CHILTON'S
AUTO REPAIR MANUAL
1993-97

Publisher & Editor-In-Chief	Kerry A. Freeman, S.A.E.
Executive Editors	Dean F. Morgantini, S.A.E., W. Calvin Settle Jr., S.A.E.
Managing Editor	Nick D'Andrea
Senior Editors	Jacques Gordon, Michael L. Grady, Kevin M. G. Maher, Debra McCall, Richard J. Rivele, S.A.E., Richard T. Smith, Jim Taylor, Ron Webb
Project Managers	Larry Braun, S.A.E., A.S.C., Thomas P. Browne III, Joseph L. DeFrancesco, A.S.E., Robert E. Doughten, Ben Greisler, S.A.E., Martin J. Gunther, Craig P. Nangle, A.S.E., S.A.E., Ernest H. Ralph, A.S.E., S.A.E., Richard Schwartz
Editorial Staff	Jaffer A. Ahmad, Robert Chabot, William Cottman, A.S.E., Leonard Davis, A.S.E., Michael DiFurio Jr., S.A.E., Sam Fiorani, Matthew E. Frederick, William C. Friedauer, Edward Giacomucci, A.S.E., S.A.E., Al Gibbs, Herbert Guie Jr, George B. Heinrich III, Dawn M. Hoch, Daniel Howells, A.S.E., David E. Jester, A.S.E., Lori L. Johnson, A.S.E., Will Kessler, A.S.E., Kenneth F. Konzelman, Neil J. Leonard, A.S.E., James R. Marotta, Robert McAnally, Thomas A. Mellon, Raymond K. Moore, A.S.E., Norman D. Norville, A.S.E., Christine L. Nuckowski, Eric S. Peterson, A.S.E., Charles Ramsey, A.S.E., William L. Renn, A.S.E., Roy Ripple, A.S.E., George E. Ritter, Robert Saxton, A.S.E., S.A.E., Paul Shanahan, Larry E. Stiles, Gordon L. Tobias, S.A.E., Albert A. Wood, A.S.E.
Production Manager	Andrea M. Steiger
Assistant Production Manager	Marsha Park-Herman
Production Specialists	Christina Davis, Kimberly T. Hayes, Joseph C. McGinty, Liz Thompson
Director of Manufacturing	Mike D'Imperio
Manufacturing Manager	Robin Norman
OFFICERS	
Senior Vice President	Ronald A. Hoxter

CHILTON BOOK COMPANY

ONE OF THE **DIVERSIFIED PUBLISHING COMPANIES,**
A PART OF **CAPITAL CITIES/ABC, INC.**

Manufactured in
© 1996 Chilton Book Company
Chilton Way, Radnor, PA 19089
ISBN 0-8019-7919-6
ISSN

CAR MODELS

Table of Contents

HOW TO USE THIS MANUAL

HOW TO USE THIS MANUAL

Car Section

Car sections are grouped by manufacturer and arranged in alphabetical order. The text and illustrations that comprise the service procedures in each Car Section are arranged in the following order of systems and components: Firing Orders, Engine Electrical, Chassis Electrical, Engine Cooling, Fuel System, Emission Controls, Engine Mechanical, Engine Lubrication, Transmission/Transaxle, Drive Axle, Steering, Front Suspension, Rear Suspension, Brakes.

All illustrations are located as close as possible to the pertinent text. Procedures are for all models in the particular section unless specifically noted otherwise.

Locating Information

The Table of Contents, at the front of the book, lists the beginning of each Car Section in the manual.

To find where a particular Car Section is located in the book, you need only look in the Table of Contents. Once you have found the proper section, you may wish to find where specific procedures are located in that section. Turn to the Index at the front of the section. At the upper left-hand side is a listing of the main topics within the section and the page number they will be found on. Following the main topics is an alphabetical listing of all the procedures within the section and their page numbers.

Safety Notice

Proper service and repair procedures are vital to the safe, reliable operation of all motor vehicles, as well as the personal safety of those performing repairs. This manual outlines procedures for servicing and repairing vehicles using safe effective methods. The procedures contain many NOTES and CAUTIONS which should be followed along with standard safety procedures to eliminate the possibility of personal injury or improper service which could damage the vehicle or compromise its safety.

It is important to note that repair procedures and techniques, tools and parts for servicing vehicles, as well as the skill and experience of the individual performing the work vary widely. It is not possible to anticipate all of the conceivable ways or conditions under which vehicles may be serviced, or to provide cautions as to all of the possible hazards that may result. Standard and accepted safety precautions and equipment should be used when handling toxic or flammable fluids, and safety goggles or other protection should be used during cutting, grinding, chiseling, prying, or any other process that can cause material removal or projectiles.

Some procedures require the use of tools specially designed for a specific purpose. Before substituting another tool or procedure, you must be completely satisfied that neither your personal safety, nor the performance of the vehicle will be endangered.

Part Numbers

Part numbers listed in this book are not recommendations by Chilton for any product by brand name. They are references that can be used with interchange manuals and aftermarket supplier catalogs to locate each brand supplier's discrete part number.

Although information in this manual is based on industry sources and is as complete as possible at the time of publication, the possibility exists that some car manufacturers made later changes which could not be included here. Information on very late models may not be available in some circumstances. While striving for total accuracy, Chilton Book Company cannot assume responsibility for any errors, changes, or omissions that may occur in the compilation of this data.

Copyright Notice

CHRYSLER CORP.

Front Wheel Drive

CHRYSLER-LeBaron **DODGE**-Daytona • Shadow • Spirit
PLYMOUTH-Acclaim • Sundance

FIRING ORDERS

NOTE: To avoid confusion, always replace spark plug wires one at a time.

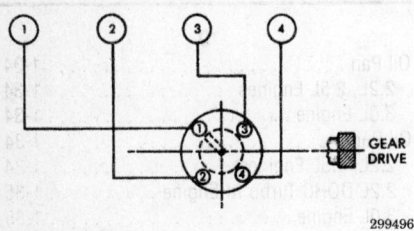

299496

2.2L and 2.5L Engines
Engine Firing Order: 1-3-4-2
Distributor Rotation: Clockwise

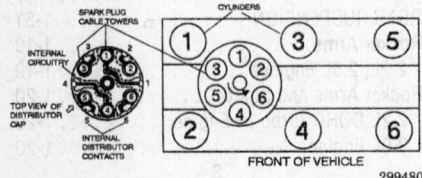

299480

3.0L Engine
Engine Firing Order: 1-2-3-4-5-6
Distributor Rotation: Counterclockwise

ENGINE ELECTRICAL

NOTE: Disconnecting the negative battery cable on some vehicles may interfere with the functions of the on board computer systems and may require the computer to undergo a relearning process, once the negative battery cable is reconnected.

Distributor

REMOVAL AND INSTALLATION

1. Disconnect the negative battery cable.
2. Disconnect the distributor pickup lead wires. Remove the splash shield, if equipped.

3. Unscrew the distributor cap hold-down screws and lift off the distributor cap with all ignition wires still connected.
4. Remove the coil wire, if necessary.
5. Matchmark the rotor to the distributor housing and the distributor housing to the engine.

NOTE: Do not crank the engine during this procedure. If the engine is cranked, the matchmark must be disregarded.

6. Remove the hold-down bolt and clamp or nut.
7. Remove the distributor from the engine.

To install:

Timing Not Disturbed

1. Install a new distributor housing O-ring.
2. Install the distributor in the engine so the rotor is aligned with the matchmark on the housing and the housing is aligned with the matchmark on the engine. Make sure the distributor is fully seated and the distributor shaft is fully engaged.
3. Install the hold-down clamp and snug the fastener.
4. Connect the distributor harness connector. Install the splash shield, if equipped.
5. Install the distributor cap and secure retainers.
6. Connect the negative battery cable.
7. Adjust the ignition timing and secure the hold-down.

Timing Disturbed

1. Install a new distributor housing O-ring.
2. Position the engine so the No. 1 piston is at TDC of the compression stroke and the mark on the vibration damper is aligned with **0** on the timing indicator.

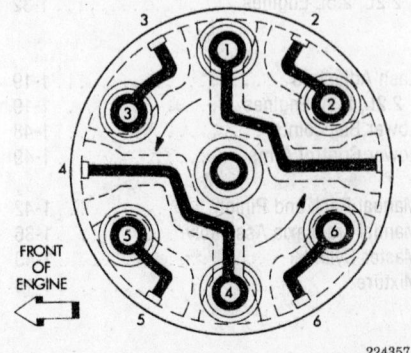

224357

Distributor cap terminal routing — 3.0L engine

3. Install the distributor in the engine so the rotor is aligned with the position of the No. 1 ignition wire on the distributor cap and the housing is aligned with the matchmark on the engine. Make sure the distributor is fully seated and the distributor shaft is fully engaged.

NOTE: There are distributor cap runners inside the cap on 3.0L engine. Make sure the rotor is pointing to where the No. 1 runner originates inside the cap and not where the No. 1 ignition wire plugs into the cap.

4. Install the hold-down clamp and snug retainer.
5. Connect the distributor wire harness connector. Install the splash shield, if equipped.
6. Install the distributor cap and tighten the screws.
7. Connect the negative battery cable.
8. Adjust the ignition timing and tighten the hold-down bolt.

Ignition Timing

ADJUSTMENT

1. Start the engine, set the parking brake and run the engine until at normal operating temperature. Be sure to turn all lights and accessories **OFF**.
2. If a magnetic timing unit is available, insert the probe into the receptacle near the timing scale. The scale is located on top of the bellhousing on 2.2L and 2.5L engines or near the crankshaft pulley on 3.0L engine.
3. If a magnetic timing unit is not available, connect a conventional power timing light to the No. 1 cylinder spark plug wire.
4. If a Diagnostic Readout Box II (DRB II, or equivalent scan tool) is available, access the Basic Timing Mode.
5. If the DRB II is not available, disconnect the coolant sensor. This sensor is located on the side of the thermostat housing on 2.2L and 2.5L engines or between the distributor or thermostat housing on 3.0L engine. The Check Engine light on the instrument panel must be ON.
6. Aim the timing light at the timing scale or read the magnetic timing unit.
7. Loosen the distributor hold-down bolt or nut enough so the distributor can be rotated.
8. Turn the distributor in the proper direction until the specified

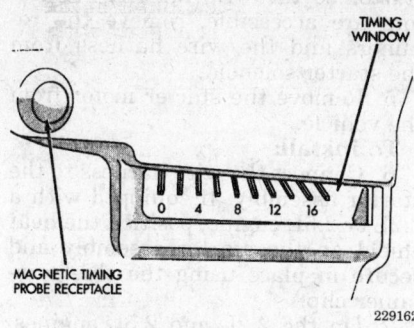

Timing scale — 2.2L and 2.5L engines

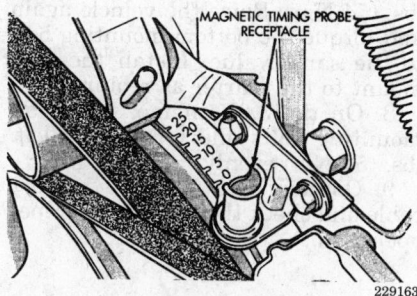

Timing scale — 3.0L engine

timing according to the VECI label is reached. Tighten the hold-down bolt or nut and recheck the timing.

9. Turn the engine **OFF**. Connect the coolant sensor and check to make sure the Check Engine light does not come ON when the vehicle is restarted. Disconnect the timing apparatus.

10. If the coolant temperature sensor was disconnected, erase the created fault code using the Erase Fault Code mode on the DRB II (or equivalent scan tool).

Alternator

PRECAUTIONS

Several precautions must be observed with alternator equipped vehicles to avoid damage to the unit.

• If the battery is removed for any reason, make sure it is reconnected with the correct polarity. Reversing the battery connections may result in damage to the 1-way rectifiers.

• When utilizing a booster battery as a starting aid, always connect the positive to positive terminals and the negative terminal from the booster

battery to a good engine ground on the vehicle being started.

• Never use a fast charger as a booster to start vehicles.

• Disconnect the battery cables when charging the battery with a fast charger.

• Never attempt to polarize the alternator.

• Do not use test lights of more than 12 volts when checking diode continuity.

• Do not short across or ground any of the alternator terminals.

• The polarity of the battery, alternator and regulator must be matched and considered before making any electrical connections within the system.

• Never separate the alternator on an open circuit. Make sure all connections within the circuit are clean and tight.

• Disconnect the battery ground terminal when performing any service on electrical components.

• Disconnect the battery if arc welding is to be done on the vehicle.

REMOVAL AND INSTALLATION

Most Chrysler products have alternators that are not intended to be disassembled for service. If defective, the alternator must be replaced as an assembly. Several different manufacturers supply alternators to Chrysler including Bosch and Nippondenso. Use care to get the correct replacement unit.

Chryslers use the Powertrain Control Module to regulate the alternator output. These vehicles, equipped with On-Board Diagnostics (OBD) may output a Diagnostic Code (seen as the Check Engine Lamp flashing) which helps when troubleshooting an alternator problem. On many applications, Code 41 indicates a problem with the alternator field not switching properly. Code 46 indicates the charging system voltage is too high. Code 47 indicates the charging system voltage is too low. The most accurate device to retrieve diagnostic codes is a Diagnostic Readout Box (DRB) or scan tool. A diagnostic scanner plugged into the data link connector (on vehicles so equipped) located near the battery, will display condition descriptions, test various circuits and component functions.

As always, check the traditional causes of vehicle electrical system problems first. Check the battery's condition and the terminals and cables for corrosion. Repair as neces-

sary. Check for a loose drive belt and tighten if necessary.

NOTE: To perform the alternator adjustment on some models, it may be necessary to raise the vehicle and remove the splash shield.

1. Disconnect the negative battery cable.

2. Disconnect and tag the wires from the alternator.

3. Loosen the belt tension adjusting bolt, disconnect the drive belt(s) and remove the supporting nuts and bolts. Remove the alternator from the engine.

To install:

4. Align alternator with pivot hole in mounting bracket and install pivot bolt. Do not fully tighten bolt at this time.

5. Check the clearance between the alternator housing and the mount. If more than 0.008 in. clearance exists, install shims and connect adjusting bolt.

6. Route belt(s) around pulleys and connect wiring.

7. Adjust belt for proper tension and tighten all mounting hardware.

8. Connect the negative battery cable and check alternator for proper charging.

Drive Belt

REMOVAL AND INSTALLATION

2.2L and 2.5L Engines

Excessive belt tension will cause damage to the alternator and water pump pulley bearings. Loose belt tension will produce slip and premature wear on the belt. Be sure to adjust the belt tension to the proper level.

To adjust the tension on a drive belt, loosen the adjusting bolt or fixing bolt locknut on the desired component, bracket or tension pulley. Then move the component or turn the adjusting bolt to adjust belt tension. Once the desired value is reached, secure the bolt or locknut and recheck tension.

Belt replacement is similar to adjustment, with the exception the belt will have to be properly routed around the pulleys. It is important to note the routing of the belt before removal. For individual belt replacement, start with the outer most belt. If a removed belt is to be reused, be certain to mark the direction of rotation on the belt, to extend belt life.

Accessory Drive Belt		Gauge	Deflection	Torque
Air Conditioning	New	105 lb.	8mm (5/16 in.)	54 N·m (40 ft. lbs.)
Compressor	Used	80 lb.	11mm (7/16 in.)	41 N·m (30 ft. lbs.)
Alternator/Water Pump	New	115 lb.	3mm (1/8 in.)	149 N·m (110 ft. lbs.)
Poly "V"	Used	80 lb.	6mm (1/4 in.)	108 N·m (80 ft. lbs.)
Power Steering Pump	New	105 lb.	6mm (1/4 in.)	102 N·m (75 ft. lbs.)
	Used	80 lb.	11mm (7/16 in.)	75 N·m (55 ft. lbs.)

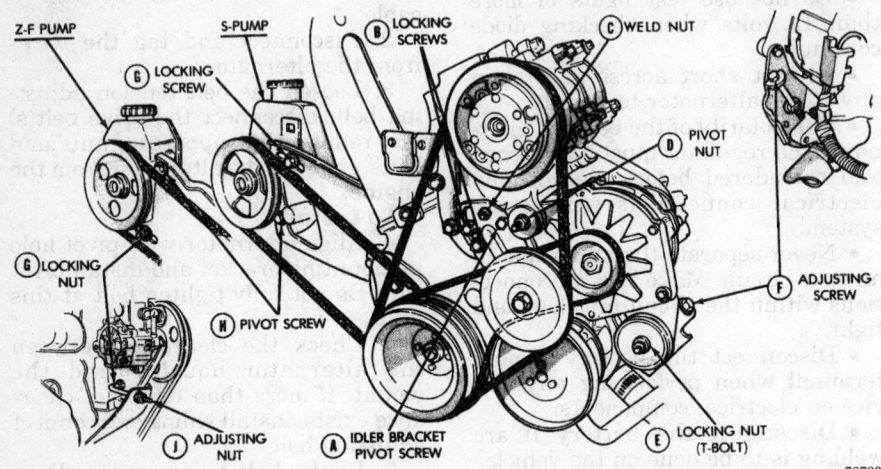

Engine accessory adjustment points — 2.2L and 2.5L engines

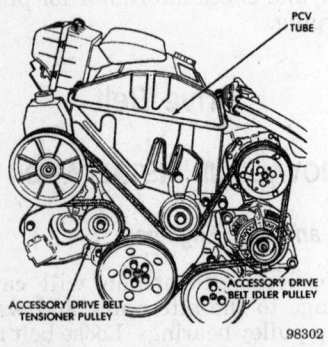

Accessory drive belt — 2.2L Turbo III engine

3.0L Engine

Excessive belt tension will cause damage to the alternator and water pump pulley bearings. Loose belt tension will produce slip and premature wear on the belt. Be sure to adjust the belt tension to the proper level.

To adjust the tension on a drive belt, loosen the adjusting bolt or fixing bolt locknut on the desired component, bracket or tension pulley. Then move the component or turn the adjusting bolt to adjust belt tension. Once the desired value is reached, secure the bolt or locknut and recheck tension.

Belt replacement is similar to adjustment, with the exception the belt will have to be properly routed around the pulleys. It is important to note the routing of the belt before removal. For individual belt replacement, start with the outer most belt. If a removed belt is to be reused, be certain to mark the direction of rotation on the belt, to extend belt life.

Serpentine belt replacement requires the use of a ½ inch breaker bar inserted into the square drive of the self adjusting tensioner. Rotate the spring loaded tensioner clockwise to release tension.

Starter

REMOVAL AND INSTALLATION

1. Disconnect the negative battery cable.
2. On 2.2L and 2.5L engines, remove the attaching nut and bolt at the top of the bellhousing. Raise the vehicle and support safely.
3. Remove the rear mount from the starter, if equipped. If equipped with 2.2L or 2.5L engine, remove the heat shield retainer clip and the shield from the starter.
4. Unbolt the starter and remove from the bellhousing. Position the

starter so the wire harness connectors are accessible, remove the retainers and the wire harness from the starter solenoid.
5. Remove the starter motor from the vehicle.

To install:
6. Connect the wire harness to the starter assembly. If equipped with a 2.2L or 2.5L engine, position the heat shield on the starter assembly and secure in place using the shield retainer clip.
7. On the 2.2L and 2.5L engines, install the lower bolt loosely, then lower the vehicle and install the uppermost nut and bolt from above. Torque starter retainer bolts to 40 ft. lbs. (54 Nm). Raise the vehicle again and torque the bottom mounting bolt to the same value. Install the rear mount to the starter as required.
8. On the 3.0L engine, install all mounting bolts and torque to 40 ft. lbs. (54 Nm) evenly.
9. Connect the negative battery cable and check the starter for proper operation.

CHASSIS ELECTRICAL

Blower Motor

REMOVAL AND INSTALLATION

Both the heater and the heater/air conditioning systems share many of the same functioning components. All vehicles are equipped with a common A/C heater unit housing assembly. On heater only systems, the evaporator and recirculating air door are omitted. A conventional blower motor resistance block supplies the motor with varied voltage (low and middle speeds) or battery voltage (high speed). Use care when working around the resistor block. Stay clear of the blower motor resistor block if the system has been recently run. The resistor block becomes very hot. In addition, do not operate the blower motor with the resistor block removed from the heater-A/C housing.

1. Disconnect the negative battery cable.
2. Remove the glove box and if equipped, the lower right side instrument panel trim cover.
3. Remove the right cowl panel trim cover.

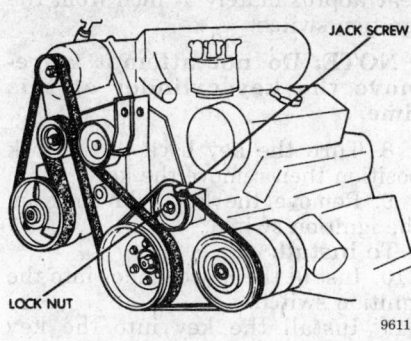

JACK SCREW

LOCK NUT

96111

A/C belt adjustment/replacement — 3.0L engine

4. Disconnect the blower motor ground lead at the right cowl panel and unplug the power lead connector.

5. Locate and disconnect the blower motor cooling tube at the heater A/C housing.

6. Remove the recirculating air door actuator and position it out of the removal path of the recirculation housing.

7. Remove the 5 attaching screws from the rear facing of the recirculation housing and 2 screws from the top (7 in all), allow the recirculation housing to drop downward and remove it from the vehicle.

8. With a pair of pliers, remove the blower wheel center securing spring clamp and remove the wheel from the motor shaft by pulling it evenly and firmly away from the motor.

9. Remove the 3 blower motor attaching screws and separate the motor from the heater A/C housing.

To install:

10. Join the blower motor to the heater A/C housing and install the 3 attaching screws.

11. Replace the blower motor wheel to the motor shaft and secure it with the spring clamp.

12. Install the recirculation housing and secure it with the 2 attaching screws on top and the 5 attaching screws on the rear.

13. Replace the recirculating air door actuator.

14. Connect the blower motor cooling tube to the A/C heater housing.

15. Connect the blower motor ground lead at the right cowl panel and plug in the power lead connector.

16. Replace the right cowl panel trim cover.

17. Install the glove box and if removed the lower right side instrument panel trim cover.

18. Connect the negative battery cable.

Windshield Wiper Motor

REMOVAL AND INSTALLATION

Whenever a wiper motor malfunction occurs, first verify that the wiper motor wire harness is properly connected to all connectors before starting normal diagnosis and repair procedures. The system is fused; always check the fuse in fuse block.

1. Disconnect the negative battery cable.

2. Remove the wiper arms, blades and the plastic cowl top cover.

3. Remove the attaching screws from each pivot assembly.

4. Remove the motor mounting bracket retainer bolts and disconnect the wiper motor harness connector.

5. Remove the wiper motor, pivot and links from the vehicle as an assembly.

6. Secure the wiper motor in a vise and remove the nut from the end of the motor shaft. Do not allow the shaft of the motor to turn from the PARK position.

To install:

7. Assemble the linkage to the motor. Make sure the crank fits over the **D** slot on the motor shaft. Tighten the motor shaft nut to 90 inch lbs. (10 Nm).

8. Make sure the motor is still in the PARK position prior to installing the wiper linkage. If not, temporarily connect the motor to the wiring harness and operate the switch to position the motor in the PARK position. Connect the linkage to the motor.

9. Install the wiper motor, pivot and links to the vehicle as an assembly.

10. Secure the mounting bracket retainer bolts to 70 inch lbs. (8 Nm). Attach the wiper motor harness.

11. Cycle the switch and turn OFF to assure motor is in the PARK position. Install the cowl top plastic cover and wiper arms tightening the retaining nuts to 150 inch lbs. (17 Nm).

12. Connect the negative battery cable and check for proper operation of the wipers.

Headlight Switch

REMOVAL AND INSTALLATION

Except 1993–94 LeBaron and 1993 Daytona

1. Disconnect the negative battery cable.

2. Remove the headlight switch bezel or cluster bezel, as required.

3. Remove the screws securing the headlight switch mounting plate to the instrument panel. Pull the assembly out and disconnect the connectors from the switch.

4. Depress the spring button and remove the headlight switch knob and stem.

5. Remove the escutcheon, if equipped, and remove the nut that attaches the switch to the mounting plate.

6. The installation is the reverse of the removal procedure.

7. Connect the negative battery cable and check the switch for proper operation.

1993–94 LeBaron and 1993 Daytona

1. Disconnect the negative battery cable.

2. Remove the panel vent grille above the switch pod assembly and remove the 2 revealed pod mounting screws.

3. Remove the 2 remaining screws under the pod and pull the pod out to disconnect the wiring harnesses. Remove the pod from the instrument panel.

4. Remove the turn signal switch lever by pulling it straight out of the pod.

5. Remove the inner panel from the pod. Remove the turn signal switch in order to gain access to the headlight switch retainers.

6. Remove the switch mounting screws.

7. Disconnect the switch linkage from the buttons by pulling the linkage straight up. Remove the switch.

To install:

8. Latch the switch linkage in the up position. Insert the dimmer shaft into the dimmer knob while aligning the switch to the pod assembly.

9. Install the switch attaching screws.

10. Unlatch linkage and install onto push buttons. Operate the switch to assure correct installation.

11. Reconnect the wiring for the turn signal switch, if disconnected, and install switch to its original position. Make sure switch wiring is properly clipped into position.

12. Place together the inner and the outer bezel sections and install the inner switch pod panel retainer screws from underneath the switch pod.

13. Install the turn signal lever by pushing straight into the switch assembly.

14. Install the switch pod assembly into the instrument panel.

15. Reconnect the negative battery terminal and check for proper system function.

Combination Switch

REMOVAL AND INSTALLATION

These vehicles use a multi-function switch (also called a combination switch) to control a variety of tasks. This switch contains the electrical circuitry for the turn signals, the hazard warning switch, headlamp beam select switch, headlamp optical horn, windshield wiper switch, pulse wipe and windshield washer switching.

This integrated switch assembly is mounted to the left side of the steering column. Should any function of the switch fail, the entire switch assembly must be replaced.

─── **CAUTION** ───
The Supplemental Inflatable Restraint (SIR) system must be disarmed before removing the combination switch. Failure to do so may cause accidental deployment of the air bag, resulting in unnecessary SIR system repairs and/or personal injury.

1. Disconnect negative battery cable.
2. Properly disarm the air bag system.
3. Tilt the steering column up and remove 6 screws holding the lower column shroud.
4. Tilt the steering column down and remove the upper column shroud.
5. Tilt the column up and remove the tilt lever.
6. Using a screwdriver in place of the lever, tilt the column down.
7. Remove the 2 mounting screws on the switch and remove the switch from the column, disconnecting the 2 electrical connectors.
To install:
8. Install the switch to the column and secure using the mounting screws. Tighten the combination switch-to-column mounting screws to 17 inch lbs. (2 Nm).
9. Install the upper and the lower steering column shrouds tightening the retainers to 17 inch lbs. (2 Nm).
10. Install the tilt lever.
11. Properly enable the SIR system.
12. Connect the negative battery cable. Check all functions of the combination switch.

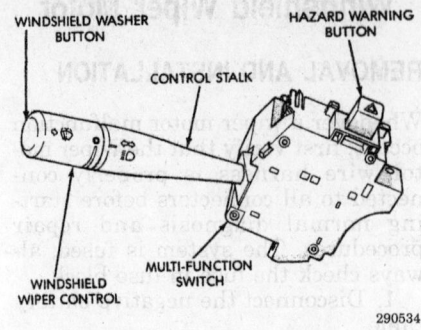

Combination switch assembly

Ignition Lock Cylinder

REMOVAL AND INSTALLATION

─── **CAUTION** ───
The Supplemental Inflatable Restraint (SIR) system must be disarmed before working around the steering column. Failure to do so may cause accidental deployment of the of the air bag, resulting in unnecessary SIR system repairs and/or personal injury.

1. Place the gear selector into the **P** position.
2. Disconnect the negative battery cable. Disarm the air bag system, if equipped.
3. If equipped with tilt steering, remove the tilt lever by turning it counterclockwise.
4. Remove the upper and lower steering column covers.
5. Insert the key and turn to the lock position.
6. Using a small screwdriver depress the key cylinder retaining pin until it is flush with the key cylinder surface.
7. Turn the key clockwise to the off position. the key cylinder will un-

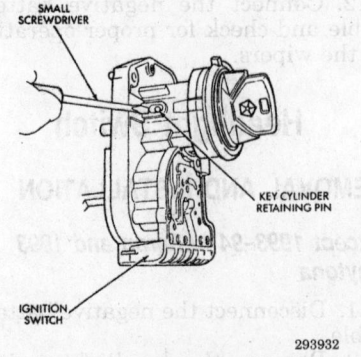

Key cylinder retaining pin

seat approximately ⅛ inch from the ignition switch.

NOTE: Do not attempt to remove the key cylinder at this time.

8. Turn the key back to the lock position then remove the key.
9. Remove the key cylinder from the ignition switch.
To install:
10. Install the key cylinder into the ignition switch.
11. Install the key into the key cylinder.
12. While gently pushing inward rotate the key clockwise until it reaches the end of its travel.
13. Install the upper and lower steering column covers.
14. Reconnect the negative battery cable.
15. Check for proper operation.

Ignition Switch

REMOVAL AND INSTALLATION

The Acustar column can be identified by the halo light around the ignition key cylinder.

1. Disconnect the negative battery cable. Disarm the air bag, if equipped.

─── **CAUTION** ───
The Supplemental Inflatable Restraint (SIR) system must be disarmed before working around the steering column. Failure to do so may result in accidental deployment of the air bag, resulting in unnecessary SIR system repairs and/or personal injury.

2. Remove the tilt lever by turning counterclockwise, if equipped.
3. Remove the 3 Torx T–20 screws and remove the upper and lower column covers.
4. Remove the 3 ignition switch tamper-resistant screws using Torx tool APEX 440–TX20H or equivalent.
5. Pull the switch away from the column. Release the connector locks on the 2 wiring connectors and disconnect them from the switch.
6. Remove the key lock cylinder from the ignition switch as follows:
   a. Insert the key and turn the switch to the **LOCK** position. Using a small tool, depress the key cylinder retaining pin until flush with the key cylinder surface.
   b. Rotate the key clockwise to the **OFF** position to unseat the key cylinder from the ignition switch assembly. The cylinder bezel

should be about ⅛ inch above the ignition switch halo light ring. Do not attempt to remove the key cylinder at this point.

c. With the key cylinder unseated, rotate the key and cylinder counterclockwise to the **LOCK** position and remove the key.

d. Remove the key cylinder from the ignition switch.

To install:

7. If equipped with floor mounted gear shifter, place the selector in the **P** position.

8. Connect the electrical harness to the ignition switch making sure the locking tabs are fully seated in the wiring connectors.

9. Place the gear shift lever in the **P** position. Mount the ignition switch to the column as follows:

a. Place the ignition switch in the **LOCK** position. The switch is in the lock position when the column lock flag is parallel to the ignition switch terminals.

b. Position the ignition switch lock dowel pin so it will engage the steering column park lock slider linkage.

c. Apply a light coat of grease to the column lock flag and the park lock dowel pin. Place the ignition against the lock housing opening on the steering column. Ensure ignition switch park lock dowel pin enters the slot in the park lock slider linkage in the steering column.

d. Install the ignition switch mounting screws and torque to 17 inch lbs. (2 Nm).

10. Install the steering column covers. If equipped with tilt wheel, install the tilt lever.

11. Install the ignition key to the lock cylinder as follows:

a. With the key cylinder and ignition switch in the **LOCK** position, key not in cylinder, gently insert the key cylinder into the ignition switch until it bottoms.

b. Insert the ignition key into the ignition cylinder. Simultaneously, push in on the cylinder and rotate the key to the end of travel. The ignition cylinder should now be fully seated in the ignition switch.

12. Connect the negative battery cable.

13. Check the push-to-lock and park lock functions, halo lighting and all ignition switch positions for proper operation.

Park/Neutral Safety Switch

REMOVAL AND INSTALLATION

1. Disconnect the negative battery cable.

2. Raise and safely support vehicle.

3. Remove the electrical connector from the Park/Neutral safety switch.

4. Position a drain pan under the transaxle and remove the switch.

To install:

5. Using a new seal, screw the switch in the transaxle case and tighten to 24 ft. lbs. (33 Nm).

6. Replace the connector plug and lower vehicle.

7. Run engine and refill the transaxle assembly. Check switch operation. Engine should start only in **P** or **N**.

Powertrain Control Module

REMOVAL AND INSTALLATION

Engine control operations such as ignition and fuel injection are controlled by an on-board computer. This computer has gone under several names in past years such as Engine Control Module, Single Board Engine Controller and others. Among car manufacturers, this computer is now known as the Powertrain Control Module (PCM) which more accurately describes its role since transmission operation and traction control and more are often included in the computer's requirements. The PCM is an expensive part and should be handled with care. Always make sure the ignition key is in the **OFF** position and the battery negative cable has been disconnected before attempting to remove the PCM. The PCM is located underhood, next to the battery.

1. Disconnect the negative battery cable.

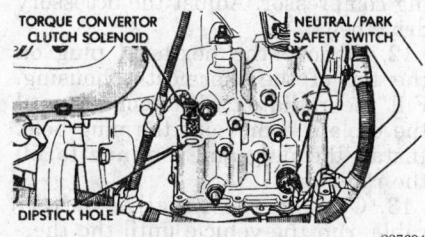

TORQUE CONVERTOR CLUTCH SOLENOID   NEUTRAL/PARK SAFETY SWITCH

DIPSTICK HOLE

227694

Park/Neutral switch location

2. Remove the air cleaner duct or air cleaner assembly.

3. Remove the battery assembly.

4. Disconnect and remove the 60-way pin connector from the PCM.

5. Remove the PCM assembly.

To install:

6. Connect the 60-way PCM connector to the PCM.

7. Install the PCM and tighten 2 mounting screws.

8. Install the battery and air cleaner assemblies.

ENGINE COOLING

Radiator

REMOVAL AND INSTALLATION

The radiators used on these vehicles are a crossflow type (horizontal tubes). Plastic tanks, while stronger than brass, are subject to damage by impact, such as slipped wrenches. Use care when working on the radiator. Some quick checks are, **with the engine cold,** idle the engine with the radiator cap removed. When the thermostat opens, you should be able to observe coolant flow while looking down the filler neck. Once flow is detected, install the radiator pressure cap. With engine idling and at normal operating temperature, feel the upper radiator hose. If it is hot, coolant is circulating.

1. Disconnect the negative battery cable.

2. With the engine and cooling system cold, drain the coolant.

3. Remove the upper hose and coolant reserve tank hose from the radiator.

4. Remove the electric cooling fan.

5. Raise the vehicle and support safely. Remove the lower hose from the radiator.

6. Disconnect the automatic transaxle cooler hoses, if equipped, and plug them. Lower the vehicle.

7. Remove the mounting brackets and carefully lift the radiator out of the engine compartment. Use care not to damage the fins.

To install:

8. Inspect the radiator hoses. Hardened, cracked, swollen or restricted hoses should be replaced. Use care not to damage radiator inlet and outlet when servicing hoses. Radiator hoses should be routed without any kinks and indexed as designed.

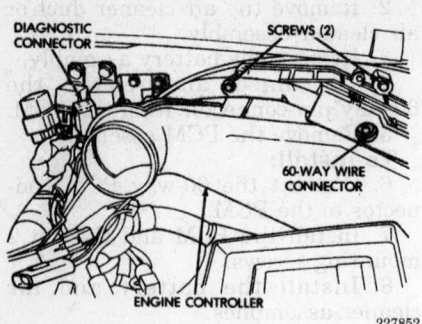

Removing the engine controller assembly

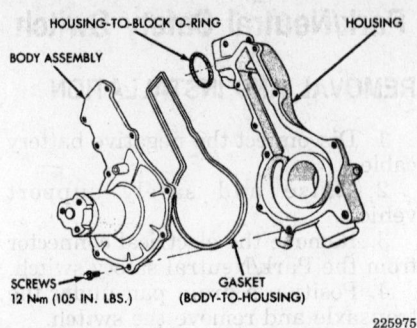

Water pump components — 2.2L and 2.5L engines

The use of molded hoses is recommended. Spring type hose clamps are used in most applications. If replacement is required, replace with original style spring type clamps.

9. Lower the radiator into position.

10. Install the mounting brackets.

11. Raise the vehicle, if necessary, and support safely. Connect the automatic transaxle cooler lines, if equipped.

12. Lower the vehicle and connect the lower hose.

13. Install the electric cooling fan.

14. Connect the upper hose and coolant reserve tank hose.

15. Fill the system with coolant.

16. Connect the negative battery cable, run the vehicle until the thermostat opens, fill the radiator completely and check the automatic transaxle fluid level, if equipped.

17. Once the vehicle has cooled, recheck the coolant level.

Water Pump

REMOVAL AND INSTALLATION

2.2L and 2.5L Engines

1. Disconnect the negative battery cable.

2. Drain the cooling system.

3. If equipped with air conditioning, remove the compressor from the bracket and position it aside. It is not necessary to discharge the air conditioning system.

4. Remove the alternator and bracket from the engine. Have a drain pan under the side mounting stud because the stud screws into a water jacket and coolant will spill out when it is removed. Remove the pulley and belt from the water pump.

5. Disconnect the lower radiator hose and heater hose from the water pump.

6. Remove the water pump housing attaching screws and remove the assembly from the vehicle. Discard the O-ring. The 2.2L Turbo III engine is equipped with a spacer (coolant deflector) between the pump housing and block on the lower mounting stud.

NOTE: Use care when removing the coolant deflector because it will be reused.

7. Remove the water pump from the housing. The 2.2L Turbo III engine is equipped with a coolant deflector that will be re-used.

To install:

8. Using a new gasket or silicone sealer, install the water pump to the housing.

9. On 2.2L Turbo III engine, install the coolant deflector to the block and install the spacer to the lower stud. Install a new O-ring to the housing and install to the engine. Torque the 3 upper bolts to 21 ft. lbs. (30 Nm) and the lower nut to 50 ft. lbs. (68 Nm).

10. Install the water pump pulley and torque the bolts to 21 ft. lbs. (30 Nm). Connect the radiator hose and heater hose to the water pump.

11. Install the alternator and compressor bracket to the engine. Install the alternator and the air conditioning compressor. Adjust the accessory drive belts.

12. Remove the hex-head plug on the top of the thermostat housing. Fill the radiator with coolant until the coolant comes out the plug hole. Install the plug and continue to fill the radiator.

13. Connect the negative battery cable, run the vehicle until the thermostat opens, fill the radiator completely and check for leaks.

14. Once engine has cooled, recheck the coolant level and add as required.

3.0L Engine

A quick test to tell whether or not the water pump is working is to see if the heater warms properly. A defective water pump will not be able to circulate heated coolant through the long heater hose.

The 3.0L water pump bolts directly to the engine block using a gasket for pump-to-block sealing. The pump is not to be repaired, but serviced as a unit. The water pump is driven by the timing belt which must be removed to service the pump. Timing belt service requires care to see that all timing marks are aligned or the engine will be severely damaged. The water pump can be replaced without discharging the air conditioning system.

1. Disconnect the negative battery cable.

2. Drain the cooling system.

3. Remove the timing cover. If the same timing belt will be reused, mark the direction of the timing belt's rotation for installation in the same direction. Make sure the engine is positioned so the No. 1 cylinder is at the TDC of its compression stroke and the sprocket's timing marks are aligned with the engine's timing mark indicators.

4. Loosen the timing belt tensioner bolt and remove the belt. Position the tensioner as far away from the center of the engine as possible and tighten the bolt. Remove the water pump mounting bolts, separate the pump from the water inlet pipe and remove the pump from the engine.

To install:

5. Inspect the water pump. Replace it, if it has any damage or cracks in the pump body, if the impeller rubs the inside of the pump or there is excessive looseness or rough turning in the bearing. If the shaft seal is leaking coolant as evidenced by coolant leaking from the vent hole, the pump should be replaced.

6. Clean all gasket and O-ring surfaces on the pump and water pipe inlet tube. Install a new O-ring on the water inlet pipe. Wet the O-ring with water (never oil or grease) to ease assembly.

7. Install the water pump to the engine with new gasket in place. Torque the water pump mounting bolts to 20 ft. lbs. (27 Nm).

8. If not already done, position both camshafts so the marks align with those on the alternator bracket (rear bank) and inner timing cover (front bank). Rotate the crankshaft so

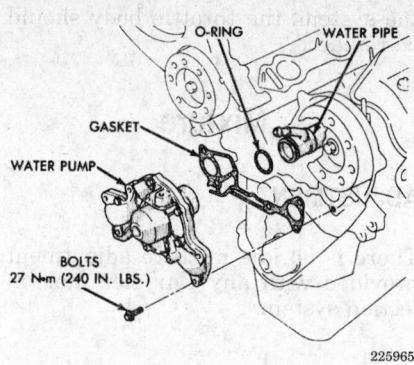

Water pump assembly — 3.0L engine

the timing mark aligns with the mark on the oil pump.

9. Install the timing belt on the crankshaft sprocket and while keeping the belt tight on the tension side (right side), install the belt on the front camshaft sprocket.

10. Install the belt on the water pump pulley, then the rear camshaft sprocket and the tensioner.

11. Rotate the front camshaft counterclockwise to tension the belt between the front camshaft and the crankshaft. If the timing marks are not aligned, repeat the procedure.

12. Install the crankshaft sprocket flange.

13. Loosen the tensioner bolt and allow the spring to tension the belt.

14. Turn the crankshaft 2 full turns in the clockwise direction only until the timing marks are aligned and torque the tensioner lock bolt to 21 ft. lbs. (29 Nm).

15. Refill the cooling system. This system uses a self-bleeding thermostat, so there is no need to bleed the system. Connect the negative battery cable, road test the vehicle and check for leaks.

Thermostat

REMOVAL AND INSTALLATION

The 2.2L and 2.5L engine thermostats are located on the front of the engine (radiator side) in what the factory calls the "water box" which is the part of the cylinder head casting which receives the thermostat. These thermostats do not have an air bleed notch.

The 3.0L engine thermostat is located in a water box formed in the timing belt end of the intake manifold. This thermostat has an air bleed valve located in the thermostat flange.

1. Disconnect the negative battery cable. Drain the coolant down to thermostat level or below.

2. Remove the thermostat housing.

3. Remove the thermostat and discard the gasket.

To install:

4. Clean all parts well especially the housing mating surfaces. Retainer bolts that thread into the cooling jacket are subject to rust and corrosion. Take care to clean these bolts well or the threaded holes in the engine can be damaged. Wet a new gasket with coolant and install.

5. Install the thermostat into the housing and position the housing on the engine. Tighten bolts to 250 inch. lbs. for 2.2L and 2.5L engines or 135 inch lbs. for 3.0L engines.

6. Add coolant to the proper level.

7. On 2.2L and 2.5L engines, remove the plug on top of the thermostat housing. On 2.2L Turbo III engine, remove the coolant temperature sensor on top of the housing. Fill the radiator with coolant until the coolant comes out the hole. Install the plug or sensor and continue to fill the radiator. The 3.0L engine thermostat is self-bleeding.

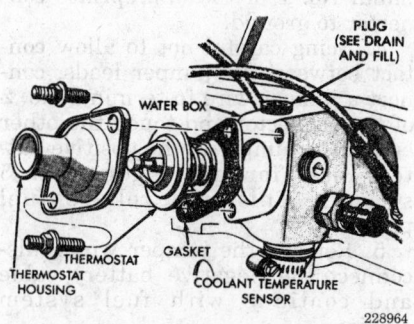

Thermostat, housing and water box — 2.2L and 2.5L engines

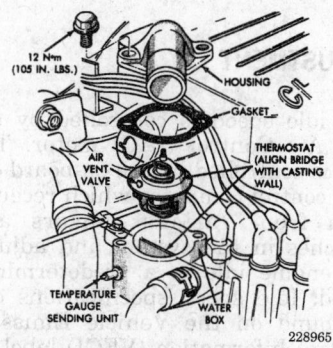

Thermostat, housing and water box — 3.0L engine

8. Connect the negative battery cable. With the radiator cap removed, run the vehicle until the thermostat opens, adding coolant as required to fill the radiator completely. Watch the temperature gauge (if equipped) for signs of problems elsewhere in the cooling system. Check for leaks. Install the radiator cap.

9. Shut the engine OFF and allow to cool. Once the vehicle has cooled, recheck the coolant level in both the radiator and the overflow tank.

Electric Cooling Fan

TESTING

————— **CAUTION** —————
*Make sure the key in in the **OFF** position when checking the electric cooling fan. If not, the fan could turn On at any time, causing serious personal injury.*

1. Unplug the fan connector.

2. Using a jumper wire, connect the terminals of the fan connector to a good 12 volt source observing correct polarity. The female terminal on the fan motor is normally the negative terminal.

3. The fan should come ON with the circuit completed and should run smoothly and free of vibrations.

4. If not, the fan is defective and should be replaced.

REMOVAL AND INSTALLATION

1. Disconnect the negative battery cable.

2. Unplug the connector.

3. Remove the mounting screws.

4. Remove the fan assembly from the vehicle.

5. The installation is the reverse of the removal procedure.

Cooling System Bleeding

To bleed air from the 2.2L and 2.5L engines, remove the plug or sensor on the top of the thermostat housing. Fill the radiator with coolant until the coolant comes out of the hole. Install the plug and continue to fill the radiator. This will vent all trapped air from the engine.

The thermostat in the 3.0L engine is equipped with a small air vent valve that allows trapped air to bleed from the system during refilling. This valve eliminates the need for cooling system bleeding.

FUEL SYSTEM

Fuel System Service Precautions

Safety is the most important factor when performing not only fuel system maintenance but any type of maintenance. Failure to conduct maintenance and repairs in a safe manner may result in serious personal injury or death. Maintenance and testing of the vehicle's fuel system components can be accomplished safely and effectively by adhering to the following rules and guidelines.

• To avoid the possibility of fire and personal injury, always disconnect the negative battery cable unless the repair or test procedure requires that battery voltage be applied.

• Always relieve the fuel system pressure prior to disconnecting any fuel system component (injector, fuel rail, pressure regulator, etc.), fitting or fuel line connection. Exercise extreme caution whenever relieving fuel system pressure to avoid exposing skin, face and eyes to fuel spray. Please be advised that fuel under pressure may penetrate the skin or any part of the body that it contacts.

• Always place a shop towel or cloth around the fitting or connection prior to loosening to absorb any excess fuel due to spillage. Ensure that all fuel spillage (should it occur) is quickly removed from engine surfaces. Ensure that all fuel soaked cloths or towels are deposited into a suitable waste container.

• Always keep a dry chemical (Class B) fire extinguisher near the work area.

• Do not allow fuel spray or fuel vapors to come into contact with a spark or open flame.

• Always use a backup wrench when loosening and tightening fuel line connection fittings. This will prevent unnecessary stress and torsion to fuel line piping. Always follow the proper torque specifications.

• Always replace worn fuel fitting O-rings with new. Do not substitute fuel hose or equivalent, where fuel pipe is installed.

Fuel System Pressure

RELIEVING

Engines With Test Port on Fuel Rail

1. Disconnect the negative battery terminal.
2. Remove the fuel filler cap.
3. Remove protective cap from the fuel test port on the fuel rail.
4. Place the open end of fuel pressure release hose tool C-4799-1 or equivalent, into an approved safety container.

NOTE: Fuel pressure test gauge tool C-4799-A contains pressure release hose C-4799-1.

5. Connect the other end of release hose to the fuel pressure test port.
6. Fuel pressure will bleed off into safety container.

Engines Without Test Port on Fuel Rail

1. Loosen the fuel filler cap to release fuel tank pressure.
2. Locate and disconnect the fuel injector harness connector.
3. Connect a jumper wire from terminal No. 1 of the appropriate connector to ground.
4. Being careful not to allow contact between the jumper leads, connect a jumper wire to terminal No. 2 of the connector and touch the other end of the jumper to the positive battery post for no longer than 5 seconds. This will relieve fuel pressure.
5. Remove the jumper wires, disconnect the negative battery cable and continue with fuel system service.

Idle Speed

ADJUSTMENT

The idle speed is controlled by the Idle Air Control (IAC) motor. The IAC is controlled by the on-board engine control computer which receives data from various sensors and switches in the system and adjusts the engine idle to a predetermined speed. Idle speed specifications can be found on the Vehicle Emission Control Information (VECI) label located in the engine compartment. If the idle speed is not within specifications and there are no problems with

the system, the throttle body should be replaced.

Mixture

ADJUSTMENT

There is no idle mixture adjustment provided with any Chrysler fuel injection system.

Fuel Filter

REMOVAL AND INSTALLATION

There were some Chrysler Flexible Fuel Vehicles (FFV) available in specific model production. Although many of the components are similar they cannot be interchanged. These vehicles can operate on a fuel mixture of gasoline and methanol up to 85 percent methanol. Components for Flexible Fuel Vehicles (FFV) can be identified by a green coloring or have a green label or tag attached. When servicing a Flexible Fuel Vehicle (FFV), use only specified FFV replacement parts. In addition, fuel filters designed for gasoline only vehicles cannot be used on Flexible Fuel vehicles.

─────── **CAUTION** ───────
Do not use conventional fuel filters, hoses or clamps when servicing this fuel system. They are not compatible with the injection system and could fail, causing personal injury or damage to the vehicle. Use only hoses and clamps specifically designed for fuel injection.

1. Disconnect the negative battery cable.
2. Relieve the fuel pressure.

─────── **CAUTION** ───────
Fuel injection systems remain under pressure after the engine has been turned OFF. Properly relieve fuel pressure before disconnecting any fuel lines. Failure to do so may result in fire or personal injury.

3. The filter is located on the frame rail toward the rear of the vehicle. Raise the vehicle and support safely. Remove the filter retaining screw and remove the filter assembly from the mounting plate.

— CAUTION —

Do not allow fuel spray or vapors to come in contact with a spark or open flame. Keep a fire extinguisher nearby. Never store fuel in an open container due to risk of fire or explosion.

4. Disconnect the quick connect fittings at the fuel filter and the fuel supply tube. Wrap a shop towel around the hoses prior to removing to absorb fuel that may leak from the connection. Remove the filter from the vehicle.

5. Loosen the outlet hose clamp on the filter and inlet hose clamp on the rear fuel tube. Wrap a shop towel around the hoses to absorb fuel that may leak from the connection. Remove the hoses from the filter and fuel tube and discard the clamps and the filter.

To install:

6. Install the inlet hose on the fuel tube and secure. Replace old clamp with new and tighten to 10 inch lbs. (1 Nm).

7. Install the outlet hose on the filter outlet fitting and secure. Tighten new clamp to 10 inch lbs. (1 Nm).

8. Position the filter assembly on the mounting plate and tighten the mounting screw to 75 inch lbs. (8 Nm).

9. Connect the negative battery cable, start the engine and check for leaks.

Fuel Pump

REMOVAL AND INSTALLATION

1. Disconnect the negative battery cable.

2. Relieve the fuel system pressure.

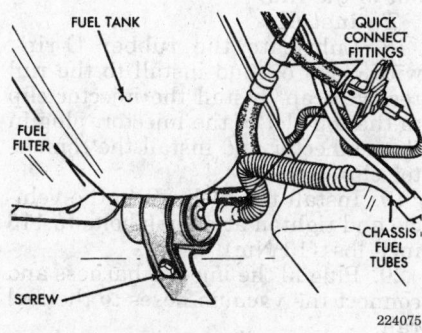

Fuel filter assembly

224075

— CAUTION —

Fuel injection systems remain under pressure after the engine has been turned OFF. The fuel system pressure must be relieved before disconnecting any fuel lines. Failure to do so may result in fire and/or personal injury.

3. Raise the vehicle and support safely.

— CAUTION —

Observe all applicable safety precautions when working around fuel. Do not allow fuel spray or fuel vapors to come into contact with a spark or open flame. Keep a dry chemical (Class B) fire extinguisher near the work area. Never drain or store fuel in an open container due to the possibility of fire or explosion.

4. Using the proper equipment, drain the fuel tank using the following procedure.

   a. Locate the drain cap on the sending unit at the rear of the tank. Remove the rubber cap from the drain tube.

   b. Connect either a portable holding tank or a siphon hose to the drain tube.

   c. Drain the fuel tank into an approved holding tank.

5. Remove the screws that hold the filler neck to the quarter panel.

6. Disconnect the wiring and hoses from the tank.

7. Place a transmission jack under the center of the tank and apply slight pressure. Remove the tank straps.

8. Remove the filler tube from the tank.

9. Lower the tank and disconnect the vapor separator rollover valve hose. Remove the fuel tank from the vehicle.

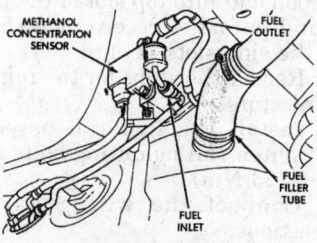

Methanol Concentration Sensor

306207

Some Chrysler produced vehicles were designated as "Flexible Fuel" and have a methanol concentration sensor located near the fuel pump module

10. Using a hammer and a brass drift, tap the lock ring counterclockwise to release the pump.

11. If necessary, partially pull the pump assembly out of the tank until the return line hose connection is visible at the of the pump assembly.

12. Disconnect the fuel fitting by pressing in on the ears.

13. Remove the pump from the tank with the O-ring. Discard the O-ring, pump inlet filter and inlet seal. Disassemble as required.

To install:

14. Install a new inlet seal, filter and strainer O-ring onto the pump. When installing strainer onto the pump reservoir body, make sure the locking tabs on the reservoir body lock over the locking tangs on the strainer.

15. Install the pump into the tank so the fuel return hose is not kinked.

16. Install the lock ring with a hammer and brass punch turning the ring clockwise. Overtightening of the lock ring may result in a leak.

17. Install the fuel tank and remaining components into position.

18. Connect the negative battery cable, start the engine and check the fuel system for leaks.

Fuel Injector

REMOVAL AND INSTALLATION

2.2L and 2.5L Engines

Fuel injection systems remain under pressure, even after the engine has been turned OFF. The fuel system pressure must be relieved before disconnecting any fuel lines. Failure to do so may result in fire and/or personal injury.

1. Disconnect the negative battery cable.

2. Remove the air cleaner assembly.

3. Relieve the fuel pressure using the recommended procedure. Loosen the fuel fill cap to prevent pressure build-up in the tank.

4. Remove the injector hold-down Torx screw and the hold-down.

5. Using a pair of small flat-tipped tools, lift the cap off the injector.

6. Gently pry the injector from its pod.

7. Remove the lower O-ring from the pod.

To install:

8. Install the new lower O-ring on the injector.

9. Align the injector terminal housing with the locating socket in the injector cap.

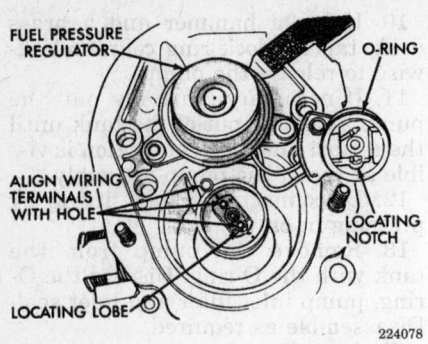

Fuel injector assembly — single point fuel injection (TBI) — 2.2L and 2.5L engines

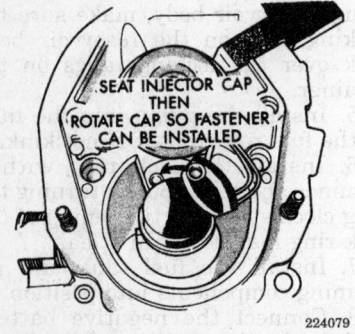

Installing fuel injector and cap — single point fuel injection (TBI) — 2.2L and 2.5L engines

10. Press the injector cap so the upper O-ring flange is flush with the lower surface of the cap.

11. Spray the inner surfaces of the injector pod with a carburetor parts cleaner to remove residual varnish and gasoline.

12. Lubricate the O-rings sparingly with petroleum jelly.

13. Place the injector and cap into the injector pod and align the cap locating pin with the locating hole in the casting.

14. Press firmly on the injector cap until it is flush with the casting surface.

15. Align the hole in the hold-down with the pin on the cap and install.

16. Push down on the cap, install the screw and torque to 35 inch lbs. (4 Nm).

17. Connect the negative battery cable and check for leaks using the DRB II or equivalent scan toll to activate the fuel pump and pressurize the fuel system.

18. Install the air cleaner.

2.2L Turbocharged Engine

1. Disconnect the negative battery cable.

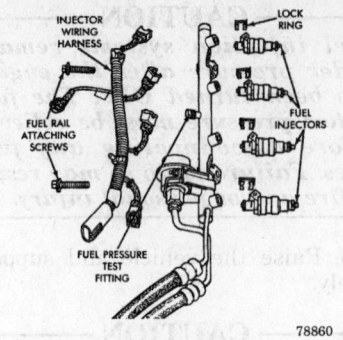

Fuel rail and injector assembly — 2.2L turbocharged engine

2. Relieve the fuel system pressure.

3. If necessary, remove the intake manifold support braces.

4. Disconnect the vacuum and vapor hoses from the fuel rail and valve cover.

5. Disconnect the injector wiring harness connections.

6. If necessary, disconnect the detonation sensor.

7. Unbolt the fuel rail from the engine.

8. Pull straight up on the injector rail and remove the rail with injectors connected.

9. Remove the injector lock clip from the fuel rail and injector. Pull the injector straight out of the fuel rail receiver cup.

10. Check the injector O-ring for damage. If the O-ring is damaged, replace it. If the injector is being reused, install a protective cap on the injector tip to prevent damage.

11. Repeat the procedure for the remaining injectors.

To install:

12. Before installing an injector the rubber O-ring should be lubricated with a drop of clean engine oil to aid in installation.

13. Install injector top end into fuel rail receiver cup.

14. Install injector clip by sliding the open end into top slot of the injector and onto the receiver cup ridge into the side slots of clip.

15. Repeat the steps for the remaining injectors.

16. Install the fuel rail assembly and tighten rail mounting bolts to 17 ft. lbs. (23 Nm).

17. Connect the wiring harness connections.

18. Connect the vacuum and vapor hoses to the fuel rail and valve cover.

19. If removed, install the intake manifold support braces.

20. Connect the negative battery cable and check for leaks using the DRB II to activate the fuel pump.

3.0L Engine

1. Disconnect the negative battery cable.

2. Relieve the fuel system pressure.

3. Remove the air cleaner to throttle body hose.

4. Disconnect the throttle cable from the throttle body and disconnect the kickdown linkage. Remove the throttle cable bracket attaching bolts.

5. Disconnect the harness connectors from the Idle Air Control (IAC) motor and the Throttle Position Sensor (TPS) on the throttle body.

6. Matchmark and carefully remove the vacuum hoses from the throttle body.

7. Remove the PCV and brake booster hoses from the air intake plenum.

8. Remove the ignition coil from the intake plenum, if mounted there.

9. Remove the EGR tube flange from the intake plenum, if equipped.

10. Unplug the coolant temperature sensor and charge temperature sensor, if equipped.

11. Remove the vacuum connection from the air intake plenum vacuum connector.

12. Remove the fuel hoses from the fuel rail and plug them.

13. Remove the air intake plenum to intake manifold bolts and remove the plenum and gaskets. Cover the intake manifold openings to prevent debris from entering the engine.

14. Remove the vacuum hoses from the fuel rail.

15. Label and disconnect the fuel injector wiring harness from each injector.

16. Remove the fuel rail attaching bolts and remove the fuel rail with the wiring harness from the vehicle. Position the rail on the bench upside down so the injectors are easily accessible.

17. Remove the retainer clip from the slot on the fuel injector and remove by pulling the injector straight out of the rail.

To install:

18. Lubricate the rubber O-ring with clean oil and install to the rail receiver cap. Install the injector clip to the top slot of the injector, plug in the connector and install the connector clip.

19. Install the fuel rail to the vehicle and tighten attaching bolts to 115 inch lbs. (13 Nm).

20. Plug in the injector harness and connect the vacuum hoses to the fuel rail.

21. Install new intake plenum gaskets with the beaded sealer side up

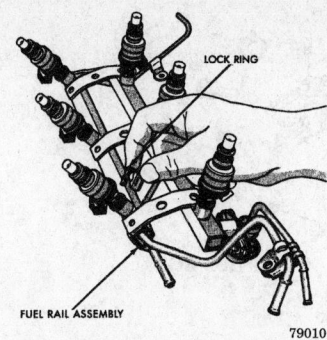

Fuel injector and rail assembly —
3.0L engine

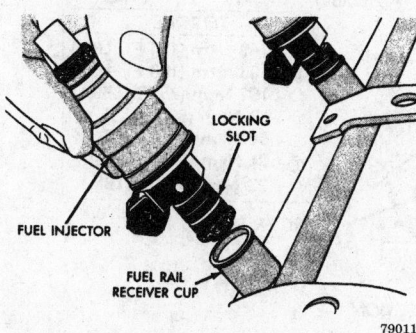

Servicing the fuel injector assembly — 3.0L
engine

and install the intake plenum. Torque the attaching bolts and nuts to 115 inch lbs. (13 Nm).

22. Reconnect the EGR tube flange and tighten to 200 inch lbs. (22 Nm).

23. Install the fuel hoses to the fuel rail. Tighten hold-down bolts to 95 inch lbs. (10 Nm).

24. Connect remaining items that were attached to the intake plenum and throttle body.

25. Connect the negative battery cable and check for leaks using the DRB II to activate the fuel pump.

ENGINE MECHANICAL

Engine Assembly

REMOVAL AND INSTALLATION

2.2L and 2.5L Engines

—————— CAUTION ——————
Fuel injection systems remain under pressure, even after the engine has been turned OFF. The fuel system pressure must be relieved before disconnecting any fuel lines. Failure to do so may result fire and/or personal injury.

1. Disconnect the negative battery cable and all engine ground straps.
2. Relieve the fuel system pressure.
3. Mark the hood hinge outline on the hood and remove the hood.
4. Drain the cooling system. Remove the radiator hoses, fan assembly, radiator shroud and radiator.
5. Remove the air cleaner, duct hoses and oil filter.
6. If equipped, remove the A/C compressor and position it out of the way. It should not be necessary to disconnect the A/C hoses from the compressor.
7. If equipped, remove the power steering pump mounting bolts and set the pump aside, without disconnecting any fluid lines.
8. Disconnect and tag the electrical connectors from the engine, alternator and fuel injection system.
9. Disconnect the fuel line(s), heater hoses and accelerator linkage.
10. If equipped, disconnect the air pump lines and remove the pump.
11. Remove the alternator from the engine.
12. Disconnect the shift linkage(s), clutch (M/T only) and speedometer cables.
13. Raise and support the front of the vehicle safely.
14. If equipped with a manual transaxle, perform the following procedures:
   a. Disconnect the clutch cable.
   b. Remove the lower cover from the transaxle case.
   c. Remove the exhaust pipe to exhaust manifold bolts. Separate the pipe from the manifold.
   d. Remove the starter and support it out of the way.

   e. If equipped, remove the anti-roll strut or damper (used on turbocharged engine) from the transaxle.
   f. Remove the manual transaxle assembly.
15. If equipped with an automatic transaxle, perform the following procedures:
   a. Remove the lower cover from the transaxle case.
   b. Remove the exhaust pipe to exhaust manifold bolts. Separate the pipe from the manifold.
   c. Remove the starter and position aside.
   d. Mark the flexplate to the torque converter, for installation purposes.
   e. Remove the torque converter to flexplate bolts. Separate the converter from the flexplate.
   f. Using a C-clamp, secure the bottom of the torque converter to the transaxle so it will not fall out.
   g. Using a transmission holding tool, secure it to the transaxle.
16. Attach an engine lifting fixture.

NOTE: If removing the engine mount insulator-to-rail screws, first mark the position of the insulator on the side rail to insure proper alignment during reinstallation.

17. Remove the transaxle to engine bolts, front engine mount nut/bolt and the left insulator through bolt (from inside the wheelhouse) or the insulator bracket-to-transaxle bolts.
18. Lift the engine from the vehicle.
   To install:
19. Lower the engine into the engine compartment. Make sure the lifting device is supporting the full weight of the engine and loosely install all of the engine mounting bolts until all are threaded.
20. Tighten front mounting bolt, left side block-to-engine bolts and right side block-to-engine bolts.
21. Remove the lifting device.
22. Raise the vehicle and support safely.
23. If equipped with a manual transaxle, install the transaxle.
24. If equipped with an automatic transaxle, align the mating marks and install the torque converter bolts. Torque bolts to 55 ft. lbs. (75 Nm). Install the torque converter inspection plate and the starter.
25. Connect the exhaust pipe. Lower the vehicle.
26. Install the alternator, power steering pump and air conditioning compressor, if equipped.
27. Connect the fuel lines and heater hoses.

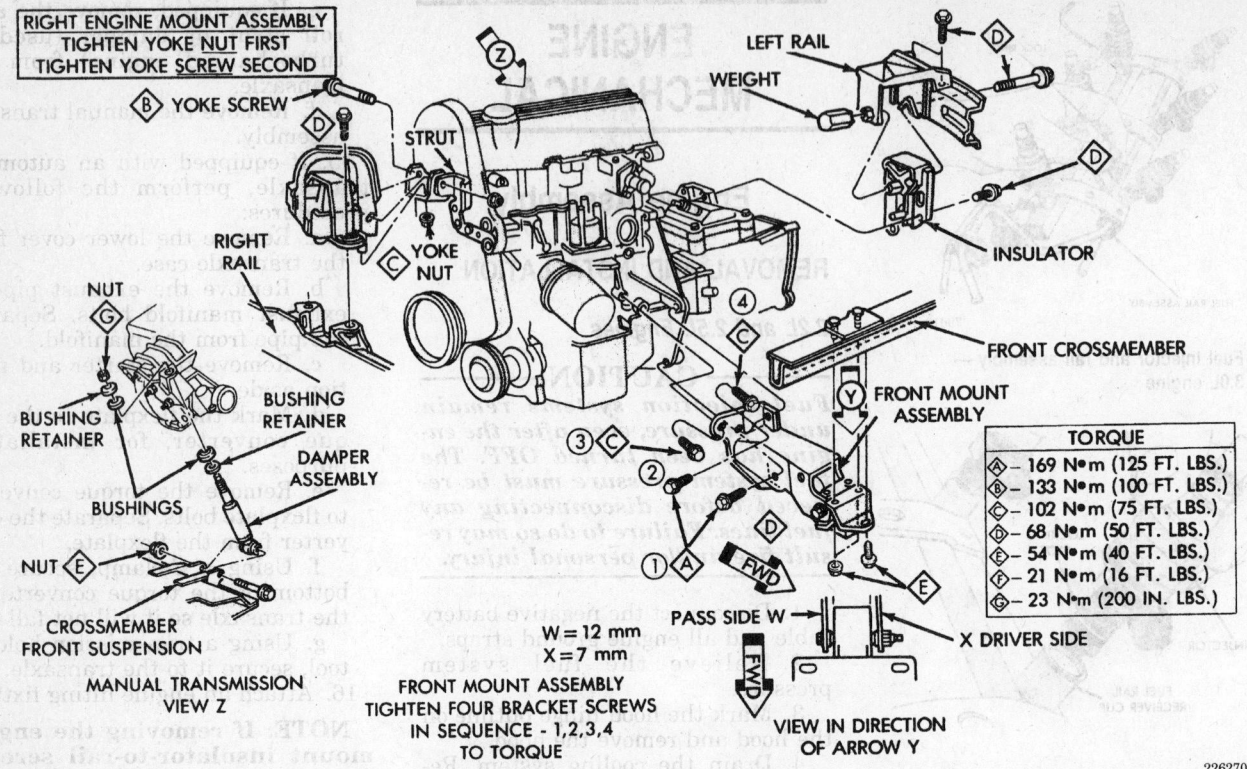

RIGHT ENGINE MOUNT ASSEMBLY
TIGHTEN YOKE **NUT** FIRST
TIGHTEN YOKE **SCREW** SECOND

B YOKE SCREW

STRUT

RIGHT RAIL

YOKE NUT

NUT **G**

BUSHING RETAINER

BUSHING RETAINER

DAMPER ASSEMBLY

BUSHINGS

NUT **E**

CROSSMEMBER FRONT SUSPENSION

MANUAL TRANSMISSION VIEW Z

LEFT RAIL

WEIGHT

INSULATOR

FRONT CROSSMEMBER

FRONT MOUNT ASSEMBLY

TORQUE	
A –	169 N•m (125 FT. LBS.)
B –	133 N•m (100 FT. LBS.)
C –	102 N•m (75 FT. LBS.)
D –	68 N•m (50 FT. LBS.)
E –	54 N•m (40 FT. LBS.)
F –	21 N•m (16 FT. LBS.)
G –	23 N•m (200 IN. LBS.)

W = 12 mm
X = 7 mm
FRONT MOUNT ASSEMBLY
TIGHTEN FOUR BRACKET SCREWS
IN SEQUENCE - 1,2,3,4
TO TORQUE

PASS SIDE W

X DRIVER SIDE

VIEW IN DIRECTION OF ARROW Y

FWD

226270

Engine mounting — 2.2L and 2.5L engines

28. Connect the throttle linkage.

29. Connect all remaining electrical connectors.

30. Install the air cleaner assembly and oil filter.

31. Install the radiator, fan assembly, hoses and intercooler, if equipped.

32. Fill the engine with the proper amount of engine oil and install a new oil filter. Connect the negative battery cable.

33. Refill the cooling system. With the radiator cap off so coolant can be added to the radiator, start the engine, allow it to reach normal operating temperature. Check for leaks and unusual noises.

34. Check the ignition timing and adjust if necessary.

35. Shut the engine down and allow to cool. Check all fluid levels.

36. Install the hood aligning the matchmarks made during the removal procedure.

3.0L Engine

⊳ CAUTION

Fuel injection systems remain under pressure, even after the engine has been turned OFF. The fuel system pressure must be relieved before disconnecting any fuel lines. Failure to do so may result in fire and/or personal injury.

1. Disconnect the negative battery cable. Relieve the fuel pressure.

2. Matchmark the hinge-to-hood position and remove the hood.

3. Drain the cooling system. Disconnect and label all engine electrical connections.

4. Remove the coolant hoses from the radiator and engine. Remove the radiator and cooling fan assembly.

5. Remove the air cleaner assembly. Disconnect the fuel lines from the engine. Disconnect the accelerator cable from the throttle body.

6. Raise the vehicle and support safely. Drain the engine oil.

7. Remove the engine drive belts.

8. Remove the air conditioning compressor mounting bolts, the drive belts and position the compressor aside. Disconnect the exhaust pipe from the exhaust manifold.

9. Remove the transaxle inspection cover, matchmark the converter to the flexplate, and remove the torque converter bolts.

10. Remove the power steering pump mounting bolts and set the pump aside, upright, with the fluid lines attached.

11. Remove the lower bellhousing bolts. Disconnect and label the starter motor wiring and remove the starter motor from the engine.

12. Lower the vehicle. Disconnect and label all electrical connectors from the engine, alternator and fuel injection system, vacuum hoses, and engine ground straps.

13. Support the transaxle with a floor jack or equivalent. Attach an engine lifting device to the engine.

14. Remove the upper transaxle-to-engine bolts.

15. To separate the engine mounts from the insulators, perform the following procedure:

   a. Mark the right insulator-to-right frame support and remove the mounting bolts.

   b. Remove the front engine mount through bolt.

   c. Remove the left insulator through bolt from inside the wheel housing.

   d. Remove the insulator bracket-to-transaxle bolts.

16. Lift and remove the engine from the vehicle.

   To install:

17. Lower the engine into the engine compartment. Align the engine mounts and install the bolts; do not tighten the bolts until all bolts have

TORQUE	
Ⓐ — 54 N•m (40 FT. LBS.)	
Ⓑ — 40 N•m (30 FT. LBS.)	
Ⓒ — 28 N•m (250 IN. LBS.)	

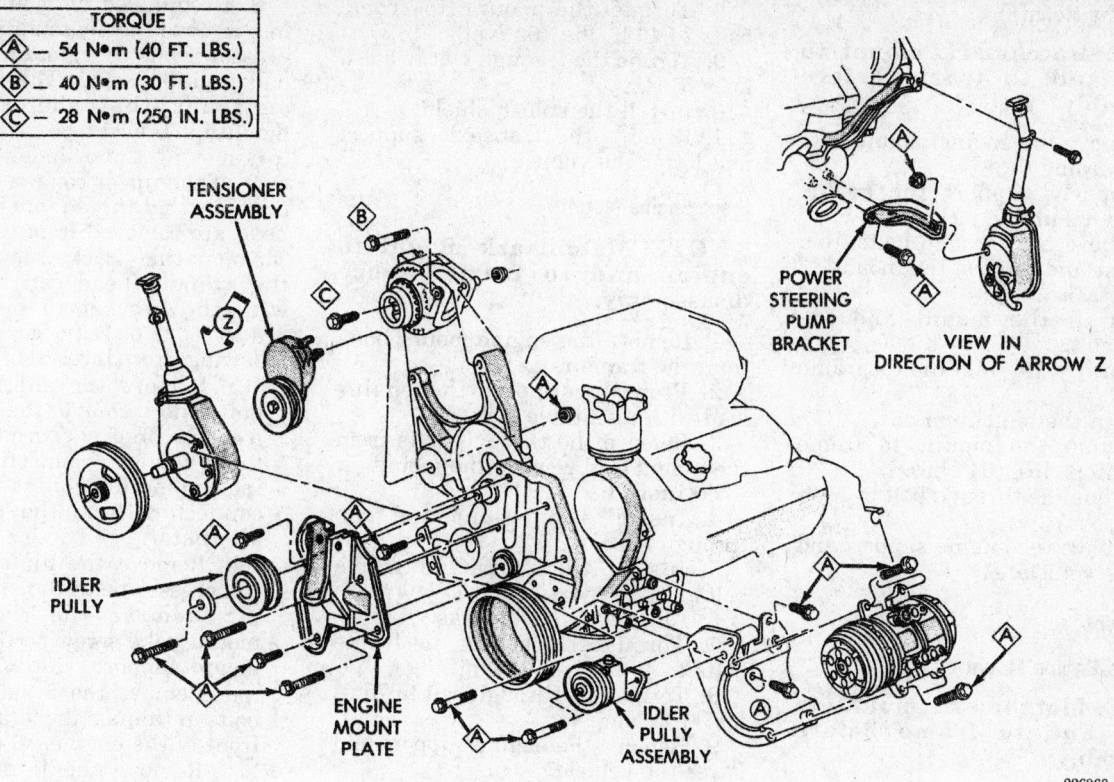

Accessories mounting brackets — 3.0L engine

been installed. Torque the through bolts to 75 ft. lbs. (102 Nm).

18. Install the upper transaxle-to-engine mounting bolts and torque to 75 ft. lbs. (102 Nm). Remove the engine lifting fixture from the engine.

19. Raise the vehicle and support safely.

20. Align the converter marks and install the torque converter bolts. Tighten the converter bolts to 55 ft. lbs. (75 Nm) and install the transaxle inspection cover.

21. Connect the exhaust pipe to the exhaust manifold. Install the starter motor and connect the wiring.

22. Install the power steering pump and air conditioning compressor. Adjust the drive belt tension, if necessary.

23. Lower the vehicle. Reconnect all vacuum hoses and electrical connections to the engine.

24. Connect the fuel lines and accelerator cable.

25. Install the radiator and fan assembly. Connect the fan motor wiring. Connect the radiator hoses and refill the cooling system.

26. Refill the engine with the proper oil to the correct level. A new filter is recommended.

27. Connect the engine ground straps. Install the hood aligning the

matchmarks made during removal. Connect the negative battery cable.

28. With the radiator cap off so coolant can be added, start and run the engine until normal operating temperature is reached. Check for fluid leaks and abnormal noises. Adjust the transaxle linkage, if necessary.

29. Shut the engine down, allow to cool and check all fluid levels.

Engine Mounts

REMOVAL AND INSTALLATION

2.2L and 2.5L Engines

Please note that these engines were used in many vehicle models for a number of years. There may be some variation in procedure due to differences in model years and engine/vehicle model combinations.

Right Side Engine Mount

1. Disconnect the negative battery cable.

2. Matchmark the engine mount to its frame mounting location.

3. Remove the right engine mount insulator vertical fasteners from the frame rail.

4. Remove the load on the engine motor mounts by carefully supporting the engine and transaxle assembly with a floor jack.

5. Remove the through-bolt from the insulator assembly and remove the insulator.

6. Installation is the reverse of the removal procedure. Make sure the matchmarks are aligned before tightening bolts.

7. Tighten the lower yoke nut first, then the through-bolt nut, then the body mounting bolts.

Left Side Engine Mount

1. Disconnect the negative battery cable.

2. Matchmark the engine mount to its frame mounting location.

3. Remove the load on the engine motor mounts by carefully supporting the engine and transaxle assembly with a floor jack.

4. Remove the bolt from the insulator and front crossmember bracket.

5. Remove the front engine mount bracket to front crossmember screws and nuts. Remove the insulator assembly.

6. Installation is the reverse of the removal procedure.

Front Engine Mount

NOTE: Matchmark mount to engine and to frame before disassembly.

1. Remove the engine mount bolts from the frame rail.
2. Properly support the engine load with a suitable jack.
3. Remove the through bolts from the mount and remove the mount.

To install:

4. Install the mount and the through bolt.
5. Install the mount to frame bolts.
6. Align the matchmarks.
7. Torque the mount to frame bolts to 40 ft. lbs. (54 Nm).
8. Torque the through bolt to 75 ft. lbs. (102 Nm).
9. Remove the engine support and lower the vehicle.

3.0L Engine

Right Side Engine Mount

NOTE: Matchmark mount to engine and to frame before disassembly.

1. Remove the engine mount bolts from the frame rail.
2. Properly support the engine load with a suitable jack.
3. Remove the through bolts from the mount and remove the mount.

To install:

4. Install the mount and the through bolt.
5. Install the mount to frame bolts.
6. Align the matchmarks.
7. Torque the mount to frame bolts to 50 ft. lbs. (68 Nm).
8. Torque the through bolt to 100 ft. lbs. (133 Nm).
9. Remove the engine support and lower the vehicle.

Left Side Engine Mount

NOTE: Matchmark mount to engine and to frame before disassembly.

1. Remove the splash shield.
2. Remove the engine mount bolts from the frame rail.
3. Properly support the transaxle load with a suitable jack.
4. Remove the through bolts from the mount and remove the mount.

To install:

5. Install the mount and the through bolt.
6. Install the mount to frame bolts.
7. Align the matchmarks.

8. Torque the mount to frame bolts to 40 ft. lbs. (54 Nm).
9. Torque the through bolt to 55 ft. lbs. (75 Nm).
10. Install the splash shield.
11. Remove the transaxle support and lower the vehicle.

Front Engine Mount

NOTE: Matchmark mount to engine and to frame before disassembly.

1. Remove the engine mount bolts from the frame rail.
2. Properly support the engine load with a suitable jack.
3. Remove the through bolts from the mount and remove the mount.

To install:

4. Install the mount and the through bolt.
5. Install the mount to frame bolts.
6. Align the matchmarks.
7. Torque the mount to frame bolts to 40 ft. lbs. (54 Nm).
8. Torque the through bolt to 75 ft. lbs. (102 Nm).
9. Remove the engine support and lower the vehicle.

Cylinder Head

REMOVAL AND INSTALLATION

2.2L and 2.5L SOHC Engines

1. Disconnect the negative battery cable from the battery and cylinder head. Relieve the fuel pressure. Drain the cooling system.
2. Rotate the engine by hand and position the engine at TDC of its compression stroke.
3. Remove the dipstick bracket nut from the thermostat housing and remove the ignition coil from the thermostat housing if installed there.
4. Remove the air cleaner assembly. Remove the upper radiator hose and disconnect the heater hoses.
5. Disconnect and label the vacuum lines, hoses and wiring connectors from the manifold(s), throttle body and from the cylinder head.
6. Disconnect all linkages and the fuel line from the throttle body. Unbolt the cable bracket. Remove the ground strap attaching screw from the firewall.
7. Remove the power steering pump assembly and position aside. It is not necessary to disconnect power steering pump supply and return hoses.

8. If equipped with air conditioning, remove the air conditioning compressor from the mounting bracket and position aside. The factory recommends that the compressor mounting bracket be removed prior to removing the cylinder head, however, if the upper compressor mounting bolts that thread into the cylinder head are removed from the compressor mounting bracket, in most cases, the cylinder head can be removed with the bracket in place. If the bracket is to be removed, perform the following procedure:

   a. Remove the alternator pivot bolt and remove the alternator from the bracket. Turn the alternator so the wire connections are facing up and disconnect the harness connectors from the rear of the alternator.

   b. Remove the air conditioning compressor belt idler.

   c. Remove the right engine mount yoke screw securing engine mount support strut to the engine.

   d. Remove the 5 side mounting bolts retaining the bracket to the front of the engine.

   e. Remove the front mounting nut. Remove the front bolt and strut and rotate the solid mount bracket away from the engine. Slide the bracket on the stud until free of the mounting studs and remove from the engine.

9. Remove the upper timing belt cover. Raise the vehicle and support safely. Disconnect the exhaust pipe from the exhaust manifold. Disconnect the water hose and oil drain from the turbocharger, if equipped.
10. If the engine is not positioned at TDC, rotate the engine by hand until the timing marks align. The No. 1 piston should be at TDC of its compression stroke. Lower the vehicle.
11. With the timing marks aligned, remove the camshaft sprocket. The camshaft sprocket can be suspended to keep the timing intact. Remove the spark plug wires from the spark plugs.
12. Remove the valve cover and curtain. Remove the cylinder head bolts and washers, starting from the outside and working inward.
13. Remove the cylinder head from the engine.
14. Clean the cylinder head gasket mating surfaces. Clean and inspect all cylinder head bolt threads for necking or stretching. It is recommended that all head bolts be replaced with new prior to installation of the cylinder head.

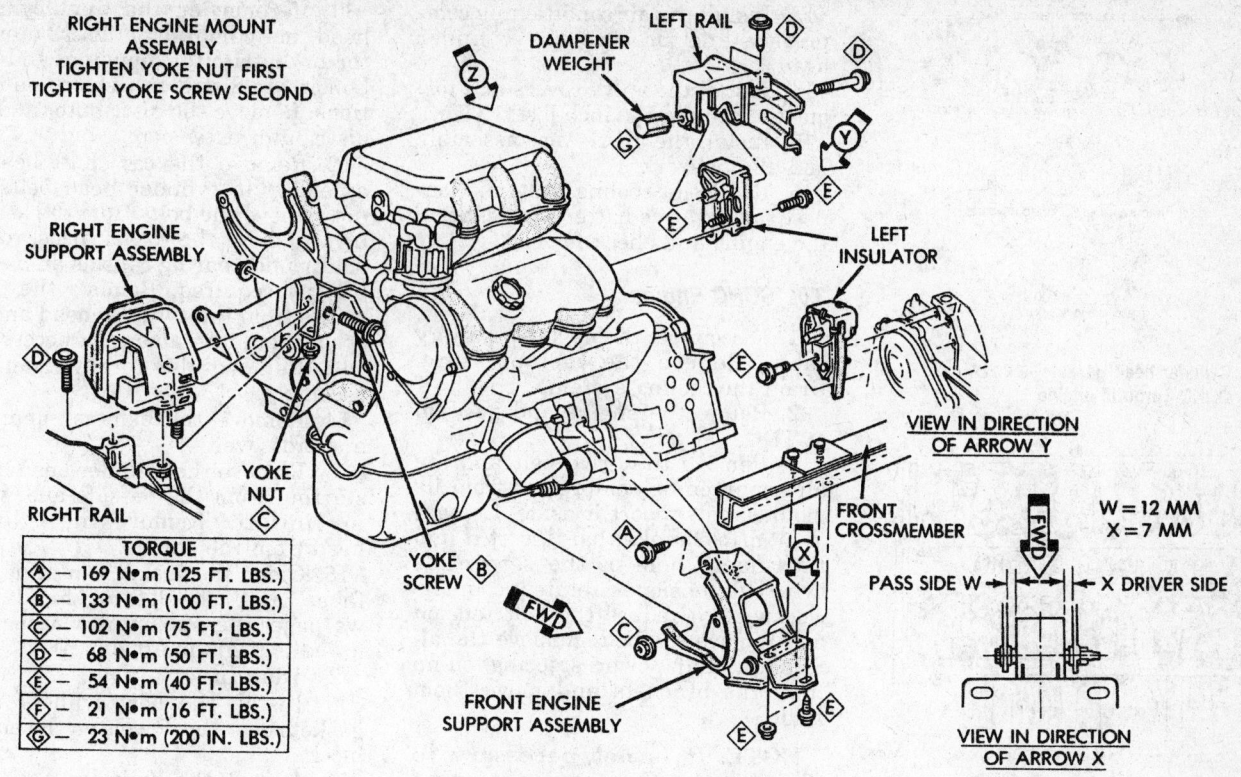

RIGHT ENGINE MOUNT
ASSEMBLY
TIGHTEN YOKE NUT FIRST
TIGHTEN YOKE SCREW SECOND

DAMPENER
WEIGHT

LEFT RAIL

LEFT
INSULATOR

RIGHT ENGINE
SUPPORT ASSEMBLY

RIGHT RAIL

YOKE
NUT

YOKE
SCREW

TORQUE	
A —	169 N•m (125 FT. LBS.)
B —	133 N•m (100 FT. LBS.)
C —	102 N•m (75 FT. LBS.)
D —	68 N•m (50 FT. LBS.)
E —	54 N•m (40 FT. LBS.)
F —	21 N•m (16 FT. LBS.)
G —	23 N•m (200 IN. LBS.)

FRONT
CROSSMEMBER

VIEW IN DIRECTION
OF ARROW Y

W = 12 MM
X = 7 MM

PASS SIDE W   X DRIVER SIDE

VIEW IN DIRECTION
OF ARROW X

FRONT ENGINE
SUPPORT ASSEMBLY

226304

Engine mounting — 3.0L engine

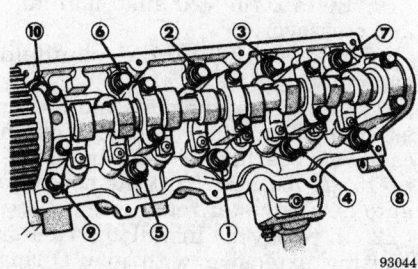

93044

Cylinder head torque sequence — 2.2L and 2.5L
SOHC engines

To install:

NOTE: Head bolt diameter is 11mm. These bolts are identified with the No. 11 on the head of the bolt. The 10mm bolts used on previous vehicles will thread into an 11mm bolt hole, but will permanently damage the cylinder block. Make sure the correct bolts are used when replacing head bolts.

15. Using new gaskets and seals, install the head to the engine.

NOTE: When torquing cylinder head bolts, tighten bolts in 3 progressive steps.

16. Using new head bolts of the correct diameter assembled with the old washers, torque the cylinder head bolts in sequence to 65 ft. lbs. (88 Nm).

17. With the cylinder head bolts at required specification turn each bolt an additional ¼ turn.

18. Install the timing belt and covers. Install the solid mount compressor bracket, if removed.

19. Install the upper air conditioning compressor mounting bracket bolts, if removed. Install the air conditioning compressor to the mounting bracket and secure with the mounting nuts.

20. Raise and safely support the vehicle. Reconnect the exhaust pipe to the manifold using a new gasket as required.

21. Connect the remaining hoses, linkage and electrical harness connectors disconnected during the removal procedure.

22. Refill the cooling system. Connect the negative battery cable. Start the engine and check for leaks. Adjust the timing, as required.

2.2L DOHC Turbo III Engine

1. Disconnect the negative battery cable at the battery terminal and un-

bolt it from the head. Relieve the fuel pressure. Drain the cooling system.

2. Remove the air cleaner assembly with all ductwork.

3. Remove the timing belt covers. Rotate the engine by hand until the timing marks align (No. 1 piston at TDC). Remove the timing belt.

4. Remove the air conditioning compressor and bracket from the cylinder head. It is not necessary to disconnect A/C hoses.

5. Disconnect the turbocharger coolant lines and separate the intake and exhaust manifolds from the cylinder head.

6. Remove the ignition cable cover and valve covers. Disconnect and label all wiring connectors, hoses and ignition wires from the cylinder head.

7. Remove the cylinder head and gasket from the engine.

8. Clean the cylinder head gasket mating surfaces. Clean and inspect all cylinder head bolt threads for necking. If necking has occurred, the threads on the bolts will not be uniform or straight when held up against a straight-edge, and will require replacement. It is recommended that all head bolts be replaced with new prior to installation of the cylinder head.

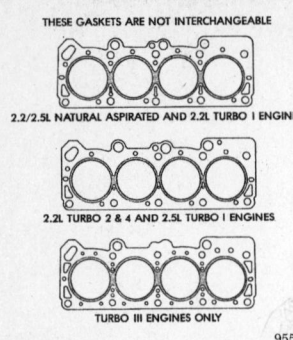

THESE GASKETS ARE NOT INTERCHANGEABLE

2.2/2.5L NATURAL ASPIRATED AND 2.2L TURBO I ENGINES

2.2L TURBO 2 & 4 AND 2.5L TURBO I ENGINES

TURBO III ENGINES ONLY

95501

Cylinder head gaskets — 2.2L DOHC Turbo III engine

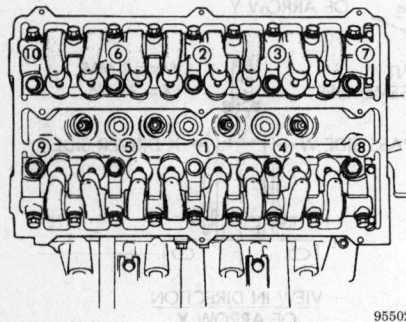

95502

Cylinder head bolt installation — 2.2L DOHC Turbo III engine

To install:

NOTE: The head gasket used on the Turbo III engine is unique to that engine. Make sure the replacement head gasket is identical to the original gasket before installing. Head bolt diameter is 11mm and the head bolts are unique to this engine. These bolts are identified with the No. 11 on the head of the bolt and are not interchangeable with other engines. Make sure the correct bolts are used when replacing head bolts.

9. Using new gaskets and seals, install the head to the engine. Using new head bolts assembled with the old washers, torque the cylinder head bolts in sequence, to 45 ft. lbs. (61 Nm). Repeating the sequence, torque the bolts to 65 ft. lbs. (88 Nm). With the bolts at 65 ft. lbs., turn each bolt an additional ¼ turn. Final torque must be over 90 ft. lbs. (122 Nm).

10. Install the timing belt and rotate the engine by hand 1 revolution until the timing marks align (No. 1 piston at TDC). Check alignment of timing marks.

11. Install the timing belt covers.

12. Install the intake and exhaust manifolds.

13. Install the air conditioning compressor and bracket the cylinder head.

14. Install the valve covers and torque the bolts to 105 inch lbs. (12 Nm).

15. Install the air cleaner assembly and all ductwork.

16. Refill the cooling system. Connect the negative battery cable. Start the engine and check for leaks.

3.0L SOHC Engine

1. Disconnect the negative battery cable. Relieve the fuel pressure. Drain the cooling system.

2. Rotate the engine and position at TDC.

3. Remove the drive belt and the air conditioning compressor from its mount and support it aside. Using a ½ in. drive breaker bar, insert it into the square hole of the serpentine drive belt tensioner, rotate it counterclockwise to reduce the belt tension and remove the belt. Remove the alternator and power steering pump from the brackets and move them aside.

NOTE: It is not necessary to disconnect A/C or power steering hoses.

4. Raise the vehicle and support safely. Remove the right front wheel assembly and the right inner splash shield.

5. Remove the crankshaft pulleys and the torsional damper.

6. Lower the vehicle. Using a floor jack and a block of wood positioned under the oil pan, raise the engine slightly. Remove the engine mount bracket from the timing cover end of the engine and the timing belt covers.

7. To remove the timing belt, perform the following procedures:

   a. Rotate the crankshaft to position the No. 1 cylinder on the TDC of its compression stroke if not already done. The crankshaft sprocket timing mark should align with the oil pan timing indicator and the camshaft sprocket's timing marks (triangles) should align with the rear timing belt cover's timing marks.

   b. Mark the timing belt in the direction of rotation for reinstallation purposes.

   c. Loosen the timing belt tensioner and remove the timing belt.

8. Remove the air cleaner assembly. Label and disconnect the spark plug wires and the vacuum hoses.

9. Remove the valve covers

10. Install auto lash adjuster retainer tools MD998443 or equivalent, on the rocker arms.

11. If removing the front cylinder head, matchmark the distributor rotor-to-distributor housing and the housing-to-distributor extension locations. Remove the distributor and the distributor extension.

12. Remove the camshaft bearing assembly to cylinder head bolts (do not remove the bolts from the assembly). Remove the rocker arms, rocker shafts and bearing caps as an assembly, as required. Remove the camshafts from the cylinder head and inspect them for damage, if necessary.

13. Remove the intake manifold assembly.

14. Remove the exhaust manifold and crossover.

15. Remove the cylinder head bolts, starting from the outside and working inward. Remove the cylinder head from the engine.

16. Clean the gasket mounting surfaces and check the heads for warpage; the maximum warpage allowed is 0.008 in. (0.20mm).

To install:

17. Install the new cylinder head gaskets over the dowels on the engine block.

18. Install the cylinder heads on the engine and torque the cylinder head bolts in sequence using 3 even steps to 80 ft. lbs. (108 Nm).

19. Install the exhaust manifold and crossover.

20. Install the intake manifold assembly.

21. Install the camshaft and rocker assembly to cylinder head. Tighten camshaft bearing cap bolts in 3 steps to 15 ft. lbs. (20 Nm). Remove the auto lash adjuster retainers.

22. If removed, install distributor housing extension with new O-ring and tighten bolts to 130 inch lbs. (15 Nm).

23. Using new gaskets, install the valve covers and tighten mounting bolts to 88 inch lbs. (10 Nm).

24. With engine at TDC and camshaft sprockets properly aligned, install the timing belt assembly.

25. When installing the timing belt over the camshaft sprocket, use care not to allow the belt to slip off the opposite camshaft sprocket.

26. Make sure the timing belt is installed on the camshaft sprocket in the same direction as when removed.

27. Install the engine mount bracket to the front of engine and install the timing belt covers.

28. Install the crankshaft pulleys and the torsional damper.

29. Install the right front wheel assembly and the right inner splash shield.

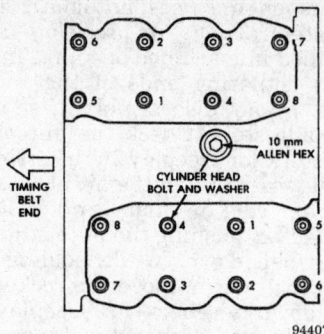

Cylinder head bolt torque
sequence — 3.0L SOHC engine

30. Install the alternator and power steering pump to the mounting brackets.

31. Using a ½ in. drive breaker bar, insert it into the square hole of the serpentine drive belt tensioner, rotate it counterclockwise to reduce the belt tension and install the belt.

32. Install the air conditioning compressor to its mount and install drive belt.

33. Refill the cooling system. Connect the negative battery cable. Start the engine and check for leaks. Adjust the ignition timing as required.

Lash Adjusters

BLEEDING

2.2L and 2.5L Engines

The hydraulic lash adjusters do not need to be bled. If the lash adjuster is not functioning properly it must be replaced. When installing a new lash adjuster soak the adjuster in oil until the adjuster has little or no plunger travel when depressed.

Valve Lash

ADJUSTMENT

The automatic lash adjuster controls the valve clearance. There is no manual adjustment. In the event of noise in the valve train check for worn parts or low oil pressure.

Rocker Arms

REMOVAL AND INSTALLATION

2.2L and 2.5L Engines

These engines use an overhead camshaft where the lifter lobes bear against a rocker arm to actuate the

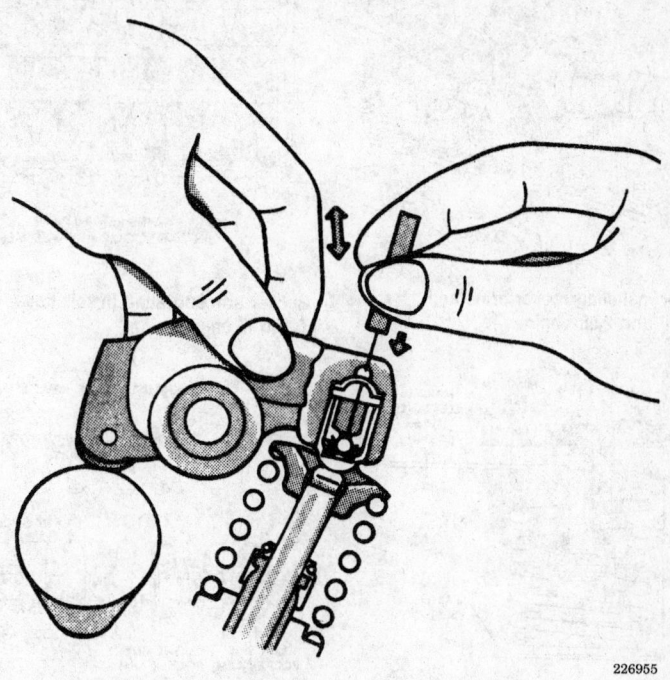

Checking the auto-lash adjuster — 3.0L engine

valves. A roller is provided on the rocker arm to reduce camshaft lobe wear. In this design engine, the rocker arm is sometimes called a cam follower. A lash adjuster (sometimes also called a tappet or lifter) set into the cylinder head provides automatic lash or clearance adjustment.

1. Disconnect the negative battery cable.

2. Remove the cylinder head cover and curtain (oil baffle).

3. Rotate the crankshaft until the low point (called the base circle) of the desired cam lobe is contacting the rocker arm.

4. Using tool C-4682A valve spring compressor tool or equivalent, depress the valve spring without dislodging the keepers and slide the rocker arm out. Work carefully so the valve spring keepers are not loosened. Do not push the valve spring down any further than necessary to slide out the rocker arm.

NOTE: If any valve train components are to be reused, keep them in order. They must be returned to their original locations.

5. If necessary, remove the valve lash adjuster(s) from the cylinder head.

To install:

6. If removed, lubricate and install the lash adjuster(s). Make sure the adjusters are at least partially full of oil. This is indicated by little or no plunger travel when the lash adjuster is depressed.

7. Rotate the camshaft until the base circle is in a position where it would contact the rocker arm. Coat the rocker arm with clean engine oil or oil supplement and, while pushing the valve spring and retainer down, slip the rocker arm into place under the camshaft lobe's base circle. Work carefully so the valve spring keepers are not loosened. Do not push the valve spring down any further than necessary to slide in the rocker arm. It is possible for the valve spring retainer locks to become dislodged which means when the engine is started, the valve could drop down, seriously damaging the engine.

NOTE: Always check that the valve locks are in their proper positions. Repeat this step for each rocker arm to be installed.

8. Install the valve curtain (oil baffle) and cylinder head cover assembly. Tighten the cylinder head cover bolts to 105 inch lbs. (12 Nm).

9. An oil and filter change is recommended. If new parts were in-

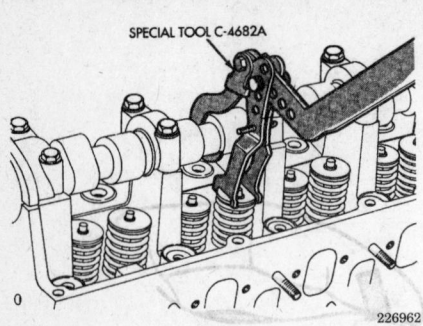

SPECIAL TOOL C-4682A

226962

Removing or installing rocker arms and lifters — 2.2L and 2.5L engines

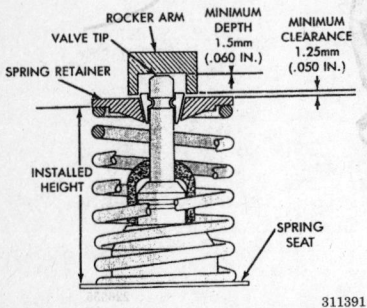

VALVE TIP
ROCKER ARM
SPRING RETAINER
MINIMUM DEPTH 1.5mm (.060 IN.)
MINIMUM CLEARANCE 1.25mm (.050 IN.)
INSTALLED HEIGHT
SPRING SEAT
311391

Checking valve spring installed height and spring retainer clearance — 2.2L and 2.5L engines

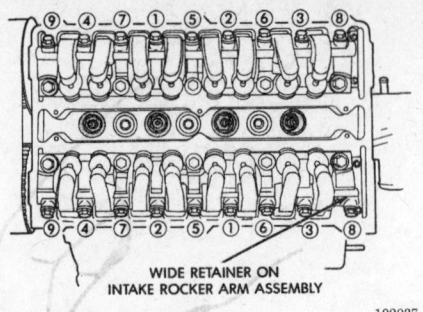

WIDE RETAINER ON INTAKE ROCKER ARM ASSEMBLY

102037

Rocker arm and shaft installation — 2.2L DOHC Turbo III engine

EXHAUST ROCKER ARM ASSEMBLY

TIMING BELT

WIDE RETAINER WITH SPACER RING

LEFT SIDE ROCKER ARM   RIGHT SIDE ROCKER ARM   INTAKE ROCKER ARM ASSEMBLY

102038

Intake and exhaust rocker assemblies — 2.2L DOHC Turbo III engine

stalled, a can of engine oil supplement may aid parts break-in.

10. Connect the negative battery cable.

11. Start the engine and allow to idle. Any valve "tappet noise" should go away as the oil pressure builds and oil reaches the engine overhead components.

Rocker Arms and Shafts

REMOVAL AND INSTALLATION

2.2L DOHC Turbo III Engine

1. Disconnect the negative battery cable.

2. Remove the valve cover(s).

3. Remove the rocker arm retaining bolts in the proper removal sequence.

4. Remove the rocker shaft assembly from the cylinder head.

5. Keep all parts in order and disassemble as required. Inspect the lash adjusters carefully.

To install:

6. Lubricate and assemble the rocker arms to the shaft.

7. Make sure the lash adjusters are at least partially full of oil. This is indicated by little or no plunger travel when depressing. Install to the rocker arms.

8. Install the assembly and tighten the bolts in the proper sequence to 18 ft. lbs. (24 Nm).

9. Install the valve cover(s).

10. Connect the negative battery cable.

3.0L Engine

The 3.0L engine uses a single overhead camshaft on each bank of cylinders. The rocker arm shaft retainers are also the upper camshaft bearing caps. When removing the camshaft bearing caps do not remove the bolts from the bearing caps. Remove the rocker arm, rocker shafts **and** the bearing caps as an assembly. This means the camshaft(s) will be loose when the rocker arm shaft retainers are loosened, which, in turn, means the timing belt must be removed and, at assembly, the engine will have to be re-timed. Work carefully on overhead camshaft engines. If the timing is off even a small amount, the engine can be seriously damaged.

This engine uses automatic lash adjusters. They are precision units installed in machined openings in the valve actuating ends of the rocker arms. Do not disassemble these auto lash adjusters. Check the auto lash adjusters for freeplay by inserting a small wire through the air bleed hole in the rocker arm and **VERY LIGHTLY** pushing the auto adjuster check ball down. While holding the check ball down, move the rocker up and down to check for freeplay. If there is no freeplay, replace the adjuster.

1. Disconnect the negative battery cable.

2. Remove the valve cover. Install lash adjuster retainer tools MD998443 or equivalent. These are small clips designed to prevent the auto-lash adjusters from falling out of the rocker arms when the rocker arms and shaft assembly is removed.

3. Rotate the engine clockwise and position at TDC, No. 1 cylinder, compression stroke (firing position). This should align all timing marks and be a point of reference for assembly.

4. Remove the distributor adapter housing.

5. Remove the timing belt assembly.

6. Loosen rocker arm and shaft assembly evenly in several steps. Remove the rocker arm and shaft assembly as a complete unit. Watch that the lash adjusters do not drop out of the rocker arms. There is 1 adjuster for each rocker arm.

To install:

7. Immerse the lash adjusters in clean diesel fuel. Using a small wire, move the plunger of the lash adjuster up and down 4 or 5 times while pushing down lightly on the check ball in order to bleed out the air. Install the lash adjusters in the rocker arms.

8. If the shaft and rockers were disassembled, clean all parts well and assemble. The springs are the same and can be used at all locations on the rocker arm shafts. Insert bolts in the number 4 bearing cap to retain the assembly.

9. Make sure the rocker arm lash adjusters are properly installed and retainer so they won't fall out.

10. Install the rocker arm and shaft assembly into the front bearing cap (No. 1) making sure the notches in the rocker shafts are facing up. Insert the hold-down bolt but do not tighten at this point.

11. Apply RTV sealer at the bearing cap ends. Lubricate the camshaft and rocker shaft with heavy engine oil or

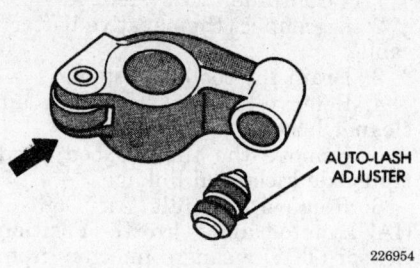

Rocker arm and lash adjuster — 3.0L engine

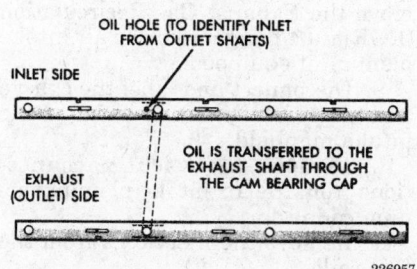

Rocker shaft identification — 3.0L engine

a quality engine oil supplement and position on the cylinder head.

NOTE: Make sure the arrow marks on the front and rear assemblies are opposite each other. The intake rocker shaft has an extra oil hole at the bottom side of shaft for identification

12. Install the remaining bearing cap bolts and tighten evenly and gradually to 15 ft. lbs. (20 Nm). Tighten the front cap retaining bolts to 17 ft. lbs. (24 Nm). Remove the lash adjuster retainers.

13. Install the distributor extension, if removed.

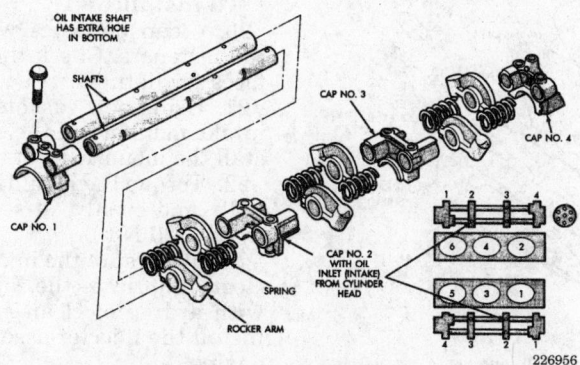

Rocker arm and shaft assembly — 3.0L engine

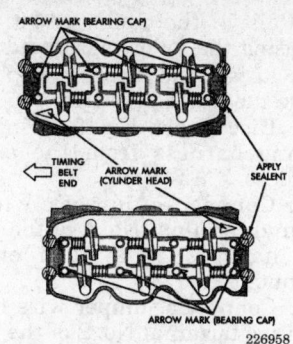

Bearing cap positioning — 3.0L engine

14. Install timing belt using the recommended procedure. Verify that all timing marks are properly aligned. This step is very important.

15. Install the valve cover with a new gasket and tighten to 5 ft. lbs. (7 Nm).

16. Connect the negative battery cable.

17. Start the engine and verify no valve noise or oil leaks.

Intake Manifold

REMOVAL AND INSTALLATION

2.2L DOHC Turbo III Engine

1. Disconnect the negative battery cable. Relieve the fuel system pressure. Drain the cooling system.

2. Remove the fresh air duct from the air filter housing. Remove the inlet hose from the intercooler.

3. Remove the radiator hose from the thermostat housing.

4. Remove the DIS ignition coil from the intake manifold.

5. Disconnect the throttle and speed control cables from the throttle body.

6. Disconnect the intercooler-to-throttle body outlet hose. Disconnect the vacuum hoses from the throttle

body and carefully remove the harness.

7. Disconnect the Idle Air Control (IAC) motor and Throttle Position Sensor (TPS) wiring connectors.

8. Remove the PCV breather/separator box and vacuum harness assembly. Remove the brake booster hose, vacuum vapor harness and fuel pressure regulator from the intake manifold.

9. Disconnect the fuel injector wiring harness and charge temperature sensor.

10. Wrap shop towels around the fittings and disconnect the fuel supply and return fuel lines.

11. Remove the intake manifold retaining bolts and remove the manifold from the cylinder head.

To install:

12. Inspect the manifold for damage of any kind. Thoroughly clean and dry the mating surfaces.

13. Install the new gasket and manifold to the cylinder head. Starting at the center and working outwards, torque the bolts gradually and evenly to 17 ft. lbs. (23 Nm).

14. Lubricate the quick-connect fuel fittings with oil and connect to the chassis tubes. Ensure they are locked by pulling on them.

15. Install the PCV breather/separator box and vacuum harness assembly. Connect the brake booster hose, vacuum vapor harness and fuel pressure regulator to the intake manifold.

16. Connect the fuel injector wiring harness and charge temperature sensor. Connect the AIS motor and TPS wiring connectors.

17. Connect the vacuum hoses from the throttle body and carefully remove the harness. Connect the intercooler-to-throttle body outlet hose.

18. Connect the throttle and speed control cables from the throttle body.

19. Install the DIS ignition coil to the intake manifold.

20. Connect the radiator hose to the thermostat housing.

21. Install the inlet hose to the intercooler. Install the fresh air duct to the air filter housing.

22. Refill and bleed the cooling system. Connect the negative battery cable. Start the engine and check for leaks.

3.0L Engine

The 3.0L engine intake manifold is a 2 piece die-cast aluminum component with an upper part called a plenum and a lower manifold which bolts to the cylinder heads. The lower manifold is also machined for the fuel rail

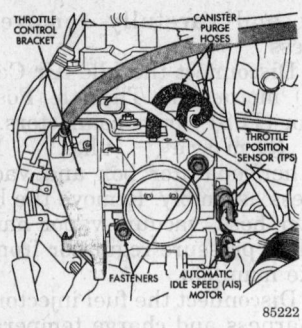

Idle air control and throttle position sensor — 2.2L DOHC Turbo III engine

attachment and injector installation. The throttle body is installed on the upper manifold (plenum). Use care when working with these light alloy parts.

— CAUTION —

Fuel injection systems remain under pressure, even after the engine has been turned OFF. The fuel system pressure must be relieved before disconnecting any fuel lines. Failure to do so may result in fire and/or personal injury.

1. Relieve the fuel system pressure using the following procedure.

a. Loosen the fuel filler cap to release fuel tank pressure.

b. Disconnect the fuel injector wiring harness from the engine harness.

c. Connect a jumper wire to ground terminal No. 1 of the injector harness to a good engine ground.

d. Connect a jumper wire to the positive terminal No. 2 of the injector harness and touch the battery positive post for no longer than 5 seconds. This releases system pressure.

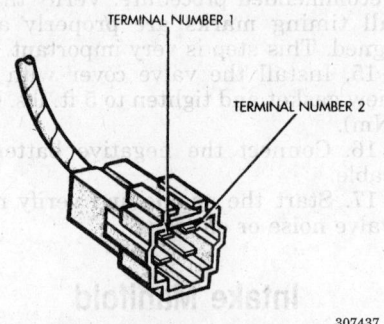

Use these terminals for fuel pressure relief — 3.0L engine

e. Remove the jumper wires.

f. Continue fuel system service.

2. Disconnect the negative battery cable.

3. Drain the cooling system.

4. Remove the throttle body to air cleaner hose.

5. Remove the throttle body and transaxle kickdown linkage.

6. Remove the Idle Air Control (IAC) motor and Throttle Position Sensor (TPS) wiring connectors from the throttle body.

7. Remove and label the vacuum hose harness from the throttle body.

8. Remove the Positive Crankcase Ventilation (PCV) and brake booster hoses from the intake plenum. Remove the Exhaust Gas Recirculation (EGR) tube flange from the air intake plenum, if equipped.

9. Disconnect and label the charge and temperature sensor wiring at the intake manifold.

10. Remove the vacuum connections from the air intake plenum vacuum connector.

11. Remove the fuel hoses from the fuel rail.

12. Remove the air intake plenum mounting bolts and remove the plenum.

13. Remove the vacuum hoses from the fuel rail and pressure regulator.

14. Disconnect the fuel injector wiring harness from the engine wiring harness.

15. Remove the fuel pressure regulator mounting bolts and remove the regulator from the fuel rail.

16. Remove the fuel rail mounting bolts and remove the fuel rail from the intake manifold.

17. Separate the radiator hose from the thermostat housing and heater hoses from the heater pipe.

18. Remove the intake manifold mounting bolts and remove the manifold from the engine.

19. Clean the gasket mounting surfaces on the engine and intake manifold.

To install:

20. Clean all parts well especially the fasteners. Check the sealing surfaces for flatness.

21. Using new gaskets, position the intake manifold on the engine and install the mounting nuts and washers.

22. Torque the mounting nuts gradually and evenly, in sequence, to 15 ft. lbs. (20 Nm).

23. Make sure the injector holes are clean. Lubricate the injector O-rings with a drop of clean engine oil and install the injector assembly onto the engine.

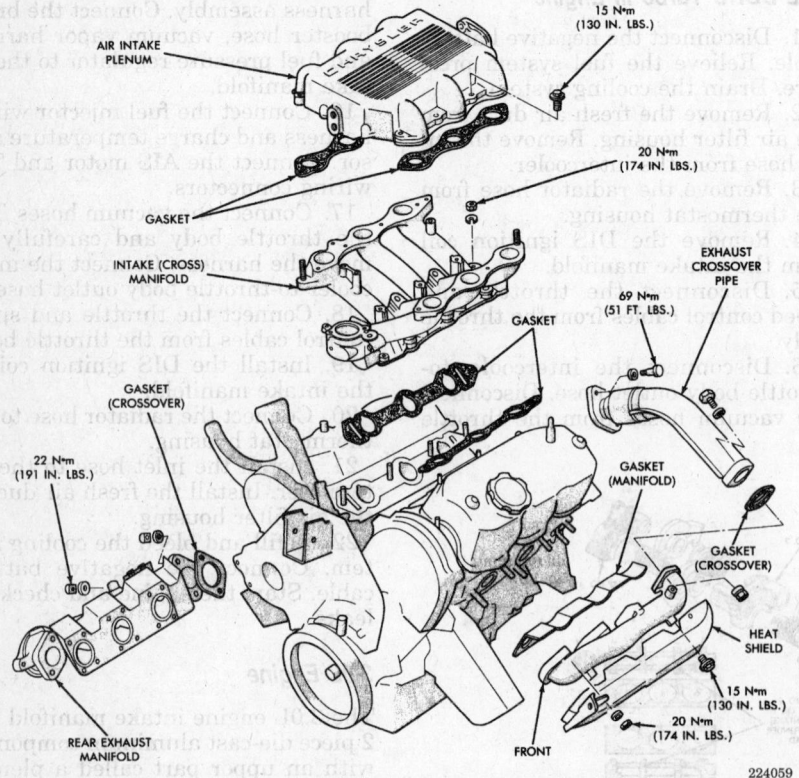

Intake and exhaust manifolds and related components — 3.0L engine

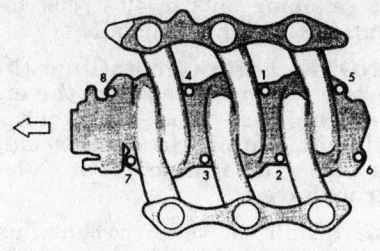

Tightening Sequence for Intake (Cross) Manifold

307441

Nut tightening sequence for the lower or cross intake manifold — 3.0L engine

24. Install and torque the fuel rail mounting bolts to 10 ft. lbs. (14 Nm).

25. Install the fuel pressure regulator onto the fuel rail.

26. Install the fuel supply and return tube and the vacuum crossover hold-down bolt.

27. Connect the fuel injection wiring harness to the engine wiring harness.

28. Connect the vacuum harness to the fuel pressure regulator and fuel rail assembly.

29. Remove the cover from the lower intake manifold and clean the mating surface.

30. Place the intake plenum gasket with the beaded sealant side up, on the intake manifold. Install the air intake plenum and torque the mounting bolts gradually and evenly, in sequence, to 10 ft. lbs. (14 Nm).

31. Connect or install all remaining items that were disconnected or removed during the removal procedure.

32. Refill the cooling system. Connect the negative battery cable and check for leaks using the DRB II to activate the fuel pump.

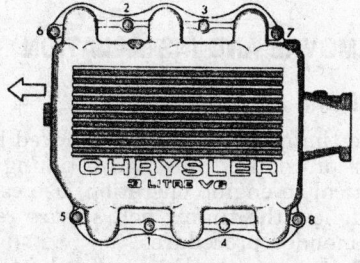

Intake Plenum Tightening Sequence

307442

Tightening sequence for the upper or air intake plenum — 3.0L engine

Exhaust Manifold

REMOVAL AND INSTALLATION

2.2L Turbo III Engine

1. Disconnect the negative battery cable.

2. Remove the turbocharger assembly, if equipped.

3. Remove the coolant tube from the cylinder head.

4. Remove the exhaust manifold retaining nuts and remove the manifold.

5. Clean the gasket mounting surfaces. Inspect the manifolds for cracks, flatness and/or damage.

To install:

6. Install a new exhaust manifold gasket. Do not use sealer of any kind.

7. Position the manifold on the studs and install the retaining nuts. Starting at the center and working outwards, torque the nuts gradually and evenly to 17 ft. lbs. (23 Nm).

8. Using a new gasket, connect the coolant tube to the cylinder head.

9. Install the turbocharger assembly, if removed.

10. Start the engine and check for exhaust leaks.

3.0L Engine

1. Disconnect the negative battery cable. Raise and safely support the vehicle.

2. Disconnect the exhaust pipe from the rear exhaust manifold, at the articulated joint.

3. Disconnect the EGR tube from the rear manifold, if equipped and disconnect the oxygen sensor wire.

4. Remove the crossover pipe to manifold bolts.

5. Remove the rear manifold to cylinder head nuts and the manifold.

6. Lower the vehicle and remove the heat shield from the front manifold.

7. Remove the bolts fastening the crossover pipe to front exhaust manifold. Remove the front manifold-to-cylinder head nuts and remove the exhaust manifold.

8. Clean the gasket mounting surfaces. Inspect the manifolds for cracks, flatness and/or damage.

To install:

NOTE: Install the gasket with the numbers 1-3-5 embossed on the top on the rear bank and those with the numbers 2-4-6 on the front (radiator side) bank.

9. Raise and safely support the vehicle.

10. Install the new gasket and rear manifold. Tighten the manifold-to-cylinder head nuts to 175 inch lbs. (20 Nm).

11. Attach the exhaust pipe to the manifold and tighten the shoulder bolt to 250 inch lbs. (28 Nm)

12. Attach the crossover pipe to the exhaust manifold and tighten the bolts to 51 ft. lbs. (69 Nm).

13. Connect the EGR tube to the rear manifold, if removed, and connect the oxygen sensor wire.

14. Lower the vehicle.

15. Install the front exhaust manifold and attach the exhaust crossover.

16. Install the front manifold heat shield and tighten the screws to 130 inch lbs. (15 Nm).

17. Reconnect the negative battery cable. Start the engine and check for exhaust leaks.

Combination Manifold

REMOVAL AND INSTALLATION

2.2L and 2.5L Engines

NOTE: On some vehicles, some of the manifold attaching bolts are not accessible or too heavily sealed from the factory and cannot be removed on the vehicle. Removal of the head would be necessary in these situations.

—— **CAUTION** ——
Fuel injection systems remain under pressure, even after the engine has been turned OFF. The fuel system pressure must be relieved before disconnecting any any fuel lines. Failure to do so may result in fire and/or personal injury.

1. Disconnect the negative battery cable.

2. Drain the cooling system. Properly relieve the fuel system pressure.

3. Remove the air cleaner. Label and disconnect all vacuum lines, electrical wiring and fuel lines from the throttle body.

4. Disconnect the throttle linkage.

5. Loosen the power steering pump and remove the power steering drive belt, if necessary.

6. Remove the power brake vacuum hose from the intake manifold.

7. Remove the EGR tube from the intake manifold. Remove the water hoses from the water crossover.

8. Raise and safely support the vehicle. Disconnect the exhaust pipe from the exhaust manifold.

9. Remove the power steering pump and set aside.

10. Remove the intake manifold support bracket, if equipped.

11. On Canadian models, remove the air injection tube bolts and the air injection tube assembly.

12. Remove the intake manifold screws/bolts.

13. Lower the vehicle and remove the intake manifold.

14. Remove the exhaust manifold nuts and remove the exhaust manifold.

To install:

15. Clean the mating surfaces and install a new combination manifold gasket. Apply a light coat of gasket sealer to the manifold side of the new combination manifold gasket.

16. Install the exhaust manifold. Install the mounting nuts and tighten them to 17 ft. lbs. (23 Nm) starting from the middle and working outward.

17. Install the intake manifold.

18. Raise and safely support the vehicle. Install the mounting screws/bolts and tighten them to 17 ft. lbs. (23 Nm) starting from the middle and working outward.

19. Connect the exhaust pipe to the exhaust manifold.

20. Install the air injection tube assembly, if equipped.

21. Install the EGR tube.

22. Install the intake support bracket.

23. Install the power steering pump and drive belt. Adjust the drive belt to the proper tension.

24. Lower the vehicle and install the water hoses to the water crossover.

25. Install the power brake vacuum hose to the intake manifold.

26. Connect the throttle linkage.

27. Install all vacuum lines, electrical wiring and fuel lines to the throttle body.

28. Install the air cleaner assembly.

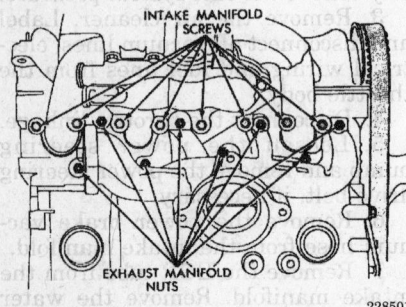

Combination intake and exhaust manifold removal and installation — 2.5L engine shown, 2.2L engine similar

29. Refill the cooling system to the proper level.

30. Reconnect the negative battery cable, run engine and check manifolds for leaks. Check coolant level and top off, if necessary.

Turbocharger

REMOVAL AND INSTALLATION

2.2L and 2.5L SOHC Engines

1. Disconnect the negative battery cable.

NOTE: On some vehicles, some of the turbocharger to exhaust manifold nuts are not accessible enough to loosen and cannot be removed from the vehicle. Head removal is necessary in these situations.

2. From above the vehicle, perform the following removal procedures:

   a. Remove the front engine mount through bolt and rotate the top of the engine forward away from the cowl.

   b. Separate the coolant line from the water box and turbocharger housing.

   c. Separate the oil feed line from the turbocharger housing.

   d. Remove the wastegate rod-to-gate retainer clip. Remove the 3 upper and 1 lower driver's side nuts retaining the turbocharger to the manifold.

   e. Disconnect the vacuum lines and the electrical lead from the oxygen sensor.

3. Raise and safely support vehicle.

4. Remove the right front wheel and tire assembly.

5. Remove the right halfshaft assembly.

6. Remove the turbocharger-to-block support bracket. Separate the oil drain back tube fitting from the turbocharger housing and remove the fitting and hose.

7. Remove the remaining turbocharger-to-manifold retaining nuts.

8. Disconnect the articulated exhaust pipe joint from the turbocharger housing.

9. Remove the lower coolant line and the turbocharger inlet fitting.

10. Lift the turbocharger off the manifold mounting studs and lower assembly from the vehicle.

To install:

11. Position the turbocharger on the exhaust manifold. Apply an anti-seize compound, Loctite™ 771-64 or equivalent, to the threads and torque

the retaining nuts to 40 ft. lbs. (54 Nm). Connect the vacuum hose.

NOTE: Before installing the turbocharger assembly to the engine, be sure it is 1st charged with oil. Failure to do this may cause damage to the turbocharger.

12. Install the lower coolant line. Install the oil drain back tube into the turbocharger housing with new gasket in place.

13. Install and tighten turbocharger to block support bracket finger-tight. First, tighten the block screw to 40 ft. lbs. (54 Nm), then tighten screw-to-turbocharger housing to 20 ft. lbs. (27 Nm).

14. Reposition exhaust pipe to the manifold and secure with the retainer bolts. Torque the shouldered bolts to 20 ft. lbs. (28 Nm).

15. Install the right driveshaft and the wheel and tire assembly to the vehicle.

16. Lower the vehicle and perform the following installation procedures:

   a. Install the 3 turbocharger to manifold retainer nuts torquing to 40 ft. lbs. (54 Nm).

   b. Reconnect the oxygen sensor lead and the vacuum harness if still disconnected.

   c. Attach the oil feed line to the turbocharger bearing housing and tighten fitting to 10 ft. lbs. (14 Nm).

   d. Apply thread sealant to the water box and turbocharger return coolant line end fittings. Install the coolant line fittings and tighten to 30 ft. lbs. (41 Nm).

   e. Align the front engine mount in the crossmember bracket.

   f. Install the through bolt and tighten to 40 ft. lbs. (54 Nm).

17. Refill the cooling system. Connect the negative battery cable and check the turbocharger for proper operation.

Front Crankshaft Seal

REMOVAL AND INSTALLATION

2.2L and 2.5L Engines

The timing belt must be removed for this procedure. Since valve timing is critical to engine operation, use care to follow the timing belt service recommended procedures.

1. Disconnect the negative battery cable.

2. It is good practice to turn the engine crankshaft to TDC No. 1 cylinder compression stroke (firing posi-

tion) before beginning timing belt removal. This should align all timing marks and serves as a point of reference for all the work that follows.

3. Remove the timing belt assembly using the recommended procedure.

4. Remove the crankshaft timing belt drive sprocket using proper puller tools. Use care if using substitutes.

5. Remove the crankshaft seal using seal puller tool 6341 or equivalent. If the camshaft and intermediate shaft seals are being serviced, their timing belt drive sprockets will have to be removed. Remove the camshaft or intermediate shaft seals using seal puller tool C-4679 or equivalent. Use care if using substitutes.

6. Clean sealing surfaces and shaft assembly.

To install:

7. Install the crankshaft seal into retainer using installation tool 6342 and 6343 or equivalent driver tool. Install the camshaft or intermediate shaft seal into retainer using tool C-4680 or equivalent driver tool.

8. Install timing belt drive sprocket(s) as required.

9. Install the timing belt using the recommended procedures. Verify that the timing marks are correctly aligned or the engine will be seriously damaged when started.

10. Connect the negative battery cable.

11. Check the oil level. If satisfactory, start engine and check for leaks to verify repair and check engine performance.

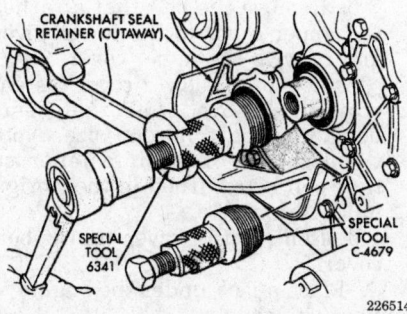

Removing oil seal — 2.2L and 2.5L engines

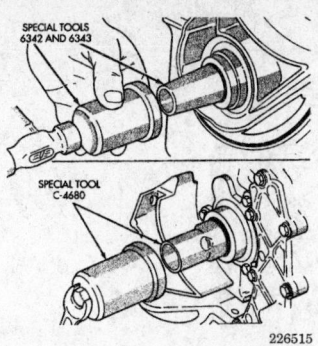

Installing oil seal — 2.2L and 2.5L engines

Timing Belt

REMOVAL AND INSTALLATION

2.2L DOHC Turbo III Engine

1. Disconnect the negative battery cable.

2. Remove the timing belt covers.

3. Install appropriate engine support tool and lift the engine slightly. Separate the right motor mount.

4. Remove the air cleaner fresh air duct, ignition cable cover, spark plugs and valve covers.

5. Raise the vehicle and support safely. Remove the lower accessory drive belt idler pulley bracket assembly.

6. Loosen the timing belt tensioner and remove the timing belt and idler pulley.

To install:

7. Loosen the rocker arm retaining bolts about 3 turns in the proper sequence. Check all lash adjusters and replace any that are damaged.

8. Align and pin both camshaft sprockets with $3/32$ in. drill bits or pin punches.

9. Install a dial indicator so the plunger is in the No. 1 spark plug hole. Rotate the crankshaft until the No. 1 piston is at TDC. Matchmark

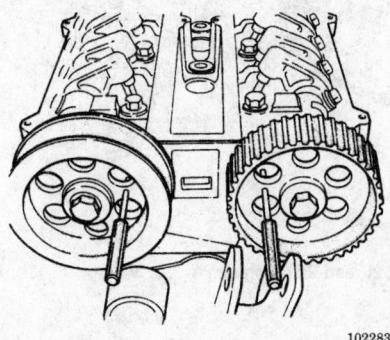

Camshaft pinned into position — 2.2L DOHC Turbo III engine

the crankshaft sprocket to the engine block for reference. The intermediate shaft sprocket does not need to be timed.

10. Install the timing belt and idler pulley starting at the crankshaft and working counterclockwise. Make sure there is no slack between sprockets when installing.

11. Install a belt tension gauge on the timing belt between the camshaft sprockets. Remove the pins from the camshaft sprockets.

12. Rotate the tensioner clockwise to adjust the belt tension to read on a belt tension gauge: 110 lbs. (445 N) for new belt or 70 lbs. (311 N) for used belt. Torque the tensioner bolt 39 ft. lbs. (53 Nm).

13. Rotate the crankshaft clockwise 2 revolutions and recheck the timing and tension. Adjust as required.

14. Torque the rocker arm bolts in sequence to 18 ft. lbs. (24 Nm).

15. Install engine mount and timing belt covers.

16. Install the spark plugs, valve covers, ignition cable cover and air duct.

17. Connect the negative battery cable.

Timing Belt, Sprockets, Tensioner and Front Cover

REMOVAL AND INSTALLATION

2.2L and 2.5L Engines

These overhead camshaft engines use a timing belt to drive the camshaft and intermediate shaft. The intermediate shaft provides distributor and oil pump drive.

If a loss of performance is noticed, ignition timing should be checked. If ignition timing is retarded by 9, 18 or 27 degrees, it may indicate that 1, 2 or 3 timing belt teeth may have skipped. The camshaft and accessory shaft timing with the crankshaft sprocket should be checked.

1. It is good practice to set the engine to TDC No. 1 cylinder compression stroke (firing position) before disassembly. This should align all timing marks (if the belt hasn't broken or jumped time) and serves as a point of reference for all work that follows. Most all body styles will have an Access Plug located in the right inner fender shield. This plug can be removed and the proper size socket, extension and ratchet inserted to rotate the crankshaft as necessary.

2. Disconnect the negative battery cable.

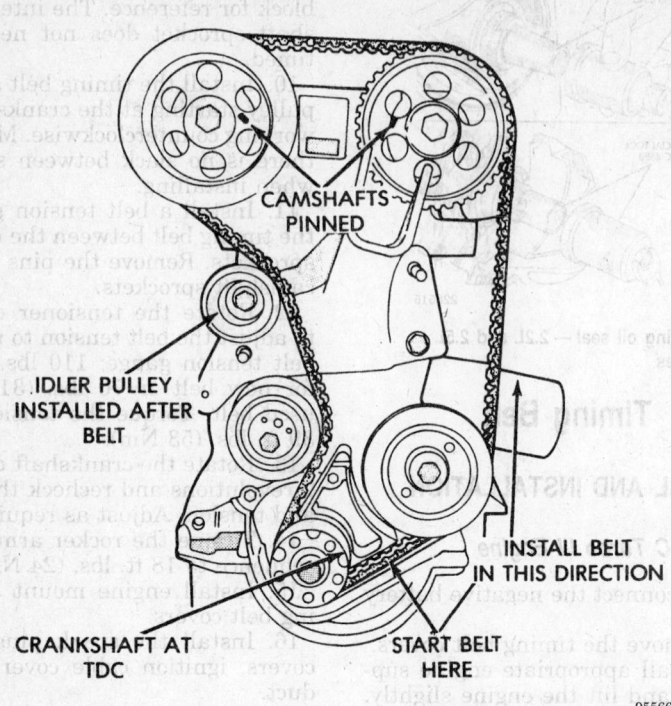

CAMSHAFTS PINNED

IDLER PULLEY INSTALLED AFTER BELT

CRANKSHAFT AT TDC

START BELT HERE

INSTALL BELT IN THIS DIRECTION

95566

Camshaft and crankshaft timing marks — 2.2L DOHC Turbo III engine

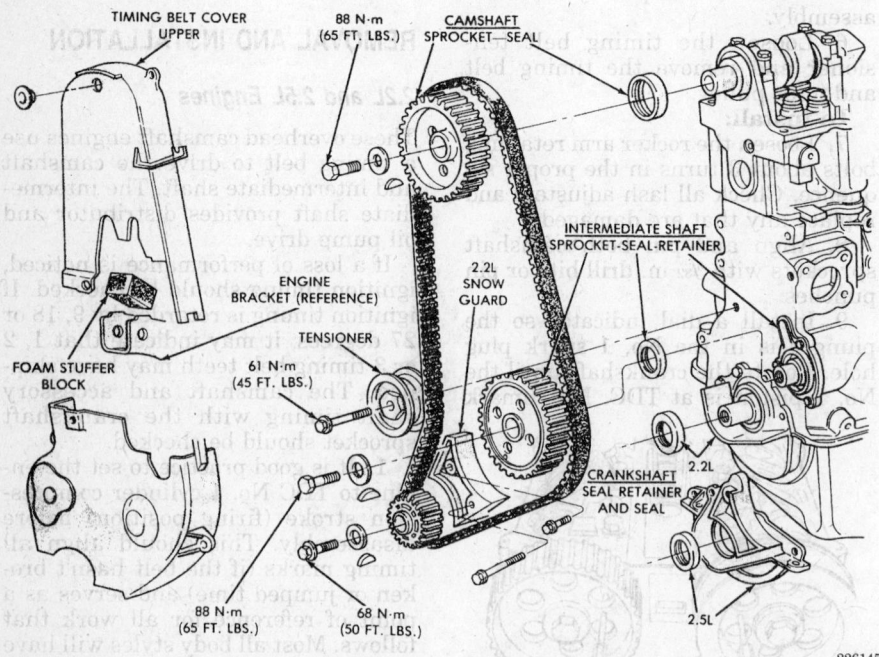

TIMING BELT COVER UPPER

88 N·m (65 FT. LBS.)

CAMSHAFT SPROCKET—SEAL

ENGINE BRACKET (REFERENCE)

TENSIONER

FOAM STUFFER BLOCK

61 N·m (45 FT. LBS.)

2.2L SNOW GUARD

INTERMEDIATE SHAFT SPROCKET-SEAL-RETAINER

CRANKSHAFT SEAL RETAINER AND SEAL

2.2L

2.5L

88 N·m (65 FT. LBS.)

68 N·m (50 FT. LBS.)

226145

Timing system and seals — 2.2L and 2.5L engines

3. Raise and safely support vehicle. Remove the right inner splash shield. Lower the vehicle.

4. The "solid mount" air conditioning compressor bracket must be removed. This requires some care. When service procedures require solid mount bracket removal and installation, it is important that bracket fasteners No. 1 through 7 be removed and installed in sequence.

a. Remove the accessory drive belts (water pump, etc.)

b. Remove the air conditioning compressor. Leave the refrigerant hoses attached and set the compressor aside.

c. Remove the alternator pivot bolt and remove the alternator. Turn the unit wire side up and disconnect, then rotate the alternator, pulley end towards the engine, and remove.

d. Remove the air conditioning compressor belt idler.

e. Remove the right engine mount yoke screw that secures the engine mount support strut to the engine mount bracket.

f. Remove the 5 side mounting bolts 1, 4, 5, 6 and 7.

g. Remove the front mounting nut, 2, and remove front bolt 3.

h. Remove the front mounting bolt and strut, rotate the solid mount bracket away from the engine and slide the bracket on the stud until nut 2 mounting stud is free. Remove the spacer from the stud.

i. Please note that installation must be done in proper sequence at assembly.

5. Remove the bolts holding the water pump pulley and crankshaft pulley and lay pulleys aside.

6. Remove both halves of the timing belt cover using the following procedure:

a. Remove the nuts and bolts that attach the upper timing belt cover to the valve cover, engine block and/or cylinder head.

b. Remove the bolt that attaches the upper cover to the lower cover and remove the upper cover.

c. Locate the lower cover fasteners and remove. Note that some bolts may be different size. Note the location of each fastener so they can be returned to their original locations.

d. Remove the lower timing belt cover.

7. Place a jack under the engine to support its weight.

8. Separate the right engine mount and raise the engine slightly.

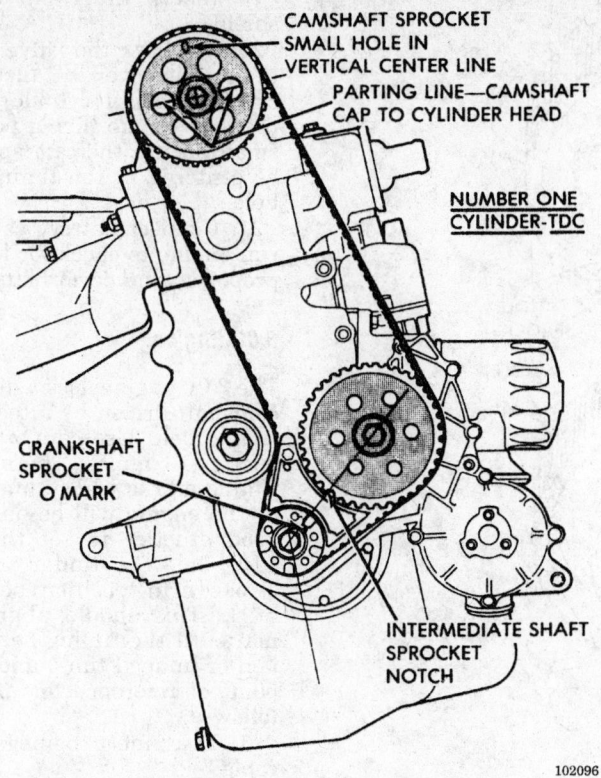

CAMSHAFT SPROCKET
SMALL HOLE IN
VERTICAL CENTER LINE

PARTING LINE—CAMSHAFT
CAP TO CYLINDER HEAD

NUMBER ONE
CYLINDER-TDC

CRANKSHAFT
SPROCKET
O MARK

INTERMEDIATE SHAFT
SPROCKET
NOTCH

102096

Timing belt and sprocket alignment — 2.2L and 2.5L Engines

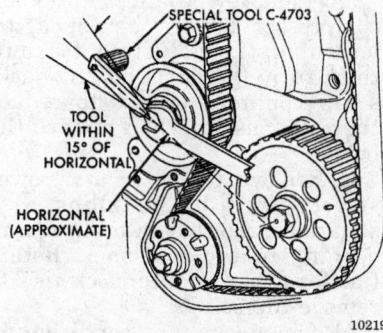

SPECIAL TOOL C-4703

TOOL
WITHIN
15° OF
HORIZONTAL

HORIZONTAL
(APPROXIMATE)

102198

Adjusting timing belt tension — 2.2L and 2.5L Engines

9. Loosen the timing belt tensioner and remove the timing belt.

10. With the timing belt removed, remove the crankshaft sprocket bolt. Use a puller to draw the crankshaft sprocket from the front of the crankshaft.

NOTE: The camshaft and intermediate shaft sprockets are unique for the 2.5L engine. They have an offset hub to provide chain clearance and are identified with a six-hole pattern. All 2.2L sprockets have a four-hole pattern.

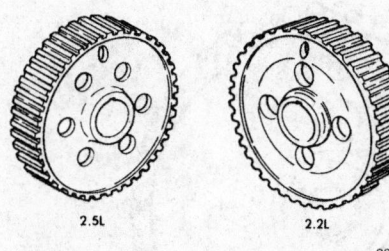

2.5L   2.2L

226152

2.5L sprocket has six-hole pattern — 2.2L has four-hole pattern

11. To remove the sprockets, a special holding tool is recommended. If the sprocket is to be reused, take precautions not to damage the sprocket. At this time, the oil seals may be replaced if necessary, using the recommended procedure.

To install:

12. If necessary, turn the crankshaft and intermediate shaft until the markings on the sprockets are in line. Removing the spark plugs makes it easier to turn the crankshaft.

13. Turn the camshaft until the arrows on the hub are in line with the No. 1 camshaft cap-to-cylinder head

mating line. The small hole in the top of the sprocket must be in vertical with the center line.

14. Install the timing belt. Turn the intermediate shaft counterclockwise slightly, if necessary, to engage the belt with the intermediate shaft sprocket. Hold the belt against the intermediate shaft sprocket and turn clockwise to take up all tension. If the timing marks are out of alignment, repeat until the alignment is correct.

15. Adjust the timing belt tension. A special tool is available that puts pressure on the pulley as the tensioner is tightened. Put this tension tool horizontally on the large hex of the timing belt tensioner pulley so the weight is at about the 9 o'clock position (parallel to the ground, hanging toward the rear of the vehicle) ± 15 degrees and loosen the tensioner locknut. Reset the belt tension tool if necessary, to have the axis within 15 degrees of horizontal. Turn the engine clockwise from TDC 2 full revolutions (remove the spark plugs, if necessary). Do not reverse the crankshaft or attempt to rotate the engine using the cam or accessory shaft attaching screw. Tighten the locknut on the tensioner holding the weighted tool in position. Do not get the tension too tight or the belt will howl and possibly break. When the timing belt is correctly tensioned, turn the crankshaft 2 full revolutions and recheck the cam timing. ALL timing marks must be aligned or the engine will be damaged.

16. Install the timing belt covers, the engine mount and crankshaft and water pump pulleys.

17. Install the solid mount bracket using the following procedure:
   a. Put the spacer onto the stud then install the bracket on the front nut mounting stud. Slide the bracket over the timing belt cover into position.
   b. Loosely install the bracket to engine fasteners.
   c. Please note that these fasteners must be tightened in sequence and to specified torque.
   d. First, torque Bolt 1 to 30 inch lbs.
   e. Second, torque Nut 2 to 40 ft. lbs.
   f. Third, torque Bolts 1 (second tightening), 4 and 5 to 40 ft. lbs.
   g. Fourth, torque bolts 6 and 7 to 40 ft. lbs.
   h. Install the alternator and air conditioning compressor. Tighten the compressor mounting bracket bolts to 40 ft. lbs.

18. Install the accessory drive belts.

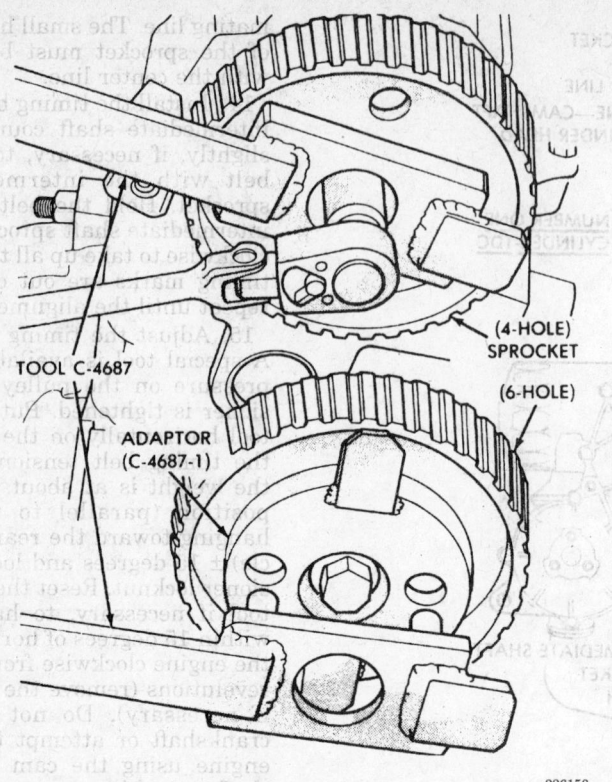

Removing or installing sprocket — 2.2L and 2.5L engines

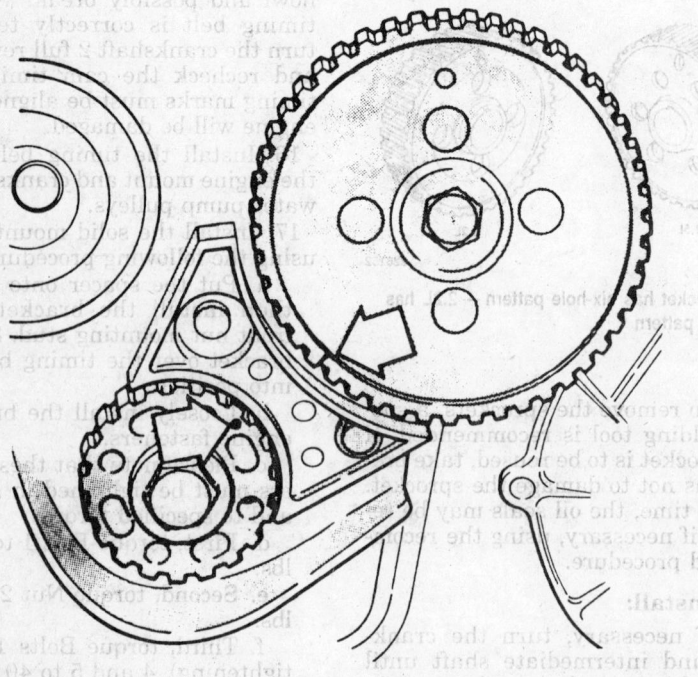

Align crankshaft and intermediate shaft timing marks — 2.2L and 2.5L engines

19. Install the right inner splash shield.

20. To check the valve timing with the timing cover installed, rotate the crank to No. 1 cylinder TDC, compression stroke (firing position). The small hole in the cam sprocket must be centered in the timing belt cover hole.

21. Connect battery cable and test run engine to check for leaks and for proper engine performance.

3.0L Engine

The 3.0L engine uses a dual overhead camshaft driven by a toothed rubber timing belt. Use care when working on these engines. There are several timing marks which must be aligned or the engine will be damaged. It is good practice to set the engine to TDC No. 1 cylinder, compression stroke (firing position) before starting work. This should align all timing marks (if the timing belt hasn't broken or jumped time) and serves as a point of reference for all work that follows.

1. Disconnect the negative battery cable.

2. Remove the accessory drive belts.

   a. Remove the air conditioning compressor belt, if equipped, by loosening the adjusting locknut, then turning the adjusting jackscrew counterclockwise to reduce the belt tension. Then remove the belt.

   b. Remove the alternator/power steering belt by installing a ½ square breaker bar into the square opening in the tensioner. Rotate the tensioner counterclockwise to remove the belt.

3. Remove the air conditioning compressor-to-mounting bracket screws and lay the compressor aside.

4. Remove the screws attaching the air conditioning compressor mounting bracket and adjustable drive belt tensioner from the block and engine mounting bracket. Remove both assemblies.

5. Remove the steering pump/alternator belt tensioner mounting bolt and remove the automatic belt tensioner. Remove the 2 steering pump-to-engine mounting bracket screws and the 1 rear support locknut. Lay the power steering pump aside.

6. Raise and safely support the vehicle. Remove the inner right splash shield.

7. Remove the crankshaft drive pulleys.

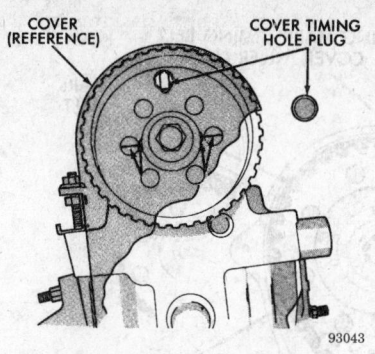

At assembly, line up camshaft timing marks — 2.2L and 2.5L engines

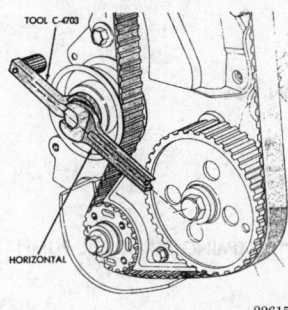

Adjusting timing belt tension with special tool to properly tension belt — 2.2L and 2.5L engines

8. Lower the vehicle and place a jack under the engine.

9. Separate the engine mount insulator from the engine mount bracket. Raise the engine slightly and remove the engine mount bracket.

10. Remove the timing belt upper front outer, upper rear and lower timing belt covers. Note that different size fasteners are used. Use care to note the location of each fastener so it can be returned to its original location.

11. If the belt is to be reused, mark its running direction for reinstallation.

12. Loosen the timing belt tensioner bolt and remove the timing belt. Remove the crankshaft flange shield.

13. If the camshaft sprockets are to be removed, a special tool is available to hold the cam sprockets in position while the center bolt is loosened (or tightened). Use care when if using substitutes.

14. If the crankshaft sprocket is to be removed, remove the crankshaft center bolt and remove the crankshaft drive sprocket. Use a suitable puller, if required.

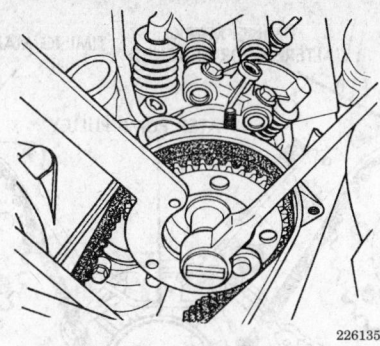

Camshaft sprocket holding tool in use — 3.0L engine

To install:

15. Clean all parts well. Because of the importance of correct valve timing, it may be helpful to apply a small amount of yellow or white paint on the timing marks on the sprockets. This should make them easier to see with the low-light conditions often found underhood.

16. Inspect the timing belt. If hardened, frayed, cracked or shows any signs of wear or distress, the belt should be replaced.

17. If removed, install the camshaft sprockets and while holding the cam sprockets, torque to 70 ft. lbs. (95 Nm). If removed, install the crankshaft sprocket.

18. If removed, install the timing belt tensioner and tensioner spring. Hook the spring upper end to the water pump pin and lower end to the tensioner bracket with the hook facing "out".

19. Turn the timing belt tensioner counterclockwise full travel in the adjustment slot and tighten the bolt to temporarily hold this position.

20. Install the timing belt on the crankshaft sprocket first, and while keeping the belt tight on the tension side install the belt on the front (radiator side) camshaft sprocket.

21. Install the belt on the water pump pulley and on the rear camshaft sprocket and finally on the timing belt tensioner.

22. Apply rotating force to the front camshaft sprocket in the opposite direction to tension the belt tension side and check that all timing marks are lined up.

23. Install the crankshaft sprocket flange.

24. Loosen the tensioner bolt and allow the spring to tension the timing belt.

25. Turn the crankshaft 2 full turns in the clockwise direction. Turn smoothly and in the clockwise direction. Verify that all timing marks are

aligned. When satisfied that the valve timing is correct, torque the tensioner locking bolt to 23 ft. lbs. (31 Nm).

26. Install the timing belt covers taking care to install the fasteners in their correct locations.

27. Reassemble the engine brackets, crankshaft pulleys and accessories and drive belts as required.

28. Connect the negative battery cable.

29. Test run vehicle to check engine performance.

Camshaft

REMOVAL AND INSTALLATION

2.2L and 2.5L Engines

The 2.2L, 2.5L engines use a Single Over Head Camshaft (SOHC) with 5 bearing journals. Flanges at the rear journal control camshaft endplay. A sintered iron timing belt sprocket is mounted on the cams nose and an oil seal is used for oil control at the front of the camshaft.

The camshaft can be replaced with the engine still in the vehicle. Camshaft service requires that the camshaft sprocket be removed from the camshaft. It is important that valve timing not be lost during this operation. To maintain camshaft, intermediate shaft and crankshaft timing during cam service, the timing belt is left indexed on the sprockets while the assembly is suspended under light tension. When removing the sprocket from the camshaft, you must maintain adequate tension on the camshaft sprocket and belt assembly to prevent the belt from disengaging with the intermediate or crankshaft timing sprockets. If the timing belt slips off the sprockets, engine timing will be lost, the engine must be retimed.

1. Disconnect the negative battery cable.

2. It is good practice to turn the crankshaft to TDC No. 1 cylinder compression stroke (firing position) before beginning work. This aligns all timing marks and serves as a point of reference for all work that follows. Turn the crankshaft so the No. 1 piston is at the TDC of the compression stroke. Remove the upper timing belt cover.

3. Remove the camshaft sprocket bolt and the sprocket and suspend tightly so the belt does not lose tension. If it does, the belt timing will have to be reset.

4. Remove the valve cover.

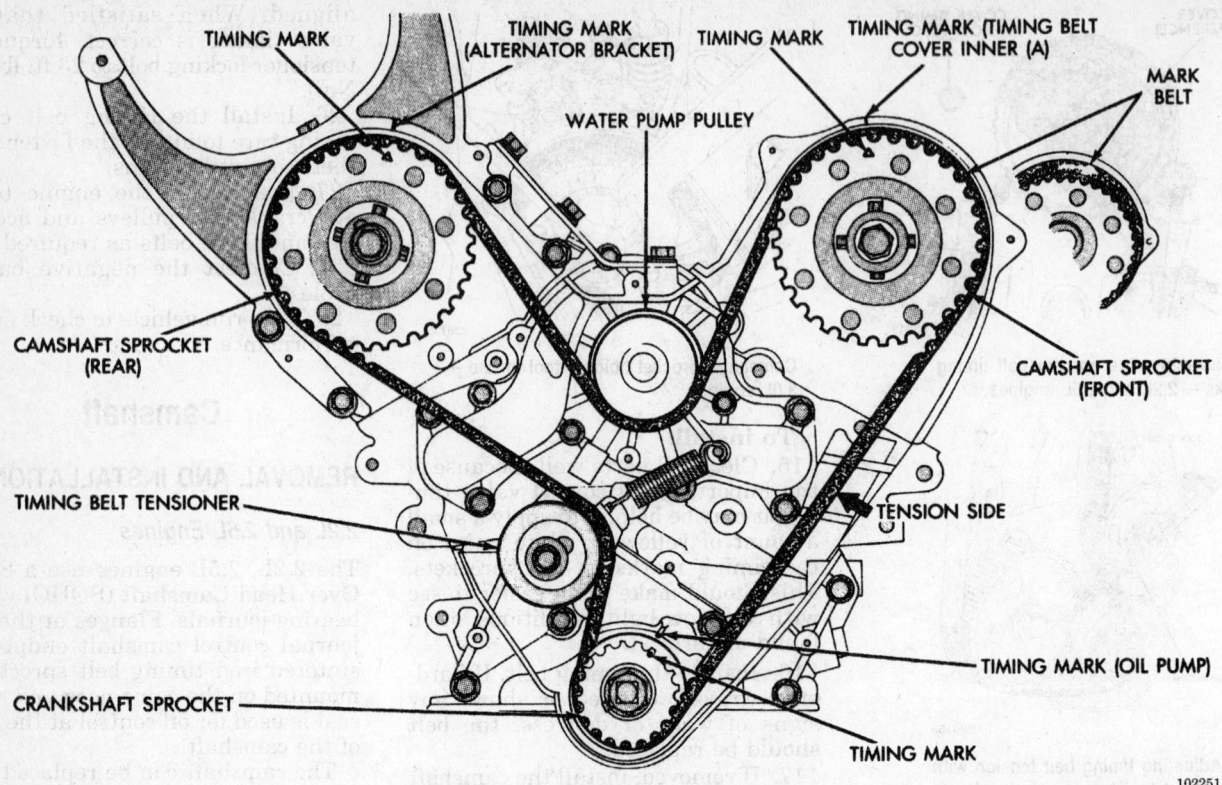

Timing belt and sprocket alignment — 3.0L engine

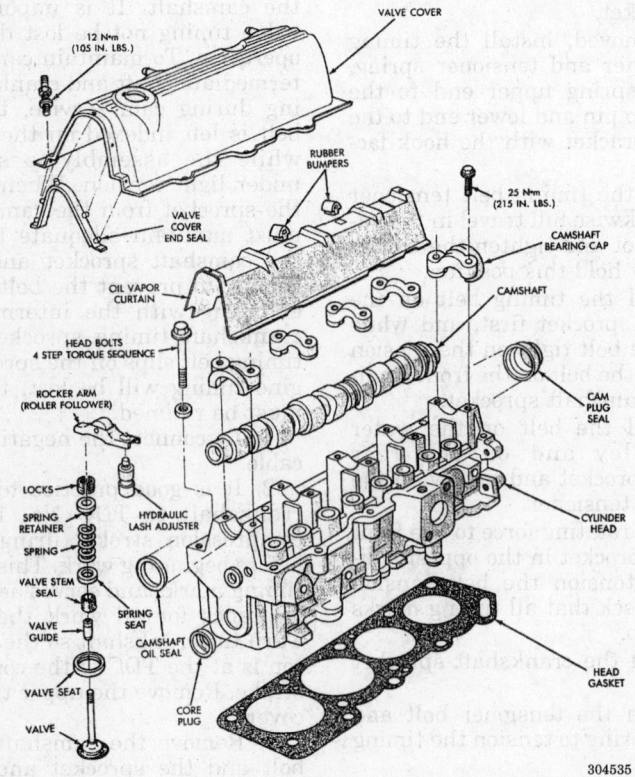

Exploded view of the cylinder head assembly — 2.2L and 2.5L engines

5. If the rocker arms are being reused, mark them for installation identification and loosen the camshaft bearing bolts, evenly and gradually.

6. Using a soft mallet, tap the rear of the camshaft a few times to break the bearing caps loose.

7. Remove the bolts, bearing caps and the camshaft with seals.

NOTE: Take note of the color of the paint stripe on the rear camshaft seal. These stripes differentiate seal sizes. If a seal with a different color stripe is installed, a severe leak will develop if the seal is too small or the cap will not be able to be fully installed if the seal is too big. Also, oversized components can be identified as follows: the top of the bearing caps are painted green and O/SJ is stamped behind the oil galley plug on the end of the head. The barrel of an oversized camshaft is also painted green and O/SJ is stamped on the end of the shaft. If normal sized parts are installed in place of oversized ones, oil pressure will be significantly reduced.

8. Check the oil passages for blockages and check parts for wear or

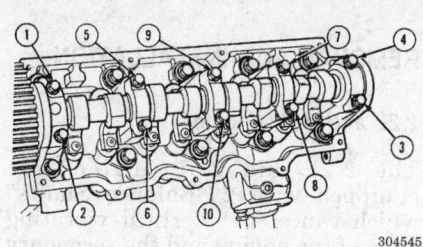

Camshaft cap removal sequence — 2.2L and 2.5L engines

damage. Clean the gasket mounting surfaces.

9. Inspect the camshaft bearing journals and lobes for wear. Lobe wear should not exceed 0.010 inch. Measure at an unworn position of the lobe, measure again in the center of the lobe and compare the figures. The difference is cam lobe wear. In actual practice, camshaft failure is usually obvious with 1 or more camshaft lobe "wiped" or rounded off. This is generally due to lubrication failure of a high mileage or neglected engine. This means the oil passages must receive extra attention when cleaning since they may become packed with sludge and metal particles. Camshaft endplay should be 0.005–0.013 inch.

To install:

10. Clean all parts well. Transfer the sprocket key to the new camshaft. New rocker arms and a new camshaft sprocket bolt are normally included with the camshaft package. Never install a new camshaft with used rocker arms or the new camshaft will quickly fail. Install the rocker arms, lubricate the camshaft with engine oil supplement or special camshaft lube, if available. Install the camshaft with end seals installed.

11. Place the bearing caps with No. 1 at the timing belt end and No. 5 at the transaxle end. The camshaft

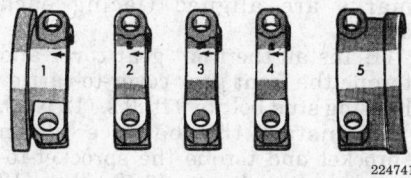

Camshaft bearing cap installation — 2.2L and 2.5L engines

bearing caps are numbered and have arrows facing forward. Torque the camshaft bearing bolts evenly and gradually to 18 ft. lbs. (24 Nm).

NOTE: Apply RTV silicone gasket material to the No. 1 and 5 bearing caps. Install the bearing caps before the seals are installed.

12. Mount a dial indicator to the front of the engine and check the camshaft endplay. Play should be 0.005–0.013 inch and not to exceed 0.020 inch

13. Install the camshaft sprocket and the new bolt.

14. Reinstall timing belt.

15. Verify that all timing marks are properly aligned. This is most important or the engine will be damaged.

16. Change the engine oil and filter. An engine oil supplement is recommended to help break-in a new camshaft.

17. Install the valve cover with a new gasket.

18. Connect the negative battery cable. Start the engine. Check ignition timing and adjust as required. Check for leaks.

19. Road test vehicle to verify performance.

2.2L DOHC Turbo III Engine

1. Disconnect the negative battery cable.

2. Remove the cylinder head.

3. Remove the rocker shaft assemblies.

4. The thrust plates in the rear of the head are not interchangeable. The intake camshaft uses a wider plate. Identify the plates and remove them.

5. To remove the cam seal, push the cam toward the seal end and the seal will be pushed out of its bore in the head.

6. Carefully pull the camshaft from the head. The intake and ex-

OIL FEED SLOT   FRONT CAM TOWER CAP

1mm (.06 IN.) DIAMETER BEAD
MOPAR GASKET MAKER   OIL FEED SLOT

REAR CAM TOWER CAP

Cam tower cap sealing — 2.2L and 2.5L engines

haust camshafts are not interchangeable. If both are being removed, identify them for installation purposes.

To install:

7. Inspect the camshaft for wear and replace any parts that are damaged.

8. Lubricate the journals with fresh engine oil and insert the camshaft into the head.

9. Install the thrust plates and tighten the retaining nuts to 55–70 inch lbs. (6–8 Nm).

10. Install new camshaft seals flush with the head surface using installation tool C–4680 or equivalent.

11. Move the camshaft as far rearward as possible. Use a dial indicator and measure the endplay. Endplay specification is 0.001–0.008 in. (0.026–0.206mm).

12. Install the rocker shaft assemblies.

13. Install the cylinder head.

14. Connect the negative battery cable.

3.0L Engine

The 3.0L engine uses 2 overhead camshafts. The front camshaft (radiator side) has a distributor drive and in longer. Both camshafts have 4 bearing journals. On the front camshaft, thrust is taken at journal 2 while rear camshaft controls thrust at journal three. Camshaft sprockets are interchangeable although in actual practice it is best to reassemble parts in their original locations. The cams and sprockets as well as the water pump are driven by a single notched timing belt.

1. Disconnect the negative battery cable.

2. Remove the air cleaner assembly and valve covers.

3. It is good practice to set the engine to TDC Cylinder No. 1 compression stroke (firing position) before starting work. This should align all timing marks and is a point of reference for all work that follows.

4. Install auto lash adjuster retainer tools MD998443 or equivalent, on the rocker arms. This is because when the rocker arm and shaft assemblies are removed, the lash adjusters cam fall out of the rocker arms and are easily damaged. The automatic lash adjusters are precision units installed machined openings in the rocker arms. Take care to keep them clean. Do not attempt to disassemble lash adjusters.

5. Remove the timing belt covers, loosen timing belt tensioner and remove the timing belt from the camshaft sprockets.

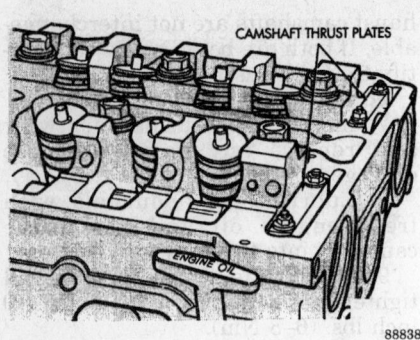

CAMSHAFT THRUST PLATES

ENGINE OIL

88838

Camshaft thrust plate — 2.2L DOHC Turbo III engine

6. If removing the front camshaft, remove the distributor extension.

7. Loosen the camshaft bearing caps but do not remove the bolts from the caps.

8. Remove the rocker arms, rocker shafts and bearing caps, as an assembly.

9. Remove the camshaft from the cylinder head.

10. Inspect the bearing journals on the camshaft, cylinder head and bearing caps.

To install:

11. Clean all parts well. Inspect the camshaft bearing journals and lobes for wear. Unworn cam lobe height should be 1.624 inch with a wear limit of 1.604 inch. In actual practice, camshaft failure is usually obvious with 1 or more camshaft lobes "wiped" or rounded off. This is generally due to lubrication failure of a high mileage or neglected engine. This means the oil passages must receive extra attention when cleaning since they may become packed with sludge and metal particles.

12. Lubricate the camshaft journals and camshaft with clean engine oil or engine oil supplement and install the camshaft in the cylinder head.

13. Align the rocker arm shaft/camshaft bearing caps with the arrow mark (depending on cylinder numbers) and in numerical order.

14. Apply sealer at the ends of the bearing caps and install the assembly.

15. Torque the bearing cap bolts, in the following sequence: No. 3, No. 2, No. 1 and No. 4 to 85 inch lbs. (10 Nm).

16. Repeat the sequence increasing the torque to 175–180 inch lbs. (18–20 Nm).

17. Install timing belt assembly using the recommended procedure. Verify that all timing marks are properly aligned before proceeding.

18. Install the distributor extension, if removed.

19. Install the valve cover and all related parts.

20. Connect the negative battery cable.

21. An oil and filter change is recommended. A can of engine oil supplement may be helpful to aid break-in a new camshaft.

22. Start the engine and look for leaks and unusual noises. Road test vehicle to check engine performance.

Intermediate Shaft

REMOVAL AND INSTALLATION

2.2L and 2.5L Engines

1. Disconnect the negative battery cable.

2. Rotate the engine so the No. 1 piston is at TDC of its compression stroke. Remove the timing belt covers to confirm that all timing marks are aligned.

3. Remove the distributor, if equipped. Looking down at the oil pump, the slot in the shaft must be parallel with the center line of the crankshaft. Remove the oil pump.

4. Remove the timing belt and the intermediate shaft sprocket.

5. Remove the intermediate shaft retainer bolts and remove the retainer from the block.

6. Remove the intermediate shaft from the engine.

7. If necessary, remove the front bushing using tool C-4697-2 and the rear bushing using tool C-4686-2.

To install:

8. Install the front bushing using tool C-4697-1 until the tool is flush with the block. Install the rear bushing using tool C-4688-1 until the tool is flush with the block.

9. Lubricate the distributor drive gear, if equipped, and install the intermediate shaft.

10. Replace the seal in the retainer and apply silicone sealer to the mating surface of the retainer. Install the retainer to the block and torque the bolts to 10 ft. lbs. (12 Nm).

11. Install the intermediate shaft sprocket and the timing belt.

12. With the timing belt properly installed, install the oil pump so the slot is parallel to the center line of the crankshaft. If equipped, install the distributor so the rotor is aligned with the No. 1 spark plug wire tower on the cap.

13. Connect the negative battery cable, check for leaks and adjust the ignition timing, as required.

Balance Shaft

REMOVAL AND INSTALLATION

2.2L and 2.5L Engines

The 2.2L and 2.5L engines are equipped with 2 "Balance Shafts" which cancel the vertical vibrating force of the engine and the secondary vibrating forces, which include the sideways rocking of the engine due to the turning direction of the crankshaft and other rolling parts. The timing sprockets are linked by a duplex chain.

1. Disconnect the negative battery cable. Raise the vehicle and support safely.

2. Remove the timing belt. Remove the oil pan, the oil pickup, the crankshaft belt sprocket and the front crankshaft oil seal retainer.

3. Remove the balance shaft chain cover, chain guide and the tensioner.

4. Remove the balance shaft sprocket-to-shaft bolt, the gear cover to balance shaft bolt and the crankshaft sprocket-to-crankshaft bolts, then the sprockets with the balance shaft chain.

5. Remove the front gear cover-to-carrier housing stud, the gear cover and the balance shaft drive gears.

6. Remove the rear gear cover-to-carrier housing bolts, the rear cover and the balance shafts from the rear of the carrier.

7. If necessary, remove the carrier housing to crankcase bolts and the housing.

To install:

8. If the carrier housing is being installed, torque the carrier housing-to-crankcase bolts to 40 ft. lbs. (54 Nm).

9. Rotate the balance shafts until the keyways are facing upward, parallel to the vertical centerline of the engine.

10. Install the short hub gear on the sprocket driven shaft and the long hub gear on the gear driven shaft. Make sure the gear timing marks are aligned (facing each other).

11. Install the front gear cover and torque the front gear cover-to-carrier housing stud bolt to 9 ft. lbs. (12 Nm).

12. Install the balance chain sprocket and torque the sprocket-to-crankshaft bolts to 11 ft. lbs. (13 Nm).

13. Rotate the crankshaft to position the No. 1 cylinder on the TDC of the compression stroke. The timing marks on the chain sprocket should

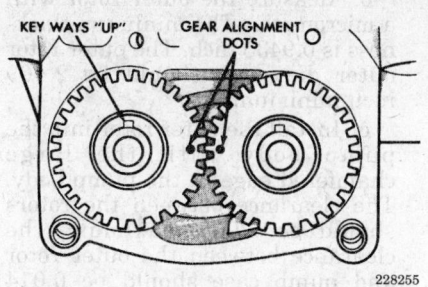

Balance shaft gear alignment — 2.2L and 2.5L engines

align with the parting line on the left side of the No. 1 main bearing cap.

14. Position the balance shaft sprocket into the balance chain so the sprocket (yellow dot) timing mark mates with the yellow link on the chain.

15. Install the balance chain/sprocket assembly onto the crankshaft and the balance shaft. Torque the sprocket-to-shaft bolts to 21 ft. lbs. (28 Nm). If necessary to secure the crankshaft while tightening the bolts, place a block of wood between the crankcase and the crankshaft counterbalance.

16. Loosely install the chain tensioners and place a shim (0.039 in. x 2.75 in.) between the chain and the tensioner. Apply firm pressure, to reduce the chain slack, to the tensioner shoe. Torque the tensioner upper bolt first, and then the lower pivot bolt to 9 ft. lbs. (12 Nm). Remove the shim from the tensioner.

17. Install the chain cover and the rear cover to the carrier housing and torque the bolts to 9 ft. lbs. (12 Nm).

18. Replace the crankshaft retainer seal, apply silicone sealer to the mating surface and install the retainer.

19. Install the oil pickup and oil pan.

20. Install the crankshaft sprocket, timing belt and related components.

21. Connect the negative battery cable, correct all engine fluid levels and road test the vehicle.

Piston and Connecting Rod

POSITIONING

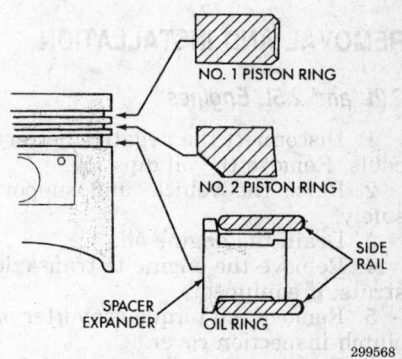

Piston ring installation

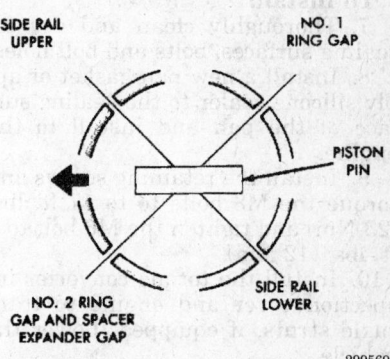

Piston ring end-gap position — 3.0L engine

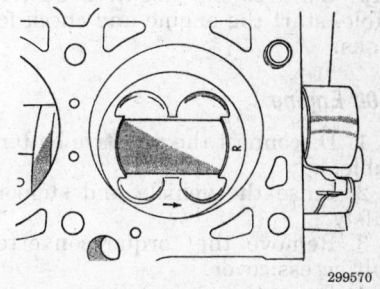

Pistons are not interchangeable. Pistons marked "R" and arrow toward the front are for cylinders 1, 3, 5. Pistons marked "L" and arrow toward front are for 2, 4, 6 — 3.0L engine

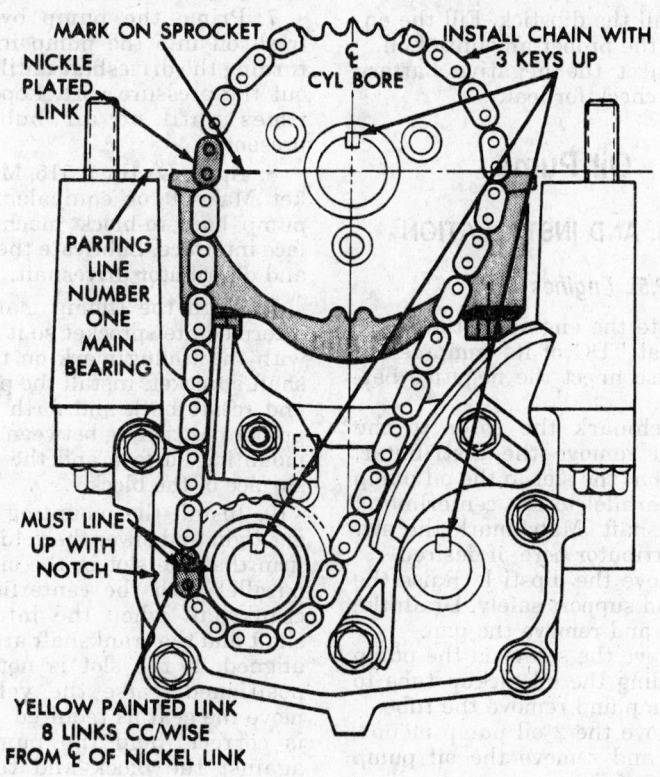

Balance shaft timing — 2.2L and 2.5L engines

ENGINE LUBRICATION

Oil Pan

REMOVAL AND INSTALLATION

2.2L and 2.5L Engines

1. Disconnect the negative battery cable. Remove the oil dipstick.
2. Raise the vehicle and support safely.
3. Drain the engine oil.
4. Remove the engine to transaxle struts, if equipped.
5. Remove the torque converter or clutch inspection cover.
6. Remove the oil pan retaining screws, oil pan and side seals.

To install:

7. Thoroughly clean and dry all sealing surfaces, bolts and bolt holes.
8. Install a new pan gasket or apply silicone sealer to the sealing surface of the pan and install to the engine.
9. Install the retaining screws and torque the M8 bolts to to 17 ft. lbs. (23 Nm) and tighten the M6 bolt to 9 ft. lbs. (12 Nm).
10. Install the torque converter inspection cover and engine to transaxle struts, if equipped. Lower the vehicle.
11. Install the dipstick. Fill the engine with the proper amount of oil.
12. Connect the negative battery cable, start the engine and check for leaks.

3.0L Engine

1. Disconnect the negative battery cable.
2. Raise the vehicle and support safely.
3. Remove the torque converter bolt access cover.
4. Drain the engine oil.
5. Remove the oil pan retaining screws and remove the oil pan and gasket.

To install:

6. Clean all parts well. Thoroughly clean and dry all sealing surfaces, bolts and bolt holes. Check that the oil pan gasket rails are flat. Use a block of wood and a mallet to correct minor distortions.
7. Apply silicone sealer to the chain cover-to-block mating seam and the rear main seal retainer-to-block seam, if equipped.

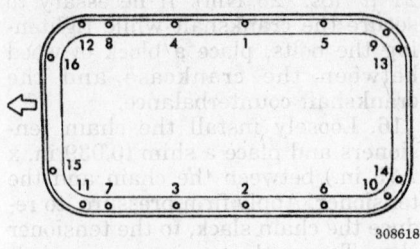

Oil pan bolt tightening sequence — 3.0L engine

308618

8. Install a new pan gasket or apply silicone sealer to the sealing surface of the pan and install to the engine.
9. Install the retaining screws and torque in sequence to 50 inch lbs. (6 Nm).
10. Install the torque converter bolt access cover, if equipped. Lower the vehicle.
11. Install the dipstick. Fill the engine with the proper amount of oil.
12. Connect the negative battery cable and check for leaks.

Oil Pump

REMOVAL AND INSTALLATION

2.2L and 2.5L Engines

1. Rotate the engine so the No. 1 piston is at TDC of its compression stroke. Disconnect the negative battery cable.
2. Matchmark the rotor to the block and remove the distributor. Confirm that the slot in the oil pump shaft is parallel to the centerline of the crankshaft. Matchmark the slot to the distributor bore, if desired.
3. Remove the dipstick. Raise the vehicle and support safely. Drain the engine oil and remove the pan.
4. Remove the screw on the pump cover holding the oil pickup tube to the oil pump and remove the tube.
5. Remove the 2 oil pump mounting bolts and remove the oil pump from the engine.

To install:

6. The pump can be inspected for wear although in most cases an oil pressure complaint results in installation of a new oil pump. If the pump is to be disassembled for wear checks, use the following guides:

a. Check the rotor end clearance with a straight-edge and feeler gauges. Limits are 0.001 inch minimum to 0.0035 inch maximum.

b. Measure the outer rotor with a micrometer. The minimum thickness is 0.9435 inch. The outer rotor outer diameter should be 2.469 inch minimum.

c. Install the outer rotor into the pump body with the large chamfered edge in the pump body. The clearance between the rotors should be 0.008 maximum, the clearance between the outer rotor and pump case should be 0.014 inch maximum.

d. Check the pump cover with a straight-edge and feeler gauge. Clearance should be 0.003 inch maximum.

e. Measure the inner rotor thickness with a micrometer. Thickness should be 0.9435 inch minimum.

f. If the oil pressure relief spring was removed to clean the relief valve, the spring free length should be 1.95 inch. Use care to assemble the relief valve and spring in the proper order.

7. Prime the pump by pouring fresh oil into the pump intake and turning the driveshaft until oil comes out the pressure port. Repeat a few times until no air bubbles are present.

8. Apply Loctite® 515, Mopar Gasket Maker® or equivalent, on the pump body-to-block machined surface interface. Lubricate the oil pump and distributor driveshaft.

9. Align the timing mark on the intermediate sprocket so it is aligned with the timing mark on the crankshaft sprocket. Install the pump fully and rotate back and forth to ensure proper positioning between the pump mounting surface and the machined surface of the block.

10. Install the mounting bolts finger-tight and lower the vehicle to confirm that the slot in the oil pump is parallel with the centerline of the crankshaft when the intermediate shaft and the crankshaft are properly aligned. If the slot is not properly positioned, raise the vehicle and move the gear as required. If the slot is correct, hold the pump firmly against the block and torque the mounting bolts to 17 ft. lbs. (23 Nm).

11. Clean out the oil pickup or replace, as required. Replace the oil pickup O-ring and install the pickup to the pump.

12. Install the oil pan using new gaskets. Lower the vehicle.

13. Install the distributor.

14. Install the oil dipstick. Fill the engine with the proper amount of oil. Install a new oil filter.

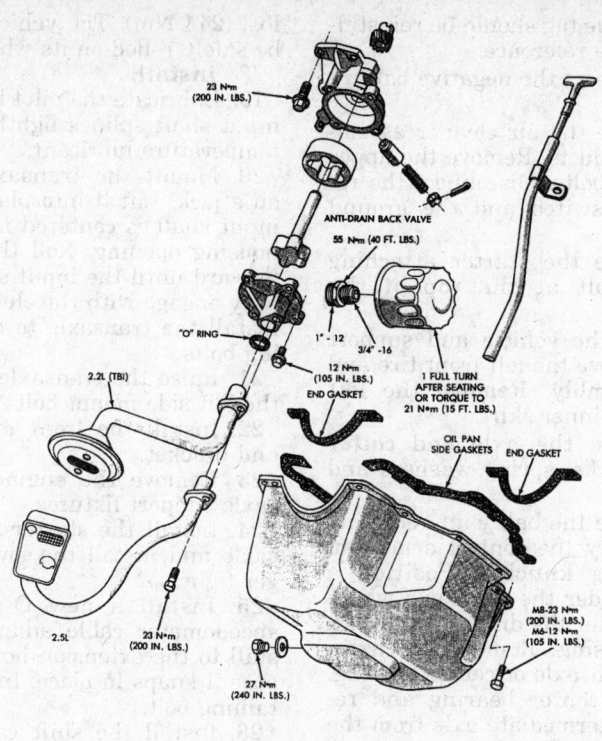

23 N•m
(200 IN. LBS.)

ANTI-DRAIN BACK VALVE

55 N•m (40 FT. LBS.)

"O" RING

2.2L (TBI)

2.5L

1" - 12   3/4" -16

12 N•m
(105 IN. LBS.)
END GASKET

1 FULL TURN
AFTER SEATING
OR TORQUE TO
21 N•m (15 FT. LBS.)

OIL PAN
SIDE GASKETS   END GASKET

2.5L

23 N•m
(200 IN. LBS.)

27 N•m
(240 IN. LBS.)

M8-23 N•m
(200 IN. LBS.)
M6-12 N•m
(105 IN. LBS.)

Engine Lubrication Components

308678

Engine lubrication components — 2.2L and 2.5L engines

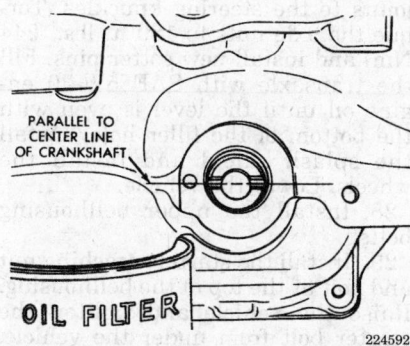

PARALLEL TO
CENTER LINE
OF CRANKSHAFT

OIL FILTER

224592

Oil pump shaft alignment — 2.2L and 2.5L engines

15. Connect the negative battery cable, refill crankcase and start the engine. It is good practice to install a mechanical oil pressure gauge so after oil pump service when the engine is first started, oil pressure can be verified immediately. Oil pressure at curb idle should be 4 psi minimum with a warm engine at idle. There should be a minimum of 25–80 psi at 3000 rpm. If, at idle, no oil pressure is seen on the gauge, shut down the engine and determine the problem.

WARNING

If oil pressure is zero at idle, DO NOT run the engine at 3000 rpm in an attempt to raise oil pressure. High engine speeds with no oil pressure will fail the engine and cause extensive internal damage.

---

16. When satisfied with the engine oil pressure. check the ignition timing and check around the oil pan for oil leaks.

2.2L DOHC Turbo III Engine

1. Rotate the engine so the No. 1 piston is at TDC of its compression stroke. Disconnect the negative battery cable.
2. Remove the dipstick. Raise the vehicle and support safely. Drain the engine oil and remove the pan.
3. Remove the screw on the pump cover holding the oil pickup tube to the oil pump and remove the tube.
4. Remove the 2 oil pump mounting bolts and remove the oil pump from the engine.
To install:
5. Prime the pump by pouring fresh oil into the pump intake and turning the driveshaft until oil comes out the pressure port. Repeat a few

times until no air bubbles are present.
6. Apply Loctite® 515, Mopar Gasket Maker® or equivalent, on the pump body-to-block machined surface interface. Lubricate the oil pump and distributor driveshaft.
7. Align the timing mark on the intermediate sprocket so it is aligned with the timing mark on the crankshaft sprocket. Install the pump fully and rotate back and forth to ensure proper positioning between the pump mounting surface and the machined surface of the block.
8. Install the mounting bolts and torque the mounting bolts to 17 ft. lbs. (23 Nm).
9. Clean out the oil pickup or replace, as required. Replace the oil pickup O-ring and install the pickup to the pump.
10. Install the oil pan using new gaskets. Lower the vehicle.
11. Install the oil dipstick. Fill the engine with the proper amount of oil.
12. Connect the negative battery cable, run engine and check the oil pressure.

3.0L Engine

1. Disconnect the negative battery cable.
2. Remove all accessory drive belts. This procedure should include removing the crankshaft drive pulleys (V-Belt and Serpentine Belt) as well as the crankshaft torsional damper.
3. Remove the timing belt using the recommended procedure. This includes removing the crankshaft timing belt drive sprocket.
4. Remove the 5 bolts attaching the oil pump to the engine block. Please note that the bolts are different lengths. Use care to mark them so they can be reinstalled in their original locations.
5. Remove the oil pump assembly.
To install:
6. Clean all parts well. The oil pump can be disassembled to check for wear or damage to the case. Remove the pump rotors to check for wear. High-mileage engines or engines that have not seen regular maintenance can experience wear in this area. Measure the clearance between the case and inner rotor. If over 0.006 inch, replace the oil pump assembly.
7. Insert the rotor into the oil pump case and measure the clearances with a feeler gauge. The clearance between the outer rotor and case should be 0.004–0.007 inch. The rotor end clearance should be

0.0015–0.0035 inch. Replace the oil pump if any part is out of specification.

8. The oil relief plunger should also be checked, especially when troubleshooting an oil pressure complaint. A stuck plunger or broken relief spring can cause improper oil pressure and filtering. Remove the plug from the underside of the oil pump body and remove the spring and plunger. Clean the plunger and bore with suitable solvent. Make sure the plunger slides smoothly in its bore. Assemble the plunger to the pump body first, then the spring and then the sealing plug. Torque the plug to 36 ft. lbs.

9. Make sure the gasket surfaces are clean. Using a new gasket, install the pump.

10. Torque the pump bolts to 120–130 inch lbs. (13–15 Nm).

11. Install the timing belt using the recommended procedure and taking all precautions to make sure the valve timing is correct.

12. Install the accessory drive belts.

13. Refill the crankcase with fresh oil. A filter change is recommended.

14. Reconnect the negative battery cable.

15. It is recommended that after oil pump service, a pressure gauge be installed and monitored to verify pressure as the engine is started and test run. Test run the engine to verify proper oil pump operation and no leaks.

TRANSAXLE

Manual Transaxle Assembly

REMOVAL AND INSTALLATION

If the vehicle is going to be rolled on its wheels while the transaxle is out of the vehicle, obtain 2 outer CV-joints to install to the hubs. If the vehicle is rolled without the proper torque applied to the front wheel bearings, the bearings will no longer be usable. Different transaxles are used according to application. It is important to use the round identification tag screwed to the top of the case when obtaining parts for exact parts

matching. The tag should be reinstalled for future reference.

1. Disconnect the negative battery cable.

2. Remove the air cleaner assembly with all ducts. Remove the upper bellhousing bolts. Disconnect the reverse light switch and the ground wire.

3. Remove the starter attaching nut and bolt at the top of the bellhousing.

4. Raise the vehicle and support safely. Remove the left front tire and wheel assembly. Remove the left front fender inner skirt.

5. Remove the axle end cotter pins, nut locks, spring washers and axle nuts.

6. Remove the ball joint retaining bolts and pry the control arm from the steering knuckle. Position a drainpan under the transaxle where the axles enter the differential or extension housing. Remove the axles from the transaxle or center bearing. Unbolt the center bearing and remove the intermediate axle from the transaxle, if equipped.

7. Remove the anti-rotation link from the crossmember. Disconnect the shifter cables from the transaxle and unbolt the cable bracket.

8. Remove the speedometer gear adaptor bolt and remove the adaptor from the transaxle.

9. Remove the rear mount from the starter, unbolt the starter and position aside.

10. Using the proper equipment, support the weight of the engine.

11. Remove the front motor mount and bracket.

12. Position a transaxle jack under the transaxle assembly.

13. Remove the lower bellhousing bolts.

14. Remove the left side splash shield. Remove the transaxle mount bolts.

15. Carefully pry the transaxle from the engine.

16. Slide the transaxle rearward until the input shaft clears the clutch disc.

17. Pull the transaxle completely away from the clutch housing and remove it from the vehicle.

18. To prepare the vehicle for rolling, support the engine with a suitable support or reinstall the front motor mount to the engine. Then reinstall the ball joints to the steering knuckle and install the retaining bolt. Install the obtained outer CV-joints to the hubs, install the washers and torque the axle nuts to 180 ft.

lbs. (244 Nm). The vehicle may now be safely rolled on its wheels.

To install:

19. Lubricate the pilot bushing and input shaft splines lightly with high temperature lubricant.

20. Mount the transaxle securely on a jack. Lift it into place until the input shaft is centered in the clutch housing opening. Roll the transaxle forward until the input shaft splines fully engage with the clutch disc and install the transaxle to clutch housing bolts.

21. Raise the transaxle and install the left side mount bolts.

22. Install the front motor mount and bracket.

23. Remove the engine and transaxle support fixtures.

24. Install the starter to the transaxle and install the lower bolt finger-tight.

25. Install a new O-ring to the speedometer cable adaptor and install to the extension housing; make sure it snaps in place. Install the retaining bolt.

26. Install the shift cable bracket and snap the cable ends in place. Install the anti-rotation link.

27. Install the axles and center bearing, if equipped. Install the ball joints to the steering knuckles. Torque the axle nuts to 180 ft. lbs. (244 Nm) and install new cotter pins. Fill the transaxle with SAE 5W-30 engine oil until the level is even with the bottom of the filler hole. Install the splash shield and install the wheels. Lower the vehicle.

28. Install the upper bellhousing bolts.

29. Install the starter attaching nut and bolt at the top of the bellhousing. Raise the vehicle and tighten the starter bolt from under the vehicle. Lower the vehicle.

30. Connect the reverse light switch and the ground wire.

31. Install the air cleaner assembly.

32. Connect the negative battery cable and check the transaxle for proper operation. Make sure the reverse lights are ON when the transaxle is in **R**.

Shift Linkage Adjustment

1. Working over the left front fender, remove the lock pin from the transaxle selector shaft housing.

2. Reverse the lock pin so the long end is down and insert it into the same threaded hole while pushing the selector shaft into the selector housing. A hole in the selector shaft will align with the lock pin, allowing the lock pin to be screwed into the

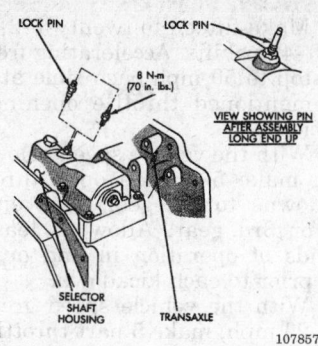

Pinning the manual transaxle in neutral

housing. This operation locks the selector shaft in the neutral position between 3rd and 4th gears.

3. Remove the gearshift knob, the retaining nut and the pull-up ring from the gearshift lever.

4. If necessary, remove the shift lever boot and console to expose the gearshift linkage. The selector cable is not adjustable.

5. Loosen the crossover cable adjusting screw and allow the cable to move in the slot. Tighten the screw to 70 inch lbs. (8 Nm).

6. Remove the lock pin from the selector shaft housing and reinstall the lock pin, with the long end up, in the selector shaft housing. Torque the lock pin to 10 ft. lbs. (12 Nm).

7. Check the 1st/reverse shifting and blockout into reverse.

8. Reinstall the console, boot, pull-up ring, retaining nut and knob.

Clutch Assembly

REMOVAL AND INSTALLATION

1. Disconnect the negative battery cable. Remove the transaxle.

2. Matchmark the pressure plate cover to the flywheel.

NOTE: To avoid the clutch disc from falling and becoming damaged during removal, insert a clutch aligning tool into the pressure plate and through the clutch disc and into the pilot bearing.

3. Loosen the flywheel to pressure plate bolts gradually and evenly to avoid warpage.

4. Remove the pressure plate and clutch assembly from the flywheel.

5. Inspect and replace the pilot bearing if necessary.

6. Sand or replace the flywheel if scoring, cracks or heat damage is present.

7. Sparingly apply anti-seize compound to the pilot bearing, input

shaft and clutch disc splines. Install a new release bearing.

To install:

8. Install the clutch disc assembly to the flywheel and, using a clutch disc alignment tool, align the disc on the flywheel and cover. Be sure to position clutch disc so the raised hub of the disc is facing away from the flywheel.

9. Torque the pressure plate/clutch assembly mounting bolts to the flywheel gradually and evenly to 21 ft. lbs. (28 Nm).

10. Install the transaxle to the vehicle.

11. Connect the negative battery cable, correct fluid levels as required and check the clutch operation and reverse lights.

Pedal Free-Play Adjustment

All vehicles are equipped with a self-adjusting cable operated mechanism and no adjustment is provided. The mechanism is located above the clutch pedal, where the cable and pivot points may be lubricated.

After installation, push and lift the clutch pedal 2 or 3 times to reset the self-adjuster.

Clutch Cable

REMOVAL AND INSTALLATION

1. Disconnect the negative battery cable.

2. Remove the retainer from the clutch release lever at the transaxle by pulling on the tail of the ball stud.

3. Pry out the ball end of the cable from the positioner adjuster on the back of the brake pedal and remove the cable, passing it through the hoop in the shock tower mounting bracket.

4. Installation is the reverse of the removal procedure.

ADJUSTMENT

After installation, push and lift the clutch pedal 2 or 3 times to allow the mechanism to adjust the cable.

Automatic Transaxle Assembly

REMOVAL AND INSTALLATION

If the vehicle is going to be rolled on its wheels while the transaxle is out of the vehicle, obtain 2 outer CV-joints to install to the hubs. If the vehicle is rolled without the proper

torque applied to the front wheel bearings, the bearings will no longer be usable.

1. Disconnect the negative battery cable.

2. If equipped, drain the coolant and remove the coolant return extension. Remove the dipstick.

3. Remove the air cleaner assembly if it is preventing access to the upper bellhousing bolts. Remove the upper bellhousing bolts and water tube, where applicable. Note locations and unplug all electrical connectors from the transaxle.

4. If equipped with a 2.2L or 2.5L engine, remove the starter attaching nut and bolt at the top of the bellhousing.

5. Disconnect the transaxle control cable at the transaxle. Raise the vehicle and support safely.

6. Remove the tire and wheel assemblies. Remove the axle end cotter pins, nut locks, spring washers and axle nuts.

7. Remove the ball joint retaining bolts and pry the control arm from the steering knuckle. Position a drain pan under the transaxle where the axles enter the differential or extension housing. Remove the axles from the transaxle or center bearing. Unbolt the center bearing and remove the intermediate axle from the transaxle, if equipped.

8. Drain the transaxle. Disconnect and plug the fluid cooler hoses.

9. If equipped with Direct Ignition System (DIS), disconnect the harness connector and remove the crankshaft position sensor from the transaxle bellhousing.

10. Remove the speedometer cable adaptor bolt and remove the adaptor from the transaxle.

11. Remove the starter. Remove the torque converter inspection cover, matchmark the torque converter to the flexplate and remove the torque converter bolts.

12. Using the proper equipment, support the weight of the engine. Remove the front motor mount and bracket.

13. Position a transaxle jack under the transaxle.

14. Remove the lower bellhousing bolts.

15. Remove the left side splash shield. Remove the transaxle mount bolts.

NOTE: The torque converter can become disengaged from the transaxle. Keep the front of transaxle slightly raised during removal.

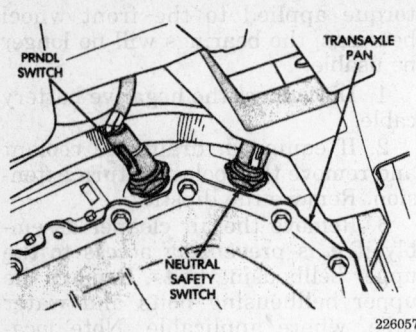

PRNDL switch and neutral safety switch — automatic transaxle

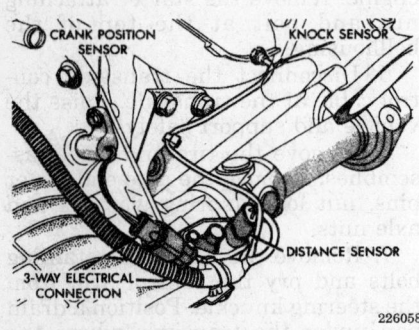

Removing crankshaft position sensor — automatic transaxle

16. Carefully pry the transaxle from the engine.

17. Slide the transaxle rearward until dowels disengage from the mating holes in the transaxle case.

18. Pull the transaxle completely away from the engine and remove it from the vehicle.

19. To prepare the vehicle for rolling, support the engine with a suitable support or reinstall the front motor mount to the engine. Then reinstall the ball joints to the steering knuckle and install the retaining bolt. Install the obtained outer CV-joints to the hubs, install the washers and torque the axle nuts to 180 ft. lbs. (244 Nm). The vehicle may now be safely rolled.

To install:

20. Install the transaxle securely on transmission jack. Rotate the converter so it will align with the positioning of the flexplate.

21. Apply a coating of high temperature grease to the torque converter pilot hub.

22. Raise the transaxle into place and push it forward until the dowels engage and the bellhousing is flush with the block. Install the transaxle to bellhousing bolts.

23. Raise the transaxle and install the left side mount bolts. Install the

torque converter bolts and torque to 55 ft. lbs. (74 Nm).

24. Install the front motor mount and bracket. Remove the engine and transaxle support fixtures.

25. Install the starter to the transaxle. Install the bolt finger-tight if equipped with a 2.2L or 2.5L engine.

26. Install a new O-ring to the speedometer cable adaptor and install to the extension housing. Make sure it snaps in place. Install the retaining bolt.

27. Connect the shifter and kickdown linkage to the transaxle, if equipped.

28. Install the axles and center bearing, if equipped. Install the ball joints to the steering knuckles. Torque the axle nuts to 180 ft. lbs. (244 Nm) and install new cotter pins. Install the splash shield and wheels. Lower the vehicle. Install the dipstick.

29. Install the upper bellhousing bolts and water pipe, if removed.

30. If equipped with 2.2L or 2.5L engine, install the starter attaching nut and bolt at the top of the bellhousing. Raise the vehicle again and tighten the starter bolt from under the vehicle. Lower the vehicle.

31. Connect all electrical wiring to the transaxle.

32. Install the air cleaner assembly, if removed. Fill the transaxle with the proper amount of Mopar ATF Plus Type 7176 or conventional Dexron®II.

33. Connect the negative battery cable and check the transaxle for proper operation.

Upshift and Kickdown Learning Procedure

The A–604 4-speed, electronic transaxle has fully adaptive controls. The controls perform their functions based on real time feedback sensor information. Although the transaxle is conventional in design, functions are controlled by its ECM.

Since the A–604 is equipped with a learning function, each time the battery cable is disconnected, the ECM memory is lost. In operation, the transaxle must be shifted many times for the learned memory to be re-input to the ECM. During this period, the vehicle will experience rough operation. The transaxle must be at normal operating temperature when learning occurs.

1. Maintain constant throttle opening during shifts. Do not move the accelerator pedal during upshifts.

2. Accelerate the vehicle with the throttle $1/8$–$1/2$ open.

3. Make fifteen to twenty 1–2, 2–3 and 3–4 upshifts. Accelerating from a full stop to 50 mph each time at the aforementioned throttle opening is sufficient.

4. With the vehicle speed below 25 mph, make 5–8 wide open throttle kickdowns to 1st gear from either 2nd or 3rd gear. Allow at least 5 seconds of operation in 2nd or 3rd gear prior to each kickdown.

5. With the vehicle speed greater than 25 mph, make 5 part throttle to wide open throttle kickdowns to either 3rd or 2nd gear from 4th gear. Allow at least 5 seconds of operation in 4th gear, preferably at road load throttle prior to performing the kickdown.

Throttle Valve Cable

ADJUSTMENT

Shift Linkage

1. Apply the parking brake. Place the shifter in the **P** detent.

2. Loosen the clamp bolt on the gearshift cable bracket.

3. Pull the shift lever all the way to the front detent position and tighten the lock screw.

4. Check for proper neutral safety switch operation.

Throttle Pressure Cable Adjustment

1. Run the engine until it reaches normal operating temperature.

2. Loosen the cable mounting bracket lock screw.

3. Position the bracket so both alignment tabs are touching the transaxle case surface and tighten the lock screws.

4. Release the cross lock on the cable assembly by pulling the cross lock up.

5. To ensure proper adjustment, the cable must be free to slide all the way toward the engine against its stop after the cross lock is released.

6. Move the transaxle throttle control lever fully clockwise and press the cross lock down until it snaps into position.

7. Road test the vehicle and check the shift points.

Throttle Pressure Rod Adjustment

1. Run the engine until it reaches normal operating temperature.

2. Loosen the adjustment swivel lock screw.

3. To ensure proper adjustment, the swivel must be free to slide along the flat end of the throttle rod. Disas-

sembly, clean and lubricate as required.

4. Hold the transaxle throttle control lever firmly toward the engine and tighten the swivel screw.

5. Road test the vehicle and check the shift points.

DRIVE AXLE

Halfshaft

REMOVAL AND INSTALLATION

1. Disconnect the negative battery cable.

2. Remove the cotter pin, nut lock and spring washer from the end of the halfshaft. Apply the brakes and loosen the hub nut while the vehicle is on the floor.

3. Raise the vehicle and support safely. Remove the tire and wheel assembly.

4. Remove the axle nut and washer.

5. Remove the ball joint retaining bolt and pry the control arm down to release the ball stud from the steering knuckle.

6. If removing the right halfshaft, remove the speedometer pinion retainer nut from the extension on the right side of the transaxle and remove the pinion.

7. Position a drainpan under the transaxle where the halfshaft enters the differential or extension housing. Remove the halfshaft from the steering knuckle and then the transaxle or center bearing by pulling on the inner joint. Unbolt the center bearing from the block and remove the intermediate shaft from the transaxle, if equipped.

To install:

8. Install the halfshaft or intermediate shaft to the transaxle, being careful not to damage the side seals. Make sure the inner joint clicks into place inside the differential. Install the center bearing retaining bolts if equipped, then install the outer shaft to the center bearing.

9. Pull the front strut out and insert the outer joint into the front hub.

10. If necessary, turn the ball joint stud to position the bolt retaining indent to the inside of the vehicle. Install the ball joint stud into the steering knuckle. Install the retaining bolt and nut. Make absolutely sure the bolt passes through the groove in the ball joint stud so the ball joint stud is locked in place in the steering knuckle. Torque to 70 ft. lbs. (95 Nm). Please note that this nut and bolt combination is unique to this application and should not be replaced with conventional hardware. Use original equipment parts if replacing.

11. Install the speedometer pinion to the extension on the transaxle and secure using the mounting bolt.

12. Install the axle nut washer and nut and torque the nut to 180 ft. lbs. (244 Nm). Install the spring washer, nut lock and a new cotter pin.

13. Install the tire and wheel assembly.

CV-Joint Boot

REPLACEMENT

NOTE: The inner tripod joint boots used on the Chrysler LH vehicles use no internal retention in the tripod housing to keep the joint assembly retained in the housing. Therefore, do not pull on the interconnecting shaft to separate the tripod housing from the transaxle stub shaft. This will damage the inner boot.

Inner Tripod Boot

1. Disconnect the negative battery cable.

2. Raise and support the vehicle safely. Remove the appropriate wheel.

3. Remove the halfshaft requiring boot replacement.

4. Remove the large boot clamp which retains the inner Tripod joint boot to the Tripod joint housing. Remove the small clamp which retains the inner Tripod joint boot to the interconnecting shaft. Discard the clamps.

5. Remove the boot from the Tripod joint housing and slide it down the shaft.

6. Slide the interconnecting shaft and spider assembly out of the Tripod joint housing.

NOTE: When removing the spider joint from the tripod housing, be careful not to allow the rollers and needle bearings to fall off of the spider assembly.

7. Remove the snapring which retains the spider assembly to the shaft. Slide the spider assembly off the shaft. If the spider assembly will not come off the shaft, tap on the spider body using a brass drift. Do not hit the outer Tripod bearings in an attempt to remove the spider assembly.

8. Slide the boot from the interconnecting shaft.

9. Thoroughly clean and inspect the spider assembly, Tripod joint housing and interconnecting shaft for excess wear.

NOTE: If any parts show excess wear, the halfshaft assembly will require replacement. Component parts of the halfshaft assemblies are not serviceable. The inner Tripod joint boot is made of 2 types of material depending on the application. The high temperature application is made from silicone rubber, which is soft. The standard temperature application boot is composed of hytrel plastic which is hard. The replacement boot must be of the same composition as the original boot.

To install:

10. Slide the new boot retainer clamp onto the interconnecting shaft followed by the new boot.

11. Install the spider assembly onto the shaft and install snapring. Make sure the snapring is fully seated in the groove of the shaft.

12. Distribute ½ of the grease provided in the boot service kit into the Tripod housing. Put the remaining grease into the CV-boot.

13. Slide the spider assembly and the interconnecting shaft into the Tripod joint housing. Position the sealing boot over the retaining groove on the interconnecting shaft. Be sure only the thinnest groove on the interconnecting shaft is visible.

14. Install the boot retaining clamp in position over the boot and crimp closed using tool C-4975 or equivalent.

CAUTION

Before crimping the CV-boot-to-Tripod housing clamp, the inner joint must be at the correct stroke position. This is required to ensure that the proper amount of air is inside the boot before the clamp is crimped. Failure to do this will result in inner CV-boot failure.

---

15. Position the sealing boot into the Tripod housing retainer groove. Install clamp on boot. Make sure the inner Tripod joint is in at the correct position. The distance between the inner clamp and the end of the housing is 8.5 inch (216mm) for standard application joint or 7.8 inch (198mm) on high temperature applications.

With the distance at the correct value, crimp the sealing boot onto the Tripod housing using tool C-4975 or equivalent.

16. Install the halfshaft into the vehicle.

17. Install the wheel and lug nuts. Torque the lug nuts, in a star pattern sequence, to 95–100 ft. lbs. (129–135 Nm).

18. Lower the vehicle. Reconnect the negative battery cable.

Outer CV-Joint Boot

1. Disconnect the negative battery cable.

2. Raise and support the vehicle safely. Remove the appropriate wheel.

3. Remove the halfshaft requiring boot replacement.

4. Remove the large boot clamp which retains the CV-joint boot to the CV-joint housing. Remove the small clamp which retains the CV-boot to the interconnecting shaft. Discard the CV-joint boot clamps.

5. Remove the boot from the joint housing and slide it down the shaft.

6. Wipe any grease away to expose the outer CV-joint-to-shaft retaining snapring. Using a pair of snapring pliers, spread the snapring and remove the CV-joint assembly off the end of the shaft.

7. Slide boot off the shaft. Thoroughly clean and inspect the CV-joint assembly and interconnecting shaft for damage or excess wear.

NOTE: If any parts show signs of excess wear, the halfshaft assembly will require replacement. Component parts for the halfshaft assemblies are not serviceable.

To install:

8. Slide a new seal boot-to-interconnecting shaft retainer clamp, followed by the new boot, onto the shaft. Install the outer CV-joint assembly onto the interconnecting shaft pushing on the shaft until the retaining snapring is seated in the groove on the shaft. Be sure the snapring is fully seated.

9. Distribute ½ of the grease provided in the boot service kit into the joint housing. Put the remaining grease into the CV-boot. Do NOT use any other type of grease.

10. Position the sealing boot over the retaining groove on the interconnecting shaft. Install the boot retaining clamp in position over the boot

and crimp closed using tool C-4975 or equivalent.

11. Position the sealing boot into the boot retaining groove on the CV-joint housing. Install clamp on boot and crimp using tool C-4975 or equivalent.

NOTE: The seal must not be dimpled, stretched or out of shape in any way. If the seal is not shaped correctly, equalize the pressure in the seal and shape it by hand.

12. Install the halfshaft into the vehicle.

13. Install the wheel and lug nuts. Torque the lug nuts, in a star pattern sequence, to 95–100 ft. lbs. (129–135 Nm).

14. Lower the vehicle. Reconnect the negative battery cable.

STEERING

Air Bag

—— CAUTION ——

Some vehicles are equipped with an air bag system, also known as the Supplemental Infaltable Restraint (SIR) or Supplemental Restraint System (SRS). The system must be disabled before performing service on or around components, steering column, instrument panel components, wiring and sensors. Failure to follow safety and disabling procedures could result in accidental air bag deployment, possible personal injury and unnecessary system repairs.

PRECAUTIONS

Several precautions must be observed when handling the inflator module to avoid accidental deployment and possible personal injury.

• Never carry the inflator module by the wires or connector on the underside of the module.

• When carrying a live inflator module, hold securely with both hands, and ensure that the bag and trim cover are pointed away.

• Place the inflator module on a bench or other surface with the bag and trim cover facing up.

• With the inflator module on the bench, never place anything on or close to the module which may be thrown in the event of an accidental deployment.

DISARMING

—— CAUTION ——

The Supplemental Restraint System (SRS) system (also called a Supplemental Inflatable Restraint or SIR system) must be disarmed before repair and/or removal of any component in its immediate area including the air bag itself. Failure to do so may cause accidental deployment of the air bag, resulting in unnecessary SIR system repairs and/or personal injury.

Except 1995–97 LeBaron, Spirit, Acclaim

To properly disarm the SRS or air bag system, disconnect the negative battery cable. It is good practice to isolate the cable end by wrapping with quality electrical tape to avoid accidental connection to the battery. Allow the system capacitor at least 2 minutes to discharge before removing air bag system components.

1995–97 LeBaron, Spirit, Acclaim

1. Disconnect the negative battery cable and isolate the cable using an appropriate insulator (wrap with quality electrical tape).

—— CAUTION ——

Always wear safety goggles when working with, or around, the air bag system.

2. Allow the system capacitor to discharge for 2 minutes before starting any repair on any air bag system or related components. This will disable the air bag system.

—— CAUTION ——

When carrying a live air bag, make sure the bag and trim cover are pointed away from the body. In the unlikely event of an accidental deployment, the bag will then deploy with minimal chance of injury. When placing a live air bag on a bench or other surface, always face the bag and trim cover up, away from the surface. This will reduce the motion of the module if it is accidentally deployed.

Steering Wheel

REMOVAL AND INSTALLATION

CAUTION

If equipped with the air bag system, the system must be disarmed prior to removing the steering wheel. Failure to disarm the air bag system may result in accidental deployment of the air bag module and possible personal injury.

Without Airbag

1. Disconnect the negative battery cable.
2. Straighten the steering wheel so the front tires are pointing straight-ahead.
3. Remove the horn pad.
4. Remove the steering wheel hold-down nut and remove the damper, if equipped. Matchmark the steering wheel to the shaft.
5. Using a steering wheel puller, pull the steering wheel off the shaft.
6. The installation is the reverse of the removal procedure. Torque the hold-down nut to 45 ft. lbs. (60 Nm).

With Airbag

1. Disconnect the negative battery cable and isolate using an appropriate insulator. Allow the system capacitor to discharge for 2 minutes prior to starting repairs on the vehicle.
2. Straighten the steering wheel so the front tires are pointing straight-ahead.
3. Remove the 4 nuts located on the back side of the steering wheel that attach the airbag module to the steering wheel.
4. Lift the module and disconnect the connectors. Remove the speed control switch, if equipped.
5. If equipped with the setscrew, place it in the clockspring to ensure

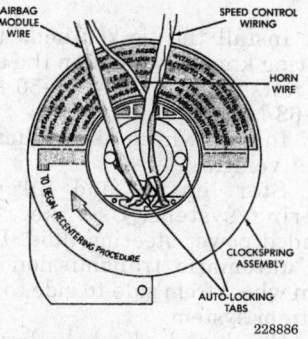

Clockspring (auto-locking) — with air bag system

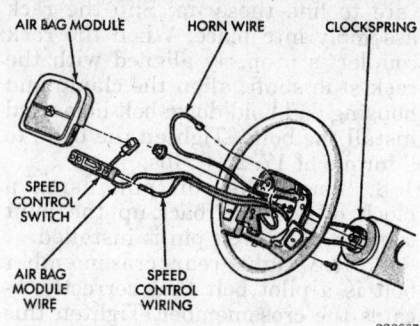

Steering wheel wiring — with air bag system

proper positioning when the steering wheel is removed.
6. Remove the steering wheel hold-down nut and remove the damper, if equipped. Matchmark the steering wheel to the shaft.
7. Using a steering wheel puller, pull the steering wheel off the shaft.

To install:

8. Position the steering wheel on the steering column. Make sure the flats on the hub of the steering wheel are aligned with the formations on the clockspring.
9. Pull the airbag and speed control connectors through the lower, larger hole in the steering wheel and pull the horn wire through the smaller hole at the top. Make sure the wires are not pinched anywhere.
10. Install the damper, if equipped.
11. Install the hold-down nut and torque to 45 ft. lbs. (60 Nm).
12. If equipped with a clockspring setscrew, remove the screw and place it in its storage location on the steering wheel.
13. Connect the horn wire.
14. Connect the speed control wire and install the speed control switch.
15. Connect the clockspring lead wire to the airbag module and install module to steering wheel.

NOTE: Do not allow anyone to enter the vehicle from this point on, until this procedure is completed.

16. Connect the DRB II to the Airbag System Diagnostic Module (ASDM) connector located to the right of the console.
17. From the passenger side of the vehicle, turn the key to the **ON** position and exit the vehicle.
18. Check to make sure no one has entered the vehicle. Connect the negative battery cable.
19. Using the DRB II, read and record any active fault data or stored codes.

20. If any active fault codes are present, perform the proper diagnostic procedures before continuing.
21. If there are no active fault codes, erase the stored fault codes. If there are active codes, the stored codes will not erase.
22. From the passenger side of the vehicle, turn the key **OFF**, then **ON** and observe the instrument cluster airbag warning light. It should come ON for 6–8 seconds, then go out, indicating the system is functioning normally. If the warning light either fails to come ON or stays lit, there is a system malfunction and further diagnostics are needed.

Tie Rod Ends

REMOVAL AND INSTALLATION

While the rack and pinion steering assembly should not be disassembled for service, the outer tie rod ends can be replaced when worn out. The rack and pinion boots can also be serviced but the factory recommends that the rack and pinion assembly be removed from the vehicle to service these boots. To remove the outer tie rod ends, use the following procedure.

1. Raise the vehicle and support safely.
2. Loosen tie rod jam nut. These underbody components are subject to rust and corrosion. It is good practice to wire brush the threaded portion of the inner tie rod (use eye protection) and apply a generous coat of penetrating oil before beginning service.
3. Remove the cotter pin and nut from the tie rod end.
4. Using a puller, remove the tie rod from the steering knuckle.
5. Mark the outer tie rod-to-inner tie rod location for approximate reassembly. Some technicians will count the turns it takes to remove the tie rod end so the replacement can be installed in the same approximate location. This saves time when resetting the toe angle after service. Remove the outer tie rod end from the inner tie rod end.

To install:

6. Clean all parts well. A light coat of chassis grease on the threaded portion of the inner tie rod end may help cut down on rust and make future service easier.
7. Install the jam nut and outer tie rod end to the inner tie rod. Do not tighten the jam nut yet. Many replacement steering parts come with a grease fitting. If so, lubricate with a quality chassis grease.

8. Install the tie rod end to the steering knuckle and tighten castle nut to 38 ft. lbs. (52 Nm) and install cotter pin.

9. Perform a front end alignment as required. It may be necessary work a small amount of silicone grease into the inner tie rod boot groove (small clamp end) to allow the inner tie rod to be turned during toe set procedures so the boot does not twist. Do not allow the boot to become twisted during front end alignment.

10. Tighten tie rod jam nut to 55 ft. lbs. (75 Nm).

11. Road test to verify that the front end alignment is correct.

Manual Rack and Pinion

REMOVAL AND INSTALLATION

1. Raise and safely support the vehicle.

2. Remove front road wheels.

3. Remove the tie rod ends using a tie rod puller C-3894-A, or equivalent.

4. Drive out the lower roll pin attaching the pinion shaft to the lower universal joint. Use a block of wood to back-up the lower universal joint to prevent damage as the roll pin is driven out. Remove the 2 rear nuts attaching the crossmember to the frame. Loosen, but do not remove the 2 front bolts attaching the crossmember to the frame.

—————— CAUTION ——————
The Supplemental Inflatable Restraint (SIR) system must be disarmed before working around the steering column. Failure to do so may cause accidental deployment to the air bag, resulting in unnecessary SIR system repairs and/or personal injury.

5. Lower the crossmember slightly. Remove the 4 rack-to-crossmember bolts and work the rack free of the steering column. If there is difficulty disengaging the rack, it may be necessary to lower the steering column from the instrument panel and pull back slightly to disengage the shaft.

6. Remove the steering rack from the left side of the vehicle.

To install:

7. Note that manual rack and pinion assemblies have a master serration on the stub shaft that must match a groove in the coupler. Use care to line these up. Slip the rack assembly into place. When the rack coupler is properly aligned with the rack stub shaft, align the clamp and housing pad hold-down bolt holes and install the bolts. Tighten the bolts to a torque of 17–25 ft. lbs.

8. Install the roll pin, using a block of wood to back-up the joint coupler as the roll pin is installed.

9. The right rear crossmember bolt is a pilot bolt that correctly locates the crossmember. Tighten this bolt first. Tighten all 4 crossmember bolts to a torque of 90 ft. lbs.

NOTE: Whenever the vehicle sub-frame (crossmember) is removed, loosened or lowered, the front wheel alignment should be checked.

10. Install the tie rod ends to the steering knuckles. Tighten the tie rod end nuts to a torque of 25–50 ft. lbs.

11. Install the wheel assemblies and lower the vehicle.

12. Check and adjust the front end alignment as required.

Power Rack and Pinion

REMOVAL AND INSTALLATION

1. Raise and safely support the vehicle.

2. Remove front wheels.

3. Remove the tie rod ends using a suitable puller.

4. Drive out the lower roll pin attaching the pinion shaft to the lower universal joint. Use a block of wood to back-up the joint to prevent damage to the universal joint.

5. Remove the 2 rear nuts attaching the crossmember to the frame. Loosen, but do not remove the 2 front bolts attaching the crossmember to the frame.

—————— CAUTION ——————
The Supplemental Inflatable Restraint (SIR) system must be disarmed before working around the steering column. Failure to do so may cause accidental deployment to the air bag, resulting in unnecessary SIR system repairs and/or personal injury.

6. Lower the crossmember slightly. Remove the 2 rack-to-crossmember bolts and work the rack free of the steering column. If there is difficulty disengaging the rack, it may be necessary to lower the steering column from the instrument panel and pull back slightly to disengage the shaft.

7. Remove any necessary splash shields. Disconnect the power steering fluid lines using the proper flare nut wrenches. Cap ends to hoses and gear ports to keep out dirt.

8. Remove any fasteners from the hose locating bracket attachment points. Disconnect the hose at opening nearest gear and drain oil from pump through the open end of the hose. Disconnect the other end of hose and remove. Discard tube end O-rings.

9. Remove the steering rack from the left side of the vehicle.

To install:

10. Slip the rack assembly into place. When the rack coupler is properly aligned with the rack stub shaft, align the clamp and housing pad hold-down bolt holes and install the bolts. Tighten the bolts to a torque of 17–25 ft. lbs.

11. Wipe the fluid hose ends and pump and gear ports clean. Install new O-rings seals on hose tube ends and lubricate with power steering fluid. Attach hose at the proper connections. Route hose smoothly in their correct location avoiding bends and kinking. Hoses must remain away from exhaust system. Do not bend tube ends.

12. Install the roll pin, using a block of wood to back-up the joint to protect the universal joint coupler.

13. The right rear crossmember bolt is a pilot bolt that correctly locates the crossmember. Tighten this bolt first. Tighten all 4 crossmember bolts to a torque of 90 ft. lbs. (122 Nm). Proper torque of these bolts is very important.

NOTE: Whenever the vehicle sub-frame (crossmember) is removed, loosened or lowered, the front wheel alignment should be checked.

14. Install the tie rod ends to the steering knuckles. Tighten the tie rod end nuts to a torque of 25–50 ft. lbs. (34–68 Nm).

15. Install wheel assemblies and lower vehicle.

16. Start engine and fill power steering system. Use the recommended power steering fluid. Do not use automatic transmission fluid. Turn wheel from side-to-side to purge air from system.

17. Check and adjust the front end alignment as required. Road test and check for leaks.

Power Steering Pump

BLEEDING

1. Check the power steering fluid level. Use MOPAR ATF PLUS automatic transmission fluid or equivalent in the power steering system.
2. Raise and safely support the vehicle to lift the front wheels off the ground.
3. Manually turn the pump pulley a few times.
4. Turn the steering wheel all the way to the left and to the right 5–6 times.
5. Disconnect the ignition high tension cable and, while operating the starter motor intermittently, turn the steering wheel all the way to the left and right 5–6 times for 15–20 seconds. During bleeding, make sure the fluid in the reservoir never falls below the lower position of the filter. If bleeding is attempted with the engine running, the air will be absorbed in the fluid. Bleed only while cranking.
6. Connect ignition high tension cable, start the engine and allow it to idle.
7. Turn the steering wheel left and right until there are no air bubbles in the reservoir. Confirm that the fluid is not milky and the level is up to the specified position on the gauge. Confirm that there is is very little change in the fluid level when the steering wheel is turned. If the fluid level changes more than 0.2 inches (about a ¼ inch) the air has not been completely bled. Repeat the process.

REMOVAL AND INSTALLATION

1. Disconnect the negative battery cable.
2. Position a drain pan under the power steering pump. Raise and safely support the vehicle.
3. Disconnect the fluid hoses from the pump and plug them.
4. If equipped, remove the tube and dipstick assembly from the pump.
5. Remove the front bracket attaching bolts and remove the belt from the pulley.
6. On 3.0L engine, disconnect the front exhaust pipe from the exhaust manifold and position aside. This is required for clearance to remove the pump.
7. Loosen the rear pump-to-bracket nut. Remove the bolt attach-

ing the pulley side of the power steering pump to the mounting bracket.
8. If necessary, remove the nut holding the power steering pump rear support bracket to the pump. Remove the 2 bolts mounting the power steering pump support bracket to the engine and remove the bracket.
9. Lower the vehicle. Remove the remaining retaining bolts and the rear mounting nut. Remove the power steering pump from the vehicle.
10. Remove the pulley from the pump with the proper puller. Install the pulley on the new pump using the special installation tools.

To install:
11. Install the pump to the engine making sure that the stud on the back of the pump is in the slotted hole in the bracket.
12. Install the mounting screws. Install the tube and dipstick assembly on the pump, if equipped.
13. Install the exhaust pipe to the manifold using a new gasket where required.
14. Install the power steering pump drive belt and adjust the tension as required.
15. Refill the pump using the correct fluid and bleed the system.

BRAKES

Anti–Lock Brake System Service

PRECAUTIONS

- Certain components within the Anti-Lock Brake System (ABS) are not intended to be serviced or repaired individually. Only those components with removal and installation procedures should be serviced.
- Do not use rubber hoses or other parts not specifically specified for and ABS system. When using repair kits, replace all parts included in the kit. Partial or incorrect repair may lead to functional problems and require the replacement of components.
- Lubricate rubber parts with clean, fresh brake fluid to ease assembly. Do not use lubricated shop air to clean parts; damage to rubber components may result.
- Use only specified brake fluid from an unopened container.

- If any hydraulic component or line is removed or replaced, it may be necessary to bleed the entire system.
- A clean repair area is essential. Always clean the reservoir and cap thoroughly before removing the cap. The slightest amount of dirt in the fluid may plug an orifice and impair the system function. Perform repairs after components have been thoroughly cleaned; use only denatured alcohol to clean components. Do not allow ABS components to come into contact with any substance containing mineral oil; this includes used shop rags.
- The Anti-Lock control unit is a microprocessor similar to other computer units in the vehicle. Ensure that the ignition switch is **OFF** before removing or installing controller harnesses. Avoid static electricity discharge at or near the controller.
- If any arc welding is to be done on the vehicle, the control unit should be unplugged before welding operations begin.

Master Cylinder

REMOVAL AND INSTALLATION

Except 1994–95

————— **WARNING** —————
Use caution when working with brake fluid. It will attack paint. Service this vehicle with DOT 3 brake fluid.

1. Disconnect the negative battery cable.
2. Disconnect brake lines from the master cylinder and plug master cylinder fluid outlets.
3. Remove the nuts attaching the master cylinder to the power booster.
4. Disconnect the electrical connector from the master cylinder, if equipped. Remove the master cylinder from the mounting studs.
5. Remove the fluid reservoir from the cylinder as required.

To install:
6. Bench bleed the master cylinder as follows:
   a. Mount master cylinder in a vise. Don't clamp tightly or the aluminum body will become distorted causing the internal components to bind.
   b. Attach tube to the fluid outlets on the master cylinder and bend tube so the outlet end of the tubes will be below the surface of brake fluid in each reservoir.

c. Fill both reservoirs with brake fluid conforming to DOT 3 specifications.

d. Slowly depress the piston and then allow the piston to return to the released position. Repeat this procedure until no bubbles are present in the fluid exiting the tubes.

e. Remove the tubes from the master cylinder and refill the reservoir with fluid.

7. Install the master cylinder to the mounting studs and install the retainer nuts. Tighten mounting nuts to 250 inch lbs. (28 Nm).

8. Install the brake lines to the master cylinder loosely. Have a helper slowly depress the brake pedal from inside the vehicle. While the pedal is being depressed, tighten the fluid lines to the master cylinder.

9. Connect the negative battery cable and check the brakes for proper operation. It may be necessary to bleed the entire system if the brake pedal is not firm.

1994–95

1. Disconnect the negative battery cable.

2. Disconnect brake lines from the master cylinder and plug master cylinder fluid outlets.

3. Remove the nuts attaching the master cylinder to the power booster.

4. Disconnect the electrical connector from the master cylinder, if equipped. Remove the master cylinder from the mounting studs.

5. Remove the fluid reservoir from the cylinder as required.

To install:

6. Bench bleed the master cylinder as follows:

a. Mount master cylinder in a vise.

b. Attach tube to the fluid outlets on the master cylinder and bend tube so the outlet end of the tubes will be below the surface of brake fluid in each reservoir.

c. Fill both reservoirs with brake fluid conforming to DOT 3 specifications.

d. Slowly depress the piston and then allow the piston to return to the released position. Repeat this procedure until no bubbles are present in the fluid exiting the tubes.

e. Remove the tubes from the master cylinder and refill the reservoir with fluid.

7. Install the master cylinder to the mounting studs and install the retainer nuts. Tighten mounting nuts to 250 inch lbs. (28 Nm).

8. Install the brake lines to the master cylinder loosely. Have a helper slowly depress the brake pedal from inside the vehicle. While the pedal is being depressed, tighten the fluid lines to the master cylinder.

9. Connect the negative battery cable and check the brakes for proper operation.

Brake Caliper

REMOVAL AND INSTALLATION

1. Raise the vehicle and support safely.

2. Remove the tire and wheel assembly.

3. Remove the caliper mounting bolts.

4. Lift the caliper off the adapter and away from the brake rotor disc. Remove the brake pads from the caliper.

5. If further work is required and the caliper must be removed from the vehicle, disconnect the brake hose retaining bolt from the caliper and remove the caliper from the vehicle. Note that new copper sealing washers will be required at installation.

To install:

6. Clean all parts well. Make sure the caliper bracket is clean of rust and corrosion. Wire brush the machined areas as necessary so the caliper can slide properly when in service. Lightly lubricate with chassis grease. Inspect the caliper guide pin bolts. Light corrosion can be removed with crocus cloth but if the corrosion is extensive, the caliper guide pins should be replaced. Inspect the bushings and guide sleeves. These should also be replaced if deteriorated. Lightly lubricate rubber parts with silicone grease.

7. If the caliper was disconnected from the brake hose, install the brake hose to the caliper using new copper sealing washers. Make sure the brake hose is not twisted after installation.

8. Install the brake pads to the caliper.

9. Position the caliper over the rotor so the caliper engages the adaptor correctly. Install the mounting bolts. Install the hold-down spring, if equipped.

10. If the caliper was separated from the brake hose, check the brake fluid in the master cylinder and refill as required. Bleed the brakes. Note that if the caliper WAS NOT disconnected from the brake hose, as in normal brake relining work, it should not be necessary to bleed the brake system.

11. Carefully test the brakes. When satisfied with the brake pedal feel, test drive the vehicle to verify correct braking operation.

Disc Brake Pads

REMOVAL AND INSTALLATION

1. Remove some of the fluid from the master cylinder so the master cylinder is ½ full. This is because as lining wears, the master cylinder reservoir brake fluid level will go down. If brake fluid has been added to the reservoir, when the caliper piston is pushed back into the caliper to make room for the new, thicker brake pads, the reservoir could overflow if some fluid is not removed.

2. Raise the vehicle and support safely. Remove the tire and wheel assemblies.

3. Remove the hold-down spring by prying away from the caliper hold-down bracket. Remove the caliper retainer pins and lift the caliper from the bracket. Do not allow the caliper to hang by the brake hose. Suspend the caliper assembly on piece of stiff wire.

4. Remove the outer pad from the caliper.

5. It may be helpful to remove the brake rotor for easier access to the inboard pad, if required. The rotor should simply pull off the front hub.

6. Remove the inner pad from the caliper hold-down bracket.

To install:

7. Use a large C-clamp to compress the piston back into the caliper bore. Be careful not to damage the piston. Most front wheel drive Chrysler products use a 54mm phenolic plastic caliper piston which can be damaged if carelessly handled.

8. Install the inner pad to the caliper hold-down and install the outer disc brake pad into caliper.

9. Install the brake rotor if removed previously.

10. Position the caliper over the rotor so the caliper engages the adaptor correctly and install the retainer pin(s). Use care, the threads will strip with the pins are not installed straight. A light coat of silicone grease (NOT chassis or wheel bearing grease) along the length of the pins may make installation easier.

11. Torque the retainer pins to 200 inch lbs.

12. Install the hold-down spring onto the hold-down bracket.

13. Depress the brake pedal several times. This is to seat the brakes.

------ **CAUTION** ------

Before the vehicle is moved after any brake service work, be sure to obtain a firm brake pedal.

14. Refill the master cylinder reservoir with clean DOT 3 brake fluid as required.
15. Install the wheel and tire and torque the lug nuts.
16. Lower the vehicle.
17. Test drive and check for proper brake operation.

Brake Rotor

REMOVAL AND INSTALLATION

1. Raise the vehicle and support safely. Remove the tire and wheel assembly.
2. Remove the caliper and brake pads.
3. Remove the factory installed clips, if equipped. It is not necessary to reinstall these clips.
4. Remove the rotor from the hub.
To install:
5. Install the rotor to the hub.
6. Install the brake caliper.
7. On front wheel disc brakes torque the caliper bolts to 30 ft. lbs. (35.5 Nm).
8. On rear wheel disc brakes torque the caliper bolts to 200 inch lbs. (22 Nm).
9. Install the wheel and tire.
10. Tighten the stud nuts in sequence until all nuts are torqued to ½ specification.
11. Torque the wheel stud nuts to 95 ft. lbs. (129 Nm).

Brake Drums

REMOVAL AND INSTALLATION

1. Raise the vehicle and support safely.
2. Remove the wheel and tire assembly.
3. Remove the dust cap.
4. Remove the cotter pin and nut lock.
5. Remove the wheel bearing nut and washer from the spindle.
6. Remove the outer wheel bearing.
7. Remove the drum with the inner wheel bearing from the spindle. If the drum is difficult to remove, remove the plug from the rear of the backing plate and push the self adjuster lever away from the star wheel. Rotate the star wheel with an upward motion to retract the shoes

and remove the drum. Remove the grease seal. Note that on this design drum brake arrangement, the star wheel adjuster is towards the top of the brake assembly, just below the wheel cylinder.
To install:
8. Lubricate and install the inner wheel bearing with a quality wheel bearing grease. Install a new grease seal.
9. Install the drum to the spindle.
10. Lubricate and install the outer wheel bearing, washer and nut. Tighten the wheel bearing adjusting nut to 240–300 inch lbs. while rotating the hub or drum assembly. This seats the bearing.
11. Back off (loosen) the adjusting nut ¼ turn then tighten the adjusting nut only finger-tight. Position the nut lock over the bearing adjusting nut with 1 pair of slots in line with the cotter pin hole in the stub axle and install the cotter pin.
12. Install the grease cap and wheel and tire assembly. Torque the wheel stud nuts to 95 ft. lbs. Install the wheel covers. Road test the vehicle to verify no wheel bearing noise and correct brake operation.

Wheel Cylinder

REMOVAL AND INSTALLATION

1. Raise the vehicle and support safely.
2. Remove the wheel, drum and brake shoes.
3. Disconnect and plug the brake line from the wheel cylinder.
4. Remove the wheel cylinder bolts and remove the cylinder from the backing plate.
To install:
5. Apply a very thin coating of silicone sealer to the cylinder mounting surface, install the cylinder to the backing plate and install the retaining bolts.
6. Connect the brake line to the wheel cylinder.
7. If the wheel cylinder was removed to correct a fluid leakage complaint, carefully clean and inspect all components. Brake shoe linings contaminated by brake fluid must not be reused. Always install new brake shoes as an axle set, both rear wheels at once. New brake hardware and return springs are always recommended. Make sure the adjuster is free and operates correctly. This is also a good time to check the parking brake cable assemblies and operation. Frozen or sticking brake cables

must be replaced. Install all brake components that were removed, using new replacement parts as required. Inspect the inside of the brake drum. Most brake drum wear can be removed by proper machining. If the drum inside diameter is beyond limits, usually marked on the drum, a replacement drum must be installed.
8. Install the tire and wheel assembly.
9. Bleed the brakes and adjust shoes.

Brake Shoes

REMOVAL AND INSTALLATION

1. Raise the vehicle and support safely. Remove the wheel and tire assemblies and the drums.
2. Remove the automatic adjuster spring and lever.
3. Rotate the automatic adjuster star wheel enough so both shoes move out far enough to be free of the wheel cylinder boots.
4. Disconnect the parking brake cable from the actuating lever.
5. Remove the lower shoe to shoe or shoe to anchor spring(s).
6. With the shoes held together by the upper shoe to shoe spring, remove them from the backing plate.
To install:
7. Thoroughly clean and dry the backing plate. To prepare the backing plate, lubricate the bosses, anchor pin and parking brake actuating lever pivot surface lightly with lithium based grease.
8. Remove, clean and dry all parts still on the old shoes. Lubricate the star wheel shaft threads with antiseize lubricant and transfer all parts to their proper locations on the new shoes.
9. Install the lower spring(s).
10. Connect the parking brake cable.
11. Install the automatic adjuster lever and spring.
12. Adjust the star wheel.
13. Remove any grease from the linings and install the drum.
14. Complete the brake adjustment with the wheels installed.

Parking Brake Cable

ADJUSTMENT

Except Daytona and LeBaron

1. Release the parking brakes fully.

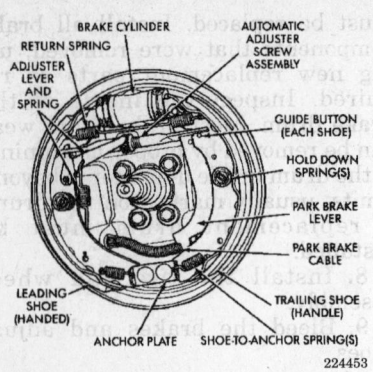

Varga left rear wheel brake assembly

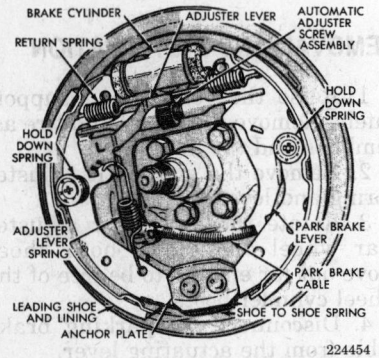

Kelsey Hayes rear brake assembly

2. Raise the vehicle and support safely.

3. Adjust the rear brakes.

4. Loosen the adjusting nut until there is slack in all the cables.

5. Rotate the rear wheels and tighten the cable adjusting nut until there is a slight drag at the wheels.

6. Continue to rotate the rear wheels and loosen the nut until all drag is eliminated.

7. Back off the nut an additional 2 turns.

8. Apply and release the parking brake several times. Upon the least release, verify there is no drag at the wheels.

9. To check the operation, make sure the parking brake holds on an incline.

Daytona and LeBaron

The parking brake hand lever contains a self-adjusting loaded clockspring feature. Routine parking brake adjustment is not required.

REMOVAL AND INSTALLATION

Front Cable

Except Daytona and LeBaron

1. Loosen the adjusting nut from under the vehicle.

2. Lift the carpet and floor matting and remove the floor pan seal.

3. Pull the cable end forward and disconnect from the clevis.

4. Pull the cable through the hole and remove.

5. The installation is the reverse of the removal procedure. Adjust cable as required.

6. Connect the negative battery cable and check the parking brakes for proper operation.

Daytona and LeBaron

> **CAUTION**
> *The parking brake hand lever contains a self-adjusting clockspring loaded to about 30 lbs. Care must be taken when handling components in the vicinity of the hand lever or serious personal injury may result.*

1. Disconnect the negative battery cable.

2. Disengage the cable from the equalizer bracket in the console.

3. Lift the carpet and floor matting and remove the floor pan seal.

4. Separate the cable from the rear parking brake lever.

5. Pull the cable through the hole and remove.

To install:

6. Install the cable and connect to the rear shoes and equalizer bracket. Install the floor pan seal and position the crapet.

7. To reload, lockout and adjust the system:

   a. Pull on the equalizer output cable with at least 30 lbs. pressure to wind up the spring. Continue until the self-adjuster lockout pawl is positioned about midway between the self-adjuster sector.

   b. Rotate the lockout pawl into the self-adjuster sector by turning the Allen screw clockwise. This action requires very little effort; do not force the screw.

   c. Adjust the rear drum-in-hat parking brake shoes.

   d. Turn the Allen screw counterclockwise about 15 degrees. When turning the lockout device, self-adjuster release is a snapping noise followed by a detent that should be felt. Very light effort is required to seat the lockout device into the de-

tent. Make sure to follow through into the detent.

   e. Cycle the negative battery cable and check the parking brakes for proper operation.

8. Connect the negative battery cable and check the parking brakes for proper operation.

Rear Cable

Rear Drum Brakes

1. Raise and support the vehicle safely. Loosen the cable adjusting nut to provide slack in the cable.

2. Remove the tire and wheel assembly.

3. Remove the brake drums. Disconnect the cable from the actuating lever on the rear brake shoe assembly.

4. Remove the retaining clip from the cable at the support bracket and pull the cable from the trailing arm assembly.

5. The installation is the reverse of the removal procedures.

Rear Disc Brakes

1. Raise and safely support the vehicle. Loosen the cable adjusting nut to provide slack in the cable.

2. Remove the tire and wheel assembly. Remove the disc brake caliper and rotor from the rear hub.

3. Disconnect the cable from the actuating lever on the rear brake shoe assembly.

4. Remove the retaining clip from the cable at the support bracket and pull the cable from the trailing arm assembly.

5. The installation is the reverse of the removal procedure.

Brake System Bleeding

Except Anti-Lock Brakes

NOTE: If using a pressure bleeder, follow the instructions furnished with the unit and choose the correct adaptor for the application. Do not substitute an adapter that almost fits as it will not work and could be dangerous.

Master Cylinder

If the master cylinder is off the vehicle, it can be bench bled.

1. Connect 2 short pieces of brake line to the outlet fittings, bend them until the free end is below the fluid level in the master cylinder reservoirs.

2. Fill the reservoir with fresh brake fluid. Pump the piston slowly until no more air bubbles appear in the reservoirs.

3. Disconnect the 2 short lines, refill the master cylinder and securely install the cylinder caps.

4. If the master cylinder is on the vehicle, it can still be bled, using a flare nut wrench.

5. Open the brake lines slightly with the flare nut wrench while pressure is applied to the brake pedal by a helper inside the vehicle.

6. Be sure to tighten the line before the brake pedal is released.

7. Repeat the process with both lines until no air bubbles come out.

8. Bleed the complete brake system.

Calipers and Wheel Cylinders

NOTE: If a pressure bleeder is not available, a good brake fluid flow can be obtained by manually bleeding the brake hydraulic system outlined in this procedure.

1. Fill the master cylinder with fresh brake fluid. Check the level often during the procedure.

2. Remove the protective caps from all 4 bleeder screws. Save for reuse. Clean the bleeder screws.

─────── **CAUTION** ───────
When bleeding the brakes, keep face away from the brake area but still wear a good pair of safety goggles. Spewing fluid may cause facial and/or visual damage. Do not allow brake fluid to spill on the car's finish. It will remove the paint.
─────────────────────

3. Connect a clear, tight fitting hose to the bleeder screw at the right rear wheel and place the hose into a clear container containing fresh brake fluid.

4. Pump the brake pedal 3 or 4 times and hold down before opening the bleeder screw.

5. Open the bleeder screw at least 1 full turn. With the bleeder screw open, the brake pedal will drop all the way down to the floor.

NOTE: If the pedal is pumped rapidly, the fluid will churn and create small air bubbles, which are difficult to remove from the system. These air bubbles will eventually congregate and a spongy pedal will result.

6. Hold down the brake pedal and release only after the bleeder screw has been closed.

7. Repeat this procedure on the remaining brake bleeder screws in the following sequence:
   a. Left rear
   b. Right front
   c. Left front

8. Be sure all trapped air has been expelled from the hydraulic system and that there is a short stroke and solid brake pedal feel. Check the brake fluid reservoir and keep filled to the proper level to ensure that no air re-enters the system through the master cylinder.

9. Hydraulic brake systems must be totally flushed if the fluid becomes contaminated with water, dirt or other corrosive chemicals. To flush, bleed the entire system until all fluid has been replaced with new fluid.

10. Install the bleeder cap(s) on the bleeder screws to keep dirt out. Always road test the vehicle, after brake work of any kind is performed.

Anti-Lock Brakes

The bleeding procedure is a 3 step process, 1 of which will require use of the DRB II scan tool or its equivalent. Bleed the system as follows:

1. Connect a pressure bleeder to the master cylinder.

2. Fully bleed the brakes hydraulic system using the conventional method in the following sequence:
   a. Right rear wheel
   b. Left rear wheel
   c. Right front wheel
   d. Left front wheel

3. Locate the diagnostic connector under the dash panel to the right of the steering column.

4. Connect the DRB II scan tool to the connector. Install cartridge for the Teves Mark IV or Teves Mark IV-G Anti-Lock Brake system. Perform the bleeding procedure using the DRB II according to the procedure outlined in the scan tool literature.

5. Once bleeding with the scan tool is complete, repeat the bleed procedure outlined in Step 2. Remove the pressure bleeder from the master cylinder.

6. Fill the master cylinder reservoir to the proper level.

7. Road test the vehicle to check for proper brake system operation.

NOTE: If using a pressure bleeder, follow the instructions furnished with the unit and choose the correct adaptor for the application. Do not substitute

an adapter that almost fits as it will not work and could be dangerous.

Master Cylinder

If the master cylinder is off the vehicle, it can be bench bled.

1. Connect 2 short pieces of brake line to the outlet fittings, bend them until the free end is below the fluid level in the master cylinder reservoirs.

2. Fill the reservoir with fresh brake fluid. Pump the piston slowly until no more air bubbles appear in the reservoirs.

3. Disconnect the 2 short lines, refill the master cylinder and securely install the cylinder caps.

4. If the master cylinder is on the vehicle, it can still be bled, using a flare nut wrench.

5. Open the brake lines slightly with the flare nut wrench while pressure is applied to the brake pedal by a helper inside the vehicle.

6. Be sure to tighten the line before the brake pedal is released.

7. Repeat the process with both lines until no air bubbles come out.

8. Bleed the complete brake system.

Calipers and Wheel Cylinders

NOTE: If a pressure bleeder is not available, a good brake fluid flow can be obtained by manually bleeding the brake hydraulic system outlined in this procedure.

1. Fill the master cylinder with fresh brake fluid. Check the level often during the procedure.

2. Remove the protective caps from all 4 bleeder screws. Save for reuse. Clean the bleeder screws.

─────── **CAUTION** ───────
When bleeding the brakes, keep face away from the brake area but still wear a good pair of safety goggles. Spewing fluid may cause facial and/or visual damage. Do not allow brake fluid to spill on the car's finish. It will remove the paint.
─────────────────────

3. Connect a clear, tight fitting hose to the bleeder screw at the right rear wheel and place the hose into a clear container containing fresh brake fluid.

4. Pump the brake pedal 3 or 4 times and hold down before opening the bleeder screw.

5. Open the bleeder screw at least 1 full turn. With the bleeder screw

open, the brake pedal will drop all the way down to the floor.

NOTE: If the pedal is pumped rapidly, the fluid will churn and create small air bubbles, which are difficult to remove from the system. These air bubbles will eventually congregate and a spongy pedal will result.

6. Hold down the brake pedal and release only after the bleeder screw has been closed.

7. Repeat this procedure on the remaining brake bleeder screws in the following sequence:
   a. Left rear
   b. Right front
   c. Left front

8. Be sure all trapped air has been expelled from the hydraulic system and that there is a short stroke and solid brake pedal feel. Check the brake fluid reservoir and keep filled to the proper level to ensure that no air re-enters the system through the master cylinder.

9. Hydraulic brake systems must be totally flushed if the fluid becomes contaminated with water, dirt or other corrosive chemicals. To flush, bleed the entire system until all fluid has been replaced with new fluid.

10. Install the bleeder cap(s) on the bleeder screws to keep dirt out. Always road test the vehicle, after brake work of any kind is performed.

Wheel Speed Sensor

REMOVAL AND INSTALLATION

1. Raise and safely support the vehicle.

2. Remove the wheel and tire assembly.

3. Remove the screw from the grommet retainer clip that holds the grommet into the fender shield.

4. Remove the 2 screws that hold the sensor routing tube to the frame rail.

5. Carefully pull the sensor assembly grommet from the fender shield.

6. Disconnect the speed sensor from the vehicle wiring harness.

7. Remove the sensor assembly grommets from the retainer brackets.

8. Remove the sensor head screw.

9. Carefully remove the sensor from the steering knuckle. If the sensor has seized due to corrosion, Do not use pliers on the sensor head. Use a hammer and a punch and tap the edge of sensor ear, rocking the sensor from side to side until free.

To install:

10. Install the sensor head and the sensor head screw.

11. Install the sensor assembly grommets.

12. Connect the speed sensor wiring and install the grommet into the fender shield.

13. Install the grommet retainer clip and screw.

14. Install the sensor routing tube.

15. Install the tire and wheel assembly.

16. Check for proper operation.

FRONT SUSPENSION

Front Strut

REMOVAL AND INSTALLATION

1. Remove the 3 mounting nuts from the strut tower under the hood.

2. Raise the vehicle and support safely.

3. Remove the brake hose bracket screw from the strut.

NOTE: When service procedure includes assembly of the original strut to the original steering knuckle, mark the cam adjusting bolt on vehicles with the eccentric camber adjusting bolt. Mark the outline of the strut on the knuckle on all other vehicles.

4. Matchmark the lower strut mount to the knuckle or camber adjuster bolt, as equipped, and remove the strut to knuckle bolts, nuts and nut plate.

--- CAUTION ---
Use extreme care when disassembling struts. Coil springs must be compressed into a loaded position for strut removal. Never remove the strut cap nut without using proper spring compression tools or equipment.

5. Position the strut assembly into spring compression tool C-4838 or equivalent.

6. Compress spring only enough to free top spring cap. Secure the strut body and remove the self-locking nut.

7. Note position and location when removing the top insulator, upper spring seat with rubber insulator and bump rubber with dust shield.

8. Remove the strut cartridge body and replace only as a complete assembly.

9. Inspect all parts for wear or damage, be sure to check top spring support bearing for smooth operation. Replace any parts found to be worn or defective.

To install:

10. Position strut assembly in the vertical position and install strut cartridge body into spring assembly.

11. Install dust boot, spring seat with rubber spring pad and top spring insulator.

12. Install the strut shaft retaining nut. Use tool L-4558 to torque the retaining nut to 55 ft. lbs. (75 Nm) plus 1/4 turn.

13. Release spring compression tool while seating spring against perch.

14. Position the strut assembly in vehicle and connect mounting hardware. Torque the upper mounting nuts to 20 ft. lbs. (27 Nm). Do not tighten the lower mounting bolts until the front end alignment has been completed.

15. Connect the brake hose to the bracket assembly.

16. Perform a front end alignment.

17. Torque the strut to knuckle nuts to 75 ft. lbs. (100 Nm) plus 1/4 turn.

Lower Ball Joints

REMOVAL AND INSTALLATION

1. Raise the vehicle and support safely. Remove the tire and wheel assembly.

2. Remove the lower control arm from the vehicle.

3. Pry off the ball joint seal. Position the receiver cup tool C-4699 or equivalent, to support the lower control arm while receiving the ball joint assembly.

4. Press against the ball joint upper housing to remove the ball joint from the lower control arm.

To install:

5. By hand, position the ball joint assembly into the bore in the lower control arm. Be sure the ball joint is not cocked in the bore of the control arm.

6. Position arm assembly in press with installer tool C-4699 or equivalent, supporting the control arm.

7. Apply pressure against the ball joint assembly until the joint is fully seated against the bottom of the control arm. Do not apply excessive pressure against the control arm.

8. Position a new seal over the stud of the ball joint so it is against the ball joint housing, and using a 1½ inch socket, press the seal onto ball joint housing until it is seated against the top surface of the control arm.

9. Install the control arm on the vehicle.

10. Install the tire and wheel assembly.

Lower Control Arms

REMOVAL AND INSTALLATION

1. Raise the vehicle and support safely. Remove the tire and wheel assembly.

2. Remove the sway bar to lower control arm retainer on both sides of the vehicle. Rotate the bar down away from the control arm.

3. Remove the ball joint stud retaining bolt and nut.

4. Pry the lower control arm from the steering knuckle.

5. Remove the control arm to crossmember bolts, nuts, bushings and retainers.

6. Remove the control arm from the vehicle.

To install:

7. Transfer all reusable parts to the new control arm and lubricate.

8. Position the control arm onto the vehicle and install the attaching bolts. Loosely assemble the nuts to the attaching bolts.

9. Install the ball joint to the steering knuckle and tighten the retaining nut and bolt to 105 ft. lbs. (145 Nm).

10. Position the sway bar against the lower control arm and install the retainers and torque to 50 ft. lbs. (70 Nm). Install the tire and wheel assembly. Lower the vehicle so the suspension is supporting the weight of the vehicle. Tighten the lower crossmember to control arm mounting bolts to 125 ft. lbs. (169 Nm).

12. Align the front suspension.

Front Sway Bar

REMOVAL AND INSTALLATION

NOTE: When attempting to eliminate a squeak, use only special lubricants designed specifically to lubricate the sway bar bushings (usually silicone grease or spray). Oil based lubricants will deteriorate the rubber and only provide a temporary fix.

1. Raise the vehicle and support safely.

2. Remove the front sway bar brackets and retainers.

3. Remove the sway bar support brackets and bushings from the lower control arm. Remove the sway bar from the vehicle.

To install:

4. Inspect for broken or distorted clamps, retainers and bushings. If bushing replacement is required, the inner bushing can be removed by opening the split. The outer bushing must be cut or hammered off the bar. If replaced, the outer sway bar bushings should be forced on (lubricate with soap and water or silicone grease) so about ½ inch of the sway bar protrudes. The sway bar to crossmember bushings should be positioned toward the rear of the vehicle. On most vehicles the control arm retainers are symmetric and bend slightly upon installation.

5. Install new bushings if required. Position the sway bar to crossmember bushings on the sway bar with the external rib up and the void in the bushing facing the rear of the vehicle. Lift the sway bar assembly into the crossmember and install the lower clamps and bolts. The center offset in the sway bar should be oriented toward the front of the vehicle.

6. Position the bushing retainers on the lower control arms and install the bolts.

7. With the lower control arms raised to design height, tighten all attaching bolts to 50 ft. lbs.

8. Lower the vehicle. Road test to confirm no noise from the sway bar.

Front Wheel Bearings

REMOVAL AND INSTALLATION

Pressed-In (Two-Piece) Hub and Bearing

NOTE: Some hub and bearing replacement packages include the one-piece unit. If this is the case, follow the installation steps for one-piece unit instead of for the two-piece unit.

1. Loosen the hub nut while the vehicle is on the floor and the brakes are applied. Raise and safely support the vehicle.

2. Remove the tire and wheel assembly. Remove the brake caliper from the caliper mount and remove the mount. Remove the brake disc.

3. Remove the hub nut and the washer from the stub shaft.

4. Disconnect the tie rod end from the steering arm using the appropriate puller.

5. Remove the clamp bolt securing the ball joint stud into the steering knuckle and separate.

6. Matchmark the lower strut mount to the knuckle. Remove the 2 strut clamp bolts and remove the knuckle from the vehicle.

7. Attach the hub removal tool C–4811 or equivalent, and the triangular adapter, to the 3 rear threaded holes of the steering knuckle housing with the thrust button inside the hub bore.

8. Tighten the bolt in the center of the tool, to press the hub from the steering knuckle. Remove the disassembly tools.

9. Remove the bolts and bearing retainer from the outside the steering knuckle.

10. Carefully pry the bearing seal from the machined recess of the steering knuckle and clean the recess.

11. Insert tool C–4811 or equivalent through the hub bearing and install bearing removal adapter to the outside of the steering knuckle. Tighten the tool to press the hub bearing from the steering knuckle. Discard the bearing and the seal.

NOTE: When service procedure includes assembly of the original strut to the original steering knuckle, mark the cam adjusting bolt on vehicles with the eccentric camber adjusting bolt. Mark the outline of the strut on the knuckle on all other vehicles.

12. If the steering knuckle is to be removed, matchmark the lower strut mount to the knuckle or camber adjusting bolt, as equipped, and remove the strut to knuckle bolts, nuts and washer plate.

To install:

13. Use tool C–4811 or equivalent, and the bearing installation adapter to press in the hub bearing into the steering knuckle.

14. Install a new seal, the bearing retainer and the bolts to the steering knuckle. Torque the bearing retainer bolts to 20 ft. lbs. (27 Nm).

15. Use the tool C–4811 or equivalent, and the hub installation adapter, to press the hub into the hub bearing.

16. Using the bearing installation tool C–4698 or equivalent, drive the new dust seal into the rear of the

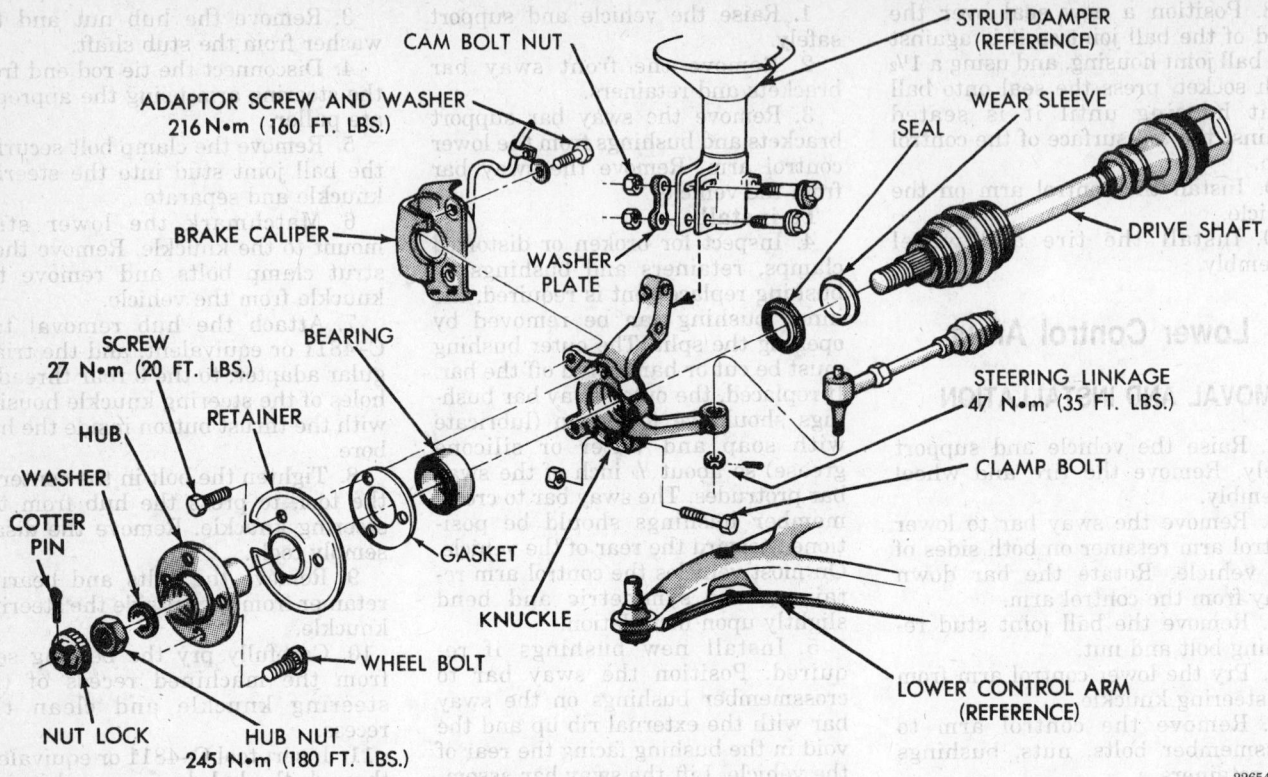

CAM BOLT NUT

ADAPTOR SCREW AND WASHER
216 N•m (160 FT. LBS.)

BRAKE CALIPER

WASHER
PLATE

SCREW
27 N•m (20 FT. LBS.)

BEARING

HUB

RETAINER

WASHER

COTTER
PIN

GASKET

KNUCKLE

WHEEL BOLT

NUT LOCK

HUB NUT
245 N•m (180 FT. LBS.)

STRUT DAMPER
(REFERENCE)

WEAR SLEEVE

SEAL

DRIVE SHAFT

STEERING LINKAGE
47 N•m (35 FT. LBS.)

CLAMP BOLT

LOWER CONTROL ARM
(REFERENCE)

226544

Pressed-in (two piece) front hub and bearing assembly

steering the hub and bearing from the knuckle as required.

17. Install the steering knuckle onto the vehicle guiding the halfshaft through the hub and install the 2 strut bolts. Verify that the steering knuckle to strut bolts, washer plate and nuts are properly installed. Torque the bolts to 75 ft. lbs. (100 Nm) plus ¼ turn. The vehicle will require a front end alignment.

NOTE: Be sure the cam bolt is installed in the same location from which it was removed.

18. Install the ball joint stud into the steering knuckle and secure with the original knuckle clamp bolt tightened to 105 ft. lbs. (145 Nm) torque.

19. Install the tie rod end to the steering knuckle. Tighten the attaching nut to 35 ft. lbs. (47 Nm) and install new cotter pin to the castle nut.

20. Install the caliper mount and caliper to the knuckle assembly.

21. Clean all foreign material from the threads of the axle stub shaft and install the washer and hub nut. With the brakes applied, torque the nut to 180 ft. lbs. (244 Nm).

22. Install the spring washer, nut lock and new cotter pin. Wrap the prongs of the cotter pin tightly around the nut lock.

23. Install the tire and wheel assembly and tighten the lug nuts to 95 ft. lbs. (129 Nm).

24. Align the front end of the vehicle if the steering knuckle was removed and replaced. Road test to verify proper front end alignment.

Bolt-In (One-Piece) Hub and Bearing

NOTE: Knuckle removal is not necessary for bearing and hub replacement. If the hub and bearing assembly requires replacement, it is to be replaced as an assembly.

1. Loosen the hub nut while the vehicle is on the floor and the brakes are applied. Raise and safely support the vehicle.

2. Remove the tire and wheel assembly from the vehicle.

3. Remove the hub nut and the washer from the stub shaft.

4. Disconnect the tie rod end from the steering arm using the appropriate puller.

5. Remove the clamp bolt securing the ball joint stud into the steering knuckle and separate.

6. Remove the caliper guide pin bolts and separate the caliper assembly from the braking disc. Support the caliper with wire hook and not by the hydraulic hose.

7. Separate the steering knuckle assembly from the ball joint stud. Pull the knuckle assembly out and away from the halfshaft.

NOTE: Care must be taken when separating the halfshaft from the knuckle, do not separate the inner CV-joint during this operation. Do not allow the halfshaft to hang by the inner CV-joint, it must be supported.

8. Remove the 4 hub and bearing assembly mounting bolts from the rear of the knuckle and remove the assembly from the knuckle.

9. Carefully pry the bearing seal from the machined recess of the steering knuckle and clean the recess.

10. Thoroughly clean and dry the knuckle and bearing mating surfaces and the seal installation area.

NOTE: When service procedure includes assembly of the original strut to the original steering knuckle, mark the cam adjusting bolt on vehicles with the eccentric camber adjusting bolt. Mark the outline of the strut on the knuckle on all other vehicles.

11. If the steering knuckle is to be removed, matchmark the lower strut

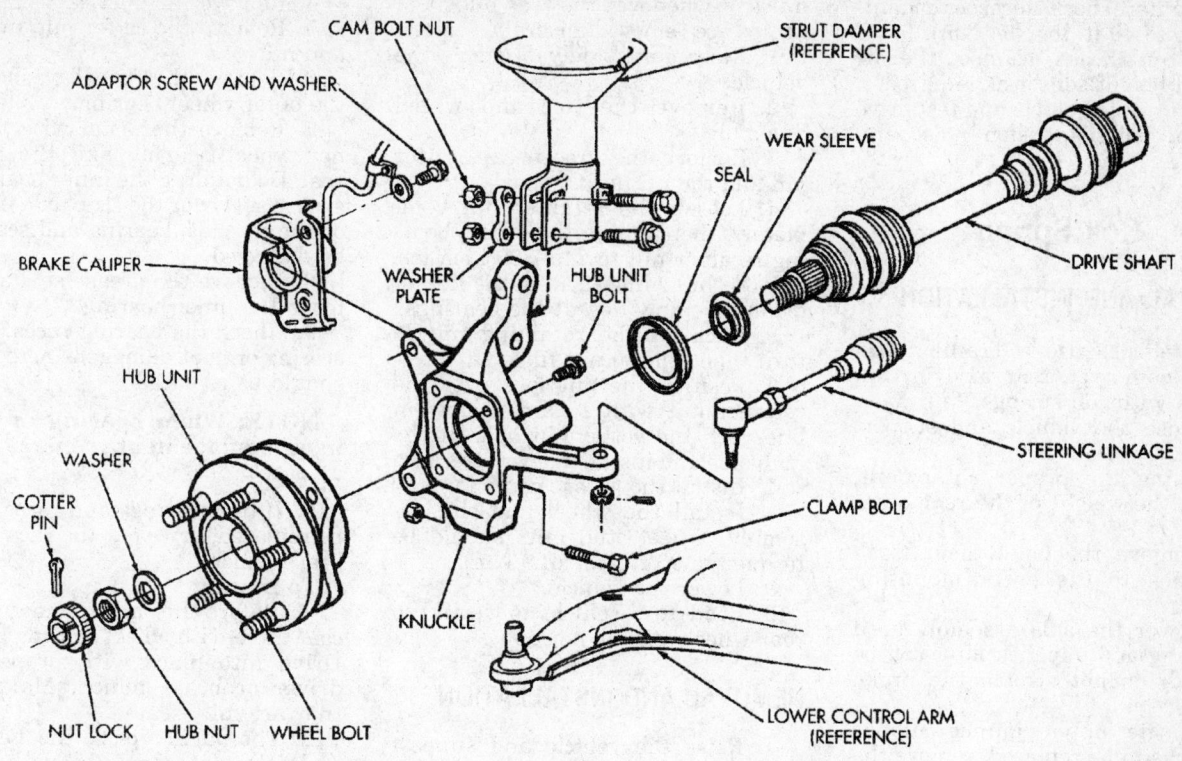

CAM BOLT NUT

ADAPTOR SCREW AND WASHER

STRUT DAMPER (REFERENCE)

WEAR SLEEVE

SEAL

BRAKE CALIPER

WASHER PLATE

HUB UNIT BOLT

DRIVE SHAFT

HUB UNIT

STEERING LINKAGE

WASHER

COTTER PIN

CLAMP BOLT

KNUCKLE

NUT LOCK   HUB NUT   WHEEL BOLT

LOWER CONTROL ARM (REFERENCE)

226543

Bolt-in (one piece) front hub and bearing assembly

mount to the knuckle or camber adjusting bolt, as equipped, and remove the strut to knuckle bolts, nuts and washer plate.

To install:

12. Clean all parts well. If the knuckle was removed, install the steering knuckle onto the vehicle guiding the halfshaft through the hub. Install the 2 strut retainer bolts and nuts loosely, aligning marks made at removal.

13. Install the hub and bearing assembly to the knuckle and torque the bolts in a criss-cross pattern to 45 ft. lbs. (65 Nm).

14. Install a new seal and wear sleeve. Lubricate the circumferences of the seal and sleeve liberally with grease.

15. Install the ball joint stud to the steering knuckle assembly and secure with the original knuckle clamp bolt tightened to 105 ft. lbs. (145 Nm) torque.

16. Install the tie rod end to the steering knuckle. Tighten the attaching nut to 35 ft. lbs. (47 Nm) and install new cotter pin.

17. Verify that the steering knuckle to strut bolts, washer plate and nuts are properly installed. Torque the bolts to 75 ft. lbs. (100 Nm) plus ¼

turn. The vehicle will require a front end alignment.

NOTE: Be sure the cam bolt is installed in the same location from which it was removed.

18. Install the brake disc and caliper to the knuckle assembly.

19. Clean all foreign material from the threads of the axle stub shaft and install the washer and hub nut. With the brakes applied, torque the nut to 180 ft. lbs. (244 Nm).

20. Install the spring washer, nut lock and new cotter pin. Wrap the prongs of the cotter pin tightly around the nut lock.

21. Install the tire and wheel assembly and tighten the lug nuts to 95 ft. lbs. (129 Nm).

22. Align the front end of the vehicle if the steering knuckle was removed and replaced. Road test to verify proper front end alignment.

REAR SUSPENSION

Shock Absorber

REMOVAL AND INSTALLATION

1. Raise the vehicle and support safely.

2. If equipped with air shocks, disconnect the air lines from the shock.

3. If removing the right rear shock on vehicle equipped with air suspension, disconnect the height sensor connector located on the right rear frame rail.

4. Support the trailing arm and remove the upper and lower shock attaching bolts.

5. Remove the shock from the vehicle.

To install:

6. Inspect the bolts. If badly rusted or if the threads are damaged, replace with new bolts of the same grade and strength.

7. Rubber bushings on the shock absorber can be lubricated with suitable silicone grease or spray.

8. Install the shock absorber. Inset the bolts and start the retainer nuts.

9. Tighten shock absorber mounting nuts to 40 ft. lbs. (54 Nm).

10. If air shocks, connect the air line and height sensor as required.

11. Lower the vehicle and test drive to check for suspension noise and handling.

Coil Spring

REMOVAL AND INSTALLATION

These vehicles use a Trailing Arm Twist Beam type rear axle in conjunction with coil springs.

1. Raise the vehicle and support safely.

2. Using the proper equipment, support the weight of the rear beam axle.

3. Remove the bolts that attach the shock to the lower mounting bracket.

4. Lower the axle assembly until the spring and upper isolator can be removed; do not stretch the brake hose.

5. If the upper jounce rubber bumper is to be replaced, remove the 2 screws holding the cup to the rail and remove the assembly.

To install:

6. If removed, position the cup and jounce bumper assembly to the rail and install the 2 attaching screws.

7. Install the isolator over the jounce bumper and install the spring.

8. Raise the axle and loosely assemble both shock absorber attaching bolts. Remove the rear axle support and lower the vehicle.

9. With the suspension supporting the weight of the vehicle, tighten both lower shock bolts to 45 ft. lbs. (61 Nm).

Wheel Bearings

ADJUSTMENT

In normal service, the lubricant and adjustment of the rear wheel bearings should be checked every 30,000

miles or whenever the rear hubs and drums are removed for brake service.

1. Raise and safely support the vehicle.

2. Remove the tire and wheel assembly.

3. Remove the grease cap, cotter pin and the nut lock.

4. Loosen the retainer nut completely, then torque the wheel bearing retainer nut to 240–300 inch lbs. (27–34 Nm) while turning the hub or drum assembly to seat the bearings.

5. Back off the retaining nut ¼ turn then tighten nut finger-tight.

6. Position the nut lock over the retainer nut with 1 pair of slots in line with the cotter pin hole in the stub axle and install a new cotter pin.

7. Install the grease cap.

8. Install the wheel and tire assembly. Wheel stud nuts should be torqued to 95 ft. lbs. (129 Nm).

9. Lower the vehicle.

10. Road test vehicle to verify no rear wheel bearing noise.

REMOVAL AND INSTALLATION

1. Raise the vehicle and support safely.

2. Remove the rear tire and wheel assembly.

3. On disc brake equipped vehicles, remove the caliper and rotor. Support the caliper out of way. Do not allow the caliper to hang by the hydraulic hose. Support with a stiff piece of wire.

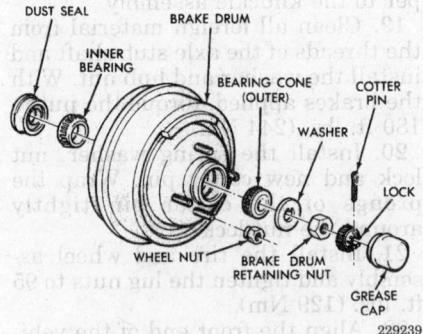

Rear wheel bearing assembly

4. Remove the dust cap.

5. Remove the cotter pin, nut lock and nut.

6. Remove the thrust washer and the outer wheel bearing.

7. Remove the drum with the inner wheel bearing and the grease seal. Do not drag the inner bearing or grease seal over the stub axle threads or the threads, bearing and seal may be damaged.

8. Remove the grease seal and remove the inner bearing.

9. Check the bearing races. If any scoring or heat damage is noted, they should be replaced.

NOTE: When bearing or races need replacement, replace them as a set.

10. If the bearings and races are to be replaced, drive out the race with a brass punch.

To install:

11. Before installing new races, coat them with wheel bearing grease. Drive into place with proper size driver or brass punch. Make sure they are fully seated.

12. Thoroughly pack the bearings and lubricate the hubs with wheel bearing grease. Install the inner bearing and coat the lip and rim of the grease seal with grease. Drive the seal into place with a seal driver.

13. Install the drum (or disc brake hub, if equipped) to the vehicle.

14. Lubricate and install the outer wheel bearing to the spindle.

15. Install the thrust washer.

16. Install and tighten the wheel bearing nut to 20–25 ft. lbs. (27–34 Nm) while rotating the drum/hub. This seats the bearings

17. Back off the adjusting nut ¼ turn then tighten it finger-tight.

18. Position the nut lock over the bearing adjusting nut with 1 pair of slots in line with the cotter pin hole in the stub axle and install the cotter pin.

19. Install the grease caps. Install the rotor and caliper if equipped with disc brakes.

20. Install the wheel and tire assemblies.

CHRYSLER CORP.

Front Wheel Drive

CHRYSLER-Concorde • New Yorker • LHS
DODGE-Intrepid EAGLE-Vision

Firing Orders

NOTE: To avoid confusion, always replace spark plug wires one at a time.

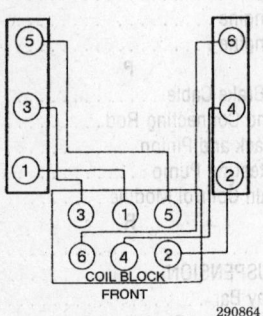

290864

**3.3L and 3.5L Engines
Engine Firing Order:
1-2-3-4-5-6
Distributorless Ignition System**

ENGINE ELECTRICAL

NOTE: Disconnecting the negative battery cable on some vehicles may interfere with the functions of the on board computer systems and may require the computer to undergo a relearning process, once the negative battery cable is reconnected.

Ignition Timing

ADJUSTMENT

The 3.3L and 3.5L engines use a distributorless ignition system referred to as Direct Ignition System (DIS). It is a fixed ignition timing system which means that basic ignition timing cannot be adjusted. All spark advance is permanently set by the Powertrain Control Module (PCM).

Alternator

PRECAUTIONS

Several precautions must be observed with alternator equipped vehicles to avoid damage to the unit.

• If the battery is removed for any reason, make sure it is reconnected with the correct polarity. Reversing the battery connections may result in damage to the 1-way rectifiers.

• When utilizing a booster battery as a starting aid, always connect the positive to positive terminals and the negative terminal from the booster battery to a good engine ground on the vehicle being started.

• Never use a fast charger as a booster to start vehicles.

• Disconnect the battery cables when charging the battery with a fast charger.

• Never attempt to polarize the alternator.

• Do not use test lights of more than 12 volts when checking diode continuity.

• Do not short across or ground any of the alternator terminals.

• The polarity of the battery, alternator and regulator must be matched and considered before making any electrical connections within the system.

• Never separate the alternator on an open circuit. Make sure all connections within the circuit are clean and tight.

• Disconnect the battery ground terminal when performing any service on electrical components.

• Disconnect the battery if arc welding is to be done on the vehicle.

REMOVAL AND INSTALLATION

3.3L (VIN T) Engine

1. Disconnect the negative battery cable.

2. Disconnect the generator field circuit plug. Remove the B+ terminal nut and wire.

3. Loosen the adjusting T-bolt and the pivot bolts. Do not remove at this time.

4. Loosen the pivot bolt but do not remove. Loosen the adjusting bolt to allow removal of the alternator drive belt. Remove the alternator drive belt.

5. Remove the adjusting T-bolt and the pivot bolt. Be careful not to loose the spacer from the pivot bolt.

6. Remove the alternator from the engine.

To install:

7. Install the alternator to the mounting bracket and secure with the pivot bolt. Make sure the spacer is properly installed on bolt prior to installation.

8. Install the adjusting T-bolt. Install the alternator drive belt and adjust tension.

9. Reconnect the B+ terminal wire to the alternator and secure with terminal nut tightened to 75 inch lbs. (9 Nm).

10. Reconnect the generator field circuit plug and the negative battery cable.

11. Start the engine and inspect the charging system for proper alternator operation.

3.5L (VIN F) Engine

1. Disconnect the negative battery cable.

2. Loosen the lower mounting bolt and the pivot bolt. Do not remove the bolts at this time.

3. Loosen the belt adjustment bolt and remove the drive belt.

4. Remove the bracket and the lower bolt. Remove the pivot bolt and the alternator from the mounting bracket.

5. Disconnect the generator field circuit plug. Remove the B+ terminal nut and wire.

6. Remove the alternator from the engine compartment.

To install:

7. Install the alternator to the engine compartment and reconnect the generator field circuit plug.

8. Reconnect the B+ terminal wire to the alternator and secure with terminal nut torqued to 75 inch lbs. (9 Nm).

9. Position the alternator on the engine bracket, install the pivot bolt and the lower mounting bolts.

10. Install the alternator drive belt and adjust the tension.

11. Reconnect the negative battery cable. Start the engine and inspect the charging system for proper alternator operation.

Drive Belt

REMOVAL AND INSTALLATION

3.3L (VIN T) Engine

1. Disconnect the negative battery cable.

2. Mark the engine drive belts for reinstallation purposes.

3. Loosen the adjusting bolt locknut and loosen the adjusting bolt

on the tensioner pulley so that removal of the drive belt is possible.

4. loosen alternator adjusting and pivot bolts.

5. Remove the engine drive belts.

To install:

6. Install the engine drive belts and tighten using adjusting bolts. Be sure that the drive belts are firmly and correctly seated in the pulley grooves.

7. Inspect belt tension with the use of belt tensioning tool kit C-4162 or equivalent, making sure tension reading agrees with the desired tension of 120 pounds (534 N) for used belt or 140–160 pounds (623–711 N) for a new belt.

8. Adjust belt tension by turning the adjusting bolt on the alternator to maintain the proper tension.

9. Torque the air conditioning belt pulley nut to 40 ft. lbs. (54 Nm).

3.5L (VIN F) Engine

1. Disconnect the negative battery cable.

2. Mark the engine drive belts for reinstallation purposes.

3. Loosen the adjusting bolt locknut and loosen the adjusting bolt

on the tensioner pulley so that removal of the drive belt is possible.

4. Remove the engine drive belts.

To install:

5. Install the engine drive belts and tighten using adjusting bolts. Be sure that the drive belts are firmly and correctly seated in the pulley grooves.

6. Inspect belt tension with the use of belt tensioning tool kit C-4162 or equivalent, making sure tension reading agrees with the desired tension of 120 pounds (534 N) for used belt or 140–160 pounds (623–711 N) for a new belt.

7. Adjust belt tension by adjusting the tensioner pulley located on the timing belt cover to maintain the proper tension.

8. Torque the air conditioning belt pulley nut to 40 ft. lbs. (54 Nm).

Starter

REMOVAL AND INSTALLATION

3.3L (VIN T) Engine

1. Disconnect the negative battery cable.

ADJUSTING BOLT   LOCKING NUT

299044

Accessory drive belts — 3.3L engine

2. Raise and safely support the vehicle.

3. Remove the three starter attaching bolts from the engine and transaxle assembly.

4. Remove the starter assembly from the engine. Position the starter to gain access to the wire harness connectors.

5. Remove the positive battery cable nut and wiring terminal nut from the starter solenoid.

6. Disconnect the push connector on the starter solenoid.

7. Remove the starter from the vehicle.

To install:

8. Clean corrosion and dirt from the wire terminals and install onto the starter solenoid. Install the starter assembly in position and secure using mounting bolts. Make sure the ground wire is installed on the lower mounting bolt during installation.

9. Torque the three mounting bolts to 40 ft. lbs. (54 Nm).

10. Lower the vehicle to the ground.

11. Reconnect the negative battery terminal.

3.5L (VIN F) Engine

1. Disconnect the negative battery cable.

2. Raise and safely support the vehicle.

3. Remove the three starter attaching bolts from the engine and transaxle assembly.

4. Remove the starter assembly from the engine. Position the starter to gain access to the wire harness connectors.

5. Remove the positive battery cable nut and wiring terminal nut from the starter solenoid.

6. Disconnect the push-on connector on the starter solenoid.

7. Remove the starter from the vehicle.

To install:

8. Clean corrosion and dirt from the wire terminals and install onto the starter solenoid. Install the starter assembly in position and secure using mounting bolts. Make sure the ground wire is installed on the lower mounting bolt during installation.

9. Torque the three mounting bolts to 40 ft. lbs. (54 Nm).

10. Lower the vehicle to the ground.

11. Reconnect the negative battery terminal.

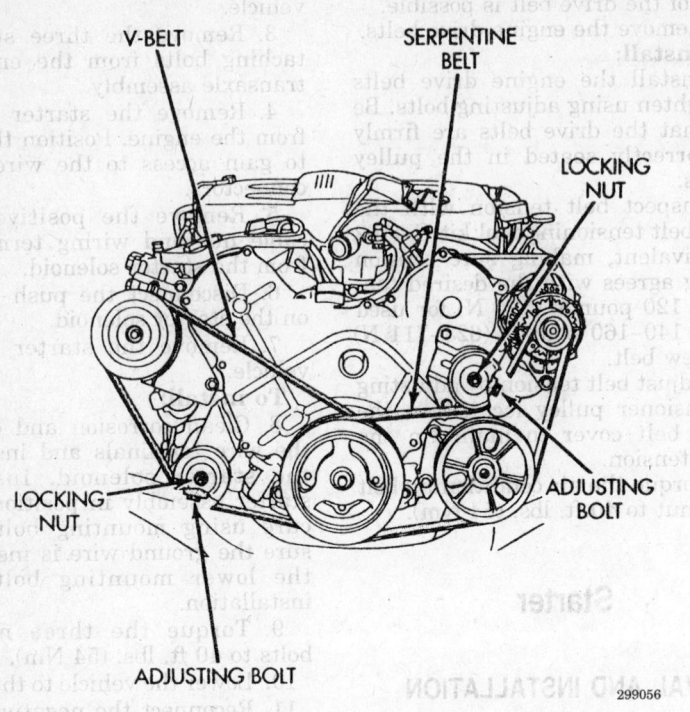

V-BELT

SERPENTINE BELT

LOCKING NUT

LOCKING NUT

ADJUSTING BOLT

ADJUSTING BOLT

Accessory drive belts — 3.5L engine

299056

CHASSIS ELECTRICAL

Blower Motor

REMOVAL AND INSTALLATION

The blower motor is located on the right side of the heater housing. All service to the blower motor is made inside the vehicle under the right side of the instrument panel. The blower motor and the blower motor wheel (fan) must be replaced as an assembly.

1. Disconnect the negative battery cable.
2. Remove the lower right side under panel silencer duct.
3. Remove the blower motor connector from the resistor block.
4. Squeeze the blower motor wiring grommet and push the grommet through the blower motor housing cover.
5. Remove the blower motor housing cover.
6. Remove the blower motor retaining screws.

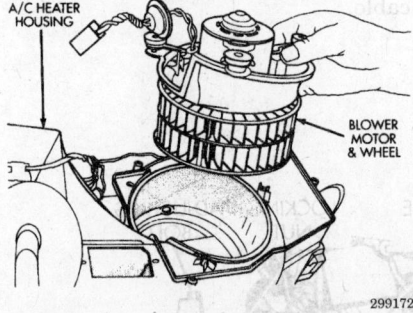

A/C HEATER HOUSING

BLOWER MOTOR & WHEEL

299172

Blower motor is replaced with the blower wheel (fan) as an assembly

7. Lower the blower motor from the housing.

To install:

8. Install the blower motor to the housing and install the retaining screws.
9. Install the blower motor housing cover and install the retaining screws.
10. Install the blower motor wiring grommet in the grommet hole on the housing cover.
11. Connect the blower motor connector to the resistor block.
12. Install the lower right side under panel silencer duct.

13. Connect the negative battery cable.
14. Check the blower motor for proper operation.

Some vehicles may exhibit a whistling noise from the heater housing unit. This noise is most noticeable with the fan in the mid-range speed and the blend air door in the high temperature position. If this noise occurs, remove the blower motor and wheel assembly. Apply strip caulking to all seam areas around the blower wheel housing and evaporator housing. The seams can be reached from the blower motor opening. Do not add an excessive amount of sealer which will restrict airflow.

Windshield Wiper Motor

REMOVAL AND INSTALLATION

1. With the ignition **OFF**, disconnect the negative battery cable.
2. Be sure that the windshield wiper arms/blades are in the **PARK** position.
3. Lift the wiper arm up to raise the wiper blades off of the windshield. Move the retainer tab to hold up the wiper arm.
4. Using a slight rocking motion, remove the wiper arm and blade assembly from the wiper arm pivot. Disconnect the washer hoses from the inline connectors. Be careful not to damage the connectors.
5. Remove the cowl top plastic screen and disconnect the washer hose at the inline connector. Make sure the connector is not lost.
6. Disconnect the motor connector at the back side of the housing. Remove the four wiper housing module mounting screws then remove the housing.
7. Remove the nut and disconnect the wiper drive link from the motor crank.
8. Remove the three motor mounting screws and lift the motor and mounting plate out of the housing.
9. Disconnect the motor harness grommet from the housing.

To install:

10. Reconnect the motor harness grommet to the housing.
11. Position the motor and mounting plate in the housing and install three mounting screws. Torque the mounting screws to 89–106 inch lbs. (11–12 Nm). Reconnect the wiper drive link to the motor crank and install the retainer nut. Torque the crank nut to 89–124 inch lbs. (11–14 Nm).

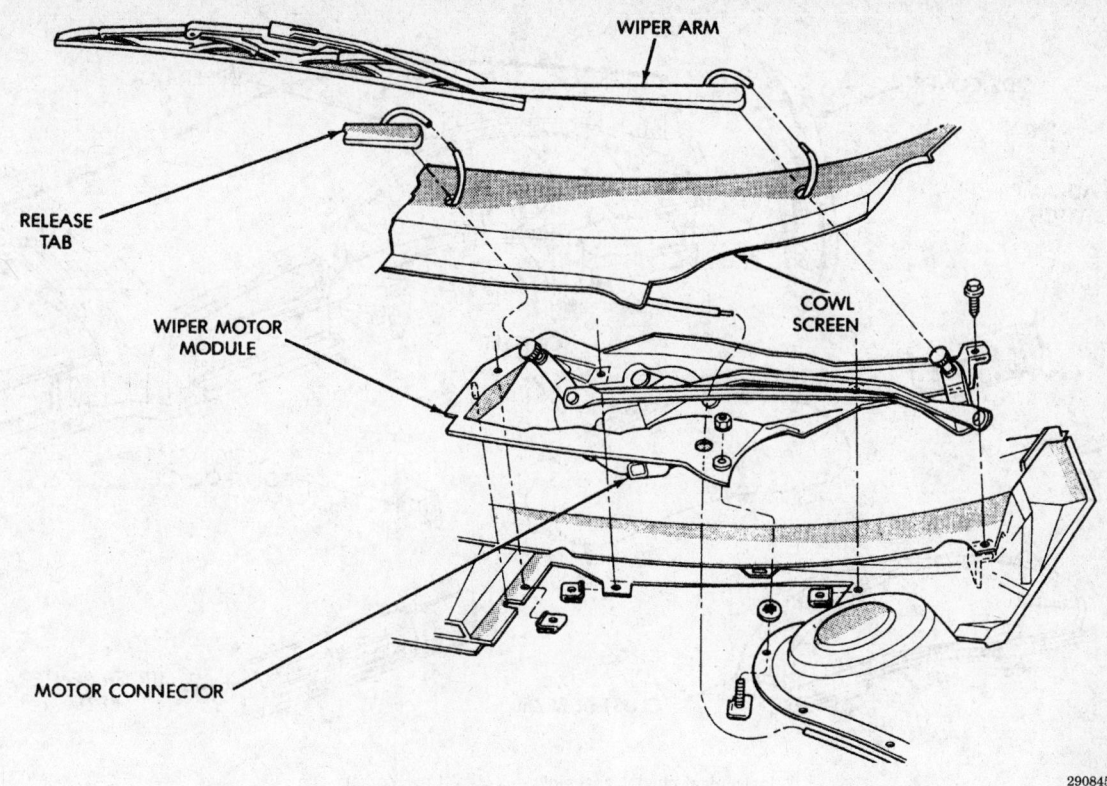

RELEASE TAB

WIPER ARM

WIPER MOTOR MODULE

COWL SCREEN

MOTOR CONNECTOR

290845

Wiper motor and linkage assembly

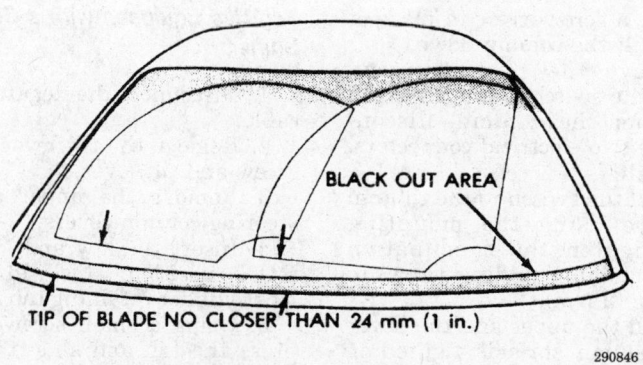

BLACK OUT AREA

TIP OF BLADE NO CLOSER THAN 24 mm (1 in.)

290846

Wiper arm adjustment

12. Install the wiper housing and secure using four mounting screws. Reconnect the motor connector at the back side of the housing.

13. Reconnect the washer hose at the inline connector and install the cowl top plastic screen.

14. Install the wiper arms and the blades. Reconnect the hose to the inline connector at the base of each arm.

15. Move the retainer tab to allow the wiper arm and blade assembly to lower onto the windshield.

16. Connect the negative battery cable. Test the wiper motor for proper operation.

17. Place the wiper arm/blades into the **PARK** position, and be sure that the tip of the wiper blades rest in the blackout area of the windshield but are no closer than 1.0 inch (24 mm) from the bottom edge of the windshield to the blade. Remove the wiper arm from the pivot and reposition the arm, if necessary.

Headlight Switch

REMOVAL AND INSTALLATION

1. Disconnect negative battery cable.

2. Open the left front door and remove the instrument panel left end cover.

3. Remove the screw from the left end of the instrument panel and pull the headlight switch bezel rearward to disengage locking clips.

4. With the bezel removed, remove the three screws on the headlight switch. Pull the switch out and disconnect the two wiring connectors. Remove the switch from the instrument panel.

To install:

5. Reconnect the two wiring connectors to the headlight switch and install the switch into the dashboard. Install the three retaining screws.

6. Install the headlight switch bezel. Be sure to engage the locking clips and secure with the retaining screw.

7. Install the instrument panel left end cover.

8. Reconnect the negative battery cable. Test the headlight switch for proper operation.

Combination Switch

REMOVAL AND INSTALLATION

These vehicles use a multi-function switch (also called a combination

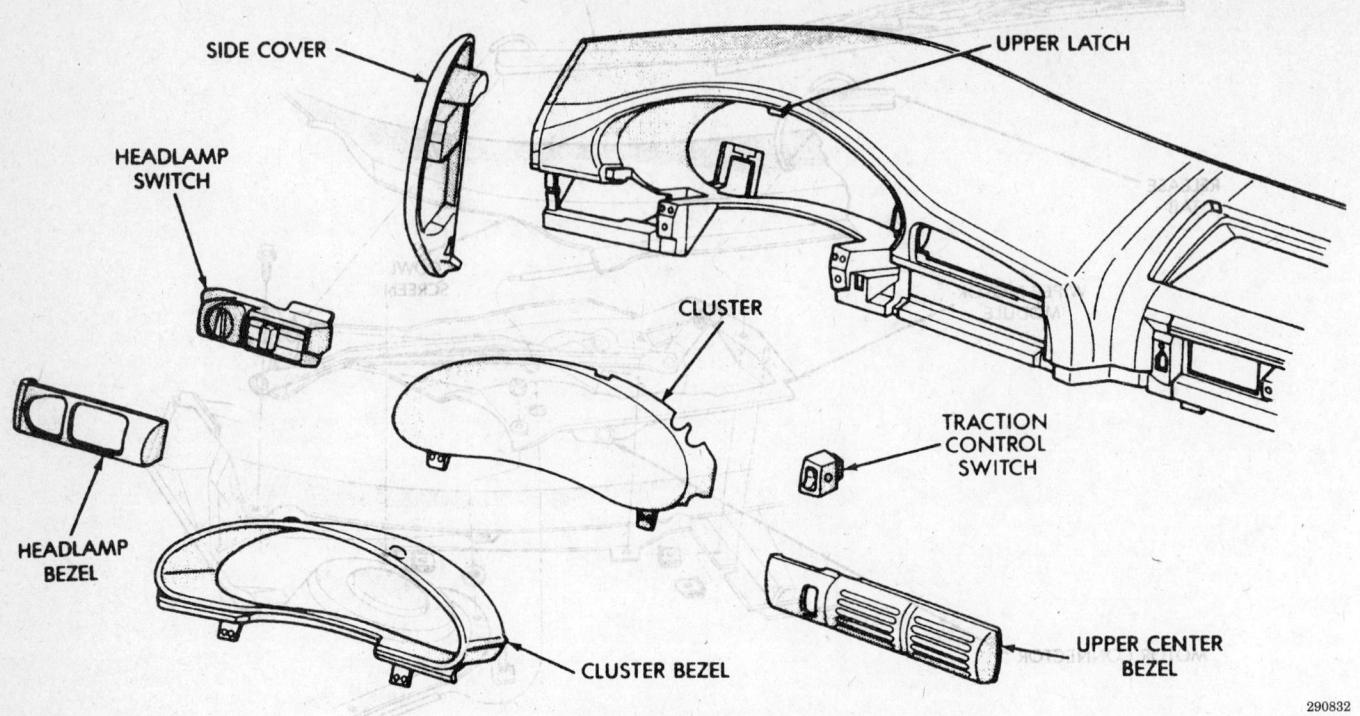

SIDE COVER

HEADLAMP SWITCH

HEADLAMP BEZEL

UPPER LATCH

CLUSTER

TRACTION CONTROL SWITCH

CLUSTER BEZEL

UPPER CENTER BEZEL

290832

Instrument cluster assembly — except New Yorker

switch) to control a variety of tasks. This switch contains the electrical circuitry for the turn signals, the hazard warning switch, headlamp beam select switch, headlamp optical horn, windshield wiper switch, pulse wipe and windshield washer switching.

This integrated switch assembly is mounted to the left side of the steering column. Should any function of the switch fail, the entire switch assembly must be replaced.

CAUTION

The Air Bag system must be disarmed before removing the combination switch. Failure to do so may cause accidental deployment of the air bag, resulting in unnecessary system repairs and/or personal injury.

1. Disconnect negative battery cable.
2. Properly disarm the air bag system.
3. Tilt the steering column up and remove six screws holding the lower column shroud.
4. Tilt the steering column down and remove the upper column shroud.
5. Tilt the column up and remove the tilt lever.

6. Using a screwdriver in place of the lever, tilt the column down.
7. Remove the two mounting screws on the switch and remove the switch from the column, disconnecting the two electrical connectors.

To install:

8. Install the switch to the column and secure using the mounting screws. Tighten the combination switch-to-column mounting screws to 17 inch lbs. (2 Nm).
9. Install the upper and the lower steering column shrouds tightening the retainers to 17 inch lbs. (2 Nm).
10. Install the tilt lever.
11. Properly enable the air bag system.
12. Connect the negative battery cable. Check all functions of the combination switch.

Ignition Lock Cylinder

REMOVAL AND INSTALLATION

NOTE: When replacing the ignition lock cylinder on a vehicle with a column shifter, a new interlock cassette must be installed. If the vehicle has a floor shift, the interlock cable at the floor shifter must be adjusted.

Vehicles Equipped with a Column Shifter

1. Disconnect the negative battery cable.
2. Remove the tilt lever mounting screw and tilt lever.
3. Remove the upper and lower steering column covers.
4. Insert the key and turn to the **RUN** position. This will cause the lock cylinder retaining tab to depress.
5. Using a small screwdriver, depress the tab and slide the ignition lock cylinder out from the housing.
6. Remove the column shift interlock cassette as follows:
   a. Push down the tab on the top of the interlock cassette.
   b. Slide the shift interlock cassette out of the housing.
   c. Disconnect the cable from the locking arm on the column shifter mechanism.

To install:

7. Install the key into the ignition lock cylinder. Turn the key to the **RUN** position. The retaining tab on the lock cylinder can be depressed using a small screwdriver.
8. The shaft at the end of the ignition key lock cylinder aligns with the socket in the end of the housing. To ensure proper alignment, the socket must be in the **RUN** position.

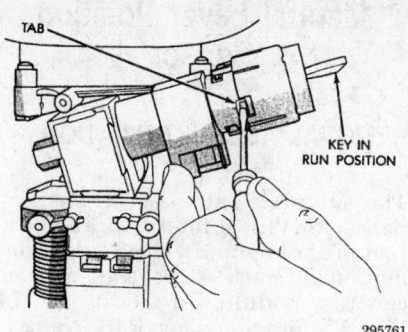

295761

Ignition lock cylinder removal

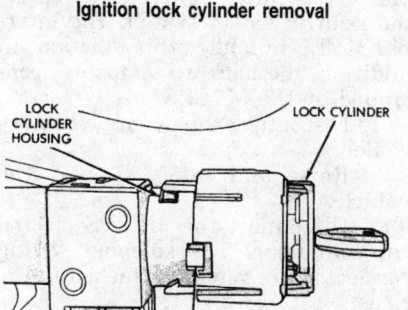

295926

Ignition lock cylinder installation

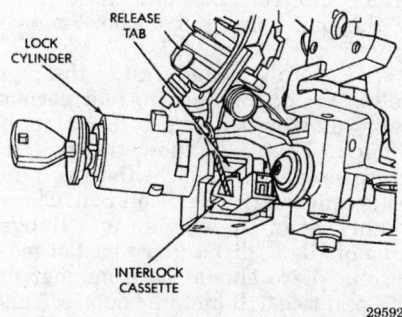

295927

Column shift ignition interlock cassette

9. Align the ignition lock cylinder with the grooves in the housing. Slide the cylinder into the housing until the tab sticks through the opening in the housing.

10. Turn the ignition key to the **OFF** position. Remove the ignition key.

NOTE: An adjustment to the column shift interlock system is performed only after a new interlock cassette has been installed. It cannot be adjusted more than once. If the interlock system op-

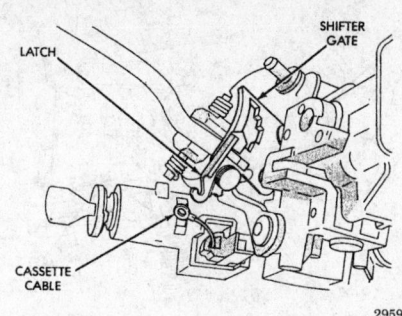

295928

Latch and shifter gate

erates incorrectly, install and adjust a new shift interlock cassette.**

11. Install and adjust a new interlock cassette as follows:

a. Be sure that the latch rotates freely on the shifter gate.

b. With the key removed and the shifter in the **PARK** position, install the cable over the hook on the locking arm of the column shifter mechanism.

c. Push the interlock cassette into the housing until it locks into place.

d. Adjust the interlock system by pushing in the adjustment tab until it stops. The adjustment tab will click as it moves into position. Be sure that the tab is fully depressed.

12. Install the upper and lower steering column covers.

13. Install the tilt lever and tighten tilt lever attaching screw.

14. Reconnect the negative battery cable.

15. Check the system for proper operation.

Vehicles Equipped with a Floor Shifter

1. Disconnect the negative battery cable.

2. Remove the tilt lever mounting screw and tilt lever.

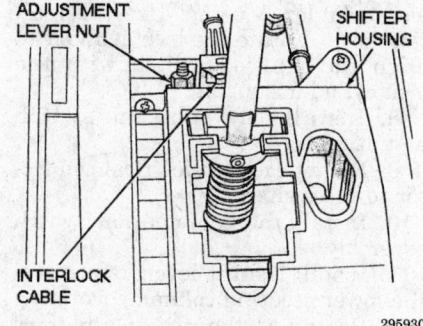

295930

Interlock cable

3. Remove the upper and lower steering column covers.

4. Insert the key and turn to the **RUN** position. This will cause the lock cylinder retaining tab to depress.

5. Using a small screwdriver, depress the tab and slide the ignition lock cylinder out from the housing.

To install:

6. Install the key into the ignition lock cylinder. Turn the key to the **RUN** position. The retaining tab on the lock cylinder can be depressed using a small screwdriver.

7. The shaft at the end of the ignition key lock cylinder aligns with the socket in the end of the housing. To ensure proper alignment, the socket must be in the **RUN** position.

8. Align the ignition lock cylinder with the grooves in the housing. Slide the cylinder into the housing until the tab sticks through the opening in the housing.

9. Turn the ignition key to the **OFF** position. Remove the ignition key.

10. Adjust the floor shift interlock cable as follows:

a. Remove the floor shifter handle.

b. Remove the floor console bezel and loosen the interlock lever adjustment nut.

c. Install the ignition key and turn it to the **RUN** position.

d. Remove interlock cable from the floor shifter housing. Slide the cable out of the groove in the shift interlock lever.

e. Inspect the interlock cable for the following:

• With the ignition key removed and the lock cylinder in the **OFF (lock)** position, the cable core wire should not move when it is pulled. If movement occurs, the cable is installed improperly or it is kinked.

• With the ignition key in the **RUN** position, the cable core wire should slide freely when it is pulled and return to the bottomed out position when it is released. If no movement occurs while in the **RUN** position, the cable is installed improperly or the wire is kinked.

f. Place the floor shifter in the **PARK** position. Slide the cable core wire into the groove on the adjustment lever and be sure that the cable end seats in the groove.

g. Slip the cable into the housing until it snaps into place. Be sure that the floor shifter remains in the **PARK** position.

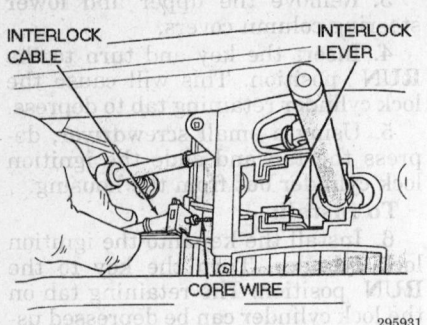

Floor shift interlock cable

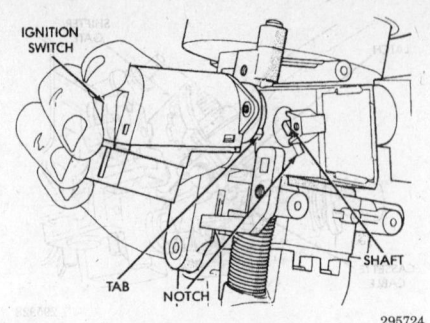

Ignition switch alignment

h. Remove the ignition key (be sure that the switch is in the **OFF** position).

i. With the adjustment nut on the interlock lever loosened, the cable should index itself to the correct position. Once this has happened, torque the adjustment nut to 50 in. lbs. (6 Nm).

j. Check the interlock adjustment as follows:

• With the key in the **OFF** position, the floor shifter should be locked in the **PARK** position. If not, re-adjust the interlock and tighten adjustment nut.

• **Without starting the engine,** turn the key to the **RUN** position. Place the shifter in the **Reverse** position. The key should not be able to be removed from the lock cylinder. If it can, re-adjust the interlock and tighten the adjustment nut.

• Place the floor shifter in the **PARK** position. Turn the ignition key to the **OFF** position. The ignition key should be able to be removed. If this is not possible, re-adjust the interlock and tighten the adjustment nut.

k. Install the bezel on the floor shifter console.

l. Install the floor shifter handle.

11. Install the upper and lower steering column covers.

12. Reconnect the negative battery cable.

13. Check the system for proper operation.

Ignition Switch

REMOVAL AND INSTALLATION

The ignition switch attaches to the lock cylinder housing opposite the lock cylinder at the top of the steer-

ing column, under the steering wheel. The steering wheel does not have to be removed but the air bag system must be disarmed.

CAUTION

The Air Bag system must be disarmed before working around the steering column. Failure to do so may cause accidental deployment of the air bag, resulting in unnecessary system repairs and/or personal injury.

1. Disconnect negative battery cable. Disarm the air bag system using the recommended procedure.

2. Remove the tilt lever attaching screws and the tilt lever from the steering column.

3. Remove the upper and the lower covers from the upper steering column.

4. Remove the combination switch assembly.

5. Disconnect the electrical connector from the ignition switch.

6. Remove the ignition switch mounting screws and the switch.

To install:

7. Position the ignition switch so the tab on the switch is aligned to the notch in the cylinder housing. Also, a slot in the end of the ignition switch fits over the shaft in the end of the lock cylinder housing. Use the ignition key to rotate the lock cylinder to align the ignition switch with the lock cylinder housing.

8. Install the ignition switch mounting screws.

9. Reconnect the electrical connector to the switch.

10. Install the combination switch assembly.

11. Install the tilt lever, upper and the lower steering column covers.

12. Reconnect the negative battery cable. Verify correct operation of the ignition switch.

Manual Lever Position Sensor

REMOVAL AND INSTALLATION

The Manual Valve Lever Position Sensor (MVLPS) interprets the position of the manual valve. This information is sent to the transmission control module. In addition, the MVLPS provides for Park/Neutral starter operation. There is no external neutral safety switch. The internal MVLPS handles this function. In addition, the back-up lamps are controlled by the MVLPS.

1. Disconnect the negative battery cable.

2. Raise and safely support the vehicle.

3. Disconnect the MVLPS electrical connector. The solenoid wiring connector can remain attached to the case.

4. Disconnect the shift cable from the shift lever at the transaxle.

5. Move the shift lever clockwise as far as it will go. This should be one position past the **L** position. Then remove the shift lever.

6. Place a large oil drain pan under the transaxle pan. Loosen all of the pan bolts, but do not remove any.

7. Carefully remove all of the pan bolts, except for the two rear corner bolts allowing the pan to hang, as if hinged. This will allow the pan to drain properly. Remove the two pan bolts and the transaxle oil pan. Thoroughly clean the oil pan and flange area of all oil, dirt and old gasket material. There should be a magnet in the pan to catch metallic debris. This should be cleaned and reinstalled.

8. Remove the oil filter from the valve body. It is held in place by two clips.

NOTE: The 42LE transaxle oil filter is not interchangeable with the 41TE transaxle filter. Installation of the 41TE oil filter in a 42LE may cause transaxle damage. Make absolutely sure the correct replacement part is installed.

9. Remove the valve body bolts and carefully remove the valve body from the transaxle.

NOTE: Use care. The overdrive and underdrive accumulator springs may fall out when removing the valve body.

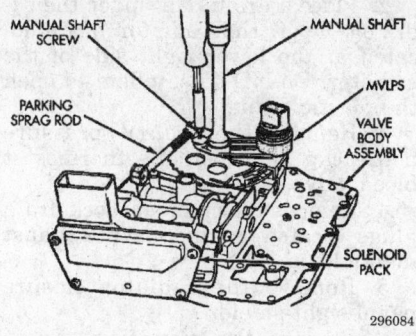

Manual valve lever position sensor removal and installation

10. Remove the manual shaft retainer screw and slide the MVLPS from the shaft.

To install:

11. Slide the MVLPS onto the manual valve shaft and tighten the retaining screw.

12. Install the valve body assembly into the transaxle. Torque the valve body screws to 40 inch lbs. (5 Nm). Be sure that the overdrive and underdrive accumulator springs have **NOT** fallen out of the valve body but are in their proper places.

13. Place the new transaxle oil filter in correct position against the valve body and install the two retaining clips.

14. Apply an ⅛-inch bead of silicone sealer around the pan flange and install the transaxle oil pan into position on the transaxle case. Install the oil pan-to-transaxle case screws and torque to 14–17 ft. lbs. (19–23 Nm).

15. Install the shift lever in the transaxle. Reconnect the shift cable to the shift lever on the transaxle.

16. Reconnect the MVLPS electrical connector.

17. Lower the vehicle. Reconnect the negative battery cable.

18. Refill the transaxle with fresh clean Mopar ATF Plus Automatic Transmission Fluid Type 7176 or DEXRON IIE Automatic Transmission Fluid only. Refill to the correct level as measured by the transaxle dipstick.

19. Start the vehicle and allow it to idle with the shifter in the **PARK** position for at least 1 minute.

20. Fully apply the parking brake. Place the shift lever in each gear position momentarily, then return it to the **PARK** position. Verify that the back-up lights work and that the engine will start only in the **PARK** position and the **NEUTRAL** position.

21. Check the fluid level while the vehicle is still idling. Add transaxle fluid to reach the correct level, if necessary.

PCM Computer

REMOVAL AND INSTALLATION

The engine management system for these vehicles is an on-board computer. It was formerly known as the Engine Control Module (ECM). It is now called the Powertrain Control Module (PCM) which more accurately reflects its expanded role. It is located under the hood. Use care handling the PCM and always disconnect the negative battery cable before servicing the PCM.

1993–95 Vehicles

1. Disconnect the negative battery cable.

2. Remove the air cleaner assembly.

3. Remove the stud bolt and push pin that secures the PCM to the body. The push pin can be removed by pulling up on the push pin center lock.

4. Lift up the PCM. Disconnect the 60-way connector to the PCM.

5. Remove the PCM from the vehicle.

To install:

6. Connect the 60-way connector to the PCM.

7. Tighten the 60-way connector retaining screw. Torque to 35 inch lbs. (4 Nm).

8. Place the PCM back to its original position in the engine compartment. Install the stud bolt and torque the bolt to 7 ft. lbs. (10 Nm). Install the push pin. Push down on the push pin center lock.

9. Install the air cleaner assembly.

10. Connect the negative battery cable. If a new PCM is being installed, program the mileage into the PCM using a DRB or equivalent scan tool.

1996 Vehicles

1. Disconnect the negative battery cable.

2. Remove the air cleaner assembly.

3. Disconnect the two 40-way connectors to the Powertrain Control Module (PCM).

4. Remove the 2 mounting bolts from the PCM.

5. Remove the PCM from the vehicle.

To install:

6. Place the PCM back to its original position in the engine compartment. Install the 2 PCM mounting bolts and torque the bolts to 35 inch lbs. (4 Nm).

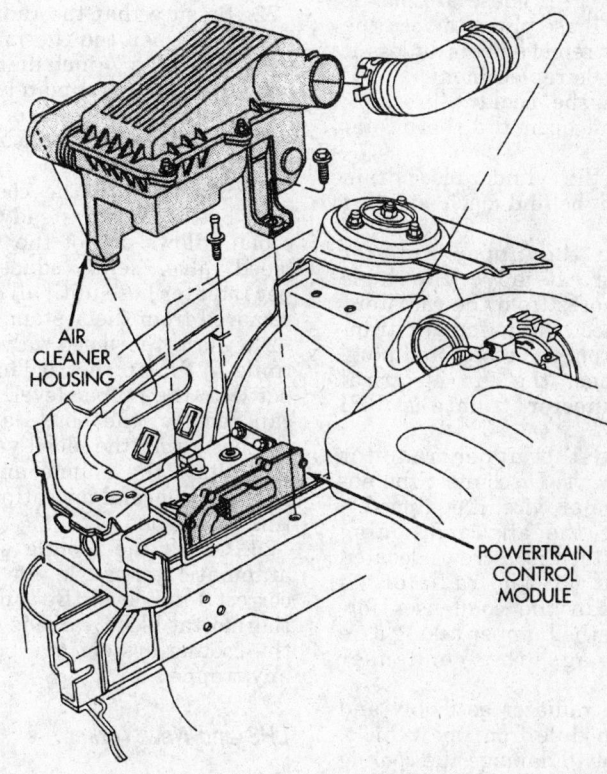

Removing the air cleaner assembly to access the PCM

7. Connect the two 40-way connectors to the PCM.

8. Install the air cleaner assembly.

9. Connect the negative battery cable. If a new PCM is being installed, program the mileage into the PCM using a DRB or equivalent scan tool.

ENGINE COOLING

Radiator

REMOVAL AND INSTALLATION

—— CAUTION ——

Do not remove the cylinder block plug or the radiator drain with the system hot and under pressure or serious burns from coolant may occur.

Concorde, Intrepid and Vision

1. Disconnect negative battery cable.

2. Place a drain pan under the radiator. Open the radiator drain located at the lower right side of the radiator. Use care. These are plastic parts. Do **NOT** use pliers to open the drain. Do not remove drain unless it leaks and needs replacement.

3. Remove the coolant pressure bottle cap and open the thermostat bleed valve.

4. Remove the cylinder block drain plugs located behind each exhaust manifold.

5. Remove the upper radiator crossmember. Remove the hose clamps and hoses from the radiator.

6. Disconnect the automatic transaxle hoses from the cooler and plug.

7. Disconnect the fan electrical harness connector from the RFI module.

8. Remove the upper radiator mounting screws. Disconnect the engine block heater wire, if equipped.

9. Remove the air conditioning condenser attaching screws located at the front of the radiator, if equipped. Lean the condenser forward against the bumper taking care not to damage the condenser assembly.

10. Lift the radiator assembly and cooling fan module from the vehicle. Take care not to damage the cooling fins or the water tubes on the radiator during removal.

11. Remove the cooling fan assembly from the radiator.

To install:

12. Attach the cooling fan assembly to the radiator.

13. Slide the radiator and the fan module down into position, seat the radiator assembly lower rubber isolator in the mount holes provided.

14. Attach the air conditioning condenser to the radiator, if equipped. Torque the mounting screws to 45 inch lbs. (5 Nm).

15. Install and torque the radiator mounting bolts to 123 inch lbs. (14 Nm).

16. Reconnect the engine block heater wire, if equipped.

17. Reconnect the lower radiator hose and clamp, then the automatic transaxle cooler hoses. Torque the radiator hose clamps to 22 inch lbs. (2.5 Nm).

18. Install the upper radiator hose and align so it will not interfere with the hood, accessory drive belt or engine.

19. Install the upper radiator crossmember. Torque the crossmember bolts to 21 ft. lbs. (28 Nm).

20. Install the cylinder block drain plugs on both sides of the engine block.

21. Reconnect the fan motor electrical connector and reconnect the negative battery cable.

22. Be sure that the radiator drain is closed. Open the thermostat bleed valve. Install a ¼-inch diameter clear hose that is approximately 48 inches in length, to the end of the bleed valve and the other end into a clean container.

23. Slowly refill the coolant pressure bottle until a steady stream of coolant flows out of the thermostat bleed valve. Gently squeeze the upper radiator hose until all of the air is removed from the system.

24. Close the bleed valve and continue to fill up the coolant pressure bottle to the proper level. Install the cap back on the bottle and remove the hose from the bleed valve.

25. Start the engine and allow to run until normal operating temperature is reached.

26. Check the cooling system and automatic transaxle for leaks and correct fluid level. Be sure that the thermostat bleed valve is closed once the cooling system has been bled of any trapped air.

LHS and New Yorker

1. Disconnect the negative battery cable.

2. Place a drain pan under the radiator. Open the radiator drain located at the lower right side of the radiator. Do **NOT** use pliers to open the plastic drain.

3. Remove the coolant pressure bottle cap and open the thermostat bleed valve.

4. Remove the cylinder block drain plugs located behind each exhaust manifold.

5. Remove the radiator closure panel sight shield.

6. Remove the mounting screws that secure the headlamp module to the mounting adapter.

7. Remove the headlamp module from the mounting adapter.

8. Disconnect the wiring connectors from the headlamp and fog lamp assemblies.

9. Remove the headlamp module from the vehicle.

10. Discharge the air conditioning system using a recovery machine.

11. Remove the upper radiator crossmember. Remove the hose clamps and the hoses from the radiator.

12. Disconnect and plug the automatic transaxle cooler hoses.

13. Disconnect the fan electrical harness connector from the RFI module.

14. Remove the upper radiator mounting screws and disconnect the engine block heater wire, if equipped.

15. Disconnect and plug the air conditioning system hoses from the condenser unit.

16. Remove the radiator/condenser/cooling fan assembly out of the vehicle. Be careful not to damage the radiator or condenser cooling fins or water tubes.

17. Separate the cooling fan assembly from the radiator. Separate the radiator from the air conditioning condenser.

To install:

18. Install the condenser unit to the radiator. Install the cooling fan assembly to the radiator and condenser.

19. Install the radiator/condenser/cooling fan assembly into position in the vehicle seating the lower rubber isolators in the mounting holes provided.

20. Reconnect the air conditioning system hoses to the condenser unit.

21. Install and torque the radiator mounting bolts to 123 inch lbs. (14 Nm).

22. Reconnect the engine block heater wire, if equipped.

23. Install the lower radiator hose and clamp, then reconnect the auto-

matic transaxle cooler hoses. Tighten the lower hose clamp.

24. Install the upper radiator hose and align so it will not interfere with the hood, accessory drive belt or engine.

25. Install the upper radiator crossmember. Torque the crossmember bolts to 21 ft. lbs. (28 Nm).

26. Install the cylinder block drain plugs on both sides of the engine block.

27. Reconnect the fan motor electrical connector and reconnect the negative battery cable.

28. Be sure that the radiator drain cock is closed. Open the thermostat bleed valve. Install a ¼-inch (6.35mm) diameter clear hose that is approximately 48 inch (1219mm) in length, to the end of the bleed valve and the other end into a clean container. The intent is to keep coolant off the drive belt. Be careful not to allow coolant to get on the drive belt.

29. Slowly refill the coolant pressure bottle until a steady stream of coolant flows out of the thermostat bleed valve. Gently squeeze the upper radiator hose until all of the air is removed from the system.

30. Close the bleed valve and continue to fill up the coolant pressure bottle to the proper level. Install the cap back on the bottle and remove the hose from the bleed valve.

31. Start the engine and allow it to run until normal operating temperature is reached.

32. Check the cooling system and automatic transaxle for leaks and correct fluid level. Be sure that the thermostat bleed valve is closed once you have bled the cooling system of all trapped air.

33. Reconnect the wiring connector to the headlamp and fog lamp assemblies.

34. Place the headlamp module into proper position on the mounting adapter. Tighten the retaining screws securing the headlamp module to the mounting adapter.

35. Install the radiator closure panel sight shield. Recharge the air conditioning system.

Water Pump

REMOVAL AND INSTALLATION

3.3L (VIN T) Engine

1. Disconnect the negative battery cable.

CAUTION

Do not remove the radiator drain with the system hot and under pressure or serious burns from coolant may occur.

---

2. Place a drain pan under the radiator. Open the radiator drain located at the lower right side of the radiator. Do **NOT** use pliers to open the plastic drain.

3. Remove the coolant pressure bottle cap and open the thermostat bleed valve.

4. Remove the serpentine belt. If necessary, remove the right front lower fender shield.

5. Remove the water pump pulley bolts and pulley.

6. Remove the water pump mounting bolts and pump. Discard the O-ring seal.

7. Clean the gasket sealing surfaces. Do not scratch the aluminum surfaces.

To install:

8. Install a new O-ring into the O-ring groove and the water pump to the timing chain case. Be sure to keep the O-ring free of any oil or grease.

9. Install the retaining bolts and torque to 105 inch lbs. (12 Nm).

10. Rotate the pump and check for freedom of movement.

11. Install the pump pulley and torque the bolts to 250 inch lbs. (30 Nm).

12. Install the serpentine belt and right lower fender shield.

13. Be sure that that the radiator drain is closed. Open the thermostat bleed valve. Install a ¼-inch clear hose about 48 inches long, to the end of the bleed valve and the other end into a clean container. The intent is to keep coolant off the drive belt(s).

14. Slowly refill the coolant pressure bottle until a steady stream of coolant flows out of the thermostat bleed valve. Gently squeeze the upper radiator hose until all of the air is removed from the system.

15. Close the bleed valve and continue to fill up the coolant pressure bottle to the proper level. Install the cap back on the bottle and remove the hose from the bleed valve.

16. Reconnect the negative battery cable. Start the engine and allow to run until normal operating temperature is reached.

17. Check the cooling system for leaks and correct coolant level. Be sure that the thermostat bleed valve is closed once the cooling system has been bled of any trapped air.

3.5L (VIN F) Engine

1. Disconnect the negative battery cable.

CAUTION

Do not open the radiator drain or the coolant pressure bottle cap with the system hot and under pressure or serious burns from coolant may occur.

---

2. Place a drain pan under the radiator. Open the radiator drain located at the lower right side of the radiator. Do **NOT** use pliers to open the plastic drain .

3. Remove the coolant pressure bottle cap and open the thermostat bleed valve.

4. Remove the timing belt using the recommended procedure. It is good practice to turn the crankshaft until the engine is at TDC No. 1 cylinder, compression stroke (firing position). This aligns all the timing marks and serves as a reference point for all the work that follows..

5. Remove the water pump mounting bolts and pump. Discard the O-ring seal.

6. Clean the gasket sealing surfaces. Do not scratch the aluminum surfaces.

To install:

7. Install a new O-ring and wet with clean coolant. Be sure to keep the new O-ring free of any oil or grease.

8. Install the water pump and O-ring to the engine block.

9. Install the retaining bolts and torque to 105 inch lbs. (12 Nm).

10. Rotate the pump and check for freedom of movement.

11. Install the timing belt using the recommended procedure. Verify that all valve timing marks align. This is most important. An engine out-of-time will be seriously damaged when first started.

12. Be sure that the radiator drain is closed. Open the thermostat bleed valve. Install a ¼-inch clear hose about 48 inches long to the end of the bleed valve and the other end into a clean container. The intent is to keep coolant off of the drive belt(s).

13. Slowly refill the coolant pressure bottle until a steady stream of coolant flows out of the thermostat bleed valve. Gently squeeze the upper radiator hose until all of the air is removed from the system.

14. Close the bleed valve and continue to fill up the coolant pressure bottle to the proper level. Install the cap back on the bottle and remove the hose from the bleed valve.

15. Reconnect the negative battery cable. Start the engine and allow to run until normal operating temperature is reached.

16. Check the cooling system for leaks and correct coolant level. Be sure that the thermostat bleed valve is closed once the cooling system has been bled of any trapped air.

Thermostat

REMOVAL AND INSTALLATION

1. Disconnect negative battery cable.

— **CAUTION** —

Do not remove the radiator pressure cap with the cooling system hot and under pressure. Serious burns from hot, pressurized coolant can result.

2. Allow cooling system to cool completely. Place a drain pan under the radiator drain, located at the lower right side of the radiator. Open the drain to drain the cooling system to a level just below the thermostat. Do **NOT** drain the cooling system completely, unless the coolant is being changed. Close the drain.

3. Loosen the radiator hose clamp from the upper radiator hose at the thermostat housing and disconnect the hose from the housing.

4. Remove the thermostat bolts and housing.

5. Remove the thermostat and discard the gasket. Thoroughly clean both gasket sealing surfaces.

To install:

6. Clean the thermostat housing bolts well. On any engine, bolts that thread into the water jacket are subject to rust and corrosion. A bolt with rust or debris in the threads can easily strip out the threads in an aluminum manifold requiring a time-consuming thread repair operation. Clean the bolts well and lightly lubricate before installation.

7. Install the thermostat into the recess in the intake manifold. Place a new gasket, moistened with water, on the water inlet mating surface on the intake manifold.

8. Install the thermostat housing over the gasket and thermostat and tighten the mounting bolts to 21 ft. lbs. (28 Nm).

9. Connect the upper radiator hose to the thermostat housing and tighten the radiator hose clamp.

10. Refill and bleed the cooling system.

Electric Cooling Fan

REMOVAL AND INSTALLATION

1. Disconnect negative battery cable.

2. Disconnect the fan electrical harness connector from the RFI module.

3. Remove the retaining screws and clips that secure the cooling fan assembly to the radiator.

4. Remove the cooling fan assembly from the radiator.

5. If it is necessary to separate the fans from the motor shafts, remove the retaining clip or nut from the motor shaft and gently, but firmly, force the fan off of the shaft.

6. If necessary, separate the fan motors from the cooling fan shroud assembly by removing the mounting screws.

To install:

7. Place the fan motor in position on the cooling fan shroud and install the mounting screws. Torque the right fan motor mounting screws to 25 inch lbs. (3 Nm) and the left fan motor mounting screws to 45 inch lbs. (5 Nm).

8. Install the fans on the motor shafts by firmly sliding the fan over the shaft and installing the retaining clip or retaining nut.

9. Install the cooling fan assembly to the radiator. Install the retaining clips and torque the shroud-to-radiator retaining screws to 45 inch lbs. (5 Nm).

10. Connect the fan motor electrical connector to the RFI module.

11. Connect the negative battery cable. Start the vehicle and road test or allow to idle until the engine reaches operating temperature to allow cooling fans to turn on.

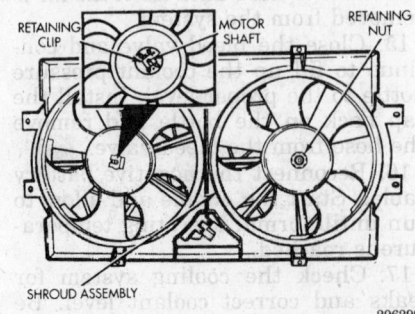

RETAINING CLIP   MOTOR SHAFT   RETAINING NUT

SHROUD ASSEMBLY

296299

Cooling fan-to-motor shaft retaining clip and retaining nut

Cooling System Bleeding

— **WARNING** —

Do not use well water or the water supply that may already be in the cooling system. It is strongly recommended that a 50/50 mixture of distilled water and ethylene glycol coolant be used.

1. Be sure that the radiator drain cock is closed. Hand tightened only.

2. If removed, install the cylinder block drain plugs.

3. Open the thermostat bleed valve. Install a ¼-inch diameter clear hose about 48 inches long, to the end of the bleed valve and the other end into a clean container.

— **WARNING** —

When installing the drain tube onto the air bleed valve of the thermostat housing, route the drain hose away from the accessory drive belt, belt pulleys and cooling fan motors.

4. Slowly refill the coolant pressure bottle until a steady stream of coolant flows out of the thermostat bleed valve. Gently squeeze the upper radiator hose until all of the air is removed from the system.

5. Close the bleed valve and continue to fill up the coolant pressure bottle to the proper level. Install the cap back on the bottle and remove the hose from the bleed valve.

6. Start the engine and allow to run until normal operating temperature is reached.

7. Check the cooling system for leaks and correct fluid level. Be sure that the thermostat bleed valve is closed once the cooling system has been bled of any trapped air.

FUEL SYSTEM

Fuel System Service Precautions

Safety is the most important factor when performing not only fuel system maintenance but any type of maintenance. Failure to conduct maintenance and repairs in a safe manner may result in serious personal injury or death. Maintenance and testing of the vehicle's fuel system components can be accomplished

safely and effectively by adhering to the following rules and guidelines.

- To avoid the possibility of fire and personal injury, always disconnect the negative battery cable unless the repair or test procedure requires that battery voltage be applied.

- Always relieve the fuel system pressure prior to disconnecting any fuel system component (injector, fuel rail, pressure regulator, etc.), fitting or fuel line connection. Exercise extreme caution whenever relieving fuel system pressure to avoid exposing skin, face and eyes to fuel spray. Please be advised that fuel under pressure may penetrate the skin or any part of the body that it contacts.

- Always place a shop towel or cloth around the fitting or connection prior to loosening to absorb any excess fuel due to spillage. Ensure that all fuel spillage (should it occur) is quickly removed from engine surfaces. Ensure that all fuel soaked cloths or towels are deposited into a suitable waste container.

- Always keep a dry chemical (Class B) fire extinguisher near the work area.

- Do not allow fuel spray or fuel vapors to come into contact with a spark or open flame.

- Always use a backup wrench when loosening and tightening fuel line connection fittings. This will prevent unnecessary stress and torsion to fuel line piping. Always follow the proper torque specifications.

- Always replace worn fuel fitting O-rings with new. Do not substitute fuel hose or equivalent, where fuel pipe is installed.

Fuel System Pressure

RELIEVING

— CAUTION —

Fuel injection systems remain under pressure, even after the engine has been turned OFF. The fuel system pressure must be relieved before disconnecting any fuel lines. Failure to do so may result in fire and/or personal injury.

1. Disconnect the negative battery cable.
2. Remove the fuel filler cap.
3. Remove the safety cap from the fuel pressure test port located on the fuel rail.
4. Place the open end of the fuel pressure release hose tool C-4799-1

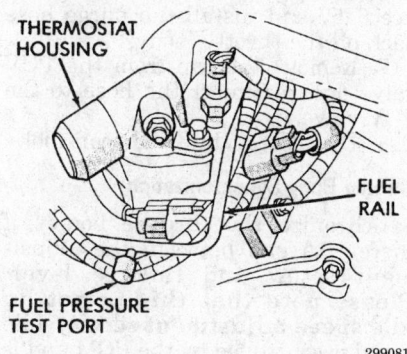

Fuel pressure test port — 3.3L engine

Fuel pressure test port — 3.5L engine

or equivalent, into a proper gasoline container. Connect the other end of the hose to the fuel pressure test port. The fuel pressure will bleed off through the hose into the container.

Idle Speed

ADJUSTMENT

3.3L (VIN T) Engine

The 3.3L engine has a fixed idle speed. The Powertrain Control Module (PCM) regulates the idle speed. Idle Speed on these vehicles is not adjustable. The minimum air flow idle, however, can be checked with a DRB scan tool as follows:

1. Warm up the engine in **PARK** or **NEUTRAL** until the cooling fan has turned on and off at least one time.
2. All accessories must be turned off.
3. Shut off the engine.
4. Disconnect the PCV valve hose from the intake manifold and cap the PCV valve.
5. Disconnect the purge hose from the throttle body.

6. Use a piece of hose to connect Air Metering Orifice 6457 (0.125 inch orifice) to the purge nipple on the throttle body.
7. Connect the DRB scan tool to the data link connector inside the passenger compartment of the vehicle and restart the engine.
8. Allow the engine to idle for at least one minute.
9. Access the minimum air flow idle speed using the DRB scan tool.
10. The following should then occur:
    a. The idle air control motor will fully close.
    b. The idle spark advance will become fixed.
    c. The DRB scan tool displays the engine RPM.
11. The minimum air flow idle specifications are as follows:
    - 540–840 RPM — under 1000 miles odometer reading
    - 600–840 RPM — over 1000 miles odometer reading
12. If the RPM is within specifications, throttle body minimum air flow is set correctly.
13. If the idle RPM is above specifications, use the DRB scan tool to check the idle air control motor operation.
14. If the idle RPM is below specifications, turn off the engine and clean the throttle body as follows:
    a. Remove the throttle body from the vehicle.
    b. Hold the throttle open and spray the entire throttle body bore, as well as the manifold side of the throttle plate with Mopar® Parts Cleaner, or equivalent.
    c. Clean the entire throttle bore, the edges and manifold side of the throttle plate using a soft scuff pad. This would also include the portion of the throttle bore that is closest to the throttle plate when it is closed.
    d. Dry the throttle body using compressed air and inspect the throttle body for any foreign material.
    e. Install the throttle body onto the manifold and repeat the idle air flow inspection procedure. If the minimum air flow idle is still out of specification, the problem is not caused by the throttle body.
15. Turn off the engine.
16. Remove the Air Metering Orifice 6457 and install the purge hose back on the throttle body.
17. Remove the cap from the PCV valve and reconnect the hose to the PCV valve.
18. Disconnect the DRB scan tool.

3.5L (VIN F) Engine

The 3.5L engine has a fixed idle speed. The Powertrain Control Module (PCM) regulates the idle speed. Idle Speed on these vehicles is not adjustable, however, the minimum air flow idle can be checked with a DRB scan tool using the following procedure.

Minimum Air Flow Setting

1. Warm up the engine in **PARK** or **NEUTRAL** until the cooling fan has turned on and off at least one time.
2. All accessories must be turned off.
3. Shut off the engine.
4. Disconnect the PCV valve hose from the intake manifold and cap the PCV valve.
5. Disconnect the purge hose from the throttle body.
6. Use a piece of hose to connect Air Metering Orifice 6457 (0.125 inch orifice) to the purge nipple on the throttle body.
7. Connect the DRB scan tool to the data link connector inside the passenger compartment of the vehicle and restart the engine.
8. Allow the engine to idle for at least one minute.
9. Access the minimum air flow idle speed using the DRB scan tool.
10. The following should then occur:
   a. The idle air control motor will fully close.
   b. The idle spark advance will become fixed.
   c. The DRB scan tool displays the engine RPM.
11. The minimum air flow idle specifications are as follows:
   • 700–1000 RPM — under 1000 miles odometer reading
   • 750–1100 RPM — over 1000 miles odometer reading
12. If the RPM is within specifications, throttle body minimum air flow is set correctly.
13. If the idle RPM is above specifications, use the DRB scan tool to check the idle air control motor operation.
14. If the idle RPM is below specifications, turn off the engine and clean the throttle body as follows:
   a. Remove the throttle body from the vehicle.
   b. Hold the throttle open and spray the entire throttle body bore, as well as the manifold side of the throttle plate with Mopar® Parts Cleaner, or equivalent.
   c. Clean the entire throttle bore, the edges and manifold side of the throttle plate using a soft scuff pad. This would also include the portion of the throttle bore that is closest to the throttle plate when it is closed.
   d. Dry the throttle body using compressed air and inspect the throttle body for any foreign material.
   e. Install the throttle body onto the manifold, start the engine and repeat the idle air flow inspection procedure. If the minimum air flow idle is still out of specification, the problem is not caused by the throttle body.
15. Turn off the engine.
16. Remove the Air Metering Orifice 6457 and install the purge hose back on the throttle body.
17. Remove the cap from the PCV valve and reconnect the hose to the PCV valve.
18. Disconnect the DRB scan tool.

Throttle Body Synchronization

Synchronize the throttle bodies if there is a gap between the adjustment screw and linkage lever. **Please note that this is not an idle speed adjustment screw.** Idle speed is controlled by the PCM and is not service adjustable. Since the 3.5L engine uses two throttle bodies, this adjustment allows synchronization of both throttle bodies. The throttle bodies must be synchronized after removing and installing the throttle bodies. Use the following procedure.

1. Install a 0.004 inch feeler gauge between the idle speed screw and throttle lever on the right side throttle body.
2. Loosen the adjustment screw locknut on the synchronization shaft.
3. Turn the adjustment screw (use the proper size Allen wrench) until it just contacts the lever on the synchronization shaft. Tighten the locknut to 36 inch lbs. (4 Nm).
4. Check the clearance between the throttle lever and idle speed screw. The 0.004 inch feeler gauge should drag when removed. Also, a 0.006 inch feeler gauge should not fit between the throttle lever and idle speed screw.

Mixture

ADJUSTMENT

The Powertrain Control Module (PCM) operates the Multi-Port Fuel Injection System. The PCM provides precise air/fuel ratios for all driving conditions and is therefore not adjustable.

Under most driving conditions, the PCM maintains an air/fuel ratio of 14.7 to 1 by constantly adjusting the fuel injector pulse width. There are no service adjustments that can be made to alter the air/fuel mixture. It is regulated by the PCM programming.

Fuel Filter

REMOVAL AND INSTALLATION

The fuel filter is a replaceable in-line filter. The filter attaches to a bracket located on the frame rail in front of the fuel tank. The inlet and outlet ends of the filter are marked for correct installation.

— CAUTION —

Fuel injection systems remain under pressure, even after the engine has been turned OFF. The fuel system pressure must be relieved before disconnecting any fuel lines. Failure to do so may result in fire and/or personal injury.

1. Disconnect the negative battery cable.
2. Release the fuel system pressure.
3. Remove any loose dirt from the quick-connect fittings on the inlet and outlet side of the fuel filter.
4. Wrap shop towels around the inlet and outlet hoses to catch any fuel spillage.
5. Disconnect the metal quick-connect fitting at the outlet side of the fuel filter as follows:
   a. Push the quick connect fitting toward the fuel tube while depressing the built in release tool on the tube side of the fitting using Special Quick-Disconnect Fitting Tool 6751 or equivalent. Slightly twist the fitting and pull it off the fuel tube.
6. Disconnect the plastic quick-connect fitting at the inlet side of the fuel filter as follows:
   a. Press the retainer tabs together and slide the quick-connect fitting off of the fuel tube nipple. The retainer remains on the fuel tube.
7. Cover both fittings to prevent contamination of the fuel system.
8. Remove the fuel filter mounting bracket and remove the filter.
   To install:
9. The inlet and the outlet side of the filter are marked for correct installation. Install the filter with the inlet side of the filter toward the

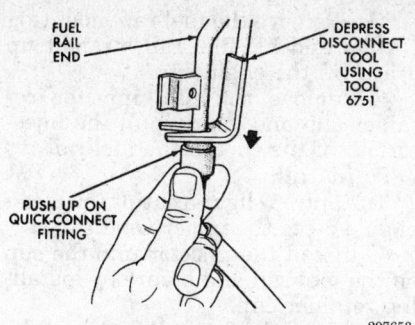

FUEL RAIL END

DEPRESS DISCONNECT TOOL USING TOOL 6751

PUSH UP ON QUICK-CONNECT FITTING

297653

Disconnecting quick-connect fitting — fuel rail shown, filter connection similar

tank. Place the filter in the mounting and install to the frame rail. Tighten the mounting bolts to approximately 105-110 inch lbs. (12 Nm).

10. Apply a light coating of clean 30 wt. engine oil to the fuel filter nipples. Install the fuel tubes to the filter, with new O-rings installed on the quick-connect fittings.

11. Install the plastic quick-connect fitting as follows:

   a. Push the quick-connect fitting over the fuel tube until the retainer locks into place and a click is heard. Be sure to check that the retainer locking ears and the fuel tube shoulder are visible through the window of the quick-connect fitting.

   b. Verify the connection by pulling back on the quick-connect fitting. If the fitting locks in place, the connection is secure.

12. Install the metal quick-connect fitting as follows:

   a. Place the Special Disconnect Tool 6751 or equivalent below the largest diameter of the quick-connect fitting.

   b. Pull disconnect tool toward the fuel tube nipple until quick-connect fitting clicks into place.

   c. Position disconnect tool between the shoulder of the disconnect tool and the top of the quick-disconnect fitting. Check the connection by pushing down using disconnect tool.

   d. Check for proper connection by pulling back on the quick connect fitting. The tube should lock in place. If the connection is not complete, make sure the black plastic ring is not causing the locking retainer to jam in the release position.

13. Lower the vehicle. Install the filler cap and reconnect the negative battery cable.

14. With the ignition in the **ON** position, connect the DRB II, or equivalent scan tool, and access the

ASD Fuel System Test to pressurize the fuel system. Check for leaks.

NOTE: When using the ASD Fuel System Test, the ASD relay will remain energized for 7 minutes or until the test is stopped or until the ignition key is turned to the OFF position.

Fuel Pump

REMOVAL AND INSTALLATION

The fuel pump module contains the fuel pump, fuel reservoir, level sensor, inlet strainer and pressure relief/rollover valve. The inlet strainer, level sensor and pressure relief/rollover valve are the only serviceable items. If the fuel pump requires service, replace the entire fuel pump module. Use care when ordering parts. Although they operate the same, flexible fuel LH-platform vehicles use different fuel pump modules than gasoline-only LH-platform vehicles. The trunk has an access cover in the floor panel for servicing the fuel pump module.

— CAUTION —
Fuel injection systems remain under pressure, even after the engine has been turned OFF. The fuel system pressure must be relieved before disconnecting any fuel lines. Failure to do so may result in fire and/or personal injury.

1. If possible, drain the fuel tank using the following procedure.

   a. If the fuel pump still operates, energize the pump and drain fuel into an approved portable fuel siphoning tank. Use the DRB ASD Fuel System test to energize the fuel pump.

   b. If the pump does not operate, remove the pump module as de-

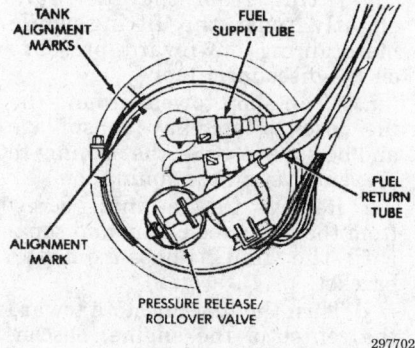

TANK ALIGNMENT MARKS

FUEL SUPPLY TUBE

FUEL RETURN TUBE

ALIGNMENT MARK

PRESSURE RELEASE/ ROLLOVER VALVE

297702

Fuel pump module alignment

scribed below and drain the tank into an approved portable fuel siphoning tank.

2. Remove the fuel tank filler cap. Relieve fuel system pressure.

3. Disconnect the negative battery cable.

4. Open the trunk lid. Remove the trunk liner and the fuel pump module access panel fasteners.

5. Remove the access panel and the gasket from the base of the trunk. Inspect the gasket and replace, if necessary.

6. Disconnect the electrical connector from the top of the fuel pump module.

7. Place shop rags around the area of the fuel supply and return lines so that it can absorb any fuel spillage. Disconnect the fuel supply and return lines from the module by squeezing the retainer tabs together and carefully sliding the quick-connect fitting assembly off of the fuel tube nipple. The retainer must remain on the fuel tube.

8. Disconnect the hose from the pressure relief/rollover valve.

9. Loosen the band clamp until the fuel pump module rises up from the tank.

— CAUTION —
The fuel pump reservoir may still contain fuel. Do not spill any fuel while removing the fuel pump module.

10. To absorb possible fuel spillage, place shop towels around access opening. Without removing module, tip the fuel pump module backwards so any excess fuel can run down the side of the module back into the tank.

11. The float arm of the level sensor will catch on the side of the inside of the tank while removing the module. Tilt the module to one side and remove from the tank.

12. Slowly and with extreme care, remove the fuel pump module and the gasket from the tank. Drain the remainder of fuel from the fuel pump module reservoir before servicing the module.

To install:

13. The fuel pump module and fuel tank have alignment marks. The tank has 2 molded lines at the 10 o'clock position. The fuel pump module has a triangular alignment mark. Carefully install the module straight into the tank with new gasket in place. Align the triangular alignment mark so it is pointing between the 2 lines on the tank.

14. Seat the module in the tank by pushing the top down. Make sure the

gasket does not become dislodged or move out of position.

15. While holding down on the module, install and tighten the clamp over the edge of the pump module and the lip of the tank. Tighten the band clamp to 31 inch lbs. (4 Nm). Be very careful not to overtighten the clamp.

16. Install the fuel tubes over the fuel return and supply nipples on the module. Push the quick-connect fitting over the fuel tube until the retainer locks into place and a click is heard. Be sure to check that the retainer locking ears and the fuel tube shoulder are visible through the window of the quick-connect fitting.

17. Verify the connection by pulling back on the quick-connect fitting. If the fitting locks in place, the connection is secure.

18. Reconnect the vent line to the pressure relief/rollover valve.

19. Reconnect the electrical connector to the fuel pump module. Reconnect the negative battery cable.

20. Turn the ignition **ON** but do not start the engine. Using the DRB II or equivalent scan tool, access the ASD Fuel System Test. This test will activate the fuel pump and pressurize the fuel system. Check for leaks.

21. Install the access cover and gasket. Tighten the fasteners.

22. Install the trunk liner and fuel tank filler cap.

Fuel Injector

REMOVAL AND INSTALLATION

3.3L (VIN T) Engine

—————— **CAUTION** ——————
Fuel injection systems remain under pressure, even after the engine has been turned OFF. The fuel system pressure must be relieved before disconnecting any

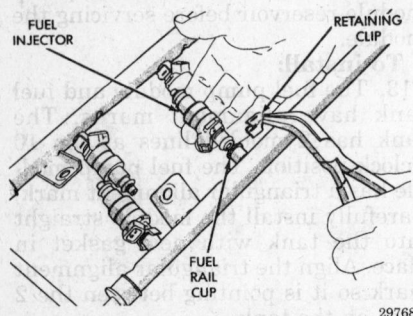

Fuel injector removal/installation — 3.3L engine

fuel lines. Failure to do so may result in fire and/or personal injury.

1. Disconnect negative battery cable.

2. Release the fuel system pressure.

3. There are numerous electrical connections to the intake manifold area. Identify and tag as required. This should save time at assembly. Remove the fuel rail from the intake manifold using the following procedure:

a. Disconnect the air plenum from the air cleaner and the throttle body.

b. With the throttle held in the wide open position, disconnect the throttle linkage and the cruise control linkage from the throttle shaft. Compress the locking tabs on both cables and remove cables from mounting bracket.

c. Disconnect the electrical connector from the solenoid on the EGR valve transducer and the MAP Sensor.

d. Disconnect the vacuum hose from the PCV valve and the fuel pressure regulator. Disconnect the brake booster hose at the rear of the intake manifold.

e. Disconnect the throttle body purge hose. Disconnect the electrical connector to the Throttle Position Sensor (TPS) and the Idle Air Control (IAC) motor.

f. Remove the EGR tube mounting screws at the intake manifold plenum (upper intake manifold). Remove the intake manifold plenum mounting bolts and the plenum from the engine.

g. At the fuel rail, push the quick-disconnect fitting toward the fuel tube while depressing the built-in disconnect tool with Special Quick-Connect Fitting Tool 6751 or equivalent. To disconnect the fitting from the fuel rail, slightly twist the fitting while maintaining downward pressure on the disconnect tool.

h. Wrap shop towels around the fuel hoses to catch any fuel spillage and be sure to cover the opening to prevent system contamination.

i. Remove the retaining screw from the fuel tube clamp and separate the fuel tubes from the bracket.

j. Turn the fuel injectors toward the center of the engine. Disconnect and label the electrical harness from the fuel injectors.

k. Remove the fuel rail mounting bolts and lift fuel rail straight up and off the engine.

4. Remove the fuel injector retainer clip and gently pull the injector out of the cup on the fuel rail.

To install:

5. Apply a light coating of clean engine oil to the upper O-ring.

6. Install the injector into the cup on the fuel rail and secure by installing retainer clip.

7. Install the fuel rail to the intake manifold as follows:

a. Apply a light coat of clean engine oil to the O-ring on the nozzle end of each injector.

b. Insert the fuel injector nozzles into the openings in the intake manifold. Seat the injectors in place and install the fuel rail mounting bolts. Torque the mounting bolts to 16 ft. lbs. (22 Nm).

c. Reconnect the electrical connectors to each fuel injector. Turn the injectors toward the cylinder head covers.

d. Reconnect the fuel supply and return tubes to the fuel rail. Be sure that the black plastic release ring to the quick-connect fitting is in the **OUT** position. Place Disconnect Special Tool 6751 or equivalent under the largest diameter of the quick-connect fitting.

e. Pull the disconnect tool toward the fuel rail until the quick-connect fitting clicks into place. Place the Special Tool between the shoulder of the built-in disconnect tool and top of the quick-connect fitting, then inspect the security of the fitting by applying a slight downward force against the fitting. It should be locked in place.

f. Install the clamp over the fuel tubes and tighten the retaining screw.

g. Install the intake plenum with new gasket onto the intake manifold. Loosely install the mounting bolts.

h. Install the EGR tube to the manifold with new gasket in place. Loosely install the mounting screws.

i. Torque the intake manifold plenum mounting bolts in sequence to 21 ft. lbs. (28 Nm).

j. Tighten the EGR tube mounting bolts.

k. Reconnect the vacuum hoses to the PCV valve and the brake booster.

l. Reconnect the electrical connectors to the idle air control motor, EGR transducer solenoid, throttle position and MAP sensors.

m. Install the throttle cable and speed control cable to the mounting bracket and reconnect to the throttle body lever while holding lever in the wide-open position.

n. Reconnect the purge hose to the throttle body. Reconnect the air plenum to the air cleaner and the throttle body.

8. Reconnect the negative battery cable.

3.5L (VIN F) Engine

CAUTION

Fuel injection systems remain under pressure, even after the engine has been turned OFF. The fuel system pressure must be relieved before disconnecting any fuel lines. Failure to do so may result in fire and/or personal injury.

1. Disconnect negative battery cable.
2. Release the fuel system pressure.
3. Remove the engine cover from the top of the intake manifold plenum.
4. Disconnect the accelerator cable and speed control cable from the throttle arm.

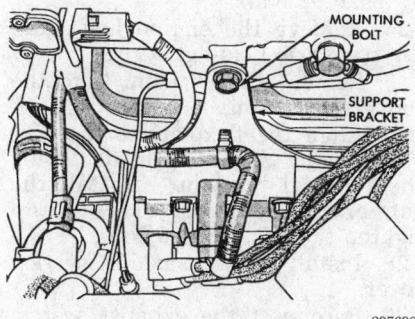

Intake manifold plenum support (right side shown) — 3.5L engine

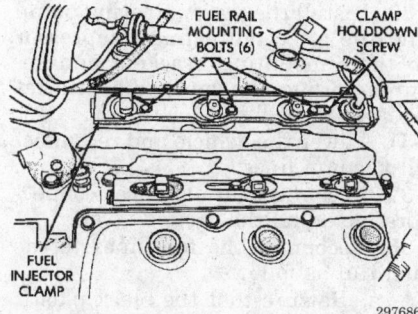

Fuel injector clamp mounting bolts — 3.5L engine

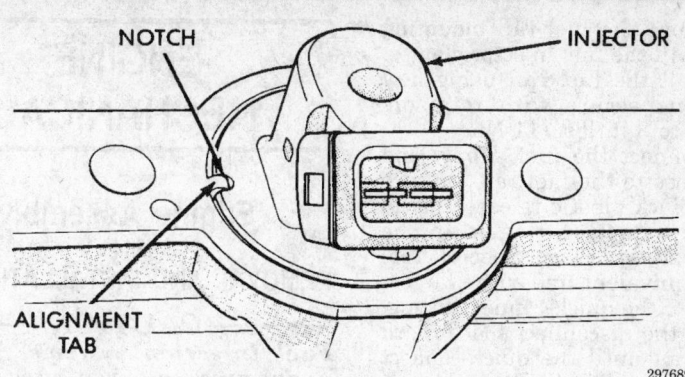

Fuel injector alignment — 3.5L engine

5. Disconnect the power brake booster hose and remove the throttle cable bracket from the manifold.
6. Disconnect the electrical connectors from the idle air control motor and the intake air temperature sensor.
7. Disconnect the vacuum hose from the Manifold Tuning Valve (MTV).
8. Disconnect the electrical connectors from the Manifold Absolute Pressure (MAP) sensor and the throttle position sensor.
9. Disconnect the throttle body purge hose, PCV make-up air hose and idle air control motor supply hose (part of the air inlet plenum). Remove the air inlet plenum from behind the manifold.
10. Disconnect the PCV hose and vacuum hoses from the intake manifold.
11. Remove the EGR tube mounting bolts at the intake manifold plenum and discard the gaskets. Remove the lower EGR mounting bolts located at the EGR valve.
12. Remove the support bracket mounting bolts on each side of the plenum. Remove the intake plenum mounting bolts.

NOTE: The intake manifold plenum uses 2 different length bolts. Take note of their position and make sure they are installed in the same location during installation.

13. Remove the intake manifold plenum from the intake manifold. Discard the old gasket. Cover the intake manifold openings.
14. At the fuel rail, push the quick-connect fitting toward the fuel tube while pushing in the built-in disconnect tool using Special Quick-Connect Fitting Tool 6751 or equivalent. Disconnect the fitting from the fuel rail by slightly turning the fitting while applying downward pressure

on Tool 6751. Be sure to wrap shop towels around the hoses to catch any fuel spillage.
15. Connect Special Fuel Gauge Adapter 6631 or exact equivalent, to the fuel supply tube end of the fuel rail. Connect fuel hose 6668 or exact equivalent, to the fuel return tube end of the fuel rail.
16. Place the other end of fuel hose 6668 into an approved gasoline container.
17. Drain the gasoline from the fuel rail. To purge the fuel from the fuel rail, spray a maximum of 55 psi of compressed air into the end of adapter 6631.

WARNING

The fuel rail must be void of gasoline prior to injector removal. If the fuel rail is not drained, the gasoline in the rail will enter the engine cylinders when the injectors are removed.

---

18. Tag all injector harness connectors noting cylinder location for reference during assembly. Disconnect the harness connectors from each injector.
19. Remove the vacuum hose from the fuel pressure regulator.
20. Remove the fuel rail mounting bolts and the injector clamp screw. Slide the injector clamp toward the rear of the engine, then lift the clamp off the rail.
21. Install the fuel rail mounting bolts finger-tight only.
22. Using a flat tipped tool, pry the fuel injector out of the fuel rail. Ensure the upper and the lower O-rings were removed with the injector. If not, remove them from the injector well in the fuel rail.

To install:

23. Lightly lubricate the O-rings with clean engine oil. Install the injector making sure to align the notch on the injector with the alignment tab on the fuel rail.

24. Remove the fuel rail mounting bolts. Install the fuel injector clamps.

25. Install the fuel rail using new gasket, and secure with retainers tightened to 8 ft. lbs. (11 Nm).

26. Reconnect the fuel supply and return tubes to the fuel rail. Be sure that the black plastic release ring to the quick-connect fitting is in the **OUT** position. Place Special Tool 6751 or equivalent under the largest diameter of the quick-connect fitting.

27. Pull the disconnect tool toward the fuel rail until the quick-connect fitting clicks into place. Place the special tool between the shoulder of the built-in disconnect tool and top of the quick-connect fitting, then inspect the security of the fitting by applying a slight downward force against the fitting. It should be locked in place.

28. Attach the electrical harness connectors to the injectors making sure of correct location.

29. Install the intake manifold plenum with a new gasket in place. Tighten mounting bolts in proper sequence to 21 ft. lbs. (28 Nm).

30. Install and tighten the support bracket bolts.

31. Attach the electrical connectors to the manifold absolute pressure sensor, throttle position sensor, idle air control motor and intake air temperature sensor.

32. Reconnect the vacuum hose to the manifold tuning valve motor.

33. Install the EGR tube. Rotate the throttle lever to the wide-open position and reconnect the speed control and throttle cables.

34. Reconnect the PCV valve hose.

35. Install the air cleaner plenum and reconnect plenum hose. Reconnect the power brake booster hose to the fitting on the intake manifold plenum.

36. Reconnect the purge tubes to the fittings on the throttle body.

37. Install the cover on the intake manifold plenum.

38. Reconnect the negative battery cable.

ENGINE MECHANICAL

Engine Assembly

REMOVAL AND INSTALLATION

—————— CAUTION ——————
Fuel injection systems remain under pressure, even after the engine has been turned OFF. The fuel system pressure must be relieved before disconnecting any fuel lines. Failure to do so may result in fire and/or personal injury.

1. Disconnect negative battery cable.

2. Release the fuel system pressure using the recommended procedure.

3. Matchmark the hood and hinges and remove the hood assembly.

4. Drain the cooling system.

5. Remove the radiator and cooling fan assemblies.

6. Label and disconnect all electrical connections.

7. Remove the coolant hoses from the engine.

8. Disconnect the fuel lines using the following procedure:

    a. At the fuel rail, push the quick-connect fitting toward the fuel tube while depressing the built-in disconnect tool with Quick-Connect Fitting Tool 6751 or equivalent.

    b. To disconnect the fitting from the fuel rail, slightly twist the fitting while maintaining downward pressure on tool 6751.

    c. Wrap shop towels around the fuel hoses to absorb any fuel spillage and be sure to cover the openings to prevent system contamination.

9. Disconnect the accelerator and cruise control cables from the throttle body.

10. Remove the air cleaner assembly. Raise and safely support the vehicle.

11. Place a drain pan under the vehicle and drain the engine oil.

12. Remove the air conditioning compressor mounting bolts and position the compressor aside. It should not be necessary to disconnect the refrigerant lines from the compressor.

13. Disconnect the exhaust pipe from the exhaust manifold.

14. Remove the transaxle inspection cover and mark the flexplate for reference during installation.

15. Remove the screws holding the torque converter to the flexplate. Attach a C-clamp or some other restraint on the converter housing to prevent the torque converter from falling out during removal of the engine.

16. Remove the power steering pump mounting bolts and position the pump aside.

17. Remove the 2 lower transaxle-to-block screws. Remove the starter motor from the transaxle housing.

18. Lower the vehicle and disconnect the the vacuum lines and ground strap. Support the transaxle with a floor jack.

19. Attach an engine lifting hoist to the engine and support.

20. Remove the upper transaxle mounting bolts.

21. Remove the insulator mounting nuts from the engine mounts.

22. Lift the engine from the vehicle.

To install:

23. Lower the engine into the engine compartment. Align the engine mounts and install all nuts. Once all mount bolts are installed, tighten the fasteners to 45 ft. lbs. (61 Nm).

24. Install the transaxle to the engine block and tighten the bolts to 75 ft. lbs. (102 Nm).

25. Remove the engine hoist and the transaxle holding fixture.

26. Remove the C-clamp from the converter housing, if installed.

27. Align the flexplate to the converter using the marks made during the removal procedure. Install the converter mounting screws and tighten to 55 ft. lbs. (75 Nm).

28. Install the transaxle inspection cover.

29. Reconnect the exhaust system to the engine exhaust manifold and install the starter. Torque the exhaust pipe-to-manifold bolts to 25 ft. lbs. (34 Nm).

30. Install the power steering pump and the air conditioning compressor to their mounting brackets and secure. Torque the mounting bracket bolts to 30 ft. lbs. (41 Nm).

31. Lower the vehicle and reconnect all vacuum lines.

32. Reconnect all electrical connectors including the ground strap.

33. Reconnect the fuel lines to the fuel rail as follows:

    a. Be sure that the black plastic release ring to the quick-connect fitting is in the OUT position. Place Special Tool 6751 under the largest diameter of the quick-connect fitting.

b. Pull Tool 6751 toward the fuel rail until the quick-connect fitting clicks into place.

c. Place the special tool between the shoulder of the built-in disconnect tool and the top of the quick-connect fitting, then inspect the security of the fitting by applying a slight downward force against the fitting. It should be locked in place.

34. Reconnect the accelerator and cruise control cables to the throttle lever.

35. Install the radiator and cooling fan assemblies.

36. Refill the crankcase with the proper amount of engine oil and replace the oil filter, if necessary.

37. Refill and bleed the cooling system.

38. Install the hood onto the vehicle aligning the marks made during removal.

39. Check to be sure that all hoses, wiring connectors, cables, vacuum and fluid lines are reconnected.

40. Reconnect the negative battery cable. Check all fluid levels.

41. Start the engine and allow to idle until normal operating temperature is reached. Inspect all fluid systems for leaks and correct level.

42. Road test the vehicle. Adjust the transaxle linkage, as necessary.

Engine Cradle

REMOVAL AND INSTALLATION

Several procedures throughout this chapter require the removal of the engine cradle assembly for access. When it becomes necessary to remove the cradle assembly from the vehicle, special tools will be required to support the engine and transaxle assembly. It will be necessary to support the engine and transaxle assembly using a suitable engine support de-

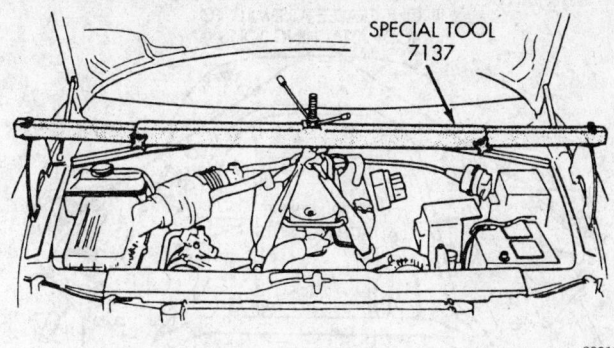

299110

Engine supporting fixture installed — 3.3L engine

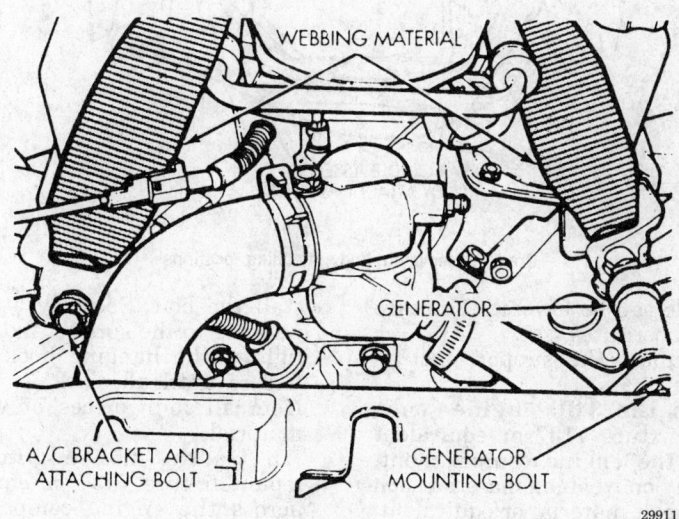

299111

Front engine support attaching locations — 3.3L engine

vice. **DO NOT** attempt to support or lift the engine using the intake manifold or any other location on the engine except as specified.

1. Disconnect the negative battery cable.

2. Support the engine and transaxle assembly using special tools such as Chrysler's Engine Support Fixture Tool 7137 or equivalent. This tool spans the engine compartment, resting on the inner fender seam flanges. Nylon straps similar to seat belt webbing hang from the support fixture and attach to engine hardpoints. The engine's weight is supported by this fixture so the cra-

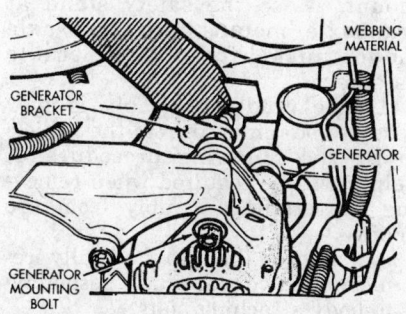

299114

Front engine support attachment to alternator bracket — 3.5L engine

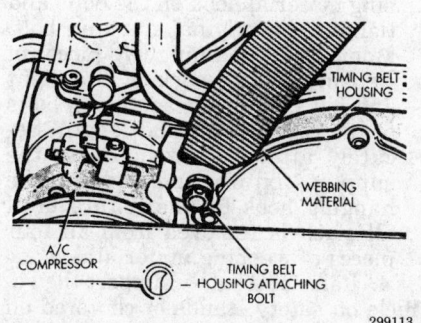

299113

Front engine support attachment to timing belt housing — 3.5L engine

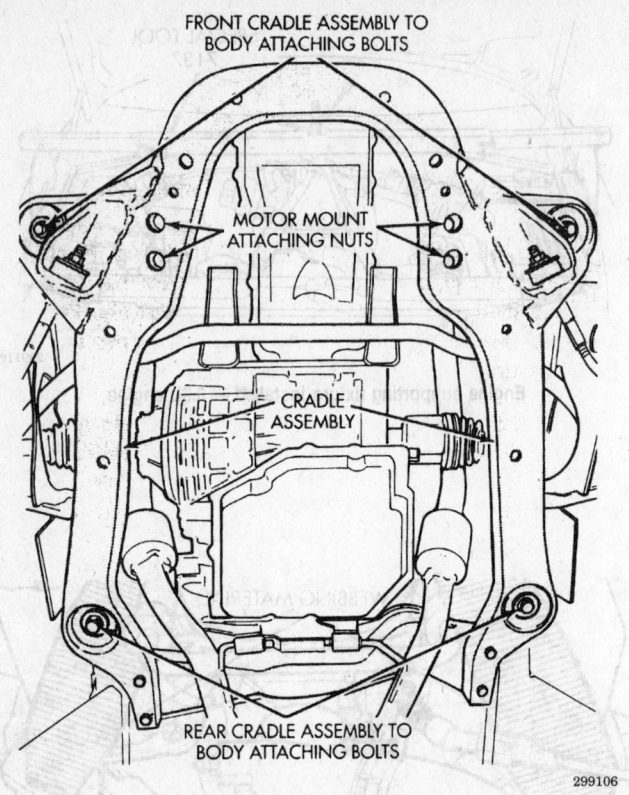

FRONT CRADLE ASSEMBLY TO
BODY ATTACHING BOLTS

MOTOR MOUNT
ATTACHING NUTS

CRADLE
ASSEMBLY

REAR CRADLE ASSEMBLY TO
BODY ATTACHING BOLTS

299106

Cradle assembly-to-body attaching locations

dle module can be lowered. Use care if using substitutes.

3. Connect the support tool as follows:

a. **On the 3.3L engine** only, mount fixture 7137 or equivalent across the engine compartment. Using nylon webbing material such as seat belt material or equivalent, attach by removing the air conditioning compressor bracket to front engine bolt, install the webbing hook on the bolt and retighten the bolt. Loosen but do not remove the alternator to engine mounting bolt, remove the spacer between the alternator and engine and install the webbing material hook between the alternator and engine, tightening the original bolt. Remove the electronic ignition coil and mounting bracket from the rear of the engine and install the webbing material hook on the bracket bolt and reinstall the bolt. **Route the webbing material between the fuel injector rail and valve cover, not between the fuel injector rail and the intake manifold.** Mount the fourth piece of webbing to the left cylinder head using a threaded hole in the back of the cylinder head. Using a bolt of correct size and length, install the webbing hook on the bolt and rein-

stall the bolt. Securely attach all webbing to the support fixture and tighten the hanging hook on the fixture until all slack is removed from all four pieces of webbing material.

b. **On the 3.5L engine** only, mount fixture 7137 or equivalent across the engine compartment. Using nylon webbing material such as seat belt material or equivalent, attach by removing the bolt attaching the timing belt housing to the front of the engine. Install the webbing material hook on the timing belt housing attaching bolt and tighten the bolt. Remove the alternator mounting bracket attaching nut and bolt and install the webbing material hook on the bolts and tighten the original nut and bolt. Mount the remaining two pieces of webbing to the back of the right and left cylinder heads using bolts of the correct size and length. Securely attach all webbing to the support fixture and tighten the hanging hook on the fixture until all slack is removed from all four pieces of webbing material.

4. Raise and safely support the vehicle on safety stands or centered on a frame contact type hoist.

5. Remove both front wheels from the vehicle.

6. Remove both left and right ball joint stud-to-steering knuckle clamp nuts and bolts. Carefully insert a prybar between the lower control arm and the steering knuckle and separate ball joint from knuckle. Make sure ball joint seal does not get damaged during separation. Use caution when separating the ball joint stud from the steering knuckle so the ball joint seal does not get cut. When the lower control arm is separated from the steering knuckle, do not let the ball joint seal hit up against the steering knuckle. If the ball joint seal hits the steering knuckle, seal damage may occur. If the ball joint seal becomes torn, replace the seal before assembling the lower control arm to the knuckle.

NOTE: Pulling the steering knuckle out from the vehicle after releasing from the ball joint can separate the inner Tripod joint. Do not separate the inner Tripod joint.

7. Remove the ground strap, located on the right side of the cradle below the halfshaft from the cradle assembly.

8. Remove the four nuts attaching the motor mounts to cradle assembly. Remove the four bolts attaching the transaxle mount to the rear of the cradle assembly.

9. Remove the bolts attaching the stabilizer bushing retainer and bushing to the cradle assembly.

───── **CAUTION** ─────

Safety stands are required to support the cradle assembly and transaxle assembly during cradle assembly removal from the vehicle. Do not attempt to remove the cradle from the vehicle without using safety stands to support components.

10. Position a safety stand under front of cradle and at the center of the transaxle to cradle assembly mount. Raise the safety stand at transaxle mount until transaxle mount just lifts off the cradle assembly.

11. The cradle assembly is now ready to be removed from the vehicle using the following procedure. A helper will be required when removing the cradle assembly from the vehicle.

a. Loosen but do not fully remove the two rear cradle assembly-to-body attaching bolts.

b. Loosen and remove the two front cradle assembly-to-body attaching bolts.

c. With a helper supporting the rear of the cradle assembly and the safety stand supporting the transaxle, remove the two rear cradle assembly-to-body attaching bolts.

d. Slowly lower the front safety stand until weight of engine is supported by engine support fixture and motor mounts bolts are clear of the cradle assembly. With a helper at the rear of the cradle, lift front of cradle assembly off the safety stand and remove from the vehicle.

To install:

12. Tie the stabilizer bar up against the two transaxle to engine block brackets. This will hold bar out of the way during cradle installation.

13. With the aid of a helper, raise the cradle into vehicle resting front of the cradle on a safety stand. Then use the following procedure to install the cradle assembly back into the vehicle.

a. With a helper, raise the rear of the cradle assembly up far enough by hand to start the two rear cradle assembly to body attaching bolts. Install the bolts far enough to securely hold the cradle assembly in place but **DO NOT TIGHTEN.**

b. Using safety stands, raise the front of the cradle up against the bottom of the motor mounts, making sure all four mount studs come through the holes in the cradle assembly.

c. Continue to raise the cradle and engine assembly safety stands until the two front cradle to body attaching bolts can be started.

d. Lower the transaxle and align the transaxle mount with the four transaxle mount attaching holes in the cradle. Install but **DO NOT TIGHTEN** the four transaxle mount to cradle assembly attaching bolts.

NOTE: The two long bolts go through the front cradle assembly to transaxle mount holes. Before tightening the cradle assembly to body attaching bolts, check that all four cradle assembly to body mounting bolts are installed straight into the mounting plates in frame rails and that the mounting plates are not cocked inside the frame rails.

e. Using a crisscross pattern, tighten all four cradle assembly to body attaching bolts until cradle is seated up against the body. Then repeating the crisscross pattern, torque all 4 bolts to 115 ft. lbs. (155 Nm).

14. Tighten the four transaxle mount-to-cradle assembly bolts.

WARNING

Check that the lower part of the stabilizer bar is centered in the middle of the cradle assembly. Failure to do so may cause the stabilizer bar to come into contact with other suspension components.

15. Untie the stabilizer bar from the brackets and position it on cradle assembly. Align the bar bushing retainers with the mounting holes in the cradle assembly. Install but do not tighten the four bushing retainers to the cradle mounting bolts. The bolts will be tightened when the vehicle is lowered to the ground.

16. Install the four motor mount to cradle assembly attaching nuts and tighten to 45 ft. lbs. (61 Nm).

17. Install the ground strap to the cradle assembly. Be sure the ground strap screw is tightened securely.

18. Install the lower ball joint stud into the steering knuckle. Install the retainer nut and tighten to 40 ft. lbs. (55 Nm).

19. Install the wheels and lug nuts. Torque the lug nuts to 95 ft. lbs. (129 Nm).

20. Lower the vehicle to the ground. With the full weight of the vehicle supported by the suspension, tighten the four stabilizer bar bushing retainer to cradle assembly attaching bolts to 40 ft. lbs. (55 Nm).

21. Remove the engine support fixture.

22. Reconnect the negative battery cable.

23. Position the vehicle on an alignment rack and check that the front suspension toe setting is correct. Adjust the toe setting as required.

Engine Mounts

REMOVAL AND INSTALLATION

NOTE: Engine hydro-mounts may show surface cracks. This will not affect performance and the engine mount should not be replaced. Only replace the engine hydro-mounts when leaking fluid.

Side Mounts

1. Disconnect negative battery cable.

2. Remove the insulator attaching nut from the top of the mounting bracket.

3. Raise and safely support the vehicle.

4. Support the engine using an appropriate jack and a block of wood across the full width of the oil pan.

5. Remove the lower attaching nuts from the bottom of the insulator to the frame.

6. Raise and carefully support the engine. Remove the insulator from its mount.

To install:

7. Install the insulator to the frame. Lower the engine onto the insulator mount and install the insulator to frame nuts. Tighten nuts to 45 ft. lbs. (61 Nm).

8. Remove the engine supporting jack out from under the vehicle.

9. Lower the vehicle and install the upper attaching nuts to mount. Tighten nuts to 45 ft. lbs. (61 Nm).

10. Reconnect the negative battery cable.

Rear Mount

1. Disconnect negative battery cable.

2. Raise and safely support the vehicle.

3. Support the transaxle with a transmission jack. Remove the insulator through bolt from the mount.

4. Remove the transaxle mount fasteners and remove the mount from the vehicle.

To install:

5. Install the rear mount and fasteners to the transaxle.

6. Install the insulator through bolt into the mount. Torque the mounting bolts and the through bolt nut to 45 ft. lbs. (65 Nm).

7. Remove the transmission jack from the vehicle.

8. Lower the vehicle. Reconnect the negative battery cable.

9. Test drive the vehicle.

Cylinder Head

REMOVAL AND INSTALLATION

3.3L (VIN T) Engine

1. Release the fuel system pressure using the recommended procedure.

2. Disconnect negative battery cable.

3. Drain the cooling system.

4. Remove the intake manifold and throttle body assemblies using the recommended procedure. Use care when handling the intake manifold gasket. It is made of very thin

metal and could cause cuts if carelessly handled.

5. Disconnect the coil wires, sending unit wire, heater hoses and by-pass hose.

6. Remove the evaporation control system, closed ventilation system and the cylinder head covers.

7. Remove the exhaust manifold(s) from the engine.

8. Remove rocker arm and shaft assemblies. Remove pushrods and identify to assure installation in original location.

9. Remove the nine head bolts from the cylinder head and remove the head from the block.

To install:

10. Thoroughly clean and dry the mating surfaces of the head and block. Check the cylinder head for cracks, damage or engine coolant leakage. Remove scale, sealing compound and carbon. Clean the oil passages thoroughly. Check the head for flatness. End to end, the head should be within 0.002 inch (0.051 mm) normally with 0.008 inch (0.203 mm) the maximum allowed out of true. The total thickness allowed to be removed from the head and block is 0.008 inch (0.203 mm) maximum.

11. Place a new head gasket on the cylinder block with the identification marks facing upward. Do not use sealer on factory type gasket.

12. Inspect the cylinder head bolts for necking (stretching) by holding a straight edge against the threads of each bolt. If all of the threads are not contacting the the scale, the bolt should be replaced. New head bolts are recommended.

NOTE: Due to the cylinder head bolt torque method used, it is imperative that the threads of the bolts be inspected for necking (stretching) prior to installation. If the threads are necked down, the bolt should be replaced. Failure to do so may result in parts failure or damage.

13. Install the cylinder head bolts. Torque bolts Nos. 1 through 8 following the proper sequence as listed below:

Step 1 — 45 ft. lbs. (61 Nm)
Step 2 — 65 ft. lbs. (88 Nm)
Step 3 — (again) 65 ft. lbs. (88 Nm)
Step 4 — additional ¼-turn

NOTE: Do not use a torque wrench for Step 4. Inspect the bolt torque after tightening. The torque should be over 90 ft. lbs. (122 Nm). If not, replace the cylinder head bolt.

14. Tighten bolt No. 9 to 25 ft. lbs. (33 Nm) only after bolts 1–8 have been torqued to specification.

15. Inspect the pushrods and replace worn or bent rods. Install the pushrods, rocker arm and shaft assemblies with the stamped steel retainers in the forward positions. Tighten the rocker shaft retainers to 250 inch lbs. (28 Nm).

NOTE: The rocker arm shaft should be torqued down slowly, starting with the centermost bolts. Allow 20 minutes tappet bleed down time after installation of the rocker shafts before engine operation.

16. Install the cylinder head covers with new gaskets in place. Tighten the retainers to 105 inch lbs. (12 Nm).

NOTE: The factory type intake manifold gasket is made of very thin metal and is very sharp. Handle with care or personal injury may occur.

17. Using all new gaskets, install the intake manifold, throttle body, air intake plenum and exhaust manifold, using the recommended procedure and following the proper torque sequences.

18. Connect all exhaust and fuel connections.

19. Reconnect the coil wires, sending unit wire, heater hoses and bypass hose.

20. Install the air intake hose.

21. Change the engine oil and oil filter.

22. Refill and bleed the cooling system.

23. Reconnect the negative battery cable and run the vehicle with the radiator cap off so coolant can be added as required until the thermostat opens. Watch for leaks and unusual engine noises that might indicate a problem. Fill the radiator completely.

24. Once the vehicle has cooled, recheck the coolant and oil level.

3.5L (VIN F) Engine

—— CAUTION ——
Fuel injection systems remain under pressure, even after the engine has been turned OFF. The fuel system pressure must be relieved before disconnecting any fuel lines. Failure to do so may result in fire and/or personal injury.

1. Release the fuel system pressure using the recommended procedure.

2. Disconnect negative battery cable.

3. Drain the cooling system.

4. Remove the radiator and cooling fan assemblies.

5. Remove the air cleaner assembly and the intake manifold plenum. Cover the lower intake manifold during service.

6. Remove the accessory drive belts.

7. Remove the crankshaft damper using the proper puller.

8. Remove the engine valve covers as follows:

a. Disconnect and relocate the spark plug wires. Label each spark plug wire to the correct cylinder location.

b. Loosen the air conditioning compressor mounting bracket and pull away from the cylinder head.

c. Remove the spark plugs, spark plug tube nut and O-ring.

d. Remove the cylinder head cover screws and the cylinder head cover from the cylinder head.

9. Remove the timing belt covers.

10. Mark the timing belt running direction for installation. Align the camshaft sprockets with the marks on the rear covers. Remove the timing belt and tensioner.

11. Pre-load the timing belt tensioner as follows:

a. Place tensioner in a vise the same way it is mounted on the engine.

b. Slowly compress the plunger into the tensioner body.

c. When the plunger is compressed into the tensioner body, install a pin through the body and plunger to retain plunger in place until the tensioner is installed.

12. Hold the camshaft sprocket with 36 mm box wrench, loosen and remove the sprocket retaining bolt and washer.

NOTE: To remove the camshaft sprocket retainer bolt while the engine is in the vehicle, it may be necessary to raise that side of the engine due to the length of the retainer bolt. The right bolt is 8.370 inches (212.6 mm) long, while the left bolt is 10.0 inches (253 mm) long. These bolts are not interchangeable and their original location during removal should be noted.

13. Remove the camshaft sprocket from the camshaft. The camshaft sprockets are not interchangeable from side to side.

14. Remove the intake manifold assembly using the recommended procedure.

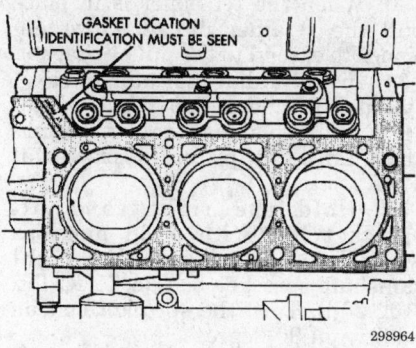

Head gasket installation — 3.3L engine

15. Remove the rear timing belt cover to cylinder head fasteners. If the right timing belt cover is to be removed, there are O-rings located behind it for the water pump passages.

16. Remove the cylinder head mounting bolts and the cylinder head from the vehicle.

To install:

17. Thoroughly clean and dry the mating surfaces of the head and block.

NOTE: When cleaning the cylinder head and block mating surfaces, do not use a metal scraper because the soft aluminum surfaces could be cut or damaged. Instead, use a scraper made of wood or plastic.

18. Check the cylinder head for cracks, damage or engine coolant leakage. Check the head for flatness. End to end, the head should be within 0.002 inch (0.051 mm) normally with 0.008 inch (0.203 mm) the maximum allowed out of true. The resurface limit is 0.008 inch (0.203 mm) maximum, the combined total dimension of stock removal from the cylinder head if any and block top surface.

19. Place a new head gaskets on the cylinder block locating dowels being sure the gasket is on the correct side.

20. Inspect the cylinder head bolts for necking by holding a straight edge against the threads of each bolt. If all of the threads are not contacting the the scale, the bolt should be replaced.

NOTE: Due to the cylinder head bolt torque method used, it is imperative that the threads of the bolts be inspected for necking prior to installation. If the threads are necked down, the bolt should be replaced. Failure to do so may result in parts failure or damage. New bolts are always recommended.

21. Install the cylinder head into position on the engine block and over the dowels. Install the cylinder head bolts, lubricating the threads with clean engine oil prior to installation. Torque bolts following the proper sequence as listed below:

Step 1 — 45 ft. lbs. (61 Nm)
Step 2 — 65 ft. lbs. (88 Nm)
Step 3 — again, 65 ft. lbs. (88 Nm)
Step 4 — additional ¼-turn

NOTE: Do not use a torque wrench for Step 4. Inspect the bolt torque after tightening. The torque should be over 90 ft. lbs. (122 Nm). If not, replace the cylinder head bolt.

22. Install the rear timing belt cover bolts and tighten as follows:

M6 bolts — 105 inch lbs. (12 Nm)
M8 bolts — 21 ft. lbs. (28 Nm)
M10 bolts — 40 ft. lbs. (54 Nm)

23. Install the intake manifold assembly and torque bolts following the proper sequence to 21 ft. lbs. (28 Nm).

NOTE: The following procedure can only be used when the camshaft sprockets have been loosened or removed from the shafts.

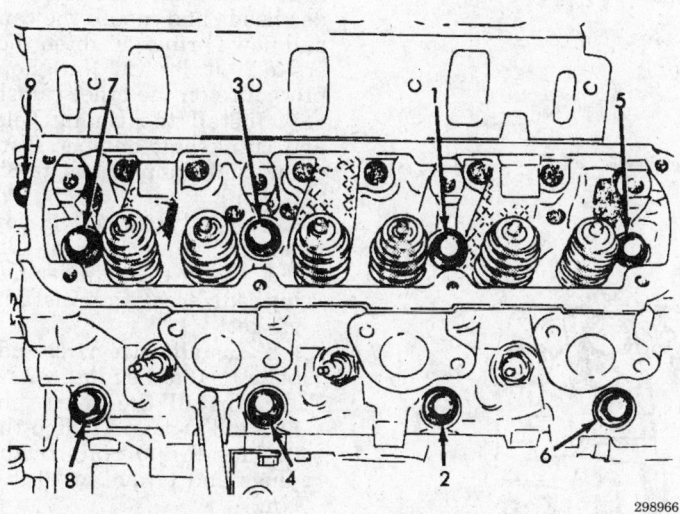

Cylinder head bolt location and torque sequence. Note location of bolt no. 9 — 3.3L engine

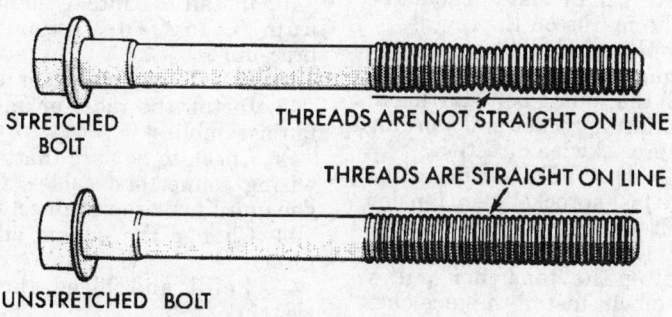

STRETCHED BOLT

THREADS ARE NOT STRAIGHT ON LINE

THREADS ARE STRAIGHT ON LINE

UNSTRETCHED BOLT

Inspecting cylinder head bolts for stretching

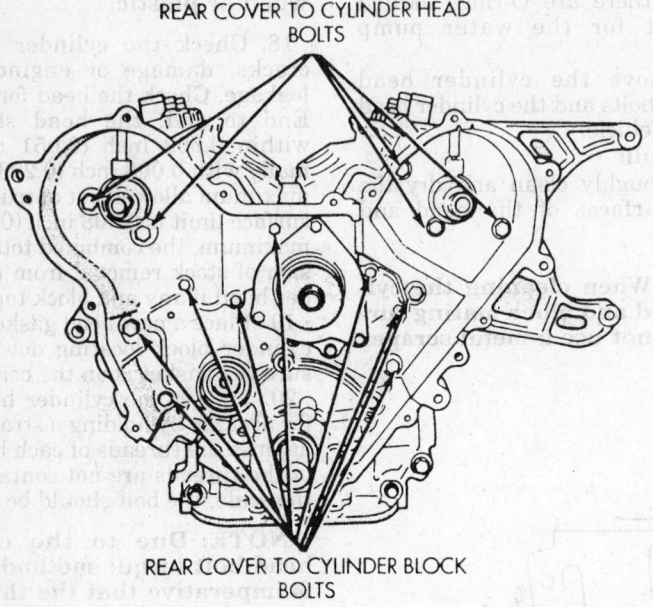

REAR COVER TO CYLINDER HEAD BOLTS

REAR COVER TO CYLINDER BLOCK BOLTS

298987

Rear timing belt cover bolts — 3.5L engine

298982

Head bolt torque sequence — 3.5L engine

24. When the camshaft sprockets are loosened or removed, the camshafts must be timed to the engine. Install the Camshaft Alignment Tools 6642-A or exact equivalent, to the rear of the cylinder heads.

25. Install both camshaft sprockets to the appropriate shafts. The left camshaft sprocket has the DIS pickup as part of the sprocket.

26. Apply threadlocking compound to the threads of the camshaft sprocket retainer bolts and install to the appropriate shafts. The right bolt is 8.380 inch (213.00 mm) long, while the left bolt is 10.0 inch (254.00 mm) long. These bolts are not interchange-

able. Do not tighten the bolts at this time. The camshaft marks should be between the marks on the cover.

27. Place the crankshaft sprocket to the TDC mark on the oil pump housing. Install the timing belt starting at the crankshaft sprocket and working in a counterclockwise direction.

28. After the belt is installed around the last sprocket keep tension on the belt until it is past the tensioner pulley.

29. Holding the tensioner pulley against the belt, install the tensioner housing and tighten to 250 inch lbs. (28 Nm)

30. When the tensioner is in place pull the retainer pin to allow tensioner to extend to the pulley bracket.

31. Install a dial indicator in No. 1 cylinder to check Top Dead Center (TDC) of the piston. Rotate the crankshaft until the piston is exactly at TDC.

32. Hold the right camshaft sprocket hex with a 36 mm box wrench and torque the right camshaft sprocket bolt to 75 ft. lbs. (102 Nm). Turn the sprocket bolt an additional 90 degrees.

33. Hold the left camshaft sprocket hex with a 36 mm box wrench and torque the left camshaft sprocket bolt to 85 ft. lbs. (115 Nm). Turn the sprocket bolt an additional 90 degrees.

34. Remove the dial indicator.

35. Remove the camshaft alignment tools from the back of the cylinder heads and install the cam covers and new O-rings. Tighten the fasteners to 20 ft. lbs. (27 Nm). Repeat this procedure on the other camshaft.

36. Install the timing belt covers and crankshaft damper. Torque the crankshaft damper bolt to 85 ft. lbs. (115 Nm).

37. Install the valve covers as follows:

a. Clean the cylinder head mating surfaces and install a new gasket.

b. Install valve cover and tighten bolts to 105 inch lbs. (12 Nm).

c. Install the spark plug tube nut and O-ring. Tighten the nut to 60 inch lbs. (7 Nm). Install spark plugs and torque to 20 ft. lbs. (28 Nm).

d. Install the air conditioning compressor to the mounting bracket. Torque the mounting bracket bolts to 30 ft. lbs. (41 Nm).

38. Reconnect the spark plug wires to the correct spark plugs.

39. Install the accessory drive belts. Adjust to the proper tension.

40. Install the intake manifold plenum using the recommended procedure.

41. Install the air cleaner assembly.

42. Install the radiator and cooling fan assemblies.

43. Check to be sure that all hoses, wiring connectors, cables, fluid and vacuum lines are reconnected.

44. Change the engine oil and oil filter.

45. Refill and bleed the cooling system.

46. Reconnect the negative battery cable, run the vehicle with the radiator cap off so coolant can be added as required until the thermostat opens.

Watch for leaks and for unusual engine noises. Fill the radiator completely as required.

47. Once the vehicle has cooled, recheck the coolant and oil level.

Valve Lifters

BLEEDING

3.3L (VIN T) Engine

The hydraulic valve lifters do not need to be bled. If the lifter is not functioning properly it must be replaced. When installing a new lifter soak the adjuster in oil until the adjuster has little or no plunger travel when depressed.

REMOVAL AND INSTALLATION

3.3L (VIN T) Engine

———— CAUTION ————
Fuel injection systems remain under pressure, even after the engine has been turned OFF . The fuel system pressure must be relieved before disconnecting any fuel lines. Failure to do so may result in fire and/or personal injury.

1. Disconnect negative battery cable. Release the fuel system pressure.
2. Drain the cooling system. Disconnect the air tube from the air cleaner and the throttle body.
3. Hold the throttle lever in the wide-open position and remove the throttle cable and the speed control cable from the lever. Compress the locking tabs on the cables and remove from the mounting brackets.
4. Disconnect the electrical connector from the solenoid on the EGR valve transducer, MAP sensor, throttle position sensor and the idle air control motor.
5. Disconnect the PCV valve hose at the rear of the intake manifold plenum.
6. Disconnect the vacuum hoses from the power brake booster at the intake manifold nipple. Disconnect the vacuum line at the fuel pressure regulator.
7. Disconnect the purge hose from the throttle body.
8. Remove the EGR tube mounting screws (under throttle body opening) at the intake manifold plenum.
9. The intake system has a large air intake plenum of aluminum alloy and a cross type intake manifold. Remove the intake manifold plenum mounting bolts and remove the plenum up off the engine. Cover the intake manifold to prevent foreign material from entering.
10. Disconnect the fuel supply and return tubes at the rear of the intake manifold. Remove the screw from the fuel clamp and separate the fuel tubes from the bracket.
11. Tag and disconnect the electrical harness from the injectors and turn toward the center of the engine.
12. Remove the four fuel rail mounting bolts and lift fuel rail straight up and off the engine.
13. Remove the upper radiator hose, bypass hose and the rear intake manifold hose.
14. Remove the intake manifold bolts and the manifold from the engine. Remove the intake manifold seal retainer screws and remove the intake manifold gasket.
15. Remove the cylinder head covers from the cylinder head. Tag and disconnect any remaining wiring such as coil wires, sending unit wires, etc.
16. Remove the exhaust manifolds by disconnecting the head pipe, EGR tube and disconnect the oxygen sensor wire. Remove the heat shield and unbolt exhaust manifold.
17. Remove the rocker arm and shaft assemblies. Remove and label the pushrods to insure installation in the original locations.
18. Remove the 9 cylinder head bolts from the cylinder head and remove the cylinder head from the engine.
19. Remove the tappet yoke retainer(s) and aligning yokes. Using tool C-4129 or equivalent, remove the tappets from their bores. If all tappets are to be removed, identify tappets to insure installation in their original location.
20. Check the tappet and bore and if they are scored scuffed or signs of sticking, ream out the bore to the next oversize and replace with an oversized tappet.

NOTE: Tappet bodies and plungers are not interchangeable, mixed parts are not compatible. Work on one tappet at a time.

To install:
21. Lubricate tappets and install in their original location. Install the aligning yokes and retainers and tighten to 105 inch lbs. (12 Nm).
22. Clean the cylinder block and head mating surfaces and install a new head gasket. Be sure the gasket is positioned correctly.

23. Install the cylinder heads. Inspect the head bolts for stretching. New bolts are recommended. Torque bolts one through eight in four steps. First, all to 45 ft. lbs. (61 NM). Second, all to 65 ft. lbs. (88 Nm). Third, all (again) to 65 ft. lbs. (88 Nm). Fourth, plus one-quarter turn (do not use a torque wrench for this step. Bolt torque after the one-quarter turn should be over 90 ft. lbs. (122 Nm). If not, replace the bolt. Tighten bolt nine to 25 ft. lbs. (33 Nm) after head bolts 1–8 have been tightened to specification.
24. Install the pushrods, rocker arm and shaft assemblies with the stamped steel retainers in the proper four positions and tighten to 250 inch lbs. (28 Nm).
25. Install intake manifold. Use new gaskets and torque bolts, working from the center, outward as follows:
   • Step 1— 10 inch lbs. (1 Nm)
   • Step 2— 200 inch lbs. (22 Nm)
   • Step 3— 200 inch lbs. (22 Nm)
26. Install the fuel rail and injectors.
27. Install the cylinder head cover and new gasket on cylinder head. Torque the cylinder head cover bolts to 105 inch lbs. (12 Nm).
28. Install the exhaust manifolds and head pipe. Manifold bolts are torqued to 200 inch lbs. (23 Nm).
29. Reconnect the electrical harness connectors and the vacuum hoses disconnected during the removal procedure.
30. Install the EGR tube to the air intake plenum.
31. Install the cruise control cable and the accelerator cable to the mounting bracket. Hold the throttle lever in the wide-open position and connect the throttle cable and the speed control cable to the lever at the throttle body.
32. Reconnect the air tube to the air cleaner and the throttle body.
33. Check the engine to be sure that all hoses, wiring connectors, cables, vacuum and fluid lines have been reconnected.
34. Refill and bleed the cooling system. An oil and filter change is recommended.
35. Raise and safely support the vehicle. Drain and replace the engine oil and the oil filter.
36. Reconnect the negative battery cable, run the vehicle until the thermostat opens and fill the radiator completely.
37. Once the vehicle has cooled, recheck the coolant level.

Lash Adjusters

BLEEDING

3.5L (VIN F) Engine

The hydraulic lash adjusters are precision units installed in machined openings in the valve actuating ends of the rocker arms, The rocker arm/lash adjuster assembly is not to be disassembled for any reason or damage to the assembly could result. The lash adjuster and rocker arm are serviced as an assembly. Lash adjuster bleeding is not possible.

REMOVAL AND INSTALLATION

3.5L (VIN F) Engine

The hydraulic lash adjusters are precision units installed in the machined openings in the valve actuating ends of the rocker arms and are serviced as an assembly. Please refer to Rocker Arm and Shaft Removal and Installation.

NOTE: Do not disassemble the hydraulic lash adjusters from the rocker arm assembly or damage to the adjuster or the rocker arm may occur.

Valve Lash

ADJUSTMENT

3.3L (VIN T) Engine

The valve train utilizes hydraulic roller valve lifters. There is no valve clearance adjustment.

3.5L (VIN F) Engine

The lash adjusters are precision units installed in machined openings in the valve actuating ends of the rocker arms and are hydraulic, The lash adjuster and rocker arm are serviced as an assembly. Valve clearance adjustments are not performed.

Rocker Arm and Shaft

REMOVAL AND INSTALLATION

3.3L (VIN T) Engine

Rocker arms are installed on a rocker arm shaft attached to the cylinder head with four bolts and retainers. Use care when working with valvetrain components. Any parts that are to be reused must be installed in their original locations.

1. Disconnect negative battery cable.

---- **CAUTION** ----
Fuel injection systems remain under pressure, even after the engine has been turned OFF . The fuel system pressure must be relieved before disconnecting any fuel lines. Failure to do so may result in fire and/or personal injury.

2. Relieve the fuel system pressure.

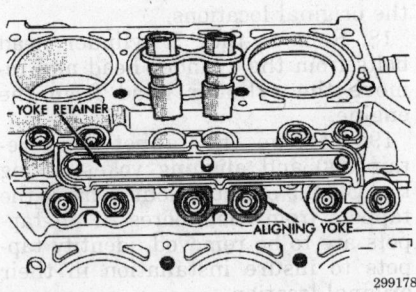

Roller tappets aligning yoke and retainer — 3.3L engine

3. Remove the upper intake manifold assembly.

4. Disconnect and label the spark plug wires from the plugs. Remove by pulling on the boot in a straight out in line with the spark plug.

5. Disconnect the closed crankcase ventilation system and the evaporative control system from the cylinder head cover.

6. Remove the cylinder head cover and gasket from the engine.

7. Remove the four rocker shaft bolts and retainers. Remove the rocker arms and shafts from the engine.

8. Inspect rocker arm and components for wear or damage and replace as required. If the rocker shaft is disassembled for cleaning or replacement, make sure to install components in their original location.

To install:

9. Install rocker arms and shaft assemblies with the stamped steel retainers in the four positions. Torque the retainer bolts slowly to 21 ft. lbs. (28 Nm), in three even progressions starting at the centermost bolts and working outward.

NOTE: After installation, allow 20 minutes tappet bleed down time before operating the engine.

10. Clean the mating surfaces of the cylinder head cover gasket. Inspect the cylinder head cover and straighten out if distorted.

11. Install the cylinder head cover with new gasket in place. Tighten fasteners to 105 inch lbs. (12 Nm).

12. Reconnect the closed crankcase ventilation system and the evaporative control system.

13. Reconnect the spark plug wires making sure that each wire is connected to the correct spark plug.

14. Install the upper intake manifold assembly and reconnect the negative battery cable.

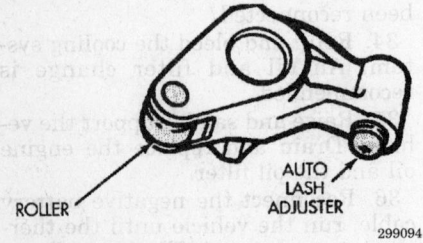

Rocker arm and lash adjuster — 3.5L engine

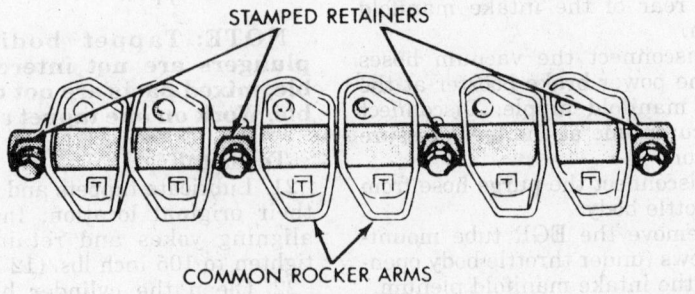

Rocker arm shaft retainers — 3.3L engine

3.5L (VIN F) Engine

NOTE: The intake and exhaust rocker arms are different and they should be identified before disassembly. The hydraulic valve lash adjusters are inside the rocker arms. DO NOT disassemble the hydraulic lash adjuster from the rocker arm assembly. Damage to the adjuster or rocker arm will occur.

1. Disconnect negative battery cable.

— **CAUTION** —

Fuel injection systems remain under pressure even after the engine has been turned OFF . The fuel system pressure must be relieved before disconnecting any fuel lines. Failure to do so may result in fire and/or personal injury.

2. Relieve the fuel system pressure.
3. Remove the air cleaner assembly and the intake manifold plenum. Cover the lower intake manifold during service.
4. Remove the cylinder head covers as follows:
   a. Disconnect, label and relocate the spark plug wires.
   b. Loosen and disengage the air conditioning compressor drive belt. Loosen the air conditioning compressor mounting bracket and pull away from the cylinder head.
   c. Remove the spark plug tube nut and O-ring.
   d. Remove the cylinder head cover screws and the cylinder head cover from the cylinder head.
5. Remove the rocker arm assembly mounting bolts and remove the assembly from the engine.
6. Inspect the rocker arms for wear or damage. Inspect the roller for

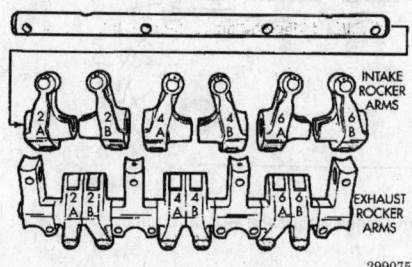

Rocker arm and shaft identification — 3.5L engine

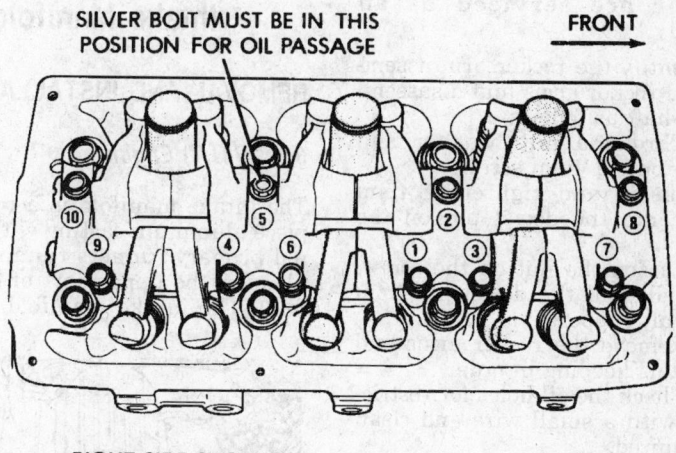

RIGHT SIDE SHOWN

299078

Rocker arm shaft assembly torque sequence, right side — 3.5L engine

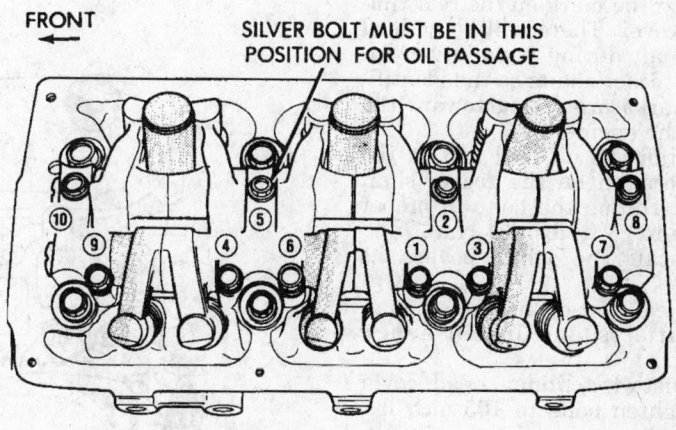

LEFT SIDE SHOWN

299079

Rocker arm shaft assembly torque sequence, left side — 3.5L engine

scuffing or wear. Replace assembly as necessary.

NOTE: Do not remove the lash adjusters from the rocker arm assembly. The rocker arm and the adjuster are serviced as an assembly.

7. Identify the rocker arm assemblies and rocker arms and disassemble the shaft as follows:

a. Thread a nut, washer and spacer onto a 4mm screw.

b. Insert and tighten a 4mm screw into the dowel pin on the shaft.

c. Loosen the nut on the screw. This will pull the dowel pin from the shaft support.

d. Remove the rocker arms and pedestals keeping in order.

e. Check the oil holes for restrictions with a small wire and clean as required.

To install:

8. Assembly the rocker shaft as follows:

a. Install the rocker arms and pedestals onto the shaft keeping in original order.

b. Press the dowel pins into the pedestals until they bottom out in the pedestals.

9. Position the camshaft so that the timing mark on the right camshaft timing belt sprocket aligns with the timing mark on the rear timing belt cover and the timing mark on the left sprocket is 45 degrees from the mark on the rear timing belt cover. There will be no load on the shaft during installation. Install the rocker shafts so the identification marks are facing toward the front of the engine.

10. Install the oil feed bolt in the correct location on the rocker shaft retainer. Torque the bolts in proper sequence to 23 ft. lbs. (31 Nm).

11. Install the valve covers as follows:

a. Clean the cylinder head mating surfaces and install a new gasket.

b. Install cylinder head cover and tighten bolts to 105 inch lbs. (12 Nm).

c. Install the spark plug tube nut and O-ring. Tighten the nut to 60 inch lbs. (7 Nm). Install the spark plug and tighten to 20 ft. lbs. (28 Nm).

d. Install the air conditioning compressor to the mounting bracket. Torque the mounting bracket bolts to 30 ft. lbs. (41 Nm).

e. Install the air conditioning compressor drive belt and adjust to the proper tension.

12. Install the intake manifold plenum and the air cleaner assembly.

13. Reconnect the negative battery cable.

Intake Manifold

REMOVAL AND INSTALLATION

3.3L (VIN T) Engine

The intake manifold is a tuned two-piece aluminum casting with individual primary runners running from a plenum (the upper part of the manifold). The intake manifold assembly

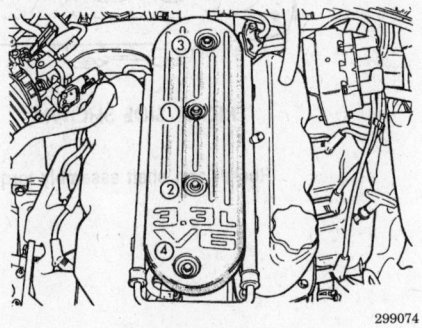

Upper intake manifold (plenum) torque sequence — 3.3L engines

is cored with upper level EGR passages for balanced cylinder-to-cylinder EGR distribution. These openings should be inspected and cleaned if necessary whenever the manifold is removed for service. Use care when working with light alloy parts. Do not over-tighten any fasteners or the casting may be damaged.

— **CAUTION** —

Fuel injection systems remain under pressure, even after the engine has been turned OFF. The fuel system pressure must be relieved before disconnecting any fuel lines. Failure to do so may result in fire and/or personal injury.

1. Disconnect negative battery cable.

2. Remove the fuel filler cap. Release the fuel system pressure using the recommended procedure.

— **CAUTION** —

Do not open the radiator drain with the system hot and under pressure or serious burns from coolant may occur.

3. Drain the cooling system using the following procedure:

a. Place a drain pan under the radiator.

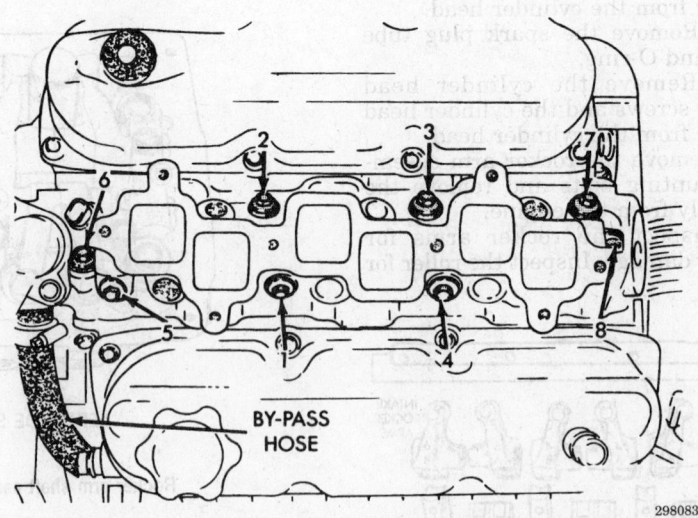

Intake manifold removal and installation torque sequence — 3.3L engines

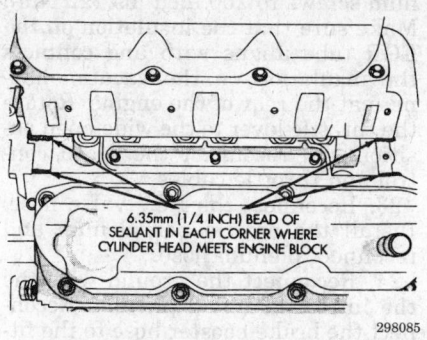

Intake manifold gasket sealing — 3.3L engines

b. Open the radiator drain located at the lower right side of the radiator. Do **NOT** use pliers to open the plastic drain fitting.

c. Remove the coolant pressure bottle cap.

4. Disconnect the air tube from the air cleaner and the throttle body.

5. Hold the throttle lever in the wide-open position and remove the throttle cable and the speed control cable from the lever. Compress the locking tabs on the cables and remove from the mounting brackets.

6. Disconnect the electrical connector from the solenoid on the EGR valve transducer, MAP sensor, throttle position sensor add the idle air control motor.

7. Disconnect the vacuum hose from the PCV valve as well as the power brake booster at the intake manifold nipple. Disconnect the vacuum line at the fuel pressure regulator.

8. Disconnect the purge hose from the throttle body.

9. Remove the EGR tube mounting screws at the intake manifold plenum.

10. Remove the intake manifold plenum (upper part of the manifold) mounting bolts and remove the plenum from the engine. Cover the lower part of the intake manifold to prevent foreign material from entering the engine.

11. Disconnect the fuel supply and return tubes to the fuel rail at the rear of the intake manifold by pushing the quick-connect fitting toward the fuel tube while depressing the built-in disconnect tool with Special Quick-Connect Fitting Tool 6751. To disconnect the fitting from the fuel rail, slightly twist the fitting while maintaining downward pressure on tool 6751. Wrap shop towels around the fuel hoses to absorb any fuel spillage.

12. Cover the fuel line openings to prevent system contamination.

13. Remove the screw from the fuel clamp and separate the fuel tubes from the bracket.

14. Tag each connector for identification, then disconnect the electrical harness from the injectors and turn toward the center of the engine.

15. Remove the fuel rail mounting bolts and lift fuel rail with the injectors attached straight up and off the engine. Cover the injector openings.

16. Remove the upper radiator hose, heater hose and the rear intake manifold hose.

17. Remove the intake manifold bolts and the manifold from the engine.

18. Remove the intake manifold seal retainers screws and remove the intake manifold gasket. Clean all mating surfaces.

19. Inspect the manifold for damage, cracks or clogged passages. Repair, clean or replace the manifold as required.

To install:

20. Verify that all intake manifold and cylinder head sealing surfaces are clean. Place a drop of sealant onto each of the four corners of the intake manifold gasket, where the cylinder head meets the engine block. Carefully install the intake manifold gasket and tighten the seal retainers to 105 inch lbs. (12 Nm).

NOTE: The intake manifold gasket is made of very thin metal and can cause cuts if handled carelessly.

21. Install the intake manifold and eight mounting bolts. Snug down evenly to just 10 inch lbs. (1 Nm).

22. Tighten intake manifold bolts in the proper sequence to 16 ft. lbs. (22 Nm). Once all bolts are torqued, repeat the torquing sequence again tightening the bolts to 16 ft. lbs. (22 Nm). Inspect to make sure all seals are still in place.

23. Apply a light coat of clean engine oil to the O-ring on the nozzle end of each injector.

24. Insert the fuel injector nozzles into the openings in the intake manifold. Seat the injectors in place and install the fuel rail mounting bolts, tightening to 16 ft. lbs. (22 Nm).

25. Attach the electrical connectors to each fuel injector. Rotate the injectors toward the cylinder head covers.

26. Reconnect the fuel supply and return tubes to the fuel rail. Be sure that the black plastic release ring to the quick-connect fitting is in the **OUT** position. Place Special Tool

6751 under the largest diameter of the quick-connect fitting.

27. Pull Tool 6751 toward the fuel rail until the quick-connect fitting clicks into place. Place the Special Tool between the shoulder of the built-in disconnect tool and top of the quick-connect fitting, then inspect the security of the fitting by applying a slight downward force against the fitting. It should be locked in place.

28. Install the intake plenum with new gasket onto the intake manifold. Loosely install the mounting bolts.

29. Install the EGR tube to the manifold with new gasket in place. Loosely install the mounting screws.

30. Tighten the intake manifold plenum mounting bolts to 21 ft. lbs. (28 Nm) following the outlined sequence.

31. Tighten the EGR tube mounting bolts.

32. Reconnect the PCV valve hose and power brake booster hose.

33. Reconnect the electrical connectors to the EGR transducer solenoid, idle air control motor, Map and throttle position sensors.

34. Install the throttle cable and speed control cable to the mounting bracket and connect to the throttle body lever while holding lever in the wide-open position.

35. Reconnect the purge hose to the throttle body. Reconnect the air tube to the air cleaner and the throttle body.

36. Drain and replace the engine oil and the oil filter.

37. Reconnect the negative battery cable. Refill the cooling system. Run the vehicle with the radiator cap removed until the thermostat opens, adding coolant as required. Watch for fuel and coolant leaks and for correct engine operation.

38. Once the vehicle has cooled, recheck the coolant level and add, if necessary.

3.5L (VIN F) Engine

————— **CAUTION** —————
Fuel injection systems remain under pressure, even after the engine has been turned OFF. The fuel system pressure must be relieved before disconnecting any fuel lines. Failure to do so may result in fire and/or personal injury.

1. Disconnect negative battery cable.

2. Remove the fuel filler cap. Release the fuel system pressure using the recommended procedure.

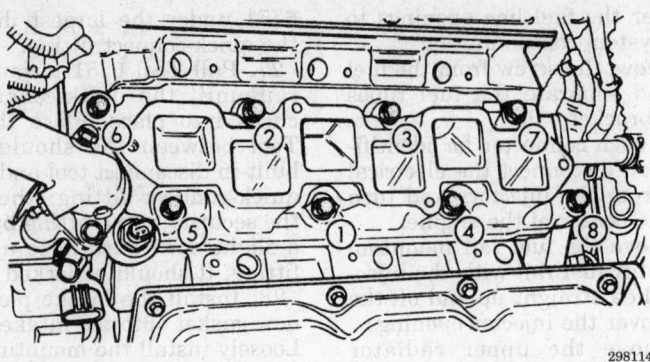

Intake manifold removal and installation torque sequence — 3.5 Engine

CAUTION

Do not open the radiator drain with the system hot and under pressure or serious burns from coolant may occur.

3. Drain the cooling system as follows:

 a. Place a drain pan under the radiator.

 b. Open the radiator drain located at the lower right side of the radiator. Do **NOT** use pliers to open the plastic drain.

 c. Remove the coolant pressure bottle cap.

4. Remove the engine cover from the top of the intake manifold.

5. Remove the accelerator and the speed control cable from the throttle lever.

6. Disconnect the electrical connector from the idle air control motor, intake air temperature sensor and the Manifold Absolute Pressure (MAP) sensor.

7. Remove the ground screw from the intake manifold. Disconnect the electrical connector to the throttle position sensor.

8. Disconnect the vacuum hoses from the manifold tuning valve, PCV make-up air hose, idle air control motor supply hose and the purge hose from the throttle bodies.

9. Disconnect the brake booster hose, PCV hose and the remaining vacuum hoses from the intake manifold. If required, label for proper installation.

10. Remove the mounting bolts for the EGR tube at the intake manifold plenum.

11. Remove the plenum support bracket mounting bolts on each side of the plenum. Remove the intake plenum mounting bolts.

NOTE: The intake manifold plenum (upper half of the intake manifold assembly) uses two different length bolts. Take note of their position and make sure they are installed in the same location during installation.

12. Remove the intake manifold plenum from the intake manifold. Discard the old gasket. Cover the intake manifold openings with tape to keep debris from entering the engine.

13. Remove the upper radiator hose from the thermostat housing and the heater hose from the rear of the intake manifold.

14. Remove the lower intake manifold retaining bolts and the manifold from the engine. Clean all gasket mating surfaces and inspect for distortion with a good straightedge.

To install:

15. Verify that all intake manifold and cylinder head sealing surfaces are clean. Carefully install the intake manifold gasket, then the lower manifold. Tighten bolts in proper sequence to 250 inch lbs. (28 Nm).

16. Reconnect the upper radiator hose to the thermostat housing and the heater hose to the rear of the intake manifold.

17. Ensure the ignition cables are routed out of the way of the intake plenum. Install the intake manifold plenum with new gasket in place and tighten in sequence to 250 inch lbs. (28 Nm).

18. Install the intake manifold plenum with a new gasket in place. Tighten mounting bolts working from the center outward, to 250 inch lbs. (28 Nm). Do not overtorque bolts when working with light alloys.

19. Install and tighten the support bracket bolts.

20. Reconnect the electrical connectors to the manifold absolute pressure sensor, throttle position sensor, idle air control motor and intake air temperature sensor.

21. Reconnect the vacuum hose to the manifold tuning valve.

22. Install the EGR tube. Tighten the EGR tube to intake manifold ple-

num screws to 200 inch lbs. (22 Nm). Make sure that the insulation on the EGR tube aligns with and contacts the insulation on the vacuum harness at the rear of the engine. Rotate the throttle lever to the wide-open position and reconnect the speed control and throttle cables.

23. Reconnect the PCV valve hose. Install the air cleaner plenum and reconnect plenum hose.

24. Reconnect the ground wire the the intake manifold plenum. Reconnect the brake booster hose to the fitting on the intake manifold plenum.

25. Reconnect the throttle body purge tubes.

26. Install the cover on the intake manifold plenum.

27. Reconnect the negative battery cable. Fill and bleed the cooling system.

28. Raise and safely support the vehicle. Change the engine oil and filter.

29. Test run the engine, check for fuel and coolant leaks and verify correct engine operation.

Exhaust Manifold

REMOVAL AND INSTALLATION

3.3L (VIN T) Engine

1. Disconnect the negative battery cable.

2. Raise and safely support the vehicle. Disconnect the exhaust pipe from the exhaust manifold.

3. Disconnect the heated oxygen sensor lead wire and remove the EGR tube from the exhaust manifold.

4. Remove the screws attaching the heat shield to the exhaust manifold. Lower the vehicle, if necessary.

5. Remove the manifold attaching bolts and remove the manifold from the cylinder head.

6. These manifolds are thin-wall designs to save weight. Inspect the manifold carefully for cracks or other damage. Check for distortion against a straightedge or thickness gauge. Replace manifold if required.

7. Remove all traces of the old manifold gasket and clean both gasket mating surfaces.

To install:

8. Install the exhaust manifold and a new manifold gasket to the cylinder head. Install the retainer bolts. Torque the retainer bolts to 17 ft. lbs. (23 Nm).

9. Reconnect the exhaust pipe to the exhaust manifold and torque the nuts to 21 ft. lbs. (28 Nm).

10. Install the EGR tube back onto the manifold.

11. Install the heat shield to the manifold. Lower the vehicle, if necessary.

12. Reconnect the electrical connector at the oxygen sensor.

13. Reconnect the negative battery cable. Operate the vehicle and inspect for exhaust leaks.

3.5L (VIN F) Engine

1. Disconnect the negative battery cable. Raise and safely support the vehicle.

2. Disconnect the exhaust pipes from the exhaust manifold.

3. Disconnect the heated oxygen sensor electrical connector.

4. Lower the vehicle and remove the screws attaching the heat shield to the exhaust manifold.

5. Remove the manifold attaching bolts and remove the manifold from the cylinder head.

6. Inspect the manifold for damage or cracks. Check for distortion against a straightedge or thickness gauge. Replace manifold if required.

7. Remove all traces of the old manifold gasket and clean both gasket mating surfaces.

To install:

8. Install the new manifold gasket and exhaust manifold to the cylinder head. Install the retainer bolts. Torque the retainer bolts to 17 ft. lbs. (23 Nm).

9. Raise and safely support the vehicle.

10. Reconnect the exhaust pipe to the exhaust manifold and torque nuts to 21 ft. lbs. (28 Nm).

11. Lower the vehicle. Install the heat shield to the manifold and tighten the retaining screws to 11 ft. lbs. (15 Nm).

12. Reconnect the electrical connector on the oxygen sensor.

13. Reconnect the negative battery cable. Operate the vehicle and inspect for exhaust leaks.

Front Cover Seal

REMOVAL AND INSTALLATION

3.3L (VIN T) Engine

This engine uses an aluminum front timing chain cover which also carries the front cover crankshaft oil seal. Use care when working with light alloy parts.

1. Disconnect negative battery cable.

2. Remove the cooling fan module assembly using the following procedure.

a. Disconnect the electric lead to the fan RFI module (connects both fan harnesses together).

b. Remove the fan module to radiator fasteners and retaining clips.

c. Remove the fan module assembly from the vehicle.

3. Remove the accessory drive belts.

4. Hold the crankshaft damper in place to keep it from turning while removing the crankshaft bolt. Using a puller, remove the crankshaft damper.

5. Using tool C-4991 or equivalent, remove the front case oil seal. Be careful not to damage the seal surface of the timing chain cover during removal.

To install:

6. Install new seal using tool C-4992 or equivalent, into the front cover. Make sure the seal spring is facing in toward the engine. Install seal until it is flush with the case cover.

7. Hold the crankshaft damper in place to keep it from turning while installing and tightening the crankshaft bolt. Torque to 40 ft. lbs. (54 Nm).

8. Install the accessory drive belts. Adjust the belts to the proper tension.

9. Install the cooling fan module assembly.

10. Reconnect the negative battery cable.

Front Crankshaft Seal

REMOVAL AND INSTALLATION

3.5L (VIN F) Engine

Note that the timing belt must be removed from the vehicle to perform this service. Use care to make sure all valve timing marks are carefully aligned both before removing the belt and after belt installation and all service has been completed. It may be good practice to set the engine to TDC No. 1 cylinder compression stroke (firing position) and aligning all timing marks before removing the timing belt. This serves as a reference for all work that follows.

1. Disconnect negative battery cable.

CAUTION

Fuel injection systems remain under pressure, even after the engine has been turned OFF. The fuel system pressure must be relieved before disconnecting any fuel lines. Failure to do so may result in in fire and/or personal injury.

---

2. Release the fuel system pressure using the recommended procedure.

3. Remove the radiator and cooling fan module assembly.

4. Remove the accessory drive belts.

5. Hold the crankshaft from turning and remove the crankshaft damper bolt. Use a balancer puller and remove the crankshaft damper.

6. Remove the stamped steel timing belt front cover and timing belt and tensioner using the recommended procedure. The sealer on the timing belt front cover may be reusable and should not be removed. Use silicone rubber adhesive sealant to replace any missing sealer. Remove the cast cover.

7. Remove the timing belt sprocket at the crankshaft using puller L-4407A or equivalent puller.

8. Locate the small dowel pin in the crankshaft. With a small punch, carefully tap out the dowel from the end of the crankshaft.

9. Remove the crankshaft seal using tool 6341A or equivalent seal puller, taking care not to nick the shaft seal surface or seal bore during removal.

To install:

10. Inspect the crankshaft seal lip surface for varnish and polish using 400 grit paper to remove as necessary.

11. Install crankshaft seal using seal installer tool 6342 or equivalent seal driver.

12. Install the rear lower timing belt cover.

13. Install the dowel into the crankshaft to 0.047 inch (1.2 mm).

14. Install the timing belt sprocket at the crankshaft using tool C-4685C1, thrust bearing, washer and 12mm bolt or equivalent setup to pull the sprocket onto crankshaft. Do not hammer on the sprocket.

15. Install the timing belt and tensioner using the recommended procedure. Verify that all valve timing marks are aligned. Rotate the crankshaft two complete turns and recheck the timing marks on the camshafts and crankshaft. The marks must line up with their respective locations. If

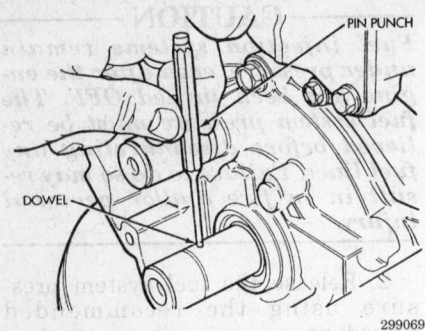

Removing the timing belt sprocket dowel pin on the crankshaft — 3.5L engine

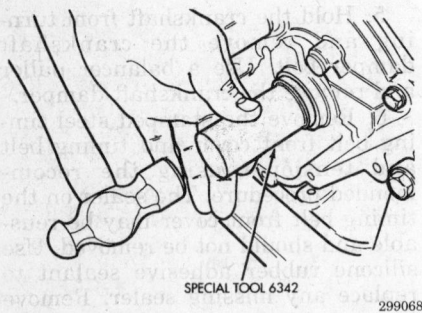

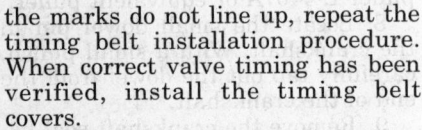

Installing the crankshaft oil seal — 3.5L engine

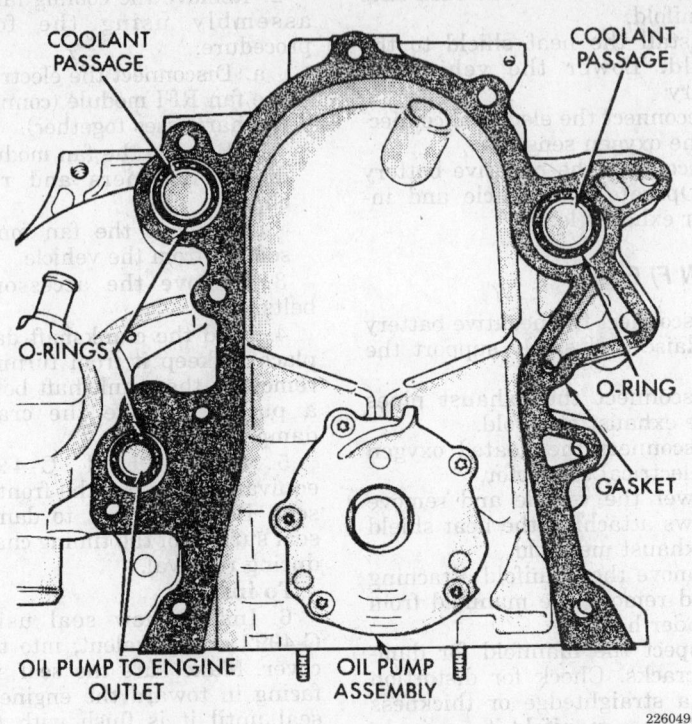

Timing case cover gaskets and O-rings — 3.3L engine

the marks do not line up, repeat the timing belt installation procedure. When correct valve timing has been verified, install the timing belt covers.

16. Install the crankshaft damper. Hold the crankshaft damper using tool L-3281 or equivalent and torque the crankshaft bolt to 85 ft. lbs. (115 Nm).

17. Install the accessory drive belts and adjust to the proper tension.

18. Install the radiator and cooling fan assemblies.

19. Refill and bleed the cooling system.

20. Reconnect the negative battery cable.

Front Cover

REMOVAL AND INSTALLATION

3.3L (VIN T) Engine

These engines use a cast aluminum cover which also carries a front crankshaft seal, oil pump, provides front oil pan closure and water pump

mounting. Use care when working with light alloy components.

1. Drain the cooling system. Remove the radiator and cooling fan assemblies.

2. Raise and safely support the vehicle. Place a drain pan under the engine oil pan and drain the engine oil. It will likely be necessary to disconnect the sway bar and place it to the rear of the vehicle to gain access to the oil pan.

3. Remove the transaxle supports brackets and inspection cover. On Imperial, it may be necessary to remove the right engine mount.

4. Remove the engine oil pan and oil pump pickup.

5. Remove the accessory drive belt(s).

6. Remove the power steering pump and set aside. On some vehicles, it may be necessary to also remove the air conditioning compressor and set aside.

7. If necessary, remove the front wheel and inner fender splash shield. Remove the crankshaft damper using a puller to draw the damper from the crankshaft.

8. Remove the tensioner pulley bracket.

9. Remove the cam sensor from the chain case cover.

10. Remove the timing chain front cover and remove the front cover oil seal.

To install:

11. Clean all parts well. Use care to remove all old sealer and gasket material from the timing chain cover.

NOTE: The crankshaft oil seal must be removed to insure correct oil pump engagement.

12. Remove the old oil seal from the timing case cover. Be sure that the mating surfaces for the timing chain cover gasket are clean and free of any burrs. Rotate the crankshaft so that the oil pump drive flats are vertical. Position the oil pump inner rotor so the mating flats are in the same position as the crankshaft drive flats. Install the timing chain front cover using new gasket and O-rings. Make sure the oil pump is engaged on the crankshaft correctly or severe damage may result.

13. Install the chain case cover screws. Snug down the two bottom screws and the top center screw. Make sure the cover is seated on the block then torque all the other screws to 20 ft. lbs. (27 Nm).

14. Install a new front cover oil seal.

15. Install the crankshaft damper and torque the center bolt to 40 ft.

lbs. (54 Nm). Install the inner splash shield if removed.

16. Install the tensioner pulley bracket.

17. Install the cam position sensor.

18. Install the accessory drive belt(s) and adjust to the proper tension.

19. Install the oil pump pickup tube and the oil pan.

20. Install the transaxle supports brackets and inspection cover.

21. Install the power steering pump and air conditioning compressor, if removed.

22. Connect the front sway bar.

23. Install the radiator and cooling fan assemblies.

24. Refill the engine with the correct amount of clean SAE 5W–30 **OR** SAE 10W–30 engine oil only. Do not mix the 2 grades of oil.

25. Refill and bleed the cooling system.

26. Reconnect the negative battery cable.

27. Test run the engine and check for leaks.

Timing Chain and Sprockets

REMOVAL AND INSTALLATION

3.3L (VIN T) Engine

The camshaft is driven by a silent chain enclosed by a cast aluminum cover which also carries a front crankshaft seal, oil pump, provides front oil pan closure and water pump mounting. Use care when working with light alloy components.

1. Rotate the engine to TDC on No. 1 cylinder. This provides a reference point.

2. Disconnect negative battery cable.

CAUTION

Fuel injection systems remain under pressure even after the engine has been turned OFF. The fuel system pressure must relieved before disconnecting any fuel lines. Failure to do so may result in fire and/or personal injury.

---

3. Relieve the fuel system pressure using the recommended procedure.

4. Drain the cooling system. Remove the radiator and cooling fan assemblies.

5. Raise and safely support the vehicle. Place a drain pan under the engine oil pan and drain the engine oil.

6. Disconnect the sway bar and place it to the rear of the vehicle to gain access to the oil pan.

7. Remove the transaxle supports brackets and inspection cover.

8. Remove the engine oil pan and oil pump pickup.

9. Remove the accessory drive belt(s).

10. Remove the power steering pump and set aside. It may be necessary to also remove the air conditioning compressor and set aside.

11. Remove the crankshaft damper using a puller to draw the damper from the crankshaft.

12. Remove the tensioner pulley bracket.

13. Remove the cam sensor from the chain case cover.

14. Remove the timing chain front cover and remove the front cover oil seal.

15. Chain stretch can be measured before chain removal. Use the following procedure.

   a. Place a small steel scale (ruler) next to the timing chain so that any movement of the chain may be measured.

   b. Place a torque wrench and socket on the camshaft sprocket attaching bolt and apply torque (ap-

---

proximately 30 ft. lbs. or 41 Nm) in the direction of crankshaft rotation to take up any slack. This is with the cylinder heads installed. If this check is done with the cylinder heads off, use approximately 15 ft. lbs. or 20 Nm. With a torque applied to the camshaft sprocket bolt, the crankshaft should not be permitted to move. It may be necessary to block the crankshaft to prevent rotation.

   c. Hold the steel scale even and reading along the edge of the chain links, apply the same amount of torque in the reverse direction. Check the amount of chain movement. Install a new timing chain if the movement exceeds $1/8$-inch (3.175 mm).

   d. If the chain is not satisfactory, it must be replaced.

16. Remove the camshaft sprocket attaching bolts and remove the timing chain with the camshaft sprocket.

17. Using a suitable puller, draw the crankshaft sprocket from the crankshaft. Use care not to damage the crankshaft surface.

To install:

18. Clean all parts well. Use care to remove all old sealer and gasket material from the timing chain cover.

19. Position a new crankshaft sprocket on the shaft and install using a properly sized driver and mallet.

20. If necessary, rotate the crankshaft so the timing arrow is at the 12 o'clock position.

21. Examine the chain. Factory replacement chains should have two links a different color than the others. Position the camshaft sprocket so the timing arrow is at the 6 o'clock position. Place the timing chain around the camshaft sprocket aligning the dark colored link of the chain with the dot on the camshaft sprocket.

22. Place the timing chain around the crankshaft sprocket aligning the dot on the crankshaft sprocket with the dark colored link on the chain. Install the camshaft sprocket in position on the shaft.

23. Using a straightedge, check the alignment of the timing arrows. Install the camshaft bolt and washer and tighten to 40 ft. lbs. (54 Nm).

24. Rotate the crankshaft 2 revolutions in the direction of engine rotation. Check the alignment of the timing arrows, which should line up with each other. If they do not align, remove the camshaft sprocket and retime the engine. Rotate the crankshaft 2 revolutions in the direction of

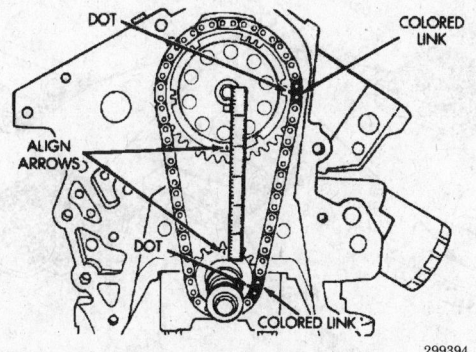

Alignment of timing marks — 3.3L engine

rotation, and confirm alignment of the timing marks.

25. Check the camshaft end-play. With new thrust plate the specification is 0.005 to 0.012 inch (0.0127 to 0.3040mm) or 0.012 inch (0.3040mm) for a used thrust plate. If not within specifications, replace thrust plate.

NOTE: The crankshaft oil seal must be removed to insure correct oil pump engagement.

26. Remove the old oil seal from the timing case cover. Be sure that the mating surfaces for the timing chain cover gasket are clean and free of any burrs. Rotate the crankshaft so that the oil pump drive flats are vertical. Position the oil pump inner rotor so the mating flats are in the same position as the crankshaft drive flats. Install the timing chain front cover using new gasket and O-rings. Make sure the oil pump is engaged on the crankshaft correctly or severe damage may result.

27. Install the chain case cover screws. Snug down the two bottom screws and the top center screw. Make sure the cover is seated on the block then torque all the other screws to 20 ft. lbs. (27 Nm).

28. Install a new front cover oil seal.

29. Install the crankshaft damper and torque the center bolt to 40 ft. lbs. (54 Nm).

30. Install the tensioner pulley bracket.

31. Install the cam position sensor.

32. Install the accessory drive belt(s) and adjust to the proper tension.

33. Install the oil pump pickup tube and the oil pan.

34. Install the transaxle supports brackets and inspection cover.

35. Install the power steering pump and air conditioning compressor, if removed.

36. Connect the front sway bar.

37. Install the radiator and cooling fan assemblies.

38. Refill the engine with the correct amount of clean SAE 5W–30 **OR** SAE 10W–30 engine oil only. Do not mix the 2 grades of oil. A filter change is recommended.

39. Refill and bleed the cooling system.

40. Reconnect the negative battery cable.

41. Test run the engine and check for leaks.

Timing Belt Front Cover

REMOVAL AND INSTALLATION

3.5L (VIN F) Engine

1. Disconnect negative battery cable.

2. Drain the cooling system. Remove radiator and cooling fan assemblies for access to timing belt covers.

3. If necessary, remove the upper radiator hose. Remove the accessory drive belts.

4. Remove the crankshaft damper with a quality puller gripping the inside of the pulley.

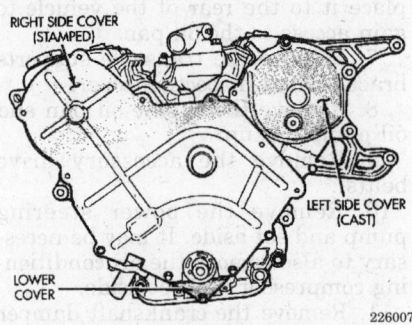

Timing belt covers — 3.5L engine

5. Remove the stamped steel cover. Do not remove the sealer on the cover. It may be reusable.

6. Remove the cast cover. If necessary, remove the lower belt cover located behind the crankshaft damper.

To install:

7. Inspect the timing belt. Replace if required.

8. Before installing, inspect the sealer on the stamped steel cover. Is some sealer is missing, use Mopar Silicone Rubber Adhesive Sealant or equivalent to replace the missing sealer.

9. Install the lower belt cover behind the crankshaft damper, if necessary.

10. Install the stamped steel cover and the cast cover. Tighten the 6mm bolts to 105 inch. lbs. (12 Nm), the 8mm bolts to 250 inch lbs. (28 Nm) and the 10mm bolts to 40 ft. lbs. (54 Nm).

11. Install the crankshaft damper using special tool L-4524, a 5.9 inch long bolt, thrust bearing and washer or equivalent damper installation tools. Torque the center bolt to 85 ft. lbs. (115 Nm).

12. Install accessory drive belts and adjust to the proper tension.

13. Install the radiator and cooling fan assemblies. Refill and bleed the cooling system.

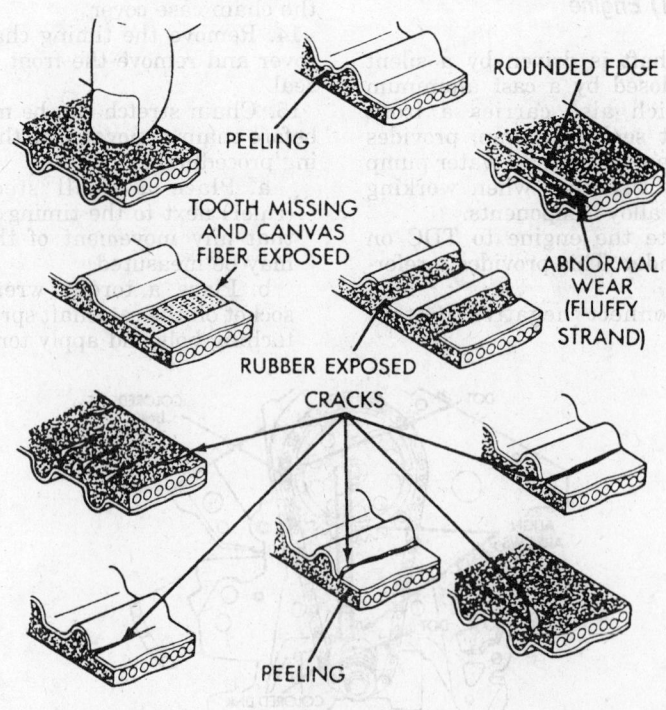

Timing belt inspection — 3.5L engine

14. Reconnect the negative battery cable and test run engine to check for leaks.

Timing Belt and Tensioner

REMOVAL AND INSTALLATION

3.5L (VIN F) Engine

Use care when servicing a timing belt. Valve timing is absolutely critical to engine performance. If the valve timing marks on all drive sprockets are not properly aligned, engine damage will result. If only the belt and tensioner are being serviced, do not loosen the camshaft drive sprockets unless they are to be replaced. The sprockets have oversize openings and can be rotated several degrees in each direction on their shafts. This means the sprockets must be retimed, requiring some special tools.

───── **CAUTION** ─────

Fuel injection systems remain under pressure, even after the engine has been turned OFF. The fuel system pressure must be relieved before disconnecting any fuel lines. Failure to do so may result in fire and/or personal injury.

1. Rotate the engine to Top Dead Center (TDC) and disconnect negative battery cable.
2. Release the fuel system pressure using the recommended procedure.
3. Drain the coolant and remove the radiator and cooling fan assemblies.
4. Remove the accessory drive belts.
5. Remove the upper radiator hose.

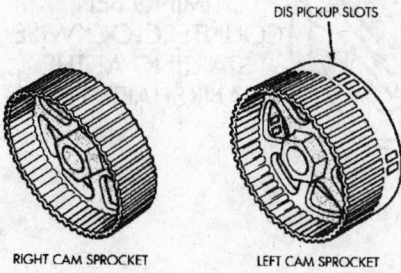

DIS PICKUP SLOTS

RIGHT CAM SPROCKET   LEFT CAM SPROCKET

299915

Camshaft sprocket identification — 3.5L engine

6. Remove the crankshaft damper with a quality puller tool gripping the inside of the pulley.
7. Remove the stamped steel cover. Do not remove the sealer on the cover. It may be reusable.
8. Remove the cast cover. If necessary, remove the lower belt cover located behind the crankshaft damper.
9. If the timing belt is to be reused, mark the timing belt running direction for installation. Align the camshaft sprockets with the marks on the rear covers.
10. Remove the timing belt and tensioner. Inspect the timing belt for excessive wear, damage and/or deterioration and replace with new belt, if necessary.
11. If it is necessary to service the camshaft sprockets, use the following procedure.

a. Hold the camshaft sprocket with a 36 mm box wrench, loosen and remove the sprocket retaining bolt and washer.

NOTE: To remove the camshaft sprocket retainer bolt while the engine is in the vehicle, it may be necessary to raise that side of the engine due to the length of the retainer bolt. The right bolt is 8 ³⁄₈-inch (213.00 mm) long, while the left bolt is 10.0 inch (254.00 mm) long. These bolts are not interchangeable and their original location during removal should be noted.

b. Remove the camshaft sprocket from the camshaft. The camshaft sprockets are not interchangeable from side to side.

c. Remove the crankshaft sprocket using puller L-4407A or equivalent.

To install:

12. If the it was necessary to remove the the camshaft sprockets, use the following procedure.

NOTE: This procedure can only be used when the camshaft sprockets have been loosened or removed from the camshafts. Each sprocket has a D-shaped hole that allows it to be rotated several degrees in each direction on its shaft. This design must be timed with the engine to ensure proper performance.

a. Install the crankshaft sprocket using tool C-4685-C1, thrust bearing, washer and 12 mm bolt.

b. When the camshaft sprockets are loosened or removed, the camshafts must be timed to the engine. Install the camshaft alignment

tools 6642-A or exact equivalent, to the rear of the cylinder heads. These tools lock the camshafts in the proper position.

13. Pre-load the belt tensioner as follows:

a. Place the tensioner in a vise the same way it is mounted on the engine.

b. Slowly compress the plunger into the tensioner body.

c. When the plunger is compressed into the tensioner body install a pin through the body and plunger to retain plunger in place until tensioner is installed.

14. Install both camshaft sprockets to the appropriate shafts. The left camshaft sprocket has the DIS pickup as part of the sprocket.

15. Apply Loctite® 271 or equivalent, to the threads of the camshaft sprocket retainer bolts and install to the appropriate shafts. The right bolt is 8 ³⁄₈-inch (213.00mm) long, while the left bolt is 10.0 inch (254.00 mm) long. These bolts are not interchangeable. Do not tighten the bolts at this time. The camshaft marks should be between the marks on the cover.

16. Align the camshaft sprockets between the marks on the covers.

17. Align the crankshaft sprocket with the TDC mark on the oil pump cover.

18. Install the timing belt starting at the crankshaft sprocket and going in a counterclockwise direction. After the belt is installed on the right sprocket, keep tension on the belt until it is past the tensioner pulley.

19. Holding the tensioner pulley against the belt, install the timing belt tensioner into the housing and tighten to 21 ft. lbs. (28 Nm).

20. When the tensioner is in place pull the retainer pin to allow tensioner to extend to the pulley bracket.

21. Remove the spark plug in the No. 1 cylinder and install a dial indicator to check Top Dead Center (TDC) of the piston. Rotate the crankshaft until the piston is exactly at TDC. Hold the camshaft sprocket hex with a 36 mm wrench and torque the right camshaft sprocket bolt to 75 ft. lbs. (102 Nm) plus an additional 90 degree turn. Torque the left camshaft sprocket bolt to 85 ft. lbs. (115 Nm) plus an additional 90 degree turn.

22. Remove the dial indicator. Install the spark plug and tighten to 20 ft. lbs. (28 Nm).

23. Remove the camshaft alignment tools from the back of the cylinder heads and install the cam covers and new O-rings. Tighten the fasten-

ers to 20 ft. lbs. (27 Nm). Repeat this procedure on the other camshaft.

24. Rotate the crankshaft sprocket two revolutions and check for proper alignment of the timing marks on the camshaft and the crankshaft. If the timing marks do not line up, repeat the procedure.

25. Before installing, inspect the sealer on the stamped steel cover. Is some sealer is missing, use MOPAR Silicone Rubber Adhesive sealant or equivalent to replace the missing sealer.

26. Install the lower belt cover behind the crankshaft damper, if necessary.

27. Install the stamped steel cover and the cast cover. Tighten the 6mm bolts to 105 inch. lbs. (12 Nm), the 8mm bolts to 250 inch lbs. (28 Nm) and the 10mm bolts to 40 ft. lbs. (54 Nm).

28. Install the crankshaft damper using special tool L-4524, a 5.9 inch long bolt, thrust bearing and washer or equivalent damper installation tools. Torque the center bolt to 85 ft. lbs. (115 Nm).

29. Install the upper radiator hose.

30. Install the accessory drive belts and adjust them to the proper tension.

31. Install the radiator and cooling fan assemblies.

32. Refill and bleed the cooling system.

33. Reconnect the negative battery cable. With the radiator cap off so coolant can be added, run the engine. Watch for leaks or unusual engine noises. Add coolant as the engine warms.

Timing Belt Sprockets

REMOVAL AND INSTALLATION

3.5L (VIN F) Engine

Camshaft Sprocket

NOTE: When the camshaft sprockets are loosened or removed, the camshafts must be timed to the engine. The following procedure can only be used when the camshaft sprockets have been loosened or removed from the shafts.

1. Rotate the engine to TDC and disconnect negative battery cable.

> ## ── CAUTION ──
> *Fuel injection systems remain under pressure, even after the engine has been turned OFF. The fuel system pressure must be re-*

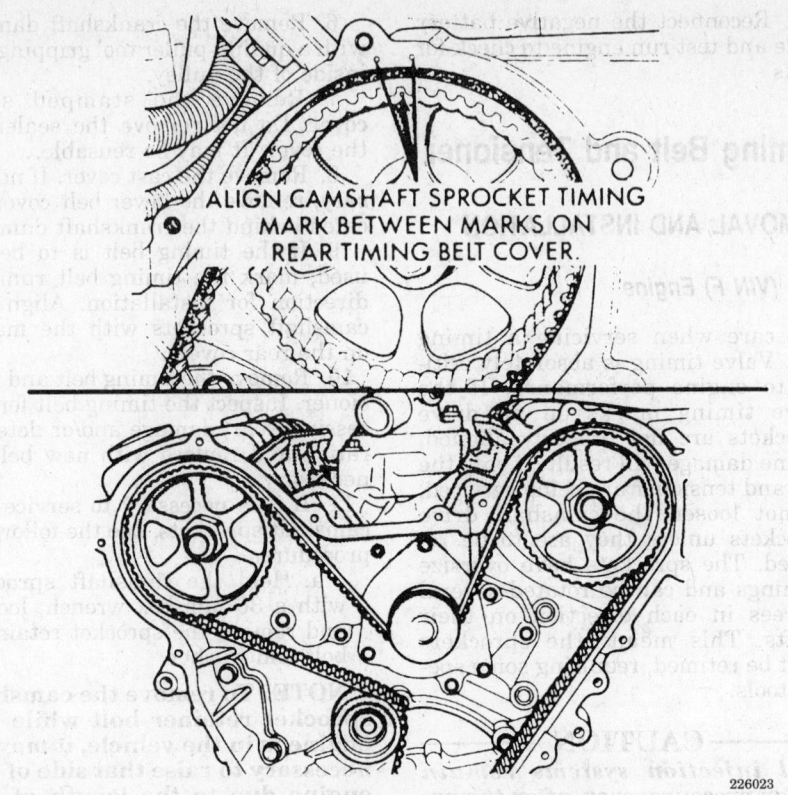

Camshaft sprocket timing marks — 3.5L engine

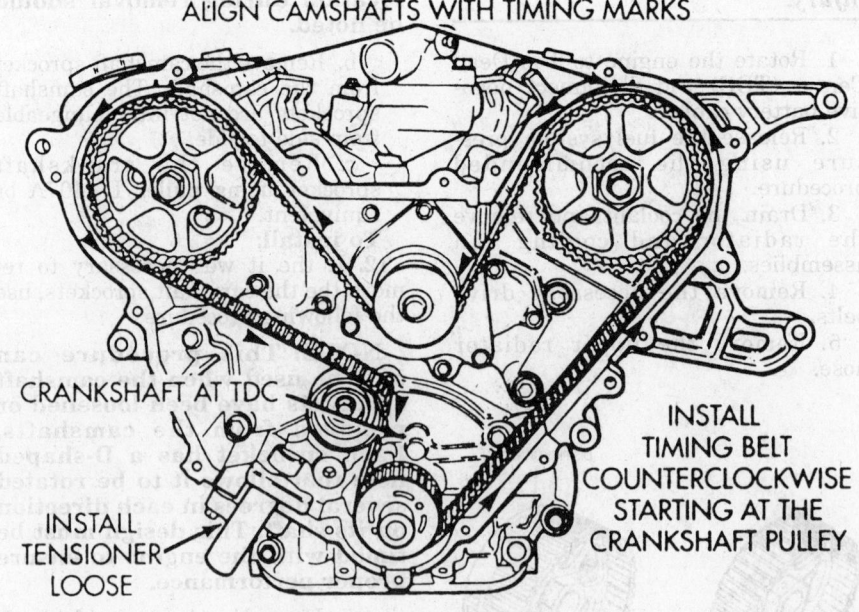

Timing belt installation — 3.5L engine

lieved before disconnecting any fuel lines. Failure to do so may result in fire and/or personal injury.

2. Release the fuel system pressure using the recommended procedure.

3. Drain the cooling system.

4. Remove the radiator and cooling fan assemblies.

5. Remove the upper radiator hose and the accessory drive belts.

6. Remove the crankshaft damper using a damper pulling tool.

7. Remove the timing belt covers. Do not remove the sealer on the cover. It should be reusable.

8. Mark the timing belt running direction for installation. Align the camshaft sprockets with the marks on the rear covers.

9. Remove the timing belt and tensioner.

10. Hold the camshaft sprocket with a 36 mm box wrench, loosen and remove the sprocket retaining bolt and washer.

NOTE: **To remove the camshaft sprocket retainer bolt while the engine is in the vehicle, it may be necessary to raise that side of the engine due to the length of the retainer bolt. The right bolt is 8 ³/₈-inch (213.00 mm) long, while the left bolt is 10.0 inch (254.00 mm) long. These bolts are not interchangeable and their original location during removal should be noted.**

11. Remove the camshaft sprocket from the camshaft. The camshaft sprockets are not interchangeable from side to side.

To install:

NOTE: **The following procedure can only be used when the camshaft sprockets have been loosened or removed from the shafts.**

12. Place the crankshaft sprocket to the TDC mark on the oil pump housing.

13. When the camshaft sprockets are loosened or removed, the camshafts must be timed to the engine. Install the camshaft alignment tools 6642-A or exact equivalent, to the rear of the cylinder heads. These tools lock the camshafts in the proper position.

14. Pre-load the belt tensioner as follows:

a. Place the tensioner in a vise the same way it is mounted on the engine.

b. Slowly compress the plunger into the tensioner body.

c. When the plunger is compressed into the tensioner body install a pin through the body and plunger to retain plunger in place until tensioner is installed.

15. Install both camshaft sprockets to the appropriate shafts. The left camshaft sprocket has the DIS pickup as part of the sprocket.

16. Apply Loctite® 271 or equivalent, to the threads of the camshaft sprocket retainer bolts and install to the appropriate shafts. The right bolt is 8 ³/₈-inch (213.00mm) long, while the left bolt is 10.0 inch (254.00 mm) long. These bolts are not interchangeable. Do not tighten the bolts at this time. The camshaft marks should be between the marks on the cover.

17. Install the timing belt starting at the crankshaft sprocket and working in a counterclockwise direction. Keep tension on the belt until it is past the tensioner pulley.

18. Holding the tensioner pulley against the belt, install the belt tensioner housing and tighten to 250 inch lbs. (28 Nm)

19. When the tensioner is in place pull the retainer pin to allow tensioner to extend to the pulley bracket.

20. Remove the spark plug in the No. 1 cylinder and install a dial indicator to check Top Dead Center (TDC) of the piston. Rotate the crankshaft until the piston is exactly at TDC. Hold the camshaft sprocket hex with a 36 mm wrench and torque the right camshaft sprocket bolt to 75 ft. lbs. (102 Nm) plus an additional 90 degree turn. Torque the left camshaft sprocket bolt to 85 ft. lbs. (115 Nm) plus an additional 90 degree turn.

21. Remove the dial indicator. Install the spark plug and tighten to 20 ft. lbs. (28 Nm).

22. Remove the camshaft alignment tools from the back of the cylin-

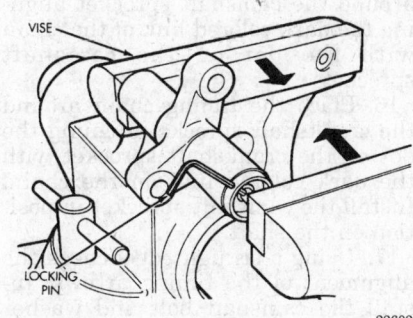

VISE

LOCKING PIN

226026

Compressing timing belt tensioner — 3.5L engine

der heads and install the cam covers and new O-rings. Tighten the fasteners to 20 ft. lbs. (27 Nm). Repeat this procedure on the other camshaft.

23. Install the timing belt covers and crankshaft damper using special tool L-4524, 5.9 in. bolt, thrust bearing and washer.

24. Install the accessory drive belts and adjust to the proper tension.

25. Install the upper radiator hose.

26. Install the radiator and cooling fan assemblies.

27. Refill and bleed the cooling system.

28. Reconnect the negative battery cable. With the radiator cap removed, run the vehicle until the thermostat opens. Add coolant as required.

29. Once the vehicle has cooled, recheck the coolant level.

Crankshaft Sprocket

--- CAUTION ---
Fuel injection systems remain under pressure, even after the engine has been turned OFF. The fuel system pressure must be relieved before disconnecting any fuel lines. Failure to do so may result in fire and/or personal injury.

1. Disconnect negative battery cable.

2. Release the fuel system pressure using the recommended procedure.

3. Remove the radiator and cooling fan assemblies.

4. Remove the accessory drive belts.

5. Remove the crankshaft damper using a damper pulling tool.

6. Remove the timing belt covers.

7. Remove the timing belt and tensioner.

8. Remove the crankshaft sprocket using puller L-4407A or equivalent.

To install:

9. Install the crankshaft sprocket using tool C-4685C1, thrust bearing, washer and 12mm bolt.

10. Install the timing belt and belt tensioner and set the engine timing. Install the timing belt covers.

11. Install the crankshaft damper and accessory drive belts. Adjust the drive belts to the proper tension.

12. Install the radiator and cooling fan assemblies. Refill the cooling system.

13. Reconnect the negative battery cable. With the radiator cap removed, run the vehicle until the thermostat opens. Add coolant as required.

Camshaft

REMOVAL AND INSTALLATION

3.3L (VIN T) Engine

NOTE: To remove and replace the camshaft on this engine and in this vehicle, the engine assembly must be removed from the vehicle.

1. Disconnect the battery negative cable.
2. Raise and safely support the vehicle. Drain the engine oil and remove the oil filter.

—————— CAUTION ——————
Fuel injection systems remain under pressure, even after the engine has been turned OFF. The fuel system pressure must be relieved before disconnecting any fuel lines. Failure to do so may result in fire and/or personal injury.

3. Relieve the fuel system pressure using the recommended procedure.

—————— CAUTION ——————
Do not open the radiator drain cock with the system hot and under pressure or serious burns from coolant may occur.

4. Remove the radiator and cooling fan assemblies.
5. Remove the engine from the vehicle using the recommended procedure.
6. With the engine removed from the vehicle, remove the cylinder head covers, rocker arm and rocker arm shaft assemblies.
7. Remove the intake manifold assembly and cylinder heads from the engine.

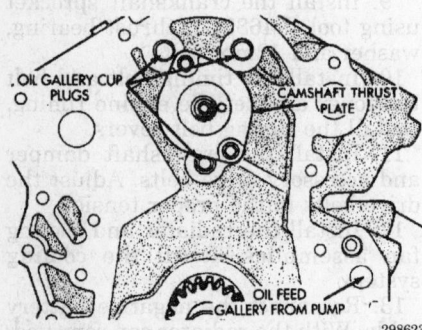

Camshaft thrust plate assembly — 3.3L engine

298623

8. Remove the harmonic balancer. Remove the timing chain case cover and timing chain from the engine.
9. Remove the pushrods and tappets. If any valvetrain components are to be reused, identify each part and its location so each can be installed in its original location.

NOTE: If the camshaft is being removed to replace with a new one, new valve lifters MUST be installed. Installing used lifters on a new camshaft will quickly fail the camshaft.

10. Remove the camshaft thrust plate. Install a long bolt into the front of the camshaft to act as a handle and aid in removal. Remove the camshaft being careful not to damage the cam bearings with the cam lobes.
11. Inspect the bearing journals and the lobes on the shaft for damage and replace the camshaft is required. New lifters must be installed on a new camshaft.

To install:
12. Lubricate the camshaft lobes and the camshaft bearing journals. Inspect the bearing journals on the camshaft and install the shaft within 2 inches of its final position in the cylinder block.

NOTE: Chrysler recommends the addition of 1 pint of Chrysler Crankcase Conditioner or equivalent, be added to the crankcase when the camshaft has been replaced. This will aid in break-in. Leave the oil mixture in the engine for a minimum of 500 miles and drain at the next normal oil change.

13. Install the camshaft thrust plate with the 2 screws and tighten to 105 inch lbs. (12 Nm).
14. Rotate the crankshaft so the timing arrow is in the 12 o'clock position.
15. Position the camshaft sprocket so the timing arrow is at the 6 o'clock position. Place the timing chain around the camshaft sprocket aligning the dark colored link of the chain with the dot on the camshaft sprocket.
16. Place the timing chain around the crankshaft sprocket aligning the dot on the crankshaft sprocket with the dark colored link on the chain. Install the camshaft sprocket in position on the shaft.
17. Using a straightedge, check the alignment of the timing arrows. Install the camshaft bolt and washer and tighten to 40 ft. lbs. (54 Nm).
18. Rotate the crankshaft 2 revolutions in the direction of engine rota-

tion. Check the alignment of the timing arrows, which should line up with each other.
　a. If they do not align, remove the camshaft sprocket and re-time the engine.
　b. Again, rotate the crankshaft 2 revolutions in the direction of rotation, and confirm alignment of the timing marks.
19. Check the camshaft end-play. With new thrust plate the specification is 0.005–0.012 inch (0.0127–0.3040 mm) or 0.012 inch (0.3040 mm) for old thrust plate. If not within specifications, replace thrust plate.
20. Lubricate and install the valve lifters (tappets) in their original position. If the camshaft was replaced, all lifters must be replaced with new parts.
21. Install the timing chain front cover using new seals and O-rings.
22. Install the cylinder heads and intake manifold assemblies onto the engine using the recommended procedures.
23. Lubricate and install the pushrods in their original position.
24. Install the rocker arm and rocker arm shaft assemblies. Install the cylinder head covers.
25. Tighten the oil pan drain plug and install a new oil filter.
26. Install the engine into the vehicle using the recommended procedure.
27. Install the radiator and cooling fan assemblies.
28. Be sure that all fluid lines, cables, hoses, and electrical connectors are reconnected and secured.
29. Refill the engine with the correct amount of clean SAE 5W–30 **OR** SAE 10W–30 engine oil only. Do not mix the 2 grades of oil.
30. Refill and bleed the cooling system.
31. Reconnect the negative battery cable. Start the engine and inspect for leaks. Test drive the vehicle.
32. Check engine fluid levels and top off if necessary.

3.5L (VIN F) Engine

1. Disconnect the negative battery cable.
2. Release the fuel system pressure. Drain the cooling system.
3. Remove the radiator/cooling fan assemblies and the accessory drive belts.
4. Remove the crankshaft damper and the timing belt covers.
5. Mark the timing belt rotation direction for installation. Align the timing belt sprockets with marks on

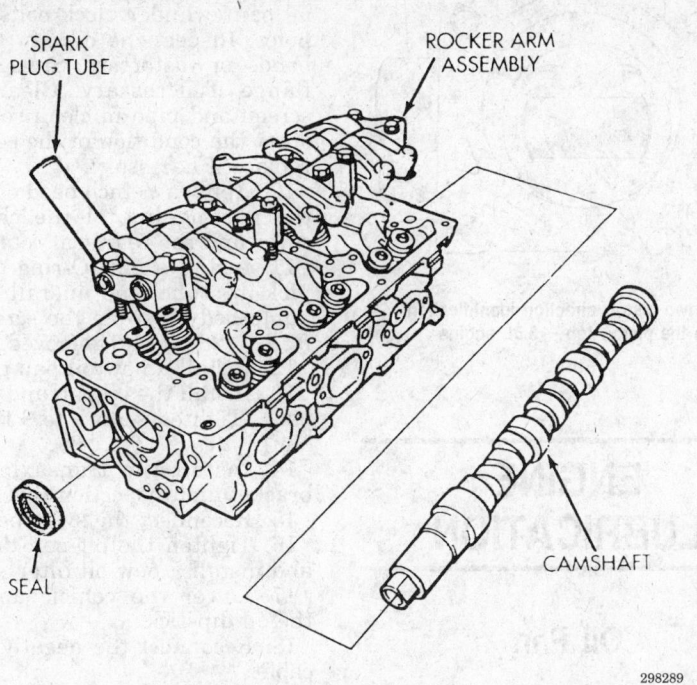

Cylinder head, camshaft and rocker arm assembly — 3.5L engine

298289

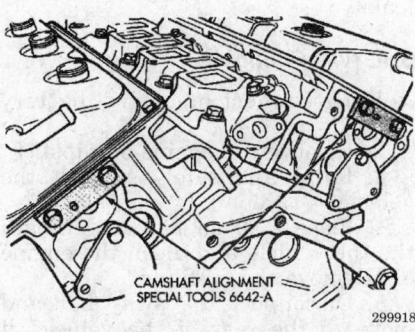

Special camshaft alignment tools — 3.5L engine

299918

the rear timing belt covers before removing the timing belt.

6. Remove the timing belt tensioner and timing belt.

7. Remove the timing belt sprockets at each camshaft.

8. Remove the intake manifold assembly using the recommended procedure.

9. Separate the exhaust manifold from the cylinder head assembly. Be sure to clean the gasket mating surfaces between the exhaust manifold and the cylinder head.

10. The rear timing belt cover must be removed to remove the cylinder heads. Remove the rear timing belt cover-to-cylinder head bolts. Remove the rear timing belt covers. The right-hand side timing belt cover has O-rings located behind it for the water pump passages.

11. Remove the cylinder head bolts and remove the cylinder head from the engine.

12. Mark the rocker arm assembly to note component locations prior to disassembly. Remove the rocker arm and shaft assemblies from the cylinder head.

13. Remove the rear camshaft cover and O-ring from the head.

14. Carefully remove the camshaft from the rear of the head taking care not to nick or scratch the journals when removing.

15. Inspect camshaft journals for wear or damage. If wear is present, inspect the cylinder head for damage. Inspect the head oil holes for clogging. Replace the camshaft as required.

16. Measure the height of the cam using a micrometer. Measure in two places; the unworn area and in the wear zone. Subtract the figures to get

cam wear. The standard specification is 0.001 inch (0.0254 mm) with the wear limit being 0.010 inch (0.254 mm). Replace the camshaft if it is worn beyond this specification

To install:

17. Lubricate the camshaft journals and cam with clean engine oil. Install camshaft into cylinder head.

18. Install the camshaft cover and O-ring to the head and tighten to 21 ft. lbs. (28 Nm).

19. Install the rocker arm assemblies in their original location.

20. Install the cylinder head assembly to the engine block. New head bolts are recommended.

21. Install the rear timing belt covers and tighten the rear cover-to-cylinder head bolts.

22. Install the timing belt sprocket, timing belt and timing belt tensioner using the recommended procedure. Be sure that once they are all installed, the timing of the camshaft(s) is accurate.

23. Install the exhaust manifold, with a new manifold gasket, to the cylinder head.

24. Install the intake manifold assembly using the recommended procedure.

25. Install the timing belt covers and crankshaft damper. Install the accessory drive belts and set them to the proper tension.

26. Install the radiator and cooling fan assembly.

27. Refill and bleed the cooling system. An oil and filter change is recommended.

28. Reconnect the negative battery cable.

Piston and Connecting Rod

POSITIONING

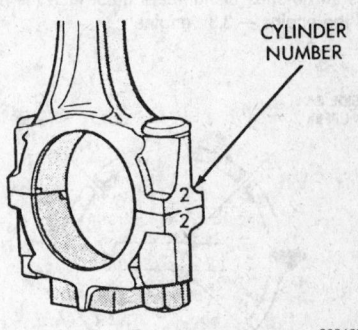

303462

Numbering the connecting rod and bearing cap to the cylinder — 3.3L engine

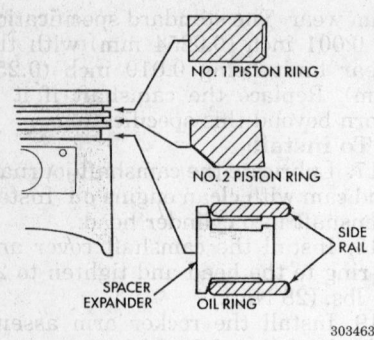

Piston ring installation — 3.3L and 3.5L Engines

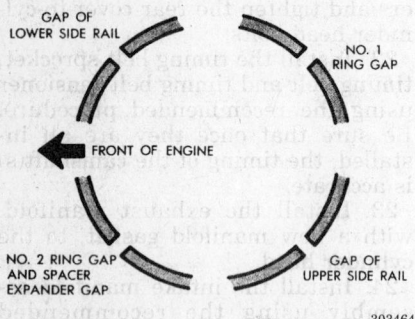

Piston ring end gap position — 3.3L engine

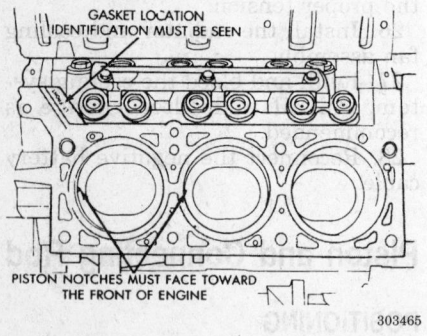

Piston notches or numbers must face the front of the engine — 3.3L engine

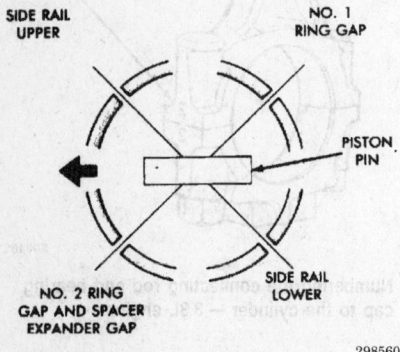

Piston ring end gap position — 3.5L engine

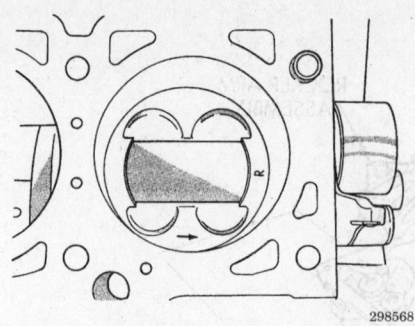

Note the two piston direction identification marks on the piston top — 3.5L engine

ENGINE LUBRICATION

Oil Pan

REMOVAL AND INSTALLATION

3.3L (VIN T) Engine

1. Disconnect negative battery cable.
2. Remove the engine oil dipstick.
3. Raise and safely support the vehicle.
4. Place a drain pan underneath the the vehicle and drain the engine oil. Remove the oil filter.
5. Disconnect the sway bar and move to the rear of the vehicle, if necessary.
6. Remove the transaxle support bracket and inspection cover.
7. Remove the oil pan screws and remove the oil pan.
8. Remove the oil pick-up tube, if necessary. Discard the old oil pick-up tube O-ring.

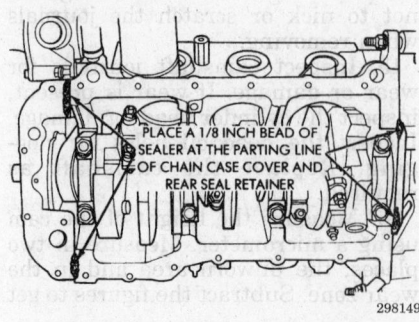

Oil pan sealing points — 3.3L engine

To install:
9. Thoroughly clean and dry the oil pan, cylinder block bolts and bolt holes. Inspect the oil pan flange for bends or distortion. Straighten the flange if necessary. Clean the oil screen and pipe in clean solvent. Inspect the condition of the screen and replace if necessary.
10. Apply a 1/8-inch bead of sealer at the parting line of the chain case cover and the rear seal retainer.
11. Install a new O-ring on the oil pick-up tube and install into the pump body. Torque the screws to 20 ft. lbs. (28 Nm), if removed.
12. Install a new oil pan gasket.
13. Install the oil pan and retaining bolts. Tighten screws to 9 ft. lbs. (12 Nm).
14. Install the transaxle support bracket and inspection cover.
15. Reconnect the sway bar.
16. Tighten the oil pan drain plug and install a new oil filter.
17. Lower the vehicle and install the oil dipstick.
18. Reconnect the negative battery cable.
19. Refill the engine with the proper amount of clean SAE 5W-30 OR SAE 10W-30 engine oil only. Do not mix the 2 grades of oil.
20. Start the engine and check for leaks.

3.5L (VIN F) Engine

1. Disconnect negative battery cable.
2. Remove the engine oil dipstick.
3. Raise and safely support the vehicle.
4. Place a drain pan underneath the the vehicle and drain the engine oil. Remove the oil filter.
5. Disconnect the sway bar and move to the rear of the vehicle, if necessary.
6. Remove the transaxle support bracket and inspection cover.
7. Remove the oil pan screws and remove the oil pan.
8. Remove the oil pick-up tube, if necessary and the windage tray/oil pan gaskets. The windage tray and oil pan gasket are integral. The silicone rubber gaskets are bonded directly to both sides of the windage tray. This assembly is reusable if it is not damaged upon removal. Discard the old oil pick-up tube O-ring.

NOTE: Any old sealant must be carefully removed if the gasket is going to be used again.

To install:
9. Thoroughly clean and dry the oil pan, cylinder block bolts and bolt

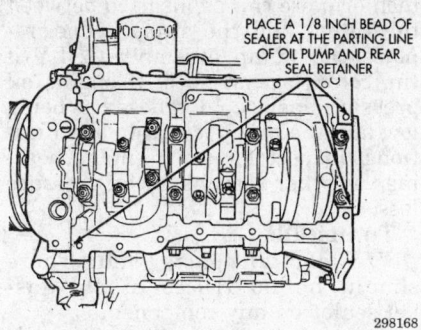

Oil pan sealing points — 3.5L engine

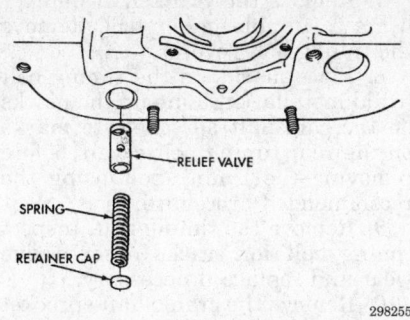

Engine oil pressure relief valve — 3.3L engine

holes. Inspect the oil pan flange for bends or distortion. Straighten flange if necessary. Clean the oil screen and pipe in clean solvent. Inspect the condition of the screen and replace if necessary.

10. Apply a ⅛-inch bead of sealer at the parting line of the oil pump body and the rear seal retainer.

11. Install a new O-ring on the oil pick-up tube and install into the pump body. Torque the screws to 20 ft. lbs. (28 Nm), if removed.

12. Install the windage tray/oil pan gasket.

13. Install the oil pan and retaining bolts. Tighten screws to 9 ft. lbs. (12 Nm).

14. Install the transaxle support bracket and inspection cover.

15. Reconnect the sway bar.

16. Tighten the oil pan drain plug and install a new oil filter.

17. Lower the vehicle and install the oil dipstick.

18. Reconnect the negative battery cable.

19. Refill the engine with the proper amount of clean SAE 5W-30 **OR** SAE 10W-30 engine oil only. Do not mix the 2 grades of oil.

20. Start the engine and check for leaks.

Oil Pump

REMOVAL AND INSTALLATION

3.3L (VIN T) Engine

1. Disconnect negative battery cable and drain the cooling system.

2. Remove the radiator assembly.

3. Raise and safely support the vehicle.

4. Drain the engine oil. Remove the oil filter.

5. Disconnect the sway bar and place it to the rear of the vehicle to gain access to the oil pan.

6. Remove the transmission support brackets and inspection cover.

7. Remove the oil pan and the oil pump pick-up tube.

8. Lower the vehicle.

9. Remove the accessory drive belts and tensioner pulley bracket.

10. Remove the power steering pump and set aside. Remove the air compressor mounting bolts and set compressor aside. Remove the compressor mounting bracket. It is not necessary to disconnect the refrigerant lines or evacuate the refrigerant from the air conditioning system.

11. Using an appropriate puller, remove the crankshaft pulley.

12. Remove the tensioner pulley bracket.

13. Remove the camshaft sensor from the chain case cover.

14. Remove the timing chain case cover mounting bolts and the cover from the front of the engine.

15. Clean the gasket material from the mating surfaces of the cover and the block.

16. Remove the oil pump cover retaining screws from the timing cover. Lift off the oil pump cover.

17. Remove the pump rotors. Wash all parts in solvent and inspect carefully for damage or wear.

18. To remove the relief valve, drill a ⅛-inch hole into the relief valve retainer cap and insert a self-threading sheet metal screw into the cap. Clamp the screw in a vise and while supporting the Chain Case Cover, remove the cap by tapping the case using a soft hammer. Discard the retainer cap and remove the spring and relief valve. Valve spring free length specification should be approximately 1.95 inches (49.5 mm) and the plunger should be free of scratches and marks.

19. Place a straightedge across the surface of the pump cover. Replace the cover if a 0.003 in. feeler gauge

can be inserted between the cover and the straightedge.

20. Measure the thickness and the diameter of the outer rotor. If the thickness measures 0.301 inches or less, or the diameter measures 3.148 inches or less, replace the outer rotor. The same thickness measurement applies to the inner rotor.

21. Insert the outer rotor into the Chain Case Cover (CCC), press it to one side and measure the clearance between the rotor and the CCC. If the measurement is 0.015 inches or more, replace the CCC only if the outer rotor is within specifications.

22. Insert the inner rotor and if the clearance between both rotors is 0.008 inches or more, replace both rotors.

23. Lay a straightedge across the face of the CCC between the bolt holes. Replace the pump assembly if a 0.004 in. feeler gauge or more, can be inserted between the rotors and the straightedge, **ONLY** if the rotors are to specifications.

To install:

24. Clean all parts well. Assemble the oil pump with new parts as required. Install the inner rotor with chamfer facing the cast iron oil pump cover.

25. Install the pump cover and tighten the fasteners to 9 ft. lbs. (12 Nm).

26. Prime the oil pump prior to installation by filling the rotor cavity with clean engine oil.

27. Remove the crankshaft oil seal from the front cover.

28. Install a new cover gasket and O-ring onto the cover.

29. Rotate the crankshaft so the oil pump drive flats are vertical. Position the oil pump inner rotor so the mating flats are in the same position as the crankshaft drive flats.

30. Install the front cover making sure the pump is correctly engaged on the crankshaft, or severe damage may result.

31. Install the chain case cover screws and snug the two bottom screws and the top center screw. Ensure the cover is seated to the block then torque all screws to 20 ft. lbs. (27 Nm).

32. Install the oil seal and the crankshaft damper.

33. Install the tensioner pulley bracket and the cam sensor.

34. Install the air conditioning compressor to the mounting bracket.

35. Install the accessory drive belt.

36. Raise and safely support the vehicle. Tighten the oil pan drain plug and install a new oil filter.

37. Install the oil pump pick-up tube and oil pan. Install the transaxle inspection cover, if removed.

38. Fill the crankcase with clean engine oil to the proper level. Install a new oil filter.

39. Install radiator assembly. Check condition of radiator hoses. Fill and bleed the cooling system.

40. Reconnect the negative battery cable. Run engine and check for leaks. Verify correct oil pressure with a gauge.

3.5L (VIN F) Engine

NOTE: The timing belt must be removed to access the oil pump behind the crankshaft drive sprocket. It is good practice to turn the crankshaft to TDC No. 1 cylinder compression stroke (firing position) before starting disassembly. This should align all timing marks and be a good point of reference for all work to follow.

1. Disconnect the negative battery cable.

— **CAUTION** —

Do not open the radiator drain or the coolant pressure bottle cap with the system hot and under pressure or serious burns from coolant may occur.

2. The cooling system must be drained and the radiator removed for access to the timing belt covers and vibration damper. Use the following procedure.

   a. Place a drain pan under the radiator. Open the radiator drain fitting located at the lower right side of the radiator. Do **NOT** use pliers to open the plastic drain fitting.

   b. Remove the coolant pressure bottle cap and open the thermostat bleed valve.

   c. Remove the radiator hoses and radiator for access to the timing belt covers and crankshaft damper.

3. Remove the accessory drive belts.

4. Raise and safely support the vehicle. Place a drain pan under the vehicle and drain the engine oil.

5. Remove the oil filter. Remove the oil pan, oil pump pick-up tube and windage tray/oil pan gasket.

6. Lower the vehicle.

7. Remove the crankshaft damper using a suitable puller tool. Remove the timing belt covers.

8. Place marks on the timing belt to aid installation. Line up the marks on the camshaft sprockets to marks on the rear timing belt covers before removing the timing belt using the recommended procedure.

9. Remove the timing belt. Inspect timing belt for cracks or excessive wear and replace if necessary.

10. Remove the crankshaft sprocket using a suitable puller tool.

11. Remove the oil pump mounting screws and the pump from the engine.

12. Remove the oil pump cover retaining screws and lift off the oil pump cover.

13. Remove the pump rotors. Wash all parts in solvent and inspect carefully for damage or wear.

14. If the relief valve is to be removed, remove the cotter pin and drill a 1/8-inch hole in the relief valve retainer cap, then insert a self-threading sheet metal screw into the cap. Clamp the screw in a vise and while supporting the oil pump body, remove the cap by tapping the oil pump body with a soft hammer. Discard the retainer cap and remove the spring and relief valve. The relief spring free length should be 49.5 mm (1.95 inch) and the relief valve free of scoring. Small marks on the plunger and in the bore may be removed with 400-grit wet or dry sandpaper.

15. Inspect the oil pump carefully. Lay a straightedge across the pump cover surface. If a 0.003 inch feeler gauge can be inserted between the cover and straight edge, the cover should be replaced.

16. Measure thickness and diameter of outer rotor. If outer rotor thickness measures 0.3695 inch or less or if the diameter of the rotor is 3.141 inch or less, replace the rotor. If the inner rotor thickness measures 0.3695 inch or less, replace the inner rotor. Slide the outer rotor into the body, press to one side with fingers and measure clearance between rotor and body. If measurement is 0.015 inch or more, replace the body only if the outer rotor is in specification. Install inner rotor into pump body. If clearance between inner and outer rotors is 0.008 inch or more, replace both rotors. Place a straightedge across the face of body between the bolt holes. If a feeler gauge of 0.004

inch or more can be inserted between the rotors and the straightedge, replace the pump assembly ONLY if the rotors are in spec. If engine oil pressure has been diagnosed as being low and the pump is within specifications, inspect for worn engine bearings or other reasons for oil pressure loss.

To install:

17. Clean all parts well. There should be no traces of old gasket/sealer on any components.

18. Assemble the oil pump with new parts as required.

19. Install the pump cover and tighten the fasteners to 9 ft. lbs. (12 Nm).

20. Prime the oil pump prior to installation by filling the rotor cavity with clean engine oil.

21. Install the oil pump over the crankshaft and carefully into position. Torque the retaining screws as follows:

   M8 screws — 21 ft. lbs. (28 Nm).
   M10 screws — 40 ft. lbs. (55 Nm).

22. Raise and safely support the vehicle. Tighten the oil pan drain plug and install a new oil filter.

23. Install the oil pump pick-up tube, windage tray/oil pan gasket and the oil pan. Torque the oil pan fasteners to 9 ft. lbs. (12 Nm). Pay attention to sealing the oil pan gasket and its integral windage tray.

24. Lower the vehicle.

25. Install the crankshaft sprocket using tool C-4685C1, thrust bearing, washer and 12mm bolt to draw the sprocket onto the crankshaft.

26. Install the timing belt and set to the correct tension using the recommended procedure.

27. Install the timing belt covers and install the vibration damper using tool L-4524, thrust bearing and washer plate or equivalent damper installation tools.

28. Install the accessory drive belts and adjust to the correct tension.

29. Install the radiator into the vehicle. Install the radiator hoses.

30. Refill and bleed the cooling system.

31. Refill the engine with the correct amount of clean SAE 5W–30 **OR** SAE 10W–30 engine oil only. Do not mix the 2 grades of oil.

32. Reconnect the negative battery cable.

33. Run the engine. Check for leaks and proper oil pressure.

TRANSAXLE

Automatic Transaxle Assembly

REMOVAL AND INSTALLATION

1. Disconnect negative battery cable.

2. Remove the engine air inlet tube.

3. The crankshaft position sensor is located on the upper right side of the transaxle bellhousing. Disconnect the crankshaft position sensor connector and remove sensor.

4. Disconnect the transaxle wiring connector block located on the right shock tower. To free the connector from the harness, remove the wire ties.

5. Raise and safely support the vehicle.

6. Remove the front wheels.

7. Remove the strut to steering knuckle bolts on both sides of the vehicle. Disconnect the tie-rod ends if required.

8. Remove the Anti-lock Brake System (ABS) wheel speed sensor, if equipped.

9. Remove the halfshafts from the transfer case by inserting a prybar between the halfshaft and the transaxle case and pry the shaft from the transaxle housing. Swing the shafts out of the way keeping the joints straight and suspend using wire. Be careful not to damage the halfshaft seals.

NOTE: Do not let the halfshafts or CV-joints hang unsupported. Internal joint damage may result if allowed to hang free.

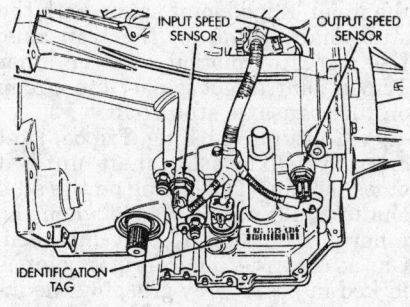

296334

Identification tag location — 42LE transaxle

10. Remove the engine-to-transaxle brackets and the transaxle bellhousing cover.

11. Mark the flexplate to the torque converter and remove the torque converter bolts. The flexplate-to-torque converter bolts are not to be reused.

12. Unbolt and remove the starter assembly from the bellhousing and allow the starter motor to sit between the engine and the frame.

13. Disconnect the oil cooler lines from the transaxle and plug to prevent excess fluid leakage.

14. Remove the transaxle dipstick.

15. Disconnect the gear selector cable from the transaxle.

16. Disconnect the exhaust pipe from the exhaust manifold and position out of the way. If the clearance will not allow for transaxle removal, remove the exhaust system from the vehicle.

17. Support the transaxle using a transmission jack. Raise the transaxle slightly to relieve the weight off the rear transaxle mount.

18. Remove the engine-to-transaxle brackets and the transaxle mount through bolt.

19. Remove the rear crossmember mounting bolts. Pry the transaxle mount rearward to separate the mount from the transaxle. Remove the rear crossmember.

20. Lower the rear of the transaxle to gain access to the bellhousing bolts. Remove the bellhousing bolts.

21. Place a drain pan under the dipstick in the transaxle to catch transaxle fluid that will drain out of the case. Remove the dipstick tube from the transaxle and plug hole.

22. Remove the engine-to-transaxle bolts and lower the transaxle from the vehicle.

NOTE: The driveplate-to-torque converter bolts and the driveplate-to-crankshaft bolts must not be reused. Install new bolts when ever these bolts are removed.

23. Inspect the driveplate for cracks. If cracks are present, replace the driveplate.

To install:

NOTE: Apply a light coating of grease to the pilot hole of the crankshaft if the torque converter is being replaced.

——————— **WARNING** ———————
When installing the transaxle unit into the vehicle, be careful that the fuel tubes at the rear of the engine do not contact the following:

• Tie rod attachment plate at the power steering rack
• Exhaust Gas Recirculation (EGR) tube
• Transaxle wiring harness

24. Install the driveplate to the engine and secure using new fasteners. Tighten the fastener to 75 ft. lbs. (101 Nm).

25. Install the transaxle into the vehicle and install the engine-to-transaxle case mounting bolts. Tighten the bolts to 75 ft. lbs. (101 Nm).

26. Install the rear transaxle case mount and the rear crossmember in position and secure all fasteners.

27. Install the dipstick tube.

28. Reconnect the exhaust pipe to the engine exhaust manifold.

29. Reconnect the gear selector cable to the transaxle. Reconnect the transaxle oil cooler lines.

30. Install the starter assembly and secure with the mounting bolts tightened to 40 ft. lbs. (54 Nm). Be sure that the starter ground strap is installed correctly.

31. Position the torque converter so matchmarks made during disassembly are in alignment. Install new torque converter to driveplate bolts and tighten to 60 ft. lbs. (81 Nm).

32. Install the transaxle bellhousing cover. Install the engine to transaxle brackets.

33. While pulling the top of the steering knuckle outward, install the inner CV-joint with new retainer clip in place, into the transaxle.

34. Install the ABS wheel sensor, if removed. Install the strut-to-steering knuckle bolts and secure.

35. Install the front wheels and lug nuts. Torque the lug nuts, in a star pattern sequence, to 95–100 ft. lbs. (129–135 Nm).

36. Lower the vehicle to the ground. Install the transaxle dipstick.

37. Reconnect the transaxle wiring harness connector on the right shock tower.

38. Install and reconnect the crankshaft position sensor.

39. Install the air inlet tube and reconnect the negative battery cable.

40. Start the engine and allow to idle for two minutes. Apply parking brake and move selector through each gear position, ending in **N**. Recheck fluid level and add if necessary. Make sure the vehicle is level when refilling the transaxle. Use Mopar Type 7176 Automatic Transmission Fluid only. Do not substitute transaxle fluid. If the differential

sump requires fluid, use 80W-90 petroleum based Hypoid gear lubricant.

41. Check the transaxle or proper operation. Adjust the shift linkage, if necessary. Make sure the reverse lamps come on when in reverse.

DRIVELINE

Halfshaft

REMOVAL AND INSTALLATION

— WARNING —

Allowing the CV-joint assemblies to dangle unsupported or pulling or pushing the ends can damage boots or CV-joints. Always support both ends of the halfshaft to prevent damage or disengagement of the Tripod joint.

1. Disconnect the negative battery cable.
2. Raise and support the vehicle safely.
3. Remove the front wheels.
4. Remove the front caliper assembly from the steering knuckle.

5. Remove the front brake rotor from the hub by pulling it straight off wheel mounting studs.
6. Remove the speed sensor cable routing bracket from the strut assembly.
7. Remove the hub and bearing-to-stub axle retainer nut.
8. Install a puller tool onto the hub and bearing assembly and secure it into place using the wheel lug nuts.
9. Protect wheel stud threads by installing a wheel lug nut onto a wheel stud. Install a flat blade prying tool to prevent the hub from turning. Using the puller tool, force the halfshaft outer stub axle from the hub and bearing assembly.
10. Dislodge the inner Tripod joint from the stub shaft retaining snapring on the transaxle. To do this, insert a prybar between the transaxle case and the inner Tripod joint and pry on Tripod joint.

NOTE: Do not try to remove the inner Tripod joint from the transaxle stub shaft at this time. Only disengage the inner Tripod joint from the retainer snapring.

11. Remove the strut assembly-to-steering knuckle attaching bolts from the strut assembly.

— WARNING —

The strut assembly to steering knuckle bolts are serrated (toothed) where they go through the strut assembly and steering knuckle. When removing the bolts, turn the nuts off the bolt. Do not turn the bolts in the steering knuckle or damage to the steering knuckle will result.

12. Separate the top of the steering knuckle from the lower end of the strut.
13. Hold the outer joint assembly with one hand. Grasp the steering knuckle with the other hand and rotate it out and to the rear of the vehicle, until the outer CV-joint clears the hub and bearing assembly.

— WARNING —

When removing the outer CV-joint from the hub and bearing assembly, do not allow the flange disc on the hub and bearing assembly to become damaged. If this happens, dirt and water can enter the bearing which will cause premature bearing failure.

14. Remove the halfshaft inner joint from the transaxle stub shaft by grasping the inner Tripod joint and the interconnecting shaft and pulling both pieces at the same time. Take care not to pull on the interconnecting shaft to remove or separation of the spider assembly will occur.

To install:

15. Replace the inner Tripod joint retaining circlip and O-ring seal on the transaxle stub shaft. These components are not reusable and must be replaced whenever the halfshaft is removed.
16. Apply an even coat of grease on the splines of the inner Tripod joint, where the O-ring seats against the Tripod joint.
17. Install the halfshaft through the hole in the splash shield. Grasp the inner joint in 1 hand and interconnecting shaft in the other. Align the inner Tripod joint spline with the stub shaft spline on the transaxle. Use a rocking motion with the inner Tripod joint to get it past the circlip on the transaxle stub shaft.
18. Continue pushing Tripod joint onto transaxle stub shaft until it stops moving. The O-ring on the stub shaft should not be visible when the inner Tripod joint is fully installed. Check that the inner Tripod joint is locked in position by grasping the inner joint and pulling. If locked in position, the joint will not move on the stub shaft.

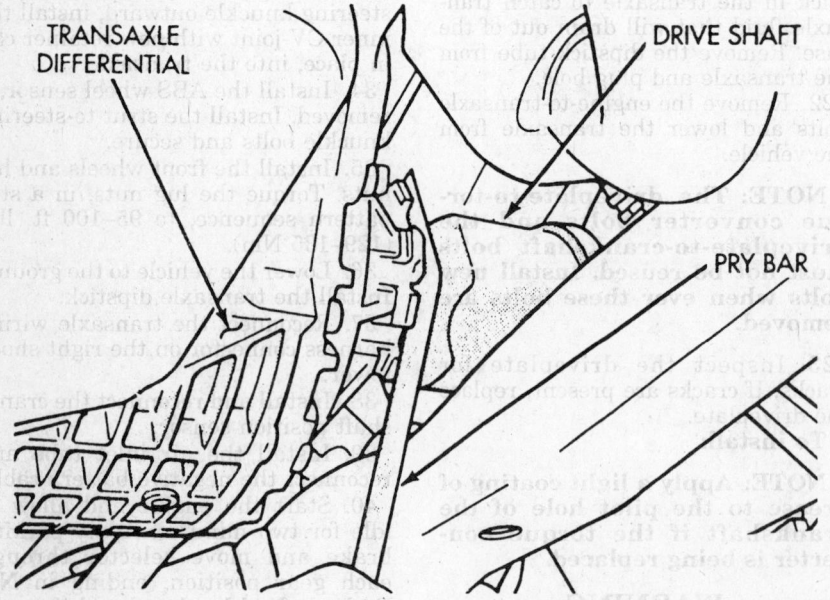

TRANSAXLE DIFFERENTIAL

DRIVE SHAFT

PRY BAR

296342

Pry as shown to remove halfshaft — 42LE transaxle

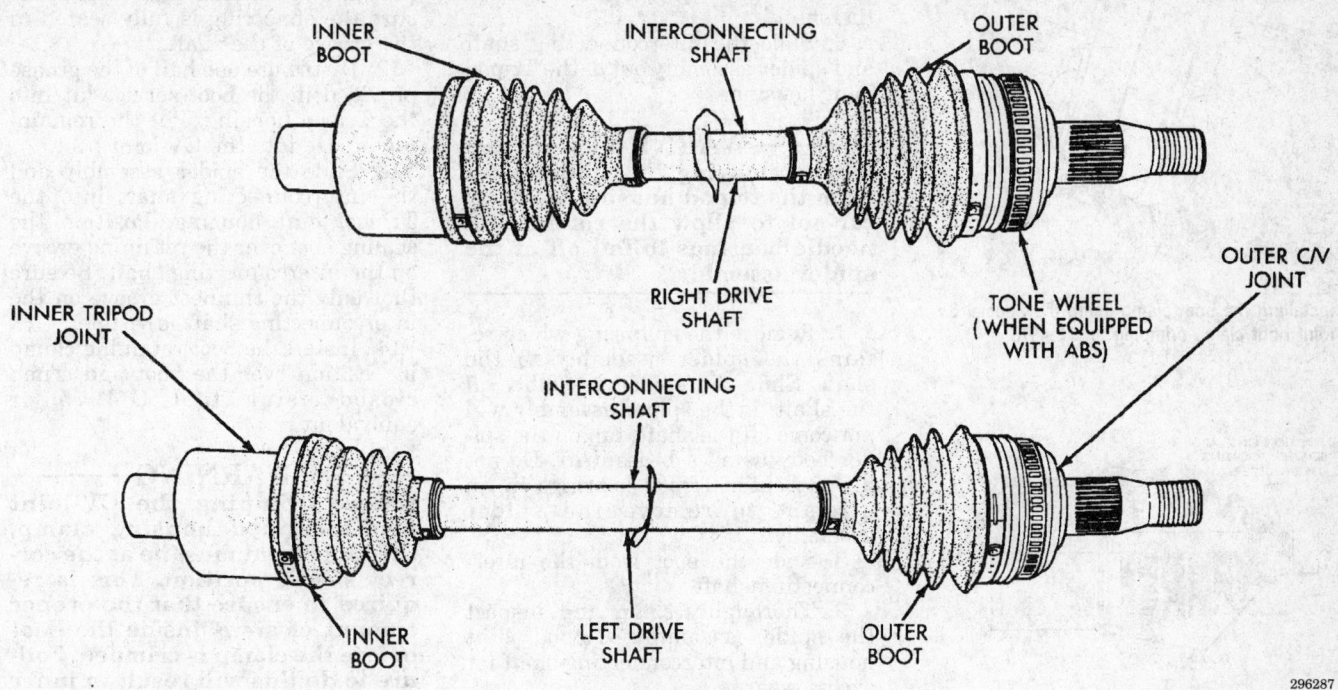

INNER BOOT

INTERCONNECTING SHAFT

OUTER BOOT

RIGHT DRIVE SHAFT

TONE WHEEL (WHEN EQUIPPED WITH ABS)

OUTER CV JOINT

INNER TRIPOD JOINT

INTERCONNECTING SHAFT

INNER BOOT

LEFT DRIVE SHAFT

OUTER BOOT

296287

Halfshaft identification

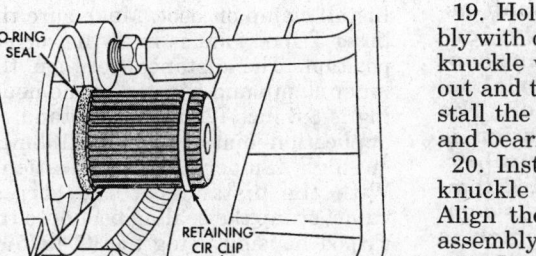

O-RING SEAL

RETAINING CIR CLIP

STUB SHAFT

296293

The Tripod joint retaining circlip and O-ring seal on the transaxle stub shaft

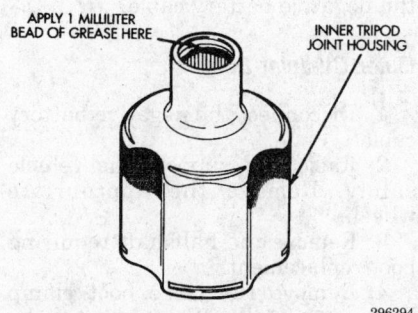

APPLY 1 MILLILITER BEAD OF GREASE HERE

INNER TRIPOD JOINT HOUSING

296294

Grease applied to the inner Tripod joint housing spline

19. Hold the outer CV-joint assembly with one hand. Grasp the steering knuckle with the other and rotate it out and to the rear of the vehicle. Install the outer CV-joint into the hub and bearing assembly.

20. Install the top of the steering knuckle into the strut assembly. Align the steering knuckle to strut assembly mounting holes.

21. Install the strut assembly-to-steering knuckle attaching bolts. Install the nuts to the attaching bolts and while holding the bolt heads, tighten nuts to 125 ft. lbs. (170 Nm). Turn the nuts on the bolts. Do **NOT** turn the bolts.

22. Install a new hub and bearing assembly-to-stub shaft retainer nut. Tighten but do not torque the nut at this time.

23. Install the speed sensor cable routing bracket and secure attaching screw.

24. Install the brake rotor and the caliper assembly. Install the caliper guide pin bolts to steering knuckle and tighten to 30 ft. lbs. (41 Nm).

25. Install the front wheels and lug nuts. Torque the lug nuts, in sequence, to 95–100 ft. lbs. (129–135 Nm). Lower the vehicle to the ground. Pump the brakes until a firm pedal is obtained.

26. Apply the brakes and torque the new stub shaft-to-hub and bearing assembly retainer nut to 120 ft. lbs. (163 Nm).

— **WARNING** —
When tightening the stub shaft retaining nut, be careful not to exceed the maximum torque specification of 120 ft. lbs. (163 Nm). If this specification is exceeded, failure of the halfshaft could result.

27. Reconnect the negative battery cable. Road test vehicle to check for noise or vibration.

CV-Joint Boot

REPLACEMENT

NOTE: The inner tripod joint boots used on the Chrysler LH vehicles use no internal retention in the tripod housing to keep the joint assembly retained in the housing. Therefore, do not pull on the interconnecting shaft to separate the tripod housing from the transaxle stub shaft. This will damage the inner boot.

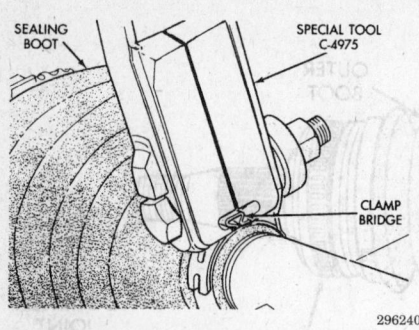

Installing the boot clamp using the Special CV-joint boot clamp crimping tool C-4975

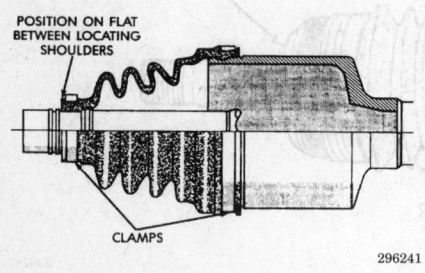

Boot and clamp positioning

Inner Tripod Boot

1. Disconnect the negative battery cable.

2. Raise and support the vehicle safely. Remove the appropriate wheel.

3. Remove the halfshaft requiring boot replacement.

4. Remove the large boot clamp which retains the inner Tripod joint boot to the Tripod joint housing. Remove the small clamp which retains the inner Tripod joint boot to the interconnecting shaft. Discard the clamps.

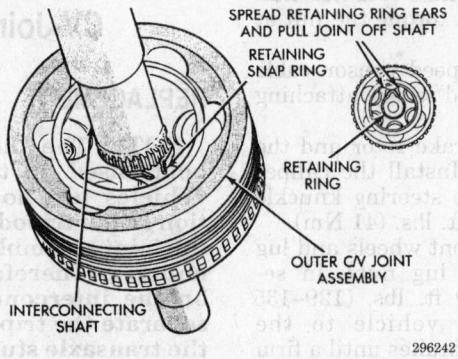

Removal of the outer CV-joint from the interconnecting shaft

5. Remove the boot from the Tripod joint housing and slide it down the shaft.

6. Slide the interconnecting shaft and spider assembly out of the Tripod joint housing.

--- WARNING ---

When removing the spider joint from the tripod housing, be careful not to allow the rollers and needle bearings to fall off of the spider assembly.

7. Remove the snap ring which retains the spider assembly to the shaft. Slide the spider assembly off the shaft. If the spider assembly will not come off the shaft, tap on the spider body using a brass drift. Do not hit the outer Tripod bearings in an attempt to remove the spider assembly.

8. Slide the boot from the interconnecting shaft.

9. Thoroughly clean and inspect the spider assembly, Tripod joint housing and interconnecting shaft for excess wear.

NOTE: If any parts show excess wear, the halfshaft assembly will require replacement. Component parts of the halfshaft assemblies are not serviceable. The inner Tripod joint boot is made of 2 types of material depending on the application. The high temperature application is made from silicone rubber, which is soft. The standard temperature application boot is composed of hytrel plastic which is hard. The replacement boot must be of the same composition as the original boot.

To install:

10. Slide the new boot retainer clamp onto the interconnecting shaft followed by the new boot.

11. Install the spider assembly onto the shaft and install snap ring. Make sure the snap ring is fully seated in the groove of the shaft.

12. Distribute one half of the grease provided in the boot service kit into the Tripod housing. Put the remaining grease into the CV-joint boot.

13. Slide the spider assembly and the interconnecting shaft into the Tripod joint housing. Position the sealing boot over the retaining groove on the interconnecting shaft. Be sure that only the thinnest groove on the interconnecting shaft is visible.

14. Install the boot retaining clamp in position over the boot and crimp closed using tool C-4975 or equivalent.

--- WARNING ---

Before crimping the CV-joint boot-to-Tripod housing clamp, the inner joint must be at the correct stroke position. This is required to ensure that the proper amount of air is inside the boot before the clamp is crimped. Failure to do this will result in inner CV-joint boot failure.

15. Position the sealing boot into the Tripod housing retainer groove. Install clamp on boot. Make sure the inner Tripod joint is in at the correct position. The distance between the inner clamp and the end of the housing is 8.5 inch (216mm) for standard application joint or 7.8 inch (198mm) on high temperature applications. With the distance at the correct value, crimp the sealing boot onto the Tripod housing using tool C-4975 or equivalent.

16. Install the halfshaft into the vehicle.

17. Install the wheel and lug nuts. Torque the lug nuts, in a star pattern sequence, to 95–100 ft. lbs. (129–135 Nm).

18. Lower the vehicle. Reconnect the negative battery cable.

Outer CV-Joint Boot

1. Disconnect the negative battery cable.

2. Raise and support the vehicle safely. Remove the appropriate wheel.

3. Remove the halfshaft requiring boot replacement.

4. Remove the large boot clamp which retains the CV-joint boot to the CV-joint housing. Remove the small clamp which retains the CV-joint boot to the interconnecting shaft. Discard the CV-joint boot clamps.

5. Remove the boot from the joint housing and slide it down the shaft.

6. Wipe any grease away to expose the outer CV-joint-to-shaft retaining snap ring. Using a pair of snap ring pliers, spread the snap ring and remove the CV-joint assembly off the end of the shaft.

7. Slide boot off the shaft. Thoroughly clean and inspect the CV-joint assembly and interconnecting shaft for damage or excess wear.

NOTE: If any parts show signs of excess wear, the halfshaft assembly will require replacement. Component parts for the half-shaft assemblies are not serviceable.

To install:

8. Slide a new seal boot-to-interconnecting shaft retainer clamp, followed by the new boot, onto the shaft. Install the outer CV-joint assembly onto the interconnecting shaft pushing on the shaft until the retaining snap ring is seated in the groove on the shaft. Be sure the snap ring is fully seated.

9. Distribute one half of the grease provided in the boot service kit into the joint housing. Put the remaining grease into the CV-joint boot. Do **NOT** use any other type of grease.

10. Position the sealing boot over the retaining groove on the interconnecting shaft. Install the boot retaining clamp in position over the boot and crimp closed using tool C-4975 or equivalent.

11. Position the sealing boot into the boot retaining groove on the CV-joint housing. Install clamp on boot and crimp using tool C-4975 or equivalent.

---- **WARNING** ----

The seal must not be dimpled, stretched or out of shape in any way. If the seal is not shaped correctly, equalize the pressure in the seal and shape it by hand.

12. Install the halfshaft into the vehicle.

13. Install the wheel and lug nuts. Torque the lug nuts, in a star pattern sequence, to 95–100 ft. lbs. (129–135 Nm).

14. Lower the vehicle. Reconnect the negative battery cable.

STEERING

Air Bag

---- **CAUTION** ----

Some vehicles are equipped with an air bag system, also known as the Supplemental Inflatable Restraint (SIR) or Supplemental Restraint System (SRS). The system must be disabled before performing service on or around system components, steering column, instrument panel components, wiring and sensors. Failure to follow safety and disabling procedures could result in accidental air bag deployment, possible personal injury and unnecessary system repairs.

PRECAUTIONS

Several precautions must be observed when handling the inflator module to avoid accidental deployment and possible personal injury.

• Never carry the inflator module by the wires or connector on the underside of the module.

• When carrying a live inflator module, hold securely with both hands, and ensure that the bag and trim cover are pointed away.

• Place the inflator module on a bench or other surface with the bag and trim cover facing up.

• With the inflator module on the bench, never place anything on or close to the module which may be thrown in the event of an accidental deployment.

DISARMING

---- **CAUTION** ----

The Air Bag system must be disarmed before repair and/or removal of any component in its immediate area including the air bag itself. Failure to do so may cause accidental deployment of the air bag, resulting in unnecessary system repairs and/or personal injury.

1. Disconnect the negative battery cable and isolate the cable using an appropriate insulator (wrap with quality electrical tape).

2. Allow the system capacitor to discharge for 2 minutes before starting any repair on any air bag system or related components. This will disable the air bag system.

---- **CAUTION** ----

Always wear safety goggles when working with, or around, the air bag system. When carrying a live air bag, make sure the bag and trim cover are pointed away from the body. In the unlikely event of an accidental deployment, the bag will then deploy with minimal chance of injury. When placing a live air bag on a bench or other surface, always face the bag and trim cover up, away from the surface. This will reduce the motion of the module if it is accidentally deployed.

Steering Wheel

REMOVAL AND INSTALLATION

---- **CAUTION** ----

The Air Bag system must be disarmed before removing the steering wheel. Failure to do so may cause accidental deployment of the air bag, resulting in unnecessary system repairs and/or personal injury. In addition, the fasteners, screws and bolts originally used for the air bag components have a special coating on them specifically designed for use in this system. They must never be replaced with any substitutes. Anytime new fasteners are needed, replace with the correct fasteners.

1. Position the front wheels in the straight ahead position, then turn the steering wheel clockwise (to the right) 180 degrees and lock the steering column with the ignition cylinder lock. This is to protect the air bag clock spring assembly.

2. Disarm the air bags as follows:

a. Disconnect the negative battery cable and isolate the negative terminal end.

b. Wait two minutes for the reserve capacitor to discharge before removing a non-deployed air bag module.

3. Remove both speed control switches from the steering wheel. If not equipped with cruise control, pry off the covers on the side of the steering wheel.

4. Remove the two bolts attaching the air bag module to the steering wheel.

5. Lift the air bag module from the steering wheel and disconnect the air bag and horn electrical harness connectors. Remove the module from the vehicle.

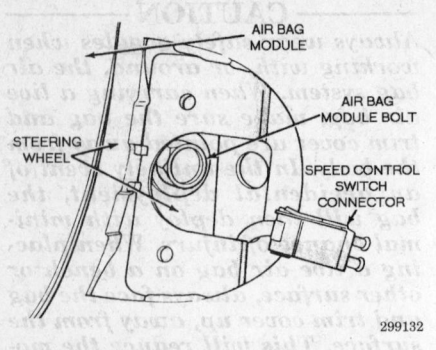

Air bag module mounting bolts

299132

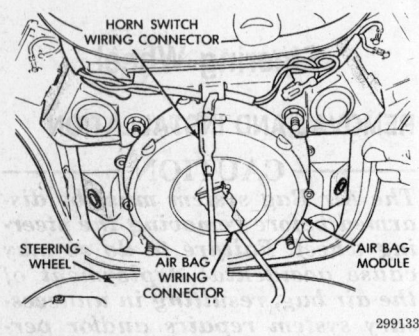

Wiring harness connectors to the air bag module

299133

---- **CAUTION** ----

When carrying a live air bag, make sure the bag and trim cover are pointed away from the body. In the unlikely event of an accidental deployment, the bag will then deploy with minimal chance of injury. When placing a live air bag on a bench or other surface, always face the bag and trim cover up, away from the surface. This will reduce the motion of the module if it is accidentally deployed.

6. Remove the steering wheel retainer nut. Use a steering wheel puller tool to remove the steering wheel from the steering shaft. While removing the steering wheel, be very careful feeding the wires carefully through the the holes in the clockspring armature. Do not hammer on steering wheel to remove.

To install:

7. Before installing the steering wheel, be sure that:

   a. The steering wheel is locked at a 180 degree turn to the right.

   b. The turn signal lever is in the neutral position.

8. Align the master splines on the steering wheel and the steering shaft and install the steering wheel to the shaft. Ensure that the flats on the steering wheel hub align with the clockspring. Install wheel retainer nut and tighten to 45 ft. lbs. (61 Nm).

9. Reconnect the horn connector.

10. Reconnect the electrical wires from the body harness to the air bag module and position module so it is at the center of the steering spokes. Install the two bolts attaching the air bag module to the steering wheel and tighten to 8 ft. lbs. (10 Nm). Be careful not to pinch any wires during installation.

11. Reconnect both harness connectors to the speed control switches and install to the steering wheel, if equipped. If not equipped with cruise control, install the covers on the side of the steering wheel.

12. Reconnect the negative battery cable. Check operation of all components disturbed during this procedure.

Tie Rod Ends

REMOVAL AND INSTALLATION

1. Raise and safely support the vehicle.

2. Remove the tire and wheel assembly.

3. Loosen the outer tie rod to adjustment sleeve jam nut.

4. Loosen the outer tie rod adjustment sleeve jam nut. Loosen but do not remove the outer tie rod to strut assembly steering arm attaching nut. Then remove the outer tie rod from the steering arm using puller tool MB-990635 or equivalent.

5. Remove the outer tie rod from the adjustment sleeve.

To install:

6. Install replacement outer rod into the adjustment sleeve. Make sure the jam nut is on the outer tie rod end. Do not tighten jam nut yet.

7. Install the outer tie rod into the steering arm on front strut assembly. Install the tie rod to steering arm attaching nut and torque to 27 ft. lbs. (37 Nm).

8. Install the tire and wheel assembly. Wheel mounting stud nuts torque specification is 95 ft. lbs. (129 Nm). Check the front wheel toe setting on the vehicle and adjust as required. Make sure the maximum number of allowable threads exposed past the edge of the tie rod jam nut is a maximum of 20 mm on either side or a combined maximum of 35 mm on both sides. Distance is measured from outside edge of jam nut to end of tie rod threads.

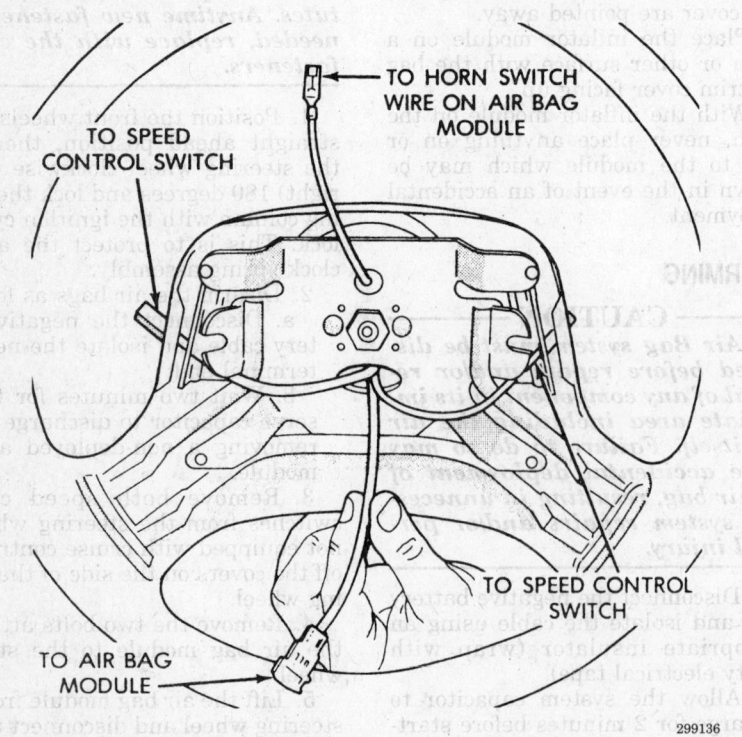

TO HORN SWITCH WIRE ON AIR BAG MODULE

TO SPEED CONTROL SWITCH

TO SPEED CONTROL SWITCH

TO AIR BAG MODULE

299136

Wire routing through the steering wheel

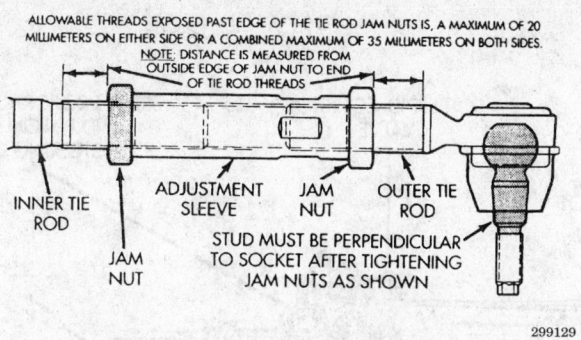

ALLOWABLE THREADS EXPOSED PAST EDGE OF THE TIE ROD JAM NUTS IS, A MAXIMUM OF 20 MILLIMETERS ON EITHER SIDE OR A COMBINED MAXIMUM OF 35 MILLIMETERS ON BOTH SIDES.
NOTE: DISTANCE IS MEASURED FROM OUTSIDE EDGE OF JAM NUT TO END OF TIE ROD THREADS

INNER TIE ROD
JAM NUT
ADJUSTMENT SLEEVE
JAM NUT
OUTER TIE ROD
STUD MUST BE PERPENDICULAR TO SOCKET AFTER TIGHTENING JAM NUTS AS SHOWN

299129

Tie rod thread engagement requirements

CAUTION

When torquing adjustment sleeve nut, the following procedure must be followed to ensure adequate torquing and retention of the adjustment sleeve jam nuts.

9. Install a 23 mm wrench of sufficient size on the flat of the adjustment sleeve to keep sleeve from turning while resisting the tightening torque of the jam nut.

10. While holding the adjustment sleeve from turning, torque the inner tie rod to adjustment sleeve jam nut to 55 ft. lbs. (75 Nm). When the outer tie rod jam nut is correctly torqued, the outer tie rod stud must be perpendicular within the tie rod end.

NOTE: When torquing the jam nut, be sure to use a 27 mm crow's foot and a clicker type torque wrench to achieve the required torque.

11. While holding the adjustment sleeve from turning, torque the outer tie rod adjustment sleeve jam nut to 55 ft. lbs. (75 Nm).
12. Lower the vehicle.
13. Adjust the front suspension toe setting.

Outer Tie Rod Socket End-Play Measurement

When measuring tie rod socket end play, the total weight of the vehicle must be supported by the tires and suspension of the vehicle. A drive-on hoist or alignment rack must be used.

1. Position the vehicle on a drive-on hoist or alignment rack so that the tires are supporting the total weight of the vehicle.
2. Raise the hoist or alignment rack so that the tie rod is positioned at eye level.
3. Mount a dial indicator to the front strut assembly.
4. Install a U-bolt (muffler clamp or similar) with an inside dimension of 2 inches and length of about 3 in-

ches on the outer tie rod end and the steering arm of the strut. Hand tighten the nuts on the U-bolt enough to just hold the U-bolt in place.
5. Position the dial indicator in the center of the tie rod. Zero the indicator.
6. Equally tighten both nuts on the U-bolt until a torque of about 24 inch lbs. is achieved on each nut.
    a. A new tie rod end should have no more than 0.100 inch end play.
    b. A tie rod end in service should no more than 0.190 inch end play.
7. If the tie rod end exceeds these specifications, it should be replaced.

Power Rack and Pinion

REMOVAL AND INSTALLATION

CAUTION

The Air Bag system must be disarmed before removing the rack and pinion steering gear. Failure to do so may cause accidental deployment of the air bag, resulting in unnecessary system repairs and/or personal injury.

---

1. Disconnect negative battery cable. Disarm the air bag system.
2. Raise and safely support the vehicle.
3. Remove the gear shift cable from the shifter lever on the transaxle.
4. Loosen the bolt at the gear shift cable to transaxle mount. Remove the cable from the transaxle.
5. Lower the vehicle.
6. If necessary, disconnect the throttle cable from the throttle body and remove throttle cable bracket.
7. Remove both wiper arm assemblies from the wiper arm pivots. Remove the cowl closure panel and weatherstrip as an assembly from the cowl.
8. Disconnect the wiper module wiring harness connector from the

vehicle wiring harness. Remove the wiper module assembly from the vehicle cowl panel.
9. Disconnect the air plenum from the throttle body, PCV make up air tube and the idle air control motor. Remove he plenum from the right side of the vehicle through the wiper module area.
10. Disconnect the vacuum connector from the power brake booster at the intake manifold.
11. Turn the front wheels to the full left position. Then turn the wheels back in the other direction until the roll pin in the lower steering coupler is accessible. Turn the ignition key switch to the LOCK position to keep the steering column from rotating after the coupler is removed from the steering gear. If the steering column shaft rotates beyond the normal number of turns in either direction, the air bag clock spring will be damaged.
12. Using paint, mark the steering coupling and steering gear shaft for orientation. Using the correct size punch, remove the roll pin from the steering coupling.
13. If equipped with a brake pedal travel sensor, remove the pedal travel sensor from the brake booster as follows:
    a. Pump the brake pedal approximately 20 times. This will bleed the vacuum stored in the booster.
    b. Remove the wiring harness connector from the sensor.
    c. Using a small flat tipped tool, lift the retainer ring from the notch. Then remove the retaining ring from the grommet.
    d. Remove the pedal travel sensor from the brake booster by carefully pulling it straight out of its mounting grommet. Do not twist the sensor.
14. Loosen and remove the two nuts attaching the master cylinder to the brake booster. Remove the master cylinder with the brake lines connected, and position aside.
15. Remove the power steering pressure hose and return hose from the power steering gear.
16. Bend back the retaining tabs on bolt attaching the tie rods to the steering gear and remove the bolts.
17. Lay the tie rods, bolts and plate as an assembly on the bellhousing of the transaxle.
18. If the rack and pinion steering gear unit being removed is a speed proportional steering gear, disconnect the vehicle wiring harness connector from the solenoid control module.

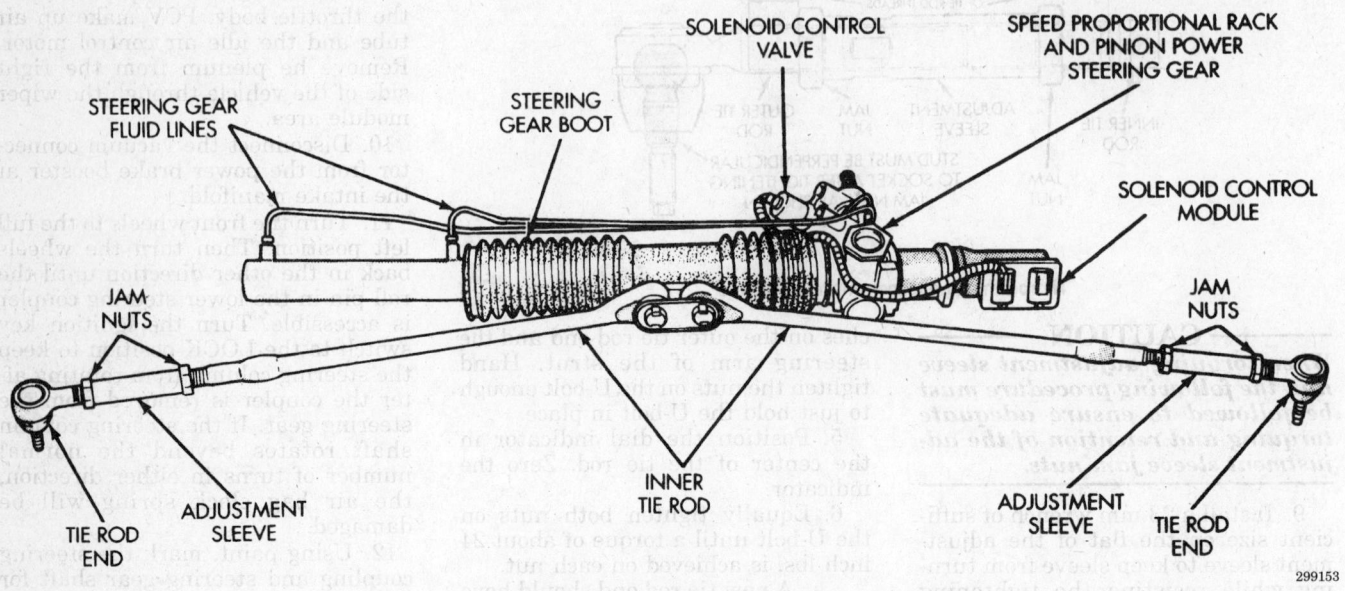

Standard power steering rack and pinion steering gear assembly

Labels in figure: SOLENOID CONTROL VALVE; SPEED PROPORTIONAL RACK AND PINION POWER STEERING GEAR; STEERING GEAR FLUID LINES; STEERING GEAR BOOT; SOLENOID CONTROL MODULE; JAM NUTS; JAM NUTS; TIE ROD END; ADJUSTMENT SLEEVE; INNER TIE ROD; ADJUSTMENT SLEEVE; TIE ROD END

299153

19. Remove the four bolts attaching the steering gear assembly to the crossmember. Slide the steering gear forward in the vehicle to disengage steering coupler from the steering gear shaft. After gear is disengaged, do not rotate the steering gear shaft.

20. Remove the steering gear assembly from the vehicle through the area in the cowl from which the windshield wiper module was previously removed.

To install:

21. If a replacement rack is being installed, grasp the shaft of the steering gear and rotate until steering gear center take off is in a full left turn position. Install the steering gear into the vehicle through the wiper module opening in cowl.

22. If the original gear is being installed, align the paint mark on the steering coupler with the mark on the steering gear shaft and install the steering gear shaft into the steering gear coupler.

23. If a replacement rack is being installed, the steering gear shaft and steering coupler must be aligned. Rotate the steering gear shaft back from the full left turn position until the master spline on the steering gear shaft is aligned with the master spline on the steering coupler. At this

point, install the steering gear into the coupler.

24. Align the steering gear with the mounting holes in the crossmember and install bolts. Be sure the brake line routing clip is installed under the left steering gear mounting bracket. Tighten the mounting bolts to 50 ft. lbs. (68 Nm).

25. Install the steering coupler to steering gear shaft retaining roll pin until it is flush with the top edge of the steering coupler.

26. If equipped with 3.5L engine, correct orientation of the power steering pressure hose at the power steering pump must be maintained. Be sure the power steering hose is installed in orientation clip at the power steering pump prior to tightening tube fitting. Attach the power steering pressure and return lines onto the proper ports of the power steering gear. Torque both fittings to 23 ft. lbs. (31 Nm).

27. Align the center take off on the steering gear with the tie rod assemblies. Install the tie rod attaching bolts and washers into the steering gear assembly. Be sure the washers are installed between the tie rods and the steering gear. Torque the tie rod to steering gear attaching bolts to 55 ft. lbs. (75 Nm). Bend the retaining tabs against the heads of the bolts.

28. Install the pedal travel sensor retainer ring on the travel sensor grommet in the vacuum booster. The tab on the retaining ring should be located in top notch of the mounting grommet.

29. Sparingly lubricate pedal travel sensor O-ring with fresh brake fluid. Install the pedal travel O-ring into the pedal travel sensor mounting grommet. Coat the end of the sensor with fresh brake fluid and install by pushing straight into the mounting grommet on the brake booster until the tab on the sensor is past the retaining ring on grommet.

30. Install the wire harness connector to the pedal travel sensor.

31. Install the master cylinder and tighten nuts to 250 inch lbs. (28 Nm).

32. Install the power booster vacuum hose to the intake manifold. Install the windshield wiper module to the cowl panel. Reconnect the electrical harness to the module.

33. If removed, install the air intake plenum and reconnect to the idle air control motor, PCV make up air tube and throttle body.

34. Install the windshield wiper module assembly into the vehicle cowl area and reconnect the wiring harness connector from the wiper module to the vehicle wiring harness.

35. Install the cowl closure panel and tighten the six mounting screws. Install the weather strip on shock towers.

36. Install the windshield washer hoses on the wiper arms, then install arms on the windshield wiper pivots.

37. If removed, install the throttle cable to the bracket and install to the throttle body.

38. Reconnect the wiring harness connector from the solenoid control valve onto the solenoid control module. Be sure that the harness connector seal is in good condition before installation.

39. Raise and safely support the vehicle.

40. Install the gear shift cable onto the shift lever on transaxle. Install gear shift cable on cable mounting bracket of transaxle and securely tighten bolt.

41. Lower the vehicle. Reconnect the negative battery cable.

42. Refill the pump reservoir to the correct lever with Mopar Power Steering Fluid or equivalent. Do not use any type of automatic transmission fluid. Start the engine and turn the steering wheel several times from stop to stop to bleed the air from the fluid in the system. Check and add fluid as required.

43. Adjust the front suspension toe setting.

Power Steering Pump

BLEEDING

1. Check the power steering fluid level. Use MOPAR Power Steering Fluid or equivalent in the power steering system. Do not use any type of automatic transmission fluid.

2. Start the engine and allow it to idle.

3. Turn the steering wheel several times from left stop to right stop until all air has been bled from the system.

4. Turn off the engine and confirm that the fluid is not milky and that the power steering fluid reservoir is filled to the proper level. Inspect the system for leaks.

5. Add approved power steering fluid to correct the reservoir fluid level, if necessary.

REMOVAL AND INSTALLATION

All LH platform vehicles with all available engine options use the Saginaw T/C style power steering pump with a remote mounted reservoir for the power steering fluid. The fluid reservoir is mounted to the left frame rail of the vehicle just rearward on the battery tray. No repair procedures are to be done on the internal components of the Saginaw power steering pumps. Repair of power steering fluid leaks from areas of the power steering pump sealed by O-rings is allowed. However, power steering pump shaft seal leakage will require replacement of the power steering pump.

1. Disconnect negative battery cable and isolate from the battery.

2. Loosen the power steering drive belt by loosening the generator mounting pivot bolts and turning the adjuster bolt or loosen the power steering drive belt by loosening the adjuster pulley locking nut and then loosening the adjuster bolt. Remove the drive belt from the pump pulley.

3. Raise and safely support the vehicle.

4. Position an oil drain pan under the vehicle to catch leaking power steering fluid. Remove the hose clamps from the power steering fluid inlet hose at the pump. Remove the hose at the pump.

5. Loosen and remove the power steering pressure hose from the power steering pump discharge fitting.

6. Loosen and remove the 3 bolts attaching the pump to the power steering pump mounting bracket. Access to the pump mounting bolts is through the holes in the pump pulley.

7. Remove the power steering pump and drive pulley as an assembly out the bottom of the engine compartment.

8. Remove the power steering pump pulley as follows:

   a. Mount the pump in a vise using the mounting bosses.

   b. Remove the power steering pump pulley from the shaft using special puller tool C-4333, or equivalent.

NOTE: Do not press or hammer on the shaft of the pump. This will cause internal pump damage.

9. Transfer parts to replacement pump, as required.

To install:

10. Install the power steering pump pulley as follows:

   a. Place the pulley onto the shaft and make sure it is installed squarely.

   b. Install the spacer provided with the replacement pump into the hub of the pulley.

   c. Insert the pulley installer tool C-4063 (without adapters) through the hole in the spacer. Thread the tool into the pump shaft and tighten the tool into the shaft.

   d. Hold the installer with one wrench so it does not rotate. Turn the hex down threaded rod of installer to push the pulley onto the shaft. Make sure the pulley does not become cocked during installation.

   e. Continue to push the pulley onto the shaft until the tool will not turn.

   f. Remove the installer tool and spacer. Turn the pulley and make sure it does not wobble. If it does, remove the pulley and check for a bent pump shaft, bent pulley or other malfunction.

11. Install the power steering pump back into the vehicle from below the engine compartment. Install mounting bolts and tighten to 40 ft. lbs. (54 Nm).

12. Correct orientation of the power steering pressure hose at the power steering pump must be maintained. Be sure the power steering hose is installed in the routing clip at the alternator (if equipped) prior to tightening tube fitting.

13. Attach the power steering pressure hose on the outlet port of the pump and tighten fitting to 25 ft. lbs. (34 Nm).

14. Install the hose from the remote fluid reservoir to the low pressure port on the power steering pump.

15. Install the accessory drive belt over the power steering pulley. Lower the vehicle.

16. Adjust the belt tension and reconnect the negative battery cable.

17. Refill the pump reservoir to the correct level with Mopar Power Steering Fluid, or equivalent. Do not use any type of automatic transmission fluid. Start the engine and turn the steering wheel several times from stop to stop to bleed the air from the fluid in the system. Check and adjust fluid level as required.

18. Test drive vehicle to verify repair.

BRAKES

Anti-Lock Brake System Service

PRECAUTIONS

- Certain components within the Anti-Lock Brake System (ABS) are not intended to be serviced or repaired individually. Only those components with removal and installation procedures should be serviced.
- Do not use rubber hoses or other parts not specifically specified for and ABS system. When using repair kits, replace all parts included in the kit. Partial or incorrect repair may lead to functional problems and require the replacement of components.
- Lubricate rubber parts with clean, fresh brake fluid to ease assembly. Do not use lubricated shop air to clean parts; damage to rubber components may result.
- Use only specified brake fluid from an unopened container.
- If any hydraulic component or line is removed or replaced, it may be necessary to bleed the entire system.
- A clean repair area is essential. Always clean the reservoir and cap thoroughly before removing the cap. The slightest amount of dirt in the fluid may plug an orifice and impair the system function. Perform repairs after components have been thoroughly cleaned; use only denatured alcohol to clean components. Do not allow ABS components to come into contact with any substance containing mineral oil; this includes used shop rags.
- The Anti-Lock control unit is a microprocessor similar to other computer units in the vehicle. Ensure that the ignition switch is **OFF** before removing or installing controller harnesses. Avoid static electricity discharge at or near the controller.
- If any arc welding is to be done on the vehicle, the control unit should be unplugged before welding operations begin.

Master Cylinder

REMOVAL AND INSTALLATION

Without ABS

Service this vehicle with DOT 3 brake fluid from a tightly sealed container. Do not use brake fluid with a lower boiling point than DOT 3 as brake failure could result during prolonged hard braking. Do not use any petroleum-based fluid because seal damage in the brake system will result.

1. Disconnect the negative battery cable.
2. Disconnect the brake fluid level sensor connector.

NOTE: Be sure to clean the master cylinder, brake fluid reservoir and surrounding area so that the hydraulic system does not become contaminated during the removal procedure. Be very careful not to allow brake fluid to accidentally spill out of the reservoir. Brake fluid damages painted surfaces.

3. Disconnect the brake lines from the master cylinder. Install plugs at the brake line outlets so that dirt or moisture does not contaminate the hydraulic system.
4. Remove the two mounting nuts securing the master cylinder to the booster and remove the cylinder from the mounting studs.
5. Remove the master cylinder straight out from the brake booster and out of the vehicle.
6. Remove the brake reservoir as follows:
   a. Clean the master cylinder housing and brake reservoir.
   b. Remove the reservoir caps and empty the reservoir of fluid.
   c. Position the master cylinder securely in a bench vise.
   d. Remove the two reservoir retaining pins using a hammer and a small punch.
   e. Rock the reservoir from side to side and remove the reservoir from the rubber grommets. Do not pry the reservoir off with any tool, or damage to the reservoir will result.
   f. Remove the rubber master cylinder-to-reservoir grommets.

To install:

7. Install the brake reservoir as follows:
   a. Install new grommets into the master cylinder housing.
   b. Lubricate the grommets with clean brake fluid.
   c. Install the reservoir by rocking back and forth until it is fully seated.
   d. Install the reservoir retaining pins.

NOTE: The master cylinder used on this vehicle has outlet ports with ISO style flares and metric threads. Be sure that the bleeding tubes used to bleed the master cylinder are of the correct style flares and metric tube nuts.

8. Bench bleed the master cylinder as follows:
   a. With the master cylinder secured in a bench vise, connect bleeding tubes to the ports of the master cylinder. Be sure to position the tubes so the tube outlets will be below the surface of the brake fluid when the reservoir is filled to the proper level.
   b. Fill up the reservoir with clean, fresh DOT 3 type brake fluid.
   c. Using a wooden dowel, or equivalent, slowly push in the master cylinder pistons then slowly release the pistons allowing them to return to the released position. Repeat this step several times until all the air bubbles have been expelled from the master cylinder.
   d. Remove the bleeding tubes from the master cylinder ports, plug the outlets and install cap.

NOTE: Bleeding the entire hydraulic system is not necessary after replacing the master cylinder if the master cylinder has been adequately bled and refilled upon installation.

9. Install master cylinder to the mounting studs, aligning the push rod with the master cylinder piston. Install the mounting nuts and tighten to 21 ft. lbs. (28 Nm).
10. Connect the brake lines to the ports on the side of the master cylinder. Torque the brake line fittings at the master cylinder to 13 ft. lbs. (17 Nm).
11. Reconnect the brake fluid level sensor connector. Reconnect the negative battery cable.
12. Check brake fluid level in reservoir and top off with clean, fresh DOT 3 type brake fluid as required.
13. Road test vehicle to verify correct braking operation and no leaks.

With ABS

1. Disconnect the negative battery cable.
2. Disconnect brake lines from the master cylinder and plug master cylinder fluid outlets.
3. Remove the nuts attaching the master cylinder to the power booster.
4. Disconnect the electrical connector from the master cylinder, if equipped. Remove the master cylinder from the mounting studs.
5. Remove the fluid reservoir from the cylinder as required.

To install:

6. Bench bleed the master cylinder as follows:

a. Mount master cylinder in a vise.

b. Attach tube to the fluid outlets on the master cylinder and bend tube so the outlet end of the tubes will be below the surface of brake fluid in each reservoir.

c. Fill both reservoirs with brake fluid conforming to DOT 3 specifications.

d. Slowly depress the piston and then allow the piston to return to the released position. Repeat this procedure until no bubbles are present in the fluid exiting the tubes.

e. Remove the tubes from the master cylinder and refill the reservoir with fluid.

7. Install the master cylinder to the mounting studs and install the retainer nuts. Tighten mounting nuts to 250 inch lbs. (28 Nm).

8. Install the brake lines to the master cylinder loosely. Have a helper slowly depress the brake pedal from inside the vehicle. While the pedal is being depressed, tighten the fluid lines to the master cylinder.

9. Connect the negative battery cable and check the brakes for proper operation.

Brake Caliper

REMOVAL AND INSTALLATION

The brake calipers may be inspected on while on the vehicle. Check the piston seals for any signs of visible leaks. Inspect the boot seals for any ruptures. Inspect the pins and slides for excessive rust/dirt build up, which may prevent proper movement.

Front Brake Calipers

1. Disconnect the negative battery cable.

2. Raise and support the vehicle safely.

3. Remove the appropriate wheel and tire assemblies.

4. Remove the two caliper guide pin bolts and remove the caliper assembly. If the caliper is not being removed from the vehicle as during brake pad renewal, simply hang the caliper with a piece of wire to take the weight off the brake hose. If the caliper is being removed for rebuild or replacement, continue to Step 5.

5. Remove the bolt retaining the brake hose to the caliper. Be sure to plug the end of the brake hose or cover it with a plastic bag to prevent contamination from entering the hydraulic system. Remove the caliper from the vehicle.

To install:

NOTE: Before installing the caliper assembly, be sure to clean the machined mounting surfaces of the steering knuckle of any grease or dirt.

6. Reconnect the brake hose to the caliper, if removed, using new gasket.

7. If new linings are being installed, the caliper pistons must be pushed back into their bore to accommodate the thickness of the new lining. Special tools are available for pushing the piston back although a large C-clamp can often be used. It is good practice to remove some (one-third to one-half) brake fluid from the master cylinder reservoir. This prevents overflow caused by brake fluid being forced through the lines as the piston is pushed back.

8. Install the brake shoes into the caliper.

9. Install the caliper to the steering knuckle in the correct position. Lubricate the machined areas that support the caliper with high-temperature grease.

10. Install and tighten the caliper guide pin bolts. Use care not to cross the threads of the caliper pin bolts. Torque the guide pin bolts to 30 ft. lbs. (41 Nm).

11. Torque the brake hose fittings to 35 ft. lbs. (48 Nm).

12. Be sure to bleed the brake system.

13. Install the wheels and lug nuts. Torque the wheel lug nuts in a star pattern sequence to 95–100 ft. lbs. (129–135 Nm). Lower the vehicle.

14. Before attempting to move the vehicle, pump the brake pedal to seat the pads against the rotors. Make sure the vehicle has a firm brake pedal. Check the level of the brake fluid and add DOT 3 brake fluid, if necessary.

15. Road test the vehicle and make several stops to wear off any foreign material on the brakes and to seat the brake linings, if replaced.

Rear Brake Calipers

1. Disconnect the negative battery cable.

2. Raise and support the vehicle safely.

3. Remove the appropriate wheel and tire assemblies.

4. Remove the two caliper guide pin bolts and remove the caliper assembly. If the caliper is not being removed from the vehicle as during brake pad renewal, simply hang the caliper with a piece of wire to take the weight off the brake hose. If the caliper is being removed for rebuild or replacement, continue to Step 5.

5. Remove the bolt retaining the brake hose to the caliper. Be sure to plug the end of the brake hose or cover it with a plastic bag to prevent contamination from entering the hydraulic system. Remove the caliper from the vehicle.

To install:

NOTE: Before installing the caliper assembly, be sure to clean the machined mounting surfaces of the rear caliper adapter of any grease or dirt.

6. Reconnect the brake hose to the caliper, if removed, using new gasket.

7. If new linings are being installed, the caliper pistons must be pushed back into their bore to accommodate the thickness of the new lining. Special tools are available for this although a large C-clamp can often be used. It is good practice to remove some brake fluid from the master cylinder reservoir. This prevents overflow caused by brake fluid being forced through the lines as the piston is pushed back.

8. Install the brake shoes into the caliper.

9. Install the brake caliper to the caliper adapter. Lubricate the machined areas that support the caliper with high-temperature grease.

10. Install and tighten the caliper guide pin bolts. Use care not to cross the threads of the caliper pin bolts. Torque the guide pin bolts to 17 ft. lbs. (22 Nm).

11. Torque the brake hose fittings to 35 ft. lbs. (48 Nm).

12. Be sure to bleed the brake system.

13. Install the wheels and lug nuts. Torque the wheel lug nuts in a star pattern sequence to 95–100 ft. lbs. (129–135 Nm). Lower the vehicle.

14. Before attempting to move the vehicle, pump the brake pedal to seat the pads against the rotors. Make sure the vehicle has a firm brake pedal. Check the level of the brake fluid and add DOT 3 brake fluid, if necessary.

15. Road test the vehicle and make several stops to wear off any foreign material on the brakes and to seat the brake linings, if replaced.

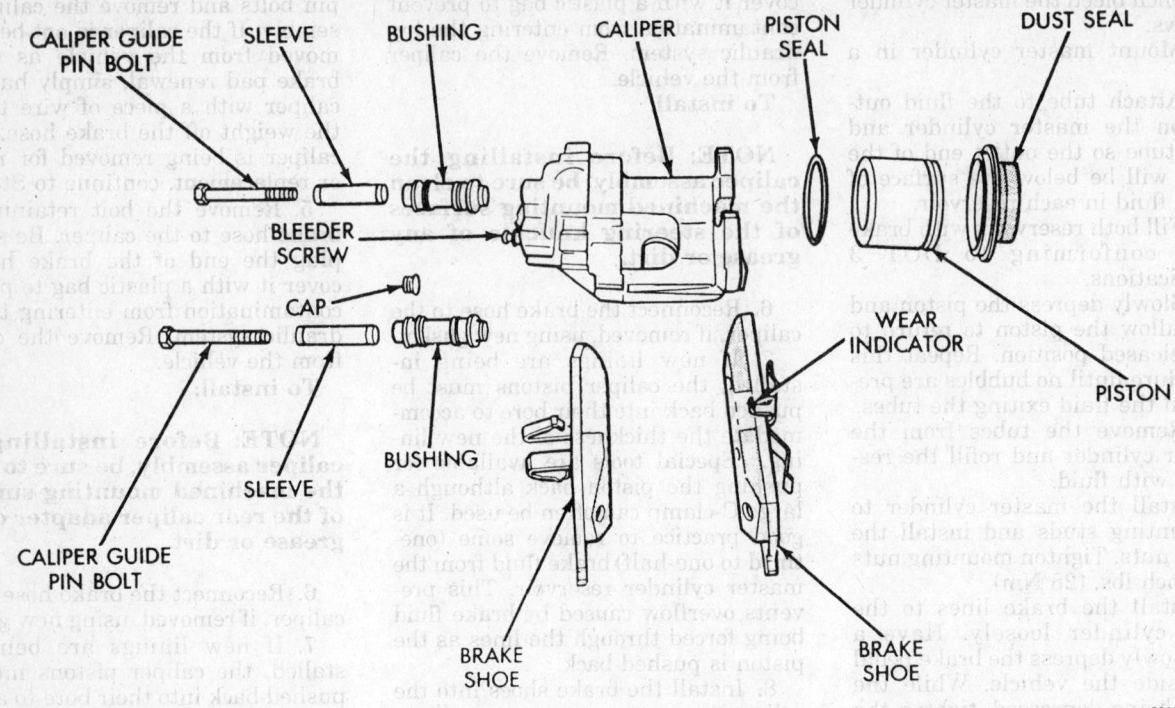

Front brake caliper and related components

Labels: CALIPER GUIDE PIN BOLT, SLEEVE, BUSHING, CALIPER, PISTON SEAL, DUST SEAL, BLEEDER SCREW, CAP, BUSHING, SLEEVE, CALIPER GUIDE PIN BOLT, WEAR INDICATOR, PISTON, BRAKE SHOE, BRAKE SHOE

291947

Disc Brake Pads

REMOVAL AND INSTALLATION

1. Remove some of the fluid from the master cylinder.

2. Raise the vehicle and support safely. Remove the appropriate wheels.

3. Remove the two caliper guide pin bolts. Remove the caliper assembly by swinging the top part of the caliper away from the brake rotor edge, then lift the caliper assembly up.

4. Prevent strain or other damage to the brake hose by supporting the caliper assembly with a strong piece of wire hanging from the strut.

5. Remove the outboard brake pad by prying the brake pad retaining clip over raised area on the caliper. Then slide the pad down and off the caliper.

6. Before removing the inboard brake pad, use a large C-clamp to press the piston back into the caliper. This will prevent possible damage to the caliper piston. It is good practice to remove some (one-third to one-half) of the brake fluid from the reservoir. This is because as the caliper piston is pushed back into the caliper, brake fluid will be pushed back

through the lines, back into the master cylinder and fluid reservoir, possibly causing the reservoir to overflow. Remove the inboard brake pad by pulling away from piston until the retainer clip is free from the cavity in the piston.

To install:

7. Lubricate both the caliper mating surface and the machined abutment surfaces with multi-purpose lubricant.

8. Before brake pad installation, be sure to lightly coat the outer backing plate surface of the new brake pads with a disc brake pad antisqueal lubricant, usually a gel-like material that deadens any high-frequency vibration that can be the source of disc brake squeal.

9. Install brake pads into the caliper assembly making sure both pads are seated securely onto the caliper.

10. Install the caliper assembly back into position over the brake rotor and install the caliper guide pin bolts. Torque the front caliper guide pin bolts to 30 ft. lbs. (41 Nm). Torque the rear caliper guide pin bolts to 17 ft. lbs. (22 Nm).

11. Install the wheels and lug nuts. Torque the lug nuts, in a star pattern sequence, to 95–100 ft. lbs. (129–135 Nm). Lower the vehicle and top off the master cylinder to the appropri-

ate level, using Dot 3 type brake fluid only.

12. Before moving the vehicle, pump the brakes until a firm pedal is obtained. Road test the vehicle to make sure the brake operation is normal.

Brake Rotor

REMOVAL AND INSTALLATION

1. Raise and safely support the vehicle. Remove the appropriate wheels.

2. Remove the caliper guide pin bolts and remove the caliper/brake pad assembly. Do not disconnect the brake line. Support the caliper assembly by hanging it off of the steering knuckle or strut with a strong piece of wire to prevent strain on the brake hose.

3. Remove the factory installed clips, if equipped. These need not be reinstalled.

4. Remove the rotor from the hub assembly by pulling straight off wheel mounting studs.

5. Inspect the brake rotor for the maximum allowable runout using Special Dial Indicator Tool C-3339 and Special Adapter Tool SP-1910 or equivalent. Runout of the rotor

should not exceed 0.003 inch (0.08 mm). Inspect the brake rotor for excessive lining deposits or corrosion. Resurface or replace the rotor if any of these conditions apply.

To install:

6. Clean both sides of the brake rotor with a brake cleaning solvent. Install the brake rotor onto the hub assembly.

7. Install the brake caliper. If installing a new rotor, compress the caliper piston back into the caliper bore to provide clearance for the rotor. It is good practice to remove some (one-third to one-half) of the brake fluid from the master cylinder reservoir to avoid fluid overflow as the piston pushed fluid back through the brake lines into the master cylinder and reservoir.

8. Tighten the caliper guide pin bolts. Torque the front caliper guide pin bolts to 30 ft. lbs. (41 Nm). Torque the rear caliper guide pin bolts to 17 ft. lbs. (22 Nm).

9. Install the wheels and lug nuts. Torque the lug nuts in a star pattern sequence to 95–100 ft. lbs. (129–135 Nm).

10. Lower the vehicle to the ground. Pump the brakes several times to seat the brake pads against the brake rotors before attempting to move the vehicle.

11. Road test the vehicle to verify good brake performance.

Brake Drums

REMOVAL AND INSTALLATION

1. Raise the vehicle and support safely.

2. Remove the rear wheels.

3. Remove the brake drum from the rear hub and bearing assembly. If the drum is difficult to remove, increase the clearance between the brake shoes and the drum as follows:

  a. Remove the rubber plug from the top of the brake support plate.

  b. Rotate the automatic shoe adjuster screw with an upward motion using a medium size flat tipped tool.

4. Remove the brake drum from the rear hub and bearing.

5. Inspect the brake shoe linings and drums for wear, contamination and scoring.

To install:

6. Install the rear brake drum onto the rear hub and bearing assembly. Adjust the brake shoes.

7. Install the rear wheels and lug nuts. Torque the lug nuts in a star

pattern sequence to 95–100 ft. lbs. (129–135 Nm).

Wheel Cylinder

REMOVAL AND INSTALLATION

1. Raise the vehicle and support safely. Remove the rear wheel and brake drum.

2. Remove the rear hub and bearing unit dust cap, cotter pin, nut retainer and wave washer. Discard the old cotter pin.

3. Remove the hub and bearing assembly retaining nut and washer from the rear spindle.

4. Remove the hub and bearing assembly from the spindle.

5. Remove the brake shoes assembly from the rear brake support plate.

6. Remove the brake hose bracket from the support plate.

NOTE: Often the brake line connection to the wheel cylinder is rusty and seized. The line often breaks the line when attempts are made to disconnect it from the wheel cylinder. Take time to wire brush the brake line tube and fitting nut at the back of the wheel cylinder. Some emery cloth can be used to clean the

rust where the brake line goes into the fitting nut. Apply a generous amount of penetrating oil before attempting to loosen the brake line fitting nut.

7. Using a line wrench, disconnect the brake line fitting nut from the wheel cylinder. Plug the open end of the brake line to prevent fluid loss or system contamination.

8. Remove the rear wheel cylinder attaching bolts and pull the wheel cylinder off the brake support plate.

9. Clean all parts well. If the wheel cylinder was removed to correct a fluid leakage complaint, carefully clean and inspect all components. Brake shoe linings contaminated by brake fluid must not be reused. Always install new brake shoes as an axle set, both rear wheels at once. New brake hardware and return springs are always recommended. Make sure the adjuster is free and operates correctly. This is also a good time to check the parking brake cable assemblies and operation. Frozen or sticking brake cables must be replaced (one at a time, or connection to equalizer will be difficult). Install all brake components that were removed, using new replacement parts as required. Inspect the inside of the brake drum. Most

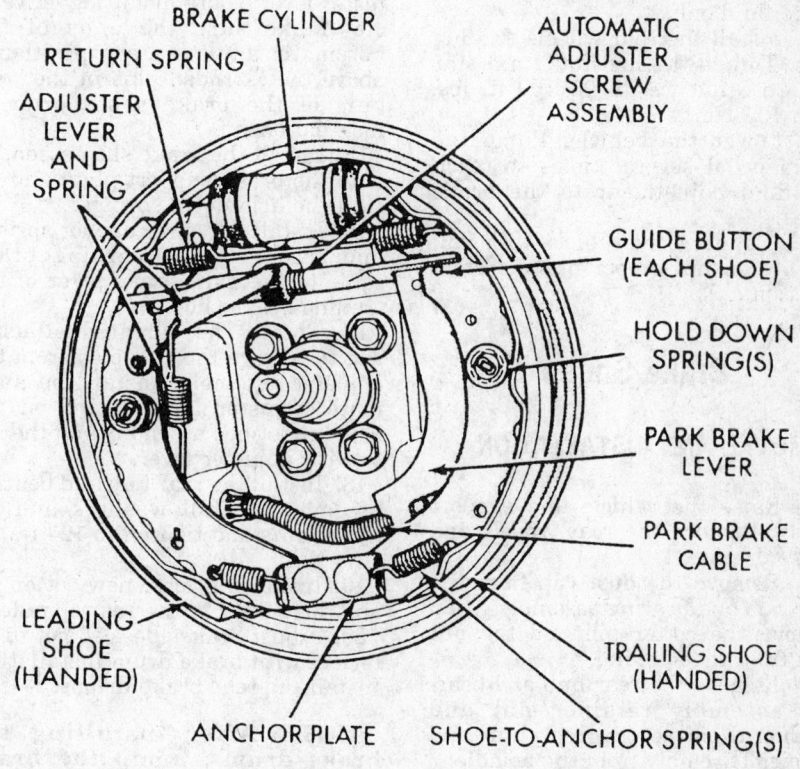

Chrysler LH platform VARGA® brand brake shoe assembly

291979

brake drum wear can be removed by proper machining. If the drum inside diameter is beyond maximum limits, usually marked on the drum, a replacement drum must be installed.

To install:

10. Install the wheel cylinder O-ring onto the wheel cylinder.

11. Install the wheel cylinder onto the brake support plate. Torque the wheel cylinder mounting bolts to 6 ft. lbs. (8 Nm).

12. Reconnect the brake hose fitting to the wheel cylinder, but do not tighten.

13. Install the brake hose bracket to the support plate. Torque the brake hose fitting, at the wheel cylinder, to 13 ft. lbs. (17 Nm).

14. Install the brake shoe assembly back onto the support plate. Install the brake drum.

15. Install the hub and bearing assembly onto the rear spindle.

16. Install the washer and retaining nut. Torque the retaining nut to 124 ft. lbs. (168 Nm).

17. Install the wave washer, nut retainer and a new cotter pin onto the spindle. Install the hub and bearing assembly dust cap.

18. Install the brake drum onto hub.

19. Bleed the entire brake system and refill the master cylinder to the appropriate level using DOT 3 type brake fluid only.

20. Install the rear wheels and lug nuts. Torque the lug nuts, in a star pattern sequence, to 95–100 ft. lbs. (129–135 Nm).

21. Lower the vehicle. Pump the brake pedal several times to reach the final adjustment to the brake shoes.

22. Test drive the vehicle, with caution, to ensure proper brake system function.

Brake Shoes

REMOVAL AND INSTALLATION

1. Raise the vehicle and support safely. Remove the rear wheels and brake drums.

2. Remove the dust cap from the rear hub and bearing assembly. Then remove the cotter pin and the nut lock from the spindle.

3. Remove the rear hub and bearing assembly retainer nut and washer. Remove the rear hub and bearing assembly from the spindle.

4. Remove the automatic adjuster spring from the adjuster lever.

5. Rotate the automatic adjuster star wheel enough so both shoes move out far enough to be free of the wheel cylinder boots.

6. Disconnect the parking brake cable from the actuating lever. Disconnect parking brake cable one side at a time.

7. Remove the both lower brake shoe to anchor springs.

8. Remove the 2 brake shoe hold-down springs from the brake shoes.

9. Remove the brake shoes, upper shoe-to-shoe return spring, automatic adjuster and automatic adjuster lever from the backing plate as an assembly.

10. Separate the brake shoes from the automatic adjuster mechanism.

11. Remove the brake shoe automatic adjuster lever from the leading brake shoe.

To install:

12. Thoroughly clean and dry the backing plate. To prepare the backing plate, lubricate the bosses, anchor pin and parking brake actuating lever pivot surface lightly with lithium based grease.

13. Remove, clean and dry all parts still on the old shoes. Lubricate the star wheel shaft threads with anti-seize lubricant.

14. Assemble both brake shoes, the top shoe to shoe return spring, automatic adjuster and automatic adjuster lever before mounting on vehicle. Make sure the ends of the automatic adjusters are positioned above the extruded pins in the webbing of the brake shoes prior to installation.

15. Install the brake shoe assembly onto the brake support plate and install the hold-down springs.

16. Install the lower anchor springs and reconnect the parking brake cable to the park brake lever of the trailing brake shoe.

17. Rotate the serrated adjuster nut to remove the free-play from the adjuster assembly. Install the automatic adjuster lever spring on the lead brake shoe assembly and the automatic adjuster lever.

18. Install the rear hub and bearing assembly. Install washers and retainer nuts and tighten to 124 ft. lbs. (168 Nm).

19. Install nut lock, new cotter pin and dust cover to the rear spindle.

20. Adjust brake shoes so not to interfere with brake drum installation. Install the rear brake drums.

NOTE: After installing the brake drums, pump the brake pedal several times to partially adjust the brake shoes. To verify

proper operation of the self-adjusting parking brake, be sure that both rear brakes are not dragging when the parking brake pedal is released.

21. Install the rear wheels and lug nuts. Torque the lug nuts, in a star pattern sequence, to 95 ft. lbs. (129 Nm).

22. Road test the vehicle. The automatic adjusters will continue brake adjustment during the road test of the vehicle.

Parking Brake Cable

CAUTION

The self-adjusting feature of this parking brake lever assembly contains a clock spring loaded to approximately 8 pounds. Care must be taken to prevent excessive jarring of the assembly. Do not release the self-adjuster lockout device before installing cables into the equalizer. Keep hands out of the self-adjuster sector and pawl area. Failure to observe this warning in handling this mechanism could lead to serious injury.

ADJUSTMENT

The parking brake foot lever assembly contains a self-adjuster. Routine parking brake cable adjustment is not required. However, adjusting of the parking brake cable on vehicles equipped with rear drum brakes relies on proper adjustment of the brake shoes. The park brake system on vehicles equipped with rear disc brakes must have the drum-in-hat brake shoes properly adjusted to 6.75 inches (171 mm) diameter.

NOTE: When repairs to the foot lever assembly or front cable are required the self-adjuster must be reloaded and locked out.

1. Be sure that the front and both rear parking brake cables are properly installed to the equalizer bracket before you adjust the cable.

2. Be sure that the cable is correctly seated on the cam surface of the parking brake mechanism.

NOTE: The parking brake pedal must be in the fully released position when releasing the parking brake self adjuster.

3. Using a pair of pliers, firmly grasp the lock pin previously installed in the parking brake pedal mechanism.

4. Remove the lock pin from the parking brake pedal mechanism by pulling it firmly and rapidly from the park brake mechanism.

5. Apply and release the parking brake pedal several times. The rear wheels should rotate freely without dragging.

REMOVAL AND INSTALLATION

The parking brake mechanism on vehicles with rear disc brake applications, consists of a small duo-servo brake which mounted to an adapter. The hat (center) section of the rear rotor serves as the braking surface (drum) for the parking brakes.

On rear wheel drum brake applications, the rear wheel service brakes also act as parking brakes. The rear drum brake shoes are mechanically operated by an internal lever and strut connected to a flexible steel cable. The wheel brake cables are joined at an equalizer which is attached to the front cable leading to the foot lever.

The parking brake foot lever assembly used on all LH platform vehicles with and without rear disc brakes contains a self-adjuster for the cable system. Routine parking brake cable adjustments on all LH platform vehicles is not required.

Front Parking Brake Cable Service

NOTE: When repairs to the foot lever assembly or front cable are required the self-adjuster must be reloaded and locked out.

——— CAUTION ———
The Air Bag system must be disarmed before working around the steering column. Failure to do so may cause accidental deployment of the air bag, resulting in unnecessary system repairs and/or personal injury.

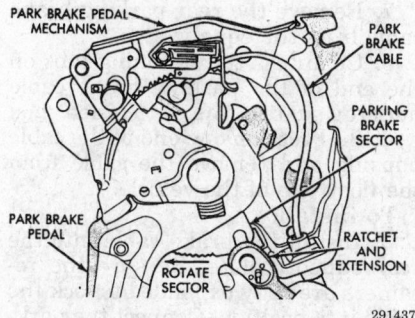

PARK BRAKE PEDAL MECHANISM — PARK BRAKE CABLE — PARKING BRAKE SECTOR — RATCHET AND EXTENSION — PARK BRAKE PEDAL — ROTATE SECTOR

291437

Winding the parking brake sector spring

1. Disconnect the negative battery cable.

2. Disarm the air bag system.

3. The driver's seat assembly must be removed from the vehicle. Use the following procedure:

a. Position the front seat far enough forward to gain access to the front mounting bolts on the floor. Remove the bolts.

b. Disengage the front seat wiring harness connector from the body connector.

c. Remove the bolts securing the front of the seat track to the floor.

d. Remove the bolts securing the rear of the seat track to the floor.

e. Separate and remove the seat from the vehicle.

4. On vehicles equipped with bucket seats:

a. Remove the shifter knob from the shift lever by removing the 3/32 inch Allen® head setscrew located on the driver side of the knob and pulling the knob straight up.

b. Remove the center console ashtray and remove the center console bezel retaining screw.

c. Disengage the holding clips of the center console bezel and remove the bezel from the console.

d. Remove the mounting screw that holds the forward crossmember to the floor mounting bracket.

e. Remove the rear mounting bracket screw located at the bottom of the glove compartment.

f. Remove the center console from the vehicle.

g. Remove the rear mounting bracket, for the center console, from the reaction bracket of the parking brake cable.

h. Remove the heat duct for the rear passenger compartment.

5. On vehicles equipped with bench seat:

a. Pull the lower rear seat cushion upward at each end of the cushion to release the loops from the cups in the floor.

b. Remove the rear seat cushion from the vehicle.

c. Remove the lower rear door opening sill moldings on both sides of the vehicle and fold the carpeting out of the way of the parking brake cable.

d. Remove the equalizer cover from the reaction bracket.

6. Remove the driver's side door opening sill molding. Remove the driver's side cowl kick molding to expose the parking brake pedal mechanism.

7. Remove the throttle pedal and bracket from the dash panel and fold the carpet back to expose the front brake cable, routing clips, reaction bracket and equalizer.

——— CAUTION ———
The parking brake pedal mechanism self-adjuster must be reloaded and locked, to remove spring tension, before attempting to remove the front parking brake cable from the parking brake pedal or equalizer. Failure to do so will make assembly extremely difficult and could lead to serious injury.

8. Reload the parking brake self-adjuster assembly as follows:

a. Insert a ¼-inch drive ratchet and extension into the hole located in the sector part of the parking brake pedal mechanism.

b. Pull on the ratchet and extension rotating the sector toward the rear of the vehicle until the ratchet extension contacts the back of the brake pedal mechanism. This will wind the sector spring.

c. Insert a pin or drill bit ⅛-inch diameter by at least two inches long into the parking brake lever mechanism to lock the sector shaft in place. Make sure the pin or drill bit is long enough to go through both sides of the parking brake lever mechanism.

9. Remove the front parking brake cable from the parking brake cable equalizer. Remove the retaining clip from the cable at the reaction bracket.

10. Remove the front parking brake cable from the reacting bracket. Remove the screw attaching the parking brake cable routing clip to the floor pan. Remove the parking brake cable from the mounting clips at the crossmember.

11. Remove the bolt attaching the parking brake mechanism to the left front door frame. Loosen but do not remove the two bolts attaching the mechanism to the cowl panel. Disconnect the electrical connector from the switch on the parking brake mechanism.

12. Remove the parking brake mechanism and the front cable from the vehicle as an assembly.

13. Mount the parking brake assembly in a vise and remove the front parking brake cable-to-mechanism retainer clip.

14. Rotate the cable to align with the notch in the cable attaching hole on the parking brake mechanism. Lift the lead of the cable out of the

retaining hole in the parking brake mechanism, then remove the cable from the mechanism.

To install:

15. Install the parking brake cable into the parking brake mechanism and rotate into position. Be sure cable is properly seated in the cam surface tract of the foot lever assembly. Install the cable retaining clip.

16. Install the parking brake mechanism to the two loosely installed mounting bolts on the cowl panel. Install the retainer bolt into the door frame. Torque the three mounting bolts to 19 ft. lbs. (27 Nm).

17. Reconnect the brake warning light connector to the switch.

18. Route the cable along the floor of the vehicle. Install the end of the cable into the hole in the reaction bracket and install retainer clip. Tap the clip with a hammer until fully seated on the cable. Install the front cable to the equalizer.

19. Install all parking brake cable routing clips and retainers. Torque the routing clip-to-floorpan mounting screws to 35 inch lbs. (4 Nm).

20. Ensure the cable is correctly routed in cam surface tract of the parking brake mechanism. Using pliers, firmly grasp the lock pin previously installed in the park brake mechanism. Remove the lock pin using a firm and rapid pull. When removed, the parking brake cables will automatically adjust. Activate and release the parking brake mechanism to assure proper operation.

21. Place the carpeting, in the left front of the vehicle, back into its proper position. Install the throttle pedal and bracket back onto the dash panel studs and tighten the mounting nuts.

22. Install the driver's side cowl kick molding and driver's side door opening sill molding.

23. On vehicles equipped with bucket seats:

a. Install the console heat duct and rear console mounting bracket. Torque the rear console mounting bracket bolts to 88 inch lbs. (10 Nm).

b. Install the center console assembly into position.

c. Install the mounting screws that hold the bottom of the glove compartment to the rear mounting bracket and the mounting screw that holds the forward cross-member to the floor mounting bracket.

d. Install the center console bezel and ashtray.

e. Install the shift knob onto the shift lever.

24. On vehicles equipped with bench seat:

a. Install the parking brake equalizer cover back on the reaction bracket. Torque the equalizer cover mounting bolts to 88 inch lbs. (10 Nm).

b. Place the rear carpeting back to its original position inside the vehicle.

c. Install the lower rear door sill moldings for both sides of the vehicle.

d. Install the rear seat cushion into the vehicle and engage the retainer loops into the retainer cup on the floor. Engage the retainers by pushing downward at each end of the rear seat.

25. Install the left front seat into the vehicle and bolt the rear of the tract to the floor.

26. Install bolts to hold front of the seat tract to the floor.

27. Reconnect the front seat wire connector to the body connector.

28. Test the left front seat for proper operation.

29. Reconnect the negative battery cable. Adjust the rear brakes, if necessary, then adjust the parking brake.

Rear Parking Brake Cable Service

Rear Drum Brakes

NOTE: When servicing the rear parking brake cable, remove one rear brake cable from the rear brake shoes at a time. Failure to do so will result in high efforts required to connect the parking brake cables to the equalizer or parking brake lever.

1. Raise and safely support the vehicle. Remove the rear wheels and rear brake drums.

2. Disconnect the parking brake cable from the parking brake lever on the rear brake assembly.

3. Position a ½-inch box wrench over the retainer on the end of the parking brake cable. Compress the cable housing retainer fingers and start the housing out of the support plate.

4. Remove the retainer clips holding the cable to the floor pan and vehicle frame.

5. On vehicles equipped with bucket seats:

a. Remove the shifter knob from the floor shifter by removing the ³/₃₂-inch Allen® head setscrew located on the driver side of the knob and pulling the knob straight up.

b. Remove center console ashtray.

c. Remove the center console bezel retaining screw and disengage the holding clips of the bezel.

d. Remove the bezel from the console.

e. Remove the mounting screw that holds the forward cross-member to the floor mounting bracket.

f. Remove the rear mounting bracket screw located at the bottom of the glove compartment.

g. Remove the center console from the vehicle.

h. Remove the rear mounting bracket for the center console from the reaction bracket of the parking brake cable.

i. Remove the heat duct for the rear passenger compartment.

6. On vehicles equipped with bench seat:

a. Move the right front seat to its furthest forward position.

b. Move the left front seat to its furthest forward position and remove the rear seat floor mounting bolts.

c. Disconnect the front seat wiring harness connector from the body harness connector.

d. Remove the mounting bolts holding the front of the seat track to the floor.

e. Remove the mounting bolts holding the rear of the seat track to the floor.

f. Separate and remove the seat from the vehicle.

g. Pull the rear seat cushion upward at each end of the cushion to release the loops from the cups in the floor.

h. Remove the rear seat cushion from the vehicle.

i. Remove the lower rear door opening sill moldings on both sides of the vehicle and fold the carpeting forward.

j. Remove the equalizer cover from the reaction bracket.

7. Remove the rear parking brake cable from the equalizer.

8. Compress the retaining tabs on the end of the parking brake cable housing using the a ½-inch box wrench. Press on one end of the cable housing and remove the cable from the floor pan of the vehicle.

To install:

9. Install the brake cable into the rear support plate. Be sure the retainers are fully expanded to lock the cable into position. Connect the parking brake cable to the parking brake shoe lever.

10. Install the parking brake cable into the mounting hole in the floor pan. Be sure that the cable retainers are expanded around the mounting hole in the floor pan of the vehicle.

11. Install the four routing clips securing the rear parking brake cable to the vehicle frame and floor pan.

12. Connect the cable to the parking brake equalizer. The equalizer can be moved rearward to aid in cable connection.

13. On vehicles equipped with bucket seats:

a. Install the console mounted heater duct and rear console mounting bracket.

b. Install the center console assembly into position.

c. Install the mounting screws that hold the bottom of the glove compartment to the rear mounting bracket and the mounting screw that holds the forward cross-member to the floor mounting bracket.

d. Install the center console bezel and ashtray.

e. Install the shift knob onto the shift lever.

14. On vehicles equipped with bench seat:

a. Install the parking brake equalizer cover back on the reaction bracket. Torque the equalizer cover mounting bolts to 88 inch lbs. (10 Nm).

b. Place the rear carpeting back to its original position inside the vehicle.

c. Install the lower rear door sill moldings for both sides of the vehicle.

d. Install the rear seat cushion into the vehicle and engage the retainer loops into the retainer cup on the floor. Engage the retainers by pushing downward at each end of the rear seat.

e. Install the left front seat back into the vehicle. Tighten the mounting bolts and reconnect the seat wiring harness connector to the body harness connector.

f. Test the left front seat for proper operation.

15. Install the brake drums and the rear wheels. Torque the wheel lug nuts in sequence to 95 ft. lbs. (129 Nm).

16. Reconnect the negative battery cable.

17. Enable the air bag system.

18. Adjust the rear brakes, then adjust the parking brake.

Rear Disc Brakes

NOTE: When servicing the rear parking brake cable, remove one rear brake cable from the rear brake shoes at a time. Failure to do so will result in high efforts required to connect the parking brake cables to the equalizer or parking brake lever.

1. Raise and safely support the vehicle. Remove the rear wheels.

2. Remove the rear disc brake caliper from the adapter. Remove the brake rotor from the vehicle.

3. Remove the parking brake shoes as follows:

a. Remove the dust cap from the rear hub. Remove the cotter pin and nut lock from the spindle.

b. Remove the hub and bearing retainer nut and washer. Pull the hub and bearing from the spindle.

c. Remove the rear brake shoe hold-down clip. Turn the parking brake adjuster wheel to its shortest length.

d. Remove the parking brake shoe adjuster assembly from the shoes. Remove the lower shoe-to-shoe spring. Pull the shoes away from the anchor and remove the shoe assembly.

e. Remove the front parking brake shoe hold-down clip and remove shoe.

4. Disconnect the parking brake cable from the brake actuator lever.

5. Remove the parking brake cable from the adapter by compressing the tangs on the cable. Remove the four routing clips securing the rear parking brake cable to the floor pan and frame of the vehicle.

6. On vehicles equipped with bucket seats:

a. Remove the shifter knob from the floor shifter, by removing the 3/32-inch Allen® head setscrew located on the left side of the knob, and pulling straight up.

b. Remove the center console ashtray.

c. Remove the center console bezel retaining screw and disengage the holding clips of the bezel.

d. Remove the bezel from the console.

e. Remove the mounting screw that holds the forward cross-member to the floor mounting bracket.

f. Remove the rear mounting bracket screw located at the bottom of the glove compartment.

g. Remove the center console from the vehicle.

h. Remove the rear mounting bracket for the center console from

the reaction bracket of the parking brake cable.

i. Remove the heat duct for the rear passenger compartment.

7. On vehicles equipped with bench seat:

a. Move the right front seat up to the full forward position.

b. Position the left front seat far enough forward to gain access to the front mounting bolts on the floor. Remove the mounting bolts.

c. Disengage the front seat wiring harness connector from the body connector.

d. Remove the mounting bolts holding the front of the seat track to the floor.

e. Remove the mounting bolts holding the rear of the seat track to the floor.

f. Separate and remove the seat from the vehicle.

g. Pull the lower rear seat cushion upward at each end of the cushion to release the loops from the cups in the floor.

h. Remove the rear seat cushion from the vehicle.

i. Remove the lower rear door opening sill moldings on both sides of the vehicle and fold the rear carpeting forward.

j. Remove the equalizer cover from the reaction bracket.

8. Remove the rear parking brake cable from the equalizer.

9. Compress the retaining tabs on the end of the parking brake cable housing using a 1/2-inch box wrench. Press on one end of the cable housing and remove the cable from the floor pan of the vehicle.

To install:

10. Install cable end into the rear disc brake adapter cable mounting hole. Be sure the cable retainer tabs are expanded around the hole opening and lock the cable in position.

11. Connect the parking brake cable end to the brake shoe assembly actuator lever.

12. Install the parking brake cable into the cable mounting hole in floor pan. Be sure the cable tabs are expanded in the opening and lock the cable in position. Install the four routing clips securing cable to floor pan and frame of the vehicle.

13. Connect the park brake cable to the brake equalizer. The equalizer can be moved toward the rear of the vehicle to aid in cable connection.

14. Install the parking brake shoes as follows:

a. Install the front parking brake shoe and the hold-down clip.

b. Install the rear parking brake shoe and the upper parking brake shoe-to-shoe return spring.

c. Move the rear parking brake shoe over the anchor block until it is correctly positioned on the adapter.

d. Install the lower parking brake shoe-to-shoe return spring. Install the adjuster assembly with the star wheel rearward.

e. Install rear parking brake shoe hold-down clip.

f. Adjust the parking brake shoes to a diameter of 6.75 inch (171.5mm).

g. Install the rear bearing and hub assembly. Install the washer and retaining nut of the hub and bearing assembly. Torque the hub retainer nut to 124 ft. lbs. (168 Nm). Install the nut lock and new cotter pin. Install the dust cap.

15. Install rear brake rotor and caliper assembly.

16. Install the rear wheels and lug nuts. Torque the lug nuts in sequence to 95 ft. lbs. (129 Nm).

17. On vehicles equipped with bucket seats:

a. Install the heater duct and console mounting bracket.

b. Install the center console assembly into position.

c. Install the mounting screws that hold the bottom of the glove compartment to the rear mounting bracket and the mounting screw that holds the forward cross-member to the floor mounting bracket.

d. Install the center console bezel and ashtray.

e. Install the shift knob onto the shift lever.

18. On vehicles equipped with bench seat:

a. Install the parking brake equalizer cover back on the reaction bracket. Torque the equalizer cover mounting bolts to 88 inch lbs. (10 Nm).

b. Place the rear carpeting back into position and reinstall the door opening sill molding.

c. Install the rear seat cushion into the vehicle and engage the retainer loops into the retainer cup on the floor. Engage the retainers by pushing downward at each end of the rear seat.

d. Install the left front seat back into the vehicle. Tighten the mounting bolts and reconnect the seat wiring harness connector to the body harness connector.

e. Test the left front seat for proper operation.

19. Reconnect the negative battery cable. Adjust the rear brakes, then adjust the parking brake.

Brake System Bleeding

NOTE: If using a pressure bleeder, follow the instructions furnished with the unit and choose the correct adaptor for the application. Do not substitute an adapter that almost fits as it will not work and could be dangerous.

NON ABS SYSTEMS

Master Cylinder

If the master cylinder is off the vehicle, it can be bench bled.

1. Connect two short pieces of brake line to the outlet fittings, bend them until the free end is below the fluid level in the master cylinder reservoirs.

2. Fill the reservoir with fresh brake fluid. Pump the piston slowly until no more air bubbles appear in the reservoirs.

3. Disconnect the two short lines, refill the master cylinder and securely install the cylinder caps.

4. If the master cylinder is on the vehicle, it can still be bled, using a flare nut wrench.

5. Open the brake lines slightly with the flare nut wrench while pressure is applied to the brake pedal by a helper inside the vehicle.

6. Be sure to tighten the line before the brake pedal is released.

7. Repeat the process with both lines until no air bubbles come out.

8. Bleed the complete brake system.

Calipers and Wheel Cylinders

NOTE: If a pressure bleeder is not available, a good brake fluid flow can be obtained by manually bleeding the brake hydraulic system outlined in this procedure.

1. Fill the master cylinder with fresh brake fluid. Check the level often during the procedure.

2. Remove the protective caps from all four bleeder screws. Save for reuse. Clean the bleeder screws.

------- CAUTION -------
When bleeding the brakes, keep face away from the brake area but still wear a good pair of safety goggles. Spewing fluid may cause facial and/or visual injury. Do not allow brake fluid to spill on the car's finish. It will remove the paint.

3. Connect a clear, tight fitting hose to the bleeder screw at the right rear wheel and place the hose into a clear container containing fresh brake fluid.

4. Pump the brake pedal three or four times and hold down before opening the bleeder screw.

5. Open the bleeder screw at least one full turn. With the bleeder screw open, the brake pedal will drop all the way down to the floor.

NOTE: If the pedal is pumped rapidly, the fluid will churn and create small air bubbles, which are difficult to remove from the system. These air bubbles will eventually congregate and a spongy pedal will result.

6. Hold down the brake pedal and release only **after** the bleeder screw has been closed.

7. Repeat this procedure on the remaining brake bleeder screws in the following sequence:

a. Left rear
b. Right front
c. Left front

8. Be sure that all trapped air has been expelled from the hydraulic system and that there is a short stroke and solid brake pedal feel. Check the brake fluid reservoir and keep filled to the proper level to ensure that no air re-enters the system through the master cylinder.

9. Hydraulic brake systems must be totally flushed if the fluid becomes contaminated with water, dirt or other corrosive chemicals. To flush, bleed the entire system until all fluid has been replaced with new fluid.

10. Install the bleeder cap(s) on the bleeder screws to keep dirt out. Always road test the vehicle, after brake work of any kind is performed.

ANTI-LOCK BRAKE SYSTEMS

The bleeding procedure is a three step process, one of which will require use of the DRB II scan tool or its equivalent. Bleed the system as follows:

1. Connect a pressure bleeder to the master cylinder.

2. Fully bleed the brakes hydraulic system using the conventional method in the following sequence:

a. Right rear wheel
b. Left rear wheel
c. Right front wheel
d. Left front wheel

3. Locate the diagnostic connector under the dash panel to the right of the steering column.

4. Connect the DRB II scan tool to the connector. Install cartridge for the Teves Mark IV or Teves Mark IV-G Anti-Lock Brake system. Check to make sure that the CAB does not have any fault codes stored in it. If it does, remove them using the DRB II scan tool.

5. Perform the bleeding procedure using the DRB II according to the procedure outlined in the scan tool literature.

6. Once bleeding with the scan tool is complete, repeat the bleed procedure outlined in Step 2. Remove the pressure bleeder from the master cylinder.

7. Fill the master cylinder reservoir to the proper level.

8. Road test the vehicle to check for proper brake system operation.

Wheel Speed Sensor

REMOVAL AND INSTALLATION

One of the primary inputs to the ABS system is from the wheel speed sensors. There is a sensor at each wheel that reads magnetic impulses from a toothed gear-like tone wheel. The sensors are easily damaged and must be handled with care. Make sure the wheel sensor surfaces are clean since they are magnetic and attract metal chips and debris. Use care when removing, installing and routing the sensor wiring. The wiring must be correctly installed to avoid ABS problems later.

Front Wheel ABS Speed Sensor

1. Disconnect negative battery cable.

2. Carefully raise and safely support the vehicle. Remove the front wheels.

3. Remove the attaching screw that mounts the grommet retaining clip to the inner fender shield.

4. Gently remove the speed sensor cable grommet from the fender shield.

5. Disconnect the vehicle wiring harness connector from the speed sensor connector.

6. Remove the routing bracket, for the speed sensor cable, from the front strut assembly.

7. Remove the screw that mounts the speed sensor head to the steering knuckle.

8. Gently remove the speed sensor from the steering knuckle.

——— **WARNING** ———
Do not remove the speed sensor with pliers for any reason. If the speed sensor has seized, due to corrosion, remove it with a small mallet and punch. Lightly tap the edge of the sensor ear, rocking the sensor from side to side until it is free.

To install:

NOTE: Coat the head of the speed sensor with some High Temperature Multi-Purpose E.P. Grease (disc brake-rated wheel bearing grease) before installing into the steering knuckle.

9. Install the speed sensor head into the steering knuckle. Install the mounting screw and torque to 60 inch lbs. (7 Nm).

10. Be sure to check the wheel speed sensor air gap clearance. The allowable clearance range is 0.020–0.065 inch (0.52–1.64 mm).

NOTE: Correct system operation depends on the wheel speed sensor cables being installed properly. Be sure that the sensor cables are installed in the retainers. Failure to do this could result in cable over extension and/or contact with moving parts. This could result in an open circuit and/or false sensor readings.

11. Route the sensor cable correctly up to the strut assembly.

12. Install the routing bracket for the sensor cable onto the strut assembly. Install and tighten the routing bracket mounting screw.

13. Connect the wheel speed sensor connector to the vehicle wiring harness connector. Be sure to securely latch the connector locking tab.

14. Push the sensor cable grommet into hole in the inner fender shield. Install the grommet retaining clip and screw.

15. Install the front wheels and lug nuts. Tighten the wheel lug nuts in a star pattern sequence and torque the nuts to half specification. Then repeat the lug nut torquing sequence to full specified torque of 95–100 ft. lbs. (129–135 Nm).

16. Lower the vehicle. Reconnect the negative battery cable.

Rear Wheel ABS Speed Sensor

1. Disengage the rear seat lower cushion retainer loops from the cups in the floor and lift upward at each end of the rear seat lower cushion.

Remove the rear seat lower cushion from the vehicle.

2. If equipped with a child safety seat:

a. Open up child safety seat and remove seat lining.

b. Remove mounting bolts securing the seat back frame to trunk closure panel through the access holes in the child seat module.

3. Remove the mounting nuts that secure the rear seat back and seat belts to the floor.

4. Lift the rear seat back upward to disengage the hooks at the top of the seat back and remove from vehicle.

5. Pull up the edge of the sound insulation on the rear bulkhead. Disconnect the wheel speed sensor connector from the vehicle wiring harness connector.

6. Carefully raise and safely support the vehicle. Remove the rear wheels.

7. Remove the sensor cable grommet retaining clip from the rear inner fender. Gently pull the sensor cable and grommet out of the hole in the fender well.

8. Remove the speed sensor head mounting bolt from the rear adapter. Remove the wheel speed sensor head from the adapter.

——— **WARNING** ———
Do not remove the wheel speed sensor with pliers for any reason. If the speed sensor has seized, due to corrosion, remove it with a hammer and punch. Lightly tap the edge of the sensor ear, rocking the sensor from side to side until it is free.

9. Remove fasteners mounting speed sensor cable mounting bracket to inner side of trailing arm.

10. Remove the sensor cable routing brackets from the rear fender well and trailing arm bracket at the rear knuckle.

11. Remove the wheel speed sensor from the vehicle.

To install:

NOTE: Coat the wheel speed sensor, where it slides into the adapter, with High Temperature Multi-Purpose E.P. Grease (disc brake-rated wheel bearing grease) before installing it into the adapter.

12. Install the wheel speed sensor head into the adapter. Install the attaching screw and torque to 60 inch lbs. (7 Nm).

13. Be sure to check the wheel speed sensor air gap clearance. The

allowable clearance range is 0.017–0.047 inch (0.45–1.21 mm).

14. Install the sensor cable mounting bracket onto the trailing arm. Install the sensor cable routing tube-to-trailing arm mounting screws.

15. Install the sensor cable routing brackets at the rear fender well and rear knuckle.

16. Install the sensor cable and grommet back into the hole in rear inner fender. Install the cable grommet retaining bracket and tighten the attaching screw.

17. Install the wheels and lug nuts. Tighten the wheel lug nuts in a star pattern sequence and torque the nuts to half specification. Then repeat the lug nut tightening sequence to full specified torque of 95–100 ft. lbs. (129–135 Nm).

18. Lower the vehicle to the ground.

19. Connect the speed sensor cable connector to the vehicle wiring harness. Be sure that the connector locking tab is securely latched.

20. Install the rear seat back cushion into the vehicle in its proper position.

21. Engage the retainer hooks at the top of the rear seat back cushion.

22. Install the mounting nuts that secure the rear seat back and seat belts to the floor.

23. If equipped with a child safety seat:

    a. Install and tighten mounting bolts to secure seat back frame to trunk closure panel through the access holes in the child seat module.

    b. Install the child safety seat liner.

24. Install the rear lower seat cushion in proper position into the vehicle.

25. Engage the retainer loops into cup on the floor kick-up.

26. To engage the retainers, be sure to push downward at each end of the lower seat cushion.

27. Reconnect the negative battery cable.

FRONT SUSPENSION

Strut

REMOVAL AND INSTALLATION

1. Disconnect the negative battery cable.

2. Raise and safely support vehicle. Do not support vehicle by placing supports under the suspension arms. The suspension arms must hang freely.

3. Remove the front wheel(s).

4. Remove the stabilizer bar attaching link at the strut assembly.

5. Loosen but do not remove the outer tie rod end to strut assembly steering arm attaching nut. Then remove the outer tie rod end from the steering arm using puller MB-990635 or equivalent.

6. If equipped with ABS, remove the speed sensor wire harness mounting bracket from the strut.

7. Remove the brake caliper assembly. Support the caliper assembly from the vehicle frame with a strong piece of wire. Do not allow the assembly to hang by the brake hose. Remove the front brake rotor disc.

NOTE: The strut assembly to steering knuckle bolts are serrated where they go through the strut and steering knuckle. Do not turn the bolts during removal. If bolts are turned, damage to the steering knuckle will result.

8. The strut assembly to steering knuckle bolts must not be turned during strut removal. Hold the bolt head with a wrench and turn the nuts off the bolts.

9. Remove the three strut assembly upper mount to shock tower mounting nuts and washers. Remove the strut from the vehicle.

— **CAUTION** —
Do NOT remove the center retaining nut. The strut assembly is held together with this nut and must only be removed following the proper strut disassembly procedure.

10. Inspect the strut assembly for signs of leakage. Actual leakage will be a stream of fluid running down the side and dripping off the lower end of the strut. A slight amount of seepage between the strut rod and strut shaft seal is not unusual and does not affect performance of the strut assembly.

To install:

11. Install the front strut into the shock tower and install the three upper mount nuts and washers. Tighten mounting nuts to 25 ft. lbs. (33 Nm).

12. Position the steering knuckle neck into the strut assembly. Install the strut assembly to steering knuckle bolts. Install the nuts onto the attaching bolts and tighten to 125 ft. lbs. (169 Nm). Do not turn the ser-

rated bolt heads during installation. Turn only the nuts.

13. Install the brake rotor and the caliper assembly to the adapter. Tighten the caliper mounting bolts 14 ft. lbs. (19 Nm).

14. Install the front speed sensor cable routing bracket onto the front strut, if equipped.

15. Install the outer tie rod on steering arm and tighten attaching nut to 27 ft. lbs. (37 Nm).

16. Install the stabilizer link assembly onto the strut assembly and tighten the attaching nut to 70 ft. lbs. (95 Nm).

17. Install the front wheel and lug nuts. Torque the lug nuts, in sequence, to 95–100 ft. lbs. (129–135 Nm).

18. Lower the vehicle to the ground. Reconnect the negative battery cable.

Coil Spring

REMOVAL AND INSTALLATION

NOTE: Service of the coil spring requires the use of a coil spring compressor tool. It is required that 5 coils be captured within the jaws of the compressor tool.

1. Disconnect the negative battery cable.

2. Raise and safely support the vehicle.

3. Remove the appropriate wheel(s).

4. Remove the strut assembly from the vehicle.

5. Securely mount the strut assembly into a vice. Using paint, mark the strut unit, lower spring isolator, spring and upper strut mount for indexing of the parts at assembly.

6. Position the spring compressor tool onto the strut. Compress the coil spring until all load is off the upper strut mount assembly.

7. Install strut rod socket tool L-4558A on the strut shaft nut and a 10 mm socket on the end of the strut shaft to prevent it from turning. Remove the strut shaft nut.

8. Remove the upper mount assembly, jounce bumper and seat bearing and dust shield as an assembly.

9. Remove the coil spring and compressor as an assembly from the strut. Remove the lower spring isolator from the strut assembly lower spring seat.

10. Inspect all components for abnormal wear, oil leakage or failure. Replace parts as required.

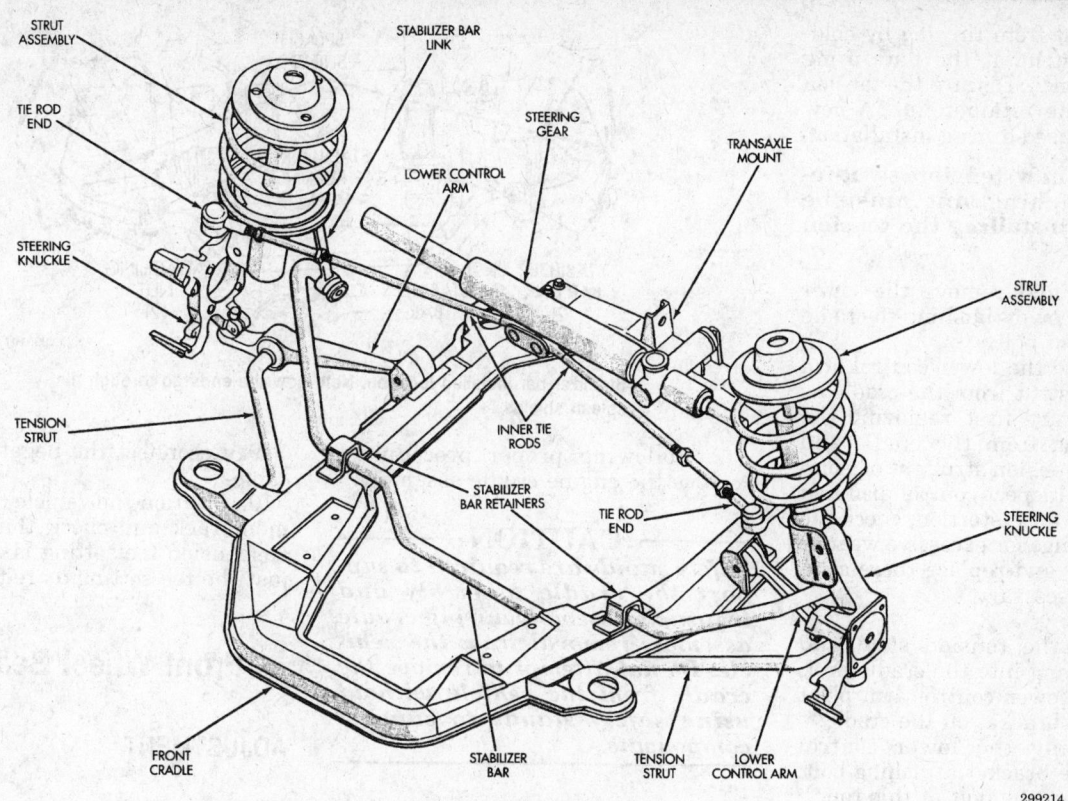

Front suspension assembly

To install:

11. Install the lower spring isolator on the strut unit. Install the compressed coil spring onto the strut assembly aligning the paint marks made during removal.

12. Install the strut bearing into the bearing seat. Bearing must be installed into the seat with notches on the bearings facing down.

13. Lower the seat bearing and dust shield onto the strut and spring assembly. Align the paint marks made during removal.

14. Install jounce bumper and upper mount on the strut shaft aligning the paint marks.

15. Install the strut mount-to-shaft retainer nut. Inspect all alignment marks made during removal and align as required. While holding the strut shaft from turning with a 10 mm socket, tighten the strut shaft nut to 70 ft. lbs. (94 Nm).

16. Equally loosen the spring compressor tool until all tension is released. Remove the spring compressor tool.

17. Install the strut assembly back into the vehicle.

18. Install the wheels and lug nuts. Torque the lug nuts, in sequence, to 95 ft. lbs. (129 Nm).

19. Lower the vehicle. Reconnect the negative battery cable.

Lower Ball Joints

REMOVAL AND INSTALLATION

NOTE: The lower ball joints on this vehicle are not serviced separately. The lower ball joints operate with no free play. If defective, the entire lower control arm must be replaced.

To inspect for ball joint wear:

1. Raise the front of the vehicle using jack stands or a frame contact type hoist until the front suspension is in full rebound and the tires are not in contact with the ground

2. Grasp the tire at the top and bottom and apply an in-and-out motion on the wheel and tire.

3. While applying force to the tire, look for any movement between the lower ball joint and the lower control arm.

4. If any movement is evident the lower ball joint is worn and the lower control arm requires replacement. Do not attempt any type of repair on the ball joint assembly.

5. Although the ball joint is not replaceable (except for replacing the entire lower control arm), the ball joint seal may be replaced. The control arm will have to be removed first. Mount the arm in a suitable press.

Then use an inch and a quarter socket as a driver and press the replacement ball joint seal onto the ball joint housing until it is squarely seated against the top of the control arm.

Lower Control Arm and Tension Strut

REMOVAL AND INSTALLATION

1. Raise and safely support the vehicle.

2. Remove the front wheel(s).

3. Remove the ball joint stud to steering knuckle clamp nut and bolt.

4. Carefully insert a prybar between the lower control arm and the steering knuckle and separate ball joint from knuckle. Make sure ball joint seal does not get damaged during separation.

NOTE: Pulling the steering knuckle out from the vehicle after releasing from the ball joint can separate the inner CV-joint. Do not separate the inner CV-joint or it can be damaged.

5. Remove tension strut to cradle attaching nut and washer from end of tension strut. When removing nut,

keep the strut from turning by holding tension strut at the flats using open end wrench. Discard the tension strut to cradle retainer nut. A new nut must be used during installation.

NOTE: A new tension strut-to-cradle attaching nut must be used when installing the tension strut.

6. Loosen and remove the lower control arm pivot bushing-to-cradle assembly pivot bolt.

7. Separate the lower control arm and tension strut from the cradle as an assembly by first removing the pivot bushing from the cradle and then sliding tension strut out of isolator bushing. Inspect control arm and tension strut for distortion, check the rubber bushings for excessive wear or deterioration and replace these components, if necessary.

To install:

8. Install the tension strut and isolator bushing into the cradle first, then install lower control arm pivot bushing into bracket on the cradle.

9. Installing the lower control arm-to-cradle bracket attaching bolt. Do not tighten the bolt at this time.

10. Install the washer and new nut on end of tension strut. Torque the tension strut-to-cradle bracket retainer nut to 110 ft. lbs. (150 Nm), while holding the tension strut flat with an open end wrench.

11. Inspect the ball joint seal and replace if damaged. Install the lower ball joint stud into the steering knuckle and install the clamp bolt and nut. Tighten the bolt to 40 ft. lbs. (55 Nm).

12. Install the front wheel and lug nuts. Torque the lug nuts, in sequence, to 95–100 ft. lbs. (129–135 Nm). Lower the vehicle so the suspension is supporting the weight of the vehicle.

13. Torque the lower control arm pivot bushing-to-cradle bracket attaching bolt to 90 ft. lbs. (123 Nm).

Sway Bar

REMOVAL AND INSTALLATION

NOTE: To access the sway bar assembly it will be necessary to remove the engine cradle assembly. Make certain that the engine is properly supported, when removing the engine cradle assembly

1. Raise and safely support the vehicle on safety stands or centered on a frame contact type hoist.

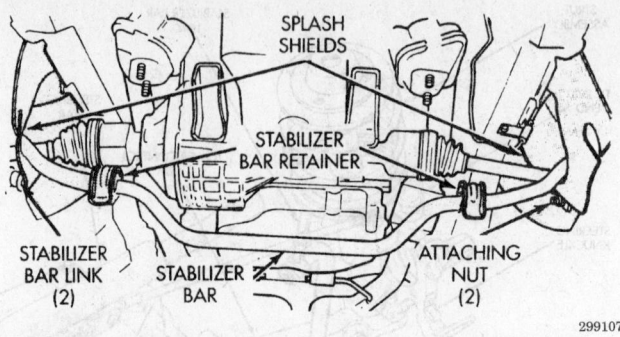

Stablizer bar installed location. Note how the ends go through the splash shields

2. Following proper procedures, remove the engine cradle assembly.

------ CAUTION ------
Safety stands are required to support the cradle assembly and transaxle assembly during cradle assembly removal from the vehicle. Do not attempt to remove the cradle from the vehicle without using safety stands to support components.

3. After the cradle is removed from the vehicle, remove the two stabilizer bar to stabilizer bar link attaching nuts and remove the bar from the vehicle.

To install:

4. Inspect for broken or distorted retainers and bushings. If bushing replacement is required, bushings can be removed by opening the slit in the bushing and removing the bushing from around the stabilizer bar. The stabilizer bar to cradle assembly bushings should be positioned on the bar so the slit in the bushing is positioned toward the front of the vehicle.

5. Install the stabilizer bar, isolator bushings and retainers back into the vehicle as an assembly. Be sure stabilizer bar is installed through the openings in the splash shields. Install the stabilizer bar link to stabilizer bar attaching nut and tighten to 70 ft. lbs. (95 Nm).

6. Tie the stabilizer bar up against the two transaxle to engine block brackets. This will hold bar out of the way during cradle installation.

7. Following the procedure, install the cradle assembly back into the vehicle. Lower the vehicle to the ground. With the full weight of the vehicle supported by the suspension, tighten the four stabilizer bar bushing retainer to cradle assembly attaching bolts to 40 ft. lbs. (55 Nm).

8. Remove the engine support fixture.

9. Reconnect the negative battery cable.

10. Position the vehicle on an alignment rack and check that the front suspension toe setting is correct. Adjust the toe setting as required.

Front Wheel Bearings

ADJUSTMENT

The front hub wheel bearing is designed for the life of the vehicle and requires no type of adjustment or periodic maintenance. The bearing is a sealed unit with the wheel hub and can only be removed and/or replaced as one unit.

REMOVAL AND INSTALLATION

On this vehicle, the front wheel hub and wheel bearing is serviced only as a complete assembly. The front wheel bearings can be neither lubricated or adjusted. The hub and bearings assembly is mounted to the steering knuckle by three mounting bolts that are removed from the rear of the steering knuckle. In addition, a new replacement hub nut must be used.

1. Raise and support the vehicle safely.

2. Remove the front wheel.

3. Remove the front caliper assembly from the steering knuckle.

4. Remove the front brake rotor from the hub by pulling it straight off of the wheel mounting stud.

5. Remove the hub and bearing to stub axle retainer nut. This hub nut is a torque prevailing retaining nut and can not be reused. A NEW retaining nut MUST be used when assembling the hub.

6. Remove the three attaching bolts that mount the hub and bearing

assembly to the steering knuckle assembly.

NOTE: If the metal seal on the hub and bearing assembly is seized to the steering knuckle and becomes dislodged on the hub and bearing during removal, the hub and bearing must be replaced. If the flinger disc becomes damaged during the removal procedure, the hub and bearing assembly must be replaced.

7. Remove the hub and bearing assembly from the steering knuckle by sliding it straight out of the knuckle and off the ends of the stub shaft. Gently pry the assembly out with a pry bar or tap it out with a soft face hammer, if necessary. Be very careful not to damage the hub and bearing assembly.

To install:

8. Clean the hub and bearing mounting surfaces of dirt and make sure there are no nicks present. Install the hub and bearing to the stub shaft and the steering knuckle. Install the the bearing assembly mounting bolts and tighten equally until the bearing assembly is seated squarely against the front of the steering knuckle. At this point, tighten the the mounting bolts to 80 ft. lbs. (110 Nm).

9. Install a new hub and bearing assembly-to-stub shaft retainer nut. A NEW retaining nut MUST be used when assembling the hub. Tighten but do not torque the nut at this time.

10. Install the brake rotor and the caliper assembly. Install the caliper-to-steering knuckle retainer bolts and tighten to 14 ft. lbs. (19 Nm).

11. Install the wheel and lug nuts. Torque the lug nuts in sequence to 95–100 ft. lbs. (129–135 Nm). Lower the vehicle to the ground. Pump the brakes until a firm pedal is obtained.

12. With the weight on the vehicle on its wheels, apply the brakes to keep the vehicle from moving. Tighten the hub and bearing assembly to stub shaft retaining nut to 120 ft. lbs. (163 Nm).

——— WARNING ———
When torquing the hub and bearing assembly to stub shaft retaining nut, do not exceed the maximum torque of 120 ft. lbs. (163 Nm). If the maximum torque is exceeded this may result in a failure of the halfshaft.

13. Inspect the toe setting on the vehicle and adjust, if necessary.

REAR SUSPENSION

Strut

REMOVAL AND INSTALLATION

1. Raise and safely support the vehicle.

2. Remove the rear wheel and tire assembly.

3. If equipped with rear disc brakes, remove the caliper assembly and rotor from the hub. If equipped with rear drum brakes, disconnect the brake flex hose from the support bracket and wheel cylinder. Plug the brake flex hose to prevent system contamination. Do not allow the rear caliper to hang down by the brake hose. Support the caliper off of the frame with a strong piece of wire.

4. If equipped with rear disc brakes, remove the speed sensor cable routing bracket and tube.

5. Remove the bolts attaching the lateral links to the rear spindle assembly.

6. Remove the rear strut assembly to stabilizer bar attaching link at the stabilizer bar. Hold the hex on the attaching link stud while breaking nut loose. The attaching link does not have to be removed from the strut.

7. Remove the rear spindle to strut assembly pinch bolt. Install a center punch in hole on spindle and tap punch into hole until jammed. This will spread spindle casting allowing it to be removed from strut.

8. Using a hammer, tap on the top surface of the spindle, driving spindle down and off the end of the strut assembly. Let the spindle and assembled components hang from the trailing arm while the strut is being serviced.

9. Lower the vehicle. From inside the trunk of the vehicle, remove the three upper strut mounting bolts and remove the strut from the vehicle.

——— CAUTION ———
Do NOT remove the center retaining nut. The strut assembly is held together with this nut and must only be removed following the proper strut disassembly procedure.

10. Inspect all components for abnormal wear, oil leakage or failure. Replace parts as required.

To install:

11. Position the strut in vehicle and install the three upper mounting nuts tightening to 20 ft. lbs. (28 Nm).

12. Install the spindle assembly onto bottom of strut. Push or tap spindle assembly onto strut until notch in spindle is tightly seated against locating tap on strut assembly. Remove the center punch from the hole in the spindle.

13. Install the strut to spindle pinch bolt and tighten to 40 ft. lbs. (55 Nm).

14. Install the lateral link to the spindle attaching bolt and tighten to 105 ft. lbs. (140 Nm).

15. Install the stabilizer bar attaching link onto the stabilizer bar and install stabilizer link to stabilizer bar attaching nut. Tighten attaching nut to 70 ft. lbs. (95 Nm), while holding the stabilizer link stud at hex with wrench.

16. If equipped with rear disc brakes, mount rear speed sensor cable routing tube and bracket in position. Install rear disc brake rotor and caliper assembly to the adapter plate. Tighten caliper mounting bolts to 16 ft. lbs. (22 Nm).

17. If equipped with rear drum brakes, install rear brake flex hose to the wheel cylinder and support plate. The brake system will require bleeding.

18. Install the rear wheel(s) and lug nuts. Torque the lug nuts, in sequence, to 95 ft. lbs. (129 Nm). Lower the vehicle. Bleed the brake system, if equipped with rear drum brakes.

19. Check and reset the rear wheel toe to specifications. Acceptable rear toe range should be 0.2 degrees OUT to 0.4 degrees IN with 0.1 degree IN the preferred setting

Coil Spring

REMOVAL AND INSTALLATION

1. Disconnect the negative battery cable.

2. Raise and safely support the vehicle. Remove the rear wheel(s).

3. Remove the rear strut assembly from the vehicle.

4. Securely mount the strut assembly into a vice. Using paint, mark the strut assembly, lower spring isolator, spring and upper strut mount for indexing of the parts at reassembly.

5. Position a spring compressor tool onto the coil spring. Compress the coil spring until all load is off of the upper strut mount assembly.

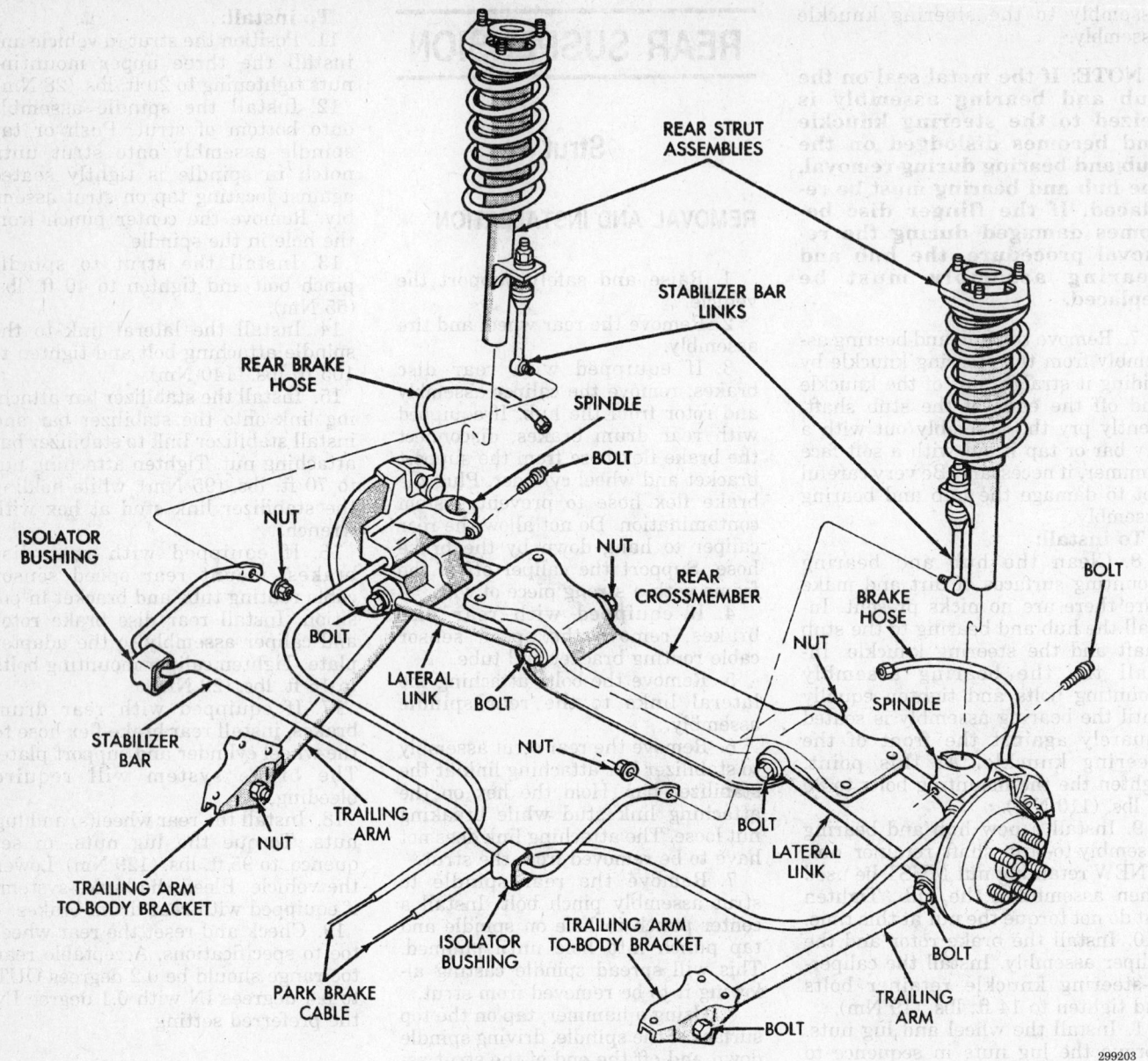

Rear suspension assembly

Labels in diagram:
- REAR STRUT ASSEMBLIES
- STABILIZER BAR LINKS
- REAR BRAKE HOSE
- SPINDLE
- BOLT
- NUT
- NUT
- REAR CROSSMEMBER
- BRAKE HOSE
- NUT
- SPINDLE
- BOLT
- ISOLATOR BUSHING
- NUT
- BOLT
- LATERAL LINK
- BOLT
- NUT
- BOLT
- LATERAL LINK
- STABILIZER BAR
- TRAILING ARM
- NUT
- TRAILING ARM TO-BODY BRACKET
- ISOLATOR BUSHING
- TRAILING ARM TO-BODY BRACKET
- TRAILING ARM
- PARK BRAKE CABLE
- BOLT
- BOLT
- 299203

6. Install strut rod socket tool L-4558 on the strut shaft nut and an 8 mm Allen® wrench on the end of the strut shaft to prevent it from turning. Remove the strut shaft nut.

7. Remove the upper strut mount assembly off of the strut shaft. Remove the coil spring and compressor tool as an assembly from the strut.

8. Remove the plate, dust shield and jounce bumper off of the strut unit.

9. Inspect all components for abnormal wear, oil leakage or failure. Replace parts as required.

To install:

10. Install the lower spring isolator on the strut unit. if it is the original isolator, align the paint marks.

11. Install the jounce bumper into the dust shield. Install the plate on top of the dust shield and into jounce bumper.

12. Install the dust shield, jounce bumper and the top plate onto the strut unit as an assembly.

13. Install the coil spring and compressor tool onto the strut unit and align the paint marks on the spring to that of the strut unit.

14. Install the upper strut mount assembly onto the strut shaft. Align the paint marks and install the strut shaft retaining nut.

15. Using the strut rod socket tool L-4558 and the 8 mm Allen® wrench to prevent the strut shaft from turning, torque the strut shaft nut to 70 ft. lbs. (95 Nm).

16. Equally loosen the spring compressor tool until all tension is released. Remove the spring compressor tool.

17. Install the strut assembly back into the vehicle.

18. Install the rear wheel(s) and lug nuts. Torque the lug nuts, in sequence, to 95 ft. lbs. (129 Nm).

19. Lower the vehicle. Reconnect the negative battery cable.

20. Check and adjust the rear wheel toe to specifications, if necessary.

Lower Control Arms

REMOVAL AND INSTALLATION

NOTE: Only frame contact or wheel lift hoisting equipment can be used on vehicles having a fully independent rear suspension. Vehicles with independent rear suspension can not be hoisted using equipment designed to lift a vehicle by the rear axle. If this type of hoisting equipment is used, damage to the rear suspension components will occur.

Left Side

1. Raise and safely support the vehicle. Remove the left rear wheel.
2. Remove the nut and bolt attaching the left lateral links to the spindle.
3. Remove the nut and bolt attaching the left lateral link to the crossmember and remove the link from the vehicle.

To install:
4. Install the solid lateral link on the front of the crossmember.
5. Install the adjustable lateral link on the rear of the crossmember with its adjusting link closer to the spindle.
6. Install the link attaching bolts at the spindle and crossmember. The bolt heads must face the front of the vehicle. Torque the link attaching bolts to 105 ft. lbs. (140 Nm).
7. Install the left rear wheel and lug nuts. Torque the lug nuts, in sequence, to 95 ft. lbs. (129 Nm).
8. Lower the vehicle to the ground.
9. Check and reset rear wheel toe to specifications.

Right Side

1. Raise and safely support the vehicle. Remove the right rear wheel.
2. Remove the nut and bolt attaching the right lateral links to the spindle.
3. Position a transmission jack under the fuel tank just forward of the crossmember to help support the fuel tank when crossmember is lowered.
4. Remove the four crossmember to frame rail attaching bolts and lower the crossmember far enough so right lateral link to crossmember attaching bolt will clear the fuel tank.
5. Remove the nut and bolt attaching the right lateral links to the rear crossmember. Remove the lateral links.

To install:
6. Install replacement lateral links to crossmember as follows:
   a. Solid lateral link is installed on crossmember toward the front of the vehicle.
   b. The adjustable lateral link is installed on crossmember toward rear of vehicle, with the adjustable link positioned toward the spindle.
7. Install the right lateral links attaching bolts at the crossmember with the heads of the bolts facing toward the front of the vehicle.
8. Position the crossmember on the frame rail and install the four attaching bolts. Tighten attaching bolts to 70 ft. lbs. (95 Nm). Remove the transmission jack from under fuel tank.
9. Torque the lateral link to crossmember attaching bolt to 105 ft. lbs. (140 Nm).
10. Align the lateral link with the spindle and install the lateral link to spindle attaching bolts. Tighten the bolts to 105 ft. lbs. (140 Nm).
11. Install the right rear wheel and lug nuts. Torque the lug nuts, in sequence, to 95 ft. lbs. (129 Nm).
12. Lower the vehicle and check the toe setting of the rear wheels. Adjust the toe to specifications as required.

Sway Bar

REMOVAL AND INSTALLATION

———— **CAUTION** ————
Fuel injection systems remain under pressure, even after the engine has been turned OFF. The fuel system pressure must be relieved before disconnecting any fuel lines. Failure to do so may result in fire and/or personal injury.

1. Disconnect the negative battery cable. Relieve the fuel system pressure.
2. Raise and safely support the vehicle.
3. Remove both rear wheels.
4. Position a transmission jack under the fuel tank just forward of the crossmember to help support the fuel tank when crossmember is removed.
5. Remove the four crossmember to frame rail attaching bolts. Remove the fuel tank.
6. Remove the stabilizer bar to link assembly attaching nuts and remove the bar and isolator bushings as an assembly from the vehicle. In-

spect the isolator bushings for damage or excessive wear and replace, if necessary.

To install:
7. Inspect for broken or distorted retainers and bushings. If bushing replacement is required, replacement bushings can be installed by locating the split in the bushing, prying open and removing the defective bushing from around the stabilizer bar.
8. Install the stabilizer bar and isolator bushings back into the vehicle as an assembly making sure bar is centered in vehicle so it doesn't contact other suspension components.
9. Install stabilizer bar attaching link onto stabilizer bar. Install new link to bar attaching nuts and tighten to 70 ft. lbs. (95 Nm).

NOTE: Replace the sway bar bracket bolt with new after loosening or removing them. Only use original equipment bolts as replacements.

10. Install the fuel tank back onto the vehicle.
11. Position the crossmember on frame rails and install four mounting bolts. Torque attaching bolts to 70 ft. lbs. (95 Nm).
12. Remove transmission jack from under the fuel tank.
13. Install the rear wheels and lug nuts. Torque the lug nuts, in a star pattern sequence, to 95–100 ft. lbs. (129–135 Nm).
14. Lower the vehicle. Reconnect the negative battery cable.
15. Pressurize the fuel system and check for leaks.
16. Check and reset the rear wheel toe to specifications as required.

Wheel Bearings

ADJUSTMENT

The rear hub and wheel bearing assembly is designed for the life of the vehicle and requires no type of adjustment or periodic maintenance. The bearing is a sealed unit with the wheel hub and can only be removed and/or replaced as one unit.

The following procedure may be used for evaluation of bearing condition.
1. Raise and safely support the vehicle.
2. Remove the rear wheels and brake drums.
3. Turn the hub flange carefully. Excessive roughness, lateral play or resistance to rotation may indicate dirt intrusion or bearing failure.

4. If the rear wheel bearings exhibit the conditions during inspection, the hub and bearing assembly should be replaced.

5. Damaged nearing seals and resulting excessive grease loss may also require bearing replacement. Moderate grease loss from the bearing is considered normal and should not require replacement of the hub and bearing assembly.

REMOVAL AND INSTALLATION

The rear wheel bearing/hub assembly is permanently lubricated and sealed at the factory. No periodic lubrication or maintenance is required. The bearing is a sealed unit with the wheel hub and can only be removed and/or replaced as one unit.

1. Raise the vehicle and support safely. Remove the rear wheel.

2. Remove the brake caliper and rotor if equipped with rear disc brakes. Remove the brake drum if equipped with drum brakes.

3. Remove the bearing dust cap using a suitable prybar.

4. Remove the cotter pin, nut retainer, nut, washer and bearing/hub assembly from the spindle.

To install:

5. Install the bearing/hub assembly. Install the bearing/hub assembly washer and retaining nut. Torque the nut to 124 ft. lbs. (168 Nm). Install the nut retainer and new cotter pin. Bend over the cotter pin.

6. Install the dust cap.

7. Install the brake drum or rotor and caliper assembly.

8. Install the rear wheel and torque the lug nuts, in sequence, to 95–100 ft. lbs. (129–135 Nm). Lower the vehicle safely.

9. Road test vehicle to verify no excessive noise from rear wheel bearing area.

FIRING ORDERS

NOTE: To avoid confusion, always replace spark plug wires one at a time.

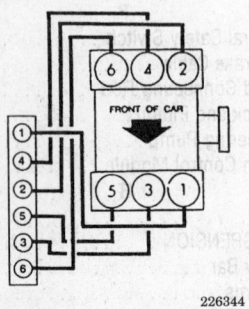

3.0L (VIN J and K) DOHC
Engines
Engine Firing Order:
1-2-3-4-5-6
Distributorless Ignition
System

226344

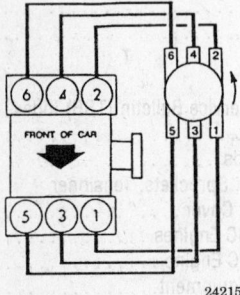

242158

3.0L (VIN H) SOHC Engine
Engine Firing Order:
1-2-3-4-5-6
Distributor Rotation:
Counterclockwise

ENGINE ELECTRICAL

NOTE: Disconnecting the negative battery cable on some vehicles may interfere with the functions of the on board computer systems and may require the computer to undergo a relearning process, once the negative battery cable is reconnected.

Distributor

REMOVAL

1. Disconnect the negative battery cable.

— CAUTION —
Wait at least 90 seconds after the negative (-) battery cable is disconnected to prevent possible deployment of the air bag.

2. Disconnect the distributor harness electrical connectors.
3. Unscrew the distributor cap hold-down screws or release the clips and lift off the distributor cap with all ignition wires connected. Remove the coil wire, if necessary.
4. Matchmark the rotor to the distributor housing and the distributor housing to the engine.

NOTE: Do not crank the engine during this procedure. If the engine is cranked, the matchmark must be disregarded.

5. Remove the hold-down nut.
6. Carefully remove the distributor from the engine.

INSTALLATION

NOTE: Some engines may be sensitive to the routing of the distributor sensor wires. If routed near the high-voltage coil wire or the spark plug wires, the electromagnetic field surrounding the high voltage wires could generate an occasional disruption of the ignition system operation.

Timing Not Disturbed

1. Install a new distributor housing O-ring and lubricate the distributor drive gear and O-ring with clean oil.
2. Install the distributor in the engine so the rotor is aligned with the matchmark on the housing and the housing is aligned with the matchmark on the engine. Make sure the distributor is fully seated and the distributor shaft is fully engaged.
3. Install the hold-down.
4. Connect the distributor harness connectors.
5. Make sure the sealing O-ring is in place, install the distributor cap and tighten the screws or secure the clips.
6. Connect the negative battery cable.
7. Adjust the ignition timing and tighten the hold-down nut to 11 ft. lbs. (14 Nm).

Timing Disturbed

1. Install a new distributor housing O-ring and lubricate with clean oil.
2. Position the engine so the No. 1 piston is at TDC of its compression stroke and the mark on the vibration damper is aligned with **0** on the timing indicator.
3. Align the distributor housing and gear mating marks. Install the distributor in engine so the slot or groove of the distributor's installation flange aligns with the distributor installation stud in the engine block. Make sure the distributor is fully seated. Inspect alignment of the distributor rotor making sure the rotor is aligned with the position of the No. 1 ignition wire in the distributor cap.

NOTE: Make sure the rotor is pointing to where the No. 1 terminal originates inside the cap, and not where the No. 1 ignition wire plugs into the cap.

4. Install the hold-down nut.
5. Connect the distributor harness connectors.
6. Make sure the sealing O-ring is in place, install the distributor cap and tighten the screws or secure the clips.
7. Connect the negative battery cable.
8. Adjust the ignition timing and tighten the hold-down bolt to 11 ft. lbs. (14 Nm).

Ignition Timing

ADJUSTMENT

1. Set the parking brake, start and run the engine until normal operating temperature is obtained. Keep all lights and accessories OFF and the front wheels straight-ahead. Place the transaxle in **P** for automatic transaxle or neutral for manual transaxle.
2. Connect a tachometer to the single pin connector terminal under the hood. Check the idle speed and if not at specification, set the idle speed to the correct level.

NOTE: The tachometer reading will be ⅓ of the actual engine speed. Multiply the reading by 3 to figure the actual engine speed.

3. Turn the engine **OFF**. Remove the water-proof cover from the ignition timing inspecting connector. This connector is located on the firewall just behind the battery.

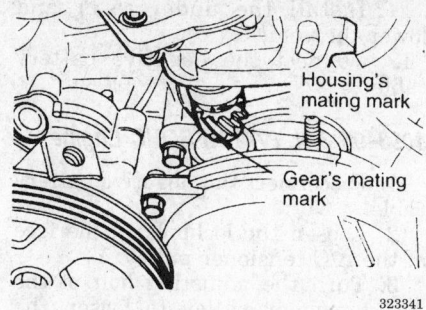

Aligning the distributor housing and gear mating marks

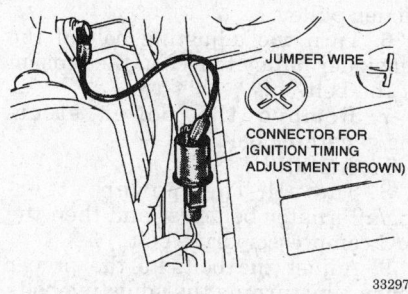

Ignition timing connector

4. Connect a conventional powered timing light to the No. 1 cylinder spark plug wire. Start the engine and run at idle.

5. Aim the timing light at the timing scale located near the crankshaft pulley. If timing marks do not align, the ignition timing is not within specification.

NOTE: The ignition timing is not adjustable; if timing is not within specification, check the Engine Control System operation and perform the required service.

6. Remove the jumper wire from the ignition timing inspection and terminal and install the water-proof cover.

7. Start the engine and check the actual timing (the timing without the terminal grounded). This reading should be approximately 5 degrees more than the basic timing. Actual timing may increase according to altitude. Also, actual timing may fluctuate because of slight variation accomplished by the ECU. As long as the basic timing is correct, the engine is timed correctly.

8. Turn the engine **OFF**. Disconnect the timing apparatus and tachometer.

Alternator

PRECAUTIONS

Several precautions must be observed with alternator equipped vehicles to avoid damage to the unit.

• If the battery is removed for any reason, make sure it is reconnected with the correct polarity. Reversing the battery connections may result in damage to the 1-way rectifiers.

• When utilizing a booster battery as a starting aid, always connect the positive to positive terminals and the negative terminal from the booster battery to a good engine ground on the vehicle being started.

• Never use a fast charger as a booster to start vehicles.

• Disconnect the battery cables when charging the battery with a fast charger.

• Never attempt to polarize the alternator.

• Do not use test lights of more than 12 volts when checking diode continuity.

• Do not short across or ground any of the alternator terminals.

• The polarity of the battery, alternator and regulator must be matched and considered before making any electrical connections within the system.

• Never separate the alternator on an open circuit. Make sure all connections within the circuit are clean and tight.

• Disconnect the battery ground terminal when performing any service on electrical components.

• Disconnect the battery if arc welding is to be done on the vehicle.

REMOVAL AND INSTALLATION

3.0L (VIN J and K) DOHC Engines

1. Disconnect the negative battery cable.

— **CAUTION** —
Wait at least 90 seconds after the negative (-) battery cable is disconnected to prevent possible deployment of the air bag.

2. On turbocharged models, remove the air intake pipe and hoses.

3. Remove the radiator surge tank.

4. If equipped with air conditioning, remove the clamp nut that secures the air conditioning hose.

5. Raise the air conditioning suction hose and suspend it from the engine hood using a cord.

6. Loosen the tensioner pulley and remove the alternator drive belt.

7. Disconnect the oxygen sensor connector.

8. Disconnect the alternator wiring harness.

9. Remove the alternator bracket to engine block mounting bolts and remove the bracket and alternator as an assembly.

10. Separate the alternator from the mounting bracket on a workbench.

To install:

11. Install the alternator onto the bracket and install bracket assembly to the engine.

12. Connect the oxygen sensor connector.

13. Connect wiring harness to the rear of the alternator.

14. Install the drive belt and adjust to proper tension using the tensioner pulley.

15. Install air conditioning suction hose to its original position and secure using clamp nut.

16. Install the radiator surge tank.

17. Install the air intake delivery hoses and air pipe.

18. Reconnect the negative battery cable and check the charging system for proper operation.

3.0L (VIN H) SOHC Engine

1. Disconnect the negative battery cable.

— **CAUTION** —
Wait at least 90 seconds after the negative (-) battery cable is disconnected to prevent possible deployment of the air bag.

2. Loosen the tensioner pulley and remove the alternator drive belt.

3. Remove the accelerator cable from the intake plenum extension.

4. Remove the brake booster vacuum hose.

5. On California models equipped with an EGR valve, unbolt the valve and remove it.

6. Disconnect the alternator wiring harness connectors.

7. Remove the alternator upper and lower mounting bolts. Remove the alternator from behind the surge tank at the center of the vehicle.

To install:

8. Position the alternator on the lower mounting fixture and install and tighten the bolts.

9. Install the EGR valve using a new gasket, if removed. Connect the vacuum hose connection at the brake booster.

10. Install the accelerator cable to the intake plenum extension. Check the accelerator cable adjustment as follows:

a. Turn the ignition key **ON** but do not start the engine. With the ignition left in this condition wait 15 seconds.

b. Check to insure that the throttle lever is in contact with the fixed Speed Adjusting Screw (SAS).

c. Check that the inner cable play is within specifications. For manual transaxle, the desired value is 0.04–0.08 in. (1–2mm). If equipped with automatic transaxle, the desired value is 0.12–0.20 in. (3–5mm).

d. If not within the desired value, loosen the adjusting bolts and slide plate so play at the inner cable will fall within the desired value. Retighten the adjusting bolts.

11. Reinstall the drive belt and adjust the tensioner until the proper belt tension is achieved.

12. Connect the negative battery cable and check the charging system for proper operation.

Drive Belt

REMOVAL AND INSTALLATION

3.0L (VIN J and K) DOHC and 1996–97 3.0L (VIN H) SOHC Engines

Alternator and A/C Compressor Belt

1. Disconnect the negative battery cable.

— CAUTION —
Wait at least 90 seconds after the negative (-) battery cable is disconnected to prevent possible deployment of the air bag.

2. Raise and safely support the vehicle and remove the front undercover.

3. Loosen the tension pulley fixing nut and relieve the tension on the belt by turning the adjusting bolt.

4. Remove the belt.

To install:

5. Install the belt on the crankshaft and alternator pulleys.

6. Using the adjusting bolt on the tensioner, tighten the belt to the desired tension.

7. Tighten the fixing nut to hold the adjustment.

8. Install the undercover and lower the vehicle to the floor.

9. Connect the negative battery cable.

Power Steering Belt

1. Disconnect the negative battery cable.

— CAUTION —
Wait at least 90 seconds after the negative (-) battery cable is disconnected to prevent possible deployment of the air bag.

2. Raise and safely support the vehicle and remove the undercover.

3. Remove the alternator and A/C compressor belt.

4. Lower the vehicle and remove the cruise control pump link assembly.

5. Place the power steering hose under the oil reservoir.

6. Loosen the tension pulley fixing bolts and remove the power steering pump drive belt.

To install:

7. Install the power steering pump drive belt.

8. Insert an extension bar or equivalent into the opening at the end of the tension pulley bracket and pivot the pulley to apply tension to the belt.

9. Tighten the fixing bolts.

10. Raise the vehicle and install the alternator and A/C compressor belt.

<SOHC>

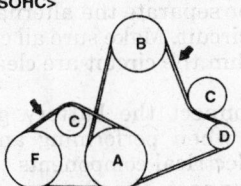

<DOHC without air conditioning>

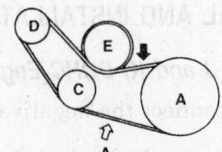

<DOHC with air conditioning>

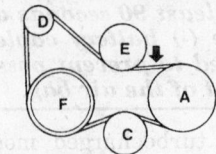

A: Crankshaft pulley
B: Power steering pump pulley
C: Tension pulley
D: Generator pulley
E: Idler pulley
F: Air conditioning compressor pulley

323658

Drive belts — 3.0L (VIN J and K) DOHC and 1996–97 3.0L (VIN H) SOHC engines

11. Install the under cover and lower the vehicle.

12. Connect the negative battery cable.

1993–95 3.0L (VIN H) SOHC Engine

1. Disconnect the negative battery cable.

2. Loosen the lockbolt on the face of the A/C tensioner pulley.

3. Turn the adjusting bolt of the A/C tensioner pulley to loosen the tension of the A/C belt.

4. Remove the A/C compressor belt.

5. Loosen the locknut on the face of the power steering/generator tensioner pulley.

6. Turn the adjusting bolt of the tensioner pulley to loosen the tension of the belt.

7. Remove the power steering/generator belt.

To install:

8. Install the power steering/alternator belt first and then the A/C compressor drive belt.

9. Adjust the belts to the proper tension by turning the adjusting bolts and tighten pulley fixing nut/bolt.

10. Tighten the mounting nut of the power steering/generator tensioner pulley to 36 ft. lbs. (50 Nm).

NOTE: The manufacturer does not provide a torque specification for the bolt that secures A/C tensioner pulley.

11. Connect the negative battery cable.

Starter

REMOVAL AND INSTALLATION

1. Disconnect the negative battery cable.

— CAUTION —
Wait at least 90 seconds after the negative (-) battery cable is disconnected to prevent possible deployment of the air bag.

2. Raise the vehicle and support safely.

3. Remove the engine undercover.

4. Disconnect the wiring from the starter.

5. Remove the mounting bolts and remove the starter from the vehicle.

To install:

6. Position the starter and install the mounting bolts.

7. Tighten starter mounting bolts to 20–25 ft. lbs. (27–34 Nm.)

Items		Check value	Adjustment value	
			New belt	Used belt
For generator and P/S pump	Tension N (lbs.)	350–600 (77–132)	700–900 (155–198)	450–600 (99–132)
	Deflection mm (in.) (Reference value)	6.0–9.0 (.24–.35)	4.0–5.0 (.16–.20)	6.0–8.0 (.24–.32)
For A/C compressor	Tension N (lbs.)	250–500 (55–110)	500–600 (110–132)	320–400 (70–88)
	Deflection mm (in.) (Reference value)	7.5–9.5 (.28–.37)	6.5–7.0 (.26–.28)	7.5–8.5 (.28–.34)

323201

Drive belt tension adjustments — 1993–95 3.0L (VIN H) SOHC engine

8. Connect the wiring for the starter.

9. Connect the negative battery cable and check the starter for proper operation.

CHASSIS ELECTRICAL

Blower Motor

REMOVAL AND INSTALLATION

1. Disconnect battery negative cable.

2. Remove the instrument panel under cover.

3. Remove the glove box and the glove box outer case assembly.

4. Remove the molded hose from the blower assembly.

5. Remove the blower motor assembly.

6. Remove the packing seal.

7. Remove the fan retaining nut and fan in order to replace the motor.

To install:

8. Check that the blower motor shaft is not bent and that the packing (sealing material) is in good condition. Clean all parts of dust, etc.

9. Assemble the motor and fan. Install the blower motor then connect the connector.

10. Install the molded hose. Install the duct or under cover.

11. Install the glove box outer case and the glove box.

12. Connect the negative battery cable and check the entire climate control system for proper operation.

Windshield Wiper Motor

REMOVAL AND INSTALLATION

Front

1. Disconnect the negative battery cable.

——— **CAUTION** ———

Wait at least 90 seconds after the negative (-) battery cable is disconnected to prevent possible deployment of the air bag.

2. Remove the windshield wiper arms by unscrewing the cap nuts and lifting the arms from the linkage posts.

3. Remove the front deck garnish assembly.

4. Remove the air inlet cover.

5. Remove the access hole cover.

6. Remove the wiper motor mounting bolts.

7. Detach the motor crank arm from the wiper linkage and remove the motor.

NOTE: The installation angle of the crank arm and motor has been factory set. Do not remove them unless necessary. If they must be removed, remove them only after marking their mounting positions.

To install:

8. Install the windshield wiper motor and connect the linkage.

9. Install the access hole cover.

10. Reinstall the wiper blades. Note that the driver's side wiper arm should be marked **D** and the passenger's side wiper arm should be marked **A**. The identification marks should be located at the base of the arm, near the pivot. Install the arms so the blades are parallel to the garnish molding when parked.

11. Connect the negative battery cable and check the wiper system for proper operation.

Rear

1. Disconnect the negative battery cable.

——— **CAUTION** ———

Wait at least 90 seconds after the negative (-) battery cable is disconnected to prevent possible deployment of the air bag.

2. Remove the liftgate lower trim. Remove the clips that hold the trim by using the following procedure:

a. Remove the clip by pressing down on the center pin with a blunt pointed tool. Press down a little more than $1/16$ in. (2mm). This releases the clip. Pull the clip outward to remove it.

b. Do not push the pin inward more than necessary because it may damage the grommet or if pushed too far, the pin may fall in. Once the clips are removed, use a plastic trim stick to pry the trim cover loose.

3. Remove the rear spoiler, center brace and center brake light.

4. Lift the small cover, remove the retaining nut and remove the wiper arm and spacer.

5. Remove the mounting bolts and remove the wiper motor.

To install:

6. Install the motor and install the retaining bolts.

7. Install the spacer, wiper arm and retaining nut. The arm should be positioned so the upper tip points to the upper left corner of the rear window when parked. Connect the battery and check the operation of the motor before proceeding. If satisfactory, disconnect the cable and proceed.

8. Install the rear spoiler and related parts.

9. Install the interior trim piece.

10. Connect the negative battery cable and recheck the system for proper operation.

Combination Switch

REMOVAL AND INSTALLATION

NOTE: The headlights, turn signals and dimmer switch are all built into one multi-function combination switch that is mounted on the left side of the steering column. The wiper and washer combination switch is mounted on the right side of the steering column.

1. Disconnect the negative battery cable.
2. If equipped with an air bag, disarm as follows:
   a. Position the front wheels in the straight-ahead position and place the key in the **LOCK** position. Remove the key from the ignition lock cylinder.
   b. Disconnect the negative battery cable and insulate the cable end with high-quality electrical tape or similar non-conductive wrapping.
   c. Wait at least one minute before working on the vehicle. The air bag system is designed to retain enough voltage to deploy the air bag for a short period of time even after the battery has been disconnected.

NOTE: If equipped with an air bag, be sure to disarm it before entering the vehicle. Failure to do so could result in personal injury.

3. Remove the steering wheel as follows:
   a. Remove the air bag module mounting nut from behind the steering wheel.
   b. To disconnect the connector of the clockspring from the air bag module, press the air bag's lock toward the module to spread the lock open. While holding lock in this position, use a small tipped prying tool to gently pry the connector from the module.
   c. Store the air bag module in a clean, dry place with the pad cover facing up.
   d. Remove the steering wheel retaining nut and use a steering wheel puller to remove the wheel. Do not use a hammer or the collapsible mechanism in the column could be damaged.
4. Remove the hood lock release handle.
5. Remove the switches from the knee protector below the steering column and remove the exposed retain-

ing screws. Then remove the knee protector.
6. Remove the steering column upper and lower covers.
7. Remove lap cooler and foot blower duct work as necessary. Gently disconnect the combination switch connectors.
8. Remove the retaining screws from the combination switch and remove the combination switch assembly from the steering column.
To install:
9. Install the switch to the steering column and connect the harness connections.
10. Install any removed duct work.
11. Install the column covers.
12. Install the knee protector and switches.
13. Install the hood release handle.
14. Confirm that the front wheels are in a straight-ahead position. Center the clockspring by aligning the **NEUTRAL** mark on the clockspring with the mating mark on the casing. Then install the steering wheel and torque the retaining nut to 29 ft. lbs. (40 Nm).
15. Attach air bag module wiring connector to clockspring connection. Install air bag module and tighten mounting nuts to 43 inch lbs. (5 Nm).
16. Connect the negative battery cable and check all functions of the combination switch and SRS warning light for proper operation.

Ignition Lock Cylinder

REMOVAL AND INSTALLATION

───────── **CAUTION** ─────────
The Supplemental Inflatable Restraint (SIR) system must be disarmed before working around the steering column. Failure to do so may cause accidental deployment of the of the air bag, resulting in unnecessary SIR system repairs and/or personal injury.

1. Place the gear selector into the PARK position.
2. Disconnect the negative battery cable. Disarm the air bag system, if equipped.
3. If equipped with tilt steering, remove the tilt lever by turning it counterclockwise.
4. Remove the upper and lower steering column covers.
5. Insert the key and turn to the lock position.
6. Using a small prybar, depress the key cylinder retaining pin until it is flush with the key cylinder surface.

7. Turn the key clockwise to the OFF position. the key cylinder will unseat approximately 1/8 inch from the ignition switch.

NOTE: Do not attempt to remove the key cylinder at this time.

8. Turn the key back to the LOCK position then remove the key.
9. Remove the key cylinder from the ignition switch.
To install:
10. Install the key cylinder into the ignition switch.
11. Install the key into the key cylinder.
12. While gently pushing inward rotate the key clockwise until it reaches the end of its travel.
13. Install the upper and lower steering column covers.
14. Reconnect the negative battery cable.
15. Check for proper operation.

Ignition Switch

REMOVAL AND INSTALLATION

1. Disconnect the negative battery cable.

───────── **CAUTION** ─────────
Wait at least 90 seconds after the negative (-) battery cable is disconnected to prevent possible deployment of the air bag.

2. Remove the steering column upper and lower covers. Use care removing covers to prevent breakage of alignment tabs.
3. Remove the knee protector.
4. Disconnect the wiring connector from the combination switch.
5. Remove the ignition switch.
To install:
6. Install the ignition switch into the interlock housing.
7. Connect the harness connections.
8. Install the knee protector.
9. Install the upper and lower steering column covers.
10. Connect the negative battery cable and check all functions of column-mounted switches and the ignition switch for proper operation.

Park/Neutral Safety Switch

REMOVAL AND INSTALLATION

1. Disconnect the negative battery cable.

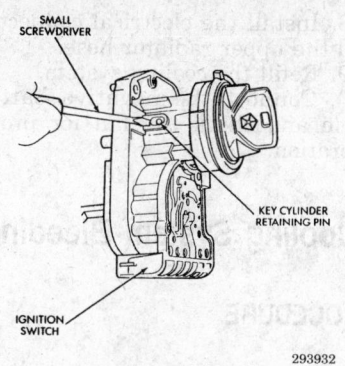

Key cylinder retaining pin

—————— **CAUTION** ——————

Wait at least 90 seconds after the negative (-) battery cable is disconnected to prevent possible deployment of the air bag.

2. Disconnect the selector cable from the lever.

3. Remove the 2 retaining screws and lift off the switch.

To install:

4. Install the lever, tighten the bolts only hand tight.

5. Rotate switch body so the manual control lever 0.20 inch (5mm) hole and the switch body 0.20 inch (5mm) holes are aligned.

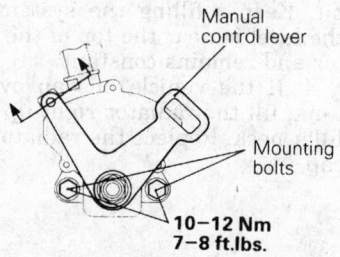

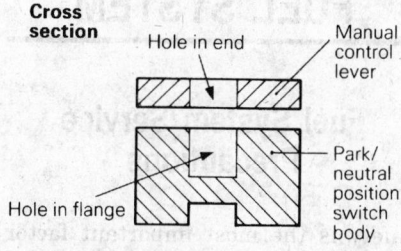

Park/Neutral switch adjustment and alignment points

6. Tighten the mounting bolts to 7–8 ft. lbs. (10–12 Nm).

7. Connect the selector cable to the lever.

8. Connect the negative battery cable.

9. After installation and adjustment make sure the engine only starts in the **P** and **N** selections. Also check that the reverse lights operate only in the **R** selection.

Powertrain Control Module

REMOVAL AND INSTALLATION

1. Disconnect the negative battery cable and wait at least 1 minute for the air bag to disarm.

2. Remove the center console left side cover to access the electronic control unit.

3. Disconnect the electrical connectors to the ECU and remove the mounting bolts.

4. Remove the ECU.

To install:

5. Position the ECU and install the mounting bolts.

6. Connect the electrical connectors to the ECU.

7. Install the side cover to the console.

8. Connect the negative battery cable.

ENGINE COOLING

Radiator

REMOVAL AND INSTALLATION

1. Disconnect the negative battery cable.

—————— **CAUTION** ——————

Wait at least 90 seconds after the negative (-) battery cable is disconnected to prevent possible deployment of the air bag.

—————— **CAUTION** ——————

Allow the cooling system to completely cool before attempting any repair or draining the system. Injury from scalding could result if radiator cap or hose connections are removed while system is hot.

2. Drain the cooling system.

3. Disconnect the overflow tube. Some vehicles may also require removal of the overflow tank.

4. Disconnect the upper and lower radiator hoses.

5. Disconnect the electrical connectors for the cooling fan and air conditioning condenser fan, if equipped. Remove the fan assembly.

6. Disconnect thermo sensor wires.

7. Disconnect and plug the automatic transaxle cooler lines, if equipped with automatic transaxle.

8. Remove the upper radiator mounts and lift out the radiator assembly.

9. Service the lower mounts, as required.

To install:

10. Install the radiator and fan assembly, if removed as an assembly.

11. Connect the automatic transaxle cooler lines, if disconnected.

12. Connect the thermo wires.

13. Install the fan if removed separately.

14. Install the radiator hoses.

15. Install the air cleaner support bracket, if removed.

16. Install the overflow tube and reservoir, if removed.

17. Fill the system with coolant.

18. Connect the negative battery cable, run the vehicle until the thermostat opens, fill the radiator completely and check the automatic transaxle fluid level, if equipped.

19. Once the vehicle has cooled, recheck the coolant level.

Water Pump

REMOVAL AND INSTALLATION

1. Disconnect the negative battery cable.

2. Drain the cooling system.

3. Remove the engine undercover.

4. Disconnect the clamp bolt from the power steering hose.

5. Support the engine with the appropriate equipment and remove the engine mount bracket.

6. Remove the timing belt(s) from the front of the engine.

7. Disconnect the coolant hoses from the pump, if equipped.

8. Remove the alternator brace.

9. Remove the water pump, gasket and O-ring where the water inlet pipe(s) joins the pump.

To install:

10. Thoroughly clean and dry both gasket surfaces of the water pump and block.

11. Install a new O-ring into the groove on the front end of the water inlet pipe. Do not apply oils or grease to the O-ring. Wet with water only.

12. Install the gasket and pump assembly and tighten the bolts to 17 ft. lbs. (24 Nm).

13. Connect the hoses to the pump.

14. Reinstall the timing belt and related parts.

15. Install the engine drive belts and adjust.

16. Fill the system with coolant.

17. Connect the negative battery cable, run the vehicle until the thermostat opens and fill the radiator completely.

18. Once the vehicle has cooled, recheck the coolant level.

Thermostat

REMOVAL AND INSTALLATION

1. Disconnect the negative battery cable.

— **CAUTION** —

Work must be started after 90 seconds from the time the ignition switch is turned to the LOCK position and the negative (-) battery cable is disconnected.

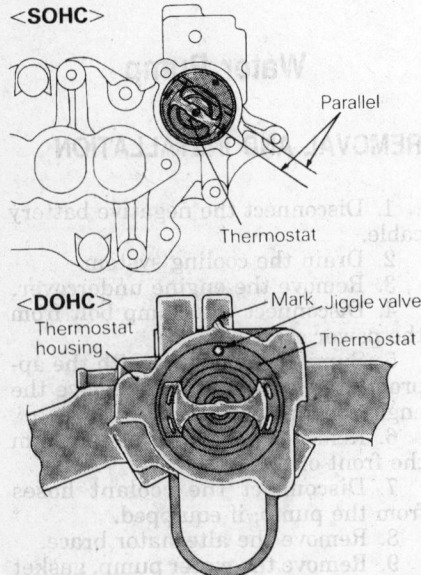

Thermostat mounting

2. Drain the cooling system.

3. Remove necessary air intake plumbing.

4. Disconnect the upper radiator hose and overflow hose from the thermostat housing.

5. Remove the thermostat housing and gasket.

6. Remove the thermostat taking note of its original position in the housing or intake manifold.

To install:

7. Install the thermostat so its flange seats tightly in the machined groove in the intake manifold or thermostat case. Refer to its location prior to removal. Align the jiggle valve with the alignment mark on the thermostat housing.

8. Use a new gasket and reinstall the thermostat housing. Torque the housing mounting bolts to 12–14 ft. lbs. (17–20 Nm).

9. Fill the system with coolant.

10. Install the removed air intake plumbing.

11. Connect the negative battery cable, run the vehicle until the thermostat opens and fill the radiator completely.

12. Once the vehicle has cooled, recheck the coolant level.

Electric Cooling Fan

REMOVAL AND INSTALLATION

1. Disconnect the negative battery cable.

— **CAUTION** —

Wait at least 90 seconds after the negative (-) battery cable is disconnected to prevent possible deployment of the air bag.

2. Drain the cooling system only when the radiator and the engine are at safe temperatures.

3. Unplug the cooling fan and radiator sensor connector(s). Most of these connectors employ a waterproof connector. When disconnecting, make sure all parts of the connector remain intact.

4. Disconnect the upper radiator hose from the radiator and remove overflow tank.

5. Remove the fan mounting screws. The radiator and condenser cooling fans are separately removable.

6. Remove the fan assembly and disassemble as required.

To install:

7. Position the fan and install the mounting screws.

8. Install the electrical connectors and the upper radiator hose.

9. Refill the cooling system.

10. Connect the negative battery cable and check the fan for proper operation.

Cooling System Bleeding

PROCEDURE

After working on the cooling system, even to replace the thermostat, the system must be bled. Air trapped in the system will prevent proper filling and leave the radiator coolant level low, causing a risk of overheating.

To bleed the system, start with the system cool, the radiator cap off and the radiator filled to about an inch below the filler neck.

1. Start the engine and run it at slightly above normal idle speed. This will insure adequate circulation. If air bubbles appear and the coolant level drops, fill the system with an antifreeze/water mixture to bring the level back to the proper level.

2. Run the engine this way until the thermostat opens. When this happens, coolant will move abruptly across the top of the radiator and the temperature of the radiator will suddenly rise.

3. At this point, air is often expelled and the level may drop quite a bit. Keep refilling the system until the level is near the top of the radiator and remains constant.

4. If the vehicle has an overflow tank, fill the radiator right up to the filler neck. Replace the radiator filler cap.

FUEL SYSTEM

Fuel System Service Precautions

Safety is the most important factor when performing not only fuel system maintenance but any type of maintenance. Failure to conduct maintenance and repairs in a safe manner may result in serious personal injury or death. Maintenance and testing of the vehicle's fuel system components can be accomplished

safely and effectively by adhering to the following rules and guidelines.

• To avoid the possibility of fire and personal injury, always disconnect the negative battery cable unless the repair or test procedure requires that battery voltage be applied.

• Always relieve the fuel system pressure prior to disconnecting any fuel system component (injector, fuel rail, pressure regulator, etc.), fitting or fuel line connection. Exercise extreme caution whenever relieving fuel system pressure to avoid exposing skin, face and eyes to fuel spray. Please be advised that fuel under pressure may penetrate the skin or any part of the body that it contacts.

• Always place a shop towel or cloth around the fitting or connection prior to loosening to absorb any excess fuel due to spillage. Ensure that all fuel spillage (should it occur) is quickly removed from engine surfaces. Ensure that all fuel soaked cloths or towels are deposited into a suitable waste container.

• Always keep a dry chemical (Class B) fire extinguisher near the work area.

• Do not allow fuel spray or fuel vapors to come into contact with a spark or open flame.

• Always use a backup wrench when loosening and tightening fuel line connection fittings. This will prevent unnecessary stress and torsion to fuel line piping. Always follow the proper torque specifications.

• Always replace worn fuel fitting O-rings with new. Do not substitute fuel hose or equivalent, where fuel pipe is installed.

Fuel System Pressure

RELIEVING

Modern Fuel Injection systems use a relatively high fuel pressure. Fuel is supplied to the system and injectors with much more pressure than that used in carbureted engines. Because of the dangers inherent in having gasoline under pressure in an automotive environment (operating or undergoing repair), it is essential that the pressure be released prior to loosening any connections where fuel flows. If the pressure is not released and a line or fitting is loosened, fuel will leak out the line under pressure,

possibly causing serious personal as well as property damage.

---— CAUTION —---
Fuel injection systems remain under pressure after the engine has been turned OFF. Properly relieve fuel pressure before disconnecting any fuel lines. Failure to do so may result in fire or personal injury.

---

1. Loosen the fuel filler cap to release fuel tank pressure.
2. Remove the fuel system access cover in the luggage compartment to gain access to the connector and disconnect the fuel pump harness connector.
3. Start the vehicle and allow it to run until it stalls from lack of fuel. Turn the key to the **OFF** position.
4. Disconnect the negative battery cable, then reconnect the fuel pump connector.
5. Wrap shop towels around the fitting that is being disconnected to absorb residual fuel in the lines.
6. Place shop towels into proper safety container.

Idle Speed

ADJUSTMENT

1993 Models

1. Warm the engine to operating temperature, leave lights, electric cooling fan and accessories **OFF**. The transaxle should be in **N**. The steering wheel in a neutral position for vehicles with power steering.
2. Check the ignition timing and adjust, if necessary. Be sure to ground ignition timing adjustment connector.
3. Connect a tachometer to the single pin connector terminal under the hood.

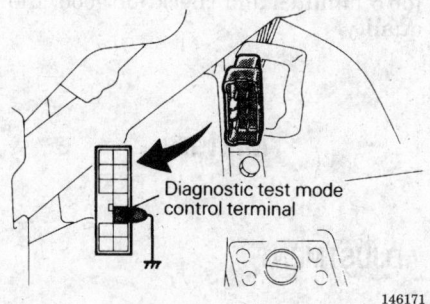

Diagnostic test mode control terminal

146171

Diagnostic terminal identification — 1993 models

4. Run the engine for more than 10 seconds at 2000–3000 rpm. Allow the engine to idle for 2 minutes. Check the idle rpm. Curb idle should be 750 rpm.
5. If adjustment is required, disconnect the waterproof female connector used for ignition timing adjustment. Connect this terminal to ground using a jumper wire.
6. Locate the self-diagnosis terminal under the dashboard and connect terminal No. **10** to ground with a jumper wire.
7. Start the engine and allow to idle. Check that the basic idle speed is at specification. The tachometer reading will be ⅓ of the actual engine speed. Multiply the reading by 3 to figure the actual engine speed. If the idle speed deviates from this speed, check the following:

    a. A new engine will idle more slowly. Break-in should take approximately 300 miles.

    b. If the vehicle stalls or has a very low idle speed, suspect a deposit buildup on the throttle valve which must be cleaned.

    c. If the idle speed is high even though the speed adjusting screw is fully closed, check that the idle position switch (fixed speed adjusting screw) position has changed. If so, adjust the idle position switch.

    d. If after all these checks the idle is still out of specification, it is probable that there is leakage resulting from deterioration of the Fast-Idle Air Valve (FIAV) and the throttle body will need to be replaced.

8. Turn the ignition switch **OFF** and stop the engine. Disconnect the jumper wire from the diagnosis connector, disconnect the jumper wire from the ignition timing connector and reconnect the waterproof connector. Disconnect the tachometer.
9. Restart the engine, allow to run for 5 minutes and check for good idle quality.

1994–97 Models

1. Warm the engine to operating temperature, leave lights, electric cooling fan and accessories **OFF**. The transaxle should be in **N**. The steering wheel in a neutral position for vehicles with power steering.
2. Check the ignition timing and adjust, if necessary. Be sure to ground ignition timing adjustment connector.
3. Connect a tachometer to the single pin connector terminal under the hood.

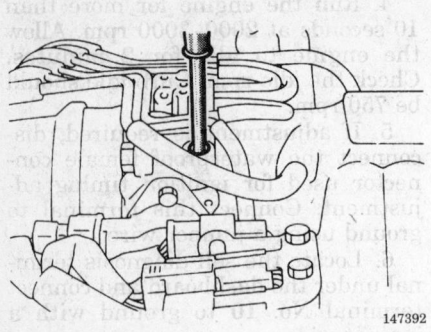

Idle speed screw location — all models

147392

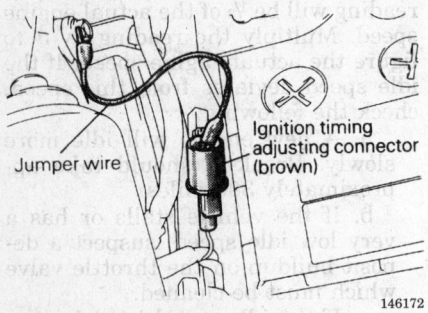

Ignition timing adjusting connector identification — all models

Jumper wire

Ignition timing adjusting connector (brown)

146172

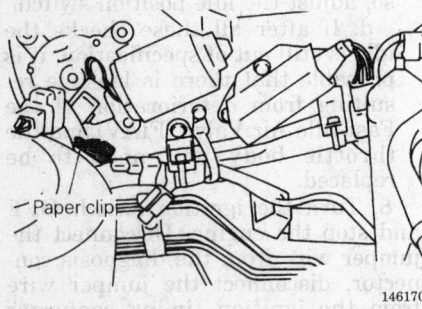

Paper clip

Tachometer connector location — all models

146170

4. Run the engine for more than 10 seconds at 2000–3000 rpm. Allow the engine to idle for 2 minutes. Check the idle rpm. Curb idle should be 750 rpm.

5. If adjustment is required, disconnect the waterproof female connector used for ignition timing adjustment. Connect this terminal to ground using a jumper wire.

6. Locate the self-diagnosis terminal under the dashboard and connect terminal No. **1** to ground with a jumper wire.

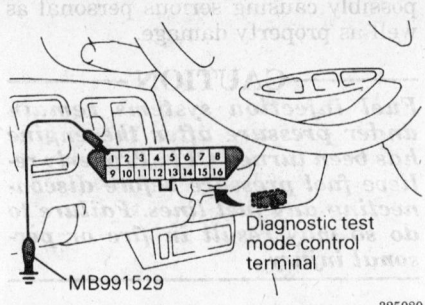

Diagnostic test mode control terminal

MB991529

325080

Diagnostic terminal identification — 1994–97 models

7. Start the engine and allow to idle. Check that the basic idle speed is at specification. The tachometer reading will be ⅓ of the actual engine speed. Multiply the reading by 3 to figure the actual engine speed. If the idle speed deviates from this speed, check the following:

a. A new engine will idle more slowly. Break-in should take approximately 300 miles.

b. If the vehicle stalls or has a very low idle speed, suspect a deposit buildup on the throttle valve which must be cleaned.

c. If the idle speed is high even though the speed adjusting screw is fully closed, check that the idle position switch (fixed speed adjusting screw) position has changed. If so, adjust the idle position switch.

d. If after all these checks the idle is still out of specification, it is probable that there is leakage resulting from deterioration of the Fast-Idle Air Valve (FIAV) and the throttle body will need to be replaced.

8. Turn the ignition switch **OFF** and stop the engine. Disconnect the jumper wire from the diagnosis connector, disconnect the jumper wire from the ignition timing connector and reconnect the waterproof connector. Disconnect the tachometer.

9. Restart the engine, allow to run for 5 minutes and check for good idle quality.

Mixture

ADJUSTMENT

Air/Fuel mixture is controlled by the ECU and is not adjustable.

Fuel Filter

REMOVAL AND INSTALLATION

The filter is located in the engine compartment, mounted on the inner fender panel.

— CAUTION —

Do not use conventional fuel filters, hoses or clamps when servicing fuel injection systems. They are not compatible with the injection system and could fail, causing personal injury or damage to the vehicle. Use only hoses and clamps specifically designed for fuel injection.

1. Relieve the fuel pressure.
2. Disconnect the negative battery cable.

— CAUTION —

Wait at least 90 seconds after the negative (-) battery cable is disconnected to prevent possible deployment of the air bag.

3. Remove the air cleaner assembly and intake hoses. Remove the battery and battery tray with washer tank.

4. Separate the flare nut connection at the line.

5. Hold the fuel filter nut securely with a backup or spanner wrench. Cover the hoses with shop towels and remove the eye bolts. Discard the gaskets.

6. Remove the mounting bolts and remove the fuel filter from the vehicle.

To install:

7. Install a new gaskets or O-rings whenever fuel connections have been disassembled.

8. Install the filter to its bracket only finger-tight. Movement of the filter will ease attachment of the fuel lines.

9. Install new gaskets and connect the high pressure hose and eye bolt, then the main pipe and eye bolt. While holding the fuel filter nut, tighten the eye bolts to 22 ft. lbs. (30 Nm). Tighten the flare nut to 25 ft. lbs. (35 Nm).

10. Tighten the mounting bolts fully.

11. Install the air cleaner assembly, battery and battery tray with washer tank, if removed.

12. Connect the negative battery cable, install the fuel filler cap, turn the key to the **ON** position to pressurize the fuel system and check for

leaks. Release the fuel pressure and repair leaks as required.

Fuel Pump

REMOVAL AND INSTALLATION

1. Relieve fuel system pressure. Remove the fuel filler cap.

———— **CAUTION** ————

Fuel injection systems remain under pressure after the engine has been turned OFF. Properly relieve fuel pressure before disconnecting any fuel lines. Failure to do so may result in fire or personal injury.

2. Disconnect the negative battery cable.

———— **CAUTION** ————

Wait at least 90 seconds after the negative (-) battery cable is disconnected to prevent possible deployment of the air bag.

3. The fuel pump is located in the fuel tank. Drain the fuel from the fuel tank.
4. Remove the fuel gauge cover located in the rear floor pan.
5. Remove the fuel pump and gauge electrical connector. Remove the overfill limiter (two-way valve).
6. Disconnect both sides of the high pressure fuel hose. When disconnecting the fuel pump side of the hose, hold the pump side nut with a wrench while turning the nut on the hose side. This will prevent any damage that will occur to the fittings and the hoses if 2 wrenches are not used.
7. Remove the fuel pump and gauge assembly from the tank.
To install:
8. Align the 3 projections on the packing with the holes on the fuel pump and the nipples on the pump facing the same direction as before removal.
9. Temporarily tighten the flare nut on the high pressure hose by hand. Making sure the hose does not twist, tighten body side nut to 22 ft. lbs. (30 Nm) and the fuel pump side nut to 25 ft. lbs. (35 Nm).
10. Install the overfill limiter (2-way valve) with the long shouldered side of the valve facing the canister.
11. Connect the electrical connector to the pump assembly.

12. Reconnect the negative battery cable and check the entire system for leaks.
13. Install sealer to the rear floor pan and install the cover into place.

Fuel Injector

REMOVAL AND INSTALLATION

1. Relieve the fuel system pressure.

———— **CAUTION** ————

Fuel injection systems remain under pressure after the engine has been turned OFF. Properly relieve fuel pressure before disconnecting any fuel lines. Failure to do so may result in fire or personal injury.

———— **CAUTION** ————

Do not allow fuel spray or fuel vapors to come in contact with spark or open flame. Keep a dry chemical fire extinguisher nearby. Never store fuel in an open container due to risk of fire or explosion.

2. Disconnect the negative battery cable.

———— **CAUTION** ————

Work MUST NOT be started until at least 90 seconds after the ignition switch is turned to the LOCK position and the negative battery cable is disconnected from the battery. This will allow time for the air bag system backup power supply to deplete its stored energy preventing accidental air bag deployment which could result in unnecessary air bag system repairs and/or personal injury.

3. Drain the cooling system.
4. Disconnect all components from the air intake plenum and remove the plenum from the intake manifold. Discard the gaskets.
5. Wrap the connection with a shop towel and disconnect the high pressure fuel line at the fuel rail.
6. Disconnect the fuel return hose and remove the O-ring.
7. Disconnect the vacuum hose from the fuel pressure regulator. Remove the fuel pressure regulator and O-ring.
8. Disconnect the electrical connectors from each injector.
9. Remove the fuel pipe connecting the fuel rails. Remove the injector

rail retaining bolts. Make sure the rubber mounting bushings do not get lost.
10. Lift the rail assemblies up and away from the engine.
11. Remove the injectors from the rail by pulling gently. Discard the lower insulator. Check the resistance through the injector. The specification for 3.0L turbocharged engine is 2–3 ohms at 68°F (20°C). The specification for non-turbocharged 3.0L engine is 13–16 ohms at 68°F (20°C).
To install:

NOTE: Some of the vehicles may have a clip that secures the the injector to the fuel rail. Be sure to remove or install the injector clip where necessary.

12. Install a new grommet and O-ring to the injector. Coat the O-ring with light oil.
13. Install the injector to the fuel rail.
14. Replace the seats in the intake manifold. Install the fuel rails and injectors to the manifold. Make sure the rubber bushings are in place before tightening the mounting bolts.
15. Tighten the retaining bolts to 7–9 ft. lbs. (10–13 Nm). Install the fuel pipe with new gasket.
16. Connect the electrical connectors to the injectors.
17. Replace the O-ring, lightly lubricate it and connect the fuel pressure regulator.
18. Connect the fuel return hose.
19. Replace the O-ring, lightly lubricate it and connect the high pressure fuel line.
20. Using new gaskets, install the intake plenum and all related items. Torque the plenum mounting bolts to 13 ft. lbs. (18 Nm).
21. Fill the cooling system.
22. Connect the negative battery cable and check the entire system for proper operation and leaks.

ENGINE MECHANICAL

Engine Assembly

REMOVAL AND INSTALLATION

1. Relieve fuel system pressure.

CAUTION

Fuel injection systems remain under pressure after the engine has been turned OFF. Properly relieve fuel pressure before disconnecting any fuel lines. Failure to do so may result in fire or personal injury.

---

2. Disconnect the negative battery cable.

CAUTION

Wait at least 90 seconds after the negative (-) battery cable is disconnected to prevent possible deployment of the air bag.

---

3. Matchmark the hood and hinges and remove the hood assembly. Remove the air cleaner assembly and all adjoining air intake duct work.

4. Disconnect and remove the cruise control linkage and actuator assemblies.

5. Drain the engine coolant and remove the radiator assembly, coolant reservoir and intercooler.

6. Disconnect the heated oxygen sensor connection at the front exhaust pipe.

7. Unbolt and remove the front exhaust pipe assembly, discard gaskets.

8. Remove the transaxle assembly.

9. Disconnect the accelerator cable, breather hose and heater hose connections from the engine

10. Note locations and remove the vacuum hoses from the engine. Be sure to disconnect the brake booster vacuum supply.

11. Disconnect the fuel feed and return hoses.

12. Remove the solenoid valve assembly and disconnect ground cable.

13. Disconnect the purge hose and EGR temperature sensor, if equipped.

14. Remove the air conditioning and power steering drive belts.

15. Unbolt and remove the air conditioning compressor and the power steering pump assemblies.

NOTE: When removing the power steering pump and a/c compressor, it is not necessary to disconnect the hoses. Position the units aside and use rope or wire to secure.

16. Disconnect the harness connections for the idle speed control, motor position sensor and throttle position sensor.

17. Disconnect the EGR temperature sensor (California).

18. If equipped with a turbocharger, disconnect the following:

   a. Connection for the booster vacuum hose.

   b. Connections for the oil cooler lines and discard the sealing rings.

   c. Connection for the oxygen sensor.

19. Disconnect the wiring at the oil pressure switch and oil pressure gauge unit.

20. Disconnect the fuel injection wiring harness plug.

21. Disconnect the wiring from the knock sensor and the crankshaft angle sensor.

22. Disconnect the coolant temperature switch, coolant temperature sensor and the coolant temperature gauge unit connections.

23. Disconnect the wiring to the ignition coil, condenser and the power transistor.

24. Disconnect the variable induction motor connection.

25. Open the cover of relay box and disconnect the alternator wiring.

26. Attach a hoist to the engine and support the engine weight. Remove the engine mount bracket.

27. Remove the front and rear roll stopper bracket mounting bolts.

28. Remove the engine assembly from the vehicle.

To install:

29. Install the engine and secure into position. Secure the engine mount bracket to block and tighten bolts to 72–87 ft. lbs. (100–120 Nm). Install through bolt and tighten bolt to 51 ft. lbs. (70 Nm).

30. Install the front and rear roll stopper through bolt and tighten to 36–43 ft. lbs. (50–60 Nm).

31. Open the cover of relay box and connect alternator wiring.

32. Connect the variable induction motor connection.

33. Connect the fuel feed and return hoses. Using a new sealing ring, tighten pressure hose connection to 4 ft. lbs. (5 Nm).

34. Connect the wiring to the ignition coil, condenser and power transistor.

35. Connect the coolant temperature switch, coolant temperature sensor and the coolant temperature gauge unit connections.

36. Connect the wiring from the knock sensor and the crankshaft angle sensor.

37. Connect the fuel injection wiring harness plug.

38. Connect the wiring at the oil pressure switch and oil pressure gauge unit.

39. If equipped with a turbocharger, connec the following:

   a. Connection for booster vacuum hose.

   b. Connections for oil cooler lines using new sealing rings. Tighten fittings to 29–33 ft. lbs. (40–49 Nm).

   c. Connection for oxygen sensor.

40. Connect EGR temperature sensor (California).

41. Connect harness connections for the idle speed control, motor position sensor and throttle position sensor.

42. Install the air conditioning compressor and the power steering pump assemblies.

43. Install the engine drive belts.

44. Connect the purge hose and the EGR temperature sensor, if equipped.

45. Install the solenoid valve assembly and connect ground cable to engine block.

46. Reconnect the vacuum hoses to the engine. Be sure to connect the brake booster vacuum supply.

47. Connect the accelerator cable, breather hose and heater hose connections to the engine.

48. Install the transaxle assembly.

49. Install the front exhaust pipe assembly, using new gaskets. Tighten manifold mounting bolts to 36 ft. lbs. (50 Nm).

50. Connect the heated oxygen sensor connection at the front exhaust pipe.

51. Replace the radiator assembly, coolant reservoir and intercooler. Refill the cooling system.

52. Install and connect the cruise control linkage and the actuator assemblies.

53. Install the hood assembly, air cleaner assembly and all adjoining air intake duct work.

54. Connect the negative battery cable and run engine.

55. Inspect all connections and check all fluid levels.

Engine Mounts

REMOVAL AND INSTALLATION

1. Disconnect the negative battery cable.

CAUTION

Except for the 1993–95 3.0L (VIN K) DOHC engine, wait at least 90 seconds after the negative (-) battery cable is disconnected to prevent possible deployment of the air bag.

---

2. Raise and safely support the engine so it is not resting on the engine mount. Use care not to bend or damage any components.

3. For the 1993–95 3.0L (VIN K) DOHC engine, remove the air hose to gain access to the mount.

4. Remove the cruise control mounting nuts and place the actuator where it will not interfere with the work.

5. Remove the engine mount bracket to body bolt.

6. Remove the engine mount bracket from the engine block and remove the engine mount.

7. Except for the 1993–95 3.0L (VIN K) DOHC engine, remove the dynamic damper.

To install:

8. Except for the 1993–95 3.0L (VIN K) DOHC engine, install the dynamic damper.

9. Install the engine mount to the engine block.

10. Install the mounting stopper and the bracket to body bolt and remove the floor jack from under the engine.

11. Install the through bolt and torque to front mount to 36–43 ft. lbs. (50–60 Nm) and the rear mount to 51–58 ft. lbs. (70–80 Nm) except for 3.0L (VIN H) SOHC engine or to 36–43 ft. lbs. (50–60 Nm) for 3.0L (VIN H) SOHC engine.

NOTE: It may be necessary to move the engine up or down to align the bolt holes.

12. Install the cruise control actuator assembly.

13. For the 1993–95 3.0L (VIN K) DOHC engine, reconnect the air hose.

14. Reconnect the negative battery cable.

Cylinder Head

REMOVAL AND INSTALLATION

3.0L (VIN J and K) DOHC Engines

— CAUTION —
Fuel injection systems remain under pressure, even after the engine has been turned OFF. The fuel system pressure must be relieved before disconnecting any fuel lines. Failure to do so may result in fire and/or personal injury.

1. Properly relieve fuel system pressure.
2. Disconnect the negative battery cable.

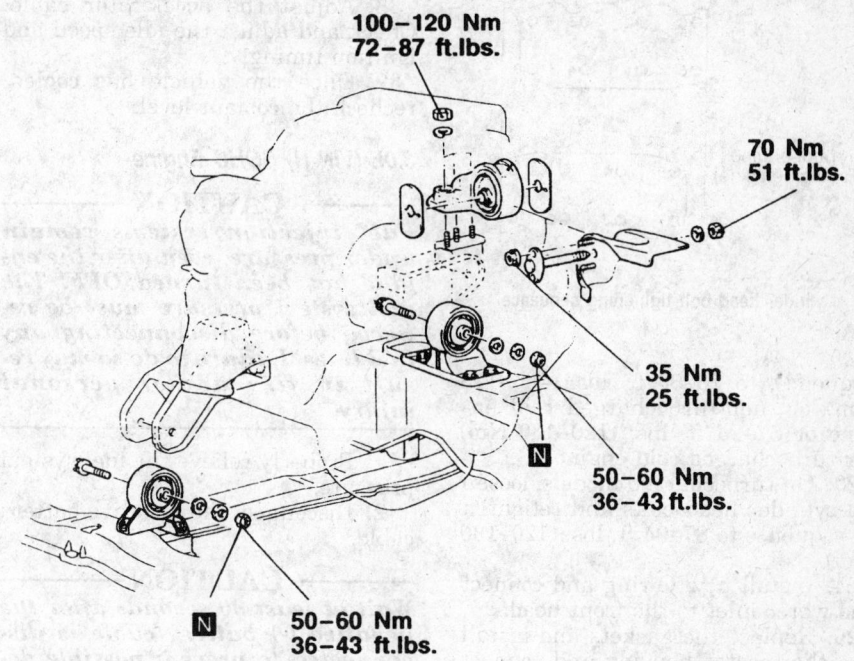

Engine mount components — 3.0L (VIN H) SOHC engine, others similar

— CAUTION —
On the 1995–97 models, wait at least 90 seconds after the negative (-) battery cable is disconnected to prevent possible deployment of the air bag.

3. Drain the cooling system.
4. Remove the air intake hoses.
5. Remove air intake plenum and intake manifold.
6. Remove the turbocharger, if equipped.
7. Remove the exhaust manifold.
8. Remove the timing belt.
9. Remove the triple pipe assembly across the top of the engine.
10. Remove the breather hose.
11. Remove the spark plug cable center cover and remove the spark plug cables.
12. When removing the valve cover, note that bolts for the front head are black and bolts for the rear head are green. Also, all bolts are 10mm long except the 1 closest to the sprockets on the rear head which is 20mm long.
13. To remove the intake camshaft sprocket, hold the camshaft with a wrench on the hexagon near the end of the camshaft and remove the bolt.
14. Remove the rear timing belt cover.
15. Remove the ignition coil.

16. Disconnect all water hoses from the thermostat housing and remove the housing.
17. Disconnect the water inlet from the front head and discard O-ring.
18. Loosen the cylinder head mounting bolts in 3 steps, starting from the outside and working inward. Lift off the cylinder head assembly and remove the head gasket.

To install:
19. Thoroughly clean and dry the mating surfaces of the head and block. Check the cylinder head for cracks, damage or engine coolant leakage. Remove scale, sealing compound and carbon. Clean oil passages thoroughly.
20. Check the cylinder head for flatness using a feeler gauge and straightedge. End to end, the head should be within 0.002 in. normally with 0.008 in. the maximum allowed out of true. The total thickness allowed to be removed from the head and block is 0.008 in. maximum.
21. Place a new head gasket on the cylinder block with the identification marks in the front top (upward) position. Do not use sealer on the gasket.
22. Carefully install the cylinder head on the block. Make sure the head bolt washers are installed with the chamfered edge upward. Using 3 even steps, torque the head bolts in

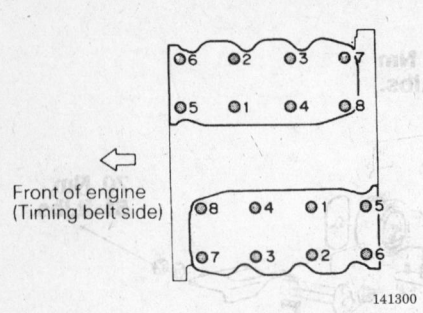

Front of engine
(Timing belt side)

141300

Cylinder head bolt tightening sequence

sequence, to 76–83 ft. lbs. (105–115 Nm) for non-turbocharged cold engine or 87–94 ft. lbs. (120–130 Nm) for turbocharged cold engine.

23. On turbocharged models, loosen all cylinder head bolts and retighten in sequence to 87–94 ft. lbs. (120–130 Nm).

24. Install new O-ring and connect the water inlet to the front head.

25. Replace the gaskets and install the thermostat housing and connect the hoses.

26. Install the ignition coil and center rear timing belt cover.

27. Install the intake camshaft sprocket. Use hex flange on camshaft to secure and torque the retaining bolt to 65 ft. lbs. (90 Nm).

28. Apply sealer to the lower edges of the half-round portions of the belt-side of the new gasket and install the valve cover. Make sure green bolts are installed on the rear head and black bolts are installed on the front head. Also, make sure the longest bolt is installed in its proper location closest to the sprockets on the rear head. Tighten the bolts in the proper sequence to 26 inch lbs. (3 Nm). Then retighten bolts No. 1–6 to 35 inch lbs. (4 Nm).

29. Connect the spark plug cables and install the center cover.

30. Install the breather hose.

31. Install the triple pipe assembly across the top of the engine and torque the retaining bolts to 7 ft. lbs. (10 Nm).

32. Install the timing belt and all related items.

33. Using all new gaskets, install the intake manifold, air intake plenum, turbocharger and exhaust manifold, following the proper torque sequences.

34. Install the air intake hoses.

35. Change the engine oil and oil filter.

36. Fill the system with coolant.

37. Connect the negative battery cable, run the vehicle until the thermostat opens, fill the radiator completely.

38. Adjust the accelerator cable. Check and adjust the idle speed and ignition timing.

39. Once the vehicle has cooled, recheck the coolant level.

3.0L (VIN H) SOHC Engine

— **CAUTION** —

Fuel injection systems remain under pressure, even after the engine has been turned OFF. The fuel system pressure must be relieved before disconnecting any fuel lines. Failure to do so may result in fire and/or personal injury.

1. Properly relieve the fuel system pressure.

2. Disconnect the negative battery cable.

— **CAUTION** —

Wait at least 90 seconds after the negative (-) battery cable is disconnected to prevent possible deployment of the air bag.

3. Drain the cooling system.

4. Remove the air intake hose.

5. Remove the exhaust manifold.

6. Remove the air intake plenum and intake manifold.

7. Remove the timing belt.

8. Remove the camshaft sprocket and rear timing belt cover.

9. Remove the power steering pump bracket. If removing the rear head, remove the alternator brace.

10. Disconnect the water inlet pipe.

11. Remove the purge pipe assembly.

12. Remove the valve cover.

13. Loosen the cylinder head mounting bolts in 3 steps, starting from the outside and working inward. Lift off the cylinder head assembly and remove the head gasket.

To install:

14. Thoroughly clean and dry the mating surfaces of the head and block. Check the cylinder head for cracks, damage or engine coolant leakage. Remove scale, sealing compound and carbon. Clean oil passages thoroughly.

15. Check the cylinder head for flatness using a feeler gauge and straightedge. End to end, the head should be within 0.002 in. normally with 0.008 in. the maximum allowed out of true. The total thickness allowed to be removed from the head and block is 0.008 in. maximum.

16. Place a new head gasket on the cylinder block making sure the identification mark on the cylinder head gasket is in the front top (upward) location. Do not use sealer on the gasket. Make sure the gasket has the proper identification mark for the engine.

17. Carefully install the cylinder head on the block. Make sure the head bolt washers are installed with the chamfered edge upward. Using 3 even steps, torque the head bolts in sequence, to 80 ft. lbs. (110 Nm). This torque specification is for a cold engine.

18. Apply sealer to the lower edges of the half-round portions of the belt-side of the new gasket and install the valve cover. Tighten valve cover bolts to 7 ft. lbs. (9 Nm).

19. Install the purge pipe assembly.

20. Connect the water inlet pipe.

21. Install the power steering pump bracket and alternator brace.

22. Install the rear timing belt cover and cam sprocket. Torque the retaining bolt to 65 ft. lbs. (90 Nm).

23. Install the timing belt and all related items.

24. Using all new gaskets, install the intake manifold, air intake plenum and exhaust manifold, following the proper torque sequences.

25. Install the air intake hose.

26. Change the engine oil and oil filter.

27. Fill the system with coolant.

28. Connect the negative battery cable, run the vehicle until the thermostat opens, fill the radiator completely.

29. Check and adjust the idle speed and ignition timing.

30. Once the vehicle has cooled, recheck the coolant level.

Lash Adjusters

BLEEDING

1. Remove the lash adjusters with the rocker arms.

2. Immerse the lash adjusters in clean diesel fuel. Using a small wire, move the plunger of the lash adjuster up and down 4 or 5 times while pushing down lightly on the check ball in order to bleed out the air.

3. Install the lash adjusters and rocker arms in the cylinder head.

Valve Lash

ADJUSTMENT

The automatic valve lash adjusters do not require manual valve adjust-

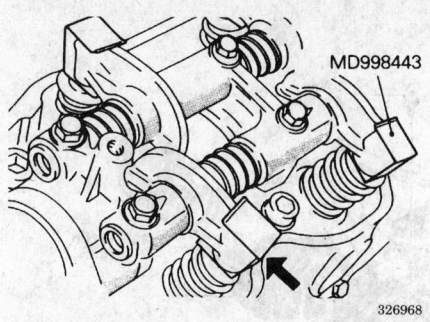

Auto lash adjuster holding tool — 3.0L (VIN H) SOHC engine

326968

ment. If the valves are making excessive noise, the problem is either with the adjusters needing to be bled or in another component. Inspect the other valve train components for damage or excessive wear.

Rocker Arms

REMOVAL AND INSTALLATION

3.0L (VIN J and K) DOHC Engines

1. Relieve the fuel system pressure.

— CAUTION —

Fuel injection systems remain under pressure after the engine has been turned OFF. Properly relieve fuel pressure before disconnecting any fuel lines. Failure to do so may result in fire or personal injury.

2. Disconnect battery negative cable.

— CAUTION —

Wait at least 90 seconds after the negative (-) battery cable is disconnected to prevent possible deployment of the air bag.

3. Remove the timing belt cover and timing belt.
4. Remove the center cover, breather and PCV hoses, and spark plug cables.
5. Remove the rocker cover, semicircular packing, throttle body stay, both camshaft sprockets, and oil seals.
6. Remove the crank angle sensor and adaptor.
7. Remove the intake and exhaust camshafts.
8. Remove rocker arms and lash adjusters from the head. It is recommended that all lash adjusters and rockers be replaced as a complete set.

To install:
9. Bleed the lash adjusters.

NOTE: Do not confuse the intake camshaft with the exhaust camshaft. The intake camshaft has a J stamped on the hexagon and the exhaust camshaft has a K or N.

10. Make sure the dowel pins on both camshaft sprocket ends are positioned properly.
11. Install the bearing caps. Tighten the caps in sequence and in 2 or 3 steps. Caps 2, 3 and 4 have a front mark. Install with the mark aligned with the front mark on the cylinder head. Intake caps have **I** stamped on the cap and exhaust caps have **E**. Also, make sure the rocker arm is correctly mounted on the lash adjuster and the valve stem end. Torque the front and rear retaining cap bolts to 15 ft. lbs. (20 Nm) and tighten the center 3 retaining cap bolts to 8 ft. lbs. (11 Nm).

NOTE: If installing the camshaft to a cylinder head that is positioned on a workbench, the valves will protrude.

12. Apply a coating of engine oil to the oil seals and install.
13. Install the timing belt, valve cover and all related parts.
14. Connect the negative battery cable and check for leaks.

Rocker Arm Shaft

REMOVAL AND INSTALLATION

3.0L (VIN H) SOHC Engine

The hydraulic lash adjusters are built into the rocker arms. If service is required, simply remove the lash adjuster from the bore in the rocker arm. It is recommended that all of the rocker arms and lash adjusters are replaced at the same time.

1. Disconnect the negative battery cable.
2. Remove the valve cover. Install lash adjuster retainer tools MD998443 or equivalent, to prevent the auto-lash adjuster from falling out of the rocker arm.
3. Rotate the engine clockwise and position No. 1 cylinder at TDC compression stroke.
4. If necessary, remove the distributor adapter housing.
5. Remove the timing belt assembly.
6. Loosen rocker arm and shaft assembly evenly in several steps. Re-

move the rocker arm and shaft assembly as a complete unit.
7. Remove the rear camshaft bearing cap and slide rocker arms, springs and washers from shaft. Note location and positioning of all rocker shaft components.
8. Visually inspect the rocker arm roller and replace if damage or seizure is evident. Check the roller for smooth rotation. Replace if excess play or binding is present. Also, inspect valve contact surface for possible damage or seizure. It is recommended that all rocker arms and lash adjusters be replaced together.

To install:
9. Bleed the lash adjusters.
10. Using a light coat of engine oil, assembly the rocker arms to the shaft. Install the rear camshaft bearing cap.
11. Lubricate the camshaft and rocker shaft with heavy engine oil and position on the cylinder head.
12. Apply a drop of sealant to the rear edges of the end caps.
13. Install the assembly making sure the notches in the rocker shafts are facing up. Insert the bolts but do not tighten at this point.
14. Install the remaining cap bolts and tighten evenly and gradually to 14 ft. lbs. (20 Nm). Remove the lash adjuster retainers.
15. Install the distributor extension, if removed.
16. Install the valve cover with a new gasket and tighten to 84 inch lbs. (9 Nm).
17. Connect the negative battery cable.

Intake Manifold

REMOVAL AND INSTALLATION

1. Relieve the fuel system pressure.
2. Disconnect battery negative cable and drain the cooling system.

— CAUTION —

Wait at least 90 seconds after the negative (-) battery cable is disconnected to prevent possible deployment of the air bag.

3. Remove the air intake hose(s).
4. Disconnect the accelerator control cables from the throttle body.
5. Matchmark and disconnect the vacuum hoses including the brake booster hose.
6. Disconnect the clutch booster vacuum hose connection, if equipped.
7. Disconnect all harness connectors.

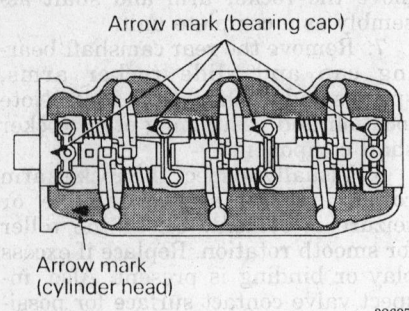

Arrow mark (bearing cap)

Arrow mark (cylinder head)

326970

Rocker arm shaft installation — 3.0L (VIN H) SOHC engine

8. Disconnect EGR components on California vehicles.

9. Remove the plenum retaining bracket.

10. Remove the plenum retaining nuts and bolts and remove the air intake plenum. Discard the gasket.

11. Disconnect the high pressure and return fuel hoses.

12. Matchmark and disconnect the vacuum hoses.

13. Disconnect the wire harness connectors.

14. Remove the fuel rail with the injectors attached.

15. On SOHC engines, disconnect the water hoses. On DOHC engines, remove the timing belt upper cover.

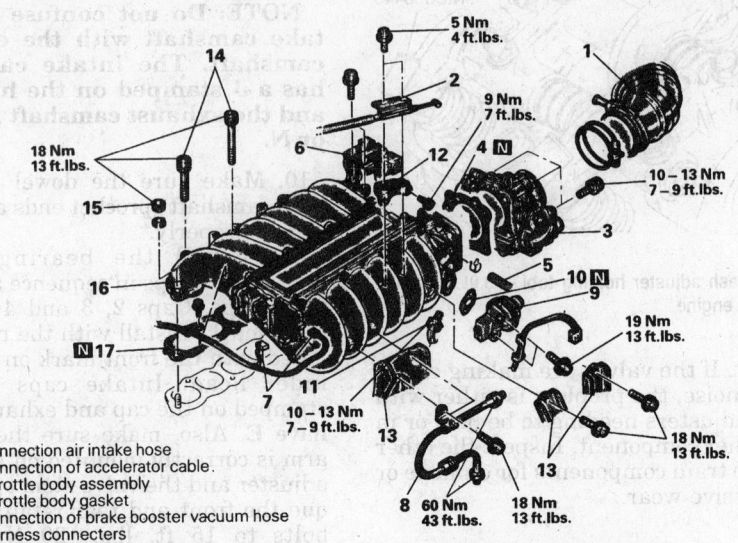

5 Nm 4 ft.lbs.

9 Nm 7 ft.lbs.

18 Nm 13 ft.lbs.

10 – 13 Nm 7 – 9 ft.lbs.

19 Nm 13 ft.lbs.

18 Nm 13 ft.lbs.

10 – 13 Nm 7 – 9 ft.lbs.

60 Nm 43 ft.lbs.

18 Nm 13 ft.lbs.

1. Connection air intake hose
2. Connection of accelerator cable.
3. Throttle body assembly
4. Throttle body gasket
5. Connection of brake booster vacuum hose
6. Harness connecters
7. Connection of VIC servo motor
8. EGR pipe
9. EGR valve
10. EGR valve gasket
11. EGR temperature sensor ⎫ <Vehicles for
12. Accelerator cable bracket ⎭ California>
13. Connection of air intake plenum stay
14. Air intake plenum installation bolts
15. Air intake plenum installation nuts
16. Air intake plenum
17. Air intake plenum gasket

323448

Air intake plenum — 3.0L (VIN J) DOHC engine

5 Nm 4 ft.lbs.

18 Nm 13 ft.lbs.

18 Nm 13 ft.lbs.

10 – 13 Nm 7 – 9 ft.lbs.

10 – 12 Nm 7 – 9 ft.lbs.

19 Nm 13 ft.lbs.

18 Nm 13.ft.lbs.

1. Connection of air intake hose
2. Connection of accelerator cable
3. Throttle body assembly
4. Throttle body gasket
5. Connection of vacuum hose
6. Connection of brake booster vacuum hose
7. Harness connector
8. EGR temperature sensor ⎫ <Vehicles for
9. EGR valve   ⎭ California>
10. EGR valve gasket
11. EGR pipe installation bolts ⎫ <Vehicles for
12. EGR pipe gasket   ⎭ California>
13. Connection of air intake plenum stay
14. Air intake plenum installation bolts
15. Air intake plenum installation nuts
16. Air intake plenum
17. Air intake plenum gasket

323447

Air intake plenum — 3.0L (VIN H) SOHC engine

16. Remove the intake manifold mounting nuts; turbocharged engines have cone disc springs under some of the nuts which should be removed. Remove the intake manifold and discard the gaskets.

To install:

17. Check all items for cracks, clogging and warpage. Maximum warpage is 0.008 in. (0.2mm). Replace any questionable parts.

18. Thoroughly clean and dry the mating surfaces of the heads, intake manifold and air intake plenum.

19. Install new intake manifold gaskets to the heads with the adhesive side facing up.

20. Place the manifold on the heads and install the cone disc springs and/or lock washers.

21. For DOHC: Lubricate the studs lightly with oil, then install the nuts following this procedure:

a. Tighten the nuts on the front bank to 4–6 ft. lbs. (5–8 Nm).

b. Tighten the nuts on the rear bank to 14–17 ft. lbs. (20–23 Nm).

c. Tighten the nuts on the front bank to 14–17 ft. lbs. (20–23 Nm).

d. Repeat Steps B and C.

22. For SOHC: Tighten the nuts to 13 ft. lbs. (18 Nm).

23. On SOHC engines, connect the water hoses. On DOHC engines, install the timing belt upper cover.

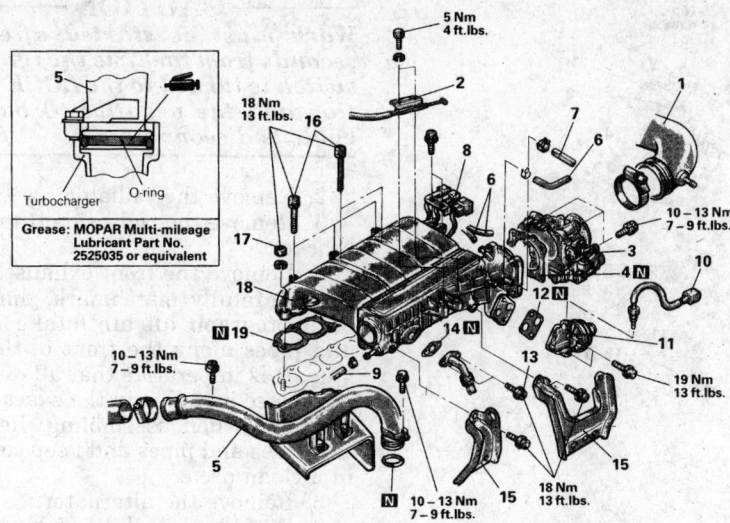

1. Connection air hose A
2. Connection of accelerator cable
3. Throttle body assembly
4. Throttle body gasket
5. Air pipe A
6. Connection of vacuum hose
7. Connection of brake booster vacuum hose
8. Harness connecter
9. Connection of clutch booster vacuum hose
10. EGR temperature sensor <Vehicles for California>

11. EGR valve
12. EGR valve gasket
13. EGR pipe installation bolts
14. EGR pipe gasket
15. Connection of air intake plenum stay
16. Air intake plenum installation bolts
17. Air intake plenum installation nuts
18. Air intake plenum
19. Air intake plenum gasket

323449

Air intake plenum — 3.0L (VIN K) DOHC engine

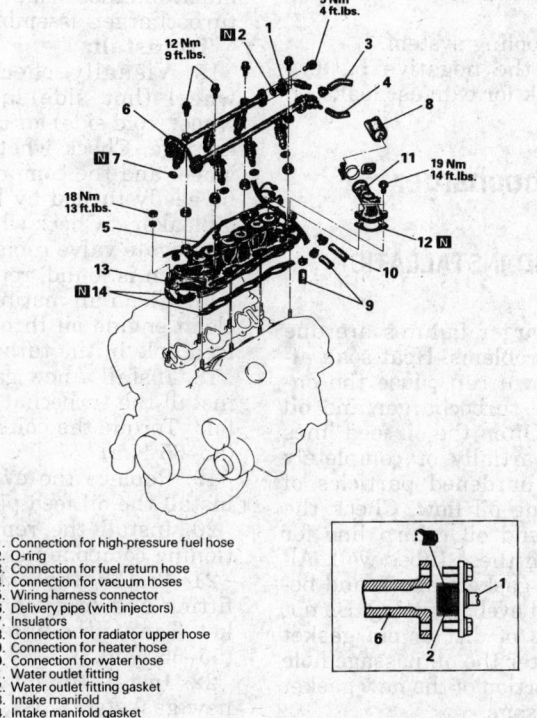

1. Connection for high-pressure fuel hose
2. O-ring
3. Connection for fuel return hose
4. Connection for vacuum hoses
5. Wiring harness connector
6. Delivery pipe (with injectors)
7. Insulators
8. Connection for radiator upper hose
9. Connection for heater hose
10. Connection for water hose
11. Water outlet fitting
12. Water outlet fitting gasket
13. Intake manifold
14. Intake manifold gasket

323450

Intake manifold — 3.0L (VIN H) SOHC engine

24. Install the fuel rail assembly.
25. Connect the harness connector and vacuum hoses.
26. Replace the O-ring and connect the fuel hoses.
27. Install a new intake air plenum gasket and install the plenum. Tighten the retaining nuts and bolts evenly and gradually to 13 ft. lbs. (18 Nm).
28. Install the retaining bracket.
29. Connect EGR components on California vehicles.
30. Connect the harness connectors and vacuum hoses.
31. Connect and adjust the accelerator cables.
32. Install the air intake hose(s).
33. Fill the system with coolant.
34. Connect the negative battery cable, run the vehicle until the thermostat opens, fill the radiator completely.
35. Check and adjust the idle speed and ignition timing.
36. Once the vehicle has cooled, recheck the coolant level.

Exhaust Manifold

REMOVAL AND INSTALLATION

Non-Turbocharged Models

1. Disconnect battery negative cable.
2. Raise the vehicle and support safely.
3. Remove the exhaust pipe to exhaust manifold nuts and separate exhaust pipe. Discard gasket.
4. Lower vehicle.
5. Remove electric cooling fan assembly, if necessary. If removing the front manifold, remove the dipstick tube. If removing the front manifold from 3.0L DOHC engine, remove the alternator.
6. Disconnect necessary EGR components.
7. Disconnect the electrical connector and remove the oxygen sensor.
8. Remove the exhaust manifold mounting bolts, the inner heat shield and the exhaust manifold.
To install:
9. Clean all gasket material from the mating surfaces and check the manifold for damage.
10. Install a new gasket and install the manifold.
11. Install the heat shields.
12. Connect EGR components.
13. Install the oxygen sensor.
14. Install the electric cooling fan assembly, dipstick tube and alternator, as required.

3-17

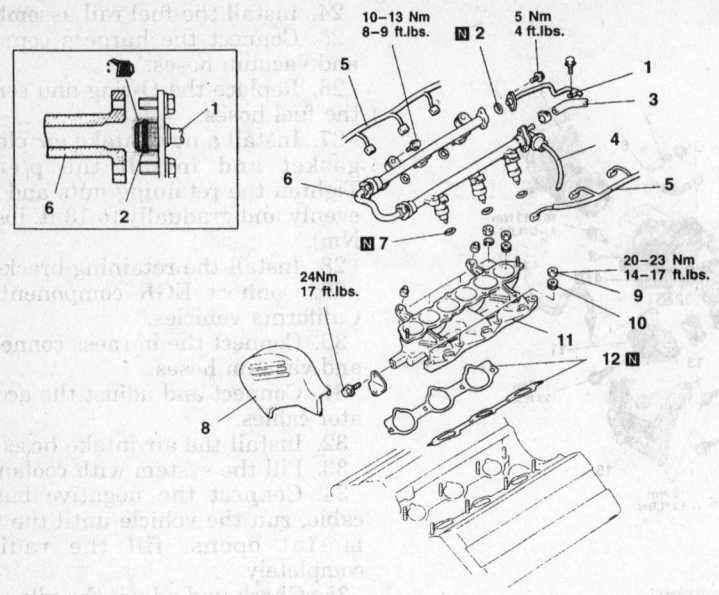

10–13 Nm
8–9 ft.lbs.

5 Nm
4 ft.lbs.

10–13 Nm 8–9 ft.lbs.

20–23 Nm
14–17 ft.lbs.

24Nm
17 ft.lbs.

1. High-pressure fuel hose connection
2. O-ring
3. Fuel return hose connection
4. Vacuum hoses connection
5. Injector connector
6. Fuel rail (with injectors)

7. Insulators
8. Timing belt upper cover
9. Intake manifold mounting nut <turbo>
10. Cone disc spring <turbo>
11. Intake manifold
12. Intake manifold gasket

323451

Intake manifold — 3.0L (VIN J and K) DOHC engines

15. Install a new flange gasket and connect the exhaust pipe.

16. Connect the negative battery cable and check for exhaust leaks.

Turbocharged Models

1. Disconnect the negative battery cable.

2. Drain the engine coolant.

3. Remove the turbocharger assembly.

4. Remove the heat shield.

5. Remove the mounting nuts and remove the exhaust manifold. Note that the cone disc springs are installed at all lower mounting points.

To install:

6. Clean all gasket material from the mating surfaces and check the manifold for damage.

7. Install new gaskets and install the manifold. Make sure all cone disc springs are in their original locations with the grooved side facing the nut. Tighten the manifold nuts using the following procedure:

   a. Tighten all but the outer 2 nuts to 22 ft. lbs. (30 Nm).

   b. Tighten the outer 2 nuts to 34–38 ft. lbs. (47–53 Nm).

   c. Loosen the outer 2 nuts, then torque them to 22 ft. lbs. (30 Nm).

8. Install the heat shield.

9. Install the turbocharger assembly.

10. Fill the cooling system.

11. Connect the negative battery cable and check for exhaust leaks.

Turbocharger

REMOVAL AND INSTALLATION

Many turbocharger failures are due to oil supply problems. Heat soak after hot shutdown can cause the engine oil in the turbocharger and oil lines to coke. Often the oil feed lines will become partially or completely blocked with hardened particles of carbon, blocking oil flow. Check the oil feed pipe and oil return line for clogging. Clean these tubes well. Always use new gaskets above and below the oil feed eyebolt fitting. Do not allow particles of dirt or old gasket material to enter the oil passage hole and that no portion of the new gasket blocks the passage.

Right Side (Front) Turbocharger

1. Disconnect the negative battery cable.

2. Remove the radiator.

3. Remove the right side transaxle bracket.

4. Remove the front exhaust pipe.

5. Carefully matchmark, diagram or photograph all air intake hoses and pipes along the front of the engine. It is imperative that all of these pieces are installed in the exact same positions when assembling. Remove the hoses and pipes and keep covered in a clean area.

6. Remove the alternator.

7. Remove the oil dipstick tube.

8. Remove the turbocharger heat protector.

9. Remove the water feed pipes.

10. Remove the oxygen sensor.

11. Remove the oil return line.

12. Remove the exhaust extension fitting and bracket.

13. Remove all air conditioning components preventing removal of the turbocharger.

14. Remove the oil feed tube.

15. Remove the turbocharger to exhaust manifold bolts and remove the turbocharger assembly.

To install:

16. Visually check the turbine wheel (hot side) and compressor wheel (cold side) for cracking or other damage. Check whether the turbine wheel and the compressor wheel can be easily turned by hand. Check for oil leakage. Check whether or not the wastegate valve remains open. If any problem is found, replace the part.

17. Clean all mating surfaces. Pour clean engine oil through the oil pipe feed hole in the turbocharger.

18. Install a new gasket and ring a install the turbocharger to the manifold. Torque the bolts to 40–47 ft. lbs. (55–65 Nm).

19. Replace the eye-bolt rings and install the oil feed pipe.

20. Install the removed air conditioning components.

21. Install the exhaust extension fitting and bracket with a new gasket. Torque the nuts to 40–47 ft. lbs. (55–65 Nm).

22. Install the oil return line with new gaskets.

23. Install the oxygen sensor.

24. Replace the eye-bolt rings and install the water feed pipes.

25. Install the turbocharger heat protector.

26. Install the dipstick tube.

27. Install the alternator.

28. Install all air intake hoses and pipes along the front of the engine. Make sure all are in their proper positions.

29. Install a new gasket and connect the front exhaust pipe.

30. Install the right side transaxle bracket.

31. Install the radiator.

32. Fill the system with coolant.

33. Connect the negative battery cable and check for exhaust leaks.

Left Side (Rear) Turbocharger

1. Remove the battery.

—————— **CAUTION** ——————
Work must be started after 90 seconds from the time the ignition switch is turned to the LOCK position and the negative (-) battery cable is disconnected.

2. Drain the coolant.

3. Remove the front exhaust pipe.

4. Disconnect the accelerator cable from the throttle body.

5. Remove the intake air hose, the air pipe across the top of the engine and its heat shield.

6. Remove the clutch booster vacuum hose and disconnect the accelerator cable from the pedal.

7. Remove the air intake hoses coming from the air cleaner box.

8. Remove the oxygen sensor and the turbocharger heat protector.

9. Remove the EGR pipe, if equipped.

10. Remove the oil feed pipe.

11. Remove the EGR valve, if equipped.

12. Remove the water feed pipes.

13. Remove the exhaust extension fitting and bracket.

14. Remove the inner heat protector.

15. Remove the oil return tube.

16. Remove the turbocharger to exhaust manifold nuts and remove the turbocharger assembly.

To install:

17. Visually check the turbine wheel (hot side) and compressor wheel (cold side) for cracking or other damage. Check whether the turbine wheel and the compressor wheel can be easily turned by hand. Check for oil leakage. Check whether or not the wastegate valve remains open. If any problem is found, replace the part.

18. Clean all mating surfaces. Pour clean engine oil through the oil pipe feed hole in the turbocharger.

19. Install a new gasket and ring a install the turbocharger to the mani-

fold. Torque the nuts to 40–47 ft. lbs. (55–65 Nm).

20. Install the oil return line with new gaskets.

21. Install the inner heat protector.

22. Install the exhaust extension fitting and bracket with a new gasket. Torque the nuts to 40–47 ft. lbs. (55–65 Nm).

23. Replace the eye-bolt rings and install the water feed pipes.

24. Install the EGR valve, if equipped.

25. Replace the eye-bolt rings and install the oil feed pipe.

26. Install the EGR pipe if equipped.

27. Install the turbocharger heat protector and oxygen sensor.

28. Install the air intake hoses coming from the air cleaner box. Make sure the triangular aligning marks are engaged.

29. Connect the accelerator cable to from the pedal and install the clutch booster vacuum hose.

30. Install the heat shield, the air pipe across the top of the engine and the air intake hose.

31. Connect the accelerator cable to the throttle body.

32. Install a new gasket and connect the front exhaust pipe.

33. Fill the system with coolant.

34. Install the battery.

35. Connect the negative battery cable and check for exhaust leaks.

Front Crankshaft Seal

REMOVAL AND INSTALLATION

1. Disconnect the negative battery cable.

—————— **CAUTION** ——————
Wait at least 90 seconds after the negative (-) battery cable is disconnected to prevent possible deployment of the air bag.

2. Remove the undercover.

3. Remove the timing belt.

4. Remove the crankshaft pulley.

5. Remove the crankshaft position sensor.

6. Remove the crankshaft sprocket, sensing blade, spacer and the key.

7. Pry out the crankshaft seal using a suitable tool.

NOTE: Be careful not to damage the crankshaft and the oil pump case.

To install:

8. Using a driver, install the new crank shaft seal.

9. Install the crankshaft key, spacer sensing blade and the sprocket.

10. Install the crankshaft pulley and torque to:

- DOHC: 130–137 ft. lbs. (180–190 Nm)
- SOHC: 108–116 ft. lbs. (150–160 Nm)

11. Install the timing belt.

12. Install the under cover and the negative battery cable.

Timing Belt, Sprockets, Tensioner and Front Cover

REMOVAL AND INSTALLATION

3.0L (VIN J and K) DOHC Engines

1. Position the engine so the No. 1 cylinder is at TDC of its compression stroke.

2. Disconnect the negative battery cable.

—————— **CAUTION** ——————
Wait at least 90 seconds after the negative (-) battery cable is disconnected to prevent possible deployment of the air bag.

3. Remove the engine under cover.

4. Remove the front under cover panel.

5. Remove the cruise control pump and the link assembly.

6. Remove the alternator.

7. Raise and suspend the engine so force is not applied to the engine mount.

8. Remove the timing covers from the engine.

9. If the same timing belt will be reused, mark the direction of the timing belt's rotation for installation in the same direction. Make sure the engine is positioned so the No. 1 cylinder is at the TDC of its compression stroke and the timing marks are aligned with the engine's timing mark indicators on the valve covers or head.

10. Loosen the center bolt of tensioner pulley and unbolt auto-tensioner assembly. The auto-tensioner assembly must be reset to correctly adjust belt tension. Remove the timing belt.

11. Using a wrench, hold the camshaft at its hexagon and remove the camshaft sprocket bolt.

12. Remove and position the auto-tensioner into a vise with soft jaws. The plug at the rear of tensioner protrudes, be sure to use a washer as a spacer to protect the plug from contacting vise jaws.

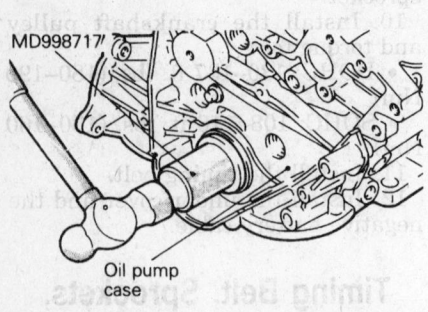

Crankshaft seal installation

MD998717

Oil pump case

MD998717 Crankshaft

Guide

Oil seal

323758

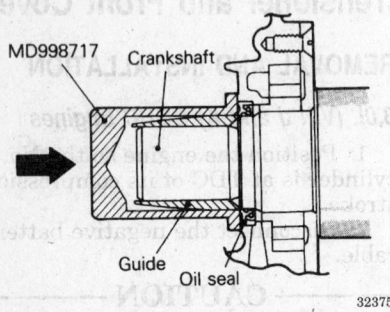

13. Slowly push the rod into the tensioner until the set hole in rod is aligned with set hole in the auto-tensioner.

14. Insert a 0.055 in. (1.4mm) wire into the aligned set holes. Unclamp the tensioner from the vise and install it on the engine. Tighten tensioner to 17 ft. lbs. (24 Nm).

15. Clean and inspect both auto tensioner mounting bolts. Coat the threads of the old bolts with thread sealer. If new bolts are installed, inspect the heads of the new bolts. If there is white paint on the bolt head, no sealer is required. If there is no paint on the head of the bolt, apply a

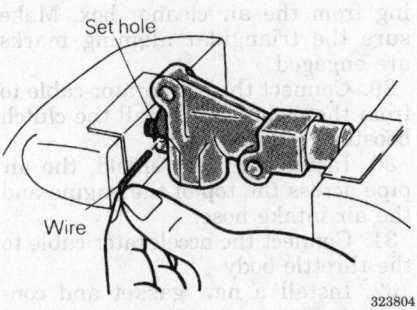

Set hole

Wire

323804

Timing belt tensioner — 3.0L (VIN J and K) DOHC engines

coat of thread sealer to the bolt. Install both bolts and torque to 17 ft. lbs. (24 Nm).

To install:

WARNING

Turning the camshaft sprocket when the timing belt is removed could cause the valves to interfere with the pistons.

---

16. Using a wrench, hold the camshaft at its hexagon and tighten the bolt holding the camshaft sprocket to 65 ft. lbs. (90 Nm).

17. Align the mark on the crankshaft sprocket with the mark on the front case. Then move the sprocket 3 teeth clockwise to lower the piston so the valves do not touch the piston if the camshafts are being moved.

18. Turn each camshaft sprocket 1 at a time to align the timing marks with the mark on the valve cover or head. If the intake and exhaust valves of the same cylinder are opened simultaneously, they could interfere with each other. Therefore, if any resistance is felt, turn the other camshaft to move the valve.

19. Using large paper clips to secure the timing belt to sprockets, install the timing belt in the following order. Be sure camshafts to cylinder heads and crankshaft to front cover timing marks are aligned.

   a. Exhaust camshaft sprocket (front bank).

   b. Intake camshaft sprocket (front bank).

   c. Water pump pulley.

   d. Intake camshaft sprocket (rear bank).

   e. Exhaust camshaft sprocket (rear bank).

   f. Idler pulley.

   g. Crankshaft pulley.

   h. Tensioner pulley.

NOTE: Since the camshaft sprockets turn easily, secure them with box wrenches to install timing belt.

20. Align all timing marks and raise tensioner pulley against belt to remove slack, snug tensioner bolt.

21. Loosen the center bolt on the tensioner pulley. Using tool MD998767 or equivalent and a torque wrench, apply a torque of 7.2 ft. lbs. (10 Nm). Tighten the tensioner bolt to 42 ft. lbs. (58 Nm) and make sure the tensioner does not rotate with the bolt.

22. Remove the set wire attached to the auto tensioner.

23. Rotate the crankshaft 2 complete turns clockwise and let it sit for approximately 5 minutes. Then,

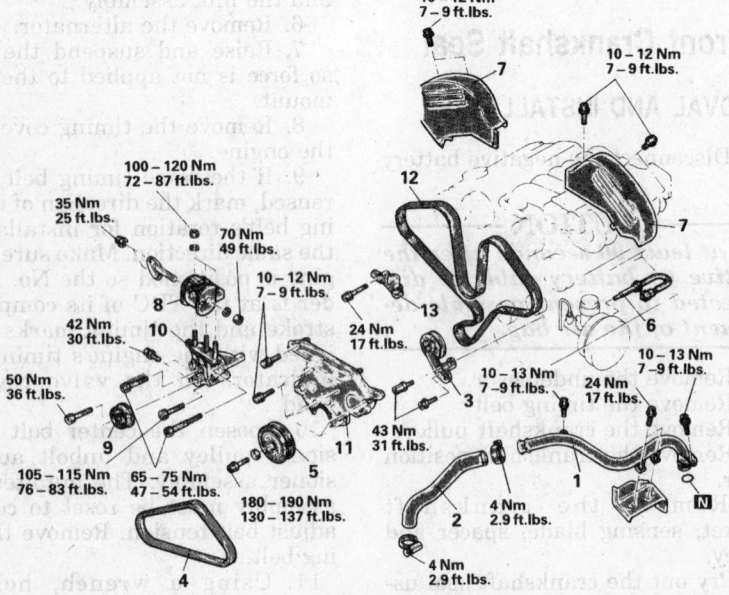

100 – 120 Nm
72 – 87 ft. lbs.

35 Nm
25 ft. lbs.

70 Nm
49 ft. lbs.

10 – 12 Nm
7 – 9 ft. lbs.

42 Nm
30 ft. lbs.

50 Nm
36 ft. lbs.

105 – 115 Nm
76 – 83 ft. lbs.

65 – 75 Nm
47 – 54 ft. lbs.

180 – 190 Nm
130 – 137 ft. lbs.

10 – 12 Nm
7 – 9 ft. lbs.

10 – 12 Nm
7 – 9 ft. lbs.

24 Nm
17 ft. lbs.

10 – 13 Nm
7 – 9 ft. lbs.

24 Nm
17 ft. lbs.

10 – 13 Nm
7 – 9 ft. lbs.

43 Nm
31 ft. lbs.

4 Nm
2.9 ft. lbs.

4 Nm
2.9 ft. lbs.

1. Air hose
2. Air pipe
3. Tensioner assembly
4. Drive belt (power steering)
5. Crankshaft pulley
6. Brake fluid level sensor
7. Timing belt upper cover
8. Engine mount bracket
9. Idler pulley (alternator/air conditioner)
10. Engine support bracket
11. Timing belt lower cover
12. Timing belt
13. Auto tensioner

adjustment of timing belt tension

323803

Engine timing belt assembly — 3.0L (VIN J and K) DOHC engines

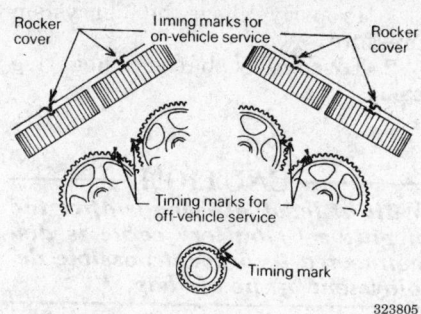

Timing marks — 3.0L (VIN J and K) DOHC engines

323805

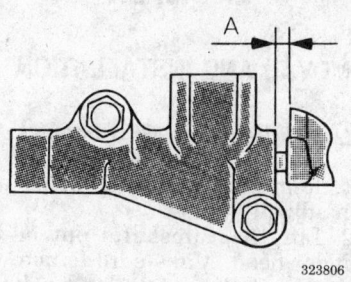

Auto tensioner protrusion — 3.0L (VIN J and K) DOHC engines

323806

check that the set pin can easily be inserted and removed from the hole in the auto tensioner.

NOTE: Even if the set pin cannot be easily inserted, the auto tensioner is normal if its rod protrusion is within specification.

24. Measure the auto tensioner protrusion (the distance between the tensioner arm and auto tensioner body) to ensure that it is within 0.15–0.18 in. (3.8–4.5mm). If out of specification, repeat adjustment procedure until the specified value is obtained.
25. Check again that the timing marks on all sprockets are in proper alignment.
26. Lower the engine so the weight is again applied to the engine mount.
27. Make any necessary engine adjustments.
28. Install the alternator.
29. Install the cruise control pump and the link assembly.
30. Install the timing belt covers and all related items.
31. Install the front under cover panel.
32. Install the engine under cover.
33. Connect the negative battery cable.
34. Start the engine and check the timing.

3.0L (VIN H) SOHC Engine

1. Position the engine so the No. 1 cylinder is at TDC of its compression stroke.
2. Disconnect the negative battery cable.

— **CAUTION** —
Wait at least 90 seconds after the negative (-) battery cable is disconnected to prevent possible deployment of the air bag.

3. Remove the engine under cover.
4. Remove the cruise control pump and the link assembly.

5. Raise and suspend the engine so force is not applied to the engine mount.
6. Remove the engine mount and support bracket from the engine.
7. Remove the timing covers from the engine.

NOTE: It is recommended to change the timing belt at 60,000 mile intervals.

8. If the same timing belt will be reused, mark the direction of the timing belt's rotation for installation in the same direction. Make sure the engine is positioned so the No. 1 cylinder is at the TDC of its compression stroke and the timing marks are al-

igned with the engine's timing mark indicators.
9. Loosen the timing belt tensioner bolt and remove the belt. If the tensioner is not being removed, position it as far away from the center of the engine as possible and tighten the bolt.
10. If the tensioner is being removed, mark the outside of the spring to ensure that it is not installed backwards. Unbolt the tensioner and remove it along with the spring.

— **WARNING** —
Turning the camshaft sprocket when the timing belt is removed could cause the valves to interfere with the pistons thus causing severe internal engine damage.

11. Using sprocket holding tool SST MB990767 or equivalent, hold the camshaft sprockets from turning and remove the camshaft sprocket bolts.
12. Note the positioning and remove the camshaft sprockets from the camshafts without disturbing the positioning of the camshafts.
   To install:
13. Install the camshaft sprocket and torque to 66 ft. lbs. (90 Nm). Be

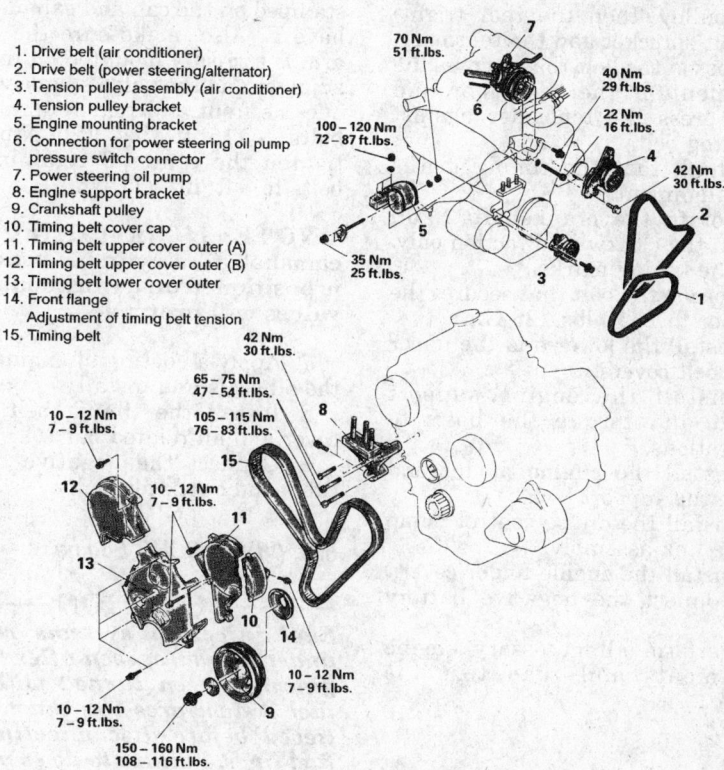

1. Drive belt (air conditioner)
2. Drive belt (power steering/alternator)
3. Tension pulley assembly (air conditioner)
4. Tension pulley bracket
5. Engine mounting bracket
6. Connection for power steering oil pump pressure switch connector
7. Power steering oil pump
8. Engine support bracket
9. Crankshaft pulley
10. Timing belt cover cap
11. Timing belt upper cover outer (A)
12. Timing belt upper cover outer (B)
13. Timing belt lower cover outer
14. Front flange
   Adjustment of timing belt tension
15. Timing belt

Engine timing belt assembly — 3.0L (VIN H) SOHC engine

327479

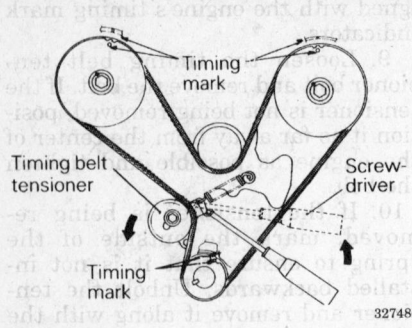

Timing mark

Timing belt tensioner

Screw-driver

Timing mark

327480

Timing marks and belt removal — 3.0L (VIN H) SOHC engine

sure to prevent the camshafts from turning.

14. Install the tensioner, if removed, and hook the upper end of the spring to the water pump pin and the lower end to the tensioner in exactly the same position as originally installed.

15. Position both camshafts so the marks align with those on the rear timing covers. Rotate the crankshaft so the timing mark aligns with the mark on the front cover.

16. Install the timing belt on the crankshaft sprocket and while keeping the belt tight on the tension side, install the belt on the front (left) camshaft sprocket.

17. Install the belt on the water pump pulley, then the rear (right) camshaft sprocket and the tensioner.

18. Loosen the bolt that secures the adjustment of the tensioner and lightly press the tensioner against the timing belt.

19. Check that the timing marks are in alignment.

20. Rotate the crankshaft 2 full turns in the clockwise direction only. Align the timing marks.

21. Torque the bolt that secures the tensioner to 19 ft. lbs. (26 Nm).

22. Install the lower and the upper timing belt covers.

23. Install the engine support bracket and torque the bolts to specifications.

24. Install the engine mount and remove the support jack.

25. Install the cruise control pump and the link assembly.

26. Install the engine under cover.

27. Connect the negative battery cable.

28. Perform all necessary engine adjustments and road test the vehicle.

Camshaft

REMOVAL AND INSTALLATION

3.0L (VIN J and K) DOHC Engines

1. Remove the rocker arms and lash adjusters.

2. Lift the camshafts out of the cylinder head. Make sure to note the positions of the intake and exhaust camshafts.

To install:

3. Bleed the lash adjusters.

4. Lubricate the camshafts with heavy engine oil and position the camshafts on the cylinder head.

NOTE: Do not confuse the intake camshaft with the exhaust camshaft. The intake camshaft has a J stamped on the hexagon and the exhaust camshaft has a K or N.

5. Make sure the dowel pins on both camshaft sprocket ends are positioned properly.

6. Install the bearing caps. Tighten the caps in sequence and in 2 or 3 steps. Caps 2, 3 and 4 have a front mark. Install with the mark aligned with the front mark on the cylinder head. Intake caps have **I** stamped on the cap and exhaust caps have **E**. Also, make sure the rocker arm is correctly mounted on the lash adjuster and the valve stem end. Torque the front and rear retaining cap bolts to 15 ft. lbs. (20 Nm) and tighten the center 3 retaining cap bolts to 8 ft. lbs. (11 Nm).

NOTE: If installing the camshaft to a cylinder head that is positioned on a workbench, the valves will protrude.

7. Apply a coating of engine oil to the oil seals and install.

8. Install the timing belt, valve cover and all related parts.

9. Connect the negative battery cable and check for leaks.

3.0L (VIN H) SOHC Engine

1. Properly relieve the fuel system pressure.

2. Disconnect battery negative cable.

3. Remove the timing belt cover and timing belt.

4. If removing the rear camshaft, remove the distributor extension.

5. Remove the valve covers.

6. Remove the rocker arms, rocker shafts and the bearing caps as an assembly.

7. Remove the crank angle sensor and adaptor.

8. Remove the camshaft from the cylinder head.

To install:

9. Immerse the lash adjusters in clean diesel fuel. Using a small wire, move the plunger of the lash adjuster up and down 4 or 5 times while pushing down lightly on the check ball in order to bleed out the air. Install the lash adjusters in the cylinder head.

10. Lubricate the camshafts with heavy engine oil and position the camshafts on the cylinder head.

11. Make sure the dowel pin on both camshaft sprocket ends are positioned properly.

12. Install the bearing caps. Tighten the caps in sequence and in 2 or 3 steps. Cap Nos. 2, 3 and 4 have a front mark. Install with the mark aligned with the front mark on the cylinder head. Torque the bearing cap bolts in the following sequence: 3, 2, 1, 4. Torque to 7 ft. lbs. (10 Nm) at first, then torque a final time to 15 ft. lbs. (20 Nm).

NOTE: If installing the camshaft to a cylinder head that is positioned on a workbench, the valves will protrude.

13. Apply a coating of engine oil to the oil seals and install them.

14. Install the distributor extension, if removed.

15. Install the timing belt, valve cover and all related parts.

16. Install the valve covers.

17. Connect the negative battery cable and check for leaks.

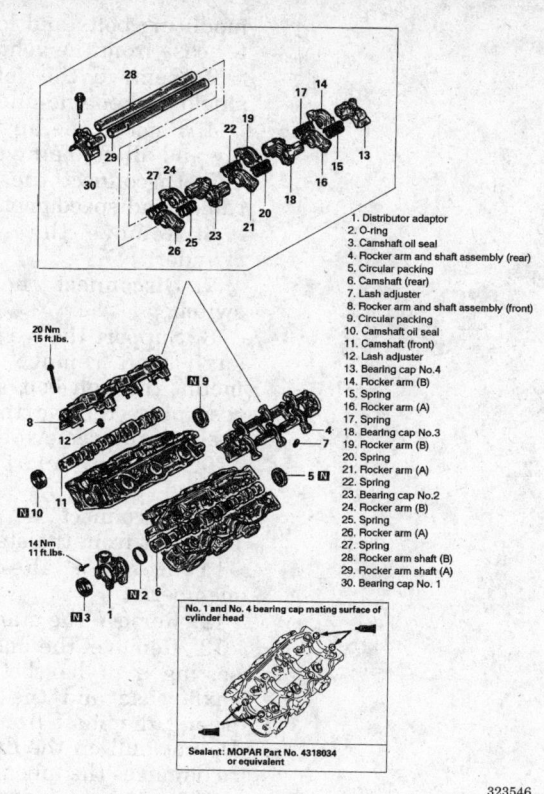

20 Nm
15 ft.lbs.

14 Nm
11 ft.lbs.

1. Distributor adaptor
2. O-ring
3. Camshaft oil seal
4. Rocker arm and shaft assembly (rear)
5. Circular packing
6. Camshaft (rear)
7. Lash adjuster
8. Rocker arm and shaft assembly (front)
9. Circular packing
10. Camshaft oil seal
11. Camshaft (front)
12. Lash adjuster
13. Bearing cap No.4
14. Rocker arm (B)
15. Spring
16. Rocker arm (A)
17. Spring
18. Bearing cap No.3
19. Rocker arm (B)
20. Spring
21. Rocker arm (A)
22. Spring
23. Bearing cap No.2
24. Rocker arm (B)
25. Spring
26. Rocker arm (A)
27. Spring
28. Rocker arm shaft (B)
29. Rocker arm shaft (A)
30. Bearing cap No. 1

No. 1 and No. 4 bearing cap mating surface of cylinder head

Sealant: MOPAR Part No. 4318034 or equivalent

323546

Camshaft assembly — 3.0L (VIN H) SOHC engine

Piston and Connecting Rod

POSITIONING

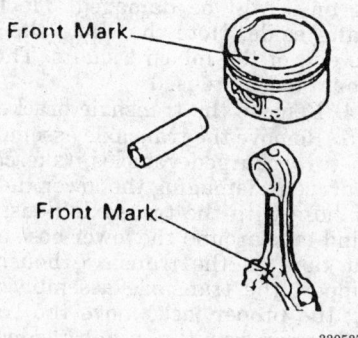

330537

Aligning the piston and the connecting rod

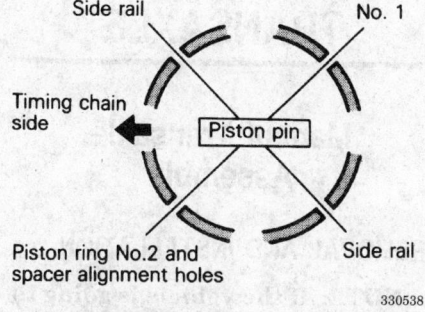

Side rail   No. 1

Timing chain side

Piston pin

Piston ring No.2 and spacer alignment holes   Side rail

330538

Arrange the piston ring and oil ring gaps (side rail and spacer) as shown

---

ENGINE LUBRICATION

Oil Pan

REMOVAL AND INSTALLATION

1. Disconnect the negative battery cable.
2. Raise the vehicle and support safely.

3. Remove the oil pan drain plug and drain the engine oil.
4. If equipped with AWD, remove the transfer assembly.
5. Disconnect and lower the exhaust pipe. On turbocharged engines, disconnect the return pipe for the turbocharger from the side of the oil pan.
6. Remove the oil pan mounting bolts.
7. Using the special tool, separate and remove the engine oil pan.

To install:

8. Thoroughly clean and dry the oil pan, cylinder block bolts and bolt holes.
9. Apply a thin bead of sealer around the surface of the oil pan.
10. Assemble the oil pan to the cylinder block within 15 minutes after applying the sealant.
11. Install the oil pan mounting bolts and torque to 4–6 ft. lbs. (6–8 Nm).
12. Fill the engine with the proper amount of oil.
13. Connect the negative battery cable and check for leaks.
14. Safely lower the vehicle to the floor.

Oil Pump

REMOVAL AND INSTALLATION

NOTE: **Whenever the oil pump is disassembled or the cover removed, the gear cavity must be filled with petroleum jelly to seal the pump and act as a prime. This allows the pump to draw oil as soon as the engine starts. Do not use grease.**

1. Disconnect the negative battery cable.

--- **CAUTION** ---
Wait at least 90 seconds after the negative (-) battery cable is disconnected to prevent possible deployment of the air bag.
---

2. Remove the front engine mount bracket and accessory drive belts.
3. Remove timing belt upper and lower covers.
4. Remove the timing belt and crankshaft sprocket.
5. Remove the oil pan.
6. Remove the oil screen and gasket.
7. Remove and tag the front cover mounting bolts. Note the lengths of the mounting bolts as they are removed for proper installation.

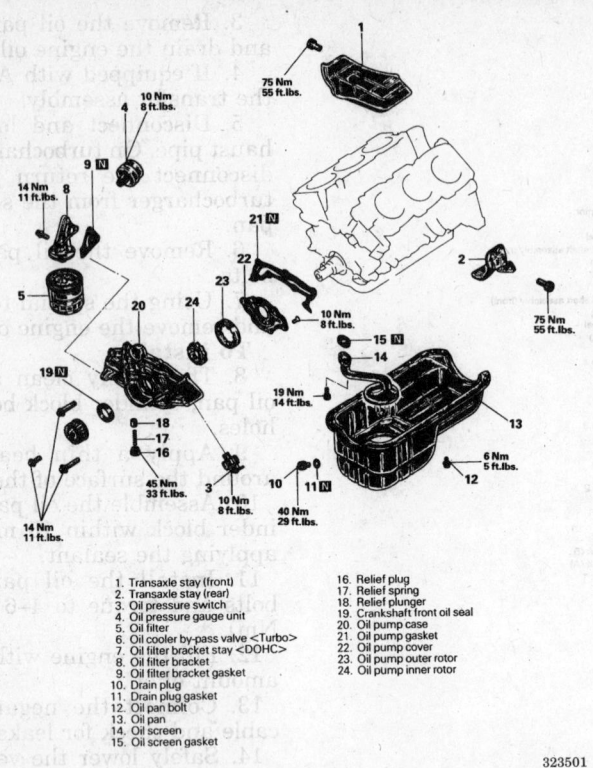

1. Transaxle stay (front)
2. Transaxle stay (rear)
3. Oil pressure switch
4. Oil pressure gauge unit
5. Oil filter
6. Oil cooler by-pass valve <Turbo>
7. Oil filter bracket stay <DOHC>
8. Oil filter bracket
9. Oil filter bracket gasket
10. Drain plug
11. Drain plug gasket
12. Oil pan bolt
13. Oil pan
14. Oil screen
15. Oil screen gasket
16. Relief plug
17. Relief spring
18. Relief plunger
19. Crankshaft front oil seal
20. Oil pump case
21. Oil pump gasket
22. Oil pump cover
23. Oil pump outer rotor
24. Oil pump inner rotor

323501

Oil pan and pump assembly — 3.0L (VIN H) SOHC engines, DOHC engines are similar

8. Remove the front case cover and oil pump assembly. Disassemble as required.

To install:

9. Thoroughly clean all gasket material from all mounting surfaces.

10. Apply engine oil to the entire surface of the gears or rotors.

11. Assemble the front case cover and oil pump assembly to the engine block using a new gasket.

12. Install the oil screen with new gasket.

13. Install the oil pan and timing belts.

14. Install the timing belt covers.

15. Install the drive belts and the front engine mount bracket.

16. Connect the negative battery cable, refill the crankcase and check for adequate oil pressure.

TRANSAXLE

Manual Transaxle Assembly

REMOVAL AND INSTALLATION

NOTE: If the vehicle is going to be rolled on its wheels while the halfshafts are out of the vehicle, obtain 2 outer CV-joints or proper equivalent tools and install to the hubs. If the vehicle is rolled without the proper torque applied to the front wheel bearings, the bearings will no longer be usable.

CAUTION
Wait at least 90 seconds after the negative (-) battery cable is disconnected to prevent possible deployment of the air bag.

---

1. Remove the battery and battery tray. Raise the vehicle and support safely. Drain the transaxle oil and the oil from the transfer case.

2. If equipped with AWD, disconnect the exhaust pipe. Remove the mounting bolts and lower the transfer case from the vehicle.

3. Remove the left side splash shield and engine under cover.

4. Remove the air cleaner assembly and all adjoining duct work.

5. Disconnect the shifter control cables and speedometer connector.

6. Remove the clutch release cylinder.

7. Disconnect the reverse light switch.

8. Support the weight of the transaxle and remove the transaxle mount through bolt. Remove the access plug, remove the bolts for the bracket and remove the brackets.

9. Disconnect the transaxle ground cable.

10. Disconnect the tie rod end and ball joint from the steering knuckle.

11. Remove the right frame member.

12. Remove the starter motor.

13. Remove the halfshafts by inserting a prybar between the transaxle case and the driveshaft and prying the shaft from the transaxle. Do not pull on the driveshaft. Doing so damages the inboard joint. Use the prybar. Do not insert the prybar so far the oil seal in the case is damaged. On AWD, remove the right side shaft as just described. The left side shaft can be removed by tapping with a plastic hammer. Remove the shaft with the hub and knuckle as an assembly. Don't tap on the center bearing or it will be damaged. Tie the shafts aside. Note the circle clip on the end of the inboard shafts. These should not be reused.

14. Remove the transaxle brackets.

15. Remove the transaxle assembly. On turbocharged vehicles, take care to prevent damaging the lower radiator hose with the transaxle housing. Wind tape around the lower hose and put tape on the transaxle housing. Support the transaxle assembly using the proper jack, move the transaxle away from the engine and lower it.

To install:

16. Install the transaxle to the engine and install the mounting bolts.

17. When installing the halfshafts, use new circlips on the axle ends. Take care to get the inboard joint parts straight, not bent relative to the axle. Care must be taken to ensure that the oil seal lip of the transaxle is not damaged by the serrated part of the driveshaft.

18. Install the starter motor and cover.

19. Install the right side frame member.

20. Install the ball joint and tie rod to the steering knuckle.

21. Connect the transaxle ground cable.

22. Install the side mount brackets and install the access plug.

23. Connect the reverse light switch.

24. Install the clutch release cylinder.

25. Connect the shifter control cables and speedometer connector.

26. Install the transfer case and related items on AWD vehicles.

27. Install the air cleaner assembly and all adjoining duct work.

28. Install the left side splash shield.

29. Install the battery tray and battery.

30. Make sure the vehicle is level when refilling the transaxle. Use Hypoid gear oil or equivalent, GL-4 or higher.

31. Connect the negative battery cable and check the transaxle and transfer case for proper operation. Make sure the reverse lamps come ON when in reverse.

Clutch Assembly

REMOVAL AND INSTALLATION

1. Disconnect the negative battery cable.

----- CAUTION -----
Wait at least 90 seconds after the negative (-) battery cable is disconnected to prevent possible deployment of the air bag.

2. Raise and safely support the vehicle.

3. Remove the transaxle assembly from the vehicle.

4. Remove the pressure plate attaching bolts. If the pressure plate is to be reused, loosen the bolts in succession, one or 2 turns at a time to prevent warping the the cover flange.

5. Remove the pressure plate release bearing assembly and the clutch disc. Do not use solvent to clean the bearing.

6. Inspect the condition of the clutch components and replace any worn parts.

To install:

7. Inspect the flywheel for heat damage or cracks. Resurface or replace the flywheel as required, using new bolts.

8. Using the proper alignment tool, install the clutch disc to the flywheel. Install the pressure plate assembly and tighten the pressure plate bolts evenly to 13 ft. lbs. (18 Nm). Remove the alignment tool.

9. Apply a very light coat of high temperature grease to the clutch fork at the ball pivot and where the fork contacts the bearing. Also a little bit of grease can be applied to end of the release cylinder's pushrod and to the pushrod hole on the fork. Apply a light coat of grease on the transaxle input shaft splines.

10. Install a new clutch release bearing. Pack its inner surface with high temperature grease.

11. Install the transaxle assembly.

12. Lower the vehicle and connect the negative battery cable.

13. Check the clutch for proper operation.

Clutch Master Cylinder

REMOVAL AND INSTALLATION

1. Disconnect the negative battery cable.

2. Remove the power brake booster and any other underhood components necessary in order to gain access to the clutch master cylinder.

3. Loosen the line at the cylinder and allow the fluid to drain. Use care, brake fluid damages paint.

4. Remove the clevis pin retainer at the clutch pedal and remove the washer and clevis pin. AWD vehicles have a clutch pedal booster which directly activates the master cylinder.

5. Remove the 2 nuts and pull the cylinder from the firewall. A seal should be between the mounting flange and firewall. This seal should be replaced.

To install:

6. Using a new seal, mount the clutch master cylinder to the firewall.

7. Install the clevis pin on FWD vehicles.

8. Lubricate all pivot points with grease.

9. Install the power brake booster.

10. Bleed the system at the slave cylinder using DOT 3 brake fluid and check the adjustment of the clutch pedal.

Clutch Slave Cylinder

REMOVAL AND INSTALLATION

1. Disconnect the negative battery cable.

----- CAUTION -----
Wait at least 90 seconds after the negative (-) battery cable is disconnected to prevent possible deployment of the air bag.

2. Remove necessary underhood components in order to gain access to the clutch release cylinder.

3. Remove the hydraulic line and allow the system to drain.

4. Remove the bolts and pull the cylinder from the transaxle housing.

To install:

5. Apply grease to the contact point on the release fork.

6. Install the slave cylinder to the transaxle housing.

7. Connect the hydraulic line to the slave cylinder.

8. Bleed the system using DOT 3 brake fluid.

9. Connect the negative battery cable.

Hydraulic Clutch System Bleeding

PROCEDURE

1. Fill the reservoir with brake fluid.

2. Loosen the bleed screw, have the clutch pedal pressed to the floor.

3. Tighten the bleed screw and release the clutch pedal.

4. Repeat the procedure until the fluid is free of air bubbles.

NOTE: Attach a hose to the bleeder and place the other end into a container at least ½ full of brake fluid during bleeding. Do not allow the reservoir to run out of fluid during bleeding.

Automatic Transaxle Assembly

REMOVAL AND INSTALLATION

NOTE: If the vehicle is going to be rolled on its wheels while the halfshafts are out of the vehicle, obtain 2 outer CV-joints or proper equivalent tools and install to the hubs. If the vehicle is rolled without the proper torque applied to the front wheel bearings, the bearings will no longer be usable.

325078

Clutch slave cylinder bleeder screw

----- CAUTION -----

Wait at least 90 seconds after the negative (-) battery cable is disconnected to prevent possible deployment of the air bag.

1. Disarm the air bag, if equipped. Remove the battery, battery tray and washer tank.

2. Remove the air cleaner assembly and adjoining duct work.

3. Disconnect the shifter control cable.

4. Disconnect and plug the oil cooler hoses.

5. Disconnect the inhibitor switch, kickdown servo switch, pulse generator, oil temperature sensor, shift control solenoid valve, and ground cable.

6. Disconnect the speedometer cable.

7. Raise the vehicle and support safely. Remove the undercovers.

8. Support the weight of the transaxle and remove the mount bracket. Remove the upper bell housing bolts.

9. Disconnect the tie rod end and ball joint from the steering knuckle.

10. Remove the right frame member.

11. Remove the starter.

12. Remove the halfshafts by inserting a prybar between the transaxle case and the driveshaft and prying the shaft from the transaxle. Do not pull on the driveshaft. Doing so damages the inboard joint. Use the prybar. Do not insert the prybar so far the oil seal in the case is damaged. Tie the halfshafts aside.

13. Remove the remaining mounting brackets.

14. Remove the bell housing cover plate.

15. Remove the special bolts holding the flexplate to the torque converter.

16. After removing the bolts, push the torque converter toward the transaxle so it doesn't stay on the engine

side and allow oil to pour out the converter hub.

17. Remove the lower transaxle to engine bolts and remove the transaxle assembly.

To install:

18. After the torque converter has been mounted on the transaxle, install the transaxle assembly on the engine. Tighten the driveplate bolts to 34–38 ft. lbs. (46–53 Nm). Install the bell housing cover.

19. Install the mounting brackets.

20. Replace the circlips and install the halfshafts to the transaxle.

21. Install the starter and frame member.

22. Install the tie rods and ball joint to the steering arm.

23. Install the upper bell housing bolts.

24. Install the transaxle mounting bracket.

25. Install the undercovers.

26. Connect the speedometer cable.

27. Connect the inhibitor switch, kickdown servo switch, pulse generator, oil temperature sensor, shift control solenoid valve, and ground cable.

28. Connect the oil cooler hoses.

29. Connect the shifter control cable.

30. Install the air cleaner assembly and adjoining duct work.

31. Install the washer tank, battery tray and battery.

32. Refill with Dexron II, Mopar ATF Plus type 7176 or equivalent, automatic transaxle fluid.

33. Start the engine and allow it to idle for 2 minutes. Apply parking brake and move selector through each gear position, ending in **N**. Recheck fluid level and add if necessary. Fluid level should be between the marks in the **HOT** range.

Transfer Case Assembly

REMOVAL AND INSTALLATION

With Manual Transaxles

1. Disconnect the negative battery cable.

----- CAUTION -----

Wait at least 90 seconds after the negative (-) battery cable is disconnected to prevent possible deployment of the air bag.

2. Raise the vehicle and support safely. Drain the transfer assembly oil.

3. Remove necessary front bumper components.

4. Disconnect the front exhaust pipe.

5. Unbolt the transfer case assembly and remove by sliding it off the rear driveshaft. Be careful not to damage the oil seal in the transfer case output housing. Do not let the rear driveshaft hang; suspend it from a frame piece. Cover the opening in the transaxle and transfer case to keep oil from dripping and to keep dirt out.

To install:

6. Lubricate the driveshaft sleeve yoke and oil seal lip on the transfer extension housing. Install the transfer case assembly to the transaxle. Use care when installing the rear driveshaft to the transfer case output shaft.

7. Tighten the transfer case to transaxle bolts to:
• 1993: 61– 65 ft. lbs. (85–90 Nm)
• 1994-97: 18–22 ft. lbs. (25–29 Nm)

8. Install the exhaust pipe using a new gasket. Install removed bumper components.

9. Refill the transfer case and check the oil levels in the transaxle and transfer case.

10. Safely lower the vehicle and connect the negative battery cable.

DRIVE AXLE

Driveshaft

REMOVAL AND INSTALLATION

1. Disconnect the negative battery cable. Raise the vehicle and support safely.

2. The rear driveshaft is a 3-piece unit, with a front, center and rear propeller shaft. Remove the nuts and insulators from the center support bearing. Work carefully. There will be a number of spacers which will differ from vehicle to vehicle. Check the number of spacers and note their locations for reference during reassembly.

3. Matchmark the rear differential companion flange and the rear driveshaft flange yoke. Remove the companion shaft bolts and remove the driveshaft, keeping it as straight as possible so as to ensure that the boot is not damaged or pinched. Use care to keep from damaging the oil

seal in the output housing of the transfer case.

NOTE: Damage to the boot can be avoided and work will be easier if a piece of cloth or similar material is inserted in the boot.

4. Do not lower the rear of the vehicle or oil will flow from the transfer case. Cover the opening to keep dirt out.

To install:

5. Install the driveshaft to the vehicle and align the matchmarks at the rear yoke. Install the bolts and torque to 36–43 ft. lbs. (50–60 Nm).

6. Install the center support bearing with all spacers in place. Torque the retaining nuts to 22–25 ft. lbs. (30–35 Nm).

7. Check the fluid levels in the transfer case and rear differential case.

U-Joints

REMOVAL AND INSTALLATION

1. Remove the driveshaft from the vehicle.

2. Matchmark the yoke and the driveshaft.

3. Using a brass bar and hammer, slightly tap in the bearing outer race.

4. Remove the 4 snaprings from the bearings.

5. Push out the bearing from the flange.

6. Clamp the bearing outer race in a vise and tap off the flange with a hammer.

7. Repeat Steps 3, 4, and 5 for the other bearings.

8. Check for worn or damaged parts. Inspect the bearing journal surfaces for wear.

To install:

9. Install the bearing cups, seals, and O-rings on the spider.

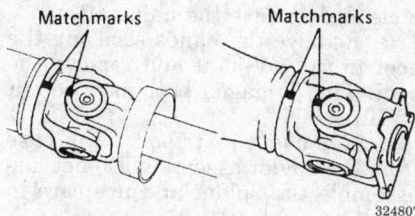

Matchmarks   Matchmarks

324807

Matchmarking the shaft and yoke

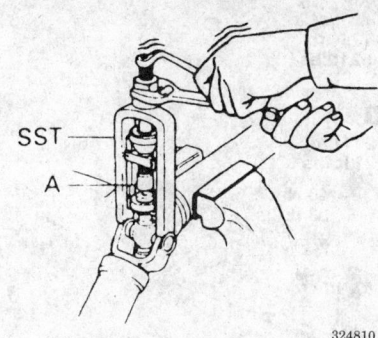

SST

A

324810

Removing the bearing from the flange

10. Grease the spider and the bearings.

NOTE: Be sure to hold the bearing caps while greasing the U-joints. The grease will force the bearing caps off the spider when they are not secured in the driveshaft yoke.

11. Remove the bearing cups from the spider.

12. Position the spider in the yoke.

13. Start the bearings in the yoke and then press them into place, using a vise. Install the snaprings to hold the bearing cups in place.

14. Make sure the bearings and snaprings are fully seated by lightly tapping on the yoke with a hammer.

15. If the axial play of the spider is greater than 0.024 inch (0.06mm), select snaprings which will provide the correct play. Be sure the snaprings are the same size on both sides or driveshaft noise and vibration will result.

16. Install the driveshaft in the vehicle.

Halfshaft

REMOVAL AND INSTALLATION

NOTE: If the vehicle is going to be rolled on its wheels while the halfshafts are out of the vehicle, obtain 2 outer CV-joints or proper equivalent tools and install to the hubs. If the vehicle is rolled without the proper torque applied to the front wheel bearings, the bearings will no longer be usable.

Front Halfshaft

1. Disconnect the negative battery cable.

2. With the vehicle on the floor and the brakes applied, remove the cotter pin, halfshaft nut and the washer.

3. Raise the vehicle and support safely. Remove the lower ball joint and the tie rod end from the steering knuckle.

4. If equipped with an inner shaft, perform the following:

   a. Remove the center support bearing bracket bolts and washers.

   b. Remove the halfshaft by setting up a puller on the outside wheel hub and pushing the halfshaft from the front hub. Then tap the shaft union at the joint case with a plastic hammer to remove the halfshaft shaft and inner shaft from the transaxle.

5. If not equipped with an inner shaft, perform the following procedures:

   a. Remove the halfshaft by setting up a puller on the outside wheel hub and pushing the halfshaft from the front hub.

   b. After pressing the outer shaft, insert a prybar between the transaxle case and the halfshaft and pry the shaft from the transaxle. Do not pull on the shaft. Doing so damages the inboard joint. Do not insert the prybar too far or the oil seal in the case may be damaged.

To install:

6. Inspect the halfshaft boot for damage or deterioration. Check the ball joints and splines for wear.

7. Replace the circlips on the ends of the halfshafts.

8. Insert the halfshaft into the transaxle. Make sure it is fully seated.

9. Pull the strut assembly out and install the other end to the hub.

10. Install the center bearing bracket bolts and tighten to 33 ft. lbs. (45 Nm).

11. Install the washer so the chamfered edge faces outward. Install the nut and tighten temporarily.

12. Install the tie rod end and ball joint.

13. Install the wheel and lower the vehicle to the floor. Tighten the axle nut with the brakes applied. Tighten the nut to a maximum torque of 188 ft. lbs. (260 Nm). Install the cotter pin and bend to secure.

Rear Halfshaft

NOTE: On vehicles with Limited Slip Differential, the right and left halfshafts are not the same. If both halfshafts are to be removed, be sure to mark one of the halfshafts (left or right) for proper installation.

1. Disconnect the negative battery cable. Raise the vehicle and support it safely.

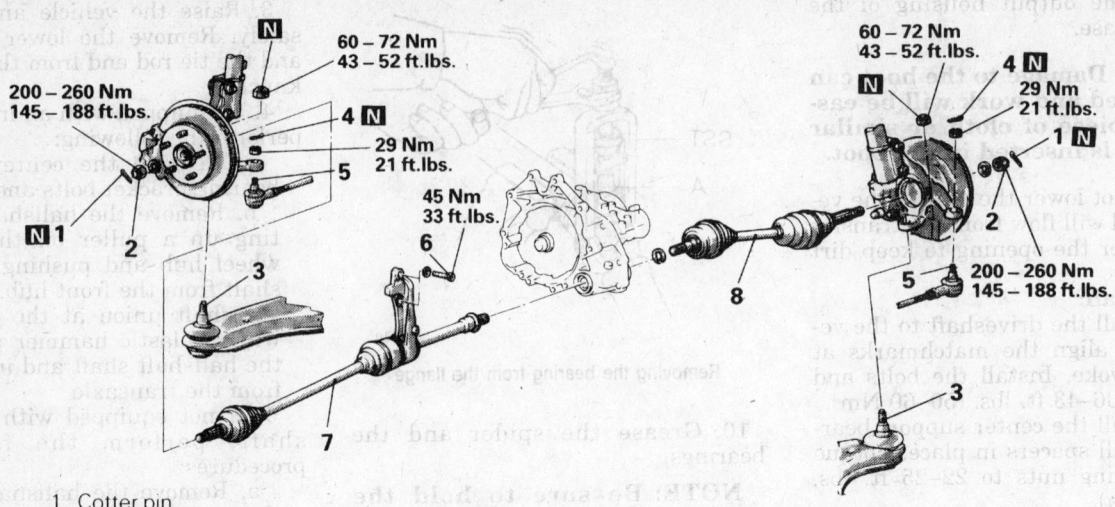

200 – 260 Nm
145 – 188 ft.lbs.

60 – 72 Nm
43 – 52 ft.lbs.

4

29 Nm
21 ft.lbs.

5

1

2

3

45 Nm
33 ft.lbs.

6

7

8

60 – 72 Nm
43 – 52 ft.lbs.

4

29 Nm
21 ft.lbs.

1

2

3

5

200 – 260 Nm
145 – 188 ft.lbs.

8

1. Cotter pin
2. Drive shaft nut
3. Lower arm ball joint connection
4. Cotter pin
5. Tie rod end connection
6. Center bearing bracket installation bolt
7. Drive shaft and inner shaft assembly (L.H.)
8. Drive shaft (R.H.)
9. Circlip

Caution
In the case of AWD-vehicles with A.B.S., take care not to damage the rotor for A.B.S. installed to the B.J. outer race.

Front halfshafts

55 – 65 Nm
40 – 47 ft.lbs.

4

3

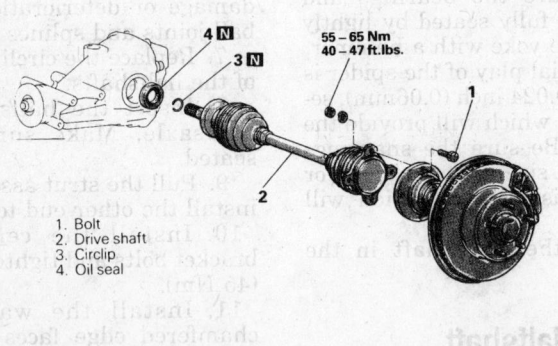

2

1

1. Bolt
2. Drive shaft
3. Circlip
4. Oil seal

324726

Rear halfshaft

2. Matchmark the halfshaft and the companion flange.

3. Remove the bolts that attach the rear halfshaft to the companion flange.

4. Use a prybar to pry the inner shaft out of the differential case. Don't insert the prybar too far or the seal could be damaged.

5. Remove the rear halfshaft from the vehicle.
To install:

6. Install a new circlip on the halfshaft and install it into the differential. Make sure it is fully seated.

7. Align the matchmarks and attach the halfshaft to the companion flange.

8. Safely lower the vehicle and connect the negative battery cable.

CV-Joint Boot

REPLACEMENT

The 2 types of joints used are the Birfield Joint, (BJ), the Tripod Joint (TJ). In addition, some left side shafts will have a center bearing bracket installed on the shaft. Special grease is generally used with these joints and is often supplied with the replacement joint and/or boot. Do not use regular chassis grease.

NOTE: If the Birfield joint boot is being replaced, the Tripod joint will have to be disassembled and the new boot can then be installed from the tripod joint side of the shaft. The Birfield joint cannot be disassembled.

1. Disconnect the negative battery cable.

— **CAUTION** —
Wait at least 90 seconds after the negative (-) battery cable is disconnected to prevent possible deployment of the air bag.

2. Raise and safely support the vehicle and remove the halfshaft.

3. Remove the bands securing the boot to the CV-joint and remove the boot. The damaged boot may be cut off if needed.

4. Remove the tripod joint case from the spider assembly. Do not disassemble the spider and use care in handling.

5. Remove the snapring and the spider assembly from the shaft.

6. If the boot is be reused, wrap vinyl tape around the spline part of

the shaft so the boot will not be damaged when removed. Remove the dynamic damper, if used, and boots from the shaft.

7. Thoroughly clean the old grease out of the CV-joint and blow dry it with compressed air.

To install:

8. Double check that the correct replacement parts are being installed. Wrap vinyl tape around the splines to protect the boot and install the boots and damper, if used, in the correct order.

9. Fill the inside of the boot with the specified grease. Often the grease supplied in the replacement parts kit is meant to be divided in half, with half being used to lubricate the joint and half being used inside the boot. Keep grease off the rubber part of the dynamic damper, if used.

10. Secure the boot bands with the halfshaft in a horizontal position.

11. Install the halfshaft.

12. Install the wheel and tire assembly if removed.

13. Safely lower the vehicle and connect the negative battery cable.

STEERING

Air Bag

─── **CAUTION** ───
Some vehicles are equipped with an air bag system, also known as the Supplemental Inflatable Restraint (SIR) or Supplemental Restraint System (SRS). The system must be disabled before performing service on or around system components, steering column, instrument panel components, wiring and sensors. Failure to follow safety and disabling procedures could result in accidental air bag deployment, possible personal injury and unnecessary system repairs.

PRECAUTIONS

Several precautions must be observed when handling the inflator

module to avoid accidental deployment and possible personal injury.

• Never carry the inflator module by the wires or connector on the underside of the module.

• When carrying a live inflator module, hold securely with both hands, and ensure that the bag and trim cover are pointed away.

• Place the inflator module on a bench or other surface with the bag and trim cover facing up.

• With the inflator module on the bench, never place anything on or close to the module which may be thrown in the event of an accidental deployment.

DISARMING

1. Position the front wheels in the straight-ahead position and place the key in the **LOCK** position. Remove the key from the ignition lock cylinder.

2. Disconnect the negative battery cable and insulate the cable end with high-quality electrical tape or similar non-conductive wrapping.

3. Wait at least one minute before working on the vehicle. The air bag system is designed to retain enough voltage to deploy the air bag for a short period of time after the battery has been disconnected.

ARMING

1. Connect the negative battery cable.

2. Turn the ignition switch to the **ON** position and check the SRS warning light for proper operation.

Steering Wheel

REMOVAL AND INSTALLATION

─── **CAUTION** ───
If equipped with an air bag, be sure to disarm it before starting repairs on the vehicle. Failure to do so could result in personal injury or death.

1. Disconnect the negative battery cable.

2. Remove the air bag module mounting nut from behind the steering wheel. Matchmark the steering wheel.

3. Disconnect the connector of the clockspring from the air bag module, press the air bag's lock towards the module to spread the lock open.

While holding lock in this position, use a small tipped prying tool to gently pry the connector from the module.

4. Store the air bag module in a clean, dry place with the pad cover facing up.

5. Remove the steering wheel retaining nut. Matchmark the steering wheel to the shaft. Use a steering wheel puller to remove the wheel. Do not use a hammer or the collapsible mechanism in the column could be damaged.

To install:

6. Confirm that the front wheels are in a straight-ahead position. Center the clockspring by aligning the **NEUTRAL** mark on the clockspring with the mating mark on the casing.

7. Line up and install the steering wheel. Torque the retaining nut to 29 ft. lbs. (40 Nm).

8. Install the air-bag module.

9. Connect the negative battery cable and check the SRS warning light operation.

Tie Rod Ends

REMOVAL AND INSTALLATION

1. Disconnect the battery negative cable.

2. Raise the vehicle and support it safely.

3. Wire brush the threads on the tie rod shaft and lubricate with penetrating oil. Loosen the locknut.

4. Remove the cotter pin and nut and press the tie rod end from the steering knuckle or trailing arm.

5. Hold the tie rod shaft with locking pliers and turn the tie rod end off, counting the number of turns for installation purposes.

To install:

6. Thread on the new tie rod end the same number of turns required to remove the old one.

7. Install the tie rod stud into the steering knuckle or trailing arm and torque the nut as follows:

Front Tie Rod End
1993 w/FWD: 29 ft. lbs. (40 Nm)
1994–97 w/FWD: 21 ft. lbs. (29 Nm)
1993–97 w/AWD: 36 ft. lbs. (50 Nm)
Rear Tie Rod End: 42 ft. lbs. (58 Nm)

8. Install a new cotter pin.

9. Tighten the locknut

10. Perform a AWD alignment.

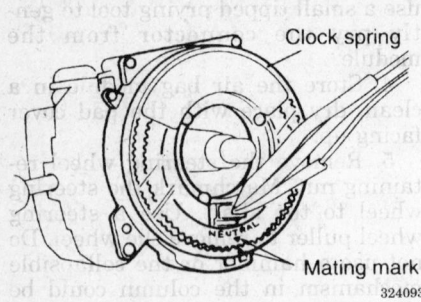

Clock spring

Mating mark
324093

Clockspring mating marks

Power Rack and Pinion

REMOVAL AND INSTALLATION

NOTE: Prior to removal of the steering gear box, center the front wheels and remove the ignition key. Failure to do so may damage the SRS clockspring and render SRS system inoperative, risking serious driver injury.

1. Disconnect the negative battery cable. Disarm the air bag.

———— CAUTION ————
Work must be started after 90 seconds from the time the ignition switch is turned to the LOCK position and the negative (-) battery cable is disconnected.

2. Disconnect the front exhaust pipe.
3. If equipped with AWD, remove the transfer case assembly.
4. Remove the bolt holding the lower steering column joint to the rack and pinion input shaft.
5. Remove the cotter pins and disconnect the tie rod ends.
6. Remove the left and right frame members.
7. Remove the stabilizer bar bracket.
8. If equipped with four-wheel steering, disconnect the lines going to the rear pump.
9. Remove the rack and pinion steering assembly and its rubber mounts. Move the rack to the right to remove it from the crossmember. Use caution to avoid damaging the boots.
To install:
10. Install the rack and install the mounting bolts, tightening bolts to 51 ft. lbs. (70 Nm). When installing the rubber rack mounts, align the projection of the mounting rubber with the

indentation in the crossmember. Install the pinch bolt.
11. Connect the lines going to the four-wheel steering rear pump and to the rack itself.
12. Install the frame members and torque the bolts to 43–51 ft. lbs. (60–70 Nm).
13. Connect the tie rods and Install new cotter pins.
14. Install the transfer case and front exhaust pipe.
15. Refill the reservoir and bleed the system.
16. Perform a front end alignment.

Power Steering Pump

BLEEDING

Front

1. Raise the vehicle and support it safely.
2. Manually turn the pump pulley a few times.
3. Turn the steering wheel all the way to the left and to the right 5–6 times.

NOTE: If bleeding is attempted with the engine running, the air will be absorbed in the fluid. Bleed only while cranking.

4. Disconnect the ignition high tension cable, and, while operating the starter motor intermittently, turn the steering wheel all the way to the left and right 5–6 times for 15-20 seconds. During bleeding, make sure the fluid in the reservoir never falls below the lower position of the filter.
5. Connect the ignition high tension cable, start the engine and allow to idle.
6. Turn the steering wheel left and right until there are no air bubbles in the reservoir. Confirm that the fluid is not milky and the level is up to the specified position on the gauge. Confirm that there is is very little change in the fluid level when the steering wheel is turned. If the fluid level changes more than 0.2 in., the air has not been completely bled. Repeat the process.

Rear

1. Bleed the front steering system.
2. Start the engine and let it idle.
3. Loosen the bleeder screw on the left side of the control valve and install a plastic tube on the bleeder.
4. Turn the steering wheel all the way to the left, then immediately

turn it half way back. Confirm that air has discharged with the fluid.
5. Repeat Step 4 2–3 times as required, to remove all of the air from the rear system. Stop the engine.
6. Loosen the power cylinder (rear steering gear) bleeder screw about ⅛ turn and install the same special tool with the rotation prevention metal fixtures to prevent the bleeder from opening more.
7. Start the engine and run to 50 mph to circulate the fluid.
8. Maintain a speed of 20 mph and turn the steering wheel back and forth. Air should be discharged through the tube of the special tool and into the oil reservoir.
9. Repeat until all air is removed from the power cylinder.

REMOVAL AND INSTALLATION

Front

1. Disconnect the negative battery cable.

———— CAUTION ————
Wait at least 90 seconds after the negative (-) battery cable is disconnected to prevent possible deployment of the air bag.

2. Remove the pressure switch connector from the side of the pump.
3. If the alternator is located under the oil pump, cover it with a shop towel to protect it from oil.
4. Disconnect the return fluid line. Remove the reservoir cap and allow the return line to drain the fluid from the reservoir. If the fluid is contaminated, disconnect the ignition high tension cable and crank the engine several times to drain the fluid from the gearbox.
5. Disconnect the pressure line.
6. Remove the pump drive belt and unbolt the pump from its bracket.
To install:
7. Install the pump, wrap the belt around the pulley and tighten the bolts.
8. Replace the O-rings and connect the pressure line. Connect the pressure line so the notch in the fitting aligns and contacts the pump's guide bracket.
9. Connect the return line.
10. Connect the pressure switch connector.
11. Adjust the belt tension and tighten the adjusting bolts.
12. Refill the reservoir and bleed the system.

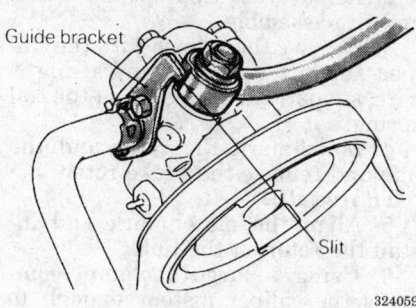

Guide bracket

Slit

324059

Front pressure hose installation

Rear

1. Disconnect the negative battery cable. Raise the vehicle and support safely.
2. Drain the power steering fluid.
3. Remove the main muffler assembly.
4. Remove the rear shock absorber lower mounting bolts.
5. Remove the 2 small cross-member brackets.
6. Using the proper equipment, support the weight of the rear differential. Remove the large self-locking crossmember mounting nuts on the differential side.
7. Disconnect the pressure and suction hoses from the fittings on the pump.
8. Remove the pump retaining bolt and remove the pump from the rear differential assembly. Do not attempt to disassemble the pump; it is not serviceable.
   To install:
9. Replace the O-ring and install the pump assembly to the differential. Make sure the housing is fully seated and the gear is fully engaged. Install the retaining bolt.
10. Replace the O-ring and connect the fluid lines to the pump.
11. Install the large self-locking crossmember mounting nuts on the

differential side. Torque to 80-94 ft. lbs. (110-130 Nm). Remove the support equipment.
12. Install the 2 small crossmember brackets.
13. Install the shock mounting bolts.
14. Install the muffler assembly.
15. Refill the reservoir and bleed the system.

— CAUTION —
Extreme caution should be taken when testing the rear steering pump. Ensure that the vehicle is supported safely and that all components are torqued to specification prior be testing.

16. To check and see if the system is functioning:
   a. Raise the vehicle safely so all 4 wheels turn freely.
   b. Run the vehicle at 50 mph.
   c. Turn the steering wheel quickly to the left and right and make sure the rear wheels steer in the same direction as the front wheels.

BRAKES

Anti-Lock Brake System Service

PRECAUTIONS

• Certain components within the Anti-Lock Brake System (ABS) are not intended to be serviced or repaired individually. Only those components with removal and installation procedures should be serviced.
• Do not use rubber hoses or other parts not specifically specified for and

ABS system. When using repair kits, replace all parts included in the kit. Partial or incorrect repair may lead to functional problems and require the replacement of components.
• Lubricate rubber parts with clean, fresh brake fluid to ease assembly. Do not use lubricated shop air to clean parts; damage to rubber components may result.
• Use only specified brake fluid from an unopened container.
• If any hydraulic component or line is removed or replaced, it may be necessary to bleed the entire system.
• A clean repair area is essential. Always clean the reservoir and cap thoroughly before removing the cap. The slightest amount of dirt in the fluid may plug an orifice and impair the system function. Perform repairs after components have been thoroughly cleaned; use only denatured alcohol to clean components. Do not allow ABS components to come into contact with any substance containing mineral oil; this includes used shop rags.
• The Anti-Lock control unit is a microprocessor similar to other computer units in the vehicle. Ensure that the ignition switch is **OFF** before removing or installing controller harnesses. Avoid static electricity discharge at or near the controller.
• If any arc welding is to be done on the vehicle, the control unit should be unplugged before welding operations begin.

DEPRESSURIZING

The ABS system is a low pressure system, and does not require depressurizing before brake system service.

Master Cylinder

REMOVAL AND INSTALLATION

1. Disconnect the negative battery cable.

— CAUTION —
Work must be started after 90 seconds from the time the ignition switch is turned to the LOCK position and the negative (-) battery cable is disconnected.

2. Disconnect the fluid level sensor connector.
3. Disconnect the brake lines from the master cylinder and plug the lines to prevent drainage.
4. Disconnect the low pressure hose.

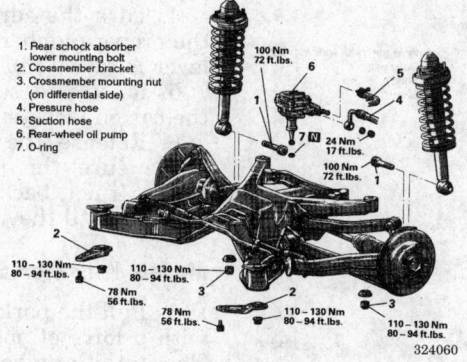

1. Rear schock absorber lower mounting bolt
2. Crossmember bracket
3. Crossmember mounting nut (on differential side)
4. Pressure hose
5. Suction hose
6. Rear-wheel oil pump
7. O-ring

100 Nm 72 ft.lbs.
24 Nm 17 ft.lbs.
100 Nm 72 ft.lbs.
110 – 130 Nm 80 – 94 ft.lbs.
78 Nm 56 ft.lbs.
110 – 130 Nm 80 – 94 ft.lbs.
78 Nm 56 ft.lbs.
110 – 130 Nm 80 – 94 ft.lbs.
110 – 130 Nm 80 – 94 ft.lbs.

324060

Rear power steering pump

5. Remove the 2 nuts securing the master cylinder to the brake booster and remove the master cylinder.

To install:

6. Bench bleed the master cylinder.

7. Install the master cylinder to the studs and install the nuts. Tighten the mounting nuts to 7 ft. lbs. (10 Nm).

8. Connect the fluid level sensor connector.

9. Install the brake lines and the low pressure hose to the master cylinder. Bleed brake system starting at the master cylinder. If air remains in the system continue bleeding the entire system.

10. Connect the negative battery cable and check the brakes for proper operation.

Brake Caliper

REMOVAL AND INSTALLATION

1. Disconnect the negative battery cable.

— **CAUTION** —

Work must be started after 90 seconds from the time the ignition switch is turned to the LOCK position and the negative (-) battery cable is disconnected.

2. Raise the vehicle and support safely. Remove appropriate wheel assembly.

3. To disconnect the front brake hose, hold the nut on the brake hose side and loosen the flared brake line nut.

4. Remove the caliper lock pins holding the caliper to the caliper support and remove the caliper.

To install:

5. Install the brake caliper to the caliper support and torque the bolts to 65 ft. lbs. (90 Nm).

6. Connect the brake hose and torque to 11 ft. lbs. (15 Nm).

7. Make sure the brake hose is not twisted after installation. Refill the brake fluid as required and bleed the brakes.

Disc Brake Pads

REMOVAL AND INSTALLATION

1. Remove approximately ⅓ of the brake fluid from the master cylinder to prevent overflow of the fluid when the caliper pistons are compressed.

2. Disconnect the negative battery cable.

— **CAUTION** —

Work must be started after 90 seconds from the time the ignition switch is turned to the LOCK position and the negative (-) battery cable is disconnected.

3. Raise the vehicle and support it safely.

4. Remove the appropriate wheel assembly.

5. On the front of All Wheel Drive (AWD) vehicles, remove the pad retaining pins and pull the pads out of the caliper body.

6. On others, remove the lock pin and swing the caliper upward and support it with a wire. Take note of the clips, pins, anti-squeal shims and other parts for reference at assembly.

To install:

7. Use a large C-clamp to compress the piston(s) back into the caliper bore.

8. Install the pads and all other small parts.

9. On AWD front disc brakes, apply a small amount of multi-purpose grease to the inner pad shims.

10. Install the caliper to the caliper support. Make sure the brake hose is not twisted after installation if both guide pins were removed.

11. Install the tire and wheel assembly and torque to 87–101 ft. lbs. (120–140 Nm).

12. Connect the negative battery cable.

13. Refill the master cylinder if needed and pump the brake pedal until firm before putting transaxle in gear or moving vehicle.

Brake Rotor

REMOVAL AND INSTALLATION

1. Raise and safely support the vehicle.

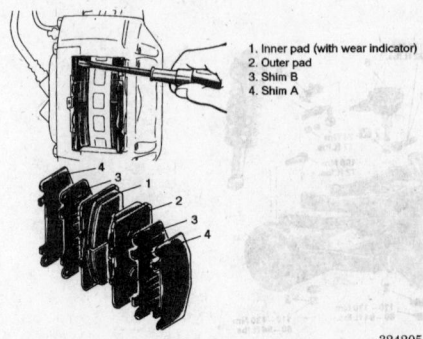

1. Inner pad (with wear indicator)
2. Outer pad
3. Shim B
4. Shim A

324205

Front disc brake pads — all wheel drive models

2. Remove the appropriate wheel and tire assembly.

3. Remove the caliper with the caliper support attached and using a wire, support the caliper from the coil spring.

4. Matchmark the rotor and the hub and remove the brake rotor.

To install:

5. Align the matchmark and install the rotor on the hub.

6. Using a large C-clamp, compress the caliper pistons enough to allow the rotor to fit between the pads and install the caliper assembly. Torque the bolts to 65 ft. lbs. (90 Nm).

7. Install the wheel and tire assembly and torque the lug nuts to 87–101 ft. lbs. (120–140 Nm).

— **CAUTION** —

Apply the brake pedal several times to insure good brake operation before moving the vehicle.

Parking Brake Cable

ADJUSTMENT

Models With Drum Brakes

NOTE: Make certain that the brake shoes are properly adjusted before attempting to adjust the parking brake.

1. Pull the parking brake lever up with a force of about 45 lbs. If that value cannot be determined, just pull it up as far as possible. The total number of clicks heard should be 5–7.

2. If the number of clicks was not within that range, release the lever and back off the cable adjuster locknut at the base of the lever and tighten the adjusting nut until there is no more slack in the cable.

3. Operate the lever and brake pedal several times, until no more clicks are heard from the automatic adjuster.

4. Turn the adjusting nut to give the proper number of clicks when the lever is raised full travel.

5. Raise and support the rear of the car on jackstands.

6. Release the brake lever and make sure the rear wheels turn freely. If not, back off on the adjusting nut until they do.

Models With Disc Brakes

1. Pull the parking brake lever up with a force of about 45 lbs. (61 N). The total number of clicks heard should be 3–5. If the number of clicks

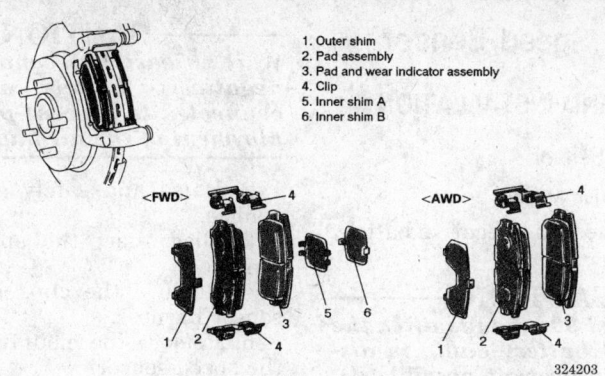

1. Outer shim
2. Pad assembly
3. Pad and wear indicator assembly
4. Clip
5. Inner shim A
6. Inner shim B

\<FWD\>   \<AWD\>

324203

Rear disc brake pads

was not within that range, system requires adjustment.

NOTE: The parking brake shoes must be adjusted before attempting to adjust the cable mechanism

2. To adjust the parking brake shoes perform the following steps:

a. remove the floor console, release the lever and back off the cable adjuster locknut at the base of the lever.

b. Raise the vehicle, support safely and remove the wheel. Remove the hole plug in the brake rotor.

c. Remove the brake caliper and hang out of the way with wire.

d. Use a suitable prybar to pry up on the self-adjuster wheel until the rotor will not turn.

e. Return the adjuster 5 notches in the opposite direction. Make sure the rotor turns freely with a slight drag.

f. Install the caliper and check operation.

3. Once the parking brake shoes have been properly adjusted, adjust the cable mechanism, by performing the following steps:

a. Turn the adjusting nut to give the proper number of clicks when the lever is raised full travel.

b. Raise and support the rear of the car on jackstands.

c. Release the brake lever and make sure the rear wheels turn freely. If not, back off on the adjusting nut until they do.

REMOVAL AND INSTALLATION

Unlike conventional rear disc brake systems, the parking brake operation is **not** incorporated into the brake caliper. This system, uses a separate set of brake shoes, located behind the brake rotor.

1. Disconnect the negative battery cable.

—————— CAUTION ——————
Work must be started after 90 seconds from the time the ignition switch is turned to the LOCK position and the negative (-) battery cable is disconnected.
————————————————

NOTE: If equipped with an air bag, be sure to disarm it before starting repairs on the vehicle.

2. Remove the front and rear floor consoles, by prying out the coin holder, box tray and remote mirror switch, if equipped, or the cover. Remove the small cover around the seat belt from the console side. Remove the screws from the center section and remove the rear part of the console.

NOTE: If equipped with SRS, when removing the floor console, don't allow any impact or shock to the SRS diagnostic unit.

3. Loosen the cable adjuster nut and then remove the parking brake cable, by pulling it from the passenger compartment.

4. Raise the vehicle and support it safely.

5. At the rear wheel, remove the brake caliper and rotor.

6. Remove the parking brake shoes, following the same procedures as conventional drum brake shoes.

7. Disconnect the cable end from the parking brake strut lever. Compress the retaining strips to remove the cable from the backing plate.

8. Unfasten any other frame retainers and remove the cables.

To install:

9. Install the cable to the rear actuator. Secure in place with the parking brake cable clip and retainer spring.

10. Install the parking brake shoes.

11. Install the brake rotor and caliper assembly.

12. Position the cable in the under the vehicle and install retainers loose.

13. Reattach the parking brake cables to the actuator inside the vehicle. Tighten the adjusting nut until the proper tension is placed on the cable. Adjust the parking brake stroke using appropriate method.

14. Secure all cable retainers. Apply and release the parking brake a number of times once all adjustments have been made.

15. Assemble the interior components which were removed.

16. Adjust the parking brake shoes and parking brake cables.

17. Connect the negative battery cable and check the rear wheels to confirm that the rear brakes are not dragging.

18. Check that the parking brake holds the vehicle on an incline.

Brake System Bleeding

PROCEDURE

NOTE: If using a pressure bleeder, follow the instructions furnished with the unit and choose the correct adaptor for the application. Do not substitute an adapter that almost fits as it will not work and could be dangerous.

Master Cylinder

If the master cylinder is off the vehicle, it can be bench bled.

1. Connect 2 short pieces of brake line to the outlet fittings, bend them until the free end is below the fluid level in the master cylinder reservoir.

2. Fill the reservoir with fresh brake fluid. Pump the piston slowly until no more air bubbles appear in the reservoirs.

3. Disconnect the 2 short lines, refill the master cylinder and securely install the cylinder caps.

4. If the master cylinder is on the vehicle, it can still be bled, using a flare nut wrench.

5. Open the brake lines slightly with the flare nut wrench while pressure is applied to the brake pedal by a helper inside the vehicle.

6. Be sure to tighten the line before the brake pedal is released.

7. Repeat the process with both lines until no air bubbles come out.

8. Refill master cylinder and always bleed the complete brake system.

Calipers

1. Fill the master cylinder with fresh brake fluid. Check the level often during the procedure.

2. Starting with the wheel farthest from the master cylinder, remove the protective cap from the bleeder and place it where it will not be lost. Clean the bleeder screw.

——— CAUTION ———
When bleeding the brakes, keep face away from the brake area. Spewing fluid may cause facial and/or visual damage. Do not allow brake fluid to spill on the car's finish. It will remove the paint.

3. If the system is empty, the most efficient way to get fluid down to the wheel is to use a pressure bleeder tool. Open the bleeder until brake fluid flows without signs of air bubbles, close bleeder.

NOTE: If the pedal is pumped rapidly, the fluid will churn and create small air bubbles, which are almost impossible to remove from the system. These air bubbles will accumulate and a spongy pedal will result.

4. If the manual procedure is to be used to pump brake fluid to the caliper or wheel cylinder, open the bleed screw and have an assistant press the brake pedal to the floor.

5. Close the bleeder bleeder screw before releasing the brake pedal, have the helper slowly release the pedal. Wait 15 seconds and repeat the procedure until no more air comes out of the bleeder upon application of the brake pedal. Remember to close the bleeder before the pedal is released inside the vehicle each time the bleeder is opened. If not, air will be introduced into the system.

6. Repeat the procedure on remaining wheel cylinders in the following order:
• Right rear caliper.
• Left front caliper.
• Left rear caliper.
• Right front caliper.

7. Hydraulic brake systems must be totally flushed if the fluid becomes contaminated with water, dirt or other corrosive chemicals. To flush, bleed the entire system until all fluid has been replaced with the correct type of new fluid.

8. Install the bleeder cap on the bleeder to keep dirt out and refill master cylinder. Always road test the vehicle after brake work of any kind is done.

Wheel Speed Sensor

REMOVAL AND INSTALLATION

Front Speed Sensor

Front Wheel Drive Models

1. Disconnect the negative battery cable.

——— CAUTION ———
Wait at least 90 seconds after the negative (-) battery cable is disconnected to prevent possible deployment of the air bag.

2. Raise and safely support the vehicle.

3. Remove the splash shield.

4. Disconnect the speed sensor connector.

5. Remove the clips holding the sensor harness.

6. Remove the speed sensor bracket.

To install:

7. Install the speed sensor bracket and torque to 9 ft. lbs. (12 Nm).

8. Install the clips holding the sensor harness.

9. Connect the speed sensor connector.

10. Connect the negative battery cable.

All Wheel Drive Models

1. Disconnect the negative battery cable.

——— CAUTION ———
Work must be started after 90 seconds from the time the ignition switch is turned to the LOCK position and the negative (-) battery cable is disconnected.

2. Raise and safely support the vehicle.

3. Remove the splash shield.

4. Disconnect the speed sensor connector.

5. Remove the clips holding the sensor harness.

6. Remove the speed sensor.

To install:

7. Install the speed sensor and torque to 9 ft. lbs. (12 Nm).

8. Install the clips holding the sensor harness.

9. Connect the speed sensor connector.

10. Connect the negative battery cable.

Rear Speed Sensor

Front Wheel Drive Models

1. Disconnect the negative battery cable.

——— CAUTION ———
Wait at least 90 seconds after the negative (-) battery cable is disconnected to prevent possible deployment of the air bag.

2. Raise and safely support the vehicle.

3. Disconnect the speed sensor connector.

4. Remove the clips holding the sensor harness.

5. Remove the mounting bolt and the speed sensor.

To install:

6. Install the speed sensor.

7. Install the clips holding the sensor harness.

8. Install the sensor wire harness and connect the connector.

9. Connect the negative battery cable.

All Wheel Drive Models

1. Disconnect the negative battery cable.

——— CAUTION ———
Work must be started after 90 seconds from the time the ignition switch is turned to the LOCK position and the negative (-) battery cable is disconnected.

2. Raise and safely support the vehicle.

3. Disconnect the speed sensor connector.

4. Remove the clips holding the sensor harness.

5. Remove the cable band.

6. Remove the mounting bolt and the speed sensor with the O-ring.

To install:

7. Install the speed sensor with the O-ring and torque to 9 ft. lbs. (12 Nm).

8. Install the clips holding the sensor harness.

9. Install the cable band.

10. Install the sensor wire harness and connect the connector.

11. Connect the negative battery cable.

FRONT SUSPENSION

Strut

REMOVAL AND INSTALLATION

1. Disconnect the negative battery cable.

<FWD>

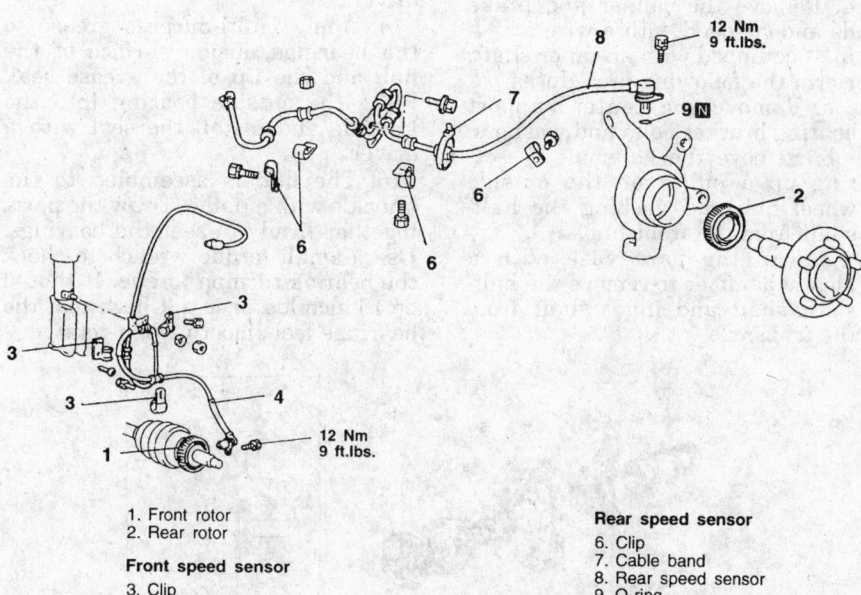

1. Front rotor
2. Rear rotor

Front speed sensor
3. Clip
4. Front speed sensor
5. Front speed sensor bracket

Rear speed sensor
6. Clip
8. Rear speed sensor

330660

Speed sensor — front wheel drive models

<AWD>

1. Front rotor
2. Rear rotor

Front speed sensor
3. Clip
4. Front speed sensor

Rear speed sensor
6. Clip
7. Cable band
8. Rear speed sensor
9. O-ring

330661

Speed sensor — all wheel drive models

CAUTION

Work must be started after 90 seconds from the time the ignition switch is turned to the LOCK position and the negative (-) battery cable is disconnected.

2. Raise and safely support vehicle.

3. Remove the brake hose and tube bracket. Do not pry the brake hose and tube clamp away when removing it.

4. If equipped with ABS, disconnect the front speed sensor mounting clamp from the strut.

5. Support the lower arm and remove the strut to knuckle bolts. Use a piece of wire to suspend the knuckle to keep the weight off the brake hose.

6. Disconnect the ECS connector.

7. Before removing the top bolts, make matchmarks on the body and the strut insulator for proper reassembly. If this plate is installed improperly, the wheel alignment will be wrong. Remove the strut upper bolts and remove the strut assembly from the vehicle.

To install:

8. Install the strut to the vehicle and install the top bolts.

9. Install the strut to the knuckle and install the bolts.

10. Connect the ECS connector.

11. Install the brake hose bracket and the ABS clamp.

12. Install the daytime running lamp delay and control unit to the mounting bracket located on top of the left strut tower.

13. Install the auto-cruise control actuator.

14. Install the wheel and tire assembly.

15. Connect the negative battery cable.

16. Perform a front end alignment.

Lower Ball Joints

REMOVAL AND INSTALLATION

The lower ball joints are integral to the lower control arms and cannot be replaced separately from the arms.

Lower Control Arms

REMOVAL AND INSTALLATION

1. Disconnect the negative battery cable.

2. Raise the vehicle and support safely allowing wheels and suspension to hang freely.

3. Remove the sway bar links from the lower control arm.

4. Disconnect the ball joint stud from the steering knuckle.

5. Remove the inner mounting frame through bolt and nut.

6. Remove the rear mount bolts. Remove the clamp if equipped.

7. Remove the rear rod bushing if servicing.

To install:

8. Assemble the control arm and bushing.

9. Install the control arm to the vehicle and install the through bolt. Replace the nut and snug temporarily.

10. Install the rear mount clamp, bolts and replacement nuts. Torque the bolts to 72–87 ft. lbs. (100–120 Nm). Torque the nuts to 29 ft. lbs. (40 Nm).

11. Connect the ball joint stud to the knuckle. Install a new nut and torque to 43–52 ft. lbs. (60–72 Nm).

12. Install the sway bar and links.

13. Lower the vehicle to the floor for the final torquing of the frame mount through bolt.

14. Once the full weight of the vehicle is on the floor, torque the frame mount through bolt nuts to 75–90 ft. lbs. (102–122 Nm).

15. Connect the negative battery cable.

16. Check the wheel alignment and adjust if necessary.

Sway Bar

REMOVAL AND INSTALLATION

1. Disconnect the negative battery cable.

— CAUTION —
Wait at least 90 seconds after the negative (-) battery cable is disconnected to prevent possible deployment of the air bag.

2. Raise the vehicle and support safely.

3. Remove the front exhaust pipe and engine undercover.

4. Remove the left and right frame members.

5. On AWD vehicles with automatic transaxle, remove the transfer case bracket and transfer case.

6. Remove the sway bar link.

7. Remove the sway bar brackets and remove the sway bar from the vehicle.

To install:

8. Note that the bar brackets are marked left and right. Lubricate all rubber parts and install the bushings, the sway bar and brackets.

9. Install the sway bar link and torque to 29 ft. lbs. (40 Nm).

10. Install the transfer case and bracket.

11. Install the frame members and torque to 29 ft. lbs. (40 Nm).

12. Install the engine undercover and exhaust pipe.

13. Connect the negative battery cable.

Front Wheel Bearings

REMOVAL AND INSTALLATION

1. Disconnect the negative battery cable.

— CAUTION —
Work must be started after 90 seconds from the time the ignition switch is turned to the LOCK position and the negative (-) battery cable is disconnected.

2. Remove the cotter pin, halfshaft nut and washer.

3. Raise the vehicle and support safely. If equipped with ABS, remove the front wheel speed sensor. Remove the ball joint and tie rod end from the steering knuckle.

4. Remove the caliper and brake pads and suspend with a wire.

5. If equipped with an inner shaft, perform the following procedures:

   a. Remove the center support bearing bracket bolts and washers.

   b. Remove the halfshaft by setting up a puller on the outside wheel hub and pushing the halfshaft from the front hub.

   c. Tap the joint case with a plastic hammer to remove the halfshaft shaft and inner shaft from the transaxle.

6. If not equipped with an inner shaft, perform the following procedures:

   a. Remove the halfshaft by setting up a puller on the outside wheel hub and pushing the halfshaft from the front hub.

   b. After pressing the outer shaft, insert a prybar between the transaxle case and the halfshaft and pry the shaft from the transaxle.

7. If equipped with AWD, the front hub/bearing assembly can be serviced at this point as a unit. If the knuckle is being removed, proceed.

8. Unbolt the lower end of the strut and remove the hub and steering knuckle assembly.

9. Set up a puller with the knuckle/hub in a vise and pull the hub from the knuckle. Do not use a hammer to accomplish this or the bearing will be damaged.

10. Once the hub and outer bearing inner race are removed with a puller, the bearing outer races can be removed by tapping out with a brass drift pin and a hammer.

To install:

11. Assemble the hub/knuckle assembly with pressing tools, using new parts as required.

12. Install the knuckle assembly to the vehicle and install the strut bolts.

13. Apply a thin coat of grease to the outside of the outer races and install into the hub with a bearing driver.

14. Apply multi-purpose grease to the bearings, inside surface of the hub and the lip of the grease seal. Place the outside bearing into the knuckle and install the seal with a driver.

15. The hub is assembled to the knuckle with a puller. Draw the parts together firmly to seat the bearings. Use a small torque wrench to check the bearing turning torque. It should be 16 inch lbs. or less. Check that the bearings feel smooth when rotated.

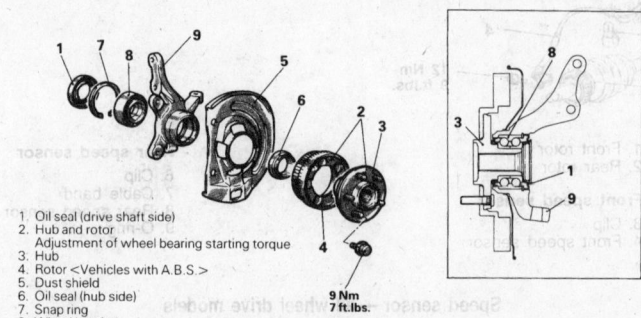

1. Oil seal (drive shaft side)
2. Hub and rotor
   Adjustment of wheel bearing starting torque
3. Hub
4. Rotor <Vehicles with A.B.S.>
5. Dust shield
6. Oil seal (hub side)
7. Snap ring
8. Wheel bearing
9. Knuckle

9 Nm
7 ft.lbs.

324364

Steering knuckle and hub assembly — front wheel drive models

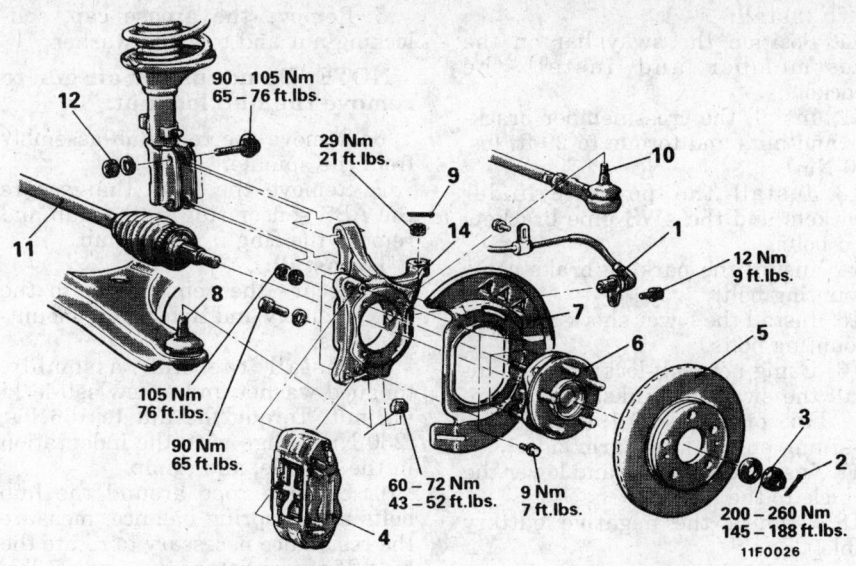

90 – 105 Nm
65 – 76 ft.lbs.

29 Nm
21 ft.lbs.

105 Nm
76 ft.lbs.

90 Nm
65 ft.lbs.

60 – 72 Nm
43 – 52 ft.lbs.

9 Nm
7 ft.lbs.

12 Nm
9 ft.lbs.

200 – 260 Nm
145 – 188 ft.lbs.

11F0026

1. Front speed sensor connection
   <Vehicles with A.B.S.>
2. Cotter pin
3. Drive shaft nut
4. Caliper assembly
5. Brake disc
6. Front hub unit bearing
7. Dust shield
8. Lower arm ball joint connection
9. Cotter pin
10. Tie rod end connection
11. Drive shaft
12. Front strut mounting bolt
13. Hub and knuckle
14. Hub

324365

Steering knuckle and hub assembly — all wheel drive models

16. Apply a thin coat of grease to the lip of the halfshaft side axle seal and drive into place until it contacts the inner bearing outer race.

17. Replace the circlips on the ends of the halfshafts.

18. Insert the halfshaft into the transaxle. Make sure it is fully seated.

19. Pull the strut assembly out and install the other end to the hub.

20. Install the center bearing bracket bolts and tighten to 33 ft. lbs. (45 Nm).

21. Install the washer so the chamfered edge faces outward. Install the nut and tighten temporarily.

22. Install the tie rod end and ball joint.

23. Install the wheel and lower the vehicle to the floor. Tighten the axle nut with the brakes applied. Tighten the nut to 166 ft. lbs. (230 Nm)

REAR SUSPENSION

Strut

REMOVAL AND INSTALLATION

1. Disconnect the negative battery cable and wait one minute for the air bag to disarm before working on the vehicle.

2. Raise and safely support vehicle.

3. Remove the rear side trim in the luggage compartment and remove the ECS connector and cap.

4. Support the suspension and remove the upper strut mounting bolts.

5. Remove the wheel and tire assembly and the lower strut mounting bolt.

6. Remove the strut from the vehicle.

To install:

7. Position the strut in the trailing arm and torque the lower mounting bolt to 72 ft. lbs. (100 Nm).

8. Guide the upper mounting studs through the body and torque the upper mounting nuts to 33 ft. lbs. (45 Nm).

9. Install the cap and the ECS connector.

10. Install the wheel and tire assembly and connect the negative battery cable.

11. Lower the vehicle to the floor.

Upper Control Arms

REMOVAL AND INSTALLATION

1. Disconnect the negative battery cable and wait one minute for the air bag to disarm before working on the vehicle.

2. Raise and safely support the vehicle.

3. Remove the wheel and tire assembly.

4. Remove the self locking nut on the upper control arm and using the special tool, separate the ball stud from the knuckle.

5. Remove the upper arm mounting bolt and nut.

6. Remove the upper control arm from the vehicle.

To install:

7. Position the upper arm on the vehicle and torque the mounting bolt to the knuckle to 54–64 ft. lbs. (75–89 Nm) and the bolt to the body to 101–116 ft. lbs. (140–160 Nm).

NOTE: Do not tighten the mounting bolt until the suspension is at the normal riding height. Lower the vehicle to the floor or on an alignment rack and then tighten the mounting bolt to specification.

8. Lower the vehicle to the floor and connect the negative battery cable.

Lower Control Arms

REMOVAL AND INSTALLATION

1. Disconnect the negative battery cable and wait at least one minute before working on the vehicle to allow time for the airbag to disarm.

2. Raise and safely support vehicle.

3. Remove the rear strut assembly and the brake line clamp bolt.

4. Matchmark and remove the lower arm mounting bolt and nut.

5. Remove the self-locking nut on the lower control arm ball stud and using the special tool, remove the ball stud from the trailing arm.

6. Remove the lower control arm from the vehicle.

To install:

NOTE: Replace all self-locking nuts. Do not torque the inboard pivot nuts until the full weight of the vehicle is on the ground.

7. Position the lower control arm and install the mounting bolt and nut. After the vehicle is on the floor, torque the nut to 101–116 ft. lbs. (140–160 Nm).

8. Install the ball stud into the trailing arm and torque the new nut to 54–64 ft. lbs. (75–89 Nm).

9. Install the rear strut and the brake line clamp bolt.

10. Lower the vehicle to the floor and torque the lower control arm pivot bolt to specifications.

11. Connect the negative battery cable.

12. Perform a rear wheel alignment.

Sway Bar

REMOVAL AND INSTALLATION

1. Disconnect the negative battery cable and wait one minute for the air bag to disarm before working on the vehicle.

2. Remove both wheel and tire assemblies.

3. Raise and safely support the vehicle.

4. Remove the self-locking nuts securing the sway bar link to the sway bar and the control arm.

5. Disconnect the power cylinder tie rod ends from the trailing arms.

6. Remove the lower shock absorber mounting bolts.

7. Remove the parking brake cable mounting bolt.

8. Remove the Four Wheel Steering (4WS) pipe bracket bolts and the power cylinder brackets.

9. Support the rear suspension assembly and remove the crossmember brackets and mounting nuts.

10. Remove the sway bar brackets and remove the sway bar.

To install:

11. Position the sway bar on the crossmember and install the brackets.

12. Install the crossmember brackets and nuts and torque to 29 ft. lbs. (40 Nm).

13. Install the power cylinder brackets and the 4WS pipe brackets and bolts.

14. Install the parking brake cable mounting bolt.

15. Install the lower shock absorber mounting bolts.

16. Using new self-locking nuts, install the sway bar links. Make sure the link protrudes 0.197–0.276 in. (5–7mm) on the lower arm side.

17. Install the wheels and lower the vehicle to the floor.

18. Connect the negative battery cable.

Wheel Bearings

REMOVAL AND INSTALLATION

NOTE: The hub assembly is not repairable, if defective replacement is the only option. If the hub is removed for any reason it must be replaced.

1. Raise and support vehicle safely.

2. Remove the both of the rear wheels.

3. Remove the caliper and the brake disc. Support the caliper with wire to prevent stress to the brake hose.

4. If equipped with ABS, remove the bolt holding the speed sensor to the trailing arm and remove the sensor.

NOTE: The speed sensor has a pole piece projecting from it. This exposed tip must be protected from impact or scratches. Do not allow the pole piece to contact the toothed wheel during removal or installation.

5. Remove the grease cap, self-locking nut and tongued washer.

NOTE: Do not use an air gun to remove the hub locknut.

6. Remove the rear hub assembly from the spindle.

7. Remove the bolts that secure the ABS sensor ring to the hub and remove the ring from the hub.

To install:

8. Secure the sensor ring to the hub assembly and tighten the mounting bolts.

9. Install the hub assembly, tongued washer and a new self-locking nut. Torque the nut to 166 lbs. (230 Nm), align with the indentation in the spindle, and crimp.

10. Using a rope around the hub bolts and a spring balance, measure the resistance necessary to rotate the hub. If the resistance exceeds 7 lbs. (31 N), loosen and retighten the locknut. If the resistance still exceeds the specification, the hub must be replaced.

11. Using a dial indicator, measure the hub endplay. The endplay should be 0.002 in. (0.05mm) or less.

12. Install the brake rotor and caliper assembly.

13. Install the speed sensor to the knuckle.

NOTE: Route the speed sensor cable correctly. Improper installation may cause cable damage and system failure. Use the white stripe on the outer insulation to keep the sensor harness properly positioned.

14. Use a brass or other non-magnetic feeler gauge to check the air gap between the tip of the pole piece and the toothed wheel. Correct gap is 0.008–0.028 in. (0.2–0.7mm). Tighten the sensor bracket nut with the sensor located so the gap is the same at several points on the toothed wheel. If the gap is incorrect, it is likely that the toothed wheel is worn or improperly installed.

15. Bleed the brake system and install the rear wheels.

CHRYSLER CORP.

Front Wheel Drive

CHRYSLER-Sebring Coupe DODGE-Avenger

FIRING ORDERS

NOTE: To avoid confusion, always replace spark plug wires one at a time.

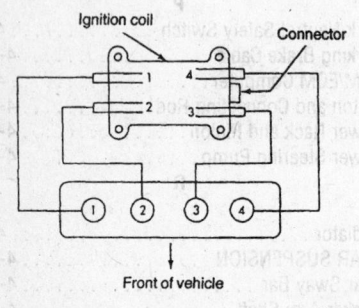

2.0L Engine
Engine Firing Order 1–3–4–2
Distributorless Ignition System

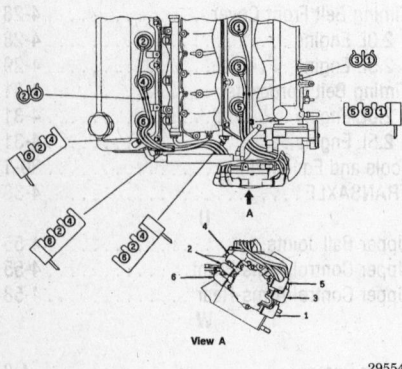

2.5L Engine
Engine Firing Order 1–2–3–4–5
Distributor Rotation: Counterclockwise

ENGINE ELECTRICAL

NOTE: Disconnecting the negative battery cable on some vehicles may interfere with the functions of the on board computer systems and may require the computer to undergo a relearning process, once the negative battery cable is reconnected.

Distributor

REMOVAL AND INSTALLATION

2.5L (VIN N) Engine

NOTE: This operation requires removal of the intake manifold plenum.

1. Disconnect the negative battery cable.
2. Disconnect the distributor harness electrical connectors.
3. Unscrew the distributor cap hold-down screws and lift off the distributor cap with all ignition spark plug wires connected.
4. Position the engine (turn the crankshaft clockwise) so that the No. 1 piston is at TDC of its compression stroke and the mark on the vibration damper is aligned with **0** on the timing indicator.
5. Remove the hold-down nuts.
6. Carefully remove the distributor from the engine.

To install:

7. Clean all parts well. Inspect for cracks or carbon tracks in the cap. Check the condition of all electrodes in both the cap and on the rotor. Replace any parts that are questionable.
8. Install a new distributor housing O-ring.

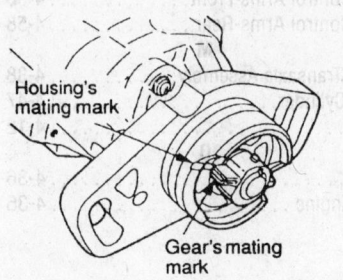

Aligning the distributor's mating marks — 2.5L engine

9. Align the housing's mating mark with the gear's mating mark and install the distributor on the cylinder head. Make sure the distributor is fully seated and the distributor shaft is fully engaged.
10. Install the hold-down nuts and tighten them to 10 ft. lbs. (13 Nm). The basic ignition timing is not adjustable.
11. Connect the distributor harness connectors.
12. Make sure the sealing O-ring (sometimes called "packing") is in place on the distributor housing and install the distributor cap and tighten the screws.
13. Verify that the spark plug cables are properly installed.
14. Connect the negative battery cable and check for proper engine operation.

Ignition Timing

ADJUSTMENT

The ignition timing is determined by the PCM from the engine speed, intake air volume, engine coolant temperature and atmospheric pressure. Since the ignition timing is controlled by the PCM basic ignition timing is not adjustable on either the 2.0L or the 2.5L engines.

Alternator

PRECAUTIONS

Several precautions must be observed with alternator equipped vehicles to avoid damage to the unit.

- If the battery is removed for any reason, make sure it is reconnected with the correct polarity. Reversing the battery connections may result in damage to the 1-way rectifiers.
- When utilizing a booster battery as a starting aid, always connect the positive to positive terminals and the negative terminal from the booster battery to a good engine ground on the vehicle being started.
- Never use a fast charger as a booster to start vehicles.
- Disconnect the battery cables when charging the battery with a fast charger.
- Never attempt to polarize the alternator.
- Do not use test lights of more than 12 volts when checking diode continuity.
- Do not short across or ground any of the alternator terminals.

• The polarity of the battery, alternator and regulator must be matched and considered before making any electrical connections within the system.

• Never separate the alternator on an open circuit. Make sure all connections within the circuit are clean and tight.

• Disconnect the battery ground terminal when performing any service on electrical components.

• Disconnect the battery if arc welding is to be done on the vehicle.

REMOVAL AND INSTALLATION

2.0L(VIN Y) Engine

1. Disconnect the negative battery cable.
2. Remove the right side under cover.
3. Remove the speed control vacuum reservoir and related components as required for alternator access.
4. Remove the alternator drive belt.
5. Remove the alternator mounting bolts.
6. Remove the alternator top brace from the engine.
7. Disconnect the alternator wiring and remove the alternator from the vehicle.
To install:
8. Install the alternator in position and connect the electrical harness.
9. Install the alternator top brace to the engine.
10. Install the alternator mounting bolts loosely.
11. Install the drive belt and adjust until the proper tension is achieved. Secure the lower alternator through bolt nut to 45 ft. lbs. (61 Nm) and the upper alternator lock bolt to 40 ft. lbs. (54 Nm).
12. Install the speed control vacuum reservoir and whatever related

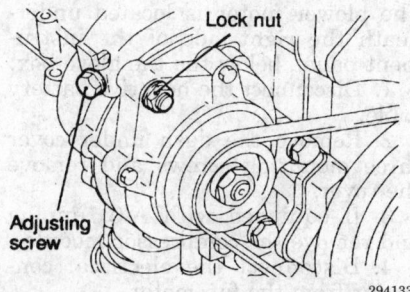

Alternator belt tension adjustment — 2.0L engine

components were removed for alternator access.
13. Install the right side undercover.
14. Connect the negative battery cable. Start the engine and check the alternator for proper operation.

2.5L(VIN N) Engine

1. Disconnect the negative battery cable.
2. Remove the right side under cover.
3. Remove the speed control vacuum reservoir and related components as required for alternator access.
4. Remove the power steering pump cover.
5. Remove the A/C compressor drive belt.
6. Remove the alternator/power steering pump drive belt.
7. Remove the intake manifold plenum.
8. Remove the upper alternator bracket.
9. Remove the intake manifold plenum stay.
10. Remove the electrical connections at the alternator.
11. Remove the upper and lower bolts and remove the alternator. Be

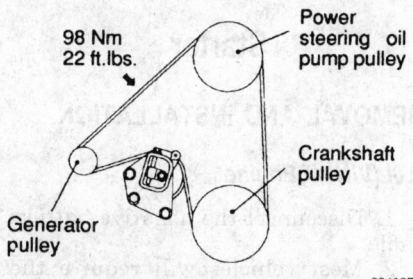

Measure the alternator/power steering pump belt tension at the point illustrated — 2.5L engine

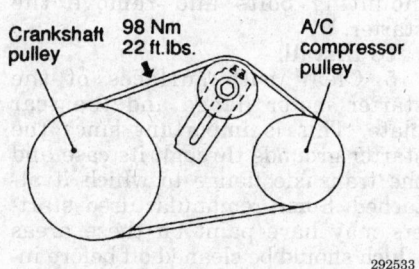

Measure the A/C belt tension at the point illustrated — 2.5L engine

very careful not to damage neighboring components while removing the alternator.
To install:
12. Connect the wiring to the alternator.
13. Place the alternator on its lower bracket and install the upper and lower bolts finger tight.
14. Install the upper alternator bracket.
15. Install the alternator/power steering pump belt, making sure it is properly seated.
16. Adjust the drive belt to the correct tension using the belt adjuster and tighten the upper mounting bolt to 16 ft. lbs. (22 Nm) and the lower mounting bolt to 30 ft. lbs. (41 Nm).
17. Install the intake manifold plenum.
18. Install the intake manifold plenum stay.
19. Install the A/C compressor drive belt.
20. Install the power steering pump cover.
21. Connect the negative battery cable.
22. Check the charging system operation.

Drive Belt

REMOVAL AND INSTALLATION

2.0L(VIN Y) Engine

When replacing or adjusting the engine drive belts, it is important that the belts are adjusted with the correct amount of tension. Excessive belt tension will cause damage to the alternator and power steering pump pulley bearings, while loose belt tension will produce slip and premature wear on the belt.

Alternator Belt

1. Place a straightedge along the bottom edge of the belt and across the 2 pulleys. Allow both ends of the straightedge to rest on the bottom of each pulley for support.
2. Measure the deflection of the belt from the straightedge with a force of about 22 lbs. applied midway between the 2 pulleys. Deflection should be 0.35–0.45 inch (9.0–11.5mm)
3. To adjust the tension on the alternator drive belt, loosen the adjusting bolt and the pivot locknut, at the alternator. Then move the alternator, by turning the adjusting bolt. Once the desired value is reached, secure the bolt and locknut. Recheck the belt tension.

Power Steering Pump Belt (vehicles without A/C)

1. Press the belt in, at about the center, between the power steering pump pulley and the crankshaft pulley. With reasonable pressure applied (about 22 lbs.) the belt should deflect about 0.43–0.55 inches (11–14 mm).

2. Adjustment can be made by loosening the 3 bolts that hold the pump. Place a suitable bar or lever between the body of the pump and gently pry to get the desired tension.

3. Retighten the 3 bolts to 29 ft. lbs. (39 Nm) and check belt tension again.

Power Steering Pump and A/C Belt

NOTE: Fix the power steering pump closest to the front of the vehicle.

1. Press the belt in, at about the center between the power steering pump pulley and the crankshaft pulley. With reasonable pressure applied (about 22 lbs.) the belt should deflect about 0.39–0.43 inches (10–11 mm).

2. Adjustment can be made by loosening the tensioner pulley nut

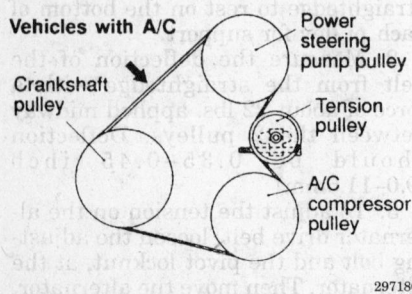

Apply force as shown and measure the alternator (generator) belt deflection — 2.0L engine

Apply force as shown and measure the power steering pump belt deflection (with A/C) — 2.0L engine

and turning the adjuster bolt until the desired tension is attained.

3. Tighten the pulley nut and check the belt tension again.

2.5L(VIN N) Engine

Alternator and Power Steering Pump Belt

1. Press the belt in, at about the center, between the power steering pump pulley and the alternator pulley. With reasonable pressure applied (about 22 lbs.) the belt should deflect about 0.43–0.55 inches (11–14 mm).

2. Adjustment can be made by loosening the tensioner pulley nut and turning the adjuster bolt until the desired tension is attained.

3. Tighten the pulley nut and check the belt tension again.

Air Conditioning Compressor Belt

1. Press the belt in, at about the center, between the crankshaft pulley and the tensioner pulley. With reasonable pressure applied (about 22 lbs.) the belt should deflect about 0.32–0.35 inches (8–9 mm).

2. Adjustment can be made by loosening the tensioner pulley nut and turning the adjuster bolt until the desired tension is attained.

3. Tighten the pulley nut and check the belt tension again.

Starter

REMOVAL AND INSTALLATION

2.0L(VIN Y) Engine

1. Disconnect the negative battery cable.

2. Most vehicles will require the removal of the front exhaust pipe. Use penetrating oil on the fasteners to ease removal.

3. Disconnect the starter motor electrical connections.

4. Remove the starter motor mounting bolts and remove the starter.

To install:

5. Clean both surfaces of the starter motor flange and the rear plate. This is important since the starter grounds through its case and the transaxle flange to which it attached. Some remanufactured starters may have paint on these areas which should be cleaned off before installation. Install the starter motor onto the transaxle and secure with the retaining bolts. Tighten the bolts to 40 ft. lbs. (34 Nm).

6. Connect the electrical harness connectors to the starter.

7. Connect the negative battery cable and check the starter for proper operation.

2.5L(VIN N) Engine

1. Disconnect the negative battery cable.

2. Most vehicles will require the removal of the front exhaust pipe. Use penetrating oil on the fasteners to ease removal.

3. Disconnect the starter motor electrical connections.

4. Remove the starter motor mounting bolts and remove the starter.

To install:

5. Clean both surfaces of the starter motor flange and the rear plate. This is important since the starter grounds through its case and the transaxle flange to which it attached. Some remanufactured starters may have paint on these areas which should be cleaned off before installation. Install the starter motor onto the transaxle and secure with the retaining bolts. Tighten the bolts to 20–25 ft. lbs. (26–33 Nm).

6. Connect the electrical harness connectors to the starter.

7. Connect the negative battery cable and check the starter for proper operation.

CHASSIS ELECTRICAL

Blower Motor

REMOVAL AND INSTALLATION

The blower motor is located underneath the right side of the instrument panel, below the the glove box.

1. Disconnect the negative battery cable.

2. Remove the dash under cover three mounting screws and remove the cover.

3. If equipped with A/C, unplug and remove the compressor module.

4. Disconnect the electrical connector from the fan motor.

5. Remove the three small bolts holding the motor to the housing and remove the motor and fan.

To install:

NOTE: Check the inside of the case carefully. Any debris can snag the fan and cause noise or poor airflow.

6. Install the blower motor in the blower case and secure with the three mounting bolts.
7. Connect the blower motor electrical connector.
8. Install the compressor module, if removed.
9. Install the undercover, taking care to insure it is in place and all the fasteners are secure.
10. Connect the negative battery cable.
11. Test system operation.

Windshield Wiper Motor

REMOVAL AND INSTALLATION

1. Disconnect the negative battery cable.
2. Remove the windshield wiper arms by unscrewing the cap nuts and lifting the arms from the linkage posts. Scribe matchmarks on the arms and wiper shafts to aid in installation.
3. Remove the front deck garnish panel.

4. Disconnect the electrical harness from the wiper motor.
5. Loosen the wiper motor assembly mounting bolts and remove the windshield wiper motor. Disconnect the linkage from the motor assembly. If necessary, remove the linkage from the vehicle.

To install:

6. Install the windshield wiper motor and connect the linkage. Connect the electrical harness to the motor.
7. Install the front deck garnish panel.
8. Align the matchmarks and install the wiper arms. Tighten the retaining nuts to 9 ft. lbs. (13 Nm).
9. Connect the negative battery cable and check the wiper system for proper operation.

Combination Switch

REMOVAL AND INSTALLATION

1. Set the steering wheel and the front wheels to the straight ahead position, and then remove the ignition key.
2. Disconnect the negative battery cable and insulate the cable end with high-quality electrical tape or similar non-conductive wrapping.

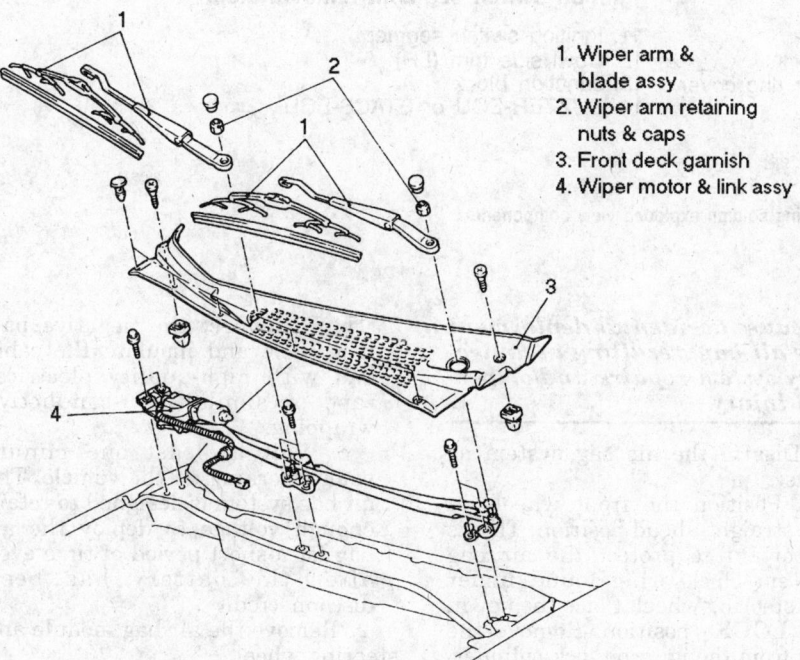

1. Wiper arm & blade assy
2. Wiper arm retaining nuts & caps
3. Front deck garnish
4. Wiper motor & link assy

289264

Exploded view of wiper assembly and related components

CAUTION

The Air Bag system must be disarmed before removing the steering wheel. Failure to do so may cause accidental deployment of the air bag, resulting in unnecessary system repairs and/or personal injury.

3. Wait at least 1 minute before working on the vehicle. The air bag system is designed to retain enough voltage to deploy for a short period of time even after the battery has been disconnected.
4. Remove the bolts securing the air bag module and remove the electrical connectors and the module.

CAUTION

When carrying a live air bag, make sure the bag and trim cover are pointed away from the body. In the unlikely event of an accidental deployment, the bag will then deploy with minimal chance of injury. When placing a live air bag on a bench or other surface, always face the bag and trim cover up, away from the surface. This will reduce the motion of the module if it is accidentally deployed.

5. Remove the lock nut and the steering wheel using a MB990803 or equivalent puller tool.
6. Remove the lower steering column cover and the column pad.
7. Remove the upper steering column cover.
8. Remove the clock spring and combination switch assembly retaining screws and electrical connectors and remove the complete assembly.
9. Remove the 4 retaining screws and separate the clock spring from the combination switch.

To install:

10. Attach the clock spring to the combination switch with the 4 retaining screws.
11. Install the electrical connectors and the clock spring and combination switch assembly and secure it with the 2 retaining screws
12. Install the upper steering column cover.
13. Install the column pad and the lower steering column cover.
14. Install the steering wheel and secure it with a new lock nut. Tighten the lock nut to 30 ft. lbs. (41 Nm).
15. Install the air bag module and tighten the retaining bolts to 4 ft. lbs. (6 Nm).
16. Connect the negative battery cable.

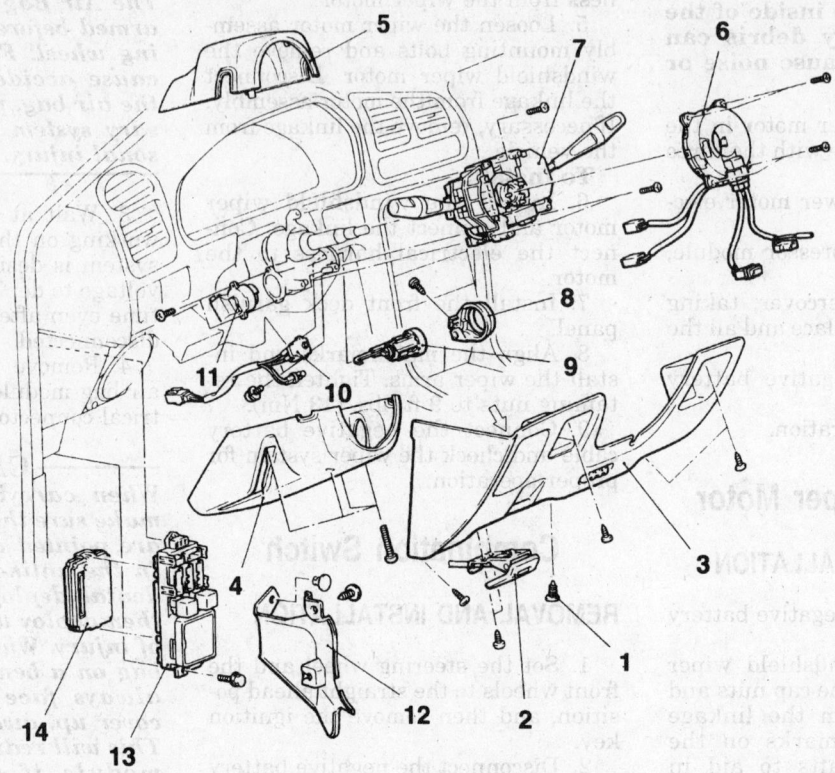

1. Clip
2. Hood lock release handle
3. Instrument under cover
4. Column cover lower
5. Column cover upper
6. Clock spring
7. Column switch
8. Ignition key illumination ring or ring cover
9. Steering lock cylinder
10. Key reminder switch segment or key hole illumination light

Ignition switch segment removal steps

11. Ignition switch segment
12. Cowl side trim (LH)
13. Junction block
14. BUZZER-ECU or ETACS-ECU

291971

Steering column exploded view components

17. Check the switch for proper operation.

Ignition Switch/Lock Cylinder

REMOVAL AND INSTALLATION

—————— CAUTION ——————
The Air Bag system must be disarmed before working around the steering column. Failure to do so

may cause accidental deployment of the air bag, resulting in unnecessary system repairs and/or personal injury.

1. Disarm the air bag system as follows:

a. Position the front wheels in the straight-ahead position. This is important to protect the air bag system clockspring found under the steering wheel. Place the key in the **LOCK** position. Remove the key from the ignition lock cylinder.

b. Disconnect the negative battery cable and insulate the cable end with high-quality electrical tape or similar non-conductive wrapping.

c. Wait at least one minute before working on the vehicle. The air bag system is designed to retain enough voltage to deploy the air bag for a short period of time even after the battery has been disconnected.

2. Remove the air bag module and steering wheel.

3. Remove the hood lock release handle.

4. Remove the knee protector.

5. Remove the steering column upper and lower covers. Use care removing covers to prevent breakage of alignment tabs.

6. Disconnect the windshield wiper, combination switch and ignition switch harness connectors.

7. Remove the retaining screws and remove the entire column switch/clockspring assembly from the left side of the steering column.

8. Remove mounting screws from the ignition switch and pull the switch from the interlock cylinder.

9. To remove the lock cylinder, insert the key and place in the **ACC** position. With a small pointed tool, push the lock pin of the steering lock cylinder inward and pull the lock cylinder out.

To install:

10. Install the lock cylinder into the interlock housing. Be sure the lock pin snaps into place.

11. Install ignition switch into interlock housing. Align keyway of ignition switch with lock cylinder and secure with mounting screws.

12. Install the column switch/clockspring assembly to the steering column and connect the harness connections.

13. Install the knee protector and the hood release handle.

14. Install the clockspring, steering wheel and air bag module.

15. Connect the negative battery cable and check all functions of column-mounted switches and the ignition switch for proper operation.

Park/Neutral Safety Switch

REMOVAL AND INSTALLATION

On vehicles equipped with automatic transaxles, the back-up (reverse) light switch is integrated within the park/neutral position switch. The

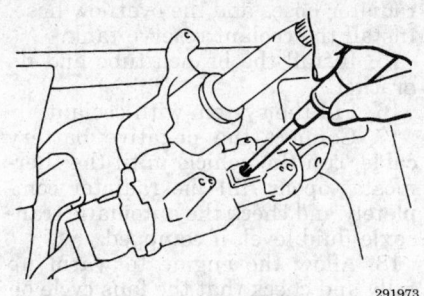

291973

Removal of the ignition switch lock cylinder

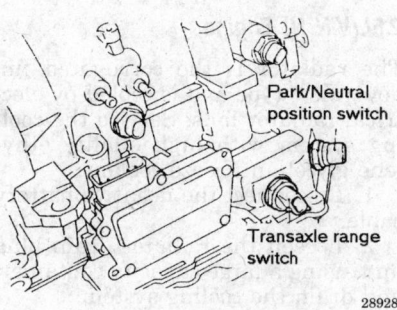

Park/Neutral position switch

Transaxle range switch

289285

Park/Neutral switch location

switch is located on the transaxle next to the electronic transaxle range control switch. Do not confuse the two. The park/neutral position switch is not adjustable.

1. Disconnect the negative battery cable.

2. Raise and safely support the vehicle.

3. Disconnect the electrical connector from the switch.

4. Remove the switch from the vehicle.

To install:

5. Install the switch and connect the electrical connector.

6. Connect the negative battery cable.

7. Check for proper switch operation.

——— CAUTION ———

*Verify that the engine will only start in the **P** or **N** positions.*

PCM/ECM Computer

REMOVAL AND INSTALLATION

The Powertrain Control Module (PCM) and the Electronically Con-

trolled Automatic Transaxle Electronic Control Module (EATX/ECM) are located in the engine compartment on the inner fender on the driver's side behind the air cleaner assembly.

Powertrain Control Module

The powertrain control module is mounted closest to the air cleaner assembly.

1. Disconnect the negative battery cable.

2. Remove the retaining screw and disconnect the electrical connector from the PCM.

3. Remove the retaining bolts and the PCM.

To install:

4. Install the PCM and secure it with the retaining bolts.

5. Connect the electrical connector and secure it with the retaining screw.

6. Connect the negative battery cable.

Electronic Transaxle Control Module (Automatic Transaxles Only)

The EATX/ECM is located behind the Powertrain Control Module.

1. Disconnect the negative battery cable.

2. Remove the retaining screws and disconnect the electrical connector from the ECM.

3. Remove the retaining bolts and the ECM.

To install:

4. Install the ECM and secure it with the retaining bolts.

5. Connect the electrical connector and secure it with the retaining screws.

6. Connect the negative battery cable.

ENGINE COOLING

Radiator

REMOVAL AND INSTALLATION

2.0L(VIN Y) Engine

The radiator is the corrugated fin, down flow type and is cooled by electrical radiator fans. Service the cooling system with high quality ethylene glycol antifreeze coolant.

1. Disconnect the negative battery cable.

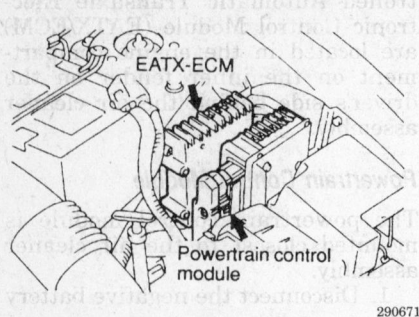

Location of the PCM and EATX\ECM

2. Loosen the radiator drain plug and, using a large capacity container, drain the cooling system.

3. Remove the bracket and plastic branch tube running from the air cleaner.

4. Disconnect the overflow tube and remove the coolant reserve tank.

5. Disconnect the upper radiator hose.

NOTE: It is recommended that each clamp be matchmarked to the hose. Observe the marks and reinstall the clamps in exactly the same position when reinstalling the radiator.

6. Label and disconnect the wiring to the thermosensors and the electric fan assemblies.

7. For vehicles with automatic transaxles, disconnect the oil cooler lines at the radiator. Plug the transaxle ports and the hose ends to contain the fluid and prevent contamination.

8. Remove the lower radiator hose.

9. Remove the bolts holding the upper mounting brackets to the support member. Remove the radiator, with the cooling fans as an assembly.

To install:

10. If the fan and shroud assemblies were removed with the radiator, they must be reinstalled before installing the radiator. The mounting bolts for the fans should be tightened to 10 ft. lbs. If the thermosensors were removed, they should be reinstalled and tightened to 10 ft. lbs.

11. Reinstall the radiator in position, making certain all the mounts and bushings are correctly installed. Tighten the mounting bolts to 10 ft. lbs. Double check the drain plug to make sure it is closed.

12. Connect the oil cooler lines and attach the brackets.

13. Connect the wiring to the electrical components, making sure each is correctly located and securely fastened.

14. Connect the upper and lower radiator hoses and the overflow hose. Install the coolant reserve tank.

15. Install the branch tube and its bracket.

16. Fill the system with coolant.

17. Connect the negative battery cable, run the vehicle until the thermostat opens, fill the radiator completely and check the automatic transaxle fluid level, if equipped.

18. Allow the engine to warm up fully and check that the fans cycle on and off correctly. Watch the coolant level carefully in the overflow tank.

19. Once the vehicle has cooled, recheck the coolant level.

2.5L(VIN N) Engine

The radiator is the corrugated fin, down flow type and is cooled by electrical radiator fans. Service the cooling system with high quality ethylene glycol antifreeze coolant.

1. Disconnect the negative battery cable.

2. Loosen the radiator drain plug and, using a large capacity container, and drain the cooling system.

3. Remove the bracket and plastic branch tube running from the air cleaner.

4. Disconnect the overflow tube and remove the coolant reserve tank.

5. Disconnect the upper radiator hose.

NOTE: It is recommended that each clamp be matchmarked to the hose. Observe the marks and reinstall the clamps in exactly the same position when reinstalling the radiator.

6. Label and disconnect the wiring to the thermosensors and the electric fan assemblies.

7. For vehicles with automatic transaxles, disconnect the oil cooler lines at the radiator. Plug the transaxle ports and the hose ends to contain the fluid and prevent contamination.

8. Remove the lower radiator hose.

9. Remove the bolts holding the upper mounting brackets to the support member. Remove the radiator, with the cooling fans as an assembly.

To install:

10. If the fan and shroud assemblies were removed with the radiator, they must be reinstalled before installing the radiator. The mounting bolts for the fans should be tightened to 10 ft. lbs. If the thermosensors were removed, they should be reinstalled and tightened to 10 ft. lbs.

11. Reinstall the radiator in position, making certain all the mounts

and bushings are correctly installed. Tighten the mounting bolts to 10 ft. lbs. Double check the drain plug to make sure it is closed.

12. Connect the oil cooler lines and attach the brackets.

13. Connect the wiring to the electrical components, making sure each is correctly located and securely fastened.

14. Connect the upper and lower radiator hoses and the overflow hose. Install the coolant reserve tank.

15. Install the branch tube and its bracket.

16. Fill the system with coolant.

17. Connect the negative battery cable, run the vehicle until the thermostat opens, fill the radiator completely and check the automatic transaxle fluid level, if equipped.

18. Allow the engine to warm up fully and check that the fans cycle on and off correctly.

19. Once the vehicle has cooled, recheck the coolant level.

Water Pump

REMOVAL AND INSTALLATION

The water pump available on both the 2.0L and 2.5L engines, is driven by the timing belt from the crankshaft. It is good practice to turn the engine crankshaft by hand (clockwise) to set the engine to TDC No. 1 cylinder compression stroke (firing position) before starting work. This should align all timing marks and serve as a reference point for later work.

2.0L(VIN Y) Engine

1. Disconnect the negative battery cable.

2. Drain the cooling system.

3. Remove the accessory drive belts.

4. Using a C 3281 or equivalent crankshaft holding tool remove the crankshaft retaining bolt.

5. Using puller tools 1026 and 6827 or equivalent remove the crankshaft pulley.

6. Remove the power steering pump with the hose attached and position it aside.

7. Remove the power steering pump bracket.

8. Place a floor jack under the engine oil pan, with a block of wood in between and jack up the engine so that the weight of the engine is no longer being applied to the engine mount bracket.

22. Using a torque wrench on the tensioner pulley, apply 21 ft. lbs. (28 Nm) of torque to the pulley.

23. With the torque being applied to the tensioner pulley, move the tensioner up against the tensioner pulley bracket and tighten the retaining bolts to 23 ft. lbs. (31 Nm).

24. Remove the tensioner plunger pin. Pretension is correct when the pin can be removed and installed.

25. Rotate the crankshaft 2 revolutions and check the timing marks. If the timing marks are not properly aligned remove the belt and repeat Steps 19 through 24.

26. Install the front timing belt cover.

27. Lower the engine enough to install the engine mount bracket.

28. Install the bracket and remove the floor jack.

29. Install the power steering pump bracket and pump.

30. Install the crankshaft pulley using a C 4685 C or equivalent pulley installer.

31. Install the accessory drive belts.

32. Properly fill the cooling system.

33. Connect the negative battery cable.

34. Check for leaks and proper engine operation.

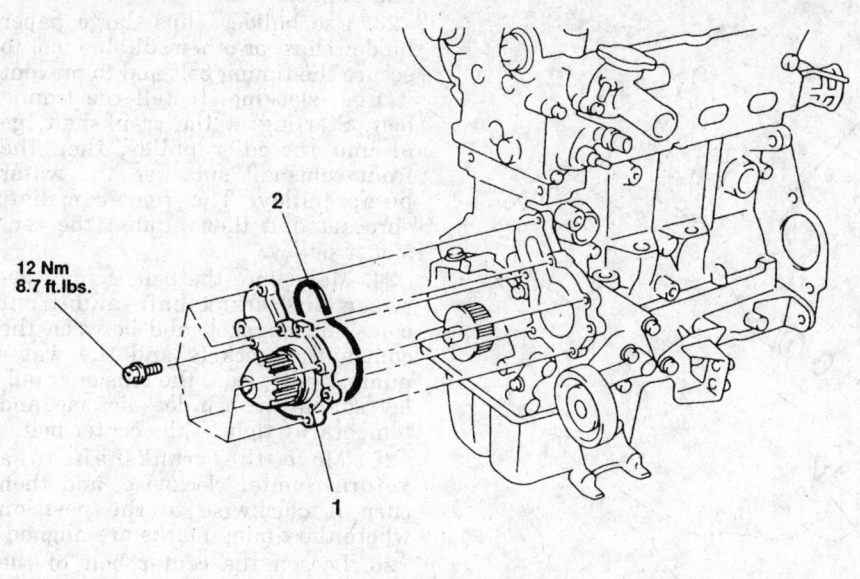

1. Water pump
2. O-ring

298802

Water pump and O-ring, frontal view — 2.0L engine

12 Nm
8.7 ft.lbs.

9. Remove the upper engine mount and the engine mounting bracket.

10. Remove the front timing belt cover.

11. Align the timing marks. Loosen the timing belt tensioner and remove the belt.

WARNING

Do not rotate the crankshaft or camshafts after removing the timing belt, or valve train components may be damaged. Always align the timing marks before removing the timing belt.

12. If the timing belt tensioner is to be replaced, remove the retaining bolts and remove the timing belt tensioner. When the timing belt tensioner is removed from the engine it is necessary to compress the plunger into the tensioner body.

13. Place the tensioner in a vise and slowly compress the plunger.

NOTE: Position the tensioner in the vise the same way it will be installed on the engine. This is to ensure proper pin orientation for when the tensioner is installed on the engine.

14. When the plunger is compressed into the tensioner body, install a pin through the body and plunger to hold the plunger in place until the tensioner is installed.

15. Remove the water pump retaining bolts and remove the water pump and O-ring.

To install:

16. Thoroughly clean all mating surfaces of the water pump and engine block.

17. Lubricate the O-ring with water or antifreeze then install into the groove in the water pump. Never use any type of oil based lube or grease on rubber parts.

18. Do not allow oil or other grease to get on the O-ring. Install the water pump. When inserting the water pump, check that there is no sand, dirt, etc. on its inner surface. Tighten the water pump retaining bolts to 8.7 ft. lbs. (12 Nm).

19. Check that all timing marks are still aligned. Bring the crankshaft sprocket to ½-notch before TDC.

20. Install the timing belt. Starting at the crankshaft, go around the water pump sprocket, idler pulley, camshaft sprockets and then around the tensioner pulley.

21. Move the crankshaft to TDC to take up the slack in the belt. Install the tensioner to the block but do not tighten the retaining bolts.

2.5L(VIN N) Engine

1. Disconnect the negative battery cable.

2. Drain the cooling system.

3. Remove the serpentine drive belt.

4. Using MB990767 and MB998754 or equivalent crankshaft holding tools remove the crankshaft bolt and remove the pulley.

5. Remove the heated oxygen sensor connection.

6. Remove the power steering pump with the hose attached and position it aside.

7. Remove the power steering pump bracket.

8. Place a floor jack under the engine oil pan, with a block of wood in between and jack up the engine so that the weight of the engine is no longer being applied to the engine support bracket.

9. Remove the upper engine mount. Spraying lubricant, slowly remove the reamer (alignment) bolt and remaining bolts and remove the engine support bracket.

NOTE: The reamer bolt may be heat seized on the engine support bracket.

10. Remove the front timing belt covers.

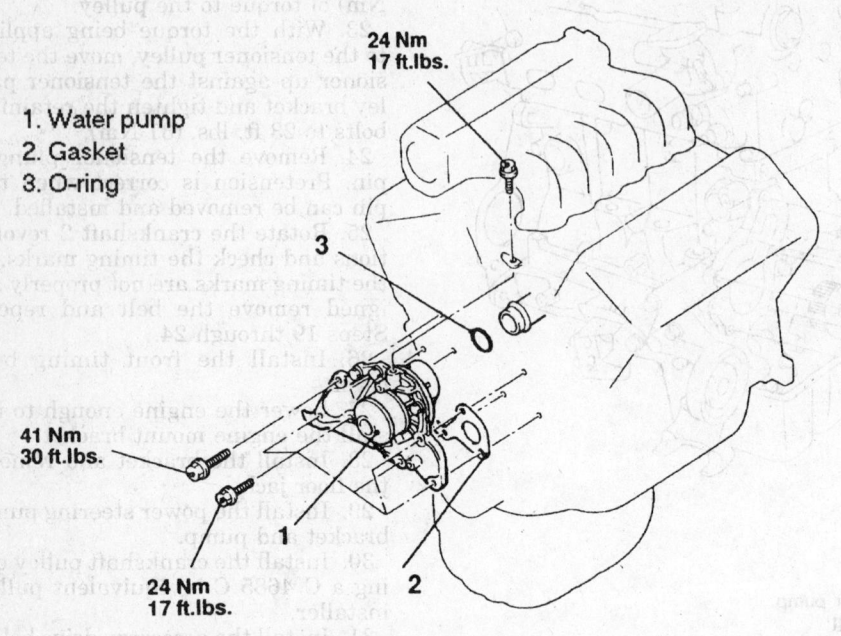

24 Nm
17 ft.lbs.

1. Water pump
2. Gasket
3. O-ring

41 Nm
30 ft.lbs.

24 Nm
17 ft.lbs.

291850

Water pump and related components — 2.5L engine

11. If the timing belt is to be re-used, chalk an arrow indicating the direction of rotation on the back of the belt for reinstallation.

12. If not done so earlier, align the timing marks by turning the crankshaft with MD998769 or equivalent crankshaft turning tool. Loosen the center bolt on the timing belt tensioner pulley and remove the belt.

─────── **WARNING** ───────
Do not rotate the crankshaft or camshaft after removing the timing belt, or valve train components may be damaged. Always align the timing marks before removing the timing belt.
────────────────────

13. Check the belt tensioner for leaks and check the push rod for cracks.

14. If the timing belt tensioner is to be replaced, remove the retaining bolts and remove the timing belt tensioner. When the timing belt tensioner is removed from the engine it is necessary to compress the plunger into the tensioner body.

15. Place the tensioner in a vise and slowly compress the plunger. Take care not to damage the pushrod.

NOTE: Position the tensioner in the vise the same way it will be installed on the engine. This is to

ensure proper pin orientation for when the tensioner is installed on the engine.

16. When the plunger is compressed into the tensioner body, install a pin through the body and plunger to hold the plunger in place until the tensioner is installed.

17. Remove the water pump retaining bolts and remove the water pump, gasket and O-ring.

To install:

18. Thoroughly clean all mating surfaces of the water pump and engine block.

19. Insert the O-ring onto the water inlet pipe. Lubricate the O-ring with water or coolant to make pump installation easier. Do not use any type of oil-base lubricant or grease.

20. Do not allow oil or other grease to get on the O-ring. Install the water pump. When inserting the water pump, check that there is no sand, dirt, etc. on its inner surface. With the exception of the long bolt (1.5 inch), tighten the water pump retaining bolts to 17 ft. lbs. (24 Nm). Tighten the long bolt to 30 ft. lbs. (41 Nm).

21. Install the timing belt tensioner and tighten the retaining bolts to 17 ft. lbs. (24 Nm), but do not remove the pin at this time.

22. Check that all timing marks are still aligned.

23. Use bulldog clips (large paper binder clips) or other suitable tool to secure the timing belt and to prevent it from slacking. Install the timing belt. Starting at the crankshaft, go around the idler pulley, then the front camshaft sprocket, the water pump pulley, The rear camshaft sprocket and then around the tensioner pulley.

24. Make sure the belt is tight between the crankshaft and front camshaft sprocket. and between the camshaft sprockets and the water pump. Gently raise the tensioner pulley, so that the belt does not sag, and temporarily tighten the center bolt.

25. Move the crankshaft to a ¼-turn counter clockwise, and then turn it clockwise to the position where the timing marks are aligned.

26. Loosen the center bolt of the tensioner pulley. Using MD998767 or equivalent tensioner tool and a torque wrench apply 3.3 ft. lbs. (4.4 Nm) tensional torque to the timing belt and tighten the center bolt to 35 ft. lbs. (48 Nm). When tightening the bolt, make sure that the tensioner pulley shaft does not rotate with the bolt.

27. Remove the tensioner plunger pin. Pretension is correct when the pin can be removed and installed easily. If the pin cannot be easily removed and installed it still satisfactory as long as it is within its standard value.

28. Check that the tensioner push rod is within the standard value. When the tensioner is engaged the pushrod should measure 0.149-0.177 inch (3.8-4.5 mm).

29. Rotate the crankshaft 2 revolutions and check the timing marks. If the timing marks are not properly aligned remove the belt and repeat Steps 20 through 26.

30. Install the timing belt covers.

31. Install the engine mounting bracket.

32. Lower the engine enough to install the engine mount onto bracket and remove the floor jack.

33. Install the power steering pump bracket and pump.

34. Install the crankshaft pulley and tighten the retaining bolt to 13 ft. lbs. (18 Nm).

35. Install the accessory drive belts.

36. Properly fill the cooling system.

37. Connect the negative battery cable.

38. Check for leaks and proper engine and cooling system operation.

Thermostat

REMOVAL AND INSTALLATION

2.0L(VIN Y) Engine

1. Disconnect the negative battery cable.
2. Drain the cooling system.
3. Disconnect the upper radiator hose from the water outlet.
4. Remove the thermostat housing.
5. Remove the thermostat taking note of its original position in the housing.
   To install:
6. Thoroughly clean the thermostat housing and cylinder head mating surfaces.
7. Install the thermostat so its flange seats tightly in the machined recess in the thermostat housing. Refer to its position prior to removal.
8. Clean the bolt threads well. Bolts that thread into openings exposed to the coolant may have a build-up of rust and corrosion. Clean bolt threads are necessary for accurate torque. Install the water outlet to the thermostat housing with a new gasket. Torque the housing mounting bolts to 16 ft. lbs. (22 Nm). Do not over-tighten or the thermostat housing and/or the water outlet may crack.
9. Connect the lower hose and fill the system with coolant.
10. Connect the negative battery cable. With the radiator cap off, start engine and allow to run until the thermostat opens. Add coolant as necessary to fill the radiator completely. Watch the coolant temperature gauge (if equipped) for signs of overheating.
11. Once the vehicle has cooled, recheck the coolant level in the radiator and the coolant overflow tank.

2.5L(VIN N) Engine

1. Disconnect the negative battery cable.
2. Drain the cooling system.
3. Disconnect the lower radiator hose from the water inlet.
4. Remove the thermostat housing.
5. Remove the thermostat taking note of its original position in the housing.
   To install:
6. Thoroughly clean the thermostat housing and cylinder head mating surfaces.
7. Install the thermostat so its flange seats tightly in the machined recess in the thermostat housing. Position the thermostat with the jiggle valve in the straight up position. Refer to its position prior to removal.
8. Install the water outlet to the thermostat housing with a new gasket. Clean the bolt threads well. Bolts that thread into openings exposed to the coolant may have a build-up of rust and corrosion. Clean bolt threads are necessary for accurate torque. Torque the housing mounting bolts to 14 ft. lbs. (20 Nm). Do not over-tighten or the thermostat housing and/or water outlet may crack.
9. Connect the lower hose and fill the system with coolant.
10. Connect the negative battery cable, run the vehicle with the radiator cap off, until the thermostat opens (cooling fan comes on), checking and filling the radiator as required. Watch the coolant temperature gauge (if equipped) for signs of overheating.
11. Install the radiator cap and shut off the engine. Once the vehicle has cooled, recheck the coolant level in the radiator and the coolant overflow tank.

Electric Cooling Fan

REMOVAL AND INSTALLATION

2.0L(VIN Y) Engine

1. Disconnect the negative battery cable.
2. Loosen the radiator drain plug and drain the cooling system beneath the level of the upper radiator hose.
3. Disconnect the upper radiator hose to allow clearance for removal of the fan and shroud assembly.

NOTE: It is recommended that each clamp be matchmarked to the hose. Observe the marks and reinstall the clamps in exactly the same position when reinstalling the radiator hose.

4. Disconnect the electrical connector from the cooling fan motor.
5. Remove the mounting bolts, fan and shroud assembly from the vehicle.
6. Remove the fan blade retainer nut from the shaft on the fan motor and separate the fan from the motor.
7. Remove the motor to shroud attaching screws and remove the motor from the shroud.
   To install:
8. Install the motor to the shroud and secure it with the mounting bolts.
9. Install the fan to the motor shaft and secure it with the retainer nut.
10. Install the fan and shroud assembly into the engine compartment and secure the assembly to the radiator. Reconnect the fan motor electrical connector.
11. Install the upper radiator hose and properly fill the cooling system.
12. Connect the negative battery cable and check the cooling fan for proper operation.

2.5L(VIN N) Engine

1. Disconnect the negative battery cable.
2. Loosen the radiator drain plug and drain the cooling system beneath the level of the upper radiator hose.
3. Disconnect the upper radiator hose to allow clearance for removal of the fan and shroud assembly.

NOTE: It is recommended that each clamp be matchmarked to the hose. Observe the marks and reinstall the clamps in exactly the same position when reinstalling the radiator hose.

4. Disconnect the electrical connector from the cooling fan motor.
5. Remove the mounting bolts, fan and shroud assembly from the vehicle.
6. Remove the fan blade retainer nut from the shaft on the fan motor and separate the fan from the motor.
7. Remove the motor to shroud attaching screws and remove the motor from the shroud.
   To install:
8. Install the motor to the shroud and secure it with the mounting bolts.
9. Install the fan to the motor shaft and secure it with the retainer nut.
10. Install the fan and shroud assembly into the engine compartment and secure the assembly to the radiator. Reconnect the fan motor electrical connector.
11. Install the upper radiator hose and properly fill the cooling system.
12. Connect the negative battery cable and check the cooling fan for proper operation.

FUEL SYSTEM

Fuel System Service Precautions

Safety is the most important factor when performing not only fuel system maintenance but any type of maintenance. Failure to conduct maintenance and repairs in a safe manner may result in serious personal injury or death. Maintenance and testing of the vehicle's fuel system components can be accomplished safely and effectively by adhering to the following rules and guidelines.

• To avoid the possibility of fire and personal injury, always disconnect the negative battery cable unless the repair or test procedure requires that battery voltage be applied.

• Always relieve the fuel system pressure prior to disconnecting any fuel system component (injector, fuel rail, pressure regulator, etc.), fitting or fuel line connection. Exercise extreme caution whenever relieving fuel system pressure to avoid exposing skin, face and eyes to fuel spray. Please be advised that fuel under pressure may penetrate the skin or any part of the body that it contacts.

• Always place a shop towel or cloth around the fitting or connection prior to loosening to absorb any excess fuel due to spillage. Ensure that all fuel spillage (should it occur) is quickly removed from engine surfaces. Ensure that all fuel soaked cloths or towels are deposited into a suitable waste container.

• Always keep a dry chemical (Class B) fire extinguisher near the work area.

• Do not allow fuel spray or fuel vapors to come into contact with a spark or open flame.

• Always use a backup wrench when loosening and tightening fuel line connection fittings. This will prevent unnecessary stress and torsion to fuel line piping. Always follow the proper torque specifications.

• Always replace worn fuel fitting O-rings with new. Do not substitute fuel hose or equivalent, where fuel pipe is installed.

Fuel System Pressure

RELIEVING

— **CAUTION** —
Fuel injection systems remain under pressure even after the engine has been turned OFF. The fuel system pressure must be relieved before disconnecting any fuel lines. Failure to do so may result in fire and/or personal injury.

1. Remove the fuel filler cap to release fuel tank pressure.
2. Remove the rear seat cushion.
3. At the fuel tank, disconnect the fuel pump harness connector.
4. Start the vehicle and allow it to run until it stalls from lack of fuel. Turn the key to the **OFF** position.
5. Disconnect the negative battery cable, then reconnect the fuel pump connector.
6. Install the rear seat cushion and the fuel filler cap.

— **CAUTION** —
Always wrap shop towels around a fitting that is being disconnected to absorb residual fuel in the lines.

Idle Speed

ADJUSTMENT

The idle speed is kept at the correct speed by controlling the amount of air that bypasses the throttle valve in accordance with changes in idling conditions and engine load during idling. The Powertrain Control Module (PCM) drives the Idle Air Control (IAC) motor to keep the engine running at the pre-set idle target speed in accordance with the engine coolant temperature and air conditioning load. In addition, when the air conditioning switch is turned off and on while the engine is idling, the IAC motor operates to adjust the throttle valve bypass air amount in accordance with the engine load conditions in order to avoid fluctuations in engine speed. Since the PCM regulates the idle speed, basic idle speed is not adjustable.

Mixture

ADJUSTMENT

The Multi-Port Fuel Injection (MFI) system consists of sensors which detect engine conditions and the Powertrain Control Module (PCM) which controls the system based on sensor signals and actuators which operate under the control of the PCM. The injector drive times and injector timing are PCM controlled for optimum air/fuel mixture. The PCM provides a richer air/fuel mixture in "open-loop" operation when the engine is cold or under extremely high load. When warm or operating normally, the PCM controls the air/fuel mixture by using the heated oxygen sensor signal to carry out "closed-loop" control for best emission controlling mixture. Since the PCM regulates the air/fuel mixture under all conditions, **the mixture is NOT adjustable** by the service technician.

Fuel Filter

REMOVAL AND INSTALLATION

A replaceable fuel filter is located in the engine compartment, on the bulkhead, next to the brake booster.

— **CAUTION** —
Fuel injection systems remain under pressure, even after the engine has been turned OFF. The fuel system pressure must be relieved before disconnecting any fuel lines. Failure to do so may result in fire and/or personal injury.

1. Following proper procedures, relieve the fuel system residual pressure.

NOTE: Wrap shop towels around the fitting that is being disconnected to absorb residual fuel in the lines.

2. Disconnect the negative battery cable.
3. Remove the air intake hose for access.
4. Hold the fuel filter housing securely with a wrench. Cover the hoses with shop towels and remove the eye bolt. Discard the gaskets.
5. Separate the flare nut connection at the bottom of the filter.

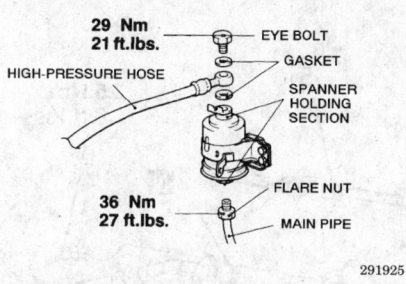

291925

Connecting fuel lines to the fuel filter

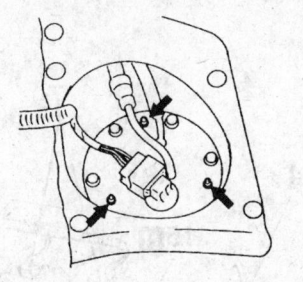

291937

Aligning the fuel pump for installation

6. Remove the mounting bolts and the fuel filter from the vehicle.

NOTE: Do not use conventional fuel filters, hoses or clamps when servicing fuel injection systems. They are not compatible with the injection system and the high pressures in fuel injection systems and could cause substandard parts to fail, causing personal injury or damage to the vehicle. Use only hoses and clamps specifically designed for fuel injection.

To install:

7. Tighten the flare nut fitting by hand before mounting the filter to the bracket.

8. Install the filter to its bracket only finger-tight. Movement of the filter will ease attachment of the fuel lines.

9. Using new gaskets, connect the high pressure hose and eye bolt. While holding the fuel filter housing, tighten the eye bolt to 21 ft. lbs. (29 Nm). Tighten the flare nut to 27 ft. lbs. (36 Nm).

10. Tighten the filter mounting bolts fully.

11. Install the intake air hose.

12. Connect the negative battery cable, turn the key to the **ON** position to pressurize the fuel system and check for leaks.

13. If necessary, release the fuel pressure and repair leaks.

Fuel Pump

REMOVAL AND INSTALLATION

— **CAUTION** —

Fuel injection systems remain under pressure, even after the engine has been turned OFF. The fuel system pressure must be relieved before disconnecting any fuel lines, Failure to do so may result in fire and/or personal injury.

Do not use conventional fuel filters, hoses or clamps when servicing fuel injection systems. They are not compatible with the injection system and could fail, causing personal injury or damage to the vehicle. Use only hoses and clamps specifically designed for fuel injection.

1. Relieve the fuel system pressure, using proper procedures. Disconnect negative battery cable.

NOTE: The rear seat cushion must be removed in order to gain access to the fuel pump.

2. Remove the rear seat cushion by pulling the seat stopper outward and lifting the lower cushion upward. There are two access covers underneath the seat. The panel on the far right side is for the fuel pump.

3. Remove the access cover.

4. Disconnect the fuel pump wiring.

5. Disconnect the return hose and the high pressure fuel hose.

— **CAUTION** —

Observe all applicable safety precautions when working around fuel. Do not allow fuel spray or fuel vapors to come into contact with a spark or open flame. Keep a dry chemical (Class B) fire extinguisher near the work area. Never drain or store fuel in an open container due to the possibility of fire or explosion. Cover all fuel hose connections with a shop towel, prior to disconnecting, to prevent splash of fuel that could be caused by residual pressure remaining in the fuel line.

6. Remove the pump mounting nuts and remove the pump assembly.

To install:

7. Align the seal position projections with the holes in the fuel pump assembly and install the assembly in the tank. Tighten the retaining nuts to 22 inch lbs. (2.5 Nm).

8. Connect the high pressure hose, return hose and the fuel pump wiring.

9. Connect the negative battery cable.

10. Check the fuel pump for proper pressure and inspect the entire system for leaks.

11. Apply sealant to the access cover and install the cover.

12. Install the rear seat cushion.

13. Pressurize the fuel system by turning the ignition key to the **ON** position. Check for leaks. Start the engine to verify proper fuel pump performance.

Fuel Injector

REMOVAL AND INSTALLATION

2.0L(VIN Y) Engine

— **CAUTION** —

Fuel injection systems remain under pressure, even after the engine has been turned OFF. The fuel system pressure must be relieved before disconnecting any fuel lines. Failure to do so may result in fire and/or personal injury.

Do not use conventional fuel filters, hoses or clamps when servicing fuel injection systems. They are not compatible with the injection system and could fail, causing personal injury or damage to the vehicle. Use only hoses and clamps specifically designed for fuel injection.

1. Relieve the fuel system pressure following proper procedures.

2. Disconnect the negative battery cable.

3. Remove the bolts holding the high pressure fuel line to the fuel rail and disconnect the line. Be prepared to contain fuel spillage. Plug the line to keep out dirt and debris.

4. Remove the vacuum hose from the fuel pressure regulator.

5. Disconnect the fuel return hose from the fuel rail. Remove the snap ring securing the fuel pressure regulator and remove the regulator from the fuel rail.

6. Label and disconnect the electrical connector from each injector.

7. Remove the bolts holding the fuel rail to the manifold. Carefully lift the rail up and remove it with the

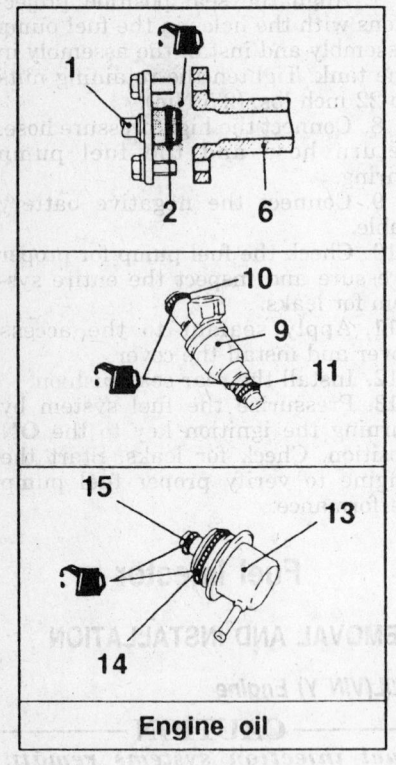

Engine oil

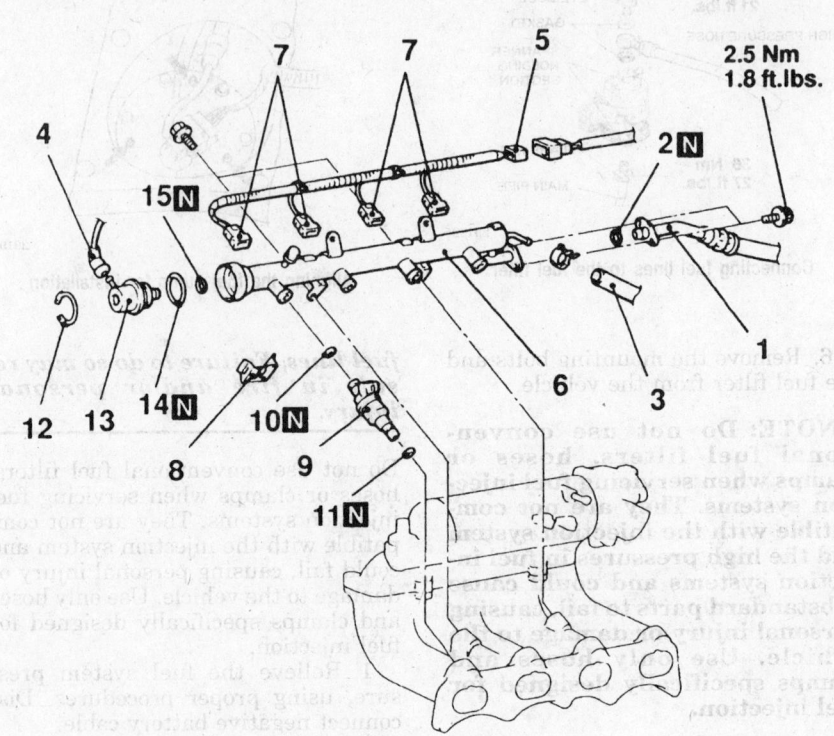

2.5 Nm
1.8 ft.lbs.

1. High-pressure fuel hose connection
2. O-ring
3. Fuel return hose connection
4. Vacuum hose connection
5. Injector harness connector
6. Fuel rail
7. Injector connectors
8. Retainers
9. Injectors
10. O-rings
11. O-rings
12. Snap ring
13. Fuel pressure regulator
14. O-ring
15. O-ring
N. Replace with new components

291929

Fuel injection system exploded view — 2.0L engine

injectors attached. Take care not to drop an injector. Place the rail and injectors in a safe location on the workbench. Protect the tips of the injectors from dirt and/or impact.

8. Remove and discard the injector insulators from the intake manifold. The insulators are not reusable.

9. Remove the injectors from the fuel rail by pulling gently in a straight outward motion. Make certain the grommet and O-ring come off with the injector.

To install:

10. Install a new insulator in each injector port in the manifold.

11. Remove the old grommet and O-ring from each injector. Install a new

grommet and O-ring. Coat the O-ring lightly with clean, thin oil.

12. If the fuel pressure regulator was removed, replace the O-rings with new ones and coat them lightly with clean, thin oil. Insert the regulator straight into the rail, then check that it can be rotated freely. If it does not rotate smoothly, remove it and inspect the O-rings for deformation or jamming. Secure the regulator with the snap ring.

13. Install the injector into the fuel rail, constantly turning the injector left and right during installation. When fully installed, the injector should still turn freely in the rail. If it does not, remove the injector and in-

spect the O-ring for deformation or damage.

14. Install the delivery pipe and injectors to the engine. Make certain that each injector fits correctly into its port and that the rubber insulators for the fuel rail mounts are in position.

15. Install the fuel rail retaining bolts.

16. Connect the wiring harnesses to the appropriate injectors.

17. Connect the vacuum hose to the pressure regulator.

18. Replace the O-ring on the high pressure fuel line, coat the O-ring lightly with clean, thin oil and install the line to the fuel rail. Tighten the

mounting bolts to 1.8 ft. lbs. (2.5 Nm).

19. Connect the negative battery cable. Pressurize the fuel system and inspect all connections for leaks.

2.5L(VIN N) Engine

----------- **CAUTION** -----------

Fuel injection systems remain under pressure, even after the engine has been turned OFF. The fuel system pressure must be relieved before disconnecting any fuel lines. Failure to do so may result in fire and/or personal injury.

Do not use conventional fuel filters, hoses or clamps when servicing fuel injection systems. They are not compatible with the injection system and could fail, causing personal injury or damage to the vehicle. Use only hoses and clamps specifically designed for fuel injection.

Front Bank Side

1. Relieve the fuel system pressure following proper procedures.
2. Disconnect the negative battery cable.
3. Disconnect the power steering pressure switch connector.
4. Disconnect the oxygen sensor connector.
5. Remove the bolts holding the fuel pipe to the fuel rail and disconnect the fuel pipe. Be prepared to contain fuel spillage. Plug the openings to keep out dirt and debris.
6. Remove the vacuum hose from the fuel pressure regulator.
7. Disconnect the fuel return hose from the pressure regulator. Remove the fuel pressure regulator mounting bolts and remove the regulator from the fuel rail.
8. Label and disconnect the electrical connectors from each injector.
9. Remove the bolts holding the fuel rail to the manifold. Carefully lift the rail up and remove it with the injectors attached. Take care not to drop an injector. Place the rail and injectors in a safe location on the workbench. Protect the tips of the injectors from dirt and/or impact.
10. Remove and discard the injector insulators from the intake manifold. The insulators are not reusable.
11. Remove the injectors from the fuel rail by pulling gently in a straight outward motion. Make cer-

tain the grommet and O-ring come off with the injector.

To install:

12. Install a new insulator in each injector port in the manifold.
13. Remove the old grommet and O-ring from each injector. Install a new grommet and O-ring. Coat the O-ring lightly with clean, thin oil.
14. If the fuel pressure regulator was removed, replace the O-ring with a new one and coat it lightly with clean, thin oil. Insert the regulator straight into the rail, then check that it can be rotated freely. If it does not rotate smoothly, remove it and inspect the O-ring for deformation or jamming. When properly installed, align the mounting holes and tighten the retaining bolts to 6.5 ft. lbs. (8.8 Nm).
15. Install the injector into the fuel rail, constantly turning the injector left and right during installation. When fully installed, the injector should still turn freely in the rail. If it does not, remove the injector and inspect the O-ring for deformation or damage.
16. Install the delivery pipe and injectors to the engine. Make certain that each injector fits correctly into its port and that the rubber insulators for the fuel rail mounts are in position.
17. Install the fuel rail retaining bolts and tighten them to 9 ft. lbs. (12 Nm).
18. Connect the wiring harnesses to the appropriate injectors.
19. Connect the vacuum hose to the pressure regulator.
20. Replace the O-ring on the fuel pipe, coat the O-ring lightly with clean, thin oil and install the pipe to the fuel rail. Tighten the mounting bolts to 6.5 ft. lbs. (8.8 Nm).
21. Connect the negative battery cable. Pressurize the fuel system and inspect all connections for leaks.

Rear Bank Side

1. Relieve the fuel system pressure following proper procedures.
2. Disconnect the negative battery cable.
3. Remove the intake manifold plenum.
4. Remove the bolts holding the high pressure line to the fuel rail and disconnect the line. Be prepared to contain fuel spillage. Plug the line to keep out dirt and debris.

5. Remove the bolts holding the fuel pipe to the fuel rail and disconnect the fuel pipe. Plug the openings to keep out dirt and debris.
6. Label and disconnect the electrical connectors from each injector.
7. Remove the bolts holding the fuel rail to the manifold. Carefully lift the rail up and remove it with the injectors attached. Take care not to drop an injector. Place the rail and injectors in a safe location on the workbench. Protect the tips of the injectors from dirt and/or impact.
8. Remove and discard the injector insulators from the intake manifold. The insulators are not reusable.
9. Remove the injectors from the fuel rail by pulling gently in a straight outward motion. Make certain the grommet and O-ring come off with the injector.

To install:

10. Install a new insulator in each injector port in the manifold.
11. Remove the old grommet and O-ring from each injector. Install a new grommet and O-ring. Coat the O-ring lightly with clean, thin oil.
12. Install the injector into the fuel rail, constantly turning the injector left and right during installation. When fully installed, the injector should still turn freely in the rail. If it does not, remove the injector and inspect the O-ring for deformation or damage.
13. Install the delivery pipe and injectors to the engine. Make certain that each injector fits correctly into its port and that the rubber insulators for the fuel rail mounts are in position.
14. Install the fuel rail retaining bolts and tighten them to 9 ft. lbs. (12 Nm).
15. Connect the wiring harnesses to the appropriate injectors.
16. Replace the O-ring on the high pressure fuel line, coat the O-ring lightly with clean, thin oil and install the line to the fuel rail. Tighten the mounting bolts to 6.5 ft. lbs. (8.8 Nm).
17. Replace the O-ring on the fuel pipe, coat the O-ring lightly with clean, thin oil and install the pipe to the fuel rail. Tighten the mounting bolts to 6.5 ft. lbs. (8.8 Nm).
18. Install the intake manifold plenum.
19. Connect the negative battery cable. Pressurize the fuel system and inspect all connections for leaks.

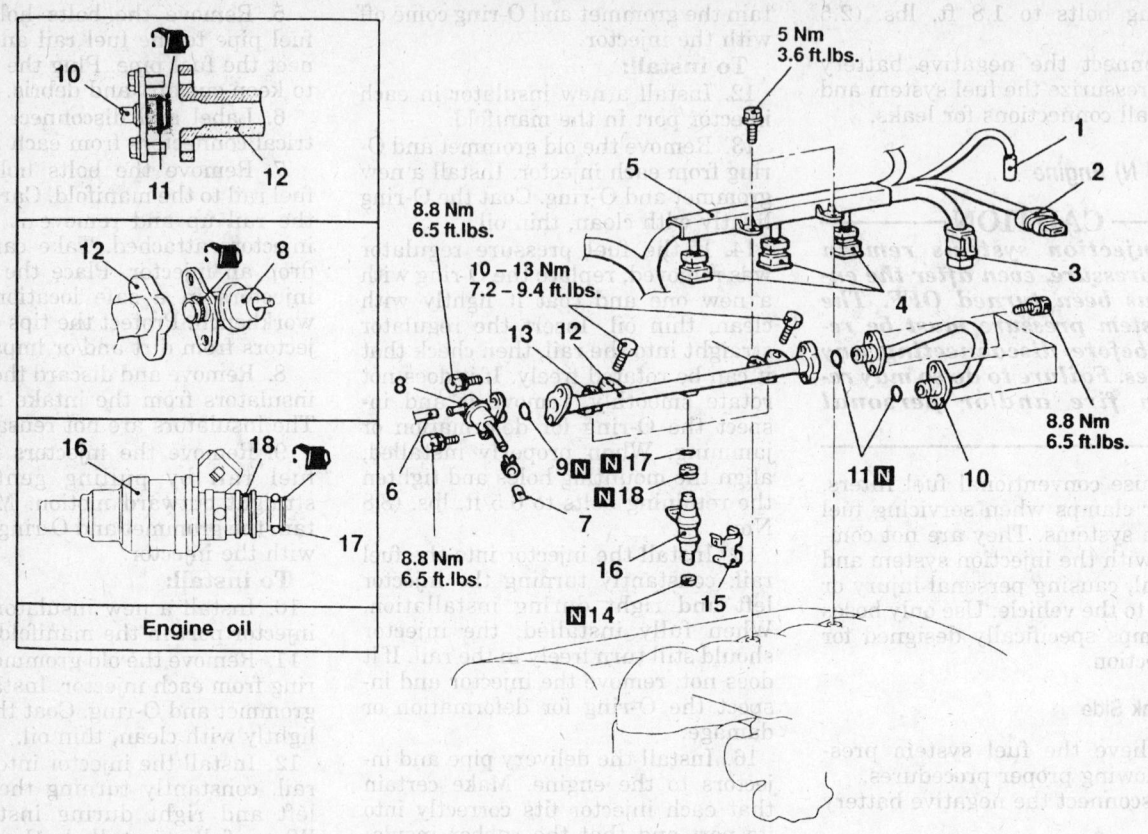

Exploded view of the front bank side of the fuel injection system — 2.5L engine

1. Power steering oil pressure switch connector
2. Heated oxygen sensor connector
3. Intake air temperature sensor connector
4. Injector connectors
5. Control wiring harness
6. Vacuum hose connection
7. Fuel return hose connection
8. Fuel pressure regulator
9. O-ring
10. Fuel pipe
11. O-rings
12. Fuel rail
13. Insulators
14. Insulators
15. Injector supports
16. Injectors
17. O-rings
18. Grommets

ENGINE MECHANICAL

Engine Assembly

REMOVAL AND INSTALLATION

2.0L(VIN Y) Engine

CAUTION
Fuel injection systems remain under pressure, even after the engine has been turned OFF. The fuel system pressure must be relieved before disconnecting any fuel lines. Failure to do so may result in fire and/or personal injury.

The transaxle must be removed before removing the engine. They will not come out as a unit.

1. Disconnect the negative battery cable.
2. Drain the engine coolant.
3. Drain the engine oil and the transmission oil.
4. Safely relieve the pressure within the fuel injection system.
5. Matchmark the hood to the hinges and remove the hood.

6. Remove the engine under covering.
7. Remove the transaxle assembly using the recommended procedure.
8. Remove the radiator, disconnecting the hoses at the engine.
9. Disconnect the accelerator cable and remove the bracket.
10. Disconnect the heater hoses.
11. Disconnect the brake booster vacuum hose at the engine.
12. Label and disconnect the vacuum hoses running to the bulkhead.
13. Disconnect the high pressure fuel line and discard the O-ring. It is not reusable.
14. Remove the fuel return hose.

15. Disconnect the electrical connectors to the engine components. All wires and connectors should be labeled at the time of removal. This should save much time at assembly. Identify, tag, then disconnect the:
- oxygen sensor
- coolant temperature sender (gauge)
- coolant temperature sensor
- idle speed control
- EGR temperature sensor
- engine knock sensor
- each injector
- power transistor
- condenser
- Throttle Position Sensor (TPS)
- Motor Position Sensor (MPS)
- distributor connector
- control harness
- alternator
- oil pressure switch
- coolant temperature switch
- crankshaft angle sensor
- body ground
- power steering oil pressure switch

16. Loosen the power steering/air conditioning compressor drive belt and remove it. Remove the bolts holding the pump to its bracket and hang the pump out of the way. Do not disconnect the hoses and do not allow the pump to hang by the hoses.

17. Remove the air conditioning compressor from its mount and hang it from a stiff wire out of the way. Note that the hoses should be left still attached. Do not loosen them or discharge the system.

18. Remove the self locking nuts and bolt at the exhaust system joint just below the manifold. Separate the exhaust pipes. Discard the gasket and the two nuts.

19. Raise and safely support the vehicle. Install the engine hoist equipment and make certain the attaching points on the engine are secure. Draw tension on the hoist just enough to support the engine's weight but no more. Do not disturb the placement of the vehicle on the stands.

20. Remove the through-bolt from the rear (bulkhead side) roll stopper. Remove the through-bolt from the front engine roll stopper.

21. Remove the nuts and bolts holding the upper (right side) engine mount to the engine. Remove the through-bolt and remove the mount assembly. Also remove the support bracket below the mount.

22. Double check for any remaining cables, wires or hoses running to the engine. Elevate the hoist and remove the engine from the vehicle. Immediately place it on an engine stand or

support it with wooden blocks. Do not allow it to rest on the oil pan or lie on its side. Do not leave the engine hanging from the hoist.

To install:
After repairs, make certain the engine is fully reassembled before installation. All components removed with the engine out of the vehicle should be in place before reinstallation.

23. Install the engine into the vehicle and lower it until the bolt holes for the mounts and roll stoppers align with the brackets. Install the through-bolts and new self-locking nuts, tightening them just snug; they will be final tightened later.

24. Install the upper (right side) mount to the engine, tightening the nuts and bolts snug. Install the through-bolt and tighten it snug.

25. Slowly release the tension on the hoist, allowing the weight of the engine to bear fully on the mounts. Once the hoist is slack, remove the lifting apparatus from the engine.

26. Connect the exhaust system to the manifold, using a new gasket and new locking nuts. Tighten the nuts and the small bolt to 33 ft. lbs. (44 Nm).

27. Final tighten the engine mount nuts and bolts. Correct torque values are:
- Nut and bolt holding the right side mount to engine: 63 ft. lbs. (86 Nm)
- Right side mount through-bolt: 71–85 ft. lbs. (98–118 Nm)
- Rear roll stopper through-bolt: 32 ft. lbs. (44 Nm)
- Front roll stopper through-bolt: 41 ft. lbs. (56 Nm)

NOTE: Allow the mounts to support the engine weight before final tightening the front roll stopper through-bolt.

28. Install the air conditioning compressor.

29. Install the power steering pump, tightening the bolts to 29 ft. lbs. (39 Nm). Install and adjust the belt.

30. Connect the wiring and harness connectors to the engine. Make certain each terminal is clean and the connector is firmly seated to its mate. Do not route wires near hot surfaces or moving parts. Connect the:
- oxygen sensor
- coolant temperature sender (gauge)
- coolant temperature sensor
- idle speed control
- EGR temperature sensor
- engine knock sensor

- each injector
- power transistor
- condenser
- throttle position sensor (TPS)
- motor position sensor (MPS)
- distributor connector
- control harness
- alternator
- oil pressure switch
- coolant temperature switch
- crankshaft angle sensor
- body ground
- power steering oil pressure switch

31. Install the fuel return hose.

32. Using a new O-ring lightly lubricated with clean engine oil, connect the high pressure fuel line and tighten the bolts to 1.8 ft. lbs. (2.5 Nm).

33. Connect the vacuum lines running to the bulkhead. Install the brake booster vacuum hose to the engine.

34. Connect the heater hoses.

35. Install the accelerator cable bracket, and connect the accelerator cable.

36. Install the radiator and connect the hoses.

37. Install the transaxle.

38. Install the engine under covering.

39. Check the engine oil drain plug and secure it if necessary. Install the proper amount of engine oil.

40. Check the transaxle drain plug, tightening it if needed, and install the proper amount of transmission oil.

41. Check the radiator and engine drain cocks, closing them if necessary and refill the coolant system.

42. Double check all installation items, paying particular attention to loose hoses or hanging wires, loosened nuts, poor routing of hoses and wires (too tight or rubbing) and tools left in the engine area.

43. Connect the negative battery cable. Start the engine and check for leaks.

44. Attend to all leaks immediately, remembering that fluids and metal surfaces may be hot. Adjust the drive belts to the correct tension. Adjust all cables (transmission, throttle, shift selector) and check the fluid levels. Check the operation of all gauges and dashboard lights.

45. With the help of an assistant, install the hood and align it for proper body fit and latching.

46. In a safe location at low speed, road test the vehicle for correct operation of steering brakes, transaxle, clutch and speedometer.

2.5L (VIN N) Engine

CAUTION

Fuel injection systems remain under pressure, even after the engine has been turned OFF. The fuel system pressure must be relieved before disconnecting any fuel lines. Failure to do so may result in fire and/or personal injury.

The transaxle must be removed before removing the engine. They will not come out as a unit.

1. Disconnect the negative battery cable.
2. Drain the engine coolant.
3. Drain the engine oil and the transmission oil.
4. Safely relieve the pressure within the fuel injection system.
5. Matchmark the hood to the hinges and remove the hood.
6. Remove the transaxle assembly.
7. Remove the radiator, disconnecting the hoses at the engine.
8. Disconnect the accelerator cable and remove the bracket.
9. Disconnect the heater hoses.
10. Disconnect the brake booster vacuum hose at the engine.
11. Label and disconnect the vacuum hoses running to the bulkhead.
12. Disconnect the high pressure fuel line and discard the O-ring. It is not reusable.
13. Remove the fuel return hose.
14. Disconnect the electrical connectors to the engine components. All wires and connectors should be labeled at the time of removal. This should save much time at assembly. Disconnect the:
- oxygen sensor
- coolant temperature sender (gauge)
- coolant temperature sensor
- idle speed control
- EGR temperature sensor
- engine knock sensor
- each injector

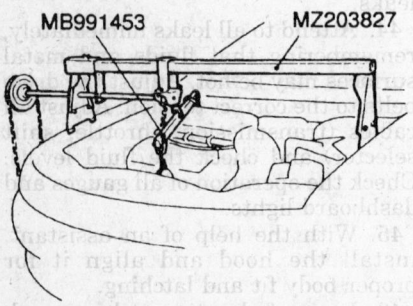

Support the engine with the proper tools to remove the engine mounts — 2.5L engine

292772

- power transistor
- condenser
- Throttle Position Sensor (TPS)
- Motor Position Sensor (MPS)
- distributor connector
- control harness
- alternator
- oil pressure switch
- coolant temperature switch
- crankshaft angle sensor
- body ground
- relay box and generator wiring harness connection
- power steering oil pressure switch

15. Remove the accessory drive belts. Remove the bolts holding the power steering pump to its bracket and hang the pump out of the way. Do not disconnect the hoses and do not allow the pump to hang by the hoses. Remove the power steering pump bracket.
16. Remove the air conditioning compressor from its mount and hang it from a stiff wire out of the way. Note that the hoses are still attached. Do not loosen them or discharge the system.
17. Remove the bolts at the exhaust system joint just below the manifold. Separate the exhaust pipes and discard the gasket.
18. Raise and safely support the vehicle. Install the engine hoist equipment and make certain the attaching points on the engine are secure. Draw tension on the hoist just enough to support the engine's weight but no more. Do not disturb the placement of the vehicle on the stands.
19. Remove the through-bolt from the rear (bulkhead side) roll stopper. Remove the through-bolt from the front engine roll stopper.
20. Remove the nuts and bolts holding the upper (right side) engine mount to the engine. Remove the through-bolt and remove the mount assembly. Also remove the support bracket below the mount.
21. Double check for any remaining cables, wires or hoses running to the engine. Elevate the hoist and remove the engine from the vehicle. Immediately place it on an engine stand or support it with wooden blocks. Do not allow it to rest on the oil pan or lie on its side. Do not leave the engine hanging from the hoist.

To install:

After repairs, make certain the engine is fully reassembled before installation. All components removed with the engine out of the vehicle should be in place before reinstallation.

22. Install the engine into the vehicle and lower it until the bolt holes for the mounts and roll stoppers align with the brackets. Install the through-bolts and new self-locking nuts, tightening them just snug; they will be final tightened later.
23. Install the upper (right side) mount to the engine, tightening the nuts and bolts snug. Install the through-bolt and tighten it snug.
24. Slowly release the tension on the hoist, allowing the weight of the engine to bear fully on the mounts. Once the hoist is slack, remove the lifting apparatus from the engine.
25. Connect the exhaust system to the manifold, using a new gasket. Tighten the bolts to 33 ft. lbs. (44 Nm).
26. Final tighten the engine mount nuts and bolts. Correct torque values are:
- Nut and bolt holding the right side mount to engine: 63 ft. lbs. (86 Nm)
- Right side mount through-bolt: 71–85 ft. lbs. (98–118 Nm)
- Rear roll stopper through-bolt: 32 ft. lbs. (44 Nm)
- Front roll stopper through-bolt: 41 ft. lbs. (56 Nm)

NOTE: Allow the mounts to support the engine weight before final tightening the front roll stopper through-bolt.

27. Install the air conditioning compressor.
28. Install the power steering pump bracket tightening the bolts to 16 ft. lbs. (22 Nm). Install the pump and adjust the belt.
29. Connect the wiring and harness connectors to the engine. Make certain each terminal is clean and the connector is firmly seated to its mate. Do not route wires near hot surfaces or moving parts. Connect the:
- oxygen sensor
- coolant temperature sender (gauge)
- coolant temperature sensor
- idle speed control
- EGR temperature sensor
- engine knock sensor
- each injector
- power transistor
- condenser
- Throttle Position Sensor (TPS)
- Motor Position Sensor (MPS)
- distributor connector
- control harness
- alternator
- oil pressure switch
- coolant temperature switch
- crankshaft angle sensor

- body ground
- relay box and generator wiring harness connection
- power steering oil pressure switch

30. Install the fuel return hose.

31. Using a new O-ring, install the high pressure fuel line.

32. Connect the vacuum lines running to the bulkhead. Install the brake booster vacuum hose to the engine.

33. Connect the heater hoses.

34. Install the accelerator cable bracket, and connect the accelerator cable.

35. Install the radiator and connect the hoses.

36. Install the transaxle.

37. Check the engine oil drain plug and secure it if necessary. Install the proper amount of engine oil.

38. Check the transaxle drain plug, tightening it if needed, and install the proper amount of transmission oil.

39. Check the radiator and engine drain cocks, closing them if necessary and refill the coolant system.

40. Double check all installation items, paying particular attention to loose hoses or hanging wires, loosened nuts, poor routing of hoses and wires (too tight or rubbing) and tools left in the engine area.

41. Connect the negative battery cable. Start the engine and check for leaks.

42. Attend to all leaks immediately, remembering that fluids and metal surfaces may be hot. Adjust the drive belts to the correct tension. Adjust all cables (transmission, throttle, shift selector) and check the fluid levels. Check the operation of all gauges and dashboard lights.

43. With the help of an assistant, install the hood and align it for proper body fit and latching.

44. In a safe location at low speed, road test the vehicle for correct operation of steering brakes, transaxle, clutch and speedometer.

Engine Mounts

REMOVAL AND INSTALLATION

2.0L(VIN Y) Engine

Upper Mount

1. Disconnect the negative battery cable. Remove the air cleaner and all necessary air duct work.

2. Using a 7137 or a C-4852 engine lift or equivalent, raise and

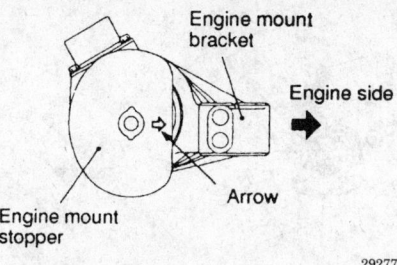

292777

Align the right side upper mount as shown on both engines

safely support the engine so it is not resting on the engine mounts.

3. Remove the engine mount bracket and body connection through bolt. Take note of the position of the arrow on the oval shaped mounting stopper plate. This is important.

4. Remove the engine mounting bracket and stopper plate.

To install:

5. Install the engine mounting bracket and stopper plate. Note the arrows on the stopper plates and make sure they are installed properly. On most engines the arrows will face towards the center of the engine. Torque the upper mount to engine nut and bolt to 63 ft. lbs. (86 Nm) and

the upper mount through bolt nut to 71–85 ft. lbs. (98–118 Nm).

6. Install the air cleaner and all air duct work previously removed.

7. Connect the negative battery cable.

Engine Roll Stopper (Lower Rear Mount)

1. Disconnect the negative battery cable.

2. Using a 7137 or a C-4852 engine lift or equivalent, raise and safely support the engine so it is not resting on the engine mounts. Use care not to bend or damage any components.

3. Remove the stopper through bolt.

4. Remove the stopper mount frame bolts and pry the mount out.

To install

5. Position the lower rear roll stopper so the part of the bracket with the hole in it is facing the front of the vehicle. Install the frame mounting bolts and tighten to 32 ft. lbs. (44 Nm).

6. Install the through bolt and torque the nut to 32 ft. lbs. (44 Nm).

7. Connect the negative battery cable.

Engine Roll Stopper (Lower Front Mount)

1. Disconnect the negative battery cable.

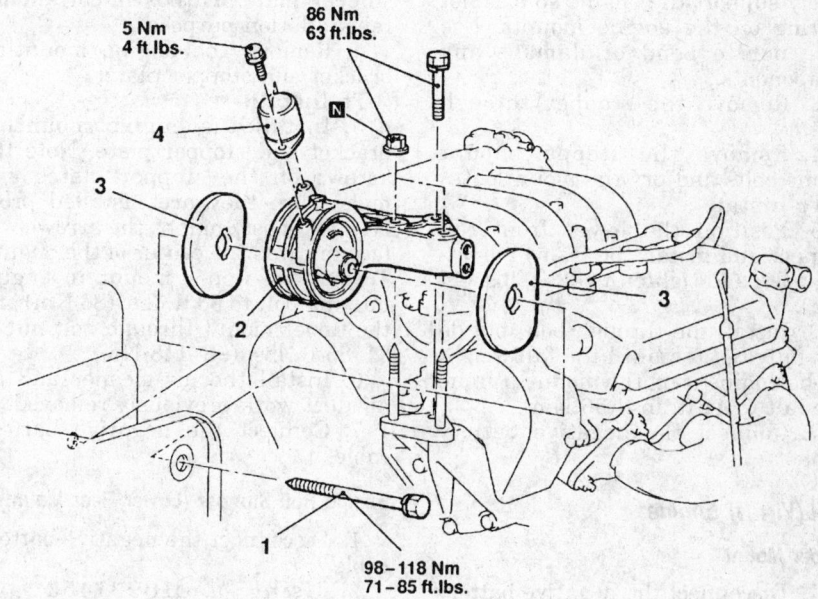

1. Engine mount insulator mounting bolt
2. Engine mount bracket
3. Engine mount stopper
4. Dynamic damper

292956

Right side upper engine mount — 2.5L engine

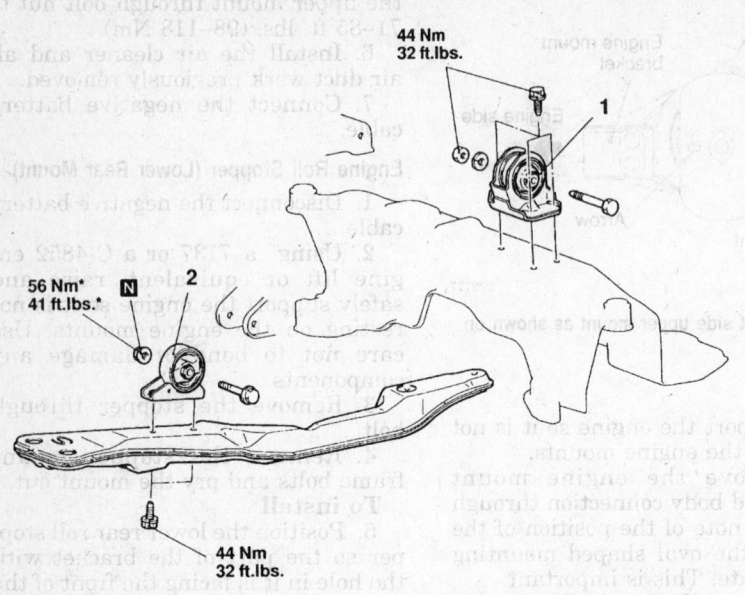

44 Nm
32 ft.lbs.

1

56 Nm*
41 ft.lbs.*

N

2

44 Nm
32 ft.lbs.

1. Rear roll stopper bracket assembly
2. Front roll stopper bracket assembly

N Replace with a new nut

Caution
*: Indicates parts which should be temporarily tightened, and then fully tightened with the vehicle on the ground in the unladen condition.

292959

Front and rear engine roll stoppers

2. Using a 7137 or a C-4852 engine lift or equivalent, raise and safely support the engine so it is not resting on the engine mounts. Use care not to bend or damage any components.

3. Remove the stopper through bolt.

4. Remove the stopper mount frame bolts and pry the mount out.

To install

5. Position the lower front roll stopper and install the frame mounting bolts and tighten to 32 ft. lbs. (44 Nm).

6. Install the through bolt, but do not fully tighten until the full weight of the engine is on the mount. Torque the nut to 41 ft. lbs. (56 Nm).

7. Connect the negative battery cable.

2.5L(VIN N) Engine

Upper Mount

1. Disconnect the negative battery cable. Remove the air cleaner and all necessary air duct work.

2. Using a MB991453 and MZ203827 engine lifting assembly or equivalent, raise and safely support the engine so it is not resting on the engine mounts.

3. Remove the engine mount bracket and body connection through

bolt. Take note of the position of the arrow on the oval shaped mounting stopper plate. This is important for reinstallation purposes.

4. Remove the engine mounting bracket and stopper plate.

To install:

5. Install the engine mounting bracket and stopper plate. Note the arrows on the stopper plates and make sure they are installed properly. On most engines the arrows will face towards the center of the engine. Torque the upper mount to engine nut and bolt to 63 ft. lbs. (86 Nm) and the upper mount through bolt nut to 71–85 ft. lbs. (98–118 Nm).

6. Install the air cleaner and all air duct work previously removed.

7. Connect the negative battery cable.

Engine Roll Stopper (Lower Rear Mount)

1. Disconnect the negative battery cable.

2. Using a MB991453 and MZ203827 engine lifting assembly or equivalent, raise and safely support the engine so it is not resting on the engine mounts. Use care not to bend or damage any components.

3. Remove the stopper through bolt.

4. Remove the stopper mount frame bolts and pry the mount out.

To install

5. Position the lower rear roll stopper so the part of the bracket with the hole in it is facing the front of the vehicle. Install the frame mounting bolts and tighten to 32 ft. lbs. (44 Nm).

6. Install the through bolt and torque the nut to 32 ft. lbs. (44 Nm).

7. Connect the negative battery cable.

Engine Roll Stopper (Lower Front Mount)

1. Disconnect the negative battery cable.

2. Using a MB991453 and MZ203827 engine lifting assembly or equivalent, raise and safely support the engine so it is not resting on the engine mounts. Use care not to bend or damage any components.

3. Remove the stopper through bolt.

4. Remove the stopper mount frame bolts and pry the mount out.

To install

5. Position the lower front roll stopper and install the frame mounting bolts and tighten to 32 ft. lbs. (44 Nm).

6. Install the through bolt, but do not fully tighten until the full weight of the engine is on the mount. Torque the nut to 41 ft. lbs. (56 Nm).

7. Connect the negative battery cable.

Cylinder Head

REMOVAL AND INSTALLATION

2.0L(VIN Y) Engine

—— CAUTION ——
Fuel injection systems remain under pressure, even after the engine has been turned OFF. The fuel system pressure must be relieved before disconnecting any fuel lines. Failure to do so may result in fire and/or personal injury.

1. Following proper procedure, relieve the fuel system pressure.

2. Disconnect the negative battery cable.

3. Remove the air cleaner with all air intake hoses.

4. Drain the cooling system and the engine oil.

5. Disconnect the accelerator cable.

6. Disconnect the small vacuum hose at the pressure regulator.

7. Disconnect the oxygen sensor, engine coolant temperature sensor, the engine coolant temperature

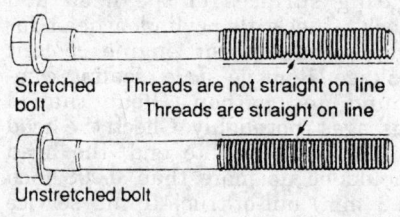

Checking the cylinder head bolts for necking (stretching) — 2.0L engine

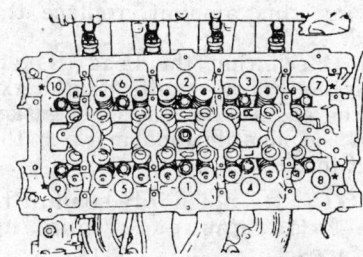

*Location of 110 mm (4.330 in.) short bolts.

Cylinder head torque sequence — 2.0L engine

gauge unit and the engine coolant temperature switch on vehicles with air conditioning.

8. Disconnect the throttle position sensor, crankshaft angle sensor, fuel injectors, ignition coil pack, EGR temperature sensor, ground cable and engine control wiring harness.

9. Remove the spark plug cable center cover and remove the spark plug cables.

10. Disconnect and plug the high pressure fuel line.

11. Remove the heater hose and the water hose connections.

12. Remove the PCV hose.

13. Disconnect and plug the fuel return hose.

14. Disconnect the brake booster vacuum hose.

15. Remove the timing belt.

16. Remove the valve cover and the half-round seal.

17. Remove the exhaust pipe self-locking nuts and separate the exhaust pipe from the exhaust manifold. Discard the gasket.

18. Remove the intake and exhaust camshafts.

19. Loosen the cylinder head mounting bolts and lift off the cylinder head assembly complete with the exhaust and intake manifolds attached.

20. Place the assembly on a suitable workbench and remove the intake and exhaust manifolds.

21. Remove the head gasket.

To install:

22. Thoroughly clean and dry the mating surfaces of the head and block. Check the cylinder head for cracks, damage or engine coolant leakage. Remove scale, sealing compound and carbon. Clean the oil passages thoroughly. Check the head for flatness. End to end, the head should be no more than 0.004 inch (0.1 mm) out-of-true.

23. Place a new head gasket on the cylinder block with the identification marks at the front top (upward) position. Make sure the gasket has the proper identification mark for the engine. Do not use sealer on the gasket.

24. New head bolts are recommended. Inspect the cylinder head bolts prior to installation. If the threads are necked down (stretched), the bolts should be replaced. Necking can be checked by holding a straight edge against the threads. If all the threads do not contact the straightedge, the bolt should be replaced.

25. Carefully install the cylinder head on the block. Before installing the bolts, the threads should be oiled with clean engine oil.

26. Tighten the cylinder head bolts in sequence as follows:

   a. Tighten the center bolts (1 through 6) to 25 ft. lbs. (33 Nm). Then tighten the outer bolts (7 through 10) to 20 ft. lbs. (27 Nm).

   b. Tighten the center bolts (1 through 6) to 50 ft. lbs. (67 Nm). Then tighten the outer bolts (7 through 10) to 20 ft. lbs. (27 Nm).

   c. Tighten the center bolts (1 through 6) to 50 ft. lbs. (67 Nm). Then tighten the outer bolts (7 through 10) to 20 ft. lbs. (27 Nm).

   d. Tighten all 10 bolts an additional 1/4 turn (90 degrees).

27. Install the new exhaust pipe gasket and connect the exhaust pipe to the manifold. Tighten the self-locking bolts to 33 ft. lbs. (44 Nm).

28. Apply sealer to the perimeter of the half-round seal and to the lower edges of the half-round portions of the belt-side of the new gasket. Install the valve cover.

29. Install the timing belt and all related items.

30. Connect or install all previously disconnected hoses, cables and electrical connections. Adjust the throttle cable(s).

31. Install the spark plug cable center cover.

32. Replace the O-rings and connect the fuel lines.

33. Install the air cleaner and intake hose. Connect the breather hose.

34. Change the engine oil and oil filter.

35. Install the radiator and fill the system with coolant.

36. Connect the negative battery cable, with the radiator cap off, run the vehicle until the thermostat opens checking the coolant level to fill the radiator completely.

37. Once the vehicle has cooled, recheck the coolant level.

2.5L (VIN N) Engine

——————— **CAUTION** ———————
Fuel injection systems remain under pressure, even after the engine has been turned OFF. The fuel system pressure must be relieved before disconnecting any fuel lines. Failure to do so may result in fire and/or personal injury.

1. Safely discharge the residual pressure from the fuel system.

2. Disconnect the negative battery cable.

3. Drain the cooling system.

4. Remove the air intake plenum.

5. Disconnect the upper radiator hose from the thermostat housing.

6. Identify and label and disconnect the various connectors along the engine control harness. This should save time at assembly. Position the harness and wiring leads out of the way.

7. Remove the intake manifold with the injectors and fuel rail attached.

8. Label and disconnect the wiring to the spark plugs corresponding to the head being removed.

9. Disconnect the oxygen sensor wiring.

10. Remove the self-locking nuts holding each (front and rear) exhaust pipe to the exhaust manifolds. Remove the small bracket bolts and separate the exhaust system from the engine.

11. Remove the dipstick and tube from the engine block (front head only).

12. Remove the heat shield from the exhaust manifold and remove the manifold from the head.

13. Remove the exhaust manifold gasket.

14. Remove the bolt holding the high pressure hose to the engine.

15. Disconnect the pressure switch wiring from the power steering pump.

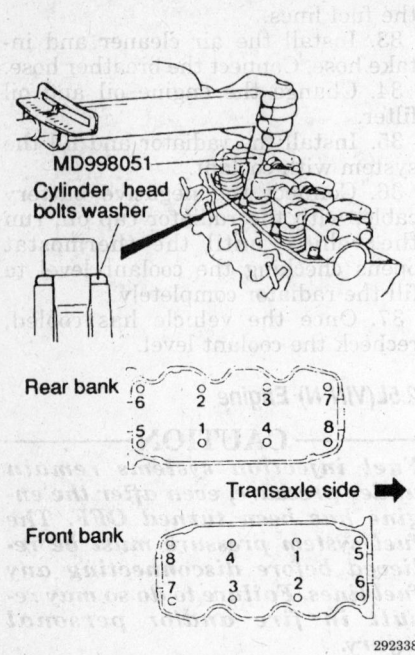

MD998051
Cylinder head bolts washer

Rear bank

Transaxle side ➡

Front bank

292338

Cylinder head bolts tightening sequence — 2.5L engine

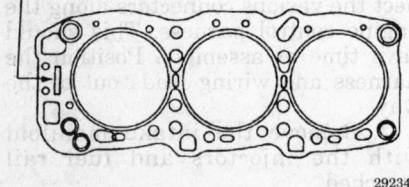

Identification mark

292343

Install the head gasket with the identification mark at the front facing upward — 2.5L engine

16. Disconnect the high pressure line from the power steering pump.

17. Place a floor jack and a broad piece of lumber under the oil pan. Jack the engine until the engine is supported, but not raised.

18. Remove the through-bolt for the left engine mount.

19. Remove the four nuts holding the upper part of the engine mount to the lower bracket.

20. Remove the air conditioning compressor drive belt, then remove the belt tensioner assembly.

21. Remove the power steering pump and the alternator. Position them out of the way.

22. Remove the belt tensioner assembly and its bracket.

23. Remove the upper outer timing belt covers. Remove the gaskets and keep them with the cover.

NOTE: The bolts holding the belt covers are of different lengths and must be correctly placed during reinstallation. Label or diagram them as they are removed.

24. Remove the lower portion of the engine support bracket. Remove the bolts in the order shown in the figure. Use spray lubricant to assist in removing the reamer (alignment) bolt. Keep in mind that the reamer bolt may be heat-seized on the bracket. Soak it in penetrating oil and remove it slowly.

25. Remove the small end cap from the rear bank belt cover, then remove the upper outer belt covers. Make certain all the gaskets are removed and placed with the covers.

26. Remove the splash shield. Turn the crankshaft clockwise until all the timing marks align. This will set the engine to TDC/compression on No. 1 cylinder.

27. Install the special counterholding tools and carefully remove the crankshaft pulley without moving the engine out of position.

28. Remove the front flange from the crank sprocket.

29. Remove the lower timing belt cover with its gasket.

30. Loosen the timing belt tensioner bolt and turn the tensioner counterclockwise along the elongated hole. This will relax the tension on the belt.

31. If the timing belt is to be reused, make a chalk arrow on the belt showing the direction of rotation so that it may be reinstalled correctly.

32. Carefully slide the belt off of the sprockets. Place the belt in a clean, dry, protected location away from the work area. if the tensioner is to be removed, disconnect the spring and remove the retaining bolt.

33. Remove the valve cover and gasket.

34. Use special hex wrench MB 998051 or equivalent, and loosen the head bolts.

35. Rock the head gently to break it loose. If tapping is necessary, do so with a rubber or wooden mallet at the corners of the head. DO NOT pry the head up by wedging tools between the head and the block or damage could result.

36. Lift the head free of the engine. Support the head assembly on wooden blocks on the workbench.

To install:

37. Thoroughly clean and dry the mating surfaces of the head and block. Check the cylinder head for cracks, damage or engine coolant leakage. Remove scale, sealing compound and carbon. Clean the oil passages thoroughly. Check the head for flatness. End to end, the head should be no more than 0.008 inch (0.2 mm) out-of-true. If the service limit is exceeded, correct to meet specifications. Note that the maximum amount from stock allowed to be removed from the cylinder head and mating cylinder block is 0.0079 inch (0.2 mm). If the cylinder head cannot be made serviceable by removing this amount, replace the head.

38. Check that the head gaskets have the proper identification marks for the engine. Lay the head gasket with the identification mark at the front top.

NOTE: Do not apply sealant to the head gasket or mating surfaces.

39. Inspect the cylinder head bolts prior to installation. If the threads are necked down (stretched), the bolts should be replaced. Necking can be checked by holding a straight edge against the threads. If all of the threads do not contact the straight-edge, the bolt should be replaced. All new head bolts are recommended.

40. Install the head straight down onto the block. Try to eliminate most of the side-to-side adjustments, as this may move the gasket out of position or damage the gasket. Before installing the bolts, the threads should be oiled with clean engine oil. Install the bolts and the special washers by hand and just start each bolt 1 or 2 turns on the threads.

NOTE: The washers must be installed correctly. The rounded shoulder of the washer denotes the face in contact with the bolt. The flat face contacts the head.

41. Correct tightening of the head bolts requires 3 steps:
   a. Following the tightening sequence, draw each bolt to 62 ft. lbs. (84 Nm).
   b. Follow the tightening sequence and tighten each bolt to 70 ft. lbs. (95 Nm).
   c. Follow the tightening sequence and tighten each bolt to 76–83 ft. lbs. (103–113 Nm).

42. Install the valve cover and gasket.

43. Check the sprockets and tensioner for wear. The sprocket teeth

should be well defined, not rounded and the valleys between the teeth should be clean. The tensioners should spin freely with no binding or unusual noise. Replace the tensioner if there is any sign of grease leaking from the seal. Clean everything with a clean, dry cloth.

NOTE: Do not spray or immerse the sprockets or tensioners in cleaning solvent. The sprocket may absorb the solvent and transfer it to the belt. The tensioners are internally lubricated and the solvent will dilute or dissolve the lubricant.

44. Observing the direction of rotation marks made earlier, install the timing belt and set to proper tension.

45. Making certain all the gaskets are correctly in place, install the lower timing belt cover and tighten the bolts to 8 ft. lbs. (11 Nm)

46. Install the crankshaft pulley. Use the special tools or equivalent to counterhold it and tighten the bolt to 134 ft. lbs. (181 Nm).

47. Install the upper outer cover on the rear bank camshaft sprocket and install the smaller end cap. Tighten the bolts to 8 ft. lbs. (11 Nm).

48. Install the cover and gaskets for the front bank camshaft sprocket.

49. Install the tensioner and bracket for the ribbed belt.

50. Install the power steering pump and the alternator.

51. Install the ribbed belt and adjust it as needed. Make certain it is properly seated on all the pulleys.

52. Install the tensioner and bracket for the air conditioner drive belt and install the belt, adjusting it as necessary.

53. Adjust the jack so that the bushing of the engine mount aligns with the bodywork bracket. Install the through-bolt and tighten the nuts snug.

54. Carefully lower the jack, allowing the full weight of the engine to bear on the mount and tighten the mounting bolt and nut to 63 ft. lbs. (86 Nm). Tighten the through-bolt to 71–85 ft. lbs. (98–118 Nm).

55. Install the exhaust manifold with a new gasket and install the heat shield.

56. Install the pressure hose to the power steering pump and tighten the nut to 32 ft. lbs. (43 Nm).

57. Connect the wiring to the power steering pump and reinstall the bracket for the high pressure hose.

58. Install the dipstick and dipstick tube, using a new O-ring on the tube and coating it with oil before installation.

59. Using new exhaust gaskets and new self-locking nuts; attach the exhaust pipes to each exhaust manifold. Tighten the nuts to 33 ft. lbs. (44 Nm). Connect the oxygen sensor wiring.

60. Connect the spark plug wires to the spark plugs.

61. Position new intake manifold gaskets and install the intake manifold.

62. Observe the labels made previously and connect each lead of the engine control harness to its proper component.

63. Connect the upper radiator hose to the thermostat housing.

64. Install the air intake plenum with a new gasket.

NOTE: Before proceeding, double check all installation items, paying particular attention to loose hoses or hanging wires, nuts not properly tightened, poor routing of hoses and wires (too tight or rubbing) and tools left in the engine area.

65. Fill the cooling system with coolant. Changing the oil and filter is recommended to eliminate pollutants such as coolant in the oil.

66. Connect the negative battery cable. With the radiator cap off, start the engine and check for leaks of fuel, vacuum, oil or coolant. Check the operation of all engine electrical systems as well as dashboard gauges and lights. Add coolant as the engine warms.

67. Perform necessary adjustments to the accelerator cable and drive belts. Allow the engine to cool and once again check and adjust the coolant level.

Lash Adjusters

BLEEDING

2.5L(VIN N) Engine

——— **WARNING** ———
The lash adjuster is a precision part. Keep it free from dust and other foreign matters. Do NOT disassemble the lash adjusters. When cleaning the lash adjusters use ONLY clean diesel fuel. Improper handling of the lash adjusters could cause engine damage.

1. Immerse the adjuster in clean diesel fuel.

2. While lightly pushing down the inner steel ball using the air bleed wire (MD998442), move the plunger up and down 4 or 5 times to bleed the air. Use of the retainer tool (MD998441) helps to facilitate the air bleeding.

3. Remove the air bleed wire and press the plunger. If the plunger is hard to push in, the lash adjuster is normal. If the plunger can be pushed in all the way readily, bleed the lash adjuster again and retest. If the lash adjuster is still loose, replace it.

NOTE: Upon completion of air bleeding, hold the lash adjusters upright to prevent the diesel fuel from spilling.

4. After air bleeding, set the lash adjuster on the leak down tester (MD998440).

5. After the plunger has gone down somewhat 0.008–0.020 inch (0.2–0.5 mm), measure the time taken for it to go down 0.040 inch (1.0 mm). Replace it if the measured time is out of specification. Standard value is 4–20 seconds/0.040 inch (1 mm). Diesel fuel should be at 50–68° F (15–20° C).

ADJUSTMENT

The 2.0L and 2.5L engines in these vehicles are equipped with hydraulic lash adjusters which are precision units installed in machined openings in the valve actuating ends of the rocker arms. Valve clearance adjustments are not performed.

Rocker Arm Shaft

REMOVAL AND INSTALLATION

——— **CAUTION** ———
Fuel injection systems remain under pressure, even after the engine has been turned OFF. The fuel system pressure must be relieved before disconnecting any fuel lines. Failure to do so may result in fire and/or personal injury.

1. Disconnect the negative battery cable.

2. Drain the coolant and remove the thermostat case assembly.

3. Remove the valve covers.

4. Remove the relay box bracket assembly, the control module and bracket assembly.

5. Install the auto lash adjuster retainer tools MD998443 or equivalent, on the rocker arms.

6. Remove the rocker arms and rocker shafts as an assembly.

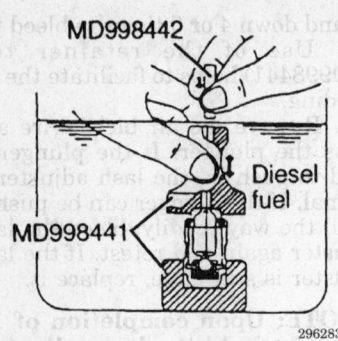

Bleeding the valve lash adjuster — 2.5L engine

7. Remove the lash adjuster retaining tools to remove the valve lash adjusters. Be sure to note where each lash adjuster was removed so that they may be reinstalled in the same rocker arm. Clean and inspect the lash adjusters for proper operation, replace as necessary.

To install:

8. Lubricate the camshaft lobes.

9. After the lash adjusters have been cleaned and inspected or replaced and properly bled, install them in their original positions. Use the retaining tools to hold the lash adjusters in position.

10. Install the rocker arm and rocker shaft assembly.

11. Remove the auto lash adjuster retainer tools from the rocker arms.

12. Install the valve cover and all related parts.

13. Install the relay box bracket assembly, the control module and bracket assembly.

14. Install the thermostat case assembly.

15. Properly fill the cooling system.

16. Connect the negative battery cable.

17. Start the engine and check for proper operation and leaks.

Intake Manifold

REMOVAL AND INSTALLATION

2.0L(VIN Y) Engine

This engine uses a two-piece aluminum intake manifold. The upper half of the manifold (also called a plenum) mounts the throttle body. The lower half of the manifold contains the fuel rail and injectors. A nonreusable gasket joins the two halves. Use care when working with light alloy parts.

> **CAUTION**
>
> *Fuel injection systems remain under pressure, even after the engine has been turned OFF. The fuel system pressure must be relieved before disconnecting any fuel lines. Failure to do so may result in fire and/or personal injury.*

1. Relieve the fuel system pressure.

2. Disconnect the negative battery cable and drain the cooling system.

3. Disconnect the accelerator cable, breather hose and air intake hose.

4. Disconnect the vacuum connection at the power brake booster and the PCV valve. Disconnect all remaining vacuum hoses and pipes, as necessary. Tag for identification, if necessary, to save time at assembly.

5. Disconnect the high pressure fuel line, fuel return hose and remove the throttle control cable and brackets.

> **CAUTION**
>
> *Do not use conventional fuel filters, hoses or clamps when servicing fuel injection systems. They are not compatible with the injection system and could fail, causing personal injury or damage to the vehicle. Use only hoses and clamps specifically designed for fuel injection.*

6. Disconnect the alternator wiring harness connection.

7. Disconnect the MAP sensor and the intake air temperature sensor connectors.

8. Disconnect the TPS connector and position the engine wiring harness aside.

9. Disconnect the EGR pipe connection.

10. Remove the intake manifold stay and the engine hanger. Disconnect the fuel injector connectors.

11. Remove the throttle body assembly.

12. Remove the mounting bolts and remove the intake manifold plenum and gasket.

13. Remove the fuel rail, injector and pressure regulator assembly. Use care since the fuel injectors can drop out of the fuel rail as it is being removed.

14. Remove the mounting bolts and remove the intake manifold and gasket from the engine.

To install:

15. Clean all gasket material from the cylinder head and intake manifold assembly. Check both surfaces for cracks or other damage. Check the intake manifold water passages and air passages for clogging. Clean if necessary. Check the gasket surface of the intake manifold for flatness using a straight edge and feeler gauge. It should be 0.006 inch or less. The limit is 0.008 inch.

16. Install a new intake manifold gasket to the head and install the manifold. Torque the manifold in a crisscross pattern, starting from the inside and working outwards to 17 ft. lbs. (23 Nm).

17. Apply a thin coat of clean engine oil to the fuel injector O-rings. Install the fuel rail, injector and pressure regulator assembly to the lower intake manifold.

18. Thoroughly clean the mating surfaces and install the intake manifold plenum with a new gasket.

19. Install the throttle body assembly.

20. Install the intake manifold stay and the engine hanger. Connect the fuel injector connectors.

21. Connect the EGR pipe connection.

22. Connect the engine control electrical connectors.

23. Connect the generator wiring harness connection.

24. Connect the high pressure fuel line, fuel return hose and the throttle control cable brackets.

25. Connect the vacuum connection at the power brake booster and the PCV valve. Connect all remaining vacuum hoses and pipes.

26. Connect the accelerator cable, breather hose and air intake hose.

27. Connect the negative battery cable.

28. Start the engine and check for proper operation.

2.5L(VIN N) Engine

This engine uses a two-piece intake manifold. The upper half of the manifold (also called a plenum) mounts the throttle body. The lower half of the manifold contains the fuel rail and injectors. A nonreusable gasket joins the two halves. The intake manifold and plenum care constructed of

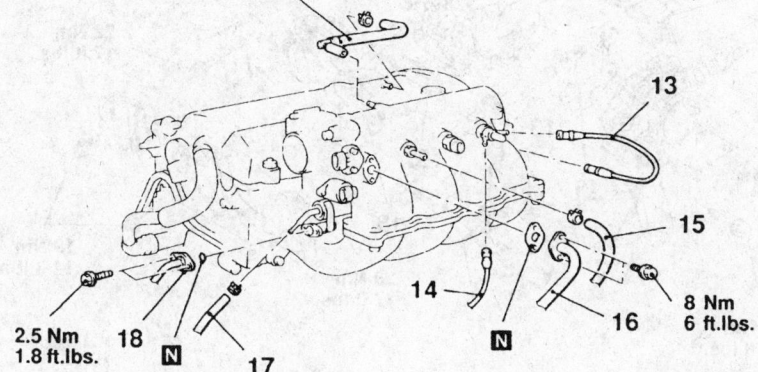

1. Air intake hose
2. Breather hose
3. Accelerator cable connection
4. Clip
5. MAP sensor connector
6. Intake air temperature sensor connector
7. Vacuum hose connection
8. TPS connector
9. Idle air control motor connector
10. Control wiring harness
11. Generator wiring harness connection
12. PCV hose assembly
13. Vacuum hose
14. Vacuum hose connection
15. Brake booster vacuum hose connection
16. EGR pipe connection
17. Fuel return hose connection
18. High-pressure fuel hose connection
Ⓝ. use new components

298384

Intake manifold assembly and related components — 2.0L engine

aluminum. Use care when working with light alloy parts.

— **CAUTION** —

Fuel injection systems remain under pressure, even after the engine has been turned OFF. The fuel system pressure must be relieved before disconnecting any fuel lines. Failure to do so may result in fire and/or personal injury.

1. Relieve the fuel system pressure.
2. Disconnect the negative battery cable.

3. Remove the breather hose and air intake hose.
4. Disconnect the vacuum connection at the power brake booster and the PCV valve. Disconnect all remaining vacuum hoses and pipes, as necessary. Tag for identification, if necessary, to save time at assembly.
5. Disconnect the high pressure fuel line, fuel return hose and remove the throttle control cable and brackets.

— **CAUTION** —

Do not use conventional fuel filters, hoses or clamps when servicing fuel injection systems. They are not compatible with the injec- *tion system and could fail, causing personal injury or damage to the vehicle. Use only hoses and clamps specifically designed for fuel injection.*

6. Remove the throttle body assembly.
7. Disconnect the MAP sensor and the intake air temperature sensor connectors.
8. Disconnect the power steering pressure switch connector.
9. Remove the oxygen sensor connector and harness.
10. Remove the IAC motor connector.

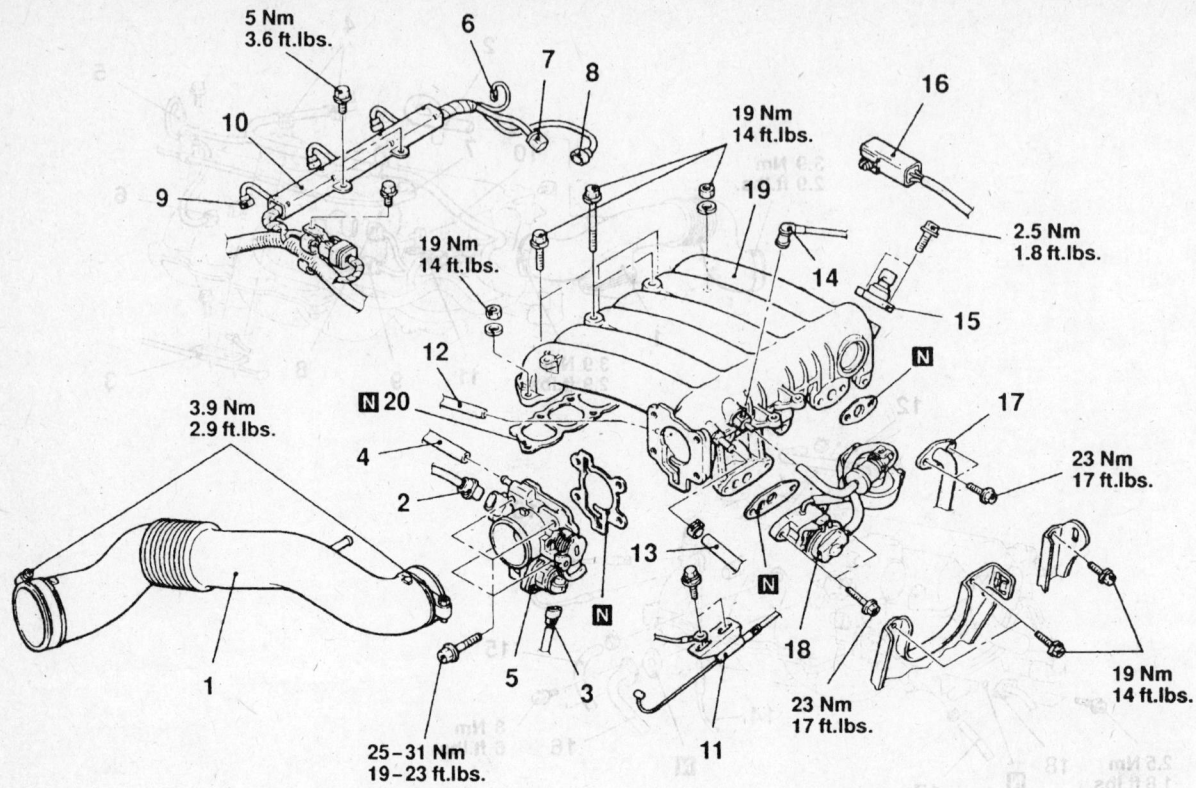

5 Nm
3.6 ft.lbs.

3.9 Nm
2.9 ft.lbs.

19 Nm
14 ft.lbs.

19 Nm
14 ft.lbs.

2.5 Nm
1.8 ft.lbs.

23 Nm
17 ft.lbs.

19 Nm
14 ft.lbs.

23 Nm
17 ft.lbs.

25–31 Nm
19–23 ft.lbs.

1. Air intake hose
2. TPS connector
3. Idle air control motor connector
4. Vacuum hose connection
5. Throttle body assembly
6. Power steering oil pressure switch connector
7. Heated oxygen sensor connector
8. Intake air temperature sensor connector
9. Injector connector
10. Control wiring harness
11. Accelerator cable connection
12. Vacuum hose connection
13. Brake booster vacuum hose connection
14. Vacuum hose connection
15. MAP sensor
16. Heated oxygen sensor harness
17. EGR pipe connection
18. EGR valve and EGR transducer assembly
19. Intake manifold plenum
20. Intake manifold plenum gasket

293034

Intake manifold plenum and related components — 2.5L engine

11. Disconnect the TPS connector and position the engine wiring harness aside.

12. Disconnect the EGR pipe connection.

13. Remove the EGR valve and the EGR transducer assembly.

14. Disconnect the fuel injector connectors.

15. Remove the mounting bolts and remove the intake manifold plenum and gasket.

16. Remove the fuel rail, injector and pressure regulator assembly. Use care since the fuel injectors can drop out of the fuel rail as it is being removed.

17. Remove the mounting bolts and remove the intake manifold and gaskets from the engine.

To install:

18. Clean all gasket material from the cylinder head and intake manifold assembly. Check both surfaces for cracks or other damage. Check the intake manifold air passages for clogging. Clean if necessary. Check the gasket surface of the intake manifold for flatness using a straight edge and feeler gauge. It should be 0.006 inch or less. The limit is 0.008 inch.

19. Properly position the new intake manifold gaskets to the heads and install the manifold. Tighten the

manifold one bank after the other by following this procedure.

a. Tighten the nuts in the front bank to 5 ft. lbs. (7 Nm).

b. Tighten the nuts in the rear bank to 14 to 17 ft. lbs. (20 to 23 Nm).

c. Tighten the nuts in the front bank to 14 to 17 ft. lbs. (20 to 23 Nm).

d. Repeat Steps **b** and **c** again.

20. Apply a thin coat of clean engine oil to the fuel injector O-rings. Install the fuel rail, injector and pressure regulator assembly.

21. Thoroughly clean the mating surfaces and install the intake manifold plenum with a new gasket. Make

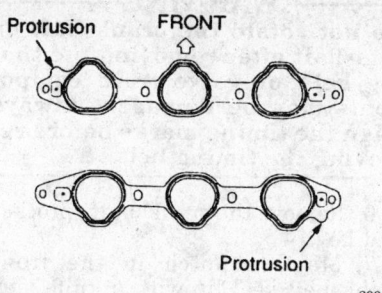

Install the intake manifold gaskets with the protrusions in the positions illustrated — 2.5L engine

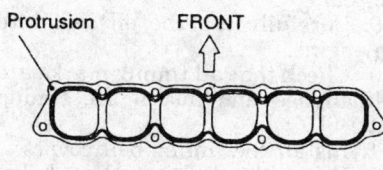

Install the plenum gasket with the protrusions in the position illustrated — 2.5L engine

sure the gasket is properly positioned. Tighten the retaining bolts to 14 ft. lbs. (19 Nm)

22. Install the throttle body assembly. Tighten the retaining bolts to 21 ft. lbs. (28 Nm).

23. Connect the fuel injector connectors.

24. Install the EGR valve and the EGR transducer assembly.

25. Connect the EGR pipe connection.

26. Connect the remaining engine control system electrical connectors.

27. Connect the high pressure fuel line, fuel return hose and the throttle control cable and brackets.

28. Connect the vacuum connection at the power brake booster and the PCV valve. Connect all remaining vacuum hoses and pipes.

29. Connect the breather hose and air intake hose.

30. Connect the negative battery cable.

31. Start the engine and check for proper operation.

Exhaust Manifold

REMOVAL AND INSTALLATION

2.0L(VIN Y) Engine

1. Disconnect the negative battery cable.

2. Remove the air intake hose and the small air hose connection.

3. Properly drain the engine coolant.

4. Disconnect the upper radiator hose from the thermostat housing.

5. Disconnect the control wiring harness connection.

6. Remove the water pipe assembly and the engine oil level gauge.

7. Remove the heat shield and the engine hanger.

8. Remove the pulsed secondary air injection valve, if equipped.

9. Raise and safely support the vehicle.

10. Remove the exhaust pipe to exhaust manifold lock nuts and separate the exhaust pipe. Discard the gasket.

11. Lower the vehicle.

12. Remove the exhaust manifold mounting bolts, and the exhaust manifold.

To install:

13. Clean all gasket material from the mating surfaces and check the manifold for cracks or warpage.

14. Install a new gasket and install the manifold. Tighten the nuts, in a crisscross pattern to 17 ft. lbs. (23 Nm).

15. Raise and safely support the vehicle.

16. Install the exhaust pipe to the exhaust manifold with a new gasket and new lock nuts. Tighten the nuts to 33 ft. lbs. (44 Nm).

17. Lower the vehicle.

18. Install the pulsed secondary air injection valve, if equipped.

19. Install the heat shield and the engine hanger.

20. Connect the control wiring harness connection.

21. Connect the upper radiator hose to the thermostat housing.

22. Properly fill the engine cooling system.

23. Install the air intake hose and the small air hose connection.

24. Connect the negative battery cable and check for exhaust leaks.

2.5L(VIN N) Engine

Front Bank Side

1. Disconnect the negative battery cable.

2. Remove the condenser fan motor assembly.

3. Remove the engine oil level dipstick and tube.

4. Remove the engine hanger.

5. Raise and safely support the vehicle.

6. Remove the exhaust pipe to exhaust manifold lock nuts and separate the exhaust pipe. Discard the gasket.

7. Lower the vehicle.

8. Remove the exhaust manifold mounting bolts, the exhaust manifold stay (brace), the exhaust manifold and the gasket.

To install:

9. Clean all gasket material from the mating surfaces and check the manifold for cracks or warpage.

10. Install a new gasket and install the manifold and manifold stay (brace). Tighten the nuts, in a crisscross pattern to 22 ft. lbs. (29 Nm).

11. Raise and safely support the vehicle.

12. Install the exhaust pipe to the exhaust manifold with a new gasket. Tighten the nuts to 33 ft. lbs. (44 Nm).

13. Lower the vehicle.

14. Install the engine hanger.

15. Install the engine oil level dipstick and tube.

16. Install the condenser fan motor assembly.

17. Connect the negative battery cable and check for exhaust leaks.

Rear Bank Side

1. Disconnect the negative battery cable.

2. Remove the intake manifold plenum and the plenum stay.

3. Remove the heated oxygen sensor.

4. Remove the generator.

5. Remove the EGR pipe assembly.

6. Raise and safely support the vehicle.

7. Remove the exhaust pipe to exhaust manifold lock nuts and separate the exhaust pipe. Discard the gasket.

8. Lower the vehicle.

9. Remove the exhaust manifold mounting bolts, the exhaust manifold stay (brace), the exhaust manifold and the gasket.

To install:

10. Clean all gasket material from the mating surfaces and check the manifold for cracks or warpage.

11. Install a new gasket and install the manifold and manifold stay (brace). Tighten the nuts, in a crisscross pattern to 22 ft. lbs. (29 Nm).

12. Raise and safely support the vehicle.

13. Install the exhaust pipe to the exhaust manifold with a new gasket. Tighten the nuts to 33 ft. lbs. (44 Nm).

14. Lower the vehicle.

15. Install the EGR pipe assembly.

16. Install the alternator and the oxygen sensor.

17. Install the intake manifold plenum and the plenum stay (brace).

18. Connect the negative battery cable and check for exhaust leaks.

Front Crankshaft Seal

REMOVAL AND INSTALLATION

2.0L(VIN Y) Engine

1. Disconnect the negative battery cable.

2. Turn the crankshaft clockwise until the engine is at TDC No. 1 cylinder compression stroke (firing position). Remove the timing belt using the recommended procedure.

3. Using a suitable puller, draw the crankshaft timing belt drive sprocket from the front of the crankshaft.

4. Drain the engine oil. Remove the oil pan.

5. Remove the oil screen feed pipe and O-ring that seals it to the pump/block assembly.

6. The front cover/oil pump mounting bolts are different sizes and must be reinstalled in their original locations. Remove and tag the front cover mounting bolts.

7. Remove the oil pump assembly.

8. Using special tool 6771 or equivalent seal puller, remove the front crankshaft oil seal. Take care not to damage the seal surface of the cover.

To install

NOTE: Whenever the oil pump is disassembled or the rear cover is removed, the gear cavity must be filled with petroleum jelly or clean engine oil. This seals the pump, acts like a prime and allows the pump to draw engine oil as soon as it starts to turn. Do not neglect this step. Do not use grease.

9. Assemble the pump using new parts as required. **Install the inner rotor with the chamfer facing the cast iron oil pump rear cover.**

10. Thoroughly clean all old gasket/sealer material from all mounting surfaces. Apply MOPAR gasket maker or equivalent sealer sparingly

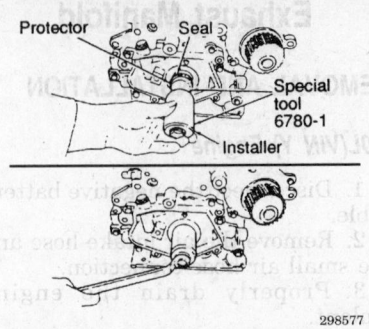

Installing a new seal using a 6780-1 seal installer — 2.0L engine

to the cover surface on the oil pump body. Install the cover and tighten the screws to 105 inch lbs.

11. Install the relief valve plunger, then the spring, gasket and cap. Tighten the cover cap to 30 ft. lbs. Install a new O-ring into the counter bore on the oil pump body discharge passage.

12. Install the oil pump carefully onto the front of the crankshaft until it is seated to the engine block. Tighten the retaining bolts to 17 ft. lbs. (23 Nm).

13. Using tool 6780-1 or equivalent, install a new front crankshaft oil seal.

14. Install the oil screen and pick-up tube assembly using a new O-ring.

15. Install the crankshaft timing belt drive sprocket. Install the timing belt following the recommended procedure. Take care that all timing marks are properly aligned or engine damage will result.

16. Install the oil pan and drain plug. Fill the crankcase with the proper quantity and type of engine oil. A filter change is recommended.

17. Connect the negative battery cable.

2.5L (VIN N) Engine

1. Disconnect the negative battery cable.

2. Remove the accessory drive belts.

3. Remove the front timing belt covers.

4. Align the timing marks by turning the crankshaft with a MD998769 crankshaft turning tool. Loosen the center bolt on the timing belt tensioner pulley and remove the belt.

5. Remove the timing belt.

NOTE: If the timing belt is to be reused, chalk an arrow indicating the direction of rotation on the back of the belt for reinstallation.

WARNING

Do not rotate the crankshaft or camshaft after removing the timing belt, or valve train components may be damaged. Always align the timing marks before removing the timing belt.

6. Remove the crankshaft sprocket and key.

7. Make a notch in the front crankshaft seal lip with a knife, etc. Cover the end of a flat tipped screwdriver with a shop towel and insert it into the notched section of the seal. Carefully pry the seal out.

To install:

8. Apply a small amount of oil to the new oil seal lip and then insert the seal.

9. Carefully tap the oil seal into the engine.

10. Check that all timing marks are still aligned and install the timing belt.

11. Install the timing belt covers.

12. Install the accessory drive belts.

13. Properly fill the cooling system.

14. Connect the negative battery cable.

15. Start the engine and adjust the timing, as necessary.

16. Check for leaks and proper engine and cooling system operation.

Timing Belt Front Cover

REMOVAL AND INSTALLATION

2.0L(VIN Y) Engine

1. Disconnect the negative battery cable.

2. Remove the accessory drive belts.

3. Using a C 3281 or equivalent crankshaft holding tool remove the crankshaft retaining bolt.

4. Using puller tools 1026 and 6827 or equivalent, remove the crankshaft pulley.

5. Remove the power steering pump with the hose attached and position it aside.

6. Remove the power steering pump bracket.

7. Place a floor jack under the engine oil pan, with a block of wood in between and jack up the engine so that the weight of the engine is no longer being applied to the engine mount bracket.

8. Remove the upper engine mount and the engine mounting bracket.

9. Remove the front timing belt cover.

To install:

10. Install the front timing belt cover.

11. Lower the engine enough to install the engine mount bracket.

12. Install the bracket and remove the floor jack.

13. Install the power steering pump bracket and pump.

14. Install the crankshaft pulley using special tool C 4685 C or equivalent pulley installer.

15. Install the accessory drive belts.

16. Connect the negative battery cable.

17. Check for leaks and proper engine operation.

2.5L(VIN N) Engine

1. Disconnect the negative battery cable.

2. Remove the accessory drive belts.

3. Using MB990767 and MB998754 or equivalent crankshaft holding tools, remove the crankshaft bolt and remove the pulley.

4. Remove the heated oxygen sensor connection.

5. Remove the power steering pump with the hose attached and position it aside.

6. Remove the power steering pump bracket.

7. Place a floor jack under the engine oil pan, with a block of wood in between and jack up the engine so that the weight of the engine is no longer being applied to the engine support bracket.

8. Remove the upper engine mount. Spraying lubricant, slowly remove the reamer (alignment) bolt and remaining bolts and remove the engine support bracket.

NOTE: The reamer (alignment) bolt may be heat seized on the engine support bracket.

9. Remove the timing belt front covers.

To install:

10. Install the timing belt front covers.

11. Install the engine mounting bracket.

12. Lower the engine enough to install the engine mount onto bracket and remove the floor jack.

13. Install the power steering pump bracket and pump.

14. Install the crankshaft pulley and tighten the retaining bolt to 13 ft. lbs. (18 Nm).

15. Install the drive belts.

16. Properly fill the cooling system.

17. Connect the negative battery cable.

18. Start the engine and check for leaks. Verify proper engine and cooling system operation.

Timing Belt and Tensioner

REMOVAL AND INSTALLATION

Valve timing is critical to engine operation. Use care when servicing the timing belt. There are a number of timing marks that must be properly aligned or engine damage will result. If the timing belt has not broken, or jumped teeth, it is recommended that the crankshaft be turned by hand (clockwise) to TDC No.1 cylinder compression stroke (firing position) before beginning work. This should align all the timing marks and serve as a reference for later work. Some technicians will apply a small amount of white paint to all timing marks. This helps make them more visible under the low-light conditions found underhood.

2.0L(VIN Y) Engine

1. Disconnect the negative battery cable.

2. Remove the accessory drive belts.

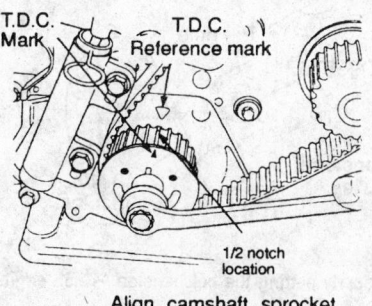

Align camshaft sprocket timing marks together

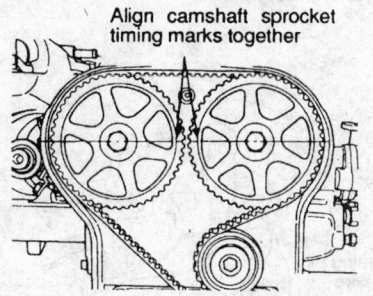

293860

Timing mark locations — 2.0L engine

3. Using a C 3281 or equivalent crankshaft holding tool remove the crankshaft pulley center retaining bolt.

4. Using puller tools 1026 and 6827 or equivalent, remove the crankshaft pulley.

5. Remove the power steering pump with the hose attached and position it aside.

6. Remove the power steering pump bracket.

7. Place a floor jack under the engine oil pan, with a block of wood in between and jack up the engine slightly so that the weight of the engine is no longer on the engine mount bracket.

8. Remove the upper engine mount and the engine mounting bracket.

9. Remove the front timing belt cover.

10. If not done so previously, align the timing marks. Loosen the timing belt tensioner and remove the belt.

— **WARNING** —

Do not rotate the crankshaft or camshafts after removing the timing belt or valve train components may be damaged. Always align the timing marks before removing the timing belt.

11. If the timing belt tensioner is to be replaced, remove the retaining bolts and remove the timing belt tensioner. When the timing belt tensioner is removed from the engine it is necessary to compress the plunger into the tensioner body.

12. Place the tensioner in a vise and slowly compress the plunger.

NOTE: Position the tensioner in the vise the same way it will be installed on the engine. This is to ensure proper pin orientation for when the tensioner is installed on the engine.

13. When the plunger is compressed into the tensioner body, install a pin through the body and plunger to hold the plunger in place until the tensioner is installed.

To install:

14. Check that all timing marks are still aligned. Bring the crankshaft sprocket to ½ a notch before TDC.

15. Install the timing belt. Starting at the crankshaft, go around the water pump sprocket, idler pulley, camshaft sprockets and then around the tensioner pulley.

16. Move the crankshaft to TDC to take up the slack in the belt. Install the tensioner to the block but do not tighten the retaining bolts.

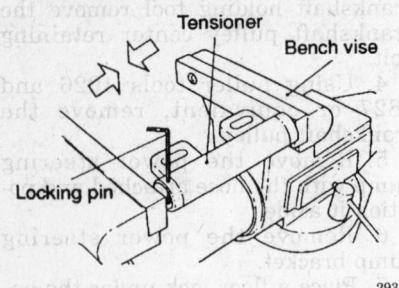

Resetting the belt tensioner — 2.0L engine

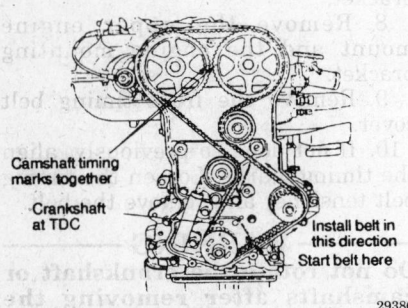

Installing the timing belt — 2.0L engine

17. Using a torque wrench on the tensioner pulley, apply 21 ft. lbs. (28 Nm) of torque to the pulley.

18. With the torque being applied to the tensioner pulley, move the tensioner up against the tensioner pulley bracket and tighten the retaining bolts to 23 ft. lbs. (31 Nm).

19. Remove the tensioner plunger pin. Pretension is correct when the pin can be removed and installed.

20. Rotate the crankshaft two revolutions and check the timing marks. If the timing marks are not properly aligned remove the belt and repeat Steps 11 through 20.

21. Install the front timing belt cover.

22. Lower the engine enough to install the engine mount bracket.

23. Install the bracket and remove the floor jack.

24. Install the power steering pump bracket and pump.

25. Install the crankshaft pulley using a C 4685 C or equivalent pulley installer.

26. Install the accessory drive belts.

27. Connect the negative battery cable.

28. Start the engine. Check for leaks and proper engine operation.

2.5L (VIN N) Engine

1. Disconnect the negative battery cable.

2. Remove the accessory drive belts.

3. Using MB990767 and MB998754 crankshaft holding tools remove the crankshaft bolt and remove the pulley.

4. Remove the heated oxygen sensor connection.

5. Remove the power steering pump with the hose attached and position it aside.

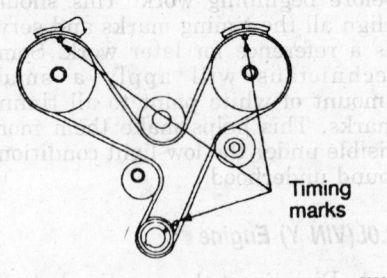

Locating the timing marks — 2.5L engine

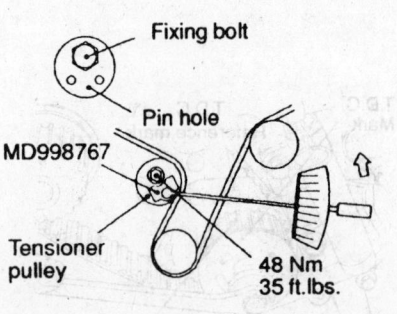

Properly setting the belt tension — 2.5L engine

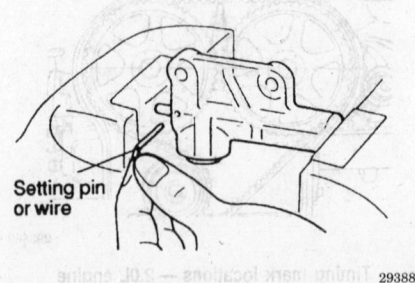

Place a pin or wire into the aligned pin holes — 2.5L engine

6. Remove the power steering pump bracket.

7. Place a floor jack under the engine oil pan, with a block of wood in between and jack up the engine so that the weight of the engine is no longer being applied to the engine support bracket.

8. Remove the upper engine mount. Spraying lubricant, slowly remove the reamer (alignment) bolt and remaining bolts and remove the engine support bracket.

NOTE: The reamer bolt is sometimes heat seized on the engine support bracket.

9. Remove the front timing belt covers.

10. If the timing belt is to be reused, chalk an arrow indicating the direction of rotation on the back of the belt for reinstallation.

11. Align the timing marks by turning the crankshaft with a MD998769 crankshaft turning tool. Loosen the center bolt on the timing belt tensioner pulley and remove the belt.

——WARNING——

Do not rotate the crankshaft or camshaft after removing the timing belt, or valve train components may be damaged. Always align the timing marks before removing the timing belt.

12. Check the belt tensioner for leaks and check the push rod for cracks.

13. If the timing belt tensioner is to be replaced, remove the retaining bolts and remove the timing belt tensioner. When the timing belt tensioner is removed from the engine it is necessary to compress the plunger into the tensioner body.

14. Place the tensioner in a vise and slowly compress the plunger. Take care not to damage the pushrod.

NOTE: Position the tensioner in the vise the same way it will be installed on the engine. This is to ensure proper pin orientation for when the tensioner is installed on the engine.

15. When the plunger is compressed into the tensioner body, install a pin through the body and plunger to hold the plunger in place until the tensioner is installed.

To install:

16. Install the timing belt tensioner and tighten the retaining bolts to 17 ft. lbs. (24 Nm), but do not remove the pin at this time.

17. Check that all timing marks are still aligned.

18. Use bulldog clips (large paper binder clips) or other suitable tool to secure the timing belt and to prevent it from slacking. Install the timing belt. Starting at the crankshaft, go around the idler pulley, then the front camshaft sprocket, the water pump pulley, The rear camshaft sprocket and then around the tensioner pulley.

19. Make sure the belt is tight between the crankshaft and front camshaft sprocket. and between the camshaft sprockets and the water pump. Gently raise the tensioner pulley, so that the belt does not sag, and temporarily tighten the center bolt.

20. Move the crankshaft to a ¼-turn counter clockwise, and then turn it clockwise to the position where the timing marks are aligned.

21. Loosen the center bolt of the tensioner pulley. Using a MD998767 or equivalent tensioner tool and a torque wrench apply 3.3 ft. lbs. (4.4 Nm) tensional torque to the timing belt and tighten the center bolt to 35 ft. lbs. (48 Nm). When tightening the bolt, make sure that the tensioner pulley shaft does not rotate with the bolt.

22. Remove the tensioner plunger pin. Pretension is correct when the pin can be removed and installed easily. If the pin cannot be easily removed and installed it still satisfactory as long as it is within its standard value.

23. Check that the tensioner push rod is within the standard value. When the tensioner is engaged the pushrod should measure 0.149-0.177 inch (3.8- 4.5 mm).

24. Rotate the crankshaft 2 revolutions and check the timing marks. If the timing marks are not properly aligned remove the belt and repeat Steps 17 through 23.

25. Install the timing belt covers.

26. Install the engine mounting bracket.

27. Lower the engine enough to install the engine mount onto bracket and remove the floor jack.

28. Install the power steering pump bracket and pump.

29. Install the crankshaft pulley and tighten the retaining bolt to 13 ft. lbs. (18 Nm).

30. Install the accessory drive belts.

31. Properly fill the cooling system.

32. Connect the negative battery cable.

33. Check for leaks and proper engine and cooling system operation.

Timing Belt Sprockets

REMOVAL AND INSTALLATION

2.0L(VIN Y) Engine

1. Disconnect the negative battery cable.

2. Remove the accessory drive belts.

3. Using special tool C 3281 or equivalent crankshaft holding tool remove the crankshaft retaining bolt.

4. Using puller tools 1026 and 6827 or equivalent, remove the crankshaft pulley.

5. Remove the power steering pump with the hose attached and position it aside.

6. Remove the power steering pump bracket.

7. Place a floor jack under the engine oil pan, with a block of wood in between and jack up the engine so that the weight of the engine is no longer being applied to the engine mount bracket.

8. Remove the upper engine mount and the engine mounting bracket.

9. Remove the front timing belt cover.

10. Align the timing marks. Loosen the timing belt tensioner and remove the belt.

WARNING

Do not rotate the crankshaft or camshafts after removing the timing belt, or valve train components may be damaged. Always align the timing marks before removing the timing belt.

11. If the timing belt tensioner is to be replaced, remove the retaining bolts and remove the timing belt tensioner. When the timing belt tensioner is removed from the engine it is necessary to compress the plunger into the tensioner body.

12. Place the tensioner in a vise and slowly compress the plunger.

NOTE: Position the tensioner in the vise the same way it will be installed on the engine. This is to ensure proper pin orientation for when the tensioner is installed on the engine.

13. When the plunger is compressed into the tensioner body, install a pin through the body and plunger to hold the plunger in place until the tensioner is installed.

14. Using special tool 6793 or equivalent puller, remove the crankshaft sprocket.

15. Using camshaft sprocket holding tools C-4687 and C-4687-1 or equivalent, remove the sprocket retaining bolts and the camshaft sprockets.

To install:

16. Install the camshaft sprockets and bolts. Using camshaft sprocket holding tools C-4687 and C-4687-1 or equivalent, hold the camshafts and tighten the retaining bolts to 75 ft. lbs. (101 Nm).

17. Using tools C-4685 and 6792 or equivalent, install the crankshaft sprocket.

18. Check that all timing marks are still aligned. Bring the crankshaft sprocket to ½ a notch before TDC.

19. Install the timing belt. Starting at the crankshaft, go around the water pump sprocket, idler pulley, camshaft sprockets and then around the tensioner pulley.

20. Move the crankshaft to TDC to take up the slack in the belt. Install the tensioner to the block but do not tighten the retaining bolts.

21. Using a torque wrench on the tensioner pulley, apply 21 ft. lbs. (28 Nm) of torque to the pulley.

22. With the torque being applied to the tensioner pulley, move the tensioner up against the tensioner pulley bracket and tighten the retaining bolts to 23 ft. lbs. (31 Nm).

23. Remove the tensioner plunger pin. Pretension is correct when the pin can be removed and installed.

24. Rotate the crankshaft 2 revolutions and check the timing marks. If the timing marks are not properly aligned remove the belt and repeat Steps 19 through 24.

25. Install the front timing belt cover.

26. Lower the engine enough to install the engine mount bracket.

27. Install the bracket and remove the floor jack.

28. Install the power steering pump bracket and pump.

29. Install the crankshaft pulley using a C 4685 C or equivalent pulley installer.

30. Install the drive belts.

31. Connect the negative battery cable.

32. Check for leaks and proper engine operation.

2.5L(VIN N) Engine

1. Disconnect the negative battery cable.

2. Remove the accessory drive belts.

3. Using MB990767 and MB998754 or equivalent crankshaft

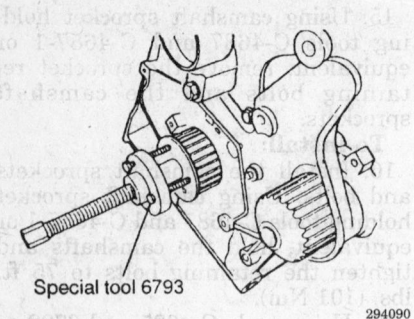

Special tool 6793

294090

Removing the crankshaft sprocket — 2.0L engine

Holder, camshaft sprocket (C-4687)

Adapter, camshaft sprocket (C-4687-1)

294091

Camshaft sprocket holding tools — 2.0L engine

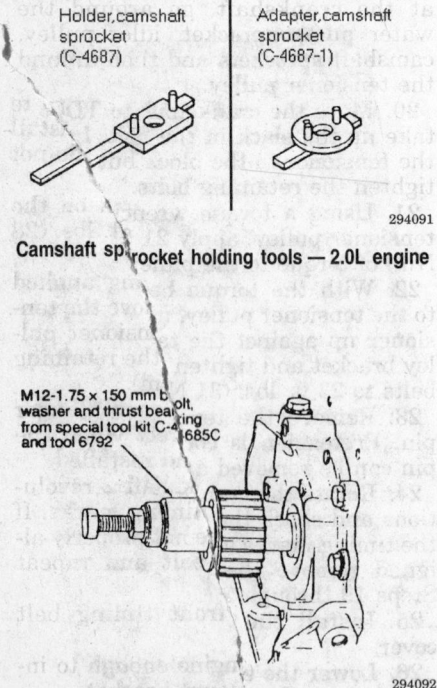

M12-1.75 x 150 mm bolt, washer and thrust bearing from special tool kit C-4685C and tool 6792

294092

Installing the crankshaft sprocket — 2.0L engine

holding tools remove the crankshaft bolt and remove the pulley.

4. Remove the heated oxygen sensor connection.

5. Remove the power steering pump with the hose attached and position it aside.

6. Remove the power steering pump bracket.

7. Place a floor jack under the engine oil pan, with a block of wood in between and jack up the engine so that the weight of the engine is no longer being applied to the engine support bracket.

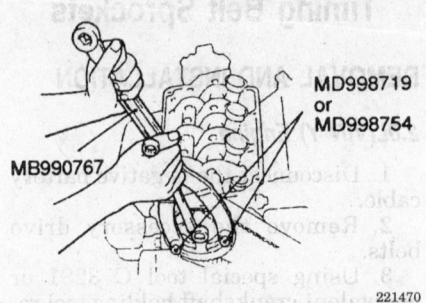

MB990767

MD998719 or MD998754

221470

Removing the camshaft sprockets — 2.5L engine

8. Remove the upper engine mount. Spraying lubricant, slowly remove the reamer (alignment) bolt and remaining bolts and remove the engine support bracket.

NOTE: Keep in mind that the reamer bolt is sometimes heat seized on the engine support bracket.

9. Remove the front timing belt covers.

10. If the timing belt is to be reused, chalk an arrow indicating the direction of rotation on the back of the belt for reinstallation.

11. Align the timing marks by turning the crankshaft with a MD998769 or equivalent crankshaft turning tool. Loosen the center bolt on the timing belt tensioner pulley and remove the belt.

WARNING
Do not rotate the crankshaft or camshaft after removing the timing belt, or valve train components may be damaged. Always align the timing marks before removing the timing belt.

12. Check the belt tensioner for leaks and check the push rod for cracks.

13. If the timing belt tensioner is to be replaced, remove the retaining bolts and remove the timing belt tensioner. When the timing belt tensioner is removed from the engine it is necessary to compress the plunger into the tensioner body.

14. Place the tensioner in a vise and slowly compress the plunger. Take care not to damage the pushrod.

NOTE: Position the tensioner in the vise the same way it will be installed on the engine. This is to ensure proper pin orientation for when the tensioner is installed on the engine.

15. When the plunger is compressed into the tensioner body, install a pin through the body and plunger to hold the plunger in place until the tensioner is installed.

16. Using camshaft sprocket holding tools MD998719 and MD990767 or equivalent, remove the sprocket retaining bolts and the camshaft sprockets.

To install:

17. Install the camshaft sprockets and bolts. Using camshaft sprocket holding tools MD998719 and MD990767 or equivalent, hold the camshafts and tighten the retaining bolts to 65 ft. lbs. (88 Nm).

18. Install the crankshaft key and sprocket.

19. Install the timing belt tensioner and tighten the retaining bolts to 17 ft. lbs. (24 Nm), but do not remove the pin at this time.

20. Check that all timing marks are still aligned.

21. Remove the crankshaft sprocket. Take care not to drop the crankshaft key.

22. Use bulldog clips or other suitable tool to secure the timing belt and to prevent it from slacking. Install the timing belt. Starting at the crankshaft, go around the idler pulley, then the front camshaft sprocket, the water pump pulley, The rear camshaft sprocket and then around the tensioner pulley.

23. Make sure the belt is tight between the crankshaft and front camshaft sprocket. Check that the belt is also tight between the camshaft sprockets and the water pump. Gently raise the tensioner pulley, so that the belt does not sag, and temporarily tighten the center bolt.

24. Move the crankshaft to a ¼-turn counterclockwise, and then turn it clockwise to the position where the timing marks are aligned.

25. Loosen the center bolt of the tensioner pulley. Using a MD998767 or equivalent tensioner tool and a torque wrench apply 3.3 ft. lbs. (4.4 Nm) tensional torque to the timing belt and tighten the center bolt to 35 ft. lbs. (48 Nm). When tightening the bolt, make sure that the tensioner pulley shaft does not rotate with the bolt.

26. Remove the tensioner plunger pin. Pretension is correct when the pin can be removed and installed easily. If the pin cannot be easily removed and installed it still satisfactory as long as it is within its standard value.

27. Check that the tensioner push rod is within the standard value.

When the tensioner is engaged the pushrod should measure 0.149-0.177 inch (3.8-4.5 mm).

28. Rotate the crankshaft 2 revolutions and check the timing marks. If the timing marks are not properly aligned remove the belt and repeat Steps 20 through 26.

29. Install the timing belt covers.

30. Install the engine mounting bracket.

31. Lower the engine enough to install the engine mount onto bracket and remove the floor jack.

32. Install the power steering pump bracket and pump.

33. Install the crankshaft pulley and tighten the retaining bolt to 13 ft. lbs. (18 Nm).

34. Install the accessory drive belts.

35. Properly fill the cooling system.

36. Connect the negative battery cable.

37. Check for leaks and proper engine and cooling system operation.

Camshaft

REMOVAL AND INSTALLATION

2.0L (VIN Y) Engine

1. Disconnect the negative battery cable.

2. Remove the cam cover.

NOTE: Always rotate the crankshaft in a clockwise direction. Make a mark on the back of the timing belt indicating the direction of rotation so it may be reassembled in the same direction if it is to be reused.

3. Rotate the crankshaft clockwise and align the timing marks so the No. 1 piston will be at TDC of the compression stroke.

4. Remove the timing belt cover.

5. Remove the timing belt.

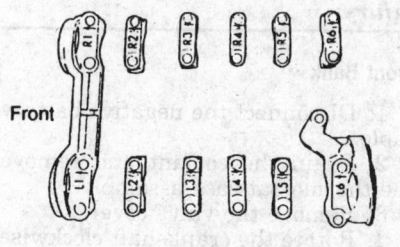

Front

320826

Identifying the camshaft bearing caps — 2.0L engine

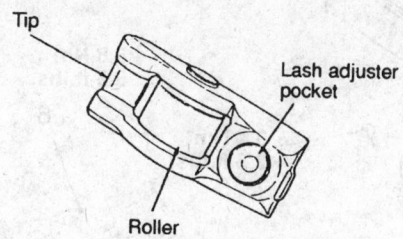

Tip

Lash adjuster pocket

Roller

320830

Camshaft follower components — 2.0L engine

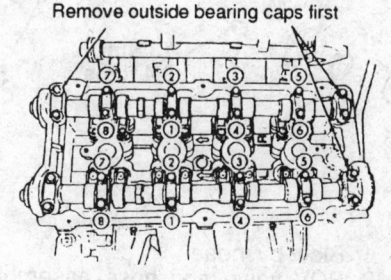

Remove outside bearing caps first

320827

Bearing cap removal sequence — 2.0L engine

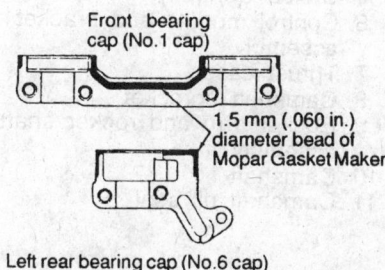

Front bearing cap (No.1 cap)

1.5 mm (.060 in.) diameter bead of Mopar Gasket Maker:

Left rear bearing cap (No.6 cap)

320835

Apply sealant as shown — 2.0L engine

6. Matchmark the crank angle sensor to the cylinder head and remove the crank angle sensor.

7. Remove both camshaft sprockets.

8. Loosen the bearing caps in sequence, one camshaft at a time. The bearing caps are identified for location. Remove the outside bearing caps first.

NOTE: If the bearing caps are difficult to remove, use a plastic hammer to gently tap the rear part of the camshaft.

9. Remove the intake and exhaust camshafts.

——— WARNING ———
The camshafts and their components are NOT interchangeable. Place an identifying mark on each component. Be sure to keep all parts organized for proper reassembly in their original positions.

To install

10. Before installation, clean the cylinder head and cover mating surfaces. Make certain that the rails are flat.

11. Check the camshaft journals on the cylinder head and the cam bearings for wear or damage. Check the cam lobes and rocker rollers for damage. Also, check the cylinder head oil holes for clogging.

12. Check camshaft end play.

13. Inspect the cam followers for wear or damage. Replace as necessary.

14. Lubricate the camshafts with heavy engine oil and position the camshafts on the cylinder head.

——— CAUTION ———
The pistons should not be at top dead center when installing the camshafts since some valves will be open depending on camshaft position.

15. Make sure the dowel pin on both camshaft sprocket ends are located on the top.

16. Install the bearing caps No. 2 through No. 5 and right No. 6 and tighten the caps in sequence to 9 ft. lbs. (12 Nm). Check the markings on the caps to identify the cap number and intake/exhaust symbol. Make sure the rocker arm is correctly mounted on the lash adjuster and the valve stem end.

17. Apply Mopar Gasket Maker® to the No. 1 and No. 6 bearing caps. Install the bearing caps and torque the retaining bolts, using the same sequence as when removed, to 20 ft. lbs. (28 Nm).

18. Apply a coating of engine oil to the oil seal. Using proper size driver, press-fit the seal into the cylinder head.

19. Install the cam sprockets. Install the timing belt using care to follow the recommended procedure. Cam timing is critical or engine damage will result. Install the timing belt cover and related components.

20. Apply Mopar® silicone rubber adhesive sealant at the camshaft cap

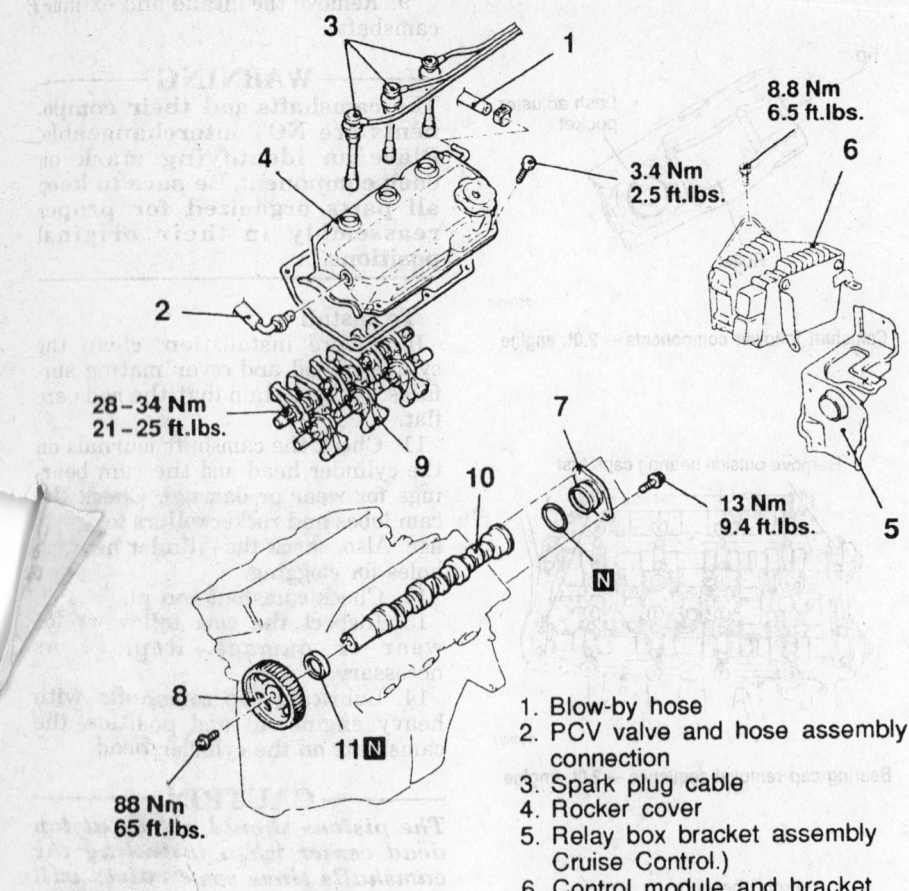

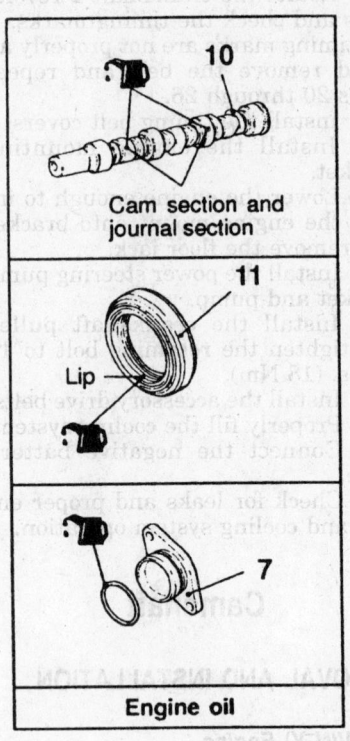

8.8 Nm
6.5 ft.lbs.

3.4 Nm
2.5 ft.lbs.

28–34 Nm
21–25 ft.lbs.

13 Nm
9.4 ft.lbs.

88 Nm
65 ft.lbs.

Cam section and journal section

Lip

Engine oil

1. Blow-by hose
2. PCV valve and hose assembly connection
3. Spark plug cable
4. Rocker cover
5. Relay box bracket assembly Cruise Control.)
6. Control module and bracket assembly
7. Thrust case
8. Camshaft sprocket
9. Rocker arm and rocker shaft assembly
10. Camshaft
11. Camshaft oil seal

Exploded view of the front bank camshaft and related components — 2.5L Engine

corners and at the top edge of the half round seal.

21. Install the cam cover using a new gasket. Tighten the cam cover retaining bolts in sequence, using a three step torque method:

   a. Tighten all bolts, in sequence to 3.3 ft. lbs. (4.5 Nm)

   b. Tighten all bolts, in sequence to 6.5 ft. lbs. (9 Nm)

   c. Tighten all bolts, in sequence to 9 ft. lbs. (12 Nm)

22. Check engine oil level. After this type of service, an oil and filter change is recommended. If a camshaft had failed, metal particles may be spread throughout the engine so an oil and filter change should be mandatory.

23. Connect the negative battery cable.

24. Start the engine and check for proper operation and leaks.

2.5L(VIN N) Engine

— **CAUTION** —
Fuel injection systems remain under pressure, even after the engine has been turned OFF. The fuel system pressure must be relieved before disconnecting any fuel lines. Failure to do so may result in fire and/or personal injury.

Front Bank

1. Disconnect the negative battery cable.

2. Drain the coolant and remove the thermostat case assembly.

3. Remove the valve cover.

4. Rotate the crankshaft clockwise and align the timing marks so the

No. 1 piston will be at TDC of the compression stroke.

NOTE: Always rotate the crankshaft in a clockwise direction. Make a mark on the back of the timing belt indicating the direction of rotation so the belt, if it is to be reused, can be reassembled in the same direction of rotation.

5. Remove the timing belt covers, timing belt and camshaft sprocket.

6. Remove the relay box bracket assembly, the control module and bracket assembly.

7. Remove the thrust case.

8. Install the auto lash adjuster retainer tools MD998443 or equivalent, on the rocker arms.

9. Remove the rocker arms and rocker shafts as an assembly.

10. Remove the camshaft from the cylinder head. Discard the oil seal.

11. Inspect the bearing journals on the camshaft.

To install:

12. Lubricate the camshaft journals and camshaft with clean engine oil. Install the camshaft in the cylinder head. Set the dowel pin on the camshaft at approximately 71 degrees before the 12 o'clock position.

13. Install the rocker arm and rocker shaft assembly.

14. Apply a thin coat of engine oil to the circumference of the camshaft (sprocket end) oil seal lip. Use seal installers such as MD 998713 and MB991559 or equivalent to install a new front seal.

15. Install the camshaft sprocket, tightening the bolt to 65 ft. lbs. (88 Nm).

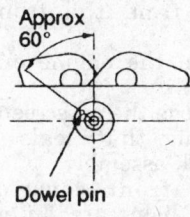

Rear bank side

Approx 60°

Dowel pin

292042

Install the rear bank camshaft as shown — 2.5L Engine

16. Install the thrust case.

17. Remove the auto lash adjuster retainer tools from the rocker arms.

18. Install the timing belt, valve cover and all related parts.

19. Install the relay box bracket assembly, the control module and bracket assembly.

20. Install the thermostat case assembly.

21. Properly fill the cooling system.

22. Connect the negative battery cable.

23. Start the engine and check for proper operation and leaks.

Rear Bank

1. Disconnect the negative battery cable.

2. Remove the front exhaust pipe.

3. Drain the coolant and remove the water inlet pipe assembly.

4. Following proper procedures, remove the intake manifold plenum.

5. Remove the valve cover.

6. Mark the position of the distributor and remove it.

7. Remove the timing belt covers, timing belt and camshaft sprocket.

8. Remove the cylinder head assembly.

9. Install auto lash adjuster retainer tools MD998443 or equivalent, on the rocker arms.

10. Remove the rocker arms and rocker shafts as an assembly.

11. Remove the camshaft from the cylinder head. Discard the oil seal.

12. Inspect the bearing journals on the camshaft.

To install:

13. Before installation, clean the cylinder head and cover mating surfaces. Make certain that the rails are flat.

14. Lubricate the camshaft journals and camshaft with clean engine oil. Install the camshaft in the cylinder head. Set the dowel pin on the camshaft at approximately 60 degrees before the 12 o'clock position.

15. Install the rocker arm and rocker shaft assembly.

16. Apply a thin coat of engine oil to the circumference of the camshaft (sprocket end) oil seal lip. Use a seal installer such as MD 998713 or equivalent to install a new front seal.

17. Install the cylinder head assembly.

18. Install the camshaft sprocket, tightening the bolt to 65 ft. lbs. (88 Nm).

19. Using the mark made earlier, install the distributor.

20. Remove the auto lash adjuster retainer tools from the rocker arms.

21. Install the timing belt, valve cover and all related parts.

22. Install the water inlet pipe assembly and properly fill the cooling system.

23. Connect the negative battery cable.

24. Drain and refill the crankcase and install a new oil filter.

25. Start the engine and check for proper operation and leaks.

Piston and Connecting Rod

POSITIONING

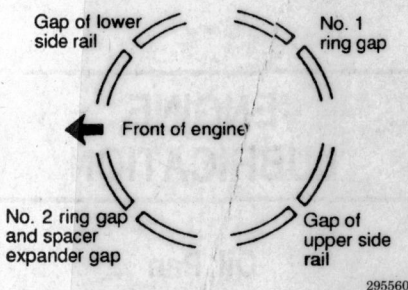

Aligning the piston ring end gaps — 2.0L engine

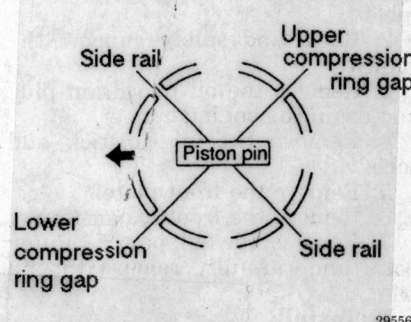

Positioning the piston ring end gaps — 2.5L engine

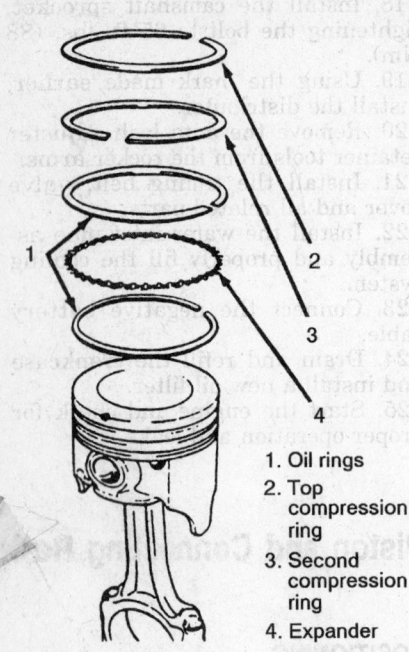

1. Oil rings
2. Top compression ring
3. Second compression ring
4. Expander

295559

Piston, connecting rod and piston rings — 2.0L and 2.5L Engines

ENGINE LUBRICATION

Oil Pan

REMOVAL AND INSTALLATION

2.0L(VIN Y) Engine

1. Disconnect the negative battery cable.
2. Raise and safely support the vehicle.
3. Remove the oil pan drain plug and drain the engine oil.
4. Remove the oil dipstick and tube.
5. Remove the front plate.
6. Remove the front exhaust pipe.
7. Remove the oil pan retaining bolts and carefully remove the oil pan.

To install:

8. Inspect the oil pan for damage and cracks. Replace if faulty. While the pan is removed, inspect the oil screen for clogging, damage and cracks. Clean and/or replace if faulty.
9. Thoroughly clean the mating surfaces of the cylinder block and the oil pan.

10. Apply sealant to the seams between the oil pump and the engine block.
11. Install the oil pan onto the cylinder block and tighten the retaining bolts to 9 ft. lbs. (12 Nm).
12. Install the front exhaust pipe.
13. Install the oil dipstick and tube.
14. Install the oil drain plug and tighten to 25 ft. lbs. (34 Nm).
15. Lower the vehicle and fill the crankcase to the proper level with clean engine oil.
16. Connect the negative battery cable. Start the engine and check for leaks.

2.5L(VIN N) Engine

1. Disconnect the negative battery cable.
2. Raise and safely support the vehicle.
3. Remove the oil pan drain plug and drain the engine oil.
4. Disconnect the oxygen sensor wiring and lower the front exhaust pipe.
5. Remove the center member.
6. Remove the front and rear plates.
7. Remove the starter.

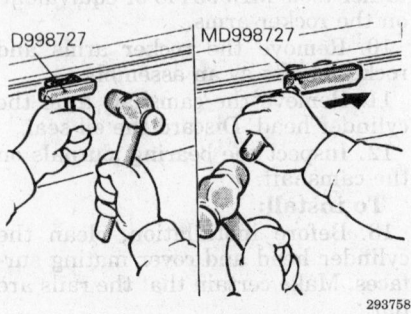

293758

Oil pan separating tool

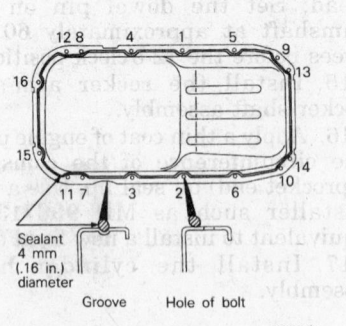

Sealant 4 mm (.16 in.) diameter

Groove   Hole of bolt

293760

Oil pan bolt tightening sequence and application of sealant to the pan — 2.5L engine

8. Remove the transaxle case lower cover.
9. Remove the oil pan mounting bolts. Using special tool MD998727 or equivalent, separate and remove the engine oil pan.

To install:

10. Thoroughly clean and dry the oil pan, cylinder block and cylinder block bolts and bolt holes.
11. Apply a thin bead of sealer around the surface of the oil pan.

NOTE: Assemble the oil pan to the cylinder block within 15 minutes after applying the sealant.

12. Install the oil pan mounting bolts. Torque the retaining bolts, in proper sequence to 4.3 ft. lbs. (5.9 Nm).
13. Install the transaxle case lower cover. install the starter and torque the bolts to 40 ft. lbs. (54 Nm).
14. Install the front and rear plates and tighten the bolts to 80 ft. lbs. (108 Nm).
15. Connect the front exhaust pipe.
16. Connect the oxygen sensor wiring.
17. Fill the engine with the proper amount of oil.
18. Connect the negative battery cable and check for leaks.

Oil Pump

REMOVAL AND INSTALLATION

2.0L(VIN Y) Engine

1. Disconnect the negative battery cable.
2. Turn the crankshaft clockwise until the engine is at TDC No. 1 cylinder compression stroke (firing position). Remove the timing belt using the recommended procedure.
3. Using a suitable puller, draw the crankshaft timing belt drive sprocket from the front of the crankshaft.
4. Drain the engine oil. Remove the oil pan.
5. Remove the oil screen feed pipe and O-ring that seals it to the pump/block assembly.
6. The front cover/oil pump mounting bolts are different sizes and must be reinstalled in their original locations. Remove and tag the front cover mounting bolts.
7. Remove the oil pump assembly.
8. Using special tool 6771 or equivalent seal puller, remove the front crankshaft oil seal. Take care not to damage the seal surface of the cover.

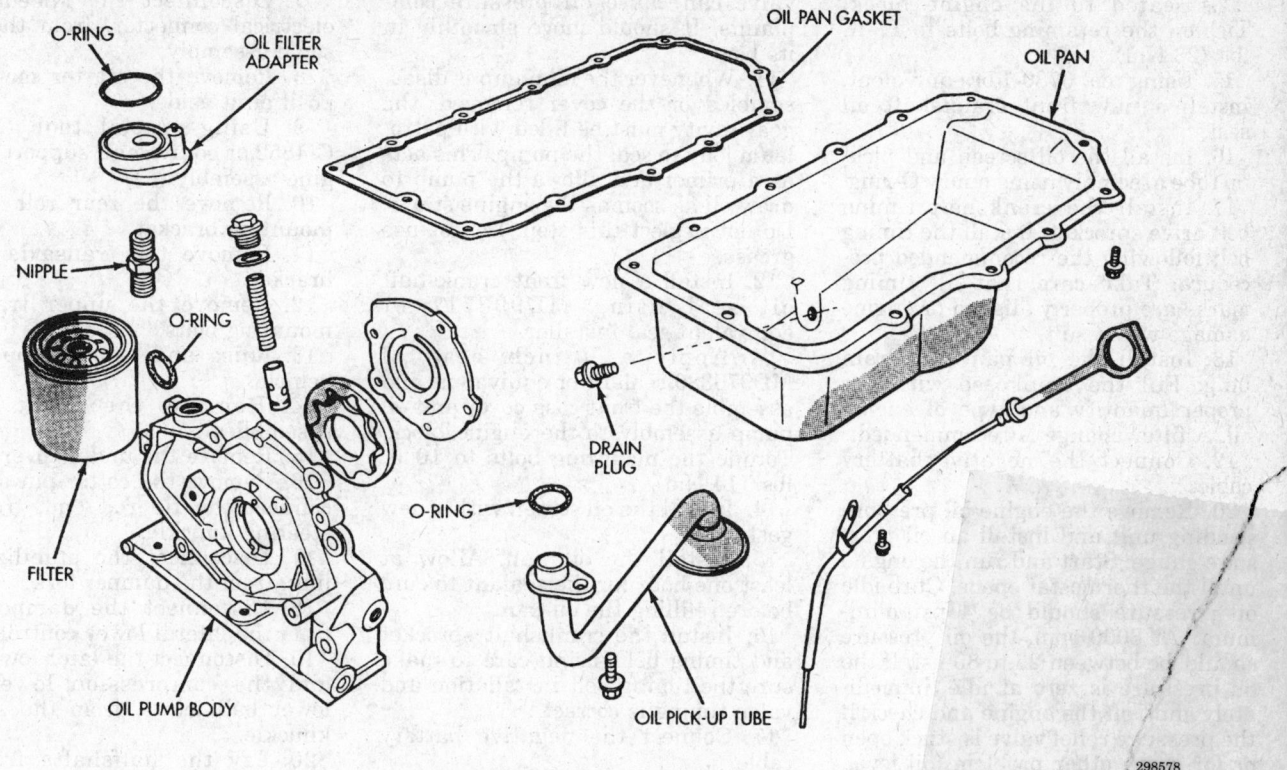

Engine lubrication components — 2.0L engine

298578

To install

9. Clean all parts well. If troubleshooting an oil pressure complaint, the oil pump internal components should be checked and carefully measured.

   a. Remove the threaded plug and gasket from the oil pump and remove the spring and relief valve. Use care to note the position of the parts. The oil pump pressure relief valve must be installed properly or the engine will be seriously damaged.

   b. Remove the oil pump cover screws and lift off the cover.

   c. Remove the pump rotors.

   d. Wash all parts in a suitable solvent, clean thoroughly and inspect carefully for wear.

   e. Make sure the mating surface of the oil pump is smooth. Replace the pump cover if scratched or grooved.

   f. Lay a straightedge across the pump cover surface. If a 0.003 inch feeler gauge can be inserted between the cover and straightedge, the cover should be replaced.

   g. Measure the thickness and diameter of the outer rotor. If the outer rotor thickness measures 0.301 inch or less, or if the diameter is 3.148 inch or less, replace the outer rotor.

   h. If the inner rotor measures 0.301 inch or less, replace the inner rotor.

   i. Slide the outer rotor into the pump housing, press to one side with fingers and measure the clearance between the rotor and housing. If measurement is 0.015 inch or more, replace the housing only if the outer rotor is in specification.

   j. Install the inner rotor into the pump housing. If clearance between the inner and outer rotors is 0.008 inch or more, replace both rotors.

   k. Place a straightedge across the face of the pump housing between bolt holes. If a 0.004 inch or more feeler gauge can be inserted between the rotors and straightedge, replace the pump assembly ONLY if the rotors are in spec.

   l. Inspect the oil relief valve plunger for scoring and free operation in its bore. Varnish or sludge can hang the plunger resulting in improper oiling. Small marks may be removed with 400 grit wet-or-dry sandpaper.

   m. The relief spring has a free length of 2.390 inches and should test between 18 and 19 pounds when compressed to 1.600 inches.

Replace the spring if it fails to meet spec.

   n. If engine oil pressure is low and the pump is within specification, inspect for worn engine bearings or other reasons for oil pressure loss.

10. Whenever the oil pump is disassembled or the rear cover is removed, the gear cavity must be filled with petroleum jelly or clean engine oil. This seals the pump, acts like a prime and allows the pump to draw engine oil as soon as it starts to turn. Do not neglect this step. Do not use grease.

11. Assemble the pump using new parts as required. **Install the inner rotor with the chamfer facing the cast iron oil pump rear cover.**

12. Thoroughly clean all old gasket/sealer material from all mounting surfaces. Apply MOPAR gasket maker or equivalent sealer sparingly to the cover surface on the oil pump body. Install the cover and tighten the screws to 105 inch lbs.

13. Install the relief valve plunger, then the spring, gasket and cap. Tighten the cover cap to 30 ft. lbs. Install a new O-ring into the counter bore on the oil pump body discharge passage.

14. Install the oil pump carefully onto the front of the crankshaft until

it is seated to the engine block. Tighten the retaining bolts to 17 ft. lbs. (23 Nm).

15. Using tool 6780-1 or equivalent, install a new front crankshaft oil seal.

16. Install the oil screen and pick-up tube assembly using a new O-ring.

17. Install the crankshaft timing belt drive sprocket. Install the timing belt following the recommended procedure. Take care that all timing marks are properly aligned or engine damage will result.

18. Install the oil pan and drain plug. Fill the crankcase with the proper quantity and type of engine oil. A filter change is recommended.

19. Connect the negative battery cable.

20. Remove the engine oil pressure sending unit and install an oil pressure gauge. Start and run the engine until the thermostat opens. Curb idle oil pressure should be 4 psi minimum. At 3000 rpm, the oil pressure should be between 25 to 80 psi. If the oil pressure is zero at idle, immediately shut off the engine and check if the pressure relief valve is stuck open or for some other problem (oil level, oil type, loose filter, etc.).

WARNING
If the oil pressure is zero at idle DO NOT run the engine at 3000 rpm looking for a pressure increase or the engine could be destroyed.

2.5L(VIN N) Engine

1. Disconnect the negative battery cable.

2. Remove the timing belt and crankshaft sprocket.

3. Remove the oil filter, oil pressure gauge unit, oil pressure switch and the oil filter bracket.

4. Remove the oil pan.

5. Remove the oil screen and gasket.

6. Remove the front cover mounting bolts. Note the lengths of the mounting bolts as they are removed for proper reinstallation.

7. Remove the front crankshaft oil seal.

8. Remove the front case cover and oil pump assembly.

To install:

9. Thoroughly clean all gasket material from all mounting surfaces.

10. The pump should be checked for wear. Matchmark the rotors before disassembly for cleaning and checking. The oil pressure relief valve can be removed by unscrewing the plug at the bottom of the pump. A stuck

valve can cause oil pressure complaints. It should move smoothly in its bore.

11. Whenever the oil pump is disassembled or the cover removed, the gear cavity must be filled with petroleum jelly to seal the pump. This acts as a primer and allows the pump to draw oil as soon as the engine starts. Do not neglect this step. Do not use grease.

12. Install a new front crankshaft oil seal using MD998717 or equivalent seal installer.

13. Apply a $\frac{1}{8}$-inch bead of MD970389 sealant or equivalent and assemble the front case cover and oil pump assembly to the engine block. Torque the mounting bolts to 10 ft. lbs. (14 Nm).

14. Install the oil screen with a new gasket.

15. Install the oil pan. Allow at least one hour for the sealant to cure before refilling the oil pan.

16. Install the crankshaft sprocket and timing belt. using care to make sure the timing belt installation and valve timing is correct.

17. Connect the negative battery cable.

18. Properly fill the crankcase and check for correct oil pressure. There should be at least 4 psi at idle. If the oil pressure is below 4 psi at idle, DO NOT rev the engine to check pressure at higher rpms. Shut the engine off immediately and begin troubleshooting for low oil pressure indications.

TRANSAXLE

Manual Transmission Assembly

REMOVAL AND INSTALLATION

1. Disconnect both battery cables, negative side first. Remove the battery and battery tray.

2. Remove the battery stay (brace).

3. Remove the air cleaner and intake hoses.

4. Drain the transaxle into a suitable waste container.

5. Remove the cotter pin securing the select and shift cables and remove the cable ends from the transaxle.

6. Disconnect the backup light switch harness and position it aside.

7. Disconnect the speedometer electrical connector, from the transaxle assembly.

8. Remove the starter motor and position it aside.

9. Using special tool 7137 or C-4852 or equivalent, support the engine assembly.

10. Remove the rear roll stopper mounting bracket.

11. Remove the transaxle mount bracket.

12. Remove the upper transaxle mounting bolts.

13. Raise and safely support the vehicle.

14. Remove the front wheel assemblies.

15. Remove the under cover.

16. Remove the cotter pin and disconnect the tie rod end, from the steering knuckle.

17. Disconnect the stabilizer bar link, from the damper fork.

18. Disconnect the damper fork, from the lateral lower control arm.

19. Disconnect the later lower arm, and the compression lower arm, lower ball joints, from the steering knuckle.

20. Pry the halfshafts from the transaxle, and secure aside.

21. Remove the connection for the clutch release cylinder and without disconnecting the hydraulic line, secure aside.

22. Remove the cover from the transaxle bellhousing.

23. Remove the engine front roll stopper through bolt.

24. Remove the centermember.

25. Support the transaxle, using a transmission jack, and remove the transaxle lower coupling bolt.

NOTE: The coupling bolt threads from the engine side, into the transaxle, and is located just above the halfshaft opening.

26. Slide the transaxle rearward and carefully lower it from the vehicle.

To install:

27. Install the transaxle to the engine and install the mounting bolts and tighten to 70 ft. lbs (95 Nm).

28. Install the cover to the transaxle bellhousing and tighten the mounting bolts to 7 ft. lbs. (9 Nm).

29. Install the centermember and tighten the front mounting bolts to 65 ft. lbs. (88 Nm) and the rear bolt to 54 ft. lbs. (73 Nm). Install the front engine roll stopper through bolt and lightly tighten. Once the full weight of the engine is on the mounts, tighten the bolt to 42 ft. lbs. (57 Nm).

30. Connect the clutch release cylinder.

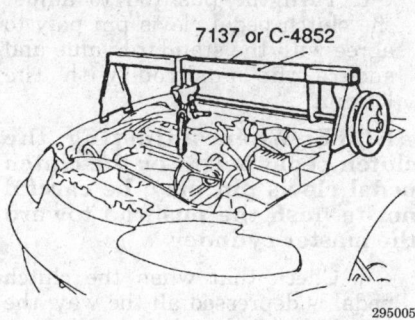

7137 or C-4852

295005

Properly support the engine assembly

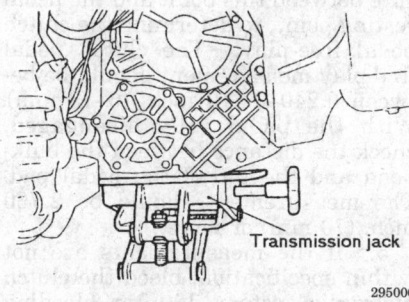

Transmission jack

295006

Use a transmission jack to support the transaxle assembly

31. Install the halfshafts, using new circlips on the axle ends.

———— WARNING ————
When installing the halfshaft, keep the inboard joint straight in relation to the axle, to avoid damaging the oil seal lip of the transaxle, with the serrated part of the halfshaft.

32. Connect the tie rod and ball joints to the steering knuckle. Tighten the ball joint self-locking nuts to 48 ft. lbs. (65 Nm). Tighten the tie rod end nut to 21 ft. lbs. (28 Nm) and secure with a new cotter pin.

33. Connect the damper fork to the lower control arm and tighten the through bolt to 65 ft. lbs. (88 Nm).

34. Connect the stabilizer link to the damper fork, and tighten the self-locking nut to 29 ft. lbs. (39 Nm).

35. Install the under cover.

36. Install wheels and lower vehicle.

37. Install the transaxle mount bracket, to the transaxle, and tighten the mounting nuts to 32 ft. lbs. (43 Nm).

38. Install the rear roll stopper mounting bracket.

39. Remove the engine support. Tighten the transaxle mount through bolt to 51 ft. lbs. (69 Nm) and tighten the front engine roll stopper through bolt.

40. Install the upper transaxle mounting bolts and tighten to 35 ft. lbs. (48 Nm).

41. Install the starter motor.

42. Connect the backup light switch and the speedometer connector.

43. Connect the select and shift cables and install new cotter pins.

44. Install the air cleaner and the air intake hose.

45. Install the battery tray and battery.

46. Install the battery stay.

47. Make sure the vehicle is level, and refill the transaxle.

48. Check the transaxle for proper operation. Make sure the reverse lights come on when in reverse.

Clutch Assembly

REMOVAL AND INSTALLATION

1. Disconnect the negative battery cable.

2. Raise and safely support the vehicle.

3. Remove the transaxle assembly from the vehicle using the recommended procedure.

4. Remove the pressure plate attaching bolts, pressure plate and clutch disc. If the pressure plate is to be reused, loosen the bolts in a diagonal pattern, 1 or 2 turns at a time. This will prevent warping the clutch cover assembly.

5. Remove the return clip and the pressure plate release bearing. Do not use solvent to clean the bearing.

6. Inspect the clutch release fork and fulcrum for damage or wear. If necessary, remove the release fork and the fulcrum from the transaxle.

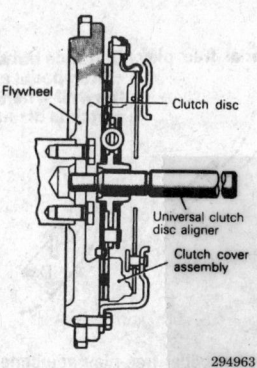

Flywheel — Clutch disc

Universal clutch disc aligner

Clutch cover assembly

294963

Clutch assembly alignment

7. Carefully inspect the condition of the clutch components and replace any worn or damaged parts.

To install:

8. Inspect the flywheel for heat damage or cracks. Resurface or replace the flywheel as required. Install the flywheel using new bolts.

9. Install the fulcrum, if removed, and tighten. Install the release fork. Apply a coating of multi-purpose grease to the point of contact with the fulcrum and the point of contact with the release bearing. Apply a coating of multi-purpose grease to the end of the release cylinder's push rod and the push rod hole in the release fork.

NOTE: When installing the clutch, apply grease to each part, but be careful not to apply excessive grease. Excessive grease will cause clutch slippage and shudder.

10. Apply multi-purpose grease to the clutch release bearing. Pack the bearing inner surface and the groove with grease. Do not apply grease to the resin portion of the bearing. Place the bearing in position and install the return clip.

11. Apply a coating of grease to the clutch disc splines and then use a brush to rub it in the grooves. Using a universal clutch disc alignment tool, position the clutch disc on the flywheel. Install the retainer bolts and tighten a little at a time, in a diagonal sequence.

12. Install the transaxle assembly using the recommended procedure and check the fluid level.

13. Verify proper clutch operation.

Clutch Master Cylinder

REMOVAL AND INSTALLATION

1. Disconnect the negative battery cable.

2. Remove necessary underhood components in order to gain access to the clutch master cylinder.

———— WARNING ————
The clutch hydraulic system uses DOT 3 or DOT 4 brake fluid. Use care when servicing since brake fluid is harmful to painted surfaces.

3. Loosen the clutch fluid line at the cylinder and allow the fluid to drain.

4. Remove the clevis pin retainer at the clutch pedal and remove the washer and clevis pin.

5. From inside the passenger compartment, remove the nut securing the master cylinder to the bulkhead.

6. From under the hood, remove the nut and pull the cylinder from the bulkhead. A seal should be between the mounting flange and bulkhead. This seal should be replaced.

To install

7. Mount master cylinder on the studs, using new seal, and torque both nuts to 10 ft. lbs. (13 Nm).

8. Lubricate all pivot points with grease and install the clevis pin.

9. Connect hydraulic line. With an assistant pressing on the clutch pedal, bleed the system at the slave cylinder. Keep the reservoir filled with fresh DOT 3 or DOT 4 brake fluid.

10. Check the adjustment of the clutch pedal for proper free-play.

11. Connect the negative battery cable. Verify correct shifting and transaxle operation.

Clutch Slave Cylinder

REMOVAL AND INSTALLATION

1. Disconnect the negative battery cable.

2. Remove the necessary underhood components in order to gain access to the clutch slave cylinder (also sometimes called a release cylinder or actuator).

3. Disconnect the hydraulic line and allow the system to drain.

4. Remove the bolts and pull the cylinder from the transaxle housing.

To install

5. Lubricate all pivot points with grease.

6. Mount the slave cylinder to the transaxle and tighten the bolts to 13 ft. lbs. (18 Nm).

7. Connect the hydraulic line and tighten to 11 ft. lbs. (15 Nm).

8. Fill the system with clean brake fluid meeting DOT 3 or DOT 4 specifications.

9. Bleed the clutch hydraulic system.

10. Check and adjust the clutch pedal height, as necessary.

Hydraulic Clutch System Bleeding

—— WARNING ——
The clutch hydraulic system uses DOT 3 or DOT 4 brake fluid. Use care. Brake fluid is harmful to painted surfaces.

1. Fill the reservoir with clean DOT 3 or DOT 4 brake fluid.

2. Loosen the bleed screw, have the clutch pedal pressed to the floor.

3. Tighten the bleed screw and release the clutch pedal.

4. Repeat the procedure until the fluid is free of air bubbles.

NOTE: It is suggested that a hose be attached to the bleeder with the other end immersed in a container at least half full of brake fluid during the bleeding operation. Do not allow the reservoir to run out of fluid during bleeding.

5. Refill the reservoir with clean brake fluid.

6. Check the clutch for proper operation.

CLUTCH PEDAL FREE-PLAY/HEIGHT ADJUSTMENT

1. Measure the clutch pedal height from the face of the pedal pad to the bulkhead. Compare the measured value with the desired distance of 7.0–7.09 inch (175–180 mm).

2. Measure the clutch pedal clevis pin play at the face of the pedal pad. Press the pedal lightly until resistance is met, and measure this distance. The clutch pedal clevis pin play should be within 0.040–0.120 inch. (1–3 mm).

3. If the clutch pedal height or clevis pin play are not within the standard values, adjust as follows:

a. For vehicles without cruise control, turn and adjust the stop bolt so the pedal height is the standard value, then tighten the locknut.

b. Vehicles with auto-cruise control system, disconnect the clutch switch connector and turn the switch to obtain the standard clutch pedal height. Then, lock by tightening the locknut.

c. Turn the pushrod to adjust the clutch pedal clevis pin play to agree with the standard value and secure the pushrod with the locknut.

NOTE: When adjusting the clutch pedal height or the clutch pedal clevis pin play, be careful not to push the pushrod toward the master cylinder.

d. Check that when the clutch pedal is depressed all the way, the interlock switch switches over from **ON** to **OFF**.

4. Move the clutch pedal until the resistance begins to increase; measure between this point and the pedal resting point, to determine the clutch pedal free-play. The clutch pedal free-play measurement should be between 0.240–0.510 inch. (6–13 mm). With the pedal fully disengaged, check the distance between the bulkhead and the top of the pedal pad. The measurement should be 2.760 inch. (70 mm) or more.

5. If the measurements are not within specification, bleed the clutch hydraulic system. If after bleeding the measurements are still not within specified range, there is a faulty component in the system, which must be replaced.

Automatic Transaxle Assembly

REMOVAL AND INSTALLATION

1. Disconnect both battery cables, negative side first. Remove the battery and battery tray.

2. Remove the battery stay (brace).

3. Remove the air cleaner and intake hoses.

4. Drain the transaxle fluid into a suitable waste container.

5. Remove the nut securing the shifter lever to the transaxle. Remove the cable retaining clip and remove the cable from the transaxle.

6. Remove the shifter cable mounting bracket.

7. Disconnect and tag the electrical connectors for the speedometer, solenoid, neutral safety switch (inhibitor switch), the pulse generator, kickdown servo switch, oil temperature sensor.

8. Disconnect and tag the oil cooler lines, at the transaxle.

9. Remove the bolt securing the fluid dipstick tube, to the transaxle. Remove the dipstick and tube from the transaxle.

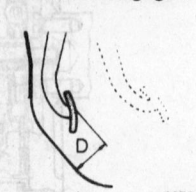

Clutch pedal free play	Distance between the clutch pedal and the firewall when the clutch is disengaged

295244

Clutch pedal free-play adjustment

10. Remove the starter motor and position aside.

11. Using special tool 7137 or C-4852 or equivalent, support the engine assembly.

12. Remove the rear roll stopper mounting bracket.

13. Remove the transaxle mount bracket.

14. Remove the upper transaxle mounting bolts.

15. Raise and safely support the vehicle.

16. Remove the front wheels.

17. Remove the left side undercover.

18. Remove the cotter pin and disconnect the tie rod end, from the steering knuckle.

19. Disconnect the stabilizer bar link, from the damper fork.

20. Disconnect the damper fork, from the lateral lower control arm.

21. Disconnect the later lower arm, and the compression lower arm, lower ball joints, from the steering knuckle.

22. Pry the halfshafts from the transaxle, and secure aside.

23. Remove the cover from the transaxle bellhousing.

24. Remove the engine front roll stopper through bolt.

25. Remove the centermember.

26. Remove the bolts holding the flexplate to the torque converter with a box wrench. Rotate the crankshaft to bring the bolts into a position for removal, one at a time.

27. To make installation easier, use chalk or paint to make matchmarks on the torque converter and flex plate. These marks will be used at assembly to realign the assembly, keeping these parts in balance. After removing the bolts, push the torque converter toward the transaxle. This will prevent the converter from remaining in contact with the engine, possibly damaging the converter.

28. Support the transaxle using a transmission jack (at the side of the case, NOT at the pan), and remove the transaxle lower coupling bolt.

NOTE: The coupling bolt threads from the engine side, into the transaxle, and is located just above the halfshaft opening.

29. Slide the transaxle rearward and carefully lower it from the vehicle.

To install:

30. After the torque converter has been mounted on the transaxle, install the transaxle assembly to the engine. Install the mounting bolts and tighten to 70 ft. lbs (95 Nm).

31. Align the balance matchmarks made at disassembly, connect the torque converter to the flexplate and tighten the bolts to 55 ft. lbs. (75 Nm).

32. Install the cover to the transaxle bellhousing and tighten the mounting bolts to 9 ft. lbs. (12 Nm).

33. Install the centermember and tighten the front mounting bolts to 65 ft. lbs. (88 Nm) and the rear bolts to 51–58 ft. lbs. (69–78 Nm). Install the front engine roll stopper through bolt and lightly tighten. Once the full weight of the engine is on the mounts, tighten the bolt to 42 ft. lbs. (56 Nm).

34. Install the halfshafts, using new circlips on the axle ends.

WARNING

When installing the halfshaft, keep the inboard joint straight in relation to the axle, to avoid damaging the oil seal lip of the transaxle, with the serrated part of the halfshaft.

---

35. Connect the tie rod and ball joints to the steering knuckle. Tighten the ball joint self-locking nuts to 48 ft. lbs. (65 Nm). Tighten the tie rod end nut to 21 ft. lbs. (28 Nm) and secure with a new cotter pin.

36. Connect the damper fork to the lower control arm and tighten the through bolt to 65 ft. lbs. (88 Nm).

37. Connect the stabilizer link to the damper fork, and tighten the self-locking nut to 29 ft. lbs. (39 Nm).

38. Install the left side undercover.

39. Install the wheels and lower the vehicle.

40. Install the transaxle mount bracket, to the transaxle, and tighten the mounting nuts to 32 ft. lbs. (43 Nm).

41. Install the rear roll stopper mounting bracket.

42. Remove the engine support. Tighten the transaxle mount through bolt to 51 ft. lbs. (69 Nm) and tighten the front engine roll stopper through bolt.

43. Install the upper transaxle mounting bolts and tighten to 35 ft. lbs. (48 Nm).

44. Install the starter motor.

45. Install the dipstick tube and the dipstick.

46. Install the shifter cable mounting bracket.

47. Connect the shifter lever and tighten the retaining nut to 14 ft. lbs. (19 Nm).

48. Connect the oil cooler lines and secure with clamps.

49. Connect the electrical connectors for the speedometer, solenoid, neutral safety switch (inhibitor switch), the pulse generator, kickdown servo switch and oil temperature sensor.

50. Install the air cleaner and the air intake hose.

51. Install the battery tray and battery.

52. Make sure the vehicle is level, and refill the transaxle with MOPAR ATF PLUS or equivalent transmission fluid. Start the engine and allow it to idle for 2 minutes. Apply the parking brake and move the selector through each gear position, ending in **N**. Recheck fluid level and add if necessary. Fluid level should be between the marks in the **HOT** range.

53. Check the transaxle for proper operation. Make sure the reverse lights come on when in reverse and the engine starts only in **P** or **N**.

DRIVE AXLE

Halfshaft

REMOVAL AND INSTALLATION

NOTE: If the vehicle is going to be rolled while the halfshafts are out of the vehicle, obtain 2 outer CV-joints or proper equivalent tools and install to the hubs. If the vehicle is rolled without the proper torque applied to the front wheel bearings, the bearings will no longer be usable.

1. Disconnect the negative battery cable.

2. Remove the cotter pin, halfshaft nut and washer.

3. Raise and safely support the vehicle.

4. Remove the wheel.

5. Using joint separation tool MB991113 or equivalent, disconnect the tie rod end from the steering knuckle.

CAUTION

Use of improper methods of joint separation can result in damage to the joint, leading to possible failure.

---

6. Disconnect the sway bar link from the damper fork.

7. Remove the damper fork lower through bolts and upper pinch bolt. Remove the damper fork assembly.

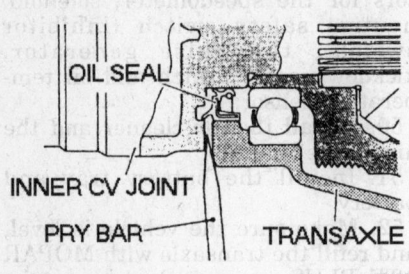

OIL SEAL

INNER CV JOINT

PRY BAR   TRANSAXLE

290558

Removing the halfshaft from the transaxle

8. Using a joint separation tool, disconnect the lateral arm and the compression arm from the steering knuckle.

9. Remove the halfshaft from the hub/knuckle by setting up a puller on the outside wheel hub and pushing the halfshaft from the front hub. After pressing the outer shaft, insert a prybar between the transaxle case and the halfshaft and pry the shaft from the transaxle.

NOTE: Do not pull on the shaft. Doing so damages the inboard joint. Do not insert the prybar too far or the oil seal in the case may be damaged.

To install:

10. Inspect the halfshaft boot for damage or deterioration. Check the ball joints and splines for wear.

11. Replace the circlips on the ends of the halfshaft(s).

12. Insert the halfshaft into the transaxle. Make sure it is fully seated.

13. Pull the knuckle assembly outward and install the other end of the halfshaft into the hub.

14. Install the washer so the chamfered edge faces outward. Install the halfshaft nut and tighten temporarily.

15. Connect the lateral arm and the compression arm to the steering knuckle. Tighten the self-locking nuts to 43–52 ft. lbs. (59–71 Nm).

16. Install the damper fork. Tighten the lower through bolt/nut to 65 ft. lbs. (88 Nm) and the upper pinch bolt to 76 ft. lbs. (103 Nm).

17. Connect the tie rod end to the steering knuckle. Tighten the retaining nut to 17–25 ft. lbs. (24–33 Nm) and install a new cotter pin.

18. Connect the sway bar link to the damper fork and tighten the link nut to 29 ft. lbs. (39 Nm).

19. Install the lock washer and axle nut. Tighten the axle nut with the special tool MB990767 to hold the hub from turning. Tighten the nut to a torque of 145–188 ft. lbs. (200–260 Nm).

NOTE: Before securely tightening the axle nut make sure there is no load on the wheel bearings.

20. Install a new cotter pin and bend to secure.

21. Install the wheel.

22. Check the transaxle fluid level and top off, if necessary.

23. Connect the negative battery cable.

24. Test drive the vehicle and check for proper operation.

CV-Joint Boot

REPLACEMENT

The driveshaft (halfshaft) has a tripod joint on the transaxle side and a birfield joint on the wheel side. A center bearing and an inner shaft is used on some models. Since the outer CV-Joint assembly is not serviceable and should not be disassembled, both boots are removed from the inner CV-Joint side of the halfshaft.

1. Raise and safely support the vehicle. Remove the halfshaft.

2. Remove the bands from the inner boot. Side cutter pliers can be used to cut off the metal retaining bands. Remove the inner CV-Joint boot and case from the halfshaft.

3. Remove the snapring next to the tripod joint spider assembly from the halfshaft with snapring pliers and remove the spider assembly from the shaft. Do not disassemble the spider assembly. Use care when handling the spider assembly.

NOTE: Both of the halfshaft boots are going to be removed from the inner CV-Joint case side of the halfshaft.

4. If the boot is to be reused, wrap vinyl tape around the spline part of the shaft so the boot will not be damaged when removed. Remove the dynamic damper, if used, and both boots from the shaft.

To install:

5. Clean old grease from joint with solvent and blow dry.

6. Double check that the correct replacement parts are being installed. Wrap vinyl tape around the splines to protect the boot and install the boots and damper, if used, in the correct order.

7. Fill the inside of the boot with the specified grease. Often the grease supplied in the replacement parts kit is meant to be divided in half, with half being used to lubricate the joint and half being used inside the boot. Keep grease off of the rubber part of the dynamic damper (if used).

8. Secure the boot bands with the halfshaft in a horizontal position. Make sure the boot span on the halfshaft is 3.15 ± 0.12 inch (80 ± 3mm) in length.

9. Check dynamic damper for proper positioning and adjust if necessary.

10. Install the halfshaft.

STEERING

Air Bag

— CAUTION —

Some vehicles are equipped with an air bag system, also known as the Supplemental Inflatable Restraint (SIR) system or Supplemental Restraint System (SRS). The system must be disabled before performing service on or around system components, steering column, instrument panel components, wiring and sensors. Failure to follow safety and disabling procedures could result in accidental air bag deployment, possible personal injury and unnecessary system repairs.

PRECAUTIONS

Several precautions must be observed when handling the inflator module to avoid accidental deployment and possible personal injury.

• Never carry the inflator module by the wires or connector on the underside of the module.

• When carrying a live inflator module, hold securely with both hands, and ensure that the bag and trim cover are pointed away.

• Place the inflator module on a bench or other surface with the bag and trim cover facing up.

• With the inflator module on the bench, never place anything on or close to the module which may be thrown in the event of an accidental deployment.

1. DO NOT use any electrical test equipment on or near any SRS components except those specified by Chrysler corporation:

a. Use a digital multi-meter for which the maximum test current is

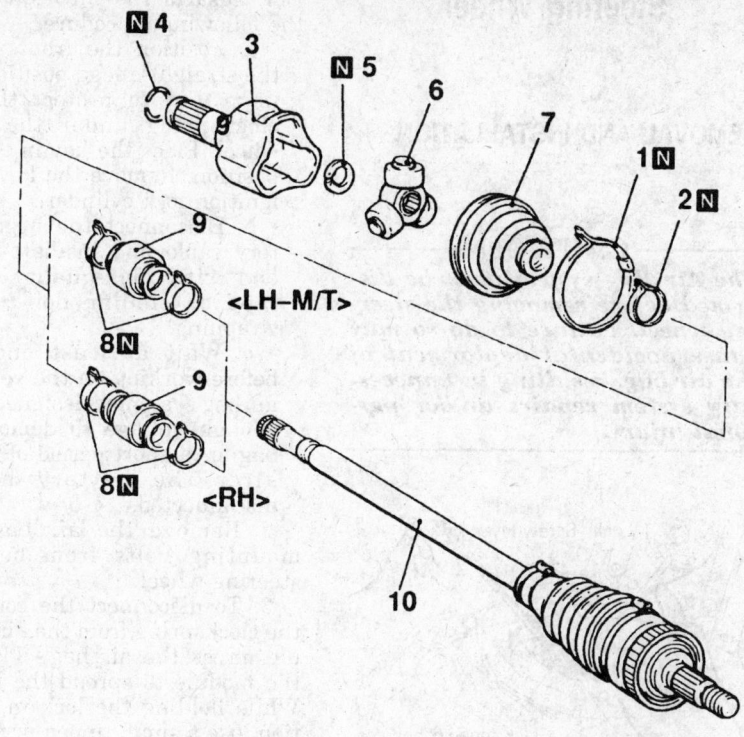

1. Inner CV joint boot band
2. Inner CV joint boot band
3. Inner CV joint case
4. Circlip
5. Snap ring
6. Spider assy
7. Inner CV joint boot
8. Damper band
9. Dynamic damper
10. Outer joint & shaft assy

Halfshaft assembly exploded view — 2.5L engine application shown

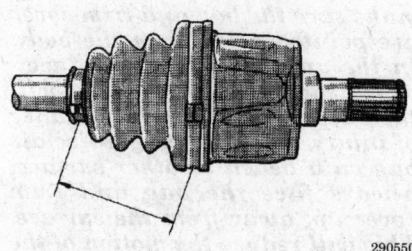

Checking the installed length of the CV-joint boot

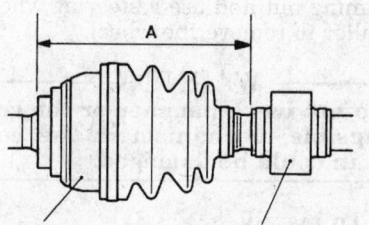

Dynamic damper installation. A=14.72 inch (374mm) for right side; A=7.64 inch (194mm) for left side, M/T

2mA or less at the minimum range of resistance measurement for use with the Chrysler SRS Check Harness when checking the SRS electrical circuitry.

b. Chrysler special tool MB991613 SRS Check Harness acts like a "break-out box" for checking SRS wiring. There are other factory special tool wiring adapters that are available and may be used.

c. DRB III or equivalent scan tool for reading and erasing air bag diagnostic codes.

2. NEVER ATTEMPT TO REPAIR THE FOLLOWING COMPONENTS:

a. Air Bag Control Unit (SRS-ECU)

b. Clock Spring under steering wheel.

c. Air Bag Modules

d. If any of these components are diagnosed as faulty, they should only be replaced.

3. Do not attempt to repair any of the air bag system wiring harness connectors. If any of the connectors or wires are faulty, replace that harness.

4. Air bag components should not be subjected to heat over 200°F. Remove the SRS-ECU, the air bag modules themselves and the clock spring

before drying or baking the vehicle after painting.

5. After air bag system service, check the SRS warning light operation to make sure that the system functions properly.

6. Make certain that the ignition switch is in the **OFF** position when a scan tool is connected or disconnected.

DISARMING

The Air Bag system (\Chrysler also calls it the Supplemental Restraint System or SRS) is designed to supplement the driver's and passenger's seat belts to help reduce the risk or severity of injury to the driver and front passenger by activating and deploying both air bags in certain frontal collisions. The system consists of two air bag modules, one located in the center of the steering wheel and another located above the glove box, which contains the folded air bag and an inflator unit. The air bag electronic control unit (SRS-ECU) located under the floor console assembly monitors the system and which contains a safing G sensor and analog G sensor. An SRS warning light is located on the instrument panel which indicates the status of the air bag system. A clock spring interconnection is located within the steering column.

To deploy the air bags, the SRS-ECU must respond to the output signal from the analog G sensor and the safing G sensor must be ON. The SRS-ECU then causes the air bag modules to ignite and deploy.

Service technicians should use care when working around any vehicle equipped with an air bag system, to avoid injury to the technician by inadvertent deployment of the air bag or to the driver by rendering the air bag system inoperative.

The SRS-ECU not only controls the air bag system, it can provide diagnostic information. The SRS-ECU monitors the air bag system and stores data concerning any detected faults in the system. When the ignition key is turned to the **ON** or **START** position, the SRS warning light should illuminate for about 7 seconds and then turn off. That indicates that the SRS system is in operating condition. If the SRS warning light does not illuminate as described or stays on for more than 7 seconds or if the SRS light illuminates while driving, immediate inspection is required. If the vehicle's SRS warning light is in any of these three conditions, the SRS system must be inspected, diagnosed and serviced.

To avoid injury from accidental deployment of the air bag during vehicle servicing, the following service precautions must be observed.

—— **CAUTION** ——
The Air Bag system must be disarmed before removing many components. Failure to do so may cause accidental deployment of the air bag, resulting in unnecessary system repairs and/or personal injury.

1. Disarm the air bag system using the following procedure:
   a. Position the front wheels in the straight-ahead position and place the key in the **LOCK** position. Remove the key from the ignition lock cylinder.
   b. Disconnect the negative battery cable and insulate the cable end with high-quality electrical tape or similar non-conductive wrapping.
   c. Wait at least one minute before working on the vehicle. The air bag system is designed to retain enough voltage to deploy the air bag for a short period of time even after the battery has been disconnected.

Steering Wheel

REMOVAL AND INSTALLATION

—— **CAUTION** ——
The Air Bag system must be disarmed before removing the steering wheel. Failure to do so may cause accidental deployment of the air bag, resulting in unnecessary system repairs and/or personal injury.

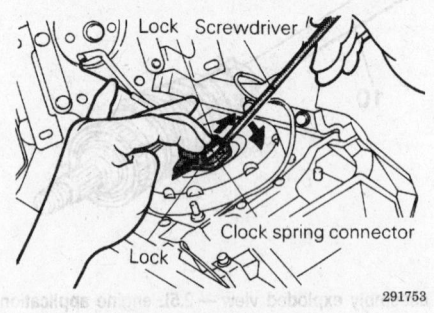

Airbag module connection

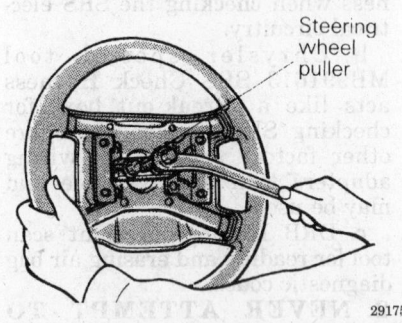

Steering wheel removal

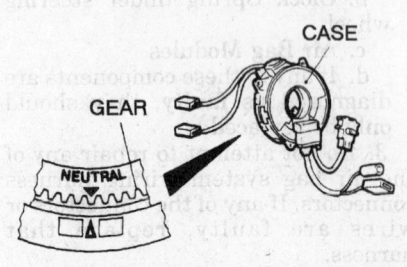

Clock spring mating mark alignment

1. Disarm the SRS system using the following procedure:
   a. Position the front wheels in the straight-ahead position. This is important to protect the airbag clockspring under the steering wheel. Place the key in the **LOCK** position. Remove the key from the ignition lock cylinder.
   b. Disconnect the negative battery cable and insulate the cable end with high-quality electrical tape or similar non-conductive wrapping.
   c. Wait at least one minute before working on the vehicle. The air bag system is designed to retain enough voltage to deploy the air bag for a short period of time even after the battery has been disconnected.
2. Remove the air bag module mounting bolts from behind the steering wheel.
3. To disconnect the connector of the clockspring from the air bag module, press the air bag's lock toward the module to spread the lock open. While holding the lock in this position, use a small tipped prying tool to gently pry the connector from the module.

—— **CAUTION** ——
When carrying a live air bag, make sure the bag and trim cover are pointed away from the body. In the unlikely event of an accidental deployment, the bag will then deploy with minimal chance of injury. When placing a live air bag on a bench or other surface, always face the bag and trim cover up, away from the surface. This will reduce the motion of the module if it is accidentally deployed.

4. Store the air bag module in a clean, dry place with the pad cover facing up. Do not place anything on top of the air bag module.
5. Remove the steering wheel retaining nut and use a steering wheel puller to remove the wheel.

—— **WARNING** ——
Do not use a hammer or the collapsible mechanism in the column could be damaged.

To install:
6. Confirm that the front wheels are in a straight-ahead position. Center the clockspring by aligning the **NEUTRAL** mark on the clockspring with the mating mark on the casing.

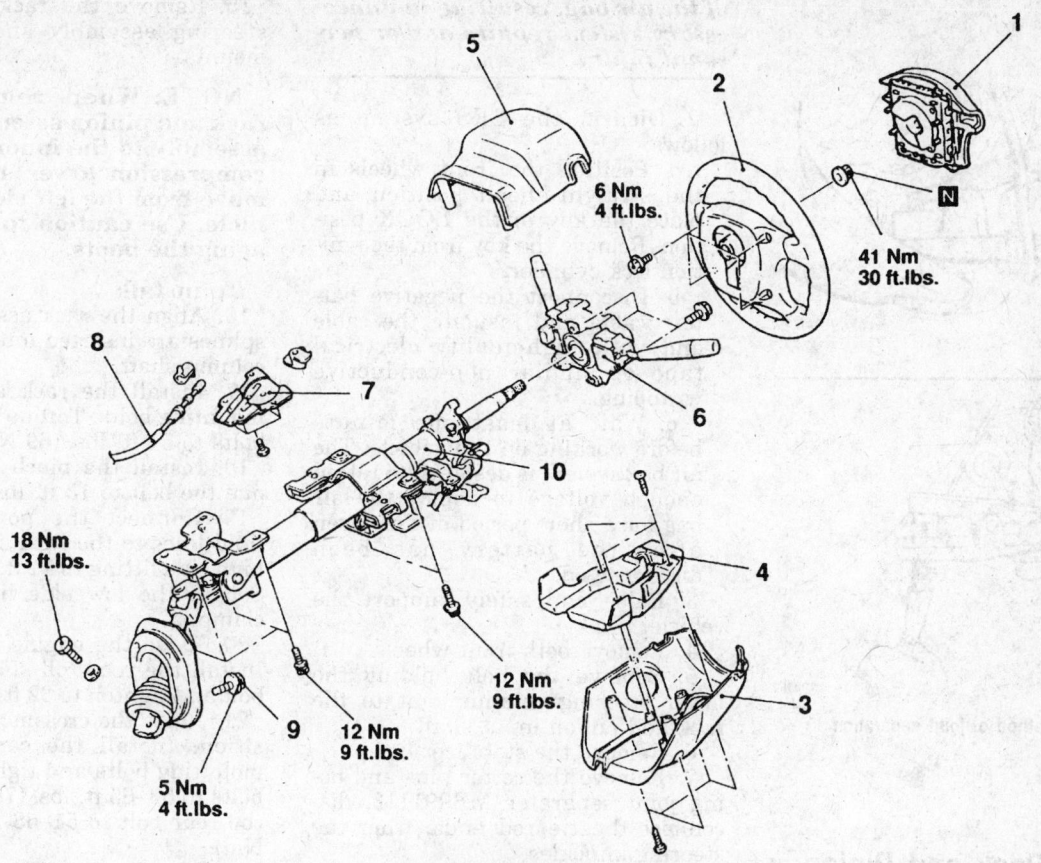

Removal steps
1. Air bag module
2. Steering wheel
3. Lower column cover
4. Column pad
5. Upper column cover
6. Clock spring and column switch assembly
7. Cover <A/T>
8. Key interlock cable <A/T>
9. Retainer attachment bolt
10. Steering column assembly

291752

Exploded view of the steering wheel and related components

7. Install the steering wheel and torque the retaining nut to 29 ft. lbs. (40 Nm).

8. Attach the air bag module wiring connector to clockspring connection. Install the air bag module and tighten the mounting bolts to 43 inch lbs. (5 Nm) and install the side covers.

9. Connect the negative battery cable and check all functions of the combination switch and air bag system for proper operation.

Tie Rod Ends

REMOVAL AND INSTALLATION

1. Raise and safely support the vehicle .

2. Remove the appropriate wheel.

3. Wire brush the threads on the tie rod shaft and lubricate them with penetrating oil. Loosen the locknut.

4. Remove the cotter pin and nut and press the tie rod end from the steering knuckle with the proper tie rod removal tool.

5. Hold the tie rod shaft with locking pliers and turn the tie rod end off.

Counting the number of turns should make installation of the replacement tie rod end close to previous alignment.

To install:

6. Install the tie rod end the same number of turns that it took to remove the old one. Tighten the locknut to 36–40 ft. lbs. (50–55 Nm).

7. Install the tie rod stud into the steering knuckle and install the nut. Tighten the nut to 17–25 ft. lbs. (24–34 Nm) and install a new cotter pin.

8. Perform a front end alignment.

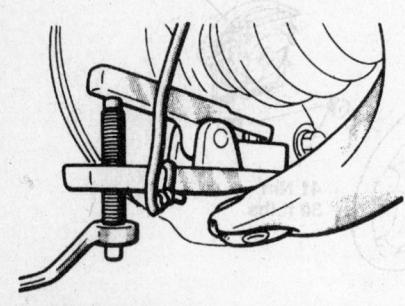

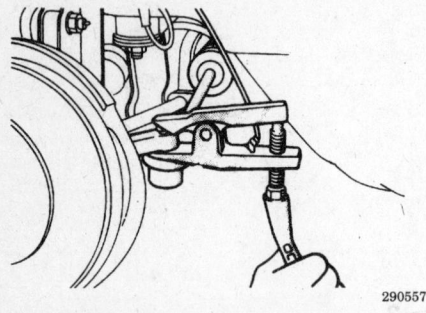

Proper method of joint separation

Power Rack and Pinion

REMOVAL AND INSTALLATION

——— **CAUTION** ———
Prior to removal of the steering rack and pinion unit, center the front wheels and remove the ignition key. Failure to do so may damage the SRS (air bag system) clock spring under the steering wheel and render SRS system inoperative, risking serious driver injury.

1. Drain the power steering fluid using the following procedure:
   a. Disconnect the power steering return (low side) hose.
   b. Connect a suitable container to the hose.
   c. Properly disable the ignition system.
   d. While cranking the engine, turn the wheels, several times, from side to side, until the fluid is removed.

——— **CAUTION** ———
The Air Bag system must be disarmed before working around the steering column. Failure to do so may cause accidental deployment

of the air bag, resulting in unnecessary system repairs and/or personal injury.

2. Disarm the SRS system as follows:
   a. Position the front wheels in the straight-ahead position and place the key in the **LOCK** position. Remove the key from the ignition lock cylinder.
   b. Disconnect the negative battery cable and insulate the cable end with high-quality electrical tape or similar non-conductive wrapping.
   c. Wait at least one minute before working on the vehicle. The air bag system is designed to retain enough voltage to deploy the air bag for a short period of time even after the battery has been disconnected.
3. Raise and safely support the vehicle.
4. Remove both front wheels.
5. Remove the bolt holding the lower steering column joint to the rack and pinion input shaft.
6. Remove the stabilizer bar.
7. Remove the cotter pins and using joint separator MB991113, disconnect the tie rod ends, from the steering knuckles.
8. On vehicles equipped with Electronic Control Power steering (EPS), disconnect the wiring harness, from the solenoid connector.
9. Locate the two triangular braces near the crossmember and remove both.
10. Support the center crossmember. Remove the through bolt from the front round roll stopper and remove the three bolts securing the center crossmember.
11. Remove the center crossmember.
12. Properly support the engine and remove the rear roll stopper through bolt. Lower the engine slightly.

——— **WARNING** ———
In order to prevent damage to the engine, when supporting and jacking the engine, place a block of wood between the jack and the oil pan.

13. Disconnect the power steering fluid pressure pipe and return hose from the rack fittings. Plug the fittings to prevent excessive fluid leakage.
14. Remove the clamp bolts and the two bolts securing the rack assembly to the chassis.

15. Remove the rack and pinion steering assembly and its rubber mounts.

NOTE: When removing the rack and pinion assembly, tilt the assembly to the inner side of the compression lower arm, and remove from the left side of the vehicle. Use caution to avoid damaging the boots.

To install:
16. Align the rack assembly so the splines are inserted into the steering column shaft.
17. Install the rack and with the mounting bolts. Torque the mounting bolts to 51 ft. lbs. (69 Nm).
18. Install the pinch bolt and torque the bolt to 13 ft. lbs. (18 Nm).
19. Connect the power steering fluid lines to the rack and tighten to high side fitting to 11 ft. lbs. (15 Nm). Secure the low side hose with the clamp.
20. Raise the engine into position. Install the rear roll stopper through bolt and tighten to 32 ft. lbs. (43 Nm).
21. Raise the crossmember into position. Install the center member mounting bolts and tighten the front bolts to 58–65 ft. lbs. (78–88 Nm) and the rear bolt to 51–58 ft. lbs. (69–78 Nm).
22. Install the front roll stopper bolt and tighten the nut to 32 ft. lbs. (43 Nm).
23. Install the two triangular braces and tighten the mounting bolts to 50–56 ft. lbs. (69–78 Nm).
24. Install the stabilizer bar.
25. Connect the tie rod ends and tighten the nuts to 20 ft. lbs. (27 Nm).
26. On vehicles equipped with EPS, connect the wiring harness to the solenoid connector.
27. Install the wheels and lower the vehicle.
28. Refill the reservoir with power steering fluid and properly bleed the power steering system.
29. Perform a front end alignment.

Power Steering Pump

BLEEDING

1. Check the power steering fluid level. Use MOPAR ATF PLUS automatic transmission fluid or equivalent in the power steering system.
2. Raise and safely support the vehicle to lift the front wheels off the ground.
3. Manually turn the pump pulley a few times.

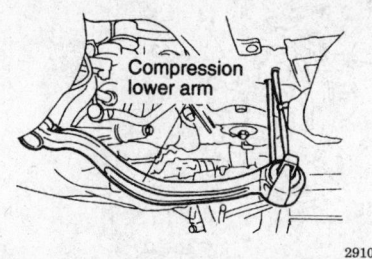

Removal of the rack and pinion assembly past the compression lower arm

4. Turn the steering wheel all the way to the left and to the right 5 or 6 times.

5. Disconnect the ignition high tension cable and, while operating the starter motor intermittently, turn the steering wheel all the way to the left and right 5-6 times for 15- 20 seconds. During bleeding, make sure the fluid in the reservoir never falls below the lower position of the filter. If bleeding is attempted with the engine running, the air will be absorbed in the fluid. Bleed only while cranking.

6. Connect ignition high tension cable, start the engine and allow it to idle.

7. Turn the steering wheel left and right until there are no air bubbles in the reservoir. Confirm that the fluid is not milky and the level is up to the specified position on the gauge. Confirm that there is very little change in the fluid level when the steering wheel is turned. If the fluid level changes more than 0.2 inches (about a ¼-inch) the air has not been completely bled. Repeat the process.

REMOVAL AND INSTALLATION

Service this vehicle's power steering system with MOPAR ATF PLUS automatic transmission fluid type 7176 or DEXRON II automatic transmission fluid.

1. Disconnect the negative battery cable.

2. Remove (drain, suction pump, etc.) as much power steering fluid as possible.

3. Disconnect the return fluid line. Remove the reservoir cap and allow the return line to drain the fluid from the reservoir. If the fluid is contaminated, disconnect the ignition high tension cable and crank the engine

several times to drain the fluid from the gearbox.

NOTE: **Cover any components located underneath the power steering pump with a shop towel to protect them from damage due to power steering fluid spillage. For example, the A/C compressor or alternator, depending on vehicle and engine, is below the power steering pump, so cover the A/C compressor or alternator with a shop towel before removing any hoses.**

4. Loosen the power steering pump mounting bolts and remove the drive belt.

5. Remove the pressure switch connector from the side of the pump.

6. Disconnect the pressure line.

7. Unbolt and remove the pump from the mounting bracket.

To install:

8. Clean all parts well. Inspect the pump pulley for cracks. Check the hoses carefully for cracks or signs of weakness.

9. Install the pump, wrap the belt around the pulley and lightly tighten the mounting bolts.

10. Replace the O-rings and connect the pressure line. Connect the pressure line so the notch in the fitting aligns and contacts the pump's guide bracket. Tighten the fitting to 13 ft. lbs. (18 Nm).

11. Connect the return line and secure with the clamp.

12. Connect the pressure switch connector.

13. Adjust the power steering belt for proper tension and tighten the adjusting bolts.

14. Refill the reservoir and bleed the power steering system system.

BRAKES

Anti-Lock Brake System Service

PRECAUTIONS

• Certain components within the Anti-Lock Brake System (ABS) are not intended to be serviced or repaired individually. Only those components with removal and installation procedures should be serviced.

• Do not use rubber hoses or other parts not specifically specified for and ABS system. When using repair kits,

replace all parts included in the kit. Partial or incorrect repair may lead to functional problems and require the replacement of components.

• Lubricate rubber parts with clean, fresh brake fluid to ease assembly. Do not use lubricated shop air to clean parts; damage to rubber components may result.

• Use only specified brake fluid from an unopened container.

• If any hydraulic component or line is removed or replaced, it may be necessary to bleed the entire system.

• A clean repair area is essential. Always clean the reservoir and cap thoroughly before removing the cap. The slightest amount of dirt in the fluid may plug an orifice and impair the system function. Perform repairs after components have been thoroughly cleaned; use only denatured alcohol to clean components. Do not allow ABS components to come into contact with any substance containing mineral oil; this includes used shop rags.

• The Anti-Lock control unit is a microprocessor similar to other computer units in the vehicle. Ensure that the ignition switch is **OFF** before removing or installing controller harnesses. Avoid static electricity discharge at or near the controller.

• If any arc welding is to be done on the vehicle, the control unit should be unplugged before welding operations begin.

Master Cylinder

REMOVAL AND INSTALLATION

—— **WARNING** ——
Use care when working with brake fluid. Brake fluid is extremely harmful to painted surfaces.

1. Disconnect the negative battery cable.

2. Disconnect the fluid level sensor connector, if equipped.

3. Disconnect the hoses from the master cylinder to the fluid reservoir. Plug the hoses to prevent drainage.

4. Disconnect the brake lines from the master cylinder.

5. Remove the two nuts securing the master cylinder to the brake booster and remove the master cylinder.

To install:

6. Install master cylinder to the booster mounting studs. Install the mounting nuts and tighten to 7 ft. lbs. (10 Nm).

7. Connect reservoir hoses to master cylinder and secure with clamps.

8. Connect the brake lines to the master cylinder.

9. Fill the reservoir to the proper level with clean DOT 3 or DOT 4 brake fluid. Bleed the master cylinder.

10. Apply the brake pedal and check for firmness. If the pedal is spongy, air is present in the system and bleeding of the entire system is required.

11. Check the brakes for proper operation and leaks.

Brake Caliper

REMOVAL AND INSTALLATION

Front Brake Calipers

NOTE: Do not allow the master cylinder reservoir to empty. An empty reservoir will allow air to enter the brake system and complete system bleeding will be required.

1. Remove about half of the brake fluid from the master cylinder.

2. Raise and safely support the vehicle and remove the wheel assembly.

3. Position a C-clamp, or other suitable tool, over the caliper. Smoothly apply pressure, forcing the caliper piston into the caliper bore until it bottoms. Remove the C-clamp, if used.

4. If the caliper is to be completely removed from the vehicle, remove the brake hose attaching bolt, disconnect the brake hose from the caliper and plug the hose to prevent fluid contamination or loss.

5. Remove the caliper mounting bolts and lift the caliper off of the support bracket.

6. Remove the caliper from the vehicle. If the caliper is only removed for access to other components, support the caliper, with the brake hose attached, so that there is no strain on the brake hose.

To install:

7. Position the caliper to the steering knuckle. Lubricate and install the mounting bolts. Tighten the bolts to 54 ft. lbs. (74 Nm).

8. If removed, unplug and install the brake line hose to the caliper and tighten the inlet fitting bolt to 22 ft. lbs. (29 Nm).

9. Fill the master cylinder with fresh brake fluid and, if the brake

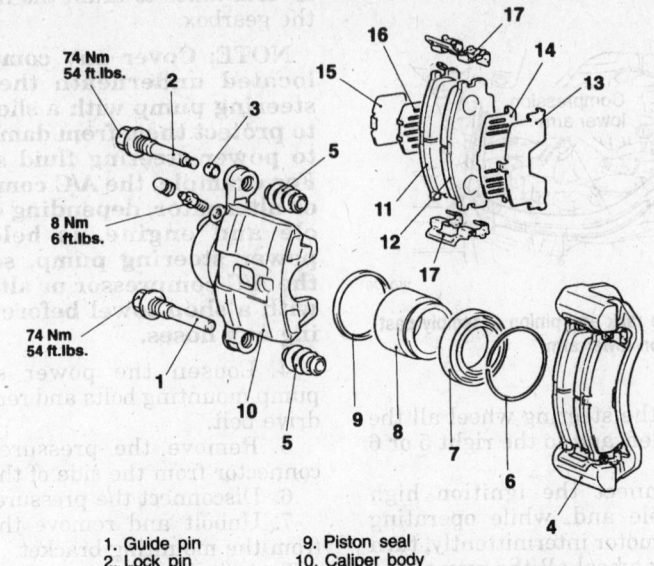

74 Nm 54 ft.lbs.
8 Nm 6 ft.lbs.
74 Nm 54 ft.lbs.

1. Guide pin
2. Lock pin
3. Bushing
4. Caliper support (pad, clip, shim)
5. Boot
6. Boot ring
7. Piston boot
8. Piston
9. Piston seal
10. Caliper body
11. Pad and wear indicator assembly
12. Pad assembly
13. Outer shim (stainless)
14. Outer shim (coated with rubber)
15. Inner shim (stainless)
16. Inner shim (coated with rubber)
17. Clip

289790

Exploded view of the front brake pads and related components

hose was removed, bleed the brake system

10. Install the wheel assembly and lower the vehicle.

11. Depress the brake pedal 3–4 times to seat the brake linings and to restore pressure in the system.

— **CAUTION** —
Do not move the vehicle until a firm pedal is obtained.

Rear Brake Calipers

Unlike many rear disc brake designs, this system does not incorporate the parking brake system, into the rear brake caliper. The rear brake system is serviced the same as the front system.

1. Remove about half of the brake fluid from the master cylinder.

2. Raise and safely support the vehicle and remove the wheel assembly.

3. Position a C-clamp, or other suitable tool, over the caliper. Smoothly apply pressure, forcing the caliper piston into the caliper bore until it bottoms. Remove the C-clamp, if used.

4. If the caliper is to be completely removed from the vehicle, remove the brake hose attaching bolt, disconnect the brake hose from the caliper and

plug the hose to prevent fluid contamination or loss.

5. Remove the caliper mounting bolts and lift the caliper off of the support bracket.

6. Remove the caliper from the vehicle. If the caliper is only removed for access to other components, support the caliper, with the brake hose attached, so that there is no strain on the brake hose.

To install:

7. Position the caliper on the support bracket, lubricate and install the mounting bolts. Tighten the bolts to 54 ft. lbs. (74 Nm).

8. If removed, unplug and install the brake line hose to the caliper and tighten the inlet fitting bolt to 22 ft. lbs. (29 Nm).

9. Fill the master cylinder with fresh brake fluid and, if the brake hose was removed, bleed the brake system

10. Install the wheel assembly and lower the vehicle.

11. Depress the brake pedal 3–4 times to seat the brake linings and to restore pressure in the system.

— **CAUTION** —
Do not move the vehicle until a firm pedal is obtained.

Disc Brake Pads

REMOVAL AND INSTALLATION

Front Disc Brake Pads

1. Remove some of the brake fluid from the master cylinder reservoir. The reservoir should be no more than full. When the pistons are depressed into the calipers, excess fluid will flow up into the reservoir.
2. Raise and safely support the vehicle.
3. Remove the appropriate tire and wheel assemblies.
4. Remove the caliper guide and lock pins and lift the caliper assembly from the caliper support. Tie the caliper out of the way using wire. Do not allow the caliper to hang by the brake line.

NOTE: On some models, the caliper can be flipped up by leaving the upper pin in place and using it as a pivot point.

5. Remove the brake pads, spring clip and shims. Take note of positioning to aid installation.
6. Install the wheel lug nuts onto the studs and lightly tighten. This is done to hold the disc on the hub.
To install:
7. Use a large C-clamp to compress the piston(s) back into the caliper bore.
8. Lubricate the slide points and install the brake pads, shims and spring clip onto the caliper support. Install the caliper over the brake pads.

NOTE: Be careful that the piston boot does not become caught when lowering the caliper onto the support. Do not twist the brake hose during caliper installation.

9. Lubricate and install the caliper guide and lock pins in their original positions. Tighten guide and locking pins to 54 ft. lbs. (74 Nm).
10. Install the tire and wheel assemblies. Lower the vehicle.

— **CAUTION** —
Pump brake pedal several times, until firm, before attempting to move the vehicle.

11. Road test the vehicle and check brakes for proper operation.

Rear Disc Brake Pads

Unlike many rear disc brake designs, this system does not incorporate the parking brake system, into the rear brake caliper, therefore, the rear brake system is serviced the same as the front system.

1. Remove some of the brake fluid from the master cylinder reservoir. The reservoir should be no more than full. When the pistons are depressed into the calipers, excess fluid will flow up into the reservoir.
2. Raise and safely support the vehicle.
3. Remove the appropriate tire and wheel assemblies.
4. Remove the caliper guide and lock pins and lift the caliper assembly from the caliper support. Tie the caliper out of the way using wire. Do not allow the caliper to hang by the brake hose.

NOTE: On some models, the caliper can be flipped up by leaving the upper pin in place and using it as a pivot point.

5. Remove the brake pads, spring clip and shims. Take note of positioning to aid installation.
6. Install the wheel lug nuts onto the studs and lightly tighten. This is done to hold the brake disc on the hub.
To install:
7. Use a large C-clamp to compress the piston(s) back into the caliper bore.
8. Lubricate the slide points and install the brake pads, shims and spring clip onto the caliper support. Install the caliper over the brake pads.

NOTE: Be careful that the piston boot does not become caught when lowering the caliper onto the support. Do not twist the brake hose during caliper installation.

9. Lubricate and install the caliper guide and lock pins in their original positions. Tighten guide and locking pins to 54 ft. lbs. (74 Nm).
10. Install the tire and wheel assemblies. Lower the vehicle.

— **CAUTION** —
Pump brake pedal several times, until firm, before attempting to move the vehicle.

11. Road test the vehicle and check brakes for proper operation.

Brake Rotor

REMOVAL AND INSTALLATION

Front Brake Rotors

1. Remove about half of the brake fluid from the master cylinder.
2. Raise and safely support the vehicle and remove the wheel assembly.
3. Position a C-clamp, or other suitable tool, over the caliper. Smoothly apply pressure, forcing the caliper piston into the caliper bore until it bottoms. Remove the C-clamp, if used.
4. If the caliper is to be completely removed from the vehicle, remove the brake hose attaching bolt, disconnect the brake hose from the caliper and plug the hose to prevent fluid contamination or loss.
5. Remove the caliper mounting bolts and lift the caliper off of the support bracket.
6. If the caliper is only removed for access to other components, support the caliper, with the brake hose attached, so that there is no strain on the brake hose.
7. Remove the caliper support bracket and slide the rotor off of the hub.
To install:
8. Inspect the rotor for cracks and excessive wear. Front brake rotor wear limit is 22.4mm (.880 inch). Rear brake rotor wear limit is 8.4mm (.330 inch). On all rotors, runout limit is 0.08mm (.0031 inch).
9. Slide the rotor onto the hub and install the caliper support bracket. Tighten the support bracket bolts to 65 ft. lbs. (88 Nm).
10. Position the caliper to the steering knuckle. Lubricate and install the mounting bolts. Tighten the bolts to 54 ft. lbs. (74 Nm).
11. If removed, unplug and install the brake line hose to the caliper and tighten the inlet fitting bolt to 22 ft. lbs. (29 Nm).
12. Fill the master cylinder with fresh brake fluid and, if the brake hose was removed, bleed the brake system.
13. Install the wheel assembly and lower the vehicle.
14. Depress the brake pedal 3–4 times to seat the brake linings and to restore pressure in the system.

— **CAUTION** —
Do not move the vehicle until a firm pedal is obtained.

Rear Brake Rotors

Unlike many rear disc brake designs, this system does not incorporate the parking brake system, into the rear brake caliper. The rear brake system is serviced the same as the front system.

1. Remove about half of the brake fluid from the master cylinder.
2. Raise and safely support the vehicle and remove the wheel assembly.
3. Position a C-clamp, or other suitable tool, over the caliper. Smoothly apply pressure, forcing the caliper piston into the caliper bore until it bottoms. Remove the C-clamp, if used.
4. If the caliper is to be completely removed from the vehicle, remove the brake hose attaching bolt, disconnect the brake hose from the caliper and plug the hose to prevent fluid contamination or loss.
5. Remove the caliper mounting bolts and lift the caliper off of the support bracket.
6. Remove the caliper from the vehicle. If the caliper is only removed for access to other components, support the caliper, with the brake hose attached, so that there is no strain on the brake hose.
7. Remove the caliper support bracket and slide the rotor off of the hub.

To install:

8. Inspect the rotor for cracks and excessive wear. Rear brake rotor wear limit is 8.4mm (.330 inch). On all rotors, runout limit is 0.08mm (.0031 inch).
9. Slide the rotor onto the hub and install the caliper support bracket. Tighten the support bracket bolts to 65 ft. lbs. (88 Nm).
10. Position the caliper on the support bracket, lubricate and install the mounting bolts. Tighten the bolts to 54 ft. lbs. (74 Nm).
11. If removed, unplug and install the brake line hose to the caliper and tighten the inlet fitting bolt to 22 ft. lbs. (29 Nm).
12. Fill the master cylinder with fresh brake fluid and, if the brake hose was removed, bleed the brake system
13. Install the wheel assembly and lower the vehicle.
14. Depress the brake pedal 3–4 times to seat the brake linings and to restore pressure in the system.

--- **CAUTION** ---
Do not move the vehicle until a firm pedal is obtained.

Brake Drums

REMOVAL AND INSTALLATION

1. Raise and safely support the vehicle .
2. Remove the wheel assembly.
3. Remove the brake drum detent (retaining) screw and remove the drum from the axle.
4. If difficulty is encountered in removing the drum:
   a. Verify the parking brake is released.
   b. Loosen the parking brake cable.
   c. Remove the access hole plug from the backing plate and move the parking brake lever until the lever stop rests on the brake shoe.
   d. If necessary, turn the adjuster so that it draws in to allow more clearance between the shoe and drum.

To install:

5. Inspect all parts. Check the drum for cracks or excessive wear. Wear limit on the brake drum inside diameter is 230.6mm (9.0 inches).
6. Turn the adjuster until it is drawn all the way in to the stop. Check that the adjuster turns freely. The nut must NOT lock at the end of the adjuster. Check that the parking brake lever stops are against the edge of the shoe web.
7. Install the brake drum and detent screw.
8. Install the wheel assembly, then lower the vehicle.
9. Apply the foot brake at least 10 times until clicking of the adjustment actuator can no longer be heard. This procedure will automatically adjust the clearance between the shoe and drum.

Wheel Cylinder

REMOVAL AND INSTALLATION

NOTE: It is important to NOT let the master cylinder reservoir run dry, at any time during this procedure, or the entire brake system will have to be bled.

1. Raise and safely support the vehicle .
2. Remove the wheel and the brake drum.
3. Remove the shoe-to-lever spring and the upper shoe-to-shoe spring. Spread the upper portion of the brake shoes slightly.
4. Remove and plug the brake line from the wheel cylinder.

5. Remove the wheel cylinder retaining bolts and remove the cylinder from the backing plate.

To install:

6. Apply a very thin coating of silicone sealer to the cylinder mounting surface.
7. Install the cylinder to the backing plate and tighten the retaining bolts to 7 ft. lbs. (10 Nm).
8. Connect the brake line to the wheel cylinder and tighten the fitting to 11 ft. lbs. (15 Nm).
9. Install the brake springs and the brake drum.
10. Install the tire and wheel assembly.
11. Fill the system with clean brake fluid, adjust and bleed the rear brakes.

Brake Shoes

REMOVAL AND INSTALLATION

1. Raise and safely support the vehicle and remove the wheel assembly.
2. Remove the detent screw and brake drum. If difficulty is encountered in removing the drum:
   a. Verify the parking brake is released.
   b. Loosen the parking brake cable.
   c. Remove the access hole plug from the backing plate and move the parking brake lever until the lever stop rests on the brake shoe.
   d. If necessary, turn the adjuster so that it draws in to allow more clearance between the shoe and drum.

NOTE: Note the location of all springs and clips for proper reassembly.

3. Remove the shoe-to-lever spring and remove the adjuster lever.
4. Remove the auto adjuster assembly.
5. Remove the retainer spring.
6. Remove the hold-down springs, washers and pins.
7. Remove the shoe-to-shoe spring.
8. Remove the brake shoes from the backing plate.
9. Using a flat-tipped tool, open up the parking brake lever retaining clip. Remove the clip and washer from the pin on the shoe assembly and remove the shoe from the lever assembly.

To install:

10. Thoroughly clean and dry the backing plate. Lubricate the backing plate at the brake shoe contact points.

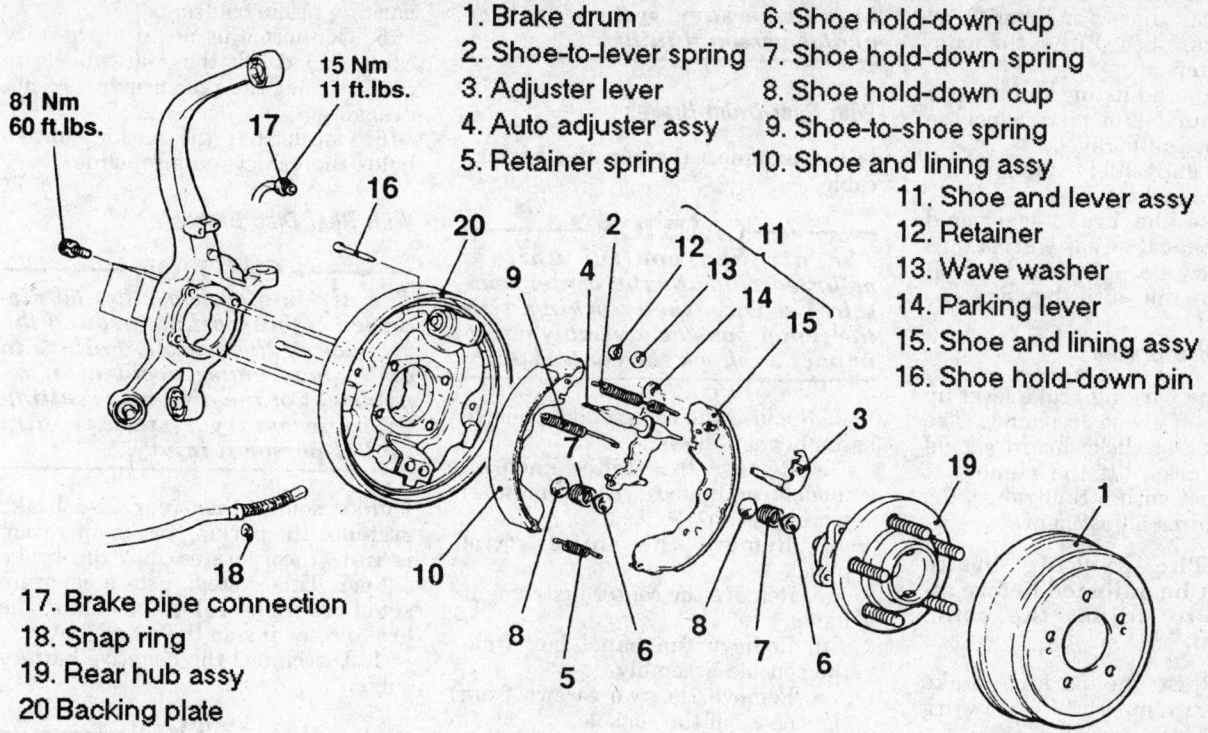

1. Brake drum
2. Shoe-to-lever spring
3. Adjuster lever
4. Auto adjuster assy
5. Retainer spring
6. Shoe hold-down cup
7. Shoe hold-down spring
8. Shoe hold-down cup
9. Shoe-to-shoe spring
10. Shoe and lining assy
11. Shoe and lever assy
12. Retainer
13. Wave washer
14. Parking lever
15. Shoe and lining assy
16. Shoe hold-down pin

81 Nm
60 ft.lbs.

15 Nm
11 ft.lbs.

17. Brake pipe connection
18. Snap ring
19. Rear hub assy
20 Backing plate

Exploded view of the drum brake assembly

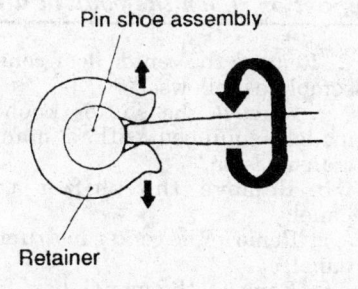

Pin shoe assembly

Retainer

Opening the retainer clip

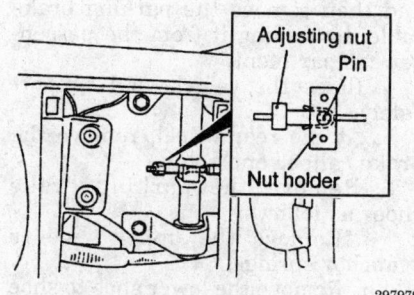

Adjusting nut
Pin

Nut holder

Adjusting nut and nut holder

11. Lubricate backing plate bosses, anchor pin, and parking brake actuating mechanism with a lithium-based grease.

12. Install the parking brake lever assembly on the lever pin. Install the wave washer and a new retaining clip. Use pliers or the like to install the retainer on the pin. If removed, connect the parking brake lever to the parking brake cable and verify that the cable is properly routed.

13. Clean and lubricate the adjuster assembly. Make sure the nut-adjuster is drawn all the way to the stop, but the nut must NOT lock firmly at the end of the assembly.

14. Install the brake shoes on the backing plate with the hold-down springs, washers and pins.

15. Install the shoe-to-shoe spring.

16. Install the retainer spring.

17. Install the auto adjuster assembly and install the adjuster lever and the shoe-to-lever spring.

18. Pre-adjust the shoes so the drum slides on with a light drag and install the brake drum.

19. Adjust the rear brake shoes and install the rear wheels.

20. Adjust the parking brake cable.

21. Lower the vehicle and check for proper brake operation.

Parking Brake Cable

ADJUSTMENT

With Rear Drum Brakes

NOTE: Make certain that the brake shoes are properly adjusted before attempting to adjust the parking brake.

1. Pull the parking brake lever up with a force of about 45 pounds. The total number of clicks heard should be 5 to 7.

2. If the number of clicks was not within that range, release the lever. Uncover the inner compartment mat of the floor console.

3. Loosen the adjusting nut to the end of the cable rod at the base of the lever, freeing the parking brake cable.

4. With the engine idling, forcefully depress the brake pedal five or six times and confirm that the pedal stroke stops changing. If the pedal stroke stops changing, the automatic adjustment mechanism is working correctly and the clearance between the shoe and drum is correct.

5. After verifying that the brake adjustment is correct, tighten the adjusting nut until there is no more slack in the cable.

6. Operate the lever and brake pedal several times, and verify no more clicks are heard from the automatic adjuster.

7. Turn the adjusting nut to give the proper number of clicks when the lever is raised full travel.

8. Raise and safely support the rear of the vehicle.

9. Release the brake lever and make sure that the rear wheels turn freely. If they do not, then back off the adjusting nut until they do.

With Rear Disc Brakes

1. Pull the parking brake lever up with a force of about 45 pounds. The total number of clicks heard should be 3 to 5 clicks. If the number of clicks was not within that range, the system requires adjustment.

NOTE: The parking brake shoes must be adjusted before attempting to adjust the cable mechanism

2. To adjust the parking brake shoes perform the following procedure.

a. remove the floor console, release the lever and back off the cable adjuster locknut at the base of the lever.

b. Raise and safely support the vehicle and remove the wheel. Remove the hole plug in the brake rotor.

c. Use a suitable adjuster tool to turn the adjuster wheel to expand the brake shoes until the rotor will not turn.

d. Return the adjuster 5 notches in the opposite direction. Make sure the rotor turns freely with a slight drag.

3. Once the parking brake shoes have been properly adjusted, adjust the cable mechanism, by performing the following steps:

a. Turn the adjusting nut to give the proper number of clicks when the lever is raised full travel.

b. Raise and safely support the rear of the vehicle.

c. Release the brake lever and make sure that the rear wheels turn freely. If they do not, then back off on the adjusting nut until they do.

REMOVAL AND INSTALLATION

— CAUTION —
The Air Bag system must be disarmed before working around the interior of the vehicle. Failure to do so may cause accidental deployment of the air bag, resulting in unnecessary system repairs and/or personal injury.

With Rear Drum Brakes

1. Disconnect the negative battery cable.

— CAUTION —
The air bag control unit is mounted beneath the center console. Use care when working with the center console assembly not to impact or shock the control unit.

2. Remove the center floor console assembly as follows:

a. Remove the shifter knob on models equipped with a manual transmission.

b. Remove the shifter trim panel.

c. Remove the center instrument panel.

d. Remove the panel box from the console assembly.

e. Remove the two screws from the center of the console.

f. Remove the four side panel screws and remove the console from the vehicle.

3. Loosen the cable adjuster nut and then remove the parking brake cable, by pulling it from the passenger compartment.

4. Raise and safely support vehicle.

5. Remove the brake drum and shoes.

6. Disconnect the cable end from the parking brake strut lever. Compress the retaining strips to remove the cable from the backing plate.

7. Unfasten any other frame retainers and remove the cables.

To install

8. Install the cable to the rear actuator. Secure in place with the parking brake cable clip and retainer spring.

9. Install the brake shoes and drum.

10. Position the cable in the under the vehicle and install retainers loose.

11. Reattach the parking brake cables to the actuator inside the vehicle. Tighten the adjusting nut until the proper tension is placed on the cable. Adjust the parking brake stroke.

12. Secure all cable retainers. Apply and release the parking brake a number of times once all adjustments have been made.

13. Assemble the interior components which were removed.

14. Adjust the rear brakes and parking brake cables.

15. Connect the negative battery cable and check the rear wheels to confirm that the rear brakes are not dragging.

16. Check that the parking brake holds the vehicle on an incline.

With Rear Disc Brakes

— CAUTION —
The Air Bag system must be disarmed before working around the interior of the vehicle. Failure to do so may cause accidental deployment of the air bag, resulting in unnecessary system repairs and/or personal injury.

Unlike some other rear disc brake systems, the parking brake operation is **not** incorporated into the brake caliper. This system, uses a separate set of brake shoes, located behind the brake rotor inside the rotor "hat".

1. Disconnect the negative battery cable.

— CAUTION —
The air bag control unit is mounted beneath the center console. Use care when working with the center console assembly not to impact or shock the control unit.

2. Remove the center floor console assembly as follows:

a. Remove the shifter knob on models equipped with a manual transmission.

b. Remove the shifter trim panel.

c. Remove the center instrument panel.

d. Remove the panel box from the console assembly.

e. Remove the two screws from the center of the console.

f. Remove the four side panel screws and remove the console from the vehicle.

3. Loosen the cable adjuster nut and then remove the parking brake cable, by pulling it from the passenger compartment.

4. Raise the vehicle and support safely.

5. At the rear wheel, remove the brake caliper and rotor.

6. Remove the parking brake shoes as follows:

a. Remove the upper shoe to anchor springs.

b. Remove the lower shoe to shoe spring.

c. Remove the brake shoe hold-down springs.

d. Disconnect the parking brake cable from the actuating lever.

7. Disconnect the cable end from the parking brake strut lever. Compress the retaining strips to remove the cable from the backing plate.

8. Unfasten any other frame retainers and remove the cables.

To install

9. Install the cable to the rear actuator. Secure in place with the parking brake cable clip and retainer spring.

10. Install the parking brake shoes.

11. Install the brake rotor and caliper assembly.

12. Position the cable in the under the vehicle and install retainers loose.

13. Reattach the parking brake cables to the actuator inside the vehicle. Tighten the adjusting nut until the proper tension is placed on the cable. Adjust the parking brake stroke.

14. Secure all cable retainers. Apply and release the parking brake a number of times once all adjustments have been made.

15. Assemble the interior components which were removed.

16. Adjust the parking brake shoes and parking brake cables.

17. Connect the negative battery cable and check the rear wheels to confirm that the rear brakes are not dragging.

18. Check that the parking brake holds the vehicle on an incline.

Brake System Bleeding

NOTE: If using a pressure bleeder, follow the instructions furnished with the unit and choose the correct adapter for the application. Do not substitute an adapter that does fit exactly as it will not work and could be dangerous.

MASTER CYLINDER

1. If the master cylinder is off of the vehicle it can be bench bled.

a. Connect two short pieces of brake line to the outlet fittings, bend them until the free end is below the fluid level in the master cylinder reservoir.

b. Fill the reservoir with fresh brake fluid. Pump the piston slowly until no more air bubbles appear in the reservoirs.

c. Disconnect the two short lines, refill the master cylinder and securely install the cylinder caps.

2. If the master cylinder is on the vehicle, it can still be bled, using a flare nut wrench.

a. Open the brake lines slightly with the flare nut wrench while pressure is applied to the brake pedal by a helper inside the vehicle.

b. Be sure to tighten the line before the brake pedal is released.

c. Repeat the process with both lines until no air bubbles come out.

3. Refill master cylinder and always bleed the complete brake system.

CALIPERS AND WHEEL CYLINDERS

1. Fill the master cylinder with fresh DOT 3 or DOT 4 brake fluid. Check the level often during the procedure.

2. Starting with the wheel furthest from the master cylinder, remove the protective cap from the bleeder and place it where it will not be lost. Clean the bleeder screw.

---— CAUTION —---
When bleeding the brakes, keep face away from the brake area. Spewing fluid may cause skin and/or eye damage. Do not allow brake fluid to spill on the car's finish. It will remove the paint.

---

3. If the system is empty, the most efficient way to get fluid down to the wheel is to use a pressure bleeder tool. Open the bleeder until brake fluid flows without signs of air bubbles, close bleeder.

NOTE: Stroke the brake pedal slowly. If the pedal is pumped rapidly, the fluid will churn and create small air bubbles, which are almost impossible to remove from the system. These air bubbles will accumulate and a spongy pedal will result.

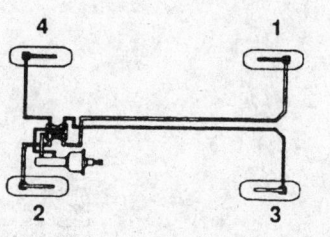

Brake system bleeding sequence

295224

4. If the manual procedure is to be used to pump brake fluid to the caliper or wheel cylinder, open the bleed screw and have an assistant press the brake pedal to the floor.

5. Close the bleeder screw before releasing the brake pedal, have the helper slowly release the pedal. Wait 15 seconds and repeat the procedure until no more air comes out of the bleeder upon application of the brake pedal. Remember to close the bleeder before the pedal is released inside the vehicle each time the bleeder is opened. If not, air will be introduced into the system.

6. Repeat the procedure on remaining wheel cylinders/calipers in the following order:
- Right rear wheel cylinder/caliper.
- Left front caliper.
- Left rear wheel cylinder/caliper.
- Right front caliper.

7. Hydraulic brake systems must be totally flushed if the fluid becomes contaminated with water, dirt or other corrosive chemicals. To flush, bleed the entire system until all fluid has been replaced with the correct type of new fluid.

8. Install the bleeder cap on the bleeder to keep dirt out and refill master cylinder. Always road test the vehicle after brake work of any kind is done to verify correct repair.

Wheel Speed Sensor

REMOVAL AND INSTALLATION

Front ABS Wheel Speed Sensor

1. Disconnect the negative battery cable.

2. Remove the clips securing the speed sensor lead.

3. Remove the retaining screw and remove the speed sensor.

To install:

4. Attach the wheel speed sensor with the retaining screw.

5. Secure the speed sensor lead with the clips.

---— WARNING —---
Make sure the lead wire is properly routed in its original position. If the lead wire is improperly installed it could be damaged by moving parts.

---

6. Connect the negative battery cable and check the ABS system for proper operation.

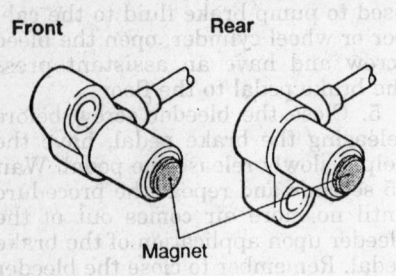

Front and rear ABS wheel speed sensors

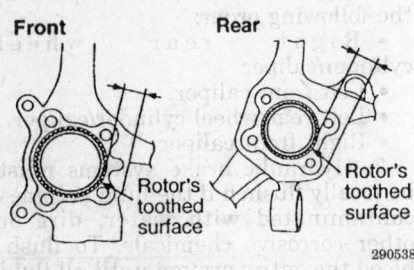

Front and rear wheel speed sensor clearance measurement

Rear ABS Wheel Speed Sensor

1. Disconnect the negative battery cable.
2. Remove the clips securing the speed sensor lead.
3. Remove the retaining screw and remove the speed sensor.

To install:

4. Attach the wheel speed sensor with the retaining screw.
5. Secure the speed sensor lead with the clips.

—— **WARNING** ——
Make sure the lead wire is properly routed in its original position. If the lead wire is improperly installed it could be damaged by moving parts.

6. Connect the negative battery cable and check the ABS system for proper operation.

FRONT SUSPENSION

Strut

REMOVAL AND INSTALLATION

1. Disconnect the negative battery cable.
2. Raise and safely support the vehicle.
3. Remove the appropriate wheel assembly.

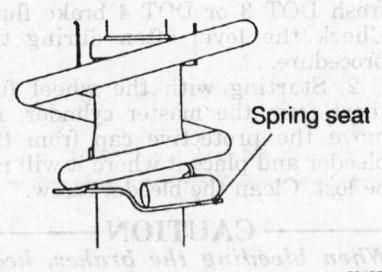

Make sure the spring is properly seated

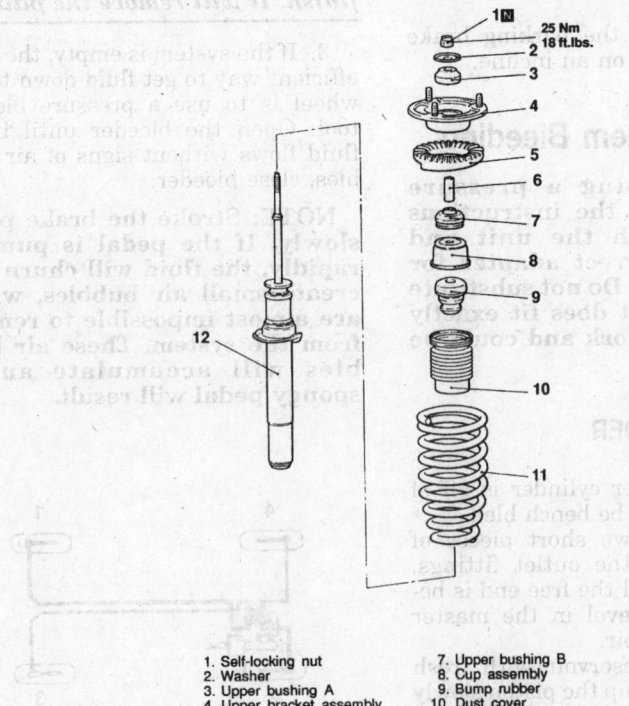

1. Self-locking nut
2. Washer
3. Upper bushing A
4. Upper bracket assembly
5. Upper spring pad
6. Collar
7. Upper bushing B
8. Cup assembly
9. Bump rubber
10. Dust cover
11. Coil spring
12. Shock absorber assembly

Strut assembly exploded view

4. Disconnect the sway bar link from the damper fork.
5. Remove the damper fork lower through bolt and upper pinch bolt. Remove the damper fork assembly.
6. Remove the strut upper nuts and remove the strut assembly from the vehicle. Do NOT remove the large center nut.

To install:

7. Install the strut to the vehicle and tighten the upper mounting nuts to 32 ft. lbs. (44 Nm).
8. Align the strut to the damper fork and install the damper fork. Tighten the lower through bolt/nut to 65 ft. lbs. (88 Nm) and the upper pinch bolt to 76 ft. lbs. (103 Nm).
9. Connect the sway bar link to the damper fork and tighten the link nut to 29 ft. lbs. (39 Nm).
10. Install the wheel and tire assembly.
11. Perform a front end alignment.

Coil Spring

REMOVAL AND INSTALLATION

1. Place the strut assembly in a MB991237 and MB991238 spring compressor assembly or equivalent. Tighten the compressor and compress the spring slowly. Make certain

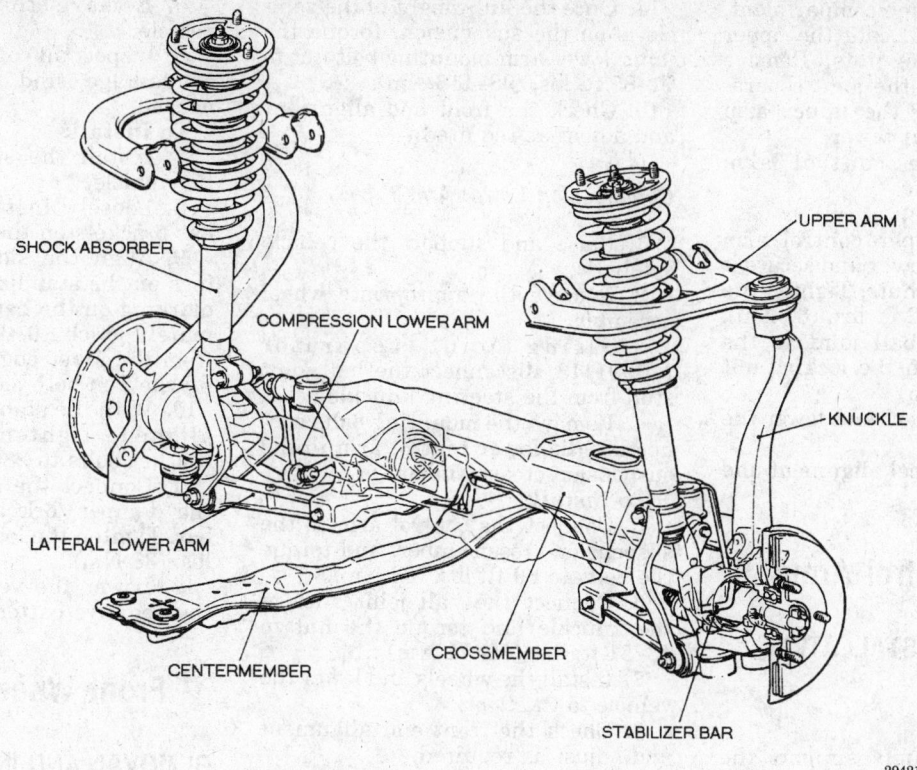

SHOCK ABSORBER

COMPRESSION LOWER ARM

UPPER ARM

KNUCKLE

LATERAL LOWER ARM

CENTERMEMBER

CROSSMEMBER

STABILIZER BAR

294218

Front suspension component identification

the compressor is properly engaged before tightening.

2. After tension has been removed from the strut assembly and strut plate, remove the piston rod nut and washer from the top of the strut assembly.

3. If the spring is to be replaced, slowly release the tension on the spring compressor. Allow the spring to expand fully. If only the strut is being replaced the spring may remain in the compressor assembly.

4. By hand, remove the upper strut bearing, washer, mount and strut shield. Remove the upper insulator ring, the spring, the bumper and lower insulator from the spring. Take notice to each components location for proper reassembly.

To install:

5. Install the lower insulator, the strut bumper and the uncompressed spring. Install the upper spring insulator, strut shield, mount and washer.

6. Install or align the spring compressor. Make certain the spring is correctly positioned relative to the upper and lower insulator rings. Smoothly compress the spring.

7. Install the washer and piston rod nut. Tighten the nut to 18 ft. lbs. (25 Nm). Install the dust cap.

8. Carefully release the spring compressor, watching the spring position as it seats. When the spring is properly seated, release/remove the compressor tools.

9. Reinstall the strut assembly.

Upper Ball Joints

REMOVAL AND INSTALLATION

The upper ball joint is an integrated part of the upper control arm assembly, and can not be serviced separately. A worn or damaged ball joint requires replacement of upper control arm assembly.

Lower Ball Joints

REMOVAL AND INSTALLATION

The front suspension is called a Multi-Link Suspension. There are two lower arms used in this front suspension; a curved arm called the Compression Lower Arm and also a straight arm called the Lateral Lower Arm. Both arms contain lower ball joints since there are two sockets in the steering knuckle. Ball joints and lower arms are removed and replaced as an assembly. A front end

alignment is required after these procedures.

Upper Control Arms

REMOVAL AND INSTALLATION

1. Raise and safely support the vehicle.

2. Remove the appropriate wheel.

3. Using the joint separation tool, MB991113 or equivalent, disconnect the ball joint stud from the steering knuckle.

4. The ball joint can be checked using the following procedure.

   a. An adapter (MB 990326) is available that fits onto the ball joint stud and adapts to an inch-pound torque wrench. If this tool is not available, a shop-made substitute can be fabricated.

   b. Turn the ball joint stud with the torque wrench. The factory standard for breakaway torque is 3 to 13 inch lbs.

   c. If the ball joint stud is out of specification (turns too easily or is too stiff), continue with the ball joint/control arm assembly replacement. Note that the boot over the ball joint can be removed, and the joint greased. A new replacement ball joint boot is recommended.

5. Inside the engine compartment, at the strut tower, locate the upper control arm mounting nuts. Remove the nuts and using the joint separation tool, separate the upper arm shafts from the strut tower.

6. Remove the control arm assembly.

To install:

7. Align the upper control arm shafts to the strut tower and secure it with the mounting nuts. Tighten the mounting nuts to 62 ft. lbs. (86 Nm).

8. Connect the ball joint to the knuckle and tighten the locking nut to 20 ft. lbs. (28 Nm).

9. Install the wheel and lower the vehicle.

10. Check the wheel alignment and adjust if necessary.

Lower Control Arms

REMOVAL AND INSTALLATION

Lateral Lower Arm

1. Raise and safely support the vehicle.

2. Remove the appropriate wheel assembly.

3. Remove the stay bracket from the crossmember.

4. Using joint separator MB991113, disconnect the ball joint stud from the steering knuckle.

5. Remove the through bolt, connecting the damper fork to the lower control arm.

6. Remove the mounting bolt connecting the lower control arm to the suspension crossmember.

7. Remove the lower control arm from the vehicle.

To install:

8. When installing the control arm, temporarily tighten the nuts and/or bolts securing the control arm to the suspension crossmember. Tighten them fully only after the vehicle is sitting on its wheels.

9. Connect the damper fork to the lower control arm and tighten the through bolt to 64 ft. lbs. (88 Nm).

10. Connect the ball joint stud to the knuckle and torque the nut to 65–80 ft. lbs. (88–108 Nm). Install a new cotter pin.

11. Connect the tension rod to the control arm and torque the nuts to 80–94 ft. lbs. (108–127 Nm).

12. Connect the stay bracket to the crossmember and tighten the mounting bolts to 50–56 ft. lbs. (69–78 Nm).

13. Install the wheels and lower the vehicle to the floor.

14. Once the full weight of the vehicle is on the suspension, torque the inner lower arm mounting bolt nut to 71–85 ft. lbs. (98–118 Nm).

15. Check the front end alignment and adjust as required.

Compression Lower Arm

1. Raise and support the vehicle safely.

2. Remove the appropriate wheel assembly.

3. Using joint separator MB991113, disconnect the ball joint stud from the steering knuckle.

4. Remove the mounting bolts connecting the lower control arm to the suspension crossmember.

To install:

5. Connect the control arm to the suspension crossmember, and torque the bolts to 60 ft. lbs. (83 Nm).

6. Connect the ball joint stud to the knuckle and torque the nut to 43–51 ft. lbs. (59–71 Nm).

7. Install the wheels and lower the vehicle to the floor.

8. Check the front end alignment and adjust as required.

Sway Bar

REMOVAL AND INSTALLATION

The front stabilizer bar (commonly called a sway bar) runs across the width of the front crossmember and uses two ball joint type links to connect the ends of the bar to the lower strut forks.

1. Disconnect the negative battery cable.

2. Raise and safely support the vehicle.

3. Disconnect the stabilizer links by removing the self-locking nuts.

4. Remove the stabilizer bar mounting brackets and bushings.

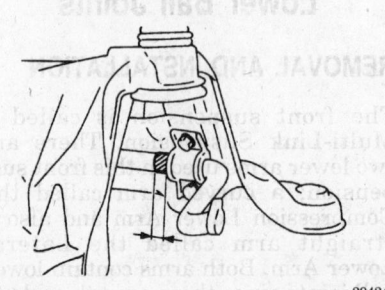

Installation of the stabilizer bar. Shaded area represents approximately 3/8-inch (10mm)

5. Remove the bar from the vehicle.

6. Inspect all components for wear or damage, and replace parts as needed.

To install:

7. Install the stabilizer bar into the vehicle.

8. Loosely install the stabilizer bar brackets on the vehicle.

9. Align the side locating markings on the stabilizer bar so that the marking on the bar extends approximately -inch (0.40 inch or 10mm) from the inner edge of the mounting bracket, on both sides.

10. With the stabilizer bar properly aligned, tighten the mounting bracket bolts to 28 ft. lbs. (39 Nm).

11. Connect the stabilizer links to the damper fork and the stabilizer bar. Tighten the locking nuts to 28 ft. lbs. (38 Nm).

12. Lower the vehicle and connect the negative battery cable.

Front Wheel Bearings

REMOVAL AND INSTALLATION

To check hub and bearing assembly end play, remove the caliper and rotor. Rig a dial indicator to bear against the hub flange near the center ridge. Wiggle the hub back and forth. If end play exceeds 0.002 inch, replace the front hub and bearing assembly.

1. Remove the cotter pin, halfshaft nut and washer.

2. Raise and safely support the vehicle.

3. Remove the appropriate wheel assembly.

4. If equipped with ABS, remove the vehicle speed sensor.

5. Remove the caliper and brake pads. Support the caliper out of the way using wire.

6. Remove the brake rotor from the hub assembly.

7. Disconnect the upper ball joint from the steering knuckle using a press type tool and pull the knuckle outward.

— CAUTION —

Use of improper methods of joint separation can result in damage to joint, leading to possible failure. Never use wedge-type tools or the ball joint can be damaged.

8. From the back of the knuckle, remove the four bolts securing the hub to the knuckle.

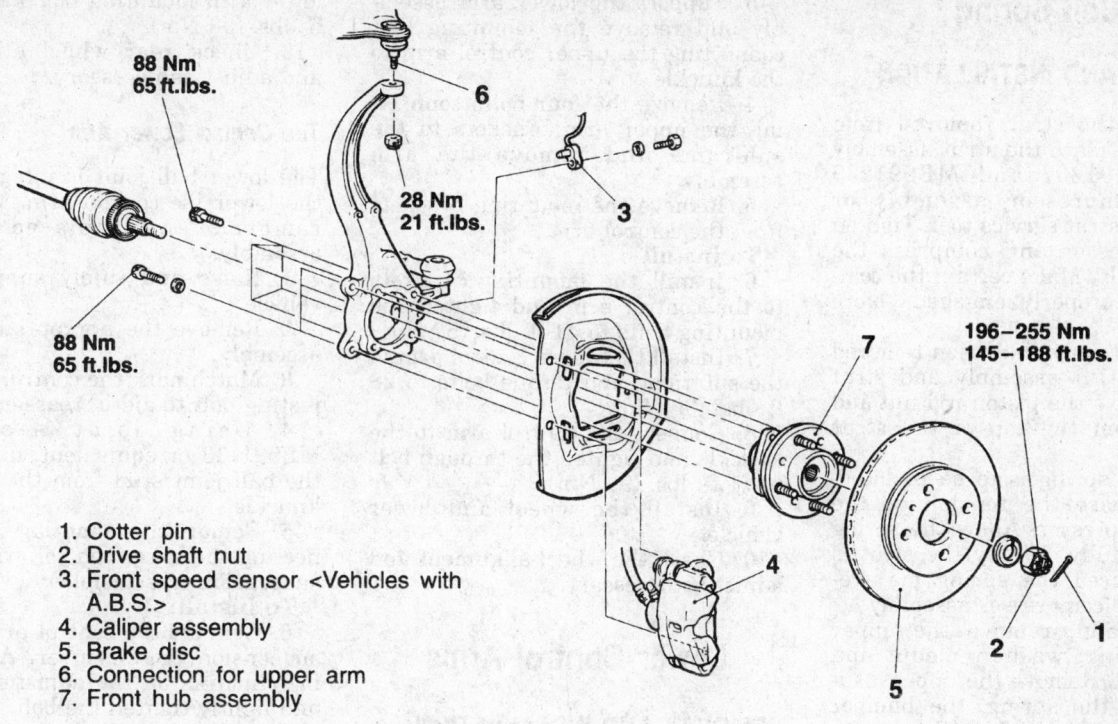

1. Cotter pin
2. Drive shaft nut
3. Front speed sensor <Vehicles with A.B.S.>
4. Caliper assembly
5. Brake disc
6. Connection for upper arm
7. Front hub assembly

88 Nm
65 ft.lbs.

28 Nm
21 ft.lbs.

88 Nm
65 ft.lbs.

196–255 Nm
145–188 ft.lbs.

294258

Front hub assembly and related components

9. Remove the hub and bearing assembly from the knuckle.

NOTE: The hub and wheel bearing assembly is not serviceable and should not be disassembled.

To install
10. Install the hub to the steering knuckle and tighten the mounting bolts to 65 ft. lbs. (88 Nm).
11. Connect the upper ball joint to the steering knuckle and tighten the self-locking nut to 21 ft. lbs. (28 Nm).
12. Position the rotor on the hub. Install a couple of lug nuts and lightly tighten to hold rotor on hub.
13. Install the caliper holder and place brake pads in holder. Slide caliper over brake pads and install guide pins. Once caliper is secured, lug nuts can be removed.
14. If equipped with ABS, install the vehicle speed sensor.
15. Install the wheel assembly and lower the vehicle.
16. Install the wheel and lower the vehicle to the floor. Examine the drive shaft (halfshaft) hub washer. Locate the chamfered side. This side is installed outward, away from the hub. Install the washer and hub nut. Tighten the axle nut with the brakes

applied. Tighten the nut to torque of 145–188 ft. lbs. (200–260 Nm).
17. Install a new cotter pin and bend to secure.

——— **CAUTION** ———
Pump the brake pedal until hard, before attempting to move the vehicle.

REAR SUSPENSION

Strut

REMOVAL AND INSTALLATION

NOTE: The strut assembly is a load bearing component, therefore the vehicle chassis and axle weight must be supported separately, requiring the use of two separate lifting devices.

1. The rear package shelf front cover(s) must be removed to access the top mounting nuts. Most connections are plastic clips. Use care when

removing these components to avoid unnecessary damage.
   a. Remove the rear shelf speaker covers.
   b. Remove the rear shelf top assembly.
   c. Remove the front cover(s) to access the strut top mounting nuts.
2. Raise and support vehicle chassis.
3. Raise and support lower control arm assembly slightly.
4. Remove the strut upper mounting nuts.
5. Remove the strut lower mounting bolt and remove the assembly from the vehicle.
   To install
6. Position the strut assembly so that the lower mounting bolt can be installed and lightly tightened.
7. Use a jack to raise or lower the lower control arm, so that top strut plate studs aligns through the body. Raise the jack to hold the strut assembly in position.
8. Install top plate nuts on studs and tighten the mounting nuts to 32 ft. lbs. (44 Nm).
9. Tighten the lower mounting bolt to 71 ft. lbs. (98 Nm).
10. Install the interior trim pieces to complete strut installation.

Coil Spring

REMOVAL AND INSTALLATION

1. With the strut removed from the vehicle, place the strut assembly in a MB991237 and MB991239 spring compressor assembly or equivalent strut service tool. Tighten the compressor and compress the spring slowly. Make certain the compressor is properly engaged before tightening.

2. After tension has been removed from the strut assembly and strut plate, remove the piston rod nut and washer from the top of the strut assembly.

3. If the spring is to be replaced, slowly release the tension on the spring compressor. Allow the spring to expand fully. If only the strut is being replaced the spring may remain in the compressor assembly.

4. By hand, remove the upper strut bearing, washer, mount and strut shield. Remove the upper insulator ring, the spring, the bumper and lower insulator from the spring. Take notice to each components location for proper reassembly.

To install:

5. Install the lower insulator, the strut bumper and the uncompressed spring. Install the upper spring insulator, strut shield, mount and washer.

6. Install or align the spring compressor. Make certain the spring is correctly positioned relative to the upper and lower insulator rings. Smoothly compress the spring.

7. Install the washer and piston rod nut. Tighten the nut to 16 ft. lbs. (22 Nm). Install the dust cap.

8. Carefully release the spring compressor, watching the spring position as it seats. When the spring is properly seated, release/remove the compressor tools.

9. Reinstall the strut assembly.

Upper Control Arms

REMOVAL AND INSTALLATION

NOTE: The following procedures include the removal of load bearing components, therefore the vehicle chassis and axle weight must be supported separately, requiring the use of two separate lifting devices.

1. Raise and safely support the vehicle.

2. Remove the appropriate wheel assembly.

3. Support the lower arm assembly and remove the mounting bolt connecting the upper control arm to the knuckle.

4. Remove the four bolts connecting the upper arm brackets to the subframe and remove the arm assembly.

5. Remove the mounting brackets from the control arm.

To install

6. Install the mounting brackets to the control arm and tighten the mounting bolts to 41 ft. lbs. (57 Nm).

7. Install the upper control arm to the subframe and torque bolts to 28 ft. lbs. (39 Nm).

8. Connect the control arm to the knuckle and tighten the through bolt to 71 ft. lbs. (98 Nm).

9. Install the wheel and lower vehicle.

10. Check the wheel alignment and adjust as necessary.

Lower Control Arms

REMOVAL AND INSTALLATION

Lower Control Arm

1. Raise and safely support the vehicle.

2. Remove the appropriate wheel assembly.

3. If equipped with ABS, disconnect the speed sensor harness brackets, from the lower control arm.

4. Disconnect the stabilizer bar link from the lower control arm.

5. Remove the through bolt, connecting the knuckle assembly to the lower control arm.

6. Remove the mounting bolt connecting the lower control arm to the suspension crossmember.

7. Remove the lower control arm from the vehicle.

To install:

NOTE: The control arm mounting bolts must not be fully tightened until the suspension is bearing the full weight of the vehicle.

8. Install the control arm to the suspension crossmember and temporarily tighten the mounting bolt.

9. Connect the knuckle to the lower control arm and lightly tighten the through bolt.

10. Connect the stabilizer bar link to the control arm and torque the nut to 28 ft. lbs. (39 Nm).

11. Install the wheels and lower the vehicle to the floor.

12. Once the full weight of the vehicle is on the suspension, torque the lower arm mounting bolt nuts to 71 ft. lbs. (98 Nm).

13. Check rear wheel alignment and adjust as necessary.

Toe Control Lower Arm

The lower ball joint is integral with the lower toe control arm. They are removed and replaced as an assembly.

1. Raise and safely support the vehicle.

2. Remove the appropriate wheel assembly.

3. Matchmark the control arm adjusting bolt to aid in reassembly.

4. Using joint separator MB991113 or equivalent, disconnect the ball joint stud from the steering knuckle.

5. Remove the mounting bolts connecting the lower control arm to the suspension crossmember.

To install:

6. Connect the control arm to the suspension crossmember. Align the matchmarks on the adjustment bolt and lightly tighten the bolt.

7. Connect the ball joint stud to the knuckle and torque the nut to 20 ft. lbs. (28 Nm).

8. Install the wheels and lower the vehicle to the floor.

9. With the full weight of the vehicle is on the ground, torque the control arm through bolt to 50–56 ft. lbs. (69–78 Nm).

10. Check rear wheel alignment and adjust as necessary.

Trailing Arm Assembly

The trailing arm is the round link that locates the suspension fore and aft. It connects to the support knuckle on the outboard end and the body on its inboard end. Note that during installation, the retaining nuts should be temporarily tightened, the fully torqued with the vehicle on the ground.

1. Remove the knuckle and trailing arm assembly connecting bolt.

2. Remove the grommet, or cover over the body to trailing arm bolt.

3. Remove the body to trailing arm bolt and the two round plates on either side of the rubber bushed end of the trailing arm.

4. Remove the arm from vehicle.

To install:

5. Install the arm to the vehicle and install the body to trailing arm bolt along with the round plates. Temporarily tighten the bolt.

6. Install the trailing arm to the knuckle, install the bolt and but and temporarily tighten the bolt.

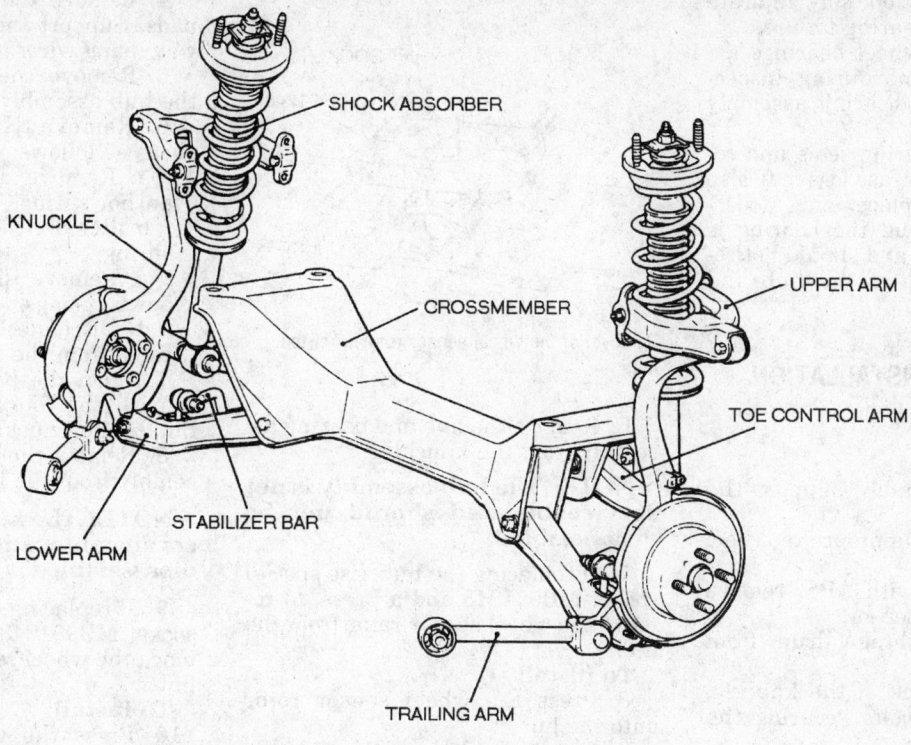

SHOCK ABSORBER

KNUCKLE

CROSSMEMBER

UPPER ARM

TOE CONTROL ARM

STABILIZER BAR

LOWER ARM

TRAILING ARM

294922

Multi-link rear suspension component identification

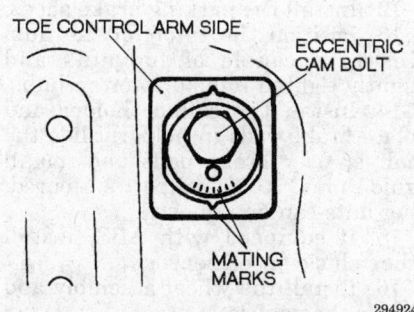

TOE CONTROL ARM SIDE

ECCENTRIC CAM BOLT

MATING MARKS

294924

Removal and installation of the toe control arm

7. Lower the vehicle until there is load on the suspension. Final torque on the knuckle and trailing arm assembly connecting bolt is 85-100 ft. lbs. Final torque on the body to trailing arm bolt is 100-115 ft. lbs.

8. Check rear wheel alignment and adjust as necessary.

Sway Bar

REMOVAL AND INSTALLATION

1. Raise and safely support the vehicle.

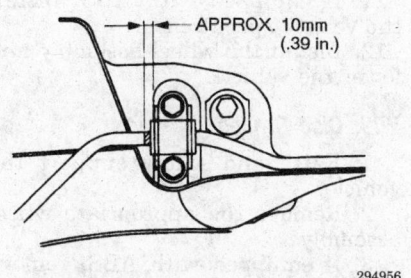

APPROX. 10mm
(.39 in.)

294956

Installation of the stabilizer bar

2. Disconnect the stabilizer links by removing the self-locking nuts.

3. Remove the stabilizer bar mounting brackets and bushings.

4. Remove the bar from the vehicle.

5. Inspect all components for wear or damage, and replace parts as needed.

To install:

6. Install the stabilizer bar into the vehicle.

7. Loosely install the stabilizer bar brackets on the vehicle.

8. Align the side locating markings on the stabilizer bar, so that the marking on the bar, extends approxi-

mately 0.39 inches (10 mm) from the outer edge of the mounting bracket, on both sides.

9. With the stabilizer bar properly aligned, tighten the mounting bracket bolts to 28 ft. lbs. (39 Nm).

10. Connect the stabilizer links to the damper fork and the stabilizer bar. Tighten the locking nuts to 28 ft. lbs. (38 Nm).

11. Lower the vehicle and road test to check for noise.

Wheel Bearings

ADJUSTMENT

The rear hub and wheel bearing assembly is designed for the life of the vehicle and requires no type of adjustment or periodic maintenance. The bearing is a sealed unit with the wheel hub and can only be removed and/or replaced as one unit.

The following procedure may be used for evaluation of bearing condition.

1. Raise and safely support the vehicle.

2. Remove the rear wheels and brake drums.

3. Turn the hub flange carefully. Excessive roughness, lateral play or

resistance to rotation may indicate dirt intrusion or bearing failure.

4. If the rear wheel bearings exhibit the conditions during inspection, the hub and bearing assembly should be replaced.

5. Damaged bearing seals and resulting excessive grease loss may also require bearing replacement. Moderate grease loss from the bearing is considered normal and should not require replacement of the hub and bearing assembly.

REMOVAL AND INSTALLATION

With Drum Brakes

1. Raise and safely support the vehicle.
2. Remove the appropriate wheel assembly.
3. If equipped with ABS, remove the vehicle speed sensor.
4. Remove the brake drum from the hub assembly.
5. From the back of the knuckle, remove the four bolts securing the hub to the knuckle.

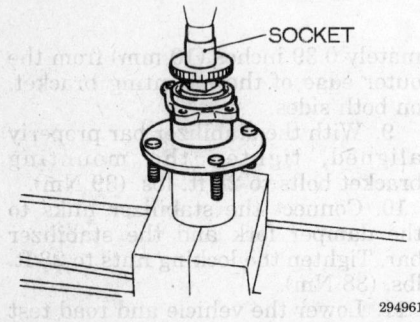

Installation of the wheel sensor rotor to the hub

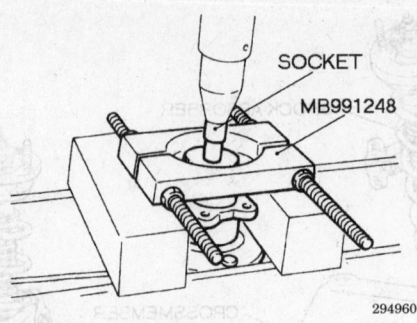

Removal of the wheel sensor rotor from the hub

6. Remove the hub and bearing assembly from the knuckle.

NOTE: The hub assembly is not serviceable and should not be disassembled.

7. If replacing the hub, use special socket MB991248 and a press, to remove the wheel sensor rotor from the hub.

To install

8. Press the wheel sensor rotor onto the hub.
9. Install the hub to the knuckle and tighten the mounting bolts to 54–65 ft. lbs. (74–88 Nm).
10. Install the brake drum on the hub.
11. If equipped with ABS, install the vehicle speed sensor.
12. Install the wheel assembly and lower the vehicle.

With Disc Brakes

1. Raise and safely support the vehicle.
2. Remove the appropriate wheel assembly.
3. If equipped with ABS, remove the vehicle speed sensor.

4. Remove the caliper and brake pads. Support the caliper out of the way using wire.
5. Remove the brake rotor from the hub assembly.
6. Remove the parking brake shoes as follows:
   a. Remove the upper shoe to anchor springs.
   b. Remove the lower shoe to shoe spring.
   c. Remove the brake shoe hold-down springs.
   d. Disconnect the parking brake cable from the actuating lever.
7. From the back of the knuckle, remove the four bolts securing the hub to the knuckle.
8. Remove the hub and bearing assembly from the knuckle.

NOTE: The hub assembly is not serviceable and should not be disassembled.

9. If replacing the hub, use special socket MB991248 and a press, to remove the wheel sensor rotor from the hub.

To install

10. Press the wheel sensor rotor onto the hub.
11. Install the hub to the knuckle and tighten the mounting bolts to 54–65 ft. lbs. (74–88 Nm).
12. Install the parking brake shoes.
13. Position the rotor on the hub. Install a couple of lug nuts and lightly tighten to hold rotor on hub.
14. Install the caliper holder and place brake pads in holder. Slide the caliper over brake pads and install guide pins. Once caliper is secured, lug nuts can be removed.
15. If equipped with ABS, install the vehicle speed sensor.
16. Install the wheel assembly and lower the vehicle.

CHRYSLER CORP.

Front Wheel Drive

EAGLE-Talon PLYMOUTH-Laser

FIRING ORDERS

NOTE: To avoid confusion, always replace spark plug wires one at a time.

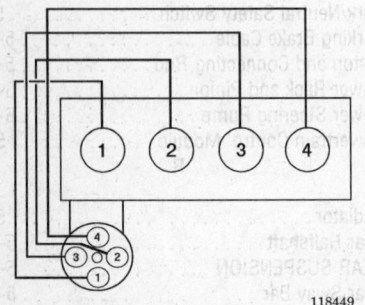

1.8L (VIN B) SOHC Engine
Engine Firing Order: 1–3–4–2
Distributor Rotation: Clockwise

118449

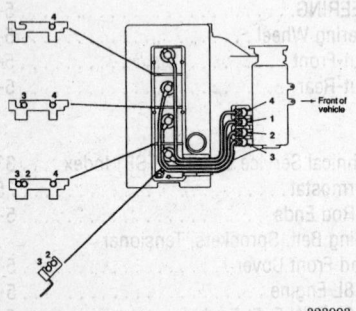

2.0L (VIN E and VIN F) DOHC Engines
Engine Firing Order: 1–3–4–2
Distributorless Ignition System

323993

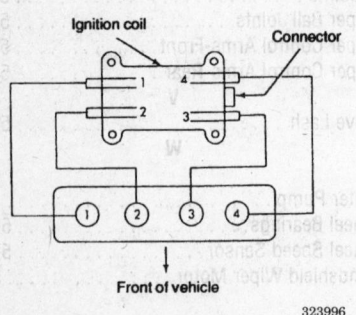

2.0L (VIN Y) DOHC Engine
Engine Firing Order: 1–3–4–2
Distributorless Ignition System

323996

ENGINE ELECTRICAL

NOTE: Disconnecting the negative battery cable on some vehicles may interfere with the functions of the on board computer systems and may require the computer to undergo a relearning process, once the negative battery cable is reconnected.

Distributor

REMOVAL AND INSTALLATION

1.8L (VIN B) Engines

1. Disconnect the negative battery cable. Remove the ignition wire cover, if equipped.
2. Disconnect the distributor harness electrical connector.
3. Unscrew the distributor cap hold-down screws or release the clips and lift off the distributor cap with all ignition wires still connected. Remove the coil wire, if necessary.
4. Matchmark the rotor to the distributor housing and the distributor housing to the engine.

NOTE: Do not crank the engine during this procedure. If the engine is cranked, the matchmark must be disregarded.

5. Remove the hold-down nut.
6. Carefully remove the distributor from the engine.

NOTE: Some engines may be sensitive to the routing of the distributor sensor wires. If routed near the high-voltage coil wire or the spark plug wires, the electromagnetic field surrounding the high voltage wires could generate an occasional disruption of the ignition system operation.

Installation — Timing Not Disturbed

1. Install a new distributor housing O-ring and lubricate with clean oil.
2. Install the distributor in the engine so the rotor is aligned with the housing and the housing is aligned with the engine. Make sure the distributor is fully seated and the distributor shaft is fully engaged.
3. Install the hold-down nut.
4. Connect the distributor harness connectors.
5. Make sure the sealing O-ring is in place, install the distributor cap

and tighten the screws or secure the clips.
6. Connect the negative battery cable.
7. Adjust the ignition timing and tighten the hold-down nut.

Installation — Timing Disturbed

1. Install a new distributor housing O-ring and lubricate with clean oil.
2. Position the engine so the No. 1 piston is at TDC of its compression stroke and the mark on the vibration damper is aligned with **0** on the timing indicator.
3. Align the distributor housing and gear mating marks. Install the distributor in engine so the slot or groove of the distributor's installation flange aligns with the distributor installation stud in the engine block. Make sure the distributor is fully seated. Inspect alignment of the distributor rotor making sure the rotor is aligned with the position of the No. 1 ignition wire in the distributor cap.

NOTE: Make sure the rotor is pointing to where the No. 1 runner originates inside the cap, if equipped, and not where the No. 1 ignition wire plugs into the cap.

4. Install the hold-down nut.
5. Connect the distributor harness connectors.
6. Make sure the sealing O-ring is in place, install the distributor cap and tighten the screws or secure the clips.
7. Connect the negative battery cable.
8. Adjust the ignition timing and tighten the hold-down bolt.

2.0L (VIN E, F and Y) Engines

These vehicles are equipped with a distributorless ignition system.

Ignition Timing

ADJUSTMENT

1.8L (VIN B) Engines

1. Apply the parking brake and block the wheels. Run the engine until the coolant reaches normal operating temperature.
2. Make certain all accessories are **OFF**.
3. Position the steering wheel in the straight ahead position and the gear selector lever in **P** or **N**.
4. Connect a timing light to the engine.

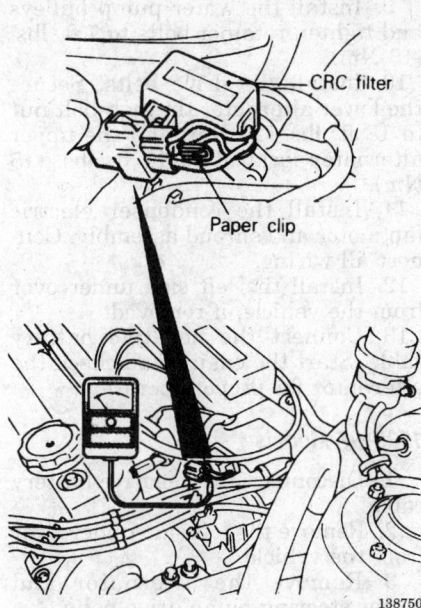

Making tachometer connection — 1.8L (VIN B) engine

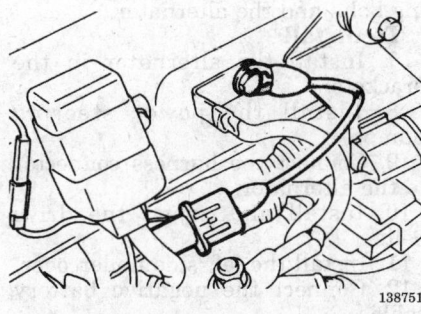

Timing connector identification — 1.8L (VIN B) engine

5. Insert a paper clip into the CRC filter connector (3-pole connector), located in the engine compartment.

6. Connect a tachometer to the inserted clip.

NOTE: During installation of the paper clip, do not separate the connector.

7. Check the curb idle speed. It should be 600–800 rpm.

8. Turn the engine **OFF**. Connect a jumper wire to the terminal for ignition-timing adjustment (located in the engine compartment), and ground it.

9. Start and run the engine at curb idle speed.

10. Check the basic ignition timing and adjust, if necessary. Basic ignition timing should be 5 degrees BTDC.

NOTE: Ignition timing can vary depending on equipment and options. Always check and use underhood decal for specification when available.

11. If the timing is not within specifications, loosen the distributor hold-down bolt and turn the distributor. Turning the distributor to the right retards timing while to the left will advance timing.

12. Tighten the hold-down bolt after adjustment. Recheck the timing and adjust if necessary.

13. Stop the engine and remove the ground for the ignition timing connector.

NOTE: Actual ignition timing may vary, depending on the control mode of the engine control unit. In such case, recheck the basic ignition timing. If there is no deviation, the ignition timing is functioning normally.

14. Start the engine and run at curb idle. Check the actual ignition timing. Actual ignition timing should be 10 degrees BTDC.

NOTE: At altitudes, more than approximately 2,300 ft. (701m) above sea level, the actual ignition timing is further advanced to ensure good combustion.

2.0L (VIN E and F) Engines

1. Apply the parking brake and block the wheels. Run the engine until normal operating temperature is reached.

2. Make certain all accessories are **OFF**.

3. Position the steering wheel in the straight ahead position and the gear selector in **P** or **N**.

4. Connect a timing light to the engine following the manufacturers instructions.

5. Insert a paper clip into the engine revolution speed detection terminal and connect a tachometer to the inserted clip.

6. On 1995–97 models, insert a paper clip from the harness side into the No. 1 pin connector (blue). Connect a primary-voltage-detection type tachometer to the paper clip.

NOTE: Do not use the scan tool. If tested with the scan tool connected to the data link connector, the ignition timing will not be the basic timing.

7. Check the curb idle speed. It should be 650–850 rpm.

NOTE: The tachometer will be reading ½ of the actual engine speed. Multiply the reading by 2 to figure actual engine speed.

8. Stop the engine and connect a jumper wire to the terminal for the ignition timing adjustment to ground.

9. On 1995–97 models, remove the waterproof connector from the ignition timing adjustment connector (brown). Connect the jumper wire with the clip to the ignition timing adjustment terminal, and ground this to the body.

10. Start and run the engine at curb idle speed.

11. Check and adjust the basic ignition timing and adjust, if necessary. Basic timing should be 5 degrees BTDC.

NOTE: Always refer to the underhood label for timing specifications.

12. If the timing was not within specifications, loosen the crank angle sensor retaining nut and turn the sensor until the timing is correct.

13. On 1995–97 models, if the basic ignition timing is not within specifications, inspect the MFI components.

14. Tighten the nut after adjusting. Recheck the timing and adjust if necessary.

15. Stop the engine and remove the ground from the timing connector.

16. Start the engine and run at curb idle. Check the actual ignition timing. It should be 8 degrees BTDC.

2.0L (VIN Y) Engine

It is not necessary to check ignition timing using a timing light, because the crankshaft position is detected directly and ignition timing is controlled electronically.

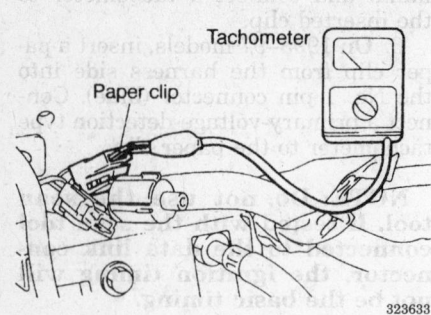

Connecting a tachometer — 1995–97 2.0L VIN F

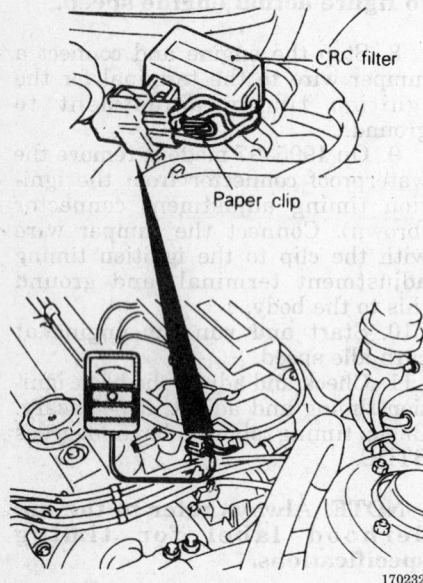

Making tachometer connection — 1993–94 VIN E and F

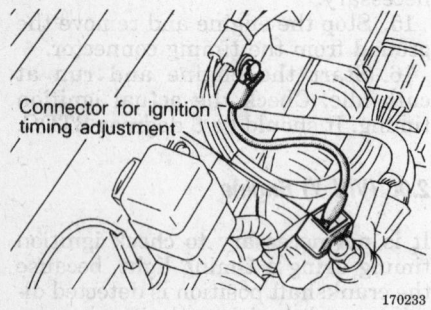

Ignition timing connector location — 1993–94 VIN E and F

Alternator

PRECAUTIONS

Several precautions must be observed with alternator equipped vehicles to avoid damage to the unit.

• If the battery is removed for any reason, make sure it is reconnected with the correct polarity. Reversing the battery connections may result in damage to the 1-way rectifiers.

• When utilizing a booster battery as a starting aid, always connect the positive to positive terminals and the negative terminal from the booster battery to a good engine ground on the vehicle being started.

• Never use a fast charger as a booster to start vehicles.

• Disconnect the battery cables when charging the battery with a fast charger.

• Never attempt to polarize the alternator.

• Do not use test lights of more than 12 volts when checking diode continuity.

• Do not short across or ground any of the alternator terminals.

• The polarity of the battery, alternator and regulator must be matched and considered before making any electrical connections within the system.

• Never separate the alternator on an open circuit. Make sure all connections within the circuit are clean and tight.

• Disconnect the battery ground terminal when performing any service on electrical components.

• Disconnect the battery if arc welding is to be done on the vehicle.

REMOVAL AND INSTALLATION

1993–94 Models

1. Disconnect the negative battery cable. Remove the left side undercover from the vehicle.
2. If equipped with A/C, remove the condenser electric fan motor and shroud assembly.
3. Remove the drive belts.
4. Remove both water pump pulleys and the alternator top brace.
5. Disconnect the alternator wiring and remove the alternator from the vehicle.

To install:

6. Install the alternator in position and connect the electrical harness.
7. Install the alternator top brace and tighten the mounting bolt to 20 ft. lbs. (27 Nm).

8. Install the alternator mounting bolt loosely.
9. Install the water pump pulleys and tighten retainer bolts to 7 ft. lbs. (10 Nm).
10. Install the drive belts. Secure the lower alternator through-bolt nut to 18 ft. lbs. (25 Nm) and the upper alternator lockbolt to 11 ft. lbs. (15 Nm).
11. Install the condenser electric fan motor and shroud assembly. Connect all wiring.
12. Install the left side undercover from the vehicle, if removed.
13. Connect the negative battery cable. Start the engine and check the alternator for proper operation.

1995–97 Models

1. Disconnect the negative battery cable.
2. Remove the left side undercover from the vehicle.
3. Remove the alternator and power steering pump drive belts.
4. Remove the harness connector from the alternator.
5. If necessary, remove the power steering pump and secure the pump above the engine mount bracket. It is not necessary to disconnect the hoses.
6. Remove the alternator mounting bolts and the alternator.

To install:

7. Install the alternator in the brackets.
8. Install the power steering pump.
9. Connect the harness connector to the alternator.
10. Install and adjust the drive belts.
11. Install the left side undercover.
12. Connect the negative battery cable.

Drive Belt

REMOVAL AND INSTALLATION

1.8L (VIN B) Engine

To adjust the tension on a drive belt, loosen the adjusting bolt or fixing bolt locknut on the desired component, bracket or tension pulley. Move the component or turn the adjusting bolt to adjust the belt tension. Once the desired value is reached, secure the bolt or locknut and recheck tension.

Belt replacement is similar to adjustment, with the exception the belt will have to be properly routed around the pulleys. It is important to note the routing of the belt before re-

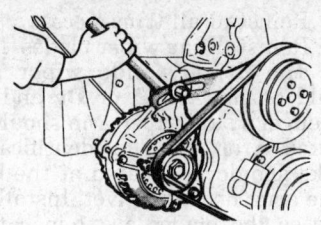

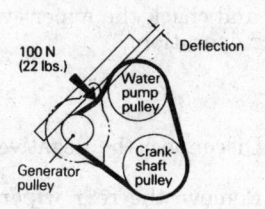

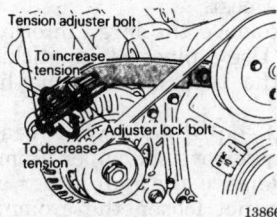

Belt adjustment procedure — 1.8L (VIN B) engine

moval. For individual belt replacement, start with the outer most belt. If a removed belt is to be reused, be certain to mark the direction of rotation on the belt, to extend belt life.

2.0L (VIN E and F) Engine

Alternator and Water Pump Belt

1. Disconnect the negative battery cable.
2. Loosen the lockbolt.
3. Loosen the pivot bolt.
4. Turn the adjusting bolt to loosen the tension on the belt.
5. Remove the belt.
To install:
6. Install the belt on the pulleys.
7. Turn the adjusting bolt until a tension of 110.2–154.3 lbs.(490–686 Nm) for a new belt or 88.2 lbs. (392 Nm) for a used belt is indicated on the tension gauge. Another method is to apply a force of 22 lbs. (98 Nm) to the belt midway between the water pump and the generator and measure the belt deflection. Belt deflection should be 0.30–0.35 in. (7.5–9.0mm) for a new belt or 0.39 in. (10.0mm) for a used belt.
8. Tighten the pivot bolt to 15–18 ft. lbs. (20–25 Nm).
9. Tighten the lockbolt to 8.7–11 ft. lbs. (12–15 Nm).

10. Now tighten the adjusting bolt to 7.2 ft. lbs. (9.8 Nm).
11. Connect the negative battery cable.

Power Steering Belt

1. Disconnect the negative battery cable.
2. Remove the alternator belt.
3. Loosen the four power steering pump mounting bolts.
4. Pivot the power steering pump towards the water pump and remove the belt.
To install:
5. Install the belt on the proper pulleys.
6. Pry the power steering pump away from the water pump pulley until a tension of 110.2–154.3 lbs.(490–686 Nm) for a new belt or 77.2–99.2 lbs. (343–441 Nm) for a used belt is indicated on the tension gauge. Another method is to apply a force of 22 lbs. (98 Nm) to the belt midway between the water pump and the power steering pulley and measure the belt deflection. Belt deflection should be 0.18–0.22 in. (4.5–5.5mm) for a new belt or 0.24–0.28 in. (6.0–7.0mm) for a used belt.
7. Tighten mounting bolt A to 21 ft. lbs. (28 Nm).
8. Tighten mounting bolts B and D to 21 ft. lbs. (28 Nm).
9. Tighten mounting bolt C to 16 ft. lbs. (22 Nm).
10. Install the alternator belt.
11. Connect the negative battery cable.

A/C Compressor Belt

1. Disconnect the negative battery cable.
2. Remove the generator belt.
3. Loosen the lockbolt on the tension pulley.
4. Turn the adjusting bolt to loosen the belt.
5. Remove the belt.
To install:
6. Install the belt on the proper pulleys.
7. Turn the adjusting bolt until a tension of 86.0–99.2 lbs.(382–411 Nm) for a new belt or 57.3–75.0 lbs. (255–333 Nm) for a used belt is indicated on the tension gauge. Another method is to apply a force of 22 lbs. (98 Nm) to the belt midway between the A/C compressor and the crankshaft pulley or between the A/C compressor and the tension pulley and measure the belt deflection. Belt deflection should be 0.22–0.24 in. (5.5–6.0mm) for a new belt or 0.26–0.30 in. (6.5–7.5mm) for a used belt.

WARNING

Measure the belt tension only after one full rotation of the crankshaft in the forward direction (right).

8. Tighten the lockbolt to 17–20 ft. lbs. (23–26 Nm).
9. Install the generator belt.
10. Connect the negative battery cable.

2.0L (VIN Y) Engine

Alternator Belt

1. Disconnect the negative battery cable.
2. Remove the power steering pump and A/C compressor belt.
3. Loosen the lock nut.
4. Loosen the pivot bolt.
5. Turn the adjusting bolt to loosen the tension on the belt.
6. Remove the belt.
To install:
7. Install the belt on the proper pulleys.
8. Turn the adjusting bolt until a tension of 110–160 lbs. (490–712 Nm) for a new belt or 90–110 lbs. (400–490 Nm) for a used belt is indicated on the tension gauge. Another method is to apply a force of 22 lbs. (98 Nm) to the belt midway between the water pump and the generator and measure the belt deflection. Belt deflection should be 0.30–0.41 in. (7.5–10.5mm) for a new belt or 0.35–0.47 in. (9.0–12.0mm) for a used belt.
9. Tighten the pivot bolt to 40 ft. lbs. (54 Nm).
10. Tighten the lock nut to 45 ft. lbs. (61 Nm).
11. Install and adjust the power steering pump and A/C compressor belt.
12. Connect the negative battery cable.

Power Steering Pump and A/C Compressor Belt

1. Disconnect the negative battery cable.
2. Loosen the nut in the center of the tension pulley.
3. Turn the adjusting bolt to loosen the belt.
4. Remove the belt.
To install:
5. Install the belt on the proper pulleys.
6. Turn the adjusting bolt until a tension of 136.7–158.7 lbs. (608–706 Nm) for a new belt or 92.62–114.6 lbs. (412–510 Nm) for a used belt is indicated on the tension gauge. Another method is to apply a force of 22 lbs. (98 Nm) to the belt midway between

the crankshaft pulley and the power steering pulley and measure the belt deflection. Belt deflection should be 0.32–0.35 in. (8.0–9.0mm) for a new belt or 0.39–0.43 in. (10.0–11.0mm) for a used belt.

—— **WARNING** ——

Make sure the power steering pump is mounted at the most forward position in the mounting bracket.

7. Tighten the tension pulley nut.

Starter

REMOVAL AND INSTALLATION

1. Disconnect the negative battery cable.
2. Remove the battery and tray from the engine compartment.
3. Disconnect the speedometer cable at the transaxle.
4. If equipped with 1.8L engine, remove the bracket on the lower side if the intake manifold.
5. Disconnect the starter motor electrical connections.
6. Remove the starter motor mounting bolts and remove the starter.

To install:

7. Clean the starter and engine mating surfaces. Install the starter motor and secure with the retainer bolts. Tighten to 25 ft. lbs. (34 Nm).
8. Connect the electrical harness to the starter.
9. Install the intake manifold stay, if removed. Tighten the retainers to 18 ft. lbs. (25 Nm).
10. Install the speedometer cable at the transaxle. Install the tray and battery.
11. Start the vehicle to test the starter.

CHASSIS ELECTRICAL

Blower Motor

REMOVAL AND INSTALLATION

1. Disconnect the negative battery cable.
2. Remove the right side duct, if equipped.

3. On 1995–97 vehicles with A/C and non-turbo engines, disconnect and remove the automatic compressor ECM.
4. Remove the molded hose from the blower assembly.
5. Remove the blower motor assembly.
6. Remove the packing seal.
7. Remove the retaining nut and fan.

To install:

8. Check that the blower motor shaft is not bent and that the packing (sealing material) is in good condition. Clean all parts of dust, etc.
9. Assemble the motor and fan. Install the blower motor, then connect the motor terminals to the battery voltage. Check that the blower motor operates smoothly. Then, reverse the polarity and check that the blower motor operates smoothly in the reverse direction.
10. On 1995–97 vehicles with A/C and non-turbo engines, install and connect the automatic compressor ECM.
11. Install the molded hose and duct, if removed.
12. Connect the negative battery cable and check the entire climate control system for proper operation.

Windshield Wiper Motor

REMOVAL AND INSTALLATION

Front

1. Disconnect the negative battery cable.

NOTE: Disconnect the linkage at the motor.

2. Remove the windshield wiper arms by unscrewing the cap nuts and lifting the arms from the linkage post.
3. Remove the front deck garnish panel.
4. Remove the air inlet trim pieces.
5. Remove the hole cover.
6. Remove the wiper motor by loosening the mounting bolts, removing the motor assembly, then disconnecting the linkage.

NOTE: The installation angle of the crank arm and motor has been factory set. Do not remove unless necessary. If removal is required, remove only after marking the mounting positions.

To install:

7. Install the windshield wiper motor and connect the linkage.

8. Reinstall all trim pieces.
9. Reinstall the wiper blades. Note that the driver's side wiper arm should be marked **D** or **Dr** and the passenger's side wiper arm should be marked **A** or **As**. The identification marks should be located at the base of the arm, near the pivot. Install the arms so the blades are 1 inch from the garnish molding when parked.
10. Connect the negative battery cable and check the wiper system for proper operation.

Rear

1. Disconnect the negative battery cable.
2. Remove the rear wiper arm by removing the cover, unscrewing the nut and lifting the arm from the linkage post.
3. Remove the large interior trim panel. Use a plastic trim stick to unhook the trim clips of the liftgate trim.
4. If equipped with rear air spoiler, remove the wiper grommet.
5. Remove the rear wiper assembly. Do not loosen the grommet for the wiper post.

To install:

6. Install the motor and grommet. Mount the grommet so the arrow on the grommet is pointing upward.
7. Install the wiper arm.
8. Connect the negative battery cable and check the rear wiper for proper operation.

Combination Switch

REMOVAL AND INSTALLATION

1993–94 Models

1. Disconnect the negative battery cable.
2. Remove the knee protector panel under the steering column, then the upper and lower column cover.
3. Remove the horn pad attaching screw on the under side of the steering wheel and remove the horn pad by pushing the pad upward.
4. Matchmark and remove the steering wheel.

NOTE: Do not hammer on the steering wheel to remove it or the collapsible mechanism may be damaged.

5. Remove the lap cooler ducts.
6. Remove the band retaining the switch wiring.

7. Disconnect all connectors, remove the wiring clip and remove the switch assembly.

To install:

8. Install the switch assembly and secure the clip. Make sure no wires are pinched or out of place.

9. Install the lap cooler ducts.

10. Install the column covers and knee protector.

11. Install the steering wheel.

12. Connect the negative battery cable and check all functions of the combination switch for proper operation.

1995–97 Models

— **CAUTION** —
The Air Bag system must be disarmed before removing the Air Bag module. Failure to do so may cause accidental deployment, property damage or personal injury.

1. Turn the front wheels to the straight ahead position and center the steering wheel. Turn the ignition switch to the lock position.

2. Disconnect the negative battery cable and wait at least 90 seconds before performing any work.

3. Remove the Air Bag module.

4. Remove the steering wheel.

5. Remove the lower trim panel and the lap duct.

6. Remove the upper and lower steering column cover and pad.

NOTE: Do not remove the clockspring from the combination switch unless absolutely necessary.

7. Disconnect the harness from the combination switch and the clockspring assembly.

8. Remove the combination switch mounting bolts.

9. Remove the combination switch with the clockspring assembly.

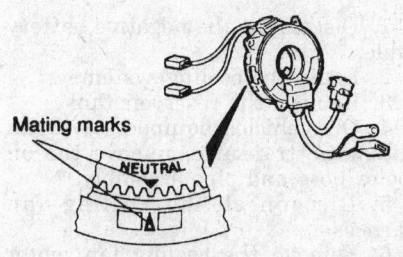

Mating marks

327592

Align the clockspring mating marks

To install:

10. Make sure the mating marks on the clockspring are aligned and install the combination switch and clockspring assembly.

11. Connect the harness to the clockspring and combination switch.

12. Install the steering column covers.

13. Install the lap duct and the lower trim panel.

14. Install the steering wheel.

15. Install the Air Bag module.

16. Connect the negative battery cable and turn the ignition switch **ON**. The Air Bag warning light should light for about seven seconds and then go off. If the light does not go off, check for trouble codes.

Ignition Switch and Lock Cylinder

REMOVAL AND INSTALLATION

— **CAUTION** —
The Air Bag system must be disarmed before removing the steering wheel. Failure to do so may cause accidental deployment, property damage or personal injury.

1. Disconnect the negative battery cable.

2. Remove the lower instrument panel knee protector.

3. Remove the steering wheel assembly.

NOTE: Use proper steering wheel puller equipment when removing the steering wheel. The use of a hammer for removal could damage the collapsible mechanism within the column.

4. Remove the upper and lower steering column cover.

5. Remove the clip that holds the wiring harness against the steering column.

6. Insert the key into the steering lock cylinder and turn to the **ACC** position.

7. With a small pointed tool, push the lock pin of the steering lock cylinder inward and pull the lock cylinder out.

8. Remove the key reminder switch, if equipped.

9. Unplug the ignition switch harness connector. Remove the ignition switch mounting screws and pull the

switch from the steering lock cylinder.

NOTE: To remove the steering lock, use a hacksaw or equivalent to cut the special bolts through the bracket. The bracket and bolts must be replaced with new ones.

To install:

10. Install the lock cylinder in the lock cylinder bracket. Make sure the cylinder operates properly before breaking off the heads of the special bolts if the bracket was replaced.

11. Install the ignition switch onto the rear of the lock cylinder housing. Be sure to align the keyway of the ignition switch with lock cylinder.

12. Connect harness connections and install the wiring clip.

13. Install the steering column upper and lower covers.

14. Install the knee protector.

15. Connect the negative battery cable and check the ignition switch and lock for proper operation.

Park/Neutral Safety Switch

REMOVAL AND INSTALLATION

1. Disconnect the negative battery cable.

2. Disconnect the selector cable from the lever.

3. Remove the two retaining screws and lift off the switch.

To install

4. Mount and position the switch. Do not tighten the bolts until the switch is adjusted.

5. Connect the selector cable and adjust switch.

6. After installation and adjustment make sure the engine only starts in the **P** and **N** selections. Also check that the reverse lights operate only in the **R** selection.

Powertrain Control Module

REMOVAL AND INSTALLATION

1. Disconnect negative battery cable.

— **WARNING** —
A grounded wrist strap should be used to prevent static discharge to the PCM. Static discharge can easily destroy the electronic components inside the PCM.

2. Remove both center console side panels.

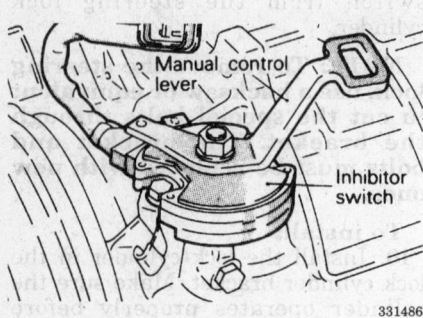

Park/Neutral switch location

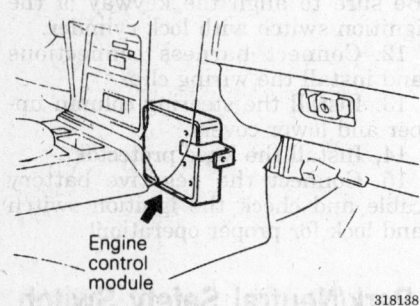

Powertrain control module location

3. Unplug the wiring connector and remove the mounting hardware. Slide the control unit out the side.

To install:

4. Slide the PCM into the side of the center console, and connect the harness.

5. Install the side panels. Connect the negative battery cable.

ENGINE COOLING

Radiator

REMOVAL AND INSTALLATION

1. Disconnect the negative battery cable.

2. Drain the cooling system.

3. Disconnect the overflow tube. Some vehicles may also require removal of the overflow tank.

4. Disconnect upper and lower radiator hoses. Matchmark the upper radiator hose to assure a proper installation.

5. Disconnect the electrical connectors for the cooling fan and A/C condenser fan, if equipped. Remove the fan assembly from the engine compartment.

6. Disconnect the thermosensor wires.

7. Disconnect and plug the automatic transaxle cooler lines, if equipped.

8. Remove the upper radiator mounts and lift out the radiator assembly.

To install:

9. Install the radiator and fan assembly.

10. Connect the automatic transaxle cooler lines, if disconnected.

11. Connect the thermowires.

12. Install the radiator hoses. Make sure to position the upper radiator hose correctly.

13. Install the overflow tube and reservoir, if removed.

14. Fill the system with coolant.

15. Connect the negative battery cable, run the vehicle until the thermostat opens, fill the radiator completely and check the automatic transaxle fluid level, if equipped.

16. Once the vehicle has cooled, recheck the coolant level.

Water Pump

REMOVAL AND INSTALLATION

1. Disconnect the negative battery cable.

2. Drain the cooling system. Remove the drive belt.

3. Remove the engine undercover.

4. Disconnect the clamp from the power steering hose. Remove the tensioner pulley bracket.

5. Support the engine and remove the engine mount bracket.

6. Remove both the outer and the inner timing belts. If reusing old belt, mark the direction of rotation on the outer belt surface.

7. Remove the alternator brace from the front of the water pump.

8. Remove the water pump and gasket from the engine.

9. Remove the O-ring from the inlet pipe. Clean all mating surfaces and inspect for cracks or other damage.

To install:

10. Install a new O-ring into the front end of the water inlet pipe. Wet with clean antifreeze only.

11. Install the gasket and pump assembly and tighten the bolts. Note the marks on the bolt heads. Those

marked **4** should be tightened to 9–11 ft. lbs. (12–15 Nm). Those bolts marked **7** should be tightened from 15–19 ft. lbs. (20–27 Nm).

12. Connect the hoses to the pump.

13. Install the timing belts.

14. Install the engine undercover.

15. Fill the system with coolant.

16. Connect the negative battery cable, run the vehicle until the thermostat opens and fill the radiator completely.

17. Recheck the coolant level.

Thermostat

REMOVAL AND INSTALLATION

1. Disconnect the negative battery cable.

2. Drain the cooling system.

3. Disconnect the upper radiator and overflow hoses from the thermostat housing.

4. Remove the housing and gasket.

5. Remove the thermostat taking note of the original position in the housing.

To install:

NOTE: In order to prevent leakage, make sure both mating surfaces are clean and free of all old gasket material.

6. Install the thermostat so the flange seats tightly.

7. Use a new gasket and install the thermostat housing. Tighten the bolts to 12–14 ft. lbs. (17–20 Nm).

8. Connect the hoses and fill the system with coolant.

9. Connect the negative battery cable, run the vehicle until the thermostat opens and fill the radiator completely.

10. Recheck the coolant level.

Electric Cooling Fan

REMOVAL AND INSTALLATION

1. Disconnect the negative battery cable.

2. Drain the cooling system.

3. Remove the reservoir tank.

4. On vehicles equipped with an automatic transaxle, remove the oil cooler hose and pipe assembly.

5. Disconnect the cooling fan harness.

6. Remove the cooling fan motor and shroud.

7. Remove the fan from the motor, then remove the motor from the shroud.

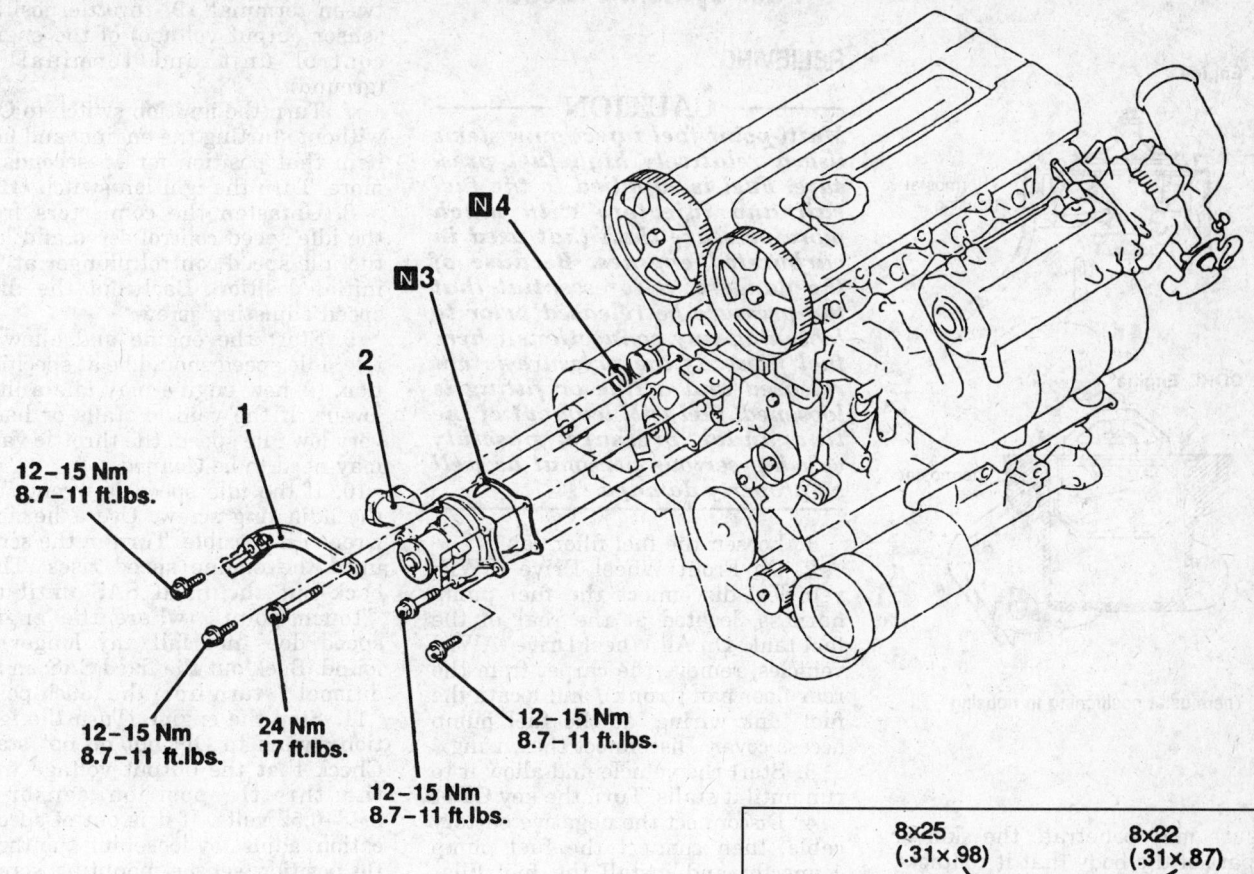

Removal steps

1. Generator brace
2. Water pump
3. Water pump gasket
4. O-ring

12–15 Nm
8.7–11 ft.lbs.

12–15 Nm
8.7–11 ft.lbs.

12–15 Nm
8.7–11 ft.lbs.

24 Nm
17 ft.lbs.

12–15 Nm
8.7–11 ft.lbs.

12–15 Nm
8.7–11 ft.lbs.

8×25
(.31×.98)

8×22
(.31×.87)

8×65
(.31×2.56)

8×22
(.31×.87)

8×14
(.31×.55)

Bolt diameter × length: mm (in.)

319006

Water pump mounting — 2.0L (VIN F and E) engines

To install:

8. Install the cooling fan motor to the fan shroud.

9. Install the fan to the motor.

10. install the shroud to the radiator.

11. Connect the fan motor harness.

12. On vehicles equipped with an automatic transaxle, connect the oil cooler pipe and hose assembly.

13. Install the radiator reservoir tank and tighten the screws to 9 ft. lbs. (12 Nm).

14. Connect the negative battery cable.

15. Refill the cooling system and bleed.

FUEL SYSTEM

Fuel System Service Precautions

Safety is the most important factor when performing not only fuel system maintenance but any type of maintenance. Failure to conduct maintenance and repairs in a safe manner may result in serious personal injury or death. Maintenance and testing of the vehicle's fuel system components can be accomplished safely and effectively by adhering to the following rules and guidelines.

• To avoid the possibility of fire and personal injury, always disconnect the negative battery cable unless the repair or test procedure requires that battery voltage be applied.

• Always relieve the fuel system pressure prior to disconnecting any fuel system component (injector, fuel rail, pressure regulator, etc.), fitting or fuel line connection. Exercise extreme caution whenever relieving fuel system pressure to avoid exposing skin, face and eyes to fuel spray. Please be advised that fuel under

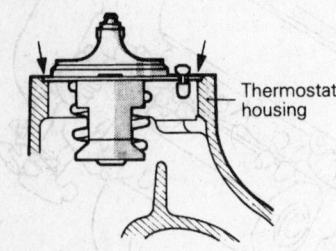

1.8L Engine

Thermostat housing

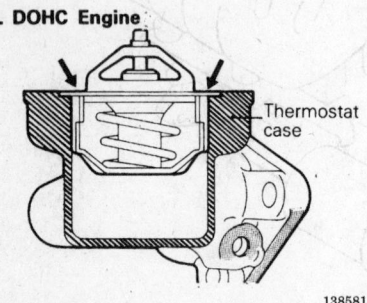

2.0L DOHC Engine

Thermostat case

138581

Thermostat positioning in housing

pressure may penetrate the skin or any part of the body that it contacts.

• Always place a shop towel or cloth around the fitting or connection prior to loosening to absorb any excess fuel due to spillage. Ensure that all fuel spillage (should it occur) is quickly removed from engine surfaces. Ensure that all fuel soaked cloths or towels are deposited into a suitable waste container.

• Always keep a dry chemical (Class B) fire extinguisher near the work area.

• Do not allow fuel spray or fuel vapors to come into contact with a spark or open flame.

• Always use a backup wrench when loosening and tightening fuel line connection fittings. This will prevent unnecessary stress and torsion to fuel line piping. Always follow the proper torque specifications.

• Always replace worn fuel fitting O-rings with new. Do not substitute fuel hose or equivalent, where fuel pipe is installed.

Fuel System Pressure

RELIEVING

--- CAUTION ---

Multi-point fuel injection systems use a relatively high fuel pressure. Fuel is supplied to the fuel rail and injectors with much more pressure than that used in carbureted engines. Because of the dangers, it is essential that the pressure be released prior to loosening any connections where fuel flows. If the pressure is not released and a line or fitting is loosened, fuel will leak out of the line under pressure, possibly causing serious personal as well as property damage.

1. Loosen the fuel filler cap.
2. On Front Wheel Drive (FWD) vehicles, disconnect the fuel pump harness, located at the rear of the fuel tank. On All Wheel Drive (AWD) vehicles, remove the carpet from the rear floor pan (trunk), and locate the fuel tank wiring, at the fuel pump access cover. Disconnect the wiring.
3. Start the vehicle and allow it to run until it stalls. Turn the key **OFF**.
4. Disconnect the negative battery cable, then connect the fuel pump connector and install the fuel filler cap.

--- WARNING ---

Always wrap shop towels around a fitting that is being disconnected to absorb residual fuel in the lines.

Idle Speed

ADJUSTMENT

1.8L (VIN B) Engines

1. Warm the engine to normal operating temperature. Make sure all accessories are **OFF**. The transaxle should be in **N** or **P**, and the steering wheel should be straight ahead.
2. Check the ignition timing and adjust, if necessary.
3. Connect a tachometer to the CRC filter connector. Use a paper clip for a tach adapter.
4. Run the engine for more than 10 seconds at 2000–3000 rpm. Allow the engine to idle for 2 minutes. Check the idle rpm. Curb idle should be 750 rpm.
5. If adjustment is required, slacken the accelerator cable.

6. Connect a digital voltmeter between terminal **19** (throttle position sensor output voltage) of the engine control unit and terminal **24** (ground).
7. Turn the ignition switch to **ON**, without starting the engine, and hold it in that position for 15 seconds or more. Turn the ignition switch **OFF**.
8. Unfasten the connectors from the idle speed control servo and lock the idle speed control plunger at the initial position. Back out the fixed speed adjusting screw.
9. Start the engine and allow to idle. Idle speed should be at specification. (A new engine may idle a little lower). If the vehicle stalls or has a very low idle speed, the throttle valve may need to be cleaned.
10. If the idle speed is wrong, use the adjusting screw. Use a hexagon wrench if possible. Turn in the screw until the engine speed rises. Then back out the fixed SAS until the ""touch point" (where the engine speed does not fall any longer) is found. Back out the fixed SAS an additional ½ turn from the touch point.
11. Stop the engine. Turn the ignition switch to **ON** but do not start. Check that the output voltage from the throttle position sensor is 0.48–0.52 volts. If it is out of specification, adjust by loosening the throttle position sensor mounting screws and rotating the throttle position sensor. Turning the throttle position sensor clockwise increases the output voltage. After adjustment, tighten screws firmly.
12. Turn the ignition switch **OFF**.
13. Adjust the free-play of the accelerator cable, reconnect the harness to the idle speed control servo. Remove the voltmeter.
14. Start the engine and check the curb idle. It should be 700 rpm.
15. Turn the ignition switch to **OFF**, disconnect the negative battery cable for more than 10 seconds and reconnect. This clears any trouble codes introduced during testing.
16. Restart the engine, allow to run for 5 minutes and check the idle quality.

2.0L (VIN E and F) Engines

1. Warm the engine to normal operating temperature. Leave all accessories **OFF**. The transaxle should be in **N**. The steering wheel should be straight ahead.
2. Check the ignition timing and adjust, if necessary.
3. Connect a tachometer to the single pin connector terminal under the hood.

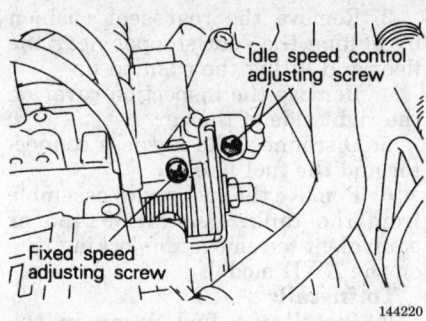

Idle speed adjusting screw location — 1.8L (VIN B) engines

4. Run the engine for more than 10 seconds at 2000–3000 rpm. Allow the engine to idle for 2 minutes. Check the idle rpm. Curb idle should be 750 rpm.

5. If adjustment is required, disconnect the waterproof female connector used for ignition timing adjustment. Connect this terminal to ground using a jumper wire.

6. Locate the self-diagnosis terminal under the dashboard and connect terminal No. 10 to ground with a jumper wire.

7. Start the engine and allow to idle. Check the idle speed. If the idle speed deviates, check the following:

  a. On new engines, break-in should take approximately 300 miles.

  b. A deposit buildup on the throttle valve which must be cleaned.

  c. The idle position switch (fixed speed adjusting screw) position has changed. Adjust the switch.

  d. Leakage resulting from deterioration of the Fast-Idle Air Valve (FIAV). The throttle body will have o be replaced.

8. Turn the ignition switch **OFF**. Disconnect the jumper wires and tachometer.

9. Restart the engine, allow to run for 5 minutes and check for the idle quality.

2.0L (VIN F) Engines

The engine idle speed adjusting screw has been set at the factory. No subsequent adjustment should be necessary.

Any adjustment should be made only after confirming that the spark plugs, injectors, idle air servo control, cylinder compression, and all engine systems affecting idle speed, are all normal.

1. Make sure the engine is at normal operating temperature, and all accessories are turned **OFF** The transaxle should be in **P** or **N**.

2. Connect a scan tool to the 16-pin data link connector.

3. If a scan tool is not used, connect a primary voltage detection tachometer to the 1-pin (blue) engine speed detection connector with a paper clip and ground terminal 1 of the 16-pin data link connector.

4. Remove the waterproof female connector from the ignition timing adjustment connector and ground the terminal with a jumper wire.

5. Start the vehicle and let it idle. Correct idle speed should be 700–800 rpm.

NOTE: Engine speed may be 20–100 rpm less than specification on new vehicles driven less than 300 miles (500km).

If the engine stalls or the idle speed is low, check the throttle valve for deposits and clean if necessary.

6. Adjust the idle speed using the engine speed adjusting screw.

NOTE: If the idle speed is high, even after the screw is fully closed, check the fixed SAS for any indication that it has been moved. If it has been moved, adjust it. If the idle speed is still high, check the fast idle air valve (FIAV) for deterioration. If the valve is deteriorated, replace the throttle body.

7. Turn the ignition **OFF**.

8. Disconnect the jumper wire from the adjustment terminal and install the connector to the original condition.

9. Start the engine, let it run for about 10 minutes and confirm proper idle speed.

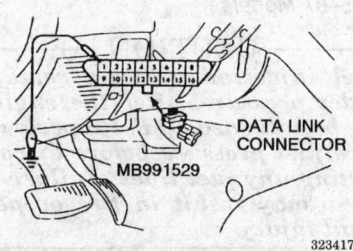

Data link connector location — 2.0L (VIN F) engines

2.0L (VIN Y) Engines

Idle speed is not adjustable on these engines.

Fuel Filter

REMOVAL AND INSTALLATION

A replaceable fuel filter is located in the engine compartment, on the firewall.

————— **CAUTION** —————
Do not use conventional fuel filters, hoses or clamps when servicing fuel injection systems. They are not compatible with the injection system and could fail, causing personal injury or damage to the vehicle. Use only hoses and clamps specifically designed for fuel injection.
————————————————

1. Relieve the fuel system pressure.

NOTE: Wrap shop towels around the fitting that is being disconnected to absorb residual fuel in the lines.

2. On turbo models, remove the battery and the air intake hose.

3. Hold the fuel filter nut securely with a backup wrench. Cover the hoses with shop towels and remove the eye bolt. Discard the gaskets.

4. Separate the flare nut connection at the filter. Discard the gaskets.

5. Remove the mounting bolts and fuel filter from the vehicle.

To install:

6. If equipped with flare fitting, install a new O-ring and tighten the fitting by hand before installing the filter to the vehicle.

7. Install the filter to the bracket finger-tight. Movement of the filter will ease attachment of the fuel lines.

8. Install new O-rings and connect the high pressure hose and eye bolt, then the main pipe and eye bolt. While holding the fuel filter nut, tighten the eye bolts to 22 ft. lbs. (30 Nm). Tighten the flare nut to 25 ft. lbs. (35 Nm).

9. Tighten the filter mounting bolts fully.

10. On turbo models, reinstall the battery and the air intake hose.

11. Connect the negative battery cable, install the fuel filler cap, turn the key **ON** to pressurize the fuel system and check for leaks.

Fuel Pump

REMOVAL AND INSTALLATION

1993-94 Models

FWD Models

1. Relieve the fuel system pressure. Disconnect negative battery cable.
2. Raise and safely support the vehicle.
3. Drain the fuel from the fuel tank.

Remove the electrical connectors at the fuel pump. Make sure there is enough slack in the electrical harness of the fuel gauge unit to allow for the fuel tank to be lowered slightly. If not, label and disconnect the electrical harness at the fuel gauge unit.

—— CAUTION ——
Cover the high pressure fuel hose with rags to prevent splash of fuel caused by residual pressure in the fuel pipe line.

4. Disconnect the high pressure line at the pump.
5. Loosen the self-locking nuts on the tank support straps to the end of the stud bolts.
6. Remove the right side lateral rod attaching bolt and disconnect the arm from the right body coupling. Lower the lateral rod and suspend from the axle beam using wire.
7. Remove the holding bolt and gasket from the base of the tank.
8. Remove the fuel pump assembly.
To install:
9. Align the 3 projections on the packing with the holes on the fuel pump. Have the nipples on the pump facing the same direction as before removal.
10. Install the holding bolt through the bottom of the tank. Make sure the gasket on the bolt is replaced and is not pinched during installation. Tighten to 10 ft. lbs. (14 Nm).
11. Install the right side lateral rod and attaching bolt into the body coupling. Tighten loosely only.
12. Tighten the self-locking nuts on the tank support straps until the tank is seated fully. Tighten nuts to 22 ft. lbs. (31 Nm).
13. Install the high pressure fuel hose connector and tighten to 29 ft. lbs. (40 Nm).
14. Install the electrical connectors onto the fuel pump and gauge unit assemblies.

15. Lower the vehicle. Tighten the lateral rod attaching bolt to 58–72 ft. lbs. (80–100 Nm).
16. Refill the fuel tank.
17. Connect the negative battery cable and check the system for proper operation and leaks.

AWD Models

1. Relieve the fuel system pressure. Disconnect the negative battery cable.
2. The fuel pump is located in the fuel tank. Remove the hole cover located in the rear floor pan.
3. Partially drain the fuel tank.
4. Remove the electrical connector from the fuel pump.
5. Remove the overfill limiter (two-way valve), as required.

—— CAUTION ——
Cover the hose connection with a shop towel to prevent any splash of fuel due to residual pressure in the fuel pipe.

6. Remove the high pressure fuel hose connector.
7. Remove the fuel pump and gauge assembly from the tank. Note the position of pump prior to removal.
To install:
8. Align the 3 projections on the packing with the holes on the fuel pump. Have the nipples on the pump facing the same direction as before removal. Install the retainers and tighten to 2 ft. lbs. (3 Nm).
9. Install the high pressure hose connection and tighten to 29 ft. lbs. (40 Nm).
10. Install the overfill limiter (two-way valve) and the electrical connector to the fuel pump.
11. Fill the fuel tank.
12. Reconnect the negative battery cable and check the entire system for leaks.
13. Install MOPAR Rope Caulk Sealer part 4026044 or equivalent, to the rear floor pan and install the cover into place.

1995-97 Models

—— CAUTION ——
Fuel injection systems remain under pressure, after the engine has been turned OFF. Properly relieve fuel pressure before disconnecting any fuel lines. Failure to do so may result in fire or personal injury.

1. Relieve the fuel system pressure.
2. Disconnect the negative battery cable.

3. Remove the rear seat cushion by pulling the seat stopper near the floor and lifting the cushion up.
4. Remove the inspection cover on the right side of the car.
5. Disconnect the harness connector and the fuel lines.
6. Remove the fuel pump assemble from the tank. Use MB991480 or equivalent to remove the locking ring on the AWD model.
To install:
7. Install the fuel pump in the tank.
8. Connect the hoses and the harness connector.
9. Install the inspection cover.
10. Install the rear seat.
11. Connect the negative battery cable.

Fuel Injector

REMOVAL AND INSTALLATION

1. Relieve the fuel system pressure.
2. Disconnect the PCV hose from the valve cover and the breather hose at the opposite end of the valve cover.
3. Remove the bolts holding the high pressure fuel line to the fuel rail. Plug the line to keep out dirt and debris.
4. Remove the vacuum hose from the fuel pressure regulator.
5. Disconnect the fuel return hose from the pressure regulator. Remove the fuel pressure regulator mounting bolts and remove from the fuel rail.
6. Disconnect the electrical connector from each injector.
7. Remove the bolt(s) holding the fuel rail to the manifold. Carefully lift the rail up and remove with the injectors attached. Place the rail and injectors in a safe location on the workbench; protect the tips from dirt and/or impact.
8. Remove and discard the injector insulators from the manifold. The insulators are not reusable.
9. Remove the injectors from the fuel rail by pulling straight out. Make certain the grommet and O-ring come off with the injector.
To install:
10. Install a new insulator in each injector port.
11. Install a new grommet and O-ring coated with clean oil.
12. If the fuel pressure regulator was removed, replace the O-ring with a new one and coat with clean oil. Insert the regulator into the rail, check that it can be rotated freely. If it does not, remove it and inspect the

O-ring for deformation or jamming. Align the mounting holes and tighten the bolts to 8 ft. lbs. (11 Nm).

13. Install the injectors into the fuel rail, constantly turning the injector left and right during installation. The injector should turn freely in the rail. If it does not, remove the injector and inspect the O-ring for deformation or damage.

14. Install the delivery pipe and injectors. Install the fuel rail retaining bolts and tighten them to 8 ft. lbs. (11 Nm).

15. Connect the wiring harnesses to the injectors.

16. Connect the fuel return hose to the pressure regulator, then connect the vacuum hose.

17. Replace the O-ring on the high pressure fuel line. Coat the O-ring lightly with clean, oil and install the line to the fuel rail. Tighten the mounting bolts to 4 ft. lbs. (6 Nm).

18. Connect the PCV hose and the breather hose.

19. Connect the negative battery cable. Pressurize the fuel system and inspect for leaks.

ENGINE MECHANICAL

Engine Assembly

REMOVAL AND INSTALLATION

1.8L (VIN B) Engine

1. Relieve fuel system pressure.
2. Remove the engine under cover, if equipped.
3. Matchmark the hood and hinges and remove the hood assembly. Remove the air cleaner and intake duct work.
4. Drain the engine coolant, remove the radiator, coolant reservoir and intercooler.
5. Remove the transaxle and transfer case, if equipped with AWD.
6. Disconnect the accelerator cable, heater hoses, brake vacuum hose, vacuum hoses, high pressure fuel line, fuel return line, oxygen sensor, coolant temperature gauge, coolant temperature sensor, thermo switch sensor, the idle speed control, the motor position sensor, the throttle position sensor the EGR temperature sensor (California vehicles), the fuel injector, the power transistor,

the ignition coil, the condenser and noise filter, the distributor and control harness, and the alternator and oil pressure switch wires.

7. Remove the A/C drive belt and compressor. Leave the hoses attached and secure the compressor aside.
8. Remove the power steering pump and secure aside.
9. Remove the exhaust manifold-to-head pipe nuts. Discard the gasket.
10. Attach a hoist to the engine and take up the engine weight. Remove the engine mount bracket. Remove the torque control brackets (roll stoppers). Note that some engine mounts have arrows on them for proper assembly. Lift the engine slowly from the engine compartment.

To install:

11. Install the engine and secure in the mounts. The front lower mount through-bolt nut should not be tightened until the full weight of the engine is on the mount. Tighten the engine mount bolts as follows:
  • Upper mount-to-engine nuts and bolts — 36–47 ft. lbs. (50–65 Nm)
  • Upper mount through-bolt nut — 43–58 ft. lbs. (60–80 Nm)
  • Lower mount through-bolt nut — 33–47 ft. lbs. (45–65 Nm)
12. Install the exhaust pipe, power steering pump and A/C compressor.
13. Connect the electrical and vacuum connections.
14. Install the transaxle, if removed earlier. Install the starter.
15. Install the radiator assembly and intercooler.
16. Install the air cleaner assembly.
17. Fill the engine with the proper amount of engine oil. Connect the negative battery cable.
18. Refill the cooling system. Start the engine, allow it to reach normal operating temperature. Check for leaks.
19. Check the ignition timing and adjust, if necessary.
20. Install the hood making sure to align the matchmarks made during disassembly.
21. Road test the vehicle and check all functions for proper operation.

2.0L (VIN E, F and Y) Engines

1. Relieve the fuel system pressure.
2. Disconnect the negative battery cable.
3. Matchmark the hood to the hinges and remove the hood.
4. Remove the intake air duct.
5. Drain the engine coolant.
6. Remove the radiator.
7. Remove the engine undercover.

8. Attach an engine lifting fixture to the engine and remove the transaxle assembly.
9. Disconnect the power steering pressure, oil pressure switch, oil pressure gauge sender and the alternator wiring connectors.
10. Remove the alternator.
11. Remove the power steering pump from the bracket and position aside. It is not necessary to disconnect the fluid lines.
12. Remove the A/C compressor from the bracket and position it aside. Do not disconnect the hoses.
13. Disconnect the accelerator cable from the throttle body and mounting bracket.
14. Disconnect the following:
  • Idle air control motor
  • Knock sensor
  • Oxygen sensor
  • Engine coolant temperature gauge sender
  • Engine coolant temperature sensor
  • Ignition module (power transistor)
  • Throttle position sensor
  • Condenser
  • Manifold differential pressure sensor
  • Injectors
  • Ignition coil
  • Camshaft position sensor
  • Crankshaft position sensor
  • A/C compressor connector
  • Engine control wiring harness
15. Disconnect the brake booster vacuum hose.

— **CAUTION** —
Fuel injection systems remain under pressure, after the engine has been turned OFF. Properly relieve fuel pressure before disconnecting any fuel lines. Failure to do so may result in fire or personal injury.

16. Disconnect the fuel lines from the fuel supply rail.
17. Disconnect the water hose connections.
18. Disconnect the vacuum hoses.
19. Disconnect the front exhaust pipe from the turbocharger.
20. Support the engine under the oil pan with a floor jack and a piece of wood.
21. Remove the engine support fixture and replace it with a hoist.
22. Remove the engine mount bracket.
23. Raise the engine up slowly out of the engine compartment.

To install:

24. Slowly lower the engine assembly into the vehicle.

25. Position the floor jack under the oil pan with a piece of wood in between. Use the floor jack to adjust the height of the engine while installing the engine mount bracket.

26. Remove the chain hoist and install an engine support fixture.

27. Connect the front exhaust pipe to the turbocharger.

28. Connect the vacuum hoses.

29. Connect the water hoses.

30. Install a new O-ring and connect the fuel lines to the fuel supply rail.

31. Connect the brake booster vacuum hose.

32. Attach the following connectors:
- Idle air control motor
- Knock sensor
- Oxygen sensor
- Engine coolant temperature gauge sender
- Engine coolant temperature sensor
- Ignition module (power transistor)
- Throttle position sensor
- Condenser
- Manifold differential pressure sensor
- Injectors
- Ignition coil
- Camshaft position sensor
- Crankshaft position sensor
- A/C compressor connector
- Engine control wiring harness

33. Install the A/C compressor and power steering pump.

34. Install the generator.

35. Connect the oil pressure gauge sender, oil pressure and power steering pressure switch connectors.

36. Install the transaxle. Remove the engine support fixture.

37. Install the engine undercover.

38. Install the radiator and hoses.

39. Install the intake air duct.

40. Refill the engine with coolant.

41. Align the matchmarks and install the hood.

42. Connect the negative battery cable.

Engine Mounts

REMOVAL AND INSTALLATION

1993–94 Models

Upper Mounts

1. Disconnect the negative battery cable. Remove the air cleaner and air duct work.

2. Raise and safely support the engine so it is not resting on the engine mount. Use care not to bend or damage any components.

3. Remove the retainer bolt from the clamp securing the power steering pressure hose and A/C low pressure hose.

4. Remove the engine mount bracket-to-body through-bolt. Take note of the position of the arrow on the oval shaped mounting stopper plate.

5. Remove the engine mounting bracket and stopper plate.

To install:

6. Install the engine mounting bracket and stopper plate, aligning the arrows properly. On most engines the arrows will face the towards the center of the engine. Tighten upper mount-to-engine nuts and bolts to 36–47 ft. lbs. (50–65 Nm) and the upper mount through-bolt nut to 43–58 ft. lbs. (60–80 Nm).

Lower Mounts

1. Disconnect the negative battery cable.

2. Raise and safely support the engine so it is not resting on the engine mount.

3. Remove the stopper through-bolt.

4. Remove the stopper mount-to-frame bolts and pry the mount out.

To install

5. Position the lower front roll stopper with the hole in it facing the front of the vehicle. Install the frame mounting bolts and tighten.

6. Install the front lower mount through-bolt nut, but do not tighten until the full weight of the engine is on the mount. Tighten the lower mount through-bolt nut to 36–47 ft. lbs. (50–65 Nm).

1995–97 Models

Engine Mount

1. Disconnect the negative battery cable.

2. Jack up the engine under the oil pan to take the weight off of the engine mount.

3. Remove the engine mount through-bolt.

4. Remove the bolt and two nuts securing the mount to the engine and remove.

To install:

5. Install the engine mount so the arrow is pointing toward the engine.

6. Line up the holes and install the through-bolt.

7. Remove the jack from under the oil pan.

Transaxle Mount

1. Disconnect the negative battery cable.

2. Remove the air cleaner.

3. Remove the battery and tray.

4. Support the weight of the engine and remove the engine roll stopper.

5. Remove the engine mount.

6. Remove the charcoal canister and bracket.

7. Remove the four nuts and transaxle mount bracket.

8. Remove the through-bolt and the transaxle mount.

To install:

9. Install the transaxle mount.

10. Install the charcoal canister and bracket.

11. Install the engine mount and the roll stopper.

12. Install the battery tray and battery.

13. Install the air cleaner.

14. Connect the negative battery cable.

Engine Roll Stopper

1. Disconnect the negative battery cable.

2. Safely raise and support the vehicle.

3. Remove the engine undercovers.

4. Raise the engine to take the weight off of the roll stopper.

5. Remove the roll stopper through-bolt.

6. Remove the roll stopper mounting bolts.

7. Remove the roll stopper.

To install:

8. Install the roll stopper and mounting bolts. Lower the engine and tighten the through-bolt.

NOTE: Measure the distance from the center member to the through-bolt on the front roll stopper. If the dimension is greater than 1.57–181 in. (40–46mm), replace the front roll stopper bracket assembly.

9. Install the engine undercovers.

10. Lower the vehicle to the floor.

11. Connect the negative battery cable.

Cylinder Head

REMOVAL AND INSTALLATION

1.8L (VIN B) Engines

1. Relieve the fuel system pressure.

2. Drain the cooling system.

3. Remove the air intake and breather hose.

4. Disconnect the accelerator cable. There will be 2 cables, if equipped with cruise control.

Front of engine (Timing belt side)

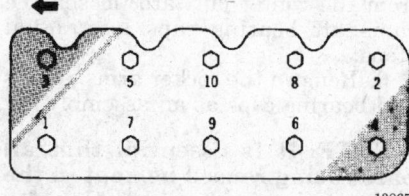

Cylinder head bolt removal sequence — 1.8L (VIN B)

138657

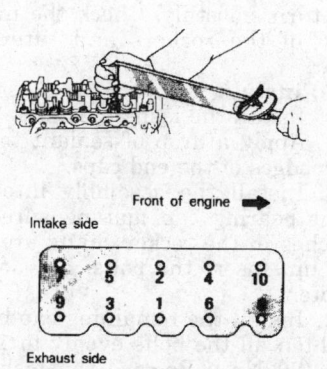

Cylinder head torque sequence — 1.8L (VIN B)

138656

5. Place a shop towel around the high pressure fuel line and disconnect the fuel line.

6. Remove the upper radiator hose, water breather hose, the water bypass hose and the heater hose.

7. Disconnect the PCV hose.

8. Remove the spark plug cables. Make sure not to pull on the cable.

9. Disconnect and plug the fuel return line.

10. Disconnect the vacuum line for the brake booster.

11. Disconnect the oxygen sensor, engine coolant temperature gauge unit and the water temperature sen-

sor, idle speed control motor, throttle position sensor, distributor, motor position sensor, fuel injectors, EGR temperature sensor (California vehicles), power transistor, condenser and engine ground cable.

12. Remove the clamp that holds the power steering pressure hose to the engine mounting bracket.

13. Lift the engine enough to take the weight off the engine mounting bracket. Remove the bracket.

14. Remove the valve cover, gasket and half-round seal. Remove the timing belt.

15. Remove the exhaust pipe self-locking nuts and separate the exhaust pipe from the exhaust manifold. Discard the gasket.

16. Loosen the cylinder head mounting bolts in 3 steps, starting from the outside and working inward. Lift off the cylinder head assembly and remove the head gasket.

To install:

17. Thoroughly clean and dry the mating surfaces. Check the cylinder head for cracks or damage. Check the head for flatness. End-to-end, the head should be within 0.002 in. normally with 0.008 in. the maximum allowed out of true. The total thickness allowed to be removed from the head and block is 0.008 in. maximum.

18. Place a new head gasket on the cylinder block with the identification marks facing upward. Make sure the gasket has the proper identification mark for the engine. Do not use sealer on the gasket.

19. Install the cylinder head on the block. Using 3 even steps, tighten the head bolts in sequence to 51–54 ft. lbs. (70–75 Nm).

20. Install a new exhaust pipe gasket and connect the pipe to the manifold.

21. Install the timing belt. Apply sealer to the perimeter of the half-round seal. Install a new valve cover gasket. Install the valve cover.

22. Install the engine mount bracket. Once secure, remove the jack.

23. Install the clamp that holds the power steering pressure hose to the engine bracket.

24. Connect or install all disconnected hoses, cables and electrical connections. Adjust the throttle cable(s).

25. Replace the O-rings and connect the fuel lines.

26. Install the air intake and breather hose.

27. Change the engine oil. Fill the cooling system.

28. Connect the negative battery cable. Run the vehicle until the thermostat opens. Fill the radiator completely.

29. Check and adjust the idle speed and ignition timing.

30. Once the vehicle has cooled, recheck the coolant level.

2.0L (VIN E, F and Y) Engines

— **CAUTION** —

Fuel injection systems remain under pressure, after the engine has been turned OFF. Properly relieve fuel pressure before disconnecting any fuel lines. Failure to do so may result in fire or personal injury.

1. Relieve the fuel system pressure.

2. Disconnect the negative battery cable.

3. Drain the engine coolant and oil.

4. Remove the air cleaner and intake duct.

5. Disconnect the following connectors:
- A/C compressor
- Power steering pressure switch
- Heated oxygen sensor
- Engine coolant temperature gauge sender
- Engine coolant temperature sensor
- MAP sensor
- Intake air temperature sensor
- Throttle position sensor
- Idle air control motor
- Injector harness
- Ignition coil
- Camshaft position sensor
- EGR solenoid valve

6. Disconnect the accelerator cable from the throttle body.

7. Disconnect the heater hoses from the rear of the engine.

8. Disconnect and plug the fuel lines from the fuel supply rail.

9. Disconnect the purge hose and the brake booster vacuum hose connections.

10. Disconnect the overflow tube connection.

11. Disconnect the upper radiator hose and the water hose connections.

12. Remove the timing belt.

13. Remove the intake manifold stay.

14. Remove the intake and exhaust camshafts.

15. Disconnect the exhaust pipe connection from the exhaust manifold.

16. Remove the 10 cylinder head mounting bolts and lift off the cylinder head.

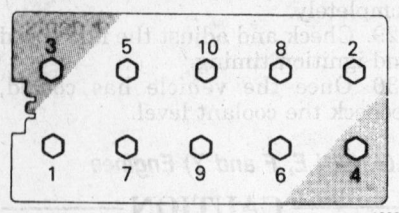

Cylinder head bolt removal sequence — 2.0L (VIN E, F and Y) engines

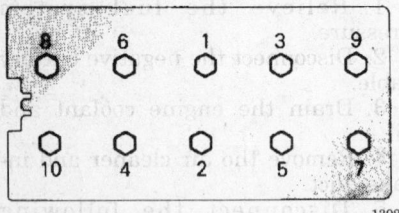

Cylinder head bolt torque sequence — 2.0L (VIN E, F and Y) engines

To install:

17. Thoroughly clean the cylinder head and engine block sealing surfaces. Check the deck and the cylinder head for warpage.

18. Place a new head gasket on the engine block and position the cylinder head on the block.

19. Coat the threads of the bolts with clean engine oil and install the bolts finger-tight. The short bolts go in the corners.

20. Tighten the cylinder head bolts in the following sequence:

• Tighten the center bolts 1 through 6 to 25 ft. lbs. (33 Nm) then tighten the outer bolts 7 through 10 to 20 ft. lbs. (27 Nm)

• Tighten the center bolts 1 through 6 to 50 ft. lbs. (67 Nm) then tighten the outer bolts 7 through 10 to 20 ft. lbs. (27 Nm)

• Turn all fasteners 1 through 10 1/4 turn (90°) more in sequence. Do not use a torque wrench for this step.

21. Use a new gasket and connect the front exhaust pipe to the exhaust manifold.

22. Install the camshafts.

23. Install the timing belts.

24. Install the intake manifold stay.

25. Connect the upper radiator hose and clamp.

26. Connect the water hose to the water pipe.

27. Connect the overflow tube.

28. Connect the brake booster vacuum hose and purge air hose connection.

29. Use a new O-ring and connect the fuel lines to the fuel supply rail.

30. Connect the heater hose.

31. Attach the following connectors:

• A/C compressor
• Power steering pressure switch
• Heated oxygen sensor
• Engine coolant temperature gauge sender
• Engine coolant temperature sensor
• MAP sensor
• Intake air temperature sensor
• Throttle position sensor
• Idle air control motor
• Injector harness
• Ignition coil
• Camshaft position sensor
• EGR solenoid valve

32. Install and adjust the accelerator cable.

33. Install the air intake duct and air cleaner assembly.

34. Refill the engine with oil and coolant.

35. Turn the ignition to the **ON** position and check for fuel leaks. Then start the engine and check for coolant leaks and proper operation.

Lash Adjusters

Bleeding

NOTE: The hydraulic lash adjuster is a precision component that relies on a clean operating environment. When handling the lash adjuster, make certain that no dirt or foreign particles are allowed to get inside the unit. DO NOT try to take the unit apart. When cleaning the unit, only use clean diesel fuel.

1. Mount the adjuster into special bleeding tool 09246–32100 or equivalent and immerse the tool and adjuster in a container of clean diesel fuel.

2. With air bleed wire 09246–32200 or equivalent inserted in the adjuster, lightly press down on the steel ball and compress the plunger four or five times.

3. Remove the air bleed wire from the adjuster and push down firmly on the plunger. If the plunger moves even slightly, repeat Steps 1 and 2 until the plunger stops moving. If the plunger continues to move, replace it.

REMOVAL AND INSTALLATION

1.8L (VIN B) Engines

1. Disconnect the negative battery cable.

2. Remove the valve cover. Install lash adjuster retainer tools MD998443 or equivalent, to the rocker arm.

3. Remove the distributor extension, if necessary.

4. Remove the timing belt.

5. Working in a crisscross pattern from the center outward, loosen the camshaft bearing caps in gradual steps.

6. Remove the rocker arms, shafts and bearing caps as an assembly.

NOTE: It is essential that all parts being reused be kept in the same order and orientation for reinstallation.

7. Remove the lash adjusters.

8. Inspect the roller surfaces of the rockers. Replace if there are any signs of damage or if the roller does not turn smoothly. Check the inside bore of the rockers and lifter for wear.

To install:

9. Install the lash adjusters.

10. Apply a drop of sealant to the rear edges of the end caps.

11. Install the assembly into the front bearing cap making sure the notches in the rocker shafts are facing up. Insert the bolts, but do not tighten.

12. Install the remaining cap bolts. Tighten all the bolts evenly to 15 ft. lbs. (20 Nm). Remove the lash adjuster retainers.

13. Install the timing belt.

14. Install the distributor extension, if removed.

15. Install the valve cover, using a new gasket and semi-circular packing in place.

16. Connect the negative battery cable.

17. Run engine and check the ignition timing.

2.0L (VIN E, F and Y) Engines

1. Relieve the fuel system pressure following proper procedure. Disconnect negative battery cable.

2. Disconnect the accelerator cable, PCV hoses, breather hoses, spark plug cables and the remove the valve cover.

3. Remove the timing belt.

NOTE: Always rotate the crankshaft in a clockwise direction. Make a mark on the back of the timing belt indicating the direction of rotation so it may be reassembled in the same direction if it is to be reused.

4. Remove the crank angle sensor.

NOTE: It is essential that all parts be kept in the same order and orientation for reinstallation. In order to prevent confusion during installation, be sure to mark and separate all parts.

5. Remove the camshafts, rocker arms and lash adjusters.
6. Visually inspect the rocker arm roller and replace if damage or seizure is evident. Check the roller for smooth rotation. Replace if excess play or binding is present. Also, inspect valve contact surface for possible damage or seizure.

To install:

7. Install the lash adjusters and rocker arms into the cylinder head. Lubricate lightly with clean oil prior to installation.
8. Apply engine oil to the lobes and journals of each camshaft. Install the camshafts into the cylinder head. Align the shafts so the dowel pins on the camshaft sprocket end are located on the top.
9. Install and tighten the camshaft bearing caps in the proper sequence torquing to specification.
10. Replace the camshaft oil seals and install the sprockets.
11. Align the punch mark on the crank angle sensor housing with the notch on the sensor plate. Install the crank angle sensor into the cylinder head.
12. Install the timing belt, cover and related components.
13. Install the valve cover using a new gasket.
14. Connect the negative battery cable.

Valve Lash

ADJUSTMENT

Valve clearance is not adjustable. Valve clearance is accomplished by the hydraulic lash adjuster.

Rocker Arms and Shafts

REMOVAL AND INSTALLATION

1.8L (VIN B) Engine

1. Disconnect the negative battery cable.
2. Disconnect the accelerator cable, PCV and breather hoses, spark plug cables and valve cover.
3. Remove the valve cover. Install lash adjuster retainer tools MD998443 or equivalent, to the rocker arm.
4. Remove the distributor extension, if necessary.
5. Remove the timing belt.
6. Working in a crisscross pattern from the center outward, loosen the camshaft bearing caps in gradual steps.
7. Remove the camshaft, rocker arms, shafts and bearing caps as an assembly.

NOTE: It is essential that all parts be kept in the same order and orientation for reinstallation. Be sure to mark and separate parts to keep them from getting mixed. This will aid assembly.

8. Disassemble the rocker shaft assembly. Starting at rear bearing cap, slide each piece off the shaft.

NOTE: Inspect the roller surface of the rockers. Replace if there are any signs of damage or if the roller does not turn smoothly.

To install:

9. Apply a drop of sealant to the rear edges of the end caps.
10. Apply engine oil to the lobes and journals of each camshaft. Install the camshafts into the cylinder head taking care not to confuse the intake and the exhaust camshaft; the intake camshaft has a slit on its rear end for driving the crank angle sensor. Align shafts so dowel pins on camshaft sprocket end are located on the top.
11. Install and tighten the camshaft bearing caps in the proper sequence torquing to specifications in three even steps.
12. Replace the camshaft oil seals and install the sprockets.
13. Locate the dowel pin on the sprocket end of the intake camshaft at the top position, if not already done.
14. Align the punch mark on the crank angle sensor housing with the notch on the sensor plate. Install the

crank angle sensor into the cylinder head.
15. Install the timing belt, covers and related components.
16. Install the valve cover using new gasket. Reconnect all related components.
17. Reconnect the negative battery cable.

2.0L (VIN E, F and Y) Engines

1. Apply engine oil to the lobes and journals of each camshaft. Install the camshafts into the cylinder head taking care not to confuse the intake and the exhaust camshaft; the intake camshaft has a slit on its rear end for driving the crank angle sensor. Align shafts so dowel pins on camshaft sprocket end are located on the top.
2. Install and tighten the camshaft bearing caps in the proper sequence torquing to specifications in three even steps.
3. Replace the camshaft oil seals and install the sprockets.
4. Locate the dowel pin on the sprocket end of the intake camshaft at the top position, if not already done.
5. Align the punch mark on the crank angle sensor housing with the notch on the sensor plate. Install the crank angle sensor into the cylinder head.
6. Install the timing belt, covers and related components.
7. Install the valve cover using new gasket. Reconnect all related components.
8. Reconnect the negative battery cable.

Intake Manifold

REMOVAL AND INSTALLATION

1.8L (VIN B) and 2.0L (VIN E and F) Engines

1. Relieve the fuel system pressure.
2. Disconnect the negative battery cable.
3. Drain the cooling system.
4. Disconnect the accelerator cable, breather hose and air intake hose.
5. Disconnect the upper radiator hose, heater hose and water bypass hose.
6. Remove all vacuum hoses and pipes as necessary, including the brake booster vacuum line.
7. Disconnect the high pressure fuel line and the fuel return hose.

8. Tag and disconnect the electrical connectors from the oxygen sensor, coolant temperature sensor, thermo switch, idle speed control, EGR temperature sensor and spark plug wires.

9. Remove the fuel rail, fuel injectors, pressure regulator and insulators from the engine.

10. Remove the intake manifold bracket.

11. Disconnect the water hose at the throttle body, and the water inlet and heater connections.

12. Remove the thermostat housing if necessary.

13. Disconnect the vacuum line from the power brake booster and PCV valve if connected.

14. Remove the intake manifold mounting bolts and remove the manifold.

To install:

15. Clean the gasket material from the intake mounting surface and manifold assembly. Check both surfaces for cracks or other damage. Check the intake manifold water passages and air passages for clogging. Clean if necessary.

16. Install a new gasket to the head, and install the manifold. Tighten the manifold in a crisscross pattern, starting from the inside and working outwards to 11–14 ft. lbs. (15–19 Nm).

17. Install the fuel delivery pipe, injectors and pressure regulator, lubricating all seals lightly with oil. Tighten the retaining bolts to 7–9 ft. lbs. (10–13 Nm).

18. Install the thermostat housing, if removed, intake manifold brace bracket, distributor (if removed) and throttle body bracket.

19. Connect or install all hoses, cables and electrical connectors that were removed or disconnected during the removal procedure.

20. Fill the system with coolant.

21. Connect the negative battery cable. Run the vehicle until the thermostat opens, fill the radiator completely. Check for leaks.

22. Adjust the accelerator cable. Check and adjust the ignition timing. Once the vehicle has cooled, recheck the coolant level.

2.0L (VIN Y) Engines

———— **CAUTION** ————

Fuel injection systems remain under pressure, after the engine has been turned OFF. Properly relieve fuel pressure before discon-

necting any fuel lines. Failure to do so may result in fire or personal injury.

1. Disconnect the negative battery cable.

2. Drain the engine coolant.

3. Remove the vacuum reservoir if equipped with cruise control.

4. Remove the air intake and breather hoses.

5. Remove the accelerator cable from the bracket.

6. Disconnect the engine harness retaining clips.

7. Disconnect the MAP sensor connector.

8. Disconnect the charge temperature sensor connector.

9. Disconnect the TPS and the AIS motor connectors.

10. Position the engine control wiring harness out of the way.

11. Disconnect the alternator wiring harness.

12. Remove the PCV hose assembly.

13. Label and disconnect the vacuum hoses.

14. Disconnect the EGR pipe connection.

15. Disconnect the fuel lines from the fuel rail.

16. Remove the intake manifold stay and the engine hanger.

17. Remove the throttle body.

18. Remove the intake manifold plenum and gasket.

19. Disconnect the injector connectors.

20. Remove the fuel rail with the injectors.

21. Remove the intake manifold.

To install:

22. Use a new gasket and install the intake manifold.

23. Install the fuel rail assembly and connect the injectors.

24. Use a new gasket and install the intake plenum.

25. Use a new gasket and install the throttle body.

26. Install the intake manifold stay and the engine hanger.

27. Use a new O-ring and connect the fuel lines to the fuel rail.

28. Connect the EGR pipe.

29. Connect the vacuum hoses and the PCV hose assembly.

30. Connect the generator wiring harness connector.

31. Reposition the engine control wiring harness and install the brackets and clips.

32. Connect the AIS motor and the TPS sensor connectors.

33. Connect the vacuum hose to the throttle body.

34. Connect the MAP and the charge temperature sensor connectors.

35. Install the accelerator cable in the bracket and connect it to the throttle body.

36. Connect the breather hose and the air intake hose.

37. Install the vacuum reservoir, if equipped.

38. Connect the negative battery cable.

39. Refill the engine with coolant.

40. Adjust the accelerator cable.

Exhaust Manifold

REMOVAL AND INSTALLATION

1.8L (VIN B) and 2.0L (VIN E and F) Engines

1. Disconnect the battery negative cable. Drain the cooling system.

2. If equipped with A/C, remove the condenser cooling fan. Remove the power steering pump and place aside.

3. Disconnect the oxygen sensor harness.

4. Raise the vehicle and support safely.

5. Drain the oil and remove the dipstick tube.

6. If turbo equipped, remove the exhaust pipe-to-turbocharger nuts and separate the exhaust pipe. Discard the gasket.

7. Lower the vehicle. Remove the air intake and vacuum hose connections.

8. Remove the exhaust manifold and turbocharger (if equipped) heat shields.

9. Remove the exhaust manifold-to-turbocharger attaching bolts and nut.

10. If applicable, remove the engine hanger, water and oil lines from the turbo.

11. Remove the exhaust manifold mounting nuts. Remove the exhaust manifold and gasket from the engine.

To install:

12. Clean all gasket material from the mating surfaces and check the manifold for damage.

13. Install new gaskets and install the manifold. On 1.8L engines, tighten the attaching nuts in a crisscross pattern to 11–14 ft. lbs. (15–20 Nm). On 2.0L (VIN E and F) engines, tighten the manifold-to-head nuts in a crisscross pattern to 18–22 ft. lbs. (25–30 Nm). Tighten the manifold-to-turbo nut and bolts to 40–47 ft. lbs. (55–65 Nm).

14. Install the engine hanger, water and oil lines to the turbocharger.

15. Install the heat shields.

16. Install the new gasket and connect the exhaust pipe.

17. Install the condenser cooling fan and power steering pump. Connect the oxygen sensor harness.

18. Install the oil level indicator and tube replacing the O-ring as required.

19. Fill the crankcase with clean oil and refill the cooling system.

20. Connect the negative battery cable and run engine until thermostat opens.

21. Check for fluid and exhaust leaks. Top off engine coolant.

2.0L (VIN Y) Engines

1. Disconnect the negative battery cable.

2. Drain the engine coolant.

3. Remove the air intake hose.

4. Disconnect the upper radiator hose from the water outlet.

5. Disconnect the air hose connection.

6. Remove the engine control wiring harness from the rear of the engine.

7. Remove the water pipe assembly.

8. Remove the oil dipstick.

9. Remove the upper heat shield.

10. Remove the engine hanger.

11. Disconnect the pulsed secondary air injection (check valve) valve from the exhaust pipe (manual transaxle only).

12. Disconnect the front exhaust pipe from the manifold.

13. Remove the lower heat shield.

14. Remove the exhaust manifold and gasket.

To install:

15. Use a new gasket and install the exhaust manifold. Tighten the nuts and bolts to 17 ft. lbs. (23 Nm).

16. Install the lower heat shield.

17. Use a new gasket and connect the front exhaust pipe to the manifold.

18. On vehicles with manual transaxles, connect the pulsed secondary air injection valve to the exhaust pipe.

19. Install the engine hanger.

20. Install the upper heat shield.

21. Install the dipstick and the water pipe.

22. Attach the engine wiring harness to the rear of the engine.

23. Connect the air hose and the upper radiator hose.

24. Install the air intake hose.

25. Connect the negative battery cable.

26. Refill the engine with coolant, start the engine and check for leaks.

Turbocharger

REMOVAL AND INSTALLATION

1. Disconnect the negative battery cable.

2. Drain the engine coolant.

3. Remove the condenser fan motor, if equipped with A/C.

4. Remove the heated oxygen sensor.

5. Remove the oil dipstick and tube.

6. Remove the air cleaner and intake hose assembly.

7. Disconnect the air intake hose from the turbocharger.

8. Disconnect the engine coolant hoses from the turbocharger.

9. Disconnect the oil supply pipe connection. Do not let dirt of foreign particles enter the oil pipe.

10. Remove the heat shields.

11. Remove the engine hanger.

12. Disconnect the front exhaust pipe from the turbocharger.

13. Remove the oil return pipe and gaskets.

14. Remove the flange bolts and nut that attach the turbo to the exhaust manifold. Take note of the positions of the coned disc springs and the washers.

15. Remove the turbocharger, gasket and ring.

To install:

16. Use a new gasket and install the turbo to the exhaust manifold. Make sure the coned disc spring and the washers are installed in their original positions. Tighten the bolts and nut to 20–23 ft. lbs. (27–31 Nm). Tighten the bolts and nuts an additional 60°– 70°.

17. Use a new gasket and connect the exhaust pipe to the turbo. Tighten the mounting bolts to 40–47 ft. lbs. (54–64 Nm).

18. Using new gaskets, install the oil return pipe.

19. Install the engine hanger.

20. Install the heat shields.

21. Connect the oil supply pipe. Tighten the flare nut fittings to 14 ft. lbs. (19 Nm).

22. Connect the engine coolant hoses to the turbo.

23. Connect the air hose.

24. Install the air cleaner and duct asscmbly.

25. Position a new O-ring on the dipstick tube and install.

26. Install the heated oxygen sensor.

27. Install the condenser fan assembly.

28. Change the engine oil and filter.

29. Connect the negative battery cable and refill the engine with coolant.

30. Start the engine and let it idle. Check for leaks.

Front Crankshaft Seal

REMOVAL AND INSTALLATION

1.8L (VIN B) and 2.0L (VIN E and F) Engines

1. Disconnect the negative battery cable.

2. Remove the timing belts.

NOTE: When reusing the timing belts, be sure to mark the direction of rotation on the belt. This will extend belt life, ensuring the same direction of rotation.

3. Remove the crankshaft pulley.

4. Remove the crankshaft sprocket. If the sprocket is difficult to remove, an appropriate puller may be used.

5. On 1.8L engines, remove the flange and inner sprocket.

NOTE: When removing the inner sprocket, use care not to loose the woodruff key.

6. Pry the seal from the bore using the proper tools.

To install:

7. Using a proper sized driver, install a new seal.

8. On 1.8L engines, install the woodruff key and inner sprocket. Install the flange.

9. Install the crankshaft sprocket.

10. Install the timing belt and remaining components.

11. Install the engine undercover. Connect the negative battery cable, start the engine and check for leaks.

2.0L (VIN Y) Engines

1. Remove the timing belt.

2. Using tool MB995027 or equivalent, remove the crankshaft sprocket.

3. Using tool MB995020 or equivalent, remove the oil seal. Be careful not the scratch the oil seal bore or the crankshaft sealing surface.

To install:

4. Apply clean engine oil to the oil seal. Using tool MB995022 or equivalent, install the oil seal.

5. Using tool MB995035 and MB995026 or equivalent, install the crankshaft sprocket.

6. Install the timing belt.

Timing Belt, Sprockets, Tensioner and Front Cover

REMOVAL AND INSTALLATION

1.8L (VIN B) Engines

1. Position the engine so the No. 1 piston is at TDC.

2. Disconnect the negative battery cable.

3. Remove the timing belt covers.

4. Loosen the timing (outer) belt tensioner. Move the tensioner toward the water pump and lightly tighten the bolt to hold the position. Remove the outer timing belt.

NOTE: If timing belts are going to be reused, mark the direction of rotation on the belt. This will ensure the belt is installed in same direction.

5. Remove the outer crankshaft sprocket and flange.

6. Remove the crankshaft sprocket retainer bolt and washer from the sprocket and remove. If the sprocket

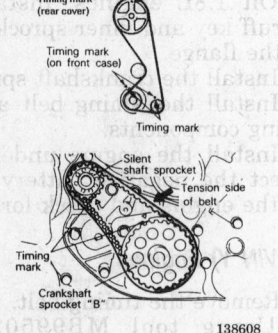

Timing mark locations and alignment — 1.8L (VIN B) engines

is difficult to remove, an appropriate puller may be used.

7. Loosen the silent shaft (inner) belt tensioner and remove the belt.

8. If necessary, remove the camshaft sprocket using a spanner wrench. Hold the shaft in position while removing the bolt.

To install:

9. Install the sprockets to their appropriate shafts. Install the retainer bolts and tighten the camshaft bolt to 58–72 ft. lbs. (80–100 Nm), crankshaft bolt to 80–94 ft. lbs. (110–130 Nm).

10. Turn both tensioner pulleys and check for any signs of bearing wear.

11. Align the timing marks of the silent shaft sprockets and the crankshaft sprocket with the timing marks on the front case. Wrap the timing belt around the sprockets so there is no slack in the upper portion of the belt.

12. Install the tensioner pulley and move the pulley by hand so the long side of the belt deflects about ¼in.

13. Hold the pulley firm and tighten the bolt to 15 ft. lbs. (20 Nm) and recheck the deflection amount.

14. Install the timing belt tensioner fully toward the water pump and tighten the bolts. Place the upper end of the spring against the water pump body.

15. Align the timing marks of the camshaft, crankshaft and oil pump sprockets with the corresponding marks on the case.

16. Before installing the timing belt, ensure that the left side (rear) silent shaft (oil pump sprocket) is in the correct position as follows:

a. Remove the plug from the rear side of the block and insert a tool with shaft diameter of 0.31 in. (8mm) into the hole.

b. With the timing marks still aligned, the shaft of the tool must be able to go in at least 2 ½ in. If the tool can only go in about 1 in., the shaft is not in the correct orientation and will cause a vibration during engine operation. Remove the tool from the hole and turn the oil pump sprocket 1 complete revolution. Realign the timing marks and insert the tool. The shaft of the tool must go in at least 2 ½.

c. Recheck and realign the timing mark.

d. Leave the tool in place to hold the silent shaft while continuing.

17. Install the belt around the crankshaft sprocket, oil pump sprocket, then camshaft sprocket. While doing so, make sure there is no

slack between the sprockets except where the tensioner is installed.

18. Recheck the timing mark alignment. If all are aligned, loosen the tensioner mounting bolt and allow tension to the belt.

19. Remove the tool that is holding the silent shaft and rotate the crankshaft a distance equal to 2 teeth on the camshaft sprocket. This will allow the tensioner to automatically apply the proper tension on the belt. Do not manually overtighten the belt or it will howl.

20. Tighten the lower mounting bolt first, then the upper spacer bolt.

21. To verify correct belt tension, check that the deflection at the longest span of the belt is about ½ in.

22. Install the timing belt covers and all related items.

23. Connect the negative battery cable.

24. Run the engine until the thermostat opens. Check and adjust ignition timing.

2.0L (VIN E and F) Engines

1. Disconnect the negative battery cable.

2. Remove the timing belt upper and lower covers.

3. Rotate the crankshaft clockwise and align the timing marks so No. 1 piston will be at TDC of the compression stroke. At this time the timing marks on the camshaft sprocket and the upper surface of the cylinder head should coincide. The dowel pin of the camshaft sprocket should be at the upper side.

NOTE: Always rotate the crankshaft in a clockwise direction. Make a mark on the back of the timing belt indicating the direction of rotation so it may be reassembled in the same direction if it is to be reused.

4. Remove the auto tensioner and remove the outer timing belt.

5. Remove the timing belt tensioner, tensioner arm, idler pulley, oil pump sprocket, special washer, flange and spacer.

6. Remove the silent shaft (inner) belt tensioner and remove the belt.

7. If necessary, remove the crankshaft sprocket bolt and washer from the sprocket and remove sprocket. If the sprocket is difficult to remove, use a gear puller.

8. If necessary to remove the camshaft sprockets, hold the camshaft stationary using the hexagon cast between journals No. 2 and 3. Remove the retainer bolt and remove the sprocket from the camshaft.

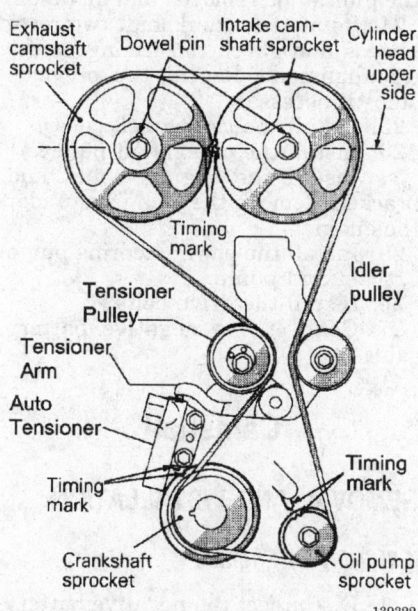

Timing belt routing and mark identification — 2.0L (VIN E and F) engines

139322

To install:

9. Check both the tensioner and the idler pulley for bearing wear, and replace if needed.

10. Install the sprockets to the appropriate shafts. Install the retainer bolts and tighten the camshaft sprocket bolt to 65 ft. lbs. (90 Nm). Install the crankshaft sprocket and tighten the retaining bolt to 87 ft. lbs. (120 Nm).

11. Align the timing marks on the crankshaft sprocket and the silent shaft sprocket. Fit the inner timing belt over the crankshaft and silent shaft sprocket. Make sure there is no slack in the belt.

12. While holding the inner timing belt tensioner, adjust the timing belt tension by applying a force towards the center of the belt, until the tension side of the belt is taut. Tighten the tensioner bolt.

NOTE: When tightening the bolt of the tensioner, ensure that the tensioner pulley shaft does not rotate with the bolt.

13. Check belt for proper tension by depressing on the long side with your finger and noting the deflection. The desired reading is 0.20–0.28 in. (5–7mm). If tension is not correct, readjust.

14. Install the flange, crankshaft and washer. The flange on the crankshaft sprocket must be installed towards the inner timing belt sprocket. Tighten bolt to 80–94 ft. lbs. (110–130 Nm).

15. To install the oil pump sprocket, insert a Phillips screwdriver with a shaft 0.31 in. (8mm) in diameter into the plug hole in the left side of the cylinder block to hold the left silent shaft. Tighten the nut to 36–43 ft. lbs. (50–60 Nm).

16. Using a wrench, hold the camshaft at the hexagon between journal No. 2 and 3 and tighten the sprocket bolt, if removed, to 58–72 ft. lbs. (80–100 Nm). If no hexagon is present between journal No. 2 and 3, hold the sprocket stationary with a spanner wrench while tightening the sprocket retainer bolt.

17. Carefully push the auto tensioner rod in until the set hole in the rod aligns with the hole in the cylinder. Place a wire into the hole to retain.

18. Install the tensioner pulley onto the arm. Locate the pinhole in the tensioner pulley shaft to the left of the center bolt. Then, tighten the center bolt finger-tight.

19. When installing the timing belt, turn the camshaft sprockets so the dowel pins are located on top. Align the timing marks facing each other, and the top surface of the cylinder head. When you let go of the exhaust camshaft sprocket, it will rotate 1 tooth in the counterclockwise direction. This should be taken into account when installing the timing belts on the sprocket.

NOTE: Both camshaft sprockets are used for the intake and exhaust camshafts and are provided with 2 timing marks. When the sprocket is mounted on the exhaust camshaft, use the timing mark on the right with the dowel pin hole on top. For the intake camshaft sprocket, use the 1 on the left with the dowel pin hole on top.

20. Align the crankshaft sprocket and oil pump sprocket timing marks.

21. Remove the plug on the cylinder block and insert a Phillips screwdriver with a shaft diameter of 0.31 in. (8mm) through the hole. If the shaft can be inserted 2.4 in. deep, the silent shaft is in the correct position. If the shaft of can only be inserted 0.8–1.0 in. (20–25mm) deep, turn the oil pump sprocket 1 turn and realign the marks. Reinsert the tool making sure it is inserted 2.4 in. deep. Keep

the tool inserted in hole for the remainder of this procedure.

NOTE: The above step assures that the oil pump socket is in correct orientation to the silent shafts. This step must not be skipped or a vibration will develop during engine operation.

22. Install the timing belt as follows:

a. Install the timing belt around the intake camshaft sprocket and retain it with 2 spring clips or binder clips.

b. Install the timing belt around the exhaust sprocket, aligning the timing marks with the cylinder head top surface using 2 wrenches. Retain the belt with 2 spring clips.

c. Install the timing belt around the idler pulley, oil pump sprocket, crankshaft sprocket and the tensioner pulley. Remove the 2 spring clips.

d. Lift upward on the tensioner pulley in a clockwise direction and tighten the center bolt. Make sure all the timing marks are aligned.

e. Rotate the crankshaft ¼ turn counterclockwise. Then turn clockwise until the timing marks are aligned again.

23. To adjust the timing (outer) belt, turn the crankshaft ¼ turn counterclockwise. Then turn clockwise to move No. 1 cylinder to TDC.

24. Loosen the center bolt. Using tool MD998738 or equivalent and a torque wrench, apply 1.88–2.03 ft. lbs. (2.6–2.8 Nm) of torque. Tighten the center bolt.

25. Screw the special tool into the engine left support bracket until the end makes contact with the tensioner arm. At this point, screw the special tool in more and remove the set wire attached to the auto tensioner, if the wire was not previously removed. Then remove the special tool.

26. Rotate the crankshaft 2 complete turns clockwise and let it sit for approximately 15 minutes. Then, measure the auto tensioner protrusion (the distance between the tensioner arm and auto tensioner body) to ensure that it is within 0.15–0.18 in. (3.8–4.5mm). If out of specification, repeat Step 1–4 until the specified value is obtained.

27. If the timing belt tension adjustment is being performed with the engine mounted in the vehicle, and clearance between the tensioner arm and the auto tensioner body cannot be measured, the following alternative method can be used:

a. Screw in special tool MD998738 or equivalent, until its

end makes contact with the tensioner arm.

b. After the special tool makes contact with the arm, screw it in some more to retract the auto tensioner pushrod while counting the number of turns the tool makes until the tensioner arm is brought into contact with the auto tensioner body. Make sure the number of turns the special tool makes conforms with the standard value of 2 1/2 to 3 turns.

c. Install the rubber plug to the timing belt rear cover.

28. Install the timing belt covers and all related items.

29. Connect the negative battery cable.

30. Run engine until thermostat opens. Check ignition timing and adjust as necessary.

2.0L (VIN Y) Engines

1. Disconnect the negative battery cable.

2. Remove the drive belts.

3. Remove the power steering pump from the bracket and position it out of the way. Do not disconnect the hoses.

4. Remove the power steering pump bracket from the engine.

5. Use a floor jack with a piece of wood on it and jack up the engine to take the weight off of the engine mount.

6. Remove the engine mount and bracket.

7. Remove the crankshaft pulley.

8. Remove the front timing belt cover.

NOTE: If the timing belt is going to be reused, mark the direction of rotation on the belt with an arrow. Install the belt in the same direction.

9. Rotate the crankshaft sprocket clockwise until the timing marks are aligned. Loosen the timing belt tensioner and remove the timing belt.

10. If necessary, use tool MB995027 or equivalent to remove the crankshaft sprocket.

——— WARNING ———
Do not rotate the crankshaft or the camshafts while the belt is removed.

11. If necessary, use tool MB990767 or equivalent to hold the camshaft sprocket while removing the mounting bolt in the center.

To install:

12. Using a vise, slowly compress the plunger into the body of the ten-

sioner and install a pin through the body of the tensioner to retain the plunger.

13. Use the special tool to hold the camshaft sprocket and install the center bolt. Tighten the bolt to 75 ft. lbs. (101 Nm).

14. Using tool MB995035 and MB995026 or equivalent, install the crankshaft sprocket.

15. Make sure the timing marks are still aligned, if not, align the camshaft sprocket timing marks facing each other. Align the crankshaft sprocket timing mark with the mark on the oil pump housing, then turn the crankshaft sprocket backward 1/2 notch.

16. Install the timing belt starting at the crankshaft, go around the water pump sprocket, idler pulley, camshaft sprockets and then around the tensioner pulley.

17. Turn the crankshaft sprocket 1/2 notch to TDC to take up the slack in the belt.

18. Install the tensioner on the engine but do not tighten the bolts.

19. Place a torque wrench on the tensioner pulley and apply 21 ft. lbs. (28 Nm) of torque in the direction of the water pump. Push the tensioner up against the tensioner pulley and tighten the mounting bolts to 23 ft. lbs. (31 Nm).

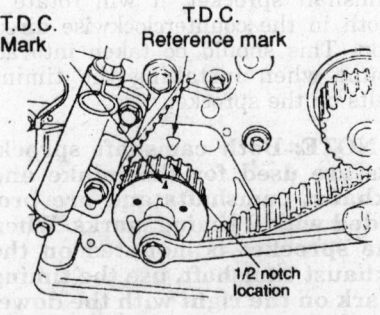

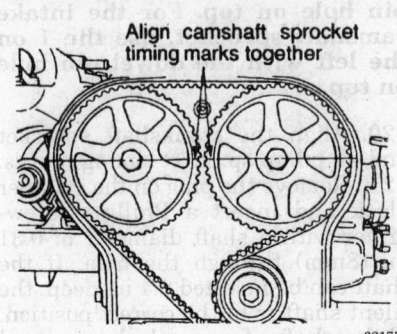

Align camshaft sprocket timing marks together

321757

Engine timing marks — 2.0L (VIN Y) engine

20. Pull the pin out of the tensioner. Belt tension is correct when the pin can be removed and installed.

21. Rotate the crankshaft two revolutions and check the timing marks for alignment. Repeat the previous steps if necessary.

22. Install the timing belt cover.

23. Install the crankshaft pulley.

24. Install the engine mount and bracket. Remove the jack from under the engine.

25. Install the power steering pump bracket and pump.

26. Install the drive belts.

27. Connect the negative battery cable.

Camshaft

REMOVAL AND INSTALLATION

1.8L (VIN B) Engines

1. Disconnect the negative battery cable. Remove the air intake hose and the PCV hose.

2. Remove the valve cover.

3. Remove distributor extension housing if necessary.

4. Remove the outer timing belt.

5. Install auto lash adjuster retainer tools MD998443 or equivalent, on the rocker arm.

6. Remove the camshaft bearing caps, in the reverse order of installation, but do not remove the bolts from the carrier.

7. Remove the rocker arms, rocker shafts and bearing caps from the engine as an assembly.

8. Remove the camshaft from the cylinder head.

9. Inspect the bearing journals on the camshaft for excess wear or damage. Measure the cam lobe height and compare to the desired readings. Inspect the bearing surfaces in the cylinder head. Replace any components that is damaged or shows signs of excess wear.

To install:

10. Lubricate the camshaft journals and camshaft with clean engine oil and install the camshaft in the cylinder head.

11. Align the camshaft bearing caps with the arrow mark depending on cylinder numbers and install in numerical order.

12. Apply sealer at the ends of the bearing caps and install the assembly.

13. Tighten the bearing cap bolts in the following sequence: No. 3, No. 2, No. 1 and No. 4 to 85 inch lbs. (10 Nm).

14. Repeat the sequence increasing the torque to 15 ft. lbs. (20 Nm).

15. Install the distributor extension if it was removed.

16. Install the timing belt, valve cover and all related parts.

17. Connect the negative battery cable and check for leaks.

2.0L (VIN E) Engine

1. Relieve the fuel system pressure following proper procedure. Disconnect the negative battery cable.

2. Disconnect the accelerator cable, PCV hoses, breather hoses, spark plug cables and remove the valve cover.

NOTE: Always rotate the crankshaft in a clockwise direction. Make a mark on the back of the timing belt indicating the direction of rotation so it may be reassembled in the same direction if it is to be reused.

3. Remove the timing belt.

4. Matchmark the crank angle sensor to the cylinder head and remove the crank angle sensor.

5. Remove the rocker cover, semi-circular packing, throttle body stay, both camshaft sprockets, and oil seal.

6. Loosen the bearing cap bolts in 2–3 steps in the reverse order of installation. Label and remove all camshaft bearing caps.

NOTE: If the bearing caps are difficult to remove, use a plastic hammer to gently tap the rear part of the camshaft.

7. Remove the intake and exhaust camshafts.

8. Check the camshaft journals for wear or damage. Check the cam lobes and rocker rollers for damage. Also, check the cylinder head oil holes for clogging.

To install:

9. Lubricate the camshafts with heavy engine oil and position the camshafts on the cylinder head.

NOTE: Do not confuse the intake camshaft with the exhaust camshaft. The intake camshaft has a split on its rear end for driving the crank angle sensor.

10. Make sure the dowel pin on both camshaft sprocket ends are located on the top.

11. Install the bearing caps. Tighten the caps in sequence and in 2 or 3 steps. No. 2 and 5 caps are of the same shape. Check the markings on the caps to identify the cap number and intake/exhaust symbol. Only **L**

(intake) or **R** (exhaust) is stamped on No. 1 bearing cap. Also, make sure the rocker arm is correctly mounted on the lash adjuster and the valve stem end. Tighten the retaining bolts to 15 ft. lbs. (20 Nm).

12. Apply a coating of engine oil to the oil seal. Using a proper sized driver, press-fit the seal into the cylinder head.

13. Install the camshaft sprockets. While holding the camshaft at its hexagon, between number 2 and 3 journals tighten sprocket bolts to 58–72 ft. lbs. (80–100 Nm).

14. Align the punch mark on the crank angle sensor housing with the notch in the plate. With the dowel pin on the sprocket side of the intake camshaft at top, install the crank angle sensor on the cylinder head.

NOTE: Do not position the crank angle sensor with the punch mark positioned opposite the notch; this position will result in incorrect fuel injection and ignition timing.

15. Install the timing belt, covers and related components.

16. Install the valve cover using new gasket. Reconnect all related components.

17. Reconnect the negative battery cable.

2.0L (VIN F) Engines

1. Disconnect the negative battery cable.

2. Disconnect the accelerator cable from the throttle body and remove the cable bracket from the intake plenum.

3. Remove the engine center cover.

4. Disconnect the spark plug cables from the spark plugs. Label them if necessary.

5. Disconnect the breather hose and the PCV hose from the rocker cover.

6. Remove the rocker cover.

7. Position the No. 1 cylinder at TDC on the compression stroke.

8. Remove the timing belt.

9. Remove the camshaft sprockets.

10. Loosen the bearing cap bolts in two or three steps and remove the bearing caps. If the bearing caps are hard to remove, tap the rear of the camshaft with a plastic hammer.

11. Remove the camshaft(s) and the oil seals.

To install:

12. Apply engine oil or assembly lube to the camshafts and install them on the cylinder head.

WARNING

If new camshaft(s) are being installed, remove the rocker arms and install the camshaft(s) and the bearing caps. Make sure the camshaft(s) can be turned by hand. After checking, remove the camshafts and install the rocker arms.

NOTE: Bearing caps and rocker arms must be installed in the same location that they were remove from.

13. Install the bearing caps and tighten the bolts evenly in two or three steps to specifications.

14. Apply engine oil to the lip of the seal. Using MB998713, install the front oil seal.

15. Install the camshaft sprockets.

16. Install the timing belt.

17. Apply sealant to the semi-circular packing and install it in the cylinder head.

18. Apply sealant to the lower part of the front and rear bearing caps where they meet the cylinder head. Use a new gasket and install the rocker cover.

19. Connect the PCV hose and the breather hose.

20. Connect the spark plug wires.

21. Install the center cover.

22. Install and adjust the accelerator cable.

23. Connect the negative battery cable.

2.0L (VIN Y) Engines

1. Disconnect the negative battery cable.

2. Remove the ignition coil pack.

3. Disconnect the PCV hose and the breather hose from the cylinder head cover.

4. Remove the semi-circular packing from the rear of the head.

5. Remove the camshaft position sensor.

6. Remove the timing belt.

7. Use tool MB990767 and MB998719 or equivalent to hold the camshaft sprockets and remove the sprocket mounting bolt and the sprocket.

8. Remove the bracket and the rear timing belt cover.

9. Remove the outside camshaft bearing cap.

10. Gradually loosen the camshaft bearing caps in sequence, one

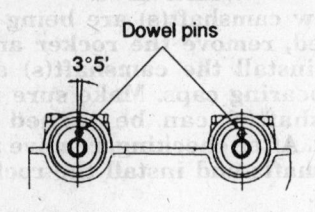

Dowel pins

Exhaust side   **Intake side**

320862

Position the camshafts with the dowels facing up — 2.0L (VIN F) engine

camshaft at a time and remove the bearing caps.

NOTE: Keep the bearing caps in order. They must be installed in the location that they were removed from.

11. Mark the camshafts for later identification and remove the camshafts. The camshafts are not interchangeable.

To install:

12. Apply engine oil or assembly lube to the camshaft and install the camshafts.

13. Install the bearing caps. Tighten the bolts evenly and in sequence.

14. Apply Loctite 518® to the outside camshaft bearing caps and install them.

15. Install the camshaft oil seal.

16. Install the rear timing belt cover and the bracket.

17. Use the special tools and install the camshaft sprockets.

18. Install the timing belt.

19. Apply Loctite 5699® or equivalent to the semi-circular packing and install it in the rear of the cylinder head.

20. Install the camshaft position sensor.

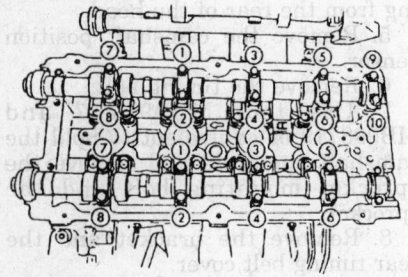

320878

Camshaft bearing cap bolt installation sequence — 2.0L (VIN Y) engine

21. Install the cylinder head cover. Tighten the bolts evenly in three steps in the proper sequence.

22. Install the air, breather and PCV hoses.

23. Install the coil pack.

24. Connect the negative battery cable.

Balance Shaft

REMOVAL AND INSTALLATION

1.8L (VIN B) Engines

NOTE: A special oil seal guide MD998285 or equivalent, is needed to complete this operation.

1. Disconnect the negative battery cable.

2. Remove the oil filter, oil pressure switch, oil gauge sending unit, oil filter mounting bracket and gasket.

3. Raise and safely support the vehicle. Drain the engine oil. Remove the engine oil pan, oil screen and gasket.

4. Lower the vehicle. Remove the timing belts.

5. Remove the front engine cover. Different length bolts are used. Take note of their locations. If the cover sticks to the block, look for a special slot provided and pry with a flat bladed tool. Discard the shaft seal and gasket.

6. Remove the oil pump driven gear flange bolt. When loosening this bolt, first insert a tool approximately 3/8 in. diameter into the plug hole on the left side of the cylinder block to hold the silent shaft. Remove the oil pump gears and remove the front case assembly. Remove the threaded plug, the oil pressure relief spring and plunger.

7. Remove the silent shaft oil seals, the crankshaft oil seal and front case gasket.

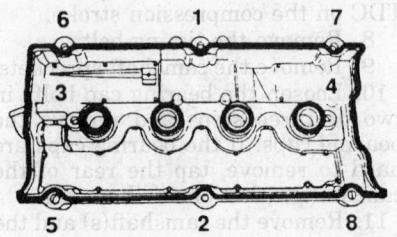

320879

Cylinder head cover bolt installation sequence — 2.0L (VIN Y) engine

8. Remove the silent shafts and inspect as follows:

a. Check the oil holes in the shaft for clogging.

b. Check journals of the shaft for seizure, damage and contact with bearing. If there is anything wrong with the journal, replace the silent shaft bearing, silent shaft or front case.

c. Check the silent shaft oil clearance. If the clearance is beyond the specifications, replace the silent shaft bearing, silent shaft or front case. The specifications for oil clearances are as follows:

- **Right shaft**
- Front — 0.0008–0.0024 in. (0.02–0.06mm)
- Rear — 0.0020–0.0036 in. (0.05–0.09mm)
- **Left shaft**
- Front — 0.0008–0.0021 in. (0.02–0.05mm)
- Rear — 0.0020–0.0036 in. (0.05–0.09mm)

To install:

NOTE: Whenever the oil pump is disassembled or the cover is removed, the gear cavity must be filled with petroleum jelly. This seals the pump and acts as a prime so the oil pump will draw oil as soon as the engine begins to turn. Do not use grease.

9. Lubricate the bearing surface of the shaft and the bearing journals with clean engine oil. Carefully install the silent shafts to the block.

10. Clean the gasket material from the mating surface of the cylinder block and the engine front cover. Install new gasket in place.

11. Using a proper sized driver, install the crankshaft oil seal into the front engine cover.

12. Using the proper sized socket wrench, press in the silent shaft oil seal into the front case.

13. Place pilot tool MD998285 or equivalent, onto the nose of the crankshaft. Apply clean engine oil to the outer circumference of the pilot tool.

14. Install the front case onto the engine block and install the retainer bolts in their original positions. Tighten retainers evenly to 12 ft. lbs. (17 Nm).

15. Install the oil pump relief plunger and spring into the bore in the front case and tighten to 33 ft. lbs. (45 Nm). Make sure a new gasket is in place.

16. Install the oil pimp drive gear and driven gear to the front case, lining up the timing marks. Lubricate the gears with clean engine oil.

17. Insert the Phillips screwdriver into the hole on the side of the engine block, to hold the silent shaft. Install and tighten the flange bolt to 27 ft. lbs. (37 Nm).

18. Install a new oil pump cover gasket in the groove of the front case. When installing the gasket, make sure the round side of the gasket is towards the oil pump cover.

19. Install the oil pump seal into the oil pump cover, making sure the lip is facing the correct direction. The lip of the seal should be installed against the oil it is to stop.

NOTE: The timing of the oil pump sprocket and connected silent shaft can be incorrect, even with the timing mark aligned. Incorrect orientation of the silent shaft will result in engine vibration during operation. Follow the alignment procedure under timing belt.

20. Install the timing belts and all related items. Make sure the timing and the orientation of the silent shafts is correct, using alignment tool in the hole in the left side of the engine block.

21. Install the oil pan, oil filter mounting bracket, oil switches and new oil filter to the engine. Fill the crankcase to the proper level with clean engine oil.

22. Connect the negative battery cable and start the engine. Check for proper timing and inspect for leaks.

2.0L (VIN E and F) Engines

NOTE: A special oil seal guide MD998285 and a plug cap socket tool MD998162 or exact equivalents are needed to complete this operation.

1. Disconnect the negative battery cable.

2. Remove the oil filter, oil pressure switch, oil gauge sending unit, oil filter mounting bracket and gasket. Remove the oil cooler bolt and oil cooler from the oil filter bracket.

3. Raise and safely support the vehicle.

4. Drain the engine oil. Remove engine oil pan, oil screen and gasket. Remove the relief plug, gasket, relief spring and relief plunger.

5. Lower the vehicle. Using the proper equipment, support the weight of the engine.

6. Remove the front engine mount bracket and accessory drive belt.

7. Using special tool MD998162, remove the plug cap in the engine front cover.

8. Remove the plug on the side of the engine block. Insert a Phillips screwdriver with a shank diameter of 0.32 in. (8mm) into the plug hole. This will hold the silent shaft.

9. Remove the driven gear bolt that secures the oil pump driven gear to the silent shaft.

10. Remove and tag the front cover mounting bolts. Note the lengths of the mounting bolts as they are removed for proper installation.

11. Remove the front case cover and oil pump assembly. If necessary, the silent shaft can come out with the cover assembly.

12. Remove the silent shaft oil seals, the crankshaft oil seal and front case gasket.

13. Remove the silent shafts and inspect as follows:

   a. Check the oil holes in the shaft for clogging.

   b. Check journals of the shaft for seizure, damage and contact with bearing. If there is anything wrong with the journal, replace the silent shaft bearing, silent shaft or front case.

   c. Check the silent shaft oil clearance. If the clearance is beyond the specifications, replace the silent shaft bearing, silent shaft or front case. The specifications for oil clearances are as follows:

   • **Right shaft**
   • Front — 0.0012–0.0024 in. (0.03–0.06mm)
   • Rear — 0.0008–0.0021 in. (0.02–0.05mm)
   • **Left shaft**
   • Front — 0.0020–0.0036 in. (0.05–0.09mm)
   • Rear — 0.0017–0.0033 in. (0.04–0.08mm)

To install:

14. Lubricate the bearing surface of the shaft and the bearing journals with clean engine oil. Carefully install the silent shafts to the block.

15. Clean the gasket material from the mating surface of the cylinder block and the engine front cover. Install new gasket in place.

16. Install the oil pump drive gear and driven gear to the front case, lining up the timing marks. Lubricate the gears with clean engine oil. Install the oil pump cover, with a new gasket in place and tighten the mounting bolts to 13 ft. lbs. (18 Nm).

17. Using a proper sized driver, install the crankshaft oil seal into the front engine case.

18. Using the proper sized socket wrench, press in the silent shaft oil seal into the front case.

19. Place pilot tool MD998285 or equivalent, onto the nose of the crankshaft. Apply clean engine oil to the outer circumference of the pilot tool.

20. Install the front case onto the engine block and and temporarily tighten the flange bolts (other than those for tightening the filter bracket). Mount the oil filter bracket with new gasket in place. Install the 4 bolts with washers and tighten to 16 ft. lbs. (22 Nm).

21. Insert the Phillips screwdriver into the hole on the side of the engine block. Secure the oil pump driven gear onto the left silent shaft by tightening the driven gear flange bolt to 29 ft. lbs. (40 Nm).

22. Install a new O-ring onto the groove in the front case. Using special socket tool, install and tighten the plug cap to 20 ft. lbs. (27 Nm).

23. Install the oil pump relief plunger and spring into the bore in the oil filter bracket and tighten to 36 ft. lbs. (50 Nm). Make sure a new gasket is in place.

24. Clean both mating surfaces of the oil pan and the cylinder block. Apply sealant in the groove in the oil pan flange, keeping towards the inside of the bolt holes. The width of the sealant bead applied is to be about 0.16 in. (4mm) wide.

NOTE: After applying sealant to the oil pan, do not exceed 15 minutes before installing the oil pan.

25. Install the oil pan to the engine and secure with the retainers. Tighten bolts to 6 ft. lbs. (8 Nm).

26. Install the oil pressure gauge unit and the oil pressure switch. Connect the electrical harness connector.

27. Install the oil cooler and secure with the bolt. Tighten to 33 ft. lbs. (45 Nm).

28. Install a new oil filter and fill the engine with clean engine oil.

29. Install the timing belts and all related items.

NOTE: The timing of the oil pump sprocket and connected silent shaft can be incorrect, even with the timing mark aligned. Make certain that all special timing belt installation procedures are followed to ensure proper orientation of the silent shafts.

30. Install any remaining components removed during disassembly.

31. Connect the negative battery cable and start the engine. Check for proper timing and inspect for leaks.

Piston and Connecting Rod

POSITIONING

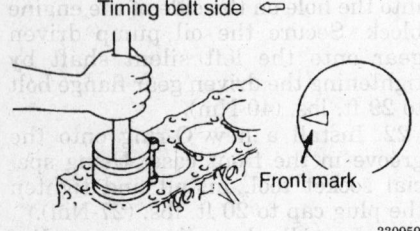

Piston position in the cylinder block

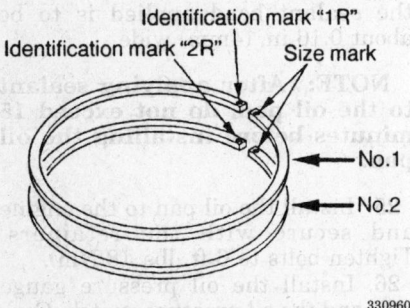

Piston ring identification

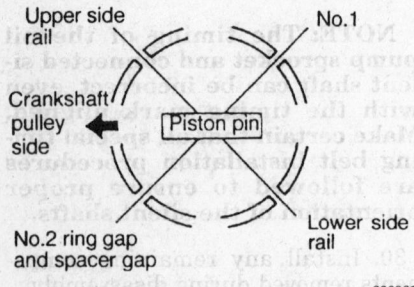

Piston ring end gap

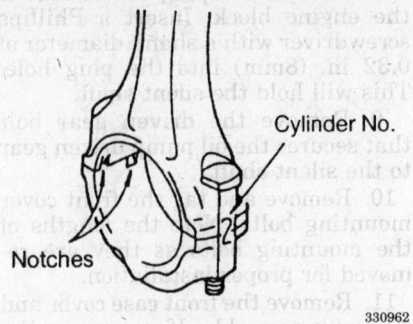

Correct connecting rod installation

ENGINE LUBRICATION

Oil Pan

REMOVAL AND INSTALLATION

1.8L (VIN B) and 2.0L (VIN E) Engines

1. Disconnect the negative battery cable.
2. Raise the vehicle and support safely.
3. Drain the engine oil.
4. Disconnect and lower the exhaust pipe from the manifold.
5. If AWD equipped, remove the transfer assembly and right driveshaft.
6. Support the weight of the engine. Remove the retainer bolts and the center crossmember.
7. Remove the oil pan bolts, followed by the pan.

NOTE: Do not use a chisel, screwdriver or similar tool when removing the oil pan. Damage to components may occur.

8. Inspect the oil pan for damage and cracks. Replace if faulty. Inspect the oil screen for clogging, damage and cracks. Replace if faulty.

To install:

9. Remove all gasket material from the mating surfaces of the cylinder block and oil pan.
10. Apply sealant around the gasket surfaces of pan in such a manner that all the bolt holes are circled and

there is a continuous bead around the entire perimeter of the oil pan.

NOTE: The continuous bead of sealer should be applied in a bead approximately 0.16 in. (4mm) in diameter.

11. Install the oil pan onto the cylinder block and install the fasteners. Tighten the bolts to 4–6 ft. lbs. (6–8 Nm).
12. Install the crossmember and tighten the crossmember bolts to 72 ft. lbs. (100 Nm).
13. If AWD equipped, install the transfer assembly and right driveshaft, if removed.
14. Connect the exhaust pipe to the manifold with new gasket in place. Tighten the flange nuts to 29 ft. lbs. (40 Nm). Install and tighten the support bolt to 29 ft. lbs. (40 Nm).
15. Lower the vehicle and fill the crankcase to the proper level with clean engine oil.
16. Connect the negative battery cable. Start the engine and check for leaks.

2.0L (VIN F) Engines

FWD Models

1. Disconnect the negative battery cable.
2. Raise the vehicle and support safely.
3. Remove the oil pan drain plug and drain the engine oil.
4. Disconnect and lower the exhaust pipe from the engine manifold.
5. Using the appropriate equipment, support the weight of the engine. Remove the retainer bolts and the center crossmember.
6. Disconnect the return pipe for the turbocharger from the side of the oil pan.
7. Remove the oil pan bolts. Tap in thin prybar between the engine block and the oil pan.

NOTE: Do not use a chisel, screwdriver or similar tool when removing the oil pan. Damage to engine components may occur.

8. Inspect the oil pan for damage and cracks. Replace if faulty. While the pan is removed, inspect the oil screen for clogging, damage and cracks. Replace if faulty.

To install:

9. Using a wire brush or other tool, scrape clean all gasket surfaces of the cylinder block and the oil pan so that all loose material is removed. Clean sealing surfaces of all dirt and oil.
10. Apply sealant around the gasket surfaces of the oil pan in such a

manner that all bolt holes are circled and there is a continuous bead of sealer around the entire perimeter of the oil pan.

NOTE: The continuous bead of sealer should be applied in a bead approximately 0.16 in. (4mm) in diameter.

11. Install the oil pan onto the cylinder block within 15 minutes after applying sealant. Install the fasteners and tighten to 4–6 ft. lbs. (6–8 Nm).

12. Install the oil return pipe using a new gasket. Tighten retainers to 5–7 ft. lbs. (7–10 Nm).

13. Install the crossmember and tighten the mounting bolts to 72 ft. lbs. (100 Nm).

14. Connect the exhaust pipe from the engine manifold with a new gasket in place. Tighten the exhaust pipe-to-manifold flange nuts to 43 ft. lbs. (60 Nm).

15. Install the oil drain plug and tighten to 33 ft. lbs. (42 Nm).

16. Lower the vehicle and fill the crankcase to the proper level with clean engine oil.

17. Connect the negative battery cable. Start the engine and check for leaks.

AWD Models

1. Disconnect the negative battery cable.

2. Raise the vehicle and support safely.

3. Remove the oil pan drain plug and drain the engine oil.

4. Disconnect and lower the exhaust pipe from the engine manifold.

5. Remove the transfer assembly and right driveshaft.

6. Using the appropriate equipment, support the weight of the engine and remove the left member.

7. Disconnect the return pipe for the turbocharger from the side of the oil pan.

8. Remove the oil pan bolts. Tap in thin prybar between the engine block and the oil pan.

NOTE: Do not use a chisel, screwdriver or similar tool when removing the oil pan. Damage to engine components may occur.

9. Inspect the oil pan for damage and cracks. Replace if faulty. While the pan is removed, inspect the oil screen for clogging, damage and cracks. Replace if faulty.

To install:

10. Using a wire brush or other tool, scrape clean all gasket surfaces of the cylinder block and the oil pan so that all loose material is removed.

Clean sealing surfaces of all dirt and oil.

11. Apply sealant around the gasket surfaces of the oil pan in such a manner that all bolt holes are circled and there is a continuous bead of sealer around the entire perimeter of the oil pan.

NOTE: The continuous bead of sealer should be applied in a bead approximately 0.16 in. (4mm) in diameter.

12. Install the oil pan onto the cylinder block within 15 minutes after applying sealant. Install the fasteners and tighten to 4–6 ft. lbs. (6–8 Nm).

13. Install the oil return pipe using a new gasket, if removed. Tighten retainers to 5–7 ft. lbs. (7–10 Nm).

14. Install the left member and tighten the forward retainer bolts to 72 ft. lbs. (100 Nm). Tighten the rearward left member bolts to 58 ft. lbs. (80 Nm).

15. Install the transfer assembly and right driveshaft.

16. Connect the exhaust pipe from the engine manifold with a new gasket in place. Tighten the exhaust pipe-to-manifold flange nuts to 43 ft. lbs. (60 Nm).

17. Install the oil drain plug and tighten to 33 ft. lbs. (42 Nm).

18. Lower the vehicle and fill the crankcase to the proper level with clean engine oil.

19. Connect the negative battery cable. Start the engine and check for leaks.

2.0L (VIN Y) Engines

1. Disconnect the negative battery cable.

2. Raise and safely support the vehicle.

3. Drain the engine oil.

4. Remove the front exhaust pipe.

5. Remove the dipstick and tube assembly.

6. Remove the front plate.

7. Remove the oil pan mounting bolts.

8. Remove the oil pan and gasket.

To install:

9. Apply sealant at the point where the engine block meets the oil pump.

10. Use a new gasket and install the oil pan. Tighten the mounting bolts to 8.9 ft. lbs. (12 Nm).

11. Install the front plate.

12. Install the front exhaust pipe.

13. Install the dipstick and tube assembly.

14. Safely lower the vehicle to the floor.

15. Refill the crankcase with oil to the proper level.

16. Connect the negative battery cable.

17. Start the engine and check for leaks.

Oil Pump

REMOVAL AND INSTALLATION

1.8L (VIN B) Engines

1. Disconnect the negative battery cable.

2. Remove the front engine mount bracket and accessory drive belts. Make sure to support the engine using the proper equipment prior to removing the front engine mount.

3. Remove timing belt upper and lower covers.

4. Remove the timing belt and crankshaft sprocket.

5. Drain the engine oil and remove the engine oil pan.

6. Remove the oil screen and gasket.

7. Remove the retainer bolts, from the oil pump cover. Be sure to notate bolt location, because length differs according to location. Remove the oil pump cover from the front engine case cover.

8. Remove the flange bolt and the oil pump drive and oil pump driven gears.

9. After disassembling the oil pump, clean all components and remove all mating surfaces of gasket material.

To install:

NOTE: Whenever the oil pump is disassembled or the cover removed, the gear cavity must be filled with petroleum jelly for priming purposes. Do not use grease.

10. Align the timing mark on the oil pump drive gear with that on the driven gear and install them into the engine front cover. Apply engine oil to the gears.

11. Insert a Phillips screwdriver with a shank diameter of 0.320 in. (8mm) into the plug hole on the left side of the cylinder to hold the silent shaft. Tighten the flange bolt to 29 ft. lbs. (40 Nm).

12. Install a new oil pump gasket into the groove in the front case. When installing the gasket, face the round side to the oil pump cover.

13. Install a new oil seal into the oil pump cover, making sure the lip of the seal is in the correct direction.

14. Install the oil pump cover onto the engine front case and tighten the bolts to 13 ft. lbs. (18 Nm).

15. Install the oil screen in position with new gasket in place.

16. Clean both mating surfaces of the oil pan and the cylinder block. Apply sealant in the groove in the oil pan flange, keeping towards the inside of the bolt holes. The width of the sealant bead applied is to be about 0.160 in. (4mm) wide.

NOTE: After applying sealant to the oil pan, do not exceed 15 minutes before installing the oil pan.

17. Install the oil pan to the engine and secure with the retainers. Tighten bolts to 6 ft. lbs. (8 Nm).

18. Install the oil filter bracket to the engine with new gasket in place. Tighten the retainer bolts to 15 ft. lbs. (22 Nm).

19. Install the oil pressure sending unit as follows:

  a. Apply a thin bead of sealant to the threaded portion of the oil pressure sensor. Do not allow sealer to contact the end of the threaded portion of the sensor.

  b. Install the sensor and tighten to 8 ft. lbs. (12 Nm). Do not overtighten the sensor.

  c. Connect the electrical harness connector to the sensor.

20. Lubricate the sealing ring on the oil filter with a small amount of clean engine oil. Install new oil filter, filled with clean oil, onto the filter bracket.

21. Fill the engine to the correct level with clean engine oil.

22. Connect the negative battery cable and start the engine. Check the oil pressure, making sure it is at the correct reading. Inspect for leaks.

2.0L (VIN E and F) Engines

NOTE: Whenever the oil pump is disassembled or the cover removed, the gear cavity must be filled with petroleum jelly. This seals the pump and acts like a primer so the oil pump draws oil as soon as the engine turns. Do not use grease.

1. Disconnect the negative battery cable. Rotate the engine so No. 1 cylinder is on Top Dead Center (TDC) of its compression stroke. The timing marks should be aligned at this point.

2. Raise and safely support the vehicle.

3. Drain the engine oil. Lower the vehicle.

4. Using the proper equipment, support the weight of the engine. Remove the front engine mount bracket and accessory drive belts.

5. Remove timing belt upper and lower covers.

6. Remove the timing belt and crankshaft sprocket.

7. Disconnect the electrical connector from the oil pressure sending unit and remove the oil pressure sensor. Remove the oil filter and the oil filter bracket.

8. Remove the oil pan, oil screen and gasket.

9. Using special tool MD998162, remove the plug cap in the engine front cover.

10. Remove the plug on the side of the engine block. Insert a Phillips screwdriver with a shank diameter of 0.32 in. (8mm) into the plug hole. This will hold the silent shaft.

11. Remove the driven gear bolt that secures the oil pump driven gear to the silent shaft.

12. Remove and tag the front cover mounting bolts. Note the lengths of the mounting bolts as they are removed for proper installation.

13. Remove the front case cover and oil pump assembly. If necessary, the silent shaft can come out with the cover assembly.

14. Remove the oil pump cover, located on the back of the engine front cover. Remove the oil pump drive and driven gears.

15. After disassembling the oil pump, clean all components and remove gasket material from mating surfaces.

16. Assemble the oil pump gears into the front case and rotate it to ensure smooth rotation and no looseness. Make sure there is no ridge wear on the contact surface between the front case and the gear surface of the oil pump front cover.

To install:

17. Align the timing mark on the oil pump drive gear with that on the driven gear and install them into the engine front case. Apply engine oil to the gears.

18. Install the oil pump cover and tighten the retainer bolts to 13 ft. lbs. (18 Nm).

19. Using the appropriate driver, install a new crankshaft seal into the front case.

20. Position new front case gasket in place. Set seal guide tool MD998285 on the front end of the crankshaft to protect the seal from damage. Apply a thin coat of oil to the outer circumference of the seal pilot tool.

21. Install the front case assembly through a new front case gasket and temporarily tighten the flange bolts.

22. Mount the oil filter on the bracket with new oil filter bracket gasket in place. Install the 4 bolts with washers and tighten to 25 ft. lbs. (34 Nm).

23. Insert a Phillips screwdriver into a hole in the left side of the engine block to lock the silent shaft in place.

24. Secure the oil pump drive gear onto the left silent shaft by installing and tightening the driven gear bolt to 29 ft. lbs. (40 Nm).

25. Install a new O-ring to the groove in the front case and install the plug cap. Using the special tool MD998162, tighten the cap to 20 ft. lbs. (27 Nm).

26. Install the oil screen in position with new gasket in place.

27. Clean both mating surfaces of the oil pan and the cylinder block. Apply sealant in the groove in the oil pan flange, keeping towards the inside of the bolt holes. The width of the sealant bead applied is to be about 0.016 in. (4mm) wide.

NOTE: After applying sealant to the oil pan, do not exceed 15 minutes before installing the oil pan.

28. Install the oil pan to the engine and secure with the retainers. Tighten bolts to 9 ft. lbs. (12 Nm).

29. Install the oil pressure gauge unit and the oil pressure switch. Connect the electrical harness connector.

30. Install the oil cooler. Secure with oil cooler bolt tightened to 33 ft. lbs. (45 Nm).

31. Refill the crankcase. Install new oil filter.

32. Connect the negative battery cable and start the engine. Verify oil pressure. Inspect for leaks.

2.0L (VIN Y) Engines

1. Disconnect the negative battery cable.

2. Raise the safely support the vehicle.

3. Drain the engine oil.

4. Remove the rear plate.

5. Remove the oil filter and adapter.

6. Remove the oil pan.

7. Remove the oil pick-up tube.

8. Remove the timing belt.

9. Using tool MB995027 or equivalent, remove the crankshaft sprocket.

WARNING

Do not nick the crankshaft sealing surface or the seal bore.

10. Using tool MB995020 or equivalent, remove the crankshaft oil seal.

11. Remove the oil pump mounting bolts.

12. Remove the oil pump.

To install:

13. Apply a bead of the specified sealant to the sealing surface of the oil pump and install a new O-ring into the counter bore on the oil pump discharge passage.

14. Carefully install the oil pump on the crankshaft until seated to the engine block. Tighten the bolts to 17 ft. lbs. (23 Nm).

15. Install a new crankshaft oil seal in the oil pump.

16. Install the crankshaft sprocket using the proper installation tools.

17. Install the timing belt and related components.

18. Install the oil pick-up tube.

19. Apply Loctite®18718 or equivalent at the point where the oil pump meets the engine block.

20. Install the oil pan using a new gasket. Tighten the mounting bolts to 9 ft. lbs. (12 Nm).

21. Use a new O-ring and install the oil filter adapter to the engine. Made sure the roll pin aligns with the hole. Tighten the assembly to 40 ft. lbs. (55 Nm).

22. Install a new oil filter.

23. Install the rear plate.

24. Safely lower the vehicle to the floor.

25. Refill the engine with the proper amount of oil.

26. Start the engine and check for leaks.

TRANSAXLE

Manual Transaxle Assembly

REMOVAL AND INSTALLATION

1993–94 Models

NOTE: If the vehicle is going to be rolled on its wheels while the halfshafts are out of the vehicle, obtain two outer CV-joints or proper equivalent tools and install to the hubs. If the vehicle is rolled without the proper torque applied to the front wheel bearings, the bearings will no longer be usable.

1. Remove the battery and the air intake hoses.

2. Remove the auto-cruise actuator and underhood bracket, located on the passenger side inner fender wall.

3. Drain the transaxle and transfer case fluid, if equipped, into a suitable waste container.

4. Remove the retainer bolt and pull the speedometer cable from the transaxle assembly.

5. Remove the cotter pin securing the select and shift cables and remove the cable ends from the transaxle.

6. Remove the connection for the clutch release cylinder and without disconnecting the hydraulic line, secure aside.

7. Disconnect the backup light switch harness and position aside.

8. Disconnect the starter electrical connections, if necessary, remove the starter motor and position aside.

9. Remove the transaxle mount bracket. Remove the upper transaxle mounting bolts.

10. Raise the vehicle and support safely on jackstands. Remove the undercover and the front wheels.

11. Remove the cotter pin and disconnect the tie rod end from the steering knuckle.

12. Remove the self-locking nut from the halfshafts. Disconnect the lower arm ball joint from the steering knuckle.

13. Remove the halfshafts from the transaxle.

14. On AWD vehicles, disconnect the front exhaust pipe.

15. On AWD vehicles, remove the transfer case by removing the attaching bolts, moving the transfer case to the left and lowering the front side. Remove it from the rear driveshaft. Be careful of the oil seal. Do not allow the driveshaft to hang; once the front is removed from the transfer, tie it up. Cover the transfer case openings to keep out dirt.

16. Remove the cover from the transaxle bellhousing. On AWD models, also remove the crossmember and the triangular gusset.

17. Remove the transaxle lower coupling bolt. It is just above the halfshaft opening on FWD models or transfer case opening on AWD models.

18. Support the weight of the engine from above (chain hoist). Support the transaxle using a transmission jack and remove the remaining lower mounting bolts.

19. On turbocharged vehicles, be careful not to damage the lower radiator hose with the transaxle housing during removal. Wrap tape on both the lower hose and the transaxle housing to prevent damage. Move the transaxle assembly to the right and carefully lower it from the vehicle.

To install:

20. Install the transaxle to the engine and install the mounting bolts. Install the transaxle lower coupling bolt.

21. Install the underpan, crossmember and the triangular gusset.

22. Install the transfer case on AWD vehicles and connect the exhaust pipe.

23. Install the halfshafts, using new circlips on the axle ends. Try to keep the inboard joint straight in relation to the axle. Be careful not to damage the oil seal lip of the transaxle with the serrated part of the halfshaft.

24. Connect the tie rod and ball joint to the steering knuckle.

25. Install the transaxle mount bracket.

26. Install wheels and lower vehicle. Tighten axle shaft nuts to 145–188 ft.lbs.

27. Install the starter motor.

28. Connect the backup light switch and the speedometer cable.

29. Install the clutch release cylinder.

30. Connect the select and shift cables and install new cotter pins.

31. Install the air intake hose.

32. Install the auto-cruise actuator and bracket.

33. Install the battery.

34. Make sure the vehicle is level when refilling the transaxle. Use Hypoid gear oil or equivalent, GL-4 or higher.

35. Check the transaxle and transfer case for proper operation. Make sure the reverse lights come on when in reverse.

1995–97 Models

1. Remove the battery and the air intake hoses.

2. Remove the battery tray and support.

3. If equipped with cruise control, remove the auto-cruise actuator and bracket.

4. Drain the transaxle and transfer case fluid, if equipped, into a suitable container.

5. Remove the charcoal canister and bracket.

6. Disconnect the shift and select cables from the transaxle.

7. Disconnect the back-up light switch and the vehicle speed sensor connectors.

8. Remove the starter assembly.

9. Attach an engine support fixture to the engine and remove the transaxle mounting bolts.

10. Remove the rear roll stopper bracket mounting bolts.

11. Remove the transaxle mounting bracket mounting nuts.

12. Raise the vehicle and remove the engine undercovers.

13. If equipped with all wheel drive, remove the transfer case assembly.

---WARNING---

Do not remove or install the axle shaft nut when the vehicle is on the floor or damage to the bearings will occur.

14. Remove the axle shafts.

15. Remove the slave cylinder from the bell housing but do not disconnect the fluid line. Position it out of the way.

16. Remove the bell housing cover and the right hand center member stay (support).

17. Remove the center member.

18. Place a transmission jack under the transaxle and remove the transaxle mounting bolt.

19. Remove the transaxle mounting and lower the transaxle.

To install:

20. Raise the transaxle into position and install the transaxle mounting. Tighten the through-bolt to 50 ft. lbs. (69 Nm).

21. Install the transaxle assembly mounting bolt. Tighten the bolt to 22–25 ft. lbs. (30–34 Nm).

22. Install the center member assembly and the right hand stay.

23. Install the bell housing cover and the slave cylinder.

24. Install the axle shafts. Make sure to install the washer in the proper direction.

25. Install the engine undercovers and lower the vehicle.

26. Install the transfer case assembly if removed.

27. Install the transaxle mounting bracket mounting nuts.

28. Install the rear roll stopper bracket mounting bolts.

29. Install the transaxle assembly mounting bolts. Tighten the mounting bolts to 35 ft. lbs. (48 Nm).

30. Remove the engine support fixture.

31. Install the starter assembly.

32. Connect the vehicle speed sensor and the back-up light connectors.

33. Install the cruise control actuator if removed.

34. Install the battery tray support and the tray.

35. Install the charcoal canister bracket and the canister.

36. Install the air duct and the air cleaner assembly.

37. Refill the transaxle and the transfer case if equipped with oil.

Clutch Assembly

REMOVAL AND INSTALLATION

1. Remove the transaxle assembly from the vehicle.

2. Remove the pressure plate attaching bolts, pressure plate and clutch disc. If the pressure plate is to be reused, loosen the bolts in a diagonal pattern, 1 or 2 turns at a time. This will prevent warping the the clutch cover assembly.

3. Remove the return clip and pressure plate release bearing. Do not use solvent to clean the bearing.

4. Inspect the clutch release fork and fulcrum for damage or wear. If necessary, remove the release fork and unthread the fulcrum from the transaxle.

5. Carefully inspect the condition of the clutch components and replace any worn or damaged parts.

To install:

6. Inspect the flywheel for heat damage or cracks. Resurface or replace the flywheel as required. Install the flywheel using new bolts.

7. Install the fulcrum and tighten to 25 ft. lbs. (35 Nm). Install the release fork. Apply a coating of multipurpose grease to the point of contact with the fulcrum and the point of contact with the release bearing. Apply a coating of multi-purpose grease to the end of the release cylinder's push rod and the push rod hole in the release fork.

NOTE: When installing the clutch, apply grease to each part, but be careful not to apply excessive grease. Excessive grease will cause clutch slippage and shudder.

8. Apply multi-purpose grease to the clutch release bearing. Pack the bearing inner surface and the groove with grease. Do not apply grease to the resin portion of the bearing. Place the bearing in position and install return clip.

9. Apply a coating of grease to the clutch disc splines and then use a brush to rub it in the grooves. Using a universal clutch disc alignment tool, position the clutch disc on the flywheel. Install the retainer bolts and tighten a little at a time, in a diagonal sequence. Tighten them to a final torque of 16 ft. lbs. (22 Nm). Remove the aligning tool.

10. Install the transaxle assembly and check fluid level.

11. Check for proper clutch operation.

Clutch Master Cylinder

REMOVAL AND INSTALLATION

1. Disconnect the negative battery cable.

2. Remove necessary underhood components in order to gain access to the clutch master cylinder.

---WARNING---

Clutch hydraulic system uses brake fluid. Use care; brake fluid is harmful to painted surfaces.

3. Loosen the clutch fluid line at the cylinder and allow the fluid to drain.

4. Remove the clevis pin retainer at the clutch pedal and remove the washer and clevis pin.

5. Remove the 2 nuts and pull the cylinder from the firewall. A seal should be between the mounting flange and firewall. This seal should be replaced.

To install:

6. Mount master cylinder on firewall studs, using a new seal, and tighten the nuts to 7–11 ft. lbs. (10–15 Nm).

7. Lubricate all pivot points with grease and install the clevis pin.

8. Connect hydraulic line and bleed the system at the slave cylinder using fresh DOT 3 brake fluid.

9. Check the adjustment of the clutch pedal for proper free-play.

10. Connect the negative battery cable.

Clutch Slave Cylinder

REMOVAL AND INSTALLATION

1. Disconnect the negative battery cable. Remove necessary underhood components in order to gain access to the clutch release cylinder.

2. Remove the hydraulic line and allow the system to drain.

3. Remove the bolts and pull the cylinder from the transaxle housing.

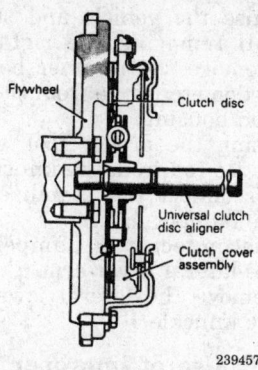

Clutch assembly alignment

239457

To install:

4. Lubricate all pivot points with grease.

5. Mount slave cylinder to transaxle and tighten bolts to 11–16 ft. lbs.

6. Connect hydraulic line and fill the system with clean brake fluid meeting DOT 3 specifications.

7. Bleed the system and adjust the clutch pedal height and the clevis pin play.

Clutch System Bleeding Procedure

——— CAUTION ———
The clutch hydraulic system uses brake fluid. Use care; brake fluid is harmful to painted surfaces.

1. Fill the reservoir with clean brake fluid meeting DOT 3 specifications.

2. Press the clutch pedal to the floor then open the bleeder screw on the slave cylinder.

3. Tighten the bleed screw and release the clutch pedal.

4. Repeat the procedure until the fluid is free of air bubbles.

NOTE: It is suggested that a hose be attached to the bleeder with the other end immersed in a container at least half full of brake fluid during the bleeding operation. Do not allow the reservoir to run out of fluid during bleeding.

Automatic Transaxle Assembly

REMOVAL AND INSTALLATION

1. Disconnect the negative battery cable. Remove the battery and battery tray.

2. On vehicles equipped with auto-cruise, remove the control actuator and bracket.

3. Drain the transaxle fluid.

4. Remove the air cleaner assembly, intercooler and air hose, as required.

5. Mark the shift cable. Remove the adjusting nut and disconnect the shift cable.

6. Remove the dipstick and tube assembly.

7. Disconnect the electrical connectors for the solenoid, neutral safety switch (inhibitor switch), the pulse generator kickdown servo switch and oil temperature sensor.

8. Disconnect the speedometer cable and oil cooler lines.

9. Disconnect the wires to the starter motor and remove the starter.

10. Remove the upper transaxle to engine bolts.

11. Support the transaxle and remove the transaxle mounting bracket.

12. Raise the vehicle and support safely. Remove the sheet metal under guard.

13. Disconnect the tie rod ends and ball joints from the steering knuckle.

14. Remove the halfshafts.

15. On AWD vehicles, disconnect the exhaust pipe and remove the transfer case.

16. Remove the lower bellhousing cover and remove the special bolts holding the flexplate to the torque converter. To remove, turn the engine crankshaft with a box wrench and bring the bolts into a position appropriate for removal, one at a time. After removing the bolts, push the torque converter toward the transaxle so it doesn't stay on the engine allowing oil to pour out the converter hub or cause damage to the converter.

17. Remove the lower transaxle-to-engine bolts and remove the transaxle assembly.

To install:

18. After the torque converter has been mounted on the transaxle, install the transaxle assembly on the engine. Tighten the driveplate bolts to 34–38 ft. lbs. (46–53 Nm). Install the bellhousing cover.

19. On AWD models, install the transfer case and frame pieces. Connect the exhaust pipe using a new gasket.

20. Replace the circlips and install the halfshafts.

21. Install the tie rods and ball joint.

22. Install the transaxle mounting bracket.

23. Install the underguard.

24. Install the starter.

25. Connect the speedometer cable and oil cooler lines.

26. Connect the solenoid, neutral safety switch (inhibitor switch), the pulse generator kickdown servo switch and oil temperature sensor.

27. Install the remaining components.

28. Refill with Dexron II, Mopar ATF Plus type 7176, Mitsubishi Plus ATF or equivalent, automatic transaxle fluid. If the vehicle is AWD check and fill the transfer case.

29. Start the engine and allow to idle for 2 minutes. Apply parking brake and move selector through each gear position, ending in **N**. Recheck fluid level and add if necessary. Fluid level should be between the marks in the **HOT**range.

Transfer Case Assembly

REMOVAL AND INSTALLATION

1. Disconnect the battery negative cable.

2. Raise and properly support vehicle. Drain the transfer assembly.

3. Disconnect the front exhaust pipe, if it interferes with removal.

4. Unbolt the transfer case assembly and remove by sliding it off the rear propeller shaft. Be careful not to damage the oil seal in the transfer case output housing. Do not let the rear propeller shaft hang; suspend it from a frame piece. Cover the opening in the transaxle and transfer case to keep oil from dripping and to keep dirt out.

To install:

5. Lubricate the driveshaft sleeve yoke and oil seal lip on the transfer extension housing. Install the transfer case assembly to the transaxle. Use care when installing the rear propeller shaft to the transfer case output shaft.

6. Tighten the transfer case to transaxle bolts to 40–43 ft. lbs. (55–60 Nm) on manual transaxle vehicles or 43–58 ft. lbs. (60–80 Nm) on automatic transaxle vehicles.

7. Install the exhaust pipe using a new gasket.

8. Refill the transfer case with gear oil of classification GL-4 or higher, SAE 75W-85W or 75W-90. Check fluid level in transaxle and add as required.

DRIVE AXLE

Driveshaft

REMOVAL AND INSTALLATION

1. Disconnect the negative battery cable. Raise the vehicle and support safely.

2. The rear driveshaft is a 3-piece unit, with a front, center and rear propeller shaft. Remove the nuts and insulators from the center support bearing. Work carefully. There will be a number of spacers which will differ from vehicle to vehicle. Check the number of spacers and write down their locations for reference during reassembly.

3. Matchmark the rear differential companion flange and the rear driveshaft flange yoke. Remove the companion shaft bolts and remove the driveshaft, keeping it as straight as possible so as to ensure that the boot is not damaged or pinched. Use care to keep from damaging the oil seal in the output housing of the transfer case.

NOTE: Damage to the boot can be avoided and work will be easier if a piece of cloth or similar material is inserted in the boot.

4. Do not lower the rear of the vehicle or oil will flow from the transfer case. Cover the opening to keep dirt out.

To install:

5. Install the driveshaft to the vehicle and align the matchmarks at the rear yoke.

6. Install the bolts at the rear differential flange and tighten to 22–25 ft. lbs. (30–35 Nm).

7. Install the center support bearing with all spacers in place. Tighten the retaining nuts to 22–25 ft. lbs. (30–35 Nm).

8. Check the fluid levels in the transfer case and rear differential case.

U-Joints

REMOVAL AND INSTALLATION

1. Raise and properly support vehicle. Matchmark the driveshaft (propeller shaft) to the companion flange and remove driveshaft and secure it in a vise.

2. Make mating marks on the yoke and the universal joint that is to be

disassembled. Remove the snaprings from the yoke with snapring pliers.

3. Force out the bearing journals from the yoke using a large C-clamp. Install a collar on the fixed side of the C-clamp. Press the journal bearing into the collar by applying pressure with the C-clamp, on the opposite side.

4. Pull the journal bearing from the yoke.

NOTE: If the journal bearing is hard to remove, strike the yoke with a plastic hammer.

5. Press the journal shaft using C-clamp or similar tool, to remove the remaining bearings.

6. Once all bearings are removed, remove the journal.

To install:

7. Apply multi-purpose grease to the shafts, grease sumps, dust seal lips and needle roller bearings of the replacement U-joint. Do not apply excessive grease. Otherwise, it may be difficult to install the bearing caps and errors in selection of snaprings may result.

8. Press fit the journal bearings to the yoke using a C-clamp as follows:

   a. Install a solid base onto the bottom of the C-clamp.

   b. Insert both bearings into the yoke. Hold and press fit them by tightening the C-clamp.

   c. Install snaprings of the same thickness onto both sides of each yoke.

9. Press the U-joint to one side of the yoke and measure the clearance between the snapring and the groove wall (bearing cap) with a feeler gauge. If the clearance exceeds 0.0004–0.0012 in. (0.01–0.03mm), the snaprings should be replaced with thicker snaprings.

10. Align the matchmark and install the driveshaft in the vehicle.

Front Halfshaft

REMOVAL AND INSTALLATION

NOTE: If the vehicle is going to be rolled while the halfshafts are out of the vehicle, obtain 2 outer CV-joints or proper equivalent tools and install to the hubs. If the vehicle is rolled without the proper torque applied to the front wheel bearings, the bearings will no longer be usable.

1. Disconnect the negative battery cable.

2. Remove the cotter pin, halfshaft nut and washer.

3. Raise the vehicle and support safely. If removing the right halfshaft, remove the retainer bolt and the speedometer drive from the right extension housing.

4. Using the proper tool, disconnect the tie rod from the knuckle.

5. Disconnect the stabilizer link from the damper fork.

6. Disconnect the damper fork from the lateral lower arm.

7. Remove the lateral lower arm from the knuckle.

NOTE: Use of improper methods of joint separation can result in damage to joint, leading to possible failure.

8. On halfshaft with an inner shaft (AWD vehicles), remove the center support bearing bracket bolts and washers. Then remove the halfshaft by setting up a puller on the outside wheel hub and pushing the halfshaft from the front hub. Then tap the shaft union at the joint case with a plastic hammer to remove the halfshaft and inner shaft from the transaxle.

9. On one piece halfshafts (FWD vehicles), remove the halfshaft from the hub/knuckle by setting up a puller on the outside wheel hub and pushing the halfshaft from the front hub. After pressing the outer shaft, insert a prybar between the transaxle case and the halfshaft and pry the shaft from the transaxle.

NOTE: Do not pull on the shaft. Doing so damages the inboard joint. Do not insert the prybar too far or the oil seal in the case may be damaged.

To install:

10. Inspect the halfshaft boot for damage or deterioration. Check the ball joints and splines for wear.

11. Replace the circlips on the ends of the halfshafts.

12. Insert the halfshaft into the transaxle. Make sure it is fully seated.

13. Pull the strut assembly outward and install the other end of the halfshaft into the hub.

14. Install the center bearing bracket bolts and tighten to 33 ft. lbs. (45 Nm), if equipped.

15. Install the washer so the chamfered edge faces outward. Install the halfshaft nut and tighten temporarily.

16. Install the tie rod end and ball joint to the steering knuckle.

17. Install the wheel and lower the vehicle to the floor. Tighten the axle nut with the brakes applied. Tighten

the nut to torque of 145–188 ft. lbs. (200–260 Nm).

18. Install a new cotter pin and bend to secure.

Rear Halfshaft

REMOVAL AND INSTALLATION

1. Raise and safely support the vehicle.
2. Remove the rear wheel(s).
3. If equipped with ABS, remove the wheel speed sensor.
4. Remove the brake caliper and rotor or brake drum.
5. Remove the parking brake shoes or the shoe and lever assembly.
6. Remove the parking brake cable from the backing plate.
7. If equipped with drum brakes, disconnect the brake line from the wheel cylinder.
8. Disconnect the shock from the knuckle.
9. Disconnect the trailing arm and the lower arm from the knuckle.
10. Disconnect the toe control arm from the knuckle.
11. Remove the cotter pin, nut and washer from the axle shaft.
12. Remove the differential mount support.
13. Pull the lower end of the knuckle outward and pry the axle shaft out of the differential housing.
14. Remove the axle shaft from the hub assembly.

To install:
15. Install the axle shaft in the hub assembly.
16. Install a new circlip on the inner shaft and install the shaft in the differential.
17. Install the differential mount support.
18. Install the washer on the axle shaft in the correct direction.
19. Install the nut and torque to specification. If the hole for the cotter pin does not line up, tighten the nut up to 188 ft. lbs. (255 Nm). and install the cotter pin in the first hole that lines up.
20. Connect the toe control arm, trailing arm and the lower arm to the knuckle.
21. Install the lower shock mount to the knuckle.
22. Connect the brake line to the wheel cylinder if removed.
23. Install the parking brake cable through the backing plate.
24. Assembly the remaining brake components.
25. Install the ABS wheel speed sensor if removed.

26. Install the wheel.
27. If equipped with drum brakes, bleed the brake system.

CV-Joint Boot

REPLACEMENT

FWD Models

These vehicles use different types of CV-joints. Engine size, transaxle type, whether the joint is an inboard or outboard joint, even which side of the vehicle is being serviced could make a difference in joint type. Be sure to properly identify the joint before attempting joint or boot replacement. Look for identification numbers at the large end of the boots and/or on the end of the metal retainer bands.

The 2 types of joints used are the Birfield Joint, (B.J.) and the Tripod Joint (T.J.). Special grease and clamps are used with these joints and is normally supplied with the replacement joint and/or boot kit. Do not use regular chassis grease.

Correct installation of the CV-boot is essential for its longevity. A specification is given for the distance between the large and small boot bands. This is so the boot will not be installed either too loose or too tight, which could cause early wear and cracking, allowing the grease to get out and water and dirt in, leading to early joint failure.

Both of the halfshaft boots are going to be removed from the T.J. case side of the halfshaft.

1. Disconnect the negative battery cable. Remove the halfshaft from the vehicle.
2. Remove the T.J. boot bands from the boot. Side cutter pliers can be used to cut off the metal retaining bands. Remove the T.J. case from the halfshaft.
3. Remove the snapring next to the tripod joint spider assembly from the halfshaft with snapring pliers. Remove the spider assembly from the shaft.

NOTE: Do not disassemble the spider and use care in handling.

4. If the boot is be reused, wrap vinyl tape around the spline part of the shaft so the boot will not be damaged when removed. Remove the dynamic damper, if used, and boots from the shaft.

To install:
5. Clean old grease from joint with solvent and blow dry.

6. Double check that the correct replacement parts are being installed. Wrap vinyl tape around the splines to protect the boot and install the boots and damper, if used, in the correct order.
7. Fill the inside of the boot with the specified grease. Often the grease supplied in the replacement parts kit is meant to be divided in half, with half being used to lubricate the joint and half being used inside the boot. Keep grease off the rubber part of the dynamic damper (if used).
8. Secure the boot bands with the halfshaft in a horizontal position. Make sure the boot span on the halfshaft is 3.35 ± 0.12 in. (85 ± 3mm) in length.
9. Install halfshaft into vehicle.
10. Check transaxle fluid level and fill if necessary.

AWD Front Boots

1. Raise and properly support vehicle. Remove the halfshaft from the vehicle.
2. Remove the T.J. large and small boot bands. Remove the T.J. case from inner shaft assembly.
3. Remove the snapring next to the tripod joint spider assembly from the halfshaft with snapring pliers. Remove the spider assembly from the shaft.

NOTE: Do not disassemble the spider and use care in handling.

4. If the boot is be reused, wrap vinyl tape around the spline part of the shaft so the boot will not be damaged when removed.
5. Remove the inner and the outer dust seals from the center support bearing assembly. Remove the center bearing from the shaft.
6. Remove the inner shaft assembly, together with the seal plate, from the T.J. case. Using puller tool, remove the inner shaft from the center bearing bracket.

To install:
7. Clean old grease from joint(s) with solvent and blow dry.
8. Apply multi-purpose grease to the center bearing and inside the the center bearing bracket. Using proper size driver, press fit the center bearing into the center bearing bracket.
9. Apply multi-purpose grease to the rear surfaces of both dust seals and install. Use a pipe to hold the inner race of the center bearing and force the inner shaft into place.
10. Install the boots in place. Apply grease to the inner shaft splines, then press fit it into the T.J. case. Press the seal plate into the T.J. case.

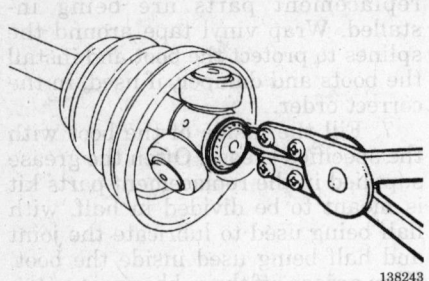

138243

Front inner joint disassembly

11. Fill the joint and the boot with the specified grease, enclosed in the repair kit. Divide the grease in half between the joint and the boot. Keep grease off the rubber part of the dynamic damper (if used).

12. Secure the boot bands with the halfshaft in a horizontal position. Make sure the boot span on the halfshaft is 3.35 ± 0.12 in. (85 ± 3mm) in length.

13. Install the halfshaft into vehicle.

14. Check transaxle fluid level and fill if necessary.

AWD Rear Boots

Both boots are removed from the DOJ (outer) side of the halfshaft.

1. Raise and properly support vehicle. Remove the halfshaft from the vehicle.

2. Remove the outer joint large and small boot bands, and slide boot back out of way.

3. Remove outer case circlip and slide case from inner shaft assembly.

4. Remove the inner snapring, located on the end of the axle shaft. Slide ball and cage assembly from the halfshaft.

NOTE: Do not disassemble the cage and ball assembly.

5. Remove inner boot clamps and slide boot from axle.

To install:

6. Clean old grease from joint(s) with solvent and blow dry.

7. Pack joint(s) with specified grease, enclosed in repair kit. Divide the grease in half between the joint and the boot. Slide boots and clamps onto axle shaft.

8. Reinstall the ball and cage assembly onto the axleshaft. Secure with the snapring.

9. Slide outer cage assemble onto shaft and secure with circlip.

10. Tighten the boot bands with the halfshaft in a horizontal position.

Make sure the boot span on the halfshaft is 3.35 ± 0.12 in. (85 ± 3mm) in length.

11. Install halfshaft into vehicle.

12. Check the differential fluid level and fill if necessary.

STEERING

Air Bag

CAUTION
Some vehicles are equipped with an Air Bag system, also known as the Supplemental Inflatable Restraint (SIR) or Supplemental Restraint System (SRS). The system must be disabled before performing service on or around system components, steering column, instrument panel components, wiring and sensors. Failure to follow safety and disabling procedures could result in accidental Air Bag deployment, possible personal injury and unnecessary system repairs.

PRECAUTIONS

Several precautions must be observed when handling the inflator module to avoid accidental deployment and possible personal injury.

• Never carry the inflator module by the wires or connector on the underside of the module.

• When carrying a live inflator module, hold securely with both hands, and ensure that the bag and trim cover are pointed away.

• Place the inflator module on a bench or other surface with the bag and trim cover facing up.

• With the inflator module on the bench, never place anything on or close to the module which may be thrown in the event of an accidental deployment.

DISARMING

1. Position the front wheels in the straight-ahead position and place the key in the **LOCK** position. Remove the key from the ignition lock cylinder.

2. Disconnect the negative battery cable and insulate the cable end with high-quality electrical tape or similar non-conductive wrapping.

3. Wait at least one minute before working on the vehicle. The Air Bag

system is designed to retain enough voltage to deploy the Air Bag for a short period of time after the battery has been disconnected.

To arm:

4. Connect the negative battery cable, turn the ignition switch to the **ON** position and check the Air Bag warning light for proper operation.

Steering Wheel

REMOVAL AND INSTALLATION

1993–94 Models

1. Disconnect the negative battery cable.

2. Remove the horn pad from the steering wheel. Remove the retainers in the horn pad. Push the pad upward to remove. Disconnect horn button connector.

3. Remove the steering wheel retaining nut.

4. Matchmark the steering wheel to the shaft.

5. Use a steering wheel puller to remove the steering wheel.

WARNING
Do not hammer on steering wheel to remove it. The collapsible column mechanism may be damaged.

To install:

6. Line up the matchmarks and install the steering wheel to the shaft.

7. Tighten the steering wheel attaching nut to 33 ft. lbs. (45 Nm).

8. Reconnect the horn connector and install the horn pad.

9. Connect the negative battery cable.

1995–97 Models

CAUTION
The Air Bag system must be disarmed before removing the steering wheel. Failure to do so may cause accidental deployment, property damage or personal injury.

1. Disconnect the negative battery cable.

2. Remove the lap cooler duct.

3. Remove the Air Bag module and disconnect the clockspring from the steering wheel.

4. Disconnect the horn connector.

5. Remove steering wheel retaining nut.

6. Matchmark the steering wheel to the shaft.

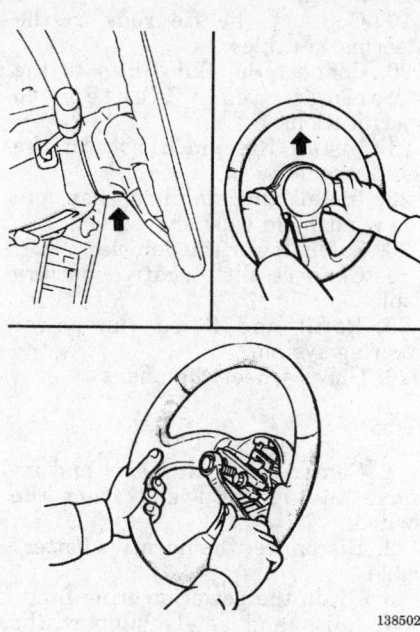

138502

Steering wheel removal

7. Use a steering wheel puller to remove the steering wheel.

———— WARNING ————
Do not hammer on steering wheel to remove it. The collapsible column mechanism may be damaged.

To install:

8. Line up the matchmarks and install the steering wheel to the shaft.

9. Tighten the steering wheel attaching nut to 30 ft. lbs. (41 Nm).

10. Reconnect the horn connector.

11. Install the Air Bag module and reconnect the clockspring.

12. Reinstall the lap cooler duct.

13. Connect the negative battery cable.

Tie Rod Ends

REMOVAL AND INSTALLATION

Outer

1. Raise the front of the vehicle and support it on jackstands. Remove the wheel.

2. Remove the cotter pin and the tie rod ball joint stud nut. Note the position of the steering linkage.

3. Wire brush the threads on the tie rod shaft and lubricate with penetrating oil.

4. Using a suitable ball joint separator tool, remove the tie rod ball joint from the steering knuckle.

5. Loosen the locknut and remove the outer tie rod end from the tie rod. Count the number of complete turns it takes to completely remove it.

To install:

6. Install the new tie rod end, turning it in exactly as many turns as it was to remove the old one. Make sure it is correctly positioned in relationship to the steering linkage.

7. Connect the outer tie rod end to the steering knuckle and install the castle nut. Tighten the nut to 25 ft. lbs. (34 Nm).

8. Install a new cotter pin to the castle nut.

9. Tighten the tie rod end locking nut to 30 ft. lbs. (42 Nm).

10. Install the wheel and tire assembly.

11. Lower the vehicle and perform a front end alignment.

Inner

1. Raise the front of the vehicle and support it on jackstands. Remove the wheel.

2. Remove the cotter pin and the outer tie rod ball joint stud nut. Note the position of the steering linkage.

3. Wire brush the threads on the tie rod shaft and lubricate with penetrating oil.

4. Using a suitable ball joint separator tool, remove the tie rod ball joint from the steering knuckle.

5. Loosen the locknut and remove the tie rod end from the tie rod. Count the number of complete turns it takes to completely remove it.

6. Remove the tie rod-to-steering gear locknut.

7. Remove the clamps that secure the flexible boot to the steering gear.

8. Slide the boot from the inner tie rod and remove the boot.

9. Bend the lock plate tabs from the inner tie rod end nut.

10. Loosen the inner tie rod end nut from the steering gear and remove the inner tie rod end.

To install:

11. Using a new lockplate, install the tie rod end and tighten the tie rod to 65 ft. lbs. (90 Nm).

12. Bend the tabs of the new lock plate to secure the inner tie rod end.

13. Slide the boot onto the steering gear and secure it with new clamps.

14. Install the outer tie rod end to the steering gear locknut.

15. Install the outer tie rod end, turning it in exactly as many turns as it was to remove the old one. Make sure it is correctly positioned in relationship to the steering linkage.

16. Connect the outer tie rod end to the steering knuckle and install the castle nut. Tighten the nut to 25 ft. lbs. (34 Nm).

17. Install a new cotter pin to the castle nut.

18. Tighten the tie rod end locking nut to 30 ft. lbs. (42 Nm).

19. Install the wheel and tire assembly.

20. Lower the vehicle and perform a front end alignment.

Manual Rack and Pinion

REMOVAL AND INSTALLATION

1. Position the wheels in a straight ahead position. Disconnect the negative battery cable. Raise the vehicle and support safely.

2. Remove the bolt holding lower steering column joint to the rack and pinion input shaft.

3. Remove the cotter pins and, using the proper separating tools, disconnect the tie rod ends from the knuckle.

4. Locate the triangular brace near the stabilizer bar brackets on the crossmember and remove both the brace and the stabilizer bar bracket.

5. Place a jack under the center member. Remove the through-bolt from the round roll stopper. Remove the rear bolts from the center crossmember.

6. Disconnect the front exhaust pipe and tie out of the way. Lower the center member slightly.

7. Remove the rack and pinion steering assembly and its rubber mounts. Move the rack to the right to remove from the crossmember. While tilting downward, remove the rack assembly from the left side of the vehicle. Use caution to avoid damaging the boots.

To install:

8. Install the rack and mounting bolts, torquing bolts to 43–58 ft. lbs. (60–80 Nm). When installing the rubber rack mounts, align the projection of the mounting rubber with the indentation in the crossmember.

9. Raise the center member using the jack and install the center support rear bolts. Tighten to 72 ft. lbs. (100 Nm).

10. Install the roll stopper bolt and new nut. Tighten nut to 47 ft. lbs. (65

Nm). Remove the jack supporting the center member.

11. Install the joint assembly and gear box connecting bolt and tighten to 14 ft. lbs. (20 Nm).

12. Reposition the exhaust pipe and connect to the manifold.

13. Install the stabilizer bar brackets and brace.

14. Connect the tie rod ends to the steering knuckles. Install the retaining nuts.

15. Perform a front end alignment.

Power Rack and Pinion

REMOVAL AND INSTALLATION

1993–94 Models

1. Disconnect the negative battery cable. Drain the power steering fluid. Raise the vehicle and support safely.

2. Remove the bolt holding lower steering column joint to the rack and pinion input shaft.

3. Remove the transfer case, if equipped.

4. Remove the cotter pins and using the proper tools, separate the tie rod ends from the steering knuckle.

5. Locate the triangular brace near the stabilizer bar brackets on the crossmember and remove both the brace and the stabilizer bar bracket.

6. Support the center crossmember. Remove the through-bolt from the round roll stopper and remove the rear bolts from the center crossmember.

7. Disconnect the front exhaust pipe, if equipped with FWD.

8. Disconnect the power steering fluid pressure pipe and return hose from the rack fittings. Plug the fittings to prevent excess fluid leakage.

9. Lower the crossmember slightly. Remove the rack and pinion steering assembly and its rubber mounts. Move the rack to the right to remove from the crossmember. Tilt the assembly downward and remove from the left side of the vehicle. Use caution to avoid damaging the boots.

To install:

10. Install the rack and install the mounting bolts. Tighten the mounting bolts to 43–58 ft. lbs. (60–80 Nm). When installing the rubber rack mounts, align the projection of the mounting rubber with the indentation in the crossmember.

11. Connect the power steering fluid lines to the rack.

12. Connect the exhaust pipe, if removed.

13. Raise the crossmember into position. Install the center member mounting bolts and tighten to 72 ft. lbs. (100 Nm). Install the roll stopper bolt and new nut. Tighten nut to 47 ft. lbs. (65 Nm).

14. Install the stabilizer bar brackets and brace.

15. Connect the tie rod ends and tighten nuts to 25 ft. lbs. (34 Nm).

16. Install the transfer case, if removed. Check and fill fluid.

17. Refill the reservoir with power steering fluid and bleed the system.

18. Perform a front end alignment.

1995–97 Models

— **CAUTION** —
The Air Bag system must be disarmed before removing the rack and pinion. Failure to do so may cause accidental deployment, property damage or personal injury.

Non-Turbo

1. Center the front wheel and remove the ignition key from the switch.

2. Disconnect the negative battery cable.

3. Drain the power steering fluid.

4. Raise and safely support the vehicle.

5. Remove the stabilizer bar.

6. Remove the windshield washer reservoir.

7. Remove the pinch bolt from the joint assembly.

8. Disconnect the fluid lines from the steering rack.

9. Using the proper tools, disconnect the tie rod ends from the steering knuckles.

10. Remove the left and right stays (supports).

11. Support the engine and remove the center member.

12. Remove the clamp and the mounting bolts.

13. Disconnect the left lower compression arm from the body side of the vehicle and support it with wire or string.

14. Disconnect the steering rack from the joint assembly and remove the rack from the left side of the vehicle.

To install:

15. Position the steering rack in the vehicle and install the clamp and the mounting bolts. Make sure the rack is centered before connecting it to the joint assembly.

16. Install the left lower compression arm to the body.

17. Install the center member.

18. Install the left and right stays and remove the engine support fixture or jack.

19. Connect the tie rods to the steering knuckles.

20. Connect the fluid lines to the steering rack. Tighten to specifications.

21. Install the pinch bolt in the joint assembly.

22. Install the stabilizer bar and the windshield washer reservoir.

23. Safely lower the vehicle.

24. Connect the negative battery cable.

25. Refill and bleed the power steering system.

26. Check wheel alignment.

Turbo

1. Center the front wheel and remove the ignition key from the switch.

2. Disconnect the negative battery cable.

3. Drain the power steering fluid.

4. Raise and safely support the vehicle.

5. Remove the stabilizer bar.

6. Disconnect the fluid level sensor and remove the brake fluid reservoir and position it out of the way. Do not disconnect the brake hose.

7. Disconnect the electrical connector from the A/C compressor.

8. Remove the A/C compressor from the bracket and position it out of the way. Do not disconnect the hoses.

9. Remove the pinch bolt from the joint assembly.

10. Disconnect the fluid lines from the steering rack.

11. Using the proper tools, disconnect the tie rod ends from the steering knuckles.

12. Remove the left and right stays (supports).

13. Support the engine and remove the center member assembly.

14. Remove the clamp and the mounting bolts.

15. Disconnect the left lower compression arm from the body side of the vehicle and support it with wire or string.

16. Disconnect the steering rack from the joint assembly and remove the rack from the left side of the vehicle.

To install:

17. Position the steering rack in the vehicle and install the clamp and the mounting bolts. Make sure the rack is centered before connecting it to the joint assembly.

18. Install the left lower compression arm to the body.

19. Install the center member assembly.

20. Install the left and right stays.

21. Connect the tie rod ends to the steering knuckles.

22. Connect the fluid lines to the steering rack.

23. Install the pinch bolt in the joint assembly.

24. Install the stabilizer bar.

25. Safely lower the vehicle.

26. Install the A/C compressor and connect the harness connector.

27. Install the brake fluid reservoir and connect the fluid level sensor.

28. Connect the negative battery cable.

29. Refill and bleed the power steering system.

30. Check wheel alignment.

Power Steering Pump

REMOVAL AND INSTALLATION

1. Disconnect the battery negative cable.

2. Remove the pressure switch connector from the side of the pump.

3. If the alternator is located under the oil pump, cover it with a shop towel to protect it from oil.

4. Disconnect the return fluid line. Remove the reservoir cap and allow the return line to drain the fluid from the reservoir. If the fluid is contaminated, disconnect the ignition high tension cable and crank the engine several times to drain the fluid from the gearbox.

5. Disconnect the pressure line.

6. Remove the pump drive belt and unbolt the pump from its bracket.

To install:

7. Install the pump, wrap the belt around the pulley and tighten the mounting bolts.

8. Replace the O-rings and connect the pressure line. Connect the pressure line so the notch in the fitting aligns and contacts the pump's guide bracket.

9. Connect the return line. Connect the pressure switch connector.

10. Adjust the belt tension and tighten the adjusting bolts.

11. Refill the reservoir and bleed the system.

BLEEDING

1. Raise the vehicle and support safely.

2. Manually turn the pump pulley a few times.

3. Turn the steering wheel all the way to the left and to the right 5 or 6 times.

4. Disable the ignition system and, while operating the starter motor intermittently, turn the steering wheel all the way to the left and right 5–6 times for 15–20 seconds. During bleeding, make sure the fluid in the reservoir never falls below the lower position of the filter. If bleeding is attempted with the engine running, the air will be absorbed in the fluid. Bleed only while cranking.

5. Enable the ignition, start engine and allow to idle.

6. Turn the steering wheel left and right until there are no air bubbles in the reservoir. Confirm that the fluid is not milky and the level is up to the specified position on the gauge. Confirm that there is is very little change in the fluid level when the steering wheel is turned. If the fluid level changes more than 0.2 in., the air has not been completely bled. Repeat the process.

BRAKES

Anti-Lock Brake System Service

PRECAUTIONS

• Certain components within the Anti-Lock Brake System (ABS) are not intended to be serviced or repaired individually. Only those components with removal and installation procedures should be serviced.

• Do not use rubber hoses or other parts not specifically specified for and ABS system. When using repair kits, replace all parts included in the kit. Partial or incorrect repair may lead to functional problems and require the replacement of components.

• Lubricate rubber parts with clean, fresh brake fluid to ease assembly. Do not use lubricated shop air to clean parts; damage to rubber components may result.

• Use only specified brake fluid from an unopened container.

• If any hydraulic component or line is removed or replaced, it may be necessary to bleed the entire system.

• A clean repair area is essential. Always clean the reservoir and cap thoroughly before removing the cap. The slightest amount of dirt in the

fluid may plug an orifice and impair the system function. Perform repairs after components have been thoroughly cleaned; use only denatured alcohol to clean components. Do not allow ABS components to come into contact with any substance containing mineral oil; this includes used shop rags.

• The Anti-Lock control unit is a microprocessor similar to other computer units in the vehicle. Ensure that the ignition switch is **OFF** before removing or installing controller harnesses. Avoid static electricity discharge at or near the controller.

• If any arc welding is to be done on the vehicle, the control unit should be unplugged before welding operations begin.

Master Cylinder

REMOVAL AND INSTALLATION

1. Disconnect the negative battery cable.

2. Disconnect the fluid level sensor connector, if equipped.

3. If equipped with M/T, remove the clutch fluid reservoir bracket.

4. If equipped with a 2.0L non-turbo engine, remove the following:

   a. The battery.

   b. The relay assembly mounting bolts.

   c. The washer tank mounting bolts.

5. If equipped with the 2.0L turbo engine, remove the following:

   a. The center member assembly mounting bolts and the engine center member roll stopper.

   b. The engine mount bracket and engine mount.

   c. The A/C compressor mounting bolts, the A/C compressor and the tensioner pulley.

   d. The A/C high pressure hose clamp and mounting bolt.

   e. The power steering pressure hose and return hose clamp mounting bolts.

6. Disconnect the brake lines from the master cylinder. A separate fluid reservoir is used. Plug the lines to prevent drainage.

7. Remove the two nuts securing the master cylinder to the brake booster and remove the master cylinder.

To install:

8. Install master cylinder to the mounting studs and install the mounting nuts. Tighten mounting nuts to 7 ft. lbs. (10 Nm).

9. Connect reservoir hoses to master cylinder and secure with clamps.

10. Fill the reservoir to the proper level with clean DOT 3 brake fluid. Bleed the master cylinder.

11. Connect the brake lines to master cylinder.

12. If equipped with the 2.0L non-turbo engine install the following:

a. The battery.

b. The relay assembly mounting bolts.

c. The washer tank mounting bolts.

13. If equipped with the 2.0L turbo engine, install the following:

a. The center member assembly mounting bolts and the engine center member roll stopper.

b. The engine mount bracket and engine mount.

c. The A/C compressor mounting bolts, the A/C compressor and the tensioner pulley.

d. The A/C high pressure hose clamp and mounting bolt.

e. The power steering pressure hose and return hose clamp mounting bolts.

14. Apply the brake pedal and check for firmness. If the pedal is spongy, air is present in the system. If air remains in the system, bleeding the entire system is required.

15. Check the brakes for proper operation and leaks.

Brake Caliper

REMOVAL AND INSTALLATION

Front

1. Raise the vehicle and support safely.

2. Remove the appropriate tire and wheel assembly.

3. To disconnect the front brake hose, hold the nut on the brake hose side and loosen the flared brake line nut. Remove the brake hose from the caliper.

4. Remove the caliper guide and lock pins and lift the caliper assembly from the caliper support.

To install:

5. Position caliper onto the caliper support. Install the guide pin and lock pin. Tighten guide and locking pins to 23 ft. lbs. (32 Nm) on vehicles built up to May, 1989. On vehicles built during and after May, 1989,

tighten caliper guide and locking pins to 54 ft. lbs. (75 Nm).

NOTE: On some vehicles markings are located on guide pin and the locking pin bolt heads

6. Reconnect the brake hose.

NOTE: Use caution not to twist brake hose during installation.

7. Bleed the brake system.

8. Apply brake pedal and inspect system. Ensure proper operation and no leakage.

9. Install tire and wheel assembly. Tighten lug nuts to 87–101 ft. lbs. (120–140 Nm).

Rear

1. Disconnect the battery negative cable.

2. Raise the vehicle and support safely.

3. Remove the appropriate tire and wheel assemblies. Loosen the parking brake cable adjustment from inside the vehicle.

4. Disconnect the parking brake cable end installed to the rear brake caliper assembly.

5. Remove the caliper lock and guide pins. Lift the caliper assembly from the caliper support.

6. Remove the rear brake hose from the caliper. Remove the caliper from the vehicle.

To install:

7. Install the rear brake hose onto the caliper with new washers in place. If equipped with brake hose retainer bolt, tighten bolt to 25 ft. lbs. (35 Nm) torque. If no bolt is used, tighten the brake hose fitting to 12 ft. lbs. (17 Nm).

NOTE: Do not twist the brake hose during installation.

8. Install the caliper over the brake pads. Lubricate and install the

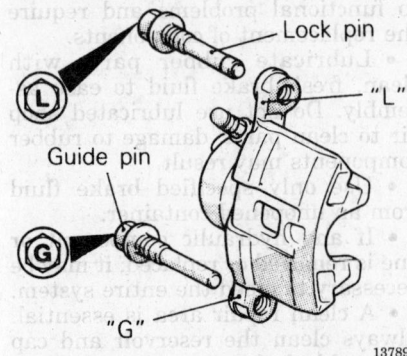

Caliper locking pin and guide pin description

lock pin and tighten to 23 ft. lbs. (32 Nm). Install the guide pin and tighten to 23 ft. lbs. (32 Nm).

9. Bleed the brake system.

10. Inspect the brake system for leaks and ensure proper operation.

11. Install tire and wheel assemblies. Tighten wheels to 87–101 ft. lbs. (120–140 Nm).

12. Properly adjust parking brake cable.

Disc Brake Pads

REMOVAL AND INSTALLATION

── **CAUTION** ──

Brake pads and shoes may contain asbestos, which has been determined to be a cancer causing agent. Never clean the brake surfaces with compressed air! Avoid inhaling any dust from brake surfaces! When cleaning brakes, use commercially available brake cleaning fluids.

Front

1. Remove some of the brake fluid from the master cylinder reservoir. The reservoir should be no more than half full. When the pistons are pressed into the calipers, excess fluid will flow up into the reservoir.

2. Raise the vehicle and support safely.

3. Remove the appropriate tire and wheel assemblies.

4. Remove the caliper guide and lock pins and lift the caliper assembly from the caliper support. Tie the caliper out of the way using wire. Do not allow the caliper to hang by the brake line.

NOTE: On some vehicles, the caliper can be flipped up by leaving the upper pin in place and using it as a pivot point.

5. Remove the brake pads, spring clip and shims. Take note of positioning to aid installation.

6. Install the wheel lug nuts onto the studs and lightly tighten. This is done to hold the disc on the hub.

To install:

7. Use a large C-clamp to compress piston(s) back into caliper bore. On two piston calipers both pistons will have to be retracted together.

8. Lubricate slide points and install the brake pads, shims and spring clip onto the caliper support.

Install the caliper over the brake pads.

NOTE: Be careful that the piston boot does not become caught when lowering the caliper onto the support. Do not twist the brake hose during caliper installation.

9. Lubricate and install the caliper guide and lock pins in their original positions. Tighten guide and locking pins to 23 ft. lbs. (32 Nm) on vehicles built up to May, 1989. On vehicles built during and after May, 1989, tighten caliper guide and locking pins to 54 ft. lbs. (75 Nm).
10. Install the tire and wheel assemblies. Lower the vehicle.

NOTE: Pump brake pedal several times, until firm, before attempting to move vehicle.

11. Road test the vehicle and check brakes for proper operation.

Rear

1. Remove some of the brake fluid from the master cylinder reservoir. The reservoir should be no more than half full. When the pistons are depressed into the calipers, excess fluid will flow up into the reservoir.
2. Raise the vehicle and support safely.
3. Remove the appropriate tire and wheel assemblies. Loosen the parking brake cable adjustment from inside the vehicle.
4. Disconnect the parking brake cable end installed to the rear brake caliper assembly.
5. Remove the caliper lock and guide pins and lift the caliper assembly from the caliper support. Tie the caliper out of the way using wire. Do not allow the caliper to hang by the brake line.
6. Remove the outer shim, brake pads and spring clips from the caliper support. Take note of positioning of each to aid in installation.
7. Install the wheel lug nuts onto the studs and lightly tighten. This is done to hold the disc on the hub.
8. Clean the caliper piston. Using rear disc brake driver tool MB990652 or equivalent, thread the piston into the caliper bore. Be sure at this point, that the stopper groove of the piston correctly fits into the projection on the replacement brake pads rear surface.
To install:
9. Lubricate all sliding and pivot points. Install the brake pads, shims and spring clip to the caliper support.

Install the caliper over the brake pads.

NOTE: Be careful that the piston boot does not become caught when lowering the caliper onto the support. Do not twist the brake hose during caliper installation.

10. Lubricate and install the caliper guide and lock pins. Tighten the pins to 23 ft. lbs. (32 Nm). Attach the parking brake cable to the rear brake assembly.
11. Start the engine and forcefully depress the brake pedal 5–6 times. Apply the parking brake and make sure the adjustment is within specifications. Adjust the parking brake cable, as required.
12. Install the tire and wheel assemblies. Lower the vehicle.
13. Test the brakes for proper operation.

Brake Rotor

REMOVAL AND INSTALLATION

Front

1. Raise the vehicle and support safely. Remove appropriate wheel assembly.
2. Remove the caliper and brake pads. Support the caliper out of the way using wire.
3. The rotor on most models is held to the hub by two small threaded screws. Remove screws, if equipped, and pull off the rotor.
4. Installation is the reverse of the removal process.

Rear

1. Raise the vehicle and support safely. Remove appropriate wheel assembly.
2. Disconnect the parking brake connection at the rear caliper assembly.
3. Remove the caliper and brake pads. Support the caliper out of the way using wire.
4. Remove the brake rotor (disc) from the rear hub assembly.
5. Installation is the reverse of the removal procedure.

Parking Brake Cable

ADJUSTMENT

1993–94 Models

1. Make sure the parking brake cable is free and is not frozen or stick-

ing. With the engine running, forcefully depress the brake pedal 5–6 times.
2. Apply the parking brake while counting the number of notches. Desired parking brake stroke should be 5–7 notches.
3. If adjustment is required, remove the carpeting from inside the floor console. This will expose the adjusting nut within the console.
4. Loosen the locknut on the cable rod. Rotate the adjusting nut to adjust the parking brake stroke to the 5–7 notch setting. After making the adjustment, check there is no looseness between the adjusting nut and the parking brake lever, then tighten the locknut.

NOTE: Do not adjust the parking brake too tight. If the number of notches is less than specification, the cable has been pulled too much and the automatic adjuster will fail or the brakes will drag.

5. After adjusting the lever stroke, raise the rear of the vehicle and safely support. With the parking brake lever in the released position, turn the rear wheels to confirm that the rear brakes are not dragging.
6. Check that the parking brake holds the vehicle on an incline.

1995–97 Models

1. Pull the parking brake lever with a force of approx. 45 lbs. (196 Nm) and count the number of notches. Standard value is 3–5 notches.

—————— **CAUTION** ——————
The 45 lbs. (196 Nm) force of the parking brake lever must be strictly observed.
————————————————

2. If the parking brake lever is not the standard value, adjust in the following manner:
   a. Remove the inner compartment mat of the floor console
   b. Loosen the adjusting nut at the end of the cable rod, freeing the parking brake.
   c. Remove the adjustment hole plug, and then with a suitable tool, turn the adjuster to expand the shoes against the brake drum so that the rotor will not turn. Return the adjuster five notches in the opposite direction.
   d. Turn the parking brake adjusting nut and adjust the parking brake lever stroke to within the standard value.

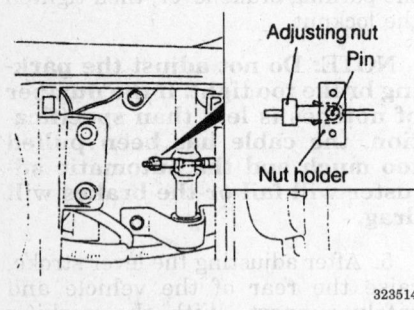

Parking brake shoe adjustment

Adjusting nut and pin

── **CAUTION** ──

If the number of brake lever notches engaged is less than the standard value, the cable has been pulled excessively. Be sure to adjust it within the standard value.

e. After making the adjustment, check to be sure that there is no play between the adjusting nut and the pin. Also check that the adjusting nut is securely held at the nut holder.

f. After adjusting the parking brake lever stroke, safely raise and support the rear of the vehicle and with the parking brake lever in the released position, turn the rear wheels to confirm that there is no brake drag.

REMOVAL AND INSTALLATION

1. Disconnect the negative battery cable.
2. Remove the floor console from the vehicle as follows:

a. Remove the screw plugs in the side covers. Remove the retainer

screws and the side covers from the vehicle.

b. Remove the front mounting screw cover from the floor console. Remove the manual transaxle shift lever knob.

c. Remove the cup holder and the carpet inserts from the floor console assembly.

d. Label and disconnect the electrical wire harness connections for the floor console.

e. Remove the mounting bolts and the floor console from the vehicle.

3. Loosen the cable adjusting nut and disconnect the rear brake cables from the actuator. Remove the center cable clamp and grommet.

4. Raise the vehicle and support safely. Remove the parking brake cable clip and retainer spring. Disconnect the cable end from the parking brake assembly.

5. Unfasten any remaining frame retainers and remove the cables from the vehicle.

To install:

6. The parking brake cables may be color coded to indicate side. Check the parking brake cables for an identification mark. If present, position the cables as follows:

a. AWD vehicle — yellow cable goes on left side

b. AWD vehicle — orange cable goes on right side

c. FWD vehicle — white cable goes on right side

d. FWD vehicle — no color marking goes on left side

7. Install the cable to the rear actuator. Secure in place with the parking brake cable clip and retainer spring.

8. Position the cable in the under the vehicle and install retainers loose.

9. Reattach the parking brake cables to the actuator inside the vehicle. Tighten the adjusting nut until the proper tension is placed on the cable. Adjust the parking brake stroke using appropriate method.

10. Secure all cable retainers. Apply and release the parking brake a number of times once all adjustments have been made. With the rear wheels raised, make sure the parking brake is not causing excess drag on the rear wheels.

11. Install the floor console assembly as follows:

a. Install the floor console in position in the vehicle. Position the seat belts as required. Install the console retainer bolts.

b. Reconnect the electrical harness connectors to the vehicle body harness.

c. Install the carpet and the cup holder to the console assembly. Install the manual shift knob.

d. Install the side covers and retainers. Cover retainer screws with plugs.

e. Connect the negative battery cable and check console electrical components for proper operation.

12. Road test the vehicle and check for proper parking brake operation. Check that the parking brake holds the vehicle on an incline.

Brake System

Bleeding Procedure

Bleeding the brake system is required anytime the normally closed system has been opened to the atmosphere. When bleeding the system, keep the brake fluid level in the master cylinder reservoir above half full. If the reservoir is empty, air will be pushed through the system. Hydraulic brake systems must be totally flushed if the fluid becomes contaminated with water, dirt or other corrosive chemicals. To flush, bleed the entire system until all fluid has been replaced with the correct type of new fluid.

NOTE: If using a pressure bleeder, follow the instructions furnished with the unit and choose the correct adaptor for the application. Do not substitute an adapter that '"almost fits" as it will not work and could be dangerous.

Master Cylinder

Due to the location of the fluid reservoir, bench bleeding of the master cylinder is not recommended. The master cylinder is to be bled while mounted on the brake booster. If the fluid reservoir runs dry, bleeding of the entire system will be necessary. Two people will be required to bleed the brake system.

1. Fill the brake fluid reservoir with clean brake fluid. Disconnect the brake tube from the master cylinder.

2. Have a helper slowly depress the brake pedal. Once depressed, hold it in that position. Brake fluid will be expelled from the master cylinder.

3. While the pedal is held down, use a finger to close the outlet port of the master cylinder. While the port is closed, have the helper release the brake pedal.

4. Repeat this procedure until all air is bled from the master cylinder. Check the brake fluid in the reservoir every 4–5 times, making sure the reservoir does not run dry. Add clean DOT 3 brake fluid to the reservoir as needed. All air is bled from the master cylinder when the fluid expelled from the port is free of bubbles.

5. Connect the brake tube to the port on the master cylinder and tighten to 10 ft. lbs. (14 Nm). Add clean fluid to fill the reservoir to the appropriate level.

6. Pressurize the system and check for leaks. On ABS cars, turn the key **ON** until the system warning light goes off. Then the remainder of the system can be bleed with the key **OFF**, following normal bleeding procedure.

Calipers

1. Fill the master cylinder with fresh brake fluid. Check the level often during this procedure. Raise and safely support the vehicle.

NOTE: On ABS cars, system must be bleed with the key in the OFF position.

2. Starting with the wheel farthest from the master cylinder, remove the protective cap from the bleeder and place where it will not be lost. Clean the bleeder screw.

3. Start the engine and run at idle.

4. If the system is empty, the most efficient way to get fluid down to the wheel is to loosen the bleeder about ½–¾ turn, place a finger firmly over the bleeder and have a helper pump the brakes slowly until fluid comes out the bleeder. Once fluid is at the bleeder, close it before the pedal is released.

NOTE: If the pedal is pumped rapidly, the fluid will churn and create small air bubbles, which are almost impossible to remove from the system. These air bubbles will accumulate and a spongy pedal will result. Also note, it is important not to exceed normal pedal travel during bleeding procedure. This will prevent possible master cylinder piston(s) damage, due to build-up on the bore walls.

5. Once fluid has been pumped to the caliper, open the bleed screw again, have a helper press the brake pedal, lock the bleeder and have the helper slowly release the pedal. Wait 15 seconds and repeat the procedure (including the 15 second wait) until no more air comes out of the bleeder upon application of the brake pedal. Remember to close the bleeder before the pedal is released inside the vehicle each time the bleeder is opened. If not, air will be introduced into the system.

6. If a helper is not available, connect a small hose to the bleeder, place the end in a container of brake fluid and proceed to pump the pedal from inside the vehicle until no more air comes out the bleeder. The hose will prevent air from entering the system.

7. Repeat the procedure on the remaining calipers in the following order:
   a. Left front caliper
   b. Left rear caliper
   c. Right front caliper

8. Pressurize the system and check for fluid leaks. Install the bleeder cap on the bleeder to keep dirt out.

9. Always road test the vehicle after brake work of any kind is done.

Wheel Speed Sensor

REMOVAL AND INSTALLATION

Front

1. Disconnect the negative battery cable. Wait at least 90 seconds before performing any work.

2. Raise and safely support the vehicle. Remove the necessary tire and wheel assembly.

3. Remove the fender splash shield.

4. Disconnect the ABS speed sensor connector.

5. Remove the sensor harness clamp bolts and clamps.

6. Remove the ABS speed sensor mounting bolt and the sensor.

To install:

7. Install the ABS speed sensor with its mounting bolt.

NOTE: The clearance between the wheel speed sensor and the rotor's toothed surface is not adjustable, but measure the distance between the sensor installation surface and the rotor's toothed surface. Standard value is: 1.11–1.12 in. If not within specifications, replace the speed sensor or the toothed rotor.

8. Reinstall the sensor harness with its clamps and bolts.

9. Reconnect the speed sensor connector.

10. Install the fender splash shield.

11. Reinstall the tire and wheel, safely lower the vehicle, and reconnect the negative battery cable.

Rear

1. Disconnect the negative battery cable. Wait at least 90 seconds before performing any work.

2. Raise and safely support the vehicle. Remove the necessary tire and wheel assembly.

3. Disconnect the ABS speed sensor connector.

4. Remove the sensor harness clamp bolts and clamps.

5. Remove the ABS speed sensor mounting bolt and the sensor.

To install:

6. Install the ABS speed sensor with its mounting bolt.

NOTE: The clearance between the wheel speed sensor and the rotor's toothed surface is not adjustable, but measure the distance between the sensor installation surface and the rotor's toothed surface. Standard value is: 1.11–1.12 in. If not within specifications, replace the speed sensor or the toothed rotor.

7. Reinstall the sensor harness with its clamps and bolts.

8. Reconnect the speed sensor connector.

9. Reinstall the tire and wheel, safely lower the vehicle, and reconnect the negative battery cable.

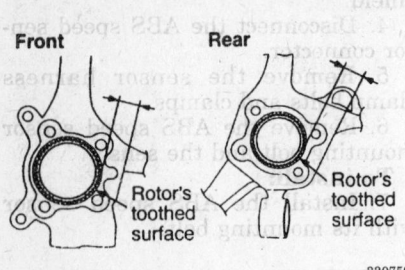

ABS speed sensor clearance

330759

Front   Rear

Rotor's toothed surface

Rotor's toothed surface

FRONT SUSPENSION

Strut

REMOVAL AND INSTALLATION

1. Disconnect the negative battery cable.
2. Raise and safely support vehicle.
3. Remove the brake hose and tube bracket retainer bolt and bracket from the front strut. Do not pry the brake hose and tube clamp away when removing.
4. If equipped with ABS, disconnect the front speed sensor mounting clamp from the strut.
5. Support the lower arm using floor jack or equivalent. Remove the lower strut-to-knuckle bolts. Once the mounting bolts have been removed, jack up the lower arm. Use a piece of wire to attach the brake hose, tube and driveshaft to the knuckle and to help keep the weight off. These components are not to be pulled.
6. Before removing the top bolts, make matchmarks on the body and the strut insulator for proper reassembly. If this plate is installed improperly, the wheel alignment will be wrong. Remove the strut upper mounting bolts. Remove the strut assembly from the vehicle.

To install:

7. Install the dust cover, bump rubber, cup assembly, upper bushing, collar, upper spring pad and bracket assembly on the strut.
8. Install the upper bushing and washer on the piston rod.
9. Install a new self locking nut on the piston rod. Temporarily tighten the nut.

10. Carefully remove the spring compressor from the spring. Tighten the self locking nut to 16 ft. lbs. (25 Nm).
11. Position the strut assembly in the damper fork and install the mounting bolt.
12. Pass the studs in the upper bracket assembly through the holes in the inner fender and install the three mounting nuts.
13. Connect the stabilizer link to the damper fork.
14. Install the wheel assembly.
15. Safely lower the vehicle to the floor.
16. Check and adjust the front wheel alignment if necessary.

Upper Ball Joints

REMOVAL AND INSTALLATION

The upper ball joint is an integral part of the upper control arm. If the upper ball joint is to be serviced, the upper control arm will have to be replaced.

Lower Ball Joints

REMOVAL AND INSTALLATION

The lower ball joint is an integral part of the lower control arm assembly, and cannot be serviced separately. A worn or damaged ball joint, requires replacement of lower control arm assembly.

Upper Control Arms

REMOVAL AND INSTALLATION

1. Raise and safely support the vehicle.
2. Remove the front wheel(s).
3. Disconnect the upper ball joint from the steering knuckle.
4. Remove the upper shaft mounting nuts from the body.
5. Remove the upper arm.
6. Remove the through-bolts that attach the upper arm to the shafts.

To install:

7. Assembly the upper arm to the shafts at the proper angle. Tighten the through-bolts and nuts to 41 ft. lbs. (57 Nm). The proper angle is 84–86°. After the arm and the shafts are connected at the right angle, measure dimensions A and B to insure correct assembly.

- A — 11.8 in. (299.9mm)
- B — 9.2 in. (234.0mm)

8. Install the control arm assembly to the body with new self locking nuts. Tighten the self locking nuts to 62 ft. lbs. (86 Nm).
9. Connect the upper ball joint to the steering knuckle with a new self locking nut. Tighten the locking nut to 20 ft. lbs. (28 Nm).
10. Install the front wheel(s).
11. Perform front wheel alignment and adjust if necessary.
12. Safely lower the vehicle to the floor.

Lower Control Arms

REMOVAL AND INSTALLATION

1993–94 Models

1. Raise the vehicle and support safely.
2. Remove the sway bar links from the lower control arm. Remove the joint cups and bushings, if equipped.
3. Disconnect the ball joint stud from the steering knuckle.
4. Remove the inner lower arm mounting bolts and nut.
5. Remove the rear mounting bolts. Remove the rear retainer clamp if equipped.
6. Remove the arm from the vehicle.
7. Remove the rear rod bushing, if service is required.

To install:

8. Assemble the control arm and bushing. Install the control arm to the vehicle and install the inner mounting bolts. Install new nut and snug temporarily.
9. Install the rear mount clamp, bolts and replacement nuts. Tighten the clamp mounting nuts to 34 ft. lbs. (47 Nm). Temporarily tighten the clamp mounting bolt. Once the weight of the vehicle is on the suspension, the bolt will be tightened to 72 ft. lbs. (100 Nm).
10. Connect the ball joint stud to the knuckle. Install a new nut and tighten to 43–52 ft. lbs. (60–72 Nm).
11. Install the sway bar links.
12. Lower the vehicle to the floor for the final torquing of the inner frame mount bolt.
13. Once the full weight of the vehicle is on the suspension, tighten the inner lower arm mounting bolt nuts to 87 ft. lbs. (120 Nm). Tighten the inner clamp mounting bolt to 72 ft. lbs. (100 Nm).
14. Inspect all suspension bolts, making sure they all have been fully tightened.

1995–97 Models

The compression lower arm ball joint and the lateral lower arm ball joint are integral parts of the arms. If the ball joints are to be replaced, the arms must be replaced.

Compression Lower Arm

1. Raise and safely support the vehicle.
2. Remove the wheel.
3. Disconnect the compression lower arm ball joint from the steering knuckle.
4. Remove the two mounting bolts and remove the compression lower arm.
   To install:
5. Install the compression lower arm. Tighten the two mounting bolts to 60 ft. lbs. (81 Nm).
6. Using a new self locking nut, connect the ball joint to the knuckle assembly. Tighten the nut to 43–51 ft. lbs. (59–71 Nm).
7. Install the wheel, and lower the vehicle to the floor.
8. Check and adjust the front wheel alignment if necessary.

Lateral Lower Arm

1. Raise and safely support the vehicle.
2. Remove the wheel.
3. Remove the stay (bracket).
4. Remove the strut lower mounting bolts.
5. Disconnect the lateral lower arm from the knuckle assembly.
6. Remove the lateral lower arm mounting bolts and the lateral arm.
   To install:
7. Install the lateral lower arm and temporarily install the mounting bolts. Do not tighten the bolt until the vehicle is on the floor at normal riding height.
8. Use a new self locking nut and connect the ball joint to the knuckle assembly. Tighten the self locking nut to 43–51 ft. lbs. (59–71 Nm).
9. Install the strut lower mounting bolts, secure with the nut and tighten to 64 ft. lbs. (88 Nm).
10. Install the stay with the bolts and tighten them to 51–58 ft. lbs. (69–78 Nm).
11. Install the wheel assembly and lower the vehicle to the floor.
12. Tighten the lateral lower arm through-bolt and nut to 71–85 ft. lbs. (98–118 Nm).
13. Check and adjust the front wheel alignment if necessary.

Sway Bar

REMOVAL AND INSTALLATION

1993–94 Models

FWD Vehicle

1. Disconnect the negative battery cable.
2. Raise and safely support vehicle. Remove the front exhaust pipe and gasket from the manifold and using wire, tie it down and out of the way.

NOTE: When relocating the front exhaust pipe, make sure the flexible joint is not bent more than a few degrees or damage to the pipe joint may occur.

3. Remove the center crossmember rear installation bolts.
4. Remove the stabilizer link bolts. On the pillow-ball type joint, hold the ball stud with a hex wrench and remove the self-locking nut with a box wrench.
5. Remove the stabilizer bar bolts and mounts.
6. Remove the bar from the vehicle.
   a. Pull both ends of the stabilizer bar toward the rear of the vehicle.
   b. Move the right stabilizer bar end until the it clears the lower arm.
   c. Remove the stabilizer bar out the right side of the vehicle.
7. Inspect all bushings for wear and deterioration and replace as required. Check the stabilizer bar for damage, and replace as required.
   To install:
8. Install the stabilizer bar into the vehicle.
9. Install the stabilizer bar brackets on the vehicle, following any side locating markings on the brackets. Temporarily tighten the stabilizer bar bracket. Align the bushing end with the marked part of the stabilizer bar and then fully tighten the stabilizer bar bracket.
10. If equipped with the pillow-ball type mounting, install the stabilizer bar links and link mounting nuts. Using a wrench, secure the ball studs at both ends of the stabilizer link while tightening the nuts. Tighten the nuts on the stabilizer bar bolt so that the distance of bolt protrusion above the top of the nut is 0.63 to 0.70 in. (16 to 18mm).
11. Install the front exhaust pipe with a new gasket in place. Tighten the new self-locking nuts to 29 ft. lbs. (40 Nm).

12. Connect the negative battery cable.

AWD Vehicle

1. Disconnect the negative battery cable.
2. Remove the front exhaust pipe.
3. Remove the center gusset and transfer assembly.
4. Using a wrench to secure the ball studs at both ends of the stabilizer link, remove the link mounting nuts. Remove the stabilizer link.
5. Remove the stabilizer bar bracket installation bolt and bar bracket and bushing.
6. Disconnect the stabilizer bar coupling at the right lower control arm. Pull out the left side stabilizer edge, pulling it out between the drive shaft and the lower arm. Pull out the right side bar below the lower arm.
   To install:
7. Install the bar into the vehicle.
8. Temporarily tighten the stabilizer bar bracket. Align the bushing end with the marked part of the stabilizer bar and then fully tighten the stabilizer bar bracket.
9. Install and tighten the stabilizer bar bracket bolt.
10. Install the stabilizer bar links and link nuts. Using a wrench, secure the ball studs at both ends of the stabilizer link while tightening the nuts. Tighten the nuts on the stabilizer bar bolt so that the distance of bolt protrusion above the top of the nut is 0.63 to 0.70 in. (16 to 18mm).
11. Install the transfer assembly and gusset.
12. Install the left crossmember. Tighten the rear mounting bolts to 58 ft. lbs. (80 Nm) and the front mounting bolts to 72 ft. lbs. (100 Nm).

1995–97 Models

1. Raise and safely support the vehicle.
2. Remove the front wheels.
3. Disconnect the stabilizer link from the bar.
4. Remove the stabilizer bar bracket mounting bolts.
5. Remove the stabilizer bar.
   To install:
6. Position the stabilizer bar so the distance from the marking on the bar and the edge of the bracket is 0.39 in. (10mm). Tighten the mounting bolts to specifications.
7. Connect the links to the stabilizer bar.
8. Install the front wheels.
9. Safely lower the vehicle to the floor.

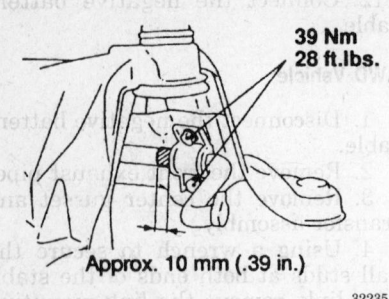

**39 Nm
28 ft.lbs.**

Approx. 10 mm (.39 in.)

322351

Correct position of the stabilizer bar in the bracket

Front Wheel Bearings

REMOVAL AND INSTALLATION

1993–94 Models

1. Disconnect the negative battery cable.
2. Remove the cotter pin from the driveshaft nut and loosen the half-shaft nut.
3. Raise the vehicle and support safely. Remove the halfshaft nut.
4. If equipped with ABS, remove the front wheel speed sensor.
5. If equipped with Active-ECS, disconnect the height sensor from the lower control arm.
6. Remove the caliper assembly and brake pads. Suspend the caliper with a wire.
7. Disconnect the ball joint and tie rod end from the steering knuckle.
8. Remove the halfshaft.
9. Unbolt the lower end of the strut and remove the hub and steering knuckle assembly from the vehicle.
10. Install the hub/knuckle assembly in a vise. Using puller MB991056 or equivalent, remove the hub from the knuckle.

NOTE: Do not use a hammer to accomplish this or the bearing will be damaged.

11. Remove the oil seal from the axle side of the knuckle using a small prying tool.
12. Remove the wheel bearing inner race from the front hub using a puller.

NOTE: Be careful that the front hub does not fall when the inner race is removed.

13. Remove the snapring from the axle side of the knuckle. Remove the bearing from the knuckle using a puller.

14. Once the bearing is removed, the bearing outer race can be removed by tapping out with a brass drift pin and a hammer.

To install:

15. Fill the wheel bearing with multi-purpose grease. Apply a thin coating of multipurpose grease to the knuckle and bearing contact surfaces.
16. Press the wheel bearing into the knuckle using appropriate pressing tool. Once the bearing is installed, install the inner race using the proper driving tool.
17. Drive the oil seal into the knuckle by using the proper size driver. Drive seal into knuckle until it is flush with the knuckle end surface.
18. Using pressing tool MB990998 or equivalent, mount the front hub assembly into the knuckle. Tighten the nut of the pressing tool to 144–188 ft. lbs. (200–260 Nm). Rotate the hub to seat the bearing.
19. Mount the knuckle assembly in a vise. Check the hub assembly turning torque and end-play as follows:

   a. Using a torque wrench and socket MB990998 or equivalent, turn the hub in the knuckle assembly. Note the reading on the torque wrench and compare to the desired reading of 16 inch lbs. (1.8 Nm) or less. This is known as the break-away torque.

   b. Check for roughness when turning the bearing.

   c. Mount a dial indicator on the hub so the pointer contacts the machined surface on the hub.

   d. Check the end-play.

   e. Compare the reading to the limit of 0.002 in. (0.05mm).

20. If the starting torque or the hub end-play are not within specifications while the nut is tightened to 144–188 ft. lbs. (200–260 Nm), the bearing, hub or knuckle have probably not been installed correctly. Repeat the disassembly and assembly procedure and recheck the starting torque and end-play.
21. Install the hub and knuckle assembly onto the vehicle. Install the lower ball joint stud into the steering knuckle and install new nut. Tighten to 52 ft. lbs. (72 Nm).
22. Install the halfshaft.
23. Install the two front strut lower mounting bolts and tighten to 80–94 ft. lbs. (110–130 Nm).
24. Connect the tie rod end. Install new cotter pin.
25. Install the brake disc and caliper assembly.

26. If equipped with Active-ECS, connect the height sensor and tighten the mounting bolt to 15 ft. lbs. (20 Nm).
27. Install the front speed sensor, if removed.

NOTE: When installing front speed sensor, make sure harness is routed in the original position and that it is not twisted.

28. Install the washer and new locknut to the end of the halfshaft. Tighten the locknut snugly.
29. Install the wheel. Lower the vehicle to the ground.
30. With the weight of the vehicle on the ground and the brakes applied, tighten the locknut to 144–188 ft. lbs. (200–260 Nm).
31. Install the cotter pin in the first matching holes and bend it securely.

1995–97 Models

1. Raise and safely support the vehicle.
2. Remove the front wheel.
3. Use tool MB990767 or equivalent to hold the hub assembly while removing the axle nut.
4. On vehicles with ABS, remove the wheel speed sensor.
5. Remove the caliper and suspend it out of the way with wire or string.
6. Remove the brake rotor.
7. Disconnect the steering knuckle from the upper arm.
8. Pull the knuckle away from the vehicle to access the hub mounting bolts on the inboard side of the hub. Be careful not to damage the ball joint boot or the ABS rotor if equipped.
9. Remove the mounting bolts and the front hub assembly.

NOTE: Do not disassemble the hub assembly. If binding or damaged, it must be replaced as a unit.

To install:

10. Install the hub to the knuckle. Tighten the mounting bolts to 65 ft. lbs. (88 Nm).
11. Connect the knuckle to the upper arm.
12. Install the brake rotor and the caliper.
13. Install the wheel speed sensor if removed.
14. Install the axle nut washer in the proper direction. Install the axle nut and tighten to 145–188 ft. lbs. (196–255 Nm).
15. Install the wheel and lower the vehicle to the floor.

REAR SUSPENSION

Strut

REMOVAL AND INSTALLATION

1993–94 Models

NOTE: The strut assembly is a load bearing component, therefore the vehicle chassis and axle weight must be supported separately, requiring the use of two separate lifting devices.

1. Disconnect the negative battery cable.
2. Raise and support vehicle chassis.
3. Raise and support the torsion axle and arm assembly slightly. Make sure the jack does not contact the lateral rod.

NOTE: Always use a wooden block between the jack receptacle and the axle beam. Place the jack at the center of the axle beam.

4. Remove the trunk interior trim to gain access to the top mounting nuts.
5. Remove the top cap and upper strut mounting nuts.
6. Remove the brake tube bracket bolt.
7. Remove the strut lower mounting bolt and remove the assembly from the vehicle.

To install:
8. Position the strut assembly so that the lower mounting bolt can be installed and lightly tightened.
9. Use a jack to raise or lower the axle assembly so that the top strut plate studs aligns through the body. Raise jack to hold the strut assembly in position.
10. Install the top plate nuts on the studs. Tighten the upper shock mounting nuts to 29 ft. lbs. (40 Nm) and the lower mounting bolt to 72 ft. lbs. (100 Nm).
11. Connect brake bracket and lower vehicle.
12. Install top cap and interior trim.

1995–97 Models

1. Remove the service lid in the luggage compartment.
2. Remove the cap and flange nuts securing the upper mounting bracket to the body. Do not remove the larger nut in the center of the strut assembly.
3. Raise and safely support the vehicle.
4. Remove the bolt attaching the lower end of the strut to the knuckle and remove the strut from the vehicle.
5. Use a coil spring compressor to compress the spring.
6. While holding the piston rod, remove the self locking nut.
7. Remove the upper bracket assembly and spring pad.
8. Remove the collar, upper bushing, cup assembly, bump rubber and dust cover.
9. Remove the coil spring from the strut.

To install:
10. Align the end of the coil spring with the stepped part of the spring seat and install the compressed coil spring on the strut.
11. Install the dust cover, bump rubber, cup assembly, upper bushing, collar, upper spring pad and bracket assembly on the strut.
12. Install the upper bushing and washer on the piston rod.
13. Install a new self locking nut on the piston rod. Temporarily tighten the nut.
14. Carefully remove the spring compressor from the spring. Tighten the self locking nut to 16 ft. lbs. (25 Nm).
15. Install the upper bracket of the shock to the vehicle. Tighten the mounting nuts to 32 ft. lbs (44 Nm).
16. Raise the suspension up with a jack or adjustable stand to align the shock absorber lower mounting holes.
17. Install the lower mounting bolt. Tighten the bolt to 71 ft. lbs.
18. Remove the jack or stand and safely lower the vehicle to the floor.
19. Install the cap and service lid.

Upper Control Arms

REMOVAL AND INSTALLATION

1993–94 Models

1. Disconnect the negative battery cable. Raise and safely support vehicle. Remove the tire and wheel assembly.
2. Support the lower control arm. Remove the brake line clamp bolt.
3. Remove the nut and separate the upper ball joint, from the steering knuckle.
4. Matchmark the eccentric on the upper installation bolt and remove.

5. Remove the upper arm from the vehicle.

To install:
6. Install the arm to the vehicle and install the upper arm installation bolt. Align the matchmarks and tighten the nut snugly only.
7. Install the upper ball joint.
8. Install the wheel assembly.
9. Lower the vehicle until the suspension supports its weight. Tighten the upper arm installation bolt to 116 ft. lbs. (160 Nm).
10. Check the rear wheel alignment.

1995–97 Models

1. Raise and safely support the vehicle.
2. Remove the wheel and tire assembly.
3. Remove the through-bolt securing the upper arm to the knuckle.
4. Remove the four bolts securing the upper arm to the vehicle.
5. Remove the upper arm assembly.
6. Remove the through-bolts and nuts and remove the upper arm to body brackets.

To install:
7. Install the upper arm to body brackets to the upper arm. Tighten the bolts and nuts to 41 ft. lbs. (57 Nm).
8. Install the upper arm to the vehicle and tighten the four bolts to 28 ft. lbs. (39 Nm).
9. Install the through-bolt securing the upper arm to the knuckle. Do not tighten the nut until the vehicle is on the floor at normal riding height.
10. Install the wheel.
11. Safely lower the vehicle to the floor and tighten the nut for the through-bolt to 71 ft. lbs. (98 Nm).
12. Check and adjust wheel alignment.

Lower Control Arms

REMOVAL AND INSTALLATION

1993–94 Models

1. Disconnect the negative battery cable.
2. Raise the vehicle and support safely.
3. Remove sway bar links or mounting nuts and bolts from lower control arm. Remove the joint cups and bushings, if equipped.

4. Disconnect the ball joint stud from the steering knuckle, using tool MB990635 or equivalent.

NOTE: It is important to use proper method when separating joints. Damage to joint could occur, resulting in possible failure.

5. Remove the inner lower arm mounting bolts and nut.
6. Remove the rear mount bolts. Remove the rear retainer clamp if equipped.
7. Remove the arm from the vehicle.
8. Remove the rear rod bushing, if service is required.

To install:
9. Assemble the control arm and bushing. Install the control arm to the vehicle and install the inner mounting bolts. Install new nut and snug temporarily.
10. Install the rear mount clamp, bolts and replacement nuts. Tighten the clamp mounting nuts to 34 ft. lbs. (47 Nm). Temporarily tighten the clamp mounting bolts. Once the weight of the vehicle is on the suspension, tightened the bolt to 72 ft. lbs. (100 Nm).
11. Connect the ball joint stud to the knuckle. Install a new nut and tighten to 43–52 ft. lbs. (60–72 Nm).
12. Install the sway bar and links.
13. Lower the vehicle to the floor for the final torquing of the inner frame mount bolt.
14. Once the full weight of the vehicle is on the suspension, tighten the inner lower arm mounting bolt nuts to 87 ft. lbs. (120 Nm). Tighten the inner clamp mounting bolt to 72 ft. lbs. (100 Nm).
15. Inspect all suspension bolts, making sure they all have been fully tightened.
16. Connect the negative battery cable.

1995–97 Models

1. Raise and safely support the vehicle.
2. Remove the wheel and tire assembly.
3. If equipped with a lower arm cover, remove the cover.
4. Disconnect the stabilizer link from the lower arm.
5. If equipped with ABS, remove the wheel speed sensor clamp bolts.
6. Remove the through-bolt securing the lower arm to the knuckle.
7. Remove the through-bolt securing the lower arm to the suspension crossmember.
8. Remove the lower control arm.

To install:
9. Install the lower arm to the suspension crossmember and the knuckle but do not tighten the through-bolts until the vehicle is on the floor at normal riding height.
10. Install the ABS wheel speed sensor clamp bolts to the lower arm, if removed.
11. Connect the stabilizer link to the lower arm.
12. Install the lower arm cover if removed.
13. Install the wheel and tire assembly.
14. Safely lower the vehicle to the floor and tighten the through-bolts to 71 ft. lbs. (98 Nm).

Sway Bar

REMOVAL AND INSTALLATION

1993–94 Models

1. Raise and support the vehicle safely.
2. Place a jack under the rear axle and suspension assembly.
3. Remove the self-locking nuts and crossmember bracket.
4. Remove the retainer bolts and the stabilizer bar brackets. Remove the bushing.
5. Hold the stabilizer bar with a wrench. Remove the self-locking nut.
6. Once the stabilizer bar nut is removed, remove the joint cups and stabilizer rubber bushing.
7. Hold the stabilizer link with a wrench and remove the self-locking nuts. Remove the stabilizer link.
8. Lower the jack supporting the rear axle slightly. Maintain a slight gap between the rear suspension and the body of the vehicle.
9. Remove the stabilizer bar.
10. Inspect the bar for damage, wear and deterioration and replace as required.

To install:
11. Install the stabilizer bar into the vehicle. Raise the rear axle and suspension into place.
12. Install the stabilizer link into the stabilizer bar and install a new self-locking nut. Tighten the nut to 33 ft. lbs. (45 Nm).
13. Install the joint cups and stabilizer rubber to the link. Install a new self-locking nut onto the link. While holding the stabilizer link ball studs with a wrench, tighten the self-locking nut so the protrusion of the stabilizer link is within 0.354–0.433 in. (9–11mm).

14. Install the center stabilizer bar bushings, brackets and bolts. Tighten the bolts to 10 ft. lbs. (14 Nm).
15. Install the parking brake cable and rear speed sensor installation bolt.
16. Install the crossmember bracket and tighten the bolt to 61 ft. lbs. (85 Nm). Tighten the crossmember bracket mounting nut to 94 ft. lbs. (130 Nm).
17. Install the rubber insulators and new self-locking nuts onto the crossmember brackets. Tighten the nuts to 80–94 ft. lbs. (110–130 Nm).
18. Lower the vehicle.

1995–97 Models

1. Raise and safely support the vehicle.
2. Remove the rear wheels.
3. If equipped with lower control arm covers, remove them.
4. Remove the stabilizer link mounting nuts and remove the link.
5. Remove the bolts securing the stabilizer bar brackets.
6. Remove the stabilizer bar from the vehicle.

To install:
7. Install the stabilizer bar so the identification mark is on the left and install the bushing and brackets. Tighten the mounting bolts to 7–10 ft. lbs. (9–14 Nm).
8. Install the stabilizer links and tighten the nuts to 28 ft. lbs. (39 Nm).
9. Install the lower control arm cover if removed.
10. Install the wheels and lower the vehicle to the floor.

Wheel Bearings

ADJUSTMENT

1993–94 Models

1. Raise and properly support vehicle. Remove appropriate tire and wheel assembly.
2. Remove the locknut hub cap, brake caliper and rotor.
3. Set up dial indicator on hub surface; move hub assembly in and out to measure total end-play.

Limit: 0.004 in. (0.01mm) or less on FWD and 0.031 in. (0.80mm) or less on AWD.

4. If end-play is not in specified limit, attempt to adjust before replacing bearing assembly.

To adjust:

NOTE: On AWD vehicles it will be necessary to partially remove axle shaft to gain access to hub nut.

5. Securely hold hub assembly, with special tool C-3281 or equivalent, loosen hub locknut.

6. While holding hub assembly tighten locknut to 144–188 ft. lbs. (200–260 Nm).

7. Recheck end-play, if still not within specifications, replace bearing assembly.

1995-97 Models

The wheel bearing play is not adjustable. If the wheel bearing play is not within specifications, the hub assembly must be replaced.

FWD Vehicles

1. Release the parking brake and remove the brake drum.

2. If equipped with rear disc brakes, remove the caliper assembly and the brake disc (rotor).

3. Place a dial gauge against the hub surface; then move the hub in the axial direction and check whether or not there is end-play. Specification is 0.002 in (0.05mm).

4. To check for rotary sliding resistance: turn the hub a few times to seat the bearing. Wind a rope around the hub bolts and turn the hub by pulling at a 90° angle with a spring balance. Measure to determine whether or not the rotary sliding resistance of the rear hub is at the limit value. Specification is: 3.9 lbs. (18 Nm) or less.

AWD Vehicles

1. Remove the caliper assembly and the brake disc (rotor).

2. Place a dial gauge against the hub surface; then move the hub in the axial direction and check whether or not there is end-play. Specification is 0.002 in (0.05mm).

REMOVAL AND INSTALLATION

1993-94 Models

FWD vehicles

1. Raise the vehicle and support safely.

2. Remove the tire and wheel assembly.

3. Remove the bolt(s) holding the speed sensor bracket to the knuckle and remove the assembly from the vehicle.

NOTE: The speed sensor has a pole piece projecting from it. This exposed tip must be protected from impact or scratches. Do not allow the pole piece to contact the toothed wheel during removal or installation.

4. Remove the caliper from the brake disc and suspend with a wire.

5. Remove the brake disc.

6. Remove the grease cap, self-locking nut and tongued washer.

7. Remove the rear hub and bearing assembly.

NOTE: The rear hub assembly can not be disassembled. If bearing replacement is required, replace the assembly as a unit.

To install:

8. Install the hub and bearing assembly.

9. Install the tongued washer and a new self-locking nut.

10. Hold hub assembly securely with special tool C-3281 or equivalent, tighten the nut to 144–188 ft. lbs. (200–260 Nm). Align with the indentation in the spindle and crimp nut down to lock in place.

11. Set up a dial indicator and measure the end-play while moving the hub in and out.

12. If the end-play exceeds 0.004 in. (0.01mm), retorque the nut and remeasure the end-play. If still beyond the limit, replace the hub and bearing unit.

13. Install the grease cap and brake parts.

14. Temporarily install the speed sensor to the knuckle; tighten the bolts only finger-tight.

15. Route the speed sensor cable correctly and loosely install the clips and retainers. All clips must be in their original position and the sensor cable must not be twisted. Improper installation may cause cable damage or system failure.

NOTE: The wiring in the harness is easily damaged by twisting and flexing. Use the white stripe on the outer insulation as a guide to keep the sensor harness properly placed.

16. Use a brass or other non-magnetic feeler gauge to check the air gap between the tip of the pole piece and the toothed wheel. Correct gap is 0.012–0.035 in. (0.3–0.9mm). Tighten the 2 sensor bracket bolts to 10 ft. lbs. (14 Nm) with the sensor located so the gap is the same at several points on the toothed wheel. If the gap is incorrect, it is likely that the toothed wheel is worn or improperly installed.

17. Install the tire and wheel assembly. Be sure to pump brake pedal until firm before moving vehicle

AWD vehicles

1. Disconnect the negative battery cable.

2. Raise and support the vehicle safely.

3. Remove the tire and wheel assembly from the vehicle.

4. If equipped with ABS, remove the rear wheel speed sensor.

NOTE: Be cautious to ensure that the tip of the pole piece on the rear speed sensor does not come in contact with other parts during removal. Sensor damage could occur.

5. Remove the rear caliper and support assembly out of the way. Remove the brake disc.

6. Remove the driveshaft and companion flange installation bolts, nuts and washers. Move the end of shaft slightly to access the self-locking nut.

7. Using axle holding tool C-3281 or equivalent, secure the rear axle shaft in position, then remove the self-locking nut.

8. Using a slide hammer with hub adapter, remove the rear axle shaft from the trailing arm.

9. If equipped with ABS, remove the rear rotor from the axle assembly using collar and press. The rotor is a press fit.

10. Remove the outer bearing and dust cover concurrently from the axle shaft using a press.

11. Using puller, remove the oil seal and inner bearing from the trailing arm.

12. Inspect the companion flange and axle shaft for wear or damage. Inspect the dust cover for deformation or damage. Inspect the bearings for burning or declaration. Replace components as required.

To install:

13. Using the proper driver, press fit the inner bearing onto the trailing arm. Press fit the oil seal onto the trailing arm with the depression in the oil seal facing upward, and until it contacts the shoulder on the inner arm.

NOTE: When tapping the oil seal in, use a plastic hammer to lightly tap the top and circumference of the seal installation tool, press fitting gradually and evenly.

14. Press fit the dust covers onto the axle until it contacts the axle shaft shoulder. Install the innermost cover so the depression is facing upward.

15. Apply multi-purpose grease around the entire circumference of the inner side of the outer bearing seal lip. Press fit the outer bearing to the axle shaft so that the bearing seal

lip surface is facing towards the axle shaft flange.

16. Press fit the rear rotor to the axle shaft with the rear rotor groove surface towards the axle shaft flange.

17. Install the rear axle shaft to the trailing arm temporarily. Install the companion flange to the rear axle shaft, then install a new self-locking nut.

18. While holding the rear axle shaft in position using holding tool, tighten a new self-locking nut to 159 ft. lbs. (220 Nm).

19. Install the driveshaft nuts, washers and bolts. Tighten to 47 ft. lbs. (65 Nm).

20. Install the rear brake disc, caliper assembly and parking brake.

21. Install the tire and wheel assembly and lower the vehicle. Check the parking brake stroke and adjust as required.

22. Before moving the vehicle, pump the brakes until a firm pedal is achieved.

1996–97 Models

The rear wheel bearing is not serviceable. If the wheel bearing must be replaced for any reason, the hub assembly must be replaced.

1. Disconnect the negative battery cable.

2. Raise and safely support the vehicle.

3. Remove the wheel and tire assembly.

4. Remove the rear wheel speed sensor if equipped with ABS.

5. Remove the caliper assembly and rotor, or drum. Suspend the caliper out of the way with wire.

6. On vehicles with rear disc brakes, remove the parking brake shoes.

7. On vehicles equipped with AWD, remove the axle shaft locking nut, and using a suitable tool, separate the hub from the axle shaft.

8. Remove the hub mounting bolts from behind the backing plate and remove the hub.

NOTE: The rotor for the ABS must be removed and installed using a press.

9. Remove the through-bolt, lockwasher and nut, and disconnect the trailing arm.

10. Remove the bolt, washer, and locknut, and disconnect the lower control arm from the knuckle.

11. Remove the locknut and the toe control arm ball joint from the knuckle.

12. Remove the lower strut mounting bolt and disconnect the strut from the knuckle.

13. Remove the through-bolt, washer and locknut and disconnect the upper control arm from the knuckle. Remove the knuckle assembly.

To install:

14. Install the knuckle assembly to the upper control arm with the through-bolt, washer and locknut. Torque is 71 ft. lbs. (98 Nm).

15. Connect the lower strut mount to the knuckle and tighten the bolt to 71 ft. lbs. (98 Nm).

16. Install the toe control arm ball joint to the knuckle and tighten the mounting locknut to 20 ft. lbs. (28 Nm).

17. Reconnect the lower control arm to the knuckle with the through-bolt, washer, and locknut. Tighten the bolt and nut to 71 ft. lbs.

18. Install the lower trailing arm to the knuckle with the bolt, washer, and locknut. Tighten the nut and bolt to 85–99 ft. lbs. (118–137 Nm).

19. Press the rotor (ABS) on the hub.

20. On AWD vehicles, engage the splines on the axle shaft with the hub and tighten the shaft lock nut to 145–188 ft. lbs. (196–225 Nm).

21. Install the hub and tighten the mounting bolts to 54–65 ft. lbs. (74–88 Nm).

22. The remainder of installation is the reverse pf removal.

CHRYSLER CORP.

Front Wheel Drive

DODGE/PLYMOUTH-Neon

FIRING ORDERS

NOTE: To avoid confusion, always replace spark plug wires one at a time.

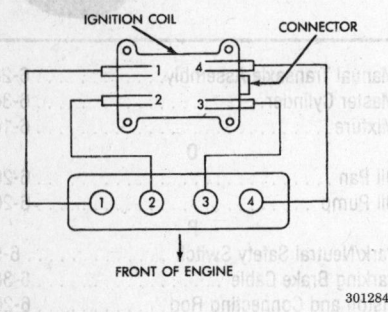

2.0L Engine
Engine Firing Order: 1–3–4–2
Distributorless Ignition System

ENGINE ELECTRICAL

NOTE: Disconnecting the negative battery cable on some vehicles may interfere with the functions of the on board computer systems and may require the computer to undergo a relearning process, once the negative battery cable is reconnected.

Ignition Timing

ADJUSTMENT

Ignition timing is controlled by the Powertrain Control Module (PCM). No adjustment is necessary or possible.

Alternator

PRECAUTIONS

Several precautions must be observed with alternator equipped vehicles to avoid damage to the unit.
• If the battery is removed for any reason, make sure it is reconnected with the correct polarity. Reversing the battery connections may result in damage to the 1-way rectifiers.
• When utilizing a booster battery as a starting aid, always connect the positive to positive terminals and the negative terminal from the booster battery to a good engine ground on the vehicle being started.
• Never use a fast charger as a booster to start vehicles.
• Disconnect the battery cables when charging the battery with a fast charger.
• Never attempt to polarize the alternator.
• Do not use test lights of more than 12 volts when checking diode continuity.
• Do not short across or ground any of the alternator terminals.
• The polarity of the battery, alternator and regulator must be matched and considered before making any electrical connections within the system.
• Never separate the alternator on an open circuit. Make sure all connections within the circuit are clean and tight.
• Disconnect the battery ground terminal when performing any service on electrical components.
• Disconnect the battery if arc welding is to be done on the vehicle.

REMOVAL AND INSTALLATION

1. Disconnect the negative battery cable from the left strut tower. The ground cable is equipped with a insulator grommet which should be placed on the stud to prevent the negative battery cable from accidentally grounding.
2. Unplug the field circuit from the alternator.
3. Remove the B+ terminal cover from by spreading the cover with a flat blade tool.
4. Remove the B+ nut and wire.
5. Loosen the adjusting bolt.
6. Loosen the pivot bolt and the adjusting bolt until the belt can be removed.
7. Remove the adjusting bolt and pivot bolt.
8. Remove the alternator by moving it toward the head lamp bucket.
   To install:
9. Position the alternator into the bracket.
10. Connect the B+ wire and torque the nut to 75 inch lbs. (9 Nm).
11. Reinstall the B+ terminal cover.
12. Reconnect the field circuit to the alternator.
13. Install the pivot bolt.
14. Install the adjusting bolt.
15. Install the drive belt.
16. Adjust the drive belt and tighten the adjusting bolt to 40 ft. lbs. (54 Nm).
17. Tighten the pivot bolt to 40 ft. lbs. (54 Nm).
18. Reconnect the negative battery cable to the strut tower.

Drive Belts

REMOVAL AND INSTALLATION

When replacing or adjusting the engine drive belts, it is important that the belts are adjusted with the correct amount of tension. Excessive belt tension will cause damage to the alternator and power steering pump pulley bearings, while loose belt tension will produce slip and premature wear on the belt.

Belt replacement is similar to adjustment, with the exception that the belt will have to be properly routed around the pulleys. Note the routing of the belt before removal, if possible. For individual belt replacement, start with the outer most belt. If a removed belt is to be reused, be certain to mark the direction of rotation on the belt to extend belt life.

Alternator Belt

1. Place a straightedge along the bottom edge of the belt and across the 2 pulleys. Allow both ends of the straightedge to rest on the bottom of each pulley for support.
2. Measure the deflection of the belt from the straightedge with a force of about 22 lbs. applied midway between the 2 pulleys. Deflection should be 0.35–0.45 inch (9.0–11.5mm)
3. To adjust the tension on the alternator drive belt, loosen the adjusting bolt and the pivot locknut, at the alternator. Then, move the alternator, by turning the adjusting bolt. Once the desired value is reached, secure the bolt and locknut. Recheck the belt tension.

Power Steering Pump Belt (Vehicles Without A/C)

1. Press the belt in, at about the center, between the power steering pump pulley and the crankshaft pulley. With reasonable pressure applied (about 22 lbs.) the belt should deflect about 0.43–0.55 inches (11–14mm).
2. Adjustment can be made by loosening the 3 bolts that hold the pump. Place a suitable bar or lever between the body of the pump and gently pry to get the desired tension.
3. Retighten the 3 bolts to 29 ft. lbs. (39 Nm) and check belt tension again.

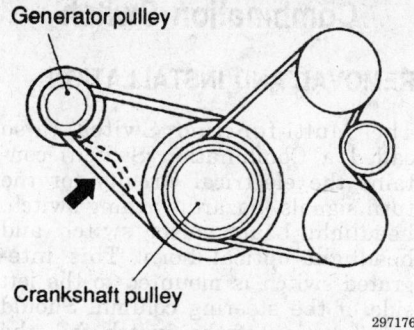

Generator pulley

Crankshaft pulley

297176

Apply force as shown and measure the alternator (generator) belt deflection

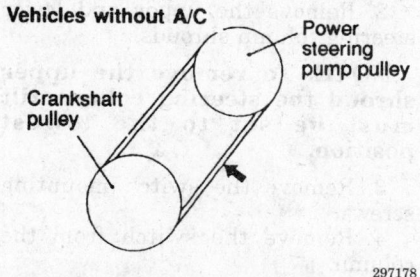

Vehicles without A/C

Power steering pump pulley

Crankshaft pulley

297178

Apply force as shown and measure the power steering pump belt deflection (without A/C)

Power Steering Pump and A/C Belt

NOTE: Fix the power steering pump closest to the front of the vehicle.

1. Press the belt in, at about the center between the power steering pump pulley and the crankshaft pulley. With reasonable pressure applied (about 22 lbs.) the belt should deflect about 0.39–0.43 inches (10–11mm).
2. Adjustment can be made by loosening the tensioner pulley nut

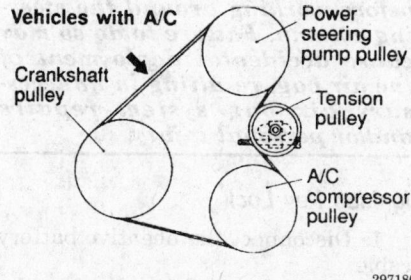

Vehicles with A/C

Power steering pump pulley

Crankshaft pulley

Tension pulley

A/C compressor pulley

297180

Apply force as shown and measure the power steering pump belt deflection (with A/C)

and turning the adjuster bolt until the desired tension is attained.

3. Tighten the pulley nut and check the belt tension again.

Starter

REMOVAL AND INSTALLATION

1. Disconnect the negative battery cable from the left strut tower.

NOTE: The ground cable is equipped with a insulator grommet which should be placed on the stud to prevent the negative battery cable from accidentally grounding.

2. Remove the air cleaner resonator.
3. Remove the battery cable nut and disconnect the wiring at the starter.
4. Disconnect the push on connector.
5. Remove the 2 bolts that attach the starter to the transaxle.
6. Remove the starter from the vehicle.

To install:

7. Install the starter and the attaching bolts.
8. Torque the attaching bolts to 40 ft. lbs. (54 Nm).
9. Reconnect the starter wiring and torque the nut to 90 inch lbs. (10 Nm).
10. Reconnect the push on connector.
11. Install the air cleaner resonator.
12. Reconnect the negative battery cable.

CHASSIS ELECTRICAL

Blower Motor

REMOVAL AND INSTALLATION

This vehicle uses 2 different blower motors. The one for vehicles equipped with A/C uses 3 retaining screws. The one for vehicles with heat only uses a tab and a twist lock mechanism. The blower motor is located on the bottom right side of the unit housing. The blower motor can be removed from the vehicle without having to remove

the unit housing assembly. The blower motor wheel is only serviced with the blower motor. The wheel and motor are balanced as an assembly. If the blower motor wheel requires replacement, the blower motor must be replaced.

To test the blower motor, disconnect the blower motor wiring. Connect a jumper from 12V to one of the blower motor terminals. Connect a jumper from a good ground source to the other blower motor connection. If the blower motor does not run, the motor is likely defective.

—— CAUTION ——
Do not allow the 2 jumpers to touch during this test. This could cause a spark or fire.

With A/C

1. Disconnect the negative battery cable.
2. Remove the right side scuff plate in the passenger compartment.
3. Pull back the carpet.
4. Cut the wheel housing silencer in line with the blower motor wiring.
5. Disconnect the blower motor wiring.
6. Remove the 3 blower motor retaining screws.
7. Lower the blower motor from the heater box.

To install:

8. Install the blower motor into the heater box.
9. Install the blower motor attaching screws.
10. Reconnect the blower motor wiring.
11. Tape the silencer into position.
12. Install the carpet and the scuff plate.
13. Connect the negative battery cable.

Without A/C

1. Disconnect the negative battery cable.
2. Detach the blower motor wiring.
3. Depress the release tab and turn the blower motor approximately 1/8 turn counterclockwise.
4. Remove the blower motor from the heater box.

To install:

5. Install the blower motor into the heater box.
6. Turn the blower motor approximately 1/8 turn clockwise until the retaining tab locks into place.
7. Reconnect the negative battery cable.

Windshield Wiper Motor

REMOVAL AND INSTALLATION

The windshield wipers are controlled by the right side lever on the steering column mounted multi-function switch. All wiper and washer functions are contained in this switch. The wipers operate only when the ignition switch is in the **ACCESSORY** or **IGNITION** positions. Fuse 15, located in the fuse block protects the wiper/washer system. The wiper motor also has an internal non-serviceable circuit breaker to provide protection against motor stall conditions. If the wiper motor does not run, check the 20 amp fuse in cavity 15 of the fuse block. Also check the ground strap on the wiper motor. On the intermittent wiper system, a ground is dedicated to the wiper switch. This is ground G201 located in the instrument panel at the rear of the center stack behind the radio. The following items are on that ground: Radio, DRL module, combination flasher and the heated rear window switch. Check these items for proper operation.

The wiper motor and linkage assembly is referred to as the wiper motor and linkage module. To remove the windshield wiper and linkage assembly module use the following procedure.

1. Disconnect and isolate the negative battery cable.
2. Lift off the wiper arm caps. Remove the wiper arm nuts at the base of the wiper arms and remove both wiper arms from their pivots with a rocking motion.
3. Remove the rear hood seal along with the cowl top plastic screen.
4. Disconnect the wiring harness at the front plenum wall.
5. Remove the wiper motor mounting screws.
6. Remove the wiper motor and linkage assembly module.
7. To remove the wiper motor from the linkage assembly, insert a suitable prytool between the ball cap and linkage, then twist the prytool and lift straight up on the linkage.
To install:
8. Connect the linkage to the wiper motor. Align the link ball cap over the ball and gently press fit against the shoulder of the ball cap to lock the ball cap into position.
9. Install the wiper motor assembly into the vehicle.

10. Reconnect the wiring harness. Make sure the motor connector seal is properly positioned.
11. Install the wiper motor mounting screws and torque to 60–80 inch lbs. (7–9 Nm).
12. Reinstall the cowl screen.
13. Install the wiper arms.
14. Connect the negative battery cable.
15. Wet the windshield to prevent wiper blade drag and test the wiper system in all modes.

Headlight Switch

REMOVAL AND INSTALLATION

──────── **CAUTION** ────────
The Supplemental Restraint System (SRS) must be disarmed before working around the steering column. Failure to do so may cause accidental deployment of the air bag, resulting in unnecessary SRS system repairs and/or personal injury.
────────────────────────

1. Disconnect the negative battery cable.
2. Remove the steering column cover and liner.

NOTE: The upper steering column cover can be removed by placing the tilt steering to the lowest position.

3. Remove the 3 screws attaching the headlight switch mounting plate to the instrument panel.
4. Pull the switch away from the panel opening.
5. Disconnect both connectors from the switch.
6. Remove the switch knob by pressing the release button and pulling the knob out of the switch.
7. Pull the switch bezel out of the mounting plate.
8. Remove the headlight switch retaining nut.
To install:
9. Install the headlight switch and the retaining nut.
10. Install the switch bezel.
11. Reconnect both wiring connectors.
12. Install the switch knob.
13. Install the mounting plate into the instrument panel.
14. Install the mounting plate screws.
15. Install the steering column liner and column cover.
16. Reconnect the negative battery cable.

Combination Switch

REMOVAL AND INSTALLATION

The Multi-function Switch (also called a Combination Switch) contains the electrical circuitry for the turn signals, hazard warning switch, headlight beam select switch and headlight optical horn. This integrated switch is mounted to the left side of the steering column. Should any function of the switch fail, the entire switch assembly must be replaced.

1. Disconnect the negative battery cable.
2. Remove the upper and lower steering column shrouds.

NOTE: To remove the upper shroud the steering column tilt must be set to the lowest position.

3. Remove the switch mounting screws.
4. Remove the switch from the column.
To install:
5. Install the switch into the column.
6. Install the mounting screws and torque to 17 inch lbs. (2 Nm).
7. Install the column shrouds.
8. Torque the shroud screws to 17 inch lbs. (2 Nm).
9. Reconnect the negative battery cable.
10. Test all switch functions.

Ignition Lock and Switch

REMOVAL AND INSTALLATION

The ignition switch (electrical component) attaches to the lock cylinder housing on the end opposite the lock cylinder (mechanical lock).

──────── **CAUTION** ────────
The Supplemental Restraint System (SRS) must be disarmed before working around the steering column. Failure to do so may cause accidental deployment of the air bag, resulting in unnecessary air bag system repairs and/or personal injury.
────────────────────────

Ignition Key Lock

1. Disconnect the negative battery cable.
2. Turn the ignition key to the **RUN** position.
3. Insert a small prytool into the tab access hole and depress the tab.

4. Pull the ignition lock from the steering column.
To install:
5. Insert the ignition lock into the steering column.
6. Lightly pull on the ignition lock to make sure it engaged the retaining tab.
7. Turn the key to the **OFF** position.
8. Reconnect the negative battery cable.

Ignition Switch

1. Disconnect the negative battery cable.
2. Place the ignition into the **RUN** position.
3. Insert a small prytool into the hole in the lower shroud and depress the cylinder release tab.
4. Remove the lock cylinder from the steering column.
5. Remove the upper and lower steering column shrouds.
6. Disconnect the electrical connectors from the switch.
7. Using a No. 10 Torx® bit, remove the retaining bolt.
8. Depress the retaining tabs and remove the switch.
To install:
9. Place the ignition switch in the **RUN** position.
10. Install the switch into the retaining tabs.
11. Install the retaining bolt.
12. Reconnect the wiring to the ignition switch.
13. Install the lower and upper shrouds.
14. Install the key lock cylinder.
15. Reconnect the negative battery cable.

Park/Neutral Safety Switch

REMOVAL AND INSTALLATION

1. Raise and safely support the vehicle.
2. Disconnect the switch located on the left side of the transaxle.
3. Using a box wrench, remove the switch from the transaxle.
To install:
4. Using a box wrench, install the switch into the transaxle.
5. Reconnect the electrical connection to the switch.
6. Lower the vehicle and check for proper operation. Vehicle should start in **P** and **N** only. Verify that the back-up lights work in **R**.

PCM/ECM Computer

REMOVAL AND INSTALLATION

The 2.0L engine uses a sequential Multi-Port Electronic Fuel Injection (MFI) system. The MFI is computer regulated and provides precise air/fuel ratios for all driving conditions. The Powertrain Control Module (PCM) operates the fuel injection system as well as many other vehicle powertrain functions. Due to its many functions and wide range of vehicle systems controlled by its outputs, the term Powertrain Control Module replaces and more accurately describes the function of the on-board computer formerly called the Electronic Control Module or Engine Control Module (ECM).

The PCM is located underhood on the inner fender panel next to the washer fluid reservoir on the passenger side.
1. Disconnect the negative battery cable.
2. Remove the PCM attaching screws.
3. Lift the PCM up and disconnect the 60–way connector from the PCM.
4. Remove the PCM from the vehicle.
To install:
5. Install the PCM into the vehicle.
6. Connect the 60–way connector and tighten the retainer screw to 40 inch lbs. (4.5 Nm).
7. Install the PCM attaching screws and tighten to 80 inch lbs. (9 Nm).
8. Connect the negative battery cable.

ENGINE COOLING

Radiator

REMOVAL AND INSTALLATION

1. Disconnect the negative battery cable.
2. Loosen the radiator drain plug and drain the cooling system.
3. Unfasten the hose clamps and disconnect hoses from the radiator.
4. If equipped with an automatic transaxle, disconnect the automatic transaxle hoses and plug them to keep out dirt.

5. Remove the radiator-to-battery strut brace.
6. Remove the fan module assembly by disconnecting the fan motor electrical connector. Remove the fan shroud retaining screws located at the top of the shroud. Lift the shroud up and out of the bottom shroud attachment clips, separating the shroud from the radiator. For dual fan applications, the left fan module may be removed first, then the right side module. Use care not to damage the fans.
7. Remove the upper radiator isolator bracket mounting screws. If equipped, disconnect the engine block heater wire.
8. If equipped, remove the air conditioning condenser attaching screws located at the front of the radiator. Lean the condenser forward. It is not necessary to discharge the air conditioning system to remove the radiator.
9. The radiator can now be lifted free from the engine compartment.

NOTE: Care should be taken not to damage the radiator cooling fins or water tubes during removal.

To install:
10. Slide the radiator down into position behind the radiator support (yoke).
11. If equipped with air conditioning, attach the air conditioning condenser-to-radiator with 4 mounting screws and tighten to 50 inch lbs. (5.6 Nm). Then, seat the assembly lower rubber isolators into the mounting holes provided in the lower crossmember.
12. Tighten the radiator isolator mounting bracket screws to 65 inch lbs. (7.3 Nm). The radiator should have clearance to move up approximately ¼ inch after assembly.
13. If equipped with automatic transaxle, unplug and connect the cooler hoses, then tighten the clamps to 35 inch lbs. (3.9 Nm).
14. Slide the cooling fan module down into the clip(s) on the lower radiator flange. For dual fan applications, install the right fan module first and then the left fan module. Install the retaining screws and tighten to 50 inch lbs. (5.6 Nm).
15. Install the radiator-to-battery strut brace.
16. Attach the radiator hoses and coolant reserve hose. Align and the position the hose clamps so they will not interfere with the engine or hood.
17. Connect the fan motor electrical connection and connect the negative battery cable.

18. Refill the cooling system with 50/50 mix of approved anti-freeze and water.

19. Operate the engine until it reaches normal operating temperature. Check the cooling system reserve reservoir for proper coolant level. Add as required. Check the automatic transaxle fluid level, if equipped.

Water Pump

REMOVAL AND INSTALLATION

2.0L (VIN Y) DOHC Engine

These vehicles use a conventional water-cooled pressure forced circulation type cooling system in which the water pump pressurizes coolant and circulates it through the engine. The water pump is the centrifugal type and is driven by the timing belt from the crankshaft. This means that timing belt removal is required, a lengthy process requiring careful work. It is good practice to turn the engine crankshaft by hand (clockwise) to set the engine to TDC No. 1 cylinder compression stroke (firing position) before starting work. This should align all timing marks and serve as a reference point for later work.

1. Disconnect the negative battery cable.
2. Drain the cooling system.
3. Remove the accessory drive belts.
4. Using C-3281 or equivalent crankshaft holding tool, remove the crankshaft retaining bolt.
5. Using puller tools 1026 and 6827 or equivalent, remove the crankshaft pulley.
6. Remove the power steering pump with the hose attached, and position the assembly aside.
7. Remove the power steering pump bracket.
8. Place a floor jack under the engine oil pan, with a block of wood in between and raise the engine so the weight of the engine is no longer being applied to the engine mount bracket.
9. Remove the upper engine mount and the engine mounting bracket.
10. Remove the front timing belt cover.
11. Align the timing marks. Loosen the timing belt tensioner and remove the belt.

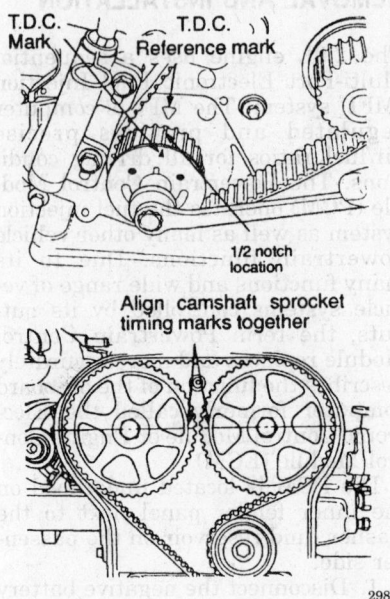

Setting the timing marks — 2.0L DOHC engine

WARNING

Do not rotate the crankshaft or camshafts after removing the timing belt, or valve train components may be damaged. Always align the timing marks before removing the timing belt.

---

12. If the timing belt tensioner is to be replaced, remove the retaining bolts and remove the timing belt tensioner. When the timing belt tensioner is removed from the engine it is necessary to compress the plunger into the tensioner body.

13. Place the tensioner in a vise and slowly compress the plunger.

NOTE: Position the tensioner in the vise the same way it will be installed on the engine. This is to ensure proper pin orientation for when the tensioner is installed on the engine.

14. When the plunger is compressed into the tensioner body, insert a pin through the body and plunger to hold the plunger in place until the tensioner is installed.

15. Unfasten the water pump retaining bolts, then remove the water pump and O-ring.

To install:

16. Thoroughly clean all mating surfaces of the water pump and engine block.

17. Lubricate the O-ring with water or antifreeze then install into the groove in the water pump. Never use any type of oil based lube or grease on rubber parts.

18. Do not allow oil or other grease to get on the O-ring. Install the water pump. When inserting the water pump, check that there is no sand, dirt, etc. on its inner surface. Tighten the water pump retaining bolts to 8.7 ft. lbs. (12 Nm).

19. Check that all timing marks are still aligned. Bring the crankshaft sprocket to 1/2-notch before TDC.

20. Install the timing belt. Starting at the crankshaft, go around the water pump sprocket, idler pulley, camshaft sprockets and then around the tensioner pulley.

21. Move the crankshaft to TDC to take up the slack in the belt. Install the tensioner to the block, but do not tighten the retaining bolts.

22. Using a torque wrench on the tensioner pulley, apply 21 ft. lbs. (28 Nm) of torque to the pulley.

23. With the torque being applied to the tensioner pulley, move the tensioner up against the tensioner pulley bracket and tighten the retaining bolts to 23 ft. lbs. (31 Nm).

24. Remove the tensioner plunger pin. Pretension is correct when the pin can be removed and installed.

25. Rotate the crankshaft 2 revolutions and check the timing marks. If the timing marks are not properly aligned remove the belt and repeat Steps 19 through 24.

26. Install the front timing belt cover.

27. Lower the engine enough to install the engine mount bracket.

28. Install the bracket and remove the floor jack.

29. Install the power steering pump bracket and pump.

30. Install the crankshaft pulley using tool C-4685-C, or equivalent pulley installer.

31. Install the accessory drive belts.

32. Properly fill the cooling system.

33. Connect the negative battery cable.

34. Check for leaks and proper engine operation.

2.0L (VIN C) SOHC Engine

The water pump has a diecast aluminum body and housing with a stamped steel impeller. The water pump bolts directly to the block. The cylinder block-to-water pump sealing

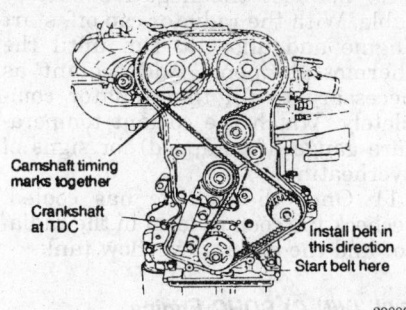

Installing the timing belt — 2.0L DOHC engine

is provided by a large rubber O-ring. The water pump is driven by the timing belt which must be removed to service the water pump.

This procedure requires removing the engine timing belt and the auto tensioner. To help assure proper alignment at assembly, it may be helpful to set the engine at TDC on No. 1 cylinder, compression stroke (firing position). This should align all timing marks on the crankshaft sprocket and the camshaft sprocket. This serves as a reference for all the work that follows.

1. Disconnect the negative battery cable remote connection located on the left strut tower.

2. Turn the crankshaft until the engine is at TDC on No. 1 cylinder, compression stroke (firing position).

3. Raise and safely support the vehicle.

4. Remove the right inner splash shield.

5. Remove the drive belts and power steering pump.

6. Drain the cooling system.

7. Support the engine and remove the right motor mount.

8. Remove the power steering pump bracket.

9. Remove the right engine mount bracket.

10. Remove the timing belt and the inner timing belt cover using the following procedure:

   a. Remove the front timing belt cover.

   b. Verify that the valvetrain timing marks are aligned.

   c. Loosen the timing belt tensioner screws, then remove the tensioner and timing belt.

WARNING

With the timing belt removed, DO NOT rotate the camshaft or crankshaft or damage to the engine could occur.

---

   d. With the timing belt removed, remove the camshaft sprocket bolt. Do not allow the camshaft to turn when the camshaft sprocket is being removed.

   e. Remove the rear timing belt cover to access the water pump.

11. Unfasten the water pump attaching bolts, then remove the water pump.

To install:

12. Clean all parts well. Replace the water pump if there are any cracks, signs of coolant leakage from the shaft seal, loose or rough turning bearing, damaged impeller or sprocket or sprocket flange loose or damaged.

13. Clean all sealing surfaces. Install a new rubber O-ring into the water pump.

NOTE: Make sure the O-ring is properly seated in the water pump groove before tightening the screws. An improperly located O-ring may cause damage to the O-ring and cause a coolant leak.

14. Position the water pump to the engine, then install the retaining bolts. Torque the bolts to 105 inch lbs. (12 Nm).

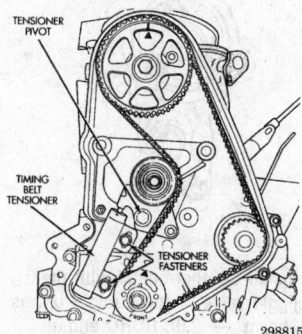

Timing belt tensioner — 2.0L SOHC engine

15. Pressurize the cooling system to 15 psi and check for leaks. If okay, release the pressure and continue the engine assembly process.

16. Install the inner timing belt cover.

17. Install the timing belt using the following procedure:

   a. Place the timing belt tensioner in a soft-jaw vise in the same relative position as it is installed in the engine. Compress the plunger into the tensioner by closing the vise jaws. When the plunger is compressed all the way, insert a pin (a $5/64$ inch Allen wrench will work) through the body and plunger to hold the plunger in place until the tensioner is installed.

   b. If not done so at disassembly, set the crankshaft sprocket to TDC by aligning the sprocket with the arrow mark on the oil pump housing, them back off 3 notches before TDC.

   c. Move the crankshaft sprocket to $1/2$ mark before TDC for belt installation.

   d. Install the timing belt starting at the crankshaft sprocket, going around the water pump sprocket and then around the camshaft sprocket.

   e. Move the crankshaft sprocket to TDC to take up belt slack. Install the tensioner, but do not tighten the fasteners. Using a torque wrench on the tensioner pulley, apply 250 inch lbs. (28 Nm) torque. With the torque applied to the pulley, move the tensioner pulley bracket and tighten the fasteners to 275 inch lbs. (31 Nm). Pull out the tensioner plunger retaining pin. Pretension is correct when the pin can be removed and installed.

   f. Rotate the crankshaft 2 full revolutions clockwise and check the alignment of the timing marks. If correct, install the timing belt front cover.

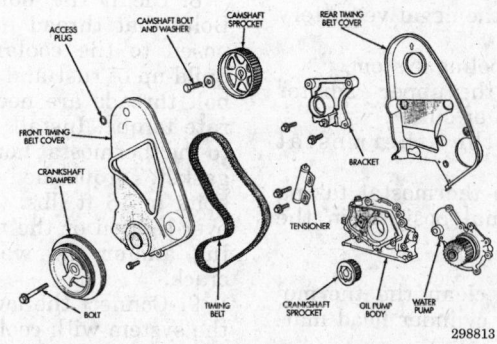

Timing belt, covers and water pump arrangement — 2.0L SOHC engine

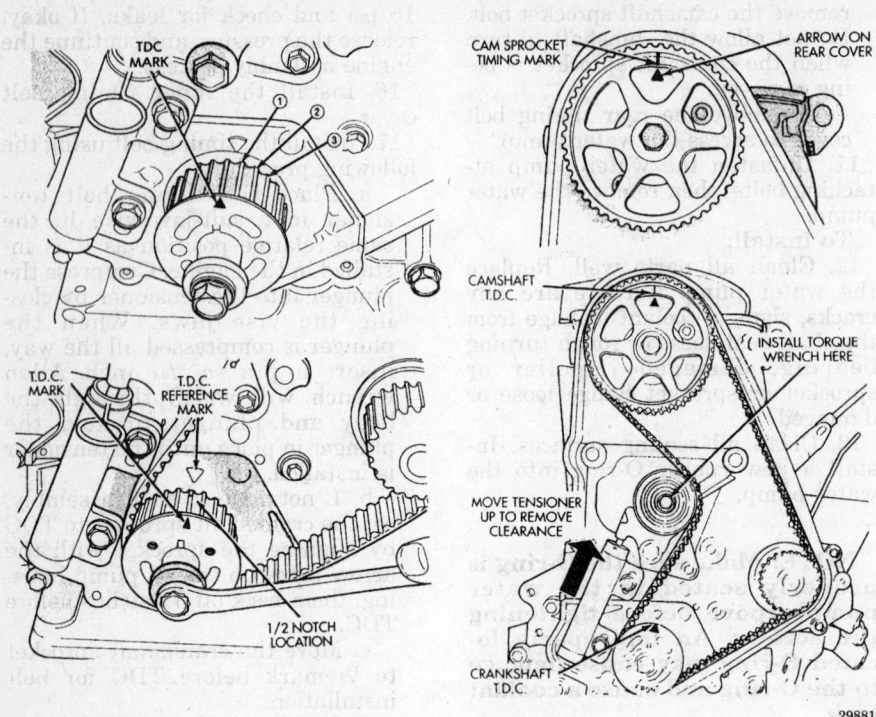

Timing marks — 2.0L SOHC engine

18. Install the right engine mount bracket and engine mount.
19. Lower the vehicle.
20. Install the power steering pump bracket and pump.
21. Install the drive belts.
22. Fill and bleed the cooling system.
23. Start the engine and check for leaks and/or proper operation.

Thermostat

REMOVAL AND INSTALLATION

2.0L (VIN Y) DOHC Engine

1. Disconnect the negative battery cable.
2. Drain the cooling system.
3. Disconnect the upper radiator hose from the water outlet.
4. Remove the thermostat housing.
5. Remove the thermostat taking note of its original position in the housing.
To install:
6. Thoroughly clean the thermostat housing and cylinder head mating surfaces.
7. Install the thermostat so its flange seats tightly in the machined

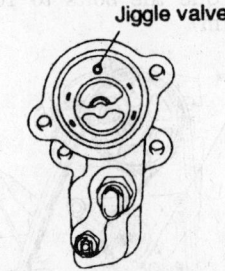

Install the thermostat with the jiggle valve facing straight up, as shown — 2.0L DOHC engine

recess in the thermostat housing. Refer to its position prior to removal.
8. Clean the bolt threads well. Bolts that thread into openings exposed to the coolant may have a build-up of rust and corrosion. Clean bolt threads are necessary for accurate torque. Install the water outlet to the thermostat housing with a new gasket. Torque the housing mounting bolts to 16 ft. lbs. (22 Nm). Do not over-tighten or the thermostat housing and/or the water outlet may crack.
9. Connect the lower hose and fill the system with coolant.

10. Connect the negative battery cable. With the radiator cap off, start engine and allow to run until the thermostat opens. Add coolant as necessary to fill the radiator completely. Watch the coolant temperature gauge (if equipped) for signs of overheating.
11. Once the vehicle has cooled, recheck the coolant level in the radiator and the coolant overflow tank.

2.0L (VIN C) SOHC Engine

1. Disconnect the negative battery cable at the remote terminal at the shock tower.
2. Drain the coolant to below the thermostat level.
3. Disconnect the coolant recovery hose.
4. Remove the thermostat housing bolts.
To install:
5. Clean all gasket surfaces.
6. Install the thermostat and align the air bleed with the notch on the cylinder head. Install the housing using a new gasket.
7. Install the thermostat housing bolts and torque to 110 inch lbs. (12.5 Nm).
8. Reconnect the coolant recovery hose.
9. Connect the negative battery cable to the remote terminal at the shock tower.
10. Fill and bleed the engine cooling system.
11. Pressure test for leaks.

Electric Cooling Fan

REMOVAL AND INSTALLATION

1. Disconnect the negative battery cable.
2. Disconnect the fan motor(s) electrical connector.
3. It may be necessary to remove the radiator-to-battery strut brace.
4. Remove the fan shroud retaining screws located at the top of the shroud. Lift the shroud up and out of the bottom shroud attachment clips, separating the shroud from the radiator. For dual fan applications, perform the following procedures:
   a. Remove the left fan module first.
   b. Remove the right side module. Use care not to damage the fans.
5. There are no repairs that are to be made to the fan. If the fan blades are warped, cracked or otherwise damaged, the fan blade MUST be re-

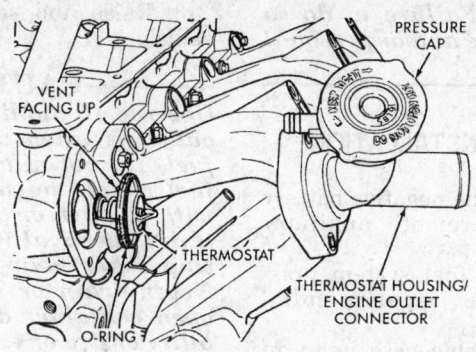

Thermostat and engine outlet connector — 2.0L SOHC engine

placed. To replace the fan blade, use the following procedure.

a. To remove the fan from the motor shaft, bench support the motor and motor shaft while removing the fan retaining clip. Use care. Do not damage the motor shaft by excessive force.

b. To install a replacement fan blade, slide the fan over the shaft. Support the motor and shaft while installing the fan retaining clip.

c. Do not disassemble the fan motor from the support bracket. The electric fan motor is serviced as an assembly with the fan module.

NOTE: Surface burr removal may be required to remove the fan from the motor shaft.

To install:

6. Slide the cooling fan module down into the clip(s) on the lower radiator flange. For dual fan applications, install the right fan module first and then the left fan module. Install the retaining screws and tighten to 50 inch lbs.

7. If removed, install the radiator-to-battery strut brace.

8. Connect the fan motor electrical connection and connect the negative battery cable.

Cooling System Bleeding

Do not use straight antifreeze as engine coolant. Inadequate engine running temperatures can result. Do not operate the vehicle without the proper concentration (50% antifreeze and 50% water) of recommended ethylene glycol coolant or high temperatures and cooling system corrosion can result. The use of aluminum cylinder heads, intake manifolds and water pumps requires special corrosion protection. Chrysler Corporation recommends their Mopar Antifreeze,

Prestone II, Peak or antifreeze containing Alugard 340-2 or their equivalent to protect the aluminum parts in these engines.

Coolant level should be checked whenever other engine compartment service is performed or when a coolant leak is suspected. Coolant recovery tank level should read between the ADD and FULL marks, located on the side of the the recovery tank when the engine is at normal operating temperature. Cooling system freeze protection should be tested at the onset of the winter season or every 12 months. Service is required if coolant is low, contaminated, rusty or if freeze protection is inadequate.

The following procedure is to be used whenever any repair work is done to the engine cooling system.

1. Fill the cooling system to the filler neck with a 50/50 solution of antifreeze and water.

2. Start the engine, with the radiator cap off, and run until warm.

3. Add coolant as necessary to maintain the level.

4. Install the radiator cap.

FUEL SYSTEM

Fuel System Service Precautions

Safety is the most important factor when performing not only fuel system maintenance but any type of maintenance. Failure to conduct maintenance and repairs in a safe manner may result in serious personal injury or death. Maintenance and testing of the vehicle's fuel system components can be accomplished

safely and effectively by adhering to the following rules and guidelines.

• To avoid the possibility of fire and personal injury, always disconnect the negative battery cable unless the repair or test procedure requires that battery voltage be applied.

• Always relieve the fuel system pressure prior to disconnecting any fuel system component (injector, fuel rail, pressure regulator, etc.), fitting or fuel line connection. Exercise extreme caution whenever relieving fuel system pressure to avoid exposing skin, face and eyes to fuel spray. Please be advised that fuel under pressure may penetrate the skin or any part of the body that it contacts.

• Always place a shop towel or cloth around the fitting or connection prior to loosening to absorb any excess fuel due to spillage. Ensure that all fuel spillage (should it occur) is quickly removed from engine surfaces. Ensure that all fuel soaked cloths or towels are deposited into a suitable waste container.

• Always keep a dry chemical (Class B) fire extinguisher near the work area.

• Do not allow fuel spray or fuel vapors to come into contact with a spark or open flame.

• Always use a backup wrench when loosening and tightening fuel line connection fittings. This will prevent unnecessary stress and torsion to fuel line piping. Always follow the proper torque specifications.

• Always replace worn fuel fitting O-rings with new. Do not substitute fuel hose or equivalent, where fuel pipe is installed.

Fuel System Pressure

RELIEVING

——— **CAUTION** ———

Fuel injection systems remain under pressure, even after the engine has been turned OFF. The fuel system pressure must be relieved before disconnecting any fuel lines. Failure to do so may result in fire and/or personal injury.

1. Disconnect the negative battery cable from the remote connection located on the left strut tower.

2. Remove the fuel filler cap.

3. Remove the cap on the fuel pressure test port on the fuel rail.

4. Place the open end of fuel pressure release hose special tool number C-4799-1 or equivalent into an ap-

proved gasoline container. Connect the other end of hose C-4799-1 or equivalent to the fuel pressure test port. Fuel pressure will bleed off through the hose onto the gasoline container.

Idle Speed

ADJUSTMENT

Idle speed is maintained by the Powertrain Control Module. Adjustment is neither necessary nor possible.

Mixture

ADJUSTMENT

Mixture is maintained by the Powertrain Control Module (PCM). Adjustment is neither necessary nor possible.

Fuel Filter

The fuel delivery system contains a replaceable in-line filter. The fuel filter mounts to the frame above the rear of the fuel tank. The inlet and outlet tubes are permanently attached to the filter. Please note that the fuel system pressure must be relieved before servicing fuel system components. In addition, Chrysler uses Quick Connect fittings on fuel line connections. Specific instructions are given at the end of this procedure.

——— CAUTION ———
Fuel injection systems remain under pressure, even after the engine has been turned OFF. The fuel system pressure must be properly relieved before disconnecting

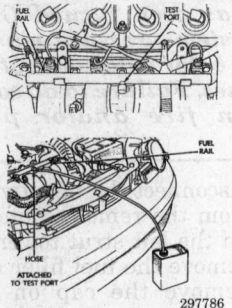

297786

Fuel pressure test port location and pressure relief hose installation

any fuel lines. Failure to do so may result in fire and/or personal injury.

REMOVAL AND INSTALLATION

1. Disconnect the negative battery cable from the remote auxiliary jumper terminal.
2. Release the fuel system pressure using the recommended procedure.
3. Place a suitable hose in to an approved gasoline container and connect other end to the fuel test port.
4. Remove the fuel cap slowly to release tank pressure.
5. Raise and safely support the vehicle.
6. Disconnect the fuel lines from the fuel filter. These are quick connect fittings. Depress the releasing tabs on the fuel filter hose connections and disconnect the fuel lines.
7. Remove the fuel filter.

To install:

8. Apply a light coat of clean 30W engine oil to the fuel filter nipples. Install the fuel tubes.
9. Install the fuel filter and reconnect the fuel lines.
10. Lower the vehicle.
11. Connect the negative battery cable.

Fuel Pump

REMOVAL AND INSTALLATION

1995 Vehicles

The fuel pump module is installed in the fuel tank. The fuel pump module contains the electric fuel pump, fuel pump reservoir, level sensor, inlet strainer and pressure regulator. The inlet strainer, fuel pressure regulator and fuel level sensor are the only serviceable items. If the fuel pump requires service, replace the fuel pump module.

1. Relieve the fuel system pressure.

——— CAUTION ———
Fuel injection systems remain under pressure, even after the engine has been turned OFF. The fuel system pressure must be relieved before disconnecting any fuel lines. Failure to do so may result in fire and/or personal injury.

2. Disconnect the negative battery cable.

3. Raise and safely support the vehicle.

——— CAUTION ———
Observe all applicable safety precautions when working around fuel. Do not allow fuel spray or fuel vapors to come into contact with a spark or open flame. Keep a dry chemical (Class B) fire extinguisher near the work area. Never drain or store fuel in an open container due to the possibility of fire or explosion.

4. Locate and remove the rubber cap from the drain tube. The tube is located on the rear of the fuel tank. Connect either a portable holding tank or siphon hose to the drain tube.
5. Drain the fuel tank into a an approved safety gasoline container.
6. Disconnect the fuel lines from the fuel pump module by depressing the quick connect retainers with thumb and fore-finger.
7. Using a hammer and a brass drift, carefully tap locking ring counterclockwise to release the pump.

——— WARNING ———
The fuel reservoir of the fuel pump module does not empty out when the tank is drained. The fuel in the reservoir will spill out when the module is removed.

8. Remove the pump from the tank and O-ring from the tank.

To install:

9. It is good practice to replace the inlet strainer whenever a pump module is removed.
10. On 1995, the fuel pump module must be removed to service the pressure regulator. On 1996, the pressure regulator is moved to the top of the module.
11. Clean the seal area of the tank.
12. Install a new O-ring in position on the pump and install the fuel pump.
13. Install the lockring using a hammer and brass drift. Drive the lockring clockwise to lock the pump in place.

NOTE: Do not over tighten the lockring. This may cause a leak.

14. Connect the fuel lines onto the fuel pump module and make sure the quick-connect fittings are secure.
15. Lower the vehicle.
16. Refill the fuel tank, as required.
17. Reconnect the negative battery cable.
18. Pressurize the fuel system and check for leaks.

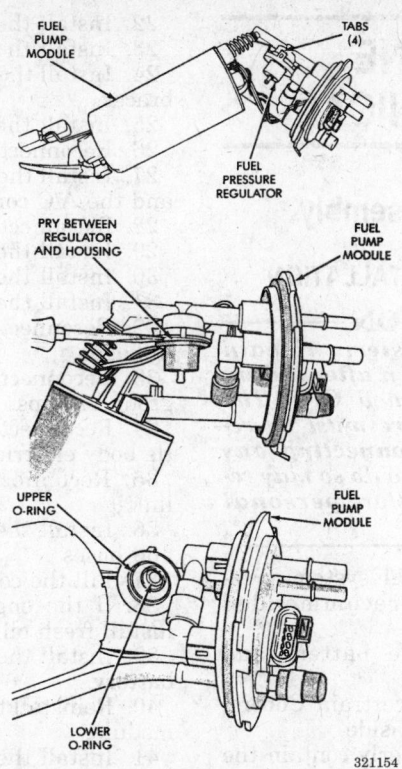

The fuel pressure regulator can be serviced
once the fuel pump module is removed —
1995 vehicles

1996–97 Vehicles

The fuel pump module is installed in
the top of the fuel tank. The fuel
pump module contains the following:
electric fuel pump. fuel pump reservoir,
inlet strainer, fuel filter/pressure
regulator, fuel gauge
sending unit and fuel supply line connection.
The fuel pump has a maximum
deadhead pressure output of
approximately 130 psi. The regulator
adjusts fuel system pressure to approximately
49 psi.

The inlet strainer, fuel pressure
regulator and fuel level sensor are
the only serviceable items. If the fuel
pump requires service, replace the
fuel pump module. The fuel tank
must be lowered for this procedure.

> **CAUTION**
> *Fuel injection systems remain
> under pressure, even after the engine
> has been turned OFF. The
> fuel system pressure must be relieved
> before disconnecting any
> fuel lines. Failure to do so may result
> in fire and/or personal
> injury.*

1. Relieve the fuel system
pressure.
2. Disconnect the negative battery
cable.

> **CAUTION**
> *Observe all applicable safety precautions
> when working around
> fuel. Do not allow fuel spray or
> fuel vapors to come into contact
> with a spark or open flame. Keep
> a dry chemical (Class B) fire extinguisher
> near the work area.
> Never drain or store fuel in an
> open container due to the possibility
> of fire or explosion.*

3. Remove the fuel filler cap to
make sure there is no pressure in the
tank.
4. Raise and safely support the
vehicle.
5. Locate and remove the quick
connect cap from the drain port. The
drain port is located on the rear top of
the fuel tank. Connect siphon hose
into the drain tube.
6. Drain the fuel tank into a an
approved safety gasoline container.
7. Disconnect the fuel lines from
the fuel pump module by depressing
the quick connect retainers with
thumb and fore-finger.
8. Disconnect the electrical connection
from the fuel pump module
by pushing down on the connector retainer
and pulling the connector off
the module.

9. Use a transmission jack to support
the fuel tank. Remove the bolts
from the fuel tank straps.
10. Lower the tank slightly.
11. Use special tool 6856 or
equivalent adjustable spanner-type
tool to grip the fuel pump module
locknut. Use care if using a substitute
tool. Turn counterclockwise to
remove.
12. Remove the old fuel pump module
and O-ring seal from the fuel
tank. Discard the old seal.
   To install:
13. Wipe the seal area of the tank
clean and place a new seal in position
in the tank opening.
14. Position the fuel pump module
in the tank. Make sure the alignment
tab on the underside of the fuel pump
module flange sits in the notch on the
fuel tank.
15. Position the locknut over the
fuel pump module and using special
tool 6856 or equivalent adjustable
spanner-type tool to grip the fuel
pump module locknut, turn clockwise
to tighten to 40 ft. lbs. (55 Nm).

> **NOTE: Use care. Overtightening
> the fuel pump lockring may
> result in a leak.**

16. Connect the electrical connector
and fuel quick-connect fittings.
17. Raise the tank back into position
and secure the retainer straps.
18. Lower the vehicle.
19. Refill the tank. Connect the
negative battery cable. Pressurize
the fuel system and check for leaks.

Fuel Injector

REMOVAL AND INSTALLATION

1. Disconnect the negative battery
remote cable located on the left strut
tower.
2. Release the fuel system pressure
using the recommended
procedure.

> **CAUTION**
> *Fuel injection systems remain
> under pressure, even after the engine
> has been turned OFF. The
> fuel system pressure must be relieved
> before disconnecting any
> fuel lines. Failure to do so may result
> in fire and/or personal
> injury.*

3. Disconnect the fuel supply line
from the fuel rail. This is a quick connect
fitting. Squeeze the fitting retainer
tabs together and pull the
quick connect fitting assembly apart.

CAUTION

Wrap shop towels around the hose connection to catch any gasoline spillage.

4. Disconnect the fuel injector electrical connectors.

5. Remove the fuel rail attaching screws.

6. Lift the fuel rail and injectors off the intake manifold. Cover the fuel injector openings in the intake manifold.

7. Remove the fuel injector clip and pull the injector from the fuel rail. Note that whenever a fuel injector is removed, the O-rings must be replaced.

To install:

8. Install the injectors into the fuel rail and install the retaining clips.

9. Apply a light coating of clean engine oil to the O-ring on the nozzle end of each injector.

10. Insert the fuel injector nozzles into the openings in the intake manifold. Seat the injectors and tighten the fuel rail mounting screws to 200 ± 30 inch lbs. (22.5 ± 3 Nm).

11. Reconnect the electrical connectors to the fuel injectors.

12. Lightly oil the tube end, then reconnect the fuel supply line quick connect fitting to the fuel rail.

13. Connect the negative battery remote cable to the left strut tower.

14. Start engine and check for fuel leaks.

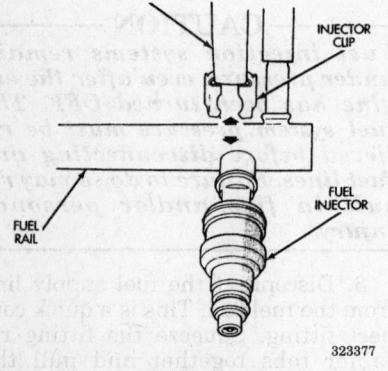

Fuel injector clip

ENGINE MECHANICAL

Engine Assembly

REMOVAL AND INSTALLATION

CAUTION

Fuel injection systems remain under pressure, even after the engine has been turned OFF. The fuel system pressure must be relieved before disconnecting any fuel lines. Failure to do so may result in fire and/or personal injury.

1. Release the fuel system pressure using the recommended procedure.

2. Disconnect the battery and tray.

3. Unbolt the Powertrain Control Module and move it aside.

4. Drain and properly contain the coolant from the engine.

5. Remove the upper and lower radiator hose, radiator and cooling fan.

6. Disconnect the clutch cable and transaxle shift linkage, if equipped.

7. Disconnect the throttle body linkage and the engine wiring harness.

8. Disconnect the heater hoses.

9. Discharge and properly contain the refrigerant.

10. Raise and safely support the vehicle.

11. Remove the drive belts.

12. Disconnect the exhaust pipe from the exhaust manifold.

13. Remove the front and rear engine mount brackets from the body.

14. Lower the vehicle and remove the air cleaner.

15. Remove the power steering pump and the A/C compressor.

16. Disconnect the ground straps from the engine.

17. Raise the vehicle and install a engine dolly under the vehicle and support engine.

18. Remove the transaxle and engine mount through bolts.

19. Raise the vehicle slowly allowing the engine and transaxle assembly to remain on the dolly.

To install:

20. Position the engine and the transaxle under the vehicle and lower the vehicle onto the engine assembly.

21. Align the engine mounts and install the right and left mount bolts.

22. Install the transaxle mount.

23. Install the axle shafts.

24. Install the transaxle and engine braces.

25. Install the splash shields.

26. Reconnect the exhaust pipe.

27. Install the power steering pump and the A/C compressor.

28. Reconnect the heater hoses.

29. Install the A/C compressor.

30. Install the inner splash shield.

31. Install the wheels and tires.

32. Reconnect the clutch cable and linkages.

33. Reconnect the fuel lines and the ground straps.

34. Reconnect the engine end throttle body electrical harnesses.

35. Reconnect the throttle body linkage.

36. Install the radiator, cooling fan and hoses.

37. Fill the cooling system.

38. If the engine oil was drained, install fresh oil and a new oil filter.

39. Install the battery tray and the battery.

40. Reinstall the powertrain control module.

41. Install the air cleaner.

42. Start the engine and run until operating temperature.

43. Check for leaks and proper operation.

Engine Mounts

REMOVAL AND INSTALLATION

Upper Mount

1. Disconnect the negative battery cable. Remove the air cleaner and all necessary air duct work.

2. Using a 7137 or a C-4852 engine lift or equivalent, raise and safely support the engine so it is not resting on the engine mounts.

3. Remove the engine mount bracket and body connection through bolt. Take note of the position of the arrow on the oval shaped mounting stopper plate; this is important.

4. Remove the engine mounting bracket and stopper plate.

To install:

5. Install the engine mounting bracket and stopper plate. Note the arrows on the stopper plates and make sure they are installed properly. On most engines, the arrows will face towards the center of the engine. Torque the upper mount to engine nut and bolt to 63 ft. lbs. (86 Nm) and the upper mount through bolt nut to 71–85 ft. lbs. (98–118 Nm).

6. Install the air cleaner and all air duct work previously removed.

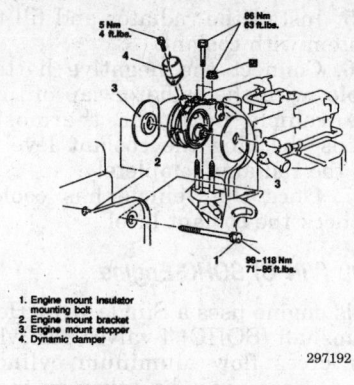

1. Engine mount insulator
   mounting bolt
2. Engine mount bracket
3. Engine mount stopper
4. Dynamic damper

297192

Right side upper engine mount

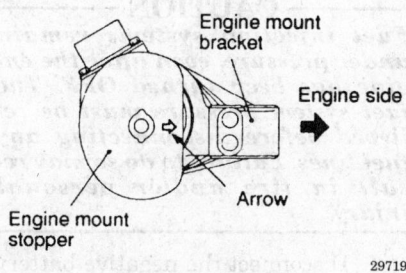

297194

Align the right side upper mount as shown

7. Connect the negative battery cable.

Engine Roll Stopper (Lower Rear Mount)

1. Disconnect the negative battery cable.

2. Using a 7137 or a C-4852 engine lift or equivalent, raise and safely support the engine so it is not resting on the engine mounts. Use care not to bend or damage any components.

3. Remove the stopper through bolt.

4. Remove the stopper mount frame bolts and pry the mount out.

To install:

5. Position the lower rear roll stopper so the part of the bracket with the hole in it is facing the front of the vehicle. Install the frame mounting bolts and tighten to 32 ft. lbs. (44 Nm).

6. Install the through bolt and torque the nut to 32 ft. lbs. (44 Nm).

7. Connect the negative battery cable.

Engine Roll Stopper (Lower Front Mount)

1. Disconnect the negative battery cable.

2. Using a 7137 or a C-4852 engine lift or equivalent, raise and safely support the engine so it is not resting on the engine mounts. Use care not to bend or damage any components.

3. Remove the stopper through bolt.

4. Remove the stopper mount frame bolts and pry the mount out.

To install:

5. Position the lower front roll stopper and install the frame mounting bolts and tighten to 32 ft. lbs. (44 Nm).

6. Install the through bolt, but do not fully tighten until the full weight of the engine is on the mount. Torque the nut to 41 ft. lbs. (56 Nm).

7. Connect the negative battery cable.

Cylinder Head

REMOVAL AND INSTALLATION

2.0L (VIN Y) DOHC Engine

This engine uses a Dual Over Head Camshaft (DOHC) aluminum cylinder head. Care must be taken to make sure all valve timing marks align after cylinder head service.

---**CAUTION**---

Fuel injection systems remain under pressure, even after the engine has been turned OFF. The fuel system pressure must be relieved before disconnecting any fuel lines. Failure to do so may result in fire and/or personal injury.

---

1. Following proper procedure, relieve the fuel system pressure.

2. Disconnect the negative battery cable.

3. Remove the air cleaner with all air intake hoses.

4. Drain the cooling system and the engine oil.

5. Disconnect the accelerator cable.

6. Disconnect the small vacuum hose at the pressure regulator.

7. Disconnect the oxygen sensor, engine coolant temperature sensor, the engine coolant temperature gauge unit and the engine coolant temperature switch on vehicles with air conditioning.

8. Disconnect the throttle position sensor, crankshaft angle sensor, fuel injectors, ignition coil pack, EGR temperature sensor, ground cable and engine control wiring harness.

9. Remove the spark plug cable center cover and remove the spark plug cables.

10. Disconnect and plug the high pressure fuel line.

11. Remove the heater hose and the water hose connections.

12. Remove the PCV hose.

13. Disconnect and plug the fuel return hose.

14. Disconnect the brake booster vacuum hose.

15. Remove the timing belt.

16. Remove the valve cover and the half-round seal.

17. Remove the exhaust pipe self-locking nuts and separate the exhaust pipe from the exhaust manifold. Discard the gasket.

18. Remove the intake and exhaust camshafts.

19. Loosen the cylinder head mounting bolts and lift off the cylinder head assembly complete with the exhaust and intake manifolds attached.

20. Place the assembly on a suitable workbench and remove the intake and exhaust manifolds.

21. Remove the head gasket.

To install:

22. Thoroughly clean and dry the mating surfaces of the head and block. Check the cylinder head for cracks, damage or engine coolant leakage. Remove scale, sealing compound and carbon. Clean the oil passages thoroughly. Check the head for flatness. End to end, the head should be no more than 0.004 inch (0.1mm) out-of-true.

23. Place a new head gasket on the cylinder block with the identification marks at the front top (upward) position. Make sure the gasket has the proper identification mark for the engine. Do not use sealer on the gasket.

24. New head bolts are recommended. Inspect the cylinder head bolts prior to installation. If the threads are necked down (stretched), the bolts should be replaced. Necking can be checked by holding a straight edge against the threads. If all the threads do not contact the straight-edge, the bolt should be replaced.

25. Carefully install the cylinder head on the block. Before installing the bolts, the threads should be oiled with clean engine oil.

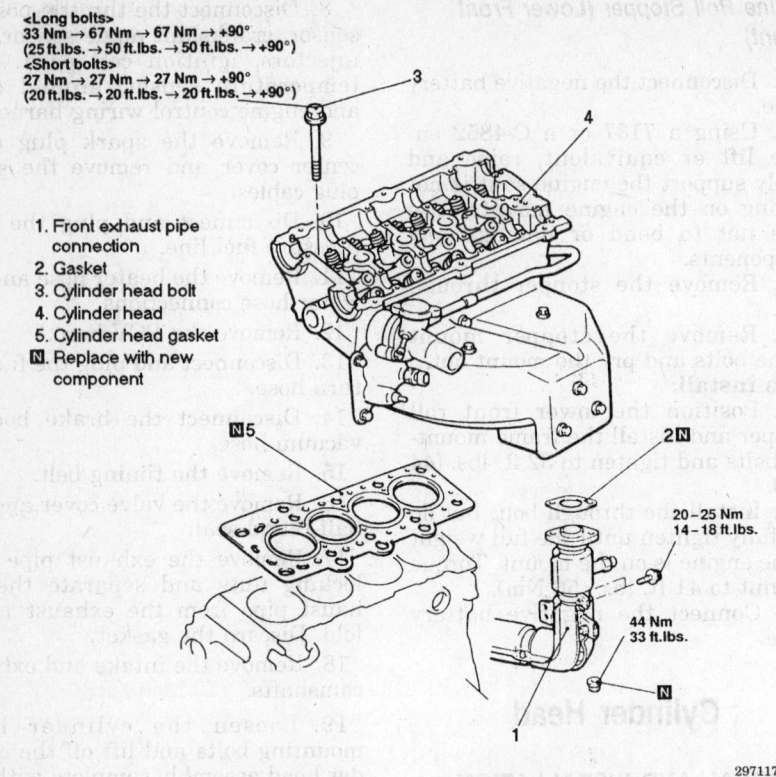

<Long bolts>
33 Nm → 67 Nm → 67 Nm → +90°
(25 ft.lbs. → 50 ft.lbs. → 50 ft.lbs. → +90°)
<Short bolts>
27 Nm → 27 Nm → 27 Nm → +90°
(20 ft.lbs. → 20 ft.lbs. → 20 ft.lbs. → +90°)

1. Front exhaust pipe connection
2. Gasket
3. Cylinder head bolt
4. Cylinder head
5. Cylinder head gasket
N. Replace with new component

20 – 25 Nm
14 – 18 ft.lbs.

44 Nm
33 ft.lbs.

297117

Cylinder head exploded view — 2.0L DOHC engine

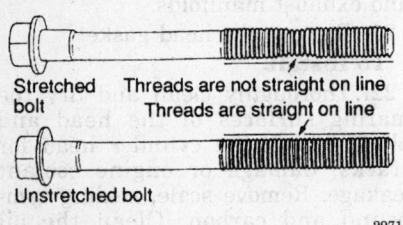

Stretched bolt — Threads are not straight on line
Threads are straight on line
Unstretched bolt

297118

Checking the cylinder head bolts for necking (stretching)

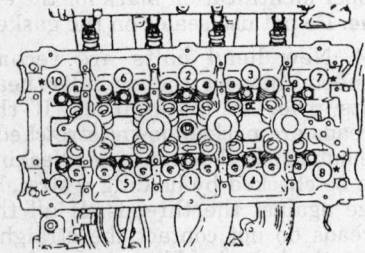

*Location of 110 mm (4.330 in.) short bolts.

297119

Cylinder head torque sequence — 2.0L DOHC engine

26. Tighten the cylinder head bolts in sequence as follows:
   a. Tighten the center bolts (1 through 6) to 25 ft. lbs. (33 Nm). Then, tighten the outer bolts (7 through 10) to 20 ft. lbs. (27 Nm).
   b. Tighten the center bolts (1 through 6) to 50 ft. lbs. (67 Nm). Then, tighten the outer bolts (7 through 10) to 20 ft. lbs. (27 Nm).
   c. Tighten all 10 bolts an additional ¼ turn (90 degrees).
27. Install the new exhaust pipe gasket and connect the exhaust pipe to the manifold. Tighten the self-locking bolts to 33 ft. lbs. (44 Nm).
28. Apply sealer to the perimeter of the half-round seal and to the lower edges of the half-round portions of the belt-side of the new gasket. Install the valve cover.
29. Install the timing belt and all related items.
30. Connect or install all previously disconnected hoses, cables and electrical connections. Adjust the throttle cable(s).
31. Install the spark plug cable center cover.
32. Replace the O-rings and connect the fuel lines.
33. Install the air cleaner and intake hose. Connect the breather hose.
34. Change the engine oil and oil filter.

35. Install the radiator and fill the system with coolant.
36. Connect the negative battery cable, with the radiator cap off, run the vehicle until the thermostat opens checking the coolant level to fill the radiator completely.
37. Once the vehicle has cooled, recheck the coolant level.

2.0L (VIN C) SOHC Engine

This engine uses a Single Over Head Camshaft (SOHC) 4-valves per cylinder cross flow aluminum cylinder head. Care must be taken to make sure all valve timing marks align after cylinder head service.

-------- CAUTION --------
Fuel injection systems remain under pressure, even after the engine has been turned OFF. The fuel system pressure must be relieved before disconnecting any fuel lines. Failure to do so may result in fire and/or personal injury.

1. Disconnect the negative battery cable remote connection located on the left strut tower.
2. Relieve the fuel system pressure using the recommended procedure.
3. Remove the air cleaner.
4. Drain and properly contain the engine coolant.
5. Disconnect and tag all vacuum hoses and all electrical connections from the throttle body.
6. Remove the throttle linkage.
7. Remove the accessory drive belt(s).
8. Disconnect the power brake booster vacuum hose.
9. Raise and safely support the vehicle.
10. Disconnect the exhaust pipe from the exhaust manifold.
11. Remove the power steering pump and move aside.
12. Remove the coil pack and bracket from the engine.
13. Disconnect the cam sensor and fuel injector wiring.
14. Remove the timing belt the recommended procedure.
15. Remove the valve cover.
16. Remove the rocker arm shafts.
17. Remove the cylinder head bolts and remove the cylinder head.

To install:

NOTE: The cylinder head bolts should be checked for stretching before reuse. If the thread area of the bolt is necked down the bolts must be replaced with new. New head bolts are recommended.

18. Clean all parts well. Clean all gasket material from the cylinder head and the engine block. Use care not to scratch the aluminum cylinder head sealing surface. Check the cylinder head for flatness using a feeler gauge and a straightedge. The cylinder head must be flat within 0.004 inch (0.1mm).

19. Check the cylinder head for cracks or other damage.

20. Install a new gasket and the cylinder head to the engine block.

21. Install the cylinder head bolts, the 4 short bolts are to be installed in positions 7, 8, 9 and 10. Torque the bolts in sequence.

22. Torque the bolts in 4 steps as follows:

    a. First: all bolts to 25 ft. lbs. (34 Nm).

    b. Second: all bolts to 50 ft. lbs. (68 Nm).

    c. Third: all bolts again to 50 ft. lbs. (68 Nm).

    d. Fourth: all bolts an additional ¼ turn.

NOTE: Do not use a torque wrench for the fourth step.

23. Set the crankshaft to 3 notches BTDC before installing the rocker arm shafts. Install the rocker arm shafts with the notches in the shafts pointing up and toward the timing belt side of the engine. Torque the retainer bolts to 200 inch lbs. (23 Nm).

24. Install the valve cover.

25. Install the timing belt using the recommended procedure to make sure the timing marks are properly aligned. Failure to do so will cause engine damage.

26. Reconnect the cam sensor and the fuel injector wiring.

27. Install the coil pack and the bracket.

28. Install the power steering pump.

29. Raise the vehicle and reconnect the exhaust pipe to the exhaust manifold.

30. Reconnect the brake booster vacuum line.

31. Reconnect the throttle linkage.

32. Reconnect all vacuum hoses and wiring.

33. Refill the cooling system. An oil and filter change are recommended.

34. Reconnect the negative battery cable remote connection.

35. Start the engine and check for leaks. Run the engine with the radiator cap off so as the engine warms and the thermostat opens, coolant can be added to the radiator. Test drive vehicle to check for proper operation.

Lash Adjusters

REMOVAL AND INSTALLATION

2.0L (VIN Y) DOHC Engine

This engine is called a Dual Over Head Camshaft (DOHC) engine since it uses 2 camshafts. Care must be taken to make sure all valve timing marks align after any cylinder head and valve train service.

1. Disconnect the negative battery cable.

2. Remove the cam cover.

NOTE: Always rotate the crankshaft in a clockwise direction. Make a mark on the back of the timing belt indicating the direction of rotation so it may be reassembled in the same direction if it is to be reused.

3. Rotate the crankshaft clockwise and align the timing marks so the No. 1 piston will be at TDC of the compression stroke.

4. Remove the timing belt cover.

5. Remove the timing belt.

6. Matchmark the crank angle sensor to the cylinder head and remove the crank angle sensor.

7. Remove both camshaft sprockets.

8. Loosen the bearing caps in sequence, 1 camshaft at a time. The bearing caps are identified for location. Remove the outside bearing caps first.

NOTE: If the bearing caps are difficult to remove, use a plastic hammer to gently tap the rear part of the camshaft.

9. Remove the intake and exhaust camshafts.

10. Remove the cam follower assemblies from the cylinder head. Keep the cam followers in the order they have been removed from the head for reassembly.

11. Mark the lash adjusters for reassembly in their original positions, then remove them.

WARNING

The camshafts and their components are NOT interchangeable. Place an identifying mark on each component. Be sure to keep all parts organized for proper reassembly in their original positions.

To install:

12. Before installation, clean the cylinder head and cover mating surfaces. Make certain that the rails are flat.

13. Install the lash adjuster assembly making sure the adjusters are at least partially full of oil. This is indicated by little or no plunger travel when the adjuster is depressed.

14. Lubricate the cam followers with clean engine oil and install the cam followers in their original position on the lash adjuster and valve stem.

15. Check the camshaft journals on the cylinder head and the cam bearings for wear or damage. Check the cam lobes and rocker rollers for damage. Also, check the cylinder head oil holes for clogging.

16. Check camshaft end-play.

17. Inspect the cam followers for wear or damage. Replace as necessary.

18. Lubricate the camshafts with heavy engine oil and position the camshafts on the cylinder head.

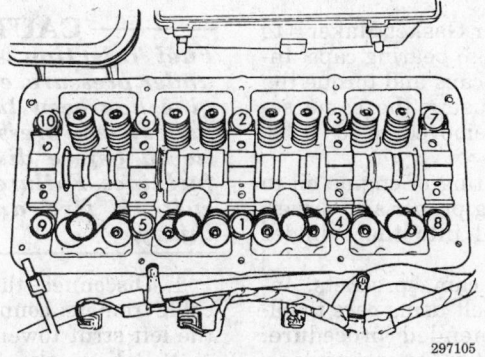

297105

Cylinder head tightening sequence — 2.0L SOHC engine

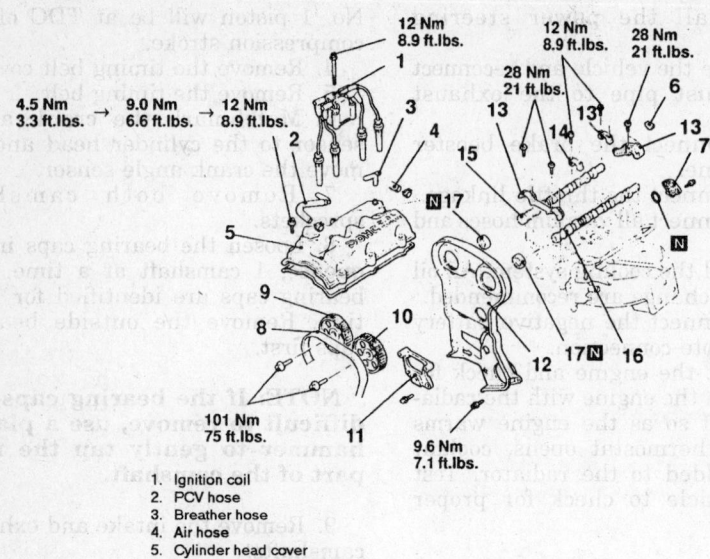

1. Ignition coil
2. PCV hose
3. Breather hose
4. Air hose
5. Cylinder head cover
6. Semi--circular packing
7. Camshaft position sensor
8. Timing belt
9. Intake camshaft sprocket
10. Exhaust camshaft sprocket
11. Bracket
12. Rear timing belt cover
13. Outside timing belt cover
14. Camshaft bearing cap
15. Intake camshaft
16. Exhaust camshaft

Exploded view of the camshafts and related components — 2.0L DOHC engine

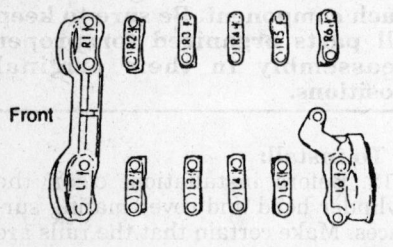

Identifying the camshaft bearing caps — 2.0L DOHC engine

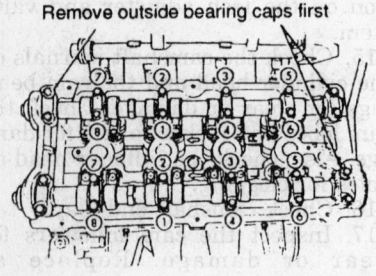

Bearing cap removal sequence — 2.0L DOHC engine

WARNING
The pistons should not be at top dead center when installing the camshafts since some valves will be open depending on camshaft position.

19. Make sure the dowel pin on both camshaft sprocket ends are located on the top.

20. Install the bearing caps No. 2 through No. 5 and right No. 6 and tighten the caps in sequence to 9 ft. lbs. (12 Nm). Check the markings on the caps to identify the cap number and intake/exhaust symbol. Make sure the rocker arm is correctly mounted on the lash adjuster and the valve stem end.

21. Apply Mopar Gasket Maker® to the No. 1 and No. 6 bearing caps. Install the bearing caps and torque the retaining bolts, using the same sequence as when removed, to 20 ft. lbs. (28 Nm).

22. Apply a coating of engine oil to the oil seal. Using proper size driver, press-fit the seal into the cylinder head.

23. Install the cam sprockets. Install the timing belt using care to follow the recommended procedure. Cam timing is critical or engine damage will result. Install the timing belt cover and related components.

---

24. Apply Mopar® silicone rubber adhesive sealant at the camshaft cap corners and at the top edge of the half round seal.

25. Install the cam cover using a new gasket. Tighten the cam cover retaining bolts in sequence, using a 3 step torque method:

   a. Tighten all bolts, in sequence to 3.3 ft. lbs. (4.5 Nm).

   b. Tighten all bolts, in sequence to 6.5 ft. lbs. (9 Nm).

   c. Tighten all bolts, in sequence to 9 ft. lbs. (12 Nm)

26. Check engine oil level. After this type of service, an oil and filter change is recommended. If a camshaft had failed, metal particles may be spread throughout the engine so an oil and filter change should be mandatory.

27. Connect the negative battery cable.

28. Start the engine and check for proper operation and leaks.

Valve Lash

ADJUSTMENT

The engines in these vehicles do not require periodic valve lash adjustment.

Rocker Arms/Shaft

REMOVAL AND INSTALLATION

2.0L (VIN C) SOHC Engine

This engine uses a Single Over Head Camshaft (SOHC) running in an aluminum cylinder head. Rocker arm shafts mount directly to the cylinder head. Care must be taken to make sure all valve timing marks align after cylinder head and valve train service. Please note that the cylinder head must be removed from the vehicle to service the camshaft.

CAUTION
Fuel injection systems remain under pressure, even after the engine has been turned OFF. The fuel system pressure must be relieved before disconnecting any fuel lines. Failure to do so may result in fire and/or personal injury.

---

1. Disconnect the negative battery cable remote connection located on the left strut tower.

2. Relieve the fuel system pressure using the recommended procedure.

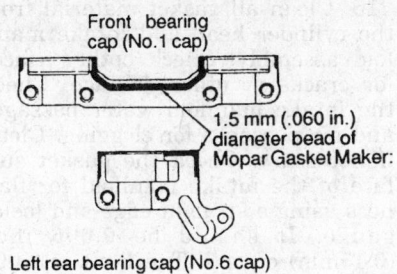

Apply sealant as shown — 2.0L DOHC engine

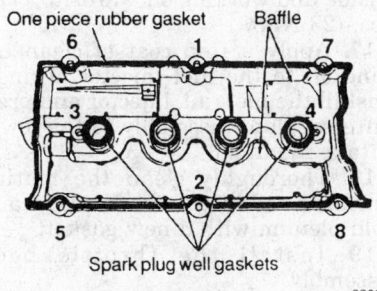

Cam cover tightening sequence — 2.0L DOHC engine

3. Remove the cylinder head cover using the recommended procedure.

4. Mark the rocker arm shaft assemblies to identify them for later installation.

5. Unfasten the rocker arm shaft bolts and remove the rocker arm shaft assemblies from the cylinder head.

To install:

6. If the rocker arms and shaft are to be serviced, mark the rocker arms so any that are to be returned to service will be installed in their original locations. Slide the rocker arms off the shaft. Keep the spacers and

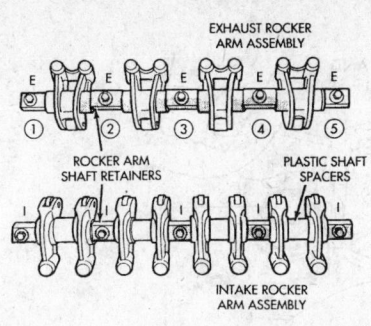

Rocker arm shaft identification — 2.0L SOHC engine

rocker arms in the same location for reassembly.

a. Inspect the rocker arms for scoring, wear on the roller or damage to the shaft. Replace parts as necessary.

b. The rocker arm shaft is hollow and used as a lubrication oil duct. Check that the shaft is clean inside and out.

c. Check all oil holes for clogging with a small wire and clean and required.

d. To assemble, thoroughly lubricate all rocker arm components and spacers and install on the rocker arm shaft in the original locations.

7. If the vehicle exhibited a tappet-like noise, the valve lash adjusters built into the rocker arms should be cleaned and checked. Lash adjusters removed from a rocker arm should be returned to their original locations. Replace worn or defective lash adjusters. To install a lash adjuster, use the following procedure.

a. Lubricate the lash adjuster thoroughly with clean engine oil.

b. Install the adjuster into the rocker arm making sure the adjuster is at least partially filled with oil.

c. Place the rocker arm in clean engine oil and pump the plunger

until the lash adjuster travel is taken up. If travel is not reduced, replace the adjuster.

d. Install the rocker arm back on the rocker arm shaft.

8. Before installing the rocker arm and shaft assemblies, set the crankshaft to 3 notches before TDC on the crankshaft sprocket.

9. Install the rocker arm and shaft assemblies with the small notches in the rocker shafts pointing up and toward the timing belt side of the engine. Install the retainers in their original positions on the exhaust and intake shafts. Torque the bolts to 200 inch lbs. (23 Nm).

10. Install the cylinder head cover.

11. Connect all electrical, vacuum and fluid connections as required.

12. Refill the cooling system. An oil and filter change are recommended.

13. Reconnect the negative battery cable remote connection.

14. Start the engine and check for leaks. Run the engine with the radiator cap off so as the engine warms and the thermostat opens, coolant can be added to the radiator. Test drive vehicle to check for proper operation.

Intake Manifold

REMOVAL AND INSTALLATION

2.0L (VIN Y) DOHC Engine

This engine uses a two-piece aluminum intake manifold. The upper half of the manifold (also called a plenum) mounts the throttle body. The lower half of the manifold contains the fuel rail and injectors. A nonreusable gasket joins the 2 halves. Use care when working with light alloy parts.

—— **CAUTION** ——

Fuel injection systems remain under pressure, even after the engine has been turned OFF. The fuel system pressure must be relieved before disconnecting any fuel lines. Failure to do so may result in fire and/or personal injury.

1. Relieve the fuel system pressure.

2. Disconnect the negative battery cable and drain the cooling system.

3. Disconnect the accelerator cable, breather hose and air intake hose.

4. Disconnect the vacuum connection at the power brake booster and the PCV valve. Disconnect all remaining vacuum hoses and pipes, as

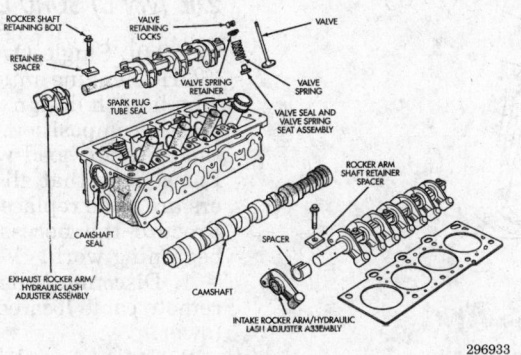

Cylinder head and valve assembly — 2.0L SOHC engine

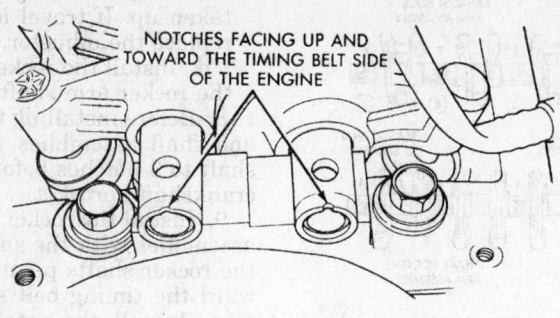

NOTCHES FACING UP AND TOWARD THE TIMING BELT SIDE OF THE ENGINE

296936

Rocker arm shaft notch location — 2.0L SOHC engine

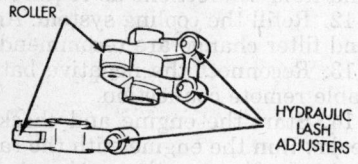

EXHAUST ROCKER ARM

ROLLER

HYDRAULIC LASH ADJUSTERS

INTAKE ROCKER ARM

296938

Rocker arm/hydraulic lash adjuster assemblies — 2.0L engine

necessary. Tag for identification, if necessary, to save time at assembly.

5. Disconnect the high pressure fuel line, fuel return hose and remove the throttle control cable and brackets.

CAUTION

Do not use conventional fuel filters, hoses or clamps when servicing fuel injection systems. They are not compatible with the injection system and could fail, causing personal injury or damage to

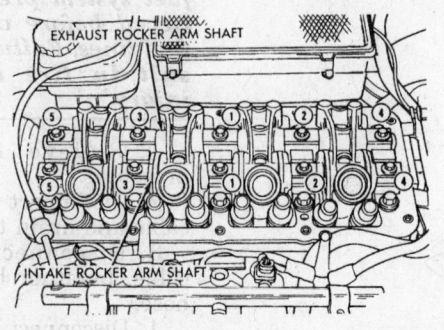

EXHAUST ROCKER ARM SHAFT

INTAKE ROCKER ARM SHAFT

296939

Rocker arm shaft tightening sequence — 2.0L SOHC engine

the vehicle. Use only hoses and clamps specifically designed for fuel injection.

6. Disconnect the alternator wiring harness connection.

7. Disconnect the MAP sensor and the intake air temperature sensor connectors.

8. Disconnect the TPS connector and position the engine wiring harness aside.

9. Disconnect the EGR pipe connection.

10. Remove the intake manifold stay and the engine hanger. Disconnect the fuel injector connectors.

11. Remove the throttle body assembly.

12. Remove the mounting bolts and remove the intake manifold plenum and gasket.

13. Remove the fuel rail, injector and pressure regulator assembly. Use care since the fuel injectors can drop out of the fuel rail as it is being removed.

14. Remove the mounting bolts and remove the intake manifold and gasket from the engine.

To install:

15. Clean all gasket material from the cylinder head and intake manifold assembly. Check both surfaces for cracks or other damage. Check the intake manifold water passages and air passages for clogging. Clean if necessary. Check the gasket surface of the intake manifold for flatness using a straight edge and feeler gauge. It should be 0.006 inch (0.15mm) or less. The limit is 0.008 inch (0.20mm).

16. Install a new intake manifold gasket to the head and install the manifold. Torque the manifold in a criss-cross pattern, starting from the inside and working outwards to 17 ft. lbs. (23 Nm).

17. Apply a thin coat of clean engine oil to the fuel injector O-rings. Install the fuel rail, injector and pressure regulator assembly to the lower intake manifold.

18. Thoroughly clean the mating surfaces and install the intake manifold plenum with a new gasket.

19. Install the throttle body assembly.

20. Install the intake manifold stay and the engine hanger. Connect the fuel injector connectors.

21. Connect the EGR pipe connection.

22. Connect the engine control electrical connectors.

23. Connect the generator wiring harness connection.

24. Connect the high pressure fuel line, fuel return hose and the throttle control cable brackets.

25. Connect the vacuum connection at the power brake booster and the PCV valve. Connect all remaining vacuum hoses and pipes.

26. Connect the accelerator cable, breather hose and air intake hose.

27. Connect the negative battery cable.

28. Start the engine and check for proper operation.

2.0L (VIN C) SOHC Engine

The 2.0L Single Overhead Camshaft (SOHC) engine intake manifold is a long branch design made of a molded plastic composition. It is attached to the cylinder head with 10 fasteners. Please note that all seals and fasteners are to be replaced for installation. Procure the necessary parts before beginning work.

1. Disconnect the negative battery remote cable located on the left strut tower.

2. Release the fuel system pressure.

19. Install the intake manifold. Tighten the fasteners in sequence to 105 inch lbs. (11.9 Nm).

20. Apply a light coating of engine oil to the fuel injector O–rings. Remove the covers from the fuel injector openings and install the fuel injectors into the engine. Seat the injectors in place and tighten the fuel rail bolts to 200 inch lbs. (23 Nm).

21. Connect the electrical connectors to the fuel injectors.

22. Lubricate the quick-connect fittings with clean engine oil. Connect the fuel supply line to the fuel rail. Check the connection by pulling on the connector to insure it is locked into position.

23. Connect the PCV and the brake booster hoses.

24. Install the throttle body and torque to 200 inch lbs. (23 Nm). Install the transaxle to throttle body support bracket and tighten to 105 inch lbs. (11.9 Nm) at the throttle body first. Next tighten the bracket at the transaxle.

25. Reconnect the MAP sensor and the air temperature sensor wiring connectors.

26. Connect the knock sensor electrical and starter relay connectors. Connect the wiring harness to the intake manifold tab.

27. Reconnect the IAC and TPS wiring connectors.

28. Reconnect the throttle body vacuum hoses.

29. Install the accelerator, kickdown and speed control cables to their bracket and connect then to the throttle lever.

30. Loosely assemble the EGR tube onto the valve and intake manifold finger tight. Tighten the tube fasteners at the EGR valve first to 95 inch lbs. (11 Nm), then tighten the intake manifold side fasteners to 95 inch lbs. (11 Nm).

31. Install the fresh air duct to the air filter housing.

32. Connect the negative battery cable, pressurize the fuel system and check for leaks.

Exhaust Manifold

REMOVAL AND INSTALLATION

2.0L (VIN Y) DOHC Engine

1. Disconnect the negative battery cable.

2. Remove the air intake hose and the small air hose connection.

3. Properly drain the engine coolant.

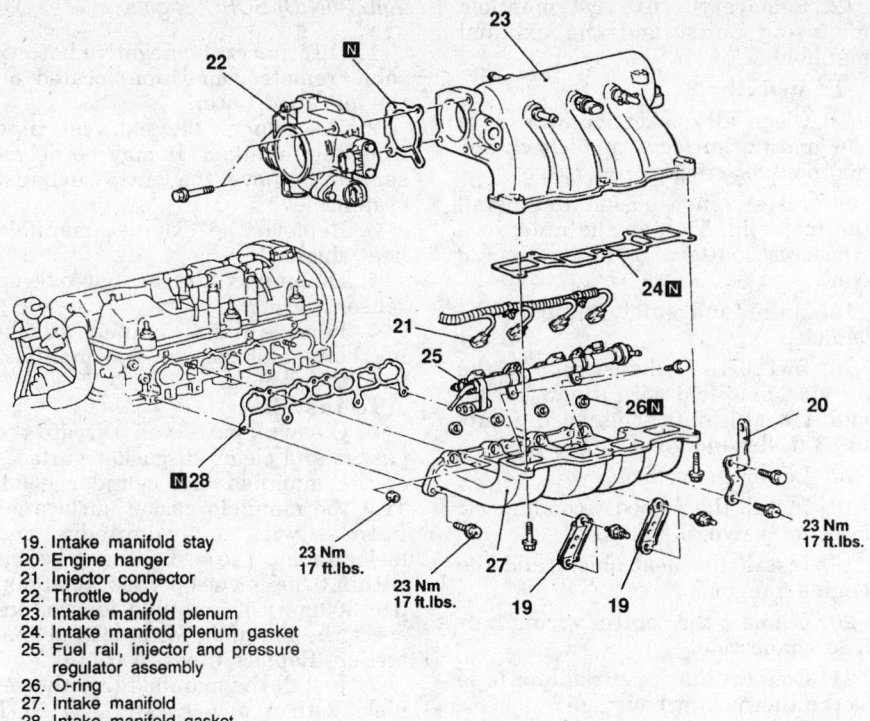

19. Intake manifold stay
20. Engine hanger
21. Injector connector
22. Throttle body
23. Intake manifold plenum
24. Intake manifold plenum gasket
25. Fuel rail, injector and pressure regulator assembly
26. O-ring
27. Intake manifold
28. Intake manifold gasket

298386

Exploded view of the intake manifold assembly — 2.0L DOHC engine

——— **CAUTION** ———
Fuel injection systems remain under pressure after the engine has been turned OFF. The fuel system pressure must be relieved before disconnecting any fuel lines. Failure to do so may result fire and/or personal injury.

3. Disconnect the fuel supply line from the fuel rail.

4. Remove the fuel rail attaching screws and remove the fuel rail. Use care when handling the fuel injectors. Do not set them on their tips. Cover the fuel injector openings after fuel rail removal.

5. Remove the fresh air duct and upper air filter housing.

6. Remove the accelerator, kickdown and speed control cables from the throttle lever and bracket.

7. Disconnect the Throttle Position Sensor (TPS) and the Idle Air Control (IAC) motor electrical connections.

8. Disconnect the vacuum hoses from the throttle body.

9. Disconnect the connectors from the Manifold Absolute Pressure (MAP) sensor and the intake air temperature sensors.

10. Disconnect the vapor and brake booster hoses.

11. Disconnect the knock sensor electrical connector and disconnect the wiring harness from the tab located on the intake manifold.

12. Disconnect the electrical connector from the starter relay.

13. Remove the transaxle to throttle body support bracket fasteners at the throttle body and loosen the fastener at the transaxle end.

14. Remove the throttle body.

15. Remove the EGR tube bolts at the valve and at the intake manifold. Remove the tube from the engine.

16. Remove the intake manifold to inlet water tube support fastener.

17. Remove the 9 intake manifold screws and washer assemblies and the 1 nut and washer assembly Discard the fasteners. At assembly, they should be replaced with new fasteners. Remove the intake manifold.

To install:

18. Clean all gasket surfaces. Check upper and lower manifold gasket surfaces for flatness with a straight edge. Surface must be flat within 0.006 inch (0.15mm) per foot.

NOTE: All seals are to be replaced with new seals and all fasteners are to be replaced with new fasteners.

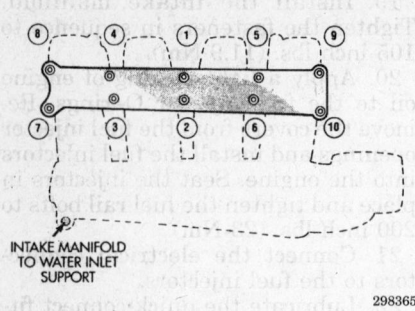

Intake manifold torque sequence — 2.0L SOHC engine

298365

4. Disconnect the upper radiator hose from the thermostat housing.

5. Disconnect the control wiring harness connection.

6. Remove the water pipe assembly and the engine oil level gauge.

7. Remove the heat shield and the engine hanger.

8. Remove the pulsed secondary air injection valve, if equipped.

9. Raise and safely support the vehicle.

10. Remove the exhaust pipe to exhaust manifold lock nuts and separate the exhaust pipe. Discard the gasket.

11. Lower the vehicle.

12. Remove the exhaust manifold mounting bolts, and the exhaust manifold.

To install:

13. Clean all gasket material from the mating surfaces and check the manifold for cracks or warpage.

14. Install a new gasket and install the manifold. Tighten the nuts, in a criss-cross pattern to 17 ft. lbs. (23 Nm).

15. Raise and safely support the vehicle.

16. Install the exhaust pipe to the exhaust manifold with a new gasket and new lock nuts. Tighten the nuts to 33 ft. lbs. (44 Nm).

17. Lower the vehicle.

18. Install the pulsed secondary air injection valve, if equipped.

19. Install the heat shield and the engine hanger.

20. Connect the control wiring harness connection.

21. Connect the upper radiator hose to the thermostat housing.

22. Properly fill the engine cooling system.

23. Install the air intake hose and the small air hose connection.

24. Connect the negative battery cable and check for exhaust leaks.

2.0L (VIN C) SOHC Engine

1. Disconnect the negative battery cable remote connection located on the left strut tower.

2. Disconnect the exhaust pipe from the manifold. It may be necessary to remove the entire exhaust system.

3. Remove the exhaust manifold heat shield.

4. Disconnect the heated oxygen sensor.

5. Remove the 8 manifold attaching bolts and remove the manifold from the vehicle.

To install:

6. Clean all parts well. Discard the gasket and clean all gasket surfaces of the manifold and cylinder head. Test the manifold gasket surface for flatness with a straightedge and feeler gauge. The surface must be flat within 0.006 inches per foot (0.15mm per 300mm) of manifold length. Inspect the manifold for cracks or distortion. Replace if necessary.

7. Install the manifold into the vehicle with a new gasket. DO NOT APPLY SEALER.

8. Install the 8 manifold bolts and tighten starting at the center and working outward. Torque to 200 inch lbs. (23 Nm).

9. Reconnect the heated oxygen sensor.

10. Install the heat shield.

11. Install the exhaust pipe and torque fasteners to 250 inch lbs. (28 Nm).

Front Crankshaft Seal

REMOVAL AND INSTALLATION

1. Disconnect the negative battery cable.

2. Turn the crankshaft clockwise until the engine is at TDC No. 1 cylinder compression stroke (firing position). Remove the timing belt using the recommended procedure.

3. Using a suitable puller, draw the crankshaft timing belt drive sprocket from the front of the crankshaft.

4. Drain the engine oil. Remove the oil pan.

5. Remove the oil screen feed pipe and O-ring that seals it to the pump/block assembly.

6. The front cover/oil pump mounting bolts are different sizes and must be reinstalled in their original locations. Remove and tag the front cover mounting bolts.

7. Remove the oil pump assembly.

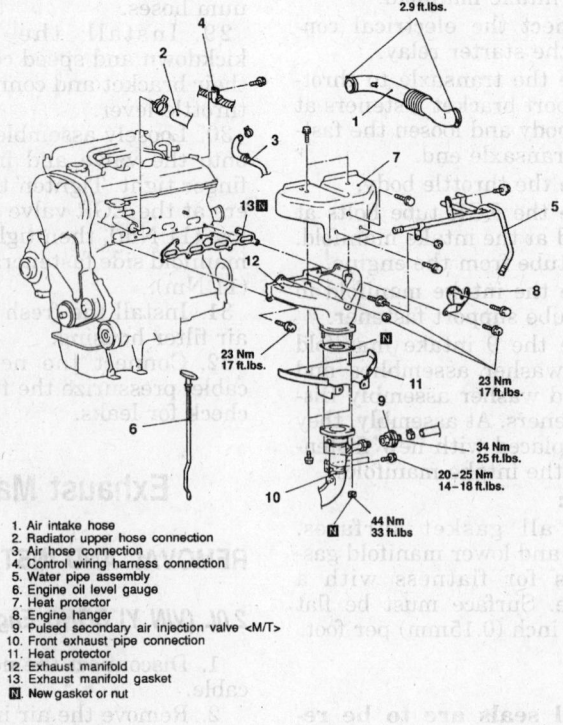

1. Air intake hose
2. Radiator upper hose connection
3. Air hose connection
4. Control wiring harness connection
5. Water pipe assembly
6. Engine oil level gauge
7. Heat protector
8. Engine hanger
9. Pulsed secondary air injection valve <M/T>
10. Front exhaust pipe connection
11. Heat protector
12. Exhaust manifold
13. Exhaust manifold gasket
N. New gasket or nut

Exhaust manifold and related components — 2.0L DOHC engine

297218

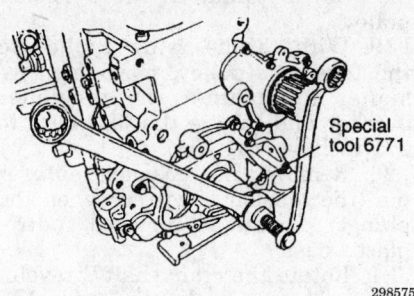

298575

Removing the front crankshaft oil seal — 2.0L engine

8. Using special tool 6771 or equivalent seal puller, remove the front crankshaft oil seal. Take care not to damage the seal surface of the cover.

To install:

9. Using tool 6780-1 or equivalent, install a new front crankshaft oil seal.

10. Install the oil screen and pick-up tube assembly using a new O-ring.

11. Install the crankshaft timing belt drive sprocket. Install the timing belt following the recommended procedure. Take care that all timing marks are properly aligned or engine damage will result.

12. Install the oil pan and drain plug. Fill the crankcase with the proper quantity and type of engine oil. A filter change is recommended.

13. Connect the negative battery cable.

14. Remove the engine oil pressure sending unit and install an oil pressure gauge. Start and run the engine until the thermostat opens. Curb idle oil pressure should be 4 psi minimum. At 3000 rpm, the oil pressure should be between 25 to 80 psi. If the oil pressure is zero at idle, immediately shut OFF the engine and check if the pressure relief valve is stuck open or for some other problem (oil level, oil type, loose filter, etc.).

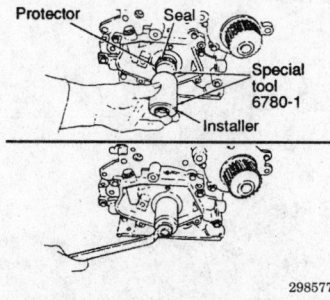

298577

Installing a new seal using a 6780-1 seal installer. Use care if using substitute tools — 2.0L engine

WARNING

If the oil pressure is zero at idle DO NOT run the engine at 3000 rpm looking for a pressure increase or the engine could be destroyed.

Front Camshaft Seal(s)

REMOVAL AND INSTALLATION

1. Disconnect the negative battery cable.

WARNING

Before removing the timing belt, set the crankshaft sprocket 3 notches BTDC, as this will prevent possible engine damage.

2. Remove the timing belt cover, then remove the timing belt.

3. Hold the camshaft sprocket with C-4687 with adaptor C-4687-1 or equivalent while removing the bolts.

4. Use a prybar to carefully remove the crankshaft seal(s). Be sure not to nock or damage the camshaft seal surface or cylinder head seal retaining bore.

5. Shaft seal lip surface must be free of varnish, dirt or nicks. Polish with 400 grit paper if necessary.

To install:

6. Install camshaft seal into the cylinder head using tool MD 998713 until flush with the head.

7. Install camshaft sprockets and tighten the attaching bolts to 75 ft. lbs. (101 Nm).

8. Install the timing belt and the timing belt cover.

Timing Belt, Sprockets, Tensioner and Front Cover

REMOVAL AND INSTALLATION

2.0L (VIN Y) DOHC Engine

1. Disconnect the negative battery cable remote connection located on the left strut tower.

2. Remove the drive belts and accessories.

3. Remove the crankshaft damper.

4. Remove the right engine mount

5. Place a support under the engine.

6. Remove the engine mount bracket

7. Remove the timing belt cover.

8. Loosen the timing belt tensioner bolts.

9. Remove the timing belt and the tensioner.

10. Place the tensioner into a soft jawed vise to compress the tensioner.

11. After compressing the tensioner place a pin ($^5/_{64}$ Allen wrench will work) into the plunger side hole to retain the plunger until installation.

To install:

12. Set the crankshaft sprocket to Top Dead Center (TDC) by aligning the notch on the sprocket with the arrow on the oil pump housing.

13. Set the camshafts to align the timing marks.

14. Move the crankshaft to ½ notch before TDC.

15. Install the timing belt starting at the crankshaft then around the water pump and around the camshafts last.

16. Move the crankshaft to TDC to take up the belt slack.

17. Install the tensioner to the block but do not tighten.

18. Using a torque wrench on the tensioner pulley apply 250 inch lbs. (28 Nm) of torque to the tensioner pulley.

19. With torque being applied to the tensioner pulley, move the tensioner up against the tensioner bracket and torque the fasteners to 275 inch lbs. (31 Nm).

20. Remove the tensioner plunger pin, the tension is correct when the plunger pin can be removed and replaced easily.

21. Rotate the crankshaft 2 revolutions and recheck the timing marks.

22. Install the timing belt cover.

23. Install the engine mount bracket.

24. Install the right engine mount.

25. Remove the engine support.

26. Install the crankshaft damper and torque to 105 ft. lbs. (142 Nm).

27. Install the drive belts and accessories.

28. Install the right inner splash shield.

29. Perform the crankshaft and camshaft relearn alignment procedure using the DRB scan tool or equivalent.

2.0L (VIN C) SOHC Engine

1. Disconnect the negative battery cable remote connection located on the left strut tower.

2. Remove the drive belts and accessories.

3. Remove the crankshaft damper.

4. Remove the right engine mount

5. Place a support under the engine.

6. Remove the engine mount bracket

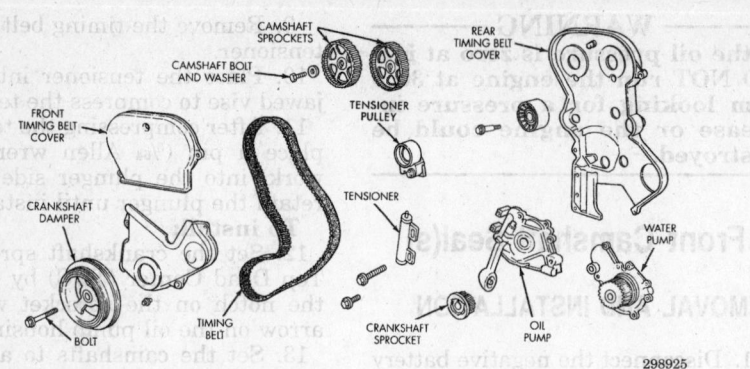

Timing belt assembly — 2.0L DOHC engine

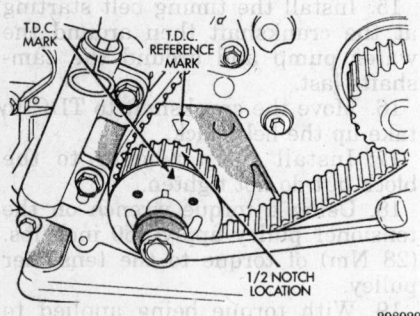

Crankshaft position — 2.0L DOHC engine

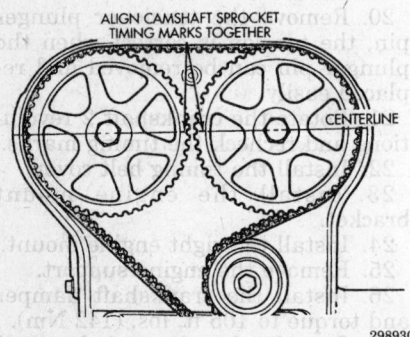

Camshaft position — 2.0L DOHC engine

7. Remove the timing belt cover.

8. Loosen the timing belt tensioner bolts.

9. Remove the timing belt and the tensioner.

10. Place the tensioner into a soft jawed vise to compress the tensioner.

11. After compressing the tensioner place a pin (⁵/₆₄ inch Allen wrench will work) into the plunger side hole to retain the plunger until installation.

To install:

12. Set the crankshaft sprocket to Top Dead Center (TDC) by aligning the notch on the sprocket with the arrow on the oil pump housing then

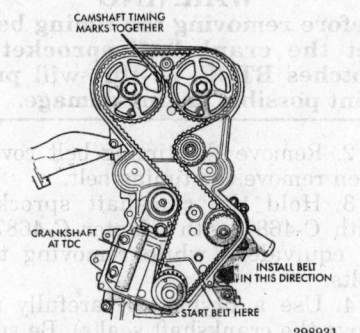

Timing belt proper position — 2.0L DOHC engine

back off the sprocket 3 notches before TDC.

13. Set the camshaft to align the timing marks.

14. Move the crankshaft to ½ notch before TDC.

15. Install the timing belt starting at the crankshaft then around the water pump and around the camshaft last.

16. Move the crankshaft to TDC to take up the belt slack.

17. Install the tensioner to the block but do not tighten.

18. Using a torque wrench on the tensioner pulley apply 250 inch lbs

(28 Nm) of torque to the tensioner pulley.

19. With torque being applied to the tensioner pulley, move the tensioner up against the tensioner bracket and torque the fasteners to 275 inch lbs. (31 Nm).

20. Remove the tensioner plunger pin, the tension is correct when the plunger pin can be removed and replaced easily.

21. Rotate the crankshaft 2 revolutions and recheck the timing marks.

22. Install the timing belt cover.

23. Install the engine mount bracket.

24. Install the right engine mount.

25. Remove the engine support.

26. Install the crankshaft damper and torque to 105 ft. lbs. (142 Nm).

27. Install the drive belts and accessories.

28. Install the right inner splash shield.

29. Perform the crankshaft and camshaft relearn alignment procedure using the DRB scan tool or equivalent.

Camshaft

REMOVAL AND INSTALLATION

2.0L (VIN Y) DOHC Engine

This engine is called a Dual Over Head Camshaft (DOHC) engine since it uses 2 camshafts. Care must be taken to make sure all valve timing marks align after any cylinder head and valve train service.

1. Disconnect the negative battery cable.

2. Remove the cam cover.

NOTE: Always rotate the crankshaft in a clockwise direction. Make a mark on the back of the timing belt indicating the direction of rotation so it may be reassembled in the same direction if it is to be reused.

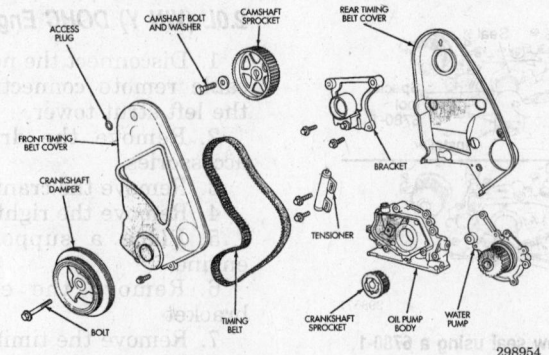

Timing belt assembly — 2.0L SOHC engine

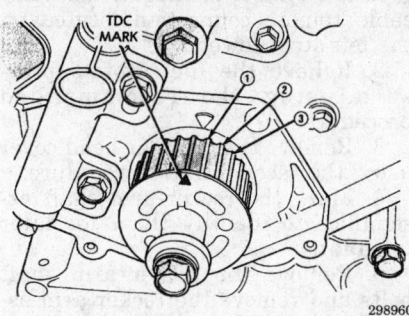

Crankshaft position — 2.0L SOHC engine

298960

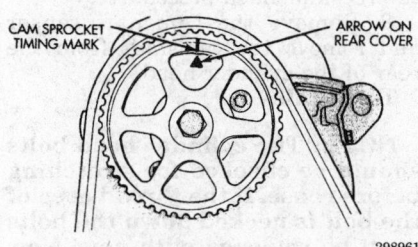

CAM SPROCKET TIMING MARK

ARROW ON REAR COVER

Camshaft position — 2.0L SOHC engine

298961

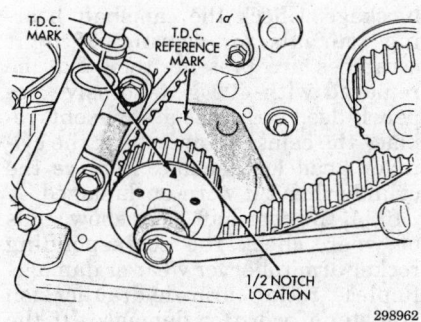

T.D.C. MARK

T.D.C. REFERENCE MARK

1/2 NOTCH LOCATION

Adjusting crankshaft position — 2.0L SOHC engine

298962

3. Rotate the crankshaft clockwise and align the timing marks so the No. 1 piston will be at TDC of the compression stroke.
4. Remove the timing belt cover.
5. Remove the timing belt.
6. Matchmark the crank angle sensor to the cylinder head and remove the crank angle sensor.
7. Remove both camshaft sprockets.
8. Loosen the bearing caps in sequence, 1 camshaft at a time. The bearing caps are identified for loca-

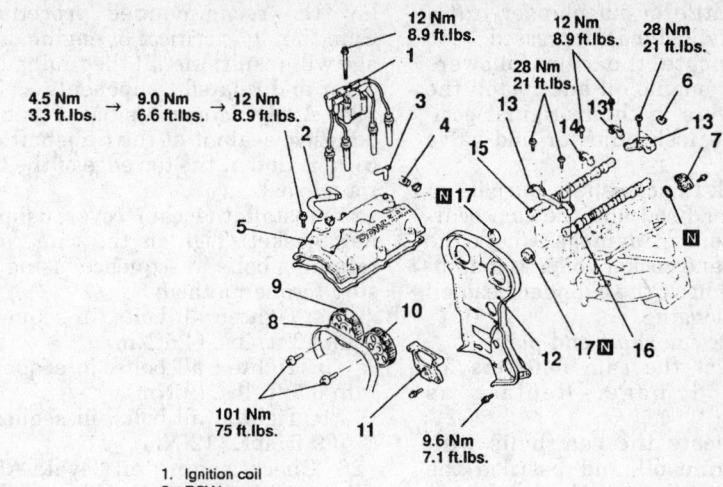

12 Nm 8.9 ft.lbs.

12 Nm 8.9 ft.lbs.

28 Nm 21 ft.lbs.

4.5 Nm 3.3 ft.lbs. → 9.0 Nm 6.6 ft.lbs. → 12 Nm 8.9 ft.lbs.

28 Nm 21 ft.lbs.

101 Nm 75 ft.lbs.

9.6 Nm 7.1 ft.lbs.

1. Ignition coil
2. PCV hose
3. Breather hose
4. Air hose
5. Cylinder head cover
6. Semi-circular packing
7. Camshaft position sensor
8. Timing belt
9. Intake camshaft sprocket
10. Exhaust camshaft sprocket
11. Bracket
12. Rear timing belt cover
13. Outside timing belt cover
14. Camshaft bearing cap
15. Intake camshaft
16. Exhaust camshaft

320825

Exploded view of the camshafts and related components — 2.0L DOHC engine

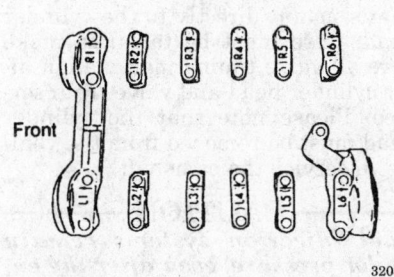

Front

Identifying the camshaft bearing caps — 2.0L DOHC engine

320826

Remove outside bearing caps first

Bearing cap removal sequence — 2.0L DOHC engine

320827

tion. Remove the outside bearing caps first.

NOTE: If the bearing caps are difficult to remove, use a plastic hammer to gently tap the rear part of the camshaft.

9. Remove the intake and exhaust camshafts.
10. Remove the cam follower assemblies from the cylinder head. Keep the cam followers in the order they have been removed from the head for reassembly.
11. Mark the lash adjusters for reassembly in their original positions.

—— WARNING ——
The camshafts and their components are NOT interchangeable. Place an identifying mark on each component. Be sure to keep all parts organized for proper reassembly in their original positions.

To install:
12. Before installation, clean the cylinder head and cover mating surfaces. Make certain that the rails are flat.
13. Install the lash adjuster assembly making sure the adjusters are at least partially full of oil. This is indi-

cated by little or no plunger travel when the adjuster is depressed.

14. Lubricate the cam followers with clean engine oil and install the cam followers in their original position on the lash adjuster and valve stem.

15. Check the camshaft journals on the cylinder head and the cam bearings for wear or damage. Check the cam lobes and rocker rollers for damage. Also, check the cylinder head oil holes for clogging.

16. Check camshaft end-play.

17. Inspect the cam followers for wear or damage. Replace as necessary.

18. Lubricate the camshafts with heavy engine oil and position the camshafts on the cylinder head.

——— WARNING ———
The pistons should not be at top dead center when installing the camshafts since some valves will be open depending on camshaft position.

19. Make sure the dowel pin on both camshaft sprocket ends are located on the top.

20. Install the bearing caps No. 2 through No. 5 and right No. 6 and tighten the caps in sequence to 9 ft. lbs. (12 Nm). Check the markings on the caps to identify the cap number and intake/exhaust symbol. Make sure the rocker arm is correctly mounted on the lash adjuster and the valve stem end.

21. Apply Mopar Gasket Maker® to the No. 1 and No. 6 bearing caps. Install the bearing caps and torque the retaining bolts, using the same sequence as when removed, to 20 ft. lbs. (28 Nm).

22. Apply a coating of engine oil to the oil seal. Using proper size driver, press-fit the seal into the cylinder head.

23. Install the cam sprockets. Install the timing belt using care to fol-

low the recommended procedure. Cam timing is critical or engine damage will result. Install the timing belt cover and related components.

24. Apply Mopar® silicone rubber adhesive sealant at the camshaft cap corners and at the top edge of the half round seal.

25. Install the cam cover using a new gasket. Tighten the cam cover retaining bolts in sequence, using a 3 step torque method:

   a. Tighten all bolts, in sequence to 3.3 ft. lbs. (4.5 Nm)

   b. Tighten all bolts, in sequence to 6.5 ft. lbs. (9 Nm)

   c. Tighten all bolts, in sequence to 9 ft. lbs. (12 Nm)

26. Check engine oil level. After this type of service, an oil and filter change is recommended. If a camshaft had failed, metal particles may be spread throughout the engine so an oil and filter change should be mandatory.

27. Connect the negative battery cable.

28. Start the engine and check for proper operation and leaks.

2.0L (VIN C) SOHC Engine

This engine uses a Single Over Head Camshaft (SOHC) running in an aluminum cylinder head. Rocker arm shafts mount directly to the cylinder head. Care must be taken to make sure all valve timing marks align after cylinder head and valve train service. Please note that the cylinder head must be removed from the vehicle to service the camshaft.

——— CAUTION ———
Fuel injection systems remain under pressure, even after the engine has been turned OFF. The fuel system pressure must be relieved before disconnecting any fuel lines. Failure to do so may result in fire and/or personal injury.

1. Disconnect the negative battery cable remote connection located on the left strut tower.

2. Relieve the fuel system pressure using the recommended procedure.

3. Remove the cylinder head cover using the recommended procedure.

4. Mark the rocker arm shaft assemblies to identify them for later installation.

5. Remove the rocker arm shaft bolts and remove the rocker arm assemblies from the cylinder head.

6. Remove the timing belt and camshaft sprocket using the recommended procedure.

7. Remove the cylinder head using the recommended procedure.

8. Remove the camshaft sensor and remove the camshaft from the rear of the cylinder head.

   To install:

   NOTE: The cylinder head bolts should be checked for stretching before reuse. If the thread area of the bolt is necked down the bolts must be replaced with new. New head bolts are recommended.

9. Clean all parts well. Inspect the camshaft journals for scoring. Check the oil feed holes in the head for blockage. Check the camshaft bearing journals for scoring. If light scratches are present, they may be removed with 400 grit abrasive paper. If deep scratches are present, replace the camshaft and check the cylinder head for damage. Replace the cylinder head if worn or damaged.

10. If the camshaft lobes show signs of wear, check the corresponding rocker arm roller for wear or damage. Replace rocker arms/hydraulic lash adjuster if worn or damaged. If the camshaft lobes show signs of pitting on the nose, flank or base circle, replace the camshaft.

11. If the rocker arms and shaft are to be serviced, mark the rocker arms so any that are to be returned to service will be installed in their original locations. Slide the rocker arms off the shaft. Keep the spacers and rocker arms in the same location for reassembly.

   a. Inspect the rocker arms for scoring, wear on the roller or damage to the shaft. Replace parts as necessary.

   b. The rocker arm shaft is hollow and used as a lubrication oil duct. Check that the shaft is clean inside and out.

   c. Check all oil holes for clogging with a small wire and clean and required.

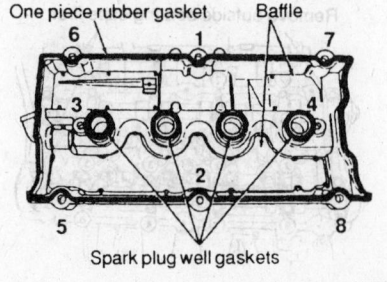

Front bearing cap (No.1 cap)

1.5 mm (.060 in.) diameter bead of Mopar Gasket Maker:

Left rear bearing cap (No.6 cap)

320835

Apply sealant as shown — 2.0L DOHC engine

One piece rubber gasket — Baffle

Spark plug well gaskets

320839

Cam cover tightening sequence — 2.0L DOHC engine

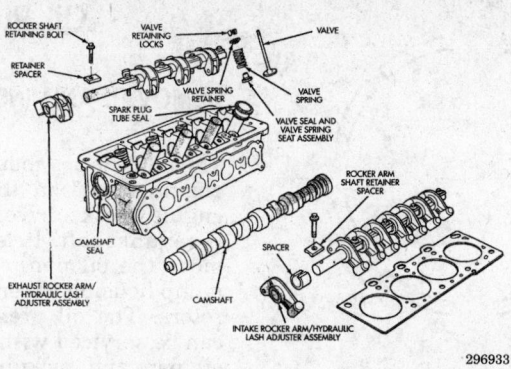

Cylinder head and valve assembly — 2.0L SOHC engine

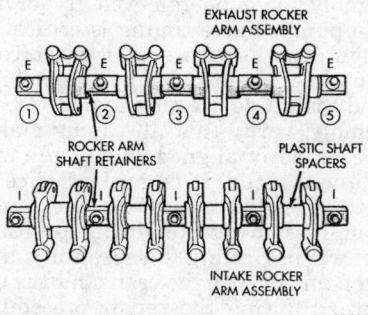

Rocker arm shaft identification — 2.0L SOHC engine

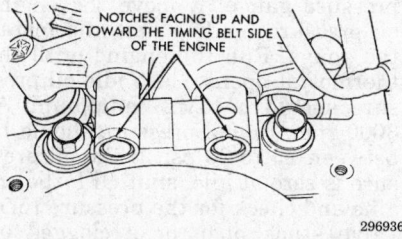

Rocker arm shaft notch location — 2.0L SOHC engine

d. To assemble, thoroughly lubricate all rocker arm components and spacers and install on the rocker arm shaft in the original locations.

12. If the vehicle exhibited a tappet-like noise, the valve lash adjusters built into the rocker arms should be cleaned and checked. Lash adjusters removed from a rocker arm should be returned to their original locations. Replace worn or defective lash adjusters. To install a lash adjuster, use the following procedure.

a. Lubricate the lash adjuster thoroughly with clean engine oil.

b. Install the adjuster into the rocker arm making sure the adjuster is at least partially filled with oil.

c. Place the rocker arm in clean engine oil and pump the plunger until the lash adjuster travel is taken up. If travel is not reduced, replace the adjuster.

d. Install the rocker arm back on the rocker arm shaft.

13. Camshaft end-play can be checked. Use the following procedure.

a. Oil the camshaft journals and install the camshaft without the rocker arm assemblies. Install the cam sensor and tighten the screws to 85 inch lbs. (9.6 Nm).

b. Setup a dial indicator to touch on the nose of the camshaft.

c. Using a suitable tool, move the camshaft as far rearward as it will go. Make sure the dial indicator probe is in contact with the camshaft.

d. Zero the dial indicator.

e. Move the camshaft as far forward as it will go.

f. Read the end-play on the dial indicator. Specification is 0.005–0.013 inch (0.13–0.33mm).

14. To install the camshaft, lubricate the bearing journals thoroughly. Install the camshaft into the cylinder head carefully. Make sure it turns freely. If the camshaft installation is satisfactory, install the cam sensor and tighten the screws to 85 inch lbs. (9.6 Nm).

15. Install the camshaft seal. The camshaft must be installed before the camshaft seal is installed. The seal should be flush with the cylinder head after installation.

16. Install the camshaft sprocket and tighten to the bolt to 85 ft. lbs. (115 Nm).

17. Install the cylinder head using the recommended procedure.

18. Before installing the rocker arm and shaft assemblies, set the crankshaft to 3 notches before TDC on the crankshaft sprocket.

19. Install the rocker arm and shaft assemblies with the small notches in the rocker shafts pointing up and toward the timing belt side of the engine. Install the retainers in their original positions on the exhaust and intake shafts. Torque the bolts to 200 inch lbs. (23 Nm).

20. Install the timing belt using care to align all valve timing marks, using the recommended procedure.

21. Connect all electrical, vacuum and fluid connections as required.

22. Refill the cooling system. An oil and filter change are recommended.

23. Reconnect the negative battery cable remote connection.

24. Start the engine and check for leaks. Run the engine with the radiator cap off so as the engine warms and the thermostat opens, coolant can be added to the radiator. Test drive vehicle to check for proper operation.

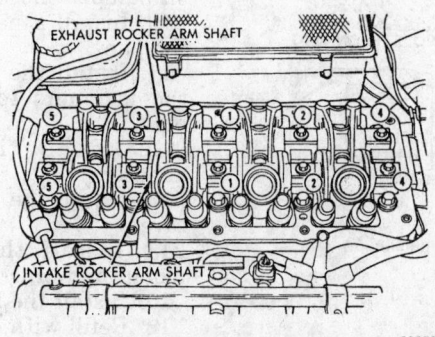

Rocker arm shaft tightening sequence — 2.0L SOHC engine

Piston and Connecting Rod

POSITIONING

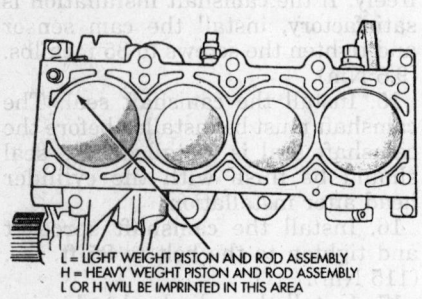

L = LIGHT WEIGHT PISTON AND ROD ASSEMBLY
H = HEAVY WEIGHT PISTON AND ROD ASSEMBLY
L OR H WILL BE IMPRINTED IN THIS AREA

302414

Markings for correct piston installation

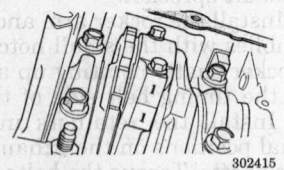

302415

Connecting rod to end cap identification

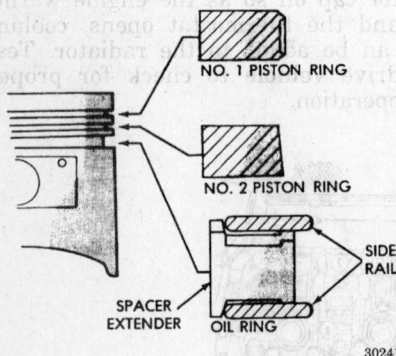

302416

Piston ring installation

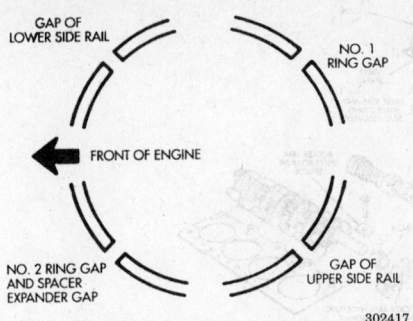

302417

Piston ring end-gap position

ENGINE LUBRICATION

Oil Pan

REMOVAL AND INSTALLATION

1. Disconnect the negative battery cable.
2. Drain the engine oil.
3. Remove the transaxle bracket.
4. Safely support the engine's weight. Remove the front engine mount and bracket.
5. Remove the transaxle inspection cover.
6. If equipped with air conditioning, remove the oil filter and adapter.
7. Remove the oil pan.
8. Clean the oil pan and all gasket surfaces.
   To install:
9. Apply MOPAR Silicone Rubber Adhesive Sealant or equivalent at the oil pump-to-block parting line.
10. Install a new oil pan gasket to the pan.
11. Install the oil pan and tighten the fasteners to 105 inch lbs. (12 Nm).
12. Install the oil filter and adapter.
13. Install the transaxle inspection cover.
14. Install the front transaxle mount and bracket.
15. Install the transaxle bracket.
16. Refill with engine oil. A new oil filter is recommended.
17. Connect the negative battery cable.
18. Test run the engine to check for leaks.

Oil Pump

REMOVAL AND INSTALLATION

The oil pump is mounted in the front engine cover, on the outside of the engine block, driven by the nose of the crankshaft. It is necessary to remove the oil pan, oil pickup and oil pump housing to service the oil pump rotors. The oil pressure relief valve can be serviced without removing the oil pan and pickup tube. Since the crankshaft timing belt drive sprocket must be removed, the timing belt assembly must be removed to service the pump. Valve timing is critical to engine performance and great care must be exercised when removing and installing the timing belt. At assembly, verify that all timing marks are properly aligned or engine damage will occur. It is good practice to set the engine to TDC No. 1 cylinder compression stroke (firing position) before beginning disassembly and then, make sure the crankshaft is not turned during the repair procedure. This should show all timing marks in alignment and can be used as a reference for all later work.

If troubleshooting an oil pressure complaint, check the engine oil pressure with a reliable mechanical oil pressure gauge. Remove the engine oil pressure sending unit and install the gauge. Run the engine until the thermostat opens. Curb idle oil pressure should be 4 psi minimum. At 3000 rpm, the oil pressure should be between 25 to 80 psi. If the oil pressure is zero at idle, shut OFF the engine and check for the pressure relief valve stuck open or a clogged oil pickup screen.

--- **WARNING** ---

If the oil pressure is zero at idle DO NOT run the engine at 3000 rpm looking for a pressure increase or the engine could be destroyed.

1. Disconnect the negative battery cable.
2. Turn the crankshaft clockwise until the engine is at TDC No. 1 cylinder compression stroke (firing position). Remove the timing belt using the recommended procedure.
3. Using a suitable puller, draw the crankshaft timing belt drive sprocket from the front of the crankshaft.
4. Drain the engine oil. Remove the oil pan.

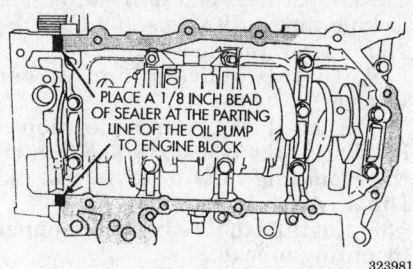

Apply sealer as indicated — 2.0L engine

5. Remove the oil screen feed pipe and O-ring that seals it to the pump/block assembly.

6. The front cover/oil pump mounting bolts are different sizes and must be reinstalled in their original locations. Remove and tag the front cover mounting bolts.

7. Remove the oil pump assembly.

8. Using special tool 6771 or equivalent seal puller, remove the front crankshaft oil seal. Take care not to damage the seal surface of the cover.

To install:

9. Clean all parts well. If troubleshooting an oil pressure complaint, the oil pump internal components should be checked and carefully measured.

a. Remove the threaded plug and gasket from the oil pump and remove the spring and relief valve. Use care to note the position of the parts. The oil pump pressure relief valve must be installed properly or the engine will be seriously damaged.

b. Remove the oil pump cover screws and lift off the cover.

c. Remove the pump rotors.

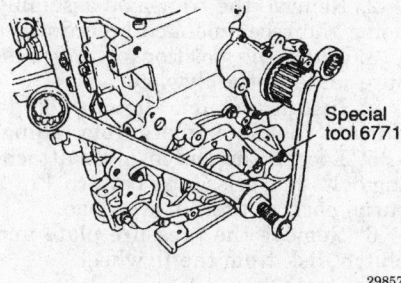

Removing the front crankshaft oil seal — 2.0L engine

d. Wash all parts in a suitable solvent, clean thoroughly and inspect carefully for wear.

e. Make sure the mating surface of the oil pump is smooth. Replace the pump cover if scratched or grooved.

f. Lay a straightedge across the pump cover surface. If a 0.003 inch feeler gauge can be inserted between the cover and straightedge, the cover should be replaced.

g. Measure the thickness and diameter of the outer rotor. If the outer rotor thickness measures 0.301 inch or less, or if the diameter is 3.148 inch or less, replace the outer rotor.

h. If the inner rotor measures 0.301 inch or less, replace the inner rotor.

i. Slide the outer rotor into the pump housing, press to one side with fingers and measure the clearance between the rotor and housing. If measurement is 0.015 inch or more, replace the housing only if the outer rotor is in specification.

j. Install the inner rotor into the pump housing. If clearance between the inner and outer rotors is 0.008 inch or more, replace both rotors.

k. Place a straightedge across the face of the pump housing between bolt holes. If a 0.004 inch or more feeler gauge can be inserted between the rotors and straightedge, replace the pump assembly ONLY if the rotors are in specification.

l. Inspect the oil relief valve plunger for scoring and free operation in its bore. Varnish or sludge can hang the plunger resulting in improper oiling. Small marks may be removed with 400 grit wet-or-dry sandpaper.

m. The relief spring has a free length of 2.390 inches and should test between 18–19 lbs. when compressed to 1.600 inches. Replace the spring if it fails to meet specification.

n. If engine oil pressure is low and the pump is within specification, inspect for worn engine bearings or other reasons for oil pressure loss.

10. Whenever the oil pump is disassembled or the rear cover is removed, the gear cavity must be filled with petroleum jelly or clean engine oil. This seals the pump, acts like a

prime and allows the pump to draw engine oil as soon as it starts to turn. Do not neglect this step. Do not use grease.

11. Assemble the pump using new parts as required.

NOTE: Install the inner rotor with the chamfer facing the cast iron oil pump rear cover.

12. Thoroughly clean all old gasket/sealer material from all mounting surfaces. Apply MOPAR gasket maker or equivalent sealer sparingly to the cover surface on the oil pump body. Install the cover and tighten the screws to 105 inch lbs.

13. Install the relief valve plunger, then the spring, gasket and cap. Tighten the cover cap to 30 ft. lbs. Install a new O-ring into the counter bore on the oil pump body discharge passage.

14. Install the oil pump carefully onto the front of the crankshaft until it is seated to the engine block. Tighten the retaining bolts to 17 ft. lbs. (23 Nm).

15. Using tool 6780-1 or equivalent, install a new front crankshaft oil seal.

16. Install the oil screen and pick-up tube assembly using a new O-ring.

17. Install the crankshaft timing belt drive sprocket. Install the timing belt following the recommended procedure. Take care that all timing marks are properly aligned or engine damage will result.

18. Install the oil pan and drain plug. Fill the crankcase with the proper quantity and type of engine oil. A filter change is recommended.

19. Connect the negative battery cable.

20. Remove the engine oil pressure sending unit and install an oil pressure gauge. Start and run the engine until the thermostat opens. Curb idle oil pressure should be 4 psi minimum. At 3000 rpm, the oil pressure should be between 25 to 80 psi. If the oil pressure is zero at idle, immediately shut off the engine and check if the pressure relief valve is stuck open or for some other problem (oil level, oil type, loose filter, etc.).

— **WARNING** —

If the oil pressure is zero at idle DO NOT run the engine at 3000 rpm looking for a pressure increase or the engine could be destroyed.

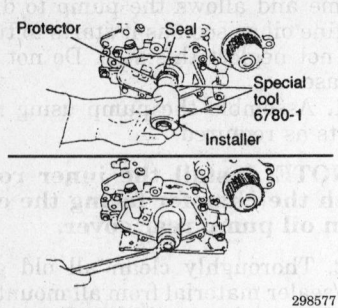

298577

Installing a new seal using a 6780-1 seal installer. Use care if using substitute tools — 2.0L engine

TRANSAXLE

Manual Transaxle Assembly

REMOVAL AND INSTALLATION

1. Disconnect both battery cables, negative side first. Remove the battery and battery tray.

2. Remove the battery stay (brace).

3. Remove the air cleaner and intake hoses.

4. Drain the transaxle into a suitable waste container.

5. Remove the cotter pin securing the select and shift cables and remove the cable ends from the transaxle.

6. Disconnect the backup light switch harness and position it aside.

7. Disconnect the speedometer electrical connector, from the transaxle assembly.

8. Remove the starter motor and position it aside.

9. Using special tool 7137 or C-4852 or equivalent, support the engine assembly.

10. Remove the rear roll stopper mounting bracket.

11. Remove the transaxle mount bracket.

12. Remove the upper transaxle mounting bolts.

13. Raise and safely support the vehicle.

14. Remove the front wheel assemblies.

15. Remove the under cover.

16. Remove the cotter pin and disconnect the tie rod end, from the steering knuckle.

17. Disconnect the stabilizer bar link, from the damper fork.

18. Disconnect the damper fork, from the lateral lower control arm.

19. Disconnect the later lower arm, and the compression lower arm, lower ball joints, from the steering knuckle.

20. Pry the halfshafts from the transaxle, and secure aside.

21. Remove the connection for the clutch release cylinder and without disconnecting the hydraulic line, secure aside.

22. Remove the cover from the transaxle bellhousing.

23. Remove the engine front roll stopper through bolt.

24. Remove the centermember.

25. Support the transaxle, using a transmission jack, and remove the transaxle lower coupling bolt.

NOTE: The coupling bolt threads from the engine side, into the transaxle, and is located just above the halfshaft opening.

26. Slide the transaxle rearward and carefully lower it from the vehicle.

To install:

27. Install the transaxle to the engine and install the mounting bolts and tighten to 70 ft. lbs. (95 Nm).

28. Install the cover to the transaxle bellhousing and tighten the mounting bolts to 7 ft. lbs. (9 Nm).

29. Install the centermember and tighten the front mounting bolts to 65 ft. lbs. (88 Nm) and the rear bolt to 54 ft. lbs. (73 Nm). Install the front engine roll stopper through bolt and lightly tighten. Once the full weight of the engine is on the mounts, tighten the bolt to 42 ft. lbs. (57 Nm).

30. Connect the clutch release cylinder.

31. Install the halfshafts, using new circlips on the axle ends.

— WARNING —
When installing the halfshaft, keep the inboard joint straight in relation to the axle, to avoid damaging the oil seal lip of the transaxle, with the serrated part of the halfshaft.

32. Connect the tie rod and ball joints to the steering knuckle. Tighten the ball joint self-locking nuts to 48 ft. lbs. (65 Nm). Tighten the tie rod end nut to 21 ft. lbs. (28 Nm) and secure with a new cotter pin.

33. Connect the damper fork to the lower control arm and tighten the through bolt to 65 ft. lbs. (88 Nm).

34. Connect the stabilizer link to the damper fork, and tighten the self-locking nut to 29 ft. lbs. (39 Nm).

35. Install the under cover.

36. Install wheels and lower vehicle.

37. Install the transaxle mount bracket, to the transaxle, and tighten the mounting nuts to 32 ft. lbs. (43 Nm).

38. Install the rear roll stopper mounting bracket.

39. Remove the engine support. Tighten the transaxle mount through bolt to 51 ft. lbs. (69 Nm) and tighten the front engine roll stopper through bolt.

40. Install the upper transaxle mounting bolts and tighten to 35 ft. lbs. (48 Nm).

41. Install the starter motor.

42. Connect the backup light switch and the speedometer connector.

43. Connect the select and shift cables and install new cotter pins.

44. Install the air cleaner and the air intake hose.

45. Install the battery tray and battery.

46. Install the battery stay.

47. Make sure the vehicle is level, and refill the transaxle.

48. Check the transaxle for proper operation. Make sure the reverse lights come ON when in reverse.

Clutch Assembly

REMOVAL AND INSTALLATION

Please note that the entire transaxle assembly must be removed from the vehicle to service the clutch assembly.

1. Disconnect the negative battery cable.

2. Remove the transaxle assembly using the recommended procedure.

3. Mark the position of the pressure plate to the flywheel.

4. Install a clutch aligning tool to prevent the clutch plate from falling.

5. Loosen the clutch cover attaching bolts in a crisscross pattern 1 or 2 turns each to prevent warpage.

6. Remove the pressure plate and clutch disk from the flywheel.

To install:

7. Inspect the flywheel for any heat cracks, glazing or scoring. If any of these conditions are present, replace the flywheel.

8. If the flywheel has been removed, torque the flywheel bolts to 70 ft. lbs. (95 Nm).

9. Install the clutch disk and the pressure plate using a suitable alignment tool.

NOTE: Be sure to align the marks previously made during removal.

10. Tighten the pressure plate bolts in a crisscross pattern 1 to 2 turns at a time.

11. Torque the pressure plate bolts to 250 inch lbs. (28 Nm).

12. Install a new throwout bearing into the release fork in the transaxle.

13. Install the transaxle using the recommended procedure.

14. Reconnect the negative battery cable.

Clutch Cable

ADJUSTMENT

The manual transaxle clutch release system has a unique self-adjusting mechanism to compensate for clutch disc wear. This adjuster mechanism is located with the clutch cable assembly. The preload spring maintains tension on the cable. This tension keeps the clutch release bearing continuously loaded against the fingers of the clutch cover assembly. No manual adjustment is obtainable.

When servicing this vehicle or if removing and installing the clutch cable, do not pull on the clutch cable housing to remove it from the dash panel. Damage to the cable self-adjuster may occur.

To check the function of the adjuster mechanism, use the following procedure:

1. With slight pressure, pull the clutch release lever end of the cable to draw the cable taut.

2. Push the clutch cable housing toward the dash panel. With less than 25 pounds of effort the cable housing should move 1.2–2.0 in. (30–50mm). This indicates proper adjuster mechanism function.

3. If the cable does not adjust, determine if the mechanism is properly seated on the bracket.

Automatic Transaxle Assembly

REMOVAL AND INSTALLATION

1. Disconnect both battery cables, negative side first. Remove the battery and battery tray.

2. Remove the battery stay (brace).

3. Remove the air cleaner and intake hoses.

4. Drain the transaxle fluid into a suitable waste container.

5. Remove the nut securing the shifter lever to the transaxle. Remove the cable retaining clip and remove the cable from the transaxle.

6. Remove the shifter cable mounting bracket.

7. Disconnect and tag the electrical connectors for the speedometer, solenoid, neutral safety switch (inhibitor switch), the pulse generator, kickdown servo switch, oil temperature sensor.

8. Disconnect and tag the oil cooler lines, at the transaxle.

9. Remove the bolt securing the fluid dipstick tube, to the transaxle. Remove the dipstick and tube from the transaxle.

10. Remove the starter motor and position aside.

11. Using special tool 7137 or C-4852 or equivalent, support the engine assembly.

12. Remove the rear roll stopper mounting bracket.

13. Remove the transaxle mount bracket.

14. Remove the upper transaxle mounting bolts.

15. Raise and safely support the vehicle.

16. Remove the front wheels.

17. Remove the left side undercover.

18. Remove the cotter pin and disconnect the tie rod end, from the steering knuckle.

19. Disconnect the stabilizer bar link, from the damper fork.

20. Disconnect the damper fork, from the lateral lower control arm.

21. Disconnect the later lower arm, and the compression lower arm, lower ball joints, from the steering knuckle.

22. Pry the halfshafts from the transaxle, and secure aside.

23. Remove the cover from the transaxle bellhousing.

24. Remove the engine front roll stopper through bolt.

25. Remove the centermember.

26. Remove the bolts holding the flexplate to the torque converter with a box wrench. Rotate the crankshaft to bring the bolts into a position for removal, one at a time.

27. To make installation easier, use chalk or paint to make matchmarks on the torque converter and flex plate. These marks will be used at assembly to realign the assembly, keeping these parts in balance. After removing the bolts, push the torque converter toward the transaxle. This will prevent the converter from remaining in contact with the engine, possibly damaging the converter.

28. Support the transaxle using a transmission jack (at the side of the case, NOT at the pan), and remove the transaxle lower coupling bolt.

NOTE: The coupling bolt threads from the engine side, into the transaxle, and is located just above the halfshaft opening.

29. Slide the transaxle rearward and carefully lower it from the vehicle.

To install:

30. After the torque converter has been mounted on the transaxle, install the transaxle assembly to the engine. Install the mounting bolts and tighten to 70 ft. lbs. (95 Nm).

31. Align the balance matchmarks made at disassembly, connect the torque converter to the flexplate and tighten the bolts to 55 ft. lbs. (75 Nm).

32. Install the cover to the transaxle bellhousing and tighten the mounting bolts to 9 ft. lbs. (12 Nm).

33. Install the centermember and tighten the front mounting bolts to 65 ft. lbs. (88 Nm) and the rear bolts to 51–58 ft. lbs. (69–78 Nm). Install the front engine roll stopper through bolt and lightly tighten. Once the full weight of the engine is on the mounts, tighten the bolt to 42 ft. lbs. (56 Nm).

34. Install the halfshafts, using new circlips on the axle ends.

WARNING

When installing the halfshaft, keep the inboard joint straight in relation to the axle, to avoid damaging the oil seal lip of the transaxle, with the serrated part of the halfshaft.

35. Connect the tie rod and ball joints to the steering knuckle. Tighten the ball joint self-locking nuts to 48 ft. lbs. (65 Nm). Tighten the tie rod end nut to 21 ft. lbs. (28 Nm) and secure with a new cotter pin.

36. Connect the damper fork to the lower control arm and tighten the through bolt to 65 ft. lbs. (88 Nm).

37. Connect the stabilizer link to the damper fork, and tighten the self-locking nut to 29 ft. lbs. (39 Nm).

38. Install the left side undercover.

39. Install the wheels and lower the vehicle.

40. Install the transaxle mount bracket, to the transaxle, and tighten

the mounting nuts to 32 ft. lbs. (43 Nm).

41. Install the rear roll stopper mounting bracket.

42. Remove the engine support. Tighten the transaxle mount through bolt to 51 ft. lbs. (69 Nm) and tighten the front engine roll stopper through bolt.

43. Install the upper transaxle mounting bolts and tighten to 35 ft. lbs. (48 Nm).

44. Install the starter motor.

45. Install the dipstick tube and the dipstick.

46. Install the shifter cable mounting bracket.

47. Connect the shifter lever and tighten the retaining nut to 14 ft. lbs. (19 Nm).

48. Connect the oil cooler lines and secure with clamps.

49. Connect the electrical connectors for the speedometer, solenoid, neutral safety switch (inhibitor switch), the pulse generator, kickdown servo switch and oil temperature sensor.

50. Install the air cleaner and the air intake hose.

51. Install the battery tray and battery.

52. Make sure the vehicle is level, and refill the transaxle with MOPAR ATF PLUS or equivalent transaxle fluid. Start the engine and allow it to idle for 2 minutes. Apply the parking brake and move the selector through each gear position, ending in **N**. Recheck fluid level and add if necessary. Fluid level should be between the marks in the **HOT** range.

53. Check the transaxle for proper operation. Make sure the reverse lights come ON when in **R** and the engine starts only in **P** or **N**.

Throttle Valve (Pressure) Cable

ADJUSTMENT

1. Make the the engine is at normal operating temperature.

2. Release the cross-lock on the cable assembly by pulling the cross-lock upward.

3. To ensure proper adjustment, the cable must be free to slide all the way toward the engine, against its stop, after the cross-lock is released.

4. Move the transaxle throttle control lever fully clockwise, against its internal stop, and press the cross-lock downward into a locked position.

The adjustment is complete and transaxle throttle cable backlash was automatically removed.

5. Test the cable freedom of operation by moving the transaxle throttle lever forward (counterclockwise). then slowly release it to confirm it will return fully rearward (clockwise).

NOTE: No lubrication is required for any component of the throttle cable system.

DRIVE AXLE

Halfshaft

REMOVAL AND INSTALLATION

1. Raise and safely support the vehicle.

2. Remove the tire and wheel.

3. Remove the cotter pin, nut lock and washer from the halfshaft.

4. Remove the halfshaft bearing nut.

5. Remove the brake caliper from the steering knuckle. Hang it out of the way on a piece of wire. Do not allow the caliper to hang by the brake hose.

6. Remove the brake rotor.

7. Disconnect the tie rod end.

   a. Remove the tie rod end nut by holding the tie rod end stud with an $^{11}/_{32}$ inch socket while loosening the nut.

   b. Remove the tie rod end using a puller. Do not hammer on the tie rod end. Do not use a wedge type tool to hammer the joint apart.

8. Remove the pinch bolt that retains the lower control arm ball joint to the steering knuckle. Use a suitable prybar push down on the lower control arm to release the steering knuckle from the ball joint stud.

9. Pull the steering knuckle away from the halfshaft. Support the outer end of the halfshaft assembly.

WARNING
Care must be taken not to separate the inner CV-joint during this operation. Do not allow the driveshaft to hang by the inner CV-joint of the halfshaft assembly.

10. Remove the halfshaft from the transaxle.

   a. Depending on the model year and halfshaft type used, it may be

possible to insert a prybar between the inner joint and the transaxle case and carefully lever the inner joint free. Pry against the inner joint until the retaining snapring is disengaged from the transaxle side gear.

   b. Another method is to remove the inner tripod joint from the side gears of the transaxle by using a brass punch (drift pin) to dislodge the inner tripod joint retaining ring from the transaxle side gear.

   c. If removing the right side inner tripod joint, position the punch against the inner tripod joint. Strike the punch sharply with a hammer to dislodge the right inner joint from the side gear.

   d. If removing the left side inner tripod joint, position the punch in the groove of the inner tripod joint. Strike the punch sharply with a hammer to dislodge the left inner tripod joint from the transaxle side gear.

   e. Removal of the inner joints is made easier if you apply outward pressure on the joint as you strike the punch with the hammer.

   f. Hold the inner tripod joint and interconnecting shaft of the halfshaft assembly. Remove the inner tripod joint from the transaxle by pulling it straight out of the transaxle side gear and transaxle oil seal. Use care when removing the tripod joint. Do not let the spline or snapring drag across the sealing lip of the transaxle seal.

WARNING
The halfshaft, when installed, acts as a bolt and secures the front/hub bearing assembly. If the vehicle is to be supported or moved on its wheels with a halfshaft removed, install a proper sized bolt and nut through the front hub. Tighten the bolt and nut to 135 ft. lbs. (183 Nm). This will ensure that the hub bearing cannot loosen.

To install:
11. Clean all parts well. Inspect the sealing boots on the joints and replace if necessary. The boots are the only serviceable components of the halfshaft.

12. Thoroughly clean the spline and oil seal sealing surfaces on the tripod joint. Lightly lubricate the oil seal sealing surfaces with fresh transaxle fluid. Holding the halfshaft by the tripod joint and interconnecting shaft, install the tripod joint into the transaxle side gear as far as it will go.

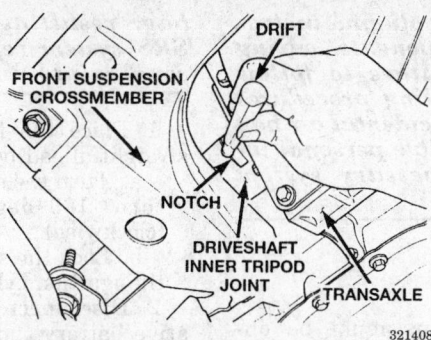

Disengaging the left inner tripod joint from the transaxle using a brass punch as a driver, positioned in the groove of the joint — 1996 shown

13. Carefully align the tripod joint with the transaxle side gears. Then, grasp the halfshaft interconnecting shaft and push the tripod joint into the transaxle side gear until it is fully seated.

NOTE: Test that the snapring is fully engaged with the side gear by attempting to remove the tripod joint from the transaxle by hand. If the snapring is fully engaged with the side gear, the tripod joint will not be removable by hand.

14. Thoroughly clean the splines in the steering knuckle hub/bearing assembly. Make sure the outer CV-joint which fits into the steering knuckle is free of debris and moisture before assembling into the steering knuckle.

15. Slide the halfshaft stub axle end back into the front hub. Install the steering knuckle onto the ball joint stud.

16. Install a NEW steering knuckle to ball joint stud bolt and nut. Tighten the nut and bolt to 70 ft. lbs. (95 Nm).

17. Install the tie rod end into the steering knuckle. Start the tie rod end-to-steering knuckle nut onto the stud of the tie rod end. Tighten the tie rod end nut by holding the tie rod end stud with an $^{11}/_{32}$ inch socket while tightening the nut with a box wrench. Then, using a "crowsfoot" and torque wrench, tighten the tie rod nut to 45 ft. lbs. (61 Nm) while still holding the stud stationary with the $^{11}/_{32}$ inch socket.

18. Install the disc brake caliper on the hub and bearing assembly.

19. Install the disc brake caliper assembly on the steering knuckle. The caliper is installed by first sliding the top of the caliper under the top abutment on the steering knuckle. Then, install the bottom of the caliper against the bottom abutment of the steering knuckle. Tighten the caliper-to-knuckle bolts to 23 ft. lbs. (31 Nm).

20. Clean all foreign matter and dirt from the threads of the outer CV-joint stub axle. Install the hub nut and washer onto the threads and tighten the nut. With the vehicle brakes firmly applied to keep the halfshaft from turning, tighten the hub nut to 135 ft. lbs. (183 Nm). Install the spring washer, nut lock and a new cotter pin to the stub axle.

21. Install the front tire and wheel assembly and torque the lug nuts to 100 ft. lbs. (135 Nm).

22. Check for correct fluid level in the transaxle. Lower the vehicle and road test to verify no front halfshaft noise.

CV-Joint Boot

REPLACEMENT

─────── **WARNING** ───────
Use only clamps provided with the replacement package when servicing. Plastic wire ties and other straps will not clamp tightly enough and grease will leak, causing damage to the joint.

Inner Joint

1. Remove the halfshaft from the vehicle using the recommended procedure.

2. If cutting the boot away, mark and note the boot positioning on the shaft relative to the raised shoulders. Remove the boot.

3. Separate the housing from the tripod.

NOTE: Always hold the rollers in place when removing the housing from the tripod or the needle bearings may fall out.

4. Remove the snapring from the end of the shaft and remove the tripod.

5. If not already done, mark the boot positioning on the shaft, relative to the raised shoulders and remove the boot from the shaft.

6. Remove as much old grease as possible from the joint. Inspect all parts for wear or damage.

NOTE: Do not use petroleum-based solvents on the joints, shaft or boot to clean. It will ruin hidden rubber seals within the joint. Use only chlorine based cleaner or hot soapy water to clean the joint, if necessary. Make sure the joint it completely dry before assembling.

To install:

7. If equipped, slide a new rubber washer seal over the stub shaft and down into the groove provided.

8. If the clamping device is not a straight strap, install it on the shaft first, then install the boot to the shaft in the proper position. Using the proper tool, C–4124 or equivalent for crimping with rubber boot or C–4653 or equivalent for clamping a strap, secure the clamp.

9. Slide the tripod on the shaft.

10. Install the snapring into its groove on the shaft to lock the tripod in position.

11. Pack the grease provided in the housing.

12. Install the tripod joint into the housing.

13. Position the larger end of the boot over the housing.

14. Using the proper tool for crimping the rubber boot clamp, secure the clamp.

15. Install the halfshaft to the vehicle.

16. Fill the transaxle if fluid was lost when removing the halfshaft.

17. Road test the vehicle.

Outer Joint

1. Remove the halfshaft from the vehicle using the recommended procedure.

2. If cutting the boot away, mark and note the boot positioning on the shaft relative to the raised shoulders. Remove the boot.

3. Support the halfshaft in a vise equipped with protective caps on the jaws of the vise to prevent damage to the shaft. Using a soft faced hammer give a sharp hit to the end of the CV-joint housing to dislodge the internal clip on the shaft.

4. If damaged, remove the wear sleeve from the CV-joint machined ledge.

5. Remove the circlip from the groove.

6. Remove as much old grease as possible from the joint. Inspect all parts for wear or damage.

NOTE: Do not use petroleum based solvents on the joints, shaft or boot to clean. It will ruin hidden rubber seals within the joint. Use only chlorine based cleaner or hot soapy water to clean the joint, if necessary. Make sure the joint it completely dry before assembling.

To install:

7. If the clamping device is not a straight strap, install it on the shaft first, then install the boot to the shaft in the proper position. Using the proper tool, C–4124 or equivalent for crimping with rubber boot or C–4653 or equivalent for clamping a strap, secure the clamp. Install a new circlip if provided in the replacement package.

8. Fill the boot with the proper amount of grease according to the instructions provided with the package.

9. Pack the joint with the remaining grease. Position the outer joint on the shaft with hub nut installed, engage the splines and strike sharply with a soft-face hammer to install. Make sure the circlip did not become dislodged.

10. Position the larger end of the boot over the housing.

11. Purge the air from the boot.

12. Using the proper tool, C–4124 or equivalent for crimping with rubber boot or C–4653 or equivalent for clamping a strap, secure the clamp.

13. Install the halfshaft in the vehicle.

14. Fill the transaxle if fluid was lost when removing the halfshaft.

15. Road test the vehicle.

STEERING

Air Bag

—————— CAUTION ——————
Some vehicles are equipped with an air bag system, also known as the Supplemental Restraint System (SRS). The system must be disabled before performing service on or around system compo- nents, *steering column, instrument panel components, wiring and sensors. Failure to follow safety and disabling procedures could result in accidental air bag deployment, possible personal injury and unnecessary system repairs.*

PRECAUTIONS

Several precautions must be observed when handling the inflator module to avoid accidental deployment and possible personal injury.

• Never carry the inflator module by the wires or connector on the underside of the module.

• When carrying a live inflator module, hold securely with both hands, and ensure that the bag and trim cover are pointed away.

• Place the inflator module on a bench or other surface with the bag and trim cover facing up.

• With the inflator module on the bench, never place anything on or close to the module which may be thrown in the event of an accidental deployment.

DISARMING

Proper SRS disarming can be obtained by disconnecting and isolating the negative battery cable remote connection. The battery is hidden inside the left fenderwell. The negative battery cable runs from the battery to a remote terminal on one of the shock towers, which helps when jump-starting the vehicle. The negative battery cable terminal has a plastic "ear" on it that can be placed on a stud near the body side negative terminal. This isolates the negative battery cable, and helps keep it from grounding against the body, accidentally completing the ground circuit. Before beginning service work, disconnect the negative battery cable at this remote terminal. Allow the air bag system capacitor at least 2 minutes to discharge before removing air bag system components.

Steering Wheel

REMOVAL AND INSTALLATION

—————— CAUTION ——————
The Supplemental Restraint System (SRS) must be disarmed before removing the steering wheel. Failure to do so may cause accidental deployment of the air bag, *resulting in unnecessary SRS system repairs and/or personal injury.*

1. Place the front wheels in the straight ahead position then:

a. Turn the steering wheel a half turn (180 degrees) to the right (clockwise).

b. Lock the steering column with the ignition cylinder lock.

2. Disconnect and isolate the negative battery cable. Allow a minium of 2 minutes for the system capacitor to discharge before removal of any air bag components.

3. Remove the speed control switches and connections.

4. Remove the driver's air bag attaching bolts from the back of the steering wheel.

5. Disconnect the air bag and horn wiring connectors.

—————— CAUTION ——————
When carrying a live air bag, make sure the bag and trim cover are pointed away from the body. In the unlikely event of an accidental deployment, the bag will then deploy with minimal chance of injury. When placing a live air bag on a bench or other surface, always face the bag and trim cover up, away from the surface. This will reduce the motion of the module if it is accidently deployed.

6. Remove the steering wheel retaining nut and vibration damper if equipped.

7. Remove the steering wheel using a suitable puller.

NOTE: When removing the steering wheel, carefully feed the wiring through the holes in the clockspring armature.

To install:

8. Check that the turn signal switch is in the neutral position and the steering wheel position has not changed from the removal process (still a half turn to the right and locked with the ignition cylinder lock.

9. Install the steering wheel ensuring that the flats on the hub align with the clock spring.

10. Pull the horn lead, air bag and speed control wiring through the larger slot in the hub taking care not to pinch the wiring.

11. Install the retaining nut and torque to 45 ft. lbs. (61 Nm).

12. Position the speed control wiring under and behind the speed control mounting flanges.

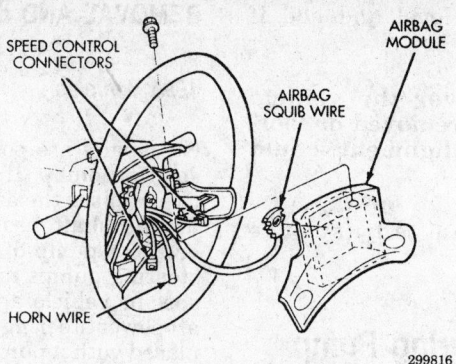

Air bag module

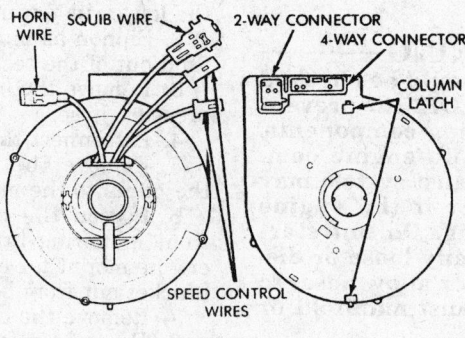

Clock spring assembly

13. Install the horn wiring to the horn switches. Connect the air bag wiring. Make the air bag connection by pressing straight in on the connector. The connector should be fully seated. Feel for a positive snap to assure positive connection.

14. Install the air bag module and torque the bolts to 95 inch lbs. (10.5 Nm).

15. Connect and install the speed control switches.

16. Torque the screws to 20 inch lbs. (2 Nm).

17. Connect a DRB scan tool or equivalent to the diagnostic connector. Make sure the proper programming cartridge (if used) is in place.

18. Turn the ignition key to the **ON** position and exit the vehicle.

19. Reconnect the negative battery cable.

20. Using the scan tool, read and record any active diagnostic codes. Read and record any stored data.

21. Erase stored data. Clear any trouble codes. Codes will not erase if a problem exists.

22. Turn the key to **OFF** then **ON**. The air bag light should stay lit for 6–8 seconds indicating that the system is functioning properly.

23. If the air bag light does not come ON or the light stays ON there is a malfunction in the air bag system.

Tie Rod Ends

REMOVAL AND INSTALLATION

1. Raise and safely support the vehicle.
2. Remove the wheel and tire.
3. Loosen the outer tie rod to inner tie rod jam nut.
4. Remove the tie rod to steering knuckle attaching nut. Hold the tie rod stud with an $^{11}/_{32}$ inch socket. Loosen the nut with a wrench.
5. Separate the tie rod from the steering knuckle using a suitable press-type separating tool.
6. Unscrew the tie rod from the steering rack.
To install:
7. Make sure the jam nut is on the inner tie rod. Thread the tie rod end onto the steering rack inner tie rod.
8. Install the outer tie rod end into the steering knuckle. Start the tie rod end to steering knuckle attaching nut onto the stud of the tie rod end. While holding the stud of the tie rod end stationary, tighten the tie rod to steering knuckle nut. Then, using a crowsfoot and an $^{11}/_{32}$ inch socket, torque the tie rod end attaching nut to 45 ft. lbs. (61 Nm).

9. Check and adjust the Toe alignment.
10. Tighten the jam nut to 55 ft. lbs. (75 Nm). Make sure the tie rod boots are not twisted. Correct as necessary.
11. Install the wheel and tire.

Manual/Power Rack and Pinion

REMOVAL AND INSTALLATION

1. Disconnect the negative battery cable.
2. Remove the retaining pin from the intermediate shaft coupler pinch bolt and remove the pinch bolt.
3. Raise and safely support the vehicle.
4. Remove the front wheels.
5. Disconnect the dampener from the crossmember.
6. Disconnect the electrical connections from the steering rack.
7. Disconnect the tie rod ends by holding the tie rod end stud with an $^{11}/_{32}$ inch socket and loosen the retaining nut with a wrench.
8. Separate the tie rod end using a suitable tool.

WARNING
Before removing the front suspension crossmember from the vehicle you must scribe the front suspension crossmember and the vehicle body. This must be done to retain the proper alignment. The caster and camber are not adjustable.

9. Scribe a line on the body and on the crossmember on all 4 sides.
10. Disconnect the stabilizer from the body.
11. If equipped with antilock brakes, remove the 3 bolts attaching it to the crossmember and tie the antilock brake controller to the vehicle body.
12. Place a suitable lifting device under the front suspension crossmember.
13. Remove the 2 front bolts attaching the crossmember to the frame rails.
14. Loosen but do not remove the 2 rear bolts attaching the crossmember to the frame rails.
15. Lower the lifting device enough to gain access to the steering rack.
16. Disconnect the power steering lines and drain the fluid if equipped.

17. Remove the 2 steering rack attaching bolts.

18. Remove the steering rack from the vehicle.

To install:

19. Install the steering rack into the crossmember.

20. Install the attaching bolts. Torque the bolts to 50 ft. lbs. (68 Nm).

21. If equipped with power steering install the pressure and return lines. Torque the lines to 275 inch lbs. (31 Nm).

22. Raise the crossmember against the frame rails and install the 2 front bolts.

23. Tighten all 4 bolts until the crossmember contacts the body.

24. Torque the bolts to just 20 inch lbs. (2 Nm).

25. Using a soft faced hammer tap the crossmember into position.

NOTE: Be sure to align the scribed marks on the crossmember.

26. Starting with the rear bolts, final torque the crossmember bolts to 120 ft. lbs. (163 Nm).

27. Reconnect the transaxle dampener.

28. Reconnect the power steering pressure switch

29. Install the antilock brake control unit.

30. Torque the antilock brake bolts to 250 inch lbs. (28 Nm).

31. Reinstall the heat shield on the tie rod ends.

32. Install the tie rod ends and torque to 45 ft. lbs. (61 Nm).

33. Reinstall the wheels and torque to 95 ft. lbs. (129 Nm).

34. Lower the vehicle.

35. Reinstall the intermediate shaft pinch bolt and retaining pin.

36. Torque the pinch bolt to 240 inch lbs. (27 Nm).

37. Reconnect the negative battery cable.

38. Fill the power steering system to the cold level with approved fluid. Do not use automatic transaxle fluid.

39. Start the engine and allow it to run for a few minutes.

40. Shut OFF the engine and check the power steering fluid.

41. Add power steering fluid if necessary.

42. Raise the front wheels off the ground.

43. Start the engine and turn the wheel from stop to stop to bleed any air from the system.

44. Check fluid level and add if necessary.

NOTE: Whenever the vehicle crossmember is removed or loosened, the wheel alignment should be checked.

45. Check and adjust the alignment.

Power Steering Pump

BLEEDING

───────── **WARNING** ─────────
The fluid level should be checked with the engine OFF to prevent injury from moving components. Power steering oil, engine components and exhaust system may be extremely hot if the engine has been running. Do not start the engine with any loose or disconnected hoses or allow hoses to touch a hot exhaust manifold or catalyst.

NOTE: In all power steering pumps, use only MOPAR Power Steering Fluid or equivalent. Do not use any type of automatic transaxle fluid in the power steering system.

Wipe the filler cap clean, then check the fluid level. The dipstick should indicate FULL COLD when the fluid is at normal room temperature of approximately 70–80°F (21–27°C).

1. Fill the power steering pump fluid reservoir to the proper level.

2. Start the engine and let run for a few seconds. Turn the engine OFF.

3. Add fluid if necessary. Repeat this procedure until the fluid level remains constant after running the engine.

4. Raise the front wheels of the vehicle off the ground.

5. Start the engine. Slowly turn the steering wheel right and left, lightly contacting the wheel stops. Then, turn the engine **OFF**.

6. Add power steering fluid if necessary.

7. Lower the vehicle and turn the steering wheel slowly from lock to lock.

8. Check the fluid level and refill as required.

9. If the fluid is extremely foamy, allow the vehicle to stand a few minutes and repeat the above procedure.

REMOVAL AND INSTALLATION

1995 Vehicles

No repairs are possible to the power steering pump. If the pump is defective, replace the assembly. Because of unique shaft bearings, flow control levels or pump displacements, power steering pumps may be used only on specific vehicle applications. Be sure all power steering pumps are only replaced with a pump that is correct for that application.

1. Disconnect the negative battery cable remote connection located on the left strut tower.

2. Siphon as much power steering fluid out of the reservoir as possible.

3. Remove the banjo bolt from the pressure hose.

4. Disconnect the supply hose.

5. Remove the 2 bolts attaching the pump to the mounting bracket.

6. Loosen the bolt attaching the front mounting bracket to the front engine mount far enough to slide the bracket out from behind the bolt.

7. Remove the drive belt.

8. Remove the pump and the bracket as an assembly.

To install:

9. Install the pump and bracket.

10. Install the 2 pump attaching bolts.

11. Install the drive belt.

12. Adjust the tension on the belt.

13. Tighten all 3 mounting bolts to 40 ft. lbs. (54 Nm).

14. Connect the supply hose.

15. Reconnect the banjo bolt using new O–rings.

16. Fill the power steering reservoir with fluid.

17. Reconnect the negative battery cable. Bleed the power steering system and check for leaks.

NOTE: Fluid level should be checked with the engine off to prevent injury from moving parts. Do NOT use any type of automatic transaxle fluid in the power steering system. Use only Mopar Power Steering Fluid or equivalent. Do not overfill.

1996–97 Vehicles

The power steering pump removal procedure and pump and bracket fastener locations are the same for both engine applications. The front power steering pump bracket must be removed as an assembly with the power steering pump and removed

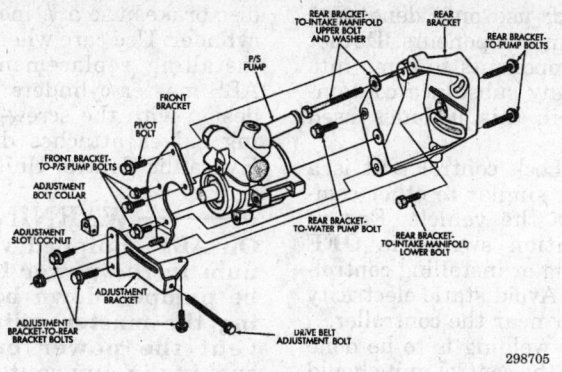

Power steering pump and brackets — 1995 vehicles

from the pump after removing the pulley from the pump.

— CAUTION —

Power steering oil, engine components and the exhaust system may be extremely hot if the engine has been running. Do not start the engine with any loose or disconnected hoses or allow hoses to touch a hot exhaust manifold or catalyst.

1. Disconnect the negative battery cable.
2. Remove the Banjo Bolt and power steering fluid pressure hose from the pressure fitting on the power steering pump.
3. Discard all used O-rings on the power steering pressure hose fitting and bolt.
4. Remove the hose clamp attaching the power steering fluid supply hose to the power steering pump fitting and remove the fluid supply hose from the pump.
5. Raise and safely support the vehicle.
6. If equipped with the Dual Over Head Cam (DOHC) engine, remove

the bolt attaching the coolant tube to the bottom of the intake manifold.

NOTE: The bolt requires removal to allow the coolant tube to be moved out of the way for access to the power steering pump mounting bolt. The coolant tube does not need to be removed or the cooling system drained.

7. Remove the 2 power steering pump-to-cast bracket mounting and adjustment bolts.

NOTE: The power steering pump front mounting bracket is slotted at the bolt attaching it to the front engine mount. This bolt only needs to be loosened to remove the mounting bracket from the engine.

8. Loosen the bolt attaching the power steering pump front mounting to the front engine mount only far enough to slide the bracket out from under the bolt.
9. Remove the power steering pump drive belt from the pump pulley.
10. Remove the power steering pump and front mounting bracket as an assembly from the engine.

Power steering hose attachment to the power steering pump. Note locating pin position — 1996 shown

11. Transfer the required parts from the removed power steering pump to the replacement pump.

To install:

12. Install the power steering pump and mounting bracket as an assembly back on the engine.
13. Slide the front power steering pump bracket between the bracket mounting bolt and front engine mount.

NOTE: Be sure the washer on the bolt is between the head of the bolt and bracket and does not get trapped between the bracket and engine mount.

14. Install the 2 power steering pump-to-cast mounting bracket attaching bolts; do not tighten the bolts at this time.
15. Install the power steering pump drive belt on the pulley.
16. Install a ½ inch breaker bar in the square hole in the front power steering pump mounting bracket. Then, rotate the pump inwards to obtain the correct drive belt tension. When the belt is correctly adjusted, torque the 2 bolts at the cast bracket to 40 ft. lbs. (54 Nm). Then, torque the front power steering pump bracket bolts to 40 ft. lbs. (54 Nm).
17. Install the power steering supply hose on the power steering pump suction fitting. Install the hose clamp on the hose being sure the hose clamp is installed on the hose past the upset bead on the pump tube.
18. Using a lint-free towel, wipe clean all open power steering hose ends and pump fittings.
19. Install new O-rings, 1 on the end of the pressure hose banjo fitting and another on the bolt. Lubricate both O-rings with fresh power steering fluid. Install the banjo bolt into the fitting and attach the pressure hose to the power steering pump.
20. Position the locating pin on the power steering pressure hose banjo fitting so it is against the pump mounting bracket. While holding the locating pin against the pump bracket, torque the banjo bolt to 25 ft. lbs. (34 Nm).

— WARNING —

Do not use automatic transaxle fluid in this power steering system. Service this vehicle with MOPAR Power Steering Fluid or equivalent.

21. Fill the power steering fluid reservoir to the correct fluid level.
22. Connect the negative battery cable. Start the engine and allow to run for a few seconds, then shut OFF.

Add fluid if necessary. Repeat this procedure until the fluid level remains constant after running the engine.

23. Raise the front wheels off the ground.

24. Start the engine then slowly turn the steering wheel right and left several times until lightly contacting the wheel stops. Then, turn the engine OFF.

25. Add power steering fluid if necessary.

26. Lower the vehicle. Start the engine again and tun the steering wheel slowly lock-to-lock. Stop the engine and check the fluid level and refill as required. If the fluid is extremely foamy, allow the vehicle to stand a few minutes and repeat the above procedure. Finally, check all hose connections and fittings for leaks.

BRAKES

Anti-Lock Brake System Service

PRECAUTIONS

• Certain components within the Anti-lock Brake System (ABS) are not intended to be serviced or repaired individually. Only those components with removal and installation procedures should be serviced.

• Do not use rubber hoses or other parts not specifically specified for and ABS system. When using repair kits, replace all parts included in the kit. Partial or incorrect repair may lead to functional problems and require the replacement of components.

• Lubricate rubber parts with clean, fresh brake fluid to ease assembly. Do not use lubricated shop air to clean parts; damage to rubber components may result.

• Use only specified brake fluid from an unopened container.

• If any hydraulic component or line is removed or replaced, it may be necessary to bleed the entire system.

• A clean repair area is essential. Always clean the reservoir and cap thoroughly before removing the cap. The slightest amount of dirt in the fluid may plug an orifice and impair the system function. Perform repairs after components have been thor-

oughly cleaned; use only denatured alcohol to clean components. Do not allow ABS components to come into contact with any substance containing mineral oil; this includes used shop rags.

• The Anti-Lock control unit is a microprocessor similar to other computer units in the vehicle. Ensure that the ignition switch is **OFF** before removing or installing controller harnesses. Avoid static electricity discharge at or near the controller.

• If any arc welding is to be done on the vehicle, the control unit should be unplugged before welding operations begin.

Master Cylinder

REMOVAL AND INSTALLATION

Vehicles Without ABS

Vehicles not equipped with ABS use a standard compensating port design master cylinder. The master cylinder is an anodized aluminum casting with a see-through plastic reservoir.

1. With the ignition OFF, pump the brake pedal until a firm pedal is achieved.

2. Disconnect the brake fluid level sensor electrical connector.

3. Disconnect the hydraulic lines and plug to prevent dirt from entering the open system.

4. Clean the area around the mounting of the master cylinder.

5. Remove the master cylinder mounting bolts.

6. Remove the master cylinder.

To install:

7. Install the master cylinder onto the brake booster.

8. Tighten the master cylinder mounting bolts to 250 inch lbs. (28 Nm).

9. Connect the hydraulic lines to the master cylinder.

10. Torque the hydraulic lines to 145 inch lbs. (17 Nm).

11. Reconnect the brake fluid level sensor electrical connector.

12. Bleed the brake hydraulic system.

13. Road test and check for proper operation.

Vehicles With ABS

Four different master cylinder assemblies are available. Vehicles equipped with rear drum brakes use a master cylinder with a 21mm diameter bore while vehicles with rear

disc brakes use a ⅞ inch bore master cylinder. Use care when ordering and installing replacement parts. The ABS master cylinders are a 2 outlet design with the screw-in proportioning valves attaches directly to the Hydraulic Control Unit (HCU).

— **WARNING** —
On ABS equipped vehicles, vacuum in the power booster must be pumped down before removing the master cylinder to prevent the power booster from sucking in any contamination.

1. With the ignition OFF, pump the brake pedal until a firm pedal is achieved. This depletes the vacuum from the booster.

2. Disconnect the brake fluid level sensor electrical connector.

3. Disconnect the hydraulic lines and plug to prevent dirt from entering the open system.

4. Clean the area around the mounting of the master cylinder.

5. Remove the master cylinder mounting bolts.

6. Remove the master cylinder.

NOTE: On ABS equipped vehicles, the master cylinder is used to create the seal for holding vacuum in the power booster. The seal in the front of the power booster must be replaced when ever the master cylinder is removed.

7. Insert a small prybar between the seal and the master cylinder pushrod and carefully pry out the seal.

To install:

8. Lubricate the master cylinder pushrod with silicone lubricant.

9. Install the vacuum seal on the pushrod with the seal notches pointing toward the master cylinder.

10. Slide the seal onto the pushrod until it is seated against the master cylinder.

11. Install the master cylinder onto the brake booster.

12. Tighten the master cylinder mounting bolts to 250 inch lbs. (28 Nm).

13. Connect the hydraulic lines to the master cylinder.

14. Torque the hydraulic lines to 145 inch lbs. (17 Nm).

15. Reconnect the brake fluid level sensor electrical connector.

16. Bleed the brake hydraulic system.

17. Road test and check for proper operation.

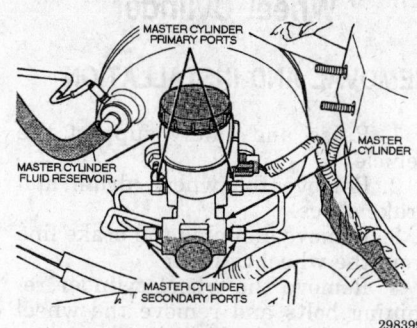

Master cylinder assembly

Brake Caliper

REMOVAL AND INSTALLATION

1. Raise and safely support the vehicle.
2. Remove the wheel and tire.
3. Remove the 2 caliper slide pin bolts.
4. Disconnect the brake hose from the caliper.
5. Remove the caliper from the steering spindle.

To install:

6. Connect the brake hose to the caliper and torque to 35 ft. lbs. (48 Nm).
7. Install the caliper to the steering spindle.
8. Install the caliper slide pin bolts and torque to 16 ft. lbs. (22 Nm).
9. Properly bleed the brake system.
10. Install the wheel and tire and torque the lug nuts to 95 ft. lbs. (129 Nm).
11. Pump the brake pedal to seat the front brake pads before moving the vehicle.
12. Road test the vehicle and check for proper operation.

Disc Brake Pads

REMOVAL AND INSTALLATION

1. Raise and safely support the vehicle.
2. Remove the tire and wheel assembly.
3. Remove the 2 caliper to steering knuckle guide pin bolts.
4. Lift the caliper away from the steering knuckle by first rotating the free end of the caliper away from the steering knuckle. Then, slide the opposite end of the caliper out from

under the machined end of the steering knuckle.
5. Support the caliper from the upper control arm to prevent the weight of the caliper from being supported by the brake flex hose which will damage the hose.
6. Remove the brake pads from the caliper. Remove the outboard brake pad by prying the pad retaining clip over the raised area on the caliper. Then, slide the pad down and off the caliper. Pull the inboard brake pad away from the piston until the retaining clip is free from the cavity in the piston.
7. If required, the rotor can be removed by pulling it straight off the wheel mounting studs.

To install:

8. Clean all parts well. Inspect the caliper for piston seal leaks (brake fluid in and around the boot area and inboard lining) and for any ruptures of the piston dust boot. If the boot is damaged or fluid leak is visible, disassemble the caliper and install a new seal and boot (and piston, if scored).
9. Inspect the caliper pin bushings. Replace if damaged, dry or brittle.
10. Completely retract the piston into the caliper using a large C-clamp or other suitable tool.
11. Lubricate the area on the steering knuckle where the caliper slides with high temperature grease.
12. Reinstall the rotor if removed.
13. Install the brake pads into the caliper. Note that the inboard and outboard pads are different. Make sure the inboard brake shoe assembly is positioned squarely against the face of the caliper piston.

NOTE: Be sure to remove the noise suppression gasket paper cover if the pads come so equipped.

14. Carefully position the caliper and brake shoe assemblies over the rotor by hooking the lower end of the caliper over the steering knuckle. Then, rotate the caliper into position at the top of the steering knuckle. Make sure the caliper guide pin bolts, bushings and sleeves are clear of the steering knuckle bosses.
15. Install the caliper guide pin bolts and tighten to 168 inch lbs. (19 Nm).
16. Reinstall the wheel and tire. Tighten the stud nuts in 2 steps, in a criss-cross pattern to 95 ft. lbs. (129 Nm).
17. Lower the vehicle.

18. Pump the brake pedal until the brake pads are seated and a firm pedal is achieved before attempting to move the vehicle.
19. Road test the vehicle for proper operation.

Brake Rotor (Disc)

REMOVAL AND INSTALLATION

1. Raise and safely support the vehicle.
2. Remove the tire and wheel assembly.
3. Remove the 2 caliper to steering knuckle guide pin bolts.
4. Lift the caliper away from the steering knuckle by first rotating the free end of the caliper away from the steering knuckle. Then, slide the opposite end of the caliper out from under the machined end of the steering knuckle.
5. Support the caliper from the upper control arm to prevent the weight of the caliper from being supported by the brake flex hose which will damage the hose.
6. Remove the rotor by pulling it straight off the wheel mounting studs.

To install:

7. Install the rotor by positioning it over the wheel mounting studs.
8. Carefully position the caliper and brake shoe assemblies over the rotor by hooking the lower end of the caliper over the steering knuckle. Then, rotate the caliper into position at the top of the steering knuckle. Make sure the caliper guide pin bolts, bushings and sleeves are clear of the steering knuckle bosses.
9. Install the caliper guide pin bolts and tighten to 168 inch lbs. (19 Nm).
10. Reinstall the wheel and tire. Tighten the stud nuts in 2 steps, in a criss-cross pattern to 95 ft. lbs. (129 Nm).
11. Lower the vehicle.
12. Pump the brake pedal until the brake pads are seated and a firm pedal is achieved before attempting to move the vehicle.
13. Road test the vehicle for proper operation.

Brake Drums

REMOVAL AND INSTALLATION

1. Raise and safely support the vehicle.

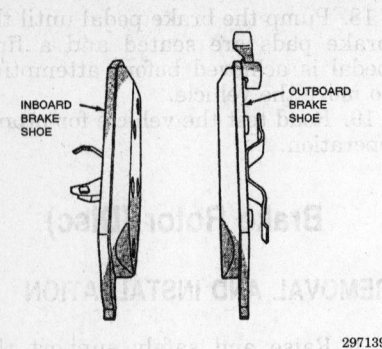

INBOARD BRAKE SHOE

OUTBOARD BRAKE SHOE

297139

Disc brake pad identification

2. Remove the rear wheel assembly.

NOTE: If the vehicle has high mileage, the brake drums may have a ridge worn in them by the brake shoes. This ridge causes the brake drum to interfere with the brake shoes preventing drum removal. Clearance can be obtained by backing off the brake's automatic self-adjuster mechanism, using the following procedure.

a. Locate and remove the rubber plug from the top of the brake support plate (backing plate).

b. Insert a small prybar through the automatic adjuster access hole and engage the teeth on the adjuster wheel. Rotate the adjuster wheel so it is moved toward the front of the vehicle. This will back off the adjustment of the rear brake shoes.

c. Continue moving the adjuster wheel toward the front of the vehicle until it stops moving.

3. Remove the rear brake drum from the hub assembly.

To install:

4. Inspect the brake drums for cracks or signs of overheating. Measure the drum runout and diameter. If not to specification, resurface the drum. Runout should not exceed 0.006 inch. The diameter variation (oval shape) of the drum braking surface must not exceed either 0.0025 inch in 30 degrees or 0.0035 inch in 360 degrees. All brake drums are marked with the maximum allowable brake drum diameter on the face of the drum.

5. Install the rear brake drum onto the hub assembly.

6. Install the wheel and tire. Tighten the lug nuts in a crisscross pattern to about 45 ft. lbs. then repeat the pattern and final torque to 95 ft. lbs. (129 Nm).

7. Properly adjust the rear brakes.

8. Lower the vehicle.

9. Road test vehicle to check brake operation.

Brake Shoes

REMOVAL AND INSTALLATION

1. Raise and safely support the vehicle. Remove the wheel and tire assemblies and the drums.

2. Remove the automatic adjuster spring and lever.

3. Remove the hold-down clips and pins.

4. Rotate the automatic adjuster star wheel enough so both shoes move out far enough to be free of the wheel cylinder boots.

5. Disconnect the parking brake cable from the actuating lever.

6. Remove the lower shoe to shoe spring.

7. With the shoes held together by the upper shoe to shoe spring, remove them from the backing plate.

To install:

8. Thoroughly clean and dry the backing plate. To prepare the backing plate, lubricate the bosses, anchor pin and parking brake actuating lever pivot surface lightly with lithium based grease.

9. Remove, clean and dry all brake components. Lubricate the star wheel shaft threads with anti-seize lubricant and transfer all parts to their proper locations on the new shoes.

10. Install the lower spring.

11. Connect the parking brake cable.

12. Install the automatic adjuster lever and spring.

13. Adjust the star wheel.

14. Remove any grease from the linings and install the drum.

15. Complete the brake adjustment with the wheels installed.

16. Check for proper brake system operation.

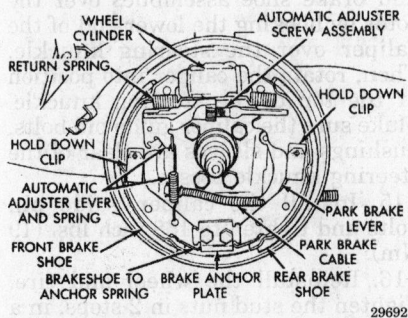

WHEEL CYLINDER

RETURN SPRING

AUTOMATIC ADJUSTER SCREW ASSEMBLY

HOLD DOWN CLIP

HOLD DOWN CLIP

AUTOMATIC ADJUSTER LEVER AND SPRING

FRONT BRAKE SHOE

BRAKESHOE TO ANCHOR SPRING

BRAKE ANCHOR PLATE

PARK BRAKE LEVER

PARK BRAKE CABLE

REAR BRAKE SHOE

296921

Kelsey Hayes rear brake assembly (left side shown)

Wheel Cylinder

REMOVAL AND INSTALLATION

1. Raise and safely support the vehicle.

2. Remove the wheel, drum and brake shoes.

3. Remove and plug the brake line from the wheel cylinder.

4. Remove the wheel cylinder retaining bolts and remove the wheel cylinder from the backing plate.

To install:

5. Apply a very thin coating of silicone sealer to the wheel cylinder mounting surface, install the cylinder to the backing plate and install the retaining bolts. Tighten the retaining bolts to 75 inch lbs. (8 Nm).

6. Connect the brake line to the wheel cylinder. Tighten the rear brake flex hose tube fitting to 145 inch lbs. (17 Nm).

7. If the wheel cylinder was removed to correct a fluid leakage complaint, carefully clean and inspect all components. Brake shoe linings contaminated by brake fluid must not be reused. Always install new brake shoes as an axle set, both rear wheels at once. New brake hardware and return springs are always recommended. Make sure the adjuster is free and operates correctly. This is also a good time to check the parking brake cable assemblies for proper operation. Frozen or sticking brake cables must be replaced. Install all brake components that were removed, using new replacement parts as required. Inspect the inside of the brake drum. Most brake drum wear can be removed by proper machining. If the drum inside diameter is beyond the limit stamped on it, a replacement drum must be installed.

8. Install the tire and wheel assembly.

9. Bleed the brake hydraulic system and adjust the brake shoes.

Parking Brake Cable

ADJUSTMENT

With Rear Drum Brakes

Due to the auto adjust feature of the parking brake lever, adjustment of the parking brake system on vehicles equipped with rear drum brakes relies on proper drum brake show adjustment.

1. Raise and safely support the vehicle so all wheels are free to turn.

2. Remove the tire and wheel assembly.

3. Remove the rear brake adjusting hole rubber plug from the rear brake show support plate. Be sure the parking brake lever is fully released.

4. Insert Brake Adjuster C-3784 or equivalent, through the adjusting hole in the support plate and against the star wheel of the adjusting screw. Move the handle of the tool downward until a slight drag is felt when the road wheel is rotated.

5. Insert a thin prytool or piece of welding rod onto the brake adjusting hole. Push the adjusting lever out of engagement with the star wheel. Be careful not to bend the adjusting lever or distort the lever spring. While holding the adjusting lever out of engagement with the star wheel, back off the star wheel to ensure a free wheel with no brake shoe drag.

6. Repeat the above adjustment at the other rear wheel, Install the adjusting hole rubber plugs in the rear brake supports.

7. Install the tire and wheel assemblies, then carefully lower the vehicle.

8. Apply and release the parking brake lever 1 time after the wheel brake adjustment.

REMOVAL AND INSTALLATION

—————— CAUTION ——————
The Supplemental Restraint System (SRS) must be disarmed before working around the interior of the vehicle. Failure to do so may cause accidental deployment of the air bag, resulting in unnecessary SRS system repairs and/or personal injury.

With Rear Drum Brakes

1. Disconnect the negative battery cable.

—————— WARNING ——————
The air bag control unit is mounted beneath the center console. Use care when working with the center console assembly not to impact or shock the control unit.

2. Remove the center floor console assembly as follows:
   a. Remove the shifter knob, if equipped with a manual transaxle.
   b. Remove the shifter trim panel.
   c. Remove the center instrument panel.

d. Remove the panel box from the console assembly.
   e. Remove the 2 screws from the center of the console.
   f. Remove the 4 side panel screws and remove the console from the vehicle.

3. Loosen the cable adjuster nut and then remove the parking brake cable, by pulling it from the passenger compartment.

4. Raise and safely support the vehicle.

5. Remove the brake drum and shoes.

6. Disconnect the cable end from the parking brake strut lever. Compress the retaining strips to remove the cable from the backing plate.

7. Unfasten any other frame retainers and remove the cables.

To install:

8. Install the cable to the rear actuator. Secure in place with the parking brake cable clip and retainer spring.

9. Install the brake shoes and drum.

10. Position the cable in the under the vehicle and install retainers loose.

11. Reattach the parking brake cables to the actuator inside the vehicle. Tighten the adjusting nut until the proper tension is placed on the cable. Adjust the parking brake stroke.

12. Secure all cable retainers. Apply and release the parking brake a number of times once all adjustments have been made.

13. Assemble the interior components which were removed.

14. Adjust the rear brakes and parking brake cables.

15. Connect the negative battery cable and check the rear wheels to confirm that the rear brakes are not dragging.

16. Check that the parking brake holds the vehicle on an incline.

With Rear Disc Brakes

—————— CAUTION ——————
The Supplemental Restraint System (SRS) must be disarmed before working around the interior of the vehicle. Failure to do so may cause accidental deployment of the air bag, resulting in unnecessary SRS system repairs and/or personal injury.

Unlike some other rear disc brake systems, the parking brake operation is **not** incorporated into the brake caliper. This system, uses a separate

set of brake shoes, located behind the brake rotor inside the rotor "hat".

1. Disconnect the negative battery cable.

—————— WARNING ——————
The air bag control unit is mounted beneath the center console. Use care when working with the center console assembly not to impact or shock the control unit.

2. Remove the center floor console assembly as follows:
   a. Remove the shifter knob, if equipped with a manual transaxle.
   b. Remove the shifter trim panel.
   c. Remove the center instrument panel.
   d. Remove the panel box from the console assembly.
   e. Remove the 2 screws from the center of the console.
   f. Remove the 4 side panel screws and remove the console from the vehicle.

3. Loosen the cable adjuster nut and then remove the parking brake cable, by pulling it from the passenger compartment.

4. Raise the vehicle and support safely.

5. At the rear wheel, remove the brake caliper and rotor.

6. Remove the parking brake shoes as follows:
   a. Remove the upper shoe to anchor springs.
   b. Remove the lower shoe to shoe spring.
   c. Remove the brake shoe hold-down springs.
   d. Disconnect the parking brake cable from the actuating lever.

7. Disconnect the cable end from the parking brake strut lever. Compress the retaining strips to remove the cable from the backing plate.

8. Unfasten any other frame retainers and remove the cables.

To install:

9. Install the cable to the rear actuator. Secure in place with the parking brake cable clip and retainer spring.

10. Install the parking brake shoes.

11. Install the brake rotor and caliper assembly.

12. Position the cable in the under the vehicle and install retainers loose.

13. Reattach the parking brake cables to the actuator inside the vehicle. Tighten the adjusting nut until the proper tension is placed on the cable. Adjust the parking brake stroke.

14. Secure all cable retainers. Apply and release the parking brake a number of times once all adjustments have been made.

15. Assemble the interior components which were removed.

16. Adjust the parking brake shoes and parking brake cables.

17. Connect the negative battery cable and check the rear wheels to confirm that the rear brakes are not dragging.

18. Check that the parking brake holds the vehicle on an incline.

Brake System Bleeding

VEHICLES WITHOUT ABS

Pressure Bleeding

1. Use bleeder tank special tool C-3496-B with the required adapter for the master cylinder reservoir to pressurize the hydraulic system for bleeding. Connect the tool(s) according to the manufacturers instructions.

When bleeding the brake system, some air may be trapped in the brake lines or valves far upstream, as much as 10 feet from the bleeder screw. Therefore is it vital to have a fast flow of a large volume of brake fluid when bleeding the brakes to ensure all the air gets out.

The following wheel sequence for bleeding the hydraulic brake system should be used to ensure adequate removal of all trapped air from the hydraulic system:
- Left rear wheel
- Right front wheel
- Right rear wheel
- Left front wheel

2. Attach a clear plastic hose to the bleeder screw starting at the right rear wheel and feed the hose into a clear jar containing enough fresh brake fluid to submerge the end of the hose.

3. Open the bleeder screw at least 1 full turn or more to obtain a steady stream of brake fluid.

4. After 4–8 ounces of fluid has been bled through the brake and an air-free flow if maintained in the clear plastic hose and jar, close the bleeder screw.

5. Repeat the process at all the other retaining bleeder screws. Check the pedal for travel. If the pedal travel is excessive or has not been improved, enough fluid has not passed through the system to expel

all the trapped air. Be sure to monitor the fluid level in the pressure bleeder. It must stay at the proper level so air will not be allowed to reenter the brake system through the master cylinder reservoir.

Manual Bleeding

The following wheel sequence for bleeding the hydraulic brake system should be sued to ensure adequate removal of all trapped air from the hydraulic system:
- Left rear wheel
- Right front wheel
- Right rear wheel
- Left front wheel

1. Attach a clear plastic hose to the bleeder screw starting at the right rear wheel and feed the hose into a clear jar containing enough fresh brake fluid to submerge the end of the hose.

2. Have an assistant pump the brake pedal 3 or 4 times and hold it down before the bleeder screw is opened.

3. Open the bleeder screw at least 1 full turn. When the bleeder screw opens, the brake pedal will drop all the way to the floor.

4. Close the bleeder screw. Release the brake pedal AFTER the screw is closed.

5. Repeat the procedure 4 or 5 times at each bleeder screw. Check the pedal for travel. If the pedal travel is excessive or has not been improved, enough fluid has not passed through the system to expel all the trapped air. Be sure to monitor the fluid level in the pressure bleeder. It must stay at the proper level so air will not be allowed to reenter the brake system through the master cylinder reservoir.

6. Test drive the vehicle to be sure the brakes are operating correctly and that the pedal is solid.

Master Cylinder Bleeding

It is not necessary to bleed the entire hydraulic system after replacing the master cylinder, but the master cylinder must have been bled and filled upon installation.

1. Remove the master cylinder, then clamp it in a suitable vise. Attach Bleeding Tubes 6802 to the master cylinder. Position the tubes so the outlets of the tubes will be below the surface of the brake fluid when the reservoir is filled to the proper level.

2. Fill the brake fluid reservoir with brake fluid conforming to DOT 3 specifications such as Mopar or equivalent.

3. Use a wooden dowel to depress the pushrod slowly, then allow the pistons to return the the released position. Repeat several times until all of the air bubbles are expelled.

4. Remove the bleeding tubes from the master cylinder outlet ports, plug the ports, then install the fill cap on the reservoir.

5. Remove the master cylinder from the vise, then install on the vehicle.

VEHICLES WITH ABS

When bleeding the ABS system, the following bleeding sequence must be followed to insure complete and adequate bleeding. The ABS system can be bled using a manual bleeding procedure or standard pressure bleeding.

1. If pressure bleeding, use bleeder tank special tool C-3496-B with the required adapter for the master cylinder reservoir to pressurize the hydraulic system for bleeding. Connect the tool(s) according to the manufacturers instructions.

When bleeding the brake system, some air may be trapped in the brake lines or valves far upstream, as much as 10 feet from the bleeder screw. Therefore is it vital to have a fast flow of a large volume of brake fluid when bleeding the brakes to ensure all the air gets out.

The following wheel sequence for bleeding the hydraulic brake system should be used to ensure adequate removal of all trapped air from the hydraulic system:
- Left rear wheel
- Right front wheel
- Right rear wheel
- Left front wheel

2. Assemble and install all brake system components on the vehicle, making sure all hydraulic fluid lines are installed and properly torqued.

3. Connect the DRB Diagnostics Tester to the diagnostics connector, located under the steering column cover, directly below the steering column.

4. Using the DRB, check to make sure the CAB (ABS control module) does not have any fault codes stored. If it does, remove them using the DRB.

5. Bleed the brake system.

6. Have an assistant apply the brake firmly, then initiate the "Bleed ABS" cycle, on the DRB, one time. Release the brake pedal.

7. Bleed the brake system, as in step 5 above.

8. Repeat steps 6 and 7 above until the brake fluid flows clear and free of bubbles. Check the brake fluid level in the reservoir periodically to prevent reservoir from running low on brake fluid.

9. Test drive the vehicle to be sure the brakes are operating properly and the pedal is solid.

Wheel Speed Sensor

REMOVAL AND INSTALLATION

Front Wheel Speed Sensor

1. Disconnect the negative battery cable.

2. Raise and safely support the vehicle.

3. Remove the tire and wheel assembly.

4. Detach the speed sensor connector from the vehicle wiring harness. Remove the clip attaching the speed sensor cable connector to the vehicle body.

5. Remove the wheel speed sensor head-to-steering knuckle attaching bolt.

6. Carefully remove the sensor head from the steering knuckle. If the sensor has seized due to corro-

sion, do NOT use pliers on the sensor head. Use a hammer and a punch to gently tap the edge of the sensor ear, rocking the sensor side to side, until its free.

7. Remove the speed sensor cable assembly grommets from the retaining bracket. Remove the speed sensor cable routing clip from the frame of the vehicle.

To install:

8. Attach the wheel speed sensor cable connector to the vehicle wiring harness.

9. Install the speed sensor cable assembly grommets into the retaining bracket. Install the speed sensor cable routing clip onto the frame of the vehicle.

10. Fasten the speed sensor-to-steering knuckle attaching screw. Torque the screw to 60 inch lbs. (7 Nm).

11. Install the wheel and tire assembly.

12. Carefully lower the vehicle, then connect the negative battery cable.

Rear Wheel Speed Sensor

1. Disconnect the negative battery cable.

2. Raise and safely support the vehicle, then remove the tire and wheel assembly.

3. Detach the speed sensor cable connector from the vehicle wiring harness. Remove the clip attaching the speed sensor cable connector to the vehicle body.

4. Remove the speed sensor cable routing bracket from the rear brake flex hose mounting bracket. Remove the speed sensor cable from the routing clips on the rear brake flex hose and chassis brake tube.

5. Depending upon application, remove the bolt attaching the speed sensor head to the drum brake support plate or disc brake adapter. Remove the bolt attaching the speed sensor cable routing bracket to the rear strut assembly.

6. Remove the sensor head from the support plate or disc brake adapter assembly, depending upon application. If the sensor has seized due to corrosion, do NOT use pliers to remove it.

To install:

7. Install the speed sensor head into the bracket support plate or disc brake adapter, as applicable.

8. Install the wheel speed sensor attaching bolt, then torque to 60 inch lbs. (70 Nm).

9. If equipped with rear drum brakes, install the speed sensor cable routing bracket onto the brake flex hose bracket. If with rear disc brakes, mount the brake flex hose and speed sensor cable routing bracket onto the rear strut bracket.

10. Fasten the speed sensor cable onto the routing clips on the rear brake flex hose and chassis brake tube.

11. Attach the speed sensor cable connector into the vehicle wiring harness. Install the clip attaching the speed sensor cable connector to the vehicle body.

12. Install the tire and wheel assembly.

13. Carefully lower the vehicle, then connect the negative battery cable.

FRONT SUSPENSION

Strut

REMOVAL AND INSTALLATION

The front strut assemblies may be inspected on the vehicle. Inspect for damaged or broken coil springs. Inspect for torn or damaged strut assembly dust boots. Lift the dust boot and check for evidence of fluid running from the upper end of the fluid reservoir. Actual leakage will be fluid running down the side and dripping off the lower end of the strut. A slight amount of seepage between the strut rod and strut shaft seal is not unusual and does not affect performance of the strut assembly. Also check the jounce bumpers for signs of deterioration.

1. Raise and safely support the vehicle.

2. Remove the wheel and tire.

3. Remove the bolt attaching the brake hose to the strut assembly.

4. If equipped with antilock brakes, disconnect the speed sensor wiring.

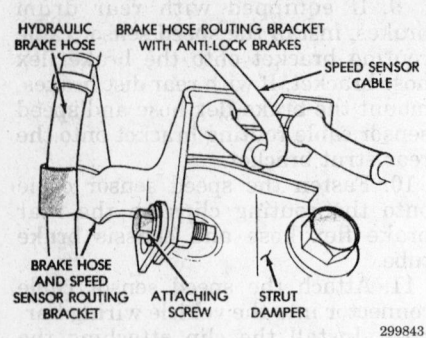

HYDRAULIC BRAKE HOSE

BRAKE HOSE ROUTING BRACKET WITH ANTI-LOCK BRAKES

SPEED SENSOR CABLE

BRAKE HOSE AND SPEED SENSOR ROUTING BRACKET

ATTACHING SCREW

STRUT DAMPER

299843

ABS speed sensor cable

WARNING

The pinch bolts are serrated and must not be turned during removal or installation. Remove the pinch bolt nut while holding the bolt still then remove the bolt by pushing it straight out of the strut bracket.

---

5. Remove the pinch bolts attaching the steering knuckle to the lower control arm.

6. Remove the 3 nuts attaching the strut to the strut tower.

7. Remove the strut assembly. If both struts are being removed, mark the strut assemblies right or left according to which side of the vehicle they were removed from.

To install:

8. Install the strut assembly into the strut tower.

9. Install the 3 upper strut mounting nuts.

10. Torque the nuts to 23 ft. lbs. (31 Nm).

11. Install the steering knuckle onto the strut.

12. Reinstall the steering knuckle pinch bolts.

NOTE: The pinch bolts must be installed with the nuts facing the front of the vehicle.

13. Torque the steering knuckle to strut pinch bolts to 40 ft. lbs. (53 Nm). Remember that the pinch bolts are serrated and must not be turned during installation. Turn the nuts only.

14. Install the wheel and tire.

15. Lower the vehicle.

16. The front end alignment should be checked and adjusted as necessary.

Lower Ball Joints

REMOVAL AND INSTALLATION

The front suspension ball joints operate with no free-play. The ball joints are replaceable ONLY as an assembly. Do not attempt any type of repair on the ball joint assembly. The ball joint is a press fit into the lower control arm with the joint stud retained in the steering knuckle by the clamp bolt. To check the ball joint, with the weight of the vehicle resting on the road wheels, grasp the grease fitting and without using any tools, attempt to move the grease fitting. If the ball joint is worn the grease fitting will move easily. If movement is noted, replacement of the ball joint is recommended.

1. Raise and safely support the vehicle.

2. Remove the lower control arm.

3. Pry the seal boot off the ball joint.

4. Using a suitable press remove the ball joint from the lower control arm.

To install:

5. Install the ball joint into the lower control arm with the notch in the ball joint stud facing the front lower control arm bushing.

6. Using a suitable press, press the ball joint into the lower control arm.

7. Install the ball joint boot seal using a suitable driver such as a large socket or suitable sized piece of pipe. Do not use a shop press as was used to install the ball joint as a press exerts too much force.

8. Install the lower control arm.

9. Lower the vehicle and check for proper operation. A front end alignment check is recommended.

Lower Control Arms

REMOVAL AND INSTALLATION

A gas pressurized MacPherson strut front suspension design is used. A cast lower control arm assembly is attached to the front suspension crossmember using 2 rubber isolator bushings and to the steering knuckle by means of a replaceable ball joint.

If damaged, the lower control arm casting is serviced as a complete component. Inspect the lower control arm for signs of damage from contact with the ground or road debris. If the lower control arm shows any signs of damage, inspect the arm for distortion. Do not attempt to repair or straighten a broken or bent lower control arm.

1. Raise and safely support the vehicle.

2. Remove the front wheels and tires.

3. Remove the steering knuckle to ball joint stud clamping nut.

4. Remove the sway bar attaching links (bolts) from both control arms.

5. Loosen, but do not remove, the sway bar to crossmember attaching bolts and rotate the sway bar away from the control arm.

6. Using a prybar, separate the lower control arm from the steering knuckle. Use care not to damage the ball joint boot.

WARNING

Pulling the steering knuckle out from the vehicle after releasing the ball joint can separate the inner CV-joint. Use care.

---

7. Remove the 2 lower control arm bushing to crossmember attaching nuts and bolts.

8. Remove the control arm.

To install:

9. Install the control arm into the vehicle.

10. Install the 2 control arm attaching bolts. Install the rear bolt first but do not tighten yet.

11. Torque the front control arm attaching bolt first. Tighten to 120 ft. lbs. (163 Nm). Then, torque the rear bolt to same specification.

12. Install the control arm to the steering knuckle and torque the ball joint stud clamping nut and bolt to 70 ft. lbs. (95 Nm).

13. Rotate the sway bar up to the control arms.

14. Install the sway bar attaching bolts but do not tighten yet.

15. Install the wheels and tires.

16. Lower the vehicle so the vehicle's weight is on the floor.

17. Torque the sway bar to lower control arm bolts to 21 ft. lbs. (28 Nm).

18. Torque the sway bar bushing retainer to crossmember attaching bolts to 21 ft. lbs. (28 Nm).

Sway Bar

REMOVAL AND INSTALLATION

The sway bar interconnects the front lower control arms of the vehicle and attaches to the crossmember. Jounce and rebound movements affecting 1 wheel are partially transmitted to the opposite wheel to restrict body roll. Attachment of the sway bar to the front crossmember is through 2 rubber bushings. The sway bar is connected to the lower control arms using rubber isolated attaching links. The bushings are split for easy replacement. The split in the sway bar bushing should be positioned toward the front of the vehicle.

1. Raise and safely support the vehicle.

2. Remove the nuts and link and sway bar attaching link assemblies from the lower control arms.

3. Remove the bolts from the front crossmember retainer bushings.

4. Remove the sway bar.

To install:

5. Install the sway bar and the front crossmember retainer bushings. Align the cut outs in the bushings with the raised bead in the crossmember.

6. Install the sway bar to the lower control arms.

7. Torque the link assemblies to 21 ft. lbs. (28 Nm).

8. Torque the bushing attaching bolts to 21 ft. lbs. (28 Nm).

Front Wheel Bearings

ADJUSTMENT

The hub and bearing assemblies are non-serviceable. If the assembly is damaged, the complete unit must be replaced.

REMOVAL AND INSTALLATION

This vehicle uses a sealed for life front hub and bearing assembly attached to the front steering knuckle. The outer CV-joint assembly is spined to the front hub and bearing assembly. The front wheel bearing is called a cartridge bearing. The wheel bearing can be serviced separately from the front steering knuckle and hub assembly. Installation and retention of the front wheel bearing in the steering knuckle is by an interference fit and retained by a snapring. If the front wheel bearing requires replacement, the hub must be removed from the original wheel bearing and transferred to the replacement bearing.

Note that the Neon does not use a rubber lip seal as on past front wheel drive cars to prevent contamination of the front wheel bearing. On this vehicle, the face of the outer CV-joint fits deeply into the steering knuckle using a close fit. This design deters direct water splash on the bearing seal while allowing any water that gets in, to run out the bottom. It is important to thoroughly clean the outer CV-joint and the wheel bearing area in the steering knuckle before it is assembled after servicing.

The steering knuckle MUST be removed to replace both the hub and the front wheel bearing.

1. Remove the front hub cotter pin, nut lock and spring washer.

WARNING

Wheel bearing damage will result if, after loosening the hub nut, the vehicle is rolled on the ground or the weight of the vehicle is allowed to be supported by the tires.

---

2. Loosen the hub nut while the vehicle is on the floor with the brakes applied. The front hub and halfshaft are splined together through the knuckle (bearing) and retained by the hub nut. The front wheel bearing supports the front hub and weight of the vehicle.

3. Raise and safely support the vehicle.

4. Remove the wheel and tire.

5. Remove the front disc brake caliper from the steering knuckle. The caliper is removed by first lifting the bottom of the caliper away from the steering knuckle and then removing the top of the caliper out from under the steering knuckle. Support the caliper using wire. Do not allow the caliper to hand by the brake hose.

6. Remove the brake rotor from the front hub/bearing assembly.

7. Remove the nut attaching the outer tie rod end to the steering knuckle.

   a. Hold the tie rod end stud with an $^{11}/_{32}$ inch socket while loosening and removing the nut with the wrench.

   b. Remove the tie rod end from the steering knuckle using a puller. Do not hammer wedge-type tools or the steering tie rod end joint will be damaged.

8. Locate and remove the lower control arm ball joint clamping nut and bolt and separate the ball joint stud from the steering knuckle by prying down on the lower control arm. Use care not to damage the steering tie rod end or seal. In addition, use care not to allow the halfshaft to become overextended as the steering knuckle is removed. Do not allow the halfshaft to hang by the inner CV-joint boot. The halfshaft must be supported.

NOTE: The steering knuckle to strut assembly attaching bolts are serrated and must NOT be turned during removal. Remove the nuts while holding the bolts stationary in the steering knuckle.

9. With the steering knuckle removed from the vehicle, use a suitable hydraulic press to remove the wheel bearing from the steering knuckle. Use care to jig the knuckle level in the press bed and press the wheel hub and bearing slowly from the knuckle. One bearing race may come out with the hub when the hub is removed. Remove the knuckle from the press and remove the snapring retaining the hub bearing in the steering knuckle. Reposition the knuckle in the press and press the hub bearing from its bore.

To install:

10. Clean all parts well. Again, use care to jig the steering knuckle level in the press bed. Place the new hub bearing into the bore of the steering knuckle so it is square with the bore. Place a bearing driver on the outer race of the hub bearing and press the hub bearing into the steering knuckle until it is fully seated in the bottom of its bore. Install the hub bearing retaining snapring into its groove in the knuckle bore. Be sure it is fully seated in its groove. Use care not to damage the just-installed bearing seal when installing the snapring.

11. Again, place the knuckle assembly with the hub bearing installed on the press bed using care to align and level the assembly. Use suitable drivers and arbors to support the hub bearing on its inner race. Place the wheel hub in the bearing using care to align it square with the bearing. Using suitable drivers, press the hub into the bearing until it bottoms in the hub bearing.

12. Install the steering knuckle/hub/wheel bearing assembly back into the front strut and install the through bolts. The steering knuckle-to-strut bolts are serrated (toothed) and must not be turned in the steering knuckle during installation. Torque the nuts (do not turn the bolt heads) to 40 ft. lbs. (54 Nm) plus an additional 1/4 turn after the specified torque is met.

13. Slide the halfshaft back into the front hub and bearing assembly. Then, install the steering knuckle onto the ball joint stud.

14. Install a NEW steering knuckle-to-ball joint stud, clamp bolt and nut. Torque the clamp bolt to 75 ft. lbs. (100 Nm).

15. Install the tie rod end into the steering knuckle. Start the tie rod end-to-steering knuckle attaching

nut onto the stud of the tie rod end. While holding the stud of the tie rod end stationary, tighten the tie rod end nut. Using a "crow's foot" and $^{11}/_{32}$ inch socket, torque the tie rod end nut to 40 ft. lbs. (55 Nm).

16. Install the brake rotor and the caliper onto the steering knuckle.

a. The caliper is installed by first sliding the top of the caliper under the top abutment on the steering knuckle, then installing the bottom of the caliper against the bottom abutment of the steering knuckle.

b. Install the caliper attaching bolts and torque to 23 ft. lbs. (31 Nm).

17. Clean all foreign matter from the threads of the outer CV-joint stub axle. Install the hub nut onto the threads of the stub axle and tighten the nut.

18. With the vehicle's brakes applied to keep the brake rotor from turning, tighten the hub nut to 135 ft. lbs. (183 Nm).

19. Install the wheel and tire. Torque the lug nuts to 100 ft. lbs. (135 Nm).

20. Lower the vehicle.

21. Install the spring washer, hub nut lock and a new cotter pin. Wrap the cotter pin prongs tightly around the hub nut lock.

22. Check the front end alignment and adjust Toe as required.

REAR SUSPENSION

Strut

REMOVAL AND INSTALLATION

NOTE: The strut assembly is a load bearing component, therefore the vehicle chassis and axle weight must be supported separately, requiring the use of 2 separate lifting devices.

1. Raise and support vehicle chassis.

2. Raise and support lower control arm assembly slightly.

3. Remove the strut upper mounting nuts.

4. Remove the shock lower mounting bolts and remove the assembly from the vehicle.

To install:

5. Position the strut assembly so the lower mounting bolts can be installed and lightly tightened.

6. Use a jack to raise or lower the lower control arm, so top strut plate studs aligns through the body. Raise the jack to hold the strut assembly in position.

7. Install top plate nuts on studs and tighten the mounting nuts to 275 inch lbs. (31 Nm).

8. Tighten the lower mounting bolt to 70 ft. lbs. (95 Nm).

Coil Spring

REMOVAL AND INSTALLATION

1. Place the strut assembly in a MB991237 and MB991239 spring compressor assembly or equivalent. Tighten the compressor and compress the spring slowly. Make certain the compressor is properly engaged before tightening.

2. After tension has been removed from the strut assembly and strut plate, remove the piston rod nut and washer from the top of the strut assembly.

3. If the spring is to be replaced, slowly release the tension on the spring compressor. Allow the spring to expand fully. If only the strut is being replaced, the spring may remain in the compressor assembly.

4. By hand, remove the upper strut bearing, washer, mount and strut shield. Remove the upper insulator ring, the spring, the bumper and lower insulator from the spring. Take notice to each components location for proper reassembly.

To install:

5. Install the lower insulator, the strut bumper and the uncompressed spring. Install the upper spring insulator, strut shield, mount and washer.

6. Install or align the spring compressor. Make certain the spring is correctly positioned relative to the upper and lower insulator rings. Smoothly compress the spring.

7. Install the washer and piston rod nut. Tighten the nut to 16 ft. lbs. (22 Nm). Install the dust cap.

8. Carefully release the spring compressor, watching the spring position as it seats. When the spring is properly seated, release/remove the compressor tools.

9. Reinstall the strut assembly.

Wheel Bearings

ADJUSTMENT

The hub and bearing assemblies are non-serviceable. If the assembly is damaged, the complete unit must be replaced.

REMOVAL AND INSTALLATION

The vehicle uses a rear hub and wheel bearing assembly that is designed for the life of the vehicle and should require no maintenance. To inspect, with the wheel and brake drum removed, rotate the flanged outer ring of the hub. Excessive roughness or resistance to rotation may indicate dirt in the bearing or bearing failure. If the rear wheel bearings exhibit these conditions, the hub and bearing assembly should be replaced. Damaged bearing seals and resulting excessive grease loss may also require bearing replacement. Moderate grease loss from the bearing is considered normal and should not require replacement of the hub and bearing assembly.

1. Raise and safely support the vehicle.

2. Remove the wheel and tire.

3. Remove the brake drum or brake rotor.

4. Remove the hub dust cap.

5. Loosen and remove the bearing nut.

6. Remove the hub and bearing assembly.

To install:

7. Install the hub and bearing assembly.

8. Install the bearing retaining nut and torque to 124 ft. lbs. (168 Nm).

9. Install the dust cap.

10. Install the brake drum or brake rotor.

11. Install the wheel and tire.

12. Torque the lug nuts to 100 ft. lbs. (135 Nm).

13. Lower the vehicle and check for proper operation.

CHRYSLER CORP.

Front Wheel Drive

CHRYSLER-Cirrus • Sebring Convertible DODGE-Stratus
PLYMOUTH-Breeze

FIRING ORDERS

NOTE: To avoid confusion, always replace spark plug wires one at a time.

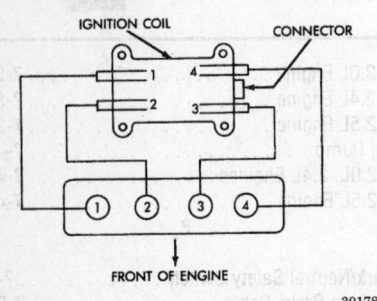

2.0L and 2.4L Engines
Engine Firing Order: 1–3–4–2
Distributorless Ignition System

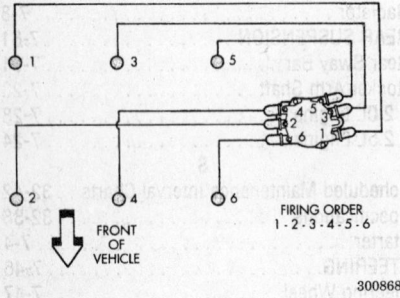

2.5L Engine
Engine Firing Order: 1–2–3–4–5–6
Distributor Rotation: Counterclockwise

ENGINE ELECTRICAL

NOTE: Disconnecting the negative battery cable on some vehicles may interfere with the functions of the on board computer systems and may require the computer to undergo a relearning process, once the negative battery cable is reconnected.

Distributor

REMOVAL

2.5L Engine

The 2.5L engine is equipped with a camshaft driven mechanical distributor. This engine uses a fixed ignition timing system. The basic ignition timing is not adjustable. The Powertrain Control Module (PCM) determines spark advance. The crankshaft position sensor and camshaft position sensor are Hall Effect devices. The crankshaft sensor is mounted remotely from the distributor while the camshaft position sensor is mounted inside the distributor housing. Both sensors generate pulses that are inputs to the PCM. The PCM determines crankshaft position from these sensors. The PCM calculates injector sequence and ignition timing. There is a resistor built into the distributor cap. An ohmmeter connected between the center button and ignition coil terminal should read 5000 ohms.

1. Disconnect the negative battery remote cable located on the left strut tower.
2. Remove the bolt attaching the air inlet resonator to the intake manifold.
3. Loosen the clamps holding the air cleaner cover to the air cleaner housing.
4. Remove the PCV make-up air hose from the air inlet tube.
5. Remove the EGR tube.
6. Mark for identification, if necessary and remove the spark plug wires from the distributor cap.
7. Remove the distributor cap.
8. Mark the rotor position. A scribe mark indicates where to position the rotor when reinstalling the distributor. Remove the rotor.
9. Disconnect the 2 electrical harness connections from the distributor.
10. Remove the 2 distributor hold-down nuts and washers.
11. Remove the spark plug cable mounting bracket.
12. Remove the transaxle dipstick tube.
13. Carefully remove the distributor from the engine.

INSTALLATION

Timing Not Disturbed

1. Inspect the rotor for cracks or burned electrode. Replace if defective. Install the rotor onto the distributor.
2. Inspect the O-ring seal. If nicked or cracked, replace with a new one. Make sure the O-ring is properly seated on the distributor.
3. Carefully engage the distributor drive with the slotted end of the camshaft. When the distributor is installed properly, the rotor will be in line with the previously made mark.
4. Verify proper rotor alignment with mark made at disassembly.
5. Install the distributor hold-down nuts and washers. Torque the nuts to 9 ft. lbs. (13 Nm).
6. Install the spark plug cable bracket.
7. Reconnect the 2 distributor wiring connectors.
8. Install the distributor cap.
9. Install the spark plug cables following the identification marks made at disassembly.
10. Install the transaxle dipstick tube.
11. Reinstall the EGR tube and torque the bolts to 95 inch lbs. (11 Nm).
12. Reinstall the PCV hose.
13. Install the air cleaner and tighten the hose clamps.
14. Install the air inlet resonator.
15. Reconnect the negative battery remote cable.

Timing Disturbed

1. Rotate the crankshaft until No. 1 piston is at Top Dead Center of the compression stroke.
2. Rotate the rotor to the No. 1 terminal position on the distributor cap.
3. Lower the distributor into place, engaging the distributor drive with the drive on the camshaft. With the distributor fully seated on the engine, the rotor should be under the No. 1 terminal.
4. Verify proper rotor alignment with mark made at disassembly.
5. Install the distributor hold-down nuts and washers. Torque the nuts to 9 ft. lbs. (13 Nm).
6. Install the spark plug cable bracket.
7. Reconnect the 2 distributor wiring connectors.
8. Install the distributor cap.
9. Install the spark plug cables following the identification marks made at disassembly.
10. Install the transaxle dipstick tube.
11. Reinstall the EGR tube and torque the bolts to 95 inch lbs. (11 Nm).
12. Reinstall the PCV hose.
13. Install the air cleaner and tighten the hose clamps.
14. Install the air inlet resonator.
15. Reconnect the negative battery remote cable.

Ignition Timing

ADJUSTMENT

These engines use a fixed ignition system. The Powertrain Control Module (PCM) regulates the ignition timing. Basic ignition timing is not adjustable.

Alternator

PRECAUTIONS

Several precautions must be observed with alternator equipped vehicles to avoid damage to the unit.

• If the battery is removed for any reason, make sure it is reconnected with the correct polarity. Reversing the battery connections may result in damage to the 1-way rectifiers.

• When utilizing a booster battery as a starting aid, always connect the positive to positive terminals and the negative terminal from the booster battery to a good engine ground on the vehicle being started.

• Never use a fast charger as a booster to start vehicles.

• Disconnect the battery cables when charging the battery with a fast charger.

• Never attempt to polarize the alternator.

• Do not use test lights of more than 12 volts when checking diode continuity.

• Do not short across or ground any of the alternator terminals.

• The polarity of the battery, alternator and regulator must be matched and considered before making any electrical connections within the system.

• Never separate the alternator on an open circuit. Make sure all connections within the circuit are clean and tight.

• Disconnect the battery ground terminal when performing any service on electrical components.

• Disconnect the battery if arc welding is to be done on the vehicle.

REMOVAL AND INSTALLATION

2.0L Engine

The 2.0L engine uses a Nippondenso, 90 amp alternator.

1. Disconnect the negative battery cable from the left strut tower. The ground cable is equipped with a insulator grommet which should be placed on the stud to prevent the negative battery cable from accidentally grounding.

2. Unplug the field circuit from the alternator.

3. Remove the B+ terminal cover by spreading the cover with a small flat blade tool.

4. Remove the B+ nut and wire.

5. Loosen the adjusting bolt, but do not remove.

6. Loosen the pivot bolt and the adjusting bolt until the drive belt can be removed.

7. Remove the adjusting bolt and pivot bolt, but do not drop the spacer.

8. Remove the alternator by moving it toward the headlight bucket.

To install:

9. Install the alternator into the bracket on the engine.

10. Install the pivot bolt, but do not tighten.

11. Install the adjusting bolt, but do not tighten.

12. Connect the B+ wire and install the retaining nut. Torque the nut to 75 inch lbs. (9 Nm).

13. Reinstall the B+ terminal cover.

14. Reconnect the field circuit to the alternator.

15. Install the drive belt, making sure it is correctly routed and seated on the alternator pulley. Do not tension the drive belt at this time.

16. Adjust the drive belt and tighten the adjusting bolt to 40 ft. lbs. (54 Nm).

17. Tighten the pivot bolt to 40 ft. lbs. (54 Nm).

18. Reconnect the negative battery cable to the strut tower.

2.4L Engine

The 2.4L engine is equipped with a Nippondenso, 90 amp alternator.

1. Disconnect the negative battery cable from the left strut tower. The ground cable is equipped with a insulator grommet which should be placed on the stud to prevent the negative battery cable from accidentally grounding.

2. Unplug the field circuit from the alternator.

3. Remove the B+ terminal cover from by spreading the cover with a flat blade tool.

4. Remove the B+ nut and wire.

5. Loosen the adjusting bolt, but do not remove.

6. Loosen the pivot bolt and the adjusting bolt until the drive belt can be removed. Remove the accessory drive belt.

7. Remove the adjusting bolt and pivot bolt.

8. Remove the ABS braking unit by removing the 2 lower plate mounting bolts. Leave all lines connected.

9. Remove the coolant overflow bottle.

10. Remove the alternator by sliding the alternator under the air conditioner lines towards the passenger side of the vehicle.

To install:

11. Install the alternator into the bracket on the engine.

12. Install the adjusting and pivot bolts, but do not tighten at this time.

13. Connect the B+ wire and torque the nut to 75 inch lbs. (9 Nm).

14. Reinstall the B+ terminal cover.

15. Reconnect the field circuit to the alternator.

16. Install the accessory drive belt. Be sure the drive belt is correctly routed on the engine and correctly seated on the alternator pulley.

17. Adjust the drive belt and tighten the adjusting bolt to 40 ft. lbs. (54 Nm).

18. Tighten the pivot bolt to 40 ft. lbs. (54 Nm).

19. Reinstall the ABS braking unit. Install and tighten the 2 lower plate mounting bolts.

20. Reinstall the coolant overflow bottle.

21. Reconnect the negative battery cable to the strut tower.

2.5L Engine

A Melco 90 amp alternator is used and is not intended to be disassembled for service. If defective, it must be replaced as an assembly.

1. Disconnect the negative battery cable from the left strut tower. The ground cable is equipped with an insulator grommet which should be placed on the stud to prevent the negative battery cable from accidentally grounding.

2. Unplug the field circuit from the alternator.

3. Remove the B+ terminal nut and wire.

4. Loosen the top mounting ear bolt.

5. Loosen, but do not remove, the pivot bolt. Use care not to loose the nut.

6. Loosen the adjusting bolt on the idler to allow removal of the alternator drive belt.

7. Remove the pivot bolt using care not to loose the spacer.

8. Remove the top mounting ear bolt.

9. Remove the alternator upper bracket.

10. Remove the alternator from the vehicle.

To install:

11. Install the alternator into the bracket on the engine.

12. Install the alternator upper bracket and tighten mounting bolts.

13. Install the pivot bolt, but do not tighten at this time.

14. Install the top mounting ear bolt, but do not tighten at this time.

15. Connect the B+ terminal wire and torque the nut to 75 inch lbs. (9 Nm).

16. Reconnect the field circuit to the alternator.

17. Install the drive belt. Be sure the accessory drive belt is correctly routed on the engine and properly seated on the alternator pulley.

18. Adjust the drive belt and tighten the idler pulley bolt to 40 ft. lbs. (54 Nm).

19. Tighten the pivot bolt and top mounting ear bolt to 40 ft. lbs. (54 Nm).

20. Reconnect the negative battery cable to the strut tower.

Drive Belt

REMOVAL AND INSTALLATION

When replacing or adjusting the engine drive belts, it is important that the belts are adjusted with the correct amount of tension. Excessive belt tension will cause damage to the alternator and power steering pump pulley bearings, while loose belt tension will produce slip and premature wear on the belt.

Belt replacement is similar to adjustment, with the exception the belt will have to be properly routed around the pulleys. Note the routing of the belt before removal, if possible. For individual belt replacement, start with the outer belt. If a removed belt is to be reused, be certain to mark the direction of rotation on the belt, to extend belt life.

2.0L and 2.4L Engines

Alternator and Air Conditioning Compressor Belt

1. Disconnect the negative battery cable remote connection located on the left strut tower.

2. Loosen the locking nut and the pivot bolt located on the top and bottom sides of the alternator.

3. Rotate the adjusting bolt in the direction that loosens the drive belt enough to slide it off the pulleys and remove it from the engine.

To install:

4. Install the drive belt and properly route it around the alternator, idler, A/C compressor and crankshaft damper pulleys. Be sure the belt is seated correctly in each pulley otherwise, the belt could jump out of position.

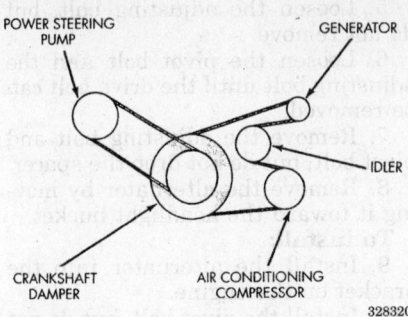

Accessory drive belt routing — 2.0L, 2.4L and 2.5L engines

5. Rotate the adjusting bolt at the alternator until tension starts to increase on the drive belt. The proper belt tension is as follows:

  a. **NEW** A/C compressor-alternator belt: 150 lbs.

  b. **USED** A/C compressor-alternator belt: 80 lbs.

6. Tighten the locking nut and pivot bolt to 40 ft. lbs. (54 Nm) and check the belt tension again. Adjust if necessary.

7. Reconnect the negative battery cable. Start engine and check for proper belt operation.

Power Steering Pump Belt

1. Disconnect the negative battery cable remote location located on the left strut tower.

2. Loosen the power steering pump pivot bolt located above the power steering pump.

3. Loosen the locking bolts located in front and behind the power steering pump located at the bottom of the pump.

4. Pivot the power steering pump closer to the engine to release the drive belt from the pump pulley and crankshaft damper pulley. Remove the drive belt from the engine.

To install:

5. Install the drive belt and properly route it around the power steering pump and crankshaft damper pulleys. Be sure the belt is properly seated in each pulley otherwise, the belt could jump out of position.

6. Using a ½ in. breaker bar, affixed to the square opening on the adjusting bracket, apply enough force on the adjusting bracket to set the drive belt to the proper tension. The proper belt tension is as follows:

  a. **NEW** power steering pump belt: 130 lbs.

  b. **USED** power steering pump belt: 80 lbs.

7. On the 2.0L and 2.4L engines, tighten the locking nuts and pivot bolt to 40 ft. lbs. (54 Nm). Check the belt tension again.

8. On the 2.5L engine, tighten the rear locking bolt to 21 ft. lbs. (28 Nm). Tighten the front locking bolt and pivot bolt to 40 ft. lbs. (54 Nm). Check the belt tension again.

9. Reconnect the negative battery cable. Start the engine and check for proper belt operation.

Starter

REMOVAL AND INSTALLATION

2.0L Engine

A Bosch starter unit is used.

1. Disconnect the negative battery cable remote connection from the left strut tower.

NOTE: The ground cable is equipped with a insulator grommet which should be placed on the stud to prevent the negative battery cable from accidentally grounding.

2. Remove the air cleaner resonator.

3. Remove the positive battery cable retaining nut from the starter.

4. Disconnect the positive battery cable and alternator output wire from the starter.

5. Disconnect the push-on solenoid connector from the starter.

6. Remove the 2 bolts that attach the starter to the transaxle.

7. Remove the starter from the vehicle.

To install:

8. Install the starter and the attaching bolts to the transaxle assembly.

9. Torque the attaching bolts to 40 ft. lbs. (54 Nm).

NOTE: Clean all dirt and/or corrosion from the wire terminals before reconnecting the wiring to the solenoid.

10. Reconnect the push-on solenoid connector to the starter.

11. Reconnect the alternator output wire and positive battery cable to the starter and torque the retaining nut to 90 inch lbs. (10 Nm).

12. Install the air cleaner resonator.

13. Reconnect the negative battery cable.

2.4L Engine

A Nippondenso starter is used.

1. Disconnect the negative battery cable remote connection from the left strut tower.

NOTE: The ground cable is equipped with a insulator grommet which should be placed on the stud to prevent the negative battery cable from accidentally grounding.

2. Remove the air cleaner resonator.

3. Remove the 3 bolts attaching the Transmission Control Module (TCM). Do not disconnect the TCM wiring. Move the TCM to gain access to the upper starter mounting bolt.

4. Remove the upper starter mounting bolt.

5. Raise and safely support the vehicle.

6. Remove the positive battery cable nut and disconnect the cable from the starter.

7. Disconnect the push-on solenoid connector.

8. Remove the lower mounting bolt that attaches the starter to the transaxle.

9. Remove the starter from the vehicle.

To install:

10. Install the starter onto the transaxle and the lower mounting bolt.

11. Torque the mounting bolts to 40 ft. lbs. (54 Nm).

NOTE: Before reconnecting the wiring to the starter solenoid, be sure to clean the wiring of any dirt or corrosion.

12. Reconnect the positive battery cable to the solenoid post and torque the retaining nut to 90 inch lbs. (10 Nm).

13. Reconnect the push-on solenoid connector.

14. Lower the vehicle.

15. Install the upper attaching bolt and torque to 40 ft. lbs. (54 Nm).

16. Install the TCM to its original location and install the mounting screws.

17. Install the air cleaner resonator.

18. Reconnect the negative battery cable.

2.5L Engine

A Melco starter unit is used.

1. Disconnect the negative battery cable from remote negative terminal located on the left strut tower.

NOTE: The ground cable is equipped with a insulator grommet which should be placed on the stud to prevent the negative battery cable from accidentally grounding.

2. Raise and safely support vehicle.

3. Remove the oil filter. Place a drain pan under the oil filter to prevent small oil spillage.

4. Remove the positive battery cable retaining nut and battery cable from the starter.

5. Disconnect the push-on solenoid connector.

6. Remove the 3 bolts that attach the starter unit to the transaxle.

7. Remove the starter unit from the vehicle.

To install:

8. Install the starter unit into the vehicle onto the transaxle and install the 3 mounting bolts.

9. Torque the mounting bolts to 40 ft. lbs. (54 Nm).

NOTE: Before reconnecting any wiring to the starter solenoid, clean the wire terminals of any dirt or corrosion.

10. Reconnect the positive battery terminal and retaining nut to the starter solenoid post. Torque the retaining nut to 90 inch lbs. (10 Nm).

11. Reconnect the push-on solenoid connector.

12. Install the oil filter.

13. Lower the vehicle.

14. Reconnect the negative battery cable.

CHASSIS ELECTRICAL

Blower Motor

REMOVAL AND INSTALLATION

The blower motor is located on the right side of the heater housing.

1. Disconnect the remote negative battery cable connection located at the left strut tower.

2. Remove the lower right under panel silencer duct.

3. Disconnect the blower motor wiring connector from the resistor block.

4. Remove the blower motor case attaching screws.

5. Remove blower motor case from heater housing.

6. Remove the fan scroll from the blower motor shaft.

7. Remove the blower motor unit from the blower motor case.

To install:

8. Install the blower motor unit into the blower motor case and install the fan scroll onto the shaft of the blower motor.

9. Install the blower motor assembly into the heater housing.

10. Install the blower motor attaching screws.

11. Reconnect the blower motor wiring connector to the resistor block.

12. Reinstall the lower right under panel silencer duct.

13. Reconnect the remote negative battery cable and test blower motor for proper operation.

Windshield Wiper Motor

REMOVAL AND INSTALLATION

The windshield wipers are controlled by multi-function switch. All wiper and washer functions are contained in this switch. The wipers operate only when the ignition switch is in the **ACCESSORY** or **IGNITION** positions. Fuse 15, located in the junction block and fuses 8 and 14 in the Power Distribution Center block, protect the wiper/washer system. The wiper motor also has an internal non-serviceable circuit breaker to provide protection against motor stall conditions.

The wiper and washer system switch located on the steering column provides inputs to the Body Control Module (BCM) which in turn operates 2 relays. One relay turns the wipers ON/OFF and the other changes the HIGH/LOW speeds. Intermittent delay functions are also controlled by the BCM.

The wiper motor and linkage assembly is referred to as the wiper motor and linkage module.

1. Disconnect the negative battery cable at the remote terminal on the shock tower. Be sure the wiper arm/blades are in the **PARK** position.

2. Lift the cover at the base of the wiper arm and remove the wiper arm securing nut.

3. Remove both wiper arms by gently rocking them from side to side until they slide off the pivot.

4. Remove the cowl screen.

5. Remove the 4 wiper motor mounting screws.

6. Lift the wiper motor to gain access to the harness clip.

7. Disconnect the harness clip and wiring harness.

8. Remove the wiper motor and linkage assembly module.

9. To remove the wiper motor from the linkage assembly, disconnect the drive linkage from the motor output crank. Carefully separate the ball cap from the ball.

To install:

10. Connect the linkage to the wiper motor. Align the link ball cap over the ball and gently press fit against the shoulder of the cap to lock the cap into position.

11. If removed, install the wiper motor output crank nut and torque to 17 ft. lbs. (22 Nm). Torque the wiper motor mounting nuts to 18 ft. lbs. (25 Nm).

12. Install the wiper motor assembly into the vehicle.

13. Reconnect the wiring harness and the harness clip. Make sure the motor connector seal is properly positioned.

14. Install the 4 wiper motor mounting screws and torque to 89–106 inch lbs. (10–12 Nm).

15. Reinstall the cowl screen.

16. Install the wiper arm/blade assemblies as follows:

   a. Install the wiper arm onto the pivot shaft, lining up the wiper arm to the keyway while pressing down on the arm to start on the pivot shaft.

   b. Start the wiper arm retainer nut.

   c. Raise the wiper arm/blade off windshield while tightening the retaining nut. Torque the retaining nut to 27–32 ft. lbs. (37–43 Nm).

   d. Install cover at base of wiper arm.

17. Connect the negative battery cable at the remote terminal on the shock tower.

18. Wet the windshield to prevent wiper blade drag and test the wiper system in all modes. With the wiper arm/blades in the **PARK** position, measure the distance from the tip of the wiper blade to the edge of the cowl screen. The distance measurement should be 0.75–1.60 in. (18–42mm).

Combination Switch

REMOVAL AND INSTALLATION

The multi-function switch (often also called the combination switch) is mounted center over the steering column. There are 2 levers, one on each side of the steering column. The left side controls the signaling and lighting. The right side controls the windshield wiper and washer system. An intermediate detent in the turn signal switch allows the usual "lane-change" feature, the turn signal switch returning to the OFF position as soon as the lever is released. In addition, hazard warning, windshield wiper and wash system, exterior lamp control and instrument panel dimmer functions are all built into this switch.

Should any function of the switch fail, the entire switch assembly must be replaced.

--- **CAUTION** ---
The Supplemental Inflatable Restraint (SIR) system must be disarmed before working around the steering column. Failure to do so may cause accidental deployment of the air bag, resulting in unnecessary SIR system repairs and/or personal injury.

1. Disconnect and isolate the battery negative remote cable before beginning any work around the steering column. This will disable the air bag system. Allow the system capacitor to discharge for 2 full minutes before working around the steering column.

2. Remove the upper steering column cover as follows:

   a. Remove the steering column lower cover retaining screws.

   b. Loosen the lower section of the instrument cluster hood for clearness as necessary.

   c. Remove the steering column upper cover.

3. Remove the combination switch mounting screws.

4. Disconnect the wiring connectors.

5. Lift the switch upwards to remove.

To install:

NOTE: The turn signal flasher and the hazard warning flasher are combined into one unit called a combination flasher (combo-flasher). An inoperative or incomplete turn signal circuit will result in an increase in flasher

speed. The flasher is mounted to the back of the multi-function (combination) switch. The flasher is serviced separately from the switch. The flasher is black in color for ease of identification.

6. Install the combination switch.

7. Reconnect the wiring connectors.

8. Install the combination switch mounting screws and torque to 20 inch lbs. (2.3 Nm).

9. Install the steering column cover as follows:

   a. Install upper cover onto steering column.

   b. Tighten the lower part of the instrument cluster hood.

   c. Install the lower steering column cover retaining screws and torque to 17 inch lbs. (2 Nm).

10. Reconnect the negative battery cable. Test switch functions.

Ignition Lock Cylinder

REMOVAL AND INSTALLATION

--- **CAUTION** ---
The Supplemental Inflatable Restraint (SIR) system must be disarmed before working around the steering column. Failure to do so may cause accidental deployment of the air bag, resulting in unnecessary SIR system repairs and/or personal injury.

1. Disconnect the negative battery cable. Disable the air bag system.

2. Remove the upper steering column shroud.

3. Pull down lower steering column shroud enough to access lock cylinder retaining tab.

4. Turn the ignition key to the **RUN** position.

5. Insert a small prybar into the tab access hole and depress the tab.

6. Pull the ignition lock cylinder from the steering column.

To install:

7. With the ignition key in the ignition lock cylinder, turn the key to the **RUN** position. Depress the lock cylinder retaining tab.

8. The shaft at the end of the ignition lock cylinder lines up to the socket at the end of the lock cylinder housing. The socket must be in the **RUN** position for the socket and lock cylinder to line up.

9. Line up the lock cylinder to the grooves in the lock cylinder housing. Insert the ignition lock cylinder into the housing until the retaining tab

sticks through the opening in the housing.

10. Lightly pull on the ignition lock to make sure it engaged the retaining tab.

11. Turn the key to the **OFF** position. Remove the ignition key.

12. Install the steering column shrouds.

13. Reconnect the negative battery cable.

Ignition Switch

REMOVAL AND INSTALLATION

1995 Vehicles

——————— CAUTION ———————
The Supplemental Inflatable Restraint (SIR) system must be disarmed before working around the steering column. Failure to do so may cause accidental deployment of the air bag, resulting in unnecessary SIR system repairs and/or personal injury.

1. Disarm the air bag system. Disconnect the negative battery remote cable connection located on the left strut tower.

2. Remove the center bezel and the knee bolster.

3. Remove the lower steering column shroud.

4. Tilt the steering wheel to the full down position.

5. Remove the upper shroud.

6. Remove the screws attaching the multifunction switch to the ignition switch.

7. Turn the key to the **RUN** position.

8. Depress the lock cylinder retaining tab and remove the key cylinder.

9. Disconnect the electrical connectors from the ignition switch.

10. Remove the ignition switch mounting screws.

11. Depress the retaining tabs and remove the ignition switch.

   To install:

12. Install the lock cylinder and the retaining screws.

13. Connect the electrical connections to the ignition switch.

14. Install the multifunction switch attaching screws.

15. Install the upper and lower shrouds.

16. Install the center bezel and knee bolster.

17. Reconnect the negative battery cable and enable the air bag system.

1996–97 Vehicles

——————— CAUTION ———————
The Supplemental Inflatable Restraint (SIR) system must be disarmed before working around the steering column. Failure to do so may cause accidental deployment of the air bag, resulting in unnecessary SIR system repairs and/or personal injury.

1. Disarm the air bag system. Disconnect the negative battery remote cable located on the left strut tower.

2. Remove the left end instrument panel cover/fuse panel and remove the retaining screw holding the end of the instrument panel top cover.

3. Remove the instrument panel center bezel.

4. Remove the screws that secure the instrument panel top cover to the center of the instrument panel.

5. Lift the instrument panel top cover enough to gain access to the knee bolster attaching screws.

6. Remove the lower knee bolster attaching screws and knee bolster from the vehicle.

7. Remove the lower steering column shroud attaching screws. Pull down on the lower shroud to clear the ignition key cylinder and key release (if equipped).

8. Remove the lower steering column shroud by sliding the lower shroud forward while holding down the steering wheel tilt lever.

9. Tilt the steering wheel down to the fully lowered position and remove the upper steering column shroud.

10. Remove the screws that hold the combination switch to the ignition lock housing.

11. Place the ignition into the **RUN** position.

12. Insert a small prybar into the hole in the lower shroud and depress the cylinder release tab.

13. Remove the ignition lock cylinder from the steering column.

14. Disconnect the electrical connectors from the switch.

15. Using a No. 10 Torx® bit, remove the ignition switch mounting screw.

16. Depress the retaining tabs and remove the switch.

   To install:

17. Place the ignition switch in the **RUN** position. Be sure the actuator shaft in the lock housing is also in the **RUN** position.

18. Install the switch making sure the switch snaps over the retaining tabs.

19. Install the retaining screw.

20. Reconnect the wiring to the ignition switch.

21. Install the ignition lock cylinder.

22. Install the 2 combination switch mounting screws.

23. Install the lower and upper shrouds.

24. Install the knee bolster to the lower dash panel and tighten the lower knee bolster attaching screws.

25. Install the screws that secure the instrument panel top cover to the center of the instrument panel.

26. Install the instrument panel center bezel.

27. Install and tighten the screw holding the end of the instrument panel top cover. Install the left side instrument panel/fuse panel cover.

28. Reconnect the negative battery cable. Check for proper ignition switch and key-in warning switch operation.

Park/Neutral Safety Switch

REMOVAL AND INSTALLATION

1995 Vehicles

The transaxle range and Park/Neutral position switches send signals to the Transmission Control Module (TCM) on the position of the transaxle manual lever valve. The TCM receives the switch signals and processes the data. The TCM sends the Shift Lever Position (SLP) information to the BCM via the CCD bus. The BCM then illuminates the appropriate shifter position indicator in the instrument cluster.

1. Disconnect and isolate the negative battery cable at the left strut tower.

2. Raise and safely support the vehicle.

NOTE: The Park/Neutral position switch is a black switch located on the transaxle to the right of the range switch. Do not confuse the 2 switches. The transaxle range switch has a white electrical connector and is located on the front of the transaxle just above the oil pan. The Park/Neutral position switch has a black electrical connector.

3. Disconnect the switch wiring.

4. Loosen and remove the switch.

To install:

5. Install the switch and tighten.

NOTE: Switch seal washer must be seated properly before switch installation. Failure to do so may result in transaxle fluid leakage.

6. Reconnect the switch wiring.
7. Lower the vehicle.
8. Reconnect the negative battery cable and check for proper operation.

1996–97 Vehicles

This vehicle does not come equipped with a Park/Neutral switch or Manual Valve Lever Position Sensor. Instead, the automatic transaxle comes equipped with a Transaxle Range Sensor, which is located on top of the valve body. This sensor will provide accurate transaxle gear position measurement. This sensor is very similar to the MVLPS that is currently on the 42LE transaxle.

The Transaxle Range Sensor, if defective, must be removed from the transaxle as an assembly. The valve body must be removed to service the sensor. The Transaxle Range Sensor is mounted on the top side of the valve body. It can only be removed from the valve body after the valve body is removed from the vehicle. The valve body can be removed with the transaxle remaining in the vehicle or with the transaxle removed; the sensor is a **non-adjustable** component.

1. Disconnect the negative battery cable at the left strut tower.
2. Remove the air cleaner assembly.
3. Disconnect the gear shift cable.
4. Remove the manual valve lever.
5. Disconnect the transaxle range sensor connector.
6. Carefully raise and safely support the vehicle.
7. Place a drain pan, with a large opening, under the transaxle oil pan. Loosen the transaxle oil pan mounting bolts and tap the oil pan at one corner to break it loose allowing the fluid to drain. After the fluid has drained, remove the transaxle oil pan.
8. Remove the transaxle oil filter while allowing the residual transaxle fluid to fully drain.
9. Remove the mounting bolts for the valve body.
10. Separate the park rod from the guide bracket and remove the valve body assembly from the transaxle.
11. Place the valve body assembly on a workbench.
12. Remove the Transaxle Range Sensor (TRS) attaching screw.

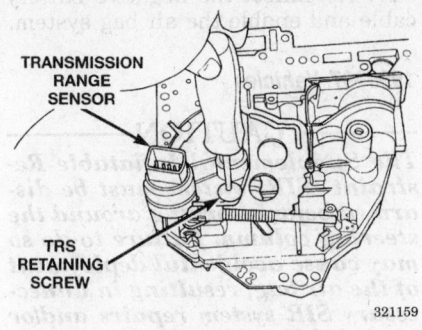

TRANSMISSION RANGE SENSOR

TRS RETAINING SCREW

321159

Removal of retaining screw — 1996–97 vehicles

13. Remove the manual shaft seal and slide the TRS up the manual shaft to remove from the valve body.

To install:

14. Install the TRS by sliding it down onto the manual shaft.
15. Install the manual shaft seal halfway down onto the manual shaft and seat it in the shaft seal groove.
16. Install and tighten the TRS retaining screw. Torque the retaining screw to 45 inch lbs. (5 Nm).
17. Install the valve body assembly up into the transaxle. Engage the park rod into the guide bracket.
18. Install and tighten the valve body mounting bolts. Torque the mounting bolts to 105 inch lbs. (12 Nm).
19. Install a new transaxle oil filter and O-ring.
20. Before installing the transaxle oil pan, be sure to thoroughly clean the gasket mating surfaces of the transaxle case and transaxle oil pan as well as the pan magnet. Then, place a light bead of RTV sealer on the oil pan gasket surface. Properly position the new pan gasket on top of the pan gasket mating surface.
21. Position the transaxle oil pan onto the transaxle case and install the pan mounting bolts. Torque the oil pan mounting bolts to 165 inch lbs. (19 Nm).
22. Lower the vehicle.
23. Reconnect the Transaxle Range Sensor connector.
24. Install the manual valve lever and reconnect the gear shift cable.
25. Install the air cleaner assembly.
26. Pour 4 quarts of MOPAR® ATF PLUS Type 7176 or equivalent into the transaxle filler tube.
27. Reconnect the negative battery cable.
28. Start the engine and allow it to idle for at least one minute. Apply both parking and service brakes. Move the gear shift selector momentarily through each gear position, ending up in the **P** or **N** position.

29. Check the fluid level, while the engine is running, and if necessary add sufficient fluid to bring the fluid level to 1/8 in. below the "ADD" mark on the dipstick. At normal operating temperature, the level should be in the "HOT" region of the dipstick.
30. Road test the vehicle.

Powertrain Control Module

REMOVAL AND INSTALLATION

The Powertrain Control Module (PCM) also known as the Electronic Control Module, regulates the ignition and fuel injection systems. The unit mounts underhood attached to a bracket between the air cleaner housing and the Power Distribution Center (PDC). It is easily identified by the 2 large (40-pin) harness connectors. Use care when handling the PCM and its connectors to avoid damage.

1. Disconnect the negative battery cable remote connection located on the left strut tower.
2. Disconnect the 2 40-pin connectors from the PCM.
3. Remove the 2 PCM retaining screws.
4. Lift the PCM and remove from the vehicle.

To install:

5. Install the PCM into the vehicle.
6. Install the 2 attaching screws.
7. Reconnect the two 40 pin connectors.
8. Reconnect the negative battery remote cable.

ENGINE COOLING

Radiator

REMOVAL AND INSTALLATION

———— CAUTION ————

Do not open the radiator draincock or remove the radiator cap when the cooling system is hot and under pressure. This can cause serious burns from hot, pressurized coolant. Allow a sufficient amount of time for the cooling system to cool down before opening up the system.

The radiator uses plastic tanks. Plastic tanks, while stronger than brass, are subject to damage by impact, such as slipped wrenches. Use care when working around these radiators.

One type of corrosion encountered with aluminum cylinder heads is aluminum hydroxide deposits. Corrosion products are carried to the radiator and deposited when cooled off. They appear as dark grey when wet and white when dry. This corrosion may be removed with a two-part cleaner (oxalic acid and neutralizer) available at auto parts outlets. Follow the manufacturer's directions for use.

1. Disconnect the negative battery cable end at the remote terminal on the left shock tower.

NOTE: The ground cable is equipped with a insulator grommet which should be placed on the stud to prevent the negative battery cable from accidentally grounding.

2. Remove the air inlet resonator.

3. Place a large drain pan under the radiator drain plug. Drain and properly contain engine coolant.

4. Remove the upper radiator crossmember as follows:

   a. Remove the push-in mounting fasteners securing the front fascia/grille unit to the radiator support crossmember.

   b. Remove the mounting bolts securing the support braces to the bottom of the crossmember.

   c. Remove the bolts securing the crossmember to the radiator closure panel.

   d. Remove the mounting nuts attaching the hood latch to the radiator crossmember. Remove the crossmember from the vehicle.

5. Disconnect the hoses from the radiator.

6. Disconnect the engine block heater wire, if equipped.

7. Disconnect and plug the transaxle cooler lines, if equipped.

8. Unplug the cooling fan wiring.

9. Remove the air conditioning condenser mounting screws. Use care when working around the air conditioning condenser. Avoid bending the condenser inlet tube. Care should be taken not to damage radiator or condenser cooling fins or water tubes during removal. It is not necessary to discharge the air conditioning system to remove the radiator.

10. Carefully remove the radiator from the vehicle. The cooling fan/shroud assembly can be separated from the radiator at this time.

To install:

11. If separated, install the cooling fan/shroud assembly to the radiator unit.

12. Lower the radiator and fan module (assembly) into position. Seat the radiator assembly lower isolators in the mount holes provided.

13. Install the air conditioning condenser mounting screws and torque to 45 inch lbs. (5 Nm).

14. Connect the radiator hoses and tighten hose clamps. Torque the hose clamps to 22 inch lbs. (2.5 Nm). Be sure the hoses do not interfere with the accessory drive belt. Be sure the upper hose clamp does not interfere with the hood liner.

15. Connect the cooling fan wiring.

16. Connect the transaxle cooler lines to the radiator, if equipped.

17. Install the upper radiator crossmember as follows:

   a. Place the radiator crossmember into the vehicle in proper position.

   b. Install the hood latch to the radiator crossmember. Install and tighten the hood latch mounting nuts.

   c. Install and tighten the crossmember-to-radiator closure panel mounting bolts.

   d. Install and tighten the mounting bolts that secure the support braces to bottom of the radiator crossmember.

   e. Install the push-in mounting fasteners holding the front fascia/grille unit to the radiator crossmember.

18. Connect the engine block heater, if equipped.

19. Install the air inlet resonator.

20. Fill the cooling system with a 50/50 mix of clean water and Mopar Antifreeze, Prestone II, Peak or other approved antifreeze containing Alugard 340-2 or their equivalents to protect the aluminum components in the engine and cooling system that come into contact with coolant.

21. Connect the negative battery cable. Start the engine and allow it to idle until it reaches full operating temperature. Check the cooling system for correct fluid level and top off if necessary.

Water Pump

REMOVAL AND INSTALLATION

2.0L and 2.4L Engines

This engine uses a die-cast aluminum body water pump with a stamped steel impeller. The water pump bolts directly to the block. The cylinder block to water pump sealing is provided by a large rubber O-ring. The water pump is driven by the timing belt which must be removed to service the water pump.

1. Disconnect the negative battery cable remote connection located on the left strut tower.

NOTE: This procedure requires removing the engine timing belt and the auto tensioner. The factory specifies that the timing marks should always be aligned before removing the timing belt. Set the engine at TDC on No. 1 compression stroke. This should align all timing marks on the crankshaft sprocket and both camshaft sprockets.

2. Raise and safely support the vehicle.

3. Remove the right inner splash shield.

4. Remove the accessory drive belts.

5. Place a drain pan under the radiator drain plug. Drain and properly contain the cooling system.

6. Support the engine and remove the right motor mount.

7. Remove the power steering pump mounting bracket bolts and place the pump/bracket assembly off to one side. Do not disconnect the power steering fluid lines.

8. Remove the right engine mount bracket.

9. Remove the front timing belt upper and lower covers.

10. Loosen the timing belt tensioner screws and remove the belt tensioner and timing belt.

————— **WARNING** —————
With the timing belt removed, DO NOT rotate the camshaft or crankshaft or damage to the engine could occur.

11. Remove the camshaft sprockets. With the timing belt removed, remove both camshaft sprocket bolts. Do not allow the camshafts to turn when the camshaft sprockets are being removed.

12. Remove the rear timing belt cover to access the water pump.

13. Remove the water pump attaching bolts.

14. Remove the water pump.

To install:

15. Clean all parts well. Replace the water pump if there are any cracks, signs of coolant leakage from the shaft seal, loose or rough tuning bearing, damaged impeller or

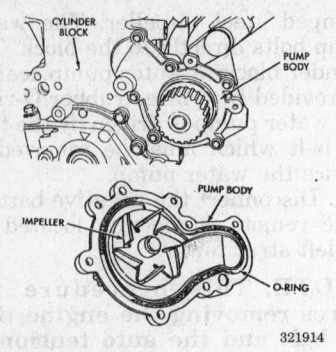

Water pump assembly — 2.0L and 2.4L engines

sprocket or sprocket flange loose or damaged.

16. Clean all sealing surfaces. Install a new rubber O-ring into the water pump.

NOTE: Make sure the O-ring is properly seated in the water pump groove before tightening the screws. An improperly located O-ring may cause damage to the O-ring and cause a coolant leak.

17. Install the water pump and torque the bolts to 105 inch lbs. (12 Nm).

18. Pressurize the cooling system to 15 psi and check for leaks. If okay, release the pressure and continue the engine assembly process.

19. Install the rear timing belt cover.

20. Install the camshaft sprockets and tighten the attaching bolts to 75 ft. lbs. (101 Nm). DO NOT allow the camshafts to turn while the sprockets bolts are being tighten to maintain timing mark alignment.

**———— WARNING ————
Do not attempt to compress the tensioner plunger with the tensioner assembly installed in the engine. This will cause damage to the tensioner and other related components. The tensioner MUST be compressed in a vise.**

21. Install the timing belt tensioner and timing belt. Be sure to properly tension the timing belt.

22. Install the front upper and lower timing belt covers.

23. Install the right engine mount bracket and engine mount.

24. Install the crankshaft damper and torque the center bolt to 105 ft. lbs. (142 Nm).

25. Install the right inner splash shield.

26. Lower the vehicle.

27. Install the power steering pump bracket and power steering pump.

Torque the bracket mounting bolts to 40 ft. lbs. (54 Nm).

28. Install the drive belts. Properly tension the drive belts.

29. Refill the cooling system using a mixture of 50/50 water and ethylene glycol antifreeze. Bleed the cooling system.

30. Start the engine and check for proper operation. Perform relearn camshaft and crankshaft alignment procedure in the Engine Miscellaneous Section of the DRB Scan Tool (or equivalent).

31. Check and top off cooling system, if necessary.

2.5L Engine

The water pump bolts directly to the engine block using a gasket for pump-to-block sealing. The pump is serviced as a unit. The 2.5L engine uses metal piping beyond the lower radiator hose to route coolant to the suction side of the water pump, located in the "V" of the cylinder banks. These pipes also have connections for thermostat bypass and heater return coolant hoses. The pipes use O-rings for sealing.

The water pump is driven by the timing belt which must be removed to service the water pump. Timing belt covers must be removed to access the timing belt.

1. Disconnect the negative battery cable remote connection located on the left strut tower.

2. Place a large drain pan under the radiator drain plug. Drain and properly contain the engine coolant.

NOTE: This procedure requires removing the engine timing belt and the auto tensioner. To help assure proper alignment at assembly, it may be helpful to set the engine at TDC on No. 1 compression stroke. This should align all timing marks on the crankshaft sprocket and both camshaft sprockets.

3. Remove the accessory drive belts and crankshaft damper.

4. Remove the right engine mount. This requires safely supporting the engine so the mount can be removed.

5. Remove the timing belt covers in this order: upper left cover, upper right cover and last, the lower cover.

6. Remove the timing belt and tensioner.

7. Remove the water pump mounting bolts.

8. Separate the water pump from the water inlet pipe and remove the pump.

To install:

9. Inspect the pump for damage or cracks, signs of coolant leakage at the vent and excessive looseness or rough turning bearing. Any problems require a new pump.

10. Clean all gasket and O-ring surfaces on the block, water pump and water inlet pipe. Install a new O-ring on the water inlet pipe. Wet the O-ring with water to make installation easier. DO NOT use oil or grease on the O-ring.

11. Install a new gasket on the water pump and fit the pump inlet opening over the water pipe. Press the assembly together to force the pipe into the water pump.

12. Install the water pump to engine bolts and torque to 20 ft. lbs. (27 Nm).

13. Install the timing belt and timing belt tensioner. Set the timing belt tension.

14. Install the timing belt covers. Install the right engine mount.

15. Install the crankshaft damper.

16. Install the accessory drive belts and set to proper tension.

17. Reconnect the negative battery cable remote connection.

18. Fill and bleed the engine cooling system.

19. Start the engine and verify proper operation.

Thermostat

REMOVAL AND INSTALLATION

2.0L and 2.4L Engines

1. Disconnect the negative battery cable at the remote terminal at the left shock tower.

2. Place a large drain pan under the radiator drain plug. Allow the cooling system to sufficiently cool down before opening the drain plug to avoid personal injury. Drain the coolant to below the thermostat level.

3. Disconnect the coolant recovery hose and radiator hose.

4. Remove the thermostat housing bolts.

5. Remove the thermostat assembly from the vehicle and discard.

To install:

6. Clean all gasket surfaces.

7. Install the thermostat and align the air bleed with the notch on the cylinder head. Install the thermostat housing using a new gasket.

8. Install the thermostat housing bolts and torque to 110 inch lbs. (12.5 Nm).

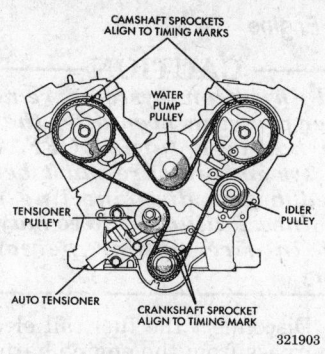

Timing belt routing and timing — 2.5L engine

9. Reconnect the coolant recovery hose and radiator hose. Tighten the radiator hose clamp.

10. Connect the negative battery cable to the remote terminal at the shock tower.

11. Fill and bleed the engine cooling system.

12. Pressure test for leaks.

2.5L Engine

1. Disconnect the negative battery cable at the remote terminal at the left shock tower.

2. Place a large drain pan under the radiator drain plug. Allow the

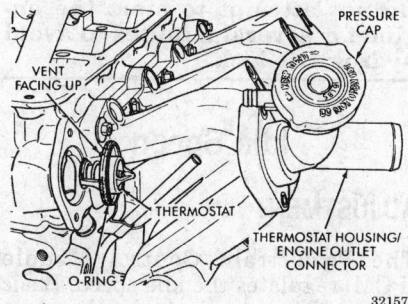

Thermostat and engine outlet connector — 2.0L engine

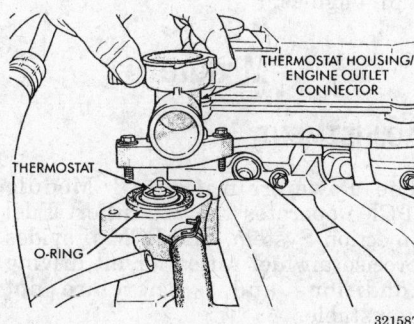

Thermostat and engine outlet connector — 2.4L engine

cooling system to sufficiently cool down before opening the drain plug to avoid personal injury. Drain the coolant to below the thermostat level.

3. Remove the inlet radiator hose hose and coolant elbow from the thermostat housing.

4. Remove the thermostat housing bolts.

5. Remove the thermostat assembly from the vehicle and discard.

To install:

6. Clean all gasket surfaces.

7. Install the thermostat into the recess of the thermostat housing. Be sure to install the new thermostat with the bleed vent hole positioned upward.

8. Install the thermostat housing using a new gasket. Install the thermostat housing bolts and torque to 133 inch lbs. (13 Nm).

9. Connect the inlet radiator hose to the thermostat housing and tighten the radiator hose clamp.

10. Connect the negative battery cable to the remote terminal on the shock tower.

11. Fill and bleed the engine cooling system.

12. Pressure test for leaks.

Electric Cooling Fan

REMOVAL AND INSTALLATION

1. Disconnect the negative battery cable from the left strut tower. The ground cable is equipped with a insulator grommet which should be placed on the stud to prevent the negative battery cable from accidentally grounding.

2. Disconnect the cooling fan electrical connections.

3. Remove the 4 cooling fan/shroud assembly mounting bolts.

4. Remove the cooling fan/shroud assembly.

5. To remove the fan blade from the fan motor on the shroud assem-

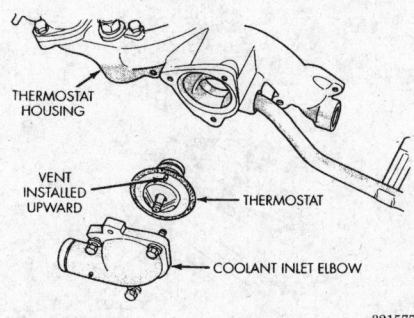

Thermostat, housing and inlet elbow — 2.5L engine

bly, first support the motor on a bench then remove the fan retaining clip from the motor shaft. Slide the fan off the motor shaft.

6. To remove the fan motor from the shroud unit, remove the fan mounting screws and remove the fan motor from the fan shroud unit.

To install:

7. Install the fan motor onto the fan shroud assembly. Install and tighten the mounting screws.

8. Install the fan blade onto the fan motor shaft and install the retaining clip.

9. Install the cooling fan/shroud assembly into the vehicle.

10. Install the cooling fan mounting bolts.

11. Torque the fan mounting bolts to 65 inch lbs. (7.5 Nm).

12. Connect the fan electrical connections.

13. Reconnect the negative battery cable to the left strut tower.

Cooling System Bleeding

Do not use straight antifreeze as engine coolant. Inadequate engine running temperatures can result. Do not operate the vehicle without the proper concentration (50 antifreeze and 50 water) of recommended ethylene glycol coolant or high temperatures and cooling system corrosion can result. The use of aluminum cylinder heads, intake manifolds and water pumps requires special corrosion protection. Chrysler Corporation recommends their Mopar Antifreeze, Prestone II, Peak or antifreeze containing Alugard 340-2 or their equivalent to protect the aluminum parts in these engines.

Coolant level should be checked whenever other engine compartment service is performed or when a coolant leak is suspected. Coolant recovery tank level should read between the ADD and FULL marks, located on the side of the the recovery tank when the engine is at normal operating temperature. Cooling system freeze protection should be tested at the onset of the winter season or every 12 months. Service is required if coolant is low, contaminated, rusty or if freeze protection is inadequate.

The following procedure is to be used whenever any repair work is done to the engine cooling system.

1. Fill the cooling system to the filler neck with a 50/50 solution of antifreeze and water.

2. Start the engine with the radiator cap off and run until engine reaches full operating temperature.

3. Add coolant as necessary to maintain the level.

NOTE: Air can only be bled from the cooling system by gathering beneath the radiator pressure cap. On the next engine heat up, the air will be pushed past the pressure cap into the coolant overflow tank due to thermal expansion of the coolant. It then escapes into the atmosphere and is replaced with solid coolant upon engine cool down.

4. Install the radiator cap.

FUEL SYSTEM

Fuel System Service Precautions

Safety is the most important factor when performing not only fuel system maintenance but any type of maintenance. Failure to conduct maintenance and repairs in a safe manner may result in serious personal injury or death. Maintenance and testing of the vehicle's fuel system components can be accomplished safely and effectively by adhering to the following rules and guidelines.

• To avoid the possibility of fire and personal injury, always disconnect the negative battery cable unless the repair or test procedure requires that battery voltage be applied.

• Always relieve the fuel system pressure prior to disconnecting any fuel system component (injector, fuel rail, pressure regulator, etc.), fitting or fuel line connection. Exercise extreme caution whenever relieving fuel system pressure to avoid exposing skin, face and eyes to fuel spray. Please be advised that fuel under pressure may penetrate the skin or any part of the body that it contacts.

• Always place a shop towel or cloth around the fitting or connection prior to loosening to absorb any excess fuel due to spillage. Ensure that all fuel spillage (should it occur) is quickly removed from engine surfaces. Ensure that all fuel soaked cloths or towels are deposited into a suitable waste container.

• Always keep a dry chemical (Class B) fire extinguisher near the work area.

• Do not allow fuel spray or fuel vapors to come into contact with a spark or open flame.

• Always use a backup wrench when loosening and tightening fuel line connection fittings. This will prevent unnecessary stress and torsion to fuel line piping. Always follow the proper torque specifications.

• Always replace worn fuel fitting O-rings with new. Do not substitute fuel hose or equivalent, where fuel pipe is installed.

Fuel System Pressure

RELIEVING

2.0L and 2.4L Engines

— **CAUTION** —

Fuel injection systems remain under pressure, even after the engine has been turned OFF. The fuel system pressure must be relieved before disconnecting any fuel lines. Failure to do so may result in fire and/or personal injury.

1. Disconnect the negative battery cable from the remote connection located on the left strut tower.
2. Remove the fuel filler cap.
3. Remove the cap on the fuel pressure test port on the fuel rail.
4. Place the open end of fuel pressure release hose special tool number C-4799-1 or equivalent into an approved gasoline container. Connect the other end of hose C-4799-1 or equivalent to the fuel pressure test port. Fuel pressure will bleed off through the hose onto the gasoline container.

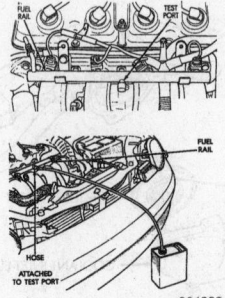

Fuel pressure test port location — 2.0L and 2.4L engines

2.5L Engine

— **CAUTION** —

Fuel injection systems remain under pressure, even after the engine has been turned OFF. The fuel system pressure must be relieved before disconnecting any fuel lines. Failure to do so may result in fire and/or personal injury.

1. Disconnect the fuel rail electrical harness from the engine harness. This is connector C165, a black plastic connector located at the right rear of the intake manifold.
2. Circuit A142 supplies voltage for the fuel injectors while the Powertrain Control Module (PCM) controls the ground for each injector. Connect a jumper wire to the terminal for Circuit A142 (18 ga. wire, Dark Green with Orange tracer, from ASD relay).
3. Connect the other end of the jumper wire to a 12 volt power source.
4. Connect one end of a second jumper wire to a ground source.
5. Momentarily ground each of the injectors by connecting the other end of the jumper wire to the injector terminal in the harness connector. Repeat this procedure for 2 or 3 injectors.

— **WARNING** —

Do not attempt to start the engine for several minutes to avoid hydrostatic lock.

Idle Speed

ADJUSTMENT

The Powertrain Control Module (PCM) regulates the idle speed. Basic idle speed is not adjustable. However, the minimum air flow idle rpm specification is 600–1300 rpm for 2.0L and 2.4L engines, or 500–1100 rpm for 2.5L engines.

Mixture

ADJUSTMENT

The Powertrain Control Module (PCM) operates the Multi-Port Fuel Injection System. The PCM provides precise air/fuel ratios for all driving conditions and is therefore not adjustable.

Under most driving conditions, the PCM maintains an air/fuel ratio of 14.7:1 by constantly adjusting the fuel injector pulse width. There are no service adjustments that can be made to alter the air/fuel mixture. It is regulated by the PCM programming.

Fuel Filter

REMOVAL AND INSTALLATION

The fuel delivery system contains a replaceable in-line filter. The fuel filter mounts to the frame above the rear of the fuel tank. The fuel tank assembly must be loosened and lowered slightly to access the filter. The inlet and outlet tubes are permanently attached to the filter. Please note that the fuel system pressure must be relieved before servicing fuel system components. In addition, Chrysler uses Quick Connect fittings on fuel line connections. Specific instructions are given at the end of this procedure.

— **CAUTION** —

Fuel injection systems remain under pressure, even after the engine has been turned OFF. The fuel system pressure must be relieved before disconnecting any fuel lines. Failure to do so may result in fire and/or personal injury.

1. Disconnect the negative battery cable from the remote auxiliary jumper terminal at the left strut tower.
2. Release the fuel system pressure using the recommended procedure.
3. From inside the trunk, disconnect the fuel pump module wiring jumper from the main body harness. The 4-pin connector is located under the trunk mat on the left side of the trunk near the base of the shock tower. Locate the body grommet for the jumper near the base of the rear seat. Push the grommet out and feed the jumper completely through the hole in the body.
4. Remove the fuel cap slowly to release tank pressure.
5. Raise and safely support the vehicle.
6. Locate the drain plug on the bottom left of the fuel tank. Place a fuel approved container with a capacity of at least 16 gallons, under the drain plug. Remove the plug and drain the fuel tank. When finished draining, install the plug since there will be one to 2 gallons of fuel remaining. Tighten the drain plug to 32 inch lbs.

— **CAUTION** —

Observe all applicable safety precautions when working around fuel. Do not allow fuel spray or fuel vapors to come in contact with a spark or open flame. Keep a dry chemical (Class B) fire extinguisher near the work area. Never drain or store fuel in an open container due to the possibility of fire or explosion.

7. Remove the driver's side fuel tank strap. Loosen, but do not remove the passenger's side fuel tank strap allowing the fuel tank neck to touch the rear suspension crossmember.

— **CAUTION** —

Wrap shop towels around the fuel hoses to catch any gasoline spillage.

8. Disconnect the fuel lines from the fuel pump module. These are quick connect fittings. Depress the releasing tabs on the fuel filter hose connections and disconnect the fuel lines.
9. Remove the fuel filter.
To install:
10. The fuel supply (to filter) tube and the return tube (to fuel pump

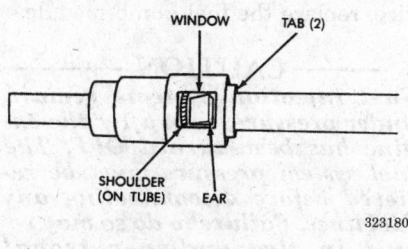

Fitting release clip

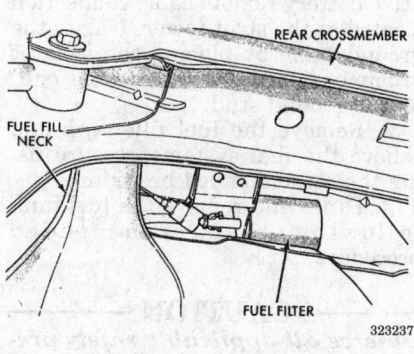

Remove fuel filter

module) are permanently attached to the fuel filter. The ends of the fuel supply and return tubes have different size quick connect fittings. The large quick connect fitting attaches to the large nipple (supply side) on the fuel pump module. The smaller quick connect fitting attaches the the small nipple (return side) on the fuel pump module. Specific Quick Connect fitting service procedures are given below.

11. Apply a light coat of clean 30W engine oil to the fuel filter nipples. Install the fuel tubes.
12. Install the fuel tank, filter and fuel tank straps. Torque the strap bolts to 250 inch lbs. (23 Nm).

NOTE: Make sure the fuel pump module electrical harness grommet is installed in the body as the tank is raised into position.

13. Lower the vehicle and connect the pump module connector.
14. Fill the tank with fuel.
15. Connect the negative battery cable to the auxiliary jumper terminal.

Quick Connect Fitting Service

Fuel line clamps are not used on any fuel system components. If there is a fitting or line that needs repair, it must be replaced. DO NOT use rubber hose and clamps to repair a fuel line. The fuel pump has a maximum deadhead pressure output of approximately 130 psi. The regulator adjusts fuel system pressure to approximately 49 psi. These high pressure systems must be respected and serviced only with approved parts.

Inspect all lines and quick connect fittings for leaks. Replace lines that rub against other vehicle components or show signs of wear. Make sure all lines are secured in their proper holders.

Fuel tubes connect fuel system components with plastic quick connect fuel fittings. The fitting contains non-serviceable O-ring seals. Note that, on this vehicle, quick connect fittings are not serviced separately. Do not attempt to repair damaged quick connect fittings or fuel tubes. Replace the complete fuel tube/quick connect fitting assembly.

The quick connect fitting consists of the O-rings, retainer and casing. When the fuel tube enters the fitting, the retainer locks the shoulder of the nipple in place and the O-rings seal the tube. When disconnecting a quick connect fitting, the retainer will remain on the fuel tube nipple.

CAUTION

Always release the fuel system pressure using the recommended procedure before attempting to disconnect quick connect fittings.

To service a quick connect fitting, use the following procedure.

1. Disconnect the negative battery cable from the auxiliary jumper terminal.

2. Perform the fuel pressure relief recommended procedure.

3. Squeeze the fitting retainer tabs together and pull the fuel tubes/quick connect fitting assembly off the fuel tube nipples. The retainer will remain on the fuel tube.

4. When installing, or connecting this type of quick connect fitting, first clean the fuel tube nipple and retainer using a clean, lint-free cloth.

5. Prior to connecting the fitting to the fuel tube, coat the fuel tube nipple with clean 30W engine oil.

CAUTION

Never install a quick connect fitting without the retainer being either on the fuel tube or already in the quick connect fitting. In either case, make sure the retainer locks securely into the quick connect fitting by firmly pulling on the fuel tube and fitting to ensure it is secured.

6. Push the quick connect fitting over the fuel tube until retainer seats and a click is heard.

7. Check the windows (openings) in the sides of the quick connect casing. When the fitting completely attaches to the fuel tube, the retainer locking ears and fuel tube shoulder are visible in the windows. If they are not visible, the retainer was not properly installed. DO NOT rely on the audible click to confirm a secure connection.

8. Make sure the locking ears on the retainer and shoulder (stop bead) on the fuel tube are completely visible in the quick connect fitting windows. DO NOT rely on the audible click to confirm a secure fitting connection. Always pull on the line and fitting to ensure that the retainer is seated.

9. Verify connection by pulling on the lines. If the fitting locks in place, the connection is secure.

10. The factory recommends using the DRB scan tool (or equivalent) Automatic Shutdown (ASD) Fuel System Test to pressurize the fuel system to check for leaks.

CAUTION

When using the ASD Fuel System Test, the Auto Shutdown (ASD) relay remains energized for either 7 minutes, until the test is stopped or until the ignition switch is turned to the OFF position.

Fuel Pump

REMOVAL AND INSTALLATION

The in-tank fuel pump module contains the fuel pump and pressure regulator which adjusts fuel system pressure to approximately 49 psi. Voltage to the fuel pump is supplied through the fuel pump relay.

The fuel pump is serviced as part of the fuel pump module. The fuel pump module is installed in the top of the fuel tank and contains the electric fuel pump, fuel pump reservoir, inlet strainer fuel gauge sending unit, fuel supply and return line connections and the pressure regulator. The inlet strainer, fuel pressure regulator and level sensor are the only serviceable items. If the fuel pump requires service, replace the fuel pump module.

CAUTION

Fuel injection systems remain under pressure, even after the engine has been turned OFF. The fuel system pressure must be relieved before disconnecting any fuel lines. Failure to do so may result in fire and/or personal injury.

1. Disconnect and isolate the negative battery remote cable connection located at the strut tower. Isolate the ground cable by placing the isolator grommet (located on the cable end) on the ground stud.

2. Remove the fuel filler cap and relieve the fuel system pressure using the recommended procedure.

3. Drain and remove the fuel tank following the recommended procedure.

CAUTION

Observe all applicable safety precautions when working around fuel. Do not allow fuel spray or

fuel vapors to come in contact with a spark or open flame. Keep a dry chemical (Class B) fire extinguisher near the work area. Never drain or store fuel in an open container due to the possibility of fire or explosion.

4. Clean the top of the tank to remove any loose dirt.

5. Disconnect fuel lines from the fuel pump module.

6. Using special tool 6856 or equivalent, remove the fuel pump locknut.

CAUTION

The fuel reservoir of the fuel pump module does not empty out when the tank is drained. The fuel in the reservoir may spill out when the module is removed.

7. Remove the fuel pump and O-ring from the tank. Discard the O-ring.

To install:

8. Clean all parts well. Wipe the seal area of the tank clean. Place a new O-ring on the ledge between the tank threads and the pump module opening.

9. Position the fuel pump module in the tank. Make sure the alignment tab on the underside of the pump module flange sits in the corresponding notch in the fuel tank.

10. While holding the fuel pump module in place install the locking ring and torque to 40–45 ft. lbs. (54–61 Nm) using special tool 6856 or equivalent spanner-type tool.

11. Install the fuel tank assembly using the recommended procedure.

12. Reconnect the negative battery remote cable connection.

13. Fill up fuel tank with clean fuel. Turn the ignition switch to the **ON** position to pressurize the system. Check the fuel system for leaks.

Fuel Injector

REMOVAL AND INSTALLATION

2.0L and 2.4L Engines

1. Disconnect the negative battery remote cable located on the left strut tower.

2. Release the fuel system pressure using the recommended procedure.

— **CAUTION** —

Fuel injection systems remain under pressure, even after the engine has been turned OFF. The fuel system pressure must be relieved before disconnecting any fuel lines. Failure to do so may result in fire and/or personal injury.

3. Disconnect the fuel supply line from the fuel rail. This is a quick connect fitting. Squeeze the fitting retainer tabs together and pull the quick connect fitting assembly apart.

— **CAUTION** —

Wrap shop towels around the hose connection to catch any gasoline spillage.

4. Disconnect the fuel injector electrical connectors.
5. Remove the fuel rail attaching screws.
6. Lift the fuel rail and injectors off the intake manifold. Cover the fuel injector openings in the intake manifold.
7. Remove the fuel injector clip and pull the injector from the fuel rail. Note that whenever a fuel injector is removed, the O-rings must be replaced.

To install:
8. Install the injectors into the fuel rail and install the retaining clips.
9. Apply a light coating of clean engine oil to the O-ring on the nozzle end of each injector.
10. Insert the fuel injector nozzles into the openings in the intake manifold. Seat the injectors and tighten the fuel rail mounting screws to 170–230 inch lbs. (19.5–23.5 Nm).
11. Reconnect the electrical connectors to the fuel injectors.
12. Lightly oil the tube end, then reconnect the fuel supply line quick connect fitting to the fuel rail. Be

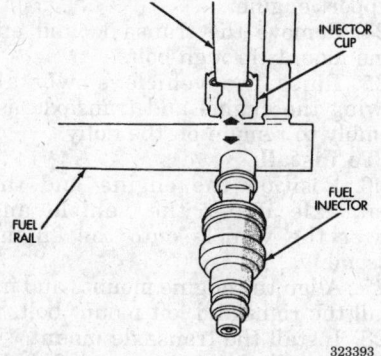

Fuel injector clip — 2.0L and 2.4L engines

sure the quick connect fittings are fully engaged.
13. Connect the negative battery remote cable to the left strut tower.
14. Start engine and check for fuel leaks.

2.5L Engine

The intake manifold assembly is composed of an upper plenum and lower manifold. This aluminum alloy manifold has long runners to improve airflow inertia. The plenum chamber absorbs air pulsations created during the suction phase of each cylinder. The lower intake manifold is machined for 6 injectors and the fuel rail mounts. The fuel injectors are mounted in 2 fuel rails which bolt to the lower half of the two-piece intake manifold assembly. The upper half (plenum) of the intake manifold must be removed to access the fuel rails for fuel injector removal.

1. Disconnect the negative battery remote cable located on the left strut tower.
2. Release the fuel system pressure.

— **CAUTION** —

Fuel injection systems remain under pressure, even after the engine has been turned OFF. The fuel system pressure must be relieved before disconnecting any fuel lines. Failure to do so may result in fire and/or personal injury.

3. Disconnect the fuel supply line from the fuel rail. This is a quick connect fitting. Squeeze the fitting retainer tabs together and separate the connection.

— **CAUTION** —

Wrap shop towels around the connection to catch any gasoline spillage.

NOTE: It may be helpful to identify and tag each sensor connector as it is being removed. This may save time at assembly.

4. Disconnect the connectors from the Manifold Absolute Pressure (MAP) sensor and the intake air temperature sensors.
5. Remove the plenum support bracket located to the rear of the MAP sensor.
6. Remove the air inlet resonator attaching bolt.
7. Loosen the throttle body air inlet hose clamp.
8. Release the snaps holding the air cleaner housing cover to the hous-

ing. Remove the air cleaner cover and inlet hoses from the engine.
9. Disconnect the Throttle Position Sensor (TPS) and the Idle Air Control (IAC) motor electrical connections.
10. Squeeze the retainer tab on the throttle cable and slide the cable out of the bracket.
11. If equipped with speed control, slide the speed control cable out of the bracket.
12. Remove the EGR tube from the engine.
13. Remove the plenum support bracket located to the rear of the EGR tube.
14. Remove the 7 bolts attaching the upper intake plenum to the intake manifold and remove the plenum.
15. Disconnect the fuel injector electrical connectors.
16. Remove the 4 bolts attaching the fuel rails to the lower intake manifold and lift the fuel rails off the engine. Use care. There are spacers under each fuel rail bolt.
17. Remove the fuel injector clip.
18. Pull the fuel injector out of the fuel rail.

To install:
19. Using new O-rings, apply a light coating of engine oil to the fuel injector O-rings on the nozzle end of each injector.
20. Install the fuel injectors into the fuel rail and secure with the fuel injector clips. Install the fuel injector/fuel rail assembly into the engine.
21. Seat the injectors in place making sure the spacers are properly located under each fuel rail mounting position and tighten the fuel rail bolts to 8 ft. lbs. (12 Nm).
22. Reconnect the electrical connectors to the fuel injectors.
23. Reconnect the fuel supply line to the fuel rail. Be sure the quick connect fittings are fully engaged.
24. Install the upper intake plenum with new gaskets.
25. Torque the plenum bolts to 13 ft. lbs. (18 Nm).
26. Install the plenum support brackets and torque to 13 ft. lbs. (18 Nm).
27. Install the EGR tube and torque the screws to 95 inch lbs. (11 Nm).
28. Reinstall the throttle cables.
29. Reconnect the TPS and IAC electrical connections.
30. Reconnect the MAP sensor and the intake air temperature sensors.
31. Install the air cleaner assembly and torque the hose clamps to 25 inch lbs. (3 Nm).

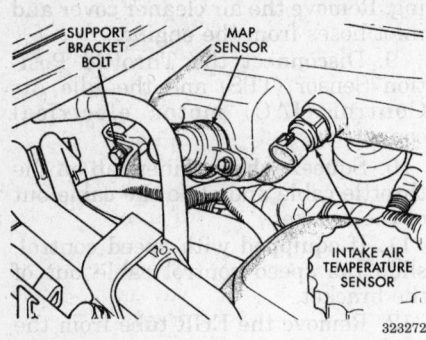

Intake manifold sensors — 2.5L engine

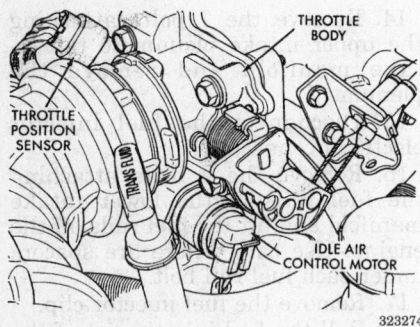

Idle Air Control (IAC) motor — 2.5L engine

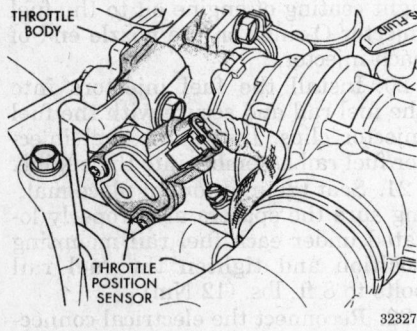

Throttle Position sensor — 2.5L engine

32. Install the air inlet resonator attaching bolt.

33. Reconnect the negative battery cable. Start the engine and check for fuel leaks.

ENGINE MECHANICAL

Engine Assembly

REMOVAL AND INSTALLATION

2.0L, 2.4L and 2.5L Engines

— CAUTION —

Fuel injection systems remain under pressure, even after the engine has been turned OFF. The fuel system pressure must be relieved before disconnecting any fuel lines. Failure to do so may result in fire and/or personal injury.

1. Release the fuel system pressure using the recommended procedure. Disconnect the fuel line quick-connect fitting from the fuel rail by squeezing the retainer tabs together and pulling the fuel tube/quick-connect fitting assembly off the fuel tube nipple.

2. Remove the battery and battery tray according to the following:

   a. Be sure the ignition switch is in the OFF unlocked position and all vehicle accessories are turned OFF.

   b. Disconnect the negative battery cable remote connection located on the left strut tower.

   c. Turn the steering wheel to the extreme left position.

   d. Release the shield by twisting the 4 plastic screws ¼ turn. Remove the shield.

   e. Disconnect the the battery blanket heater, if equipped.

   f. Remove the negative battery cable, then remove the positive battery cable.

   g. Remove the bolt securing the battery strap to the battery hold-down bracket. Remove the hold-down bracket bolt.

   h. Slide the battery to the rear of the battery tray and lift over the lip. Be careful not to tip the battery or acid will spill out.

   i. Remove the battery from the vehicle. Remove the battery blanket heater, if equipped.

   j. Remove the battery tray mounting bolts.

   k. Remove the battery tray and battery strap.

3. Remove the complete air cleaner and inlet duct assembly.

4. Unbolt the Powertrain Control Module and move it aside.

5. Drain and properly contain the coolant from the engine.

6. Remove the upper and lower radiator hose, radiator and cooling fan.

7. Disconnect and plug the automatic transaxle cooler lines, if equipped.

8. Disconnect the clutch cable and transaxle shift linkage, if equipped.

9. Disconnect the throttle body linkage and the engine wiring harness.

10. Disconnect the heater hoses.

11. Recover and properly contain the refrigerant of the A/C system with an R-134a recovery unit.

12. Raise and safely support the vehicle. Remove the front wheels.

13. Drain the engine oil, if necessary.

14. Remove the right side inner splash shield.

15. Remove the accessory drive belts.

16. Remove the right and left half-shaft assemblies.

17. Disconnect the exhaust pipe from the exhaust manifold.

18. Remove the front and rear engine mount brackets from the body.

19. Lower the vehicle.

20. Remove the power steering pump and reservoir, set them aside.

21. Remove the A/C compressor as follows:

   a. Disconnect the compressor clutch wire lead.

   b. Disconnect and plug the refrigerant lines from the compressor.

   c. Remove the compressor mounting bolts.

   d. Remove the compressor unit from the vehicle. Be sure to plug all openings in the A/C system to prevent moisture contamination.

22. Disconnect the ground straps from the engine.

23. Raise the vehicle and install an engine dolly under the vehicle and support engine.

24. Remove the transaxle and engine mount through bolts.

25. Raise the vehicle slowly allowing the engine and transaxle assembly to remain on the dolly.

To install:

26. Position the engine and the transaxle under the vehicle and lower the vehicle onto the engine assembly.

27. Align the engine mounts and install the right and left mount bolts.

28. Install the transaxle mount.

29. Install the right and left half-shaft assemblies.

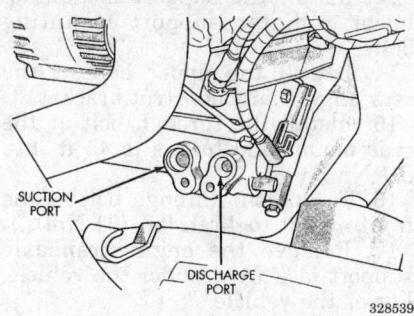

A/C refrigerant lines — 2.0L engine

328539

30. Install the transaxle and engine braces.

31. Install the splash shields.

32. Reconnect the exhaust pipe to the exhaust manifold.

33. Install the power steering pump and reservoir.

34. Install the A/C compressor on the engine as follows:

   a. Position the compressor correctly against the engine.

   b. Install the compressor mounting bolts. Torque the compressor mounting bolts to 30 ft. lbs. (41 Nm).

   c. Reconnect the A/C refrigerant hoses with new seals.

   d. Reconnect the compressor clutch wire.

35. Install the accessory drive belts and adjust.

36. Install the front engine mount.

37. Install the inner splash shield. Install the front wheels and lug nuts. Torque the lug nuts, in a star pattern sequence, to 95–100 ft. lbs. (129–135 Nm).

38. If equipped with manual transaxle, reconnect the clutch cable and linkages.

39. If equipped with automatic transaxle, reconnect the shifter and kickdown linkages.

40. Reconnect the fuel lines and heater hoses.

41. Install the ground straps.

42. Reconnect the engine end throttle body electrical harnesses and connections.

43. Reconnect the throttle body linkage.

44. Install the radiator, cooling fan/shroud assembly and hoses. Reconnect the automatic transaxle cooler lines to the radiator, if equipped.

45. Refill the cooling system with a $^{50}/_{50}$ mixture of clean water and ethylene glycol antifreeze.

46. If the engine oil was drained, install fresh oil and a new oil filter.

47. Install the battery tray and the battery as follows:

   a. Install the battery strap and battery tray into the vehicle through the left front fender well.

   b. Install and tighten the battery tray mounting bolts.

   c. Install the battery blanket heater onto the battery, if equipped.

   d. Install the battery onto the battery tray in the proper position.

   e. Install the hold-down bracket bolt and the bolt securing the battery strap to the battery hold-down bracket. Torque the battery hold-down bracket bolt to 124 inch lbs. (14 Nm).

   f. Connect the positive battery cable to the battery and then connect the negative battery cable to the battery. Torque the battery cables to 150 inch lbs. (17 Nm). Do NOT reconnect the negative battery cable remote connection to the left strut tower at this time.

   g. Reconnect the battery blanket heater, if equipped.

   h. Install shield. Turn the 4 plastic screws to secure the shield in place.

48. Reinstall the powertrain control module.

49. Install the air cleaner and inlet duct assembly.

50. Reconnect the negative battery cable remote connection located at the left strut tower.

51. Recharge the air conditioning system.

52. Check to be sure all ducts, hoses, fuel lines and wiring connectors have been properly reconnected.

53. Start the engine and run until operating temperature.

54. Check for leaks and proper operation.

Engine Mounts

REMOVAL AND INSTALLATION

The right and left engine support assemblies are slotted to allow for right and/or left drive train adjustments in relation to halfshaft assembly length. Check and adjust the right and left engine support assemblies as necessary.

Right Side Mount

NOTE: The right side engine mount is a Hydro-Mount and may show surface cracks. This will not effect its performance and should not be replaced. The Hydro-Mount should only be replaced when it is leaking fluid.

1. Disconnect the negative battery cable remote connection located on the left strut tower.

2. Raise and safely support the vehicle. Remove the right inner splash shield.

3. Remove the right engine support assembly vertical mounting bolts from the vehicle frame rail.

4. Lower the vehicle.

5. Carefully place a floor jack under the engine assembly to remove any load from the motor mount.

6. Remove the 3 bolts securing the engine support assembly to the engine mount bracket.

7. Remove the right side engine mount.

To install:

8. Install the engine support assembly to the engine mount bracket and install the 3 mounting bolts. Torque the bolts to 45 ft. lbs. (61 Nm).

9. Remove the floor jack from under the engine assembly. Raise and safely support the vehicle.

10. Install the right engine support assembly-to-vehicle frame rail mounting bolts. Torque the mounting bolts to 45 ft. lbs. (61 Nm).

11. Install the right inner splash shield. Lower the vehicle.

12. Perform engine support adjustment procedure, if necessary.

13. Reconnect the negative battery cable.

Left Side Engine Mount

NOTE: The left side engine mount is a Hydro-Mount and may show surface cracks. This will not effect its performance and should not be replaced. The Hydro-Mount should only be replaced when it is leaking fluid.

1. Disconnect the negative battery cable remote connection located on the left strut tower.

2. Carefully support the transaxle assembly with a transaxle jack.

3. Remove the 3 mounting bolts from the mount to the transaxle.

4. Remove the transaxle mount attaching bolts and remove the mount.

To install:

5. Install the transaxle support mount to the vehicle frame rail and install the attaching bolts. Torque the attaching bolts to 24 ft. lbs. (33 Nm).

6. Position the transaxle mount to the transaxle assembly and install the 3 mounting bolts. Torque the 3

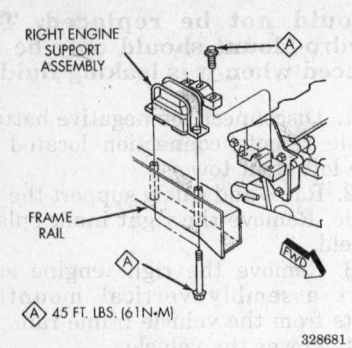

Right side engine mount — 2.0L, 2.4L and 2.5L engines

TORQUE	
Ⓐ	61 N•m (45 ft. lbs.)
Ⓑ	33 N•m (24 ft. lbs.)

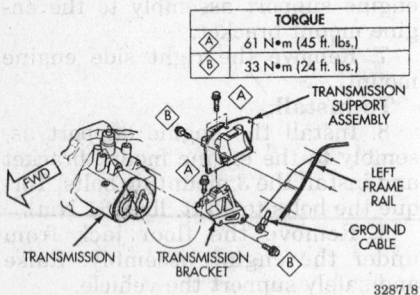

Left side engine mount — Type 1

TORQUE	
Ⓐ	61 N•m (45 ft. lbs.)
Ⓑ	33 N•m (24 ft. lbs.)

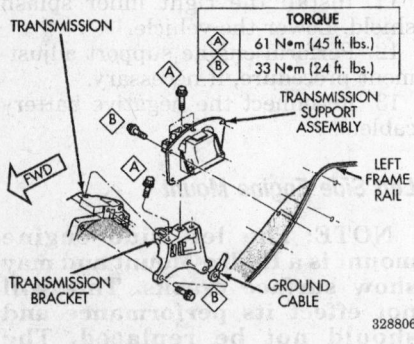

Left side engine mount — Type 2

transaxle mount-to-transaxle mounting bolts to 45 ft. lbs. (61 Nm).

7. Remove the transaxle jack from under the vehicle.

8. Perform engine support adjustment procedure, if necessary.

9. Reconnect the negative battery cable.

Engine Support Module — Front and Rear Mounts

1. Disconnect the negative battery cable remote connection located on the left strut tower.

2. Raise and safely support the vehicle. Support the engine/transaxle assembly so it will not rotate

3. Remove the through bolt at the rear mount and remove the bolts securing the module to the crossmember.

4. Remove the upper mounting bolt from the rear support strut bracket.

5. Remove the front mounting bolts from the support module to the lower radiator support.

6. Support the cooling module.

7. Remove the lower radiator support bolts, and remove the support member.

8. Remove the through bolt at the front mount and remove the engine support module.

9. If necessary, separate the front and rear engine mounting torque brackets from the engine.

To install:

10. If removed, install the front and rear engine mounting torque brackets. Install the mounting bolts and torque to 80 ft. lbs. (110 Nm).

11. Position the engine support module up under the engine. Install the through bolt at the front mount, but do not fully tighten at this time.

12. Install the lower radiator support member.

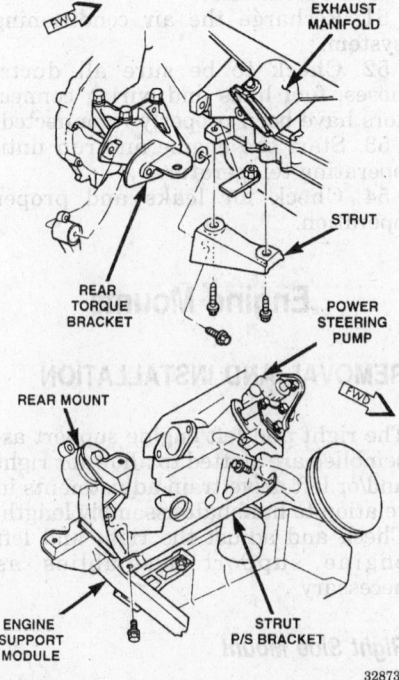

Rear engine mounting torque bracket — 2.0L engine

13. Install the support module-to-lower radiator support mounting bolts.

14. Install the upper bolt at the rear engine support strut bracket.

15. Install the through bolt at the rear mount and torque to 45 ft. lbs. (61 Nm).

16. Torque the through bolt at the front mount to 45 ft. lbs. (61 Nm).

17. Remove the engine/transaxle support jack from under the vehicle. Lower the vehicle.

18. Reconnect the negative battery cable.

NOTE: Whenever the vehicle sub-frame is removed or lowered, the wheel alignment should be checked.

Engine Support Adjustment

1. Carefully support the engine and transaxle assembly with a floor jack to remove all load off the engine mounts.

2. Loosen the vertical fasteners to the right engine support assembly.

3. Loosen the vertical bolts to the left engine support assembly.

4. Pry the engine/transaxle assembly to the right or left as required to attain the proper halfshaft length.

5. Torque the right engine support assembly vertical bolts to 45 ft. lbs. (61 Nm). Torque the left engine support assembly bolts to 45 ft. lbs. (61 Nm).

6. Recheck the halfshaft length.

Cylinder Head

REMOVAL AND INSTALLATION

2.0L Engine

This engine uses a Single Over Head Camshaft (SOHC) 4-valves per cylinder cross flow aluminum cylinder head. Care must be taken to make sure all valve timing marks align after cylinder head service.

--- **CAUTION** ---
Fuel injection systems remain under pressure, even after the engine has been turned OFF. The fuel system pressure must be relieved before disconnecting any fuel lines. Failure to do so may result in fire and/or personal injury.

1. Disconnect the negative battery cable remote connection located on the left strut tower.

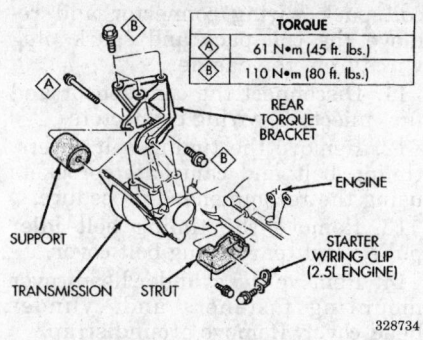

TORQUE	
Ⓐ	61 N•m (45 ft. lbs.)
Ⓑ	110 N•m (80 ft. lbs.)

328734

Rear engine mounting torque bracket — 2.4L and 2.5L engines

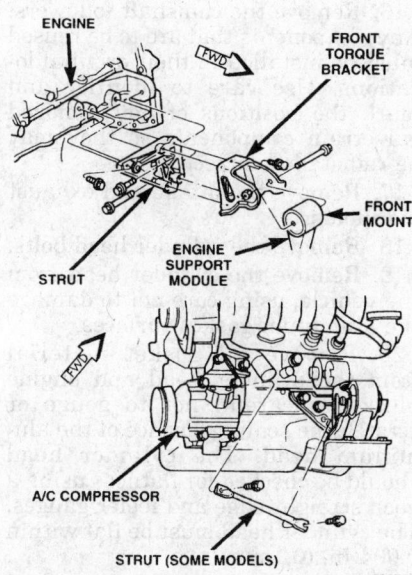

328735

Front engine mounting torque bracket — 2.0L and 2.4L engines

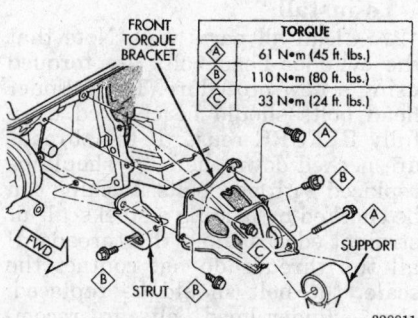

TORQUE	
Ⓐ	61 N•m (45 ft. lbs.)
Ⓑ	110 N•m (80 ft. lbs.)
Ⓒ	33 N•m (24 ft. lbs.)

328811

Front engine mounting torque bracket — 2.5L engine

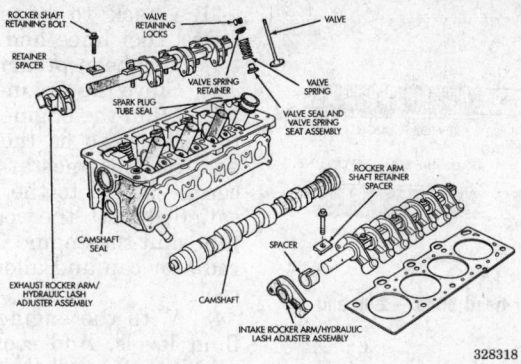

328318

Cylinder head and valvetrain assembly — 2.0L engine

2. Relieve the fuel system pressure using the recommended procedure.

3. Remove the air cleaner assembly.

4. Drain and properly contain the engine coolant.

5. Disconnect and tag all vacuum hoses and all electrical connections from the throttle body. Disconnect the fuel line quick-connect fitting to the fuel injectors by squeezing the retainer tabs together and pulling the fuel tube/quick-connect fitting assembly off the fuel tube nipple.

6. Disconnect the throttle linkage.

7. Remove the accessory drive belt(s).

8. Disconnect the power brake booster vacuum hose from the intake manifold.

9. Raise and safely support the vehicle.

10. Disconnect the exhaust pipe from the exhaust manifold. Lower the vehicle.

11. Remove the power steering pump and move aside.

12. Disconnect the the coil pack wiring connector. Disconnect the spark plug wires from the spark plugs. Remove the ignition coil pack unit from the engine.

13. Remove the cylinder head cover.

14. Disconnect the cam sensor and fuel injector wiring.

15. Remove the intake and exhaust manifolds, if necessary.

16. Remove the timing belt cover, timing belt, camshaft sprocket and rear timing belt cover using the recommended procedure.

17. Remove the rocker arm/rocker arm shaft assemblies.

18. Remove the cylinder head bolts and remove the cylinder head.

To install:

NOTE: The cylinder head bolts should be checked for stretching before reuse. If the thread area of the bolt is necked down the bolts must be replaced with new. New head bolts are recommended.

19. Clean all parts well. Clean all gasket material from the cylinder head and the engine block. Use care not to scratch the aluminum cylinder head sealing surface. Check the cylinder head for flatness using a feeler gauge and a straight-edge. The cylinder head must be flat within 0.004 in. (0.1mm).

20. Check the cylinder head for cracks or other damage.

21. Install a new gasket and the cylinder head to the engine block.

22. Be sure to oil the cylinder head bolt threads with clean engine oil. Install the cylinder head bolts, the 4 4.330 in. (110mm) short bolts are to be installed in positions 7, 8, 9 and 10. Torque the bolts in proper sequence.

23. Torque the bolts in 4 steps as follows:

    a. First: all bolts to 25 ft. lbs. (34 Nm).

    b. Second: all bolts to 50 ft. lbs. (68 Nm).

    c. Third: all bolts again to 50 ft. lbs. (68 Nm).

    d. Fourth: all bolts an additional ¼ turn.

NOTE: Do not use a torque wrench for the fourth step.

24. Set the crankshaft to 3 notches BTDC before installing the rocker arm shafts. Install the rocker arm/rocker arm shaft assemblies.

25. Install the cylinder head cover with a new cylinder head cover gasket. Be sure the cover gasket mating surfaces are clean of any dirt, oil or old gasket material. Torque the cylinder head cover mounting bolts to 105 inch lbs. (12 Nm).

26. Install the timing belt rear cover and camshaft sprocket. Install the timing belt using the recommended procedure to make sure the

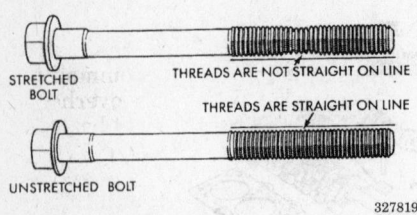

STRETCHED BOLT — THREADS ARE NOT STRAIGHT ON LINE

THREADS ARE STRAIGHT ON LINE

UNSTRETCHED BOLT

327819

Checking the cylinder head bolts — 2.0L and 2.4L engines

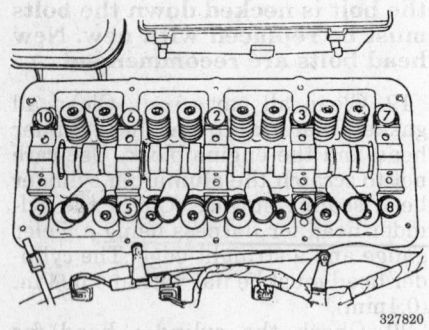

327820

Cylinder head tightening sequence — 2.0L engine

timing marks are properly aligned. Failure to do so will cause engine damage.

27. Install the timing belt cover.

28. Install the intake and exhaust manifolds, if removed.

29. Reconnect the cam sensor and the fuel injector wiring.

30. Install the ignition coil pack onto the engine. Reconnect the coil pack wiring connector and the spark plug wires to the correct spark plugs.

31. Install the power steering pump.

32. Raise the vehicle and reconnect the exhaust pipe to the exhaust manifold. Lower the vehicle.

33. Reconnect the brake booster vacuum line.

34. Install and adjust the accessory drive belts.

35. Reconnect the throttle linkage.

36. Reconnect all vacuum hoses and wiring connectors to the throttle body.

37. Reconnect the fuel line to the fuel injectors.

38. Install the air cleaner assembly.

39. Refill the cooling system with a $^{50}/_{50}$ mixture of clean antifreeze and water. A complete oil and filter change is recommended.

40. Reconnect the negative battery cable remote connection.

41. Check to be sure all ducts, hoses, fuel lines and wiring connectors have been properly reconnected.

42. Start the engine and check for leaks. Run the engine with the radiator cap off so as the engine warms and the thermostat opens, coolant can be added to the radiator. When satisfied that the cooling system is full, shut the engine **OFF**, install the radiator cap and allow the engine to cool.

43. With the engine cool, check all fluid levels. Add coolant and oil as required. Restart the engine and test drive vehicle to check for proper operation.

2.4L Engine

This engine uses a Dual Over Head Camshaft (DOHC) 4-valves per cylinder cross flow aluminum cylinder head. The valves are actuated by roller cam followers which pivot on stationary hydraulic valve adjusters. Care must be taken to make sure all valve timing marks align after cylinder head and valvetrain service.

── CAUTION ──

Fuel injection systems remain under pressure, even after the engine has been turned OFF. The fuel system pressure must be relieved before disconnecting any fuel lines. Failure to do so may result in fire and/or personal injury.

1. Relieve the fuel system pressure using the recommended procedure.

2. Disconnect the negative battery cable remote connection located on the left strut tower.

3. Place a large drain pan under the radiator drain plug. Open up the drain plug and drain the cooling system.

4. Remove the air cleaner assembly and disconnect all vacuum lines, electrical wiring and fuel lines from the throttle body.

5. Disconnect the throttle linkage.

6. Remove the accessory drive belts.

7. Disconnect the power brake vacuum hose from the intake manifold.

8. Raise and safely support the vehicle. Disconnect the exhaust pipe from the exhaust manifold.

9. Lower the vehicle as required to remove the power steering pump. Do not disconnect the fluid lines. Set the pump aside.

10. Label the spark plug wires for correct installation. Disconnect the coil pack wiring connector and remove the coil pack and spark plug wires from the engine.

11. Disconnect the cam sensor and fuel injectors' wiring connectors.

12. Remove the timing belt covers, timing belt and camshaft sprockets using the recommended procedure.

13. Remove the timing belt idler pulley and rear timing belt cover.

14. Remove the cylinder head cover mounting fasteners and cylinder head cover. Remove ground strap.

15. Identify the camshafts, if they are to be reused, for later installation. The camshafts are not interchangeable. Remove the camshaft bearing caps and remove the camshafts in the prescribed sequence.

16. Remove the camshaft followers. Any components that are to be reused must be installed in their original locations. Use care to identify and mark the positions of any removed valvetrain components so they may be reinstalled correctly.

17. Remove the intake and exhaust manifolds.

18. Remove the cylinder head bolts.

19. Remove the cylinder head from the vehicle, using care not to damage the aluminum gasket surfaces.

20. Remove all gasket material from the cylinder head and engine block. Be careful not to gouge or scratch the sealing surface of the aluminum head. The cylinder head should be checked for flatness using a good straight-edge and feeler gauges. The cylinder head must be flat within 0.004 in. (0.1mm).

21. Inspect the camshaft bearing oil feed holes in the cylinder head for clogging. Inspect the camshaft bearing journals for wear or scoring. Check the cam surface for abnormal wear and damage. A visible worn groove in the roller path or on the cam lobes is cause for replacement. Valve service may be performed at this time.

To install:

22. Clean all parts well. Note that the cylinder head bolts are torqued using a new procedure. The cylinder head bolts should be checked carefully BEFORE reuse. If the threads are necked down the bolts should be replaced with new bolts. Necking can be checked by holding a steel scale or straight-edge against the threads. If all the threads do not contact the scale, the bolt should be replaced. New cylinder head bolts are recommended for any engine rebuild, especially if it is known that the engine has been disassembled before.

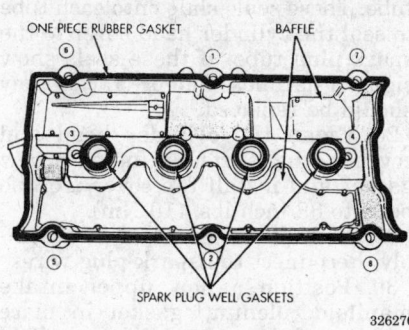

ONE PIECE RUBBER GASKET   BAFFLE

SPARK PLUG WELL GASKETS

326270

Cylinder head cover and gaskets — 2.4L engine

23. Make sure both the top of the engine block and the bottom of the cylinder head are clean. Install a new gasket making sure all holes align with the openings in the engine block. Carefully set the cylinder head in place. A helper may be required.

24. Before installing the bolts the threads should be oiled with clean engine oil. Install the bolts and torque in sequence in 4 Steps as follows:

    a. First: tighten all bolts to 25 ft. lbs. (34 Nm).

    b. Second: tighten all bolts to 50 ft. lbs. (68 Nm).

    c. Third: tighten all bolts again to 50 ft. lbs. (68 Nm).

    d. Fourth: tighten all bolts and additional ¼ turn.

NOTE: Do not use a torque wrench for the fourth step.

25. Check the camshaft end-play, then install the camshaft.

26. Apply Mopar Gasket Maker or equivalent sealer to the No. 1 and No. 6 bearing caps. Install the bearing caps and tighten the M8 fasteners to 250 inch lbs. (28 Nm). The end caps must be installed before the seals may be installed.

27. Apply a light coating of clean engine oil to the lip of the new camshaft seal. Install the camshaft

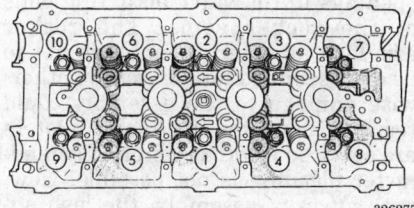

326277

Cylinder head tightening sequence — 2.4L engine

seal until it fits flush with the cylinder head.

28. Install the camshaft sprockets, if removed. Install the rear timing belt cover and timing belt using care to make sure all timing marks are properly aligned, using the recommended procedure. Install timing belt cover.

——— WARNING ———
Verify that all timing marks are correct. If the timing belt or sprockets are incorrectly installed, engine damage will occur. Take time to make sure all timing marks are correctly aligned.

29. Install the intake and exhaust manifolds.

30. Clean the cylinder head cover and cylinder head gasket rails (mating surfaces). Make certain the rails are flat.

31. Install new cylinder head cover gaskets. Use care. DO NOT allow oil or solvents to contact the timing belt as they can deteriorate the rubber and cause tooth skipping. Apply Mopar Silicone Rubber Adhesive Sealant, or equivalent, at the camshaft cap corners and at the top edge of the ½ round seal.

32. Install the cylinder head cover assembly to the head and tighten the fasteners in sequence using the following 3 Steps:

    a. First: tighten all cylinder head cover fasteners to 40 inch lbs. (4.5 Nm).

    b. Second: tighten all fasteners to 80 inch lbs. (9 Nm).

    c. Third: tighten all fasteners to 105 inch lbs. (12 Nm).

33. Install the ground strap.

34. Install the ignition coil pack and reconnect the spark plug wiring.

35. Connect the cam sensor and fuel injectors' wiring.

36. Install the power steering pump assembly.

37. Connect the exhaust pipe to the exhaust manifold.

38. Connect all vacuum lines and remaining wiring. Connect the throttle linkage and fuel lines.

39. Install and adjust the accessory drive belts.

40. Refill the cooling system. An oil and filter change is recommended since coolant can enter the oil system when a head is removed.

41. Connect the remaining air ducting. Connect the negative battery cable and test run vehicle. Check for leaks and for proper operation.

2.5L Engine

This engine uses aluminum alloy cylinder heads with 4-valves per cylinder and pressed-in cast iron valve guides. The cylinders are common to either cylinder bank. Two overhead camshafts are supported by 4 bearing journals which are part of the head with the distributor driven off the right (firewall side) cylinder head. Right and left camshaft drive sprockets are interchangeable. The sprockets and engine water pump are driven by the timing belt. Care must be taken to make sure all valve timing marks align after cylinder head and valvetrain service. Please note that for camshaft service, the cylinder head must be removed.

——— CAUTION ———
Fuel injection systems remain under pressure, even after the engine has been turned OFF. The fuel system pressure must be relieved before disconnecting any fuel lines. Failure to do so may result in fire and/or personal injury.

1. Disconnect the negative battery cable from the remote auxiliary jumper terminal.

2. Relieve the fuel system pressure using the recommended procedure.

3. Place a large drain pan under the radiator drain plug. Drain the cooling system.

4. Remove the accessory drive belts.

5. Remove the front timing belt covers and timing belt using the recommended procedure. Remove the camshaft sprockets.

6. The intake manifold is a 2-piece unit. The upper part is a large air intake plenum of aluminum alloy. Use care working with light alloy parts. Remove the air intake plenum first, then remove the lower intake manifold.

7. Label, disconnect and set aside the spark plug wires.

8. Remove the cylinder head cover screws and remove the cover.

9. Identify the rocker arm shaft assemblies before removal.

10. Install the auto lash adjuster retainers Special Tool MD 998443 or equivalent to keep the auto lash adjusters from falling out of the rocker arms when the rocker arm assembly is removed.

11. Loosen the attaching fasteners and remove the rocker arm shaft assemblies from the cylinder head.

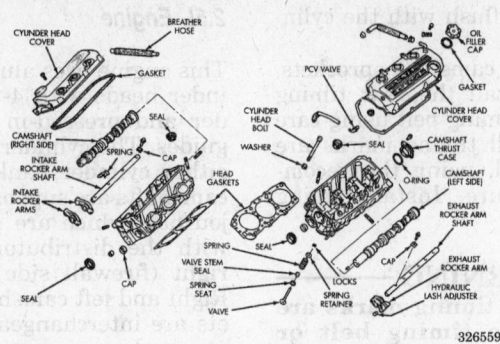

Cylinder head, camshafts and rocker assemblies — 2.5L engine

12. Remove the distributor assembly.

13. Remove the exhaust manifold and crossover.

14. Remove the cylinder head bolts and remove the cylinder head from the vehicle.

15. If the camshaft(s) are to be serviced, remove the thrust case from the left head assembly and remove the camshaft from the rear of the head. If not already done, remove the distributor from the right cylinder head and remove the camshaft from the rear of the head.

16. Valve service may be performed at this time, if required.

To install:

17. Clean all parts well. All gasket surfaces on the engine block, cylinder head(s) and both the upper and lower sections of the intake manifold must be clean. Check for cracks, signs of wear in the camshaft bores or other damage. With a straight-edge and feeler gauge, check the head for flatness. It should be within 0.0012 in. (0.03mm) along its length. The service limit is 0.008 in. (0.2mm). If the head must be resurfaced, the grinding limit is 0.008 in. (0.2mm). Note that this dimension is a combined total dimension of stock material removal from the cylinder head, if any, and the block top surface is 0.0079 in. (0.2mm).

18. Camshaft end-play can be checked. Oil the camshaft journals with clean engine oil and install (if removed) the camshaft without the rocker arm assemblies. Move the camshaft as far rearward as it will go. Mount a dial indicator to bear on the front of the camshaft. Zero the indicator. Move the camshaft as far forward as it will go. End-play should be 0.004–0.008 in. (0.1–0.2mm). Maximum allowed end-play is 0.016 in. (0.4mm).

19. If the camshafts were removed, lubricate the camshaft journals and carefully install the camshaft into the

cylinder head. Install the thrust case and torque the thrust case mounting bolts to 9 ft. lbs. (13 Nm).

20. Apply a light coating of engine oil to the lip of the camshaft seal(s). Install the camshaft seal(s). The camshaft must be installed before installing the seal. Install the camshaft sprocket(s) and tighten to 65 ft. lbs. (88 Nm).

21. Install a new head gasket over the locating dowels in the cylinder head. Install the head, using care to locate on top of the dowels. Install the 10mm Allen hex head bolts with washers. New head bolts are recommended. Torque the head bolts in proper sequence. Tighten gradually, working in 2 or 3 steps and finally tighten to 80 ft. lbs. (108 Nm).

22. Install the lower intake manifold using new gaskets.

23. Clean and install the rocker arm assemblies.

24. Install the exhaust manifold and crossover exhaust pipe.

25. Install the timing belt. This is a somewhat complex procedure. Use the recommended procedure. It is most important that all valve timing marks align properly or engine damage will result.

26. Install the timing belt covers.

27. Inspect the spark plug tube seals located on the ends of each

tube. These seals slide onto each tube to seal the cylinder head cover to the spark plug tube. If these seals show signs of hardness and/or cracks, they should be replaced.

28. Clean the cylinder head and cover mating surfaces. Install a new gasket and install the cover. Torque bolts to 88 inch lbs. (10 Nm).

29. Install the distributor assembly. Reconnect the spark plug wires.

30. Position a new upper intake manifold (plenum) gasket in place and install the upper intake manifold plenum. Torque the bolts to 13 ft. lbs. (18 Nm). Install the bolts for the plenum support brackets and connect the EGR tube.

31. Install the speed control cable (if equipped) and the throttle cable.

32. Reconnect the TPS and the idle air control motor electrical connectors.

33. Install the air cleaner cover, inlet hoses and air inlet resonator. Tighten the air tube connections.

34. Reconnect the MAP and intake air temperature sensor wiring connectors.

35. Install and adjust the accessory drive belts.

36. Refill the cooling system with a $^{50}/_{50}$ mixture of clean antifreeze and water. An oil and filter change is recommended whenever a cylinder head has been removed since coolant can get into the oil system.

37. Reconnect the negative battery cable. Check to be sure all ducts, hoses, fuel lines and wiring connections have all been properly reconnected.

38. Start the engine and check for leaks, abnormal noises and vibrations. Bleed the cooling system.

Lash Adjusters

BLEEDING

2.0L and 2.5L Engines

The hydraulic lash adjusters are precision units installed in machined openings in the valve actuating ends of the rocker arms. The rocker arm/lash adjuster assembly is not to be disassembled for any reason otherwise damage to the assembly could result. The rocker arm/lash adjusters are serviced as an assembly. Lash adjuster bleeding is not possible. However, upon re-assembly, the lash adjusters should be partially filled with oil which would be indicated by little or no plunger travel when the lash adjuster is depressed. If plunger travel is excessive, place assembly in

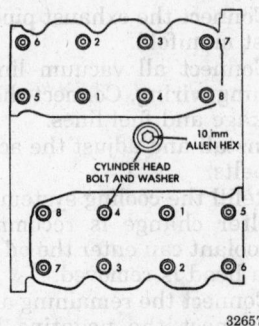

Cylinder head bolt tightening sequence — 2.5L engine

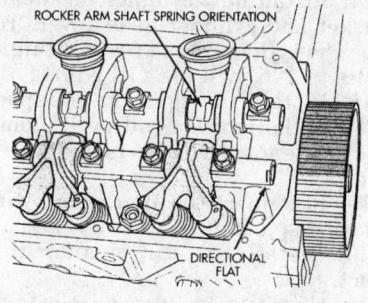

The flats on the rocker arm shaft aid proper orientation — 2.5L engine

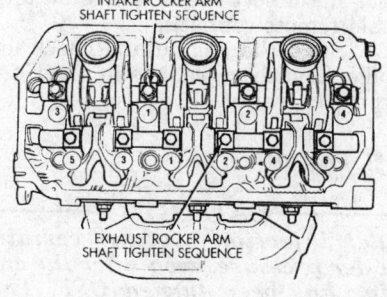

Rocker arm shaft tightening sequence — 2.5L engine

fresh oil and pump until the travel is taken up. If the travel is not reduced, the assembly must be replaced.

2.4L Engine

The hydraulic valve lash adjusters do not need to be bled. If any lash adjuster is not functioning properly it must be replaced. When installing a new lash adjuster soak the adjuster in oil until the adjuster has little or no plunger travel when depressed. However, if plunger travel is not re-

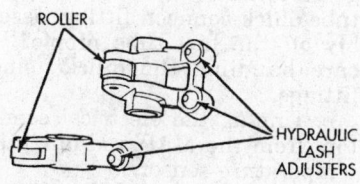

Hydraulic lash adjuster — 2.0L engine

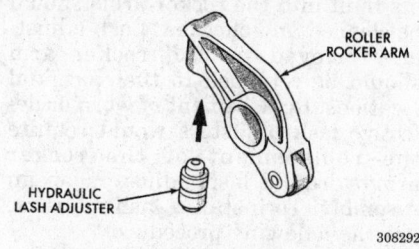

Hydraulic lash adjuster — 2.5L engine

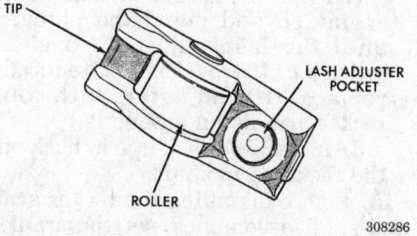

Lash adjuster in cam follower assembly — 2.4L engine

duced, the lash adjuster needs replacement.

Valve Lash

ADJUSTMENT

2.0L and 2.5L Engines

The engines are equipped with hydraulic lash adjusters which are precision units installed in machined openings in the valve actuating ends of the rocker arms. Valve clearance adjustments are not performed.

2.4L Engine

The valves are actuated by roller cam followers which pivot on stationary hydraulic lash adjusters. The hydraulic lash adjusters are precision units installed in machined openings of the cam follower. Valve clearance adjustments are not performed.

Rocker Arm Shaft

REMOVAL AND INSTALLATION

2.0L Engine

This engine uses a Single Over Head Camshaft (SOHC) running in an aluminum cylinder head. Rocker arm shafts mount directly to the cylinder head. Care must be taken to make sure all valve timing marks align after cylinder head and valvetrain service. The hydraulic lash adjusters are located in the valve actuating end of the rocker arm and are serviced as an assembly.

> ### CAUTION
> *Fuel injection systems remain under pressure, even after the engine has been turned OFF. The fuel system pressure must be relieved before disconnecting any fuel lines. Failure to do so may result in fire and/or personal injury.*

1. Disconnect the negative battery cable remote connection located on the left strut tower.
2. Relieve the fuel system pressure using the recommended procedure.
3. Label the spark plug wires to the correct spark plugs. Disconnect the spark plug wires from each spark plug.
4. Remove the inlet duct for the air cleaner.
5. Remove the ignition coil pack.
6. Remove the cylinder head cover retaining bolts and remove the cylinder head cover. Remove and discard any gasket material. Thoroughly clean the cylinder head cover and cylinder head cover gasket mating surfaces. Be sure the gasket mating surfaces are flat.
7. Mark the rocker arm shaft assemblies to identify them for later installation.
8. Remove the rocker arm shaft bolts and remove the rocker arm assemblies from the cylinder head.
9. Mark the rocker arm spacers and retainers to identify them for correct installation. Disassemble the rocker arm/shaft assemblies by removing the attaching bolts from the rocker arm shaft.
10. Slide the rocker arm/hydraulic lash adjuster assembly and rocker arm spacers off the rocker arm shaft. Be sure the rocker arms and spacers are re-assembled in the same positions they are removed from.

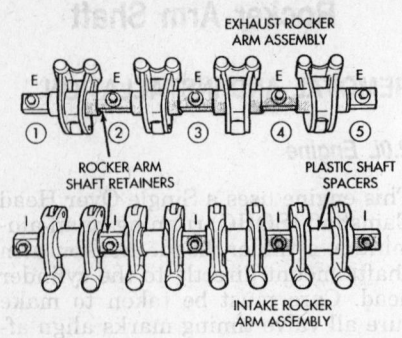

Rocker arm shaft identification — 2.0L engine

To install:

NOTE: Inspect the rocker arms and shaft for scoring and/or wear on the rollers or damage to the rocker arm. If scoring, wear or damage is present, replace the rocker arm assemblies. The rocker arm shaft is hollow and therefore, used as an oil lubrication duct. Inspect the oil holes for clogging, using a small wire and clean, if necessary. Inspect the location where the rocker arms mount to the shaft and replace if damaged or worn.

11. If the camshaft lobes show signs of wear, check the corresponding rocker arm roller for wear or damage. Replace rocker arms/hydraulic lash adjuster if worn or damaged. If the camshaft lobes show signs of pitting on the nose, flank or base circle, replace the camshaft.

12. Inspect the rocker arms for scoring, wear on the roller or damage to the shaft. Replace parts as necessary. Check that the rocker arm shaft is clean inside and out. Check all oil holes for clogging with a small wire and clean and required.

13. Thoroughly lubricate all rocker arm components and spacers and install on the rocker arm shaft in the original locations.

14. If the vehicle exhibited a tappet-like noise, the valve lash adjusters built into the rocker arms should be cleaned and checked. Lash adjusters removed from a rocker arm should be returned to their original locations. Replacement of worn or defective lash adjusters would require the replacement of the rocker arm/hydraulic lash adjusters as an assembly. To install a lash adjuster, use the following procedure.

a. Lubricate the lash adjuster thoroughly with clean engine oil.

b. Install the adjuster into the rocker arm making sure the adjuster is at least partially filled with oil.

c. Place the rocker arm in clean engine oil and pump the plunger until the lash adjuster travel is taken up. If travel is not reduced, replace the adjuster with the rocker arm as an assembly.

d. Install the rocker arm back on the rocker arm shaft.

15. Before installing the rocker arm and shaft assemblies, set the crankshaft to 3 notches before TDC on the crankshaft sprocket.

NOTE: When installing the intake rocker arm/shaft assembly, be sure the plastic rocker arm spacers do not interfere with the spark plug tubes. If there is interference, rotate the plastic spacers until they are at the proper angle. Do not rotate the spacers by forcing down on the shaft assembly or damage to the spark plug tubes will occur.

16. Install the rocker arm and shaft assemblies with the small notches in the rocker shafts pointing up and toward the timing belt side of the engine. Install the retainers in their original positions on the exhaust and intake shafts. Torque the bolts in proper sequence to 200 inch lbs. (23 Nm).

17. Install new cylinder head cover gasket and cylinder head cover. Torque the cylinder head cover retaining bolts to 105 inch lbs. (12 Nm).

18. Install the ignition coil pack. Torque the ignition coil pack mounting fasteners to 200 inch lbs. (23 Nm).

19. Reconnect the spark plug wires to each spark plug.

20. Install the air cleaner inlet duct.

21. Check to be sure all electrical, vacuum and fluid connections are reconnected as required.

22. An oil and filter change are recommended.

23. Reconnect the negative battery cable remote connection.

24. Start the engine and check for leaks. Test drive vehicle to check for proper operation.

2.5L Engine

—————— **CAUTION** ——————
Fuel injection systems remain under pressure, even after the engine has been turned OFF. The fuel system pressure must be relieved before disconnecting any fuel lines. Failure to do so may result in fire and/or personal injury.
————————————————————

1. Disconnect the negative battery cable from the remote auxiliary jumper terminal.

2. Relieve the fuel system pressure using the recommended procedure.

3. If removing the right (firewall) side rocker arm/shaft assembly, Remove the upper intake manifold (air intake plenum), which is a 2-piece unit of aluminum alloy. Use care working with light alloy parts. Remove the air intake plenum using the following procedure.

a. Verify that the fuel system pressure relief procedure has been performed. Disconnect the fuel supply tube from the fuel rail by squeezing the retainer tabs together and pulling the fuel tube/quick-connect fitting assembly off the fuel tube nipple. Use care handling the quick-connect fittings.

b. Unplug the electrical connectors from the MAP and air intake temperature sensors.

c. Remove the plenum support bracket bolt located rearward of the MAP sensor.

d. Remove the bolt holding the air inlet resonator to the intake manifold.

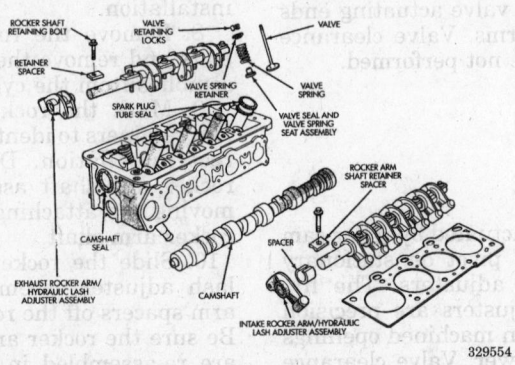

Cylinder head and valve assembly — 2.0L engine

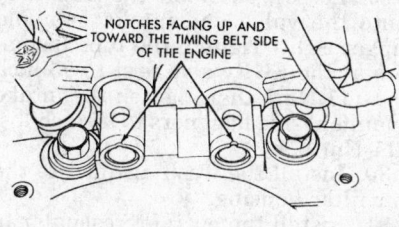

Rocker arm shaft notch location — 2.0L engine

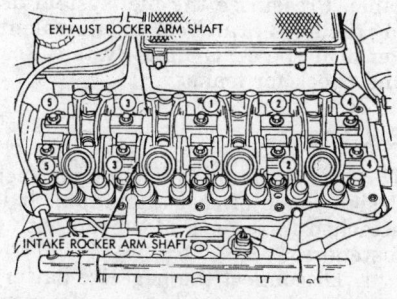

Rocker arm shaft tightening sequence — 2.0L engine

e. Loosen the throttle body air inlet hose clamp. Release the snaps holding the air cleaner housing cover to the housing. Remove the air cleaner cover and inlet hoses from the engine.

f. Unplug the TPS and idle air control motor electrical connections.

g. Squeeze the retainer tab on the throttle cable and slide the cable out of the bracket. Slide the speed control cable out of its bracket, if equipped.

h. Remove the EGR tube from the intake manifold.

i. Remove the plenum support bracket bolt located rearward of the EGR tube.

j. Remove the 7 bolts holding the upper intake plenum and remove the plenum.

4. Disconnect, label and set aside the spark plug wires.

5. Remove the cylinder head cover screws and remove the cover.

6. Identify the rocker arm shaft assemblies before removal.

7. Install the auto lash adjuster retainers Special Tool MD 998443 or equivalent to keep the auto lash adjusters from falling out of the rocker arms when the rocker arm assembly is removed.

8. Loosen the attaching fasteners and remove the rocker arm shaft assemblies from the cylinder head.

NOTE: The hydraulic automatic lash adjusters are precision units installed in the machined openings in the rocker arm units. Do not disassemble the auto lash adjusters from the rocker arms.

To install:

9. The rocker arm shafts are hollow and used as a lubrication oil duct. Make sure all valvetrain parts are clean. Check the rocker arm mounting portion of the shafts for wear or damage. Replace if necessary. Check all oil holes for clogging with a small wire and clean as required. If any rockers were removed, lubricate and install on the shafts in their original positions.

10. Install the rocker arm and shaft assemblies with the FLAT in the rocker arm shafts facing toward the timing belt side of the engine for the right cylinder head. For the left cylinder head install the rocker arm and shaft assembly with the FLAT in the rocker arm shaft facing toward the transaxle side of the engine. Install the retainers and spring clips in their original positions on the exhaust and intake shafts. Torque the retainer bolts to 276 inch lbs. (31 Nm) working from the center, outward. Remove the valve lash retainer tools that should have been installed at disassembly.

11. Inspect the spark plug tube seals located on the ends of each tube. These seals slide onto each tube to seal the cylinder head cover to the spark plug tube. If these seals show signs of hardness and/or cracks, they should be replaced.

12. Clean the cylinder head and cover mating surfaces. Install a new gasket and install the cover. Torque bolts to 88 inch lbs. (10 Nm). Reconnect the spark plug wires.

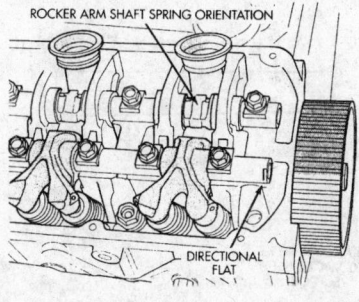

The flats on the rocker arm shaft aid proper orientation — 2.5L engine

13. Position a new upper intake manifold (plenum) gasket in place and install the upper manifold. Torque the bolts to 13 ft. lbs. (18 Nm). Install the bolts for the plenum support brackets and connect the EGR tube.

14. Reconnect the throttle and speed control cables.

15. Reconnect the TPS and idle air control motor electrical connectors.

16. Reconnect the MAP and intake air temperature sensors.

17. Install the air inlet resonator, air inlet hose and the air cleaner housing cover.

18. Check to be sure all remaining electrical connectors have been reconnected. Tighten the air tube connections.

19. An oil and filter change is recommended.

20. Connect the negative battery cable. Start the engine and check for leaks, abnormal noises and vibrations.

Intake Manifold

REMOVAL AND INSTALLATION

2.0L Engine

The intake manifold is a long branch design made of a molded plastic composition. It is attached to the cylinder head with 10 fasteners. Please note that all seals are to be replaced with new seals and all fasteners are to be replaced with new fasteners. Procure the necessary parts before beginning work.

1. Disconnect the negative battery remote cable located on the left strut tower.

—— **CAUTION** ——

Fuel injection systems remain under pressure after the engine has been turned OFF. The fuel system pressure must be relieved before disconnecting any fuel lines. Failure to do so may result fire and/or personal injury.

2. Release the fuel system pressure using the recommended procedure.

3. Remove the air inlet resonator as follows:

a. Loosen the screw securing the air inlet resonator to the throttle body.

b. Loosen the clamp holding the air inlet resonator to the air inlet tube. Remove the resonator.

4. Disconnect the fuel supply line quick connect fitting from the fuel

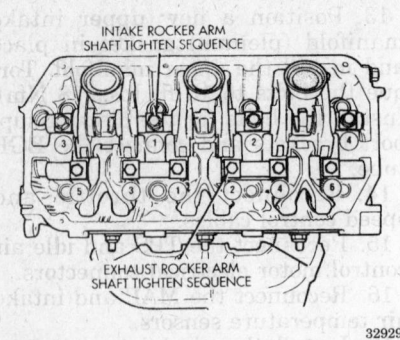

329290

Rocker arm shaft tightening sequence — 2.5L engine

rail by squeezing the retainer tabs together and pulling the fuel tube/quick connect fitting from the fuel tube nipple. The retainer will remain on the fuel tube. Wrap shop towels around the fuel line openings to catch any spilling fuel.

5. Remove the fuel rail attaching screws and remove the fuel rail. Use care when handling the fuel injectors. Do not set them on their tips. Cover the fuel injector openings after fuel rail removal.

6. Remove the accelerator, kickdown and speed control cables from the throttle lever and bracket.

7. Disconnect the Throttle Position Sensor (TPS) and the Idle Air Control (IAC) motor electrical connections.

8. Disconnect the vacuum hoses from the throttle body.

9. Disconnect the connectors from the Manifold Absolute Pressure (MAP) sensor and the intake air temperature sensors.

10. Disconnect the vapor and brake booster hoses.

11. Disconnect the knock sensor electrical connector, starter relay connector and the wiring harness from the tab located on the intake manifold.

12. Remove the transaxle to throttle body support bracket fasteners at the throttle body and loosen the fastener at the transaxle end.

13. Remove the throttle body assembly.

14. Remove the EGR tube bolts at the valve and at the intake manifold. Remove the tube from the engine.

15. Remove the intake manifold to inlet water tube support fastener.

16. Remove the 9 intake manifold screws and washer assemblies and the one nut and washer assembly. Discard the fasteners. At assembly, they should be replaced with new fasteners. Remove the intake manifold from the vehicle.

To install:

17. Clean all gasket surfaces. Check upper and lower manifold gasket surfaces for flatness with a straight-edge. Surface must be flat within 0.006 in. (0.15mm) per foot.

NOTE: All seals are to be replaced with new seals and all fasteners are to be replaced with new fasteners.

18. Install the intake manifold with new O-ring seals. Tighten the fasteners in proper sequence to 105 inch lbs. (12 Nm).

19. Apply a light coating of engine oil to the fuel injector O-rings. Remove the covers from the fuel injector openings and install the fuel injectors into the engine. Seat the injectors in place and tighten the fuel rail bolts to 200 inch lbs. (23 Nm).

20. Connect the electrical connectors to the fuel injectors.

21. Lubricate the quick-connect fittings with clean 30W engine oil. Connect the fuel supply line to the fuel rail. Check the connection by pulling on the connector to insure it is locked into position.

22. Connect the PCV and the brake booster hoses.

23. Install the throttle body and torque to 200 inch lbs. (23 Nm). Install the transaxle to throttle body support bracket and tighten to 105 inch lbs. (11.9 Nm) at the throttle body first. Next tighten the bracket at the transaxle.

24. Reconnect the MAP sensor and the air temperature sensor wiring connectors.

25. Connect the knock sensor electrical and starter relay connectors. Connect the wiring harness to the intake manifold tab.

26. Reconnect the IAC and TPS wiring connectors.

27. Reconnect the throttle body vacuum hoses.

28. Install the accelerator, kickdown and speed control cables to

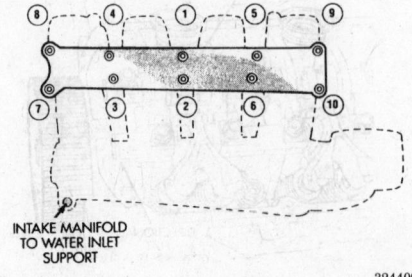

INTAKE MANIFOLD TO WATER INLET SUPPORT

324409

Intake manifold torque sequence — 2.0L engine

their bracket and connect then to the throttle lever.

29. Loosely assemble the EGR tube onto the valve and intake manifold finger tight. Tighten the tube fasteners at the EGR valve first to 95 inch lbs. (11 Nm), then tighten the intake manifold side fasteners to 95 inch lbs. (11 Nm).

30. Install the fresh air duct to the air filter housing.

31. Install the air inlet resonator to the throttle body. Connect the air inlet tube to the resonator and torque the clamps to 20–30 inch lbs. (2–3 Nm).

32. Connect the negative battery cable. Pressurize the fuel system using the DRB Scan Tool, or equivalent. Perform the ASD Fuel System Test and check for leaks.

2.4L Engine

The intake manifold is a long branch design made of cast aluminum. It is attached to the cylinder head with 8 fasteners.

1. Disconnect the negative battery remote cable located on the left strut tower.

— CAUTION —

Fuel injection systems remain under pressure even after the engine has been turned OFF. The fuel system pressure must be relieved before disconnecting any fuel lines. Failure to do so may result in fire and/or personal injury.

2. Release the fuel system pressure using the recommended procedure.

3. Remove the air inlet resonator as follows:

   a. Remove the 2 mounting bolts that secure the air inlet resonator to the intake manifold.

   b. Loosen the screw securing the resonator to the throttle body.

   c. Loosen the clamp securing the air inlet resonator to the air inlet tube. Remove the resonator.

4. Disconnect the fuel supply line quick connect at the fuel tube assembly by squeezing the retainer tabs together and pulling the fuel tube/quick connect fitting assembly from the fuel tube nipple. The retainer will remain on the fuel tube. Use shop towels to catch any dripping fuel.

5. Remove the fuel rail assembly attaching screws and remove the fuel rail assembly from the engine. Use care when handling the fuel injectors. Do not set them on their tips. Cover

the fuel injector openings after fuel rail removal.

6. Remove the accelerator, kickdown and speed control cables from the throttle lever and bracket.

7. Disconnect the Idle Air Control (IAC) motor and Throttle Position Sensor (TPS) wiring connectors.

8. Disconnect the vacuum hoses from the throttle body.

9. Disconnect the Manifold Absolute Pressure (MAP) and Intake Air Temperature (IAT) electrical connectors. Disconnect the vapor and brake booster hoses.

10. Disconnect the knock sensor electrical connector and disconnect the wiring harness from the tab located on the intake manifold.

11. Remove the transaxle to throttle body support bracket fasteners at the throttle body and loosen the fastener at the transaxle end. Remove the throttle body from the intake manifold.

12. Remove the EGR tube bolts at the valve end and at the intake manifold. Remove the tube from the engine.

13. Remove the intake manifold support bracket. Remove the 8 intake manifold fasteners and washers. Remove the intake manifold from the engine.

To install:

14. Clean all parts well. Check the mating surfaces for cracks or distortion.

15. Install a new intake manifold gasket and position the manifold on the cylinder head. Torque the fasteners to 200 inch lbs. (23 Nm) in correct sequence starting from the center and working out.

16. Remove the covering from the fuel injector openings and make sure the openings are clean. Install the fuel rail assembly to the intake manifold. Tighten the retainer screws to 200 inch lbs. (23 Nm).

17. Connect the PCV and brake booster hoses.

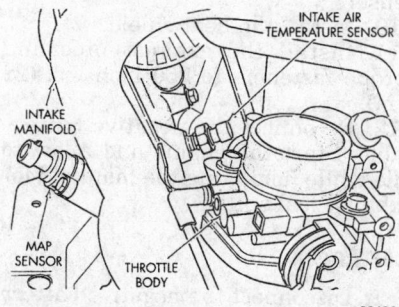

Manifold Absolute Pressure (MAP) and Intake Air Temperature (IAT) sensor locations — 2.4L engine

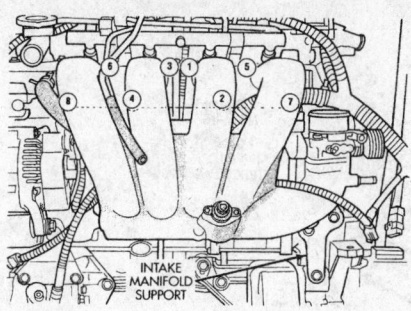

Intake manifold bolt torque sequence — 2.4L engine

18. Inspect the quick connect fittings for damage and repair as required. Lube the fuel tube with clean 30W engine oil. Connect the fuel supply tube hose to the fuel rail assembly. Check the connection by pulling on the connector to insure it is locked in position.

19. Install the throttle body. Tighten the fasteners to 200 inch lbs. (23 Nm). Install the transaxle to throttle body support bracket and tighten to 105 inch lbs. (11.9 Nm) at the throttle body first, then tighten the bracket at the transaxle.

20. Connect the MAP and IAT wiring connectors.

21. Connect the knock sensor electrical connector and connect the wiring harness to the tab located on the intake manifold.

22. Connect the IAC and TPS wiring connectors.

23. Connect the remaining vacuum hoses to the throttle body.

24. Connect the accelerator, kickdown and speed control cables to the throttle lever and bracket.

25. Loosely assemble the EGR tube onto the valve and intake manifold finger tight. Tighten the tube fasteners at the EGR valve first to 95 inch lbs. (11 Nm) then tighten the intake manifold side fasteners to 95 inch lbs. (11 Nm).

26. Install the air inlet resonator to the throttle body, then install the air inlet tube to the resonator. Torque the clamps to 20–30 inch lbs. (2.5–3.5 Nm). Tighten the 2 air inlet resonator-to-intake manifold mounting bolts.

27. Connect the negative battery cable. Pressurize the fuel system using the DRB scan tool, or equivalent. Perform the ASD Fuel System Test and check for leaks.

2.5L Engine

The intake manifold assembly is composed of an upper plenum and lower manifold. This aluminum alloy manifold has long runners to improve airflow inertia. The plenum chamber absorbs air pulsations created during the suction phase of each cylinder. The lower intake manifold is machined for 6 injectors and the fuel rail mounts.

1. Disconnect the negative battery remote cable located on the left strut tower.

2. Release the fuel system pressure.

CAUTION

Fuel injection systems remain under pressure, even after the engine has been turned OFF. The fuel system pressure must be relieved before disconnecting any fuel lines. Failure to do so may result in fire and/or personal injury.

---

3. Disconnect the fuel supply line from the fuel rail. This is a quick connect fitting. Squeeze the fitting retainer tabs together and separate the connection.

CAUTION

Wrap shop towels around the connection to catch any gasoline spillage.

NOTE: It may be helpful to identify and tag each sensor connector as it is being removed. This may save time at assembly.

4. Disconnect the connectors from the Manifold Absolute Pressure (MAP) sensor and the intake air temperature sensors.

5. Remove the plenum support bracket located to the rear of the MAP sensor.

6. Remove the air inlet resonator attaching bolt.

7. Loosen the throttle body air inlet hose clamp.

8. Release the snaps holding the air cleaner housing cover to the housing. Remove the air cleaner cover and inlet hoses from the engine.

9. Disconnect the Throttle Position Sensor (TPS) and the Idle Air Control (IAC) motor electrical connections.

10. Squeeze the retainer tab on the throttle cable and slide the cable out of the bracket.

11. If equipped with speed control, slide the speed control cable out of the bracket.

12. Remove the EGR tube from the engine intake manifold.

13. Remove the plenum support bracket located to the rear of the EGR tube.

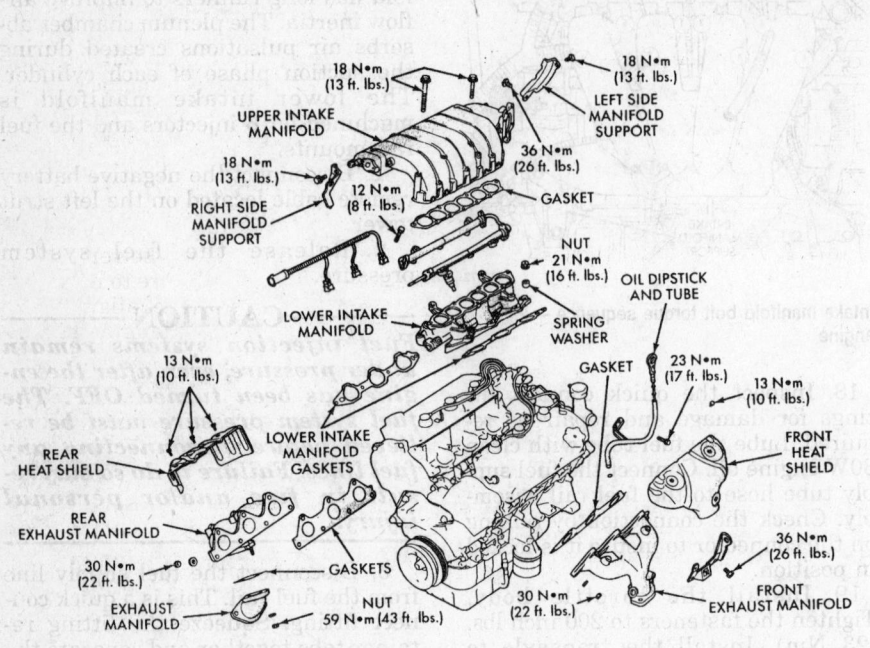

Intake and exhaust manifolds and related components — 2.5L engine

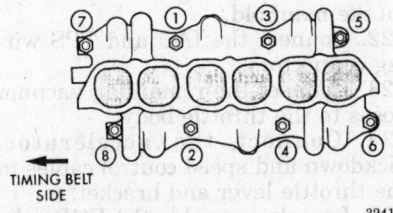

Intake manifold torque sequence — 2.5L engine

14. Remove the 7 bolts attaching the upper intake plenum to the intake manifold and remove plenum.

15. Disconnect the fuel injector electrical connectors.

16. Remove the 4 bolts attaching the fuel rail to the intake manifold. Use care. There are spacers under each fuel rail bolt. Remove the fuel rail.

17. Remove the lower intake manifold attaching bolts.

18. Remove the intake manifold and discard the old gaskets.

To install:

19. Clean all gasket surfaces.

20. Check upper and lower manifold gasket surfaces for flatness with a straight-edge.

21. Surface must be flat within 0.006 in. (0.15mm) per 12 in. of manifold length.

22. Install the lower intake manifold with new gaskets.

23. Tighten the bolts in correct sequence to 185 inch lbs. (21 Nm).

24. Apply a light coating of engine oil to the fuel injector O-rings.

25. Install the fuel injectors into the engine.

26. Seat the injectors in place and tighten the fuel rail bolts to 8 ft. lbs. (12 Nm).

27. Reconnect the electrical connectors to the fuel injectors.

28. Reconnect the fuel supply line to the fuel rail.

29. Install the upper intake plenum with new gaskets.

30. Torque the plenum bolts to 13 ft. lbs. (18 Nm).

31. Install the plenum support brackets and torque to 13 ft. lbs. (18 Nm).

32. Install the EGR tube and torque the screws to 95 inch lbs. (11 Nm).

33. Reinstall the throttle cables.

34. Reconnect the TPS and IAC electrical connections.

35. Reconnect the MAP sensor and the intake air temperature sensor connectors.

36. Install the air cleaner assembly and torque the hose clamps to 25 inch lbs. (3 Nm).

37. Install the air inlet resonator attaching bolt.

38. Connect the negative battery remote cable at the left strut tower.

39. Start the engine and check for leaks.

Exhaust Manifold

REMOVAL AND INSTALLATION

2.0L and 2.4L Engines

1. Disconnect the negative battery cable remote connection located on the left strut tower.

2. Disconnect the exhaust pipe from the exhaust manifold. Apply penetrating oil on the exhaust manifold-to-exhaust pipe flange bolts to aid in removal. It may be necessary to remove the entire exhaust system.

3. Remove the exhaust manifold heat shield.

4. Disconnect the heated oxygen sensor, if necessary.

5. Remove the 8 manifold attaching bolts and remove the manifold from the vehicle.

To install:

6. Clean all parts well. Discard the gasket and clean all gasket surfaces of the manifold and cylinder head. Test the manifold gasket surface for flatness with a straight-edge and feeler gauge. The surface must be flat within 0.006 inches per foot (0.15mm per 300mm) of manifold length. Inspect the manifold for cracks or distortion. Replace if necessary.

7. Install the manifold into the vehicle with a new gasket. DO NOT APPLY SEALER.

8. Install the 8 manifold bolts and tighten starting at the center and working outward in both directions. Torque to 200 inch lbs. (23 Nm).

9. Reconnect the heated oxygen sensor.

10. Install the heat shield.

11. Install the exhaust pipe and torque fasteners to 250 inch lbs. (28 Nm).

12. Reconnect the negative battery cable. Start the engine and allow to idle while inspecting the manifold for exhaust leaks.

2.5L Engine

1. Disconnect the negative battery cable remote connection located at the left strut tower.

2. Raise and safely support the vehicle.

3. Disconnect the exhaust pipe connection to the rear (cowl side) exhaust manifold at the flex joint.

NOTE: It may be necessary to remove the whole exhaust system.

4. Remove the bolts attaching the cross-over pipe to the manifolds and remove the cross-over pipe assembly.

5. Disconnect the oxygen sensor lead wire at the rear manifold. Remove the oxygen sensor at the rear exhaust manifold.

6. Remove the power steering bracket.

7. Remove the rear exhaust manifold heat shield.

8. Remove the rear manifold attaching nuts and remove the rear manifold.

9. Lower the vehicle and disconnect the front heated oxygen sensor wiring connector. Remove the front heated oxygen sensor.

10. Remove the front manifold heat shield.

11. Remove the front manifold securing nuts and remove the front manifold.

To install:

12. Clean all parts well. Inspect the exhaust manifolds for damage or cracks and check for distortion of the cylinder head mounting surface and exhaust crossover mounting surface with a straight-edge and thickness gauge.

13. Install a new front manifold gasket.

14. Install the front manifold and torque the nuts to 22 ft. lbs. (30 Nm).

15. Install the front exhaust manifold heat shield and torque the heat shield mounting screws to 115 inch lbs. (13 Nm).

16. Install and tighten the front heated oxygen sensor. Reconnect the oxygen sensor wiring connector.

17. Raise the vehicle.

18. Install a new rear exhaust manifold gasket. Install the rear exhaust manifold.

19. Torque the manifold nuts to 22 ft. lbs. (30 Nm).

20. Install the power steering bracket.

21. Install the crossover pipe and torque the nuts to 22 ft. lbs. (30 Nm).

22. Install and tighten the rear heated oxygen sensor. Reconnect the rear heated oxygen sensor lead.

23. Reconnect the exhaust pipe to the rear manifold. Torque the exhaust pipe-to-rear exhaust manifold flange mounting bolts to 21 ft. lbs. (28 Nm).

24. Lower the vehicle. Reconnect the negative battery cable. Start the engine and allow the engine to idle while inspecting the vehicle for exhaust leaks at the manifold.

Front Crankshaft Seal

REMOVAL AND INSTALLATION

2.0L and 2.4L Engines

The timing belt must be removed for this procedure. Use care that all timing marks are aligned after installation or then engine will be damaged.

1. Disconnect the negative battery cable remote location located on the left strut tower.

2. Remove the accessory drive belts.

3. Raise and safely support the vehicle. Drain the engine oil.

4. Remove the crankshaft damper/pulley using a jaw puller tool.

5. Remove the timing belt cover and timing belt using the recommended procedures.

6. Remove the crankshaft timing belt sprocket using special tool No. 6793.

―――― **CAUTION** ――――
Do not nick the seal surface of the crankshaft or the seal bore.

7. Remove the front crankshaft seal using tool No. 6771 or equivalent seal puller. Be careful not to damage the seal contact area of the crankshaft.

To install:

8. Apply a light coating of clean engine oil to the lip of the new oil seal. Install the new front crankshaft oil seal by using oil seal installer tool No. 6780-1 or equivalent seal tool.

9. Place new oil seal into the opening with the seal spring facing the inside of the engine. Be sure the oil seal is installed flush with the front cover.

10. Install the crankshaft timing belt sprocket using tool No. 6792.

NOTE: Be sure the word "FRONT" on the timing belt sprocket is facing you.

11. Install the timing belt and timing belt cover.

12. Install the crankshaft damper/pulley onto the crankshaft. Use thrust bearing/washer and 12M-1.75 x 150mm bolt from special tool No. 6792. Install the crankshaft damper/pulley retaining bolt and torque to 105 ft. lbs. (142 Nm).

13. Lower the vehicle.

14. Install the accessory drive belts. Adjust the belts to the proper tension.

15. Refill the engine with the correct amount of clean engine oil.

16. Reconnect the negative battery cable. Start the engine and check for leaks.

2.5L Engine

The timing belt must be removed for this procedure. Use care to make sure all timing marks are aligned after this service or the engine will be damaged.

1. Disconnect the negative battery cable remote location located on the left strut tower.

2. Remove the accessory drive belts.

3. Raise and safely support the vehicle. Drain the engine oil.

4. Remove the right inner splash shield.

5. Remove the crankshaft damper/pulley.

6. Remove the timing belt covers and timing belt.

7. Remove the crankshaft timing belt sprocket and key.

8. Remove the front crankshaft seal by prying it out with a flat tipped prytool. Be sure to cover the end of the prytool tip with a shop towel.

―――― **CAUTION** ――――
Do not nick the seal surface of the crankshaft or the seal bore.

To install:

9. Apply a light coating of clean engine oil to the lip of the new oil seal. Install the new front crankshaft oil seal into the oil pump housing by using oil seal installer tool No. MD998717 or equivalent seal installer. Be sure the oil seal is installed flush with the oil pump cover.

10. Install the crankshaft timing belt sprocket and key.

11. Install the timing belt and timing belt covers using the recommended procedure. Verify that all timing marks are correctly aligned or the engine will be damaged.

12. Install the crankshaft damper/pulley onto the crankshaft. Install the crankshaft damper/pulley retaining bolt and torque to 134 ft. lbs. (182 Nm).

13. Install the right inner splash shield.

14. Lower the vehicle.

15. Install the accessory drive belts. Adjust the belts to the proper tension.

16. Refill the engine with the correct amount of clean engine oil.

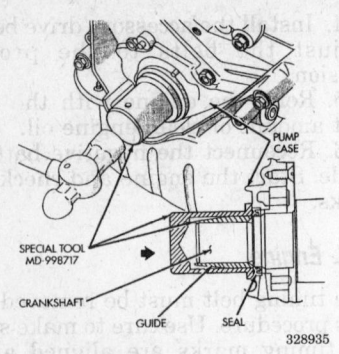

Front crankshaft oil seal installation — 2.5L engine

17. Reconnect the negative battery cable. Start the engine and check for leaks.

Timing Belt, Sprockets, Tensioner and Front Cover

ADJUSTMENT

2.0L and 2.4L Engines

Once the timing belt has been properly installed onto the engine, it can be adjusted as follows:

1. Move the crankshaft to TDC to take up the belt slack.
2. Install the tensioner to the block but do not tighten.
3. Using a torque wrench on the tensioner pulley apply 250 inch lbs. (28 Nm) of torque to the tensioner pulley.
4. With torque being applied to the tensioner pulley, move the tensioner up against the tensioner bracket and torque the fasteners to 275 inch lbs. (31 Nm).
5. Remove the tensioner plunger pin, the tension is correct when the plunger pin can be removed and replaced easily.
6. Rotate the crankshaft 2 revolutions and recheck the timing marks.
7. Install the timing belt cover.

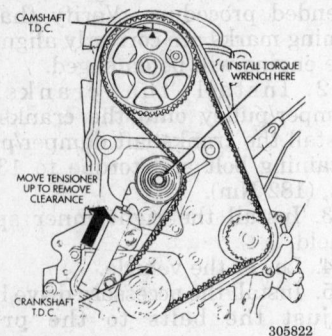

Adjusting timing belt tension — 2.0L engine

2.5L Engine

Once the timing belt has been installed onto the engine, it is adjusted as follows:

1. Install a binder clip on the belt to the sprocket so it will not slip out of position.
2. Be sure to keep the belt taught in position.
3. Install a binder clip on the front sprocket and belt.
4. Rotate the crankshaft to TDC.
5. Rotate the crankshaft sprocket in the clockwise direction to tension the belt. Be sure all timing marks are still in alignment.
6. Using tool number MD998767 and a torque wrench on the tensioner pulley, apply 39 inch lbs. (4.4 Nm) of torque to the tensioner pulley. Torque the tensioner pulley bolt to 35 ft. lbs. (48 Nm).
7. With torque being applied to the tensioner pulley, install the tensioner to the tensioner pulley bracket and torque the fasteners to 205 inch lbs. (23 Nm).
8. Remove the tensioner plunger pin, the tension is correct when the plunger pin can be removed and replaced easily.
9. Remove the timing belt binder clips.
10. Rotate the crankshaft 2 revolutions and recheck the timing marks.
11. Install the timing belt cover.

REMOVAL AND INSTALLATION

2.0L Engine

1. Disconnect the negative battery cable remote connection located on the left strut tower.
2. Remove the drive belts and accessories.
3. Remove the crankshaft damper.
4. Remove the right engine mount
5. Place a support under the engine.

6. Remove the engine mount bracket
7. Remove the timing belt cover.
8. Loosen the timing belt tensioner bolts.
9. Remove the timing belt and the tensioner.
10. Place the tensioner into a soft jawed vise to compress the tensioner.
11. After compressing the tensioner place a pin (5/64 in. Allen wrench will work) into the plunger side hole to retain the plunger until installation.

To install:

12. Set the crankshaft sprocket to Top Dead Center (TDC) by aligning the notch on the sprocket with the arrow on the oil pump housing then back off the sprocket 3 notches before TDC.
13. Set the camshaft to align the timing marks.
14. Move the crankshaft to 1/2 notch before TDC.
15. Install the timing belt starting at the crankshaft then around the water pump and around the camshaft last.
16. Move the crankshaft to TDC to take up the belt slack.
17. Install the tensioner to the block but do not tighten.
18. Using a torque wrench on the tensioner pulley apply 250 inch lbs. (28 Nm) of torque to the tensioner pulley.
19. With torque being applied to the tensioner pulley, move the tensioner up against the tensioner bracket and torque the fasteners to 275 inch lbs. (31 Nm).
20. Remove the tensioner plunger pin, the tension is correct when the plunger pin can be removed and replaced easily.
21. Rotate the crankshaft 2 revolutions and recheck the timing marks.
22. Install the timing belt cover.
23. Install the engine mount bracket.
24. Install the right engine mount.
25. Remove the engine support.

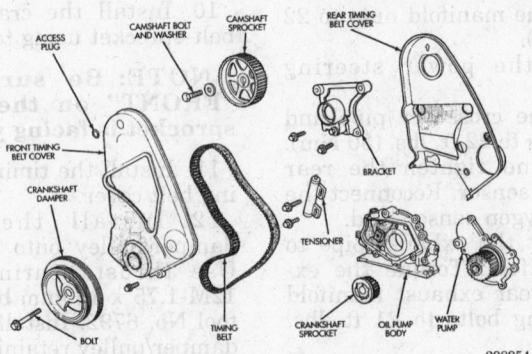

Timing belt assembly — 2.0L engine

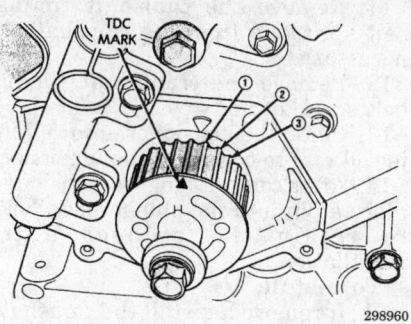

Crankshaft position — 2.0L engine

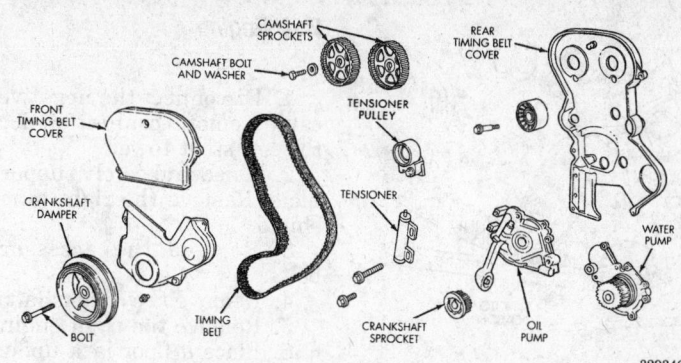

Timing belt assembly — 2.4L engine

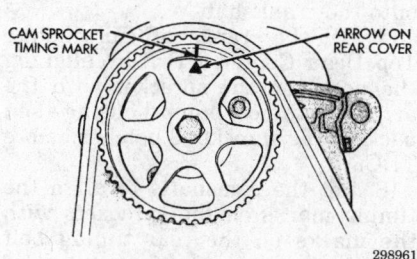

Camshaft position — 2.0L engine

26. Install the crankshaft damper and torque to 105 ft. lbs. (142 Nm).
27. Install the drive belts and accessories.
28. Install the right inner splash shield.
29. Perform the crankshaft and camshaft relearn alignment procedure using the DRB scan tool or equivalent.

2.4L Engine

1. Disconnect the negative battery cable remote connection located on the left strut tower.

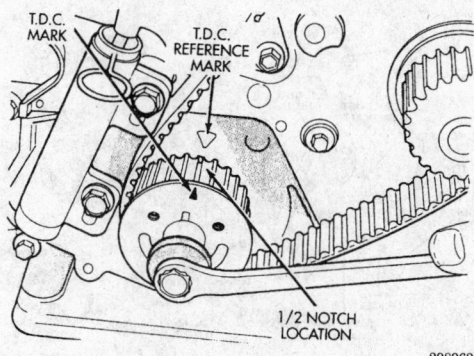

Adjusting crankshaft position — 2.0L engine

2. Remove the right inner splash shield.
3. Remove the accessory drive belts.
4. Remove the crankshaft damper.
5. Remove the right engine mount.
6. Place a floor jack under the vehicle to support the engine.
7. Remove the engine mount bracket
8. Remove the timing belt cover.

NOTE: Do not rotate the crankshaft or the camshafts after the timing belt has been removed. Damage to the valve components

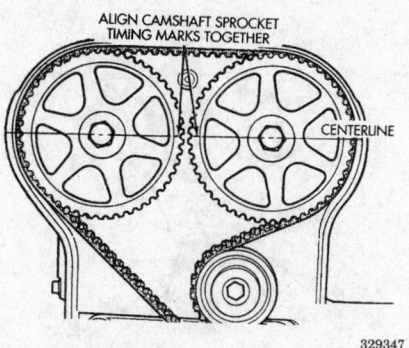

Camshaft alignment — 2.4L engine

may occur. Before removing the timing belt, always align the timing marks.

9. Align the timing marks of the timing belt sprockets to the timing marks on the rear timing belt cover and oil pump cover. Loosen the timing belt tensioner bolts.
10. Remove the timing belt and the tensioner.
11. Remove the camshaft timing belt sprockets.
12. Remove the crankshaft timing belt sprocket using special removal tool No. 6793 or equivalent.
13. Place the tensioner into a soft jawed vise to compress the tensioner.
14. After compressing the tensioner place a pin ($\frac{5}{64}$ in. Allen wrench will work) into the plunger side hole to retain the plunger until installation.
   To install:
15. Using special tool No. 6792, install the crankshaft timing belt sprocket onto the crankshaft.
16. Install the camshaft sprockets onto the camshafts. Install and torque the camshaft sprocket bolts to 75 ft. lbs. (101 Nm).
17. Set the crankshaft sprocket to Top Dead Center (TDC) by aligning the notch on the sprocket with the arrow on the oil pump housing.
18. Set the camshafts to align the timing marks on the sprockets.
19. Move the crankshaft to ½ notch before TDC.
20. Install the timing belt starting at the crankshaft then around the water pump sprocket, idler pulley, camshaft sprockets and then around the tensioner pulley.
21. Move the crankshaft sprocket to TDC to take up the belt slack.
22. Install the tensioner to the block but do not tighten.
23. Using a torque wrench on the tensioner pulley apply 250 inch lbs. (28 Nm) of torque to the tensioner pulley.

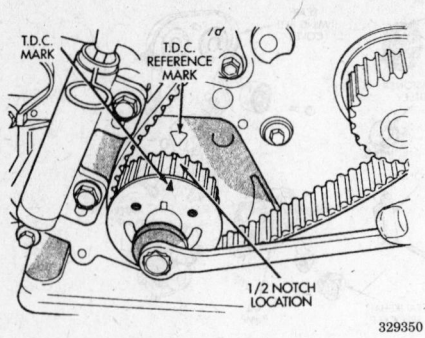

Crankshaft position — 2.4L engine

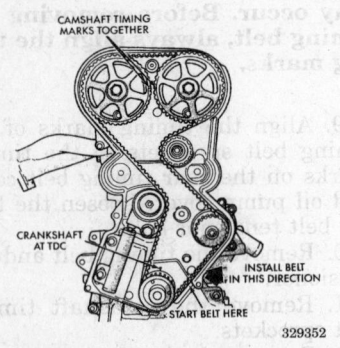

Timing belt proper position — 2.4L engine

24. With torque being applied to the tensioner pulley, move the tensioner up against the tensioner pulley bracket and torque the fasteners to 275 inch lbs. (31 Nm).

25. Remove the tensioner plunger pin, the tension is correct when the plunger pin can be removed and replaced easily.

26. Rotate the crankshaft 2 revolutions and recheck the timing marks. Wait several minutes and then recheck that the plunger pin can easily be removed and installed.

27. Install the front timing belt cover.

28. Install the engine mount bracket.

29. Install the right engine mount.

30. Remove the floor from under the vehicle.

31. Install the crankshaft damper and torque to 105 ft. lbs. (142 Nm).

32. Install the accessory drive belts and adjust to the proper tension.

33. Install the right inner splash shield.

34. Reconnect the negative battery cable.

35. Perform the crankshaft and camshaft relearn alignment procedure using the DRB scan tool or equivalent.

2.5L Engine

1. Disconnect the negative battery cable remote connection located on the left strut tower.

2. Raise and safely support the vehicle. Remove the right inner splash shield.

3. Remove the accessory drive belts.

4. Remove the crankshaft damper.

5. Remove the right engine mount.

6. Place a floor jack under the vehicle to support the engine.

7. Remove the right engine mount bracket

8. Remove the timing belt upper left cover, upper right cover and lower cover.

9. Loosen the timing belt tensioner bolts.

NOTE: Before removing timing belt, be sure to align the sprocket timing marks to the timing marks on the rear timing belt cover.

10. If the present timing belt is going to be reused, mark the running direction of the timing belt for installation. Remove the timing belt and the tensioner.

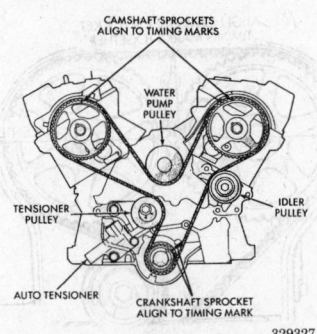

Timing belt routing and timing mark relationship — 2.5L engine

11. Remove the camshaft timing belt sprockets from the camshaft, if necessary.

12. Remove the crankshaft timing belt sprocket and key.

13. Place the tensioner into a soft jawed vise to compress the tensioner.

14. After compressing the tensioner place a pin into the plunger side hole to retain the plunger until installation.

To install:

15. If removed, install the camshaft sprockets onto the camshaft. Install the camshaft sprocket bolt and torque to 65 ft. lbs. (88 Nm).

16. If removed, install the crankshaft timing belt sprocket and key onto the crankshaft.

17. Set the crankshaft sprocket to Top Dead Center (TDC) by aligning the notch on the sprocket with the arrow on the oil pump housing then back off the sprocket 3 notches before TDC.

18. Set the camshafts to align the timing marks on the sprockets with the marks on the rear timing belt cover.

19. Install the belt on the rear camshaft sprocket first.

20. Install a binder clip on the belt to the sprocket so it won't slip out of position.

21. Keeping the belt taunt, install it under the water pump pulley and around the front camshaft sprocket.

22. Install a binder on the front sprocket and belt.

23. Rotate the crankshaft to TDC.

24. Continue routing the belt by the idler pulley and around the crankshaft sprocket to the tensioner pulley.

25. Move the crankshaft sprocket clockwise to TDC to take up the belt slack. Check that all timing marks are in alignment.

26. Install the tensioner to the block but do not tighten.

27. Using special tool No. MD998767 and a torque wrench on

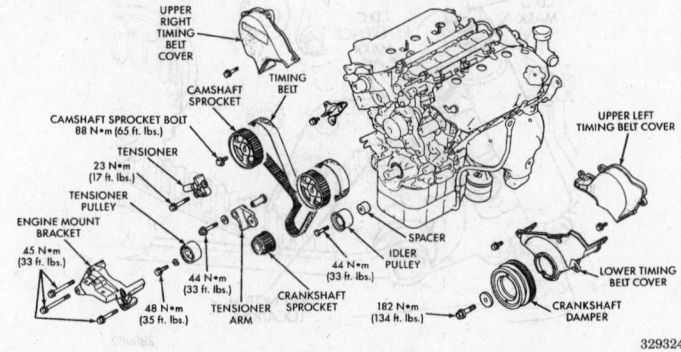

Timing belt and related components — 2.5L engine

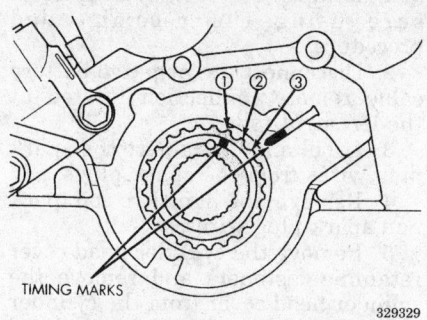

Crankshaft position — 2.5L engine

RIGHT CAM SPROCKET   ALIGN TIMING MARKS   LEFT CAM SPROCKET

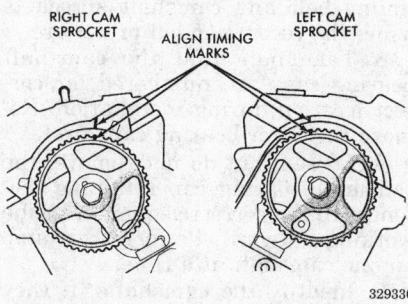

Camshaft sprocket timing mark alignment — 2.5L engine

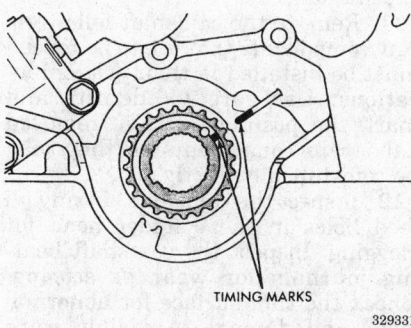

TIMING MARKS

Adjusting crankshaft position — 2.5L engine

the tensioner pulley, apply 39 inch lbs. (4.4 Nm) of torque to the tensioner. Torque the tensioner pulley bolt to 35 ft. lbs. (48 Nm).

28. With torque being applied to the tensioner pulley, move the tensioner up against the tensioner bracket and torque the fasteners to 17 ft. lbs. (23 Nm).

29. Remove the tensioner plunger pin, the tension is correct when the plunger pin can be removed and replaced easily.

30. Rotate the crankshaft 2 revolutions clockwise and recheck the tim-

ing marks. Check to make sure the tensioner plunger pin can be easily installed and removed. If the pin does not remove and install easily, perform the procedure again.

31. Install the timing belt cover.

32. Install the engine mount bracket.

33. Install the right engine mount.

34. Remove the engine support.

35. Install the crankshaft damper and torque to 134 ft. lbs. (182 Nm).

36. Install the accessory drive belts and adjust to proper tension.

37. Install the right inner splash shield.

38. Perform the crankshaft and camshaft relearn alignment procedure using the DRB scan tool or equivalent.

Camshaft

REMOVAL AND INSTALLATION

2.0L Engine

This engine uses a Single Over Head Camshaft (SOHC) running in an aluminum cylinder head. Rocker arm shafts mount directly to the cylinder head. Care must be taken to make sure all valve timing marks align after cylinder head and valvetrain service. Please note that the cylinder head must be removed from the vehicle to service the camshaft.

— **CAUTION** —
Fuel injection systems remain under pressure, even after the engine has been turned OFF. The fuel system pressure must be relieved before disconnecting any fuel lines. Failure to do so may result in fire and/or personal injury.

1. Disconnect the negative battery cable remote connection located on the left strut tower.

2. Make sure the engine is cool before starting cylinder head removal.

3. Relieve the fuel system pressure using the recommended procedure.

4. Place a large drain pan under the vehicle radiator drain plug. Open the drain plug and drain out the engine coolant.

5. Remove the complete air cleaner assembly.

6. Label spark plug wires to the correct spark plugs. Disconnect the spark plug wires from each spark plug.

7. Remove the ignition coil pack.

8. Remove the cylinder head cover retaining bolts and remove the cylinder head cover. Be sure to remove and discard the old gasket material. Clean the cylinder head and cover gasket mating surfaces. Inspect the gasket mating surfaces for flatness.

9. Mark the rocker arm shaft assemblies to identify them for later installation.

10. Remove the rocker arm shaft bolts and remove the rocker arm assemblies from the cylinder head.

11. Remove the timing belt and camshaft sprocket using the recommended procedure.

12. Remove the cylinder head using the recommended procedure.

13. Remove the camshaft sensor and remove the camshaft from the rear of the cylinder head.

To install:

NOTE: The cylinder head bolts should be checked for stretching before reuse. If the thread area of the bolt is necked down the bolts must be replaced with new. New head bolts are recommended.

14. Clean all parts well. Inspect the camshaft journals for scoring. Check the oil feed holes in the head for blockage. Check the camshaft bearing journals for scoring. If light scratches are present, they may be removed with 400 grit abrasive paper. If deep scratches are present, replace the camshaft and check the cylinder head for damage. Replace the cylinder head if worn or damaged.

15. If the camshaft lobes show signs of wear, check the corresponding rocker arm roller for wear or damage. Replace rocker arms/hydraulic lash adjuster if worn or damaged. If the camshaft lobes show signs of pitting on the nose, flank or base circle, replace the camshaft.

16. If the rocker arms and shaft are to be serviced, mark the rocker arms so any that are to be returned to service will be installed in their original locations.

17. Install the rocker arm back on the rocker arm shaft.

18. To install the camshaft, lubricate the bearing journals thoroughly. Install the camshaft into the cylinder head carefully. Make sure it turns freely. If the camshaft installation is satisfactory, install the cam sensor and tighten the screws to 85 inch lbs. (9.6 Nm).

19. Camshaft end-play can be checked. Use the following procedure.

a. Oil the camshaft journals and install the camshaft without the rocker arm assemblies. Install the

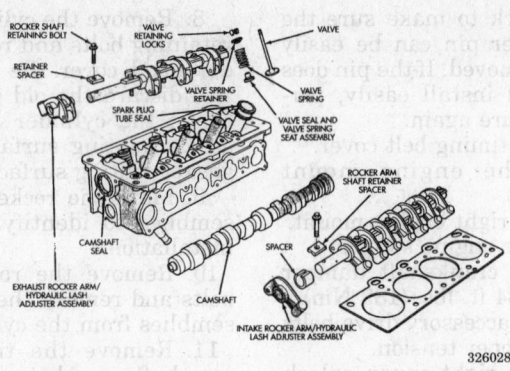

Cylinder head and valve assembly — 2.0L engine

cam sensor and tighten the screws to 85 inch lbs. (9.6 Nm).

b. Setup a dial indicator to touch on the nose of the camshaft.

c. Using a suitable tool, move the camshaft as far rearward as it will go. Make sure the dial indicator probe is in contact with the camshaft.

d. Zero the dial indicator.

e. Move the camshaft as far forward as it will go.

f. Read the end-play on the dial indicator. Specification is 0.005–0.013 in. (0.13–0.33mm).

20. Install the camshaft seal. The camshaft must be installed before the camshaft seal is installed. The seal should be flush with the cylinder head after installation.

21. Install the camshaft sprocket and tighten to the bolt to 85 ft. lbs. (115 Nm).

22. Install the cylinder head using the recommended procedure. Be sure to use new cylinder head mounting bolts.

23. Before installing the rocker arm and shaft assemblies, set the crankshaft to 3 notches before TDC on the crankshaft sprocket.

24. Install the rocker arm and shaft assemblies.

25. Install the camshaft sprocket and timing belt using care to align all

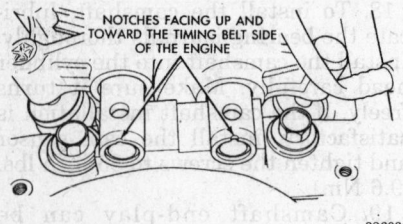

Rocker arm shaft notch location — 2.0L engine

valve timing marks, using the recommended procedure.

26. Install the cylinder head cover along with a new gasket. Torque the cylinder head cover retaining bolts to 105 inch lbs. (12 Nm).

27. Install the ignition coil pack and torque the retaining fasteners to 200 inch lbs. (23 Nm).

28. Install the complete air cleaner assembly.

29. Reconnect the spark plug wires to the correct spark plugs.

30. Check to be sure all electrical, vacuum and fluid connections have been reconnected properly.

31. Refill the cooling system. An oil and filter change are recommended.

32. Reconnect the negative battery cable remote connection.

33. Start the engine and check for leaks. Run the engine with the radiator cap off so as the engine warms and the thermostat opens, coolant can be added to the radiator.

34. Shut down the engine and allow to cool. Verify correct fluid levels. Test drive vehicle to check for proper operation.

2.4L Engine

This engine uses a Dual Over Head Camshaft (DOHC) 4-valves per cylinder cross flow aluminum cylinder head. The valves are actuated by roller cam followers which pivot on stationary hydraulic valve adjusters. Care must be taken to make sure all valve timing marks align after cylinder head and valvetrain service.

CAUTION

Fuel injection systems remain under pressure, even after the engine has been turned OFF. The fuel system pressure must be relieved before disconnecting any fuel lines. Failure to do so may result in fire and/or personal injury.

1. Relieve the fuel system pressure using the recommended procedure.

2. Disconnect the negative battery cable remote connection located at the left strut tower.

3. Label and disconnect the spark plug wires from the spark plugs.

4. Remove the ignition coil pack and spark plug wires.

5. Remove the cylinder head cover retaining fasteners and remove the cylinder head cover from the cylinder head. Discard the old cylinder head cover gasket.

6. Remove the ground strap.

7. Remove the timing belt covers, timing belt and camshaft sprockets using the recommended procedure.

8. Take note that the camshaft bearing caps are numbered for correct location during installation. Remove the outer bearing caps first.

9. Loosen, but do not remove, the camshaft bearing cap retaining fasteners in the correct sequence, inside working outward. Perform this step on one camshaft at a time.

10. Identify the camshafts, if they are to be reused, for later installation. The camshafts are not interchangeable. Remove the camshaft bearing caps and remove the camshafts.

11. Remove the camshaft followers. Any components that are to be reused must be installed in their original locations. Use care to identify and mark the positions of any removed valvetrain components so they may be reinstalled correctly.

12. Inspect the camshaft bearing oil feed holes in the cylinder head for clogging. Inspect the camshaft bearing journals for wear or scoring. Check the cam surface for abnormal wear and damage. A visible worn groove in the roller path or on the cam lobes is cause for replacement.

To install:

13. Thoroughly clean all camshaft and related parts.

14. The camshaft end-play should be checked using the following procedure:

a. Oil the camshaft journals and install the camshaft **WITHOUT** the cam follower assemblies. Install the rear cam caps and torque to 250 inch lbs. (28 Nm).

b. Carefully push the camshaft as far rearward as it will go.

c. Set up a dial indicator to bear against the front of the camshaft (the sprocket end). Zero the indicator.

d. Move the camshaft forward as far as it will go. Read the dial indi-

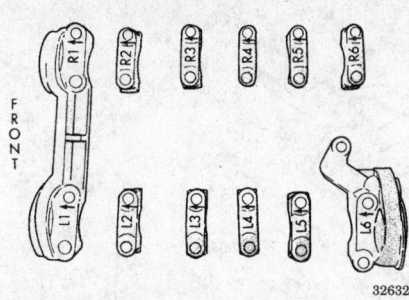

Camshaft bearing cap identification — 2.4L engine

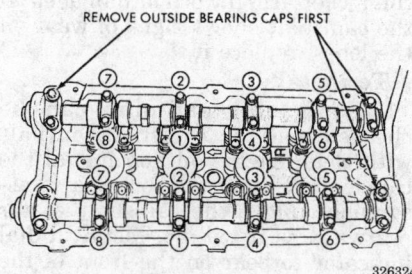

Camshaft bearing cap removal — 2.4L engine

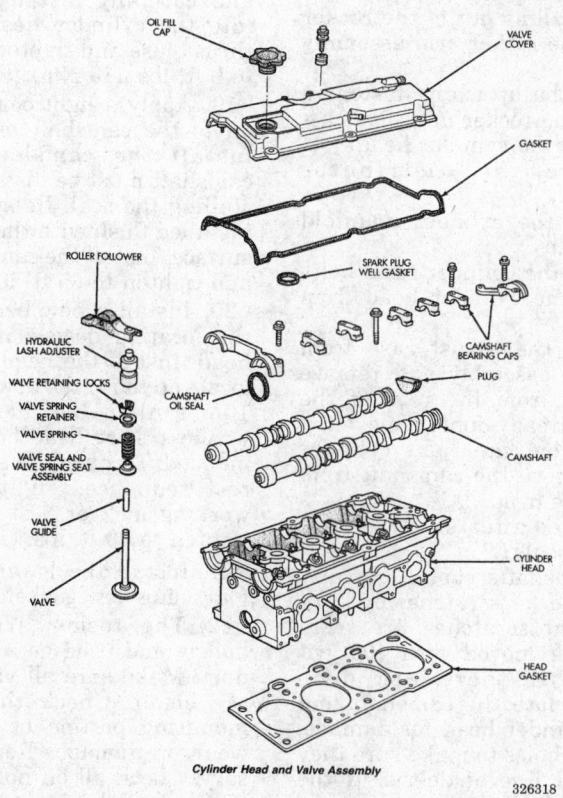

Cylinder Head and Valve Assembly

326318

Cylinder head, valves and related components — 2.4L engine

cator. End-play specification is 0.002–0.010 in. (0.05–0.15mm).

e. If excessive end-play is present, inspect the cylinder head and camshaft for wear; replace if necessary.

15. If satisfied with the fit and condition of the camshafts, remove the camshafts for installation of the cam followers.

16. The hydraulic valve lash adjusters are inside the roller cam followers. Make sure they are clean, well-lubricated with clean engine oil and properly positioned. Install the cam followers in their original positions on the hydraulic adjuster and valve stem.

WARNING

Make sure NONE of the pistons are at Top Dead Center when installing the camshafts.

17. Lubricate the camshaft bearing journals and cam followers with clean engine oil and install the camshafts. Install right and left camshaft bearing caps No. 2 through No. 5 and right side No. 6. Tighten the M6 fasteners to 105 inch lbs. (12 Nm) in correct sequence.

18. Apply Mopar® Gasket Maker or equivalent sealer to the No. 1 and left side No. 6 bearing caps. Install the bearing caps and tighten the M8 fasteners to 250 inch lbs. (28 Nm). The end caps must be installed before the seals may be installed.

19. Install the camshaft end seals.

20. Install the camshaft sprockets, if removed. Install the timing belt using care to make sure all timing marks are properly aligned, using the recommended procedure. Install the timing belt covers.

WARNING

Verify that all timing marks are correct. If the timing belt or sprockets are incorrectly installed, engine damage will occur. Take time to make sure all timing marks are correctly aligned.

21. Clean the cylinder head cover and cylinder head gasket rails (mating surfaces). Make certain the rails are flat.

22. Install new cylinder head cover gaskets. Use care. DO NOT allow oil or solvents to contact the timing belt as they can deteriorate the rubber and cause tooth skipping. Apply Mopar Silicone Rubber Adhesive Sealant, or equivalent, at the camshaft cap corners and at the top edge of the ½ round seal.

NOTE: Inspect the spark plug well seals for cracking and/or swelling and replace if necessary.

23. Install the cylinder head cover assembly to the head and tighten the fasteners in sequence using the following 3 Steps:

a. First: tighten all cylinder head cover fasteners to 40 inch lbs. (4.5 Nm).

b. Second: tighten all fasteners to 80 inch lbs. (9 Nm).

c. Third: tighten all fasteners to 105 inch lbs. (12 Nm).

24. Install the ignition coil pack and connect the spark plug wiring to the correct spark plugs. Torque the coil pack retaining fasteners to 105 inch lbs. (12 Nm).

25. Reconnect the ground strap.

26. Check to be sure all vacuum lines and remaining wiring have been reconnected.

27. An oil and filter change is recommended.

28. Connect the negative battery cable and test run vehicle. Check for leaks and for proper operation.

2.5L Engine

This engine uses aluminum alloy cylinder heads with 4-valves per cylinder and pressed-in cast iron valve guides. The cylinders are common to

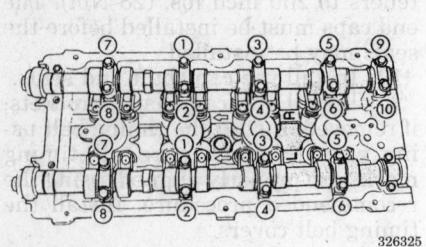

Camshaft bearing cap tightening sequence —
2.4L engine

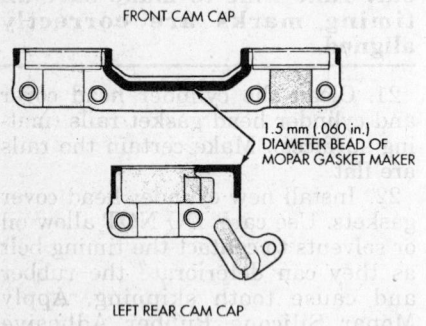

FRONT CAM CAP

1.5 mm (.060 in.)
DIAMETER BEAD OF
MOPAR GASKET MAKER

LEFT REAR CAM CAP

Camshaft bearing cap sealing — 2.4L engine

either cylinder bank. Two overhead camshafts are supported by 4 bearing journals which are part of the head with the distributor driven off the right (firewall side) cylinder head. Right and left camshaft drive sprockets are interchangeable. The sprockets and engine water pump are driven by the timing belt. Care must be taken to make sure all valve timing marks align after cylinder head and valvetrain service. Please note that for camshaft service, the cylinder head must be removed.

CAUTION

Fuel injection systems remain under pressure, even after the engine has been turned OFF. The fuel system pressure must be relieved before disconnecting any fuel lines. Failure to do so may result in fire and/or personal injury.

---

1. Disconnect the negative battery cable from the remote auxiliary jumper terminal.

2. Relieve the fuel system pressure using the recommended procedure.

3. Place a large drain pan under the radiator drain plug. Drain the cooling system.

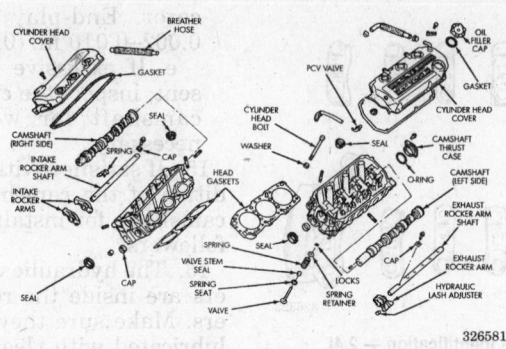

Cylinder head, camshafts and rocker assemblies — 2.5L engine

4. Remove the timing belt covers, timing belt and camshaft sprockets using the recommended procedure.

5. The intake manifold is a two-piece unit. The upper part is a large air intake plenum of aluminum alloy. Use care working with light alloy parts. Remove the air intake plenum first, then remove the lower intake manifold.

6. Disconnect, label and set aside the spark plug wires.

7. Remove the cylinder head cover screws and remove the cover.

8. Identify the rocker arm shaft assemblies before removal.

9. Install the auto lash adjuster retainers Special Tool MD 998443 or equivalent to keep the auto lash adjusters from falling out of the rocker arms when the rocker arm assembly is removed.

10. Loosen the attaching fasteners and remove the rocker arm shaft assemblies from the cylinder head.

11. Remove the distributor assembly.

12. Remove the exhaust manifold and crossover.

13. Remove the cylinder head bolts and remove the cylinder head from the vehicle.

14. Remove the thrust case from the left head assembly and remove the camshaft from the rear of the head. If not already done, remove the distributor from the right cylinder head and remove the camshaft from the rear of the head.

15. Remove and discard the camshaft oil seals.

16. The camshafts should be carefully inspected for scratches or worn areas. If light scratches are seen, they may be removed with 400 grit sandpaper. If there are deep scratches, replace the camshaft and check the cylinder head for damage. Check the oil holes to make sure they are open and free of debris. If the camshaft lobes show signs of wear, check the corresponding rocker arm

roller for wear or damage. Replace the rocker arm if worn or damaged. If the camshaft shows signs of wear on the lobes, replace it.

To install:

17. Camshaft end-play can be checked. Oil the camshaft journals with clean engine oil and install the camshaft without the rocker arm assemblies. Move the camshaft as far rearward as it will go. Mount a dial indicator to bear on the front of the camshaft. Zero the indicator. Move the camshaft as far forward as it will go. End-play should be 0.004–0.008 in. (0.1–0.2mm). Maximum allowed end-play is 0.016 in. (0.4mm).

18. Lubricate the camshaft journals and carefully install the camshaft into the cylinder head. Install the thrust case and tighten the fasteners to 9 ft. lbs. (13 Nm).

19. Apply a light coating of engine oil to the camshaft oil seal lip and install the camshaft seal. The camshaft must be installed before installing the seal. Be sure the seal is installed flush with the cylinder head surface. Install the camshaft sprocket and tighten to 65 ft. lbs. (88 Nm).

20. Install a new head gasket over the locating dowels in the cylinder head. Install the head, using care to locate on top of the dowels. Install the 10mm Allen hex head bolts with washers. New head bolts are recommended. Torque the head bolts in correct sequence. Tighten gradually, working in 2 or 3 steps and finally tighten to 80 ft. lbs. (108 Nm).

21. Install the lower intake manifold using new gaskets.

22. The rocker arm shafts are hollow and used as a lubrication oil duct. Make sure all valvetrain parts are clean. Check the rocker arm mounting portion of the shafts for wear or damage. Replace if necessary. Check all oil holes for clogging with a small wire and clean as required. If any rockers were removed,

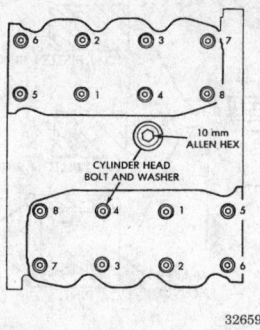

Cylinder head bolt tightening sequence — 2.5L engine

326592

lubricate and install on the shafts in their original positions.

23. Install the rocker arm and shaft assemblies.

24. Install the timing belt. This is a somewhat complex procedure. Use the recommended procedure. It is most important that all valve timing marks align properly or engine damage will result.

25. Inspect the spark plug tube seals located on the ends of each tube. These seals slide onto each tube to seal the cylinder head cover to the spark plug tube. If these seals show signs of hardness and/or cracks, they should be replaced.

26. Clean the cylinder head and cover mating surfaces. Install a new gasket and install the cover. Torque bolts to 88 inch lbs. (10 Nm). Reconnect the spark plug wires.

27. Position a new upper intake manifold (plenum) gasket in place and install the upper manifold. Torque the bolts to 13 ft. lbs. (18 Nm). Install the bolts for the plenum support brackets and connect the EGR tube.

28. Reconnect the throttle and speed control cables.

29. Reconnect the TPS and idle air control motor electrical connectors.

30. Reconnect the MAP and intake air temperature sensors.

31. Install the air inlet resonator, air inlet hose and the air cleaner housing cover.

32. Check to be sure all remaining electrical connectors have been reconnected. Tighten the air tube connections.

33. Refill the cooling system. An oil and filter change is recommended whenever a cylinder head has been removed since coolant can get into the oil system.

34. Connect the negative battery cable. Start the engine and check for leaks, abnormal noises and vibrations. Bleed the cooling system.

Balance Shaft

REMOVAL AND INSTALLATION

2.4L Engine

The 2.4L engine is equipped with 2 balance shafts installed in a carrier mounted to the lower crankcase. These balance shafts interconnect through gears to rotate in opposite directions. The gears are powered by a short, crankshaft-driven chain and rotate at 2 times the speed of the crankshaft. This will counterbalance certain reciprocating masses of the engine.

An oil passage from the No. 1 main bearing cap through the balance shaft carrier support leg provides lubrication to the balance shafts. This passage directly supplies engine oil to the front bearings and internal machined passages in the shafts that routes engine oil from the front to the rear shaft bearing journals.

Please note that this procedure requires removal of the timing belt. Valvetrain timing is critical to engine performance and to prevent engine damage. Work carefully and verify all timing marks as the engine is reassembled.

1. Disconnect the negative battery cable remote connection located on the left strut tower.

2. Remove the accessory drive belts.

3. Remove the timing belt cover and timing belt.

4. Raise and safely support the vehicle.

5. Place a large drain pan under the oil pan drain plug and drain the engine oil.

6. Remove the oil pump and oil pan.

7. Remove the balance shaft drive chain cover.

8. Remove the drive chain guide and drive chain tensioner.

9. Remove the gear cover retaining stud (double ended to also mount drive chain guide).

10. Remove the balance shaft gear and chain sprocket retaining screws.

11. Remove the drive chain and chain sprocket assembly by using 2 prybars to work the sprocket back and forth until it is removed from the crankshaft.

12. Remove the gear cover and balance shafts.

13. Remove the 4 balance shaft carrier-to-crankcase mounting bolts and separate the carrier from the engine bedplate.

To install:

14. Install the balance shafts into the carrier and place the carrier into proper position against the engine bedplate. Install and torque the 4 mounting bolts to 40 ft. lbs. (54 Nm).

15. Rotate the balance shafts until both balance shaft key ways are pointed up parallel to the vertical centerline of the engine.

16. Install the short hub drive gear onto the sprocket driven balance shaft and the long hub gear onto the chain driven shaft. Once the gears are installed onto the shafts, the gear and balance shaft key ways must be up and gear alignment dots properly meshed.

17. Install the gear cover and retaining stud fastener. Torque the double ended retaining stud fastener to 105 inch lbs. (12 Nm).

18. Install the drive chain crankshaft sprocket. Be sure to align the flat on the sprocket to the flat on the crankshaft with the sprocket facing front.

19. Rotate the crankshaft until the No. 1 cylinder is at Top Dead Center (TDC). The timing marks on the chain sprocket should align with the parting line on the left side of No. 1 main bearing cap.

20. Place the drive chain around the crankshaft sprocket so the nickel plated link of the chain is on the No. 1 cylinder timing mark of the crankshaft sprocket.

21. Install the balance shaft sprocket into the drive chain. Be sure the timing mark on the balance shaft sprocket (yellow dot) lines up with lower nickel plated link on the chain.

22. With the key ways of the balance shaft pointing up in the 12 o'clock position, slide the balance shaft drive chain sprocket onto the end of the balance shaft. If necessary to allow for clearance, the balance shaft may be pushed in slightly.

NOTE: The lower nickel plated link, timing mark on the balance shaft sprocket and arrow on the side of the gear cover should all line up when the balance shafts are properly timed.

23. If the sprockets are timed correctly, install the balance shaft bolts and torque to 250 inch lbs. (28 Nm). It may be necessary to place a wooden block between the crankcase and crankshaft counterbalance to prevent crankshaft and gear rotation.

24. Install the drive chain tensioner but keep it loose at this point.

25. Position the drive chain guide onto the double ended stud fastener. Be sure the tab on the guide fits into

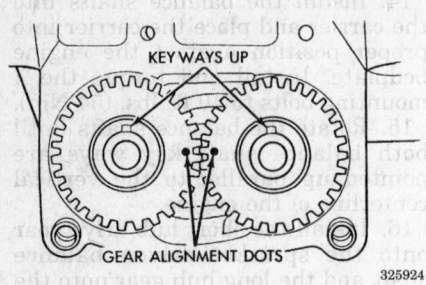

Balance shaft gear timing — 2.4L engine

325924

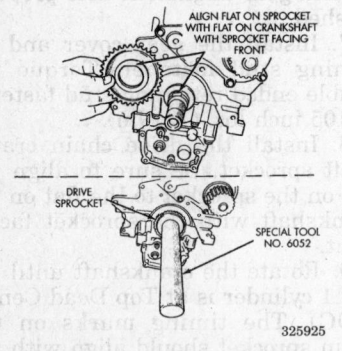

Crankshaft sprocket installation — 2.4L engine

325925

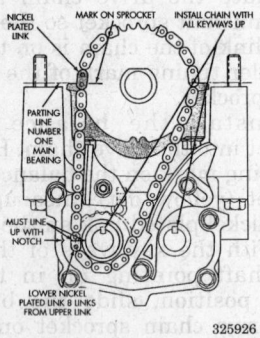

Balance shaft timing — 2.4L engine

325926

the slot on the gear cover. Install the nut/washer assembly and torque to 105 inch lbs. (12 Nm).

26. Place a shim 0.039 in. x 2.75 in. (1mm x 70mm) long between the tensioner and chain. Push the tensioner and shim up against the chain. Apply pressure of 5.5–6.6 lbs. directly behind the adjustment slot to take up the slack. Be sure the chain makes shoe radius contact.

27. With the pressure applied, tighten the top tensioner adjustment bolt first, then the lower pivot bolt.

Torque the bolts to 105 inch lbs. (12 Nm). Remove the shim after torquing the bolts.

28. Install the chain cover and torque the screws to 105 inch lbs. (12 Nm).

29. Install oil pump and oil pan.

30. Install the oil pan drain plug and gasket. Torque the drain plug to 25 ft. lbs. (34 Nm).

31. Lower the vehicle.

32. Install the timing belt and timing belt cover. Verify that all timing marks are correct. This is absolutely critical.

33. Install the accessory drive belts and adjust as necessary.

34. Refill the engine with new, clean engine oil to the proper level. An oil filter change is also recommended.

35. Reconnect the negative battery cable. Start the engine and check for leaks.

Piston and Connecting Rod

POSITIONING

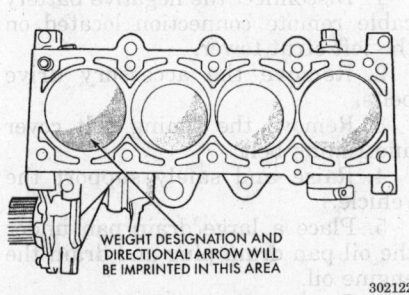

Markings for correct piston installation — 2.0L, 2.4L engines

302122

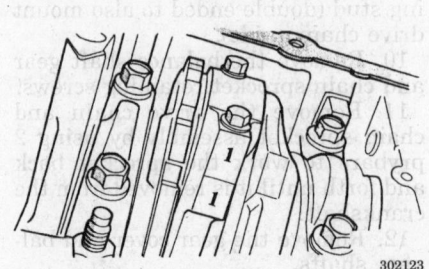

Connecting rod-to-end cap identification — 2.0L, 2.4L and 2.5L engines

302123

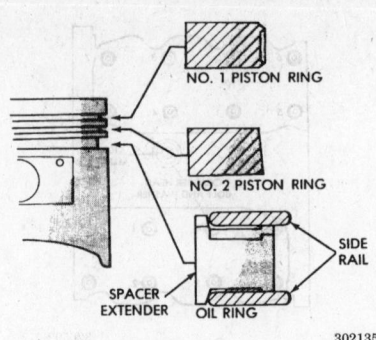

Piston ring installation — 2.0L, 2.4L, 2.5L engines

302135

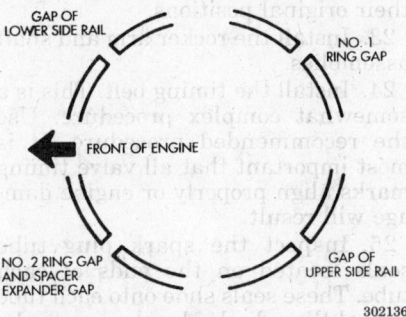

Piston ring end gap position — 2.0L, 2.4L engines

302136

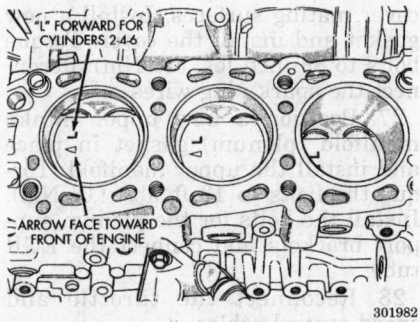

Markings on pistons to note direction of installation — 2.5L engine

301982

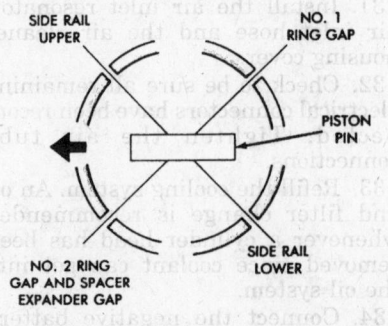

Piston ring end gap positioning — 2.5L engine

301984

ENGINE LUBRICATION

Oil Pan

REMOVAL AND INSTALLATION

2.0L Engine

1. Disconnect the negative battery cable remote connection located on the left strut tower.
2. Raise and safely support the vehicle.
3. Place a large oil pan under the oil pan drain plug. Drain the oil from the engine.
4. Remove the transaxle bending bracket.
5. Remove the front engine mount and bracket.
6. Remove the transaxle inspection cover.
7. Remove the oil filter and oil filter adapter.
8. Remove the oil pan attaching bolts.
9. Remove the oil pan.
10. Clean the oil pan as well as the oil pan gasket mating surfaces.

To install:

11. Using a suitable rubber adhesive gasket sealant, apply a ⅛ in. bead at the oil pump-to-engine block parting line.
12. Install the new oil pan gasket by positioning it properly onto the oil pan.

NOTE: If a gasket is not available, use a ⅛ in. bead of silicone gasket maker.

13. Install the oil pan to the engine.
14. Torque the oil pan attaching bolts to 105 inch lbs. (12 Nm).

15. Install the oil filter adapter as follows:
    a. Be sure the O-ring seal is seated in the groove on the adapter.
    b. Align the locating roll pin into the engine block.
    c. Torque the retaining fastener to 60 ft. lbs. (80 Nm).
16. Install a new oil filter as follows:
    a. Lubricate the contact surface of the rubber oil filter gasket with a light bead of clean engine oil.
    b. Be sure the gasket contact surface on the oil filter adapter is smooth, flat and clean of any debris or old pieces of rubber.
    c. Rotate the oil filter clockwise until the gasket contacts the adapter base. Tighten the filter to 15 ft. lbs. (21 Nm).
17. Install the transaxle inspection cover.
18. Install the front engine mount and engine mount bracket.
19. Install the transaxle bending bracket.
20. Install the oil pan drain plug and gasket. Torque the drain plug to 25 ft. lbs. (34 Nm).
21. Lower the vehicle.
22. Fill the engine with fresh oil to the proper level:
    a. With an oil filter change — 4.5 qts. (4.25L)
    b. Without an oil filter change — 4.0 qts. (3.8L)
23. Reconnect the negative battery cable. Start the engine and check for leaks.

2.4L Engine

1. Disconnect the negative battery cable remote connection located on the left strut tower.
2. Raise and safely support the vehicle.
3. Place a large drain pan under the oil pan drain plug and drain the oil from the engine.

4. If necessary, remove the transaxle bending bracket.
5. Remove the front engine mount and bracket.
6. Remove the oil pan attaching bolts.
7. Remove the oil pan.

To install:

8. Using a suitable gasket sealant apply a ⅛ in. bead at the oil pump to engine block parting line.
9. Install the new gasket.

NOTE: If a gasket is not available, use a ⅛ in. bead of silicone gasket maker.

10. Install the oil pan to the engine.
11. Torque the oil pan attaching bolts to 105 inch lbs. (12 Nm).
12. Install the front engine mount and engine mount bracket.
13. Install the transaxle bending bracket, if necessary.
14. Install the oil pan drain plug and gasket. Torque the drain plug to 25 ft. lbs. (34 Nm).
15. Lower the vehicle.
16. Fill the engine with fresh oil to the proper level. A filter change is recommended.
    a. With oil filter — 5.0 qts. (4.7L)
    b. Without oil filter — 4.5 qts. (4.3L)
17. Reconnect the negative battery cable. Start the engine and check for leaks.

2.5L Engine

1. Disconnect the negative battery cable remote connection at the left strut tower.
2. Raise and safely support the vehicle.
3. Place a large drain pan under the oil pan drain plug and drain the oil from the engine.
4. Remove the engine support module as follows:
    a. Place a support jack underneath the engine/transaxle assembly at the transaxle to prevent it from rotating.
    b. Remove the through bolt at the rear mount and remove the bolts securing the support module to the crossmember.
    c. Remove the upper mounting bolt from the rear support strut bracket.
    d. Remove the front mounting bolts from the support module to the lower radiator support member.
    e. Support the radiator/cooling fan assembly. Remove the lower radiator support member.

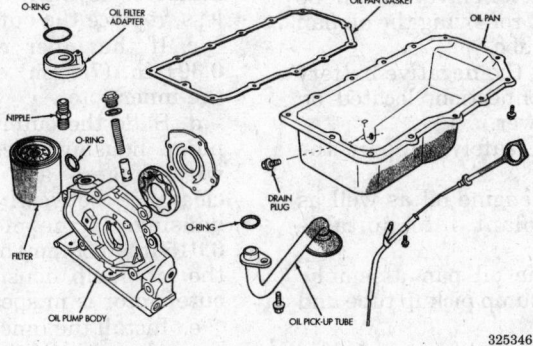

O-RING   OIL FILTER ADAPTER   OIL PAN GASKET   OIL PAN   NIPPLE   O-RING   DRAIN PLUG   O-RING   FILTER   OIL PUMP BODY   OIL PICK-UP TUBE

325346

Oil pan and pump assembly — 2.0L engine

f. Remove the through bolt at the front engine mount and remove the engine support module.

5. Remove the engine oil dipstick tube and dipstick.

6. Remove the starter motor.

7. Remove the engine-to-transaxle struts.

8. Remove the transaxle inspection cover.

9. Remove the oil pan attaching bolts.

10. Remove the oil pan.

To install:

NOTE: Oil pan-to-engine block sealing is provided by using MOPAR® Silicone Rubber Adhesive Sealant or equivalent gasket material. The gasket material should be applied in a continuous bead approximately 0.157 in. (4mm) in diameter with all mounting holes circled. Sealant that is not cured can be removed with a shop towel. Components should be torqued in place while the gasket sealer is still wet to the touch (within a 10 minute period). It is recommended that locating dowels be used during assembly to prevent smearing the gasket material from its location.

11. Apply a continuous 0.157 in. (4mm) bead of MOPAR® Silicone Adhesive Sealant or equivalent to the oil pan gasket surface. Be sure to circle all mounting bolt holes as well. Install the oil pan within a 10 minute period of applying the gasket material to ensure proper sealing.

12. Install the oil pan to the engine.

13. Torque the oil pan attaching bolts to 53 inch lbs. (6 Nm).

14. Install transaxle inspection cover.

15. Install the engine-to-transaxle struts.

16. Install the starter motor.

17. Install the engine oil dipstick tube and dipstick.

18. Install the engine support module as follows:

a. Place the engine support module in proper position under the engine/transaxle assembly.

b. Install the through bolt at the front mount. Do not torque the through bolt at this time.

c. Install the lower radiator support.

d. Install the mounting bolts from the engine support module to the lower radiator support.

e. Install the upper mounting bolt to the rear support strut bracket.

f. Install the through bolt to the rear mount and torque to 45 ft. lbs. (61 Nm).

g. Torque the through bolt at the front mount to 45 ft. lbs. (61 Nm).

h. Remove the support jack holding the engine/transaxle assembly in place.

19. Install the oil pan drain plug and gasket. Torque the drain plug to 29 ft. lbs. (40 Nm).

20. Lower the vehicle.

21. Fill the engine with fresh oil to the proper level. An oil filter change is recommended.

a. With oil filter — 4.5 qts. (4.3L)

b. Without oil filter — 4.0 qts. (3.8L)

22. Reconnect the negative battery cable. Start the engine and check for leaks.

NOTE: Whenever the vehicle sub-frame is removed or lowered, the wheel alignment should be checked.

Oil Pump

REMOVAL AND INSTALLATION

2.0L and 2.4L Engines

The oil drawn up through the pickup tube is pressurized by the pump and routed through the full flow filter to the main oil gallery running the length of the cylinder block. The oil pickup, pump and check valve provide oil flow to the main oil gallery. A vertical hole at the number 5 bulkhead routes pressurized oil through a restrictor up past a cylinder head bolt to an oil gallery running the length of the cylinder head. The camshaft journals are slotted to allow pressurized oil to pass into the bearing cap cavities. Small holes in the bearing caps direct oil to the camshaft lobes.

It is necessary to remove the oil pan, oil pickup and oil pump housing to service the oil pump rotors. The oil pump pressure relief valve can be serviced without removing the oil pan and oil pickup tube.

1. Disconnect the negative battery cable remote connection, located on the left strut tower.

2. Raise and safely support the vehicle.

3. Drain the engine oil as well as the engine coolant into suitable containers.

4. Remove the oil pan assembly. Remove the oil pump pickup tube and O-ring.

5. Remove the right inner fender splash shield.

6. Lower the vehicle.

7. Remove the drive belts as required.

8. Using a puller, remove the crankshaft damper from the front of the crankshaft.

9. Take up the weight of the engine with a suitable lift and remove the right engine mount and bracket. Make sure the engine is safely supported.

10. Remove the timing belt cover.

11. Loosen the timing belt tensioner bolts and remove the tensioner and the timing belt.

12. Using a suitable puller, draw the crankshaft sprocket from the front of the crankshaft.

13. Loosen the oil pump bolts and remove. Take note of the location of each bolt for reassembly. Remove the oil pump from the face of the engine block. If necessary, tap lightly with a soft face mallet. Use care working with light alloy parts. Remove and discard the front oil seal.

14. Remove the relief valve from the pump body by removing the threaded plug and gasket and pulling out the spring and relief valve. Note the order of parts removal.

To install:

15. Clean all parts well for inspection. Remove the screws holding the back cover to the pump body. Remove the pump rotors. The mating surface of the oil pump should be smooth. Replace the pump cover if scratched or grooved.

16. The pump should be checked for wear by carefully measuring the components. Use the following procedure:

a. Lay a straight-edge across the pump cover surface. If a 0.003 in. (0.076mm) feeler gauge can be inserted between the cover and straight-edge, the cover should be replaced.

b. Measure the thickness and diameter of the outer rotor. If the outer rotor thickness measures 0.301 in. (7.6mm) or less, or if the diameter is 3.148 in. (80mm) or less, replace the outer rotor.

c. If the inner rotor measures 0.301 in. (7.6mm) or less, replace the inner rotor.

d. Slide the outer rotor into the pump housing, press to one side with fingers and measure the clearance between the rotor and housing. If the measurement is 0.015 in. (0.38mm) or more, replace the oil pump housing only if the outer rotor is in specification.

e. Install the inner rotor into the pump housing. If the clearance between the inner and outer rotors is

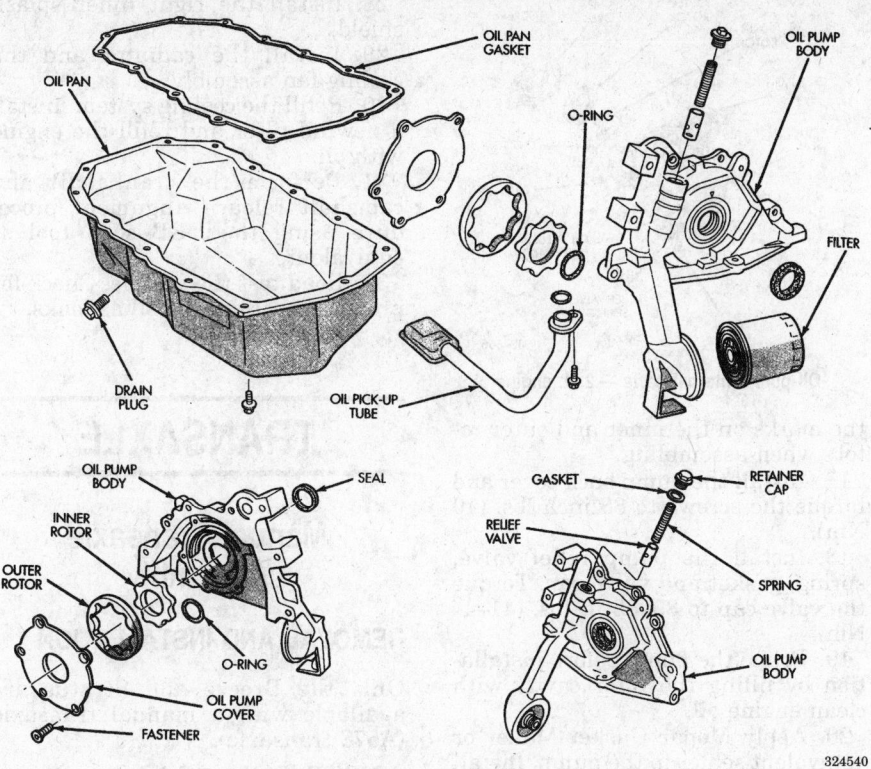

Lubrication system components — 2.4L engine

0.008 in. (0.20mm) or more, replace both rotors.

f. Place a straight-edge across the face of the pump housing between the bolt holes. If a feeler gauge of 0.004 in. (0.10mm) or more can be inserted between the rotors and straight-edge, replace the pump assembly only if the rotors are within specification.

g. Inspect the oil pressure relief valve plunger for scoring and free operation in its bore. Small marks may be removed with 400 grit wet or dry sandpaper.

h. The relief valve spring has a free length of approximately 2.39 in. (60.7mm). It should test between 18–19 lbs. (80–85mm) when compressed to 1.60 in. (40.6mm). Replace the spring if weak, damaged or fails to meet specifications.

NOTE: If oil pressure is low and the pump is within specifications, inspect for worn engine bearings, clogged oil filter, pressure relief valve stuck open, damaged or missing oil pickup tube O-ring, clogged oil pickup tube screen or other reasons for oil pressure loss.

17. Clean all oil pump parts in suitable solvent before assembly. Assemble the pump with new parts as required. Install the inner rotor with the chamfer facing the cast iron oil pump cover (back of the pump). Tighten the cover screws to 105 inch lbs. (12 Nm).

18. Install the relief valve first, then the spring, gasket and cover cap into the pump body. Note that installing the spring first will seriously damage the engine. The relief valve goes in first. Tighten the cover cap to 30 ft. lbs. (41 Nm) for 2.0L engine or 40 ft. lbs. (55 Nm) for 2.4L engine.

19. Prime the oil pump before installation by filling the rotor cavity with clean engine oil.

20. Insert a new oil ring seal to the oil pump counter bore on the pump body discharge passage. Apply Mopar Gasket Maker or equivalent anaerobic type gasket sealer to the oil pump body flange. This material cures in the absence of air when squeezed between 2 flat machined metal surfaces. For this reason, the mating surfaces of both the pump body and the engine block must be spotlessly clean so all air will be expelled when the parts are bolted together and torqued. Install the pump slowly onto the crankshaft aligning the oil pump rotor flats with the flats on the crankshaft until seated to the engine block. Tighten the fasteners to 250 inch lbs. (28 Nm).

21. Install a new front oil seal. Install the seal with the spring side towards the inside of the engine. Tap the seal into place until flush with the cover.

22. Install the crankshaft sprocket. A special tool is used to draw the sprocket onto the end of the crankshaft. Use care if using substitutes.

23. Install the timing belt and covers using the recommended procedures. Use care to make sure all valve timing marks are aligned. This is most important. Failure to properly align the timing marks will result in severe engine damage.

24. Install the crankshaft damper. A special tool making use of a 12mm x 1.75 x 150mm bolt is used to draw the crankshaft damper onto the end of the crankshaft. Use care if using substitutes. Torque the center bolt to 105 ft. lbs. (142 Nm).

25. Raise and safely support the vehicle.

26. Install the oil pump pickup tube and O-ring. Torque the oil pump pickup tube mounting screw to 21 ft. lbs. (28 Nm).

27. Clean the oil pan well and make sure the gasket rails are in good condition. Use Mopar Silicone Rubber Adhesive Sealant or equivalent sealer at the oil pump to engine block parting line. Use a new oil pan gasket, install the oil pan and torque the 13 oil pan bolts to 105 inch lbs. (12 Nm).

28. Install a new oil filter.

29. Install the right inner fender splash shield.

30. Lower the vehicle.

31. Install the drive belts and accessories as required. Adjust the accessory drive belts.

32. Install the engine mount and bracket as required.

33. Refill the engine with new, clean engine oil and coolant.

34. Test run vehicle to check for leaks. An oil pressure gauge should be installed to verify proper engine oil pressure.

2.5L Engine

The oil pump assembly is mounted on the timing belt end of the cylinder block with the inner pump rotor indexed and installed on the crankshaft nose. The oil pump case also retains the crankshaft front oil seal and provides oil pan front end closure.

Oil pressure can be checked with a mechanical oil pressure gauge installed at the oil switch location. Oil pressure should be 6 psi at idle and 35–75 psi at 3000 rpm, engine at operating temperature. If an oil pressure problem is suspected or if oil pressure is zero at idle, Do not run the engine up to 3000 rpm in an attempt to raise oil pressure or the engine will be severely damaged.

Because the oil pump is driven off the nose of the crankshaft, the timing belt must be removed to access the pump. This is a lengthy process. Use care to properly align all valve timing marks.

1. Disconnect the negative battery cable remote connection at the left strut tower.

2. Remove the drive belts and accessories.

3. Drain the coolant and remove the radiator and cooling fan.

4. Raise and safely support the vehicle. Drain the engine oil.

5. Remove the right inner splash shield.

6. Remove the crankshaft damper.

7. Support the engine and remove the right engine mount and the engine mount bracket

8. Remove the timing belt upper and lower covers.

9. Loosen the timing belt tensioner bolts and remove the timing belt and the tensioner.

10. Using a suitable puller, draw the crankshaft sprocket from the crankshaft.

11. Remove the 5 bolts that attach the oil pump to the block and remove the oil pump.

12. Inspect the oil pump case for damage and remove the rear cover.

13. Remove the pump rotors and inspect the inside of the case for excessive wear.

14. Check that the oil relief plunger slides smoothly and check for a broken spring.

To install:

15. Clean all parts well. Make sure the block and pump surfaces are clean and free of old sealer.

16. Assemble the pump using new parts as required with clean oil. Align

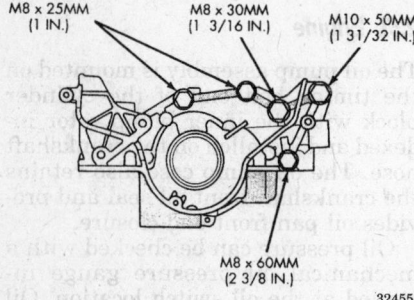

Oil pump assembly. Note the different bolt lengths — 2.5L engine

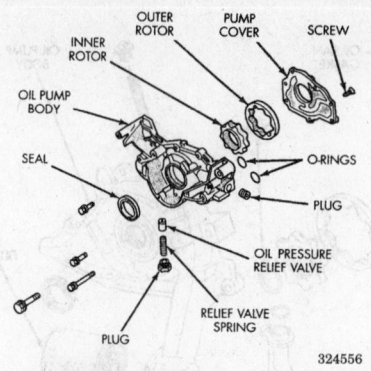

Oil pump internal parts — 2.5L engine

the marks on the inner and outer rotors when assembling.

17. Install the pump back cover and torque the screws to 88 inch lbs. (10 Nm).

18. Install the pump relief valve, spring, gasket and valve cap. Torque the valve cap to 30–33 ft. lbs. (41–44 Nm).

19. Prime the pump before installation by filling the rotor cavity with clean engine oil.

20. Apply Mopar Gasket Maker or equivalent sealer to the pump. Install the O-ring into the counter bore on the pump body discharge passage. Position the pump onto the crankshaft until seated on the block. Torque the size M8 fasteners to 10 ft. lbs. (14 Nm) and size M10 fasteners to 30 ft. lbs. (41 Nm).

21. Install the timing belt and crankshaft sprocket using the recommended procedure and following all cautions. Valve timing MUST be correct or engine damage will result.

22. Install the timing belt cover.

23. Install the engine mount bracket.

24. Install the right engine mount.

25. Remove the engine support.

26. Install the crankshaft damper and torque to 134 ft. lbs. (182 Nm).

27. Install the drive belts and accessories.

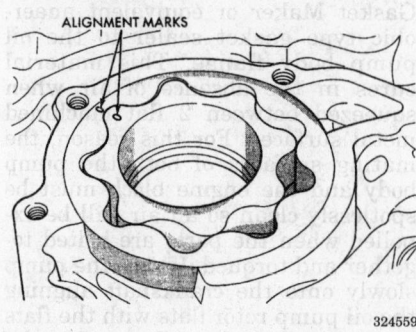

Matching the inner and outer rotor alignment marks — 2.5L engine

28. Install the right inner splash shield.

29. Install the radiator and the cooling fan assembly.

30. Refill the cooling system. Install a new oil filter and refill the engine with oil.

31. Perform the crankshaft and camshaft relearn alignment procedure using the DRB scan tool or equivalent.

32. Road test the vehicle. Check for proper operation as well as leaks.

TRANSAXLE

Manual Transaxle Assembly

REMOVAL AND INSTALLATION

Only the Breeze and Stratus are available with a manual transaxle (A578 transaxle).

NOTE: If the vehicle is going to be rolled on its wheels while the transaxle is out of the vehicle, obtain 2 outer CV-joints to install to the hubs. If the vehicle is rolled without the proper torque applied to the front wheel bearings, the bearings will no longer be usable. Different transaxles are used according to application. It is important to use the round identification tag screwed to the top of the case when obtaining parts for exact parts matching. The tag should be reinstalled for future reference.

1. Disconnect the negative battery cable at the strut tower.

2. Remove the air cleaner assembly with all ducts. Remove the clutch housing vent cap, exposing the clutch cable and clutch release lever.

3. Disconnect the clutch cable from the transaxle bell housing.

4. Disconnect the selector lever cable from the transaxle.

5. Disconnect the crossover lever cable from the transaxle and remove the shift cable mounting bracket.

6. Disconnect the accelerator cables from the throttle body and remove the accelerator cable bracket from throttle body.

7. Remove the starter upper mounting bolt and upper stud at the bell housing.

8. Remove support bracket for the throttle body.

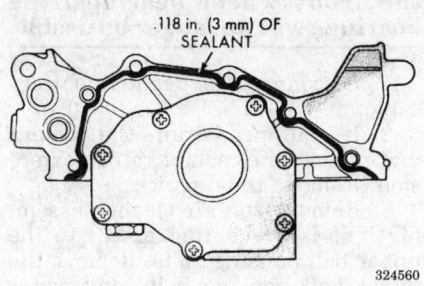

.118 in. (3 mm) OF SEALANT

324560

Oil pump sealing — 2.5L engine

9. Remove the upper bolts to the left transaxle mount.

10. Remove the upper bell housing bolts and vehicle speed sensor.

11. Disconnect the back-up lamp wiring at the transaxle.

12. Support the engine assembly using an engine support fixture.

13. Raise the vehicle and support safely. Remove the front wheels.

14. Drain fluid from the transaxle.

15. Remove both front halfshaft assemblies.

16. Remove the lower left inner fender splash shield/battery cover.

17. Remove the lower bracket bolts for the left transaxle mounts.

18. Remove the engine lower crossbar retaining bolts.

19. Remove the front steel engine mount bracket.

20. Remove the 3 mounting bolts for the aluminum front engine mount bracket.

21. Remove the lower mounting bolt for the starter motor.

22. Remove the transaxle rear mount bracket.

23. Remove the transaxle to rear lateral bending strut from engine and transaxle.

24. Remove screw to the lower dust shield.

25. Position a transaxle jack under the transaxle assembly.

26. Rotate the engine clockwise to access the clutch driveplate bolts. Remove the clutch driveplate bolts.

27. Remove the lower bell housing bolts.

28. Carefully pry the transaxle from the engine.

29. Slide the transaxle rearward until the input shaft clears the clutch disc.

30. Pull the transaxle completely away from the clutch housing and remove it from the vehicle.

31. To prepare the vehicle for rolling, support the engine with a suitable support or reinstall the front mo-

tor mount to the engine. Then reinstall the ball joints to the steering knuckle and install the retaining bolt. Install the obtained outer CV-joints to the hubs, install the washers and torque the axle nuts to 180 ft. lbs. (244 Nm). The vehicle may now be safely rolled on its wheels.

To install:

32. Lubricate the pilot bushing and input shaft splines lightly with high temperature lubricant.

33. Mount the transaxle securely on a jack. Lift it into place until the input shaft is centered in the clutch housing opening. Roll the transaxle forward until the input shaft splines fully engage with the clutch disc and install the transaxle to clutch housing bolts.

34. Raise the transaxle and install the lower transaxle-to-engine bell housing bolts.

35. Install screw to lower dust shield. Install the transaxle to rear lateral bending strut to engine and transaxle.

36. Install the rear transaxle mount bracket.

37. Install the lower mounting bolt for the starter motor.

38. Install the front steel engine mount bracket.

39. Install the aluminum front engine mount bracket and tighten the 3 mounting bolts.

40. Install retaining bolts for engine lower crossbar.

41. Install the lower bracket bolts to the left transaxle mount.

42. Install the battery cover/lower left inner fender splash shield.

43. Install both front halfshaft assemblies. Be sure to use new halfshaft retaining clips during installation.

44. Fill up transaxle, to the bottom of fill hole, with MOPAR® Type M.S. 9417 Manual Transaxle Fluid. This must be performed before the vehicle is lowered to the ground.

45. Install the front wheels and lug nuts. Torque the lug nuts, in a star pattern sequence, to 95–100 ft. lbs. (129–135 Nm).

46. Lower the vehicle to the ground.

47. Remove the engine/transaxle support fixture.

48. Connect the back-up lamp wiring at the transaxle.

49. Install the vehicle speed sensor.

50. Install the upper bell housing bolts.

51. Install upper bolts to the left transaxle mount.

52. Install the upper mounting bolt to the starter. Install the throttle body support bracket.

53. Install the upper bell housing stud nut at the bell housing.

54. Install the accelerator cable bracket at the throttle body.

55. Reconnect the accelerator cable ends to the throttle body.

56. Install the shift cable mounting bracket.

57. Install the crossover lever cable at the transaxle.

58. Install the gear selector lever cable at the transaxle.

59. Install the clutch housing vent cap.

60. Install the air cleaner assembly.

61. Connect the negative battery cable and check the transaxle for proper operation. Make sure the vehicles back-up lights function properly when the transaxle is shifted into reverse. Check the speedometer and make sure it functions properly.

62. Gear shifter crossover cable adjustment is required.

63. Road test vehicle.

Clutch Assembly

REMOVAL AND INSTALLATION

Breeze and Stratus

NOTE: The transaxle assembly must be removed to service the clutch assembly.

1. Disconnect the negative battery cable remote connection located on the left strut tower.

2. Raise and safely support the vehicle.

3. Disconnect the starter wiring and remove the starter assembly.

4. Remove the rear and front transaxle support brackets.

5. Remove the clutch inspection cover.

6. Remove the bolts attaching the modular clutch to the flywheel.

7. Remove the transaxle assembly with the clutch as an assembly.

8. Remove the clutch assembly from input shaft of the transaxle.

To install:

9. Clean all parts well. Inspect for oil leakage through the engine rear crankshaft oil seal and transaxle input shaft seal. If leakage is noted, it should be corrected at this time.

10. Examine the throwout or clutch release bearing. It is prelubricated and sealed and should not be washed in solvent. The bearing should turn smoothly when held in the hand with a light thrust load. A light drag caused by the lubricant fill is normal. If the bearing is noisy, rough or dry, replace the complete bearing assem-

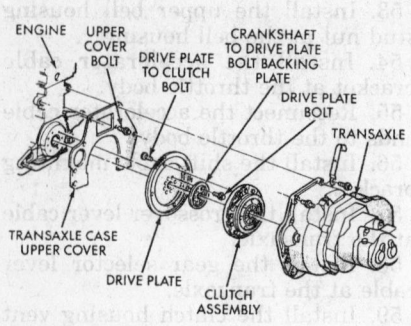

ENGINE UPPER COVER BOLT

DRIVE PLATE TO CLUTCH BOLT

CRANKSHAFT TO DRIVE PLATE BOLT BACKING PLATE

DRIVE PLATE BOLT

TRANSAXLE

TRANSAXLE CASE UPPER COVER

DRIVE PLATE

CLUTCH ASSEMBLY

310498

Clutch assembly — A578 manual transaxle

bly. In most cases where a clutch is being serviced, the complete clutch assembly and release bearing are usually replaced together.

11. Check the condition of the stud pivot spring clips on the back side of the clutch fork. If the clips are broken or distorted, replace the clutch fork. The pivot ball pocket in the fork is Teflon® coated and should be installed WITHOUT any lubricant such as grease which will break down the Teflon® coating. Make sure the ball stud and fork pocket are clean of contamination and dirt. When assembling the fork to the bearing, the small pegs on the bearing must go over the fork arms.

12. Check the flywheel for cracks, glazing or grooves. If any of these conditions exist, machine (reface) or replace the flywheel to prevent clutch chatter and premature clutch wear.

NOTE: The manual transaxle is equipped with a reverse brake. It functions as a synchronizer, but only if the vehicle is not moving. When the clutch pedal is depressed to the floor and held for 3 seconds, and the transaxle shifts to reverse, no gear clash should be present. If there is, the input shaft should be checked. When the transaxle is removed for clutch service, check the input clutch shaft, clutch disc splines and release bearing for dry rust. If present, clean rust off and apply a light coat of high temperature bearing grease to the input shaft splines. Apply grease on the input shaft splines only where the clutch disc slides. Verify that the clutch disc slides freely along the input shaft splines.

13. Install the modular clutch assembly onto the input shaft of the transaxle.

14. Install the transaxle assembly using the recommended procedure.

15. Install new clutch-to-driveplate (flywheel) bolts. Tighten the bolts in a crisscross pattern a few turns at a time to prevent distortion of the flywheel. Torque the bolts to 55 ft. lbs. (75 Nm).

16. Install the clutch inspection cover.

17. Install the transaxle lower support brackets.

18. Install the starter assembly.

19. Lower the vehicle. Reconnect the negative battery cable.

20. Road test vehicle to check for proper clutch operation.

Clutch Cable

ADJUSTMENT

The manual transaxle clutch release system has a unique self-adjusting mechanism to compensate for clutch disc wear. This adjuster mechanism is located with the clutch cable assembly. The preload spring maintains tension on the cable. This tension keeps the clutch release bearing continuously loaded against the fingers of the clutch cover assembly. No manual adjustment is obtainable.

When servicing this vehicle or if removing and installing the clutch cable, do not pull on the clutch cable housing to remove it from the dash panel. Damage to the cable self-adjuster may occur.

To check the function of the adjuster mechanism, use the following procedure:

1. With slight pressure, pull the clutch release lever end of the cable to draw the cable taut.

2. Push the clutch cable housing toward the dash panel. With less than 25 lbs. (111 N) of effort the cable housing should move 1.2–2.0 in. (30–50mm). This indicates proper adjuster mechanism function.

3. If the cable does not adjust, determine if the mechanism is properly seated on the bracket.

Automatic Transaxle Assembly

REMOVAL AND INSTALLATION

41TE Transaxle

— WARNING —
If the vehicle is going to be rolled on its wheels while the transaxle is out of the vehicle, obtain 2 outer CV-joints to install to the hubs. If the vehicle is rolled without the proper torque applied to the front wheel bearings, the bearings will no longer be usable.

1. Disconnect the negative battery cable.

2. If equipped, drain the coolant and remove the coolant return extension. Remove the dipstick.

3. Remove the air cleaner assembly if it is preventing access to the upper bell housing bolts. Remove the upper bell housing bolts and water tube, where applicable. Note locations and unplug all electrical connectors from the transaxle.

4. If equipped with a 2.2L or 2.5L engine, remove the starter attaching nut and bolt at the top of the bell housing.

5. Disconnect the transaxle control cable at the transaxle. Raise the vehicle and support safely.

6. Remove the tire and wheel assemblies. Remove the axle end cotter pins, nut locks, spring washers and axle nuts.

7. Remove the ball joint retaining bolts and pry the control arm from the steering knuckle. Position a drain pan under the transaxle where the axles enter the differential or extension housing. Remove the axles from the transaxle or center bearing. Unbolt the center bearing and remove the intermediate axle from the transaxle, if equipped.

8. Drain the transaxle. Disconnect and plug the fluid cooler hoses.

9. If equipped with Direct Ignition System (DIS), disconnect the harness connector and remove the crankshaft position sensor from the transaxle bell housing.

10. Remove the speedometer cable adaptor bolt and remove the adaptor from the transaxle.

11. Remove the starter. Remove the torque converter inspection cover, matchmark the torque converter to the flexplate and remove the torque converter bolts.

12. Using the proper equipment, support the weight of the engine. Remove the front motor mount and bracket.

13. Position a transaxle jack under the transaxle.

14. Remove the lower bell housing bolts.

15. Remove the left side splash shield. Remove the transaxle mount bolts.

NOTE: The torque converter can become disengaged from the transaxle. Keep the front of transaxle slightly raised during removal.

16. Carefully pry the transaxle from the engine.

17. Slide the transaxle rearward until dowels disengage from the mating holes in the transaxle case.

18. Pull the transaxle completely away from the engine and remove it from the vehicle.

19. To prepare the vehicle for rolling, support the engine with a suitable support or reinstall the front motor mount to the engine. Then reinstall the ball joints to the steering knuckle and install the retaining bolt. Install the obtained outer CV-joints to the hubs, install the washers and torque the axle nuts to 180 ft. lbs. (244 Nm). The vehicle may now be safely rolled.

To install:

20. Install the transaxle securely on transmission jack. Rotate the converter so it will align with the positioning of the flexplate.

21. Apply a coating of high temperature grease to the torque converter pilot hub.

22. Raise the transaxle into place and push it forward until the dowels engage and the bell housing is flush with the block. Install the transaxle to bell housing bolts.

23. Raise the transaxle and install the left side mount bolts. Install the torque converter bolts and torque to 55 ft. lbs. (74 Nm).

24. Install the front motor mount and bracket. Remove the engine and transaxle support fixtures.

25. Install the starter to the transaxle. Install the bolt finger-tight if equipped with a 2.2L or 2.5L engine.

26. Install a new O-ring to the speedometer cable adaptor and install to the extension housing. Make sure it snaps in place. Install the retaining bolt.

27. Connect the shifter and kickdown linkage to the transaxle, if equipped.

28. Install the axles and center bearing, if equipped. Install the ball joints to the steering knuckles. Torque the axle nuts to 180 ft. lbs. (244 Nm) and install new cotter pins. Install the splash shield and wheels. Lower the vehicle. Install the dipstick.

29. Install the upper bell housing bolts and water pipe, if removed.

30. If equipped with 2.2L or 2.5L engine, install the starter attaching nut and bolt at the top of the bell housing. Raise the vehicle again and tighten the starter bolt from under the vehicle. Lower the vehicle.

31. Connect all electrical wiring to the transaxle.

32. Install the air cleaner assembly, if removed. Fill the transaxle with the proper amount of Mopar ATF Plus Type 7176 or conventional Dexron®II.

33. Connect the negative battery cable and check the transaxle for proper operation.

Upshift and Kickdown Learning Procedure

The 41TE 4-speed, electronic transaxle has fully adaptive controls. The controls perform their functions based on real time feedback sensor information. Although the transaxle is conventional in design, functions are controlled by its ECM.

Since the 41TE is equipped with a learning function, each time the battery cable is disconnected, the ECM memory is lost. In operation, the transaxle must be shifted many times for the learned memory to be re-input to the ECM. During this period, the vehicle may experience rough operation. The transaxle must be at normal operating temperature when learning occurs.

1. Maintain constant throttle opening during shifts. Do not move the accelerator pedal during upshifts.

2. Accelerate the vehicle with the throttle ⅛–½ open.

3. Make 15–20 upshifts: 1st to 2nd, 2nd to 3rd and 3rd to 4th gear. Accelerating from a full stop to 50 mph (80 km/h) each time, at the aforementioned throttle opening, is sufficient.

4. With the vehicle speed below 25 mph (40 km/h), make 5–8 wide open throttle kickdowns to 1st gear from either 2nd or 3rd gear. Allow at least 5 seconds of operation in 2nd or 3rd gear prior to each kickdown.

5. With the vehicle speed greater than 25 mph (40 km/h), make 5 part throttle to wide open throttle kickdowns to either 3rd or 2nd gear from 4th gear. Allow at least 5 seconds of operation in 4th gear, preferably at road load throttle prior to performing the kickdown.

DRIVE AXLE

Halfshaft

REMOVAL AND INSTALLATION

1. Remove the cotter pin, nut lock and spring washer from the end of the outer CV-joint stub axle shaft. Discard the cotter pin.

2. Loosen, but do not remove, the wheel hub nut. Loosen the nut with the vehicle on the ground with the brakes applied.

3. Raise and safely support the vehicle.

4. Remove the front wheel and tire.

5. Remove the hub nut.

6. Remove the caliper assembly from the steering knuckle. Support the caliper assembly from the vehicle strut with a strong piece of wire. Do not disconnect the brake line from the caliper. Do not allow the caliper to hang from the brake hose.

7. Remove the brake rotor from the hub.

8. Disconnect the tie rod end from the steering knuckle.

9. If equipped with ABS, remove the wheel speed sensor and cable routing bracket from the steering knuckle.

10. Disconnect the lower ball joint from the steering knuckle using the following procedure:

   a. Remove the cotter pin and castle nut from the lower ball joint at the steering knuckle. Discard the cotter pin.

   b. Turn the steering knuckle so the front of the steering knuckle is facing as far outboard as possible in the wheel well.

   c. Strike only the steering knuckle boss until the steering knuckle separates from the lower ball joint stud. Do NOT strike the lower control arm or ball joint grease seal.

11. Pull the steering knuckle out and away from the outer CV-joint. Support the outer CV-joint preventing it from hanging off the inner joint.

12. Insert a prybar between the halfshaft inner CV-joint and the transaxle.

13. Pry carefully on the CV-joint until the snapring disengages. Be careful not to damage the transaxle case oil seal.

14. Remove the halfshaft assembly. Do not allow the splines or snapring

of the shaft to drag across and damage the transaxle oil seal when removing the halfshaft assembly.

To install:

15. Clean the shaft spline and oil seal sealing surface of the tripod joint. Using clean transaxle fluid, lightly lubricate the oil seal sealing surface of the tripod joint.

16. Install the halfshaft into the transaxle as far as it will go.

17. Forcefully push on the inner CV-joint until the snapring in engaged.

18. Be sure the CV-joint, stub shaft and steering knuckle are clean of any debris or moisture before installing the halfshaft into the steering knuckle. Install the halfshaft into the steering knuckle assembly.

19. Install the steering knuckle onto the ball joint stud.

20. Install the ball joint castle nut and torque the nut to 70 ft. lbs. (95 Nm). Install a new cotter pin.

21. If equipped ABS, install the wheel speed sensor and routing bracket to the steering knuckle. Tighten the mounting bolt.

22. Reconnect the tie rod end. Torque the nut to 45 ft. lbs. (61 Nm).

23. Install the brake rotor onto the wheel hub.

24. Install the brake caliper assembly onto the brake rotor and steering knuckle.

25. Install the halfshaft-to-steering knuckle hub nut.

26. With the brakes applied, torque the hub nut to 180 ft. lbs. (244 Nm).

27. Install the spring washer, nut lock and a new cotter pin to the end of the stub axle. Wrap the prongs of the cotter pin tightly around the hub nut lock.

28. Install the front wheels and lug nuts. Torque the lug nuts, in a star sequence, to 95–100 ft. lbs. (129–135 Nm).

29. Lower the vehicle.

30. Check the transaxle fluid level and add fluid, if necessary.

31. Check and set front toe, if necessary.

CV-Joint Boot

REPLACEMENT

1. Raise and safely support the vehicle.

2. Remove the halfshaft requiring boot replacement from the vehicle.

NOTE: The outer CV-joint is not removable from the halfshaft. If the outer boot is to be replaced, **the halfshaft assembly, minus the inner boot and inner tripod joint, must be replaced as one unit.**

Inner CV-Joint Boot

1. Remove the 2 boot clamps (one large, one small diameter clamp) and discard.

2. Slide the CV-boot down the shaft

3. Slide the tripod joint housing off the spider assembly while holding the spider bearings to prevent them from falling off the spider assembly.

4. Remove the spider assembly retaining snapring.

5. Using a brass drift near the shaft, tap the spider assembly toward the end of the shaft to remove.

6. Remove boot from the shaft.

7. Thoroughly clean and inspect the spider assembly, interconnecting shaft and tripod housing for indications of excessive wear. Replace the halfshaft assembly will require replacement if there are signs of excessive wear. Component parts of these halfshaft assemblies are not serviceable separately.

To install:

NOTE: The inner CV-joint boots are made from 2 different types of material. Silicone rubber is used for high temperature applications, whereas hytrel plastic is used for standard temperature applications. The silicone rubber boots are soft and pliable, while the hytrel boots are stiff and rigid. The new CV-boot MUST BE of the same material as the CV-boot which was removed.

8. Install the small inner boot retaining clamp onto the halfshaft. slide the new boot onto the shaft. Be sure the boot is positioned onto the shaft so the raised bead on the inside of the boot is seated on the groove on the shaft.

9. Install the spider assembly.

10. Using a brass drift, tap the spider assembly onto the shaft until the retaining snapring can be installed.

11. Fill the tripod joint housing with ½ the amount of grease provided in the seal boot service package. Fill the CV-joint boot with the remaining ½ amount of grease. Do NOT use any other type of grease in this procedure.

12. Installed the tripod housing.

13. Slide the boot onto the tripod housing and seat it on the tripod housing retaining groove.

14. Correctly position the small diameter sealing clamp over the boot. Using a suitable CV-boot clamp crimping tool, crimp the CV-boot clamp. Be sure the seal is shaped correctly and if not, shape it by hand.

15. Insert a flat plastic stick between the boot and the joint to vent the joint .

16. Lift the stick slightly and press the joint inward to release any excess air from the boot.

17. Remove the plastic stick.

18. Install the tripod boot clamp and crimp it using the CV-boot clamp crimping tool.

19. Install the halfshaft assembly back into the vehicle.

Outer CV-Joint Boot

The outer CV-joint boot is not a serviceable component of the halfshaft assembly. CV-joint and/or CV-joint boot failure will require the replacement of the outer CV-joint/boot-interconnecting shaft-vibration damper as one assembly. However, the inner CV-joint boot and inner tripod joint assembly are still serviceable components of the halfshaft assembly.

STEERING

Air Bag

─── **CAUTION** ───

Some vehicles are equipped with an air bag system, also known as the Supplemental Inflatable Restraint (SIR) or Supplemental Restraint System (SRS). The system must be disabled before performing service on or around system components, steering column, instrument panel components, wiring and sensors. Failure to follow safety and disabling procedures could result in accidental air bag deployment, possible personal injury and unnecessary system repairs.

PRECAUTIONS

Several precautions must be observed when handling the inflator module to avoid accidental deployment and possible personal injury.

• Never carry the inflator module by the wires or connector on the underside of the module.

• When carrying a live inflator module, hold securely with both hands, and ensure that the bag and trim cover are pointed away.

• Place the inflator module on a bench or other surface with the bag and trim cover facing up.

• With the inflator module on the bench, never place anything on or close to the module which may be thrown in the event of an accidental deployment.

DISARMING

This air bag system, is a sensitive, complex, electromechanical unit. Proper SRS (also called Supplemental Inflatable Restraint, or SIR, or air bag system) disarming can be obtained by disconnecting and isolating the negative battery cable. Failure to disconnect the battery could result in accidental air bag deployment and possible personal injury. Before beginning service work, allow the system capacitor 2 minutes to discharge.

Steering Wheel

REMOVAL AND INSTALLATION

─────── **CAUTION** ───────
The Supplemental Inflatable Restraint (SIR) system must be disarmed before removing the steering wheel. Failure to do so may cause accidental deployment of the air bag, resulting in unnecessary SIR system repairs and/or personal injury.

1. Disconnect the negative battery cable end at the left strut tower.

NOTE: The ground cable is equipped with a insulator grommet which should be placed on the stud to prevent the negative battery cable from accidentally grounding.

2. Disconnect the negative battery cable at the remote terminal on the shock tower. This will disable the air bag system. Allow a minium of 2 minutes for the system capacitor to discharge before removal of any air bag components.

3. Place the wheels in the straight position.

4. Remove the driver's air bag attaching bolts from the back of the steering wheel.

5. Lift the air bag module. While holding the module lift the secondary latch and disconnect the wiring connector.

NOTE: Never use a metal tool to pry on the air bag wiring connector.

6. Remove the speed control screws from the back of the steering wheel, lift out the pods and disconnect the wiring.

7. Disconnect the horn wire from the air bag bracket. Remove the speed control wires from under the bracket.

8. Remove the steering wheel retaining nut.

9. Remove the steering wheel using a suitable pulling tool.

To install:

10. Install the steering wheel onto the column.

11. Install the steering wheel retaining nut and torque to 45 ft. lbs. (61 Nm).

12. Reconnect the horn wire and the speed control wires.

13. Install the speed control pods and reattach the speed control unit to the steering wheel. Torque the screws to 15 inch lbs. (1.7 Nm).

14. Reconnect the air bag lead and push the secondary latch into place.

15. Install the air bag module and torque the screws to 85 inch lbs. (9.6 Nm).

16. Connect a scan tool to the data link connector located on the left side kick panel just above the hood release.

17. Turn the ignition key to the **ON** position.

18. Make sure no one is inside the vehicle.

19. Connect the negative battery cable.

20. Check for diagnostic codes and disconnect the scan tool.

Tie Rod Ends

REMOVAL AND INSTALLATION

1. Raise and safely support the vehicle.

2. Remove the wheel and tire.

3. Loosen the outer tie rod to inner tie rod jam nut.

4. Remove the tie rod to steering knuckle attaching nut. Hold the tie rod stud with an $^{11}/_{32}$ in. socket. Loosen the nut with a wrench.

5. Separate the tie rod from the steering knuckle using a suitable press-type separating tool.

6. Unscrew the tie rod from the steering rack while counting the amount of turns required to remove.

To install:

7. Make sure the jam nut is on the inner tie rod. Thread the tie rod end onto the steering rack inner tie rod.

8. Install the outer tie rod end into the steering knuckle. Rotate the tie

rod end the same amount of turns required to remove it. Do not tighten the jam nut at this time.

9. Install the tie rod end seal boot heat shield onto the tie rod end.

10. Install the tie rod end stud into the steering knuckle. Start the tie rod end to steering knuckle attaching nut onto the stud of the tie rod end. While holding the stud of the tie rod end stationary, tighten the tie rod to steering knuckle nut. Then using a crows foot and an socket, torque the tie rod end attaching nut to 45 ft. lbs. (61 Nm).

11. Check and adjust the Toe alignment setting.

12. Tighten the jam nut to 55 ft. lbs. (75 Nm). Make sure the steering rack-to-inner tie rod boots are not twisted. Correct as necessary.

13. Install the wheel and lug nuts. Torque the lug nuts, in a star pattern sequence, to 95–100 ft. lbs. (129–135 Nm).

Power Rack and Pinion

REMOVAL AND INSTALLATION

1. Disconnect the negative battery cable remote connection from the strut tower and properly isolate the cable.

NOTE: The negative battery is equipped with a insulated grommet that is to be placed on the strut tower connection to prevent accidental grounding.

2. Remove the retaining pin from the intermediate shaft coupler pin bolt and remove the pin bolt.

3. Raise and safely support the vehicle.

4. Remove the front wheels.

5. Disconnect the tie rod ends by holding the tie rod end stud with a $^{11}/_{32}$ in. socket and loosen the retaining nut with a wrench.

6. Separate the tie rod end using a suitable tool.

─────── **CAUTION** ───────
Before removing the front suspension crossmember from the vehicle you must scribe the front suspension crossmember and the vehicle body. This must be done to retain the proper alignment. The caster and camber are not adjustable.

7. Scribe a line on the body and on the crossmember on all 4 sides.

8. Disconnect the stabilizer from the body.

9. If equipped with anti-lock brakes remove the 3 bolts attaching it to the crossmember and tie the anti-lock brake controller to the vehicle body.

10. Disconnect the shock absorber clevis from the lower control arm.

11. Remove the 2 bolts attaching the engine support bracket to the crossmember.

12. Remove the bolt attaching the engine support bracket to the transaxle mounting bracket.

13. Place a suitable lifting device under the front suspension crossmember.

14. Remove the 8 bolts attaching the crossmember to the body of the vehicle.

15. Lower the lifting device enough to gain access to the steering rack.

16. Disconnect the power steering lines and drain the fluid if equipped.

17. Disconnect the power steering pressure switch wiring.

18. If equipped with speed proportional steering, disconnect the solenoid control module wiring.

19. Remove the 2 steering rack isolator attaching bolts.

20. Remove the 2 steering rack saddle bracket attaching bolts.

21. Remove the steering from the vehicle.

To install:

22. Install the steering rack into the crossmember.

23. Install the isolator and saddle bracket bolts. Torque the bolts to 50 ft. lbs. (68 Nm).

24. If equipped with power steering install the pressure and return lines. Torque the lines to 275 inch lbs. (31 Nm).

25. Raise the crossmember against the frame rails and install the 2 rear bolts.

26. Install the 2 front bolts.

27. Tighten all 4 bolts until the crossmember contacts the body.

28. Torque the bolts to 20 inch lbs. (2 Nm).

29. Using a soft faced hammer tap the crossmember into position.

NOTE: Be sure to align the scribed marks on the crossmember.

30. Starting with the rear bolts, torque the crossmember bolts to 120 ft. lbs. (163 Nm).

31. Install the engine support bracket.

32. Install the engine support bracket bolts to crossmember.

33. Install the engine support bracket bolt to the transaxle bracket.

34. Torque the 3 bolts to 55 ft. lbs. (75 Nm).

35. Reconnect the power steering pressure switch

36. Install the anti-lock brake control unit.

37. Torque the anti-lock brake bolts to 21 ft. lbs. (28 Nm).

38. Reinstall the heat shield on the tie rod ends.

39. Connect the shock clevis to the lower control arm.

40. Install the tie rod ends and torque to 45 ft. lbs. (61 Nm).

41. Reinstall and tighten the 2 stabilizer clamps.

42. Torque the shock absorber clevis bolt to 68 ft. lbs. (92 Nm).

43. Reinstall the wheels and torque to 95 ft. lbs. (129 Nm).

44. Lower the vehicle.

45. Reinstall the intermediate shaft pin bolt and retaining pin.

46. Torque the pin bolt to 240 inch lbs. (27 Nm).

47. Reconnect the negative battery cable.

48. Fill the power steering system to the cold level with approved fluid.

49. Start the engine and allow it to run for a few minutes.

50. Shut **OFF** the engine and check the power steering fluid.

51. Add power steering fluid if necessary.

52. Raise the front wheels off the ground.

53. Start the engine and turn the wheel from stop-to-stop to bleed any air from the system.

54. Check fluid level and add if necessary.

55. Check and adjust the alignment.

Power Steering Pump

BLEEDING

--- **CAUTION** ---

The power steering fluid level should be checked with the engine OFF to prevent injury from moving components. Power steering oil, engine components and exhaust system may be extremely hot if the engine has been running. Do not start the engine with any loose or disconnected hoses or allow hoses to touch a hot exhaust manifold or catalyst.

NOTE: In all power steering pumps, use only MOPAR® Power Steering Fluid or equivalent. DO NOT use any type of automatic transaxle fluid in the power steering system.

Wipe the filler cap clean, then check the fluid level. The dipstick should indicate FULL COLD when the fluid is at normal room temperature of approximately 70–80°F (21–27°C).

1. Fill the power steering pump fluid reservoir to the proper level. Allow the fluid to settle for at least 2 minutes.

2. Start the engine and let run for a few seconds. Turn the engine **OFF**.

3. Add fluid if necessary. Repeat this procedure until the fluid level remains constant after running the engine.

4. Raise the front wheels of the vehicle off the ground.

5. Start the engine. Slowly turn the steering wheel right and left, lightly contacting the wheel stops; then, turn the engine **OFF**.

6. Add power steering fluid if necessary.

7. Lower the vehicle and turn the steering wheel slowly from lock-to-lock.

8. Check the fluid level and refill as required.

9. If the fluid is extremely foamy, allow the vehicle to stand a few minutes and repeat the above procedure.

REMOVAL AND INSTALLATION

1. Disconnect the negative battery cable remote connection located on the left strut tower.

2. Siphon as much power steering fluid out of the reservoir as possible.

3. Raise and safely support the vehicle.

4. Remove the right front tire and wheel.

5. Remove the splash shield from the right front wheel well.

6. Disconnect the power steering pressure hose from the pump.

7. Remove the hose connection on the power steering pump.

8. Remove the power steering adjusting bolt.

9. Remove the power steering pump rear attaching bolt.

10. Remove the anti-lock brake control unit heat shield.

11. Remove the wheel speed sensor retainer bracket from the right inner fender.

12. Remove the wheel speed sensor sealing grommet from the right inner fender.

13. Disconnect the speed sensor wiring.

14. Push the wiring through the hole in the inner fender.

NOTE: If not equipped with anti-lock brakes the hole will just have a sealing plug.

15. Remove the bolt attaching the power steering front bracket to the mounting bracket. Access to the bolt is gained through the hole for the speed sensor wiring.

16. Remove the power steering pump drive belt.

17. Remove the power steering pump and the front bracket as an assembly.

To install:

18. Install the power steering pump and bracket.

19. Install the bolt at the adjusting slot but do not tighten.

20. Install the bolt mounting the power steering pump to the rear mounting bracket but do not tighten.

21. Install the power steering pump top bolt but do not tighten.

22. Connect the power steering hoses.

NOTE: Use a new O-ring when reinstalling the power steering pressure hose.

23. Install the drive belt.

24. Adjust the drive belt and torque the power pump bolts to 40 ft. lbs. (54 Nm).

25. Install the splash shield.

26. Install the tire and wheel.

27. Lower the vehicle.

28. Reconnect the negative battery cable.

29. Check the fluid level in the reservoir. Add fluid as necessary. Bleed the power steering system as required.

BRAKES

Anti-Lock Brake System Service

PRECAUTIONS

• Certain components within the Anti-Lock Brake System (ABS) are not intended to be serviced or repaired individually. Only those components with removal and installation procedures should be serviced.

• Do not use rubber hoses or other parts not specifically specified for and ABS system. When using repair kits, replace all parts included in the kit. Partial or incorrect repair may lead to functional problems and require the replacement of components.

• Lubricate rubber parts with clean, fresh brake fluid to ease as-

sembly. Do not use lubricated shop air to clean parts; damage to rubber components may result.

• Use only specified brake fluid from an unopened container.

• If any hydraulic component or line is removed or replaced, it may be necessary to bleed the entire system.

• A clean repair area is essential. Always clean the reservoir and cap thoroughly before removing the cap. The slightest amount of dirt in the fluid may plug an orifice and impair the system function. Perform repairs after components have been thoroughly cleaned; use only denatured alcohol to clean components. Do not allow ABS components to come into contact with any substance containing mineral oil; this includes used shop rags.

• The Anti-Lock control unit is a microprocessor similar to other computer units in the vehicle. Ensure that the ignition switch is **OFF** before removing or installing controller harnesses. Avoid static electricity discharge at or near the controller.

• If any arc welding is to be done on the vehicle, the control unit should be unplugged before welding operations begin.

Master Cylinder

REMOVAL AND INSTALLATION

Vehicles Without ABS

1. With the ignition switch **OFF**, pump the brake pedal until a firm pedal is achieved.

2. Disconnect the brake fluid level sensor electrical connector, located on the side of the master cylinder brake fluid reservoir.

3. Disconnect the hydraulic brake fluid lines from the master cylinder and plug off openings to prevent dirt from contaminating the hydraulic system.

4. Clean the area around the mounting of the master cylinder using Brake Parts Cleaner, or equivalent.

5. Remove the master cylinder mounting bolts.

6. Remove the master cylinder.

To install:

7. Be sure to bleed the master cylinder before installation into the vehicle.

8. Install the master cylinder onto the brake booster.

9. Install and tighten the master cylinder mounting bolts to 250 inch lbs. (28 Nm).

10. Connect the hydraulic lines to the master cylinder.

11. Torque the hydraulic lines to 145 inch lbs. (17 Nm).

12. Reconnect the brake fluid level sensor electrical connector.

NOTE: Bleeding the entire brake system after replacing the master cylinder is not necessary. However, the master cylinder must have been properly bled and filled upon installation.

13. Road test and check for proper operation.

Vehicles With ABS

1. With the ignition switch in the **OFF** position, pump the brake pedal until a firm pedal is achieved.

2. Disconnect the brake fluid level sensor electrical connector.

3. Disconnect the hydraulic brake fluid lines from the master cylinder and plug off the openings to prevent dirt from contaminating the hydraulic system.

4. Clean the area around the mounting of the master cylinder with Brake Parts Cleaner, or equivalent.

5. Remove the master cylinder mounting bolts.

6. Remove the routing clip and chassis brake lines, as an assembly, from the inboard mounting stud for the master cylinder. Be careful not to bend or kink the chassis brake lines.

7. Remove the master cylinder.

To install:

8. Bench bleed the master cylinder before installing it into the vehicle.

9. Install the master cylinder onto the brake booster.

10. Install the routing clip and chassis brake lines on the inboard mounting stud for the master cylinder.

11. Tighten the master cylinder mounting bolts to 250 inch lbs. (28 Nm).

12. Connect the hydraulic lines to the master cylinder.

13. Torque the hydraulic lines to 145 inch lbs. (17 Nm).

14. Reconnect the brake fluid level sensor electrical connector.

NOTE: It is not necessary to the bleed the entire brake system after replacing the master cylinder. However, the master cylinder must have been properly bled and filled upon installation.

15. Verify a firm brake pedal before attempting to move the vehicle. Carefully road test and check for proper operation.

Brake Caliper

REMOVAL AND INSTALLATION

1. Raise and safely support the vehicle.
2. Remove the front wheels.
3. Remove the 2 caliper-to-steering knuckle guide pin bolts.
4. If the caliper is to be removed from the vehicle completely, for example, for overhaul perform the following procedure.
   a. Disconnect the brake hose from the caliper.
   b. Cover the opening of the brake hose so the hydraulic system does not become contaminated.
5. Remove the caliper from the steering knuckle.

To install:

6. Clean and lubricate both steering knuckle abutments with a coating of multipurpose grease.
7. Position the caliper and brake pad assembly over the brake rotor. Be sure to properly install the caliper assembly into the abutments of the steering knuckle. Be sure the caliper guide pin bolts, rubber bushings and sleeves are clear of the steering knuckle bosses.
8. Install the caliper guide pin bolts and torque to 16 ft. lbs. (22 Nm).
9. If removed, connect the brake hose to the caliper and torque to 35 ft. lbs. (48 Nm).
10. Properly bleed the brake system.
11. Install the wheel and tire and torque the lug nuts in a star pattern sequence to half specification. Repeat the tightening procedure to full specified torque of 95–100 ft. lbs. (129–135 Nm).
12. Pump the brake pedal to seat the front brake pads before moving the vehicle.
13. Road test the vehicle and check for proper operation.

Disc Brake Pads

REMOVAL AND INSTALLATION

1. Raise and safely support the vehicle.
2. Remove the front wheels.
3. Remove the 2 caliper-to-steering knuckle guide pin bolts.
4. Lift the caliper away from the steering knuckle by first rotating the free end of the caliper away from the steering knuckle. Then slide the opposite end of the caliper out from

under the machined end of the steering knuckle.
5. Support the caliper from the upper control arm to prevent the weight of the caliper from being supported by the brake flex hose which will damage the hose.
6. Remove the brake pads from the caliper. Remove the outboard brake pad by prying the pad retaining clip over the raised area on the caliper. Then slide the pad down and off the caliper. Pull the inboard brake pad away from the piston until the retaining clip is free from the cavity in the piston.

To install:

7. Clean all parts well. Inspect the caliper for piston seal leaks (brake fluid in and around the boot area and inboard lining) and for any ruptures of the piston dust boot. If the boot is damaged or fluid leak is visible, disassemble the caliper and install a new seal and boot (and piston, if scored).
8. Inspect the caliper pin bushings. Replace if damaged, dry or brittle.
9. Completely retract the piston into the caliper using a large C-clamp or other suitable tool.
10. Lubricate the area on the steering knuckle where the caliper slides with high temperature grease.
11. Install the new inboard brake pad into the caliper piston by firmly pressing into the piston bore. Install the brake pads into the caliper. Note that the inboard and outboard pads are different. Make sure the inboard brake shoe assembly is positioned squarely against the face of the caliper piston.

NOTE: Be sure to remove the noise suppression gasket paper cover if the pads come so equipped.

12. Install the new outboard brake pad onto the caliper assembly.

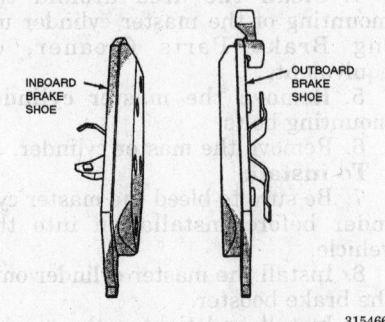

INBOARD BRAKE SHOE   OUTBOARD BRAKE SHOE

315466

Disc brake pad identification

13. Carefully position the caliper and brake pad assemblies over the rotor by hooking the lower end of the caliper over the steering knuckle. Then rotate the caliper into position at the top of the steering knuckle. Make sure the caliper guide pin bolts, bushings and sleeves are clear of the steering knuckle bosses.
14. Install the caliper guide pin bolts and tighten to 168 inch lbs. (19 Nm).
15. Reinstall the wheel and tire. Tighten the lug nuts in 2 steps, in a star pattern sequence to 95–100 ft. lbs. (129–135 Nm).
16. Lower the vehicle.
17. Pump the brake pedal until the brake pads are seated and a firm pedal is achieved before attempting to move the vehicle.
18. Road test the vehicle for proper operation.

Brake Rotor

REMOVAL AND INSTALLATION

1. Raise and safely support the vehicle.
2. Remove the front wheels.
3. Remove the 2 caliper to steering knuckle guide pin bolts.
4. Lift the caliper away from the steering knuckle by first rotating the free end of the caliper away from the steering knuckle. Then slide the opposite end of the caliper out from under the machined end of the steering knuckle.
5. Support the caliper from the upper control arm to prevent the weight of the caliper from being supported by the brake flex hose which will damage the hose.
6. Remove the brake rotor by pulling it straight off the wheel mounting studs.
7. Inspect the brake rotor surface. Resurface the rotor if deep scoring or warpage is evident. Replace the rotor if cracks or burned spots are evident.
8. Check the rotor for run-out using a dial indicator. If rotor run-out exceeds 0.005 in. (0.13mm), the rotor must be resurfaced or replaced.
9. Measure the thickness of the brake rotor at 12 equal points on the rotor at approximately 1 in. from the edge of the rotor using a micrometer. If the thickness variation is more than 0.0005 in. (0.013mm), the rotor must be resurfaced or replaced.
10. Using a micrometer, measure the rotor for minimum thickness. If the thickness measurement falls below 0.843 in. (21.4mm), replace the

rotor. If resurfacing the brake rotor, be sure to remove equal amounts from both sides of the rotor and remove as little as necessary from each rotor side.

To install:

11. Clean all parts well. Inspect the caliper for piston seal leaks (brake fluid in and around the boot area and inboard lining) and for any ruptures of the piston dust boot. If the boot is damaged or fluid leak is visible, disassemble the caliper and install a new seal and boot (and piston, if scored).

12. Inspect the caliper pin bushings. Replace if damaged, dry or brittle.

13. Completely retract the piston into the caliper using a large C-clamp or other suitable tool.

14. Lubricate the area on the steering knuckle where the caliper slides with high temperature grease.

15. Install the brake rotor onto the wheel hub.

16. Carefully position the caliper and brake shoe assemblies over the rotor by hooking the lower end of the caliper over the steering knuckle. Then rotate the caliper into position at the top of the steering knuckle. Make sure the caliper guide pin bolts, bushings and sleeves are clear of the steering knuckle bosses.

17. Install the caliper guide pin bolts and tighten to 168 inch lbs. (19 Nm).

18. Reinstall the front wheels. Tighten the stud nuts in 2 steps, in a star pattern sequence to 95–100 ft. lbs. (129–135 Nm).

19. Lower the vehicle.

20. Pump the brake pedal until the brake pads are seated and a firm pedal is achieved before attempting to move the vehicle.

21. Road test the vehicle for proper operation.

Brake Drums

REMOVAL AND INSTALLATION

All vehicles except Sebring Convertible, are equipped with rear wheel, 2 shoe leading/trailing, internal expanding type of drum brakes with automatic self-adjuster mechanisms. The automatic self-adjuster mechanisms used on these vehicles are new designs and function differently than the screw type adjusters used in the

past. These new self-adjusters are still actuated each time the vehicle's service brakes are applied. The new adjusters are located directly below the wheel cylinders.

The Sebring Convertible's rear wheel drum brake is a 2 shoe leading/trailing internal expanding type with an automatic self-adjuster mechanism. The automatic self-adjuster mechanism used on this vehicle is the screw type adjuster. The self-adjuster mechanism is actuated each time the vehicle service brakes are applied. Generally, drum brakes with a self adjusting mechanism do not require manual brake shoe adjustment. Although, in the event that the brake shoes are replaced, it is advisable to make the initial adjustment manually to speed up the initial adjustment time. The initial adjustment procedure must be done prior to driving the vehicle.

1. Raise and safely support the vehicle.

2. Remove the rear wheel assembly.

NOTE: If the vehicle has high mileage, the brake drums may have a ridge worn in them by the brake shoes. This ridge causes the brake drum to interfere with the brake shoes preventing drum removal. Clearance can be obtained by backing off the brake's automatic self-adjuster mechanism, using the following procedures.

3. For all vehicles, except Sebring Convertible, use the following procedure:

    a. Locate and remove the rubber plug from the brake support plate (backing plate).

    b. Insert a brake adjuster tool or similarly-shaped prytool through the automatic adjuster access hole and engage the teeth on the adjuster wheel. Rotate the adjuster wheel so it is moved toward the front of the vehicle. Continue moving the adjuster until it stops; this will back off the adjustment of the rear brake shoes.

4. For the Sebring Convertible, use the following procedure for releasing the self-adjusting mechanism:

    a. Locate and remove the rubber plug from the brake support plate (backing plate).

    b. Insert a brake adjuster tool or similarly shaped prytool through

the automatic adjuster access hole and carefully push the adjuster actuating lever out of engagement with the adjuster star wheel. While holding the lever away from the star wheel, insert a second prytool through the access hole and engage the teeth on the adjuster wheel. Rotate the adjuster wheel upward away from the ground; this will back off the adjustment of the rear brake shoes.

5. Remove the rear brake drum from the hub assembly.

To install:

6. Inspect the brake drums for cracks or signs of overheating. Measure the drum run-out and diameter. If not to specification, resurface the drum. Run-out should not exceed 0.006 in. (0.15mm). The diameter variation (oval shape) of the drum braking surface must not exceed either 0.0025 in. (0.064mm) in 30 degrees rotation, or 0.0035 in. (0.089mm) in 360 degrees rotation. All brake drums are marked with the maximum allowable brake drum diameter on the face of the drum.

7. Install the rear brake drum onto the hub assembly.

8. Install the wheel and tire. Tighten the lug nuts in a star pattern sequence to about 45 ft. lbs. (61 Nm); then, repeat the pattern and final torque to 95–100 ft. lbs. (129–135 Nm).

9. Properly adjust the rear brakes.

10. Lower the vehicle.

11. Road test vehicle to check brake operation.

Brake Shoes

REMOVAL AND INSTALLATION

Except Sebring Convertible

All vehicles except Sebring Convertible, are equipped with rear wheel, 2 shoe leading/trailing, internal expanding type of drum brakes with automatic self-adjuster mechanisms. The automatic self-adjuster mechanisms used on these vehicles are new designs and function differently than the screw type adjusters used in the past. These new self-adjusters are still actuated each time the vehicles' service brakes are applied. The new adjusters are located directly below the wheel cylinders.

1. Raise and safely support the vehicle.

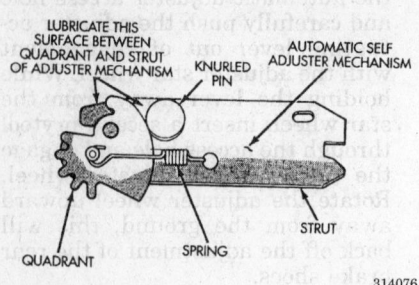

Automatic self-adjuster mechanism — except Sebring Convertible

314076

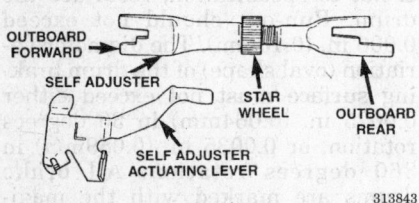

Rear brake shoe automatic self-adjuster mechanism and actuating lever — Sebring Convertible

313843

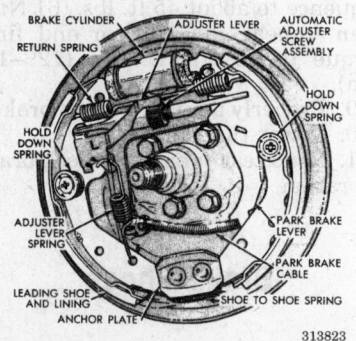

Kelsey Hayes rear brake assembly — Sebring Convertible

313823

2. Remove the rear wheel assembly.

NOTE: If the vehicle has high mileage, the brake drums may have a ridge worn in them by the brake shoes. This ridge causes the brake drum to interfere with the brake shoes preventing drum removal. Clearance can be obtained by backing off the brake's automatic self-adjuster mechanism, using the following procedure.

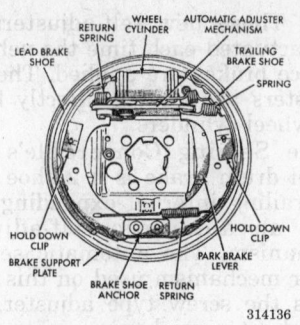

Varga rear wheel brake assembly (left side shown) — except Sebring Convertible

314136

a. Locate and remove the rubber plug from the top of the brake support plate (backing plate).

b. Insert a brake adjuster tool or similarly shaped prytool through the automatic adjuster access hole and engage the teeth on the adjuster quadrant. Then rotate the quadrant so the teeth of the quadrant are moved toward the front of the vehicle. This will back off the adjustment of the rear brake shoes.

c. Continue moving the quadrant toward the front of the vehicle until it stops moving.

3. Remove the drum from the hub assembly.

4. Remove the actuating spring from the adjuster mechanism and trailing brake shoe.

5. Remove the upper return spring from the brake shoes.

6. Remove the lower return spring from the brake shoes.

7. Remove the brake shoe retainer and pin attaching the leading brake shoe assembly to the brake support plate.

8. Remove the leading brake shoe and the adjuster mechanism as an assembly from the rear brake support plate. The adjuster mechanism cannot be separated from the leading brake shoe until the brake shoe and the adjuster mechanism is removed from the support plate.

9. Remove the trailing brake shoe retainer and pin attaching the trailing brake shoe assembly to the brake support plate. Remove the trailing brake shoe assembly.

NOTE: On this vehicle, the park brake actuating lever is permanently attached to the trailing brake shoe assembly. Do not attempt to remove it from the original brake shoe assembly or reuse the original actuating lever on a replacement brake shoe assembly. All replacement brake shoe assemblies for this vehicle must

have the actuating lever as part of the trailing brake shoe assembly.

10. Remove the parking brake cable from the parking brake lever. Do not remove the lever from the brake shoe.

11. Remove the automatic adjuster mechanism from the brake shoe by fully extending the adjuster and rotate the adjuster out to release from the brake shoe.

To inspect:

12. Clean all parts well. The brake lining should show contact across the entire width and from heel to toe; otherwise, replace. Clean and inspect the brake support plate and the automatic adjuster mechanism. Be sure the quadrant (toothed part) of the adjuster is free to rotate throughout its entire tooth contact range and is free to slide the full length of its mounting slot. Check the knurled pin. It should be securely attached to the adjuster mechanism and its teeth should be in good condition. If the adjuster is worn or damaged, replace it. If the adjuster is serviceable, lubricate lightly with high-temperature grease between the strut and the quadrant. Check the brake springs. Overheating indications are paint discoloration or distorted end coils. Replace parts as required.

13. Inspect the brake drums for cracks or signs of overheating. Measure the drum run-out and diameter. If not to specification, reface the drum. Run-out should not exceed 0.006 in. (0.15mm). The diameter variation (oval shape) of the drum braking surface must not exceed either 0.0025 in. (0.064mm) in 30 degrees rotation, or 0.0035 in. (0.089mm) in 360 degrees rotation. All brake drums are marked with the maximum allowable brake drum diameter on the face of the drum.

To install:

14. Lubricate the 8 brake shoe contact points with high-temperature grease.

NOTE: The trailing brake shoe assemblies used on the rear brakes of this vehicle are unique (handed) for the left and right side of the vehicle. Care must be taken to ensure the brake shoes are properly installed in their correct side of the vehicle. When the trailing shoes are properly installed on their correct side of the vehicle, the park brake actuating lever will be positioned under the brake shoe web.

15. Install the parking brake cable onto the parking brake lever and in-

stall the trailing brake shoe and attaching pin.

16. Install the automatic self-adjuster on the leading brake shoe by rotating it inward to attach. Install the leading shoe and adjuster assembly to the brake support plate.

17. Make sure the leading brake shoe is squarely seated on the brake support plate shoe contact areas and install the brake retainer on the retainer pin.

18. Install the lower return spring.

NOTE: The upper brake shoe return spring and adjuster mechanism actuating spring are unique for the side of the vehicle they are used on. The springs are colored for identification. The left side springs are green and the right side springs are blue.

19. Install the upper return spring (blue, right side; green, left side) on the leading brake shoe first, then on the trailing brake shoe.

20. Install the self-adjuster spring on the trailing brake shoe first then attach it to the adjuster.

21. Install the rear brake drums.

22. Install the wheel and tire. Tighten the lug nuts in a star pattern sequence to about 45 ft. lbs. (61 Nm), then repeat the pattern and final torque to 95–100 ft. lbs. (129–135 Nm).

23. Lower the vehicle.

24. Adjust the rear brakes by depressing the brake pedal. Brake shoe adjustment will occur the first time the brake pedal is depressed, pushing the rear brake shoes against the braking surface of the rear brake drums. Brake shoes should now be correctly adjusted and will not require any type of manual adjustment.

25. Road test vehicle to check brake operation.

Sebring Convertible

The Sebring Convertible's rear wheel drum brake is a 2 shoe leading/trailing internal expanding type with an automatic self-adjuster mechanism. The automatic self-adjuster mechanism used on this vehicle is the screw type adjuster. The self-adjuster mechanism is actuated each time the vehicle service brakes are applied. Generally, drum brakes with a self adjusting mechanism do not require manual brake shoe adjustment. Although, in the event that the brake shoes are replaced, it is advisable to make the initial adjustment manually to speed up the initial adjustment time. The initial adjust-

ment procedure must be done prior to driving the vehicle.

NOTE: When removing the rear brake shoes, replace the brake shoes from only one side of the vehicle at a time. This is due to the automatic adjustment feature of the parking brake system. If the brake shoes are removed from both sides of the vehicle at the same time, the automatic adjuster will remove all slack from the parking brake cables, which will make brake shoe installation extremely difficult.

1. Raise and safely support the vehicle.

2. Remove the rear wheel assembly.

NOTE: If the vehicle has high mileage, the brake drums may have a ridge worn in them by the brake shoes. This ridge causes the brake drum to interfere with the brake shoes preventing drum removal. Clearance can be obtained by backing off the brake's automatic self-adjuster mechanism, using the following procedure.

a. Locate and remove the rubber plug from the brake support plate (backing plate).

b. Insert a brake adjuster tool or similarly shaped prytool through the automatic adjuster access hole and carefully push the adjuster actuating lever out of engagement with the adjuster star wheel. While holding the lever away from the star wheel, insert a second prytool through the access hole and engage the teeth on the adjuster wheel. Rotate the adjuster wheel upward away from the ground. This will back off the adjustment of the rear brake shoes.

3. Remove the drum from the hub assembly.

4. Remove the adjusting lever actuating spring from the leading brake shoe. Remove the automatic adjuster actuating lever from the leading brake shoe.

5. Thread the adjuster star wheel all the way into the adjuster, which will remove all tension from the adjuster.

6. Remove the upper and lower return springs from the brake shoes.

7. Remove the brake shoe hold-down spring and pin attaching the leading brake shoe assembly to the brake support plate.

8. Remove the leading brake shoe from the support plate.

9. Remove the automatic adjuster from the parking brake actuating lever and trailing brake shoe.

10. Remove the retaining clip securing the parking brake actuating lever to the trailing brake shoe.

11. Remove the trailing brake shoe hold-down spring and pin attaching the trailing brake shoe assembly to the brake support plate.

12. Remove the trailing brake shoe from the brake support plate and separate the shoe from the parking brake actuating lever.

To inspect:

13. Clean all parts well. The brake lining should show contact across the entire width and from heel to toe; otherwise, replace. Clean and inspect the brake support plate and the automatic adjuster mechanism. Be sure the adjuster is free to rotate throughout its entire range. If the adjuster is worn or damaged, replace it. If the adjuster is serviceable, lightly lubricate the threaded portion with high-temperature grease. Check the brake springs. Overheating indications are paint discoloration or distorted end coils. Replace parts as required.

14. Inspect the brake drums for cracks or signs of overheating. Measure the drum run-out and diameter. If not to specification, reface the drum. Run-out should not exceed 0.006 in. (0.15mm). The diameter variation (oval shape) of the drum braking surface must not exceed either 0.0025 in. (0.064mm) in 30 degrees rotation, or 0.0035 in. (0.089mm) in 360 degrees rotation. All brake drums are marked with the maximum allowable brake drum diameter on the face of the drum.

To install:

15. Lubricate the 6 brake shoe contact points and the brake shoe anchor points with high-temperature grease.

16. Install the wave washer on the pivot pin of the parking brake actuating lever.

17. Install the trailing brake shoe onto the attaching pin of the parking brake actuating lever.

18. Position the trailing brake shoe onto the brake support plate and be sure the trailing brake shoe is squarely seated on the support plate shoe contact areas and install the brake shoe hold-down spring on the hold-down pin.

19. Install the parking brake actuating lever-to-trailing brake shoe retaining clip.

20. Install the automatic adjuster on the trailing brake shoe and the parking brake actuating lever.

21. Place the leading brake shoe onto the brake support plate in proper position and install the attaching pin and hold-down spring.

22. Install the lower and upper return springs.

23. Install the automatic adjuster actuating lever and spring onto the leading brake shoe.

24. Manually adjust the brake shoes to the furthest adjusted position but not so far as to interfere with the installation of the brake drum.

25. Install the rear brake drums. Check and adjust the brake shoes as necessary.

26. Install the wheel and tire. Tighten the lug nuts in a star pattern sequence to about 45 ft. lbs. (61 Nm), then repeat the pattern and final torque to 100 ft. lbs. (135 Nm).

27. Lower the vehicle.

28. Road test vehicle to check brake operation.

Wheel Cylinder

REMOVAL AND INSTALLATION

The hydraulic brake system is diagonally split on both the Non-ABS and ABS braking system. The rear drum brakes use wheel cylinders with piston boots of the push-on type. Note that wheel cylinders with cup expanders must have cup expanders after any service procedures (reconditioning or replacement).

With brake drums removed, inspect the wheel cylinder dust boots for evidence of a brake fluid leak. Visually check the boots for cuts, tears or heat cracks. If any of these conditions exist, the wheel cylinders should be completely cleaned, inspected and new parts installed.

To remove the wheel cylinders for repair or replacement, use the following procedure.

1. Raise and safely support the vehicle.

2. Remove the rear wheels.

3. Remove the brake drums.

4. In case of a leak, remove the brake shoes. Replace if soaked with grease or brake fluid.

5. Disconnect the rear brake flex hose from the wheel cylinder.

6. Remove the 2 wheel cylinder attaching bolts.

7. Remove the wheel cylinder from the backing plate.

To inspect:

8. Disassemble the wheel cylinders by prying the outer dust boots off.

9. Press IN on one piston to force out the opposite piston, cup and spring (with cup expanders). Using a soft tool such as a wood dowel, press out the remaining cup and piston.

10. Wash the wheel cylinder, pistons and spring in clean brake fluid or alcohol; do not use any petroleum base solvents. Clean all parts well and dry. Inspect the cylinder bore and piston for scoring and pitting. Do not use a rag to clean these parts as lint will stick to the bore surface. Wheel cylinder bores and pistons that are badly scored or pitted should be replaced. Cylinder walls that have light scratches or show signs of corrosion can usually be cleaned with crocus cloth using a circular motion. Black stains on the cylinder walls are caused by the piston cups and will not impair operation of the cylinder.

11. Before assembling the pistons and new cups in the wheel cylinder, dip them in clean brake fluid. If the boots are cracked, deteriorated or do not fit tightly on the pistons or wheel cylinder casting, install new boots. Coat the cylinder bore with clean brake fluid and install the expansion spring with cup expanders. Install the rubber cups in each end of the cylinder with the open ends of the cups facing each other.

12. Install the piston in each end of the cylinder with the flat face of each piston contacting the flat face of each cup already installed.

13. Install a dust boot over each end of the wheel cylinder. Be careful not to damage these boots during installation.

To install:

14. Apply a small bead of silicone sealer around the mating surface of the backing plate and the wheel cylinder.

15. Position the wheel cylinder on the backing plate and install the 2 wheel cylinder attaching bolts. Torque to 97 inch lbs. (11 Nm).

16. Hand start the brake line to the wheel cylinder. Torque the brake line tube nut to 145 inch lbs. (17 Nm).

17. Install the rear brake shoes. Install the rear brake drum onto the wheel hub.

18. Install the tire and wheel. Torque the lug nuts in a star pattern sequence in 2 steps with final torque of 95–100 ft. lbs. (129–135 Nm).

19. Adjust the rear brakes.

20. Bleed the entire brake hydraulic system.

21. Lower vehicle and road test.

Parking Brake Cable

ADJUSTMENT

Except Sebring Convertible

This vehicle uses a "bent nail" type parking brake tension cable equalizer. The tension equalizer can only be used one time to set the parking brake cable tension. If the parking brake cables require adjustment during the life of the vehicle, a NEW tension equalizer MUST be installed before the adjustment is made.

1. Remove the screws attaching the rear of the center console assembly to the console bracket.

2. Remove the 2 screws attaching the front of the center console to the forward console. For access to the 2 screws attaching the front of the center console to the forward console, the shifter boot or PRNDL plate must be removed from the center console by disengaging the retaining clips.

3. Raise the parking brake hand lever assembly as high as it will go for required clearance to remove the center console. Remove the center console from the vehicle.

4. Lower the park brake handle.

5. Loosen the brake cable adjusting nut on the parking brake cable output cable. This will take tension off the output cable, allowing it to be easily removed from the tension equalizer.

CAUTION

Discard the output cable retaining clip after removing it from the parking brake cable tension equalizer. Retainer is not to be reused. A new retainer is to be installed when attaching output cable to the tensioner equalizer.

---

6. Using a flat bladed prytool, unlatch the parking brake output cable retainer. Then remove the cable retainer from the parking brake cable tension equalizer. Remove the equalizer from the cables.

7. Install a NEW tension equalizer and NEW retaining clip. The cable retainer must be closed and securely latched.

8. Adjust the cable tension using the following steps:

 a. Position the parking brake lever so it is in the fully released position.

 b. Tighten the adjusting nut on the parking brake lever output cable until 12mm of thread is out past the top edge of the adjustment nut.

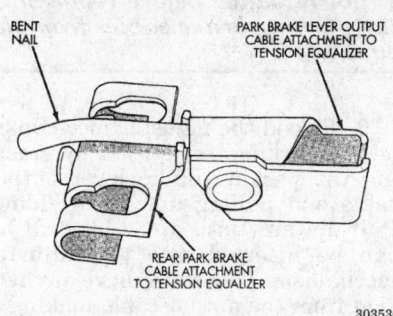

BENT NAIL

PARK BRAKE LEVER OUTPUT CABLE ATTACHMENT TO TENSION EQUALIZER

REAR PARK BRAKE CABLE ATTACHMENT TO TENSION EQUALIZER

303530

"Bent nail" parking brake cable tensioner equalizer — except Sebring Convertible

c. Actuate the parking brake lever to its fully applied position (15 clicks) one time and then release.

d. Actuating the parking brake lever to its fully applied position one time after tightening the adjustment nut will stretch the bent nail portion of the tension equalizer about ¼ in. (6mm). This process will correctly set the parking brake cable tension.

9. Check the rear wheels of the vehicle. They should rotate freely without dragging.

10. After the parking brake cable tension has been properly adjusted, check for free-play in the parking brake lever. The parking brake hand lever should feel firm at all clicks with a maximum of 15 clicks of lever travel possible.

11. Install the console components and shifter boot or PRNDL plate as required.

Sebring Convertible

Manual adjustment of the parking brake is not required due to the automatic adjustment feature used on this vehicles parking brake system. Proper adjustment of the parking brake on this vehicle depends on the rear drum brake shoes being adjusted properly.

REMOVAL AND INSTALLATION

Except Sebring Convertible

1. Disconnect the negative battery cable.

2. Remove the mounting screws securing the rear of the floor console assembly to the floor bracket.

3. If equipped with automatic transaxle, disengage the clips securing the PRNDL plate from the console and remove.

4. If equipped with manual transaxle, remove the shifter boot and knob as follows:

a. Pull down shift boot enough to expose shifter roll pin.

b. Using a flat blade tool, pry open the legs of the shift knob away from the roll pin and remove knob from shift lever.

c. Squeeze the shift boot at its base and pull up to remove.

5. Remove the 2 front floor console mounting screws. Raise the parking brake lever as high as it will it go to allow for removal clearance.

6. Remove the floor console from the vehicle. Lower the parking brake lever.

7. Loosen the output cable adjuster nut. This will relieve tension from the parking brake cables allowing for easy removal.

8. Remove the floor console rear mounting bracket. Disconnect the parking brake cable requiring replacement from the cable tension equalizer.

9. Remove the rear seat cushion from the vehicle by pulling upward at the front edge of the cushion to disengage from the retainer cups in the rear floor.

10. Carefully remove the right and left side rear scuff plates by prying scuff plate retaining clips out of the door sills.

11. Fold the rear section of carpet forward to expose the rear parking brake cables.

12. Remove the rear parking brake cables-to-floor pan routing clip.

13. Compress the parking brake cable retainer tabs at the console bracket using a ½ in. box wrench. Pull the parking brake cable straight out of the console bracket.

14. Raise and safely support the vehicle. Remove the rear wheel(s) requiring parking brake cable replacement.

15. Remove the rear brake drum.

16. Remove the dust cap from the rear hub and wheel bearing assembly.

17. Remove the rear wheel hub and bearing assembly from the rear spindle.

18. Disconnect the parking brake cable from the parking brake actuating lever on the trailing brake shoe.

19. Remove the parking brake cable from from the rear brake support plate by compressing the locking tabs on the cable retainer using a ½ in. box wrench.

20. Remove the 2 parking brake cable routing brackets located on the vehicle frame rail.

21. Remove the parking brake cable and cable sealing grommet from the vehicle floor pan.

To install:

22. Install the cable into the vehicle floor pan. Be sure the sealing grommet is installed into the floor pan as far as possible to guarantee a proper seal.

23. Install the parking brake cable into the rear brake support plate. Be sure the cable retainer locking tabs are expanded to ensure that the cable is securely locked in the brake support plate.

24. Install the 2 parking brake cable routing brackets onto the vehicle frame rail. Install and securely tighten the bracket mounting bolts.

25. Connect the parking brake cable end to the parking brake actuating lever of the trailing brake shoe.

26. Install the wheel hub and bearing assembly onto the rear spindle. Install a new rear hub/bearing assembly retaining nut and torque nut to 185 ft. lbs. (250 Nm).

27. Using a soft faced hammer, install the wheel hub and bearing dust cap.

28. Install rear brake drum.

29. Install the rear wheel(s) and lug nuts. Tighten the lug nuts in a star pattern sequence to half torque specification. Repeat the tightening procedure and tighten the lug nuts to a final torque of 95–100 ft. lbs. (129–135 Nm).

30. Lower the vehicle.

31. Grasp the park brake cable-to-floor pan sealing grommet from inside the vehicle. Pull the sealing grommet into the floor pan to ensure that it is fully seated into the floor pan.

32. Route the parking brake cable under the carpet and up to the hole in the console bracket on the floor pan. Insert the cable into the console bracket hole and engage the cable retainer locking tabs. Be sure the locking tabs are expanded to ensure that the locking tab will lock into place.

33. Install the park brake cable routing/retaining clip to the floor pan of the vehicle and tighten the mounting nut.

34. Using a suitable prytool, unseat the parking brake output cable retainer. Remove cable retainer and parking brake cable tension equalizer from the parking brake lever output cable and discard components.

35. Install a NEW parking brake cable tension equalizer on the parking brake lever output cable and rear parking brake cables.

36. Install a NEW parking brake lever output cable to tension equalizer retaining clip on tension equalizer. The cable retainer must be closed and securely latched.

37. Adjust the cable tension as follows:

a. Be sure the parking brake lever is in the fully released position.

b. Tighten the parking brake lever output cable adjusting nut until 0.5 in. (12mm) of thread is extended out past the top edge of the adjuster nut.

c. Pull the parking brake lever to its fully applied position (approximately 15 clicks) 1 time and then fully release the parking brake lever. This should stretch the bent nail portion of the tension equalizer approximately ¼ in. (6mm) which will correctly tension the parking brake cable.

38. Slightly raise and support the vehicle. Check the rear wheels with the parking brake lever fully released, to make sure they rotate freely without dragging.

39. Lower the vehicle to the ground.

40. Check the parking brake lever for free-play. The park brake lever should feel firm at all clicks. Maximum lever travel should only be 15 clicks.

41. Install the floor console into the vehicle. Install the front and rear floor console mounting screws.

42. If equipped with automatic transaxle, install the PRNDL plate onto the console and secure into place by engaging the mounting clips.

43. If equipped with manual transaxle, install the shift boot and knob as follows:

a. Slide the rubber boot down over the shift lever and squeeze at the base of the boot to engage it into position on the console.

b. Install the shift knob onto the shift lever and bend the legs of the

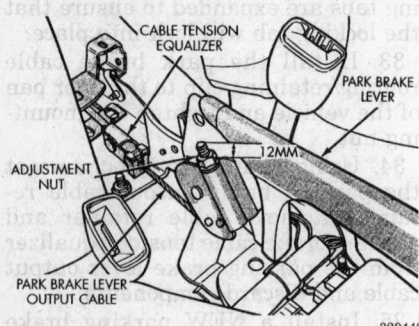

Adjustment of parking brake lever output cable — except Sebring Convertible

shift knob tightly on the shift lever over the roll pin.

44. Place the rear carpet back into its proper position in the rear of the vehicle interior.

45. Install both right and left rear door sill plate scuff moldings by snapping them into place on the rear door sills.

46. Install the rear lower seat cushion into the vehicle. Be sure to properly engage the seat cushion in the retainers on the vehicle floor pan.

47. Check and adjust the rear brakes, if necessary.

48. Connect the negative battery cable.

49. Check that the parking brake holds the vehicle on an incline.

Sebring Convertible

NOTE: To avoid extreme difficulty during the installation procedure, remove only one rear parking brake cable from the vehicle at a time.

1. Disconnect the negative battery cable.

2. Using a 2mm Allen wrench, remove the set screw securing the shifter knob to the shift lever. Remove the shifter knob by pulling it up off the shift lever.

3. Remove the 3 mounting screws securing the rear of the console to the console bracket.

4. Remove the screw hole garnish cap and the PRNDL strip from the console.

5. Remove the 2 mounting screws attaching the console to the shifter. Raise the parking brake lever to a 45 degree angle to allow for removal clearance of the console.

6. Raise the rear of the console high enough to access the console wiring harness connector. Disconnect the 8-way wiring harness connector.

7. Remove the console from the vehicle. Fully release the parking brake lever.

8. Disconnect the wiring harness from the ground switch on the parking brake lever and unclip the harness from the park brake lever.

CAUTION

The parking brake lever contains an auto adjusting feature which consists of a clock spring loaded to approximately 20 lbs. (89 N). The auto adjuster must be reloaded before releasing the parking brake cables from the equalizer. Serious injury could result if the adjuster mechanism

is not reloaded before removal of the parking brake cables from the equalizer.

9. Reload the adjuster mechanism on the parking brake lever by grasping the parking brake lever output cable and pulling upward by hand. Pull upward until a ¹⁵/₆₄ in. drill bit can be inserted into the adjuster mechanism. This will relieve any tension from the output cable making it easier to disconnect the rear parking brake cables from the equalizer.

10. Disconnect the parking brake cable requiring replacement from the cable tension equalizer.

11. Remove the rear seat cushion from the vehicle by pushing the cushion rearward and upward at one attachment point to disengage the wire loops from the retainers in the floor pan. Repeat this procedure for the other attachment point and remove the seat cushion from the vehicle. Be careful not to damage the rear seat cushion frame.

12. Carefully remove the right and left side door sill scuff plates by prying scuff plate retaining clips out of the door sills.

13. Remove the right and left side interior quarter trim panels from the vehicle as follows:

a. Lower the convertible top.

b. Pull the lower section of the rear seat back forward until the seat back brackets clear the studs on the floor pan.

c. Push the rear seat back upward to disengage the hooks that secure the seat back to the rear seat back support. Remove the rear seat back from the vehicle.

d. Remove the door sill trim panel and push-in fastener securing the quarter trim panel to the door sill panel.

e. Remove the speaker grille.

f. Remove the vertical mounting screws securing the quarter trim panel to the inner quarter panel.

g. Remove the mounting screws securing the quarter trim panel to the inner quarter panel through the speaker grille opening.

h. Remove the mounting screws securing the quarter trim panel to the inner quarter panel at the rear of the trim panel.

i. Remove the push-in fasteners securing the quarter trim panel to the inner quarter panel at the front of the trim panel.

j. Remove the quarter trim panel from the inner quarter panel. Disconnect the speaker wiring con-

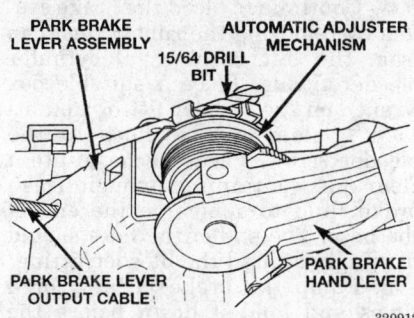

Parking brake lever properly reloaded — Sebring Convertible

nector. Remove the quarter trim panel from the vehicle.

14. Remove the 2 wiring harness routing clips from the car cross beam and the 2 clips that hold down the carpet to the car cross beam.

15. Fold the rear section of carpet forward to expose the rear parking brake cables.

16. Remove the rear parking brake cables-to-floor pan routing clip.

17. Compress the parking brake cable retainer tabs at the console bracket using a ½ in. box wrench. Pull the parking brake cable straight out of the console bracket.

18. Raise and safely support the vehicle. Remove the rear wheel(s) requiring parking brake cable replacement.

19. Remove the rear brake drum.

20. Remove the dust cap from the rear hub and wheel bearing assembly.

21. Remove the rear wheel hub and bearing assembly from the rear spindle.

22. Disconnect the parking brake cable from the parking brake actuating lever on the trailing brake shoe.

23. Remove the parking brake cable from from the rear brake support plate by compressing the locking tabs on the cable retainer using a ½ in. box wrench.

24. Remove the 2 parking brake cable routing brackets located on the vehicle frame rail.

25. Remove the parking brake cable and cable sealing grommet from the vehicle floor pan.

To install:

26. Install the cable into the vehicle floor pan. Be sure the sealing grommet is installed into the floor pan as far as possible to guarantee a proper seal.

27. Install the 2 parking brake cable routing brackets onto the vehicle frame rail. Install and securely tighten the bracket mounting bolts.

28. Install the parking brake cable into the rear brake support plate but **DO NOT** engage the cable retainer locking tabs into the brake support plate at this time.

29. Connect the parking brake cable end to the parking brake actuating lever of the trailing brake shoe. Be sure the end of the spring is under the lip on the parking brake actuating lever.

30. Fully push the parking brake cable into the rear brake support plate. Be sure the cable retainer locking tabs are securely locked into the rear brake support plate.

31. Install the wheel hub and bearing assembly onto the rear spindle. Install a new rear hub/bearing assembly retaining nut and torque nut to 185 ft. lbs. (250 Nm).

32. Using a soft faced hammer, install the wheel hub and bearing dust cap.

33. Install rear brake drum.

34. Install the rear wheel(s) and lug nuts. Tighten the lug nuts in a star pattern sequence to ½ torque specification. Repeat the tightening procedure and tighten the lug nuts to a final torque of 100 ft. lbs. (135 Nm).

35. Lower the vehicle.

36. Grasp the park brake cable-to-floor pan sealing grommet from inside the vehicle. Pull the sealing grommet into the floor pan to ensure that it is fully seated into the floor pan.

37. Route the parking brake cable under the carpet and up to the hole in the console bracket on the floor pan. Insert the cable into the console bracket hole and engage the cable retainer locking tabs. Be sure the locking tabs are expanded to ensure that the locking tab will lock into place.

38. Install the park brake cable routing/retaining clip to the floor pan of the vehicle and tighten the mounting nut.

CAUTION

This parking brake lever contains an auto adjusting feature which consists of a clock spring loaded to approximately 20 lbs. (89 N). DO NOT unload the auto adjuster mechanism using any procedure other than the one outlined in this procedure. Serious injury could result if the adjuster mechanism is unloaded using an alternative procedure.

---

39. Unload the adjuster mechanism by grasping the park brake lever output cable by hand and pulling upward on it until all tension is relieved from the drill bit. Remove the drill bit

from the clock spring of the adjuster mechanism. Then slowly release the cable until all the slack is removed from the cable.

40. Clip the wiring harness onto the parking brake lever bracket.

41. Connect the wiring harness connector to the ground switch of the parking brake lever.

42. Cycle the parking brake lever to its fully applied position and then lower it to its fully released position. This will position the parking brake cables and fully adjust them to the proper tension.

43. Slightly raise and support the vehicle. Check the rear wheels with the parking brake lever fully released, to make sure they rotate freely without dragging.

44. Lower the vehicle to the ground.

45. Raise the park brake lever to approximately a 45 degree angle to provide proper clearance for console installation.

46. Install the floor console into the vehicle. Connect the center console wiring harness connector to the vehicle wiring harness. Install the front and rear floor console mounting screws.

47. Install the PRNDL strip onto the console and screw hole garnish cap.

48. Install the shifter knob down onto the shift lever. Install and securely tighten the shifter knob set screw.

49. Place the rear carpet back into its proper position in the rear of the vehicle interior.

50. Install the 2 clips retaining the carpet to the car cross beam.

51. Install the 2 wiring harness routing clips onto the car cross beam.

52. Install the rear interior quarter trim panels as follows:

    a. Position the quarter trim panel in the vehicle and reconnect the speaker wire connector.

    b. Install the push-in fasteners securing the quarter trim panel to the inner quarter panel at the front of the quarter trim panel.

    c. Install the mounting screws securing the quarter trim panel to the inner quarter panel through the speaker opening.

    d. Install the vertical mounting screws securing the quarter trim panel to the inner quarter panel.

    e. Install the mounting screws securing quarter trim panel to the inner quarter panel at the rear of trim panel.

    f. Install the push-in fastener securing the quarter trim panel to the door sill panel.

g. Install the door sill trim panel.

h. Install the speaker grill.

i. Position the rear seat back properly into the vehicle and tilt back rearward. Lift seat back above retaining brackets on rear seat back support.

j. Lower the seat back until the center loop starts to engage.

k. Lower the rear seat back and press rearward on the outboard corners of the seat back to engage the outboard hooks to the brackets on the rear seat back support.

l. Make sure all hooks are fully engaged into the retaining brackets by pushing downward on the rear seat back.

m. Install the rear inner seat belt assembly and install the mounting nuts. Torque the mounting nuts to 350 inch lbs. (40 Nm).

53. Install both right and left door sill plate scuff moldings by snapping them into place on the rear door sills.

54. Install the rear lower seat cushion into the vehicle. Be sure to push the rear seat cushion rearward then downward to properly engage the wire loops to the retainers on the vehicle floor pan. Be sure the inner seat belts are on top of the seat cushion.

55. Check and adjust the rear brakes, if necessary.

56. Connect the negative battery cable.

57. Check that the parking brake holds the vehicle on an incline.

Brake System Bleeding

Use only MOPAR brake fluid or an equivalent from a tightly sealed container. Brake fluid must conform to DOT 3 specifications. Do not use petroleum-based fluid because seal damage in the brake system will result. On all vehicles (both non-ABS and ABS) the brake system must be bled anytime air is permitted to enter the hydraulic system. The ABS system, particularly the HCU, should only be bled when the HCU is replaced or removed from the vehicle, or if there is reason to believe the HCU has ingested air. Under most circumstances that would require brake bleeding (both non-ABS and ABS), only the base brake system needs to be bled. In cases where the ABS portion of the system must be bled, the bleeding procedure requires the use of the DRB scan tool (or equivalent). Note that correct manual bleeding of the brake hydraulic system (both non-ABS and ABS) will

require an assistant. The following wheel sequence for bleeding the brake hydraulic system should be used to ensure adequate removal of all trapped air from the hydraulic system: left rear wheel, right front wheel, right rear wheel and left front wheel.

Master Cylinder

If the master cylinder is off the vehicle, it can be bench bled.

1. Connect 2 short pieces of brake line to the outlet fittings, bend them until the free end is below the fluid level in the master cylinder reservoirs.

2. Fill the reservoir with fresh DOT 3 type brake fluid.

3. Using a wooden dowel, or equivalent, pump the piston slowly several times until no more air bubbles appear in the reservoirs.

4. Disconnect the 2 short lines, refill the master cylinder and securely install the cylinder cap.

5. If the master cylinder is on the vehicle, it can still be bled, using a flare nut wrench.

6. Open the brake lines slightly with the flare nut wrench while pressure is applied to the brake pedal by a helper inside the vehicle.

7. Be sure to tighten the line before the brake pedal is released.

8. Repeat the process with both lines until no air bubbles come out.

9. Bleed the complete brake system, if necessary.

NOTE: If the master cylinder has been thoroughly bled and filled to the proper level upon installation into the vehicle, it is not necessary to bleed the brakes entire hydraulic system.

Wheel Cylinders and Calipers

1. Clean all dirt from the master cylinder filler cap.

2. If the master cylinder is known or suspected to have air in the bore, it must be bled before any of the wheel cylinders or calipers. To bleed the master cylinder, position a shop towel under the primary (rear) outlet fitting and loosen the fitting approximately ¾ turn. Have an assistant depress the brake pedal slowly through its full travel. Close the outlet fitting and let the pedal return slowly to the fully released position. Wait 5 seconds and then repeat the operation until all air bubbles disappear.

3. Repeat Step 2 with the secondary (front) outlet fitting.

4. Continue to bleed the brake system by removing the rubber dust cap from the left rear wheel cylinder bleeder fitting. Place a suitable box wrench on the bleeder fitting and attach a clear plastic hose to the bleeder screw. Feed the hose into a clear jar containing enough fresh brake fluid to submerge the end of the hose. The end of the hose should fit snugly around the bleeder fitting.

5. Pump the brake pedal 3 or 4 times and hold it down before the bleeder is opened.

6. Open the bleeder screw at least one full turn. When the bleeder screw opens the brake pedal will drop all the way to the floor.

7. Close the bleeder screw. Release the brake pedal only after the bleeder screw is closed.

8. Repeat this procedure until no bubbles appear at the submerged end of the bleeder tube. Secure the bleeder fitting and remove the bleeder tube. Install the rubber dust cap on the bleeder fitting.

9. Repeat this procedure on the remaining wheels in the following sequence: right front, right rear, left front. Refill the master cylinder reservoir after each wheel has been bled and install the master cylinder cover and gasket. When brake bleeding is completed, the fluid level should be filled to the maximum level indicated on the reservoir.

NOTE: Never reuse brake fluid that has been drained from the hydraulic system or has been allowed to stand in an open container for an extended period of time.

10. Always make sure the disc brake pistons are returned to their normal positions by depressing the brake pedal several times until normal pedal travel is established. If the pedal feels spongy, repeat the bleeding procedure.

11. Test drive the vehicle to be sure the brakes are operating correctly and that the pedal is solid.

Anti-Lock Brakes

The bleeding procedure is a 3 step process, one of which will require use of the DRB II scan tool or its equivalent. Bleed the system as follows:

1. Locate the diagnostic connector under the dash panel next to the left kick panel.

2. Connect the DRB II scan tool to the connector. Install cartridge for the Bendix ABX-4 or Allied-Signal

ABX-4 Anti-Lock Brake systems. Check to make sure the CAB does not have any fault codes stored in it. If it does, remove them using the DRB II scan tool.

3. Bleed the base brake system using the manual or pressure bleeding method in the following sequence:
   a. Left rear wheel
   b. Right front wheel
   c. Right rear wheel
   d. Left front wheel

4. Utilizing the DRB II, go to the "Bleed ABS" routine. Firmly apply the brake pedal to initiate the "Bleed ABS" cycle one time. Release the brake pedal.

5. Using the scan tool, go on to bleed the Anti-Lock Brake System according to the scan tool literature.

6. Once bleeding with the scan tool is complete, repeat the conventional bleed procedure for the base brake system.

7. Perform this procedure until the brake fluid flows clear and free of air bubbles. Check brake fluid level periodically to prevent the reservoir from running low on fluid. Top off the master cylinder reservoir to the proper level with DOT 3 type brake fluid only.

8. Road test the vehicle to check for proper brake system operation.

Wheel Speed Sensor

REMOVAL AND INSTALLATION

Each wheel has its own wheel speed sensor and sends a small AC signal to the control module. Correct ABS operation depends on accurate wheel speed signals. The vehicles wheels and tires must all be the same size and type in order to generate accurate signals. If there is a variation between wheel and tire sizes, inaccurate wheel speed signals will be produced.

------- **CAUTION** -------
It is very critical that the wheel speed sensor(s) be installed correctly to ensure continued system operation. The sensor cables must be installed, routed and clipped properly. Failure to install the sensor(s) properly could result in contact with moving parts or over extension of sensor cables. This will cause ABS component failure and an open circuit.

Front Wheel Speed Sensor

1. Disconnect the negative battery cable at strut tower.

2. Raise and safely support the vehicle.

3. Remove the front wheel(s) of the wheel speed sensor(s) requiring removal.

4. Remove the speed sensor cable routing bracket from the steering knuckle.

5. Remove the speed sensor wiring harness sealing grommet retainer/routing bracket from the inner fender.

6. Remove the sealing grommet from the inner fender and disconnect the speed sensor cable from the vehicle wiring harness.

7. Remove the speed sensor head-to-steering knuckle retaining bolt. Remove the speed sensor head from the steering knuckle.

8. If the speed sensor head cannot be removed from the steering knuckle due to corrosion, remove the brake caliper from the brake rotor and support from the knuckle using a strong piece of wire. Then remove the brake rotor from the vehicle and insert a pin punch through the hole on the front part of the steering knuckle for the sensor head locating pin. Tap the locating pin for the wheel speed sensor out of the steering knuckle.

9. Carefully inspect the tone wheel for broken or missing teeth which can cause erratic sensor signals.

To install:

10. Connect the wheel speed sensor wiring connector to the vehicle wiring harness.

11. Install the sensor cable sealing grommet into the front inner fender. Install the sensor cable sealing grommet retainer/routing bracket on the front inner fender. Install and tighten routing bracket mounting bolt.

12. Install the speed sensor cable routing bracket onto the steering knuckle. Install the bracket mounting bolt and torque the bolt to 55 inch lbs. (6 Nm). Be sure the sensor cable is looped toward the strut. If the sensor cable is not routed in this direction, it will contact the wheel or tire, causing serious damage to the speed sensor cable.

13. Apply a light coating of MOPAR® Multi-Purpose Grease or equivalent to the locating pin of the speed sensor and install the speed sensor head onto the steering knuckle. Install the sensor head mounting screw and torque to 55 inch lbs. (6 Nm).

14. Install front wheel(s) and lug nuts. Torque the lug nuts, in a star pattern sequence, to ½ specifications. Repeat the torquing procedure to a

final torque of 95–100 ft. lbs. (129–135 Nm).

15. Lower the vehicle. Reconnect the negative battery cable. Test drive the vehicle to check for proper operation of the base and ABS systems.

Rear Wheel Speed Sensor

1. Disconnect the negative battery cable at strut tower.

2. Disconnect the speed sensor cable electrical connector from the vehicle wiring harness. The wiring connection is located in the trunk of the vehicle.

3. Raise and safely support the vehicle.

4. Remove the rear wheel(s) of the wheel speed sensor(s) requiring removal.

5. Remove the speed sensor cable sealing grommet retainer from the vehicle rear frame rail.

6. Remove the sealing grommet and sensor cable from the hole in the body of the vehicle.

7. Remove the speed sensor routing clips from the rear upper control arm and brake line flex hose routing bracket.

8. Remove the speed sensor head from the rear brake support plate.

9. Carefully inspect the tone wheel for broken or missing teeth which can cause erratic sensor signals.

To install:

10. Install the wheel speed sensor head into the brake support plate. Install the speed sensor retaining bolt and torque to 55 inch lbs. (6 Nm).

11. Install the speed sensor cable routing clips on the brake line flex hose bracket and upper control arm. Install and tighten the mounting bolts for the routing clip.

12. Install the wheel speed sensor wiring connector through the hole in the inner fender and into the vehicle trunk.

13. Install the speed sensor sealing grommet into the hole in the inner fender. Install the sensor sealing grommet retainer and attaching bolt on the rear frame rail.

14. Install front wheel(s) and lug nuts. Torque the lug nuts, in a star pattern sequence, to ½ specifications. Repeat the torquing procedure to a final torque of 95–100 ft. lbs. (129–135 Nm).

15. Lower the vehicle.

16. Connect the sensor cable connector to the vehicle wiring harness and slide the foam insulation sleeve over the wiring connection to prevent it from rattling against the body of the vehicle.

17. Reconnect the negative battery cable. Test drive the vehicle to check for proper operation of the base and ABS systems.

FRONT SUSPENSION

Strut

REMOVAL AND INSTALLATION

1. Raise and safely support the vehicle.
2. Remove the wheel and tire.
3. Remove the steering knuckle.
4. Remove the pin bolt attaching the strut to the strut clevis.
5. Remove the through bolt attaching the clevis to the lower control arm.
6. Tap the clevis with a brass drift to remove from the strut.
7. Remove the 4 bolts attaching the strut to the strut tower.
8. Remove the strut and upper control arm as an assembly.

To install:
9. Install the strut assembly into the strut tower.
10. Install the 4 upper strut mounting bolts.
11. Torque the bolts to 23 ft. lbs. (31 Nm).
12. Install the clevis onto the strut with a brass drift until the clevis is fully seated against the locating tab.
13. Reinstall the clevis pin bolt.
14. Install the clevis onto the lower control arm.
15. Install the clevis through bolt.
16. Install the steering knuckle.
17. Tighten the clevis to strut pin bolt.
18. Lower the vehicle onto a jack stand supporting the lower control arm.
19. Tighten the clevis to lower control arm mounting bolt.
20. Install the wheel and tire.
21. Lower the vehicle.

Lower Ball Joints

REMOVAL AND INSTALLATION

On all vehicles, the ball joint cannot be serviced separately. If the ball joint is defective it will require replacement of the lower control arm.

Lower Control Arms

REMOVAL AND INSTALLATION

1. Raise and safely support the vehicle.
2. Remove the front wheels and tires.
3. If equipped with 15 in. wheels, the heat shield will need to be removed before the lower control arm can be separated from the steering knuckle.
4. Remove the ball joint clamping nut.
5. Remove the sway bar attaching bolts from both control arms.
6. Disconnect the shock clevis from the lower control arm.
7. Loosen the sway bar to crossmember attaching bolts and rotate the sway bar away from the control arm.
8. Using a prybar separate the lower control arm from the steering knuckle.
9. Remove the 2 control arm attaching bolts.
10. Remove the control arm.

To install:
11. Install the control arm into the vehicle.
12. Install the 2 control arm attaching bolts.
13. Torque the control arm attaching bolts to 120 ft. lbs. (163 Nm).
14. Install the control arm to the steering knuckle and torque to 70 ft. lbs. (95 Nm).
15. Rotate the sway bar up to the control arms.
16. Install the shock absorber clevis.
17. Install the sway bar attaching bolts.
18. Torque the sway bar bolts to 21 ft. lbs. (28 Nm).
19. Torque the sway bar to crossmember attaching bolts 21 ft. lbs. (28 Nm).
20. Install the wheels and tires.

Sway Bar

REMOVAL AND INSTALLATION

The sway bar (also called a stabilizer bar) connects the front lower control arms and attaches to the front suspension crossmember and the body of the vehicle. Jounce and rebound movements affecting one wheel are partially transmitted to the opposite wheel through the stabilizer bar to restrict body roll.

Attachment of the stabilizer bar to the front crossmember and the body of the vehicle, is through 2 rubber isolator bushings. The stabilizer bar is connected to the lower control arms using ball joint type attaching links. All stabilizer bar components are serviceable as separate components and the stabilizer bar to crossmember bushings are split for easy removal and installation. The split in the stabilizer bar to crossmember bushing should be positioned toward the front of the vehicle when installed on the stabilizer bar.

1. Raise and safely support the vehicle.
2. Remove the nuts and stabilizer bar attaching link assemblies from the front lower control arms. When removing the attaching link nut keep the stud from turning by installing an Allen wrench in the end of the stud.
3. Remove the 4 bolts attaching the sway bar bushing retainers to the crossmember and body.
4. Remove the stabilizer bar bushings, bushing retainers the stabilizer bar and attaching links from the vehicle as an assembly.

To install:
5. Inspect for broken or distorted stabilizer bar bushings, bushing retainers and attaching links. If stabilizer bar to front crossmember bushing replacement is required, use the following procedure:
   a. Bend back the 4 crimp locations on the stabilizer bar bushing retainer.
   b. Separate the stabilizer bar bushing retainer.
   c. Stabilizer bar bushings are removed by opening the slit and peeling the bushing off the stabilizer bar.
   d. Install the new stabilizer bar bushings on the bar. The bushings must be installed on the sway bar with the slit in the bushing facing the front of the vehicle when the sway bar is installed.
   e. Install new bushing retainers on the stabilizer bar.
6. Install the stabilizer bar and bushings as an assembly into the vehicle.
7. Align the stabilizer bar attaching link and bushing assemblies with the attaching link mounting holes in the lower control arms. Install the attaching links into the control arms. Torque the attaching link nuts to 21 ft. lbs. (28 Nm).
8. Install the 4 sway bar retainer bushing bolts into the crossmember and torque the bolts to 21 ft. lbs. (28 Nm).
9. Install the tire and wheel.

10. Lower the vehicle and test for proper operation.

Front Wheel Bearings

ADJUSTMENT

The front hub wheel bearing is designed for the life of the vehicle and requires no type of adjustment or periodic maintenance. The bearing is a sealed unit with the wheel hub and can only be removed and/or replaced as one unit.

REMOVAL AND INSTALLATION

The front wheel bearing used on this vehicle is a bolt-in type wheel bearing.

The wheel bearing is serviced separately from the front steering knuckle and front hub assembly. Retention of the front wheel bearing into the steering knuckle is by means of 3 bolts installed from the rear of the steering knuckle. The 3 bolts attach the hub/bearing to the front surface of the steering knuckle. Removal and installation of the hub/bearing assembly from the steering knuckle must be done with the steering knuckle removed from the vehicle.

This vehicle does not use a rubber lip seal as on past front wheel drive vehicles to prevent contamination of the front wheel bearing. On this vehicle the face of the outer CV-joint has a metal bearing shield pressed on it. This design deters direct water splash on the bearing seal while allowing any water that gets in to run out the bottom of the steering knuckle. It is important to thoroughly clean the outer CV-joint and the wheel bearing area in the steering knuckle before it is assembled after servicing the front wheel bearing or driveshaft.

At no time when servicing this vehicle, can a sheetmetal screw, bolt or other metal fastener be installed in the shock tower to take the place of an original plastic clip. Also, NO holes can be drilled into the front shock tower for the installation of any metal fasteners into the shock tower. Because of the minimum clearance in this area installation of metal fasteners could damage the coil spring's protective coating and lead to corrosion failure of the spring. If a plastic clip is missing, lost or broken during servicing a vehicle, replace

only with the equivalent part listed in the Mopar parts catalog.

1. Raise and safely support the vehicle.
2. Remove the front tire and wheel.
3. Remove the steering knuckle assembly using the recommended procedure.
4. Remove the 3 bolts attaching the hub/bearing assembly to the steering knuckle.
5. Remove the hub/bearing assembly out from the front of the steering knuckle. The bolt-in front wheel bearing used on the vehicle is transferable to a replacement steering knuckle is the bearing found in serviceable condition. If the bearing will not come out of the steering knuckle, it can be tapped out using a soft-faced hammer.

To install:
6. Clean all parts well. Thoroughly clean all the hub/bearing assembly mounting surfaces on the steering knuckle.
7. Install the replacement hub/bearing assembly in the steering knuckle aligning the bolt holes in the bearing flange with the holes in the steering knuckle.
8. Install the 3 attaching bolts and tighten evenly to make sure the bearing is square to the face of the steering knuckle. Torque the attaching bolts to 80 ft. lbs. (110 Nm).
9. Install the steering knuckle. Install the tire and wheel.
10. Lower the vehicle and check for proper operation.

REAR SUSPENSION

Strut

REMOVAL AND INSTALLATION

1. Pull back the carpeting from the rear strut tower.
2. Remove the plastic cover from the top of the strut tower.
3. Remove the 2 nuts attaching the strut assembly to the body.
4. Raise and safely support the vehicle.
5. Remove the wheel and tire.
6. Remove the bolt attaching the strut to the rear knuckle.
7. Push downward on the rear suspension and tilt the top of the strut outward.

8. Remove the strut from the vehicle.

To install:
9. Install the strut into the vehicle at the rear knuckle.
10. Push downward on the rear suspension and insert the top of the strut into the vehicle.
11. Install the strut to rear knuckle attaching bolt.
12. Torque the bolt to 70 ft. lbs. (95 Nm).
13. Lower the vehicle enough to gain access to the trunk.
14. Install the strut upper mounting nuts and torque to 25 ft. lbs. (34 Nm).
15. Install the strut top cover.
16. Install the rear wheel and tire and torque to 95 ft. lbs. (125 Nm).
17. Lower vehicle and check for proper operation.

Lower Control Arms

REMOVAL AND INSTALLATION

The vehicle uses 2 lateral links instead of a control arm for the rear lower suspension.

1. Raise and safely support the vehicle.
2. Remove the wheel and tire.
3. Disconnect the sway bar bushings from the lateral link.

NOTE: The sway bar bushings are located on the front lateral link only. The rear lateral link does not have any connection to the sway bar.

4. Remove the 2 bolts attaching the lateral link to the rear knuckle and the crossmember.
5. Remove the lateral link from the vehicle.

To install:
6. Install the lateral link into the vehicle.
7. Install the 2 attaching bolts and torque to 70 ft. lbs. (95 Nm).
8. Install the sway bar bushings (front lateral link only).
9. Install the wheel and tire.
10. Align the rear suspension.
11. Lower vehicle and check for proper operation.

Sway Bar

REMOVAL AND INSTALLATION

The sway bar (also called the Stabilizer Bar) should be inspected for broken or distorted retainers and bushings. If bushing replacement is required, bushings can be removed by opening the slit in the bushing and removing the bushing from around the stabilizer bar. Note that when replacement bushings are required, the bushings must be installed with the slit positioned on the stabilizer bar so the slit will face forward in the vehicle when the stabilizer bar is installed.

1. Raise and safely support the vehicle.

2. Remove both rear wheels.

3. Remove the nuts attaching the sway bar isolator bushings to the sway bar.

4. Remove the isolator bushings from the sway bar link.

5. Remove the 4 bolts attaching the sway bar clamps to the crossmember.

6. Remove the sway bar between the crossmember and the exhaust pipe.

To install:

7. Install the sway bar into the vehicle.

NOTE: The bend in the sway bar must be positioned up in the vehicle for proper installation.

8. Install the sway bar to the sway bar links.

9. Install the isolator bushings.

10. Torque the sway bar link nuts to 40 ft. lbs. (55 Nm).

11. Install the sway bar hold-down clamps to the crossmember and center the sway bar in the vehicle.

12. Torque the hold-down clamp bolts to 250 inch lbs. (28 Nm).

13. Install the wheels and tires.

14. Torque the lug nuts to 95 ft. lbs. (129 Nm).

15. Lower the vehicle and check for proper operation.

Wheel Bearings

ADJUSTMENT

The rear hub and wheel bearing assembly is designed for the life of the vehicle and requires no type of adjustment or periodic maintenance. The bearing is a sealed unit with the wheel hub and can only be removed and/or replaced as one unit.

The following procedure may be used for evaluation of bearing condition.

1. Raise and safely support the vehicle.

2. Remove the rear wheels and brake drums.

3. Turn the hub flange carefully. Excessive roughness, lateral play or resistance to rotation may indicate dirt intrusion or bearing failure.

4. If the rear wheel bearings exhibit the conditions during inspection, the hub and bearing assembly should be replaced.

5. Damaged nearing seals and resulting excessive grease loss may also require bearing replacement. Moderate grease loss from the bearing is considered normal and should not require replacement of the hub and bearing assembly.

REMOVAL AND INSTALLATION

All vehicles are equipped with permanently lubricated and sealed for life rear wheel bearings. There is no periodic lubrication or maintenance recommended for these units.

To evaluate the condition of the rear wheel bearings, remove the wheel and brake drum and rotate the flanged outer ring of the hub. Excessive roughness or resistance to rotation may indicate dirt intrusion or wheel bearing failure. If the rear wheel bearings exhibit these conditions during inspection, the hub and bearing assembly should be replaced. Damaged bearing seals and resulting excessive grease loss may also require bearing replacement. Moderate grease loss from the bearing is considered normal and should not require replacement of the hub and bearing assembly. If service requires removal for inspection or replacement of the rear wheel bearing and hub assembly, use the following procedure.

1. Raise and safely support the vehicle.

2. Remove the wheel and tire assembly.

3. Remove the brake drum.

4. Remove the rear hub dust cap.

5. Remove the rear hub retaining nut and discard.

6. Remove the rear hub and bearing assembly by pulling straight off the spindle.

To install:

7. Install the new bearing on the rear spindle.

8. Install the hub and a new retaining nut.

9. Torque the retaining nut to 185 ft. lbs. (250 Nm). Install the dust cap by tapping on with a soft-faced mallet.

10. Install the brake drum.

11. Install the wheel and tire. Tighten the lug nuts by tightening in a crisscross pattern in 2 steps with a final torque of 95 ft. lbs. (129 Nm).

12. Lower the vehicle.

FORD MOTOR CO.

Front Wheel Drive

FORD-Tempo MERCURY-Topaz

8

FIRING ORDERS

NOTE: To avoid confusion, always replace spark plug wires one at a time.

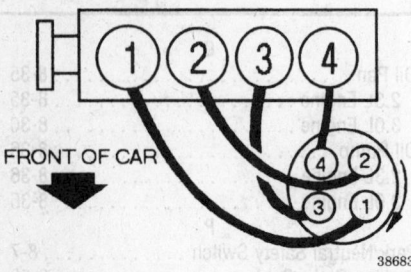

2.3L Engine
Engine Firing Order: 1-3-4-2
Distributor Rotation: Clockwise

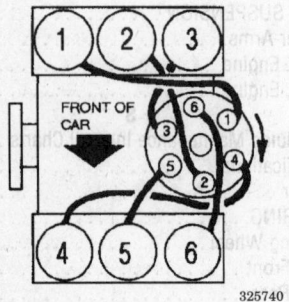

3.0L Engine
Engine Firing Order: 1-4-2-5-3-6
Distributor Rotation: Clockwise

ENGINE ELECTRICAL

NOTE: Disconnecting the negative battery cable on some vehicles may interfere with the functions of the on board computer systems and may require the computer to undergo a relearning process, once the negative battery cable is reconnected.

Distributor

REMOVAL AND INSTALLATION

2.3L Engine

1. Disconnect the negative battery cable.

2. Mark the position of the No. 1 cylinder wire tower on the distributor base.

NOTE: This reference is necessary in case the engine is disturbed while the distributor is removed.

3. Remove the distributor cap and position the cap and ignition wires to the side. Disconnect the wiring harness plug from the distributor connector.

4. Scribe a mark on the distributor body to indicate the position of the rotor tip. Scribe a mark on the distributor housing and engine block or timing cover to indicate the position of the distributor in the engine.

5. Remove the hold-down bolt and clamp located at the base of the distributor. Remove the distributor from the engine. Note the direction the rotor tip points if it moves from the No. 1 position when the drive gear disengages. For reinstallation purposes, the rotor should be at this point to insure proper gear mesh and timing.

6. Cover the distributor opening in the engine to prevent the entry of dirt or foreign material.

7. Avoid turning the crankshaft, if possible, while the distributor is removed. If the engine is disturbed, the No. 1 cylinder piston will have to be brought to TDC on the compression stroke before the distributor is installed.

To install:

NOTE: Before installing, visually inspect the distributor. The drive gear should be free of nicks, cracks and excessive wear. The distributor driveshaft should move freely, without binding. The O-ring should fit tightly and be free of cuts.

8. Position the distributor in the engine, aligning the rotor and distributor housing with the marks that were made during removal. If the distributor does not fully seat in the engine block or timing cover, it may be because the distributor is not engaging properly with the oil pump intermediate shaft. Remove the distributor and, using a suitable tool, turn the intermediate shaft until the distributor will seat properly.

9. Install the hold-down clamp and bolt. Snug the mounting bolt so the distributor can be turned for ignition timing purposes.

10. Install the distributor cap and connect the distributor to the wiring harness.

11. Connect the negative battery cable. Check and, if necessary, adjust the ignition timing.

12. After the timing has been set, tighten the distributor hold-down clamp bolt.

3.0L Engine

NOTE: If possible, position No. 1 cylinder at TDC on its compression stroke before removing the distributor. Align the timing marks on the crankshaft pulley and timing cover.

1. Disconnect the negative battery cable.

2. Disconnect the distributor wiring harness.

3. Matchmark the base of the distributor housing with a mark on the cylinder block for reference when installing the distributor.

4. Loosen the distributor cap hold-down screws and remove the distributor cap. Position the distributor cap and ignition wires aside.

5. Matchmark the tip of the rotor blade with the distributor housing.

6. Remove the distributor hold-down bolt and clamp.

7. Remove the distributor by pulling upward. Avoid rotating the engine while the distributor is removed.

8. Cover the distributor opening in the cylinder block with a shop towel to prevent contamination from entering the engine.

To install:

9. If the engine has turned from its TDC position, rotate the crankshaft as necessary to align the TDC timing marks. Make sure the No. 1 piston is at TDC on its compression stroke. Remove No. 1 spark plug if necessary.

10. Lubricate the distributor gear with engine assembly lubricant, or equivalent before installation.

11. Insert the distributor with the matchmarks on the distributor body and engine aligned and the rotor pointing a few degrees counterclockwise of the mark. While moving the distributor down into place, the rotor will turn to point at the mark. The oil pump is driven through a shaft that engages the bottom of the distributor shaft. If the distributor does not fully seat, turn the engine slightly to fully engage the shafts.

12. Return the crankshaft to TDC. If the rotor does not point at the matchmark on the distributor housing, lift the distributor out just enough to turn the shaft to the next tooth on the drive gear.

13. After the distributor has been fully seated and aligned, install the

hold-down clamp and bolt. Only snug the bolt at this time.

14. Install the distributor cap and position the ignition wires. Tighten the distributor cap hold-down screws to 18–23 inch lbs. (2.0–2.6 Nm).

15. Connect the distributor wiring harness.

16. Install the No. 1 spark plug, if removed and connect the ignition wire securely.

17. Connect the negative battery cable.

18. Check and adjust the ignition timing. Tighten the distributor hold-down clamp bolt to 17–25 ft. lbs. (23–34 Nm).

Ignition Timing

ADJUSTMENT

On the 2.3L engine, the timing marks are located on the flywheel and are visible through an access hole in the transaxle case. If equipped with a manual transaxle, the timing cover plate must be removed in order to view the timing marks and adjust the timing.

On the 3.0L engine, the timing marks are located on the crankshaft damper and timing chain cover.

1. Place the transaxle in **P** or **N** and apply the parking brake, Make sure the air conditioner and heater are OFF.

2. Open the hood, locate the timing marks and clean with a stiff brush or solvent. If equipped with 2.3L engine and manual transaxle, it will be necessary to remove the cover plate which allows access to the timing marks.

3. Using white chalk or paint, mark the specified timing mark and pointer.

4. Remove the in-line SPOUT connector or remove the shorting bar from the double wire SPOUT connector.

5. Connect a suitable inductive timing light and a tachometer according to the manufacturer's instructions.

6. Start the engine and bring to normal operating temperature.

7. Check the engine idle speed and adjust if it is not within specifications. Aim the timing light at the timing marks. If they are not aligned, loosen the distributor clamp bolt slightly and rotate the distributor body until the marks are aligned.

NOTE: To set timing correctly, a remote starter should not be used. Use the ignition key only to start the vehicle. Disconnecting the start wire at the starter relay will cause the TFI module to revert to start mode timing after the vehicle is started. Reconnecting the start wire after the vehicle is running will not correct the timing.

8. Tighten the distributor clamp bolt, then recheck the ignition timing to make sure it did not change when the bolt was tightened.

9. Shut the engine **OFF** and remove all test equipment. Reconnect the in-line SPOUT connector or reinstall the shorting bar on the double wire SPOUT connector.

10. If equipped with 2.3L engine and manual transaxle, reinstall the cover plate.

Alternator

PRECAUTIONS

Several precautions must be observed with alternator equipped vehicles to avoid damage to the unit.

- If the battery is removed for any reason, make sure it is reconnected with the correct polarity. Reversing the battery connections may result in damage to the 1-way rectifiers.
- When utilizing a booster battery as a starting aid, always connect the positive to positive terminals and the negative terminal from the booster battery to a good engine ground on the vehicle being started.
- Never use a fast charger as a booster to start vehicles.
- Disconnect the battery cables when charging the battery with a fast charger.
- Never attempt to polarize the alternator.
- Do not use test lights of more than 12 volts when checking diode continuity.
- Do not short across or ground any of the alternator terminals.
- The polarity of the battery, alternator and regulator must be matched and considered before making any electrical connections within the system.
- Never separate the alternator on an open circuit. Make sure all connections within the circuit are clean and tight.
- Disconnect the battery ground terminal when performing any service on electrical components.
- Disconnect the battery if arc welding is to be done on the vehicle.

REMOVAL AND INSTALLATION

1. Disconnect the negative battery cable.

2. Disconnect the wire harness attachments to the integral alternator/regulator assembly. Pull the 2 connectors straight out.

3. Loosen the alternator pivot bolt. Remove the adjustment arm bolt from the alternator.

4. Disengage the alternator drive belt from the alternator pulley.

5. Remove the alternator pivot bolt and alternator/regulator assembly.

6. Remove the alternator fan shield, if equipped.

To install:

7. Position the integral alternator/regulator assembly on the engine.

8. Install the alternator pivot and adjuster arm bolts. Do not tighten the bolts until the belt is properly tensioned.

9. Install the drive belt over the alternator pulley.

10. Adjust the belt tension.

11. Connect wiring harness to the alternator/regulator assembly. Push both connectors straight in.

12. Attach the alternator fan shield to the alternator, if equipped.

13. Connect the negative battery cable.

14. Check the charging system for proper operation.

Drive Belt

REMOVAL AND INSTALLATION

2.3L Engine

NOTE: An automatic tensioner maintains correct belt tension during operation. No adjustment is necessary.

The automatic belt tensioners are spring loaded devices which set and maintain the drive belt tension. The drive belt should not require tension adjustments for the life of the belt. All automatic belt tensioners have wear indicators on them. Check the indicator marks on the tensioner. If the mark is not between the indicator lines, the belt is worn or the wrong belt has been installed. Make sure the indicator mark inspection is done with the engine not running.

To install a new belt, loosen the bracket lock bolt, retract the belt tensioner by applying a counter-clockwise torque to the pulley bolt and slide the old belt off the pulleys. Slip on a new belt and release the ten-

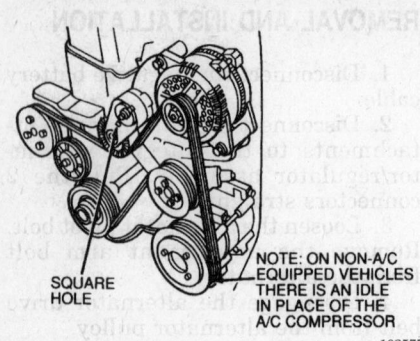

SQUARE HOLE

NOTE: ON NON-A/C EQUIPPED VEHICLES THERE IS AN IDLE IN PLACE OF THE A/C COMPRESSOR

102577

Drive belt — 2.3L engine

sioner and tighten the lock bolt. The automatic spring powered tensioner eliminates the need for periodic adjustments.

── WARNING ──

Check to make sure the V-ribbed belt is located properly in all drive pulleys before applying tensioner pressure.

3.0L Engine

NOTE: If the belt is to be re-used, mark the direction of rotation on the belt using a crayon or marking pen to ensure correct installation. Failure to do so may result in belt noise.

Accessory Drive Belt

1. Disconnect the negative battery cable.
2. Remove the plastic belt shield from the power steering pump.
3. Using a ½ in. drive breaker bar or equivalent inserted in the pulley tensioner, release the tension on the drive belt.
4. Remove the drive belt from the tensioner pulley.
5. Inspect the drive belt for cuts or wear; replace as necessary.

To install:

NOTE: If replacing the water pump drive belt, do so before installing the accessory drive belt.

6. Position the accessory drive belt on all pulleys but the tensioner pulley. If the same belt is being used, install the belt with the correct direction of rotation.
7. Place a ½ in. drive breaker bar or equivalent in the pulley tensioner and move the tensioner to allow the drive belt to be installed.
8. Position the drive belt to the pulley tensioner and slowly relax the tensioner. Remove the breaker bar or equivalent, from the tensioner.

9. Make sure all belt V-grooves make contact with the pulleys.
10. Connect the negative battery cable.

Water Pump Drive Belt

1. Disconnect the negative battery cable.
2. Remove the accessory drive belt.
3. Raise and safely support the vehicle.
4. Place a wrench on the water pump pulley tensioner. Turn the wrench clockwise to release the tension on the belt.
5. Remove the water pump drive belt.
6. Inspect the drive belt for cuts or wear; replace as necessary.

To install:

7. Position the water pump drive belt on all pulleys except at the tensioner pulley.
8. Place a wrench on the water pump pulley tensioner. Turn the wrench clockwise to move the tensioner pulley.
9. Position the water pump drive belt to the tensioner pulley and slowly relax the tensioner.
10. Remove the wrench from the water pump pulley tensioner.
11. Install the accessory drive belt.
12. Connect the negative battery cable.
13. Make sure all belt V-grooves make proper contact with the pulleys.

Starter

REMOVAL AND INSTALLATION

1. Disconnect the negative battery cable.
2. Raise and safely support the vehicle.
3. Disconnect the battery cable at the starter terminal. Disconnect the electrical connector at the solenoid.

NOTE: When disconnecting the plastic hard shell connector at the solenoid S terminal, grasp the plastic connector, depress the plastic tab and pull off the lead assembly. Do not pull on the lead wire or damage may result.

4. Remove the 2 bolts attaching the starter rear support bracket, if equipped. Remove the bracket.
5. If equipped with roll restrictor brace-to-starter studs on the transaxle housing, remove the nuts and remove the brace.
6. Remove the starter retaining bolts and remove the starter.

7. For installation, reverse the removal procedure. Tighten the attaching studs or bolts to 16–20 ft. lbs. (21–27 Nm).

CHASSIS ELECTRICAL

Blower Motor

REMOVAL AND INSTALLATION

1. Disconnect negative battery cable.
2. Remove the glove box liner.
3. Disconnect the wires at the blower motor.
4. Remove the 4 mounting screws and pull the blower motor from the heater housing.
5. Installation is in reverse of removal procedure.

NOTE: To remove blower wheel: Remove pushnuts and slide the blower wheel from the shaft.

Windshield Wiper Motor

REMOVAL AND INSTALLATION

1. Disconnect the negative battery cable.
2. Lift the water shield cover from the cowl on the passenger side.
3. Disconnect the power lead from the motor.
4. Remove the linkage retaining clip from the operating arm on the motor by lifting locking tab up and pulling clip away from pin.
5. Remove the attaching screws from the motor and bracket assembly and remove.
6. Remove the operating arm from the motor. Unscrew the 3 bolts and separate the motor from the mounting bracket.

To install:

7. Position the motor on the mounting bracket and install the retaining bolts.
8. Install the operating arm to the motor. Install the linkage retaining clip to the operating arm.
9. Connect the electrical lead to the motor.
10. Install the water shield cover to the cowl.

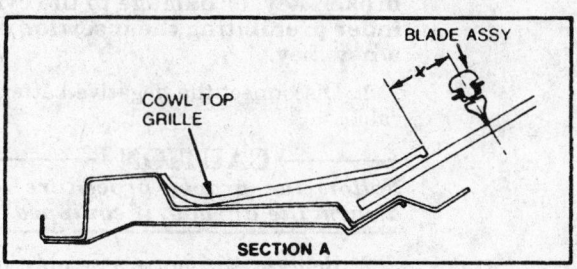

DIMENSION "X" €	
OF BLADE TO UPPER EDGE OF COWL TOP GRILLE	
DRIVER SIDE	PASSENGER SIDE
41-71mm (1.61-2 8 INCHES)	53-83mm (2.1-3.2 INCHES)

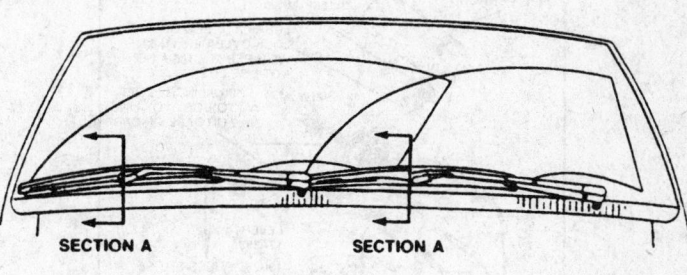

Wiper blade adjustment

90355

11. Connect the negative battery cable.

Headlight Switch

REMOVAL AND INSTALLATION

1. Disconnect the negative battery cable.

2. On vehicles without air conditioning, remove the left side air vent control cable retaining screws and let the cable hang.

3. Remove the fuse panel bracket retaining screws. Move the fuse panel assembly aside to gain access to the headlight switch.

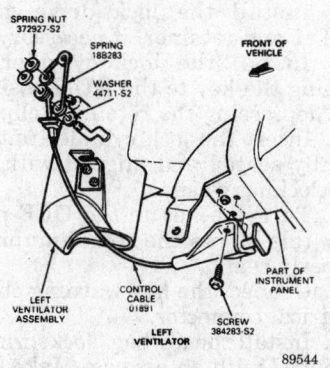

SPRING NUT 372927-S2
SPRING 18B283
WASHER 44711-S2
FRONT OF VEHICLE
LEFT VENTILATOR ASSEMBLY
CONTROL CABLE 01891
LEFT VENTILATOR
SCREW 384283-S2
PART OF INSTRUMENT PANEL

89544

Left ventilator control cable

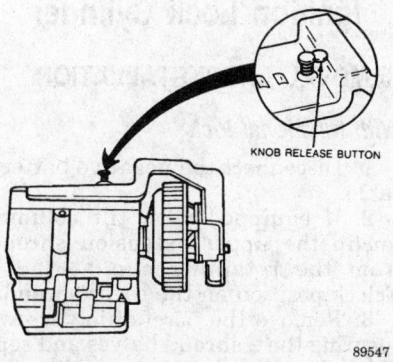

KNOB RELEASE BUTTON

89547

Headlight switch release button

4. Pull the headlight knob out to the **ON** position. Depress the headlight knob and shaft retainer button and remove the knob and shaft assembly from the switch.

5. Remove the headlight switch retaining bezel. Disconnect the multiple connector plug and remove the switch from the instrument panel.

To install:

6. Install the headlight switch into the instrument panel. Connect the multiple connector and install the headlight switch retaining bezel.

7. Install the knob and shaft assembly by inserting the shaft into the switch and gently pushing until the shaft locks in position.

8. Move the fuse panel back into position and install the fuse panel bracket with the 2 retaining screws.

9. On vehicles without air conditioning, install the left side air vent control cable and bracket.

10. Connect the negative battery cable.

Combination Switch

REMOVAL AND INSTALLATION

1. Disconnect the negative battery cable.

2. Remove the 5 column shroud screws and remove the lower column shroud.

3. Loosen the 4 steering column attaching nuts enough to allow the removal of the upper trim shroud. Remove the upper shroud.

4. Remove the turn signal switch lever by pulling the lever straight out from the switch. To make removal easier, work the outer end of the lever around with a slight rotary movement before pulling it out.

5. Peel back the foam sight shield from the turn signal switch.

6. Disconnect the turn signal switch electrical connectors.

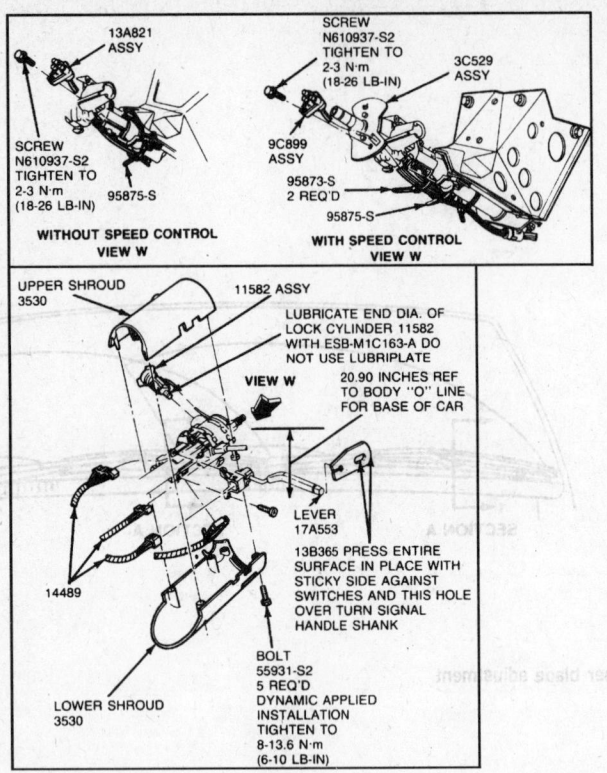

Combination switch

7. Remove the 2 self-tapping screws that attach the turn signal switch to the lock cylinder housing and disengage the switch from the housing.

To install:

8. Align the turn signal switch mounting holes with the corresponding holes in the lock cylinder housing and install 2 self-tapping screws until tight.

9. Apply the foam sight shield to the turn signal switch.

10. Install the turn signal switch lever into the switch by aligning the key on the lever with the keyway in the switch and pushing the lever toward the switch to full engagement.

11. Install the turn signal switch electrical connectors to full engagement.

12. Install the upper steering column trim shrouds.

13. Torque the steering column attaching nuts to 15–22 ft. lbs. (20–30 Nm).

14. Connect the negative battery cable.

15. Check the steering column and switch for proper operation.

Ignition Lock Cylinder

REMOVAL AND INSTALLATION

With functional lock

1. Disconnect the negative battery cable.

2. If equipped with tilt column, unclip the upper extension shroud from the retaining clip at the 9 o'clock position on the cover shroud.

3. Remove the 5 retaining screws securing the 2 shroud halves and separate the shrouds.

4. Disconnect the key warning chime connector.

5. Turn the key to the **RUN** position.

6. Place a ⅛ in. wire, pin, or small drift punch into the casting hole surrounding the lock cylinder. Depress the retaining pin while pulling the lock cylinder out of the column housing.

To install:

7. Install the lock cylinder by turning the key to the **RUN** position and depressing the retaining clip.

8. Insert the lock cylinder until it is fully seated and aligned with the interlocking washer.

9. Turn the key to the **OFF** position, allowing the retaining pin to properly seat.

10. Connect the key warning chime electrical connector.

11. Install the 2 shroud halves and retaining screws.

12. Connect the negative battery cable.

13. Check for proper starting in **P** and **N**.

Non-Functional lock

NOTE: This procedure should be used when the lock cylinder cannot be rotated due to a lost or broken key, or damage to the cylinder prohibiting the insertion of a new key.

1. Disconnect the negative battery cable.

—————— CAUTION ——————

Follow the proper procedure to disarm the air bag, if equipped.

2. Remove the horn assembly or air bag.

3. Remove the steering wheel.

4. Remove the air bag clockspring, if equipped with air bag.

5. If equipped with tilt column, unclip the upper extension shroud from the retaining clip at the 9 o'clock position on the cover shroud.

6. Remove the 5 retaining screws securing the 2 shroud halves and separate the shrouds.

7. Disconnect the key warning chime connector.

8. Using a ⅛ in. drill bit, drill out the retaining clip. Be careful not to drill deeper than ½ in.

9. Using a chisel, strike the base of the key slot until it breaks away from the lock cylinder.

10. Using a ⅜ in. drill bit, drill down the middle of the lock cylinder about 1¾ inches until the cylinder breaks away from the housing.

11. Remove the lock cylinder and clean out all metal shavings from the housing. Inspect the housing and replace if damaged.

To install:

12. Install the lock drive gear, washer and retainer, if necessary.

13. Install the lock cylinder by turning the key to the **RUN** position and depressing the retaining clip.

14. Insert the lock cylinder until it is fully seated and aligned with the interlocking washer.

15. Turn the key to the **OFF** position. this allows the retaining pin to properly seat.

16. Connect the key warning chime electrical connector.

17. Install the air bag clockspring if equipped with an air bag. Make sure the clockspring is properly aligned.

18. Install the steering wheel.

19. Install the air bag or horn assembly, as equipped.

20. Install the 2 shroud halves and retaining screws.

21. Connect the negative battery cable.

22. Check for proper starting in **P** and **N**.

Ignition Switch

REMOVAL AND INSTALLATION

1. Disconnect the negative battery cable.

2. If equipped, remove the steering column lower cover from the instrument panel by removing the 2 screws from the bottom and disengaging the snap-in retainers at the top.

3. Remove the steering column shroud self-tapping screws.

4. Remove 2 bolts and nuts holding the steering column assembly to the steering column bracket assembly and lower the steering column to the seat.

5. Remove the steering column shrouds.

6. Disconnect the electrical connector from the ignition switch.

7. Rotate ignition lock cylinder to the **RUN** position.

8. Remove 2 screws attaching the switch to the lock cylinder housing.

9. Disengage the ignition switch from the actuator pin.

To install:

10. Check to see that the actuator pin slot in the ignition switch is in the **RUN** position.

NOTE: A new switch assembly will be pre-set in the RUN position.

11. Make certain the ignition key lock cylinder is in approximately the **RUN** position to properly locate the

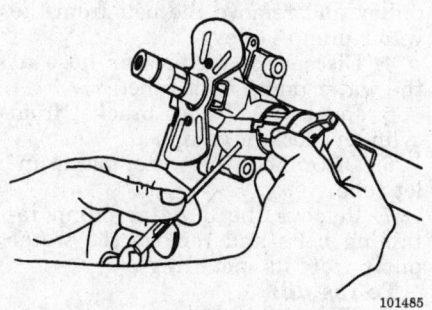

101485

Ignition switch retaining pin

lock actuator pin. The **RUN** position is achieved by rotating the key lock cylinder approximately 90 degrees from the **LOCK** position.

12. Install the ignition switch onto the actuator pin. It may be necessary to move the switch slightly back and fourth to align the switch mounting holes with the column lock housing threaded holes.

13. Install the new screws and tighten to 50–70 inch lbs. (5.6–7.9 Nm).

14. Connect the electrical connector to ignition switch.

15. Connect the negative battery cable.

16. Check the ignition switch for proper function including **START** and **ACC** positions. Also make certain the steering column is locked when in the **LOCK** position.

17. Position the top half of the shroud on the steering column.

18. Install the 2 bolts and nuts attaching the steering column assembly to the steering column bracket assembly. Tighten to 15–25 ft. lbs. (20–34 Nm).

19. Position lower shroud to upper shroud and install 5 self-tapping screws. Install the steering column lower cover on the instrument panel, if equipped.

Park/Neutral Safety Switch

REMOVAL AND INSTALLATION

The mounting location of the neutral safety switch does not provide for adjustment of the switch position when installed. If the engine will not start in **P** or **N** or if it will start in **R** or any of the **D** ranges, check the control linkage adjustment and/or replace with a known good switch.

1. Set parking brake.

2. Disconnect the battery negative cable.

3. Disconnect the wire connector from the neutral safety switch.

4. Remove the 2 retaining screws from the neutral start switch and remove the switch.

To install:

5. Place the switch on the manual shift shaft and loosely install the retaining bolts.

6. Use a No. 43 drill — 0.089 in. (2.26mm) and insert it into the switch to set the contacts.

7. Tighten the retaining screws of the switch, remove the drill and complete the assembly by reversing the removal procedure.

8. Connect negative battery cable.

9. Check the ignition switch for proper starting in **P** or **N**. Also make certain that the start circuit cannot be actuated in the **D** or **R** position and that the column is locked in the **LOCK** position.

EEC Computer

LOCATION

The Electronic Engine Control (EEC) module is located behind the center of the instrument panel.

ENGINE COOLING

Radiator

REMOVAL AND INSTALLATION

2.3L Engine

1. Disconnect the negative battery cable.

2. Place fender covers on the aprons.

3. Drain the cooling system.

4. Remove the upper hose from the radiator.

5. Remove the 2 fasteners retaining the upper end of the fan shroud to the radiator and sight shield.

NOTE: If equipped with air conditioning, remove the nut and screw retaining the upper end of the fan shroud to the radiator at the cross support and nut and screw at the inlet end of the tank.

6. Disconnect the electric cooling fan motor wires and air conditioning discharge line, if equipped, from the shroud and remove the fan shroud from the vehicle.

7. Loosen the hose clamp and disconnect the radiator lower hose from the radiator.

8. Disconnect the overflow hose from the radiator filler neck.

9. If equipped with an automatic transaxle, disconnect the oil cooler hoses at the transaxle using a quick-disconnect tool. Cap the oil tubes and plug the oil cooler hoses.

10. Remove the 2 nuts retaining the top of the radiator to the radiator support. If the stud loosens, make sure it is tightened before the radiator is installed. Tilt the top of the radiator rearward to allow clearance with the upper mounting stud and

lift the radiator from the vehicle. Make sure the mounts do not stick to the radiator lower mounting brackets.

To install:

11. Make sure the lower radiator isomounts (rubber supports) are installed over the bolts on the radiator support.

12. Position the radiator to the radiator support making sure the radiator lower brackets are positioned properly on the lower mounts.

13. Position the top of the radiator to the mounting studs on the radiator support and install 2 retaining nuts. Tighten to 5–7 ft. lbs. (7–10 Nm).

14. Connect the radiator lower hose to the engine water pump inlet tube. Install the constant tension hose clamp between alignment marks on the hose.

15. Check to make sure the radiator lower hose is properly positioned on the outlet tank and install the constant tension hose clamp. The stripe on the lower hose should be indexed with the rib on the tank outlet.

16. Connect the oil cooler hoses to the automatic transaxle oil cooler lines, if equipped. Use an appropriate oil resistant sealer.

17. Position the fan shroud to the radiator lower mounting bosses. On vehicles with air conditioning, insert the lower edge of the shroud into the clip at the lower center of the radiator. Install 2 nuts and bolts retaining the upper end of the fan shroud to the radiator. Tighten the nuts to 35–41 inch lbs. (4–5 Nm). Do not overtighten.

18. Connect the electric cooling fan motor wires to the wire harness.

19. Connect the upper hose to the radiator inlet tank fitting and install the constant tension hose clamp.

20. Connect the overflow hose to the nipple just below the radiator filler neck.

21. Install the air intake tube or sight shield.

22. Connect the negative battery cable.

23. Refill the cooling system. Start the engine and allow it to come to normal operating temperature.

24. Check for leaks and confirm the operation of the electric cooling fan.

3.0L Engine

—— CAUTION ——

Never remove the radiator cap while the engine is hot. Wait for the engine to cool before opening the cooling system to prevent possible injury and/or damage to the cooling system or engine.

1. Disconnect the negative battery cable.

2. Place fender covers on the aprons.

3. Drain the cooling system into a suitable container.

4. Remove the upper hose from the radiator.

5. Remove the 2 fasteners retaining the upper end of the fan shroud to the radiator end tank and the upper side rail.

NOTE: If equipped with air conditioning, remove the nut and screw retaining the upper end of the fan shroud to the radiator at the cross support and nut and screw at the inlet end of the tank.

6. Disconnect the electric cooling fan motor wires and air conditioning discharge line, if equipped, from the shroud and remove the fan shroud from the vehicle.

7. Loosen the hose clamp and disconnect the radiator lower hose from the radiator.

8. Disconnect the overflow hose from the radiator filler neck.

9. If equipped with an automatic transaxle, disconnect the oil cooler hoses at the transaxle using a quick-disconnect tool. Cap the oil tubes and plug the oil cooler hoses.

10. Remove the 2 nuts retaining the top of the radiator to the radiator support. If the stud loosens, make sure it is tightened before the radiator is installed. Tilt the top of the radiator rearward to allow clearance with the upper mounting stud and lift the radiator from the vehicle. Make sure the mounts do not stick to the radiator lower mounting brackets.

To install:

11. Make sure the lower radiator isomounts (rubber insulator supports) are installed over the bolts on the radiator support.

NOTE: Make sure the retaining screws are installed into their original positions. The screws are different lengths and if used in the wrong positions, may rub on the radiator end tanks causing coolant leaks.

12. Position the radiator to the radiator support making sure the radiator lower brackets are positioned properly on the lower mounts.

13. Position the top of the radiator to the mounting studs on the radiator support and install the 2 retaining nuts. Tighten to 5–7 ft. lbs. (7–10 Nm).

14. Connect the radiator lower hose to the engine water pump inlet tube. Install the constant tension hose clamp between alignment marks on the hose.

15. Check to make sure the radiator lower hose is properly positioned on the outlet tank and install the constant tension hose clamp. The stripe on the lower hose should be indexed with the rib on the tank outlet.

16. Connect the oil cooler hoses to the automatic transaxle oil cooler lines, if equipped. Use an appropriate oil resistant sealer.

17. Position the fan shroud to the radiator lower mounting bosses. On vehicles with air conditioning, insert the lower edge of the shroud into the clip at the lower center of the radiator. Install 2 nuts and bolts retaining the upper end of the fan shroud to the radiator. Tighten the nuts to 35–40 inch lbs. (4–5 Nm). Do not overtighten.

18. Connect the electric cooling fan motor wires to the wire harness.

19. Connect the upper hose to the radiator inlet tank fitting and install the constant tension hose clamp.

20. Connect the overflow hose to the nipple just below the radiator filler neck.

21. Connect the negative battery cable.

22. Refill the cooling system. Start the engine and allow to come to normal operating temperature.

23. Check for leaks and confirm the operation of the electric cooling fan.

Water Pump

REMOVAL AND INSTALLATION

2.3L Engine

1. Drain the cooling system.

2. Disconnect the negative battery cable.

3. Loosen the water pump idler pulley and remove the belt from the water pump pulley.

4. Disconnect the heater hose at the water pump inlet tube.

5. Remove A/C hose bracket from cylinder block, if equipped.

6. Disconnect the water pump inlet tube.

7. Remove the 3 water pump retaining bolts and remove the water pump from its mounting.

To install:

8. Thoroughly clean both gasket mating surfaces on the water pump and cylinder block.

9. Coat the new gasket on both sides with a water resistant sealer and position on the cylinder block.

10. Install the water pump retaining bolts and tighten to 15–22 ft. lbs. (20–30 Nm).

11. Connect the water pump inlet tube.

12. Attach A/C hose bracket to cylinder block, if equipped.

13. Connect the heater hose.

14. Install water pump belt on the pulley and adjust the tension.

15. Connect the negative battery cable.

16. Fill the cooling system. Operate the engine until normal operating temperature is reached. Check for leaks and recheck the coolant level.

3.0L Engine

1. Disconnect the negative battery cable and allow the engine to cool.

— **CAUTION** —

Do not open the radiator cap or draincock without allowing the engine to cool. Severe burns from hot coolant may occur if the cooling system is opened prior to the engine cooling down.

2. Remove the radiator cap. Place a drain pan under the radiator drain cock. Open the drain cock on the radiator and drain the cooling system.

3. Remove the water pump drive belt as follows:

a. Mark the direction of rotation on the accessory drive belt so it can be reinstalled in the same direction.

b. Remove the plastic belt shield from the power steering pump.

c. Using a ½ in. drive breaker bar or equivalent, inserted in the idler pulley tensioner, release the tension on the accessory drive belt and remove the belt.

d. Raise and safely support the vehicle.

e. Mark the direction of rotation on the water pump drive belt so it can be reinstalled in the same direction.

f. Use a suitable wrench to turn the water pump belt idler pulley tensioner clockwise and release the tension on the belt. Remove the water pump drive belt.

4. Lower the vehicle and remove the water pump to front cover hose.

5. Raise and safely support the vehicle.

6. Loosen and remove the retaining nut from the upper bracket and the bolt from the lower bracket. Gently grasp the tube at the water end and pull the tube out of the water pump. Remove the lower water pump tube.

7. Lower the vehicle.

8. Remove the heater hose from the rear of the water pump and remove the water pump pulley shield.

9. Remove the water pump from the bracket.

To install:

10. If replacing the water pump, transfer the pulley to the new pump.

11. Align the water pump to the bracket and install the mounting bolts. Tighten to 15–22 ft. lbs. (20–30 Nm).

12. Raise and safely support the vehicle.

13. Install the lower water pump tube. Lubricate the water pump end of the tube with compound ESE–M99B144–A or equivalent, before inserting into the water pump.

14. Install the retaining nut to the upper bracket stud bolt and tighten to 5 ft. lbs. (7 Nm). Install the lower tube bracket retaining bolt and tighten to 71–106 inch lbs. (8–12 Nm).

15. Lower the vehicle.

16. Install the water pump pulley shield and tighten the retaining nut to 7–10 ft. lbs. (9–14 Nm).

17. Install the heater hose to the rear of the pump. Make sure the hose is clamped securely.

18. Install the water pump to front cover hose. Tighten the clamps to 19–37 inch lbs. (2.1–4.1 Nm).

19. Install the water pump and accessory drive belts.

20. Fill and bleed the cooling system.

21. Connect the negative battery cable. Start and operate the engine until normal operating temperature is reached. Check for leaks.

Thermostat

REMOVAL AND INSTALLATION

2.3L Engine

1. Disconnect the negative battery cable.

2. Drain the cooling system.

3. Disconnect the wire connector at the thermostat housing thermoswitch.

4. Loosen the top radiator hose clamp. Remove the thermostat housing mounting bolts and lift up the housing.

5. Remove the thermostat by turning counterclockwise.

6. Clean the thermostat housing and engine gasket mounting surfaces.

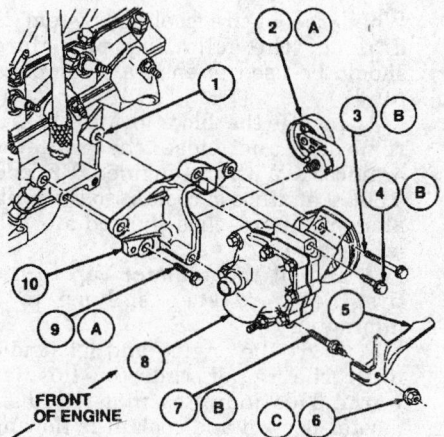

1. Cylinder block assy
2A. Tensioner assy
3B. Bolt M8 × 1.25 × 68
4B. Bolt M8 × 1.25 × 43
5. Water pump pulley shield
6C. Nut & washer assy
7B. Stud assy
8. Water pump assy
9A. Bolt M10 × 1.5 × 33.5 (3)
10. Bracket assy
A. 30–41 ft. lb.(40–55 Nm)
B. 15–22 ft. lb.(20–30 Nm)
C. 7–10 ft. lb.(9–14 Nm)

271631

Water pump assembly — 3.0L engine

To install:

7. Install new mounting gasket and fully insert the thermostat to compress the mounting gasket. Turn the thermostat clockwise to secure in housing.

8. Position the housing onto the engine. Make sure the bolt threads are clean. Install the mounting bolts and torque to 15–22 ft. lbs.

9. Refill the cooling system.

10. Connect the negative battery cable, start the engine and check for leaks.

3.0L Engine

1. Disconnect the negative battery cable. Allow the engine to cool before proceeding.

─────── **CAUTION** ───────
Do not open the radiator cap, drain cap or any cooling system component until the engine has cooled. Opening or removing a cooling system component prior to the engine cooling could result in severe burns from the engine coolant.

2. Place a suitable drain pan under the radiator.

3. Remove the radiator cap and open the drain. Drain the cooling system.

4. Remove the upper radiator hose from the thermostat housing.

5. Remove the 3 retaining bolts from the thermostat housing.

6. Remove the housing and the thermostat as an assembly.

To install:

7. Make sure all sealing surfaces are free of old gasket material.

8. Install the thermostat into the housing.

9. Make sure the bolt threads are clean. Position a new gasket onto the housing. Install the thermostat assembly and tighten the bolts to 9 ft. lbs. (12 Nm).

10. Install the upper radiator hose and tighten the clamp.

NOTE: Make sure the hose clamps are beyond the bead and placed in the center of the clamping surface of the connection. Any used hose clamps must be replaced with a new clamp to ensure proper sealing at the connection. Tighten the hose clamps to 20–30 inch lbs. (2.2–3.4 Nm).

11. Fill and bleed the cooling system. Connect the negative battery cable, start the engine and check for coolant leaks. Check the coolant level and add as required.

Electric Cooling Fan

REMOVAL AND INSTALLATION

─────── **CAUTION** ───────
Although the fans used with the 2.3L engine is similar to the 3.0L engine, they are not interchangeable. If the incorrect fan is used cooling problems may result.

1. Disconnect the negative battery cable.

2. To disconnect the wiring connector from the cooling fan motor, push down on the 2 lock fingers and pull the connector from the motor end. Disconnect wire loom from the clip on shroud.

3. Remove the 2 retaining screws securing the cooling fan and shroud assembly to the radiator.

4. Remove the cooling fan and shroud assembly from the vehicle.

5. Remove the retaining clip from the motor shaft and remove the cooling fan.

NOTE: A metal burr may be present on the motor shaft after the retaining clip has been removed. If necessary, remove the burr to facilitate fan removal.

6. Remove the 3 screws securing the cooling fan motor to the shroud and remove.

To install:

7. Position the cooling fan motor to the shroud.

8. Install 3 retaining screws and tighten to 36–84 inch lbs. (4.0–9.5 Nm).

9. Position the cooling fan on the motor shaft and install the retaining clip.

10. Install the cooling fan and shroud assembly to the vehicle.

11. Make sure the shroud slides into the clips on the bottom of the radiator end tanks.

12. Install the 2 retaining screws securing the cooling fan and shroud assembly to the radiator and tighten to 31–41 inch lbs. (3.5–4.6 Nm).

13. Connect the cooling fan motor wire loom in the clip provided on fan shroud. Connect the wiring connector to the cooling fan motor. Make sure the connector is fully seated.

14. Connect the negative battery cable.

15. Run the engine and allow to reach normal operating temperature. Make sure the cooling fan operates properly.

Cooling System

BLEEDING

When the entire cooling system is drained, the following procedure should be used to ensure a complete fill.

1. Install the block drain plug, if removed, and close the radiator draincock. With the engine **OFF**, add a 50/50 mixture of coolant to the radiator until it reaches the radiator filler neck seat.

2. Install the radiator cap to the first notch to keep spillage to a minimum.

3. Start the engine and let it idle until the upper radiator hose is warm. This indicates that the thermostat is open and coolant is flowing through the entire system.

4. Carefully remove the radiator cap and top off the radiator with water. Install the cap on the radiator securely.

5. Fill the coolant recovery reservoir with a 50/50 mixture of coolant.

6. Check for leaks at the radiator draincock and the block drain plug.

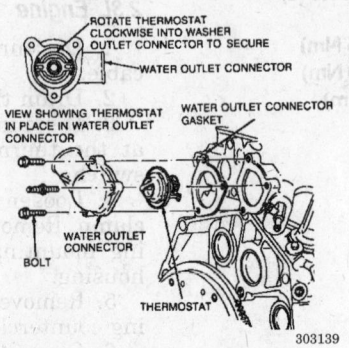

ROTATE THERMOSTAT CLOCKWISE INTO WASHER OUTLET CONNECTOR TO SECURE

WATER OUTLET CONNECTOR

VIEW SHOWING THERMOSTAT IN PLACE IN WATER OUTLET CONNECTOR

WATER OUTLET CONNECTOR GASKET

WATER OUTLET CONNECTOR

BOLT

THERMOSTAT

303139

Thermostat and related components — 3.0L engine

FUEL SYSTEM

Fuel System Service Precautions

Safety is the most important factor when performing not only fuel system maintenance but any type of maintenance. Failure to conduct maintenance and repairs in a safe manner may result in serious personal injury or death. Maintenance and testing of the vehicle's fuel system components can be accomplished safely and effectively by adhering to the following rules and guidelines.

• To avoid the possibility of fire and personal injury, always disconnect the negative battery cable unless the repair or test procedure requires that battery voltage be applied.

• Always relieve the fuel system pressure prior to disconnecting any fuel system component (injector, fuel rail, pressure regulator, etc.), fitting or fuel line connection. Exercise extreme caution whenever relieving fuel system pressure to avoid exposing skin, face and eyes to fuel spray. Please be advised that fuel under pressure may penetrate the skin or any part of the body that it contacts.

• Always place a shop towel or cloth around the fitting or connection prior to loosening to absorb any excess fuel due to spillage. Ensure that all fuel spillage (should it occur) is quickly removed from engine surfaces. Ensure that all fuel soaked cloths or towels are deposited into a suitable waste container.

• Always keep a dry chemical (Class B) fire extinguisher near the work area.

• Do not allow fuel spray or fuel vapors to come into contact with a spark or open flame.

• Always use a backup wrench when loosening and tightening fuel line connection fittings. This will prevent unnecessary stress and torsion to fuel line piping. Always follow the proper torque specifications.

• Always replace worn fuel fitting O-rings with new. Do not substitute fuel hose or equivalent, where fuel pipe is installed.

Fuel System Pressure

RELIEVING

> ——— **CAUTION** ———
> *Fuel injection systems remain under pressure, even after the engine has been turned OFF. The fuel system pressure must be relieved before disconnecting any fuel lines. Failure to do so may result in fire and/or personal injury.*

1. Disconnect the negative battery cable.
2. Remove the fuel tank cap to relieve the pressure in the fuel tank.
3. Remove the cap from the Schrader valve located on the fuel supply manifold.
4. Attach fuel pressure gauge T80L-9974-A or equivalent, to the Schrader valve and drain the fuel through the drain tube into a suitable container.
5. After the fuel system pressure is relieved, remove the fuel pressure gauge and install the cap on the Schrader valve.

Fuel Filter

REMOVAL AND INSTALLATION

> ——— **CAUTION** ———
> *Fuel injection systems remain under pressure, even after the engine has been turned OFF. The fuel system pressure must be relieved before disconnecting any fuel lines. Failure to do so may result in fire and/or personal injury.*

1. Disconnect the negative battery cable.
2. Relieve the fuel system pressure as follows:
   a. Remove the fuel tank cap to relieve the pressure in the fuel tank.
   b. Remove the cap from the Schrader valve located on the fuel supply manifold.
   c. Attach fuel pressure gauge T80L-9974-A or equivalent, to the Schrader valve and drain the fuel through the drain tube into a suitable container.
   d. After the fuel system pressure is relieved, remove the fuel pressure gauge and install the cap on the Schrader valve.
3. Remove the retainer clips at the fuel filter hose connectors and discard.

4. Remove the push connect fittings located at both ends of the fuel filter. Install new retainer clips in each connector fitting.
5. Remove the filter from the bracket by loosening the filter retaining clamp enough to allow the filter to pass through.

To install:
6. Install the fuel filter in the bracket, ensuring the proper direction of flow. Tighten the clamp to 15–25 inch lbs. (1.7–2.8 Nm).

NOTE: The flow arrow direction should be noted to ensure proper flow of fuel through the replacement filter.

7. Install the push connect fittings at both ends of the fuel filter. The fitting should be fully engaged when a distinct click sound is heard. Pull on each fuel line fitting to ensure that they are fully engaged.
8. Connect the negative battery cable.
9. Start the engine and inspect for leaks.

Fuel Pump

REMOVAL AND INSTALLATION

> ——— **CAUTION** ———
> *Fuel injection systems remain under pressure, even after the engine has been turned OFF. The fuel system pressure must be relieved before disconnecting any fuel lines. Failure to do so may result in fire and/or personal injury.*

1. Disconnect the negative battery cable.
2. Relieve the fuel system pressure as follows:
   a. Remove the fuel tank cap to relieve the pressure in the fuel tank.
   b. Remove the cap from the Schrader valve located on the fuel supply manifold.
   c. Attach fuel pressure gauge T80L-9974-A or equivalent, to the Schrader valve and drain the fuel through the drain tube into a suitable container.
   d. After the fuel system pressure is relieved, remove the fuel pressure gauge and install the cap on the Schrader valve.
3. Drain the fuel from the tank as completely as possible. This is accomplished by siphoning or pumping the fuel out through the fuel filler neck using equipment designed for this purpose.

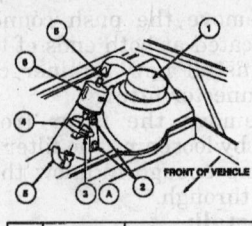

Item	Part Number	Description
1		Fender Apron
2	N802731-S56	Screw and Washer
3A		Clamp
4	9A525	Fuel Filter Bracket Assy
5		Fuel Line
6	9155	Fuel Filter
A		Tighten to 1.7-2.6 N·m (16-24 Lb-In)

298995

Fuel filter and related components

─── **CAUTION** ───

Observe all applicable safety precautions when working around fuel. Do not allow fuel spray or fuel vapors to come in contact with a spark or open flame. Keep a dry chemical (Class B) fire extinguisher near the work area. Never drain or store fuel in an open container due to the possibility of fire or explosion.

4. Raise and safely support the vehicle.

5. Remove the fuel tank.

6. Remove any dirt that has accumulated around the fuel pump retaining flange so it will not enter the fuel tank during removal and installation.

7. Turn the fuel pump locking ring counterclockwise using fuel tank sender wrench D84P-9257-A or equivalent, and remove the lock ring.

8. On all, except all-wheel drive vehicles, remove the fuel pump and bracket assembly and remove the seal gasket and discard.

9. On all-wheel drive vehicles, proceed as follows:

a. Partially lift the sender unit and disconnect the jet pump line and the electrical connector to the resistor.

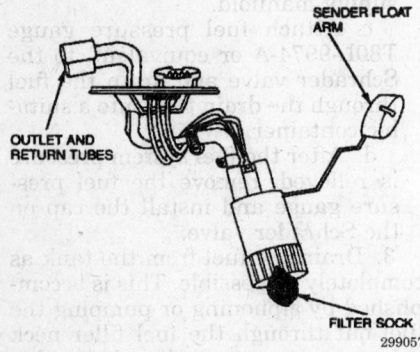

299057

Fuel pump

b. Remove the fuel pump and bracket assembly and remove the seal gasket and discard.

c. Remove the jet pump assembly attaching screw and remove the jet pump assembly.

To install:

10. Clean the fuel pump mounting flange and fuel tank mounting surface and seal ring groove.

11. Put a light coating of multi-purpose lubricant C1AZ-19590-BA or equivalent, on a new seal ring to hold it in place during assembly and install it in the fuel ring groove.

12. On all-wheel drive vehicles, install the jet pump assembly and attaching screw. Tighten the screw to 10-15 ft. lbs. (14-20 Nm).

13. Install the fuel pump and sender assembly carefully to ensure that the filter is not damaged. Make sure the locating keys are in the keyways and the seal ring remains in place.

14. On all-wheel drive vehicles, connect the jet pump line and the electrical connector to the resistor. Make sure the locating keyways and seal ring remain in place.

15. Hold the assembly in place and install the locking ring finger-tight. Make sure all locking tabs are under the tank lock ring tabs.

16. Rotate the locking ring clockwise using fuel tank sender wrench D84P-9275-A or equivalent, until the ring stops against the stops.

17. Install the fuel tank into the vehicle.

18. Lower the vehicle.

19. Add fuel to the tank and check for leaks.

20. Connect the negative battery cable.

21. Run the engine and check for leaks and proper operation.

Fuel Injector

REMOVAL AND INSTALLATION

2.3L Engine

─── **CAUTION** ───

Fuel injection systems remain under pressure, even after the engine has been turned OFF. The fuel system pressure must be relieved before disconnecting any fuel lines. Failure to do so may result in fire and/or personal injury.

1. Disconnect the negative battery cable.

2. Relieve the fuel system pressure as follows:

a. Remove the fuel tank cap to relieve the pressure in the fuel tank.

b. Remove the cap from the Schrader valve located on the fuel supply manifold.

c. Attach fuel pressure gauge T80L-9974-A or equivalent, to the Schrader valve and drain the fuel through the drain tube into a suitable container.

d. After the fuel system pressure is relieved, remove the fuel pressure gauge and install the cap on the Schrader valve.

3. Properly relieve the fuel system pressure.

4. Disconnect the engine air cleaner outlet tube from the air intake throttle body and the throttle position sensor from the wiring harness.

5. Disconnect the vacuum lines from the upper manifold and disconnect the EGR tube at the manifold connection.

6. Disconnect the air bypass valve connector, remove the accelerator and, if equipped, speed control cables and remove the manifold upper support bracket top bolt.

7. Remove the fuel supply manifold shield and the 4 upper manifold retaining bolts and 1 retaining shoulder stud.

8. Remove the upper manifold assembly and gasket and set it aside.

9. Disconnect the fuel supply and return lines and the vacuum line at the pressure regulator.

10. Disconnect the fuel injector wiring harness and disconnect the connectors from the injectors.

11. Remove the fuel supply manifold retaining bolts and remove the fuel supply manifold.

12. Grasping the injector body, pull up while gently rocking the injector from side-to-side.

13. Inspect the injector O-rings, the injector plastic hat and washer for signs of deterioration. Replace as necessary. If the hat is missing, look for it in the intake manifold.

To install:

14. Lubricate new O-rings with clean light grade oil and install 2 on each injector.

NOTE: Never use silicone grease as it will clog the injectors.

15. Install the fuel supply manifold and injectors into the intake manifold. Push the fuel rail down to make sure all the fuel injector O-rings are

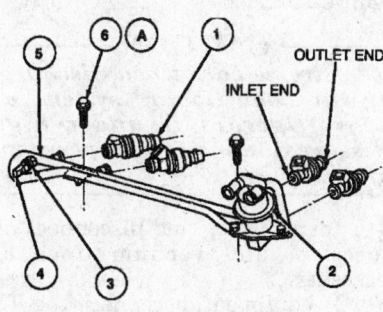

Item	Part Number	Description
1	9F593	Fuel Injector
2	9C968	Fuel Pressure Regulator
3	9H321	Fuel Pressure Relief Valve
4	9H323	Fuel Pressure Relief Valve Cap
5	9D280	Fuel Injection Supply Manifold
6A	—	Bolt
A		Tighten to 20-30 N·m (15-22 Lb-Ft)

299273

Fuel supply manifold — 2.3L Engine

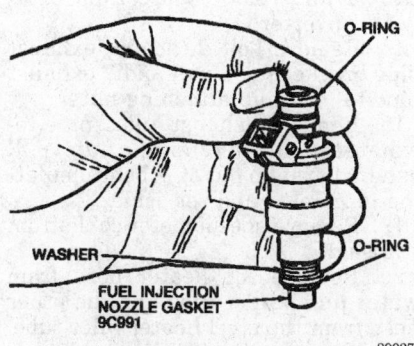

299274

Fuel injector and O-rings — 2.3L Engine

fully seated in the fuel rail cups and intake manifold.

16. Install the fuel manifold assembly retaining bolts and tighten to 15–22 ft. lbs. (20–30 Nm) while holding the assembly down.

17. Connect the fuel lines to the manifold assembly.

18. After the fuel rail assembly has been installed and before the fuel injector wiring is connected, connect the negative battery cable and turn the key to the **ON** position. This will cause the fuel pump to run for 2–3 seconds and pressurize the system.

19. Check for fuel leaks, especially where the fuel injector is installed into the fuel rail.

20. Disconnect the negative battery cable.

21. Install the upper intake manifold in the reverse order of removal and tighten the bolts to 15–22 ft. lbs. (20–30 Nm).

22. Connect the fuel injector wire connectors.

23. Connect the negative battery cable.

24. Start the engine and let it idle.

25. Turn the engine **OFF** and check for fuel leaks.

3.0L Engine

CAUTION

Fuel injection systems remain under pressure, even after the engine has been turned OFF. The fuel system pressure must be relieved before disconnecting any fuel lines. Failure to do so may result in fire and/or personal injury.

1. Disconnect the negative battery cable.

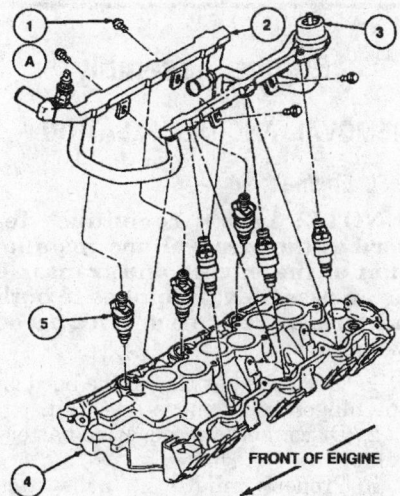

FRONT OF ENGINE

1 Bolt (4 Req'd)
2 Fuel Injection Supply Manifold
3 Fuel Pressure Regulator
4 Intake Manifold
5 Fuel Injector (6 Req'd)
A Tighten to 8-12 N·m (71-106 Lb-In)

235944

Fuel rail and injector installation — 3.0L Engine

2. Relieve the fuel system pressure as follows:

a. Remove the fuel tank cap to relieve the pressure in the fuel tank.

b. Remove the cap from the Schrader valve located on the fuel supply manifold.

c. Attach fuel pressure gauge T80L-9974-A or equivalent, to the Schrader valve and drain the fuel through the drain tube into a suitable container.

d. After the fuel system pressure is relieved, remove the fuel pressure gauge and install the cap on the Schrader valve.

3. Remove the air intake hose from the throttle body.

4. Label and disconnect all vacuum lines and electrical connectors from the throttle body.

5. Loosen the lower EGR tube nut and rotate the tube away from the valve.

6. Disconnect the accelerator and throttle valve linkage from the throttle body.

7. Disconnect the PCV hose.

8. Remove the air intake throttle body retaining bolts and remove the throttle body.

9. On some vehicles the distributor must be raised to allow the crossover tube to clear the distributor housing and lower intake manifold assembly. If the distributor needs to be removed, proceed as follows:

a. Scribe an alignment mark on the base of distributor and the intake manifold.

b. Remove the hold–down clamp.

c. Lift the distributor enough to allow the fuel injection supply manifold to clear the distributor and intake manifold.

10. Disconnect the fuel supply and fuel return lines.

11. Disconnect the wiring harness from the injectors.

12. Disconnect the vacuum line from the fuel pressure regulator valve.

13. Remove the 4 fuel injector manifold retaining bolts.

14. Carefully disengage the fuel rail assembly from the fuel injectors by lifting and gently rocking the rail.

15. Remove the injectors by lifting while gently rocking from side to side.

To install:

16. Lubricate new O-rings with clean engine oil and install 2 on each injector.

17. Make sure the injector cups are clean and undamaged.

18. Install the injectors in the fuel rail using a light twisting-pushing motion.

19. If the distributor was lifted for fuel injection manifold removal:

   a. Lift the distributor enough to allow the fuel injection supply manifold to clear the distributor and intake manifold. Position the fuel injection supply manifold.

   b. Lower the distributor into position.

   c. Install the hold–down clamp and align the scribe marks. Tighten the hold–down clamp bolt to 18 ft. lbs. (24 Nm).

20. Carefully install the rail assembly and injectors into the lower intake manifold, one side at a time. Make sure the O-rings are seated by pushing down on the fuel rail.

21. While holding the fuel rail assembly in place, install the 2 retaining bolts and tighten to 84 inch lbs. (10 Nm).

22. Connect the fuel supply and fuel return lines and the fuel pressure regulator vacuum line.

23. Before connecting the fuel injector harness, connect the negative battery cable and turn the ignition switch to the **ON** position. This will pressurize the fuel system.

24. Using a clean paper towel. check for leaks where the injector connects to the fuel rail. Turn the ignition switch **OFF** and disconnect the negative battery cable.

25. Connect the fuel injector harness.

26. Install the air intake throttle body using a new gasket. Tighten the bolts to 15–22 ft. lbs. (20–30 Nm).

27. Connect the PCV hose and connect the accelerator and throttle valve linkage.

28. Connect the EGR tube.

29. Connect all vacuum hoses and electrical connectors to their proper locations.

30. Connect the air intake hose to the throttle body.

31. Connect the negative battery cable, start the engine and let it idle for 2 minutes.

32. Using a clean paper towel, check for leaks where the injector is installed into the intake manifold.

EMISSION CONTROLS

Maintenance Light

RESETTING

1993 Vehicles

Approximately every 5000 or 7500 miles, (depending on engine application) the word **SERVICE** will appear on the electronic display for the first 1.5 miles to remind the driver that it is time for the regular vehicle service interval maintenance (i.e. oil change).

To reset the service interval reminder light for another interval, use this procedure. With the engine running, press the **TRIP** and **TRIP RESET** buttons at the same time. Hold the buttons down until 3 beeps are heard to verify that the service reminder has been reset.

ENGINE MECHANICAL

Engine Assembly

REMOVAL AND INSTALLATION

2.3L Engine

NOTE: This procedure describes the removal and installation of the engine and transaxle as an assembly. It applies to both automatic and manual transaxles.

1. Mark the position of the hood on the hinges and remove the hood.

2. Disconnect the negative battery cable.

3. Properly relieve the fuel system pressure. Remove the air cleaner.

4. Remove lower radiator hose to drain the engine coolant.

5. Remove upper radiator hose and disconnect transaxle cooler lines at rubber hoses below radiator, if equipped with automatic transaxle.

6. Disconnect the coolant fan at the electrical connection.

7. Remove radiator shroud and cooling fan as an assembly. Remove radiator.

8. Properly discharge air conditioning system, if equipped and remove pressure and suction lines from compressor.

— **CAUTION** —
Use extreme care when discharging air conditioning system, as the refrigerant is under high pressure and may cause personal injury.

9. Identify, tag and disconnect all electrical and vacuum lines as necessary.

10. If equipped, disconnect the TV linkage at the automatic transaxle. If equipped, disconnect the clutch cable from the shift lever on the transaxle.

11. Disconnect accelerator linkage and fuel lines.

12. Remove coil and brackets assembly.

13. Disconnect power steering lines at pump and remove the bracket at the cylinder head, if equipped.

14. Install 2 engine lifting eyes and install engine support tool D88L–6000–A or equivalent, to engine lifting eyes.

15. Raise and safely support the vehicle.

16. Remove battery cable from starter and remove hose from catalytic converter.

17. Remove bolt attaching exhaust pipe bracket-to-oil pan and 2 exhaust pipe-to-manifold attaching nuts.

18. Remove exhaust inlet pipe-to-exhaust manifold retaining nuts, pull exhaust system out of rubber insulating grommets and set aside.

19. Remove speedometer cable from transaxle.

20. Remove the heater hose from water pump inlet tube and the other hose from the steel heater inlet tube.

21. Remove the clamp retaining bolts or nuts at the underside of the oil pan and remove the inlet tube.

22. Remove bolts attaching control arms to body. Remove stabilizer bar brackets retaining bolts and remove brackets.

23. Remove both halfshaft assemblies. After removing the halfshafts, install transaxle plugs in the differential side gears.

— **WARNING** —
Failure to install the plugs can result in dislocation of the differential side gears. If the gears become misaligned, the differential must be removed from the transaxle to realign the gears.

24. On manual transaxle equipped vehicles, remove roll restrictor nuts

from transaxle. Pull roll restrictor from mounting bracket.

25. On manual transaxle equipped vehicles, remove shift stabilizer bar to transaxle attaching bolts. Remove shift mechanism to shift shaft attaching nut and bolt at transaxle.

26. On automatic transaxle equipped vehicles, disconnect manual shift cable clip from lever on transaxle. Remove manual shift linkage bracket bolts from transaxle and remove bracket.

27. Remove the left rear insulator mount bracket from body bracket.

28. Remove the left front insulator to transaxle mounting bolts.

29. Lower the vehicle. Install lifting equipment to the 2 lifting eyes on engine.

NOTE: Do not allow front wheels to touch floor.

30. Remove the engine support tool.

31. Remove right No. 3A insulator intermediate bracket-to-engine bracket bolts, intermediate bracket-to-insulator attaching nuts and the nut on the bottom of the double ended stud which attaches the intermediate bracket-to-engine bracket. Remove bracket.

32. Carefully lower engine and transaxle assembly to the floor.

To install:

33. Raise and safely support the vehicle.

34. Position engine and transaxle assembly directly below engine compartment.

35. Slowly lower vehicle over engine and transaxle assembly.

NOTE: Do not allow the front wheels to touch the floor.

36. Install lifting equipment to both existing engine lifting eyes on engine.

37. Raise engine and transaxle assembly up through engine compartment and position accordingly.

38. Install right side No. 3A insulator intermediate attaching nuts to intermediate bracket. Tighten to 55–75 ft. lbs. (75–100 Nm). Attach intermediate bracket-to-engine bracket bolts. Tighten to 52–70 ft. lbs. (70–95 Nm). Install nut on bottom of double-ended stud that attaches the intermediate bracket-to-engine bracket. Tighten to 60–90 ft. lbs. (80–120 Nm).

39. Install engine support tool to engine lifting eye.

40. Remove lifting equipment.

41. Raise and safely support the vehicle.

42. Position transaxle jack under engine. Raise engine and transaxle assembly into mounted position.

43. Install insulator-to-bracket nut and tighten to 45–65 ft. lbs. (61–68 Nm). Tighten the left rear No. 4 insulator bracket-to-body bracket nuts to 45–65 ft. lbs. (61–68 Nm).

44. If equipped with manual transaxle, position roll restrictor onto starter studs. Install nuts attaching roll restrictor to transaxle and tighten to 25–39 ft. lbs. (35–50 Nm).

45. Install starter cable to starter. Install water pump inlet tube and tighten the fastener to 71–97 inch lbs. (8–11 Nm).

46. Install lower radiator hose.

47. If equipped with manual transaxle, install shift stabilizer bar-to-transaxle attaching bolt. Tighten to 23–35 ft. lbs. (31–47 Nm).

48. If equipped with manual transaxle, install shift mechanism-to-input shift shaft (on transaxle) bolt and nut. Tighten to 7–10 ft. lbs. (9–13 Nm).

49. If equipped with automatic transaxle, install manual shift linkage bracket bolts to transaxle. Install cable clip to lever on transaxle.

50. Install lower radiator hose to radiator.

51. Install speedometer cable to transaxle.

52. Position exhaust system up and into insulating rubber grommets located at rear of vehicle.

53. Install exhaust pipe-to-exhaust manifold studs. Install exhaust pipe bracket-to-oil pan bolt.

54. Connect pulse air hose to catalytic converter.

55. Place stabilizer bar and control arm assembly into position. Install control arm-to-body attaching bolts. Install stabilizer bar brackets and tighten all fasteners.

56. Install the halfshaft assemblies.

57. Lower vehicle.

58. Remove engine support tool.

59. Connect any remaining electrical and vacuum lines.

60. Install heater hose.

61. Install air conditioning discharge and suction lines to compressor, if equipped. Do not charge at this time.

62. Connect fuel supply and return lines to engine.

63. Connect accelerator cable.

64. Install power steering pressure and return lines.

65. If equipped with automatic transaxle, connect TV linkage at transaxle.

66. If equipped with manual transaxle, connect clutch cable to shift lever on transaxle. Check clutch adjustment.

67. Install radiator shroud and coolant fan assembly. Connect the coolant fan electrical connector and install the upper radiator hose.

68. If equipped with automatic transaxle, connect transaxle cooler lines to rubber hoses below radiator.

69. Fill cooling system.

70. Install the coil and the air cleaner assembly.

71. Connect the negative battery cable.

72. Install the hood, aligning the marks made during the removal procedure.

73. Charge air conditioning system, if equipped.

74. Check all fluid levels.

75. Start the engine and check for leaks.

3.0L Engine

CAUTION

Fuel injection systems remain under pressure, even after the engine has been turned OFF. The fuel system pressure must be relieved before disconnecting any fuel lines. Failure to do so may result in fire and/or personal injury.

---

1. Disconnect the battery cables and remove the battery. Remove the battery tray with the air cleaner assembly attached.

2. Drain the cooling system.

3. If equipped, properly recover the refrigerant from the air conditioning system.

4. Mark the position of the hood on its hinges and remove the hood.

5. Properly relieve the fuel system pressure, then disconnect the fuel lines and position them aside.

6. Remove the upper radiator hose.

7. Tag and disconnect all necessary electrical connectors and vacuum lines.

8. Disconnect the lines from the power steering pump and remove the power steering reservoir.

9. Disconnect the air conditioning lines from the condenser, leaving the manifold lines attached to the compressor.

10. Disconnect the accelerator linkage, transaxle throttle valve linkage and cruise control cable, if equipped.

11. Disconnect the speedometer cable.

12. If equipped with automatic transaxle, disconnect the transaxle cooler lines from the radiator.

13. Remove the coolant overflow bottle and the lower radiator hose.

14. Remove the power steering lines at the rear of the engine above the transaxle.

15. Raise and safely support the vehicle.

16. Drain the engine oil and remove the heater hoses.

17. Remove the front wheel and tire assemblies.

18. Support the exhaust system and remove the exhaust Y-pipe.

19. Remove the bolt retaining the air conditioner line to the engine block.

20. Disconnect the tie rod ends from the spindles.

21. Disconnect the lower ball joints and pull down on the control arms to disengage them from the spindles.

22. Remove both halfshaft assemblies. After removing the halfshafts, install transaxle plugs T81P-1177-B or equivalent, in the differential side gears.

---------- WARNING ----------

Failure to install the plugs can result in dislocation of the differential side gears. If the gears become misaligned, the differential must be removed from the transaxle to realign the gears.

23. Lower the vehicle.

24. Remove the ignition coil bracket bolts and position the coil assembly aside.

25. Install suitable engine lifting eyes to the engine at the front of the right cylinder and at the rear of the left cylinder head. Attach suitable engine lifting equipment to the lifting eyes.

26. Remove the through bolts from the engine mounts.

27. Carefully lift the engine from the vehicle. The engine must be tilted to clear the master cylinder.

To install:

28. Carefully lower the engine into the engine compartment, being careful to clear the master cylinder.

29. Position the engine and install the through bolts in the engine mounts.

30. Remove the engine lifting equipment and the lifting eyes.

31. Position the ignition coil/bracket assembly and install the attaching bolts.

32. Raise and safely support the vehicle.

33. Remove the plugs and install the halfshaft assemblies.

34. Connect the lower ball joints and the tie rod ends to the spindles.

35. Install the bolt retaining the air conditioning line to the engine block.

36. Install the exhaust Y-pipe.

37. Install the front wheel and tire assemblies.

38. Connect the heater hoses.

39. Lower the vehicle.

40. Connect the power steering lines at the rear of the engine above the transaxle.

41. Install the lower radiator hose and the coolant overflow bottle.

42. If equipped with automatic transaxle, connect the transaxle cooler lines to the radiator.

43. Connect the speedometer cable.

44. Connect the accelerator linkage, transaxle throttle valve linkage and cruise control cable, if equipped.

45. Connect the air conditioning lines to the condenser.

46. Install the power steering fluid reservoir and connect the lines to the power steering pump.

47. Connect all vacuum lines and electrical connectors that were marked and disconnected during the removal procedure.

48. Install the upper radiator hose.

49. Connect the fuel lines.

50. Install the hood on the hinges, aligning the marks that were made during the removal procedure.

51. Install the battery tray and battery. Connect the battery cables.

52. Fill the cooling system.

53. Start the engine and bring to normal operating temperature. Check for leaks and check all fluid levels.

54. If equipped, properly evacuate and charge the air conditioning system.

Engine Mounts

REMOVAL AND INSTALLATION

2.3L Engine

Right Insulator (No. 3A)

1. Disconnect the negative battery cable.

2. Place a floor jack and a block of wood under the engine oil pan. Raise the engine approximately $\frac{1}{2}$ inch (12.7mm), or enough to take the load off the insulator.

3. Remove the insulator attaching nut from the bottom of the double-ended stud.

4. Remove the insulator lower attaching nuts through the right side front wheel opening. Remove the insulator lower retaining nuts through the engine compartment.

5. Remove the 2 bolts attaching the insulator-to-engine bracket.

6. Remove the insulator from the vehicle.

To install:

7. Position the insulator into the body opening.

8. Loosely install the retaining nuts and bolts. Tighten the nuts to 73–97 ft. lbs. (98–132 Nm) and bolts to 40–53 ft. lbs. (53–71 Nm).

9. Loosely install the retaining bolts and nut. Tighten the bolts and nut to 65–87 ft. lbs. (88–118 Nm).

10. Lower the engine and remove the jack and block of wood.

11. Connect the negative battery cable.

Left Rear Insulator (No. 4)

1. Disconnect the negative battery cable.

2. Raise and safely support the vehicle.

3. Place a transaxle jack and a block of wood under the transaxle.

4. Raise the transaxle approximately $\frac{1}{2}$ inch (12.7mm), or enough to take the load off the insulator.

5. Remove the insulator attaching nuts from the support bracket. Remove the 2 through bolts and remove the insulator from the transaxle.

To install:

6. Position the insulator over the left rear transaxle housing and support bracket studs.

7. Install the 2 insulator through bolts and tighten to 30–40 ft. lbs. (41–54 Nm).

8. Install 2 insulator-to-support bracket attaching nuts. Tighten to 73–97 ft. lbs. (98–132 Nm).

9. Remove the transaxle jack.

10. Lower the vehicle.

11. Connect the negative battery cable.

NOTE: To remove the left rear support bracket, remove the left rear engine insulator No. 4. Then remove the support bracket attaching bolts. When installing the support bracket, torque the attaching bolts to 51–67 ft. lbs. (68–92 Nm).

Left Front Insulator (No. 1)

1. Disconnect the negative battery cable.

2. Raise and safely support the vehicle.

3. Place a transmission jack and a block of wood under the transaxle. Raise the transaxle approximately $\frac{1}{2}$ inch (12.7mm), or enough to take the load off the insulator.

4. Remove the 2 insulator-to-support bracket retaining nuts.

5. Remove the stud bolt and 2 insulator-to-transaxle retaining bolts.

6. Remove the insulator from the vehicle.

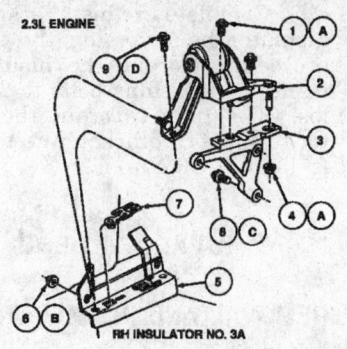

Item	Part Number	Description
1A	N605812-S100	Bolt (2 Req'd)
2	6B014	Insulator Assy
3	6030	Engine Bracket Assy
4A	N802074-S2	Nut
5	—	Body Assy
6B	N801641-S2	Nut (2 Req'd)
7	N803452-S2	Nut
8C	N806676-S100	Bolt (3 Req'd)
9D	N803529-S2	Bolt (2 Req'd)
A		Tighten to 87.5-118.5 N-m (65-87 Lb-Ft)
B		Tighten to 97.7-132.3 N-m (73-97 Lb-Ft)
C		Tighten to 68-92 N-m (51-67 Lb-Ft)
D		Tighten to 53.1-71.9 N-m (40-53 Lb-Ft)

300111

Engine mount, No. 3A — 2.3L engine

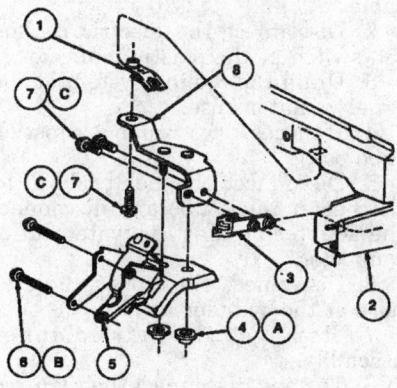

Item	Part Number	Description
1	N802397-S2	Nut
2	—	Body Assy
3	N802778-S2	Nut
4A	N801995-S101	Nuts (2 Req'd)
5	6F026	Insulator
6B	N802774-S2	Bolt (2 Req'd)
7C	N806902-S56	Bolt (3 Req'd)
8	6E042	Support Bracket Assy
A		Tighten to 97.7-132.3 N-m (73-97 Lb-Ft)
B		Tighten to 40.3-54.7 N-m (30-40 Lb-Ft)
C		Tighten to 68-92 N-m (51-67 Lb-Ft)

300112

Engine mount, No. 4 — 2.3L engine

To install:

7. Position the left front insulator to the support bracket and transaxle.

8. Install the 2 insulator-to-transaxle retaining bolts and the stud bolt. Tighten to 26–33 ft. lbs. (34–46 Nm).

9. If required, lower the transmission jack enough for the insulator to contact the support bracket.

10. Install the 2 insulator-to-support bracket retaining nuts. Tighten to 26–33 ft. lbs. (34–46 Nm).

11. Remove the transmission jack and block of wood.

12. Lower the vehicle.

Item	Part Number	Description
1	N802397-S2	Nut
2	—	Body Assy
3	N802778-S2	Nut
4A	N801995-S101	Nut (2 Req'd)
5	6F026	Insulator
6B	N802774-S2	Bolt (2 Req'd)
7C	N806902-S56	Bolt (3 Req'd)
8	6E042	Support Bracket Assy
A		Tighten to 97.7-132.3 N-m (73-97 Lb-Ft)
B		Tighten to 40.3-54.7 N-m (30-40 Lb-Ft)
C		Tighten to 68-92 N-m (51-67 Lb-Ft)

300113

Engine mount support bracket, No. 4 — 2.3L engine

13. Connect the negative battery cable.

3.0L Engine

Right Insulator (No. 3A)

1. Disconnect the negative battery cable. Place a floor jack and a block of wood under the engine oil pan. Raise the engine approximately ½ in. or enough to take the load off the insulator.

2. Remove the insulator attaching nut from the bottom of the double-ended stud.

3. Remove the insulator lower attaching nuts through the right side

front wheel opening. Remove the insulator lower retaining nuts through the engine compartment.

4. Remove the 2 nuts and 1 bolt attaching the insulator-to-engine bracket.

5. Remove the insulator from the vehicle.

To install:

6. Position the insulator into the body opening.

7. Loosely install the retaining nuts and bolts. Tighten the nuts to 73–97 ft. lbs. (98–132 Nm) and bolts to 40–53 ft. lbs. (53–71 Nm).

8. Loosely install the retaining nuts, attaching bolt and retaining lower nut. Tighten to 51–67 ft. lbs. (68–92 Nm), 22–29 ft. lbs. (30–40 Nm) and 65–87 ft. lbs. (88–118 Nm).

9. Lower the engine and remove the jack. Connect the negative battery cable.

Right Insulator (No. 2A)

1. Disconnect the negative battery cable. Place a floor jack and a block of wood under the engine oil pan. Raise the engine approximately ½ in. or enough to take the load off the insulator.

2. Remove the insulator lower nut.

3. Remove the stabilizer bar bracket bolts.

4. Remove the insulator-to-A/C bracket bolt.

5. Remove the insulator from the vehicle.

To install:

6. Position the insulator onto the A/C bracket and loosely attach the bolt. Tighten the bolt to 26–36 ft. lbs. (34–46 Nm).

7. Position the insulator onto the stabilizer bar bracket and loosely attach the nut. Loosely attach the stabilizer bar bracket bolts and tighten to 40–53 ft. lbs. (53–72 Nm). Tighten the nut to 26–36 ft. lbs. (34–46 Nm).

8. Lower the engine and remove the jack. Connect the negative battery cable.

Left Rear Insulator (No. 4)

1. Disconnect the negative battery cable. Raise the vehicle and support safely. Place a transaxle jack and a block of wood under the transaxle.

2. Raise the transaxle approximately ½ in. or enough to take the load off the insulator.

3. Remove the insulator attaching nuts from the support bracket. Remove the 2 through bolts and remove the insulator from the transaxle.

To install:

4. Install the insulator over the left rear transaxle housing and support bracket studs.

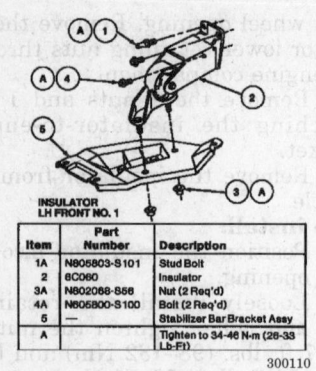

Item	Part Number	Description
1A	N805803-S101	Stud Bolt
2	6C060	Insulator
3A	N802068-S56	Nut (2 Req'd)
4A	N805800-S100	Bolt (2 Req'd)
5	—	Stabilizer Bar Bracket Assy
A		Tighten to 34-46 N·m (26-33 Lb-Ft)

Engine mount, No. 1 — 2.3L engine

3.0L Engine

Item	Part Number	Description
1A	N806677-S100	Bolt
2B	N620482-S2	Nut (2 Req'd)
3	6F012	Insulator
4A	N806810-S100	Bolt (2 Req'd)
5	6A023	Bracket Assy
6C	N802074-S2	Nut
7	N803452-S2	Nut (2 Req'd)
8	—	Body Assy
9D	N801641-S2	Nut (2 Req'd)
10E	N803529-S2	Bolt (2 Req'd)
A		Tighten to 29.7-40.3 N·m (22-29 Lb-Ft)
B		Tighten to 68-92 N·m (51-67 Lb-Ft)
C		Tighten to 87.5-118.5 N·m (65-87 Lb-Ft)
D		Tighten to 97.7-132.3 N·m (73-97 Lb-Ft)
E		Tighten to 53.1-71.9 N·m (40-53 Lb-Ft)

Engine mount, No. 3A — 3.0L engine

5. Install the 2 insulator through bolts and tighten to 30–40 ft. lbs. (41–54 Nm).

6. Install 2 insulator-to-support bracket attaching nuts. Tighten to 73–97 ft. lbs. (98–132 Nm).

7. Lower vehicle and remove floor jack. Connect negative battery cable.

NOTE: To remove the left rear support bracket, remove the left rear engine insulator No. 4. Then remove the support bracket attaching bolts. When installing the

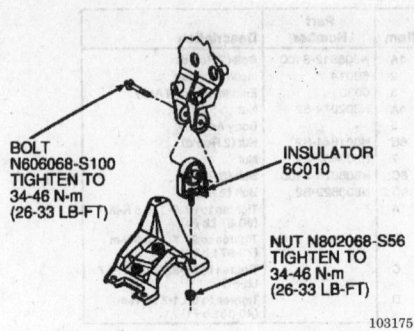

BOLT
N606068-S100
TIGHTEN TO
34-46 N·m
(26-33 LB-FT)

INSULATOR
6C010

NUT N802068-S56
TIGHTEN TO
34-46 N·m
(26-33 LB-FT)

Engine mount, No. 2A — 3.0L engine

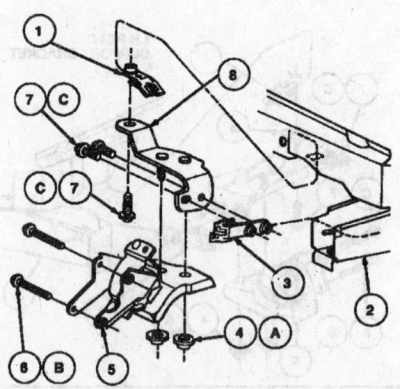

Item	Part Number	Description
1	N802397-S2	Nut
2	—	Body Assy
3	N802778-S2	Nut
4A	N801995-S101	Nuts (2 Req'd)
5	6F026	Insulator
6B	N802774-S2	Bolt (2 Req'd)
7C	N806902-S56	Bolt (3 Req'd)
8	6E042	Support Bracket Assy
A		Tighten to 97.7-132.3 N·m (73-97 Lb-Ft)
B		Tighten to 40.3-54.7 N·m (30-40 Lb-Ft)
C		Tighten to 68-92 N·m (51-67 Lb-Ft)

Engine mount, No. 4 — 3.0L engine

support bracket, torque the attaching bolts to 51–67 ft. lbs. (68–92 Nm).

Left Front Insulator (No. 1)

1. Disconnect the negative battery cable. Raise and the vehicle and support safely. Place a transaxle jack and a block of wood under the transaxle. Raise the transaxle approximately ½ in. or enough to take the load off the insulator.

2. Remove the insulator-to-support bracket attaching nut(s). Remove the insulators and transaxle attaching bolts and remove the insulator from the vehicle.

3. Complete the installation of the insulator by reversing the removal procedure. Torque the insulator to transaxle attaching bolts to 26–36 ft. lbs. (34–46 Nm). Torque the insulator-to-support bracket nut to 26–36 ft. lbs. (34–46 Nm).

Cylinder Head

REMOVAL AND INSTALLATION

2.3L Engine

1. Disconnect the negative battery cable.

2. Disconnect the electric cooling fan switch at the plastic connector.

3. Drain the cooling system at the lower radiator hose.

4. If necessary, remove dipstick tube bolt.

5. Disconnect the heater hose at the heater inlet tube and disconnect the adapter hose at the water outlet connector.

6. Disconnect the upper radiator hose at the cylinder head.

7. Remove the air cleaner assembly.

8. Tag and disconnect the required electrical connectors and vacuum hoses.

9. Remove the distributor cap and spark plug wires as an assembly. Tag the spark plug wires prior to removal.

10. Disconnect all accessory drive belts.

11. Remove the rocker arm cover and gasket.

12. Remove the rocker arm fulcrum retaining bolts and remove the fulcrum, rocker arms and pushrods. Mark the location of each rocker arm, pushrod and fulcrum in its original position.

13. Properly relieve the fuel system pressure, then disconnect the fuel supply and return lines at the fuel rail.

14. Disconnect the accelerator cable and cruise control cable, if equipped.

15. Raise and safely support the vehicle.

16. Disconnect the exhaust system at the exhaust pipe and the hose at the tube.

17. Lower the vehicle.

18. Remove the cylinder head bolts.

19. Remove the cylinder head and gasket with the exhaust and intake manifolds attached.

NOTE: Do not lay the cylinder head flat. Damage to spark plugs or gasket surfaces may result.

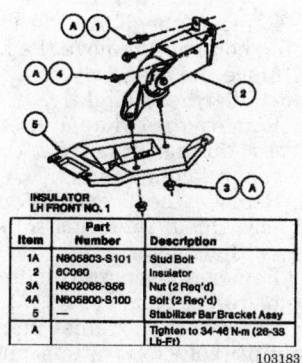

LH REAR SUPPORT BRACKET

Item	Part Number	Description
1	N802397-S2	Nut
2	—	Body Assy
3	N802778-S2	Nut
4A	N801995-S101	Nut (2 Req'd)
5	6F026	Insulator
6B	N802774-S2	Bolt (2 Req'd)
7C	N806902-S56	Bolt (3 Req'd)
8	6E042	Support Bracket Assy
A		Tighten to 97.7-132.3 N-m (73-97 Lb-Ft)
B		Tighten to 40.3-54.7 N-m (30-40 Lb-Ft)
C		Tighten to 68-92 N-m (51-67 Lb-Ft)

103177

Engine mount support bracket, No. 4 — 3.0L engine

INSULATOR LH FRONT NO. 1

Item	Part Number	Description
1A	N805803-S101	Stud Bolt
2	6C060	Insulator
3A	N802068-S56	Nut (2 Req'd)
4A	N605800-S100	Bolt (2 Req'd)
5	—	Stabilizer Bar Bracket Assy
A		Tighten to 34-46 N-m (26-33 Lb-Ft)

103183

Engine mount, No. 1 — 3.0L engine

To install:

20. Clean all gasket material from the mating surfaces of the cylinder head and block.

21. Position the head gasket on the cylinder block.

NOTE: Before installing the cylinder head, thread 2 cylinder head alignment studs T84P-6065-A or equivalent, into the block at opposite corners.

22. Install the cylinder head over the alignment studs onto the cylinder block. Start and run down several head bolts until snug. Remove the alignment studs and install the remaining head bolts. Tighten the bolts in sequence in 2 steps, first to 52-59 ft. lbs. (70-80 Nm) and then to 70-76 ft. lbs. (95-103 Nm).

23. Raise and safely support the vehicle.

24. Connect the exhaust system at the exhaust pipe and the hose to the metal tube.

25. Lower the vehicle.

26. Connect the accelerator cable and cruise control cable, if equipped.

27. Connect the fuel supply and return lines.

28. Install the fulcrums, rocker arms and pushrods in their original positions. Tighten the fulcrum bolts to 20-26 ft. lbs. (26-38 Nm).

29. Install the rocker arm cover gasket and cover.

30. Install the distributor cap and spark plug wires as an assembly.

31. Connect the accessory drive belts.

32. Connect the required electrical connectors and vacuum hoses.

33. Install the air cleaner assembly.

34. Connect the cooling fan switch at the plastic connector.

35. Connect the upper radiator hose and the heater hose.

36. If necessary, install dipstick tube bolt.

37. Fill the cooling system.

38. Connect the negative battery cable.

39. Start the engine and check for leaks.

40. After the engine has reached operating temperature, check and, if necessary, add coolant.

3.0L Engine

CAUTION

Fuel injection systems remain under pressure, even after the engine has been turned OFF. The fuel system pressure must be relieved before disconnecting any fuel lines. Failure to do so may result in fire and/or personal injury.

NOTE: Do not continue with this procedure unless new cylinder head bolts are available. The cylinder bolts removed have stretched and cannot be reused.

1. Disconnect the negative battery cable.

2. Rotate the crankshaft until No. 1 piston is at TDC on its compression stroke.

3. Drain the engine cooling system.

4. Remove the PCV hose and the air cleaner outlet tube from the rocker arm cover.

5. Remove the air cleaner outlet tube from the throttle body and mass air flow sensor.

6. Relieve the fuel system pressure as follows:

   a. Remove the fuel tank cap to relieve the pressure in the fuel tank.

   b. Remove the cap from the Schrader valve located on the fuel supply manifold.

   c. Attach fuel pressure gauge T80L-9974-A or equivalent, to the Schrader valve and drain the fuel through the drain tube into a suitable container.

   d. After the fuel system pressure is relieved, remove the fuel pressure gauge and install the cap on the Schrader valve.

7. Tag and disconnect all necessary vacuum lines.

8. Disconnect the TPS, idle air control valve, ECT, PFE, distributor, ignition coil and engine coolant temperature sending unit electrical connectors. Tag the location of each so they can be reconnected properly.

9. Disconnect the upper radiator hose from the thermostat housing.

10. Loosen the EGR tube retaining nuts and remove the tube.

11. Disconnect the throttle and TV cable from the throttle body linkage.

12. Remove the retaining nuts from the alternator brace and remove the brace.

13. Remove the 6 throttle body retaining bolts and remove the throttle body.

14. Disconnect the fuel injector harness retaining stand-offs from the inboard rocker arm cover studs. Carefully disconnect the electrical connections at each injector and remove the harness from the engine.

15. Disconnect the heater hose.

16. Tag and disconnect the ignition wires from the spark plugs, then remove the harness retaining stand-offs from the rocker arm cover studs.

17. Remove the distributor cap. Mark the position of the distributor rotor in relation to the distributor body and the position of the distributor body in relation to the engine block. Remove the distributor hold-down bolt and remove the distributor.

18. Remove the oil cooler tube assembly retaining bolt from the ignition coil bracket. Remove the ignition coil from the left cylinder head.

19. Remove the rocker arm covers.

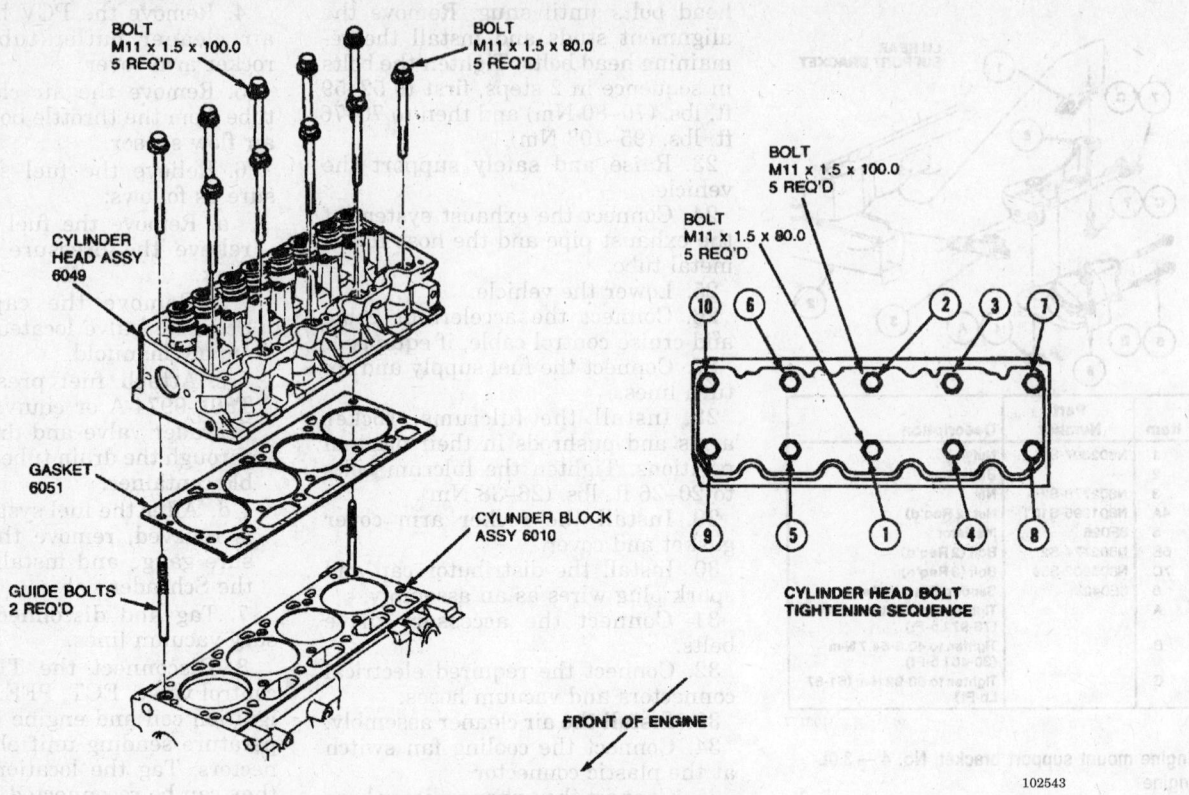

BOLT
M11 x 1.5 x 100.0
5 REQ'D

BOLT
M11 x 1.5 x 80.0
5 REQ'D

BOLT
M11 x 1.5 x 100.0
5 REQ'D

BOLT
M11 x 1.5 x 80.0
5 REQ'D

CYLINDER
HEAD ASSY
6049

GASKET
6051

CYLINDER BLOCK
ASSY 6010

GUIDE BOLTS
2 REQ'D

**CYLINDER HEAD BOLT
TIGHTENING SEQUENCE**

FRONT OF ENGINE

102543

Cylinder head bolt tightening sequence — 2.3L engine

20. Remove the rocker arms and pushrods. Keep all rocker arms, fulcrums and pushrods in order so they can be reinstalled in their original locations.

NOTE: **Regardless of the cylinder head being removed, the No. 3 cylinder intake valve rocker arm and pushrod must be removed in order to remove the intake manifold.**

21. Remove the intake manifold retaining bolts. Wedge a prybar between the manifold and engine block and pry upward to break the manifold-to-engine block seal, using the area between the thermostat and transaxle as a leverage point.

NOTE: **The intake manifold may be removed with the fuel supply manifold and injectors in place.**

22. If removing the right (rear) cylinder head, proceed as follows:

a. Remove the accessory and water pump drive belts.

b. Remove the water pump to front cover hose.

c. Raise and safely support the vehicle.

d. Remove the lower water pump tube. Loosen and remove the re-

taining nut from the upper bracket and the bolt from the lower bracket. Gently grasp the tube at the water pump end and pull the tube out of the water pump. Set the assembly aside.

e. Loosen and remove the exhaust inlet pipe flange retaining nuts from the exhaust manifold studs.

f. Lower the vehicle.

g. Remove the heater hose from the rear of the water pump.

h. Remove the water pump pulley shield. Remove the nut from the stud bolt.

i. Remove the water pump from the bracket.

j. Remove the exhaust manifold heat shield and the exhaust manifold.

23. If removing the left (front) cylinder head, proceed as follows:

a. Remove the accessory drive belt.

b. Remove the power steering pulley shield and the accessory belt tensioner.

c. Remove the 3 alternator bracket to cylinder head retaining bolts.

d. Remove the upper alternator retaining bolt.

e. Remove the 3 A/C brace retaining bolts and remove the brace.

f. Move the assembly away from the cylinder head slightly.

g. Remove the exhaust inlet pipe flange retaining nuts from the exhaust manifold studs.

h. Remove the 2 exhaust manifold heat shield retaining nuts and remove the shield.

i. Remove the engine oil dipstick tube or rotate it aside.

j. Remove the exhaust manifold retaining bolts and studs and remove the exhaust manifold.

24. Remove and discard the cylinder head bolts.

25. Remove the cylinder head(s). If the cylinder head is stuck to the gasket, place a prybar into the intake port and rock the cylinder head to break the seal.

NOTE: **When breaking the seal, be careful not to damage machined surfaces or the intake valve.**

26. Remove the cylinder head and discard the gasket.

27. If any coolant leaked into the cylinder bores from the cylinder head removal, immediately wipe the cylinder dry and apply a light coating of

engine oil to the cylinder bore surface.

NOTE: Engine coolant is corrosive to engine bearing material and piston rings.

To install:

28. Lightly oil all bolt and stud threads prior to installation. Always use new cylinder heads bolts.

29. Place shop rags in the lifter valley, cylinder bores and cylinder block coolant passages to catch any dirt or gasket material. Clean the sealing surfaces of the cylinder head, intake manifold, rocker arm covers and cylinder block.

30. If the cylinder head was removed for head gasket replacement, check the cylinder head and block for flatness using a straight-edge and feeler gauge. Warpage should not exceed 0.003 inch in 6 inch span. Replace or machine the cylinder head, as necessary. If machining, do not grind off more than 0.010 in. (0.254mm).

31. Position new head gasket(s) on the cylinder block, with the V-cut toward the front of the engine. Use dowels to align and hold the gasket in place.

NOTE: Replace any dowels that are damaged or loose.

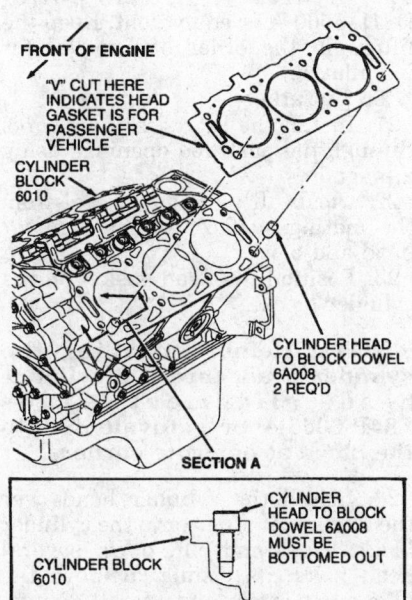

Cylinder head gasket installation — 3.0L engine

32. Install and hand-tighten the new cylinder head bolts. Tighten as follows:

   a. Tighten, in sequence, to 52–66 ft. lbs. (70–90 Nm).

   b. Back off all bolts one turn.

   c. Tighten, in sequence, to 33–41 ft. lbs. (45–55 Nm).

   d. Tighten, in sequence, to 63–73 ft. lbs. (85–99 Nm).

33. Apply a ¼ in. (6mm) drop of silicone sealer to the intersection of the cylinder block and cylinder head at the 4 corners of the lifter valley.

34. Position the intake gaskets on the cylinder heads and align the locking tabs to the cylinder head gaskets.

35. Install the front and rear intake manifold seals and secure with the retaining features.

36. Carefully lower the intake manifold into position, aligning the manifold bolt holes with the holes in the cylinder heads. Be careful not to disturb the sealer.

37. Install the No. 1, 2, 3 and 4 bolts and hand-tighten. Install the remaining bolts and tighten all bolts, in sequence, in 2 steps. First tighten to 15–22 ft. lbs. (20–30 Nm), then again in sequence, to 19–24 ft. lbs. (26–32 Nm).

38. Lubricate the distributor gear teeth and the distributor O-ring with engine oil. Install the distributor, aligning the marks that were made during the removal procedure. Install the hold-down bolt and snug.

39. Lubricate the pushrods and rocker arms with engine oil, then install them in their original locations. Snug the retaining bolts.

40. Rotate the crankshaft one turn clockwise. Tighten the rocker arm retaining bolts on the No. 1 intake valve, No. 2 exhaust valve, No. 4 intake valve and No. 5 exhaust valve to 5–11 ft. lbs. (7–15 Nm), making sure the rocker arms are seated on the pushrods and the rocker arm fulcrums are seated on the cylinder head.

41. Rotate the crankshaft 120 degrees clockwise. Tighten the remaining rocker arm retaining bolts to 5–11 ft. lbs. (7–15 Nm), making sure the rocker arms are seated on the pushrods and the rocker arm fulcrums are seated on the cylinder head.

42. Final tighten the rocker arm retaining bolts to 19–28 ft. lbs. (26–38 Nm) with the crankshaft in any position.

43. Install the rocker arm covers.

44. If the right (rear) cylinder head was removed, proceed as follows:

   a. Install the exhaust manifold and tighten the retaining bolts and studs to 15–22 ft. lbs. (20–30 Nm). Install the heat shield and tighten the retaining nuts to 12–15 ft. lbs. (16–20 Nm).

   b. Install the water pump to the bracket and tighten the retaining bolts and stud to 15–22 ft. lbs. (20–30 Nm).

   c. Install the water pump pulley shield and tighten the retaining nut to 7–10 ft. lbs. (9–14 Nm).

   d. Connect the heater hose at the fitting on the rear of the water pump and tighten the clamp.

   e. Raise and safely support the vehicle.

   f. Lubricate the water pump end of the water pump tube with soapy water and install it into the water pump. Install the retaining nut to the upper bracket stud bolt and tighten to 5 ft. lbs. (7 Nm). Install the lower tube bracket retaining bolt and tighten to 71–106 inch lbs. (8–12 Nm).

   g. Install the exhaust pipe flange nuts and tighten to 25–34 ft. lbs. (34–47 Nm).

   h. Lower the vehicle.

   i. Install the water pump to the front cover hose and tighten the clamp.

   j. Install the water pump drive belt. If the left (front) cylinder head

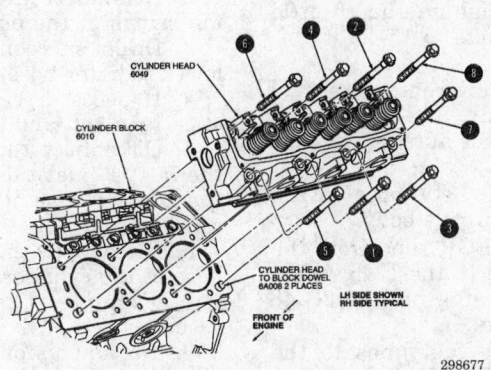

Cylinder head bolt tightening sequence — 3.0L engine

was not removed, at this time install the accessory drive belt.

45. If the left (front) cylinder head was removed, proceed as follows:

a. Install the exhaust manifold and tighten the retaining bolts and studs to 15–22 ft. lbs. (20–30 Nm).

b. Rotate into position or install the engine oil dipstick tube, as required.

c. Install the exhaust manifold heat shield and tighten the retaining nuts to 12–15 ft. lbs. (16–20 Nm).

d. Install the exhaust pipe flange nuts and tighten to 25–34 ft. lbs. (34–47 Nm).

e. Install the alternator bracket to the cylinder head and tighten the retaining bolts to 30–41 ft. lbs. (40–55 Nm).

f. Install the A/C brace and the retaining bolts and upper alternator bolt. Tighten the long bolts to 30–41 ft. lbs. (40–55 Nm) and the remaining bolt to 15–22 ft. lbs. (20–30 Nm).

g. Install the accessory belt tensioner and tighten the retaining bolt to 30–41 ft. lbs. (40–55 Nm).

h. Install the accessory drive belt.

i. Install the power steering pulley shield and tighten the retaining bolts to 6–8 ft. lbs. (8.5–11.0 Nm).

46. Install the fuel injector electrical harness to the injectors and secure the harness with the stand-offs to the inboard rocker arm cover studs.

47. Install the oil cooler tube assembly retaining bolt to the ignition coil bracket. Install the ignition coil and tighten the retaining bolts to 15–22 ft. lbs. (20–30 Nm).

48. Install the distributor cap and ignition wires. Install the wire harness stand-offs to the rocker arm cover studs and connect the wires to the spark plugs and ignition coil.

49. Install the throttle body, using a new gasket. Tighten the throttle body mounting bolts to 15–22 ft. lbs. (20–30 Nm).

50. Install the alternator brace to the throttle body and alternator bracket. Tighten the nuts to 12 ft. lbs. (16 Nm).

51. Connect the PCV hose to the tube under the throttle body.

52. Install the EGR tube from the exhaust manifold to the EGR valve. Tighten the retaining nuts to 26–48 ft. lbs. (36–65 Nm).

53. Connect the fuel lines to the fuel supply manifold. Install the fuel line safety clips.

54. Install the upper radiator hose and heater hose and tighten the clamps.

55. Connect all removed vacuum lines to their original locations as marked during the removal procedure.

56. Connect the electrical connectors at the TPS, idle air control, ECT, PFE, distributor, ignition coil and engine coolant temperature sending unit.

57. Connect the throttle and TV cables to the throttle body linkage.

58. Install the air cleaner outlet tube between the throttle body and mass air flow sensor. Tighten the clamps to 24–35 inch lbs. (2.7–4.0 Nm).

59. Install the PCV hose and air cleaner outlet tube to the rocker arm cover.

60. Drain the crankcase and fill with the proper type and quantity of engine oil.

NOTE: Engine coolant is corrosive to all engine bearing material. Changing the oil after the replacement of a coolant carrying component prevents failure later.

61. Fill and bleed the cooling system.

62. Connect the negative battery cable.

63. Start the engine and check for leaks.

64. Check, and if necessary, adjust the ignition timing.

65. Install the idle air control shield.

Valve Lifters

REMOVAL AND INSTALLATION

2.3L Engine

1. Disconnect the negative battery cable.

2. Disconnect the electric cooling fan switch at the plastic connector.

3. Drain the cooling system at the lower radiator hose.

4. If necessary, remove the dipstick tube bolt and dipstick.

5. Disconnect the heater hose at the heater inlet tube and disconnect the adapter hose at the water outlet connector.

6. Disconnect the upper radiator hose at the cylinder head.

7. Remove the air cleaner assembly.

8. Tag and disconnect the required electrical connectors and vacuum hoses.

9. Remove the distributor cap and spark plug wires as an assembly. Tag the spark plug wires prior to removal.

10. Disconnect all accessory drive belts.

11. Remove the rocker arm cover and gasket.

12. Remove the rocker arm fulcrum retaining bolts and remove the fulcrum, rocker arms and pushrods. Mark the location of each rocker arm, pushrod and fulcrum for reinstallation in its original position.

13. Properly relieve the fuel system pressure, then disconnect the fuel supply and return lines at the fuel rail.

14. Disconnect the accelerator cable and cruise control cable, if equipped.

15. Raise and safely support the vehicle.

16. Disconnect the exhaust system at the exhaust pipe and the hose at the tube.

17. Lower the vehicle.

18. Remove the cylinder head bolts.

19. Remove the cylinder head and gasket with the exhaust and intake manifolds attached.

NOTE: Do not lay the cylinder head flat. Damage to spark plugs or gasket surfaces may result.

20. Using a magnet, remove the lifters. If the lifters are stuck in their bores do to excessive gum or varnish, use hydraulic lifter puller D81L-6500-A or equivalent. Keep the lifters in the order of removal for installation.

To install:

21. Install the lifters in their bores through the push rod openings using a magnet.

22. Clean all gasket material from the mating surfaces of the cylinder head and block.

23. Position the head gasket on the cylinder block.

NOTE: Before installing the cylinder head, thread 2 cylinder head alignment studs T84P-6065-A or equivalent, into the block at opposite corners.

24. Install the cylinder head over the alignment studs onto the cylinder block. Start and run down several head bolts until snug. Remove the alignment studs and install the remaining head bolts. Tighten the bolts in sequence in 2 steps, first to 52–59 ft. lbs. (70–80 Nm) and then to 70–76 ft. lbs. (95–103 Nm).

25. Raise and safely support the vehicle.

26. Connect the exhaust system at the exhaust pipe and the hose to the metal tube.

27. Lower the vehicle.

28. Connect the accelerator cable and cruise control cable, if equipped.

29. Connect the fuel supply and return lines.

30. Install the fulcrums, rocker arms and pushrods in their original positions. Tighten the fulcrum bolts to 20–26 ft. lbs. (26–38 Nm).

31. Install the rocker arm cover gasket and cover.

32. Install the distributor cap and spark plug wires as an assembly.

33. Connect the accessory drive belts.

34. Connect the required electrical connectors and vacuum hoses.

35. Install the air cleaner assembly.

36. Connect the cooling fan switch at the plastic connector.

37. Connect the upper radiator hose and the heater hose.

38. If necessary, install dipstick tube bolt.

39. Fill the cooling system.

40. Connect the negative battery cable.

41. Start the engine and check for leaks.

42. After the engine has reached operating temperature, check and, if necessary, add coolant.

3.0L Engine

— CAUTION —

Fuel injection systems remain under pressure, even after the engine has been turned OFF. The fuel system pressure must be relieved before disconnecting any fuel lines. Failure to do so may result in fire and/or personal injury.

NOTE: Before replacing a lifter for noisy operation make sure the noise is not caused by improper valve to rocker arm clearance or by worn rocker arms or pushrods.

1. Rotate the crankshaft until the piston in No. 1 cylinder is at TDC on the compression stroke.

2. Disconnect the negative battery cable.

3. Drain the engine coolant.

4. Remove the PCV tube from from the rocker arm cover.

5. Remove the aspirator hose from air the air cleaner outlet.

6. Properly relieve the fuel system pressure.

7. Disconnect the fuel lines.

8. Tag and remove the vacuum lines.

9. Disconnect the IAT and distributor connectors. Disconnect the wiring to the throttle position sensor, idle air control and the PFE sensors. On flexible fuel vehicles disconnect the CSI and camshaft sensor.

10. Disconnect the upper radiator hose from the thermostat housing.

11. Remove the brace from the generator to throttle body stud.

12. Remove the EGR valve to exhaust manifold tube.

13. Remove the throttle body.

14. Disconnect the fuel charging wiring retainers from the valve cover studs.

15. Disconnect the heater hoses.

16. Tag and remove the spark plug wires from the spark plugs.

17. Mark the position of the distributor rotor in relation to the distributor housing and mark the position of the distributor housing in relation to the engine. Remove distributor retaining bolt and remove the distributor.

18. Remove the ignition coil from the left cylinder head.

19. Remove the valve covers.

20. Loosen cylinder No. 3 intake valve rocker arm retaining nut and rotate arm off pushrod and away from top valve stem. Remove the pushrod.

21. Remove intake manifold retaining bolts and intake manifold.

22. Loosen rocker arm fulcrum retaining bolt of lifter to be replaced enough to allow rocker arm to be lifted off the pushrod and rotated to one side.

23. Remove the pushrods. If more than one pushrod is being removed be sure to tag its original location for reinstallation.

24. Loosen the 2 roller lifter guide plate retaining bolts. Remove the guide plate retainer assembly from the lifter valley.

25. Remove the lifter guide plates from the lifters by lifting straight up.

NOTE: If the lifters are stuck in the bore, it may be necessary to use a claw–type tool to aid removal. Rotate the valve lifter back and forth to loosen it from the deposits.

26. To remove the lifter, grasp lifter and pull straight in line with bore.

To install:

NOTE: Lightly oil all retaining bolt and stud bolt threads before installation.

27. Clean mating gasket surfaces of the intake manifold and cylinder head. Use care when working with light-alloy components

28. Lubricate lifters and bore. Install the lifters into the bore.

29. Aligning the lifters flats, install the lifter guide plate. Install plate with word UP and/or button visible.

30. Install guide plate retainer assembly over the guide plates. Install the 2 retaining bolts and tighten to 8–10 ft. lbs. (10–14 Nm).

31. Apply a bead of rubber sealer to intersection of cylinder block and cylinder head assembly at 4 corners.

32. Position intake gaskets onto cylinder heads. Align intake gasket locking tabs to provisions made on the cylinder head gasket.

33. Install front and rear manifold seals.

34. Lower intake manifold into position. Install bolts and tighten in 2 steps. First tighten each bolt to 15–22 ft. lbs. (20–30 Nm). Second, tighten the bolts to 19–24 ft. lbs. (26–32 Nm).

35. Coat the distributor drive gear with engine assembly lubricant. Install the distributor, aligning the marks made during the removal procedure.

36. Install the pushrod(s), making sure they are seated in the lifters.

37. Lubricate the pushrods and rocker arms with clean engine oil. Move the rocker arms into position with the pushrods and snug the rocker arm retaining bolt.

38. Before tightening the rocker arm bolts, for each valve rotate the crankshaft to position the camshaft lobe straight down and away from the valve lifter.

39. Tighten rocker arm bolts in 2 steps. First tighten to 5–11 ft. lbs. (7–15 Nm). Second, tighten to 19–28 ft. lbs. (26–38 Nm) in any position.

40. Install valve covers.

41. Install the ignition coil.

42. Install the throttle body assembly and the upper intake manifold gaskets.

43. Install the upper intake manifold assembly.

44. Install the EGR valve to exhaust manifold tube.

45. Install fuel lines and install the fuel line safety clips.

46. Install upper radiator hose and heater hoses.

47. Reconnect the vacuum lines.

48. Connect the electrical connectors to the IAT sensor, distributor, idle air control, throttle position sensor, PFE sensor and coolant temperature sensor.

49. Fill and bleed cooling system. Refill the engine with proper grade motor oil.

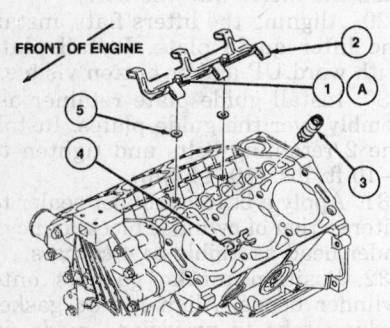

FRONT OF ENGINE

Item	Part Number	Description
1A	—	Bolt (2 Req'd)
2	6K564	Guide Plate Retainer Assembly
3	6500	Tappet (12 Req'd)
4	6K512	Guide Plate (6 Req'd)
5	—	Washer (2 Req'd)
A		Tighten to 10-14 N·m (8-10 Lb-Ft)

238539

Roller lifter guide plate retainer assembly installation — 3.0L engine

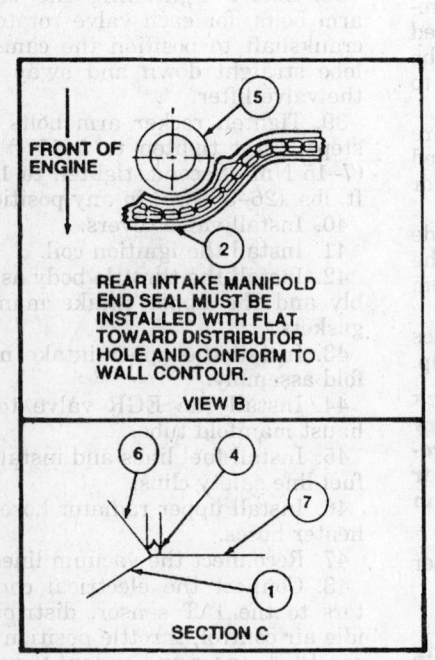

FRONT OF ENGINE

VIEW B

REAR INTAKE MANIFOLD END SEAL MUST BE INSTALLED WITH FLAT TOWARD DISTRIBUTOR HOLE AND CONFORM TO WALL CONTOUR.

VIEW B

SECTION C

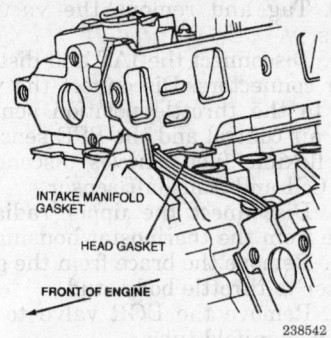

INTAKE MANIFOLD GASKET

HEAD GASKET

FRONT OF ENGINE

238542

Head gaskets and intake manifold gaskets have locking/locating tabs — 3.0L engine

50. Install air cleaner tube and engine air cleaner.
51. Reconnect the PCV hose.
52. Reconnect the negative battery cable.
53. Run the engine and check for leaks and proper operation.

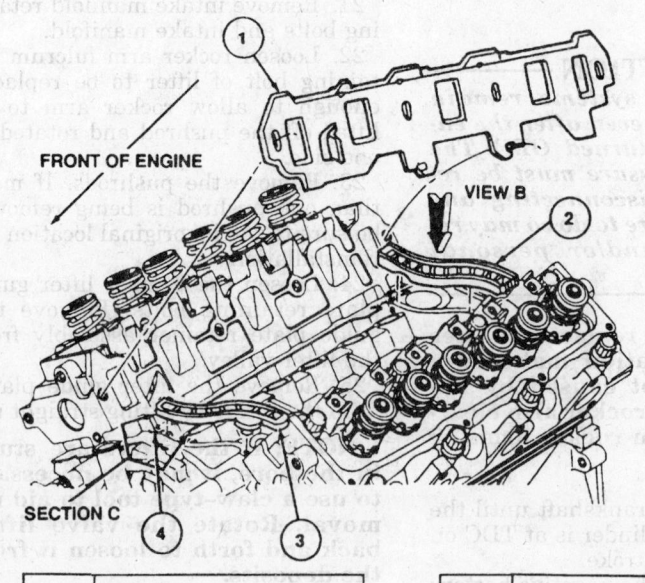

FRONT OF ENGINE

VIEW B

SECTION C

Item	Description
1	Intake Manifold Gasket (2 Req'd)
2	Rear Intake Manifold Seal
3	Front Intake Manifold Seal

Item	Description
4	Silicone Rubber
5	Distributor Hole
6	Cylinder Head Assy
7	Cylinder Block Assy

238541

Intake manifold gasket and seal installation — 3.0L engine

Valve Lash

ADJUSTMENT

2.3L Engine

NOTE: The clearance check is usually only needed when the valves, valve seats and/or cylinder head gasket surface have been installed. Clearance must be checked when the lifter is completely collapsed.

1. Disconnect the negative battery cable.
2. Remove the rocker arm cover.
3. Rotate the engine until the No. 1 cylinder is at TDC of its compression stroke. The timing marks on the camshaft and crankshaft gears will be together. Check the clearance on No. 1 intake, No. 1 exhaust, No. 2 intake and No. 3 exhaust valve.
4. To check the clearance, use lifter bleed down wrench T71P-6513-B or equivalent, to push down on the rocker arm and bleed the oil from the lifter.
5. Insert the appropriate thickness feeler gauge between the rocker arm and valve stem to check the clearance.
6. Rotate the crankshaft 1 complete turn. Check the clearance on

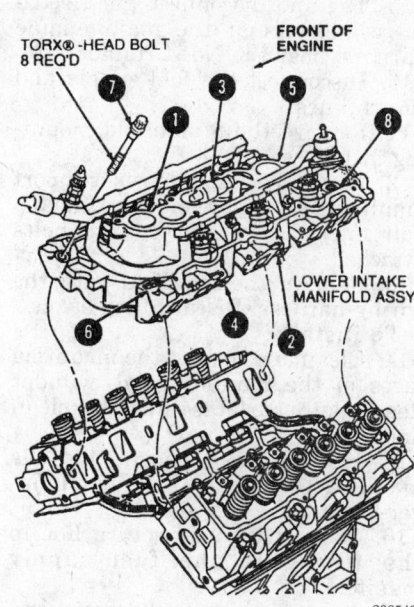

Intake manifold bolt torque sequence — 3.0L engine

238543

No. 2 exhaust, No. 3 intake, No. 4 intake and No. 4 exhaust.

7. The clearance between the rocker arm and the valve stem tip should be 0.072–0.174 in. (1.80–4.34mm) with the lifter on the base circle of the cam.

8. If the clearance is less than specified, shorter pushrods are available to correct the problem. If the clearance is greater than specified, longer pushrods are available.

3.0L Engine

NOTE: The clearance check is usually only needed when the valves, valve seats and/or cylinder head gasket surface have been installed. Clearance must be checked when the lifter is completely collapsed.

1. Disconnect the negative battery cable.

2. Remove the rocker arm covers.

3. Rotate the engine until the No. 1 cylinder is at TDC of its compression stroke and check the clearance between the rocker arm and the following valve: No. 1 intake, No. 1 exhaust, No. 2 exhaust, No. 3 intake and No. 4 exhaust and No. 6 intake.

4. To check the clearance, use lifter bleed down wrench T71P-6513-B or equivalent, to push down on the rocker arm and bleed the oil from the lifter.

5. Insert the appropriate thickness feeler gauge between the rocker arm and valve stem to check the clearance.

6. Rotate the crankshaft 360 degrees and check the clearance the rocker arm and the following valves: No. 2 intake, No. 3 exhaust, No. 4 intake, No. 5 intake, No. 5 exhaust and No. 6 exhaust.

7. The clearance should be 0.09–0.19 in. (2.3–4.8mm).

8. If the clearance is less than specified, shorter pushrods are available to correct the problem. If the clearance is greater than specified, longer pushrods are available.

Rocker Arms

REMOVAL AND INSTALLATION

2.3L Engine

1. Disconnect the negative battery cable.

2. Remove and tag all necessary vacuum hoses from the rocker cover. Remove the oil fill cap and set it aside. Disconnect the PCV hose and set it aside.

3. Remove the rocker arm cover bolts. Remove the rocker cover from the engine.

4. Remove the rocker arm bolts, fulcrums, rocker arms and fulcrum washers. Keep all parts in order so they can be reinstalled to their original position.

To install:

5. Before installation, coat the valve tips, rocker arm and fulcrum contact areas with Lubriplate® or equivalent.

6. For each valve, rotate the engine until the lifter is on the base circle of the cam (valve closed).

7. Install the rocker arm and components and torque the rocker arm bolts in 2 steps: the first to 4.5–7.5 ft. lbs. (6–10 Nm) and the second torque to 20–26 ft. lbs. (26–38 Nm). Be sure the lifter is on the base circle of the cam for each rocker arm as it is installed. For the final tightening, the camshaft may be in any position.

NOTE: If new valve train components have been installed or if the cylinder head has been machined, the valve clearance must be checked before installing the rocker arm cover. Clearance is adjusted with different length pushrods.

8. Clean the rocker cover rail on the cylinder head. The rocker arm cover has a reusable "mould in place gasket". If the gasket is damaged by a cut/nick of about 1/8 in. (maximum 2 places), the damaged area may be filled in with RTV sealer. If the gasket is damaged by cuts longer than 1/8 in. or by more than 2 cuts/nicks, replace the rocker arm cover.

9. Install the rocker arm cover with the retaining bolts and tighten to 5.9–8.5 ft. lbs. (8.0–11.5 Nm). Apply suitable threadlock adhesive to the bolts if they are being reused, to prevent leaks.

10. Install oil fill cap, all necessary vacuum hoses and the PCV hose.

11. Connect negative battery cable.

3.0L Engine

1. Disconnect the negative battery cable. Disconnect and tag the spark plug wires.

2. Remove the ignition wire/separator assembly from the rocker arm attaching bolt studs.

3. If the left rocker arm cover is being removed, remove the oil fill cap, disconnect the air cleaner closure system hose and remove the fuel injector harness from the inboard rocker arm cover studs.

4. If the right rocker arm cover is being removed, remove the throttle body as follows:

   a. Tag and disconnect the vacuum hoses at the vacuum tree.

   b. Loosen the EGR tube nuts, if equipped, at the EGR valve and exhaust manifold fitting. Remove or rotate the tube aside.

   c. Remove the PCV hose from the tube under the throttle body.

   d. Remove the air cleaner duct hose and the idle speed control solenoid shield.

   e. Disconnect the throttle and TV cables from the throttle body linkage.

   f. Tag and disconnect the electrical connectors from the air charge temperature sensor, idle speed control solenoid and throttle position sensor.

   g. Remove the alternator brace.

   h. Remove the throttle body retaining bolts and remove the throttle body. Note the location of the bolts so they can reinstalled in their original positions.

5. If the right rocker arm cover is being removed, remove the PCV valve, loosen the lower EGR tube, if equipped, retaining nut and rotate the tube aside, remove the throttle body and move the fuel injection harness aside.

6. Remove the rocker arm cover attaching screws and the covers and gaskets from the vehicle.

7. Remove the rocker arm bolts, fulcrums, rocker arms and fulcrum washers. Keep all parts in order so they can be reinstalled to their original positions.

8. Remove the pushrods, if necessary. Keep them in order so they can be reinstalled in their original positions.

9. Inspect the rocker arms, fulcrums and pushrods for wear and/or damage. Replace as necessary.

To install:

10. Install the pushrods, if removed, making sure they seat in the lifters.

11. Coat the valve and pushrod tips, rocker arm and fulcrum contact areas with Lubriplate® or equivalent. Lightly oil all the bolt and stud threads before installation.

12. Rotate the engine until the lifter is on the base circle of the cam (valve closed).

13. Install the rocker arm and components and torque the rocker arm fulcrum bolts in 2 steps: the first to 8 ft. lbs. (11 Nm) and the final to 24 ft. lbs. (32 Nm). Be sure the lifter is on the base circle of the cam for each rocker arm as it is installed.

14. Clean the cylinder head and rocker arm cover sealing surfaces of all dirt and old sealer. If not equipped with integral gaskets, make sure all old gasket material is removed.

15. Apply a bead of silicone sealant at the cylinder head to intake manifold rail step. If not equipped with integral gaskets, install a new rocker arm cover gasket.

16. Install the rocker arm cover and the bolts and studs. Tighten to 9 ft. lbs. (12 Nm) in the proper sequence.

17. Install the remaining components in the reverse order of their removal.

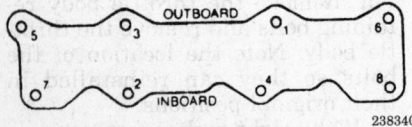

Rocker arm cover bolt tightening sequence — 3.0L engine

Intake Manifold

REMOVAL AND INSTALLATION

2.3L Engine

1. Disconnect the negative battery cable.
2. Properly relieve the fuel system pressure.
3. Remove the air duct from between the throttle body and air cleaner.
4. Disconnect the accelerator and, if equipped, cruise control cables from the mounting bracket and throttle lever.
5. Tag and disconnect the rear vacuum line to the dash panel vacuum tree, the vacuum line at the intake manifold, MAP sensor vacuum line and fuel pressure regulator vacuum line.
6. Disconnect the hoses from the PCV valve at the intake manifold.
7. Disconnect the EGR vacuum line at the EGR valve and EGR tube. Disconnect the EGR tube from the upper intake manifold by supporting the connector while loosening the compression nut.
8. Disconnect the upper support manifold bracket by removing the top

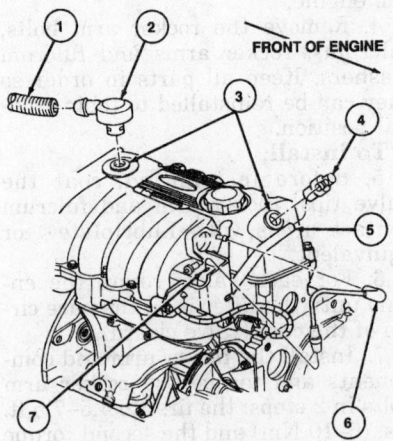

FRONT OF ENGINE

Item	Part Number	Description
1	6853	Crankcase Vent Hose Assy
2	6A768	Filter Assy
3	—	Grommet
4	6A666	PCV Valve
5	6A664	Hose
6	—	Pressure Regulator Vacuum Harness
7	9C968	Fuel Pressure Regulator

101566

Intake manifold, upper — 2.3L engine

bolt only. Leave the bottom bolts attached.

9. Tag and disconnect the electrical connectors at the main engine harness, near the No. 4 runner.
10. Disconnect the fuel supply and return lines.
11. Remove the 8 manifold mounting fasteners.
12. Disconnect the lower support manifold bracket by removing the top bolt only. Leave the bottom bolts attached.
13. Remove the manifold with the wiring harness. Discard the gasket.

To install:

14. Clean and inspect the mounting faces of the manifold and cylinder head. Both surfaces must be clean and flat.
15. Install a new gasket and the manifold assembly. Install and finger-tighten the fasteners.
16. Connect the fuel return line to the fitting in the fuel supply manifold.
17. Tighten all manifold fasteners, in sequence, to 15–22 ft. lbs. (20–30 Nm).
18. Connect the upper and lower manifold support brackets and tighten to 15–22 ft. lbs. (20–30 Nm).
19. Install the EGR tube with the oil-coated compression nut and tighten to 30–40 ft. lbs. (40–55 Nm).
20. Connect the large PCV vacuum line to the upper manifold fitting.
21. Connect the rear manifold vacuum connections at the dash panel vacuum tree and connect the vacuum line(s) to the upper manifold.
22. Connect the accelerator and, if equipped, the cruise control cables.
23. Connect the wiring harness at the electronic engine control harness.
24. Connect the fuel supply hose from the filter to the fuel supply manifold.
25. Connect the negative battery cable. Start the engine and check for fuel and/or vacuum leaks.

3.0L Engine

— **CAUTION** —
Fuel injection systems remain under pressure, even after the engine has been turned OFF. The fuel system pressure must be relieved before disconnecting any fuel lines. Failure to do so may result in fire and/or personal injury.

1. Disconnect the negative battery cable.
2. Drain the engine cooling system.

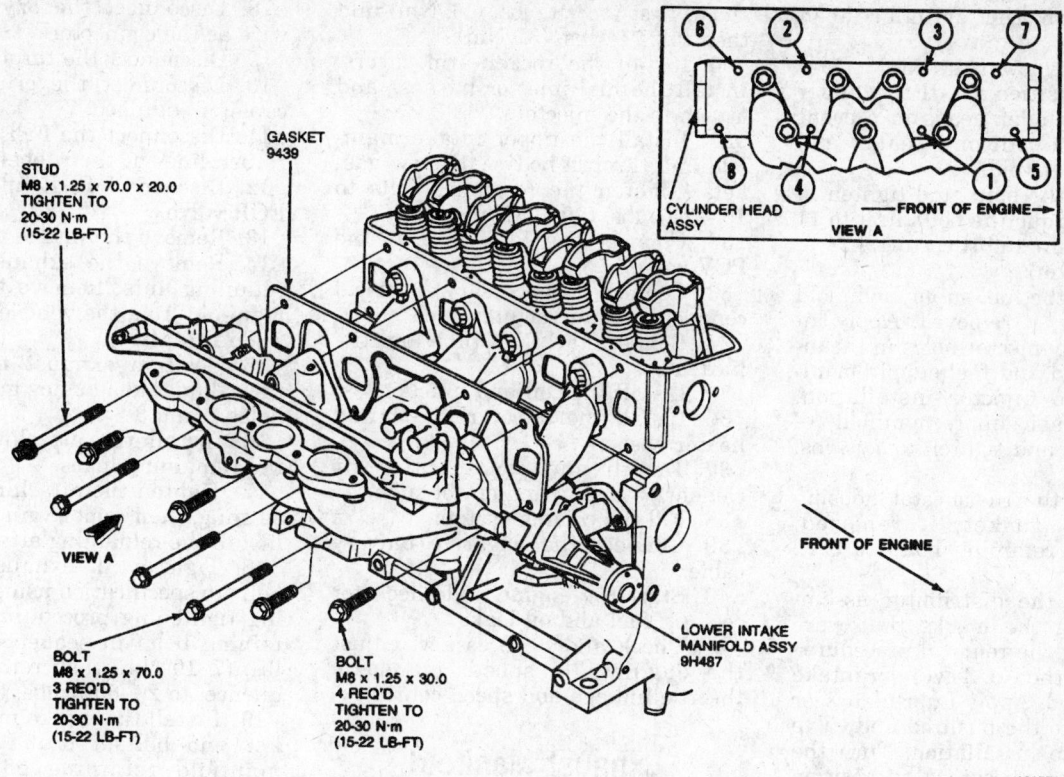

STUD
M8 x 1.25 x 70.0 x 20.0
TIGHTEN TO
20-30 N·m
(15-22 LB-FT)

GASKET
9439

VIEW A

BOLT
M8 x 1.25 x 70.0
3 REQ'D
TIGHTEN TO
20-30 N·m
(15-22 LB-FT)

BOLT
M8 x 1.25 x 30.0
4 REQ'D
TIGHTEN TO
20-30 N·m
(15-22 LB-FT)

LOWER INTAKE
MANIFOLD ASSY
9H487

FRONT OF ENGINE

CYLINDER HEAD
ASSY

FRONT OF ENGINE

VIEW A

101571

Intake manifold, lower — 2.3L engine

3. Relieve the fuel system pressure as follows:

a. Remove the fuel tank cap to relieve the pressure in the fuel tank.

b. Remove the cap from the Schrader valve located on the fuel supply manifold.

c. Attach fuel pressure gauge T80L-9974-A or equivalent, to the Schrader valve and drain the fuel through the drain tube into a suitable container.

d. After the fuel system pressure is relieved, remove the fuel pressure gauge and install the cap on the Schrader valve.

4. Loosen the hose clamp attaching the flex hose to the throttle body. Remove the air cleaner flex hose.

5. Label and disconnect the vacuum hoses from the throttle body.

6. Loosen the lower EGR tube nut and rotate the tube away from the valve.

7. Disconnect the throttle and TV cables from the throttle linkage.

8. Label and disconnect the throttle position sensor, air charge temperature sensor and idle speed control electrical connectors.

9. Disconnect the PCV hose and disconnect the alternator support brace.

10. Remove the throttle body retaining bolts and the throttle body.

11. Disconnect the fuel lines from the fuel supply manifold.

12. Label the electrical connectors and remove the fuel injection wiring harness from the engine. The intake manifold assembly can be removed with the fuel supply manifold and injectors in place.

13. Label and disconnect the spark plug wires from the spark plugs. Remove the distributor cap from the distributor.

14. Remove the rocker arm covers.

15. Disconnect the upper radiator hose and heater hoses.

16. Mark the position of the rotor in relation to the distributor housing and mark the position of the distributor housing in relation to the engine. Remove the distributor assembly.

17. Disconnect the engine coolant temperature sensor and temperature sending unit connector.

18. Loosen the intake valve rocker arm retaining bolt from cylinder No. 3 and rotate the rocker arm away from the valve stem and pushrod. Remove the pushrod.

19. Remove the intake manifold retaining bolts using a Torx® head socket. Use a suitable prybar to loosen the intake manifold. Pry upward using the area between the

thermostat and transaxle as a leverage point. Remove the manifold and old gaskets and seals.

To install:

20. Lightly oil all the attaching bolts and stud threads before installation. When using a silicone rubber sealer, assembly must occur within 15 minutes after the sealer has been applied. After this time, the sealer may start to set-up and its sealing quality may be reduced. In high temperature/humidity conditions, the sealant will start to set up in approximately 5 minutes.

21. Clean the gasket mating surfaces of the intake manifold and cylinder head. Lay a shop rag in the lifter valley to catch any gasket material. After scraping, carefully lift the cloth from the lifter valley, being careful not to let any particles enter the oil drain holes or cylinder head. If necessary, use a suitable solvent to remove old rubber sealant.

22. Apply a suitable silicone rubber sealer to the intersection of the cylinder block end rails and cylinder heads. Be careful not to let sealer that may block oil passages fall into the engine.

23. Install the front and rear intake manifold end seals in place and secure. Install the intake manifold gaskets, aligning the locking tabs to the

provisions on the cylinder head gaskets.

24. Carefully lower the intake manifold into position on the cylinder block and cylinder heads to prevent smearing the silicone sealer and causing gasket voids.

25. Install the bolts and tighten in sequence. Torque the bolts first to 11 ft. lbs. (15 Nm) and then to 19–24 ft. lbs. (26–32 Nm).

26. Install the fuel supply manifold and injectors, if removed. Apply lubricant to the injector holes in the intake manifold and fuel supply manifold prior to injector installation. Install the fuel supply manifold retaining bolts and tighten to 7 ft. lbs. (10 Nm).

27. Install the thermostat housing and a new gasket, if removed. Tighten the retaining bolts to 9 ft. lbs. (12 Nm).

28. Install the distributor assembly, aligning the marks that were made during the removal procedure.

29. Install the No. 3 cylinder intake valve pushrod. Apply Lubriplate® or equivalent, to the pushrod and valve stem prior to installation. Turn the crankshaft as necessary to position the lifter on the base circle of the camshaft (pushrod all the way down). Tighten the rocker arm bolt in 2 steps, first to 8 ft. lbs. (11 Nm) and then to 24 ft. lbs. (32 Nm).

30. Install the rocker arm covers. Install the fuel injector harness and attach to the injectors.

31. Install the upper intake manifold and throttle body with new gaskets. Tighten the retaining bolts to 15–22 ft. lbs. (20–30 Nm).

32. Connect the PCV line at the PCV valve.

33. Connect all necessary electrical connections and vacuum lines.

34. Connect the EGR tube and the fuel lines.

35. Install the coil and bracket.

36. Install the upper radiator and heater hose.

37. Install and connect the air cleaner assembly and outlet tube.

38. Fill the cooling system.

39. Connect the negative battery cable.

40. Start the engine and check for coolant, fuel and oil leaks.

41. Check and if necessary, adjust the engine idle speed, transaxle throttle linkage and speed control.

Exhaust Manifold

REMOVAL AND INSTALLATION

2.3L Engine

1. Disconnect the negative battery cable.

2. Properly relieve the fuel system pressure.

3. Drain the cooling system.

4. Remove the accelerator cable and position to the side.

5. Remove air cleaner assembly and heat stove tube at heat shield.

6. Identify, tag and disconnect all necessary vacuum lines.

7. Disconnect the exhaust pipe-to-exhaust manifold retaining nuts. Remove exhaust manifold heat shield, if equipped.

8. Disconnect the oxygen sensor wire at the connector.

9. Disconnect the throttle linkage.

10. Disconnect the cruise control cable, if equipped.

11. Disconnect the fuel supply and return lines at the rubber connector.

12. Disconnect EGR tube from the EGR valve.

13. Remove the intake manifold.

14. Remove the exhaust manifold retaining nuts. Remove the exhaust manifold from the vehicle.

To install:

15. Position exhaust manifold to the cylinder head using guide bolts in holes 2 and 3.

16. Install the attaching bolts in the remaining holes.

17. Tighten the attaching bolts until snug, then remove guide bolts and install the remaining attaching bolts.

18. Tighten all exhaust manifold bolts to specification using the following tightening procedure: torque retaining bolts in sequence to 5–7 ft. lbs. (7–10 Nm) then retorque, in sequence, to 20–30 ft. lbs. (27–41 Nm).

19. Install the intake manifold gasket and bolts. Torque the intake manifold retaining bolts, in the proper sequence to 15–22 ft. lbs. (20–30 Nm).

20. Connect the oxygen sensor wire at the connector.

21. Connect the EGR tube to EGR valve.

22. Install exhaust manifold studs.

23. Connect exhaust pipe to exhaust manifold. Install the exhaust manifold heat shield, if equipped.

24. Connect the fuel supply and return lines.

25. Install vacuum lines.

26. Install air cleaner assembly.

27. Install accelerator cable and cruise control cable, if equipped.

28. Connect the negative battery cable.

29. Fill the cooling system.

30. Start engine and check for leaks.

3.0L Engine

Left Side (Front)

1. Disconnect the negative battery cable.

2. Remove the 2 retaining nuts and remove the heat shield.

3. Remove the engine oil dipstick tube or rotate it out of the way.

4. Remove the exhaust pipe retaining nuts from the exhaust manifold studs.

5. Remove the exhaust manifold retaining bolt and stud.

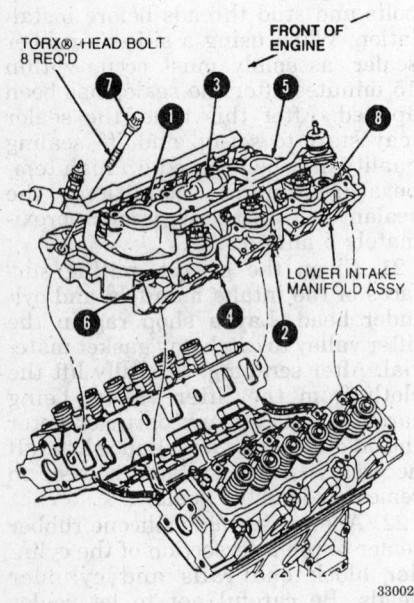

Intake manifold torque sequence — 3.0L engine

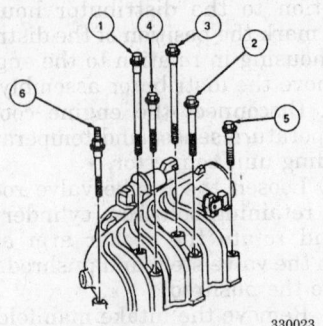

Upper intake manifold and throttle body installation — 3.0L engine

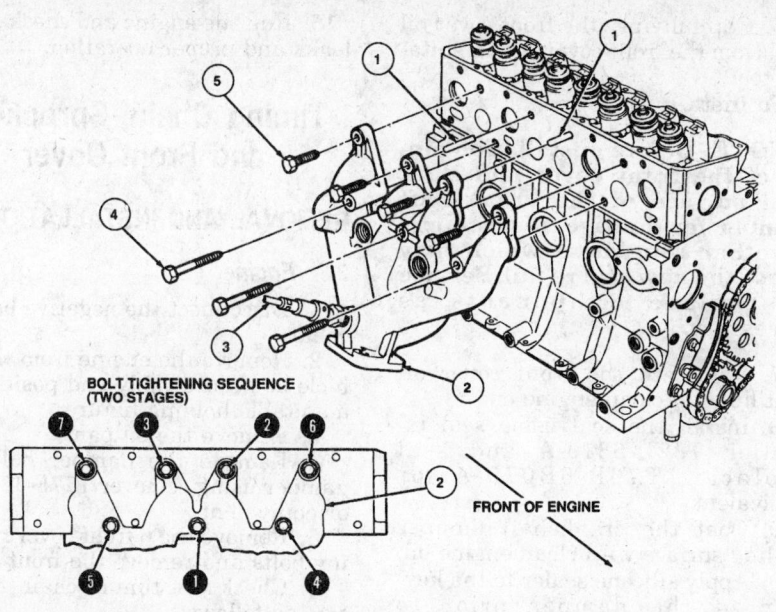

BOLT TIGHTENING SEQUENCE
(TWO STAGES)

FRONT OF ENGINE

Item	Part Number	Description
1	T84P-6065-B	Alignment Stud
2	9430	Exhaust Manifold
3	9F472	Heated Oxygen Sensor (HO2S)
4	—	Bolt M 10 x 1.5 x 8D (3 Req'd)
5	—	Bolt M 10 x 1.5 x 35 (4 Req'd)

100685

Exhaust manifold — 2.3L engine

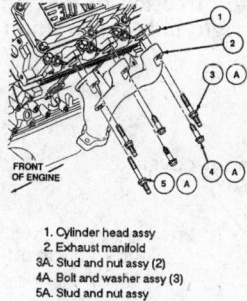

1. Cylinder head assy
2. Exhaust manifold
3A. Stud and nut assy (2)
4A. Bolt and washer assy (3)
5A. Stud and nut assy
A. 15-22 ft. lb.(20-30 Nm)

315256

Left exhaust manifold — 3.0L engine

6. Remove the manifold from the cylinder head, being careful not to damage the spark plugs.

To install:

7. Lightly oil all bolt and stud threads.

8. Clean the mating surfaces of the cylinder head, manifold and exhaust pipe.

9. Align the exhaust manifold studs with the exhaust pipe flange and install the exhaust manifold to the cylinder head. Install the retaining bolts and stud and tighten to 15–22 ft. lbs. (20–30 Nm).

10. Install the exhaust pipe retaining nuts and tighten to 25–34 ft. lbs. (34–47 Nm).

11. Rotate or install the dipstick tube bracket to the manifold retaining stud and tighten the nut to 11–14 ft. lbs. (15–20 Nm).

12. Install the heat shield and tighten the retaining nuts to 12–14 ft. lbs. (16–20 Nm).

13. Connect the negative battery cable. Start the engine and check for exhaust and oil leaks.

Right Side (Rear)

1. Disconnect the negative battery cable.

2. Drain the cooling system.

3. Disconnect the Pressure Feedback EGR (PFE) sensor hose connection to the EGR tube.

4. Loosen the EGR supply tube nuts at the manifold and EGR valve and remove the tube.

5. Remove the water pump.

6. Remove the exhaust pipe retaining nuts from the exhaust manifold studs.

7. Remove the heat shield retaining nuts and the shield.

8. Remove the exhaust manifold retaining bolts and studs. Remove the manifold from the cylinder head, being careful not to damage the spark plugs.

To install:

9. Lightly oil all bolt and stud threads.

10. Clean the mating surfaces of the cylinder head, manifold, exhaust pipe and EGR tube.

11. If installing a new manifold, install the EGR tube adapter/orfice, noting the small hole end (orfice) goes to the manifold.

12. Align the exhaust manifold studs with the exhaust pipe flange and install the exhaust manifold to the cylinder head. Install the retaining bolts and stud; then, tighten to 15–22 ft. lbs. (20–30 Nm).

13. Install the exhaust pipe retaining nuts and tighten to 25–34 ft. lbs. (34–47 Nm).

14. Install the heat shield and tighten the retaining nuts to 12–15 ft. lbs. (16–20 Nm).

15. Install the water pump.

16. Install the EGR tube and tighten the nuts to 26–48 ft. lbs. (35–65 Nm).

17. Connect the PFE hose to the EGR tube.

18. Fill and bleed the cooling system.

19. Connect the negative battery cable. Start the engine and check for coolant and exhaust leaks.

Front Cover Oil Seal

REMOVAL AND INSTALLATION

2.3L Engine

NOTE: The removal and installation of the front cover oil seal on the 2.3L engine can only be accomplished with the engine removed from the vehicle.

1. Remove the engine from the vehicle and position in a suitable holding fixture.

2. Remove bolt and washer at crankshaft pulley.

3. Remove the crankshaft pulley using a suitable puller.

4. Using a suitable tool, remove the front cover oil seal.

To install:

5. Coat a new seal with grease. Using a suitable driver, install the seal into the cover. Drive the seal in until it is fully seated. Check the seal after installation to be sure the spring is properly positioned in the seal.

6. Install crankshaft pulley, attaching bolt and washer. Torque the crankshaft pulley bolt to 140–170 ft. lbs. (190–230 Nm).

7. Install the engine in the vehicle.

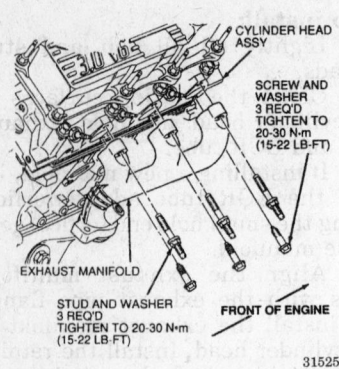

CYLINDER HEAD ASSY

SCREW AND WASHER 3 REQ'D TIGHTEN TO 20-30 N·m (15-22 LB-FT)

EXHAUST MANIFOLD

STUD AND WASHER 3 REQ'D TIGHTEN TO 20-30 N·m (15-22 LB-FT)

FRONT OF ENGINE

315259

Right exhaust manifold — 3.0L engine

3.0L Engine

1. Disconnect the negative battery cable.
2. Remove the engine from the vehicle.
3. Remove the accessory drive belts.
4. Remove the crankshaft damper retaining bolt and washer.
5. Remove the damper from the crankshaft using remover tool T58P-6316-d and adapter T82L-6316-B or equivalent.

NOTE: Use care not to damage the front cover or crankshaft.

6. Carefully pry the front cover oil seal from the front cover with a suitable tool.

To install:

NOTE: Before installation, inspect the front cover and shaft seal surface of the crankshaft damper for damage, nicks, burrs or other roughness which may cause the new seal to fail. Service or replace components as necessary.

7. Lubricate the front cover oil seal lip with clean engine oil.
8. Install the seal using seal installer T82L-6316-A and seal replacer T70P-6B070-A or equivalent.
9. Coat the crankshaft damper sealing surface with clean engine oil.
10. Apply silicone sealer to the keyway of the damper prior to installation.
11. Install the damper using seal installer T82L-6316-A or equivalent.
12. Install the damper retaining bolt and washer. Tighten to 93–121 ft. lbs. (125–165 Nm).
13. Install the accessory drive belts.
14. Install the engine into the vehicle.
15. Connect the negative battery cable.

16. Run the engine and check for oil leaks and proper operation.

Timing Chain, Sprockets and Front Cover

REMOVAL AND INSTALLATION

2.3L Engine

1. Disconnect the negative battery cable.
2. Remove the engine from the vehicle as an assembly and position on a suitable holding fixture.
3. Remove the oil pan.
4. Remove the damper bolt and damper using remover T77F-4220-B1 or equivalent.
5. Remove the 6 front cover retaining bolts and remove the front cover.
6. Check the timing chain deflection as follows:
   a. Rotate the crankshaft counterclockwise, as viewed from the front of the engine, to take up slack on the left side of chain.
   b. Make a reference mark on the block at approximately the midpoint of the chain. Measure from this point to the chain.
   c. Rotate the crankshaft in the opposite direction to take up slack on the right side of the chain. Force

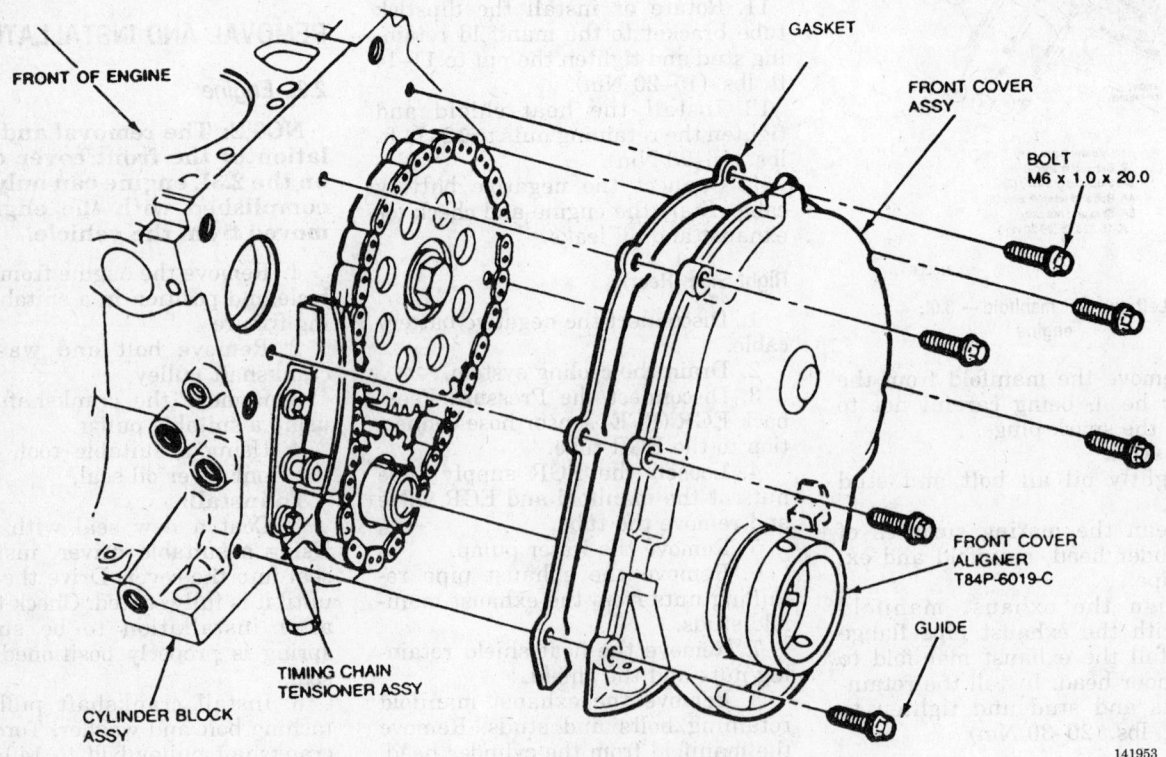

FRONT OF ENGINE

GASKET

FRONT COVER ASSY

BOLT M6 x 1.0 x 20.0

FRONT COVER ALIGNER T84P-6019-C

GUIDE

TIMING CHAIN TENSIONER ASSY

CYLINDER BLOCK ASSY

141953

With engine out of vehicle, the front cover may be removed — 2.3L engine

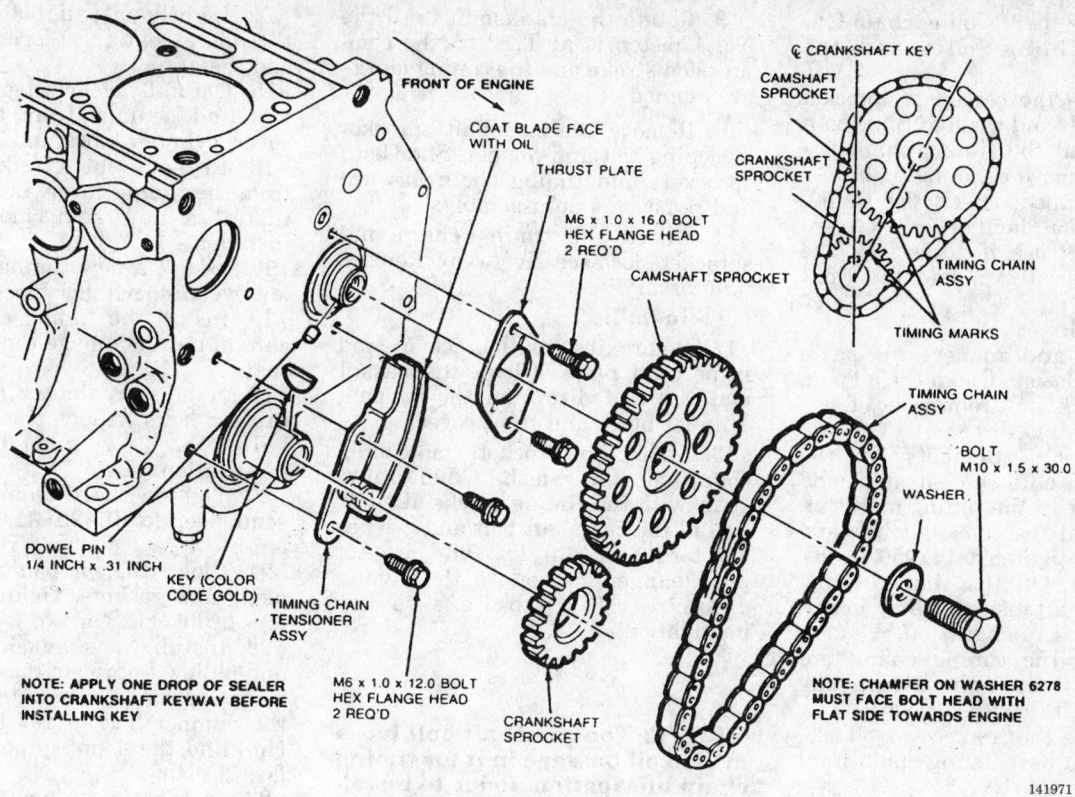

Timing chain arrangement — 2.3L engine

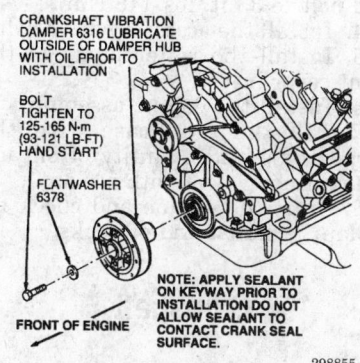

Crankshaft damper installation — 3.0L engine

the left side of the chain out with fingers and measure the distance between the reference point and the chain. The deflection is the difference between the 2 measurements.

d. If the deflection measurement exceeds 0.5 inch (12.7mm), replace the timing chain and sprockets. If wear on the tensioner face exceeds 0.06 inch (1.5mm), replace the tensioner.

7. Turn the crankshaft until the timing marks are aligned.

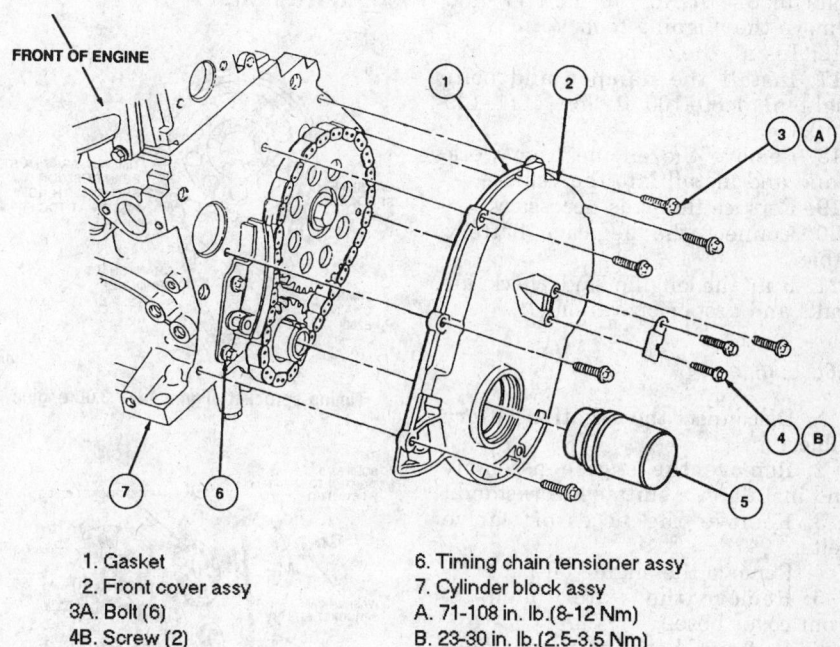

1. Gasket
2. Front cover assy
3A. Bolt (6)
4B. Screw (2)
5. Pilot tool
6. Timing chain tensioner assy
7. Cylinder block assy
A. 71-108 in. lb.(8-12 Nm)
B. 23-30 in. lb.(2.5-3.5 Nm)

Timing chain front cover and related components — 2.3L engine

8. Remove the 2 timing chain tensioner retaining bolts and the tensioner.

9. Remove the camshaft sprocket attaching bolt and washer. Slide both sprockets and the timing chain forward and remove as an assembly.

10. If equipped, check the timing chain vibration damper for excessive wear and replace if necessary. The damper is located inside the front cover.

To install:

11. Clean and inspect all parts before installation. Clean the oil pan, cylinder block and front cover of gasket material and dirt.

12. Slide both sprockets and the timing chain onto the camshaft and crankshaft with the timing marks aligned. Install the camshaft bolt and washer and tighten to 41–56 ft. lbs. (55–75 Nm). Oil the timing chain, sprockets and tensioner after installation with clean engine oil.

13. Install the timing chain tensioner and tighten the 2 retaining bolts to 6–8 ft. lbs. (8–12 Nm).

14. Replace the front cover oil seal and install a new timing chain front cover gasket.

15. Install the timing chain front cover using aligning tool T81P-6019-C. Tighten the retaining bolts to 6–8 ft. lbs. (8–12 Nm) and remove the aligning tool.

16. Install the oil pan.

17. Install the damper and bolt. Tighten to 80–100 ft. lbs. (111–139 Nm).

18. Remove the engine from work stand and install into the vehicle.

19. Replace fluids as necessary.

20. Connect the negative battery cable.

21. Run the engine and check for leaks and proper operation.

3.0L Engine

1. Disconnect the negative battery cable.

2. Remove the engine assembly and install on a suitable workstand.

3. Remove the accessory drive belts.

4. Remove the engine oil pan.

5. Remove the water pump-to-front cover hose.

6. Remove both belt tensioner assemblies.

7. Remove the vibration damper using a suitable puller.

8. Remove the front cover retaining bolts and remove the front cover. If replacing the front cover, transfer the engine mount to the cover mounting pad.

9. Rotate the crankshaft until the No. 1 piston is at TDC on the compression stroke and the timing marks are aligned.

10. Remove the camshaft sprocket attaching bolt and washer. Slide both sprockets and timing chain forward and remove as an assembly.

11. Check the timing chain and sprockets for excessive wear. Replace if necessary.

To install:

12. Before installation, clean and inspect all parts. Clean the gasket material and dirt from the oil pan, cylinder block and front cover.

13. Slide both sprockets and timing chain onto the camshaft and crankshaft with the timing marks aligned. Install the camshaft bolt and washer and torque to 46 ft. lbs. (63 Nm). Apply clean engine oil to the timing chain and sprockets after installation.

NOTE: The camshaft bolt has a drilled oil passage in it for timing chain lubrication. Prior to installation, clean the passage and make sure it is clear. Never replace the camshaft bolt with a standard bolt.

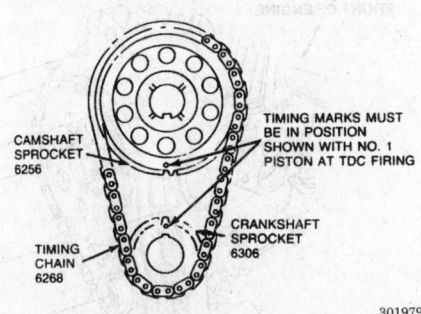

CAMSHAFT SPROCKET 6256

TIMING MARKS MUST BE IN POSITION SHOWN WITH NO. 1 PISTON AT TDC FIRING

TIMING CHAIN 6268

CRANKSHAFT SPROCKET 6306

301979

Timing sprocket alignment — 3.0L engine

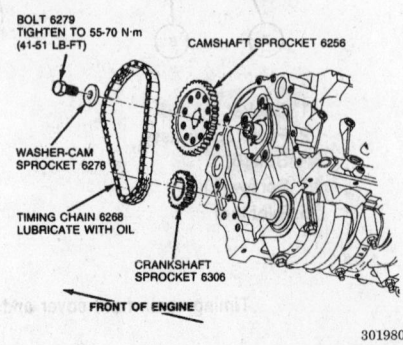

BOLT 6279 TIGHTEN TO 55-70 N·m (41-51 LB-FT)

CAMSHAFT SPROCKET 6256

WASHER-CAM SPROCKET 6278

TIMING CHAIN 6268 LUBRICATE WITH OIL

CRANKSHAFT SPROCKET 6306

FRONT OF ENGINE

301980

Timing chain and sprocket installation — 3.0L engine

14. Lightly oil all bolt and stud threads except those specifying special sealant.

15. Carefully clean all gasket material and sealant from the timing cover, cylinder block and oil pan.

16. Pry out the crankshaft seal from the timing cover. Lubricate and install a new seal, using a seal installer.

17. Install a new timing cover gasket over the cylinder block dowels.

18. Install the timing cover, being careful not to damage the crankshaft seal.

19. Hand start the timing cover retaining bolts. Apply pipe sealant to bolt No. 1, 2 and 3 prior to installation.

20. Tighten the retaining bolts, in sequence, to 15–22 ft. lbs. (20–30 Nm).

21. Clean the oil pan and install, using new gaskets. Tighten the bolts to 9 ft. lbs. (12 Nm).

22. Install the crankshaft damper and pulley. Lubricate the seal mating surface prior to installation. Tighten the damper bolt to 107 ft. lbs. (145 Nm) and the 4 pulley bolts to 26 ft. lbs. (35 Nm).

23. Install the automatic belt tensioners. Tighten the retaining nuts and bolt to 35 ft. lbs. (48 Nm).

24. Install the accessory drive belts.

25. Install the water pump to the front cover hose.

26. Install the engine assembly.

27. Fill the crankcase with the proper type and quantity of oil and the cooling system with coolant.

28. Start the engine and check for coolant, exhaust and oil leaks.

Camshaft

REMOVAL AND INSTALLATION

2.3L Engine

1. Disconnect the negative battery cable.

2. Drain the cooling system and crankcase. Properly relieve the fuel system pressure.

3. Remove the engine from the vehicle and position in a suitable holding fixture. Remove the engine oil dipstick.

4. Remove necessary drive belts and pulleys.

5. Remove the cylinder head.

6. Remove the distributor.

7. Using a magnet, remove the hydraulic lifters and label them so they can be installed in their original positions. If the lifters are stuck in the

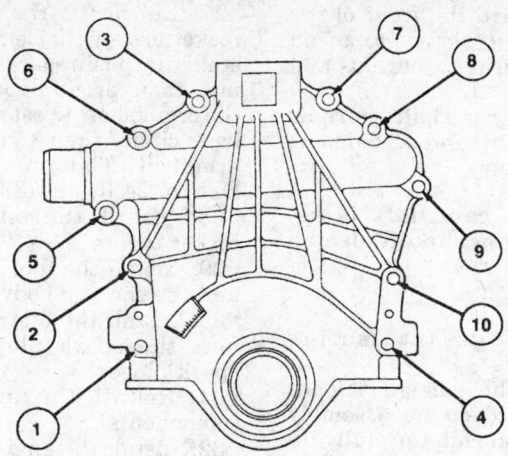

Timing chain front cover bolt identification and torque sequence — 3.0L engine

Hole No.	Bolt Size	Torque
1-2	M8 x 1.25 x 33	20-30 N·m (15-22 Lb-Ft)
3-4-5-6-7-8-9-10-11-12-13 Front Cover Assy	M8 - 1.25 x 53	20-30 N·m (15-22 Lb-Ft)

301961

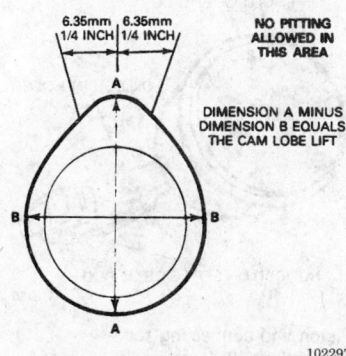

Camshaft inspection — 2.3L engine

102297

bores by excessive varnish, etc., use a suitable puller to remove them.

8. Remove the crankshaft pulley.

9. Remove the oil pan.

10. Remove the cylinder front cover and gasket.

11. Check the camshaft endplay as follows:

   a. Push the camshaft toward the rear of the engine and install a dial indicator tool, so the indicator foot is positioned on the camshaft sprocket attaching bolt.

   b. Zero the dial indicator. Position a small prybar or equivalent, between the camshaft sprocket or gear and block.

   c. Pull the camshaft forward and release it. Maximum camshaft endplay specification is 0.009 in.

   d. If the camshaft endplay is over the amount specified, replace the thrust plate.

12. Remove the timing chain, sprockets and timing chain tensioner. Remove the VRS sensor (located near base of distributor on the block).

13. Remove camshaft thrust plate. Carefully remove the camshaft by pulling it toward the front of the engine. Use caution to avoid damaging bearings, journals and lobes.

To install:

14. Clean and inspect all parts before installation.

15. Lubricate camshaft lobes and journals with heavy engine oil. Carefully slide the camshaft through the bearings in the cylinder block.

16. Install the thrust plate. Tighten attaching bolts to 6–9 ft. lbs. (8–12 Nm).

17. Install the timing chain, sprockets and timing chain tensioner. Install the VRS sensor (located near base of distributor on the block).

18. Install the cylinder front cover and crankshaft pulley.

19. Clean the oil pump inlet tube screen, oil pan and cylinder block gasket surfaces. Prime oil pump by filling the inlet opening with oil and

rotate the pump shaft until oil emerges from the outlet tube. Install oil pump, oil pump inlet tube screen and oil pan.

20. Install the accessory drive belts and pulleys.

21. Lubricate the lifters and lifter bores with heavy engine oil. Install lifters into their original bores.

22. Install cylinder head.

23. Position No. 1 piston at TDC after the compression stroke. Position distributor in the block with the rotor at the No. 1 firing position. Install distributor retaining clamp.

24. Install engine in vehicle.

25. Fill the cooling system and crankcase to the proper levels.

26. Connect negative battery cable.

27. Start the engine. Check and adjust ignition timing. Check for leaks.

3.0L Engine

— **CAUTION** —

Fuel injection systems remain under pressure, even after the engine has been turned OFF. The fuel system pressure must be relieved before disconnecting any fuel lines. Failure to do so may result in fire and/or personal injury.

1. Disconnect the negative battery cable.

2. Drain the cooling system and crankcase.

3. Relieve the fuel system pressure as follows:

   a. Remove the fuel tank cap to relieve the pressure in the fuel tank.

   b. Remove the cap from the Schrader valve located on the fuel supply manifold.

   c. Attach fuel pressure gauge T80L-9974-A or equivalent, to the Schrader valve and drain the fuel through the drain tube into a suitable container.

   d. After the fuel system pressure is relieved, remove the fuel pressure gauge and install the cap on the Schrader valve.

4. Remove the engine from the vehicle and position in a suitable holding fixture.

5. Remove the accessory drive components from the front of the engine.

6. Remove the throttle body and the fuel injector harness.

7. Label and disconnect the spark plug wires from the spark plugs.

8. Remove the distributor assembly.

9. Remove the rocker arm covers.

10. Loosen the rocker arm fulcrum nuts and position the rocker arms to the side for easy access to the pushrods. Remove the pushrods and label so they may be installed in their original positions.

11. Remove the intake manifold leaving the fuel supply manifold and injectors in place.

12. Using a suitable magnet or lifter removal tool, remove the hydraulic lifters and keep them in order so they can be installed in their original positions. If the lifters are stuck in the bores by excessive varnish, use a hydraulic lifter puller to remove the lifters.

13. Remove the crankshaft pulley and damper using a suitable removal tool.

14. Remove the oil pan.

15. Remove the front cover assembly.

16. Align the timing marks on the camshaft and crankshaft sprockets. Check the camshaft end-play as follows:

a. Push the camshaft toward the rear of the engine and install a dial indicator tool, so the indicator point is on the camshaft sprocket attaching screw.

b. Zero the dial indicator. Position a small prybar or equivalent, between the camshaft sprocket and block.

c. Pull the camshaft forward and release it. Compare the dial indicator reading with the camshaft end-play service limit specification of 0.005 in. (0.13mm).

d. If the camshaft end-play is over the amount specified, replace the thrust plate.

17. Remove the timing chain and sprockets.

18. Remove the camshaft thrust plate. Carefully remove the camshaft by pulling it toward the front of the engine. Remove it slowly to avoid damaging the bearings, journals and lobes.

19. Inspect the camshaft journals and lobes for wear and/or damage. Replace as necessary.

NOTE: If the camshaft is replaced, new lifters should also be installed.

To install:

20. Clean all gasket mating surfaces.

21. Lubricate the camshaft lobes and journals with engine assembly lube or clean engine oil. Carefully insert the camshaft through the bearings into the cylinder block.

22. Install the thrust plate. Tighten the retaining bolts to 84 inch lbs. (10 Nm).

23. Install the timing chain and sprockets. Tighten the camshaft sprocket retaining bolt to 46 ft. lbs. (63 Nm).

───── CAUTION ─────
The camshaft bolt has a drilled oil passage it it for timing chain lubrication. Make sure the passage is clean prior to bolt installation. If the bolt is damaged, do not replace the camshaft bolt with a standard bolt or engine damage may result.

24. Install the front timing cover and crankshaft damper and pulley. Tighten the crankshaft damper bolt to 107 ft. lbs. (145 Nm).

25. Lubricate the lifters and lifter bores with clean engine oil. Install the lifters into their original bores.

26. Install the intake manifold assembly.

27. Lubricate the pushrods and rocker arms with clean engine oil. Install the pushrods and rocker arms into their original positions. Rotate the crankshaft to set each lifter on its base circle, then tighten the rocker arm bolt. Tighten the rocker arm bolts to 24 ft. lbs. (32 Nm).

28. Install the oil pan and the rocker covers.

29. Install the fuel injector harness and the throttle body.

30. Install the distributor and connect the spark plug wires to the spark plugs.

31. Install the accessory drive components.

32. Install the engine assembly.

33. Restore all fluid levels.

34. Connect the negative battery cable.

35. Start the engine and check for leaks.

36. Check and adjust the ignition timing.

Piston and Connecting Rod

POSITIONING

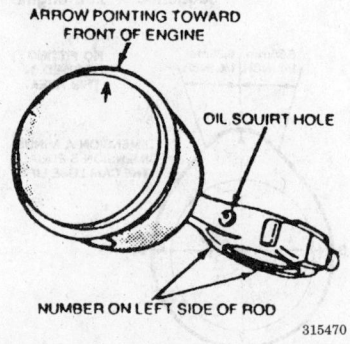

Piston and connecting rod installation — 2.3L engine

Camshaft and crankshaft sprocket alignment — 3.0L engine

CAMSHAFT SPROCKET 6256
TIMING MARKS MUST BE IN POSITION SHOWN WITH NO. 1 PISTON AT TDC FIRING
CRANKSHAFT SPROCKET 6306
TIMING CHAIN 6268

298452

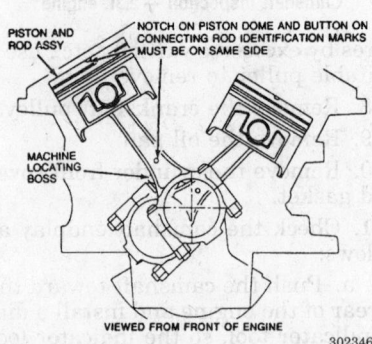

Camshaft installation — 3.0L engine

CAMSHAFT THRUST PLATE-6269 APPLY OIL TO BOTH SIDES PRIOR TO INSTALLATION
FRONT OF ENGINE
BOLT 2 REQ'D TIGHTEN TO 10 N·m (7 LB-FT)
CAMSHAFT 6250 COAT ALL CAMSHAFT JOURNALS OR CAMSHAFT BEARINGS WITH OIL PRIOR TO INSTALLATION

298453

Piston and connecting rod installation — 3.0L engine

PISTON AND ROD ASSY
NOTCH ON PISTON DOME AND BUTTON ON CONNECTING ROD IDENTIFICATION MARKS MUST BE ON SAME SIDE
MACHINE LOCATING BOSS
VIEWED FROM FRONT OF ENGINE

302346

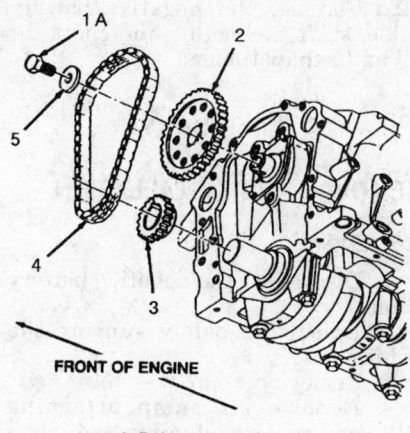

FRONT OF ENGINE

1. Bolt
2. Camshaft sprocket
3. Crankshaft sprocket
4. Timing chain
5. Washer
A. Tighten to 50-70 N.m
   (37-51 lb-ft)

298454

Timing chain and sprocket installation — 3.0L engine

NO. 3 INT
NO. 3 EXH
NO. 2 INT
NO. 1 EXH
© OF ENGINE
NO. 6 EXH
NO. 6 INT
NO. 5 INT
NO. 4 EXH
TIMING MARKS
CYLINDER HEAD
CAMSHAFT POSITION B
CYLINDER HEAD

BOLT
12 REQ'D
TIGHTEN IN
TWO STEPS:
A: 7-15 N·m (5-11 LB-FT)
B: 26-38 N·m (19-28 LB-FT)
ROCKER ARM
12 REQ'D
CYLINDER HEAD
CYLINDER HEAD
FULCRUM
12 REQ'D
NO. 1 INT
NO. 5 EXH
PUSH ROD
12 REQ'D
NO. 4 INT
TIMING MARKS
NO. 2 EXH
CAMSHAFT POSITION A
RIGHT SIDE
ROCKER ARM
SEAT AND BOLT
MUST BE FULLY SEATED
AFTER FINAL TORQUE
LEFT SIDE
2.15-4.69mm (0.085-0.185 INCH) WITH
VALVE TAPPET FULLY COLLAPSED
ON BASE CIRCLE OF CAMSHAFT
LOBE AFTER ASSEMBLY

298455

Camshaft position for rocker arm bolt tightening — 3.0L engine

ENGINE LUBRICATION

Oil Pan

REMOVAL AND INSTALLATION

2.3L Engine

1. Disconnect the negative battery cable. Raise the vehicle and support safely.
2. Drain the crankcase and drain the cooling system by removing the lower radiator hose.
3. Remove the roll restrictor on manual transaxle equipped vehicles.
4. Disconnect the starter cable and remove the starter.
5. Disconnect the exhaust pipe from oil pan.
6. Remove the engine coolant tube from the lower radiator hose, the water pump and at the tabs on the oil pan. Position air conditioner line off to the side.
7. Remove the retaining bolts and remove the oil pan.
To install:
8. Clean both mating surfaces of oil pan and cylinder block making certain all traces of RTV sealant are removed. Ensure that the block rails, front cover and rear cover retainer are also clean.
9. Remove and clean oil pump pickup tube and screen assembly. After cleaning, install tube and screen assembly.
10. Apply RTV E8AZ-19562-A Sealer or equivalent, in oil pan groove. Completely fill oil pan groove with sealer. Sealer bead should be 0.200 in. (5mm) wide and 0.080-0.150 in. (2.0-3.8mm) high, above oil pan surface, in all areas except the half-rounds. The half-rounds should have a bead 0.200 in. (5mm) wide and 0.150-0.200 in. (3.8-5.1mm) high, above the oil pan surface.

NOTE: Applying RTV in excess of the specified amount will not improve the sealing of the oil pan, and could cause the oil pickup screen to become clogged with sealer. Use adequate ventilation when applying sealer.

11. Install oil pan to cylinder block within 5 minutes to prevent skinning over. RTV needs to cure completely before coming in contact with any engine oil, about 1 hour at ambient temperature between 65-75°F.
12. Install oil pan bolts lightly until the 2 oil pan-to-transmission bolts can be installed.

NOTE: If oil pan is installed on engine outside of vehicle, a transaxle case or equivalent fixture must be bolted to the block to align the oil pan with the rear face of block.

13. Install 2 oil pan-to-transaxle bolts. Tighten to 30-39 ft. lbs. (40-54 Nm) to align oil pan with transaxle. Loosen bolts ½ turn.
14. Tighten all oil pan flange bolts to 15-22 ft. lbs. (20-30 Nm).
15. Tighten 2 oil pan-to-transmission bolts to 30-39 ft. lbs. (40-54 Nm).
16. If required, rework exhaust bracket to fit to oil pan.
17. Replace water inlet tube O-ring and install tube.
18. Install the starter. and connect the starter cable.
19. Install roll restrictor, if equipped.
20. Lower vehicle.
21. Fill the crankcase with the proper type and quantity of engine oil. Fill and bleed the cooling system.
22. Connect negative battery cable.
23. Start engine and check for coolant and oil leaks.

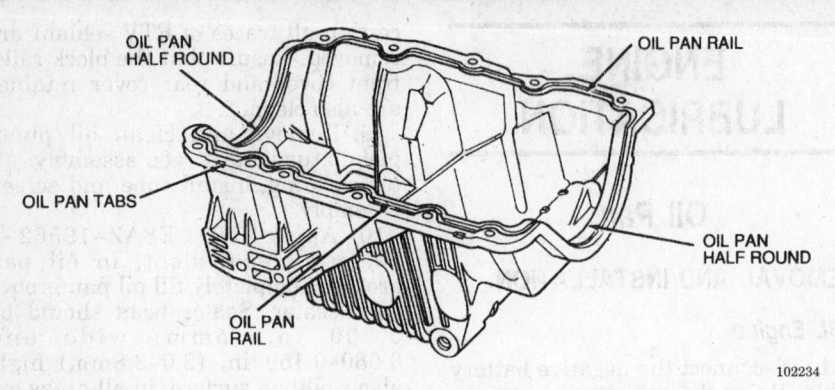

Oil pan — 2.3L engine

102234

3.0L Engine

1. Disconnect the negative battery cable.
2. Remove the engine oil dipstick.
3. Raise and safely support the vehicle.
4. If equipped, remove the low oil level sensor retainer clip and disconnect the electrical connector at the sensor.
5. Drain the engine oil from the crankcase into a suitable container.
6. Remove the starter.
7. Disconnect the exhaust gas oxygen sensors.
8. Remove the catalytic converter and exhaust pipe assembly.

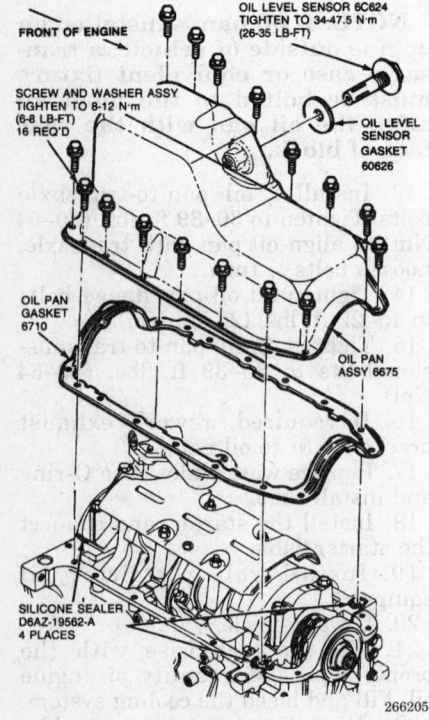

FRONT OF ENGINE

OIL LEVEL SENSOR 6C624
TIGHTEN TO 34-47.5 N·m
(26-35 LB-FT)

SCREW AND WASHER ASSY
TIGHTEN TO 8-12 N·m
(6-8 LB-FT)
16 REQ'D

OIL LEVEL
SENSOR
GASKET
60626

OIL PAN
GASKET
6710

OIL PAN
ASSY 6675

SILICONE SEALER
D6AZ-19562-A
4 PLACES

266205

Oil pan installation — 3.0L engine

9. If equipped with automatic transaxle, remove the torque converter access plate from the transaxle. If equipped with manual transaxle, remove the left and right transaxle support plates.
10. Remove the oil pan retaining bolts and remove the oil pan, making sure the internal pan baffle does not snag on the oil pump pickup tube and screen. Remove and discard the oil pan gasket.

To install:
11. Clean the oil pan and all gasket mating surfaces.
12. Install a new oil pan gasket on the cylinder block using the retaining features and a suitable gasket adhesive. Snug retaining bolts at the 4 corners and 2 middle places on the cylinder block to support the gasket until the adhesive cures.
13. Apply a $3/16$ in. (4.75mm) bead of silicone sealer to the junction of the rear main bearing cap and cylinder block and the junction of the front cover and cylinder block.

NOTE: Do not let the sealer cure longer than 4 minutes prior to oil pan installation or 7 total minutes before bolts are tightened to specification.

14. Position the oil pan and install the retaining bolts, hand tight.
15. Tighten the 4 corner bolts to 7–10 ft. lbs. (10–14 Nm), then tighten the remaining bolts to the same specification.
16. If equipped with automatic transaxle, install the torque converter access plate. If equipped with manual transaxle, install the left and right transaxle plates.
17. Install the catalytic converter and pipe assembly. Connect the oxygen sensors.
18. Install the starter.
19. If equipped, connect the low oil level sensor electrical connector and install the retainer clip.
20. Lower the vehicle.

21. Fill the crankcase with the proper type and quantity of engine oil. Install the dipstick.
22. Connect the negative battery cable, start the engine and check for oil and exhaust leaks.

Oil Pump

REMOVAL AND INSTALLATION

2.3L Engine

1. Disconnect the negative battery cable.
2. Raise and safely support the vehicle.
3. Remove oil pan.
4. Remove oil pump attaching bolts and remove oil pump and intermediate driveshaft.

To install:
5. Prime oil pump by filling inlet port with engine oil. Rotate pump shaft until oil flows from outlet port.
6. If screen and cover assembly have been removed, replace gasket. Clean screen and reinstall screen and cover assembly. Tighten attaching bolts and nut.
7. Position intermediate driveshaft into distributor socket.
8. Insert intermediate driveshaft into oil pump. Install pump and shaft as an assembly.

--- **CAUTION** ---

Do not attempt to force the pump into position if it will not seat. The shaft hex may be misaligned with the distributor shaft. To align, remove the oil pump and rotate the intermediate driveshaft into a new position.

9. Tighten the oil pump attaching bolts to 15–22 ft. lbs. (20–30 Nm).
10. Install oil pan with new gasket.
11. Connect negative battery cable.
12. Fill the crankcase with the proper type and quantity of engine oil. Start engine and check for leaks.

3.0L Engine

1. Disconnect the negative battery cable.
2. Raise and safely support the vehicle.
3. Drain the crankcase into a suitable container and remove the oil pan.
4. Remove the oil pump attaching bolts. Lift the oil pump off the engine and withdraw the oil pump driveshaft.

To install:
5. Prime the oil pump by filling either the inlet or the outlet port with

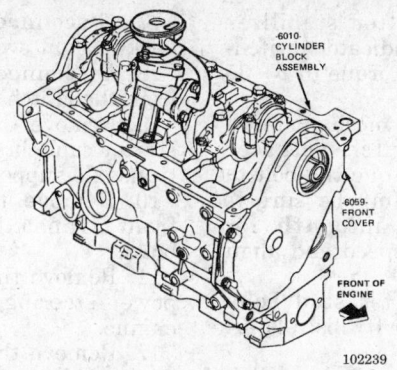

Oil pump — 2.3L engine

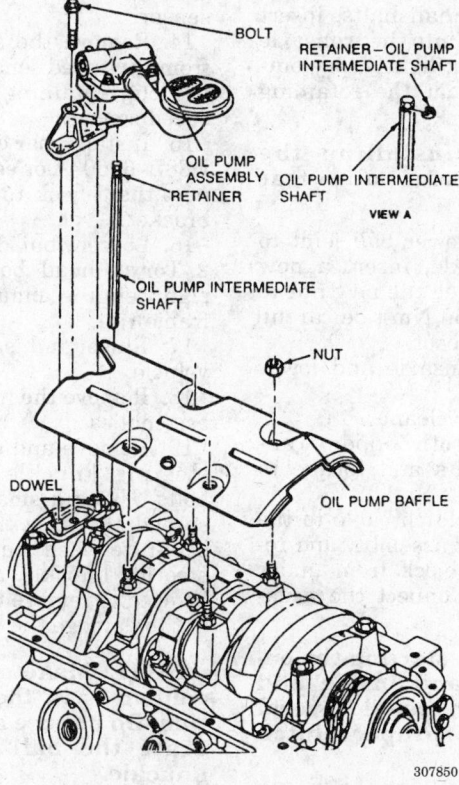

Oil pump installation — 3.0L engine

engine oil. Rotate the pump shaft to distribute the oil within the oil pump body cavity.

6. Insert the oil pump intermediate shaft assembly into the hex drive hole in the oil pump assembly until the retainer "clicks" into place. Place the oil pump in the proper position and install the retaining bolt.

7. Torque the oil pump retaining bolt to 35 ft. lbs. (48 Nm).

8. Install the oil pan with a new gasket.

9. Lower the vehicle.

10. Fill the crankcase with the proper type and quantity of engine

oil. Start engine and check for leaks and proper oil pressure.

TRANSAXLE

Manual Transaxle Assembly

REMOVAL AND INSTALLATION

2.3L Engine

1. Disconnect the negative battery cable. Wedge a 7 in. wooden block under the clutch pedal to hold the pedal up slightly beyond its normal position. Grasp the clutch cable, pull it forward and disconnect it from the clutch release shaft assembly. Remove the clutch casing from the rib on the top surface of the transaxle case.

2. Remove the upper 2 transaxle-to-engine bolts. Remove the air cleaner.

3. Raise and safely support the vehicle.

4. Remove the front stabilizer bar-to-control arm nut and washer, on the driver's side and discard the nut. Remove both front stabilizer bar mounting brackets and discard the bolts.

5. Remove the lower control arm ball joint-to-steering knuckle nut/bolt and discard the nut/bolt; repeat this procedure on the opposite side.

6. Using a large prybar, pry the lower control arm from the steering knuckle; repeat this procedure on the opposite side.

NOTE: Be careful not to damage or cut the ball joint boot and do not contact the lower arm.

7. Using a large prybar, pry the left-side inboard CV-joint assembly from the transaxle.

NOTE: Plug the seal opening (both sides), with transaxle plugs T81P–1177–B or equivalent, to prevent lubricant leakage.

8. Grasp the left-hand steering knuckle and swing it and the half-shaft outward from the transaxle; this will disconnect the inboard CV-joint from the transaxle.

NOTE: If the CV-joint assembly cannot be pried from the transaxle, insert a differential rotator tool through the left-side and tap the joint out; the tool can be used from either side of transaxle.

9. Using mechanics wire, support the halfshaft in a near level position to prevent damage to the assembly

during the remaining operations; repeat this procedure on the opposite side.

10. Disengage the locking tabs and remove the backup light switch connector from the transaxle backup light switch.

11. Remove the 3 nuts from the starter mounting studs. Remove the ground strap. Remove the starter stud bolts.

12. Remove the shift mechanism-to-shift shaft nut/bolt, the control selector indicator switch arm and the shift shaft.

13. Remove the shift mechanism stabilizer bar-to-transaxle bolt, control selector indicator switch and bracket assembly.

14. Using a crowsfoot wrench, remove the speedometer cable from the transaxle.

15. Remove both oil pan-to-clutch housing bolts.

16. Using a floor jack and a transaxle support, position it under the transaxle and secure the transaxle to it.

17. Remove the both left-hand rear No. 4 insulator-to-body bracket nuts and the left-hand front No. 1 insulator-to-body bracket bolts.

18. Lower the floor jack, until the transaxle clears the rear insulator. Support the engine by placing wood under the oil pan.

19. Remove the engine-to-transaxle bolts and lower the transaxle from the vehicle.

NOTE: On Tempo/Topaz, one of the engine-to-transaxle bolts attaches the ground strap and wiring loom stand off bracket.

To install:

20. Raise the transaxle into position and engage the input shaft with the clutch plate. Install the lower engine-to-transaxle bolts and torque to 28–31 ft. lbs. (38–42 Nm).

——— **WARNING** ———
Never attempt to start the engine prior to installing the CV-joints or differential side gear for dislocation and/or damage may occur.

21. Tighten the left front No. 1 insulator bolts to 25–35 ft. lbs. (34–47 Nm) and the left rear No. 4 insulator bolts to 35–50 ft. lbs. (47–68 Nm).

22. Remove the floor jack and adapter.

23. Using a crowsfoot wrench, install the speedometer cable; be careful not to cross-thread the cable nut.

24. Install the oil pan-to-transaxle bolts and tighten to 28–38 ft. lbs. (38–51 Nm).

25. Install the shifter stabilizer bar/control selector indicator switch-to-transaxle bolt and torque to 23–35 ft. lbs. (31–47 Nm).

26. Install the shift mechanism-to-shift shaft, the switch actuator bracket clamp and torque the bolt to 7–10 ft. lbs. (9–13 Nm); be sure to shift the transaxle into **4th** for 4-speed or **5th** for 5-speed and align the actuator.

27. Install the starter stud bolts and tighten to 30–40 ft. lbs. (41–54 Nm).

28. Install the backup light switch connector to the transaxle switch.

29. Install the new circlip onto both inner joints of the halfshafts, insert the inner CV-joints into the transaxle and fully seat them; lightly, pry outward to confirm that the retaining rings are seated.

NOTE: When installing the halfshafts, be careful not to tear the oil seals.

30. Connect the lower ball joint to the steering knuckle, insert a new pinch bolt and torque the new nut to 37–44 ft. lbs. (50–60 Nm); be careful not to damage the boot.

31. Refill the transaxle and lower the vehicle.

32. Install the air cleaner.

33. Install the both upper transaxle-to-engine bolts and torque to 28–31 ft. lbs. (38–42 Nm).

34. Connect the clutch cable to the clutch release shaft assembly and remove the wooden block from under the clutch pedal. Connect the negative battery cable.

NOTE: Prior to starting the engine, set the hand brake and pump the clutch pedal several times to ensure proper clutch adjustment.

3.0L Engine

1. Prop the clutch pedal to keep it from moving toward the floor when the clutch cable is disconnected.

2. Disconnect the negative battery cable.

3. Disconnect the mass air flow sensor and air charge temperature sensor connectors at the air cleaner.

4. Remove the air cleaner retaining bolt, loosen the outlet tube at the throttle body and remove the air cleaner assembly.

5. Remove the retaining bolts and remove the coil bracket assembly from the left cylinder head. Position the assembly aside.

6. Install engine lifting bracket tools T70P–6000 or equivalent, on the rear of the left cylinder head.

7. Disconnect and remove the backup light switch.

8. Disconnect the clutch cable from the clutch release lever.

9. Remove the nut attaching the starter cable bracket to the left front transaxle support.

10. Remove the 2 top left-hand front transaxle mount-to-engine bolts.

11. Remove the nuts attaching the power steering line bracket to the engine.

12. Remove the top 4 transaxle-to-engine bolts.

13. Disconnect the vehicle speed sensor connector from the speed sensor.

14. Remove the speedometer cable from the speed sensor. Do not remove the clip retaining the cable to the speed sensor.

15. Install 3 bar engine support tool D88L–6000–A or equivalent, and connect the J-hook to the engine lifting bracket.

16. Loosen, but do not remove, the 2 Torx® head bolts attaching the right engine mounts to the right frame rail.

17. Raise and safely support the vehicle.

18. Remove the front wheel and tire assemblies.

19. Remove and discard both lower steering knuckle ball joint pinch bolts. Using a small prybar, slightly spread the knuckle pinch joint and separate each ball joint from the steering knuckle. A drift punch may be used to remove the bolt. Be careful not to damage the ball joint boot seal.

NOTE: Make sure the steering column is in the unlocked position. Do not use a hammer to separate the ball joint from the knuckle.

20. Remove the cotter pins and the nuts from the tie-rod ends. Use a suitable tool to disconnect the tie-rod ends from the steering knuckles.

21. Use a suitable prybar to disengage the CV-joints from the transaxle. Install transaxle plugs T81P–1177–B or equivalent, to prevent transaxle fluid from leaking from the transaxle.

22. Remove the nuts attaching the halfshafts to the hubs and remove the halfshafts.

23. Remove the bolt attaching the shift lever linkage to the transaxle shift rod and position the linkage aside.

24. Remove the bolt attaching the stabilizer rod to the transaxle and position the stabilizer rod aside.

25. Disconnect the neutral sensing switch connector.

26. Remove the starter support bracket and the starter.

27. Loosen the a front retaining bolt from each side on the engine-to-transaxle support bracket.

28. Remove the 2 rear retaining bolts on the engine-to-transaxle bracket and remove the 2 bracket-to-transaxle bolts.

29. Remove the accessory drive belt tensioner and install damper tool T93P6316-A or equivalent.

30. From the right wheel well, loosen, but do not remove, 2 right engine mount retaining nuts.

31. Remove the lower retaining bolt from the left front transaxle mount. Loosen the through bolt and pivot the mount up away from the transaxle.

32. Remove the 2 retaining nuts from the rear transaxle mount.

33. Carefully lower the engine/transaxle assembly using the engine support fixture until the crankshaft damper just contacts the right frame rail.

NOTE: Do not let the weight of the engine rest on the crankshaft damper or the damper and crankshaft thrust bearings may be damaged.

34. Position a transaxle jack under the transaxle and install safety chains. Lower the transaxle.

35. Remove the 2 remaining transaxle-to-engine bolts.

36. Remove the transaxle from the vehicle. After clearing the clutch assembly, rotate the transaxle clutch housing toward the front of the vehicle to clear the suspension stabilizer bar.

NOTE: Do not move the vehicle with the wheels on the ground with the transaxle removed.

To install:

37. Position the transaxle on the transaxle jack with safety chains.

38. Raise the transaxle into position. Rotate the transaxle clutch housing to the front of the engine compartment to allow the rear of the transaxle to clear the suspension stabilizer bar.

39. Align the transaxle input shaft with the clutch splines and locating pin on the engine and seat the transaxle against the engine.

40. Install 2 transaxle-to-engine bolts.

41. Using the engine support fixture, raise the engine/transaxle assembly into position.

42. Position the engine-to-transaxle support bracket to the transaxle and install but do not tighten the 2 bolts, 1 on each side.

43. Install 2 rear engine support bracket-to-engine bolts, 1 on each side.

44. Tighten the support bracket-to-engine bolts.

45. From the right wheel well, tighten the 2 right engine mount nuts to 73–97 ft. lbs. (98–132 Nm).

46. Install the lower left front engine mount-to-engine bolt and tighten to 26–33 ft. lbs. (34–46 Nm). Tighten the transaxle mount through bolt.

47. Install 2 left rear engine mount-to-body bracket nuts and tighten to 73–97 ft. lbs. (98–132 Nm).

48. Remove the transaxle jack.

49. Remove damper tool T-93P6316-A or equivalent and install the accessory drive belt tensioner.

50. Install the starter and the starter bracket.

51. Connect the neutral sensing switch connector.

52. Install the transaxle stabilizer bar and tighten the bolt.

53. Position the shift linkage on the shift rod and install the bolt, washer and nut.

54. Position the halfshafts in the vehicle and insert the outer CV-joints through the hub assemblies. Install the retaining nuts and tighten to 180–200 ft. lbs. (244–271 Nm).

55. Remove the plugs from the transaxle that were installed during the removal procedure.

56. Install new clips on the inner CV-joint stub axles. Install the halfshafts into the transaxle. Pull on the CV-joints to make sure they are fully seated in the transaxle.

57. Install the ball joints into the steering knuckles using new bolts. Tighten the bolts to 38–45 ft. lbs. (52–60 Nm).

58. Check the transaxle fluid level and add fluid, if necessary.

59. Lower the vehicle. Remove the engine support and lifting eye.

60. Position the coil bracket and install the retaining bolts.

61. Install the speedometer cable into the speed sensor. Pull on the cable to make sure it is fully seated in the sensor. Connect the electrical connector to the speed sensor.

62. Connect the clutch cable to the clutch release lever.

63. Coat the threads of the backup light switch with pipe sealant and install the switch. Tighten to 12–15 ft. lbs. (16–20 Nm).

64. Install the 4 upper transaxle-to-engine bolts and tighten to 25–34 ft. lbs. (34–47 Nm).

65. Position the power steering line bracket to the upper transaxle-to-engine stud bolts and install the 2 nuts.

66. Install the 2 bolts in the left front transaxle mount.

67. Position the starter cable bracket and install the attaching nut.

68. Install the air cleaner and outlet tube.

69. Connect the mass air flow sensor and air charge temperature sensor connectors at the air cleaner.

70. Tighten the top 2 right mount-to-body Torx® bolts to 40–52 ft. lbs. (54–71 Nm).

71. Connect the negative battery cable.

72. Remove the prop from the clutch pedal. Road test the vehicle.

Clutch Assembly

REMOVAL AND INSTALLATION

1. Disconnect the negative battery cable.

2. Raise and safely support the vehicle.

3. Remove the transaxle assembly.

4. If the clutch assembly is to be reused, matchmark the pressure plate and the flywheel so they can be assembled in the same position.

5. Loosen the pressure plate-to-flywheel bolts one turn at a time, in a criss-cross pattern, until the spring tension is relieved, to prevent pressure plate cover distortion.

6. Support the pressure plate and remove the bolts. Remove the pressure plate and clutch disc from the flywheel.

7. Inspect the flywheel, clutch disc, pressure plate, release bearing, pilot bearing and the clutch fork for wear; replace parts, as required.

NOTE: If the flywheel shows any signs of overheating (blue discoloration) or if it is badly grooved or scored, it should be refaced or replaced.

To install:

8. If removed, install a new pilot bearing using a suitable installation tool.

9. If removed, install the flywheel. Make sure the flywheel and crankshaft flange mating surfaces are clean. Tighten the flywheel bolts to 54–64 ft. lbs. (73–86 Nm) on the 2.3L engine or 59 ft. lbs. (80 Nm) on the 3.0L engine.

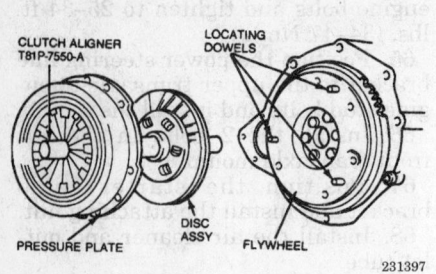

Clutch disc and pressure plate

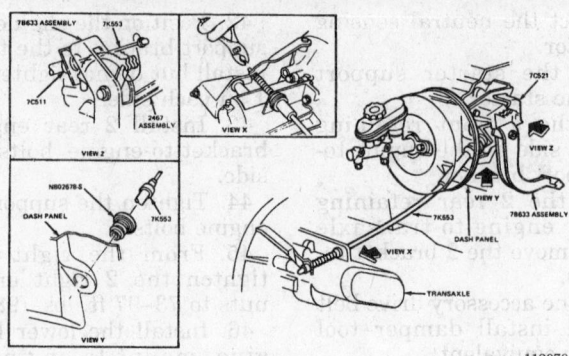

Clutch cable

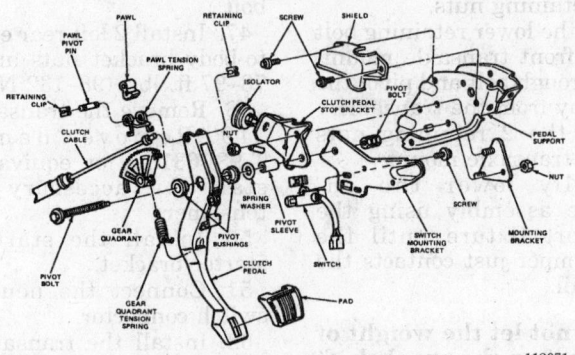

Clutch pedal assembly

10. Clean the pressure plate and flywheel surfaces thoroughly. Position the clutch disc and pressure plate into the installed position and support them with a dummy shaft or clutch aligning tool. If the clutch assembly is being reused, align the matchmarks that were made during the removal procedure.

11. Install the pressure plate-to-flywheel bolts. Tighten them gradually in a criss-cross pattern to 13–20 ft. lbs. (18–26 Nm). Remove the alignment tool.

12. If the release bearing was removed, lubricate the release fork where it contacts the bearing and install the bearing in the fork.

13. Install the transaxle assembly.

14. Lower the vehicle.

15. Connect the negative battery cable.

16. Bleed the hydraulic clutch system, if needed.

17. Check the clutch for proper operation.

Clutch Cable

ADJUSTMENT

The clutch cable is self-adjusting by pedal motion. After proper installation of the cable, depress the clutch pedal several times. Adjustment is completed by pulling the clutch pedal to its upward most position.

Automatic Transaxle Assembly

REMOVAL AND INSTALLATION

NOTE: On all, except the 3.0L engine, the right-side halfshaft assembly must be removed first.

The differential rotator tool T81P-4026–A or equivalent, is then inserted into the transaxle to drive the left-side inboard CV-joint assembly from the transaxle.

1. Disconnect the negative battery cable.

2. Remove the air cleaner assembly.

3. Disconnect the electrical harness connector from the neutral safety switch.

4. Disconnect the throttle valve linkage, throttle cable if equipped with 3.0L engine, and the manual lever cable from their levers.

NOTE: Failure to disconnect the linkage or cable and allowing the transaxle to hang, will fracture the throttle valve camshaft joint, which is located under the transaxle cover.

5. To prevent contamination, cover the timing window in the converter housing. If equipped, remove the bolts retaining the thermactor hoses and position out of the way.

6. If equipped, remove the ground strap, located above the upper engine mount, and the coil and bracket assembly.

NOTE: If equipped with 3.0L engine, be careful not to damage the TV cable while accessing the upper retaining bolts.

7. Remove both transaxle-to-engine upper bolts; the bolts are located below and on both sides of the distributor. Install bar engine support D88L–6000–A or equivalent.

8. Raise and safely support the vehicle. Remove the front wheel and tire assemblies.

9. Remove the nut from the control arm-to-steering knuckle attaching bolt at the ball joint. Using a hammer and a punch, drive the bolt from the steering knuckle; repeat this step on the other side. Discard the nut and bolt.

NOTE: Be careful not to damage or cut ball joint boot. The prybar must not contact lower arm.

10. Using a prybar, disengage the control arm from the steering

knuckle; repeat this step on the other side.

NOTE: Do not hammer on the knuckle to remove the ball joints. The plastic shield installed behind the rotor contains a molded pocket into which the lower control arm ball joint fits. When disengaging the control arm from the knuckle, clearance for the ball joint can be provided by bending the shield back toward the rotor. Failure to provide clearance for the ball joint can result in damage to the shield.

11. Remove the stabilizer bar bracket-to-frame rail bolts and discard the bolts; repeat this step on the other side.

12. Remove the stabilizer bar-to-control arm nut/washer and discard the nut; repeat this step on the other side.

13. Pull the stabilizer bar from the control arms.

14. Remove the brake hose clip-to-strut bracket bolts.

15. Remove the tie rod end nuts and disengage the tie rods from the steering knuckles.

16. Using a halfshaft removal tool, pry the halfshaft from the right side of the transaxle and support the end of the shaft with mechanics wire.

NOTE: It is normal for some fluid to leak from the transaxle when the halfshaft is removed. Be careful not to damage the transaxle case or the bottom of the oil pan flange when prying the halfshaft out.

17. Using differential rotator tool T81P–4026–A or equivalent, drive the left-side halfshaft from the differential side gear.

18. Pull the halfshaft from the transaxle and support the end of the shaft with mechanics wire.

NOTE: Do not allow the shaft to hang unsupported, as damage to the outboard CV-joint may result.

19. Install transaxle plugs T81P–1177–B or equivalent, into the differential seals.

20. Remove the starter support bracket, if equipped. Disconnect the starter cable. Remove the starter bolts and the starter. If equipped, remove the hose and bracket bolts on the starter and a bolt at the converter and disconnect the hoses.

21. If equipped, remove the transaxle support bracket. Remove the dust cover from the torque converter housing.

22. Remove the torque converter-to-flywheel nuts by turning the crankshaft pulley bolt to bring the nuts into position.

23. Position a suitable transmission jack under the transaxle and remove the rear support bracket nuts.

24. Remove the left front insulator-to-body bracket nuts, the bracket-to-body bolts and the bracket.

25. Disconnect the transaxle cooler lines.

26. Remove the manual lever bracket-to-transaxle case bolts.

27. Support the engine. Make sure the transaxle is supported and remove the remaining transaxle-to-engine bolts.

28. Make sure the torque converter studs will be clear the flywheel. Insert a prybar between the flywheel and the converter, then pry the transaxle and converter away from the engine. When the converter studs are clear of the flywheel, lower the transaxle about 2–3 in. (51–76mm).

29. Disconnect the speedometer cable and lower the transaxle.

NOTE: When moving the transaxle away from the engine, watch the No. 1 insulator. If it contacts the body before the converter studs clear the flywheel, remove the insulator.

To install:

30. Raise the transaxle and align it with the engine and flywheel. Install the No. 1 insulator, if removed. Torque the transaxle-to-engine bolts to 25–33 ft. lbs. (34–45 Nm) on 2.3L engine or 34–47 ft. lbs. (46–63 Nm) on 3.0L engine. Tighten the torque converter-to-flywheel bolts to 23–39 ft. lbs. (31–53 Nm).

31. Install the manual lever bracket-to-transaxle case bolts and connect the transaxle cooler lines.

32. Install the left front insulator-to-body bracket nuts and torque the nuts to 40–50 ft. lbs. (55–70 Nm). Install the bracket-to-body and torque the bolts to 55–70 ft. lbs. (75–90 Nm).

33. Install the transaxle support bracket and the dust cover to the torque converter housing.

34. If equipped, install the hose and bracket bolts on the starter and a bolt to the converter and connect the hoses. Install the starter and the support bracket; torque the starter-to-engine bolts to 30–40 ft. lbs. (41–54 Nm). Connect the starter cable.

35. Remove the seal plugs from the differential seals and install the half-

shaft by performing the following procedures:

a. Prior to installing the halfshaft in the transaxle, install a new circlip onto the CV-joint stub.

b. Install the halfshaft in the transaxle by carefully aligning the CV-joint splines with the differential side gears. Be sure to push the CV-joint into the differential until the circlip is felt to seat in the differential side gear. Use care to prevent damage to the differential oil seal.

c. Attach the lower ball joint to the steering knuckle, taking care not to damage or cut the ball joint boot. Insert a new pinch bolt and a new nut. While holding the bolt with a 2nd wrench, torque the nut to 40–54 ft. lbs. (54–74 Nm).

36. Engage the tie rod with the steering knuckle and torque the nut to 23–35 ft. lbs. (31–47 Nm).

37. Install the brake hose clip-to-strut bracket and torque the bolt to 8 ft. lbs. (11 Nm).

38. Install the stabilizer bar to control arm and using a new nut, torque it to 98–125 ft. lbs. (133–169 Nm).

39. Install the stabilizer bar bracket-to-frame rail bolts and using new bolts, torque them to 60–70 ft. lbs. (81–95 Nm).

40. Install the wheel and tire assemblies and lower the vehicle. Install the upper transaxle-to-engine bolts and torque to 25–33 ft. lbs. (34–45 Nm) on 2.3L engine or 34–47 ft. lbs. (46–63 Nm) on 3.0L engine.

41. If equipped, install the ground strap, located above the upper engine mount, and the coil and bracket assembly.

42. If equipped, install the bolts retaining the thermactor hoses. Uncover the timing window in the converter housing.

43. Connect the throttle valve linkage or cable and the manual lever cable to their levers.

44. Connect the electrical harness connector from the neutral safety switch.

45. Install the air cleaner assembly.

46. Connect the negative battery cable and road test the vehicle.

Throttle Valve Linkage

ADJUSTMENT

2.3L Engine

1. Disconnect the negative battery cable.

2. Remove the splash shield from the cable retainer bracket.

3. Loosen the trunnion bolt on the throttle valve rod.

4. Install plastic clip using TV linkage adjustment tool T91P-7000-A or equivalent, to bottom of throttle valve rod; be sure the clip keeps the rod from telescoping.

5. Be sure the return spring is connected between the throttle valve rod and the retaining bracket to hold the transaxle throttle vale lever at it's idle position.

6. Make sure the throttle lever is resting on the throttle return control screw.

7. Tighten the throttle valve rod trunnion bolt and remove the plastic clip.

8. Install the splash shield. Connect the negative battery cable and check the vehicle's operation.

3.0L Engine

1. Remove the splash shield from the cable retainer bracket.

2. Unsnap the white adjuster locking clip at the cable retainer bracket.

3. Hold the transaxle lever in the idle position against the idle stop.

4. Make sure the throttle lever adjusting screw is resting against the idle stop.

5. Snap the white adjuster locking clip into the lock position.

6. Install the splash shield.

7. Check the linkage for proper operation.

DRIVE AXLE

Halfshaft

REMOVAL AND INSTALLATION

Front Halfshaft

NOTE: On the automatic transaxle, the right side halfshaft must be removed before removing the left halfshaft. Special tool T81P-4026-A or an equivalent differential rotator tool is required.

— CAUTION —

The front hub nut torque is very high. Tighten and loosen the hub nut only with the vehicle sitting on all 4 wheels.

1. Set the parking brake. Remove the cap from the hub and loosen the hub nut. The nut must be loosened

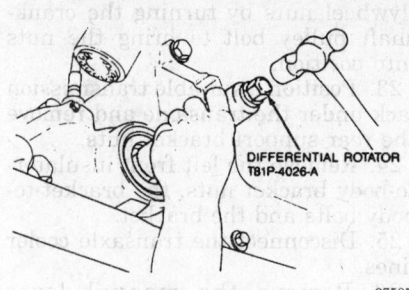

97585

Differential rotator

without unstaking; the use of a chisel or similar tool may damage the spindle thread.

2. Raise and safely support the vehicle. Remove the wheel and tire assembly. Remove the hub nut/washer and discard the nut.

3. Remove the brake hose routing clip-to-strut bolt.

4. Remove the nut from the ball joint-to-steering knuckle bolt. Drive the bolt from the steering knuckle with a punch and discard the bolt/nut. This bolt cannot be re-used.

5. To separate the ball joint from the steering knuckle, pry down against the stabilizer bar. Make sure the bar does not damage any bushings or the CV-joint boot. Carefully bend the brake disc shield away from the ball joint as needed while prying.

6. On automatic transaxles, remove the right side halfshaft first. Position the prybar between the differential housing and the CV-joint assembly and carefully pry the halfshaft away from the differential housing. Be careful not to damage the differential oil seal, case, CV-joint boot or the transaxle.

7. Hang the halfshaft from a convenient underbody component. Do not allow the shaft to hang unsupported or the outboard CV-joint may be pulled apart.

NOTE: If removing both halfshafts at the same tine, shipping plugs T81P-1177-B or equivalent must be installed in the differential housing or the differential side gears will move out of place. If the gears become misaligned, the differential will have to be removed from the transaxle to realign the gears.

8. To remove the left halfshaft from a manual transaxle, carefully pry it out as before.

9. To remove the left side halfshaft from an automatic transaxle, insert special tool T81P-4026-A or

an equivalent differential rotator tool into the right side CV-joint opening. Carefully drive the left side inner CV-joint out of the transaxle.

10. Use a wheel puller or a front hub removal tool to press the outboard CV-joint stub shaft out of the hub. Do not drive the stub shaft out with a hammer or the CV-joint internal components will be damaged.

To install:

11. Install a new circlip onto the inboard CV-joint stub shaft. To install the circlip properly, start one end in the groove and work the circlip over the stub shaft end and into the groove; this will avoid over expanding the circlip.

12. Lubricate the seal and align the splines of the inboard CV-joint stub shaft with the splines in the differential. Push the CV-joint into the differential until the circlip is seated in the differential side gear. A plastic mallet may be used to aid in seating the circlip into the differential side gear groove. If so, tap only on the outboard CV-joint stub shaft.

13. Carefully align the outboard CV-joint stub shaft splines with the hub splines and push the shaft into the hub as far as possible. If necessary, use the front hub replacer tool to firmly press the halfshaft into the hub.

14. Connect the control arm-to-steering knuckle and torque the new nut/bolt to 40–54 ft. lbs. (54–74 Nm).

15. Position the brake hose routing clip on the suspension strut and torque the bolt to 96 inch lbs. (11 Nm).

16. Install the hub nut washer and a new hub nut.

17. Install the wheel/tire assembly and torque the lug nuts to 80–105 ft. lbs. (108–144 Nm). Lower the vehicle so it is sitting on all 4 wheels.

18. Torque the hub nut to 180–200 ft. lbs. (244–271 Nm). The new nut will click, indicating the locking tabs are ratcheting into the slot in the stub axle. Make sure one of the tabs is in the slot after the final tightening.

19. Refill the transaxle and road test.

Rear Halfshaft

1. Raise and safely support the vehicle. Remove the rear suspension control arm bolt.

2. Remove the outboard U-joint retaining bolts and straps. Remove the inboard U-joint retaining bolts and straps.

3. Slide the shafts together; do not allow the splined shafts to contact with excessive force. Remove the

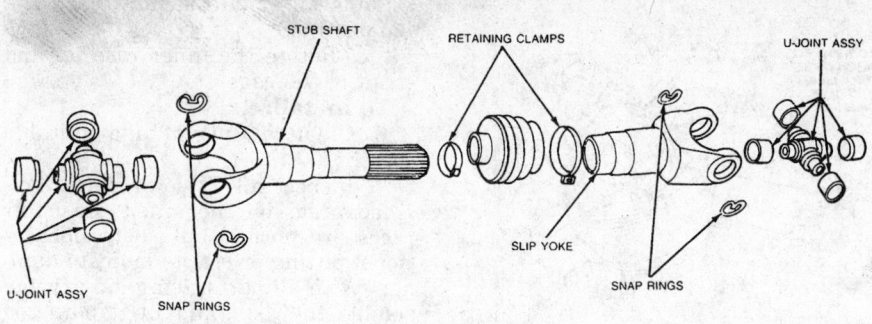

Rear halfshaft

halfshafts; do not drop the halfshafts as the impact may cause damage to the U-joint bearing cups.

4. Retain the bearing cups. Inspect the U-joint assemblies for wear or damage, replace the U-joint if necessary.

To install:

5. Install the halfshaft at the inboard U-joint; the inboard shaft has a larger diameter than the outboard shaft. Install the U-joint retaining caps and bolts and torque them to 15–17 ft. lbs. (21–23 Nm).

NOTE: Be sure to apply Loctite® to the U-joint bolts.

6. Install the halfshaft at the outboard U-joint. Install the U-joint retaining caps and bolts and torque them to 15–17 ft. lbs. (21–23 Nm).

7. Install the rear suspension control arm and torque the bolt to 60–86 ft. lbs. (82–116 Nm).

CV-Joint Boot

REPLACEMENT

Inner CV-Joint

1. Remove halfshaft assembly from vehicle and clamp it in a soft jaw vise. Do not allow vise jaws to contact the boot or its clamp.

2. Cut the large boot clamp using side cutters and peel it away from the boot. After removing the clamp, roll the boot back over the shaft.

3. Remove the wire ring or roll crimp ball retainer from the outer race.

4. Carefully pull the outer race off and be ready to catch the balls if they fall out.

5. Pull the inner race assembly out until it rests on the circlip. Using snapring pliers, spread the stop ring and move it back on the shaft.

6. Slide the inner race assembly down the shaft to allow access to the circlip and remove the circlip.

7. Remove the inner race assembly and the stop ring and pull the boot off the halfshaft.

NOTE: Circlips must not be reused. Replace with new circlips before assembly.

8. When replacing damaged boots, the grease should be checked for contamination. If the CV-joints were operating satisfactorily and the grease does not appear to be contaminated, add grease and replace the boot. If the lubricant appears contaminated, disassemble the CV-joint for inspection.

   a. Remove the balls by carefully prying them from the cage. Be careful not to scratch the inner race or cage.

   b. Rotate the inner race to align the lands with the cage windows.

   c. Lift the inner race out through the wider end of the cage.

To install:

9. Clean all parts in a suitable solvent.

10. Inspect all CV-joint parts. Polished areas on the inner and outer races are normal but rust, indentations, pitting or cracks indicate damage. With all parts clean and dry, assemble the joint and fit it onto the halfshaft splines. If the outer race can be twisted and clicks loudly when twisting, it is worn and the joint should be replaced.

NOTE: CV-joint components cannot be interchanged. If inspection reveals damage or excess wear, the entire joint must be replaced as an assembly.

11. Install a new circlip, supplied with the service kit, in the groove at the end of the shaft. Do not over-expand or twist the circlip during installation.

12. Install the inner race in the cage. The race is installed through the large end of the cage with the circlip counterbore facing the large end of the cage.

13. With the cage and inner race properly aligned, install the balls by pressing them through the cage windows by hand.

14. Push the inner race and cage assembly into the outer race. The inner race chamfer must be facing out. Install the ball retainer.

15. Place the inner boot clamp onto the halfshaft.

16. Wrap smooth black tape around the halfshaft spline and push the boot onto the shaft.

17. Tighten clamp securely but not to the point where the clamp bridge is cut or the boot is damaged.

18. Position the stop ring and new circlip onto the halfshaft.

19. Fill the CV-joint outer race with 3.2 oz. (90 grams) of grease, then spread 1.4 oz. (40 grams) of grease evenly inside boot for a total combined fill of 4.6 oz. (130 grams). It is important that the boot be lubricated to prevent early failure.

20. Push the boot back and fit the CV-joint onto the splines. Using a plastic hammer, one sharp blow should drive the CV-joint over the circlip and against the stop ring.

21. Move CV-joint in or out as necessary to fit the boot onto the joint.

NOTE: Insert a suitable tool between the boot and the outer bearing race and allow the trapped air to escape from the boot.

22. Ensure the boot is seated in its groove and install the clamp. Tighten it securely but not to the point where the clamp bridge is cut or the boot is damaged.

23. Wipe away all excess grease from the CV-joint external surfaces.

24. Install the halfshaft assembly.

Outer CV-Joint

1. Remove halfshaft assembly from vehicle and clamp it in a soft jaw vise. Do not allow vise jaws to contact the boot or its clamp.

2. Cut the large boot clamp using side cutters and peel it away from the boot. After removing the clamp, roll the boot back over the shaft.

3. Support the halfshaft in the vise with the inner bearing race angled down.

4. Using a brass drift and hammer, give a sharp tap to the inner bearing race to drive it off the internal circlip and separate the CV-joint

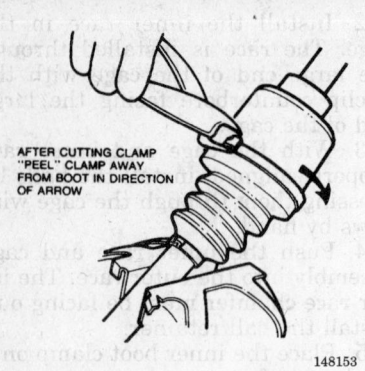

AFTER CUTTING CLAMP "PEEL" CLAMP AWAY FROM BOOT IN DIRECTION OF ARROW

148153

CV-joint boot clamp removal

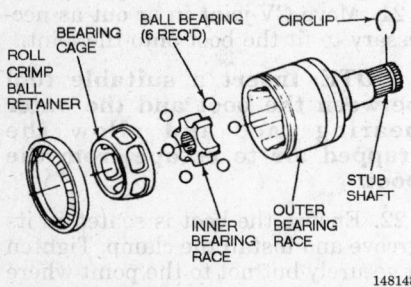

WIRE RING BALL RETAINER — BALL CAGE — BALL 6 REQ'D — OUTER RACE — CIRCLIP — INNER RACE — STUB SHAFT

148154

Inner CV-joint with retainer ring

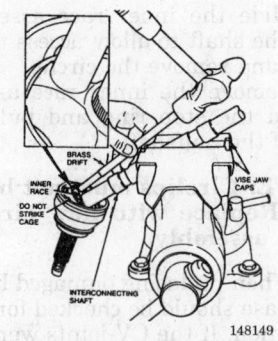

ROLL CRIMP BALL RETAINER — BEARING CAGE — BALL BEARING (6 REQ'D) — CIRCLIP — INNER BEARING RACE — OUTER BEARING RACE — STUB SHAFT

148148

Inner CV-joint with roll crimp retainer

from the interconnecting shaft. Take care not to drop the CV-joint at separation.

5. When replacing damaged boots, the grease should be checked for contamination. If the CV-joints were operating satisfactorily and the grease does not appear to be contaminated, add grease and replace the boot. If the lubricant appears contaminated, disassemble the CV-joint for inspection.

 a. Clamp the CV-joint stub shaft in a soft jawed vise with the outer

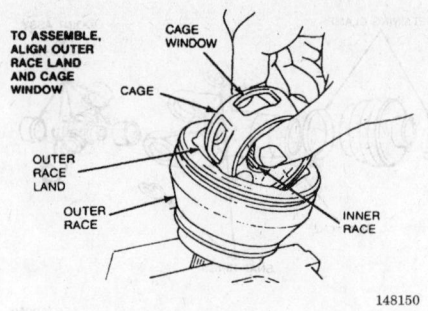

TO ASSEMBLE, ALIGN OUTER RACE LAND AND CAGE WINDOW — CAGE WINDOW — CAGE — OUTER RACE LAND — OUTER RACE — INNER RACE

148150

Installing inner race and ball cage

BRASS DRIFT — INNER RACE DO NOT STRIKE CAGE — VISE JAW CAPS — INTERCONNECTING SHAFT

148149

One sharp tap will disengage the CV-joint from the circlip

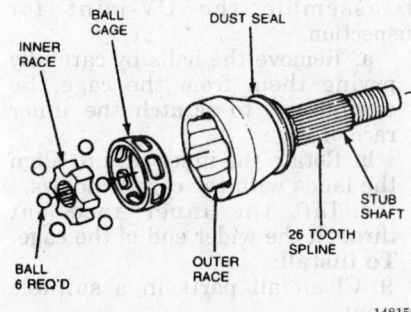

INNER RACE — BALL CAGE — DUST SEAL — BALL 6 REQ'D — OUTER RACE — 26 TOOTH SPLINE — STUB SHAFT

148152

Outer CV-joint with dust seal

face facing up. Care should be taken not to damage the dust seal.

 b. Press down on the inner race to tilt is as far as possible. A tight assembly can be tilted by tapping the inner race with wooden dowel and hammer. Do not hit the cage.

 c. With the cage sufficiently tilted, remove a ball from cage. Remove all 6 balls in this manner.

 d. Pivot the cage and inner race assembly until it is straight up and down in the outer race. Align the cage windows with the outer race

lands and lift assembly from the outer race.

 e. Rotate the inner race up and out of the cage.

To install:

6. Clean all parts in a suitable solvent.

7. Inspect all CV-joint parts. Polished areas on the inner and outer races are normal but rust, indentations, pitting or cracks indicate damage. With all parts clean and dry, assemble the joint and fit it onto the halfshaft splines. If the outer race can be twisted and clicks loudly when twisting, it is worn and the joint should be replaced.

NOTE: CV-joint components cannot be interchanged. If inspection reveals damage or excess wear, the entire joint must be replaced as an assembly.

8. Place the inner boot clamp onto the halfshaft.

9. Wrap smooth black tape around the halfshaft spline and push the boot onto the shaft.

10. Tighten clamp securely but not to the point where the clamp bridge is cut or the boot is damaged.

11. Position the stop ring and new circlip onto the halfshaft.

12. Fill the CV-joint outer race with 3.2 oz. (90 grams) of grease, then spread 1.4 oz. (40 grams) of grease evenly inside boot for a total combined fill of 4.6 oz. (130 grams). It is important that the boot be lubricated to prevent early failure.

13. Push the boot back and fit the CV-joint onto the splines. Using a plastic hammer, one sharp blow should seat the CV-joint onto the circlip.

14. Move CV-joint in or out as necessary to fit the boot onto the joint.

NOTE: Insert a suitable tool between the boot and the outer bearing race and allow the trapped air to escape from the boot.

15. Ensure the boot is seated in its groove and install the clamp. Tighten it securely but not to the point where the clamp bridge is cut or the boot is damaged.

16. Wipe away all excess grease from the CV-joint external surfaces.

17. Install the halfshaft assembly.

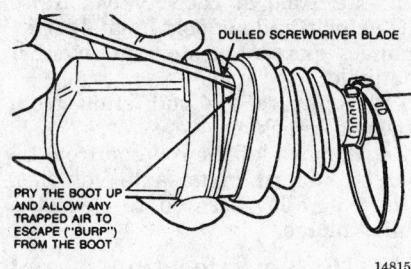

Release the air before installing the large clamp

STEERING

Air Bag

CAUTION

Some vehicles are equipped with an air bag system, also known as the Supplemental Inflatable Restraint (SIR) or Supplemental Restraint System (SRS). The system must be disabled before performing service on or around components, steering column, instrument panel components, wiring and sensors. Failure to follow safety and disabling procedures could result in accidental air bag deployment, possible personal injury and unnecessary system repairs.

PRECAUTIONS

Several precautions must be observed when handling the inflator module to avoid accidental deployment and possible personal injury.

• Never carry the inflator module by the wires or connector on the underside of the module.

• When carrying a live inflator module, hold securely with both hands, and ensure that the bag and trim cover are pointed away.

• Place the inflator module on a bench or other surface with the bag and trim cover facing up.

• With the inflator module on the bench, never place anything on or close to the module which may be thrown in the event of an accidental deployment.

DISARMING

1. Disconnect the negative battery cable.

2. Wait 1 minute for the backup power supply to deplete its stored energy.

3. Remove the 4 nut and washer assemblies retaining the air bag to the steering wheel.

4. Disconnect the air bag electrical connector.

5. Attach air bag simulator tool 105-00008 or equivalent on the clockspring to simulate the air bag.

6. Connect the positive battery cable.

Steering Wheel

REMOVAL AND INSTALLATION

CAUTION

If equipped with an air bag, the negative battery cable and backup power supply must be disconnected before working on the system. Failure to do so may result in deployment of the air bag and possible personal injury.

WARNING

Always carry the air bag module with soft face away from your body to prevent injury in case of an accidental deployment.

WARNING

Always place the driver's side air bag module soft face up on a bench to prevent injury in case of an accidental deployment.

1. Disconnect the negative battery cable. On air bag equipped vehicles, disconnect the backup power supply as follows:

a. Remove 2 screws retaining the steering column opening cover to the instrument panel and remove the cover.

b. Remove the 4 bolts retaining the bolster and remove the bolster.

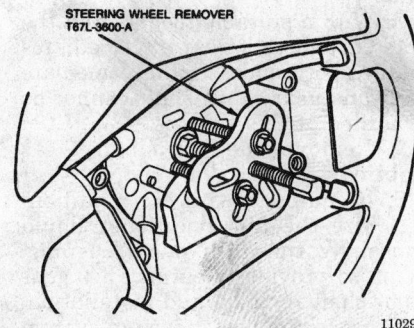

Steering wheel removal

c. Disconnect the connector from the backup power supply.

2. Remove the horn pad cover by removing the retaining screws from the steering wheel assembly.

NOTE: The emblem assembly is removed, after the horn pad cover is removed, by pushing it out from the backside of the emblem.

3. Remove the energy absorbing foam from the wheel assembly, if equipped. Remember, the energy absorbing foam must be installed when the steering wheel is assembled. Disconnect the horn pad wiring connector.

4. If equipped with an air bag restraint system, remove the 4 nuts located on the back of the steering wheel which hold the air bag module to the steering wheel.

5. Lift the air bag module from the wheel and disconnect the air bag module-to-clockspring connector.

6. Loosen the steering wheel retaining bolt 4–6 turns but do not remove. On air bag equipped vehicles, remove the bolt completely to remove the vibration damper, then reinstall the bolt loosely on the shaft.

7. Remove the steering wheel with a suitable puller. Do not use a knockoff type puller, because it will cause damage to the collapsible steering column. Loosen the retaining bolt, grasp the rim of the steering wheel, and pull the steering wheel from the upper shaft.

To install:

8. Install the steering wheel assembly on the steering column, making sure the alignment marks are correct.

9. Install a new retaining bolt. Torque the bolt to 23–33 ft. lbs. (31–45 Nm). On air bag equipped vehicles, install the vibration damper before installing the bolt.

10. If equipped with air bag, connect the air bag module wire to the clockspring connector, and place the module on the steering wheel with the 4 attaching nuts, torque the nuts to 35–53 inch lbs. (4–6 Nm).

11. On vehicles without an air bag, connect the horn pad wiring connector and, if equipped, install the energy absorbing foam. Install the horn pad cover and torque the retaining screws to 8–10 inch lbs. (0.9–1.1 Nm).

12. On air bag equipped vehicles, connect the backup power supply connector and reinstall the bolster and steering column opening cover.

13. Reconnect the negative battery cable and check the steering wheel for proper operation.

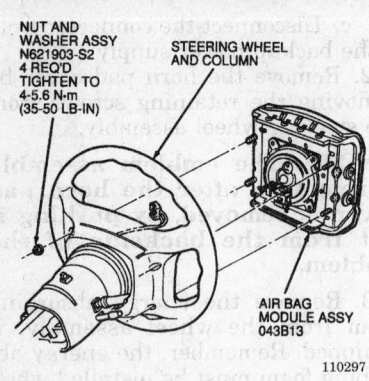

NUT AND
WASHER ASSY
N621903-S2
4 REQ'D
TIGHTEN TO
4-5.6 N·m
(35-50 LB-IN)

STEERING WHEEL
AND COLUMN

AIR BAG
MODULE ASSY
043B13

110297

Air bag module

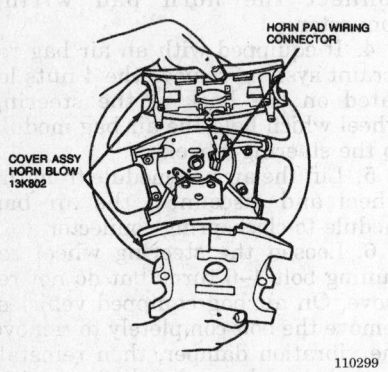

HORN PAD WIRING
CONNECTOR

COVER ASSY
HORN BLOW
13K802

110299

Horn wiring connector

Tie Rod Ends

REMOVAL AND INSTALLATION

1. Disconnect the negative battery cable.
2. Raise and safely support the vehicle.
3. Remove the wheel.
4. Remove and discard the cotter pin from the castellated tie rod end nut.
5. Loosen the tie rod end jam nut.
6. Separate the tie rod end from the steering knuckle, using tie rod

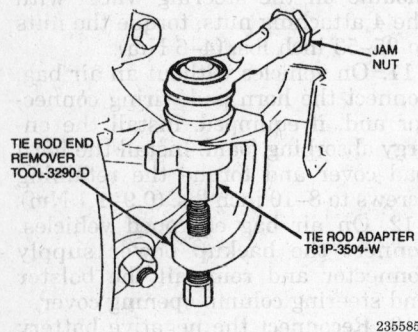

JAM
NUT

TIE ROD END
REMOVER
TOOL-3290-D

TIE ROD ADAPTER
T81P-3504-W

235585

Tie rod end removal

end remover tool 3290-D and adapter T81P-3504-W or equivalent.

7. Count and record the number of turns to unscrew the tie rod end from the inner tie rod or use the jam nut as a marker for reassembly.

To install:

8. Clean the inner tie rod threads. Apply a light coating of suitable grease to the threads.
9. If removed, install the jam nut.
10. Thread the new tie rod end onto the inner tie rod to the same depth as the removed tie rod end. Tighten the jam nut to 25–29 ft. lbs. (34–49 Nm).
11. Place the tie rod end stud into the steering knuckle using a good dust boot.
12. Install the castellated nut onto the tie rod end stud. Tighten the nut to 25–34 ft. lbs. (34–46 Nm) and align the next slot with the cotter pin hole in the tie rod end stud. Install a new cotter pin.
13. Install the wheel and torque the lug nuts to 65–87 ft. lbs. (88–118 Nm).
14. Lower the vehicle.
15. Check the front wheel alignment and adjust as necessary.
16. Tighten the tie rod jam nut to 25–29 ft. lbs. (34–49 Nm).
17. Road test the vehicle.

Power Rack and Pinion

REMOVAL AND INSTALLATION

1. Disconnect the negative battery cable.
2. Turn the ignition key to the **RUN** position.
3. Remove access panel from dash below the steering column.
4. Remove screws from steering column boot at the dash panel and slide boot up intermediate shaft.
5. Remove intermediate shaft bolt at gear input shaft and loosen the bolt at the steering column shaft joint.
6. With a suitable tool, spread the slots enough to loosen intermediates shaft at both ends. The intermediate shaft and gear input shaft cannot be separated at this time.
7. Remove the air cleaner, as required.
8. If equipped with air conditioning, wire the air conditioner liquid line above the dash panel opening. Doing so provides clearance for gear input shaft removal and installation.
9. Separate the steering system pressure and return lines at intermediate connections.

10. Disconnect the exhaust secondary air tube at check valve. Raise the vehicle and support it safely. Disconnect exhaust system at exhaust manifold.
11. Separate tie rod ends from steering knuckles.
12. Remove left tie rod end from tie rod on manual transaxle vehicles. This will allow tie rod to clear the shift linkage.

NOTE: Mark location of rod end prior to removal.

13. If equipped with an automatic transaxle:
    a. Disconnect speedometer cable at transaxle and remove the vehicle speed sensor.
    b. Remove transaxle shift cable assembly at transaxle.
14. Turn steering wheel to full left turn stop for easier gear removal.
15. Remove screws holding the heater water tube to shake brace below the oil pan.
16. Remove nut from the lower of 2 bolts holding engine mount support bracket to transaxle housing. Tap bolt out as far as it will go.
17. Remove the gear mounting brackets and insulators.
18. Drape cloth towel over both apron opening edges to protect bellows during gear removal.
19. Separate gear from intermediate shaft by either pushing up on shaft with a bar from underneath the vehicle while pulling the gear down or with an assistant removing the shaft from inside the vehicle.
20. Rotate gear forward and down to clear the input shaft through the dash panel opening.
21. Make sure input shaft is in full left turn position. Move gear through the right (passenger) side apron opening until left tie rod clears left apron opening and other parts so it may be lowered. Guide the power steering hoses around the nearby components as the gear is being removed.
22. Lower the left side of the gear and remove the gear out of the vehicle. Use care not to tear the bellows.

To install:

23. Rotate the input shaft to a full left turn stop. Position the right road wheel to a full left turn.
24. Start the right side of the gear through the opening in the right apron. Move the gear in until the left tie rod clears all parts so it may be raised up to the left apron opening.
25. Raise the gear and insert the left side through the apron opening. Move the power steering hoses into their proper position at the same

time. Rotate the gear so the joint shaft enters the dash panel opening.

26. With an assistant guiding the intermediate shaft from the inside of the vehicle, insert the input shaft into the intermediate shaft coupling. Insert the intermediate shaft clamp bolts finger-tight. Do not tighten at this time.

NOTE: The right and left side insulators and brackets are not interchangeable side to side.

27. Install the gear mounting insulators and brackets in their proper places. Ensure the flat in the left mounting area is parallel to the dash panel. Tighten the bracket bolts to 40–55 ft. lbs. (54–75 Nm) in the sequence as described below:

　a. Tighten the left (driver's side) upper bolt halfway.

　b. Tighten the left side lower bolt.

　c. Tighten the left side upper bolt.

　d. Tighten the right side bolts.

28. Attach the tie rod ends to the steering knuckles. Tighten the castellated nuts to 27–32 ft. lbs. (36–43 Nm), then tighten as required to insert new cotter pins.

29. Install the engine mount nut.

30. Install the heater water tube to the shake brace.

31. Install the exhaust system. If removed, install the speedometer cable, the vehicle speed sensor and the transaxle shift cable.

32. Connect the secondary air tube at the check valve. Connect the pressure and return lines at the intermediate connections or steering gear.

33. Install the air cleaner.

34. Tighten the gear input shaft to intermediate shaft coupling clamp bolt first. Then, tighten the upper intermediate shaft clamp bolt. Tighten to 20–30 ft. lbs. (27–40 Nm).

35. Install the access panel below the steering column. Turn the ignition key to the **OFF** position.

36. Fill the system. Check and adjust the toe. Tighten the tie rod end jam nuts to 40–50 ft. lbs. (54–68 Nm), check for twisted bellows.

37. Connect negative battery cable.

Power Steering Pump

BLEEDING

If air bubbles are present in the power steering fluid, bleed the system by performing the following:

1. Fill the reservoir to the proper level.

2. Operate the engine until the fluid reaches normal operating temperature of 165–175°F (73.8–79.4°C).

3. Turn the steering wheel all the way to the left then all the way to the right several ties. Do not hold the steering wheel in the far left or far right position stops.

4. check the fluid level and recheck the fluid for the presence of trapped air. If apparnet that air is still in the system, fabricate or obtain a vacuum tester and purge the system as follows:

　a. Remove the pump dipstick cap assembly.

　b. Check and fill the pump reservoir with fluid to the **COLD FULL** mark on the dipstick.

　c. Disconnect the ignition coil wire of the coil pack electrical connector if equipped with distributorless ignition, and raise the front of the vehicle and support safely.

　d. Crank the engine with the starter and check the fluid level. Do not turn the steering wheel at this time.

　e. Fill the pump reservoir to the **COLD FULL** mark on the dipstick. Crank the engine with the starter while cycling the steering wheel lock-to-lock. Check the fluid level.

　f. Tightly insert a suitable size rubber stopper and air evacuator pump into the reservoir fill neck. Connect the ignition coil wire or coil pack electrical connector.

　g. With the engine idling, apply 15 in. Hg vacuum to the reservoir for 3 minutes. As air is purged from the system, the vacuum will drop off. Maintain the vacuum on the system as required throughout the 3 minutes.

　h. Remove the vacuum source. Fill the reservoir to the **COLD FULL** mark on the dipstick.

　i. With the engine idling, re-apply 15 in. Hg vacuum to the reservoir. Slowly cycle the steering wheel to lock-to-lock stops for approximately 5 minutes. Do not hold the steering wheel on the stops during cycling. Maintain the vacuum as required.

　j. Release the vacuum and disconnect the vacuum source. Add fluid, as required.

　k. Start the enigne and cycle the wheel slowly and check for leaks at all connections.

　l. Lower the front wheels.

5. In case of severe aeration, repeat the bleeding procedure.

REMOVAL AND INSTALLATION

2.3L Engine

1. Disconnect the negative battery cable.

2. Disconnect the fluid return line at the remote reservoir and drain the power steering fluid into a suitable container.

3. Disconnect the pressure hose from the pump outlet and drain the fluid into a suitable container.

4. Loosen the tensioner and remove the drive belt from the pump pulley.

5. Remove the 4 bolts from the pump pulley and remove the pulley.

6. Remove the 3 pump retaining bolts and remove the pump.

To install:

7. Position the pump on its bracket and install the retaining bolts.

8. Install the pulley and secure with the 4 bolts.

9. Connect the pressure line to the pump but do not overtighten the fitting. Swivel and/or endplay of the fitting is normal and does not indicate a loose fitting.

10. Connect the inlet hose to the pump and secure with the hose clamp.

11. Fill the reservoir with the proper type of fluid.

12. Connect the negative battery cable and bleed the system. Check for leaks.

3.0L Engine

1. Disconnect the negative battery cable.

2. Disconnect the fluid return hose from the pump inlet and drain the fluid into a suitable container.

3. Remove the pressure line from the pump outlet and drain the fluid into a suitable container.

4. Remove the plastic pulley guard.

5. Loosen the tensioner and remove the drive belt from the pulley.

6. Remove the pulley-to-pump shaft bolt or nut and remove the pulley from the pump shaft.

7. Remove the pump-to-bracket bolts and remove the pump.

To install:

8. Install the pump onto the pump bracket and install the retaining screws. Tighten to 47–63 inch lbs. (5.2–7.2 Nm).

9. Install the pulley on the pump shaft and secure with the bolt or nut.

10. Install the drive belt on the pulley.

11. Install the plastic pulley guard.

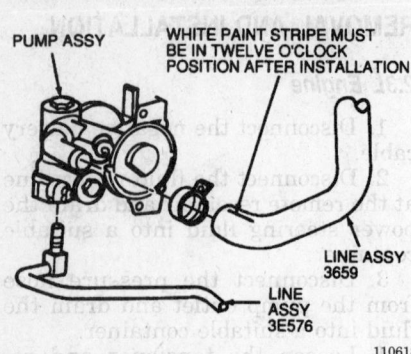

PUMP ASSY

WHITE PAINT STRIPE MUST BE IN TWELVE O'CLOCK POSITION AFTER INSTALLATION

LINE ASSY 3659

LINE ASSY 3E576

110618

Steering pump — 2.3L engine

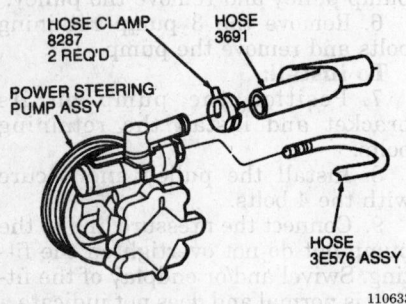

HOSE CLAMP 8287 2 REQ'D

HOSE 3691

POWER STEERING PUMP ASSY

HOSE 3E576 ASSY

110633

Steering pump, 3.0L engine

12. Connect the pressure line to the pump but do not overtighten the fitting. Swivel and/or endplay of the fitting is normal and does not indicate a loose fitting.

13. Connect the inlet hose to the pump and secure with the hose clamp.

14. Fill the reservoir with the proper type of fluid.

15. Connect the negative battery cable and bleed the system. Check for leaks.

BRAKES

Anti-Lock Brake System Service

PRECAUTIONS

• Certain components within the Anti-Lock Brake System (ABS) system are not intended to be serviced or repaired individually. Only those components with removal and installation procedures should be serviced.

• Do not use rubber hoses or other parts not specifically specified for and ABS system. When using repair kits, replace all parts included in the kit. Partial or incorrect repair may lead to functional problems and require the replacement of components.

• Lubricate rubber parts with clean, fresh brake fluid to ease assembly. Do not use lubricated shop air to clean parts; damage to rubber components may result.

• Use only DOT 3 brake fluid from an unopened container.

• If any hydraulic component or line is removed or replaced, it may be necessary to bleed the entire system.

• A clean repair area is essential. Always clean the reservoir and cap thoroughly before removing the cap. The slightest amount of dirt in the fluid may plug an orifice and impair the system function. Perform repairs after components have been thoroughly cleaned; use only denatured alcohol to clean components. Do not allow ABS components to come into contact with any substance containing mineral oil; this includes used shop rags.

• The Anti-Lock control unit is a microprocessor similar to other computer units in the vehicle. Ensure that the ignition switch is **OFF** before removing or installing controller harnesses. Avoid static electricity discharge at or near the controller.

• If any arc welding is to be done on the vehicle, the ALCU connectors should be unplugged before welding operations begin.

Master Cylinder

REMOVAL AND INSTALLATION

1. Disconnect the negative battery cable.

2. Remove the brake lines from the primary and secondary outlet ports of the master cylinder and pressure control valves. Plug the brake lines to prevent contamination.

3. Remove 2 nuts retaining the master cylinder to the brake booster assembly. Disconnect the brake warning light electrical connector.

4. Slide the master cylinder forward and upward from the vehicle.

To install:

5. Before installation, bench bleed the new master cylinder as follows:

   a. Mount the new master cylinder in a suitable holding fixture. Be careful not to damage the housing.

   b. Fit short lengths of brake lines to the master cylinder ports

positioned so they are directed into the master cylinder reservoir and submerged by brake fluid.

   c. Fill the master cylinder reservoir with clean DOT 3 or equivalent, brake fluid.

   d. Using a suitable tool inserted into the booster pushrod cavity, push the master cylinder piston in slowly and allow it to return.

   e. Repeat the procedure until clear fluid only (no bubbles) is expelled into the master cylinder reservoir.

   f. Remove the short brake lines and plug the outlet ports. Remove the master cylinder from the holding fixture.

6. Position the master cylinder over the booster pushrod and booster mounting studs. Install the nuts and tighten to 16–21 ft. lbs. (21–29 Nm).

7. Remove the plugs and connect the brake lines. Tighten the fittings.

8. Make sure the master cylinder reservoir is full. To finish bleeding the master cylinder:

   a. Have an assistant push down on the brake pedal.

   b. When the pedal is down, crack open the brake line fittings at the master cylinder one at a time, to expel any remaining air in the master cylinder.

   c. With the pedal still down, tighten the fitting, then allow the brake pedal to return.

   d. Repeat until all air is expelled from the master cylinder. Final tighten the brake line fittings to 10–18 ft. lbs. (14–24 Nm).

9. Make sure the master cylinder reservoir is full. To finish bleeding the brake system, proceed as follows:

   a. Have an assistant push down on the brake pedal.

   b. When the pedal is all the way down, crack open the brake bleeder screws, one at a time beginning at the rear wheels, to expel any remaining air in the hydraulic brake system.

   c. With the pedal still down, tighten the bleeder screw, then allow the brake pedal to return.

   d. Repeat until all air is expelled from the brake lines. Tighten the brake bleeder screws.

   e. If the brake pedal feels spongy, repeat the brake bleeding procedure.

10. Connect the brake warning light electrical connector.

11. Make sure the master cylinder reservoir is full.

12. Connect the negative battery cable.

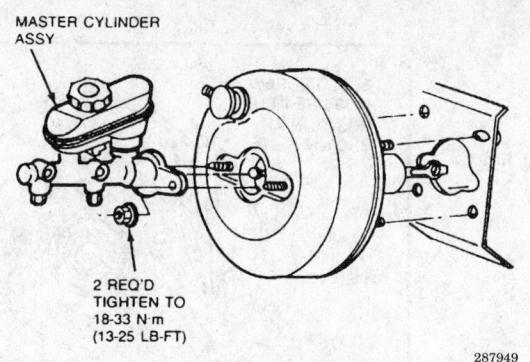

MASTER CYLINDER
ASSY

2 REQ'D
TIGHTEN TO
18-33 N·m
(13-25 LB-FT)

287949

Master cylinder

13. Check for fluid leaks and check for proper operation.

Brake Caliper

REMOVAL AND INSTALLATION

1. Raise and safely support the vehicle.
2. Remove the wheel and tire assembly.
3. Disconnect the brake hose from the disc brake caliper by removing the banjo bolt and copper sealing washers. Plug the brake hose end and discard the washers.
4. Remove the 2 disc brake caliper locating pins using a T40 Torx® drive bit or equivalent.
5. Lift the caliper off the disc brake rotor and the steering knuckle and anchor plate, using a rotating motion.

NOTE: If the brake hose is not to be removed from the disc brake caliper, make sure to hang the caliper with wire to prevent damage to the brake hose.

To install:
6. Retract the caliper piston fully into its bore using a retracting tool or C-clamp and an old disc brake pad or block of wood.

NOTE: Do not press directly against the plastic piston or damage to the piston may result.

7. Position the caliper assembly above the disc brake rotor with the anti-rattle spring under the upper arm of the steering knuckle. Install the caliper over the rotor with a rotating motion. Ensure the inner disc brake pad is properly positioned.

NOTE: Verify the correct caliper assembly is installed on the correct steering knuckle. The cal-iper bleed screw should be positioned on top of the caliper when the assembled is installed.

8. Lubricate the 2 caliper locating pins and the inside of the insulators with silicone grease or equivalent. Install the locating pins through the caliper insulators and into the steering knuckle attaching holes. The caliper locating pins must be inserted and threads started by hand.
9. Using a T40 Torx® drive bit or equivalent, tighten the caliper locating pins to 18–25 ft. lbs. (24–34 Nm).
10. Remove the plug from the brake hose end and install 2 new copper sealing washers to the banjo bolt, 1 on each side of the brake hose end, and fit the brake hose to the disc brake caliper. Tighten the banjo bolt to 44 ft. lbs. (60 Nm).
11. Bleed the brake system of any air, filling the master cylinder as required using clean DOT 3 or equivalent brake fluid.
12. Install the wheel and tire assembly. Tighten the lug nuts to 80–105 ft. lbs. (115–142 Nm).
13. Lower the vehicle.
14. Pump the brake pedal several times prior to moving the vehicle to position the brake pads to the rotor.
15. Road test the vehicle and check for leaks and proper brake operation.

Disc Brake Pads

REMOVAL AND INSTALLATION

1. Remove the master cylinder reservoir cap and check the fluid level in the reservoir. Remove enough brake fluid until the reservoir is ½ full. Discard the removed fluid.
2. Raise and safely support the vehicle.
3. Remove the wheel and tire assembly.
4. Remove the 2 disc brake caliper locating pins.

5. Lift the caliper assembly from the steering knuckle and rotor using a rotating motion. Suspend the caliper with wire to prevent damage to the bake hose. Do not pry directly against the plastic caliper piston or damage will occur.
6. Remove both disc brake pads.
7. Inspect the brake rotor braking surfaces. Machine or replace the brake rotor as required.
To install:
8. Seat the caliper piston in its bore using an appropriate tool or a C-clamp and block of wood.

NOTE: Extra care must be taken during this procedure to prevent damage to the caliper piston. Metal or sharp objects cannot come into direct contact with the piston surface or damage will result.

9. Remove all rust buildup from the inside of the caliper where the outer disc brake pad makes contact.
10. Install the inner disc brake pad to the caliper and caliper piston. Do not bend the anti-rattle clips during the installation of the pad.
11. Install the correct outer disc brake pad. Ensure the clips are properly seated.
12. Install the caliper over the rotor and steering knuckle. Install the 2 locating pins and tighten to 18–25 ft. lbs. (24–34 Nm).
13. Install the wheel and tire assembly. Tighten the lug nuts to 80–105 ft. lbs. (109–142 Nm).
14. Lower the vehicle.
15. Pump the brake pedal prior to moving the vehicle to position the brake pads and piston.
16. Check the fluid level in the master cylinder and add as required.
17. Road test the vehicle and check for proper brake operation.

Brake Rotor

REMOVAL AND INSTALLATION

1. Raise and safely support the vehicle.
2. Remove the wheel and tire assembly.
3. Remove the 2 disc brake caliper locating pins.
4. Lift the caliper assembly from the steering knuckle and rotor using a rotating motion. Do not pry directly against the plastic piston or damage will occur.
5. Position the caliper aside and support it with a length of wire to avoid damaging the brake hose.

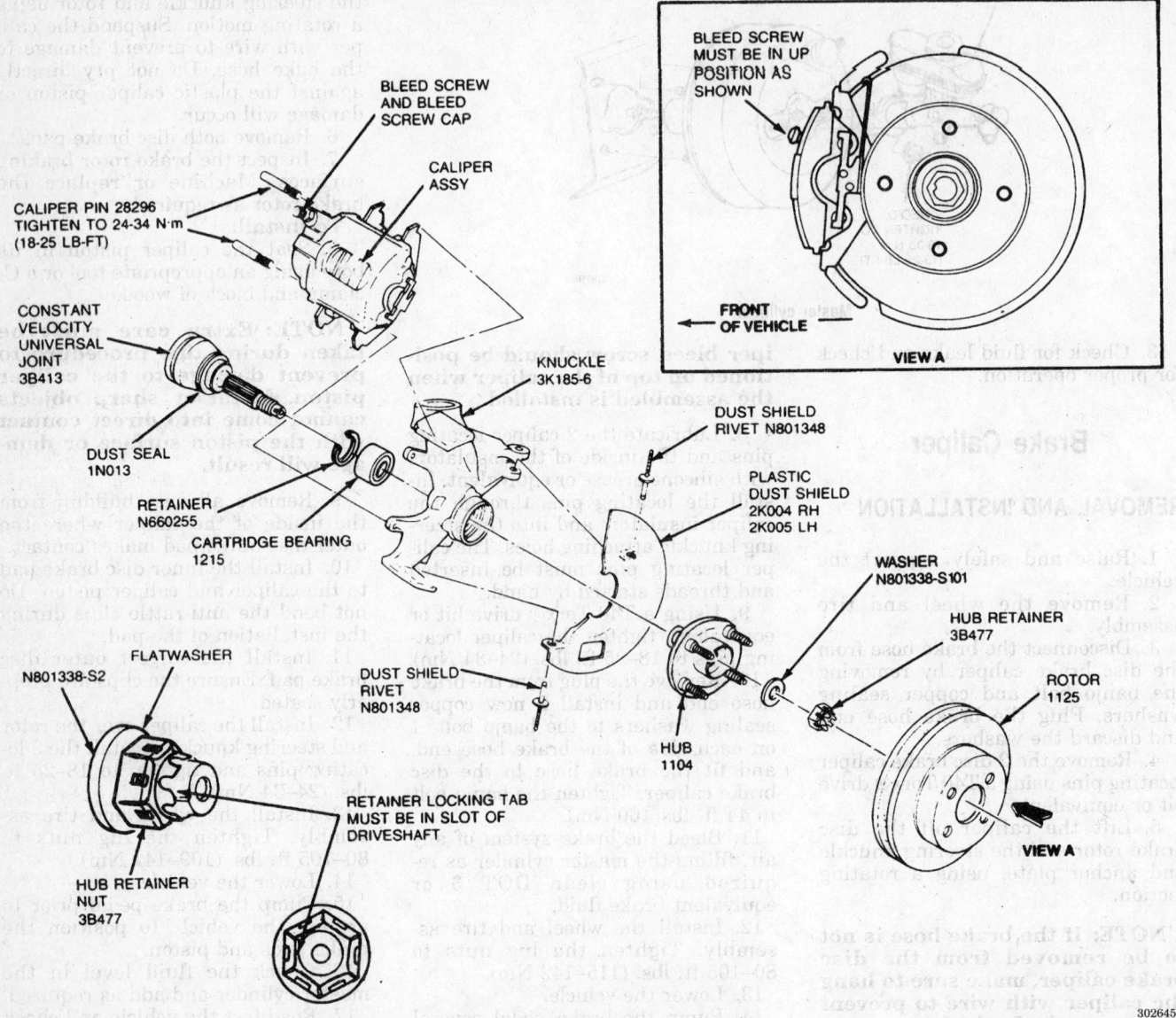

BLEED SCREW
AND BLEED
SCREW CAP

CALIPER PIN 28296
TIGHTEN TO 24-34 N·m
(18-25 LB-FT)

CALIPER
ASSY

CONSTANT
VELOCITY
UNIVERSAL
JOINT
3B413

KNUCKLE
3K185-6

DUST SEAL
1N013

DUST SHIELD
RIVET N801348

RETAINER
N660255

PLASTIC
DUST SHIELD
2K004 RH
2K005 LH

CARTRIDGE BEARING
1215

WASHER
N801338-S101

FLATWASHER

HUB RETAINER
3B477

N801338-S2

DUST SHIELD
RIVET
N801348

ROTOR
1125

HUB
1104

HUB RETAINER
NUT
3B477

RETAINER LOCKING TAB
MUST BE IN SLOT OF
DRIVESHAFT

VIEW A

302645

BLEED SCREW
MUST BE IN UP
POSITION AS
SHOWN

FRONT
OF VEHICLE

VIEW A

Brake rotor and related components

6. Remove the rotor from the hub assembly by pulling it off the hub studs. Inspect the rotor and refinish or replace, as necessary.

To install:

7. If the disc brake rotor is being replaced, remove the protective coating from the new rotor with carburetor cleaner or similar solvent. If the original rotor is being installed, make sure the rotor braking and mounting surfaces are clean.

8. Install the rotor on the hub assembly.

9. Install the caliper assembly.

10. Install the wheel and tire assembly. Tighten the lug nuts to 80–105 ft. lbs. (109–142 Nm).

11. Lower the vehicle.

12. Pump the brake pedal several times prior to moving the vehicle to position the disc brake pads.

13. Road test the vehicle and check for proper brake operation.

Brake Drums

REMOVAL AND INSTALLATION

1. Raise and safely support the vehicle.

2. Remove wheel and tire assembly.

3. Remove grease cap from hub. Remove cotter pin, nut lock, adjusting nut and keyed flat washer from spindle. Remove outer bearing.

4. Remove hub and drum assembly as a unit.

NOTE: If the hub/drum assembly will not come off, pry the rubber plug from the backing plate inspection hole. On vehicles with 7 in. brakes, insert a suitable tool in the hole until it contacts the adjuster assembly pivot. Apply side pressure on this pivot point to allow the adjuster quadrant to ratchet and release the brake adjustment. On vehicles with 8 in. brakes, remove the brake line-to-axle retention bracket. This will allow sufficient room for insertion of suitable tools to disengage the adjusting lever and back-off the adjusting screw.

5. Inspect the brake drum and refinish or replace, as necessary. If refinishing, check the maximum inside diameter specification.

To install:

6. Inspect and lubricate bearings, as necessary. Replace grease seal if any damage is visible.
7. Clean spindle stem and apply a thin coat of wheel bearing grease.
8. Install hub and drum assembly on spindle.
9. Install outer bearing into hub on spindle.
10. Install keyed flat washer and adjusting nut. Tighten nut finger-tight.
11. Adjust wheel bearing. Install nut retainer and a new cotter pin.
12. Install grease cap.
13. Install wheel and tire assembly. Tighten wheel nuts to 80–105 ft. lbs. (109–142 Nm).
14. Pump brake pedal prior to moving vehicle to position brake linings.
15. Connect negative battery cable.
16. Road test vehicle.

Wheel Cylinder

REMOVAL AND INSTALLATION

1. Raise and safely support the vehicle. Remove wheel/tire and hub/drum assemblies.
2. Remove brake shoe assembly.
3. Disconnect hydraulic fitting from wheel cylinder.
4. Remove wheel cylinder attaching bolts and remove wheel cylinder.

NOTE: Use caution to prevent brake fluid from contacting brake linings and drum braking surface. Contaminated linings must be replaced.

To install:

5. Ensure ends of hydraulic fittings are free of foreign matter before making connections.
6. Position wheel cylinder and foam seal on backing plate and finger-tighten the hydraulic fitting to cylinder.
7. Secure cylinder to backing plate by installing attaching bolts. Tighten bolts to 8–10 ft. lbs. (10–14 Nm).
8. Tighten hydraulic fitting.
9. Install and adjust brakes.
10. Install hub/drum and wheel assembly.
11. Bleed brake system and lower the vehicle.

Brake Shoes

REMOVAL AND INSTALLATION

7 Inch Brake Drum

1. Raise and safely support the vehicle.
2. Remove wheel and tire assembly.
3. Remove hub and drum assembly.
4. Remove hold-down spring and pins.
5. Lift brake shoe and adjuster assembly up and away from anchor block and shoe guide. Do not damage the boots when rotating shoes off the wheel cylinder.
6. Remove lower shoe-to-shoe spring from leading and trailing shoe slots.
7. Hold brake shoe/adjuster assembly, remove leading shoe-to-adjuster strut retracting spring. This can be done by rotating shoe over adjuster quadrant until spring is slack and then disconnecting spring. The leading shoe should now be free.
8. Remove trailing shoe-to-parking brake strut retracting spring by pivoting strut downward until it disengages from trailing shoe.
9. Disassemble adjuster, if necessary, by pulling quadrant away from knurled pin in strut and rotating quadrant in either direction until quadrant teeth are no longer meshed with pin. Remove spring and slide quadrant out of slot. Do not overstress spring during disassembly.
10. Remove parking brake lever from trailing shoe and lining assembly by removing horseshoe retaining clip and spring washer, and lifting lever off pin on brake shoe.

To install:

11. Apply light coating of high temperature grease at points where brake shoes contact the backing plate.
12. Apply light uniform coating of multi-purpose lubricant to strut at contact surface between strut and adjuster quadrant.
13. Install adjuster quadrant pin into slot in strut and install adjuster spring. Pivot quadrant until it meshes with knurled pin in third and fourth notch of outboard end of quadrant.
14. Assemble parking brake lever to trailing shoe. Install spring washer and new horseshoe clip. Crimp clip until lever is securely fastened.
15. Install trailing shoe-to-parking brake strut retracting spring by attaching spring to slots in each part and pivoting strut into position to tension spring. Make sure the end of the spring, with the hook that is parallel to the center line of the coils, is installed in hole in shoe web. Installed spring should be flat against shoe web and parallel to strut.
16. Install lower shoe-to-shoe retracting spring between leading and trailing shoes. The spring hook with the longest straight section fits into hole in trailing shoe.
17. Install leading shoe-to-adjuster/strut retracting spring by installing spring to both parts and pivoting leading shoe over quadrant into position to tension spring.
18. Expand shoe and strut assembly to fit over anchor plate and wheel cylinder piston inserts.
19. Attach parking brake cable to parking brake lever.
20. Install hold-down pins and springs on each shoe and lining assembly.
21. Set brake shoe diameter using a suitable brake adjusting gauge.
22. Install hub/drum and wheel and tire assemblies.
23. Adjust wheel bearings.
24. Lower vehicle and check brake operation.

8 Inch Brake Drum

1. Raise and safely support the vehicle.
2. Remove the wheel, tire, and hub and drum assembly.
3. Remove 2 shoe hold-down springs and pins.
4. Lift the brake shoes, springs and adjuster assembly off backing plate and wheel cylinder assembly. Be careful not to bend adjusting lever during removal.
5. Remove the parking brake cable from the parking brake lever.
6. Remove the retracting springs from the lower brake shoe attach-

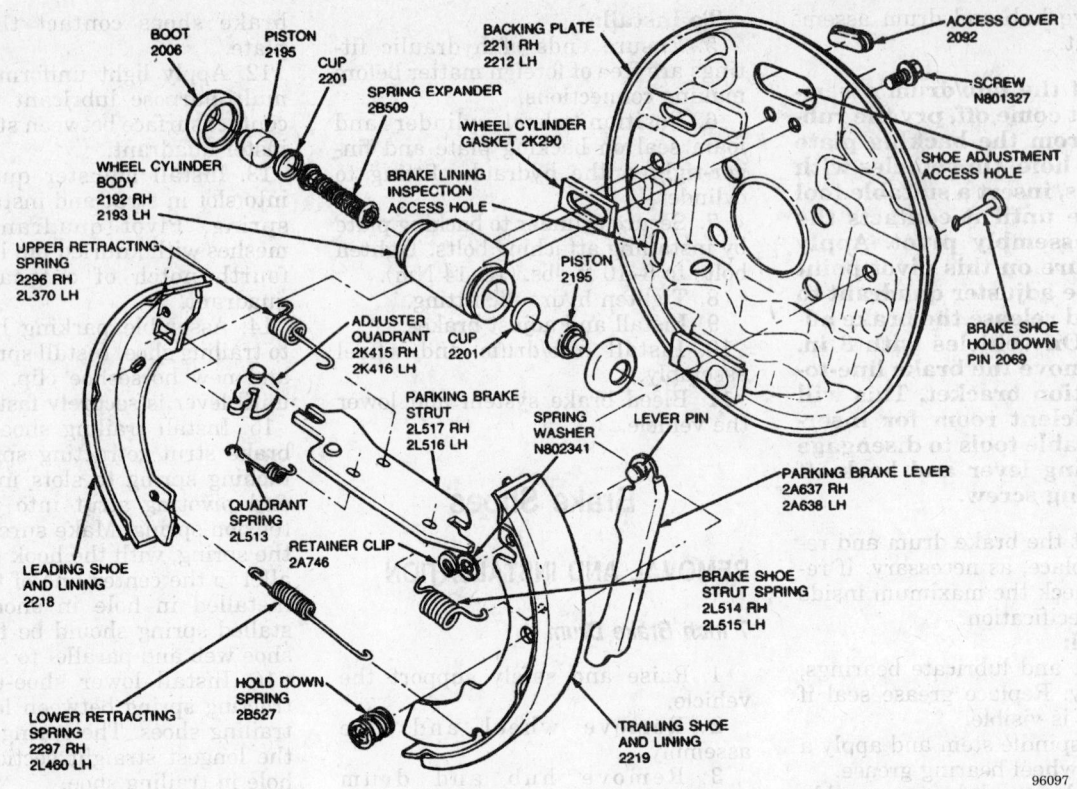

Drum brake, 7 inch

BOOT 2006
PISTON 2195
CUP 2201
SPRING EXPANDER 2B509
WHEEL CYLINDER BODY 2192 RH 2193 LH
BRAKE LINING INSPECTION ACCESS HOLE
BACKING PLATE 2211 RH 2212 LH
WHEEL CYLINDER GASKET 2K290
ACCESS COVER 2092
SCREW N801327
SHOE ADJUSTMENT ACCESS HOLE
UPPER RETRACTING SPRING 2296 RH 2L370 LH
ADJUSTER QUADRANT 2K415 RH 2K416 LH
CUP 2201
PISTON 2195
BRAKE SHOE HOLD DOWN PIN 2069
PARKING BRAKE STRUT 2L517 RH 2L516 LH
SPRING WASHER N802341
LEVER PIN
PARKING BRAKE LEVER 2A637 RH 2A638 LH
QUADRANT SPRING 2L513
RETAINER CLIP 2A746
LEADING SHOE AND LINING 2218
BRAKE SHOE STRUT SPRING 2L514 RH 2L515 LH
LOWER RETRACTING SPRING 2297 RH 2L460 LH
HOLD DOWN SPRING 2B527
TRAILING SHOE AND LINING 2219

96097

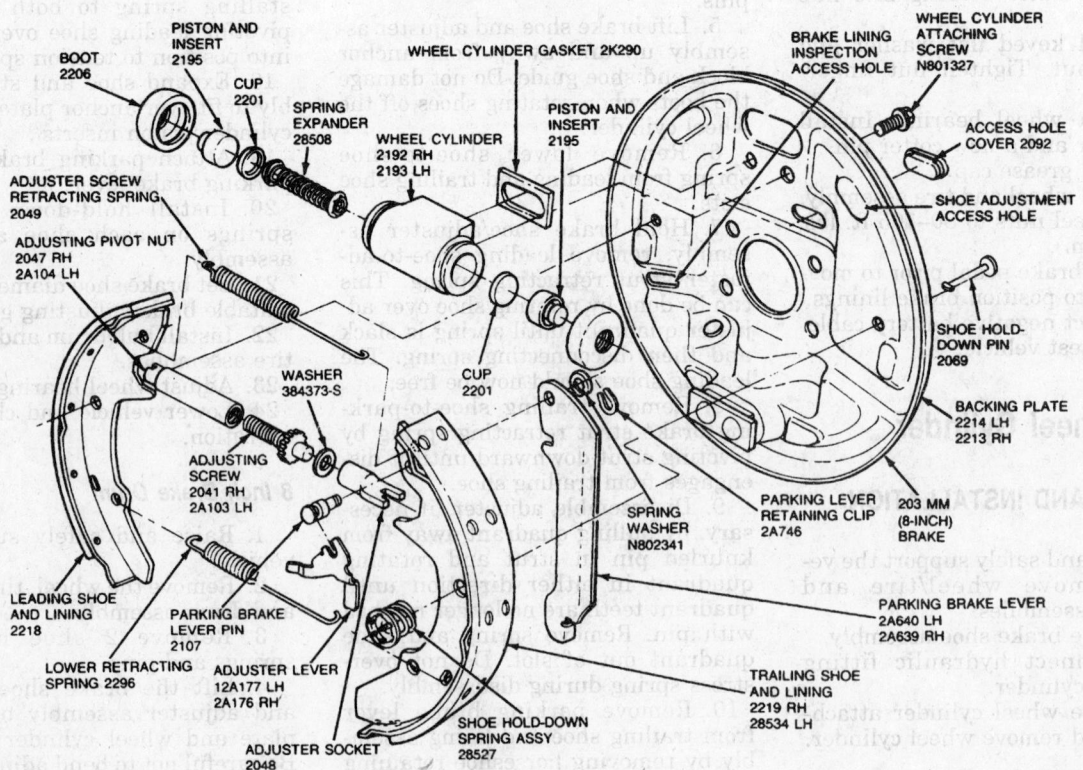

Drum brake, 8 inch

BOOT 2206
PISTON AND INSERT 2195
CUP 2201
SPRING EXPANDER 28508
WHEEL CYLINDER 2192 RH 2193 LH
WHEEL CYLINDER GASKET 2K290
BRAKE LINING INSPECTION ACCESS HOLE
PISTON AND INSERT 2195
BRAKE LINING INSPECTION ACCESS HOLE
WHEEL CYLINDER ATTACHING SCREW N801327
ACCESS HOLE COVER 2092
SHOE ADJUSTMENT ACCESS HOLE
ADJUSTER SCREW RETRACTING SPRING 2049
ADJUSTING PIVOT NUT 2047 RH 2A104 LH
WASHER 384373-S
CUP 2201
ADJUSTING SCREW 2041 RH 2A103 LH
LEADING SHOE AND LINING 2218
PARKING BRAKE LEVER PIN 2107
LOWER RETRACTING SPRING 2296
ADJUSTER LEVER 12A177 LH 2A176 RH
ADJUSTER SOCKET 2048
SPRING WASHER N802341
SHOE HOLD-DOWN SPRING ASSY 28527
SHOE HOLD-DOWN PIN 2069
BACKING PLATE 2214 LH 2213 RH
PARKING LEVER RETAINING CLIP 2A746
203 MM (8-INCH) BRAKE
PARKING BRAKE LEVER 2A640 LH 2A639 RH
TRAILING SHOE AND LINING 2219 RH 28534 LH

96098

ments and upper shoe-to-adjusting lever attachment points. This will separate the brake shoes and disengage the adjuster mechanism.

7. Remove the horseshoe retaining clip and spring washer and slide the lever off the parking brake lever pin on the trailing shoe.

To install:

8. Apply a light coating of high temperature grease at the points where the brake shoes contact the backing plate.

9. Apply a light coating of lubricant to the adjuster screw threads and the socket end of the adjusting screw. Install the stainless steel washer over the socket end of the adjusting screw and install the socket. Turn the adjusting screw into the adjusting pivot nut to the limit of the threads and then back-off ½ turn.

10. Assemble the parking brake lever to the trailing shoe by installing the spring washer and a new horseshoe retaining clip. Crimp the clip until it retains the lever to the shoe securely.

11. Attach the parking brake cable to the parking brake lever.

12. Attach the lower shoe retracting spring to the leading and trailing shoe assemblies and install to backing plate. It will be necessary to stretch the retracting spring as the shoes are installed downward over the anchor plate to inside of shoe retaining plate.

13. Install the adjuster screw assembly between the leading shoe slot and the slot in the trailing shoe and parking brake lever. The adjuster socket end slot must fit into the trailing shoe and parking brake lever.

NOTE: The adjuster socket blade is marked R or L for the right or left brake assemblies. The adjuster blade must be installed with the letter R or L in the upright position, facing the wheel cylinder. The deeper of the 2 slots in the adjuster socket fits into the parking brake lever.

14. Assemble the adjuster lever in the groove located in the parking brake lever pin and into the slot of the adjuster socket that fits into the trailing shoe web.

15. Attach the upper retracting spring to the leading shoe slot. Using a suitable spring tool, stretch the other end of the spring into the notch on the adjuster lever. If the adjuster lever does not contact the star wheel after installing the spring, it is possible that the adjuster socket is installed incorrectly.

16. Set the brake shoe diameter using a suitable brake adjusting gauge.

17. Install the hub/drum and wheel/tire assemblies and adjust the wheel bearings.

18. Lower the vehicle and check brake operation.

Parking Brake Cable

ADJUSTMENT

NOTE: The rear brake shoes should be properly adjusted before adjusting the parking brake.

1. With the engine running, apply approximately 100 lbs. pedal effort to the hydraulic service brake 3 times before adjusting the parking brake. Stop the engine.

2. Block the front wheels and place the transaxle in **N**. Raise and safely support the rear of the vehicle just enough to rotate the wheels.

3. Place the parking brake control assembly in the 12th notch position, 2 notches from full application. Tighten the adjusting nut until approximately 1 in. (25mm) of threaded rod is exposed beyond the nut. Release the parking brake control and rotate the rear wheels by hand. There should be no brake drag.

4. If the brakes drag when the control assembly is fully released, or the handle travels too far on full apply, repeat the procedure and adjust the nut accordingly.

REMOVAL AND INSTALLATION

Control Assembly

1. Disconnect the negative battery cable.

2. Place the control assembly in the seventh notch position and remove the adjusting nut. Completely release the control assembly.

3. Remove the 2 bolts that attach the control assembly to the floor pan.

4. Disconnect the brake light and ground wire from the control assembly.

5. Remove the control assembly from the vehicle.

To install:

6. Install the adjusting rod into the control assembly clevis and position the control assembly on the floor pan.

7. Install the brake light and ground wire.

8. Install the adjusting nut and adjust the parking brake.

9. Connect the negative battery cable.

Cable Removal and Installation

1. Place control assembly in seventh notch position and loosen adjusting nut. Completely release control assembly.

2. Raise and safely support the vehicle. Remove rear parking brake cable from equalizer.

3. Remove hairpin clip holding cable to floor pan tunnel bracket.

4. Remove wire retainer holding cable to fuel tank mounting bracket. Remove cable from wire retainer. Remove cable and clip from the fuel pump bracket.

5. Remove screw holding cable retaining clip to rear sidemember. Remove cable from clip.

6. Remove the wheel and tire assembly and rear brake drum.

7. Disengage cable end from brake assembly parking brake lever. Depress cable prongs holding cable to backing plate. Remove cable through hole in backing plate.

To install:

8. Insert cable through hole in backing plate. Attach cable end to rear brake assembly parking brake lever.

9. Insert conduit end fitting into backing plate. Ensure retention prongs are locked into place.

10. Insert cable into rear attaching clip and attach clip to rear sidemember with screw.

11. Route cable through bracket in floorpan tunnel and install hairpin retaining clip.

12. Install cable end into equalizer.

13. Insert cable into wire retainer and snap retainer into hole in fuel tank mounting bracket. Insert cable and install clip into suspension torque box bracket.

14. Install rear drum, wheel and tire assembly and wheel cover.

15. Lower vehicle.

16. Adjust parking brake.

Brake System Bleeding

1. Clean all the dirt from around the master cylinder filler cap.

2. Fill the reservoir with barke fluid. The reservoir must be at least ³/₄ full throughout the bleeding procedure.

3. If the master cylinder is known or suspected to have air in the bore, it must be bled before any wheel cylinders or calipers.

4. To bleed the master cylinder, loosen 2 outlet fitting approximately

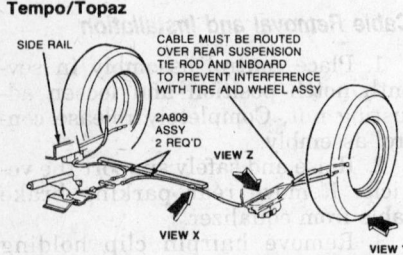

Tempo/Topaz

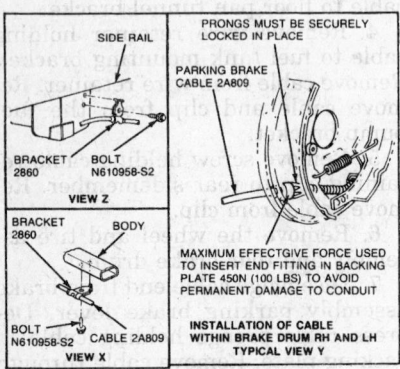

Parking brake cable

³/₄ turn. Have an assistant push the brake pedal down slowly through full travel. Close the outlet fitting, then return the pedal slowly to the full released position. Wait 5 seconds, then repeat the operation until the air bubbles cease to appear.

5. Loosen the other outlet fitting approximately ³/₄ turn and repeat Step 4.

6. To continue to bleed the system, remove the rubbler dust cap from the wheel cylinder bleeder fitting or caliper fitting. Check to make sure the bleeder fitting is positioned at the upper half on the front of the caliper; if not, the caliper is located on the wrong side.

7. Attach a suitable length of rubber hose to the fitting. Submerge the free end of the hose in a container partially filled with clean brake fluid and loosen the bleeder fitting approximately ³/₄ turn.

8. Have the assistant push brake pedal down slowly through full travel. Close the bleeder fitting, then return the pedal to the full released position. Wait 5 seconds, then repeat this operation until the air bubbles cease to appear at the submerged end of the bleeder hose.

9. When the fluid is completely free of air bubbles, properly tighten

the bleeder fitting and reinstall the rubber dust cap. Repeat this process on the opposite diagonal system. Refill the master cylinder reservoir after each wheel cylinder or caliper is bled and reinstall the master cylider cap.

NOTE: If all wheels are to be bled, proceed as follows: right rear, left front, left rear and right front.

10. When the bleeding operation is complete, the fluid level should be filled to the maximum fill level indicated on the reservoir. Always ensure the disc brake pistons are returned to their normal positions by depressing the brake pedal several times until the normal pedal travel is established. Check the pedal feel. If the pedal feels spongy, repeat the bleeding procedure.

FRONT SUSPENSION

Strut

REMOVAL AND INSTALLATION

NOTE: All vehicles are equipped with gas pressurized shock absorbers which will extend unassisted. Do not apply heat or flame to the shock strut tube during removal.

1. Loosen but do not remove the 2 top mount-to-shock tower nuts.
2. Raise and safely support the vehicle. Raise vehicle to a point where it is possible to reach the 2 top mount-to-shock tower nuts and the strut-to-knuckle pinch bolt.
3. Remove the wheel and tire assembly.
4. Remove the brake flex line-to-strut bolt.
5. Remove the strut-to-knuckle pinch bolt.
6. Using a suitable tool, spread the knuckle-to-strut pinch joint slightly.
7. Using a suitable bar, pry down on the knuckle until the strut separates from the knuckle. Be careful not to pinch the brake hose. Do not pry against the brake caliper or brake hose bracket.
8. Remove the 2 top mount-to-shock tower nuts and remove the strut from the vehicle.

9. To disassemble the strut:
a. Install the strut into a suitable spring compressor and compress the spring.
b. Place a deep 18mm socket on the strut shaft nut. Insert an 8mm deep socket with a ¼ inch drive wrench. Remove the top shaft mounting nut from the shaft while holding the ¼ inch drive socket with a suitable extension.
c. Loosen the spring compressor tool and remove the top mount bracket assembly, bearing, insulator and spring.

NOTE: Do not attempt to remove the shaft nut by turning the shaft and holding the nut. The nut must be turned and the shaft held to avoid possible damage to the shaft.

To install:
10. To assemble the strut:
a. Install the replacement strut in the spring compressor.
b. Install the spring, insulator, bearing and top mount bracket assembly. Make sure these parts are assembled in the correct sequence and proper position. If the bearing and seal assembly are out of position, the bearing will be damaged.
c. Install the top shaft mounting nut while holding the shaft with a ¼ drive 8mm deep socket and extension. Tighten the nut to 35–50 ft. lbs. (48–68 Nm).
11. Install the strut assembly in the vehicle. Install 2 top mount-to-shock tower nuts. Tighten to 25–30 ft. lbs. (37–41 Nm).
12. Slide the strut mounting flange onto the knuckle.
13. Install the strut-to-knuckle pinch bolt. Tighten to 68–80 ft. lbs. (92–110 Nm).
14. Install the brake flex line-to-strut bolt.
15. Install the wheel and tire assembly.
16. Lower the vehicle.
17. Check the alignment.

Lower Ball Joints

REMOVAL AND INSTALLATION

The lower ball joint is integral to the lower control assembly and connot be serviced individually. Any movement of the lower ball joint detected as a result of inspection requires replacement of the lower control assembly.

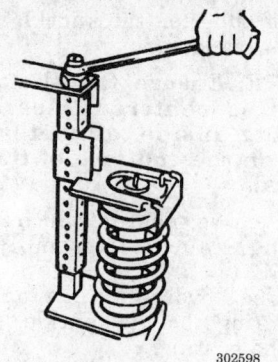

Spring compressor

302598

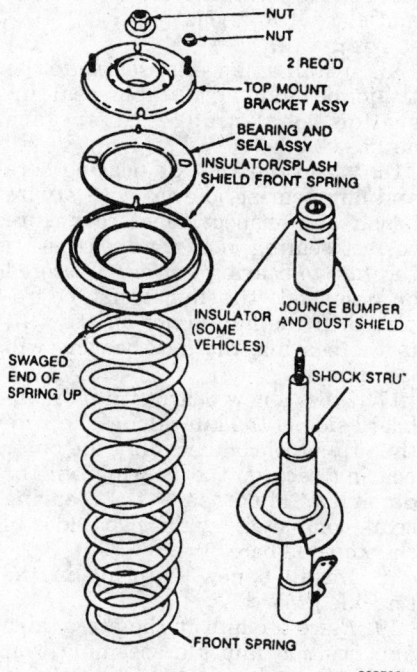

NUT
NUT
2 REQ'D
TOP MOUNT BRACKET ASSY
BEARING AND SEAL ASSY
INSULATOR/SPLASH SHIELD FRONT SPRING
JOUNCE BUMPER AND DUST SHIELD
INSULATOR (SOME VEHICLES)
SWAGED END OF SPRING UP
SHOCK STRUT
FRONT SPRING

302599

Strut and spring assembly

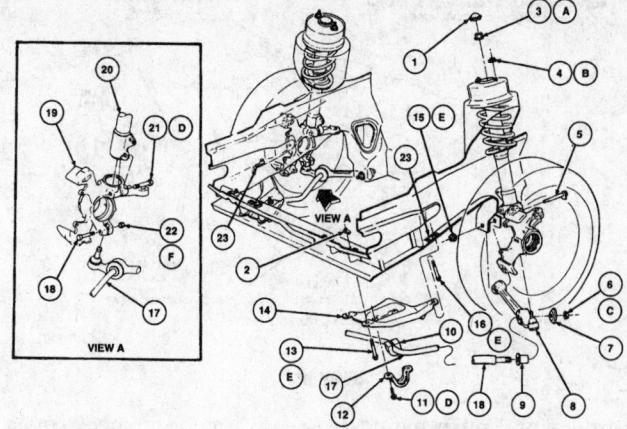

Front suspension, exploded view

111068

Item	Part Number	Description
1	18A179	Cap (2 Req'd)
2	N807050-S56	Nut (4 Req'd)
3A	N802613-S100	Nut (2 Req'd)
4B	N801310-S100	Nut (4 Req'd)
5	N803336-S150	Bolt (2 Req'd)
6C	N801311-S101	Nut (2 Req'd)
7	N801649-S5	Washer (2 Req'd)
8	3051 LH / 3042 RH	Control Arm
9	5490	Spacer (2 Req'd)
10	5484	Bushing (2 Req'd)
11D	N806734-S60	Bolt (4 Req'd)
12	5486	U-Bracket (2 Req'd)
13E	N806731-S60	Bolt (4 Req'd)
14	5486	Stabilizer Bar Bracket
15	N800937-S160	Nut (2 Req'd)
16E	N804522-S151	Bolt (2 Req'd)

(Continued)

Item	Part Number	Description
17	5494	Stabilizer Bar
18	N801588-S100	Bolt (2 Req'd)
19	3K185 RH / 3K186 LH	Knuckle Assy
20	18045	Strut Assy
21D	N801942-S100	Bolt (2 Req'd)
22F	N801308-S100	Nut (2 Req'd)
23	N802642-S161	Nut (2 Req'd)
A		Tighten to 47-63 N-m (35-46 Lb-Ft)
B		Tighten to 30-40 N-m (23-29 Lb-Ft)
C		Tighten to 133-153 N-m (99-112 Lb-Ft)
D		Tighten to 91-104 N-m (67-77 Lb-Ft)
E		Tighten to 63-72 N-m (47-53 Lb-Ft)
F		Tighten to 48-55 N-m (35-41 Lb-Ft)

Lower Control Arms

REMOVAL AND INSTALLATION

NOTE: The ball joint is part of the control arm and cannot be replaces separately.

1. Raise and safely support the vehicle. Make sure steering column is unlocked.
2. Remove nut from stabilizer bar end. Pull off large dished washer.
3. Remove lower control arm inner pivot nut and bolt.
4. Remove lower control arm ball joint pinch bolt. Using a suitable tool, slightly spread knuckle pinch joint and separate control arm from steering knuckle. A drift punch may be used to remove bolt but do not use a hammer to separate ball joint from knuckle.

— CAUTION —

Do not allow the steering knuckle/halfshaft to move outward. Over extension of the tripod CV-joint could result in separation of internal parts, causing failure of the CV-joint.

5. Remove stabilizer bar spacer from the arm bushing.

To install:

6. Assemble lower control arm ball joint stud to the steering knuckle, ensuring that the ball stud groove is properly positioned.
7. Insert a new pinch bolt and nut. Tighten to 38–40 ft. lbs. (52–55 Nm).
8. Insert stabilizer bar spacer into arm bushing with flange facing forward.
9. Clean stabilizer bar threads to remove dirt and contamination.
10. Position lower control arm onto stabilizer bar and position lower control arm to the inner underbody mounting. Install a new nut and bolt. Tighten to 48–55 ft. lbs. (65–74 Nm).

11. Assemble stabilizer bar, dished washer and a new nut to stabilizer. Tighten nut to 98–115 ft. lbs. (132–156 Nm).
12. Lower vehicle.

Sway Bar

REMOVAL AND INSTALLATION

1. Raise and safely support the vehicle.
2. Remove nut from stabilizer bar at each lower control arm and pull off large dished washer. Discard nuts.
3. Remove stabilizer bar insulator U-bracket bolts and U-brackets and remove stabilizer bar assembly. Discard bolts.

NOTE: Stabilizer bar U-bracket insulators can be serviced without removing the stabilizer bar assembly.

To install:

4. Slide new insulators onto the stabilizer bar and position them in the approximate location.
5. Clean stabilizer bar threads to remove dirt and contamination.
6. Install spacers into the control arm bushings from forward side of control arm so washer end of spacer will seat against stabilizer bar

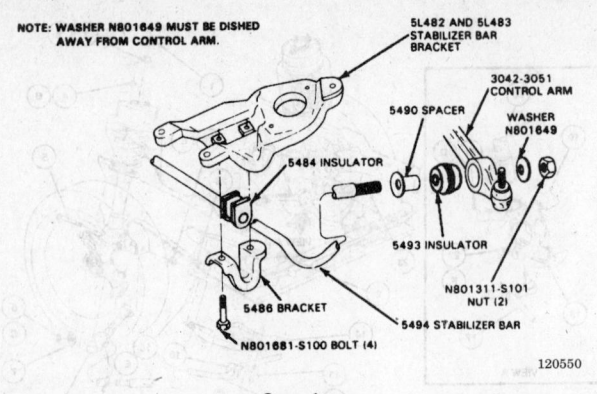

NOTE: WASHER N801649 MUST BE DISHED AWAY FROM CONTROL ARM.

5L482 AND 5L483 STABILIZER BAR BRACKET

3042-3051 CONTROL ARM

5490 SPACER

WASHER N801649

5484 INSULATOR

5493 INSULATOR

N801311-S101 NUT (2)

5486 BRACKET

5494 STABILIZER BAR

N801681-S100 BOLT (4)

120550

Sway bar

machined shoulder and push mounting brackets over insulators.

7. Insert end of stabilizer bar into the lower control arms. Using new bolts, attach the stabilizer bar and the insulator U-brackets to the bracket assemblies. Hand start all 4 U-bracket bolts. Tighten all bolts halfway, then tighten bolts to 82–88 ft. lbs. (110–120 Nm).

8. Using new nuts and the original dished washers (dished side away from bushing), attach the stabilizer bar to the lower control arm. Tighten nuts to 99–112 ft. lbs. (133–153 Nm).

9. Lower vehicle.

Front Wheel Bearings

ADJUSTMENT

There is no adjustment for the front wheel bearings due to the cartridge design which contains both inner and outer bearings. The cartridge bearings are sealed and permanently lubricated.

REMOVAL AND INSTALLATION

NOTE: New wheel bearings must be installed any time they are removed.

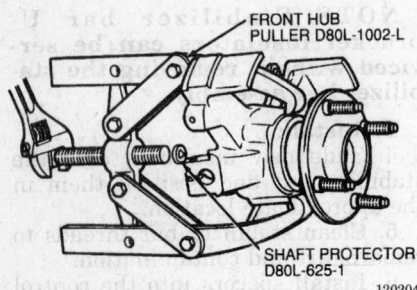

FRONT HUB PULLER D80L-1002-L

SHAFT PROTECTOR D80L-625-1

120204

Hub and knuckle separation

1. Remove wheel cover/hub cover and loosen the hub nut. The hub nut must be discarded after removal.

2. Raise and safely support the vehicle. Remove wheel and tire assembly.

3. Without disconnecting the brake hose, remove brake caliper and support it with a length of wire. Do not allow caliper assembly to hang from brake hose.

4. Remove rotor from hub by pulling it off hub bolts. If rotor is difficult to remove from hub, strike rotor sharply between studs with a rubber or plastic hammer. If rotor will not pull off, apply rust penetrator to inboard and outboard rotor hub mating surfaces. Install a 3 jaw puller and remove rotor by pulling on rotor outside diameter and pushing on hub center.

NOTE: If excessive force is required for rotor removal, check rotor for lateral runout. Lateral runout must be checked with wheel nuts clamping that section of rotor.

5. Remove rotor splash shield.

6. Disconnect lower control arm and tie rod from knuckle (leave strut attached).

7. Loosen the 2 strut top mount-to-apron nuts.

8. Install a suitable hub removal tool and remove hub/bearing/knuckle assembly by pushing out CV-joint outer shaft until it is free of assembly.

9. Support knuckle with a length of wire, remove strut bolt and slide hub/knuckle assembly off strut.

10. Carefully remove support wire and transfer hub/bearing/knuckle assembly to bench.

11. Install front hub puller D80L–1002–L and shaft protector D80L–625–1 or equivalents, with

jaws of puller on the knuckle bosses and remove hub.

NOTE: Ensure the shaft protector is centered, clears the bearing inside diameter and rests on the end face of the hub journal.

12. Remove snapring which retains bearing knuckle assembly and discard.

13. Use a hydraulic press to remove the bearing from the knuckle and discard the bearing.

14. At the outer CV-joint, remove bearing dust seal by uniformly tapping on outer edge with a light-duty hammer and a small prybar. It may be necessary to remove the halfshaft and place it in a vise.

To install:

15. Install a new dust seal onto the CV-joint using a suitable seal installer. Seal flange must face outboard.

16. Clean the knuckle bearing bore and hub bearing journal with a wire brush and inspect them to ensure correct seating of a new bearing. If the hub-to-bearing surface is scored or damaged, the hub must be replaced. If the hub is not an exact fit in the bearing, the new bearing will fail quickly.

17. Press a new bearing into the inboard side of the knuckle. Make sure the press tool contacts only the outer bearing race or the bearing will be damaged. Make sure the bearing seats completely against shoulder of the knuckle bore.

18. Install a new snapring in the knuckle groove.

19. Place the hub on the press table and position knuckle assembly over the hub barrel with the outboard side down. Press the knuckle onto the hub. Make sure the press tool contacts only the inner bearing race or the bearing will be destroyed. Make sure the hub rotates freely in the knuckle after installation.

20. Install the hub/knuckle assembly on vehicle and attach strut loosely to knuckle. Lubricate CV-joint stub shaft splines with SAE 30 weight motor oil and insert shaft into hub splines as far as possible using hand pressure only. Check that splines are properly engaged.

21. Complete installation of front suspension components.

22. Install disc brake rotor to hub assembly.

23. Install disc brake caliper over rotor.

24. Ensure outer brake shoe spring end is seated under upper arm of knuckle.

25. Install wheel and tire assembly, tightening wheel nuts finger-tight.

26. Lower vehicle and block wheels to prevent vehicle from rolling.

27. Tighten wheel nuts to 85–105 ft. lbs. (115–142 Nm).

28. Manually thread the new hub nut onto the axle shaft as far as possible using a 30mm or 1³/₁₆ inch socket. Using hand tools only, torque the nut to 188–236 ft. lbs. (255–320 Nm). Do not use power or impact tools to tighten the hub nut. Do not move the vehicle before the nut is tightened.

NOTE: During tightening, an audible click will indicate proper ratchet function of the hub nut retainer. As the hub nut retainer tightens, ensure that one of the 3 locking tabs is in the slot of the CV-joint shaft. If the hub nut retainer is damaged, or more than 1 locking tab is broken, replace the hub nut retainer.

29. Install wheel cover or hub cover.

30. Remove wheel blocks.

REAR SUSPENSION

Strut

REMOVAL AND INSTALLATION

NOTE: All vehicles are equipped with gas-pressurized shock absorbers which will extend unassisted. Do not apply heat or flame to the shock strut during removal.

1. Open the luggage compartment and loosen but do not remove the 2 nuts retaining the upper strut mount to body.

2. Raise and safely support the vehicle.

3. Remove the wheel and tire assembly.

4. Place a jackstand under the control arms to support the suspension.

NOTE: Care should be taken when removing the strut that the rear brake flex hose is not stretched or the steel brake tube is not bent.

5. Remove the bolt attaching brake hose bracket to strut and move it aside.

6. Remove the 2 bolts attaching the strut to the spindle.

7. Remove the 2 upper mount-to-body nuts.

8. Remove the strut from vehicle.

9. Place the strut, spring and upper mount assembly in a suitable spring compressor.

— CAUTION —
Attempting to remove the spring from the strut without first compressing the spring with a tool designed for that purpose could cause bodily injury.

10. With the spring compressed, remove the strut shaft-to-mount retaining nut and then remove the spring, strut and mount from the compressor tool.

NOTE: Do not attempt to remove shaft nut by turning shaft and holding nut. Nut must be turned and shaft held to avoid possible fracture of shaft at base of hex.

To install:

11. With the spring compressed, install the spring, spring insulator, top mount and upper washer on the strut shaft.

12. Ensure spring is properly located in the upper and lower spring seats. The spring end must be within ¹³/₃₂ inch (10mm) of the step in the spring seat.

13. Tighten the shaft nut to 35–46 ft. lbs. (47–63 Nm). Use an 18mm deep socket to turn the nut and a ¼ inch drive 8mm deep socket to hold the shaft so it will not turn while tightening the nut.

14. Insert 2 upper mount studs into the strut tower and hand start 2 new nuts. Do not tighten at this time.

15. Position the spindle into the lower strut mount and install 2 new bolts. Tighten to 85–96 ft. lbs. (115–130 Nm).

16. Install the brake flex-hose bracket on the strut.

17. Install the wheel and tire assembly.

18. Remove the jackstand.

19. Lower the vehicle.

20. Tighten the 2 top mount-to-body nuts to 23–29 ft. lbs. (30–40 Nm).

21. Road test and check for proper operation.

Lower Control Arms

REMOVAL AND INSTALLATION

1. Raise and safely support the vehicle.

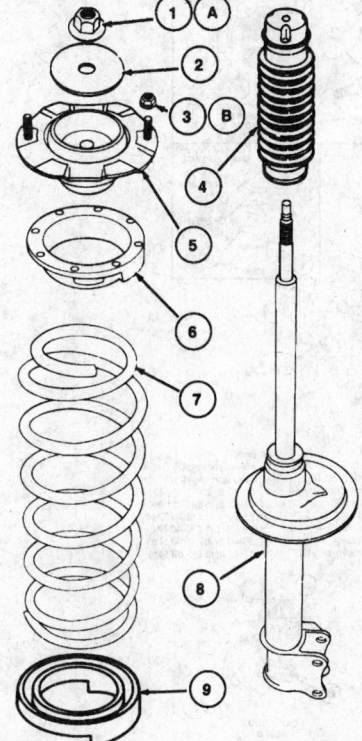

Item	Part Number	Description
1A	N802074-S150	Nut
2	N802552-S2	Washer
3B	N801310-S100	Nut (2 Req'd)
4	18K005	Jounce Bumper and Dust Shield
5	18169	Top Mount
6	5536	Insulator
7	5560	Rear Spring
8	18080	Shock Strut
9	5K617	Lower Insulator
A		Tighten to 47-63 N·m (35-46 Lb-Ft)
B		Tighten to 30-40 N·m (23-29 Lb-Ft)

302612

Rear strut and related components

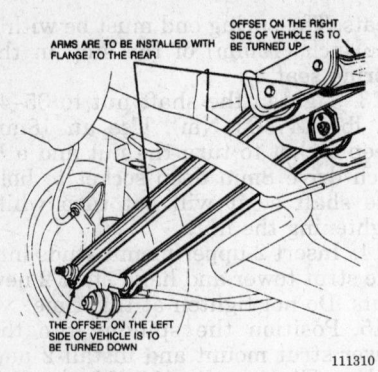

Rear suspension arms

2. Remove and discard arm-to-spindle bolt and nut.

3. Remove and discard center retaining bolt and nut.

4. Remove arm from vehicle.

To install:

NOTE: When installing new control arms, the bushing with the 0.39 in. (10mm) hole is installed to the center of the vehicle and the bushing with the 0.48 in. (12mm) hole is installed to the spindle. The offset on the arm must face up on the right side of the vehicle and down on the left

side of the vehicle. The flange edge of the arm stamping must also face the rear of the vehicle.

5. Position arm at center of vehicle and insert new bolt and nut. Do not tighten at this time.

6. Move arm end up to spindle and insert new bolt, washer and nut. Ensure bolt engages both arms and spindle.

7. Tighten arm-to-body bolt to 41–46 ft. lbs. (55–63 Nm).

8. Tighten arm-to-spindle nut to 60–80 ft. lbs. (81–109 Nm).

9. Lower vehicle.

Sway Bar

REMOVAL AND INSTALLATION

1. Remove nut and washer assembly from attaching stud and washer assembly on the stabilizer bar end.

2. Remove the U-bracket-to-body bolts.

3. Remove the sway bar from the vehicle.

4. Installation is the reverse of removal. Tighten the U-bracket-to-body bolts to 18–22 ft. lbs. (25–29 Nm). Tighten the nut and washer assembly to 6–17 ft. lbs. (8–24 Nm).

Wheel Bearings

ADJUSTMENT

Front Wheel Drive

1. Raise and safely support the vehicle.

2. Remove the wheel cover or ornament and nut covers. Revome grease cap from the hub.

3. Remove the cotter pin and nut retainer. Discard the cotter pin.

4. Back-off adjusting nut 1 full turn. Ensure nut turns freely on spindle threads. Correct any binding condition.

5. Tighten adjusting nut to 17–25 ft. lbs. (23–34 Nm) while rotating hub and drum assembly to seat bearings. Loosen adjusting nut ½ turn and tighten adjusting nut to 24–28 inch lbs. (2.7–3.2 Nm) using inch lb. torque wrench.

6. Position adjusting nut retainer over adjusting nut so slots in nut retainer flange are aligned with cotter pin hole in spline.

7. Install a new cotter pin and bend ends around retainer flange.

8. Check hub rotation. If hub rotates freely, install grease cap. If not, check bearing for damage and replace as necessary.

9. Install wheel and tire assembly, wheel cover or ornaments, and nut covers as required.

10. Lower vehicle.

All Wheel Drive

Bearings on AWD vehicles are not adjustable.

REMOVAL AND INSTALLATION

Front Wheel Drive

1. Raise and safely support the vehicle.

2. Remove wheel and tire assembly. Remove grease cap from hub.

3. Remove cotter pin, nut retainer, adjusting nut and flat washer from spindle. Discard cotter pin.

4. Pull hub and drum assembly off spindle, being careful not to drop outer bearing assembly.

5. Remove outer bearing assembly.

6. Using seal remover, remove and discard grease seal. Remove inner bearing assembly from hub.

7. Wipe all lubricant from spindle and inside of hub. Cover spindle with a clean cloth and vacuum all loose dust and dirt from brake assembly. Carefully remove cloth to prevent dirt from falling on spindle.

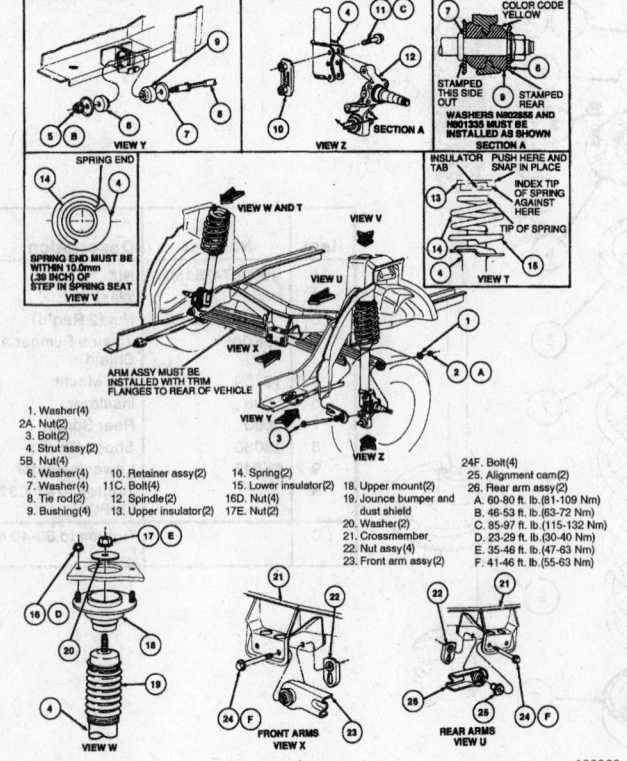

Rear suspension

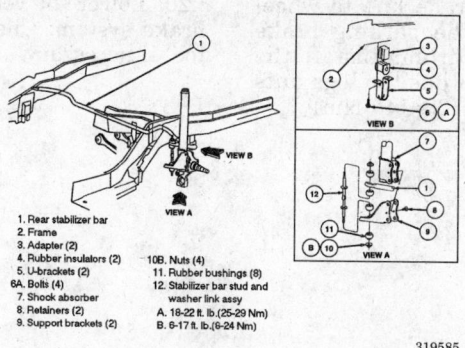

1. Rear stabilizer bar
2. Frame
3. Adapter (2)
4. Rubber insulators (2)
5. U-brackets (2)
6A. Bolts (4)
7. Shock absorber
8. Retainers (2)
9. Support brackets (2)
10B. Nuts (4)
11. Rubber bushings (8)
12. Stabilizer bar stud and washer link assy
A. 18-22 ft. lb.(25-29 Nm)
B. 6-17 ft. lb.(6-24 Nm)

319585

Sway bar

8. Clean both bearing assemblies and cups using solvent. Inspect bearing assemblies and cups for excessive wear, scratches, pits or other damage. Replace all worn or damaged parts as required.

NOTE: Allow solvent to dry before repacking bearings. Do not spin-dry bearings with air pressure.

9. If bearings must be replaced, remove the cups with wheel hub cup remover T73T–1217–A (outer bearing cup) or T77F–1102–A (inner bearing cup) or equivalent.

To install:

10. If inner or outer bearing cups were removed, install replacement cups using driver handle T80T–4000–W and bearing cup replacers T77F–1202–A and T73T–1217–A or equivalent. Support drum hub on wood block to prevent damage. Insure cups are properly seated in hub.

11. Ensure all spindle and bearing surfaces are clean.

12. Using a bearing packer, pack bearing assemblies with a suitable wheel bearing grease. If a packer is not available, work in as much grease as possible between the rollers and the cages. Grease the cup surfaces.

13. Place inner bearing cone and roller assembly in inner cup. Apply light film of grease to lips of a new grease seal and install seal with rear hub seal replacer T81P–1249–A or equivalent. Ensure retainer flange is seated all around.

14. Apply light film of grease on spindle shaft bearing surfaces.

15. Install hub and drum assembly on spindle. Keep hub centered on spindle to prevent damage to grease seal and spindle threads.

16. Install outer bearing assembly and keyed flat washer on spindle. Install adjusting nut finger-tight.

17. To adjust the wheel bearings:

a. Make sure the brakes are not dragging.

b. Back-off adjusting nut 1 full turn. Ensure nut turns freely on spindle threads. Correct any binding condition.

c. Tighten adjusting nut to 17–25 ft. lbs. (23–34 Nm) while rotating hub and drum assembly to seat bearings.

d. Loosen adjusting nut ½ turn, then tighten to 24–28 inch lbs. (2.7–3.2 Nm) using inch lb. torque wrench.

e. Position adjusting nut retainer over adjusting nut so slots in nut retainer flange are in line with cotter pin hole in spline.

f. Install a new cotter pin and bend ends around retainer flange.

18. Check hub rotation. If hub rotates freely, install grease cap. If not, check bearing adjustment again.

19. Install wheel and tire on drum.

20. Lower vehicle.

All Wheel Drive

1. Raise and support the vehicle safely. Remove the tire and wheel assembly.

2. Remove the brake drum. Remove the parking brake cable from the brake backing plate.

3. Remove the brake line from the wheel cylinder. Remove the outboard U-joint retaining bolts. Remove the outboard end of the halfshaft from the wheel stub shaft yoke and wire it to the control arm.

4. Remove and discard the control arm to spindle bolt, washer and nut. Remove the tie rod nut, bushing and washer and discard the nut.

5. Remove and discard the 2 bolts retaining the spindle to the strut. Remove the spindle from the vehicle. Mount the spindle and backing plate assembly in a suitable vise.

6. Remove the cotter pin and nut attaching the stub shaft yoke to the stub shaft. Discard the cotter pin.

7. Remove the spindle and backing plate assembly from the vise. Remove the stub shaft yoke using a 2 jaw puller and shaft protector. After removing end yoke from spindle assembly, inspect the nylon bushing and replace, as necessary.

8. Position the spindle and backing plate assembly into a vise and remove the wheel stub shaft.

9. Remove the snapring retaining the bearing. Remove the bolts retaining the spindle to the backing plate and remove the backing plate.

10. Remove the spindle from the vise and mount it into a suitable press. With the spindle side facing upward, carefully press out the bearing from the spindle, using a driver handle and bearing cup driver. Discard the bearing after removal.

To install:

11. Mount the spindle in a press, spindle side facing down. Position a new bearing in the outboard side of the spindle and carefully press in the new bearing using a driver handle and bearing installer.

NOTE: Make sure the press tool contacts only the outer bearing race or the bearing will be damaged.

12. Remove the spindle from the press and mount it in a vise. Install the snapring retaining the bearing. Position the backing plate to the spindle and install the retaining bolts.

13. Install the wheel stub shaft. Install the stub shaft yoke and attaching nut. Torque the nut to 120–150 ft. lbs. (163–204 Nm) install a new cotter pin.

14. Remove the spindle and backing plate assembly from the vise. Position the spindle onto the tie rod and then into the strut lower bracket. Insert 2 new strut-to-spindle bolts. Do not tighten at this time.

15. Install the tie rod bushing washer and new nut. Install the new control arm to spindle bolt, washers and nut. Do not tighten them at this time.

16. Install a jackstand to support the suspension at the normal curb height before tightening the fasteners.

17. Torque the spindle to strut bolts to 70–96 ft. lbs. (95–130 Nm). Torque the tie rod nut to 52–74 ft. lbs. (71–101 Nm). Torque the control arm to spindle nut to 60–86 ft. lbs. (82–117 Nm).

18. Position the outboard end of the halfshaft to the wheel stub shaft yoke. Install the retaining caps and bolts and torque them to 15–17 ft. lbs.

19. Install the brake line to wheel cylinder. Install the parking brake cable and brake drum. Install the wheel assembly, torque the lugs nuts to 80–105 ft. lbs. (109–140 Nm).

20. Lower the vehicle and bleed the brake system. Check and adjust the toe, if necessary.

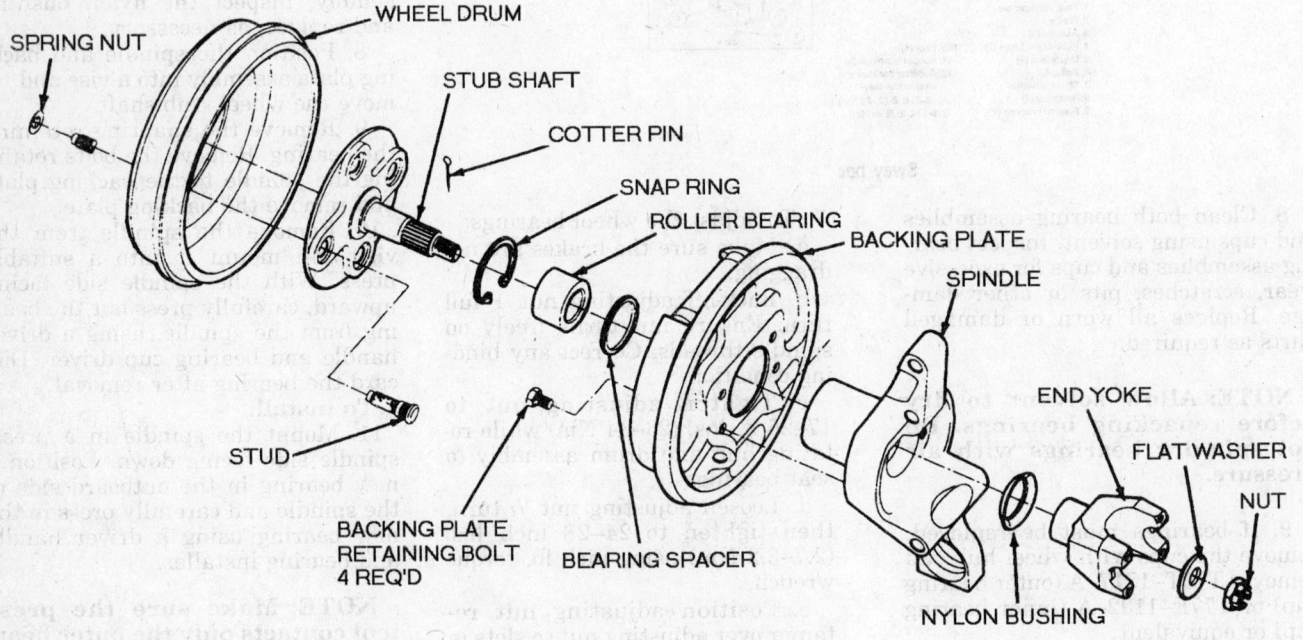

Spindle and bearing assembly on all wheel drive models

SPRING NUT

WHEEL DRUM

STUB SHAFT

COTTER PIN

SNAP RING

ROLLER BEARING

BACKING PLATE

SPINDLE

END YOKE

FLAT WASHER

NUT

NYLON BUSHING

STUD

BACKING PLATE RETAINING BOLT 4 REQ'D

BEARING SPACER

175552

FORD MOTOR CO.

Front Wheel Drive

FORD-Aspire

FIRING ORDER

NOTE: To avoid confusion, always replace spark plug wires one at a time.

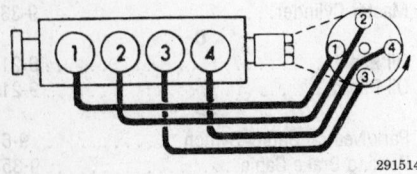

1.3L Engine
Firing Order: 1-3-4-2
Distributor Rotation: Counterclockwise

291514

ENGINE ELECTRICAL

Distributor

REMOVAL AND INSTALLATION

NOTE: The ignition coil and ignition module are integral components of the distributor. If the ignition coil or ignition module are defective, the entire distributor must be replaced. If replacing the distributor cap, mark the distributor cap towers with the cylinder numbers before removing the spark plug wires, for installation reference.

1. Disconnect the negative battery cable.
2. Remove the 2 distributor cap screws, pull off the distributor cap and position it aside. If the ignition wires are to be removed from the distributor cap, tag their location on the cap and use a twisting, pulling motion to remove.
3. Disconnect the 2 electrical harness connectors at the distributor.
4. Scribe a reference mark across the distributor mounting flange and cylinder head surface to ensure that the distributor will be installed without altering the timing. Note the position of the rotor.
5. Remove the 2 distributor mounting bolts and remove the distributor assembly from the cylinder head.
6. Remove the distributor O-ring and inspect for damage and replace as required. Coat the O-ring with clean engine oil and install.

To install:

7. Install the distributor, making sure the offset drive tang engages with the camshaft slot. With the offset drive tang, the distributor can only be installed one way regardless of camshaft position.
8. After the distributor is engaged with the camshaft, align the timing reference marks scribed across the flange base and cylinder head. Install the 2 mounting bolts but do not tighten at this time.
9. Connect the 2 electrical harness connectors.
10. Install the distributor cap. If the spark plug wires were removed, connect them to the correct distributor cap towers.
11. Connect the negative battery cable.
12. Start the engine and check the ignition timing. When the timing is set, tighten the distributor mounting bolts to 14–19 ft. lbs. (19–25 Nm).

Ignition Timing

ADJUSTMENT

1. Start the engine and allow it to warm up to normal operating temperature.
2. Turn **OFF** all accessories.
3. Connect a timing light, and ground the Self-Test Input (STI) test connector terminal at the Data Link Connector.
4. Check the base ignition timing. The white ignition timing mark should line up with the pointer on the timing belt cover.
5. If the timing marks do not line up, loosen the distributor bolts, and rotate the distributor until the timing

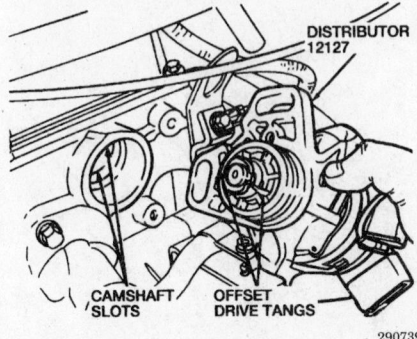

DISTRIBUTOR
12127

CAMSHAFT
SLOTS

OFFSET
DRIVE TANGS

290739

Distributor

marks are properly aligned. Tighten the distributor bolts to 14–19 ft. lbs. (19–25 Nm.).
6. Check the timing marks to ensure they did not move when the distributor bolts were tightened.
7. Remove the STI connector and the timing light.

Alternator

PRECAUTIONS

Several precautions must be observed with alternator equipped vehicles to avoid damage to the unit.
• If the battery is removed for any reason, make sure it is reconnected with the correct polarity. Reversing the battery connections may result in damage to the 1-way rectifiers.
• When utilizing a booster battery as a starting aid, always connect the positive to positive terminals and the negative terminal from the booster battery to a good engine ground on the vehicle being started.
• Never use a fast charger as a booster to start vehicles.
• Disconnect the battery cables when charging the battery with a fast charger.
• Never attempt to polarize the alternator.
• Do not use test lights of more than 12 volts when checking diode continuity.
• Do not short across or ground any of the alternator terminals.
• The polarity of the battery, alternator and regulator must be matched and considered before making any electrical connections within the system.
• Never separate the alternator on an open circuit. Make sure all connections within the circuit are clean and tight.
• Disconnect the battery ground terminal when performing any service on electrical components.
• Disconnect the battery if arc welding is to be done on the vehicle.

REMOVAL AND INSTALLATION

1. Disconnect the negative battery cable.
2. If equipped, position the rubber boot away from the **B** terminal to expose the terminal nut. Remove the nut and electrical lead from the terminal post.
3. Disconnect the alternator electrical connector.
4. Remove the alternator adjustment bolt.

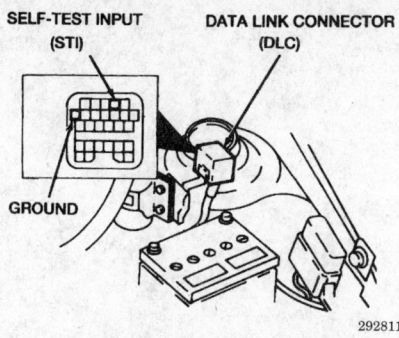

Self-Test Input (STI) ground

292811

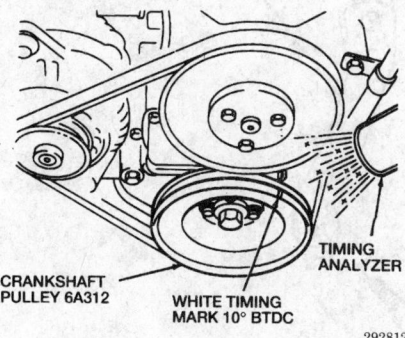

CRANKSHAFT PULLEY 6A312 WHITE TIMING MARK 10° BTDC TIMING ANALYZER

292812

Timing marks

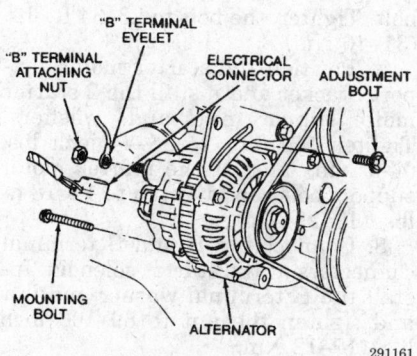

"B" TERMINAL EYELET "B" TERMINAL ATTACHING NUT ELECTRICAL CONNECTOR ADJUSTMENT BOLT MOUNTING BOLT ALTERNATOR

291161

Alternator installation

5. Raise and safely support the vehicle.

6. Remove the alternator lower bolt.

7. Remove the drive belt from the alternator pulley.

8. Remove the alternator from the vehicle. If necessary, bend the catalytic converter shield brace enough to allow clearance for removal.

To install:

9. Position the alternator and fit the drive belt to the alternator pulley.

10. Install the alternator lower bolt and the adjustment bolt, hand tight.

11. If bent, restore the catalytic converter shield brace to its original position.

12. Lower the vehicle.

13. Connect the alternator electrical connector.

14. Connect the **B** terminal lead and nut. Tighten the nut securely and fit the rubber boot.

15. Properly adjust the alternator drive belt as follows:

• NEW ALTERNATOR BELT: Tighten the belt and tension to 86–103 lbs. (39–47 kg) using Offset Belt Tension Gauge 021-0028A or equivalent. Run the engine for 10 minutes and readjust the belt.

• USED ALTERNATOR BELT: Tighten the belt and tension to 68–86 lbs. (31–39 kg) using Offset Belt Tension Gauge 021-0028A or equivalent.

NOTE: An alternate method of belt adjustment is to check belt deflection. With the engine cold, apply 22 lbs. (98N) at a point midway between the alternator and crankshaft pulleys. Belt deflection should be 0.31–0.35 inch (8–9 mm) for a new belt and 0.35–0.39 inch (9–10 mm) for a used belt.

16. Tighten the alternator adjustment bolt to 14–19 ft. lbs. (19–25 Nm).

17. Raise and safely support the vehicle.

18. Tighten the alternator lower bolt to 27–38 ft. lbs. (37–52 Nm).

19. Lower the vehicle.

20. Connect the negative battery cable.

21. Check the alternator/charging circuit for proper operation.

Drive Belt

REMOVAL AND INSTALLATION

1. Disconnect the negative battery cable.

2. Remove the A/C compressor drive belt, if equipped.

3. Remove the power steering pump drive belt, if equipped.

4. Loosen the alternator adjustment bolt.

5. Raise and safely support the vehicle.

6. Loosen the lower alternator mounting bolt.

7. Lower the vehicle.

8. Remove the alternator drive belt.

To install:

9. Position the new alternator drive belt over the pulleys. Verify

that the belt is seated properly on all of the pulleys.

10. Position a suitable prybar between the engine and the alternator. Position the bar against the alternator in an area around a case bolt. Do not pry on the stator frame.

11. Adjust the belt tension by prying with the bar. Measure the belt tension using a belt tension gauge or by using the deflection method.

12. If using a belt tension gauge, position the gauge on the longest accessible belt span. The belt tension should be 86–103 lbs. (39–47 Kg) for a new belt or 68–86 lbs. (31–39 Kg) for a used belt (more than 10 minutes running time).

13. If using the deflection method, apply approximately 22 lbs. (10 Kg) of pressure to the middle of the longest accessible belt span. The deflection should be 0.31–0.35 in. (8–9mm) for a new belt or 0.35–0.39 in. (9–10mm) for a used belt (more than 10 minutes running time).

14. When the belt tension is as specified, tighten the adjustment bolt to 14–19 ft. lbs. (19–25 Nm).

15. Raise and safely support the vehicle.

16. Tighten the lower mounting bolt to 27–38 ft. lbs. (37–52 Nm).

17. Lower the vehicle.

18. If equipped, install and adjust the A/C and/or power steering belts.

19. Connect the negative battery cable.

20. Run the engine for several minutes and recheck the alternator drive belt tension.

Starter

REMOVAL AND INSTALLATION

With Automatic Transaxle

1. Disconnect the negative battery cable.

2. Remove the 2 upper starter mounting bolts.

3. Raise and safely support the vehicle.

4. Remove the 2 bolts that secure the intake manifold support bracket, then remove the bracket.

5. Remove the bolt that secures the mounting bracket to the support bracket and remove the support bracket.

6. Remove the 2 nuts and washers that secure the mounting bracket to the starter and remove the mounting bracket.

7. Remove the **B** terminal washer and nut. Disconnect the **B** and **S** ter-

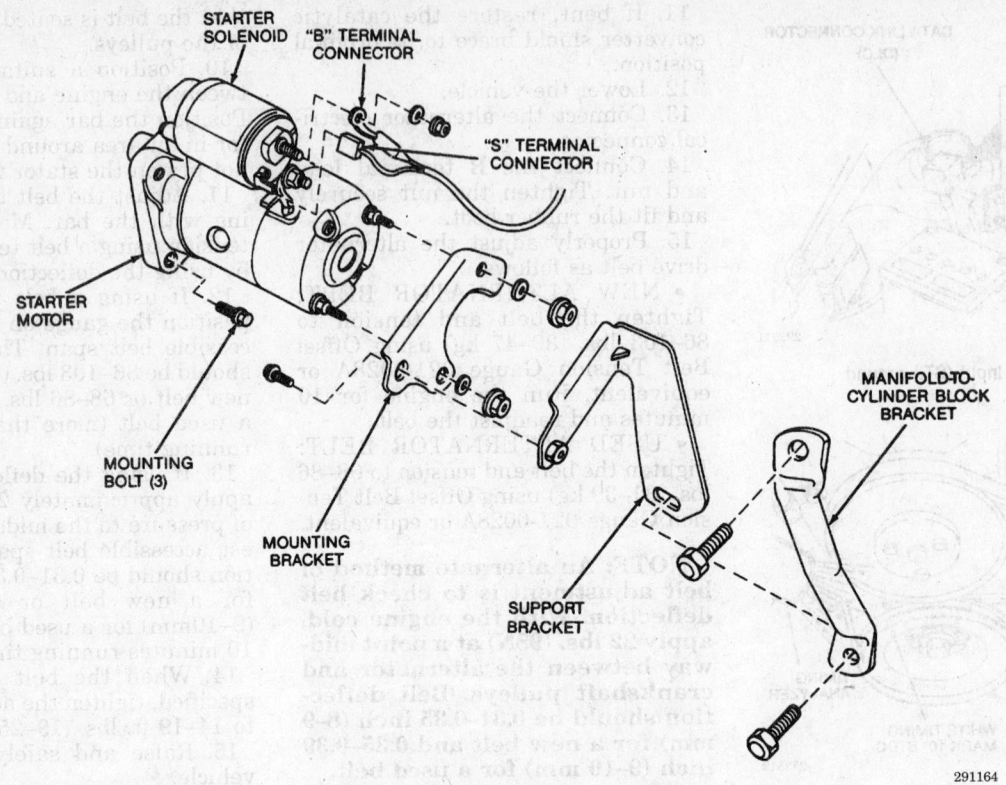

Starter motor with automatic transaxle

minal connectors at the starter solenoid.

8. Remove the lower starter mounting bolt and remove the starter.

To install:

9. Position the starter and install the lower starter mounting bolt. Tighten the bolt to 23–34 ft. lbs. (31–46 Nm).

10. Connect the **B** and **S** terminal connectors at the starter solenoid. Install the **B** terminal washer and nut and tighten the nut to 89–106 inch lbs. (10–12 Nm).

11. Position the mounting bracket and install the washers and nuts. Tighten the nuts to 35–44 inch lbs. (4–5 Nm).

12. Position the support bracket and install the bolt.

13. Position the intake manifold support bracket and install the bolts. Tighten the bolts to 12–14 ft. lbs. (16–22 Nm).

14. Lower the vehicle.

15. Install the 2 upper starter mounting bolts and tighten the bolts to 23–34 ft. lbs. (31–46 Nm).

16. Connect the negative battery cable.

17. Check the starter motor for proper operation.

With Manual Transaxle

1. Disconnect the negative battery cable.

2. Remove the **B** terminal washer and nut. Disconnect the **B** and **S** terminal connectors at the starter solenoid.

3. Remove the 2 bolts that secure the starter support bracket to the transaxle.

4. Remove the 2 starter motor support nuts, washers and the starter motor support.

5. Remove the starter mounting bolts and remove the starter.

To install:

6. Position the starter motor and install the lower starter mounting

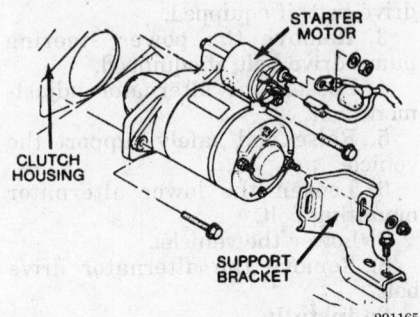

Starter motor - manual transaxle installation

bolt. Tighten the bolt to 23–34 ft. lbs. (31–46 Nm).

7. Position the starter motor support bracket and install the 2 starter motor support nuts and washers. Tighten the nuts to 35–44 inch lbs. (4–5 Nm). Install the starter motor support bolts and tighten to 14–18 ft. lbs. (19–25 Nm).

8. Connect the **B** and **S** terminal connectors at the starter solenoid. Install the **B** terminal washer and nut and tighten the nut to 89–106 inch lbs. (10–12 Nm).

9. Connect the negative battery cable.

10. Check the starter motor for proper operation.

CHASSIS ELECTRICAL

Blower Motor

REMOVAL AND INSTALLATION

1. Disconnect the negative battery cable.

2. Remove the 2 top blower motor housing nuts.

3. Remove the left-hand lower nut and pull the blower motor housing out from behind the evaporator case.

4. Disconnect the blower motor wiring.

5. Remove the 3 blower motor mounting screws and the blower motor from the case.

6. Remove the blower wheel circlip and the blower wheel.

7. Remove the 3 blower motor cover screws and remove the blower motor cover.

To install:

8. Position the blower motor cover onto the new blower motor and install the 3 cover screws.

9. Slide the blower wheel onto the motor shaft and install the circlip.

10. Position the blower motor to the blower motor housing and install the 3 retaining screws.

11. Connect the blower motor wiring.

12. Install the blower motor housing to the evaporator case and secure with 3 blower motor housing nuts.

13. Connect the negative battery cable.

14. Check the blower motor for proper operation.

Wiper Motor

REMOVAL AND INSTALLATION

Front Wiper Motor

1. Disconnect the negative battery cable.

2. Disconnect the windshield wiper motor electrical connector.

3. Remove the 2 EGR solenoid vacuum valve bracket nuts and slide the vacuum outlet fitting and cap off the access plate.

4. Remove the 2 access panel nuts.

5. Loosen the 4 windshield wiper motor bolts.

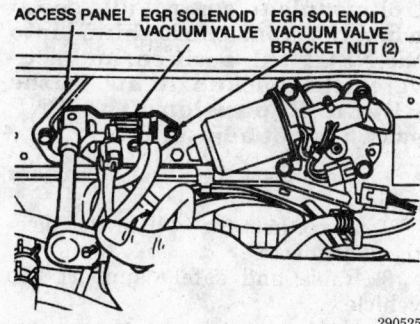

ACCESS PANEL   EGR SOLENOID   EGR SOLENOID
VACUUM VALVE   VACUUM VALVE —
BRACKET NUT (2)

290525

Removing the solenoid bracket to access the front wiper motor

6. Pull the windshield wiper motor and the access panel away from the bulkhead.

7. Disengage the windshield wiper linkage pivot from the wiper motor output arm.

8. Separate the wiper motor from the access panel.

To install:

9. Attach the wiper motor to the access panel.

10. Install the windshield wiper linkage pivot onto the wiper motor output arm.

11. Install the windshield wiper motor and the access panel onto the bulkhead.

12. Install the 4 windshield wiper motor bolts. Ensure the ground wire is installed with the top left windshield wiper motor bolt. Tighten the bolts to 62–88 inch lbs. (7–10 Nm).

13. Install the 2 access panel nuts.

14. Slide the vacuum outlet fitting and cap back onto the access plate and install the two EGR solenoid vacuum valve bracket nuts.

15. Reconnect the windshield wiper motor electrical connector.

16. Connect the negative battery cable.

17. Check for proper wiper motor operation.

Rear Wiper Motor

1. Disconnect the negative battery cable.

2. Lift the rear wiper arm cover and remove the wiper arm pivot nut.

3. Carefully pry on the wiper pivot arm to disengage it from the splined wiper motor shaft and remove.

4. Remove the pivot shaft mounting cap from the wiper motor shaft.

5. Remove the 2 pivot shaft cowl nuts.

6. Remove the pivot shaft mounting spacer.

7. Remove the 3 wiper motor cover screws.

8. Remove the rear wiper motor cover.

9. Remove the liftgate trim panel.

10. Disconnect the wiper motor electrical connector.

11. Remove the 4 wiper motor attaching bolts and the wiper motor.

12. If necessary, remove the inner bushing and O-ring from the motor shaft.

To install:

NOTE: Ensure the locating tab on the inner bushing engages the alignment tab on the brush lead cover tab.

13. If removed, replace the O-ring and inner bushing on the motor shaft.

14. Position the wiper motor and install the 4 wiper motor attaching bolts. Tighten the bolts to 72–96 inch lbs. (8–11 Nm).

15. Connect the wiper motor electrical connector.

16. Install the liftgate trim panel.

17. Install the wiper motor cover.

18. Install the 3 wiper motor cover screws.

19. Install the pivot shaft mounting spacer.

20. Install the 2 pivot shaft cowl nuts. Tighten the nuts to 27–44 inch lbs. (3–5 Nm).

21. Install the pivot shaft mounting cap onto the wiper motor shaft.

22. Connect the negative battery cable.

23. Operate the rear wiper motor through several cycles then turn the rear wiper switch to the **OFF** position.

24. Install the rear wiper pivot arm so that the tip of the wiper blade is 3.15 inches (80mm) from the edge of the back window glass weatherstrip.

25. Install the rear wiper pivot arm nut and tighten to 54–79 inch lbs. (6–9 Nm).

26. Check for proper rear wiper motor operation.

Combination Switch

REMOVAL AND INSTALLATION

—————— CAUTION ——————

The Supplemental Inflatable Restraint (SIR) system must be disarmed before performing service on or around SIR system components or SIR system wiring. Failure to do so may cause accidental deployment of the air bag, resulting in unnecessary SIR system repairs and/or personal injury.

1. Disconnect the negative battery cable.

2. Before proceeding further, wait 1 minute for the air bag backup power supply to deplete its stored energy.

3. Remove the steering wheel.

4. Remove the 3 steering column shroud screws.

5. Separate the upper and lower portions of the steering column shroud.

6. Remove the steering column shroud from the steering column.

7. Apply 2 strips of tape across the air bag sliding contact to prevent accidental rotation.

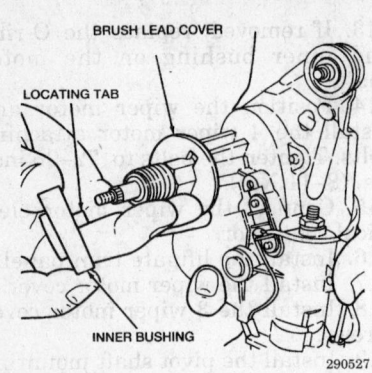

Rear wiper motor inner bushing installation

8. Remove the 3 air bag sliding contact screws and pull the air bag sliding contact off the steering column.

9. Remove the air bag sliding contact ground wire screw.

10. Disconnect the air bag sliding contact electrical connector.

11. Remove the air bag sliding contact.

12. Remove the 3 combination switch retaining screws.

13. Disconnect the combination switch electrical connectors.

14. Slide the combination switch off the steering column.

To install:

15. Slide the combination switch onto the steering column.

16. Reconnect the combination switch electrical connectors.

17. Install the 3 combination switch retaining screws and tighten.

18. Reconnect the air bag sliding contact electrical connector.

19. Install and tighten the air bag sliding contact ground wire screw.

20. Place the air bag sliding contact onto the steering column and tighten the 3 air bag sliding contact screws to 18–26 inch lbs. (2–3 Nm).

21. Remove the 2 strips of tape from the air bag sliding contact.

NOTE: If the air bag sliding contact has been accidentally rotated, the air bag sliding contact alignment must be adjusted.

22. If required, adjust the air bag sliding contact as follows:

   a. Center the front wheels to the straight ahead position.

   b. Turn the air bag sliding contact clockwise to its lock position and then back it off 2¾ turns.

   c. Align the arrows on the air bag sliding contact and the air bag sliding contact housing.

23. Place the steering column shroud onto the steering column.

24. Install and tighten the 3 steering column shroud screws.

25. Install the steering wheel.

26. Connect the negative battery cable.

27. Check all functions of the combination switch for proper operation.

28. Turn the ignition switch to the **RUN** position and visually check that the air bag warning indicator lights on the dash momentarily to prove out the air bag system.

Ignition Lock and Switch

REMOVAL AND INSTALLATION

1. Disconnect the negative battery cable.

2. Remove the 3 screws retaining the steering column shroud.

3. Separate and remove the upper and lower steering column shrouds.

4. Remove the ignition switch attaching screw and separate the ignition switch from the steering column.

5. Disconnect the ignition switch electrical connector.

6. Disconnect the ignition key reminder switch electrical connector.

7. If necessary, remove the 2 ignition key reminder switch screws and the ignition key reminder switch.

To install:

8. If removed, position the ignition key reminder switch and install the 2 ignition key reminder switch screws.

9. Connect the ignition key reminder switch electrical connector.

10. Connect the ignition switch electrical connector.

11. Position the ignition switch in the lock cylinder housing and install with the attaching screw.

12. Install the upper and lower steering column shrouds.

13. Connect the negative battery cable.

14. Check the ignition switch operation.

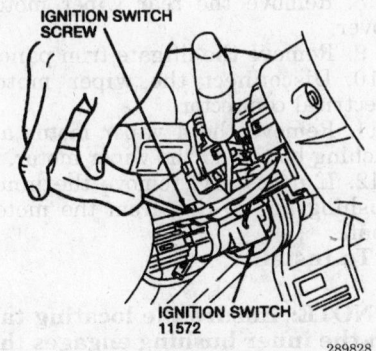

Ignition switch removal

Park/Neutral Safety Switch

REMOVAL AND INSTALLATION

NOTE: The Park/Neutral Position (PNP) switch on vehicles equipped with the automatic transaxle, allows the engine to start only in the PARK or NEUTRAL positions, activates the backup lights in the REVERSE position and signals gear position to the powertrain control module for engine idle control. The PNP switch is not adjustable.

With Automatic Transaxle

1. Disconnect the negative battery cable.

2. Disconnect the PNP switch electrical connector.

3. Place a drain pan under the transaxle to catch any lost transaxle fluid.

4. Using an extension and a crow's foot wrench, remove the PNP switch from the front left-hand side of the transaxle.

To install:

5. Apply a silicone sealer to the threads of the PNP switch.

6. Install the PNP switch to the transaxle case and tighten to 14–19 ft. lbs. (19–26 Nm).

7. Connect the PNP switch electrical connector.

8. Reconnect the negative battery cable.

9. Check the PNP switch operation. Make sure that the engine will only crank in the **PARK** or **NEUTRAL** positions and that the backup lights operate when the transaxle is in the **REVERSE** position.

With Manual Transaxle

NOTE: The Park/Neutral Position (PNP) switch on vehicles equipped with the manual transaxle, signals the powertrain control module to control idle speed when the transaxle is shifted into the NEUTRAL position, and activates the backup lights in the REVERSE position. The PNP switch is not adjustable.

1. Disconnect the negative battery cable.

2. Disconnect the PNP switch electrical connector.

3. Raise and safely support the vehicle.

4. Using an open-end wrench, remove the PNP switch from the flywheel housing.

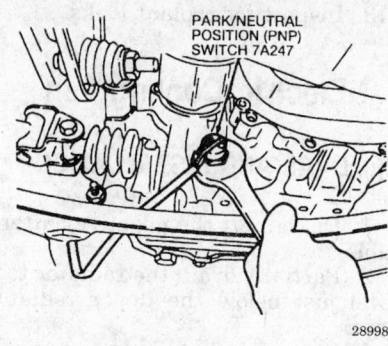

PARK/NEUTRAL POSITION (PNP) SWITCH 7A247

289985

Park/Neutral position switch — with manual transaxle

To install:

5. Apply a silicone sealer to the threads of the PNP switch.

6. Install the PNP switch to the flywheel housing and tighten to 15–21 ft. lbs. (20–29 Nm).

7. Lower the vehicle.

8. Connect the PNP switch electrical connector.

9. Connect the negative battery cable.

10. Check the PNP switch operation. Make sure that the backup lights operate when the transaxle is in the **REVERSE** position.

Powertrain Control Module

REMOVAL AND INSTALLATION

—————— **CAUTION** ——————
Electronic modules are sensitive to static electrical charges. Wear a static discharge strap to prevent damage or exposure to these charges.

1. Disconnect the negative battery cable.

2. Reach under the left-hand side of the instrument panel and remove the 3 Powertrain Control Module (PCM) nuts.

3. If equipped, move the anti-lock brake module aside.

4. Pull the PCM down to access the electrical connectors.

5. Disconnect the PCM electrical connectors.

6. Remove the PCM.

To install:

7. Position the PCM and connect the electrical connectors to the PCM.

8. Install the anti-lock brake module, if equipped.

9. Install the 3 PCM nuts.

10. Reconnect the negative battery cable.

ENGINE COOLING

Radiator

REMOVAL AND INSTALLATION

1. Disconnect the negative battery cable.

2. Make sure the engine is cool. Remove the radiator pressure cap from the filler neck.

3. Drain the cooling system.

4. Disconnect the coolant recovery hose from the filler neck.

5. Loosen the retaining clamp and disconnect the upper radiator hose from the radiator.

6. Disconnect the cooling fan wiring harness connector. Disengage the wiring harness from the routing clamps on the cooling fan shroud.

7. Loosen the retaining clamp and disconnect the lower radiator hose.

8. On vehicles with automatic transaxles, disconnect the oil cooler hoses and plug. Remove the lower bolt and cooler hose bracket.

9. Remove the 4 bolts attaching the radiator upper tank brackets to the vehicle body and remove the radiator/cooling fan assembly. Separate the fan and shroud assembly from the radiator, if necessary.

To install:

10. If removed, install the fan and shroud assembly on the radiator.

11. Lower the radiator/cooling fan assembly into the engine compartment, making sure the mounting insulators engage with their supports. Attach the radiator to the mounting brackets with the four bolts. Tighten the 4 bracket bolts to 71–89 inch lbs. (8–10 Nm).

12. Connect the cooling fan wiring and position the wiring harness in the routing clips on the fan shroud.

13. On vehicles with automatic transaxles, unplug and connect the oil cooler hoses. Install the lower bolt and cooler hose bracket.

14. Connect the coolant recovery hose and the upper and lower radiator hoses.

15. Connect the negative battery cable.

16. Fill and bleed the cooling system.

17. Install the radiator pressure cap on the radiator filler neck.

18. Allow the engine to reach normal operation temperature and check for leaks.

Water Pump

REMOVAL AND INSTALLATION

1. Disconnect the negative battery cable.

2. Remove the timing belt.

3. Drain the cooling system into a suitable container.

4. Remove the 2 bolts attaching the inlet tube to the water pump housing. Remove the inlet tube and gasket.

5. Remove the 4 water pump-to-cylinder block retaining bolts and remove the water pump.

6. Remove all existing gasket material from the cylinder block and inlet tube gasket surfaces.

To install:

7. Coat both sides of the new water pump and inlet tube gaskets with a suitable water resistant sealer. Apply the gaskets to the engine and inlet tube surfaces. Make certain the gasket holes are aligned with the bolt holes.

8. Position the water pump against the gasket. Make sure the holes in the water pump are aligned with the gasket holes and that the pump does not shift the position of the gasket.

9. Install the 4 water pump retaining bolts and torque to 14–19 ft. lbs. (19–26 Nm).

10. Position the inlet tube and gasket against the water pump housing and install the attaching bolts. Torque the bolts to 14–22 ft. lbs. (19–30 Nm).

11. Install the timing belt.

12. Fill the cooling system to the proper level.

13. Connect the negative battery cable.

14. Start the engine and allow it to reach normal operating temperature. Check for coolant leaks and proper operation.

Thermostat

REMOVAL AND INSTALLATION

1. Disconnect the negative battery cable.

2. Drain the cooling system to a level below the radiator upper hose. Disconnect the radiator upper hose from the thermostat housing.

3. Remove the thermostat housing-to-cylinder head attaching nut and bolt. Remove the thermostat housing and housing gasket.

4. Remove the thermostat.

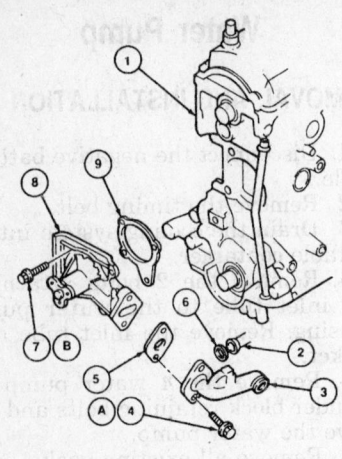

1 Cylinder Block
2 Heater Water Hose
3 Water Inlet Connection
4 Hot Water Heater Elbow
  Connector Bolt (2 Req'd)
5 Water Pump Inlet Gasket
6 O-Ring
7 Water Pump Bolt (4 Req'd)
8 Water Pump
9 Water Pump Housing Gasket
A Tighten to 19-30 N-m (14-22 Lb-Ft)
B Tighten to 19-26 N-m (14-19 Lb-Ft)

Water pump and related components

5. Remove all gasket material from the thermostat housing and cylinder block surfaces.

To install:

6. Install the thermostat in the cylinder head, with the valve end first and the sub-valve at the top.

7. Coat a new thermostat housing gasket with a suitable water resistant sealer. Apply the gasket to the cylinder block surface making sure the gasket and cylinder block holes are aligned.

8. Position the thermostat housing onto the cylinder head making sure the bolt holes are aligned and the gasket does not shift. Install the thermostat housing retaining nut and bolt. Before tightening the nut and bolt, ensure that the thermostat flange is properly seated against the recess of the housing. Torque the nut and bolt to 14–19 ft. lbs. (19–26 Nm).

9. Connect the upper radiator hose to the thermostat housing.

10. Fill the cooling system to the proper level and install the radiator cap.

11. Connect the negative battery cable.

12. Start the engine and check for proper thermostat operation by allowing the engine to reach normal operating temperature.

13. Inspect for coolant leaks.

Electric Cooling Fan

REMOVAL AND INSTALLATION

1. Disconnect the negative battery cable.

2. Partially drain the radiator to a level just below the upper radiator hose.

3. Loosen the retaining clamp and disconnect the upper radiator hose at the radiator.

4. Disconnect the cooling fan wiring harness connector and disengage the wiring harnesses from the routing clamps on the cooling fan shroud.

5. Remove the 2 bolts attaching the top of the fan shroud to the radiator.

6. Support the fan/shroud assembly and loosen the 2 bolts attaching the bottom of the fan shroud to the radiator. Remove the fan/shroud assembly from the vehicle.

7. Remove the fan blade retaining nut and the fan from the motor shaft.

8. Remove the wiring harness routing strap.

9. Remove the 3 retaining screws and remove the cooling fan motor from the fan shroud.

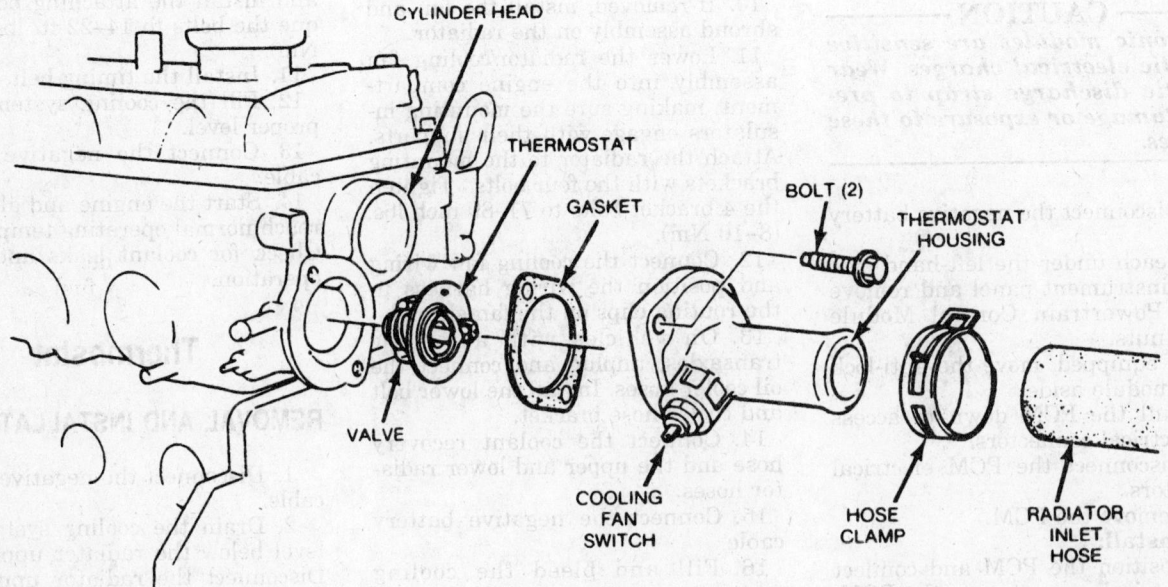

Thermostat and related components

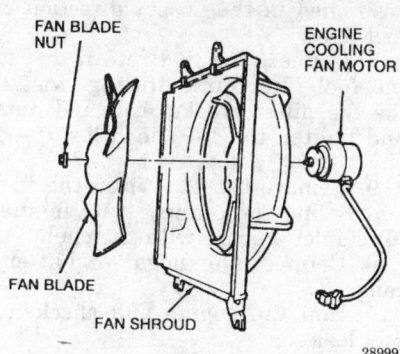

Cooling fan assembly

To install:

10. Position the cooling fan motor to the shroud and install the retaining screws.

11. Install the wiring harness routing strap.

12. Install the fan blade to the cooling fan motor shaft and secure with the nut.

13. Install the fan/shroud assembly into the vehicle making sure to align the lower fan shroud tabs to the radiator bolts.

14. Install the bolts attaching the top of the fan shroud to the radiator.

15. Tighten the lower fan shroud radiator bolts.

16. Connect the cooling fan wiring harness connector and engage the wiring harnesses to the routing clamps on the cooling fan shroud.

17. Connect the upper radiator hose and tighten the retaining clamp.

18. Refill the cooling system to the proper level.

19. Connect the negative battery cable.

20. Allow the engine to reach normal operating temperature and check the engine cooling fan for proper operation.

Cooling System

BLEEDING

─── **CAUTION** ───
Never remove the coolant pressure relief cap or draincock while the engine is running. Allow a hot engine to cool down before cap or draincock removal. When removing the pressure relief cap, always place a heavy shop towel around the cap and slowly turn the cap until coolant pressure begins to release. Once the pressure has been released, push down on the cap and finish removal.

When the entire cooling system is drained, the following procedure should be used to ensure a complete fill.

1. Install the block drain plug if removed, and close the draincock. With the engine off, add anti-freeze to the radiator to a level of 50 percent of the total cooling system capacity. Then add water until it reaches the radiator filler neck seat.

2. Install the radiator cap to the first notch to keep spillage to a minimum.

3. Start the engine and let it idle until the upper radiator hose is warm. This indicates that the thermostat is open and coolant is flowing through the entire system.

4. Carefully remove the radiator cap and top off the radiator with water. Install the cap on the radiator securely.

5. Fill the coolant recovery reservoir to the **FULL COLD** mark with anti-freeze, then add water to the **FULL HOT** mark. This will ensure that a proper mixture is in the coolant recovery bottle.

6. Check the cooling system for leaks.

FUEL SYSTEM

Fuel System Service Precautions

Safety is the most important factor when performing not only fuel system maintenance but any type of maintenance. Failure to conduct maintenance and repairs in a safe manner may result in serious personal injury or death. Maintenance and testing of the vehicle's fuel system components can be accomplished safely and effectively by adhering to the following rules and guidelines.

• To avoid the possibility of fire and personal injury, always disconnect the negative battery cable unless the repair or test procedure requires that battery voltage be applied.

• Always relieve the fuel system pressure prior to disconnecting any fuel system component (injector, fuel rail, pressure regulator, etc.), fitting or fuel line connection. Exercise extreme caution whenever relieving fuel system pressure to avoid exposing skin, face and eyes to fuel spray.

Please be advised that fuel under pressure may penetrate the skin or any part of the body that it contacts.

• Always place a shop towel or cloth around the fitting or connection prior to loosening to absorb any excess fuel due to spillage. Ensure that all fuel spillage (should it occur) is quickly removed from engine surfaces. Ensure that all fuel soaked cloths or towels are deposited into a suitable waste container.

• Always keep a dry chemical (Class B) fire extinguisher near the work area.

• Do not allow fuel spray or fuel vapors to come into contact with a spark or open flame.

• Always use a backup wrench when loosening and tightening fuel line connection fittings. This will prevent unnecessary stress and torsion to fuel line piping. Always follow the proper torque specifications.

• Always replace worn fuel fitting O-rings with new. Do not substitute fuel hose or equivalent, where fuel pipe is installed.

Fuel System Pressure

RELIEVING

─── **CAUTION** ───
Fuel injection systems remain under pressure, even after the engine has been turned OFF. The fuel system pressure must be relieved before disconnecting any fuel lines. Failure to do so may result in fire and/or personal injury.

1. Remove the fuel tank filler cap and relieve the fuel tank pressure.

2. Locate the fuel pump relay behind the left-hand side of the instrument panel.

3. Start the engine.

4. Disconnect the fuel pump relay electrical connector and wait for the engine to stall.

5. After the engine stalls, turn the ignition key to the **OFF** position.

6. Connect the fuel pump relay electrical connector.

Idle Speed

ADJUSTMENT

NOTE: Before adjusting the idle speed, make sure the ignition timing is adjusted to specifications. Turn off all the lamps and electrical accessories. Disable the cooling fan.

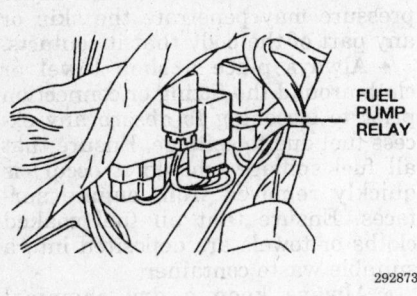

Fuel pump relay location

292873

1. Connect a Rotunda 88 Digital Multimeter 105–00053 or equivalent, with an inductive pickup attached to the No. 1 spark plug.

2. Ground the PCM STI 10–pin at the Data Link Connector (DLC).

3. Start the engine.

4. Run the engine until it reaches normal operating temperature.

5. Check the idle speed.

6. If adjustment is necessary, turn the idle speed adjustment screw until

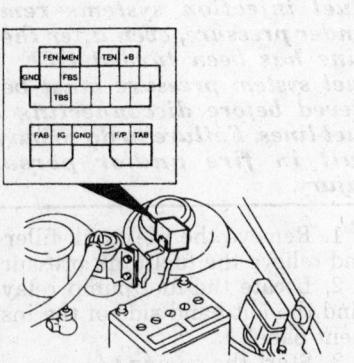

Data link connector

292780

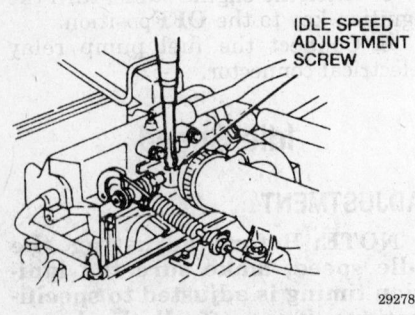

Idle speed adjustment screw

292781

the correct idle speed is attained. Idle speed specifications are as follows:

• 650–750 rpm (Manual transaxle in NEUTRAL)

• 700–800 rpm (automatic transaxle in PARK)

7. Turn the engine off and allow to cool.

8. Remove the ground from the PCM STI 10–pin at the DCL.

9. After the engine has cooled, start and run the engine until it reaches normal operation temperature and make sure the idle speed is set to specifications.

Air/Fuel Mixture

ADJUSTMENT

The air/fuel mixture adjustment is controlled by the Engine Control Module (ECM) and cannot be altered. The ECM receives inputs from various engine sensors to calculate the required fuel flow rate necessary to maintain the prescribed air/fuel ratio. The ECM then sends signals based on the input information to the fuel injectors to meter the required quantity of fuel necessary for proper combustion.

Fuel Filter

REMOVAL AND INSTALLATION

— **CAUTION** —
Fuel injection systems remain under pressure, even after the engine has been turned OFF. The fuel system pressure must be relieved before disconnecting any fuel lines. Failure to do so may result in fire and/or personal injury.

1. Properly relieve the fuel system pressure.

2. Disconnect the negative battery cable.

3. Loosen the clamp and disconnect the fuel supply line at the inlet of the fuel filter. Plug the end to prevent spillage.

4. Loosen the clamp and disconnect the fuel return line at the fuel filter outlet fitting.

5. Remove the filter bracket attaching bolt and nuts from the fuel filter.

6. Remove the fuel filter from the bracket and properly dispose.

To install:

7. Install the new fuel filter into the bracket. Make sure the filter is

positioned in the proper direction of fuel flow.

8. Connect the fuel return line to the fuel filter outlet fitting. Install the fuel filter bracket bolt and nuts and tighten to 71–97 inch lbs. (8–11 Nm).

9. Remove the plug from the fuel supply line and connect it to the fuel filter inlet. Secure with the clamp.

10. Connect the negative battery cable.

11. Run the engine and check for fuel leaks.

Fuel Pump

REMOVAL AND INSTALLATION

— **CAUTION** —
Fuel injection systems remain under pressure, even after the engine has been turned OFF. The fuel system pressure must be relieved before disconnecting any fuel lines. Failure to do so may result in fire and/or personal injury.

1. Properly relieve the fuel system pressure as follows:

a. Start the engine.

b. Disconnect the fuel pump relay electrical connector.

c. After the engine stalls, turn the ignition to the **OFF** position.

d. Connect the fuel pump relay electrical connector.

2. Disconnect the negative battery cable.

3. Remove the rear seat cushion and cover.

4. Remove the 3 luggage compartment floor cover hold-down pins and fold the luggage compartment floor cover forward until the sending unit access plate is visible.

5. On 2-door vehicles, remove the 6 inner rear floor filler cover bolts and the 2 inner rear floor filler cover nuts.

6. Remove the inner rear floor filler cover.

7. Remove the 4 fuel pump assembly access plate screws, the ground lead, and the sending unit access plate from the chassis.

8. Disconnect the fuel pump and sending unit electrical connector.

9. On 4-door vehicles, remove the 2-way check valve bolt and the 2-way check valve from the bracket on the fuel pump and sending unit housing.

10. Loosen the 2 hose clamps and disconnect the 2 fuel pump hoses from the fuel pump hose fittings on the fuel pump and sending unit housing.

11. Remove the 4 screws (2-door) or 8 screws (4-door) and lift the fuel pump and sending unit housing from the fuel tank.

12. Disconnect the fuel tank sending unit electrical connector.

13. Remove the two fuel tank sending unit washers and nuts and the fuel tank sending unit.

14. Remove the fuel pump screw, fuel pump grommet and the fuel pump bracket from the bottom of the fuel pump.

15. Remove the fuel filter bracket, fuel filter and the retainer from the fuel pump.

16. Disconnect the fuel pump electrical connector.

17. Remove the clamp and the fuel pump from the housing.

To install:

18. Position the fuel pump on the housing and secure with the clamp.

19. Install the fuel tank sender filter, bracket and the retainer on the fuel pump.

20. Clip the fuel tank sender bracket on to the fuel pump.

21. Install the fuel pump grommet, fuel pump screw and the fuel pump bracket onto the bottom of the fuel pump.

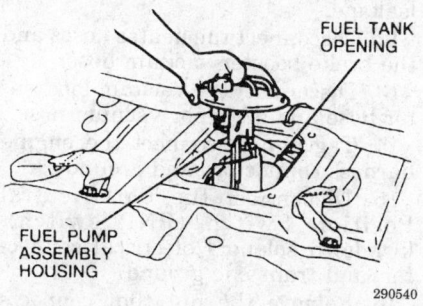

Removing the fuel pump housing from the fuel tank

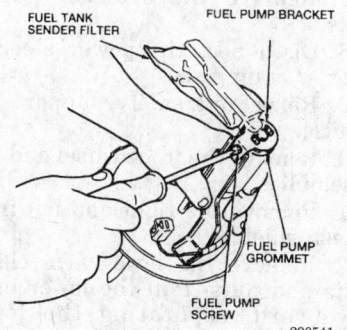

Removing the fuel pump from the fuel pump housing

22. Position the fuel tank sending unit on the housing and install the sending unit washers and nuts.

23. Connect the fuel tank sending unit electrical connector.

24. Place the fuel pump and sending unit housing into the fuel tank opening and install the 4 screws (2-door) or 8 screws (4-door).

25. Install the 2 fuel hoses on the 2 sending unit fuel pump hose fittings and tighten the 2 hose clamps.

26. On 4-door vehicles, position the 2-way check valve on the fuel pump and sending unit housing and secure with the bolt.

27. Connect the fuel pump and sending unit electrical connector.

28. Connect the negative battery cable.

29. Start the engine and check for leaks at the fuel line connections and proper fuel pump operation. Turn the engine **OFF**.

30. Install the 4 fuel pump and sending unit access plate screws, the ground lead and the fuel pump and sending unit access plate.

31. Fold the luggage compartment floor cover over the fuel pump and sending unit access plate and install the 3 hold-down pins.

32. Install the rear seat cushion and cover.

Fuel Injector

REMOVAL AND INSTALLATION

CAUTION

Fuel injection systems remain under pressure, even after the engine has been turned OFF. The fuel system pressure must be relieved before disconnecting any fuel lines. Failure to do so may result in fire and/or personal injury.

NOTE: Before preforming this procedure, make sure to have replacement insulators and injector O-rings.

1. Properly relieve the fuel system pressure as follows:
   a. Start the engine.
   b. Disconnect the fuel pump relay electrical connector.
   c. After the engine stalls, turn the ignition to the **OFF** position.
   d. Connect the fuel pump relay electrical connector.

2. Disconnect the negative battery cable.

3. Remove the upper intake manifold.

4. Remove the vacuum line from the fuel pressure regulator.

5. Disconnect and plug the fuel inlet and return lines from the fuel injection supply manifold.

6. Disconnect the 4 electrical connectors at the injectors.

7. Remove the 2 fuel injection supply manifold retaining bolts.

8. Remove the fuel injection supply manifold with the fuel injectors.

9. Remove the injectors from the fuel injection supply manifold.

10. Remove and discard the fuel injector insulators and O-rings from the fuel injectors.

To install:

11. Install new O-rings and insulators onto the fuel injectors and lubricate the O-rings with clean engine oil.

12. Position the injectors onto the fuel injection supply manifold and then fit the assembly to the cylinder head.

13. Install the 2 fuel injection supply manifold retaining bolts and tighten to 14–17 ft. lbs. (19–23 Nm.).

14. Connect the 4 fuel injector electrical connectors.

15. Remove the plugs from the fuel inlet and return lines and install to the fuel injection supply manifold. Install the clamps.

16. Connect the vacuum line to the fuel pressure regulator.

17. Install the upper intake manifold.

18. Connect the negative battery cable.

19. Run the engine and check for fuel leaks and proper engine operation.

EMISSION CONTROLS

Service Interval Lamp

RESETTING

Approximately every 5000 or 7500 miles, (depending on engine application) the word **SERVICE** will appear on the electronic display for the first 1.5 miles to remind the driver that it is time for the regular vehicle service interval maintenance (i.e. oil change).

To reset the service interval reminder light for another interval: With the engine running, press the

1. Fuel charging wiring
2. Fuel injection supply manifold bolts (2)
3. Fuel tube hose
4. Fuel pressure regulator bolts (2)
5. Fuel hose
6. Fuel pressure regulator
7. Fuel injector
8. Fuel injector o-rings
9. O-rings
10. Fuel injector insulators
11. Fuel injection supply manifold
   A. 14-17 ft. lb.(19-23 Nm)
   B. 71-97 in. lb.(8-11 Nm)

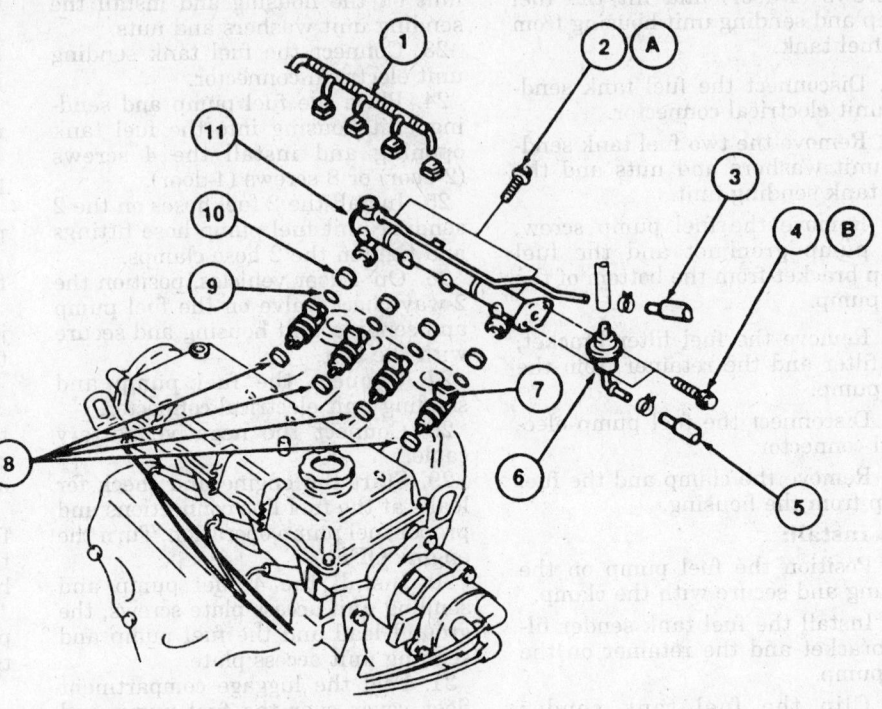

Fuel injector and related components

TRIP and TRIP RESET buttons at the same time. Hold the buttons down until 3 beeps are heard to verify that the service reminder has been reset.

ENGINE MECHANICAL

Engine Assembly

REMOVAL AND INSTALLATION

NOTE: The engine and transaxle are removed together by lifting them out of the vehicle from above.

— CAUTION —

Fuel injection systems remain under pressure, even after the engine has been turned OFF. The fuel system pressure must be relieved before disconnecting any fuel lines. Failure to do so may result in fire and/or personal injury.

With Automatic Transaxle

1. Relieve the fuel system pressure as follows:
   a. Start the engine.
   b. Disconnect the fuel pump relay electrical connector.
   c. After the engine stalls, turn the ignition to the OFF position.
   d. Connect the fuel pump electrical connector.
2. Disconnect the battery cables from the battery, negative cable first.
3. Remove the battery and battery tray.
4. Mark the hinge location and remove the hood.
5. Remove the engine air cleaner and air cleaner to intake manifold tube.
6. Drain the radiator coolant, engine oil, transaxle fluid and, if equipped, the power steering fluid into suitable containers.
7. Properly recover the refrigerant from the A/C system, if equipped.
8. Remove the radiator and cooling fan as an assembly.
9. Disconnect the accelerator cable from the mounting bracket and throttle lever. Remove the two accelerator shaft bracket bolts from the valve cover, and remove the bracket.
10. Disconnect the speedometer cable from the transaxle.

11. Disconnect the fuel hoses. Plug or cover the hose openings to prevent dirt from entering and to avoid fuel leakage.
12. Disconnect the heater hoses and the brake booster vacuum hose.
13. Disconnect the vacuum modulator hose and governor vacuum hose.
14. Tag and disconnect the engine harness connectors and grounds.
15. Disconnect the Park/Neutral Position Switch (PNP switch), kickdown solenoid electrical connector, and transaxle ground.
16. Remove the nut that connects the shift lever to the manual shaft assembly. Remove the shift cable and bracket from the transaxle.
17. Remove the accessory drive belts.
18. Disconnect the power steering lines, if equipped.
19. Raise and safely support the vehicle.
20. Remove the front wheel and tire assemblies.
21. Remove the right and left front splash shields.
22. Remove the lower arm clamp bolts and nuts. Pull the lower arms downward, separating the lower arms from the knuckles.
23. Remove the halfshafts and install differential plugs T87C-7025-C.

or equivalent, between the differential side gears.

24. If equipped, remove the 2 power steering pump bolts and separate the power steering pump from the engine.

25. If equipped, remove the A/C compressor.

26. Remove the exhaust inlet pipe.

27. Remove the starter motor.

28. Remove the 2 front and 2 rear transaxle support insulator nuts.

29. Remove the 2 muffler pipe bracket bolts, and the bracket.

30. Remove the 6 transaxle case to cylinder block front and rear bracket bolts, and brackets.

31. Remove the engine rear plate bolt, and the 4 flywheel-to-torque converter nuts.

32. Lower the vehicle.

33. Disconnect the vacuum lines from intake manifold vacuum outlet fitting and cap and fuel vapor canister.

34. Attach lifting hooks to the engine lifting eyes and remove any slack in the chains.

35. Remove the engine support insulator through bolt and nut from engine support insulator.

36. Carefully raise and remove the engine and transaxle as an assembly.

37. Remove the 4 engine-to-transaxle bolts and separate the transaxle from the engine.

38. Remove the flywheel.

39. Mount the engine on an engine stand, and remove the lifting hooks from the engine.

To install:

40. Attach lifting hooks to the engine lifting eyes and remove any slack in the chains.

41. Remove the engine from the engine stand.

42. Install the flywheel.

43. Mount the transaxle to the engine. Install the engine-to-transaxle bolts and tighten to 41–59 ft. lbs. (55–80 Nm).

44. Position the engine and transaxle assembly in the engine compartment.

45. Install the engine support insulator through bolt and nut into the engine support insulator, and tighten to 39–47 ft. lbs. (53–64 Nm).

46. Remove the lifting cables from the engine lifting eyes.

47. Connect the vacuum lines from intake manifold vacuum outlet fitting and cap and fuel vapor canister.

48. Raise and safely support the vehicle.

49. Install the 4 flywheel-to-torque converter nuts and tighten to 25–36 ft. lbs. (34–49 Nm).

50. Install the engine rear plate and tighten the bolts to 61–87 inch lbs. (7–10 Nm).

51. Install the 6 transaxle case to cylinder block front and rear bracket bolts, and brackets. Tighten the bolts to 27–38 ft. lbs. (37–52 Nm).

52. Install the muffler pipe bracket, and tighten the bolts to 28–41 ft. lbs. (38–56 Nm).

53. Install the 2 rear transaxle support insulator nuts, and tighten to 21–34 ft. lbs. (28–46 Nm).

54. Install the 2 front transaxle support insulator nuts, and tighten to 27–38 ft. lbs. (37–52 Nm).

55. Install the starter motor.

56. Install the exhaust inlet pipe and tighten the nuts to 23–34 ft. lbs. (31–46 Nm).

57. If equipped, install the A/C compressor.

58. If equipped, install the power steering pump and tighten the bolts to 27–40 ft. lbs. (36–54 Nm).

59. Remove the differential plugs and install the halfshafts. Install the lower arm ball joint to the knuckle and tighten the clamp nut and bolt to 32–40 ft. lbs. (43–54 Nm).

60. Install the right and left front splash shields.

61. Install the front wheel and tire assemblies.

62. Lower the vehicle.

63. If equipped, connect the power steering lines.

64. Install the accessory drive belts.

65. Connect the shift cable and bracket onto the transaxle.

66. Connect the Park/Neutral Position Switch (PNP switch), kickdown solenoid electrical connector, and transaxle ground.

67. Connect all engine harness connectors and grounds.

68. Connect the vacuum modulator hose and governor vacuum hose.

69. Connect the brake booster vacuum hose, the heater hoses and the fuel lines.

70. Connect the speedometer cable. Connect the accelerator cable and bracket to the throttle lever and mounting bracket.

71. Install the radiator and cooling fan.

72. Install the engine air cleaner, and the air cleaner to intake manifold tube.

73. Install the hood, aligning the marks that were made during the removal procedure.

74. Install the battery tray and the battery. Connect the battery cables, negative cable last.

75. Add the proper types and quantities of engine oil, transaxle fluid and coolant.

76. If equipped, add power steering fluid to the reservoir.

77. If equipped, evacuate, recharge and leak test the air conditioning system.

78. Start the engine and check for leaks and proper fluid levels.

79. Road test the vehicle and recheck all fluid levels.

With Manual Transaxle

1. Relieve the fuel system pressure as follows:

a. Start the engine.

b. Disconnect the fuel pump relay electrical connector.

c. After the engine stalls, turn the ignition to the **OFF** position.

d. Connect the fuel pump electrical connector.

2. Disconnect the battery cables from the battery, negative cable first.

3. Remove the battery and battery tray.

4. Mark the hinge location and remove the hood.

5. Remove the engine air cleaner and air cleaner to intake manifold tube.

6. Drain the radiator coolant, engine oil, transaxle fluid and, if equipped, the power steering fluid into suitable containers.

7. Properly recover the refrigerant from the A/C system, if equipped.

8. Remove the radiator and cooling fan as an assembly.

9. Disconnect the accelerator cable from the mounting bracket and throttle lever. Remove the 2 accelerator shaft bracket bolts from the valve cover, and remove the bracket.

10. Disconnect the speedometer cable from the transaxle.

11. Disconnect the fuel hoses. Plug or cover the hose openings to prevent dirt from entering and to avoid fuel leakage.

12. Disconnect the heater hoses and the brake booster vacuum hose.

13. Tag and disconnect the engine harness connectors and grounds.

14. Disconnect the backup lamp switch, Park/Neutral Position Switch (PNP switch), and transaxle ground.

15. Remove the starter motor.

16. Disconnect the clutch release cable.

17. Remove the accessory drive belts.

18. Disconnect the power steering lines, if equipped.

19. Raise and safely support the vehicle.

20. Remove the front wheel and tire assemblies.

21. Remove the right and left front splash shields.

22. Remove the lower arm clamp bolts and nuts. Pull the lower arms downward, separating the lower arms from the knuckles.

23. Remove the halfshafts and install differential plugs T87C-7025-C or equivalent, between the differential side gears.

24. If equipped, remove the 2 power steering pump bolts and separate the power steering pump from the engine.

25. If equipped, remove the A/C compressor.

26. Disconnect the transmission gearshift rod and clevis and gearshift stabilizer bar from the transaxle.

27. Remove the exhaust inlet pipe.

28. Remove the 2 front and 2 rear transaxle support insulator nuts.

29. Lower the vehicle.

30. Disconnect the vacuum lines from intake manifold vacuum outlet fitting and cap and fuel vapor canister.

31. Attach lifting hooks to the engine lifting eyes and remove any slack in the chains.

32. Remove the engine support insulator through bolt and nut from engine support insulator.

33. Carefully raise and remove the engine and transaxle as an assembly.

34. Remove the 2 muffler pipe bracket bolts and bracket.

35. Remove the 5 transaxle case to cylinder block front and rear bracket bolts, and brackets.

36. Remove the 4 engine-to-transaxle bolts and separate the transaxle from the engine.

37. Remove the flywheel.

38. Mount the engine on an engine stand, and remove the lifting hooks from the engine.

To install:

39. Attach lifting hooks to the engine lifting eyes and remove any slack in the chains.

40. Remove the engine from the engine stand.

41. Install the flywheel.

42. Mount the transaxle to the engine. Install the engine-to-transaxle bolts and tighten to 41–59 ft. lbs. (55–80 Nm).

43. Install the starter motor.

44. Install the transaxle case to cylinder block front and rear bracket bolts, and brackets. Tighten the bolts to 27–38 ft. lbs. (37–52 Nm).

45. Install the muffler pipe bracket, and tighten the bolts to 28–41 ft. lbs. (38–56 Nm).

46. Position the engine and transaxle assembly in the engine compartment.

47. Install the engine support insulator through bolt and nut into the engine support insulator, and tighten to 39–47 ft. lbs. (53–64 Nm).

48. Remove the lifting cables from the engine lifting eyes.

49. Connect the vacuum lines from intake manifold vacuum outlet fitting and cap and fuel vapor canister.

50. Raise and safely support the vehicle.

51. Install the 2 rear transaxle support insulator nuts, and tighten to 21–34 ft. lbs. (28–46 Nm).

52. Install the 2 front transaxle support insulator nuts, and tighten to 27–38 ft. lbs. (37–52 Nm).

53. Install the exhaust inlet pipe and tighten the nuts to 23–34 ft. lbs. (31–46 Nm).

54. If equipped, install the A/C compressor.

55. If equipped, install the power steering pump and tighten the bolts to 27–40 ft. lbs. (36–54 Nm).

56. Remove the differential plugs and install the halfshafts. Install the lower arm ball joint to the knuckle and tighten the clamp nut and bolt to 32–40 ft. lbs. (43–54 Nm).

57. Install the right and left front splash shields.

58. Install the front wheel and tire assemblies and lower the vehicle.

59. If equipped, connect the power steering lines.

60. Install the accessory drive belts.

61. Connect the clutch release cable.

62. Connect the starter motor electrical connectors.

63. Connect the backup lamp switch, Park/Neutral Position Switch (PNP switch), and transaxle ground.

64. Connect all engine harness connectors and grounds.

65. Connect the brake booster vacuum hose, the heater hoses and the fuel lines.

66. Connect the speedometer cable. Connect the accelerator cable and bracket to the throttle lever and mounting bracket.

67. Install the radiator and cooling fan.

68. Install the engine air cleaner, and the air cleaner to intake manifold tube.

69. Install the hood, aligning the marks that were made during the removal procedure.

70. Install the battery tray and the battery. Connect the battery cables, negative cable last.

71. Add the proper types and quantities of engine oil, transaxle fluid and coolant.

72. If equipped, add power steering fluid to the reservoir.

73. If equipped, evacuate, recharge and leak test the air conditioning system.

74. Start the engine. Check for leaks and proper fluid levels.

75. Road test the vehicle and recheck all fluid levels.

Engine Mounts

REMOVAL AND INSTALLATION

NOTE: Procedures are given for engine mounts (support insulators) as well as transaxle mounts (support insulators). Make sure to properly support the engine and/or transaxle when replacing mounts.

Right Front Mount

1. Disconnect the negative battery cable.

2. Remove the engine air cleaner.

3. Support the engine with a suitable jack.

4. Remove the 3 engine mount (support insulator) nuts and washers.

5. Remove the engine mount through bolt and nut.

6. Remove the engine mount.

To install:

7. Position the engine mount and install the engine mount nuts. Tighten the nuts to 39–47 ft. lbs. (53–64 Nm).

8. Install the through bolt and nut. Tighten to 48–69 ft. lbs. (67–93 Nm).

9. Remove the jack from under the engine.

10. Install the engine air cleaner.

11. Connect the negative battery cable.

Front and Rear Transaxle Mounts

1. Disconnect the negative battery cable.

2. Support the engine with a Three-Bar Engine Support or equivalent.

3. Raise and safely support the vehicle.

4. Remove the 2 front mount and the 2 rear mount nuts from rear engine support crossmember.

5. Remove the 4 engine support rebound insulator bolts from the crossmember and remove the crossmember.

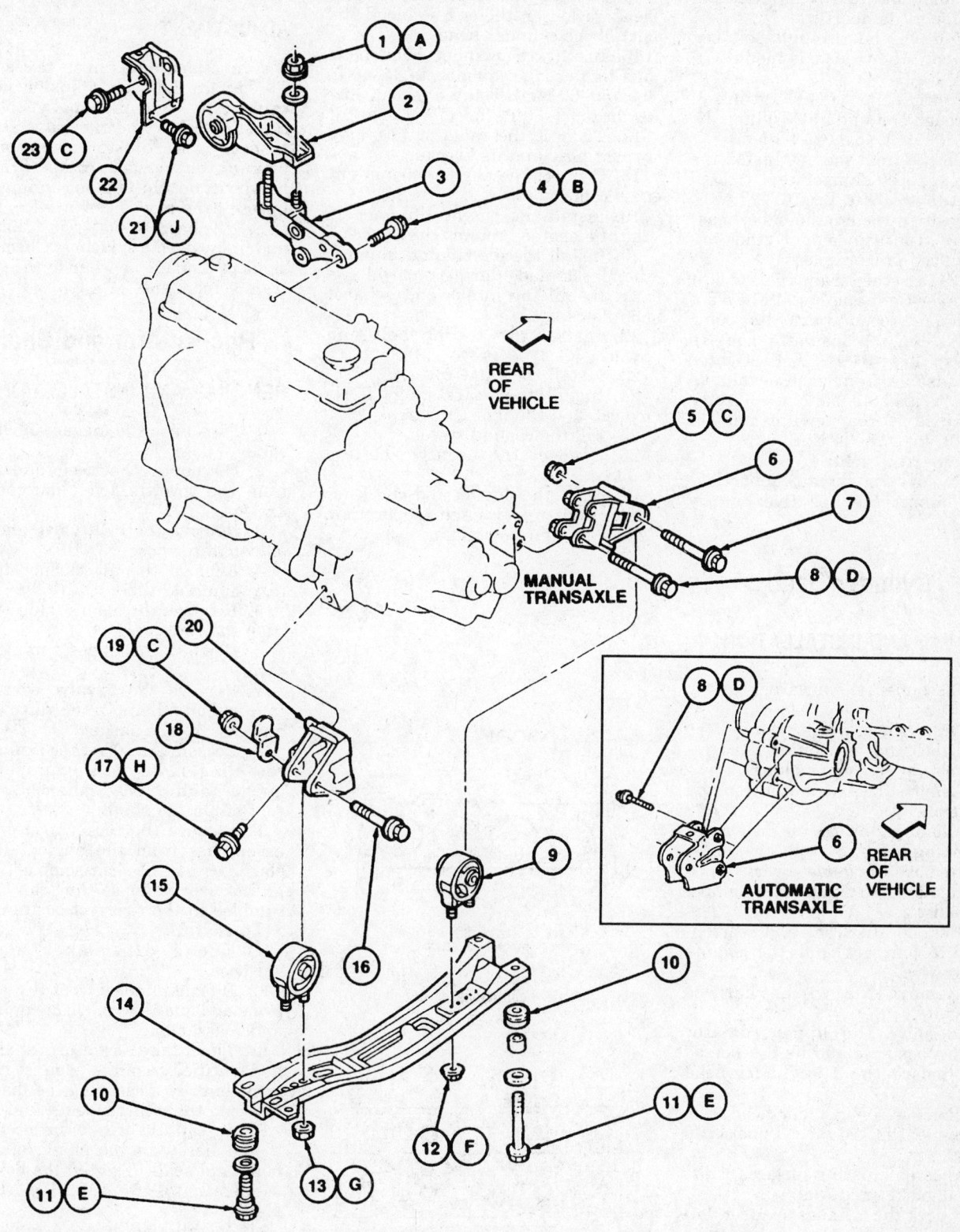

REAR
OF
VEHICLE

MANUAL
TRANSAXLE

AUTOMATIC
TRANSAXLE

REAR
OF
VEHICLE

295410

Engine and transaxle mounting

6. On the front mount, remove the front mount nut and remove the engine support damper.

7. Remove the through bolt nut and remove the transaxle mount.

To install:

8. Position the mount and install the through bolt and nut. Tighten the nut to 48–69 ft. lbs. (67–93 Nm).

9. On the front mount, Install the engine support damper and tighten the nut to 48–69 ft. lbs. (67–93 Nm).

10. Position the crossmember making sure transaxle mount studs are through the proper holes.

11. Install the transaxle mount nuts finger-tight and install the 4 engine support rebound insulator bolts.

12. Tighten the insulator bolts to 47–66 ft. lbs. (64–89 Nm). Tighten the transaxle front mount nuts to 32–34 ft. lbs. (43–52 Nm). Tighten the transaxle rear mount nuts to 21–34 ft. lbs. (28–34 Nm).

13. Lower the vehicle.

14. Remove the engine support.

15. Connect the negative battery cable.

Cylinder Head

REMOVAL AND INSTALLATION

1. Disconnect the negative battery cable.

2. Drain the cooling system.

3. Label and disconnect the ignition wires at the spark plugs.

4. Remove the distributor from the cylinder head.

5. Remove the timing belt cover and timing belt.

6. Remove the valve cover.

7. Remove the exhaust and intake manifolds.

8. Remove the front and rear engine lift hangers and the engine ground wire.

9. Remove the wiring harness connectors.

10. Remove the upper radiator hose, bypass hose and its bracket.

11. Remove the 10 cylinder head bolts.

12. Remove the cylinder head from the engine. Discard the cylinder head gasket.

13. Clean all mating surfaces of dirt and old gasket material.

14. Check the cylinder head for flatness using a feeler gauge and straightedge. The cylinder head should be within 0.006 inch (0.15mm) over the entire cylinder head area. Machine or replace the cylinder head, as necessary.

To install:

15. Properly position the cylinder head gasket on the engine block and install the cylinder head.

16. Install 10 cylinder head bolts and tighten, in sequence, to 35–40 ft. lbs. (50–60 Nm). Tighten in sequence again to 56–60 ft. lbs. (75–81 Nm).

17. Connect the radiator hose and bypass hose and its bracket.

18. Connect the engine wiring harness connectors.

19. Install the engine lift hangers and the engine ground wire.

20. Install the distributor, aligning the marks made during removal.

21. Install the intake and exhaust manifolds.

22. Install the timing belt and cover.

23. Install the valve cover.

24. Install the spark plugs if removed, and the ignition wires.

25. Fill the cooling system.

26. Connect the negative battery cable.

27. Start the engine and check for leaks and proper engine operation. Check the ignition timing.

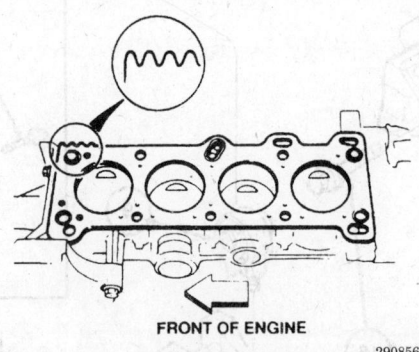

FRONT OF ENGINE

290856

Cylinder head gasket positioning

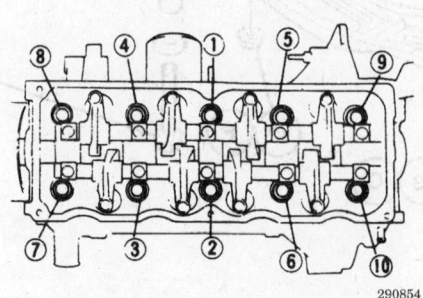

290854

Cylinder head bolt torque sequence

Valve Lash

ADJUSTMENT

No valve lash adjustment necessary. The hydraulic lash adjuster maintains a 0 clearance between the camshaft lobes and the valve stems. Inspect lash adjuster operation by pushing down each rocker arm by hand without the engine running. If a rocker arm moves down with little force, the lash adjuster is collapsed and may need replacing. Check for signs of damage or wear to the valve train components.

Rocker Arm and Shaft

REMOVAL AND INSTALLATION

1. Disconnect the negative battery cable.

2. Disconnect the accelerator cable from the throttle lever and routing bracket.

3. Remove the PCV valve and the oil separator hose.

4. Remove the air cleaner to intake manifold tube.

5. Remove the spark plug wires from the routing clips.

6. Remove the upper timing belt cover.

7. Remove the 6 valve cover retaining bolts. Remove the valve cover and discard the gasket.

8. Loosen the 10 rocker arm shaft retaining bolts in the proper sequence and remove the bolts and seats from the shafts.

9. Remove the rocker arms/shafts assemblies from the engine. If the shafts are to be disassembled, keep all parts in order so they can be assembled in their correct positions.

To install:

10. Clean all gasket mating surfaces.

11. If disassembled, coat the rocker arms and shafts with clean engine oil and reassemble.

12. The intake rocker arm shaft can be identified from the exhaust rocker arm shaft by measuring the distance between the oiling holes **A** and **B**.

13. Install the rocker arms/shafts assemblies with the shaft retaining bolts and seats. Tighten the bolts, in sequence, to 16–21 ft. lbs. (22–28 Nm).

14. Install the valve cover with a new gasket. Tighten the 6 valve cover retaining bolts to 44–79 inch lbs. (5–9 Nm).

15. Install the upper timing belt cover.

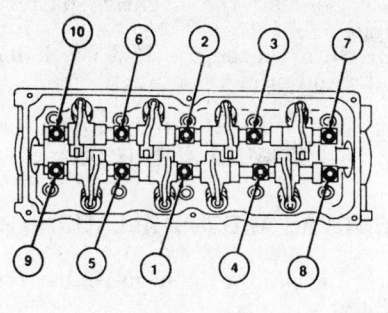

Rocker arm and shaft retaining bolt removal sequence

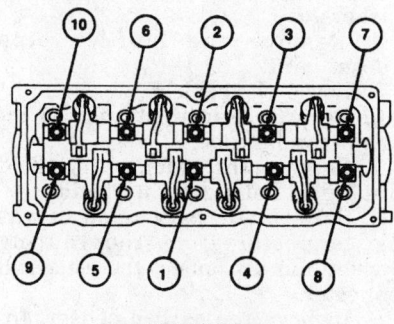

TIGHTENING SEQUENCE

Rocker arm shaft bolt torque sequence

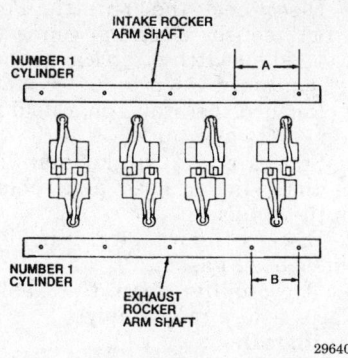

Method of identifying the rocker arm shafts

16. Install the spark plug wires onto the routing clips.

17. Install the PCV valve and the oil separator hose onto the valve cover.

18. Install the accelerator cable.

19. Install the air cleaner to intake manifold tube.

20. Connect the negative battery cable.

21. Run the engine and check for leaks and proper engine operation.

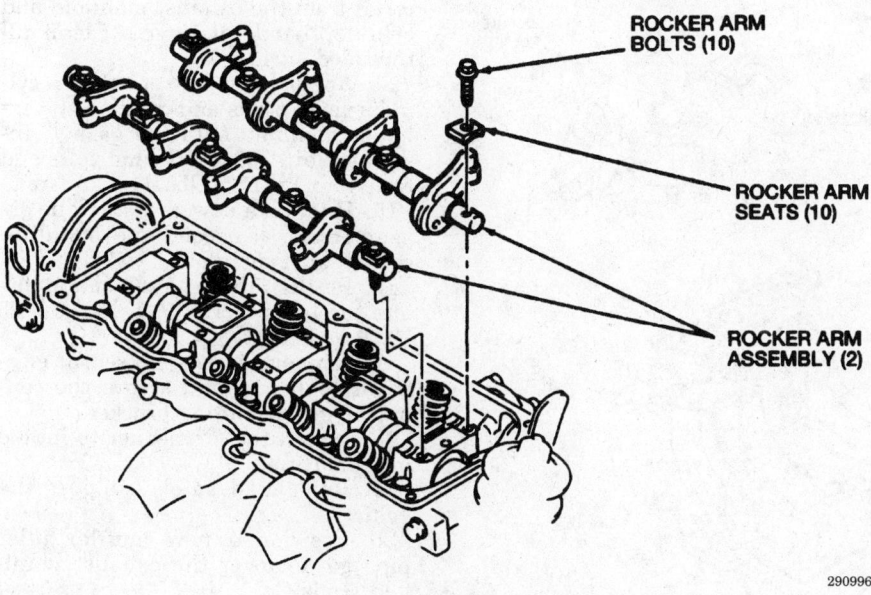

Rocker arm and shaft assembly

Intake Manifold

REMOVAL AND INSTALLATION

—— **CAUTION** ——

Fuel injection systems remain under pressure, even after the engine has been turned OFF. The fuel system pressure must be relieved before disconnecting any fuel lines. Failure to do so may result in fire and/or personal injury.

NOTE: The intake manifold removal and installation procedure includes removal of the intake plenum as one assembly. If servicing the fuel rail or fuel injectors, only remove the intake plenum.

Intake Plenum Only

1. Properly relieve the fuel system pressure as follows:

   a. Start the engine.

   b. Disconnect the fuel pump relay electrical connector.

   c. After the engine stalls, turn the ignition to the **OFF** position.

   d. Disconnect the negative battery cable.

   e. Connect the fuel pump relay electrical connector.

2. Drain the cooling system.

3. Disconnect the accelerator cable from the throttle body.

4. Remove the air duct from the throttle body.

5. Identify, tag and disconnect the necessary vacuum and coolant hoses and remove.

6. Disconnect the Throttle Position (TP) sensor electrical connector and idle switch electrical connector, if equipped.

7. Remove the intake plenum retaining bolts and nuts.

8. Separate the intake plenum from the intake manifold and remove from the vehicle.

9. Discard the intake plenum-to-intake manifold gasket.

10. Thoroughly clean the gasket surfaces before reassembly.

   To install:

11. Install a new intake plenum-to-intake manifold gasket.

12. Position the intake plenum to the intake manifold and install the retaining nuts and bolts. Torque the retaining bolts and nuts to 14–20 ft. lbs. (19–26 Nm).

13. Connect the vacuum and coolant hoses as removed.

14. Connect the TP sensor electrical connector and the idle switch electrical connector, if equipped.

15. Install the air duct to the throttle body.

16. Install the accelerator cable to the throttle body.

17. Refill the cooling system to the proper level.

18. Connect the negative battery cable.

19. Run the engine and check for leaks and proper operation.

Intake Manifold and Plenum

1. Properly relieve the fuel system pressure as follows:

  a. Start the engine.

  b. Disconnect the fuel pump relay electrical connector.

  c. After the engine stalls, turn the ignition to the **OFF** position.

  d. Disconnect the negative battery cable.

  e. Connect the fuel pump relay electrical connector.

2. Drain the cooling system.

3. Remove the intake manifold bracket retaining bolts and the bracket.

4. Disconnect the accelerator cable from the throttle body.

5. Remove the air duct from the throttle body.

6. Identify, tag and disconnect the necessary vacuum and coolant hoses and remove.

7. Disconnect the Throttle Position (TP) sensor electrical connector and the idle switch electrical connector, if equipped.

8. Remove the intake manifold retaining bolts and nuts.

9. Separate the intake manifold from the cylinder head and remove from the vehicle.

10. Discard the intake manifold-to-cylinder head gasket.

11. Thoroughly clean the gasket surfaces before reassembly.

To install:

12. Install a new intake manifold-to-cylinder head gasket.

13. Position the intake manifold to the cylinder head and install the retaining nuts and bolts. Torque the retaining bolts and nuts to 14–20 ft. lbs. (19–26 Nm).

14. Connect the vacuum and coolant hoses as removed.

15. Connect the TP sensor electrical connector and the idle switch electrical connector, if equipped.

16. Install the air duct to the throttle body.

17. Install the accelerator cable to the throttle body.

18. Install the intake manifold bracket and retaining bolts. Tighten the bolts to 22–34 ft. lbs. (31–46 Nm).

19. Refill the cooling system to the proper level.

20. Connect the negative battery cable.

21. Run the engine and check for leaks and proper operation.

Exhaust Manifold

REMOVAL AND INSTALLATION

1. Disconnect the negative battery cable.

2. Raise and safely support the vehicle.

3. Remove the exhaust inlet pipe-to-exhaust manifold nuts and washers.

4. Remove the muffler pipe bracket bolts.

5. Lower the vehicle.

6. Remove the air cleaner to intake manifold tube.

7. Remove the exhaust manifold heat shield bolts and the shield.

8. Separate the oxygen sensor wiring connector from the routing bracket and disconnect the electrical connector.

9. Remove the oxygen sensor. Inspect the sensor gasket for damage and replace if necessary.

10. Remove the 4 exhaust manifold retaining bolts and 3 nuts.

11. Remove the exhaust manifold from the vehicle.

12. Remove the inlet pipe and exhaust manifold gaskets and discard.

To install:

13. Remove all existing gasket material from the exhaust manifold and cylinder head inlet pipe. Clean all threaded surfaces.

14. Apply a new gasket onto the cylinder head studs and position the exhaust manifold onto the gasket. Install the attaching nuts and bolts and torque to 12–17 ft. lbs. (16–23 Nm).

15. Position a new gasket on the oxygen sensor if needed and install to the exhaust manifold.

16. Position the heat shield and torque the bolts to 12–17 ft. lbs. (16–23 Nm).

17. Connect the oxygen sensor electrical connector and secure the connector in the routing bracket.

18. Install the air cleaner to intake manifold tube.

19. Raise and safely support the vehicle.

20. Position a new muffler inlet pipe gasket over the exhaust manifold studs.

21. Raise the exhaust inlet pipe into position on the exhaust manifold and support by hand. Install the attaching nuts and washers and torque to 23–34 ft. lbs. (31–46 Nm).

FUEL RAIL

PRESSURE REGULATOR

INTAKE PLENUM

INTAKE MANIFOLD

302326

Intake plenum and intake manifold assembly

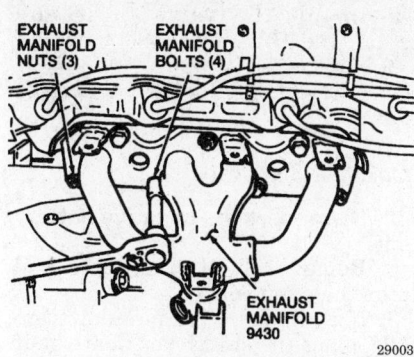

Exhaust manifold installation

22. Position the muffler pipe bracket and install the bolts. Tighten to 28–47 ft. lbs. (38–56 Nm).

23. Lower the vehicle.

24. Connect the negative battery cable.

25. Start the engine and inspect for exhaust leaks.

Front Cover Seal

REMOVAL AND INSTALLATION

NOTE: The timing belt must be removed for this procedure. The crankshaft seal is located behind the crankshaft sprocket in the oil pump front housing.

1. Disconnect the negative battery cable.

2. Remove the accessory drive belts.

3. Remove the timing belt cover and the timing belt.

4. Remove the crankshaft sprocket and crankshaft key.

5. Using seal remover T92C-6700-CH or equivalent, remove the crankshaft front seal from the oil pump housing.

To install:

6. Lubricate the lip of the new seal and the crankshaft seal surface with clean engine oil.

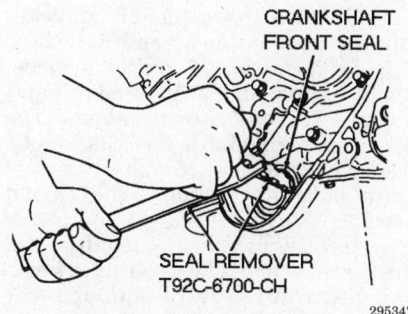

Removing the front crankshaft seal

7. Using front seal installer T87C-6019-A or equivalent, draw the front crankshaft seal into the oil pump housing.

8. Install the crankshaft sprocket and key.

9. Install the timing belt and timing belt cover.

10. Install the accessory drive belts.

11. Connect the negative battery cable.

12. Run the engine and check for oil leaks and proper engine operation.

Timing Belt

REMOVAL AND INSTALLATION

1. Disconnect the negative battery cable.

2. Remove the accessory drive belts.

3. Remove the 3 water pump pulley attaching bolts and remove the water pump pulley.

4. Raise and safely support the vehicle.

5. Remove the right front wheel and tire assembly and the right inner fender panel.

6. Remove the 4 attaching bolts and the screws from the crankshaft pulley. Remove the spacer and outer pulley, if equipped. Remove the inner spacer, inner pulley and the baffle or guide plates, as required.

7. Remove the attaching bolts and the upper and lower covers.

8. Rotate the crankshaft until the sprocket timing marks are aligned.

9. Remove the timing belt tensioner spring, spring cover and timing belt tensioner bolt. Remove the timing belt.

NOTE: If the timing belt is to be reused, mark the direction of rotation on the belt, using a crayon, so the belt can be reinstalled in the same direction.

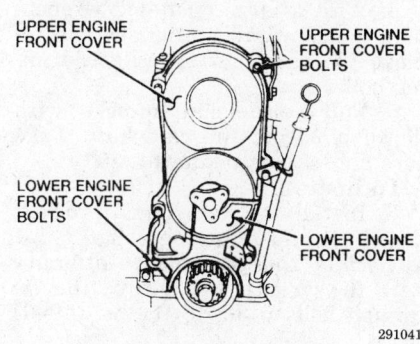

Upper and lower front covers

10. If the camshaft sprocket requires removal, proceed as follows:

a. Hold the camshaft stationary with an open end wrench and remove the camshaft sprocket retaining bolt.

b. Pull the camshaft sprocket with dowel pin off of the camshaft. Use care not to drop the dowel pin.

11. If the crankshaft sprocket requires removal, proceed as follows:

a. Remove the crankshaft pulley retaining bolt.

b. Pull the crankshaft pulley hub, sprocket and key from the crankshaft. Make sure not to drop the crankshaft key.

To install:

12. If removed, install the crankshaft sprocket as follows:

a. Install the sprocket with the key onto the crankshaft.

b. Install the crankshaft pulley hub.

c. Clean the threads of the crankshaft pulley bolt and coat with a non-hardening sealer.

d. Install the bolt and tighten to 80–85 ft. lbs. (108–118 Nm).

13. If removed, install the camshaft sprocket as follows:

a. Position the sprocket and dowel pin to the camshaft and install the retaining bolt.

b. Hold the camshaft stationary with an open end wrench and torque the retaining bolt to 36–45 ft. lbs. (49–61 Nm).

14. Align the camshaft and crankshaft timing marks with the marks located on the cylinder head and oil pump housing.

15. If reusing the original timing belt, install the timing belt with the mark made indicating the direction of rotation.

16. Install the timing belt tensioner spring and cover on the tensioner. Position the tensioner and spring assembly on the engine and install the attaching bolt. Do not tighten the bolt at this time.

17. Rotate the crankshaft 2 turns in the direction of normal rotation and align the timing marks. Ensure all marks are still correctly aligned.

18. Reconnect the free end of the spring to the spring anchor. Torque the tensioner bolt to 14–19 ft. lbs. (19–26 Nm).

19. Install the upper and lower covers. Install the attaching bolts and tighten to 71–97 inch lbs. (8–11 Nm).

20. Install the crankshaft pulley baffle with the curved lip facing outward or install the large guide plate and then the small guide plate, as required.

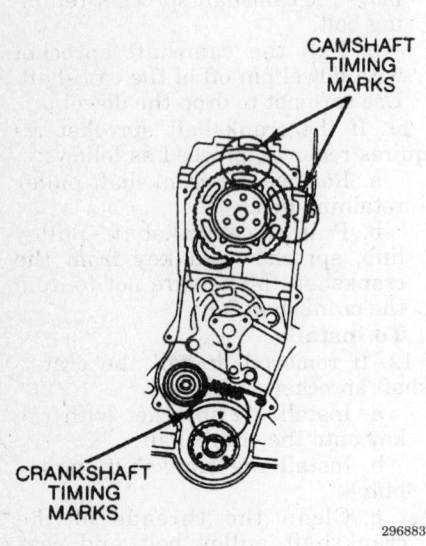

Timing marks

296883

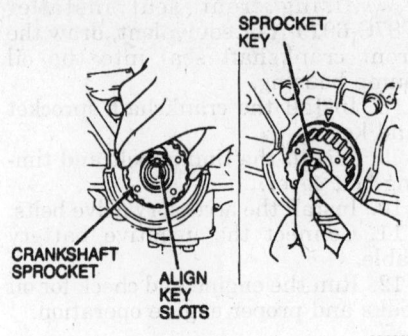

Crankshaft sprocket removal

296887

21. Install the inner pulley with the deep recess facing outward. Install the spacer and then the outer pulley, spacer and screws. Install the pulley bolts and tighten to 109–152 inch lbs. (12–17 Nm).

22. Install the inner fender panel.

23. Install the wheel and tire assembly. Torque the lug bolts to 65–87 ft. lbs. (88–118 Nm).

24. Lower the vehicle.

25. Install the water pump pulley and tighten the bolts to 36–45 ft. lbs. (49–61 Nm).

26. Install the accessory drive belts.

27. Connect the negative battery cable.

28. Run the engine and check for proper operation.

Timing Belt, Sprockets, Tensioner and Front Cover

REMOVAL AND INSTALLATION

Camshaft Sprocket

1. Disconnect the negative battery cable.

2. Remove the timing belt and timing belt tensioner.

3. Remove the valve cover.

4. With a large open-end wrench, hold the camshaft stationary and remove the camshaft sprocket retaining bolt.

5. Pull the camshaft sprocket with dowel pin from the camshaft. Take care not to lose the dowel pin.

To install:

6. Install the camshaft sprocket, dowel pin and retaining bolt.

7. Hold the camshaft stationary with the wrench and torque the retaining bolt to 36–45 ft. lbs. (49–61 Nm).

8. Install the timing belt and tensioner.

9. Install the valve cover and tighten the bolts to 44–80 inch lbs. (5–9 Nm).

10. Connect the negative battery cable.

Crankshaft Sprocket

1. Disconnect the negative battery cable.

2. Remove the timing belt and timing belt tensioner.

3. If equipped with manual transaxle, place the shift lever in 4th gear and apply the parking brake. If equipped with automatic transaxle vehicle, install flywheel holding tool T84P-6375-A or equivalent.

4. Remove the crankshaft pulleys, pulley hub and retaining bolt.

5. Pull the crankshaft sprocket, timing belt guide and key from the crankshaft. Make certain not to lose the key when removing the crankshaft sprocket. Replace the key if worn or damaged.

To install:

6. Position the crankshaft sprocket onto the crankshaft and align the keyways. Install the key.

7. Install the timing belt guide

8. Coat the threads of the retaining bolt with non-hardening sealer. Install the pulley hub and retaining bolt and torque to 80–85 ft. lbs. (108–118 Nm).

9. Install the crankshaft pulleys

10. Remove the flywheel holding tool.

11. Install the timing belt.

12. Connect the negative battery cable.

Camshaft

REMOVAL AND INSTALLATION

1. Disconnect the battery cables, negative cable first.

2. Remove the battery.

3. Remove the timing belt and valve cover.

4. Remove the camshaft sprocket and the distributor assembly.

5. Remove the 10 rocker arm assembly bolts and seats and remove the two rocker arm assemblies. Tag the rocker arm shafts for reassembly.

6. Remove the camshaft thrust plate bolt and remove the thrust plate.

7. Gently pull the camshaft out of the flywheel side of the cylinder head, being careful not to damage the camshaft bearing surfaces.

8. Inspect the camshaft seal and replace as necessary.

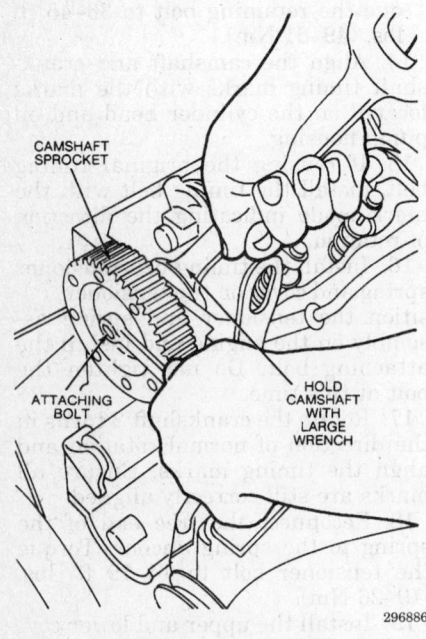

Camshaft sprocket removal

296886

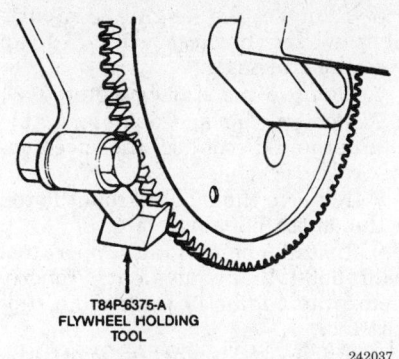

T84P-6375-A
FLYWHEEL HOLDING
TOOL

242037

Flywheel Holding Tool

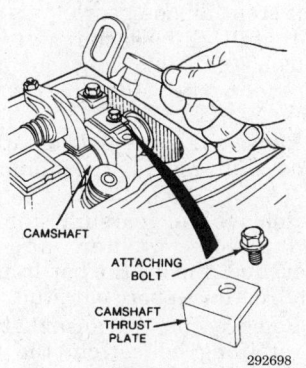

CAMSHAFT

ATTACHING
BOLT

CAMSHAFT
THRUST
PLATE

292698

Camshaft thrust plate

9. Inspect the camshaft journals and lobes for wear and/or damage.

To install:

10. Clean all gasket mating surfaces. Thoroughly clean the camshaft and cylinder head.

11. Coat the camshaft with clean engine oil and install it through the flywheel side of the cylinder head. Be careful not to damage the bearing surfaces in the cylinder head.

12. Install the camshaft thrust plate and bolt. Tighten the bolt to 71–88 inch lbs. (8–10 Nm).

13. Install the two rocker arm assemblies, bolts and seats. Tighten the bolts to 16–21 ft.lbs. (22–28 Nm) in the correct sequence.

NOTE: While tightening the rocker arm assembly bolts, pull back on the rocker springs to prevent them from being pinched between the rocker shaft and the cylinder head pedestal.

14. Install the distributor and the camshaft sprocket.

15. Install the valve cover and timing belt.

16. Install the battery and connect the battery cables, negative cable last.

17. Run the engine and check for leaks and proper operation.

18. Check and adjust the ignition timing, as required.

Piston and Connecting Rod

POSITIONING

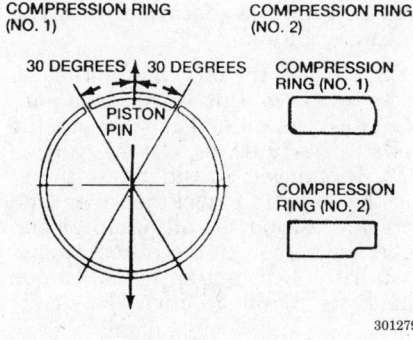

COMPRESSION RING (NO. 1)
COMPRESSION RING (NO. 2)

30 DEGREES 30 DEGREES

PISTON PIN

COMPRESSION RING (NO. 1)

COMPRESSION RING (NO. 2)

301279

Piston ring location

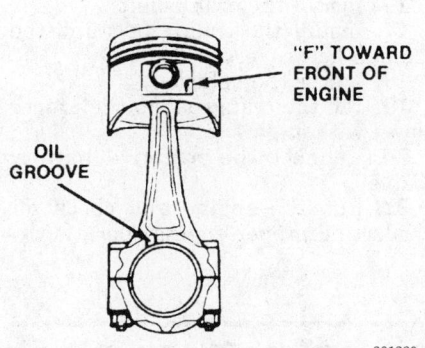

"F" TOWARD FRONT OF ENGINE

OIL GROOVE

301280

Piston Location

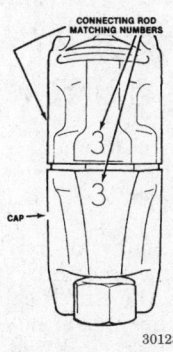

CONNECTING ROD MATCHING NUMBERS

3

3

CAP

301281

Connecting rod cap to rod matchmarks

ENGINE LUBRICATION

Oil Pan

REMOVAL AND INSTALLATION

1. Disconnect the negative battery cable.

2. Raise and safely support the vehicle.

3. Drain the engine oil into a suitable container.

4. Remove the exhaust inlet pipe.

5. Remove the oil pan-to-cylinder block retaining nuts and bolts. Lower the oil pan and discard the old gasket.

To install:

6. Clean the oil pan and cylinder block sealing surfaces to remove all traces of existing gasket material. Thoroughly clean the oil pan.

7. Apply a suitable oil resistant sealant to the joint lines formed at the cylinder block and front and rear engine covers.

8. Apply a new gasket to the oil pan.

9. Position the oil pan and gasket to the cylinder block and install the retaining nuts and bolts. Torque the oil pan nuts and bolts in an alternating pattern to 69–78 inch lbs. (8–9 Nm). Do not over tighten the bolts or the gasket will split.

10. Install the oil pan drain plug and tighten to 22–30 ft. lbs. (29–41 Nm).

11. Lower the vehicle.

12. Install the exhaust inlet pipe.

13. Fill the crankcase with the proper amount of clean engine oil.

14. Connect the negative battery cable.

15. Run the engine and check for oil leaks.

Oil Pump

REMOVAL AND INSTALLATION

1. Disconnect the negative battery cable.

2. Raise and safely support the vehicle.

3. Remove the timing belt covers and the timing belt.

4. Remove the crankshaft sprocket.

5. Drain the engine oil and remove the oil pan.

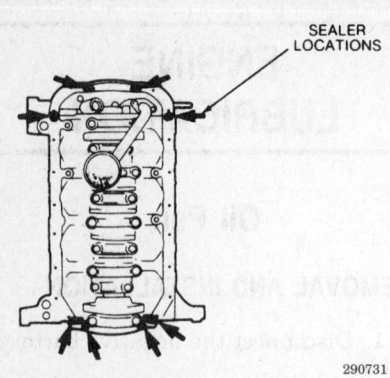

Apply sealer at these locations

6. If equipped, remove the crankshaft position sensor (CKP) bracket bolt and move the sensor and bracket aside.

7. Remove the 6 oil pump assembly retaining bolts and remove the oil pump assembly and gasket from the engine. Discard the gasket.

8. Remove the oil pump pickup tube and screen, if required.

9. Disassembly the oil pump and inspect all components as necessary. Replace the oil pump if needed.

To install:

10. Clean all gasket mating surfaces. If disassembled, clean the oil

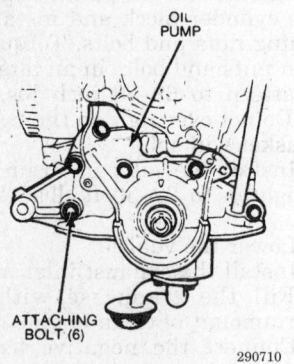

Oil pump attaching bolt locations

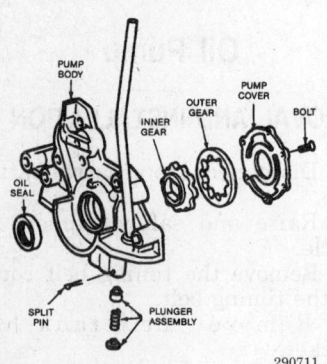

Oil pump components

pump housing and components with a suitable solvent and allow to dry. Reassemble the oil pump, lubricating all components with clean engine oil. Install a new front crankshaft seal using a suitable seal installer.

11. Carefully coat both sides of the new oil pump gasket with a suitable sealant compound. Apply the gasket to the oil pump and remove any excess sealant.

NOTE: Do not allow the sealant compound to enter the oil pump discharge opening once the gasket is in place. This opening must be free and clear before the oil pump is installed onto the cylinder block.

12. Position the oil pump to the cylinder block mating surface and install the 6 retaining bolts. Torque the bolts to 14–19 ft. lbs. (19–25 Nm).

13. If removed, install a new gasket on the oil pump pickup tuber and screen. Position the oil pump pickup tube and screen to the oil pump and install the 2 retaining bolts. Torque the bolts to 69–95 inch lbs. (8–11 Nm).

14. If removed, install the (CKP) and tighten the bracket bolt to 25 inch lbs. (3 Nm).

15. Install the oil pan.

16. Install the crankshaft.

17. Install the timing belt and the timing belt covers.

18. Lower the vehicle.

19. Fill the crankcase to the proper level with engine oil.

20. Connect the negative battery cable.

21. Run the engine and check for leaks and proper engine operation.

TRANSAXLE

Manual Transaxle Assembly

REMOVAL AND INSTALLATION

1. Disconnect the negative battery cable.

2. Disconnect the 2 backup light switch wiring connectors.

3. Disconnect the park/neutral switch wiring connector.

4. Remove the clutch cable adjusting nut and disengage the cable from the release lever. Pull the clutch release cable through the clutch cable bracket.

5. Remove the engine compartment wiring harness ground strap from the transaxle.

6. Remove the starter motor.

7. Loosen the speedometer cable retainer and disconnect the speedometer cable.

8. Remove the 2 bolts from the top of the clutch housing.

9. Install 3 bar engine support tool D88L-6000-A, or equivalent. Properly secure the engine to the engine support tool.

10. Raise and safely support the vehicle.

11. Disengage the halfshafts from the differential side gears.

12. Install differential side gear plug tool T87C-7025-C or equivalent, to prevent the side gears from moving.

13. Remove the nut and bolt attaching the shift rod to the input shift rail.

14. Remove the gearshift stabilizer bar nut, lock washer, and flat washer, and remove the bar from the control rod-to-support bar stud.

15. Remove the 3 transaxle-to-engine retaining bolts from the transaxle case rear bracket and remove the bracket.

16. Remove the 3 transaxle-to-engine retaining bolts from the transaxle case front bracket and remove the bracket.

17. Remove the 2 rear and 2 front transaxle support insulator nuts from the rear engine support.

18. Remove the 4 rear engine support rebound insulator bolts and remove the rear engine support.

19. Position a suitable transmission jack under the transaxle and secure it with a safety chain or strap.

20. Remove the 4 flywheel reinforcing plate bolts.

21. Remove the 2 remaining transaxle-to-engine block retaining bolts.

22. Carefully separate the transaxle from the engine and lower the transaxle from the vehicle.

To install:

23. Raise the transaxle into position and seat it against the rear of the engine.

24. Install 4 flywheel reinforcing plate bolts and tighten to 62–86 inch lbs. (7–10 Nm).

25. Install 2 lower transaxle retaining bolts and tighten to 47–66 ft. lbs. (64–89 Nm).

26. Remove the transmission jack.

27. Position the rear engine support. Install 2 rear transaxle support insulator nuts and tighten to 21–34 ft. lbs. (28–46 Nm.).

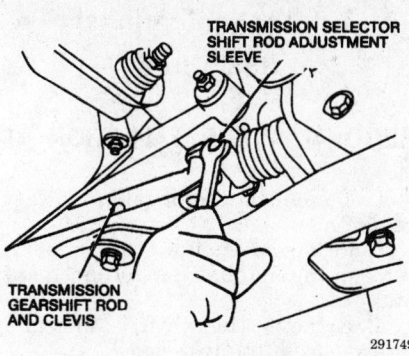

Gearshift rod and clevis

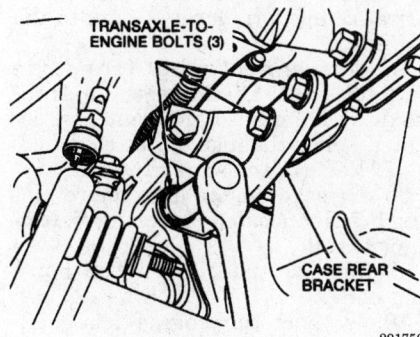

Transaxle case rear bracket

28. Install 2 front transaxle support insulator nuts and tighten to 32–38 ft. lbs. (43–52 Nm.).

29. Install the 4 rear engine support rebound insulator bolts and tighten to 47–66 ft. lbs. (64–89 Nm.).

30. Install the transaxle case to cylinder block front and rear brackets and install the 3 bolts on each. Tighten the bolts to 27–38 ft. lbs. (37–52 Nm.).

31. Install the washer and the gearshift stabilizer bar on the control rod-to-support bar stud.

32. Install the washer, lock washer, and gearshift stabilizer bar nut. Tighten the nut to 28–38 ft. lbs. (38–52 Nm.).

33. Position the gearshift rod and clevis on the main shift control shaft, and install the selector shift rod adjustment sleeve. Tighten the nut to 12–17 ft. lbs. (16–23 Nm.).

34. Route the park/neutral (PNP) switch wiring over the rear engine support.

35. Install the halfshaft and joint assemblies.

36. Check and fill the transaxle, if needed.

37. Lower the vehicle and remove the engine support bar.

38. Install 2 retaining bolts at the top of the clutch housing. The top bolt

is installed through the heater pipe bracket. Torque the bolts to 47–66 ft. lbs. (64–89 Nm.).

39. Connect the ground strap to the transaxle case.

40. Connect the speedometer cable onto the sleeve and hand tighten.

41. Install the starter motor.

42. Connect the park/neutral and backup light switch wiring connectors.

43. Connect the clutch cable to the release lever and adjust the clutch pedal free-play.

44. Connect the negative battery cable.

45. Road test the vehicle and check for proper transaxle operation.

Clutch Assembly

REMOVAL AND INSTALLATION

1. Disconnect the negative battery cable.

2. Raise and safely support the vehicle.

3. Remove the transaxle assembly.

NOTE: During the removal procedure, do not allow oil or grease to come in contact with the clutch disc facing if the disc is to be reused. Handle the disc with clean rags wrapped around the edges and do not touch the disc facing. Even a small amount of dirt or grease may cause the clutch to grab or slip.

4. If the pressure plate is to be reused, paint or scribe alignment marks on the pressure plate and flywheel for assembly reference.

5. Install an appropriate locking tool to prevent the flywheel from turning.

6. Install a clutch aligning tool to prevent the clutch plate from dropping when the retaining bolts are removed.

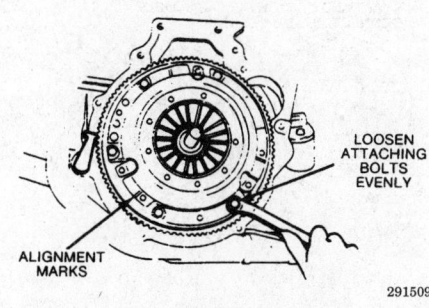

Pressure plate alignment marks

7. Loosen the 6 pressure plate retaining bolts in an alternate pattern, 1 turn at a time. This will relieve the pressure plate spring tension evenly and prevent distortion of the pressure plate. Remove the pressure plate and clutch disc once the retaining bolts are removed.

8. Inspect all clutch components including the clutch release fork and release bearing and replace as required.

9. Inspect the flywheel for scoring, cracks and heat checks. Resurface or replace the flywheel, as necessary.

10. Inspect the pilot bearing for damage. Make sure the bearing turns easily. If replacement is necessary, remove the flywheel and remove the pilot bearing.

To install:

11. If necessary, install a new pilot bearing using a suitable installation tool. Use only a driver tool that contacts the bearing outer race. A driver tool that contacts the inner race or the bearing area will damage the bearing.

12. If the flywheel was removed, clean the sealant from the flywheel retaining bolts. Coat the bolt threads with a suitable sealer compound.

13. Make sure the crankshaft flange and the back of the flywheel are clean. Position the flywheel on the crankshaft and install the 6 retaining bolts. Tighten the bolts to 71–76 ft. lbs. (96–103 Nm).

14. Position the clutch disc on the flywheel and install a clutch alignment tool to hold the disc in place.

NOTE: When installing the clutch disc, make sure the disc dampener springs are facing away from the flywheel. A new disc will be stamped FLYWHEEL to indicate the correct installation position.

15. Align the reference marks, if present, and position the pressure plate on the flywheel and install the retaining bolts. Torque the bolts evenly, in an alternate pattern, to 13–20 ft. lbs. (18–26 Nm). The bolts must be tightened in this manner to prevent distortion of the pressure plate.

16. Remove the clutch alignment tool.

17. Clean the clutch disc splines on the input shaft with a dry rag and coat the spline surfaces with a light film of clutch grease.

18. Install the transaxle.

19. Connect the negative battery cable.

20. Adjust the clutch pedal free-play.

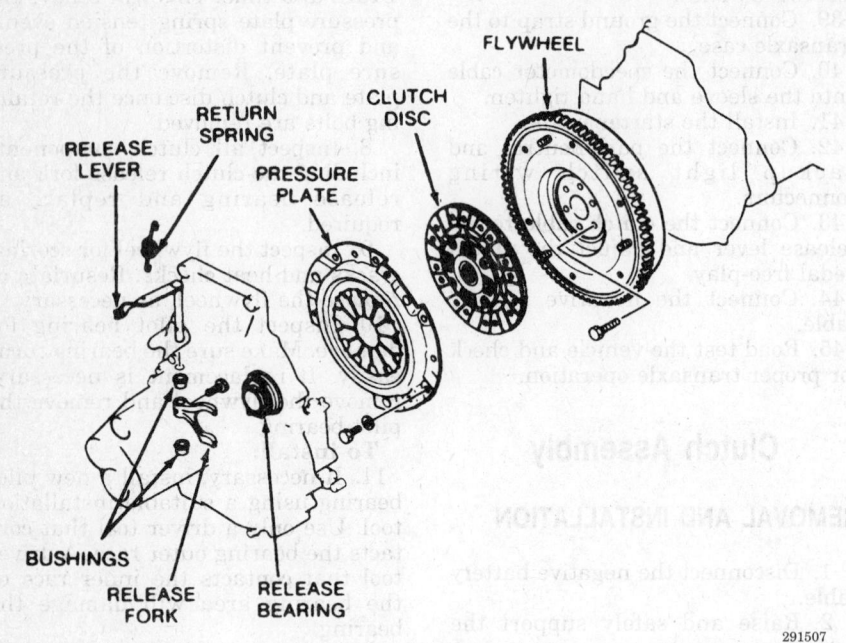

RETURN SPRING

RELEASE LEVER

FLYWHEEL

CLUTCH DISC

PRESSURE PLATE

BUSHINGS

RELEASE FORK

RELEASE BEARING

291507

Clutch assembly

21. Road test the vehicle for proper clutch operation.

Clutch Cable

ADJUSTMENTS

Free-Play Adjustment

1. Carefully move the clutch pedal back and forth and measure the amount of travel. If the clutch pedal free-play is 0.35–0.59 in. (9–15mm), no adjustment is necessary. If the free-play is not within specification, proceed to Step 2.

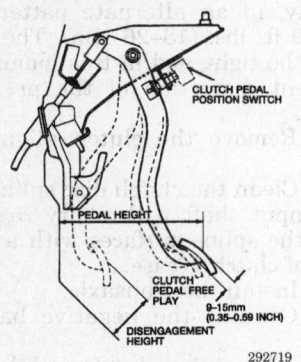

CLUTCH PEDAL POSITION SWITCH

PEDAL HEIGHT

CLUTCH PEDAL FREE PLAY

9-15mm (0.35-0.59 INCH)

DISENGAGEMENT HEIGHT

292719

Clutch pedal free-play

2. Pull back the transaxle release lever and measure the clearance between the lever and the cable connecting link. Thread the adjuster in or out until the clearance between the connecting link and the release lever is 0.06–0.10 in. (1.5–2.5mm).

3. Check the free-play at the clutch. If it is not within specification, inspect the clutch release components for a problem.

4. After adjusting the clutch, make sure the clutch disengagement height is 2.92 inches (74 mm) minimum.

CLEARANCE BETWEEN CLUTCH RELEASE SHAFT AND CLUTCH RELEASE CABLE CONNECTING LINK

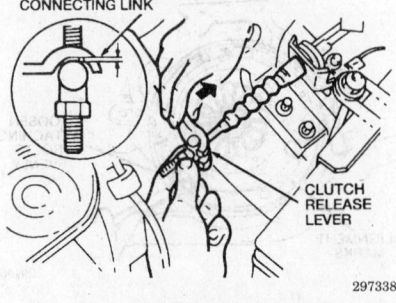

CLUTCH RELEASE LEVER

297338

Clearance check between the release lever and connecting link

Automatic Transaxle Assembly

REMOVAL AND INSTALLATION

1. Disconnect the negative battery cable.

2. From the engine compartment, remove the manual control lever nut and arm.

3. Remove the shift cable and bracket from the transaxle.

4. Disconnect the speedometer cable from the transaxle.

5. Disconnect the transaxle electrical connectors, located next to the governor.

6. Disconnect the transaxle ground wire. Disconnect the transaxle vacuum hose and vent hose located below the distributor cap.

7. Remove the starter motor.

8. Remove the coolant pipe retaining bracket, located below the distributor cap.

9. Remove the 2 upper bell housing bolts.

10. Support the engine using engine support bar tool D87L–6000–A or equivalent.

11. Raise and safely support the vehicle.

12. Remove the front wheel and tire assemblies.

13. Drain the transaxle fluid.

14. Remove the front fender splash shield.

15. Remove the front stabilizer bar.

16. Remove the lower arm clamp bolts and nuts. Pull the lower arms downward, separating the lower arms from the knuckles.

NOTE: Use care not to damage the ball joint dust boots.

17. Remove the tie rod cotter pin and nut. Disconnect the tie rod end from the knuckle and discard the cotter pin

18. Remove the halfshafts. Install differential plug tool T87C–7025–C or equivalent, between the differential side gears to prevent side gear movement.

19. Remove the front and rear transaxle support insulator nuts.

20. Remove the 4 rear engine support rebound insulator bolts and the transmission support crossmember.

21. Remove the front transaxle support insulator through bolt and nut. Remove the front transaxle support insulator.

22. Remove the 4 front transaxle support bracket bolts and the front transaxle support bracket.

23. Remove the 2 rear transaxle support bracket bolts and the rear transaxle support bracket and insulator.

24. Remove the intake manifold support.

25. Remove the 3 transaxle-to-engine bolts and remove the transaxle case rear bracket.

26. Remove the 3 transaxle-to-engine bolts from the transaxle case to cylinder block front bracket and remove the bracket.

27. Remove the flywheel cover bolts and cover.

28. Using a wrench, rotate the crankshaft pulley bolt clockwise to gain access to 4 torque converter-to-flywheel nuts and remove all nuts.

29. Make alignment marks between the oil cooler tubes and hoses. Disconnect and plug the oil cooler lines.

30. Position a transmission jack under the transaxle and secure with a chain or strap.

31. Remove the remaining engine-to-transaxle retaining bolts.

32. Carefully separate and lower the transaxle from the vehicle.

To install:

33. Raise the transaxle into position and install 2 engine-to-transaxle bolts. Make sure that the torque converter is in alignment with the flex plate. Tighten to 47–66 ft. lbs. (64–89 Nm).

34. Remove the transaxle jack.

35. Install the starter motor.

36. Install the torque converter bolts and tighten to 26–36 ft. lbs. (34–49 Nm).

37. Install the flywheel cover and tighten the bolts to 71–97 inch lbs. (8–11 Nm).

38. Position the front transaxle-to-engine support and install the 3 retaining bolts. Tighten the bolts to 27–38 ft.lbs. (37–52 Nm.).

39. Install the intake manifold support and tighten the bolts to 27–38 ft.lbs. (37–52 Nm.).

40. Install the transaxle case rear bracket and 3 retaining bolts. Tighten the bolts to 27–38 ft.lbs. (37–52 Nm.).

41. Install the front transaxle support insulator bracket and 4 retaining bolts. Tighten the bolts to 28–37 ft. lbs. (38–51 Nm.).

42. Install the front transaxle support bracket and the through bolt and nut. Do not tighten the the through bolt and nut until the rear engine support is installed.

43. Position the transmission support crossmember and install the rear engine support rebound insulator bolts. Tighten the rear engine support rebound insulator bolts to 47–66 ft. lbs. (64–89 Nm.).

44. Tighten the front transaxle support bracket through bolt and nut to 69–83 ft. lbs. (93–113 Nm.).

45. Install 2 front transaxle support insulator nuts and tighten to 32–38 ft. lbs. (43–52 Nm).

46. Install 2 rear transaxle support insulator nuts and tighten to 21–34 ft. lbs. (28–46 Nm).

47. Remove the differential plugs and install the halfshafts.

48. Align the marks made on the oil cooler lines and install. Install the hose clamps.

49. Connect the tie rod ends to the steering knuckles and tighten the attaching nuts to 26–30 ft. lbs. (35–40 Nm). Install new cotter pins.

50. Attach the lower arm ball joints to the knuckles. Tighten the lower arm clamp bolt to 40–50 ft. lbs. (54–68 Nm).

51. Install the front stabilizer bar. Tighten the retaining nuts to 43–52 ft. lbs. (43–52 Nm).

52. Install the front fender splash shield and tighten the bolts to 65–95 inch lbs. (8–10 Nm.).

53. Install the front wheel and tire assemblies. Tighten the lug bolts to 65–87 ft. lbs. (88–118 Nm).

54. Lower the vehicle.

55. Remove the engine support tool.

56. Attach the shift cable and bracket to the transaxle. Install the manual control lever arm on the manual control lever. Install the retaining nut and tighten to 34–47 ft. lbs. (44–64 Nm).

NOTE: Do not use any type of power wrench to tighten the nut. Damage to the transaxle may result.

57. Install the coolant pipe retaining bracket located below the distributor.

58. Connect the vacuum hose and vent hose located below the distributor cap.

59. Connect the transaxle electrical connectors located next to the governor.

60. Connect the ground wire to the transaxle and tighten the bolt to 65–95 inch lbs. (8–10 Nm.).

61. Connect the speedometer cable.

62. Connect the negative battery cable.

63. Fill the transaxle with the specified fluid to the proper level.

NOTE: Make sure that the gearshift lever position aligns with the manual control lever position exactly before starting the engine.

64. Start the engine. Check for leaks and proper fluid level.

65. Road test the vehicle and check for proper operation.

DRIVELINE

Halfshaft

REMOVAL AND INSTALLATION

1. Disconnect the negative battery cable.

2. With the vehicle sitting on all 4 wheels, use a small cape chisel to raise the staked portion of the front wheel hub locknut. Have an assistant apply the brakes, then loosen but do not remove the locknut.

3. Raise the vehicle and support it safely.

4. Drain the transaxle fluid.

5. Remove the front wheel and tire assembly.

6. Remove the ball joint clamp bolt and nut from the steering knuckle. Carefully pry the lower control arm down to disconnect the ball joint from the steering knuckle. Be careful not to tear or puncture the dust boot when disconnecting the ball joint.

7. Using a small prybar, separate the halfshaft from the transaxle.

NOTE: The halfshaft must be separated from the transaxle gradually. If the halfshaft is yanked out suddenly, the oil seal may be damaged.

8. Install differential plug tool T87C-7025-C or equivalent, to prevent the differential side gear from moving.

9. Remove and discard the wheel hub locknut.

10. Withdraw the halfshaft from the hub. Be careful not to damage the oil seal. If necessary, use a suitable wheel puller to push the halfshaft out of the hub.

To install:

11. Inspect the differential and wheel hub oil seals for damage and replace as required.

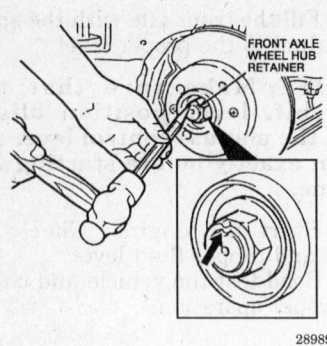

Raising the staked edge of the wheel hub retaining nut

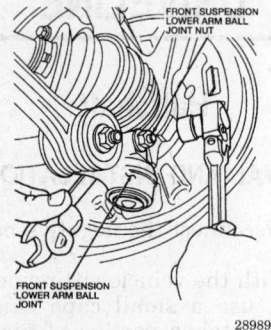

Removing the clamp bolt and nut from the steering knuckle

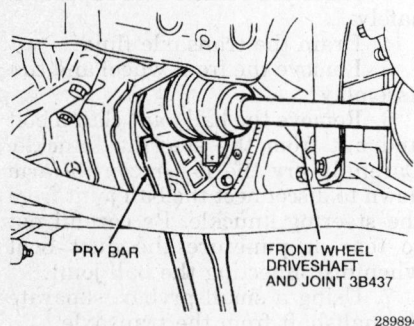

Separating the halfshaft from the transaxle

12. Remove the circlip from the inboard halfshaft spline end and replace with a new clip. Lubricate the inboard and outboard halfshaft spline ends with grease.

13. Remove the differential gear holding plug.

14. Position and install the inboard end of the halfshaft into the differential side gear. Make sure the circlip snaps into place and take care not to damage the differential oil seal.

15. Position and install the outboard end of the halfshaft into the

wheel hub. Take care not to damage the wheel hub oil seal.

16. Install the wheel hub locknut onto the halfshaft and tighten by hand.

17. Raise the lower control arm and connect the ball joint. Take care not to damage the ball joint dust boot. Install the clamp nut and bolt and torque the nut to 32–40 ft. lbs. (43–54 Nm).

18. Install the wheel and tire assembly. Torque the lug bolts to 65–87 ft. lbs. (88–118 Nm).

19. Install and tighten the transaxle drain plug.

20. Lower the vehicle.

21. With the vehicle sitting on all four wheels, have an assistant apply the brakes. Torque the wheel hub locknut to 116–174 ft. lbs. (157–235 Nm). Stake the nut using a suitable tool.

NOTE: Do not stake the locking tab with a pointed tool. Make sure the locking tab is depressed at least 0.16 in. (4mm) into the locknut slot to ensure proper locking capability.

22. Raise and safely support the vehicle.

23. Grasp the wheel hub and pull on it to ensure that the halfshaft is installed properly. Rotate the wheel hub by hand to check that the hub and halfshaft assembly turns smoothly.

24. Fill the transaxle with the proper grade and type fluid to specification.

25. Lower the vehicle.

26. Connect the negative battery cable.

27. Road test the vehicle and check for transaxle leaks and proper operation.

CV-Joint Boot

REMOVAL AND INSTALLATION

NOTE: On all halfshafts, the outboard CV-joint (Birfield-type) is permanently fitted onto the halfshaft and cannot be removed. To replace the outboard CV-joint boot, the inner CV-joint (Tripot-type) must first be removed. If a boot has failed due to age or wear, all boots should be replaced at the same time.

1. Raise and safely support the vehicle.

2. Remove the halfshaft from the vehicle and support the assembly in a vise equipped with soft jaws.

3. Remove the large boot clamp from the inboard CV-joint and roll the boot back over the shaft.

4. Matchmark the outer race, axle shaft and tripot bearing for reassembly using paint or marker. Do not use a punch or chisel.

5. Remove the wire ring bearing retainer from inside the outer race/housing and remove the outer race.

6. Matchmark the tripot bearing and the shaft. Remove the tripot bearing snap-ring and remove the tripot bearing from the shaft. It may be necessary to drive the tripot off the shaft with a brass drift.

7. Remove the small clamp and the inboard CV-joint boot from the halfshaft.

NOTE: Test the CV-joint grease for contamination by rubbing a small amount between 2 fingers. If a gritty feeling is present, the grease is contaminated and the CV-joint must be disassembled and thoroughly cleaned and inspected before adding new grease.

8. To remove the outboard CV-joint boot, remove the dynamic damper (right-hand halfshaft only), then remove the boot clamps and slide the boot off of the shaft from the inboard side. If the outboard boot is to be reused, wrap tape around the shaft splines to protect the boot during removal.

9. On vehicles equipped with ABS brakes, the wheel speed sensor can be driven off the outboard CV-joint if the joint is to be replaced.

NOTE: Do not remove the anti-lock sensor ring if it does not need to be replaced. If the sensor ring must be removed replace with a new sensor ring.

To install:

10. If not already installed, wrap smooth electrical tape around the halfshaft spline to protect and ease the installation of the CV-joint boot(s). Slide the clamps and the outboard boot onto the shaft.

11. Before positioning the boot over the CV-joint, pack the CV-joint and boot with grease. Be sure to use all of the grease in the pouch supplied with the boot kit.

12. Fit the boot into place on the CV-joint, making sure it is fully seated in the grooves in the shaft and outer race. Insert a suitable tool between the boot and the outer bearing race to allow trapped air to escape from the boot.

Automatic Transaxle Vehicles

Vehicles equipped with an automatic transaxle have tripot type CV joints.

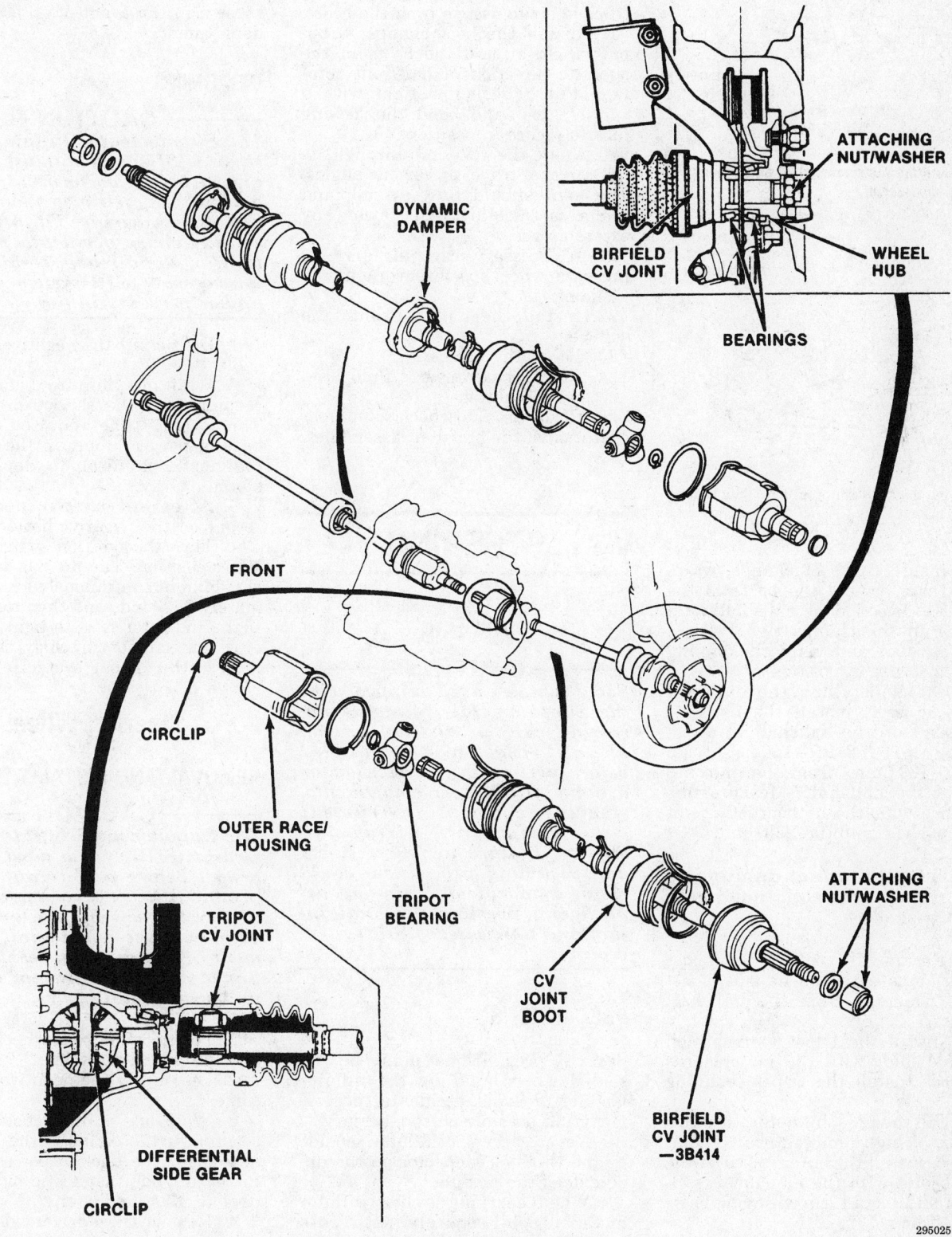

ATTACHING NUT/WASHER

BIRFIELD CV JOINT

WHEEL HUB

BEARINGS

DYNAMIC DAMPER

FRONT

CIRCLIP

OUTER RACE/ HOUSING

TRIPOT BEARING

TRIPOT CV JOINT

DIFFERENTIAL SIDE GEAR

CIRCLIP

CV JOINT BOOT

BIRFIELD CV JOINT —3B414

ATTACHING NUT/WASHER

295025

Halfshafts with Tripot CV-joints shown disassembled

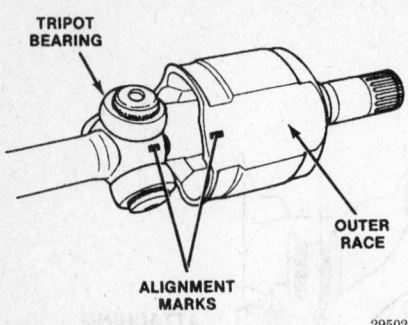

Matchmark the outer race, shaft and bearing prior to disassembly

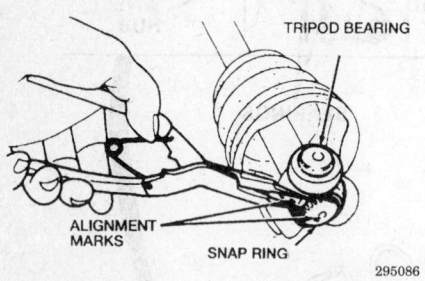

Removing the snap-ring from the Tripot bearing

13. Install the boot clamps, wrapping them around the boots in the opposite direction of halfshaft rotation. Pull the clamps tight with a suitable tool and bend the locking tabs to secure in position.

14. After installing the outboard CV-joint boot, install the dynamic damper onto the halfshaft at a distance of 18.99–19.27 inch (482.5–489.5mm) from the outboard end of the halfshaft. Measure this distance with the outboard CV-joint fully pushed onto the halfshaft.

NOTE: Dynamic damper is only used on the right-hand halfshaft assembly.

15. Fit the inboard CV-joint boot and clamps onto the halfshaft. Remove the tape from the halfshaft splines.

16. Install the tripot assembly on the halfshaft with the matchmarks aligned. Install the tripot retaining ring.

17. Fill the CV-joint outer race with 3.5 oz. of high temperature CV-joint grease. Install the outer race over the tripot joint with the matchmarks aligned and install the wire ring bearing retainer.

18. Fit the boot into place on the CV-joint, making sure it is fully

seated in the grooves in the shaft and outer race. The distance between the CV-joint boot clamp grooves will measure about 3.5 inch (90mm). Insert a suitable tool between the boot and the outer bearing race to allow trapped air to escape from the boot.

19. Install the boot clamps, wrapping them around the boots in the opposite direction of halfshaft rotation. Pull the clamps tight with a suitable tool and bend the locking tabs to secure in position.

20. Work the CV-joint through its full range of travel at various angles. The joint should flex, extend, and compress smoothly. Wipe away any excess grease.

21. If necessary, carefully drive or press the wheel speed sensor onto the CV-joint.

22. Install the halfshaft into the vehicle.

23. Lower the vehicle.

24. Check the transaxle fluid level and add if required.

25. Road test and check for proper operation of the halfshaft assemblies.

STEERING

Air Bag

———— CAUTION ————

Some vehicles are equipped with the Supplemental Inflatable Restraint (SIR) or air bag system. The SIR system must be disabled before performing service on or around SIR system components, steering column, instrument panel components, wiring and sensors. Failure to follow safety and disabling procedures could result in accidental air bag deployment, possible personal injury and unnecessary SIR system repairs.

PRECAUTIONS

Several precautions must be observed when handling the inflator module to avoid accidental deployment and possible personal injury.

• Never carry the inflator module by the wires or connector on the underside of the module.

• When carrying a live inflator module, hold securely with both hands, and ensure that the bag and trim cover are pointed away.

• Place the inflator module on a bench or other surface with the bag and trim cover facing up.

• With the inflator module on the bench, never place anything on or close to the module which may be thrown in the event of an accidental deployment.

DISARMING

———— CAUTION ————

The Supplemental Inflatable Restraint (SIR) system must be disarmed before performing service around SIR system components or SIR system wiring. Failure to do so may cause accidental deployment of the air bag, resulting in unnecessary SIR system repairs and/or personal injury.

1. Disconnect the negative battery cable.

2. Wait one minute before proceeding with the service procedure. This is the time required for the backup power supply in the air bag diagnostic monitor to deplete its stored energy.

3. After service is completed, reconnect the negative battery cable.

4. Turn the ignition switch to the **RUN** position. The air bag indicator should light continuously for approximately 6 seconds and then turn OFF. If the indicator fails to light, flashes or remains lit continuously, there is a fault in the air bag system.

Steering Wheel

REMOVAL AND INSTALLATION

———— CAUTION ————

The Supplemental Inflatable Restraint (SIR) system must be disarmed before performing service around SIR system components or SIR system wiring. Failure to do so may cause accidental deployment of the air bag, resulting in unnecessary SIR system repairs and/or personal injury.

1. Position the front wheels in the straight-ahead position.

2. Disconnect the negative battery cable.

3. Wait one minute before proceeding further. This is the time required for the backup power supply in the air bag diagnostic monitor to deplete its stored energy.

4. Remove the 4 driver side air bag module bolts and remove the module. Disconnect the 2 electrical connectors

by pressing the knobs on the connectors.

— CAUTION —

When carrying a live air bag, make sure the bag and trim cover are pointed away from the body. In the unlikely event of an accidental deployment, the bag will then deploy with minimal chance of injury. When placing a live air bag on a bench or other surface, always face the bag and trim cover up, away from the surface. This will reduce the motion of the module if it is accidently deployed.

5. Set the air bag module aside with the trim cover face up.
6. Place tape across the air bag sliding contact to prevent the contact from rotating.
7. Matchmark the steering wheel and steering column shaft for assembly reference.
8. Remove the steering wheel nut.
9. Using a steering wheel puller tool, remove the steering wheel. Do not hit the steering column with a hammer. Route the wiring harness through the steering wheel as it is lifted off the steering column shaft.

To install:
10. If the air bag sliding contact has been accidently rotated, proceed as follows:
  a. Remove the steering column shroud screws, separate and remove the shroud.

— WARNING —

Do not rotate the air bag sliding contact more than 2½ turns in either direction to avoid damaging the air bag sliding contact.

  b. Turn the air bag sliding contact clockwise to lock it, then back it off 2¾ turns.
  c. Align the arrows on the air bag sliding contact and the air bag sliding contact housing.
  d. Install the shroud and secure with the screws.
Proper alignment of the air bag sliding contact

295623

1. Position the steering wheel onto the steering column shaft and align the matchmarks. Tighten the nut to 29–36 ft. lbs. (39–49 Nm).
2. Connect the 2 electrical connectors, position the air bag module, and install the 4 bolts. Tighten the bolts to 36–53 inch lbs. (4–6 Nm.).
3. Connect the negative battery cable.

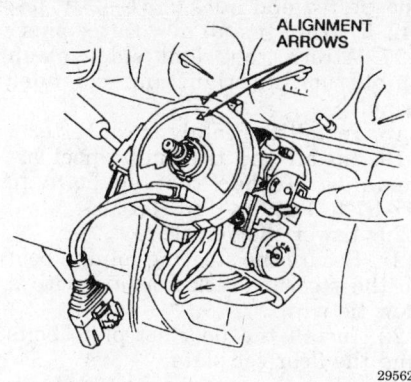

295623

4. Prove out the air bag system by turning the ignition switch to the **RUN** position while watching the air bag warning indicator. The air bag diagnostic monitor will illuminate the warning indicator for approximately 6 seconds and then turn it off indicating that the air bag warning indicator is functional. If the air bag warning indicator does not illuminate, stays on steady or flashes, a fault has been detected by the air bag diagnostic monitor.

Tie Rod Ends

REMOVAL AND INSTALLATION

1. Raise and safely support the vehicle.
2. Remove the wheel and tire assembly.
3. Remove the cotter pin and nut from the tie rod end stud. Discard the cotter pin. Examine the nut for damage and replace as required.
4. Separate the tie rod end from the steering knuckle using tie rod end remover tool T85M-3395-A or equivalent.
5. With paint or a suitable marker, mark the tie rod end, jam nut and tie rod spindle to ease assembly without changing the toe setting.

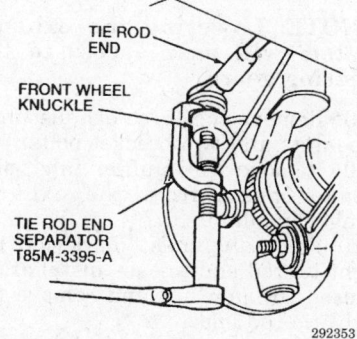

292353

Tie rod end and removal tool

6. Loosen the jam nut and unscrew the tie rod end counting the number of turns required for removal. Replace the tie rod end as required.

NOTE: **If new tie rod ends are being installed, place the old and new ends side-by-side and place alignment marks in the new end that match as closely as possible to the marks on the old end. The existing jam nut may not seat in exactly the same position on the new tie rod end and the toe setting may have to be checked and/or readjusted as a precaution.**

To install:
7. When replacing a tie rod end, install a new dust boot over the stud with a suitable adapter. A ¾ inch. socket will accomplish the task simply and effectively.
8. Thread the jam nut and tie rod end onto the tie rod and align the index marks made during the removal procedure.
9. If installing a new tie rod end, match the position of the old one as closely as possible.
10. Install the tie rod end into the knuckle and tighten the nut to 25–36 ft. lbs. (34–50 Nm). If the cotter pin does not align with stud bore, tighten (do not loosen) the nut until the castellations align with the pin bore. Install a new cotter pin.
11. Tighten the tie rod end jam nut to 31–42 ft. lbs. (42–57 Nm).
12. Install the wheel and tire assembly. Tighten the lug bolts to 65–87 ft. lbs. (88–118 Nm).
13. Lower the vehicle.
14. Check the toe setting.
15. Road test the vehicle and check for proper operation.

Manual Rack and Pinion

REMOVAL AND INSTALLATION

1. Disconnect the negative battery cable.
2. Matchmark the steering column lower universal joint and steering rack pinion for assembly reference. Remove the steering column and intermediate shaft assembly from the vehicle.
3. Remove the floor set plate bolts and the floor set plate.
4. Cut the plastic tie wrap securing the steering column boot to the steering rack.
5. Raise and safely support the vehicle.

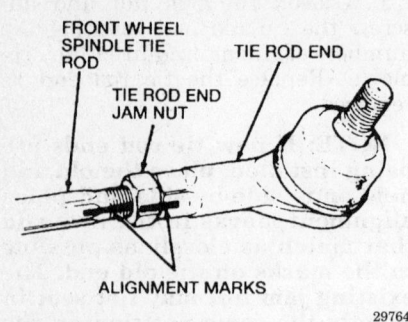

FRONT WHEEL
SPINDLE TIE
ROD

TIE ROD END

TIE ROD END
JAM NUT

ALIGNMENT MARKS

297649

Placing alignment marks on the tie rod end and spindle

6. Remove the front wheel and tire assemblies.

7. Using the proper tool, separate both tie rod ends from the steering knuckles.

8. Remove the catalytic converter.

9. Remove the plastic splash shield from the right inner fender.

10. Remove the 2 steering rack mounting bolts and lower the steering rack until it is free of the steering column boot. Slide the rack to the right, through the inner fender opening until the left tie rod is clear of the left inner fender, then lower the left end until the steering rack assembly can be withdrawn from the left side of the vehicle.

NOTE: While maneuvering the tie rod boots in and out of the inner fender openings, guide the steering rack assembly carefully to avoid cutting or nicking the boots.

11. Remove the steering column intermediate shaft coupling bolt and the coupling from the steering rack.

To install:

12. Install the steering column intermediate shaft coupling and retaining bolts to the steering rack and tighten to 13–20 ft. lbs. (18–26 Nm).

13. Position the steering rack by starting the right side tie rod end through the right inner fender opening far enough to insert the left end of the steering gear assembly into the left front fender opening. Adjust the positioning of the steering rack to the left being careful not to snag the bellows.

14. Align the intermediate shaft coupling with the steering column boot. Raise the steering rack fully into position.

15. Install the left steering rack mounting bolt followed by the right steering rack mounting bolt. Torque the bolts to 27–38 ft. lbs. (37–52 Nm).

16. Connect the tie rod ends to the steering knuckles. Install and tighten

the tie rod end nuts to 31–42 ft. lbs. (42–57 Nm). Install new cotter pins.

17. Attach the right side splash shield on the right inner fender panel.

18. Install the catalytic converter.

19. Install the tire and wheel assemblies. Torque the lug bolts to 65–87 ft. lbs. (88–118 Nm).

20. Lower the vehicle.

21. Secure the steering column boot to the steering rack housing with a new tie wrap.

22. Install the floor set plate bolts and the floor set plate.

23. Align the matchmarks made on the steering column lower universal joint and the steering rack pinion shaft. Install the steering column when the proper alignment is achieved.

24. Connect the negative battery cable.

25. Check the front end alignment.

26. Road test the vehicle and check for proper steering rack operation.

Power Rack and Pinion

REMOVAL AND INSTALLATION

1. Disconnect the negative battery cable.

2. Remove the steering column tube boot retainer and pry up the boot.

3. Remove the intermediate shaft coupling bolt.

4. Raise and safely support the vehicle.

5. Remove the front tire and wheel assemblies.

6. Remove the power steering hose bracket. Disconnect and plug the high pressure and return lines.

7. Remove the tie rod end cotter pins and nuts. Using the proper tool, separate both tie rod ends from the steering knuckles. Discard the cotter pins.

8. Remove the front fender splash shield.

NOTE: Lowering the exhaust system will ease access to the steering gear.

9. Remove the three exhaust inlet pipe nuts and two bracket bolts.

10. Remove the muffler inlet pipe hanger posts from the exhaust hanger insulators.

11. Place alignment marks on the right tie rod end to ease installation. Loosen the jam nut and remove the right tie rod end.

12. Remove the steering rack mounting bolts and lower the steer-

ing rack until it is free of the steering column boot. Slide the rack to the left and pull the right tie rod through the fender opening. Remove the steering gear by sliding it to the right.

To install:

13. Position the steering rack in its mounting location.

14. With an assistant lifting the steering gear, align the intermediate shaft with the universal joint and install the coupling bolt, but do not tighten.

15. Install the 4 steering rack bracket bolts and tighten to 27–38 ft. lbs. (37–52 Nm).

16. Tighten the intermediate shaft coupling bolt to 13–20 ft. lbs. (18–26 Nm).

17. Unplug and connect the high pressure and return lines and install the power steering hose bracket.

18. Install the muffler inlet pipe hanger posts onto the exhaust hanger insulators.

19. Raise the exhaust system and install the 3 exhaust inlet pipe nuts. Tighten the nuts to 28–38 ft. lbs. (38–53 Nm.).

20. Install the 2 exhaust inlet pipe bracket bolts.

21. Install the right tie rod end and attach the tie rod ends to the steering knuckles. Install the tie rod end nuts and tighten to 31–42 ft. lbs. (42–57 Nm). Install new cotter pins.

22. Install the small front fender splash shield.

23. Install the front wheel and tire assemblies and torque the lug bolts to 65–87 ft. lbs. (88–118 Nm).

24. Lower the vehicle.

25. Connect the negative battery cable.

26. Add power steering fluid and allow any air to bleed from the power steering system.

27. Check for leaks.

28. Adjust the toe setting by performing a front end alignment.

29. Road test the vehicle and check for proper operation.

Power Steering Pump

REMOVAL AND INSTALLATION

1. Disconnect the negative battery cable.

2. Remove the air cleaner intake tube and engine air cleaner.

3. Disconnect the electrical connector from the Power Steering Pressure (PSP) switch.

4. Disconnect and plug the reservoir hose at the power steering pump.

Manual Steering Gear Mounting and Related Components

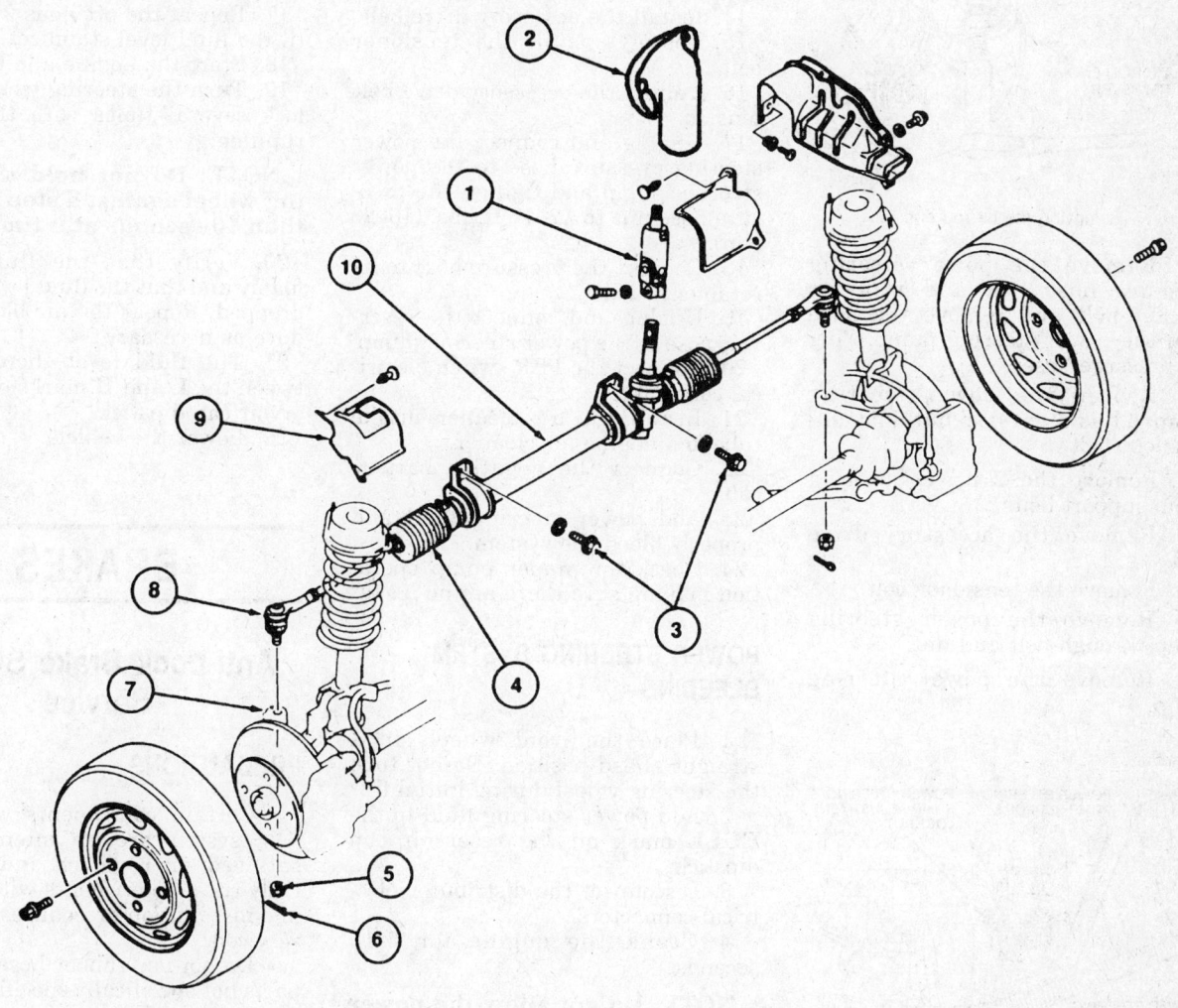

1 Steering Column Gear Input Shaft Coupling
2 Steering Column Tube Boot
3 Steering Gear Bolts
4 Front Wheel Spindle Connecting Rod Bellow
5 Front Wheel Spindle Connecting Rod End Nut
6 Front Wheel Spindle Connecting Rod End Cotter Pin
7 Front Wheel Knuckle
8 Front Wheel Spindle Connecting Rod Or End
9 Front Splash Shield
10 Steering Gear and Linkage

291363

Manual steering gear and related components

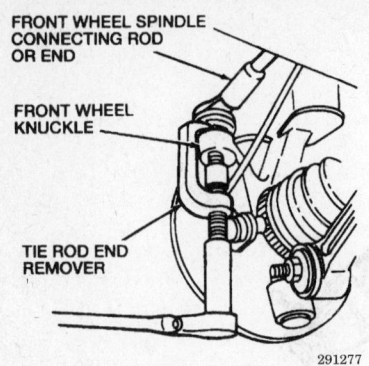

Removing the tie rod end

291277

5. Remove the power steering pressure hose nut and retainer bracket bolt and remove the hose from the power steering pump. Plug the pressure hose.

6. Loosen the power steering pump adjustment locknut and adjustment bolt.

7. Remove the 2 power steering pump support bolts.

8. Remove the accessory drive belt.

9. Remove the tensioner bolt.

10. Remove the power steering pump through-bolt and nut.

11. Remove the power steering pump.

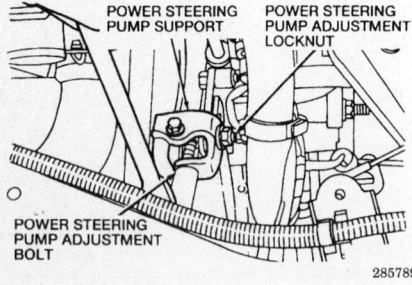

Power steering pump locknut and support

285789

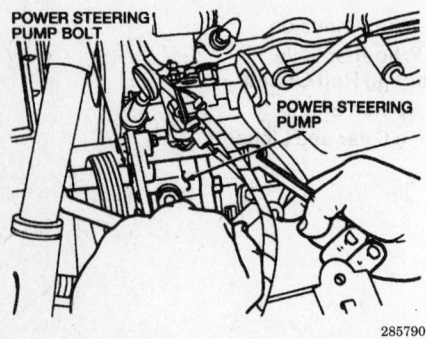

Power steering pump through-bolt

285790

To install:

12. Position the power steering pump and install the 2 power steering pump support bolts. Tighten the bolts to 22–29 ft. lbs. (29–39 Nm).

13. Loosely install the power steering pump through-bolt and nut.

14. Install the accessory drive belt.

15. Loosely install the tensioner bolt.

16. Adjust the accessory drive belt tension.

17. Unplug and connect the power steering pressure hose to the power steering pump and tighten the pressure hose nut to 12–17 ft. lbs. (16–23 Nm).

18. Connect the pressure hose to its retainer bracket.

19. Unplug and connect the reservoir hose to the power steering pump.

20. Connect the PSP switch electrical connector.

21. Install the air cleaner intake tube and engine air cleaner.

22. Connect the negative battery cable.

23. Add power steering fluid and properly bleed the system.

24. Check for proper pump operation making sure there are no leaks.

POWER STEERING SYSTEM BLEEDING

1. Place the front wheels in a straight ahead position. Do not turn the steering wheel during initial fill.

2. Add power steering fluid to the **FULL** mark on the reservoir cap dipstick.

3. Disconnect the distributor electrical connectors.

4. Crank the engine for 5-10 seconds.

NOTE: Do not allow the power steering reservoir to run dry.

5. Refill the power steering reservoir after cranking.

6. Repeat the cranking procedure until the fluid level in the reservoir remains constant.

7. Connect the distributor electrical connectors.

8. Start the engine and allow it to idle for several minutes.

9. Turn the steering wheel lock to lock several times.

10. Turn off the engine and check the fluid level. Add fluid if necessary.

11. If noise or aeration is present, the system must be purged of air.

12. Make sure that the power steering reservoir is full.

13. Raise and safely support the vehicle so that the front wheels are off the ground.

14. Turn the ignition key to the **ON** position.

15. Turn the steering wheel lock to lock several times with the engine not running.

16. Recheck the fluid and add if needed.

17. Repeat the previous 2 steps until the fluid level stabilizes.

18. Start the engine and let it idle.

19. Turn the steering wheel lock to lock several times with the engine running.

NOTE: Do not hold the steering wheel against a stop for more than 10 seconds at a time.

20. Verify that the fluid is not foamy and that the fluid level has not dropped. Repeat the air bleed procedure as necessary.

21. The fluid level should be between the **L** and **H** marks on the reservoir cap dipstick.

22. Lower the vehicle.

BRAKES

Anti-Lock Brake System Service

PRECAUTIONS

• Certain components within the ABS system are not intended to be serviced or repaired individually. Only those components with removal and installation procedures should be serviced.

• Do not use rubber hoses or other parts not specifically specified for and ABS system. When using repair kits, replace all parts included in the kit. Partial or incorrect repair may lead to functional problems and require the replacement of components.

• Lubricate rubber parts with clean, fresh brake fluid to ease assembly. Do not use lubricated shop air to clean parts; damage to rubber components may result.

• Use only DOT 3 brake fluid from an unopened container.

• If any hydraulic component or line is removed or replaced, it may be necessary to bleed the entire system.

• A clean repair area is essential. Always clean the reservoir and cap thoroughly before removing the cap. The slightest amount of dirt in the fluid may plug an orifice and impair the system function. Perform repairs after components have been thoroughly cleaned; use only denatured alcohol to clean components. Do not allow ABS components to come into contact with any substance containing mineral oil; this includes used shop rags.

• The Anti-Lock control unit is a microprocessor similar to other computer units in the vehicle. Ensure that the ignition switch is **OFF** before removing or installing controller harnesses. Avoid static electricity discharge at or near the controller.

• If any arc welding is to be done on the vehicle, the ALCU connectors should be disconnected before welding operations begin.

Master Cylinder

REMOVAL AND INSTALLATION

1. Disconnect the negative battery cable.
2. Disconnect the low fluid level sensor connector.
3. Disconnect the brake lines from the master cylinder connections. Cap the brake lines and plug the master cylinder ports.
4. Remove the 2 master cylinder retaining nuts and washers.
5. Remove the master cylinder from the vehicle.

To install:

6. If a new master cylinder is being installed, check the pushrod length adjustment as follows:
   a. Position master cylinder gauge T87C-2500-A or equivalent, on the end of the master cylinder, loosen the set screw and push the gauge plunger against the bottom of the primary piston.
   b. While holding the gauge in position, tighten the set screw.
   c. Invert the master cylinder gauge and place it over the brake booster pushrod.
   d. If the clearance is not zero, loosen the pushrod locknut and adjust the pushrod.

NOTE: Proper pushrod length adjustment is critical. If the pushrod is adjusted too long, the brakes will drag. If the pushrod is adjusted too short, the brake pedal will be low.

7. Before installation, bench bleed a new master cylinder as follows:
   a. Mount the new master cylinder in a suitable holding fixture. Be careful not to damage the housing.

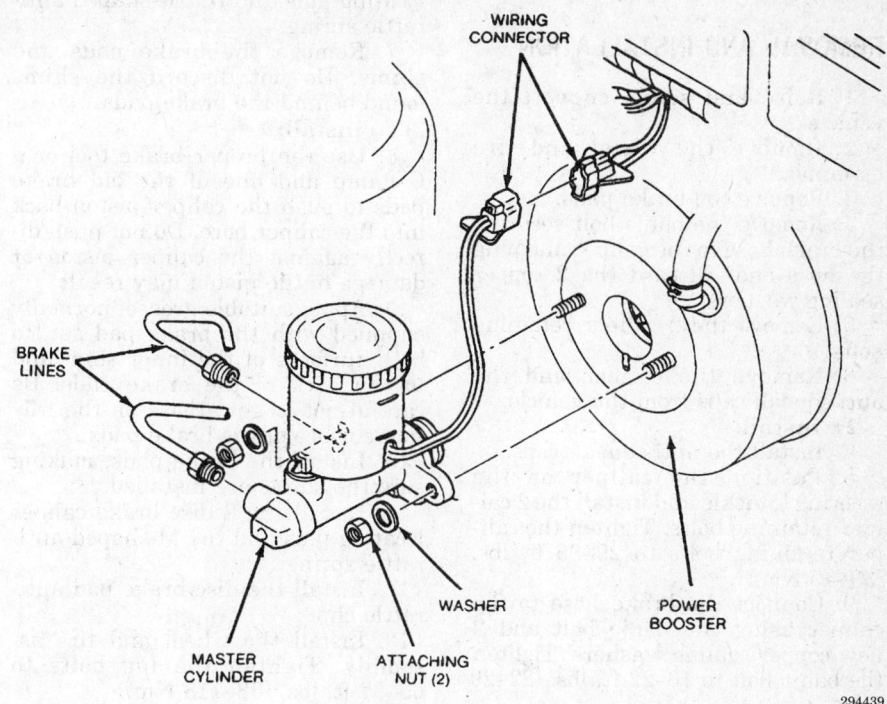

Master cylinder assembly

294439

   b. Fit short lengths of brake lines to the master cylinder ports positioned so that they are directed into the master cylinder reservoir and submerged by brake fluid.
   c. Fill the master cylinder reservoir with clean DOT 3 or equivalent, brake fluid.
   d. Using a suitable tool inserted into the booster pushrod cavity, push the master cylinder piston in slowly and allow it to return.
   e. Repeat the procedure until clear fluid only (no bubbles) is expelled into the master cylinder reservoir.
   f. Remove the short brake lines and plug the outlet ports. Remove the master cylinder from the holding fixture.

8. Position the master cylinder and install the attaching washers and nuts. Torque the nuts to 7–12 ft. lbs. (10–16 Nm).
9. Connect the brake lines to the master cylinder ports and tighten.
10. Have an assistant push down on the brake pedal.
11. When the pedal is all the way down, crack open the brake line fittings at the master cylinder, one at a time, to expel any remaining air at the master cylinder.

12. With the pedal still down, tighten the brake line fitting, then allow the brake pedal to return.
13. Repeat until all air is expelled. Final tighten the brake line fittings to 10–15 ft. lbs. (13–21 Nm).
14. Make sure the master cylinder reservoir is full. To finish bleeding the brake system, proceed as follows:
    a. Have an assistant push down on the brake pedal.
    b. When the pedal is all the way down, crack open the brake bleeder screws, one at a time, to expel any remaining air in the hydraulic brake system.
    c. With the pedal still down, tighten the bleeder screw, then allow the brake pedal to return.
    d. Repeat until all air is expelled from the brake lines. Securely tighten the brake bleeder screws.
    e. If the brake pedal feels spongy, repeat the brake bleeding procedure.
15. Connect the low fluid level sensor.
16. Make sure the master cylinder reservoir is full.
17. Connect the negative battery cable.
18. Check for leaks and proper brake operation.

Brake Caliper

REMOVAL AND INSTALLATION

1. Raise and safely support the vehicle.
2. Remove the wheel and tire assembly.
3. Remove the brake pads.
4. Remove the banjo bolt securing the brake hose to the caliper and plug the hose end. Discard the 2 copper sealing washers.
5. Remove the 2 caliper retaining bolts.
6. Remove the caliper and the anti-squeak caps from the vehicle.
To install:
7. Install the anti-squeak caps.
8. Position the caliper on the steering knuckle and install the 2 caliper retaining bolts. Tighten the caliper retaining bolts to 29–36 ft. lbs. (39–49 Nm).
9. Connect the brake hose to the caliper using the banjo bolt and 2 new copper sealing washers. Tighten the banjo bolt to 16–22 ft. lbs. (22–29 Nm).
10. Install the brake pads.
11. Bleed the brake system.
12. Install the wheel and tire assembly.
13. Lower the vehicle.
14. Apply the brake pedal several times to position the brake pads, before attempting to move the vehicle.
15. Check for proper brake operation.

Disc Brake Pads

REMOVAL AND INSTALLATION

1. Remove brake fluid from the master cylinder reservoir to lower the level by approximately 1/3 preventing brake fluid overflow when the caliper piston is pressed back into its bore.
2. Raise and safely support the vehicle.
3. Remove the tire and wheel assembly.
4. Use the proper brake tool or a C-clamp to move the caliper piston into its bore approximately 1/8 in. (3mm) to allow removal of the disc brake pads.

NOTE: Do not use a screwdriver or similar tool to pry the piston away from the rotor.

5. Remove the brake pad anti-rattle clip. Disengage the M-shaped anti-rattle spring from the brake pads.

6. Remove the 2 disc brake caliper locating pins and the M-shaped anti-rattle spring.
7. Remove the brake pads and shims. Do not discard the shims found behind the brake pads.
To install:
8. Use the proper brake tool or a C-clamp and one of the old brake pads to push the caliper piston back into the caliper bore. Do not push directly against the caliper piston or damage to the piston may result.
9. Apply suitable grease, normally supplied with the brake pad set, to both surfaces of the inner shim and to the back of the brake pads. Be careful not to get grease on the friction surface of the brake pads.
10. Install the brake pads, making sure the shims are installed.
11. Install the 2 disc brake caliper locating pins and the M-shaped anti-rattle spring.
12. Install the disc brake pad anti-rattle clip.
13. Install the wheel and tire assembly. Tighten the lug bolts to 65–87 ft. lbs. (88–118 Nm).
14. Lower the vehicle.
15. Apply the brake pedal several times to seat the pads, before moving the vehicle. Check the brake fluid level in the master cylinder and add fluid as necessary.
16. Check for proper brake operation.

Brake Rotor

REMOVAL AND INSTALLATION

1. Raise and safely support the vehicle.
2. Remove the tire and wheel assembly.
3. Remove the disc brake pads and shims.
4. Remove the 2 brake caliper retaining bolts. Lift the caliper assembly from the steering knuckle.

NOTE: Do not allow the caliper to hang by the brake hose. Support the caliper by a length of wire attached to the strut.

5. Remove the two front disc brake rotor retaining screws and remove the disc brake rotor from the vehicle.
To install:
6. Position the disc brake rotor to the wheel hub and install the 2 brake rotor retaining screws. Tighten securely.
7. Install the brake caliper to the steering knuckle and secure with the 2 brake caliper retaining bolts. Tighten the brake caliper retaining bolts to 29–36 ft. lbs. (39–49 Nm).

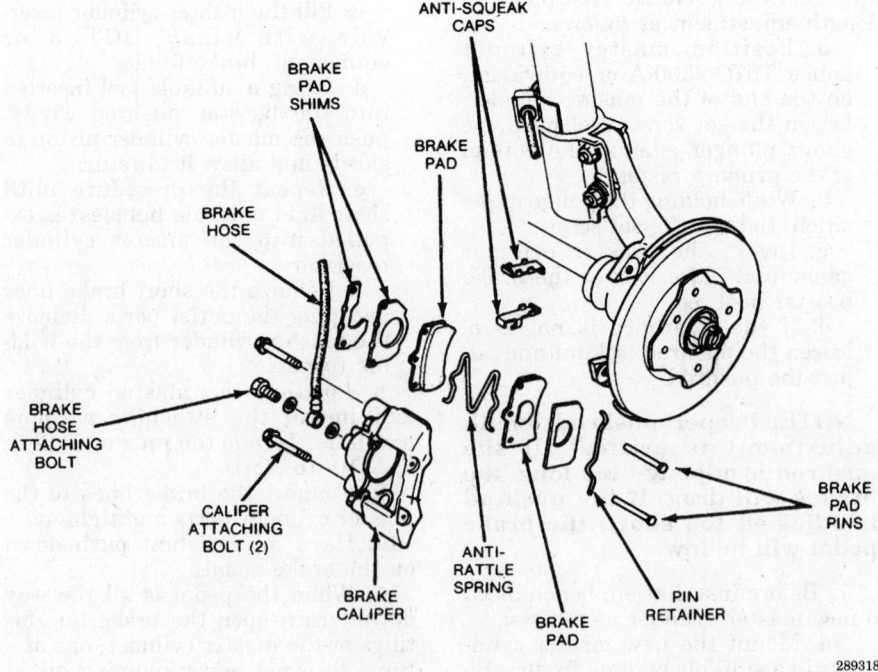

Disc brake assembly

8. Install the disc brake pads and shims.

9. Install the tire and wheel assembly. Tighten the lug bolts to 65–87 ft. lbs. (88–118 Nm).

10. Lower the vehicle.

11. Apply the brake pedal several times to seat the pads, before moving the vehicle.

12. Check the brake fluid level in the master cylinder and add fluid as necessary.

13. Check for proper brake operation.

Brake Drums

REMOVAL AND INSTALLATION

1. Raise and safely support the vehicle.

2. Remove the tire and wheel assembly.

3. Remove the hub grease cap.

4. Remove the cotter pin and wheel bearing nut cover. Discard the cotter pin.

5. Remove the wheel bearing nut.

NOTE: A left-hand threaded locknut is used on the vehicles right rear wheel spindle. Turn this locknut clockwise to loosen.

6. Remove the brake drum, washer and bearings as an assembly. Be careful not to let the outer wheel bearing fall out of the hub during removal.

7. If the brake drum is to be machined or replaced, remove the inner wheel bearing and grease seal.

To install:

8. If removed, install the inner wheel bearing and a new grease seal.

9. Make sure the bearings and hub contain an adequate amount of clean wheel bearing grease.

10. Adjust the distance between the brake shoes to match the inner diameter of the brake drum if the brake drum has been machined or replaced.

11. Position the brake drum on the spindle. Keep the drum centered on the spindle to prevent damage to the grease seal and spindle threads.

12. Install the outer wheel bearing, washer and wheel bearing nut.

13. Properly adjust the wheel bearing preload.

14. Install the wheel bearing nut cover and a new cotter pin.

15. Install the tire and wheel assembly. Torque the lug bolts to 65–87 ft. lbs. (88–118 Nm).

16. Lower the vehicle.

17. Check the brake system for proper operation.

Brake Shoes

REMOVAL AND INSTALLATION

1. Raise and safely support the vehicle.

2. Remove the wheel and tire assembly.

3. Remove the brake drum.

4. Remove the brake shoe hold-down springs and pins.

5. Remove the brake shoe retracting springs and the right hand anti-rattle spring.

6. Pull the brake shoes away from the backing plate and remove.

To install:

7. Clean the brake backing plate.

8. Lubricate the backing plate shoe pads with a suitable high temperature grease.

9. Install the brake shoe upper retracting spring on the primary brake shoe. Position the primary brake shoe on the backing plate and install the hold-down pin and spring.

10. Connect the upper retracting spring to the secondary brake shoe and position the shoe against the backing plate. Install the secondary brake shoe hold-down pin and spring.

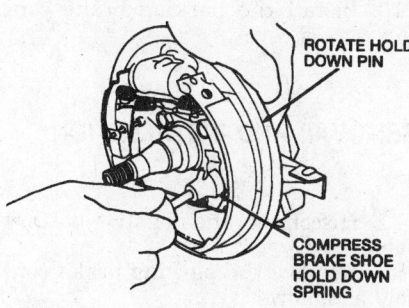

Brake shoe hold-down spring and pin removal

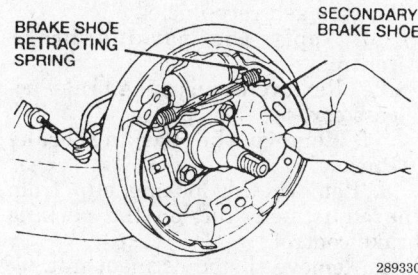

Removing the secondary brake shoe

11. Install the right hand anti-rattle spring and the lower brake shoe retracting spring.

12. Set the self adjuster to the fully released position. Place a suitable tool against the adjuster cam and push it to the released position.

13. Install the brake drum and properly adjust the wheel bearing preload.

14. Lower the vehicle.

15. Push the brake pedal several times to set the self adjuster.

16. Check the brake system for proper operation.

Wheel Cylinder

REMOVAL AND INSTALLATION

1. Raise and safely support the vehicle.

2. Remove the wheel and tire assembly.

3. Remove the brake drum and discard the cotter pin.

NOTE: The right-hand wheel bearing nut uses left-hand threads.

4. Remove the brake shoes.

5. Disconnect the brake line from the wheel cylinder. Plug or cover the brake line opening to prevent the entry of dirt or grease.

6. Remove the 2 wheel cylinder retaining bolts and remove the wheel cylinder from the backing plate.

To install:

7. Position the wheel cylinder onto the backing plate and hand start the brake line flare nut.

8. Install the 2 retaining bolts and torque to 7–9 ft. lbs. (10–13 Nm).

9. Tighten the flare nut to 10–15 ft. lbs. (13–21 Nm).

10. Install the rear brake shoes.

11. Install the brake drum and adjust the wheel bearing preload.

12. Install the wheel and tire assembly. Torque the lug bolts to 65–87 ft. lbs. (88–118 Nm).

13. Lower the vehicle.

14. Bleed the brake system and adjust the rear brakes as required.

15. Check the brake system for leaks and proper operation.

Parking Brake Cable

ADJUSTMENT

1. Make sure the parking brake is fully released.

2. Remove the parking brake console access cover.

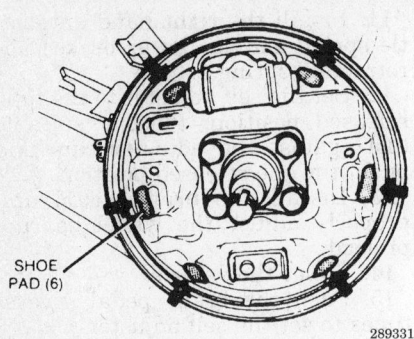

Lubricate the brake shoe pads with high temperature grease

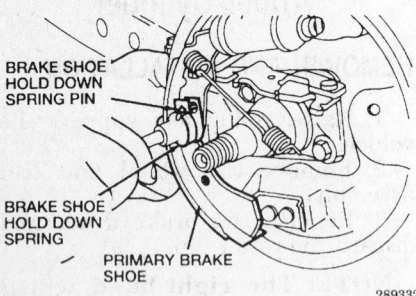

Securing the primary brake shoe with the hold-down pin and spring

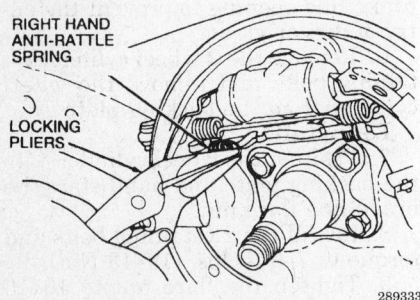

Installing the anti-rattle spring

3. Remove the locking clip from the cable adjuster nut.

4. Raise and safely support the vehicle.

5. Make sure the rear wheels turn freely.

6. Tighten the cable adjuster nut until there is a slight brake drag when the rear wheels are rotated.

7. Back off on the adjuster nut until the brake drag disappears.

8. Check the operation of the parking brake. The rear brakes should be fully applied when the brake lever is pulled upward 11–16 notches.

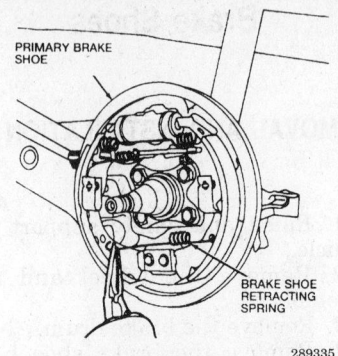

Installing the lower brake shoe return spring

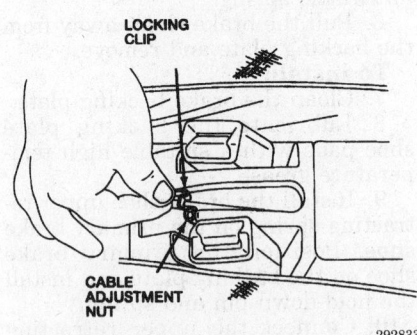

Parking brake cable adjustment nut location

9. Install the locking clip onto the cable adjuster nut.

10. Install the parking brake console access cover.

REMOVAL AND INSTALLATION

1. Disconnect the negative battery cable.

2. Remove the parking brake console as follows:

   a. Slide both front seats all the way forward.

   b. Remove the ash receptacle from the console by lifting the cover and pulling straight up.

   c. Remove the screw located below the ash receptacle.

   d. Apply the parking brake control.

   e. Remove the parking brake access cover.

   f. Remove the parking brake console.

3. Remove the locking clip from the adjustment nut on the parking brake control.

4. Remove the adjustment nut.

5. Disconnect the parking brake switch electrical connector from the switch and bracket.

6. Remove the 2 parking brake control retaining bolts and the parking brake control.

7. Remove the attaching screws and parking brake console mounting bracket.

8. Remove the bolts attaching the lower half of the rear seat hinge to the floor pan.

9. Fold the rear seat forward and remove the bolts attaching the upper half of the rear seat hinge to the floor pan.

10. Remove the rear seat.

11. Remove the rear carpet push retainers and carefully pull the carpeting forward to expose the parking brake cable cover.

12. Remove the 2 parking brake cable cover screws and remove the cable cover.

13. Raise and safely support the vehicle.

14. Remove the rear wheel and tire assemblies.

15. Remove the 2 cotter pins and clevis pins attaching the parking brake cable ends to the rear brake levers. Discard the cotter pins.

16. Remove the parking brake cable retaining clips.

17. Disengage the parking brake routing sleeves from the routing brackets.

18. Remove the nut and bolt attaching the parking brake routing bracket to the fuel tank.

19. Remove the parking brake cable equalizer attaching bolts.

20. Withdraw the lever end of the cable through the body opening and remove from the vehicle.

To install:

21. Position the lever end of the cable through the body opening.

22. Position the cable routing bracket on the fuel tank and install the attaching bolt and nut.

23. Make sure the cable seal is properly positioned in the floor pan.

24. Position the cable equalizer and install the attaching bolts. Make sure the equalizer spacers are in position before tightening the attaching bolts.

25. Route the cable ends through the body brackets and install the retaining clips.

26. Seat the cable sleeves in the routing brackets.

27. Attach the cable ends to the brake levers using the clevis pins and new cotter pins.

28. Install the rear wheel and tire assemblies. Tighten the wheel lug bolts to 65–87 ft. lbs. (88–118 Nm).

29. Lower the vehicle.

30. Route the end of the cable through the park brake lever.

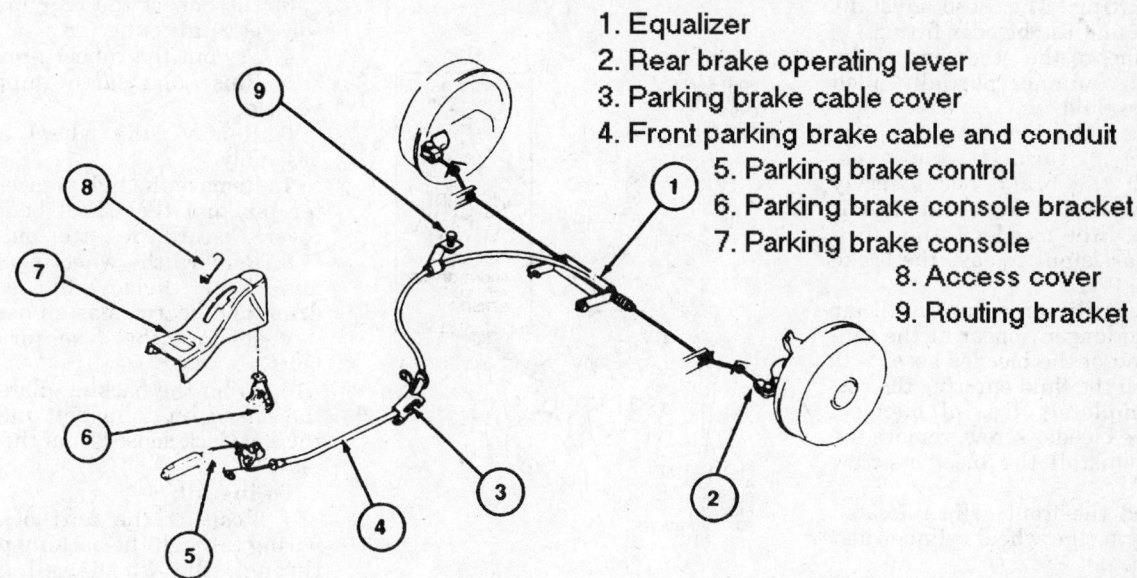

1. Equalizer
2. Rear brake operating lever
3. Parking brake cable cover
4. Front parking brake cable and conduit
5. Parking brake control
6. Parking brake console bracket
7. Parking brake console
8. Access cover
9. Routing bracket

289717

Parking brake cable and conduit

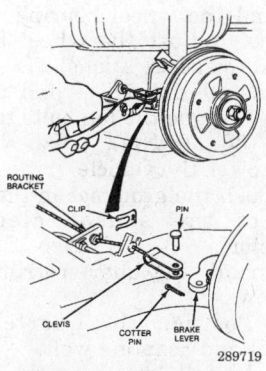

Cable attachment at rear brakes

289719

31. Position the parking brake cable cover and secure with the attaching screws.

32. Position the carpet and install the luggage compartment floor cover hold–down pins.

33. Install the rear seat cushion and cover.

34. Position the console mounting bracket and install the retaining screws.

35. Position the parking brake control and install the 2 retaining bolts. Tighten the bolts to 14–19 ft. lbs. (19–26 Nm).

36. Connect the parking brake switch electrical connector to the switch and bracket.

37. With the end of the parking brake cable properly routed at the parking brake control, install the adjustment nut.

38. Adjust the parking brake cable.

39. Install the locking clip to the adjustment nut.

40. Install the console as follows:

    a. Install the parking brake access cover.

    b. Apply the parking brake control.

    c. Install the retaining screw below the ash tray receptacle.

    d. Install the ash tray receptacle.

    e. Reposition both seats.

41. Connect the negative battery cable.

42. Road test the vehicle and verify proper operation of the parking brake system.

BRAKE SYSTEM BLEEDING

The brake hydraulic circuits form a split diagonal hydraulic system. The left front and right rear form one circuit while the right front and left rear form the other circuit. When bleeding one of these circuits, bleed the rear wheel first and then the front wheel at the opposite corner.

NOTE: Do not allow the master cylinder to run dry during the bleeding procedure. Only use fresh DOT 3 or equivalent brake fluid from a closed container.

1. Clean all dirt from the master cylinder filler cap.

2. Fill the master cylinder with DOT 3 brake fluid.

3. If the master cylinder is known or suspected to contain air, it must be bled before the wheel cylinders or caliper. Bleed the master cylinder as follows:

    a. Loosen the front line fitting and have an assistant push the brake pedal slowly through it's full travel.

    b. While the assistant holds the pedal down, tighten the brake line fitting. After the line fitting is tightened, the assistant may release the brake pedal.

    c. Repeat the procedure on the rear brake line.

    d. Repeat the entire process several times to make sure all air has been removed from the master cylinder.

4. Remove the bleeder screw cap from the appropriate rear wheel cyl-

inder. Position a box end wrench on the bleeder fitting.

5. Attach a rubber hose to the bleeder fitting. The hose must fit snugly around the bleeder fitting.

6. Submerge the other end of the hose in a container partially filled with brake fluid.

7. Loosen the bleeder fitting approximately ¾ turn. Have an assistant push the brake pedal slowly through it's full travel and hold it there. Close the bleeder fitting, then have the assistant release the brake pedal.

8. Repeat the procedure until air bubbles no longer appear at the submerged end of the bleeder hose.

9. When the fluid entering the bottle is completely free of bubbles, tighten the bleeder screw, remove the hose and install the bleeder screw cap.

10. Bleed the front caliper located diagonally to the wheel cylinder just completed.

11. Check the master cylinder fluid level and add, if necessary.

12. Bleed the other diagonal circuit in the same manner.

13. Check the pedal feel. If the pedal is still spongy, repeat the bleeding procedure.

14. Road test the vehicle and check for proper brake system operation.

ABS Speed Sensor

REMOVAL AND INSTALLATION

Front

1. Disconnect the negative battery cable.

2. Disconnect the front anti-lock sensor electrical connector and remove the grommet.

3. Raise and safely support the vehicle.

4. Remove the wheel and tire assembly.

5. Remove the 2 sensor wiring harness support bolts.

6. Remove the 2 bolts retaining the wheel sensor from the knuckle and remove the sensor from the knuckle.

To install:

7. Position the wheel sensor in the knuckle and install the 2 retaining bolts. Tighten the bolts to 12–16 ft. lbs. (16–23 Nm).

8. Install the 2 sensor wiring harness support bolts.

9. Install the wheel and tire assembly. Torque the lug bolts to 65–87 ft. lbs. (88–118 Nm).

10. Lower the vehicle.

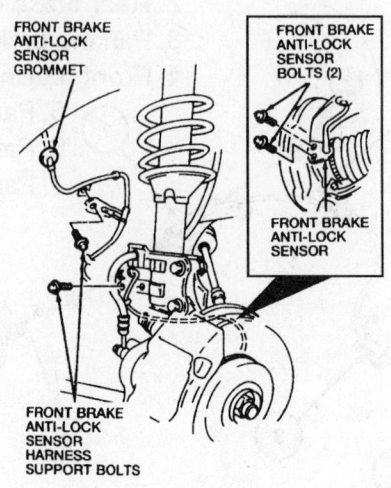

FRONT BRAKE ANTI-LOCK SENSOR GROMMET

FRONT BRAKE ANTI-LOCK SENSOR BOLTS (2)

FRONT BRAKE ANTI-LOCK SENSOR

FRONT BRAKE ANTI-LOCK SENSOR HARNESS SUPPORT BOLTS

289355

Front brake wheel sensor

11. Push the rubber grommet into place and connect the electrical connector.

12. Connect the negative battery cable.

13. Prove out the ABS system by turning the ignition key to the **KEY ON-ENGINE OFF** position while watching the ABS warning indicator. When the ABS system is operating properly, the indicator will illuminate while in the **KEY ON-ENGINE OFF** position and will go out after the engine has started with a delay of up to 60 seconds.

Rear

1. Disconnect the negative battery cable.

2. Remove the quarter trim panel as follows:

   a. Remove the luggage compartment cover.

   b. Remove the rear seat.

   c. Remove the screws and push-pins from the package tray. Disconnect the radio speaker electrical connector.

   d. Remove the rear safety belt anchor bolt.

   e. Remove the push-pins and the luggage compartment side cover.

   f. Remove the rear door scuff plate.

   g. Pull the seaming welt away from the quarter trim panel and remove the panel.

3. Disconnect the rear brake sensor electrical connector.

4. Pry out the rubber grommet.

5. Raise and safely support the vehicle.

6. Remove the wheel and tire assembly.

7. Remove the hub grease cap, cotter pin and the wheel bearing nut cover. Discard the cotter pin.

8. Remove the wheel bearing nut and washer. Remove the rear brake drum and bearings as an assembly.

9. Remove the 4 sensor harness bolts.

10. From the backing plate remove the sensor retaining bolt and remove the anti-lock sensor from the backing plate.

To install:

11. Position the anti-lock sensor wiring through the backing plate and through the wheel well grommet hole.

12. Install the sensor and retaining bolt and tighten the bolt to 12–16 ft. lbs. (16–23 Nm).

13. Install the 4 sensor harness bolts.

14. Install the brake drum, bearings and the wheel bearing washer and nut. Adjust the wheel bearing preload to specifications.

15. Install the wheel and tire assembly. Torque the lug bolts to 65–87 ft. lbs. (88–118 Nm).

16. Lower the vehicle.

17. Push the grommet in place and connect the sensor electrical connector.

18. Install the quarter trim panel as follows:

   a. Position the quarter trim panel and seaming welt.

   b. Install the rear door scuff plate.

   c. Install the luggage compartment side cover using the push-pins.

   d. Install the rear safety belt and anchor pin. Torque the anchor bolt to 28–58 ft. lbs. (38–788 Nm).

   e. Connect the speaker electrical connector.

   f. Install the push-pins in the package tray.

   g. Install the rear seat.

   h. Install the luggage compartment cover.

19. Connect the negative battery cable.

20. Prove out the ABS system by turning the ignition key to the **KEY ON-ENGINE OFF** position while watching the ABS warning indicator.

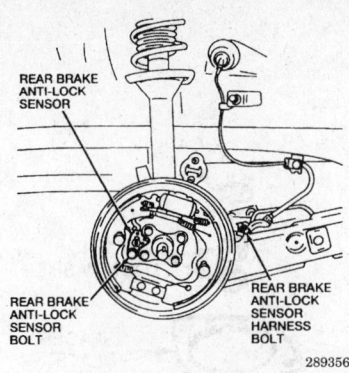

REAR BRAKE
ANTI-LOCK
SENSOR

REAR BRAKE
ANTI-LOCK
SENSOR
BOLT

REAR BRAKE
ANTI-LOCK
SENSOR
HARNESS
BOLT

289356

Rear brake wheel sensor

When the ABS system is operating properly, the indicator will illuminate while in the **KEY ON-ENGINE OFF** position and will go out after the engine has started with a delay of up to 60 seconds.

FRONT SUSPENSION

Strut and Spring

REMOVAL AND INSTALLATION

1. Raise and safely support the vehicle.
2. Remove the wheel and tire assembly.
3. Remove the brake line clip from the strut lower mounting bracket and disengage the brake line.
4. Remove the 2 nuts and bolts securing the strut lower bracket to the steering knuckle.
5. Working in the engine compartment, remove the 2 nuts securing the strut mounting block in the strut tower.
6. Disengage the strut lower bracket from the steering knuckle and lower the strut clear of the wheel well.
7. To separate the coil spring from the strut, attach spring compressor tool T81P-5310-A or equivalent, and compress the coil spring.
8. Pry out the mounting block cap and remove the strut upper nut and lockwasher.
9. Remove the strut mounting block and spacer plate. Remove the washer, bearing seal and bearing from the strut rod.

10. Remove the upper spring seat, seat insulator and spring. Slide the jounce bumper/shield off the strut.

NOTE: If replacing the spring, release the spring compressor progressively to prevent spring arching. Open the compressor jaws wide enough to grip the new spring in the same position and tighten the compressor screws progressively, compressing the spring until the strut can be assembled without interference.

To install:
11. Check the condition of the jounce bumper and spring seat insulator and replace, as necessary. Make sure the bearing operates smoothly. Check the spring for uniform coil spacing, for nicks or burrs and compare the spring length with a new spring to check for excessive spring set; replace as necessary.
12. Slide the jounce bumper/shield onto the strut rod and over the body. Install the compressed spring, upper spring seat insulator and upper seat, positioning the spring ends against the steps in the seats.
13. Install the bearing, seal and plain washer on the strut rod. Install the strut mounting block with the white alignment spot on the same side of the strut as the steering knuckle mounting bracket.
14. Install the spacer plate. Install the lockwasher and nut and tighten to 40–50 ft. lbs. (54–67 Nm). Release and remove the spring compressor.
15. Place the strut assembly with spacer plate in the strut tower with the white alignment mark facing outward.
16. Install the 2 upper mounting block stud nuts and torque to 34–46 ft. lbs. (46–63 Nm).
17. Engage the steering knuckle in the strut tower lower bracket and install the 2 retaining bolts and nuts. Torque to 69–86 ft. lbs. (93–117 Nm).
18. Position the brake line into the strut lower mounting bracket cutout and install the retaining clip.
19. Install the wheel and tire assembly and torque the lug bolts to 65–87 ft. lbs. (88–118 Nm).
20. Lower the vehicle.
21. Check the front wheel alignment.
22. Road test the vehicle and check for proper operation.

Lower Ball Joints

REMOVAL AND INSTALLATION

The lower ball joint is an integral component of the lower control arm. If the lower ball joint is defective, the entire lower control arm must be replaced.

Lower Control Arm

REMOVAL AND INSTALLATION

NOTE: The ball joint is an integral component of the lower control arm and cannot be serviced separately.

1. Raise and safely support the vehicle.
2. Remove the front wheel and tire assembly.
3. Remove the lower control arm-to-chassis bolt and washer at the frame bracket.
4. Remove the ball joint clamp bolt and and nut from the steering knuckle assembly.
5. Remove the cotter pin and stabilizer bar bushing nut from the rear of the control arm and remove the rear bushing washer and bushing. Discard the cotter pin.
6. Lower the control arm, prying the ball joint stud out of the steering knuckle if necessary. Disengage and remove the control arm from the stabilizer end and remove from the vehicle.
7. If the lower control arm is to be reused, inspect the control arm bushings for damage or excessive wear. Verify that the ball joint swivels freely but is not loose.
8. If the lower control arm bushing are to be replaced, remove the old bushing with C-frame tool T74P-3044-A1, bushing tool T81P-5493-B2 and receiver cup tool T88C-5493E or equivalents. Center the new bushing in the center of the control arm eye and install using the same tools.
9. If the ball joint boot is damaged or deteriorated, pry the boot off with a small cold chisel. Install the new boot onto the ball joint using a suitable adapter such as a ¾ inch socket to properly seat the boot.
10. Replace the lower control arm as required.
To install:
11. Position the front bushing washer and bushing onto the stabilizer end. Engage the lower control arm with the stabilizer.

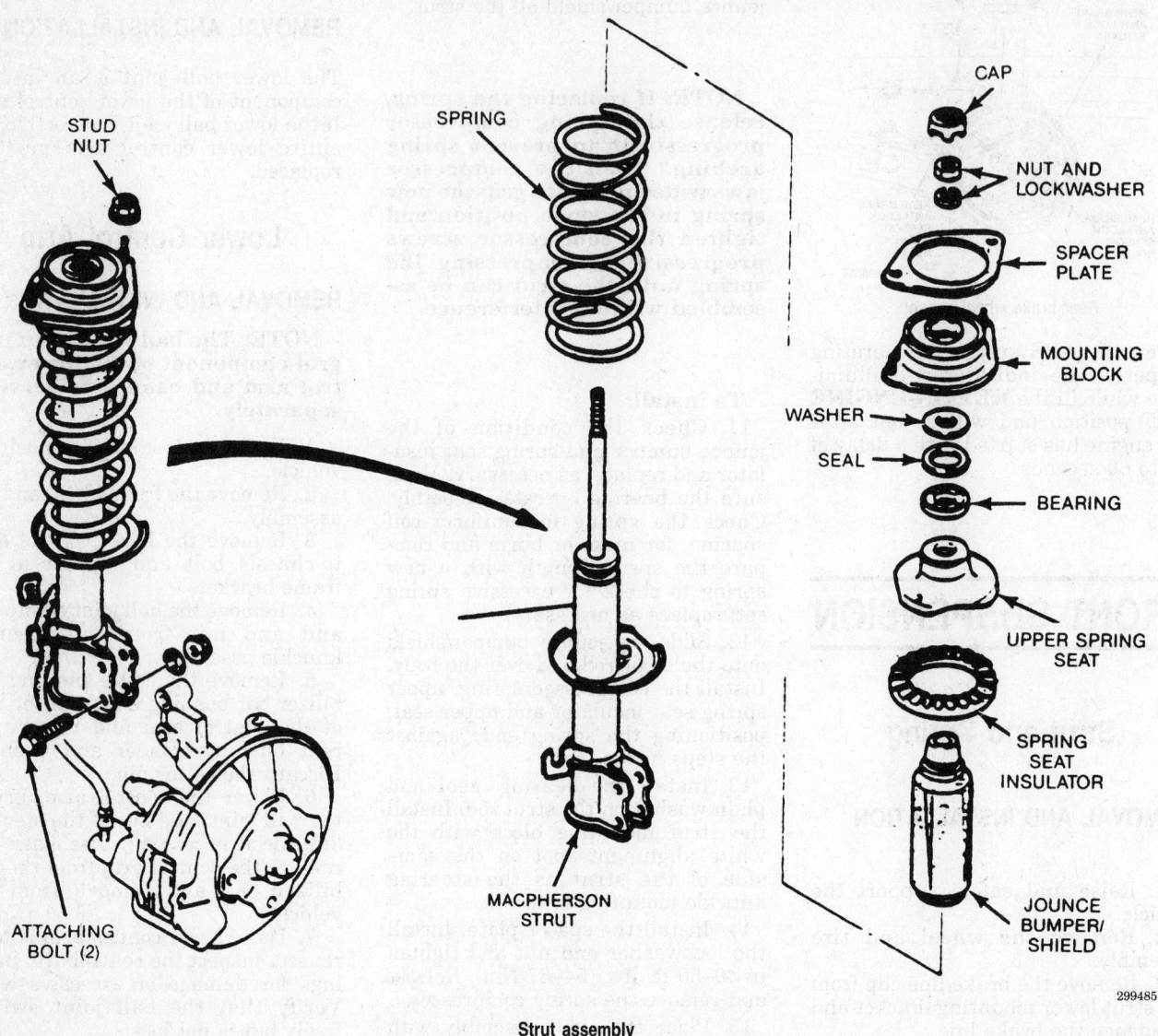

Strut assembly

12. Raise the control arm inner end into the pivot bracket on the frame and start the pivot bolt to hold the control arm in place. Do not completely tighten the bolt at this time.

13. Engage the control arm ball joint stud with the clamp bore in the steering knuckle and install the clamp bolt and nut. Do not tighten yet.

14. Install the stabilizer rear bushing and washer onto the stabilizer end with the retaining nut. Torque the retaining nut to 47–57 ft. lbs. (64–77 Nm). Install a new cotter pin.

15. Torque the lower arm-to- chassis bolt at the frame bracket to 32–40 ft. lbs. (43–54 Nm).

16. Hold the clamp bolt stationary and torque the clamp nut to 32–40 ft. lbs. (43–54 Nm).

17. Install the wheel and tire assembly. Torque the lug bolts to 65–87 ft. lbs. (88–118 Nm).

18. Lower the vehicle.

19. Check the front wheel alignment.

Sway Bar

REMOVAL AND INSTALLATION

1. Raise and safely support the vehicle.

2. Remove the 4 stabilizer bar (sway bar) bracket nuts and 2 stabilizer bar brackets.

3. Remove the split bushings from the stabilizer bar. Replace deteriorated or worn bushings as required.

4. Remove the cotter pins and front stabilizer bar retaining nuts at the lower control arms and remove the rear washers and bushings. Discard the cotter pins.

5. Pull the stabilizer bar forward to disengage it from the lower control arms. Remove the front bushings and washers. Replace deteriorated or worn bushings as required.

Lower Control Arms
Removal

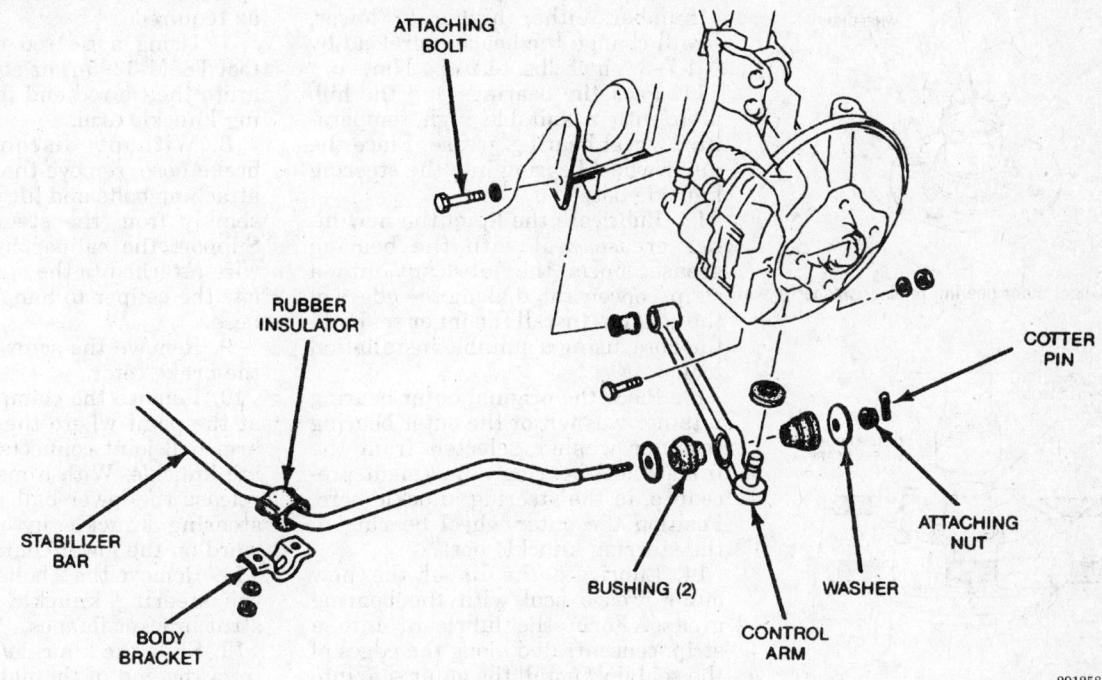

Lower control arm and related components

To install:

6. Install the control arm bushing washers on the ends of the stabilizer bar and install the control arm front bushings.

7. Support the stabilizer bar by hand and insert the ends of the bar into the lower control arms. Install the rear control arm bushings and washers with the retaining nuts. Make the retaining nuts finger-tight.

8. Install the 2 split bushings on the the stabilizer bar with the split side forward and position them next to the white alignment marks on the bar.

9. Install the 2 stabilizer bar brackets. Torque the 4 bracket retaining nuts to 40–50 ft. lbs. (54–68 Nm).

10. Torque the front stabilizer bar retaining nuts to 47–57 ft. lbs. (64–77 Nm) and install new cotter pins.

11. Lower the vehicle.

Front Wheel Bearings

ADJUSTMENT

NOTE: Wheel bearing adjustment (bearing preload) is normally not required except when the wheel bearing(s) and/or steering knuckle are replaced.

The adjustment procedure requires removal of the steering knuckle and disassembly of the wheel bearings from the knuckle.

1. Disconnect the negative battery cable.

2. Remove the wheel and tire assembly.

3. Remove the steering knuckle/hub assembly.

4. Separate the hub from the steering knuckle.

5. Remove the outer bearing retainer washer.

NOTE: The outer bearing retainer washer is preselected to yield the correct bearing preload. Save the washer for use during assembly.

6. Remove the outer wheel bearing from the hub using a bearing pulling attachment, driver and a press.

7. Remove the grease seals from the hub and steering knuckle bore and discard. Remove the inner wheel bearing.

8. If the bearings are to be replaced, remove the bearing races from the steering knuckle using a suitable puller and slide hammer.

9. Thoroughly clean the hub and knuckle. Inspect the hub and knuckle for wear and/or damage. Replace as necessary.

To install:

10. To check wheel bearing preload, proceed as follows:

a. Install the outer bearing races in the steering knuckle using suitable bearing cup installation tools.

b. Lubricate the bearing races and bearing with a thin film of clean grease. Install the bearings in the steering knuckle.

c. Install spacer selection tool T87C-1104-B or equivalent, and clamp the bolt head in a vise.

d. Tighten the center bolt in increments, to 36, 72, 108 and 145 ft. lbs. (49, 98, 147 and 196 Nm). After tightening the center bolt to a specified increment, seat the bearings by rotating the steering knuckle.

e. Remove the tool/steering knuckle from the vise. Remount the assembly in the vise, clamping it where the strut mounts.

f. Measure the amount of torque required to rotate the spacer selector tool, using an inch pound torque wrench. The torque wrench reading must be taken just as the tool starts to rotate.

g. If the torque wrench indicates 2.2–10.4 inch lbs. (0.25–1.80 Nm), the outer bearing retainer washer is the correct thickness. If the torque wrench indicates less than 2.2 inch lbs. (0.25 Nm), a thinner outer

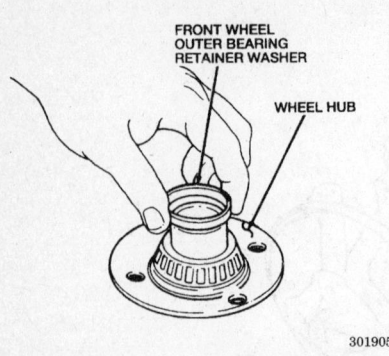

Front wheel outer bearing retainer washer

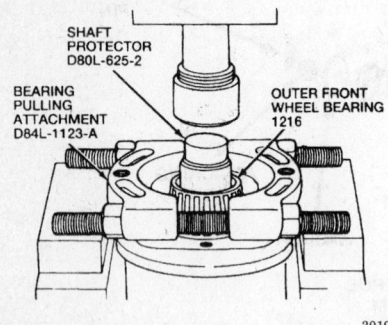

Removing the outer wheel bearing from the hub

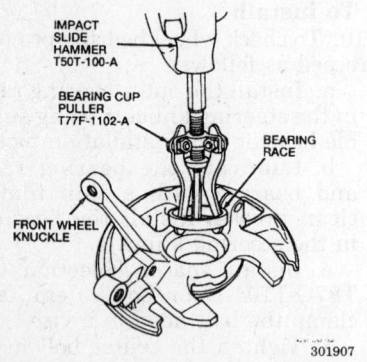

Removing the bearing races from the knuckle

bearing retainer washer must be installed. If the torque wrench indicates more than 10.4 inch lbs. (1.8 Nm), a thicker outer bearing retainer washer must be installed.

h. Each outer bearing retainer washer has a numerical code that identifies its thickness, which is stamped onto the outer diameter of the washer. The numbers range from 1–21, with 1 being the thinnest washer. If the number stamped on the washer is not legible, measure the washer with a micrometer and compare it to the

thickness chart to determine the number.

i. Changing the outer bearing retainer washer thickness by 1 number, either higher or lower, will change the bearing preload by 1.7–3.5 inch lbs. (0.2–0.4 Nm).

11. Pack the bearings and the hub area with a suitable high temperature wheel bearing grease. Place the inner wheel bearing into the steering knuckle bore.

12. Lubricate the lip of the new inner grease seal with the bearing grease. Form the lubricant into a strip, concentrated along the edges of the seal lip. Install the inner seal into the bore, using a suitable installation tool.

13. Place the original outer bearing retainer washer, or the outer bearing retainer washer selected from the front wheel bearing adjustment procedure, in the steering knuckle bore. Position the outer wheel bearing in the steering knuckle bore.

14. Lubricate the lip of the new outer grease seal with the bearing grease. Form the lubricant into a strip, concentrated along the edges of the seal lip. Install the outer seal into the bore, using a suitable installation tool.

15. Install the hub to the steering knuckle.

16. Install the steering knuckle/hub assembly in the vehicle.

17. Install the wheel and tire assembly. Tighten the lug bolts to 65–87 ft. lbs. (88–118 Nm).

18. Lower the vehicle.

19. Road test the vehicle and check for proper operation.

REMOVAL AND INSTALLATION

NOTE: If the bearings are being replaced, a new bearing preload spacer must be selected. This requires spacer selection tool T87C-1104-B or equivalent and both inch pound and foot pound torque wrenches. Do not attempt bearing replacement without these tools.

1. With the vehicle sitting on all 4 wheels, use a small chisel to straighten the staked edge of the wheel hub retaining nut. Take care not to damage the halfshaft threads.

2. Remove and discard the wheel hub retaining nut.

3. Raise and safely support the vehicle.

4. Remove the wheel.

5. Remove the retaining clip securing the caliper hose to the strut bracket.

6. Remove the cotter pin and tie rod end attaching nut. Discard the cotter pin and set the nut aside. Inspect the nut for damage and replace as required.

7. Using a tie rod end separator tool T85M-3395-A or equivalent, separate the tie rod end from the steering knuckle arm.

8. Without disconnecting the brake hose, remove the brake caliper attaching bolts and lift the caliper assembly from the steering knuckle. Support the caliper by a length of wire attached to the strut. Do not allow the caliper to hang by the brake hose.

9. Remove the screws and remove the brake rotor.

10. Remove the clamp bolt and nut at the point where the lower control arm ball joint connects to the steering knuckle. With a medium prybar, release the lower ball joint from the steering knuckle by prying downward on the lower control arm.

11. Remove the 2 bolts that position the steering knuckle between the strut bracket flanges.

12. Slide the knuckle/hub assembly from the end of the halfshaft. If necessary, use a wheel puller to press the shaft out of the hub and remove from the vehicle.

13. If necessary, use a chisel to separate the disc brake rotor shield from the knuckle/hub assembly. Discard the shield.

NOTE: Shield replacement is not a requirement for normal bearing service.

14. Using puller tool T87C-1104-A and hub/bearing remover adapter T92C-1104-AH or equivalents, separate the hub from the knuckle.

15. Remove the outer bearing retainer washer.

NOTE: The outer bearing retainer washer is preselected to yield the correct bearing preload. Save the washer for use during assembly.

16. Remove the outer bearing from the wheel hub using a bearing pulling attachment, driver and a press.

17. Remove the grease seals from the hub and steering knuckle bore and discard. Remove the inner wheel bearing.

18. If the bearings are to be replaced, remove the bearing races from the steering knuckle using a suitable puller and slide hammer.

19. Thoroughly clean the hub and knuckle. Inspect the hub and knuckle for wear and/or damage. Replace as necessary.

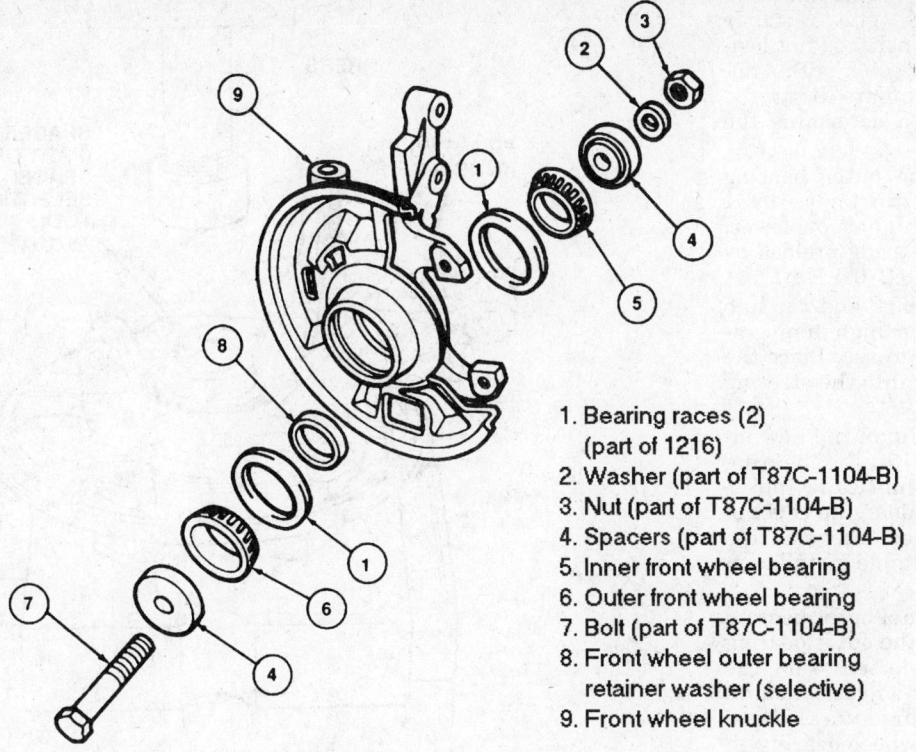

1. Bearing races (2)
   (part of 1216)
2. Washer (part of T87C-1104-B)
3. Nut (part of T87C-1104-B)
4. Spacers (part of T87C-1104-B)
5. Inner front wheel bearing
6. Outer front wheel bearing
7. Bolt (part of T87C-1104-B)
8. Front wheel outer bearing
   retainer washer (selective)
9. Front wheel knuckle

301912

Tools required for wheel bearing adjustment

Stamped mark	Thickness
1	6.285 mm (0.2474 in)
2	6.325 mm (0.2490 in)
3	6.365 mm (0.2506 in)
4	6.405 mm (0.2522 in)
5	6.445 mm (0.2538 in)
6	6.485 mm (0.2554 in)
7	6.525 mm (0.2570 in)
8	6.565 mm (0.2586 in)
9	6.605 mm (0.2602 in)
10	6.645 mm (0.2618 in)
11	6.685 mm (0.2634 in)
12	6.725 mm (0.2650 in)
13	6.765 mm (0.2666 in)
14	6.805 mm (0.2682 in)
15	6.845 mm (0.2698 in)
16	6.885 mm (0.2714 in)
17	6.925 mm (0.2730 in)
18	6.965 mm (0.2746 in)
19	7.005 mm (0.2762 in)
20	7.045 mm (0.2778 in)
21	7.085 mm (0.2794 in)

301913

Front wheel outer bearing retainer washer thickness chart

To install:

20. If the brake rotor shield was removed, install a new one using installation tools T80T-4000-W and T94C-1175-B or equivalents.

21. If the wheel bearings or steering knuckle are being replaced, bearing preload must be checked before assembly as follows:

a. Install the outer bearing races in the steering knuckle using suitable bearing cup installation tools.

b. Lubricate the bearing races and bearing with a thin film of clean grease. Install the bearings in the steering knuckle.

FRONT DISC BRAKE ROTOR SHIELD

DRIVER HANDLE T80T-4000-W

DUST SHIELD REPLACER T94C-1175-B

320540

Installing the disc brake rotor shield

c. Install spacer selection tool T87C-1104-B or equivalent, and clamp the bolt head in a vise.

d. Tighten the center bolt in increments, to 36, 72, 108 and 145 ft. lbs. (49, 98, 147 and 196 Nm). After tightening the center bolt to a specified increment, seat the bearings by rotating the steering knuckle.

e. Remove the tool/steering knuckle from the vise. Remount the assembly in the vise, clamping it where the strut mounts.

f. Measure the amount of torque required to rotate the spacer selector tool, using an inch pound torque wrench. The torque wrench reading must be taken just as the tool starts to rotate.

g. If the torque wrench indicates 2.2–10.4 inch lbs. (0.25–1.80 Nm), the outer bearing retainer washer is the correct thickness. If the torque wrench indicates less than 2.2 inch lbs. (0.25 Nm), a thinner outer bearing retainer washer must be installed. If the torque wrench indicates more than 10.4 inch lbs. (1.8 Nm), a thicker outer bearing retainer washer must be installed.

h. Each outer bearing retainer washer has a numerical code that identifies its thickness, which is stamped onto the outer diameter of the washer. The numbers range from 1–21, with 1 being the thinnest washer. If the number stamped on the washer is not legible, measure the washer with a micrometer and compare it to the thickness chart to determine the number.

i. Changing the outer bearing retainer washer thickness by 1 number, either higher or lower, will change the bearing preload by 1.7–3.5 inch lbs. (0.2–0.4 Nm).

22. Pack the bearings and the hub area with a suitable high temperature wheel bearing grease. Place the inner wheel bearing into the steering knuckle bore.

23. Lubricate the lip of the new inner grease seal with the bearing grease. Form the lubricant into a strip, concentrated along the edges of the seal lip. Install the inner seal into the bore, using a suitable installation tool.

24. Place the original outer bearing retainer washer, or the outer bearing retainer washer selected from the front wheel bearing adjustment procedure, in the steering knuckle bore. Position the outer wheel bearing in the steering knuckle bore.

25. Lubricate the lip of the new outer grease seal with the bearing grease. Form the lubricant into a strip, concentrated along the edges of the seal lip. Install the outer seal into the bore, using a suitable installation tool.

26. Position the hub in the steering knuckle bore and press it into position using a suitable driver.

27. Install the steering knuckle/hub assembly in the vehicle.

28. Clean the halfshaft spline end and lubricate with a coating of wheel bearing grease. Apply a thin film of clean SAE 30 weight oil to the steering knuckle/hub assembly up to the point where the uppermost arm of the steering knuckle seats into the strut bracket. Guide the steering knuckle/hub assembly onto the halfshaft and the strut.

29. Install the strut-to-steering knuckle bolts and attaching nuts. Tighten the nuts to 69–86 ft. lbs. (93–117 Nm).

30. Position the lower control arm ball joint in the steering knuckle. Install the lower control arm pinch bolt and attaching nut. Tighten the nut to 32–40 ft. lbs. (43–54 Nm).

31. Install the brake rotor.

32. Position the caliper on the steering knuckle and install the attaching bolts. Tighten the bolts to 29–36 ft. lbs. (39–49 Nm). Position

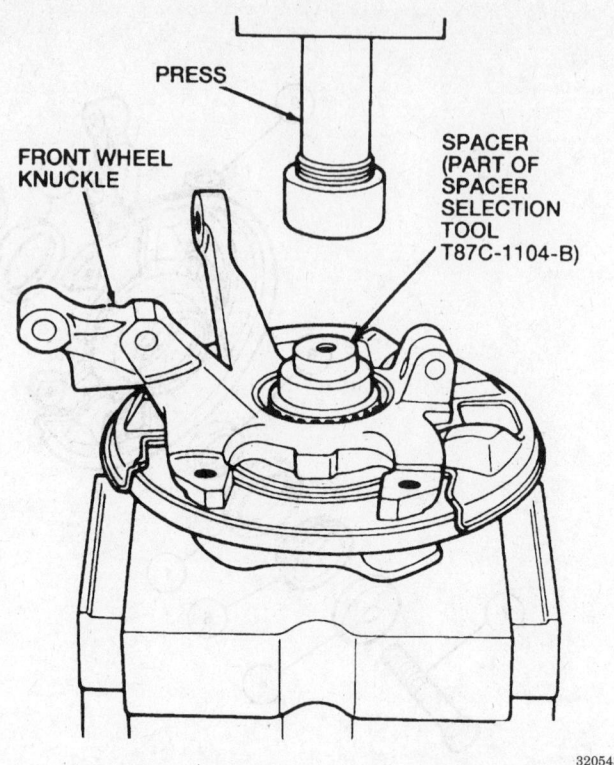

PRESS

FRONT WHEEL KNUCKLE

SPACER (PART OF SPACER SELECTION TOOL T87C-1104-B)

320543

Installing the hub into the steering knuckle

the caliper hose in the strut routing bracket and install the retaining clip.

33. Install a new wheel hub retaining nut and tighten by hand.

34. Connect the tie rod end to the steering knuckle and install the attaching nut. Tighten the attaching nut to 22–33 ft. lbs. (29–44 Nm). Install a new cotter pin through the nut and ball stud. If the openings in the nut and the hole in the ball stud are not aligned, tighten the nut to the point of alignment. Never loosen the nut.

35. Install the wheel and tire assembly. Tighten the lug bolts to 65–87 ft. lbs. (88–118 Nm).

36. Lower the vehicle.

37. Tighten the wheel hub retaining nut to 116–174 ft. lbs. (157–235 Nm). After installation, the wheel hub assembly must rotate freely by hand. Stake the halfshaft attaching nut into the shaft groove.

NOTE: Do not use a pointed tool to stake the nut. If the nut cracks even slightly during staking, replace it with another new one.

38. Check the front wheel alignment.

39. Road test the vehicle and check for proper operation.

REAR SUSPENSION

Strut and Spring

REMOVAL AND INSTALLATION

— **WARNING** —
Do not attempt to remove both left and right spring and strut assemblies at the same time. Do one side at a time to prevent damage to the rear suspension.

1. From the cargo compartment, remove the side cover.

2. Remove the quarter trim panel as follows:

a. Remove the luggage compartment cover.

b. Remove the rear seat.

c. Remove the screws and push-pins from the package tray. Disconnect the radio speaker electrical connector.

d. Remove the rear safety belt anchor bolt.

e. Remove the push pins and the luggage compartment side cover.

f. Remove the rear door scuff plate.

g. Pull the seaming welt away from the quarter trim panel and remove the panel.

3. Remove the strut cap, jam nut and flanged nut from the strut rod and remove the bushing washer and upper bushing.

4. Raise and safely support the vehicle.

NOTE: Raising the vehicle will release any tension left on the coil spring.

5. Remove the rear wheel and tire assembly.

6. Remove the lower strut mounting bolt from the torsion beam.

7. Remove the strut assembly from the vehicle and separate it from the spring and seat insulator.

8. Inspect the condition of the spring, spring seat insulator and strut. Replace any damaged or deteriorated components, as required.

To install:

9. If the upper spring seat insulator is replaced, install the new insulator on the spring upper end, seating the end of the coil against the step in the insulator. Position the spring on the strut, making sure the end of the coil seats against the step in the strut spring seat.

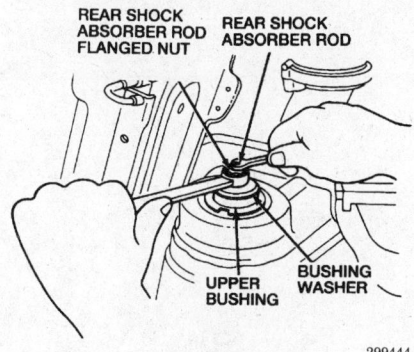

Upper strut mounting

299444

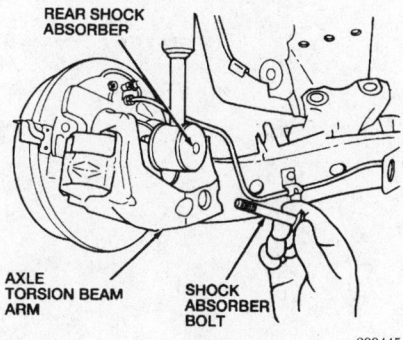

Lower strut bolt

299445

10. Guide the strut into the upper strut mounting hole through the wheel well.

11. Align the strut lower end with the mounting hole in the torsion beam. Start the mounting bolt in by hand to hold the strut in position.

12. Install the wheel and tire assembly. Torque the lug bolts to 65–87 ft. lbs. (88–118 Nm).

13. Lower the vehicle.

NOTE: Make sure that the coil spring is positioned properly on the spring seat insulator.

14. From the cargo compartment, install the rod upper end bushing, bushing washer and flanged nut. Torque the flanged nut to 12–18 ft. lbs. (16–24 Nm). Hold the flanged nut stationary and tighten the locknut.

15. Install the jam nut and the strut cap.

16. Install the side cover in the cargo compartment.

17. Raise and safely support the vehicle.

18. Torque the lower strut mounting bolt to 50–60 ft. lbs. (68–81 Nm).

19. Lower the vehicle.

20. Check the rear wheel alignment.

Wheel Bearings

ADJUSTMENT

1. Make sure the parking brake is fully released.

2. Raise and safely support the vehicle.

3. Remove the wheel and tire assembly.

4. Remove the grease cap.

5. Rotate the brake drum to make sure there is no brake drag.

6. Remove the cotter pin, wheel bearing nut cover. Discard the cotter pin.

7. To seat the bearings, torque the wheel bearing nut to 18–22 ft. lbs. (25–29 Nm). Rotate the brake drum by hand while tightening the nut.

8. Loosen the wheel bearing nut until it can be turned by hand.

9. Before the bearing preload can be set, the amount of seal drag must be measured and added to the the required preload.

10. To measure the seal drag proceed as follows:

a. Install a lug bolt and rotate the brake drum until the stud is in the 12 o'clock position.

b. Place an inch pound torque wrench onto the bolt to measure

the amount of force required to rotate the break drum.

c. Pull the torque wrench and note and record the torque reading when rotation begins.

11. Add the oil seal drag value obtained in Step 9 to the specified value of 0.6–1.9 lbs. (2.6–8.5 N). This is the standard bearing preload.

12. Loosely tighten the wheel bearing nut. Rotate the brake drum until the nut and wheel are returned to the 12 o'clock position. Position the inch pound torque wrench onto the nut and measure the amount of pull required to rotate the brake drum. Tighten the wheel bearing nut until the torque shown on the torque wrench is within the range calculated previously.

13. Turn the wheel bearing nut slowly to adjust to the standard bearing preload.

14. Install the nut retaining cap and a new cotter pin.

15. Install the grease cap.

16. Install the wheel and tire assembly. Torque the lug bolts to 65–87 ft. lbs. (88–118 Nm).

17. Lower the vehicle.

18. Road test the vehicle and check for proper operation.

REMOVAL AND INSTALLATION

1. Make sure the parking brake is fully released.

2. Raise and safely support the vehicle.

3. Remove the wheel and tire assembly.

4. Remove the hub grease cap.

5. Remove the cotter pin, nut cover and the nut. Discard the cotter pin.

6. Pull the brake drum bearings and hub assembly away from the spindle shaft. Take care not to damage the spindle shaft threads.

7. Remove the outer wheel bearing assembly and washer.

8. With a small roll head prybar or equivalent, remove the bearing grease seal from the bearing hub. Discard the seal regardless of condition.

9. Remove the inner wheel bearing assembly from the bearing hub. If the bearings are to be reused, identify and tag each bearing for installation reference.

10. Thoroughly clean the wheel bearings and hub using suitable solvent and allow to dry. Inspect the bearings and bearing races for scoring, pitting, wear or other damage and replace as necessary.

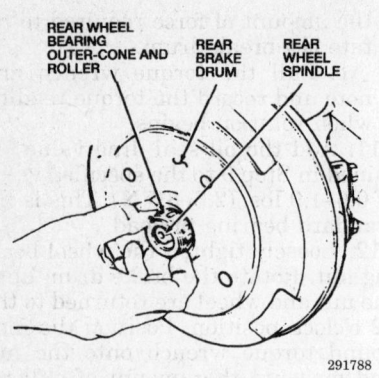

REAR WHEEL BEARING OUTER-CONE AND ROLLER — REAR BRAKE DRUM — REAR WHEEL SPINDLE

291788

Removing the outer wheel bearing

11. If replacing the bearings, the bearing races must also be replaced.

Remove the bearing races from the hub using a brass drift.

To install:

12. If replacing the bearing races, install new races in the hub using suitable installation tools.

13. Pack the bearings and the drum hub area with high temperature wheel bearing grease. Do not fill the entire hub with grease.

14. Position the inner bearing in the hub. Install and seat a new grease seal with a suitable driving tool. Lubricate the lip of the seal with wheel bearing grease.

15. Position the brake drum and hub assembly on the spindle. Keep the hub centered during positioning

to prevent damage to the new grease seal and the spindle threads.

16. Install the outer wheel bearing, washer and nut.

17. Adjust the bearing preload.

18. Install the wheel bearing nut cover and a new cotter pin.

19. Install the hub grease cap.

20. Install the wheel and tire assembly. Torque the lug bolts to 65–87 ft. lbs. (88–118 Nm).

21. Lower the vehicle.

22. Check and adjust the brakes as required.

23. Check for proper brake operation.

FORD MOTOR CO.

Front Wheel Drive

FORD-Probe

10

FIRING ORDERS

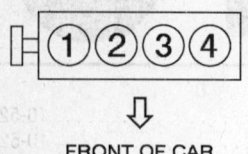

FRONT OF CAR

314511

2.0L (VIN A) Engine
Firing Order: 1-3-4-2
Distributor Rotation: Clockwise

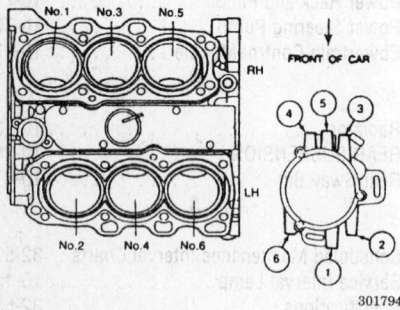

301794

2.5L (VIN B) Engine
Firing Order: 1-2-3-4-5-6
Distributor Rotation: Counterclockwise

ENGINE ELECTRICAL

NOTE: Disconnecting the negative battery cable on some vehicles may interfere with the functions of the on board computer systems and may require the computer to undergo a relearning process, once the negative battery cable is reconnected.

Distributor

REMOVAL AND INSTALLATION

1993 2.0L (VIN A) Engine

1. Disconnect the negative battery cable.
2. Remove the distributor cap and position aside, leaving the spark plug wires connected. Before removing the distributor, mark the position of the distributor cap No. 1 spark plug wire tower on the distributor housing.
3. Disconnect the distributor electrical connector. If equipped with automatic transaxle, disconnect the coil connector.
4. Using a wrench on the crankshaft pulley, rotate the crankshaft to position the No. 1 piston at TDC on the compression stroke. The crankshaft pulley notch should align with the timing plate indicator and the distributor rotor should be pointing to the No. 1 spark plug tower position on the distributor cap.
5. Using chalk or paint, mark the position of the distributor housing on the cylinder head.
6. Remove the distributor hold-down bolt(s) and remove the distributor.
7. Inspect the O-ring on the distributor housing and replace it if damaged or worn.

Installation — Timing Not Disturbed

1. Using clean engine oil, lubricate the distributor O-ring.
2. Align the marks on the distributor shaft and housing.
3. Install the distributor making sure to engage the drive gear into the camshaft distributor drive gear.
4. Make sure the distributor rotor aligns with the No. 1 spark plug tower position on the distributor cap and the distributor housing mark aligns with the cylinder head or cylinder block mark.

NOTE: There are existing marks on the distributor shaft and housing, which when aligned, indicate the No. 1 spark plug wire tower position.

5. Install and loosely tighten the distributor hold-down bolt(s).
6. Connect the electrical connectors to their original locations. Install the distributor cap.

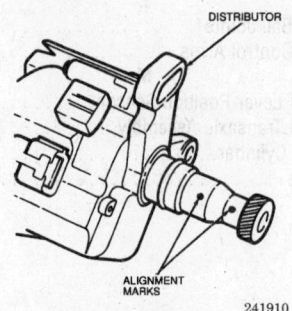

DISTRIBUTOR

ALIGNMENT MARKS

241910

Distributor assembly for automatic transaxle (manual transaxle similar) — 1993 2.0L (VIN A) engine

7. Connect the negative battery cable. Start the engine and check or adjust the ignition timing.

Installation — Timing Disturbed

1. Using clean engine oil, lubricate the distributor O-ring.
2. Disconnect the spark plug wire from the No. 1 cylinder spark plug. Remove the spark plug from the No. 1 cylinder and press a thumb over the spark plug hole.
3. Using a wrench on the crankshaft pulley, rotate the crankshaft until pressure is felt at the spark plug hole, indicating the piston is approaching TDC on the compression stroke. Continue rotating the crankshaft until the crankshaft pulley mark aligns with the timing cover indicator.
4. Position the distributor rotor so it aligns with the No. 1 spark plug wire tower on the distributor cap.
5. Install the distributor. Align the mark that was made on the distributor housing with the mark that was made on the cylinder block. Loosely tighten the distributor hold-down bolts.
6. Connect the electrical connectors to their original locations. Install the distributor cap.
7. Install the spark plug in the No. 1 cylinder and connect the spark plug wire.
8. Connect the negative battery cable.
9. Run the engine and check the ignition timing. Adjust the ignition timing if required.

1994–97 2.0L (VIN A) Engine

Removal

1. Disconnect the negative battery cable.
2. Remove the distributor cap and position aside, leaving the spark plug wires connected. Before removing the distributor, mark the position of the distributor cap No. 1 spark plug wire tower on the distributor.
3. Disconnect the distributor electrical connector.
4. Using a wrench on the crankshaft pulley, rotate the crankshaft to position the No. 1 piston at TDC on the compression stroke. The crankshaft pulley notch should align with the timing plate indicator and the distributor rotor should be pointing to the No. 1 spark plug tower position on the distributor cap.
5. Using chalk or paint, mark the position of the distributor housing on the cylinder head.

6. Remove the distributor hold-down bolt and remove the distributor.

7. Inspect the O-ring on the distributor housing and replace it if damaged or worn.

Installation — Timing Not Disturbed

1. Using clean engine oil, lubricate the distributor O-ring.

2. Align marks on the distributor shaft and the distributor housing.

3. Install the distributor. Make sure the distributor rotor aligns with the No. 1 spark plug tower position on the distributor cap and the distributor housing mark aligns with the cylinder head or cylinder block mark.

NOTE: There are existing marks on the distributor shaft and housing, which when aligned, indicate the No. 1 position.

4. Install the distributor hold-down bolt.

5. Connect the electrical connectors to their original locations. Install the distributor cap.

6. Connect the negative battery cable.

7. Run the engine and check the ignition timing. Adjust the ignition timing if required.

Installation — Timing Disturbed

1. Using clean engine oil, lubricate the distributor O-ring.

2. Disconnect the spark plug wire from the No. 1 cylinder spark plug. Remove the spark plug from the No. 1 cylinder and press a thumb over the spark plug hole.

3. Using a wrench on the crankshaft pulley, rotate the crankshaft until pressure is felt at the spark plug hole, indicating the piston is approaching TDC on the compression stroke. Continue rotating the crankshaft until the crankshaft pulley mark aligns with the timing cover indicator.

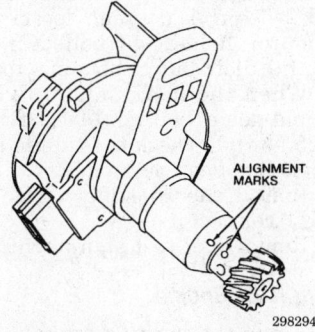

ALIGNMENT MARKS

298294

Distributor assembly — 1994–97 2.0L (VIN A) engine

4. Position the distributor rotor so it aligns with the No. 1 spark plug wire tower on the distributor cap.

5. Install the distributor. Align the mark that was made on the distributor housing with the mark that was made on the cylinder block. Loosely tighten the distributor hold-down bolts.

6. Connect the electrical connectors to their original locations. Install the distributor cap.

7. Install the spark plug in the No. 1 cylinder and connect the spark plug wire.

8. Connect the negative battery cable.

9. Run the engine and check the ignition timing. Adjust the ignition timing if required.

2.5L (VIN B) Engine

1. Disconnect the negative battery cable.

2. Remove the air cleaner intake tube nuts. Loosen the spring clamp at the front of the engine air cleaner assembly and slide it forward. Remove the engine air cleaner intake tube.

3. Loosen the clamp on the front of the Volume Air Flow (VAF) meter and disconnect the air duct. Disconnect the VAF meter electrical connector at the left side of the air cleaner.

4. Disconnect the evaporative emission canister vacuum hose from the routing clip on the front of the air cleaner.

5. Remove the fuel pressure regulator control solenoid from the air cleaner and position aside.

6. Remove the 2 engine air cleaner nuts and bolt and remove the air cleaner assembly.

7. Tag and disconnect the spark plug wires from the distributor cap. Disconnect the 2 electrical connectors from the top of the distributor.

8. Using chalk or paint, mark the position of the distributor housing on the cylinder head. Remove the 2 distributor hold-down bolts and remove the distributor.

To install:

9. Align the distributor shaft with the camshaft end and install the distributor.

NOTE: The tangs on the distributor shaft are different sizes, allowing the distributor to be installed in only one position.

10. Install the distributor hold-down bolts. Align the mark that was made on the distributor housing with the mark that was made on the cylinder head and loosely tighten the bolts.

11. Connect the electrical connectors to the distributor and the spark plug wires to the distributor cap.

12. Install the engine air cleaner assembly and tighten the 2 nuts and bolt to 14–18 ft. lbs. (19–25 Nm).

13. Install the fuel pressure regulator solenoid and connect the evaporative emission canister vacuum hose to the routing clip.

14. Connect the VAF meter electrical connector. Connect the air duct and tighten the clamp.

15. Align the engine air cleaner intake tube and install it to the engine air cleaner assembly. Loosen the spring clamp and slide it into position. Install the engine air cleaner intake tube nuts and tighten to 71–88 inch lbs. (8–10 Nm).

16. Connect the negative battery cable.

17. Run the engine and check the ignition timing. Adjust the ignition timing if required.

Ignition Timing

ADJUSTMENT

1993 2.0L (VIN A) Engine

Manaul Transaxle

1. Apply the parking brake.

2. Make sure that the transmission is in **NEUTRAL**.

3. Locate the timing marks on the crankshaft pulley and the timing indicator scale on the engine front cover. If the marks are hard to see, clean them with degreaser and a stiff brush.

4. Start the engine and bring to normal operating temperature.

5. Make sure all accessories are OFF.

6. Connect a tachometer and timing light to the engine according to the manufacturer's instructions.

7. Remove the shorting bar from the SPOUT connector.

8. Verify that the idle speed is 700 ± 50 rpm; adjust if necessary.

9. Aim the timing light at the timing marks. The mark on the crankshaft pulley should line up with the 10 degrees BTDC ± 1 degree timing mark on the timing indicator on the front engine cover.

10. If the timing marks are not aligned (within 1 degree), loosen the distributor hold-down bolts (2) and turn the distributor housing to adjust. When the marks align, tighten the hold-down bolt to 14–19 ft. lbs. (19–25 Nm). Recheck the timing after the bolt has been tightened.

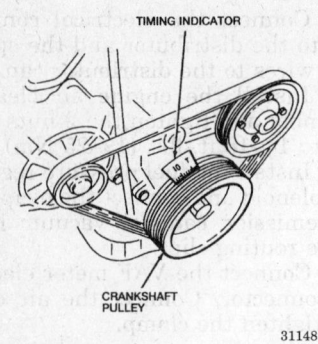

Location of timing marks — all 2.0L (VIN A) engine

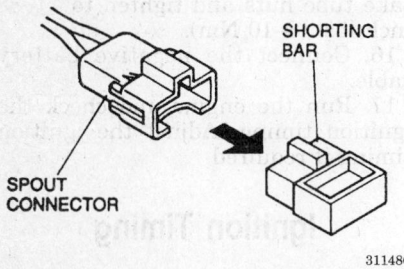

Removing the shorting bar from the SPOUT connector — all 2.0L (VIN A) engine

11. Install the shorting bar to the SPOUT connector.

12. Remove all test equipment.

Automatic Transaxle

1. Apply the parking brake. Place the shift lever in **PARK**.

2. Locate the timing marks on the crankshaft pulley and the timing indicator scale on the engine front cover. If the marks are hard to see, clean them with degreaser and a stiff brush.

3. Start the engine and bring to normal operating temperature.

4. Make sure all accessories are OFF.

5. Connect a tachometer and timing light to the engine according to the manufacturer's instructions.

6. Connect a jumper wire between the **STI (TEN)** terminal and **GROUND** terminal on the Data Link Connector (DLC) located next to the battery.

7. Check the idle speed. If necessary adjust the idle speed to 700 ± 50 rpm.

8. Aim the timing light at the timing marks. The mark on the crankshaft pulley should line up with the 12 degrees BTDC ± 1 degree timing mark on the timing indicator on the front engine cover.

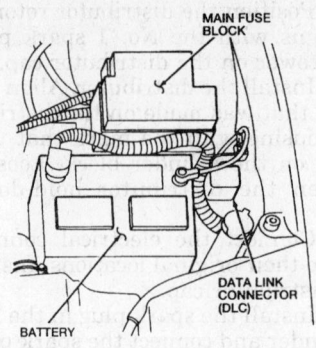

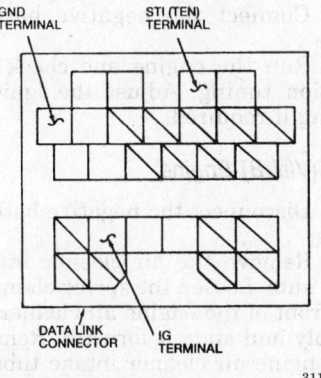

Data Link Connector (DLC) and terminal locations — 1993 2.0L (VIN A) engine

9. If the timing marks are not aligned (within 1 degree), loosen the distributor hold-down bolts (2) and turn the distributor housing to adjust. When the marks align, tighten the hold-down bolt to 14–18 ft. lbs. (19–25 Nm).

10. Recheck the timing after the bolt has been tightened.

11. Remove the jumper wire from the DLC.

12. Remove the tachometer and the timing light.

1994 2.0L (VIN A) Engine

1. Apply the parking brake.

2. Make sure that the transmission is in **NEUTRAL** if equipped with manual transaxle or in **PARK** if equipped with automatic transaxle.

3. Locate the timing marks on the crankshaft pulley and the timing indicator scale on the engine front cover. If the marks are hard to see, clean them with degreaser and a stiff brush.

4. Start the engine and bring to normal operating temperature.

5. Make sure all accessories are off.

6. Connect a tachometer and timing light to the engine according to the manufacturer's instructions.

7. Remove the shorting bar from the SPOUT connector.

8. Verify that the idle speed is 700 ± 50 rpm; adjust if necessary.

9. Aim the timing light at the timing marks. The mark on the crankshaft pulley should line up with the 12 degrees BTDC ± 1 degree timing mark on the timing indicator on the front engine cover.

10. If the timing marks are not aligned (within 1 degree), loosen the distributor hold-down bolts (2) and turn the distributor housing to adjust. When the marks align, tighten the hold-down bolt to 14–19 ft. lbs. (19–25 Nm). Recheck the timing after the bolt has been tightened.

11. Install the shorting bar to the SPOUT connector.

12. Remove all test equipment.

1995–97 2.0L (VIN A) Engine

1. Apply the parking brake.

2. Make sure that the transmission is in **NEUTRAL** if equipped with manual transaxle or in **PARK** if equipped with an automatic transaxle.

3. Locate the timing marks on the crankshaft pulley and the timing indicator scale on the engine front cover. If the marks are hard to see, clean them with degreaser and a stiff brush.

4. Start the engine and bring to normal operating temperature.

5. Make sure all accessories are OFF.

6. Connect a tachometer and timing light to the engine according to the manufacturer's instructions.

7. Remove the shorting bar from the SPOUT connector.

8. Verify that the idle speed is 700 ± 50 rpm; adjust if necessary.

9. Aim the timing light at the timing marks. The mark on the crankshaft pulley should line up with the 10 degrees BTDC ± 1 degree timing mark on the timing indicator on the front engine cover.

10. If the timing marks are not aligned (within 1 degree), loosen the distributor hold-down bolts (2) and turn the distributor housing to adjust. When the marks align, tighten the hold-down bolt to 14–19 ft. lbs. (19–25 Nm). Recheck the timing after the bolt has been tightened.

11. Install the shorting bar to the SPOUT connector.

12. Remove all test equipment.

2.5L (VIN B) Engine

1. Apply the parking brake.

2. If equipped with a manual transaxle, place the shift lever in

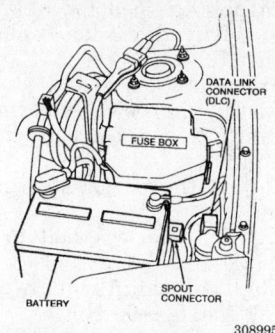

Location of the SPOUT connector — 1995–97 2.0L (VIN A) engine

308995

NEUTRAL. If equipped with an automatic transaxle, place the shift lever in **PARK**.

3. Locate the timing marks on the crankshaft pulley and timing belt cover. If the marks are hard to see, clean them off with some degreasing cleaner and a wire brush.

4. Start the engine and allow it to come to normal operating temperature.

5. Make sure all accessories are **OFF**.

6. Connect a tachometer and a timing light according to the manufacturer's instructions.

7. Connect terminals **STI (TEN)** and **GND** on the Data Link Connector (DLC) with a jumper wire.

8. Check the idle speed and verify that it is 650 ± 50 rpm. Adjust the idle speed if necessary.

9. Aim the timing light at the timing marks. The mark on the crankshaft pulley should line up with the 10 degrees BTDC ± 1 degree timing mark on the timing indicator on the front engine cover.

NOTE: Do not pinch the ignition coil-to-distributor high tension wiring when turning the distributor.

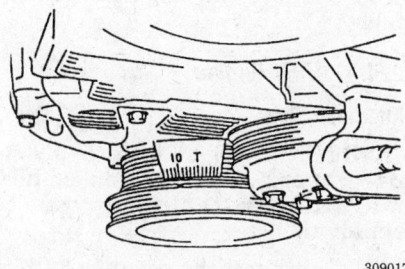

Timing mark location — 2.5L (VIN B) engine

309017

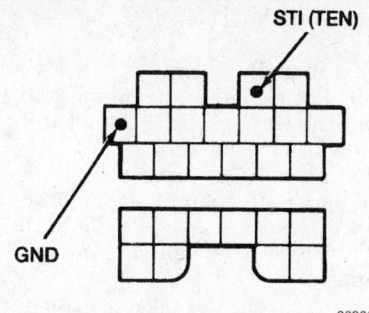

Data Link Connector terminal locations — 2.5L (VIN B) engine

309018

10. If the marks are not aligned, loosen the distributor hold-down bolts (2) just enough to turn the distributor housing. While aiming the timing light at the timing marks, turn the distributor until the marks are aligned. Tighten the distributor hold-down bolts to 14–19 ft. lbs. (19–25 Nm) and recheck the timing.

11. Stop the engine.

12. Remove the jumper wire and test equipment.

Alternator

PRECAUTIONS

Several precautions must be observed with alternator equipped vehicles to avoid damage to the unit.

• If the battery is removed for any reason, make sure it is reconnected with the correct polarity. Reversing the battery connections may result in damage to the 1-way rectifiers.

• When utilizing a booster battery as a starting aid, always connect the positive to positive terminals and the negative terminal from the booster battery to a good engine ground on the vehicle being started.

• Never use a fast charger as a booster to start vehicles.

• Disconnect the battery cables when charging the battery with a fast charger.

• Never attempt to polarize the alternator.

• Do not use test lights of more than 12 volts when checking diode continuity.

• Do not short across or ground any of the alternator terminals.

• The polarity of the battery, alternator and regulator must be matched and considered before making any electrical connections within the system.

• Never separate the alternator on an open circuit. Make sure all connec-

tions within the circuit are clean and tight.

• Disconnect the battery ground terminal when performing any service on electrical components.

• Disconnect the battery if arc welding is to be done on the vehicle.

REMOVAL AND INSTALLATION

2.0L (VIN A) Engine

1. Disconnect the negative battery cable.

2. Remove the alternator upper mounting bolt.

3. Loosen the alternator adjusting arm and remove the drive belt from the alternator pulley.

4. Raise and safely support the vehicle.

5. Remove the 6 bolts and remove the crossmember.

6. Disconnect the electrical connectors from the alternator.

7. If more clearance is required, remove the front exhaust pipe as follows:

a. Support the exhaust system at the catalytic converter with a jack.

b. Disconnect the oxygen sensor electrical connector and remove the sensor using sensor wrench T79P-9472-A or equivalent.

c. Remove the 3 exhaust manifold flange nuts and remove the clamp from the hold-down bracket.

d. Remove the exhaust pipe-to-converter nuts and pry the rubber hangers from the mounting hooks. Remove the pipe.

8. Remove the alternator lower through bolt and remove the alternator.

To install:

9. Install the alternator with the lower through bolt. Do not tighten the bolt at this time.

10. If removed, install the exhaust pipe, using new gaskets. Tighten the pipe-to-converter nuts to 66 ft. lbs. (89 Nm) and the exhaust manifold flange nuts to 38 ft. lbs. (52 Nm). Tighten the exhaust clamp nuts to 34 ft. lbs. (47 Nm).

11. If removed, install the oxygen sensor, using sensor wrench T79P-9472-A or equivalent, and tighten to 36 ft. lbs. (49 Nm). Connect the oxygen sensor electrical connector.

12. Connect the alternator electrical connectors.

13. Install the crossmember and tighten the 6 bolts to 69–96 ft. lbs. (94–131 Nm).

14. Lower the vehicle.

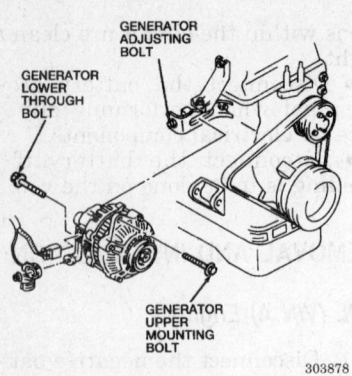

Alternator bolt locations — 2.0L (VIN A) engine

15. Install the drive belt and upper mounting bolt. Adjust the belt tension.

16. Tighten the lower through bolt to 27–38 ft. lbs. (37–52 Nm) and the upper mounting bolt to 8–10 ft. lbs. (10–15 Nm).

17. Connect the negative battery cable.

18. Check that the charging system is operating properly.

2.5L (VIN B) Engine

1. Disconnect the negative battery cable.

2. Disconnect the electrical connectors from the alternator.

3. Loosen the belt tensioner locknut and tension adjusting bolt. Remove the alternator upper mounting bolt.

4. Raise and safely support the vehicle.

5. Remove the right splash shield.

6. Remove the drive belt from the alternator pulley.

7. If equipped with A/C, remove the A/C compressor mounting bolts and and support the compressor aside. Leave the refrigerant lines connected.

8. Loosen the alternator lower through bolt and remove the alternator.

To install:

9. Position the alternator and install the through bolt. Do not tighten the bolt at this time.

10. Lower the vehicle and install the upper alternator bolt. Tighten the bolt to 14–18 ft. lbs. (19–25 Nm).

11. Raise the vehicle and safely support.

12. Tighten the lower alternator through bolt to 28–38 ft. lbs. (37–51 Nm).

13. If equipped, install the A/C compressor and tighten the A/C compressor mounting bolts to 18–26 ft. lbs. (24–35 Nm).

14. Install the drive belt.

15. Install the right splash shield.

16. Lower the vehicle.

17. Adjust the alternator drive belt tension and tighten the belt tensioner locknut and tension adjusting bolt.

18. Connect the electrical connectors to the alternator.

19. Connect the negative battery cable.

20. Run the engine and check charging system operation.

Drive Belts

REMOVAL AND INSTALLATION

2.0L (VIN A) Engine

Power Steering Pump Drive Belt

1. Disconnect the negative battery cable.

2. Remove the power steering pump belt shield.

3. Loosen the adjusting bolt, lock bolt, and through-bolt.

4. Remove the power steering belt. If the belt is to be reused, mark the belt for its direction of rotation.

5. Inspect the belt for wear, glazing and frayed cords; replace as necessary.

To install:

6. Install the power steering belt and make sure it is correctly lined up on the pulleys.

7. Adjust the power steering belt tension.

8. Tighten the lock bolt to 24–34 ft. lbs. (32–46 Nm).

9. Tighten the through-bolt to 32–44 ft. lbs. (44–60 Nm).

10. Install the belt shield and tighten the attaching bolts to 61–86 inch lbs. (7–9 Nm).

11. Connect the negative battery cable.

Alternator Drive Belt

1. Disconnect the negative battery cable.

2. Remove the power steering belt.

3. Loosen the alternator adjusting bolt and upper mounting bolt.

4. Raise and safely support the vehicle.

5. Remove the RH front splash shield.

6. Loosen the lower through-bolt.

7. Lower the vehicle and remove the alternator belt. If the belt is to be reused, mark the belt for its direction of rotation.

8. Inspect the belt for wear, glazing and frayed cords; replace as necessary.

To install:

9. Install the alternator belt and make sure it is correctly lined up on the pulleys.

10. Adjust the alternator belt tension.

11. Tighten the upper mounting bolt to 14–18 ft. lbs. (19–25 Nm).

12. Raise and safely support the vehicle.

13. Tighten the lower through-bolt to 28–38 ft. lbs. (38–51 Nm).

14. Install the RH splash shield and tighten the bolts to 71–88 inch lbs. (8–10 Nm).

15. Lower the vehicle.

16. Install the power steering belt.

17. Connect the negative battery cable.

2.5L (VIN B) Engine

Alternator Belt

NOTE: If the belt is to be reused, mark the direction of normal belt rotation prior to removal.

1. Disconnect the negative battery cable.

2. Loosen the tensioner locknut on the alternator belt tensioner.

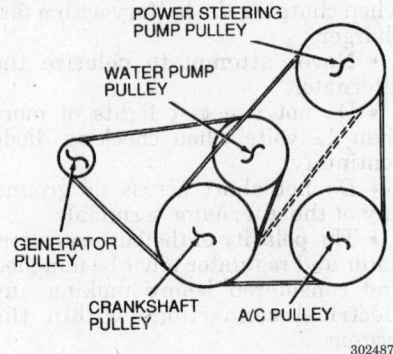

Alternator bolt locations — 2.5L (VIN B) engine

Drive belt routing — 2.0L (VIN A) engine

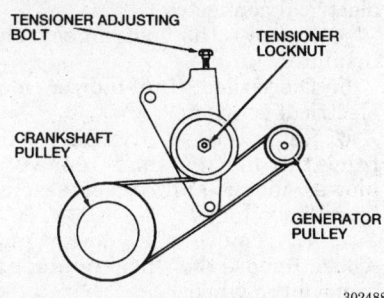

Without A/C

Alternator drive belt routing, Without A/C —
2.5L (VIN B) engine

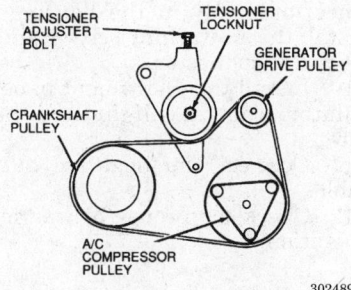

With A/C

Alternator drive belt routing, With A/C —
2.5L (VIN B) engine

3. Loosen the tensioner adjusting bolt until there is enough slack to remove the belt.

4. Remove the belt from each of the pulleys and remove from the vehicle.

5. Inspect the belt for wear, glazing, fraying or chunking; replace as necessary.

To install:

NOTE: Make sure that the belt is properly seated on all of the pulleys before adjusting the tension.

6. Route the alternator belt around the alternator and crankshaft pulleys and, if equipped, the A/C pulley.

7. Adjust the alternator belt to the proper tension and tighten the tensioner locknut to 24–34 ft. lbs. (32–46 Nm).

8. Connect the negative battery cable.

Power Steering and Water Pump Belt

NOTE: If the belt is to be reused, mark the direction of normal belt rotation prior to removal.

1. Disconnect the negative battery cable.

2. Remove the alternator belt.

3. Loosen the tensioner locknut on the power steering and water pump belt tensioner.

4. Loosen the tensioner adjusting bolt until there is enough slack to remove the belt.

5. Remove the belt from each of the pulleys and remove from the vehicle.

6. Inspect the belt for wear, glazing, fraying or chunking; replace as necessary.

To install:

NOTE: Make sure that the belt is properly seated on all of the pulleys before adjusting the tension.

7. Route the power steering and water pump drive belt around the 3 pulleys.

8. Adjust the belt to the proper tension and tighten the tensioner locknut to 24–34 ft. lbs. (32–46 Nm).

9. Install the alternator belt.

10. Connect the negative battery cable.

Starter

REMOVAL AND INSTALLATION

2.0L (VIN A) Engine

1. Disconnect the negative battery cable.

2. Remove the air cleaner and the air cleaner-to-intake manifold tube.

3. Remove the 2 upper starter mounting bolts.

4. Raise and safely support the vehicle.

5. Remove the intake manifold support bracket bolts and the bracket.

6. Disconnect the electrical connectors from the starter solenoid.

7. Remove the lower starter mounting bolt and remove the starter.

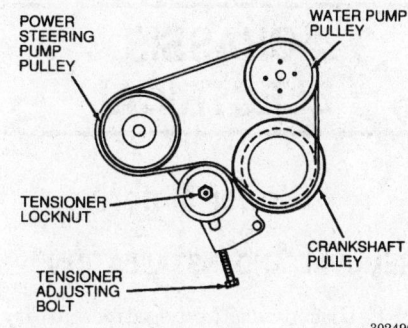

Power steering and water pump belt routing —
2.5L (VIN B) engine

To install:

8. Position the starter and loosely tighten the lower starter mounting bolt.

9. Connect the electrical connectors to the starter solenoid.

10. Install the intake manifold support bracket and bolts.

11. Lower the vehicle.

12. Install the upper starter mounting bolts and tighten to 24–34 ft. lbs. (31–46 Nm).

13. Tighten the lower starter mounting bolt to 24–34 ft. lbs. (31–46 Nm).

14. Install the air cleaner-to-intake manifold tube and the air cleaner.

15. Connect the negative battery cable.

16. Check the starter motor for proper operation.

2.5L (VIN B) Engine

Manual Transaxle

1. Disconnect the negative battery cable.

2. Remove the air cleaner intake tube and the engine air cleaner.

3. Remove the S-terminal wire from the starter solenoid.

4. Remove the nut and the B-terminal wire from the starter solenoid.

5. Remove the 3 starter mounting bolts and remove the starter.

To install:

6. Position the starter motor and install the 3 mounting bolts.

7. Tighten the starter motor mounting bolts to 24–33 ft. lbs. (32–46 Nm).

8. Install the B-terminal wire and nut to the starter solenoid and tighten to 12–16 ft. lbs. (16–22 Nm).

9. Install the S-terminal wire to the starter solenoid.

10. Install the engine air cleaner and air cleaner intake tube.

11. Connect the negative battery cable.

12. Check for proper starter motor operation.

Automatic Transaxle

1. Disconnect the negative battery cable.

2. Remove the air cleaner intake tube and the engine air cleaner.

3. Remove the transmission shift cable from the selector lever using a screwdriver or suitable tool.

4. Squeeze the lock tabs on the shift cable and remove the cable from the bracket.

5. Label and disconnect the following electrical connectors:
   a. Knock sensor
   b. Throttle position Sensor
   c. Fuel injector harness

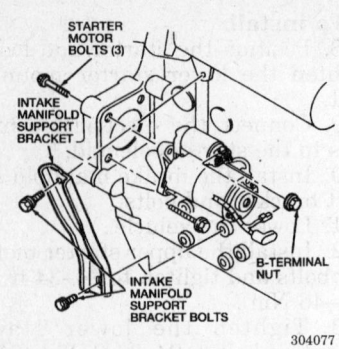

Starter motor bolt locations — 2.0L
(VIN A) engine

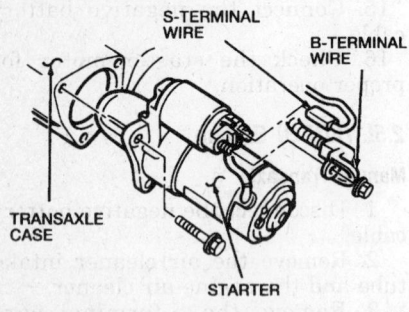

Starter motor assembly, With manual
transaxle — 2.5L (VIN B) engine

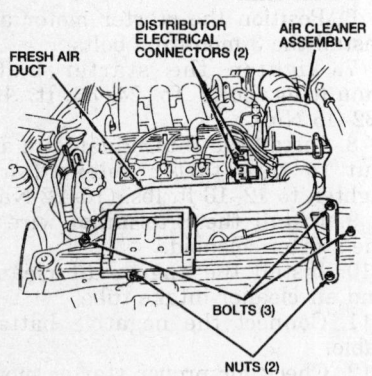

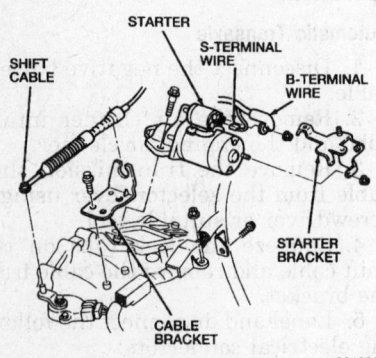

Starter motor assembly, With automatic
transaxle — 2.5L (VIN B) engine

d. Ignition distributor
e. Transmission range selector
f. Transaxle connector
g. Other wiring harness connectors in the work area

6. Position the wiring harness out of the way.

7. Remove the 2 selector cable bracket mounting bolts and the bracket.

8. Remove the two nuts and the bolt from the starter bracket and remove the bracket.

9. Remove the S-terminal wire from the starter solenoid.

10. Remove the nut and the B-terminal wire from the starter solenoid.

11. Remove the 3 starter mounting bolts and remove the starter.

To install:

12. Position the starter and tighten the starter mounting bolts to 24–33 ft. lbs. (32–46 Nm).

13. Install the B-terminal wire and nut. Tighten the nut to 12–16 ft. lbs. (16–22 Nm).

14. Install the S-terminal wire to the solenoid.

15. Position the starter bracket and install the bolt and nuts.

16. Install the selector cable bracket and tighten the bolts to 60–84 inch lbs. (7–9 Nm).

17. Connect the following electrical connectors:
  a. Knock sensor
  b. Throttle position Sensor
  c. Fuel injector harness
  d. Ignition distributor
  e. Transmission range selector
  f. Transaxle connector

18. Install the shift cable into the cable bracket and into the selector lever.

19. Install the engine air cleaner and the air cleaner intake tube.

20. Connect the negative battery cable.

21. Check for proper starter motor operation.

CHASSIS ELECTRICAL

Blower Motor

REMOVAL AND INSTALLATION

1. Disconnect the negative battery cable.

2. Remove the 2 instrument panel insulator screws.

3. Disconnect the courtesy lamp electrical connector.

4. Remove the instrument panel insulator.

5. Disconnect the blower motor electrical connector.

6. Remove the 4 attaching screws from the housing and remove the blower motor and wheel assembly from the vehicle.

7. To remove the blower motor wheel, remove the wheel retainer and remove the wheel.

To install:

8. If removed, install the blower motor wheel onto the blower motor shaft and install the retainer.

9. Position the blower motor and wheel assembly in the housing and install the 4 attaching screws and the electrical connector.

10. Install the instrument panel insulator, courtesy light and panel screws.

11. Connect the negative battery cable.

12. Check for proper blower motor operation.

Windshield Wiper Motor

REMOVAL AND INSTALLATION

Front Wiper Motor

1. Disconnect the negative battery cable.

2. Remove the wiper arm cover cap and attaching nut and remove the wiper arm and blade assemblies.

3. Remove the 7 screw covers and screws securing the cowl vent grille. Remove the cowl vent grille.

4. Remove the 5 cowl deflector clips and the the 3 cowl deflector screws. Remove the cowl deflector.

5. Use a small prybar to separate the wiper mounting arm and pivot shaft from the wiper motor output arm.

6. Remove the 4 pivot shaft retaining bolts from the pivot shaft brackets.

7. Remove the wiper mounting arm and pivot shaft from the vehicle.

8. Remove the wiring harness bracket from the wiper motor mounting bracket.

9. Disconnect the wiper motor ground and the electrical connector.

10. Remove the wiper motor bracket bolts and remove the wiper motor from the vehicle.

11. Disconnect the wiper motor ground and electrical connector and remove the wiper motor.

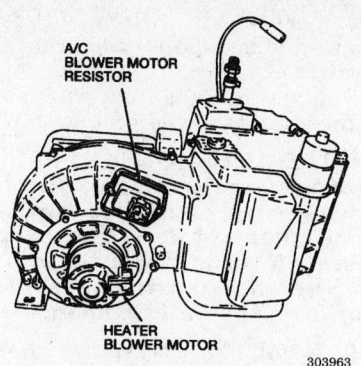

Blower motor assembly

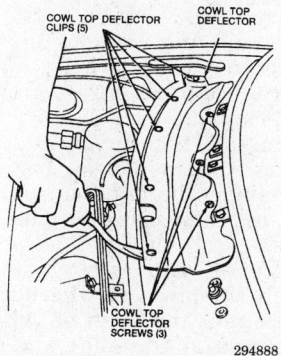

Cowl vent grille screw locations

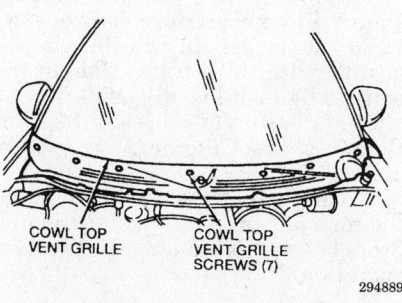

Cowl deflector screw and clip locations

To install:

12. Position the wiper motor and secure with the wiper motor bracket bolts. Tighten the wiper motor bolts to 61–87 inch lbs. (7–9 Nm).

13. Connect the wiper motor electrical connector and ground.

14. Install the wiring harness bracket into the wiper motor mounting bracket.

15. Connect the negative battery cable. Run the wiper motor, then turn the wiper switch **OFF** and allow the motor to stop in the park position.

16. Position the wiper mounting arm and pivot shaft. Tighten the 4

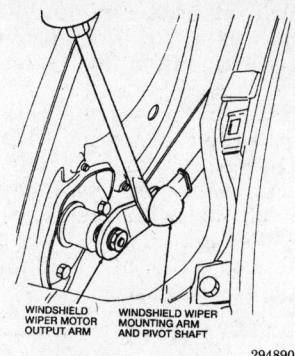

Separating the pivot shaft

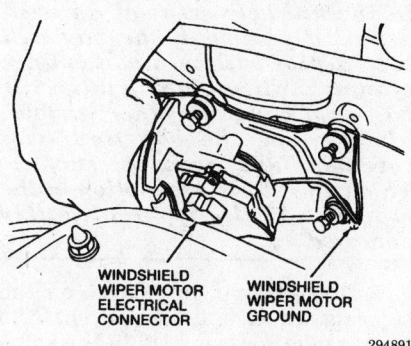

Front wiper motor

pivot shaft retaining bolts to 61–87 inch lbs. (7–9 Nm).

17. Connect the wiper mounting arm and pivot shaft to the wiper motor output arm.

18. Install the cowl deflector and secure with 3 screws and 5 clips.

19. Install the cowl vent grille, screws and screw covers.

20. Install the wiper arm and blade assemblies so the tips of the blades are 1.12–1.28 in. (28–32mm) from the top of the cowl grille. Tighten the wiper arm nuts to 86–121 inch lbs. (10–15 Nm).

NOTE: The drivers side wiper arm and blade assembly is marked D while the passenger side arm is marked PL.

21. Connect the negative battery cable.

22. Check for proper operation.

Rear Wiper Motor

1. Disconnect the negative battery cable.

2. Lift the wiper arm nut cover and remove the nut. Remove the wiper arm and blade assembly.

3. Remove the cover and remove the wiper motor shaft support nut and bezel.

4. Raise the liftgate and remove the liftgate lower trim.

5. Disconnect the wiper motor electrical connector.

6. Remove the 3 wiper motor mounting bolts and disconnect the ground wire. Remove the wiper motor.

To install:

7. Position the wiper motor and install the 3 mounting bolts making certain the ground wire is connected. Tighten the wiper motor mounting bolts to 61–87 inch lbs. (7–9 Nm).

8. Connect the wiper motor electrical connector and connect the negative battery cable.

9. Run the wiper motor, then turn the wiper switch **OFF** and allow the motor to stop in the park position.

10. Install the liftgate lower trim and lower the liftgate.

11. Install the wiper motor shaft support nut and bezel and tighten to 27–52 inch lbs. (3–5 Nm). Install the cover.

12. Install the wiper arm and blade assembly so the tip of the wiper blade is 1.0–1.6 in. (25–40mm) from the shaded glass area. Tighten the wiper arm nut to 87 inch lbs. (9 Nm).

13. Connect the negative battery cable.

14. Check for proper operation.

Concealed Headlights

MANUAL OPERATION

If the power headlight door system becomes inoperative, the vehicle has a manual retractor system that allows the headlights to be raised manually (the headlights will not turn on).

1. Disconnect the negative battery cable.

2. Open the hood and remove the cover for the main fuse panel located on the right inner fender panel.

—————— CAUTION ——————
Always remove the RETRA (20) amp fuse before manually operating a headlight retractor or attempting to remove anything from the headlight. Failure to remove the RETRA fuse could cause injury to a hand or fingers.
——————————————————

3. Remove the RETRA (20) amp fuse.

4. Locate the retractor motor behind each headlight.

5. Remove the cap and turn the knob. Each headlight must be raised separately.

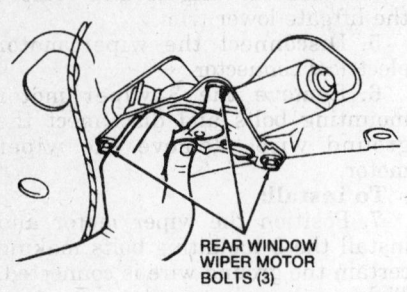

REAR WINDOW
WIPER MOTOR
BOLTS (3)

294892

Rear wiper motor bolt locations

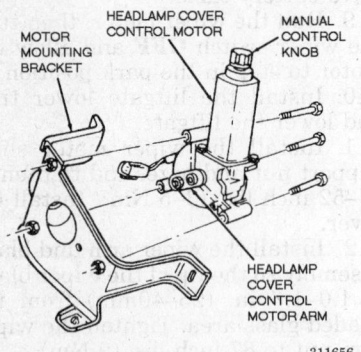

MOTOR
MOUNTING
BRACKET

HEADLAMP COVER
CONTROL MOTOR

MANUAL
CONTROL
KNOB

HEADLAMP
COVER
CONTROL
MOTOR ARM

311656

Headlight retractor motor

--- CAUTION ---

When reinstalling the RETRA fuse, make sure nothing is near the headlight retractor. It may move suddenly while the fuse is being inserted, causing injury to hands and fingers.

6. Install the RETRA (20) amp fuse and replace the main fuse panel cover.

7. Connect the negative battery cable.

Combination Switch

REMOVAL AND INSTALLATION

--- CAUTION ---

The Supplemental Restraint System (SRS) must be disarmed before removing the air bag module. Failure to do so may cause accidental deployment of the air bag, resulting in unnecessary SRS repairs and/or personal injury.

1. Center the wheels to the straight-ahead position.

2. Disconnect the negative battery cable.

3. Wait at least 1 minute for the air bag backup power supply in the diagnostic monitor to deplete its stored energy.

4. Remove the 4 bolts retaining the air bag to the steering wheel.

5. Disconnect the air bag/horn electrical connector. Disconnect the cruise control switch electrical connector, if equipped.

6. Remove the air bag module from the steering wheel.

--- CAUTION ---

When carrying a live air bag, make sure the bag and trim cover are pointed away from the body. In the unlikely event of an accidental deployment, the bag will then deploy with minimal chance of injury. When placing a live air bag on a bench or other surface, always face the bag and trim cover up, away from the surface. This will reduce the motion of the module if it is accidently deployed.

7. Make alignment marks on the steering wheel and column shaft so they can be reassembled in the same position.

8. Remove the steering wheel nut and remove the steering wheel with a suitable puller. Route the wiring harness through the steering wheel as the wheel is lifted off the shaft. Do not try to remove the steering wheel by hitting the column shaft with a hammer, as the column shaft will collapse.

9. Remove the 4 lower steering column panel screws and separate the upper and lower steering column panels.

10. Remove the lock cylinder illumination bulb from the lower steering column panel and remove the upper and lower steering column panels.

11. Apply 2 strips of tape across the clockspring and housing to prevent accidental rotation.

12. Remove the 3 clockspring assembly screws and pull the clockspring assembly off the steering column shaft.

13. Disconnect the ground wire and electrical connector and remove the clockspring assembly.

14. Remove the cancel cam and spring. Remove the 3 combination switch screws.

15. Disconnect the electrical connectors and remove the combination switch.

To install:

16. Slide the combination switch over the column shaft and connect the electrical connectors.

17. Secure the switch with the screws and install the cancel cam and spring.

18. Make sure the front wheels are in the straight-ahead position.

NOTE: If the clockspring has been accidently rotated, turn the clockspring clockwise until it stops, then rotate counterclockwise 2.75 turns. Align the marks on the clockspring with the marks on the outer housing.

19. Install the clockspring ground wire screw and position the clockspring assembly on the column shaft. Connect the clockspring electrical connector.

20. Install the clockspring screws and tighten to 26 inch lbs. (3 Nm). Remove the tape strips from the clockspring and housing.

21. Install the upper and lower steering column panels. Install the lock cylinder illumination bulb and the 4 lower steering column panel screws.

22. Route the wiring harness through the steering wheel opening and position the steering wheel on the column shaft. Align the marks that were made during removal.

23. Install the steering wheel nut and tighten to 36 ft. lbs. (49 Nm).

24. Connect the air bag module and, if equipped, cruise control electrical connectors. Install the air bag module with the 4 bolts. Tighten the bolts to 54 inch lbs. (6 Nm).

25. Make sure no one is in the vehicle and connect the negative battery cable.

26. Check that the air bag light in the instrument cluster is functioning properly to prove-out the system operation.

Ignition Lock Cylinder

REMOVAL AND INSTALLATION

NOTE: The following procedure requires new ignition lock cylinder bolts.

--- CAUTION ---

The Supplemental Restraint System (SRS) must be disarmed before performing service around SRS components or wiring. Failure to do so may cause accidental deployment of the airbag, resulting in unnecessary SRS repairs and/or personal injury.

1. Position the vehicle with the front wheels in a straight ahead position.

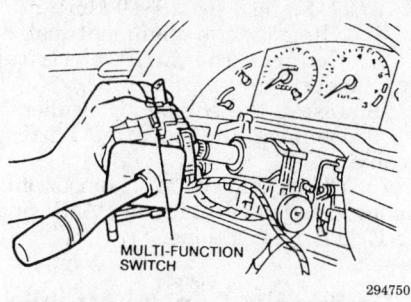

Combination switch (multi-function switch)

2. Disconnect the negative battery cable.

3. Disconnect the positive battery cable.

4. Wait at least 1 minute for the air bag backup power supply to drain before continuing.

5. Remove the steering wheel.

6. Remove the 4 lower steering column shroud screws and separate the upper and lower shrouds.

7. Remove the lock cylinder illumination bulb from the steering column shroud, then remove the steering column shrouds.

8. Remove the combination switch.

9. If equipped with automatic transaxle, remove the ignition/shifter interlock cable screw and disconnect the ignition/shifter interlock cable from the lock cylinder.

10. Using a suitable chisel, remove the 2 lock cylinder bolts, then remove the lock cylinder from the steering column.

11. Disconnect the ignition key reminder switch wiring connector.

12. If necessary, remove the ignition key reminder switch.

To install:

13. If removed, install the ignition key reminder switch.

14. Connect the key reminder switch electrical connector.

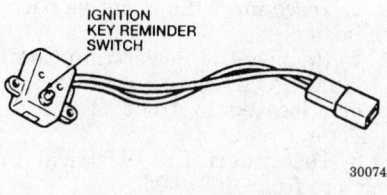

Ignition key reminder switch

15. Install the lock cylinder using new bolts. Tighten the bolts until the bolt heads break off.

16. If equipped with automatic transaxle, connect the ignition/shifter interlock cable to the lock cylinder and install the screw.

17. Install the combination switch.

18. Install the lock cylinder illumination bulb to the lower steering column shroud. Install the shrouds and secure with the screws.

19. Install the steering wheel.

20. Connect the negative battery cable. Check for proper ignition switch and lock operation.

Ignition Switch

REMOVAL AND INSTALLATION

——— **CAUTION** ———
The Supplemental Restraint System (SRS) must be disarmed before performing service around SRS wiring or SRS components. Failure to do so may cause accidental deployment of the air bag, resulting in unnecessary SRS repairs and/or personal injury.

1. Disconnect both battery cables from the battery, negative cable first. Wait at least 1 minute for the air bag backup power supply to drain.

2. Remove the 4 lower steering column panel screws and separate the upper and lower steering column panels.

3. Remove the lock cylinder illumination bulb from the lower steering column panel and remove the upper and lower steering column panels.

4. Disconnect the ignition switch electrical connector and remove the retaining screw.

5. Remove the ignition switch from the steering column.

To install:

6. Position the ignition switch and install the retaining screw.

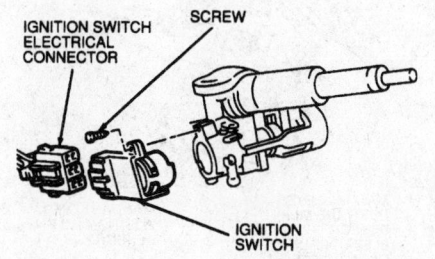

Ignition switch

7. Install the ignition switch electrical connector.

8. Position the upper and lower steering column panels and install the lock cylinder illumination bulb to the lower steering column panel.

9. Install the 4 steering column panel retaining screws and secure the panels.

10. Connect the battery cables, negative cable last.

11. Check for proper ignition switch operation.

Park/Neutral Safety Switch

REMOVAL AND INSTALLATION

1. Disconnect the negative battery cable.

2. Raise and safely support the vehicle.

3. Drain the transmission fluid into a suitable container.

4. Disconnect the Park/Neutral Position (PNP) switch electrical connector.

5. Remove the PNP switch.

To install:

6. Install the PNP switch using a new gasket and tighten the switch to 15–21 ft. lbs. (20–29 Nm).

7. Connect the PNP switch electrical connector.

8. Fill the transaxle with the correct amount of transmission fluid.

9. Lower the vehicle.

10. Connect the negative battery cable.

11. Check that the PNP switch will only allow engine cranking in the **PARK** and **NEUTRAL** positions.

NOTE: The PNP switch is not adjustable and if it does not function properly, it must be replaced.

Manual Lever Position Sensor

REMOVAL AND INSTALLATION

4EAT Transaxle

1. Disconnect the negative battery cable.

2. Remove the air cleaner assembly.

3. Pry the shift cable from the manual control shift outer lever, using a small prybar.

4. Disconnect the MLP sensor, also known as the Transmission Range (TR) switch, electrical connector.

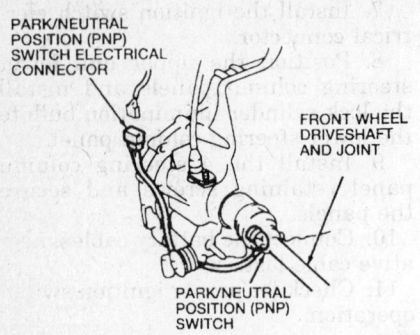

Park/neutral position switch

241697

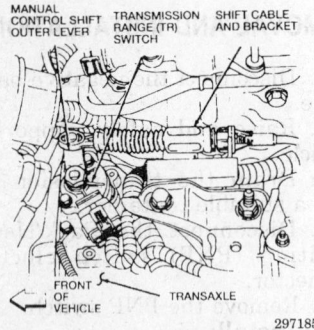

Manual lever position sensor location — 4EAT transaxle

297185

5. Remove the sensor mounting bolts and remove the sensor.

To install:

6. Install the MLP sensor and loosely secure the mounting bolts.

7. To adjust the MLP sensor:

a. Rotate the sensor shaft to the **NEUTRAL** mark.

b. Connect a suitable volt/ohmmeter to the sensor and check for continuity to make sure the sensor is closed.

c. Tighten the sensor mounting bolts to 71–88 inch lbs. (8–10 Nm). Make sure the sensor shaft is still aligned with the neutral mark with the MLP sensor closed.

8. Connect the MLP sensor electrical connector.

9. Install the shift cable onto the manual control shift outer lever.

10. Install the air cleaner assembly and connect the negative battery cable.

11. Check that the MLP sensor only allows engine cranking in **PARK** and **NEUTRAL** positions.

CD4E Transaxle

1. Disconnect the negative battery cable.

2. Remove the engine air cleaner.

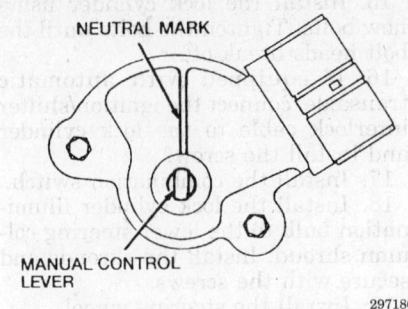

Manual lever position sensor adjustment — 4EAT transaxle

297186

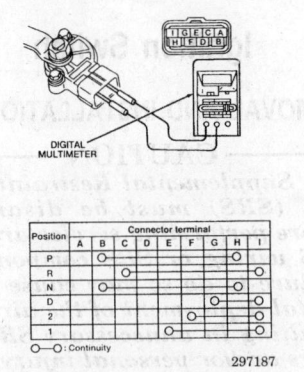

Continuity check of the MLP sensor

297187

3. Disconnect the MLP, also known as the Transmission Range (TR) switch, electrical connector.

4. Remove the 2 MLP mounting bolts and remove the MLP sensor.

To install:

5. Install the MLP sensor to the transaxle leaving the mounting bolts loose.

6. Adjust the sensor as follows:

a. Place Transmission Range Alignment Tool T945P-70010-AH or equivalent, onto the MLP sensor and align the manual valve detent lever shaft.

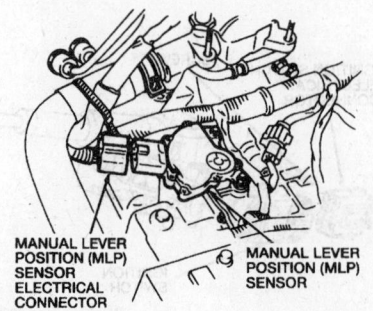

Manual lever position sensor location — CD4E transaxle

297188

b. Tighten the 2 mounting bolts to 71–88 inch lbs. (8–10 Nm).

c. Remove the alignment tool.

7. Connect the MLP electrical connector.

8. Install the engine air cleaner.

9. Connect the negative battery cable.

10. Check that the MLP sensor only allows engine cranking in **PARK** and **NEUTRAL** positions.

Powertain Control Module

REMOVAL AND INSTALLATION

2.0L (VIN A) Engine

— **CAUTION** —

Electronic modules are sensitive to static electrical charges. Wear a static discharge strap to prevent damage or exposure to these charges.

1. Disconnect the negative battery cable.

2. Remove the floor console and instrument panel insulator.

3. Disconnect the engine control sensor wiring connectors from the PCM.

4. Remove the 2 PCM screws.

5. Remove the PCM from the vehicle.

To install:

6. Position the PCM in the dash and install the 2 screws.

7. Tighten the screws to 32 inch lbs. (3.7 Nm).

8. Connect the engine control sensor wiring connectors to the PCM.

9. Install the floor console and instrument panel insulator.

10. Connect the negative battery cable.

2.5L (VIN B) Engine

— **CAUTION** —

Electronic modules are sensitive to static electrical charges. Wear a static discharge strap to prevent damage or exposure to these charges.

1. Disconnect the negative battery cable.

2. Remove the Powertrain Control Module (PCM) center console access panel, located in front of the gear selector.

3. Disconnect the electrical connectors from the PCM.

4. Remove the 2 PCM retaining bolts.

5. Remove the PCM from the console panel.

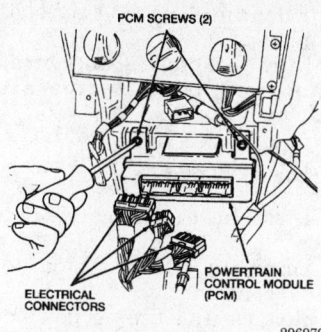

Powertrain control module attaching screw locations — 2.0L (VIN A) engine

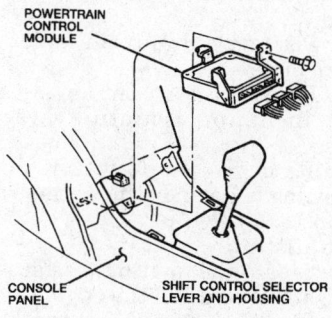

Powertrain control module attaching screw locations — 2.5L (VIN B) engine

To install:

6. Position the PCM in the console panel and install the 2 retaining bolts.

7. Connect the electrical connectors to the PCM.

8. Install the center console access panel.

9. Connect the negative battery cable.

ENGINE COOLING

Radiator

REMOVAL AND INSTALLATION

2.0L (VIN A) Engine

—————— **CAUTION** ——————
Never remove the radiator cap or filler cap while the engine is hot. Wait for the engine to cool before opening the cooling system to prevent possible injury and/or damage to the cooling system or engine.

1. Disconnect the negative battery cable.

2. Disconnect the cooling fan wiring harness connectors.

3. Remove the fresh air duct.

4. Make sure the engine is cool. Press the radiator cap down with a heavy cloth and open enough to relieve the cooling system pressure. Remove the radiator pressure cap from the filler neck.

5. Position a pan under the radiator drain valve and drain the cooling system.

6. Disconnect the upper and lower radiator hoses.

7. Disconnect the overflow hose from the filler neck.

8. Disconnect and plug the oil cooler lines, if equipped with automatic transaxle.

9. Remove the radiator upper mounting brackets.

10. Remove the radiator and cooling fan as an assembly.

11. If necessary, remove the fan and shroud assembly from the radiator.

To install:

12. If removed, install the fan and shroud assembly. Tighten the mounting bolts to 88 inch lbs. (10 Nm).

13. Install the radiator, making sure the bottom of the radiator engages the insulators.

14. Install the upper brackets and tighten the retaining bolts to 14–18 ft. lbs. (19–25 Nm).

15. Unplug and connect the oil cooler lines, if equipped.

16. Install the upper and lower radiator hoses to the radiator.

17. Connect the overflow hose and connect the cooling fan wiring connector.

18. Close the radiator drain valve and fill the system with coolant.

19. Connect the negative battery cable.

20. Run the engine to operating temperature to pressurize the system and check for leaks.

21. Recheck the coolant level and refill if necessary.

2.5L (VIN B) Engine

—————— **CAUTION** ——————
Never remove the radiator cap or filler cap while the engine is hot. Wait for the engine to cool before opening the cooling system to prevent possible injury and/or damage to the cooling system or engine.

1. Disconnect the negative battery cable.

2. Remove the fresh air duct.

3. Make sure the engine is cool. Press the radiator cap down with a heavy cloth and open enough to relieve the cooling system pressure. Remove the radiator pressure cap from the filler neck.

4. Position a drain pan under the radiator drain valve and drain the cooling system.

5. Remove the expansion tank upper and lower hoses from the radiator.

6. Disconnect the overflow hose from the expansion tank.

7. Remove the 2 expansion tank bolts. Remove the expansion tank.

8. Disconnect the electrical connectors from the electric cooling fan and the A/C condenser cooling fan.

9. Remove the upper radiator hose from the radiator.

10. Disconnect and plug the oil cooler lines, if equipped with automatic transaxle.

11. Remove the lower radiator hose from the radiator.

12. Remove the radiator upper mounting brackets.

NOTE: If required for added clearance, move the air bag electrical harness aside.

13. Remove the radiator and cooling fan(s) as an assembly.

14. If necessary, remove the fan and shroud assembly from the radiator.

To install:

15. If removed, install the fan and shroud assembly to the radiator. Tighten the mounting bolts to 88 inch lbs. (10 Nm).

16. Install the radiator and shroud assembly into the vehicle, making sure the bottom of the radiator engages the insulators.

17. Install the upper brackets and tighten the retaining bolts to 14–18 ft. lbs. (19–25 Nm).

18. Install the lower radiator hose.

19. Unplug and connect the oil cooler lines, if equipped.

20. Install the upper radiator hose to the radiator.

21. Connect the electric cooling fan and the A/C condenser cooling fan electrical connectors.

22. Install the expansion tank and the 2 retaining bolts.

23. Connect the overflow hose to the expansion tank.

24. Connect the expansion tank upper and lower hoses to the radiator.

25. Close the radiator drain valve and fill the system with coolant.

26. Connect the negative battery cable.

27. Run the engine to operating temperature to pressurize the system and check for leaks.

28. Recheck the coolant level and refill if necessary.

Water Pump

REMOVAL AND INSTALLATION

2.0L (VIN A) Engine

1. Disconnect the negative battery cable.

2. Drain the cooling system into a suitable container.

3. Remove the accessory drive belts.

4. Disconnect the Power Steering Pressure (PSP) switch electrical connector.

5. Remove the power steering pump drive belt idler pulley shield bolts and remove the shield.

6. Remove the power steering pump through bolt and lock bolt, and position it out of the way.

7. Loosen the cylinder head cover bolts in 2–3 steps in the proper sequence. Remove the cylinder head cover.

8. Raise and safely support the vehicle.

9. Remove the water pump pulley using pulley tool T92C-6312-AH or equivalent, to hold the pulley while the bolts are removed.

10. Remove the splash shields.

11. Remove the timing belt.

12. Remove the power steering pump lower bracket from the water pump.

13. Remove the 5 water pump mounting bolts and remove the water pump.

To install:

14. Clean all gasket mating surfaces.

15. Install a new gasket on the water pump and install the water

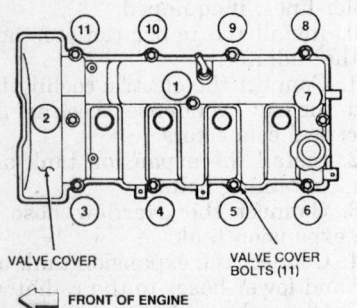

Cylinder head cover bolt removal sequence — 2.0L (VIN A) engine

297414

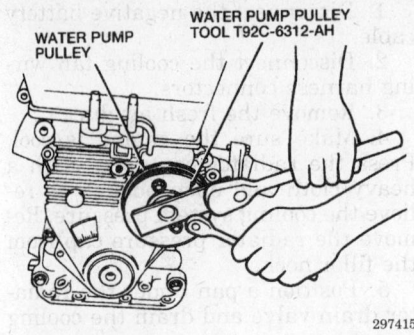

Water Pump Pulley Tool T92C-6312-AH — 2.0L (VIN A) engine

297415

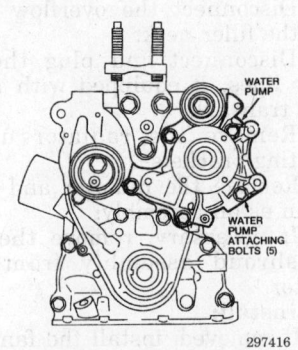

Water pump attaching bolt locations — 2.0L (VIN A) engine

297416

pump on the engine. Install the mounting bolts and tighten to 14–19 ft. lbs. (19–25 Nm).

16. Install the power steering pump lower bracket from the water pump.

17. Install the water pump pulley and bolts. Hold the pulley with the tool and tighten the bolts to 71–88 inch lbs. (8–10 Nm).

18. Install the timing belt.

19. Install the splash shields and tighten the bolts to 71–88 inch lbs. (8–10 Nm).

20. Lower the vehicle and install the cylinder head cover. Tighten the bolts in 2–3 steps to 52–69 inch lbs. (6–7 Nm) in the proper sequence.

21. Place the power steering pump in position. Install the through bolt and tighten to 32–45 ft. lbs. (43–61 Nm). Install the lock bolt and tighten to 23–34 ft. lbs. (31–46 Nm).

22. Connect the PSP switch electrical connector.

23. Install the accessory drive belts and adjust the tension.

24. Install the steering idler pulley shield and tighten the bolts to 61–86 ft. lbs. (7–9 Nm).

25. Connect the negative battery cable.

26. Fill and bleed the cooling system.

27. Run the engine and bring to normal operating temperature. Check for leaks.

2.5L (VIN B) Engine

1. Disconnect the negative battery cable.

2. Drain the cooling system into a suitable container.

3. Remove the timing belt covers and the timing belt.

4. Use pulley removal tool T92C-6312-AH or equivalent, to hold the water pump pulley and remove the bolts. Remove the water pump pulley.

5. Position a drain pan under the water pump.

6. Remove the 3 front engine support insulator mounting bracket bolts.

7. Remove the 5 water pump mounting bolts and remove the water pump.

To install:

8. Clean the mating surfaces of the water pump and the engine block.

9. Install a new O-ring onto the water pump.

10. Install the water pump and torque the bolts 14–18 ft. lbs. (19–25 Nm).

11. Install the 3 engine support mounting bracket bolts.

12. Install the water pump pulley with the bolts. Hold the pulley with the tool and tighten the bolts to 71–88 inch lbs. (8–10 Nm).

13. Install the timing belt and timing belt covers.

14. Connect the negative battery cable.

15. Fill and bleed the cooling system.

16. Run the engine and bring to normal operating temperature. Check for leaks.

Thermostat

REMOVAL AND INSTALLATION

2.0L (VIN A) Engine

1. Disconnect the negative battery cable.

2. Drain the cooling system into a suitable container.

3. Remove the lower radiator hose from the thermostat housing.

4. Remove the 2 thermostat housing mounting bolts and remove the thermostat housing.

5. Remove the thermostat.

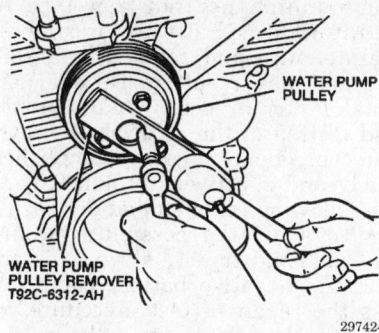

Water Pump Pulley Tool T92C-6312-AH — 2.5L (VIN B) engine

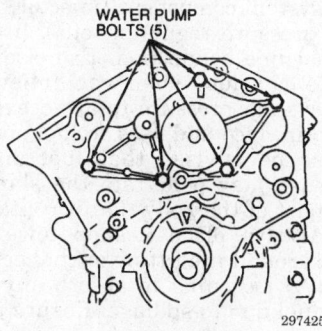

Water pump attaching bolt locations — 2.5L (VIN B) engine

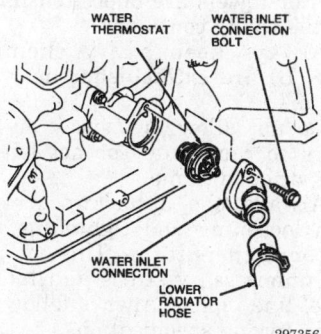

Thermostat and related components — 2.0L (VIN A) engine

To install:

6. Clean the thermostat housing and engine block thermostat housing mating surfaces.

7. Install the thermostat, aligning the tab on the thermostat with the tab on the engine block thermostat housing.

8. Install the thermostat housing and tighten the bolts to 14–18 ft. lbs. (19–25 Nm).

9. Connect the lower radiator hose.

10. Connect the negative battery cable.

11. Fill and bleed the cooling system.

12. Run the engine and bring to normal operating temperature. Check for leaks.

2.5L (VIN B) Engine

1. Disconnect the negative battery cable.

2. Drain the cooling system into a suitable container.

3. Remove the fresh air duct and air cleaner assembly.

4. Remove the lower radiator hose from the coolant inlet pipe.

5. Remove the coolant inlet pipe mounting bolt and pull the coolant inlet pipe away from the thermostat housing.

6. Remove the 2 thermostat housing bolts and remove the thermostat housing. Discard the O-ring.

7. Remove the thermostat.

To install:

8. Clean the thermostat housing and engine block thermostat housing mating surfaces.

9. Install the thermostat, aligning the tab on the thermostat with the tab on the engine block thermostat housing.

10. Install the thermostat housing and tighten the bolts to 14–18 ft. lbs. (19–25 Nm).

11. Install a new thermostat housing O-ring and connect the coolant inlet pipe to the thermostat housing.

12. Install the coolant inlet pipe mounting bolt and tighten to 14–18 ft. lbs. (19–25 Nm).

13. Connect the lower radiator hose to the coolant inlet pipe.

14. Connect the negative battery cable.

15. Fill and bleed the cooling system.

16. Run the engine and bring to normal operating temperature. Check for leaks.

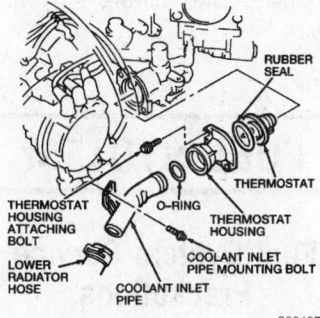

Thermostat and related components — 2.5L (VIN B) engine

Electric Cooling Fan

REMOVAL AND INSTALLATION

2.0L (VIN A) Engine

1. Disconnect the negative battery cable.

2. Drain the cooling system and remove the radiator and fan shroud assembly.

3. On vehicles equipped with automatic transaxles, remove the oil cooler lines.

4. Remove the 4 fan shroud retaining bolts and remove the fan shroud assembly from the radiator.

5. Remove the cooling fan blade retaining clip.

6. Remove the fan shroud insert and remove the cooling fan blade.

7. Remove the 3 cooling fan motor mounting bolts and remove the cooling fan motor.

To install:

8. Position the cooling fan motor and install the 3 motor mounting bolts. Tighten the motor mounting bolts to 16–19 inch lbs. (1.8–2.1 Nm).

9. Position the cooling fan blade and install the fan clip.

10. Install the fan shroud insert.

11. Position the fan shroud to the radiator and tighten the 4 fan shroud attaching bolts to 71–88 inch lbs. (8–10 Nm).

12. If equipped with an automatic transaxle, install the oil cooler tubes.

13. Install the radiator and fill the cooling system.

14. Connect the negative battery cable.

15. Run the engine and check for leaks and proper cooling fan operation.

2.5L (VIN B) Engine

1. Disconnect the negative battery cable.

2. Drain the cooling system and remove the radiator and fan shroud assembly.

3. Remove the 4 fan shroud retaining bolts and remove the fan shroud assembly from the radiator.

4. Remove the cooling fan blade clip and remove the cooling fan blade.

5. Remove the 3 cooling fan motor mounting bolts and remove the cooling fan motor.

6. Repeat the fan removal procedure for the A/C condenser cooling fan blade and motor, if required.

To install:

7. Position the cooling fan motor and install the 3 motor mounting bolts. Tighten the motor mounting bolts to 16–19 inch lbs. (1.8–2.1 Nm).

8. Position the cooling fan blade and install the fan clip.

9. Repeat the installation procedure for the A/C condenser cooling fan motor and blade, if required,

10. Position the fan shroud to the radiator and tighten the 4 fan shroud attaching bolts to 71–88 inch lbs. (8–10 Nm).

11. Install the radiator and fill the cooling system.

12. Connect the negative battery cable.

13. Run the engine and check for coolant leaks.

Cooling System

BLEEDING

2.0L (VIN A) Engine

When the cooling system has been drained, the following procedure should be used to ensure a complete fill.

CAUTION

Never remove the radiator cap when the engine is operating or damage to the engine and/or personal injury may result. Before removing the radiator cap, wait for the radiator to cool down. Wrap a thick cloth around the radiator cap and turn it slowly to the first stop. Stay clear while the pressure is being released. Once all of the pressure is released, finish removing the radiator cap.

1. Check all hose clamps and close the radiator draincock.

2. Place the climate control assembly in the maximum heat position.

3. Fill the radiator to the filler neck seat with a 55/45 mixture of coolant and water.

NOTE: Before operating the engine with the hood open, inspect the engine cooling fan blades for possible cracks or separation.

4. Leave the radiator cap off and run the engine until the thermostat opens and hot coolant is running through the upper radiator hose.

5. Turn the engine off.

NOTE: Coolant level cannot be accurately checked while the engine is running.

6. Add coolant to the radiator as required and install the radiator cap.

7. Add a 55/45 mixture of coolant and water to the F mark on the radiator overflow container.

8. Check the cooling system for leaks.

2.5L (VIN B) Engine

When the cooling system has been drained, the following procedure should be used to ensure a complete fill.

CAUTION

Never remove the radiator cap or the filler cap when the engine is operating or damage to the engine and/or personal injury may result. Before removing a radiator cap or filler cap, wait for the radiator to cool down. Wrap a thick cloth around the radiator cap or filler cap and turn it slowly to the first stop. Stay clear while the pressure is being released. Once all of the pressure is released, finish removing the radiator cap or filler cap.

1. Check all hose clamps and close the radiator draincock.

2. Place the climate control assembly in the maximum heat position.

3. Fill the cooling system to the filler neck seat with a 55/45 mixture of coolant and water.

NOTE: Before operating the engine with the hood open, inspect the engine cooling fan blades for possible cracks or separation.

4. Leave the filler cap off and run the engine until the lower radiator hose feels warm indicating that the thermostat has opened.

5. With the engine idling, add coolant until it reaches the top of the filler neck seat.

6. Turn off the engine.

7. Add a 55/45 mixture of coolant and water to raise the coolant level to the top of the filler neck.

8. Fully install the filler cap and fill the radiator coolant recovery reservoir until it reaches the F mark on the coolant level dipstick.

9. Check the cooling system for leaks.

FUEL SYSTEM

Fuel System Service Precautions

Safety is the most important factor when performing not only fuel system maintenance but any type of maintenance. Failure to conduct maintenance and repairs in a safe manner may result in serious personal injury or death. Maintenance and testing of the vehicle's fuel system components can be accomplished safely and effectively by adhering to the following rules and guidelines.

• To avoid the possibility of fire and personal injury, always disconnect the negative battery cable unless the repair or test procedure requires that battery voltage be applied.

• Always relieve the fuel system pressure prior to disconnecting any fuel system component (injector, fuel rail, pressure regulator, etc.), fitting or fuel line connection. Exercise extreme caution whenever relieving fuel system pressure to avoid exposing skin, face and eyes to fuel spray. Please be advised that fuel under pressure may penetrate the skin or any part of the body that it contacts.

• Always place a shop towel or cloth around the fitting or connection prior to loosening to absorb any excess fuel due to spillage. Ensure that all fuel spillage (should it occur) is quickly removed from engine surfaces. Ensure that all fuel soaked cloths or towels are deposited into a suitable waste container.

• Always keep a dry chemical (Class B) fire extinguisher near the work area.

• Do not allow fuel spray or fuel vapors to come into contact with a spark or open flame.

• Always use a backup wrench when loosening and tightening fuel line connection fittings. This will prevent unnecessary stress and torsion to fuel line piping. Always follow the proper torque specifications.

• Always replace worn fuel fitting O-rings with new. Do not substitute fuel hose or equivalent, where fuel pipe is installed.

Fuel System Pressure

RELIEVING

CAUTION

Fuel injection systems remain under pressure even after the engine has been turned OFF. The fuel system pressure must be relieved before disconnecting any fuel lines. Failure to do so may result in fire and/or personal injury.

1. Relieve the fuel system pressure as follows:

a. Start the engine and let it idle.

b. Locate and remove the fuel pump relay from the main fuse junction panel located next to the left-hand front strut tower.

c. After the engine stalls, turn **OFF** the ignition switch.

d. Install the fuel pump relay.

2. Disconnect the negative battery cable.

Idle Speed

ADJUSTMENT

1993 2.0L (VIN A) Engine

Automatic Transaxle

1. Place the gearshift lever in **PARK** and apply the parking brake.

2. Start the engine and bring to normal operating temperature.

3. Turn **OFF** all electrical loads and accessories.

4. Using a jumper wire, connect the **GND** terminal to the **STI (TEN)** terminal on the Data Link Connector (DLC), located next to the battery.

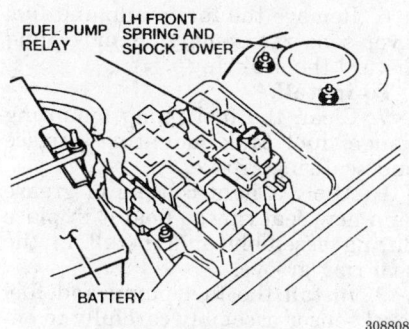

Fuel pump relay location

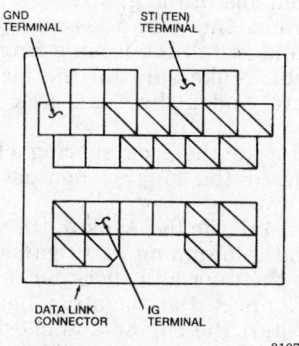

Data link connector terminal locations

5. Connect a suitable tachometer according to the manufacturer's instructions.

6. Observe the idle speed; it should be 650–750 rpm.

NOTE: Do not check the idle speed while the cooling fan is running.

7. If the idle speed is not as specified, adjust it by turning the idle speed adjusting screw.

8. Remove the tachometer and the jumper wire .

Manual Transaxle

1. Set the parking brake and make sure that the transaxle is in **NEUTRAL**.

2. Start the engine and bring to normal operating temperature.

3. Turn the engine **OFF**.

4. Turn **OFF** all accessories.

5. Connect a suitable tachometer according to the manufacturer's instructions.

6. Disconnect the Idle Air Control Bypass Air (IAC BPA) valve connector.

7. Start and run the engine at 2500 rpm for 30 seconds.

8. Let the engine idle and make a note of the idle speed.

9. Turn the idle speed adjusting screw until the idle speed is 650–750 rpm.

10. Turn the engine **OFF**, then start it again and rerun the test to make sure the idle speed is correct.

11. Connect the IAC BPA valve connector and remove the tachometer.

1994–97 2.0L (VIN A) Engine

1. Place the gearshift lever in **P** or **N** and apply the parking brake.

2. Start the engine and bring to normal operating temperature.

3. Turn the ignition switch to the **OFF** position and turn OFF all electrical loads and accessories.

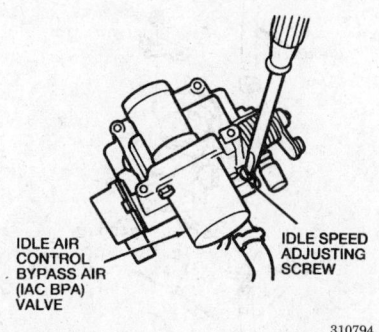

Idle air adjust screw location — 2.0L (VIN A) engine

4. Connect a tachometer following manufacturers instructions.

5. For vehicles equipped with the CD4E transaxle, connect a jumper wire between the **GND** terminal to the **STI (TEN)** terminal on the Data Link Connector (DLC), located next to the battery.

6. Disconnect the Idle Air Control Bypass Air (IAC BPA) valve electrical connector.

7. Start and run the engine at 2500 rpm for 30 seconds.

8. Let the engine idle and make a note of the idle speed.

9. Turn the idle adjusting screw to set the idle between 650–750 rpm.

NOTE: Do not check the idle speed while the cooling fan is running.

10. Turn OFF the engine and rerun the test to verify that the idle speed is correct.

11. Connect the IAC BPA valve.

12. Remove the jumper wire and the tachometer.

2.5L (VIN B) Engine

1. Apply the parking brake.

2. If equipped with a manual transaxle, place the gearshift lever in **NEUTRAL**. If equipped with an automatic transaxle, place the gearshift lever in **PARK**.

3. Start the engine and bring to normal operating temperature.

4. Make sure all electrical loads are turned **OFF**.

5. Connect a suitable tachometer according to the manufacturer's instructions.

6. Connect a jumper wire between the **GROUND** and **STI (TEN)** terminals of the Data Link Connector (DLC), located next to the battery.

7. Observe the idle speed; it should be 650 ± 50 rpm.

NOTE: Do not check the idle speed while the cooling fan is operating.

8. If the idle speed is not as specified, proceed with a check of the ignition timing.

9. Connect a suitable timing light according to the manufacturer's instructions.

10. Aim the timing light at the marks on the crankshaft pulley and timing belt cover and make sure the timing is 10 degrees BTDC ± 1 degree.

11. If the timing is not within specification, loosen the distributor bolts and turn the distributor until the timing is correct. Tighten the bolts to

14–18 ft. lbs. (19–25 Nm) and verify the timing.

12. Remove the jumper wire from between the **GND** and **STI (TEN)** terminals.

13. Verify that the ignition timing is 6–18 degrees BTDC.

14. Reconnect the jumper wire between the **GND** and **STI (TEN)** terminals.

15. Turn the idle speed adjusting screw to set the idle speed to 650 ± 50 rpm.

16. Turn the engine **OFF**.

17. Remove the jumper wire, tachometer and timing light.

Mixture

ADJUSTMENT

2.0L (VIN A) Engine

The air/fuel mixture is not adjustable. The air/fuel mixture is controlled by the engine control system, which consists of various sensors, switches and the Powertrain Control Module (PCM). The various sensors and switches send signals to the PCM regarding engine operating conditions; the PCM then uses the data to determine the timing and opening duration of the fuel injectors.

Fuel Filter

REMOVAL AND INSTALLATION

— CAUTION —
Fuel injection systems remain under pressure even after the engine has been turned OFF. The fuel system pressure must be relieved before disconnecting any fuel lines. Failure to do so may result in fire and/or personal injury.

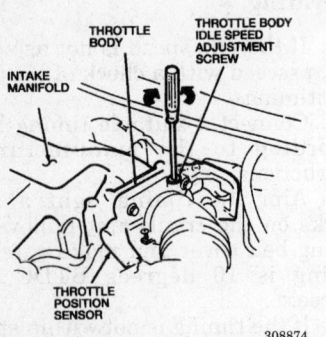

Idle air adjust screw location — 2.5L (VIN B) engine

1. Properly relieve the fuel system pressure as follows:

a. Start the engine and let it idle.

b. Locate and remove the fuel pump relay from the main fuse junction panel located next to the left-hand front strut tower.

c. After the engine stalls, turn **OFF** the ignition switch.

d. Install the fuel pump relay.

2. Disconnect the negative battery cable.

3. If equipped with cruise control, remove the 3 speed control servo nuts and position the servo aside.

4. Remove the 2 fuel filter bracket nuts and remove the fuel tube clamps.

5. Disconnect the fuel lines from both ends of the fuel filter. Plug the lines to prevent leakage.

6. Remove the filter from the mounting bracket.

To install:

7. Position the fuel filter in the mounting bracket.

8. Unplug the fuel lines and place new fuel line clamps onto each fuel line.

9. Install the fuel lines to the fuel filter and position the clamps.

10. Install the 2 bracket nuts and tighten to 71–97 inch lbs. (8–11 Nm).

11. If equipped with cruise control, position the speed control servo and install the 3 mounting nuts.

12. Connect the negative battery cable.

13. Run the engine and check for fuel leaks.

Fuel Pump

REMOVAL AND INSTALLATION

— CAUTION —
Fuel injection systems remain under pressure even after the engine has been turned OFF. The

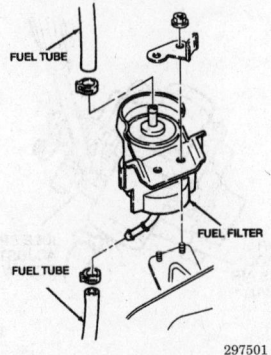

Fuel filter assembly

fuel system pressure must be relieved before disconnecting any fuel lines. Failure to do so may result in fire and/or personal injury.

1. Relieve the fuel system pressure as follows:

a. Start the engine and let it idle.

b. Locate and remove the fuel pump relay from the main fuse junction panel located next to the left-hand front strut tower.

c. After the engine stalls, turn **OFF** the ignition switch.

d. Install the fuel pump relay.

2. Disconnect the negative battery cable.

3. Remove the fuel tank and place it on a bench.

4. Remove any dirt that has accumulated around the fuel pump retaining flange so it will not enter the tank during pump removal and installation.

5. Turn the fuel pump locking ring counterclockwise and remove the locking ring.

— CAUTION —
If the locking ring is tapped around, be sure to use a brass drift and mallet to prevent sparks.

6. Remove the fuel pump and fuel level sensor assembly. Remove and discard the seal ring.

To install:

7. Clean the fuel pump mounting flange, fuel tank mounting surface and seal ring groove.

8. Apply a light coating of grease on a new seal ring to hold it in place during assembly and install in the seal ring groove.

9. Install the fuel pump and fuel level sensor assembly carefully to ensure the filter and sensor arm are not damaged. Make sure the locating keys are in the keyways and the seal ring remains in the groove.

10. Hold the pump assembly in place and install the locking ring finger-tight. Make sure all the locking tabs are under the tank lock ring tabs.

11. Rotate the locking ring clockwise until the ring is against the stops.

12. Install the fuel tank in the vehicle. Add a minimum of 10 gallons of fuel to the tank and check for leaks.

13. Connect the negative battery cable, start the engine and check for proper system operation and for fuel leaks.

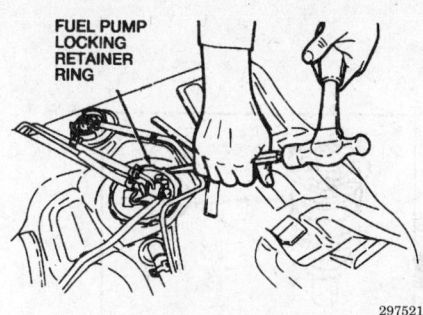

Fuel pump lock ring

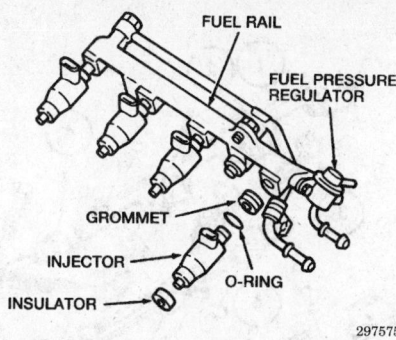

Fuel rail and injector assembly — 2.0L (VIN A) engine

Fuel Injector

REMOVAL AND INSTALLATION

2.0L (VIN A) Engine

— **CAUTION** —

Fuel injection systems remain under pressure even after the engine has been turned OFF. The fuel system pressure must be relieved before disconnecting any fuel lines. Failure to do so may result in fire and/or personal injury.

1. Relieve the fuel system pressure as follows:

    a. Start the engine and let it idle.

    b. Locate and remove the fuel pump relay from the main fuse junction panel located next to the left-hand front strut tower.

    c. After the engine stalls, turn **OFF** the ignition switch.

    d. Install the fuel pump relay.

2. Disconnect the negative battery cable.

3. Label and disconnect the fuel injector wiring harness.

4. Disconnect and plug the fuel lines at the fuel rail.

5. Disconnect the vacuum hose from the fuel pressure regulator.

6. Remove the fuel line mounting bracket bolt.

7. Remove the fuel rail mounting bolts, spacers, insulators and the fuel rail, with the injectors attached.

8. Remove the fuel injectors, grommets and O-rings from the fuel rail. Remove the O-rings from the fuel injectors.

To install:

9. Apply a small amount of clean engine oil to new O-rings and install them and the grommets on the fuel injectors.

10. Install the insulators and injectors on the intake manifold.

11. Install the grommets and the fuel rail onto the injectors.

12. Install the fuel rail attaching bolts and tighten to 14–18 ft. lbs. (19–25 Nm).

13. Connect the vacuum hose to the fuel pressure regulator and the fuel lines to the fuel rail.

14. Install the fuel line mounting bracket and tighten the bolt to 71–97 inch lbs. (8–11 Nm).

15. Connect the fuel injector wiring harness.

16. Connect the negative battery cable and turn the ignition switch **ON** to pressurize the fuel system. Check for leaks and correct as necessary, before starting the engine.

2.5L (VIN B) Engine

— **CAUTION** —

Fuel injection systems remain under pressure even after the engine has been turned OFF. The fuel system pressure must be relieved before disconnecting any fuel lines. Failure to do so may result in fire and/or personal injury.

1. Relieve the fuel system pressure as follows:

    a. Start the engine and let it idle.

    b. Locate and remove the fuel pump relay from the main fuse junction panel located next to the left-hand front strut tower.

    c. After the engine stalls, turn **OFF** the ignition switch.

    d. Install the fuel pump relay.

2. Disconnect the negative battery cable.

3. Remove the air cleaner housing and air ducts.

4. Label and disconnect the fuel injector electrical connectors.

5. Disconnect and plug the fuel supply line. Discard the copper crush washer.

6. Remove the vacuum hose and the fuel return line from the fuel pressure regulator.

7. Remove the fuel pressure regulator mounting bolts and the fuel pressure regulator.

8. Remove the 2 fuel rail (fuel injection supply manifold) mounting bolts and the fuel rails.

9. Remove the 6 fuel injector O-rings.

10. Remove the 7 fuel injector wiring harness attaching screws and remove the fuel injector wiring harness from the fuel rails.

11. Remove and discard the spacer from the top of each fuel injector.

12. Remove the fuel injectors from the fuel rails by rotating back and forth while pulling straight up.

To install:

13. Apply clean engine oil to new O-rings and install them on the injectors. Install the injectors into the fuel rails.

14. Install new spacers on the injectors, then install the fuel injector wiring harness with the screws. Tighten the screws to 31 inch lbs. (3 Nm).

15. Install 6 new insulators and the fuel rails. Install the fuel rail mounting bolts and tighten to 14–18 ft. lbs. (19–25 Nm).

16. Install the fuel pressure regulator and tighten the bolts to 71–97 inch lbs. (8–11 Nm).

17. Install the fuel return line and vacuum hose to the fuel pressure regulator.

18. Using a new copper crush washer, connect the fuel supply line.

19. Connect the fuel injector electrical connectors.

20. Install the air ducts and air cleaner housing.

21. Connect the negative battery cable. Turn the ignition switch **ON** to pressurize the fuel system. Check for fuel leaks and correct as necessary before starting the engine.

EMISSION CONTROLS

Service Interval Lamp

RESETTING

Approximately every 5000 or 7500 miles, (depending on engine application) the word **SERVICE** will appear on the electronic display for the first

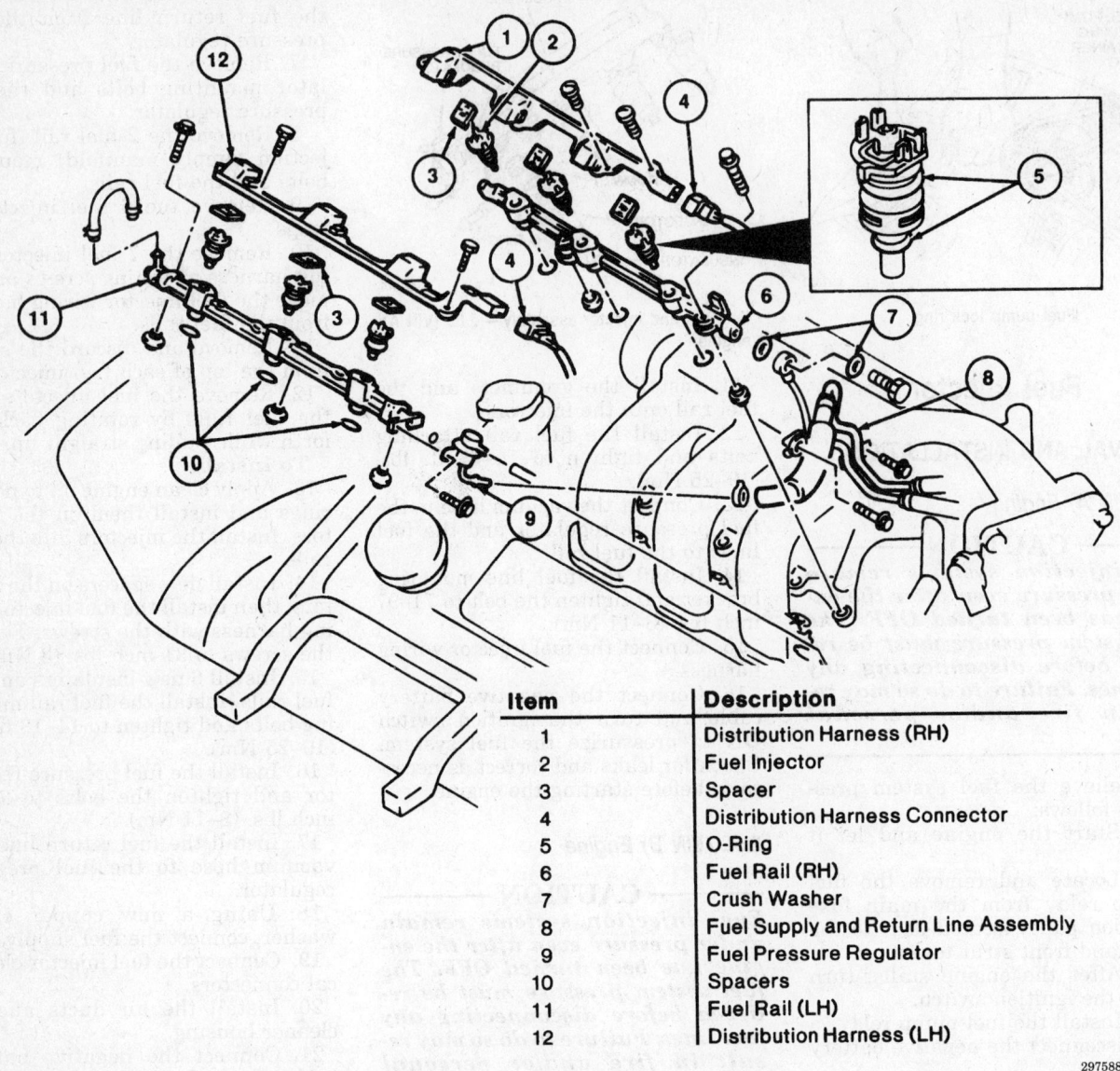

Item	Description
1	Distribution Harness (RH)
2	Fuel Injector
3	Spacer
4	Distribution Harness Connector
5	O-Ring
6	Fuel Rail (RH)
7	Crush Washer
8	Fuel Supply and Return Line Assembly
9	Fuel Pressure Regulator
10	Spacers
11	Fuel Rail (LH)
12	Distribution Harness (LH)

297588

Fuel rail and injector assembly — 2.5L (VIN B) engine

1.5 miles to remind the driver that it is time for the regular vehicle service interval maintenance (i.e. oil change).

To reset the service interval reminder light, with the engine running, press and hold the **SERVICE RESET** button located on the speed alarm keyboard. Hold the button until the **SERVICE** light disappears from the display and 3 audible beeps are heard to verify that the service reminder has been reset.

NOTE: On cars not equipped with the speed alarm keyboard, locate the 3/16 in. hole in the over- head console. Use a small rod or other tool to depress the reset button located behind the hole.

ENGINE MECHANICAL

Engine Assembly

REMOVAL AND INSTALLATION

2.0L (VIN A) Engine

Automatic Transaxle

NOTE: The engine is lifted from the engine compartment, leaving the transaxle in the vehicle.

CAUTION
Fuel injection systems remain under pressure even after the engine has been turned OFF. The fuel system pressure must be relieved before disconnecting any fuel lines. Failure to do so may result in fire and/or personal injury.

1. Relieve the fuel system pressure as follows:
   a. Start the engine and let it idle.
   b. Locate and remove the fuel pump relay from the main fuse junction panel located next to the left-hand front strut tower.
   c. After the engine stalls, turn **OFF** the ignition switch.
   d. Install the fuel pump relay.
2. Disconnect the battery cables, negative cable first. Remove the battery and battery tray.
3. Mark the position of the hood on its hinges and carefully remove the hood.
4. Drain the cooling system and the engine oil into suitable containers. Reinstall the engine oil pan plug.
5. Remove the air intake system.
6. If equipped, remove the A/C compressor and position aside, leaving the refrigerant lines attached. Support the compressor with suitable wire.
7. Label, disconnect and plug the fuel lines at the fuel rail.
8. Label and disconnect the electrical connectors from the distributor, engine coolant temperature sensor, cooling fan temperature sensor (if equipped), water temperature indicator sender unit, throttle position sensor, idle air control valve, idle switch (if equipped), fuel injector harness, EGR solenoid vacuum valve and alternator.

9. Remove the sensor harness tie wrap and retainer clip from the cylinder head cover bracket.
10. Remove the power steering pump drive belt pulley shield. Loosen the power steering pump adjusting bolt, lock bolt and through-bolt. Remove the power steering drive belt. Remove the power steering hose brackets from the cylinder head cover.
11. Remove the power steering pump support and disconnect the Power Steering Pressure (PSP) switch electrical connector. Remove the power steering pump through-bolt and the power steering pump and position aside, leaving the hoses connected.
12. Loosen the alternator adjusting bolt and remove the upper lock bolt.
13. Remove the upper and lower radiator hoses. If equipped, disconnect the cruise control vacuum hose from the back right-hand side of the intake manifold.
14. Disconnect the vacuum hose connecting the Evaporative Emission (EVAP) canister to the metal vacuum tube. If equipped, disconnect the EGR temperature sensor connector.
15. Disconnect the accelerator cable.
16. Disconnect the brake power booster vacuum line from the back left-hand side of the intake manifold.
17. Disconnect the heater hoses at the bulkhead.
18. Remove the upper starter motor mounting bolts.
19. Raise and safely support the vehicle.
20. Remove the splash shields.
21. Remove the intake manifold support bracket.
22. Remove the starter motor. Remove the torque converter access plug.
23. Remove the halfshaft support bearing bracket bolts. Disconnect the oil pressure switch electrical connector.
24. Remove the 4 torque converter-to-flexplate nuts.
25. Remove the 3 engine-to-transaxle bolts and the 2 transaxle-to-engine mounting bolts.
26. Disconnect the Heated Oxygen Sensor (HO2S) electrical connector.
27. Remove and discard the exhaust inlet pipe-to-catalytic converter nuts.
28. Remove the exhaust pipe bracket bolts. Remove and discard the exhaust inlet pipe-to-exhaust manifold nuts and remove the exhaust pipe. Support the remaining exhaust system with mechanics wire.

29. Label and disconnect the remaining alternator wiring. Remove the wiring harness bracket from the back of the alternator, remove the through-bolt and remove the alternator.
30. Disconnect the fuel injector wiring harness from the bottom of the intake manifold.
31. Use crankshaft pulley holder tool T92C-6316-AH or equivalent, to hold the crankshaft pulley and remove the pulley bolt. Remove the crankshaft pulley.
32. Lower the vehicle.
33. Attach an engine sling to the engine lifting eyes and a suitable hoist. Raise the engine slightly and remove the right-hand engine support insulator.
34. Remove the remaining transaxle-to-engine mounting bolts and carefully remove the engine from the vehicle.
35. Remove the flexplate from the crankshaft and mount the engine on a workstand, if required.
To install:
36. Remove the engine from the workstand. Remove the old sealant from the flexplate mounting bolts and bolt holes.
37. If reusing the flexplate bolts, apply silicone sealant to the bolt threads. Install the flexplate and loosely install the bolts.

NOTE: New flexplate mounting bolts come with sealant already on them.

38. Tighten the 6 flexplate bolts in 2–3 steps to 70–75 ft. lbs. (96–103 Nm) in a crisscross pattern.
39. Carefully lower the engine into the vehicle and align it to the transaxle. Make sure that the torque converter studs are aligned with the holes in the flexplate.
40. Install the upper 4 transaxle-to-engine bolts and tighten mounting bolts **A** to 50–73 ft. lbs. (68–99 Nm) for vehicles equipped with the 4EAT transaxle and 66–86 ft. lbs. (90–116 Nm) for vehicles equipped with the CD4E transaxle.
41. Raise the engine slightly and install the right-hand engine support insulator. Tighten the support insulator through-bolt to 63–86 ft. lbs. (86–116 Nm) and support insulator nuts to 54–75 ft. lbs. (74–103 Nm).
42. Remove the engine lifting equipment.
43. Raise and safely support the vehicle.
44. Install the 4 torque converter-to-flexplate nuts and tighten to 32–45 ft. lbs. (44–60 Nm). Rotate the flexplate, as necessary, to gain access to

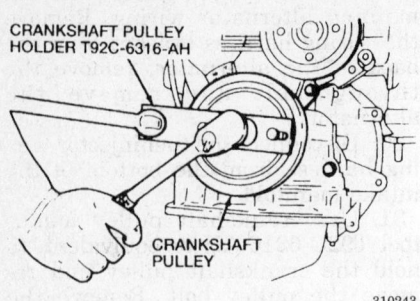

Crankshaft pulley holder — 2.0L (VIN A) engine

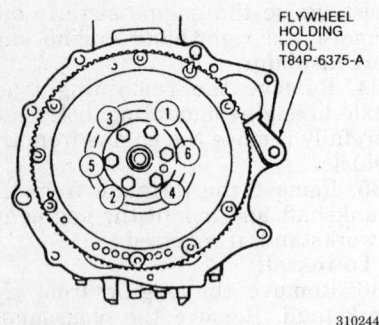

Flexplate bolt torque sequence — 2.0L (VIN A) engine

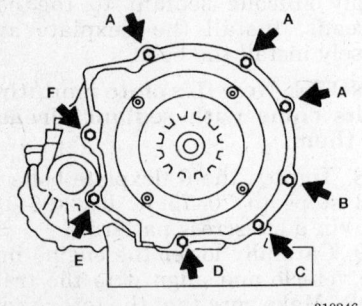

Transaxle and engine mounting bolt identification, 4EAT transaxle — 2.0L (VIN A) engine

all of the nuts. Install the torque converter access plug.

45. Install the remaining transaxle-to-engine mounting bolts. Tighten mounting bolts **B** to 50–73 ft. lbs. (68–99 Nm) for vehicles equipped with the 4EAT transaxle and 28–38 ft. lbs. (38–51 Nm) for vehicles equipped with the CD4E transaxle. Tighten mounting bolt **C** to 28–38 ft. lbs. (38–51 Nm) for vehicles equipped with the 4EAT transaxle and 14–18 ft. lbs. (19–25 Nm) for vehicles

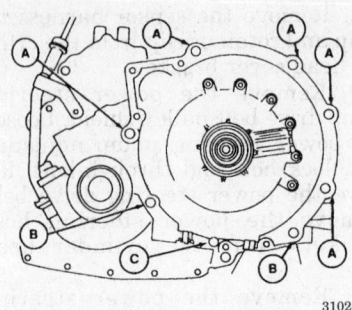

Transaxle and engine mounting bolt identification, CD4E transaxle — 2.0L (VIN A) engine

equipped with the CD4E transaxle. On vehicles equipped with the 4EAT transaxle, install and tighten mounting bolt **D** to 14–18 ft. lbs. (19–25 Nm), **E** to 28–38 ft. lbs. (38–51 Nm) and **F** to 50–73 ft. lbs. (68–99 Nm).

46. Install the alternator and loosely install the through bolt. Connect the alternator wiring and install the harness bracket to the back of the alternator.

47. Install the starter motor and tighten the bolts to 23–34 ft. lbs. (31–46 Nm). Install the intake manifold support bracket and tighten the bolts to 27–38 ft. lbs. (37–52 Nm).

48. Install the halfshaft bracket bearing bolts and tighten, in sequence, to 32–45 ft. lbs. (43–61 Nm).

49. Connect the fuel injector wiring harness to the bottom of the intake manifold.

50. Connect the oil pressure switch electrical connector. If equipped, install the A/C compressor on the mounting bracket and tighten the bolts to 26 ft. lbs. (35 Nm).

51. Install the crankshaft pulley and hold it with a suitable tool. Tighten the pulley bolt to 116–123 ft. lbs. (157–167 Nm).

52. Install the exhaust pipe to the catalytic converter and tighten the new nuts to 27–38 ft. lbs. (37–52 Nm). Attach the exhaust pipe support bracket to the engine and tighten the bolts to 27–38 ft. lbs. (37–52 Nm).

53. Install the new exhaust inlet pipe-to-exhaust manifold nuts and tighten to 27–38 ft. lbs. (37–52 Nm). Install the exhaust hanger insulators and remove the mechanics wire.

54. Connect the HO2S electrical connector.

55. Lower the vehicle.

56. Loosely attach the alternator to the alternator adjuster arm. Install the alternator belt and adjust the tension. Tighten the alternator upper

mounting bolt to 14–18 ft. lbs. (19–25 Nm).

57. Raise and safely support the vehicle.

58. Tighten the alternator through-bolt to 27–38 ft. lbs. (37–52 Nm).

59. Install the splash shields.

60. Lower the vehicle.

61. Install the power steering pump support and loosely install the power steering pump through-bolt and lock bolt. Connect the PSP switch electrical connector and install the power steering pump drive belt.

62. Adjust the power steering drive belt tension, then tighten the through-bolt to 27–38 ft. lbs. (37–52 Nm) and the lock bolt to 14–18 ft. lbs. (19–25 Nm).

63. Install the power steering pump belt shield and tighten the bolts to 71–88 inch lbs. (8–10 Nm). Install the power steering hose brackets to the cylinder head cover and tighten the bolts to 71–88 inch lbs. (8–10 Nm).

64. Connect the heater hoses. If equipped, connect the cruise control vacuum line to the back right-hand side of the intake manifold.

65. Connect the vacuum line connecting the EVAP canister to the metal vacuum tube.

66. Connect the power brake booster vacuum line to the back left-hand side of the intake manifold.

67. Unplug and connect the fuel lines to the fuel rail and all remaining electrical connectors.

68. Connect the sensor harness retainer clip to the cylinder head cover and the tie wrap to the water bypass hose.

69. Install the accelerator cable and the upper and lower radiator hoses.

70. Install the air intake system.

71. Carefully install the hood, aligning the marks that were made during removal.

72. Install the battery tray and battery.

73. Connect the battery cables, negative cable first.

74. Fill the engine with the proper type and quantity of oil.

75. Fill and bleed the cooling system.

76. Run the engine and bring to normal operating temperature. Check for leaks and proper engine operation.

Manual Transaxle

NOTE: The engine and transaxle are lifted from the engine compartment as an assembly.

— CAUTION —

Fuel injection systems remain under pressure even after the engine has been turned OFF. The fuel system pressure must be relieved before disconnecting any fuel lines. Failure to do so may result in fire and/or personal injury.

1. Relieve the fuel system pressure as follows:

   a. Start the engine and let it idle.

   b. Locate and remove the fuel pump relay from the main fuse junction panel located next to the left-hand front strut tower.

   c. After the engine stalls, turn **OFF** the ignition switch.

   d. Install the fuel pump relay.

2. Disconnect the battery cables, negative cable first. Remove the battery and battery tray.

3. Mark the position of the hood on its hinges and carefully remove the hood.

4. Drain the cooling system and the engine oil into suitable containers. Reinstall the engine oil pan plug.

5. Remove the air intake system.

6. Remove the upper and lower radiator hoses and remove the radiator.

7. If equipped, remove the A/C compressor and position aside, leaving the refrigerant lines attached. Support the compressor with suitable wire.

8. Label, disconnect and plug the fuel lines at the fuel rail and set the fuel lines aside.

9. Label and disconnect the electrical connectors from the distributor, coil, engine coolant temperature sensor, coolant temperature gauge sensor, throttle position sensor, idle air control valve, fuel injectors, EGR solenoid vacuum valve, EGR temperature sensor and the alternator.

10. Remove the power steering pump drive belt pulley shield and the power steering pump drive belt. Remove the power steering hose brackets from the cylinder head cover.

11. Loosen the power steering pump drive belt adjusting bolt, lock bolt and through-bolt. Remove the power steering pump support.

12. Disconnect the Power Steering Pressure (PSP) switch electrical connector.

13. Remove the power steering hose bracket bolts and bracket from the cylinder head cover. Remove the power steering pump and position aside, leaving the hoses connected.

14. Loosen the alternator adjusting bolt, remove the upper lock bolt.

15. If equipped, disconnect the cruise control vacuum hose from the back right-hand side of the intake manifold.

16. Disconnect the vacuum line connecting the Evaporative Emission (EVAP) canister to the metal vacuum tube. If equipped, disconnect the EGR temperature sensor connector.

17. Disconnect the accelerator cable. Disconnect the power booster vacuum hose from the back left-hand side of the intake manifold.

18. Disconnect the heater hoses at the bulkhead and remove the upper starter motor mounting bolts. If equipped, disconnect the speed control electrical connector, remove the 2 speed control servo mounting nuts and position the speed control servo aside.

19. Remove the ignition coil.

20. Remove the 2 fuel filter bracket bolts and position the filter and bracket aside.

21. Remove the ignition control module.

22. Remove the ground wire bracket from between the transaxle and rear transaxle support insulator (mount).

23. Remove the rear transaxle support insulator through-bolt and remove the transaxle ground from the top rear of the transaxle.

24. Label and disconnect the Brake On/Off (BOO) switch and Vehicle Speed Sensor (VSS) electrical connectors from the rear of the transaxle.

25. Disconnect and plug the lower clutch slave cylinder tube fitting at the slave cylinder. Pull the spring clips from the slave cylinder line mounting brackets, then remove the hydraulic clutch hose from the lower clutch slave cylinder tube.

26. Label and disconnect the park/neutral position switch from the front of the transaxle.

27. Raise and safely support the vehicle.

28. Remove the splash shields and the front wheels.

29. Remove the 6 crossmember bolts and the crossmember.

30. Remove the 2 lower transaxle support insulator bolts and the lower transaxle support insulator.

31. Remove the 6 rear engine support nuts and 2 bolts and remove the rear engine support.

32. Remove the halfshafts.

33. Install transaxle plug tools T88C-7025-AH or equivalent, into the differential side gears.

NOTE: If the plugs are not installed, the differential side gears may become mispositioned. If the

gears are mispositioned, the differential may have to be removed to reposition them.

34. Remove the intake manifold support bolts and the support. Remove the 3 rear transaxle support bracket bolts and remove the rear transaxle support bracket.

35. Remove the starter.

36. Label and disconnect the oil pressure switch and Heated Oxygen Sensor (HO2S) electrical connectors.

37. Remove and discard the exhaust inlet pipe-to-catalytic converter nuts. Remove the exhaust support bolts. Remove and discard the exhaust inlet pipe-to-exhaust manifold nuts and remove the exhaust pipe. Support the exhaust system with mechanics wire.

38. Remove the control rod-to-support bar stud nut, then disengage the bar from the transaxle. Remove the transaxle shift rod adjustment sleeve through-bolt and nut, then disengage the linkage from the transaxle.

39. Remove the alternator drive belt if not already done. Remove the wiring harness bracket from the rear of the alternator and remove the alternator through-bolt. Label and disconnect the remaining alternator wiring and remove the alternator.

40. Hold the crankshaft pulley with crankshaft pulley holder T92C-6316-AH or equivalent tool and remove the pulley bolt. Remove the crankshaft pulley.

41. If equipped with A/C, remove the A/C compressor bolts and position the compressor aside by supporting the compressor with mechanics wire. Do not remove the A/C lines.

42. Lower the vehicle.

43. Attach an engine sling to the engine lifting eyes and a suitable hoist to the sling. Raise the engine slightly and remove the right-hand engine support insulator.

44. Remove the left-hand engine support insulator nuts and bolt and through-bolt and remove the left-hand engine support insulator.

45. Carefully raise and remove the engine/transaxle assembly from the vehicle.

46. Remove the transaxle-to-engine bolts and the engine-to-transaxle bolts. Separate the transaxle from the engine.

47. Remove the clutch assembly, flywheel and crankshaft rear cover plate. Mount the engine on a workstand.

To install:

48. Remove the engine from the workstand using an engine sling and hoist. Install the crankshaft rear

cover plate and tighten the bolt to 71–88 inch lbs. (8–10 Nm).

49. Install the flywheel and clutch assembly.

50. Install the transaxle on the engine. Install the transaxle-to-engine bolts. Tighten bolts **A** to 66–86 ft. lbs. (90–116 Nm), **B** to 28–38 ft. lbs. (38–51 Nm) and **C** to 14–18 ft. lbs. (19–25 Nm). Install the engine-to-transaxle bolts. Tighten bolt **D** to 28–38 ft. lbs. (38–51 Nm) and bolt **E** to 66–86 ft. lbs. (90–116 Nm).

51. Carefully lower the engine/transaxle assembly into the engine compartment.

52. Install the left-hand transaxle support insulator and tighten the 2 nuts and 1 bolt to 50–68 ft. lbs. (67–93 Nm). Tighten the left-hand transaxle support insulator through-bolt to 63–86 ft. lbs. (86–116 Nm).

53. Raise the engine slightly and install the right-hand engine support insulator. Tighten the right-hand engine support insulator through-bolt to 63–86 ft. lbs. (86–116 Nm) and the right-hand engine support insulator nuts to 54–75 ft. lbs. (74–103 Nm). Remove the engine lifting equipment.

54. Raise and safely support the vehicle.

55. Install the alternator and loosely install the alternator through-bolt. Install the alternator drive belt. Connect the alternator wiring and install the wiring harness bracket to the rear of the alternator.

56. Connect the control rod-to-support bar. Tighten the nut to 28–38 ft. lbs. (38–51 Nm).

57. Connect the shift rod adjustment sleeve to the transaxle with the through-bolt and nut. Tighten the through-bolt to 14–18 ft. lbs. (19–25 Nm).

58. Install the exhaust pipe to the catalytic converter and tighten the new nuts to 27–38 ft. lbs. (37–52 Nm). Attach the exhaust pipe support

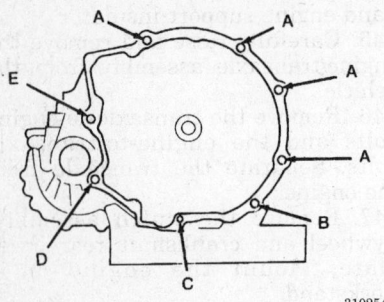

Transaxle and engine mounting bolt identification, MTX transaxle — 2.0L (VIN A) engine

310254

bracket to the engine and tighten the bolts to 27–38 ft. lbs. (37–52 Nm).

59. Install the new exhaust pipe-to-exhaust manifold nuts and tighten to 27–38 ft. lbs. (37–52 Nm).

60. Connect the HO2S and oil pressure sensor electrical connectors.

61. Install the starter motor and tighten the bolts to 23–34 ft. lbs. (31–46 Nm).

62. Install the rear transaxle support insulator and tighten the 3 bolts to 50–68 ft. lbs. (67–93 Nm).

63. Install the intake manifold support bracket and tighten the bolts to 27–38 ft. lbs. (38–52 Nm).

64. Remove the plugs from the differential side gears and install the halfshafts.

65. Install the lower transaxle support insulator and tighten the bolts to 41–59 ft. lbs. (55–80 Nm).

66. Install the rear engine support. Tighten bolts and nuts **B** to 50–68 ft. lbs. (67–93 Nm), nuts **A** to 55–77 ft. lbs. (75–104 Nm) and nuts **C** to 32–44 ft. lbs. (44–60 Nm).

67. Install the crossmember and tighten the 6 bolts to 68–96 ft. lbs. (94–131 Nm).

68. Install the crankshaft pulley. Hold the crankshaft pulley with a suitable tool. Install the pulley bolt and tighten to 116–123 ft. lbs. (157–167 Nm).

69. If equipped, install the A/C compressor and tighten the bolts to 18–26 ft. lbs. (24–35 Nm).

70. Lower the vehicle.

71. Loosely attach the alternator to the alternator adjuster arm. Adjust the alternator drive belt tension. Tighten the alternator upper mounting bolt to 14–18 ft. lbs. (19–25 Nm).

72. Raise and safely support the vehicle.

73. Tighten the alternator through-bolt to 27–38 ft. lbs. (38–52 Nm).

74. Install the splash shields and the wheels. Torque the lug nuts to 65–86 ft. lbs. (88–118 Nm).

75. Lower the vehicle.

76. Connect the park/neutral position switch electrical connector.

77. Remove the plug and install the hydraulic clutch hose to the clutch slave cylinder metal tube. Install the clips to the clutch cylinder hose brackets. Install the hydraulic line fitting on the slave cylinder.

78. Connect the VSS and BOO switch electrical connectors at the rear of the transaxle.

79. Install the ground wire bracket between the transaxle and the rear transaxle support insulator. Install the rear transaxle support insulator through-bolt and tighten to 50–68 ft.

lbs. (67–93 Nm). Install the transaxle ground at the top rear of the transaxle.

80. Install the ignition control module.

81. Install the fuel filter and bracket and tighten the bolts to 71–97 inch lbs. (8–11 Nm).

82. Install the ignition coil.

83. If equipped, install the speed control servo and tighten the nuts. Connect the speed control servo electrical connector.

84. Install the upper starter motor mounting bolts and tighten to 23–34 ft. lbs. (31–46 Nm).

85. Connect the heater hoses and connect the power brake booster vacuum line to the back left-hand side of the intake manifold.

86. Connect the accelerator cable. Connect the EGR temperature sensor, if equipped.

87. Connect the vacuum line between the EVAP canister and the metal vacuum tube. If equipped, connect the cruise control vacuum line to the back right-hand side of the intake manifold.

88. Loosely install the power steering pump through-bolt and lock bolt. Connect the PSP switch electrical connector and install the power steering drive belt.

89. Adjust the power steering drive belt tension, then tighten the through-bolt to 32–45 ft. lbs. (43–61 Nm) and the lock bolt to 23–34 ft. lbs. (31–46 Nm).

90. Install the power steering pump drive belt pulley shield and tighten the bolts to 61–86 inch lbs. (7–9 Nm). Install the power steering hose brackets to the cylinder head cover and tighten the bolts to 71–88 inch lbs. (8–10 Nm).

91. Connect all remaining electrical connectors.

92. Unplug and connect the fuel lines.

93. Install the radiator and the upper and lower radiator hoses.

94. Install the air intake system.

95. Install the battery tray and battery.

96. Install the hood, aligning the marks that were made during removal.

97. Connect the battery cables, negative cable last.

98. Fill the engine with the proper type and quantity of oil.

99. Fill and bleed the cooling system.

100. Bleed the clutch hydraulic system.

101. Run the engine and bring to normal operating temperature.

Check for leaks and proper engine operation.

2.5L (VIN B) Engine

Automatic Transaxle

NOTE: The engine and transaxle are lifted from the engine compartment as an assembly.

——— **CAUTION** ———
Fuel injection systems remain under pressure even after the engine has been turned OFF. The fuel system pressure must be relieved before disconnecting any fuel lines. Failure to do so may result in fire and/or personal injury.

1. Relieve the fuel system pressure as follows:
    a. Start the engine and let it idle.
    b. Locate and remove the fuel pump relay from the main fuse junction panel located next to the left-hand front strut tower.
    c. After the engine stalls, turn **OFF** the ignition switch.
    d. Install the fuel pump relay.
2. Disconnect the battery cables, negative cable first. Remove the battery and battery tray.
3. Mark the position of the hood on its hinges and carefully remove the hood.
4. Drain the cooling system and the engine oil into suitable containers. Reinstall the engine oil pan plug.
5. Remove the engine air intake system.
6. Loosen the A/C and alternator belt tensioner locknut and adjuster bolt and remove the drive belt.
7. Raise and safely support the vehicle.
8. Remove the front wheels and the splash shields.
9. Remove the 6 crossmember bolts and remove the crossmember.
10. Disconnect the front and rear Heated Oxygen Sensor (HO2S) electrical connectors. Remove the exhaust inlet pipe-to-exhaust manifold nuts.
11. Disconnect the oil pressure switch electrical connector, located near the oil filter.
12. Remove the halfshafts.
13. Loosen the power steering and water pump drive belt tensioner locknut and adjuster bolt and remove the drive belt.
14. Remove the 3 power steering pump mounting bolts through the holes in the pump pulley. Remove the power steering hose bracket-to-power steering pump bolt and the pump rear bracket bolt. Secure the pump aside with mechanics wire, leaving the hoses connected.
15. If equipped, remove the 4 A/C compressor mounting bolts and secure the compressor aside with mechanics wire, leaving the refrigerant lines attached. Do not let the compressor hang by the refrigerant lines.
16. Lower the vehicle.
17. Remove the upper and lower radiator hoses and overflow hose.
18. Disconnect the cooling fan electrical connectors. Disconnect and plug the transaxle cooler lines.
19. Remove the 2 radiator hold-down bolts and remove the radiator and cooling fan assembly.

NOTE: Use care when lifting the radiator/cooling fan assembly not to damage the radiator and condenser cooling fins.

20. Label and disconnect the wiring from the alternator and distributor. Remove the 2 A/C and alternator wiring harness retaining bolts, then disconnect the harness from the engine block.
21. Label and disconnect the electrical connectors from the fuel injector harness, vehicle speed sensor, starter motor, throttle position sensor, knock sensor, EGR, idle air control valve, EGR valve position sensor, neutral safety switch, engine coolant temperature sensor, cooling fan engine coolant temperature sensor, temperature gauge sending unit and crank position sensor, as equipped.
22. Label and disconnect the vacuum hoses from the speed control servo, EGR, throttle body, power brake booster, climate control assembly and fuel pressure regulator.
23. Remove the 2 heater hoses from the thermostat housing. Remove the wiring harness grounds.
24. If equipped, disconnect the speed control servo electrical connector. Remove the 2 nuts from the servo bracket and position the speed control servo and bracket aside.
25. Disconnect the fuel supply and return lines and discard the copper crush washer. Remove the 2 fuel line retaining bolts from the fuel line bracket.
26. Disconnect the accelerator cable from the throttle body. Remove the 2 nuts from the fuel filter bracket and position the filter aside, without disconnecting the fuel lines.
27. Remove the spring clip from the shift cable bracket and pull the cable from the switch. Remove the 2 bolts from the cooling fan relay bracket and position the bracket aside.

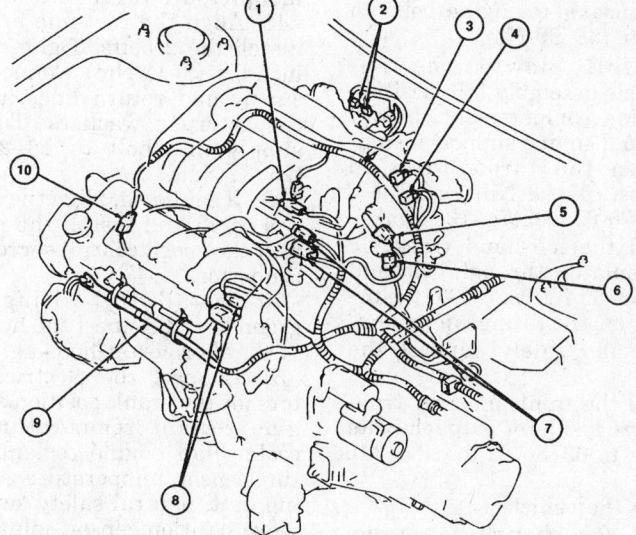

1. Knock sensor (KS)
2. Variable resonance induction system (VRIS) solenoids
3. Vent solenoid
4. EGR solenoid vacuum valve
5. Throttle position sensor
6. Idle air control valve
7. Fuel charging wiring harness
8. Distributor electrical connectors
9. A/C and generator wiring harness
10. Crankshaft position sensor (2)

307425

Engine assembly electrical connector locations — 2.5L (VIN B) engine

28. Raise and safely support the vehicle.

29. Remove the front and rear transaxle support insulator through-bolts.

30. Lower the vehicle.

31. Attach suitable lifting equipment to the engine lifting eyes and remove any slack using an engine hoist attached to the lifting cables.

32. Remove the left-hand transaxle support insulator through-bolt and the right-hand transaxle support insulator through-bolt and 2 nuts. Remove the right-hand engine support insulator from the vehicle.

33. Carefully lift the engine/transaxle assembly from the vehicle.

34. Separate the transaxle from the engine.

35. Remove the flexplate and mount the engine on a suitable engine stand.

To install:

36. Attach suitable lifting cables to the engine lifting eyes and remove any slack on the cables.

37. Remove the engine from the engine stand and install the flexplate. Tighten the flexplate bolts, in 2–3 steps, in sequence to 45–49 ft. lbs. (61–67 Nm).

38. Position the engine to the transaxle and install the retaining bolts. Tighten the engine-to-transaxle bolts and the transaxle-to-engine bolts to 50–73 ft. lbs. (68–99 Nm).

39. Carefully lower the engine/transaxle assembly into position in the engine compartment. Install the right-hand engine support insulator. Tighten the through-bolt to 50–68 ft. lbs. (67–93 Nm) and the 3 nuts to 54–76 ft. lbs. (74–103 Nm).

40. Install the left-hand transaxle support insulator through-bolt and tighten to 63–86 ft. lbs. (86–116 Nm). Remove the engine lifting equipment.

41. Raise and safely support the vehicle.

42. Install the front and rear transaxle support insulator through-bolts and tighten to 63–86 ft. lbs. (86–116 Nm).

43. Lower the vehicle.

44. Align the cooling fan relay bracket and install the 2 bolts. Tighten to 88 inch lbs. (10 Nm).

45. Install the shift cable and retain with the spring clip. Align the fuel filter and install the 2 nuts. Tighten to 71–88 inch lbs. (8–10 Nm).

46. If equipped, connect the vacuum line to the climate control assembly. Connect the vacuum line to the power brake booster.

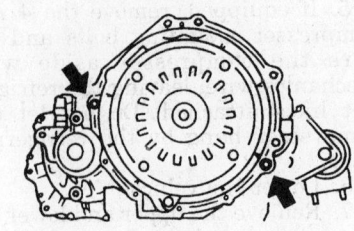

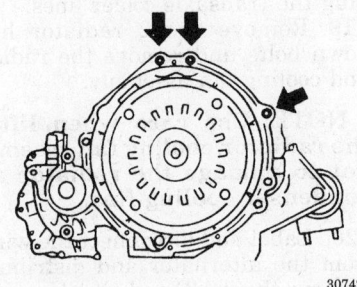

307427

Engine and transaxle retaining bolt locations — 2.5L (VIN B) engine

47. Connect the accelerator cable and vacuum lines to the throttle body. Connect the vacuum line to the fuel pressure regulator.

48. Align the fuel line bracket and install the 2 bolts. Tighten to 71–88 inch lbs. (8–10 Nm). Connect the fuel supply and return lines, using new copper crush washers. Tighten the supply line bolt to 18–25 ft. lbs. (25–34 Nm).

49. If equipped, align the speed control servo and install the nuts. Connect the speed control servo electrical connector.

50. Install the wiring harness grounds and connect the heater hoses to the thermostat housing.

51. Connect the electrical connectors for the crank position sensor, engine coolant temperature sensor, cooling fan engine coolant temperature sensor, temperature gauge sending unit, neutral safety switch, EGR valve position sensor, idle air control valve, EGR, knock sensor, throttle position sensor, starter, vehicle speed sensor and fuel injector harness, as equipped.

52. Connect the vacuum hose to the EGR and if equipped, the speed control servo.

53. Align the A/C and alternator wiring harness and install the 2 bolts. Connect the 2 electrical connectors to the top of the distributor.

54. Carefully install the radiator and cooling fan assembly and connect the cooling fan electrical connectors. Install the 2 radiator hold-down bolts and tighten to 71–88 inch lbs. (8–10 Nm).

55. Unplug and connect the transaxle oil cooler lines and install the upper and lower radiator hoses.

56. Raise and safely support the vehicle.

57. If equipped, install the A/C compressor and tighten the bolts to 28–38 ft. lbs. (38–51 Nm).

58. Position the power steering pump and install the rear bracket bolt. Tighten to 24–34 ft. lbs. (32–46 Nm). Install the power steering hose bracket bolt and tighten to 24–34 ft. lbs. (31–46 Nm).

59. Install the 3 power steering pump bolts through the pulley and tighten to 23–34 ft. lbs. (31–46 Nm). Install the power steering pump drive belt and adjust the tension.

60. Install the halfshafts.

61. Connect the oil pressure switch electrical connector.

62. Install the exhaust inlet pipe to the manifolds and tighten the nuts to 30–41 ft. lbs. (40–55 Nm). Connect the front and rear HO2S electrical connectors.

63. Position the crossmember and install the 6 bolts. Tighten to 69–93 ft. lbs. (94–126 Nm).

64. Install the splash shields and the wheels. Torque the lug nuts to 65–87 ft. lbs. (88–118 Nm).

65. Lower the vehicle.

66. Install the alternator drive belt and adjust the tension. Make sure all electrical connectors and vacuum hose are connected.

67. Install the air intake system.

68. Carefully install the hood, aligning the marks that were made during removal.

69. Install the battery tray and battery. Connect the battery cables, negative cable last.

70. Fill the engine with the proper type and quantity of oil.

71. Add transmission fluid to the transaxle if needed.

72. Fill and bleed the cooling system.

73. Run the engine and bring to normal operating temperature. Check for leaks and proper engine operation.

74. Top off all fluids.

Manual Transaxle

NOTE: The engine and transaxle are lifted from the engine compartment as an assembly.

━━━━ CAUTION ━━━━
Fuel injection systems remain under pressure even after the engine has been turned OFF. The fuel system pressure must be relieved before disconnecting any fuel lines. Failure to do so may result in fire and/or personal injury.

1. Relieve the fuel system pressure as follows:
   a. Start the engine and let it idle.
   b. Locate and remove the fuel pump relay from the main fuse junction panel located next to the left-hand front strut tower.
   c. After the engine stalls, turn **OFF** the ignition switch.
   d. Install the fuel pump relay.
2. Disconnect the battery cables, negative cable first. Remove the battery and battery tray.
3. Mark the position of the hood on its hinges and carefully remove the hood.
4. Drain the cooling system and the engine oil into suitable containers. Reinstall the engine oil pan plug.
5. Remove the air intake system.
6. Raise and safely support the vehicle.
7. Remove the front wheels and the splash shields.
8. Remove the 6 crossmember bolts and remove the crossmember.
9. Remove the 2 bolts and 6 nuts from the rear engine support and remove the rear engine support.
10. Disconnect the front and rear Heated Oxygen Sensor (HO2S) electrical connectors. Remove the exhaust inlet pipe-to-exhaust manifold nuts.
11. Remove the control rod-to-support bar stud nut, then disengage the bar from the transaxle. Remove the transaxle shift rod adjustment sleeve through-bolt and nut, then disengage the linkage from the transaxle.
12. Disconnect the A/C and oil pressure switch electrical connectors. Disconnect and plug the hydraulic line at the slave cylinder, then remove the 2 spring clips from the lower clutch slave cylinder tube.
13. Remove the halfshafts.
14. Remove the 3 bolts and 1 through-bolt from the rear transaxle support bracket and remove the rear transaxle support bracket.

15. Loosen the locknut and adjuster bolt on the power steering pump drive belt tensioner and remove the drive belt. Remove the 3 power steering pump mounting bolts working through the pulley holes.
16. Remove the rear bracket bolt from the power steering pump and secure the pump aside with mechanics wire.
17. If equipped, remove the 4 A/C compressor mounting bolts and secure the compressor aside with mechanics wire, leaving the refrigerant lines attached. Do not let the compressor hang by the refrigerant lines.
18. Remove the power steering hose bracket from the pump. Loosen the alternator drive belt tensioner locknut and adjuster bolt and remove the drive belt.
19. Remove the upper and lower radiator hoses and overflow hose. Disconnect the cooling fan electrical connectors.
20. Remove the 2 radiator hold-down bolts and remove the radiator and the cooling fan as an assembly.

NOTE: Use care when lifting the radiator/cooling fan assembly not to damage the radiator and condenser cooling fins.

21. Label and disconnect the electrical connectors at the alternator. Remove the 2 A/C and alternator wiring harness bolts, then disconnect the harness from the engine block.
22. Label and disconnect the electrical connectors from the distributor, fuel rail, vehicle speed sensor, starter motor, throttle position sensor, engine coolant temperature sensor, cooling fan engine coolant temperature sensor, temperature gauge sending unit, knock sensor, crank position sensor, EGR valve, park/neutral position switch, idle air control valve, EGR valve position sensor and, if equipped, speed control servo.
23. Label and disconnect the vacuum hoses from the speed control servo, if equipped, EGR valve and fuel pressure regulator.
24. Remove the ground-to-engine bracket bolt located near the starter. If equipped, remove the 2 nuts from the speed control servo and position aside.
25. Remove the transaxle ground and backup lamp switch electrical connector from the rear of the transaxle. Remove the starter-to-chassis ground.
26. Disconnect the heater hoses from the engine.

27. Disconnect and plug the fuel supply and return lines. Remove the 2 fuel line retaining bolts and bracket.
28. Label and disconnect the vacuum lines and the accelerator cable from the throttle body. Label and disconnect the vacuum line from the intake manifold to the climate control assembly and the power brake booster vacuum hose.
29. Remove the 2 fuel filter mounting nuts and position the filter aside, leaving the fuel lines connected.
30. Attach suitable engine lifting equipment to the engine lifting eyes and take up any slack.
31. Remove the 2 left-hand transaxle support insulator nuts and through-bolt.
32. Remove the 3 right-hand engine support insulator nuts and the through-bolt and remove the right-hand engine support insulator.
33. Carefully lift the engine/transaxle assembly from the vehicle.
34. Remove the transaxle-to-engine bolts and the engine-to-transaxle bolts. Separate the transaxle from the engine.
35. Remove the clutch assembly, flywheel and crankshaft rear cover plate. Mount the engine on a workstand.

To install:
36. Remove the engine from the workstand using an engine sling and hoist.
37. Install the flywheel and clutch assembly.
38. Align the engine with the transaxle and install the retaining bolts. Tighten the 6 transaxle-to-engine bolts to 50–73 ft. lbs. (68–99 Nm) and the 3 engine-to-transaxle bolts to 28–38 ft. lbs. (38–51 Nm).
39. Carefully lower the engine/transaxle assembly into position in the engine compartment.
40. Install the rear transaxle support insulator. Tighten the 3 nuts to 50–68 ft. lbs. (67–93 Nm) and the through-bolt to 63–86 ft. lbs. (86–116 Nm).
41. Install the left-hand transaxle support insulator. Tighten the 2 nuts to 55–77 ft. lbs. (75–104 Nm) and the through-bolt to 63–86 ft. lbs. (86–116 Nm).
42. Install the right-hand transaxle support insulator. Tighten the 2 nuts to 54–77 ft. lbs. (74–104 Nm) and the through-bolt to 50–68 ft. lbs. (67–93 Nm). Remove the engine lifting equipment.
43. Raise and safely support the vehicle.

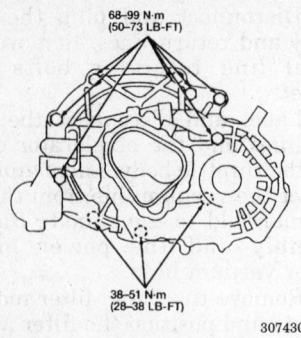

Engine and transaxle mounting bolt locations — 2.5L (VIN B) engine

44. Install the power steering pump and tighten the 3 bolts to 23–34 ft. lbs. (31–46 Nm).

45. Tighten the power steering pump rear bracket bolt to 24–34 ft. lbs. (32–46 Nm). Install the power steering pump drive belt and adjust the tension.

46. If equipped, install the A/C compressor and tighten the 4 bolts to 28–38 ft. lbs. (38–51 Nm). Install the alternator and A/C drive belt and adjust the tension.

47. Connect the control rod-to-support bar. Tighten the nut to 23–33 ft. lbs. (32–46 Nm).

48. Connect the shift rod adjustment sleeve to the transaxle with the through-bolt and nut. Tighten the through-bolt to 12–16 ft. lbs. (16–22 Nm).

49. Install the halfshafts.

50. Install the rear engine support. Tighten bolts and nuts **B** to 50–68 ft. lbs. (67–93 Nm), nuts **A** to 55–77 ft. lbs. (75–104 Nm) and nuts **C** to 32–44 ft. lbs. (44–60 Nm).

51. Install the exhaust inlet pipe to the exhaust manifolds and tighten the nuts to 30–41 ft. lbs. (40–55 Nm).

52. Connect the HO2S electrical connectors.

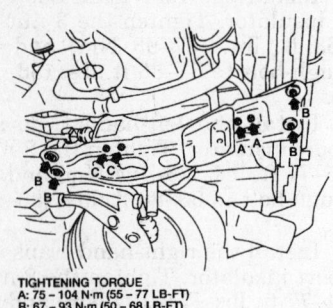

TIGHTENING TORQUE
A: 75–104 N·m (55–77 LB-FT)
B: 67–93 N·m (50–68 LB-FT)
C: 44–60 N·m (32–44 LB-FT)

307431

Rear engine support mounting nut and bolt identification — 2.5L (VIN B) engine

53. Install the crossmember and tighten the 6 bolts to 69–93 ft. lbs. (94–126 Nm).

54. Install the splash shields and the wheels. Tighten the lug nuts to 65–86 ft. lbs. (88–118 Nm).

55. Lower the vehicle.

56. Install the power steering hose bracket bolt to the pump.

57. Connect the electrical connectors to the knock sensor, engine coolant temperature sensor, cooling fan engine coolant temperature sensor, temperature gauge sending unit, crank position sensor, EGR solenoids, EGR valve position sensor, vehicle speed sensor, starter motor, backup lamp switch, fuel injectors, throttle position sensor, distributor and park/neutral position switch.

58. Connect the vacuum hoses to the climate control assembly, located in the right-hand rear of the engine compartment, EGR valve, fuel pressure regulator and, if equipped, speed control servo.

59. Install the starter-to-chassis grounds, the transaxle ground and the ground-to-engine bracket bolt.

60. Connect the heater hoses.

61. Unplug and connect the fuel supply and return lines. Tighten the 2 fuel line bracket bolts to 71–88 inch lbs. (8–10 Nm) and the fuel supply line bolt to 18–25 ft. lbs. (25–34 Nm). Make sure to use new copper crush washers.

62. Install the fuel filter to the bracket and install the 2 nuts. Install the speed control servo with the nuts and connect the electrical connector.

63. Connect the vacuum lines to the throttle body and connect the accelerator cable. Connect the power brake booster vacuum hose.

64. Connect the idle air control valve, oil pressure switch, A/C compressor and alternator electrical connectors. Connect the A/C and alternator harness bracket to the engine.

65. Connect the hydraulic line to the slave cylinder and install the line bracket spring clips.

66. Carefully install the radiator and cooling fan assembly and install the 2 radiator hold-down bolts. Tighten the 2 bolts to 71–88 inch lbs. (8–10 Nm). Connect the cooling fan electrical connectors. Install the upper and lower radiator hoses.

67. Install the air intake system.

68. Check that all vacuum and electrical connectors are installed.

69. Carefully install the hood, aligning the marks that were made during removal.

70. Install the battery tray and the battery. Connect the battery cables, negative cable last.

71. Fill and bleed the cooling system.

72. Fill the engine with the proper type and quantity of oil.

73. Fill the transaxle with the proper type and quantity of oil, if needed.

74. Bleed the clutch hydraulic system.

75. Run the engine and bring to normal operating temperature. Check for leaks and proper operation.

76. Stop the engine and check all fluid levels.

Engine Mounts

REMOVAL AND INSTALLATION

2.0L (VIN A) Engine and 2.5L (VIN B) Engine

Front Transaxle Support Insulator

1. Disconnect the negative battery cable.

2. Support the engine with a suitable engine support.

3. Raise and safely support the vehicle.

4. Remove the 6 crossmember bolts and the crossmember.

5. Remove the screws securing the front splash shields to the rear engine support.

6. Remove the 2 bolts and 6 nuts securing the rear engine support and remove the rear engine support.

7. Remove the front transaxle support insulator through-bolt and the 4 support bracket bolts.

8. If required, remove the front transaxle support bracket and insulator from the vehicle.

To install:

9. If removed, position the front transaxle support bracket and insulator into the vehicle. Install the 4 support bracket bolts and tighten to 28–38 ft. lbs. (38–51 Nm).

10. Install the front transaxle support insulator through-bolt and tighten to 63–86 ft. lbs. (86–116 Nm).

11. Position the rear engine support and install the 2 bolts and 6 nuts. Tighten the rear engine support bolts and nuts to 50–68 ft. lbs. (67–93 Nm). Tighten the rear engine support-to-front mount nuts to 55–77 ft. lbs. (75–104 Nm) and the rear engine support-to-rear mount nuts to 32–44 ft. lbs. (44–60 Nm).

12. Install the splash shield screws.

13. Position the crossmember and install the 6 retaining bolts. Tighten

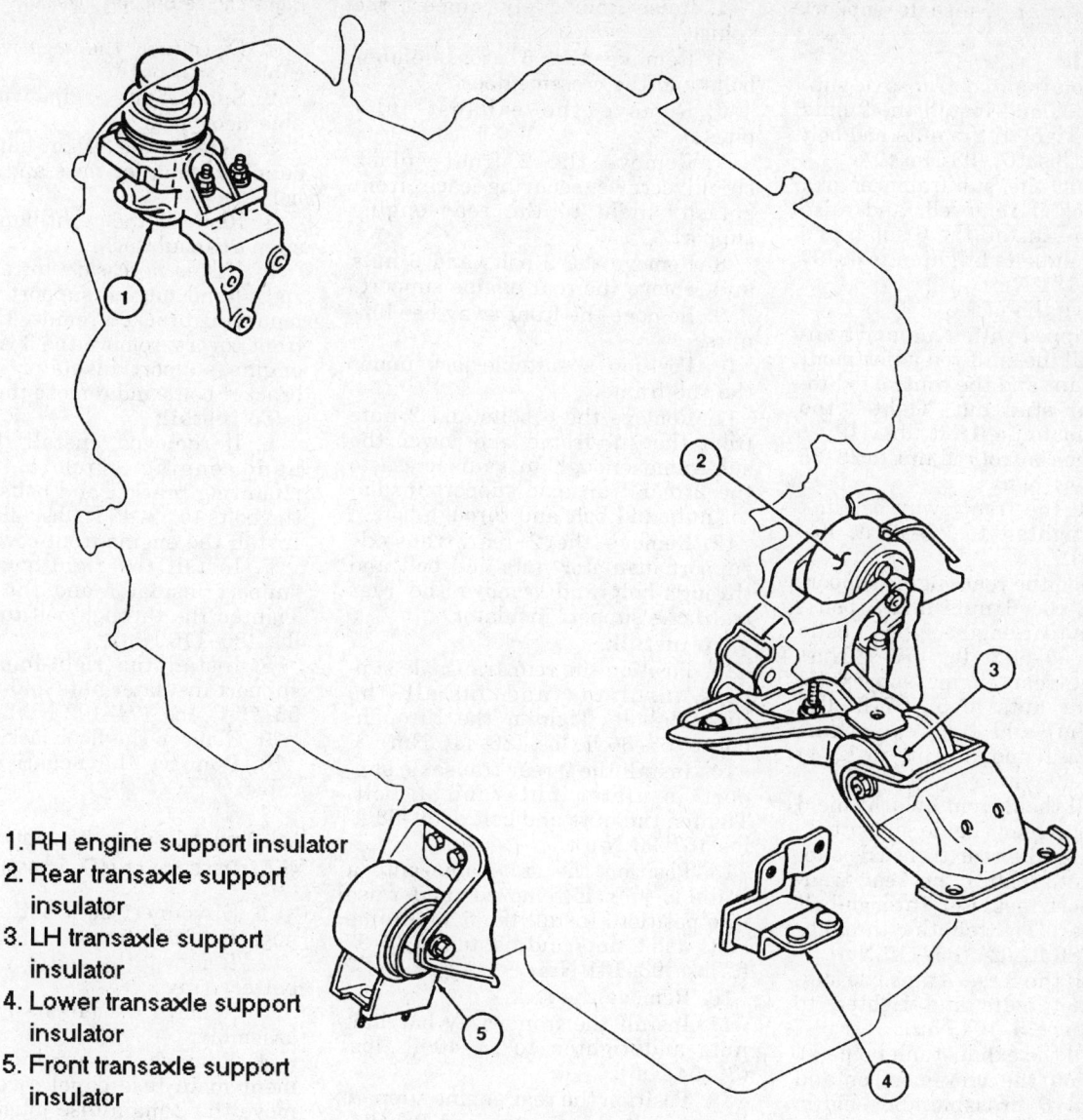

1. RH engine support insulator
2. Rear transaxle support insulator
3. LH transaxle support insulator
4. Lower transaxle support insulator
5. Front transaxle support insulator

303601

Engine and transaxle mount locations

the bolts to 69–97 ft. lbs. (93–131 Nm).

14. Lower the vehicle.

15. Remove the engine support.

16. Connect the negative battery cable.

Rear Transaxle Support Insulator with 4EAT Transaxle or MTX

1. Disconnect the negative battery cable.

2. Remove the ground wire bracket bolt.

3. Remove the lower steering column shaft-to-power steering gear input shaft bolt.

4. Support the engine with a suitable engine support.

5. Raise and safely support the vehicle.

6. Remove the 6 crossmember bolts and the crossmember.

7. Remove the exhaust inlet pipe(s).

8. Remove the 3 rear transaxle support bolts.

9. Remove the rear transaxle support insulator through-bolt and damper.

10. Remove the rear transaxle support bracket.

11. Remove the 2 front splash shield screws securing each front splash shield to the rear engine support.

12. Remove the 2 bolts and 6 nuts and remove the rear engine support.

13. Remove the front sway bar link nuts.

14. If equipped with a manual transaxle, remove the control rod-to-support bar stud nut and the shift rod adjustment sleeve and nut.

15. Position a suitable jack under the sub-frame.

16. Remove the 6 bolts and 2 nuts from the sub-frame and lower the sub-frame enough to gain access to the 2 rear transaxle support insulator nuts and bolt.

17. Remove the 2 rear transaxle support insulator nuts and bolt and

remove the rear transaxle support insulator.

To install:

18. Position the rear transaxle support insulator and install the 2 nuts and 1 bolt. Tighten the nuts and bolt to 50–68 ft. lbs. (67–93 Nm).

19. Position the sub-frame onto a suitable jack if removed, and raise into position. Install the 6 sub-frame bolts and 2 nuts and tighten to 69–97 ft. lbs. (93–131 Nm).

20. Remove the jack.

21. If equipped with a manual transaxle, install the shift rod adjustment sleeve and nut and the control rod-to-support bar stud nut. Tighten the shift rod nut to 14–18 ft. lbs. (19–25 Nm) and the control rod nut to 28–38 ft. lbs. (38–51 Nm).

22. Install the front sway bar link nuts and tighten to 27–40 ft. lbs. (36–54 Nm).

23. Position the rear engine support and install the 6 nuts and 2 bolts. Tighten the rear engine support bolts and nuts to 50–68 ft. lbs. (67–93 Nm). Tighten the rear engine support-to-front mount nuts to 55–77 ft. lbs. (75–104 Nm) and the rear engine support-to-rear mount nuts to 32–44 ft. lbs. (44–60 Nm).

24. Install the 2 front splash shield screws to the rear engine support.

25. Position the rear transaxle support bracket. Install the rear transaxle support insulator through-bolt and damper. Tighten the through-bolt to 63–86 ft. lbs. (86–116 Nm).

26. Install the 3 rear transaxle support bracket bolts and tighten to 55–77 ft. lbs. (75–104 Nm).

27. Install the exhaust inlet pipe(s).

28. Position the crossmember and install the 6 crossmember bolts. Tighten the bolts to 69–97 ft. lbs. (93–131 Nm).

29. Lower the vehicle.

30. Remove the engine support.

31. Install the lower steering column shaft-to-power steering input shaft bolt and tighten to 13–20 ft. lbs. (18–26 Nm).

32. Install the ground wire bracket bolt.

33. Connect the negative battery cable.

Rear Transaxle Support Insulator with CD4E Transaxle

1. Disconnect the negative battery cable.

2. Remove the lower steering column shaft-to-power steering gear input shaft bolt.

3. Support the engine with a suitable engine support.

4. Raise and safely support the vehicle.

5. Remove the 6 crossmember bolts and the crossmember.

6. Remove the exhaust inlet pipe(s).

7. Remove the 2 front splash shield screws securing each front splash shield to the rear engine support.

8. Remove the 2 bolts and 6 nuts and remove the rear engine support.

9. Remove the front sway bar link nuts.

10. Position a suitable jack under the sub-frame.

11. Remove the 6 bolts and 2 nuts from the sub-frame and lower the sub-frame enough to gain access to the 2 rear transaxle support insulator nuts and bolt and through-bolt.

12. Remove the 2 rear transaxle support insulator nuts and bolt and through-bolt and remove the rear transaxle support insulator.

To install:

13. Position the rear transaxle support insulator and install the through-bolt. Tighten the through-bolt to 63–86 ft. lbs. (86–116 Nm).

14. Install the 2 rear transaxle support insulator nuts and 1 bolt. Tighten the nuts and bolt to 50–68 ft. lbs. (67–93 Nm).

15. Position the sub-frame onto a suitable jack if removed, and raise into position. Install the 6 sub-frame bolts and 2 nuts and tighten to 69–97 ft. lbs. (93–131 Nm).

16. Remove the jack.

17. Install the front sway bar link nuts and tighten to 27–40 ft. lbs. (36–54 Nm).

18. Position the rear engine support and install the 6 nuts and 2 bolts. Tighten the rear engine support bolts and nuts to 50–68 ft. lbs. (67–93 Nm). Tighten the rear engine support-to-front mount nuts to 55–77 ft. lbs. (75–104 Nm) and the rear engine support-to-rear mount nuts to 32–44 ft. lbs. (44–60 Nm).

19. Install the 2 front splash shield screws to the rear engine support.

20. Install the exhaust inlet pipe(s).

21. Position the crossmember and install the 6 crossmember bolts. Tighten the bolts to 69–97 ft. lbs. (93–131 Nm).

22. Lower the vehicle.

23. Remove the engine support.

24. Install the lower steering column shaft-to-power steering input shaft bolt and tighten to 13–20 ft. lbs. (18–26 Nm).

25. Connect the negative battery cable.

Right Engine Support Insulator

1. Disconnect the negative battery cable.

2. Support the engine with a suitable floor jack.

3. Remove the 3 right-hand engine support insulator nuts and through-bolt.

4. Remove the right-hand engine support insulator.

5. If it is necessary to remove the right-hand engine support insulator mounting bracket, remove the engine front covers, remove the 3 right-hand engine support insulator mounting bracket bolts and remove the bracket.

To install:

6. If removed, install the right-hand engine support insulator mounting bracket and bolts. Tighten the bolts to 32–45 ft. lbs. (43–61 Nm). Install the engine front covers.

7. Install the right-hand engine support insulator and through-bolt. Tighten the through-bolt to 63–86 ft. lbs. (86–116 Nm).

8. Install the right-hand engine support insulator nuts and tighten to 55–75 ft. lbs. (74–102 Nm).

9. Remove the floor jack.

10. Connect the negative battery cable.

Left Transaxle Support Insulator with 4EAT Transaxle or MTX Transaxle

1. Disconnect both battery cables, negative cable first.

2. Remove the battery and the battery tray.

3. Remove the engine air cleaner assembly.

4. Remove the engine compartment main fuse panel cover and remove the 2 main fuse junction panel cable nuts and the cables.

5. Remove the 2 main fuse junction panel nuts. Pry the Data Link Connector (DLC) loose from the mounting bracket and set aside.

6. Remove the main fuse junction panel bracket nut and slide the in-line electrical connector off of the main fuse junction panel. Position the main fuse junction panel out of the way.

7. If equipped, remove the anti-lock relay bracket nut. Remove the 2 fuel filter bracket nuts.

8. Support the transaxle with a suitable floor jack.

9. Remove the engine control sensor wiring harness nut and lift the harness from the left-hand transaxle support insulator.

10. Remove the 2 left-hand transaxle support insulator nuts, bolt and through-bolt.

11. Remove the left-hand transaxle support insulator from the vehicle.

12. If it is necessary to remove the left-hand transaxle support bracket, begin by removing the 2 engine control sensor wiring harness-to-left-hand transaxle support bracket bolts. Remove the 4 left-hand transaxle support bracket bolts and remove the left-hand transaxle support bracket.

To install:

13. If removed, position the left-hand transaxle support bracket and install the 4 bolts. Tighten the bolts to 44–59 ft. lbs. (59–80 Nm). Install the 2 engine control sensor wiring harness-to-left-hand transaxle support bracket bolts and tighten securely.

14. Position the left-hand transaxle support insulator. Install the 2 left-hand transaxle support insulator nuts, bolt and through-bolt. Tighten the through-bolt to 63–86 ft. lbs. (86–116 Nm). Tighten the 2 nuts and 1 bolt to 50–68 ft. lbs. (67–93 Nm)

15. Position the engine control sensor wiring harness and nut to the left-hand transaxle support insulator and tighten.

16. Install the 2 fuel filter bracket nuts. Tighten the nuts to 71–97 inch lbs. (8–11 Nm).

17. If equipped, install the anti-lock relay bracket. Tighten the nut to 81–113 inch lbs. (9–13 Nm).

18. Position the main fuse panel and install the in-line electrical connector. Install the main fuse panel nut and tighten.

19. Place the DLC in position and secure. Install the 2 main fuse panel nuts and tighten. Install the 2 main fuse panel cables and tighten the nuts.

20. Install the fuse panel cover.

21. Install the engine air cleaner assembly.

22. Install the battery tray and the battery.

23. Connect the battery cables, negative cable last.

Left Transaxle Support Insulator with CD4E Transaxle

1. Disconnect both battery cables, negative cable first.

2. Remove the battery and the battery tray.

3. Remove the engine air cleaner assembly.

4. Remove the 2 fuel filter bracket nuts.

5. Remove the 2 ignition coil nuts and move the ignition coil aside.

6. If equipped, remove the 3 speed control servo nuts and position the speed control servo aside.

7. Remove the 3 ignition coil mounting strap bolts. Disconnect the wiring harness clips and remove the ignition coil straps.

8. Support the transaxle with a suitable floor jack.

9. Remove 2 nuts and 2 bolts from the left-hand transaxle support insulator and remove the left-hand transaxle support insulator through-bolt.

10. Remove the left-hand transaxle support insulator.

11. If required, remove the 4 left-hand transaxle support bracket bolts and remove the left-hand transaxle support bracket.

To install:

12. If removed, position the left-hand transaxle support bracket and install the 4 bolts. Tighten the bolts to 44–59 ft. lbs. (59–80 Nm).

13. Install the left-hand transaxle support insulator and through-bolt. Tighten the through-bolt to 63–86 ft. lbs. (86–116 Nm).

14. Install the 2 nuts and 2 bolts from the left-hand transaxle support insulator. Tighten the 2 nuts to 12–17 ft. lbs. (16–23 Nm) and the 2 bolts to 28–38 ft. lbs. (38–51 Nm).

15. Remove the floor jack.

16. Position the ignition coil mounting straps and connect the wiring harness clips. Install the 3 ignition coil mounting strap bolts and secure.

17. If equipped, position the speed control servo and nuts. Tighten the 3 nuts securely.

18. Position the ignition coil and 2 nuts. Tighten the nuts to 71–88 inch lbs. (8–10 Nm).

19. Install the 2 fuel filter bracket nuts and tighten to 71–97 inch lbs. (8–11 Nm).

20. Install the engine air cleaner assembly.

21. Install the battery tray and the battery.

22. Connect the battery cables, negative cable last.

Lower Transaxle Support Insulator

1. Disconnect the negative battery cable.

2. Raise and safely support the vehicle.

3. Remove the 2 lower transaxle support insulator bolts.

4. Remove the 2 lower transaxle support insulator nuts.

5. Remove the lower transaxle support insulator.

To install:

6. Position the lower transaxle support insulator and install the 2 lower transaxle support insulator nuts. Tighten the nuts to 32–44 ft. lbs. (44–60 Nm).

7. Install the 2 lower transaxle support insulator bolts and tighten to 41–44 ft. lbs. (55–80 Nm).

8. Lower the vehicle.

9. Connect the negative battery cable.

Rear Transaxle Support Bracket with CD4E Transaxle

1. Disconnect the negative battery cable.

2. Remove the transaxle.

3. Remove the 3 rear transaxle support bracket bolts and remove the rear transaxle support bracket from the transaxle.

To install:

4. Position the rear transaxle support bracket to the transaxle and install the 3 rear transaxle support bracket bolts. Tighten the bolts to 55–77 ft. lbs. (75–104 Nm).

5. Install the transaxle.

6. Connect the negative battery cable.

Front Engine Support Damper

1. Disconnect the negative battery cable.

2. Raise and safely support the vehicle.

3. Remove the 6 crossmember bolts and remove the crossmember.

4. Remove the engine damper nut and remove the front engine support damper from the crossmember.

To install:

5. Position the front engine support damper to the crossmember. Install and tighten the engine damper nut.

6. Position the crossmember and install the 6 bolts. Tighten the bolts to 69–97 ft. lbs. (93–131 Nm).

7. Lower the vehicle.

8. Connect the negative battery cable.

Cylinder Head

REMOVAL AND INSTALLATION

2.0L (VIN A) Engine

1993–94 Vehicles

NOTE: The cylinder head bolts are a torque-to-yield design and cannot be reused. Before beginning this procedure, make sure new cylinder head bolts are available.

CAUTION

Fuel injection systems remain under pressure even after the engine has been turned OFF. The fuel system pressure must be relieved before disconnecting any fuel lines. Failure to do so may result in fire and/or personal injury.

1. Relieve the fuel system pressure as follows:

   a. Start the engine and let it idle.

   b. Locate and remove the fuel pump relay from the main fuse junction panel located next to the left-hand front strut tower.

   c. After the engine stalls, turn OFF the ignition switch.

   d. Install the fuel pump relay.

2. Disconnect the negative battery cable.

3. Drain the engine cooling system into a suitable container.

4. Remove the air intake system.

5. Remove the power steering hose bracket bolts from the cylinder head cover.

6. Disconnect the Power Steering Pressure (PSP) switch.

7. Remove the 2 power steering pump idler pulley shield bolts and the shield.

8. Remove the accessory drive belts.

9. Remove the power steering pump through-bolt and lockbolt and secure the pump aside with mechanics wire, leaving the hoses attached.

10. Remove the exhaust manifold.

11. Label and disconnect the spark plug wires from the spark plugs.

12. Disconnect the hoses from the cylinder head cover and loosen the cover bolts in sequence, in 2–3 steps. Remove the cylinder head cover.

13. Remove the timing belt covers and remove the timing belt.

14. Remove the 2 intake manifold support bolts and the intake manifold support.

15. Label and disconnect the distributor/coil connectors, engine coolant temperature sensor, temperature gauge sensor and if equipped, the cooling fan temperature sensor .

16. Remove the 4 retaining bolts and the coolant temperature sensor housing from the back of the cylinder head.

17. Disconnect the fuel supply and return lines and tag for reassembly. Plug and position the fuel lines aside.

18. Label, disconnect and move aside the following electrical connectors:

- Idle switch
- Throttle position sensor
- Idle air control valve
- Fuel injector harness wiring
- EGR solenoid vacuum valve
- Alternator

19. Disconnect the vacuum hose from the vacuum fitting on the right-hand side of the intake manifold, emissions canister and the brake vacuum booster.

20. Disconnect the accelerator cable.

21. Remove the distributor.

22. Remove the camshafts.

23. Loosen the cylinder head bolts in 3 steps in the proper sequence.

24. Remove the cylinder head bolts, the cylinder head and the cylinder head gasket.

25. If needed, remove the intake manifold from the cylinder head.

26. Clean all gasket mating surfaces. Inspect the cylinder head for damage, cracks, and fluid leakage. Check the head gasket surface for distortion (warpage) using a straight-edge and feeler gauge. Maximum allowable distortion is 0.004 in. (0.10mm).

To install:

27. If removed, install the intake manifold using a new gasket.

28. Position a new cylinder head gasket on the cylinder block and carefully install the cylinder head.

29. Install new cylinder head bolts.

30. Tighten the cylinder head bolts in the proper sequence as follows:

   a. Tighten each bolt to 10 ft. lbs. (13 Nm).

   b. Tighten each bolts again to 16 ft. lbs. (22 Nm).

   c. Paint a mark on the socket or the edge of each cylinder head bolt to use as a reference.

   d. Using the same torque sequence, tighten each bolt 90 degrees ± 5 degrees.

   e. Use the same sequence and tighten each bolt an additional 90 degrees ± 5 degrees.

31. Install the valve lifters, if removed and the camshafts.

NOTE: Make sure that none of the camshaft lobes are located directly on the hydraulic valve lifters when tightening the camshaft cap bolts.

32. Install the distributor.

33. Connect the brake booster vacuum hose, emissions canister vacuum hose and the vacuum hose to the right-hand side of the intake manifold.

34. Connect the accelerator cable.

35. Connect the following engine wiring connectors:

- Idle switch
- Throttle position sensor
- Idle air control valve
- Fuel injector harness wiring
- EGR solenoid vacuum valve
- Alternator

36. Unplug and connect the fuel supply and return lines.

37. Connect the distributor/coil electrical connectors.

38. Install the timing belt and the timing belt covers.

39. Install the intake manifold support and bolts. Tighten the support bolts to 27–38 ft. lbs. (37–52 Nm).

40. Install a new cylinder head cover gasket on the cover. Apply sealant to the cylinder head surface in the area adjacent to the front camshaft caps, then install the cover. Tighten the bolts in 2 steps, in sequence to 52–69 inch lbs. (6–7 Nm).

41. Connect the hoses to the cylinder head cover.

42. Connect the spark plug wires.

43. Install the exhaust manifold.

44. Install the alternator belt and adjust the tension.

45. Loosely install the power steering pump through-bolt and lockbolt.

46. Connect the PSP switch electrical connector.

47. Install the power steering pump drive belt and adjust the tension. Tighten the pump through-bolt to 32–45 ft. lbs. (43–61 Nm) and the lockbolt to 23–34 ft. lbs. (31–46 Nm).

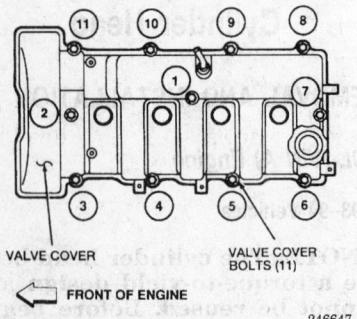

Cylinder head cover bolt removal
sequence — 2.0L (VIN A) engine

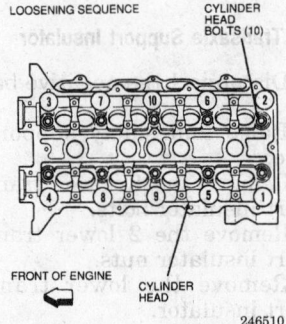

Cylinder head bolt loosening
sequence — 2.0L (VIN A) engine

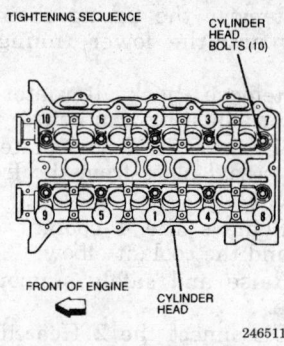

TIGHTENING SEQUENCE CYLINDER HEAD BOLTS (10)

FRONT OF ENGINE CYLINDER HEAD

246511

Cylinder head bolt tightening sequence — 2.0L (VIN A) engine

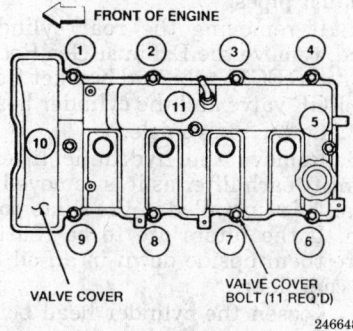

FRONT OF ENGINE

VALVE COVER VALVE COVER BOLT (11 REQ'D)

246648

Cylinder head cover bolt tightening sequence — 2.0L (VIN A) engine

48. Install the power steering pump idler pulley and retaining bolts. Tighten the shield retaining bolts to 61–86 inch lbs. (7–9 Nm).

49. Install the power steering hose brackets to the cylinder head cover. Tighten to 71–88 inch lbs. (8–10 Nm).

50. Install the coolant temperature sensor housing with a new gasket. Tighten the 4 bolts to 14–18 ft. lbs. (19–25 Nm).

51. Connect the engine coolant temperature sensor, temperature gauge sensor and if equipped, the cooling fan temperature sensor electrical connectors .

52. Install the air intake system.

53. Connect the negative battery cable.

54. Fill and bleed the cooling system.

55. Run the engine and check for leaks and proper operation.

56. Adjust the ignition timing if needed.

1995–97 Vehicles

NOTE: Before beginning this procedure, make sure new cylinder head bolts are available, if needed.

— CAUTION —

Fuel injection systems remain under pressure even after the engine has been turned OFF. The fuel system pressure must be relieved before disconnecting any fuel lines. Failure to do so may result in fire and/or personal injury.

---

1. Relieve the fuel system pressure as follows:

    a. Start the engine and let it idle.

    b. Locate and remove the fuel pump relay from the main fuse junction panel located next to the left-hand front strut tower.

    c. After the engine stalls, turn **OFF** the ignition switch.

    d. Install the fuel pump relay.

2. Disconnect the negative battery cable.

3. Drain the engine cooling system into a suitable container.

4. Remove the air intake system.

5. Remove the power steering hose bracket bolts from the cylinder head cover.

6. Disconnect the Power Steering Pressure (PSP) switch.

7. Remove the 2 power steering pump idler pulley shield bolts and the shield.

8. Remove the accessory drive belts.

9. Remove the power steering pump through-bolt and lock bolt and secure the pump aside with mechanics wire, leaving the hoses attached.

10. Remove the exhaust manifold.

11. Label and disconnect the spark plug wires from the spark plugs.

12. Disconnect the hoses from the cylinder head cover and loosen the cover bolts in sequence, in 2–3 steps. Remove the cylinder head cover.

13. Remove the timing belt covers and remove the timing belt.

14. Remove the 2 intake manifold support bolts and the intake manifold support.

15. Label and disconnect the wiring from the distributor/coil connectors, engine coolant temperature sensor and temperature gauge sensor.

16. Remove the 4 retaining bolts and the coolant temperature sensor housing from the back of the cylinder head.

17. Disconnect the fuel supply and return lines and tag for reassembly. Plug and position the fuel lines aside.

18. Label, disconnect and move aside the following electrical connectors:

    • Idle switch
    • Throttle position sensor
    • Idle air control valve
    • Fuel injector harness wiring
    • EGR solenoid vacuum valve
    • Alternator

19. Disconnect the vacuum hose from the vacuum fitting on the right-hand side of the intake manifold, emissions canister and the brake vacuum booster.

20. Disconnect the accelerator cable.

21. Remove the distributor.

22. Remove the camshafts.

23. Loosen the cylinder head bolts in 3 steps in the proper sequence.

24. Remove the cylinder head bolts, the cylinder head and the cylinder head gasket.

25. If necessary, remove the intake manifold from the cylinder head.

26. Clean all gasket mating surfaces. Inspect the cylinder head for damage, cracks, and fluid leakage. Check the head gasket surface for distortion (warpage) using a straight-edge and feeler gauge. Maximum allowable distortion is 0.004 in. (0.10mm).

To install:

27. If removed, install the intake manifold using a new gasket.

28. Position a new cylinder head gasket on the cylinder block and carefully install the cylinder head.

29. Measure the cylinder head bolts to determine if they are reusable or if they are stretched beyond use. If the cylinder head bolt is longer than 4 inches (105.5 mm), it is stretched and must be replaced. Measurements are taken from under the shoulder to the end of the threads.

30. Tighten the cylinder head bolts in the proper sequence as follows:

    a. Tighten each bolt to 10 ft. lbs. (13 Nm).

    b. Tighten each bolts again to 16 ft. lbs. (22 Nm).

    c. Paint a mark on the socket or the edge of each cylinder head bolt to use as a reference.

    d. Using the same torque sequence, tighten each bolt 90 degrees ± 5 degrees.

    e. Use the same sequence and tighten each bolt an additional 90 degrees ± 5 degrees.

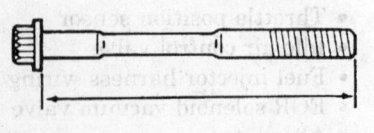

302206

Measuring the cylinder head bolt length — 1995–97 2.0L (VIN A) engine

31. Install the valve lifters, if removed and the camshafts.

NOTE: Make sure that none of the camshaft lobes are located directly on the hydraulic valve lifters when tightening the camshaft cap bolts.

32. Install the distributor.

33. Connect the brake booster vacuum hose, emissions canister vacuum hose and the vacuum hose to the right-hand side of the intake manifold.

34. Connect the accelerator cable.

35. Connect the following engine wiring connectors:
- Idle switch
- Throttle position sensor
- Idle air control valve
- Fuel injector harness wiring
- EGR solenoid vacuum valve
- Alternator

36. Unplug and connect the fuel supply and return lines.

37. Connect the distributor/coil electrical connectors.

38. Install the timing belt and the timing belt covers.

39. Install the intake manifold support and bolts. Tighten the support bolts to 28–38 ft. lbs. (38–51 Nm).

40. Install a new cylinder head cover gasket on the cover. Apply sealant to the cylinder head surface in the area adjacent to the front camshaft caps, then install the cover. Tighten the bolts in 2 steps, in sequence, to 52–69 inch lbs. (6–7 Nm).

41. Connect the hoses to the cylinder head cover.

42. Connect the spark plug wires.

43. Install the exhaust manifold.

44. Install the alternator belt and adjust the tension.

45. Loosely install the power steering pump through-bolt and lock bolt.

46. Connect the PSP switch electrical connector.

47. Install the power steering pump drive belt and adjust the tension. Tighten the pump through-bolt to

32–45 ft. lbs. (43–61 Nm) and the lock bolt to 23–34 ft. lbs. (31–46 Nm).

48. Install the power steering pump idler pulley and retaining bolts. Tighten the shield retaining bolts to 61–86 inch lbs. (7–9 Nm).

49. Install the power steering hose brackets to the cylinder head cover. Tighten to 71–88 inch lbs. (8–10 Nm).

50. Install the coolant temperature sensor housing with a new gasket. Tighten the 4 bolts to 14–18 ft. lbs. (19–25 Nm).

51. Connect the engine coolant temperature sensor and temperature gauge sensor electrical connectors .

52. Install the air intake system.

53. Connect the negative battery cable.

54. Fill and bleed the cooling system.

55. Run the engine and check for leaks and proper operation.

56. Adjust the ignition timing, if necessary.

2.5L (VIN B) Engine

1993–94 Vehicles

NOTE: The cylinder head bolts are a torque-to-yield design and cannot be reused. Before beginning this procedure, make sure new cylinder head bolts are available.

─── **CAUTION** ───

Fuel injection systems remain under pressure even after the engine has been turned OFF. The fuel system pressure must be relieved before disconnecting any fuel lines. Failure to do so may result in fire and/or personal injury.

1. Relieve the fuel system pressure as follows:
   a. Start the engine and let it idle.
   b. Locate and remove the fuel pump relay from the main fuse junction panel located next to the left-hand front strut tower.
   c. After the engine stalls, turn **OFF** the ignition switch.
   d. Install the fuel pump relay.

2. Disconnect the negative battery cable.

3. Drain the engine cooling system into a suitable container.

4. Remove the timing belt covers and the timing belt.

5. Remove the intake manifold.

6. Disconnect the ventilation pipe from the left cylinder head cover, remove the bolts and remove both cylinder head covers.

7. Remove the camshafts.

8. Remove the timing chain tensioner and the lower timing belt idler.

9. Remove the 2 alternator-to-alternator adjusting arm bolts.

10. Remove the 3 seal plate bolts and the seal plate from the front of the engine.

11. Remove the 4 coolant elbow bolts and the coolant elbow.

12. Raise and safely support the vehicle.

13. Disconnect the 2 Heated Oxygen Sensor (HO2S) electrical connectors. Remove the exhaust pipe-to-exhaust manifold nuts and lower the exhaust pipes.

14. If removing the rear cylinder head, remove the Exhaust Gas Recirculation (EGR) tube and bracket from the EGR valve and the cylinder head.

15. Lower the vehicle.

16. Remove the hydraulic lifters. Identify each lifter as it is removed so it can be reinstalled in the same position. If the lifters are to be reused, store them upside down in an oil filled container.

17. Loosen the cylinder head bolts, in 2–3 steps in the correct removal sequence.

18. Remove the cylinder head bolts and washers and remove the cylinder heads.

19. If required, remove the exhaust manifolds, manifold shields and gaskets.

20. Clean all gasket mating surfaces. Inspect the cylinder head(s) for damage, cracks, and water and oil leakage.

21. Check the cylinder head gasket surface for distortion using a straight-edge and feeler gauge. Maximum allowable distortion is 0.004 in. (0.10mm).

To install:

22. If removed, install the exhaust manifolds using new gaskets. Tighten the exhaust manifold nuts and bolts to 14–18 ft. lbs. (19–25 Nm). Tighten the manifold heat shield bolts to 71–88 inch lbs. (8–10 Nm).

23. Position new head gaskets on the cylinder block. The gaskets cannot be interchanged and are marked **R** and **L** for the right and left bank.

24. Carefully position the cylinder heads. Apply clean engine oil to the threads of new cylinder head bolts and install the washers from the old bolts. Install the cylinder head bolts.

25. Torque the new cylinder head bolts using the following procedure:
   a. Tighten the cylinder head bolts in sequence to 10 ft. lbs. (13 Nm).

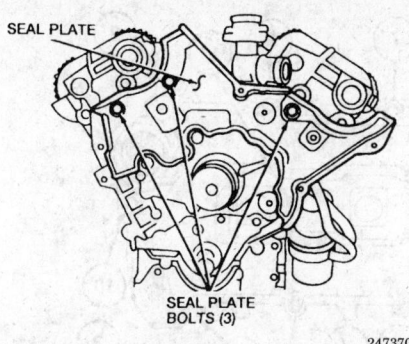

Seal plate bolt locations — 2.5L (VIN B) engine

247370

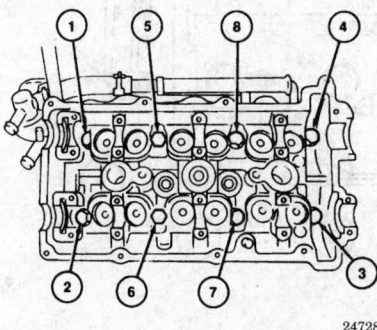

Cylinder head bolt loosening sequence
(same for both heads) — 2.5L (VIN B) engine

247282

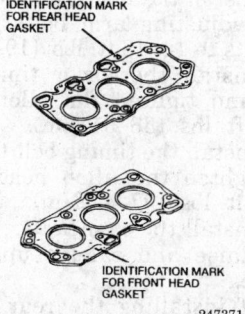

IDENTIFICATION MARK
FOR REAR HEAD
GASKET

IDENTIFICATION MARK
FOR FRONT HEAD
GASKET

247371

Identification marks on
cylinder head gaskets — 2.5L
(VIN B) engine

b. Tighten the bolts again in sequence to 19 ft. lbs. (26 Nm).

c. Paint a mark on the edge of the socket or each cylinder head bolt to use as a reference.

d. Turn each bolt in sequence 90 degrees ± 5 degrees.

e. Turn each bolt in sequence an additional 90 degrees ± 5 degrees.

26. Apply clean engine oil to the hydraulic lifters and install them in their original positions. Make sure they move freely in the bores.

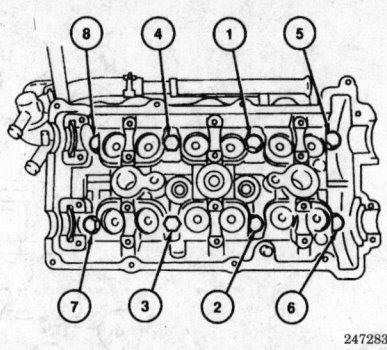

Cylinder head bolt tightening sequence
(same for both heads) — 2.5L (VIN B)
engine

247283

27. Install the seal plate and the seal plate bolts. Tighten the seal plate bolts to 71–88 inch lbs. (8–10 Nm).

28. Install the coolant elbow. Tighten the 4 bolts to 14–18 ft. lbs. (19–25 Nm).

29. Install the 2 alternator-to-alternator adjusting arm bolts. Tighten the bolts to 14–18 ft. lbs. (19–25 Nm).

30. Install the lower timing belt idler and tighten the idler bolt to 28–38 ft. lbs. (38–51 Nm).

31. Install the timing chain tensioner and tighten the Allen head bolt to 27–33 ft. lbs. (37–44 Nm).

32. Install the camshafts.

33. Raise and safely support the vehicle.

34. If installing the rear cylinder head, install the EGR tube and the EGR tube bracket.

35. Connect the exhaust inlet pipes to the manifolds and tighten the nuts to 41 ft. lbs. (55 Nm).

36. Connect the 2 heated oxygen sensor electrical connectors.

37. Lower the vehicle.

38. Apply sealant to the cylinder head surface in the area of the front and rear camshaft caps. Install new gaskets and install the cylinder head covers. Tighten the bolts in 2 steps, in sequence, to 43–78 inch lbs. (5–8 Nm).

39. Install the intake manifold.

40. Install the timing belt and timing belt covers.

41. Connect the negative battery cable.

42. Fill and bleed the cooling system.

43. Run the engine and check for leaks and proper engine operation.

1995–97 Vehicles

NOTE: The cylinder head bolts may be replaced with new, or measured and reused if they are not stretched beyond allowable

limits. Before beginning this procedure, make sure new cylinder head bolts are available, if needed.

———— **CAUTION** ————
Fuel injection systems remain under pressure even after the engine has been turned OFF. The fuel system pressure must be relieved before disconnecting any fuel lines. Failure to do so may result in fire and/or personal injury.

1. Relieve the fuel system pressure as follows:

   a. Start the engine and let it idle.

   b. Locate and remove the fuel pump relay from the main fuse junction panel located next to the left-hand front strut tower.

   c. After the engine stalls, turn **OFF** the ignition switch.

   d. Install the fuel pump relay.

2. Disconnect the negative battery cable.

3. Drain the engine cooling system into a suitable container.

4. Remove the timing belt covers and the timing belt.

5. Remove the intake manifold.

6. Disconnect the ventilation pipe from the left cylinder head cover, remove the bolts and remove both cylinder head covers.

7. Remove the camshafts.

8. Remove the timing belt tensioner and the lower timing belt idler.

9. Remove the 2 alternator-to-alternator adjusting arm bolts.

10. Remove the 3 seal plate bolts and the seal plate from the front of the engine.

11. Remove the 4 coolant elbow bolts and the coolant elbow.

12. Raise and safely support the vehicle.

13. Disconnect the 2 Heated Oxygen Sensor (HO2S) electrical connectors. Remove the exhaust pipe-to-exhaust manifold nuts and lower the exhaust pipes.

14. If removing the rear cylinder head, remove the Exhaust Gas Recirculation (EGR) tube and bracket from the EGR valve and the cylinder head.

15. Lower the vehicle.

16. Remove the hydraulic lifters. Identify each lifter as it is removed so it can be reinstalled in the same position. If the lifters are to be reused, store them upside down in an oil filled container.

17. Loosen the cylinder head bolts, in 2–3 steps in the correct removal sequence.

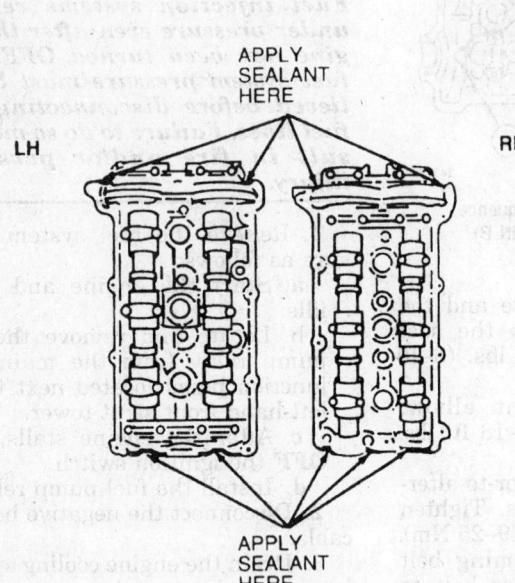

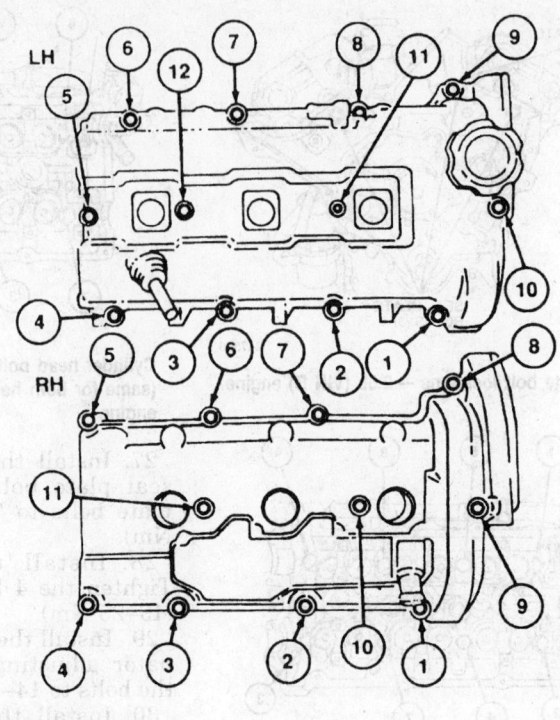

APPLY SEALANT HERE

LH

RH

APPLY SEALANT HERE

LH

RH

247373

Cylinder head cover sealant location and tightening sequence — 2.5L (VIN B) engine

18. Remove the cylinder head bolts and, if equipped, head bolt washers and remove the cylinder heads.

19. If required, remove the exhaust manifolds, manifold shields and gaskets.

20. Clean all gasket mating surfaces. Inspect the cylinder head(s) for damage, cracks, and water and oil leakage.

21. Check the cylinder head gasket surface for distortion using a straight-edge and feeler gauge. Maximum allowable distortion is 0.004 in. (0.10mm).

To install:

22. If removed, install the exhaust manifolds using new gaskets. Tighten the exhaust manifold nuts and bolts to 14–18 ft. lbs. (19–25 Nm). Tighten the manifold heat shield bolts to 71–88 inch lbs. (8–10 Nm).

23. Position new head gaskets on the cylinder block. The gaskets cannot be interchanged and are marked **R** and **L** for the right and left bank.

24. Carefully position the cylinder heads on the head gaskets.

25. Measure the cylinder head bolts from the shoulder to the end of the threads. If the length of the bolt measures more than 5.315 inches (135.0mm) on 1995 vehicles or 5.217 in. (132.5mm) on 1996–97 vehicles, it has stretched beyond its service limit

and must be replaced. Make sure that the washer is removed when measuring.

26. Apply clean engine oil to the threads of the cylinder head bolts and, if equipped, install the washers.

27. Install the cylinder head bolts.

28. Torque the new or good used cylinder head bolts using the following procedure:

a. Tighten the cylinder head bolts in sequence to 10 ft. lbs. (13 Nm).

b. Tighten the bolts again in sequence to 19 ft. lbs. (26 Nm).

c. Paint a mark on the edge of the socket or each cylinder head bolt to use as a reference.

d. Turn each bolt in sequence 90 degrees ± 5 degrees.

e. Turn each bolt in sequence an additional 90 degrees ± 5 degrees.

29. Apply clean engine oil to the hydraulic lifters and install them in their original positions. Make sure they move freely in the bores.

30. Install the seal plate and the seal plate bolts. Tighten the seal plate bolts to 71–88 inch lbs. (8–10 Nm).

31. Install the coolant elbow. Tighten the 4 bolts to 14–18 ft. lbs. (19–25 Nm).

32. Install the 2 alternator-to-alternator adjusting arm bolts. Tighten the bolts to 14–18 ft. lbs. (19–25 Nm).

33. Install the lower timing belt idler and tighten the idler bolt to 28–38 ft. lbs. (38–51 Nm).

34. Install the timing belt tensioner and tighten the Allen head bolt to 27–33 ft. lbs. (37–44 Nm).

35. Install the camshafts.

36. Raise and safely support the vehicle.

37. If installing the rear cylinder head, install the EGR tube and the EGR tube bracket.

38. Connect the exhaust inlet pipes to the manifolds and tighten the nuts to 41 ft. lbs. (55 Nm).

39. Connect the 2 heated oxygen sensor electrical connectors.

40. Lower the vehicle.

41. Apply sealant to the cylinder head surface in the area of the front and rear camshaft caps. Install new gaskets and install the cylinder head covers. Tighten the bolts in 2 steps, in sequence, to 43–78 inch lbs. (5–8 Nm).

42. Install the intake manifold.

43. Install the timing belt and timing belt covers.

44. Connect the negative battery cable.

45. Fill and bleed the cooling system.

46. Run the engine and check for leaks and proper engine operation.

Valve Lifters

REMOVAL AND INSTALLATION

2.0L (VIN A) Engine

1. Disconnect the negative battery cable.
2. Label and disconnect the spark plug wires and clips from the cylinder head cover. Remove the ignition distributor with wiring and set aside.
3. Remove the power steering hose brackets from the cylinder head cover.
4. Disconnect the breather tube and PCV valve from the cylinder head cover. Loosen the cylinder head cover bolts in 2–3 steps. Remove the cylinder head cover.
5. Remove the accessory drive belts, timing belt covers and timing belt.
6. Remove the camshaft sprockets.
7. Note the location of the numbers on top of the camshaft caps, so the caps can be reinstalled in their original positions.
8. Loosen the camshaft cap bolts in 2 steps, in the proper sequence. Remove the camshaft caps and the oil seals.
9. Remove the camshafts.
10. Number each hydraulic lifter (hydraulic lash adjuster) with a paint marker or equivalent, as they are removed, so they can be reinstalled in their original locations.
11. Remove the lifters and inspect for unusual wear.
    To install:
12. Apply clean engine oil to the lifters and install. If the original lifters are to be reused, be sure to install them in their original positions.
13. Lubricate the camshaft lobes and journals with clean engine oil and install the camshafts on the cylinder head. Make sure none of the lobes are depressing any of the hydraulic lifters.
14. Apply silicone sealant to the cylinder head on the front camshaft caps mating surface. Do not get sealant on the camshaft journals.
15. Install the camshaft bearing caps in their original locations. Install the bolts and tighten, in sequence, in 3 steps:
    Step 1: 35 inch lbs. (4 Nm)
    Step 2: 71 inch lbs. (8 Nm)
    Step 3: 100–126 inch lbs. (12–14 Nm)
16. Apply clean engine oil to the lips of 2 new camshaft front seals.

Install the seals using seal replacer T90P-6256-BH or equivalent.
17. Install the camshaft sprockets, timing belt and timing belt covers.
18. Install the accessory drive belts and adjust the tension.
19. Apply silicone sealant to a new cylinder head cover gasket and install the gasket on the cylinder head cover.
20. Apply silicone sealant to the cylinder head in the area adjacent to the front camshaft caps.
21. Install the cylinder head cover. Tighten the bolts in 2 steps, starting at the camshaft pulley and moving around the cover, to 52–69 inch lbs. (6–7 Nm).
22. Install the power steering hose brackets and tighten the bolts to 71–88 inch lbs. (8–10 Nm).
23. Connect the breather hose and PCV valve.
24. Install the ignition distributor and connect the spark plug wires and clips.
25. Connect the negative battery cable.
26. Run the engine and check for leaks and proper engine operation.
27. Check the ignition timing and adjust, if required.

2.5L (VIN B) Engine

1. Disconnect the negative battery cable.
2. Remove the intake manifold.
3. Label and disconnect the spark plug wires from the spark plugs and move aside.
4. Remove the upper timing belt cover bolts. On the left cylinder head cover, disconnect the ventilation pipe from the front of the left side (front) cylinder head cover.
5. Remove the bolts and the cylinder head covers.
6. Remove the timing belt.
7. Hold the camshafts on the hexagon casting and loosen the sprocket retaining bolts. Remove the camshaft sprockets.

NOTE: Do not remove any of the camshaft bearing caps when the camshaft lobes are depressing the valve lifters or damage to the thrust journal support or camshaft may result.

8. Turn the camshafts so the knock pins are aligned with the marks on the camshaft bearing end caps. This will reduce the pressure on the hydraulic lifters.
9. Note the markings on the camshaft bearing caps prior to removal, so they can be reinstalled in the same positions. The right bank

(rear) caps are marked with numbers and the left bank (front) caps are marked with letters.
10. Loosen the camshaft bearing end cap bolts in sequence, in 5–6 steps. Remove the camshaft bearing end caps.
11. Remove the blind caps.
12. Remove the remaining camshaft bearing cap bolts in the proper sequence. Remove the caps, being sure to remove the thrust caps last. Do not damage the cylinder head thrust bearing support.
13. Remove the camshafts and oil seals. Tag the camshafts for identification.
14. Number each hydraulic lifter (hydraulic lash adjuster) with a paint marker or equivalent, as they are removed, so they can be reinstalled in their original locations.
15. Remove the hydraulic lifters and inspect for unusual wear.
    To install:
16. Apply clean engine oil to the hydraulic lifters and install. If the original lifters are to be reused, make sure they are installed in their original positions. Make sure that the hydraulic lifters move freely in their bore.
17. Apply clean engine oil to the camshaft lobes, journals and supports.
18. Install the camshafts so the timing marks align on the camshaft gears.
19. Apply silicone sealant to the cylinder head surface in the area forward of the camshaft gear cavity on both cylinder heads and to the left (front) cylinder head on the rear exhaust camshaft end cap mating surface.
20. Install the thrust caps. Tighten the thrust cap bolts until the caps are fully seated on the cylinder head.

WARNING

Do not install any of the camshaft bearing caps when the camshaft lobes are depressing the valve lifters or damage to the thrust journal support or camshaft may result.

---

21. Install the remaining camshaft bearing caps and camshaft bearing end caps in their original positions. Tighten the caps, in sequence, in 5 equal steps, with the final step being 98–123 inch lbs. (11–14 Nm).

NOTE: The right bank (rear) camshaft bearing caps are marked with numbers and the left bank (front) camshaft bearing caps are marked with letters.

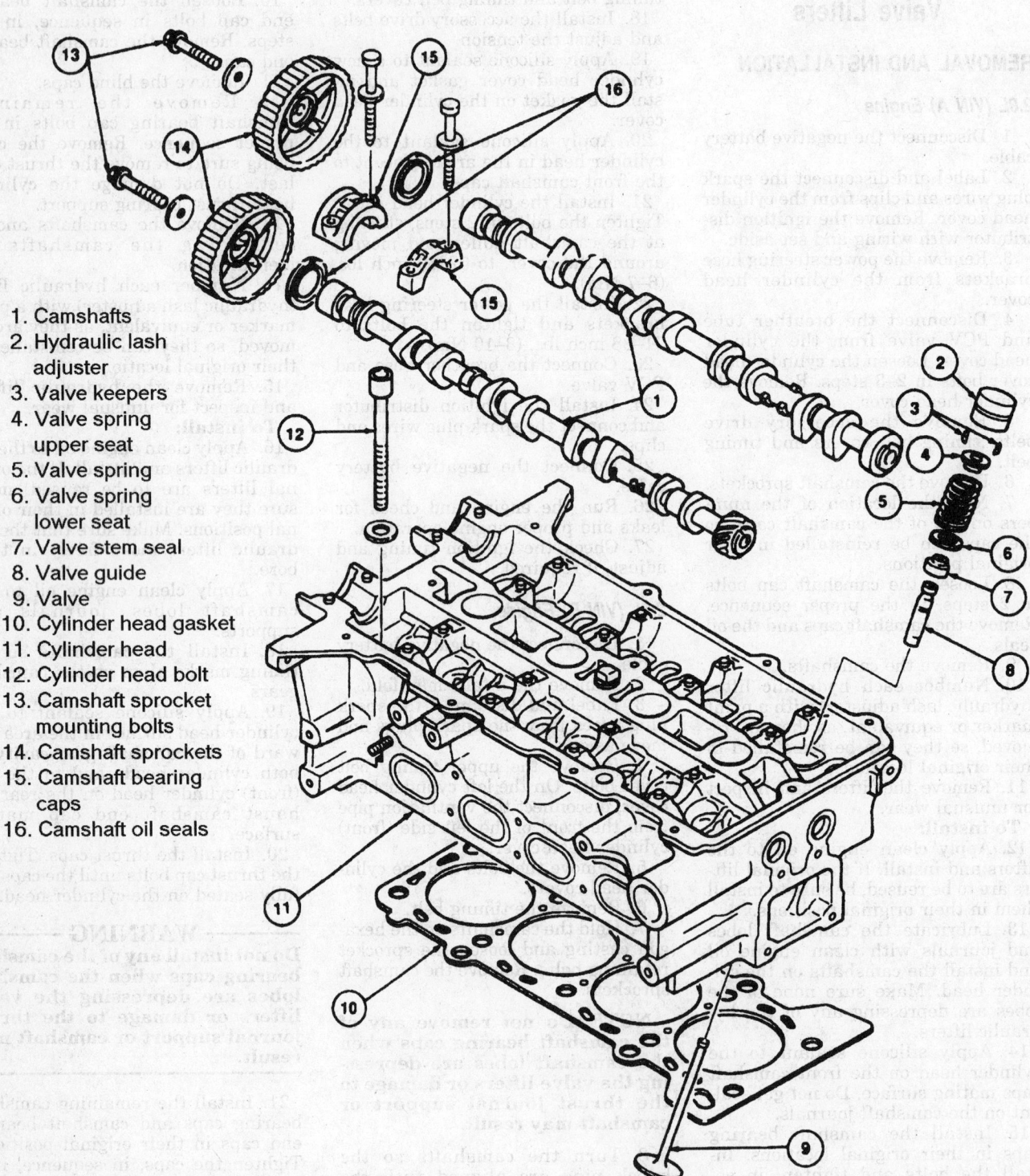

1. Camshafts
2. Hydraulic lash adjuster
3. Valve keepers
4. Valve spring upper seat
5. Valve spring
6. Valve spring lower seat
7. Valve stem seal
8. Valve guide
9. Valve
10. Cylinder head gasket
11. Cylinder head
12. Cylinder head bolt
13. Camshaft sprocket mounting bolts
14. Camshaft sprockets
15. Camshaft bearing caps
16. Camshaft oil seals

Camshaft and lifter components — 2.0L (VIN A) engine

303809

1. Front blind cap(LH)
2. Front end cap(LH)
3. Front thrust cap(LH)
4. Front camshaft caps(LH)
5. Front cylinder head cover(LH)
6. Rear cylinder head bolt(RH)
7. Front oil seal(LH)
8. Front camshaft pulley(LH)

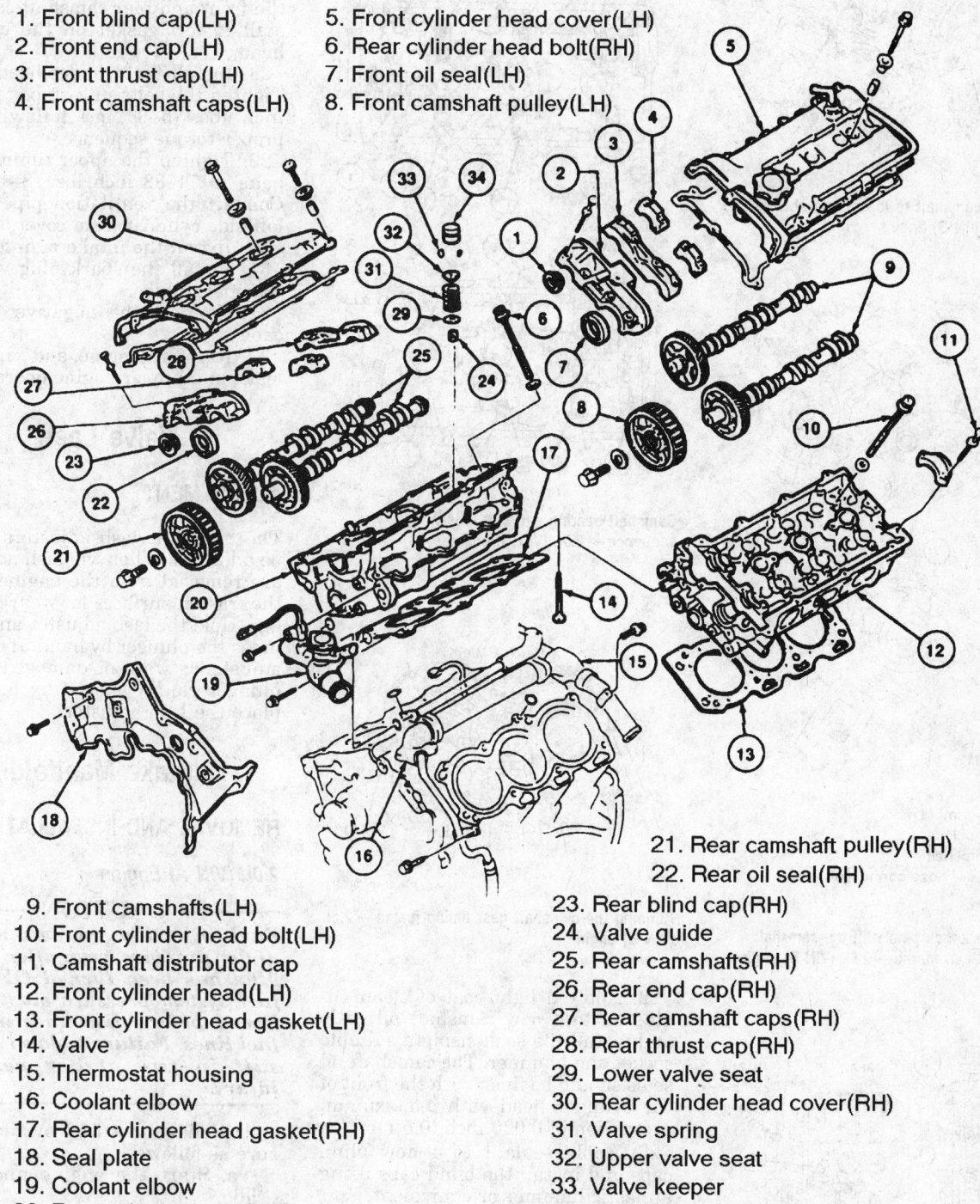

9. Front camshafts(LH)
10. Front cylinder head bolt(LH)
11. Camshaft distributor cap
12. Front cylinder head(LH)
13. Front cylinder head gasket(LH)
14. Valve
15. Thermostat housing
16. Coolant elbow
17. Rear cylinder head gasket(RH)
18. Seal plate
19. Coolant elbow
20. Rear cylinder head(RH)

21. Rear camshaft pulley(RH)
22. Rear oil seal(RH)
23. Rear blind cap(RH)
24. Valve guide
25. Rear camshafts(RH)
26. Rear end cap(RH)
27. Rear camshaft caps(RH)
28. Rear thrust cap(RH)
29. Lower valve seat
30. Rear cylinder head cover(RH)
31. Valve spring
32. Upper valve seat
33. Valve keeper
34. Hydraulic lash adjuster(HLA)

303856

Cylinder head and valve train components — 2.5L (VIN B) engine

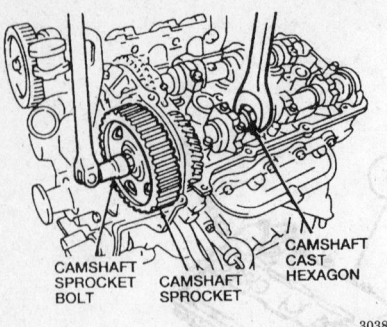

Holding the camshaft to loosen the sprocket bolt — 2.5L (VIN B) engine

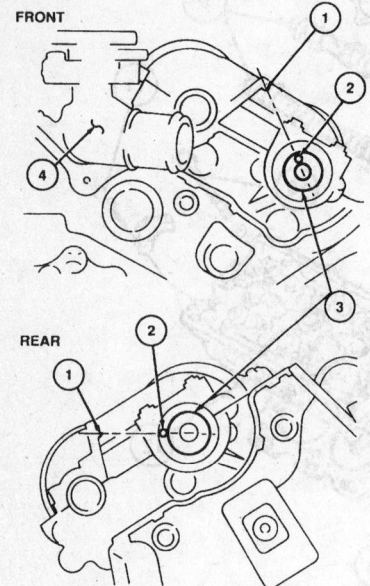

1. Alignment mark
2. Knock pin
3. Camshaft
4. Water hose connection, front

Aligning the knock pins with the camshaft bearing end cap marks — 2.5L (VIN B) engine

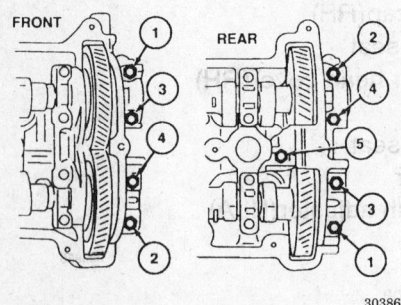

Camshaft bearing end cap bolts loosening sequence — 2.5L (VIN B) engine

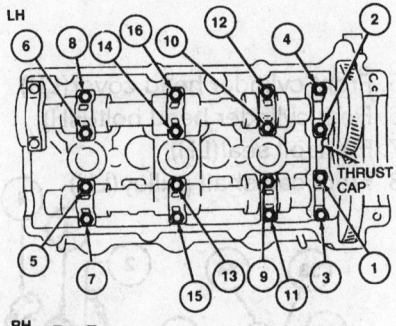

Camshaft bearing cap bolt loosening sequence — 2.5L (VIN B) engine

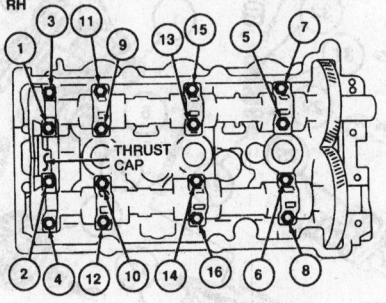

ALIGN THE MARKS

Aligning the camshaft gear timing marks — 2.5L (VIN B) engine

22. Apply a light coat of clean engine oil to 4 new camshaft oil seals and install the seals using a suitable socket and hammer. The camshaft oil seals should be flush with the front of the cylinder head with a maximum protrusion of 0.020 inch (0.5 mm).

23. Apply sealant to 4 new blind caps and install the blind caps using a plastic hammer or equivalent.

24. Install the camshaft sprockets and retaining bolts. Hold the hexagon casting on the camshafts with a suitable wrench and tighten the retaining bolts to 90–103 ft. lbs. (123–140 Nm).

25. Install the timing belt.

26. Remove any sealant and gasket material from the cylinder head cover contact surfaces.

27. Apply silicone sealant to the cylinder head in the area adjacent to the front and rear camshaft caps. Install a new gasket on the cylinder head.

28. Install the cylinder head cover. Tighten the bolts in 2 steps to 43–78 inch lbs. (5–8 Nm) following the proper torque sequence.

29. Tighten the upper timing cover bolts to 71–88 inch lbs. (8–10 Nm). Connect the ventilation pipe to the left side cylinder head cover.

30. Install the intake manifold.

31. Install the spark plug wires to the spark plugs.

32. Connect the negative battery cable.

33. Run the engine and check for leaks and proper engine operation.

Valve Lash

ADJUSTMENT

The hydraulic lash adjusters cannot be adjusted. When the lash adjusters are removed from the engine, check the friction surfaces for wear or damage. Hold the lash adjuster and try to press the plunger by hand. If the lash adjuster is worn or damaged, or the plunger can be moved by hand, replace the lash adjuster.

Intake Manifold

REMOVAL AND INSTALLATION

2.0L (VIN A) Engine

— CAUTION —

Fuel injection systems remain under pressure, even after the engine has been turned OFF. The fuel system pressure must be relieved before disconnecting any fuel lines. Failure to do so may result in fire and/or personal injury.

1. Relieve the fuel system pressure as follows:

   a. Start the engine and let it idle.

   b. Locate and remove the fuel pump relay from the main fuse junction panel located next to the left-hand front strut tower.

   c. After the engine stalls, turn OFF the ignition switch.

   d. Install the fuel pump relay.

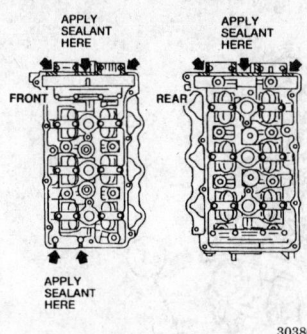

Applying sealant for the camshaft bearing end caps — 2.5L (VIN B) engine

303863

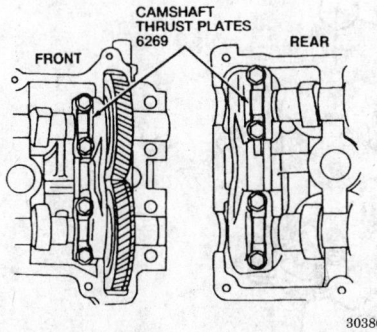

Installing the thrust caps (plates) — 2.5L (VIN B) engine

303864

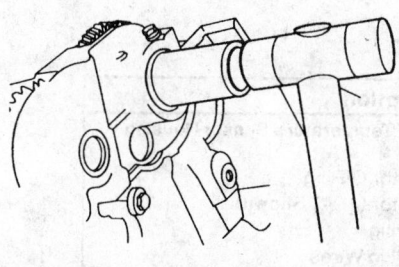

Installing the camshaft front oil seals — 2.5L (VIN B) engine

303866

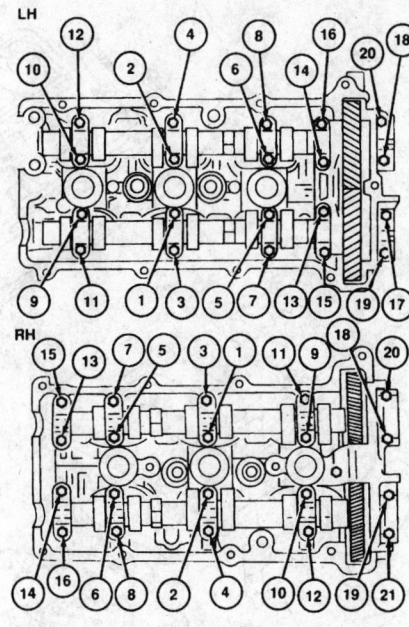

Camshaft bearing cap bolt torque sequence — 2.5L (VIN B) engine

303865

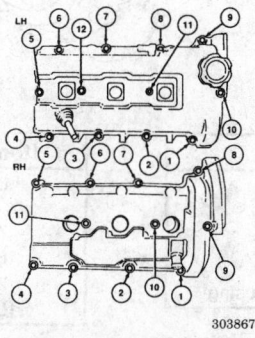

Cylinder head cover bolt torque sequence — 2.5L (VIN B) engine

303867

2. Disconnect the negative battery cable.

3. Remove the air ducts and air cleaner assembly.

4. Remove and plug the fuel supply and return lines from the intake manifold.

5. Disconnect the accelerator cable.

6. Disconnect and plug the coolant lines at the Idle Air Control Bypass Air (IAC BPA) valve and the throttle body.

7. Label and disconnect the vacuum lines to and from the throttle body, the vacuum lines for the brake booster at the left-hand side of the intake manifold and disconnect the vacuum line fitting on the right-hand side of the intake manifold.

8. Label and disconnect the electrical connectors for the Throttle Position (TP) sensor, EGR temperature sensor, if equipped, and the EGR solenoid vacuum valve.

9. Disconnect the PCV valve from the cylinder head cover.

10. Raise and safely support the vehicle.

11. Remove the intake manifold support bracket and remove the EGR pipe from the intake manifold.

12. Lower the vehicle.

13. Remove the 5 bolts and 2 nuts securing the intake manifold and remove the manifold and gasket.

To install:

14. Clean all gasket mating surfaces.

15. Install the intake manifold, using a new gasket. Tighten the nuts and bolts to 14–19 ft. lbs. (19–25 Nm) in the proper sequence.

16. Raise and safely support the vehicle.

17. Attach the EGR pipe to the manifold and install the intake manifold support bracket. Tighten the support bracket bolts to 28–38 ft. lbs. (38–51 Nm).

18. Lower the vehicle.

19. Connect the PCV valve to the cylinder head cover. Connect the electrical connectors to the EGR solenoid vacuum valve, EGR temperature sensor, if equipped and the TP sensor.

20. Connect the vacuum lines running to and from the throttle body, the brake booster and the vacuum line fitting on the right-hand side of the intake manifold.

21. Connect the coolant lines to the IAC BPA valve and the throttle body.

22. Connect the accelerator cable and the fuel lines. Install the fuel line mounting bracket and tighten the bolt to 97 inch lbs. (11 Nm).

23. Install the air cleaner assembly and ducts.

24. Connect the negative battery cable.

25. Fill and bleed the cooling system.

26. Run the engine and check for leaks.

2.5L (VIN B) Engine

— CAUTION —

Fuel injection systems remain under pressure, even after the engine has been turned OFF. The fuel system pressure must be relieved before disconnecting any fuel lines. Failure to do so may result in fire and/or personal injury.

1. Relieve the fuel system pressure as follows:

a. Start the engine and let it idle.

b. Locate and remove the fuel pump relay from the main fuse junction panel located next to the left-hand front strut tower.

c. After the engine stalls, turn **OFF** the ignition switch.

Intake Manifold, Coolant Temperature Sensor Housing and Distributor

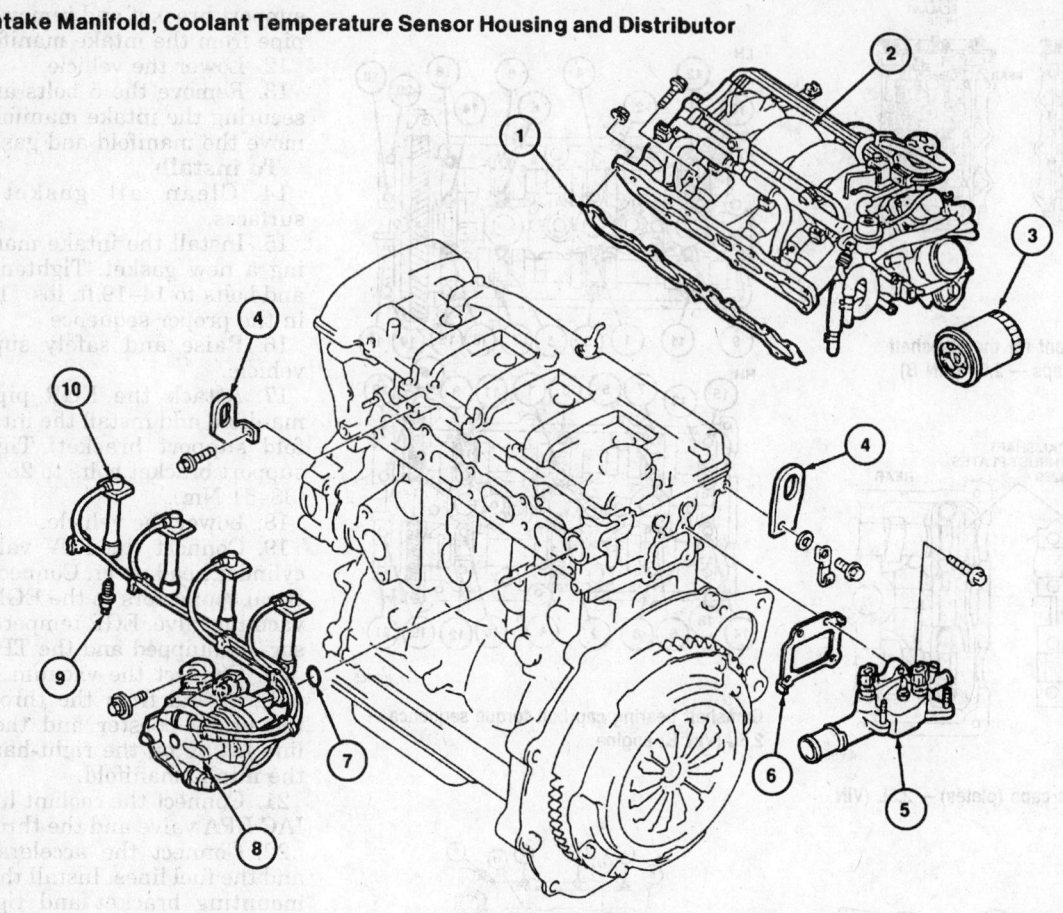

Item	Description
1	Intake Manifold Gasket
2	Intake Manifold
3	Oil Filter
4	Engine Lifting Eye
5	Coolant Temperature Sensor Housing

Item	Description
6	Coolant Temperature Sensor Housing Gasket
7	Distributor O-Ring
8	Distributor (4EAT Shown)
9	Spark Plug
10	Spark Plug Wires

298640

Intake manifold removal — 2.0L (VIN A) engine

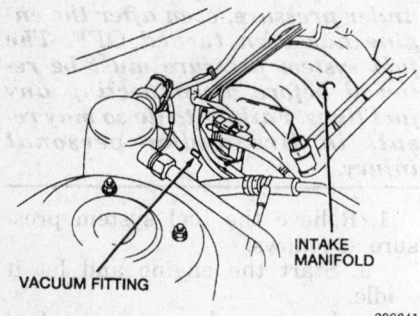

298641

Vacuum fitting at right-hand side of the intake manifold — 2.0L (VIN A) engine

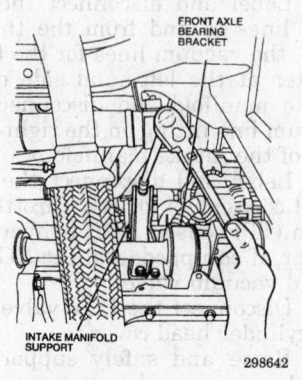

298642

Intake manifold support — 2.0L (VIN A) engine

d. Install the fuel pump relay.

2. Disconnect the negative battery cable.

3. Drain the cooling system into a suitable container.

4. Disconnect the vacuum hoses and electrical connectors from the air cleaner housing. Remove the air cleaner assembly.

5. Disconnect the Knock Sensor (KS) connector and remove the knock sensor bracket from the intake manifold. Remove the Crankshaft Position Sensor (CPS) bracket from the right side of the intake manifold.

6. Remove the right bank (rear) spark plug wires from the spark plugs and the routing clips. Remove

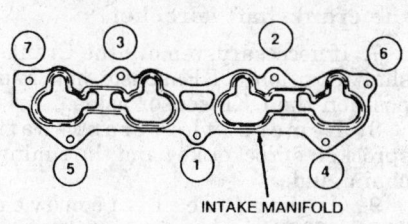

Intake manifold fastener torque sequence — 2.0L (VIN A) engine

298644

the Variable Resonance Induction System (VRIS) solenoid connector bracket from the rear of the intake manifold.

7. Label and disconnect the necessary vacuum hoses from the rear of the intake manifold and EGR valve. Disconnect the PCV valve hose from the intake manifold, near the throttle body.

8. Label and disconnect the Throttle Position (TP) sensor and fuel rail electrical connectors. Disconnect the accelerator cable from the throttle body and the vacuum hose from the evaporative canister.

9. Disconnect and plug the fuel supply line at the fuel rails and dis-

card the copper crush washers. Disconnect the fuel and vacuum lines from the fuel pressure regulator.

10. Disconnect the EGR breather tube.

11. Remove the intake manifold mounting nuts and bolts in 2–3 steps, then remove the intake manifold.

To install:

12. Clean all gasket mating surfaces.

13. Position new gaskets and install the intake manifold. Tighten the nuts and bolts in 2–3 steps to 14–18 ft. lbs. (19–25 Nm).

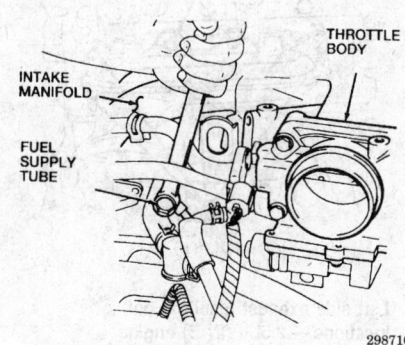

Location of fuel supply fitting — 2.5L (VIN B) engine

298710

Item	Description
1	Intake Manifold Nut (2 Req'd)
2	Throttle Body
3	Intake Manifold
4	Intake Manifold Bolt (5 Req'd)
5	Spark Plug
6	Distributor to Spark Plug Wires
A	Tighten to 19-25 N·m (14-18 Lb-Ft) Refer to the procedure in this section for tightening sequence.

298708

Intake manifold removal — 2.5L (VIN B) engine

14. Connect the EGR breather tube and connect the fuel and vacuum lines to the fuel pressure regulator.

15. Connect the fuel supply line to the fuel rail, using new copper crush washers. Tighten the fuel line fittings to 18–25 ft. lbs. (25–34 Nm).

16. Connect the vacuum hoses to the evaporative canister, intake manifold, throttle body and EGR valve.

17. Connect the TP sensor and fuel rail electrical connectors. Install the VRIS solenoid connector bracket.

18. Connect the spark plug wires to the spark plugs and routing clips. Install the CPS bracket.

19. Install the KS bracket and connect the KS electrical connector.

20. Install the air cleaner assembly and connect the vacuum hoses and electrical connectors to the air cleaner housing.

21. Connect the negative battery cable.

22. Fill and bleed the cooling system.

23. Run the engine and check for leaks.

Exhaust Manifold

REMOVAL AND INSTALLATION

2.0L (VIN A) Engine

1. Disconnect the negative battery cable.

2. Remove the 7 exhaust manifold heat shield bolts and the heat shield.

3. Disconnect the oxygen sensor electrical connector.

4. Raise and safely support the vehicle.

5. Remove and discard the exhaust inlet pipe-to-exhaust manifold nuts. Suspend the exhaust system with wire.

6. Remove the 2 exhaust inlet pipe bracket bolts.

7. Disconnect the EGR pipe from the exhaust manifold.

8. Lower the vehicle.

9. Remove the 2 nuts and 8 bolts and remove the exhaust manifold. Discard the nuts.

To install:

10. Clean all gasket mating surfaces.

11. Position a new exhaust manifold gasket over the studs and install the exhaust manifold. Tighten the 8 mounting bolts to 12–17 ft. lbs. (16–23 Nm).

12. Install 2 new manifold mounting nuts and tighten to 14–21 ft. lbs. (20–28 Nm).

13. Raise and safely support the vehicle.

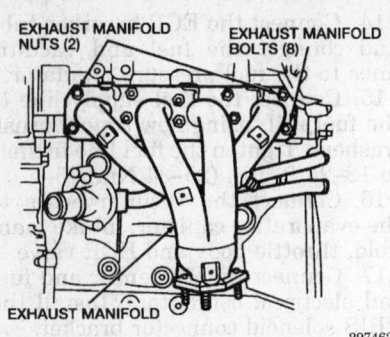

Exhaust manifold bolt and nut locations — 2.0L (VIN A) engine

14. Connect the exhaust pipe to the manifold. Install new nuts and tighten to 27–38 ft. lbs. (37–52 Nm). Connect the oxygen sensor connector.

15. Install the 2 exhaust inlet pipe bracket bolts.

16. Connect the EGR pipe to the back of the exhaust manifold and tighten to 24–34 ft. lbs. (32–47 Nm).

17. Lower the vehicle.

18. Install the exhaust manifold heat shield and tighten the bolts to 71–88 inch lbs. (8–10 Nm).

19. Connect the negative battery cable. Run the engine and check for exhaust leaks.

2.5L (VIN B) Engine

1. Disconnect the negative battery cable.

2. Raise and safely support the vehicle.

3. Disconnect the oxygen sensor connectors.

4. Remove the nuts from the front and rear exhaust pipes and lower the exhaust system. Both pipes must be disconnected, even if only one manifold is to be removed.

5. If removing the rear (right side) manifold, disconnect the EGR pipe.

6. Remove the 3 exhaust manifold shields bolts and remove the shield.

7. Remove the 2 nuts and 5 bolts and remove the exhaust manifold(s).

To install:

8. Clean all gasket mating surfaces.

9. Install the exhaust manifold, using a new gasket, and tighten the nuts and bolts to 14–18 ft. lbs. (19–25 Nm).

10. Install the exhaust manifold shield and tighten the bolts to 71–88 inch lbs. (8–10 Nm).

11. If installing the rear (right side) manifold, connect the EGR pipe.

12. Connect the exhaust pipes to the manifolds, using new gaskets and nuts, and tighten the nuts to 30–41 ft. lbs. (40–55 Nm).

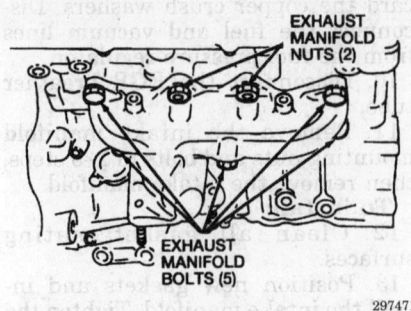

Right side exhaust manifold bolt locations — 2.5L (VIN B) engine

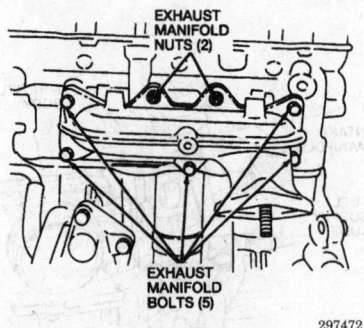

Left side exhaust manifold bolt locations — 2.5L (VIN B) engine

13. Connect the oxygen sensor connectors and the negative battery cable. Run the engine and check for exhaust leaks.

Front Crankshaft Seal

REMOVAL AND INSTALLATION

2.0L (VIN A) Engine

1. Disconnect the negative battery cable.

2. Remove the accessory drive belts.

3. Raise and safely support the vehicle.

4. Remove the front splash shields.

5. Prevent the crankshaft pulley from rotating by installing crankshaft pulley holder T92C-6316-AH or equivalent, and remove the crankshaft pulley bolt.

6. Remove the timing belt.

NOTE: Be careful not to turn either the crankshaft or camshaft sprockets after the belt has been removed. This is especially important when removing

the crankshaft damper retaining bolt and when using a puller on the crankshaft sprocket.

7. If necessary, remove the Crankshaft Position (CKP) sensor bolt and position the CKP sensor aside.

8. Remove the crankshaft sprocket, sprocket key and the timing chain guide.

9. Using seal remover T92C-6700-CH or equivalent, remove the front crankshaft seal.

NOTE: Seal extractor tool T92C-6700-CH is recommended. It hooks the seal from inside and pulls the seal outward. If the special seal extractor recommended by the vehicle manufacturer is not available, remove the oil pump to remove the seal. Be careful not to score the crankshaft or the seal seat.

To install:

10. Lubricate the seal lip with clean engine oil. Using oil seal installation tool T74P-6150-A or equivalent, press or drive the new seal into the oil pump cavity. Install the seal so it flush with the oil pump body.

11. Install the timing chain guide and crankshaft sprocket.

12. If necessary, position the crankshaft position (CKP) sensor and install the (CKP) bolt. Torque sensor bolt to 71–88 inch lbs. (8–10 Nm).

13. Install the timing belt.

14. Install the crankshaft pulley and torque the crankshaft pulley bolt to 116–123 ft. lbs. (157–166 Nm).

15. Install the front splash shields.

16. Lower the vehicle.

17. Install the accessory drive belts.

18. Connect the negative battery cable.

19. Run the engine and check for oil leaks and proper engine operation.

2.5L (VIN B) Engine

1. Disconnect the negative battery cable.

2. Remove the oil pump.

NOTE: Protect the oil pump housing with a rag while removing the crankshaft front seal.

3. Using a suitable drift or punch tool protected with a rag, knock out the front crankshaft seal from the oil pump housing.

To install:

4. Clean the oil, dirt and old sealant from all contact surfaces.

5. Lubricate the new front crankshaft seal lip with clean engine oil. Using front seal replacer T74P-6150-A or equivalent, press or drive the new seal into the oil pump

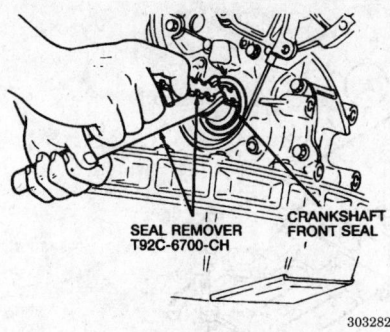

Removing the front crankshaft oil seal — 2.0L (VIN A) engine

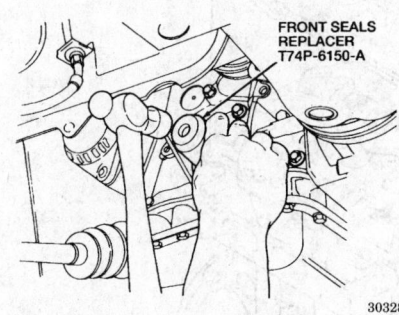

Installing the front crankshaft oil seal — 2.0L (VIN A) engine

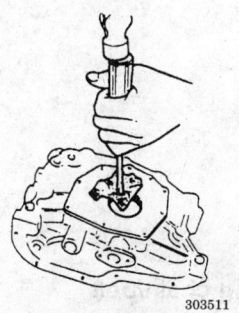

Removing the front crankshaft seal — 2.5L (VIN B) engine

housing. Install the seal so it is flush with the edge of the oil pump housing or protrudes no more than 0.021 inch (0.7 mm) from the edge of the pump body.

6. Install the oil pump.
7. Connect the negative battery cable.
8. Run the engine and check for leaks and proper engine operation.

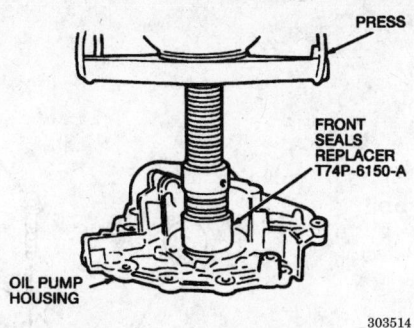

Installing the front crankshaft seal — 2.5L (VIN B) engine

Timing Belt, Tensioner and Front Cover

REMOVAL AND INSTALLATION

2.0L (VIN A) Engine

1. Disconnect the negative battery cable.
2. Label and disconnect the spark plug wires and clips from the cylinder head cover. Remove the ignition distributor with wiring and set aside.
3. Remove the power steering hose brackets from the cylinder head cover. If necessary disconnect the crankshaft position sensor.
4. Disconnect the breather tube and PCV valve from the cylinder head cover.
5. Loosen the cylinder head cover bolts in 2–3 steps. Remove the cylinder head cover.
6. Remove the power steering belt shield. Loosen the power steering adjusting bolt, lock bolt and through bolt and remove the power steering belt.
7. Loosen the alternator adjusting bolt and upper mounting bolt. Remove the alternator belt.
8. Support the engine with engine support tool 014-00750 or equivalent. Raise the engine slightly with a jack and remove the right side engine support insulator (mount).
9. Remove the oil level indicator bolt and 4 upper timing belt cover bolts and remove the upper timing belt cover.
10. Raise and safely support the vehicle.
11. Remove the splash shields. Using holder tool T92C-6316-AH or equivalent, hold the crankshaft pulley and remove the pulley bolt. Use a suitable puller to remove the pulley, then remove the guide plate.
12. Remove the 4 lower timing belt cover bolts and remove the lower timing belt cover.

13. Temporarily install the crankshaft pulley bolt.
14. Turn the crankshaft until the timing mark on the crankshaft sprocket aligns with the timing mark on the oil pump and the camshaft sprocket timing marks, **E** and **I**, line up on the camshaft sprockets.
15. Lower the vehicle.
16. Insert camshaft sprocket holding tool T92C-6256-AH or equivalent, between the camshaft sprockets.

Timing belt tensioner — 2.0L (VIN A) engine

1. Turn the timing belt tensioner with an Allen wrench and remove the tensioner spring from the tensioner spring pin.
2. If the timing belt is to be reused, mark the direction of rotation on the timing belt. Remove the timing belt.
3. If it is necessary to remove the sprockets, remove the camshaft sprocket holding tool. Hold the camshaft by placing a suitable wrench on the hexagon which is cast into the camshaft. Place another wrench onto the camshaft sprocket retaining bolt and loosen the bolt.

NOTE: Before removing the camshaft sprocket(s), make sure that the camshafts are still in alignment and tag each sprocket to the camshaft from which it was removed.

4. Remove the camshaft sprocket bolt and the camshaft sprocket from the camshaft.
5. Repeat the camshaft sprocket removal procedure for the opposite camshaft if required.
6. Raise and safely support the vehicle.
7. Slide off the crankshaft sprocket and remove the crankshaft key.

To install:

8. Install the crankshaft key and slide the crankshaft sprocket into position.
9. Lower the vehicle.
10. Install the camshaft sprocket onto the proper camshaft, making sure to align the dowel pin.
11. Make sure that the **I** and **E** are in alignment.
12. Install the camshaft sprocket bolt. Hold the hexagon on the camshaft with a suitable wrench and tighten the sprocket bolt to 35–48 ft. lbs. (47–65 Nm). Make sure the camshaft sprockets are still properly aligned and reinstall the sprocket holding tool.

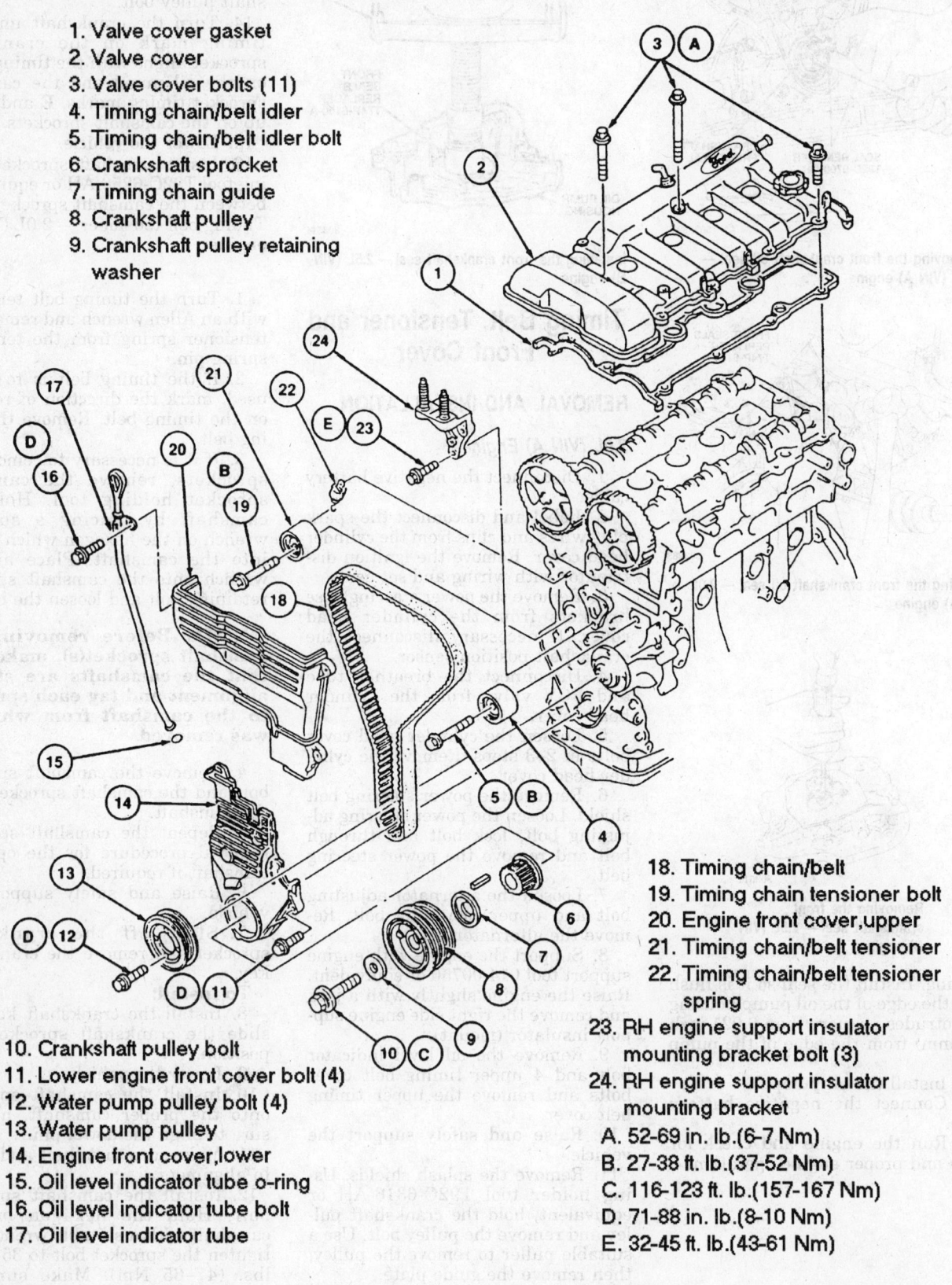

1. Valve cover gasket
2. Valve cover
3. Valve cover bolts (11)
4. Timing chain/belt idler
5. Timing chain/belt idler bolt
6. Crankshaft sprocket
7. Timing chain guide
8. Crankshaft pulley
9. Crankshaft pulley retaining washer

10. Crankshaft pulley bolt
11. Lower engine front cover bolt (4)
12. Water pump pulley bolt (4)
13. Water pump pulley
14. Engine front cover, lower
15. Oil level indicator tube o-ring
16. Oil level indicator tube bolt
17. Oil level indicator tube

18. Timing chain/belt
19. Timing chain tensioner bolt
20. Engine front cover, upper
21. Timing chain/belt tensioner
22. Timing chain/belt tensioner spring
23. RH engine support insulator mounting bracket bolt (3)
24. RH engine support insulator mounting bracket

A. 52-69 in. lb.(6-7 Nm)
B. 27-38 ft. lb.(37-52 Nm)
C. 116-123 ft. lb.(157-167 Nm)
D. 71-88 in. lb.(8-10 Nm)
E. 32-45 ft. lb.(43-61 Nm)

Timing belt and related components — 2.0L (VIN A) engine

306567

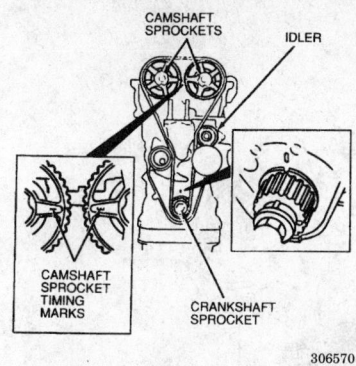

Timing mark locations — 2.0L (VIN A) engine

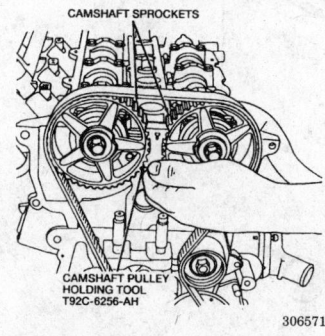

Camshaft sprocket holding tool — 2.0L (VIN A) engine

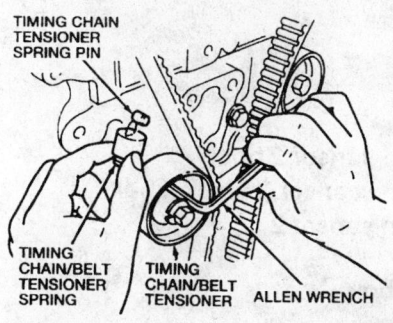

13. Make sure the timing marks on the camshaft and crankshaft sprockets are still aligned.

14. Install the timing belt. If re-using the original timing belt, make sure it is installed in the same direction of rotation.

15. Turn the tensioner clockwise with an Allen wrench and install the tensioner spring. Remove the holding tool from between the camshaft sprockets.

16. Rotate the crankshaft clockwise 2 turns and align the timing marks.

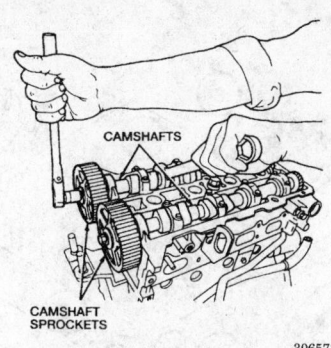

Holding the camshaft — 2.0L (VIN A) engine

Make sure all marks are still correctly aligned.

NOTE: The timing chain tensioner automatically adjust the tension on the timing belt.

17. Raise and safely support vehicle.

18. Install the timing belt lower cover and tighten the 4 bolts to 71–88 inch lbs. (8–10 Nm).

19. Install the guide plate, crankshaft pulley and pulley bolt. Hold the pulley with the holder tool and tighten the bolt to 116–123 ft. lbs. (157–167 Nm).

20. Install the splash shields and lower the vehicle.

21. Raise the engine slightly with the jack and install the right side engine mount. Tighten the mount through bolt to 63–86 ft. lbs. (86–116 Nm) and the mount attaching nuts to 54–75 ft. lbs. (74–103 Nm). Remove the engine support tool.

22. Install the upper timing belt cover and tighten the bolts to 71–88 inch lbs. (8–10 Nm).

23. Clean the cylinder head and valve cover mating surfaces thoroughly.

24. Apply silicone sealant to the cylinder head surface in the area adjacent to the front camshaft bearing caps. Apply sealant to a new gasket and install it on the cylinder head cover.

25. Install the cylinder head cover and tighten the bolts, in sequence, in 2 steps to 52–69 inch lbs. (6–7 Nm).

26. Install the power steering hose brackets and tighten the bolts to 71–88 inch lbs. (8–10 Nm). Connect the spark plug wires and wire clips. Connect the breather tube and PCV valve. If necessary connect the crankshaft position sensor.

27. Install the alternator belt and adjust the tension. Tighten the upper mounting bolt to 14–18 ft. lbs. (19–25 Nm) and the lower through bolt to 27–38 ft. lbs. (37–52 Nm).

28. Install the power steering belt and adjust the tension. Tighten the through bolt to 32–45 ft. lbs. (43–61 Nm) and the lock bolt to 23–34 ft. lbs. (31–46 Nm). Install the power steering belt shield and tighten the bolts to 61–86 inch lbs. (7–9 Nm).

29. Connect the negative battery cable.

30. Run the engine and check for leaks and proper engine operation.

2.5L (VIN B) Engine

1. Disconnect the negative battery cable.

2. Label and disconnect the electrical connectors from the coolant elbow. Label and disconnect the electrical connectors from the knock sensor (if required) and crankshaft position sensor.

3. Loosen the drive belt tensioner locknuts and adjusting bolts. Remove the accessory drive belts.

4. Raise and safely support the vehicle.

5. Remove the lower bolt from the A/C and alternator tensioner bracket.

6. Remove the right wheel and splash shields.

7. Hold the crankshaft pulley (damper) with holder tool T92C-6316-AH or equivalent, and remove the crankshaft pulley bolt. Remove the crankshaft pulley using a puller if needed.

8. Remove the 5 front timing belt cover bolts.

9. Hold the water pump pulley with holder tool T92C-6312-AH or equivalent, remove the 4 bolts and the water pump pulley.

10. Hold the power steering pump pulley with strap wrench D85L-6000-A or equivalent and remove the power steering pump pulley nut and the pulley.

11. Lower the vehicle.

12. Remove the upper bolt from the belt idler bracket and remove the bracket.

13. Remove the engine oil dipstick tube retaining bolt and the tube.

14. Remove the 8 rear timing belt cover retaining bolts and remove the timing belt covers.

15. Temporarily reinstall the crankshaft pulley bolt.

16. Remove the 3 nuts and through-bolt from the right-hand engine support insulator and remove the support insulator. Remove the support insulator bracket.

17. Raise and safely support the vehicle.

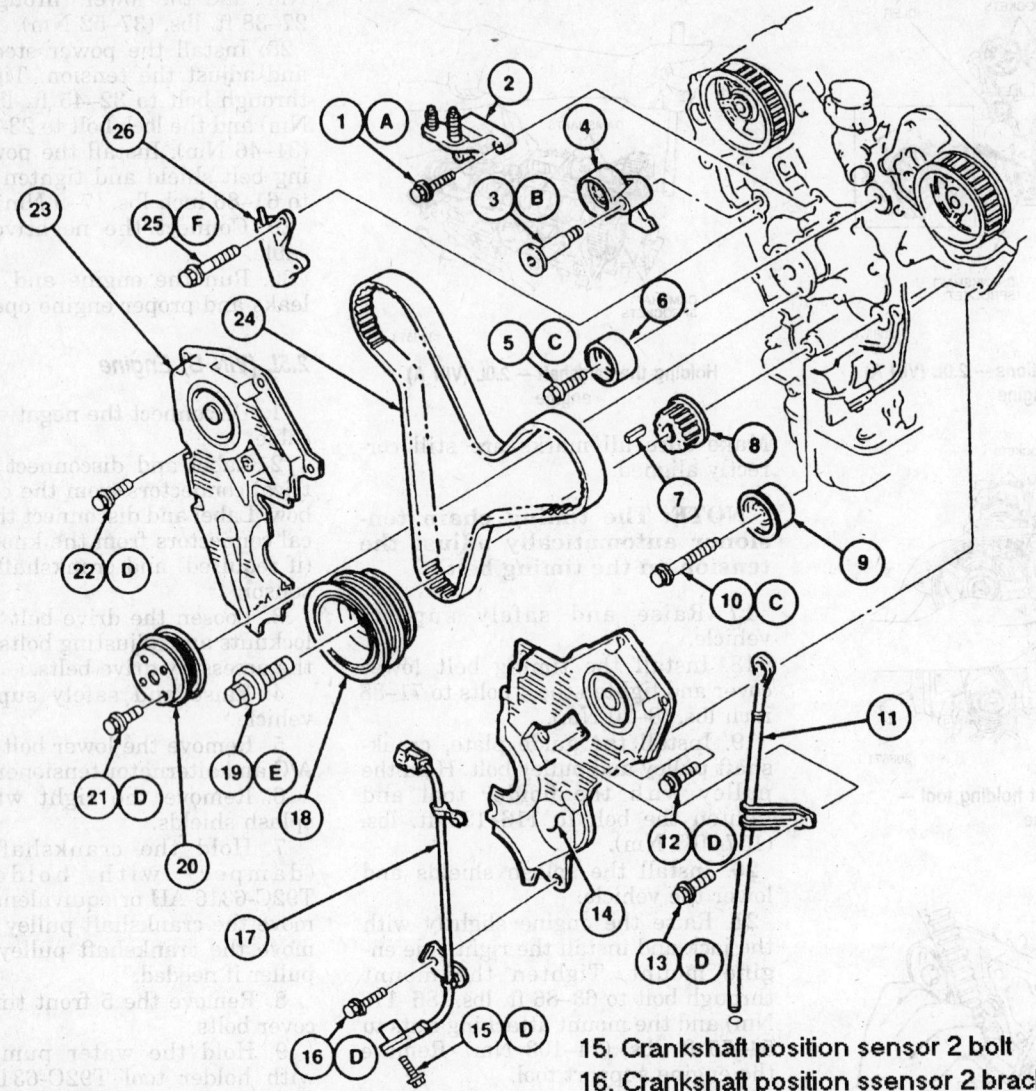

1. RH engine support insulator mounting bracket bolts (3)
2. RH engine support insulator mounting bracket
3. Timing chain/belt tensioner bolt
4. Timing chain/belt tensioner
5. Timing chain/belt idler bolt, upper
6. Timing chain/belt idler, upper
7. Crankshaft sprocket key
8. Crankshaft sprocket
9. Timing chain/belt idler, lower
10. Timing chain/belt idler bolt, lower
11. Oil level indicator tube
12. Engine front cover bolt, front (8)
13. Oil level indicator tube bracket bolt
14. Engine front cover, front

15. Crankshaft position sensor 2 bolt
16. Crankshaft position ssensor 2 bracket bolt
17. Crankshaft position sensor 2
18. Crankshaft pulley
19. Crankshaft pulley bolt
20. Water pump pulley
21. Water pump pulley bolt (4)
22. Engine front cover bolt (7)
23. Engine front cover, rear
24. Timing chain/belt
25. Timing chain tensioner arm bolt (2)
26. Timing chain tensioner arm
A. 32–44 ft. lb.(44–60 Nm)
B. 28–32 ft. lb.(38–44 Nm)
C. 28–38 ft. lb.(38–51 Nm)
D. 71–88 in. lb.(8–10 Nm)
E. 116–122 ft. lb.(157–166 Nm)
F. 14–18 ft. lb.(19–25 Nm)

303201

Timing belt front cover and related components — 2.5L (VIN B) engine

18. Turn the crankshaft to TDC No. 1 cylinder in the direction of normal rotation. Make sure that the timing mark on the crankshaft sprocket aligns with the timing mark on the oil pump.

19. Remove the 2 bolts from the timing belt tensioner arm, removing the lower bolt first.

20. Remove the timing belt tensioner arm.

21. If the timing belt is to be reused, mark the direction of rotation on the timing belt.

22. Loosen the Allen bolt on the timing belt tensioner.

23. Remove the timing belt.

24. If the timing belt sprockets are to be removed, proceed as follows:

   a. Remove the intake manifold.

   b. Label and disconnect the necessary hoses from the cylinder head covers.

   c. Label and disconnect the spark plug wires from the spark plugs.

   d. Remove the cylinder head cover retaining bolts and remove the cylinder head covers.

   e. Hold the camshaft using a suitable wrench on the hexagon cast into the camshaft. Remove the camshaft sprocket bolts and the camshaft sprockets.

   f. Use crankshaft damper puller T74P-6316-A or equivalent, to remove the crankshaft sprocket. Remove the crankshaft sprocket key.

To install:

25. If the timing belt sprockets were removed, proceed as follows:

   a. Install the crankshaft sprocket key and crankshaft sprocket.

   b. Install the camshaft sprockets on the camshafts with the retaining bolts.

   c. Hold the camshaft using a suitable wrench on the hexagon cast into the camshaft. Tighten the camshaft sprocket bolts to 90–103 ft. lbs. (123–140 Nm).

   d. Make sure the cylinder head cover and cylinder head contact surfaces are clean and free of dirt, oil and old sealant and gasket material.

   e. Apply silicone sealant to the cylinder heads in the area adjacent to the front and rear camshaft caps. Install new gaskets on the cylinder heads.

   f. Install the cylinder head covers and tighten the retaining bolts, in sequence, in 2 Steps, to 43–78 inch lbs. (5–8 Nm).

   g. Connect the spark plug wires to the spark plugs and connect the hoses to the cylinder head covers.

   h. Install the intake manifold. Cylinder head cover bolt torque sequence — 2.5L (VIN B) engine

1. Position the timing belt tensioner arm in a suitable press.

2. Compress the tensioner until the hole in the piston is aligned with the 2nd hole in the tensioner case. Insert a 0.060 in. (1.6mm) diameter wire or pin through the 2nd hole to keep the piston compressed.

3. Align the camshaft sprockets to TDC.

4. Turn the crankshaft counter-clockwise until the crankshaft sprocket is offset from TDC by 1 tooth.

5. Install the timing belt.

6. If the original belt is being reused, make sure it is installed in the same direction of rotation.

7. Turn the crankshaft in the direction of normal engine rotation until the crankshaft sprocket timing mark is at TDC. This should place all of the belt slack in the timing belt tensioner portion of the timing belt.

8. Install the timing belt tensioner arm and 2 bolts. Tighten the bolts to 14–18 ft. lbs. (19–25 Nm).

9. Remove the wire or pin from the tensioner.

NOTE: When properly timed, the crankshaft timing marks will line up and the crankshaft sprocket timing mark will no longer be 1 tooth off.

10. Rotate the crankshaft 2 complete turns in the direction of normal rotation and align the timing marks. Make sure all marks are still correctly aligned. This will also set the timing belt tension.

NOTE: The timing belt tensioner will automatically adjust the timing belt tension.

11. Tighten the timing belt tensioner Allen bolt to 28–32 ft. lbs. (35–51 Nm).

12. Install the right-hand engine support insulator. Tighten the 3 nuts to 54–76 ft. lbs. (74–103 Nm) and the through-bolt to 50–68 ft. lbs. (67–93 Nm).

13. Remove the crankshaft damper bolt.

14. Install the timing belt covers with the rear 8 bolts. Tighten to 71–88 inch lbs. (8–10 Nm).

15. Install the engine oil dipstick tube and retaining nut.

16. Install the belt idler bracket and the upper retaining bolt.

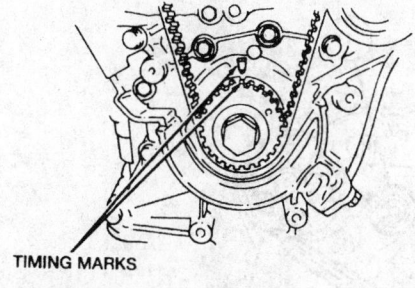

TIMING MARKS

312556

Crankshaft sprocket and oil pump timing marks — 2.5L (VIN B) engine

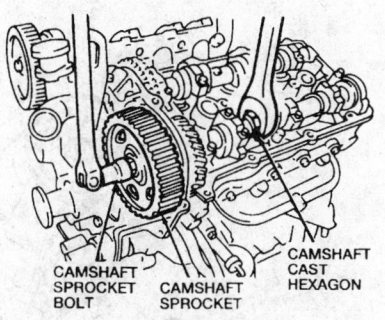

CAMSHAFT SPROCKET BOLT   CAMSHAFT SPROCKET   CAMSHAFT CAST HEXAGON

312598

Holding the camshaft — 2.5L (VIN B) engine

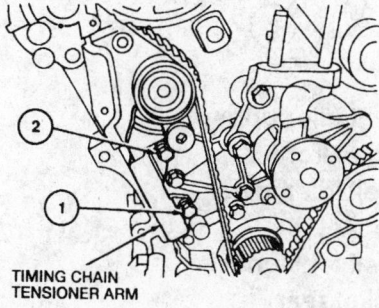

TIMING CHAIN TENSIONER ARM

312557

Sequence for removing the timing belt tensioner arm bolts — 2.5L (VIN B) engine

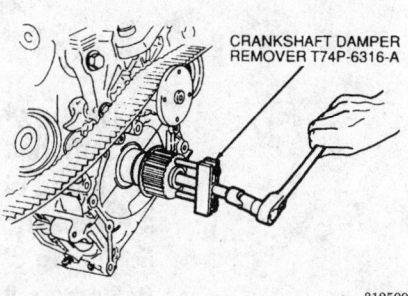

CRANKSHAFT DAMPER REMOVER T74P-6316-A

312599

Removing the crankshaft sprocket — 2.5L (VIN B) engine

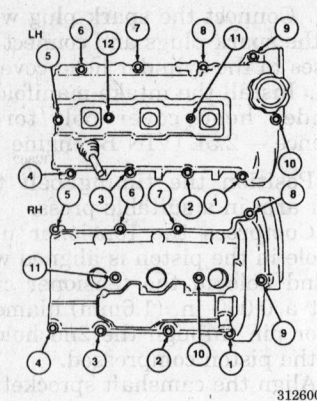

312600

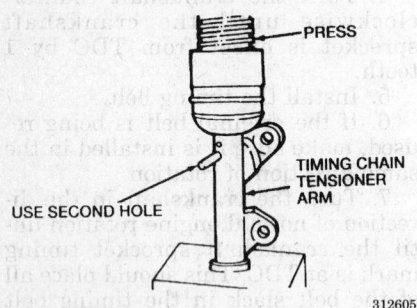

Compressing the timing belt tensioner — 2.5L (VIN B) engine

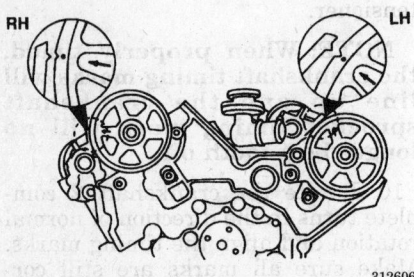

312606

Camshaft sprocket timing marks for TDC — 2.5L (VIN B) engine

17. Raise and safely support the vehicle.

18. Install the power steering pump pulley and nut. Tighten the nut to 36–43 ft. lbs. (49–59 Nm) while holding the pulley with a strap wrench.

19. Install the water pump pulley and 4 bolts. Hold the pulley with the holder tool and tighten the bolts to 71–88 inch lbs. (8–10 Nm).

20. Install the 5 front timing belt cover bolts and tighten to 71–88 inch lbs. (8–10 Nm).

21. Install the crankshaft pulley (damper) with the bolt. Hold the crankshaft pulley with the holding

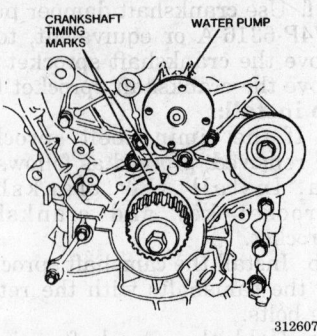

312607

Positioning the crankshaft sprocket timing mark 1 degree off of TDC — 2.5L (VIN B) engine

tool and tighten to 116–122 ft. lbs. (157–166 Nm).

22. Install the splash shields.

23. Install the wheel and torque the lug nuts to 65–87 ft. lbs. (88–118 Nm).

24. Install the lower bolt into the A/C and alternator tensioner bracket.

25. Lower the vehicle.

26. Install the accessory drive belts and adjust the tension.

27. Connect the electrical connectors to the sensors at the coolant elbow and the knock sensor and crankshaft position sensor.

28. Connect the negative battery cable.

29. Run the engine and check for leaks and proper engine operation. Make sure that the timing marks on the crankshaft sprocket align with the timing mark on the oil pump.

Camshaft

REMOVAL AND INSTALLATION

2.0L (VIN A) Engine

1. Disconnect the negative battery cable.

2. Label and disconnect the spark plug wires and clips from the cylinder head cover. Remove the ignition distributor with wiring and set aside.

3. Remove the power steering hose brackets from the cylinder head cover.

4. Disconnect the breather tube and PCV valve from the cylinder head cover. Loosen the cylinder head cover bolts in 2–3 steps. Remove the cylinder head cover.

5. Remove the accessory drive belts, timing belt covers and timing belt.

6. Remove the camshaft sprockets.

7. Note the location of the numbers on top of the camshaft caps, so the caps can be reinstalled in their original positions.

8. Loosen the camshaft cap bolts in 2 steps, in the proper sequence.

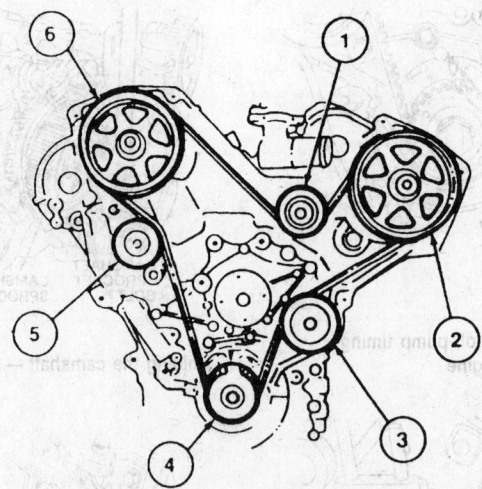

1. Timing chain/belt idler, upper
2. Camshaft sprocket, front
3. Timing chain/belt idler, lower
4. Crankshaft sprocket
5. Timing chain/belt tensioner
6. Camshaft sprocket, rear

312608

Timing belt, sprocket and idler positioning — 2.5L (VIN B) engine

Remove the camshaft caps and the oil seals.

9. Remove the camshafts.

10. If they are to be removed, number each hydraulic lifter (hydraulic lash adjuster) with a paint marker or equivalent during removal.

11. Inspect the camshafts and lifters for wear and/or damage and replace, as necessary.

To install:

12. If removed, apply clean engine oil to the lifters and install. If the original lifters are to be reused, be sure to install them in their original positions.

13. Lubricate the camshaft lobes and journals with clean engine oil and install the camshafts on the cylinder head. Make sure none of the lobes are depressing any of the hydraulic lifters.

14. Apply silicone sealant to the cylinder head on the front camshaft caps mating surface. Do not get sealant on the camshaft journals.

15. Install the camshaft bearing caps in their original locations. Install the bolts and tighten, in sequence, in 3 steps:

Step 1: 35 inch lbs. (4 Nm)
Step 2: 71 inch lbs. (8 Nm)
Step 3: 100–126 inch lbs. (12–14 Nm)

16. Apply clean engine oil to the lips of 2 new camshaft front seals. Install the seals using seal replacer T90P-6256-BH or equivalent.

17. Install the camshaft sprockets, timing belt and timing belt covers.

18. Install the accessory drive belts and adjust the tension.

19. Apply silicone sealant to a new cylinder head cover gasket and install the gasket on the cylinder head cover.

20. Apply silicone sealant to the cylinder head in the area adjacent to the front camshaft caps.

21. Install the cylinder head cover. Tighten the bolts in 2 steps, in sequence, to 52–69 inch lbs. (6–7 Nm).

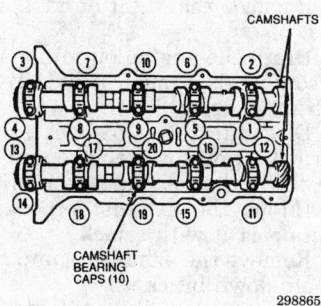

LOOSENING SEQUENCE

CAMSHAFTS

CAMSHAFT BEARING CAPS (10)

298865

Camshaft bearing cap bolt loosening sequence — 2.0L (VIN A) engine

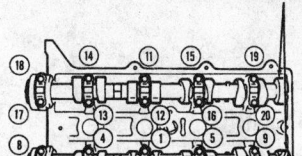

TIGHTENING SEQUENCE

CAMSHAFTS

CAMSHAFT BEARING CAPS (10)

298866

Camshaft bearing cap bolt torque sequence — 2.0L (VIN A) engine

22. Install the power steering hose brackets and tighten the bolts to 71–88 inch lbs. (8–10 Nm).

23. Connect the breather hose and PCV valve.

24. Install the ignition distributor and connect the spark plug wires and clips.

25. Connect the negative battery cable.

26. Run the engine and check for leaks and proper engine operation.

27. Check the ignition timing and adjust if required.

2.5L (VIN B) Engine

1. Disconnect the negative battery cable.

2. Remove the intake manifold.

3. Label and disconnect the spark plug wires from the spark plugs and move aside.

4. Remove the upper timing belt cover bolts. On the left cylinder head cover, disconnect the ventilation pipe from the front of the left side (front) cylinder head cover.

5. Remove the bolts and the cylinder head covers.

6. Remove the timing belt.

7. Hold the camshafts on the hexagon casting and loosen the sprocket retaining bolts. Remove the camshaft sprockets.

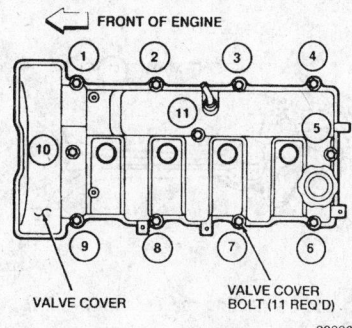

← FRONT OF ENGINE

VALVE COVER

VALVE COVER BOLT (11 REQ'D)

298867

Cylinder head cover bolt torque sequence — 2.0L (VIN A) engine

8. Turn the camshafts so the knock pins are aligned with the marks on the camshaft bearing end caps. This will reduce the pressure on the hydraulic lifters.

9. Note the markings on the camshaft bearing caps prior to removal, so they can be reinstalled in the same positions. The right bank (rear) caps are marked with numbers and the left bank (front) caps are marked with letters.

10. Loosen the camshaft bearing end cap bolts in sequence, in 5–6 steps. Remove the camshaft bearing end caps.

11. Remove the blind caps.

12. Remove the remaining camshaft bearing cap bolts, in 5–6 steps, in the proper sequence. Remove the caps, being sure to remove the thrust caps last. Do not damage the cylinder head thrust bearing support.

13. Remove the camshafts and oil seals. Tag the camshafts for identification.

14. If they are to be removed, number each hydraulic lifter (hydraulic lash adjuster) with a paint marker or equivalent during removal.

15. Inspect the camshafts and lifters for wear and/or damage; replace as necessary.

To install:

16. If removed. apply clean engine oil to the hydraulic lifters and install. If the old lifters are to be reused, be sure to install them in their original positions. Make sure that the hydraulic lifters move freely in their bore.

17. Apply clean engine oil to the camshaft lobes, journals and supports.

18. Install the camshafts so the timing marks align on the camshaft gears.

19. Apply silicone sealant to the cylinder head surface in the area forward of the camshaft gear cavity on both cylinder heads and to the left (front) cylinder head on the rear exhaust camshaft end cap mating surface.

20. Install the thrust caps. Tighten the thrust cap bolts until the caps are fully seated on the cylinder head.

WARNING

Do not install any of the camshaft bearing caps when the camshaft lobes are depressing the valve lifters or damage to the thrust journal support or camshaft may result.

21. Install the remaining camshaft bearing caps and camshaft bearing end caps in their original positions. Tighten the caps, in sequence, in 5 equal steps, with the final step being 98–123 inch lbs. (11–14 Nm).

NOTE: The right bank (rear) camshaft bearing caps are marked with numbers and the left bank (front) camshaft bearing caps are marked with letters.

22. Apply a light coat of clean engine oil to 4 new camshaft oil seals and install the seals using a suitable socket and hammer. The camshaft oil seals should be flush with the front of the cylinder head with a maximum protrusion of 0.020 inch (0.5 mm).

23. Apply sealant to 4 new blind caps and install the blind caps using a plastic hammer or equivalent.

24. Install the camshaft sprockets and retaining bolts. Hold the hexagon casting on the camshafts with a suitable wrench and tighten the retaining bolts to 90–103 ft. lbs. (123–140 Nm).

25. Install the timing belt.

26. Remove any sealant and gasket material from the cylinder head cover contact surfaces.

27. Apply silicone sealant to the cylinder head in the area adjacent to the front and rear camshaft caps. Install a new gasket on the cylinder head.

28. Install the cylinder head cover. Tighten the bolts in 2 steps to 43–78 inch lbs. (5–8 Nm) following the proper torque sequence.

29. Tighten the upper timing cover bolts to 71–88 inch lbs. (8–10 Nm). Connect the ventilation pipe to the left side cylinder head cover.

30. Install the intake manifold.

31. Connect the spark plug wires to the spark plugs.

32. Connect the negative battery cable.

33. Run the engine and check for leaks and proper engine operation.

Piston and Connecting Rod

POSITIONING

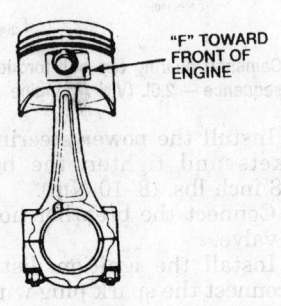

Piston and connecting rod assembly — 2.0L (VIN A) engine

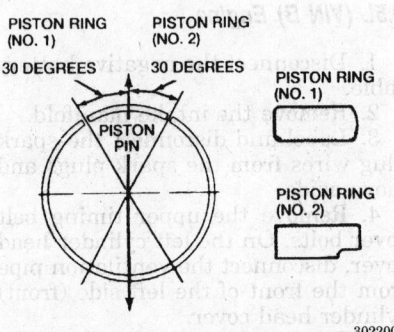

Compression ring positioning — 2.0L (VIN A) engine

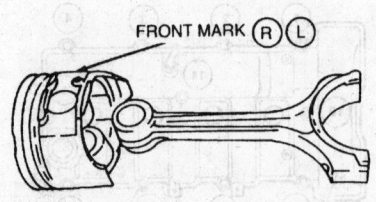

Piston and connecting rod assembly — 2.5L (VIN B) engine

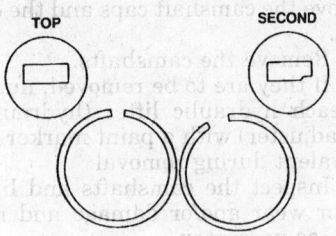

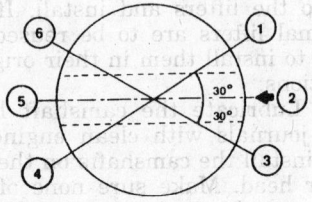

1. Top ring end gap
2. Piston pin end gap
3. Second ring end gap
4. Oil ring lower rail end gap
5. Oil ring spacer end gap
6. Oil ring upper rail end gap

Piston ring positioning — 2.5L (VIN B) engine

ENGINE LUBRICATION

Oil Pan

REMOVAL AND INSTALLATION

2.0L (VIN A) Engine

1. Disconnect the negative battery cable.

2. Raise and safely support the vehicle.

3. Remove the right-hand splash shield.

4. Drain the engine oil into a suitable container. Temporarily reinstall the drain plug.

5. Disconnect the oxygen sensor electrical connector. Remove and discard the exhaust pipe-to-manifold nuts. Move the exhaust pipe aside and support it with a jack.

6. Remove the exhaust clamp from the hold-down bracket.

7. Remove the oil pan bolts and carefully pry the oil pan from the engine stiffener.

To install:

8. Clean the oil pan. Clean all dirt, oil and old sealant from the oil pan and cylinder block and stiffener contact surfaces.

9. Apply a continuous bead of silicone sealant around the oil pan, going on the inside of the bolt holes.

10. Install the oil pan and tighten the bolts to 14–19 ft. lbs. (19–25 Nm).

11. Install a new exhaust clamp and tighten the clamp nuts to 26–34 ft. lbs. (34–47 Nm).

12. Connect the exhaust pipe to the manifold with new nuts. Tighten the nuts to 27–38 ft. lbs. (37–52 Nm).

13. Connect the oxygen sensor electrical connector.

14. Install the right-hand splash shield.

15. Tighten the oil pan drain plug to 22–30 ft. lbs. (30–41 Nm).

16. Lower the vehicle.

17. Fill the engine with the proper type and quantity of engine oil.

18. Connect the negative battery cable.

19. Run the engine and check for leaks.

2.5L (VIN B) Engine

1. Disconnect the negative battery cable.

2. Raise and safely support the vehicle.

3. Drain the engine oil into a suitable container and reinstall the drain plug.

4. Disconnect the 2 oxygen sensor electrical connectors.

5. Remove the 6 crossmember bolts and remove the crossmember.

6. Remove the exhaust pipe-to-manifold nuts (3 per side) and lower

the exhaust system to gain access to the oil pan bolts.

NOTE: The oil pan bolts are different lengths. Identify them as they are removed so they can be reinstalled in their proper locations.

7. Remove the oil pan bolts and the oil pan.

To install:

8. Clean the oil pan. Clean all dirt, oil and old sealant from the oil pan and cylinder block contact surfaces. Remove the old sealant from the threads of the oil pan bolts and the bolt holes in the block.

———— **WARNING** ————
Failure to remove the old sealant from the bolts and bolt holes may cause the block to crack.

9. Apply a continuous bead of silicone sealant along the inside of the bolt holes, overlapping the ends.

NOTE: Once the new silicone sealant is applied, the oil pan must be installed within 5 minutes.

10. Install the oil pan with the bolts. Tighten the long oil pan bolts to 14–18 ft. lbs. (19–25 Nm) and short bolts to 71–88 inch lbs. (8–10 Nm).

11. Connect the exhaust pipes to the manifolds with new gaskets and tighten the nuts to 30–41 ft. lbs. (40–55 Nm).

12. Connect the 2 oxygen sensor electrical connectors.

13. Install the crossmember and tighten the 6 retaining bolts to 69–93 ft. lbs. (94–126 Nm).

14. Tighten the oil pan drain plug to 22–30 ft. lbs. (30–41 Nm).

15. Lower the vehicle.

16. Fill the engine with the proper type and quantity of engine oil.

17. Run the engine and check for leaks.

Oil Pump

REMOVAL AND INSTALLATION

2.0L (VIN A) Engine

1. Disconnect the negative battery cable.

2. Remove the accessory drive belts.

3. Raise and safely support the vehicle.

4. Remove the engine oil pan and the oil pump pick-up tube and screen.

5. Remove the 2 rear main seal housing-to-stiffener nuts and the 12 stiffener bolts in 2 steps, in the proper sequence, and remove the stiffener.

6. Remove the engine front covers, timing belt and crankshaft sprocket.

7. If equipped, remove the A/C compressor and secure it aside, leaving the refrigerant lines attached. Remove the compressor mounting bracket.

8. Remove the 7 oil pump bolts and remove the oil pump.

To install:

9. Clean the oil, dirt and sealant from all contact surfaces.

10. Apply a bead of silicone to the oil pump-to-cylinder block contact surface, along the inside of the bolt holes. Make sure the sealer does not fall into the engine and form plugs that could block oil passages.

11. Install the oil pump and tighten the bolts to 14–19 ft. lbs. (19–25 Nm).

12. Clean the engine block contact area and stiffener. Apply a bead of silicone sealant to the perimeter of the stiffener, going on the inside of the bolt holes. Apply sealant to the 4 rear bolt holes. Make sure the sealer does not fall into the engine and form plugs that could block oil passages.

13. Install the stiffener and the mounting bolts. Tighten the bolts in 2 steps, in sequence, to 14–19 ft. lbs. (19–25 Nm). Tighten the rear main seal housing-to-stiffener nuts to 88 inch lbs. (10 Nm).

14. Install a new gasket and the oil pump pick-up tube and screen. Tighten the mounting bolts to 71–88 inch lbs. (8–10 Nm).

15. Install the oil pan.

16. If equipped, install the A/C compressor bracket and tighten the bolts to 38 ft. lbs. (52 Nm). Install the A/C compressor and tighten the bolts to 26 ft. lbs. (35 Nm).

17. Install the crankshaft sprocket, timing belt and the engine front covers.

18. Install the accessory drive belts.

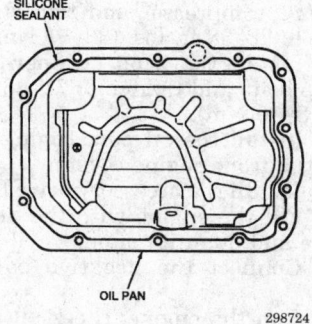

Apply sealant to the oil pan — 2.0L (VIN A) engine

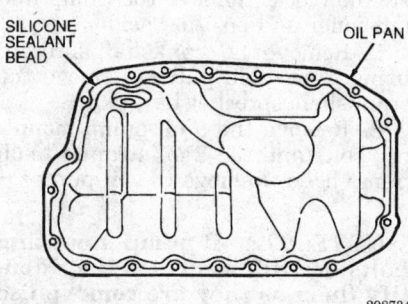

Apply sealant to the oil pan — 2.5L (VIN B) engine

10-53

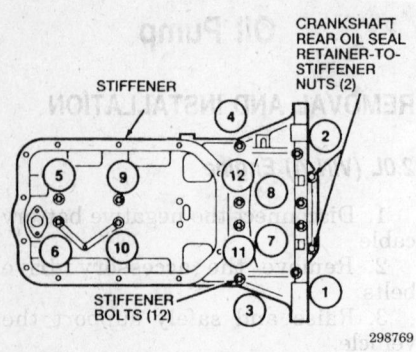

Stiffener bolt and nut loosening sequence — 2.0L (VIN A) engine

298769

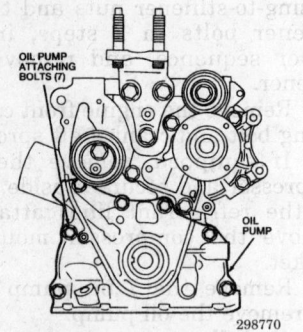

Oil pump attaching bolt locations — 2.0L (VIN A) engine

298770

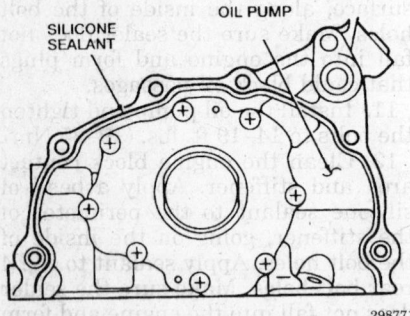

Applying sealant to the oil pump — 2.0L (VIN A) engine

298771

19. Fill the engine with the proper grade and quantity of oil.

20. Connect the negative battery cable.

21. Run the engine; check oil pressure and check for leaks.

2.5L (VIN B) Engine

The oil pump is located behind the crankshaft timing belt sprocket. Oil pump replacement is an in-vehicle service, however it requires that the timing belt covers, timing belt, and oil pan be removed. It is recom-

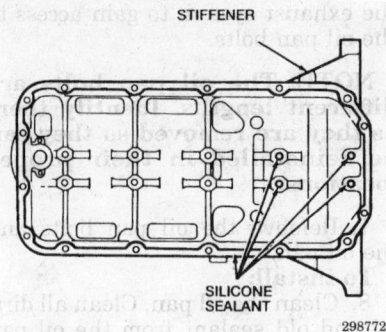

Applying sealant to the stiffener — 2.0L (VIN A) engine

298772

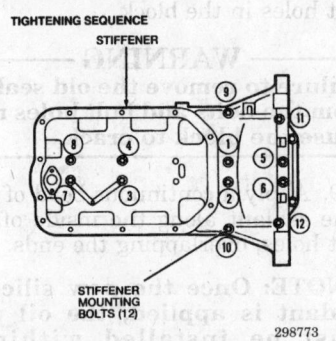

Stiffener bolt torque sequence — 2.0L (VIN A) engine

298773

mended that the front crankshaft seal be replaced whenever the oil pump is removed from the vehicle.

1. Disconnect the negative battery cable.

2. Remove the accessory drive belts.

3. Remove the engine front covers and the timing belt.

4. Remove the engine oil pan.

5. If equipped, remove the 4 A/C compressor mounting bolts and support the compressor aside, without disconnecting the refrigerant lines. Remove the 5 A/C compressor mounting bracket bolts and remove the bracket.

6. Remove the power steering pump and tensioner bolts from the engine block. Remove the pump and tensioner and position aside.

7. Remove the crankshaft sprocket using a suitable puller. Remove the crankshaft sprocket key.

8. Remove the 9 oil pump mounting bolts and the 2 oil strainer-to-oil pump bolts. Remove the oil pump.

NOTE: The oil pump mounting bolts are different lengths. Identify them as they are removed so they can be reinstalled in their proper locations.

9. Remove the oil pump O-ring. If necessary, use a small prybar or punch protected with a rag to knock out the front crankshaft seal from the oil pump housing.

NOTE: Protect the oil pump housing with a rag while removing the crankshaft front seal.

To install:

10. Clean the oil, dirt and sealant from all contact surfaces.

11. Lubricate the new front crankshaft seal lip with clean engine oil. Using front seal replacer T74P-6150-A or equivalent, press or drive the new seal into the oil pump housing. Install the seal so it is flush with the edge of the oil pump housing or protrudes no more than 0.021 inch (0.7 mm) from the edge of the pump body.

12. Install a new O-ring onto the oil pump. Apply a continuous bead of silicone sealant to the oil pump mating surface and install the pump.

13. Install the 9 oil pump mounting bolts.

NOTE: Make sure that the proper length bolts are placed into the correct positions.

14. Tighten bolts A and B to 14–18 ft. lbs. (19–25 Nm).

15. Install the crankshaft timing belt sprocket and key.

16. Install the power steering pump and tensioner. Tighten the 2 power steering belt tensioner upper bolts and the power steering pump rear bracket bolt to 24–33 ft. lbs. (32–46 Nm). Tighten the tensioner lower bolt to 14–18 ft. lbs. (19–25 Nm).

17. If equipped, install the A/C compressor bracket and tighten the bolts to 28–38 ft. lbs. (38–51 Nm). Install the A/C compressor and tighten the bolts to 28–38 ft. lbs. (38–51 Nm).

18. Install the 2 oil strainer-to-oil pump bolts and tighten to 71–88 inch lbs. (8–10 Nm).

19. Install the oil pan, timing belt and the front engine covers.

20. Install the accessory drive belts.

21. Fill the engine with the proper grade and quantity of oil.

22. Connect the negative battery cable.

23. Run the engine; check oil pressure and check for leaks.

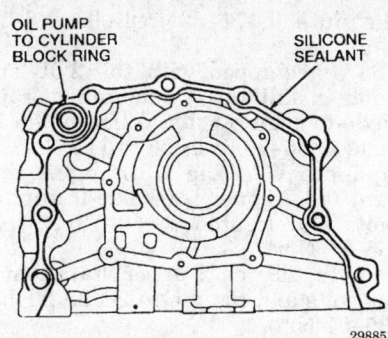

Applying sealant to the oil pump housing — 2.5L (VIN B) engine

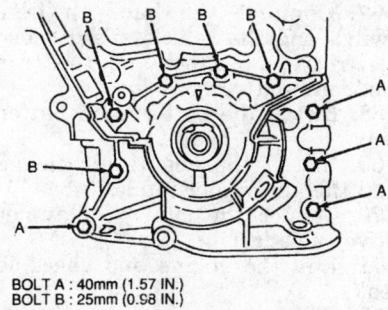

BOLT A : 40mm (1.57 IN.)
BOLT B : 25mm (0.98 IN.)

Oil pump attaching bolts are different lengths — 2.5L (VIN B) engine

TRANSAXLE

Manual Transaxle Assembly

REMOVAL AND INSTALLATION

1. Disconnect the battery cables, negative cable first, and remove the battery and battery tray.
2. Remove the air cleaner intake tube and the engine air cleaner.
3. Remove the transaxle ground strap bolts and straps.
4. Label and disconnect the vehicle speed sensor electrical connector at the top right-hand rear corner of the transaxle.
5. Label and disconnect the park/neutral position switch electrical connector from the lower front of the transaxle.
6. Label and disconnect the backup light switch electrical connector from the rear of the transaxle.
7. Disconnect the 2 spring clips from the clutch slave cylinder hydraulic line and remove the 2 clutch slave cylinder mounting bolts. Posi-tion the clutch slave cylinder aside, without disconnecting the hydraulic line.
8. Support the engine with engine support tool 014-00750 or equivalent.
9. Remove the 4 upper transaxle-to-engine mounting bolts on 2.0L engine or 7 upper transaxle-to-engine mounting bolts on 2.5L engine.
10. Remove the upper starter bolts. Remove the fuel filter mounting nuts and position the filter aside, without disconnecting the fuel lines.
11. Remove the 2 nuts and the through-bolt from the left side engine support insulator.
12. Raise and safely support the vehicle.
13. If equipped with 2.0L engine, remove the intake manifold support bolts and bracket.
14. Disconnect the wiring from the starter, remove the lower starter bolt and remove the starter.
15. Remove the drain plug and drain the transaxle fluid into a suitable container. Discard the drain plug washer.
16. Remove the front wheels.
17. Unstake the left-hand and right-hand halfshaft retaining nuts. Have an assistant apply the brakes to keep the hubs from turning, then remove the nuts and discard them.
18. Remove the lower splash shields.
19. Remove the 6 crossmember bolts and the crossmember.
20. On vehicles equipped with anti-lock brakes, remove the clips from the left-hand side anti-lock sensor and bracket. Remove the bracket nuts from the anti-lock sensor and bracket harness mount on the left-hand side of the vehicle.
21. Remove the lower ball joint pinch bolt and nut from the left side knuckle. Pry the lower ball control arm down to separate the ball joint from the knuckle. Be careful not to damage the ball joint dust boot.
22. Pull the hub/knuckle assembly outward to separate it from the half-shaft. If the halfshaft is stuck in the hub, push it out using a suitable puller.
23. Position a prybar between the transaxle case and inner CV-joint. Pry the left halfshaft from the tran-saxle case. Install transaxle plug tool T88C-7025-AH or equivalent, to keep the differential side gear from becom-ing mispositioned.
24. If equipped with 2.5L engine, disconnect the oxygen sensor connec-tors. Remove and discard the exhaust pipe-to-manifold nuts. Lower the ex-haust system enough to gain access to the right side halfshaft support bearing.
25. Remove the 3 right halfshaft support bearing bolts. If equipped with anti-lock brakes, remove the clips from the wheel speed sensor and the wheel speed sensor mounting nuts from the sensor harness mount on the right side of the vehicle.
26. Remove the lower ball joint pinch bolt and nut from the right side knuckle. Pry the lower ball control arm down to separate the ball joint from the knuckle. Be careful not to damage the ball joint dust boot.
27. Pull the hub/knuckle assembly outward to separate it from the half-shaft. If the halfshaft is stuck in the hub, push it out using a suitable puller.
28. Pull the right halfshaft from the transaxle case. Install transaxle plug tool T88C-7025-AH or equivalent, to keep the differential side gear from becoming mispositioned.
29. Remove the 6 nuts and 2 bolts from the rear engine support and re-move the rear engine support.
30. Disconnect the shift rod and control rod from the transaxle.
31. Remove the 3 rear transaxle support bracket bolts.
32. Support the transaxle with a suitable jack.
33. Remove the 3 rear transaxle support insulator bolts and the rear transaxle support insulator.
34. Remove the 2 lower transaxle-to-engine mounting bolts on the 2.0L engine or the 3 lower transaxle-to-en-gine mounting bolts on the 2.5L engine.
35. Separate the transaxle from the engine and carefully lower it from the vehicle.

To install:

36. Place the transaxle on a suita-ble jack. Apply a thin coating of mo-lybdenum grease or equivalent, to the input shaft splines. Raise the transaxle into position and align it with the engine.
37. Position the transaxle to the en-gine and loosely install the lower transaxle-to-engine bolts. Remove the transmission jack.
38. Install the 2 nuts and through-bolt to the left-hand engine support insulator. Tighten the nuts to 32–44 ft. lbs. (44–60 Nm) and the through-bolt to 63–86 ft. lbs. (86–116 Nm).
39. Install the rear transaxle sup-port insulator with the 3 bolts. Tighten the bolts to 50–68 ft. lbs. (67–93 Nm).
40. Connect the control rod and shift rod to the transaxle. Tighten the control rod nut to 28–38 ft. lbs.

(38–51 Nm) and the shift rod bolt and nut to 14–18 ft. lbs. (19–25 Nm).

41. Install the rear engine support and tighten the bolts and nuts as follows:
- Tighten **A** to 55–77 ft. lbs. (75–104 Nm)
- Tighten **B** to 50–68 ft. lbs. (67–93 Nm)
- Tighten **C** to 32–44 ft. lbs. (44–60 Nm)

42. Remove the transaxle plug from the right-hand side of the transaxle and install the right halfshaft. Pull out on the right hub/knuckle assembly and install the halfshaft into the hub.

43. Pry the lower control arm down and insert the lower ball joint stud into the knuckle. Install the pinch bolt and nut and tighten to 26–41 ft. lbs. (35–56 Nm).

44. If equipped with anti-lock brakes, install the anti-lock sensor harness mounting nuts and tighten to 71–88 inch lbs. (8–10 Nm). Install the sensor harness clips.

45. Install the 3 halfshaft support bearing bracket bolts. Tighten the bolts, in sequence, to 32–45 ft. lbs. (43–61 Nm).

46. Remove the transaxle plug from the left-hand side and install the left halfshaft. Pull out on the left

2.0L

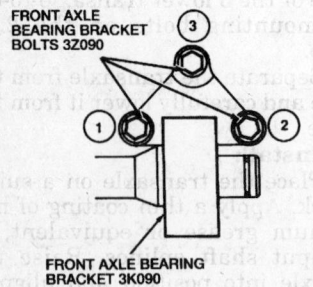

FRONT AXLE BEARING BRACKET BOLTS 3Z090

FRONT AXLE BEARING BRACKET 3K090

2.5L

2.5L MODELS

FRONT AXLE BEARING BRACKET BOLTS 3Z090

FRONT AXLE BEARING BRACKET 3K090

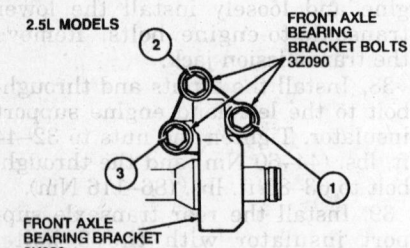

305235

Front axle bearing bracket bolt locations and torque sequence

hub/knuckle assembly and install the halfshaft into the hub.

47. Pry the lower control arm down and insert the lower ball joint stud into the knuckle. Install the pinch bolt and nut and tighten to 26–41 ft. lbs. (35–56 Nm).

48. If equipped with anti-lock brakes, install the anti-lock sensor harness mounting nuts and tighten to 71–88 inch lbs. (8–10 Nm). Install the sensor harness clips.

49. If equipped with 2.5L engine, connect the exhaust pipes to the manifolds and tighten the new nuts to 30–41 ft. lbs. (40–55 Nm). Connect the 2 oxygen sensor connectors.

50. Position the crossmember and tighten the 6 retaining bolts to 69–96 ft. lbs. (94–131 Nm).

51. Install new halfshaft retaining nuts. Have an assistant apply the brakes to lock the hubs, then tighten the nuts to 174–235 ft. lbs. (235–319 Nm). Stake the nuts using a dull chisel or similar tool.

52. Install the wheels and torque the lug nuts to 65–87 ft. lbs. (88–118 Nm).

53. Install a new washer on the transaxle drain plug. Install and tighten the drain plug to 29–43 ft. lbs. (40–58 Nm).

54. Position the starter motor and tighten the lower retaining bolt to 28–38 ft. lbs. (35–52 Nm). Connect the starter wiring.

55. If equipped with 2.0L engine, install the intake manifold support bracket and bolts. Tighten the bolts to 27–38 ft. lbs. (37–52 Nm).

56. Install the lower splash shields.

57. If equipped with the 2.0L engine, install the 2 lower transaxle-to-engine bolts. Tighten bolt **B** to 28–38 ft. lbs. (38–51 Nm) and tighten bolt **C** to 14–18 ft. lbs. (19–25 Nm). Install the 2 engine-to-transaxle bolts. Tighten bolt **D** to 28–38 ft. lbs. (38–51 Nm) and bolt **E** to 66–86 ft. lbs. (90–116 Nm).

58. If equipped with the 2.5L engine, install and tighten the 3 lower transaxle-to-engine bolts to 28–38 ft. lbs. (38–51 Nm).

59. Install the 3 rear transaxle support bracket bolts and tighten to 50–68 ft. lbs. (67–93 Nm).

60. Fill the transaxle with the proper fluid to a level even with the lower edge of the oil level plug port, with the vehicle level. Install the plug, using a new washer, and tighten to 29–43 ft. lbs. (40–58 Nm) if not already done.

61. Lower the vehicle.

62. Position the fuel filter and install the 2 retaining nuts. Tighten

the nuts to 71–88 inch lbs. (8–10 Nm).

63. If equipped with the 2.0L engine, install the 4 remaining transaxle-to-engine bolts. Tighten the **A** bolts to 66–86 ft. lbs. (90–116 Nm). If equipped with the 2.5L engine, install the 7 upper transaxle-to-engine bolts and tighten to 50–73 ft. lbs. (68–99 Nm).

64. Install the 2 upper starter motor bolts and tighten to 28–38 ft. lbs. (35–51 Nm).

65. Remove the engine support tool.

66. Install the clutch slave cylinder and the 2 retaining bolts. Tighten the bolts to 12–16 ft. lbs. (16–22 Nm).

67. Connect the back-up lamp switch and the vehicle speed sensor electrical connectors and the transaxle ground straps.

68. Install the battery and battery tray.

69. Install the engine air cleaner and the air cleaner intake tube.

70. Connect the battery cables, negative cable last.

71. Run the engine and check for leaks.

72. Check transaxle operation and bleed the clutch hydraulic system if required.

Clutch Assembly

REMOVAL AND INSTALLATION

1. Disconnect the negative battery cable.

2. Remove the transaxle assembly.

3. Position a suitable clutch alignment tool through the pressure plate, clutch disc and into the pilot bearing; this will keep the assembly from dropping when the bolts are removed.

4. Install flywheel holding tool T74P-6375-A or equivalent, to keep the flywheel from turning.

5. Remove the pressure plate-to-flywheel bolts evenly to relieve the spring pressure.

6. Remove the pressure plate, clutch disc and alignment tool.

7. Inspect the pressure plate and clutch disc for wear and/or damage and replace, as necessary.

8. Inspect the pilot bearing for excessive wear or scoring. Remove it using a suitable puller, only if replacement is necessary.

9. Inspect the flywheel for scoring, cracks, worn or broken teeth, or other damage. Remove the flywheel if machining or replacement is necessary. Use care when removing the last bolt to prevent dropping the flywheel.

2.0L

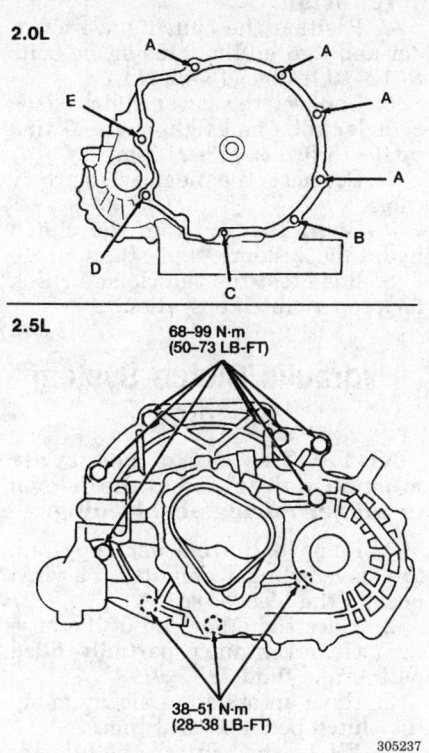

2.5L

68–99 N·m
(50–73 LB-FT)

38–51 N·m
(28–38 LB-FT)

305237

Engine and transaxle mounting bolt locations

CLUTCH ALIGNER
T74P-7137-K

CLUTCH
PRESSURE
PLATE

FLYWHEEL

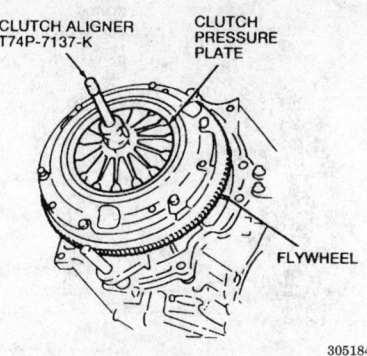

305184

Clutch aligner tool installed

10. Remove the release bearing and fork. Inspect them for wear or damage and replace as necessary

To install:

11. Apply molybdenum grease to the release bearing where it contacts the release fork. Apply molybdenum grease to the release fork at the pivot point and to the area where it contacts the release bearing.

12. Install the release fork and bearing.

13. If removed, install the flywheel. Make sure the crankshaft flange and flywheel mating surfaces are clean. Remove the old sealant from the flywheel bolts and apply stud and bearing mount sealant to them. If the old

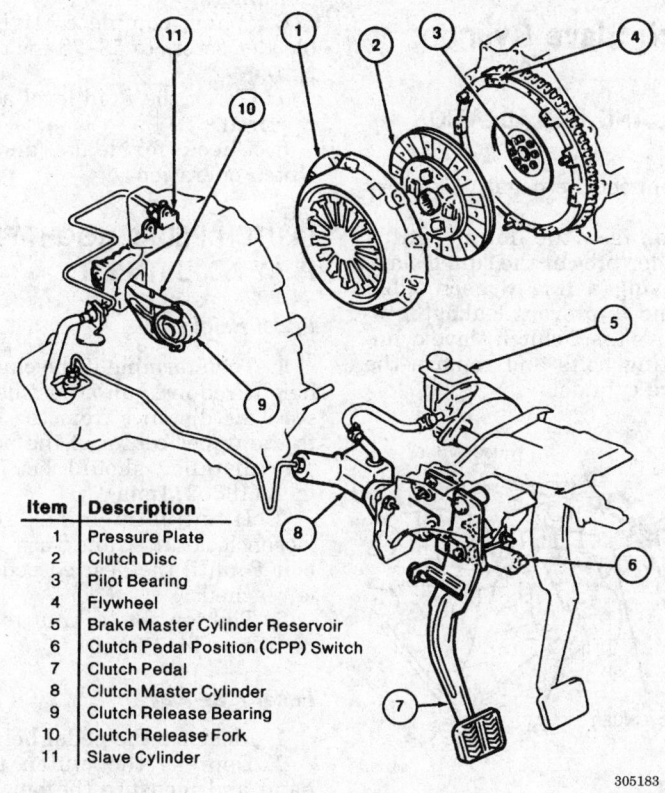

Item	Description
1	Pressure Plate
2	Clutch Disc
3	Pilot Bearing
4	Flywheel
5	Brake Master Cylinder Reservoir
6	Clutch Pedal Position (CPP) Switch
7	Clutch Pedal
8	Clutch Master Cylinder
9	Clutch Release Bearing
10	Clutch Release Fork
11	Slave Cylinder

305183

Clutch assembly

sealant cannot be removed, replace the bolts.

14. Install the flywheel holding tool. Tighten the flywheel bolts, in sequence, to 71–75 ft. lbs. (97–102 Nm) on 2.0L engine or 45–49 ft. lbs. (61–67 Nm) on 2.5L engine.

15. If removed, install a new pilot bearing using a suitable installation tool. When installed, the pilot bearing should be 0–0.016 in. (0–0.4mm) below the surface of the crankshaft flange.

16. Apply a small amount of molybdenum grease to the clutch disc and input shaft splines. Do not let grease get on the clutch face.

17. Install the clutch disc with the spring side of the disc toward the transaxle. Install the alignment tool to hold the disc in place.

18. Install the pressure plate to the flywheel. Install the pressure plate-to-flywheel bolts and tighten evenly until the bolts are seated. Torque to 13–18 ft. lbs. (18–26 Nm) in the proper sequence.

19. Install the transaxle assembly.

20. Connect the negative battery cable.

21. Check for proper clutch operation.

Clutch Master Cylinder

REMOVAL AND INSTALLATION

— **CAUTION** —
Brake fluid contains Polyglycols and Polyglycol Ethers. Do not allow brake fluid to contact skin or eyes. Do not allow brake fluid to contact vehicle painted surfaces.

1. Disconnect the negative battery cable.

2. Using pliers, slide the spring clip from the clutch master cylinder end of the reservoir hose. Remove and plug the hose.

3. Loosen and remove the clutch slave cylinder to clutch master cylinder tube at the clutch master cylinder, using a line wrench, while taking care not to twist the tube.

4. Working inside the vehicle, remove the clutch master cylinder upper retaining nut. From the engine compartment, remove the clutch master cylinder lower retaining nut.

5. Carefully pull the clutch master cylinder from the bulkhead and discard the gasket.

To install:

6. Carefully position the clutch master cylinder to the bulkhead using a new mounting gasket.

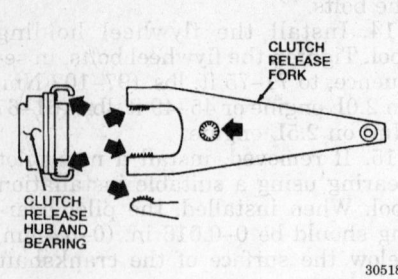

Installing the release bearing to the clutch fork

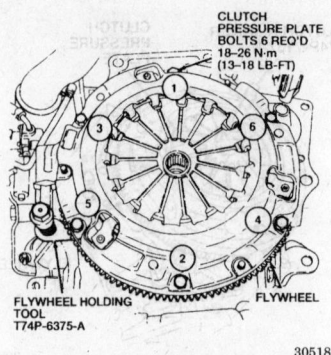

Clutch pressure plate bolt torque sequence

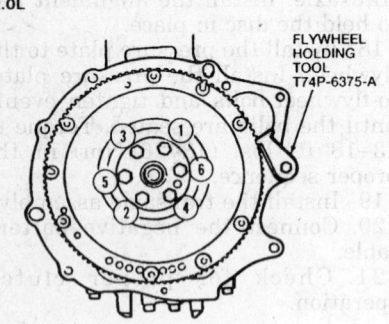

Flywheel bolt torque sequence

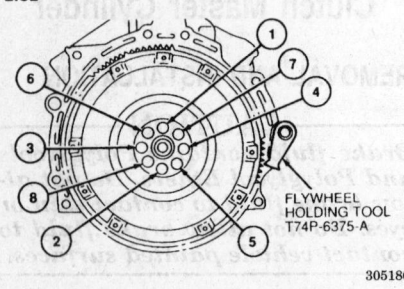

Flywheel bolt torque sequence

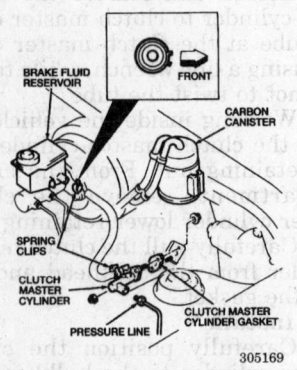

Clutch master cylinder

7. Tighten the 2 clutch master cylinder mounting nuts to 14–18 ft. lbs. (19–26 Nm).

8. Connect the clutch slave cylinder to clutch mater cylinder tube and tighten the nut to 10–15 ft. lbs. (13–21 Nm).

9. Unplug and connect the clutch master cylinder end of the reservoir hose. Properly position the spring clip.

10. Bleed the air from the clutch hydraulic system.

11. Connect the negative battery cable.

12. Road test the vehicle to check for proper clutch operation.

Clutch Slave Cylinder

REMOVAL AND INSTALLATION

1. Disconnect the negative battery cable.

2. Disconnect the lower clutch slave cylinder tube at the clutch slave cylinder using a line wrench. Plug the tube end to prevent leakage.

3. Remove the 2 clutch slave cylinder mounting bolts and remove the clutch slave cylinder.

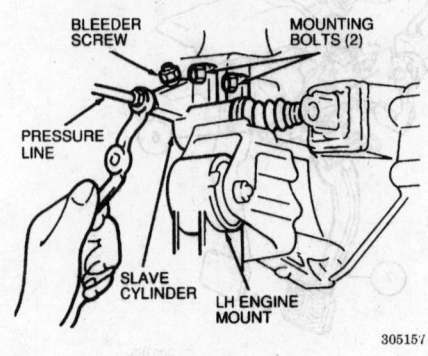

Clutch slave cylinder

To install:

4. Position the clutch slave cylinder and tighten the 2 mounting bolts to 12–16 ft. lbs. (16–22 Nm).

5. Connect the lower clutch slave cylinder tube and tighten the fitting to 10–15 ft. lbs. (13–21 Nm).

6. Connect the negative battery cable.

7. Bleed the air from the clutch hydraulic system.

8. Road test the vehicle and check for proper clutch operation.

Hydraulic Clutch System Bleeding

NOTE: The fluid reservoir must be maintained at the ¾ level or higher during air bleeding.

1. Remove the bleeder cap from the slave cylinder and attach a vinyl hose to the bleeder screw.

2. Place the other end of the hose in a clear container partially filled with brake fluid.

3. Have an assistant slowly pump the clutch pedal several times.

4. With the clutch pedal depressed, loosen the bleeder screw to release the fluid and air.

5. Tighten the bleeder screw. Repeat this procedure until there are no air bubbles in the fluid in the container.

6. When complete, tighten the bleeder screw to 53–78 inch lbs. (6–8 Nm).

7. Check the fluid level and fill as required.

8. Check for leaks and proper clutch operation.

CLUTCH PEDAL HEIGHT/FREE-PLAY ADJUSTMENT

Pedal Height

1. To determine if the clutch pedal height requires an adjustment, measure the distance from the bulkhead to the upper center of the pedal pad. The distance should be 7.32–8.31 inch (186–211mm).

2. If adjustment is required, loosen locknut **A** and turn adjusting bolt **B** until the desired pedal height is reached.

3. Tighten the locknut to 122–156 inch lbs. (14–17 Nm).

Pedal Free-Play

1. Measure the pedal height.

2. Depress the clutch pedal by hand and measure the height of the pedal when resistance is felt.

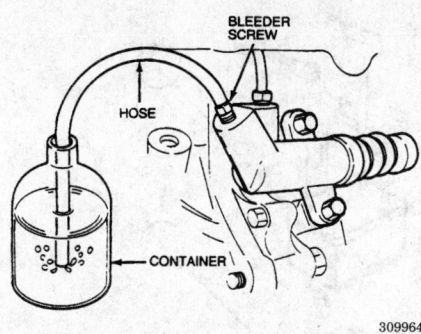

Bleeding the clutch hydraulic system

309964

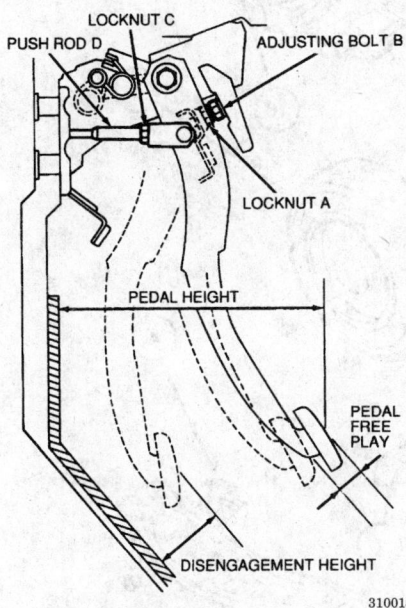

Clutch pedal adjustment

310018

3. The free-play should be 0.04–0.12 inch (1–3mm).

4. If adjustment is necessary, proceed as follows:

    a. Loosen locknut **C** and turn clutch master cylinder pushrod **D** until the pedal play is within specifications.

    b. Measure the distance from the floor to the center of the pedal pad when the pedal is fully depressed. The distance should be 2.64 inches (67 mm).

    c. Tighten the locknut to 105–147 inch lbs. (12–16 Nm).

Automatic Transaxle Assembly

REMOVAL AND INSTALLATION

1993 Vehicles

2.0L (VIN A) Engine

1. Disconnect the battery cables, negative cable first and remove the battery and battery tray.

2. Remove the air cleaner assembly.

3. Pry the shift cable from the transaxle manual lever. Remove the cable bracket lock tab retainer, press in on the lock tabs and pull the cable through the bracket.

4. Disconnect the Manual Lever Position (MLP) sensor electrical connector.

5. Disconnect the heated oxygen sensor electrical connector.

6. Disconnect the transaxle electrical connector.

7. Remove the wiring harness bracket from the cable bracket.

8. Disconnect the Vehicle Speed Sensor (VSS) electrical connector. Remove the ground wire bracket and the ground wire.

9. Remove the harness support bracket to the engine block, located at the rear transaxle mount.

10. Disconnect and plug the oil cooler inlet and outlet hoses.

11. Remove the 4 transaxle-to-engine mounting bolts.

12. Support the engine from above with engine support tool 014-00750 or equivalent.

13. Remove the 2 left-hand transaxle mount nuts and bolt and the mount through-bolt.

14. Remove the 2 fuel filter bracket nuts from the left transaxle mount. Position the filter and bracket aside without disconnecting the fuel lines.

15. Remove the left-hand transaxle mount.

16. Disconnect the pulse signal generator connector.

17. Raise and safely support the vehicle.

18. Remove the front wheels and the splash shields.

19. Remove the 6 crossmember bolts and the crossmember.

20. Remove the 6 rear engine support nuts and 2 bolts and remove the rear engine support.

21. Remove the halfshafts.

22. Remove the intake manifold support bracket and the starter.

23. Disconnect the transaxle vent hose and the dipstick tube.

24. Remove the seal rubber located next to the starter opening. Use a small prybar to hold the flexplate and reach through the opening to remove the torque converter nuts.

25. Support the transaxle with a suitable transmission jack. Secure the transaxle to the jack to keep it from falling.

26. Remove the engine-to-transaxle and transaxle-to-engine bolts.

27. Remove the 3 rear transaxle mount bolts.

28. Use a small prybar or suitable tool to separate the transaxle from the engine. Slightly tilt the transaxle and engine to ease removal.

29. Remove the transaxle from the engine and lower the transaxle from the vehicle.

To install:

NOTE: Flush the transmission oil cooler lines whenever the transaxle is being replaced or rebuilt.

30. Place the transaxle onto a suitable transmission jack and secure the transaxle to the jack.

31. Raise the transaxle into position with the engine. Align the torque converter studs with the flexplate.

32. Install the transaxle-to-engine bolts and tighten as follows:

• Bolt **B** to 50–73 ft. lbs. (68–99 Nm)

• Bolt **C** to 28–38 ft. lbs. (38–51 Nm)

33. Install the engine-to-transaxle bolts and tighten as follows:

• Bolt **D** to 14–18 ft. lbs. (19–25 Nm)

• Bolt **E** to 28–38 ft. lbs. (38–51 Nm)

• Bolt **F** to 50–73 ft. lbs. (68–99 Nm).

34. Install the 3 rear transaxle mount bolts and tighten to 50–68 ft. lbs. (67–93 Nm). Install the 4 torque converter-to-flexplate nuts and tighten to 32–45 ft. lbs. (44–60 Nm).

35. Install the intake manifold support bracket. Tighten the bolts to 27–38 ft. lbs. (37–52 Nm).

36. Install the starter motor and the starter motor wiring.

37. Connect the transaxle vent hose and install the dipstick tube. Tighten the dipstick tube mounting bolts to 71–88 inch lbs. (8–10 Nm).

38. Remove the transaxle plugs and install the halfshafts.

39. Install the transaxle lower mount and tighten the bolts to 50–68 ft. lbs. (67–93 Nm).

40. Remove the transmission jack.

41. Install the rear engine support. Tighten the rear engine support bolts and nuts to 50–68 ft. lbs. (67–93 Nm).

2.0L Models

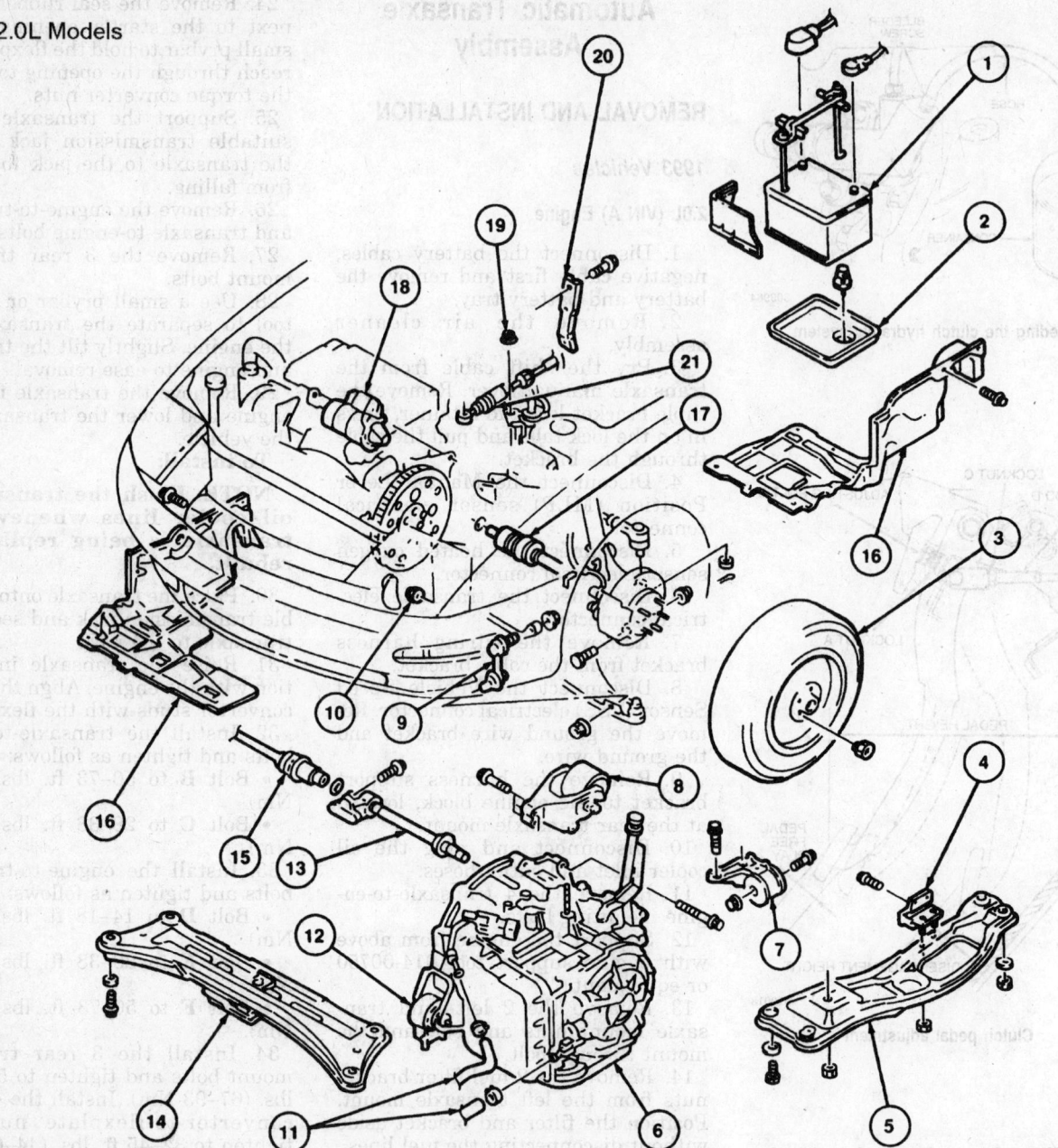

1 Battery
2 Battery tray
3 Wheel and tire
4 Lower transaxle mount
5 Transaxle cradle
6 Transaxle
7 LH transaxle mount
8 Rear transaxle mount
9 Stabilizer link
10 Torque converter nut
11 Oil cooler hoses

12 Front transaxle mount
13 Joint shaft
14 Transverse member
15 RH halfshaft
16 Splash shield
17 LH halfshaft
18 Starter
19 Fuel filter bracket mounting nut
20 Intake manifold support bracket
21 Shift cable

247170

Transaxle mounting points — 1993 2.0L (VIN A) engine

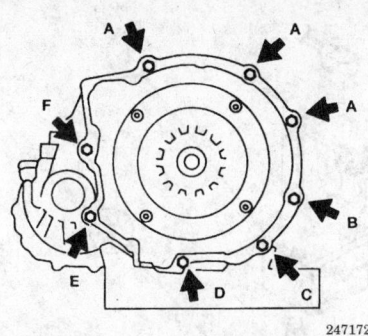

247172

**Engine-to-transaxle mounting bolt
identification — 1993 2.0L (VIN A) engine**

Tighten the rear engine support-to-front mount nuts to 55–77 ft. lbs. (75–104 Nm) and the rear engine support-to-rear mount nuts to 32–44 ft. lbs. (44–60 Nm).

42. Install the crossmember and tighten the 6 bolts to 68–96 ft. lbs. (94–131 Nm).

43. Install the splash shields and the wheels. Tighten the lug nuts to 67–86 ft. lbs. (88–118 Nm).

44. Lower the vehicle.

45. Install the 3 transaxle-to-engine bolts **A** and tighten to 50–73 ft. lbs. (68–99 Nm).

46. Connect the VSS and the pulse signal generator electrical connectors. Install the ground wire bracket and the ground wire.

47. Install the harness support bracket to the engine block located at the rear transaxle mount.

48. Install the left side transaxle mount. Tighten the 2 nuts and bolt to 50–68 ft. lbs. (67–93 Nm) and the through-bolt to 63–86 ft. lbs. (86–116 Nm).

49. Remove the engine support tool.

50. Install the fuel filter bracket and tighten the retaining nuts to 71–88 inch lbs. (8–10 Nm).

51. Connect the oil cooler inlet and outlet hoses.

52. Connect the transaxle electrical connector and the oxygen sensor electrical connector.

53. Insert the shift cable through the cable bracket and pull the cable until the lock tabs engage. Install the lock tab retainer. Connect the shift cable to the manual lever arm.

54. Connect the MLP switch electrical connector. Snap the wiring harness bracket on the cable bracket.

55. Install the air cleaner assembly.

56. Install the battery tray and battery. Connect the battery cables, negative cable last.

57. Fill the transaxle with the proper type and quantity of fluid.

58. Run the engine and check for leaks.

59. Road test and check for proper transaxle operation.

2.5L (VIN B) Engine

1. Disconnect the battery cables, negative cable first and remove the battery and battery tray.

2. Remove the engine air cleaner assembly.

3. Pry the shift cable from the manual control shift outer lever. Remove the cable bracket lock tab retainer, press in on the lock tabs to release the shift cable and pull the cable through the bracket.

4. Disconnect the Manual Lever Position (MLP) sensor electrical connector. Disconnect the 2 heated oxygen sensor electrical connectors and the transaxle electrical connector.

5. Remove the wiring harness bracket from the shift cable bracket.

6. Remove the starter motor wiring and remove the starter motor.

7. Disconnect the Vehicle Speed Sensor (VSS) electrical connector. Remove the ground wire bracket and the ground wire.

8. Remove the harness support bracket to the alternator, located at the rear engine support insulator bracket.

9. Disconnect and plug the inlet and outlet oil cooler tubes.

10. Remove the 3 transaxle-to-engine mounting bolts.

11. Support the engine from above with engine support tool 014-00750 or equivalent.

12. Remove the 2 left-hand engine support insulator nuts and bolt and the mount through-bolt.

13. Remove the 2 fuel filter bracket nuts from the left transaxle mount. Position the filter and bracket aside without disconnecting the fuel lines.

14. Remove the left-hand engine support insulator bracket.

15. Disconnect the Pulse Signal Generator (PSG) electrical connector.

16. Raise and safely support the vehicle.

17. Remove the front wheels and the splash shields.

18. Remove the 6 crossmember bolts and the crossmember.

19. Remove the 6 rear engine support nuts and 2 bolts and remove the rear engine support.

20. Remove the 2 lower transaxle support insulator bolts and remove the lower transaxle support insulator.

21. Remove the halfshafts and install transaxle plugs to prevent the differential gears from becoming misaligned.

22. Disconnect the transaxle vent hose and the dipstick tube.

23. Remove the 3 inspection cover bolts. Use a small prybar to hold the flexplate and remove the 4 torque converter nuts.

24. Support the transaxle with a suitable transmission jack. Secure the transaxle to the transmission jack to keep it from falling.

25. Remove the 2 engine-to-transaxle bolts.

26. Remove the 3 rear transaxle support bracket bolts.

27. Use a small prybar to separate the transaxle from the engine. Slightly tilt the transaxle and engine to ease removal.

28. Carefully remove the transaxle from the engine and lower the transaxle from the vehicle.

To install:

NOTE: If the transaxle was rebuilt or replaced, be sure to flush the transmission oil cooler tubes to remove contaminants before installing the oil cooler tubes to the transaxle.

29. Position the transaxle onto a suitable transmission jack and secure the transaxle to the jack.

30. Raise the transaxle into position. Align the torque converter studs with the flexplate.

31. Install the 2 engine-to-transaxle bolts and tighten to 50–73 ft. lbs. (68–99 Nm).

32. Install the 3 rear transaxle support bracket bolts and tighten to 50–68 ft. lbs. (67–93 Nm).

33. Install the 4 torque converter-to-flexplate nuts and tighten to 32–45 ft. lbs. (44–60 Nm).

34. Install the inspection cover.

35. Connect the transaxle vent hose and install the dipstick tube. Tighten the dipstick tube mounting bolts to 71–88 inch lbs. (8–10 Nm).

36. Remove the transaxle plugs and install the halfshafts.

37. Install the lower transaxle support insulator and tighten the bolts to 41–59 ft. lbs. (55–80 Nm).

38. Remove the transmission jack.

39. Install the rear engine support. Tighten the rear engine support-to-body bolts and nuts to 50–68 ft. lbs. (67–93 Nm). Tighten the rear engine support-to-front mount nuts to 55–77 ft. lbs. (75–104 Nm) and the rear engine support-to-rear mount nuts to 32–44 ft. lbs. (44–60 Nm).

40. Install the crossmember and tighten the 6 bolts to 68–96 ft. lbs. (94–131 Nm).

41. Install the splash shields and the front wheels. Torque the lug nuts to 66–86 ft. lbs. (88–118 Nm).

42. Lower the vehicle.

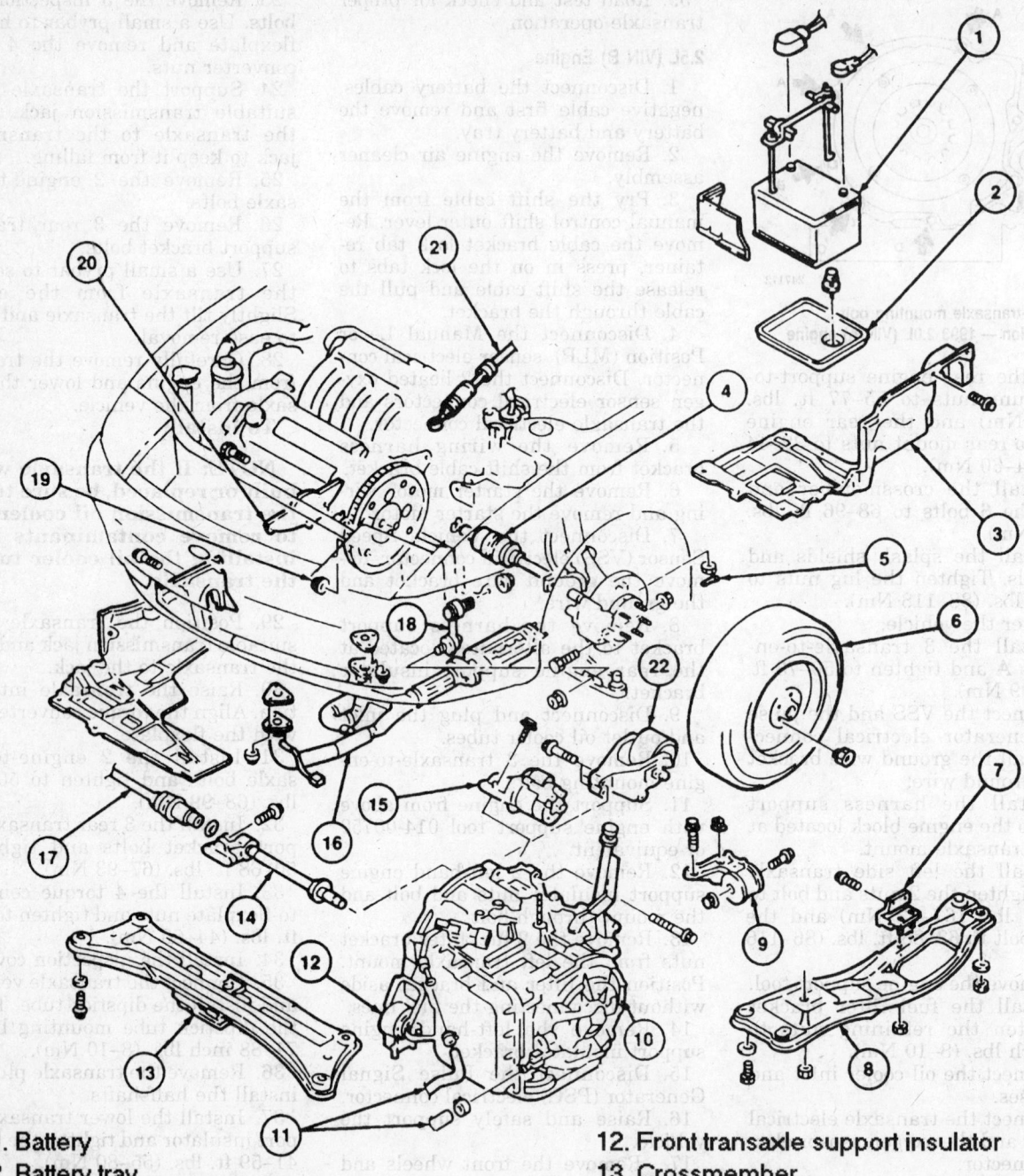

1. Battery
2. Battery tray
3. Front fender splash shield
4. LH front wheel driveshaft & joint
5. Cotter pin
6. Wheel & tire
7. Lower transaxle support insulator
8. Rear engine support
9. LH transaxle support insulator
10. Transaxle
11. Oil cooler tubes
12. Front transaxle support insulator
13. Crossmember
14. Halfshaft
15. Starter motor
16. Exhaust inlet pipe
17. RH front wheel driveshaft & joint
18. Stabilizer bar link
19. Flywheel to converter retaining nut
20. Cover plate
21. Transmission shift cable & bracket
22. Rear transaxle support insulator

247171

Transaxle mounting points — 1993 2.5L (VIN B) engine

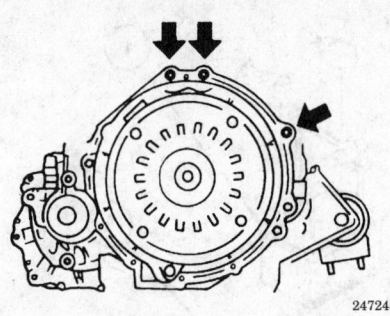

Location of the 3 transaxle-to-engine bolts — 2.5L (VIN B) engine

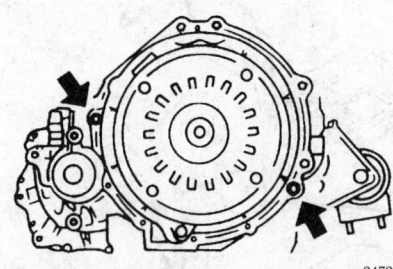

Location of the 2 engine-to-transaxle bolts — 2.5L (VIN B) engine

43. Install the 3 transaxle-to-engine support bolts and tighten to 50–73 ft. lbs. (68–99 Nm).

44. Connect the VSS and the PSG electrical connectors. Install the ground wire bracket and the ground wire.

45. Install the harness support bracket to the engine block located at the left-hand engine support insulator bracket.

46. Install the left-hand engine support insulator and 2 nuts and 1 bolt. Tighten the 2 nuts and bolt to 50–68 ft. lbs. (67–93 Nm). Install the through-bolt and tighten to 63–86 ft. lbs. (86–116 Nm).

47. Remove the engine support tool.

48. Install the fuel filter bracket and tighten the 2 nuts to 71–88 inch lbs. (8–10 Nm).

49. Unplug and connect the inlet and outlet oil cooler tubes.

50. Install the starter motor and wiring.

51. Connect the transaxle electrical connector and the 2 heated oxygen sensor electrical connectors.

52. Insert the shift cable through the cable bracket and pull the cable until the lock tabs engage. Install the lock tab retainer. Connect the shift cable to the manual control shift outer lever.

53. Connect the MLP sensor electrical connector. Snap the wiring harness bracket on the cable bracket.

54. Install the air cleaner assembly.

55. Install the battery tray and battery. Connect the battery cables, negative cable last.

56. Fill the transaxle with the proper type and quantity of fluid.

57. Run the engine and check for leaks. Road test and check for proper transaxle operation.

1994–97 Vehicles

With 4EAT Transaxle

1. Disconnect the battery cables, negative cable first, and remove the battery and battery tray.

2. Remove the engine air cleaner assembly.

3. Pry the shift cable from the manual control shift outer lever. Remove the cable bracket lock tab retainer, press in on the lock tabs to release the shift cable and pull the cable through the bracket.

4. Disconnect the Manual Lever Position (MLP) sensor electrical connector. Disconnect the 2 heated oxygen sensor electrical connectors and the transaxle electrical connector.

5. Remove the wiring harness bracket from the shift cable bracket.

6. Remove the starter motor wiring and remove the starter motor.

7. Disconnect the Vehicle Speed Sensor (VSS) electrical connector. Remove the ground wire bracket and the ground wire.

8. Remove the harness support bracket to the alternator, located at the rear engine support insulator bracket.

9. Disconnect and plug the inlet and outlet oil cooler tubes.

10. Remove the 3 transaxle-to-engine mounting bolts.

11. Support the engine from above with engine support tool 014-00750 or equivalent.

12. Remove the 2 left-hand engine support insulator nuts and bolt and the mount through-bolt.

13. Remove the 2 fuel filter bracket nuts from the left transaxle mount. Position the filter and bracket aside without disconnecting the fuel lines.

14. Remove the left-hand engine support insulator bracket.

15. Disconnect the Pulse Signal Generator (PSG) electrical connector.

16. Raise and safely support the vehicle.

17. Remove the front wheels and the splash shields.

18. Remove the 6 crossmember bolts and the crossmember.

19. Remove the 6 rear engine support nuts and 2 bolts and remove the rear engine support.

20. Remove the 2 lower transaxle support insulator bolts and remove the lower transaxle support insulator.

21. Remove the halfshafts and install transaxle plugs to prevent the differential gears from becoming misaligned.

22. Disconnect the transaxle vent hose and the dipstick tube.

23. Remove the 3 inspection cover bolts. Use a small prybar to hold the flexplate and remove the 4 torque converter nuts.

24. Support the transaxle with a suitable transmission jack. Secure the transaxle to the transmission jack to keep it from falling.

25. Remove the 2 engine-to-transaxle bolts.

26. Remove the 3 rear transaxle support bracket bolts.

27. Use a small prybar to separate the transaxle from the engine. Slightly tilt the transaxle and engine to ease removal.

28. Carefully remove the transaxle from the engine and lower the transaxle from the vehicle.

To install:

NOTE: If the transaxle was rebuilt or replaced, be sure to flush the transmission oil cooler tubes to remove contaminants before installing the oil cooler tubes to the transaxle.

29. Position the transaxle onto a suitable transmission jack and secure the transaxle to the jack.

30. Raise the transaxle into position. Align the torque converter studs with the flexplate.

31. Install the 2 engine-to-transaxle bolts and tighten to 50–73 ft. lbs. (68–99 Nm).

32. Install the 3 rear transaxle support bracket bolts and tighten to 50–68 ft. lbs. (67–93 Nm).

33. Install the 4 torque converter-to-flexplate nuts and tighten to 32–45 ft. lbs. (44–60 Nm).

34. Install the inspection cover.

35. Connect the transaxle vent hose and install the dipstick tube. Tighten the dipstick tube mounting bolts to 71–88 inch lbs. (8–10 Nm).

36. Remove the transaxle plugs and install the halfshafts.

37. Install the lower transaxle support insulator and tighten the bolts to 41–59 ft. lbs. (55–80 Nm).

38. Remove the transmission jack.

39. Install the rear engine support. Tighten the rear engine support-to-body bolts and nuts to 50–68 ft. lbs.

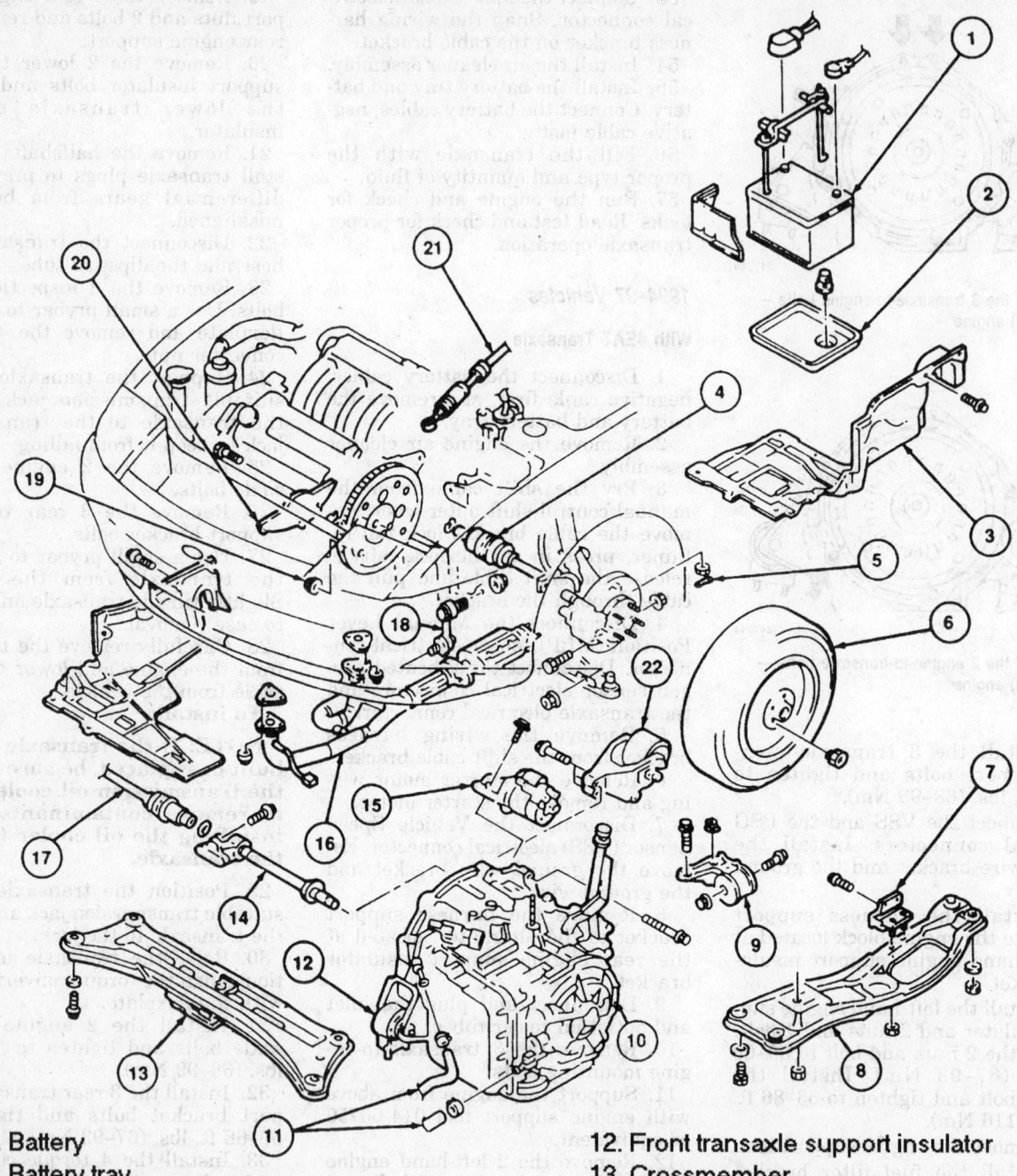

1. Battery
2. Battery tray
3. Front fender splash shield
4. LH front wheel driveshaft & joint
5. Cotter pin
6. Wheel & tire
7. Lower transaxle support insulator
8. Rear engine support
9. LH transaxle support insulator
10. Transaxle
11. Oil cooler tubes
12. Front transaxle support insulator
13. Crossmember
14. Halfshaft
15. Starter motor
16. Exhaust inlet pipe
17. RH front wheel driveshaft & joint
18. Stabilizer bar link
19. Flywheel to converter retaining nut
20. Cover plate
21. Transmission shift cable & bracket
22. Rear transaxle support insulator

305128

Transaxle assembly and related components — 1994–97 with 4EAT transaxle

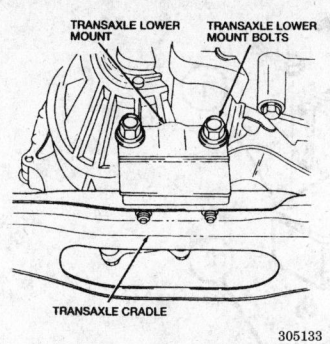

TRANSAXLE LOWER MOUNT   TRANSAXLE LOWER MOUNT BOLTS

TRANSAXLE CRADLE

305133

**Transaxle lower mount location —
1994–97 with 4EAT transaxle**

(67–93 Nm). Tighten the rear engine support-to-front mount nuts to 55–77 ft. lbs. (75–104 Nm) and the rear engine support-to-rear mount nuts to 32–44 ft. lbs. (44–60 Nm).

40. Install the crossmember and tighten the 6 bolts to 68–96 ft. lbs. (94–131 Nm).

41. Install the splash shields and the front wheels. Torque the lug nuts to 66–86 ft. lbs. (88–118 Nm).

42. Lower the vehicle.

43. Install the 3 transaxle-to-engine support bolts and tighten to 50–73 ft. lbs. (68–99 Nm).

44. Connect the VSS and the PSG electrical connectors. Install the ground wire bracket and the ground wire.

45. Install the harness support bracket to the engine block located at the left-hand engine support insulator bracket.

46. Install the left-hand engine support insulator and 2 nuts and 1 bolt. Tighten the 2 nuts and bolt to 50–68 ft. lbs. (67–93 Nm). Install the through-bolt and tighten to 63–86 ft. lbs. (86–116 Nm).

47. Remove the engine support tool.

48. Install the fuel filter bracket and tighten the 2 nuts to 71–88 inch lbs. (8–10 Nm).

49. Unplug and connect the inlet and outlet oil cooler tubes.

50. Install the starter motor and wiring.

51. Connect the transaxle electrical connector and the 2 heated oxygen sensor electrical connectors.

52. Insert the shift cable through the cable bracket and pull the cable until the lock tabs engage. Install the lock tab retainer. Connect the shift cable to the manual control shift outer lever.

53. Connect the MLP sensor electrical connector. Snap the wiring harness bracket on the cable bracket.

54. Install the air cleaner assembly.

55. Install the battery tray and battery. Connect the battery cables, negative cable last.

56. Fill the transaxle with the proper type and quantity of fluid.

57. Run the engine and check for leaks. Road test and check for proper transaxle operation.

With CD4E Transaxle

1. Disconnect the battery cables, negative cable first, and remove the battery and battery tray.

2. Remove the air cleaner assembly.

3. Disconnect the Manual Lever Position (MLP) sensor electrical connector. Remove the 2 MLP sensor bolts and the sensor.

4. Remove the ground wire bracket and the ground wire.

5. Remove the shift cable from the cable bracket. Remove the 2 cable bracket bolts and remove the bracket.

6. Disconnect the transaxle electrical connector by pushing on the retaining ring and gently pulling up on the connector.

7. Disconnect and plug the oil cooler inlet and outlet hoses.

8. Remove the two upper starter motor bolts.

9. Support the engine from above with engine support tool 014-00750 or equivalent.

10. Remove the top 3 transaxle-to-engine mounting bolts.

11. Remove the 2 fuel filter bracket nuts from the left-hand transaxle mount. Position the filter and bracket aside without disconnecting the fuel lines.

12. Remove the 2 ignition coil nuts and position the ignition coil out of the way.

13. Remove the 3 speed control servo nuts and position the speed control servo out of the way.

14. Remove the 3 ignition coil mounting strap bolts.

15. Disconnect the wire harness clips from the ignition coil mounting straps. Remove the ignition coil mounting straps.

16. Remove the 2 nut and 2 bolts from the left-hand engine support insulator. Remove the left-hand support insulator through-bolt. Remove the left-hand support insulator.

17. Raise and safely support the vehicle.

18. Remove the front wheels and the splash shields.

19. Remove the 6 crossmember bolts and the crossmember.

20. Remove the 6 rear engine support nuts and 2 bolts and remove the rear engine support.

21. Remove the halfshafts. Install transaxle plugs into the differential side gears.

22. Remove the intake manifold support bolts and the support.

23. Disconnect the starter motor wiring. Remove the lower starter motor bolt and remove the starter motor.

24. Disconnect the Transmission Speed Sensor (TSS) connector.

25. Remove the 4 torque converter-to-flywheel retaining nuts through the starter opening.

26. Remove the front transaxle support insulator through-bolt and remove the front support insulator.

27. Remove the 4 front transaxle support bracket bolts and remove the support bracket.

28. Remove the rear transaxle support insulator through-bolt.

29. Lower the vehicle.

30. Lower the transaxle by loosening the bolt on the engine support.

31. Raise and safely support the vehicle.

32. Secure the transaxle to a suitable jack using the appropriate adapters.

33. Remove the 3 engine-to-transaxle bolts.

34. Remove the 2 remaining transaxle-to-engine bolts.

35. Use a small prybar or similar tool to separate the transaxle from the engine.

36. Partially lower the transaxle from the engine.

37. Disconnect the Vehicle Speed Sensor (VSS) connector.

38. Finish lowering the transaxle and remove from the vehicle.

To install:

NOTE: If the transaxle was rebuilt or is being replaced, be sure to flush the transmission oil cooler hoses to remove contaminants before installing the oil cooler hoses to the transaxle.

39. Place the transaxle on a suitable jack using the appropriate adapters.

40. Raise the transaxle into position. Align the torque converter studs with the flexplate.

41. Install the transaxle-to-engine bolts. Tighten bolts **B** to 28–38 ft. lbs. (38–51 Nm) and bolts **C** to 14–18 ft. lbs. (19–25 Nm).

42. Remove the jack.

43. Install the rear transaxle support insulator through-bolt. Do not tighten fully at this time.

44. Install the front transaxle support bracket. Tighten the 4 front transaxle support bracket bolts to 28–38 ft. lbs. (38–51 Nm). Install the

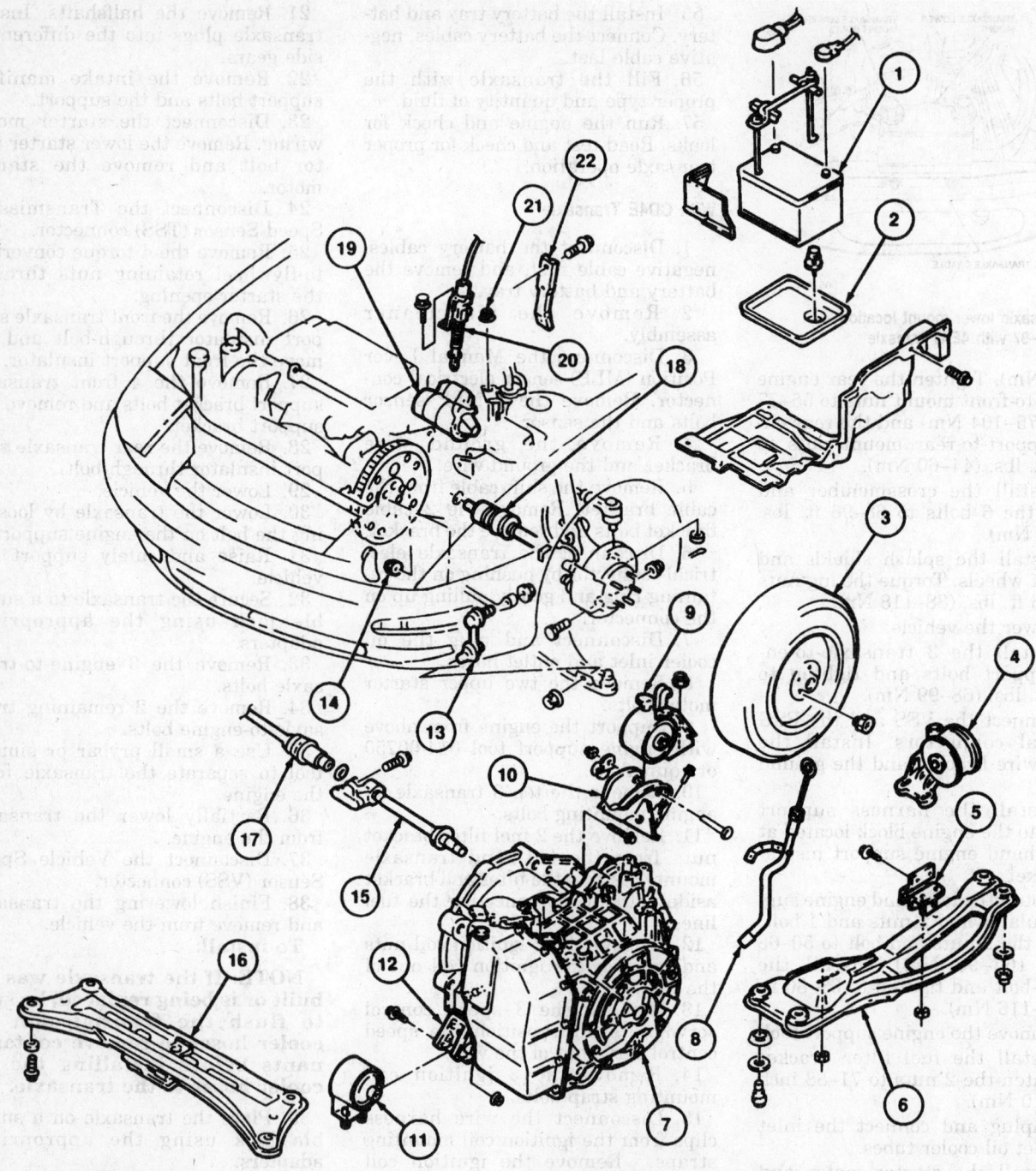

1 Battery
2 Battery tray
3 Wheel and tire
4 LH transaxle support insulator
5 Lower transaxle support insulator
6 Rear engine support
7 Transaxle
8 Oil filler tube
9 Rear transaxle support insulator
10 Rear transaxle support bracket
11 Front transaxle support insulator

12 Front transaxle support bracket
13 Stabilizer bar link
14 Torque converter nut
15 Halfshaft
16 Crossmember
17 RH front wheel driveshaft and joint
18 LH front wheel driveshaft and joint
19 Starter motor
20 Shift cable
21 Fuel filter bracket mounting nut
22 Intake manifold support

305137

Transaxle and related components — 1994–97 with CD4E transaxle

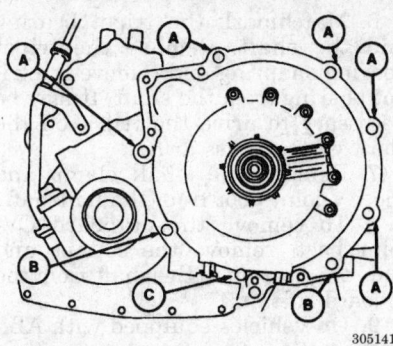

Transaxle-to-engine mounting bolts — 1994 with CD4E transaxle

front transaxle support insulator and through-bolt. Do not tighten the through-bolt fully at this time.

45. Connect the TSS and VSS connectors.

46. Install the 4 torque converter-to-flexplate nuts and tighten to 32–45 ft. lbs. (44–60 Nm) on 1994 vehicles or 24–30 ft. lbs. (33–40 Nm) on 1995–97 vehicles.

47. Install the torque converter access plug.

48. Lower the vehicle.

49. Install the left-hand engine support insulator and tighten the through-bolt to 63–86 ft. lbs. (86–116 Nm). Install the 2 left-hand engine support insulator nuts and 2 bolts. Do not tighten them fully at this time.

50. Remove the engine support.

51. Raise and safely support the vehicle.

52. Install the rear engine support. Tighten the rear engine support bolts and nuts as follows:
 • Tighten **A** to 55–77 ft. lbs. (75–104 Nm)
 • Tighten **B** to 50–68 ft. lbs. (67–93 Nm)
 • Tighten **C** to 32–44 ft. lbs. (44–60 Nm)

53. Tighten the front and rear transaxle support insulator through-bolts to 63–86 ft. lbs. (86–116 Nm).

54. Install the starter motor and the lower starter motor bolt. Tighten the lower starter motor bolt to 23–34 ft. lbs. (31–46 Nm). Connect the starter motor wiring.

55. Install the intake manifold support and bolts. Tighten the bolts to 27–38 ft. lbs. (37–52 Nm).

56. Remove the transaxle plugs and install the halfshafts.

57. Install the crossmember and 6 bolts. Tighten the crossmember bolts to 68–96 ft. lbs. (94–131 Nm).

58. Install the splash shields and the wheels. Torque the lug nuts to 67–86 ft. lbs. (88–118 Nm).

59. Lower the vehicle.

60. Tighten the 2 left-hand support insulator nuts to 12–17 ft. lbs. (16–23 Nm). Tighten the 2 left-hand engine support insulator bolts to 28–38 ft. lbs. (38–51 Nm).

61. Position the ignition coil mounting straps and install the wire harness clips and the 3 ignition coil strap bolts.

62. Position the speed control servo and install and tighten the 3 nuts.

63. Position the ignition coil and install the 2 nuts. Tighten the ignition coil nuts to 71–88 inch lbs (8–10 Nm).

64. Install the 2 fuel filter bracket nuts and tighten to 71–97 inch lbs. (8–11 Nm).

65. Install the 2 upper starter motor bolts and tighten to 23–34 ft. lbs. (31–46 Nm).

66. Install the 3 remaining transaxle-to-engine bolts. Tighten the **A** bolts to 66–86 ft. lbs. (89–117 Nm).

67. Install the shift cable bracket and 2 bolts.

68. Connect the transaxle electrical connector.

69. Install the MLP sensor and the two MLP sensor bolts. Adjust the MLP sensor and tighten the bolts to 96–117 inch lbs. (11–13 Nm). Connect the MLP sensor electrical connector

70. Connect the oil cooler inlet and outlet hoses.

71. Install the air cleaner assembly.

72. Install the battery tray and battery.

73. Install the ground wire bracket and the ground wire.

74. Connect the battery cables, negative cable last.

75. Fill the transaxle with the proper type and quantity of fluid.

76. Run the engine and check for leaks.

77. Road test and check for proper transaxle operation.

DRIVELINE

Halfshaft

REMOVAL AND INSTALLATION

1. Disconnect the negative battery cable.

2. With the vehicle sitting on all 4 wheels, use a chisel to raise the staked portion of the hub nut. Have an assistant apply the brakes to lock the wheels, then loosen but do **not** remove the hub nut.

3. Raise and safely support the vehicle.

4. Remove the wheel and the necessary inner fender splash guards.

5. Remove the stabilizer link assembly from the lower control arm.

6. Remove the ball joint clamp bolt from the lower control arm. Carefully pry the lower control arm downward to separate the ball joint from the steering knuckle.

NOTE: If removing the right halfshaft, remove the support bearing bracket from the cylinder block.

7. Separate the halfshaft from the transaxle by positioning a prybar between the halfshaft and transaxle case. Pry out the halfshaft while pulling out on the steering knuckle. Be careful not to damage the transaxle case, transaxle oil seal, CV-joint or CV-joint boot.

8. Remove and discard the hub nut. Pull the halfshaft out of the wheel hub. If necessary, use a plastic hammer to tap it out or a wheel puller to press it out. Do not use a metal hammer.

9. Support the halfshaft and slide it out of the transaxle.

10. Install transaxle plug T88C-7025-AH or equivalent, into the halfshaft opening of the transaxle case and into the differential side gear; this will keep the differential side gear from falling out of place.

To install:

NOTE: If replacing a halfshaft and joint on an ABS equipped vehicle, install a new front brake anti-lock sensor indicator.

11. On the end of the halfshaft, install a new circlip. Start one end of the clip in the groove and work the clip over the stub shaft end and into the groove. This will prevent over-expanding the clip. Make sure the end gap is positioned at the top of the splines.

12. Remove the transaxle plug and inspect the transaxle oil seal. Replace if necessary.

13. Lubricate the halfshaft splines with a suitable grease, align the splines with the differential side gear and push the halfshaft into the differential. Make sure the retaining clip clicks into the differential side gear groove.

14. Position the halfshaft through the wheel hub and install a new attaching nut. Do not tighten the nut at this time.

15. If installing the right halfshaft, install the halfshaft support bearing

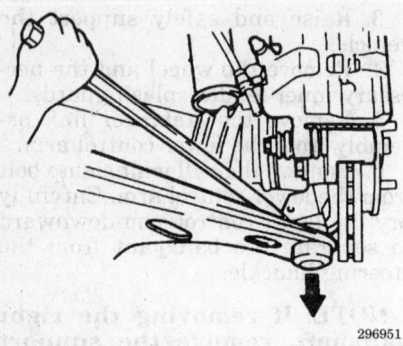

Pry the control arm down to disconnect the ball joint

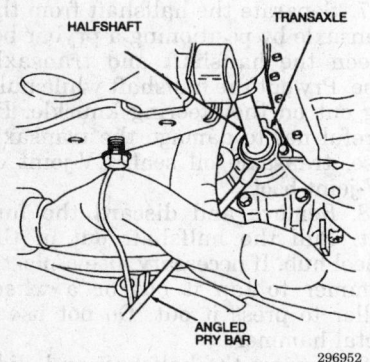

Pry the halfshaft out of the transaxle

and tighten the mounting bolts to 31–46 ft. lbs. (42–62 Nm).

NOTE: The support bearing bolts must be torqued in the proper sequence.

16. Position the ball joint in the steering knuckle and install the clamp bolt/nut. Tighten the nut to 25–42 ft. lbs. (34–57 Nm).

17. Install the stabilizer link assembly and tighten the nuts to 27–40 ft. lbs. (36–54 Nm).

18. Install the splash shields and wheel and lower the vehicle.

19. With the vehicle sitting on all 4 wheels, have an assistant apply the

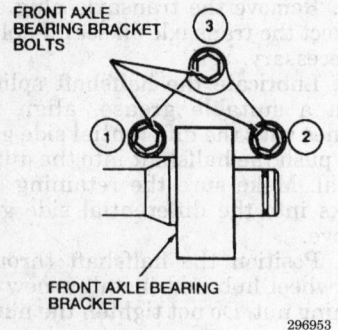

Axle bearing bracket bolt torque sequence — 2.0L engine

10-68

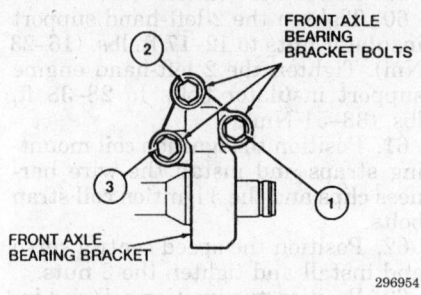

Axle bearing bracket bolt torque sequence — 2.5L engine

brakes and tighten the halfshaft attaching nut to 174–235 ft. lbs. (235–319 Nm). Stake the nut using a suitable chisel with a rounded cutting edge.

NOTE: If the nut splits or cracks after staking, it must be replaced with a new nut.

20. Connect the negative battery cable.

21. Check the transaxle fluid level. Road test the vehicle and check for leaks.

CV-Joint Boot

REPLACEMENT

NOTE: The outboard CV-joint is an integral part of the halfshaft and cannot be removed. To replace the outboard CV-joint boot, the inner CV-joint must be removed. If a boot has failed due to age or wear, all boots should be replaced at the same time. Vehicles with automatic transaxle are equipped with a Tripot type CV-joint while vehicles with manual transaxle are equipped with a Double Offset type CV-joint.

Automatic Transaxle

1. Disconnect the negative battery cable.

2. Raise and support the vehicle safely.

3. Remove the halfshaft from the vehicle. Support the assembly in a vise with soft jaws.

4. Remove the large boot clamp from the inboard CV-joint and roll the boot back over the shaft.

5. Matchmark the outer race, axle shaft and tripot bearing for reassembly. Remove the wire ring bearing retainer from inside the outer race/housing and remove the outer race.

6. Matchmark the tripot bearing and the shaft. Remove the tripot bearing snapring and remove the tripot bearing from the shaft. It may be necessary to drive the tripot off the shaft with a brass drift.

7. Remove the small clamp and the CV-joint boot from the halfshaft.

8. To remove the outboard CV-joint boot, remove the clamps and slide the boot off of the shaft from the inboard side.

9. On vehicles equipped with ABS brakes, the wheel speed sensor can be removed with a bearing puller if the halfshaft is to be replaced.

To install:

NOTE: Inspect the CV-joint boot grease for contamination. If grit can be found in the grease, the CV-joint should be disassembled, inspected for abnormal wear and cleaned before assembly.

10. Wrap smooth electrical tape around the halfshaft spline to ease installation of the boot. Slide the clamps and the outboard boot onto the shaft.

11. Before positioning the boot over the CV-joint, pack the CV-joint and boot with grease. Be sure to use all of the grease in the pouch supplied with the boot kit.

12. Fit the boot into place on the CV-joint, making sure it is fully seated in the grooves in the shaft and outer race. Insert a suitable tool between the boot and the outer bearing race to allow trapped air to escape from the boot.

13. Install the boot clamps, wrapping them around the boots in the opposite direction of halfshaft rotation. Pull the clamps tight with a suitable tool and bend the locking tabs to secure in position.

14. Fit the inboard CV-joint boot and clamps onto the halfshaft.

15. Install the tripot assembly on the halfshaft with the matchmarks aligned. Install the tripot retaining ring.

16. Fill the CV-joint outer race with 3.5 oz. of high temperature CV-joint grease. Install the outer race over the tripot joint with the matchmarks aligned and install the wire ring bearing retainer.

17. Fit the boot into place on the CV-joint, making sure it is fully seated in the grooves in the shaft and outer race. Insert a suitable tool between the boot and the outer bearing race to allow trapped air to escape from the boot.

18. Install the boot clamps, wrapping them around the boots in the

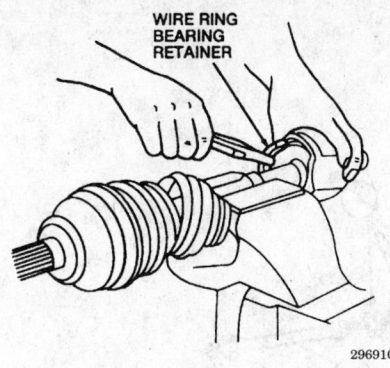

Remove the wire ring to remove the outer race

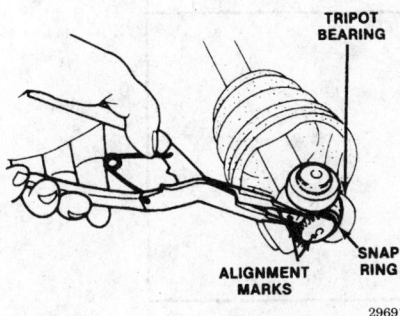

Remove the snapring to remove the tripot bearing

opposite direction of halfshaft rotation. Pull the clamps tight with a suitable tool and bend the locking tabs to secure in position. Wipe away any excess grease.

19. Work the CV-joint through its full range of travel at various angles. The joint should flex, extend, and compress smoothly.

20. Measure the total halfshaft length to check for proper assembly. The left halfshaft should be 25.4–25.93 inches (648.7–658.7mm) and the right halfshaft should be 23.57–23.96 inches (598.7–608.7mm).

21. If necessary, carefully drive or press the wheel speed sensor onto the CV-joint.

22. Install the halfshaft in the vehicle.

23. Connect the negative battery cable. Road test

Manual Transaxle

NOTE: Inspect the CV-joint boot grease for contamination. If grit can be found in the grease, the CV-joint should be disassembled, inspected for abnormal wear and cleaned before assembly.

1. Disconnect the negative battery cable.

2. Raise and safely support the vehicle.

3. Remove the halfshaft assembly and clamp it in a vise with soft jaws.

4. Remove the boot bands and peel the boot back off of the inboard CV-joint.

5. Mark the outer race and halfshaft with paint for proper positioning during assembly. Do not use a punch to make the mark.

6. Remove the wire ring retainer from the inside end of the outer race. Carefully pull the outer race off and be ready to catch the balls (they may fall out of the cage).

7. Mark the halfshaft and inner race with paint for proper positioning during assembly. Do not use a punch to make the mark.

8. Remove the snapring from the end of the halfshaft and remove the inner race and ball assembly from the halfshaft.

9. If it is necessary to disassemble the CV-joint:

   a. Mark the inner race and cage with paint for proper alignment at assembly.

   b. Use a dull tool to carefully pry the balls out of the cage. Be careful not to scratch the inner race surface.

   c. Turn the cage approximately 30 degrees to separate it from the inner race.

10. Remove the inboard CV-joint boot.

11. To remove the outboard CV-joint boot, remove the clamps and slide the boot off of the shaft from the inboard side.

12. On vehicles equipped with ABS brakes, the wheel speed sensor can removed with a bearing puller if the halfshaft is to be replaced.

To install:

13. Wrap smooth electrical tape around the halfshaft spline to ease installation of the boot. Slide the clamps and the outboard boot onto the shaft.

14. Before positioning the boot over the CV-joint, pack the CV-joint and boot with grease. Be sure to use all of the grease in the pouch supplied with the boot kit.

15. Fit the boot into place on the CV-joint, making sure it is fully seated in the grooves in the shaft and outer race. Insert a suitable tool between the boot and the CV-joint to allow trapped air to escape from the boot.

16. Install the boot clamps, wrapping them around the boots in the opposite direction of halfshaft rotation. Pull the clamps tight with a suitable tool and bend the locking tabs to secure in position.

17. Fit the inboard CV-joint boot and clamps onto the halfshaft.

18. If the inner CV-joint was disassembled, reassemble the ball cage, balls and inner ring with the matchmarks aligned.

19. Install the assembly onto the halfshaft with the matchmarks aligned and install the snapring.

20. Fill the CV-joint outer race with high temperature CV-joint grease. Install the outer race over the balls with the matchmarks aligned and install a new circlip.

21. Fit the boot into place on the CV-joint, making sure it is fully seated in the grooves in the shaft and outer race. Insert a suitable tool between the boot and the outer race to allow trapped air to escape from the boot.

22. Install the boot clamps, wrapping them around the boots in the opposite direction of halfshaft rotation. Pull the clamps tight with a suitable tool and bend the locking tabs to secure in position.

23. Work the CV-joint through its full range of travel at various angles. The joint should flex, extend, and compress smoothly. Wipe away any excess grease.

24. Measure the total halfshaft length to check for proper assembly. The left halfshaft should be 25.56–25.95 in. (649.7–659.7mm) and the right halfshaft should be 23.64–24.02 in. (600.2–610.2mm).

25. If necessary, carefully drive or press the wheel speed sensor onto the CV-joint.

26. Install the halfshaft in the vehicle.

27. Connect the negative battery cable.

STEERING

Air Bag

——— CAUTION ———

The Supplemental Restraint System (SRS) must be disarmed before removing the air bag module. Failure to do so may cause accidental deployment of the air bag, resulting in unnecessary SRS repairs and/or personal injury.

Double Offset CV Joint (MTX)
Disassembly
Disassembled View

RIGHT SIDE

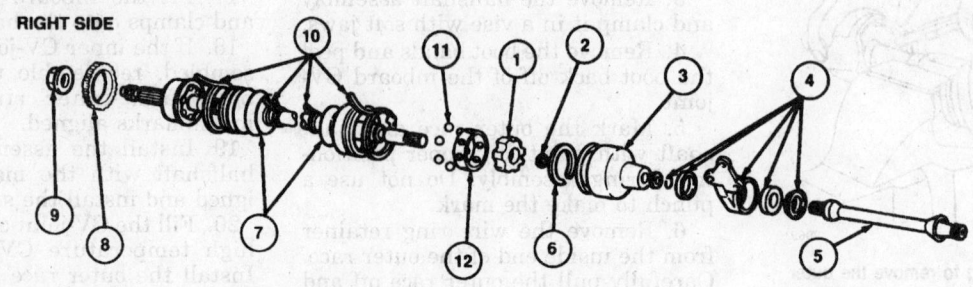

Item	Description
1	Inner Race
2	Snap Ring
3	Outer Race
4	Support Bearing Assembly
5	Joint Shaft
6	Bearing Retainer
7	Boots
8	ABS Wheel Sensor Rotor
9	Halfshaft Attaching Nut
10	Boot Clamps
11	Ball Bearing
12	Bearing

296913

Halfshaft with double offset CV-joint used on manual transaxle

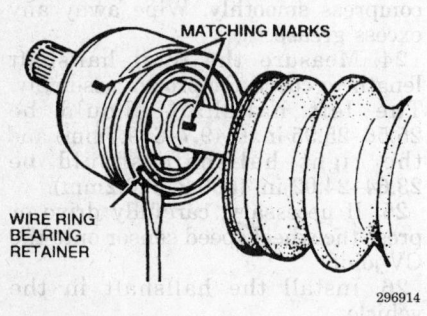

MATCHING MARKS

WIRE RING
BEARING RETAINER

296914

Remove the wire ring to pull the outer race off

WHEEL SIDE   TRANSAXLE SIDE

RIGHT-HAND SIDE

LEFT-HAND SIDE

296912

Inboard (transaxle side) and outboard CV-joint boots

PRECAUTIONS

Several precautions must be observed when handling the inflator module to avoid accidental deployment and possible personal injury.

Never carry the inflator module by the wires or connector on the underside of the module.

When carrying a live inflator module, hold securely with both hands, and ensure that the bag and trim cover are pointed away.

Place the inflator module on a bench or other surface with the bag and trim cover facing up.

With the inflator module on the bench, never place anything on or close to the module which may be thrown in the event of an accidental deployment.

DISARMING

— CAUTION —
The air bag must be disarmed before performing service around air bag components or air bag wiring. Failure to do so may cause accidental deployment of the air bag, resulting in unnecessary repairs and/or personal injury.

1. Position the vehicle with the front wheels in a straight ahead position.
2. Disconnect the negative battery cable.
3. Disconnect the positive battery cable.
4. Wait at least 1 minute for the air bag backup power supply to drain before continuing.
5. Proceed with repair.
6. Once complete, connect the battery cables, negative cable last.

Steering Wheel

REMOVAL AND INSTALLATION

— CAUTION —
The air bag must be disarmed before removing the air bag. Failure to do so may cause accidental deployment of the air bag, resulting in unnecessary system repairs and/or personal injury.

NOTE: Always wear safety glasses when servicing an air bag vehicle and when handling an air bag.

1. Center the front wheels in the straight-ahead position.
2. Disconnect both battery cables from the battery, negative cable first,

and wait at least 1 minute for the air bag backup power supply energy to be depleted.

3. Remove the 4 air bag module retaining bolts and lift the module from the steering wheel.

4. Label and disconnect the electrical connectors and remove the air bag module.

CAUTION

When carrying a live air bag, make sure the bag and trim cover are pointed away from the body. In the unlikely event of an accidental deployment, the bag will then deploy with minimal chance of personal injury. When placing a live air bag on a bench or other surface, always face the bag and trim cover up, away from the surface. This will reduce the motion of the module if it is accidentally deployed.

5. If equipped, remove the speed control switch electrical connectors.

6. Make an alignment mark on the steering wheel and steering shaft for assembly reference.

7. Remove the steering wheel nut.

8. Remove the steering wheel using a suitable puller. Route the wire

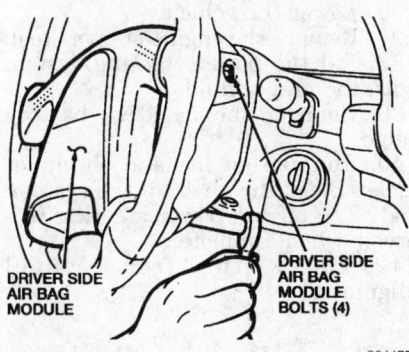

DRIVER SIDE AIR BAG MODULE

DRIVER SIDE AIR BAG MODULE BOLTS (4)

304477

Air bag module bolt locations

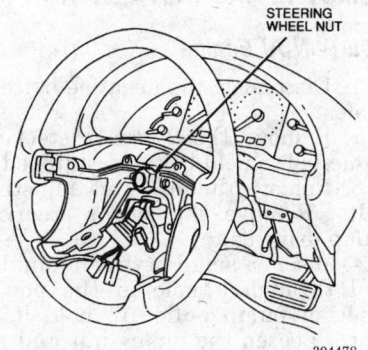

STEERING WHEEL NUT

304478

Remove the steering wheel attaching nut

harness through the steering wheel as the wheel is lifted from the shaft.

WARNING

Do not try to remove the steering wheel by hitting the steering shaft with a hammer. The steering shaft will collapse, causing the steering wheel to bind.

9. Apply 2 strips of tape across the clockspring and housing to prevent accidental rotation.

To install:

10. Make sure the wheels are in the straight-ahead position.

11. If the clockspring has been accidentally rotated, the clockspring alignment must be adjusted; proceed as follows:

a. Make sure the wheels are in the straight-ahead position.

b. Turn the clockspring clockwise until it stops. Do not apply excessive force.

c. Rotate the clockspring counterclockwise 2 ¾ turns.

d. Align the marks on the clockspring with the marks on the outer housing.

12. Remove the tape strips from the clockspring and housing.

13. Route the wire harness through the steering wheel opening and position the steering wheel on the shaft. Align the marks made during removal.

14. Install the steering wheel nut and tighten to 29–36 ft. lbs. (39–49 Nm).

15. Connect the electrical connectors to the air bag module.

16. Connect the speed control switch electrical connectors, if equipped.

17. Position the air bag module and install the 4 attaching bolts. Tighten the bolts to 36–54 inch lbs. (4–6 Nm).

18. Make sure no one is in the vehicle and connect the battery cables, negative cable last.

19. Check that the air bag proveout light in the instrument cluster goes out after a few seconds from engine start-up to show that the system is functioning properly.

Tie Rod Ends

REMOVAL AND INSTALLATION

1. Disconnect the negative battery cable.

2. Raise and safely support the vehicle.

3. Remove the wheel.

4. Remove the cotter pin and the castellated nut from the tie rod end stud.

5. Separate the tie rod end from the steering knuckle using separator tool 3290-D or equivalent. If the tie rod end does not separate easily, give the steering knuckle a sharp blow with a brass hammer or drift to shock the taper.

6. Paint or mark an alignment stripe on the tie rod end, jam nut, and tie rod. If the tie rod end is being replaced, count the number of turns it takes to remove the tie rod end and record.

7. Loosen the jam nut and remove the tie rod end.

To install:

8. Thread the jam nut and tie rod end onto the inner tie rod spindle.

9. Align the marks made during removal. If a new tie rod end is being used, screw the new tie rod end onto the inner tie rod spindle the same number of turns as required for removal of the old tie rod end. Hold the tie rod end and tighten the jam nut to 51–72 ft. lbs. (69–98 Nm).

10. Install the tie rod end in the steering knuckle. Install the nut and tighten to 22–33 ft. lbs. (29–44 Nm).

11. Install a new cotter pin. If the slots in the nut do not align with the hole in the tie rod end stud, tighten the nut for proper alignment; never loosen the nut.

12. Install the wheel and torque the lug nuts to 65–87 ft. lbs. (88–118 Nm).

13. Lower the vehicle.

14. Check the front wheel alignment.

Power Rack and Pinion

REMOVAL AND INSTALLATION

1. Disconnect the negative battery cable.

2. Support the engine with engine support tool D88L-6000-A or equivalent.

3. Raise and safely support the vehicle.

4. Remove both front wheels.

5. Remove the cotter pins and castellated nuts from the tie rod ends. Use tool 3290-D or equivalent, to separate the tie rod ends from the steering knuckles.

6. Remove the splash shields. Remove the 6 crossmember bolts and remove the crossmember.

7. Remove the 2 bolts and 6 nuts from the rear engine support and remove the support.

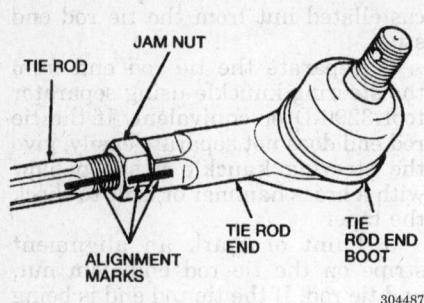

Tie rod end

304487

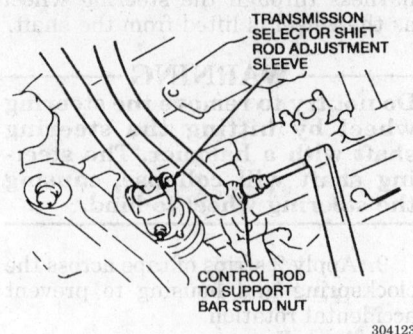

Removing the control rod nut

304123

8. If equipped with 2.5L engine, disconnect the oxygen sensor connectors. Remove the exhaust pipe-to-manifold nuts and separate the pipes from the manifolds. Move the front exhaust pipe aside.

9. Disconnect and plug the power steering pressure and return hoses.

10. Remove the steering shaft U-joint shield bolt and shield.

11. Remove the ground wire bracket from the rear transaxle support bracket (4EAT and MTX transaxles only).

12. Remove the 3 rear transaxle support bracket bolts and the transaxle insulator through-bolt. Remove the transaxle support bracket (4EAT and MTX transaxles only).

13. Remove the 4 bolts from the 2 rack and pinion assembly mounting brackets and remove the brackets.

14. Remove the steering column lower yoke-to-power steering gear input shaft and bolt.

15. If equipped with a manual transaxle, remove the control rod to support bar stud nut and position the control rod aside.

16. Position a jack under the front sub-frame. Remove the 6 sub-frame bolts and 2 nuts.

17. Remove the vent tube attached to the drivers side of the sub-frame.

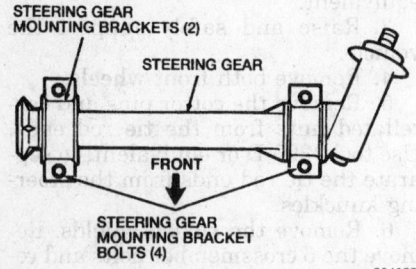

Rack and pinion assembly (Steering gear) mounting bracket bolt locations

304122

18. Remove the upper stabilizer bar link nuts.

19. Lower the sub-frame to allow removal of the rack and pinion assembly. Remove the rack and pinion assembly from the drivers side of the vehicle.

To install:

NOTE: If a new rack and pinion assembly is being installed or if the old assembly was turned, place the assembly into a soft jaw vise and rotate the steering gear input shaft counting the number of turns lock to lock. Back the steering gear input shaft up half of the number of the turns counted to center the rack and pinion. Do not damage the input shaft splines.

20. Position the rack and pinion assembly in the vehicle.

21. With the aide of an assistant, install the steering column lower yoke-to-power steering gear input shaft and bolt. Tighten the bolt to 13–20 ft. lbs. (18–26 Nm).

22. Raise the front sub-frame into position. Install the sub-frame mounting bolts and nuts and tighten to 69–97 ft. lbs. (93–131 Nm).

23. Remove the jack.

24. Install the upper stabilizer bar link nuts and tighten to 27–40 ft. lbs. (36–54 Nm). Install the vent tube.

25. Position the 2 rack and pinion assembly mounting brackets and install the 4 mounting bolts. Tighten to 28–38 ft. lbs. (38–51 Nm).

26. Position the rear transaxle support insulator and install the through-bolt, if equipped. Tighten the through-bolt to 63–86 ft. lbs. (85–117 Nm).

27. Install the 3 rear transaxle support bracket bolts if equipped, and tighten to 50–68 ft. lbs. (67–93 Nm). Install the ground wire bracket to the rear engine mount.

28. Install the steering shaft U-joint shield and bolt and tighten securely.

29. Remove the plugs and connect the power steering pressure and return lines.

30. If equipped with a manual transaxle, install the control rod to the support bar stud and install the nut. Tighten the nut to 28–38 ft. lbs. (38–51 Nm).

31. If equipped with 2.5L engine, connect the exhaust pipes to the manifolds and tighten the nuts to 38 ft. lbs. (51 Nm). Connect the oxygen sensor connectors.

32. Position the rear engine support and install the 6 nuts and 2 bolts. Tighten the rear engine support nuts and bolts as follows:

   a. Tighten **A** to 55–77 ft. lbs. (75–104 Nm)

   b. Tighten **B** to 50–68 ft. lbs. (67–93 Nm)

   c. Tighten **C** to 32–44 ft. lbs. (44–60 Nm)

33. Position the crossmember and install the 4 crossmember bolts. Tighten to 69–97 ft. lbs. (93–131 Nm).

34. Connect the tie rod ends to the steering knuckles and tighten the castellated nuts to 23–33 ft. lbs. (31–44 Nm). Install new cotter pins.

35. Install the splash shields and the front wheels. Torque the lug nuts to 65–87 ft. lbs. (88–118 Nm).

36. Lower the vehicle.

37. Remove the engine support tool.

38. Fill the power steering system with the proper fluid.

39. Connect the negative battery cable.

40. Run the engine and check for leaks. Bleed the air from the system.

41. Top off the power steering reservoir when complete.

42. Check the front wheel alignment.

Power Steering Pump

REMOVAL AND INSTALLATION

2.0L (VIN A) Engine

1. Disconnect the negative battery cable.

2. Remove the 2 power steering pump belt shield bolts, if equipped.

3. Remove the lock and adjusting bolts. Remove the power steering pump drive belt.

4. Insert a small prybar or similar tool through a hole in the power steering pump pulley to hold it in place. Loosen the pulley nut and remove the nut, washer and the power steering pump pulley.

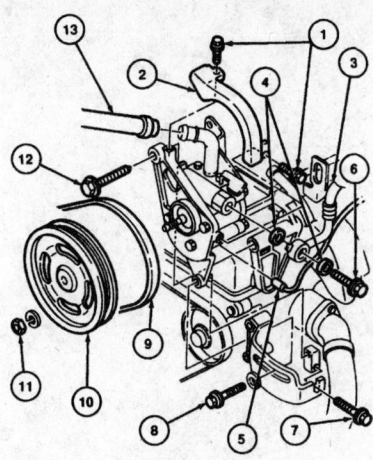

Item	Description
1	Power Steering Pump Belt Shield Bolts
2	Power Steering Pump Belt Shield
3	High Pressure Line
4	Banjo Bolt Crush Washers
5	Power Steering Pump Pressure Switch
6	High Pressure Line Banjo Bolt
7	Adjusting Bolt
8	Lock Bolt
9	Power Steering Pump Belt
10	Power Steering Pump Pulley
11	Pulley Nut
12	Power Steering Pump Through Bolt
13	Supply Line (From Reservoir)

304389

Power steering pump assembly — 2.0L (VIN A) engine

5. Remove the 2 reservoir pump hose manifold bolts and remove the pressure hose banjo bolt.

6. Disconnect the Power Steering Pressure (PSP) switch electrical connector and remove the pump through-bolt.

7. Remove the power steering pump.

To install:

8. Position the power steering pump and loosely install the through-bolt.

9. Connect the PSP switch electrical connector.

10. Install the pressure hose and banjo bolt using new washers. Tighten to 22–33 ft. lbs. (30–44 Nm).

11. Connect the reservoir pump hose manifold and install the 2 retaining bolts. Tighten the bolts to 10–13 ft. lbs. (14–18 Nm).

12. Install the power steering pump pulley, washer and the retaining nut. Insert a small prybar or similar tool through a hole in the pulley to hold it in place and torque the nut to 36–43 ft. lbs. (49–59 Nm).

13. Install the power steering pump drive belt and the lock and adjusting bolts. Adjust the belt tension.

14. Install the 2 power steering pump belt shield bolts and tighten to 63–81 inch lbs. (7–9 Nm).

15. Connect the negative battery cable.

16. Fill the power steering system with the proper fluid.

17. Bleed the air from the system.

18. Check for leaks.

19. Top off the power steering reservoir when complete.

2.5L (VIN B) Engine

1. Disconnect the negative battery cable.

2. Remove the power steering pressure hose hold-down bracket bolt.

3. Remove the power steering pressure hose banjo bolt. Dispose of the crush washers.

4. Raise and safely support the vehicle.

5. Remove the right-hand front tire and splash shield.

6. Loosen the locknut and the adjusting bolt and remove the power steering pump drive belt.

7. Insert a small prybar or similar tool through a hole in the power steering pump pulley to hold it in place. Loosen the pulley nut and remove the nut, washer and the power steering pump pulley.

8. Remove the 2 reservoir pump manifold bolts and remove the power steering pressure hose hold-down bracket nut.

9. Disconnect the Power Steering Pressure (PSP) switch electrical connector.

10. Remove the 4 power steering pump bracket-to-engine bolts and remove the power steering pump.

To install:

11. Position the power steering pump and install the 4 pump bracket-to-engine bolts. Tighten to 23–34 ft. lbs. (31–46 Nm).

12. Connect the PSP switch electrical connector.

13. Install the power steering pressure hose bracket nut and tighten to 61–86 inch lbs. (6–9 Nm). Install the power steering reservoir pump manifold and bolts and tighten to 10–13 ft. lbs. (14–18 Nm).

14. Install the power steering pump pulley, washer and the retaining nut. Insert a small prybar or similar tool through a hole in the pulley to hold it in place and torque the retaining nut to 36–43 ft. lbs. (49–59 Nm).

15. Install the power steering pump drive belt and adjust the tension. Be sure to tighten the adjusting bolt and locknut.

16. Install the splash shield and the wheel. Tighten the lug nuts to 65–87 ft. lbs. (88–118 Nm).

17. Lower the vehicle.

18. Install the power steering pressure hose banjo bolt, using new crush washers. Tighten to 22–33 ft. lbs. (30–44 Nm).

19. Install the power steering pressure hose hold-down bracket bolt and tighten to 61–86 inch lbs. (6–9 Nm).

20. Connect the negative battery cable.

21. Fill the power steering system with the proper fluid.

22. Bleed the air from the system.

23. Check for leaks.

24. Top off the power steering fluid when complete.

POWER STEERING SYSTEM BLEEDING

1. Raise and support the vehicle, safely.

2. Disconnect the coil wire. Refill the power steering pump reservoir to the specified level.

3. Crank the engine. Check and refill the reservoir.

4. Crank the engine and rotate the steering wheel from lock-to-lock.

NOTE: The front wheels must be off the ground during lock-to-lock rotation of the steering wheel.

5. Check and refill the power steering pump reservoir.

6. Connect the coil wire, start the engine and allow it to run for several minutes.

7. Rotate the steering wheel from lock-to-lock several times, until the air bubbles are eliminated from the fluid.

8. Turn the engine **OFF**. Check and/or refill the reservoir.

9. Disconnect the negative battery cable, depress the brake pedal for at least 5 seconds and reconnect the negative battery cable.

BRAKES

Anti-Lock Brake System Service

PRECAUTIONS

• Certain components within the ABS system are not intended to be serviced or repaired individually. Only those components with removal and installation procedures should be serviced.

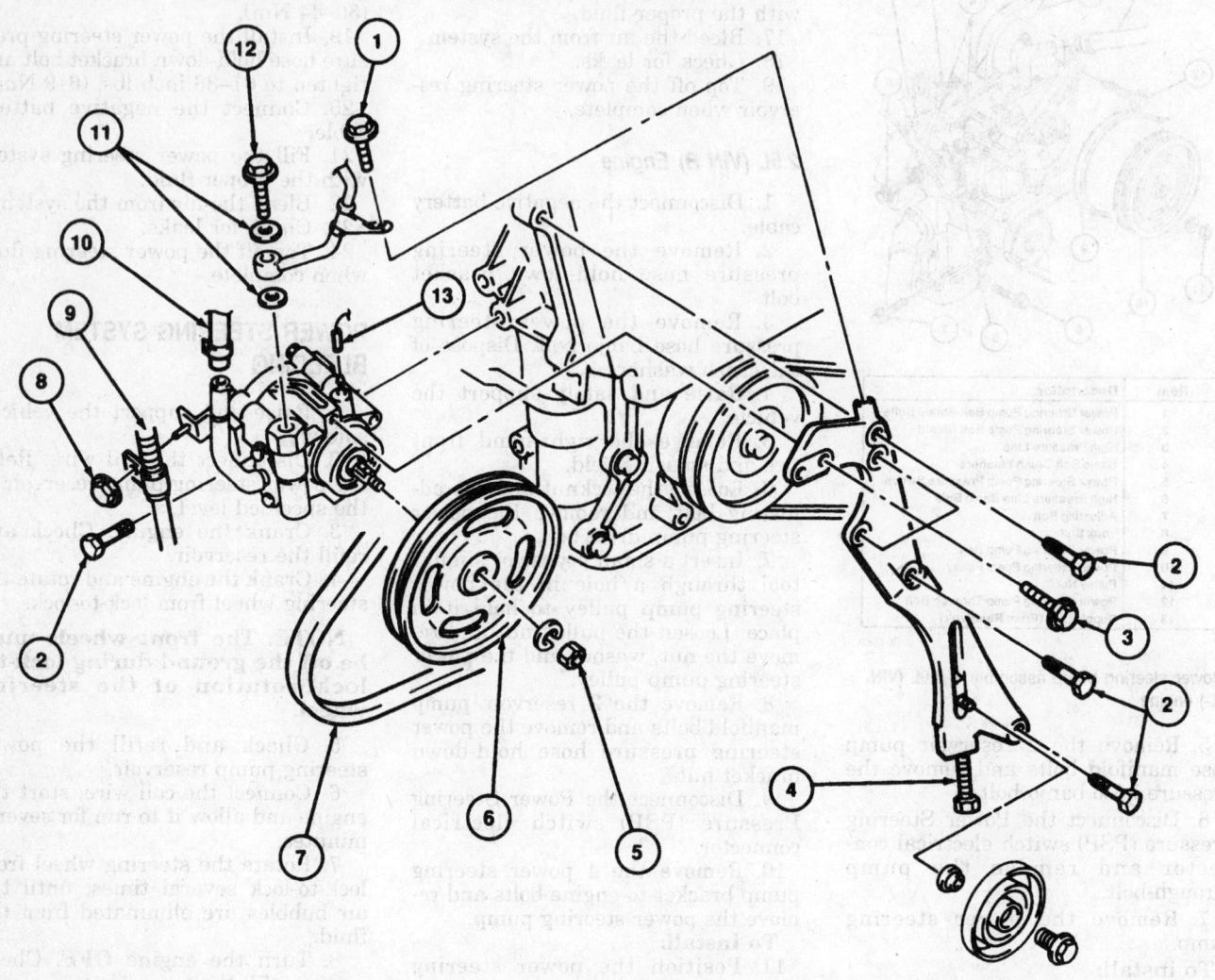

Item	Description
1	High Pressure Line Hold-Down Bracket Bolt
2	Power Steering Pump Bracket-to-Engine Block Bolts
3	Power Steering Pump Bracket-to-Power Steering Pump Bolts
4	Adjusting Bolt
5	Pulley Nut
6	Power Steering Pump Pulley
7	Power Steering Pump Belt
8	High Pressure Line Hold-Down Bracket Nut
9	High Pressure Line
10	Supply Line (From Reservoir)
11	Banjo Bolt Crush Washers
12	High Pressure Line Banjo Bolt
13	Power Steering Pump Pressure Switch

304391

Power steering pump assembly — 2.5L (VIN B) engine

- Do not use rubber hoses or other parts not specifically specified for and ABS system. When using repair kits, replace all parts included in the kit. Partial or incorrect repair may lead to functional problems and require the replacement of components.
- Lubricate rubber parts with clean, fresh brake fluid to ease assembly. Do not use lubricated shop air to clean parts; damage to rubber components may result.
- Use only DOT 3 brake fluid from an unopened container.
- If any hydraulic component or line is removed or replaced, it may be necessary to bleed the entire system.
- A clean repair area is essential. Always clean the reservoir and cap thoroughly before removing the cap. The slightest amount of dirt in the fluid may plug an orifice and impair the system function. Perform repairs after components have been thoroughly cleaned; use only denatured alcohol to clean components. Do not allow ABS components to come into contact with any substance containing mineral oil; this includes used shop rags.
- The Anti-Lock control unit is a microprocessor similar to other computer units in the vehicle. Ensure that the ignition switch is **OFF** before removing or installing controller harnesses. Avoid static electricity discharge at or near the controller.
- If any arc welding is to be done on the vehicle, the control unit should be unplugged before welding operations begin.

Master Cylinder

REMOVAL AND INSTALLATION

— **WARNING** —
Be careful not to get brake fluid on painted surfaces as it will destroy the finish.

1. Disconnect the negative battery cable.
2. Remove the cruise control actuator from its bracket, if equipped.
3. Disconnect the electrical connector from the fluid level sensor.
4. Disconnect the brake lines from the master cylinder. On vehicles with a manual transaxle, disconnect and plug the reservoir hose for the clutch master cylinder.
5. Cap the brake lines and the master cylinder ports.
6. Remove the 2 mounting nuts and remove the master cylinder.

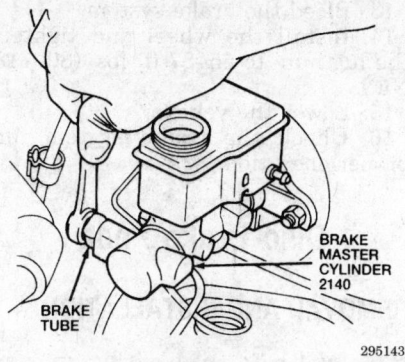

Master cylinder

To install:
7. Adjust the master cylinder pushrod.
8. Position the master cylinder on the power brake booster studs.
9. Install the master cylinder mounting nuts and tighten to 8–12 ft. lbs. (10–16 Nm).
10. Connect short lengths of brake line to the master cylinder that point back into the reservoir. Position the ends of the lines so they will be submerged in brake fluid.
11. Fill the master cylinder reservoir with DOT-3 or equivalent brake fluid and cover the reservoir with a shop towel. Slowly pump the brake pedal until clear fluid comes out of both temporary brake lines.
12. Remove the temporary brake lines and connect the brake lines to the master cylinder. Temporarily tighten the brake line fittings. Bleed the master cylinder as follows:
   a. Have an assistant pump the brake pedal 10 times and then hold firm pressure on the brake pedal.

NOTE: Position shop towels beneath the master cylinder before proceeding with the next step, to catch and absorb the brake fluid.

   b. Loosen the rearmost brake line fitting, until a stream of brake fluid comes out. Have the assistant maintain pressure on the brake pedal until the brake line fitting is tightened.
   c. Repeat the operation until clear, bubble-free fluid comes out from around the brake line fitting.
   d. Repeat the bleeding procedure at the front brake line fitting.
13. Final tighten the brake line nuts to 10–16 ft. lbs. (13–22 Nm).
14. Connect the clutch master cylinder supply hose, if equipped.
15. Connect the fluid level sensor electrical connector.
16. Connect the cruise control actuator to its bracket, if equipped.

17. Make sure the master cylinder reservoir is filled to the proper level. Bleed the brake system, if necessary.
18. Connect the negative battery cable.
19. Road test the vehicle and check for proper brake operation.

Brake Caliper

REMOVAL AND INSTALLATION

Front Disc Brake Caliper

1. Raise and safely support the vehicle.
2. Remove the wheel.
3. Remove the banjo bolt attaching the brake hose to the caliper and discard the 2 sealing washers. Plug the hose to prevent fluid leakage.
4. Remove the caliper mounting bolt and pivot the caliper upward and off the brake pads.
5. Slide the caliper from the guide pin and remove from the vehicle.
To install:
6. Remove the guide pin bushing dust boots and push out the caliper guide pin bushing.
7. Lubricate the guide pin bushings with high temperature grease and install them in the caliper. Install the guide pin bushing dust boots.
8. Slide the caliper onto the guide pin and pivot the caliper down onto the brake pads. To provide the necessary clearance, it may be necessary to pull slightly outward on the caliper.
9. Install the caliper mounting bolt and tighten to 33–36 ft. lbs. (44–49 Nm).
10. Install 2 new copper washers and the banjo bolt on the brake hose banjo fitting.
11. Position the brake hose on the caliper and install the banjo bolt. Tighten the bolt to 16–22 ft. lbs. (22–29 Nm).
12. Bleed the brake system.
13. Install the wheel and tighten the lug nuts to 65–87 ft. lbs. (80–118 Nm).
14. Lower the vehicle.
15. Check the brake system for proper operation.

Rear Disc Brake Caliper

1. Raise and safely support the vehicle.
2. Remove the wheel.
3. Remove the parking brake cable retaining clip.
4. Loosen the parking brake cable housing adjustment nut. Remove the

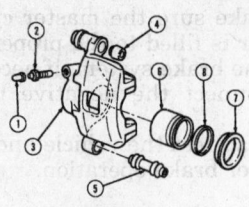

Item	Description
1	Bleeder Screw Cap
2	Bleeder Screw
3	Caliper Body
4	Caliper Guide Dust Boot / Bushing
5	Dust Boot
6	Piston
7	Dust Cover
8	Piston Seal

294935

Front brake caliper (exploded view)

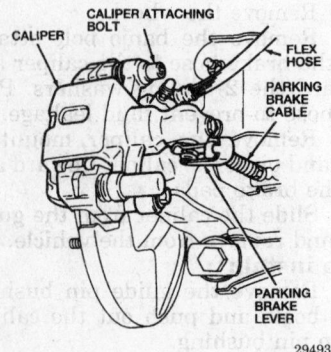

294936

Rear brake caliper

cable housing from the bracket and the parking brake lever.

5. Remove the banjo bolt mounting the brake hose to the caliper.

6. Remove and discard the copper washers from the banjo fitting.

7. Remove the caliper mounting bolt.

8. Pivot the caliper off the brake pads and slide the caliper off the guide pin.

To install:

9. Lubricate the guide pin bushings with high temperature grease. Install the caliper onto the guide pin and pivot the caliper over the brake pads. Tighten the caliper attaching bolt to 25–29 ft. lbs. (34–39 Nm).

10. Install new copper washers and the banjo bolt mounting the brake hose to the caliper. Tighten the banjo bolt to 16–22 ft. lbs. (23–29 Nm).

11. Position the parking brake cable into the parking brake lever and bracket. Install the retaining clip.

12. Adjust the parking brake cable so there is no clearance between the cable end and the parking brake lever. Tighten the parking brake cable locknut.

13. Bleed the brake system.

14. Install the wheel and tighten the lug nuts to 65–87 ft. lbs. (80–118 Nm).

15. Lower the vehicle.

16. Check the brake system for proper operation.

Disc Brake Pads

REMOVAL AND INSTALLATION

Front Disc Brake Pads

1. Remove brake fluid from the master cylinder reservoir until the reservoir is approximately ½ full. Discard the removed fluid.

2. Raise and safely support the vehicle.

3. Remove the wheel.

4. If necessary, clean the brake assembly with brake cleaner and allow to dry.

5. Using an appropriate tool, pry the caliper outboard.

6. Remove the caliper mounting bolt. Pivot the caliper upward on the fixed guide pin and secure it out of the way.

— **WARNING** —
Do not allow the caliper to hang by the brake hose or damage to the brake hose may result.

7. Remove the 2 anti-rattle springs.

8. Remove the shims. Tag the shims so they can be reinstalled in their original position.

9. Remove the brake pads and retaining clips from the caliper anchor.

10. Inspect the disc brake rotor for wear and/or damage. Machine or replace, as necessary. If machining, observe the minimum thickness specification.

To install:

11. Install the retaining clips.

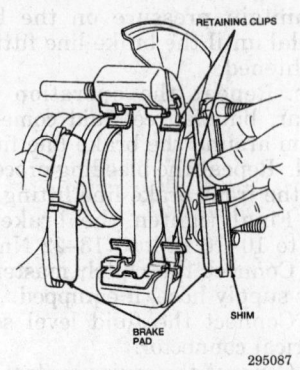

295087

Front disc brake pads

12. Install the brake pads into the caliper anchor. The pad with the wear indicator is the inboard pad.

13. Install the shims in their original position.

14. Compress the caliper piston into its bore using a large C-clamp and one of the old brake pads. Do not allow the clamp to push directly on the caliper piston.

15. Pivot the caliper down over the brake pads. Install the anti-rattle springs. Install the caliper mounting bolt and tighten the bolt to 33–36 ft. lbs. (44–49 Nm).

16. Install the wheel. Tighten the lug nuts to 65–87 ft. lbs. (80–118 Nm).

17. Lower the vehicle.

18. Pump the brake pedal several times to position the caliper piston.

19. Check the fluid level in the master cylinder reservoir and add fluid as necessary.

20. Road test the vehicle for proper brake operation.

Rear Disc Brake Pads

1. Remove brake fluid from the master cylinder reservoir until the reservoir is approximately ½ full. Discard the removed fluid.

2. Raise and safely support the vehicle.

3. Remove the wheel.

4. Remove the parking brake cable retaining clip.

5. Loosen the parking brake cable housing adjusting nut. Remove the cable housing from the bracket and the parking brake lever.

6. Insert an Allen wrench into the back of the caliper and turn the manual adjustment gear counterclockwise to pull the caliper piston inward. Turn the gear until it stops.

7. Remove the caliper mounting bolt and pivot the caliper to clear the brake pads. Remove the caliper and suspend it with wire from the rear strut.

— **WARNING** —
Do not allow the caliper to hang by the brake hose or damage to the brake hose may result.

8. Remove the anti-rattle spring from the disc brake pads. Remove the disc brake pads, the shims and retaining clips. If the brake pads and shims are to be reused, tag them so they can be installed in their original positions.

9. Inspect the disc brake rotor for wear and/or damage. Machine or replace, as necessary. If machining, ob-

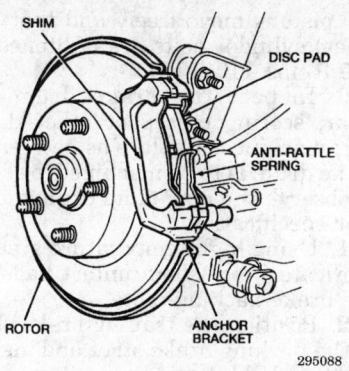

Rear disc brake pads

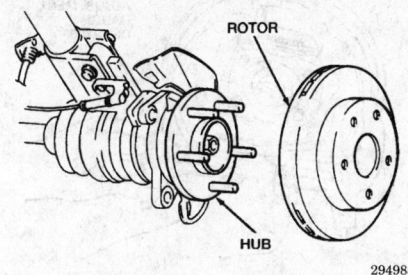

Front brake rotor

serve the minimum thickness specification.

To install:

10. Install the retaining clips. Position the shims on the disc brake pads and install the pads into the caliper anchor bracket.

11. Install the anti-rattle spring onto the disc brake pads.

12. Lightly lubricate the guide pin bushings with high temperature grease and install the caliper onto the guide pin. Pivot the caliper over the disc brake pads.

13. Install the caliper mounting bolt and tighten to 25–29 ft. lbs. (34–39 Nm).

14. Install the parking brake cable into the parking brake lever and bracket. Install the parking brake cable retaining clip.

15. Adjust the cable so there is no clearance between the cable end and the parking brake lever. Tighten the parking brake cable locknut.

16. Turn the caliper manual adjustment gear clockwise with an Allen wrench until the brake pads just touch the rotor, then back off 1/3 turn.

17. Install the wheel and tighten the lug nuts to 65–87 ft. lbs. (80–118 Nm).

18. Lower the vehicle.

19. Pump the brake pedal several times to position the caliper piston. Check the fluid level in the master cylinder reservoir and add clean brake fluid, if necessary.

20. Road test and check for proper brake operation.

Brake Rotor

REMOVAL AND INSTALLATION

Front Disc Brake Rotor

1. Raise and safely support the vehicle.

2. Remove the wheel.

3. Remove the caliper anchor bracket bolts and remove the anchor bracket and caliper as an assembly. Support the caliper assembly from the coil spring with mechanics wire or string; do not disconnect the brake hose.

NOTE: Do not let the caliper assembly hang by the brake hose.

4. Remove the disc brake rotor. Handle the rotor with care, to prevent nicking or scratching the rotor surface.

5. Inspect the disc brake rotor for wear, scoring, cracks or other damage. Machine or replace the rotor, as necessary. If machining, observe the rotor minimum thickness specification.

To install:

6. Fit the disc brake rotor onto the hub.

7. Install the caliper anchor bracket and tighten the bolts to 58–72 ft. lbs. (78–98 Nm).

8. Install the wheel and tighten the lug nuts to 65–87 ft. lbs. (80–118 Nm).

9. Lower the vehicle.

10. Before moving the vehicle, apply the brake pedal several times to make sure the caliper piston is positioned.

Rear Disc Brake Rotor

1. Raise and safely support the vehicle.

2. Remove the wheel.

3. Remove the dust cap.

4. Use a proper chisel to unstake the rear axle hub retaining nut. Remove the retaining nut and discard.

5. Remove the 2 anchor bracket bolts and remove the caliper and anchor bracket assembly. Do not disconnect the brake hose from the caliper. Support the caliper with mechanics wire from the coil spring.

NOTE: Do not let the caliper hang by the brake hose.

6. Remove the disc brake rotor from the hub.

7. Inspect the disc brake rotor for wear, scoring, cracks or other damage. Machine or replace the rotor, as necessary. If machining, observe the rotor minimum thickness specification.

To install:

8. Install the rotor on the hub assembly.

9. Install the caliper and anchor bracket assembly. Tighten the caliper anchor bracket bolts to 33–49 ft. lbs. (45–67 Nm).

10. Install a new wheel hub retaining nut and torque to 130–174 ft. lbs. (177–235 Nm).

11. Stake the retaining nut using a cold chisel with the cutting edge rounded off.

12. Install the dust cap.

13. Install the wheel and torque the lug nuts to 65–87 ft. lbs. (80–118 Nm).

14. Lower the vehicle.

15. Connect the negative battery cable.

16. Before moving the vehicle, apply the brake pedal several times to make sure the caliper piston is positioned.

Brake Drums

REMOVAL AND INSTALLATION

1. Raise and safely support the vehicle.

2. Remove the wheel.

3. Remove the hub grease cap.

4. Remove the 2 brake drum screws and remove the brake drum.

5. Inspect the brake drum for wear, scoring, cracks or other damage. Machine or replace the drum, as necessary. If machining, observe the maximum drum diameter specification.

To install:

6. If the brake drum has been machined and/or the brake shoes replaced, measure the inside diameter of the brake drum using adjustment gauge D81L-1103-A or equivalent.

7. Insert a suitable tool into the knurled quadrant of the rear brake strut and quadrant. Adjust the shoes to the same measurement as the brake drum.

8. Install the brake drum.

9. Install the brake drum screws and tighten to 89–123 inch lbs. (10–14 Nm).

10. Install the hub grease cap.

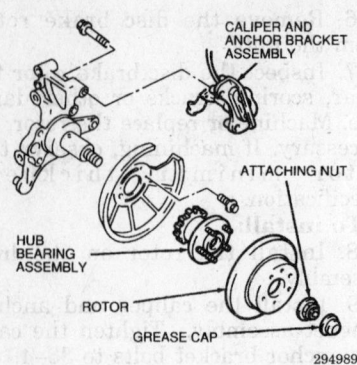

Rear brake rotor and related components

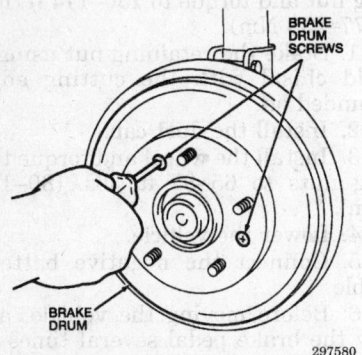

Removing the brake drum

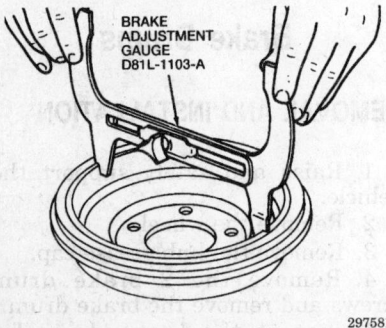

Measuring the drum with the brake adjustment gauge

11. Install the wheel and torque the lug nuts to 65–87 ft. lbs. (88–118 Nm).

12. Lower the vehicle. Complete the brake adjustment by sharply applying the brakes several times while driving the vehicle alternately forward and reverse.

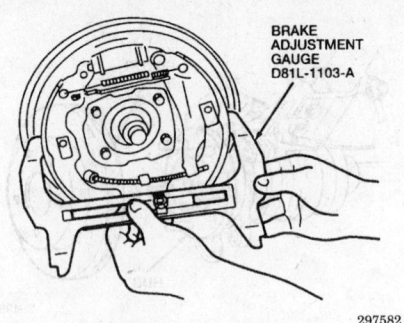

Measuring the brake shoes with the brake adjustment gauge

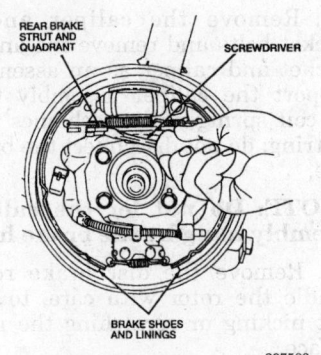

Adjusting the brake shoes

Brake Shoes

REMOVAL AND INSTALLATION

1. Raise and safely support the vehicle.

2. Remove the wheel and dust cap.

3. Unstake the hub locknut. Have an assistant apply the brakes to lock the hub, then remove the locknut.

4. Remove the brake drum and the hub.

5. Remove the brake shoe hold-down springs. Push the hold-down spring inward and twist the hold-down pin using needle-nose pliers until the head of the pin aligns with the slot in the spring. Release the spring and pin.

6. Remove the parking brake cable from the parking brake anchor plate.

7. Remove the brake shoe return springs and remove the brake shoes.

To install:

8. Clean the brake backing plate and brake hardware with suitable brake cleaner and air dry.

9. Inspect all components for wear and/or damage and replace, as necessary. Inspect the wheel cylinder for moisture which may indicate leakage. Verify that the rear wheel cylinder pistons move freely and that the wheel cylinder bolts are tightened to 7–9 ft. lbs. (10–13 Nm).

10. Inspect the brake drum for wear, scoring cracks or other damage; machine or replace as necessary. If the drum is to be machined, be sure to observe the maximum drum diameter specification.

11. Using high temperature grease, lubricate the 6 shoe contact pads on the brake backing plate.

12. Position the trailing brake shoe in the parking brake strut and install the rear hold-down pin and spring.

13. Position the leading brake shoe against the parking brake strut and backing plate and install the hold-down pin and spring.

14. Install the brake shoe return springs. Connect the parking brake cable.

15. Measure the drum inside diameter using gauge tool D81L-1103-A or equivalent. Insert a small prybar or screwdriver into the knurled quadrant of the parking brake strut and adjust the brake shoes to the same measurement as the brake drum. The brake shoes should just touch the brake drum when properly adjusted.

16. Install the hub and the brake drum. Install a new locknut and tighten to 130–174 ft. lbs. (177–235 Nm). Stake the locknut using a dull bladed chisel. Install the dust cap.

17. Install the wheel and torque the lug nuts to 65–87 ft. lbs. (88–118 Nm).

18. Lower the vehicle.

19. Complete the brake adjustment by sharply applying the brakes several times while driving the vehicle alternately forward and reverse.

Wheel Cylinder

REMOVAL AND INSTALLATION

1. Raise and safely support the vehicle.

2. Remove the wheel.

3. Remove the brake drum, hub and the brake shoes.

4. Using a tubing wrench, disconnect the brake line from the wheel cylinder.

5. Remove the 2 wheel cylinder-to-backing plate bolts.

6. Remove the wheel cylinder.

To install:

7. Install the wheel cylinder and loosely install the mounting bolts.

8. Connect the brake line to the wheel cylinder and tighten the fitting.

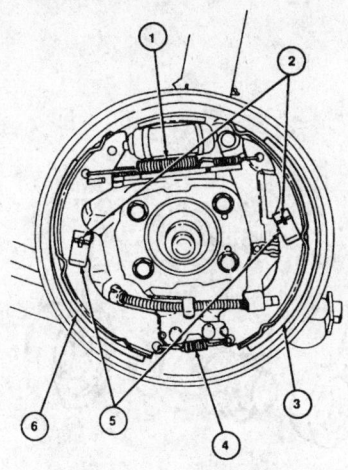

Item	Description
1	Brake Shoe Retracting Spring, Upper
2	Brake Shoe Hold Down Spring Pin
3	Trailing Brake Shoe and Lining (Part of 2200)
4	Brake Shoe Retracting Spring, Lower
5	Brake Shoe Hold-Down Spring
6	Leading Brake Shoe and Lining (Part of 2200)

295049

Brake shoes and related components

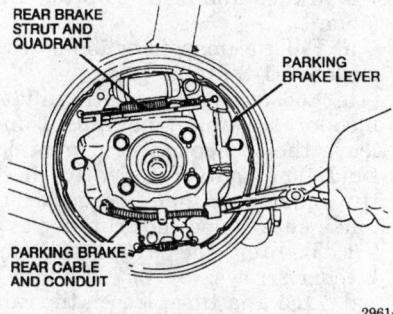

296145

Removing the parking brake cable from the parking brake lever

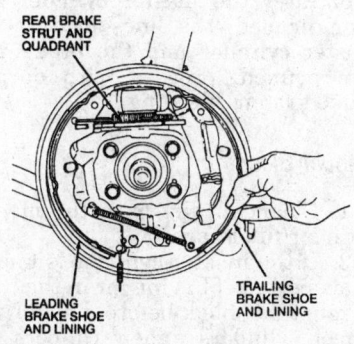

296168

Removing the trailing brake shoe

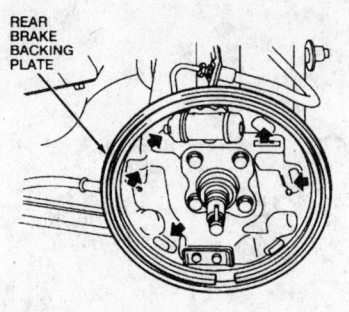

296173

Apply lubricant to the brake backing plate at these locations

9. Tighten the wheel cylinder mounting bolts to 84–108 inch lbs. (10–13 Nm).
10. Install the brake shoes, hub and brake drum.
11. Bleed the brakes.
12. Install the wheel and torque the lug nuts to 65–87 ft. lbs. (88–118 Nm).
13. Lower the vehicle.
14. Road test the vehicle and check for proper brake operation.

Parking Brake Cable

ADJUSTMENT

1. Start the engine and depress the brake pedal several times while the vehicle is moving in reverse.
2. Stop the engine.
3. Remove the center console and remove the parking brake lever cover.
4. Adjust the parking brake adjusting nut so the brakes are fully applied when the parking brake lever can be lifted 5–7 notches.
5. Make sure the brakes do not drag.
6. Reinstall the lever cover and the center console.

REMOVAL AND INSTALLATION

1. Raise and safely support the vehicle.
2. Remove the rear wheels.
3. If equipped with drum brakes, proceed as follows:
   a. Remove the brake drum and hub.
   b. Remove the parking brake cable from the parking brake anchor plate.
   c. Remove the cable from the hole in the backing plate.
   d. Disconnect the parking brake return spring from the parking brake strut.

4. If equipped with disc brakes, loosen the parking brake cable adjusting nut and remove the cable from the parking brake lever.
5. Remove the 4 cable housing clamp nuts from the rear suspension trailing arms.
6. Remove the exhaust heat shield.
7. Remove the 2 cable housing clamp nuts from the body and remove the cables from the equalizer and housing bracket.

To install:

8. Install the cable ends into the parking brake equalizer and housing brackets.
9. Apply suitable grease to the cable clamps and the cable, then install the cable housing support clamps on the body.
10. Install the cable housing brackets on each rear suspension trailing arm and tighten the nuts to 19 ft. lbs. (25 Nm).
11. Install the exhaust heat shield.
12. If equipped with drum brakes, proceed as follows:
   a. Route the parking brake cable through the hole in the backing plate and connect it to the parking brake anchor plate.
   b. Connect the parking brake return spring to the parking brake strut.
   c. Install the hub and the brake drum.
13. If equipped with disc brakes, proceed as follows:
   a. Connect the parking brake cable to the parking brake lever.
   b. Install the parking brake cable retaining clip.
   c. Tighten the parking brake cable adjusting nut.

NOTE: On rear disc brakes, there must be no clearance between the cable end and the lever.

14. Install the wheels and torque the lug nuts to 65–87 ft. lbs. (88–118 Nm).
15. Lower the vehicle.
16. Adjust the parking brake.

Brake System Bleeding

SYSTEM PRIMING

When a new master cylinder has been installed, or the brake system emptied or partially emptied, fluid may not flow from the bleeder screws during normal bleeding. It may be

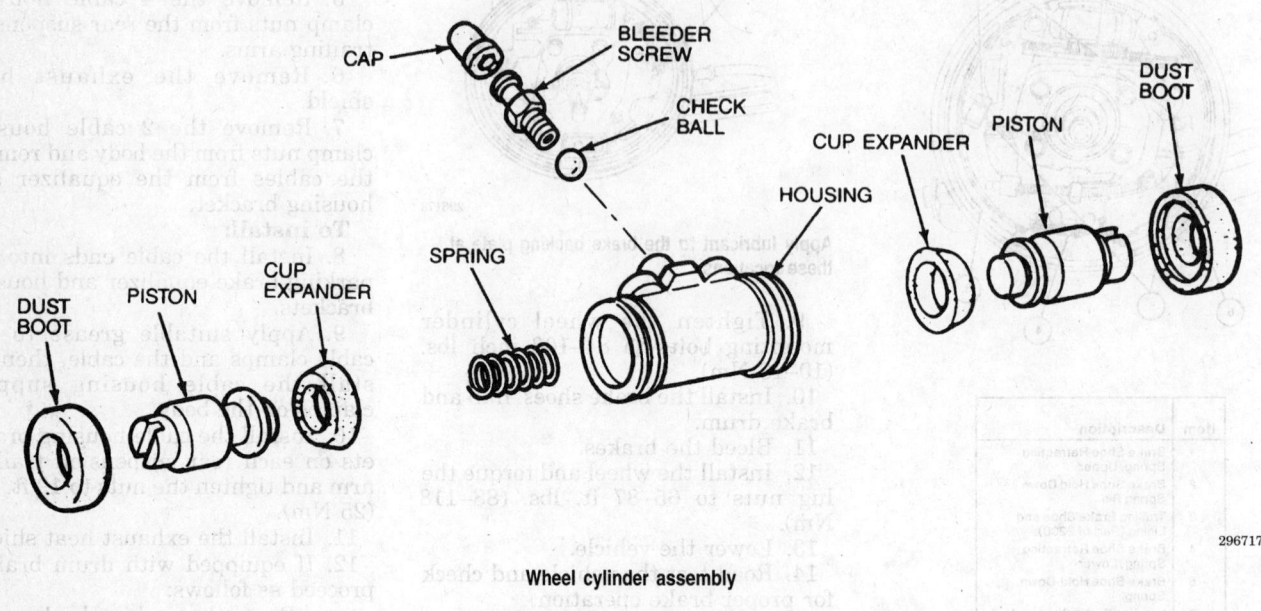

Wheel cylinder assembly

296717

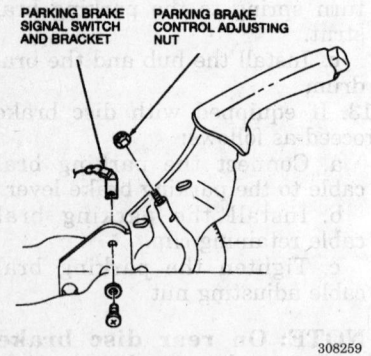

Parking brake lever

308259

necessary to prime the system using the following procedure:

1. Disconnect the brake lines from the master cylinder.

2. Install short brake lines in the master cylinder ports and position them so they loop back up into the reservoir and the ends of the lines are submerged in brake fluid.

3. Fill the reservoir with clean DOT-3 brake fluid and cover the reservoir with a shop towel.

4. Slowly pump the brake pedal until clear, bubble-free fluid comes out of both temporary brake lines.

WARNING

Do not allow brake fluid to spill on the vehicle's finish; it will remove the paint. In case of a spill, flush the area with water.

---

5. Remove the short brake lines and reconnect the vehicle brake lines to the master cylinder.

6. Bleed each brake line at the master cylinder as follows:

a. Have an assistant slowly pump the brake pedal 10 times and then hold firm pressure on the pedal.

b. Position a shop towel under the rear most brake line fitting. Open the fitting with a tubing wrench until a stream of brake fluid comes out. Have the assistant maintain pressure on the brake pedal until the brake line fitting is tightened.

c. Repeat Steps a and b until clear, bubble-free fluid comes out from around the tubing fitting.

d. Repeat the operation on the front brake line fitting.

7. If any of the brake lines, calipers, or wheel cylinders have been removed, it may be helpful to prime the system by gravity bleeding. This should be done after the master cylin-

der is primed and bled. To prime the system:

a. Fill the master cylinder with clean DOT-3 brake fluid.

b. Loosen both wheel cylinder bleeder screws, if equipped, and leave them open until clear brake fluid flows out. Frequently check the master cylinder reservoir to make sure it does not run dry.

c. Tighten the wheel cylinder bleeder screws.

d. One at a time, loosen the caliper bleeder screws and leave them open until clear fluid flows out. Frequently check the master cylinder reservoir to make sure it does not run dry.

e. Tighten the bleeder screws.

8. After the master cylinder has been primed, the lines bled at the master cylinder and the brake system primed, proceed with normal brake system bleeding.

Manual Bleeding

1. Clean all dirt from the master cylinder filler cap.

2. If the master cylinder is known or suspected of having air in the bore, it must be bled before any of the wheel cylinders and/or calipers are bled. Use the System Priming procedure.

3. Bleed the wheel cylinders and/or calipers as follows:

a. Begin at the rear bleeder screw.

NOTE: The brake system is diagonally split. If bleeding is begun at the right rear wheel, bleed the left front caliper next, followed by the left rear and right front. If bleeding is begun at the left rear wheel, bleed the right front caliper next, followed by the right rear and left front.

b. Attach a drain hose to the bleeder screw. The end of the hose should fit snugly around the end of the bleeder screw.

c. Place the other end of the hose in a container partially filled with clean brake fluid.

d. Have an assistant slowly pump the brake pedal 5–10 times and maintain pressure on the pedal after the last stroke.

e. Loosen the bleeder screw approximately 3/4 turn. Make sure your assistant keeps constant pressure on the pedal until the pedal drops all the way down and the bleeder screw is closed again. If the pedal pressure is released, air will be drawn back into the system.

f. Tighten the bleeder screw.

g. Repeat this operation until the fluid is clear and air bubbles no longer appear in the container.

h. Repeat these steps at the other wheel cylinder and calipers.

NOTE: Never reuse the brake fluid expelled from the bleeder screws during the bleeding operation.

4. After the bleeding procedure is completed, make sure the fluid level is correct in the master cylinder reservoir.

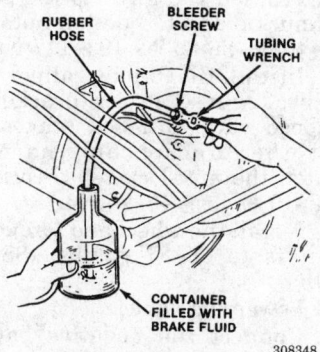

RUBBER HOSE · BLEEDER SCREW · TUBING WRENCH · CONTAINER FILLED WITH BRAKE FLUID

308348

Bleeding the brake system

Wheel Speed Sensor

REMOVAL AND INSTALLATION

Front

1. Disconnect the negative battery cable.
2. Raise and safely support the vehicle.
3. Remove the front wheel.
4. Remove the 2 speed sensor mounting bolts and remove the sensor from the steering knuckle.
5. Remove the routing brackets from the strut and inner fender well.
6. Disconnect the sensor electrical connector and remove the sensor.

To install:

NOTE: The left and right sensors are not interchangeable; L is marked on the left bracket and R is marked on the right bracket. Make sure that the wiring harness will clear all suspension components.

7. Route the sensor wiring harness in the vehicle and connect the electrical connector.
8. Install the routing bracket onto the inner fender well and tighten the nut to 71–88 inch lbs. (8–10 Nm).
9. Install the routing bracket onto the strut and tighten the bolt to 13–19 ft. lbs. (18–25 Nm).
10. Install the sensor into the steering knuckle and tighten the bolts to 12–17 ft. lbs. (16–23 Nm).
11. Make sure that the clearance between the sensor and the sensor ring on the CV-joint is 0.012–0.043 in. (0.3–1.1mm)
12. Install the wheel and lower the vehicle.
13. Connect the negative battery cable.

Rear

1. Disconnect the negative battery cable.
2. Raise and safely support the vehicle.
3. Remove the rear wheel.
4. Remove the speed sensor mounting bolt and remove the sensor from the spindle.
5. Remove the routing brackets from the strut and inner fender well.
6. Remove luggage compartment interior panels as necessary, to gain access to the wiring harness.
7. Disconnect the sensor electrical connector and remove the sensor.

To install:

NOTE: The left and right sensors are not interchangeable; L is marked on the left bracket and R is marked on the right bracket. Make sure that the wiring harness will clear all suspension components.

8. Route the sensor wiring harness in the vehicle and connect the electrical connector.
9. Install the routing bracket onto the inner fender well and tighten the bolt to 81–113 inch lbs. (9–13 Nm).
10. Install the routing bracket onto the strut and tighten the bolt to 13–19 ft. lbs. (18–25 Nm).
11. Install the sensor to the spindle and tighten the bolt to 12–17 ft. lbs. (16–23 Nm).
12. Make sure that the clearance between the sensor and sensor ring on the hub is 0.012–0.043 in. (0.3–1.1mm).
13. Install any luggage compartment interior panels that were removed.
14. Install the wheel and lower the vehicle.
15. Connect the negative battery cable.

FRONT SUSPENSION

Strut and Spring

REMOVAL AND INSTALLATION

1. Disconnect the negative battery cable.
2. Raise and support the vehicle safely.
3. Remove the wheel.
4. If equipped with anti-lock brakes, disconnect the electrical harness and remove the bracket.
5. Remove the U-clip from the brake line hose and slide it out of the strut bracket.
6. Remove the 2 strut-to-steering knuckle bolts and nuts.
7. Lower the vehicle enough to remove the 4 upper strut mounting bolts.
8. Remove the strut from the vehicle.
9. Place the strut assembly in a suitable holding fixture. Loosen, but do not remove the shock mounting nut. Compress the spring with a suitable compressor tool, then remove the shock mounting nut. Gradually

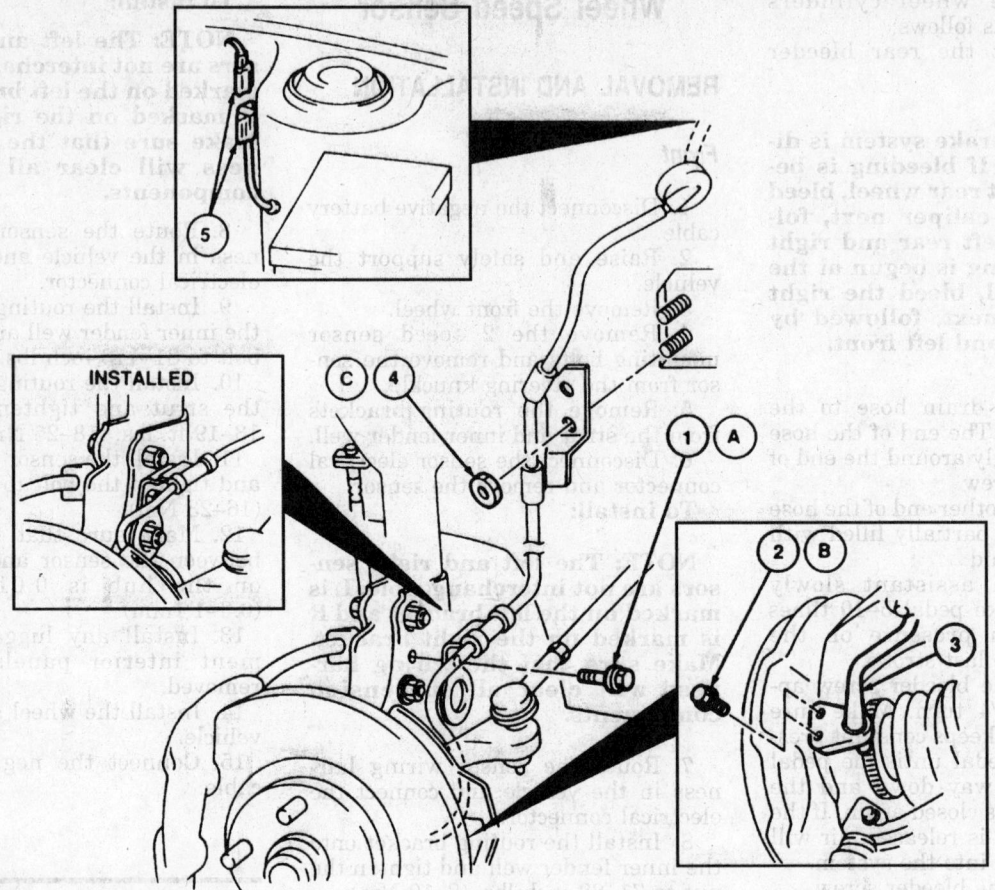

INSTALLED

1. Routing bracket bolt
2. Front brake anti-lock sensor bolts(2)
3. Front brake anti-lock sensor
4. Routing bracket nut(2)
5. Front brake anti-lock sensor electrical connector
   A. 13-19 ft. lb.(18-25 Nm)
   B. 12-17 ft. lb.(16-23 Nm)
   C. 71-88 in. lb.(8-10 Nm)

296231

Front wheel speed sensor mounting

release the spring compressor to relieve the spring tension.

10. Remove the strut mounting bracket, thrust bearing, upper spring seat with insulator, dust tube and bumper.

11. Remove the coil spring and the lower spring insulator. Replace components as required.

To install:

NOTE: Face the direction indicator on the strut mounting bracket towards the rear outboard position during reassembly. Make sure that the notch on the upper spring seat faces towards the outboard position during reassembly.

12. Compress the coil spring with the coil spring compressor and install the coil spring and the lower spring insulator onto the strut.

13. Install the bumper, dust tube, upper spring seat with insulator, thrust bearing and the strut mounting bracket.

14. Install the strut mounting nut and tighten to 66–86 ft. lbs. (89–117 Nm).

15. Gradually release the compressor tool and remove from the strut assembly.

16. Install the strut in the shock tower with the direction indicator facing the rear outboard position.

Tighten the 4 upper strut mounting nuts to 34–46 ft. lbs. (46–63 Nm).

17. Position the strut to the steering knuckle and torque the nuts and bolts to 68–86 ft. lbs. (93–117 Nm).

18. Install the brake caliper and the brake hose in its bracket. If equipped with anti-lock brakes, install the bracket and harness. Tighten the anti-lock sensor bracket to 13–19 ft. lbs. (18–25 Nm).

19. Install the wheel and torque the lug nuts to 65–87 ft. lbs. (88–118 Nm).

20. Lower the vehicle.

21. Connect the negative battery cable.

22. Check the wheel alignment.

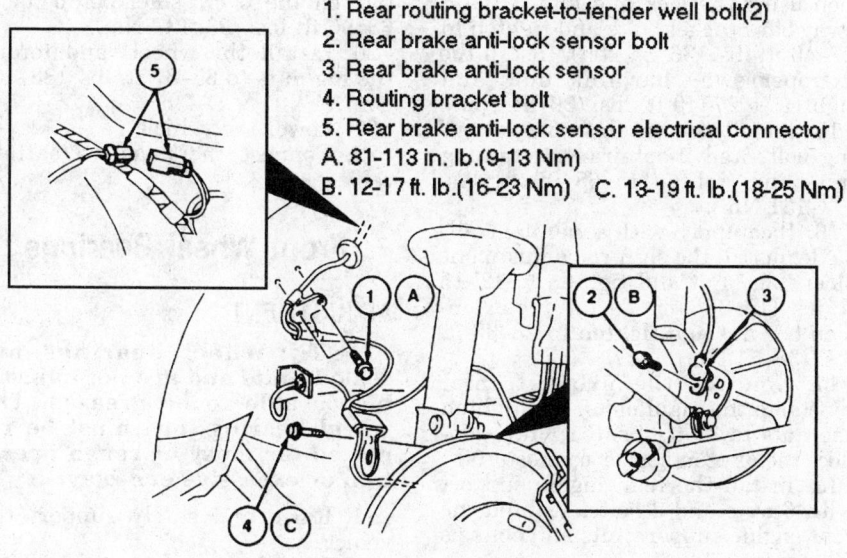

1. Rear routing bracket-to-fender well bolt(2)
2. Rear brake anti-lock sensor bolt
3. Rear brake anti-lock sensor
4. Routing bracket bolt
5. Rear brake anti-lock sensor electrical connector
A. 81-113 in. lb.(9-13 Nm)
B. 12-17 ft. lb.(16-23 Nm)   C. 13-19 ft. lb.(18-25 Nm)

296232

Rear wheel speed sensor mounting

7. Remove the 2 lower control arm rear mounting bolts.
8. Remove the lower control arm front mounting bolt and remove the lower control arm from the vehicle.
   To install:
9. Position the lower control arm, install the 2 lower control arm rear mounting bolts and tighten to 69–96 ft. lbs. (93–131 Nm).
10. Install the lower control arm front mounting bolt and tighten to 58–78 ft. lbs. (78–106 Nm).
11. Install the ball joint stud into the steering knuckle and tighten the clamp bolt to 32–40 ft. lbs. (43–54 Nm).
12. Install the stabilizer bar link nut and tighten to 27–40 ft. lbs. (36–54 Nm).
13. Install the wheel and tighten the lug nuts to 65–87 ft. lbs. (88–118 Nm).
14. Lower the vehicle.
15. Check the front suspension and steering for proper operation. Check wheel alignment.

Sway Bar

REMOVAL AND INSTALLATION

1. Disconnect the negative battery cable.
2. Raise and safely support the vehicle.
3. Remove the front wheels.
4. Remove the 6 crossmember bolts and remove the crossmember.
5. Remove the 2 bolts and 6 nuts and remove the rear engine support.
6. Disconnect the oxygen sensor connector(s).
7. Disconnect the exhaust inlet pipe(s) from the manifold(s) and position aside.
8. If equipped with manual transaxle, remove the extension bar nut and shift rod adjustment sleeve and nut.
9. Position a jack under the front sub-frame. Remove the 4 bolts and 2 nuts attaching the front sub-frame to the body.
10. Remove the 2 upper sway bar link nuts. Remove the 4 sway bar brackets bolts.
11. Lower the front sub-frame to allow removal of the sway bar.
12. Remove the sway bar from the right side of the vehicle.
    To install:

NOTE: Apply rubber grease to the inside surface of the sway bar bushings and align the bushings with the installation mark on the sway bar.

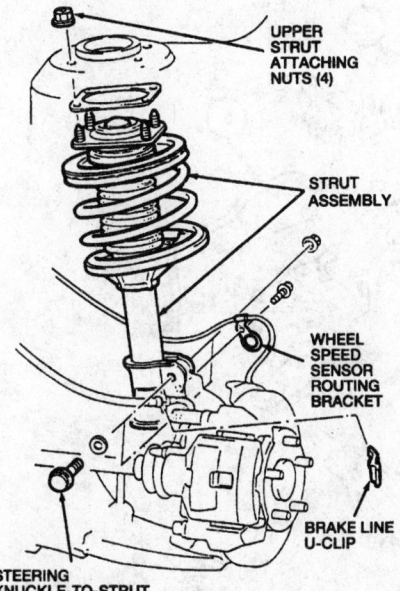

UPPER STRUT ATTACHING NUTS (4)

STRUT ASSEMBLY

WHEEL SPEED SENSOR ROUTING BRACKET

BRAKE LINE U-CLIP

STEERING KNUCKLE-TO-STRUT MOUNTING BOLTS (2)

285435

Front strut assembly

Lower Ball Joints

REMOVAL AND INSTALLATION

The lower ball joint is an integral part of the lower control arm and cannot be serviced separately. If the lower ball joint is defective, the entire lower control arm must be replaced.

Lower Control Arms

REMOVAL AND INSTALLATION

NOTE: The lower ball joint is an integral part of the lower control arm and cannot be serviced separately. If the lower ball joint is defective, the entire lower control arm must be replaced.

1. Disconnect the negative battery cable.
2. Raise and safely support the vehicle.
3. Remove the wheel.
4. Remove the ball-joint clamp bolt and nut from the steering knuckle.
5. Remove the stabilizer bar link nut.
6. Using a prybar, pry downward to separate the ball joint from the steering knuckle.

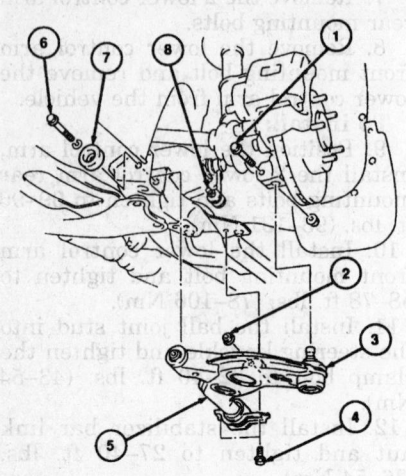

Item	Description
1	Ball Joint Clamp Bolt
2	Stabilizer Control Link-to-Lower Control Arm Nut
3	Ball Joint
4	Rear Bushing Bolt
5	Lower Control Arm
6	Front Bushing Bolt
7	Front Bushing Dynamic Damper
8	Stabilizer Control Link

304772

Lower control arm assembly

13. Install the sway bar from the right side of the vehicle and position the sway bar to the sub-frame.

14. Raise the sub-frame into position using the jack and install the 4 sway bar bracket bolts and tighten to 27–40 ft. lbs. (36–54 Nm). Install the 2 upper sway bar link nuts and tighten to 27–40 ft. lbs. (36–54 Nm).

15. Install the 4 sub-frame retaining bolts and 2 sub-frame retaining nuts and tighten to 68–96 ft. lbs. (93–131 Nm).

16. If equipped with a manual transaxle, install the shift rod adjustment sleeve and nut and tighten to 12–16 ft. lbs. (19–25 Nm). Install the extension bar nut and tighten to 28–38 ft. lbs. (38–51 Nm).

17. Connect the exhaust inlet pipe(s) to the manifold(s) and tighten the nuts to 38 ft. lbs. (51 Nm). Connect the oxygen sensor connector(s).

18. Install the rear engine support with 6 nuts and 2 bolts. Tighten the rear engine support nuts and bolts as follows:

a. Torque nuts and/or bolts in figure **A** to 55–77 ft. lbs. (75–104 Nm).

b. Torque nuts and/or bolts in figure **B** to 50–68 ft. lbs. (67–93 Nm).

c. Torque nuts and/or bolts in figure **C** to 32–44 ft. lbs. (44–60 Nm).

19. Install the crossmember and tighten the 6 crossmember bolts to 68–96 ft. lbs. (93–131 Nm).

20. Install the wheels and torque the lug nuts to 65–87 ft. lbs. (88–118 Nm).

21. Lower the vehicle.

22. Connect the negative battery cable.

Front Wheel Bearings

ADJUSTMENT

NOTE: Wheel bearings are sealed units and are not adjustable or able to be greased. The wheel bearing unit must be replaced for noisy or rough operation or excessive end-play.

1. Raise and safely support the vehicle.

2. Make sure the parking brake is fully released.

3. Remove the wheel.

4. On disc brakes, remove the disc brake caliper.

5. On rear disc brakes install the lug nuts to hold the rotor in place.

6. Rotate the drum or rotor to make sure there is no brake drag.

Item	Description
1	Rear Engine Mount
2	Front Crossmember
3	Stabilizer Bar
4	Steering Gear
5	Transaxle Cradle
6	Transverse Member
7	Converter Inlet Pipe — 2.5L
8	Converter Inlet Pipe — 2.0L

2.0L ENGINE

2.5L ENGINE

304815

Front sway bar (stabilizer bar) and related components

7. Position a suitable dial indicator.

8. Check the wheel bearing end-play. End-play should not exceed 0.002 in. (0.05mm).

9. If the end-play exceeds specification, replace the wheel bearing or hub/bearing assembly, as required.

10. When complete, install the calipers if removed.

11. Install the wheel and torque the lug nuts to 65–87 ft. lbs. (88–118 Nm).

REMOVAL AND INSTALLATION

1. Disconnect the negative battery cable.

2. Raise and safely support the vehicle.

3. Remove the front wheel.

4. Using a small cape chisel or similar tool and a hammer, raise the staked portion of the hub retaining nut. With an assistant applying the brakes to prevent the hub from turning, loosen and remove the wheel hub retaining nut. Discard the nut after removal; it must not be reused.

5. If equipped with anti-lock brakes, remove the 2 anti-lock sensor retaining bolts and the sensor and bracket.

6. At the tie rod end, remove the cotter pin and castellated nut. Using a tie rod end separator tool or equivalent, separate the tie rod end from the steering knuckle.

7. Remove the lower control arm ball joint clamp nut/bolt. Using a prybar, pry the lower control arm downward and separate the ball joint from the steering knuckle.

8. Remove the steering knuckle-to-strut attaching bolts.

9. Remove the front disc brake caliper anchor bracket and suspend the caliper assembly from the coil spring with mechanics wire.

10. Remove the disc brake rotor.

11. Pull the steering knuckle/hub assembly off of the halfshaft. If necessary, lightly tap the end of the halfshaft CV-joint with a plastic faced hammer or use a wheel puller to press the shaft out of the hub. Support the halfshaft with wire or a stand; do not let the halfshaft hang on the inner CV-joint or the joint could be pulled apart.

12. Remove the steering knuckle, hub and wheel bearing assembly from the vehicle.

13. Using a prybar, pry the grease seal from the knuckle.

14. Position the steering knuckle in a suitable fixture and remove the hub from the knuckle. If the inner race remains on the hub, grind a section of the inner race to approximately 0.020

inch (0.5mm) and use a chisel to remove it. Wear appropriate eye protection.

15. Remove the snapring from the steering knuckle.

NOTE: Wheel bearings are contained within a sealed unit and are not serviceable.

16. Position the steering knuckle in a suitable fixture and remove the wheel bearing from the knuckle.

NOTE: Unless the disc brake rotor dust shield is damaged, it should be left on the steering knuckle; it is pressed on and must be replaced if removed or damaged.

To install:

17. Inspect the steering knuckle and hub for cracks, wear and scoring. Replace parts as necessary.

18. Position the steering knuckle in a suitable fixture and press in the wheel bearing. Make sure the press tool contacts only the outer bearing race or the bearing will be damaged.

19. Install the snapring.

20. Position the steering knuckle in a suitable fixture and press the hub into the steering knuckle. Make sure the inner bearing race is supported or the bearing will be damaged.

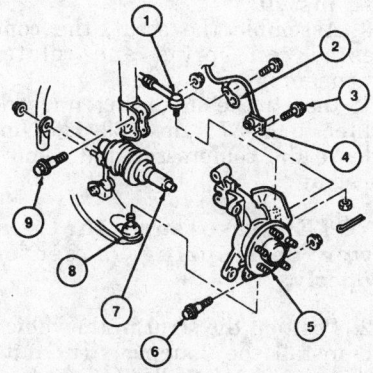

Item	Description
1	Tie Rod End
2	Speed Sensor Routing Bracket
3	Speed Sensor Bolts
4	Speed Sensor
5	Wheel Hub / Steering Knuckle Assembly
6	Ball Joint Clamp Bolt
7	Halfshaft
8	Ball Joint
9	Steering Knuckle-to-Strut Mounting Bolts

304731

Front suspension components

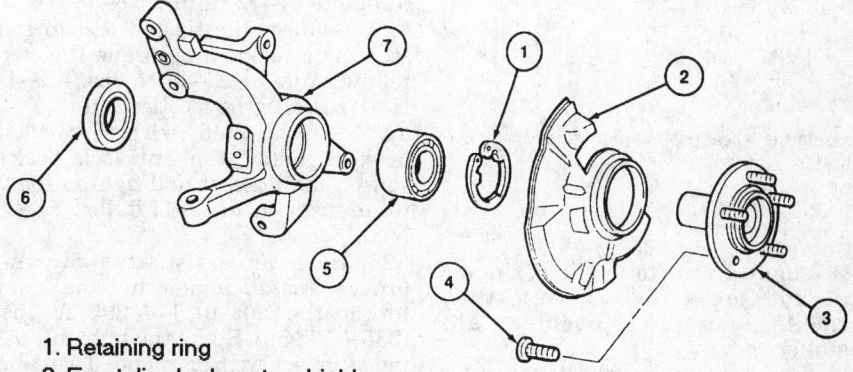

1. Retaining ring
2. Front disc brake rotor shield
3. Wheel hub
4. Wheel hub bolt
5. Front wheel bearing
6. Inner wheel bearing oil seal
7. Front wheel knuckle

304732

Steering knuckle, hub and bearing

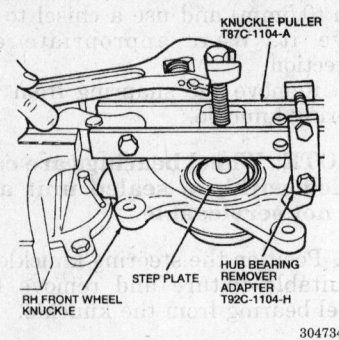

Removing the wheel bearing from the steering knuckle

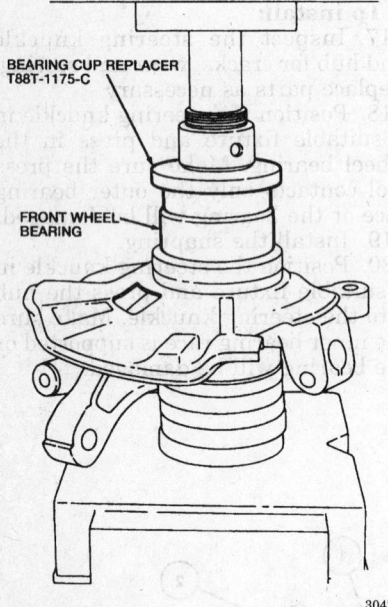

Pressing the wheel bearing into the steering knuckle

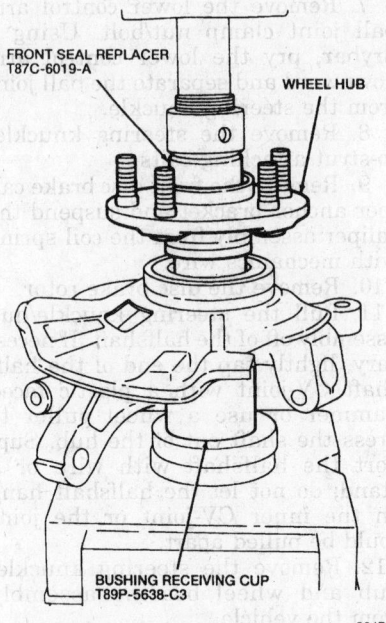

Pressing the hub into the steering knuckle

stall the clamp bolt and nut and torque to 25–42 ft. lbs. (34–57 Nm).

25. Install the disc brake rotor.

26. Position the caliper anchor bracket. Install the caliper anchor bracket-to-steering knuckle bolts and torque to 58–72 ft. lbs. (78–98 Nm).

27. Connect the tie rod end to the steering knuckle and torque the castellated nut to 22–33 ft. lbs. (29–44 Nm). Install a new cotter pin.

28. If equipped with anti-lock brakes, install the anti-lock brake sensor and bracket and tighten the 2 retaining bolts to 12–17 ft. lbs. (16–23 Nm).

29. Have an assistant apply the brakes. Install a new hub nut and torque the nut to 174–235 ft. lbs. (235–319 Nm). Stake the hub nut, using a chisel with a rounded cutting edge or similar tool.

30. Install the wheel and torque the lug nuts to 65–87 ft. lbs. (88–118 Nm).

31. Lower the vehicle.

32. Connect the negative battery cable.

33. Check the front brakes, suspension and steering for proper operation. Check the front wheel alignment.

21. Apply grease to the lip of a new seal and press the seal into the knuckle, using a suitable seal installer.

22. Grease the halfshaft splines. Slide the hub/steering knuckle onto the halfshaft and position it into the strut bracket.

23. Install the strut-to-steering knuckle bolts and nuts and torque to 68–86 ft. lbs. (93–117 Nm).

24. Push the lower control arm ball joint into the steering knuckle. In-

REAR SUSPENSION

Strut and Spring

REMOVAL AND INSTALLATION

1. Raise and safely support the vehicle.

2. Remove the rear wheel.

3. If equipped with anti-lock brakes, remove the speed sensor routing bracket.

4. Remove the brake line U-clip from the strut housing.

5. Remove the 2 spindle-to-strut mounting nuts and bolts.

6. Remove the trunk side panel to gain access to the strut assembly.

7. Remove the 3 upper strut attaching nuts and remove the strut.

— CAUTION —

Do not attempt to remove the coil spring from the strut assembly without compressing the coil spring first.

8. Use spring compressor tool D85P-7178-A or equivalent, to compress the coil spring.

9. Remove the mounting nut from the strut and disassemble the strut components.

To install:

10. Assemble the strut, the compressed coil spring and related components.

11. Install the mounting nut and tighten to 66–87 ft. lbs. (89–117 Nm). Release the compressor and remove the strut.

NOTE: Make sure that the lower coil spring is seated properly.

12. Position the strut in the vehicle and install the 3 upper strut nuts. Tighten to 34–46 ft. lbs. (46–63 Nm).

13. Install the trunk side panel.

14. Install the 2 spindle-to-strut bolts and nuts and tighten to 69–87 ft. lbs. (93–117 Nm).

15. Install the brake line U-clip.

16. Install the wheel speed sensor bracket, if equipped.

17. Install the wheel and torque the lug nuts to 65–87 ft. lbs. (88–118 Nm).

18. Lower the vehicle.

19. Check the wheel alignment.

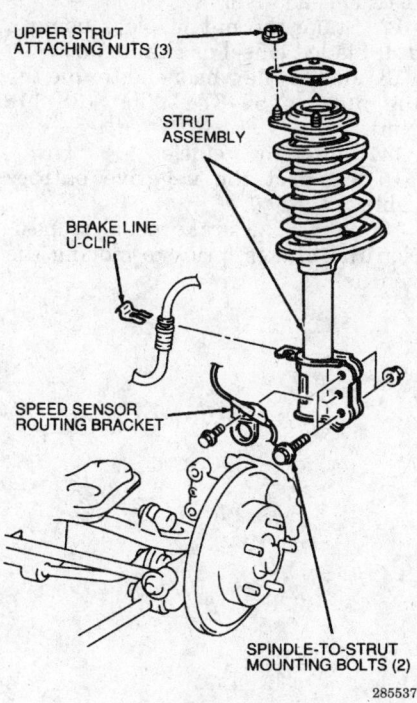

UPPER STRUT
ATTACHING NUTS (3)

STRUT
ASSEMBLY

BRAKE LINE
U-CLIP

SPEED SENSOR
ROUTING BRACKET

SPINDLE-TO-STRUT
MOUNTING BOLTS (2)

285537

Rear strut assembly

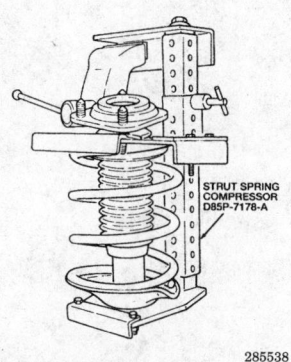

STRUT SPRING
COMPRESSOR
D85P-7178-A

285538

Compressing the strut coil spring

Sway Bar

REMOVAL AND INSTALLATION

1. Raise and safely support the vehicle.

2. Remove the 2 lower sway bar link nuts.

3. Remove the 2 sway bar bracket nuts, remove the 2 sway bar bracket bolts, and remove the sway bar.

To install:

4. Apply rubber grease to the inside surface of the sway bar insulators (bushings). Align each insulator

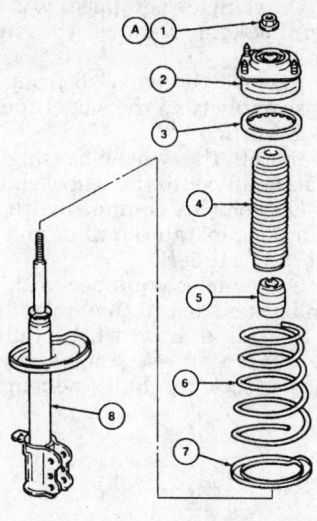

1. Shock absorber strut mounting nut
2. Rear shock absorber bracket
3. Rear spring anti-squeak insert, upper
4. Shock absorber dust tube
5. Rear shock absorber dust boot
6. Rear spring
7. Rear spring anti-squeak insert, lower
8. Rear shock absorber
A. 66-87 ft. lb. (89-117 Nm)

285539

Strut components

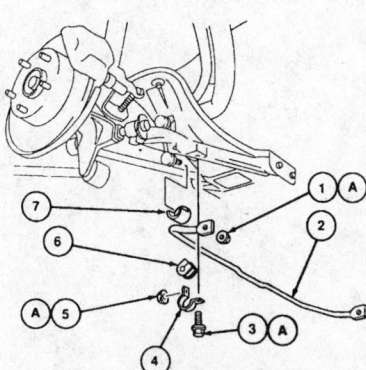

1. Lower rear stabilizer bar link nut (2)
2. Rear stabilizer bar
3. Stabilizer bar bracket bolt (2)
4. Stabilizer bar bracket
5. Stabilizer bar bracket nut (2)
6. Lower suspension arm stabilizer
   bar insulator
7. Rear stabilizer bar link protector (2)
A. 27-40 ft. lb. (36-54 Nm)

304929

Rear sway bar (stabilizer bar) assembly

with the installation mark on the bar.

5. Install the sway bar and sway bar brackets. Tighten the bracket bolts and nuts to 27–40 ft. lbs. (36–54 Nm)

6. Install the two lower sway bar link nuts. Tighten the nuts to 27–40 ft. lbs. (36–54 Nm)

7. Lower the vehicle.

Wheel Bearings

ADJUSTMENT

NOTE: Wheel bearings are sealed units and are not adjustable or able to be greased. The wheel bearing unit must be replaced for noisy or rough operation or excessive end-play.

Wheel Bearing End-Play check

1. Raise and safely support the vehicle.

2. Make sure the parking brake is fully released.

3. Remove the wheel.

4. On disc brakes, remove the disc brake caliper.

5. On rear disc brakes install the lug nuts to hold the rotor in place.

6. Rotate the drum or rotor to make sure there is no brake drag.

7. Position a suitable dial indicator.

8. Check the wheel bearing end-play. End-play should not exceed 0.002 in. (0.05mm).

9. If the end-play exceeds specification, replace the wheel bearing or hub/bearing assembly, as required.

10. When complete, install the calipers if removed.

11. Install the wheel and torque the lug nuts to 65–87 ft. lbs. (88–118 Nm).

REMOVAL AND INSTALLATION

NOTE: The wheel bearing cannot be disassembled from the hub and must be replaced as an assembly.

1. Disconnect the negative battery cable.

2. Raise and safely support the vehicle.

3. Remove the rear wheel.

4. Carefully unstake the wheel hub retainer. Have an assistant apply the brakes to lock the hub, then remove the retainer. Discard the nut.

5. On vehicles equipped with rear disc brakes, remove the disc brake caliper and rotor.

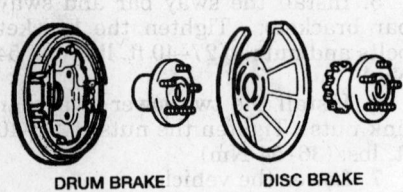

DRUM BRAKE **DISC BRAKE**

304738

Hub and wheel bearing assembly

6. On vehicles equipped with rear drum brakes, remove the brake drum.

7. Remove the wheel bearing and hub assembly from the wheel spindle.

To install:

8. Install the wheel bearing and hub assembly onto the wheel spindle.

9. On vehicles equipped with rear disc brakes, install the rotor and rear disc brake caliper.

10. On vehicles equipped with rear drum brakes, install the brake drum.

11. Install a new wheel hub retainer. Have an assistant apply the brakes to lock the hub, then tighten the wheel hub retainer to 130–174 ft. lbs. (177–235 Nm).

12. Stake the nut in place using a dull bladed chisel or similar tool.

13. Install the wheel and torque the lug nuts to 65–87 ft. lbs. (88–118 Nm).

14. Lower the vehicle.

15. Connect the negative battery cable.

16. Pump the brake pedal to position the brakes, prior to moving the vehicle.

FORD MOTOR CO.
Front Wheel Drive
MERCURY-Capri

11

FIRING ORDERS

NOTE: To avoid confusion, always replace spark plug wires one at a time.

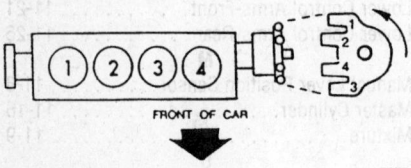

50734

1.6L (VIN 6) Engine
Engine Firing Order: 1–3–4–2
Distributor Rotation: Counterclockwise

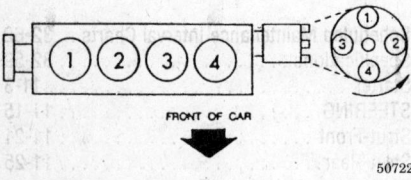

50722

1.6L (VIN Z) Engine
Firing Order: 1–3–4–2
Distributor Rotation: Counterclockwise

ENGINE ELECTRICAL

NOTE: Disconnecting the negative battery cable on some vehicles may interfere with the functions of the on board computer systems and may require the computer to undergo a relearning process, once the negative battery cable is reconnected.

Distributor

REMOVAL AND INSTALLATION

Timing Not Disturbed

1. Disconnect the negative battery cable.

2. Disconnect the vacuum hose from the vacuum control unit and disconnect the electrical connectors.

3. Disconnect the coil wire from the distributor cap, remove the 2 retaining screws from the distributor cap and position aside.

4. Mark the position of the distributor housing on the cylinder head and note or mark the position of the rotor tip to the distributor housing.

5. Remove the distributor hold-down bolts and remove the distributor. Inspect the O-ring and replace if damaged or worn.

To Install:

NOTE: The drive tang of the distributor is offset so as to allow only one installation position.

6. Set the rotor to the position noted during removal. Lubricate the O-ring seal with clean engine oil.

7. Install the distributor on the rear of the cylinder head and align the marks made during removal.

8. Install the 2 distributor hold-down bolts. Install the distributor cap and tighten the 2 retaining screws.

9. Connect the coil wire to the distributor cap, the electrical connectors and the vacuum hoses to the vacuum control unit.

10. Connect the negative battery cable.

11. Start the engine and check the ignition timing and adjust, as necessary.

Timing Disturbed

1. Disconnect the ignition wire from the No. 1 cylinder spark plug and remove the spark plug.

2. Place a finger over the spark plug hole and turn the crankshaft, in the direction of normal rotation, until compression is felt. Continue turning the crankshaft until the TDC mark on the timing belt front cover is aligned with the mark on the crankshaft damper.

3. Disconnect the negative battery cable.

4. Set the rotor to point in the direction of the No. 1 ignition wire tower on the distributor cap. Lubricate the O-ring seal with clean engine oil.

5. Install the distributor on the rear of the cylinder head and install the 2 distributor hold-down bolts. Tighten the bolts snug only, allowing distributor movement for timing purposes.

6. Install the distributor cap and tighten the 2 retaining screws. Connect the coil wire.

7. Connect the electrical connectors and the vacuum hoses. Install the No. 1 cylinder spark plug and connect the ignition wire.

8. Connect the negative battery cable.

9. Start the engine and check or adjust the ignition timing.

10. Tighten the distributor hold-down bolts after timing has been set.

Ignition Timing

ADJUSTMENT

1. Apply the parking brake. If equipped with a manual transaxle, place the shift lever in **NEUTRAL**. If equipped with an automatic transaxle, place the shift lever in **PARK**.

2. Locate the timing marks on the crankshaft pulley and timing belt cover. If the marks are hard to see, clean them off with some degreasing cleaner and a wire brush.

3. Start the engine and allow it to come to normal operating temperature. Make sure all accessories are **OFF**.

4. Connect a tachometer and inductive timing light according to the manufacturer's instructions.

5. Connect terminal 10 on the self test input (STI) connector and ground with a jumper wire.

6. Check the idle speed and adjust, if necessary; it should be 850 ± 50 rpm.

7. Aim the timing light at the timing marks. The timing should be as follows:

Non-turbocharged engines — 2 degrees BTDC ±1 degree.

Turbocharged engines — 12 degrees BTDC ±1 degree.

8. If the marks are not aligned, loosen the distributor hold-down bolts just enough to turn the distributor housing. While aiming the timing light at the timing marks, turn the distributor until the marks are aligned. Tighten the distributor hold-down bolts and then recheck the timing.

9. Stop the engine. Remove the jumper wire and test equipment.

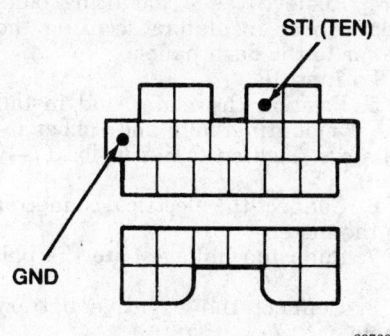

Data Link Connector terminal identification

Alternator

PRECAUTIONS

Several precautions must be observed with alternator equipped vehicles to avoid damage to the unit.

• If the battery is removed for any reason, make sure it is reconnected with the correct polarity. Reversing the battery connections may result in damage to the 1-way rectifiers.

• When utilizing a booster battery as a starting aid, always connect the positive to positive terminals and the negative terminal from the booster battery to a good engine ground on the vehicle being started.

• Never use a fast charger as a booster to start vehicles.

• Disconnect the battery cables when charging the battery with a fast charger.

• Never attempt to polarize the alternator.

• Do not use test lights of more than 12 volts when checking diode continuity.

• Do not short across or ground any of the alternator terminals.

• The polarity of the battery, alternator and regulator must be matched and considered before making any electrical connections within the system.

• Never separate the alternator on an open circuit. Make sure all connections within the circuit are clean and tight.

• Disconnect the battery ground terminal when performing any service on electrical components.

• Disconnect the battery if arc welding is to be done on the vehicle.

REMOVAL AND INSTALLATION

1. Disconnect the negative battery cable.

2. Remove the nut and eyelet connector from the **B** terminal.

3. Disconnect the electrical connector.

4. Remove the adjustment bolt from the top of the alternator.

5. Remove the pivot bolt from the bottom of the alternator and remove the alternator.

To Install:

6. Loosely install the pivot bolt into the bottom of the alternator.

7. Loosely install the upper alternator mounting bolt.

8. Install the alternator belt and position the alternator to provide proper tension on the drive belt.

9. Tighten the pivot bolt to 14–15 ft. lbs. (19–21 Nm) and the upper adjustment bolt to 28–39 ft. lbs. (38–53 Nm).

10. Connect the negative battery cable.

11. Check the charging system for proper operation.

Drive Belt

REMOVAL AND INSTALLATION

Power Steering Drive Belt

1. Loosen the drive belt locknut and adjusting bolts at the power steering pump.

2. Remove the belt.

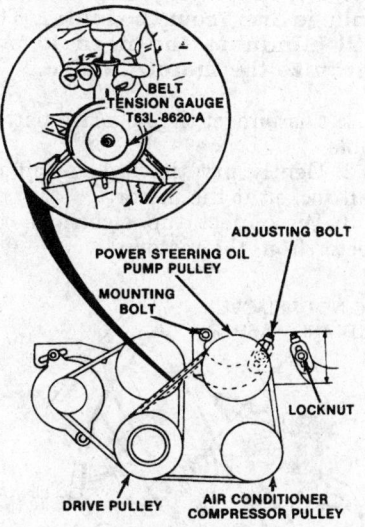

Vehicles With A/C

BELT
TENSION GAUGE
T63L-8620-A

ADJUSTING BOLT

POWER STEERING OIL
PUMP PULLEY

MOUNTING
BOLT

LOCKNUT

DRIVE PULLEY

AIR CONDITIONER
COMPRESSOR PULLEY

Typical drive belt routing

To Install:

3. Position drive belt over the pulley on the crankshaft, power steering pump and A/C compressor if equipped.

4. Adjust the tension of the belts using the deflection method or a belt tension gauge.

Alternator Drive Belt

1. Remove the power steering belt.

2. Loosen the drive belt adjusting bolts at the alternator and rotate the alternator towards the engine.

3. Remove the drive belt.

To install:

4. Position the alternator belt over the crankshaft pulley, alternator and water pump pulleys.

5. Adjust the tension of the drive belt using the deflection method or a belt tension gauge.

Starter

REMOVAL AND INSTALLATION

1. Disconnect the negative battery cable.

2. Disconnect the starter wires and remove the starter upper retaining bolts.

3. Remove the intake manifold support bracket upper retaining bolts.

4. Raise and safely support the vehicle.

5. Remove the starter support bracket-to-intake manifold support bracket retaining bolt.

6. Remove the intake manifold support bracket lower retaining bolts and remove the starter lower retaining bolt.

NOTE: Loosen the rubber exhaust hangers, if added clearance is required.

7. Remove the starter. Remove the support bracket from the starter, if required.

To install:

8. Install the support bracket to the starter, if removed. Tighten the retaining nuts to 54–70 inch lbs. (6–8 Nm).

9. Position the starter and loosely install the lower retaining bolt.

10. Position the intake manifold support bracket and loosely install the lower retaining bolts.

11. Install the starter support bracket-to-intake manifold support bracket retaining bolt. Tighten to 14–19 ft. lbs. (19–25 Nm).

12. Tighten the lower starter bolt to 23–30 ft. lbs. (31–41 Nm).

13. Lower the vehicle.

14. Install the starter upper retaining bolts and tighten to 23–30 ft. lbs. (31–41 Nm). Make sure the starter wire support bracket is secured with the rear upper starter bolt.

15. Install the intake manifold support bracket upper retaining bolts and tighten the upper and lower bolts to 23–34 ft. lbs. (31–46 Nm).

16. Connect the starter wires. Tighten the **B** terminal retaining nut to 71–106 inch lbs. (8–12 Nm).

17. Connect the negative battery cable.

18. Check the starter for proper operation.

CHASSIS ELECTRICAL

Blower Motor

REMOVAL AND INSTALLATION

1. Disconnect the negative battery cable.

2. Disconnect the blower motor electrical harness connector.

3. Remove the 3 screws retaining the motor and cover to the blower case.

4. Remove the cover, cooling tube and the blower motor from the vehicle.

5. Remove the nut retaining the blower wheel to the blower motor and remove the blower wheel.

6. Remove the gasket from the blower motor.

To Install:

7. Position the gasket onto the blower motor.

8. Install the blower wheel onto the blower motor and secure with the retaining nut.

9. Position the blower motor, cooling tube and cover into the blower case. Install and tighten the 3 retaining screws.

10. Connect the negative battery cable.

11. Check for proper blower motor operation.

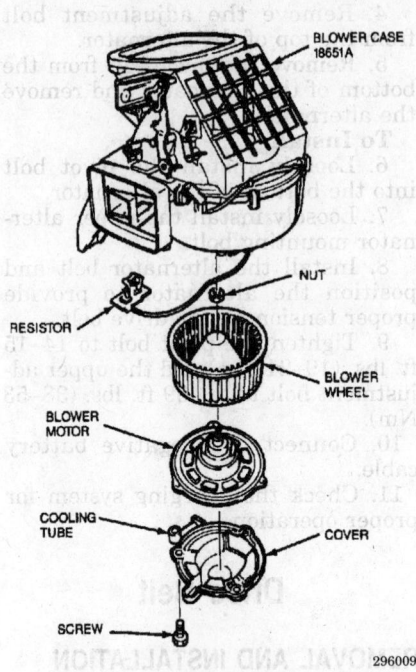

Blower motor and related components

Windshield Wiper Motor

REMOVAL AND INSTALLATION

NOTE: Disconnect the linkage from the motor at the ball socket, not by removing the nut and linkage arm from the motor. This will eliminate the need to synchronize the motor/linkage.

1. Disconnect the negative battery cable.

2. Gently pry the linkage off the ball socket at the motor.

3. Disconnect the electrical connector from the motor.

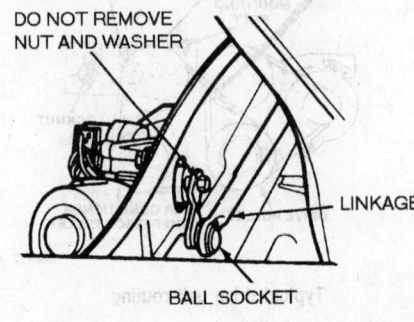

Wiper motor linkage connection point

4. Remove the 4 mounting bolts and rubber insulators securing the motor to the dash panel.

To install:

5. Position the motor and install the 4 mounting bolts and rubber insulators. Tighten to 5–7 ft. lbs. (7–10 Nm).

6. Connect the electrical connector to the motor.

7. Snap the linkage onto the ball socket.

8. Connect the negative battery cable.

9. Check the wipers for proper operation.

Concealed Headlights

MANUAL OPERATION

The headlight motor switch, located on the console, is used to raise and lower the headlights without turning the headlights on. This switch allows the headlights to be serviced without the use of the headlight switch. A manual control knob, located under a rubber boot, is provided behind each door motor. Each headlight door can be operated manually if electrical power is not available.

NOTE: Do not raise or lower the doors manually if electrical power is available.

1. Open the hood and remove the rubber boot from the manual knob.

2. Rotate the knob to raise or lower the headlight.

Headlight Switch

REMOVAL AND INSTALLATION

1. Disconnect the negative battery cable.

2. Pull out the storage compartment. Remove the 2 upper screws, 2 lower screws and heater/radio bezel.

3. Remove the trim covers located on both sides of the steering column by pulling outward.

4. Remove the retaining screws and carefully pull the instrument cluster bezel partially away from the instrument panel.

5. Disconnect the electrical connectors from the clock and the switches in the bezel.

6. Depress the tangs on both sides of the headlight switch and remove the switch from the bezel.

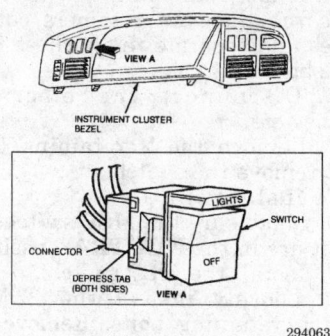

Headlight switch removal

To install:

7. Position the headlight switch into the instrument cluster bezel. Make sure the switch is fully seated.

8. Connect the electrical connector to the clock and switches.

9. Install the instrument cluster bezel and install the retaining screws.

10. Install the trim covers.

11. Install the heater/radio bezel and 4 retaining screws.

12. Install the storage compartment.

13. Connect the negative battery cable.

14. Check the headlight switch for proper operation.

Combination Switch

REMOVAL AND INSTALLATION

CAUTION

The Supplemental Air Bag System (SRS) must be disarmed before performing service around SRS components or SRS wiring. Failure to do so may cause accidental deployment of the air bag, resulting in unnecessary SRS repairs and/or personal injury.

NOTE: The combination switch incorporates the windshield wiper switch, turn signal switch, high beam switch, flash-to-pass switch and the hazard warning switch. The combination switch is serviced as an assembly.

1. Disconnect the negative battery cable.

2. Wait at least 1 minute for the backup power supply to deplete its stored energy.

3. Remove the center trim panel and access cover under the steering column.

NOTE: Removal of the combination switch does not require the removal of the steering wheel. If the steering wheel is removed for any reason, place 2 strips of tape across the air bag clockspring to prevent rotation of the clockspring and possible damage. If the clockspring is rotated it must be centered before installing the steering wheel.

4. Remove the lower steering column shroud.

5. Remove the steering column upper retaining bolts and allow the column to rest on the instrument panel brace.

NOTE: Check that no wires are pinched when lowering the column.

6. Remove the 2 combination switch retaining screws and remove the combination switch.

7. Disconnect the electrical connectors from the combination switch.

8. Grasp the combination switch and lever firmly and pull the lever out of the switch.

To install:

9. Align the key with the slot and install the lever in the new combination switch assembly.

10. Connect the electrical connectors to the switch.

11. Position the switch on the steering column and install the 2 retaining screws.

12. Make sure the column support bracket is in position. Raise the column into position and install the retaining bolts. Tighten to 17–23 ft. lbs. (23–31 Nm).

13. Install the lower column shroud. Install the access cover and trim panel.

14. Connect the negative battery cable.

15. Check the combination switch for proper operation.

Ignition Lock Cylinder

REMOVAL AND INSTALLATION

Functional Lock

1. Disconnect the negative battery cable.

2. Remove the lower steering column shroud.

3. With the ignition key installed, rotate the ignition lock cylinder while pushing the release pin with a ⅛ inch drift or similar tool.

4. Remove the lock cylinder assembly by pulling it out of the housing.

To install:

5. Install the ignition lock cylinder assembly with the ignition key installed. Make sure the cylinder is fully seated.

6. Install the lower column shroud.

7. Connect the negative battery cable.

8. Check the ignition lock for proper operation.

Non-functional Lock

The following procedure applies to vehicles in which the ignition lock is inoperative and the lock cylinder cannot be rotated due to a lost or broken lock cylinder key, the key number is not known or the lock cylinder cap is damaged and/or broken to the extent that the lock cylinder cannot be rotated.

1. Disconnect the negative battery cable.

2. Remove the lower steering column shroud.

3. Using a ⅛ inch drill, drill out the retaining pin, being cautious not to drill deeper than ½ inch.

4. Place a chisel at the base of the ignition lock cylinder cap and using a hammer, strike the chisel with sharp blows to break the cap away from the lock cylinder.

5. Using a ⅜ inch diameter drill, drill out the middle of the ignition lock key slot approximately 1 ¾ inch until the lock cylinder breaks loose from the break-away base of the lock cylinder. Remove the lock cylinder and drill shavings from the lock cylinder housing.

6. Remove the snapring, washer and steering column lock gear. Thoroughly clean all drill shavings and other foreign materials from the casting.

7. Carefully inspect the lock cylinder housing for damage. If any damage is apparent, the housing must be replaced.

To install:

8. Install the steering column lock gear, washer and snapring.

9. Install the ignition lock cylinder assembly with the ignition key installed. Make sure the cylinder is fully seated.

10. Install the lower column shroud.

11. Connect the negative battery cable.

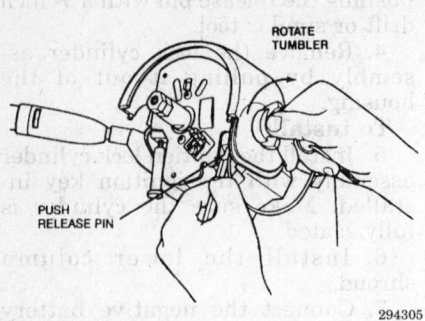

Removing the tumbler assembly

12. Check the ignition lock for proper operation.

Ignition Switch

REMOVAL AND INSTALLATION

1. Disconnect the negative battery cable.
2. Remove the lower steering column shroud. Remove the center access panel and trim cover under the steering column.
3. Remove the left side defroster connector tube.
4. Remove the steering column upper retaining bolts and allow the column to rest on the instrument panel brace.

NOTE: Make sure no wiring is pinched beneath the steering column when lowered.

5. Remove the ignition lock tumbler.
6. Remove the upper column cover and column lock shield.
7. Disconnect the ignition switch connector, remove the switch retaining screws and remove the switch.

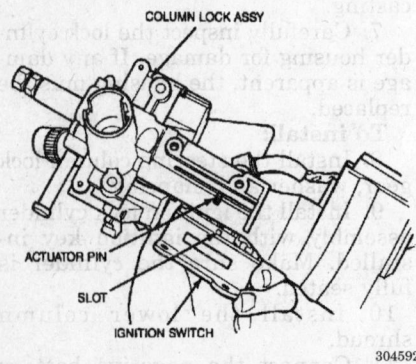

Ignition switch installation

To install:

8. Position the ignition switch to the column lock assembly. Make sure the actuator pin of the lock assembly fits into the slot in the ignition switch.
9. Install the switch retaining screws and tighten to 62–76 inch lbs. (7–9 Nm).
10. Connect the switch electrical connector.
11. Install the column lock shield. Tighten the screws and nut to 14–18 ft. lbs. (19–25 Nm).
12. Install the upper column shroud.
13. Install the lock tumbler assembly. Make sure the tumbler snaps in place.
14. Raise the column and install the upper retaining bolts. Tighten to 17–23 ft. lbs. (23–31 Nm).
15. Install the defroster connector tube and the column lower shroud.
16. Install the access panel and trim cover.
17. Connect the negative battery cable.
18. Check the ignition switch for proper operation.

Manual Lever Position Sensor

REMOVAL AND INSTALLATION

NOTE: The Manual Lever Position (MLP) sensor is only used on vehicles equipped with automatic transaxles.

1. Disconnect the negative battery cable.
2. Place the shift selector lever in the **NEUTRAL** position.
3. Remove the engine air cleaner assembly.
4. Remove the shift cable retaining nut and disconnect the cable.

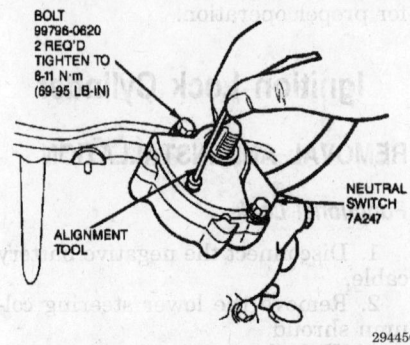

Manual lever position sensor adjustment

5. Remove the MLP switch harness from the metal retainers, cut the tapes and separate the harness from the sheathing.
6. Disconnect the electrical connectors.
7. Remove the 2 retaining bolts and remove the switch.

To install:

8. Make sure the MLP switch and shaft are in the **NEUTRAL** position.
9. Install the MLP switch.
10. Loosely install the 2 MLP switch retaining bolts. Remove the cover screw and align the internal hole with the cover screw hole. Hold the alignment by inserting a 0.079 inch (2mm) pin through the holes.
11. Remove the pin, install the cover screw and tighten to 3–6 inch lbs. (0.4–0.7 Nm).
12. Route the harness into the sheathing and secure in the metal retainers.
13. Connect the electrical connectors.
14. Install the shift cable and tighten the retaining nut to 33–47 ft. lbs. (44–64 Nm).
15. Install the engine air cleaner assembly.
16. Connect the negative battery cable.
17. Check the MLP switch for proper operation. The engine must only be able to crank when the shift selector is in the **NEUTRAL** or **PARK** positions.

Powertrain Control Module

REMOVAL AND INSTALLATION

------ **WARNING** ------
Never disconnect a Powertrain Control Module (PCM) with the battery connected. Be sure to wear a grounding device when removing or installing a PCM to prevent damage to the unit due to static electricity.

1. Disconnect the negative battery cable.
2. Pull back on the front edges of both center carpet panels and disengage the push pin retainers.
3. Remove the retaining screws and carpet panels.
4. Install a grounding device to prevent static charge.
5. Remove the screws retaining the PCM to the floor.
6. Disconnect the electrical harness connectors from the PCM.

To install:

NOTE: Carefully inspect the harness and the PCM terminal connections. Check that all terminals are clean and not corroded. Also check that no terminals are bent or deformed. Be sure that the connectors are fully seated; a poorly seated connector can cause intermittent operation or poor engine performance.

7. Install the electrical harness connectors to the PCM.

8. Install the PCM to the floorpan with the retaining screws.

9. Remove the grounding device.

10. Install the center carpet panels and push pin connectors.

11. Connect the negative battery cable.

12. Road test the vehicle and check for proper engine operation.

ENGINE COOLING

Radiator

REMOVAL AND INSTALLATION

──────── CAUTION ────────

Never remove the radiator cap while the engine is running or personal injury from scalding hot coolant or steam may result.

1. Disconnect the negative battery cable.

2. Disconnect the cooling fan wiring harness connector.

3. Remove the radiator pressure cap. If the cooling system is pressurized, be sure to wrap a cloth around the cap and slowly turn it to the first stop, to release the pressure.

4. Place a drain pan under the radiator and open the drain valve at the bottom left of the radiator. Drain the cooling system.

5. Disconnect the radiator hoses from the radiator. Disconnect the overflow tube from the filler neck.

6. Disengage the wiring harness from the routing clips attached to the cooling fan shroud.

7. If equipped with an automatic transaxle, disconnect and plug the cooler lines.

8. Remove the 6 bolts retaining the radiator upper tank brackets to the radiator core support.

9. Remove the radiator and cooling fan assembly.

10. Remove the 4 fan shroud retaining bolts and separate the fan and shroud assembly from the radiator.

To install:

11. Place the fan and shroud assembly against the rear of the radiator and secure with the 4 bolts. Tighten to 23–33 ft. lbs. (31–46 Nm).

12. Make sure the radiator insulators are positioned on the radiator supports. Position the radiator, making sure the lower tank engages the insulators.

13. Install the 6 radiator retaining bolts through the top tank mounting brackets into the core support. Make sure the insulators are aligned and tighten the bolts securely.

14. If equipped, unplug and connect the automatic transaxle oil cooler lines.

15. Secure the wiring harness in the routing clips.

16. Connect the radiator hoses to the radiator. Connect the overflow tube to the filler neck.

17. Close the drain valve and fill the cooling system. Install the pressure cap.

18. Connect the cooling fan harness connector.

19. Connect the negative battery cable.

20. Start the engine and bring to normal operating temperature.

21. Check for leaks and proper function of the cooling system.

22. Recheck the coolant level.

Water Pump

REMOVAL AND INSTALLATION

1. Disconnect the negative battery cable.

2. Drain the cooling system.

3. Remove the timing belt, timing belt tensioner and idler pulleys.

4. Remove the engine oil dipstick bracket retaining bolt.

5. Remove the power steering pump from the retaining bracket, leaving the hoses connected. Remove the power steering pump bracket and position the pump aside.

6. Remove the water pump outlet.

7. Remove the 4 water pump retaining bolts.

8. Remove the water pump and gasket.

NOTE: Raise the engine slightly with a floor jack, if required, to gain clearance for removal.

To install:

9. Clean the water pump and cylinder head gasket mating surfaces. Transfer the rubber belt cover seal to the new water pump, if required.

10. Position the water pump with a new gasket. Install the 4 retaining bolts and tighten to 14–19 ft. lbs. (19–25 Nm).

11. Install the water pump outlet with a new gasket and O-ring. Tighten the retaining bolts to 14–19 ft. lbs. (19–25 Nm).

12. Install the oil dipstick retaining bolt.

13. Install the timing belt tensioner, idler pulleys and the timing belt.

14. Install the power steering pump bracket. Tighten the nut and bolts to 35–48 ft. lbs. (47–66 Nm).

15. Install the power steering pump.

16. Fill the cooling system.

17. Connect the negative battery cable.

18. Start the engine and allow it to reach normal operating temperature.

19. Check for leaks and proper engine operation.

Thermostat

REMOVAL AND INSTALLATION

1. Disconnect the negative battery cable.

2. Disconnect the wire from the engine cooling fan switch on the thermostat housing.

──────── CAUTION ────────

Never remove the radiator cap while the engine is running or personal injury from scalding hot coolant or steam may result.

3. Remove the radiator pressure cap. If the cooling system is pressurized, be sure to wrap a cloth around the cap and slowly turn it to the first stop, to release the pressure.

4. Partially drain the cooling system.

5. Disconnect the upper radiator hose from the thermostat housing.

6. Remove the 2 bolts retaining the thermostat housing to the cylinder head and remove the housing. Remove the thermostat and gasket.

To install:

7. Clean the thermostat gasket mating surfaces.

8. Install the thermostat into the cylinder head, valve end first, with the jiggle valve at the top.

9. Coat a new housing gasket with water resistant sealer and position it

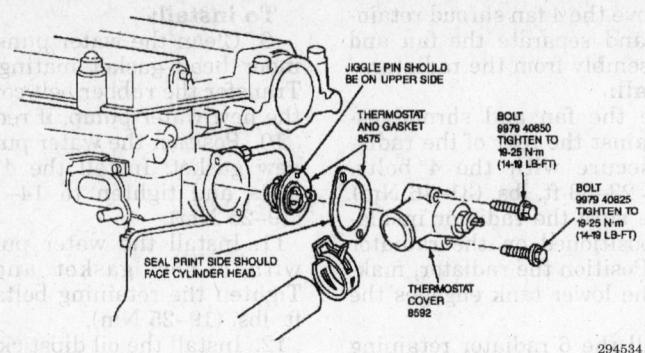

JIGGLE PIN SHOULD BE ON UPPER SIDE

THERMOSTAT AND GASKET 8575

BOLT 9979 40850 TIGHTEN TO 19-25 N·m (14-19 LB-FT)

BOLT 9979 40825 TIGHTEN TO 19-25 N·m (14-19 LB-FT)

SEAL PRINT SIDE SHOULD FACE CYLINDER HEAD

THERMOSTAT COVER 8592

294534

Thermostat and related components

on the cylinder head with the bolt holes correctly aligned.

NOTE: The painted side of the gasket must face the thermostat.

10. Carefully position the thermostat housing to align the bolt holes without shifting the gasket and install the 2 retaining bolts. Tighten to 14–19 ft. lbs. (19–25 Nm).
11. Connect the upper radiator hose to the thermostat housing and install the hose clamp.
12. Fill the cooling system and install the pressure cap.
13. Connect the wire from the engine cooling fan to the switch on the thermostat housing.
14. Connect the negative battery cable.
15. Start the engine and bring to normal operating temperature.
16. Check for leaks and proper cooling system operation.

Electric Cooling Fan

REMOVAL AND INSTALLATION

1. Disconnect the negative battery cable.
2. Disengage the fan wiring harness from the routing clamps. Separate the cooling fan wiring connector.
3. Remove the 4 screws retaining the fan shroud to the radiator and remove the fan and shroud.
4. Remove the retaining nut and washer and remove the fan from the motor shaft.
5. Remove the 3 retaining screws and washers and separate the fan motor from the shroud.
To install:
6. Position the cooling fan on the shroud and install the 3 retaining

screws and washers. Tighten to 36–48 inch lbs. (4–6 Nm).
7. Install the fan on the motor shaft and install the retaining washer and nut.
8. Position the fan and shroud and install the 4 retaining screws. Tighten to 23–34 ft. lbs. (31–46 Nm).
9. Connect the cooling fan wiring and secure in place using the routing clamps.
10. Connect the negative battery cable.
11. Allow the engine to warm to normal operating temperature and verify cooling fan operation.

Cooling System Bleeding

When the entire cooling system is drained, the following procedure should be used to ensure a complete fill.
1. Install the block drain plug, if removed, and close the radiator draincock. With the engine off, add a 50/50 mixture of coolant to the radiator until it reaches the radiator filler neck seat.
2. Install the radiator cap to the first notch to keep spillage to a minimum.
3. Start the engine and let it idle until the upper radiator hose is warm. This indicates that the thermostat is open and coolant is flowing through the entire system.
4. Carefully remove the radiator cap and top off the radiator with water. Install the cap on the radiator securely.
5. Fill the coolant recovery reservoir with a 50/50 mixture of coolant.
6. Check for leaks at the radiator draincock and the block drain plug.

FUEL SYSTEM

Fuel System Service Precautions

Safety is the most important factor when performing not only fuel system maintenance but any type of maintenance. Failure to conduct maintenance and repairs in a safe manner may result in serious personal injury or death. Maintenance and testing of the vehicle's fuel system components can be accomplished safely and effectively by adhering to the following rules and guidelines.
• To avoid the possibility of fire and personal injury, always disconnect the negative battery cable unless the repair or test procedure requires that battery voltage be applied.
• Always relieve the fuel system pressure prior to disconnecting any fuel system component (injector, fuel rail, pressure regulator, etc.), fitting or fuel line connection. Exercise extreme caution whenever relieving fuel system pressure to avoid exposing skin, face and eyes to fuel spray. Please be advised that fuel under pressure may penetrate the skin or any part of the body that it contacts.
• Always place a shop towel or cloth around the fitting or connection prior to loosening to absorb any excess fuel due to spillage. Ensure that all fuel spillage (should it occur) is quickly removed from engine surfaces. Ensure that all fuel soaked cloths or towels are deposited into a suitable waste container.
• Always keep a dry chemical (Class B) fire extinguisher near the work area.
• Do not allow fuel spray or fuel vapors to come into contact with a spark or open flame.
• Always use a backup wrench when loosening and tightening fuel line connection fittings. This will prevent unnecessary stress and torsion to fuel line piping. Always follow the proper torque specifications.
• Always replace worn fuel fitting O-rings with new. Do not substitute fuel hose or equivalent, where fuel pipe is installed.

Fuel System Pressure

RELIEVING

— **CAUTION** —

Fuel injection systems remain under pressure, even after the engine has been turned OFF. The fuel system pressure must be relieved before disconnecting any fuel lines. Failure to do so may result in fire and/or personal injury.

1. Disconnect the inertia fuel shut off switch (IFS) located on the LH side of the spare tire well in the luggage compartment.
2. Run the engine until it stalls. The fuel pressure is now relieved.
3. Disconnect the negative battery cable.

Idle Speed

ADJUSTMENT

NOTE: Before adjusting the idle speed, make sure the ignition timing is adjusted to specification. Turn off all lights and other unnecessary electrical loads. This adjustment must be done while the cooling fan motor is not operating.

1. Allow the engine to warm up to normal operating temperature.
2. Attach a suitable tachometer to the test connector (white: 1 pin).
3. Check the idle speed on the tachometer. Connect a jumper wire between the test connector (green: 1 pin) and ground and turn the air adjustment screw to obtain the correct idle speed of 800–900 rpm.
4. Remove the jumper wire and the tachometer.
5. Road test the vehicle and check for proper operation.

Mixture

ADJUSTMENT

The air/fuel mixture is electronically controlled by the Powertrain Control Module (PCM) and cannot be adjusted. The PCM continuously adjust the air/fuel ratio in response to signals received from operator controls, sensor and switch signals monitoring the engines running condition.

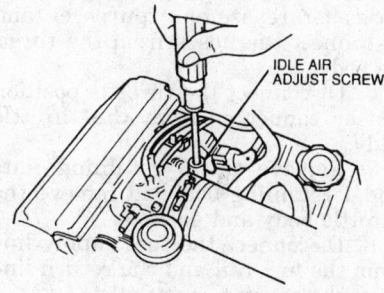

IDLE AIR ADJUST SCREW

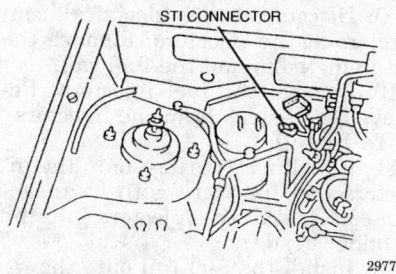

STI CONNECTOR

297756

Idle air adjustment screw and STI connector locations

Fuel Filter

REMOVAL AND INSTALLATION

— **CAUTION** —

Fuel injection systems remain under pressure, even after the engine has been turned OFF. The fuel system pressure must be relieved before disconnecting any fuel lines. Failure to do so may result in fire and/or personal injury.

1. Relieve the fuel system pressure by performing the following:
    a. Disconnect the Inertia Fuel Shutoff (IFS) switch located on the LH side of the spare tire well in the luggage compartment.
    b. Run the engine until it stalls. The fuel pressure is now relieved.
2. Disconnect the negative battery cable.
3. Reconnect the IFS switch.
4. Remove the clamp and supply line from the bottom of the fuel filter. Plug the supply line.
5. Remove the clamp and outlet line from the top of the fuel filter.
6. Remove the fuel filter from the bracket.

To install:
7. Install the fuel filter in the retaining bracket and tighten the retaining bolt securely.
8. Remove the plug from the supply line and install the supply line on the bottom of the fuel filter. Securely tighten the hose clamp.
9. Install the outlet line on the top of the fuel filter and position the squeeze clamp.
10. Connect the inertia switch wire connector.
11. Connect the negative battery cable.
12. Start the engine and check the fuel filter for leaks.

Fuel Pump

REMOVAL AND INSTALLATION

— **CAUTION** —

Fuel injection systems remain under pressure, even after the engine has been turned OFF. The fuel system pressure must be relieved before disconnecting any fuel lines. Failure to do so may result in fire and/or personal injury.

1. Relieve the fuel system pressure by performing the following:
    a. Remove the fuel filler cap to relieve fuel pressure in the fuel tank.
    b. Remove the rear seat cushion.
    c. Start the engine and disconnect the fuel pump electrical connector.
    d. Allow the engine to stall. Fuel pressure is now relieved.
2. Disconnect the negative battery cable.
3. Disconnect the fuel pump ground wire from the access cover and remove the cover.
4. Remove and plug the fuel supply and return lines. Remove the fuel pump/sending unit retaining bolts.
5. Remove the fuel pump/sending unit and gasket from the fuel tank. Cover the opening of the tank to prevent dirt from entering.
6. Remove the 2 fuel pump wires from the sending unit. Remove the retaining clamp screw and remove the clamp.
7. Remove the rubber retaining band and remove the fuel pump from the sending unit.
To install:
8. Install the fuel pump to the sending unit bracket and secure with the retaining clamp.

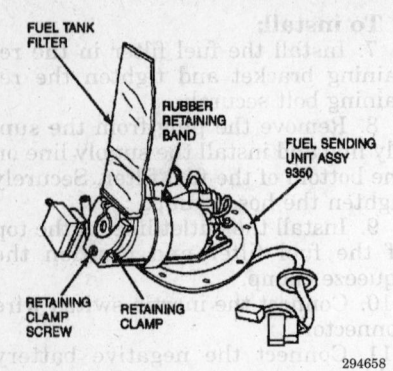

FUEL TANK
FILTER

RUBBER
RETAINING
BAND

FUEL SENDING
UNIT ASSY
9350

RETAINING
CLAMP
SCREW

RETAINING
CLAMP

294658

Fuel pump components

9. Install the rubber retaining band and connect the fuel pump wires to the sending unit.

10. Position a new gasket onto the fuel tank and install the fuel pump/sending unit with the retaining bolts.

11. Unplug and connect the supply and return lines and secure with the clamps.

12. Install the access cover and connect the fuel pump connector and ground wire.

13. Connect the negative battery cable.

14. Run the engine and check for leaks and proper engine operation.

15. Check the operation of the fuel gauge.

16. Install the rear seat cushion.

Fuel Injector

REMOVAL AND INSTALLATION

——— CAUTION ———
Fuel injection systems remain under pressure, even after the engine has been turned OFF. The fuel system pressure must be relieved before disconnecting any fuel lines. Failure to do so may result in fire and/or personal injury.

1. Relieve the fuel system pressure by performing the following:
   a. Remove the fuel filler cap to relieve fuel pressure in the fuel tank.
   b. Remove the rear seat cushion.
   c. Start the engine and disconnect the fuel pump electrical connector.
   d. Allow the engine to stall. Fuel pressure is now relieved.

2. Disconnect the negative battery cable.

3. Partially drain the cooling system.

4. Remove the air duct and the accelerator cable from the throttle body.

5. Mark all vacuum and coolant hoses for reassembly purposes and disconnect the hoses from the throttle body.

6. Disconnect the throttle position sensor connector from the throttle body.

7. Remove the 3 retaining nuts and 1 retaining bolt and remove the throttle body and gaskets.

8. Disconnect the fuel supply line from the fuel rail and the return line from the pressure regulator.

9. Disconnect the electrical connectors at the injectors. Remove the retaining bolts and the fuel rail.

10. Remove the fuel injectors. Remove the O-rings from the injectors.

To install:

11. Install new O-rings onto the injectors and lubricate with clean engine oil. Install the injectors into the cylinder head.

12. Install the fuel rail onto the injectors and install the retaining bolts. Tighten to 14–19 ft. lbs. (19–25 Nm).

13. Connect the electrical connectors to the injectors.

14. Connect the fuel return line to the pressure regulator and the supply line to the fuel rail.

15. Clean the gasket mating surfaces of the throttle body and intake manifold.

16. Install new gaskets and position the throttle body onto the intake manifold. Install the 3 retaining nuts and 1 bolt and tighten to 12–17 ft. lbs. (16–23 Nm).

17. Connect the throttle position sensor connector. Connect all vacuum and coolant hoses to the throttle body at the locations noted during removal.

18. Install the accelerator cable and the air duct.

19. Fill the cooling system.

20. Connect the negative battery cable.

21. Run the engine and check for leaks and proper operation.

ENGINE MECHANICAL

Engine Mounts

REMOVAL AND INSTALLATION

Right Hand Engine Mount

1. Disconnect the negative battery cable.

2. Support the engine assembly with a floor jack.

3. Remove the right hand mount to engine bracket retaining nuts and washers.

4. Remove the through-bolt and retaining nut.

5. Remove the bracket to body retaining bolts.

6. Remove the right hand mount and bracket.

To install:

7. Position the bracket to the body. Tighten the smaller bolts to 14–21 ft. lbs. (20–28 Nm). Tighten the larger bolt to 49–67 ft. lbs. (67–91 Nm).

8. Install the engine mount to the engine bracket. Tighten the nuts to 44–63 ft. lbs. (60–85 Nm).

9. Install the through-bolt and tighten to 33–48 ft. lbs. (45–65 Nm).

10. Remove the floor jack.

11. Connect the negative battery cable.

Front Engine Mount

1. Disconnect the negative battery cable.

2. Support the engine with engine support tool D88L-6000-A, or equivalent.

3. Raise and safely support the vehicle.

4. Remove the engine mount retaining nuts from the crossmember.

5. Remove the engine mount retaining bolts from the transaxle. Remove the through-bolt, if required.

6. Remove the front engine mount.

To install:

7. Position the front engine mount to the transaxle. Install the retaining bolts and tighten to 27–38 ft. lbs. (37–52 Nm).

8. Apply 2 drops of thread-locking compound to the studs and install the engine mount to crossmember retaining nuts. Tighten to 47–65 ft. lbs. (64–89 Nm).

9. Install the through-bolt, if removed. Tighten to 48 ft. lbs. (65 Nm).

10. Lower the vehicle.

11. Remove the engine support tool.

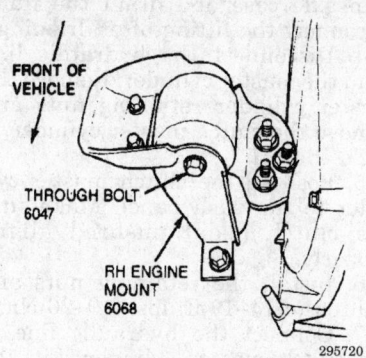

Upper right hand engine mount installation

12. Connect the negative battery cable.

Rear Engine Mount and Crossmember

1. Support the engine with engine support tool D88L–6000–A or equivalent.
2. Raise and safely support the vehicle.
3. Remove the front and rear engine mount to crossmember retaining nuts. Remove the crossmember retaining bolts, washers and nut to crossmember braces.
4. Remove the left A-arm retaining bolt and washer.
5. Support the crossmember and remove the engine support crossmember retaining bolts and washers.
6. Remove the crossmember.
7. Remove the rear engine mount retaining bolts.
8. Remove the rear engine mount.
To install:
9. Install the rear engine mount. Tighten the retaining bolts to 28–38 ft. lbs. (37–52 Nm).
10. Install the crossmember. Apply 2 drops of thread-locking compound and tighten the retaining bolts to 47–66 ft. lbs. (64–89 Nm). Tighten the rear engine retaining nut to 47–66 ft. lbs. (64–89 Nm).
11. Install the crossmember brace retaining bolts, washers and nut. Apply 2 drops of thread-locking compound and tighten the retaining nut to 26–32 ft. lbs. (35–44 Nm). Tighten the bolts to 14–19 ft. lbs. (18–26 Nm).
12. Install the left A-arm retaining bolt. Tighten to 47–66 ft. lbs. (64–89 Nm).
13. Install the front and rear engine mount to crossmember retaining nuts. Apply 2 drops of thread-locking compound and tighten to 47–66 ft. lbs. (64–89 Nm).
14. Lower the vehicle.
15. Remove the engine support fixture.

16. Connect the negative battery cable.

Valve Lash

ADJUSTMENT

Both versions of the 1.6L engine use hydraulic lash adjusters, therefore, valve clearance adjustment is not required. Inspect the lash adjuster operation by holding the bucket body while pressing the plunger by hand. If the plunger moves, replace the lash adjuster. Inspect the friction surfaces for wear and replace if necessary.

Oil Pan

REMOVAL AND INSTALLATION

1.6L (VIN Z) Engine

1. Disconnect the negative battery cable.
2. Raise and safely support the vehicle.
3. Drain the engine oil.
4. Remove the frame brace retaining bolt. Loosen the right A-arm front bolt and pivot brace downward.
5. Disconnect the front exhaust pipe from the exhaust manifold. Remove the front exhaust pipe bracket retaining bolts.
6. Loosen the rubber exhaust hangers at the catalyst. Allow the exhaust to hang supported by mechanic's wire.
7. Remove the oil pan retaining bolts. Carefully pry the oil pan loose from the cylinder block.

NOTE: Do not force a prying tool between the cylinder block and oil pan.

8. Remove the front and rear oil pan seals.

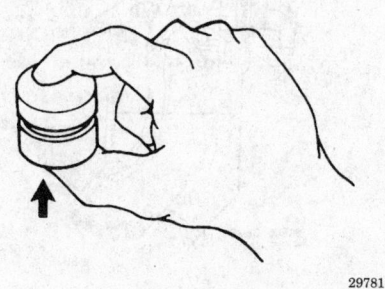

Hydraulic lash adjuster

To install:
9. Clean the oil pan and the oil pan and cylinder block gasket surfaces.
10. Apply gasket sealant to new front and rear oil pan seals. Install the seals to the cylinder block.
11. Apply gasket sealant to the oil pan gasket surface and install the oil pan. Tighten the retaining bolts to 71–97 inch lbs. (8–11 Nm).
12. Install the rubber exhaust hanger to the brackets.
13. Install a new gasket and connect the front exhaust pipe to the exhaust manifold. Tighten the retaining nuts to 23–34 ft. lbs. (31–46 Nm).
14. Install the front exhaust pipe bracket. Tighten the retaining bolts to 34 ft. lbs. (46 Nm).
15. Pivot the frame brace into position. Tighten the retaining bolt to crossmember to 26–37 ft. lbs. (35–50 Nm). Tighten the right A-arm front retaining bolt to 72–86 ft. lbs. (97–117 Nm).
16. Lower the vehicle.
17. Fill the engine with the proper type and quantity of engine oil.
18. Connect the negative battery cable.
19. Start the engine and check for leaks and proper engine operation.

1.6L (VIN 6) Engine

1. Disconnect the negative battery cable.
2. Raise and safely support the vehicle.
3. Drain the engine oil.
4. Remove the frame brace retaining bolt. Loosen the right A-arm front bolt and pivot brace downward.
5. Disconnect the front exhaust pipe from the turbocharger. Remove the front exhaust pipe bracket retaining bolts.
6. Loosen the rubber exhaust hangers at the catalytic converter. Allow the exhaust to hang supported by mechanic's wire.
7. Disconnect the turbocharger oil return hose.
8. Remove the oil pan retaining bolts. Carefully pry the oil pan loose from the cylinder block.

NOTE: Do not force a prying tool between the cylinder block and oil pan.

9. Remove the front and rear oil pan seals.
To install:
10. Clean the oil pan and the oil pan and cylinder block gasket surfaces.

11. Apply gasket sealant to new front and rear oil pan seals. Install the seals to the cylinder block.

12. Apply gasket sealant to the oil pan gasket surface and install the oil pan. Tighten the retaining bolts to 71–97 inch lbs. (8–11 Nm).

13. Connect the turbocharger oil return hose.

14. Install the rubber exhaust hanger to the brackets.

15. Install a new gasket and connect the front exhaust pipe to the turbocharger. Tighten the retaining nuts to 18–24 ft. lbs. (24–32 Nm).

16. Install the front exhaust pipe bracket. Tighten the retaining bolts to 34 ft. lbs. (46 Nm).

17. Pivot the frame brace into position. Tighten the retaining bolt to crossmember to 26–37 ft. lbs. (35–50 Nm). Tighten the right A-arm front retaining bolt to 72–86 ft. lbs. (97–117 Nm).

18. Lower the vehicle.

19. Fill the engine with the proper type and quantity of engine oil.

20. Connect the negative battery cable.

21. Start the engine and check for leaks and proper engine operation.

CLUTCH

Clutch Cable

ADJUSTMENT

Turbocharged Engine

Depress the clutch pedal lightly by hand until all free-play is removed and measure the free-play distance.

The distance should be 0.350–0.590 inch (9–15mm).

1. If adjustment is necessary, adjust the free-play at the clutch release lever as follows:

  a. Depress the clutch release lever and pull the pin away from the lever.

  b. Adjust clearance **A** to 0.06–0.100 inch (1.5–2.5mm) by turning adjusting nut **B**.

  c. After adjustment, make sure that when the clutch is disengaged, the pedal height is 8.0–8.2 inch (202–207mm).

Non-Turbocharged Engine

Depress the clutch pedal lightly by hand until all free-play is removed and measure the free-play distance. The distance should be 0.02–1.2 inch (0.6–3.0mm).

1. If adjustment is necessary, adjust the clutch pedal free-play at the clutch pedal pushrod as follows:

  a. Loosen the locknut.

  b. Turn the pushrod adjusting nut in the direction required to achieve the required clearance.

  c. Tighten the locknut to 9–12 ft. lbs. (12–17 Nm). Make sure the pedal height is 7.95–8.15 inch (202–207mm).

Clutch Master Cylinder

REMOVAL AND INSTALLATION

NOTE: The hydraulic clutch system is only used on vehicles equipped with naturally aspirated engines.

1. Disconnect the battery cables, negative cable first. Remove the battery.

2. Remove the windshield wiper motor.

3. Disconnect the hydraulic line fitting at the retaining bracket on the

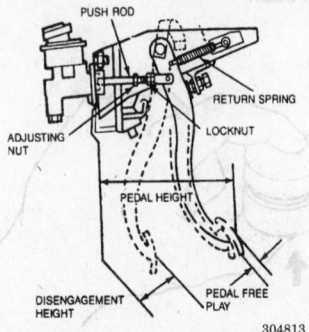

Clutch pedal free-play adjustment — non-turbocharged engine

transaxle case and drain the fluid. Reconnect the fitting after draining.

4. Disconnect the hydraulic line from the master cylinder. Remove the master cylinder retaining nuts and remove the clutch master cylinder.

To install:

5. Position the clutch master cylinder to the dash panel. Make sure the clutch pedal pushrod aligns properly.

6. Install the retaining nuts and tighten to 14–19 ft. lbs. (19–26 Nm).

7. Connect the hydraulic line to the clutch master cylinder. Fill the reservoir and bleed the clutch system.

8. Install the wiper motor.

9. Install the battery. Connect the battery cables, negative cable last.

10. Check the clutch system for leaks and proper operation.

Clutch Slave Cylinder

REMOVAL AND INSTALLATION

NOTE: The hydraulic clutch system is only used on vehicles equipped with naturally aspirated engines.

1. Disconnect and plug the hydraulic line at the clutch slave cylinder.

2. Remove the 2 bolts retaining the slave cylinder and remove the slave cylinder.

To install:

3. Position the clutch slave cylinder and install the 2 retaining bolts. Tighten to 12–16 ft. lbs. (16–23mm).

4. Unplug and connect the hydraulic line.

5. Fill the clutch master cylinder reservoir and bleed the clutch system.

Hydraulic Clutch System Bleeding

NOTE: The hydraulic clutch system is only used on vehicles equipped with naturally aspirated engines. Only use heavy-duty brake fluid C6AZ-19542-AA or equivalent.

1. Raise and safely support the vehicle.

2. Attach a hose to the bleeder valve on the clutch slave cylinder.

3. Open the bleeder valve ½ turn and watch for air bubbles in the brake fluid at the open end of the hose.

NOTE: Keep the reservoir full of brake fluid while bleeding.

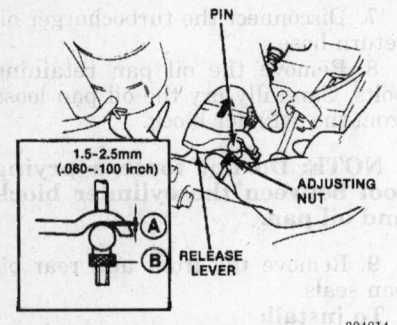

Clutch pedal free-play adjustment — turbocharged engine

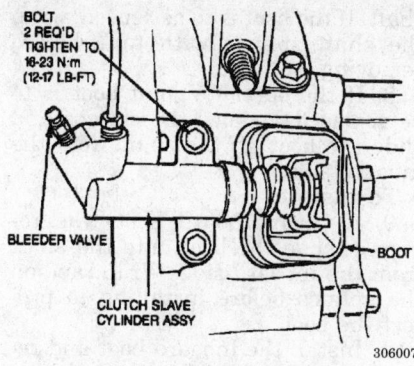

Clutch slave cylinder components

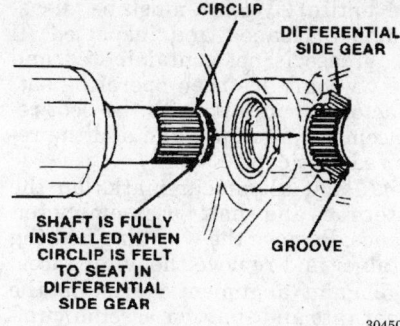

Halfshaft installation into the differential side gear

4. Close the bleeder valve when the bubbling stops.

5. Depress the clutch pedal to the floor and hold.

6. Open the bleed valve ¼ turn and push the clutch pedal down as far as it will go. Close the valve, then release the pedal.

7. Fill the clutch master cylinder reservoir.

8. Check the clutch system for proper operation.

DRIVELINE

Halfshaft

REMOVAL AND INSTALLATION

1. Disconnect the negative battery cable.

2. Carefully raise the staked portion of the halfshaft attaching nut using a small cape chisel.

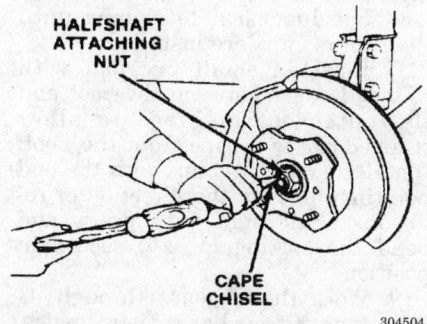

Proper method of removing the halfshaft nut

3. With the vehicle sitting on all four wheels, loosen, but do not remove the halfshaft attaching nut. If necessary, lock the hub by applying the brakes.

4. Raise and safely support the vehicle.

5. Remove the wheel and tire assemblies.

6. Remove the necessary engine compartment underbody covers.

7. Remove the stabilizer bar to control arm attaching nuts, bolt, washers and bushings.

8. Remove the lower control arm ball joint clamp bolt. Pry downward on the lower control arm to separate the ball joint from the steering knuckle.

NOTE: Extreme care must be taken to ensure the prybar does not damage the transaxle case, oil seal, CV-joint or CV-joint boot.

9. To separate the halfshaft from the transaxle, pull outward on the steering knuckle with just enough force to disengage the circlip on the inner end of the halfshaft. If the halfshaft is difficult to remove, a prybar can be inserted between the halfshaft and the transaxle case to loosen it. To prevent damage to the transaxle oil seal, do not pull the halfshaft all the way out of the transaxle.

10. Remove the halfshaft attaching nut and washer and discard the nut. Pull the halfshaft out of the wheel hub. If necessary, use a wheel puller to press the halfshaft out of the hub. Do not use a hammer or the CV-joint will be damaged.

11. Support the halfshaft and slide it out of the transaxle. Use care to prevent damage to the transaxle oil seal.

12. Install differential plugs T87C–7025–C or equivalent, to pre-

vent misalignment of the differential carrier.

To install:

13. Install a new circlip on the inboard CV-joint stub shaft. To install the circlip, start one end in the groove and work the circlip over the stub shaft end and into the groove. This will avoid over-expanding the circlip.

NOTE: The circlip must always be replaced.

14. Make sure the dynamic damper, if equipped, is positioned properly.

15. Inspect the transaxle oil seal. If it shows any signs of wear or damage that may cause a leak, replace the seal.

16. Remove the differential plugs. Make sure the circlip gap is positioned at the top of the halfshaft splines and lightly lubricate the splines with grease.

17. Carefully align the CV-joint splines with the differential side gear splines and push the halfshaft into the differential. When it seats properly, it will be possible to feel the circlip snap into the differential side gear groove. Try to pull the halfshaft out to make sure the circlip is seated.

18. Position the halfshaft through the wheel hub and install a new attaching nut. Do not tighten the nut at this time.

19. Install the lower control arm ball joint through the steering knuckle. Install the pinch bolt and attaching nut and tighten the nut to 32–40 ft. lbs. (43–54 Nm).

20. Position the stabilizer bar and install the attaching bolt, nuts, washers and bushings. Tighten the nut until 7/16 inch (10.8mm) of the bolt threads extend beyond the nut.

21. Install the removed underbody covers.

22. Install the wheel and tire assemblies.

23. Lower the vehicle.

24. Torque the new halfshaft attaching nut to 116–174 ft. lbs. (157–235 Nm). Stake the nut using a cold chisel with the cutting edge rounded.

NOTE: If the nut splits or cracks after staking, it must be replaced with a new nut.

25. Connect the negative battery cable.

26. Road test the vehicle an check for leaks and proper operation.

CV-Joint Boot

REPLACEMENT

NOTE: Three types of CV-joints are used. The turbocharged engine is equipped with Rzeppa or double offset inboard CV-joints. The non-turbocharged engine is equipped with tripot inboard CV-joints. All outboard CV-joints are the Birfield type. The Birfield outboard CV-joint cannot be disassembled. If the outboard CV-joint boot is to be replaced, it will be necessary to remove the inboard CV-joint.

Turbocharged Engine

1. Remove the halfshaft from the vehicle and clamp it in a vise equipped with jaw caps to prevent damage to the machined surfaces. Do not allow the vise to contact the boot or its clamps.
2. Remove the large boot clamp from the inboard CV-joint, using side cutters. After removing the clamp, roll the boot back over the shaft.
3. Check the grease for contamination by rubbing it between 2 fingers. Any gritty feeling indicates a contaminated CV-joint, in which case

the entire CV-joint must be disassembled, cleaned and inspected. If the grease is not contaminated and the CV-joint has been operating satisfactorily, continue with the boot replacement procedure and add the required lubricant.

4. Paint alignment marks on the outer race and shaft for assembly reference. Remove the wire ring bearing retainer and remove the outer race.
5. Paint alignment marks on the inner race and shaft for assembly reference. Remove the inner race snapring from the end of the halfshaft and remove the inner race, cage and ball bearings from the shaft as an assembly. Use care to prevent damage to the bearing surfaces and cage.
6. If only the boot is being replaced, go to the next step. If it is necessary to disassemble the CV-joint further, proceed as follows:
   a. Pry the ball bearing out of the bearing cage using a small prybar with blunted edges. Mark the inner race and the bearing cage for proper assembly.
   b. Rotate the inner race to align the bearing lands with the windows in the bearing cage. Remove the inner race through the larger end of the cage.
7. Remove the small clamp and remove the inner boot from the half-

shaft. If the boot is to be reused, wrap the shaft splines with tape before removing.

8. If the outer CV-joint boot is to be replaced, remove the clamps and slide the boot off the shaft from the inboard side.

To install:

9. If the outboard boot was removed, slide the boot onto the shaft from the inboard side. Wrap tape on the splines before installing to protect the boot.
10. Install the inboard boot and remove the tape from the shaft.
11. Lubricate the inner race, bearing cage and ball bearings with high temperature CV-joint grease.
12. Position the inner race in the bearing cage and align the matchmarks.

NOTE: Install the race with the chamfered splines facing the large end of the cage.

13. Install the ball bearings in the bearing cage. The balls can be pressed into the cage windows with the heel of the hand.
14. Install the inner race, cage and balls on the halfshaft as an assembly. Make sure the chamfer on the bearing cage faces the snapring and the paint marks made during removal line up. Install the inner race snapring.
15. Lubricate the outer race with 1.4–2.1 oz. (40–60 grams) of high temperature CV-joint grease. Install the outer race and add another 0.7–1.0 oz. (20–30 grams) of high temperature CV-joint grease to the outer race. Install the wire ring bearing retainer.
16. Grease the inside of the boot and position the CV-joint boot(s). Make sure the boot is fully seated in the grooves in the shaft and outer race.
17. Extend or compress the inner CV-joint as necessary until the distance between the boot clamp grooves measures 3.5 inch (90mm). Do not allow this dimension to change until the boot clamps are installed.
18. Insert a small pry tool with rounded edges between the boot and the outer bearing race to allow trapped air to escape from the boot. Install new boot clamps with the end pointing opposite the direction of rotation. Pull tight with pliers and bend the locking tabs to secure in position.
19. Work the CV-joint through its full range of travel at various angles. The joint should flex, extend and compress smoothly.

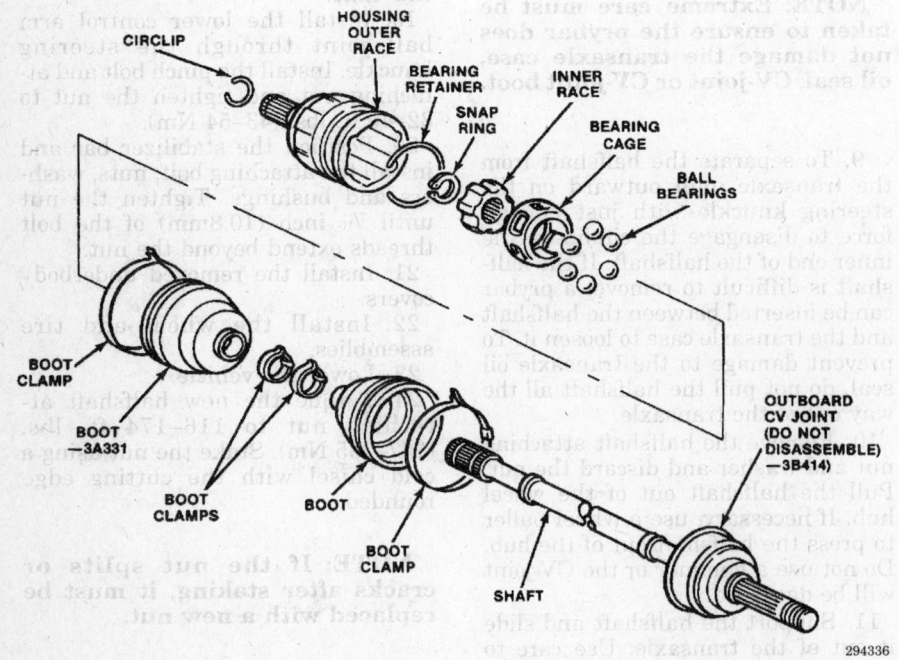

CIRCLIP   **HOUSING OUTER RACE**   **BEARING RETAINER**   **INNER RACE**   **SNAP RING**   **BEARING CAGE**   **BALL BEARINGS**   **BOOT CLAMP**   **BOOT —3A331**   **BOOT CLAMPS**   **BOOT**   **BOOT CLAMP**   **SHAFT**   **OUTBOARD CV JOINT (DO NOT DISASSEMBLE) —3B414**

294336

Exploded view of the Rzeppa CV-joint — turbocharged engines

20. Install the halfshaft into the vehicle.

21. Road test the vehicle and check for proper operation.

Non-Turbocharged Engine

1. Remove the halfshaft from the vehicle and clamp it in a vise equipped with jaw caps to prevent damage to the machined surfaces. Do not allow the vise to contact the boot or its clamps.

2. Remove the large boot clamp from the inboard CV-joint, using side cutters. After removing the clamp, roll the boot back over the shaft.

3. Check the grease for contamination by rubbing it between 2 fingers. Any gritty feeling indicates a contaminated CV-joint, in which case the entire CV-joint must be disassembled, cleaned and inspected. If the grease is not contaminated and the CV-joint has been operating satisfactorily, continue with the boot replacement procedure and add the required lubricant.

4. Paint alignment marks on the outer race and shaft for assembly reference. Remove the wire ring bearing retainer and remove the outer race.

5. Paint alignment marks on the tripot bearing and shaft for assembly reference. Remove the tripot bearing

snapring and, using a brass drift and hammer, remove the tripot bearing from the shaft.

6. Remove the small clamp and remove the inner boot from the halfshaft. If the boot is to be reused, wrap the shaft splines with tape before removing.

7. If the outer CV-joint boot is to be replaced, remove the clamps and slide the boot off the shaft from the inboard side.

To install:

8. If the outboard boot was removed, slide the boot onto the shaft from the inboard side. Wrap tape on the splines before installing to protect the boot.

9. Install the inboard boot and remove the tape from the shaft.

10. Install the tripot assembly on the halfshaft. Tap the assembly onto the shaft using a hammer and brass drift. Install the tripot assembly retaining ring.

11. Fill the CV-joint outer race with 3.5 oz. (100 grams) of high temperature CV-joint grease. Install the outer race over the tripot joint and install the wire ring bearing retainer.

12. Grease the inside of the boot and position the CV-joint boot(s). Make sure the boot is fully seated in the grooves in the shaft and outer race.

13. Extend or compress the inner CV-joint as necessary until the distance between the boot clamp grooves measures 3.5 inch (90mm). Do not allow this dimension to change until the boot clamps are installed.

14. Insert a small pry tool with rounded edges between the boot and the outer bearing race to allow trapped air to escape from the boot. Install new boot clamps.

15. Wrap the clamps around the boots with the end pointing opposite the direction of rotation. Pull tight with pliers and bend the locking tabs to secure in position.

16. Work the CV-joint through its full range of travel at various angles. The joint should flex, extend and compress smoothly.

17. Install the halfshaft into the vehicle.

18. Road test the vehicle and check for proper operation.

STEERING

Air Bag

——— **CAUTION** ———

Some vehicles are equipped with an air bag system, also known as the Supplemental Inflatable Restraint (SIR) or Supplemental Restraint System (SRS). The system must be disabled before performing service on or around system components, steering column, instrument panel components, wiring and sensors. Failure to follow safety and disabling procedures could result in accidental air bag deployment, possible personal injury and unnecessary system repairs.

PRECAUTIONS

Several precautions must be observed when handling the inflator module to avoid accidental deployment and possible personal injury.

• Never carry the inflator module by the wires or connector on the underside of the module.

• When carrying a live inflator module, hold securely with both hands, and ensure that the bag and trim cover are pointed away.

• Place the inflator module on a bench or other surface with the bag and trim cover facing up.

BOOT DYNAMIC DAMPER

BOOT —3A331

TRIPOT BEARING

OUTBOARD CV JOINT (DO NOT DISASSEMBLE)

BEARING RETAINER

CIRCLIP

BOOT CLAMP

BOOT CLAMP

SNAP RING

OUTER RACE

294339

Exploded view of the Tripot joint — non-turbocharged engines

- With the inflator module on the bench, never place anything on or close to the module which may be thrown in the event of an accidental deployment.

DISARMING

1993 Vehicles

—————— **CAUTION** ——————
The Supplemental Air Bag System (SRS) must be disarmed before performing service around SRS components or SRS wiring. Failure to do so may cause accidental deployment of the air bag, resulting in unnecessary SRS repairs and/or personal injury.

1. Disconnect both battery cables, negative cable first.
2. Lower the glove compartment door fully by depressing the stops.
3. Disconnect the electrical connector from the battery backup, which is a blue rectangular box on the outer left side of the glove compartment and attached to the instrument panel.

NOTE: The backup power supply provides air bag firing circuit power if the battery or battery cables are damaged or cut very early in an accident before the sensors can close. The battery backup contains a capacitor that takes approximately 15 minutes to discharge after the battery is disconnected.

4. After servicing is completed, connect the electrical connector to the battery backup.
5. Close the glove compartment door.
6. Connect the battery cables, negative cable last.
7. Turn the ignition switch to the **RUN** position. The air bag indicator should light continuously for approximately 6 seconds and then turn OFF. If the indicator fails to light, flashes or remains lit continuously, there is a fault in the air bag system.

1994 Vehicles

—————— **CAUTION** ——————
The Supplemental Air Bag System (SRS) must be disarmed before performing service around SRS components or SRS wiring. Failure to do so may cause accidental deployment of the air bag, resulting in unnecessary SRS repairs and/or personal injury.

1. Disconnect the negative battery cable.
2. Wait one minute before proceeding with the service procedure. This is the time required for the backup power supply in the air bag diagnostic monitor to deplete its stored energy.
3. After service is completed, reconnect the negative battery cable.
4. Turn the ignition switch to the **RUN** position. The air bag indicator should light continuously for approximately 6 seconds and then turn OFF. If the indicator fails to light, flashes or remains lit continuously, there is a fault in the air bag system.

Power Steering Pump

BLEEDING

1. Disable the ignition system.
2. Raise the vehicle until the front tires are just off of the ground and safely support. Make sure that the transaxle is not in gear.
3. Fill the power steering pump reservoir.
4. Crank the engine for 30 seconds without turning the steering wheel and recheck the fluid level. Add fluid if needed.
5. Crank the engine for 30 seconds while turning the steering wheel lock to lock. Check the fluid level and fill if needed.

—————— **WARNING** ——————
Do not hold the steering wheel against a stop for more than 5 seconds as damage to the steering pump could result.

6. Lower the vehicle.
7. Restore the ignition system.
8. Start the vehicle and check that the power steering system is free of air. If not, repeat the procedure. If required, purge the system of air using an external vacuum source following the equipment manufacturers instructions.

REMOVAL AND INSTALLATION

1. Disconnect the negative battery cable.
2. Remove the right-hand radiator support and brace. Remove the power steering pump drive belt.
3. Disconnect the intercooler outlet hose at the throttle inlet, if equipped, and position out of the way.
4. Remove the ground wire from the engine lifting eye and remove the

electrical connector from the pressure switch.
5. Place a drain pan below the power steering pump. Remove the inlet and return hoses from the pump and plug.
6. Remove the adjusting screw, nut and block from the bracket. Remove the pivot bolt.
7. Position the pump below the pump bracket in the engine compartment. Remove the pump bracket retaining nut, bolts and pump bracket.
8. Remove the power steering pump from the vehicle.
   To install:
9. Position the power steering pump in the engine compartment below the pump bracket mounting stud.
10. Install the power steering pump bracket. Tighten the retaining bolts and nut to 27–38 ft. lbs. (37–52 Nm).
11. Install the pump on the pump bracket and install the pivot bolt finger-tight.
12. Install the adjusting screw block, nut and screw finger-tight. Install the drive belt, adjust the tension and tighten the bolts.
13. Connect the pressure switch electrical connector and install the ground wire onto the engine lifting eye bracket.
14. Unplug and install the pressure and return hoses. Connect the intercooler outlet hose, if equipped.
15. Install the right-hand radiator support and brace.
16. Fill the power steering pump reservoir with power steering fluid and bleed the system.
17. Road test the vehicle and check for proper operation of the power steering system.

BRAKES

Master Cylinder

REMOVAL AND INSTALLATION

NOTE: Pump the brake pedal several times to exhaust any vacuum in the brake booster.

1. Disconnect the negative battery cable.
2. Remove the brake lines from the master cylinder using a flarenut wrench. Cap the lines and master cylinder ports.
3. Remove the vacuum valve from the brake booster and disconnect the pressure warning switch connector.

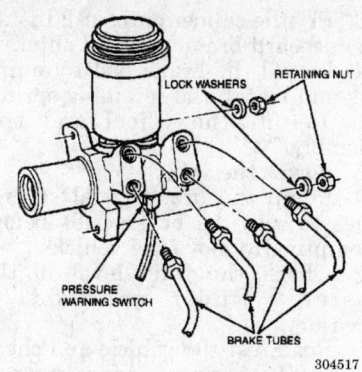

Master cylinder connection points

4. Remove the 2 nuts and lockwashers retaining the master cylinder to the brake booster. Remove the master cylinder from the booster. It may be necessary to pry between the booster and the master cylinder to free the master cylinder.

To install:

5. Position the master cylinder onto the booster assembly studs. Install the 2 lockwashers and nuts and tighten to 8–11 ft. lbs. (10–16 Nm).

6. Install short brake lines in the master cylinder ports and position them so they point back into the reservoir and the ends of the lines are submerged in brake fluid.

7. Fill the reservoir with clean DOT 3 brake fluid from a closed container and cover the reservoir with a shop towel.

8. Pump the brakes until clear, bubble-free brake fluid comes out of both brake lines. If any brake fluid spills on the paint, wash it off immediately with water.

9. Remove the short brake lines and connect the vehicle brake lines to the master cylinder. Using a flarenut wrench, torque the brake lines to 10–17 ft. lbs. (13–22 Nm). Bleed each brake line at the master cylinder using the following procedure:

a. Have an assistant pump the brake pedal 10 times and then hold firm pressure on the pedal.

b. Loosen the lower rear most brake line fitting with a flarenut wrench until a stream of brake fluid comes out. Have the assistant maintain pressure on the brake pedal until the brake line fitting is tightened again.

c. Repeat this operation until clear, bubble free fluid comes out from around the brake line fitting.

d. Repeat this bleeding operation at the upper rear, lower front and upper front brake line fittings, in that order.

10. Install the vacuum valve to the booster and connect the pressure warning switch connector.

11. Finish bleeding the brake system at each wheel.

12. Make sure the reservoir is filled to the proper level.

13. Connect the negative battery cable.

14. Check and if necessary, adjust the brake light switch.

15. Road test the vehicle and check for proper brake system operation.

Brake Caliper

REMOVAL AND INSTALLATION

Front

1. Raise and safely support the vehicle.

2. Remove the front wheel and tire assembly.

3. Remove the disc brake pads.

4. Remove the banjo bolt attaching the brake flex hose to the caliper. Remove and discard the 2 copper washers.

5. Remove the 2 caliper retaining bolts and lift the caliper off of the rotor.

To install:

6. Before installing the caliper, remove the guide pin bushing dust boots and push out the caliper guide pin bushings.

7. Lubricate the guide pin bushings with disc brake caliper slide grease and install them in the caliper. Install the guide pin bushing dust boots.

8. Position the caliper over the rotor.

NOTE: To provide the necessary clearance, it may be necessary to pull outward slightly on the caliper bushings.

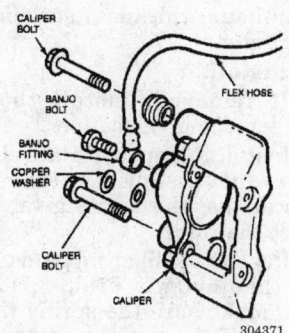

Front brake caliper retaining bolts and flex hose

9. Install the caliper retaining bolts and tighten to 29–36 ft. lbs. (39–49 Nm).

10. Install 2 new copper washers and the banjo bolt on the flex hose banjo fitting. Position the flex hose on the caliper and install the banjo bolt. Tighten to 17–21 ft. lbs. (22–29 Nm).

11. Install the brake pads and shims.

12. Bleed the brake system.

13. Install the wheel and tire assembly.

14. Lower the vehicle.

15. Pump the brake pedal several times to seat the brake pads to the rotor before attempting to move the vehicle.

16. Road test the vehicle and check for proper brake system operation.

Rear

1. Raise and safely support the vehicle.

2. Remove the rear wheel and tire assembly.

3. Remove the disc brake pads.

4. Remove the attaching clip from the brake flex hose.

5. Remove the banjo bolt attaching the brake flex hose to the caliper. Remove and discard the 2 copper washers.

6. Remove the lower caliper attaching bolt.

7. Using a cold chisel, remove the upper caliper guide pin dust cap to gain access to the Allen head on the guide pin. Using an Allen wrench, remove the upper caliper guide pin.

8. Lift the caliper off the rotor.

To install:

9. Install the disc brake pads and shims.

10. Before installing the caliper, remove the upper guide pin and the lower guide pin bushing. Remove the guide pin and guide pin bushing dust boots.

11. Lubricate the upper guide pin and lower guide pin bushing with disc brake caliper slide grease. Install the guide pin and guide pin bushing dust boots.

12. Position the caliper over the rotor. The piston is threaded onto the parking brake actuator inside the caliper. It may be necessary to rotate the piston into the caliper to provide the necessary clearance.

13. Tighten the upper guide pin with an Allen wrench and install the dust cap with a plastic hammer.

14. Install the lower caliper attaching bolt through the caliper guide pin bushing. Tighten to 29–36 ft. lbs. (39–49 Nm).

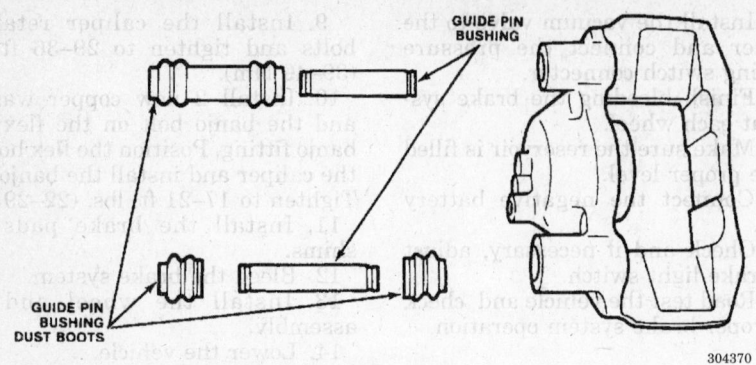

GUIDE PIN
BUSHING

GUIDE PIN
BUSHING
DUST BOOTS

304370

Front caliper guide pin bushings and boots

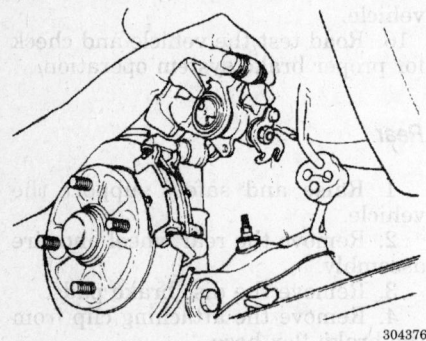

304376

Rear brake caliper removal

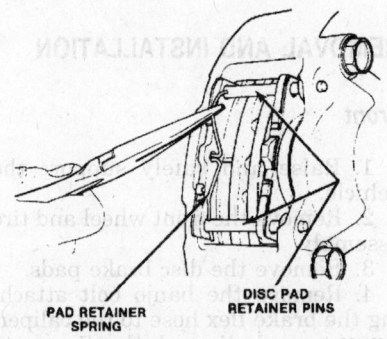

PAD RETAINER
SPRING

DISC PAD
RETAINER PINS

304432

Front brake pad retainer spring removal

15. Install 2 new copper washers and the banjo bolt on the flex hose fitting. Position the flex hose on the caliper and install the banjo bolt. Tighten the bolt to 17–21 ft. lbs. (22–29 Nm).

16. Bleed the brake system.

17. Install the wheel and tire assembly.

18. Lower the vehicle.

19. Pump the brake pedal several times to seat the brake pads to the rotor before attempting to move the vehicle.

20. Road test the vehicle and check for proper brake system operation.

Disc Brake Pads

REMOVAL AND INSTALLATION

Front

1. Remove approximately ⅔ of the brake fluid from the master cylinder.

2. Raise and safely support the vehicle.

3. Remove the front wheel and tire assembly.

4. Using needle-nose pliers, remove the pad retainer spring that locks in the disc pad retainer pins.

5. Remove the disc pad retainer pins using a hammer and pin punch.

6. Using a small prybar, pry the caliper outboard and remove the outboard brake pad and shim. Tag the shims so they can be installed in their original position.

7. Push the caliper inboard with one hand and remove the inboard brake shoe and shims with the other hand. Be careful to prevent damage to the caliper piston dust boot.

8. Remove the anchor plate clips from the caliper anchor plate. Attach tape to the anchor plate clips and label **TOP** and **BOTTOM**.

9. Inspect the disc brake rotor for scoring or wear. Replace or machine, as necessary. If machining, observe the minimum thickness specification on the rotor.

To install:

10. If removed, install the disc brake rotor and the caliper.

11. Install the anchor plate clips. If they are not installed in their original locations, the locating tabs may contact the rotor.

12. Push the caliper inboard and install the inboard brake pad and shims. Make sure the spring tabs on the back of the brake pad are properly aligned and fully seated in the caliper piston.

13. Pry the caliper outboard. Install the outboard brake pad and shim.

14. Install the brake pad retaining pins and install the retaining spring.

15. Install the wheel and tire assembly.

16. Lower the vehicle.

17. Pump the brake pedal several times to seat the brake pads before attempting to move the vehicle.

18. Check the fluid level in the master cylinder and add, if necessary.

19. Road test the vehicle and check the brake system for proper operation.

Rear

1. Remove approximately ⅔ of the brake fluid from the master cylinder.

2. Raise and safely support the vehicle.

3. Remove the rear wheel and tire assembly.

4. Using needle-nose pliers, remove the parking brake return springs at the back of the caliper.

5. Loosen the parking brake cable housing adjusting nut. Remove the cable housing from the bracket on the rear lower control arm.

6. Loosen the attaching bolt connecting the parking brake cable bracket to the rear caliper. Remove the parking brake cable from the rear caliper.

7. Loosen the lower caliper bolt. Pivot the caliper upward on the upper caliper guide pin.

8. Remove the disc pad retaining spring and remove the disc pads and shims.

9. Remove the anchor plate clips from the caliper anchor plate. Attach tape to the anchor plate clips and label **TOP** and **BOTTOM**.

10. Inspect the disc brake rotor for scoring or wear. Replace or machine, as necessary. If machining, observe the minimum thickness specification on the rotor.

To install:

11. If removed, install the disc brake rotor.

12. Install the anchor plate clips. If they are not installed in their original locations, the locating tabs may contact the rotor. Lubricate the anchor plate clips with disc brake caliper slide grease.

13. Install the shims on the backs of the pads and install in the anchor plate. Install the disc pad retaining spring.

14. The piston is threaded onto the parking brake adjuster spindle. Rotate the piston using appropriate tools to seat it fully into the bore.

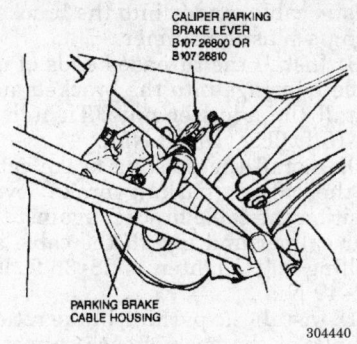

Disconnecting the parking brake cable from the caliper

15. Pivot the caliper down over the pads. Install the lower caliper bolt and tighten to 29–36 ft. lbs. (39–49 Nm).

16. Install the parking brake cable in the caliper parking brake lever. Position the cable bracket against the rear caliper and install the cable attaching bolt.

17. Install the wheel and tire assembly.

18. Pump the brake pedal several times to seat the brake pads before attempting to move the vehicle.

19. Check the fluid level in the master cylinder and add, if necessary.

20. With the wheels off the ground, spin each wheel several times to make sure the calipers are not frozen and the parking brake is not adjusted too tight.

21. Lower the vehicle.

22. Road test the vehicle and check fro proper brake system operation.

Brake Rotor

REMOVAL AND INSTALLATION

Front

1. Raise and safely support the vehicle.

2. Remove the front wheel and tire assembly.

3. Carefully raise the staked portion of the halfshaft locknut using a small cape chisel. Remove the halfshaft locknut and washer and discard the nut.

NOTE: When loosening the locknut, lock the hub by applying the brakes.

4. Remove the stabilizer bar to control arm attaching bolt, nut, washers and bushings.

5. Remove the cotter pin and tie rod end attaching nut. Separate the tie rod end from the steering knuckle arm using a suitable tool. Discard the cotter pin.

6. Remove the U-shaped retaining clip from the center section of the caliper flex hose. Remove the disc brake pads and the caliper. Suspend the caliper from the strut spring with wire to prevent damage to the brake hose.

7. Remove the lower control arm ball joint clamp bolt and nut. Pry downward on the lower control arm to separate the ball joint from the steering knuckle.

8. Remove the steering knuckle to strut retaining bolts. Slide the hub/steering knuckle assembly out of its bracket in the strut and off the end of the halfshaft. Use care to prevent damage to the grease seals.

NOTE: If the hub binds on the halfshaft splines, it can be loosened by lightly tapping with a plastic faced hammer on the end of the halfshaft. Never use a metal faced hammer as damage to the CV-joint internal components will result. If the halfshaft splines become rusted to the hub, a jaw type puller must be used to separate them.

9. Remove the hub and rotor assembly from the steering knuckle using knuckle puller T87C–1104–A, or equivalent.

10. If the rotor is to be reused, paint aligning marks on the hub and rotor assembly so they can be assembled in the same position. Remove the retaining bolts and separate the rotor from the hub. It may be helpful to mount the rotor in a soft jawed vise.

11. Inspect the rotor for scoring and wear. Replace or machine as necessary. If machining, observe the minimum thickness specification.

To install:

12. Position the hub on the rotor and install the retaining bolts. Make

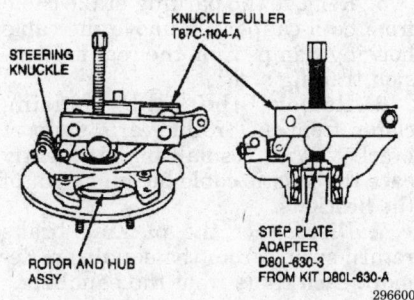

Removing the rotor from the knuckle

sure the index marks on the hub and rotor align. Tighten the retaining bolts to 33–40 ft. lbs. (44–54 Nm).

13. Install the hub and rotor assembly to the steering knuckle using a hydraulic press and suitable fixtures, or equivalent.

14. Position the front hub/steering knuckle assembly over the halfshaft and into the strut. Install the steering knuckle to strut retaining bolts and nuts. Tighten the retaining nuts to 69–86 ft. lbs. (93–117 Nm).

15. Position the lower control arm ball joint through the steering knuckle and install the clamp bolt and nut. Tighten the clamp bolt to 32–40 ft. lbs. (43–54 Nm).

16. Position the brake caliper over the rotor and install the retaining bolts. Tighten to 29–36 ft. lbs. (39–49 Nm). Install the U-clip on the caliper flex hose.

17. Install a new halfshaft locknut. Tighten to 116–174 ft. lbs. (157–235 Nm). Stake the halfshaft locknut using a cold chisel with the cutting edge rounded.

NOTE: If the nut splits or cracks after staking, it must be replaced with a new nut.

18. Connect the tie rod to the steering knuckle arm and install the retaining nut. Tighten to 22–33 ft. lbs. (29–44 Nm) and install a new cotter pin.

NOTE: If the slots in the nut do not align with the hole in the ball joint stud, tighten the nut for proper alignment. Never loosen the nut.

19. Position the stabilizer bar and install the stabilizer link assembly including the retaining bolt, nut, washers, sleeve and rubber bushings. Tighten the attaching nut until 0.43 inch (10.8mm) of the bolt threads extend beyond the nut.

20. Install the wheel and tire assembly.

21. Lower the vehicle.

22. Pump the brake pedal several times to position the brake pads to the rotor before attempting to move the vehicle.

23. Road test the vehicle and check for proper brake system operation.

Rear

1. Raise and safely support the vehicle.

2. Remove the rear wheel and tire assembly.

3. Remove the disc brake pads and caliper from the anchor plate. Support the caliper with wire from the

coil spring to prevent damage to the brake hose.

4. Remove the grease cap.

5. Carefully raise the staked portion of the spindle nut using a small cape chisel. Remove the spindle locknut and washer and outer wheel bearing. Discard the spindle locknut.

NOTE: The locknuts are threaded left and right. The left hand threaded locknut is on the right side of the vehicle. Turn this locknut clockwise to loosen. The right hand threaded locknut is turned counterclockwise to loosen.

6. Remove the rotor/hub assembly and inspect for scoring and wear. Replace or machine as necessary. If machining, observe the minimum thickness specification.

To install:

7. If the inner wheel bearing was removed, install the bearing and a new grease seal.

NOTE: Make sure that the wheel bearings and hub contains adequate lubricant.

8. Position the rotor/hub assembly on the spindle. Keep the hub centered on the spindle to prevent damage to the grease seal.

9. Install the outer bearing, washer and a new spindle locknut.

10. Properly adjust the bearing preload.

11. Install the grease cap.

12. Install the caliper and brake pads.

13. Install the wheel and tire assembly.

14. Lower the vehicle.

15. Pump the brake pedal several times to position the brake pads to the rotor before attempting to move the vehicle.

16. Road test the vehicle and check for proper brake system operation.

Parking Brake Cable

ADJUSTMENT

1. Remove the rear console as follows:

   a. Slide the front seats completely forward and remove the screws retaining the rear of the rear console.

   b. Slide the front seat completely rearward and remove the screws retaining the rear console to the front console.

   c. Raise the parking brake lever as far as it will go. Raise the rear of the rear console and pull backwards to remove.

   d. Release the parking brake lever.

2. Loosen the locknut.

3. Loosen or tighten the adjusting nut so the parking brake begins to apply when the lever is pulled up 5 notches and is fully set at 7–11 notches.

4. Using spring scale tool T74P–3504–Y or equivalent, check the force required to apply the parking brake. A properly operating system will require 44 lbs. of force to fully apply the parking brakes.

5. Tighten the locknut against the adjusting nut.

6. Make sure the brakes do not drag when the parking brake lever is released.

7. Make sure the brake warning light illuminates when the parking brake lever is raised.

8. Install the rear console as follows:

   a. Raise the parking brake lever as far as it will go.

   b. Lower the console into position.

   c. Slide the front seat completely rearward and install the screws retaining the rear console to the front console.

   d. Slide the front seats completely forward and install the screws retaining the rear of the rear console.

REMOVAL AND INSTALLATION

1. Raise and safely support the vehicle.

2. Remove the rear wheel and tire assemblies.

3. Using a pair of needle-nose pliers, remove the parking brake return springs at the back of each caliper.

4. Loosen the parking brake cable housing adjusting nut. Loosen the attaching bolt connecting the parking brake cable bracket to the rear caliper.

5. Remove the parking brake cable from both calipers. Remove the cable housing clamps from the rear suspension trailing arms.

6. Remove the cable housing clamp from the trailing arm support bracket. With a small prybar, gently ease the plastic cable bushings out of the brackets.

7. Disconnect the parking brake return spring from the equalizer. Remove each cable from the equalizer.

To install:

8. Install the 2 cable ends in the parking brake equalizer. Install the 2 plastic cable guides into the brackets using a plastic hammer.

9. Install the threaded ends of the cable housings into the brackets and install the adjuster nuts. Tighten to 12–16 ft. lbs. (16–23 Nm).

10. Install the parking brake cable in the caliper parking brake lever. Position the cable bracket against the rear caliper and install the cable attaching bolt. Tighten to 28–36 ft. lbs. (37–49 Nm).

11. Install the parking brake return springs at the back of each caliper.

12. Install the cable housing support clamps and install the equalizer return spring.

13. Install the wheel and tire assemblies.

14. Lower the vehicle.

15. Adjust the parking brake.

16. Check the parking brake system for proper operation.

Brake System Bleeding

MANUAL BLEEDING

1. Clean all dirt from the master cylinder filler cap.

2. If the master cylinder is known or suspected of having air in the bore, it must be bled before any of the calipers. Bleed the master cylinder as follows:

   a. Using a tubing wrench, remove the brake lines from the master cylinder.

   b. Install short brake lines in the master cylinder and position them so they point back into the reservoir and the ends of the lines are submerged in brake fluid.

   c. Fill the reservoir with clean DOT 3 or equivalent brake fluid and cover the reservoir with a shop towel.

   d. Pump the brakes until clear, bubble-free fluid comes out of both brake lines. If any brake fluid spills on the paint, wash it off immediately with water.

   e. Remove the short brake lines and connect the vehicle brake lines to the master cylinder.

   f. Have an assistant pump the brake pedal 10 times and then hold firm pressure on the pedal.

   g. Crack the rear most brake line fitting with a tubing wrench until a stream of brake fluid comes out. Have the assistant maintain pressure on the brake pedal until the brake line fitting is tightened again.

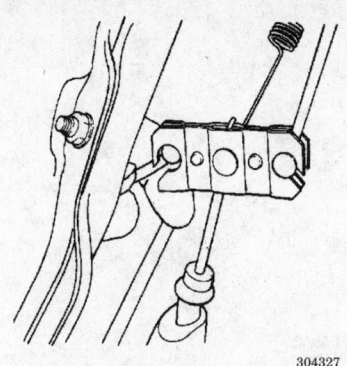

Parking brake cable equalizer assembly

h. Repeat this operation until clear, bubble free fluid comes out from around the brake line fitting.

i. Repeat this bleeding operation at the front brake line fitting.

NOTE: Never allow the master cylinder reservoir to run dry during the brake bleeding procedure.

3. To bleed the calipers, proceed as follows:

a. Attach a rubber drain hose to the right rear caliper bleeder screw. The end of the tube should fit snugly around the end of the bleeder screw.

b. Place the free end of the hose in a container partially filled with clean brake fluid.

c. Have an assistant apply and maintain pressure on the brake pedal.

d. Loosen the bleeder screw approximately ¾ turn. It is very important the helper maintain constant pressure on the pedal until the pedal drops all the way down and the bleeder screw is closed again. If the pedal pressure is released, air will be drawn back into the system.

e. Tighten the bleeder screw and release the brake pedal.

f. Repeat this operation until the fluid is clear and no more air bubbles appear at the submerged end of the hose.

g. Repeat these steps at the other calipers in the following order: left rear, then right front, then left front. Maintain proper fluid level in the reservoir at all times.

4. Top up the brake fluid when bleeding is complete. Never reuse brake fluid.

PRESSURE BLEEDING

For pressure bleeding, use only a diaphragm-type bleeder and follow the manufacturer's instructions.

FRONT SUSPENSION

Strut

REMOVAL AND INSTALLATION

—— **CAUTION** ——
Do not remove the MacPherson strut upper nut without using an approved MacPherson strut compressor, 086–00029 or equivalent. The coil spring is under extreme pressure and can cause severe bodily injury if the nut is removed without a spring compressor.

1. Raise and safely support the vehicle.
2. Remove the wheel and tire assembly.
3. Remove the brake caliper and suspend it with mechanics wire. Do not allow the caliper to hang from the brake hose.
4. Paint a white aligning mark on the inside of the strut mounting block. Loosen and remove the steering knuckle to strut retaining bolts and nuts.
5. Remove the U-clip from the brake line hose and slide it out of its bracket on the strut.
6. Remove the 4 strut mount nuts from the strut tower and remove the strut from the vehicle.

NOTE: Paint a white index mark between the strut and the strut tower if one is not visible for installation reference.

7. Compress the spring with spring compressor 086–00029 or equivalent and remove the strut rod nut. Gradually release the spring compressor.
8. Remove the mounting block, upper spring seat, bump stopper, coil spring and lower spring seat from the strut.
To install:
9. Install the lower spring seat, coil spring, bump stopper, upper spring seat and mounting block on the strut.
10. Compress the spring with the spring compressor and install the strut rod nut. Tighten to 22–27 ft. lbs.

(29–36 Nm). Gradually release the spring compressor.
11. Install the strut assembly in the strut tower. Make sure that the white mark or rubber nib is pointed inward. Install the 4 strut attaching nuts and tighten to 17–22 ft. lbs. (23–29 Nm).

NOTE: Make sure the white aligning mark faces the center of the vehicle.

12. Install the steering knuckle to the strut and install the attaching bolts and nuts. Tighten the steering knuckle-to-strut attaching bolts to 69–86 ft. lbs. (93–117 Nm).
13. Install the brake caliper and brake hose in the bracket.
14. Install the wheel and tire assembly.
15. Lower the vehicle.
16. Check for proper operation.

Lower Ball Joints

REMOVAL AND INSTALLATION

1. Raise and safely support the vehicle.
2. Remove the wheel and tire assembly.
3. Remove the ball joint clamp bolt from the steering knuckle.
4. Using a prybar, pull down on the lower control arm to separate it from the steering knuckle.
5. Remove the 2 ball joint retaining nuts from the control arm. Using a small prybar, pry the ball joint off the control arm.
To install:
6. Install the ball joint to the control arm. Tighten the bolts to 69–86 ft. lbs. (93–117 Nm).
7. Raise the lower arm and install the ball joint stud in the spindle. Install the ball joint clamp bolt and tighten to 32–40 ft. lbs. (43–54 Nm).
8. Install the wheel and tire assembly.
9. Lower the vehicle.
10. Check for proper operation.

Lower Control Arms

REMOVAL AND INSTALLATION

1. Raise and safely support the vehicle.
2. Remove the wheel and tire assembly.
3. Disconnect the stabilizer bar from the lower control arm.
4. Remove the ball joint clamp bolt.

(29–35 Nm). Gradually release the spring compressor.

11. Install the strut assembly in the strut cover. Make sure that the white mark or rubber rib is pointed inward. Install the 4 strut attaching nuts and tighten to 17–22 ft. lbs. (23–29 Nm).

NOTE: Make sure the white aligning mark faces the center of the vehicle.

12. Install the steering knuckle to the strut and install the attaching bolts and nuts. Tighten the steering knuckle-to-strut attaching bolts to 69–86 ft. lbs. (93–117 Nm).

13. Install the brake caliper and brake hose in the bracket.

14. Install the wheel and tire assembly.

15. Lower the vehicle.

16. Check for proper operation.

Lower Ball Joints

REMOVAL AND INSTALLATION

1. Raise and safely support the vehicle.

2. Remove the wheel and tire assembly.

3. Remove the ball joint clamp bolt from the knuckle.

4. Using a prybar, pull down on the lower control arm to separate it from the steering knuckle.

5. Remove the 2 ball joint retaining nuts from the control arm. Using a small prybar, pry the ball joint off the control arm.

To install:

6. Install the ball joint to the control arm. Tighten the bolts to 69–86 ft. lbs. (93–117 Nm).

7. Raise the lower arm and install the ball joint stud in the spindle. Install the ball joint clamp bolt and tighten to 32–40 ft. lbs. (43–54 Nm).

8. Install the wheel and tire assembly.

9. Lower the vehicle.

10. Check for proper operation.

Control Arms

REMOVAL AND INSTALLATION

1. Raise and safely support the vehicle.

2. Remove the wheel and tire assembly.

3. Disconnect the stabilizer bar from the lower control arm.

4. Remove the ball joint clamp bolt.

PRESSURE BLEEDING

For pressure bleeding, use only a diaphragm-type bleeder and follow the manufacturer's instructions.

b. Repeat this operation until clear bubble free fluid comes out from around the brake line fitting.

c. Repeat this bleeding operation at the front brake line.

NOTE: Never allow the master cylinder to run dry during the brake bleeding procedure.

3. To bleed the brakes, do as follows:

a. Attach a rubber hose onto the right rear caliper bleeder screw, end of the tube should be snugly around the end of the bleeder screw.

b. Place the free end of the hose in a container partially filled with clean brake fluid.

c. Have an assistant apply and maintain pressure on the brake pedal.

d. Loosen the bleeder screw approximately ½ turn. It is very important the helper maintain constant pressure on the pedal until the pedal drops all the way down.

e. Tighten the bleeder screw and have the assistant slowly release the pedal.

f. Repeat this operation until the fluid is clear and no more air bubbles appear.

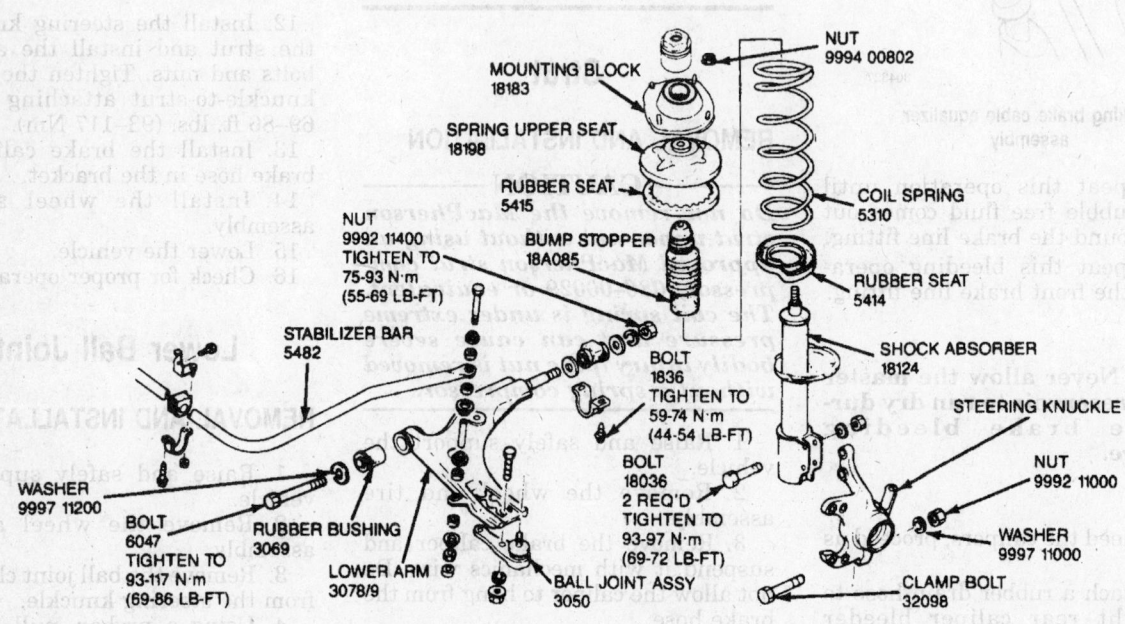

Exploded view of the front suspension components

Labels on diagram:
- MOUNTING BLOCK 18183
- SPRING UPPER SEAT 18198
- RUBBER SEAT 5415
- NUT 9992 11400 TIGHTEN TO 75–93 N·m (55–69 LB-FT)
- BUMP STOPPER 18A085
- STABILIZER BAR 5482
- WASHER 9997 11200
- BOLT 6047 TIGHTEN TO 93–117 N·m (69–86 LB-FT)
- RUBBER BUSHING 3069
- LOWER ARM 3078/9
- BALL JOINT ASSY 3050
- BOLT 1836 TIGHTEN TO 59–74 N·m (44–54 LB-FT)
- BOLT 18036 2 REQ'D TIGHTEN TO 93–97 N·m (69–72 LB-FT)
- NUT 9994 00802
- COIL SPRING 5310
- RUBBER SEAT 5414
- SHOCK ABSORBER 18124
- STEERING KNUCKLE
- NUT 9992 11000
- WASHER 9997 11000
- CLAMP BOLT 32098
- 306294

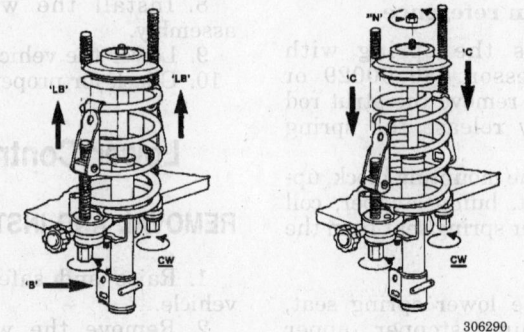

Proper method of compressing the strut coil spring for strut disassembly

306290

---

5. Remove the control arm front retaining bolt and the control arm rear bracket and retaining bolts.

6. Remove the control arm.

To install:

7. Position the control arm and loosely install the front retaining bolt.

8. Install the control arm rear retaining bracket and bolts. Tighten the bolts to 44–54 ft. lbs. (59–74 Nm).

9. Tighten the front retaining bolt to 69–86 ft. lbs. (93–117 Nm).

10. Install the ball joint to the steering knuckle. Tighten the clamp bolt to 32–40 ft. lbs. (43–54 Nm).

11. Connect the stabilizer bar to the lower control arm.

12. Install the wheel and tire assembly.

13. Lower the vehicle.

14. Check for proper operation.

Sway Bar

REMOVAL AND INSTALLATION

1. Raise and safely support the vehicle.

2. Remove the wheel and tire assembly.

3. Remove the stabilizer bar to control arm attaching bolt, nut, washers and bushings.

4. Remove the stabilizer bar bracket bolts and remove the brackets and the stabilizer bar.

To install:

5. Install the stabilizer bar with the brackets and bracket bolts. Tighten the bolts securely.

6. Install the stabilizer link assembly including the attaching bolt, nut, washers, sleeve and rubber bushings. Tighten the attaching nut until 0.43 in. (10.8mm) of the bolt threads extend beyond the nut.

7. Install the wheel and tire assembly.

8. Lower the vehicle.

9. Check for proper operation.

Front Wheel Bearings

ADJUSTMENT

The front wheel bearing adjustment is controlled by a shim located between the front wheel bearings which is selected during front hub and bearing assembly. If adjustment is required, the front hub assembly must be removed and disassembled and the correct shim installed.

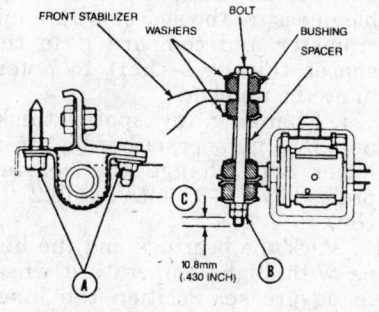

297395

Correct assembly of the front stabilizer link

REMOVAL AND INSTALLATION

1. Raise and safely support the vehicle.

2. Remove the front wheel and tire assembly.

3. Carefully raise the staked portion of the halfshaft attaching nut using a small cape chisel. Remove the halfshaft attaching nut and washer and discard the nut.

NOTE: When loosening the nut, lock the hub by applying the brakes.

4. Remove the stabilizer bar to control arm attaching bolt, nut, washers and bushings.

5. Remove the cotter pin and tie rod end attaching nut. Separate the tie rod end from the steering knuckle arm using a suitable tool. If the tie rod end does not separate easily, give the steering knuckle a sharp blow with a soft faced hammer to shock the taper. Discard the cotter pin.

6. Remove the U-shaped retaining clip from the center section of the caliper flex hose.

7. Remove the disc brake pads and the caliper. Suspend the caliper from the strut spring. Do not allow the caliper to hang from the brake hose.

8. Remove the lower control arm ball joint clamp bolt and nut. Pry downward on the lower control arm to separate the ball joint from the steering knuckle.

9. Remove the steering knuckle to strut attaching bolts. Slide the hub/steering knuckle assembly out of its bracket in the strut and off the end of the halfshaft. Use care to prevent damage to the grease seals.

NOTE: If the hub binds on the halfshaft splines, it can be loosened by lightly tapping with a plastic faced hammer on the end of the halfshaft. Never use a metal faced hammer as damage to the CV-joint internal components will result. If the halfshaft splines become rusted to the hub, a jaw type puller must be used to separate them.

10. Remove the hub and rotor assembly from the steering knuckle using knuckle puller T87C–1104–A or equivalent. Remove the bearing preload spacer from the hub and rotor assembly.

NOTE: The spacer located between the bearings determines bearing preload. It must not be discarded.

11. Paint aligning marks on the hub and rotor assembly so they can be assembled in the same position. Remove the attaching bolts and separate the rotor from the hub. It may be helpful to mount the rotor in a soft jawed vise.

12. Remove the bearing from the wheel hub using bearing puller attachment D84L–1123–A and puller D80L–927–A or equivalent. A bearing splitter and a large vibration damper puller can also be used. A spacer block will have to be used over the hub. A socket may also have to be used to finish pulling the bearing off the hub.

13. Remove the outer grease seal from the hub. Remove the inner grease seal from the steering knuckle using a small prybar.

14. Remove the bearing from the steering knuckle. Unless it has been damaged, the disc brake dust shield should be left on the steering knuckle.

15. If the bearings are to be replaced, remove the old bearing races using a brass drift and a hammer.

To install:

16. Clean and inspect all components that will be reused. Check the hub, knuckle and rotor dust shield for cracks, scoring, rusting, etc.

17. If the brake rotor dust shield was removed, install a new one using dust shield installation tool T87C–1175–B or equivalent.

18. If the original bearings and knuckle are being reused, proceed to Step 17. If the bearings or knuckle are being replaced, bearing preload must be checked as follows before assembly.

a. Install the outer bearing races in the steering knuckle using bearing cup installation tool D79P–1202–A or equivalent.

b. Lubricate the bearing races and bearing with a thin film of clean engine oil. Install the bearings and preload spacer in the steering knuckle.

c. Install spacer selection tool T87C–1104–B or equivalent, and clamp the tool in a vise.

d. Tighten the center bolt in increments, to 36, 72, 108 and 145 ft. lbs. (49, 98, 147 and 196 Nm). After tightening the center bolt to a specified increment, seat the bearings by rotating the steering knuckle. Verify the torque of the center bolt is 145 ft. lbs. (196 Nm).

e. Remove the tool/steering knuckle from the vise. Remount the assembly in the vise, clamping it where the MacPherson strut mounts.

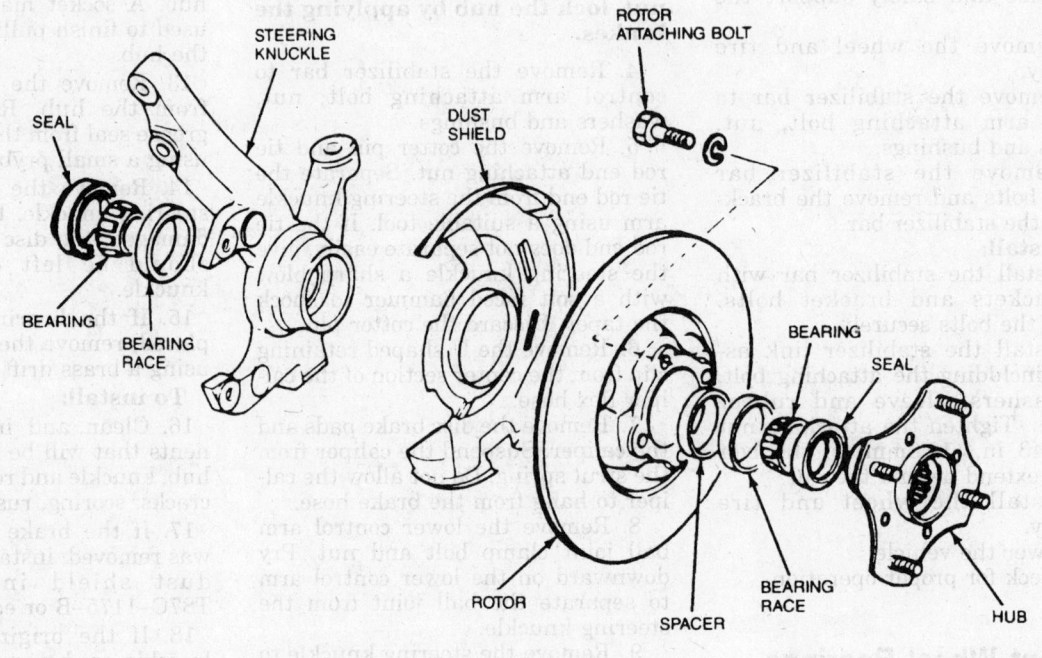

Exploded view of the hub, knuckle and bearing assembly

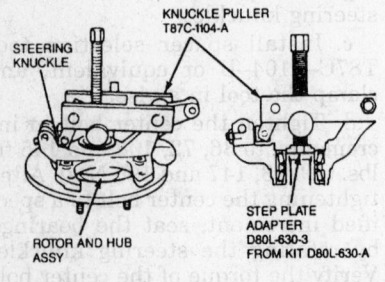

Removing the hub and rotor from the knuckle assembly

f. Measure the amount of torque required to rotate the spacer selector tool, using an inch pound torque wrench. The torque wrench reading must be taken just as the tool starts to rotate.

g. If the torque wrench indicates 2.2–10.4 inch lbs. (0.25–1.8 Nm), the spacer is the correct thickness. If the torque wrench indicates less than 2.2 inch lbs. (0.25 Nm), a thinner spacer must be installed. If the torque wrench indicates more than 10.4 inch lbs. (1.8 Nm), a thicker spacer must be installed.

h. Each bearing spacer has a numerical code that identifies it's thickness, stamped onto the outer

diameter of the spacer. The numbers range from 1–21, with 1 being the thinnest spacer. If the number stamped on the spacer is not legible, measure the spacer with a micrometer and compare it to the spacer thickness chart to determine the number.

i. Changing the spacer thickness by 1 number, either higher or lower, will change the bearing preload by 1.7–3.5 inch lbs. (0.2–0.4 Nm).

19. Pack the bearings and the hub area with high temperature wheel bearing grease. Position the inner bearing in the steering knuckle.

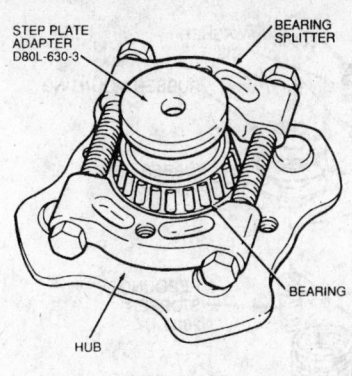

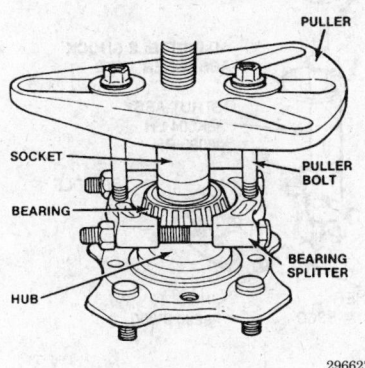

296627

Removing the bearing from the hub

Stamped Mark	Thickness
1	6.285 mm (0.2474 in)
2	6.325 mm (0.2490 in)
3	6.365 mm (0.2506 in)
4	6.405 mm (0.2522 in)
5	6.445 mm (0.2538 in)
6	6.485 mm (0.2554 in)
7	6.525 mm (0.2570 in)
8	6.565 mm (0.2586 in)
9	6.605 mm (0.2602 in)
10	6.645 mm (0.2618 in)
11	6.685 mm (0.2634 in)
12	6.725 mm (0.2650 in)
13	6.765 mm (0.2666 in)
14	6.805 mm (0.2682 in)
15	6.845 mm (0.2698 in)
16	6.885 mm (0.2714 in)
17	6.925 mm (0.2730 in)
18	6.965 mm (0.2746 in)
19	7.005 mm (0.2762 in)
20	7.045 mm (0.2778 in)
21	7.085 mm (0.2794 in)

296631

Spacer thickness chart

20. Lubricate the new grease seal lip with grease and install using seal installer T87C–1175–A or equivalent.

21. Install the bearing preload spacer and bearing in the steering knuckle.

22. Lubricate the new outer grease seal lip with grease and install using seal installer T87C–1175–A or equivalent.

23. Position the hub on the rotor and install the attaching bolts. Make sure the index marks on the hub and rotor align. Tighten the attaching bolts to 33–40 ft. lbs. (44–54 Nm).

24. Install the hub and rotor assembly in the steering knuckle using a hydraulic press and suitable fixtures.

25. Position the front hub/steering knuckle assembly over the halfshaft and into the strut. Install the steering knuckle to strut attaching bolts and nuts. Tighten the attaching nuts to 69–86 ft. lbs. (93–117 Nm).

26. Position the lower control arm ball joint through the steering knuckle and install the clamp bolt and nut. Tighten the clamp bolt to 32–40 ft. lbs. (43–54 Nm).

27. Position the brake caliper over the rotor and install the attaching bolts. Tighten to 29–36 ft. lbs. (39–49 Nm). Install the U-clip on the caliper flex line.

28. Install a new halfshaft attaching nut. Tighten to 116–174 ft. lbs. (157–235 Nm). Stake the halfshaft using a cold chisel with the cutting edge rounded.

NOTE: If the nut splits or cracks after staking, it must be replaced with a new nut.

29. Connect the tie rod to the steering knuckle arm and install the attaching nut. Tighten to 22–33 ft. lbs. (29–44 Nm). Install a new cotter pin.

NOTE: If the slots in the nut do not align with the hole in the ball joint stud, tighten the nut for proper alignment. Never loosen the nut.

30. Position the stabilizer bar and install the stabilizer link assembly including the attaching bolt, nut, washers, sleeve and rubber bushings. Tighten the attaching nut until 0.43 in. (10.8mm) of the bolt threads extend beyond the nut.

31. Install the wheel and tire assembly.

32. Lower the vehicle.

33. Pump the brake pedal several times to position the brake pads to the brake rotor before attempting to move the vehicle.

34. Road test the vehicle and check for proper operation.

REAR SUSPENSION

Strut

REMOVAL AND INSTALLATION

—————— **CAUTION** ——————
Do not remove the MacPherson strut upper nut without using an approved Spring Compressor, 086–00029 or equivalent. The coil spring is under extreme pressure and can cause severe bodily injury if the nut is removed without a spring compressor.

1. Raise and safely support the vehicle.

2. Remove the wheel and tire assembly.

3. Remove the rear disc brake caliper and rotor assembly. Hand the caliper with wire to prevent damage to the brake hose.

4. Remove the anti-moan bracket, if equipped.

5. Loosen and remove the trailing arm bolt and the spindle to strut retaining bolts.

6. Paint a white index mark on the strut rubber mounting bracket.

7. Remove the 2 upper strut mounting nuts from the strut tower.

8. Remove the strut from the vehicle.

9. Compress the coil spring using spring compressor 086–00029, or equivalent.

10. Remove the strut rod nut while the spring is compressed and remove the rubber mounting bracket, spring upper seat, lower seat and the rubber spring seat.

11. Slowly release the spring and remove the spring compressor. Remove the coil spring, dust boot and rebound bumpers.

To Install:

12. Install the rebound bumpers and dust boot on the strut.

13. Compress and install the coil spring on the strut. Lubricate the strut rod.

14. Install the rubber seat, spring upper seat with rubber mounting bracket and strut rod nut on the strut. Tighten to 40–50 ft. lbs. (55–68 Nm).

15. Release the spring compressor.

16. Install the strut into the strut tower.

17. Install the 2 upper strut mounting nuts and tighten to 17–22 ft. lbs. (23–29 Nm).

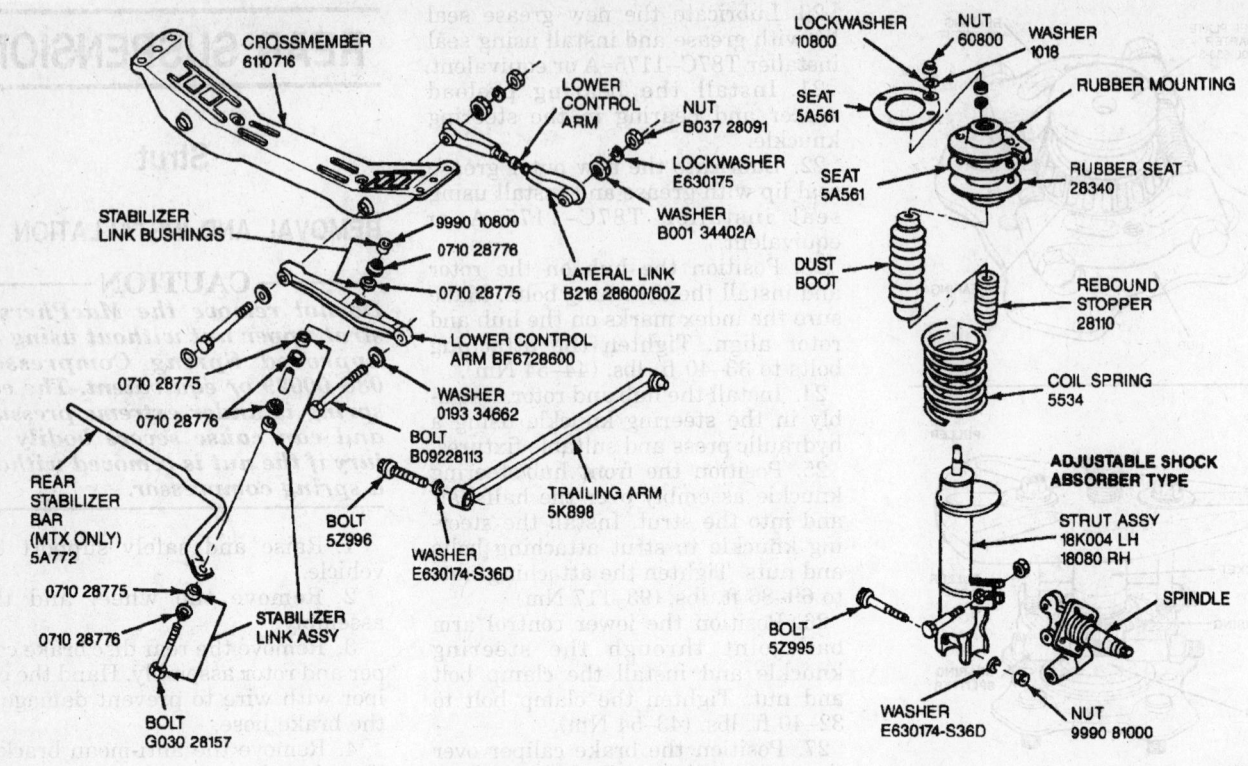

Exploded view of the rear suspension assembly

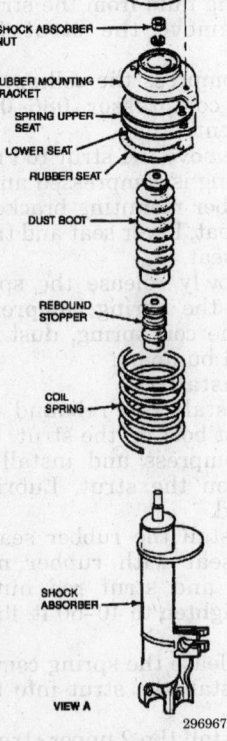

Strut, coil spring and related components

18. Install the trailing arm and spindle to strut mounting bolts. Do not tighten the bolts at this time.

19. Install the anti-moan bracket, if equipped.

20. Install the disc brake caliper and rotor assembly.

21. Install the wheel and tire assembly. Tighten the lug nuts to 67–88 ft. lbs. (90–120 Nm).

22. Lower the vehicle.

23. With the suspension loaded to normal ride height, tighten the trailing arm and spindle to strut mounting bolts to 69–86 ft. lbs. (93–117 Nm).

24. Check for proper operation.

Lower Control Arms

REMOVAL AND INSTALLATION

1. Raise and safely support the vehicle.

2. Remove the wheel and tire assembly.

3. Remove the rear disc brake caliper and rotor assembly.

4. Paint an aligning mark on each control arm and control arm bushing. Paint an aligning mark on each side of the trailing arm and crossmember.

5. Remove the stabilizer link assembly. Loosen and remove the stabilizer bar, bushings and the stabilizer.

6. Loosen both inner and outer lower control arm bolts. Loosen the spindle to strut attaching bolts.

7. Remove the parking brake attaching bolt from the rear trailing assembly.

8. Loosen the trailing arm to strut attaching bolts.

9. When all control arm and trailing arm bolts are loosened, remove all the bolts and remove both the control arms and the trailing arm.

To install:

10. Mount the control arm and trailing arm on the rear crossmember and hand-tighten the bolts. Make sure the left and right arms are in the correct position.

11. Connect both control arms with the outer control arm bolt but do not install the spindle at this time. Raise both control arms so the painted stripes align and tighten the rear control arm bolts.

12. Install the spindle in the strut. Tighten the spindle to strut attaching bolts to 69–86 ft. lbs. (93–117 Nm).

13. Install and tighten the control arm to spindle attaching bolt to 45–55 ft. lbs. (60–75 Nm).

14. Tighten the inner control arm bolt to 69–86 ft. lbs. (93–117 Nm).

15. Loosely install the rear stabilizer bar in the stabilizer bushing. Make sure the alignment stripe painted on the stabilizer bar aligns with the bushings. Do not fully tighten the bracket bolts at this time.

16. Install the stabilizer link assembly. Tighten the stabilizer bushing bracket bolts to 32–39 ft. lbs. (45–55 Nm). Tighten the stabilizer link bolt until 0.310 in. (18mm) of thread extends beyond the nut. Final tightening must be done with the suspension loaded.

17. Install the rear brake assembly.

18. Install the wheel and tire assembly.

19. Lower the vehicle.

20. Check the rear alignment if required.

21. Road test the vehicle to check for proper operation.

Sway Bar

REMOVAL AND INSTALLATION

1. Raise and safely support the vehicle.

2. Remove the wheel and tire assembly.

3. Paint an aligning mark at each sway bar bushing to aid in bushing placement during installation.

4. Loosen and remove the sway bar bracket retaining bolts, brackets and bushings.

5. Loosen and remove the sway bar link assemblies at the lower control arms.

6. Remove the sway bar.

7. Inspect all bushings and replace as required.

To Install:

8. Install the sway bar bushings, if removed.

9. Make sure the alignment stripe painted on the sway bar aligns with the bushings.

10. Install the sway bar assembly.

11. Install the sway bar brackets and retaining bolts. Tighten the bolts to 32–39 ft. lbs. (45–55 Nm).

12. Install the sway bar link assemblies to the lower control arms. Tighten the sway bar link bolt until 0.709 inch (18mm) of thread extends beyond the nut. Final tightening must be done with the suspension loaded.

13. Install the wheel and tire assembly.

14. Lower the vehicle.

15. Check for proper operation.

Wheel Bearings

ADJUSTMENT

NOTE: The wheel bearing locknut on the right side has a left hand thread. The locknuts are staked into position and must be replaced with new nuts when removed.

1. Make sure the parking brake is fully released.

2. Raise and safely support the vehicle.

3. Remove the wheel and tire assembly.

4. Remove the grease cap.

5. Rotate the brake rotor to make sure there is no brake drag. If the brakes drag, press on the inner brake pad to push the caliper piston back slightly. Bearing preload cannot be adjusted if the brake drags.

6. With a small cape chisel, carefully raise the staked portion of the locknut.

7. Remove the locknut and discard it.

8. To seat the bearings, torque the locknut to 18–22 ft. lbs. (25–29 Nm). Rotate the brake rotor by hand while tightening the locknut.

9. Loosen the locknut until it can be turned by hand.

10. Before the bearing preload can be measured, the seal drag must be measured:

   a. Install a lug nut onto the wheel stud and rotate the brake rotor until the stud is in the 12 o'clock position.

   b. Fit an inch pound torque wrench onto the nut to measure the amount of force required to rotate the brake rotor.

   c. Pull the torque wrench and record the torque reading when rotation just begins.

   d. Bearing preload without seal drag is 1.3–4.3 inch lbs. (0.15–0.49 Nm). To calculate the final bearing preload, add these figures to the seal drag measurement.

11. After the preload range is determined, tighten the locknut slightly.

12. Rotate the brake rotor until the nut and wheel are returned to the 12 o'clock position. Position the inch lb. torque wrench onto the nut and measure the amount of pull required to rotate the brake rotor. Tighten the locknut until the torque shown on the torque wrench is within the range that was calculated.

13. Using a cold chisel with the cutting edge rounded, stake the locknut in place.

NOTE: If the nut splits or cracks after staking, it must be replaced with a new nut.

14. Install the grease cap.

15. Install the wheel and tire assembly.

16. Lower the vehicle.

REMOVAL AND INSTALLATION

1. Raise and safely support the vehicle.

2. Remove the wheel and tire assembly.

3. Remove the 2 guide pin bolts from the caliper and lift the caliper clear of the disc with the inner cable and flexible hose attached. Tie the caliper to the strut spring to prevent damage to the brake hose.

4. Remove the grease cap.

5. Carefully raise the staked portion of the locknut using a small cape chisel. Remove and discard the locknut.

NOTE: The locknuts are threaded left and right. The left hand threaded locknut is on the right side of the vehicle. Turn this locknut clockwise to loosen. The right hand threaded locknut is turned counterclockwise to loosen.

6. Remove the washer and outer bearing from the bearing hub. Remove the brake rotor/bearing hub assembly.

7. Remove the bearing grease seal using a small prybar. Discard the seal.

8. Remove the inner bearing from the hub.

NOTE: If the bearings are to be reused, they should be tagged so they can be installed in their original positions.

9. If the bearings are to be replaced, remove the bearing races using a brass drift.

10. Clean the bearings with solvent and wipe out all the old grease from inside the hub.

To install:

11. If the bearing races were removed, install new ones using a brass drift.

12. Pack the bearings and the hub area with high temperature wheel bearing grease. Position the inner bearing in the hub.

13. Lubricate the lip of a new grease seal. Install the seal using

Wheel Bearings

ADJUSTMENT

DISC BRAKE

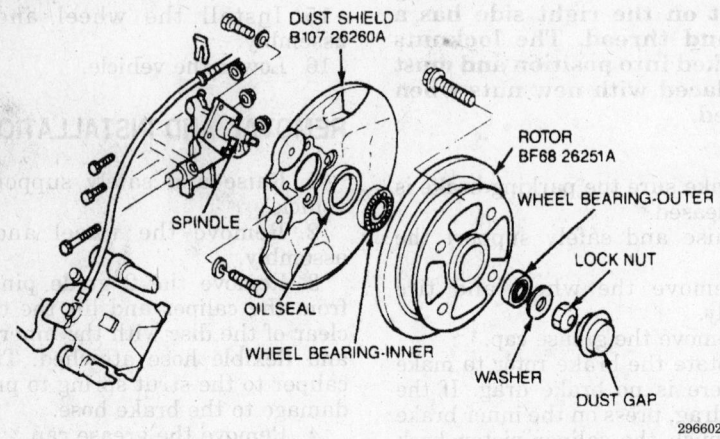

Exploded view of rear wheel bearing components

Labels: DUST SHIELD B107 26260A, SPINDLE, OIL SEAL, WHEEL BEARING-INNER, ROTOR BF68 26251A, WHEEL BEARING-OUTER, LOCK NUT, WASHER, DUST CAP, 296602

NOTE: The wheel bearing locknut on the right side has a left hand thread. The locknut must be staked into position and final be replaced with a new nut when removed.

1. Make sure the caliper is fully released.
2. Raise and safely support the vehicle.
3. Remove the wheel and tire assemble.
4. Remove the caliper.
5. Rotate the rotor to make sure there is no drag. If the brakes are dragging, the piston may need to be pressed into the caliper bore slightly. Pressing the piston back slightly, freeing the rotor, is not needed if the brake drags.
6. With a small cape chisel, carefully raise the staked portion of the locknut.
7. Remove the locknut and discard.

seal installer T87C-1175-A or equivalent.

14. Position the brake rotor/bearing hub assembly on the spindle.

NOTE: Keep the hub centered on the spindle to prevent damage to the grease seal and spindle threads.

15. Install the outer bearing, washer and a new locknut.

16. To seat the bearings, torque the locknut to 18–22 ft. lbs. (25–29 Nm). Rotate the brake rotor by hand while tightening the locknut.
17. Loosen the locknut until it can be turned by hand.
18. Before the bearing preload can be adjusted, the seal drag must be measured:

    a. Install a lug nut onto the wheel stud and rotate the brake ro-

tor until the stud is in the 12 o'clock position.

    b. Fit an inch pound torque wrench onto the nut to measure the amount of force required to rotate the brake rotor.

    c. Pull the torque wrench and record the torque reading when rotation just begins.

    d. Bearing preload without seal drag is 1.3–4.3 inch lbs. (0.15–0.49 Nm). To calculate the final bearing preload, add these figures to the seal drag measurement.

19. After the preload range is determined, tighten the locknut slightly.
20. Rotate the brake rotor until the nut and wheel are returned to the 12 o'clock position. Position the inch lb. torque wrench onto the nut and measure the amount of pull required to rotate the brake rotor. Tighten the locknut until the torque shown on the torque wrench is within the range that was calculated.
21. Using a cold chisel with the cutting edge rounded, stake the locknut in place.

NOTE: If the nut splits or cracks after staking, it must be replaced with a new nut.

22. Install the grease cap
23. Install the disc brake caliper assembly.
24. Install the wheel and tire assembly.
25. Lower the vehicle.
26. Pump the brake pedal several times to position the brake pads before attempting to move the vehicle.
27. Road test the vehicle and check for proper operation.

FORD MOTOR CO.

Front Wheel Drive

FORD-Contour **MERCURY**-Mystique

FIRING ORDERS

NOTE: To avoid confusion, always replace spark plug wires one at a time.

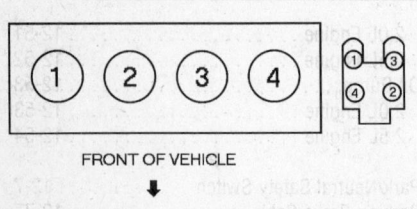

FRONT OF VEHICLE

⬇

302855

2.0L (VIN 3) Engine
Firing Order: 1-3-4-2
Distributorless Ignition System

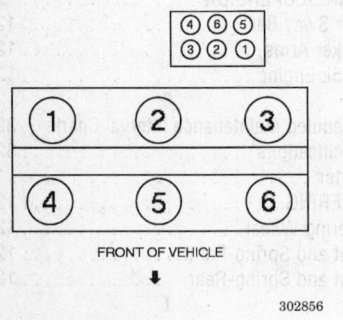

FRONT OF VEHICLE

⬇

302856

2.5L (VIN L) Engine
Firing Order: 1-4-2-5-3-6
Distributorless Ignition System

ENGINE ELECTRICAL

NOTE: Disconnecting the negative battery cable on some vehicles may interfere with the functions of the on board computer systems and may require the computer to undergo a relearning process, once the negative battery cable is reconnected.

Ignition Timing

ADJUSTMENT

The ignition timing is set at 10 degrees Before Top Dead Center (BTDC) and is not adjustable.

Alternator

PRECAUTIONS

Several precautions must be observed with alternator equipped vehicles to avoid damage to the unit.

• If the battery is removed for any reason, make sure it is reconnected with the correct polarity. Reversing the battery connections may result in damage to the 1-way rectifiers.

• When utilizing a booster battery as a starting aid, always connect the positive to positive terminals and the negative terminal from the booster battery to a good engine ground on the vehicle being started.

• Never use a fast charger as a booster to start vehicles.

• Disconnect the battery cables when charging the battery with a fast charger.

• Never attempt to polarize the alternator.

• Do not use test lights of more than 12 volts when checking diode continuity.

• Do not short across or ground any of the alternator terminals.

• The polarity of the battery, alternator and regulator must be matched and considered before making any electrical connections within the system.

• Never separate the alternator on an open circuit. Make sure all connections within the circuit are clean and tight.

• Disconnect the battery ground terminal when performing any service on electrical components.

• Disconnect the battery if arc welding is to be done on the vehicle.

REMOVAL AND INSTALLATION

2.0L (VIN 3) Engine

1. Disconnect the negative battery cable.
2. Remove the air intake resonator, mass air flow sensor and the air cleaner outlet tube assembly.
3. Disconnect the wiring harness to the alternator (generator) and regulator.
4. Raise and safely support the vehicle.
5. Relax the tension on the accessory drive belt and remove it from the alternator pulley.
6. Remove the alternator mounting bracket retaining bolts and the alternator mounting bracket.
7. Lower the vehicle.

8. Remove the power steering pressure hose bracket retaining nut and bolt.
9. Remove the power steering pressure hose bracket from the engine lifting eye.
10. Remove the alternator mounting bolts and remove the alternator from the vehicle.
   To install:
11. Place the alternator on the engine alternator bracket.
12. Install the alternator mounting bolts, finger tight.
13. Install the power steering pressure hose bracket to the engine lifting eye.
14. Install the power steering pressure hose bracket retaining nut and bolt.
15. Raise and safely support the vehicle.
16. Install the alternator mounting bracket and retaining bolts. Torque the retaining bolts to 15–22 ft. lbs. (20–30 Nm).
17. Install the accessory drive belt over the alternator pulley.
18. Lower the vehicle.
19. Torque the alternator mounting bolts to 15–22 ft. lbs. (20–30 Nm).
20. Connect the wiring harness to the alternator and regulator. Tighten the output terminal nut to 80–97 inch lbs. (9–11 Nm).
21. Install the mass air flow sensor, air cleaner outlet tube and the air intake resonator.
22. Connect the negative battery cable.
23. Check the charging system for proper operation.

2.5L (VIN L) Engine

1. Disconnect the negative battery cable.
2. Remove the accessory drive belt from the alternator (generator) pulley.
3. Raise and safely support the vehicle.
4. Remove the right front wheel and tire assembly.
5. Remove the right tie rod end from the steering knuckle. Discard the cotter pin.
6. Remove the exhaust system Y-pipe.
7. Disconnect the wiring harness to the alternator.
8. Remove the alternator brace retaining bolts and the brace from the alternator.
9. Remove the right halfshaft.
10. Remove the alternator retaining bolts from the alternator mounting bracket.

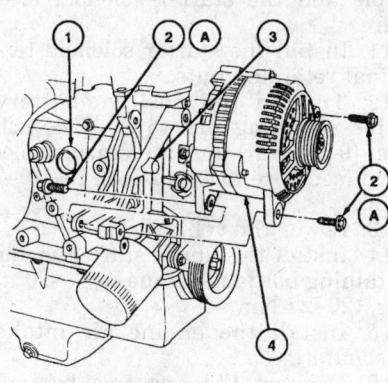

1. Cylinder block
2. Bolt
3. Generator mounting bracket
4. Generator
A. 15–22 ft. lb.(20–30 Nm)

298261

Alternator and mounting components — 2.0L (VIN 3) engine

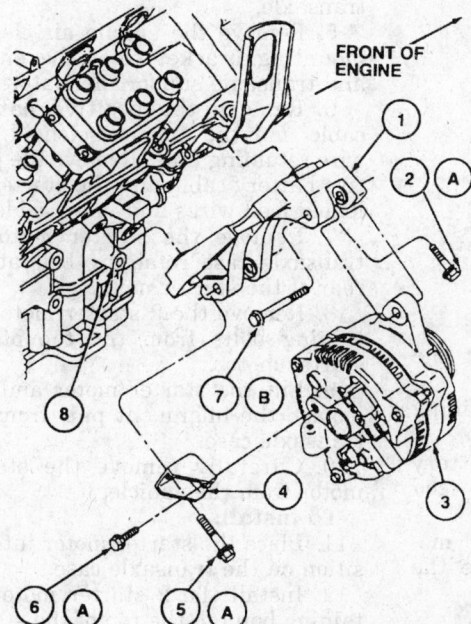

FRONT OF ENGINE

1. Generator mounting bracket bolt(2)
2. Bolt(2)
3. Generator
4. Generator brace
5. Bolt
6. Bolt
7. Bolt(2)
8. Cylinder block
A. 15–22 ft. lb.(20–30 Nm)
B. 29–40 ft. lb.(40–55 Nm)

NOTE: DURING INSTALLATION, TIGHTEN
GENERATOR BRACE RETAINING BOLT
TO GENERATOR PRIOR TO TIGHTENING
GENERATOR BRACE RETAINING BOLT
TO MOUNTING BRACKET

298268

Alternator and mounting components — 2.5L (VIN L) engine

11. Rotate the alternator and remove through the right side of the vehicle.

To install:

12. Position the alternator onto the alternator mounting bracket.

13. Install the alternator mounting bolts and torque to 29–40 ft. lbs. (40–55 Nm).

14. Install the right halfshaft.

15. Install the alternator brace and retaining bolts. Torque the bolts to 15–22 ft. lbs. (20–30 Nm).

16. Connect the wiring harness to the alternator. Tighten the output terminal nut to 80–97 inch lbs. (9–11 Nm).

17. Install the exhaust system Y-pipe.

18. Install the right outer tie rod end to the steering knuckle. Install a new cotter pin.

19. Install the wheel and tire assembly. Torque the lug nuts to 62 ft. lbs. (85 Nm).

20. Lower the vehicle.

21. Install the accessory drive belt onto the alternator pulley.

22. Connect the negative battery cable.

23. Check for proper charging system operation.

Drive Belt

REMOVAL AND INSTALLATION

2.0L (VIN 3) Engine

1. Disconnect the negative battery cable.

2. Raise and safely support the vehicle.

3. Place a 13mm wrench onto the pulley mounting bolt of the accessory drive belt tensioner.

4. Rotate the tensioner away from the accessory drive belt.

5. Lift the belt over the pulley and remove.

To install:

6. Place the accessory drive belt onto the tensioner pulley and all other pulleys but one.

7. Make sure that the belt is properly installed on each pulley.

8. Using the 13mm wrench, rotate the accessory drive belt tensioner away from the drive belt.

9. Place the drive belt onto the last pulley and release the tensioner.

10. Remove the 13mm wrench and visually check that the V-grooves in the drive belt are matched to the grooves in each pulley.

11. Lower the vehicle.

12. Connect the negative battery cable.

13. Run the engine and check for proper belt operation.

2.5L (VIN L) Engine

1. Disconnect the negative battery cable.

2. Place a ³⁄₈ inch breaker bar into the square hole in the drive belt tensioner arm.

3. Rotate the accessory drive belt tensioner away from the accessory drive belt.

4. Lift the belt over the pulley and remove.

5. If the water pump drive belt is to be replaced, remove the water pump pulley shield.

6. Rotate the water pump drive belt tensioner clockwise by hand and remove the belt.

To install:

7. If the water pump drive belt was removed, rotate the water pump drive belt tensioner clockwise by hand and install the water pump drive belt onto the pulleys.

8. Check that the V-grooves in the water pump belt are matched to the grooves in each pulley.

9. Install the water pump pulley shield.

10. Position the accessory drive belt under the tensioner pulley into the

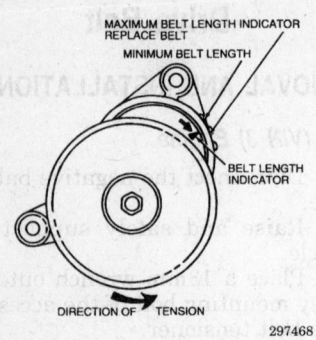

Drive belt tensioner showing belt length indicator — 2.0L (VIN 3) engine

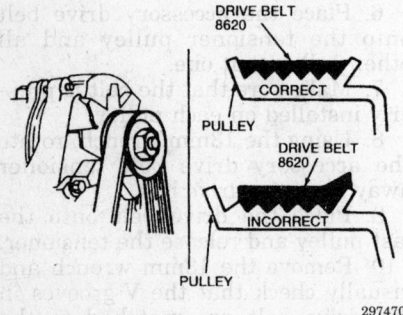

View of correct and incorrect drive belt positioning

recess in the engine front cover, and up onto the tensioner pulley. Place the belt onto all the other pulleys but one.

11. Make sure that the belt is properly installed on each pulley.

12. Insert the ³⁄₈ inch breaker bar into the accessory drive belt tensioner and rotate the tensioner clockwise.

13. Place the belt onto the last pulley and release the tensioner.

14. Check that the V-grooves in the drive belt are matched to the grooves in each pulley.

15. Connect the negative battery cable.

16. Run the engine and check for proper belt operation.

Starter

REMOVAL AND INSTALLATION

2.0L (VIN 3) Engine

1. Disconnect the negative battery cable.

2. Remove the engine air intake resonator.

3. Remove the upper starter retaining bolt.

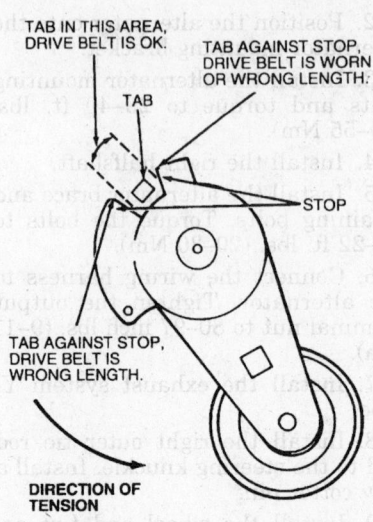

Accessory drive belt tensioner showing the belt wear indicator — 2.5L (VIN L) engine

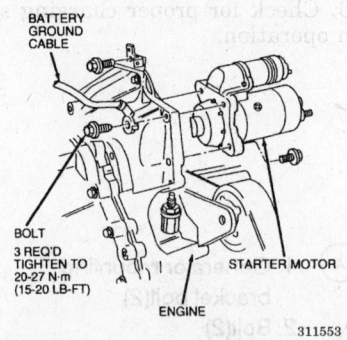

Starter motor and related components — 2.0L (VIN 3) engine

4. Raise and safely support the vehicle.

5. Remove the terminal retaining nuts from the starter solenoid.

6. Remove the positive battery cable and the starter solenoid feed wire.

7. Remove the 2 lower starter motor retaining bolts and remove the starter.

To install:

8. Place the starter in position and install the 2 lower retaining bolts.

NOTE: Be sure to install the ground cable.

9. Torque the lower retaining bolts to 15–20 ft. lbs. (20–27 Nm).

10. Install the positive battery cable and the starter solenoid feed wire.

11. Install the starter solenoid terminal retaining nuts.

12. Tighten the positive battery cable retaining nut to 80–124 inch lbs. (9–14 Nm) and the starter solenoid feed wire to 44–62 inch lbs. (5–7 Nm).

13. Lower the vehicle.

14. Install the upper starter motor retaining bolt and torque to 15–20 ft. lbs. (20–27 Nm).

15. Install the engine air intake resonator.

16. Connect the negative battery cable.

17. Check for proper starter motor operation.

2.5L (VIN L) Engine

1. Disconnect the negative battery cable.

2. Remove the engine air cleaner.

3. Remove the fuel supply and return lines from the fuel line support bracket located on the accelerator cable bracket and remove the fuel line support bracket.

4. If equipped with automatic transaxle, remove the gear selector and gear shift cable bracket from the transaxle.

5. Remove the engine air cleaner mounting bracket from the engine and transaxle support insulator.

6. Remove the positive battery cable and the starter solenoid feed wire retaining nuts. Remove the positive battery cable and the starter solenoid feed wires and move aside.

7. Remove the starter motor to transaxle case retaining bolt at the rear of the starter motor.

8. Remove the 2 starter motor retaining bolts from the top of the starter motor.

9. Lift the starter motor and disengage the alignment pins from the transaxle case.

10. Carefully remove the starter motor from the vehicle.

To install:

11. Place the starter motor into position on the transaxle case.

12. Install the 2 starter motor retaining bolts to the top of the starter motor, finger tight.

13. Install the starter motor to transaxle case retaining bolt at the rear of the starter motor, finger tight.

14. Torque all 3 starter motor retaining bolts to 15–21 ft. lbs. (21–29 Nm).

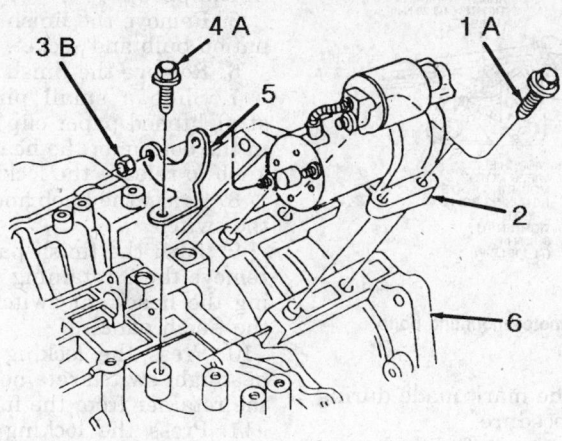

1. Bolt (2 req'd)
2. Starter motor
3. Nut (2 req'd)
4. Bolt
5. Starter bracket
6. Engine
A. Tighten to 21-29 Nm (15-21 lb. ft.)
B. Tighten to 15-20 Nm (12-14 lb. ft.)

311582

Starter motor and related components — 2.5L (VIN L) engine

15. Install the positive battery cable and the starter solenoid feed wire to their terminals. Install the positive battery cable and the starter solenoid feed wire retaining nuts.

16. Tighten the positive battery cable retaining nut to 80–124 inch lbs. (9–14 Nm).

17. Tighten the starter solenoid feed wire retaining nut to 45–62 inch lbs. (5–7 Nm).

18. Install the engine air cleaner mounting bracket to the engine and transaxle support insulator. Torque the retaining bolts to 15–22 ft. lbs. (20–30 Nm).

19. If equipped with automatic transaxle, install the gear selector cable and gear shift cable bracket to the transaxle.

20. Install the fuel line support bracket to the accelerator cable bracket.

21. Install the fuel supply and return lines to the support bracket.

22. Install the engine air cleaner.

23. Connect the negative battery cable.

24. Check for proper starter motor operation.

CHASSIS ELECTRICAL

Blower Motor

REMOVAL AND INSTALLATION

1. Disconnect the negative battery cable.

2. Working from inside the vehicle, remove the push pins and the upper footwell trim panel from the passenger side.

3. Disconnect the wiring harness connector from the blower motor.

4. Carefully lift the retaining lug on the A/C blower motor flange and rotate the blower motor counterclockwise approximately 30 degrees to disengage the blower motor from the A/C evaporator housing.

5. Pull the blower motor out of the A/C evaporator housing.

6. If the blower motor is to be replaced, remove the blower motor wheel retainer and remove the blower motor wheel.

To install:

7. If removed, install the blower motor wheel to the blower motor and secure with the retainer.

8. Install the blower motor into the A/C evaporator housing and turn the unit counterclockwise until the retaining lugs are engaged.

9. Connect the wiring harness to the blower motor.

10. Install the upper footwell trim panel.

11. Connect the negative battery cable.

12. Verify proper operation of the blower motor.

Windshield Wiper Motor

REMOVAL AND INSTALLATION

1. Disconnect the negative battery cable.

2. Raise the covers over the wiper arm retaining nuts.

3. Loosen the nut on each arm approximately 2 turns.

4. Lift each wiper arm and free it from the pivot shafts.

5. Finish removing the 2 retaining nuts and remove the wiper arms from the vehicle.

6. Remove the hood weather-strip from the cowl top extension.

7. Remove the upper fastener caps and remove the 5 screws.

8. Remove the lower screws that secure the cowl vent screens to the upper cowl panel.

9. Lift the right–hand cowl vent screen away from the upper cowl panel and remove.

10. Lift the left–hand cowl vent screen away from the upper cowl panel and remove.

11. Mark the position of the wiper motor output arm in relation to the wiper motor mounting bracket.

12. Remove the 4 bolts retaining the wiper linkage to the upper cowl panel.

13. Remove the wiper linkage/motor assembly from the upper cowl panel.

14. Disconnect the wiring harness connector at the wiper motor.

15. Remove the bolt retaining the wiper motor output arm to the wiper motor crankshaft.

16. Remove the 3 bolts retaining the wiper motor to its mounting plate.

17. Remove the wiper motor.

To install:

18. Mount the wiper motor to the mounting plate and torque the 3 bolts to 71–106 inch lbs. (8–12 Nm).

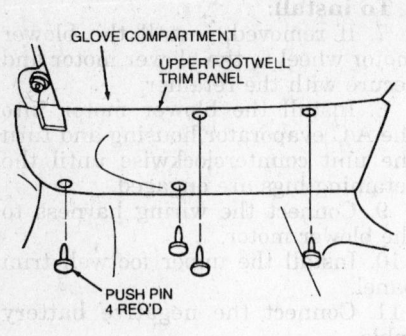

GLOVE COMPARTMENT

UPPER FOOTWELL TRIM PANEL

PUSH PIN 4 REQ'D

298272

View of the upper footwell trim panel

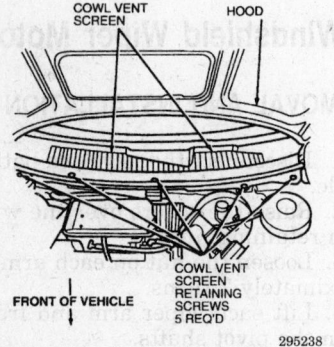

COWL VENT SCREEN

HOOD

COWL VENT SCREEN RETAINING SCREWS 5 REQ'D

FRONT OF VEHICLE

295238

Loaction of cowl vent screen retaining screws

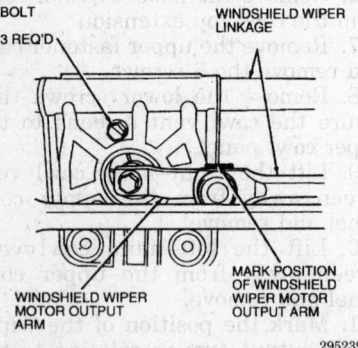

BOLT 3 REQ'D

WINDSHIELD WIPER LINKAGE

WINDSHIELD WIPER MOTOR OUTPUT ARM

MARK POSITION OF WINDSHIELD WIPER MOTOR OUTPUT ARM

295239

Marking position of wiper motor output arm

19. Position the wiper motor output arm to the wiper motor crankshaft and torque the nut to 19 ft. lbs. (26 Nm).

20. Connect the wiring harness to the wiper motor.

21. Position the wiper linkage/motor assembly to the upper cowl panel. Install the 4 retaining bolts and torque to 71 inch lbs. (8 Nm).

22. Check that the position of the wiper motor output arm is indexed

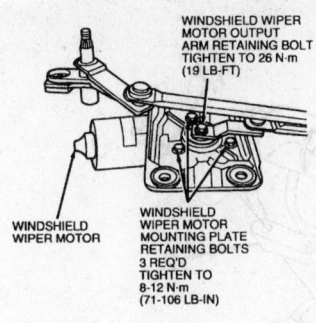

WINDSHIELD WIPER MOTOR OUTPUT ARM RETAINING BOLT TIGHTEN TO 26 N·m (19 LB-FT)

WINDSHIELD WIPER MOTOR

WINDSHIELD WIPER MOTOR MOUNTING PLATE RETAINING BOLTS 3 REQ'D TIGHTEN TO 8-12 N·m (71-106 LB-IN)

295241

View of wiper motor mounting bolts and output arm

properly with the mark made during the removal procedure.

23. Reinstall the left–hand cowl vent screen.

24. Reinstall the right–hand cowl vent screen.

25. Secure the 5 screws securing the cowl vent screens and reinstall the upper fastener caps.

26. Reinstall the lower screws retaining the cowl vent screens to the upper cowl panel.

27. Reinstall the hood weatherstrip to the cowl top extension.

28. Make sure that the wiper motor is in the park position before installing the wiper arms.

29. Install the wiper arms and retaining nuts. Only snug the nuts at this time.

30. Connect the negative battery cable.

31. Run the wiper motor at low speed and check the wiper arms for proper positioning.

32. If the wiper arm positioning is correct, continue to tighten the wiper arm retaining nuts to 18 ft. lbs. (25 Nm). If not, re–index the wiper arms, check their positioning and then tighten the retaining nuts.

33. Fasten the retaining nut covers.

34. Check the wiper motor for proper operation at all speeds.

Headlight Switch

REMOVAL AND INSTALLATION

1. Disconnect the negative battery cable.

2. Remove the 2 screws retaining the left–hand finish panel to the instrument panel.

3. Pull straight out on the finish panel to release the retaining clips.

4. Disconnect the wiring harness connectors to the headlight switch

and outside rear view mirror control (if equipped).

5. Remove the finish panel illumination bulb and socket.

6. Remove the finish panel.

7. Slide a small pin such as a straightened paper clip into the hole in the bottom of the headlight switch knob to release the locking tab.

8. Grasp the knob and pull it off of the switch.

9. Turn the finish panel over and remove the 3 retaining screws securing the headlight switch retainer to the finish panel.

10. Press the locking tabs on the headlight switch retainer and remove the retainer from the finish panel.

11. Press the locking tabs on the headlight switch and remove the switch from the headlight switch retainer.

To Install:

12. Secure the headlight switch to the headlight switch retainer.

13. Secure the headlight switch retainer to the finish panel.

14. Install the 3 retaining screws securing the headlight switch retainer to the finish panel.

15. Turn the finish panel over. Index the switch knob and push it onto the headlight switch.

16. Install the finish panel illumination bulb and socket.

17. Connect the wiring harnesses for the headlight switch and the outside rear view mirror control (if equipped).

18. Push the finish panel back onto the instrument panel, making sure that the retaining clips are properly aligned.

19. Install the 2 screws retaining the finish panel.

20. Connect the negative battery cable.

21. Check the headlight switch for proper operation.

Combination Switch

REMOVAL AND INSTALLATION

1. Disconnect the negative battery cable.

2. Remove the upper and lower steering column shrouds.

3. Disconnect the wiring harness connector from the combination switch.

4. Release the locking tab securing the combination switch to the steering column.

5. Slide the switch from the steering column.

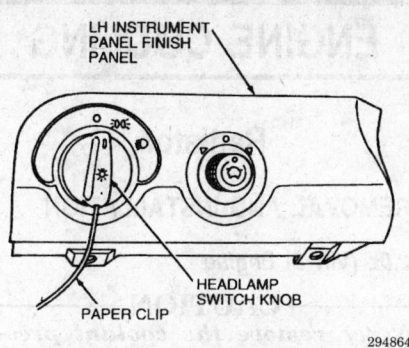

Removing the headlight switch knob

294864

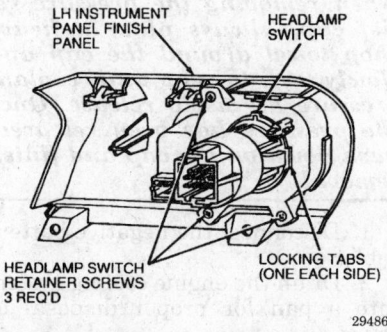

Locations of headlight switch retaining screws and tabs

294865

To Install:

6. If the combination switch is being replaced, transfer the indicator flasher from the old switch.

7. Slide the switch into position and make sure the locking tab is fully seated.

8. Connect the wiring harness connector to the combination switch.

9. Install the upper and lower steering column shrouds.

10. Reconnect the negative battery cable.

11. Check the combination switch for proper operation.

Ignition Lock Cylinder

REMOVAL AND INSTALLATION

Functional lock cylinder

1. Disconnect the negative battery cable.

2. Remove the upper and lower steering column shrouds.

3. Turn the ignition lock to the accessory position.

4. Insert a 0.125 (3.17mm) wire pin or small drift punch in the hole at the top of the cylinder housing and depress the retaining pin while pulling out the lock cylinder.

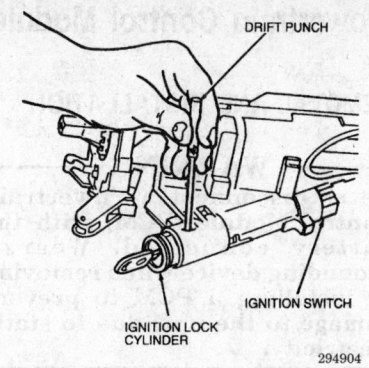

Releasing the lock cylinder retaining pin

294904

To Install:

5. Turn the new lock cylinder to accessory position and while pressing the retaining pin in, insert the lock cylinder into the housing.

6. Turn the key to the **OFF** position the release the retaining pin.

7. Try the lock cylinder operation in all positions.

8. Install the shrouds.

9. Connect the negative battery cable.

Non-Functional lock cylinder

1. Disconnect the negative battery cable.

2. Remove the upper and lower steering column shrouds.

3. Insert a 0.125 (3.17mm) drill bit in the hole at the top of the cylinder housing and drill out the retaining pin.

NOTE: Be careful not to drill into the lock cylinder housing.

4. Pull out the lock cylinder.

5. Clean out all the metal shavings and check the tube for damage. If damaged, the housing must be replaced.

To Install:

6. Turn the new lock cylinder to accessory position and while pressing the retaining pin in, insert the lock cylinder into the housing.

7. Turn the key to the **OFF** position the release the retaining pin.

8. Try the lock cylinder operation in all positions.

9. Install the shrouds.

10. Connect the negative battery cable.

Ignition Switch

REMOVAL AND INSTALLATION

1. Disconnect the negative battery cable.

2. Remove the 2 upper steering column shroud screws and remove the shroud.

NOTE: It is not necessary to remove the steering wheel for this procedure.

3. Remove the 3 lower steering column shroud retaining screws and remove the shroud.

4. Disconnect the electrical harness connector at the ignition switch.

5. Depress the locking tabs and remove the ignition switch.

To install:

6. Place the new ignition switch into position and push into place making sure that the switch is held securely.

7. Connect the electrical harness connector to the ignition switch.

8. Install the lower steering column shroud and its retaining screws.

9. Install the upper steering column shroud and its retaining screws.

10. Connect the negative battery cable.

11. Check for proper ignition switch operation.

Park/Neutral Safety Switch

REMOVAL AND INSTALLATION

NOTE: The manual lever position sensor (transmission range sensor) is only used on vehicles equipped with an automatic transaxle.

1. Disconnect the negative battery cable.

2. Place the transmission manual control lever in the **NEUTRAL** position.

3. Remove the engines air intake resonators and Mass Air Flow (MAF) sensor.

4. Disconnect the electrical harness connector from the manual lever position sensor, which is located on top of the automatic transaxle.

5. Remove the 2 retaining bolts and remove the manual lever position sensor.

To install:

6. Make sure that the transmission manual control lever is in the **NEUTRAL** position.

7. Install the manual lever position sensor and loosely install the 2 retaining bolts.

8. Align the sensor slots using Transmission Range Sensor Tool T94P–70010–AH, or equivalent.

9. Tighten the retaining bolts to 7–9 ft. lbs. (9–12 Nm).

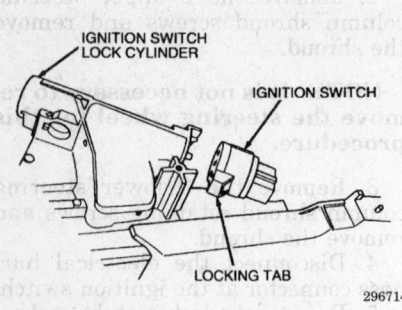

View of ignition switch and locking tab

296714

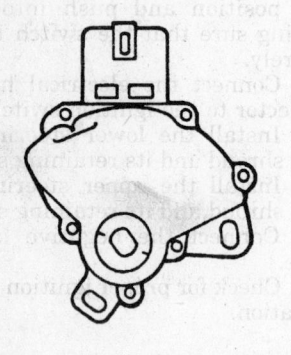

295777

View of the manual lever position sensor

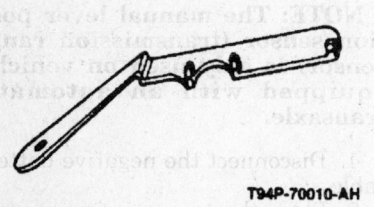

T94P-70010-AH

295778

Sensor alignment tool

10. Install the electrical harness connector to the sensor.

11. Install the MAF sensor and the air intake resonator.

12. Connect the negative battery cable.

13. Engage the parking brake and check for proper operation and adjustment of the manual lever position sensor as follows:

a. Check that the engine will only crank in the **PARK** or **NEUTRAL** positions.

b. Check that the back–up lamps only work in the **REVERSE** position.

Powertrain Control Module

REMOVAL AND INSTALLATION

— WARNING —

Never disconnect a Powertrain Control Module (PCM) with the battery connected. Wear a grounding device when removing or installing a PCM to prevent damage to the unit due to static electricity.

1. Disconnect the negative battery cable.

2. From the engine compartment, move the power steering oil reservoir to gain access to the PCM harness retainer bolt.

3. Loosen the harness retainer bolt and carefully remove the engine control sensor wiring harness connector from the PCM.

4. From inside the vehicle, loosen the PCM bracket support nut located on the right-hand lower side of the dash panel.

5. Remove the PCM bracket support.

6. Remove the PCM.

To Install:

7. Place the PCM into position.

8. Install the support bracket and nut.

9. Torque the nut to 24–32 inch lbs. (2.7–3.7 Nm).

10. From the engine compartment, connect the engine control sensor wiring harness connector to the PCM.

11. Torque the retaining bolt to 32 inch lbs. (3.7 Nm).

12. Reposition the power steering oil reservoir.

13. Connect the negative battery cable.

14. Run the engine and check for proper operation.

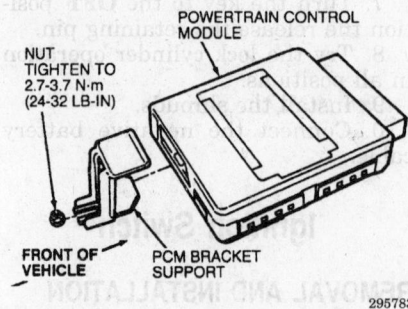

295783

Powertrain control module and bracket

ENGINE COOLING

Radiator

REMOVAL AND INSTALLATION

2.0L (VIN 3) Engine

— CAUTION —

Never remove the coolant pressure relief cap while the engine is running. Allow a hot engine to cool down before cap removal. When removing the pressure relief cap, always place a heavy shop towel around the cap and slowly turn the cap until coolant pressure begins to release. Once the pressure has been released, push down on the cap and finish removal.

1. Disconnect the negative battery cable.

2. Drain the engine cooling system into a pan for proper disposal or reuse.

3. Remove the upper radiator hose.

4. If equipped with automatic transaxle, remove the transaxle oil cooler tube from the oil cooler inlet fitting.

5. Remove the nuts securing the fan shroud to the radiator.

6. Raise and safely support the vehicle.

7. If equipped with automatic transaxle, loosen the transaxle oil cooler tubes while holding the radiator connector with a back-up wrench and remove the tubes. This will prevent damage to the radiator.

8. Remove the lower radiator hose.

9. Support the fan shroud, radiator and A/C condenser core with a suitable jack stand.

10. Remove the lower radiator supports from the subframe.

11. Remove the 2 retaining bolts securing the A/C condenser core to the brackets at the bottom of the radiator.

12. Move the jack stand aside while grasping the radiator.

13. Carefully lower the radiator and remove from the vehicle.

To install:

14. If the radiator is being replaced, remove the necessary parts (transaxle oil cooler fittings, draincock, etc.) from the old unit and install on the new one.

15. Carefully raise the radiator up into position.

16. Rest the radiator, A/C condenser core and fan shroud on a jack stand.

17. Position the A/C condenser core to the radiator and secure with 2 bolts to the brackets at the bottom of the radiator.

18. Install the lower radiator supports to the subframe.

19. Remove the jack stand.

20. Install the transaxle oil cooler tubes to the radiator. Tighten the tubes while holding the radiator connector with a back-up wrench to prevent damage to the radiator.

21. Install the lower radiator hose.

22. Lower the vehicle.

23. Install the nuts retaining the fan shroud to the radiator.

24. Connect the transaxle oil cooler tube to the oil cooler inlet fitting.

25. Install the upper radiator hose.

26. Refill the engine cooling system.

27. Connect the negative battery cable.

28. Run the engine and top off the coolant, as necessary. If equipped, check the automatic transaxle fluid level.

2.5L (VIN L) Engine

-------- **CAUTION** --------
Never remove the coolant pressure relief cap while the engine is running. Allow a hot engine to cool down before cap removal. When removing the pressure relief cap, always place a heavy shop towel around the cap and slowly turn the cap until coolant pressure begins to release. Once the pressure has been released, push down on the cap and finish removal.

1. Disconnect the negative battery cable.

2. Drain the engine cooling system into a pan for proper disposal or reuse.

3. Remove the upper radiator hose.

4. Remove the radiator overflow hose from the radiator.

5. If equipped with automatic transaxle, remove the transaxle oil cooler tube from the oil cooler inlet fitting.

6. Remove the nuts securing the fan shroud to the radiator.

7. Raise and safely support the vehicle.

8. If equipped with automatic transaxle, loosen the transaxle oil cooler tubes while holding the radiator connector with a back-up wrench and remove the tubes. This will prevent damage to the radiator.

9. Remove the lower radiator hose.

10. Support the fan shroud, radiator and A/C condenser core with a suitable jack stand.

11. Remove the lower radiator supports from the subframe.

12. Remove the 2 retaining bolts securing the A/C condenser core to the brackets at the bottom of the radiator.

13. Move the jack stand aside while grasping the radiator.

14. Carefully lower the radiator and remove from the vehicle.

To install:

15. If the radiator is being replaced, remove the necessary parts (transaxle oil cooler fittings, draincock, etc.) from the old unit and install on the new one.

16. Carefully raise the radiator up into position.

17. Rest the radiator, A/C condenser core and fan shroud on a jack stand.

18. Position the A/C condenser core to the radiator and secure with 2 bolts to the brackets at the bottom of the radiator.

19. Install the lower radiator supports to the subframe.

20. Remove the jack stand.

21. Install the transaxle oil cooler tubes to the radiator. Tighten the tubes while holding the radiator connector with a back-up wrench to prevent damage to the radiator.

22. Install the lower radiator hose.

23. Lower the vehicle.

24. Install the nuts retaining the fan shroud to the radiator.

25. Connect the transaxle oil cooler tube to the oil cooler inlet fitting.

26. Connect the radiator overflow hose to the radiator.

27. Install the upper radiator hose.

28. Refill the engine cooling system.

29. Connect the negative battery cable.

30. Run the engine and top off the coolant, as necessary. If equipped, check the automatic transaxle fluid level.

Water Pump

REMOVAL AND INSTALLATION

2.0L (VIN 3) Engine

1. Disconnect the negative battery cable.

2. Drain the engine cooling system.

3. Raise and safely support the vehicle.

4. Remove the lower radiator hose from the water pump.

5. Lower the vehicle.

6. Remove the accessory drive belt.

7. Remove the timing belt covers and the timing belt.

8. Remove the 4 water pump retaining bolts.

9. Remove the water pump.

To Install:

10. Clean the water pump and cylinder block gasket sealing surfaces as required.

11. Install a new water pump gasket and the water pump onto the cylinder block.

12. Torque the retaining bolts to 12–15 ft. lbs. (16–20 Nm).

13. Install the timing belt and the timing belt covers.

14. Install the accessory drive belt.

15. Raise and safely support the vehicle.

16. Install the lower radiator hose.

17. Lower the vehicle.

18. Fill the engine cooling system.

19. Connect the negative battery cable.

20. Run the engine and check for coolant leaks and proper operation.

2.5L (VIN L) Engine

NOTE: Before continuing with this procedure, make sure 3 new water pump retaining bolts (W701544) are available. Due to their torque-to-yield design, the bolts stretch and cannot be reused.

1. Disconnect the negative battery cable.

2. Drain the engine cooling system.

3. Remove the water pump pulley shield.

4. Remove the water pump drive belt.

5. Remove the water pump inlet and outlet hoses from the water pump.

6. Remove the 3 water pump to left-hand cylinder head retaining bolts.

7. Remove the water pump and water pump housing from the vehicle.

8. Remove the water pump to water pump housing retaining bolts and separate the water pump from the water pump housing.

To install:

9. Clean the water pump to water pump housing gasket sealing surfaces.

10. Install the water pump to the water pump housing using a new gasket and install the retaining bolts. Torque the retaining bolts to 16–18 ft. lbs. (22–25 Nm).

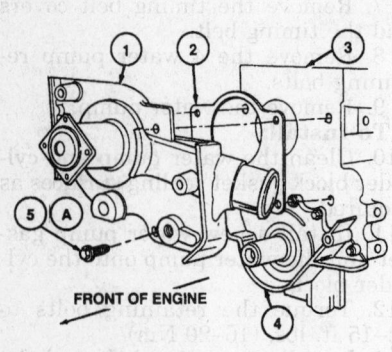

1. Water pump
2. Water pump housing gaskets
3. Cylinder block
4. Oil pump
5. Bolt(4)
A. 12-15 ft. lb.(16-20 Nm)

296174

Water pump and related components — 2.0L (VIN 3) engine

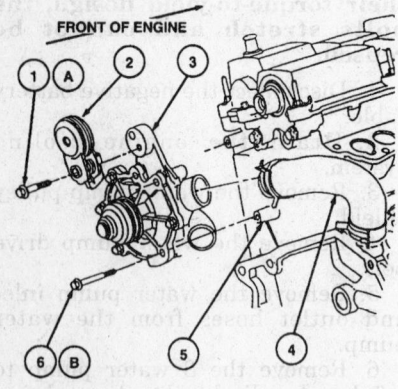

1. Bolt
2. Water pump drive belt tensioner
3. Water pump
4. Water pump outlet hose
5. LH cylinder head
6. Bolt(3)
A. 71-106 in. lb.(8-12 Nm)
B. 11-13 ft. lb.(15-18 Nm)
   then rotate 85-95°

296176

Water pump and related components — 2.5L (VIN L) engine

11. Position the water pump and water pump housing and install 3 new torque-to-yield retaining bolts into the left-hand cylinder head.

12. Torque the new retaining bolts to 11–13 ft. lbs. (15–18 Nm), then rotate the retaining bolts 85–95 degrees.

13. Install the water pump inlet and outlet hoses to the water pump.

14. Install the water pump drive belt.

15. Install the water pump shield.

16. Fill the engine cooling system.

17. Connect the negative battery cable.

18. Run the engine and check for coolant leaks and proper operation.

Thermostat

REMOVAL AND INSTALLATION

2.0L (VIN 3) Engine

------- **CAUTION** -------
Never remove the coolant pressure relief cap while the engine is running. Allow a hot engine to cool down before cap removal. When removing the pressure relief cap, always place a heavy shop towel around the cap and slowly turn the cap until coolant pressure begins to release. Once the pressure has been released, push down on the cap and finish removal.

1. Disconnect the negative battery cable.

2. Drain the engine cooling system to a level below the thermostat housing.

3. Remove the engine air intake resonators to gain access to the water hose connection.

4. Disconnect the upper radiator hose and the overflow hose from the water hose connection.

5. Remove the 3 retaining bolts from the water hose connection.

6. Remove the water hose connection from the water thermostat housing and separate the thermostat and O-ring seal.

7. Remove the thermostat and O-ring seal from the vehicle.

To install:

8. Clean all gasket mating surfaces.

9. Install the thermostat and O-ring seal into the water thermostat housing.

10. Install the water hose connection and install the 3 retaining bolts.

Torque the retaining bolts to 71–97 inch lbs. (8–11 Nm).

11. Install the upper radiator hose and the overflow hose to the water hose connection.

12. Install the engine air resonators.

13. Fill the engine cooling system.

14. Connect the negative battery cable.

15. Start the engine and allow it to reach normal operating temperature.

16. Check the cooling system for proper operation. Top off the cooling system as needed.

2.5L (VIN L) Engine

------- **CAUTION** -------
Never remove the coolant pressure relief cap while the engine is running. Allow a hot engine to cool down before cap removal. When removing the pressure relief cap, always place a heavy shop towel around the cap and slowly turn the cap until coolant pressure begins to release. Once the pressure has been released, push down on the cap and finish removal.

1. Disconnect the battery cables, negative cable first.

2. Drain the engine cooling system to a level below the thermostat housing.

3. Disconnect the radiator hoses from the thermostat housing assembly and remove the thermostat housing.

4. Remove the 2 thermostat housing retaining bolts.

5. Separate the front and rear thermostat housings and remove the O-ring and thermostat.

To install:

6. Clean all gasket mating surfaces.

7. Install a new thermostat and O-ring and position the front and rear thermostat housings together.

8. Install the 2 thermostat housing retaining bolts and alternately tighten the bolts to a final torque of 15–22 ft. lbs. (20–30 Nm).

9. Position the thermostat housing assembly and install the radiator hoses.

10. Install the battery and connect the battery cables, negative cable last.

11. Fill the engine cooling system.

12. Start the engine and allow it to reach normal operating temperature.

13. Check the cooling system for proper operation. Top off the cooling system as needed.

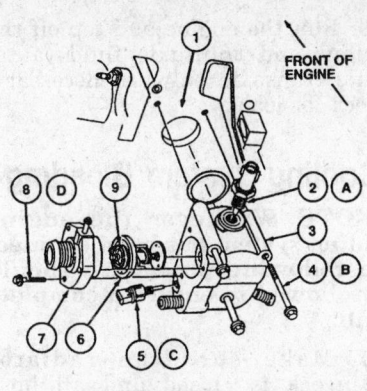

Item	Description
1	Cylinder Block
2	Engine Coolant Temperature Sensor
3	Water Thermostat Housing
4	Bolt (3 Req'd)
5	Water Temperature Indicator Sender Unit
6	O-Ring (Part of 8575)
7	Water Hose Connection
8	Bolt (3 Req'd)
9	Water Thermostat
A	Tighten to 10-14 N·m (89-124 Lb-In)
B	Tighten to 18-22 N·m (13-16 Lb-Ft)
C	Tighten to 7-10 N·m (62-89 Lb-In)
D	Tighten to 8-11 N·m (71-97 Lb-In)

295924

Thermostat and related components

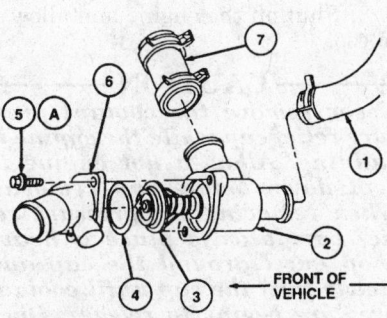

1 Water bypass hose
2 Rear water thermostat housing
3 Water thermostat
4 O-ring seal
5 Bolt (2 req'd)
6 Front water thermostat housing
7 Engine return hose
A Tighten fo 20-30 Nm (15-22 lb-ft)

307791

Thermostat and related components

Electric Cooling Fan

REMOVAL AND INSTALLATION

2.0L (VIN 3) Engine

——————— **CAUTION** ———————
Never remove the coolant pressure relief cap while the engine is running. Allow a hot engine to cool down before cap removal. When removing the pressure relief cap, always place a heavy shop towel around the cap and slowly turn the cap until coolant pressure begins to release. Once the pressure has been released, push down on the cap and finish removal.

1. Disconnect the negative battery cable.
2. Drain the engine cooling system into a pan for proper disposal or reuse.
3. Remove the wiring harness connectors and move aside.
4. Remove the cooling fan wiring ground cable from the right-hand front fender apron.
5. Remove the upper radiator hose.
6. On vehicles equipped with automatic transaxles, remove the transaxle oil cooler tube from the oil cooler inlet fitting.
7. Raise and safely support the vehicle.
8. If equipped, loosen the automatic transaxle oil cooler tubes while holding the radiator connector with a back–up wrench and remove the tubes. This will prevent damage to the radiator.
9. Remove the lower radiator hose.
10. Disconnect the A/C clutch field wiring harness connector.
11. If equipped, remove the automatic transaxle oil cooler tubes from the retaining bracket located at the front of the subframe.
12. Remove the retaining bracket from the subframe.
13. Support the fan shroud, radiator and A/C condenser core with a suitable jack stand.
14. Remove the lower radiator supports from the subframe.
15. Remove the 2 retaining bolts securing the A/C condenser core to the brackets at the bottom of the radiator.
16. Move the jack stand aside while grasping the radiator/fan shroud assembly.
17. Carefully remove the radiator/fan shroud assembly and place on a work bench.
18. Remove the fan shroud from the radiator.
19. Position the fan shroud to access the fan assembly.

NOTE: Depending on the application, some fan shrouds contain either one or two cooling fans and related brackets and wiring connectors.

20. Remove the fan blade retaining clip.
21. Remove the fan blade.
22. Disconnect the wiring harness connector from the cooling fan motor.
23. Remove the retaining screws securing the cooling fan motor.
24. Remove the cooling fan motor.
To install:
25. Position the cooling fan motor to the fan shroud.
26. Install the retaining screws and tighten.
27. Reposition the wiring harness connector to the fan shroud.
28. Install the fan blade and secure with a fan blade retaining clip.
29. Position the fan shroud to the radiator and secure.
30. Carefully raise the radiator/fan shroud assembly up into position.
31. Rest the radiator/fan shroud assembly on a jack stand.
32. Position the A/C condenser core to the radiator and secure with 2 bolts to the brackets at the bottom of the radiator.
33. Install the lower radiator supports to the subframe.
34. Remove the jack stand.
35. If equipped, install the transaxle oil cooler tubes to the bracket at the front of the subframe.
36. Install the A/C clutch field wiring harness connector.
37. Install the lower radiator hose.
38. If equipped, install the transaxle oil cooler tubes to the radiator. Tighten the tubes while holding the radiator connector with a back–up wrench to prevent damage to the radiator.
39. Lower the vehicle.
40. If equipped, connect the transaxle oil cooler tube to the oil cooler inlet fitting.
41. Install the upper radiator hose.
42. Connect the cooling fan wiring ground cable to the right–hand front fender apron.
43. Install the wiring harness connectors.
44. Refill the engine cooling system.
45. Connect the negative battery cable.
46. Run the engine and top off the coolant and transaxle fluid (automatic transaxle only) as necessary. Check for leaks.

2.5L (VIN L) Engine

— **CAUTION** —

Never remove the coolant pressure relief cap while the engine is running. Allow a hot engine to cool down before cap removal. When removing the pressure relief cap, always place a heavy shop towel around the cap and slowly turn the cap until coolant pressure begins to release. Once the pressure has been released, push down on the cap and finish removal.

1. Disconnect the negative battery cable.
2. Drain the engine cooling system into a pan for proper disposal or reuse.
3. Remove the wiring harness connector for the A/C cycling switch.
4. Remove the wiring harness connectors and move aside.
5. Remove the cooling fan wiring ground cable from the right-hand front fender apron.
6. Remove the upper radiator hose.
7. Remove the overflow hose from the radiator.
8. If equipped with automatic transaxle, remove the transaxle oil cooler tube from the oil cooler inlet fitting.
9. Raise and safely support the vehicle.
10. If equipped, loosen the automatic transaxle oil cooler tubes while holding the radiator connector with a back–up wrench and remove the tubes. This will prevent damage to the radiator.
11. Remove the lower radiator hose.
12. Disconnect the A/C clutch field wiring harness connector.
13. If equipped, remove the automatic transaxle oil cooler tubes from the retaining bracket located at the front of the subframe.
14. Remove the retaining bracket from the subframe.
15. Support the fan shroud, radiator and A/C condenser core with a suitable jack stand.
16. Remove the lower radiator supports from the subframe.
17. Remove the 2 retaining bolts securing the A/C condenser core to the brackets at the bottom of the radiator.
18. Move the jack stand aside while grasping the radiator/fan shroud assembly.
19. Carefully remove the radiator/fan shroud assembly and place on a work bench.

20. Remove the fan shroud from the radiator.
21. Position the fan shroud to access the fan assembly.

NOTE: Depending on the application, some fan shrouds contain either one or two cooling fans and related brackets and wiring connectors.

22. Remove the fan blade retaining clip.
23. Remove the fan blade.
24. Disconnect the wiring harness connector from the cooling fan motor.
25. Remove the retaining screws securing the cooling fan motor.
26. Remove the cooling fan motor.

To install:

27. Position the cooling fan motor to the fan shroud.
28. Install the retaining screws and tighten.
29. Reposition the wiring harness connector to the fan shroud.
30. Install the fan blade and secure with a fan blade retaining clip.
31. Position the fan shroud to the radiator and secure.
32. Carefully raise the radiator/fan shroud assembly up into position.
33. Rest the radiator/fan shroud assembly on a jack stand.
34. Position the A/C condenser core to the radiator and secure with 2 bolts to the brackets at the bottom of the radiator.
35. Install the lower radiator supports to the subframe.
36. Remove the jack stand.
37. If equipped, install the automatic transaxle oil cooler tubes to the bracket at the front of the subframe.
38. Install the A/C clutch field wiring harness connector.
39. Install the lower radiator hose.
40. If equipped, install the automatic transaxle oil cooler tubes to the radiator. Tighten the tubes while holding the radiator connector with a back–up wrench to prevent damage to the radiator.
41. Lower the vehicle.
42. If equipped, connect the automatic transaxle oil cooler tube to the oil cooler inlet fitting.
43. Install the upper radiator hose.
44. Install the overflow hose to the radiator.
45. Connect the cooling fan wiring ground cable to the right-hand front fender apron.
46. Install the wiring harness connector.
47. Install the wiring harness connector for the A/C cycling switch.
48. Refill the engine cooling system.
49. Connect the negative battery cable.

50. Run the engine and top off the coolant and transaxle fluid (automatic transaxle only) as necessary. Check for leaks.

Cooling System Bleeding

NOTE: Whenever the engine cooling system has been drained, the following procedure should be followed to ensure a complete refill.

1. Make sure the radiator draincock is closed and all hose clamps are tight.
2. Add a 50/50 mixture of water and cooling system fluid into the fill neck on the radiator coolant recovery reservoir until coolant reaches the **MAX** mark on the radiator coolant recovery reservoir.
3. Install the pressure relief cap to the radiator coolant recovery reservoir.
4. Select the maximum heat and blower motor settings.
5. Set the discharged air to vent through the A/C ducts in the instrument panel.
6. Start the engine and allow it to idle.
7. Feel the discharge air at the A/C vents. If the air discharged remains cool and the engine coolant temperature gauge does not rise, then the coolant level in the engine is too low.
8. Shut off the engine and allow it to cool.

— **CAUTION** —

Never remove the coolant pressure relief cap while the engine is running. Allow a hot engine to cool down before cap removal. When removing the pressure relief cap, always place a heavy shop towel around the cap and slowly turn the cap until coolant pressure begins to release. Once the pressure has been released, push down on the cap and finish removal.

9. Carefully remove the pressure relief cap from the radiator coolant recovery reservoir.
10. Add coolant to the radiator coolant recovery reservoir until it reaches the **MAX** mark.
11. Reinstall the pressure relief cap.
12. Start the engine and allow it to idle.
13. Repeat the procedure from Step 4.
14. The engine coolant temperature gauge should maintain a stabilized

reading in the middle area of the **NORMAL** range and the upper radiator hose should be hot to the touch.

15. Recheck the coolant level in the recovery reservoir and fill as necessary.

NOTE: If the engine coolant level indicator flashes, it is indicating that the coolant level is low and will require at least 1–1.5 quarts (0.946–1.416 liters) to fill the system.

FUEL SYSTEM

Fuel System Service Precautions

Safety is the most important factor when performing not only fuel system maintenance but any type of maintenance. Failure to conduct maintenance and repairs in a safe manner may result in serious personal injury or death. Maintenance and testing of the vehicle's fuel system components can be accomplished safely and effectively by adhering to the following rules and guidelines.

• To avoid the possibility of fire and personal injury, always disconnect the negative battery cable unless the repair or test procedure requires that battery voltage be applied.

• Always relieve the fuel system pressure prior to disconnecting any fuel system component (injector, fuel rail, pressure regulator, etc.), fitting or fuel line connection. Exercise extreme caution whenever relieving fuel system pressure to avoid exposing skin, face and eyes to fuel spray. Please be advised that fuel under pressure may penetrate the skin or any part of the body that it contacts.

• Always place a shop towel or cloth around the fitting or connection prior to loosening to absorb any excess fuel due to spillage. Ensure that all fuel spillage (should it occur) is quickly removed from engine surfaces. Ensure that all fuel soaked cloths or towels are deposited into a suitable waste container.

• Always keep a dry chemical (Class B) fire extinguisher near the work area.

• Do not allow fuel spray or fuel vapors to come into contact with a spark or open flame.

• Always use a backup wrench when loosening and tightening fuel line connection fittings. This will prevent unnecessary stress and torsion to fuel line piping. Always follow the proper torque specifications.

• Always replace worn fuel fitting O-rings with new. Do not substitute fuel hose or equivalent, where fuel pipe is installed.

Fuel System Pressure

RELIEVING

───── **CAUTION** ─────

Fuel injection systems remain under pressure, even after the engine has been turned OFF. The fuel system pressure must be relieved before disconnecting any fuel lines. Failure to do so may result in fire and/or personal injury.

1. Disconnect the negative battery cable.
2. Remove the engine air cleaner assembly.
3. Loosen the fuel tank filler cap to relieve pressure in the fuel tank.
4. Connect fuel pressure gauge T80L-9974-B or equivalent, to the fuel pressure relief valve located on the fuel rail.
5. Open the manual valve on the fuel pressure gauge and drain the fuel through the drain tube into a suitable container.
6. Remove the fuel pressure gauge.
7. When service on the vehicle is complete, be sure to install the engine air cleaner assembly, tighten the fuel tank filler cap and connect the negative battery cable.

Idle Speed

ADJUSTMENT

1995 Vehicles

2.0L (VIN 3) Engine

This vehicle is equipped with On Board Diagnostics I (OBD I). Engine idle speed is controlled by the Powertrain Control Module (PCM) and the Idle Air Control (IAC) valve. The throttle body and the IAC valve does not allow for adjustments and must NOT be cleaned or damage to the special "sludge tolerant" coating in the throttle body bore will be harmed. If a problem is encountered with the engine idle speed, the OBD I system

must be thoroughly inspected before a proper repair can be performed.

NOTE: The idle speed should be checked with the engine warm, in NEUTRAL or PARK and with the A/C OFF.

The idle speed should be 880 ± 100 rpm on vehicles equipped with manual transaxles and 800 ± 100 rpm on vehicles equipped with automatic transaxles.

2.5L (VIN L) Engine

This vehicle is equipped with On Board Diagnostics I (OBD I). Engine idle speed is controlled by the Powertrain Control Module (PCM) and the Idle Air Control (IAC) valve. The throttle body and the IAC valve does not allow for adjustments and must NOT be cleaned or damage to the special "sludge tolerant" coating in the bore of the throttle body will be harmed. If a problem is encountered with the engine idle speed, the OBD I system must be thoroughly inspected before a proper repair can be performed.

NOTE: The idle speed should be checked with the engine warm, in NEUTRAL or PARK and with the A/C OFF.

The idle rpm is checked using Rotunda New Generation Star (NGS) tester 007-00500 or equivalent scan tool, using the proper software as follows:

1. Activate the engine running self-test.
2. After the Diagnostic Trouble Code (DTC) slow code output is completed, unlatch and within 4 seconds, latch the STI button.
3. A single pulse code indicates the entry mode. Following the single pulse code, the NGS tool should produce a constant tone, solid light or read "STO LO" indicating that the idle rpm is within allowable parameters. If using an equivalent, scan tool, follow the manufacturers testing procedures.
4. To exit the test procedure using the NGS, unlatch the STI button then wait 4 seconds for reinitialization (after 10 minutes it will exit by itself).
5. If a beeping tone is heard or the light flashes, the idle rpm is not within allowable parameters and further diagnosis is required.

1996–97 Vehicles

This vehicle is equipped with On Board Diagnostics II (OBD II). Engine idle speed is controlled by the

Powertrain Control Module (PCM) and the Idle Air Control (IAC) valve. The throttle body does not allow for adjustments and must not be cleaned or damage to the special "sludge tolerant" coating in the throttle body bore will be harmed. If a problem is encountered with the engine idle speed, the OBD II system must be thoroughly inspected before a proper repair can be performed.

Mixture

ADJUSTMENT

The air/fuel mixture is electronically controlled by the Powertrain Control Module (PCM) and the Idle Air Control (IAC) valve, and cannot be adjusted. The PCM continuously adjust the air/fuel ratio in response to signals received from operator controls, sensor and switch signals monitoring the engines running condition.

Fuel Filter

REMOVAL AND INSTALLATION

------ CAUTION ------
Fuel injection systems remain under pressure, even after the engine has been turned OFF. The fuel system pressure must be relieved before disconnecting any fuel lines. Failure to do so may result in fire and/or personal injury.

1. Disconnect the negative battery cable.
2. Relieve the fuel system pressure as follows:
   a. Remove the air cleaner assembly.
   b. Connect fuel pressure gauge T80L-9974-B or equivalent, to the fuel pressure relief valve located on the fuel rail.
   c. Open the manual valve on the fuel pressure gauge and drain the fuel through the drain tube into a suitable container.
   d. Remove the fuel pressure gauge and reinstall the air cleaner assembly.
3. Locate the fuel filter at the fuel tank. Clean the area before disassembly.
4. Remove the retainer clips at both ends of the fuel filter in the following manner:
   a. Bend the shipping tab downward so that it will clear the body.

b. Using hands only, spread the clip legs to disengage the body and to push the legs up into the fitting.
   c. Pull on the triangular and of the clip to finish removal.

------ WARNING ------
Never use any tools to remove the retainer clips as distortion to the fittings may result causing fuel leaks.

5. Twist and pull the fuel lines off of the fuel filter. Prepare for fuel to be released from the fuel filter and fuel lines by placing a shop towel around the area to absorb the fuel being released.
6. Check the fuel line fittings for any internal parts that may have been dislodged during removal and correct.
7. Loosen the mounting clamp and remove the fuel filter.
8. Drain and then properly dispose of the fuel filter.

To install:

9. Place the new fuel filter into its mount with the arrow pointed in the direction of flow and tighten the clamp.
10. Install new retainer clips onto the fuel line fittings before placing the lines onto the fuel filter ends.

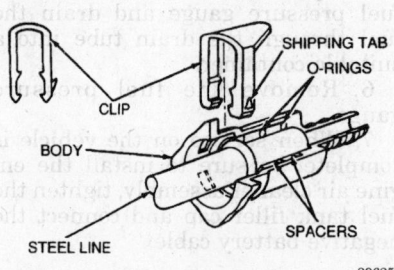

View of the retainer clip showing the shipping tab

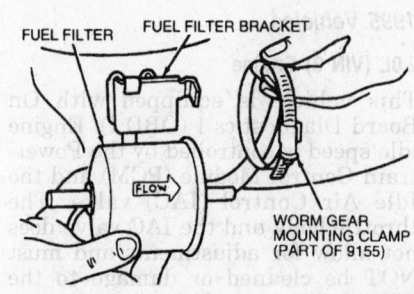

Fuel filter showing direction of fuel flow and bracket

11. Push the fuel lines onto the fuel filter until an audible click is heard. Pull on the fitting to verify a good connection.
12. Start the engine and check for fuel leaks.

Fuel Pump

REMOVAL AND INSTALLATION

------ CAUTION ------
Fuel injection systems remain under pressure, even after the engine has been turned OFF. The fuel system pressure must be relieved before disconnecting any fuel lines. Failure to do so may result in fire and/or personal injury.

1. Disconnect the negative battery cable.
2. Relieve the fuel pressure in the following manner:
   a. Open the fuel tank filler cap to vent off pressure in the tank.
   b. Remove the air cleaner assembly.
   c. Connect fuel pressure gauge T80L-9974-B or equivalent, to the fuel pressure relief valve located on the fuel rail.
   d. Open the manual valve on the fuel pressure gauge and drain the fuel through the drain tube into a suitable container.
   e. Remove the fuel pressure gauge.
   f. Secure the fuel fill cap and install the air cleaner assembly.
3. Remove the rear seat cushion.
4. Remove the plastic grommet from the floor pan.
5. Disconnect the fuel pump electrical harness connector.
6. Disconnect the fuel lines from the fuel pump by compressing the tabs on both sides of each nylon push connect fitting and easing the fuel line off of the fuel pump.
7. Using Fuel Tank Sender Wrench D84P-9275-A or equivalent, turn the fuel pump locking ring counterclockwise to loosen the ring.
8. Remove the fuel pump locking ring.
9. Remove the fuel pump being careful not to damage the fuel gauge sending unit.
10. Place a shop towel over the opening in the fuel tank to prevent dirt from contaminating the fuel.

To install:

11. Remove the shop towel over the opening in the fuel tank and clean the groove for the fuel pump seal. Be

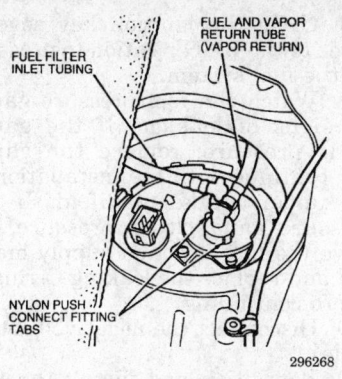

View of fuel pump fittings through the floorpan

careful not to allow dirt to enter the fuel tank.

12. Apply a light coat of grease onto a new O-ring seal and install into the groove of the fuel tank.

13. Carefully install the fuel pump into the tank to prevent damage to the fuel gauge sender or the fuel pickup filter.

NOTE: It is recommended that the in-line fuel filter be replaced whenever a fuel pump is being replaced.

14. Ensure that the flange of the fuel pump mounting plate is located properly in its keyway and that the O-ring has not shifted out of position.

15. Keep a light downward pressure on the fuel pump while installing the fuel pump locking retainer ring.

16. Install the ring ensuring that all of the locking tabs are under the fuel tank lock ring tabs. Turn the ring clockwise finger-tight.

17. Install the fuel tank sender wrench or equivalent, over the retainer ring and finish tightening until the retainer ring is resting against its stops.

18. Connect the fuel lines to the fuel pump.

19. Install the fuel pump electrical harness connector.

20. Install the plastic grommet into the floorpan.

21. Install the rear seat cushion.

22. Connect the negative battery cable.

23. Start the engine and check for leaks and proper operation.

Fuel Injector

REMOVAL AND INSTALLATION

2.0L (VIN 3) Engine

————— CAUTION —————
Fuel injection systems remain under pressure, even after the engine has been turned OFF. The fuel system pressure must be relieved before disconnecting any fuel lines. Failure to do so may result in fire and/or personal injury.

1. Disconnect the negative battery cable.

2. Remove the engine air intake resonators.

3. Relieve the fuel system pressure as follows:

   a. Loosen the fuel tank filler cap to relieve pressure in the fuel tank.

   b. Remove the air cleaner assembly.

   c. Connect fuel pressure gauge T80L–9974–B or equivalent, to the fuel pressure relief valve located on the fuel rail.

   d. Open the manual valve on the fuel pressure gauge and drain the fuel through the drain tube into a suitable container.

   e. Remove the fuel pressure gauge.

   f. Secure the fuel fill cap and install the air cleaner assembly.

4. Disconnect the wiring harness to the fuel injectors and move aside.

5. Disconnect the vacuum line at the fuel pressure regulator.

6. Using the proper spring lock coupling disconnect tools (³⁄₈ inch and ¹⁄₂ inch), remove the fuel supply and return hoses from the fuel injection supply manifold (fuel rail).

7. Remove the fuel supply and return hoses from the retaining bracket on the intake manifold and move aside.

8. Remove the 3 retaining bolts securing the fuel injection supply manifold to the intake manifold.

9. Carefully disengage the fuel injection supply manifold with the fuel injectors attached and remove from the engine.

10. Remove the fuel injectors from the manifold as required.

To install:

11. If a new fuel injector is to be installed, swap the retaining clip from the injector being replaced.

12. Inspect the fuel injector O-rings, replace as required.

13. Lightly coat the injector O-rings with clean engine oil. Install the fuel

injectors with retaining clips onto the fuel injection supply manifold using a light twisting/pushing motion.

NOTE: Never use silicone grease on the fuel injector O-rings as it will clog the injector nozzles.

14. Carefully position the fuel injection supply manifold with injectors into the intake manifold.

15. Install the 3 fuel injection supply manifold retaining bolts. Tighten the retaining bolts to 71–106 inch lbs. (8–12 Nm).

16. Install the fuel supply and return hoses to the fuel injection supply manifold.

17. Install the vacuum line to the fuel pressure regulator.

18. Install the fuel injector wiring harness.

19. Install the air intake resonators.

20. Connect the negative battery cable.

21. Run the engine and check for leaks and proper operation.

2.5L (VIN L) Engine

————— CAUTION —————
Fuel injection systems remain under pressure, even after the engine has been turned OFF. The fuel system pressure must be relieved before disconnecting any fuel lines. Failure to do so may result in fire and/or personal injury.

1. Disconnect the negative battery cable.

2. Relieve the fuel system pressure as follows:

   a. Loosen the fuel tank filler cap to relieve pressure in the fuel tank.

   b. Remove the air cleaner and outlet tube assembly.

   c. Connect fuel pressure gauge T80L–9974–B or equivalent, to the fuel pressure relief valve located on the fuel rail.

   d. Open the manual valve on the fuel pressure gauge and drain the fuel through the drain tube into a suitable container.

   e. Remove the fuel pressure gauge.

   f. Secure the fuel fill cap and install the air cleaner and outlet tube assembly.

3. Remove the upper intake manifold.

4. Disconnect the fuel injector wiring harness and move aside.

5. Disconnect the vacuum line at the fuel pressure regulator.

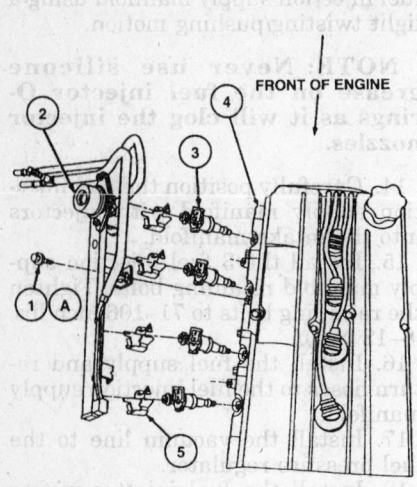

FRONT OF ENGINE

1 Bolt (3 req'd)
2 Fuel pressure regulator
3 Fuel injector (4 req'd)
4 Intake manifold
5 Retainer clip
A Tighten to 8-12 Nm (71-106 lb-in)

296264

Fuel injectors and mounting components — 2.0L (VIN 3) engine

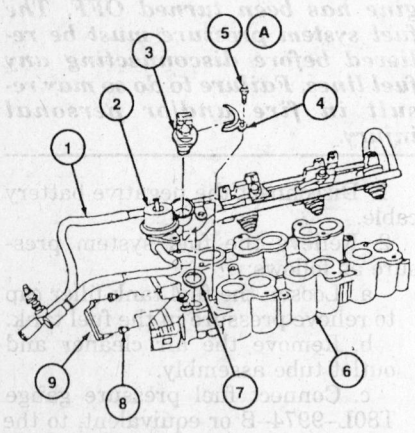

1 Fuel injection supply manifold
2 Fuel pressure regulator
3 Fuel injector (6 req'd)
4 Fuel injector retainer (6 req'd)
5 Bolt (7 req'd)
6 Lower intake manifold (IMRC)
7 Fuel injector seal (6 req'd)
8 IMRC linkage rod bushings (2 req'd)
9 IMRC linkage rod
A Tighten to 8-12 Nm (71-106 lb-in)

296266

Fuel injectors and related components — 2.5L (VIN L) engine

6. Remove the spring lock coupling retainer clips from the fuel supply and return fittings.

7. Use spring lock coupling disconnect tools (³⁄₈ inch and ¹⁄₂ inch) to disconnect the fuel supply and return hoses from the fuel injection supply manifold.

8. Remove 7 fuel injection supply manifold retaining bolts and retainers.

9. Remove the IMRC vacuum solenoid linkage rod from the lower intake manifold by carefully prying the rod with a screwdriver.

10. Carefully disengage the fuel injection supply manifold with the fuel injectors and remove as an assembly.

11. Remove the fuel injectors from the fuel injection supply manifold using Fuel Injection Remover Tool T94P-9000-AH or equivalent.

12. Remove the fuel injector O-rings from the fuel injectors that were removed.

13. Remove the fuel injector seals if any remained on the lower intake manifold.

To install:

14. Place new O-rings onto each injector and lightly coat with clean engine oil.

15. Install the fuel injectors into the fuel injection supply manifold using a light twisting/pushing motion.

16. Place new seals between the fuel injection supply manifold and the lower intake manifold.

17. Install the fuel injection supply manifold with injectors to the lower intake manifold.

18. Install the 7 fuel injector retainers and bolts.

19. While holding down on the fuel injection supply manifold, verify that the retainers are correctly positioned and tighten the retaining bolts to 71–106 inch lbs. (8–12 Nm).

20. Replace the IMRC linkage rod bushings.

21. Install the IMRC vacuum solenoid linkage rod to the lower intake manifold.

22. Install the fuel supply and return hoses to the fuel supply manifold and ensure that the spring lock couplings are correctly installed.

23. Install the retaining clips onto the spring lock couplings.

24. Connect the vacuum line to the fuel pressure regulator.

25. Temporarily connect the negative battery cable.

26. Connect the fuel pressure gauge to the fuel pressure relief valve located on the fuel injection supply manifold.

27. Cycle the ignition key several times to the **RUN** position to pressurize the fuel system.

28. Watch the fuel pressure gauge for signs of leakage. If the gauge holds pressure, remove the gauge and continue with the installation of the upper intake manifold. If the pressure gauge loses pressure, remove the fuel injection supply manifold and replace the leaking O-ring(s) before continuing.

29. Disconnect the negative battery cable.

30. Reposition and install the fuel injector wiring harness.

31. Install the upper intake manifold.

32. Connect the negative battery cable.

33. Run the engine and check for leaks and proper operation.

ENGINE MECHANICAL

Engine Assembly

REMOVAL AND INSTALLATION

2.0L (VIN 3) Engine

— **CAUTION** —

Fuel injection systems remain under pressure, even after the engine has been turned OFF. The fuel system pressure must be relieved before disconnecting any fuel lines. Failure to do so may result in fire and/or personal injury.

1. Disconnect the battery cables, negative cable first.

2. Properly relieve the fuel system pressure.

3. Remove the pinch bolt and disconnect the steering shaft and joint at the cowl, inside the vehicle.

4. Remove the engine air cleaner and the engine air intake resonators.

5. Properly recover the refrigerant from the A/C system.

6. Raise and safely support the vehicle.

7. Remove the front splash shield from between the front subframe and the body.

8. Remove the catalytic converter.

9. Drain the engine cooling system.

10. Remove the front wheel and tire assemblies.

11. Separate the left-hand and right-hand stabilizer bar links from the front stabilizer bar.

12. Separate the left-hand and right-hand outer tie rod ends from the front wheel knuckles. Discard the cotter pins.

13. Remove the pinch bolts and separate the front suspension lower control arms from the front wheel knuckles at the ball joints.

14. Remove the left-hand and right-hand wheel hub retainer nuts from the halfshaft ends and remove the halfshafts from the front wheel knuckles.

15. Remove the A/C accumulator retaining screws from the front subframe.

16. Disconnect the vehicle speed sensor wiring harness at the connector.

17. Disconnect the speedometer drive cable from the transaxle.

18. If the engine is to be separated from the transaxle after removal from the vehicle, and the transaxle is an automatic, remove the right-hand splash shield from the front fender apron.

19. Remove the access plug from the engine rear plate and remove the 4 torque converter retaining nuts.

20. Push the torque converter into the transaxle front pump support and gear.

21. Disconnect the wiring to the knock sensor and the oil pressure sensor located on the right-hand side of the cylinder block.

22. Lower the vehicle.

23. Secure the radiator and fan shroud assembly to the radiator support, using safety wire or equivalent.

24. Disconnect the accelerator cable and speed control actuator from the throttle body.

25. Remove the accelerator cable bracket.

26. Disconnect the wiring to the fuel injectors at the connector located near the fuel pressure regulator.

27. Remove the retaining screws from the engine control wiring at the intake manifold.

28. Remove the power steering pump auxiliary reservoir from the bracket and lay on top of the engine assembly using shop towels to absorb the fluid.

29. Disconnect the return hose from the power steering reservoir and plug the hose.

WARNING
Do not allow the power steering fluid to come into contact with the accessory drive belts.
---

30. Disconnect the power steering return hose from the power steering pump.

31. Disconnect the wiring from the power steering pressure switch located on the power steering pressure hose.

32. Disconnect the wiring from the alternator and the grounding strap from the alternator mounting bracket.

33. Disconnect the vacuum supply hose from the fitting on the rear of the intake manifold.

34. Disconnect the coolant hoses from the radiator coolant recovery reservoir.

35. Disconnect the hoses from the A/C compressor and plug.

36. Disconnect the fuel return and supply lines from the fuel rail and plug.

37. Disconnect the vacuum supply hose to the EGR valve, EGR pressure sensor and the EGR valve to exhaust manifold tube.

38. If equipped with automatic transaxle, pry the end of the shift cable from the stud, remove the 2 retaining bolts and remove shift cable and bracket from the transaxle. Remove the wiring to the transmission range sensor and remove the wire retainers.

39. Disconnect the grounding strap from the transaxle.

40. Disconnect the wiring from the ignition coil and the radio ignition interference capacitor. Move the wiring out of the way.

41. Remove the vacuum supply line from the power brake booster.

42. Disconnect the upper radiator hose from the radiator.

43. Disconnect the evaporative emission hose from the connector located near the radio ignition interference capacitor.

44. Disconnect the heater hose from the connector located near the EGR valve.

45. Disconnect the positive and negative battery cable retainer from the battery tray.

46. If equipped with manual transaxle, remove the retainer and disconnect the clutch hydraulic line from the clutch actuator pipe at the transaxle case.

47. If equipped, disconnect the block heater power supply wiring from the left-hand side of the radiator support.

48. If equipped with automatic transaxle, remove the transmission oil cooler lines from the transaxle. Remove the oil cooler return line from the bracket on the left-hand side of the transaxle.

49. If equipped with manual transaxle, remove the bolt from the shift rod and the nut from the stabilizer bar and remove from the transaxle.

50. Disconnect the lower radiator hose from the radiator.

51. Remove the 4 bolts retaining the lower radiator supports to the front subframe. Rotate the radiator supports forward.

52. Disconnect the wiring harness from the A/C compressor.

53. Disconnect the engine wiring from the heated oxygen sensor, engine coolant temperature sensor and the crankshaft position sensor.

54. Remove the 2 screws retaining the bumper cover braces to the left-hand and right-hand sides of the front subframe and rotate the cover braces forward.

55. Disconnect the power steering oil cooler hoses at the right-hand front subframe. Drain the fluid from the hoses.

56. Partially lower the vehicle.

57. Position and secure all lines, hoses and components that will be removed with the engine.

58. Install Powertrain and Subframe Support Bracket 134-00250 or equivalent, with Powertrain Lift (hydraulic lift) 134-00251 or equivalent, to support the powertrain assembly for removal from the vehicle.

NOTE: Make sure that the powertrain and subframe support bracket and lift are correctly positioned for safe removal of the powertrain assembly.

59. Remove the 4 subframe to body retaining bolts.

60. Remove the upper front engine support bracket and the engine and transmission support insulator retaining nuts.

61. With an assistant, carefully lower the powertrain assembly while checking for body interference.

62. With the powertrain assembly lowered from the vehicle, carefully roll the powertrain lift or equivalent away from the vehicle.

63. Using Floor Crane 014-00071 or equivalent, and Floor Crane Positioning Sling 014-00036 or equivalent, support the engine using the engine lifting eyes.

64. Remove the left-hand and the right-hand halfshafts from the transaxle.

65. Remove the left-hand and the right-hand front engine support insulators from the subframe and transaxle.

66. Using the floor crane and sling or equivalent, raise the engine and transaxle assembly and remove from the subframe.

67. Position the transaxle part of the assembly onto Transmission Jack 066-00016 or equivalent.

68. Remove the 2 starter retaining bolts and remove the starter motor.

69. Remove the battery ground cable from the engine to transaxle retaining stud bolt.

70. Remove the transaxle to engine retaining bolts and separate the engine from the transaxle.

71. If equipped with manual transaxle, remove the 6 clutch pressure plate retaining bolts and remove the pressure plate and the clutch disc.

72. Remove the 8 flywheel retainer bolts and the flywheel from the crankshaft.

73. Remove the engine rear plate.

74. Install the engine to an engine stand for further service.

To install:

75. Remove the engine from the engine stand using the floor crane and sling or equivalent, attached to the engine lifting eyes.

76. Install the engine rear plate and flywheel.

77. Torque the flywheel retaining bolts in an alternating sequence to 79–86 ft. lbs. (107–117 Nm) for automatic transaxles and 81–89 ft. lbs. (110–120 Nm) for manual transaxles.

78. If equipped with manual transaxle, install the clutch disc and the clutch pressure plate.

79. Torque the retaining bolts in an alternating sequence to 22 ft. lbs. (30 Nm).

80. Install the engine to the transaxle. If equipped with automatic transaxle, align the torque converter to the flywheel while positioning the engine to the transaxle.

81. Install the engine to transaxle retaining bolts and torque to 25–34 ft. lbs. (34–46 Nm).

82. If removed, place the subframe onto the powertrain and subframe support bracket and hydraulic lift.

83. Using the floor crane and sling, position the engine and transaxle assembly onto the subframe keeping the crane attached for support.

84. Install Powertrain Alignment Gauge T94P-6000-AH or equivalent, to the left–hand front engine support bracket and subframe. Install the through bolt. Tighten the retaining

bolts and the through bolt to 20 ft. lbs. (27 Nm).

85. Install the right-hand engine support insulator retaining bolts and through bolt to the subframe. Leave the bolts finger tight.

86. Install the battery ground cable to the engine at the transaxle stud bolt. Torque the retaining nut to 15–22 ft. lbs. (20–30 Nm).

87. Install the starter motor.

88. Install the left-hand and right-hand halfshafts into the transaxle.

89. Install Sub-frame Alignment Pin Set 94P-2100-AH or equivalent into the subframe.

90. Install the 4 subframe retaining bolts and torque to 92–100 ft. lbs. (125–135 Nm).

91. Remove the subframe alignment pins.

NOTE: Make sure that the engine and transaxle are firmly seated against the front and rear insulator brackets.

92. Using new nuts, install the upper front engine support bracket and the transmission support insulator. Torque the nuts to 7 ft. lbs. (10 Nm).

93. Remove the subframe support bracket and the hydraulic lift.

94. Raise and safely support the vehicle.

95. Tighten the right-hand front engine support insulator to the subframe bolts to 30–41 ft. lbs. (41–55 Nm).

NOTE: Check the position of the right-hand front engine support insulator. It must be centered in its bracket and in perfect alignment front to rear.

96. Lower the vehicle.

97. Torque the front engine support bracket nuts to 52–70 ft. lbs. (70–95 Nm).

98. Torque the engine and transmission support insulator nuts to 30–41 ft. lbs. (41–55 Nm) for automatic transaxles and 52–70 ft. lbs. (70–95 Nm) for manual transaxles.

99. Torque the right-hand front engine support insulator through-bolt to 75–102 ft. lbs. (103–137 Nm).

100. Remove the powertrain alignment gauge.

101. Install the left-hand front engine support insulator to the subframe. Toque the retaining bolts to 7 ft. lbs. (10 Nm).

NOTE: Check the position of the left-hand front engine support insulator to ensure perfect front to rear alignment.

102. Retorque the 2 retaining bolts to 30–41 ft. lbs. (41–55 Nm).

103. Install the left-hand front engine support insulator through-bolt. Torque the through-bolt to 75–102 ft. lbs. (103–137 Nm).

104. Connect the power steering oil cooler hoses at the right-hand side of the subframe. Tighten the hose clamps securely.

105. Connect the engine wiring to the heated oxygen sensor, engine coolant temperature sensor and the crankshaft position sensor.

106. Connect the wiring harness to the A/C compressor.

107. Install the radiator supports to the subframe bolts. Tighten to 71–97 inch lbs. (8–11 Nm).

108. Connect the lower radiator hose to the radiator.

109. If equipped with manual transaxle, install the shift rod stabilizer to the transaxle.

110. Torque the shift rod bolt to 17 ft. lbs. (23 Nm) and the stabilizer nut to 41 ft. lbs. (55 Nm).

111. If equipped with automatic transaxle, install the transmission oil cooler lines and torque the nuts to 18–22 ft. lbs. (24–31 Nm).

112. Connect the heater hose to the heater water tube located under the engine.

113. Lower the vehicle.

114. If equipped, connect the block heater power supply wiring.

115. If equipped with manual transaxle, connect the clutch hydraulic line to the clutch actuator pipe at the transaxle case and install the retainer.

116. Connect the positive and negative battery cable retainer to the battery tray.

117. Install the heater hose to the connector tube located near the EGR valve and clamp securely.

118. Connect the evaporative emission hose to the connector located near the radio ignition interference capacitor.

119. Connect the upper radiator hose to the radiator.

120. Connect the vacuum supply line to the power brake booster.

121. Connect the wiring to the ignition coil and the radio ignition interference capacitor.

122. Connect the ground strap to the transaxle case.

123. Connect the wiring harness to the transmission range sensor.

124. If equipped with automatic transaxle, install the shift cable and bracket to the transaxle. Torque the retaining bolts to 15–19 ft. lbs. (20–25 Nm).

125. Connect the vacuum hoses from the EGR pressure sensor to the EGR

valve to exhaust manifold tube and the vacuum supply hose to the EGR valve.

126. Unplug and connect the fuel supply and return lines to the fuel rail.

127. Unplug and connect the refrigerant hoses to the A/C compressor.

128. Connect the coolant hoses to the radiator coolant recovery reservoir.

129. Connect the vacuum supply hose for the body to the fitting at the rear of the intake manifold.

130. Install the ground strap on the alternator mounting bracket.

131. Connect the wiring harness to the alternator.

132. Connect the wiring to the power steering pressure switch located on the power steering pressure hose.

133. Connect the power steering return hose to the power steering pump. Clamp the hose securely.

134. Install the power steering pump auxiliary reservoir to its bracket.

135. Install the retainer screws for the engine wiring to the intake manifold.

136. Connect the wiring to the connector located near the fuel pressure regulator for the fuel injectors.

137. Connect the accelerator cable and the speed control actuator to the accelerator cable bracket and the throttle body.

138. Remove the safety wire holding the radiator and fan shroud to the radiator support.

139. Raise and safely support the vehicle.

140. Connect the wiring harness to the knock sensor and the oil pressure sensor located on the right-hand side of the cylinder block.

141. If equipped with automatic transaxle, install the torque converter to flywheel retaining nuts. Torque the retaining nuts in an alternating sequence to 54–64 ft. lbs. (73–87 Nm). Install the access plug into the engine rear plate.

142. Install the right-hand splash shield onto the front fender apron.

143. Connect the vehicle speed sensor wiring harness to the connector.

144. Install the speedometer drive cable to the transaxle.

145. Install the A/C accumulator retaining screws to the subframe.

146. Install the left-hand and right-hand halfshafts.

147. Install the left-hand and right-hand front wheel knuckles into the front suspension lower control arms at the ball joints. Torque the bolts to 37–43 ft. lbs. (50–58 Nm).

148. Install the left-hand and right-hand tie rod ends to the front wheel knuckles. Install new castellated nuts onto the tie rod end studs. Torque the castellated nuts to 21 ft. lbs. (28 Nm). Install new cotter pins.

149. Connect the left-hand and right-hand front stabilizer bar links to the front stabilizer bar. Torque the nuts to 35 ft. lbs. (48 Nm).

150. Install the front wheel and tire assemblies. Torque the lug nuts to 95 ft. lbs. (129 Nm).

151. Install the catalytic converter.

152. Install the splash shield to the front of the subframe and body.

153. Rotate the front bumper cover braces rearward. Install the 2 screws to the subframe and tighten securely.

154. Lower the vehicle.

155. Install the engine air cleaner and the engine air intake resonators.

156. Connect the steering shaft and joint inside the vehicle. Torque the bolt to 18 ft. lbs. (24 Nm).

157. If equipped with automatic transaxle, fill the transmission with the proper amount and type of fluid.

158. Fill the power steering reservoir with the proper type of fluid.

159. Fill the engine cooling system.

160. Connect the battery cables, negative cable last.

161. Check all fluid levels.

162. Evacuate and recharge the A/C system.

NOTE: Whenever the vehicle subframe is removed or lowered, the wheel alignment should be checked.

163. Run the engine and check for leaks. Check for proper engine operation.

2.5L (VIN L) Engine

—— CAUTION ——
Fuel injection systems remain under pressure, even after the engine has been turned OFF. The fuel system pressure must be relieved before disconnecting any fuel lines. Failure to do so may result in fire and/or personal injury.

1. Disconnect the battery cables, negative cable first.

2. Properly relieve the fuel system pressure.

3. Remove the water pump pulley shield.

4. Remove the pinch bolt and disconnect the steering shaft and joint at the cowl inside the vehicle.

5. Remove the engine air cleaner.

6. Properly recover the refrigerant from the A/C system.

7. Raise and safely support the vehicle.

8. Remove the exhaust crossover and the catalytic converter.

9. Drain the engine cooling system.

10. Remove the front wheel and tire assemblies.

11. Separate the left-hand and right-hand stabilizer bar links from the front stabilizer bar.

12. Separate the left-hand and right-hand outer tie rod ends from the front wheel knuckles. Discard the cotter pins.

13. Remove the pinch bolts and separate the front suspension lower control arms from the front wheel knuckles at the ball joints.

14. Remove the left-hand and right-hand wheel hub retainer nuts from the halfshaft ends and remove the halfshafts from the front wheel knuckles.

15. Remove the A/C accumulator retaining screws from the front subframe.

16. Disconnect the vehicle speed sensor wiring harness at the connector.

17. Disconnect the speedometer drive cable from the vehicle speed sensor.

18. If the engine is to be separated from the transaxle after removal from the vehicle, and the transaxle is an automatic, remove the access plug from the engine rear plate and remove the 4 torque converter retaining nuts.

19. Push the torque converter into the transaxle front pump support and gear.

20. Lower the vehicle.

21. Secure the radiator and fan shroud assembly to the radiator support, using safety wire.

22. Disconnect the accelerator cable and speed control actuator from the throttle body.

23. Remove the accelerator cable bracket from the throttle body.

24. Disconnect the 3 connectors for the engine control wiring from the bracket located on the left-front fender apron and unplug the connectors.

25. Remove the retainer for the engine control wiring from the air cleaner bracket.

26. Remove the ignition control module from the bulkhead, if equipped.

27. Remove the power steering pump auxiliary reservoir from the bracket and lay on top of the engine assembly using shop towels to absorb the fluid.

28. Disconnect the return hose from the power steering reservoir and plug the hose.

---- **WARNING** ----

Do not allow the power steering fluid to come into contact with the accessory drive belts.

29. Disconnect the power steering pressure hose from the power steering pump.

30. Disconnect the power steering pressure hose bracket from the upper front engine support bracket. Lay the hose on top of the engine.

31. Disconnect the wiring from the powertrain control module and retainer located on the right-hand side of the dash panel.

32. Remove the ground strap for the engine control wiring at the right-hand fender apron.

33. Disconnect the coolant hoses from the radiator coolant recovery reservoir.

34. Disconnect the fuel return and supply lines from the fuel rail.

35. If equipped with automatic transaxle, pry the end of the shift cable from the stud, remove the 2 retaining bolts and remove shift cable and bracket from the transaxle. Remove the wiring to the transmission range sensor and remove the wire retainers.

36. Disconnect the grounding strap from the transaxle.

37. Remove the vacuum supply line from the power brake booster.

38. Disconnect the upper radiator hose from the radiator.

39. Disconnect the positive and negative battery cable retainer from the battery tray.

40. If equipped with manual transaxle, remove the retainer and disconnect the clutch hydraulic line from the clutch actuator pipe at the transaxle case.

41. Disconnect the block heater power supply wiring from the right-hand side of the radiator support if equipped.

42. Raise and safely support the vehicle.

43. Disconnect the heater hoses from the heater core.

44. Disconnect the A/C suction hose from the A/C condenser core and plug the hose.

45. Disconnect the A/C discharge hose from the A/C accumulator and plug the hose.

46. If equipped with automatic transaxle, remove the transmission oil cooler lines from the transaxle. Remove the oil cooler return line

from the bracket on the left-hand side of the transaxle.

47. If equipped with manual transaxle, remove the bolt from the shift rod and the nut from the stabilizer bar and remove from the transaxle.

48. Disconnect the lower radiator hose from the radiator.

49. Remove the 4 bolts retaining the lower radiator supports to the front subframe. Rotate the radiator supports forward.

50. Disconnect the wiring harness from the A/C compressor.

51. Remove the 2 screws retaining the bumper cover braces to the left-hand and right-hand sides of the front subframe and rotate the cover braces forward.

52. Partially lower the vehicle.

53. Position and secure all lines, hoses and components that will be removed with the engine.

54. Install Powertrain and Subframe Support Bracket 134-00250 or equivalent with Powertrain Lift (hydraulic lift) 134-00251 or equivalent to support the powertrain assembly for removal from the vehicle.

NOTE: Make sure that the powertrain and subframe support bracket and lift are correctly positioned for safe removal of the powertrain assembly.

55. Remove the 4 subframe to body retaining bolts.

56. Remove the upper front engine support bracket and the engine and transmission support insulator retaining nuts.

57. With an assistant, carefully lower the powertrain assembly while checking for body interference.

58. With the powertrain assembly lowered from the vehicle, carefully roll the powertrain lift or equivalent away from the vehicle.

59. Using Floor Crane 014-00071 or equivalent, and Floor Crane Positioning Sling 014-00036 or equivalent, support the engine using the engine lifting eyes.

60. Remove the left-hand and the right-hand halfshafts from the transaxle.

61. Remove the left-hand and the right-hand front engine support insulators from the subframe and transaxle.

62. Using the floor crane and sling or equivalent, raise the engine and transaxle assembly and remove from the subframe.

63. Position the transaxle part of the assembly onto Transmission Jack 066-00016 or equivalent.

64. Remove the 2 starter retaining bolts and remove the starter motor.

65. Disconnect the engine control wiring connectors and retaining clips at the transaxle and move aside.

66. Remove the battery ground cable from the engine to transaxle retaining stud bolt.

67. Remove the transaxle to engine retaining bolts and separate the engine from the transaxle.

68. If equipped with manual transaxle, remove the 6 clutch pressure plate retaining bolts and remove the pressure plate and the clutch disc.

69. Remove the 8 flywheel retainer bolts and the flywheel from the crankshaft.

70. Remove the engine rear plate.

71. Install the engine to an engine stand for further service.

To install:

72. Remove the engine from the engine stand using the floor crane and sling or equivalent, attached to the engine lifting eyes.

73. Install the engine rear plate and flywheel.

74. Torque the flywheel retaining bolts in an alternating sequence to 79–86 ft. lbs. (107–117 Nm) for automatic transaxles, or 81–89 ft. lbs. (110–120 Nm) for manual transaxles.

75. If equipped with manual transaxle, install the clutch disc and the clutch pressure plate. Torque the retaining bolts in an alternating sequence to 22 ft. lbs. (30 Nm).

76. Install the engine to the transaxle. If equipped with automatic transaxle, align the torque converter to the flywheel while positioning the engine to the transaxle.

77. Install the engine to transaxle retaining bolts and torque to 25–34 ft. lbs. (34–46 Nm).

78. If removed, place the subframe onto the powertrain and subframe support bracket and hydraulic lift.

79. Using the floor crane and sling, position the engine and transaxle assembly onto the subframe keeping the crane attached for support.

80. Install Powertrain Alignment Gauge T94P-6000-AH or equivalent, to the left-hand front engine support bracket and subframe. Install the through-bolt. Tighten the retaining bolts and the through-bolt to 20 ft. lbs. (27 Nm).

81. Install the right-hand engine support insulator retaining bolts and through-bolt to the subframe. Leave the bolts finger tight.

82. Install the battery ground cable to the engine at the transaxle stud bolt. Torque the retaining nut to 15–22 ft. lbs. (20–30 Nm).

83. Position the engine control wiring across the transaxle and install the retaining clips.

84. Install the engine control wiring to the transaxle connectors.

85. Install the starter motor.

86. Install the left-hand and right-hand halfshafts into the transaxle.

87. Install Sub-frame Alignment Pin Set 94P–2100–AH or equivalent into the subframe.

88. Install the 4 subframe retaining bolts and torque to 92–100 ft. lbs. (125–135 Nm).

89. Remove the subframe alignment pins.

NOTE: Make sure that the engine and transaxle are firmly seated against the front and rear insulator brackets.

90. Using new nuts, install the upper front engine support bracket and the transmission support insulator. Torque the nuts to 7 ft. lbs. (10 Nm).

91. Remove the subframe support bracket and the hydraulic lift.

92. Raise and safely support the vehicle.

93. Tighten the right-hand front engine support insulator to the subframe bolts to 30–41 ft. lbs. (41–55 Nm).

NOTE: Check the position of the right-hand front engine support insulator. It must be centered in its bracket and in perfect alignment front to rear.

94. Lower the vehicle.

95. Torque the front engine support bracket nuts to 52–70 ft. lbs. (70–95 Nm).

96. Torque the engine and transmission support insulator nuts to 30–41 ft. lbs. (41–55 Nm) for automatic transaxles or 52–70 ft. lbs. (70–95 Nm) for manual transaxles.

97. Torque the right-hand front engine support insulator through-bolt to 75–102 ft. lbs. (103–137 Nm).

98. Remove the powertrain alignment gauge.

99. Install the left-hand front engine support insulator to the subframe. Toque the retaining bolts to 7 ft. lbs. (10 Nm).

NOTE: Check the position of the left-hand front engine support insulator to ensure perfect front to rear alignment.

100. Retorque the 2 retaining bolts to 30–41 ft. lbs. (41–55 Nm).

101. Install the left-hand front engine support insulator through-bolt. Torque the through-bolt to 75–102 ft. lbs. (103–137 Nm).

102. Connect the wiring harness to the A/C compressor.

103. Install the front bumper cover braces to the left-hand and right-hand side of the subframe.

104. Install the radiator supports to the subframe. Tighten the retaining bolts to 71–97 inch lbs. (8–11 Nm).

105. Connect the lower radiator hose to the radiator.

106. If equipped with manual transaxle, install the shift rod stabilizer to the transaxle. Torque the shift rod bolt to 17 ft. lbs. (23 Nm) and the stabilizer nut to 41 ft. lbs. (55 Nm).

107. If equipped with an automatic transaxle, install the transmission oil cooler lines and torque the nuts to 18–22 ft. lbs. (24–31 Nm).

108. Unplug and connect the A/C suction hose to the A/C condenser core.

109. Unplug and connect the A/C discharge hose to the A/C accumulator.

110. Lower the vehicle.

111. If equipped, connect the block heater power supply wiring.

112. If equipped with manual transaxle, connect the clutch hydraulic line to the clutch actuator pipe at the transaxle case and install the retainer.

113. Connect the positive and negative battery cable retainer to the battery tray.

114. Connect the upper radiator hose to the radiator.

115. Connect the vacuum supply line to the power brake booster.

116. Connect the ground strap to the transaxle case.

117. If equipped with automatic transaxle, install the shift cable and bracket to the transaxle. Torque the retaining bolts to 15–19 ft. lbs. (20–25 Nm).

118. Connect the fuel supply and return lines to the fuel rail.

119. Install the ground strap for the engine control wiring to the right-hand cylinder head.

120. Connect the wiring to the powertrain control module and position the wiring into the retainer bracket.

121. Connect the power steering pressure hose to the power steering pump.

122. Install the power steering pressure hose bracket to the upper front engine support bracket.

123. Install the power steering pump auxiliary reservoir to its bracket.

124. Install the ignition control module and bracket to the bulkhead, if equipped.

125. Install the retainer for the engine control wiring to the air cleaner bracket.

126. Connect the 3 engine control connectors and install to the bracket on the left-hand front fender apron.

127. Install the accelerator cable bracket to the throttle body and tighten to the retaining bolts to 71–106 inch lbs. (8–12 Nm).

128. Connect the accelerator cable and the speed control actuator to the accelerator cable bracket.

129. Remove the safety wire holding the radiator and fan shroud to the radiator support.

130. Raise and safely support the vehicle.

131. If equipped with an automatic transaxle, push the torque converter into the flywheel pilot and install 4 torque converter to flywheel retaining nuts. Torque the retaining nuts in an alternating sequence to 54–64 ft. lbs. (73–87 Nm). Install the access plug into the engine rear plate.

132. Connect the vehicle speed sensor wiring harness to the connector.

133. Install the speedometer drive cable to the vehicle speed sensor.

134. Install the A/C accumulator retaining screws to the subframe.

135. Install the left-hand and right-hand halfshafts into the front wheel knuckles and install the front axle wheel retaining nuts.

136. Install the left-hand and right-hand front wheel knuckles into the front suspension lower control arms at the ball joints. Torque the bolts to 37–43 ft. lbs. (50–58 Nm).

137. Install the left-hand and right-hand tie rod ends to the front wheel knuckles. Install new castellated nuts onto the tie rod end studs. Torque the castellated nuts to 21 ft. lbs. (28 Nm). Install new cotter pins.

138. Connect the left-hand and right-hand front stabilizer bar links to the front stabilizer bar. Torque the nuts to 35 ft. lbs. (48 Nm).

139. Install the front wheel and tire assemblies. Torque the lug nuts to 95 ft. lbs. (129 Nm).

140. Install the exhaust crossover and the three way catalytic converter.

141. Install the splash shield to the front of the subframe and body.

142. Rotate the front bumper cover braces rearward. Install the 2 screws to the subframe and tighten securely.

143. Lower the vehicle.

144. Install the engine air cleaner.

145. Connect the steering shaft and joint inside the vehicle. Torque the bolt to 18 ft. lbs. (24 Nm).

146. Install the water pump pulley shield.

147. If equipped with automatic transaxle, fill the transmission with the proper amount and type of fluid.

148. Fill the power steering reservoir with the proper type of fluid.

149. Fill the engine cooling system.

150. Check all fluid levels.

151. Connect the battery cables, negative cable last.

152. Evacuate and recharge the A/C system.

NOTE: Whenever the vehicle subframe is removed or lowered, the wheel alignment should be checked.

153. Run the engine and check for leaks. Check for proper engine operation.

Engine Mounts

REMOVAL AND INSTALLATION

2.0L (VIN 3) Engine

NOTE: When replacing mounts using self-locking nuts, make sure to have new replacement nuts as the originals loose their holding capabilities and must not be reused.

NOTE: When replacing the upper front engine mount, it is recommended that the left-hand front and right-hand front engine mounts be replaced at the same time.

Upper Front Engine Mount

1. Position the vehicle over a hoist but do not raise at this time.

2. Disconnect the negative battery cable.

3. Remove the engine air intake resonators.

4. Install Three Bar Engine Support D88L–6000–A, or equivalent, to the engine lifting eyes and support the engine.

5. Remove the upper front engine support retaining nuts from the engine and the upper front engine mount (front engine support insulator).

6. Remove the upper front engine support bracket from the vehicle.

7. Remove the retaining bolts for the radiator coolant recovery reservoir and position the reservoir aside.

8. Remove the power steering hose bracket and position the power steering return hose out of the way.

9. Remove the upper front engine mount retaining bolts.

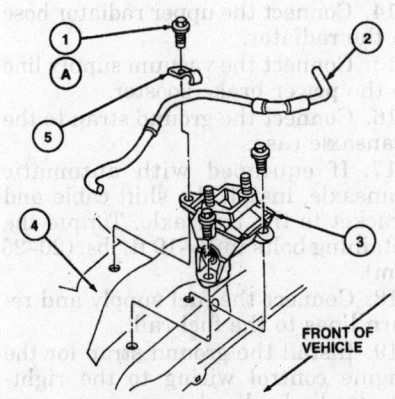

1. Nut(6)
2. Front engine support bracket(upper)
3. Generator mounting bracket
4. A/C compressor mounting bracket
5. Front engine support insulator(upper)
A. 52–70 ft. lb. (70-95 Nm)

297730

Upper front engine support bracket removal — 2.0L (VIN 3) engine

1. Bolt (3)
2. Power steering return hose
3. Front engine support insulator (upper)
4. Front fender apron (RH)
5. Power steering hose bracket
A. 52–70 ft. lb.(70–95 Nm)

297731

Upper front engine mount removal — 2.0L (VIN 3) engine

10. Remove the upper front engine mount.

To install:

NOTE: The engine and transaxle position must be realigned to the front subframe.

11. Raise and safely support the vehicle.

12. Remove the 2 retaining bolts to the front subframe.

13. Remove the through-bolt.

14. Remove the left-hand front engine mount (support insulator).

15. Install Powertrain Alignment Gauge T94P–6000–AH or equivalent, to the transaxle bracket and the front subframe.

16. Install and torque the 2 retaining bolts and then the through-bolt to 20 ft. lbs. (27 Nm).

17. Loosen the right-hand front engine mount (support insulator) through-bolt.

18. Lower the vehicle.

19. Position the power steering hose bracket and the power steering return hose onto the upper front engine support before installing the retaining bolts.

20. Install the new upper front engine mount (support insulator) and its retaining bolts to the right-hand front fender apron. Torque the retaining bolts to 52–70 ft. lbs. (70–95 Nm).

21. Position the radiator coolant recovery reservoir and install the reservoir retaining bolts.

NOTE: The upper front engine support bracket uses self–locking nuts. These nuts must always be replaced with new ones once removed.

22. Install the upper front engine support bracket onto the engine and the upper front engine mount. Install new self-locking nuts and torque the retaining nuts to 7 ft. lbs. (10 Nm).

23. Lower the vehicle.

24. Remove the three bar engine support tool, or equivalent.

25. Retorque the upper front engine support bracket nuts to 52–70 ft. lbs. (70–95 Nm).

26. Raise and safely support the vehicle.

27. Check the position of the right-hand front engine mount (support insulator). It must be centered in the transaxle bracket and in perfect front to rear alignment.

28. Torque the right-hand front engine mount through bolt to 75–102 ft. lbs. (103–137 Nm).

29. Remove the 2 retaining bolts and the through bolt securing the powertrain alignment gauge.

30. Remove the powertrain alignment gauge.

31. Install the left-hand front engine mount to the front subframe using the 2 retaining bolts. Torque the retaining bolts to 7 ft. lbs. (10 Nm).

32. Check the position of the left-hand front engine mount to ensure perfect front to rear alignment.

33. Retorque the 2 retaining bolts to 30–40 ft. lbs. (41–55 Nm).

34. Install the left–hand front engine mount through-bolt. Torque the through-bolt to 75–102 ft. lbs. (103–137 Nm).

35. Lower the vehicle.

36. Install the engine air intake resonators.

37. Fill the power steering system with the proper power steering fluid.

38. Connect the negative battery cable.

39. Run the engine and check for leaks and proper operation.

Left Front Engine Mount

1. Disconnect the negative battery cable.

2. Raise and safely support the vehicle.

3. Remove the 2 bolts retaining the left-hand front engine mount (support insulator) to the front subframe.

4. Remove the left-hand front engine mount through-bolt.

5. Remove the left-hand engine mount from the vehicle.

To install:

6. Install the left-hand front engine mount to the front subframe with 2 retaining bolts. Torque the retaining bolts to 7 ft. lbs. (10 Nm).

7. Check the position of the left-hand front engine mount to ensure perfect front to rear alignment.

8. Retorque the 2 retaining bolts to 30–40 ft. lbs. (41–55 Nm).

9. Install the left-hand front engine mount through-bolt. Torque the through-bolt to 75–102 ft. lbs. (103–137 Nm).

10. Lower the vehicle.

11. Connect the negative battery cable.

12. Check for proper operation.

Right Front Engine Mount

1. Disconnect the negative battery cable.

2. Raise and safely support the vehicle.

3. Remove the through-bolt from the right-hand front engine mount (support insulator).

4. Remove the 2 retaining bolts securing the right-hand front engine mount to the front subframe.

5. Remove the right-hand front engine mount from the vehicle.

To install:

6. Install the right-hand front engine mount to the support bracket and the front subframe.

7. Loosely install the 2 retaining bolts and the through-bolt.

8. Locate the left-hand front engine mount.

9. Remove the 2 retaining bolts to the front subframe.

10. Remove the through-bolt and remove the left-hand front engine mount.

11. Install Powertrain Alignment Gauge T94P–6000–AH, or equivalent to the transaxle bracket and the front subframe.

12. Install and torque the 2 retaining bolts and then the through-bolt to 20 ft. lbs. (27 Nm).

13. Torque the right-hand front engine mount retaining bolt to the front subframe to 7 ft. lbs. (10 Nm).

14. Check the position of the right-hand front engine mount (support insulator). It must be centered in the transaxle bracket and in perfect front to rear alignment.

15. Retorque the 2 retaining bolts to 30–40 ft. lbs. (41–55 Nm).

16. Install the right-hand front engine mount through-bolt. Torque the

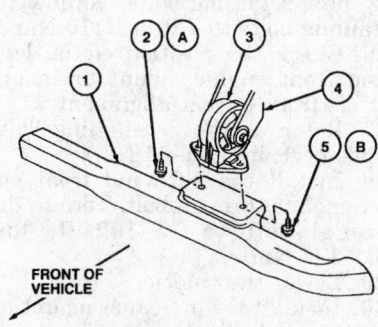

FRONT OF VEHICLE

1. Front sub-frame
2. Bolt(2)
3. Front engine support insulator(LH)
4. Front engine support bracket(LH)
5. Through bolt
A. 30-40 ft. lb.(41-55 Nm)
B. 75-102 ft. lb.(103-137 Nm)

297734

Left front engine mount — 2.0L (VIN 3) engine

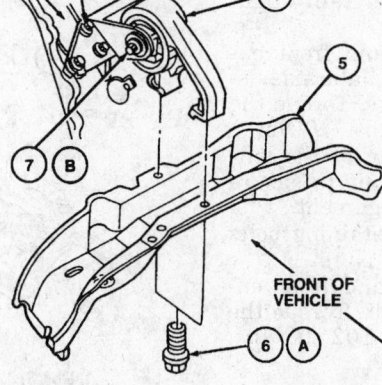

FRONT OF VEHICLE

1. Transaxle
2. Bolt(3)
3. Front engine support bracket(RH)
4. Front engine support insulator(RH)
5. Front sub-frame
6. Bolt(2)
7. Through bolt
A. 30-40 ft. lb.(41-55 Nm)
B. 75-102 ft. lb.(103-137 Nm)

297736

Right-hand front engine mount used with automatic transaxles — 2.0L (VIN 3) engine

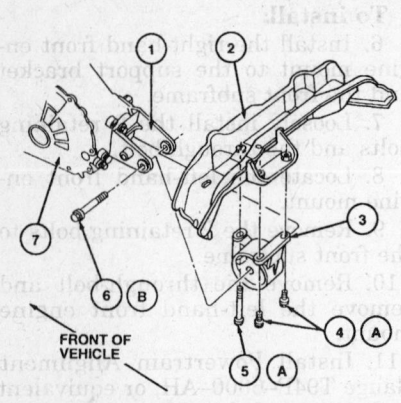

1. Front engine support bracket(RH)
2. Front sub-frame
3. Front engine support insulatorRH)
4. Bolt(2)
5. Bolt
6. Through bolt
7. Transaxle
   A. 30-40 ft. lb.(41-55 Nm)
   B. 75-102 ft. lb.(103-137 Nm)

297738

Right front engine mount used with manual transaxles — 2.0L (VIN 3) engine

through-bolt to 75–102 ft. lbs. (103–137 Nm).

17. Remove the 2 retaining bolts and the through-bolt securing the powertrain alignment gauge.

18. Remove the powertrain alignment gauge.

19. Install the left-hand front engine mount to the front subframe using the 2 retaining bolts. Torque the retaining bolts to 7 ft. lbs. (10 Nm).

20. Check the position of the left-hand front engine mount to ensure perfect front to rear alignment.

21. Retorque the 2 retaining bolts to 30–40 ft. lbs. (41–55 Nm).

22. Install the left-hand front engine mount through-bolt. Torque the through bolt to 75–102 ft. lbs. (103–137 Nm).

23. Lower the vehicle.

24. Connect the negative battery cable.

25. Check for proper operation.

Transaxle Mount

1. Disconnect the negative battery cable.

2. Remove the engine air cleaner.

3. Remove the engine air cleaner support bracket from the transaxle mount (engine and transmission support).

4. Install Three Bar Engine Support D88L–6000–A, or equivalent to the engine lifting eyes and support the engine and transaxle.

5. Remove the transaxle mount retaining bolts to the left-hand front fender apron.

6. Remove the retaining nuts from the engine and transmission support.

7. Remove the transaxle mount from the vehicle.

To install:

NOTE: The engine and transaxle position must be realigned to the front subframe.

8. Raise and safely support the vehicle.

9. Remove the left-hand front engine mount retaining bolts and through-bolt.

10. Remove the left-hand engine mount.

11. Install in its place, Powertrain alignment Gauge T94P–6000–AH, or equivalent to the transaxle bracket and the front subframe.

12. Install and torque the 2 retaining bolts and then the through-bolt to 20 ft. lbs. (27 Nm).

13. Loosen the right-hand front engine mount (support insulator) through-bolt.

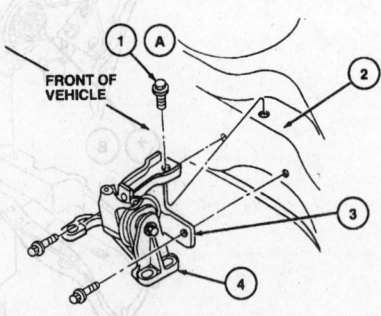

1. Bolt(3)
2. Front fender apron(LHL)
3. Engine and transmission
   support insulator bolt
4. Bolt
5. Engine and transmission
   support
   A. 52-70 ft. lb.(70-95 Nm)

297739

Transaxle mounting to the front fender apron — 2.0L (VIN 3) engine

14. Lower the vehicle.

15. Install the transaxle mount and its retaining bolts to the front fender apron. Torque the retaining bolts to 52–70 ft. lbs. (70–95 Nm).

NOTE: The transaxle mount uses self-locking nuts. These nuts must always be replaced with new ones once removed.

16. Install new retaining nuts to the transaxle mount and torque the retaining nuts to 7 ft. lbs. (10 Nm).

17. Lower the vehicle.

18. Remove the three bar engine support tool, or equivalent.

19. Retorque the transaxle mount retaining nuts to 52–70 ft. lbs. (70–95 Nm) for manual transaxles or 30–40 ft. lbs. (41–55 Nm) for automatic transaxles.

20. Raise and safely support the vehicle.

21. Check the position of the right-hand front engine mount (support insulator). It must be centered in the transaxle bracket and in perfect front to rear alignment.

22. Torque the right-hand front engine mount through-bolt to 75–102 ft. lbs. (103–137 Nm).

23. Remove the 2 retaining bolts and the through-bolt securing the powertrain alignment gauge.

24. Remove the powertrain alignment gauge.

25. Install the left–hand front engine mount to the front subframe using the 2 retaining bolts. Torque the retaining bolts to 7 ft. lbs. (10 Nm).

26. Check the position of the left-hand front engine mount to ensure perfect front to rear alignment.

27. Retorque the 2 retaining bolts to 30–40 ft. lbs. (41–55 Nm).

28. Install the left-hand front engine mount through-bolt. Torque the through-bolt to 75–102 ft. lbs. (103–137 Nm).

29. Lower the vehicle.

30. Install the air cleaner mounting bracket and the bracket retaining bolts. Torque the retaining bolts to 15–22 ft. lbs. (20–30 Nm).

31. Install the engine air cleaner.

32. Connect the negative battery cable.

33. Check for proper operation.

2.5L (VIN L) Engine

NOTE: When replacing mounts using self-locking nuts, make sure to have new replacement nuts available as the originals loose their holding capabilities and must not be reused.

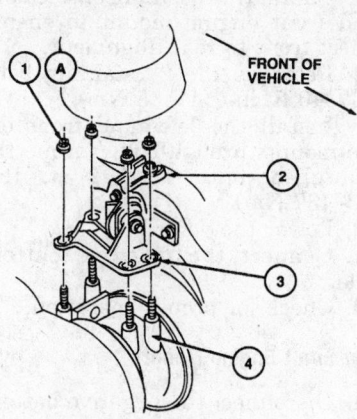

1. Nut(4)
2. Engine and transmission support insulator
3. Bolt
4. Engine and transmission support
5. Transaxle
A. 30–40 ft. lb.(41–55 Nm)

297740

Transaxle mount and components for automatic transaxle — 2.0L (VIN 3) engine

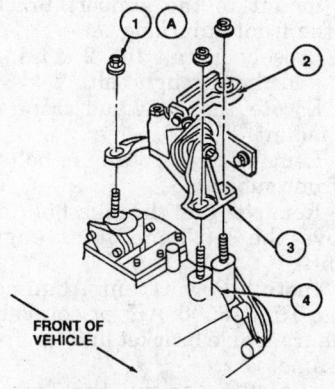

1. Nut(3)
2. Engine and transmission support insulator
3. Bolt
4. Engine and transmission support
5. Transaxle
A. 52–70 ft. lb.(70–95 Nm)

297741

Transaxle mount and components for manual transaxle — 2.0L (VIN 3) engine

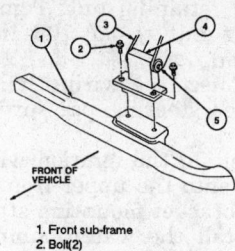

1. Front sub-frame
2. Bolt(2)
3. Front engine support bracket(LH)
4. Powertrain alignment gauge
5. Through bolt

297732

Installing the Powertrain Alignment Gauge — 2.0L (VIN 3) engine

NOTE: When replacing the upper front engine mount, it is recommended that the left-hand front and right-hand front engine mounts be replaced at the same time.

Upper Front Engine Mount

1. Position the vehicle over a hoist but do not raise at this time.
2. Disconnect the negative battery cable.
3. Remove the 4 screws securing the water pump shield and remove the shield.
4. Install Three Bar Engine Support D88L–6000–A, or equivalent to the engine lifting eyes and support the engine.
5. Disconnect the wiring harness from the power steering pump switch connector.
6. Remove the power steering pressure hose bracket retaining bolt from the upper engine support bracket.
7. Place a shop towel under the power steering pressure hose connection at the power steering pump. Remove the hose and move aside. Allow the shop towel to absorb the fluid to prevent it from spilling onto the drive belt.
8. Position the ignition wires away from the upper front engine support bracket.
9. Remove the upper front engine support retaining nuts from the engine and the upper front engine mount (front engine support insulator).
10. Remove the upper front engine support bracket from the vehicle.
11. Remove the retaining bolts for the radiator coolant recovery reservoir and position the reservoir aside.
12. Remove the upper front engine mount retaining bolts.
13. Remove the upper front engine mount.

To install:

NOTE: The engine and transaxle position must be realigned to the front subframe.

14. Raise and safely support the vehicle.
15. Remove the 2 retaining bolts to the front subframe.
16. Remove the through-bolt.
17. Remove the left-hand front engine mount (support insulator).
18. Install Powertrain Alignment Gauge T94P–6000–AH, or equivalent to the transaxle bracket and the front subframe.
19. Install and torque the 2 retaining bolts and then the through-bolt to 20 ft. lbs. (27 Nm).
20. Loosen the right-hand front engine mount (support insulator) through-bolt.
21. Lower the vehicle.
22. Install the new upper front engine mount (support insulator) and its retaining bolts to the right-hand front fender apron. Torque the retaining bolts to 52–70 ft. lbs. (70–95 Nm).
23. Position the radiator coolant recovery reservoir and install the reservoir retaining bolts.

NOTE: The upper front engine support bracket uses self–locking nuts. These nuts must always be replaced with new ones once removed.

24. Install the upper front engine support bracket onto the engine and the upper front engine mount. Install new retaining nuts and torque the retaining nuts to 7 ft. lbs. (10 Nm).
25. Lower the vehicle.
26. Remove the three bar engine support tool, or equivalent.
27. Retorque the upper front engine support bracket nuts to 52–70 ft. lbs. (70–95 Nm).
28. Raise and safely support the vehicle.
29. Check the position of the right-hand front engine mount (support insulator). It must be centered in the transaxle bracket and in perfect front to rear alignment.
30. Torque the right-hand front engine mount through-bolt to 75–102 ft. lbs. (103–137 Nm).
31. Remove the 2 retaining bolts and the through-bolt securing the powertrain alignment gauge.
32. Remove the powertrain alignment gauge.
33. Install the left-hand front engine mount to the front subframe using the 2 retaining bolts. Torque the retaining bolts to 7 ft. lbs. (10 Nm).

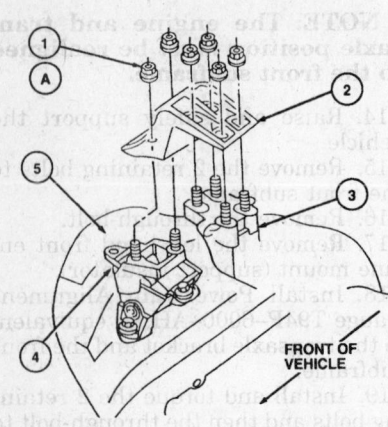

1. Nut(6)
2. Front engine support bracket(upper)
3. Power steering pump bracket
4. Front engine support insulator(upper)
5. Front fender apron(RH)
A. 52-70 ft. lb.(70-95 Nm)

297747

Upper front engine support bracket removal — 2.5L (VIN L) engine

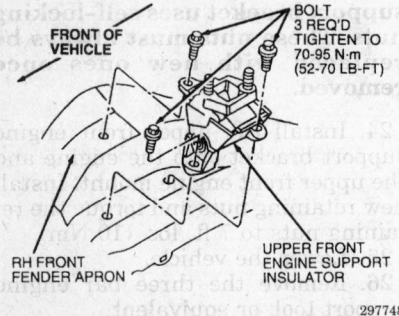

297748

Upper front engine mount removal — 2.5L (VIN L) engine

34. Check the position of the left-hand front engine mount to ensure perfect front to rear alignment.

35. Retorque the 2 retaining bolts to 30-40 ft. lbs. (41-55 Nm).

36. Install the left-hand front engine mount through-bolt. Torque the through-bolt to 75-102 ft. lbs. (103-137 Nm).

37. Lower the vehicle.

38. Install the power steering pressure hose to the power steering pump.

39. Position the power steering pressure hose bracket to the upper front engine support bracket and in-

stall the retaining nut. Tighten the retaining nut to 71-106 inch lbs. (8-12 Nm).

40. Connect the wiring harness to the power steering pressure switch connector.

41. Position the ignition wires and retainer onto the upper front engine support bracket mounting studs.

42. Install the water pump pulley shield and tighten the 4 screws securely.

43. Fill the power steering system with the proper power steering fluid.

44. Connect the negative battery cable.

45. Run the engine and check for leaks and proper operation.

Left Front Engine Mount

1. Disconnect the negative battery cable.

2. Raise and safely support the vehicle.

3. Remove the 2 bolts retaining the left-hand front engine mount (support insulator) to the front subframe.

4. Remove the left-hand front engine mount through-bolt.

5. Remove the left-hand engine mount from the vehicle.

To install:

6. Install the left-hand front engine mount to the front subframe us-

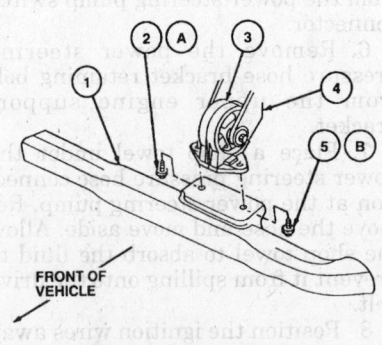

1. Front sub-frame
2. Bolt(2)
3. Front engine support insulator(LH)
4. Front engine support bracket(LH)
5. Through bolt
A. 30-40 ft. lb.(41-55 Nm)
B. 75-102 ft. lb.(103-137 Nm)

297750

Left-hand front engine mount — 2.5L (VIN L) engine

ing with 2 retaining bolts. Torque the retaining bolts to 7 ft. lbs. (10 Nm).

7. Check the position of the left-hand front engine mount to ensure perfect front to rear alignment.

8. Retorque the 2 retaining bolts to 30-40 ft. lbs. (41-55 Nm).

9. Install the left-hand front engine mount through-bolt. Torque the through-bolt to 75-102 ft. lbs. (103-137 Nm).

10. Lower the vehicle.

11. Connect the negative battery cable.

12. Check for proper operation.

Right Front Engine Mount

1. Disconnect the negative battery cable.

2. Raise and safely support the vehicle.

3. Remove the through-bolt from the right-hand front engine mount (support insulator).

4. Remove the 2 retaining bolts securing the right-hand front engine mount to the front subframe.

5. Remove the right-hand front engine mount from the vehicle through the bottom of the right-hand side of the front subframe, past the right-hand exhaust manifold and the alternator.

To install:

6. Install the right-hand front engine mount to the support bracket and the front subframe.

7. Loosely install the 2 retaining bolts and the through-bolt.

8. Locate the left-hand front engine mount.

9. Remove the 2 retaining bolts to the front subframe.

10. Remove the through-bolt and remove the left-hand front engine mount.

11. Install Powertrain Alignment Gauge T94P-6000-AH, or equivalent to the transaxle bracket and the front subframe.

12. Install and torque the 2 retaining bolts and then the through-bolt to 20 ft. lbs. (27 Nm).

13. Torque the right-hand front engine mount retaining bolt to the front subframe to 7 ft. lbs. (10 Nm).

14. Check the position of the right-hand front engine mount (support insulator). It must be centered in the transaxle bracket and in perfect front to rear alignment.

15. Retorque the 2 retaining bolts to 30-40 ft. lbs. (41-55 Nm).

16. Install the right-hand front engine mount through-bolt. Torque the through-bolt to 75-102 ft. lbs. (103-137 Nm).

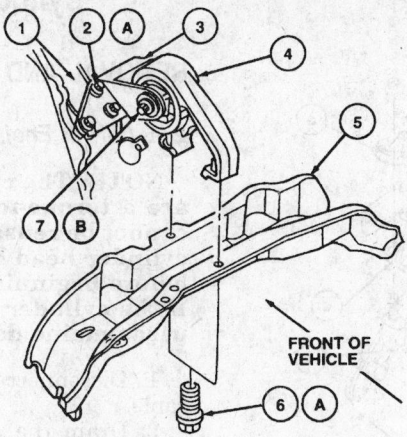

1. Transaxle
2. Bolt(3)
3. Front engine support bracket(RH)
4. Front engine support insulator(RH)
5. Front sub-frame
6. Bolt(2)
7. Through bolt
A. 30-40 ft. lb.(41-55 Nm)
B. 75-102 ft. lb.(103-137 Nm)

297751

Right-hand front engine mount used with automatic transaxles — 2.5L (VIN L) engine

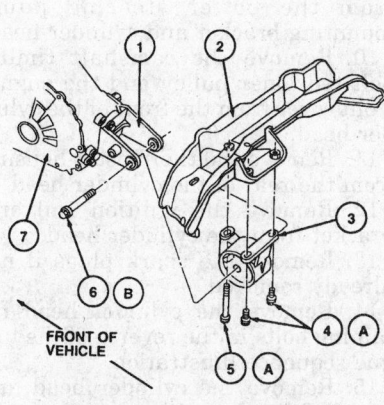

1. Front engine support bracket(RH)
2. Front sub-frame
3. Front engine support insulatorRH)
4. Bolt(2)
5. Bolt
6. Through bolt
7. Transaxle
A. 30-40 ft. lb.(41-55 Nm)
B. 75-102 ft. lb.(103-137 Nm)

297752

Right-hand front engine mount used with manual transaxles — 2.5L (VIN L) engine

17. Remove the 2 retaining bolts and the through-bolt securing the powertrain alignment gauge.
18. Remove the powertrain alignment gauge.
19. Install the left-hand front engine mount to the front subframe using the 2 retaining bolts. Torque the retaining bolts to 7 ft. lbs. (10 Nm).
20. Check the position of the left-hand front engine mount to ensure perfect front to rear alignment.
21. Retorque the 2 retaining bolts to 30–40 ft. lbs. (41–55 Nm).
22. Install the left-hand front engine mount through-bolt. Torque the through-bolt to 75–102 ft. lbs. (103–137 Nm).
23. Lower the vehicle.
24. Connect the negative battery cable.
25. Check for proper operation.

Transaxle Mount

1. Disconnect both battery cables, negative cable first.
2. Remove the engine air cleaner.
3. Remove the battery and the battery tray.
4. Remove the engine air cleaner support bracket from the transaxle mount (engine and transmission support).

5. Remove the 4 retaining screws securing the water pump shield and remove the shield.
6. Install Three Bar Engine Support D88L–6000–A, or equivalent to the engine lifting eyes and support the engine and transaxle.
7. Remove the transaxle mount retaining bolts to the left-hand front fender apron.
8. Remove the retaining nuts from the engine and transmission support.
9. Remove the transaxle mount from the vehicle.
To install:

NOTE: The engine and transaxle position must be realigned to the front subframe.

10. Raise and safely support the vehicle.
11. Remove the left-hand front engine mount retaining bolts and through-bolt.
12. Remove the left-hand engine mount.
13. Install in its place, Powertrain alignment Gauge T94P–6000–AH, or equivalent to the transaxle bracket and the front subframe.
14. Install and torque the 2 retaining bolts and then the through-bolt to 20 ft. lbs. (27 Nm).
15. Loosen the right-hand front engine mount (support insulator) through-bolt.
16. Lower the vehicle.
17. Install the transaxle mount and its retaining bolts to the front fender apron. Torque the retaining bolts to 52–70 ft. lbs. (70–95 Nm).

NOTE: The transaxle mount uses self-locking nuts. These nuts must always be replaced with new ones once removed.

18. Install new retaining nuts to the transaxle mount and torque the retaining nuts to 7 ft. lbs. (10 Nm).
19. Lower the vehicle.
20. Remove the three bar engine support tool, or equivalent.
21. Retorque the transaxle mount retaining nuts to 52–70 ft. lbs. (70–95 Nm) for manual transaxles or 30–40 ft. lbs. (41–55 Nm) for automatic transaxles.
22. Raise and safely support the vehicle.
23. Check the position of the right-hand front engine mount (support insulator). It must be centered in the transaxle bracket and in perfect front to rear alignment.
24. Torque the right-hand front engine mount through-bolt to 75–102 ft. lbs. (103–137 Nm).

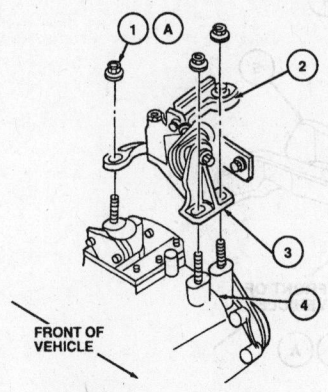

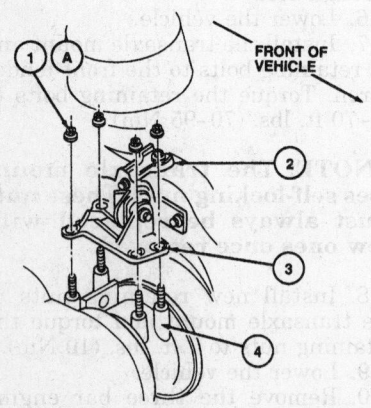

1. Bolt(3)
2. Front fender apron(LHL)
3. Engine and transmission support insulator bolt
4. Bolt
5. Engine and transmission support
A. 52–70 ft. lb.(70–95 Nm)

297753

Transaxle mounting to the front fender apron — 2.5L (VIN L) engine

1. Nut(3)
2. Engine and transmission support insulator
3. Bolt
4. Engine and transmission support
5. Transaxle
A. 52–70 ft. lb.(70–95 Nm)

297755

Transaxle mount and components for manual transaxle — 2.5L (VIN L) engine

1. Nut(4)
2. Engine and transmission support insulator
3. Bolt
4. Engine and transmission support
5. Transaxle
A. 30–40 ft. lb.(41–55 Nm)

297754

Transaxle mount and components for automatic transaxle — 2.5L (VIN L) engine

25. Remove the 2 retaining bolts and the through-bolt securing the powertrain alignment gauge.
26. Remove the powertrain alignment gauge.
27. Install the left-hand front engine mount to the front subframe using the 2 retaining bolts. Torque the retaining bolts to 7 ft. lbs. (10 Nm).
28. Check the position of the left-hand front engine mount to ensure perfect front to rear alignment.
29. Retorque the 2 retaining bolts to 30–40 ft. lbs. (41–55 Nm).
30. Install the left-hand front engine mount through-bolt. Torque the through-bolt to 75–102 ft. lbs. (103–137 Nm).
31. Lower the vehicle.
32. Install the battery tray and the battery.
33. Install the air cleaner mounting bracket and the bracket retaining bolts.
34. Torque the retaining bolts to 15–22 ft. lbs. (20–30 Nm).
35. Install the engine air cleaner.
36. Install the water pump pulley shield and the water pump shield retaining screws. Tighten the screws securely.
37. Connect both battery cables, negative cable last.
38. Check for proper operation.

Cylinder Head

REMOVAL AND INSTALLATION

2.0L (VIN 3) Engine

NOTE: The cylinder head bolts are a torque-to-yield design and cannot be reused. Make sure new cylinder head bolts are available before beginning this procedure. If the cylinder head bolts are re-used, engine damage may occur.

1. Disconnect the negative battery cable.
2. Drain the engine coolant from the radiator and the cylinder block drain plugs.
3. Remove the intake manifold.
4. Remove the exhaust manifold.
5. Remove the camshafts and valve tappets.
6. Support the engine with a wood block between the crankshaft pulley and the front subframe.
7. Remove the Three Bar Engine Support D88L-6000-A or equivalent, previously installed for the timing belt cover removal.
8. Remove the right–hand engine lifting eye retaining bolt and the lifting eye.
9. Remove the support bracket from the power steering pump mounting bracket and cylinder head.
10. Remove the camshaft timing belt tensioner pulley and the engine front cover from the front of the cylinder head.
11. Remove the thermostat housing from the rear of the cylinder head.
12. Remove the ignition coil and bracket from the cylinder head.
13. Remove the spark plugs if not already removed.
14. Remove the cylinder head retaining bolts in the reverse of the torque sequence illustration.
15. Remove the cylinder head and gasket from the engine.
16. If the cylinder head is to be serviced, remove the left–hand engine lifting eye.
To install:
17. Clean the cylinder head and cylinder block gasket surfaces and check for flatness.
18. Install a new cylinder head gasket onto the cylinder block . Make sure the head gasket is properly positioned on the dowels.

───── **WARNING** ─────
Use care when positioning the cylinder head to prevent damage to the head gasket or dowels.

19. Place a light coating of engine oil onto the threads of the new cylinder head bolts and install.

20. Torque the cylinder head bolts in sequence and in the following steps:

- Torque all bolts to 15–22 ft. lbs. (20–30 Nm)
- Torque all bolts to 30–37 ft. lbs. (40–50 Nm)
- Rotate all bolts 90–120 degrees.

21. Install the ignition coil bracket and the ignition coil.

22. Install the water thermostat housing.

23. Install the engine front cover. Tighten the retaining bolts to 71–97 inch lbs. (8–11 Nm).

24. Install the camshaft timing belt tensioner pulley and retaining bolt onto the front of the cylinder head.

25. Install the support bracket to the power steering pump mounting bracket and the cylinder head.

26. Torque the support bracket to 29–41 ft. lbs. (39–55 Nm).

27. Install the right-hand engine lifting eye to the cylinder head and the alternator mounting bracket. Torque the retaining bolts to 30–41 ft. lbs. (41–55, Nm).

28. If removed, install the left–hand engine lifting eye to the cylinder head and torque to 10–13 ft. lbs. (14–18 Nm).

29. Install the Three Bar Engine Support D88L-6000-A or equivalent to the engine lifting eyes and support the engine.

30. Remove the wood block from between the subframe and the crankshaft pulley.

31. Install the valve tappets and camshaft into their original locations.

32. Install the exhaust manifold.

33. Install the intake manifold.

34. Install the spark plugs.

35. Drain the engine oil and remove the engine oil filter.

36. Reinstall the drain plug and torque to 15–21 ft. lbs. (21–28 Nm).

37. Install a new engine oil filter and fill the crankcase with the proper amount and grade of oil.

38. Fill the engine cooling system.

39. Connect the negative battery cable.

40. Run the engine and check for oil and coolant leaks. Check for proper engine operation.

2.5L (VIN L) Engine

NOTE: The cylinder head bolts are a torque-to-yield design and cannot be reused. Make sure new cylinder head bolts are available before beginning this procedure.

1. Disconnect the negative battery cable.

2. Drain the engine coolant from the radiator and cylinder block drain plugs.

3. Close the radiator draincock and install the drain plugs into the cylinder block.

4. Remove the upper and lower intake manifolds.

5. Drain the engine oil and remove the oil pan.

6. Remove the alternator and the alternator mounting bracket.

7. Remove the heated oxygen sensor from the right-hand cylinder head exhaust manifold.

8. Remove the left-hand cylinder head exhaust manifold.

9. Remove the water pump.

10. Remove the engine front cover.

11. Install the upper front engine mount (support insulator) and the upper front engine support bracket to the front of the engine and the right-hand front fender apron.

12. Remove Three Bar Engine Support D88L-60000-A or equivalent.

13. Remove the camshafts and valve tappets from both cylinder heads.

14. Disconnect the hoses from the EGR pressure sensor at the EGR valve, to the exhaust manifold tube.

15. Disconnect the electrical connector to the EGR pressure sensor.

16. Disconnect the interior vacuum source hose from the main emission vacuum harness.

17. Disconnect the fuel vapor hose from the PCV valve.

18. Disconnect the electrical connector to the EGR transducer.

19. Remove the EGR valve to exhaust manifold tube from the right-hand exhaust manifold and remove from the vehicle.

20. Remove the wiring retaining bracket from the EGR transducer bracket.

21. Remove the engine air cleaner.

22. Remove the crankcase ventilation tube from the water crossover and oil separator.

23. Remove the water crossover retaining bolt and stud bolt from the right-hand cylinder head. Set the water crossover aside.

24. Remove the oil level dipstick from the left-hand cylinder head.

25. Remove the cylinder head retaining bolts from the cylinder heads in the reverse of the torque sequence illustration.

26. Remove the right-hand cylinder head with the exhaust manifold and the EGR transducer bracket attached.

27. If required, remove the exhaust manifold and the EGR transducer

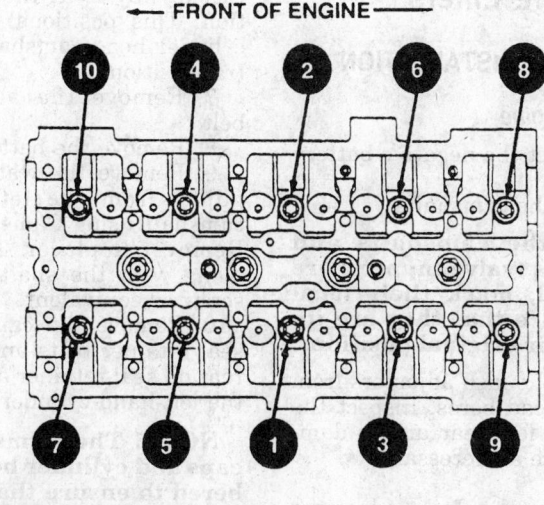

← FRONT OF ENGINE →

TIGHTEN BOLTS IN SEQUENCE SHOWN

297400

Cylinder head retaining bolt torque sequence — 2.0L (VIN 3) engine

bracket from the right-hand cylinder head.

28. Remove the left-hand cylinder head.

29. Inspect the cylinder heads and cylinder block.

To install:

30. Clean the cylinder heads, intake manifolds, valve covers and the cylinder head gasket sealing surfaces on the cylinder block.

NOTE: If the cylinder heads were removed for cylinder head gasket replacement, check the flatness of the cylinder heads and the cylinder block gasket sealing surfaces.

31. If removed, install the right-hand exhaust manifold and EGR transducer bracket to the right-hand cylinder head.

32. Install new cylinder head gaskets onto the dowels of the cylinder block.

33. Position the cylinder heads into their original positions using care not to damage the heads, block or gaskets.

34. Make sure that the cylinder heads are correctly positioned on the dowels.

35. Lightly oil the threads of the new cylinder head retaining bolts and install into the cylinder heads.

36. Tighten the new cylinder head retaining bolts as follows:
   • Torque the bolts, in sequence, to 27–32 ft. lbs. (37–43 Nm)
   • Rotate the bolts, in sequence, 85–95 degrees
   • Loosen the bolts, in sequence, a minimum of 1 full turn
   • Torque the bolts, in sequence, to 27–32 ft. lbs. (37–43 Nm)
   • Rotate the bolts, in sequence, 85–95 degrees
   • Rotate the bolts, in sequence, an additional 85–95 degrees

37. Inspect the water crossover O-rings and replace if required.

38. Install the water crossover to the cylinder heads and tighten the retaining bolts to 71–106 inch lbs. (8–12 Nm).

39. Install the crankcase ventilation tube. Tighten the tube to 44–62 inch lbs. (5–7 Nm).

40. Install the oil level dipstick to the left-hand cylinder head.

41. Install the engine air cleaner.

42. Install the wiring retaining bracket onto the EGR transducer bracket.

43. Install the EGR to exhaust manifold tube onto the right-hand exhaust manifold.

44. Connect the electrical connectors to the EGR transducer and the EGR pressure sensor.

45. Connect the fuel vapor hose to the PCV valve.

46. Connect the interior vacuum source hoses to the main emission vacuum harness.

47. Connect the hoses from the EGR pressure sensor and EGR valve to the exhaust manifold tube.

48. Install the camshafts and valve tappets.

49. Install the Three Bar Engine Support D88L-6000-A or equivalent, to the engine lifting eyes and support the engine.

50. Install the engine front cover.

51. Install the left-hand exhaust manifold.

52. Install the water pump.

53. Install the heated oxygen sensor to the right-hand exhaust manifold.

54. Install the alternator mounting bracket and the alternator.

55. Install the engine oil pan.

56. Install the lower and upper intake manifolds.

57. Replace the engine oil filter.

58. Fill the engine with the proper amount and grade of engine oil.

59. Fill the engine cooling system.

60. Connect the negative battery cable.

61. Run the engine and check for oil and coolant leaks. Check for proper engine operation.

Valve Lifters

REMOVAL AND INSTALLATION

2.0L (VIN 3) Engine

1. Disconnect the negative battery cable.

2. Remove the camshafts.

NOTE: If the camshafts and lash adjusters (valve tappets) are to be reused, mark their locations to ensure that they are installed in their original positions.

3. Remove the lash adjusters from the cylinder head bores. Inspect the lash adjusters for wear and/or damage and replace as necessary.

To install:

4. Lubricate the lash adjusters with engine assembly lubricant prior to installation.

5. If the lash adjusters are new, soak them in a container of clean engine oil and manually pump up the lash adjusters before installation.

6. Install the lash adjusters in their proper locations.

7. Install the camshafts.

8. Connect the negative battery cable.

9. Run the engine and check for leaks and proper operation.

2.5L (VIN L) Engine

1. Disconnect the negative battery cable.

2. Remove the valve covers as follows:
   a. Remove the ignition wires and spark plugs.
   b. Remove the ignition coil from the right-hand valve cover.
   c. Remove the crankcase ventilation tubes from both valve covers.
   d. Remove the wiring harness and bracket to the fuel injectors and move aside.
   e. Remove the retaining nuts and engine wiring from both valve covers and move aside.
   f. Loosen the valve cover retaining bolts and studs in reverse of the torque sequence illustration.
   g. Remove both valve covers from the engine.

3. Remove the crankshaft pulley retaining bolt.

4. Rotate the crankshaft so that the keyway is at the 11 o'clock position to locate the crankshaft at TDC for No. 1 cylinder.

5. Verify that the alignment arrows on the camshafts are aligned. If not, rotate the crankshaft one complete revolution and recheck.

6. Rotate the crankshaft so that the keyway is at the 3 o'clock position. This positions the right-hand cylinder head camshafts to the neutral position.

7. Remove the accessory drive belt.

8. Remove the battery.

9. Remove the water pump drive pulley from the left-hand intake camshaft using Camshaft Dampener Remover/Replacer T94P-6312-AH along with the shaft protector and screw or equivalent.

10. Remove the camshaft rear oil seal retainer bolts and the camshaft rear oil seal retainer and gasket from the left-hand cylinder head.

NOTE: The camshaft journal caps and cylinder heads are numbered to ensure that they are assembled in their original positions. If removed, keep the camshaft journal caps together with the cylinder head that they were removed from.

11. Remove the right-hand cylinder head camshaft journal thrust cap retaining bolts and thrust caps. Loosen

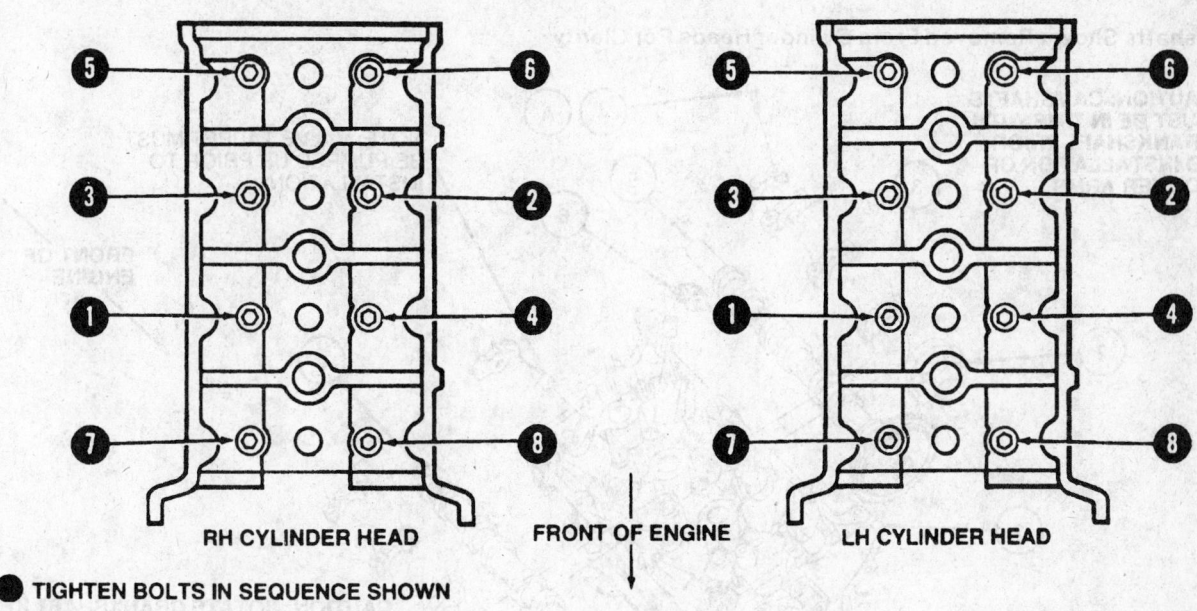

RH CYLINDER HEAD   FRONT OF ENGINE   LH CYLINDER HEAD

● **TIGHTEN BOLTS IN SEQUENCE SHOWN**

297446

Cylinder head bolt torque sequence — 2.5L (VIN L) engine

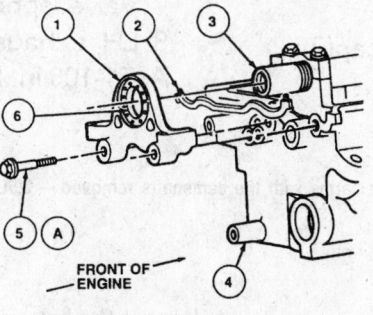

FRONT OF ENGINE

1. Valve cover
2. Water pump pulley remover/replacer
3. Water pump drive pulley
4. Screw (2)
5. Shaft protector
6. Crankshaft damper remover

298201

Removing the water pump drive pulley — 2.5L (VIN L) engine

FRONT OF ENGINE

1. Camshaft rear oil seal retainer
2. Oil seal retainer gasket
3. LH intake camshaft
4. LH cylinder head bolt(2)
5. Bolt(2)
6. Camshaft rear oil seal
A. 71–106 in. lb.(8–12 Nm)

298202

Left-hand cylinder head camshaft rear oil seal retainer — 2.5L (VIN L) engine

the bolts in reverse of the torque sequence illustration.

12. Loosen the remaining camshaft journal cap bolts in sequence, releasing the bolts several revolutions at a time by making several passes to allow the camshaft to be raised from the cylinder head evenly. Do not remove the retaining bolts completely.

NOTE: If the lash adjusters (valve tappets) and rocker arms are to be reused, mark the positions of the lash adjusters and rocker arms so that they are reassembled into their original positions.

13. With the camshafts loose, remove the rocker arms, keeping them in the order that they were removed.
14. Remove the lash adjusters from the cylinder head.
15. Rotate the crankshaft 2 revolutions and locate the crankshaft keyway at the 11 o'clock position. This will position the left-hand cylinder head camshafts to their neutral position.

Camshafts Shown Removed From Cylinder Heads For Clarity

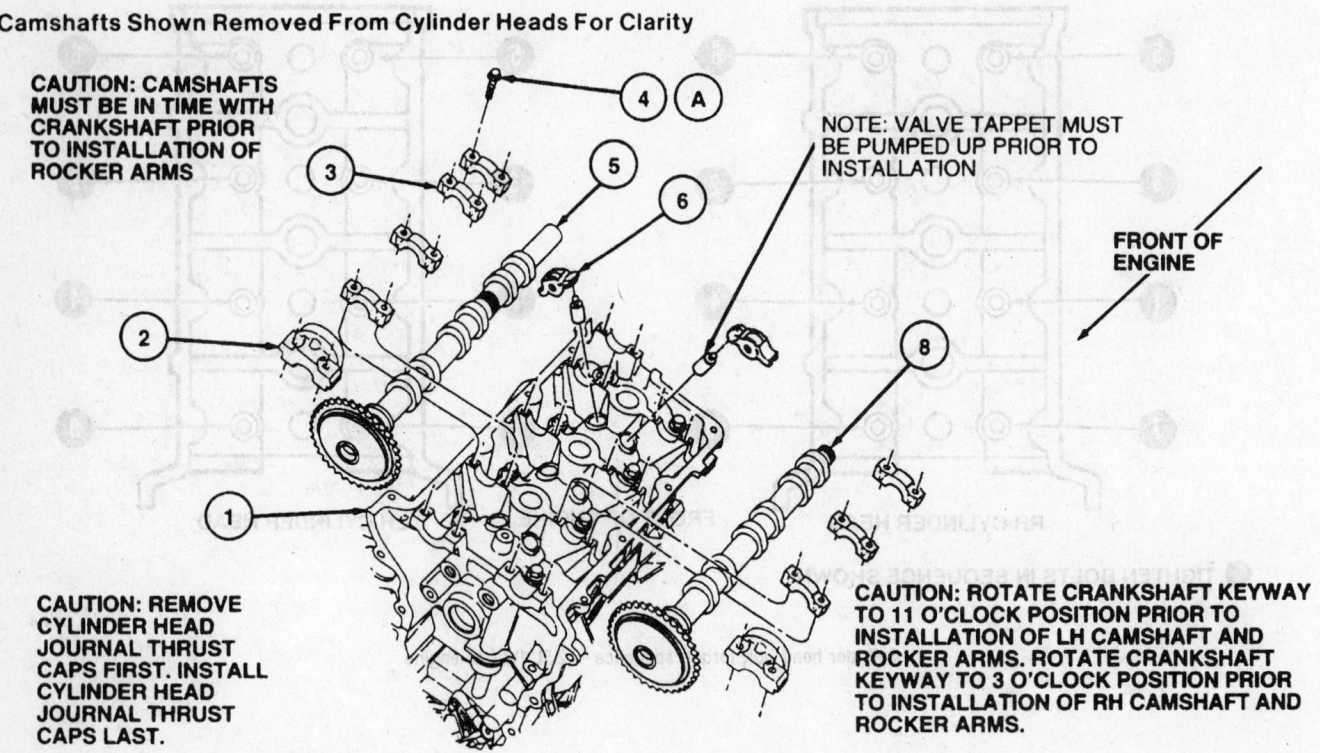

CAUTION: CAMSHAFTS MUST BE IN TIME WITH CRANKSHAFT PRIOR TO INSTALLATION OF ROCKER ARMS

NOTE: VALVE TAPPET MUST BE PUMPED UP PRIOR TO INSTALLATION

FRONT OF ENGINE

CAUTION: REMOVE CYLINDER HEAD JOURNAL THRUST CAPS FIRST. INSTALL CYLINDER HEAD JOURNAL THRUST CAPS LAST.

CAUTION: ROTATE CRANKSHAFT KEYWAY TO 11 O'CLOCK POSITION PRIOR TO INSTALLATION OF LH CAMSHAFT AND ROCKER ARMS. ROTATE CRANKSHAFT KEYWAY TO 3 O'CLOCK POSITION PRIOR TO INSTALLATION OF RH CAMSHAFT AND ROCKER ARMS.

1. Cylinder head
2. Camshaft journal thrust cap(2)
3. Camshaft journal cap(7)
4. Bolt(18)
5. LH intake camshaft
6. Rocker arm(12)
7. Valve tappet(12)
8. LH exhaust camshaft
A. 71-106 in. lb.(8-12 Nm)

View of rocker arms with the camshafts removed — 2.5L (VIN L) engine

16. Verify that the alignment arrows on the camshafts are aligned.

NOTE: The camshaft journal caps and cylinder heads are numbered to ensure that they are assembled in their original positions. If removed, keep the camshaft journal caps together with the cylinder head that they were removed from.

17. Remove the camshaft journal thrust cap retaining bolts and thrust caps from the left-hand cylinder head. Loosen the bolts in reverse of the torque sequence illustration.

18. Loosen the remaining camshaft journal cap bolts in reverse of the tor-que sequence, releasing the bolts several revolutions at a time by making several passes to allow the camshaft to be raised from the cylinder head evenly. Do not remove the retaining bolts completely.

NOTE: If the lash adjusters (valve tappets) and rocker arms are to be reused, mark the positions of the lash adjusters and rocker arms so that they are reassembled into their original positions.

19. With the camshafts loose, remove the rocker arms, keeping them in the order that they were removed.

20. Remove the lash adjusters from the cylinder head.

21. Inspect the rocker arms and lash adjusters for wear and/or damage and replace as necessary.

To install:

22. Make sure that the crankshaft keyway is at the 11 o'clock position.

23. Lubricate the left-hand cylinder head lash adjusters with engine assembly lubricant and install into their correct positions in the cylinder head.

24. If the lash adjusters are being replaced with new units, soak the lash adjusters in a container of clean engine oil and manually pump up the

lash adjusters before installing into the cylinder head.

25. If the original lash adjusters are to be installed, pump up each lash adjuster before assembly. Replace any lash adjuster that refuses to pump up.

NOTE: The lash adjusters (valve tappets) must be pumped up prior to installation.

26. Lubricate the left-hand cylinder head rocker arms with engine assembly lubricant and install the left-hand cylinder head rocker arms into their original locations.

NOTE: Do not install the camshaft journal thrust caps until the camshaft journal caps are torqued into position.

27. Tighten the left-hand cylinder head camshaft journal cap bolts in sequence making several passes to pull the camshafts down evenly. Torque the bolts to 71–106 inch lbs. (8–12 Nm).

28. Install the left–hand cylinder head thrust caps and bolts. Torque to 71–106 inch lbs. (8–12 Nm).

29. Rotate the crankshaft 2 revolutions and position the crankshaft keyway to the 3 o'clock location. This will position the right-hand cylinder head camshafts to the neutral position.

30. Lubricate the right-hand cylinder head lash adjusters with engine assembly lubricant and install into their original positions in the cylinder head.

31. If the lash adjusters are being replaced with new units, soak the lash adjusters in a container of clean engine oil and manually pump the lash adjusters up before installing into the cylinder head.

NOTE: The lash adjusters (valve tappets) must be pumped up prior to installation.

32. Lubricate the right-hand cylinder head rocker arms with engine assembly lubricant and install the right-hand cylinder head rocker arms into their original locations.

NOTE: Do not install the camshaft journal thrust caps until the camshaft journal caps are torqued into position.

33. Tighten the right-hand cylinder head camshaft journal cap bolts in sequence making several passes to pull the camshafts down evenly. Torque the bolts to 71–106 inch lbs. (8–12 Nm).

34. Install the right–hand cylinder head thrust caps and bolts. Torque to 71–106 inch lbs. (8–12 Nm).

35. Install the left-hand cylinder head camshaft rear oil seal and retainer with gasket onto the cylinder head. Tighten the retaining bolts to 71–106 inch lbs. (8–12 Nm).

36. Install the water pump drive pulley onto the left-hand intake camshaft using Power Steering Pump Pulley Replacer T91P-3A733-A, screw and replacer cup or equivalent.

37. Install the accessory drive belts.

38. Install the battery.

39. Install the crankshaft pulley retaining bolt.

40. Tighten the crankshaft pulley retaining bolt as follows:

   a. Torque to 89 ft. lbs. (120 Nm).

   b. Loosen the bolt at least 1 full turn.

   c. Torque the bolt to 35–39 ft. lbs. (47–53 Nm).

   d. Rotate the bolt 85–95 degrees.

41. Install both valve covers as follows:

   a. Clean the valve cover gasket sealing surfaces.

   b. Install new valve cover gaskets onto the valve covers.

   c. For each valve cover, place a bead of silicone sealant at 2 places on the valve cover sealing surfaces where the engine front cover and

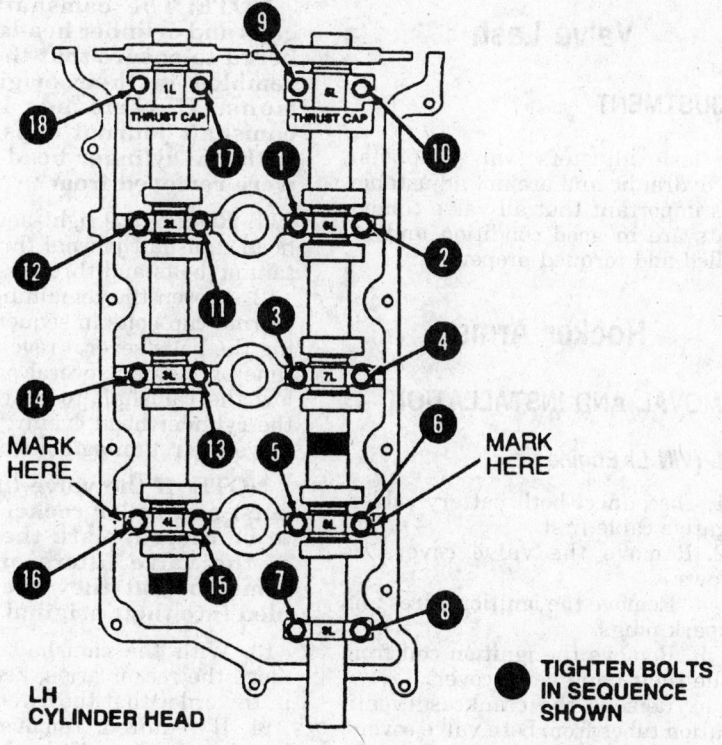

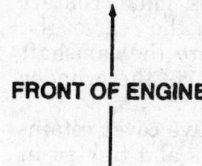

CAUTION: INSTALL CAMSHAFT JOURNAL THRUST CAPS AFTER INSTALLING OTHER CAMSHAFT JOURNAL CAPS. DAMAGE TO THE CAMSHAFT JOURNAL THRUST CAPS MAY OCCUR IF NOT INSTALLED LAST.

FRONT OF ENGINE

Torque sequence for the left-hand cylinder head camshaft journal caps — 2.5L (VIN L) engine

298207

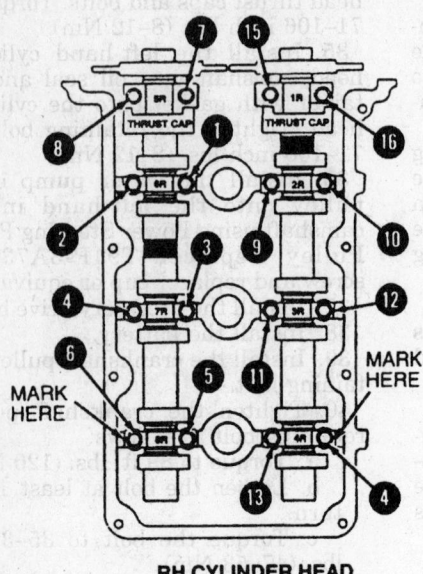

CAUTION: INSTALL CAMSHAFT JOURNAL THRUST CAPS AFTER INSTALLING OTHER CAMSHAFT JOURNAL CAPS. DAMAGE TO THE CAMSHAFT JOURNAL THRUST CAPS MAY OCCUR IF NOT INSTALLED LAST.

FRONT OF ENGINE

RH CYLINDER HEAD

TIGHTEN BOLTS IN SEQUENCE SHOWN

Torque sequence for the right-hand cylinder head camshaft journal caps — 2.5L (VIN L) engine

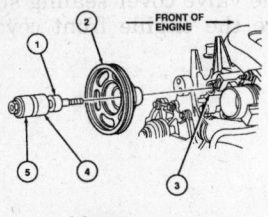

1. Screw
2. Water pump drive pulley
3. LH intake camshaft
4. Replacer cup
5. Power steering pump pulley replacer

Installing the water pump drive pulley — 2.5L (VIN L) engine

the cylinder heads make contact and at 2 places on the rear of the cylinder head where the camshaft seal retainer contacts the cylinder head.

d. Install the valve cover retaining bolts and studs and tighten in sequence to 71–106 inch lbs. (8–12 Nm).

NOTE: The valve covers must be installed and properly tightened within 6 minutes of applying the silicone sealant.

42. Connect the negative battery cable.

43. Run the engine and check for leaks and proper operation.

Valve Lash

ADJUSTMENT

The lash adjusters (valve tappets), are hydraulic and are not adjustable. It is important that all valve components are in good condition and installed and torqued properly.

Rocker Arms

REMOVAL AND INSTALLATION

2.5L (VIN L) Engine

1. Disconnect both battery cables, negative cable first.
2. Remove the valve covers as follows:
   a. Remove the ignition wires and spark plugs.
   b. Remove the ignition coil from the right-hand valve cover.
   c. Remove the crankcase ventilation tubes from both valve covers.
   d. Remove the wiring harness and bracket to the fuel injectors and move aside.

e. Remove the retaining nuts and engine wiring from both valve covers and move aside.
   f. Loosen the valve cover retaining bolts and studs in reverse of the torque sequence illustration.
   g. Remove both valve covers from the engine.
3. Remove the crankshaft pulley retaining bolt.
4. Rotate the crankshaft so that the keyway is at the 11 o'clock position to locate the crankshaft at TDC for No. 1 cylinder.
5. Verify that the alignment arrows on the camshafts are aligned. If not, rotate the crankshaft one complete revolution and recheck.
6. Rotate the crankshaft so that the keyway is at the 3 o'clock position. This positions the right-hand cylinder head camshafts to the neutral position.
7. Remove the accessory drive belts.
8. Remove the battery.
9. Remove the water pump drive pulley from the left-hand intake camshaft using Camshaft Dampener Remover/Replacer T94P-6312-AH or equivalent, along with the shaft protector and screw or equivalent.
10. Remove the camshaft rear oil seal retainer bolts and the camshaft rear oil seal retainer and gasket from the left-hand cylinder head.

NOTE: The camshaft journal caps and cylinder heads are numbered to ensure that they are assembled in their original positions. If removed, keep the camshaft journal caps together with the cylinder head that they were removed from.

11. Remove the right-hand cylinder head camshaft journal thrust cap retaining bolts and thrust caps.
12. Loosen the remaining camshaft journal cap bolts in sequence, releasing the bolts several revolutions at a time by making several passes to allow the camshaft to be raised from the cylinder head evenly. Do not remove the retaining bolts completely.

NOTE: If the valve lifters (tappets) and roller rocker arms are to be reused, mark the positions of the valve lifters and rocker arms so that they are reassembled into their original positions.

13. With the camshafts loose, remove the rocker arms, keeping them in the order that they were removed.
14. If required, remove the valve lifters from the cylinder head.
15. Rotate the crankshaft 2 revolutions and locate the crankshaft key-

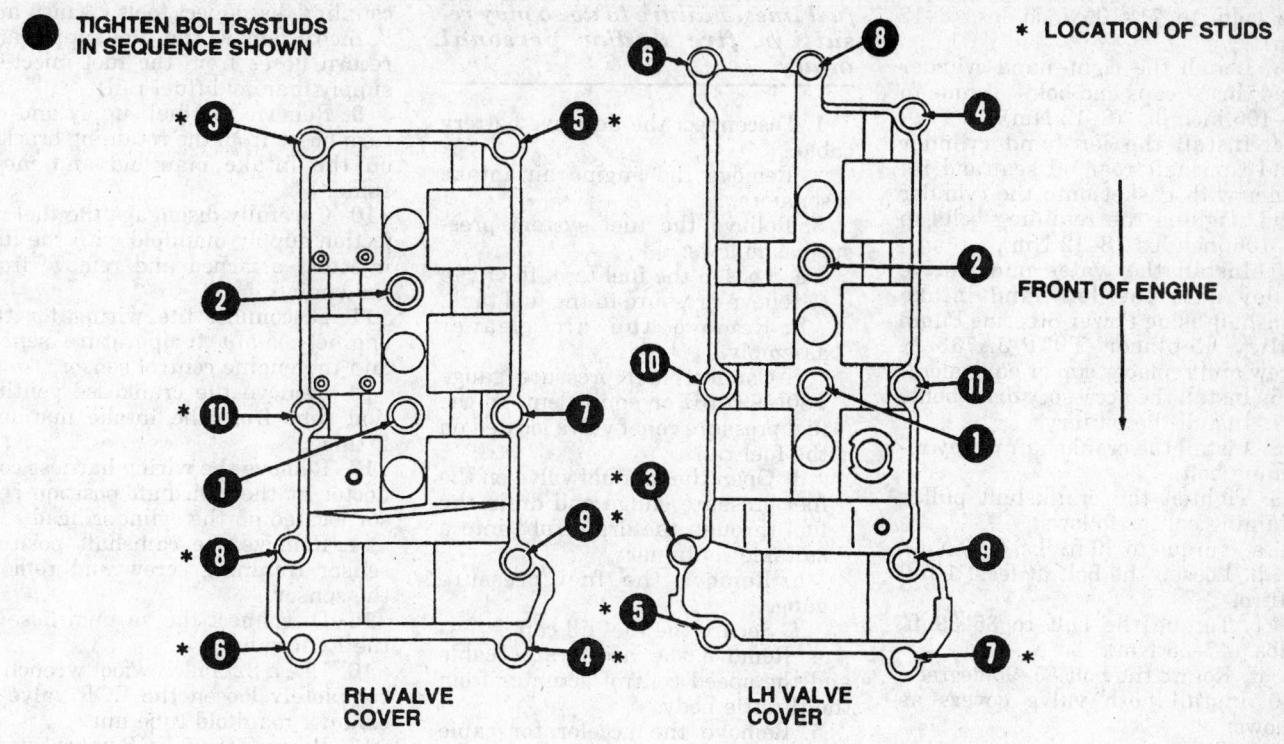

Torque sequence for the left and right valve cover installation — 2.5L (VIN L) engine

way at the 11 o'clock position. This will position the left-hand cylinder head camshafts to their neutral position.

16. Verify that the alignment arrows on the camshafts are aligned.

NOTE: The camshaft journal caps and cylinder heads are numbered to ensure that they are assembled in their original positions. If removed, keep the camshaft journal caps together with the cylinder head that they were removed from.

17. Remove the camshaft journal thrust cap retaining bolts and thrust caps from the left-hand cylinder head.

18. Loosen the remaining camshaft journal cap bolts in sequence, releasing the bolts several revolutions at a time by making several passes to allow the camshaft to be raised from the cylinder head evenly. Do not remove the retaining bolts completely.

NOTE: If the valve lifters (tappets) and roller rocker arms are to be reused, mark the positions of the valve lifters and rocker arms so that they are reassembled into their original positions.

19. With the camshafts loose, remove the rocker arms, keeping them in the order that they were removed.

20. If required, remove the valve lifters from the cylinder head.

21. Inspect the rocker arms and valve lifters for wear and/or damage and replace as necessary.

To install:

22. Make sure that the crankshaft keyway is at the 11 o'clock position.

23. Lubricate the left-hand cylinder head valve lifters with engine assembly lubricant and install into their correct positions in the cylinder head.

24. If the valve lifters are being replaced with new units, soak the lifters in a container of clean engine oil and/or manually pump up the lifters before installing into the cylinder head.

25. Lubricate the left-hand cylinder head rocker arms with engine assembly lubricant and install the left-hand cylinder head rocker arms into their original locations.

NOTE: Do not install the camshaft journal thrust caps until the camshaft journal caps are torqued into position.

26. Tighten the left-hand cylinder head camshaft journal cap bolts in sequence making several passes to pull the camshafts down evenly. Torque

the bolts to 71–106 inch lbs. (8–12 Nm).

27. Install the left-hand cylinder head thrust caps and bolts. Torque to 71–106 inch lbs. (8–12 Nm).

28. Rotate the crankshaft 2 revolutions and position the crankshaft keyway to the 3 o'clock location. This will position the right-hand cylinder head camshafts to the neutral position.

29. Lubricate the right-hand cylinder head valve lifters with engine assembly lubricant and install into their original positions in the cylinder head.

30. If the valve lifters are being replaced with new units, soak the lifters in a container of clean engine oil and/or manually pump the lifters up before installing into the cylinder head.

31. Lubricate the right-hand cylinder head rocker arms with engine assembly lubricant and install the right-hand cylinder head rocker arms into their original locations.

NOTE: Do not install the camshaft journal thrust caps until the camshaft journal caps are torqued into position.

32. Tighten the right-hand cylinder head camshaft journal cap bolts in sequence making several passes to pull

the camshafts down evenly. Torque the bolts to 71–106 inch lbs. (8–12 Nm).

33. Install the right–hand cylinder head thrust caps and bolts. Torque to 71–106 inch lbs. (8–12 Nm).

34. Install the left-hand cylinder head camshaft rear oil seal and retainer with gasket onto the cylinder head. Tighten the retaining bolts to 71–106 inch lbs. (8–12 Nm).

35. Install the water pump drive pulley onto the left-hand intake camshaft using Power Steering Pump Pulley Replacer T91P-3A733-A, screw and replacer cup or equivalent.

36. Install the accessory drive belts.

37. Install the battery.

38. Install the crankshaft pulley retaining bolt.

39. Tighten the crankshaft pulley retaining bolt as follows:

   a. Torque to 89 ft. lbs. (120 Nm).

   b. Loosen the bolt at least 1 full turn.

   c. Torque the bolt to 35–39 ft. lbs. (47–53 Nm).

   d. Rotate the bolt 85–95 degrees.

40. Install both valve covers as follows:

   a. Clean the valve cover gasket sealing surfaces.

   b. Install new valve cover gaskets onto the valve covers.

   c. For each valve cover, place a bead of silicone sealant at 2 places on the valve cover sealing surfaces where the engine front cover and the cylinder heads make contact and at 2 places on the rear of the cylinder head where the camshaft seal retainer contacts the cylinder head.

   d. Install the valve cover retaining bolts and studs and tighten in sequence to 71–106 inch lbs. (8–12 Nm).

NOTE: The valve covers must be installed and properly tightened within 6 minutes of applying the silicone sealant.

41. Connect both battery cables, negative cable last.

42. Run the engine and check for leaks and proper operation.

Intake Manifold

REMOVAL AND INSTALLATION

2.0L (VIN 3) Engine

— CAUTION —
Fuel injection systems remain under pressure, even after the engine has been turned OFF. The fuel system pressure must be re-lieved before disconnecting any fuel lines. Failure to do so may result in fire and/or personal injury.

1. Disconnect the negative battery cable.

2. Remove the engine air intake resonators.

3. Relieve the fuel system pressure as follows:

   a. Loosen the fuel tank filler cap to relieve pressure in the fuel tank.

   b. Remove the air cleaner assembly.

   c. Connect fuel pressure gauge T80L-9974-B or equivalent, to the fuel pressure relief valve located on the fuel rail.

   d. Open the manual valve on the fuel pressure gauge and drain the fuel through the drain tube into a suitable container.

   e. Remove the fuel pressure gauge.

   f. Secure the fuel fill cap.

4. Remove the accelerator cable and the speed control actuator from the throttle body.

5. Remove the accelerator cable bracket.

6. Disconnect the wiring harness to the fuel injectors and move aside.

7. Disconnect the vacuum line at the fuel pressure regulator.

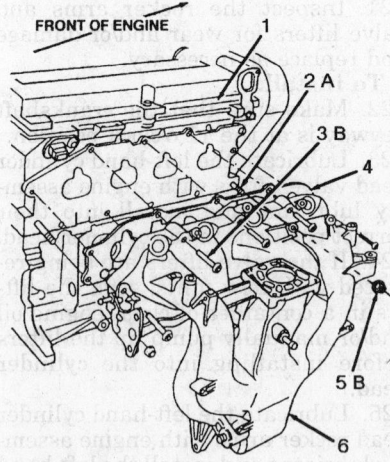

FRONT OF ENGINE

1. Cylinder head
2. Stud (2 req'd)
3. Bolt (8 req'd)
4. Intake manifold gasket
5. Nut (2 req'd)
6. Intake manifold
A. Tighten to 0-10 Nm (0-89 lb. in.)
B. Tighten to 16-20 Nm (12-15 lb. ft.)

296725

Intake manifold and related mounting components — 2.0L (VIN 3) engine

8. Using the proper spring lock coupling disconnect tools (³⁄₈ inch and ¹⁄₂ inch), remove the fuel supply and return hoses from the fuel injection supply manifold (fuel rail).

9. Remove the fuel supply and return hoses from the retaining bracket on the intake manifold and move aside.

10. Carefully disengage the fuel injection supply manifold with the fuel injectors attached and remove from the engine.

11. Disconnect the wiring for the engine coolant temperature sensor and the engine control sensor.

12. Remove the crankcase ventilation tube from the intake manifold fitting.

13. Remove the wiring harness connector at the camshaft position sensor located on the cylinder head.

14. Remove the camshaft position sensor retaining screw and remove the sensor.

15. Disconnect the vacuum hose at the EGR valve.

16. Use a 22mm crowfoot wrench to completely loosen the EGR valve to exhaust manifold tube nut.

17. Remove the 2 EGR valve retaining bolts and remove the EGR valve and gasket.

18. Remove the vacuum supply hoses for the body and brake booster from the bottom of the intake manifold.

19. Remove the retaining bracket screws for the engine control sensor wiring.

NOTE: The intake manifold can be removed without the removal of the alternator and alternator mounting bracket however their removal will make the job easier.

20. Remove the retaining bolts and nuts from the intake manifold.

21. Remove the intake manifold and gasket from the cylinder head.

   To install:

22. If the intake manifold is to be replaced, remove the throttle body and idle air control valve from the old manifold and install on the new one.

23. Clean the gasket sealing surfaces on the intake manifold and the cylinder head.

24. Install the intake manifold using a new gasket to the cylinder head.

25. Install the intake manifold retaining nuts and bolts.

26. Tighten the retaining nuts and bolts in several passes to 12–15 ft. lbs. (16–20 Nm), starting at the center and working towards the ends of the cylinder head.

27. Install the alternator mounting bracket and alternator, if removed.

28. Install the retaining screws for the engine control wiring sensor retaining bracket.

29. Install the vacuum supply hoses for the body and brake booster to the bottom of the intake manifold.

30. Install the EGR valve using a new gasket and tighten the 2 retaining bolts.

31. Install the EGR valve to exhaust manifold tube nut and tighten to 26–33 ft. lbs. (35–45 Nm).

32. Install the EGR valve vacuum hose.

33. Install the camshaft position sensor and its retaining screw and tighten to 13–17 ft. lbs. (18–23 Nm).

34. Connect the camshaft position sensor wiring harness connector.

35. Install the crankcase ventilation tube to the intake manifold fitting.

36. Connect the engine coolant temperature sensor and the engine control sensor wiring.

37. Carefully position the fuel injection supply manifold with injectors into the intake manifold.

38. Install the fuel supply and return hoses to the fuel injection supply manifold.

39. Install the vacuum line to the fuel pressure regulator.

40. Position the wiring harness for the fuel injectors and install.

41. Install the accelerator cable bracket.

42. Install the speed control actuator and the accelerator cable to the throttle body.

43. Install the air cleaner assembly and the air intake resonators.

44. Connect the negative battery cable.

45. Run the engine and check for leaks and proper operation.

2.5L (VIN L) Engine

─────── CAUTION ───────

Fuel injection systems remain under pressure, even after the engine has been turned OFF. The fuel system pressure must be relieved before disconnecting any fuel lines. Failure to do so may result in fire and/or personal injury.

Upper Intake Manifold Removal

1. Disconnect the negative battery cable.

2. Relieve the fuel system pressure as follows:

   a. Loosen the fuel tank filler cap to relieve pressure in the fuel tank.

   b. Remove the air cleaner and outlet tube assembly.

   c. Connect fuel pressure gauge T80L–9974–B or equivalent, to the fuel pressure relief valve located on the fuel rail.

   d. Open the manual valve on the fuel pressure gauge and drain the fuel through the drain tube into a suitable container.

   e. Remove the fuel pressure gauge.

3. Remove the water pump pulley shield.

4. Depress the black retainer with a screwdriver on the upper intake manifold and disconnect the main emission vacuum control connector and the brake booster vacuum connector from the upper intake manifold.

5. Remove the accelerator cable and speed control actuator from the throttle body.

6. Remove the accelerator cable bracket from the intake manifold and move aside.

7. Remove the idle air control valve fresh air supply hose from the fitting on the upper intake manifold.

8. Disconnect the wiring harnesses from the throttle position sensor, idle air control valve and the EGR vacuum regulator control.

9. Remove the vacuum supply hose from the upper intake manifold to the PCV valve at the upper intake manifold.

10. Disconnect the vacuum supply hoses to the EGR vacuum regulator control and the EGR valve.

11. Loosen and remove the EGR valve to exhaust manifold tube and move aside.

12. Remove the Intake Manifold Runner Control (IMRC) vacuum solenoid linkage rod by carefully prying with a screwdriver.

13. Remove the upper intake manifold retaining bolts in the reverse of the torque sequence illustration.

NOTE: When removing engine components such as manifolds and cylinder heads, always remove the retaining bolts in a reverse order of their tightening sequence to prevent warpage to the component.

14. Remove the upper intake manifold and gaskets from the engine.

Lower Intake Manifold Removal

1. Disconnect the fuel injector wiring harness and move aside.

2. Disconnect the vacuum line to the fuel pressure regulator and the IMRC valve and set aside.

3. Remove the spring lock coupling retainer clips from the fuel supply and return fittings.

4. Use spring lock coupling disconnect tools (³⁄₈ and ¹⁄₂) to disconnect the fuel supply and return hoses from the fuel injection supply manifold.

5. Remove the 8 lower intake manifold to cylinder head retaining bolts in reverse of the torque sequence illustration.

6. Remove the lower intake manifold and gaskets from the vehicle.

7. If the lower intake manifold is to be replaced or machined, remove the fuel injectors and the IMRC vacuum solenoid.

Lower Intake Manifold Installation

1. Install the IMRC vacuum solenoid and fuel injectors onto the lower intake manifold if removed. Use a hand vacuum pump to verify operation of the IMRC vacuum solenoid and plate operation at this time.

2. Thoroughly clean the gasket sealing areas and place 2 new intake to cylinder head gaskets into position.

3. Carefully install the lower intake manifold and install the intake manifold to cylinder head retaining bolts.

4. Tighten the retaining bolts in sequence to 71–106 inch lbs. (8–12 Nm).

5. Install the fuel supply and return hoses to the fuel supply manifold and ensure that the spring lock couplings are correctly installed.

6. Install the retaining clips onto the spring lock couplings.

7. Connect the vacuum line to the fuel pressure regulator and the IMRC vacuum solenoid.

8. Temporarily connect the negative battery cable.

9. Connect the fuel pressure gauge to the fuel pressure relief valve located on the fuel injection supply manifold.

10. Cycle the ignition key several times to the **RUN** position to pressurize the fuel system.

11. Watch the fuel pressure gauge for signs of leakage. If the gauge holds pressure, remove the gauge and continue with the installation of the upper intake manifold. If the pressure gauge loses pressure, remove the fuel injection supply manifold and replace the leaking O-ring(s) before continuing.

12. Disconnect the negative battery cable.

13. Reposition and install the fuel injector wiring harness.

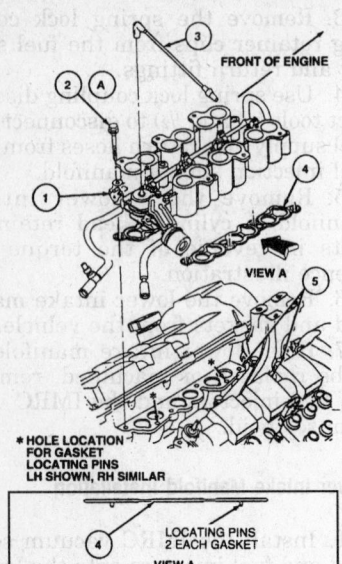

* HOLE LOCATION
FOR GASKET
LOCATING PINS
LH SHOWN, RH SIMILAR

LOCATING PINS
2 EACH GASKET
VIEW A

1. Lower intake manifold
2. Bolt (8)
3. Main emission vacuum control connector
4. Intake manifold gasket (2)
5. Cylinder head (2)
A. 71-106 in. lb. (8-12 Nm)

308703

Installing the lower intake manifold and its components — 2.5L (VIN L) engine

Installation Sequence

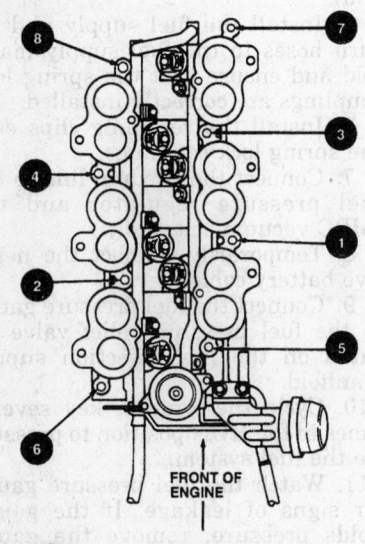

FRONT OF ENGINE

INSTALL BOLTS IN SEQUENCE SHOWN

LOWER INTAKE MANIFOLD

308704

Lower intake manifold bolt torque sequence — 2.5L (VIN L) engine

Upper Intake Manifold Installation

1. Install the upper intake manifold using 2 new gaskets onto the lower intake manifold.

2. Install the upper manifold retaining bolts and torque following the proper sequence to 71–106 inch lbs. (8–12 Nm).

3. Install new bushings for the IMRC linkage rod and install the rod.

4. Install the EGR valve to exhaust manifold tube and tighten the nut to 26–33 ft. lbs. (35–45 Nm).

5. Connect the vacuum supply hoses to the EGR vacuum regulator control and the EGR valve.

6. Connect the wiring harness connectors for the throttle position sensor, idle air control valve and the EGR vacuum regulator control.

7. Install the idle air control valve fresh air supply hose to the fitting on the upper intake manifold.

8. Install the accelerator cable bracket to the intake manifold.

9. Install the speed control actuator and the accelerator cable to the throttle body.

10. Install the main emission vacuum control connector and the brake booster vacuum connector to the upper intake manifold.

11. Install the water pump pulley shield.

Installation Sequence

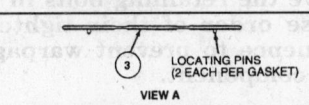

TIGHTEN BOLTS IN
SEQUENCE SHOWN
* HOLE LOCATION FOR
GASKET LOCATING PINS

FRONT OF ENGINE

LOCATING PINS
(2 EACH PER GASKET)
VIEW A

1. Bolt (6)
2. Intake manifold, upper
3. Intake manifold upper gasket
4. Intake manifold, lower
5. Isolator (6)
6. 71-106 in. lb. (8-12 Nm)

308705

Upper intake manifold bolt torque sequence — 2.5L (VIN L) engine

12. Connect the negative battery cable.

13. Run the engine and check for leaks.

Exhaust Manifold

REMOVAL AND INSTALLATION

2.0L (VIN 3) Engine

1. Disconnect the negative battery cable.

2. Remove the engine air intake resonators.

3. Disconnect the heated oxygen sensor at the wiring connector.

4. Remove the oil level indicator tube.

5. Remove the exhaust manifold heat shield retainers and the exhaust manifold heat shield.

6. Remove the heated oxygen sensor from the exhaust manifold.

7. Remove the 4 catalytic converter retaining nuts.

8. Raise and safely support the vehicle.

9. Remove the EGR valve to exhaust manifold tube, retaining bracket and clamp.

10. Remove the catalytic converter.

11. Lower the vehicle.

12. Remove the 9 exhaust manifold retaining nuts from the cylinder head studs.

13. Remove the exhaust manifold and gasket.

14. Remove the exhaust manifold from the vehicle.

15. Clean all gasket mating surfaces.

To install:

16. Position a new exhaust manifold gasket and the exhaust manifold onto the cylinder head studs.

17. Install the exhaust manifold retaining nuts and torque to 13–16 ft. lbs. (14–17 Nm).

18. Raise and safely support the vehicle.

19. Install the catalytic converter using a new exhaust converter inlet gasket.

20. Install the EGR valve to exhaust manifold tube, retaining bracket and clamp.

21. Torque the EGR valve to exhaust manifold tube nut to 44 ft. lbs. (60 Nm).

22. Lower the vehicle.

23. Install the catalytic converter to exhaust manifold retaining nuts.

24. Install the heated oxygen sensor to the exhaust manifold and torque to 44 ft. lbs. (60 Nm).

25. Install the exhaust manifold heat shield and heat shield retainers.

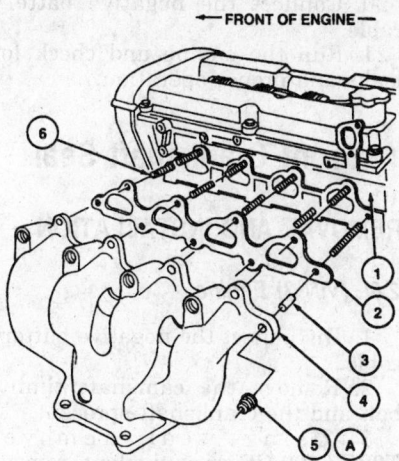

Item	Description
1	Cylinder Head
2	Exhaust Manifold Gasket
3	Spacer
4	Exhaust Manifold
5	Nut (9 Req'd)
6	Stud (9 Req'd)
A	Tighten to 14-17 N·m (13-16 Lb-Ft)

296196

Exhaust manifold and related components — 2.0L (VIN 3) engine

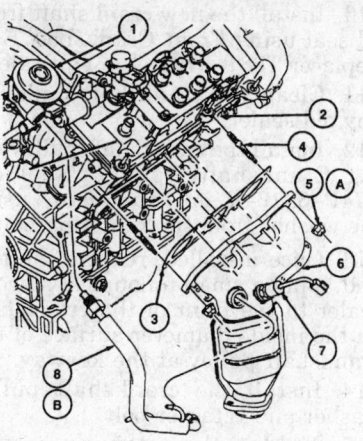

Item	Description
1	EGR Valve
2	RH Cylinder Head
3	Exhaust Manifold Gasket
4	Stud Bolt (6 Req'd)
5	Nut (6 Req'd)
6	RH Exhaust Manifold
7	Heated Oxygen Sensor
8	EGR Valve to Exhaust Manifold Tube
A	Tighten to 18-22 N·m (13-16 Lb-Ft)
B	Tighten to 35-45 N·m (26-33 Lb-Ft)

296201

Exhaust manifold and components (right-side shown) — 2.5L (VIN L) engine

26. Torque the heat shield retainers to 71–106 inch lbs. (8–11 Nm).

27. Install the oil level indicator tube.

28. Connect the heated oxygen sensor wiring harness connector.

29. Install the air intake resonators.

30. Connect the negative battery cable.

31. Run the engine and check for leaks and proper operation.

2.5L (VIN L) Engine

Right Side

1. Disconnect the negative battery cable.

2. Disconnect the wiring harness from the oxygen sensor.

3. Raise and safely support the vehicle .

4. Remove the alternator and the alternator mounting bracket.

5. Remove the retaining nuts from the outlet flange on the catalytic converter.

6. Remove the muffler and the exhaust converter outlet gasket from the catalytic converter.

7. Remove the catalytic converter retaining nuts and the exhaust pipe flange hold-down springs.

8. Remove the catalytic converter.

9. Remove the halfshaft support bearing retainer bracket from the support bearing and cylinder block.

10. Remove the oxygen sensor from the exhaust manifold.

11. Loosen the EGR valve to exhaust manifold tube nuts and remove the tube.

12. Remove the exhaust manifold retaining nuts from the cylinder head studs.

13. Remove the exhaust manifold and gasket from the engine.

14. Clean all gasket mating surfaces.

To install:

15. Install the exhaust manifold with a new exhaust manifold gasket.

16. Install the exhaust manifold retaining studs and torque to 13–16 ft. lbs. (18–22 Nm) in the proper sequence.

17. Install the EGR valve to exhaust manifold tube and torque the nuts to 26–33 ft. lbs. (35–45 Nm).

18. Install the oxygen sensor into the exhaust manifold and torque to 26–34 ft. lbs. (35–46 Nm).

19. Install the halfshaft support bearing retainer bracket to the support bearing and the cylinder block.

20. Install the catalytic converter using a new exhaust converter inlet gasket.

21. Position the muffler and a new exhaust converter outlet gasket onto the catalytic converter and loosely install the retaining nuts.

22. Align the exhaust system and tighten all nuts and bolts.

23. Torque the catalytic converter retaining nuts with the exhaust pipe flange hold-down springs to 22–30 ft. lbs. (27.9–40.3 Nm).

24. Torque the muffler inlet flange nuts to 26–33 ft. lbs. (34–46 Nm).

25. Install the alternator mounting bracket and the alternator.

26. Lower the vehicle.

27. Connect the wiring harness to the oxygen sensor.

28. Connect the negative battery cable.

29. Run the engine and check for exhaust leaks and proper operation.

Left Side

1. Disconnect the negative battery cable.

2. Disconnect the wiring connector to the oxygen sensor.

3. Raise and safely support the vehicle .

4. Remove the front and rear exhaust crossover tube flange fasteners from the exhaust manifolds.

5. Remove the stud and nut retainer from the engine oil pan.

6. Remove the 2 remaining nuts and bolts from the exhaust crossover tubes outlet connection.

7. Remove the exhaust crossover tube.

8. Remove the lower radiator hose tube retaining bracket nuts.

9. Remove the 6 exhaust manifold retaining nuts from the cylinder head studs.

10. Move the lower radiator hose tube to gain access for removal of the exhaust manifold.

11. Remove the exhaust manifold. If the manifold is being replaced, remove the oxygen sensor.

12. Clean all gasket mating surfaces.

To install:

13. If removed, install the oxygen sensor in the new manifold.

14. Place a new exhaust manifold to cylinder block gasket onto the cylinder head studs.

15. Move the lower radiator hose to allow installing the exhaust manifold.

16. Place the exhaust manifold onto the cylinder head studs and install the 6 retaining nuts.

17. Torque the retaining nuts to 13–16 ft. lbs. (18–22 Nm), in sequence.

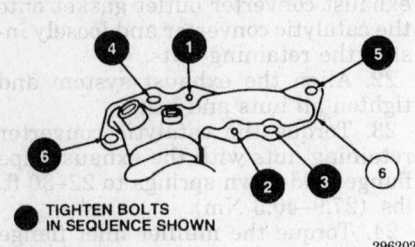

TIGHTEN BOLTS IN SEQUENCE SHOWN

296202

Proper torque pattern for exhaust manifolds — 2.5L (VIN L) engine

18. Install the lower radiator hose tube retaining bracket and torque to 71–106 inch lbs. (8–12 Nm).

19. Install the exhaust crossover tube.

20. Install 2 nuts and bolts to the exhaust crossover tube outlet connection.

21. Install the stud and nut retainer at the engine oil pan.

22. Install the front and rear exhaust crossover tube flange fasteners at the exhaust manifolds.

23. Lower the vehicle.

24. Connect the wiring harness connector to the oxygen sensor.

25. Connect the negative battery cable.

26. Run the engine and check for exhaust leaks and proper operation.

Front Cover Seal

REMOVAL AND INSTALLATION

2.5L (VIN L) Engine

1. Disconnect the negative battery cable.

2. Raise and safely support the vehicle.

3. Remove the right front wheel and tire assembly.

4. Remove the splash shield from the front fender apron.

5. Remove the accessory drive belt.

6. Remove the crankshaft pulley retaining bolt and washer from the crankshaft.

7. Using Crankshaft Damper Remover T58P-6316–D or equivalent, remove the crankshaft pulley from the crankshaft.

8. Using Seal Remover T92C-6700–CH or equivalent, remove the front cover oil seal.

To install:

9. Lubricate the crankshaft front oil seal sealing surfaces with engine assembly lubricant.

10. Install the new crankshaft front oil seal using Front Crankshaft Seal Replacer T88T-6701–A or equivalent.

11. Clean the crankshaft pulley of any old sealer.

12. Install the crankshaft pulley using Crankshaft Damper Replacer T74P-6316–B or equivalent, using the washer from the retaining bolt.

13. Once installed, remove the tool and apply a small amount of silicone sealer to the front of the crankshaft on the inside diameter surface of the crankshaft pulley at the keyway.

14. Install the crankshaft pulley washer and retainer bolt.

15. Tighten the retaining bolt as follows:

• Torque the retaining bolt to 89 ft. lbs. (120 Nm)

• Loosen the retaining bolt a minimum of 1 full turn

• Torque the retaining bolt to 35–39 ft. lbs. (47–53 Nm)

• Rotate the retaining bolt an additional 85–95 degrees

16. Install the accessory drive belt.

17. Install the splash shield to the front fender apron.

18. Install the right front wheel and tire assembly. Torque the lug nuts to 62 ft. lbs. (85 Nm).

19. Lower the vehicle.

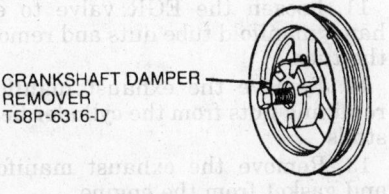

CRANKSHAFT DAMPER REMOVER T58P-6316–D

297803

Removing the crankshaft pulley — 2.5L (VIN L) engine

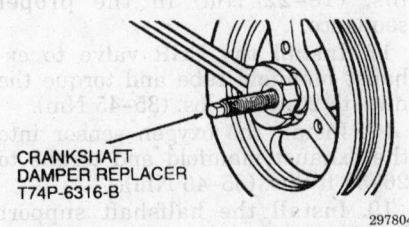

CRANKSHAFT DAMPER REPLACER T74P-6316–B

297804

Installing the crankshaft pulley — 2.5L (VIN L) engine

20. Connect the negative battery cable.

21. Run the engine and check for leaks and proper operation.

Front Crankshaft Seal

REMOVAL AND INSTALLATION

2.0L (VIN 3) Engine

1. Disconnect the negative battery cable.

2. Remove the camshaft timing belt and the crankshaft sprocket.

3. Using seal remover T92C-6700–CH or equivalent, remove the front crankshaft seal from the oil pump housing.

To install:

4. Clean and inspect the oil pump housing crankshaft front oil seal bore.

5. Lubricate the oil pump crankshaft oil seal bore and crankshaft front seal with engine assembly lubricant.

6. Use oil pump seal replacer T81P-6700–A or equivalent tool, and the crankshaft pulley retaining bolt, install the new crankshaft seal into the oil pump.

7. Install the crankshaft sprocket and the camshaft timing belt.

8. Connect the negative battery cable.

9. Run the engine and check for leaks and proper operation.

Timing Chain, Sprockets and Front Cover

REMOVAL AND INSTALLATION

2.5L (VIN L) Engine

1. Disconnect the negative battery cable.

2. Remove the upper intake manifold.

3. Remove the valve covers as follows:

a. Remove the ignition wires and spark plugs.

b. Remove the ignition coil from the right-hand valve cover.

c. Remove the crankcase ventilation tubes from both valve covers.

d. Remove the wiring harness and bracket to the fuel injectors and move aside.

e. Remove the retaining nuts and engine wiring from both valve covers and move aside.

f. Loosen the valve cover retaining bolts and studs following the proper removal sequence.

g. Remove both valve covers from the engine.

4. Install Three Bar Engine Support D88L-6000-A or equivalent to the engine lifting eyes and support the engine.

5. Remove the power steering pressure hose bracket retainer bolt from the upper front engine support bracket.

6. Disconnect the power steering pressure hose from the power steering pump and position out of the way.

NOTE: Mark the position of the upper front engine support bracket before removing.

7. Remove the upper front engine support bracket retainer nuts and remove the bracket.

8. Disconnect the low coolant level sensor connector from the wiring harness.

9. Remove the radiator coolant recovery reservoir retainers and move the recovery reservoir aside.

10. Remove the upper front engine insulator.

11. Set the radiator coolant recovery reservoir back into its position but do not secure.

12. Disconnect the three wiring harness connectors located at the inline connector bracket at the front of the right-hand cylinder head and set aside.

13. Loosen the power steering pump pulley retaining bolts but do not remove the bolts completely.

14. Remove the accessory drive belt.

15. Finish removing the power steering pump pulley retaining bolts and remove the pulley.

16. Remove the power steering pump and pump support retaining nuts and bolts and remove the power steering pump and pump support.

17. Raise and safely support the vehicle.

18. Remove the right-front wheel and tire assembly.

19. Remove the alternator from its mounting bracket and move aside.

20. Remove the alternator mounting bracket.

21. Remove the crankshaft pulley.

22. Disconnect the wiring from the Crankshaft Position (CKP) sensor and the Camshaft Position (CMP) sensor electrical connectors.

23. Remove the engine oil pan.

24. Loosen the A/C compressor retaining bolts and move the A/C compressor to gain access to the front cover retaining bolt.

25. Partially lower the vehicle.

26. Remove the bracket retainers and the bracket for the engine wiring and the A/C hose at the engine front cover.

NOTE: It may be necessary to raise and lower the vehicle several times in order to follow the engine front cover bolt removal sequence.

27. Remove the engine front cover bolts in reverse of the torque sequence illustration.

28. Remove the engine front cover and gasket from the vehicle.

29. Rotate the crankshaft so that the keyway is at the 11 o'clock position to locate the crankshaft at TDC for No. 1 cylinder.

30. Verify that the alignment arrows on the camshafts are aligned. If not, rotate the crankshaft 1 complete revolution and recheck.

31. Rotate the crankshaft so that the keyway is at the 3 o'clock position. This positions the right-hand cylinder head camshafts to the neutral position.

32. Remove the right-hand cylinder head timing chain tensioner retaining bolts and the timing chain tensioner.

NOTE: The camshaft journal caps and cylinder heads are numbered to ensure that they are assembled in their original positions. If removed, keep the camshaft journal caps together with the cylinder head that they were removed from.

33. Remove the right-hand cylinder head camshaft journal thrust cap retaining bolts and thrust caps.

34. Loosen the remaining camshaft journal cap bolts in sequence, releasing the bolts several revolutions at a time by making several passes to allow the camshaft to be raised from the cylinder head evenly. Do not remove the retaining bolts completely.

NOTE: If the valve tappets and roller rocker arms are to be reused, mark the positions of the valve tappets and rocker arms so that they are reassembled into their original positions.

35. With the camshafts loose, remove the rocker arms, keeping them in the order that they were removed.

NOTE: If the right-hand cylinder head timing chain tensioner arm and timing chain guide are to be reused, mark the position of the timing chain tensioner arm and the timing chain guide so that they are reassembled into their original positions.

36. Remove the right-hand cylinder head timing chain tensioner arm and the timing chain.

37. Remove the right-hand cylinder head timing chain guide retaining bolts and the timing chain guide.

38. If worn, replace the timing chain guide.

39. Remove the right-hand crankshaft timing chain sprocket.

40. Remove the right-hand cylinder head camshaft timing chain sprockets if they are to be replaced.

41. Rotate the crankshaft 2 revolutions and locate the crankshaft keyway at the 11 o'clock position. This will position the left-hand cylinder head camshafts to their neutral position.

42. Verify that the alignment arrows on the camshafts are aligned.

43. Remove the left-hand cylinder head timing chain tensioner retaining bolts and the timing chain tensioner.

NOTE: The camshaft journal caps and cylinder heads are numbered to ensure that they are assembled in their original positions. If removed, keep the camshaft journal caps together with the cylinder head that they were removed from.

44. Remove the camshaft journal thrust cap retaining bolts and thrust caps from the left-hand cylinder head.

45. Loosen the remaining camshaft journal cap bolts in sequence, releasing the bolts several revolutions at a time by making several passes to allow the camshaft to be raised from the cylinder head evenly. Do not remove the retaining bolts completely.

NOTE: If the valve tappets and roller rocker arms are to be reused, mark the positions of the valve tappets and rocker arms so that they are reassembled into their original positions.

46. With the camshafts loose, remove the rocker arms, keeping them in the order that they were removed.

NOTE: If the left-hand cylinder head timing chain tensioner arm and timing chain guide are to be reused, mark the position of the timing chain tensioner arm and the timing chain guide so that they are reassembled into their original positions.

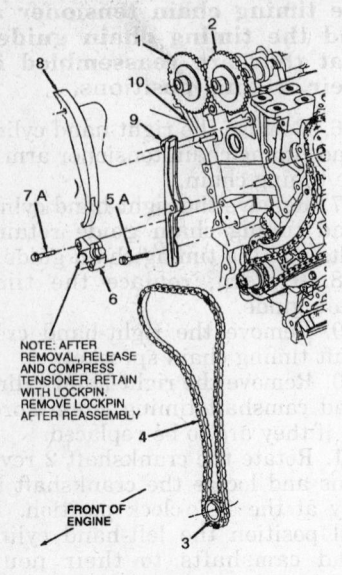

NOTE: AFTER
REMOVAL, RELEASE
AND COMPRESS
TENSIONER. RETAIN
WITH LOCKPIN.
REMOVE LOCKPIN
AFTER REASSEMBLY

FRONT OF
ENGINE

1. Exhaust camshaft, RH
2. Intake camshaft, RH
3. Timing chain crankshaft sprocket, RH
4. Timing chain, RH
5. Bolt (2 req'd)
6. Timing chain tensioner
7. Bolt (2 req'd)
8. Timing chain tensioner arm
9. Timing chain guide
10. Cylinder head, RH
A. Tighten to 20–30 Nm (15–22 lb. ft.)

297989

**Removal of the right-hand cylinder
head timing chain and components —
2.5L (VIN L) engine**

47. Remove the left-hand cylinder
head timing chain tensioner arm and
the timing chain.
48. Remove the left-hand cylinder
head timing chain guide retaining
bolts and the timing chain guide.
49. If worn, replace the timing
chain guide.
50. Remove the left-hand crankshaft timing chain sprocket.
51. If required, the left-hand cylinder head camshaft sprockets may be
removed at this time.
To install:

**NOTE: Inspect the timing
chains, tensioners, tensioner
arms, guides and sprockets for
wear or damage. If any components are to be replaced for premature wear or damage, the
camshaft damper should also be
replaced.**

52. Install or replace the camshaft
sprockets, if removed.
53. Make sure that the crankshaft
keyway is still at the 11 o'clock
position.
54. Install the left-hand timing
chain crankshaft sprocket onto the
crankshaft.
55. Install the left-hand timing
chain guide and retaining bolts to the

engine. Torque the retaining bolts to
15–22 ft. lbs. (20–30 Nm).
56. Verify that the alignment arrows on the left-hand cylinder head
camshafts are aligned before
proceeding.
57. Install the left-hand timing
chain over the left-hand crankshaft
sprocket and the left-hand camshaft
sprockets.
58. Align the timing index marks
on the left-hand cylinder head timing
chain with the timing index marks on
the crankshaft sprocket and the
camshaft sprockets.
59. Install the left-hand timing
chain tensioner arm over the alignment dowel on the left-hand cylinder
head.

**NOTE: Before installing the
timing chain tensioner, it must
be properly compressed and
locked.**

60. Using a small screwdriver, release the timing chain tensioner
ratchet/pawl mechanism through the
access hole in the timing chain tensioner as follows:
    a. Insert a small piece of wire
into the top of the piston and gently unseat the oil check ball.
    b. Compress the timing chain
tensioner by hand.
    c. With the tensioner com-
pressed, install a 0.060 inch
(1.5mm) drill bit or wire into the
small hole above the ratchet, engaging the lock groove in the rack
of the timing chain tensioner.
61. Install the compressed and
locked left-hand cylinder head timing
chain tensioner and retaining bolts
onto the cylinder block. Torque the
retaining bolts to 15–22 ft. lbs. (20–30
Nm).
62. Verify that the timing index
marks on the left-hand timing chain
are in alignment with the timing index marks on the crankshaft sprocket
and the camshaft sprockets.
63. Install or replace the camshaft
sprockets if removed.
64. Install the right-hand timing
chain crankshaft sprocket onto the
crankshaft.
65. Install the right-hand timing
chain guide and retaining bolts to the
engine. Torque the retaining bolts to
15–22 ft. lbs. (20–30 Nm).
66. Verify that the alignment arrows on the right-hand cylinder head
camshafts are aligned before
proceeding.
67. Install the right-hand timing
chain over the right-hand crankshaft
sprocket and the right-hand
camshaft sprockets.

68. Align the timing index marks
on the right-hand cylinder head timing chain with the timing index
marks on the crankshaft sprocket
and the camshaft sprockets.
69. Install the right-hand timing
chain tensioner arm over the alignment dowel on the right-hand cylinder head.

**NOTE: Before installing the
timing chain tensioner, it must
be properly compressed and
locked.**

70. Using a small screwdriver, release the timing chain tensioner
ratchet/pawl mechanism through the
access hole in the timing chain tensioner as follows:
    a. Insert a small wire into the
top of the piston and gently unseat
the oil check ball.
    b. Compress the timing chain
tensioner by hand.
    c. With the tensioner com-
pressed, install a 0.060 inch
(1.5mm) drill bit or wire into the
small hole above the ratchet, engaging the lock groove in the rack
of the timing chain tensioner.
71. Install the compressed and
locked right-hand cylinder head timing chain tensioner and retaining
bolts onto the cylinder block. Torque
the retaining bolts to 15–22 ft. lbs.
(20–30 Nm).
72. Verify that the timing index
marks on the right-hand timing
chain are in alignment with the timing index marks on the crankshaft
sprocket and the camshaft sprockets.
73. Make sure that the crankshaft
keyway is at the 11 o'clock position.
74. Lubricate the left–side rocker
arms with engine assembly lubricant
and install the left-hand cylinder
head rocker arms into their original
locations.

**NOTE: Do not install the
camshaft journal thrust caps until the camshaft journal caps are
torqued into position.**

75. Tighten the left-hand cylinder
head camshaft journal cap bolts in sequence making several passes to pull
the camshafts down evenly. Torque
the bolts to 71–106 inch lbs. (8–12
Nm).
76. Install the left-hand cylinder
head thrust caps and bolts. Torque to
71–106 inch lbs. (8–12 Nm).
77. Rotate the crankshaft 2 revolutions and position the crankshaft
keyway to the 3 o'clock location. This
will position the right-hand cylinder
head camshafts to the neutral
position.

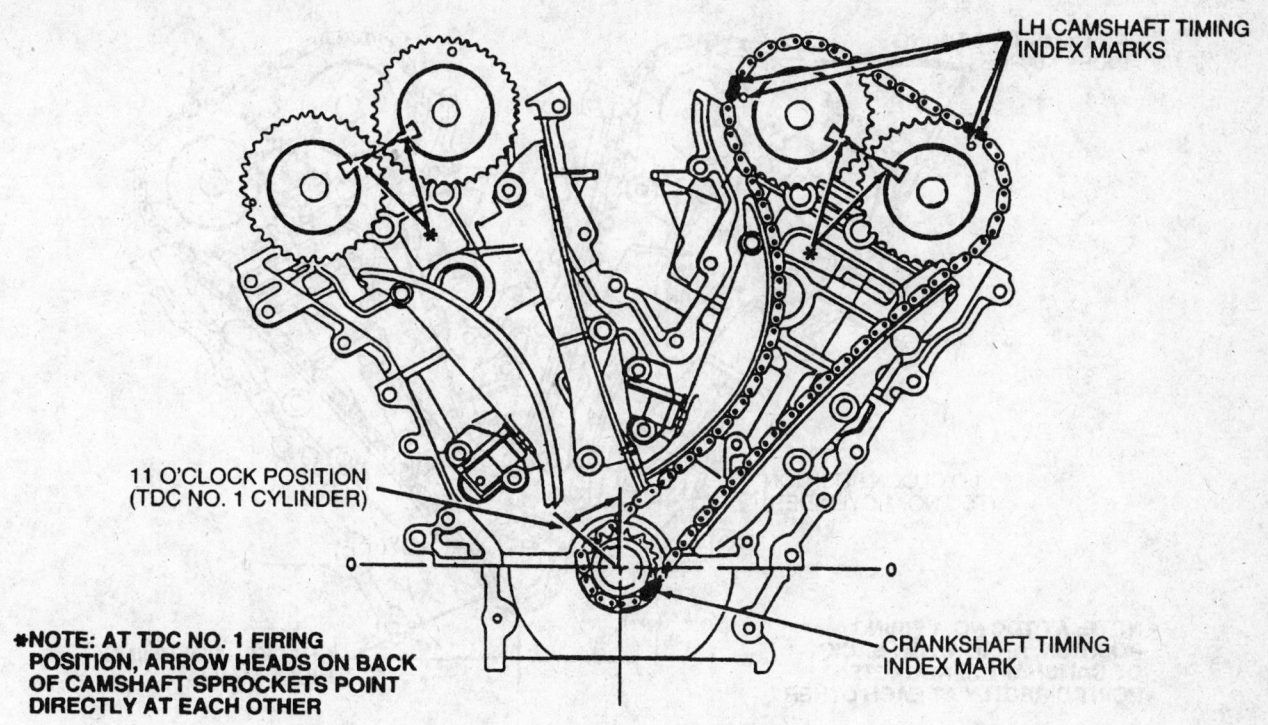

LH CAMSHAFT TIMING INDEX MARKS

11 O'CLOCK POSITION (TDC NO. 1 CYLINDER)

CRANKSHAFT TIMING INDEX MARK

✱NOTE: AT TDC NO. 1 FIRING POSITION, ARROW HEADS ON BACK OF CAMSHAFT SPROCKETS POINT DIRECTLY AT EACH OTHER

297994

Correct timing chain index marks for the left-hand cylinder head — 2.5L (VIN L) engine

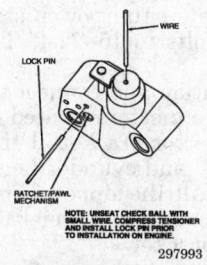

WIRE

LOCK PIN

RATCHET/PAWL MECHANISM

NOTE: UNSEAT CHECK BALL WITH SMALL WIRE. COMPRESS TENSIONER AND INSTALL LOCK PIN PRIOR TO INSTALLATION ON ENGINE.

297993

Positioning of the wire and pin to compress and lock the tensioner — 2.5L (VIN L) engine

78. Lubricate the right–side rocker arms with engine assembly lubricant and install the right-hand cylinder head rocker arms into their original locations.

NOTE: Do not install the camshaft journal thrust caps until the camshaft journal caps are torqued into position.

79. Tighten the right-hand cylinder head camshaft journal cap bolts in sequence making several passes to pull the camshafts down evenly. Torque the bolts to 71–106 inch lbs. (8–12 Nm).

80. Install the right-hand cylinder head thrust caps and bolts. Torque to 71–106 inch lbs. (8–12 Nm).
81. Remove the lock pins from the timing chain tensioners.
82. Verify that the timing index marks on the timing chains are in alignment with the timing index marks on the crankshaft sprocket and the camshaft sprockets.
83. Clean the engine front and the front cover to cylinder block gasket sealing surfaces.

NOTE: The front cover must be installed and properly tightened within 6 minutes of the application of the sealer.

84. Apply silicone sealer to the 6 critical areas shown in View **A**, to the cylinder block to prevent oil seepage.
85. Place new front cover gaskets onto the dowel pins on the cylinder block and heads.
86. Place the front cover into position by placing the front cover onto the dowel pins at the cylinder block.
87. Install the 6 front cover retaining bolts and stud bolts where the silicone sealer was applied.
88. Tighten the bolts and stud bolts until the front cover contacts the cylinder block and heads an then turn the bolts and stud bolts an additional ¼ turn.

89. Install the remaining front cover retaining bolts and stud bolts.
90. Torque all of the front cover retaining bolts and stud bolts in proper sequence to 15–22 ft. lbs. (20–30 Nm).
91. Install the bracket and bracket retainers to the front cover. Install the wiring and the A/C hose to the bracket.
92. Raise and safely support the vehicle.
93. Install the A/C compressor and its retaining bolts. Torque the retaining bolts to 15–22 ft. lbs. (20–30 Nm).
94. If required, replace the crankshaft front seal at this time.
95. Install the engine oil pan.
96. Connect the wiring harness to the CKP sensor and the CMP sensor.
97. Install the crankshaft pulley.
98. Install the alternator mounting bracket and the alternator.
99. Install the right-front wheel and tire assembly. Torque the lug nuts to 62 ft. lbs. (85 Nm).
100. Lower the vehicle on the hoist.
101. Lower the engine to its correct position.
102. Install the power steering pump support and the power steering pump to the front of the engine.
103. Install the power steering pump pulley. Loosely install the retaining bolts.
104. Install the accessory drive belt.

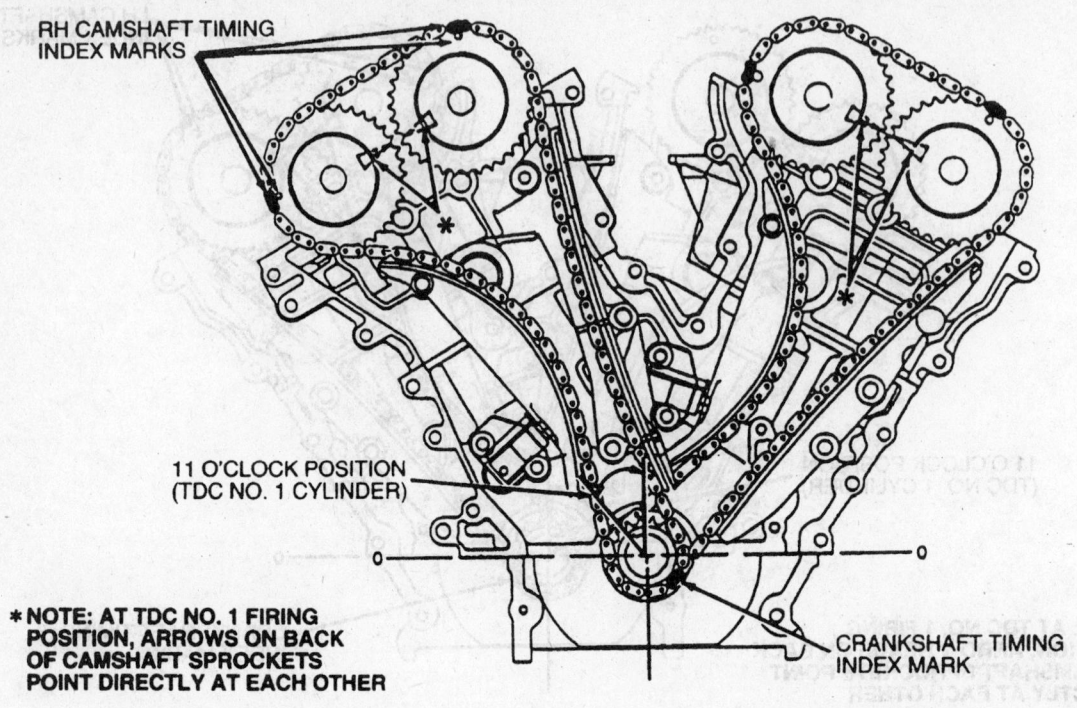

RH CAMSHAFT TIMING
INDEX MARKS

11 O'CLOCK POSITION
(TDC NO. 1 CYLINDER)

*NOTE: AT TDC NO. 1 FIRING
POSITION, ARROWS ON BACK
OF CAMSHAFT SPROCKETS
POINT DIRECTLY AT EACH OTHER

CRANKSHAFT TIMING
INDEX MARK

297996

Correct timing chain index marks for the right-side cylinder head — 2.5L (VIN L) engine

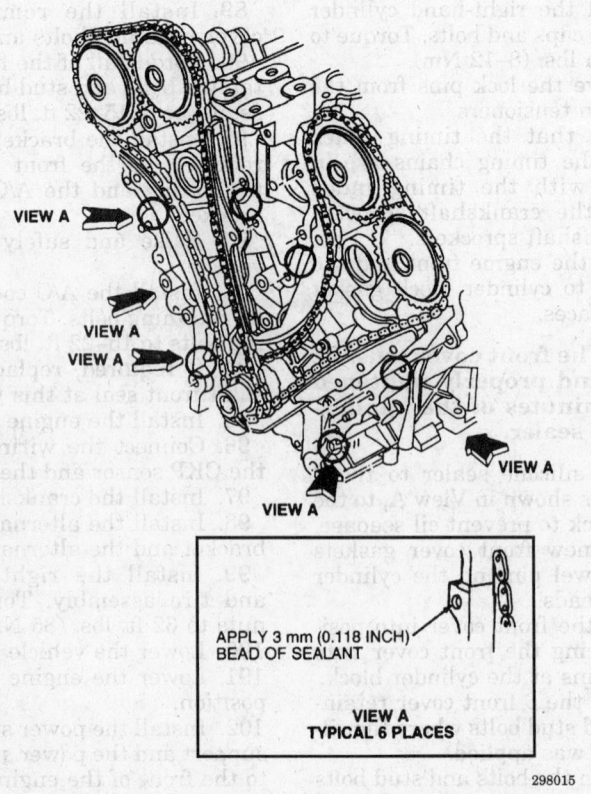

VIEW A

VIEW A

VIEW A

VIEW A

VIEW A

APPLY 3 mm (0.118 INCH)
BEAD OF SEALANT

VIEW A
TYPICAL 6 PLACES

298015

The 6 points for sealer placement — 2.5L (VIN L) engine

80. Install the right-hand cylinder head clamps and bolts. Torque to 71–106 inch lbs. (8–12 Nm).
81. Remove the lock pins from the timing chain tensioners.
82. Verify that the timing index marks on the timing chains are aligned with the timing index marks on the camshaft sprockets.
83. Clean the engine front cover to cylinder head and block sealing surfaces.

NOTE: The front cover must be installed and properly torqued within 5 minutes after application of the sealer.

84. Apply silicone sealer to the 6 areas shown in View A, to the cylinder block to cylinder head joints.
85. Place new front cover gaskets onto the dowel pins on the block and heads.
86. Place the front cover in position by placing it over the dowel pins of the cylinder block.
87. Install the front cover top bolts and stud bolts finger-tight.
88. Tighten the bolts and stud bolts.

105. Tighten the power steering retaining bolts to 15–22 ft. lbs. (20–30 Nm).
106. Position and connect the wiring to the 3 connectors located on the in-line connector bracket at the front of the right-hand cylinder head.
107. Install the upper front support insulator and the upper front engine support bracket.
108. Loosen and remove the Three Bar Engine Support or equivalent.
109. Connect the power steering pressure hose to the power steering pump.
110. Connect the power steering pressure hose bracket and retainer bolt to the upper front engine support bracket.
111. Install both valve covers as follows:
   a. Clean the valve cover gasket sealing surfaces.
   b. Install new valve cover gaskets onto the valve covers.
   c. For each valve cover, place a bead of silicone sealant at 2 places on the valve cover sealing surfaces where the engine front cover and the cylinder heads make contact and at 2 places on the rear of the cylinder head where the camshaft seal retainer contacts the cylinder head.

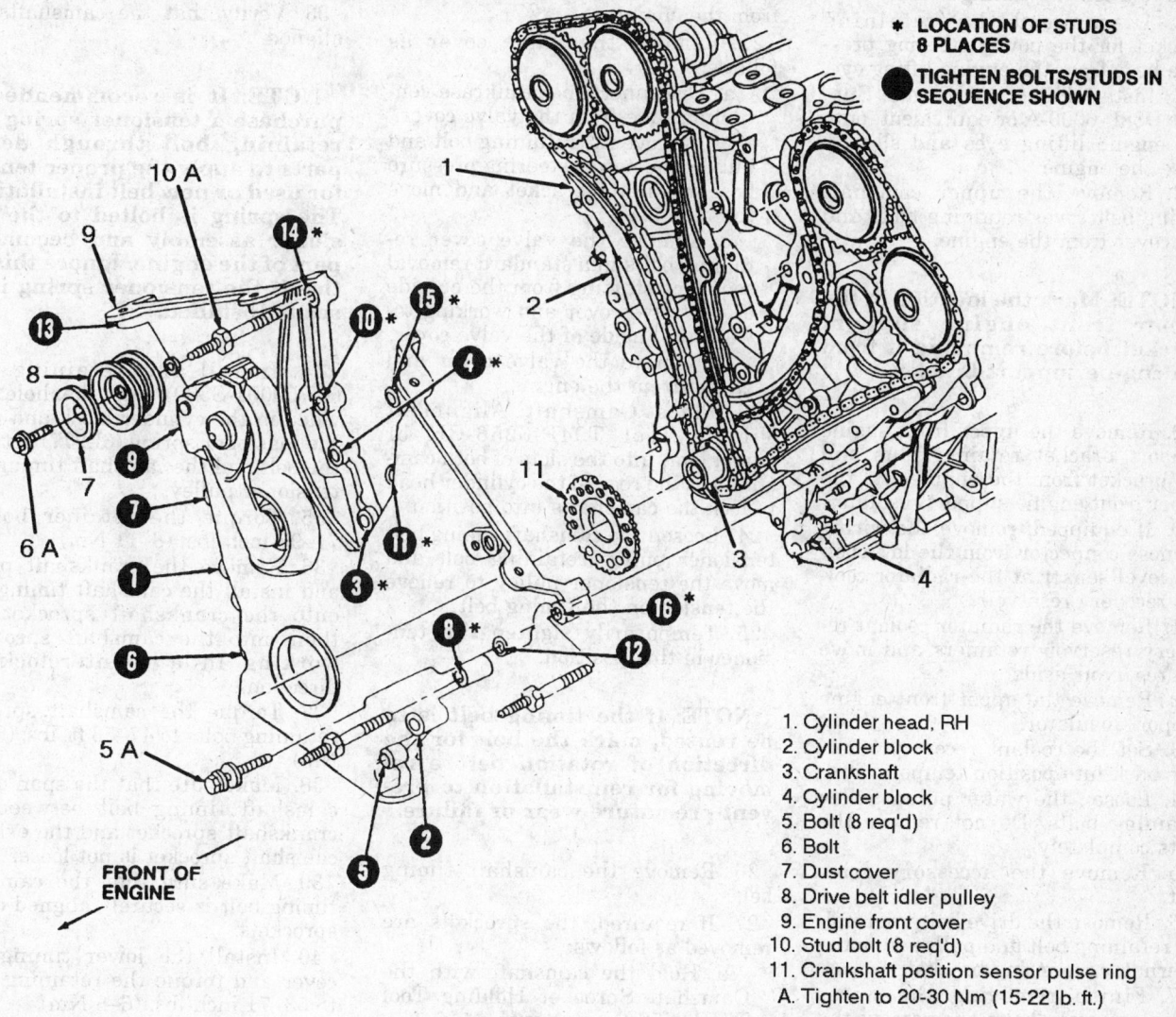

* **LOCATION OF STUDS 8 PLACES**
* **TIGHTEN BOLTS/STUDS IN SEQUENCE SHOWN**

1. Cylinder head, RH
2. Cylinder block
3. Crankshaft
4. Cylinder block
5. Bolt (8 req'd)
6. Bolt
7. Dust cover
8. Drive belt idler pulley
9. Engine front cover
10. Stud bolt (8 req'd)
11. Crankshaft position sensor pulse ring
A. Tighten to 20-30 Nm (15-22 lb. ft.)

298017

Front cover bolt location and torque sequence — 2.5L (VIN L) engine

d. Install the valve cover retaining bolts and studs and tighten in sequence to 71–106 inch lbs. (8–12 Nm).

NOTE: The valve covers must be installed and properly tightened within 6 minutes of applying the silicone sealant.

112. Install the upper intake manifold.
113. Fill the engine with the proper amount and grade of oil.
114. Replace any lost fluid to the power steering reservoir.
115. Connect the negative battery cable.

116. Run the engine and check for leaks and proper operation.
117. Recheck the fluid levels.

Timing Belt, Sprockets, Tensioner and Front Cover

ADJUSTMENT

2.0L (VIN 3) Engine

When installing a timing belt, tensioner spring (6L277) and retaining bolt (W700001-S309) must be purchased and properly installed on the engine. First check to see if these parts are already installed. The ten-

sioner spring will adjust the timing belts tension and should not require further adjustments.

REMOVAL AND INSTALLATION

2.0L (VIN 3) Engine

1. Disconnect the negative battery cable.
2. Remove the engine air intake resonators.
3. Label and remove the ignition wires from the spark plugs. Move the ignition wires aside.
4. Remove the spark plugs.
5. Manually rotate the crankshaft to Top Dead Center (TDC) for No. 1

piston on its compression stroke. Be sure to align the timing marks.

6. Disconnect the retaining bracket for the power steering pressure hose from the engine lifting eye.

7. Install Three Bar Engine Support D88L-6000-A or equivalent, onto the engine lifting eyes and slightly raise the engine.

8. Remove the upper camshaft timing belt cover retaining bolts and the cover from the engine.

NOTE: Mark the location of the upper front engine support bracket before removing it from the engine support bracket.

9. Remove the upper front engine support bracket retainer nuts and the bracket from the engine and the upper front engine support insulator.

10. If equipped, remove the wiring harness connector from the low coolant level sensor at the radiator coolant recovery reservoir.

11. Remove the radiator coolant recovery reservoir retainers and move the reservoir aside.

12. Remove the upper front engine support insulator.

13. Set the coolant recovery reservoir back into position temporarily.

14. Loosen the water pump pulley retaining bolts. Do not remove the bolts completely.

15. Remove the accessory drive belt.

16. Remove the drive belt idler pulley retaining bolt and pulley from the alternator mounting bracket.

17. Finish removing the water pump retaining bolts and remove the water pump pulley.

18. Remove the center camshaft timing belt cover retaining bolts and the cover from the engine.

19. Raise and safely support the vehicle.

20. Remove the crankshaft pulley.

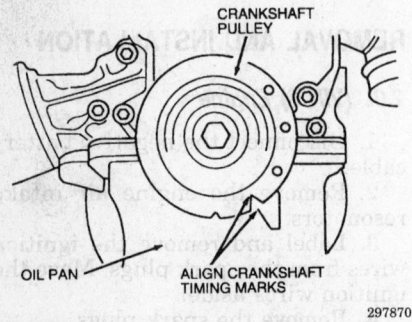

Location of the crankshaft timing marks — 2.0L (VIN 3) engine

21. Remove the lower camshaft timing belt cover bolts and the cover from the engine.

22. Remove the valve cover as follows:

a. Disconnect the crankcase ventilation tube from the valve cover.

b. Remove the retaining bolt and nut for the power steering pressure hose retaining bracket and move the hose aside.

c. Remove the valve cover retaining bolts in a standard removal sequence starting from the outside of the valve cover and working toward the inside of the valve cover.

d. Remove the valve cover and gasket from the engine.

23. Place Camshaft Alignment Timing Tool T94P-6256-CH or equivalent, into the slots of both camshafts at the rear of the cylinder head to lock the camshafts into position.

24. Loosen the camshaft timing belt tensioner pulley retaining bolt and move the tensioner pulley to relieve the tension on the timing belt.

25. Temporarily tighten the tensioner in this position.

NOTE: If the timing belt is to be reused, mark the belt for the direction of rotation before removing for reinstallation to prevent premature wear or failure.

26. Remove the camshaft timing belt.

27. If required, the sprockets are removed as follows:

a. Hold the camshaft with the Camshaft Sprocket Holding Tool T74P-6256-B or equivalent.

b. Loosen and remove the camshaft sprocket retaining bolt.

c. Remove the camshaft sprocket from the camshaft.

d. Repeat the procedure for the 2nd camshaft sprocket.

e. Remove the crankshaft sprocket.

To install:

28. Slide the crankshaft sprocket onto the crankshaft aligning the keyway.

29. Align the camshafts using the Camshaft Alignment Timing Tool T94P-6256-CH.

30. Install the sprockets onto the camshafts and loosely install the camshaft retaining bolts.

31. Torque the camshaft sprocket retaining bolts to 47–53 ft. lbs. (64–72 Nm).

32. Loosely install the crankshaft pulley to verify that the engine is at

TDC. Realign the marks if they have moved.

33. Verify that the camshafts are aligned.

NOTE: It is recommended to purchase a tensioner spring and retaining bolt through dealer parts to apply the proper tension for used or new belt installations. The spring is bolted to the tensioner assembly and becomes a part of the engine. Ignore this notice if the tensioner spring is already installed.

34. Install the retaining bolt (W700001-S309) into the hole provided in the cylinder block and place the tensioner spring (6L277) between the bolt and the camshaft timing belt tensioner pulley.

35. Torque the retainer bolt to 71–97 inch lbs. (8–11 Nm).

36. Remove the crankshaft pulley and install the camshaft timing belt onto the crankshaft sprocket and then onto the camshaft sprockets working in a counterclockwise direction.

37. Torque the camshaft sprocket retaining bolts to 47–53 ft. lbs. (64–72 Nm).

38. Make sure that the span of the camshaft timing belt between the crankshaft sprocket and the exhaust camshaft sprocket is not loose.

39. Make sure that the camshaft timing belt is securely aligned on all sprockets.

40. Install the lower timing belt cover and torque the retaining bolts to 53–71 inch lbs. (6–8 Nm).

41. Apply silicone sealer to the keyway of the crankshaft pulley and install. Torque the retaining bolt to 81–89 ft. lbs. (110–120 Nm).

42. Inspect the timing mark on the crankshaft pulley to verify that the engine is still at TDC.

43. Loosen the camshaft timing belt tensioner pulley retaining bolt and allow the tensioner spring attached to the pulley to pull the tensioner pulley against the camshaft timing belt.

44. Remove the camshaft alignment timing tool from the camshafts at the rear of the engine.

45. Turn the crankshaft 2 revolutions in a clockwise direction.

46. Tighten the camshaft timing belt tensioner pulley retaining bolt to 26–30 ft. lbs. (35–40 Nm).

47. Recheck the crankshaft timing mark is at TDC for the No. 1 piston and that both camshafts are in align-

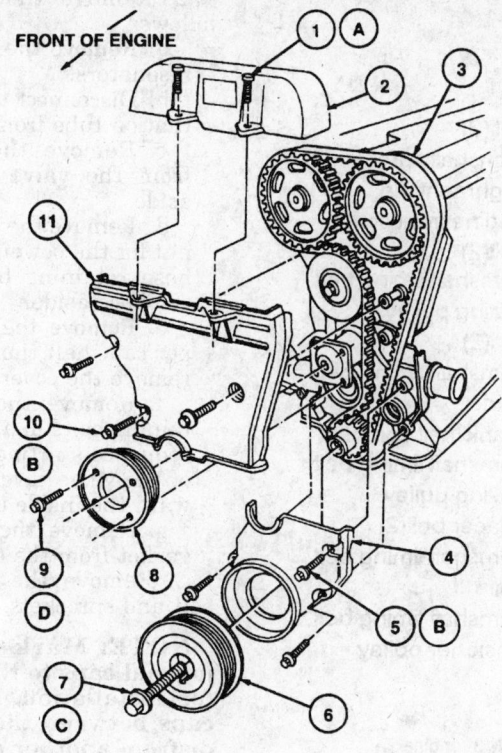

FRONT OF ENGINE

1. Bolt(2)
2. Upper camshaft timing belt cover
3. Cylinder block
4. Lower camshaft timing belt cover
5. Bolt(3)
6. Crankshaft pulley
7. Bolt
8. Water pump pulley
9. Bolt(4)
10. Bolt(3)
11. Center camshaft timing belt cover
A. 27–44 in. lb.(3–5 Nm)
B. 53–71 in. lb.(6–8 Nm)
C. 81–89 ft. lb.(110–120 Nm)
D. 89–124 in. lb.(10–14 Nm)

297871

Removal of components to access the timing belt — 2.0L (VIN 3) engine

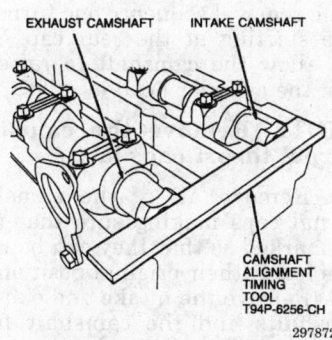

EXHAUST CAMSHAFT   INTAKE CAMSHAFT

CAMSHAFT ALIGNMENT TIMING TOOL T94P-6256-CH

297872

Rear view of the camshafts and the alignment tool — 2.0L (VIN 3) engine

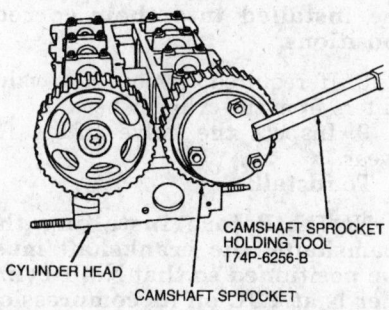

CAMSHAFT SPROCKET HOLDING TOOL T74P-6256-B

CYLINDER HEAD

CAMSHAFT SPROCKET

297955

Camshaft sprocket being held with the camshaft holding tool — 2.0L (VIN 3) engine

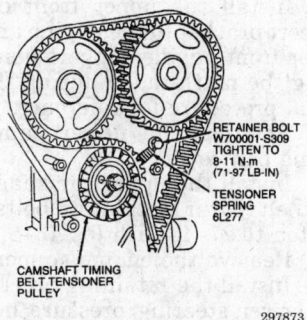

RETAINER BOLT W700001-S309 TIGHTEN TO 8-11 N-m (71-97 LB-IN)

TENSIONER SPRING 6L277

CAMSHAFT TIMING BELT TENSIONER PULLEY

297873

Timing belt installed showing the tensioner spring — 2.0L (VIN 3) engine

ment using the camshaft alignment timing tool.

NOTE: A slight adjustment of the camshafts to allow the insertion of the camshaft alignment timing tool is permissible as long as the crankshaft stays at the TDC location.

48. Camshaft sprocket holding tool T74P-6256–B can be used to move the camshaft sprocket(s) if a slight adjustment is required.

49. If a camshaft is not properly aligned, perform the following procedure:

a. Loosen the retaining bolt securing the camshaft sprocket to the camshaft while holding the camshaft sprocket from turning with the camshaft sprocket holding tool.

b. Turn the camshaft until the camshaft alignment timing tool can be installed.

c. Verify that the crankshaft timing mark is at TDC for No. 1 cylinder.

d. While holding the camshaft sprocket with the camshaft sprocket holding tool, torque the retaining bolt to 47–53 ft. lbs. (64–72 Nm).

e. Remove the tool and rotate the crankshaft 2 revolutions (clockwise).

f. Verify that the camshafts are aligned and that the crankshaft is at TDC for No. 1 cylinder.

50. Install the valve cover as follows:

a. Clean the gasket sealing surfaces.

b. Inspect the valve cover gasket and O-rings, replace as required.

c. Install the valve cover retaining bolts and tighten in a standard sequence starting from the center

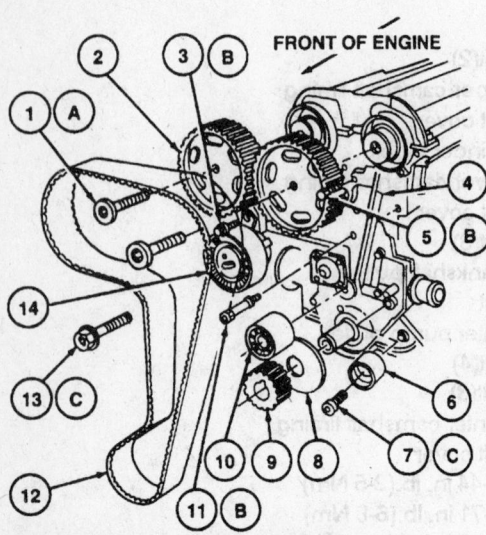

FRONT OF ENGINE

1. Bolt(2)
2. Camshaft sprocket(2)
3. Alignment pin
4. Engine front cover
5. Bolt(2)
6. Camshaft timing belt routing pulley
7. Bolt(2)
8. Camshaft timing belt guide
9. Crankshaft sprocket
10. Camshaft timing belt routing pulley
11. Spacer bolt(2)
12. Camshaft timing belt
13. Bolt
14. Camshaft timing belt tensioner pulley

297874

NOTE: FACES OF CAMSHAFTS AND CAMSHAFT SPROCKETS MUST BE FREE OF BURRS, DIRT AND DEBRIS PRIOR TO ASSEMBLY OF CAMSHAFT SPROCKETS TO CAMSHAFTS.

View of the sprockets and related components — 2.0L (VIN 3) engine

and working towards the outside of the valve cover to 53–71 inch lbs. (6–8 Nm).

d. Install the power steering hose retaining bracket and the power steering hose.

e. Install the crankcase ventilation tube to the valve cover.

51. Position the center camshaft timing belt cover.

52. Install the center camshaft timing belt cover retaining bolts and tighten to 53–71 inch lbs. (6–8 Nm).

53. Install the water pump pulley and the retaining bolts. Install the bolts, finger tight.

54. Install the drive belt idler pulley.

55. Install the drive belt idler pulley retaining bolt and torque to 35 ft. lbs. (48 Nm).

56. Install the accessory drive belt.

57. Tighten the water pump pulley retaining bolts to 89–124 inch lbs. (10–14 Nm).

58. Move the radiator coolant recovery reservoir aside.

59. Install the upper front engine support insulator.

60. Position the radiator coolant recovery reservoir and install the retainers.

61. If equipped, install the wiring harness to the low coolant level sensor on the coolant recovery reservoir.

62. Install the upper front engine support bracket to the engine and the upper front engine support insulator using the mark made during the removal procedure for reference.

63. Position the upper camshaft timing belt cover.

64. Install the upper camshaft timing belt cover retaining bolts and tighten to 27–44 inch lbs. (3–5 Nm).

65. Remove the engine support.

66. Install the retaining bracket for the power steering pressure hose to the engine lifting eye.

67. Install the spark plugs and the ignition wires.

68. Install the engine air intake resonators.

69. Replace the engine oil.

70. Connect the negative battery cable.

71. Run the engine and check for leaks and proper operation.

Camshaft

REMOVAL AND INSTALLATION

2.0L (VIN 3) Engine

1. Disconnect the negative battery cable.

2. Remove the valve cover as follows:

a. Remove the engine air intake resonators.

b. Disconnect the crankcase ventilation tube from the valve cover.

c. Remove the ignition wires from the valve cover and move aside.

d. Remove the retaining bolt and nut for the power steering pressure hose retaining bracket and move the hose aside.

e. Remove the bolts for the upper camshaft timing belt cover and remove the cover.

f. Remove the valve cover retaining bolts in a standard removal sequence starting from the outside of the valve cover and working toward the inside of the valve cover.

g. Remove the valve cover and gasket from the engine.

3. Remove the camshaft timing belt and sprockets.

NOTE: Mark the camshaft journal caps to the cylinder head for installation. Do not mix the caps between the two camshafts or from another cylinder head.

4. Loosen all of the camshaft journal cap bolts in pairs and in reverse of the removal sequence one turn at a time starting at the rear cap. This will allow the camshaft to raise up from the cylinder head evenly.

NOTE: Remove the camshaft journal thrust caps last.

5. Remove all of the camshaft journal caps making sure that they are marked so that they can be reassembled to their original positions.

6. Remove the intake and exhaust camshafts and the camshaft front seals from the cylinder head.

7. Inspect the camshafts and cylinder head for wear.

NOTE: If the valve lifters (tappets) are to be reused, mark their locations to ensure that they will be installed into their correct positions.

8. If required, remove the valve lifters from the cylinder head.

9. Inspect the valve lifters for wear.

To install:

NOTE: Before installing the camshafts, the crankshaft must be positioned so that No. 1 cylinder is at TDC on its compression stroke.

10. If removed, lubricate the valve lifters with engine assembly lubri-

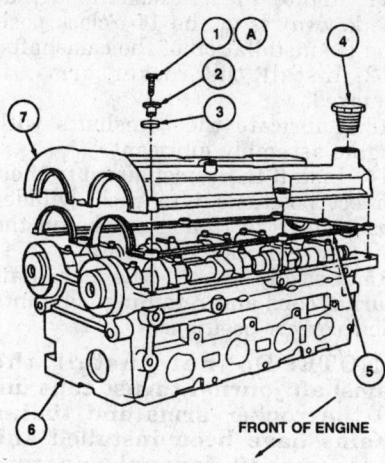

1. Bolt (10)
2. Spacer (10)
3. O-ring (10)
4. Oil filler cap
5. Valve cover gasket
6. Cylinder head
7. Valve cover
A. 53–71 in. lb.(6–8 Nm)

297125

Valve cover and related components — 2.0L (VIN 3) engine

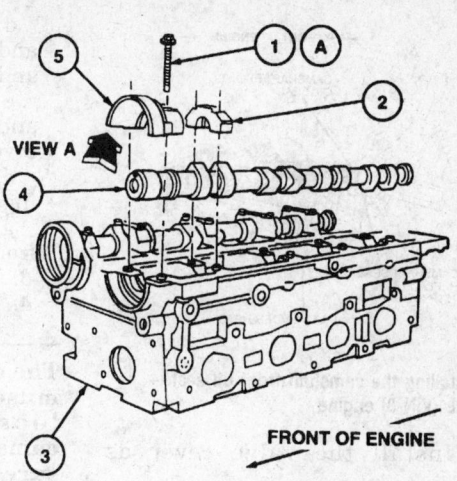

1. Bolt(20)
2. Camshaft journal cap(8)
3. Cylinder head
4. Camshaft
5. Camshaft journal thrust cap(2)
A. 13–15 ft. lb.(17–21 Nm)

VIEW A

FRONT OF ENGINE

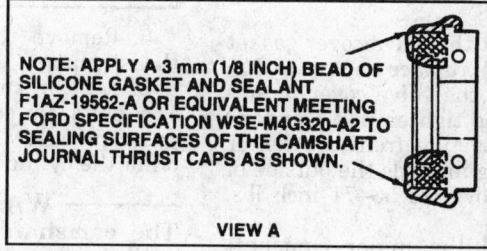

NOTE: APPLY A 3 mm (1/8 INCH) BEAD OF SILICONE GASKET AND SEALANT F1AZ-19562-A OR EQUIVALENT MEETING FORD SPECIFICATION WSE-M4G320-A2 TO SEALING SURFACES OF THE CAMSHAFT JOURNAL THRUST CAPS AS SHOWN.

VIEW A

297127

Installing the camshafts and journal caps — 2.0L (VIN 3) engine

cant and install into the lifter bores that they were removed from.

11. If the valve lifters are new, soak the lifters in a container of clean engine oil or manually pump up the lifters before installation.

12. Lubricate the camshafts with engine assembly lubricant and place the camshafts into the cylinder head.

NOTE: The intake and exhaust camshafts are marked for identification, also the intake camshaft has an extra cam lobe for the Camshaft Position Sensor (CMP).

13. Loosely install the camshaft journal caps and retaining bolts into their original positions.

14. Install the camshaft journal thrust caps and retaining bolts last. Apply a bead of silicone sealant to the sealing surfaces of the camshaft journal thrust caps.

15. Tighten all of the camshaft journal caps in several steps pulling the camshaft down evenly following the proper sequence.

16. Once the camshaft journal caps are fully seated, torque the retaining bolts to 13–15 ft. lbs. (17–21 Nm).

17. Install new camshaft front seals using the Camshaft Seal Replacer T92C–6700–CH or equivalent.

18. Install the camshaft sprockets and the camshaft timing belt.

FRONT OF ENGINE

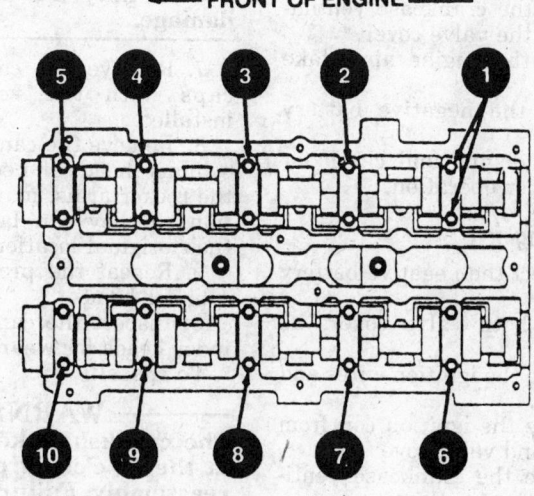

TIGHTEN BOLTS IN SEQUENCE SHOWN

297128

Tightening sequence for camshaft journal cap retaining bolts — 2.0L (VIN 3) engine

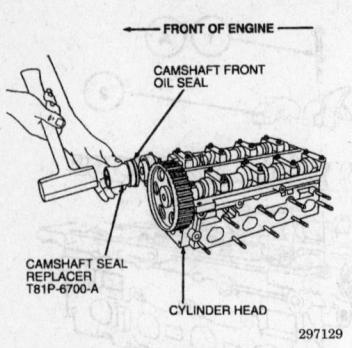

Installing the camshaft front oil seal — 2.0L (VIN 3) engine

19. Install the valve cover as follows:

a. Clean the gasket sealing surfaces.

b. Inspect the valve cover gasket and O-rings, replace as required.

c. Install the valve cover retaining bolts and tighten in a standard sequence starting from the center and working towards the outside of the valve cover to 53–71 inch lbs. (6–8 Nm).

d. Install the upper camshaft timing belt cover and tighten the retaining bolts to 27–44 inch lbs. (3–5 Nm).

e. Install the power steering hose retaining bracket and the power steering hose.

f. Install the ignition wires.

g. Install the crankcase ventilation tube to the valve cover.

h. Install the engine air intake resonators.

20. Connect the negative battery cable.

21. Run the engine and check for leaks and proper operation.

2.5L (VIN L) Engine

1. Disconnect the negative battery cable.

2. Remove both valve covers as follows:

a. Remove the ignition wires and spark plugs.

b. Remove the ignition coil from the right-hand valve cover.

c. Remove the crankcase ventilation tubes from both valve covers.

d. Remove the wiring harness and bracket to the fuel injectors and move aside.

e. Remove the retaining nuts and engine wiring from both valve covers and move aside.

f. Loosen the valve cover retaining bolts and studs, in sequence.

g. Remove both valve covers from the engine.

3. Remove the engine front cover.

4. Remove the timing chains.

——— WARNING ———

The camshaft journal thrust caps must be removed first, before loosening the remaining camshaft journal cap bolts, to ensure that the camshaft journal thrust caps are not damaged.

5. Remove the camshaft journal thrust caps.

6. Loosen the camshaft journal cap bolts in sequence, in several passes, to allow the camshaft to raise off of the cylinder head evenly.

——— WARNING ———

The camshaft journal caps and cylinder heads are numbered to ensure that they are assembled in their original positions. Keep the camshaft journal caps from each cylinder head together; do not mix them with caps from another cylinder head. Failure to do so may result in engine damage.

7. Remove the camshaft journal caps with the retaining bolts installed.

8. Remove the camshafts from the cylinder head. If necessary, remove the rocker arms, marking their position so they can be reinstalled in their original locations.

9. Repeat the procedure for both cylinder heads.

10. Inspect the camshafts and cylinder heads for wear or damage.

To install:

——— WARNING ———

The crankshaft keyway must be at the 11 o'clock position before reassembly. Failure to do so may lead to engine damage.

11. Rotate the crankshaft so that the keyway is at the 11 o'clock position for installation of the camshafts.

12. Install the rocker arms, if removed.

13. Lubricate the camshafts with engine assembly lubricant.

14. Install the camshafts into their correct positions into each cylinder head with the timing marks on the camshaft sprockets aligned.

15. Loosely install the camshaft journal caps and retaining bolts into their correct positions.

NOTE: Do not install the camshaft journal thrust caps until the rocker arms and timing chains have been installed and the camshaft journal caps are torqued into position.

16. Install the rocker arms.

17. Install the timing chains.

18. Tighten the camshaft journal cap bolts, in sequence, to 71–106 inch lbs. (8–12 Nm). Install the thrust caps and tighten the retaining bolts to 71–106 inch lbs. (8–12 Nm).

19. Install the engine front cover.

20. Install both valve covers as follows:

a. Clean the valve cover gasket sealing surfaces.

b. Install new valve cover gaskets onto the valve covers.

c. For each valve cover, place a bead of silicone sealant at 2 places on the valve cover sealing surfaces where the engine front cover and the cylinder heads make contact and at 2 places on the rear of the cylinder head where the camshaft seal retainer contacts the cylinder head.

d. Install the valve cover retaining bolts and studs and tighten in sequence to 71–106 inch lbs. (8–12 Nm).

NOTE: The valve covers must be installed and properly tightened within 6 minutes of applying the silicone sealant.

21. Connect the negative battery cable.

22. Run the engine and check for leaks and proper engine operation.

LH Side Shown, RH Side Similar

CAUTION: CAMSHAFTS MUST BE IN TIME WITH CRANKSHAFT PRIOR TO INSTALLATION OF ROCKER ARMS

NOTE: VALVE TAPPET MUST BE PUMPED UP PRIOR TO INSTALLATION

FRONT OF ENGINE

CAUTION: REMOVE CYLINDER HEAD JOURNAL THRUST CAPS FIRST. INSTALL CYLINDER HEAD JOURNAL THRUST CAPS LAST.

CAUTION: ROTATE CRANKSHAFT KEYWAY TO 11 O'CLOCK POSITION PRIOR TO INSTALLATION OF LH CAMSHAFT AND ROCKER ARMS. ROTATE CRANKSHAFT KEYWAY TO 3 O'CLOCK POSITION PRIOR TO INSTALLATION OF RH CAMSHAFT AND ROCKER ARMS.

326654

Camshaft removal — left cylinder head shown — 2.5L (VIN L) engine

Piston and Connecting Rod

POSITIONING

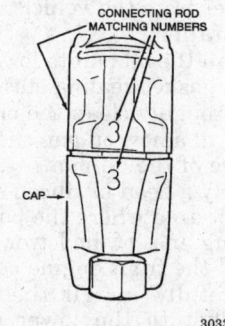

Connecting rod and cap positioning — all engines

303310

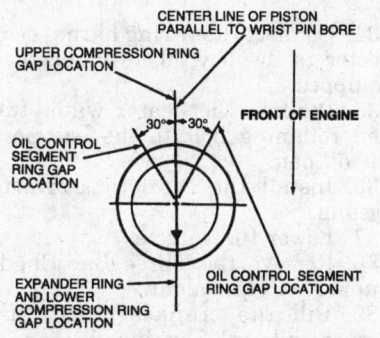

Piston ring positioning — all engines

303308

ENGINE LUBRICATION

Oil Pan

REMOVAL AND INSTALLATION

2.0L (VIN 3) Engine

1. Disconnect the negative battery cable.

2. Install Three Bar Engine Support D88L-6000-A or similar engine support, to the engine lifting eyes and support the engine.

3. Raise and safely support the vehicle.

4. Remove the catalytic converter system.

5. Disconnect the wiring to the low oil level sensor, if equipped.

6. Remove the heater water bolt from the bottom of the oil pan and position the tube out of the way.

7. Drain the engine oil.

8. Reinstall the oil pan drain plug and torque to 15–21 ft. lbs. (21–28 Nm).

9. Remove the oil pan retaining bolts from the transaxle housing.

10. Remove the lower engine rear plate.

11. Remove the through bolt for the left-hand and right-hand support insulators.

12. Lower the vehicle to work on the top side of the engine but keep on the hoist.

NOTE: Mark the location of the upper front engine support bracket before removing from the front engine support bracket.

13. Remove the upper front engine support bracket retaining nuts from

the upper front engine support insulator.

14. Raise the engine to allow room for removal of the engine oil pan, using the Three Bar Engine Support or equivalent engine brace.

15. Raise and safely support the vehicle.

16. Remove the oil pan retaining bolts from the cylinder block, working from the ends of the block toward the center.

17. Loosen and remove the oil pan and gasket.

18. Inspect the oil pump pickup tube and screen and clean or replace as necessary.

To install:

19. If removed, install the oil pump pickup tube and screen. Tighten the retaining bolts to 71–97 inch lbs. (8–11 Nm). Install a new self-locking oil pump pickup tube support nut to the crankshaft main bearing cap retaining stud bolt. Tighten to 13–15 ft. lbs. (17–21 Nm).

20. Clean the gasket sealing surfaces for the oil pan at the cylinder block.

21. Clean the oil pan thoroughly leaving no traces of gasket material, grease or solvents.

22. Apply a bead of silicone gasket sealer to the oil pump parting lines and at the crankshaft rear main seal retainer on the cylinder block.

23. Install the oil pan and a new gasket into position and hold with several oil pan retaining screws.

24. Install the rest of the retaining bolts and push the oil pan flush against the transaxle case before tightening the retaining bolts.

25. Tighten the 10 oil pan retaining bolts, in several passes, to 15–18 ft. lbs. (20–24 Nm), working from the center of the block towards the ends.

26. Install the lower engine rear plate.

27. Install the oil pan to transaxle housing retaining bolts.

28. Torque the retaining bolts to 25–34 ft. lbs. (34–46 Nm).

29. Lower the vehicle but keep on the hoist.

30. Lower the engine into position by adjusting the Three Bar Engine Support or equivalent.

31. Install the upper front engine support bracket retaining nuts onto the upper front engine support insulator.

32. Raise and safely support the vehicle.

33. Install the through bolts for the left-hand and the right-hand insulators.

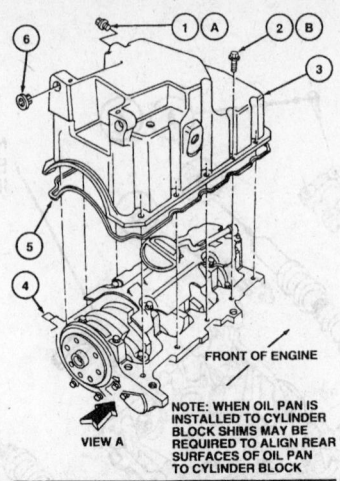

FRONT OF ENGINE

VIEW A

NOTE: WHEN OIL PAN IS INSTALLED TO CYLINDER BLOCK SHIMS MAY BE REQUIRED TO ALIGN REAR SURFACES OF OIL PAN TO CYLINDER BLOCK

3 mm (0.25 INCH)

NOTE: APPLY A 3 mm (0.25 INCH) BEAD OF SILICONE GASKET AND SEALANT F1AZ-19562-A OR EQUIVALENT MEETING FORD SPECIFICATION WSE-M4G320-A2

SEALER

VIEW A TYPICAL FOUR PLACES

1 Oil pan drain plug
2 Bolt (10 req'd)
3 Oil pan
4 Cylinder block
5 Oil pan gasket
6 Oil pan spacer (as req'd)
A Tighten to 21–28 Nm (15–21 lb-ft)
B Tighten to 20–24 Nm (15–18 lb-ft)

296810

Installing the oil pan and related components

34. Connect the wiring harness connector to the low oil level sensor, if equipped.

35. Position the heater water tube and retaining bolt to the bottom of the oil pan.

36. Install the catalytic converter system.

37. Lower the vehicle.

38. Remove the Three Bar Engine Support or equivalent.

39. Fill the crankcase with the proper amount of engine oil.

40. Connect the negative battery cable.

41. Run the engine and check for leaks and proper operation.

2.5L (VIN L) Engine

1. Disconnect the negative battery cable.

2. Remove the water pump pulley shield.

3. Install Three Bar Engine Support D88L-6000-A or equivalent, to the engine lifting eyes and support the engine.

4. Raise and safely support the vehicle.

5. Remove the exhaust crossover and the exhaust retaining bracket located on the right-hand side of the oil pan.

6. Remove the exhaust heat shield retaining nuts and the exhaust heat shields from the left-hand side of the oil pan.

7. Drain the engine oil.

8. Reinstall the oil pan drain plug using a new gasket and torque to 16–22 ft. lbs. (22–30 Nm).

9. Remove the oil pan retaining bolts from the transaxle housing.

10. If equipped with automatic transaxle, remove the access plug from the engine rear plate.

11. Remove the through bolt for the left-hand and right-hand front engine support insulators.

12. Partially lower the vehicle on the hoist.

NOTE: Mark the location of the upper front engine support bracket before it is removed.

13. Remove the upper front engine support bracket retaining nuts and remove the upper front engine support bracket.

14. Using the three bar engine support or equivalent, raise the engine to allow room for removal of the engine oil pan.

15. Raise and safely support the vehicle.

16. Remove the oil pan retaining bolts and studs from the lower cylinder block following the bolt removal sequence.

17. Remove the oil pan and the oil pan gasket from the vehicle.

To Install:

18. Clean the oil pan to lower cylinder block gasket sealing surfaces.

19. Thoroughly clean the oil pan.

20. Install a new oil pan gasket into the groove of the oil pan.

21. Apply a bead of silicon sealer to the gasket area where the pan meets the parting lines of the lower cylinder block and the front engine cover.

22. Carefully install the oil pan with gasket to the lower cylinder block.

23. Install the bolts and studs but do not tighten.

24. Push the oil pan against the transaxle case and tighten the oil pan bolts and studs.

25. Install the oil pan to transaxle case bolts and torque to 25–34 ft. lbs. (34–46 Nm).

26. Tighten the oil pan bolts and studs in the proper sequence to 15–22 ft. lbs. (20–30 Nm).

27. If equipped with automatic transaxle, install the access plug into the engine rear plate.

28. Lower the vehicle but keep it on the hoist.

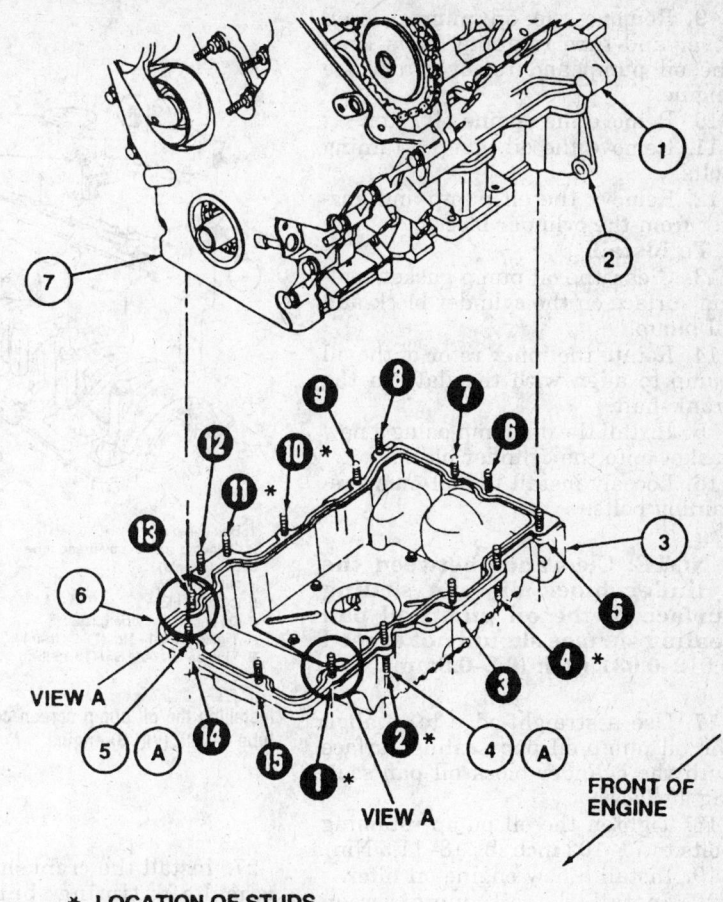

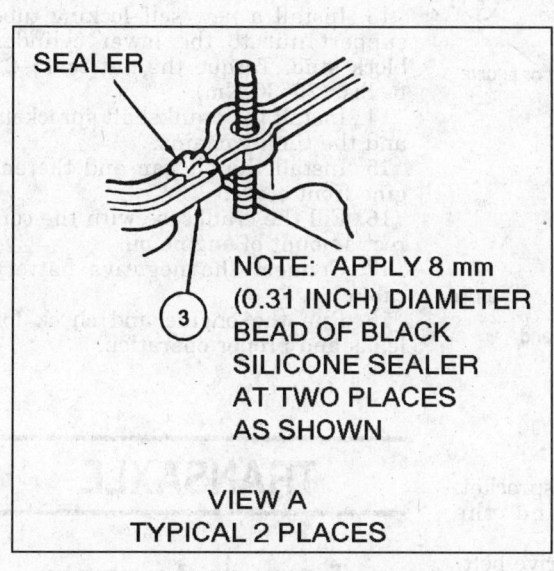

SEALER

NOTE: APPLY 8 mm
(0.31 INCH) DIAMETER
BEAD OF BLACK
SILICONE SEALER
AT TWO PLACES
AS SHOWN

VIEW A
TYPICAL 2 PLACES

* LOCATION OF STUDS
● TIGHTEN BOLTS/STUDS IN
SEQUENCE SHOWN

FRONT OF
ENGINE

1 Upper cylinder block	5	Bolt (10 req'd)
2 Lower cylinder block	6	Oil pan gasket
3 Oil pan	7	Engine front cover
4 Stud bolt (5 req'd)	A	Tighten to 20-30 Nm (15-22 lb-ft)

296839

Oil pan bolt and stud torque sequence — 2.5L (VIN L) engine

29. Using the three bar engine support or equivalent, lower the engine into its proper position.

30. Install the upper front engine support bracket and its retaining nuts onto the upper front engine support insulator.

31. Raise and safely support the vehicle.

32. Install the through bolts for the left-hand and right-hand front engine support insulators.

33. Install the exhaust crossover, exhaust retaining bracket and heat shield.

34. Replace the engine oil filter.

35. Lower the vehicle.

36. Remove the three bar engine support or equivalent.

37. Fill the crankcase with the correct amount and grade of engine oil.

38. Connect the negative battery cable.

39. Run the engine and check for leaks and proper operation.

Oil Pump

REMOVAL AND INSTALLATION

2.0L (VIN 3) Engine

1. Disconnect the negative battery cable.

2. Remove the accessory drive belt.

3. Remove the camshaft timing belt covers, camshaft timing belt and the crankshaft sprocket.

4. Install Three Bar Engine Support D88L-6000-A or similar engine support, to the engine lifting eyes and support the engine.

5. Raise and safely support the vehicle.

6. Remove the catalytic converter system.

7. Remove the oil pan.

8. Remove the oil pump screen cover and tube retaining nut from the crankshaft main bearing cap stud bolt.

9. Remove the oil pump screen cover and tube retaining bolts from the oil pump and remove from the engine.

10. Remove the engine oil filter.

11. Remove the oil pump retaining bolts.

12. Remove the oil pump and gasket from the cylinder block.

To install:

13. Clean the oil pump gasket sealing surface on the cylinder block and oil pump.

14. Rotate the inner rotor of the oil pump to align with the flats on the crankshaft.

15. Install the oil pump using a new gasket onto the cylinder block.

16. Loosely install the oil pump retaining bolts.

NOTE: Clearance between the cylinder block oil pan sealing surface to the oil pump oil pan sealing surface should not exceed 0.012–0.031 inch (0.3–0.8mm).

17. Use a straight-edge to to align the oil pump oil pan sealing surface with the cylinder block oil pan sealing surface.

18. Tighten the oil pump retaining bolts to 71–102 inch lbs. (8–11.5 Nm).

19. Install a new engine oil filter.

20. Install the oil pump screen cover and tube using a new gasket to the oil pump. Tighten the retaining bolts to 71–97 inch lbs. (8–11 Nm).

21. Install a new self–locking oil pump screen cover and tube support nut to the crankshaft main bearing cap stud bolt. Torque the retaining nut to 13–15 ft. lbs. (17–21 Nm).

22. Install the engine oil pan.

23. Install the catalytic converter system.

24. Lower the vehicle.

25. Remove the three bar engine support or equivalent.

26. If required, replace the front crankshaft seal at this time.

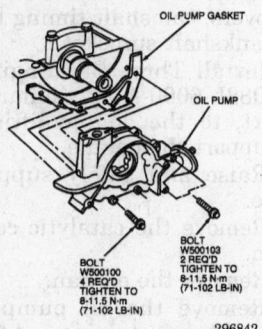

OIL PUMP GASKET

OIL PUMP

BOLT W500100 4 REQ'D TIGHTEN TO 8-11.5 N·m (71-102 LB-IN)

BOLT W500103 2 REQ'D TIGHTEN TO 8-11.5 N·m (71-102 LB-IN)

296842

View of the oil pump and gasket — 2.0L (VIN 3) engine

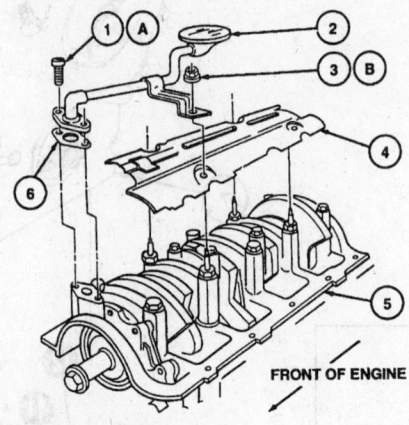

1 Bolt (2 req'd)
2 Oil pump screen cover and tube
3 Nut (4 req'd)
4 Oil pan baffle
5 Cylinder block
6 Oil pump inlet tube gasket
A Tighten to 8-11 Nm (71-97 lb-in)
B Tighten to 17-21 Nm (13-15 lb-ft)

FRONT OF ENGINE

296843

Installing the oil pump screen cover and tube — 2.0L (VIN 3) engine

27. Install the crankshaft sprocket, camshaft timing belt and the camshaft timing belt covers.

28. Install the accessory drive belt.

29. Fill the crankcase with the proper amount and grade of engine oil.

30. Connect the negative battery cable.

31. Run the engine and check for leaks and proper operation.

2.5L (VIN L) Engine

1. Disconnect the negative battery cable.

2. Remove the oil pan and the engine front cover.

3. Remove the timing chains and the crankshaft sprockets.

4. Remove the oil pump screen cover and tube retaining nut from the lower cylinder block stud bolt.

5. Remove the oil pump screen cover and tube retaining bolts from the oil pump and remove the tube from the engine.

6. Remove the 4 oil pump retaining bolts in reverse of the removal sequence.

7. Remove the oil pump from the vehicle.

To install:

8. Rotate the inner rotor of the oil pump to align with the flats on the crankshaft.

9. Install the oil pump flush to the cylinder block.

10. Install the oil pump retaining bolts and tighten in sequence to 71–106 inch lbs. (8–12 Nm).

11. Inspect the oil pump screen and tube O-ring and replace if needed.

12. Position the oil pump screen and tube with the O-ring to the oil pump. Tighten the retaining bolts to 71–106 inch lbs. (8–12 Nm).

13. Install a new self–locking tube support nut to the lower cylinder block stud. Torque the nut to 15–22 ft. lbs. (20–30 Nm).

14. Install the crankshaft sprockets and the timing chains.

15. Install the oil pan and the engine front cover.

16. Fill the crankcase with the correct amount of engine oil.

17. Connect the negative battery cable.

18. Run the engine and check for leaks and proper operation.

TRANSAXLE

Transaxle Assembly

REMOVAL AND INSTALLATION

1. Disconnect the battery cables, negative cable first.

2. Remove the battery.

3. Secure the radiator and fan shroud to the radiator support using safety wire.

4. Loosen the front strut upper retaining nuts a total of 5 turns to allow room for the removal of the half-shafts. Do not remove the nuts.

5. Remove the Mass Air Flow sensor (MAF) and the air cleaner assembly. Remove the air cleaner lower bracket.

6. Install Three Bar Engine Support D88L-6000-A or equivalent, and support the engine.

7. Disconnect the electrical connector at the backup lamp switch.

8. Remove the bolts securing the ground strap to the transaxle housing.

9. Remove the engine and transmission support insulator (mount).

10. Disconnect the hydraulic line and rubber grommet from the support insulator bracket.

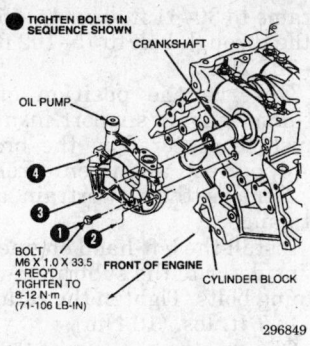

Oil pump retaining bolt torque sequence — 2.5L (VIN L) engine

11. Remove the rubber inspection cover from the transaxle clutch housing.

12. Remove the retaining clip and remove the hydraulic line fitting at the clutch slave cylinder.

13. Remove the upper transaxle to engine bolt.

14. Remove the 2 upper starter motor retaining bolts with the ground strap.

15. If equipped with the 2.0L (VIN 3) engine, remove the exhaust manifold heat shield and the catalytic converter retaining nuts at the exhaust manifold.

16. Remove the wheel and tire assemblies.

17. Remove the accessory drive belt pulley cover.

18. Raise and safely support the vehicle.

19. If equipped with the 2.0L (VIN 3) engine, remove the oil level dipstick. Remove the catalytic converter to engine bracket strap and the retaining bolts to the halfshaft bracket.

20. Remove the catalytic converter.

21. If equipped with the 2.5L (VIN L) engine, remove the water pump pulley shield. Remove the front Y-pipe nuts and the rear Y-pipe to catalytic converter nuts and remove the Y-pipe.

22. Disconnect the Vehicle Speed Sensor (VSS) electrical connector.

23. Remove the speedometer cable.

24. Remove the 9 screws securing the lower radiator air deflector and remove the deflector.

25. Push the shift rod forward and remove the shift rod pinch bolt.

26. Pull the shift rod back and remove it from the transaxle.

27. Remove the shift control stabilizer bar nut at the stud and remove the stabilizer bar.

28. Remove the shift control stabilizer bar and bracket from the right-hand engine support insulator bracket.

29. Remove the underbody heat shield from under the shift control.

30. Reposition the shift rod and stabilizer bar to allow transaxle removal.

31. Remove the 2 screws securing the A/C accumulator to the subframe.

32. Remove the halfshafts and the intermediate shaft.

33. Remove the bolts and the right-hand engine support insulator mounting nuts and remove the bracket from the transaxle.

34. Remove the left-hand engine support insulator through-bolt.

35. Lower the vehicle.

36. Adjust the three bar engine support or equivalent, to relieve tension on the right-hand front engine support bracket.

37. Remove the right-hand front engine support bracket through-bolt.

38. Raise and safely support the vehicle.

39. Disconnect the steering column from the steering gear at the pinch bolt.

40. Remove the catalytic converter.

41. Disconnect the tie rod ends from the steering knuckles and discard the cotter pins.

42. Remove the lower control arm to ball joint pinch bolts and separate the lower control arms from the ball joints.

43. Separate the sway bar (stabilizer bar) links from the sway bar.

44. Remove the splash shield at the front of the subframe.

45. Remove the through-bolt from the left-hand front engine support insulator and remove the right-hand front engine support insulator and mounting bracket.

46. Disconnect the power steering oil cooler hoses at the right-hand front of the subframe and drain the power steering fluid.

47. Remove the A/C accumulator retaining bracket screws from the subframe.

48. Remove the 4 bolts retaining the lower radiator supports to the subframe. Rotate the radiator supports forward.

49. Remove the 2 screws retaining the bumper cover braces to the left-hand and right-hand sides of the subframe. Rotate the bumper cover braces forward.

50. Position Powertrain Lift 014-00765 or equivalent, with wood blocks approximately 40 inches (1,016mm) in length secured to the lift under the subframe.

51. Remove the 4 subframe to body retaining bolts.

52. Allow the subframe to lower slightly and disconnect the power steering pressure and return hoses

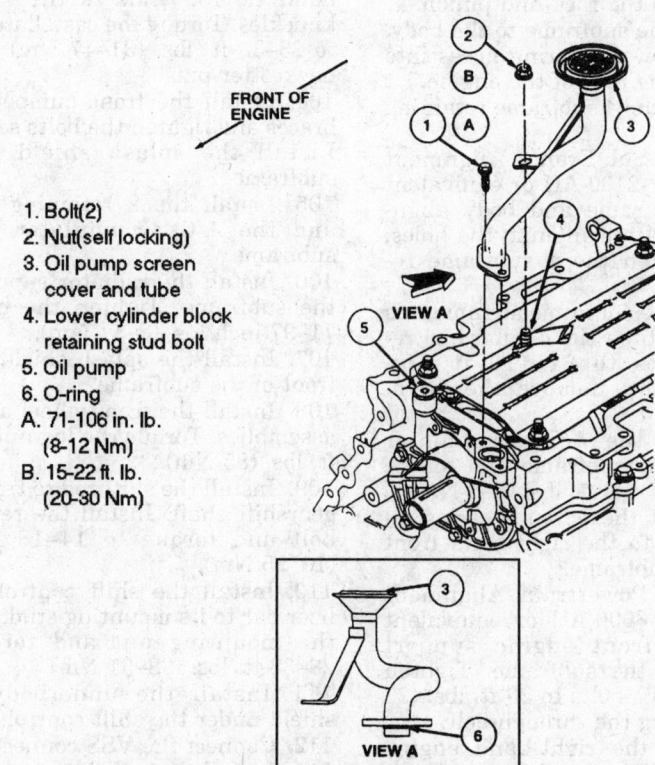

1. Bolt(2)
2. Nut(self locking)
3. Oil pump screen cover and tube
4. Lower cylinder block retaining stud bolt
5. Oil pump
6. O-ring
A. 71-106 in. lb. (8-12 Nm)
B. 15-22 ft. lb. (20-30 Nm)

Oil pump screen cover tube and components — 2.5L (VIN L) engine

from the rack and pinion (steering gear).

53. Finish lowering the subframe and move aside.

54. Lower the vehicle.

55. Loosen the front mount retaining nuts 5 turns.

56. Place a floor jack and a block of wood under the transaxle and raise the transaxle enough to release the tension on the three bar engine support, or equivalent.

57. Back off on the three bar engine support adjustment to allow downward travel of the transaxle.

58. Slowly lower the transaxle until it reaches the limits of the front engine mount movement.

59. Adjust the three bar engine support or equivalent, to hold the transaxle in this position.

60. Remove the floor jack and wood block.

61. Raise the vehicle and safely support.

62. Position Transmission Jack 014-00210 or equivalent, to the transaxle and secure.

63. Remove the last starter motor retaining bolt and hang the starter off to the side using safety wire.

64. Remove the 2 bolts retaining the engine oil pan to the transaxle.

65. Remove the remaining bolts.

66. Separate the transaxle from the engine and carefully remove from the vehicle.

To install:

67. If removed, place the transaxle on the transmission jack and secure.

68. Apply a film of grease to the input shaft splines.

69. If removed, install the right-hand engine support insulator bracket to the transaxle case. Torque the bolts to 62 ft. lbs. (84 Nm).

70. If removed, install the shift stabilizer bar mounting bracket to the right-hand transaxle support insulator bracket stud and transaxle case. Torque the nut and bolt to 28–38 ft. lbs. (38–51 Nm).

71. Carefully raise the transaxle into position with the engine.

72. If required, use an 18mm socket to rotate the engine to line up the splines on the clutch disc with the input shaft.

73. Install 2 lower side and 2 lower transaxle retaining bolts. Torque the retaining bolts to 30 ft. lbs. (40 Nm).

74. Install the starter motor and the lower bolt. Torque the starter motor lower bolt to 35 ft. lbs. (48 Nm).

75. Lower the vehicle.

76. Using the floor jack and a block of wood positioned at the engine and

transaxle mating area, raise the assembly into position.

77. Adjust the three bar engine support to maintain the engine in the correct position.

78. Remove the floor jack and the block of wood. Install the upper transaxle retaining bolts.

79. Torque the retaining bolts to 28–38 ft. lbs. (38–51 Nm).

80. Install the nuts retaining the front engine support bracket. Torque the retaining nuts to 61 ft. lbs. (83 Nm).

81. Install the upper starter motor retaining bolts. Torque the retaining bolts to 35 ft. lbs. (48 Nm).

82. Install the ground strap to the transaxle retaining bolt.

83. Connect the backup lamp switch connector.

84. Install the engine and transaxle support insulator.

85. Raise and safely support vehicle.

86. Install the left-hand and right-hand halfshafts and the intermediate shaft.

87. If removed, place the subframe on the powertrain lift, or equivalent with wood blocks approximately 40 inches (1,016mm) in length secured to the lift under the subframe.

88. Raise the subframe and connect the power steering pressure and return hoses to the rack and pinion.

89. Align the subframe to the body. Route the power steering hoses into position at the rear of the engine.

90. Install the 4 subframe retaining bolts loosely.

91. Install Sub-Frame Alignment Pin Set T94P-2100-AH or equivalent into the subframe and body alignment holes. After aligning the holes, slightly tighten the 4 subframe retaining bolts.

92. After the subframe alignment is complete, tighten the 4 subframe retaining bolts to 81–110 ft. lbs. (110–150 Nm). Remove the alignment tools.

93. Install the A/C accumulator bracket to the subframe and torque the screws to 4–6 ft. lbs. (6–8 Nm).

94. Connect the power steering oil cooler hoses to the right-hand front side of the subframe.

95. Install Powertrain Alignment Gauge T94P-6000-AH or equivalent to the left-front engine support bracket and the subframe. Tighten the 2 retaining bolts to 20 ft. lbs. (27 Nm) and snug the through-bolt.

96. Install the right-hand engine support insulator with retaining bolts to the subframe and through-bolt. Tighten the 2 retaining bolts to

subframe to 30–41 ft. lbs. (41–55 Nm) and the through-bolt to 75–102 ft. lbs. (103–137 Nm).

97. Observe the position of the right-hand engine support insulator. It must be centered in the bracket and in perfect alignment front to rear. Remove the powertrain alignment gauge.

98. Install the left-hand engine support insulator to the subframe with 2 retaining bolts. Tighten the retaining bolts to 7 ft. lbs. (10 Nm).

99. Observe the position of the left-hand engine support insulator to ensure perfect alignment front to rear. Retighten the bolts to 30–41 ft. lbs. (41–55 Nm). Install the left-hand engine support insulator through-bolt and tighten to 75–102 ft. lbs. (103–137 Nm).

100. Install the left-hand and right-hand lower control arms to the steering knuckles. Install new pinch bolts and nuts. Torque the pinch bolts to 61 ft. lbs. (83 Nm).

101. Connect the steering yoke to the steering gear shaft. Torque the steering yoke retaining bolt to 15–20 ft. lbs. (20–27 Nm).

102. Install the sway bar (stabilizer bar) links. Install the retaining nuts and tighten to 35–48 ft. lbs. (47–65 Nm).

103. Install the left-hand and right-hand tie rod ends to the steering knuckles. Torque the castellated nuts to 23–35 ft. lbs. (31–47 Nm). Install new cotter pins.

104. Install the front bumper cover braces and tighten the bolts securely. Install the splash shield to the subframe.

105. Install the 2 retaining screws and the A/C accumulator to the subframe.

106. Install the radiator supports to the subframe. Tighten the bolts to 71–97 inch lbs. (8–11 Nm).

107. Install the splash shield at the front of the subframe.

108. Install the front wheel and tire assemblies. Torque the lug nuts to 62 ft. lbs. (85 Nm).

109. Install the shift rod to transaxle gearshift shaft. Install the retaining bolt and torque to 14–18 ft. lbs. (19–25 Nm).

110. Install the shift control stabilizer bar to its mounting stud. Install the mounting nut and torque to 28–38 ft. lbs. (38–51 Nm).

111. Install the underbody heat shield under the shift control.

112. Connect the VSS connector.

113. Install the speedometer cable.

114. If equipped with 2.5L (VIN L) engine, install the Y-pipe to the ex-

haust manifolds and the catalytic converter using new gaskets. Install the water pump pulley shield.

115. If equipped with 2.0L (VIN 3) engine, install the catalytic converter between the exhaust manifold and the exhaust pipe using new gaskets. Install the exhaust manifold heat shield and the oil level dipstick. Tighten the retaining bolts to 71–106 inch lbs. (8–11 Nm).

116. Install the support bracket strap to the catalytic converter and engine.

117. Install the bracket to the catalytic converter and the intermediate halfshaft bearing bracket.

118. Install the lower radiator air deflector.

119. Install the belt pulley cover by sliding it up and under the front fender splash shield.

120. Check the transmission fluid level and add fluid as required.

121. Lower the vehicle.

122. Remove the three bar engine support, or equivalent.

123. Install the wire loom retainers removed during the transaxle removal.

124. Install the hydraulic line to the clutch slave cylinder and install the clip.

125. Install the rubber inspection cover to the clutch housing and the hydraulic line to the retaining grommet.

126. Remove the upper strut mounting nuts and replace with new locknuts. Torque the strut mounting nuts to 34 ft. lbs. (46 Nm).

127. Install the air cleaner lower bracket.

128. Install the MAF sensor and the air cleaner assembly.

129. Remove the wire retaining the radiator and fan shroud to the radiator support.

130. Install the battery and cables, negative cable last.

131. Adjust the shift linkage and bleed the hydraulic clutch system as required.

132. Road test the vehicle and check for proper operation.

Clutch Assembly

REMOVAL AND INSTALLATION

1. Disconnect the negative battery cable.

2. Raise and safely support the vehicle.

3. Remove the starter motor.

4. Disconnect the hydraulic coupling for the slave cylinder at the transaxle by sliding the sleeve on the tube towards the slave cylinder and applying a slight pulling force to the tube.

5. Remove the transaxle.

6. Mark the assembled position of the clutch and pressure plate to the flywheel if it is to be reinstalled.

NOTE: The clutch pressure plate is only held in place by the retaining bolts. No dowel pins are used, therefore the pressure plate must be supported when removing the retaining bolts.

7. Loosen the pressure plate bolts evenly until the pressure plate spring pressure is released, then finish removing the bolts while supporting the clutch and pressure plate assembly.

8. Remove the clutch and pressure plate from the vehicle.

9. Inspect the flywheel, slave cylinder and other components for wear or damage.

To install:

10. Clean the pressure plate and flywheel surfaces.

11. Install the clutch disc using Clutch Aligner T74P-7137-K or equivalent.

NOTE: If the clutch disc and pressure plate are being reused, align the marks made during disassembly.

12. Install Flywheel Holding Tool T74P-6375-A or equivalent, to hold the flywheel.

13. Install the pressure plate and start the retaining bolts.

14. Torque the retaining bolts evenly and in sequence to 13–18 ft. lbs. (18–26 Nm).

15. Remove the clutch aligner tool.

16. Install the transaxle.

17. Connect the slave cylinder tube coupling by pushing the male coupling into the slave cylinder female coupling.

18. Lower the vehicle.

19. Connect the negative battery cable.

20. Bleed the hydraulic clutch system, if required.

21. Check the clutch system for proper operation.

Clutch Master Cylinder

REMOVAL AND INSTALLATION

1. Disconnect the negative battery cable.

2. Lower the storage compartment door located to the lower left side of the steering column by depressing the locking tabs.

3. Remove the lower instrument panel by removing 3 retaining screws and 2 retaining clips.

4. Lower the fuse box and disconnect the 2 multi-plugs and remove the multi-plugs from the fuse box.

5. Unhook the fuse box from its support.

6. Unplug the 5 electrical connectors from the back of the fuse box and remove the fuse box.

7. Remove the spring clip retaining the clutch master cylinder to the clutch pedal.

8. Remove the spring clip retaining the clutch slave cylinder tube to the clutch master cylinder.

NOTE: Place a fender cover or equivalent on the floor to protect the carpet from the hydraulic fluid.

9. Disconnect the clutch slave cylinder tube and brake master cylinder fluid reservoir tube from the clutch master cylinder and plug the tubes.

10. Remove the 2 nuts retaining the clutch master cylinder to the bracket and remove the clutch master cylinder from the vehicle.

To install:

11. Install the clutch master cylinder to the bracket and install the 2 retaining nuts. Torque the retaining nuts to 7.5 ft. lbs. (10 Nm).

12. Install the clutch slave cylinder tube to the clutch master cylinder and install the spring clip. Connect the brake master cylinder fluid reservoir tube to the clutch master cylinder.

13. Install the spring clip retaining the clutch master cylinder to the clutch pedal.

14. Connect the 5 electrical connectors to the back of the fuse box and position the fuse box.

15. Connect the fuse box to its support.

16. Connect the 2 multi-plugs to the fuse box.

17. Raise the fuse box into its normal position.

18. Install the lower instrument panel using the retaining screws and clips.

19. Close the lower storage compartment door.

20. Connect the negative battery cable.

21. Bleed the hydraulic clutch system.

22. Check for proper clutch system operation.

Clutch System

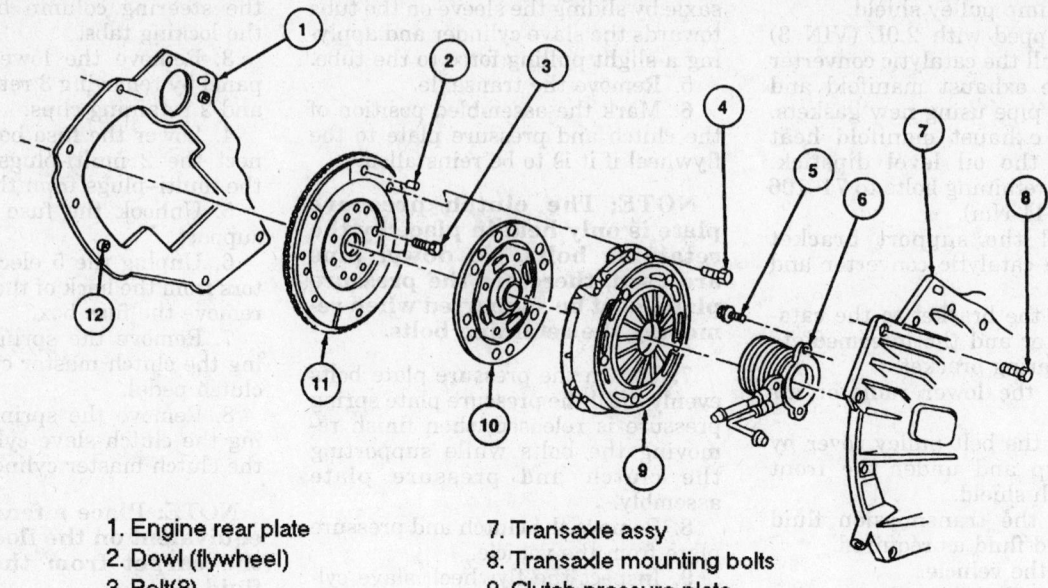

1. Engine rear plate
2. Dowl (flywheel)
3. Bolt(8)
4. Bolt(6)
5. Bolt(3)
6. Clutch slave cylinder
7. Transaxle assy
8. Transaxle mounting bolts
9. Clutch pressure plate
10. Clutch disc
11. Flywheel
12. Dowl bushing (engine plate)

Clutch assembly components

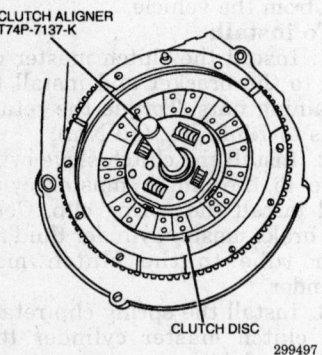

CLUTCH ALIGNER
T74P-7137-K

CLUTCH DISC

299497

Aligning the clutch disc

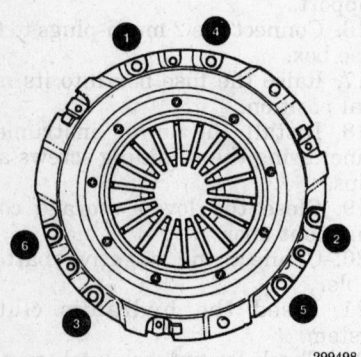

299498

Torque sequence for the pressure plate

12-58

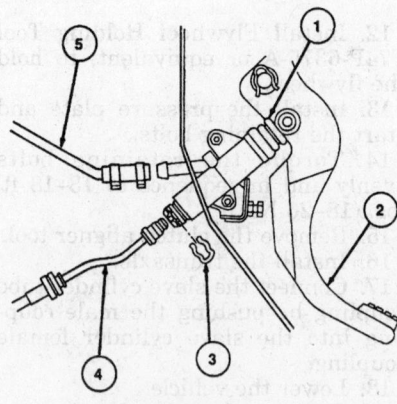

1 Clutch Master Cylinder
2 Clutch Pedal
3 Clip
4 Clutch Slave Cylinder to Clutch Master Cylinder Tube
5 Brake Reservoir-to-Clutch Master Cylinder Tube

299542

Clutch master cylinder and related components

299537

Clutch Slave Cylinder

REMOVAL AND INSTALLATION

1. Disconnect the negative battery cable.
2. Raise and safely support the vehicle.
3. Disconnect the hydraulic coupling for the clutch slave cylinder at the transaxle by removing the clip and then sliding the sleeve on the tube towards the clutch slave cylinder while applying a slight pulling force to the tube.
4. Remove the transaxle.
5. From inside the transaxles belhousing, remove the 3 clutch slave cylinder retaining bolts and remove the clutch slave cylinder.
6. Remove the clutch slave cylinder bleed tube.

To install:
7. Place the clutch slave cylinder over the input shaft splines and position the clutch slave cylinder.
8. Install the clutch slave cylinder retaining bolts. Torque the retaining bolts to 7–14 ft. lbs. (9–19 Nm).
9. Install the bleed tube. Torque the bleed tube fitting to 10.4 ft. lbs. (14 Nm).
10. Install the transaxle.

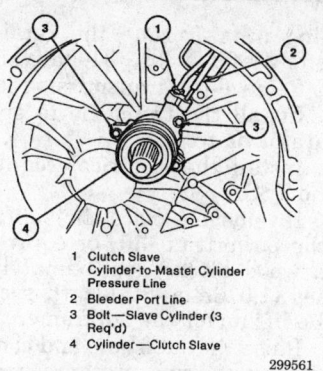

1 Clutch Slave
  Cylinder-to-Master Cylinder
  Pressure Line
2 Bleeder Port Line
3 Bolt—Slave Cylinder (3
  Req'd)
4 Cylinder—Clutch Slave

299561

Clutch slave cylinder and related components

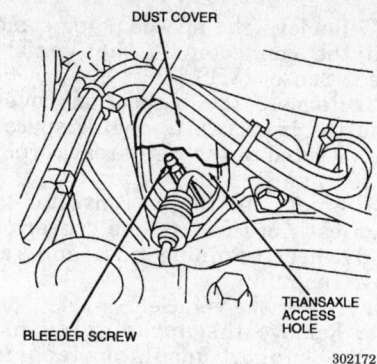

Clutch slave cylinder bleeder valve location

302172

11. Connect the slave cylinder tube coupling by pushing the male coupling into the slave cylinder female coupling and then installing the clip.

12. Lower the vehicle.

13. Connect the negative battery cable.

14. Bleed the hydraulic clutch system, if required.

15. Check for proper clutch operation.

Hydraulic Clutch System Bleeding

1. Disconnect the negative battery cable.

2. Remove the air cleaner outlet tube and the Mass Air Flow (MAF) sensor.

3. Clean the top of the brake master cylinder fluid reservoir before opening it.

NOTE: The brake master cylinder fluid reservoir is also the reservoir for the hydraulic clutch master cylinder.

4. Make sure that there is adequate fluid in the master cylinder fluid reservoir before attempting to bleed the system. Check the fluid level throughout the bleeding procedure.

5. Remove the rubber inspection cover from the bell housing.

6. Connect a hose to the bleeder valve fitting on the clutch slave cylinder. Submerge the other end of the hose into a container of clean brake fluid.

7. Push the clutch pedal down while opening the bleeder on the clutch slave cylinder. Watch for air bubbles escaping from the hydraulic system.

8. Close the bleeder before releasing the clutch pedal.

9. Repeat the procedure until no more air bubbles are seen.

10. Install the rubber inspection cover to the bell housing.

11. Top off the brake master cylinder fluid reservoir and install the diaphragm and cap securely.

12. Install the MAF sensor and air cleaner outlet tube.

13. Connect the negative battery cable.

14. Check the clutch for proper operation.

Clutch Adjustments

Because the clutch system is hydraulic, the clutch pedal free-play is self–adjusting and requires no additional maintenance.

Automatic Transaxle Assembly

REMOVAL AND INSTALLATION

1. Disconnect the negative battery cable.

2. Disconnect the positive battery cable and remove the battery.

3. On vehicles equipped with the 2.0L (VIN 3) engine, remove the oil level dipstick and the exhaust manifold heat shield.

4. On vehicles equipped with the 2.5L (VIN L) engine, remove the water pump pulley shield.

5. Secure the radiator and fan shroud with safety wire to the radiator support.

6. Remove the air cleaner assembly and mounting bracket.

7. Loosen the left-hand and right-hand upper strut mounting nuts 5 turns to allow room for removal of the halfshafts. Do not remove the nuts completely.

8. Disconnect the shift cable and remove the 2 retaining bolts securing the shift cable bracket to the transaxle case.

9. Disconnect the Transmission Range (TR) sensor connector.

10. Remove the TR sensor.

11. Disconnect the 10–pin harness connector from the transaxle.

12. Support the engine with Three Bar Engine Support D88L-6000-A or equivalent.

13. Remove the upper transaxle support insulator bracket mounting nuts.

14. Remove the 3 bolts retaining the upper transaxle support insulator to the inner fenderwell and remove the insulator.

15. Remove the upper bell housing to engine retaining bolts.

16. Raise and safely support the vehicle.

17. Remove the wheel and tire assemblies.

18. Disconnect the steering column from the rack and pinion by removing the pinch bolt.

19. Disconnect the tie rod ends from the steering knuckles. Discard the cotter pins.

20. Remove the ball joint to lower control arm pinch bolts and separate the ball joints from the lower control arms.

21. Remove the sway bar (stabilizer bar) link nuts and separate the sway bar links from the stabilizer bar.

22. Remove the splash shield at the front of the subframe.

23. Remove the through-bolts retaining the left-hand and right-hand front engine support insulators to the subframe.

24. Disconnect the power steering oil cooler hoses at the right-hand front subframe and drain the power steering system.

25. Remove the 2 screws retaining the bumper cover braces to the subframe. Rotate the bumper cover braces forward.

26. Remove the radiator air deflector.

27. Remove the 4 bolts retaining the left-hand and right-hand lower radiator support brackets to the front of the subframe. Rotate the radiator supports forward.

28. Disconnect and remove the exhaust system components necessary for transaxle removal.

29. Remove the 2 bolts retaining the A/C accumulator to the subframe, located at the drivers side front corner of the vehicle.

30. Remove the bolt retaining the transaxle cooler line bracket to the front of the subframe.

31. Position Powertrain Lift 014-00765 or equivalent with wood blocks approximately 40 inches (1,016mm) in length secured to the lift under the subframe.

32. Remove the 4 subframe to body retaining bolts. Lower the subframe slightly and disconnect the power steering pressure and return hoses from the rack and pinion.

33. Finish lowering the subframe and set aside.

34. Disconnect the Turbine Shaft Speed (TSS) sensor connector at the transaxle oil pump.

35. Drain the transaxle oil into a suitable container for recycling.

36. Disconnect the transaxle cooler inlet line at the transaxle case.

37. Remove the transaxle cooler inlet line from the bracket at the transaxle oil pump.

38. Disconnect the transaxle cooler outlet line at the transaxle case.

39. Disconnect the transaxle cooler inlet and outlet lines at the radiator and remove from the vehicle.

40. Remove the left-hand halfshaft and the right-hand halfshaft and intermediate shaft from the vehicle.

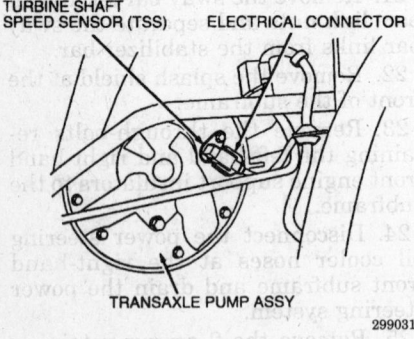

Turbine shaft speed sensor

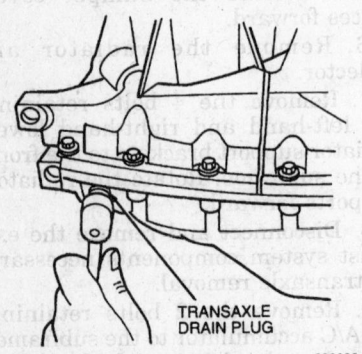

Transaxle drain plug location

41. Remove the speedometer cable and the connector to the Vehicle Speed Sensor (VSS).

42. Remove the inspection cover from the transaxle to engine spacer plate located at the right–rear corner of the engine.

43. Rotate the torque converter to align each of the 4 torque converter to flywheel retaining nuts and remove the nuts.

44. Lower the vehicle.

45. Remove the upper right-hand engine support insulator (engine mount) bracket nuts.

46. Lower the engine and transaxle assembly using the three bar engine support until the transaxle assembly is level with the left-hand frame member.

47. On vehicles equipped with A/C, lower the engine until the A/C compressor is below the right-hand frame member.

48. Raise and safely support the vehicle.

49. Support the transaxle on a transmission jack and secure the transaxle to the jack.

50. Remove the starter motor retaining nuts and remove the starter.

51. Remove the lower transaxle to engine retaining bolts.

52. Separate the transaxle from the engine.

NOTE: Use care when removing the transaxle to prevent the torque converter from falling out.

53. Carefully lower the transaxle from the vehicle.

To install:

54. Make sure that the torque converter is fully engaged in the transaxle.

55. There should be approximately a ⁷/₁₆ inch (10mm) air gap between a straightedge across the bellhousing flange and the torque converter.

56. If removed, place the transaxle on the transmission jack and secure.

NOTE: Use care not to allow the torque converter to fall out of the transaxle when tilted.

57. Raise the transaxle into position and align with the engine.

58. Align the torque converter studs with the mating holes in the flywheel.

59. Once the transaxle is fitted to the engine, install the lower transaxle to engine retaining bolts. Torque the retaining bolts to 41–50 ft. lbs. (55–68 Nm).

60. Rotate the torque converter and install the 4 converter to flywheel re-

taining nuts. Torque the retaining nuts to 23–39 ft. lbs. (31–53 Nm).

61. Remove the transmission jack.

62. Install the transaxle to engine separator plate.

63. Install the electrical connector to the TSS.

64. If removed, place the subframe on the powertrain lift, or equivalent with wood blocks approximately 40 inches (1,016mm) in length secured to the lift under the subframe.

65. Raise the subframe and connect the power steering pressure and return hoses to the rack and pinion.

66. Align the subframe. Route the power steering hoses into position at the rear of the engine.

67. Install the 4 subframe retaining bolts loosely.

68. Install Sub-Frame Alignment Pin Set T94P-2100-AH or equivalent into the subframe and body alignment holes. After aligning the holes, slightly tighten the 4 subframe retaining bolts.

69. After the subframe alignment is complete, tighten the 4 subframe retaining bolts to 81–110 ft. lbs. (110–150 Nm). Remove the alignment tools.

70. Install the A/C accumulator bracket to the subframe and torque the screws to 4–6 ft. lbs. (6–8 Nm).

71. Connect the power steering oil cooler hoses to the right-hand front side of the subframe.

72. Install Powertrain Alignment Gauge T94P-6000-AH or equivalent to the left-front engine support bracket and the subframe. Tighten the 2 retaining bolts to 20 ft. lbs. (27 Nm) and snug the through-bolt.

73. Install the right-hand engine support insulator with retaining bolts to the subframe and through-bolt. Tighten the 2 retaining bolts to subframe to 30–41 ft. lbs. (41–55 Nm) and the through-bolt to 75–102 ft. lbs. (103–137 Nm).

74. Observe the position of the right-hand engine support insulator. It must be centered in the bracket and in perfect alignment front to rear. Remove the powertrain alignment gauge.

75. Install the left-hand engine support insulator to the subframe with 2 retaining bolts. Tighten the retaining bolts to 7 ft. lbs. (10 Nm).

76. Observe the position of the left-hand engine support insulator to ensure perfect alignment front to rear. Retighten the bolts to 30–41 ft. lbs. (41–55 Nm). Install the left-hand engine support insulator through-bolt and tighten to 75–102 ft. lbs. (103–137 Nm).

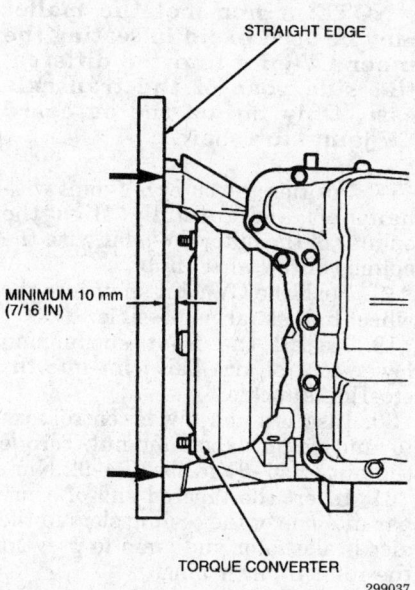

STRAIGHT EDGE

MINIMUM 10 mm
(7/16 IN)

TORQUE CONVERTER

299037

**Example of a properly installed torque
converter**

77. Install the transaxle cooler inlet and outlet lines.

78. Install the lower radiator support brackets to the subframe. Torque the bolts to 20 ft. lbs. (27 Nm).

79. Connect the speedometer cable and the electrical connector to the VSS.

80. Install the exhaust system.

81. Install the left-hand halfshaft using a new circlip.

82. Install the intermediate halfshaft and tighten the 2 retaining nuts to 20 ft. lbs. (27 Nm).

83. Install the right-side halfshaft.

84. Install the left-hand and right-hand lower control arms to the steering knuckles.

85. Install new pinch bolts and nuts. Torque the pinch bolts to 61 ft. lbs. (83 Nm).

86. Install the sway bar (stabilizer bar) links. Install the retaining nuts and tighten to 35–48 ft. lbs. (47–65 Nm).

87. If equipped, install the ABS wiring loom retainer to the sway bar link stud. Torque the retaining nuts to 35 ft. lbs. (47 Nm).

88. Install the left-hand and right-hand tie rod ends to the steering knuckles. Torque the castellated nuts to 23–35 ft. lbs. (31–47 Nm). Install new cotter pins.

89. Install the front bumper cover braces and tighten the bolts securely. Install the splash shield to the subframe.

90. Install the wheel and tire assemblies. Torque the lug nuts to 62 ft. lbs. (85 Nm).

91. Lower the vehicle.

92. Connect the steering yoke to the steering gear shaft. Torque the steering yoke retaining bolt to 15–20 ft. lbs. (20–27 Nm).

93. Install the upper transaxle to engine retaining bolts. Torque the retaining bolts to 23–39 ft. lbs. (55–68 Nm).

94. Remove the safety wire securing the radiator and fan shroud to the radiator support.

95. Reinstall all wiring retaining clips that were disturbed during transaxle removal.

96. Raise the engine and transaxle assembly into position using the three bar engine support or similar tool.

97. Install the engine and transmission support insulator to the left-hand front fender apron. Torque the bolts to 40–55 ft. lbs. (54–75 Nm).

98. Install new locknuts retaining the engine and transmission support insulator to the transaxle. Torque the locknuts to 40–55 ft. lbs. (54–75 Nm).

99. Install the front engine support insulator.

100. If equipped with 2.5L (VIN L) engine, install the power steering line retaining bracket to the front engine support insulator. Torque the new locknuts to 56–76 ft. lbs. (77–103 Nm). Install the water pump pulley shield.

101. If equipped with the 2.0L (VIN 3) engine, install the exhaust manifold heat shield and oil level dipstick. Tighten the retaining bolts to 71–106 inch lbs. (8–12 Nm).

102. Remove the three bar engine support, or equivalent.

103. Install the TR sensor and adjust.

104. Connect the 10-pin harness connector to the transaxle.

105. Install the TR sensor electrical connector.

106. Install the starter motor and retaining bolts. Torque the starter motor retaining bolts to 43–58 ft. lbs. (59–79 Nm) for the 2.5L (VIN L) engine or 15–20 ft. lbs. (20–27 Nm) for the 2.0L (VIN 3) engine.

107. Install the shift cable mounting bracket. Torque the retaining bolts to 15–19 ft. lbs. (20–25 Nm).

108. Install the shift cable to the manual lever by pressing the cable end onto the stud until a click is heard.

109. Install the battery tray, battery and the battery hold-down.

110. Install new upper strut mounting nuts. Torque the strut mounting nuts to 34 ft. lbs. (46 Nm).

111. Fill the power steering remote oil reservoir.

112. Connect the battery cables, negative cable last.

113. Fill the transaxle with the proper type and amount of transmission fluid.

114. Run the engine and check the transaxle for leaks.

115. Recheck the transmission fluid level.

NOTE: Whenever the vehicle subframe is removed or lowered, the wheel alignment should be checked.

116. Check the alignment and adjust if necessary.

117. Road test the vehicle to check for proper transmission operation.

DRIVELINE

Halfshaft

REMOVAL AND INSTALLATION

NOTE: Do not begin this procedure without a new wheel hub retaining nut, a new lower control arm to steering knuckle pinch bolt and new retainer circlips for the CV-joints. Once removed, these parts loose their torque holding or retention capabilities and must not be reused.

Left Side

1. Raise and safely support the vehicle.

2. Remove the left front wheel and tire assembly.

3. Snug 2 of the lug nuts back onto the rotor.

4. Insert the tapered end of a pry bar or steel rod into one of the cooling slots of the disc brake rotor and place the bar against the disc brake anchor plate, to keep the rotor from turning.

5. Loosen and remove the wheel hub retaining nut. Discard the nut.

6. Remove the nut of the stabilizer bar link and separate the stabilizer bar from the strut using a tie rod end removal tool.

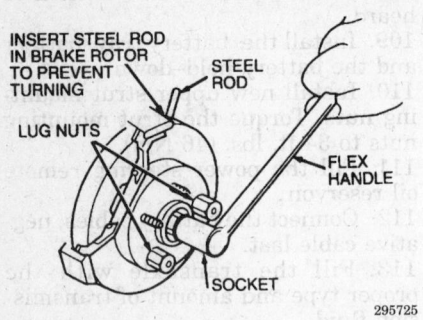

INSERT STEEL ROD
IN BRAKE ROTOR
TO PREVENT
TURNING

STEEL
ROD

LUG NUTS

FLEX
HANDLE

SOCKET

295725

Loosening the wheel hub retaining nut

7. Remove the cotter pin and castellated nut that secures the tie rod end to the steering knuckle. Discard the cotter pin.

8. Using a tie rod end removal tool, separate the tie rod end from the steering knuckle.

9. Remove the lower control arm to steering knuckle pinch bolt and nut.

10. Using a pry bar or similar tool, separate the lower control arm ball joint from the steering knuckle.

--- WARNING ---

Never use a hammer to separate the halfshaft from the front wheel hub as damage to the threads or internal components may result.

11. Separate the outer CV-joint and halfshaft from the wheel hub using Front Hub Remover/Replacer T81P–1104–C and its associated components, or equivalent.

12. Install CV-joint puller T86P–3514–A1 or equivalent, between the inner CV-joint and the transaxle case.

13. Attach the corresponding extension and slide hammer to the CV-joint puller and remove the halfshaft

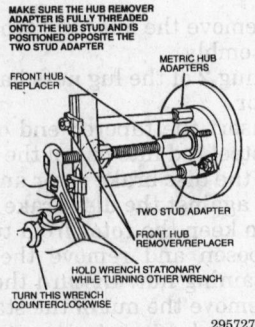

MAKE SURE THE HUB REMOVER
ADAPTER IS FULLY THREADED
ONTO THE HUB STUD AND IS
POSITIONED OPPOSITE THE
TWO STUD ADAPTER

FRONT HUB
REPLACER

METRIC HUB
ADAPTERS

TWO STUD ADAPTER

FRONT HUB
REMOVER/REPLACER

HOLD WRENCH STATIONARY
WHILE TURNING OTHER WRENCH

TURN THIS WRENCH
COUNTERCLOCKWISE

295727

Separating the outer CV-joint from the wheel hub

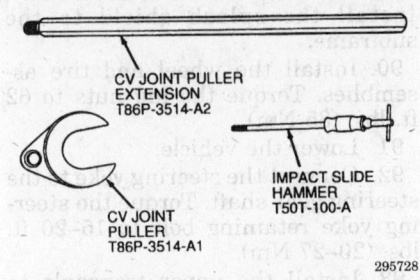

CV JOINT PULLER
EXTENSION
T86P-3514-A2

IMPACT SLIDE
HAMMER
T50T-100-A

CV JOINT
PULLER
T86P-3514-A1

295728

Tools used for inner CV-joint removal

with both CV-joints as an assembly from the transaxle case.

14. Remove the assembly from the vehicle.

To Install:

15. Replace the driveshaft bearing retainer circlip. Start one end of the circlip into the groove and work the circlip over the housing end and into the groove. This will avoid over–expanding the circlip.

16. Carefully align the splines of the inner CV-joint (install the halfshaft and both CV-joints as an assembly) with the splines in the transaxle case and push it into the

differential side gear until the circlip is felt to seat.

NOTE: A non–metallic mallet may be used to aid in seating the inner CV-joint into the differential side gear of the transaxle case. Only tap on the outboard CV-joint stub shaft.

17. Position the outer CV-joint with halfshaft and carefully align the splines of the outer CV-joint with the splines of the wheel hub.

18. Push the CV-joint shaft into the wheel hub as far as possible.

19. Install the front suspension lower control arm ball joint into the steering knuckle.

20. Install a new lower control arm to knuckle pinch bolt and nut. Torque the nut to 54–67 ft. lbs. (74–92 Nm).

21. Insert the tapered end of a pry bar into one of the cooling slots in the disc brake rotor and jamb to prevent the rotor from turning.

22. Install a new wheel hub retaining nut onto the exposed threads of the outer CV-joint and manually thread the nut on as far as possible.

23. Finish running the nut up and torque to 246 ft. lbs. (340 Nm).

24. Remove the steel rod or pry bar.

25. Install the tie rod end on the steering knuckle. Install the castel-

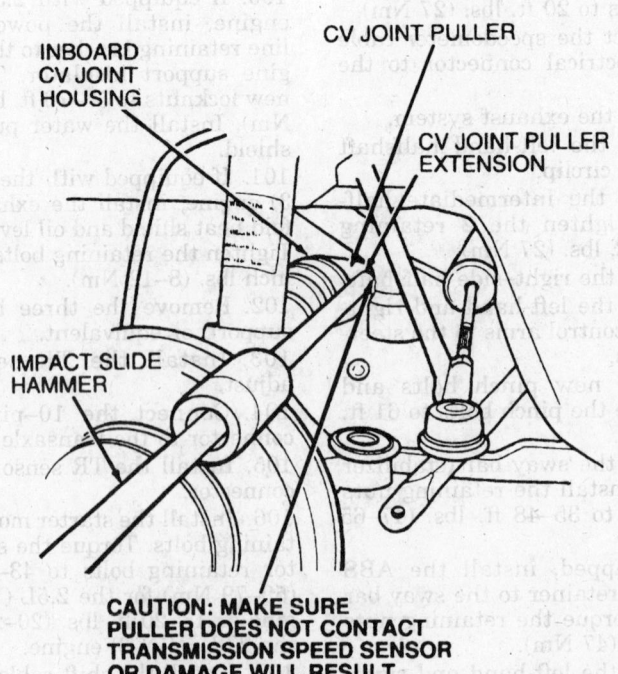

INBOARD
CV JOINT
HOUSING

CV JOINT PULLER

CV JOINT PULLER
EXTENSION

IMPACT SLIDE
HAMMER

CAUTION: MAKE SURE
PULLER DOES NOT CONTACT
TRANSMISSION SPEED SENSOR
OR DAMAGE WILL RESULT

295729

Positioning of the CV-joint puller for halfshaft removal

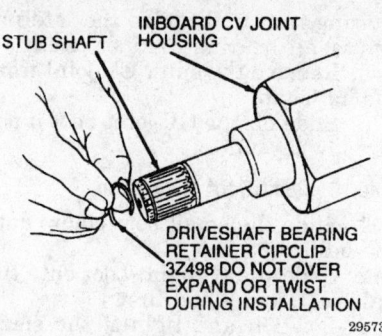

Replacing the driveshaft bearing retainer circlip on inner CV-joint

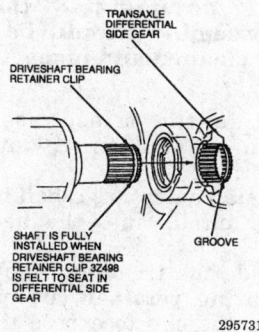

Aligning the inner CV-joint shaft with the differential side gear

lated nut and torque to 20 ft. lbs. (28 Nm). Install a new cotter pin.

26. Install the stabilizer link bar with a new nut. Torque the nut to 14–23 ft. lbs. (20–32 Nm).

27. Install the wheel and tire assembly. Torque the lug nuts to 62 ft. lbs. (85 Nm).

28. Lower the vehicle.

29. Road test the vehicle and check for proper operation.

Right Side and Intermediate

1. Raise and safely support the vehicle.

2. Remove the right front wheel and tire assembly.

3. Snug 2 of the lug nuts back onto the rotor.

4. Insert the tapered end of a pry bar into one of the cooling slots of the disc brake rotor and place the bar against the disc brake anchor plate, to prevent the rotor from turning.

5. Loosen and remove the wheel hub retaining nut. Discard the retaining nut.

6. Remove the nut of the stabilizer bar link and separate the stabilizer bar from the strut using a tie rod end removal tool.

7. Remove the cotter pin and castellated nut that secures the tie rod end to the steering knuckle. Discard the cotter pin.

8. Using a tie rod end removal tool, separate the tie rod end from the steering knuckle.

9. Remove the lower control arm to wheel spindle pinch bolt and nut.

10. Using a pry bar or similar tool, separate the lower control arm ball joint from the steering knuckle.

--- **WARNING** ---

Never use a hammer to separate the halfshaft from the front wheel hub as damage to the threads or internal components may result.

11. Separate the outer CV-joint and halfshaft from the wheel hub using Front Hub Remover/Replacer T81P–1104–C and its associated components, or an equivalent tool.

12. Install CV-joint puller T86P–3514–A1 or equivalent, between the inner CV-joint and the intermediate halfshaft.

13. Using the extension and slide hammer on the CV-joint puller, separate the right hand halfshaft with CV-joints from the intermediate halfshaft.

14. Remove the right side halfshaft from the vehicle.

15. If intermediate shaft removal is required, proceed as follows:

    a. Remove the 2 nuts securing the support bracket.

    b. Remove the intermediate shaft and bearing shield.

    c. On the 2.0L (VIN 3) engine, the exhaust clamp and 2 bolts will need to be removed to allow removal of the intermediate shaft.

    d. Remove the intermediate shaft from the vehicle.

    e. If the intermediate shaft support bracket needs to be removed, locate the 3 bolts securing the support bracket and remove the bolts.

    f. Remove the support bracket.

To install:

16. If the intermediate shaft was removed, reinstall as follows:

    a. If the support bracket was removed, reinstall with the 3 bolts and torque to 15–23 ft. lbs. (21–32 Nm).

    b. Carefully align the splines of the intermediate shaft with the splines of the differential side gears in the transaxle case.

    c. Push the shaft into the case until it is fully seated. The inner CV-joint is to be installed with the halfshaft and outer CV-joint attached as an assembly.

    d. Install the 2 nuts onto the intermediate shaft support bracket and bearing shield. Torque the nuts to 17–22 ft. lbs. (24–30 Nm).

    e. On the 2.0L (VIN 3) engine, install the 2 bolts and the exhaust clamp.

17. Replace the bearing retainer circlip on the intermediate shaft. Start one end of the circlip into the groove and work the circlip over the housing end into the groove. This will prevent over–expanding the circlip.

18. Carefully align the splines of the inner CV-joint with the splines of the intermediate shaft. The inner CV-joint is to be installed as an assembly with the right-hand halfshaft and outer CV-joint.

19. Push the inner CV-joint onto the intermediate shaft until it is fully seated.

20. Carefully align the splines of the outer CV-joint with the wheel hub and push the CV-joint into the hub as far as possible.

21. Install the lower control arm ball joint into the steering knuckle.

22. Install a new lower control arm to knuckle pinch bolt and nut. Torque to 54–67 ft. lbs. (74–92 Nm).

23. Insert the tapered end of a pry bar into one of the cooling slots in the front disc brake rotor and jamb the pry bar to prevent the rotor from turning.

24. Install a new wheel hub retaining nut onto the exposed threads of the outer CV-joint and manually thread the nut on as far as possible.

25. Torque the wheel hub retaining nut to 246 ft. lbs. (340 Nm).

26. Remove the pry bar.

27. Install the tie rod end into the steering knuckle. Install the castellated nut and torque to 20 ft. lbs. (28 Nm). Install a new cotter pin.

28. Install the stabilizer link bar and nut. Torque to 14–23 ft. lbs. (20–32 Nm).

29. Install the right front wheel and tire assembly. Torque the lug nuts to 62 ft. lbs. (85 Nm).

30. Lower the vehicle.

31. Road test the vehicle and check for proper operation.

CV-Joint Boot

REPLACEMENT

NOTE: Do not begin this procedure without a new wheel hub retainer, a new lower control arm to steering knuckle pinch bolt and new retainer circlips for the CV-joints. Once removed, these parts loose their torque holding or retention capabilities and must not be reused.

Left Side

1. Remove the left side halfshaft from the vehicle.
2. Place the halfshaft on a bench or in a soft jaw vise.

Inner CV-Joint Boot Removal

1. Cut off the 2 boot clamps.
2. Cut the CV-joint boot off or slide the boot back onto the halfshaft.
3. Slide the CV-joint stub shaft housing off to gain access to the tripod assembly.
4. Using snapring pliers, move the stop ring back onto the halfshaft.
5. Slide the tripod assembly back onto the halfshaft to allow access to the driveshaft bearing retainer circlip.

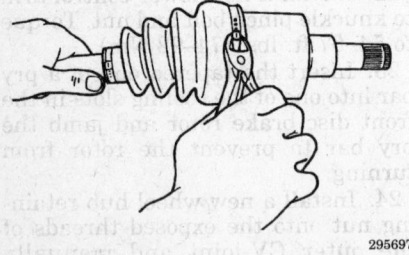

Removing the inner boot clamps with side-cutters — left side

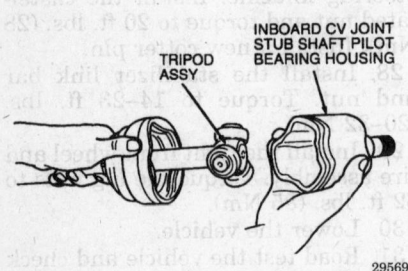

Removing the CV-joint stub shaft housing

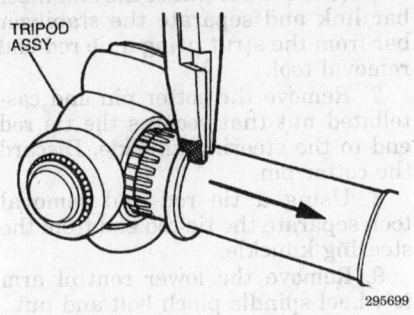

Moving the stop ring on the halfshaft — left side

6. Remove the driveshaft bearing retainer circlip from the halfshaft.
7. Slide off the tripod assembly.
8. Slide off the CV-joint boot if not already cut off.

Outer CV-Joint Boot Removal

1. Cut off the 2 CV-joint boot clamps.
2. Cut off the CV-joint boot or push back onto the halfshaft.
3. Angle the outer CV-joint away from the halfshaft to expose the inner bearing race.
4. Using a brass drift and hammer, apply a sharp tap to the inner

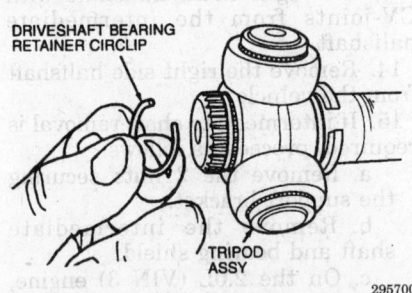

View of the driveshaft bearing retainer circlip

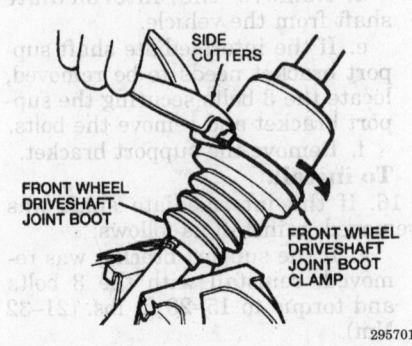

Removing the outer boot clamps using side-cutters

bearing race to dislodge the internal driveshaft bearing retainer circlip.
5. Remove the outer CV-joint from the halfshaft.
6. Slide off the CV-joint boot if not already cut off.

Inner CV-Joint Boot Installation

1. Slide the small boot clamp onto the halfshaft.
2. Slide the CV-joint boot onto the halfshaft, small end first.
3. Position and tighten the small boot clamp.

NOTE: Inspect the constant velocity joint grease for contamination. If contaminated, the CV-joint assembly should be thoroughly cleaned and inspected for wear.

4. Slide the tripod assembly (chamfered side first) onto the halfshaft.
5. Install a new driveshaft bearing retainer circlip into the halfshaft groove.
6. Slide the tripod assembly over the bearing retainer circlip compressing the clip to expose the stop ring groove at the other end.
7. Using snapring pliers, move the stop ring into its groove and make sure it is fully seated.
8. Install the Trilobe insert onto the CV-joint housing, if removed and fill the housing with the proper grease. Place the remaining grease evenly into the boot area.
9. Install the CV-joint stub shaft housing onto the tripod assembly. Make sure it is fully seated.
10. Place a screwdriver under the lip of the boot, being careful not to damage the boot, to allow trapped air to escape.
11. Place the large clamp into position and tighten the clamp.
12. Wipe off any remaining grease.

Outer CV-Joint Boot Installation

1. Install the small boot clamp.
2. Install the CV boot, small end first. Tape wrapped around the splines will aid in installing the new boot.
3. Position and tighten the small boot clamp.

NOTE: Inspect the constant velocity joint grease for contamination. If contaminated, the CV-joint assembly should be thoroughly cleaned and inspected for wear.

4. Install an new driveshaft bearing retainer circlip onto the halfshaft.
5. Fill the CV-joint with the proper grease.

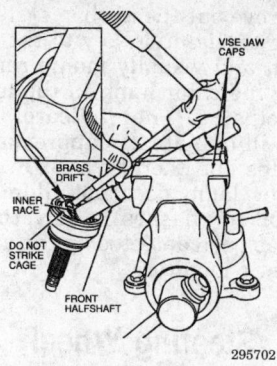

Method of removing the outer
CV-joint

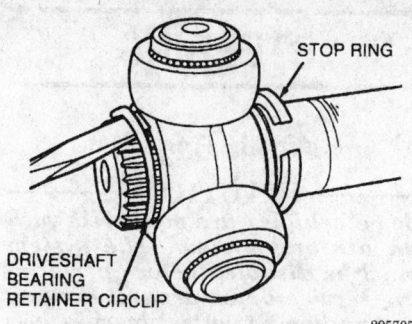

Installing a new driveshaft bearing retainer
circlip

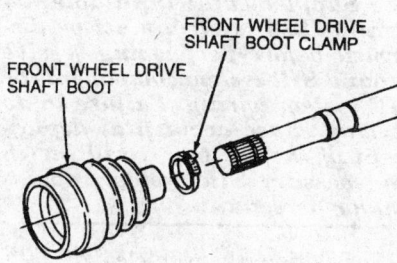

Installing the boot and clamp

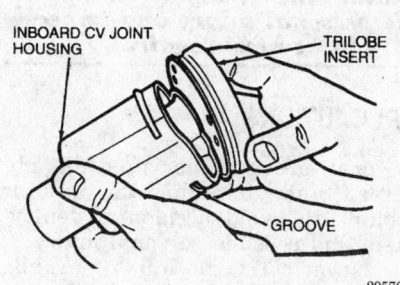

Positioning the Trilobe insert to the housing

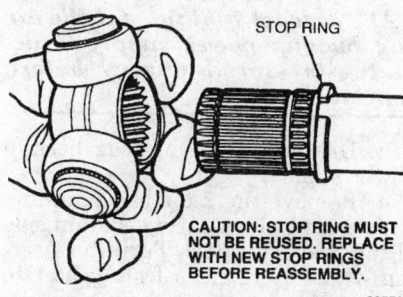

Installing the tripod onto the halfshaft

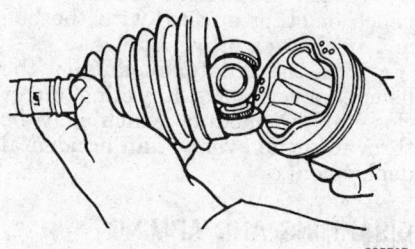

Installing the housing onto the tripod assembly

6. Place the CV-joint onto the half-shaft and align the splines.

7. Using a plastic faced hammer, tap the CV-joint onto the halfshaft until it is fully seated.

8. Wipe off any remaining grease.

9. Install the large boot clamp and tighten the clamp.

10. Install the halfshaft into the vehicle.

Right Side

1. Remove the right side halfshaft from the vehicle.

2. Place the halfshaft on a bench or in a soft jaw vise.

Inner CV-Joint Boot Removal

1. Cut off the 2 boot clamps.

2. Cut the CV-joint boot off or slide the boot back onto the halfshaft.

3. Slide the CV-joint stub shaft housing off to gain access to the tripod assembly.

4. Using snapring pliers, move the stop ring back onto the halfshaft.

5. Slide the tripod assembly back onto the halfshaft to allow access to the driveshaft bearing retainer circlip.

6. Remove the driveshaft bearing retainer circlip from the halfshaft.

7. Slide off the tripod assembly.

8. Slide off the CV-joint boot if not already cut off.

Outer CV-Joint Boot Removal

1. Clamp the halfshaft assembly in a soft jaw vise.

2. Cut off the 2 CV-joint boot clamps.

3. Cut off the CV-joint boot or push back onto the halfshaft.

4. Angle the outer CV-joint away from the halfshaft to expose the inner bearing race.

5. Using a brass drift and hammer, apply a sharp tap to the inner bearing race to dislodge the internal driveshaft bearing retainer circlip.

6. Remove the outer CV-joint from the halfshaft.

7. Slide off the CV-joint boot if not already cut off.

Inner CV-Joint Boot Installation

1. Slide the small boot clamp onto the halfshaft.

2. Slide the CV-joint boot onto the halfshaft, small end first.

3. Position and tighten the small boot clamp.

NOTE: Inspect the constant velocity joint grease for contamination. If contaminated, the CV-joint assembly should be thoroughly cleaned and inspected for wear.

4. Slide the tripod assembly (chamfered side first) onto the halfshaft.

5. Install a new driveshaft bearing retainer circlip into the halfshaft groove.

6. Slide the tripod assembly over the bearing retainer circlip compressing the clip to expose the stop ring groove at the other end.

7. Using the snapring pliers, move the stop ring into its groove and make sure it is fully seated.

8. Install the Trilobe insert onto the CV-joint housing, if removed and fill the housing with the proper grease. Place the remaining grease evenly into the boot area.

9. Install the CV-joint stub shaft housing onto the tripod assembly. Make sure it is fully seated.

10. Place a screwdriver under the lip of the boot, being careful not to damage the boot, to allow trapped air to escape.

11. Place the large clamp into position and tighten the clamp.

12. Wipe off any remaining grease.

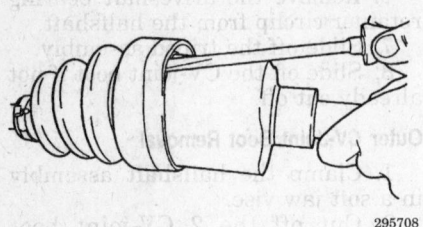

295708

Method of removing excess air pressure

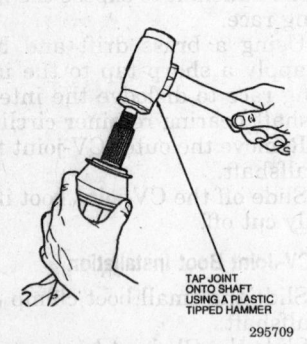

TAP JOINT
ONTO SHAFT
USING A PLASTIC
TIPPED HAMMER

295709

**Method of installing the outer
CV joint onto the halfshaft**

Outer CV-Joint Boot Installation

1. Install the small boot clamp.
2. Install the CV-joint boot, small end first. Tape wrapped around the splines will aid in installing the new boot.
3. Position and tighten the small boot clamp.

NOTE: Inspect the constant velocity joint grease for contamination. If contaminated, the CV-joint assembly should be thoroughly cleaned and inspected for wear.

4. Install a new driveshaft bearing retainer circlip onto the halfshaft.
5. Fill the CV-joint with the proper grease.
6. Place the CV-joint onto the halfshaft and align the splines.
7. Using a plastic faced hammer, tap the CV-joint onto the halfshaft until it is fully seated.
8. Wipe off any remaining grease.
9. Install the large boot clamp and tighten the clamp.
10. Install the right–side halfshaft into the vehicle.

STEERING

Air Bag

CAUTION

Some vehicles are equipped with an air bag system. The system must be disabled before performing service on or around air bag system components, steering column, instrument panel components, wiring and sensors. Failure to follow safety and disabling procedures could result in accidental air bag deployment, possible personal injury and unnecessary SIR system repairs.

PRECAUTIONS

Several precautions must be observed when handling the inflator module to avoid accidental deployment and possible personal injury.

- Never carry the inflator module by the wires or connector on the underside of the module.
- When carrying a live inflator module, hold securely with both hands, and ensure that the bag and trim cover are pointed away.
- Place the inflator module on a bench or other surface with the bag and trim cover facing up.
- With the inflator module on the bench, never place anything on or close to the module which may be thrown in the event of an accidental deployment.

DISARMING AND ARMING

CAUTION

The air bag system must be disarmed before performing service around air bag components or wiring. Failure to do so may cause accidental deployment of the air bag, resulting in unnecessary repairs and/or personal injury.

1. Position the vehicle with the front wheels in a straight ahead position.
2. Disconnect the negative battery cable.
3. Disconnect the positive battery cable.
4. Wait at least 1 minute for the air bag backup power supply to drain before continuing.
5. Proceed with repair.
6. Once complete, connect the battery cables, negative cable last.

7. Prove out the air bag system by turning the ignition key to the **RUN** position and visually monitoring the air bag indicator lamp in the instrument cluster. The indicator lamp should illuminate for approximately 6 seconds and then turn **OFF**. If the indicator lamp does not illuminate, stays on, or flashes at any time, a fault has been detected by the air bag diagnostic monitor.

Steering Wheel

REMOVAL AND INSTALLATION

CAUTION

The Supplemental Inflatable Restraint (SIR) system must be disarmed before performing service around SIR system components or SIR system wiring. Failure to do so may cause accidental deployment of the air bag, resulting in unnecessary SIR system repairs and/or personal injury.

1. Position the vehicle with the front wheels in a straight ahead position.
2. Disconnect both battery cables, negative cable last.

CAUTION

WAIT at least 1 minute for the air bag backup power supply to deplete its stored energy before continuing.

3. Disconnect the air bag backup power supply.
4. Remove the 2 air bag retaining screws from the steering column side of the steering wheel. Turn the steering wheel 90 degrees from center to remove 1 screw and then turn the steering wheel 180 degrees to gain access to the 2nd screw.

CAUTION

When carrying a live air bag, make sure the bag and trim cover are pointed away from the body. In the unlikely event of an accidental deployment, the bag will then deploy with minimal chance of injury. When placing a live air bag on a bench or other surface, always face the bag and trim cover up, away from the surface. This will reduce the motion of the module if it is accidently deployed.

5. Carefully remove the air bag module from the steering wheel and disconnect the electrical connector.

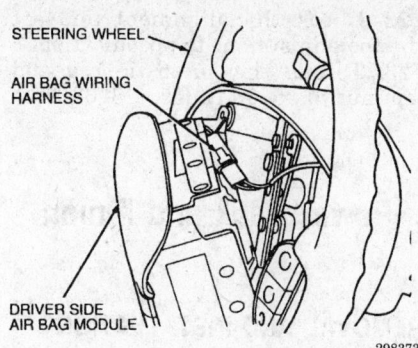

Air bag module and wiring harness

6. Remove the air bag module from the vehicle and place on a bench with the trim cover facing up.

7. Center the front wheels to the straight ahead position.

8. Disconnect the speed control wiring harness from the steering wheel.

9. Remove the steering wheel retaining bolt.

10. Carefully lift the steering wheel off of the shaft while routing the air bag sliding contact wire harness through the steering wheel opening.

11. Remove the steering wheel from the vehicle.

To install:

12. Ensure that the vehicles front wheels are in a straight ahead position.

NOTE: If the air bag was deployed due to an accident, the steering column must also be replaced.

13. Feed the air bag sliding contact wire harness through the steering wheel opening and position the steering wheel on the steering shaft.

14. Make sure that the shaft alignment marks are aligned and that the air bag contact wire is not pinched.

15. Install a new steering wheel retaining bolt.

16. Torque the retaining bolt to 37 ft. lbs. (50 Nm).

17. Connect the speed control wiring harness to the steering wheel and snap the connector assembly into the steering wheel clip.

18. Connect the air bag wiring harness to the air bag module and install the air bag to the steering wheel. Torque the air bag retaining screws to 8–10 ft. lbs. (11–13 Nm).

19. Connect the air bag backup power supply.

20. Connect both battery cables, negative cable last.

21. Prove out the air bag system by turning the ignition key to the **RUN**

position and visually monitoring the air bag indicator lamp in the instrument cluster. The indicator lamp should illuminate for approximately 6 seconds and then turn off. If the indicator lamp does not illuminate, stays on, or flashes at any time, a fault has been detected by the air bag diagnostic monitor.

Tie Rod Ends

REMOVAL AND INSTALLATION

NOTE: The outer tie rod ends may be serviced with the rack and pinion (steering gear) in the vehicle, however to service the inner tie rod ends, the rack and pinion must be removed from the vehicle. It is recommended to replace both inner tie rod ends while the rack and pinion is out of the vehicle.

Outer

1. Disconnect the negative battery cable.

2. Remove the wheel and tire assembly.

3. Remove the cotter pin and the castellated nut from the outer tie rod end. Discard the cotter pin.

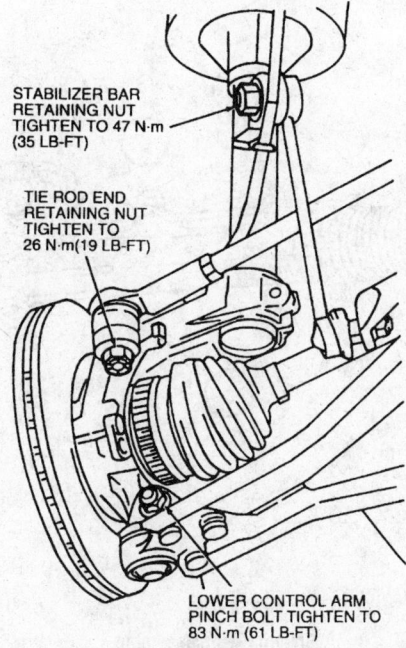

STABILIZER BAR RETAINING NUT TIGHTEN TO 47 N·m (35 LB-FT)

TIE ROD END RETAINING NUT TIGHTEN TO 26 N·m (19 LB-FT)

LOWER CONTROL ARM PINCH BOLT TIGHTEN TO 83 N·m (61 LB-FT)

299038

Front suspension and steering components

4. Separate the outer tie rod end from the steering knuckle using Tie Rod End Remover Tool 3290-D or equivalent.

5. Hold the outer tie rod end with a wrench and loosen the tie rod end jam nut.

6. Note the depth that the outer tie rod end jam nut is located.

7. Remove the outer tie rod end from the inner tie rod spindle. Count and record the number of turns required to remove the outer tie rod end.

8. Remove the outer tie rod end from the vehicle.

To install:

9. Clean the threads on the inner tie rod spindle (front wheel spindle connecting rod).

10. Thread the new outer tie rod end onto the inner tie rod. Use the same number of turns recorded during disassembly. The jam nut should also indicate that the new outer tie rod end is positioned properly.

11. Place the outer tie rod end stud into the steering knuckle. Set the front wheels in a straight ahead position.

12. Install a new castellated nut onto the outer tie rod end stud.

13. Torque the nut to 21 ft. lbs. (28 Nm).

14. Continue to tighten the castellated nut until a new cotter pin can be inserted through the hole in the stud. Install a new cotter pin.

15. If required, repeat the procedure for the opposite side.

16. Install the wheel and tire assembly. Torque the lug nuts to 62 ft. lbs. (85 Nm).

17. Connect the negative battery cable.

18. Check the alignment and set the toe adjustment to specification.

19. Torque the outer tie rod end jam nut to 35–50 ft. lbs. (48–68 Nm).

Inner

1. Remove the front subframe and the rack and pinion (steering gear).

2. Remove the rack and pinion from the front subframe and secure to a bench mounted holding fixture.

3. Working on one side, remove the outer tie rod end making sure to record the number of turn required for removal.

4. Remove the outer tie rod end jam nut.

5. Remove the outer clamp securing the inner tie rod bellows to the tie rod spindle.

6. Loosen the larger inner clamp using a wrench or screwdriver and remove the bellows.

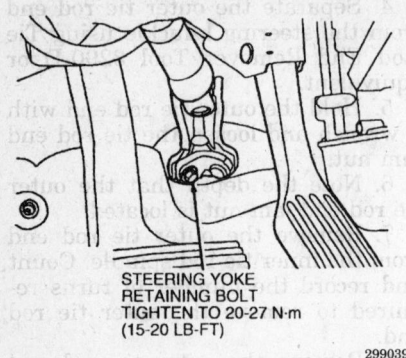

STEERING YOKE
RETAINING BOLT
TIGHTEN TO 20-27 N·m
(15-20 LB-FT)

299039

Installing the steering yoke retaining bolt

7. Position the rack and pinion so that several teeth of the rack are exposed.

8. Hold the rack with an adjustable wrench on the end teeth only while loosening the inner tie rod nut (ball joint nut) with a pipe wrench.

9. Once the inner tie rod is loose, remove by hand.

10. Remove the inner tie rod from the vehicle and inspect the rack and pinion for seal leakage. Replace the rack and pinion if there is excessive leakage from the rack seals.

To install:

11. Install a new inner tie rod assembly.

12. Turn the rack and pinion against the left-hand stop.

13. While holding the rack with an adjustable wrench nearest the rack end, tighten the inner tie rod nut (ball joint nut) using a pipe wrench.

14. Apply a small amount of grease to the lip of the bellows where it clamps to the inner tie rod spindle to allow the shaft to turn without twisting the bellows.

15. Install the bellows with the larger inner clamp and tighten with a wrench or screwdriver.

16. Install a new outer clamp using needlenose pliers.

17. Install the jam nut onto the tie rod spindle.

18. Apply a small amount of grease to the outer tie rod threads and install the outer tie rod end using the same number of threads recorded during disassembly.

19. Repeat the procedure for the opposite side.

20. Install the rack and pinion to the front subframe and install the front subframe to the vehicle.

21. Connect the negative battery cable.

NOTE: Whenever the vehicle subframe is removed or lowered, the wheel alignment should be checked.

22. Check the alignment and set the toe adjustment to specification.

23. Torque the outer tie rod end jam nut to 35–50 ft. lbs. (48–68 Nm).

Power Rack and Pinion

REMOVAL AND INSTALLATION

1. Disconnect the negative battery cable.

2. Working inside the vehicle, remove the clamp plate bolt retaining the steering column shaft to the flexible coupling.

3. Rotate the clamp plate to separate it from the shaft of the flexible coupling.

4. Remove the floor seal being careful not to damage the sealing lip.

5. Remove the pinch bolt securing the flexible coupling to the rack and pinion (steering gear) pinion shaft and remove the flexible coupling.

6. Remove as much of the power steering fluid as possible from the power steering auxiliary reservoir using a suction gun or similar method.

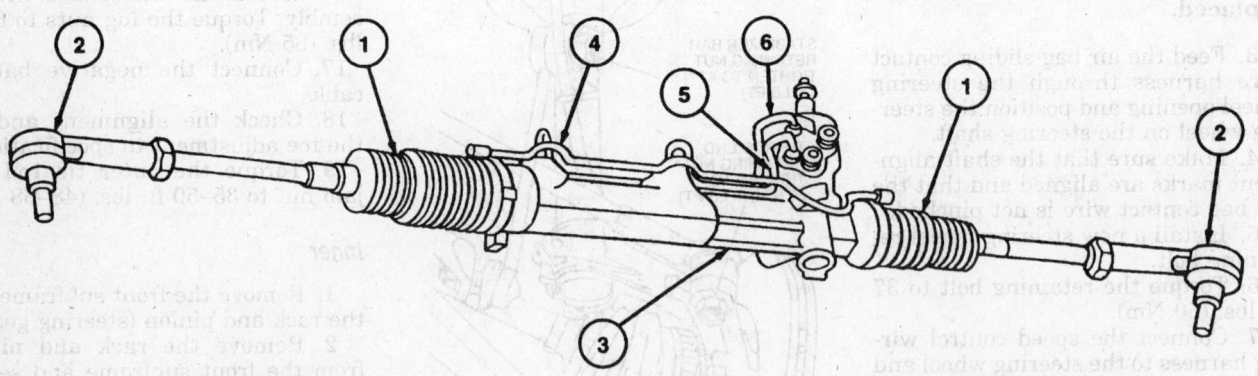

1	Front Suspension Steering Ball Dust Seal
2	Front Wheel Spindle Connecting End
3	Steering Gear
4	Power Steering Gear Rack Balance Tube
5	Power Steering Left Turn Pressure Tube
6	Power Steering Right Turn Pressure Tube

298379

Steering gear assembly

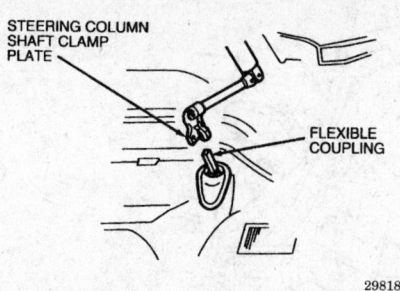

Steering column shaft removed from the flexible coupling

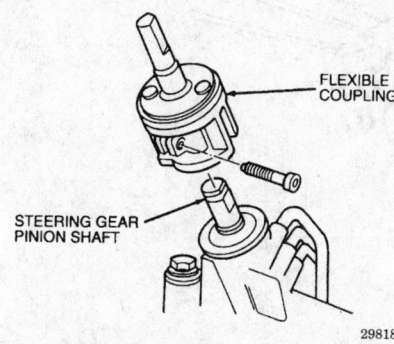

Flexible coupling removed from the steering gear pinion shaft

7. Disconnect the power steering return hose from the power steering pump auxiliary reservoir.

NOTE: The front subframe must be removed in order to allow removal of the rack and pinion (steering gear).

8. If equipped with the 2.0L (VIN 3) engine, remove the oil level dipstick and the exhaust manifold shield.

9. If equipped with the 2.5L (VIN L) engine, remove the water pump pulley shield.

10. Secure the radiator and fan shroud assembly to the radiator support using safety wire.

11. Install Three Bar Engine Support D88L–6000–A or equivalent, to the engine lifting eyes and support the engine/transaxle assembly.

12. Raise and safely support the vehicle.

13. Remove the catalytic converter.

14. Remove the front wheel and tire assemblies.

15. Separate the left-hand and right-hand stabilizer bar links from the front stabilizer bar.

16. Separate the left-hand and right-hand outer tie rod ends from the steering knuckles. Discard the cotter pins.

17. Remove the pinch bolts and separate the front suspension lower control arms from the steering knuckles at the ball joints.

18. Remove the splash shield at the front of the subframe.

19. If equipped with an automatic transaxle, remove the retaining through-bolts from the left-hand and right-hand front engine support insulators (engine mounts) to the subframe.

20. If equipped with a manual transaxle, remove the through-bolt from the left-hand front engine support insulator and remove the right-hand front engine support insulator and mounting bracket.

21. Disconnect the power steering oil cooler hoses at the right-hand front of the subframe and drain the power steering system.

22. Remove the A/C accumulator retaining screws from the front subframe.

23. Remove the 4 bolts retaining the lower radiator supports to the front subframe. Rotate the radiator supports forward.

24. Remove the 2 screws retaining the bumper cover braces to the left-hand and right-hand sides of the front subframe and rotate the cover braces forward.

25. Position Powertrain Lift (hydraulic lift) 014-00765 or equivalent, and 2 wood blocks approximately 40 inches in length attached to the subframe to support the subframe for removal from the vehicle.

NOTE: Make sure that the powertrain lift and wood blocks are correctly positioned for safe removal of the subframe.

26. Remove the 4 subframe to body retaining bolts.

27. Lower the subframe slightly and disconnect the power steering pressure and return hoses from the rack and pinion.

28. Finish lowering the subframe.

29. Remove the 6 bolts and the steering gear cover plate from the subframe.

30. Disconnect the power steering pressure and return hose unions from the steering gear.

31. Remove the 2 bolts retaining the steering gear to the subframe and remove the rack and pinion.

To install:

32. If the rack and pinion (steering gear) is being replaced, remove the inner tie rods and boots from the old unit and install on the new one, if they are in good condition.

33. Install new plastic seals on the power steering pressure and return line fittings as required.

34. Install the rack and pinion to the subframe and install the retaining bolts.

35. Torque the 2 rack and pinion retaining bolts to 101 ft. lbs. (137 Nm).

36. Connect the power steering and return hose unions to the rack and pinion.

37. Torque the unions to 23 ft. lbs. (31 Nm).

38. Install the rack and pinion cover plate and install the 6 retaining bolts.

39. Torque the retaining bolts to 37 ft. lbs. (50 Nm).

40. If lowered, raise and safely support the vehicle.

41. If removed, position the front subframe onto the powertrain lift and raise.

42. Install the power steering pressure and return hoses to the rack and pinion.

43. Position the front subframe to the body.

44. Route the power steering hoses to their correct positions.

45. Loosely install the 4 subframe to body bolts.

46. Install Sub-Frame Alignment Pin Set T94P-2100-AH or equivalent, into the front subframe to body alignment holes.

47. Slightly tighten the 4 subframe to body retaining bolts.

48. Move the subframe to complete the alignment.

49. Tighten the 4 subframe to body retaining bolts to 81–110 ft. lbs. (110–150 Nm).

50. Remove the alignment tools.

51. Install the A/C bracket retaining screw to the subframe and secure.

52. Connect the power steering oil cooler hoses to the front of the subframe.

53. Install the Powertrain Alignment Gauge T94P-6000-AH or equivalent, to the left-hand front engine support bracket and the subframe.

54. Tighten the 2 retaining bolts to 20 ft. lbs. (27 Nm) and snug the through bolt.

55. Lower the vehicle.

56. Remove the engine support from the top of the engine compartment.

57. Connect the power steering return hose to the power steering pump auxiliary reservoir.

58. Working inside the vehicle, install the flexible coupling to the steering gear pinion shaft and install the pinch bolt.

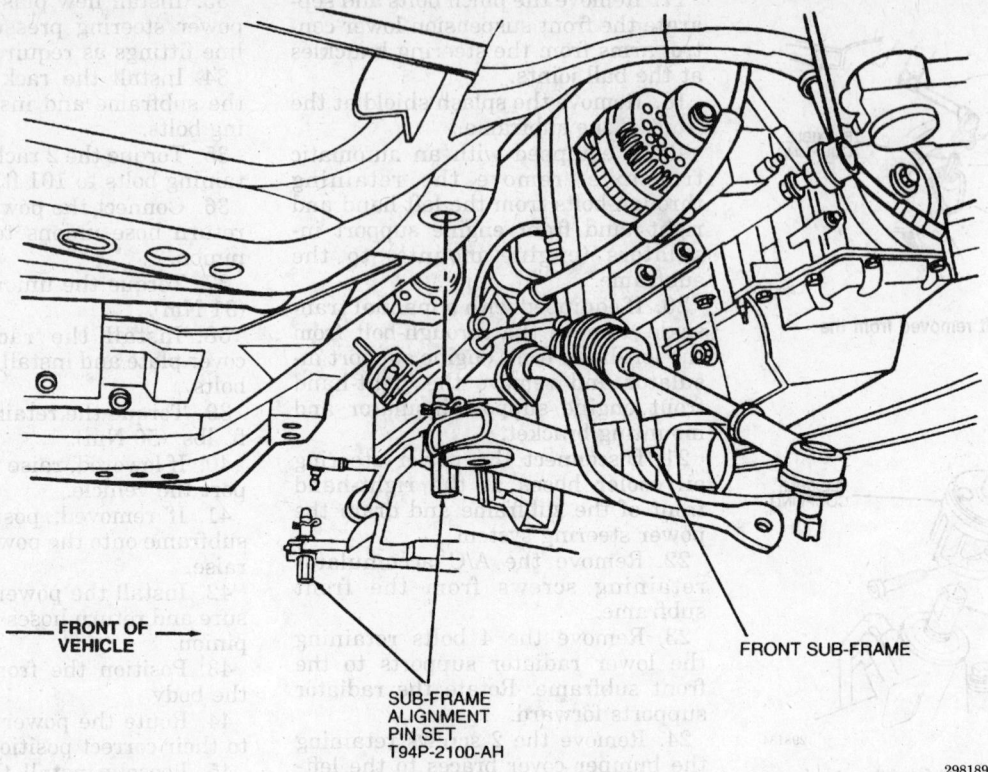

FRONT OF VEHICLE →

SUB-FRAME
ALIGNMENT
PIN SET
T94P-2100-AH

Sub-frame with alignment pins installed

FRONT SUB-FRAME

298189

59. Torque the pinch bolt to 21 ft. lbs. (28 Nm).

60. Install the floor seal.

61. Align the steering column shaft clamp plate with the flexible coupling and install the clamp plate bolt. Torque the clamp plate bolt to 18 ft. lbs. (24 Nm).

62. Partially raise and safely support the vehicle.

63. Install the right-hand front engine support insulator (engine mount). Torque the 2 retaining bolts to 30–41 ft. lbs. (41–55 Nm) and the through-bolt to 75–102 ft. lbs. (103–137 Nm).

64. Check the position of the right-hand front engine mount (support insulator). It must be centered in the transaxle bracket and in perfect front to rear alignment.

65. Remove the 2 retaining bolts and the through-bolt securing the powertrain alignment gauge and remove the powertrain alignment gauge.

66. Install the left-hand front engine mount to the front subframe using the 2 retaining bolts. Torque the retaining bolts to 7 ft. lbs. (10 Nm).

67. Check the position of the left-hand front engine mount to ensure perfect front to rear alignment.

68. Retorque the 2 retaining bolts to 30–40 ft. lbs. (41–55 Nm).

69. Install the left-hand front engine mount through-bolt. Torque the through-bolt to 75–102 ft. lbs. (103–137 Nm).

70. Install the stabilizer bar link to the front stabilizer bar. Torque the retaining nuts to 35–48 ft. lbs. (47–65 Nm).

71. Connect the left-hand and right-hand lower control arms to the ball joints and install the pinch bolts. Torque the pinch bolts to 37–43 ft. lbs. (50–58 Nm).

72. Install the catalytic converter.

73. Install both tie rod ends to the steering knuckles. Install new cotter pins.

74. Install the wheel and tire assemblies and torque the lug nuts to 63 ft. lbs. (85 Nm).

75. Install the front bumper cover braces to both sides of the subframe.

76. Install the radiator supports and the splash shield to the subframe.

77. Lower the vehicle.

78. Remove the safety wire supporting the radiator and fan shroud assembly.

79. If equipped with the 2.0L (VIN 3) engine, install the exhaust manifold shield and the oil level dipstick.

80. If equipped with the 2.5L (VIN L) engine, install the water pump pulley shield and secure.

81. Fill the power steering system with the proper fluid.

82. Connect the negative battery cable.

83. Run the engine and check for leaks and proper operation.

NOTE: Whenever the vehicle subframe is removed or lowered, the wheel alignment should be checked.

84. Bleed the power steering system, if needed.

Power Steering Pump

REMOVAL AND INSTALLATION

2.0L (VIN 3) Engine

1. Disconnect the negative battery cable.

2. Remove the exhaust manifold heat shield.

3. Remove the power steering reservoir pump hose from the power steering pump bracket and engine lifting bracket.

4. Disconnect the power steering return hose and the power steering reservoir pump hose from the power steering pump.

5. Allow the fluid to drain into an appropriate container.

6. Raise and safely support the vehicle.

7. Remove the lower drive belt guard.

8. Remove the accessory drive belt.

9. Lower the vehicle.

10. Remove the 3 bolts from the traction assist module and move aside.

11. Rotate the power steering pump pulley to gain access to the 3 power steering pump retaining bolts and remove the bolts.

12. Remove the bolt at the back of the power steering pump.

13. Remove the power steering pump from the vehicle.

14. If required, remove the power steering pump pulley using Power Steering Pump Pulley Remover T69L-10300-B or equivalent.

To install:

15. If the pulley was removed from the power steering pump, install the pulley using Power Steering Pump Pulley Replacer T91P-3A733-A or equivalent.

16. The power steering pulley must be flush or within 0.010 inch (0.25mm) of the end of the pump shaft.

17. Install the power steering pump and install the single retaining bolt at the rear of the pump. Do not tighten the bolt at this time.

18. Rotate the power steering pump pulley to install the 3 retaining bolts into the front of the power steering pump. Torque all 4 bolts to 18 ft. lbs. (25 Nm).

19. Position the traction assist module and install the 3 retaining bolts. Tighten the retaining bolts to 53 inch lbs. (6 Nm).

20. Raise and safely support the vehicle.

21. Install the accessory drive belt.

22. Install the lower drive belt guard.

23. Lower the vehicle.

24. Connect the power steering pressure hose and the power steering reservoir pump hose to the power steering pump.

25. Torque the pressure hose fitting to 48 ft. lbs. (65 Nm).

26. Install the power steering pressure hose to the power steering pump bracket and engine lifting bracket.

27. Install the exhaust manifold heat shield.

28. Connect the negative battery cable.

29. Fill the power steering system with the proper fluid.

30. Run the engine and check for leaks and proper operation.

31. Bleed the power steering system of air if necessary.

2.5L (VIN L) Engine

1. Disconnect the negative battery cable.

2. Remove the retaining bolt securing the power steering pressure hose to the upper engine mount.

3. Disconnect the power steering pressure hose from the power steering pump and allow the fluid to drain into a proper container.

4. Move the power steering pressure hose aside.

5. Move the ignition wire organizer aside.

6. Remove the front engine support insulator (engine mount). This requires the use of an engine support brace.

7. Loosen but do not remove the 4 power steering pump pulley bolts.

8. Remove the accessory drive belt.

9. Remove the 4 bolts and the power steering pump pulley.

10. Disconnect the power steering reservoir pump hose from the power steering pump.

11. Remove the power steering pressure hose clamp from the power steering pump bracket.

12. Remove the 6 retaining nuts and 5 retaining bolts from the power steering pump bracket.

13. Remove the power steering pump and bracket from the vehicle.

To install:

14. Install the power steering pump and bracket into the vehicle.

15. Install the 6 retaining nuts and 5 retaining bolts. Torque the nuts and bolts to 18 ft. lbs. (25 Nm).

16. Install the power steering pressure hose clamp to the power steering pump bracket.

17. Connect the power steering reservoir pump hose to the power steering pump.

18. Install the power steering pump pulley and the 4 retaining bolts, finger tight.

19. Install the accessory drive belt.

20. Tighten the power steering pump pulley bolts to 97 inch lbs. (11 Nm).

21. Install the front engine support insulator (engine mount) and remove the engine support brace.

22. Reposition the ignition wire organizer.

23. Connect the power steering pressure hose to the power steering pump. Torque the pressure hose fitting to 48 ft. lbs. (65 Nm).

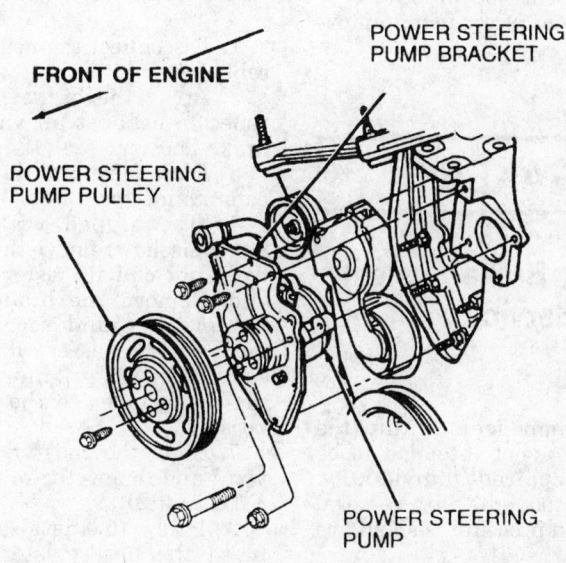

Power steering pump and related components - 2.5L (VIN L) engine

24. Install the power steering pressure hose to the front engine support insulator.

25. Connect the negative battery cable.

26. Fill the power steering system.

27. Run the engine and check for leaks and proper operation.

28. Bleed the power steering system of air if necessary.

BLEEDING

1. Disable the ignition system.

2. Raise the vehicle until the front tires are just off of the ground and safely support. Make sure that the transaxle is not in gear.

3. Fill the power steering pump auxiliary reservoir.

4. Crank the engine for 30 seconds without turning the steering wheel and recheck the fluid level. Add fluid if needed.

5. Crank the engine for 30 seconds while turning the steering wheel lock to lock. Check the fluid level and fill if needed.

------ WARNING ------

Do not hold the steering wheel against a stop for more than 5 seconds as damage to the steering pump could result.

6. Lower the vehicle.

7. Restore the ignition system.

8. Start the vehicle and check that the power steering system is free of air. If not, repeat the procedure or purge the system of air using an external vacuum source.

BRAKES

Anti-Lock Brake System Service

PRECAUTIONS

• Certain components within the ABS system are not intended to be serviced or repaired individually. Only those components with removal and installation procedures should be serviced.

• Do not use rubber hoses or other parts not specifically specified for and ABS system. When using repair kits, replace all parts included in the kit. Partial or incorrect repair may lead to functional problems and require the replacement of components.

• Lubricate rubber parts with clean, fresh brake fluid to ease assembly. Do not use lubricated shop air to clean parts; damage to rubber components may result.

• Use only DOT 3 brake fluid from an unopened container.

• If any hydraulic component or line is removed or replaced, it may be necessary to bleed the entire system.

• A clean repair area is essential. Always clean the reservoir and cap thoroughly before removing the cap. The slightest amount of dirt in the fluid may plug an orifice and impair the system function. Perform repairs after components have been thoroughly cleaned; use only denatured alcohol to clean components. Do not allow ABS components to come into contact with any substance containing mineral oil; this includes used shop rags.

• The Anti-Lock control unit is a microprocessor similar to other computer units in the vehicle. Ensure that the ignition switch is **OFF** before removing or installing controller harnesses. Avoid static electricity discharge at or near the controller.

• If any arc welding is to be done on the vehicle, the control unit should be unplugged before welding operations begin.

Master Cylinder

REMOVAL AND INSTALLATION

1. Disconnect the negative battery cable.

2. Apply the brake pedal several times to exhaust all vacuum in the brake booster.

3. Disconnect the fluid level indicator connector.

4. If equipped with a manual transmission, remove the hose for the hydraulic clutch master cylinder.

5. Remove the brake lines from the primary and secondary outlet ports on the master cylinder.

6. Remove the 2 nuts securing the master cylinder to the power brake booster.

7. Slide the master cylinder forward and remove it from the vehicle.

To install:

8. Prior to installation, bench bleed the new master cylinder as follows:

a. Secure the master cylinder in a soft-jawed vise. Be careful not to damage or distort the master cylinder housing.

b. Attach short lengths of brake tubing to the master cylinder out-let ports, positioning the ends of the tubing inside the master cylinder reservoir.

c. Fill the master cylinder reservoir with fresh clean brake fluid, making sure the ends of the brake tubing are submerged.

d. Using a suitable tool to push on the master cylinder piston, stroke the piston in the master cylinder bore until no more air bubbles are seen in the master cylinder fluid reservoir.

e. Remove the short lengths of brake tubing and install temporary plugs in the master cylinder cylinder outlet ports, to keep fluid from spilling.

f. Install the cap on the fluid reservoir and remove the master cylinder from the vise.

9. Position the master cylinder onto the studs of the power brake booster.

10. Install the 2 retaining nuts and torque to 15–18 ft. lbs. (21–29 Nm).

11. Install the primary and secondary brake lines to the master cylinder outlet ports and torque to 10–18 ft. lbs. (14–22 Nm).

12. If equipped with a manual transmission, connect the hose to the hydraulic clutch master cylinder.

13. Connect the brake warning indicator switch connector.

14. Fill the master cylinder with the proper brake fluid to just below the full line.

15. Bleed the brake system starting with the right rear wheel and working to the left front. Top off the master cylinder when complete.

16. Road test the vehicle and check for proper brake system operation.

Brake Caliper

REMOVAL AND INSTALLATION

Front Brake Caliper

1. Raise and safely support the vehicle.

2. Remove the wheel and tire assembly.

3. Remove the outer disc brake pad spring clip (anti–rattle clip).

4. Remove the 2 locator pin covers and remove the locator pins.

5. Free the hose from its mounting on the strut.

6. Lift the caliper off of the brake rotor.

7. Remove the inboard disc brake pad from the caliper.

8. If the brake caliper is to be removed from the vehicle, place a pan under the caliper to catch the brake

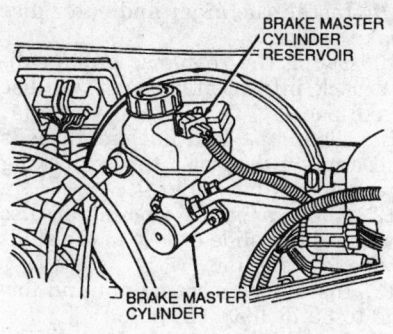

View of master cylinder and components

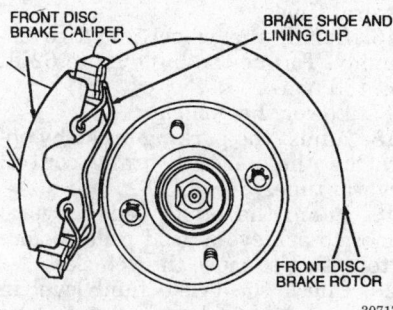

Position of disc brake pad (brake shoe and lining) spring clip

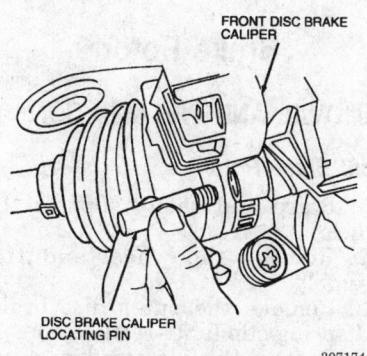

View of disc brake caliper locating pin

fluid for proper disposal. Disconnect the brake hose at the caliper and allow to drain.

9. Remove the brake caliper.
10. If the caliper is not to be serviced, tie off the caliper to prevent strain on the brake hose.

To install:

11. If removed, install the brake hose onto the caliper. Torque the fitting to 10 ft. lbs. (14 Nm).

12. Position the inboard disc brake pad into the caliper.

13. Make sure that the outboard disc brake pad is positioned properly.

14. Install the caliper over the brake rotor and position onto the caliper anchor plate.

15. Install the 2 caliper locator pins and torque to 20 ft. lbs. (28 Nm).

16. Install the caliper locator pin covers.

17. Install the outer disc brake pad spring clip.

18. Connect the brake hose to the front strut.

19. Bleed the brake system of air. Top off the master cylinder when complete.

20. If the brake pedal feels spongy, repeat the brake bleeding procedure.

21. Install the wheel and tire assembly. Torque the lug nuts to 62 ft. lbs. (85 Nm).

22. Lower the vehicle.

23. Pump the brake pedal several times to position the brake pads before attempting to move the vehicle.

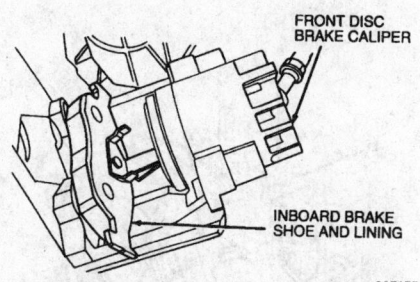

View of inboard disc brake pad (brake shoe and lining)

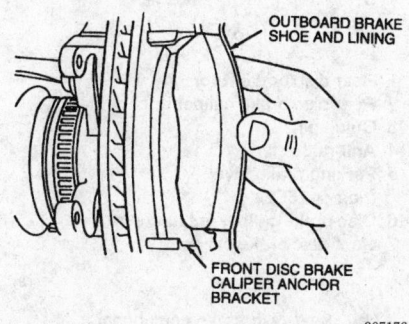

Positioning the outboard disc brake pad (brake shoe and lining)

24. Road test the vehicle and check for proper brake system operation.

Rear Brake Caliper

1. Raise and safely support the vehicle.

2. Remove the wheel and tire assembly.

3. Remove the parking brake rear cable and conduit from the parking brake lever at the disc brake caliper, using a pair of pliers.

4. Remove the cotter pin and guide pin.

5. Remove the caliper locating pin cover and remove the locating pin.

6. Lift the rear disc brake caliper off of the anchor plate.

7. Place a pan under the caliper to catch any lost brake fluid. Dispose of properly.

8. Crack open the rear brake hose fitting and allow the brake fluid to drain.

9. Finish removing the rear brake hose and washers from the caliper. Discard the washers.

10. Remove the disc brake caliper.

To install:

11. Connect the rear brake hose to the caliper using new washers.

12. Fit the rear disc brake caliper over the rotor and position onto the anchor plate.

13. Install the caliper locating pin and torque to 30 ft. lbs. (41 Nm).

14. Install the caliper locating pin cover.

15. Install the guide pin and the cotter pin.

16. Install the parking brake rear cable and conduit onto the parking brake lever.

17. Adjust the parking brake by operating the parking brake control several times.

18. Properly bleed the brake system of air. Top off the master cylinder when complete.

19. If the brake pedal feels spongy, repeat the brake bleeding procedure.

20. Install the wheel and tire assembly. Torque the lug nuts to 62 ft. lbs. (85 Nm).

21. Lower the vehicle.

22. Pump the brake pedal several times to position the brake pads before attempting to move the vehicle.

23. Road test the vehicle and check for proper brake system operation.

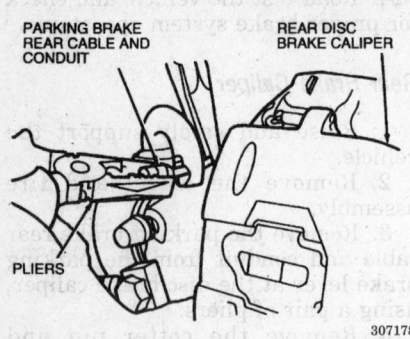

PARKING BRAKE
REAR CABLE AND
CONDUIT

REAR DISC
BRAKE CALIPER

PLIERS

307178

Method of removing the parking brake cable

Disc Brake Pads

REMOVAL AND INSTALLATION

Front Disc Brake Pads

1. Remove ½ of the brake fluid from the master cylinder reservoir.
2. Raise and safely support the vehicle.
3. Remove the wheel and tire assembly.
4. Remove the outer disc brake pad spring clip (anti–rattle clip).
5. Remove the 2 locating pin covers and remove the locating pins.
6. Free the hose from its mounting on the strut.
7. Lift the caliper off of the brake rotor and tie it off to prevent damage to the brake hose.
8. Remove the outboard disc brake pad from the anchor plate.
9. Remove the inboard disc brake pad from the brake caliper.
To install:
10. If installing new disc brake pads, use a C-clamp or similar tool to push the caliper piston into the caliper bore. This will allow room for the new pads.
11. Place the inboard disc brake pad into the caliper.
12. Place the outboard disc brake pad into position in the anchor plate.
13. Position the disc brake caliper onto the rotor.
14. Install the 2 locating pins and torque to 20 ft. lbs. (28 Nm).
15. Install the caliper locating pin covers.
16. Secure the brake hose to its support on the strut.
17. Install the disc brake pad spring clip.
18. Install the wheel and tire assembly. Torque the lug nuts to 62 ft. lbs. (85 Nm).
19. Lower the vehicle.

20. Pump the brake pedal several times to achieve a good pedal before attempting to move the vehicle.
21. Check the brake fluid level in the master cylinder fluid reservoir and add fluid as necessary.
22. Road test the vehicle and check for proper brake system operation.

Rear Disc Brake Pads

1. Remove ½ of the brake fluid from the master cylinder reservoir.
2. Raise and safely support the vehicle.
3. Remove the wheels.
4. Remove the cotter pin and guide pin.
5. Remove the caliper locating pin cover and remove the locating pin.
6. Swing the rear disc brake caliper away from the brake rotor and anchor plate. There is no need to remove the parking brake cable or brake hose.
7. Remove the inner and outer disc brake pads.
To install:
8. If installing new disc brake pads, use rear caliper piston adjuster T87P–2588–A or similar tool, to rotate the rear disc brake piston clockwise, retracting the caliper piston. This will allow room for the new brake pads.

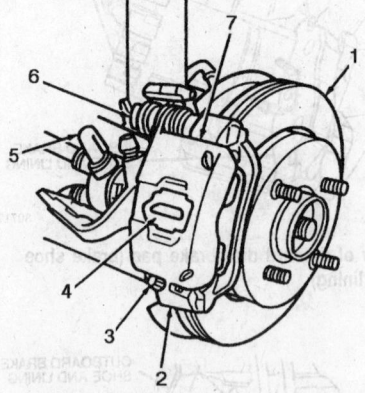

1 Rear disc brake rotor
2 Rear disc brake caliper anchor plate
3 Guide pin
4 Anti-rattle clip
5 Parking brake lever (part of 2552)
6 Disc brake caliper locating pin
7 Rear disc brake caliper

295340

Rear disc brake components

9. Install the inner and outer disc brake pads.
10. Swing the rear disc brake caliper back into position over the disc brake pads.
11. Clean the locating pin threads and apply 1 drop of a thread locking agent or similar sealer.
12. Apply a small amount of disc brake caliper slide grease to the shaft of the locating pin.
13. Install the locating pin and torque to 30 ft. lbs. (41 Nm).
14. Install the guide pin and the cotter pin.
15. Adjust the parking brake by operating the parking brake control several times.
16. Install the wheel and tire assembly. Torque the lug nuts to 62 ft. lbs. (85 Nm).
17. Lower the vehicle.
18. Adjust the parking brake by operating the parking brake control several times.
19. Pump the brake pedal several times to achieve a good pedal before attempting to move the vehicle.
20. Check the brake fluid level in the master cylinder fluid reservoir and add fluid as necessary.
21. Road test the vehicle and check for proper brake system operation.

Brake Rotor

REMOVAL AND INSTALLATION

Front Brake Rotor

1. Raise and safely support the vehicle.
2. Remove the wheel and tire assembly.
3. Remove the outer disc brake pad spring clip (anti–rattle clip).
4. Remove the 2 locator pin covers and remove the locator pins.
5. Free the brake hose from its mounting on the strut.
6. Lift the caliper off of the brake rotor and tie it off to prevent damage to the brake hose.
7. Remove the outboard disc brake pad from the anchor plate.
8. Remove the 2 bolts securing the anchor plate and remove the anchor plate.
9. Remove the disc brake rotor.
10. Inspect the brake rotor surfaces for scoring, wear or other damage. Machine or replace the brake rotor as necessary.
To install:
11. If using a new disc brake rotor, thoroughly clean the surfaces to remove any protective coating before assembly.

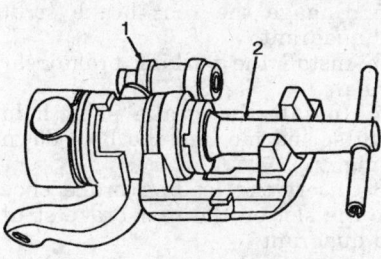

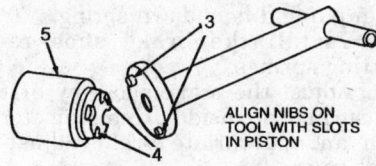

ALIGN NIBS ON
TOOL WITH SLOTS
IN PISTON

1 Rear disc brake caliper
2 Rear caliper piston adjuster
3 Nibs (part of T87P-2588-A)
4 Piston adjuster slots
  (part of 2B588)
5 Rear disc brake caliper
  piston and adjuster

295341

Diagram of rear caliper piston and the adjuster tool

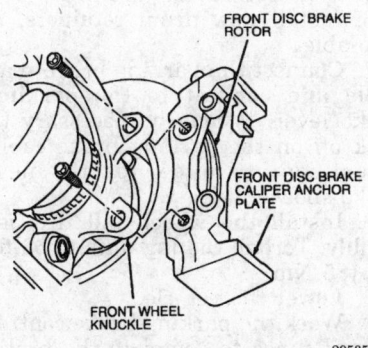

FRONT DISC BRAKE
ROTOR

FRONT DISC BRAKE
CALIPER ANCHOR
PLATE

FRONT WHEEL
KNUCKLE

295359

View of front caliper anchor plate and bolts

12. Position the disc brake rotor onto the wheel hub.

13. Install the anchor plate and secure with the 2 bolts. Torque the bolts to 88 ft. lbs. (120 Nm).

14. Set the outboard disc brake pad into position on the anchor plate.

15. Position the caliper over the brake rotor and install it to the anchor plate with the 2 locator pins. Torque the locator pins to 20 ft. lbs. (28 Nm).

16. Install the brake caliper locator pin covers.

17. Secure the brake hose to its support on the strut.

18. Install the disc brake pad spring clip.

19. Install the wheel and tire assembly. Torque the lug nuts to 62 ft. lbs. (85 Nm).

20. Lower the vehicle.

21. Pump the brake pedal several times to achieve a good pedal before attempting to move the vehicle.

22. Road test the vehicle and check the brake system for proper operation.

Rear Brake Rotor

1. Raise and safely support the vehicle.

2. Remove the wheel and tire assembly.

3. Remove the parking brake rear cable and conduit from the parking brake lever at the disc brake caliper using a pair of pliers.

4. Remove the cotter pin and guide pin.

5. Remove the caliper locating pin cover and remove the locating pin.

6. Lift the rear disc brake caliper off of the anchor plate.

7. Secure the caliper using wire to prevent damage to the brake hose.

8. Remove the inner and outer disc brake pads.

9. Remove the 2 bolts securing the anchor plate. Remove the anchor plate.

10. Remove the retainers on the wheel studs securing the brake rotor to the hub (if not already removed).

11. Remove the brake rotor.

12. Inspect the brake rotor surfaces for scoring, wear or other damage. Machine or replace the disc brake rotor as necessary.

   To Install:

13. If the rotor is being replaced, make sure that the new rotor is thoroughly cleaned of its protective coating.

14. Lubricate the area of the wheel hub that the rotor fits to with a suitable grease.

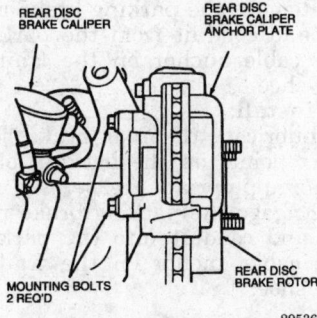

REAR DISC
BRAKE CALIPER

REAR DISC
BRAKE CALIPER
ANCHOR PLATE

MOUNTING BOLTS
2 REQ'D

REAR DISC
BRAKE ROTOR

295362

Location of rear caliper anchor plate retaining bolts

15. Install the rear disc rotor onto the wheel hub using new retainers.

16. Fit the caliper anchor plate and secure with the 2 bolts after applying 1 drop of threadlock sealer to the threads of each bolt.

17. Torque the retaining bolts to 43 ft. lbs. (59 Nm).

18. Install the inner and outer disc brake pads.

19. Install the rear disc brake caliper.

20. Fit the caliper locating pin and torque to 30 ft. lbs. (41 Nm).

21. Install the caliper locating pin cover.

22. Install the guide pin and the cotter pin.

23. Install the parking brake rear cable and conduit onto the parking brake lever.

24. Adjust the parking brake by operating the parking brake control several times.

25. Install the wheel and tire assembly. Torque the lug nuts to 62 ft. lbs. (85 Nm).

26. Lower the vehicle.

27. Pump the brake pedal several times to achieve a good pedal before attempting to move the vehicle.

28. Road test the vehicle and check the brake system for proper operation.

Brake Drums

REMOVAL AND INSTALLATION

1. Raise and safely support the vehicle.

2. Remove the wheel and tire assembly.

3. Remove the brake drum retainers, if installed.

4. Grasp the brake drum and remove.

5. If the drum will not slide off with light force, then the brake shoes will need to be backed off as follows:

   a. Remove the rubber plug on the backing plate and insert a screwdriver or small brake adjusting tool into the slot to contact the brake strut and quadrant.

   b. A forward motion of the tool will separate the quadrant from the knurled wheel and allow the brake shoes to retract.

   c. Remove the brake drum.

   To install:

6. Make sure the brake drum and shoes are clean of any oils or protective coatings.

7. Position the brake drum onto the wheel hub.

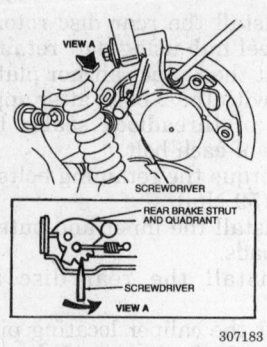

Positioning a screwdriver to release the brake shoes

8. Install new brake drum retainers if available.

9. Install the wheel and tire assembly. Torque the lug nuts to 62 ft. lbs. (85 Nm).

10. Lower the vehicle.

11. Work the parking brake control several times to adjust the rear brake shoes.

12. Pump the brake pedal several times to assure a good pedal before attempting to move the vehicle.

13. Road test the vehicle and check for proper brake system operation.

Brake Shoes

REMOVAL AND INSTALLATION

1. Raise and safely support the vehicle.

2. Remove the rear wheel and tire assembly.

3. Remove the brake drum retainers, if equipped.

4. Grasp the brake drum and remove.

5. If the drum will not slide off with light force, then the brake shoes will need to be backed off:

　a. Remove the rubber plug on the backing plate and insert a screwdriver or small brake adjusting tool into the slot to contact the brake strut and quadrant.

　b. A forward motion of the screwdriver will separate the quadrant from the knurled wheel and allow the brake shoes to retract.

　c. Remove the brake drum.

6. Remove the brake shoe hold down springs and the brake shoe hold down pins.

7. Remove the brake shoe retracting springs.

8. Disengage the parking brake cable and conduit from the parking brake lever.

9. Remove the brake shoes.

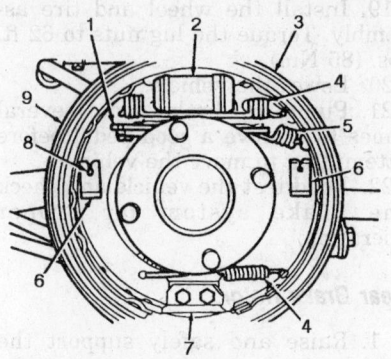

1 Rear brake strut and quadrant
2 Rear wheel cylinder
3 Rear brake shoe lining
4 Brake shoe retracting spring (part of 2A25)
5 Parking brake return spring
6 Brake shoe hold down spring (part of 2A225)
7 Anchor block (part of 2212)
8 Brake shoe hold down spring pin (part of 2A225)
9 Rear brake backing plate

Location of brake components

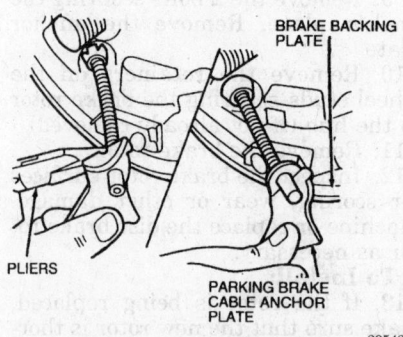

Removal of parking brake cable

10. Disengage the the rear brake strut and quadrant from the rear brake shoe.

11. Remove the parking brake rear cable and conduit from the parking brake cable anchor on the trailing brake shoe.

　To install:

12. Lubricate the rear brake shoe contact points on the backing plate with an appropriate grease.

13. Engage the parking brake rear cable and conduit into the parking brake cable anchor on the trailing brake shoe.

14. Position the trailing brake shoe on the backing plate.

15. Engage the rear brake strut and quadrant.

16. Install the parking brake rear spring.

17. Install the brake shoe hold down spring pin and the hold down spring.

18. Insert the leading brake shoe into the slot on the rear brake strut and quadrant.

19. Install the brake shoe hold down pin and hold down spring.

20. Install the brake shoe retracting springs.

21. Adjust the brake shoes by first measuring the inside drum diameter with an appropriate brake adjustment gauge.

22. Insert a small screwdriver or similar tool into the knurled quadrant of the brake strut and quadrant to adjust the brake shoes to the same measurement of the brake drum by expanding the brake strut and quadrant.

23. Trial fit the brake drum. The shoes should just contact the drum surface when properly adjusted.

24. Make sure that the brake drum and brake shoes are clean of any oils or protective coatings.

25. Install the brake drum.

26. Install new drum retainers, if available.

27. Connect the parking brake rear cable and conduit to the parking brake lever. It may be necessary to back off on the parking brake cable adjustment to allow for the new brake shoes.

28. Install the wheel and tire assembly. Torque the lug nuts to 62 ft. lbs. (85 Nm).

29. Lower the vehicle.

30. Work the parking brake control several times to complete the brake shoe adjustment and to check the parking brake adjustment as well.

31. Pump the brake pedal several times to assure a good pedal.

32. Road test the vehicle and check for proper brake system operation.

Wheel Cylinder

REMOVAL AND INSTALLATION

1. Raise and safely support the vehicle.

2. Remove the rear wheel and tire assembly.

3. Remove the brake drum retainers, if equipped.

4. Grasp the brake drum and remove.

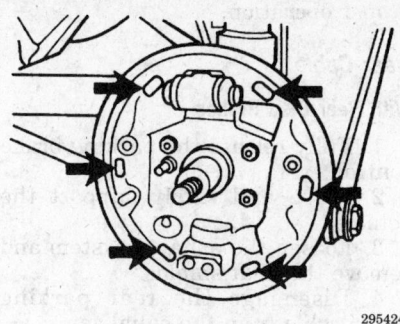

Lubricating points on back plate

295424

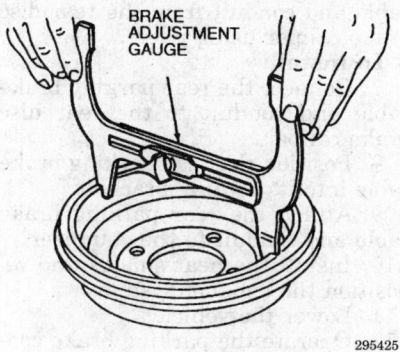

BRAKE
ADJUSTMENT
GAUGE

Measuring the brake drum inner diameter

295425

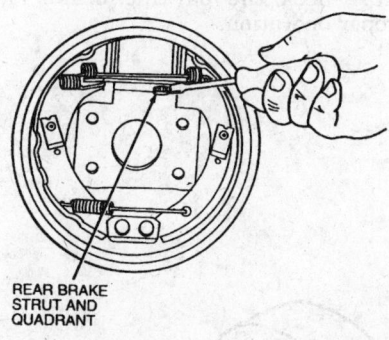

REAR BRAKE
STRUT AND
QUADRANT

295426

Adjusting the brake shoes to fit the drum

5. If the drum will not slide off with light force then the brake shoes need to be backed off.

   a. Remove the rubber plug on the backing plate and insert a screwdriver or small brake adjusting tool into the slot to contact the brake strut and quadrant.

   b. A forward motion of the screwdriver will separate the quadrant from the knurled wheel and allow the brake shoes to retract.

   c. Remove the brake drum.

6. Remove the brake shoe hold down springs and the brake shoe hold down pins.

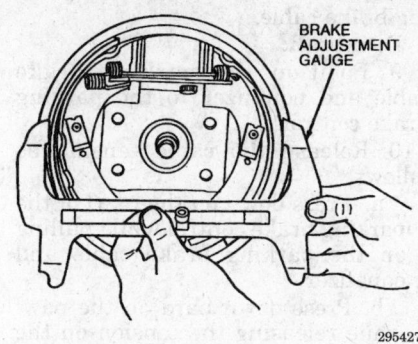

BRAKE
ADJUSTMENT
GAUGE

Measuring the brake shoe adjustment

295427

7. Remove the brake shoe retracting springs.

8. Disengage the parking brake cable and conduit from the parking brake lever.

9. Remove the brake shoes.

10. Disengage the the rear brake strut and quadrant from the rear brake shoe.

11. Remove the parking brake rear cable and conduit from the parking brake cable anchor on the training brake shoe.

12. Disconnect the brake line at the wheel cylinder.

13. Remove the 2 bolts securing the wheel cylinder to the backing plate and remove the wheel cylinder.

To install:

14. Install the wheel cylinder to the brake backing plate and install the 2 retaining bolts. Torque the retaining bolts to 7–9 ft. lbs. (10–13 Nm).

15. Connect the brake line to the wheel cylinder and torque the fitting to 10–18 ft. lbs. (14–24 Nm).

16. Lubricate the rear brake shoe contact points on the backing plate with an appropriate grease.

17. Engage the parking brake rear cable and conduit into the parking brake cable anchor on the trailing brake shoe.

18. Position the trailing brake shoe on the backing plate.

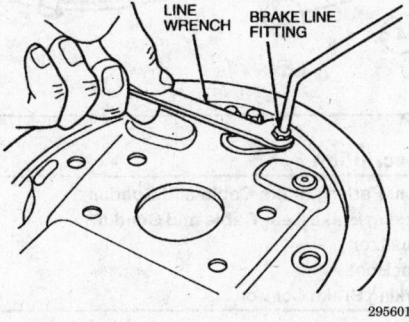

LINE
WRENCH

BRAKE LINE
FITTING

Removing the brake line from the wheel cylinder

295601

19. Engage the rear brake strut and quadrant.

20. Install the parking brake rear spring.

21. Install the brake shoe hold down spring pin and the hold down spring.

22. Insert the leading brake shoe into the slot on the rear brake strut and quadrant.

23. Install the brake shoe hold down pin and hold down spring.

24. Install the brake shoe retracting springs.

25. Adjust the brake shoes by first measuring the inside drum diameter with an appropriate brake adjustment gauge.

26. Insert a small screwdriver into the knurled quadrant of the brake strut and quadrant to adjust the brake shoes to the same measurement of the brake drum by expanding the brake strut and quadrant.

27. Trial fit the brake drum. The shoes should just contact the drum surface when properly adjusted.

28. Make sure that the brake drum and brake shoes are clean of any oils or protective coatings.

29. Install the brake drum.

30. Install new drum retainers, if available.

31. Connect the parking brake rear cable and conduit to the parking brake lever. It may be necessary to back off on the parking brake cable adjustment to allow for the new brake shoes.

32. Bleed the brake system of air until a firm pedal is achieved. Top off the brake fluid in the master cylinder.

33. Install the wheel and tire assembly. Torque the lug nuts to 62 ft. lbs. (85 Nm).

34. Lower the vehicle.

35. Work the parking brake control several times to complete the brake shoe adjustment and to check the parking brake adjustment as well.

36. Pump the brake pedal several times to assure a good pedal.

37. Road test the vehicle and check the brake system for proper operation.

Parking Brake Cable

ADJUSTMENT

The parking brake cable is adjusted by operating the parking brake control handle several times.

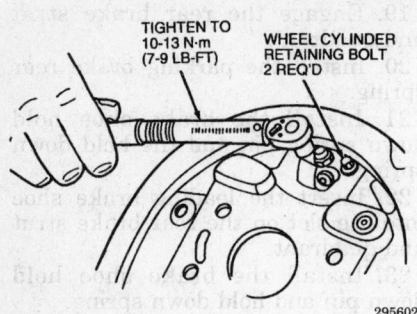

TIGHTEN TO
10-13 N·m
(7-9 LB-FT)

WHEEL CYLINDER
RETAINING BOLT
2 REQ'D

295602

Using a torque wrench to secure the wheel cylinder retaining bolts

REMOVAL AND INSTALLATION

Front Cable

1. Fully release the parking brake.
2. Raise and safely support the vehicle.
3. Loosen and lower the exhaust system and remove the heat shields.
4. Disengage the rear parking brake cables from the equalizer.
5. Lower the vehicle.
6. Remove the console for removal of the cable.
7. Route the parking brake cable and equalizer through the hole in the floor pan.
8. Disconnect the parking brake cable and equalizer from the parking

brake control and remove the parking brake cable.

To install:

9. Position the parking brake cable and equalizer to the parking brake control.
10. Release the cable tension as follows:
    a. Press down on the pawl of the parking brake control while pulling on the parking brake cable and equalizer.
    b. Press down hard on the pawl while releasing the tension on the cable. The cable will stay released as long as no tension is applied to the cable or as long as the parking brake control handle is not moved.
11. Route the parking brake cable and equalizer through the hole in the floor pan.
12. Install the grommet into the opening.
13. Raise and safely support the vehicle.
14. Connect the parking brake rear cables to the equalizer.
15. Install the heat shields and reposition the exhaust system.
16. Lower the vehicle.
17. Install the console.
18. Operate the parking brake control several times to set the parking brake adjuster.

19. Check the parking brake for proper operation.

Rear Cable

With Rear Disc Brakes

1. Fully release the parking brake control.
2. Raise and safely support the vehicle.
3. Loosen the exhaust system and remove the heat shields.
4. Disengage the rear parking brake cable from the equalizer.
5. Remove the rear parking brake cable from its routing brackets.
6. Remove the rear parking brake cable and conduit from the rear disc brake caliper using pliers.

To install:

7. Connect the rear parking brake cable and conduit to the rear disc brake caliper.
8. Position the rear parking brake cable into its routing brackets.
9. Attach the rear parking brake cable and conduit to the equalizer.
10. Install the heat shields and reposition the exhaust system.
11. Lower the vehicle.
12. Operate the parking brake control several times to adjust the parking brake cable tension.
13. Check the parking brake for proper operation.

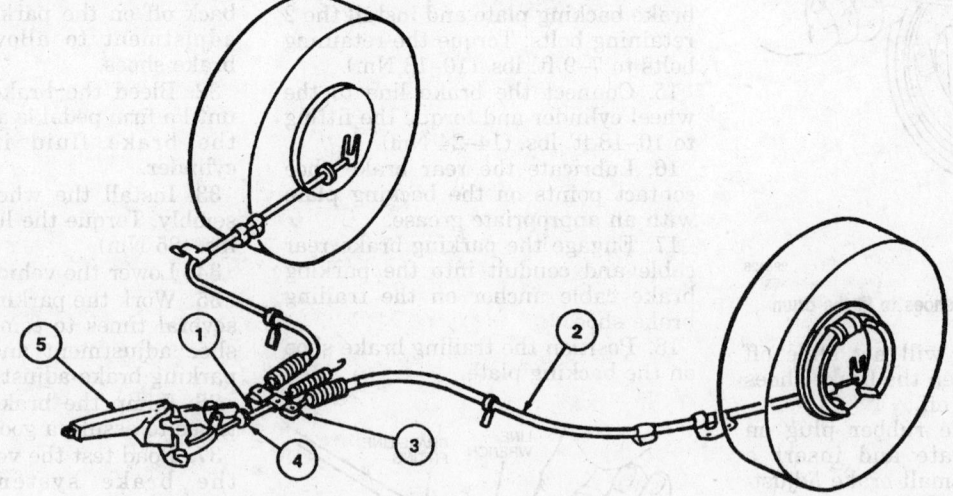

Item	Description
1	Front Parking Brake Cable and Conduit
2	Parking Brake Rear Cable and Conduit
3	Equalizer
4	Dust Boot
5	Parking Brake Control

308000

Parking brake components

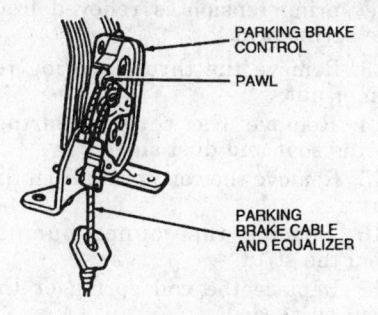

View of parking brake pawl

308002

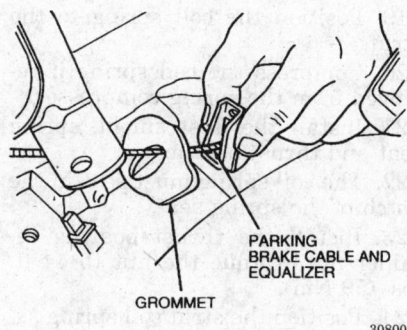

Removing front parking brake cable

308004

With Rear Drum Brakes

1. Fully release the parking brake control.
2. Raise and safely support the vehicle.
3. Loosen the exhaust system and remove the heat shields.
4. Disengage the rear parking brake cable to be removed from the equalizer.
5. Remove the rear parking brake cable from its routing brackets.
6. Remove the rear wheel of the cable being serviced.
7. Remove the brake drum to gain access to the parking brake cable anchor.
8. Disengage the cable end from the parking brake cable anchor (parking brake lever).
9. Press the prongs on the cable conduit attaching the cable to the backing plate using the box end of a ½ inch wrench.
10. With the prongs compressed, pull the cable and conduit out of the backing plate and remove the cable.
   To install:
11. Push the new cable end for the brake side into the backing plate until the prongs on the conduit are fully seated.

12. Connect the rear parking brake cable to the parking brake cable anchor (parking brake lever).
13. Install the brake drum.
14. Install the wheel and torque the lug nuts to 62 ft. lbs. (85 Nm).
15. Install the rear parking brake cable into its routing brackets.
16. Connect the parking brake cable to the equalizer.
17. Install the heat shields and reposition the exhaust system.
18. Lower the vehicle.
19. Operate the parking brake control several times to adjust the parking brake cable tension.
20. Check the parking brake for proper operation.

Brake System Bleeding

The hydraulic brake system is bled the same for ABS brake systems as well as non-ABS brake systems. The only exception is that the ABS brake system must be bled at least twice to ensure that no more air exist in the hydraulic system.
1. Clean the area around the master cylinder filler cap and remove the cap.
2. Fill the reservoir with clean brake fluid and install the cap.
3. If the master cylinder is known or suspected to have air in the bore, it must be bled before any of the wheel cylinders or calipers. To bleed the master cylinder, position a shop towel under the rear master cylinder outlet fitting and loosen the fitting approximately ¾ turn. Have an assistant depress the brake pedal slowly through its full travel. Close the outlet fitting and let the pedal return slowly to the fully released position. Wait 5 seconds, then repeat the operation until all air bubbles disappear.

— **WARNING** —
Be careful not to spill brake fluid on painted surfaces, as it can destroy the finish. If any brake fluid is spilled, rinse the area immediately with water.

4. Repeat the master cylinder bleeding procedure at the front master cylinder outlet fitting.
5. Position the correct box-end wrench on the brake bleeder on the right-rear wheel.

NOTE: Always bleed the longest line first, which is the right-rear.

6. Place a drain tube to the brake bleeder screw and the free end into a

container half full of clean brake fluid.
7. Loosen the bleeder screw while an assistant apply the brake pedal slowly through its entire travel.
8. Close the brake bleeder and then have the assistant release the brake pedal.
9. Continue bleeding the right-rear wheel until the fluid is free of air bubbles.

NOTE: DO NOT allow the master cylinder to run out of brake fluid. Only use fresh DOT 3 or equivalent brake fluid from a closed container.

10. Repeat the procedure on the left-front wheel, then the left-rear and finally the right-front wheel.
11. Throughout the brake bleeding process, continually check the brake master cylinder and top off the fluid as required.
12. If the brake pedal is spongy, repeat the brake bleeding procedure, or if available bleed the brake system using a pressure bleeder following the manufacturers recommendations.

Wheel Speed Sensor

REMOVAL AND INSTALLATION

Front brake sensor

1. Raise and safely support the vehicle.
2. Remove the sensor retaining bolt and sensor from the front wheel knuckle.
3. Disconnect the sensor wire from the strut bracket.
4. To gain access to the sensor connector, remove the 5 inner fender front splash shield push pins and 3 screws. Remove the splash shield and disconnect the sensor connector.
   To Install:
5. Connect the speed sensor to the vehicle harness. Install the splash shield.
6. Connect the sensor wire to the strut housing.
7. Install the speed sensor and tighten the retaining bolt to 7 ft. lbs. (10 Nm).
8. Lower the vehicle.
9. Road test the vehicle.

Rear wheel sensor

1. Disconnect the negative battery cable.
2. Remove the rear seat cushion by pulling upward on the front of the cushion.

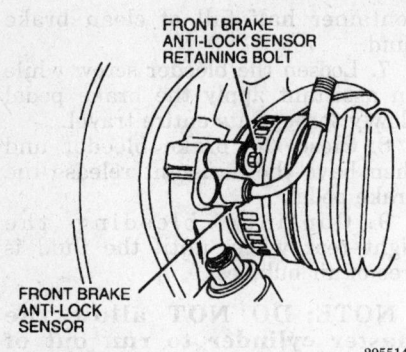

Front wheel anti-lock speed sensor

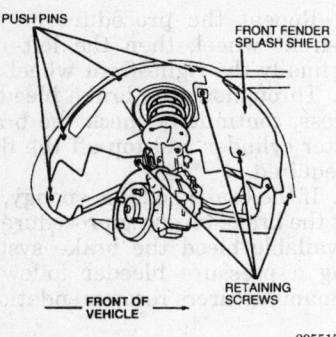

Front wheel splash shield

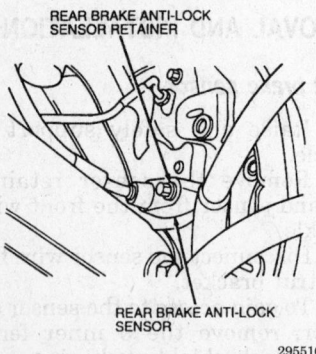

Rear wheel anti-lock wheel sensor

3. Disconnect the the sensor connector and feed the wire through the floorpan.

4. Raise and safely support the vehicle.

5. Disconnect the wire from the routing brackets.

6. Remove the sensor retaining bolt and remove the sensor.

To Install:

7. Install the speed sensor and tighten the retaining bolt 7 ft. lbs. (10 Nm).

8. Feed the sensor wire through the floorpan and connect the wire to the routing brackets.

9. Lower the vehicle.

10. Connect the wiring connector and install the rear seat cushion.

11. Connect the negative battery.

12. Road test the vehicle.

FRONT SUSPENSION

Strut and Spring

REMOVAL AND INSTALLATION

1. Disconnect the negative battery cable.

2. Raise and safely support the vehicle.

3. Remove the wheel and tire assembly.

4. Lower the vehicle enough to gain access to the strut retaining nut.

5. From inside the engine compartment, hold the strut piston with an 8mm Allen head wrench while removing the top retaining nut.

6. Raise and safely support the vehicle.

7. Disconnect the stabilizer bar link from the strut.

8. Remove the brake hose and anti–lock wiring from the strut bracket.

9. Remove the steering knuckle to strut pinch bolt.

10. Work the strut out of the steering knuckle and lower the strut out of the strut tower.

11. Remove the strut/coil spring assembly from the vehicle.

— CAUTION —

Do not attempt to remove the coil spring from the strut without first compressing the coil spring with the appropriate tool.

12. Install Spring Compressor 086–00029 or equivalent, to the coil

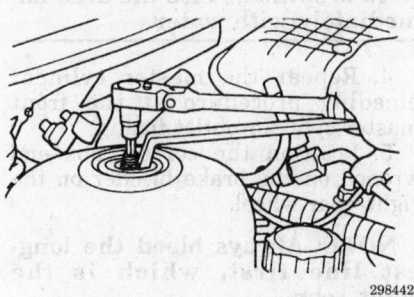

Removing the top strut retaining nut

spring and compress the spring until the spring tension is relieved from the spring seat.

13. Remove the thrust bearing retainer nut.

14. Remove the thrust bearing, spring seat and dust shield.

15. Remove the coil spring from the strut.

16. Remove the jounce bumper from the strut.

17. Replace the coil spring or the strut as needed.

To install:

18. Install the jounce bumper to the strut.

19. Position the coil spring to the strut.

20. Compress the coil spring if removed from the spring compressor.

21. Install the dust shield, spring seat and thrust bearing.

22. The coil spring must seat in the notch of the spring seat.

23. Install the thrust bearing retainer nut. Torque the nut to 44 ft. lbs. (59 Nm).

24. Position the strut/coil spring assembly into the strut tower and fit the lower portion of the strut into the steering knuckle.

25. Install the knuckle to strut pinch bolt. Do not tighten the bolt at this time.

26. Partially lower and safely support the vehicle.

27. Install the top strut mounting nut.

28. Use an 8mm Allen head wrench to prevent the strut piston rod from turning while torquing the mounting nut to 34 ft. lbs. (46 Nm).

29. Torque the knuckle to strut pinch bolt to 40 ft. lbs. (54 Nm).

30. Raise and safely support the vehicle.

31. Install the stabilizer bar link. Be careful not to damage the ball joint seal. Replace the stabilizer bar link if the seal is damaged.

32. Torque the stabilizer bar link retaining nut to 37 ft. lbs. (50 Nm).

33. Position the brake hose and the anti-lock wiring to the strut bracket.

34. Install the wheel and tire assembly. Torque the lug nuts to 63 ft. lbs. (85 Nm).

35. Lower the vehicle.

36. Connect the negative battery cable.

37. Check the front wheel alignment.

38. Road test the vehicle and check for proper operation.

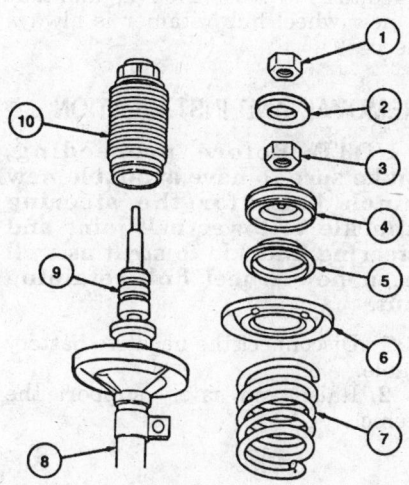

ROTUNDA SPRING COMPRESSOR
298443

Strut and coil spring assembly in spring compressor

1	Nut
2	Retainer
3	Upper Mount Retainer Nut
4	Upper Mount
5	Bearing
6	Spring Seat
7	Front Coil Spring
8	Front Shock Absorber
9	Jounce Bumper
10	Dust Shield

298444

Strut and coil spring components

Lower Ball Joints

REMOVAL AND INSTALLATION

If the lower ball joint requires replacement, the lower control arm and ball joint assembly must be replaced together as the lower ball joint is not available separately.

Lower Control Arms

REMOVAL AND INSTALLATION

Right Side

1. Disconnect the negative battery cable.
2. Raise and safely support the vehicle.
3. Remove the wheel and tire assembly.
4. Remove the lower ball joint to steering knuckle pinch bolt.
5. Separate the lower control arm ball joint from the steering knuckle.
6. Remove the 4 lower control arm bushing to front subframe nuts and bolts.
7. Remove the right side lower control arm.

To install:

8. Place the right side lower control arm to the front subframe.
9. Install the mounting bolts with the threads pointed down. Torque the mounting bolts to 96 ft. lbs. (130 Nm).
10. Install the lower ball joint stud to the steering knuckle.
11. Install the lower ball joint to steering knuckle pinch bolt. Torque the pinch bolt to 40 ft. lbs. (54 Nm).

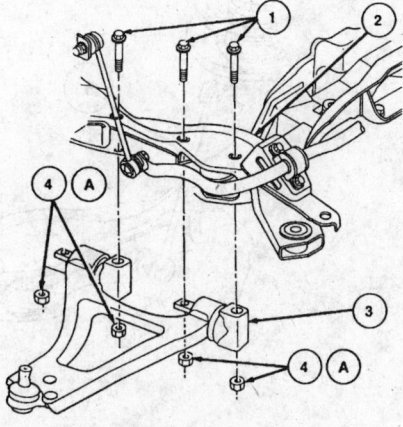

1	Bolts (4 Req'd)
2	Sub-frame
3	Front Suspension Lower Arm
4	Nuts (4 Req'd)
A	Tighten to 130 N·m (96 Lb-Ft)

298694

Lower control arm and related components (left side shown)

12. Install the wheel and tire assembly. Torque the lug nuts to 62 ft. lbs. (85 Nm).
13. Lower the vehicle.
14. Connect the negative battery cable.
15. Check the wheel alignment.
16. Road test the vehicle and check for proper operation.

Left Side

1. Disconnect the negative battery cable.
2. Support the radiator and fan shroud using safety wire to the radiator support.
3. If equipped with 2.0L (VIN 3) engine, remove the heat shield and the catalytic converter retaining nuts from the exhaust manifold.
4. Raise and safely support the vehicle.
5. Remove the catalytic converter.
6. Disconnect the steering column lower yoke.
7. If equipped with a manual transaxle, disconnect the gearshift rod and clevis.
8. Remove the front lower radiator cover.
9. Remove the radiator supports.
10. Remove the ball joint to steering knuckle pinch bolt.
11. Separate the lower control arm from the steering knuckle.
12. Disconnect the power steering cooler lines at the right–hand side of the front subframe.
13. Disconnect the stabilizer bar link from the front stabilizer bar.
14. Remove the through-bolts from the front and rear engine mounts.
15. Position Powertrain Lift 014-00765 (transmission jack) or equivalent, under the front subframe.
16. Remove the subframe mounting bolts.
17. Remove the 4 lower control arm bushing to front subframe mounting nuts.
18. Lower the front subframe enough to allow removal of the lower control arm mounting bolts.
19. Remove the mounting bolts and remove the lower control arm.

To install:

20. Position the lower control arm into the front subframe.
21. Install the mounting bolts with the threads pointed down.
22. Raise the front subframe into position.
23. Install the lower control arm to subframe mounting bolts. Torque the mounting bolts and nuts to 96 ft. lbs. (130 Nm).

24. Install the subframe mounting bolts. Torque the bolts to 81–110 ft. lbs. (110–150 Nm).

25. Remove the powertrain lift or equivalent.

26. Install the front and rear engine mount through-bolts. Torque the through-bolts to 40–55 ft. lbs. (54–75 Nm).

27. Connect the power steering cooler lines to the right–hand side of the subframe.

28. Position the lower control arm ball joint stud to the steering knuckle.

29. Install the pinch bolt and torque to 40 ft. lbs. (54 Nm).

30. Connect the stabilizer bar link to the front stabilizer bar. Torque the bar link to 37 ft. lbs. (50 Nm).

31. Install the radiator support.

32. Install the front lower radiator cover.

33. Install the catalytic converter.

34. Lower the vehicle.

35. If equipped with 2.0L (VIN 3) engine, install the heat shield and the catalytic converter retaining nuts.

36. Raise and safely support the vehicle.

37. If equipped with a manual transaxle, install the gearshift rod and clevis.

38. Connect the steering column lower yoke.

39. Install the wheel and tire assembly. Torque the lug nuts to 62 ft. lbs. (85 Nm).

40. Lower the vehicle.

41. Remove the safety wire supporting the radiator.

42. Connect the negative battery cable.

NOTE: Whenever the vehicle subframe is removed or lowered, the wheel alignment should be checked.

43. Check the wheel alignment.

44. Road test the vehicle and check for proper operation.

Sway Bar

REMOVAL AND INSTALLATION

1. Disconnect the negative battery cable.

2. Raise and safely support the vehicle.

3. Remove the front wheel and tire assemblies.

4. Remove the left-hand and right-hand sway bar (stabilizer bar) link retaining nuts and the sway bar links

from the bracket on the strut housings.

5. Remove the sway bar links from the sway bar and set aside.

6. If necessary, use a ball joint remover to separate the sway bar link from the sway bar.

7. Remove the 4 sway bar insulator bracket to subframe bolts (2 on each side).

8. Remove the sway bar from the vehicle.

To install:

9. Position the sway bar (stabilizer bar) into the vehicle.

10. Install the 4 mounting bracket to subframe bolts.

11. Torque the bolts to 37 ft. lbs. (50 Nm).

12. Attach the sway bar links to each side of the sway bar and to each strut bracket.

NOTE: If the sway bar link boot seals are damaged, the link must be replaced.

13. Install the 4 retaining nuts and torque to 37 ft. lbs. (50 Nm).

14. Install the wheel and tire assemblies. Torque the lug nuts to 62 ft. lbs. (85 Nm).

15. Lower the vehicle.

16. Connect the negative battery cable.

17. Road test the vehicle and check for proper operation.

Front Wheel Bearings

ADJUSTMENT

The front wheel bearings consist of a cartridge design and are permanently lubricated and sealed requiring no further maintenance. The bearings are preset and cannot be adjusted. If any part of a wheel bearing assembly is defective, the unit must be replaced. It is critical that the wheel hub retainer is properly torqued to 210 ft. lbs. (290 Nm) and that a new wheel hub retainer is always used.

REMOVAL AND INSTALLATION

NOTE: Before proceeding, make sure to have available new pinch bolts for the steering knuckle to lower ball joint and steering knuckle to strut as well as a new wheel hub retaining nut.

1. Disconnect the negative battery cable.

2. Raise and safely support the vehicle.

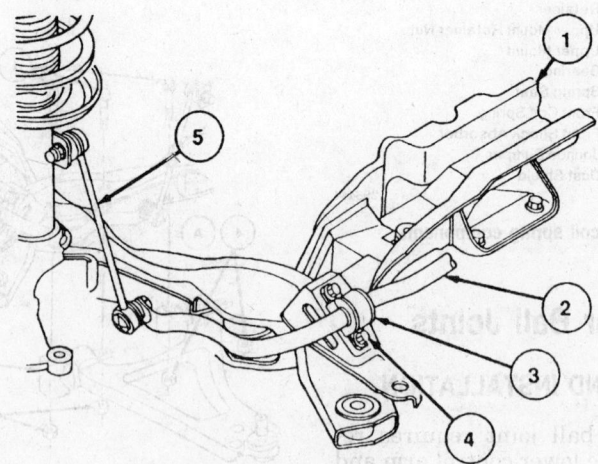

1 Front Sub-Frame
2 Stabilizer Bar
3 Stabilizer Bar Bracket
4 Stabilizer Bar Bracket Bolt (4 Req'd)
5 Stabilizer Bar Link

298706

Sway bar and related components (left side shown)

3. Remove the wheel and tire assembly.

4. Remove the disc brake caliper and rotor.

5. Support the disc brake caliper with safety wire. Do not let the caliper hang by the brake hose.

6. Remove the anti-lock brake sensor retaining bolt and remove the sensor from the steering knuckle.

7. Remove the outer tie rod end cotter pin and remove the castellated nut. Discard the cotter pin.

8. Separate the outer tie rod end from the steering knuckle using Tie Rod End Remover 3290-D or equivalent.

9. Remove the wheel hub retaining nut. Discard the nut.

10. Separate the halfshaft from the wheel hub using Front Hub Remover/Replacer T81P-1104-C or equivalent, and the related adapters.

11. Once removed, support the end of the halfshaft.

12. Remove the steering knuckle to lower ball joint pinch bolt.

13. Remove the steering knuckle to strut pinch bolt.

14. Work the steering knuckle off of the lower ball joint and out of the lower strut tube.

15. Remove the steering knuckle from the vehicle.

16. Place the steering knuckle assembly onto a suitable workbench.

17. Install Front Hub Remover/Replacer T81P-1104-C or equivalent with the appropriate adapters and separate the hub from the steering knuckle.

18. Remove the inner and outer snap rings securing the wheel bearing.

19. Remove the wheel bearing from the steering knuckle. Drive or press the old wheel bearing out as required.

To install:

20. Install the outer snap ring in the steering knuckle.

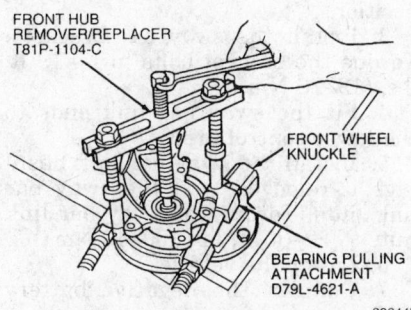

Removing the hub from the steering knuckle

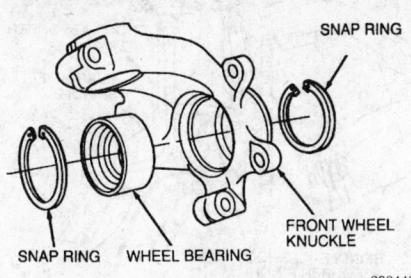

Wheel bearing and related components

21. Install the wheel bearing using a hydraulic press with Pinion Bearing Cup Replacer T80T-4000-E or equivalent.

22. Install the inner snap ring in the steering knuckle.

23. Install the hub to the steering knuckle using Threaded Drawbar T75T-1176-A or equivalent.

24. Carefully align the splines of the outer CV-joint with the splines in the hub.

25. Position the steering knuckle to the lower ball joint stud.

26. Position the steering knuckle to the lower strut tube.

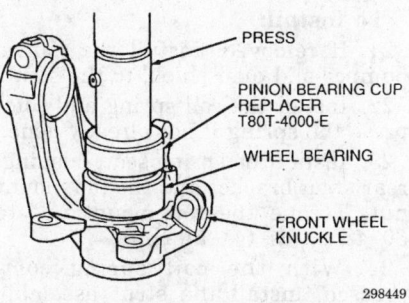

Pressing in the new wheel bearing

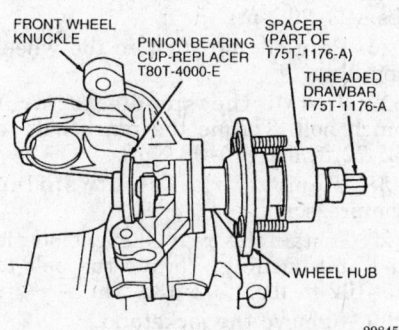

Installing the hub to the steering knuckle

27. Install a new steering knuckle to lower ball joint pinch bolt. Torque the bolt to 55–58 ft. lbs. (75–79 Nm).

28. Install a new steering knuckle to strut pinch bolt. Torque the bolt to 40 ft. lbs. (54 Nm).

29. Install the anti-lock brake sensor and retaining bolt. Torque the retaining bolt to 7 ft. lbs. (10 Nm).

30. Install the disc brake rotor and caliper assembly.

31. Install a new wheel hub retaining nut. Torque the retaining nut to 210 ft. lbs. (290 Nm).

NOTE: Do not use an impact gun to tighten the wheel hub retaining nut or damage to the wheel bearing may result.

32. Attach the tie rod end to the steering knuckle. Install the castellated nut and torque the nut to 18–22 ft. lbs. (24–30 Nm). Install a new cotter pin.

33. Install the wheel and tire assembly. Torque the lug nuts to 62 ft. lbs. (85 Nm).

34. Lower the vehicle.

35. Connect the negative battery cable.

36. Pump the brake pedal several times to position the disc brake pads before attempting to move the vehicle.

37. Road test the vehicle and check for proper operation.

REAR SUSPENSION

Strut and Spring

REMOVAL AND INSTALLATION

1. Disconnect the negative battery cable.

2. Raise and safely support the vehicle.

3. Remove the wheel and tire assembly.

4. Remove the anti–lock sensor wiring from the strut bracket.

5. Remove the anti-lock sensor mounting bolt and remove the sensor.

6. Disconnect the rear brake hose fitting from the brake tube. Plug the brake lines.

7. Remove the retainer and the rear brake hose from the strut.

8. Remove the tie strap retaining the parking brake rear cable and conduit to the rear suspension tie rod.

9. Disconnect the rear sway bar link and bushings from the rear control arm (suspension arm).

10. Disconnect the rear suspension tie rod from the wheel spindle.

NOTE: The front and rear control arms (suspension arms) must be supported prior to the removal of the upper or lower strut attachments.

11. Position a jack stand under the front and rear control arms.

12. Remove the spindle to strut pinch bolt.

13. Separate the spindle from the strut by tapping down on the wheel spindle.

14. Compress the coil spring using Strut Spring Compressor T81P-5310-A or equivalent.

15. Remove the 2 top retaining bolts and remove the strut assembly.

16. Place the strut assembly on a suitable workbench.

17. Compress the coil spring enough to relieve the tension on the spring seat.

18. Remove the top mount nut, rear strut bracket, bushing and spring seat.

19. Remove the coil spring and slowly relieve the tension on the spring if it is not to be reinstalled.

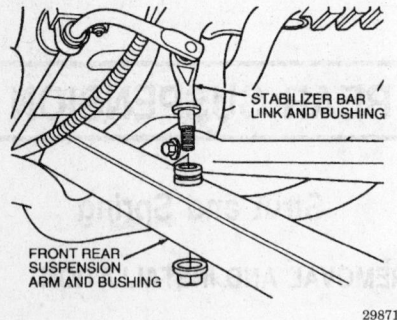

STABILIZER BAR LINK AND BUSHING

FRONT REAR SUSPENSION ARM AND BUSHING

298719

Removing the sway bar link

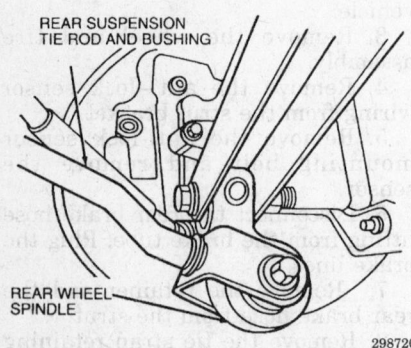

REAR SUSPENSION TIE ROD AND BUSHING

REAR WHEEL SPINDLE

298720

Rear suspension tie rod

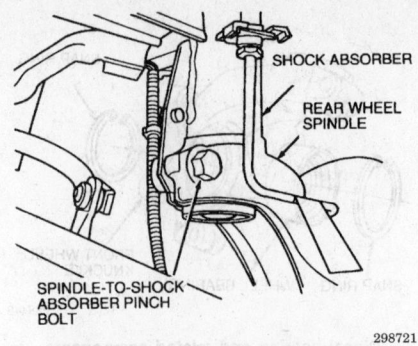

SHOCK ABSORBER

REAR WHEEL SPINDLE

SPINDLE-TO-SHOCK ABSORBER PINCH BOLT

298721

Rear strut pinch bolt and spindle

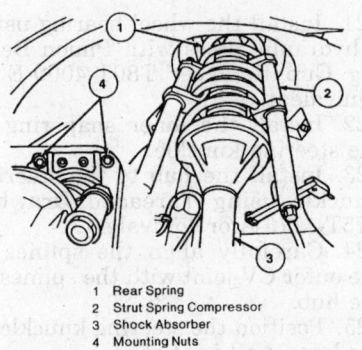

1 Rear Spring
2 Strut Spring Compressor
3 Shock Absorber
4 Mounting Nuts

298722

Compressing the coil spring prior to removal

20. Remove the dust shield and the jounce bumper if the strut is being replaced.

To install:

21. If removed, install the jounce bumper and dust shield to the strut.

22. Install the coil spring and compress the spring if not already done.

23. Install the spring seat, bushing, rear strut bracket and the top mount nut. Torque the top mount nut to 30–43 ft. lbs. (41–58 Nm).

24. With the coil spring compressed, install the strut assembly into position.

25. Install the 2 rear strut bracket mounting bolts. Torque the rear strut bracket mounting bolts to 17–22 ft. lbs. (23–30 Nm).

26. Position the strut to the wheel spindle.

27. Install the spindle to strut pinch bolt. Torque the pinch bolt to 52–72 ft. lbs. (70–98 Nm).

28. Remove the strut spring compressor.

29. Install the rear suspension tie rod and bushing. Torque the bolt to 75–102 ft. lbs. (102–138 Nm).

30. Remove the jackstand.

31. Attach the rear brake hose to the strut.

32. Install the rear brake anti–lock sensor and retaining bolt. Tighten the retaining bolt to 7 ft. lbs. (9 Nm).

33. Install the anti–lock wiring to the strut.

34. Secure the parking brake cable and conduit to the rear suspension tie rod with a tie strap.

35. Unplug and connect the rear brake hose fitting to the brake tube. Tighten the fitting securely.

36. Bleed the brake system.

37. Install the wheel and tire assembly. Torque the lug nuts to 62 ft. lbs. (85 Nm).

38. Lower the vehicle.

39. Connect the negative battery cable.

40. Check the rear wheel alignment.

41. Road test the vehicle and check for proper operation.

Sway Bar

REMOVAL AND INSTALLATION

1. Disconnect the negative battery cable.

2. Raise and safely support the vehicle.

3. Remove the sway bar (stabilizer bar) link nut from the lower control arm on each side of the sway bar.

4. Rotate the sway bar to separate the sway bar links from the lower control arms.

5. Remove the sway bar bracket bolts and remove the brackets.

6. Remove the sway bar from the vehicle.

7. If the sway bar is being replaced, remove the sway bar links by prying them off of the studs on each end of the sway bar and remove the rubber insulators from the sway bar. Sway bar and related components

298765

To install:

1. Connect the rubber insulators and the sway bar links if removed.

2. Place the sway bar into position.

3. Install the sway bar brackets. Torque the bracket bolts to 14–19 ft. lbs. (19–26 Nm).

4. Fit the sway bar link ends to the lower control arms.

5. Install the sway bar link bushings if removed, and the sway bar link nuts. Torque the sway bar link nuts to 22–30 ft. lbs. (30–40 Nm).

6. Lower the vehicle.

7. Connect the negative battery cable.

8. Road test the vehicle and check for proper operation.

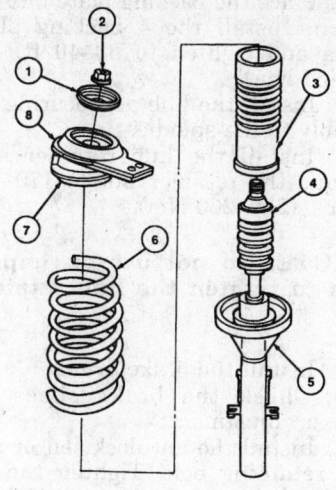

1 Rear Shock Absorber Bracket
2 Shock Absorber Mounting Nut
3 Rear Shock Absorber Dust Boot
4 Rear Suspension Jounce Bumper
5 Shock Absorber
6 Rear Spring
7 Spring Seat
8 Shock Absorber Bushing

298723

Rear strut, coil spring and related components

Wheel Bearings

ADJUSTMENT

The rear wheel bearings consist of a cartridge design and are permanently lubricated and sealed requiring no further maintenance. The bearings are preset and cannot be adjusted. If any part of a wheel bearing assembly is defective, the unit must be replaced. It is critical that the wheel hub retainer is properly torqued to 170–192 ft. lbs. (230–260 Nm) and that a new wheel hub retainer is always used.

REMOVAL AND INSTALLATION

NOTE: The wheel bearings are contained within the wheel hub and must be replaced as an assembly.

With Rear Disc Brakes

1. Disconnect the negative battery cable.
2. Raise and safely support the vehicle.
3. Remove the wheel and tire assembly.
4. Remove the anti-lock sensor retaining bolt and remove the sensor.

5. Remove the rear disc brake caliper and rotor.

NOTE: Do not use an impact gun to remove the hub retainer nut.

6. Remove the hub retainer nut.
7. Slide the hub and wheel bearing assembly off of the spindle and remove.
8. Remove the disc brake dust shield.
9. Disconnect the rear tie rod and bushing from the spindle.
10. Disconnect the rear control arm (suspension arm) from the spindle. Remove the front control arm (suspension arm) from the spindle.
11. Remove the strut-to-spindle pinch bolt.
12. Separate the spindle from the strut and remove from the vehicle.
   To install:
13. Install the spindle to the to the strut. Tighten the pinch nut to 52–72 ft. lbs. (70–98 Nm).
14. Connect the rear control arm (suspension arm) to the spindle. Connect the front control arm (suspension arm) to the spindle. Tighten the bolt, but do not torque at this time.
15. Connect the rear tie rod and bushing to the spindle. Tighten the bolt, but do not torque at this time.
16. Install the disc brake dust shield.
17. Install the hub and bearing assembly to the spindle.
18. Install the hub retainer nut. Torque the hub retainer nut to 170–192 ft. lbs. (230–260 Nm).

NOTE: Do not use an impact gun to tighten the hub retainer nut.

19. Install the rear disc brake rotor and caliper.
20. Install the anti-lock sensor. Tighten the retaining bolt for the sensor to 7–8 ft. lbs. (9–11 Nm).
21. Install the wheel and tire assembly. Tighten the wheel nuts to 62 ft. lbs. (85 Nm).
22. Lower the vehicle until the wheels are supporting the vehicles weight. This will properly load the suspension.
23. Tighten the spindle-to-control arm bolt to 52–79 ft. lbs. (70–98 Nm).
24. Tighten the tie rod-to-spindle bolt to 75–102 ft. lbs. (98 Nm).
25. Finish lowering the vehicle.
26. Road test the vehicle and check for proper operation.

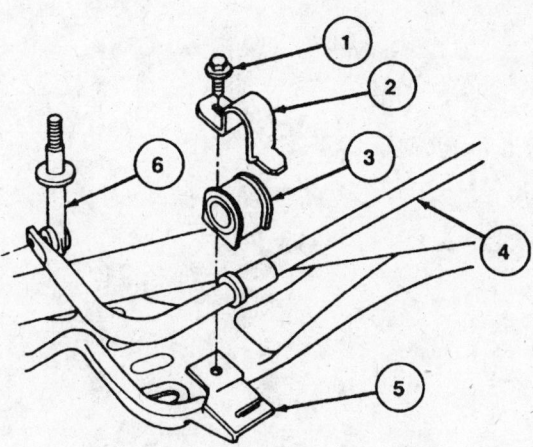

1 Bolt (4 Req'd)
2 Stabilizer Bar Bracket (2 Req'd)
3 Stabilizer Bar Insulator (2 Req'd)
4 Rear Stabilizer Bar
5 Rear Crossmember
6 Stabilizer Bar End

298765

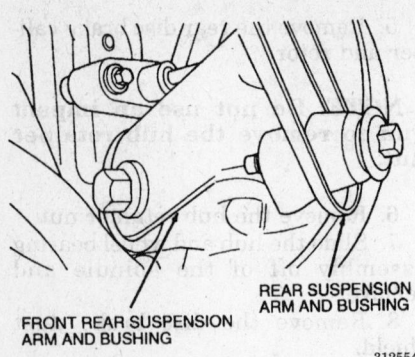

FRONT REAR SUSPENSION
ARM AND BUSHING

REAR SUSPENSION
ARM AND BUSHING

312554

Front control arm and spindle

With Rear Drum Brakes

NOTE: The wheel bearings are contained within the wheel hub and must be replaced as an assembly.

1. Disconnect the negative battery cable.
2. Raise and safely support the vehicle.
3. Remove the wheel and tire assembly.
4. Remove the anti-lock sensor retaining bolt and remove the sensor.

5. Remove the brake drum retainer and the brake drum.

NOTE: Do not use an impact gun to remove the hub retainer nut.

6. Remove the wheel hub retainer nut.
7. Slide the hub and bearing assembly off of the spindle.
8. Remove the 4 backing plate bolts and move the backing plate out of the way. Be careful not to damage the brake line to the wheel cylinder.
9. Disconnect the rear tie rod and bushing from the spindle.
10. Disconnect the rear control arm (suspension arm) from the spindle. Remove the front control arm (suspension arm) from the spindle.
11. Remove the strut-to-spindle pinch bolt.
12. Separate the spindle from the strut and remove.
To install:
13. Install the spindle to the strut. Tighten the pinch bolt to 52–72 ft. lbs. (70–98 Nm).
14. Connect the rear control arm (suspension arm) to the spindle. Connect the front control arm (suspension arm) to the spindle.
15. Connect the rear tie rod and bushing to the spindle. Do not torque the bolt at this time.

16. Place the backing plate into position. Install the 4 backing plate bolts and tighten to 33–40 ft. lbs. (45–54 Nm).
17. Install the hub and bearing assembly to the spindle.
18. Install the hub retainer nut. Torque the retainer nut to 170–192 ft. lbs. (230–260 Nm).

NOTE: Do not use an impact gun to tighten the hub retainer nut.

19. Install the brake drum.
20. Check the brake shoes for proper adjustment.
21. Install the anti-lock sensor and the retaining bolt. Tighten the retaining bolt to 7–8 ft. lbs. (9–11 Nm).
22. Install the wheel and tire assembly. Tighten the wheel nuts to 62 ft. lbs. (85 Nm).
23. Lower the vehicle until the wheels are supporting the vehicles weight. This will properly load the suspension.
24. Tighten the spindle-to-control arm bolt to 52–79 ft. lbs. (70–98 Nm).
25. Tighten the tie rod-to-spindle bolt to 75–102 ft. lbs. (98 Nm).
26. Finish lowering the vehicle.
27. Road test the vehicle and check for proper operation.

FORD MOTOR CO.

Front Wheel Drive

FORD-Taurus **LINCOLN**-Continental
MERCURY-Sable

13

FIRING ORDERS

NOTE: To avoid confusion, always replace spark plug wires one at a time.

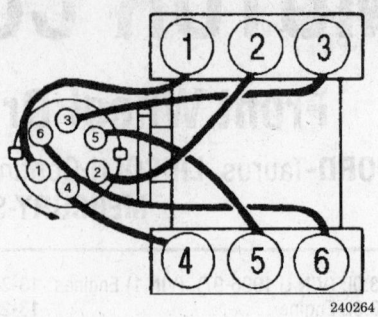

3.8L (VIN 4) Engine
Engine Firing Order: 1–4–2–5–3–6
Distributor Rotation: Counterclockwise

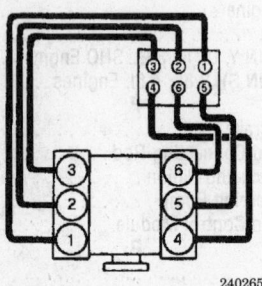
240265

**3.0L and 3.2L SHO Engines
1996–97 3.0L (VIN U and 1) Engines**
Engine Firing Order:
1–4–2–5–3–6
Distributorless Ignition System

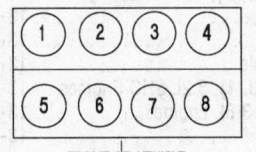
325598

4.6L Engine
Engine Firing Order: 1–3–7–2–6–5–4–8
Distributorless Ignition System

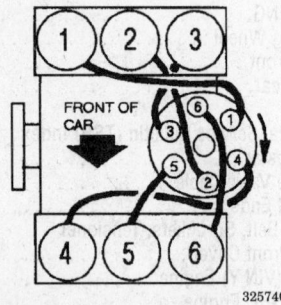
325740

1993–95 3.0L (VIN U) Engine
Engine Firing Order: 1–4–2–5–3–6
Distributor Rotation: Clockwise

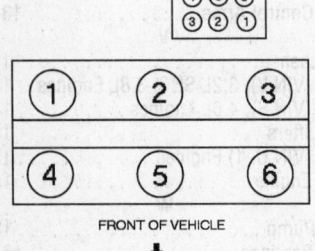
325674

3.0L (VIN S) Engine (4V)
Firing Order: 1–4–2–5–3–6
Distributorless Ignition System

ENGINE ELECTRICAL

NOTE: Disconnecting the negative battery cable on some vehicles may interfere with the functions of the on board computer systems and may require the computer to undergo a relearning process, once the negative battery cable is reconnected.

Distributor

REMOVAL AND INSTALLATION

1993–95 3.0L (VIN U) and 3.8L Engines

1. Disconnect the negative battery cable.
2. Disconnect the wiring connector from the distributor.
3. Mark the position of the distributor cap No. 1 cylinder wire tower on the distributor base.

4. Remove distributor cap and position it and the attached wires aside.
5. Mark the position of the rotor in relation to the distributor housing and mark the position of the distributor housing on the engine.
6. Remove the distributor hold-down bolt and clamp and remove the distributor.

NOTE: Before installation, inspect the distributor O-ring and drive gear for wear and/or damage. Rotate the distributor shaft to make sure it moves freely, without binding.

To install:

Timing Not Disturbed

1. Install the distributor, aligning the distributor housing and rotor with the marks that were made during the removal procedure.
2. Install the distributor hold-down bolt and clamp. Only snug the bolt at this time.
3. Connect the distributor to the wiring harness.
4. Install the distributor cap. Make sure the ignition wires are securely connected to the distributor cap and spark plugs. Tighten the distributor cap screws to 18–23 inch lbs. (2.0–2.6 Nm).
5. Connect a suitable timing light and set the initial timing.
6. Tighten the distributor hold-down bolt to 14–21 ft. lbs. (19–28 Nm) on the 3.0L engine or 20–29 ft. lbs. (27–40 Nm) on the 3.8L engine.
7. Recheck the initial timing and adjust if necessary.

Timing Disturbed

1. Disconnect the spark plug wire from the No. 1 cylinder spark plug and remove the spark plug.
2. Place a finger over the spark plug hole. Rotate the engine clockwise until compression is felt at the spark plug hole.
3. Align the timing pointer with the TDC mark on the crankshaft damper.
4. Rotate the distributor shaft so the rotor tip is pointing to the distributor cap No. 1 spark plug tower position.
5. While installing the distributor, continue rotating the rotor slightly so the leading edge of the vane is centered in the vane switch stator assembly.
6. Rotate the distributor in the block to align the leading edge of the vane and vane switch stator assembly. Make sure the rotor is pointing

to the distributor cap No. 1 spark plug tower position.

NOTE: If the vane and vane switch stator cannot be aligned by rotating the distributor in the block, remove the distributor just enough to disengage the distributor gear from the camshaft gear. Rotate the rotor enough to engage the distributor gear on another tooth of the camshaft gear. Repeat Steps 1 and 2, if necessary.

7. Install the distributor hold-down bolt and clamp. Only snug the bolt at this time.

8. Connect the distributor to the wiring harness and install the distributor cap. Tighten the distributor cap hold-down screws to 18–23 inch lbs. (2.0–2.6 Nm).

9. Install the No. 1 cylinder spark plug and connect the spark plug wire.

10. Connect a suitable timing light and set the initial timing.

11. Tighten the distributor hold-down bolt to 14–21 ft. lbs. (19–28 Nm) on the 3.0L engine or 20–29 ft. lbs. (27–40 Nm) on the 3.8L engine.

12. Recheck the initial timing and adjust if necessary.

Ignition Timing

ADJUSTMENT

1993–95 3.0L (VIN U) and 3.8L Engines

1. Apply the parking brake. If equipped with manual transaxle, place the shift lever in neutral. If equipped with automatic transaxle, place the shift lever in **P**.

2. Locate the timing marks on the crankshaft pulley and timing belt cover. If the marks are hard to see, clean them off with some degreasing cleaner and a wire brush.

3. Connect a suitable inductive timing light according to the manufacturer's instructions.

4. Disconnect the single wire in-line SPOUT connector, located near the distributor, by pulling the plug from the connector housing.

5. Start the engine and allow it to warm to operating temperature. Make sure the idle speed is correct.

NOTE: To set timing correctly, a remote starter should not be used. Use only the ignition key to start the vehicle. Disconnecting the start wire at the starter relay

will cause the TFI module to revert to start mode timing after the vehicle is started. Reconnecting the start wire after the vehicle is running will not correct the timing.

6. Aim the timing light at the timing marks. The timing should be at 10 degrees BTDC but always check the underhood vehicle emission label.

7. If the timing is incorrect, loosen the distributor hold-down bolt just enough to turn the distributor housing. While aiming the timing light at the timing marks, turn the distributor until the marks are aligned. Tighten the distributor hold-down bolt and recheck the timing.

8. Reconnect the single wire inline SPOUT connector. Check the timing advance to verify the distributor is advancing beyond the initial setting.

9. Remove the inductive timing light.

3.0L (VIN Y) and 3.2L SHO Engines

The base ignition timing is set at 10 degrees Before Top Dead Center (BTDC) and is not adjustable.

Alternator

PRECAUTIONS

Several precautions must be observed with alternator equipped vehicles to avoid damage to the unit.

- If the battery is removed for any reason, make sure it is reconnected with the correct polarity. Reversing the battery connections may result in damage to the 1-way rectifiers.
- When utilizing a booster battery as a starting aid, always connect the positive to positive terminals and the negative terminal from the booster battery to a good engine ground on the vehicle being started.
- Never use a fast charger as a booster to start vehicles.
- Disconnect the battery cables when charging the battery with a fast charger.
- Never attempt to polarize the alternator.
- Do not use test lights of more than 12 volts when checking diode continuity.
- Do not short across or ground any of the alternator terminals.
- The polarity of the battery, alternator and regulator must be matched and considered before making any electrical connections within the system.

- Never separate the alternator on an open circuit. Make sure all connections within the circuit are clean and tight.
- Disconnect the battery ground terminal when performing any service on electrical components.
- Disconnect the battery if arc welding is to be done on the vehicle.

REMOVAL AND INSTALLATION

3.0L (VIN Y) SHO Engine

1. Disconnect the battery cables and remove the battery and battery tray.

2. Tag and disconnect the wire harness from the alternator.

3. Loosen the belt tensioner and remove the alternator belt from the pulley.

4. Remove the mounting bolts and the alternator.

To install:

5. Position the alternator and install the mounting bolts.

6. Tighten the front mounting bolt to 36–53 ft. lbs. (48–72 Nm) on 1993 vehicles or 30–41 ft. lbs. (40–55 Nm) on 1994–95 vehicles. Tighten the rear mounting bolts to 25–37 ft. lbs. (34–50 Nm) on 1993 vehicles or 15–22 ft. lbs. (20–30 Nm) on 1994–95 vehicles.

7. Connect the wire harness to the alternator.

8. Install the alternator drive belt and adjust the belt tension.

9. Connect the negative battery cable.

1993–95 3.0L (VIN U) and 3.8L Engines

1. Disconnect the negative battery cable.

2. Tag and disconnect the wire harness from the alternator.

3. Rotate the belt tensioner counterclockwise and remove the drive belt from the pulley.

4. Remove the alternator mounting bolts and remove the alternator.

5. Installation is the reverse of the removal procedure. Torque the mounting brace bolt to 15–22 ft. lbs. (20–30 Nm) and the pivot bolt to 30–41 ft. lbs. (40–55 Nm).

3.2L SHO Engine

1. Disconnect the negative battery cable.

2. Tag and disconnect the wire harness connector and output terminal connector from the alternator.

CAUTION

Be extremely careful when removing and installing belts to ensure that the tool does not slip.

3. Using a 14mm socket over the bolt on the drive belt tensioner pulley, rotate clockwise (downward) to release the drive belt tension. Loosen the belt tensioner and remove the alternator belt from the pulley.

4. Remove the mounting bolt from the rear alternator bracket to alternator.

5. Remove the 2 alternator mounting brackets.

6. Remove the alternator from the vehicle.

To install:

7. Install the alternator mounting brackets. Tighten the bolts to 15–22 ft. lbs. (20–30 Nm).

8. Install the remaining components in the reverse order of removal. Tighten the upper rear bolt to 15–22 ft. lbs. (20–30 Nm) and the lower bolt to 30–41 ft. lbs. (40–55 Nm).

9. Connect the negative battery cable.

3.0L (VIN S) Engine

1. Disconnect the negative battery cable.

2. Tag and disconnect the wiring harness from the alternator.

3. Remove the accessory drive belt from the alternator pulley.

4. Remove the 2 retaining bolts and 1 stud securing the alternator to the engine.

5. Remove the alternator.

To install:

6. Place the alternator into position and install the mounting bolts. Tighten the bolts to 15–22 ft. lbs. (20–30 Nm).

7. Install the remaining components in the reverse order of removal. Tighten the wiring harness mounting nut to 11–14 ft. lbs. (15–20 Nm).

8. Connect the negative battery cable. Check the alternator for proper operation.

1996–97 3.0L (VIN 1 and U) Engines

1. Disconnect the negative battery cable.

2. Tag and disconnect the wire harness from the alternator.

3. Rotate the belt tensioner counterclockwise and remove the drive belt from the alternator pulley.

4. Remove the alternator mounting bolts and remove the brace.

5. Remove the pivot bolt and remove the alternator.

To install:

6. Place the alternator into position and install the pivot and mounting brace bolts but do not tighten.

7. Rotate the belt tensioner counterclockwise and install the drive belt to the alternator pulley.

8. Tighten the retaining nut and 2 mounting brace retaining bolts securely. Tighten the pivot bolt to 30–40 ft. lbs.

9. Connect the alternator wiring harness. Tighten the output terminal nut to 97 inch lbs. (11 Nm).

10. Connect the negative battery cable. Check the alternator for proper operation.

4.6L Engine

1. Disconnect the negative battery cable.

2. Remove 2 screws retaining the power steering pump auxiliary reservoir to the radiator coolant recovery reservoir and set the power steering pump auxiliary reservoir aside.

3. Disconnect the hoses from the radiator coolant recovery reservoir. Remove the retaining nut and bolt and remove the radiator coolant recovery reservoir.

4. Remove the accessory drive belt from the alternator pulley.

5. Tag and disconnect the wiring connectors from the rear of the alternator. To disconnect push-on type terminals, depress the lock tab and pull straight off.

6. Remove the alternator mounting bracket from the top of the alternator.

7. Remove 2 mounting bolts and remove the alternator.

To install:

8. Place the alternator in position and install 2 mounting bolts finger-tight.

9. Install the alternator mounting bracket and 2 retaining bolts. Tighten the alternator mounting bolts to 15–22 ft. lbs. (20–30 Nm). Tighten the alternator mounting bracket bolts to 6–8 ft. lbs. (8–12 Nm).

10. Install the remaining components in the reverse order of removal.

11. Connect the negative battery cable.

12. Start the engine and check the alternator/charging system for proper operation.

Drive Belt

REMOVAL AND INSTALLATION

3.0L (VIN Y) SHO Engine

Alternator and A/C Compressor Belt

1. Disconnect the negative battery cable. Loosen the nut in the center of the idler pulley.

2. Loosen the idler adjusting screw until the belt can be removed, then remove the belt.

To install:

3. Install the belt over the pulleys. Make sure the belt V-grooves make proper contact with the pulleys.

4. Position a suitable belt tension gauge over the longest accessible belt span.

5. Turn the idler adjusting screw until there is 220–265 lbs. on the tension gauge for a new belt or 148–192 lbs. for a used belt.

6. Torque the idler pulley nut to 25–37 ft. lbs. (34–50 Nm).

7. Connect the negative battery cable.

Power Steering and Water Pump Belt

1. Disconnect the negative battery cable. Remove the alternator belt.

2. Loosen the nut on the belt tensioner pulley.

3. Turn the belt adjusting screw on the tensioner counterclockwise until the belt can be removed.

To install:

4. Position the belt over the pulleys. Make sure the belt V-grooves properly contact the pulleys.

5. Position a suitable belt tension gauge over the longest accessible belt span.

6. Turn the idler adjusting screw until there is 154–198 lbs. on the tension gauge for a new belt or 112–157 lbs. or for a used belt.

7. Torque the idler pulley nut to 25–37 ft. lbs. (34–50 Nm).

8. Install the alternator belt.

3.2L SHO Engine

CAUTION

Be extremely careful when removing or installing the drive belt to ensure that the tool does not slip.

1. Disconnect the negative battery cable.

2. Using a 14mm socket or wrench on the bolt attaching the drive belt tensioner pulley, rotate the drive belt tensioner downward to release the drive belt tension.

3. Remove the drive belt from the pulleys.

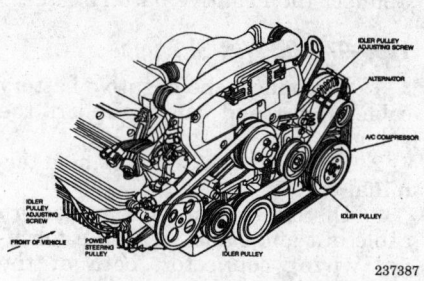

Drive belt routing — 3.0L (VIN Y) SHO engine

237387

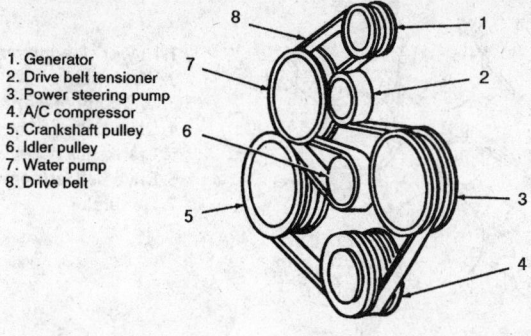

1. Generator
2. Drive belt tensioner
3. Power steering pump
4. A/C compressor
5. Crankshaft pulley
6. Idler pulley
7. Water pump
8. Drive belt

Drive belt routing — 3.0L (VIN U and 1) engines

314495

To install:

4. Install the drive belt over all the pulleys, except the power steering pump pulley.

5. Rotate the drive belt tensioner carefully and install the drive belt over the power steering pulley. Ensure that all V-grooves make proper contact with the pulleys.

6. Connect the negative battery cable.

3.0L (VIN U and 1) and 3.8L Engines

— CAUTION —
Be extremely careful when removing or installing the drive belt to ensure that the tool does not slip.

1. Disconnect the negative battery cable.

2. Using a 15mm socket or wrench on the bolt attaching the drive belt tensioner pulley, rotate the drive belt tensioner clockwise to remove the drive belt from the pulleys.

To install:

3. Install the drive belt over all the pulleys except the drive belt tensioner pulley.

4. Rotate the drive belt tensioner carefully in a clockwise direction and install the drive belt over the tensioner pulley. Ensure that all V-

3.2L SHO Accessory Drive Belt

1. Drive belt
2. Water pump
3. Idler pulley
4. Generator
5. A/C compressor
6. Idler pulley
7. Crankshaft vibration damper and pulley
8. Drive belt tensioner
9. Power steering pump

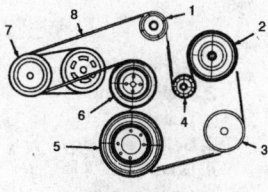

1. Generator
2. Power steering pump
3. A/C compressor
4. Idler pulley
5. Crankshaft vibration damper and pulley
6. Water pump
7. Drive belt tensioner
8. Drive belt

314496

Drive belt routing — 3.8L engine

grooves make proper contact with the pulleys.

5. Slowly release the tensioner and remove the tool. Connect the negative battery cable.

6. Start the engine and check for proper operation.

3.0L (VIN S) Engine

— CAUTION —
Be extremely careful when removing or installing the drive belt to ensure that the tool does not slip.

1. Disconnect the negative battery cable.

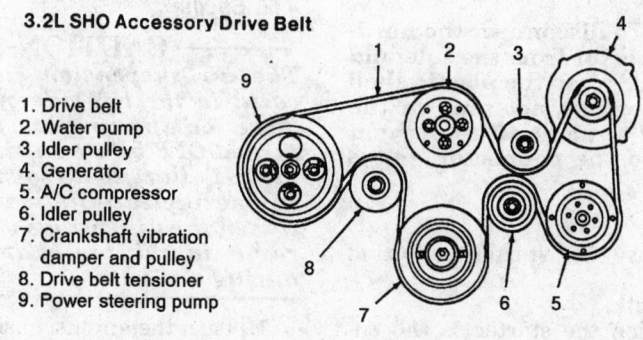

Drive belt routing — 3.2L SHO engine

256582

2. Using a 15mm socket or wrench on the bolt attaching the drive belt tensioner pulley, rotate the drive belt tensioner counterclockwise to remove the drive belt from the pulleys.

To install:

3. Install the drive belt over all the pulleys except the drive belt tensioner pulley.

4. Rotate the drive belt tensioner carefully counterclockwise and install the drive belt over the tensioner pulley. Ensure that all V-grooves make proper contact with the pulleys.

5. Connect the negative battery cable. Start the engine and check for proper operation.

4.6L Engine

— CAUTION —
Do not allow the drive belt tensioner to snap back as damage to the drive belt tensioner or personal injury could result.

1. Disconnect the negative battery cable.

2. Install a breaker bar or equivalent tool, in the 3/8 inch square hole in the drive belt tensioner.

3. Rotate the tensioner away from the belt with the breaker bar.

4. Lift the belt over the alternator pulley flange and remove it.

5. Slowly relax the tensioner and remove the breaker bar, or equivalent.

To install:

6. Place the breaker bar or equivalent, in the drive belt tensioner and rotate the tensioner to allow belt installation.

7. Install the belt over the pulleys, making sure it is properly routed. Ensure that the ribs on the belt properly contact the grooves on the pulleys.

8. Slowly relax the tensioner and remove the breaker bar.

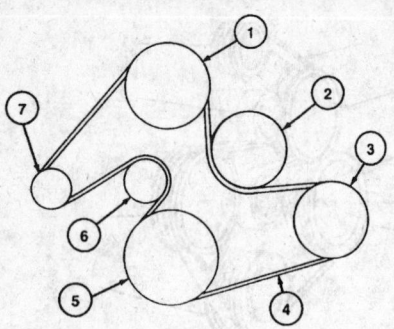

1. Power steering pump
2. Water pump
3. A/C compressor
4. Drive belt
5. Crankshaft pulley
6. Drive belt tensioner
7. Generator

314507

Drive belt pulley routing — 3.0L (VIN S) engine

Drive Belt Routing

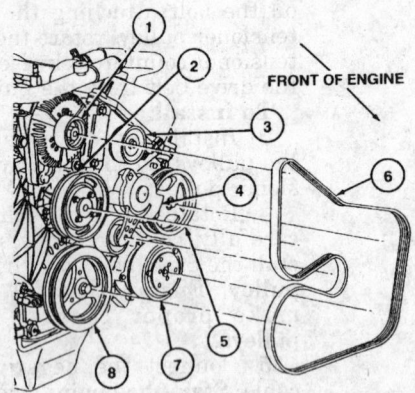

FRONT OF ENGINE

1. Generator
2. Water pump pulley
3. Belt idler pulley
4. Drive belt tensioner
5. Power steering pump
6. Drive belt
7. A/C compressor
8. Crankshaft pulley

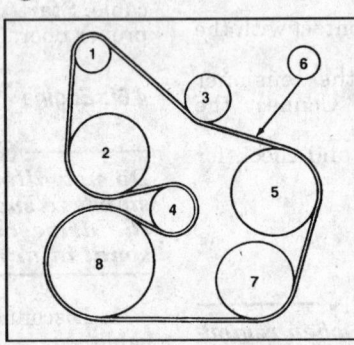

314543

Drive belt routing and tensioner location — 4.6L engine

9. Connect the negative battery cable. Start the engine and check for proper operation.

Starter

REMOVAL AND INSTALLATION

1993–95 Vehicles, Except 4.6L Engines

1. Disconnect the negative battery cable. Raise the vehicle and support it safely.
2. Disconnect the battery cable to the starter. If equipped with starter mounted solenoid, disconnect the push-on connector from the solenoid.

NOTE: To disconnect the hard-shell connector from the solenoid S terminal, grasp the plastic shell and pull off; do not pull on the wire. Pull straight off to prevent damage to the connector and S terminal.

3. Remove the starter bolts and the starter.

To install:

4. Position the starter to the engine and tighten the mounting bolts to 15–20 ft. lbs. (20–27 Nm).

5. Reconnect the electrical leads. Connect the negative battery cable.

1996–97 3.0L (VIN S) Engine

1. Disconnect the negative battery cable. Raise and safely support the vehicle.
2. Remove the starter solenoid terminal cover.
3. Disconnect the starter battery cable nut and cable and the S terminal wiring connector, both at the starter solenoid.

NOTE: To disconnect the hard-shell connector from the solenoid S terminal, grasp the plastic shell and pull off; do not pull on the wire. Pull straight off to prevent damage to the connector and S terminal.

4. Remove the ground wire retaining nut at the upper starter motor retaining bolt and move the ground wire aside.
5. Remove the upper and lower starter motor retaining bolts, then remove the starter motor from the vehicle.

To install:

6. Position the starter motor to the engine and install the 2 retaining bolts. Tighten the retaining bolts to 15–21 ft. lbs. (21–29 Nm).
7. Connect the ground cable to the upper starter motor retaining bolt and install the retaining nut. Tighten the nut to 15–21 ft. lbs. (21–29 Nm).
8. Connect the starter motor battery cable to the solenoid and install the retaining nut. Tighten the retaining nut to 80–123 inch lbs. (9–14 Nm).
9. Connect the S terminal connector being careful to push the connector on straight until it locks into position with a noticeable click or detent.
10. Install the starter solenoid terminal cover. Lower the vehicle.
11. Connect the negative battery cable. Check the starter motor for proper operation.

4.6L Engine

— CAUTION —
The air suspension switch, located in the left side of the luggage compartment, must be turned OFF before raising the vehicle. Failure to do so may result in unexpected inflation or deflation of the air springs which may result in shifting of the vehicle during service.

1. Turn the air suspension switch, located in the left side of the luggage compartment, to the **OFF** position.

2. Disconnect the negative battery cable. Raise and safely support the vehicle.

3. Remove the splash shield from the subframe and lower radiator support.

4. Remove the positive battery cable and S-terminal connector wiring retaining nuts from the starter solenoid. Remove the cables and position aside.

5. Remove the 2 starter motor bolts. Lift the starter motor and disengage the alignment pins from the transaxle case. Carefully remove the starter motor from the vehicle.

To install:

6. Position the starter motor to the transaxle case and install 2 retaining bolts. Tighten the bolts to 15–20 ft. lbs. (20–27 Nm).

7. Install the positive battery cable and S-terminal connector to the starter solenoid and install the terminal retaining nuts. Tighten the battery terminal nut to 80–124 inch lbs. (9–14 Nm). Tighten the S-terminal nut to 45–62 inch lbs. (5–7 Nm).

8. Install the red starter solenoid safety cap.

9. Install the splash shield to the subframe and lower radiator support. Tighten the retaining screws securely. Lower the vehicle.

10. Turn the air suspension switch to the **ON** position.

11. Connect the negative battery cable. Check the starter motor for proper operation.

CHASSIS ELECTRICAL

Blower Motor

REMOVAL AND INSTALLATION

1993–95 Taurus and Sable and 1993–94 Continental

1. Disconnect the negative battery cable. Open the glove compartment door, release the door retainers and lower the door.

2. Remove the screw attaching the recirculation duct support bracket to the instrument panel cowl.

3. Remove the vacuum connection to the recirculation door vacuum motor. Remove the screws attaching the recirculation duct to the heater assembly.

4. Remove the recirculation duct from the heater assembly, lowering the duct from between the instrument panel and the heater case.

5. Disconnect the blower motor electrical lead. Remove the blower motor wheel clip and remove the blower motor wheel.

6. Remove the blower motor mounting plate screws and remove the blower motor from the evaporator case.

To install:

7. Feed the A/C blower motor electrical connector through the A/C evaporator housing.

8. Position the blower motor into the A/C evaporator housing. Install the 4 retaining screws. Make sure the mounting seal is in place.

9. Place the blower motor wheel to the A/C blower motor shaft aligning the flat on the shaft with the flat on the blower motor wheel. Make sure the wheel is fully seated. Install the blower motor wheel retainer on the blower motor shaft.

10. Install the remaining components in the reverse order of removal.

11. Connect the negative battery cable, then check the operation of the blower motor.

1996–97 Taurus and Sable

1. Disconnect the negative battery cable.

2. Disengage the instrument panel insulator from the instrument panel.

3. Disconnect the wiring connector at the blower motor.

4. Remove the 3 screws retaining the blower motor to the evaporator housing and remove the blower motor and wheel assembly.

5. If required, separate the blower motor wheel from the blower motor by removing the retainer from the blower motor shaft and sliding the blower motor wheel off the blower motor shaft.

To install:

6. If the blower motor wheel was removed, align the flats on the inside diameter of the wheel hub with the flat surface on the blower motor shaft and slide the blower motor wheel onto the blower motor shaft. Install a new blower motor wheel retainer onto the shaft.

7. Install the blower motor and wheel assembly into the evaporator housing and firmly tighten the 3 retaining screws.

8. Attsch the connector to the blower motor. Install the insulator panel to the instrument panel.

9. Connect the negative battery cable. Check for proper blower motor operation.

1995–97 Continental

1. Disconnect the negative battery cable.

2. Remove 2 push pins retaining the instrument panel insulator to the instrument panel. Lower the instrument panel insulator enough to disengage the courtesy lamp and set the instrument panel insulator aside.

3. Remove the passenger door trim kick panel, pull back the carpet and reposition the main wiring harness connector and bracket to improve access to the blower motor.

4. Disconnect the wiring connector at the blower motor.

5. Remove 3 screws retaining the blower motor to the evaporator housing, and remove the blower motor and wheel assembly.

To install:

6. Install the blower motor and wheel assembly into the evaporator housing. Install and tighten 3 retaining screws. Connect the wiring connector to the blower motor.

7. Install the remaining components in the reverse order of removal.

8. Connect the negative battery cable. Check for proper blower motor operation in all speeds.

Windshield Wiper Motor

REMOVAL AND INSTALLATION

1993–95 Vehicles, Except 1995 Continental

1. Disconnect the negative battery cable. Detach the power lead from the motor.

2. Remove the left wiper arm.

3. Lift the water shield cover from the cowl on the passenger side.

4. Remove the linkage retaining clip from the operating arm on the motor.

5. Remove the attaching screws from the motor and bracket assembly and remove.

To install:

6. Position the windshield wiper motor and install the retaining bolts. Tighten to 60–80 inch lbs. (7–9 Nm).

7. Install the retaining clip on the windshield wiper mounting arm and pivot shaft.

8. Install the windshield wiper mounting arm and pivot shaft on the windshield wiper motor. Verify that the wiper mounting arm and the

pivot shaft is securely attached to the wiper motor.

9. Install the windshield wiper mounting arm and pivot shaft by pulling until the clip snaps in place.

10. Install the LH watershield.

11. Connect the battery negative cable. Check the windshield wiper motor operation through all modes.

12. Install the LH wiper blade.

1996–97 Taurus and Sable

1. Disconnect the negative battery cable. Mark the position of the wiper blade tips on the windshield using a removable marker.

2. Remove the wiper arm covers using a suitable prytool and remove both wiper arms by removing the 15mm attaching nuts.

3. Remove the 8 plastic retainers and the 6 clips holding the cowl vent screens to the to the inner panels. Turn the retainers 1/4 turn counterclockwise to remove. Remove the cowl vent screens.

4. Disconnect the electrical connector from the wiper motor and remove the 3 screws retaining the module assembly to the cowl.

5. Remove the windshield wiper module from the vehicle.

To install:

6. Position the windshield wiper motor module and install the retaining bolts. Tighten to 89–124 inch lbs. (10–14 Nm).

7. Attach the wiper motor connector. Install the cowl vent screens to the inner panels.

8. Install the windshield wiper pivot shaft, making sure the wiper blade tip is aligned with the reference mark previously made on the windshield.

9. Install the wiper arm pivot shaft nuts and tighten to 22–29 ft. lbs. (30–40 Nm). Install the wiper arm covers.

10. Connect the battery negative cable. Check the windshield wiper motor operation through all modes and check the blade positioning.

1995–97 Continental

1. Disconnect the negative battery cable. Detach the connector from the wiper motor.

2. Remove both windshield wiper arm and blades.

3. Lift the watershield cover from the cowl on the right side.

4. Remove the linkage retaining clip from the windshield wiper motor output arm by lifting the locking tab

up and pulling the clip away from the pin.

5. Remove the output arm from the motor. Remove 3 retaining bolts from the windshield wiper motor and bracket assembly and remove the wiper motor.

To install:

6. Position the windshield wiper motor assembly and install the retaining bolts. Tighten to 62–80 inch lbs. (7–9 Nm).

7. Install the wiper drive arm on the wiper motor.

8. Install the retaining clip on the windshield wiper drive arm. Make sure the windshield wiper mounting arm and pivot shaft is securely attached to the windshield wiper motor output arm. Make sure the clip snaps in place.

9. Install the left cowl vent screen. Tighten the screws to 27–44 inch lbs. (3–5 Nm). Connect the negative battery cable.

10. Check the wiper motor operation through all modes. Install the windshield wiper blades.

Headlight Switch

REMOVAL AND INSTALLATION

Taurus and Sable

1993–95 Vehicles

1. Disconnect the negative battery cable.

2. Pull off the headlight switch knob and remove the retaining nut.

3. Remove the instrument cluster finish panel as follows:

   a. Apply the parking brake.

   b. Remove the ignition lock cylinder.

   c. If equipped with a tilt column, tilt the column to the most downward position and remove the tilt lever.

   d. Remove the 4 bolts and opening cover from under the steering column.

   e. Remove the steering column trim shrouds. Disconnect all electrical connections from the steering column combination switch.

   f. Remove the 2 screws retaining the combination switch and remove the switch.

   g. Pull the gear shift lever to the full down position.

   h. Remove the cluster opening finish panel retaining screws. On Taurus there are 4 screws, on Sable there are 5 screws.

   i. Remove the finish panel by pulling it toward the driver to unsnap the snap-in retainers and disconnect the wiring from the switches, clock and warning lights.

4. Remove the 2 screws retaining the headlight switch, pull the switch out of the instrument panel and disconnect the electrical connector.

5. Installation is the reverse of the removal procedure.

1996–97 Vehicles

1. Disconnect the negative battery cable. Carefully pry the headlight switch and housing away from the instrument panel.

2. Push the release button down on the headlight switch and pull the switch assembly away from the instrument panel.

NOTE: If equipped with autolamps there will be 3 wire connectors on the headlight switch. If not equipped with autolamps, it will have 2 connectors.

3. Disconnect the wire connectors from the headlight switch and remove the switch.

4. Installation is the reverse of the removal procedure.

Continental

1993–94 Vehicles

1. Disconnect the negative battery cable. Gently pull off the headlight switch knob.

2. Snap out both mouldings, remove the 5 cluster opening finish panel retaining screws and the panel.

3. Remove the 2 screws retaining the headlight switch to the finish panel, disconnect the electrical connector and remove the switch.

4. Installation is the reverse of the removal procedure.

1995–97 Vehicles

1. Disconnect the negative battery cable.

2. Remove 3 screws from the bottom of the instrument panel steering column cover and remove the cover.

3. Remove 3 screws retaining the headlight switch assembly (lighting control module) to the instrument panel.

4. Carefully pull the switch assembly out and remove 3 wiring connectors from the switch.

5. Remove the headlight switch assembly from the vehicle.

6. Installation is the reverse of the removal procedure.

Combination Switch

REMOVAL AND INSTALLATION

1. Disconnect the negative battery cable.

2. If equipped with a tilt steering column, set the tilt column to its lowest position and remove the tilt lever.

3. Remove the ignition lock cylinder. Remove the steering column shroud screws and remove the upper and lower shrouds.

4. Remove 2 self tapping screws attaching the combination switch to the steering column and disengage the switch from the steering column tube casting.

5. Remove the wiring harness retainer, if equipped and disconnect the electrical connectors.

To install:

6. Connect the electrical connectors. Install the wiring harness retainer, if equipped.

7. Align the combination switch mounting holes with the corresponding holes in the steering column tube casting and install 2 self-tapping screws. Torque the screws to 17–26 inch lbs. (2–3 Nm).

8. Install the upper and lower steering column shroud and shroud retaining screws. Torque the screws to 6–10 inch lbs. (0.7–1.1 Nm).

9. Install the ignition lock cylinder and attach the tilt lever, if removed.

10. Connect the negative battery cable. Check the combination switch and the steering column for proper operation.

Ignition Lock Cylinder

REMOVAL AND INSTALLATION

—— **CAUTION** ——
The Supplemental Inflatable Restraint (SIR) system must be disarmed before performing service

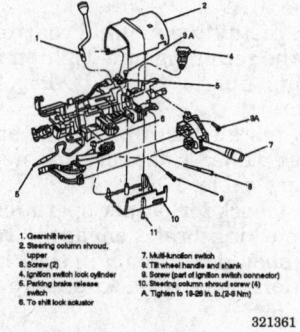

1. Gearshift lever
2. Steering column shroud, upper
3. Screw (2)
4. Ignition switch lock cylinder
5. Parking brake release switch
6. To shift lock actuator
7. Multi-function switch
8. Tilt wheel handle and shank
9. Screw (part of ignition switch connector)
10. Steering column shroud screw (4)
A. Tighten to 19–26 in. lb. (2–3 Nm)

321361

Combination switch typical installation

around SIR system components or SIR system wiring. Failure to do so may cause accidental deployment of the air bag, resulting in unnecessary SIR system repairs and/or personal injury.

Functional Lock Cylinder

1. Disconnect the negative battery cable. Turn the lock cylinder key to the **RUN** position.

2. Using a ⅛ inch (3mm) diameter wire pin or a small drift, depress the lock cylinder retaining pin through the access hole, while pulling out on the lock cylinder to remove it from the steering column housing.

To install:

3. Install the lock cylinder by turning it to the **RUN** position and depressing the retaining pin. Insert the lock cylinder into its housing in the steering column. Make sure the cylinder is fully seated and aligned in the interlocking washer before turning the key to the **OFF** position. This will permit the cylinder retaining pin to extend into the cylinder housing hole.

4. Rotate the lock cylinder using the lock cylinder key, to ensure correct mechanical operation in all positions.

5. Connect the negative battery cable. Verify proper lock cylinder operation.

Non-Functional Lock Cylinder

1. Disconnect the negative battery cable. Remove the steering wheel.

—— **CAUTION** ——
When carrying a live air bag, make sure the bag and trim cover are pointed away from the body. In the unlikely event of an accidental deployment, the bag will then deploy with minimal chance of injury. In addition, when placing a live air bag on a bench or other surface, always face the bag and trim cover up, away from the surface. This will reduce the motion of the unit if it is accidentally deployed.

2. Using channel lock or vise grip pliers, twist the lock cylinder cap until it separates from the lock cylinder.

3. Using a ⅜ inch (3mm) diameter drill bit, drill down the middle of the ignition lock key slot approximately 1¾ inch (44mm) until the lock cylinder breaks loose from the breakaway base of the lock cylinder. Remove the lock cylinder and drill shavings from

the lock cylinder housing in the steering column.

NOTE: It may be helpful to drill a small pilot hole before using a ⅜ inch drill.

4. Remove the retainer, washer, ignition switch and actuator. Thoroughly clean all drill shavings and other foreign materials from the housing.

5. Inspect the lock cylinder housing for damage from the removal operation. If the housing is damaged, it must be replaced.

To install:

6. Position the lock drive gear in the same position as that noted during the removal procedure. The position of the lock drive gear is correct if the last tooth on the drive gear is meshed with the last tooth on the rack.

7. Position the bearing retainer in the ignition switch lock cylinder housing. Insert the tip of a screwdriver or similar tool into the double-D slot of bearing and then rotate 90 degrees.

8. Press the blue plastic bearing retainer into the lock cylinder housing. Make sure the retainer is in its original position.

9. Line up the flats of the drive gear with the flats of the washer by pulling down on the column lock actuator.

10. Install the lock cylinder by turning it to the **RUN** position and depressing the retaining pin. Insert the lock cylinder into its housing. Make sure the cylinder is fully seated and aligned in the interlocking washer before turning the key to the **OFF** position. This will permit the cylinder retaining pin to extend into the cylinder housing hole.

11. Install the steering wheel. Connect the negative battery cable.

12. Check that the vehicle will start in **P** and **N** only. Also, check to ensure that the start circuit cannot be actuated in **D** or **R** positions and the column is locked in the **LOCK** position.

13. If equipped with air bag(s), prove out the air bag system by turning the ignition key to the **RUN** position and visually monitoring the air bag indicator lamp in the instrument cluster. The indicator lamp should illuminate for approximately 6 seconds and then turn OFF. If the indicator lamp does not illuminate, stays ON, or flashes at any time, a fault has been detected by the air bag diagnostic monitor.

Ignition Switch

REMOVAL AND INSTALLATION

1993–95 Vehicles, Except 1995 Continental

1. Disconnect the negative battery cable. Remove the steering column shroud by removing the self-tapping screws. Remove the tilt lever, if equipped.
2. Remove the instrument panel lower steering column cover.
3. Disconnect the ignition switch electrical connector.
4. Turn the ignition key lock cylinder to the **RUN** position.
5. Remove the 2 screws attaching the ignition switch and disengage the switch from the actuator pin.

To install:
6. Adjust the ignition switch by sliding the carrier to the switch **RUN** position. A new replacement switch assembly will already be set in the **RUN** position.
7. Make sure the ignition key lock cylinder is in the **RUN** position. The **RUN** position is achieved by rotating the key lock cylinder approximately 90 degrees from the lock position.
8. Install the ignition switch into the actuator pin. It may be necessary to move the switch slightly back and forth to align the switch mounting holes with the column lock housing threaded holes.
9. Install the attaching screws and tighten to 50–69 inch lbs. (5.6–7.9 Nm). Connect the electrical connector to the ignition switch.
10. Connect the negative battery cable. Check the ignition switch for proper function, including **START** and **ACC** positions. Make sure the

column is locked with the switch in the **LOCK** position.
11. Install the instrument panel lower steering column cover, the steering column trim shrouds and the tilt lever, if equipped.

1996–97 Taurus and Sable 1995–97 Continental

1. Disconnect the negative battery cable. Remove the I/P lower steering column cover.
2. Disconnect the ignition switch electrical connector.
3. Turn the ignition key lock cylinder to the **RUN** position.
4. Remove the 2 screws attaching the ignition switch and disengage the switch from the actuator pin.

To install:
5. Adjust the ignition switch by sliding the carrier to the switch **RUN** position. A new replacement switch assembly will already be set in the **RUN** position.
6. Make sure the ignition key lock cylinder is in the **RUN** position. The **RUN** position is achieved by rotating the key lock cylinder approximately 90 degrees from the lock position.
7. Install the ignition switch into the actuator pin. It may be necessary to move the switch slightly back and forth to align the switch mounting holes with the column lock housing threaded holes.
8. Install the attaching screws and tighten to 50–69 inch lbs. (5.6–7.9 Nm). Attach the switch connector.
9. Connect the negative battery cable. Check the ignition switch for proper function, including **START** and **ACCESSORY** positions. Make sure the column is locked with the switch in the **LOCK** position.

10. Install the instrument panel lower steering column cover.

Manual Lever Position Sensor (MLPS)

REMOVAL AND INSTALLATION

NOTE: The Manual Lever Position Sensor (MLPS) also known as the Transmission Range (TR) sensor, is located on the automatic transaxle and is responsible for providing information to the PCM on manual control lever positioning. It also permits the engine to start only in P and N and activates the backup lamps in the R position.

1. Disconnect the negative battery cable. Place the manual control lever in the **N** position.
2. Remove the air cleaner and the air cleaner outlet tube. Detach the connector from the MLPS.

NOTE: Do not pry on the electrical connector. Press the button and pull up on the harness connector to prevent damage to the connector and switch.

3. Remove the nut retaining the manual control lever to the manual control lever shaft.
4. Remove the manual control lever. Remove the 2 MLPS retaining bolts and remove the sensor.

To install:
5. Make sure the manual control lever is in the **N** position.
6. Install the MLPS and loosely install the 2 retaining bolts.
7. Align the MLPS using MLPS alignment tool T94P-70010-AH or equivalent. Tighten the 2 retaining bolts to 7–9 ft. lbs. (9–12 Nm).
8. Connect the electrical connector to the MLPS.
9. Install the manual control lever and the retaining nut. Tighten the retaining nut to 12–16 ft. lbs. (16–22 Nm).
10. Install the air cleaner and the outlet tube. Connect the negative battery cable.
11. Check for proper operation with the parking brake engaged. The engine should start only in the **N** and **P** positions. Verify backup lamp operation in **R**.

1. Gearshift lever
2. Steering column shroud
3. Ignition switch lock cylinder
4. Parking brake release switch
5. Screw(2)
6. Multi-function switch
7. Tilt wheel handle and shank
8. Screw(4)
9. Steering column shroud
10. Ignition switch
11. Parking brake release vacuum hose
12. Screw(2)
13. Gear shift lever pin
14. Parking brake release vacuum hose extension

A. 17-26 in. lb. (2-3 Nm)
B. 30-44 in. lb. (3.5-5.0 Nm)
C. 6-10 in. lb. (0.8-1.18 Nm)

235896

Ignition switch and related components — Taurus and Sable

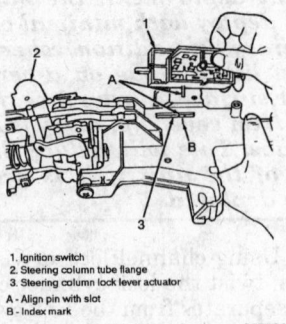

1. Ignition switch
2. Steering column tube flange
3. Steering column lever actuator
A - Align pin with slot
B - Index mark

305408

Ignition switch installation — 1995–97 Continental with 4.6L engine

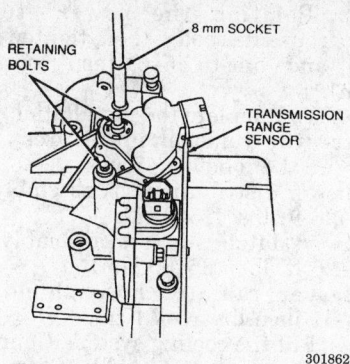

Manual lever position sensor retaining bolts

301862

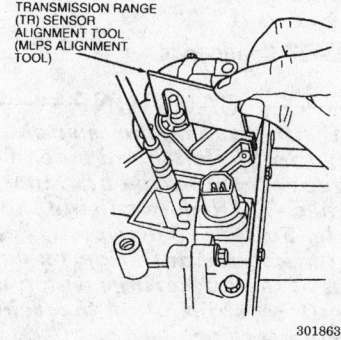

301863

Manual lever position sensor alignment tool

Powertrain Control Module

REMOVAL AND INSTALLATION

──────── **WARNING** ────────
Never disconnect a Powertrain Control Module (PCM) with the battery connected. Be sure to wear a grounding device when removing or installing a PCM to prevent damage to the unit due to static electricity.

1993–95 Vehicles, Except 1995 Continental

1. Disconnect the negative battery cable.
2. Wearing a grounding device, loosen the engine control sensor wiring harness connector retaining bolt and carefully disconnect the harness connector from the PCM.
3. Remove the PCM bracket clip from the module bracket and remove the PCM.
To install:
4. Wearing a grounding device, install the PCM to the mounting bracket and secure the bracket clip.

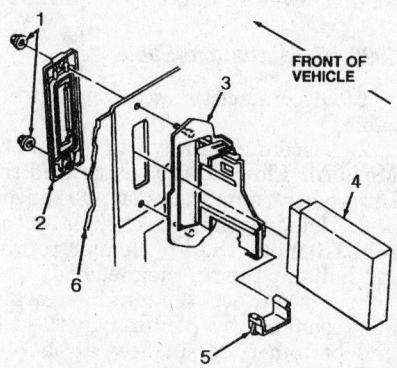

1 Nut (2 req'd)
2 Powertrain control module seal
3 Powertrain control module bracket
4 Powertrain control module
5 Powertrain control module bracket clip
6 Dash panel

327470

Powertrain control module mounting — 1993–95 vehicles, except 1995 Continental

5. Carefully fit the engine control sensor wiring harness connector to the PCM and secure with the retaining bolt. Tighten the retaining bolt to 32 inch lbs. (3.7 Nm).
6. Connect the negative battery cable. Check for proper operation.

1996–97 Taurus and Sable 1995–97 Continental

NOTE: Carefully inspect the harness and the PCM terminal connections. Check that all terminals are clean and not corroded. Also check that no terminals are bent or deformed. Be sure the connectors are fully seated; a poorly seated connector can cause intermittent operation or poor engine performance.

1. Disconnect the negative battery cable.
2. Remove the engine appearance cover or cowl deflector, as applicable.
3. For Taurus and Sable, remove the screw retaining the engine control ground cable to the dash panel.
4. Install a grounding device to prevent static charge.
5. Loosen the engine control sensor wiring harness connector retaining bolt and then disconnect the harness connector from the PCM.

6. Remove the control module insulator from the dash panel.
7. Remove the PCM from the vehicle.
To install:
8. Push the PCM firmly into the control module bracket.
9. Install the control module insulator.
10. Reconnect the powertrain control module wiring harness connector to the control module. Tighten the bolt to 79–107 inch lbs. (8.9–12.1 Nm).
11. For Taurus and Sable, install the engine control ground cable to the dash panel. Securely tighten the retaining screw.
12. Remove the grounding device.
13. Install the engine appearance cover or cowl deflector, as applicable.
14. Connect the negative battery cable Road test the vehicle and check for proper operation.

NOTE: When the battery has been disconnected, abnormal driving symptoms may occur until the PCM relearns its strategy. The vehicle may need to be driven 10 miles (18 km) or more to relearn its strategy.

ENGINE COOLING

Radiator

REMOVAL AND INSTALLATION

1993–95 Vehicles, Except 1995 Continental

──────── **CAUTION** ────────
Never remove the radiator cap under any conditions while the engine is operating. Failure to follow these instructions could result in personal injury and/or damage to the cooling system or engine.

1. Disconnect the negative battery cable. Position a suitable drain pan under the radiator.
2. Drain the cooling system by removing the radiator cap and opening the drain located at the lower rear corner of the radiator inlet tank.
3. Remove the rubber overflow tube from the coolant recovery bottle and detach it from the radiator. Disconnect the tube from the radiator and remove the recovery bottle.

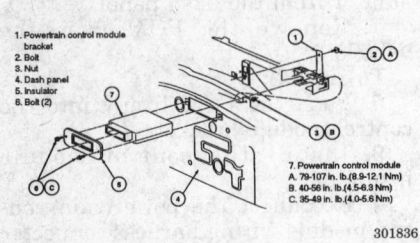

1. Powertrain control module bracket
2. Bolt
3. Nut
4. Dash panel
5. Insulator
6. Bolt (2)

7. Powertrain control module
A. 79–107 in. lb (8.9–12.1 Nm)
B. 40–56 in. lb. (4.5–6.3 Nm)
C. 35–49 in. lb. (4.0–5.6 Nm)

301836

Powertrain control module mounting — 1996–97 Taurus and Sable and 1995–97 Continental

4. On Taurus SHO, disconnect the tube from the radiator and remove the recovery bottle.

5. Remove 2 upper shroud retaining screws and lift the shroud out of the lower retaining clips.

6. Disconnect the electric cooling fan motor wires and remove the fan and shroud assembly.

7. Loosen the upper and lower hose clamps at the radiator and remove the hoses from the radiator connectors.

8. If equipped with an automatic transaxle, disconnect the transaxle oil cooling lines from the radiator fittings using tool T82L-9500-AH or equivalent.

9. Remove the upper radiator retaining screws or upper radiator support brackets, as required. Remove 2 radiator upper retaining screws.

10. Tilt the radiator rearward approximately 1 inch (25mm) and lift it directly upward, clear of the radiator support.

11. Remove the radiator lower support rubber pads, if pad replacement is necessary.

To install:

12. Position the radiator lower support rubber pads to the lower support, if removed.

13. Position the radiator into the engine compartment and to the radiator support. Insert the moulded pins at the bottom of each tank through the slotted holes in the lower support rubber pads.

14. To complete the installation, reverse the removal procedures. Position the hose on the radiator connector so the index arrow on the hose is in line with the mark on the connector. Tighten the screw clamps to 20–30 inch lbs. (2.2–3.4 Nm).

15. If equipped with automatic transaxle, connect the transaxle cooler lines using oil resistant pipe sealer.

16. Refill the cooling system. If the coolant is being replaced, refill with a 50/50 mixture of water and antifreeze. Connect the negative battery cable. Operate the engine for 15 minutes and check for leaks. Check the coolant level and add, as required.

1996–97 Taurus and Sable

1. Disconnect the negative battery cable.

2. Disconnect the wiring from the Mass Air Flow (MAF) sensor and the Intake Air Temperature (IAT) sensor.

3. Remove the air cleaner outlet tube. Remove the air cleaner retaining bolts at the air cleaner body.

4. Disconnect the engine intake air resonator by pushing in the top and bottom tube surfaces at the engine air cleaner and pulling the air cleaner outward. Lift the air cleaner up and out of the engine compartment.

5. Remove the battery and the battery tray. Raise and safely support the vehicle.

6. Drain the cooling system and remove 4 bolts retaining the lower radiator hose shield and remove the lower radiator hose shield.

7. Disconnect the lower radiator hose from the radiator.

8. Disconnect the lower transaxle oil cooler tube from the radiator using Fuel Line Disconnect tool T90T-9550-S (⅜ inch), or equivalent.

9. Remove the lower radiator mounts and lower the vehicle.

10. Open the hood and remove the 4 bolts retaining the hood latch support. Position the latch aside.

11. Remove the bumper cover and the upper radiator support.

12. Remove the 6 screws retaining the engine sensor wiring and position aside.

13. Disconnect the cooling fans from the from the control sensor wiring.

14. Remove 1 nut from the power steering/transaxle cooler bracket.

15. Remove 3 bolts and 1 stud bolt from the A/C condenser core; then, position the condenser forward and support with mechanics wire.

16. Disconnect the upper transaxle cooler tube from the radiator. Remove both lower radiator brackets and lower the radiator 2 inches. Tilt the radiator forward and remove the radiator.

To install:

17. Position the radiator in the fan shroud. Install both lower radiator brackets and tighten the nuts to 71–106 inch lbs. (8–12 Nm).

18. Position the condenser core to the radiator; then, tighten 2 bolts and 1 stud to 45–61 inch lbs. (5–7 Nm).

19. Position the power steering/transaxle cooler; then, tighten the bolt and nut to 45–61 inch lbs. (5–7 Nm).

20. To complete the installation, reverse the removal procedures and torque the following items:
Power distribution box assembly to 18–26 ft. lbs. (25–35 Nm)
Hood latch support assembly to 18–26 ft. lbs. (25–35 Nm).
Lower radiator hose shield to 45–61 inch lbs. (5–7 Nm).

21. Fill the cooling system. Connect the negative battery cable. Start the engine and allow it to reach normal operating temperature. Check for leaks and proper cooling system operation.

1995–97 Continental

― **CAUTION** ―
The air suspension switch, located in the left–hand side of the luggage compartment, must be turned OFF before raising the vehicle. Failure to do so may result in unexpected inflation or deflation of the air springs which may result in shifting of the vehicle during service.

1. Turn the air suspension switch, located in the left side of the luggage compartment, to the **OFF** position.

2. Disconnect the negative battery cable. Drain the engine cooling system.

3. Remove the engine air cleaner assembly. Remove the upper radiator hose from the water bypass tube.

4. Remove the radiator overflow hose from the radiator and fan shroud.

5. Loosen and remove the upper transaxle oil cooler tube while holding the radiator connector with a backup wrench at the inlet fitting.

6. Remove the nuts retaining the A/C condenser core to the radiator.

7. Disconnect the wiring from the cooling fan motors and the Constant Control Relay Module (CCRM).

8. Raise and safely support the vehicle. Remove the splash shield from the lower radiator support and subframe. Remove the lower radiator hose from the radiator.

9. Loosen and remove the lower transaxle oil cooler tube while holding the radiator connector with a backup wrench at the outlet fitting.

10. Remove the retaining screws for the power steering/transaxle oil cooler and position the cooler aside.

11. Support the fan shroud, radiator and A/C condenser with a suitable

jackstand. Remove the lower radiator support.

12. Remove the jackstand and carefully remove the radiator and fan shroud.

13. Remove 2 retaining bolts retaining the fan shroud to the top of the radiator and remove the shroud from the radiator.

14. Remove the upper radiator hose from the radiator.

To install:

15. Install upper radiator hose and shroud to the radiator and tighten the bolts to 24–48 inch lbs. (2.7–5.4 Nm).

16. Carefully lift the radiator and fan shroud into position and support with a suitable jackstand.

17. To complete the installation, reverse the removal procedures and torque the following items:

Lower oil cooler tube to 18–21 ft. lbs. (24.4–28.5 Nm).

Lower radiator support assembly bolts to 71–97 inch lbs. (8–11 Nm).

Upper oil cooler tube to 18–21 ft. lbs. (24.4–28.5 Nm).

18. Connect the negative battery cable. Turn the air suspension switch to the **ON** position.

19. Refill the cooling system. If the coolant is being replaced, refill with a 50/50 mixture of water and antifreeze. Operate the engine for 15 minutes and check for leaks. Check the coolant level and add, as required.

Water Pump

REMOVAL AND INSTALLATION
CAUTION
Do not remove the radiator cap or open the cooling system until the engine has cooled. Removing the radiator cap or opening the cooling system prior to the engine cooling could cause severe burns from scalding engine coolant.

3.0L (VIN Y) and 3.2L SHO Engines

1. Disconnect the battery cables, negative cable first and remove the battery and the battery tray.

2. Drain the cooling system and remove the accessory drive belts.

3. Remove the left hand drive belt idler pulley.

4. Disconnect the electrical connector from the ignition module and ground strap.

5. Loosen the clamps on the upper intake connector tube, remove the retaining bolts and remove the connector tube.

6. Remove the upper outer timing belt cover.

7. Raise and safely support the vehicle. Remove the right wheel and tire assembly. Remove the splash shield.

8. Remove the crankshaft pulley using a suitable puller.

9. Remove the lower outer timing belt cover. Disconnect the crankshaft position sensor wire harness and move it aside.

10. Remove the center timing belt cover. Remove the right hand drive belt tensioner idler pulley.

11. Remove the water pump attaching bolts and remove the water pump.

To install:

12. Lightly oil all bolt threads before installation. Clean gasket surfaces on pump and engine block.

13. Position a new gasket on the water pump and use a gasket sealer to hold the gasket in place. Install water pump and retaining bolts. Tighten to 12–17 ft. lbs. (16–23 Nm).

14. To complete the installation, reverse the removal procedures.

15. Tighten the crankshaft pulley retaining bolt to 112–127 ft. lbs. (152–172 Nm). Tighten the upper intake connector tube retaining bolts to 11–17 ft. lbs. (15–23 Nm).

16. Reconnect the battery cables.

17. Make sure draincock is closed and refill cooling system. Run the engine and check for leaks.

3.0L (VIN U and 1) Engines

1. Disconnect the negative battery cable. Allow the engine to cool. Remove the radiator cap and drain the cooling system.

2. Loosen 4 retaining bolts securing the water pump pulley to the water pump hub.

3. Remove the accessory drive belts. Remove the idler pulley or automatic tensioner, as required.

4. Disconnect and remove the heater hose from the water pump.

5. Remove the engine control sensor wiring from the locating stud bolt, if equipped.

6. Remove the water pump-to-engine retaining bolts and lift the water pump and pulley up and out of the vehicle.

To install:

7. Clean the gasket surfaces on the water pump and engine front cover. Install a new gasket on the water pump using gasket adhesive.

8. Place the water pump in position on the engine with the pulley and 4 retaining bolts loosely installed on the hub.

9. Lightly oil the retaining bolts except those requiring sealant and install the bolts in the water pump housing. Tighten the bolts designated by reference No. 1 to 15–22 ft. lbs. (20–30 Nm) and the bolts designated by reference No. 2 to 6–8 ft. lbs. (8–12 Nm).

NOTE: The bolts are of different lengths and must be installed in the correct locations.

10. Install the remaining components in the reverse order of removal. Tighten the water pump pulley bolts to 15–22 ft. lbs. (20–30 Nm).

11. Fill the cooling system. Connect the negative battery cable.

12. Start the engine and allow it to reach normal operating temperature. Check for leaks and proper operation.

3.0L (VIN S) Engine

1. Disconnect the negative battery cable. Drain the engine cooling system.

2. Remove the water pump drive belt. Remove the radiator and heater hoses from the water pump.

3. Remove the 4 nuts securing the water pump to the engine and remove the water pump.

To install:

4. Clean the water pump to engine gasket sealing surfaces.

5. Install the water pump using a new gasket and install the 4 retaining nuts. Torque the retaining nuts to 15–22 ft. lbs. (20–30 Nm).

6. Install the remaining components in the reverse order of removal.

7. Fill the engine cooling system, then connect the negative battery cable.

8. Start the engine and allow it to reach normal operating temperature, then check for coolant leaks and proper engine operation.

3.8L Taurus and Sable

1. Disconnect the negative battery cable. Allow the engine to cool before proceeding.

2. Remove the radiator cap and drain the cooling system by opening the radiator draincock.

3. Support the engine using engine support bar D88L-6000-A or equivalent. Remove the lower nut on both right engine mounts. Raise the engine.

4. Loosen the accessory drive belt idler. Remove the drive belt and water pump pulley. Remove the air suspension pump, if equipped.

5. Remove the power steering pump mounting bracket attaching bolts. Leaving hoses connected, place

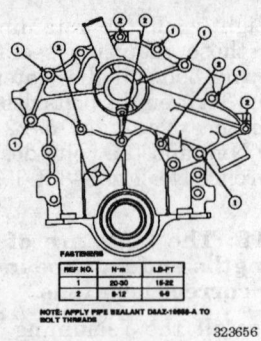

Water pump retaining bolt location and torque specifications — 3.0L (VIN U and 1) engines

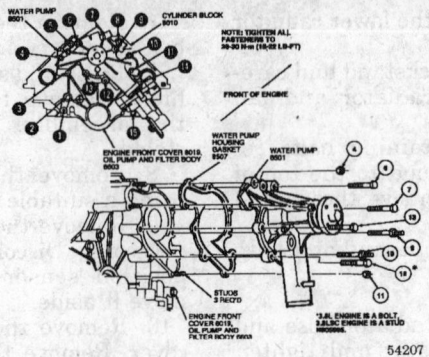

Water pump installation — 3.8L Engines

pump/bracket assembly aside in a position to prevent fluid from leaking out.

6. If equipped with air conditioning, remove the compressor front support bracket. Leave the compressor in place.

7. Disconnect coolant bypass and heater hoses at the water pump.

8. Remove the water pump-to-engine block attaching bolts and remove the pump from the vehicle. Discard the gasket and replace with new.

To install:

9. Lightly oil all bolt and stud threads before installation, except those that require sealant. Thoroughly clean the water pump and front cover gasket contact surfaces.

10. Apply a coating of contact adhesive to both surfaces of the new gasket. Position a new gasket on water pump sealing surface.

11. Position water pump on the front cover and install attaching bolts. Tighten to 15–22 ft. lbs. (20–30 Nm).

12. Install the remaining components in the reverse order of removal.

13. Remove the engine support bar. Fill cooling system to the proper level. Start engine and check for coolant leaks.

1993–94 3.8L Continental

1. Disconnect the negative battery cable. Drain the cooling system, remove the cooling fan and the shroud.

2. Release the belt tensioner and remove the accessory drive belt.

3. Remove the bolts retaining the water pump pulley to the water pump and remove the pulley.

4. Remove power steering pump pulley and remove water pump to power steering pump brace.

5. Disconnect water bypass and heater hoses at water pump; then, the lower radiator hose.

6. Remove the water pump-to the engine bolts assembly and remove the water pump.

––––––– **CAUTION** –––––––
If using a prying device do not mar mating surfaces.

To install:

7. Clean the sealing surfaces of the water pump and engine block. Position a new gasket on water pump housing using gasket adhesive.

8. Install the water pump and tighten the retaining bolts to 15–22 ft. lbs. (20–30 Nm).

NOTE: Pipe sealant should be applied to No. 1 mounting bolt.

9. Install the remaining components in the reverse order of removal. Tighten the water pump pulley bolts to 15–22 ft. lbs. (20–30 Nm).

10. Connect the negative battery cable and fill the cooling system. Run the engine and check for leaks.

4.6L Engine

1. Disconnect the negative battery cable. Drain the engine cooling system.

2. Remove the coolant recovery reservoir assembly.

3. Loosen 4 bolts retaining the water pump pulley to the water pump.

4. Release the drive belt tensioner and remove the drive belt.

5. Remove 4 bolts retaining the water pump pulley to the water pump and remove the pulley.

6. Remove 4 bolts retaining the water pump to the cylinder block and remove the water pump.

To install:

7. Replace the water pump O-ring and clean the sealing surface of the cylinder block and the water pump.

8. Lubricate the water pump O-ring with fresh coolant and install the water pump into position. Make sure the water pump is fully seated. Install the 4 water pump retaining bolts and tighten to 15–22 ft. lbs. (20–30 Nm).

9. Install the water pump pulley on the water pump with 4 retaining bolts. Tighten to 15–22 ft. lbs. (20–30 Nm).

10. Install the remaining components in the reverse order of removal.

11. Fill the cooling system to the proper level. Connect the negative battery cable, then start the engine and check for coolant leaks.

Thermostat

REMOVAL AND INSTALLATION
––––––– **CAUTION** –––––––
Do not open the radiator cap, draincock or any cooling system component unless the engine has cooled. Failure to allow the engine to cool could cause severe burns from the engine coolant.

3.0L (VIN Y) and 3.2L SHO Engines

1. Disconnect the negative battery cable. Allow the engine to cool before proceeding.

2. Place a suitable drain pan below the radiator. Remove the radiator cap and open the draincock. Partially drain the cooling system and then close the draincock.

3. Remove the air cleaner tube. Disconnect the hose from the water outlet tube.

4. Remove the water outlet tube nuts. Remove the thermostat and seal from the water outlet housing.

To install:

5. Install the seal around the outer rim of the thermostat and install the thermostat into the water outlet housing. Align the jiggle valve

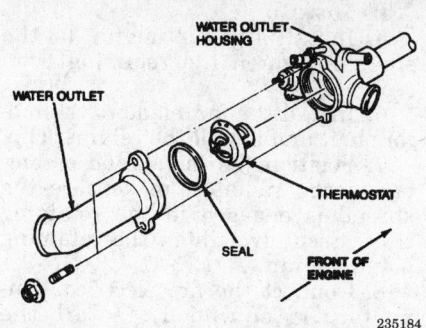

Thermostat and related components — SHO engines

of the thermostat with the upper bolt on the water outlet housing.

6. Install the water outlet tube. Tighten the 2 retaining nuts to 5–8 ft. lbs. (7–11 Nm). Install the air cleaner tube.

7. Refill the cooling system. Connect the negative battery cable. Start the engine and check for leaks. Check the coolant level and add as necessary.

3.0L (VIN 1 and U) Engines

1. Disconnect the negative battery cable. Drain the engine cooling system.

2. Remove the upper radiator hose from the thermostat housing.

3. Remove the thermostat housing bolts. Remove the housing and the thermostat as an assembly.
To install:
4. Make sure all sealing surfaces are free of old gasket material. Install the thermostat into the housing.

5. Make sure the bolt threads are clean. Position a new gasket onto the housing.

6. Install the thermostat assembly and tighten the 3 bolts to 9 ft. lbs. (12

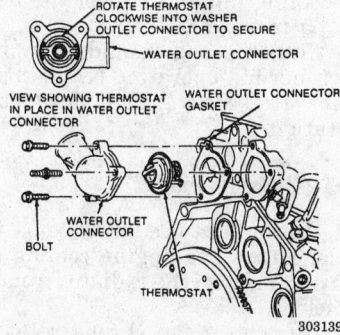

Thermostat and related components — 3.0L (VIN 1 and U) engines

Nm). Install the upper radiator hose and tighten the clamp.

NOTE: Make sure the hose clamps are beyond the bead and placed in the center of the clamping surface of the connection. Any used hose clamps must be replaced with a new clamp to ensure proper sealing at the connection. Tighten the hose clamps to 20–30 inch lbs. (2.2–3.4 Nm).

7. Fill and bleed the cooling system. Connect the negative battery cable.

8. Start the engine and allow it to reach normal operating temperature. Check for coolant leaks and proper cooling system operation. Check the coolant level and add as required.

3.0L (VIN S) Engine

1. Disconnect the negative battery cable. Drain the engine cooling system to a level below the thermostat housing.

2. Raise and safely support the vehicle. Disconnect the lower radiator hose.

3. Remove the thermostat housing bolts and the housing. Remove the O-ring and the thermostat from the housing.
To install:
4. Clean all mating surfaces.

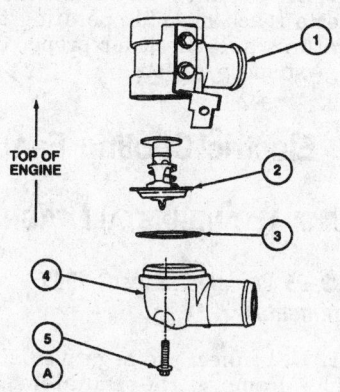

1. Upper water inlet housing
2. Water thermostat
3. O-ring seals
4. Lower water inlet housing
5. Bolt (2)
A. 71–106 in. lb.(8–12 Nm)

303181

Thermostat and related components — 3.0L (VIN S) engine

5. Install a new thermostat and O-ring and position the thermostat housings. Alternately tighten the retaining bolts to 71–106 inch lbs. (8–12 Nm).

6. Install the lower radiator hose to the water inlet connection. Compress the hose clamp tongs with suitable pliers and install the clamp at least 0.125 inch (3.2mm) from the end of the hose.

7. Lower the vehicle. Connect the negative battery cable.

8. Fill and bleed the engine cooling system. Start the engine and allow it to reach normal operating temperature. Check for leaks and proper cooling system operation.

3.8L Engine

1. Disconnect the negative battery cable. Allow the engine to cool before proceeding.

2. Drain the engine cooling system to a level below the thermostat housing. Loosen the top radiator hose clamp at the radiator and remove the hose from the radiator connection.

3. Remove 2 water outlet connection (thermostat housing) retaining bolts. Loosen and lift the water outlet and hose clear of the engine.

NOTE: Do not pry the housing off or damage to the housing or intake manifold may result.

4. Remove the thermostat from the housing by rotating it counterclockwise until the thermostat becomes free to remove.
To install:
5. Clean the gasket mating surfaces.

6. Install the thermostat into the water outlet connection by rotating it clockwise until the engaging ramps on the thermostat are secure. Make sure the bolt threads are clean. Install the water outlet connection on the intake manifold with a new gasket. Tighten 2 mounting bolts to 15–22 ft. lbs. (20–30 Nm).

7. Install the top hose to the radiator and install a new clamp. Tighten the hose clamp to 20–30 inch lbs. (2.2–3.4 Nm).

NOTE: Make sure the hose clamps are beyond the bead and placed in the center of the clamping surface of the connection. Any used hose clamps must be replaced with a new clamp to ensure proper sealing at the connection.

8. Refill the cooling system. Connect the negative battery cable.

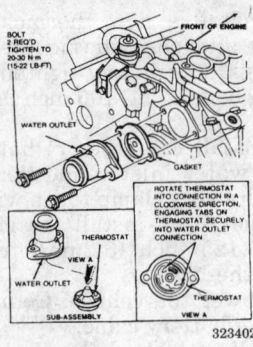

Thermostat and related
components — 3.8L engine

9. Start the engine and allow it to
reach normal operation temperature.
Check for coolant leaks and proper
cooling system operation. Check the
coolant level and add as required.

4.6L Engine

CAUTION

The air suspension switch, located in the left–hand side of the luggage compartment, must be turned OFF before raising the vehicle. Failure to do so may result in unexpected inflation or deflation of the air springs which may result in shifting of the vehicle during service.

1. Disconnect the negative battery
cable.
2. Turn the air suspension switch,
located in the left side of the luggage
compartment, to the OFF position.
3. Drain the cooling system to a
level below the thermostat.
4. Raise and safely support the
vehicle.
5. Disconnect the lower radiator
hose and the radiator coolant recovery reservoir hose from the water inlet connection.
6. Remove water inlet connection
bolts and the connection.
7. Remove the O-ring seal and the
thermostat from the thermostat
housing. Inspect the O-ring for damage and replace if necessary.

To install:
8. Install the thermostat, O-ring,
thermostat housing, and water inlet
connection; torque bolts to 15–22 ft.
lbs. (20–30 Nm).
9. Connect the lower radiator hose
and the radiator coolant recovery reservoir hose to the water inlet
connection.
10. Lower the vehicle. Connect the
negative battery cable.

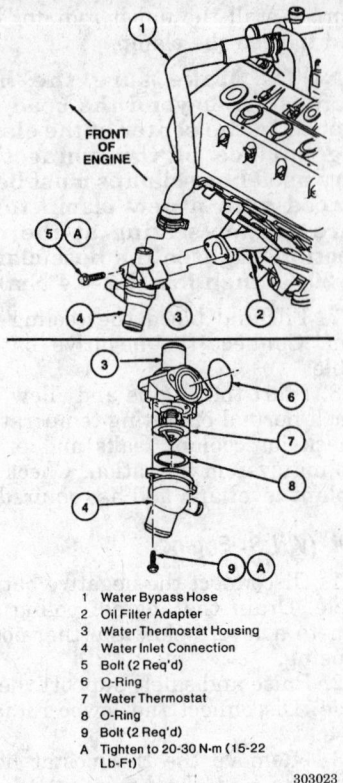

1 Water Bypass Hose
2 Oil Filter Adapter
3 Water Thermostat Housing
4 Water Inlet Connection
5 Bolt (2 Req'd)
6 O-Ring
7 Water Thermostat
8 O-Ring
9 Bolt (2 Req'd)
A Tighten to 20-30 N·m (15-22 Lb-Ft)

Thermostat and housing — 4.6L
engine

11. Turn the air suspension switch
to the **ON** position.
12. Fill the cooling system to the
proper level. Start the engine and
check for coolant leaks. Allow the engine to reach normal operating temperatures and check for proper cooling system operation.

Electric Cooling Fan

REMOVAL AND INSTALLATION

1993–95 Vehicles, Except 1995 Continental

1. Disconnect the negative battery
cable. Remove the radiator sight
shield.
2. If equipped, disconnect the electrical connector and remove the Constant Control Relay Module (CCRM) located on the radiator support.
3. Detach the fan electrical connector. If necessary, remove the air bag crash sensor. Unbolt the fan/shroud assembly from the radiator and remove.
4. Remove the retainer clip and
the fan blade from the motor shaft
and unbolt the fan motor from the
shroud.

To install:
5. Install the fan motor to the
shroud. Tighten the retaining bolts
securely.
6. Install the fan blade to the motor shaft and install the retainer clip.
7. Position the fan/shroud assembly to the radiator, make sure the shroud is engaged in the retaining clips. Securely tighten the retaining bolts and nut.
8. Connect the fan electrical connector. If equipped, install the CCRM.
9. Connect the negative battery
cable. Check the cooling fan for
proper operation.

1996–97 Taurus and Sable

NOTE: It is not necessary to remove the fan shroud to service the cooling fan and blade assembly.

1. Disconnect the negative battery
cable. Disconnect the cooling fan(s)
wiring connector.
2. Remove the bolts securing the
cooling fan motor and fan blade housing assembly to the fan shroud.
3. Move the cooling fan motor and
fan blade housing assembly to a suitable work bench.
4. Remove the fan blade retaining
clip and remove the fan blade from
the cooling fan motor.
5. Remove the cooling fan motor
screws and motor from the housing
assembly.
To install:
6. Install the cooling fan motor to
the housing assembly and tighten the
screws.
7. Slide the fan blade onto the motor shaft and secure with the retaining clip.
8. Place the cooling fan motor and
fan blade housing assembly to the fan
shroud and install the retaining
bolts. Tighten the retaining bolts to
71–106 inch lbs. (8–12 Nm).
9. Install the wiring connector to
the cooling fan(s). Connect the negative battery cable.
10. Run the engine and allow it to
reach normal operating temperature.
Check the cooling fans for proper
operation.

1995–97 Continental

1. Disconnect the negative battery
cable. Drain the engine cooling system and remove the radiator from the
vehicle.
2. Remove 2 fan shroud retaining
screws and remove the fan shroud
from the radiator.

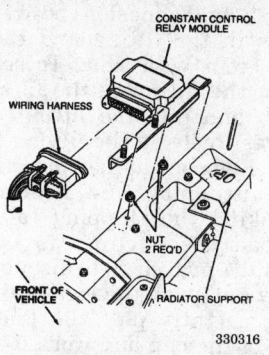

Constant control relay module
mounting — 1993–95 engines

3. Remove the cooling fan motor fan blade retaining clip and remove the fan blade from the cooling fan motor.

4. Remove the cooling fan motor screw and the motor from the fan shroud. If needed, repeat the removal for the opposite cooling fan motor and blade.

To install:

5. Install the cooling fan motor to the shroud and securely tighten the retaining screws.

6. Slide the fan blade onto the motor shaft and attach with the retaining clip. Repeat the installation for the second fan motor and blade if removed.

7. Install the fan shroud to the radiator and tighten screws to 24–48 inch lbs. (2.7–5.4 Nm).

8. Install the radiator in the vehicle. Connect the negative battery cable.

9. Fill and bleed the cooling system. Run the engine and check for leaks and proper cooling fan operation.

Cooling System Bleeding

PROCEDURE

————— **CAUTION** —————

Never remove the radiator cap under any conditions while the engine is operating. Failure to follow these instructions could result in personal injury and/or damage to the cooling system or engine.

When the entire cooling system is drained, the following procedures should be used to ensure a complete fill.

1993–95 Vehicles, Except 1995 Continental

1. Install the block drain plug, if removed and close the draincock. With the engine OFF, add anti-freeze to the radiator to a level of 50 percent of the total cooling system capacity. Then add water until it reaches the radiator filler neck seat.

2. Install the radiator cap to the first notch to keep spillage to a minimum.

3. Start the engine and let it idle until the upper radiator hose is warm. This indicates that the thermostat is open and coolant is flowing through the entire system.

4. Carefully remove the radiator cap and top off the radiator with water. Install the cap on the radiator securely.

5. Fill the coolant recovery reservoir to the **FULL COLD** mark with anti-freeze, then add water to the **FULL HOT** mark. This will ensure that a proper mixture is in the coolant recovery bottle.

6. Check for leaks at the draincock and block plug.

1996–97 Taurus and Sable 1995–97 Continental

1. On 1996–97 Taurus and Sable, carefully remove the radiator cap and top off the radiator with coolant. Install the cap on the radiator securely.

2. Make sure the radiator draincock is closed and all cylinder block drain plugs are tight. Make sure all hoses are connected and the clamps are tight.

3. Place the heater temperature selector in the maximum heat position.

4. On 1995–97 Continental, perform the following:

   a. Remove the coolant recovery reservoir pressure relief cap.

   b. Remove the cooling system fill plug from the engine water bypass tube.

————— **WARNING** —————

Do not fill the cooling system through the reservoir only, as coolant will not enter the engine and engine overheating will occur.

   c. Add a 50/50 mixture of anti-freeze and water to the fill neck on the water bypass tube until the coolant reaches the **FULL** mark on the coolant recovery reservoir.

   d. Install the pressure relief cap to the coolant recovery reservoir.

   e. Slowly continue to fill the cooling system at the water bypass

tube fill neck, allowing air to escape, until full.

5. Start the engine and let it idle.

6. On 1995–97 Continental, add coolant to the water bypass tube fill neck until full. Install the water bypass tube fill plug and tighten to 11–13 ft. lbs. (15–18 Nm).

7. Make sure the heater temperature selector is in the maximum heat position. Turn the blower fan to its highest speed setting and position the control to discharge air at the A/C registers in the instrument panel.

8. With the engine idling, feel for hot air at the A/C registers. If the air discharge remains cool and the engine temperature gauge does not move, the coolant level is low. Stop the engine, allow to cool and repeat the coolant fill procedure.

9. Start the engine and let it idle until normal operating temperature is reached. Hot air should discharge from the A/C registers, the engine temperature gauge should maintain a stabilized reading in the middle of the **NORMAL** range and the lower radiator hose should feel hot to the touch.

10. Shut the engine **OFF** and allow to cool. Check for coolant leaks.

11. On 1996–97 Taurus and Sable, check the coolant level in the coolant recovery reservoir and fill as necessary.

NOTE: When the coolant level indicator flashes, approximately 1–1.5 quarts of coolant mixture can be added to the reservoir after a proper coolant system refill.

FUEL SYSTEM

Fuel System Service Precautions

Safety is the most important factor when performing not only fuel system maintenance but any type of maintenance. Failure to conduct maintenance and repairs in a safe manner may result in serious personal injury or death. Maintenance and testing of the vehicle's fuel system components can be accomplished safely and effectively by adhering to the following rules and guidelines.

• To avoid the possibility of fire and personal injury, always disconnect the negative battery cable unless the repair or test procedure requires that battery voltage be applied.

• Always relieve the fuel system pressure prior to disconnecting any fuel system component (injector, fuel rail, pressure regulator, etc.), fitting or fuel line connection. Exercise extreme caution whenever relieving fuel system pressure to avoid exposing skin, face and eyes to fuel spray. Please be advised that fuel under pressure may penetrate the skin or any part of the body that it contacts.

• Always place a shop towel or cloth around the fitting or connection prior to loosening to absorb any excess fuel due to spillage. Ensure that all fuel spillage (should it occur) is quickly removed from engine surfaces. Ensure that all fuel soaked cloths or towels are deposited into a suitable waste container.

• Always keep a dry chemical (Class B) fire extinguisher near the work area.

• Do not allow fuel spray or fuel vapors to come into contact with a spark or open flame.

• Always use a backup wrench when loosening and tightening fuel line connection fittings. This will prevent unnecessary stress and torsion to fuel line piping. Always follow the proper torque specifications.

• Always replace worn fuel fitting O-rings with new. Do not substitute fuel hose or equivalent, where fuel pipe is installed.

Fuel System Pressure

RELIEVING

———— CAUTION ————
Fuel injection systems remain under pressure, even after the engine has been turned OFF. The fuel system pressure must be relieved before disconnecting any fuel lines. Failure to do so may result in fire and/or personal injury.

1. Disconnect the negative battery cable.
2. Remove the fuel tank fill cap to relieve the pressure in the fuel tank.
3. Remove the cap from the Schrader valve located on the fuel supply manifold (fuel rail).
4. On gasoline engines, attach Fuel Pressure Gauge T80L-9974-A or equivalent, to the valve and drain the fuel through the drain tube into a suitable container.
5. On flex-fuel engines, connect fuel pressure gauge T80L-9974-A or equivalent and fuel pressure test kit

134-R0035 or equivalent, to the Schrader valve. Drain the fuel through the drain tube into a suitable container.
6. After the fuel system pressure is relieved, remove the fuel pressure gauge and install the cap on the Schrader valve.
7. Install the fuel tank fill cap.
8. Connect the negative battery cable only after system repairs are completed.

Idle Speed

ADJUSTMENT

The idle speed is controlled by the Powertrain Control Module (PCM) and is not adjustable.

Mixture

ADJUSTMENT

The air/fuel mixture is electronically controlled by the Powertrain Control Module (PCM) and the Idle Air Control (IAC) valve, and cannot be adjusted. The PCM continuously adjust the air/fuel ratio in response to signals received from operator controls, sensor and switch signals monitoring the engines running condition.

Fuel Filter

REMOVAL AND INSTALLATION

———— CAUTION ————
Fuel injection systems remain under pressure, even after the engine has been turned OFF. The fuel system pressure must be relieved before disconnecting any fuel lines. Failure to do so may result in fire and/or personal injury.

Except 4.6L Engine

1. Disconnect the negative battery cable.
2. Properly relieve the fuel system pressure.
3. Raise and safely support the vehicle.
4. Remove the hairpin clip push connect fittings from both ends of the fuel filter as follows:
   a. Inspect the visible internal portion of the fitting for dirt accumulation. If more than a light coating of dust is present, clean the fitting before disassembly.

b. Some adhesion between the seals in the fitting and the filter will occur with time. To separate, twist the fitting on the filter, then push and pull the fitting until it moves freely on the filter.
   c. Remove the hairpin clip from the fitting by first bending and breaking the shipping tab. Next, spread the 2 clip legs by hand about 1/8 inch each, to disengage the body and push the legs into the fitting. Lightly pull the triangular end of the clip and work it clear of the filter and fitting.

NOTE: Do not use hand tools to complete this operation.

   d. Grasp the fitting and pull in an axial direction to remove the fitting from the filter. Be careful on 90 degree elbow connectors, as excessive side loading could break the connector body.
   e. After disassembly, inspect the inside of the fitting for any internal parts such as O-rings and spacers that may have been dislodged from the fitting. Replace any damaged connector.
5. Loosen the filter retaining clamp and remove the fuel filter. Note the direction of the flow arrow on the filter, so the replacement filter can be reinstalled in the same position.
   To install:
6. Install the fuel filter with the flow arrow facing the proper direction and tighten the filter retaining clamp.
7. Install the rubber insulator rings on the new filter (replace the insulator rings if the filter moves freely after the retainer is installed). Install the filter into the retainer with the flow arrow pointing out the open end of the retainer. Install the retainer on the bracket and tighten the mounting bolts to 27–44 inch lbs. (3–5 Nm).
8. Install the hairpin clip push connect fittings at both ends of the fuel filter as follows:
   a. Install a new connector if damage was found. Insert a new clip into any 2 adjacent openings with the triangular portion pointing away from the fitting opening. Install the clip until the legs of the clip are locked on the outside of the body. Piloting with an index finger is necessary.
   b. Before installing the fitting on the filter, wipe the filter end with a clean cloth. Inspect the inside of the fitting to make sure it is free of dirt and/or obstructions.

c. Apply a light coating of engine oil to the filter end. Align the fitting and filter axially and push the fitting onto the filter end. When the fitting is engaged, a definite click will be heard. Pull on the fitting to make sure it is fully engaged.

9. Lower the vehicle.

10. Connect the negative battery cable.

11. Start the engine and check for fuel leaks and proper operation.

4.6L Engine

CAUTION
The air suspension switch, located in the left side of the luggage compartment, must be turned OFF before raising the vehicle. Failure to do so may result in unexpected inflation or deflation of the air springs which may result in shifting of the vehicle during service.

1. Turn the air suspension switch, located in the left side of the luggage compartment, to the **OFF** position.

2. Disconnect the negative battery cable.

3. Properly relieve the fuel system pressure.

4. Raise and safely support the vehicle.

5. Using Fuel Line Disconnect tool T90T-9550-B or T90T-9550-C or equivalent, disconnect the push connect fittings at the fuel filter.

NOTE: Grasp the fitting and pull in an axial direction to remove the fitting from the fuel filter. Use care, as excessive side loading could break the connector body.

6. Loosen the fuel filter retaining clamp and remove the fuel filter. Note the direction of the flow arrow on the filter, so the replacement filter can be reinstalled in the same position.

To install:

7. Install the fuel filter with the flow arrow facing the proper direction. Tighten the filter retaining clamp to 15–25 inch lbs. (1.7–2.8 Nm).

8. Apply a light coating of clean engine oil to the fuel filter ends. Align the fittings and filter axially and push the fittings onto the filter ends. When the fittings are properly engaged, a definite click will be heard. Pull on the fittings to make sure they are fully engaged.

9. Lower the vehicle.

10. Connect the negative battery cable.

11. Start the engine and check for fuel leaks.

12. Turn the air suspension switch to the **ON** position.

Fuel Pump

REMOVAL AND INSTALLATION

CAUTION
Fuel injection systems remain under pressure, even after the engine has been turned OFF. The fuel system pressure must be relieved before disconnecting any fuel lines. Failure to do so may result in fire and/or personal injury.

1993–95 Vehicles, Except 1995 Continental

1. Disconnect the negative battery cable. Properly relieve the fuel system pressure.

2. Remove the fuel tank from the vehicle and place it on a bench.

3. Remove any dirt that has accumulated around the fuel pump retaining flange so it will not enter the tank during pump removal and installation.

4. Turn the fuel pump locking ring counterclockwise and remove the locking ring.

5. Remove the fuel pump/sending unit assembly. Remove and discard the seal ring.

To install:

6. Clean the fuel pump mounting flange, fuel tank mounting surface and seal ring groove.

7. Apply a light coating of grease on a new seal ring to hold it in place during assembly and install in the seal ring groove.

8. Install the fuel pump/sending unit assembly carefully to ensure the filter is not damaged. Make sure the locating keys are in the keyways and the seal ring remains in the groove.

9. Hold the pump assembly in place and install the locking ring finger-tight. Make sure all the locking tabs are under the tank lock ring tabs.

10. Rotate the locking ring clockwise until the ring is against the stops. Install the fuel tank in the vehicle. Add a minimum of 10 gallons of fuel to the tank and check for leaks.

11. Connect the negative battery cable. Turn the ignition switch to the **RUN** position several times to pressurize the fuel system. Check for fuel leaks and correct as necessary.

12. Start the engine and check for leaks. Road test the vehicle and check for proper operation.

1996–97 Taurus and Sable
1995–97 Continental

CAUTION
On 1995–97 Continental, the air suspension switch is located in the left side of the luggage compartment and must be turned OFF before raising the vehicle. Failure to do so may result in unexpected inflation or deflation of the air springs which may result in shifting of the vehicle during service.

1. On 1995–97 Continental, turn the air suspension switch, located in the left side of the luggage compartment, to the **OFF** position.

2. Disconnect the negative battery cable. Properly relieve the fuel system pressure.

3. Raise and safely support the vehicle. Remove the fuel tank and place on a suitable work bench.

4. Remove any dirt that has accumulated around the fuel pump retaining flange so it will not enter the tank during pump removal and installation.

5. Turn the fuel pump locking ring counterclockwise and remove the locking ring using Fuel Tank Sender Wrench T74P-9275-A, or equivalent.

6. Pull the fuel pump module up and out of the fuel tank until the locking tabs for the fuel pump module are accessible. Squeeze both locking tabs together and remove the fuel pump module from the fuel tank. Remove and discard the O-ring seal.

To install:

7. Clean the fuel pump mounting flange, fuel tank mounting surface and O-ring seal groove.

8. Apply a light coating of grease on a new O-ring seal to hold it in place during assembly and install the O-ring seal.

9. Install the fuel pump module carefully to ensure the filter and hoses and float rod are not damaged.

10. Align the fuel pump module and the fuel tank retainer axially and push the fuel pump module into the fuel tank retainer. When the fuel pump module is properly engaged, a definite click will be heard engaging 2 locking tabs on the outside of the fuel pump.

11. Pull on the fuel pump module to ensure that both locking tabs are properly engaged.

12. Make sure the locating keys are in the keyways and the seal ring remains in the groove.

13. Hold the fuel pump module in place and install the locking ring finger-tight. Make sure all the locking tabs are under the tank lock ring tabs.

14. Using the sender wrench or equivalent, rotate the locking ring clockwise until the ring is against the stops.

15. Install the fuel tank in the vehicle. Lower the vehicle.

16. Add a minimum of 10 gallons (38 liters) of clean fuel to the tank. Connect the negative battery cable.

17. On 1995–97 Continental, turn the air suspension switch to the **ON** position.

18. Install a suitable fuel pressure gauge to the Schrader valve on the fuel supply manifold.

19. Cycle the ignition switch from **OFF** to **ON** for 3 seconds. Repeat this procedure 5–10 times until the pressure gauge reads at least 30 psi (207 kPa). Check for fuel leaks.

20. Remove the fuel pressure gauge. Start the engine and again, check for fuel leaks. Road test the vehicle and check for proper operation.

Fuel Injector

REMOVAL AND INSTALLATION

— CAUTION —
Fuel injection systems remain under pressure, even after the engine has been turned OFF. The fuel system pressure must be relieved before disconnecting any fuel lines. Failure to do so may result in fire and/or personal injury.

3.0L (VIN Y) and 3.2L SHO Engines

1. Disconnect the negative battery cable. Properly relieve the fuel system pressure.

2. Remove the intake manifold.

3. Detach the electrical connectors at the fuel injectors.

4. Remove the fuel rail retaining bolts, then raise and slightly rotate the fuel rail assembly and remove the injectors.

To install:

5. Lubricate new O-rings with engine oil and install them on the fuel injectors.

6. Install the injectors in the fuel rail by lightly twisting and pushing the injectors into position.

7. Install the fuel rail, making sure the injectors seat properly in the cylinder head. Install the fuel rail retaining bolts and tighten to 11–17 ft. lbs. (15–23 Nm).

8. Attach the connectors at the injectors.

9. Install the intake manifold. Install the air cleaner outlet tube.

10. Connect the negative battery cable, then run the engine and check for leaks.

1993–95 3.0L (VIN U) Engine

1. Disconnect the negative battery cable. Properly relieve the fuel system pressure.

2. Remove the air intake hose from the throttle body.

3. Label and detach all vacuum lines and electrical connectors from the throttle body.

4. Loosen the lower EGR tube nut and rotate the tube away from the valve.

5. Disconnect the accelerator and throttle valve linkage from the throttle body.

6. Disconnect the PCV hose. Remove the air intake throttle body retaining bolts and remove the throttle body.

7. On some vehicles, the distributor must be raised to allow the crossover tube to clear the distributor housing and lower intake manifold assembly.

8. Disconnect the fuel supply and fuel return lines. Disconnect the wiring harness from the injectors.

9. Disconnect the vacuum line from the fuel pressure regulator valve. Remove the 4 fuel injector manifold retaining bolts.

10. Carefully disengage the fuel rail assembly from the fuel injectors by lifting and gently rocking the rail.

11. Remove the injectors by lifting while gently rocking from side to side.

To install:

12. Lubricate new O-rings with clean engine oil and install 2 on each injector. Make sure the injector cups are clean and undamaged.

13. Install the injectors in the fuel rail using a light twisting-pushing motion. If the distributor was raised for fuel injection manifold removal, reinstall it.

14. Carefully install the rail assembly and injectors into the lower intake manifold, one side at a time. Make sure the O-rings are seated by pushing down on the fuel rail.

15. While holding the fuel rail assembly in place, install the 2 retain-

ing bolts and tighten to 84 inch lbs. (10 Nm).

16. Connect the fuel supply and fuel return lines and the fuel pressure regulator vacuum line.

17. Before connecting the fuel injector harness, connect the negative battery cable and turn the ignition switch to the **ON** position. This will pressurize the fuel system.

18. Using a clean paper towel. check for leaks where the injector connects to the fuel rail. Turn the ignition switch **OFF** and disconnect the negative battery cable. Connect the fuel injector harness.

19. Install the air intake throttle body using a new gasket. Tighten the bolts to 15–22 ft. lbs. (20–30 Nm).

20. Connect the PCV hose and connect the accelerator and throttle valve linkage. Connect the EGR tube.

21. Attach all vacuum hoses and electrical connectors to their proper locations. Connect the air intake hose to the throttle body.

22. Connect the negative battery cable, start the engine and let it idle for 2 minutes.

23. Using a clean paper towel, check for leaks where the injector is installed into the intake manifold.

1996–97 3.0L (VIN S, 1 and U) Engines

— WARNING —
Do not modify flexible fuel configuration or components with parts not specially designed for use with fuel methanol and fuel ethanol. The use of different parts or materials could produce an untested configuration that could result in fire, personal injury or could cause engine damage.

1. Disconnect the negative battery cable. Properly relieve the fuel system pressure. Remove the upper intake manifold.

2. Disconnect the fuel injector wiring harness and move aside. Disconnect the vacuum line at the fuel pressure regulator.

3. If required, remove the fuel pressure regulator and the pressure relief valve from the supply manifold.

4. Remove the spring lock coupling retainer clips from the fuel supply and return fittings.

5. Use spring lock coupling disconnect tools (³⁄₈ inch and ¹⁄₂ inch) to disconnect the fuel supply and return hoses from the fuel injection supply manifold (fuel rail).

6. Remove the 4 fuel injection supply manifold retaining bolts.

7. Carefully disengage the fuel injection supply manifold with the fuel injectors and remove as an assembly.

8. Remove the fuel injector(s) from the fuel injection supply manifold using a rocking side-to-side motion while pulling up gently.

9. Remove and discard the fuel injector O-rings. Inspect the fuel injector end caps for signs of dirt and deterioration. Replace as needed.

To install:

10. Place new O-rings onto each injector and lightly coat with clean engine oil.

NOTE: On flexible fuel engines use an oil with an API designation of multi-fuel vehicles.

11. Install the fuel injector(s) into the fuel injection supply manifold using a light twisting/pushing motion.

12. Install the fuel injection supply manifold with injectors to the lower intake manifold.

13. While holding down on the fuel injection supply manifold, install the retaining bolts and tighten to 71–106 inch lbs. (8–12 Nm).

14. Install the fuel supply and return hoses to the fuel supply manifold. Ensure that the spring lock couplings are correctly installed.

15. Install the retaining clips on the spring lock couplings. Connect the vacuum line to the fuel pressure regulator. Temporarily connect the negative battery cable.

16. With the fuel injector wiring disconnected, cycle the ignition key several times to the **RUN** position to pressurize the fuel system.

17. Using a clean paper towel while wearing rubber gloves, check for leaks where the fuel injectors connect to the fuel injection supply manifold and intake manifold. If a leak is found, disconnect the negative battery cable and relieve the fuel system pressure. Remove the fuel injection supply manifold and replace the leaking O–ring(s) before continuing. If no leaks are found, continue with the next step.

18. Disconnect the negative battery cable. Connect the fuel injector wiring harness. Install the upper intake manifold.

19. Connect the negative battery cable. Run the engine and check for leaks and proper operation.

3.8L Engine

1. Disconnect the negative battery cable. Properly relieve the fuel system pressure.

2. Remove the upper intake manifold and the fuel supply manifold. Remove the injector retaining clips.

3. Disconnect the electrical connectors from the fuel injectors. To remove the injector, pull it up while gently rocking it from side-to-side.

4. Inspect the injector pintle protection cap (plastic hat) and washer for deterioration and replace, as required. If the cap is not on the injector, look for it in the manifold.

To install:

5. Lubricate new injector O-rings with clean engine oil and install 2 on each injector.

6. Install the injectors, using a light, twisting, pushing motion to install them. Install the injector retaining clips.

7. Install the fuel supply manifold, making sure the injector O-rings are fully seated in the manifold cups. Tighten the fuel supply manifold retaining bolts to 71–97 inch lbs. (8–11 Nm) and tighten the fuel pressure regulator bracket retaining bolt to 15–22 ft. lbs. (20–30 Nm).

8. Connect the fuel lines to the fuel supply manifold.

9. Connect the negative battery cable, then turn the ignition **ON** to pressurize the fuel system. Using a clean towel, check for leaks and correct as necessary.

10. Turn the ignition **OFF** and disconnect the negative battery cable. Connect the electrical harness connectors to the injectors.

11. Install the upper intake manifold. Connect the negative battery cable. Start the engine and check for fuel leaks.

4.6L Engine

1. Disconnect the negative battery cable. Properly relieve the fuel system pressure. Remove the engine appearance cover.

2. Disconnect the vacuum lines at the fuel pressure regulator and the intake manifold.

3. Detach the retainers for the engine control sensor wiring from the intake manifold studs and position the wiring aside.

4. Disconnect the engine control sensor wiring from the Intake Manifold Runner Control (IMRC) deactivation motor.

5. Remove the IMRC motor and bracket from the right cylinder head cover and intake manifold. Position the IMRC motor and bracket aside.

6. Remove the fuel coupling retainer clips from the fuel supply and return lines at the fuel injection supply manifold.

7. Using Spring Lock Coupling Disconnect tool D87L-9280-A and D87L-9280-B or equivalents, detach the fuel supply and return lines from the fuel injection supply manifold.

8. Remove 8 fuel injection supply manifold retaining bolts.

9. Carefully disengage the fuel injection supply manifold and fuel injectors and remove the fuel injection supply manifold.

10. Disconnect the electrical connectors from the fuel injectors.

11. To remove the fuel injector, pull it up while gently rocking it from side-to-side. Discard the O-rings.

To install:

12. Lubricate new O-rings with clean engine oil and install 2 on each fuel injector.

13. Install the fuel injectors, using a light, twisting, pushing motion.

14. Install the fuel injection supply manifold. Make sure all the fuel injector O-rings are fully seated in the fuel injection manifold cups and the lower intake manifolds.

15. Install 8 retaining bolts while holding down on the fuel injection supply manifold. Tighten the bolts to 40–62 inch lbs. (4.5–7 Nm).

16. Connect the electrical connectors to the fuel injectors.

17. Install the fuel supply and fuel return lines to the fuel injection supply manifold. Ensure that the fuel line connectors are properly installed. Install the fuel coupling retainer clips.

18. Install the IMRC deactivation motor and bracket. Tighten the retainers to 71–106 inch lbs. (8–12 Nm). Connect the engine control sensor wiring to the IMRC deactivation motor.

19. Install the engine control sensor wiring to the intake manifold studs.

20. Connect the vacuum lines to to the pressure regulator and intake manifold.

21. Install the fuel tank fill cap.

22. Connect the negative battery cable. Cycle the ignition switch from the **OFF** to **ON** position several times without starting the engine to pressurize the fuel system and check for fuel leaks.

23. Install the engine air cleaner outlet tube.

24. Start the engine and allow it to reach normal operating temperature and again inspect the fuel system for leaks.

25. Install the engine appearance cover. Tighten the retainers securely.

26. Road test the vehicle and check for proper engine operation.

EMISSION CONTROLS

Service Interval Lamp

RESETTING

1993 Vehicles

Approximately every 5000 or 7500 miles, (depending on engine application) the word **SERVICE** will appear on the electronic display for the first 1.5 miles to remind the driver that it is time for the regular vehicle service interval maintenance (i.e. oil change).

To reset the service interval reminder light for another interval on the Continental, with the engine running, press the **SYSTEM CHECK** and **TRIP RESET** buttons. Hold the buttons down until the **SERVICE** light disappears from the display and 3 audible beeps are heard to verify that the service reminder has been reset.

On the Sable/Taurus, with the engine running, press the **ODO SEL** and **TRIP RESET** buttons. Hold the buttons down until the **SERVICE** light disappears from the display and 3 audible beeps are heard to verify that the service reminder has been reset.

1994 Continental

The Service Interval Reminder light is used.

During the SYSTEM CHECK sequence, the SERVICE symbol comes ON and displays the miles (kilometers) to go before the next normal service is due. After service is complete, reset the service interval reminder as follows:

1. Press the **SYSTEM CHECK** button.
2. Press the **RESET** button.
3. Press the **SYSTEM CHECK** and the **RESET** button at the same time. The display now shows 7200 miles (11,580 km) and the mileage counts down from here.

ENGINE MECHANICAL

Engine Assembly

REMOVAL AND INSTALLATION

— **CAUTION** —
Fuel injection systems remain under pressure, even after the engine has been turned OFF. The fuel system pressure must be relieved before disconnecting any fuel lines. Failure to do so may result in fire and/or personal injury.

3.0L (VIN Y) and 3.2L SHO Engines

1. Disconnect the battery cables and remove the battery and battery tray. Drain the cooling system and properly relieve the fuel system pressure.
2. Detach the wiring connector retaining the under hood light, if equipped. Mark the position of the hood hinges and remove the hood.
3. Remove the oil level indicator.
4. Disconnect the alternator and voltage regulator wiring assembly.
5. Remove the radiator upper sight shield. Properly recover the refrigerant from the air conditioning system.
6. Remove the radiator coolant recovery reservoir assembly.
7. Remove the integrated relay controller, air cleaner hose assembly, upper radiator hose, electric fan and shroud assembly.
8. Remove the lower radiator hose and the radiator.
9. Disconnect the fuel inlet and return hose.
10. Remove the Barometric Air Pressure (BAP) sensor.
11. Remove the engine vibration damper and bracket assembly from the right side of the engine. Remove the engine-to-damper bracket.
12. Remove the retaining bolt from the power steering reservoir and place the reservoir aside. Disconnect the hose to the power steering cooler at the pump.
13. Disconnect the throttle linkage and disconnect and tag the vacuum hoses. Disconnect the heater hoses at the heater core.
14. Detach the electrical connectors from the harness on the rear of the engine.

15. Loosen the belt tensioner pulleys and remove the air conditioning compressor/alternator belt and the steering pump belt. Remove the lower tensioner pulley.
16. Disconnect the cycling switch on the top of the suction accumulator/drier.
17. Disconnect the air conditioning line at the dash panel and remove the accumulator and bracket assembly. Cap the openings to prevent the entry of dirt and moisture.
18. Remove the alternator assembly.
19. Disconnect the air conditioning discharge hose and remove the air conditioning compressor and bracket assembly. Cap the openings to prevent the entry of dirt and moisture.
20. Raise the vehicle and support it safely. Drain the engine oil, then remove the filter element.
21. Remove the wheel and tire assemblies. Disconnect the oil level sensor switch.
22. Disconnect the right lower ball joint, tie rod end and stabilizer bar.
23. Disconnect the center support bearing bracket and right CV-joint from the transaxle.
24. Disconnect the oxygen sensor assembly and the 4 exhaust catalyst to engine retaining bolts.
25. Remove the starter. If equipped with an automatic transaxle, remove the 4 torque converter-to-flywheel nuts. Remove the lower transaxle to engine retaining bolts.
26. Remove the engine mount to subframe nuts.
27. Remove the crankshaft pulley assembly.
28. Lower the vehicle and remove the upper transaxle to engine retaining bolts.
29. Install engine lifting eyes.
30. Position a floor jack under the transaxle.
31. Position suitable engine lifting equipment, raise the transaxle assembly slightly and remove the engine from the vehicle.

To install:
32. Position the engine assembly in the vehicle.
33. Install the upper transaxle to engine bolts and remove the floor jack and engine lifting equipment. Remove the engine lifting eyes.
34. Raise the vehicle and support it safely.
35. Install the crankshaft pulley assembly. Tighten the retaining bolt to 113–126 ft. lbs. (152–172 Nm).

36. Install the remaining components in the reverse order of removal and torque the following items:

Lower transaxle to engine bolts to 25–35 ft. lbs. (34–47 Nm)

Torque converter-to-flywheel nuts to 23–39 ft. lbs. (31–53 Nm)

Exhaust catalyst-to-engine retaining nuts 19–34 ft. lbs. (27–47 Nm)

Apply anti-seize compound to the O₂ sensor threads, then tighten to 27–33 ft. lbs. (37–45 Nm)

Oil drain plug to 15–24 ft. lbs. (20–33 Nm)

A/C compressor and bracket assembly to 27–40 ft. lbs. (36–55 Nm)

Alternator assembly to 36–53 ft. lbs. (48–72 Nm)

37. Connect the negative battery cable.

38. Fill the cooling system with the proper type and quantity of coolant and fill the crankcase with the proper type of motor oil to the required level.

39. Drain, evacuate, pressure test and recharge the air conditioning system.

40. Start the engine and check for leaks.

1993–95 3.0L (VIN U) Engines

1. Disconnect the battery cables and drain the cooling system. Mark the position of the hood on the hinges and remove the hood.

2. Properly relieve the fuel system pressure.

3. Properly recover the refrigerant from the air conditioning system.

4. Remove the air cleaner assembly. Remove the battery and the battery tray.

5. Remove the integrated relay controller, cooling fan and radiator with fan shroud.

6. Remove the engine bounce damper bracket on the strut tower.

7. Remove the evaporative emission line, upper radiator hose, starter brace and lower radiator hose.

8. Remove the exhaust pipes from both exhaust manifolds. Remove and plug the power steering pump lines. Remove the refrigerant lines from the air conditioner compressor and cap the openings to prevent contamination from dirt and moisture.

9. Remove the fuel lines and remove and tag all necessary vacuum lines.

10. Disconnect the ground strap, heater lines, accelerator cable linkage, throttle valve linkage and cruise control cable.

11. Label and detach the following wiring connectors: alternator, air conditioner compressor clutch, oxygen sensor, ignition coil, radio frequency suppressor, cooling fan voltage resistor, engine coolant temperature sensor, coolant temperature sending switch, ignition module, injector wiring harness, idle speed control motor wire, throttle position sensor, oil pressure sending switch, ground wire, block heater, if equipped, knock sensor, EGR sensor and oil level sensor.

12. Raise the vehicle and support it safely. Remove the engine mount bolts and engine mounts.

13. Remove the flywheel-to-torque converter bolts. Remove the transaxle to engine mounting bolts and transaxle brace assembly.

14. Lower the vehicle. Install a suitable engine lifting plate onto the engine and use a suitable engine hoist to remove the engine from the vehicle. Remove the main wiring harness from the engine.

To install:

15. Install the main wiring harness on the engine. Position the engine in the vehicle and remove the engine lifting plate.

16. Raise the vehicle and support it safely.

17. Install the engine mounts and bolts and tighten to 40–55 ft. lbs. (54–75 Nm). Install the transaxle brace assembly and tighten the bolts to 40–55 ft. lbs. (54–75 Nm). Install the flywheel-to-torque converter bolts.

18. Attach all wiring connectors according to their labels.

19. Install the remaining components in the reverse order of removal.

20. Fill the cooling system with the proper type and quantity of coolant. Fill the crankcase with the correct type of motor oil to the required level.

21. Install the hood, aligning the marked made during removal.

22. Connect the negative battery cable. Start the engine and check for leaks and proper engine operation.

23. Evacuate and charge the air conditioning system.

3.0L (VIN S) Engine

1. Disconnect the battery cables, negative cable first.

2. Drain the cooling system.

3. Mark the position of the hood on the hinges and remove the hood.

4. Disconnect the steering coupling at the pinch bolt joint inside the passenger compartment.

5. Remove the windshield wiper motor and then remove the cowl top inner panels.

6. Disconnect the main emission vacuum control connector at the 2 connectors located at the right side of the dash panel.

NOTE: Label all electrical connectors and vacuum hoses prior to removal so they can be reinstalled in their proper locations.

7. Disconnect the wiring from the Intake Air Temperature (IAT) sensor.

8. Remove the air cleaner outlet tube.

9. Remove the air cleaner retaining bolts at the air cleaner body.

10. Disconnect the engine intake air resonator by pushing in the top and bottom tube surfaces at the engine air cleaner and pulling the air cleaner outward. Lift the air cleaner up and out of the engine compartment.

11. Properly relieve the fuel system pressure.

12. Recover the refrigerant from the A/C system using the proper equipment.

13. Disconnect the chassis vacuum supply hose at the connection on the intake manifold. Position the hose aside.

14. Remove the ground straps from the dash panel.

15. Disconnect the control sensor wiring from the Powertrain Control Module (PCM) and position aside.

16. Disconnect the sensor wiring from the Mass Air Flow (MAF) sensor.

17. Disconnect the wiring connector from the throttle body.

18. Disconnect engine control sensor wiring from the evaporative emission canister purge valve.

19. Remove the shield and disconnect the accelerator cable and the speed control actuator from the throttle body and from the accelerator cable bracket. Position the cables aside.

20. Remove the retaining nut and disconnect the manual control lever from the manual control lever shaft at the Transmission Range (TR) sensor.

21. Remove the connectors for the engine control sensor wiring from the retaining bracket on top of the transaxle.

22. Disconnect the A/C compressor lines from the suction accumulator/drier. Cap all openings to prevent the entrance of dirt or moisture.

23. Disconnect the heater water hoses from the heater core.

24. Disconnect the upper radiator coolant recovery hose from the upper radiator hose.

25. Disconnect the upper and lower radiator hose at the water bypass tube. Position the upper hose aside.

26. Disconnect the battery ground cable from the starter mounting stud bolt.

27. Remove the power steering return hose from the power steering oil reservoir and drain.

28. Raise and safely support the vehicle. Remove the front wheel and tire assemblies.

29. Disconnect the alternator wiring harness from the alternator at the BAT terminal and stator connector plug. Remove the wiring harness retaining clip from the starter motor.

30. Disconnect the sensor wiring at the connector located at the right top of the fan shroud.

31. Remove the both front stabilizer bar links from the front stabilizer bar.

32. Separate right and left front suspension lower arms from the steering knuckles at the ball joints.

33. Separate both tie rod ends from the steering knuckles.

34. Remove both front axle wheel hub retainer nuts from the halfshaft ends.

35. Remove both halfshafts from the steering knuckles.

36. Remove the splash shield from the radiator support and front bumper.

37. Drain the engine oil.

——————— WARNING ———————
The heated oxygen sensors (HO₂S) must be removed before removing the Y-pipe or damage may occur.

38. Remove the convertor Y-pipe.

39. Disconnect the power steering pressure hose from the power steering/transaxle oil cooler connection. Position the hose aside.

40. Disconnect the fluid cooler tube from the power steering/transaxle fluid cooler connection. Position the tube aside.

41. Disconnect the fluid cooler inlet tube from the radiator fluid cooler tube. Position the tube aside.

42. Disconnect the lower radiator hose at the radiator and the radiator overflow hose.

43. Disconnect the wiring at the starter motor.

44. Disconnect the A/C compressor line from the condenser core using the proper tools.

45. Support the subframe, engine and transaxle assembly using Powertrain Lift 014-00765 and Universal Powertrain Removal Bracket 014-00766 or equivalents.

46. Remove 4 subframe retaining bolts. Lower the subframe and engine/transaxle assembly from the vehicle.

47. Disconnect the sensor wiring from the power steering gear and the secondary air injection pump.

48. Disconnect the power steering pressure hose from the power steering pump.

49. Install suitable engine lifting brackets on the engine and transaxle assembly.

50. Remove the front engine support insulator, rear engine support insulator and engine and transaxle support.

51. Lift the engine and transaxle from the subframe.

52. Lower the engine and transaxle. Support the transaxle on a level, stationary surface for transaxle storage.

53. Remove the transaxle-to-cylinder block mounting bolts and separate the engine from the transaxle/torque converter assembly.

54. Place the engine on a suitable workstand.

To install:

55. Install the transaxle/torque converter assembly to the engine. Tighten the mounting bolts to 30–44 ft. lbs. (40–60 Nm). Tighten the torque convertor nuts to 20–34 ft. lbs. (27–46 Nm).

56. Remove the engine and transaxle assembly from the workstand and position it on the subframe.

57. Install the front engine support insulator, rear engine support insulator and engine and transaxle support.

58. Install the remaining components in the reverse order of removal and torque the following:

Subframe-to-body bolts to 57–76 ft. lbs. (77–103 Nm)

Hub retainer nuts to 170–202 ft. lbs. (230–275 Nm)

Tie rod ends-to-steering knuckle nuts to 35–46 ft. lbs. (47–63 Nm)

Front suspension lower arms-to-steering knuckle nuts 50–68 ft. lbs. (68–92 Nm)

Front stabilizer bar links-to-front stabilizer bar nuts to 30–40 ft. lbs. (40–55 Nm)

Battery ground cable-to-starter motor nut to 15–22 ft. lbs. (20–30 Nm)

Manual control lever-to-manual control lever shaft to 12–16 ft. lbs. (16–22 Nm)

PCM control sensor wiring bolt to 32 inch lbs. (3.7 Nm)

Air cleaner outlet tube clamps to 12–22 inch lbs. (1.4–2.5 Nm)

59. Fill the cooling system. Fill the crankcase with the proper type of motor oil to the required level.

60. Connect the battery cables, negative cable last. Run the engine and check for leaks.

61. Evacuate and recharge the air conditioning system. Install and align the hood.

NOTE: Whenever the vehicles subframe is removed or lowered, the wheel alignment should be checked.

62. Check the front wheel alignment. Road test the vehicle and check the engine and transaxle for proper operation.

1996–97 3.0L (VIN 1 and U) Engines

1. Disconnect the battery cables, negative cable first.

2. Drain the cooling system.

3. Mark the position of the hood on the hinges and remove the hood.

4. Disconnect the steering coupling at the pinch bolt joint inside the passenger compartment.

5. Disconnect the wiring from the Mass Air Flow (MAF) sensor, and the Intake Air Temperature (IAT) sensor.

NOTE: Label all electrical connectors and vacuum hoses prior to removal so they can be reinstalled in their proper locations.

6. Remove the air cleaner outlet tube.

7. Remove the air cleaner retaining bolts at the air cleaner body.

8. Disconnect the engine intake air resonator by pushing in the top and bottom tube surfaces at the engine air cleaner and pulling the air cleaner outward. Lift the air cleaner up and out of the engine compartment.

9. Properly relieve the fuel system pressure.

10. Recover the refrigerant from the A/C system using the proper equipment.

11. Disconnect the chassis vacuum supply hose at the connection on the intake manifold. Position the hose aside.

12. Remove the ground straps from the dash panel.

13. Disconnect the control sensor wiring from the Powertrain Control Module (PCM) and position aside.

14. Remove the connectors for the engine control sensor wiring from the retaining bracket on the power brake booster. Disconnect the engine control sensor wiring at the 2 connectors.

15. Disconnect engine control sensor wiring from the evaporative emission canister purge valve.

16. Disconnect the evaporative emission hose at the crankcase vent connector and hose. Position the hose aside.

17. Remove the shield and disconnect the accelerator cable and the speed control actuator from the throttle body and from the accelerator cable bracket. Position the cables aside.

18. Remove the retaining nut and disconnect the manual control lever from the manual control lever shaft at the Transmission Range (TR) sensor.

19. Remove the connectors for the engine control sensor wiring from the retaining bracket on top of the transaxle. Disconnect the engine control sensor wiring at the 2 connectors.

20. Disconnect the wiring connector from the secondary air injection pump relay located on the retaining bracket on top of the transaxle.

21. Disconnect the main emission vacuum control connector at the connection near the fan shroud.

22. Disconnect the oil cooler inlet tube from the transaxle.

23. Disconnect the heater water hose from the water pump and water hose connection.

24. Disconnect the upper radiator hose and degas tube from the water hose connection.

25. Remove the power steering return hose from the power steering oil reservoir and drain.

26. Disconnect the alternator wiring harness from the alternator at the BAT terminal and stator connector plug. Remove the wiring harness retaining clip from the alternator mounting bracket.

27. Disconnect the retaining clips and the A/C compressor lines from the compressor. Cap all openings to prevent the entrance of dirt or moisture.

28. Raise and safely support the vehicle. Remove the front wheel and tire assemblies.

29. Remove both front stabilizer bar links from the front stabilizer bar.

30. Separate both front suspension lower arms from the steering knuckles at the ball joints.

31. Separate both tie rod ends from the front steering knuckles.

32. Remove both front axle wheel hub retainers from the halfshaft ends.

33. Remove both halfshafts from the steering knuckles.

34. Remove the splash shield from the radiator support and front bumper.

35. Drain the engine oil.

36. Remove the convertor Y-pipe.

37. Disconnect the power steering pressure hose from the power steering/transaxle oil cooler connection. Position the hose aside.

38. Disconnect the lower radiator hose at the radiator and at the radiator overflow hose.

39. Disconnect the wiring at the starter motor, and remove the starter motor.

40. Disconnect the lower cooler line from the transaxle.

41. Support the front subframe and engine/transaxle assembly using Powertrain Lift 014-00765 and Universal Powertrain Removal Bracket 014-00766 or equivalents.

42. Remove the 4 subframe retaining bolts.

43. Lower the engine/transaxle and front subframe from the vehicle.

44. Disconnect the power steering pressure hose from the power steering pump.

45. Install suitable engine lifting brackets on the engine and transaxle assembly.

46. Remove the front engine support insulator, rear engine support insulator and engine and transaxle support.

47. Lift the engine and transaxle from the subframe.

48. Lower the engine and transaxle. Support the transaxle on a level, stationary surface for transaxle storage.

49. Remove the transaxle-to-cylinder block mounting bolts and separate the engine from the transaxle/torque converter assembly.

50. Place the engine on a safe suitable workstand.

To install:

51. Install the transaxle/torque converter assembly to the engine. Tighten the mounting bolts to 30–44 ft. lbs. (40–60 Nm). Tighten the torque convertor nuts to 20–34 ft. lbs. (27–46 Nm).

52. Remove the engine and transaxle assembly from the workstand and position it on the subframe.

53. Install the front engine support insulator, rear engine support insulator and engine and transaxle support.

54. Connect the power steering pressure hose from the power steering pump.

55. Raise the engine, transaxle and subframe into position using the powertrain lifting tool.

56. Align the front subframe to the body and install the subframe-to-body bolts. Tighten the bolts to 57–76 ft. lbs. (77–103 Nm).

57. Remove the lifting equipment and move aside.

58. Install the remaining components in the reverse and torque the following:

Hub retainer nuts to 170–202 ft. lbs. (230–275 Nm)

Tie rod ends-to-steering knuckle nuts to 35–46 ft. lbs. (47–63 Nm)

Front suspension lower arms-to-front wheel knuckle nuts to 50–68 ft. lbs. (68–92 Nm)

Front stabilizer bar links-to-front stabilizer bar nuts to 30–40 ft. lbs. (40–55 Nm)

Manual control lever-to-manual control lever nut to 12–16 ft. lbs. (16–22 Nm)

Air cleaner outlet tube camps to 12–22 inch lbs. (1.4–2.5 Nm)

59. Fill the cooling system. Fill the crankcase with the proper type of motor oil to the required level.

60. Connect the battery cables, negative cable last. Run the engine and check for leaks.

61. Evacuate and recharge the air conditioning system.

62. Install and align the hood.

NOTE: Whenever the vehicles subframe is removed or lowered, the wheel alignment should be checked.

63. Check the front wheel alignment. Road test the vehicle and check the engine and transaxle for proper operation.

3.8L Engines

Except Continental

1. Drain the cooling system and disconnect the negative battery cable. Properly relieve the fuel system pressure.

2. Disconnect the underhood light wiring connector. Mark position of the hood on the hinges hinges and remove the hood.

3. Remove the oil level indicator tube.

4. Disconnect alternator to voltage regulator wiring.

5. Remove the radiator upper sight shield. Remove the engine cooling fan motor relay bolts and position the relay aside.

6. Remove the air cleaner assembly.

7. Disconnect the radiator electric fan and remove fan shroud.

8. Remove upper radiator hose.

9. Disconnect the transaxle oil cooler inlet and outlet tubes and

cover the openings to prevent the entry of dirt and grease. Disconnect the heater hoses.

10. Disconnect the power steering pressure hose.

11. Disconnect the air conditioner compressor clutch wire.

12. Properly recover the refrigerant from the air conditioning system. Disconnect the compressor-to-condenser line and cap the openings to prevent the entry of dirt and moisture.

13. Remove the radiator coolant reservoir. Remove the wiring shield.

14. Remove accelerator cable mounting bracket.

15. Disconnect fuel inlet and return lines.

16. Disconnect power steering pump pressure and return tube brackets.

17. Label and disconnect the engine control sensor wiring.

18. Identify, tag and disconnect all necessary vacuum hoses.

19. Disconnect the ground cable and remove the duct assembly.

20. Disconnect one end of the throttle control valve cable. Disconnect the bulkhead electrical connector and transaxle pressure switches.

21. Remove transaxle support bolts and remove transaxle support assembly from vehicle.

22. Raise the vehicle and support safely. Remove the wheel and tire assemblies. Drain the engine oil and remove the filter.

23. Disconnect the oxygen sensor.

24. Loosen and remove drive belts. Remove the crankshaft pulley and drive belt tensioner.

25. Remove the starter motor. Remove the converter-to-flywheel nuts.

26. Disconnect the exhaust system from the manifold and remove the catalytic converter and inlet pipe assembly.

27. Remove both front engine mount nuts.

28. Disconnect the oil level indicator sensor.

29. Disconnect the lower radiator hose.

30. Remove the engine-to-transaxle bolts and partially raise the engine.

31. Remove the water pump pulley retaining bolts and the water pump pulley.

32. Remove the distributor cap and position aside. Remove distributor rotor.

33. Remove the exhaust manifold bolt lock retaining bolts. Remove the thermactor air pump, if equipped.

34. Disconnect the oil pressure sending unit.

35. Install engine lifting eyes D81L-60001-D or equivalent, and connect suitable lifting equipment to the lifting eyes.

36. Position a suitable jack under the transaxle and raise the transaxle a small amount.

37. Carefully separate the engine from the transaxle. If equipped with an automatic transaxle, make sure the torque converter does not fall out.

38. Carefully remove the engine from the vehicle and position in a suitable holding fixture.

To install:

39. Remove the engine assembly from the work stand and position it in the vehicle. Make sure the torque converter properly mates to the flywheel.

40. Install the engine to transaxle bolts and remove the jack from under the transaxle and the engine lifting equipment. Remove the engine lifting eyes.

41. Install the remaining components in the reverse order of removal. Torque the following:
Engine to transaxle bolts to 41–50 ft. lbs. (55–68 Nm)
A/C compressor bolts to 30–45 ft. lbs. (41–61 Nm)
Converter-to-flywheel bolts to 20–34 ft. lbs. (27–46 Nm)
Crankshaft pulley bolts to 20–28 ft. lbs. (26–38 Nm)

42. Install the hood and connect the negative battery cable.

43. Fill the cooling system with the proper type and quantity of coolant and fill the crankcase with the proper type of motor oil to the required level.

44. Evacuate, pressure test and recharge the air conditioning system.

45. Start the engine and check for leaks and proper engine operation.

Continental

1. Disconnect the negative battery cable.

2. Properly relieve the fuel system pressure.

3. Drain the cooling system into a suitable container.

4. Properly recover the refrigerant from the air conditioning system.

5. Tag and disconnect the alternator-to-voltage regulator, electric cooling fan, transaxle pressure switch, air conditioning compressor clutch, electronic engine control and ground wiring.

6. Disconnect the heater hoses, power steering hoses and brackets, air conditioning discharge hose, transaxle oil cooler tubes and fuel lines. Cap all openings to prevent leakage and/or contamination.

7. Tag and disconnect the vacuum lines. Disconnect the throttle cable at the throttle valve.

8. Remove the electric cooling fan and motor assembly. Remove the fan shroud.

9. Remove the engine oil dipstick and the radiator sight shield. Remove the integrated controller relay and position aside.

10. Remove the air cleaner assembly.

11. Disconnect the upper radiator hose and remove the coolant recovery reservoir. Remove the wiring shield.

12. Remove the air suspension compressor and position aside. Remove the accelerator cable mounting bracket.

13. Remove the transaxle support assembly. Remove the air conditioning compressor.

14. Raise and safely support the vehicle.

15. Drain the engine oil and remove the oil filter. Disconnect the oxygen sensor.

16. Release the tension at the drive belts. Remove the crankshaft pulley and drive belt tensioner.

17. Remove the starter. Remove the catalytic converter housing cover and remove the converter and inlet pipe assembly from the engine.

18. Remove the nuts at the transaxle and engine mounts. Remove the torque converter-to-flywheel nuts.

19. Disconnect the oil level indicator sensor and the lower radiator hose.

20. Loosen the engine-to-transaxle bolts, leaving them loosely installed.

21. Partially lower the vehicle and remove the front wheel and tire assemblies.

22. Remove the drive belts and the water pump pulley. Remove the radiator assembly.

23. Remove the distributor cap and position aside. Remove the distributor rotor.

24. Remove the exhaust manifold lock bolts and the thermactor air pump, if equipped. Disconnect the oil pressure sending unit.

25. Install suitable engine lifting equipment and position a transaxle jack. Completely remove the engine-to-transaxle bolts.

26. Raise the transaxle assembly using the jack and lift the engine from the vehicle.

To install:

27. Position the engine assembly in the vehicle and align the engine-to-transaxle bolt bores. Install the engine-to-transaxle bolts that are accessible but do not tighten at this time.

28. Remove the transaxle jack and the engine lifting equipment.

29. Install the remaining components in the reverse order of removal. Torque the following:
A/C compressor bolts to 30–45 ft. lbs. (41–61 Nm)
Remaining transaxle-to-engine bolts to 40–50 ft. lbs. (55–68 Nm)
Torque converter-to-flywheel bolts to 20–34 ft. lbs. (27–46 Nm)
Transaxle mount retaining nuts to 50–70 ft. lbs. (68–95 Nm)
Crankshaft pulley bolts to 20–28 ft. lbs. (26–38 Nm)
Thermactor air pump mounting bolts to 30–40 ft. lbs. (40–55 Nm)

30. Lower the vehicle. Install the hood, aligning the marks that were made during the removal procedure.

31. Install the air cleaner assembly and connect the negative battery cable.

32. Fill the engine with the proper type and quantity of engine oil and coolant. Leak test, evacuate and charge the air conditioning system. Observe all safety precautions.

33. Start the engine and check for leaks.

4.6L Engine

--------- CAUTION ---------
Do not service an air spring under any circumstances when pressurized. Do not remove any suspension components supporting an air spring without first exhausting the air. The air suspension switch, located in the left–hand side of the luggage compartment, must be turned OFF before raising the vehicle. Failure to follow these instructions may result in unexpected inflation or deflation of the air springs which may result in shifting of the vehicle during service.

NOTE: **Make sure to have available a New Generation Star (NGS) tester or equivalent scan tool with the proper software to properly deflate the air springs before servicing the suspension system. Make sure to have available a new wheel hub retainer nut, lower ball-joint to lower control arm nut and a halfshaft circlip, per side.**

1. Place the vehicle over a frame contact hoist. Properly deflate the air springs.

2. Disconnect both battery cables, negative cable first. Drain the engine cooling system.

3. Mark the position of the hood on the hinges and remove the hood.

4. Disconnect the steering coupling at the pinch bolt joint inside the passenger compartment.

5. Remove the engine appearance cover from the engine. Properly relieve the fuel system pressure.

6. Recover the refrigerant from the A/C system using the proper equipment.

NOTE: **Label all electrical connectors and vacuum hoses prior to removal so they can be reinstalled in their proper locations.**

7. Disconnect the engine control sensor wiring from the Intake Air Temperature (IAT) sensor and disconnect the crankcase ventilation tube from air cleaner outlet tube.

8. Loosen the clamps on the air cleaner outlet tube to engine air cleaner and throttle body. Remove the air cleaner outlet tube.

9. Disconnect the chassis vacuum supply hose at the connection on the intake manifold. Position the hose aside.

10. Remove the ground straps from the dash panel.

11. Disconnect the engine control sensor wiring from the Powertrain Control Module (PCM) and position the engine control sensor wiring aside.

12. Remove the connectors for the engine control sensor wiring from the retaining bracket on the power brake booster. Disconnect the engine control sensor wiring at the 2 connectors.

13. Disconnect engine control sensor wiring from the Mass Air Flow (MAF) sensor.

14. Disconnect engine control sensor wiring from the evaporative emission canister purge valve.

15. Disconnect evaporative emission hose at crankcase vent connector and hose. Position the hose aside.

16. Remove the throttle cable shield and disconnect the accelerator cable and the speed control actuator from the throttle body and from the accelerator cable bracket. Position the cables aside.

17. Remove the retaining nut and disconnect the manual control lever from the manual control lever shaft at the Transmission Range (TR) sensor.

18. Remove the connectors for the engine control sensor wiring from the retaining bracket on top of the transaxle. Disconnect the engine control sensor wiring at the 2 connectors.

19. Disconnect the wiring connector from the secondary air injection

pump relay located on the retaining bracket on top of the transaxle.

20. Disconnect the main emission vacuum control connector at the connection near the fan shroud.

21. Disconnect the oil cooler inlet tube from the transaxle. Remove the oil level dipstick from the indicator tube.

22. Disconnect the heater water hose from the water bypass tube. Disconnect the heater water hose at the rear of the right cylinder head. Disconnect the upper radiator hose at the water bypass tube.

23. Remove the power steering return hose from the power steering oil reservoir and drain.

24. Disconnect the alternator wiring harness from the alternator at the BAT terminal and stator connector plug. Remove the wiring harness retaining clip from the alternator mounting bracket.

25. Partially raise and safely support the vehicle. Remove the front wheel and tire assemblies.

26. Disconnect both ride height sensor links from the lower control arms. Remove both stabilizer bar links from the front stabilizer bar.

27. Separate both lower control arms from the steering knuckles at the ball joints. Separate both tie rod ends from the steering knuckles. Discard the cotter pins.

28. Remove both wheel hub retainer nuts from the halfshaft ends. Remove both halfshafts from the steering knuckles.

29. Raise and safely support the vehicle. Remove the splash shield from the radiator support and subframe.

30. Drain the engine oil. Remove the dual convertor Y-pipe.

31. Disconnect the power steering pressure hose from the power steering/transaxle oil cooler connection. Position the hose aside.

32. Disconnect the lower radiator hose at the radiator and thermostat housing and remove the radiator.

33. Disconnect the wiring at the starter, and remove the starter. Disconnect the lower cooler line from the transaxle.

34. Disconnect the retaining clips and the A/C compressor lines from the compressor. Cap all openings to prevent the entrance of dirt or moisture.

35. Support the subframe, engine and transaxle assembly using Powertrain Lift 014-00765 and Universal Powertrain Removal Bracket 014-00766, or equivalents.

36. Remove 4 subframe-to-body retaining bolts. Slowly lower the en-

gine, transaxle and subframe from the vehicle as one assembly.

37. Disconnect the power steering pressure hose from the power steering pump.

38. Install suitable engine lifting brackets on the engine and transaxle assembly.

39. Remove the front engine support insulator, rear engine support insulator and engine and transaxle support. Lift the engine and transaxle from the subframe.

40. Lower the engine and transaxle. Support the transaxle on a level, stationary surface for transaxle storage.

41. Remove the transaxle-to-cylinder block mounting bolts and separate the engine from the transaxle/torque converter assembly. Place the engine on a suitable workstand.

To install:

42. Install the transaxle/torque converter assembly to the engine. Tighten the mounting bolts to 30–44 ft. lbs. (40–60 Nm). Tighten the torque convertor nuts to 20–34 ft. lbs. (27–46 Nm).

43. Place the engine and transaxle assembly on the subframe. Install the front engine support insulator, rear engine support insulator and engine and transaxle support.

44. Connect the power steering pressure hose from the power steering pump.

45. Raise the engine, transaxle and subframe assembly into position using the powertrain lifting tool. Loosely install 4 subframe to body bolts.

46. Align the subframe using a 0.75 inch (19mm) outside diameter pipe or similar tool installed through both alignment holes. Tighten the bolts to 55–75 ft. lbs. (75–102 Nm).

47. Remove the lifting equipment and move aside. Connect the retaining clips and the A/C compressor lines to the compressor.

48. Install the remaining components in the reverse order of removal, then torque the following:

Hub retainer nuts to 170–202 ft. lbs. (230–275 Nm)

Tie rod end-to-steering knuckle castellated nuts to 35–46 ft. lbs. (47–63 Nm)

Lower control arm-to-steering knuckle nuts to 50–68 ft. lbs. (68–92 Nm)

Stabilizer bar links-to-stabilizer bar nuts to 30–40 ft. lbs. (40–55 Nm)

Manual control lever-to-manual control lever shaft nut to 12–16 ft. lbs. (16–22 Nm)

Air cleaner outlet tube clamps to 12–22 inch lbs. (1.4–2.5 Nm)

49. Install the hood and align as needed. Install the fuel tank fill cap.

50. Fill the cooling system with the proper type and quantity of coolant and fill the crankcase with the proper type of motor oil to the required level. Check the power steering and transaxle fluid levels and fill as needed.

51. Connect both battery cables, negative cable last. Fill the air springs. Run the engine and check for leaks.

52. Evacuate, pressure test and recharge the air conditioning system.

53. Install the engine appearance cover. Tighten the retaining nuts to 61–79 inch lbs. (7–9 Nm).

NOTE: Whenever the vehicles subframe is removed or lowered, the wheel alignment should be checked.

54. Check the front wheel alignment. Road test the vehicle and check the engine for proper operation.

Engine Mounts

REMOVAL AND INSTALLATION

3.0L (VIN Y) SHO Engine

Right Front and Right Rear Mount

1. Remove the front engine mount rebound insulator retaining nuts from the right front fender apron.
2. Raise the vehicle and support it safely. Place a jack and a wood block in a suitable place under the engine.
3. Remove the roll damper to engine retaining nuts and remove the roll damper.
4. Remove the engine support insulators to front sub-frame retaining nuts.
5. Raise the engine enough to unload the right and left support insulators.
6. Remove the 2 through bolts and remove the left support insulator from the A/C compressor mounting bracket.
7. Remove the 2 through bolts and remove the right front support insulator from the power steering pump support.

To install:

8. Install the right engine support insulator to the power steering pump support with 2 through bolts. Tighten to 40–53 ft. lbs. (54–72 Nm).

9. Install the left engine support insulator to the A/C compressor mounting bracket with 2 through bolts. Tighten the bolts to 40–53 ft. lbs. (54–72 Nm).
10. Lower the engine down onto the sub-frame.
11. Install nuts retaining the left engine support insulator and the right support insulator to the front sub-frame. Tighten the nuts to 56–77 ft. lbs. (76–104 Nm).
12. Install the engine mounting damper to the front engine support insulator. Tighten the bolts to 40–53 ft. lbs. (54–72 Nm).
13. Remove jack and lower the vehicle.
14. Install nuts retaining the front engine mount rebound insulator to the front fender apron. Tighten the nuts to 21–30 ft. lbs. (28–41 Nm).

Left Engine Insulator and Support

1. Remove the bolt retaining the roll damper to the lower damper bracket and place the damper shaft aside.
2. Remove the backup light switch. Remove engine damper mounting body bracket.
3. Raise the vehicle and support it with jackstands under the vehicle body, allowing the subframe to hang.
4. Remove the left tire and wheel assembly. Place a jack and wood block under the transaxle.
5. Remove the bolts retaining the damper bracket to the engine and transaxle support insulator.
6. Remove the bolts retaining the engine and transaxle support insulator to the transaxle and front sub-frame.
7. Raise the transaxle with the jack enough to unload the insulator. Remove the engine and transaxle insulator and lower damper bracket.

To install:

8. Loosely install the engine and transaxle support insulator and damper mounting bracket to the transaxle and front sub-frame.
9. Install the damper mounting bracket to the engine and transaxle support insulator. Tighten the bolts to 40–53 ft. lbs. (54–72 Nm).
10. Install the engine and transaxle support insulator to the transaxle. Tighten the bolts to 72–97 ft. lbs. (98–132 Nm).
11. Install the engine and transaxle support insulator to the front sub-frame. Tighten the bolts to 64–88 ft. lbs. (87–119 Nm).
12. Remove the jack and the jackstands. Install the left tire and wheel. Lower the vehicle.

13. Install the engine mounting damper to the damper mounting bracket. Tighten the bolt to 40–53 ft. lbs. 54–72 Nm).

14. Install the backup light switch. Install the engine damper mounting body bracket. Tighten the retaining bolts to 21–30 ft. lbs. (28–41 Nm).

3.2L SHO Engine and 3.0L (VIN S) Engine 1996–97 3.0L (VIN 1 and U) Engines

Right Front and Right Rear Mount

1. Raise the vehicle and support it safely. Place a jack and a wood block in a suitable place under the engine.

2. Remove the nuts retaining the right front and right rear engine support insulators to front sub–frame.

3. Raise the engine enough to unload the right front and right rear support insulators.

4. Remove the 2 bolts retaining the right front support insulator and the right rear support insulator to the engine mounting brackets. Remove the right front and right rear engine support insulators.

To install:

5. On 3.2L SHO engine, install the right front engine support insulator and the right rear support insulators to the engine mounting brackets with 2 bolts. Tighten to 40–53 ft. lbs. (54–72 Nm).

6. On 3.0L (VIN S) engine, install the left engine support insulator to the engine mounting bracket with 2 bolts. Tighten the bolts to 44–59 ft. lbs. (60–80 Nm).

7. On 1996–97 3.0L (VIN 1 and U) engines, install both engine support insulators to the engine mounting brackets and secure with 2 bolts each. Tighten the retaining bolts to 44–59 ft. lbs. (60–80 Nm).

8. On 3.0L (VIN S) engine, install the right engine support insulator to the engine mounting bracket with 1 through-bolt. Tighten the through-bolt to 39–53 ft. lbs. (53–72 Nm).

9. Lower the engine down onto the sub–frame.

10. Install nuts retaining the right front engine support insulator and the right rear support insulator to the front sub–frame. Tighten the nuts to 56–77 ft. lbs. (76–104 Nm).

11. Remove jack and lower the vehicle.

Left Engine Insulator and Support

1. Raise the vehicle and support it safely. Remove left tire and wheel assembly.

2. Place a jack and wood block under the transaxle and support the transaxle.

3. Remove the nut retaining the engine and transaxle support insulator to the engine and transaxle support.

4. Remove the 2 through bolts retaining the engine and transaxle support insulator to the front sub–frame.

5. Raise the transaxle with the jack enough to unload the mount.

6. Remove the bolts retaining the support assembly to the transaxle and remove the mount and/or transaxle support assembly.

To install:

7. Loosely install the engine and transaxle support insulator.

8. Attach the engine and transaxle support to the transaxle and tighten the bolts to 40–53 ft. lbs. (54–72 Nm).

9. Attach the engine and transaxle support insulator to the front sub–frame with 2 through bolts and tighten to 64–88 ft. lbs. (87–119 Nm) for 3.2L SHO engine or to 57–76 ft. lbs. (77–103 Nm) on 3.0L (VIN S) engine and 1996–97 3.0L (VIN 1 and U) engines.

10. Lower the transaxle (enough to load the engine and transaxle support insulator).

11. Install the engine and transaxle support insulator to the engine and transaxle support retaining nut and tighten to 56–77 ft. lbs. (76–104 Nm).

12. Remove the jack. Install the left tire and wheel. Lower the vehicle.

1993–95 3.0L (VIN U) Engine

Right Side Front and Rear Insulators

1. Disconnect the negative battery cable. Raise and support the vehicle safely.

2. Place a suitable jack and a block of wood under the engine block.

3. Remove the nuts attaching the right front engine support insulator and left front support insulators to the front sub–frame.

4. Raise the engine with the jack until enough of a load is taken off the insulator.

5. Remove the insulator retaining bolts and remove the insulators from the engine support bracket.

To install:

6. Attach the left front engine support insulator and the right front engine support insulator to the engine mounting brackets with 2 bolts. Tighten the bolts to 40–53 ft. lbs. (54–72 Nm).

7. Lower the engine down onto the front sub–frame.

8. Install the nuts retaining the left front engine support insulator

and right front engine support insulator to the front sub–frame. Tighten the nuts to 56–77 ft. lbs. (76–104 Nm).

9. Remove the jack and safely lower the vehicle.

Left Insulator

1. Disconnect the negative battery cable. Raise and support the vehicle safely. Remove the wheel and tire assembly.

2. Place a suitable jack and a block of wood under the transaxle and support the transaxle.

3. Remove the nuts attaching the insulator to the support assembly. Remove the through bolts attaching the insulator to the frame.

4. Raise the transaxle with the jack enough to relieve the weight on the insulator.

5. Remove the bolts attaching the support assembly to the transaxle. Remove the insulator and/or transaxle support assembly.

To install:

6. Loosely install the mount. Attach the support assembly to the transaxle. Tighten the bolts to 40–55 ft. lbs.

7. Attach the insulator to the frame with 2 through bolts. Tighten the bolts to 60–85 ft. lbs. (81–116 Nm).

8. Lower the transaxle (enough to load insulator). Attach the insulator to the support assembly with 2 nuts. Tighten the nuts to 55–75 ft. lbs. (74–102 Nm).

9. Remove the jack. Install the left tire and wheel. Safely lower the vehicle.

3.8L Engines, Except Continental

Right Front and Left Front Insulators

1. Remove the mount upper retaining nut through the engine compartment using a long extension and an 18mm swivel socket.

2. Install 3 bar engine support D88L-6000-A or equivalent. Raise and safely support the vehicle.

3. Loosen the right rear and right front lower mount retaining nuts. Lower the vehicle.

4. Raise the engine approximately 1 inch using the engine support tool.

5. Raise and safely support the vehicle. Remove the engine mounts.

To install:

6. Install the left front engine support insulator and right front engine insulator and engine mount heat shields. Locate the anti–rotation pin to the A/C compressor mounting bracket and transaxle mounting

bracket and install the upper retaining nuts securely.

7. Lower the vehicle. Lower the engine using the 3 bar engine support D88L-6000-A or equivalent. Locate the lower studs through the front subframe.

8. Raise and safely support the vehicle.

9. Install the left front engine support insulator and right front engine insulator lower retaining nuts and tighten to 56–77 ft. lbs. (76–104 Nm).

10. Lower the vehicle. Remove the 3 bar engine support tool.

11. Tighten the left front engine support insulator upper retaining nut to 40–53 ft. lbs. (54–72 Nm) through the engine compartment between the A/C compressor and the A/C compressor mounting bracket using a long extension and a 18mm swivel socket.

12. Tighten the right front engine support insulator upper retaining nut to 40–53 ft. lbs. (54–72 Nm).

Left Engine Mount and Support Assembly

1. Raise the vehicle and support it safely. Remove the tire and wheel assembly.

2. Place a jack and wood block under the transaxle and support the transaxle.

3. Remove the nut retaining the engine and transaxle support insulator to the engine and transaxle support.

4. Remove the 2 through bolts retaining the engine and transaxle support insulator to the front sub-frame.

5. Raise the transaxle with the jack enough to unload the mount.

6. Remove the bolts retaining the support assembly to the transaxle and remove the mount and/or transaxle support assembly.

To install:

7. Attach the engine and transaxle support to the transaxle and tighten the bolts to 40–53 ft. lbs. (54–72 Nm).

8. Attach the engine and transaxle support insulator to the front sub-frame with 2 through bolts and tighten to 64–88 ft. lbs. (87–119 Nm).

9. Lower the transaxle (enough to load the engine and transaxle support insulator).

10. Install the engine and transaxle support insulator to the engine and transaxle support retaining nut and tighten to 56–77 ft. lbs. (76–104 Nm).

11. Remove the jack. Install the LH tire and wheel. Lower the vehicle.

3.8L Continental

Front

1. Disconnect the negative battery cable.

2. Remove fan shroud retaining screws. Remove the air tube to the remote air cleaner.

3. Raise and safely support the vehicle. Support engine using a jack and wood block placed under the engine.

4. Remove the engine mount through bolt. Remove shift linkage. Raise engine high enough to clear clevis brackets.

NOTE: Raise the engine carefully so as not to damage the lines and hoses at the rear of the engine.

5. Remove any accessory and oil cooler line retaining clips from the engine support brackets.

6. Remove bolts retaining the engine mount and bracket assembly to engine. Remove the mount and bracket assembly.

To install:

7. Position the engine mount and bracket assembly to the engine, install the retaining bolts and tighten to 26–34 ft. lbs. (34–47 Nm).

8. Install the accessories to the lower front engine mount support bracket stud. Tighten to 26–34 ft. lbs. (34–47 Nm).

9. Lower the engine into position and make sure the engine mounts are seated flat on the front sub frame. The left mount must seat first. Install the through-bolt and tighten to 35–50 ft. lbs. (47–68 Nm).

10. Lower the vehicle and install the air tube and the fan shroud retaining screws. Connect the negative battery cable.

Rear

1. Disconnect the negative battery cable. Raise and safely support the vehicle.

2. Support the transaxle with a jack and a wood block. Remove the rear nut attaching the mount-to-crossmember. Keep transaxle weight on the mount during nut removal.

3. Remove the 2 bolts retaining the crossmember-to-body brackets. Remove the crossmember by raising the transaxle slightly with the jack.

4. Remove the bolts retaining the rear engine mount to the transaxle. Remove the mount.

To install:

5. Position the engine mount and retainer on the transaxle. Install the 2 retaining bolts and tighten to 35–50 ft. lbs. (47–68 Nm).

6. Install the crossmember-to-body brackets. Tighten the retaining bolts or nuts to 34–47 ft. lbs. (45–65 Nm).

7. Lower the transaxle. Install the mount-to-crossmember nut. Tighten to 65–84 ft. lbs. (88–115 Nm).

8. Lower the vehicle and connect the negative battery cable.

4.6L Engine

— **CAUTION** —

The air suspension switch, located in the left-hand side of the luggage compartment, must be turned OFF before raising the vehicle. Failure to do so may result in unexpected inflation or deflation of the air springs which may result in shifting of the vehicle during service.

— **WARNING** —

Do not excessively raise the engine. Over–extension of the tripot CV-joint could result in separation of internal parts, causing failure of both halfshafts.

Left and Right Front Engine Mounts

1. Turn the air suspension switch, located in the left side of the luggage compartment, to the **OFF** position. Disconnect the negative battery cable.

2. Raise and safely support the vehicle. Support the engine using a jackstand and wood block placed under the engine.

3. Remove the engine mount (support insulator) nuts retaining the left front and right front engine support insulators to the subframe.

4. Raise the engine just high enough to unload both front engine support insulator.

— **WARNING** —

Raise the engine carefully so as not to damage the lines and hoses at the rear of the engine. If required, remove both halfshafts.

5. Remove the bolts retaining the engine mount and bracket assemblies to the engine. Remove the mount and bracket assemblies.

To install:

6. Position the engine mount and bracket assemblies to the engine, install the retaining bolts and tighten to 58–76 ft. lbs. (78–103 Nm).

7. Lower the engine into position and make sure the engine mounts are seated flat on the subframe. Install the nut retaining the insulator to the

sub–frame and tighten to 44–59 ft. lbs. (60–80 Nm).

8. Connect the negative battery cable. Lower the vehicle and turn the air suspension to the **ON** position. Road test the vehicle and check for proper operation.

Rear Transaxle Mount

1. Turn the air suspension switch, located in the left–hand side of the luggage compartment, to the **OFF** position. Disconnect the negative battery cable.

2. Raise and safely support the vehicle. Remove left front wheel and tire assembly.

3. Support the transaxle with a suitable transaxle jack. Remove the rear nut retaining the transaxle mount (support insulator) to the crossmember. Keep transaxle weight on the mount to aide nut removal.

4. Remove 2 bolts retaining the crossmember-to-body brackets. Remove the crossmember by raising the transaxle slightly with the jack.

5. Remove the bolts retaining the rear mount to the transaxle. Remove the mount.

To install:

6. Position the transaxle mount and retainer on the transaxle. Install 2 retaining bolts and tighten to 39–53 ft. lbs. (53–72 Nm).

7. Install the transaxle support insulator to the subframe with 2 through-bolts. Tighten the retaining bolts to 64–88 ft. lbs. (87–119 Nm).

8. Lower the transaxle. Install the transaxle mount-to-crossmember nut. Tighten to 56–77 ft. lbs. (76–104 Nm).

9. Install the left front wheel and tire assembly. Lower the vehicle. Connect the negative battery cable.

10. Turn air suspension switch to the **ON** position. Road test the vehicle and check for proper operation.

Cylinder Head

REMOVAL AND INSTALLATION

—————— CAUTION ——————
Fuel injection systems remain under pressure, even after the engine has been turned OFF. The fuel system pressure must be relieved before disconnecting any fuel lines. Failure to do so may result in fire and/or personal injury.

3.0L (VIN Y) and 3.2L SHO Engines

1. Disconnect the negative battery cable. Properly relieve the fuel system pressure.

2. Drain the cooling system. Remove the air cleaner outlet tube. Remove the intake manifold.

3. Loosen the accessory drive belt idlers and remove the drive belts. Remove the upper timing belt cover.

4. Remove the left idler pulley(s) and bracket assembly. Raise the vehicle and support it safely.

5. Remove the right wheel and inner fender splash shield. Remove the crankshaft damper pulley. Remove the lower timing belt cover.

6. Align both camshaft pulley timing marks with the index marks on the upper steel belt cover.

7. Release the tension on the belt by loosening the tensioner nut and rotating the tensioner with a hex head wrench. When tension is released, tighten the nut to hold the tensioner in place.

8. Disconnect the crankshaft sensor wiring assembly. Remove the center cover assembly.

9. Remove the timing belt. Note the location of the letters **KOA** on the belt. The belt must be installed to rotate in the same direction.

10. Remove the cylinder head covers. Remove the camshaft timing pulleys. Remove the upper rear and the center rear timing belt covers.

11. If the left cylinder head is being removed, remove the DIS coil bracket and the oil dipstick tube. If the right cylinder head is being removed, remove the coolant outlet hose.

12. Remove the exhaust manifold on the left cylinder head. On the right cylinder head the exhaust manifold must be removed with the head.

13. Remove the cylinder head bolts and remove the cylinder head.

To install:

14. Clean the bolt holes in the block with a tap. Lightly oil all bolt and stud bolt threads except those entering a coolant jacket Those bolts and studs must be sealed with a silicone sealer.

15. Clean the cylinder head and engine block mating surfaces of all gasket material.

16. Position the cylinder head and gasket on the engine block and align with the dowel pins.

17. Install the cylinder head bolts and tighten in sequence to 37–50 ft. lbs. (49–69 Nm). Repeat the torque sequence and tighten to 62–68 ft. lbs. (83–93 Nm).

18. When installing the left cylinder head, install the exhaust mani-

fold, DIS coil bracket and oil dipstick tube.

19. When installing the right cylinder head, install the coolant outlet hose and connect the exhaust catalyst.

20. Install the remaining components in the reverse order of removal.

21. Connect the negative battery cable.

22. Fill the engine cooling system with the proper type and quantity of coolant. Start the engine and check for coolant, fuel or oil leaks.

1993–95 3.0L (VIN U) Engines
1996–97 3.0L (VIN 1 and U) Engines

1. Rotate the crankshaft until the piston in No. 1 cylinder is at TDC on the compression stroke.

2. Disconnect the negative battery cable. Properly relieve the fuel system pressure. Drain the cooling system into a suitable container.

3. Remove the air cleaner outlet hose to throttle body. Label and disconnect the vacuum lines from the throttle body.

4. Disconnect the hoses from the EGR valve. Loosen the lower EGR tube nut and rotate the tube away from the valve.

5. Label and disconnect the wiring from the Intake Air Temperature (IAT) sensor, Throttle Position Sensor (TPS), Idle Air Control (IAC) valve and Pressure Feedback EGR (PFE) or Differential Pressure Feedback EGR (DPFE) sensors.

6. Remove the fuel line safety clips and disconnect the fuel lines from the fuel supply manifold.

7. On 1993–95 3.0L (VIN U) engine, remove the throttle body and discard the gasket.

8. On 1996–97 3.0L (VIN 1 and U) engines, remove the upper intake manifold and discard the gasket.

9. Label and disconnect the fuel injector harness from the valve cover studs and fuel injectors.

10. Remove the ignition coil and bracket from the left (front) cylinder and set aside.

11. Label and disconnect the ignition wires from the spark plugs and the valve cover studs. Disconnect the upper radiator and heater hoses.

12. On 1993–95 3.0L (VIN U) engine with non-flexible fuel, remove the distributor cap. Disconnect the wiring and remove the distributor.

13. On 1993–95 3.0L (VIN U) engine with flexible fuel, remove the Camshaft Position (CMP) sensor.

14. Disconnect the Engine Coolant Temperature (ECT) sensor and tem-

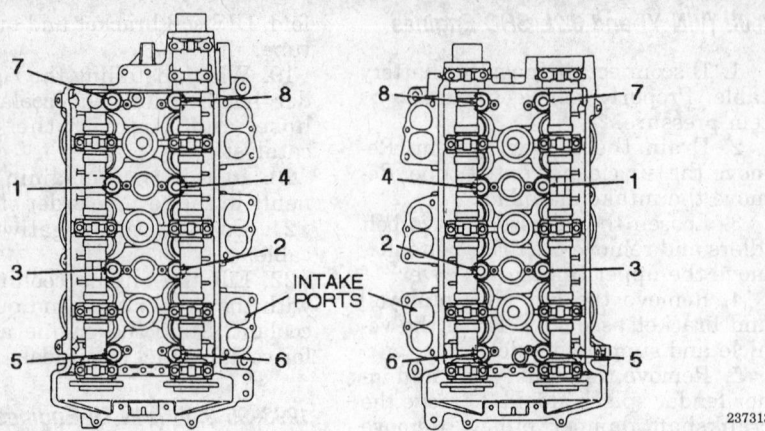

Cylinder head bolt tightening sequence — 3.0L (VIN Y) and 3.2L SHO engines

perature sending unit electrical connectors.

15. For the left (front) cylinder head, perform the following:

a. Disconnect the alternator electrical connectors.

b. Rotate the tensioner clockwise and remove the accessory drive belt.

c. Remove the automatic belt tensioner assembly.

d. Remove the alternator.

e. Remove the power steering mounting bracket retaining bolts. Leave the hoses connected and place the pump aside in a position to prevent fluid from leaking out.

f. Remove the engine oil dipstick tube from the exhaust manifold.

16. For the right (rear) cylinder head, perform the following:

a. Remove the alternator belt tensioner bracket.

b. Remove the heater supply tube retaining brackets from the exhaust manifold.

c. Remove the Vehicle Speed Sensor (VSS) cable retaining bolt.

d. Remove the throttle emission control solenoid and bracket.

17. Remove the rocker arm covers. Loosen the rocker arm fulcrum bolts and remove the rocker arms, fulcrums and bolts. Keep the assemblies in order so they can be reinstalled in their original locations.

NOTE: Regardless of the cylinder head being removed, the No. 3 cylinder intake valve pushrod must be removed to allow removal of the intake manifold.

18. Remove the pushrods and label their positions. The pushrods must be installed in their original position during reassembly.

19. Remove the intake manifold. Remove the spark plugs. Remove the exhaust manifolds.

20. Remove and discard the cylinder head bolts and remove the cylinder heads from the engine. Remove and discard the old cylinder head gaskets.

To install:

21. Clean the cylinder head bolts holes in the block with a tap. Clean the cylinder head, intake manifold, rocker arm cover and cylinder head gasket contact surfaces.

22. Position new head gaskets on the cylinder block using the dowels in the block for alignment. If the dowels are damaged, they must be replaced.

23. Position the cylinder head on the block. Install new cylinder head bolts and tighten, in sequence, to 59

Cylinder head gasket installation — 1993–95 3.0L (VIN U) and 1996–97 3.0L (VIN 1 and U) engines

ft. lbs. (80 Nm), then back off the bolts one turn.

24. Tighten the cylinder head bolts, in sequence, to 37 ft. lbs. (50 Nm). Repeat the torque sequence and tighten the bolts to 68 ft. lbs. (92 Nm).

25. Install the intake manifold. Connect the ECT and coolant temperature sending unit connectors.

26. On 1993–95 3.0L (VIN U) engine with non-flexible fuel, install the distributor.

27. Dip each pushrod end in oil conditioner or heavy engine oil. Install the pushrods in their original position.

28. Before installation, coat the valve tips, rocker arm and fulcrum contact areas with Lubriplate® or equivalent.

29. Rotate the crankshaft until the lifter is on the base circle of the cam (valve closed).

30. Install the rocker arm assemblies and torque the rocker arm fulcrum bolts to 24 ft. lbs. (32 Nm). Be sure the lifter is on the base circle of the cam for each rocker arm as it is installed.

NOTE: The fulcrums must be fully seated in the cylinder head and the pushrods must be seated in the rocker arm sockets prior to the final tightening.

31. Install the exhaust manifolds. Install the spark plugs.

32. Position the valve covers on the cylinder head and install the retaining bolts. Note the location of the ignition wire retainer stud bolts.

33. Install the fuel charging wiring to the fuel injectors and inboard valve cover stud bolts.

34. On 1993–95 3.0L (VIN U) engine, install the throttle body and intake manifold upper gasket.

35. On 1996–97 3.0L (VIN 1 and U) engines, install the upper intake manifold upper gasket and the upper intake manifold.

36. Install the ignition coil and bracket. Connect all sensor electrical connectors.

37. For the left (front) cylinder head, perform the following:

a. Connect the oil dipstick tube to the exhaust manifold stud and tighten the nut to 13 ft. lbs. (18 Nm).

b. Install the power steering support bracket and pump. Tighten the retaining bolt to 35 ft. lbs. (48 Nm).

c. Install the automatic belt tensioner and tighten the retaining nuts/bolt to 35 ft. lbs. (48 Nm).

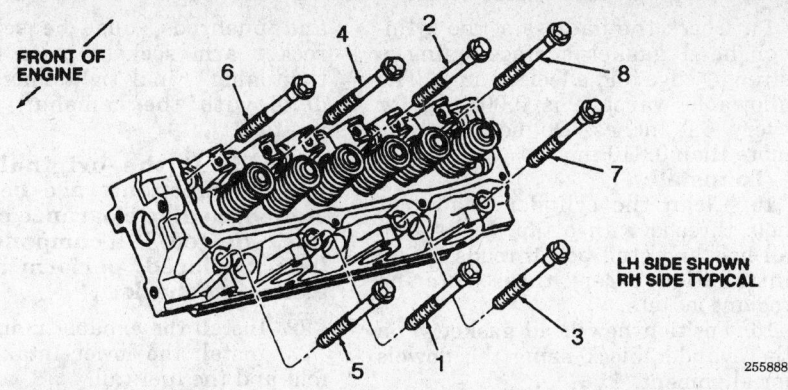

FRONT OF ENGINE

LH SIDE SHOWN
RH SIDE TYPICAL

255888

Cylinder head bolt tightening sequence — 1993–95 3.0L (VIN U) and 1996–97 3.0L (VIN 1 and U) engines

38. For the right (rear) cylinder head, perform the following:

a. Install the alternator belt tensioner bracket.

b. Install the throttle emission control solenoid and bracket. Tighten the retaining bolt to 26 ft. lbs. (35 Nm).

c. Install the heater supply tube retaining brackets to the exhaust manifold and tighten the nuts to 26 ft. lbs. (35 Nm).

d. Install the VSS cable retaining bracket.

39. On flexible fuel vehicles, install the CMP sensor.

40. Install the alternator. Install the drive belt. Connect the fuel lines.

41. Connect the upper radiator hose and heater hoses. Connect all vacuum lines to premarked locations. Change the engine oil and filter.

42. Install the air cleaner outlet tube to throttle body and engine air cleaner. Install crankcase ventilation tube to valve cover.

43. Fill and bleed the cooling system. Connect the negative battery cable. Start the engine and check for coolant, fuel, oil, vacuum and exhaust leaks.

44. On 1993–95 3.0L (VIN U) engine, with non-flexible fuel, check the ignition timing.

45. Check and if necessary, adjust the cruise control cable and the throttle valve cable.

3.0L (VIN S) Engines

NOTE: The cylinder head bolts are a torque-to-yield design and cannot be reused. Make sure new cylinder head bolts are available before beginning this procedure.

1. Disconnect the negative battery cable. Drain the engine coolant from the radiator and cylinder block drain plugs.

2. Remove the engine from the vehicle and position on a suitable workstand.

3. Remove the upper and lower intake manifolds. Remove the exhaust manifolds.

4. Drain the engine oil and remove the oil pan. Remove the engine front cover.

5. Remove the timing chains, camshafts and lash adjusters from both cylinder heads.

6. Remove the cylinder head retaining bolts from the cylinder heads in the proper removal sequence. Inspect the cylinder heads and cylinder block.

To install:

7. Clean the cylinder heads, intake manifolds, valve covers and the cylinder head gasket sealing surfaces on the cylinder block.

NOTE: If the cylinder heads were removed for cylinder head gasket replacement, check the flatness of the cylinder heads and the cylinder block gasket sealing surfaces.

8. Install new cylinder head gaskets onto the dowels of the cylinder block.

NOTE: Left and right cylinder head gaskets are not interchangeable.

9. Place the cylinder heads to their original positions using care not to damage the heads, block or gaskets.

10. Make sure the cylinder heads are correctly positioned on the dowels. Lightly oil the threads of the new cylinder head retaining bolts and install into the cylinder heads.

11. Tighten the new cylinder head retaining bolts as follows:

• Torque the bolts, in sequence, to 27–32 ft. lbs. (37–43 Nm)

• Rotate the bolts, in sequence, 85–95 degrees

• Loosen the bolts, in sequence, a minimum of 1 full turn

• Torque the bolts, in sequence, to 27–32 ft. lbs. (37–43 Nm)

• Rotate the bolts, in sequence, 85–95 degrees

• Rotate the bolts, in sequence, an additional 85–95 degrees

12. Install the EGR backpressure transducer and bracket to the rear of the right cylinder head. Tighten the retaining bolts to 71–106 inch lbs. (8–12 Nm).

13. Install the lash adjusters, camshafts and timing chains. Install the engine front cover. Install the engine oil pan.

14. Install the exhaust manifolds. Install the lower and upper intake manifolds. Replace the engine oil filter.

15. Install the engine assembly into the vehicle. Fill the engine with the proper amount and grade of engine oil. Fill the engine cooling system.

16. Connect the negative battery cable. Run the engine and check for leaks. Road test the vehicle and check for proper engine operation.

3.8L Engine

NOTE: The cylinder head bolts are a torque-to-yield design and cannot be reused. Before beginning this job, make sure new cylinder head bolts are available.

1. Disconnect the negative battery cable.

2. Relieve the fuel system pressure. Drain the cooling system.

3. Remove air cleaner assembly including the air intake duct and heat tube.

4. Loosen accessory drive belt idler. Remove drive belt.

5. For the left cylinder head, perform the following:

a. Remove oil fill cap.

b. On 1995 3.8L engine, remove the A/C mounting bracket retaining bolts. Leaving the hoses connected position the A/C compressor aside.

c. Remove the power steering pump front mounting bracket attaching bolts.

d. Remove the alternator assembly and accessory drive belt main idler.

e. Remove the power steering/pump alternator bracket retaining bolts.

f. Leaving the hoses connected, place the power steering pump/alternator bracket assembly aside in a position to prevent the fluid from leaking out.

6. For the right cylinder head, perform the following:

a. On 1993–94 3.8L engine, if equipped, disconnect the thermactor tube support bracket

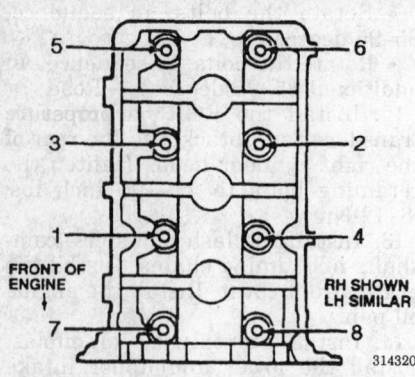

Cylinder head bolt torque sequence —
1996–97 3.0L (VIN S) engine

from the rear of the cylinder head. Remove the thermactor pump pulley and remove the pump.

b. On 1993–94 3.8L engine, if equipped, remove the air conditioner compressor belt and main drive belt.

c. On 1993–94 3.8L engine, if equipped, remove the compressor mounting bracket retaining bolts. Leave the hoses connected and position the compressor aside.

d. Remove the PCV valve.

7. Remove the upper intake manifold.

8. Remove valve rocker arm cover attaching screws. Remove the fuel rail and the lower intake manifold.

9. Remove the exhaust manifold(s).

10. Loosen the rocker arm fulcrum attaching bolts enough to allow the rocker arm to be lifted off the pushrod and rotated aside.

11. Remove the pushrods. Keep the pushrods in order because they must be installed in their original position during assembly.

12. Remove the cylinder head bolts and discard them. Remove the cylinder head(s).

13. Clean all gasket mating surfaces.

14. Check the flatness of the cylinder head gasket surface using a straight-edge and a feeler gauge. The allowable warpage is 0.003 in. for every 6.0 inches. Do not machine more than 0.010 in.

To install:

15. Clean the cylinder head bolt hole threads with a tap. Lightly oil all bolt and stud bolt threads before installation except those entering coolant jackets.

16. Position new head gasket(s) on the cylinder block using the dowels for alignment.

17. Position the cylinder head(s) on the block and install the new bolts hand tight.

18. Torque the head bolts in the proper sequence as follows:

a. 37 ft. lbs. (50 Nm)
b. 45 ft. lbs. (60 Nm)
c. 52 ft. lbs. (70 Nm)
d. 59 ft. lbs. (80 Nm)
e. Loosen each bolt one at a time in sequence 2–3 turns, then torque to 11–18 ft. lbs. (15–25 Nm).
f. Rotate each bolt in sequence an additional 90 degrees.
g. On 1995 3.8L engine, tighten each short bolt to 7–15 ft. lbs. (10–20 Nm), then rotate an additional 90 degrees.

NOTE: On 1995 3.8L engine, do not loosen more than 1 bolt at a time.

19. Lubricate each pushrod with heavy engine oil and install, in their original positions.

20. For each valve, rotate the crankshaft until the lifter rests on the base circle of the camshaft lobe (pushrod all the way down). Make sure the fulcrum is seated properly, then tighten the bolt to 43 inch lbs. (5 Nm).

21. Lubricate the rocker arm assemblies with heavy engine oil and final tighten the fulcrum bolts to 19–25 ft. lbs. (25–35 Nm). Fulcrums must be fully seated in cylinder head

and pushrods must be seated in rocker arm sockets prior to final tightening. Final tightening can be done with the camshaft in any position.

NOTE: If the original valve train components are being installed, a valve clearance check is not required. If a component has been replaced, perform a valve clearance check.

22. Install the exhaust manifold(s).

23. Install the lower intake manifold and the fuel rail.

24. Position cover and new gasket on the cylinder head and install attaching bolts. Note the location of spark plug wire routing clip stud bolts. Tighten attaching bolts to 80–106 inch lbs. (9–12 Nm).

25. Install the upper intake manifold.

26. Install the spark plugs, if removed.

27. Connect the spark plug wires to the spark plugs.

28. For the left cylinder head, perform the following:

a. Install the oil filler cap.

b. Install the alternator/power steering pump mounting bracket.

c. Install the alternator assembly.

d. On 1993–94 3.8L engine, install the A/C compressor mounting and support brackets.

e. Install the main accessory drive belt tensioner assembly.

f. Install the power steering pump assembly.

g. Install the power steering pump support bracket.

29. For the right cylinder head, perform the following:

a. Install PCV valve.

b. On 1993–94 3.8L engine, if equipped with air conditioning, install the compressor mounting and support brackets and install the compressor.

c. On 1993–94 3.8L engine, if equipped, install the thermactor pump and pump pulley.

d. On 1993–94 3.8L engine, if equipped, install the accessory drive belt idler pulley.

e. On 1993–94 3.8L engine, if equipped, install the thermactor air control valve or idle air bypass valve hose. Tighten the clamps securely to the air pump assembly.

30. Install the accessory drive belt. If equipped, attach the thermactor tube(s) support bracket to the rear of the cylinder head. Tighten attaching bolts to 30–40 ft. lbs. (40–55 Nm).

31. Connect the negative battery cable.

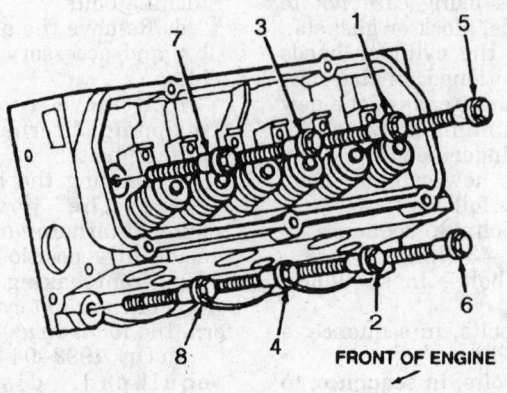

Cylinder head installation and torque sequence — 3.8L engines

32. Fill and bleed the cooling system.

33. Start engine and check for coolant, fuel and oil leaks.

34. Check and adjust the curb idle speed.

35. Install the air cleaner assembly including the air intake duct and heat tube.

4.6L Engine

NOTE: The cylinder head bolts are a torque-to-yield design and cannot be reused. Make sure new cylinder head bolts are available before performing this procedure.

1. Disconnect the negative battery cable. Remove the engine from the vehicle and position on a suitable workstand.

2. Remove both cylinder head covers. Remove the engine front cover.

3. Remove the intake manifold. Remove the Crankshaft Position (CKP) sensor pulse wheel.

4. Remove the rocker arms. Remove the exhaust manifolds. Drain the coolant from the cylinder block.

5. Rotate the engine to position the piston for No. 1 cylinder to TDC on its compression stroke.

6. Install Camshaft Positioning tool T93P-6256-A or equivalent, in the rear D-slots of the camshafts.

—————— WARNING ——————

This is not a freewheeling engine. Cam Positioning tool T93P-6256-A or equivalent, must be installed on the camshafts to prevent the camshafts from rotating. Do not rotate the camshafts or crankshaft with the timing chains removed or the valves will contact the pistons.

———————————————————

7. Remove the bolts retaining both primary timing chain tensioners to the cylinder heads and remove the timing chain tensioners.

8. Remove both primary timing chains, timing chain tensioner arms and timing chain guides. Do not loosen the camshaft timing sprocket retaining bolts.

9. Remove the outlet heater water hose retaining bolts from the rear of the right cylinder head.

10. Remove 10 bolts retaining the cylinder head to the engine block and remove the cylinder head. Loosen the bolts in the proper sequence.

11. Discard the cylinder head bolts. Remove one or both cylinder heads as required.

12. Clean all gasket mating surfaces and bolt holes. Check the cylin-

der head and cylinder block for flatness. Check the cylinder head for scratches near the coolant passage and combustion chamber that could provide leak paths.

To install:

13. Rotate the crankshaft counterclockwise 45 degrees. The crankshaft keyway should be at the 9 o'clock position as viewed from the front of the engine. This ensures that all pistons are below the top of the cylinder block deck face.

14. Position new head gaskets on the cylinder block.

15. Position the cylinder heads on the engine block dowels, being careful not to score the surface of the head face. Apply clean engine oil to the new cylinder head bolts and install them hand-tight.

16. Tighten the cylinder head bolts as follows:

   a. Tighten the bolts, in sequence, to 27–32 ft. lbs. (37–43 Nm).

   b. Rotate each bolt, in sequence, 85–95 degrees.

   c. Rotate each bolt, in sequence, an additional 85–95 degrees.

17. Position the outlet heater water hose on the right cylinder head. Install and tighten the bolts to 15–22 ft. lbs. (20–30 Nm).

18. Install the primary timing chains and set the valve timing. Install the rocker arms.

19. Install the CKP sensor pulse wheel. Install the intake manifold.

NOTE: When cleaning the sealing surfaces of the timing chain cover and oil pan-to-cylinder block joints, be careful not to damage the rubber bead of the oil pan gasket. If damaged, the oil pan gasket must be replaced.

20. Install the engine front cover. Install the cylinder head covers. Install the exhaust manifolds.

21. Install the engine in vehicle. Connect the negative battery cable.

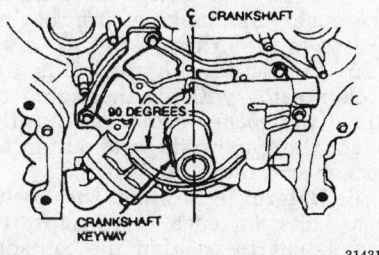

Crankshaft position before installing the cylinder heads — 4.6L engine

22. Start the engine and bring to normal operating temperature and check for leaks.

23. Check all fluid levels. Road test the vehicle and check for proper engine operation.

Lash Adjusters

BLEEDING

Lash adjusters (valve tappets), can be bled of air by placing in a container of clean engine oil, then pumping the lash adjuster plunger with the lash adjuster positioned between 2 fingers until no more air bubbles are seen. Lash adjusters must always be pumped up before installing into the cylinder head.

Valve Lifters

BLEEDING

3.0L (VIN 1 and U) and 3.8L Engines

Valve lifters (tappets), can be bled of air by using a Hydraulic Leakdown Tester or by placing in a container of clean engine oil, then pumping the valve lifter plunger with a pushrod until no more air bubbles are seen. Valve lifters must always be pumped up before installing in the engine.

REMOVAL AND INSTALLATION

1993–95 3.0L (VIN U) Engine
1996–97 3.0L (VIN 1 and U) Engines

—————— CAUTION ——————

Fuel injection systems remain under pressure, even after the engine has been turned OFF. The fuel system pressure must be relieved before disconnecting any fuel lines. Failure to do so may result in fire and/or personal injury.

———————————————————

NOTE: Before replacing a lifter for noisy operation make sure the noise is not caused by improper valve to rocker arm clearance or by worn rocker arms or pushrods.

1. Rotate the crankshaft until the piston in No. 1 cylinder is at TDC on the compression stroke. Disconnect the negative battery cable.

2. Drain the engine coolant. Remove the PCV tube from from the rocker arm cover. Remove the aspirator hose from air the air cleaner outlet.

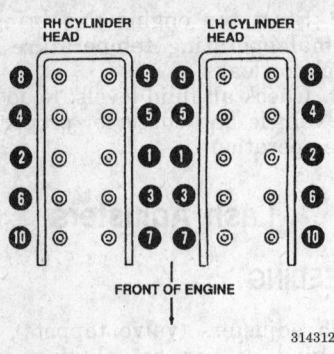

RH CYLINDER HEAD   LH CYLINDER HEAD

FRONT OF ENGINE

314312

Cylinder head bolt torque sequence — 4.6L engine

3. Properly relieve the fuel system pressure. Disconnect the fuel lines. Tag and remove the vacuum lines.

4. On 1993–95 3.0L (VIN U) engine, disconnect the IAT and distributor connectors. Disconnect the wiring to the throttle position sensor, idle air control and the PFE sensors. On flexible fuel vehicles disconnect the CSI and camshaft sensor.

5. On 1996–97 3.0L (VIN 1 and U) engine, disconnect the wiring to the Throttle Position (TP) sensor, Idle Air Control (IAC) valve, Camshaft Position (CMP) sensor, Engine Coolant Temperature (ECT) sensor, ignition coil, water temperature sender, EGR backpressure transducer, and the EGR vacuum regulator solenoid connectors.

6. Disconnect the upper radiator hose from the thermostat housing. Remove the brace from the generator to throttle body stud. Remove the EGR valve to exhaust manifold tube.

7. On 1996–97 3.0L (VIN 1 and U) engine, remove the upper intake manifold.

8. Remove the throttle body. Disconnect the fuel charging wiring retainers from the valve cover studs. Disconnect the heater hoses.

9. Tag and remove the spark plug wires from the spark plugs. Remove distributor retaining bolt and the distributor.

10. Remove the ignition coil from the left cylinder head. Remove the valve covers.

11. Loosen cylinder No. 3 intake valve rocker arm retaining nut and rotate arm off pushrod and away from top valve stem. Remove the pushrod.

12. Remove intake manifold bolts and intake manifold on 1993–95 3.0L (VIN U) engine. or remove the lower intake manifold on 1996–97 3.0L (VIN 1 and U) engine.

13. Loosen rocker arm fulcrum retaining bolt of lifter to be replaced

enough to allow rocker arm to be lifted off the pushrod and rotated to one side.

14. Remove the pushrods. If more than one pushrod is being removed be sure to tag its original location for reinstallation.

15. Loosen the 2 roller lifter guide plate retaining bolts. Remove the guide plate retainer assembly from the lifter valley.

16. Remove the lifter guide plates from the lifters by lifting straight up.

NOTE: If the lifters are stuck in the bore, it may be necessary to use a claw-type tool to aid removal. Rotate the valve lifter back and forth to loosen it from the deposits.

17. To remove the lifter, grasp lifter and pull straight in line with bore.

To install:

NOTE: Lightly oil all retaining bolt and stud bolt threads before installation.

18. Clean mating gasket surfaces of the intake manifold and cylinder head. Use care when working with light-alloy components.

19. Lubricate lifters and bore. Install the lifters into the bore. Aligning the lifters flats, install the lifter guide plate. Install plate with word UP and/or button visible.

20. Install guide plate retainer assembly over the guide plates. Install the 2 retaining bolts and tighten to 8–10 ft. lbs. (10–14 Nm).

21. Apply a bead of rubber sealer to intersection of cylinder block and cylinder head assembly at 4 corners.

22. Position intake manifold gaskets onto cylinder heads. Align the intake manifold gasket locking tabs to the tabs on the cylinder head gaskets.

23. Install front and rear manifold seals. Lower intake manifold into position. Install bolts and tighten in 2 steps. First tighten each bolt to 15–22 ft. lbs. (20–30 Nm). Second, tighten the bolts to 19–24 ft. lbs. (26–32 Nm).

24. Install the distributor. Install the pushrod(s), making sure they are seated in the lifters.

25. Lubricate the pushrods and rocker arms with clean engine oil. Move the rocker arms into position with the pushrods and snug the rocker arm retaining bolt.

26. Before tightening the rocker arm bolts, for each valve rotate the crankshaft to position the camshaft lobe straight down and away from the valve lifter.

27. Tighten rocker arm bolts in 2 steps. First tighten to 5–11 ft. lbs.

(7–15 Nm). Second, tighten to 19–28 ft. lbs. (26–38 Nm) in any position.

28. On 1993–95 3.0L (VIN U) engine, install valve covers.

29. On 1996–97 3.0L (VIN 1 and U) engine, install the fuel injector wiring harness, connect wiring to each fuel injector and secure the wiring harness to the rocker arm cover studs.

30. Install the ignition coil.

31. On the 1996–97 3.0L (VIN 1 and U) engine, install the ignition wire harness retainers to the rocker arm cover studs and connect the ignition wires to the spark plugs and the ignition coil.

32. Install the throttle body assembly and the upper intake manifold gaskets. Install the upper intake manifold assembly.

33. Install the EGR valve to exhaust manifold tube. Install fuel lines and install the fuel line safety clips.

34. Install upper radiator hose and heater hoses. Reconnect the vacuum lines.

35. On 1993–95 3.0L (VIN U) engine, connect the electrical connectors to the IAT sensor, distributor, idle air control, throttle position sensor, PFE sensor and coolant temperature sensor.

36. On 1996–97 3.0L (VIN 1 and U) engine, connect the wiring to the TP sensor, IAC valve, CMP sensor, ECT sensor, ignition coil, water temperature sender, EGR backpressure transducer, and the EGR vacuum regulator solenoid connectors.

37. Fill and bleed cooling system. Refill the engine with proper grade motor oil.

38. Install air cleaner tube and engine air cleaner. Reconnect the PCV hose.

39. Reconnect the negative battery cable. Run the engine and check for leaks and proper operation.

3.8L Engine

NOTE: Before replacing a lifter for noisy operation, be sure the noise is not caused by improper valve to rocker arm clearance or by worn rocker arms or pushrods.

1. Disconnect the negative battery cable. Tag and disconnect the spark plug wires at the spark plugs.

2. Remove plug wire routing clips from the studs on the rocker arm cover attaching bolts. Lay the plug wires with the routing clips toward the front of the engine.

3. Remove the upper intake manifold. Remove the rocker arm covers. Remove the lower intake manifold.

4. Sufficiently loosen each rocker arm fulcrum attaching bolt to allow the rocker arm to be lifted off the pushrod and rotated aside.

5. Remove the pushrods. The location of each pushrod must be identified so they can be installed in their original position.

6. Remove the 4 bolts holding the guide plate retainers in place; the bolts are held captive in the retainers. Remove the 6 guide plates from the adjacent lifters.

7. Remove the lifters using a magnet. The location of each lifter must be identified so they can be installed in their original position.

NOTE: If the lifters are stuck in the bores due to excessive varnish or gum deposits, it may be necessary to use a claw-type tool to aid removal. When using a remover tool, rotate the lifter back and forth to loosen it from gum or varnish that may have formed on the lifter.

To install:

8. Clean the rocker arm cover and cylinder head mating surfaces.

9. Install each lifter in the bore from which it was removed. If new lifters are being installed, check the new lifters for free fit in the bores.

10. Align the flats on the side of the lifters and install the 6 guide plates between the adjacent lifters. Make sure the word **UP** is showing. Install the 2 guide plate retainers and tighten the 4 captive bolts to 84–120 inch lbs. (9–14 Nm).

11. Dip each pushrod in heavy engine oil and install in its original position.

NOTE: The fulcrums must be fully seated in the cylinder head and the pushrods must be seated in the rocker arm sockets prior to final tightening.

12. For each valve, rotate the crankshaft until the lifter rests on the base circle of the camshaft lobe (pushrod all the way down). Position the rocker arms over the pushrods, install the fulcrums and tighten the bolts to 60–132 inch lbs. (7–15 Nm).

13. Lubricate all rocker arm assemblies with heavy engine oil. Tighten all the fulcrum bolts to 19–25 ft. lbs. (25–35 Nm). For final tightening the camshaft may be in any position.

14. Install the lower intake manifold and the rocker arm covers. Install the upper intake manifold.

15. Install the spark plug wire routing clips and connect the wires to the spark plugs. Connect the negative battery cable, start the engine and check for oil and coolant leaks.

Valve Lash

ADJUSTMENT

3.0L (VIN Y) and 3.2L SHO Engines

1. Disconnect the negative battery cable. Remove the intake manifold assembly. Remove the cylinder head covers.

NOTE: The cam lobes must be directed 90 degrees or more away from the lash adjusters (valve tappets). The engine must be COLD to check the valve clearance.

2. Insert a feeler gauge under the camshaft lobe at a 90 degree angle to the camshaft. Clearance for the intake valves should be 0.006–0.010 inch (0.15–0.25mm). Clearance for the exhaust valves should be 0.010–0.014 inch (0.25–0.35mm).

3. If no adjustments are required, install the cylinder head covers and intake manifold.

4. If adjustment is required, install Tappet Compressor T89P-6500-A or equivalent, under the camshaft next to the lobe and rotate the tool downward to depress the lash adjuster.

5. Install Tappet Holder T89P-6500-B or equivalent, and remove the compressor tool.

6. Using O-ring tool T71P-19703-C or equivalent, lift the adjusting spacer and remove the adjusting spacer with a magnet.

7. Determine the size of the adjusting spacer by the numbers on the bottom face of the spacer or by measuring with a micrometer.

8. Install the replacement adjusting spacer with the numbers down. Make sure the spacer is properly seated.

9. Release the tappet holder by installing the tappet compressor, then remove the tappet compressor.

10. Repeat the procedure for each valve requiring adjustment by rotating the crankshaft as necessary.

11. Check all adjusting spacers to ensure that they are properly seated.

12. Install the cylinder head covers. Install the intake manifold.

13. Connect the negative battery cable. Run the engine and check for proper operation.

3.0L (VIN 1 and U) Engines

Hydraulic valve lifters are used and no valve clearance adjustment is available. A clearance check of the rocker arm-to-valve stem gap is required when machining has been done to the cylinder heads, valves, valve seats or cylinder block head gasket surfaces or when new valve train components have been installed. The clearance check is also useful in determining loose, worn or damaged parts when there is a concern with the valve train. Clearance must be checked when the lifter is completely collapsed.

1. Disconnect the negative battery cable. Remove both rocker arm covers.

2. To check valve clearance, use Lifter Bleed Down Wrench T71P-6513-B or equivalent, to slowly push down on the pushrod end of the rocker arm and bleed the oil from the valve lifter.

3. Once the valve lifter is totally collapsed, insert the appropriate thickness feeler gauge between the rocker arm and valve stem to check the clearance. There should be 0.085–0.185 inch (2.15–4.69mm) clearance between the valve stem and rocker arm.

4. Rotate the crankshaft until No. 1 cylinder is at TDC on its compression stroke. With the engine at this position, the following valve clearances can be checked:

Intake — 1, 3 and 6
Exhaust — 1, 2 and 4

5. Rotate the crankshaft 360 degrees and check the following valves in the same manner:

Intake — 2, 4 and 5
Exhaust — 3, 5 and 6

6. If the clearance is not correct, check for loose, worn or damaged components. If no problems are found, the clearance can be adjusted by the use of shorter or longer pushrods.

3.8L Engines

The valve stem-to-rocker arm clearance should be within specification with the valve lifter completely collapsed. If the clearance is not within specifications, check for loose, worn or damaged components and repair as necessary.

1. With the crankshaft in the designated position, install Lifter Bleeder Wrench T71P-6513-B or equivalent, on the rocker arm. Slowly apply pressure to the lifter until the plunger is completely collapsed, then

use a feeler gauge to determine the valve stem-to-rocker arm clearance.

2. Rotate the engine until No. 1 piston is at TDC on its compression stroke. Check the valve stem-to-rocker arm clearance for the following valves.

Intake — 1, 3 and 6
Exhaust — 1, 2 and 4

3. Rotate the crankshaft 360 degrees and check the valve stem-to-rocker arm clearance for the following valves.

Intake — 2, 4 and 5
Exhaust — 3, 5 and 6

4. The valve stem-to-rocker arm clearance should be 0.09–0.19 inch (2.25–4.79mm) for all valves.

3.0L (VIN S) and 4.6L Engines

The lash adjusters (valve tappets), are hydraulic and are not adjustable. It is important that all valve components are in good condition and installed and torqued properly.

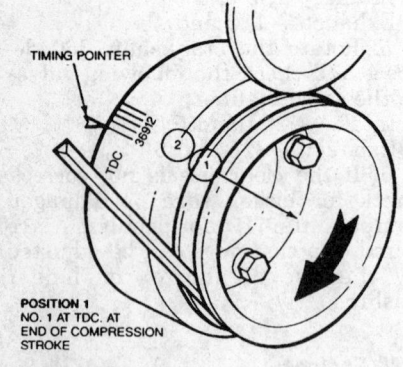

TIMING POINTER

POSITION 1
NO. 1 AT TDC. AT END OF COMPRESSION STROKE

POSITION 2
ROTATE CRANKSHAFT ONE REVOLUTION — 360 DEGREES

CYL. NO.	CRANKSHAFT POSITION	
	1	2
	SET GAP OF VALVES NOTED	
1	INT — EXH	NONE
2	EXH	INT
3	INT	EXH
4	EXH	INT
5	NONE	INT — EXH
6	INT	EXH

286203

Crankshaft positioning — 3.8L engine

Rocker Arm and Rocker Arm Shaft

REMOVAL AND INSTALLATION

1993–95 3.0L (VIN U) Engine
1996–97 3.0L (VIN 1 and U) Engines

1. Disconnect the negative battery cable. Disconnect and tag the spark plug wires.

2. Remove the ignition wire/separator assembly from the rocker arm attaching bolt studs.

3. If the left rocker arm cover is being removed, remove the oil fill cap, disconnect the air cleaner closure system hose and remove the fuel injector harness from the inboard rocker arm cover studs.

4. On 1993–95 3.0L (VIN U) engine, if the right rocker arm cover is being removed, remove the throttle body, the PCV valve, loosen the lower EGR tube, if equipped, retaining nut and rotate the tube aside and move the fuel injection harness aside.

5. On 1996–97 3.0L (VIN 1 and U) engine, if removing the right rocker arm cover, remove the upper intake manifold, tag and disconnect the vacuum hoses at the vacuum tee, loosen the EGR tube nuts at the EGR valve and exhaust manifold fitting. Remove, or rotate the tube aside, remove the PCV valve, the engine sensor wiring harness and move the wiring aside and the alternator brace.

6. Remove the rocker arm cover attaching screws and the covers and gaskets from the vehicle.

7. Remove the rocker arm bolts, fulcrums, rocker arms and fulcrum washers. Keep all parts in order so they can be reinstalled to their original positions.

8. Remove the pushrods, if necessary. Keep them in order so they can be reinstalled in their original positions.

9. Inspect the rocker arms, fulcrums and pushrods for wear and/or damage. Replace as necessary.

To install:

10. Install the pushrods, if removed, making sure they seat in the lifters.

11. Coat the valve and pushrod tips, rocker arm and fulcrum contact areas with Lubriplate® or equivalent. Lightly oil all the bolt and stud threads before installation.

12. Rotate the engine until the lifter is on the base circle of the cam (valve closed).

13. Install the rocker arm and components and torque the rocker arm fulcrum bolts in 2 steps: the first to 8 ft. lbs. (11 Nm) and the final to 24 ft. lbs. (32 Nm). Be sure the lifter is on the base circle of the cam for each rocker arm as it is installed.

14. Clean the cylinder head and rocker arm cover sealing surfaces of all dirt and old sealer. If not equipped with integral gaskets, make sure all old gasket material is removed.

15. Apply a bead of silicone sealant at the cylinder head to intake manifold rail step. If not equipped with integral gaskets, install a new rocker arm cover gasket.

16. Install the rocker arm cover and the bolts and studs. Tighten to 9 ft. lbs. (12 Nm) in the proper sequence.

17. Install the remaining components in the reverse order of their removal.

3.0L (VIN S) Engine

1. Disconnect the negative battery cable.

2. Remove both cylinder head covers.

3. Remove the crankshaft pulley retaining bolt from the front of the crankshaft allowing the keyway to be referenced.

4. Rotate the crankshaft so the keyway is at the 11 o'clock position to locate the crankshaft at TDC for No. 1 cylinder.

5. Verify that the alignment arrows on the camshafts are aligned. If not, rotate the crankshaft one complete revolution and recheck.

6. Rotate the crankshaft so the keyway is at the 3 o'clock position. This positions the right cylinder head camshafts to the neutral position.

7. Remove the right cylinder head camshaft journal thrust cap retaining bolts and thrust caps.

8. Loosen the remaining camshaft journal cap bolts in sequence, releasing the bolts several revolutions at a time by making several passes to allow the camshaft to be raised from the cylinder head evenly. Do not remove the retaining bolts completely.

9. With the camshafts loose, remove the rocker arms, keeping them in the order that they were removed.

10. If required, remove the lash adjusters from the cylinder head.

11. Rotate the crankshaft 2 revolutions and locate the crankshaft keyway at the 11 o'clock position. This will position the left cylinder head camshafts to their neutral position.

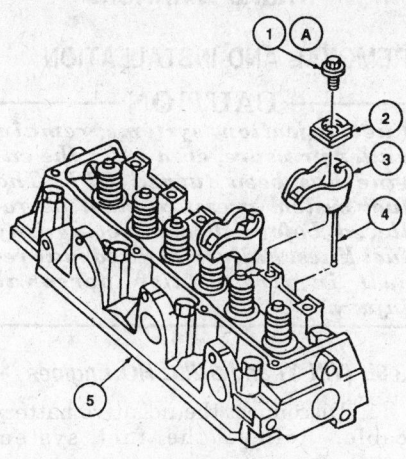

1 Bolt (12 Req'd)
2 Rocker Arm Seat (12 Req'd)
3 Rocker Arm (12 Req'd)
4 Push Rod (12 Req'd)
5 Cylinder Head (2 Req'd)
6 2.15-4.69 mm (0.085-0.185 inch)
A Tighten in Two Steps:
   7-15 N·m (5-11 Lb-Ft)
   26-38 N·m (19-28 Lb-Ft)

324358

Rocker arm assembly — 1996–97 3.0L (VIN 1 and U) engines

12. Verify that the alignment arrows on the camshafts are aligned.

NOTE: The camshaft journal caps and cylinder heads are numbered to ensure that they are assembled in their original positions. If removed, keep the camshaft journal caps together with the cylinder head that they were removed from.

13. Remove the camshaft journal thrust cap retaining bolts and thrust caps from the left cylinder head.
14. Loosen the remaining camshaft journal cap bolts in sequence, releasing the bolts several revolutions at a time by making several passes to allow the camshaft to be raised from the cylinder head evenly. Do not remove the retaining bolts completely.

NOTE: If the lash adjusters (tappets) and roller rocker arms are to be reused, mark the positions of the lash adjusters and rocker arms so they are reassembled into their original positions.

15. With the camshafts loose, remove the rocker arms, keeping them in the order that they were removed.
16. If required, remove the lash adjusters from the cylinder head.
17. Inspect the rocker arms and lash adjusters for wear and/or damage and replace as necessary.
   To install:
18. Make sure the crankshaft keyway is at the 11 o'clock position.
19. If removed, lubricate the left cylinder head lash adjusters with engine assembly lubricant and install into their correct positions in the cylinder head.
20. If the lash adjusters are being replaced with new units, soak the adjusters in a container of clean engine oil and then manually pump up the adjusters before installing into the cylinder head.
21. Lubricate the left cylinder head rocker arms with engine assembly lubricant and install the left cylinder head rocker arms into their original locations.

NOTE: Do not install the camshaft journal thrust caps until the camshaft journal caps are torqued into position.

22. Tighten the left cylinder head camshaft journal cap bolts in sequence making several passes to pull the camshafts down evenly. Torque the bolts to 71–106 inch lbs. (8–12 Nm).
23. Install the left–hand cylinder head thrust caps and bolts. Torque to 71–106 inch lbs. (8–12 Nm).

24. Rotate the crankshaft 2 revolutions and position the crankshaft keyway to the 3 o'clock location. This will position the right cylinder head camshafts to the neutral position.
25. Lubricate the right cylinder head lash adjusters with engine assembly lubricant and install into their original positions in the cylinder head.
26. If the lash adjusters are being replaced with new units, soak the adjusters in a container of clean engine oil and manually pump the adjusters up before installing into the cylinder head.
27. Lubricate the right cylinder head rocker arms with engine assembly lubricant and install the right cylinder head rocker arms into their original locations.

NOTE: Do not install the camshaft journal thrust caps until the camshaft journal caps are torqued into position.

28. Tighten the right cylinder head camshaft journal cap bolts in sequence making several passes to pull the camshafts down evenly. Torque the bolts to 71–106 inch lbs. (8–12 Nm).
29. Install the right–hand cylinder head thrust caps and bolts. Torque to 71–106 inch lbs. (8–12 Nm).
30. Install the crankshaft pulley retaining bolt.
31. Tighten the crankshaft pulley retaining bolt as follows:
   a. Torque to 89 ft. lbs. (120 Nm).
   b. Loosen the bolt at least 1 full turn.
   c. Torque the bolt to 35–39 ft. lbs. (47–53 Nm).
   d. Rotate the bolt 85–95 degrees.
32. Install both valve covers as follows:
   a. Clean the valve cover gasket sealing surfaces.
   b. Install new valve cover gaskets onto the valve covers.
   c. For each valve cover, place a bead of silicone sealant at 2 places on the valve cover sealing surfaces where the engine front cover and the cylinder heads make contact and at 2 places on the rear of the cylinder head where the camshaft seal retainer contacts the cylinder head.
   d. Install the valve cover retaining bolts and studs and tighten in sequence to 71–106 inch lbs. (8–12 Nm).

NOTE: The valve covers must be installed and properly tightened within 6 minutes of applying the silicone sealant.

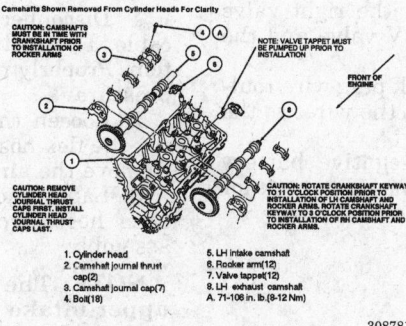

Camshafts Shown Removed From Cylinder Heads For Clarity

CAUTION: CAMSHAFTS MUST BE IN TIME WITH CRANKSHAFT PRIOR TO INSTALLATION OF ROCKER ARMS

NOTE: VALVE TAPPET MUST BE PUMPED UP PRIOR TO INSTALLATION

FRONT OF ENGINE

CAUTION: REMOVE CYLINDER HEAD JOURNAL THRUST CAPS FIRST. INSTALL CYLINDER HEAD JOURNAL THRUST CAPS LAST.

CAUTION: ROTATE CRANKSHAFT KEYWAY TO 11 O'CLOCK POSITION PRIOR TO INSTALLATION OF LH CAMSHAFT AND ROCKER ARMS. ROTATE CRANKSHAFT KEYWAY TO 3 O'CLOCK POSITION PRIOR TO INSTALLATION OF RH CAMSHAFT AND ROCKER ARMS.

1. Cylinder head
2. Camshaft journal thrust cap(2)
3. Camshaft journal cap(7)
4. Bolt(18)
5. LH intake camshaft
6. Rocker arm(12)
7. Valve tappet(12)
8. LH exhaust camshaft
A. 71-106 in. lb (8-12 Nm)

308781

View of rocker arms with the camshafts removed — 3.0L (VIN S) engine

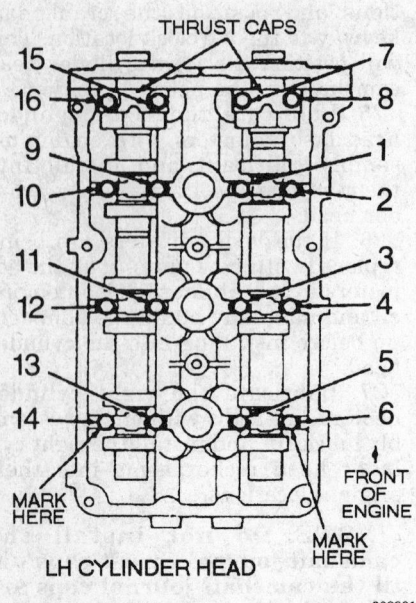

Torque sequence for the left cylinder head camshaft journal caps — 3.0L (VIN S) engine

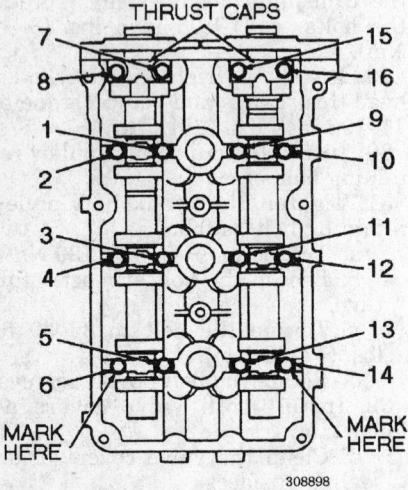

Torque sequence for the right cylinder head camshaft journal caps — 3.0L (VIN S) engine

33. Connect the negative battery cable.

34. Run the engine and check for leaks and proper operation.

3.8L Engine

1. Disconnect the negative battery cable.

2. Tag to identify, then disconnect the spark plug wires from the spark plugs. Remove the spark plug wire routing clips from the rocker arm cover attaching bolt studs.

3. To remove the left rocker arm cover, remove the oil fill cap and the crankcase vent tube.

4. To remove the right rocker arm cover, remove the PCV valve and position the air cleaner assembly aside.

5. Remove the rocker arm cover attaching screws and remove the rocker arm covers.

6. Remove the rocker arm, fulcrum and bolt assemblies. Keep each assembly together and identify the assemblies so they may be reinstalled in their original positions.

To install:

7. Clean all gasket mating surfaces on the rocker arm covers and cylinder heads. Clean the rocker arms and fulcrums and inspect for wear or damage. Replace as necessary.

8. Apply grease to the pushrod tips and valve stem tips. Lubricate the fulcrums and rocker arms with heavy engine oil and install them over the pushrods and valve stems.

9. For each valve, rotate the crankshaft until the lifter is on the base circle of the camshaft. Install the fulcrum bolt and tighten to 5–11 ft. lbs. (7–15 Nm). Make sure the pushrod and fulcrum are fully seated prior to tightening.

10. Lubricate all rocker arm assemblies with engine oil. Final tighten the fulcrum bolts to 19–25 ft. lbs. (25–35 Nm). When final tightening, the camshaft may be in any position. Make sure the pushrod and fulcrum are fully seated prior to tightening.

11. Position new gaskets on the cylinder heads and install the rocker arm covers. Tighten the attaching bolts to 80–106 inch lbs. (9–12 Nm). Note the location of the spark plug wire routing clip stud bolts prior to installation.

12. After installing the left rocker arm cover, install the oil fill cap and the crankcase vent tube.

13. After installing the right valve cover, install the PCV valve and the air cleaner assembly.

14. Install the spark plug wire routing clips and connect the wires to the spark plugs.

15. Connect the negative battery cable, start the engine and check for leaks.

Intake Manifold

REMOVAL AND INSTALLATION

— CAUTION —
Fuel injection systems remain under pressure, even after the engine has been turned OFF. The fuel system pressure must be relieved before disconnecting any fuel lines. Failure to do so may result in fire and/or personal injury.

3.0L (VIN Y) and 3.2L SHO Engines

1. Disconnect the negative battery cable. Relieve the fuel system pressure.

2. Partially drain the engine cooling system. Tag and disconnect all electrical connectors and vacuum lines from the intake assembly.

3. Remove the air cleaner tube. Disconnect the coolant lines and cables from the throttle body.

4. Remove the upper intake manifold bracket bolts. Loosen the lower bolts and remove the brackets.

5. Remove the intake manifold-to-cylinder head bolts. Remove the intake manifold and gaskets.

To install:

6. Lightly oil the attaching bolts and stud threads before installation.

NOTE: The intake gasket is reusable.

7. Position the intake manifold gasket on the cylinder heads and install the intake manifold. Tighten the bolts to 11–17 ft. lbs. (15–23 Nm). Install intake manifold support brackets. Tighten the bolts to 11–17 ft. lbs. (15–23 Nm).

8. Install the remaining components in the reverse order of removal.

9. Refill the cooling system and reconnect the negative battery cable. Run the engine and check for leaks.

1993–95 3.0L (VIN U) Engine

1. Disconnect the negative battery cable. Drain the engine cooling system. Properly relieve the fuel system pressure.

2. Loosen the hose clamp attaching the flex hose to the throttle body. Remove the air cleaner flex hose.

3. Label and disconnect the vacuum hoses from the throttle body assembly.

NOTE: The throttle body and upper intake manifold are manufactured as one assembly and will be referred to as the throttle body assembly.

4. Loosen the lower EGR tube nut and rotate the tube away from the valve. Disconnect the accelerator and TV cables from the throttle linkage.

5. Disconnect the Throttle Position (TP) sensor, Air Charge Temperature (ACT) sensor, Engine Coolant Temperature (ECT) sensor and Idle Air Control (IAC) valve electrical connectors.

6. On flex-fuel vehicles, disconnect the Camshaft Position (CMP) sensor electrical connector.

7. Disconnect the PCV hose and disconnect the alternator support brace.

8. Remove 6 throttle body retaining bolts and remove the throttle body assembly.

9. Disconnect the fuel supply and return lines at the fuel supply manifold (fuel rail).

10. Label and disconnect the fuel injection wiring harness from the engine. The manifold assembly can be removed with the fuel supply manifold and injectors in place.

11. Label and remove the ignition wires at the spark plugs.

12. Remove the rocker arm covers.

13. On non-flexible fuel vehicles, remove the distributor.

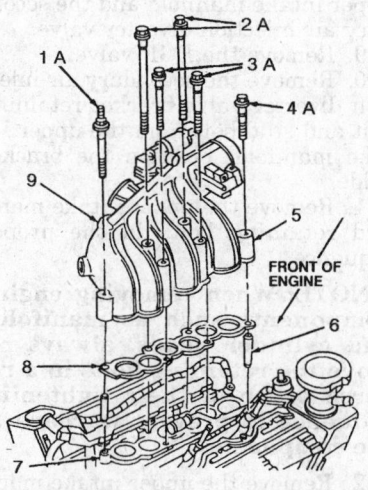

1. Stud bolt
2. Bolt (2 req'd)
3. Bolt (2 req'd)
4. Bolt
5. Throttle body
6. Guide pin (2 req'd)
7. Intake manifold
8. Intake manifold upper gasket
9. Intake manifold vacuum outlet fitting and cap
A. Tighten to 20-30 Nm (15-22 lb. ft.)

332398

Throttle body (upper intake manifold) and related components — 1993–95 3.0L (VIN U) engine

14. On flex-fuel vehicles, rotate the crankshaft until the piston for No. 1 cylinder is at TDC on its compression stroke. Note the position of the CMP electrical connector, then remove the CMP sensor housing along with the oil pump intermediate shaft.

15. Remove the ignition coil from the rear of left cylinder head.

16. Disconnect the upper radiator hose and heater hoses.

17. Loosen the intake valve rocker arm retaining bolt from No. 3 cylinder and rotate the rocker arm away from the valve stem and pushrod. Remove the pushrod.

18. Remove the intake manifold retaining bolts. Use a suitable prybar to loosen the intake manifold. Pry upward using the area between the thermostat and transaxle as a leverage point. Remove the manifold, gaskets and seals.

To install:

19. Clean the gasket mating surfaces of the intake manifold and cylinder heads. Lay a shop rag in the lifter valley to catch any gasket material. After scraping, carefully lift the cloth from the lifter valley, being careful not to let any particles enter the oil drain holes or cylinder heads. If necessary, use a suitable solvent to remove old rubber sealant.

20. Clean and lightly oil all retaining bolts and stud threads before installation.

21. Apply a suitable silicone rubber sealer to the intersection of the cylinder block end rails and cylinder heads. Be careful not to let sealer that may block oil passages fall into the engine.

NOTE: When using a silicone sealer, assembly must occur within 15 minutes after the sealer has been applied. After this time, the sealer may start to set-up and its sealing quality may be reduced. In high temperature/humidity conditions, the sealant will start to set up in approximately 5 minutes.

22. Install the front and rear intake manifold end seals in place and secure. Install the intake manifold gaskets, aligning the locking tabs to the provisions on the cylinder head gaskets.

23. Carefully lower the intake manifold into position on the cylinder block and cylinder heads to prevent smearing the silicone sealer and causing gasket voids.

24. Install the intake manifold retaining bolts and tighten the bolts starting at the center and working towards the ends. Torque the bolts in 2 steps, tighten to 15–22 lbs. (20–30 Nm). Tighten again to 19–24 ft. lbs. (26–32 Nm).

25. If removed, install the fuel supply manifold and injectors. Apply lubricant to the injector holes in the intake manifold and fuel supply manifold prior to injector installation. Install the fuel supply manifold retaining bolts and tighten to 7 ft. lbs. (10 Nm).

26. If removed, install the thermostat housing and a new gasket. Tighten the retaining bolts to 9 ft. lbs. (12 Nm).

27. On non-flexible fuel vehicles, install the distributor assembly, the distributor cap and ignition wires.

NOTE: On non–flexible fuel vehicles, coat the distributor drive gear and on flex-fuel vehicles, coat the CMP sensor drive gear with an appropriate engine assembly lubricant.

———— **WARNING** ————
For flex-fuel vehicles, Synchro Positioning tool T93P-12200-A or equivalent, must be obtained prior to installing the CMP sensor housing. Failure to follow this procedure will result in improper CMP sensor alignment. This will result in the ignition and fuel systems being out of time with the engine, possibly causing engine damage.

28. On flex-fuel vehicles, install the CMP sensor housing as follows:

a. Engage the CMP sensor housing vane into the radial slot of Synchro Positioning tool T93P-12200-A, or equivalent. Rotate the tool on the CMP sensor housing until the tool boss engages the notch in the CMP sensor.

b. Install the CMP sensor housing along with the oil pump intermediate shaft. Install the CMP sensor housing so drive gear engagement occurs when the arrow on the locator tool is pointed approximately 30 degrees counterclockwise from the rear face of the cylinder block. This step will locate the CMP sensor electrical connector in the pre-removal position.

c. Install the hold-down clamp and tighten the bolt to 15–22 ft. lbs. (20–30 Nm). Remove the synchro positioning tool.

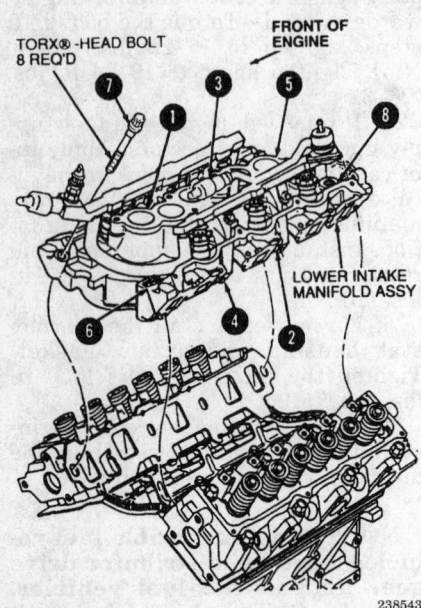

Intake manifold bolt torque sequence — 3.0L (VIN U and 1) engines

─── **WARNING** ───

If the CMP sensor electrical connector is not positioned properly, do not reposition the connector by rotating the CMP sensor housing. This will result in the ignition and fuel systems being out of time with the engine, possibly causing engine damage. Remove the housing and repeat the installation procedure.

29. Install the No. 3 cylinder intake valve pushrod. Apply engine assembly lubricant or equivalent, to the pushrod and valve stem prior to installation. Turn the crankshaft as necessary to position the lifter on the base circle of the camshaft (pushrod all the way down). Tighten the rocker arm bolt in 2 steps, first to 8 ft. lbs. (11 Nm) and then to 24 ft. lbs. (32 Nm).

30. Install the rocker arm covers.

31. Install the fuel injector harness and attach to the fuel injectors.

32. Install the ignition coil to the rear of left cylinder head. Tighten the retaining bolts to 30–40 ft. lbs. (40–55 Nm).

33. If removed, install the EGR valve on the intake manifold. Tighten the retaining bolts to 15–22 ft. lbs. (20–30 Nm).

34. Install the throttle body assembly and retaining bolts with a new gasket. Tighten the bolts in a cross-tightening sequence to 15–22 ft. lbs. (20–30 Nm).

35. Connect the rear crankcase ventilation hoses at the PCV valve and upper intake manifold.

36. If equipped with air conditioning, install the A/C compressor support bracket. Tighten the retaining nut and bolt to 15–22 ft. lbs. (20–30 Nm).

37. Install the accessory drive belt.

38. Connect the Throttle Position (TP) sensor, Air Charge Temperature (ACT) sensor, Engine Coolant Temperature (ECT) sensor and Idle Air Control (IAC) valve electrical connectors.

39. On flex-fuel vehicles, connect the Camshaft Position (CMP) sensor electrical connector.

40. Connect the necessary vacuum hoses.

41. Connect the heater water hose to the hot water heater elbow connection.

42. Position the heater tube support bracket and tighten the retaining nut to 15–22 ft. lbs. (20–30 Nm). Tighten the hose clamp at the hot water heater elbow connection securely.

43. Connect the heater water hose to the rear of the water bypass tube and tighten the hose clamp.

44. Connect the water bypass hose. Tighten the hose clamp securely.

45. Connect the upper radiator hose. Tighten the hose clamp securely.

46. Connect the fuel supply and return lines to the fuel injection supply manifold.

47. Connect the fuel line safety clips.

48. Position the accelerator cable bracket. Install and tighten the retaining bolts to 15–22 ft. lbs. (20–30 Nm).

49. Connect the speed control actuator to the throttle body assembly, if equipped.

50. Connect the accelerator cable to throttle body assembly.

51. Install the engine air cleaner and air cleaner outlet tube.

52. Install the fuel tank fill cap.

53. Connect the negative battery cable.

54. Turn the ignition switch to the **RUN** position several times without starting the engine to pressurize the fuel system and to check for fuel leaks.

55. Start the engine and check for fuel leaks and coolant leaks.

56. On non-flexible fuel vehicles, verify and if necessary, correct engine timing to 10 degrees BTDC. Tighten the distributor retaining bolt to 18 ft. lbs. (24 Nm).

57. Install the IAC valve shield.

58. Road test the vehicle and check for proper operation.

1996–97 3.0L (VIN S) Engine

Upper Intake Manifold

1. Disconnect the negative battery cable. Relieve the fuel system pressure.

2. Remove the windshield wiper motor and then remove the cowl top inner panels.

3. Remove the accelerator cable and speed control actuator from the throttle body.

4. Remove the accelerator cable bracket from the intake manifold and move aside.

5. Remove the idle air control valve fresh air supply hose from the fitting on the upper intake manifold.

6. Disconnect the wiring harnesses from the throttle position sensor and the idle air control valve.

7. Remove the vacuum supply hose from the upper intake manifold to the PCV valve at the upper intake manifold.

8. Disconnect the main emission vacuum control connector from the upper intake manifold and the secondary air injection diverter valve.

9. Remove the EGR valve.

10. Remove the secondary air injection diverter valve bracket retaining bolt and stud bolt from the upper intake manifold. Position the bracket aside.

11. Remove the upper intake manifold retaining bolts in the proper sequence.

NOTE: When removing engine components such as manifolds and cylinder heads, always remove the retaining bolts in a reverse order of their tightening sequence to prevent warpage to the component.

12. Remove the upper intake manifold and gaskets from the engine.

To install:

13. Install the upper intake manifold using 2 new gaskets onto the lower intake manifold. Install the upper manifold retaining bolts and torque following the proper sequence to 71–106 inch lbs. (8–12 Nm).

14. Install the remaining components in the reverse order of removal.

15. Connect the negative battery cable. Run the engine and check for leaks and proper engine operation.

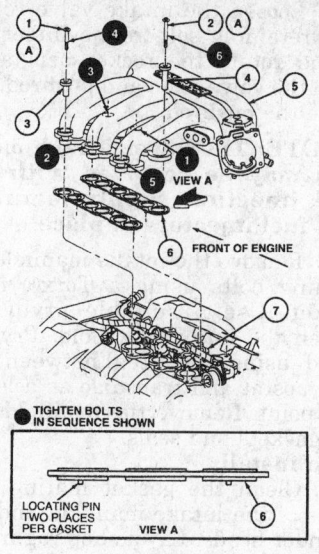

TIGHTEN BOLTS
IN SEQUENCE SHOWN

LOCATING PIN
TWO PLACES
PER GASKET   VIEW A

1 Bolt (3 Req'd)
2 Bolt (3 Req'd)
3 Isolator (3 Req'd)
4 Isolator (3 Req'd)
5 Upper Intake Manifold
6 Gasket (2 Req'd)

307195

Upper intake manifold bolt torque sequence — 3.0L (VIN S) engine

Lower Intake Manifold

1. Disconnect the fuel injector wiring harness and move aside.

2. Remove the spring lock coupling retainer clips from the fuel supply and return fittings.

3. Use spring lock coupling disconnect tools (⅜ inch and ½ inch) to disconnect the fuel supply and return hoses from the fuel injection supply manifold.

4. Disconnect the vacuum line from the fuel pressure regulator.

5. Disconnect the intake manifold runner actuator control cable from the intake manifold. Be careful not to loosen or bend the cable bracket, alignment is critical.

6. Disconnect the ignition wires from the left cylinder head and position the wires aside.

7. Remove the 8 lower intake manifold to cylinder head retaining bolts in sequence.

8. Remove the lower intake manifold and gaskets from the vehicle.

9. If the lower intake manifold is to be replaced or machined, remove the fuel injector supply manifold (fuel rail) and the fuel injectors.

To install:

10. Install the fuel injectors and the fuel supply manifold onto the lower intake manifold if removed. Verify

the operation of the manifold runner control plate.

11. Thoroughly clean the gasket sealing areas and place 2 new intake to cylinder head gaskets into position.

12. Carefully install the lower intake manifold and install the intake manifold to cylinder head retaining bolts. Tighten the retaining bolts in sequence to 71–106 inch lbs. (8–12 Nm).

13. Install the fuel supply and return hoses to the fuel supply manifold and ensure that the spring lock couplings are correctly installed.

14. Install the retaining clips onto the spring lock couplings.

15. Connect the vacuum line to the fuel pressure regulator.

16. Temporarily connect the negative battery cable.

17. Connect the fuel pressure gauge to the fuel pressure relief valve located on the fuel injection supply manifold.

18. Cycle the ignition key several times to the **RUN** position to pressurize the fuel system.

19. Watch the fuel pressure gauge for signs of leakage. If the gauge holds pressure, remove the gauge and continue with the installation of the upper intake manifold. If the pressure gauge loses pressure, re-

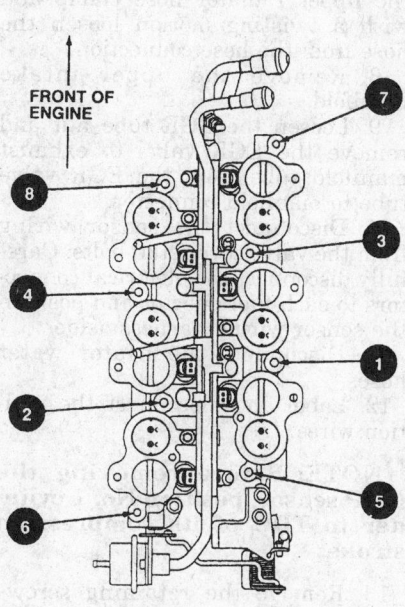

FRONT OF
ENGINE

LOWER INTAKE MANIFOLD

TIGHTEN BOLTS IN
SEQUENCE SHOWN

307194

Lower intake manifold torque sequence — 3.0L (VIN S) engine

move the fuel injection supply manifold and replace the leaking O–ring(s) before continuing.

20. Disconnect the negative battery cable.

21. Reposition and install the fuel injector wiring harness.

22. Connect the ignition wires to the left cylinder head.

23. Connect the intake manifold runner actuator control cable to the intake manifold. Be careful not to loosen or bend the cable bracket, alignment is critical.

1996–97 3.0L (VIN 1 and U) Engine

Upper Intake Manifold

1. Disconnect the negative battery cable.

2. Remove the air cleaner outlet tube.

3. Remove the accelerator cable shield from the from the cable bracket.

4. Remove the accelerator cable spring. Disconnect the accelerator and speed control cables from the throttle body.

5. Remove the 2 accelerator cable bracket retaining bolts from the side of the throttle body and position the cable bracket aside.

6. Label and disconnect the vacuum hose from the fuel pressure regulator.

7. Loosen the EGR tube nut at the EGR valve and disconnect the EGR backpressure transducer hoses from the EGR valve to exhaust manifold tube.

8. Disconnect the PCV hose, aspirator vacuum supply hose and evaporative emission return tube from the fitting underneath the upper intake manifold.

9. Disconnect the electrical connectors to the Throttle Position Sensor (TPS), Idle Air Control (IAC) valve, EGR backpressure transducer and EGR vacuum regulator solenoid.

10. Disconnect the degas tube from the radiator coolant recovery tank and lower intake manifold fitting.

11. Remove the retaining nut and bolts for the upper alternator brace and remove the brace.

12. Remove the sensor wiring bracket from the throttle body retaining stud bolt and position the wiring aside.

13. Remove the intake manifold support from the throttle body and right cylinder head.

14. Remove the upper intake manifold retaining bolts and stud bolts and note their location for installation. Remove the upper intake manifold.

15. Remove the manifold gaskets and discard.

To install:

16. Clean the gasket sealing surfaces and install the new intake manifold gaskets using locating pins as necessary to aid in gasket alignment.

17. Lightly oil all attaching bolt and stud threads before installation.

18. Position the upper intake gasket and manifold on top of the lower intake manifold. Use locating pins to secure the position of gasket between manifolds.

19. Install the retaining bolts and studs in their original locations. Tighten the stud bolts and bolts to 15–22 ft. lbs. (20–30 Nm).

20. Install the alternator brace to the upper intake manifold mounting stud and alternator mounting bracket. Tighten the nut and bolts to 9–15 ft. lbs. (12–20 Nm).

21. Install the intake manifold support to the throttle body and the right cylinder head. Tighten the top retaining bolt to 71–106 inch lbs. (8–12 Nm). Tighten the bottom bolt to 30–40 ft. lbs. (40–55 Nm).

22. Install the engine sensor wiring bracket onto the throttle body stud bolts.

23. Connect the PCV hose, aspirator vacuum supply hose and evaporative emission return tube to the fitting underneath the upper intake manifold.

24. Install the EGR tube nut to the EGR valve and tighten to 26–48 ft. lbs. (35–65 Nm).

25. Connect the vacuum hose to the fuel pressure regulator.

26. Connect the electrical connectors to the TPS, IAC, EGR backpressure transducer and EGR vacuum regulator solenoid.

27. Install the accelerator cable bracket to the side of the throttle body and install the 2 retaining bolts. Tighten the bolts to 13 ft. lbs. (17 Nm).

28. Connect the accelerator cable and speed control cable to the throttle body. Install the throttle retracting spring.

29. Install the accelerator cable shield and tighten the bolts to 13 inch lbs. (1.4 Nm).

30. Install the air cleaner outlet tube. Connect the negative battery cable.

31. Fill the cooling system.

32. Start the engine and check for leaks and proper operation.

Lower Intake Manifold

1. Disconnect the negative battery cable. Relieve the fuel system pressure.

2. Disconnect the wiring from the Mass Air Flow (MAF) sensor and the Intake Air Temperature (IAT) sensor.

3. Remove the air cleaner outlet tube. Remove the air cleaner bolts at the air cleaner body.

4. Disconnect the engine intake air resonator by pushing in the top and bottom tube surfaces at the engine air cleaner and pulling the air cleaner outward. Lift the air cleaner up and out of the engine compartment.

5. Remove the fuel line safely clips. Disconnect the fuel supply and return lines from the fuel supply manifold using the proper disconnect tools.

6. Disconnect the remaining engine wire connectors from the Camshaft Position (CMP) sensor, Throttle Position (TP) sensor, Idle Air Control (IAC) valve, Engine Coolant Temperature (ECT) sensor, ignition coil, water temperature sensor, EGR backpressure transducer and EGR vacuum regulator solenoid connector.

NOTE: Note the position of the CMP sensor electrical connector. The installation requires that the connector be located in the same location.

7. With suitable pliers, slide back the upper radiator hose clamp and with a twisting motion loosen the hose from the hose connection.

8. Remove the upper intake manifold.

9. Loosen the EGR tube nut and remove the EGR valve to exhaust manifold tube from the EGR valve tube to manifold connector.

10. Disconnect the sensor wiring from the valve cover stud bolts. Carefully disconnect the electrical connectors to each fuel injector and position the sensor wiring harness aside.

11. Disconnect the heater water hoses.

12. Label and disconnect the ignition wires.

NOTE: Before removing the CMP sensor, position No. 1 cylinder to TDC of its compression stroke.

13. Remove the retaining screws from the CMP sensor and remove the sensor from the sensor housing.

14. Remove the hold-down clamp and remove the CMP housing from the cylinder block.

15. Remove the cylinder head covers.

16. Remove the ignition coil from the rear of the left cylinder head.

17. Loosen the intake valve rocker arm retaining bolt from cylinder No. 3 and rotate the rocker arm away from the valve stem and pushrod. Remove the pushrod.

NOTE: The lower intake manifold may be removed with the fuel injection supply manifold and fuel injectors in place.

18. Remove the intake manifold attaching bolts using a Torx® head socket. Use a suitable prybar to loosen the intake manifold. Pry upward using the area between the thermostat and transaxle as a leverage point. Remove the manifold and old gaskets and seals.

To install:

19. Clean the gasket mating surfaces of the intake manifold and the cylinder heads. Lay a shop rag in the lifter valley to catch any gasket material. After scraping, carefully lift the shop rag from the lifter valley, being careful not to let any particles enter the oil drain holes or cylinder head. If necessary, use a suitable solvent to remove old rubber sealant.

20. Lightly oil all the attaching bolts and stud threads before installation. When using a silicone rubber sealer, assembly must occur within 15 minutes after the sealer has been applied. After this time, the sealer may start to set-up and its sealing quality may be reduced. In high temperature and/or humidity conditions, the sealant will start to set up in approximately 5 minutes.

21. Apply a suitable silicone rubber sealer to the intersection of the cylinder block end rails and cylinder heads. Be careful not to let sealer that may block oil passages fall into the engine.

22. Install the front and rear intake manifold end seals in place and secure. Install the intake manifold gaskets, aligning the locking tabs to the provisions on the cylinder head gaskets.

23. Carefully lower the intake manifold into position on the cylinder block and cylinder heads to prevent smearing the silicone sealer and causing gasket voids.

24. Install the bolts and tighten in sequence, starting at the center and working towards the ends. Torque the bolts in 2 steps, tighten to 15–22 lbs. (20–30 Nm). Tighten again in sequence to 19–24 ft. lbs. (26–32 Nm).

25. Install the fuel supply manifold and injectors, if removed. Apply a small amount of clean engine oil to the injector holes in the intake manifold and fuel supply manifold prior to injector installation. Install the fuel

supply manifold retaining bolts and tighten to 7 ft. lbs. (10 Nm).

WARNING

A special Synchro Positioning tool T95T-12200-A or equivalent must be used before installing the CMP sensor. If the special tool is not used the fuel system will be out of time possibly causing engine damage.

26. Attach the Synchro Position tool T95T-12200-A as follows:

    a. Engage the CMP sensor housing vane into the radial slot of the tool.

    b. Rotate the tool on the CMP sensor housing until the tool boss engages the notch in the CMP sensor housing.

27. Lube the drive gear with clean engine oil and install the CMP sensor housing so the drive gear engagement occurs when the arrow on the locator tool is pointed about 75 degrees counterclockwise from the rear face of the cylinder block. This step will locate the CMP sensor electrical connector in the same position as was noted on removal.

28. Install the hold-down clamp and tighten the bolt to 14–22 ft. lbs. (19–30 Nm).

29. Install the CMP sensor to the housing and tighten the retaining screws to 13–35 inch lbs. (1.5–4.0 Nm).

30. Install the No. 3 cylinder intake valve pushrod. Apply Lubriplate® or equivalent, to the pushrod and valve stem prior to installation. Turn the crankshaft as necessary to position the lifter on the base circle of the camshaft (pushrod all the way down). Tighten the rocker arm bolt in 2 steps, first to 8 ft. lbs. (11 Nm) and then to 24 ft. lbs. (32 Nm).

31. Install the cylinder head covers.

32. Install the fuel injector harness wiring and attach to the injectors.

33. Install the ignition coil to the rear of left cylinder head. Tighten the retaining bolts to 30–40 ft. lbs. (40–55 Nm).

34. Install the wire harness to the valve cover stud bolts and connect the ignition wires to the spark plugs and ignition coil.

35. Install the upper intake manifold.

36. Install the exhaust manifold tube to the EGR valve on the intake manifold. Tighten the upper EGR tube nut to 26–48 ft. lbs. (35–65 Nm). Tighten the lower (exhaust manifold) tube nut to 26–48 ft. lbs. (35–65 Nm).

37. Install the fuel lines and safety clips.

38. Connect the water bypass hose and the upper radiator hose and properly install the squeeze clamps.

39. Connect the vacuum hoses to their premarked positions.

40. Connect the engine sensor wiring to the CMP sensor, IAC valve, TP sensor, ECT sensor, EGR backpressure transducer, EGR vacuum regulator solenoid, ignition coil and water temperature sender.

NOTE: Make sure the CMP connector is installed in the same position as it was removed for correct operation.

41. Install engine air cleaner and air cleaner outlet tube.

42. Connect the negative battery cable.

43. Cycle the ignition switch to the **RUN** position several times without starting the engine to pressurize the fuel system and check for fuel leaks.

44. Fill the cooling system.

45. Start the engine and check for leaks and proper operation.

3.8L Engine

1. Disconnect the negative battery cable.

2. Drain the cooling system and relieve the fuel system pressure.

3. Remove the air cleaner assembly or air inlet tube.

4. Disconnect the accelerator cable at the throttle body. Disconnect the cruise control cable, if equipped.

5. If equipped with an automatic transaxle, disconnect the transaxle linkage at the upper intake manifold. Remove the retaining bolts from the accelerator cable mounting bracket and position the cables aside.

6. If equipped, disconnect the thermactor air supply hose at the check valve. The valve is located in the Y-pipe assembly.

7. Disconnect and plug the flexible fuel lines from the from steel lines over rocker arm cover.

8. Detach and plug the fuel lines at injector fuel rail assembly.

9. Disconnect the radiator hose at the thermostat housing and the coolant bypass hose at the manifold.

10. Disconnect the heater tube at the intake manifold and remove the tube support bracket retaining nut. Remove the heater hose at the rear of the heater tube. Loosen the hose clamp at the heater elbow and remove the heater tube with the hose attached. Remove the heater tube with the lines attached and set the assembly aside.

11. Tag and disconnect the vacuum lines at the fuel rail assembly and intake manifold. Tag and disconnect the necessary electrical connectors.

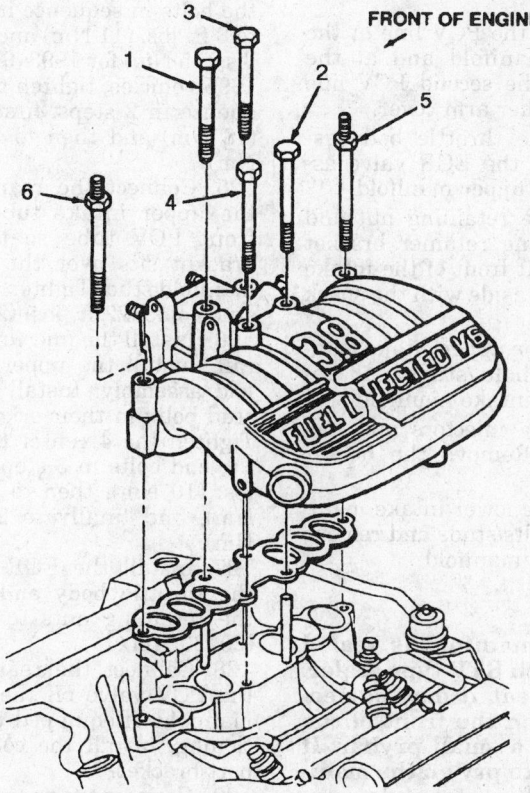

Upper intake manifold torque sequence — 3.8L engine

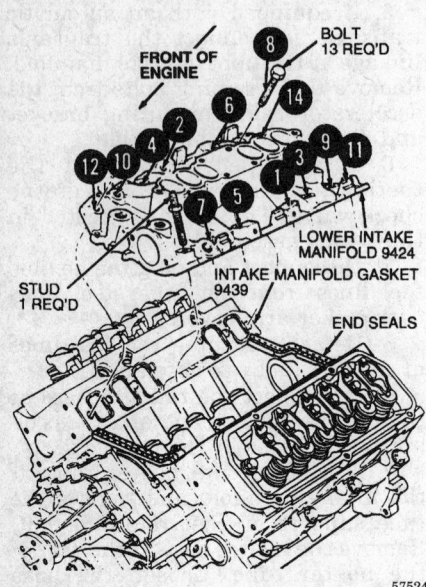

FRONT OF ENGINE · **BOLT 13 REQ'D** · 8 14 6 2 4 9 11 12 10 5 1 3 7

LOWER INTAKE MANIFOLD 9424 · **INTAKE MANIFOLD GASKET 9439** · **STUD 1 REQ'D** · **END SEALS**

57524

Lower intake manifold torque sequence — 3.8L engine

12. If equipped with air conditioning, remove the compressor support bracket.

13. Disconnect the PCV line at the upper intake manifold and at the valve. Remove the second PCV line from the left rocker arm cover.

14. Remove the throttle body assembly. Remove the EGR valve assembly from the upper manifold.

15. Remove the retaining nut and remove the wiring retainer bracket located at the left front of the intake manifold and set aside with the spark plug wires.

16. Unfasten the upper intake manifold retaining bolts/studs and remove the upper intake manifold.

17. Remove the injectors and fuel rail assembly. Remove the heater water outlet hose.

18. Remove the lower intake manifold retaining bolts/studs and remove the lower intake manifold.

NOTE: The manifold is sealed at each end with RTV-type sealer. To break the seal, it may be necessary to pry on the front of the manifold with a small prybar. If it is necessary to pry on the manifold, use care to prevent damage to the machined surfaces.

To install:

19. Clean all gasket mating surfaces. Lightly oil all retaining bolt and stud threads.

20. Apply a dab of gasket adhesive to each cylinder head mating surface. Press new intake manifold gaskets in place, using location pins as necessary to aid in installation.

21. Apply a ⅛ in. bead of silicone sealer at each corner where the cylinder head joins the cylinder block. Install the front and rear intake manifold end seals.

22. Carefully lower the intake manifold into place on the cylinder heads and cylinder block. Use locating pins as necessary to guide the manifold.

23. Install the bolts and stud bolts in their original locations and make them finger-tight.

NOTE: Bolt torque depends on the type of gasket used.

24. Graphite gaskets are usually standard. As the engine warms up, graphite gaskets allow the manifold and cylinder heads to expand at different rates without damaging the gasket and loosing the seal.

• If using the older style gasket, tighten the bolts in sequence in 3 steps: first to 8 ft. lbs. (10 Nm), then to 15 ft. lbs. (20 Nm), and finally to 24 ft. lbs. (32 Nm).

• If using graphite gaskets, tighten the bolts in sequence in 2 steps, first to 8 ft. lbs. (11 Nm) and then to 11 ft. lbs. (15 Nm) for 1993–94 vehicles. For 1995 vehicles, tighten the bolts in sequence in 2 steps, first to 13 ft. lbs. (18 Nm) and then to 16 ft. lbs. (22 Nm).

25. Connect the rear PCV line to the upper intake tube. Install the front PCV tube so the mounting bracket sits over the lower intake manifold stud. Tighten the nut on the stud to 15–22 ft. lbs. (20–30 Nm).

26. Install the injectors and the fuel rail. Install the upper intake manifold assembly. Install the bolts and stud bolts in their original locations. Tighten the 4 center bolts and then the end bolts in 3 steps, first to 8 ft. lbs. (10 Nm), then to 15 ft. lbs. (20 Nm), and finally to 24 ft. lbs. (32 Nm).

27. Install the EGR valve. Install the throttle body and cross-tighten the retaining nuts to 15–22 ft. lbs. (20–30 Nm).

28. Connect the rear PCV line at the PCV valve on the upper intake manifold. If equipped with air conditioning, install the compressor support bracket.

29. Connect the necessary electrical connectors and vacuum hoses. Con-

nect the heater tube hose to the heater elbow and position the heater tube support bracket. Tighten the retaining nut to 15–22 ft. lbs. (20–30 Nm).

30. Connect the heater hose to the heater tube and connect the coolant bypass hose and radiator upper hose.

31. Connect the fuel lines. Position the accelerator cable mounting bracket and tighten the mounting bolts to 15–22 ft. lbs. (20–30 Nm).

32. Connect the transaxle linkage at the upper intake manifold. If equipped, connect the cruise control cable.

33. Fill and bleed the cooling system. Connect the negative battery cable, start the engine and check for leaks.

34. Check and if necessary, adjust the engine idle speed, transaxle throttle linkage and cruise control.

4.6L Engine

1. Disconnect the negative battery cable. Drain the engine cooling system.

2. Remove engine appearance cover. Remove the air cleaner outlet tube.

NOTE: Label all electrical connectors and vacuum hoses as they are removed, so they can be reinstalled in their original locations.

3. Relieve the fuel system pressure. Disconnect the fuel supply and return lines from the fuel injection supply manifold.

4. Remove 8 retaining nuts on both ignition wire covers and remove the ignition wire covers.

NOTE: Do not pull on the ignition wire as the wire may separate from the connector in the ignition wire boot.

5. Tag and disconnect the ignition wires from the spark plugs with a gentle twist/pull motion on the spark plug boot.

6. Disconnect the ignition wires from the ignition coils.

7. Detach 2 center ignition wire separators from the retaining studs.

8. Remove the ignition wires.

9. Disconnect the alternator wiring harness from the alternator at the battery terminal and disconnect the stator connector plug. Remove the wiring harness retaining clip from the alternator mounting bracket.

10. Remove 2 bolts and 2 nuts attaching the alternator mounting

bracket to the intake manifold. Remove the mounting bracket.

11. Disconnect the upper radiator hose, the heater water hose and the water bypass tube to water thermostat hose at the water bypass tube.

12. Disconnect the wiring from the Engine Coolant Temperature (ECT) sensor at the water bypass tube.

13. Remove 2 retaining nuts and spacers from the water bypass tube hold-down braces to intake manifold studs and remove the water bypass tube.

14. Disconnect the accelerator cable and the speed control actuator from the throttle body and from the accelerator cable bracket. Position the accelerator cable and speed control actuator aside.

15. Disconnect the chassis vacuum supply hose from the intake manifold connector and move it aside.

16. Disconnect the vacuum harness from the fuel pressure regulator, right and left secondary air injection diverter valves, EGR control valve, EGR vacuum regulator control, and secondary air injection vacuum control solenoid and intake manifold connection. Remove the vacuum harness.

17. Disconnect the Intake Manifold Runner Control (IMRC) actuator from the IMRC assemblies at the levers. Remove the IMRC actuator retainers and the IMRC actuator.

18. Disconnect the crankcase vent connector and hose from the throttle body. Remove the PCV valve from the left side cylinder head cover. Move the crankcase vent connector and hose and PCV valve aside.

19. Disconnect the engine wiring harness retaining clips from the intake manifold studs.

20. Disconnect 8 wiring connectors from the fuel injectors.

21. Disconnect the engine control sensor wiring from the Idle Air Control (IAC) valve, Throttle Position (TP) sensor or TP sensors (if equipped with traction control). Position the wiring harness aside.

22. Remove 2 retaining bolts attaching the EGR valve to the intake manifold.

23. Loosen and remove 20 bolts and studs retaining the intake manifold to the cylinder heads in sequence.

24. Lift the engine control sensor wiring upward for intake manifold removal clearance.

25. Lift the intake manifold off the engine.

26. Remove 4 fuel injection supply manifold retaining bolts from the intake manifold. Remove the fuel injec-

tion supply manifold from the fuel injectors.

27. Remove 4 bolts retaining both lower intake manifolds (IMRC housing assemblies) to the intake manifold and remove both lower intake manifolds.

NOTE: The upper intake manifold gaskets must be replaced when the manifold retaining bolts have been loosened.

28. Remove the intake manifold upper gaskets and all load limiting spacers from both lower intake manifolds (IMRC assemblies) and discard.

29. Remove the lower intake manifold gaskets from the cylinder heads and discard.

30. Remove the EGR valve gasket and discard. Clean the EGR gasket sealing surface on the EGR valve and the intake manifold.

To install:

31. Clean and inspect all gasket mating surfaces. The upper sealing surfaces of the Intake Manifold Runner Controls (IMRC) housing assemblies and intake manifold mating surfaces should be cleaned with a suitable solvent to remove adhesive residue. Inspect the intake manifold mating surfaces and upper IMRC assembly surfaces to make sure no load limiting spacers (washers) are stuck to the sealing surfaces. The load limiting spacers are part of the intake manifold upper gasket. New load limiter spacers will be included with the replacement intake manifold upper gasket as an assembly.

32. If a new intake manifold or IMRC housing assembly is being installed, transfer all necessary components to the new intake manifold or lower intake manifold using new gaskets.

33. Install the upper intake manifold gasket on the right IMRC housing using the tapered pins at the end bolt locations for proper gasket alignment. Repeat the procedure for the left IMRC housing.

34. Install the IMRC assemblies to the intake manifold and install the 4 IMRC retaining bolts (A-D in illustration), finger-tight.

35. Install the fuel injection supply manifold to the fuel injectors and install 4 retaining bolts. Tighten to 71–106 inch lbs. (8–12 Nm).

36. Position new intake manifold gaskets onto the cylinder heads, using the gaskets integral locating pins to align the gaskets.

37. Place the intake manifold under the engine control sensor wiring and connect the fuel charging wiring to

the IAC valve and TP sensor, or sensors, depending upon application.

38. Move the intake manifold into position and install a new EGR valve gasket and the EGR retaining bolts. Finger-tighten the bolts.

39. Install the longer bolts and stud bolts in the outer bolt holes of the intake manifold, but do not tighten yet.

40. Install the shorter bolts and stud bolts in the inner bolt holes of the intake manifold. Hand–tighten all 20 fasteners.

41. Following the sequence shown, tighten fasteners 1 through 20 in numerical sequence as follows:

   a. No. 5, 7, 9 and 11 — 9–11 ft. lbs. (12–15 Nm)

   b. All others — 13–16 ft. lbs. (18–22 Nm)

   c. Then tighten all fasteners in numerical sequence, by rotating an additional 85–95 degrees.

42. Tighten 4 intake manifold to IMRC housing retaining bolts A-D to 71–89 inch lbs. (8–10 Nm). Tighten fasteners A through D, in sequence, by rotating an additional 85–95 degrees.

43. Tighten 2 EGR valve retaining bolts to 15–22 ft. lbs. (20–30 Nm).

44. Connect the wiring to the 8 fuel injectors.

45. Connect the fuel charging wiring harness retaining clips to the intake manifold studs.

46. Connect the crankcase vent connector and hose to the throttle body connection and install the PCV valve into the grommet on the right cylinder head cover.

47. Install the IMRC actuator to the IMRC assemblies at the levers. Position the actuator on the engine and install the retainers. Tighten the retainers to 71–106 inch lbs. (8–12 Nm).

48. Connect the vacuum harness to the fuel pressure regulator, right and left secondary air injection diverter valves, EGR valve, EGR vacuum regulator control and secondary air injectors vacuum control solenoids, and intake manifold vacuum rear connection.

49. Connect the chassis vacuum supply hose and spring clamp to the intake manifold vacuum connection.

50. Connect the accelerator cable and the speed control actuator to the throttle body. Connect the cables to the accelerator cable bracket.

51. Replace all O-rings on the water bypass tube. Clean the O-ring sealing surface on the cylinder heads.

52. Coat all O-rings with clean coolant.

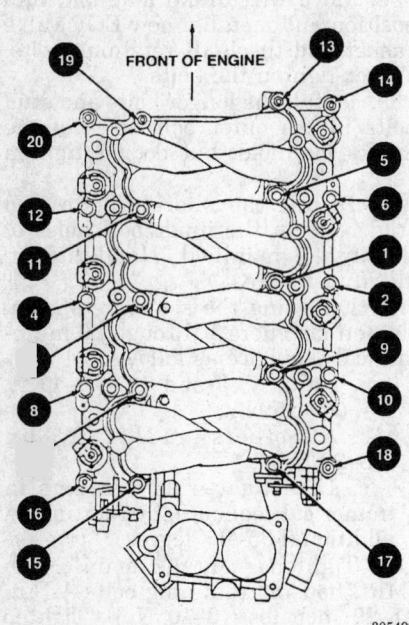

Intake manifold torque sequence — 4.6L engine

305492

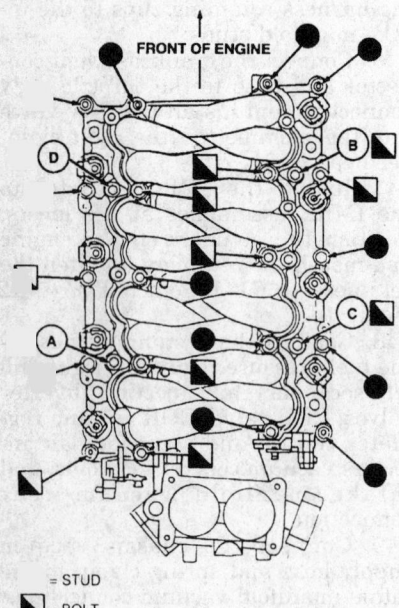

 = STUD

■ = BOLT

305493

Intake manifold-to-IMRC housing torque sequence — 4.6L engine

53. Install the water bypass tube support braces over the intake manifold inner studs and install the water bypass tube into position.

54. Install 2 water bypass tube spacers and retaining nuts. Tighten the nuts to 71–106 inch lbs. (8–12 Nm).

55. Connect the wiring connector to the ECT sensor.

56. Connect the upper radiator hose, heater water hose and water bypass tube to thermostat housing hose to the water bypass tube.

57. Position the alternator mounting bracket and install 2 front retaining bolts and 2 rear nuts. Tighten to 71–106 inch lbs. (8–12 Nm).

58. Connect the alternator wiring harness connector plug and battery terminal wire to the alternator. Install the retaining clip to the alternator mounting bracket. Tighten the retaining screw securely.

59. Position the ignition wires and connect 2 center ignition wire separators to the retaining studs on the intake manifold.

60. Connect the ignition wires to the ignition coils.

61. Connect the ignition wires to the proper spark plugs.

62. Install the ignition wires into the cylinder head cover slots and install the ignition wire covers. Install 8 retaining nuts and tighten to 18–35 inch lbs. (2–4 Nm).

63. Connect the fuel supply and return lines to the fuel injection supply manifold.

64. Install the engine air cleaner outlet tube. Tighten the air cleaner outlet tubes clamps to 12–22 inch lbs. (1.4–2.5 Nm).

65. Install the engine appearance cover. Tighten the retaining nuts to 61–79 inch lbs. (7–9 Nm).

66. Install the fuel tank fill cap.

67. Connect the negative battery cable.

68. Fill and bleed the cooling system.

69. Start the engine and check for leaks.

70. Road test the vehicle and check for proper operation.

Exhaust Manifold

REMOVAL AND INSTALLATION

3.0L (VIN Y) and 3.2L SHO Engines

Left Side

1. Disconnect the negative battery cable.

2. Remove the oil level indicator tube support bracket.

3. Remove the power steering pump pressure and return hoses.

4. Remove the manifold to exhaust pipe attaching nuts.

5. Remove the heat shield retaining bolts.

6. Remove the exhaust manifold retaining nuts and manifold.

To install:

7. Clean the mating surfaces of the exhaust manifold, cylinder head and Y-pipe.

8. Lightly oil all bolt and stud threads before installation.

9. Position the exhaust manifold on the cylinder head and install the exhaust manifold retaining nuts. Tighten to 26–38 ft. lbs. (35–52 Nm).

10. Install the heat shield retaining bolts. Tighten to 11–17 ft. lbs. (15–23 Nm).

11. Connect the Y-pipe to the exhaust manifold. Tighten the retaining nuts to 15–24 ft. lbs. (21–32 Nm).

12. Connect the power steering pressure hose and power steering return hoses.

13. Install the oil level indicator tube support bracket.

14. Connect the negative battery cable.

15. Start the engine and check for exhaust and coolant leaks.

Right Side

1. Disconnect the negative battery cable.

2. Remove the right cylinder head.

3. Remove the heat shield retaining bolts.

4. Remove the exhaust manifold retaining nuts and manifold.

To install:

5. Clean the mating surfaces of the exhaust manifold, cylinder head and Y-pipe.

6. Lightly oil all bolt and stud threads before installation.

7. Position the exhaust manifold on the cylinder head and install the exhaust manifold retaining nuts. Tighten to 26–38 ft. lbs. (35–52 Nm).

8. Install the heat shield retaining bolts. Tighten to 11–17 ft. lbs. (15–23 Nm).

9. Install the right side cylinder head.

10. Start the engine and check for exhaust and coolant leaks.

3.0L (VIN 1 and U) Engine

Left Side

1. Disconnect the negative battery cable.

2. Remove oil level indicator tube support bracket retaining nut. Re-

move the oil level indicator tube and move engine control sensor wiring aside.

3. Raise and safely support the vehicle.

4. Remove the exhaust manifold-to-exhaust pipe retaining nuts.

5. Lower the vehicle.

6. Remove the 4 exhaust manifold retaining bolts and 2 exhaust manifold stud bolts. Remove the exhaust manifold from the vehicle.

To install:

7. Clean all mating surfaces and lightly oil all bolt and stud threads prior to installation.

8. Place the exhaust manifold into position on the cylinder head using a new gasket. Tighten the 4 exhaust manifold retaining bolts and 2 stud bolts to 15–18 ft. lbs. (20–25 Nm).

9. Install the exhaust pipe to the exhaust manifold and tighten the exhaust pipe attaching nuts to 25–34 ft. lbs. (34–47 Nm).

10. Install the oil level indicator tube. Tighten the bracket nut to 12–15 ft. lbs. (16–20 Nm). Reposition the engine control sensor wiring.

11. Connect negative battery cable.

12. Run the engine and check for exhaust leaks and proper operation.

Right Side

1. Disconnect the negative battery cable.

2. Disconnect the EGR valve hoses. Remove the EGR tube from the exhaust manifold. Use a backup wrench on the lower adapter.

3. Remove the 3 retaining bolts from the exhaust manifold heat shield and remove the shield.

4. Raise and safely support the vehicle.

5. Remove the exhaust manifold-to-exhaust pipe retaining nuts and separate the exhaust pipe from the exhaust manifold.

6. Lower the vehicle.

7. Remove the 6 exhaust manifold retaining bolts and remove the exhaust manifold from the vehicle.

To install:

8. Clean all mating surfaces and lightly oil all bolt threads prior to installation.

9. Place the exhaust manifold into position on the cylinder head. Tighten the 6 exhaust manifold retaining bolts to 15–18 ft. lbs. (20–25 Nm).

10. Raise and safely support the vehicle.

11. Position the exhaust pipe and install the retaining nuts. Tighten the retaining nuts to 25–34 ft. lbs. (34–47 Nm).

12. Install the exhaust manifold heat shield and install the 3 retaining bolts. Tighten the retaining bolts to 71–106 inch lbs. (8–12 Nm).

13. Lower the vehicle.

14. Install the EGR tube to the exhaust manifold. Tighten the tube nut to 26–48 ft. lbs. (35–65 Nm). Connect the EGR valve hoses.

15. Connect the negative battery cable.

16. Run the engine and check for exhaust leaks and proper operation.

1996–97 3.0L (VIN S) Engine

1. Disconnect the negative battery cable.

2. For the right side, perform the following procedures:

 a. Remove the upper intake manifold assembly.

 b. Remove the ignition coil assembly.

 c. Loosen the EGR valve to exhaust manifold tube nuts and remove the tube.

 d. Disconnect the wiring harness from the oxygen sensor and remove the sensor.

3. Remove the secondary air injection manifold tube from the exhaust manifold.

4. Raise and safely support the vehicle.

5. Remove the dual converter Y-pipe retaining nuts from the exhaust manifolds.

6. Remove both bolt and nut retainers from the transaxle.

7. Remove the 2 remaining nuts and bolts from the dual converter Y-pipe connection. Remove the Y-pipe from the vehicle.

8. Remove the lower exhaust manifold retaining nuts from the cylinder head studs and lower the vehicle.

9. Remove the upper exhaust manifold retaining nuts from the cylinder head studs.

10. Remove the exhaust manifold and gasket from the engine.

11. Clean all gasket mating surfaces.

To install:

12. Install the exhaust manifold with a new exhaust manifold gasket.

13. Install the exhaust manifold retaining studs and torque to 13–16 ft. lbs. (18–22 Nm) in sequence.

14. Raise and safely support the vehicle.

15. Position the Y-pipe assembly using a new flange gasket and install all the retaining nuts and bolts loosely.

16. Starting at the front of the system tighten the Y-pipe to exhaust manifold nuts to 26–34 ft. lbs. (34–46 Nm). Tighten the converter to transaxle nut and bolt to 30 ft. lbs. (40.3 Nm). Tighten the converter outlet bolts to 26–34 ft. lbs. (34–46 Nm).

17. Lower the vehicle.

18. Install the oxygen sensor and torque to 26–34 ft. lbs. (35–46 Nm). Connect the electrical connector.

19. Install the secondary air injection manifold tube to the exhaust manifold and tighten the nut to 28–31 ft. lbs. (38–42 Nm).

20. For the right side, perform the following procedures:

 a. Install the EGR valve to exhaust manifold tube and torque the nuts to 26–33 ft. lbs. (35–45 Nm).

 b. Install the ignition coil assembly.

 c. Install the upper intake manifold assembly.

21. Connect the negative battery cable.

22. Run the engine and check for exhaust leaks and proper operation.

3.8L Engine

1. Disconnect the negative battery cable.

2. For the left side, perform the following procedures:

 a. Remove the oil level dipstick tube support bracket.

 b. Tag and disconnect the spark plug wires.

3. For the right side, perform the following procedures:

 a. Remove the air cleaner outlet tube assembly. If equipped, disconnect the thermactor hose from the downstream air tube check valve.

 b. Tag and disconnect the coil secondary wire from coil and the wires from spark plugs. Remove the spark plugs.

 c. Disconnect the EGR tube.

4. Raise the vehicle and support safely.

5. For the right side, remove the transaxle dipstick tube.

6. Remove the manifold-to-exhaust pipe attaching nuts.

7. Lower the vehicle.

8. Remove the exhaust manifold retaining bolts and remove the manifold from vehicle.

To install:

9. Lightly oil all bolt and stud threads before installation. Clean the mating surfaces on the exhaust manifold, cylinder head and exhaust pipe.

10. Position the exhaust manifold on the cylinder head.

11. For the left side, install the lower front bolt on No. 5 cylinder as a pilot bolt.

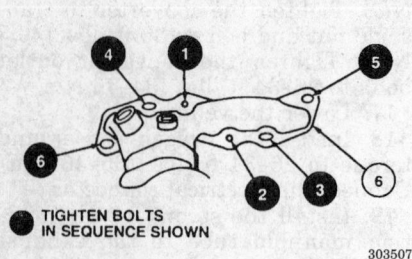

TIGHTEN BOLTS
IN SEQUENCE SHOWN

303507

Proper torque pattern for exhaust manifolds —
3.0L (VIN S) engine

CYLINDER HEAD
ASSY

FRONT OF ENGINE

STUD AND WASHER
ASSY 3 REQ'D
TIGHTEN TO
20–30 Nm
(15–22 LB-FT)

BRACKET WIRE
SUPPORT

HEX NUT
TIGHTEN TO
20–30 Nm
(15–22 LB-FT)

RH EXHAUST
MANIFOLD

SCREW AND WASHER
ASSY 3 REQ'D
TIGHTEN TO
20–30 Nm
(15–22 LB-FT)

235851

Right side exhaust manifold installation — 3.8L engine

12. For the right side, start 2 attaching bolts to align the manifold with the cylinder head.

13. Install the remaining manifold retaining bolts. Tighten the bolts to 15–22 ft. lbs. (20–30 Nm).

NOTE: On the left side, a slight warpage in the exhaust manifold may cause a misalignment between the bolt holes in the head and the manifold. Elongate the holes in the exhaust manifold as necessary to correct the misalignment, if apparent. Do not elongate the pilot hole, the lower front bolt on No. 5 cylinder.

14. Raise the vehicle and support safely.

15. Connect the exhaust pipe to the manifold. Tighten the attaching nuts to 16–24 ft. lbs. (21–32 Nm).

16. For the right side, install the transaxle dipstick tube.

17. Lower the vehicle.

18. For the left side, perform the following procedures:

   a. Connect the spark plug wires.

   b. Install dipstick tube support bracket attaching nut and tighten to 15–22 ft. lbs. (20–30 Nm).

19. For the right side, perform the following procedures:

   a. Install the outer heat shroud and tighten the retaining screws to 50–70 inch lbs. (5–8 Nm).

   b. Install the spark plugs. Connect the wires to their respective spark plugs and connect coil secondary wire to coil.

   c. Connect the EGR tube. If equipped with a thermactor hose, connect the thermactor hose to the downstream air tube and secure with clamp. Install the air cleaner outlet tube assembly.

20. Start the engine and check for exhaust leaks.

4.6L Engine

— CAUTION —

The air suspension switch, located in the left side of the luggage compartment, must be turned OFF before raising the vehicle. Failure to do so may result in unexpected inflation or deflation of the air springs which may result in shifting of the vehicle during service.

1. Turn the air suspension switch, located in the left side of the luggage compartment, to the **OFF** position.

2. Disconnect the negative battery cable.

3. Raise and safely support the vehicle.

4. For the right side, remove the wiring connectors from the Heated Oxygen Sensors (HO_2S).

5. For the left side, remove the splash shield from the lower radiator support and subframe.

6. Remove the dual converter Y-pipe from the exhaust manifolds.

7. For the right side, perform the following procedures:

   a. Remove 4 exhaust connector retaining bolts and the exhaust connector and muffler gasket from the exhaust manifold.

   b. Remove the EGR valve to exhaust manifold tube from the EGR valve tube to manifold connector using a 22mm crowsfoot wrench.

8. Loosen the retaining nut and remove the secondary air injection manifold tube from the exhaust manifold.

9. Remove 8 exhaust manifold-to-cylinder head retaining nuts.

10. Remove the exhaust manifold and gasket.

NOTE: For the right side, access may be gained through the right wheel opening area. It may be necessary to remove the EGR valve tube to manifold connector from the right exhaust manifold.

To install:

11. Clean the mating surfaces of the manifold and the cylinder head.

12. For the right side, if removed, install the EGR valve tube to manifold connector to the right exhaust manifold. Tighten to 33–48 ft. lbs. (45–65 Nm).

13. Position a new exhaust manifold gasket and the exhaust manifold to the cylinder head. Install 8 retaining nuts. Tighten the nuts in sequence to 13–16 ft. lbs. (18–22 Nm) following the same pattern for both manifolds.

14. Connect the secondary air injection manifold tube to the exhaust manifold and tighten the nut to 25–34 ft. lbs. (34–46 Nm).

15. For the right side, perform the following procedures:

   a. Connect the EGR valve to exhaust manifold tube to the EGR valve tube to manifold connector. Tighten the nut to 30–33 ft. lbs. (40–45 Nm).

   b. Install the exhaust connector with a new inlet pipe gasket.

   c. Install the inlet pipe retaining bolts and tighten to:
   1st — 13–17 ft. lbs. (17–23 Nm)
   2nd — 30–40 ft. lbs. (41–54 Nm)

16. Install the dual converter Y-pipe to the exhaust manifolds.

17. For the right side, connect the HO_2S wiring connectors.

18. For the left side, install the splash shield to the subframe and lower radiator support.

19. Lower the vehicle.

20. Connect the negative battery cable.

21. Turn the air suspension switch to the **ON** position.

22. Start the engine and check for exhaust leaks and proper engine operation.

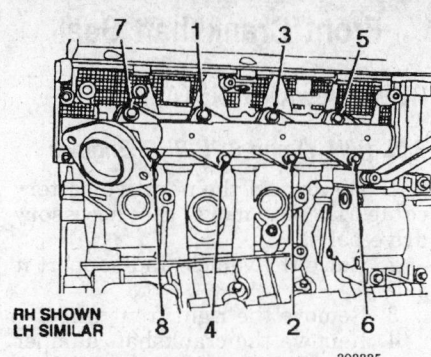

RH SHOWN LH SIMILAR

303335

Exhaust manifold torque sequence — 4.6L engine

Front Cover Seal

REMOVAL AND INSTALLATION

3.0L (VIN 1 and U) Engines

1. Disconnect the negative battery cable. Remove the accessory drive belts.

2. Raise and safely support the vehicle. Remove the right front wheel and tire assembly.

3. Remove 4 pulley-to-damper retaining bolts and remove the crankshaft pulley.

4. Remove the crankshaft damper retaining bolt and washer. Remove the damper from the crankshaft using damper removal tool T58P-6316-D or equivalent and adapter T82L-6316-B or equivalent.

5. Pry the seal from the timing cover with jet plug remover or similar tool, being careful not to damage the front cover and crankshaft.

To install:

NOTE: Before installation, inspect the front cover and shaft seal surface of the crankshaft damper for damage, nicks, burrs or other roughness which may cause the new seal to fail. Service or replace components as necessary.

6. Lubricate the seal lip with clean engine oil and install the seal using damper/seal replacer T82L-6316-A and T70P-6B070-A or equivalents.

7. Coat the crankshaft damper sealing surface with clean engine oil. Apply a silicone sealer to the keyway of the crankshaft damper just prior to installation. Install the damper using damper/seal replacer T82L-6316-A or equivalent.

8. Install the damper retaining bolt and washer. Tighten to 93–121 ft. lbs. (125–165 Nm).

9. Position the crankshaft pulley and install the 4 retaining bolts. Tighten the bolts to 30–44 ft. lbs. (40–60 Nm).

10. Install the right front wheel and tire, then lower the vehicle.

11. Install the accessory drive belts. Check the drive belts for proper routing and engagement in the pulleys.

12. Connect the negative battery cable.

13. Run the engine and check for oil leaks and proper operation.

3.0L (VIN S) Engine

1. Disconnect the negative battery cable, then raise and safely support the vehicle.

2. Remove the right-front wheel and tire assembly, then remove the right splash shield.

3. Remove the accessory drive belt.

4. Install Flywheel Holding tool T96P-6375-A or equivalent to the crankshaft damper.

NOTE: Rotating the crankshaft in a counterclockwise direction may cause the timing chains to bind causing engine damage.

5. Rotate the crankshaft pulley clockwise to remove.

NOTE: The crankshaft pulley uses left hand threading.

6. Remove 3 crankshaft pulley bracket retaining nuts from the engine front cover.

7. Remove the flywheel holding tool.

8. Remove the crankshaft pulley and bracket from the engine.

9. Remove the crankshaft damper retaining bolt and washer.

10. Using Damper Puller T58P-6316-D or equivalent, remove the damper from the crankshaft.

11. Using Seal Remover T92C-6700-CH or equivalent, remove the front cover oil seal from the front cover.

NOTE: The factory tool attaches behind the metal flange similar to a bearing splitter tool and pulls the seal outward. Be careful not to damage the crankshaft or the seal bore in the timing chain cover.

To install:

12. Lubricate the seal bore in the front cover and the seal lip area with clean engine oil.

13. Install the new front cover oil seal using Damper Replacer T74P-6316-B, or equivalent. Make

sure the seal is installed evenly and straight.

14. Thoroughly clean the front of the crankshaft and the crankshaft damper.

15. Apply silicone sealer to the front of the crankshaft at the keyway.

16. Install the crankshaft damper using Damper Replacer T74P-6316-B, or equivalent.

17. Install the crankshaft damper retaining bolt and washer. Tighten the retaining bolt as follows:

   a. Tighten the bolt to 89 ft. lbs. (120 Nm).

   b. Loosen the bolt at least 1 full turn.

   c. Tighten the bolt to 35–39 ft. lbs. (47–53 Nm).

   d. Rotate the bolt an additional 90 degrees.

18. Place the crankshaft pulley and bracket to the crankshaft and front cover.

19. Install Flywheel Holding tool T96P-6375-A or equivalent to the crankshaft damper.

20. Install 3 crankshaft pulley bracket retaining nuts and tighten to 15–22 ft. lbs. (20–30 Nm).

21. Tighten the crankshaft pulley to the crankshaft damper in a counterclockwise direction to 70–77 ft. lbs. (95–105 Nm).

22. Remove the flywheel holding tool.

23. Install the accessory drive belt.

24. Install the right splash shield.

25. Install the wheel and tire assembly.

26. Lower the vehicle.

27. Connect the negative battery cable.

28. Start the engine and check for leaks and proper operation.

3.8L Engine

1. Disconnect the negative battery cable and remove the accessory drive belts.

2. On Continental, remove the fan shroud and position it back over the fan. Remove the fan/clutch assembly and shroud.

3. Raise the vehicle and support safely.

4. Remove the pulley-to-damper attaching bolts and remove the crankshaft pulley.

5. Remove the crankshaft damper retaining bolt and washer. Remove the damper from the crankshaft using a damper removal tool.

6. Pry the seal from the timing cover with a suitable tool, being careful not to damage the front cover and crankshaft.

To install:

NOTE: Before installation, inspect the front cover and shaft seal surface of the crankshaft damper for damage, nicks, burrs or other roughness which may cause the new seal to fail. Service or replace components as necessary.

7. Lubricate the seal lip with clean engine oil and install the seal using a seal installer tool.

8. Coat the crankshaft damper sealing surface with clean engine oil. Apply RTV to the keyway of the damper prior to installation. Install the damper using a damper seal installer tool. Install the damper retaining bolt and washer. Tighten to 103–132 ft. lbs. (140–180 Nm).

9. Position the crankshaft pulley and install the attaching bolts. Tighten the attaching bolts to 19–28 ft. lbs. (26–38 Nm).

10. Position the drive belt over the crankshaft pulley. Check the drive belt for proper routing and engagement in the pulleys.

11. Reconnect the negative battery cable and start the engine and check for oil leaks.

4.6L Engine

———— CAUTION ————
The air suspension switch, located in the left side of the luggage compartment, must be turned OFF before raising the vehicle. Failure to do so may result in unexpected inflation or deflation of the air springs which may result in shifting of the vehicle during service.

1. Turn the air suspension switch, located in the left side of the luggage compartment, to the **OFF** position.

2. Disconnect the negative battery cable.

3. Release the belt tensioner and remove the accessory drive belt.

4. Raise and safely support the vehicle.

5. On the 1996–97 vehicles, perform the following procedures:

  a. Remove the front wheel and tire assemblies.

  b. Support the subframe, engine and transaxle using Powertrain Lift 014-00765 and Bracket 014-00766 or equivalents.

  c. Remove 4 subframe-to-body retaining bolts and lower the engine, transaxle and subframe approximately 2 inches (51mm). This

will allow the room needed for crankshaft damper removal.

6. Remove the crankshaft damper retaining bolt and washer.

7. Remove the crankshaft pulley from the crankshaft vibration damper using Crankshaft Damper Remover T58P-6316-D, or equivalent.

8. Use Front Cover Seal Remover T74P-6700-A, or equivalent to remove the seal from the front cover.

To install:

9. Lubricate the seal bore in the front cover and the seal lip area with clean engine oil.

10. Install the new seal using Crankshaft Seal Replacer T74P-6700-A, or equivalent. Make sure the seal is installed evenly and straight.

11. Install the crankshaft damper. Be sure to lubricate the sealing surface of the damper with clean engine oil prior to installation. Apply a suitable silicone sealer to the crankshaft keyway. Torque the crankshaft damper retaining bolt to 114–121 ft. lbs. (155–165 Nm) for the 1995 vehicles.

12. On the 1996–97 vehicles, perform the following procedures:

  a. Torque the crankshaft damper retaining bolt to 89 ft. lbs. (120 Nm) and then loosen the bolt 1 turn. Retighten the bolt to 39 ft. lbs. (53 Nm) and then rotate an additional 90 degrees clockwise.

  b. Raise the engine, transaxle and subframe into position using Rotunda Powertrain Lift 014-00765 and Bracket 014-00766 or equivalent.

  c. Align the subframe to the body and install 4 subframe-to-body bolts. Tighten the bolts to 55–75 ft. lbs. (75–102 Nm).

  d. Install the wheel and tire assemblies.

13. Lower the vehicle.

14. Relax the tensioner and install the accessory drive belt.

15. Connect the negative battery cable.

16. Turn the air suspension switch to the **ON** position.

17. Start the vehicle and check for leaks and proper operation.

NOTE: Whenever the vehicles subframe is removed or lowered, the wheel alignment should be checked.

18. Check the front end alignment.

19. Road test the vehicle and check for proper operation.

Front Crankshaft Seal

REMOVAL AND INSTALLATION

3.0L (VIN Y) and 3.2L SHO Engines

1. Disconnect the negative battery cable, then remove the accessory drive belts.

2. Raise the vehicle and support it safely.

3. Remove the right front wheel.

4. Remove the crankshaft damper attaching bolt from the crankshaft damper. Using a suitable puller, remove the crankshaft damper from the crankshaft.

5. Remove the timing belt.

6. Remove the crankshaft timing belt sprocket using a suitable puller.

NOTE: Be careful not to damage the crankshaft sensor or shutter.

7. Remove the crankshaft front oil seal using a suitable puller.

To install:

8. Inspect the oil pump and seal surface of the crankshaft for damage, nicks, burrs or other roughness which may cause the new seal to fail. Repair or replace as necessary.

9. Using suitable tools, install a new crankshaft front oil seal.

10. Install the crankshaft sprocket.

11. Install the timing belt.

12. Install the crankshaft damper. Tighten the damper attaching bolt to 113–126 ft. lbs. (152–172 Nm).

13. Install the accessory drive belts.

14. Lower the vehicle, then start the engine and check for oil leaks.

Timing Chain, Sprockets and Front Cover

REMOVAL AND INSTALLATION

3.0L (VIN U and 1) Engines

1. Disconnect the negative battery cable, then drain the engine cooling system.

2. Loosen 4 water pump pulley bolts while the accessory drive belt is in place.

3. Remove the accessory drive belts. Remove the idler pulley or automatic tensioner, as necessary.

4. Remove the lower radiator hose and the heater hose from the water pump and front cover.

5. Remove the crankshaft pulley and damper.

6. On flexible fuel vehicles, remove the Crankshaft Position (CKP) sensor.

7. Drain the engine oil and remove the oil pan.

8. If necessary, unfasten the water pump pulley retaining bolts, then remove the pulley.

9. Remove the retaining bolts from the timing cover to the cylinder block and remove the timing cover.

10. Remove the crankshaft damper and timing chain front cover.

11. Rotate the crankshaft until the No. 1 piston is at TDC on its compression stroke and the timing marks are aligned.

12. Remove the camshaft sprocket attaching bolt and washer. Slide both sprockets and timing chain forward and remove as an assembly.

13. Check the timing chain and sprockets for excessive wear. Replace if necessary.

To install:

14. Before installation, clean and inspect all parts. Clean the gasket material and dirt from the oil pan, cylinder block and front cover.

15. Slide both sprockets and timing chain onto the camshaft and crankshaft with the timing marks aligned. Install the camshaft bolt and washer and torque to 46 ft. lbs. (63 Nm). Apply clean engine oil to the timing chain and sprockets after installation.

NOTE: The camshaft bolt has a drilled oil passage in it for timing chain lubrication. Prior to installation, clean the passage and make sure it is clear. Never replace the camshaft bolt with a standard bolt.

16. Lightly oil all bolt and stud threads except bolts 1, 2 and 3 that require a suitable pipe sealant.

17. Inspect the timing cover seal for wear or damage and replace if necessary.

18. Install a new timing cover gasket over the cylinder block dowels.

19. Install the timing cover/water pump assembly onto the cylinder block with the water pump pulley loosely attached to the water pump hub.

20. Apply a non-hardening sealant to bolt numbers 1, 2 and 3 and hand start them along with the rest of the cover retaining bolts. Tighten bolts 1–10 to 19 ft. lbs. (25 Nm) and bolts 11–15 to 84 inch lbs. (10 Nm).

21. Install the engine oil pan. Tighten the retaining bolts to 108 inch lbs. (12 Nm).

22. Hand-tighten the water pump pulley retaining bolts.

23. Install the crankshaft damper and pulley. Torque the damper retaining bolt to 107 ft. lbs. (145 Nm).

24. On flexible fuel vehicles, install the CKP sensor. Tighten the retaining bolt to 44–61 inch lbs. (5–7 Nm).

25. Install the automatic belt tensioner or idler pulley, as necessary.

26. Install the water pump and accessory drive belts. Torque the water pump pulley retaining bolts to 16 ft. lbs. (21 Nm).

27. Install the lower radiator hose and the heater hose and tighten the clamps.

28. Fill the crankcase with the correct amount and type of engine oil. Fill and bleed the engine cooling system.

29. Connect the negative battery cable. Start the engine and check for coolant and oil leaks. Road test the vehicle and check for proper operation.

3.0L (VIN S) Engine

1. Disconnect the negative battery cable.

2. Remove the engine from vehicle and position on a suitable workstand.

3. Remove the upper and lower intake manifolds.

4. Remove the cylinder head covers, then remove the drive belt.

5. Remove the wire connector from the water temperature sender.

6. Remove the heater water hose from the bypass tube.

7. Remove the bypass tube bolts and the tube.

8. Remove the power steering pump bolts and the pump.

9. Remove the A/C compressor bracket to water pump brace, the mounting bolts, the compressor and bracket.

NOTE: The A/C mounting bracket bolts are torque to yield bolts and must be replaced when removed.

10. Remove the alternator and the the water pump.

11. Install Flywheel Holding tool T74P-6375-A or equivalent to keep the crankshaft from rotating.

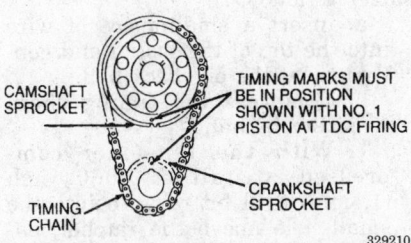

CAMSHAFT SPROCKET

TIMING MARKS MUST BE IN POSITION SHOWN WITH NO. 1 PISTON AT TDC FIRING

CRANKSHAFT SPROCKET

TIMING CHAIN

329210

Camshaft and crankshaft sprocket alignment — 3.0L (VIN U and 1) engines

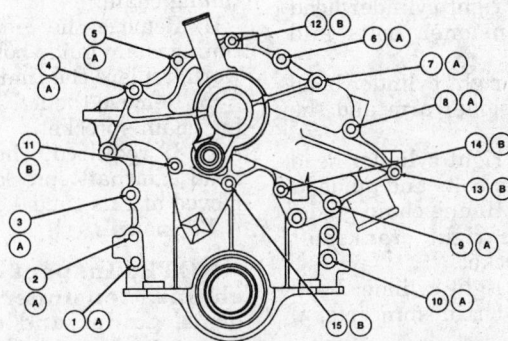

329213

Fastener And Hole No.	Fasteners		Torque Specifications	
	Size	Fastener Application	N·m	LB-FT
1A	M8 x 1.25 x 43.5	F/C TO BLOCK	20-30	15-22
2A	M8 x 1.25 x 43.5	F/C TO BLOCK	20-30	15-22
3A	M8 x 1.25 x 73	W/P & F/C TO BLOCK	20-30	15-22
4A	M8 x 1.25 x 104.3	W/P & F/C TO BLOCK	20-30	15-22
5A	M8 x 1.25 x 73	F/C TO BLOCK	20-30	15-22
6A	M8 x 1.25 x 73	W/P & F/C TO BLOCK	20-30	15-22
7A	M8 x 1.25 x 73	W/P & F/C TO BLOCK	20-30	15-22
8A	M8 x 1.25 x 104.3	W/P & F/C TO BLOCK	20-30	15-22
9A	M8 x 1.25 x 104.3	W/P & F/C TO BLOCK	20-30	15-22
10A	M8 x 1.25 x 52	F/C TO BLOCK	20-30	15-22
11B	M6 x 1 x 28.5	W/P TO F/C	8-12	71-106 (lb-in)
12B	M6 x 1 x 28.5	W/P TO F/C	8-12	71-106 (lb-in)
13B	M6 x 1 x 28.5	W/P TO F/C	8-12	71-106 (lb-in)
14B	M6 x 1 x 28.5	W/P TO F/C	8-12	71-106 (lb-in)
15B	M6 x 1 x 28.5	W/P TO F/C	8-12	71-106 (lb-in)

W/P—Water Pump
F/C—Engine Front Cover

Timing chain front cover bolt location and identification — 3.0L (VIN U and 1) engines

12. Remove the crankshaft accessory pulley and bracket.

NOTE: Rotate the pulley shaft clockwise to remove. Use a 24mm open end wrench on the inside nut to remove the pulley.

13. Remove the crankshaft pulley bolt and washer.
14. Install a suitable damper puller and remove the crankshaft damper from the crankshaft.
15. Remove the flywheel holding tool.
16. Detach the wiring from the Crankshaft Position (CKP) sensor and the Camshaft Position (CMP) sensor connectors.
17. Remove the engine oil pan, remove the retaining bolts in sequence.
18. Remove the engine front cover bolts, in sequence, the front cover and gasket.
19. Rotate the crankshaft so the keyway is at the 11 o'clock position to locate the crankshaft at TDC for No. 1 cylinder.
20. Verify that the alignment arrows on the camshafts are aligned. If not, rotate the crankshaft 1 complete revolution and recheck.
21. Rotate the crankshaft so the keyway is at the 3 o'clock position. This positions the right cylinder head camshafts to the neutral position.
22. Remove the right cylinder head timing chain tensioner bolts and tensioner.
23. Remove the right cylinder head timing chain tensioner arm and the timing chain.
24. Remove the right cylinder head timing chain guide bolts and guide; if worn, replace the timing chain guide.
25. Remove the right crankshaft timing chain sprocket.
26. Remove the right cylinder head camshaft timing chain sprockets, if being replaced.
27. Rotate the crankshaft 2 revolutions and locate the crankshaft keyway at the 11 o'clock position. This will position the left cylinder head camshafts to their neutral position.
28. Verify that the alignment arrows on the camshafts are aligned.
29. Remove the left cylinder head timing chain tensioner retaining bolts and the timing chain tensioner.

NOTE: If the left cylinder head timing chain tensioner arm and timing chain guide are to be reused, mark the position of the timing chain tensioner arm and the timing chain guide so they are reassembled into their original positions.

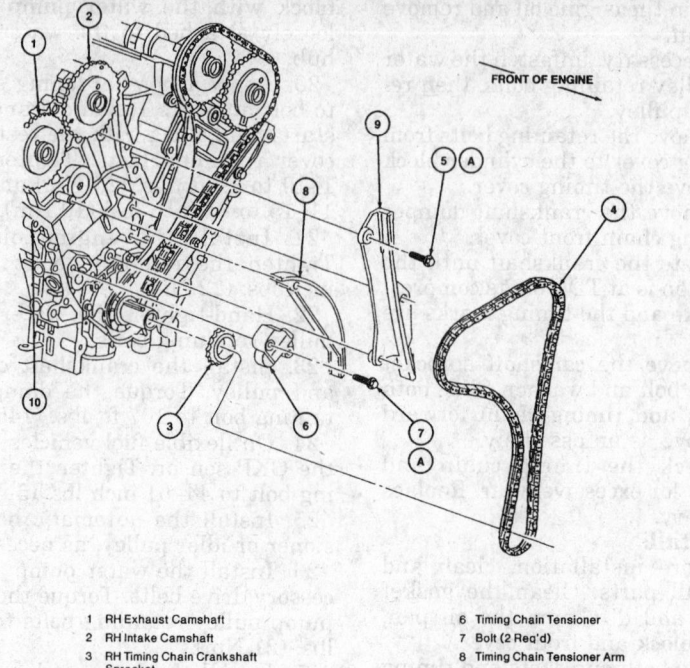

1 RH Exhaust Camshaft	6 Timing Chain Tensioner
2 RH Intake Camshaft	7 Bolt (2 Req'd)
3 RH Timing Chain Crankshaft Sprocket	8 Timing Chain Tensioner Arm
4 RH Timing Chain	9 Timing Chain Guide
5 Bolt (2 Req'd)	10 RH Cylinder Head
	A Tighten to 20–30 N·m (15–22 Lb-Ft)

326436

Right cylinder head timing chain and components — 3.0L (VIN S) engine — left side similar

30. Remove the left cylinder head timing chain tensioner arm and the timing chain.
31. Remove the left cylinder head timing chain guide bolts and guide; if worn, replace the timing chain guide.
32. Remove the left crankshaft timing chain sprocket.
33. If required, the left cylinder head camshaft sprockets may be removed at this time.

To install:

NOTE: Inspect the timing chains, tensioners, tensioner arms, guides and sprockets for wear or damage. If any components are to be replaced for premature wear or damage, the camshaft damper should also be replaced.

34. Reinstall or replace the camshaft sprockets, if removed.
35. Make sure the crankshaft keyway is still at the 11 o'clock position.
36. Install the left timing chain crankshaft sprocket onto the crankshaft.
37. Install the left timing chain guide and retaining bolts to the engine. Torque the retaining bolts to 15–22 ft. lbs. (20–30 Nm).
38. Verify that the arrows on the left cylinder head camshafts are aligned.

39. Install the left timing chain over the left crankshaft sprocket and the left camshaft sprockets.
40. Align the timing marks on the left cylinder head timing chain with the timing marks on the crankshaft sprocket and the camshaft sprockets.
41. Install the left timing chain tensioner arm over the alignment dowel on the left cylinder head.

NOTE: Before installing the timing chain tensioner, it must be properly compressed and locked.

42. Using a small prybar, release the timing chain tensioner ratchet/pawl mechanism through the access hole in the timing chain tensioner as follows:
   a. Insert a small piece of wire into the top of the piston and gently unseat the oil check ball.
   b. Compress the timing chain tensioner by hand.
   c. With the tensioner compressed, install a 0.060 inch (1.5mm) drill bit or wire into the small hole above the ratchet, engaging the lock groove in the rack of the timing chain tensioner.
43. Install the compressed and locked left cylinder head timing chain tensioner and retaining bolts onto the

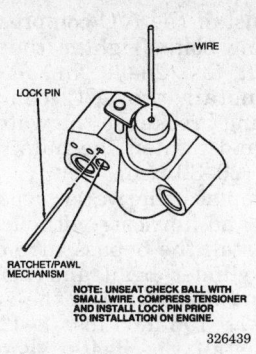

Positioning of the wire and pin to compress and lock the tensioner — 3.0L (VIN S) engine

cylinder block. Torque the retaining bolts to 15–22 ft. lbs. (20–30 Nm).

44. Verify that the timing marks on the left timing chain are in alignment with the timing marks on the crankshaft sprocket and the camshaft sprockets.

45. Reinstall or replace the camshaft sprockets, if removed.

46. Install the right timing chain crankshaft sprocket onto the crankshaft.

47. Install the right timing chain guide and retaining bolts to the engine. Torque the retaining bolts to 15–22 ft. lbs. (20–30 Nm).

48. Verify that the arrows on the right cylinder head camshafts are aligned.

49. Install the right timing chain over the right crankshaft sprocket and the right camshaft sprockets.

50. Align the timing marks on the right cylinder head timing chain with the timing marks on the crankshaft sprocket and the camshaft sprockets.

51. Install the right timing chain tensioner arm over the alignment dowel on the right cylinder head.

NOTE: Before installing the timing chain tensioner, it must be properly compressed and locked.

52. Using a small prybar, release the timing chain tensioner ratchet/pawl mechanism through the access hole in the timing chain tensioner as follows:

    a. Insert a small wire into the top of the piston and gently unseat the oil check ball.

    b. Compress the timing chain tensioner by hand.

    c. With the tensioner compressed, install a 0.060 inch (1.5mm) drill bit or wire into the small hole above the ratchet, engaging the lock groove in the rack of the timing chain tensioner.

53. Install the compressed and locked right cylinder head timing chain tensioner and retaining bolts onto the cylinder block. Torque the retaining bolts to 15–22 ft. lbs. (20–30 Nm).

54. Verify that the timing marks on the right timing chain are aligned with the timing marks on the crankshaft sprocket and the camshaft sprockets.

55. Make sure the crankshaft keyway is at the 11 o'clock position.

56. Rotate the crankshaft 2 revolutions and position the crankshaft keyway to the 3 o'clock location. This

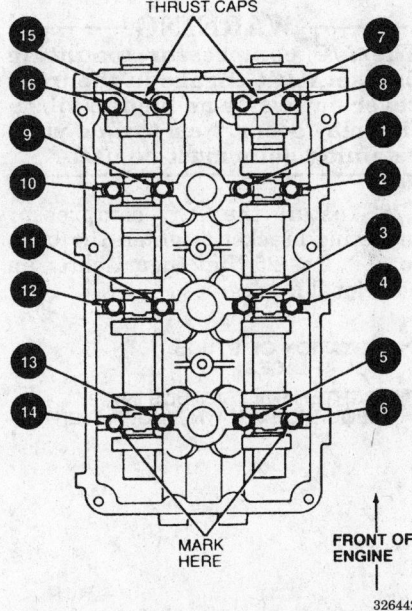

● TIGHTEN BOLTS IN SEQUENCE SHOWN

Torque sequence for the left cylinder head camshaft journal caps — 3.0L (VIN S) engine

will position the right cylinder head camshafts to the neutral position.

57. Remove the lock pins from the timing chain tensioners.

58. Verify that the timing marks on the timing chains are aligned with the timing marks on the crankshaft sprocket and the camshaft sprockets.

59. Clean the engine front and the front cover to cylinder block gasket sealing surfaces.

NOTE: The front cover must be installed and properly tightened within 6 minutes of the application of the sealer.

60. Apply silicone sealer to the 6 critical areas of the cylinder block to prevent oil seepage.

61. Place new front cover gaskets onto the dowel pins on the cylinder block and heads.

62. Place the front cover into position by placing the front cover onto the dowel pins at the cylinder block.

63. Install the 6 front cover retaining bolts and stud bolts where the silicone sealer was applied.

64. Tighten the bolts and stud bolts until the front cover contacts the cylinder block and heads; then, turn the bolts and stud bolts an additional ¼ turn.

65. Install the remaining front cover retaining bolts and stud bolts. Torque all of the front cover retaining bolts and stud bolts, in sequence, to 15–22 ft. lbs. (20–30 Nm).

66. If removed, install the belt tensioner onto the right side of the engine front cover. Tighten the bolt to 15–22 ft. lbs. (20–30 Nm).

67. Install a new oil pan gasket and apply a thin bead of silicone sealer where the timing cover meets the cylinder block.

68. Install the oil pan and and install the retaining bolts loosely.

69. With the oil pan aligned to the rear of the cylinder block, tighten the oil pan bolts and studs, in sequence, to 15–22 ft. lbs. (20–30 Nm) no more

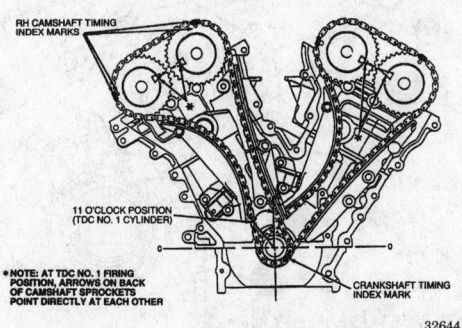

Timing chain index marks for the right cylinder head — 3.0L (VIN S) engine

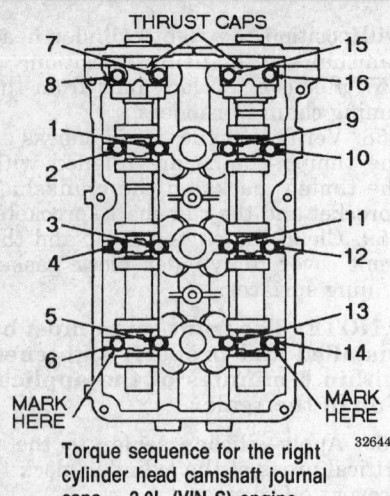

Torque sequence for the right cylinder head camshaft journal caps — 3.0L (VIN S) engine

than 6 minutes after applying the silicone sealer.

70. Connect the wiring to the CKP and CMP sensors.

71. Install Flywheel Holding tool T74P-6375-A or equivalent to keep the crankshaft from rotating.

72. Install the crankshaft damper using Crankshaft Damper Replacer T74P-6316-B, or equivalent.

73. Apply a thin coating of silicone sealer to the front of the crankshaft on the inside diameter of the damper at the keyway before installing the washer and retaining bolt.

74. Install the crankshaft damper pulley washer and bolt and tighten the bolt as follows:

    a. Tighten the bolt to 89 ft. lbs. (120 Nm) and loosen the bolt.

    b. Retighten the bolt to 39 ft. lbs. (53 Nm).

    c. Turn the retaining bolt an additional 90 degrees.

75. Install the accessory drive crankshaft pulley and bracket. Using a 34mm socket tighten the pulley counterclockwise to 70–77 ft. lbs. (95–105 Nm), tighten the bracket nuts to 15–22 ft. lbs. (20–30 Nm). Remove the flywheel holding tool.

76. Install the water pump and retaining nuts and tighten to 15–22 ft. lbs. (20–30 Nm).

77. Install the alternator and the retaining bolts. Tighten the bolts to 15–22 ft. lbs. (20–30 Nm). Connect the wiring connector.

——— WARNING ———
The A/C compressor mounting bracket must engage in the front cover dowels or an engine vibration may occur. New torque yield retaining bolts must be used.

78. Install the A/C compressor mounting bracket. Tighten the bolts to 18 ft. lbs. (25 Nm); then, tighten an additional 90 degrees.

79. Install the A/C compressor and retaining bolts. Tighten the bolts to 15–22 ft. lbs. (20–30 Nm).

80. Install the A/C compressor mounting bracket to the water pump brace and tighten the nuts to 15–22 ft. lbs. (20–30 Nm).

81. Replace the water bypass tube O-ring and lubricate with clean coolant. Install the bypass tube onto the right cylinder head and install the stud and bolt. Tighten the stud and bolt to 71–106 inch lbs. (8–12 Nm.

82. Install the heater water hose and position the hose clamp. Install the wire connector to the water temperature sender.

83. Install the power steering pump and tighten the bolts to 71–106 inch lbs. (8–12 Nm).

84. Install both cylinder head covers.

NOTE: The cylinder head covers must be installed and properly tightened within 6 minutes of applying the silicone sealant.

85. Install the lower and upper intake manifolds.

86. Release the accessory drive belt tension by rotating the tensioner clockwise and installing the drive belt.

87. Install the engine assembly into the vehicle. Restore all fluid levels. Connect the negative battery cable.

88. Run the engine and check for leaks. Road test the vehicle and check for proper operation.

3.8L Engine

1. Disconnect the negative battery cable. Drain the engine cooling system. Drain the engine oil.

2. Remove the engine air cleaner assembly and air intake duct.

3. Loosen the accessory drive belt idler. Remove the drive belt and water pump pulley.

4. Remove the power steering pump mounting bracket attaching bolts. Leaving the hoses connected, place the pump/bracket assembly in a position that will prevent the loss of power steering fluid.

5. If equipped with air conditioning, remove the compressor front support bracket. Leave the compressor in place.

6. Disconnect the coolant bypass and heater hoses at the water pump. Disconnect radiator upper hose at thermostat housing.

7. Disconnect the coil wire from distributor cap and move the cap with the ignition wires aside. Remove the distributor retaining clamp and lift distributor out of the front cover.

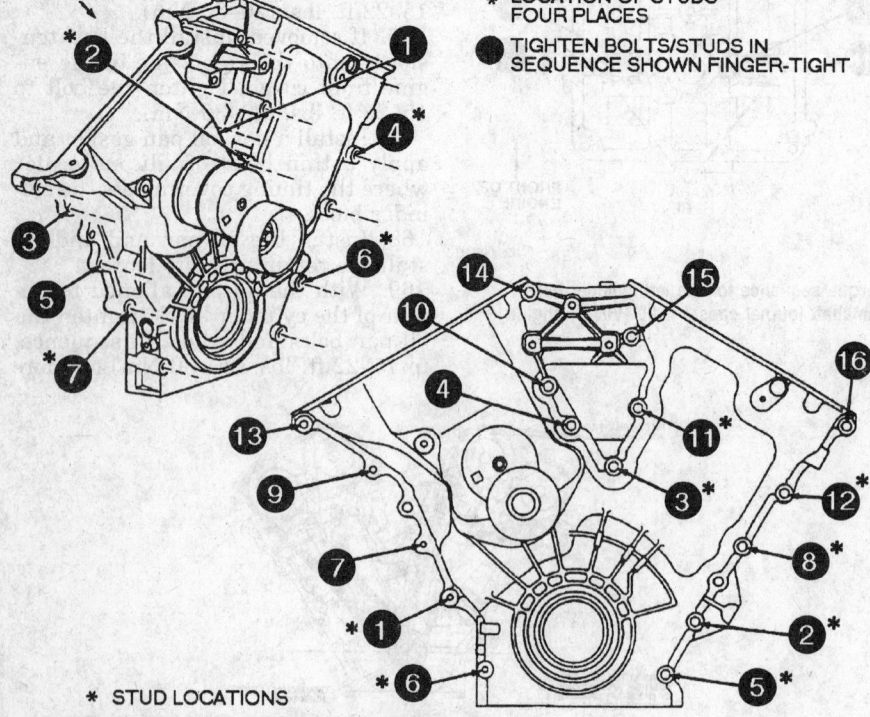

* LOCATION OF STUDS FOUR PLACES

● TIGHTEN BOLTS/STUDS IN SEQUENCE SHOWN FINGER-TIGHT

* STUD LOCATIONS

● TIGHTEN BOLTS/STUDS IN SEQUENCE SHOWN TO FINAL TORQUE SPECIFICATIONS

Front cover torque sequence — 3.0L (VIN S) engine

8. Raise and safely support the vehicle. Remove the crankshaft damper and pulley.

NOTE: If the crankshaft pulley and crankshaft vibration damper have to be separated, mark the damper and pulley so they can be reassembled in the same relative position. This is important as the damper and pulley are initially balanced as a unit. If the crankshaft damper is being replaced, check if the original damper has balance pins installed. If so, new balance pins E0SZ-6A328-A or equivalent, must be installed on the new damper in the same position as the original damper. The crankshaft pulley must also be installed in the original position.

9. Remove the engine oil filter. Disconnect the radiator lower hose at the water pump. Remove the engine oil pan.

NOTE: The front cover cannot be removed without lowering the oil pan.

10. Lower the vehicle. Remove the front cover retaining bolts.

——— **WARNING** ———
Do not overlook the front cover retaining bolt located behind the oil filter adapter. The front cover will break if all retaining bolts are not removed.

11. Remove the ignition timing indicator.

12. Remove the front cover and water pump as an assembly. Remove the cover gasket and discard.

NOTE: The front cover houses the oil pump. If a new front cover is installed, remove the water pump and oil pump from the old cover and install in the new cover.

13. Remove the camshaft retaining bolt and washer from end of the camshaft. Remove the distributor drive gear.

14. Remove the camshaft sprocket, crankshaft sprocket and timing chain as an assembly. If the crankshaft sprocket is difficult to remove, pry it off using a pair of small prybars positioned on both sides of the sprocket.

15. Pull back on the chain tensioner ratcheting mechanism and install a pin through the hole in the bracket to relieve tension. Remove 3 bolts and the chain tensioner assembly.

To install:

16. Rotate the crankshaft as necessary to position the piston for No. 1 cylinder at TDC and the crankshaft keyway at the 12 o'clock position.

17. Install the timing chain tensioner assembly. Make sure the ratcheting mechanism is in the retracted position with the pin pointing outward from the hole in the bracket assembly. Tighten the retaining bolts to 71–124 inch lbs. (8–14 Nm).

18. Lubricate the timing chain with clean engine oil. Install the camshaft sprocket, crankshaft sprocket and timing chain as an assembly.

19. Remove the pin from the tensioner assembly to load the tensioner arm against the chain. Make certain the timing marks are properly positioned across from each other.

20. Install the distributor drive gear. Install the washer and retaining bolt to the camshaft and tighten to 30–37 ft. lbs. (40–50 Nm).

21. Lightly oil all bolt and stud threads before installation. Clean all gasket surfaces on the front cover, cylinder block and fuel pump, if equipped. If reusing the front cover, replace the front cover oil seal.

22. If a new front cover is to be installed, complete the following:

    a. Pack the oil pump gear pocket with petroleum jelly and install the oil pump gears. Make sure the petroleum jelly fills the gap between the gears and the pocket. Install the oil pump cover using a new gasket and tighten the retaining bolts to 18–22 ft. lbs. (25–30 Nm).

    b. Clean the water pump gasket surface. Position a new water pump gasket on the front cover and install the water pump. Install the pump retaining bolts and tighten to 15–22 ft. lbs.

23. Install the distributor drive gear.

24. Lubricate the crankshaft front oil seal with clean engine oil.

25. Position a new front cover gasket on the cylinder block and install the front cover/water pump assembly using dowels for proper alignment. A suitable contact adhesive is recommended to hold the gasket in position while the front cover is installed.

26. Position the ignition timing indicator.

27. Install the front cover attaching bolts. Apply Loctite® or equivalent, to the threads of the bolt installed below the oil filter housing prior to installation. This bolt is to be installed and tightened last. Tighten all bolts to 15–22 ft. lbs. (20–30 Nm).

28. Raise and safely support the vehicle. Install the engine oil pan.

29. Connect the radiator lower hose. Install a new engine oil filter.

30. Coat the crankshaft damper sealing surface with clean engine oil. Apply a small amount of silicone sealant to the crankshaft keyway.

31. Position the crankshaft key in the crankshaft keyway.

32. Install the damper, washer and retaining bolt. Tighten the bolt to 104–132 ft. lbs. (140–180 Nm).

33. Install the crankshaft pulley and retaining bolts. Tighten the retaining bolts 19–28 ft. lbs. (26–28 Nm).

34. Lower the vehicle. Connect the coolant bypass hose.

35. Rotate the crankshaft, as necessary, to position the piston for No. 1 cylinder to TDC on its compression stroke. Install the distributor. Install the distributor cap and coil wire.

36. Connect the radiator upper hose at thermostat housing. Connect the heater hose.

37. If equipped with air conditioning, install the front compressor support bracket.

38. Install the power steering pump and mounting brackets. Position the accessory drive belt over the pulleys.

39. Install the water pump pulley. Position the accessory drive belt over water pump pulley and tighten the belt.

40. Fill the crankcase and cooling system to the proper levels.

41. Connect the negative battery cable. Start the engine and check for leaks.

42. Check the ignition timing and curb idle speed; adjust as required.

43. Install the engine air cleaner assembly and air intake duct. Road test the vehicle and check for proper operation.

4.6L Engine

——— **WARNING** ———
This is not a free-wheeling engine. When the timing chains are removed and the cylinder heads are installed, the crankshaft and/or camshafts must not be rotated unless as directed in this procedure. Failure to follow these instructions will result in valve and/or piston damage.

1. Disconnect the negative battery cable. Remove the engine from the vehicle and position on a suitable workstand.

2. Remove the support brackets for the engine sensor wiring from the engine front cover.

3. Loosen the water pump pulley bolts. Remove the accessory drive belt.

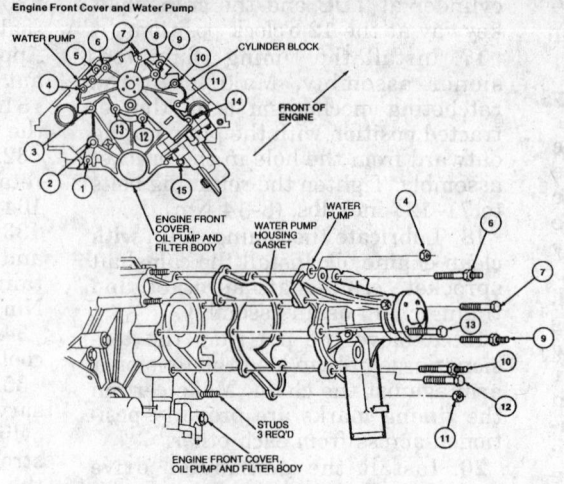

Engine Front Cover and Water Pump

Fastener and Hole No.	Hole No.		Fasteners
	Water Pump	Front Cover	Part Name
1		4	Stud
2		2	Stud
3	2	9	Stud
4	1	3	Stud
5		10	Bolt
6	9	15	Stud Bolt
7	8	16	Bolt
8		11	Bolt
9	7	17	Stud Bolt
10	6	1	Stud Bolt
11	5	7	Stud

Fastener and Hole No.	Hole No.		Fasteners
	Water Pump	Front Cover	Part Name
12	4	13	Bolt
13	3	14	Bolt
14		6	Bolt
15		5	Cap Screw
3, 4, 11	2, 1, 5	9, 8, 7	Nut

313046

Timing chain front cover and water pump assembly — 3.8L engine

4. Remove the water pump pulley and lower water pump-to-cylinder block retaining bolt for engine front cover removal clearance.

5. Using Pulley Remover T69L-10300-B or equivalent, remove the power steering pump pulley.

NOTE: The front lower bolt on the power steering pump will not come all the way out.

6. Remove the bolts retaining the power steering pump to the cylinder block and the engine front cover.

7. Position the power steering pump and reservoir aside. Remove the bolts retaining the front cover to the oil pan.

8. Remove the crankshaft pulley retaining bolt and washer from the crankshaft.

9. Install Crankshaft Damper Remover T58P-6316-D or equivalent, on the crankshaft pulley and pull the crankshaft pulley from the crankshaft.

10. Disconnect the engine control wiring from the Camshaft Position (CMP) sensor.

11. Remove the bolt retaining the belt idler pulley and remove the idler pulley.

12. Remove the drive belt tensioner retaining bolt and drive belt tensioner from the engine front cover.

13. Disconnect the Crankshaft Position (CKP) sensor and move the wiring aside.

14. Remove the engine front cover. Remove both cylinder head covers. Remove the CKP sensor tooth wheel.

15. Rotate the crankshaft to place the piston for No. 1 cylinder at Top Dead Center (TDC) on its compression stroke.

— WARNING —
Camshaft Positioning tool T93P-6256-A or equivalent, and Camshaft Holding tool T93P-6256-AH or equivalent, must be installed to prevent accidental rotation of the camshafts and possible engine damage.

16. Install Camshaft Positioning tool T93P-6256-A or equivalent, in the rear D-slots of the camshaft.

— WARNING —
Failure to use Camshaft Holding tool T93P-6256-AH or equivalent, while assembling or disassembling the timing chains will result in damage to the D-slots or Camshaft Positioning tool T93P-6256-A.

17. Install Camshaft Holding tool T93P-6256-AH or equivalent, onto the camshafts to keep the camshafts from rotating and to prevent damaging the camshaft positioning tool.

18. Remove the right primary timing chain tensioner bolts and the tensioner.

19. Remove the right tensioner arm bolt and the arm.

NOTE: The 2 bolts retaining the timing chain guide to the cylinder head are longer than the one to the block.

20. Remove the right timing chain guide bolts and guide. Remove the right primary timing chain.

21. Remove the right crankshaft sprocket. Remove the camshaft sprocket retaining bolts.

— WARNING —
The secondary timing chain tensioner plunger is spring loaded. Care must be taken to prevent the plunger from dropping out of the tensioner during disassembly.

22. Unlock and compress the secondary timing chain tensioners and lock the timing chain tensioner in the compressed position using a paper clip or stiff wire. Remove the secondary timing chains and camshaft sprockets.

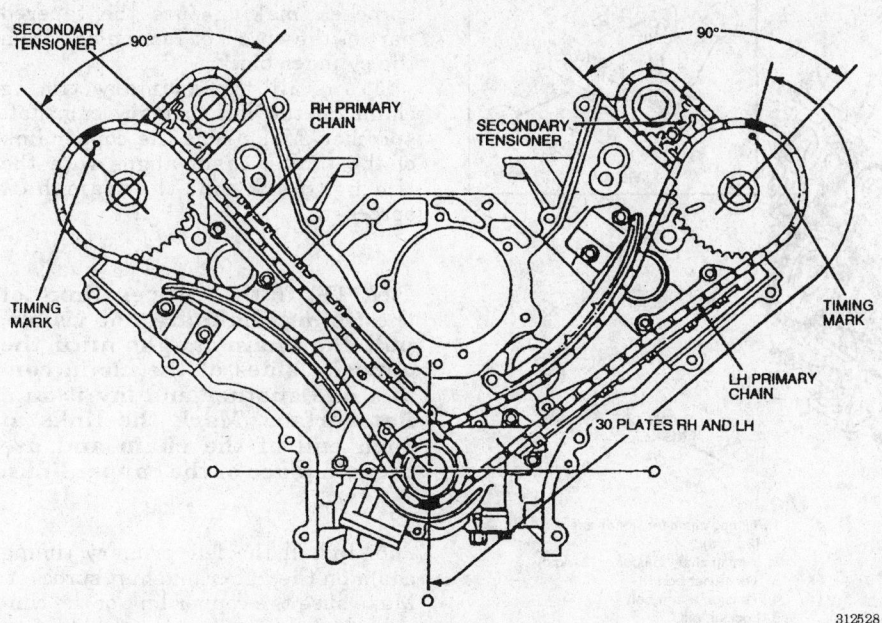

Engine positioned for timing chain removal — 4.6L engine

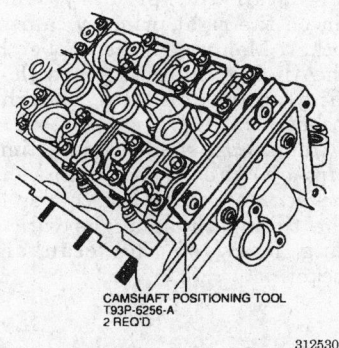

Camshaft positioning tool — 4.6L engine

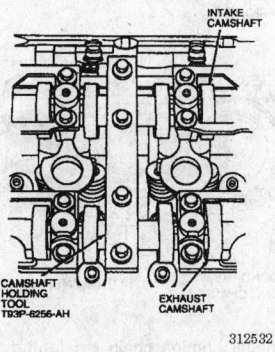

Camshaft holding tool — 4.6L engine

23. Remove the left primary timing chain tensioner bolts tensioner.

24. Remove the bolt retaining the left timing chain tensioner arm and remove the left timing chain tensioner arm.

25. Remove the left timing chain guide bolts and the guide. Remove the left primary timing chain. Remove the left crankshaft sprocket.

WARNING

Failure to use Camshaft Holding tool T93P-6256-AH or equivalent, while removing or tightening the

camshaft bolts may result in damage to the camshaft D-slots or the camshaft positioning tool.

26. Remove the camshaft sprocket retaining bolts.

WARNING

The secondary timing chain tensioner plunger is spring loaded. Care must be taken to prevent the plunger from dropping out of the tensioner during disassembly.

27. Unlock and compress the secondary chain tensioner and lock the chain tensioner in the compressed position using a paper clip or stiff wire. Remove the secondary timing chain and camshaft sprockets.

WARNING

Do not rotate the crankshaft and/or camshafts or engine damage may occur.

28. Inspect the friction material on the timing chain tensioner arms and timing chain guides. If worn or damaged, remove and clean the oil pan and replace the oil pump screen cover and tube.

To install:

29. If the engine has jumped time, make sure all service to the necessary engine components and/or valve train have been made.

30. Make sure the primary and secondary timing chain tensioners are in the collapsed position and retained with paper clips or equivalent.

31. Install Camshaft Positioning tool T93P-6256-A or equivalent, into the D-slots on the rear of the camshafts to position both intake and exhaust camshafts with the keyways pointing down towards the crankshaft. Install Camshaft Holding tool T93P-6256-AH or equivalent, on the center of the camshafts and do not remove until all parts are installed and tightened.

32. Install the left secondary timing chain tensioner and tighten the bolts to 70–106 inch lbs. (8–12 Nm).

33. Install the left secondary camshaft sprockets and secondary timing chain as an assembly. Make sure the hubs of the sprocket are facing the proper direction.

34. Install the left primary camshaft sprocket and camshaft sprocket spacer on the left camshaft.

35. Install the washers and camshaft sprocket retaining bolts, finger-tight only.

NOTE: The secondary camshaft sprockets must be free to turn.

36. Install Timing Chain Tensioning tool T93P-6256-BH or equivalent, on the left secondary timing chain tensioner.

37. Tighten the camshaft sprocket bolts to 81–95 ft. lbs. (110–130 Nm).

38. Install the right secondary timing chain tensioner and tighten the retaining bolts to 70–106 inch lbs. (8–12 Nm).

39. Install the right secondary camshaft sprockets and secondary timing chain as an assembly. Make

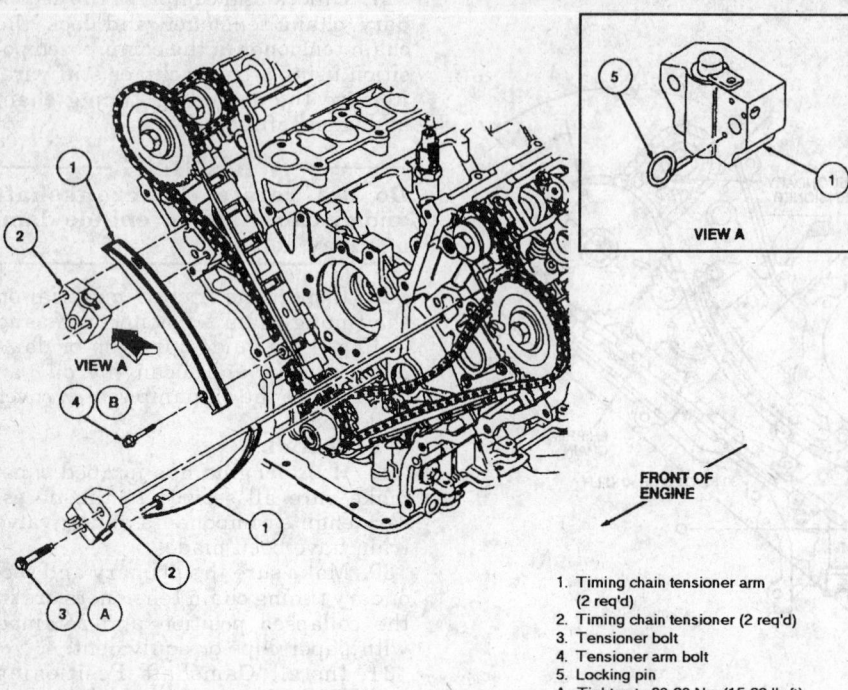

VIEW A

1. Timing chain tensioner arm (2 req'd)
2. Timing chain tensioner (2 req'd)
3. Tensioner bolt
4. Tensioner arm bolt
5. Locking pin
A. Tighten to 20-30 Nm (15-22 lb-ft)
B. Tighten to 10-15 Nm (7-11 lb-ft)

312534

Timing chain tensioners — 4.6L engine

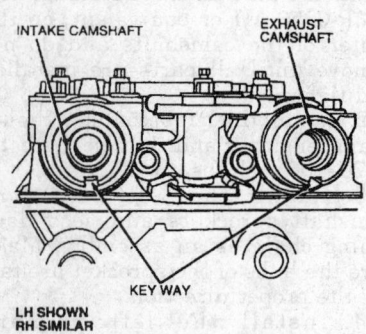

INTAKE CAMSHAFT

EXHAUST CAMSHAFT

KEY WAY

LH SHOWN
RH SIMILAR

312537

Camshaft keyway positions — 4.6L engine

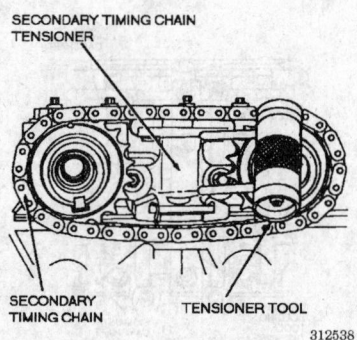

SECONDARY TIMING CHAIN TENSIONER

SECONDARY TIMING CHAIN

TENSIONER TOOL

312538

Secondary timing chain tensioning tool — 4.6L engine

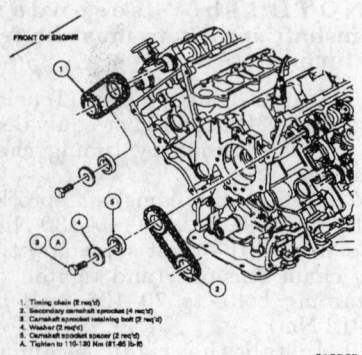

FRONT OF ENGINE

1. Timing chain (2 req'd)
2. Secondary camshaft sprocket (4 req'd)
3. Camshaft sprocket retaining bolt (2 req'd)
4. Washer (2 req'd)
5. Camshaft sprocket spacer (2 req'd)
A. Tighten to 110-130 Nm (81-95 lb-ft)

312535

Secondary timing chains — 4.6L engine

sure the hubs of the sprocket are facing the proper direction.

40. Install the right primary camshaft sprocket and camshaft sprocket spacer on the right camshaft.

41. Install the washers and camshaft sprocket retaining bolts, finger-tight only.

NOTE: The secondary camshaft sprockets must be free to turn.

42. Install Timing Chain Tensioning tool T93P-6256-BH or equivalent, on the right secondary timing chain tensioner.

43. Tighten the camshaft sprocket bolts to 81—95 ft. lbs. (110—130 Nm).

44. Install the left crankshaft sprocket, making sure the tapered part of the sprocket faces away from the cylinder block.

45. Install the primary timing chain on the left primary camshaft sprocket. Make sure the copper link of the timing chain aligns with the timing mark on the camshaft sprocket.

NOTE: If the copper links of the timing chain are not visible, pull the chain taught until the opposite sides of the chain contact one another and lay it on a flat surface. Mark the links at each end of the chain and use them in place of the copper links.

46. Install the left primary timing chain on the left crankshaft sprocket. Make sure the copper link of the timing chain aligns with the timing mark on the crankshaft sprocket.

47. Install the right crankshaft sprocket, making sure the tapered part of the sprocket faces toward the cylinder block.

48. Install the primary timing chain on the right primary camshaft sprocket. Make sure the copper link of the timing chain aligns with the timing mark on the camshaft sprocket.

49. Install the right primary timing chain on the right crankshaft sprocket. Make sure the copper link of the timing chain aligns with the timing mark on the crankshaft sprocket.

NOTE: The 2 timing chain guide bolts to the cylinder head are longer than the bolt to the cylinder block.

50. Install both timing chain guides and tighten the bolts to 71—106 inch lbs. (8—12 Nm).

51. Lubricate the timing chain tensioner arm contact surfaces with clean engine oil. Install both tensioner arms and tighten the bolts to 84—132 inch lbs. (10—15 Nm).

52. Install right and left primary timing chain tensioners and tighten the bolts to 15—22 ft. lbs. (20—30 Nm).

53. Remove the locking pins from the timing chain tensioners and make sure all timing marks are aligned. Remove the camshaft position-

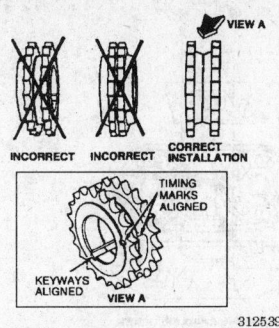

Correct installation of the
crankshaft sprockets — 4.6L engine

312539

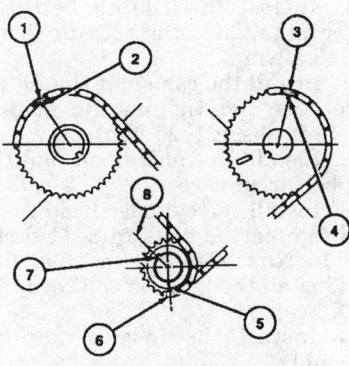

1. RH camshaft timing chain mark
2. RH camshaft sprocket mark
3. LH camshaft timing chain mark
4. RH camshaft sprocket mark
5. Camshaft sprocket mark
6. Crankshaft timing chain mark
7. Crankshaft
8. Crankshaft keyway center line

312540

Timing chain and sprocket timing marks — 4.6L
engine

ing tool and the camshaft holding
tool.

**NOTE: When cleaning the seal-
ing surfaces of the engine front
cover and oil pan-to-cylinder
block joints, be extremely careful
not to damage the rubber bead of
the oil pan gasket. If the oil pan
gasket is damaged, it must be
replaced.**

54. Clean the sealing surfaces of
the cylinder block, cylinder heads
and engine front cover; remove all
traces of oil, dirt and old sealant.

Sealing surfaces must be clean and
dry before applying sealant.

55. Replace the engine front cover
gaskets and the crankshaft front
seal. Apply silicone sealant to the
proper locations on the cylinder block
and heads.

**NOTE: The engine front cover
must be rolled into place. DO
NOT slide on the oil pan gasket.**

56. Carefully install the front cover
to the engine.

57. Install 5 studs and 10 bolts retaining the engine front cover to the
engine. Tighten, in sequence, to
15–22 ft. lbs. (20–30 Nm). Tighten
the stud bolts and bolts numbered 6,
7, 8, 9, 10, and 11 within 4 minutes
after applying the sealer.

58. Install the drive belt tensioner
and bolt. Tighten the bolt to 15–22 ft.
lbs. (20–30 Nm).

59. Install the belt idler pulley and
bolt. Tighten the bolt to 15–22 ft. lbs.
(20–30 Nm).

60. Connect the engine control sensor wiring to the CKP and CMP sensors. Install the left and the right cylinder head covers.

61. Install the crankshaft pulley on
the crankshaft, using Crankshaft
Damper Replacer T74P-6316-B or
equivalent.

62. Apply silicone sealant in the
keyway of the crankshaft pulley.

63. Install the pulley bolt and
washer. On 1995 vehicles, tighten to
114–121 ft. lbs. (155–165 Nm). On
1996–97 vehicles, tighten tighten as
follows:

 a. Tighten the bolt to 89 ft. lbs.
(120 Nm) and then loosen the bolt.

 b. Retighten the bolt to 39 ft. lbs.
(53 Nm).

 c. Turn the retaining bolt an additional 90 degrees.

64. Install 4 bolts retaining the engine front cover to the oil pan. On
1995 vehicles, tighten the bolts to
15–22 ft. lbs. (20–30 Nm). On

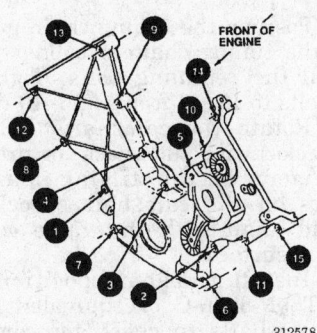

312578

Engine front cover tightening
sequence — 4.6L engine

1996–97 vehicles, tighten the bolts to
15–22 ft. lbs. (20–30 Nm) then rotate
the bolts in sequence an additional 60
degrees.

65. Position the power steering
pump on the engine and install the 4
retaining bolts. Tighten the bolts to
15–22 ft. lbs. (20–30 Nm).

66. Install the retainer brackets for
the engine control sensor wiring to
the engine front cover.

67. Using Power Steering Pump
Pulley Replacer T91P-3A733-A or
equivalent, install the power steering
pump pulley.

68. Install the water pump lower
bolt removed for engine front cover
clearance. Tighten the bolt to 15–22
ft. lbs. (20–30 Nm).

69. Install the water pump pulley
with 4 bolts. Tighten the bolts to
15–22 ft. lbs. (20–30 Nm).

70. Install the accessory drive belt.
Install the engine assembly in the vehicle. Restore all fluid levels.

71. Connect the negative battery
cable. Start the engine and check for
leaks and proper engine operation.

Camshaft Chain

REMOVAL AND INSTALLATION

3.0L (VIN Y) and 3.2L SHO Engines

— **CAUTION** —
*Fuel injection systems remain
under pressure, even after the engine has been turned OFF. The
fuel system pressure must be relieved before disconnecting any
fuel lines. Failure to do so may result in fire and/or personal
injury.*

1. Disconnect the negative battery
cable. Properly relieve the fuel system pressure.

2. Rotate the crankshaft until the
piston in No. 1 cylinder is at TDC.

3. Remove the intake manifold
assembly.

4. Remove the timing belt front
cover and timing belt.

5. If the left cylinder head cover is
being removed, remove the oil fill cap
and ignition coil plastic cover. If the
right cylinder head cover is being removed, disconnect the fuel lines. Remove the cylinder head cover(s).

6. Remove the camshaft pulleys,
noting the location of the dowel pins.

7. Remove the upper rear timing
belt cover.

8. Uniformly loosen the camshaft
bearing caps.

—————— **WARNING** ——————
If the camshaft bearing caps are not uniformly loosened, camshaft damage may result.

9. Remove the camshaft bearing caps and note their positions for installation.

10. Remove the camshaft timing chain tensioner mounting bolts.

11. Remove the camshafts together with the timing chain and tensioner.

12. Remove and discard the camshaft oil seal.

13. Remove the timing chain chain sprocket from the camshaft.

14. Inspect the camshafts for wear and/or damage and replace, as necessary.

To install:

15. Apply a thin coat of clean engine oil to the camshaft bearing surfaces on the cylinder head and camshaft journal caps.

16. Align the timing marks on the camshaft sprockets with the camshaft and install the sprockets. Tighten the bolts to 10–13 ft. lbs. (14–18 Nm).

17. Install the timing chain over the camshaft sprockets. Align the white painted link with the timing mark on the sprocket.

—————— **WARNING** ——————
Left and right timing chain tensioners are not interchangeable.

18. Rotate the camshafts 60 degrees (³/₅ turn) counterclockwise. Set the timing chain tensioner between the sprockets and install the camshafts on the cylinder head. The left and right tensioners are not interchangeable.

19. Apply a thin coat of clean engine oil to the camshaft journals and install bearing caps 2 through 5.

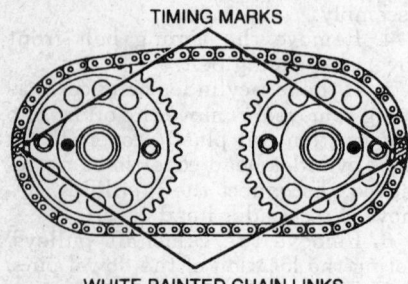

TIMING MARKS

WHITE PAINTED CHAIN LINKS

236950

Aligning the white chain link with the sprocket timing mark — SHO engine

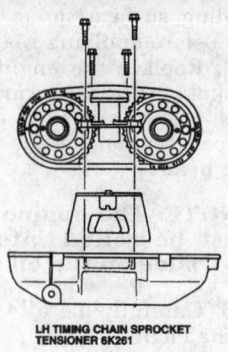 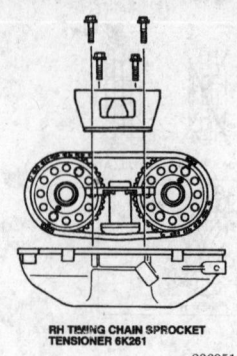

LH TIMING CHAIN SPROCKET TENSIONER 6K261   **RH TIMING CHAIN SPROCKET TENSIONER 6K261**

236951

Timing chain tensioner installation — SHO engines

Loosely install the cap retaining bolts.

NOTE: The arrows on the bearing caps point to the front of the engine when installed.

20. Apply silicone sealer to the outer diameter of the new camshaft seal and the seal seating area on the cylinder head. Install the camshaft seal using cam seal expander T89P-6256-B and cam seal replacer T89P-6256-A or equivalents.

21. Apply a 0.10 inch (2.5mm) bead of silicone sealer to the No. 1 bearing cap. Install the bearing cap while holding the camshaft seal in place with cam seal replacer T89P-6256-A or equivalent.

22. Tighten the bearing caps in sequence using a 2 step method. In the first step tighten to 71–106 inch lbs. (8–12 Nm). In the second step tighten to 12–16 ft. lbs. (16–22 Nm). For left camshaft installation, apply pressure to the timing chain tensioner to avoid damage to the bearing caps.

—————— **WARNING** ——————
The No. 5 camshaft bearing caps function as thrust bearings for the camshaft. Always tighten No. 5 camshaft bearing caps first.

23. Position the timing chain guide and the timing chain tensioner and install the retaining bolts. Tighten the bolts to 11–14 ft. lbs. (15–19 Nm).

24. Rotate the camshafts 60 degrees clockwise and check for proper alignment of the timing marks. Marks on the camshaft sprockets should align with the valve cover mating surface.

25. Install camshaft positioning tool T89P-6256-C or equivalent, on the camshafts to check for correct positioning. The flats on the tool should align with the flats on the camshaft. If the tool does not fit and/or timing marks will not line up, repeat the installation procedure from the beginning.

26. Install the timing belt rear cover and tighten the bolts to 78 inch lbs. (8.8 Nm).

27. Install the camshaft timing belt sprockets and tighten the bolts to 15–18 ft. lbs. (21–25 Nm).

28. Install the timing belt and timing belt front cover.

29. Install the cylinder head covers and tighten the bolts to 8–11 ft. lbs. (10–16 Nm). Connect the fuel lines and install the ignition coil cover and oil fill cap.

30. Install the intake manifold assembly.

31. Connect the negative battery cable. Run the engine and check for leaks and proper engine operation.

Timing Belt, Sprockets, Tensioner and Front Cover

REMOVAL AND INSTALLATION

3.0L (VIN Y) SHO Engine

1. Disconnect both battery cables, negative cable first. Remove the battery.

2. Remove the right engine roll damper. Disconnect the wiring to the ignition module.

3. Remove the intake manifold crossover tube bolts. Loosen the intake manifold tube hose clamps. Remove the intake manifold crossover tube.

4. Loosen the alternator/air conditioning belt tensioner pulley and remove the drive belt.

5. Loosen the water pump/power steering belt tensioner pulley and remove the drive belt.

6. Remove the alternator/air conditioning belt tensioner pulley and bracket assembly.

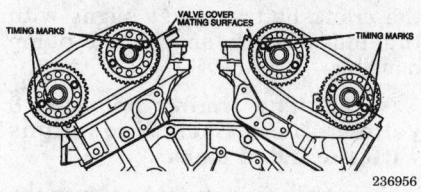

Aligning the camshaft sprockets — SHO engines

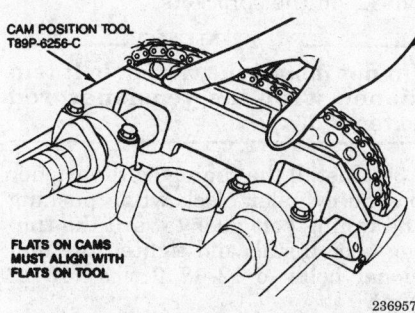

Cam position tool installation — SHO engines

7. Remove the water pump/power steering belt tensioner pulley only. Remove the upper timing belt cover.

8. Disconnect the Crankshaft Position (CKP) sensor electrical connector. Place the transaxle gear selector in N.

9. Rotate the crankshaft until the piston for No. 1 cylinder is at TDC on its compression stroke. Make sure the white mark on the crankshaft damper aligns with the 0 degree index mark on the lower timing belt cover and the marks on the intake camshaft sprockets align with the index marks on the metal timing belt cover.

10. Raise and safely support the vehicle. Remove the right front wheel and tire assembly. Loosen the fender splash shield and place it aside.

11. Remove the crankshaft damper and pulley retaining bolt. Using Puller T67L-3600-A or equivalent, remove the crankshaft damper and pulley.

12. Remove the lower timing belt cover.

13. Remove the center timing belt cover and disconnect the CKP sensor wire and grommet from the slot in the cover and the stud on the water pump.

14. Loosen the timing belt tensioner idler pulley. Rotate the idler pulley 180 degrees clockwise and tighten the tensioner nut to hold the pulley an unloaded position.

15. Lower the vehicle.

16. Remove the timing belt. If the belt is to be reused, use crayon to mark an arrow on the belt to indicate the direction of rotation, for installation reference.

17. If removing one or more camshaft sprockets, remove 2 retaining bolts securing each camshaft sprocket and remove the sprocket noting the location of the dowel pin.

18. If removing the crankshaft sprocket, install Puller T67L-3600-A, or equivalent and pull the crankshaft sprocket off the crankshaft using care not to damage the pulse wheel.

To install:

19. If the crankshaft sprocket was removed, install the crankshaft sprocket by aligning the keyway and pushing the sprocket on by hand.

20. If one or more camshaft sprockets was removed, install each camshaft sprocket by aligning the timing marks on the camshaft sprocket with the camshaft using the dowel pin as a guide. Install 2 retaining bolts and tighten to 10–13 ft. lbs. (14–18 Nm).

NOTE: Before installing the timing belt, inspect it for cracks, wear or other damage and replace, if necessary. Do not allow the timing belt to come into contact with gasoline, oil or coolant. Do not twist or turn the belt inside out.

21. Make sure the engine is at TDC for No. 1 cylinder. Check that the camshaft sprocket marks line up with the index marks on the upper steel belt cover and that the crankshaft sprocket aligns with the index mark on the oil pump housing.

NOTE: The timing belt has 3 yellow lines. Each line aligns with the index marks.

22. Install the timing belt over the crankshaft and camshaft sprockets. The lettering on the belt **KOA** should be readable from the rear of the engine (top of the lettering to the front of the engine). Make sure the yellow lines are aligned with the index marks on the sprockets.

23. Release the timing belt tensioner idler pulley locknut. Leave the nut loose. Raise and safely support the vehicle.

24. Install the center timing belt cover. Make sure the CKP sensor wiring and grommet are installed and routed properly. Tighten the mounting bolts to 60–90 inch lbs. (7–11 Nm).

25. Install the lower timing belt cover. Tighten the bolts to 60–90 inch lbs. (7–11 Nm).

26. Install the crankshaft damper and pulley using Installer T88T-6701-A, or equivalent. Install the retaining bolt and tighten to 113–126 ft. lbs. (152–172 Nm).

27. Rotate the crankshaft 2 revolutions in the clockwise direction until the yellow mark on the damper aligns with the 0 degree mark on the lower timing belt cover.

28. Remove the plastic door in the lower timing belt cover. Tighten the tensioner locknut to 25–37 ft. lbs. (33–51 Nm) and install the plastic door.

29. Rotate the crankshaft 60 degrees more in the clockwise direction until the white mark on the damper aligns with the 0 degree mark on the lower timing belt cover.

30. Lower the vehicle. Make sure the index marks on the camshaft sprockets align with the marks on the rear metal timing belt cover.

31. Route the CKP sensor wiring and connect with the engine wiring harness.

32. Install the upper timing belt cover. Tighten the bolts to 60–90 inch lbs. (7–11 Nm).

33. Install the water pump/power steering tensioner pulley. Tighten the nut to 11–17 ft. lbs. (15–23 Nm).

34. Install the alternator/air conditioning tensioner pulley and bracket assembly. Tighten the bolts to 11–17 ft. lbs. (15–23 Nm).

35. Install the water pump/power steering and alternator/air conditioning drive belts and set the tension. Tighten the idler pulley nut to 25–36 ft. lbs. (34–50 Nm).

36. Install the intake manifold crossover tube. Tighten the bolts to 11–17 ft. lbs. (15–23 Nm).

37. Install the engine roll damper. Install the battery. Connect the wiring to the ignition module.

38. Connect both battery cables, negative cable last. Raise and safely support the vehicle.

39. Install the splash shield. Install the right front wheel and tire assembly.

40. Lower the vehicle. Run the engine and check for proper operation.

3.2L SHO Engine

1. Disconnect the battery cables, negative cable first.

2. Remove the battery.

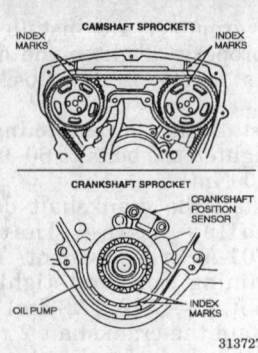

Camshaft and crankshaft sprocket index marker and alignment — SHO engines

3. Remove the right engine roll damper.

4. Disconnect the wiring to the ignition module.

5. Remove the intake manifold crossover tube bolts. Loosen the intake manifold tube hose clamps. Remove the intake manifold crossover tube.

6. Rotate the accessory drive belt tensioner clockwise to relieve belt tension. Remove the accessory drive belt.

7. Disconnect the surge tank fitting.

8. Remove the bolts retaining the upper and lower idler pulleys to the engine and remove the pulleys.

9. Using Strap Wrench D85L-6000-A or equivalent, to hold the power steering pump pulley. Remove the retaining nut and washer. Remove the power steering pulley.

10. Remove the retaining bolt from the belt tensioner and remove the tensioner.

11. Remove the upper and center timing belt covers.

12. Disconnect the Crankshaft Position (CKP) sensor electrical connector.

13. Place the transaxle selector in **N**.

14. Rotate the crankshaft until the piston for No. 1 cylinder is at TDC on its compression stroke. Make sure the white mark on the crankshaft damper aligns with the **0** degree index mark on the lower timing belt cover and the marks on the intake camshaft sprockets align with the index marks on the metal timing belt cover.

15. Raise and safely support the vehicle.

16. Remove the right front wheel and tire assembly.

17. Loosen the fender splash shield and place it aside.

18. Remove the crankshaft pulley and damper using Puller

T67L-3600-A with the appropriate adapters, or equivalent.

19. Remove the lower timing belt cover and belt guide.

20. Remove the upper timing belt tensioner bolt.

21. Slowly loosen the lower timing belt tension bolt and remove the tensioner.

22. Lower the vehicle.

23. Remove the timing belt. If the belt is to be reused, use crayon to mark an arrow on the belt to indicate the direction of rotation, for installation reference.

24. If removing one or more camshaft sprockets, remove 2 retaining bolts securing each camshaft sprocket and remove the sprocket noting the location of the dowel pin.

25. If removing the crankshaft sprocket, install Puller T67L-3600-A, or equivalent and pull the crankshaft sprocket off the crankshaft using care not to damage the pulse wheel.

To install:

26. If the crankshaft sprocket was removed, install the crankshaft sprocket by aligning the keyway and pushing the sprocket on by hand.

27. If one or more camshaft sprockets was removed, install each camshaft sprocket by aligning the timing marks on the camshaft sprocket with the camshaft using the dowel pin as a guide. Install 2 retaining bolts and tighten to 10–13 ft. lbs. (14–18 Nm).

NOTE: Before installing the timing belt, inspect it for cracks, wear or other damage and replace as necessary. Do not allow the timing belt to come into contact with gasoline, oil or coolant. Do not twist or turn the belt inside out.

28. Slowly compress the timing belt tensioner in a soft jawed vise until the hole in the tensioner housing aligns with the hole in the tensioner rod.

—————— **CAUTION** ——————
Use care when compressing the timing belt tensioner in the vise to insure that the tensioner does not slip from the vise.

29. Insert a 1/20 inch (1.5mm) hex wrench through the holes in the tensioner.

30. Release the tension from the vise.

31. If a new timing belt is being installed, loosen the timing belt idler bolt.

32. Ensure that No. 1 cylinder is at TDC on its compression stroke.

Check that the camshaft sprocket marks line up with the index marks on the upper steel belt cover and that the crankshaft sprocket aligns with the index mark on the oil pump housing.

NOTE: The timing belt has 3 yellow lines. Each line aligns with the index marks.

33. Install the timing belt over the crankshaft and camshaft sprockets. The lettering on the belt **KOB** should be readable from the rear of the engine (top of the lettering to the front of the engine). Make sure the yellow lines are aligned with the index marks on the sprockets.

—————— **WARNING** ——————
Do not install the timing belt tensioner with the tensioner rod extended.

34. Install the timing belt tensioner on the cylinder block while pushing the timing belt idler toward the timing belt. Install and tighten the tensioner bolts to 12–17 ft. lbs. (16–23 Nm).

35. Install the grommets between the timing belt tensioner and the oil pump.

36. Remove the hex wrench from the timing belt tensioner.

37. If a new timing belt is being installed, perform the following steps:

   a. Remove the hex wrench from the timing belt tensioner, if installed.

   b. Mount Timing Belt Tensioner tool T93P-6254-B or equivalent, using the holes in the power steering pump support.

   c. Hand-tighten the timing belt idler pulley bolt.

   d. Using an inch pound torque wrench with attachment T93P-6254-A or equivalent, rotate the timing belt tensioner clockwise to 4.3 inch lbs. (0.5 Nm).

   e. Tighten the timing belt idler pulley bolt to 27–37 ft. lbs. (36–50 Nm). Remove both timing belt tensioning tools.

38. Raise and safely support the vehicle.

39. Install the belt guide and lower timing belt cover. Tighten the retaining bolts to 12–17 ft. lbs. (16–23 Nm).

40. Using a suitable tool, install the crankshaft damper. Tighten the damper attaching bolt to 113–126 ft. lbs. (152–172 Nm).

41. Rotate the crankshaft 2 revolutions in the clockwise direction until the yellow mark on the damper aligns with the **0** degree mark on the lower timing belt cover.

42. Lower the vehicle.

43. Ensure that the index marks on the camshaft sprockets align with the marks on the rear metal timing belt cover.

44. Route the CKP sensor wiring and connect with the engine wiring harness.

45. Install the center and upper timing belt covers. Tighten the bolts to 12–17 ft. lbs. (16–23 Nm).

46. Install the steering pump pulley. Tighten the nut to 12–17 ft. lbs. (16–23 Nm).

47. Install the accessory drive belt while rotating accessory drive belt tensioner clockwise.

48. Install the surge tank fitting.

49. Install the intake manifold crossover tube. Tighten the bolts to 11–17 ft. lbs. (15–23 Nm).

50. Install the engine roll damper

51. Install the battery.

52. Connect the wiring to the ignition module.

53. Connect both battery cables, negative cable last.

54. Raise and safely support the vehicle.

55. Install the splash shield.

56. Install the right front wheel and tire assembly.

57. Lower the vehicle.

58. Run the engine and check for leaks and proper engine operation.

Camshaft

REMOVAL AND INSTALLATION

— **CAUTION** —

Fuel injection systems remain under pressure, even after the engine has been turned OFF. The fuel system pressure must be relieved before disconnecting any fuel lines. Failure to do so may result in fire and/or personal injury.

3.0L (VIN Y) and 3.2L SHO Engines

1. Disconnect the negative battery cable. Properly relieve the fuel system pressure.

2. Rotate the crankshaft until the piston in No. 1 cylinder is at TDC.

3. Remove the intake manifold assembly. Remove the timing belt front cover and timing belt.

4. If the left cylinder head cover is being removed, remove the oil fill cap and ignition coil plastic cover. If the right cylinder head cover is being removed, disconnect the fuel lines. Remove the cylinder head cover(s).

5. Remove the camshaft pulleys, noting the location of the dowel pins.

6. Remove the upper rear timing belt cover. Uniformly loosen the camshaft bearing caps.

— **WARNING** —

If the camshaft bearing caps are not uniformly loosened, camshaft damage may result.

7. Remove the camshaft bearing caps and note their positions for installation.

8. Remove the camshaft timing chain tensioner mounting bolts.

9. Remove the camshafts together with the timing chain and tensioner.

10. Remove and discard the camshaft oil seal.

11. Remove the timing chain chain sprocket from the camshaft.

12. Inspect the camshafts for wear and/or damage and replace, as necessary.

To install:

13. Apply a thin coat of clean engine oil to the camshaft bearing surfaces on the cylinder head and camshaft journal caps.

14. Align the timing marks on the camshaft sprockets with the camshaft and install the sprockets. Tighten the bolts to 10–13 ft. lbs. (14–18 Nm).

15. Install the timing chain over the camshaft sprockets. Align the white painted link with the timing mark on the sprocket.

— **WARNING** —

Left and right timing chain tensioners are not interchangeable.

16. Rotate the camshafts 60 degrees (³⁄₅ turn) counterclockwise. Set the timing chain tensioner between the sprockets and install the camshafts on the cylinder head. The left and right tensioners are not interchangeable.

17. Apply a thin coat of clean engine oil to the camshaft journals and install bearing caps 2 through 5. Loosely install the cap retaining bolts.

NOTE: The arrows on the bearing caps point to the front of the engine when installed.

18. Apply silicone sealer to the outer diameter of the new camshaft seal and the seal seating area on the cylinder head. Install the camshaft seal using cam seal expander T89P-6256-B and cam seal replacer T89P-6256-A or equivalents.

19. Apply a 0.10 inch (2.5mm) bead of silicone sealer to the No. 1 bearing cap. Install the bearing cap while holding the camshaft seal in place

with cam seal replacer T89P-6256-A or equivalent.

20. Tighten the bearing caps in sequence using a 2 step method. In the first step tighten to 71–106 inch lbs. (8–12 Nm). In the second step tighten to 12–16 ft. lbs. (16–22 Nm). For left camshaft installation, apply pressure to the timing chain tensioner to avoid damage to the bearing caps.

— **WARNING** —

The No. 5 camshaft bearing caps function as thrust bearings for the camshaft. Always tighten No. 5 camshaft bearing caps first.

21. Position the timing chain guide and the timing chain tensioner and install the retaining bolts. Tighten the bolts to 11–14 ft. lbs. (15–19 Nm).

22. Rotate the camshafts 60 degrees clockwise and check for proper alignment of the timing marks. Marks on the camshaft sprockets should align with the valve cover mating surface.

23. Install camshaft positioning tool T89P-6256-C or equivalent, on the camshafts to check for correct positioning. The flats on the tool should align with the flats on the camshaft. If the tool does not fit and/or timing marks will not line up, repeat the installation procedure from the beginning.

24. Install the timing belt rear cover and tighten the bolts to 78 inch lbs. (8.8 Nm).

25. Install the camshaft timing belt sprockets and tighten the bolts to 15–18 ft. lbs. (21–25 Nm).

26. Install the timing belt and timing belt front cover.

27. Install the cylinder head covers and tighten the bolts to 8–11 ft. lbs. (10–16 Nm). Connect the fuel lines and install the ignition coil cover and oil fill cap.

28. Install the intake manifold assembly.

29. Connect the negative battery cable. Run the engine and check for leaks and proper engine operation.

1993–95 3.0L (VIN U) Engine

1. Disconnect the negative battery cable. Drain the cooling system and crankcase.

2. Properly relieve the fuel system pressure.

3. Remove the engine from the vehicle and position in a suitable holding fixture.

4. Remove the accessory drive components from the front of the engine.

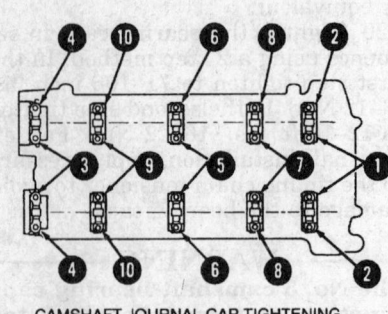

CAMSHAFT JOURNAL CAP TIGHTENING
SEQUENCE RH CYLINDER HEAD 6049

FRONT OF ENGINE

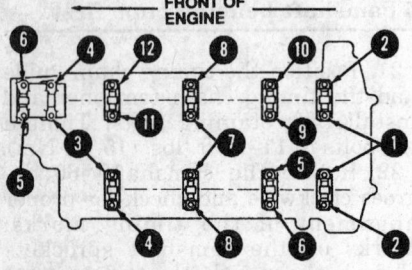

CAMSHAFT JOURNAL CAP TIGHTENING
SEQUENCE LH CYLINDER HEAD 6049

236955

Camshaft journal tightening sequence — SHO
engines

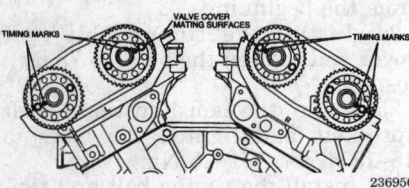

236956

Aligning the camshaft sprocket timing marks
with the valve cover mating surface — SHO
engines

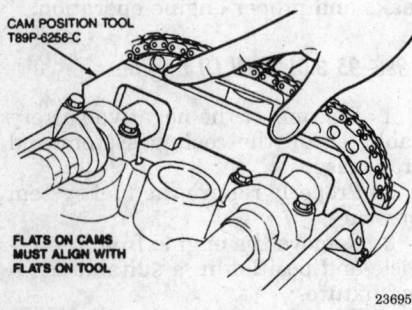

CAM POSITION TOOL
T89P-6256-C

FLATS ON CAMS
MUST ALIGN WITH
FLATS ON TOOL

236957

Cam position tool installation — SHO engines

5. Remove the throttle body and the fuel injector harness.

6. Label and disconnect the spark plug wires from the spark plugs.

7. Remove the distributor assembly.

8. Remove the rocker arm covers.

9. Loosen the rocker arm fulcrum nuts and position the rocker arms to the side for easy access to the pushrods. Remove the pushrods and label so they may be installed in their original positions.

10. Remove the intake manifold leaving the fuel supply manifold and injectors in place.

11. Using a suitable magnet or lifter removal tool, remove the hydraulic lifters and keep them in order so they can be installed in their original positions. If the lifters are stuck in the bores by excessive varnish, use a hydraulic lifter puller to remove the lifters.

12. Remove the crankshaft pulley and damper using a suitable removal tool.

13. Remove the oil pan. Remove the front cover assembly.

14. Align the timing marks on the camshaft and crankshaft sprockets. Check the camshaft end-play as follows:

    a. Push the camshaft toward the rear of the engine and install a dial indicator tool, so the indicator point is on the camshaft sprocket attaching screw.

    b. Zero the dial indicator. Position a small prybar or equivalent, between the camshaft sprocket and block.

    c. Pull the camshaft forward and release it. Compare the dial indicator reading with the camshaft end-play service limit specification of 0.005 in. (0.13mm).

    d. If the camshaft end-play is over the amount specified, replace the thrust plate.

15. Remove the timing chain and sprockets.

16. Remove the camshaft thrust plate. Carefully remove the camshaft by pulling it toward the front of the engine. Remove it slowly to avoid damaging the bearings, journals and lobes.

17. Inspect the camshaft journals and lobes for wear and/or damage. Replace as necessary.

NOTE: If the camshaft is replaced, new lifters should also be installed.

To install:

18. Clean all gasket mating surfaces. Lubricate the camshaft lobes and journals with engine assembly lube or clean engine oil. Carefully insert the camshaft through the bearings into the cylinder block.

19. Install the thrust plate. Tighten the retaining bolts to 84 inch lbs. (10 Nm).

20. Install the timing chain and sprockets. Tighten the camshaft sprocket retaining bolt to 46 ft. lbs. (63 Nm).

CAUTION

The camshaft bolt has a drilled oil passage in it for timing chain lubrication. Make sure the passage is clean prior to bolt installation. If the bolt is damaged, do not replace the camshaft bolt with a standard bolt or engine damage may result.

---

21. Install the front timing cover and crankshaft damper and pulley. Tighten the crankshaft damper bolt to 107 ft. lbs. (145 Nm).

22. Lubricate the lifters and lifter bores with clean engine oil. Install the lifters into their original bores.

23. Install the intake manifold assembly.

24. Lubricate the pushrods and rocker arms with clean engine oil. Install the pushrods and rocker arms into their original positions. Rotate the crankshaft to set each lifter on its base circle, then tighten the rocker arm bolt. Tighten the rocker arm bolts to 24 ft. lbs. (32 Nm).

25. Install the oil pan and the rocker covers.

26. Install the fuel injector harness and the throttle body.

27. Install the distributor and connect the spark plug wires to the spark plugs.

28. Install the accessory drive components.

29. Install the engine assembly.

30. Connect the negative battery cable. Restore all fluid levels.

31. Start the engine and check for leaks Check and adjust the ignition timing.

3.0L (VIN S) Engine

1. Disconnect the negative battery cable. Remove the engine assembly from the vehicle and place on an appropriate engine stand.

2. Remove the upper intake manifold. Remove both cylinder head covers.

3. Remove the engine front cover.

4. Remove the timing chains and rocker arms.

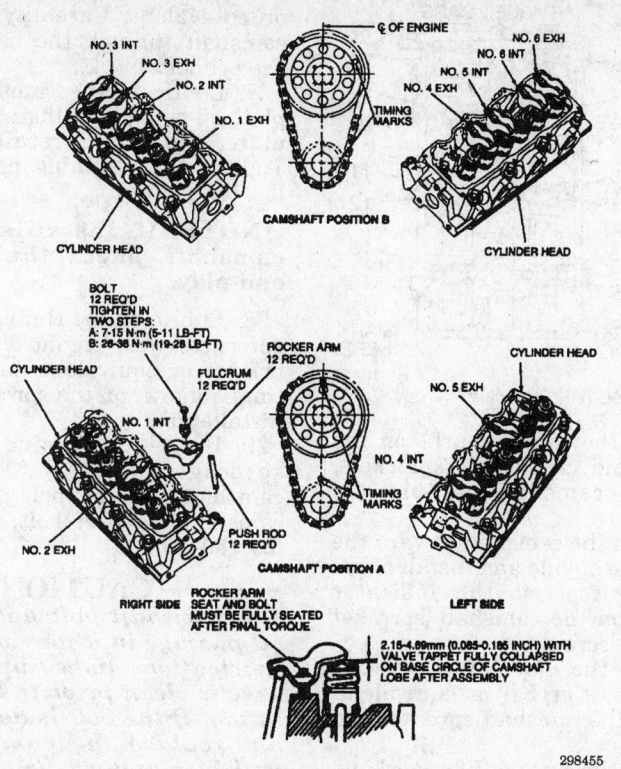

Camshaft position for rocker arm bolt tightening — 3.0L (VIN U and 1) engines

WARNING

The camshaft journal thrust caps must be removed first, before loosening the remaining camshaft journal cap bolts, to ensure that the camshaft journal thrust caps are not damaged.

5. Loosen the camshaft journal cap bolts in sequence, in several passes, to allow the camshaft to raise off the cylinder head evenly.

WARNING

The camshaft journal caps and cylinder heads are numbered to ensure that they are assembled in their original positions. Keep the camshaft journal caps from each cylinder head together. Do not mix with caps from another cylinder head. Failure to do so may result in engine damage.

6. Remove the camshaft journal caps with the retaining bolts installed.
7. Remove the camshafts from the cylinder head, then repeat the procedure for both cylinder heads.
8. Inspect the camshafts and cylinder heads for wear or damage.

To install:

WARNING

The crankshaft keyway must be at the 11 o'clock position before reassembly. Failure to do so may lead to engine damage.

9. Rotate the crankshaft so the keyway is at the 11 o'clock position for installation of the camshafts.
10. Lubricate the camshaft lobes and journals with engine assembly lubricant.
11. Install the camshafts into their correct positions into each cylinder head with the timing marks on the camshaft sprockets aligned.
12. Loosely install the camshaft journal caps and retaining bolts into their correct positions.

NOTE: Do not install the camshaft journal thrust caps until the rocker arms and timing chains have been installed and the camshaft journal caps are torqued into position.

13. Install the timing chains.
14. Tighten the camshaft journal cap bolts, in sequence, to 71–106 inch lbs. (8–12 Nm). Install the thrust caps and tighten the retaining bolts to 71–106 inch lbs. (8–12 Nm).
15. Install the engine front cover.
16. Install both cylinder head covers, then install the upper intake manifold.

NOTE: The cylinder head covers must be installed and properly torqued within 6 minutes of applying the silicone sealant.

17. Install the engine assembly into the vehicle. Connect the negative battery cable.
18. Run the engine and check for leaks and proper engine operation.

1996–97 3.0L (VIN 1 and U) Engine

1. Disconnect the negative battery cable. Remove the engine from the vehicle and position on a suitable holding fixture.
2. Rotate the crankshaft to TDC for No. 1 cylinder on its compression stroke.
3. Remove the upper intake manifold.
4. Disconnect the engine wiring harness connectors from the cylinder head cover stud bolts. Carefully disconnect and remove the fuel injector harness connectors from each fuel injector and position aside.
5. Label and disconnect the ignition wires from the spark plugs. Remove the ignition wire separators

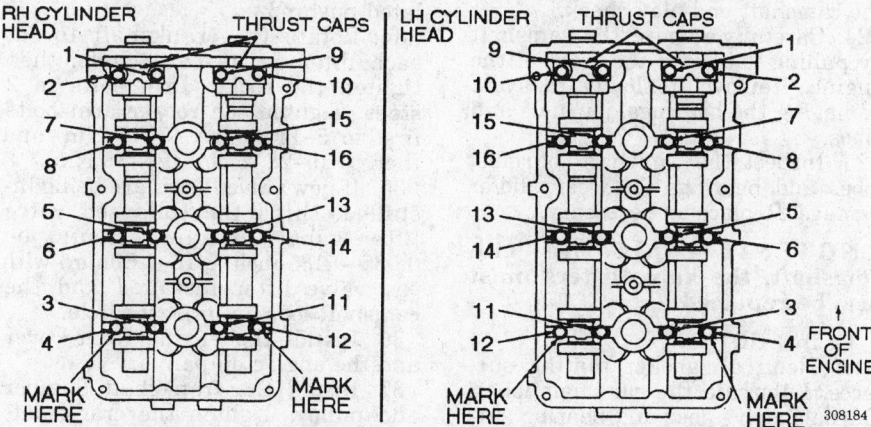

Camshaft journal cap retaining bolt removal sequence — 3.0L (VIN S) engine

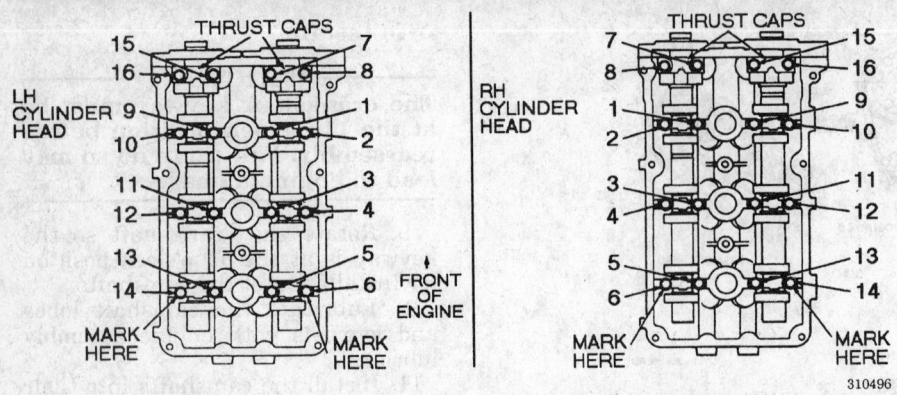

Torque sequence for both camshaft journal caps — 3.0L (VIN S) engine

from the cylinder head cover stud bolts.

6. Remove the Camshaft Position (CMP) sensor housing retainer bolt and washer and remove the CMP sensor housing.

7. Remove the ignition coil from the rear of the left cylinder head.

8. Remove the cylinder head covers.

9. Loosen the intake rocker arm fulcrum nut for No. 3 cylinder and rotate the rocker arm off the pushrod. Remove the pushrod.

10. Remove the accessory drive belt. Remove the drive belt tensioner, alternator and alternator brackets.

11. Remove the lower intake manifold leaving the fuel supply manifold (fuel rail) and fuel injectors in place.

12. Loosen the remaining rocker arm fulcrum nuts enough to allow the rocker arms to be lifted off the pushrods. Remove the remaining pushrods, identifying each pushrod for installation.

13. Remove the valve lifter guide plate retainer bolts and the valve lifter guide plate.

14. Using a suitable magnet or lifter removal tool, remove the hydraulic valve lifters and keep them in order so they can be installed in their original positions. If the valve lifters are stuck in the bores by excessive varnish, use a hydraulic lifter puller to remove the lifters.

15. Remove the crankshaft pulley retaining bolts and the crankshaft pulley.

16. Remove the crankshaft damper retaining bolt and washer. Remove the crankshaft damper using remover T58P- 6316-D or equivalent and adapter T82L-6316-B or equivalent.

17. Remove the engine oil pan.

18. Remove the engine front cover retaining bolts leaving the water pump attached. Remove the engine front cover.

19. Align the timing marks on the camshaft and crankshaft sprockets. Check the camshaft end-play as follows:

a. Push the camshaft toward the rear of the engine and install a dial indicator tool, so the indicator point is on the camshaft sprocket attaching screw.

b. Zero the dial indicator. Position a small prybar or equivalent, between the camshaft sprocket and block.

c. Pull the camshaft forward and release it. Compare the dial indicator reading with the camshaft end-play service limit specification of 0.005 inch (0.13mm).

d. If the camshaft end-play is over the amount specified, replace the thrust plate.

20. Remove the camshaft sprocket retaining bolt and washer.

21. Inspect the timing chain for excessive deflection.

22. Grasp the camshaft sprocket and the crankshaft sprocket and slide the timing chain and sprocket assembly off the engine.

23. Remove the 2 camshaft thrust plate retaining bolts and the camshaft thrust plate. Discard the thrust plate if it was found to be worn beyond specifications made during the camshaft end-play check.

24. Carefully remove the camshaft by pulling it toward the front of the engine. Remove it slowly to avoid damaging the bearings, journals and lobes.

25. Inspect the camshaft journals, lobes and bearings for wear and/or damage. Replace as necessary.

NOTE: If replacing the camshaft, the valve lifters must also be replaced.

To install:

26. Clean all gasket mating surfaces. Lubricate the camshaft lobes, journals, drive gear and bearing surfaces with engine assembly lubricant

or equivalent. Carefully insert the camshaft through the bearings into the cylinder block.

27. Lubricate the camshaft thrust plate on both sides. Install the thrust plate and the 2 retaining bolts. Tighten the retaining bolts to 7 ft. lbs. (10 Nm).

NOTE: If installing a new camshaft, check the camshaft end-play.

28. Lubricate the timing chain and sprockets with engine assembly lubricant or equivalent and align the timing marks on the sprockets before installation.

29. Install the timing chain and sprocket assembly. Install the camshaft sprocket bolt. Tighten the camshaft sprocket bolt to 37–51 ft. lbs. (50–70 Nm).

----- CAUTION -----
The camshaft bolt has a drilled oil passage in it for timing chain lubrication. Make sure the passage is clean prior to bolt installation. If the bolt is damaged, do not replace the camshaft bolt with a standard bolt or engine damage will result.
----

30. Lubricate the valve lifters and lifter bores with engine assembly lubricant or equivalent. Install the valve lifters into their original bores.

31. Align the flat surfaces on the valve lifters and install the valve lifter guide plate. The plate must be installed with the word **UP** and/or a dimple on the plate visible. Install the 2 retaining bolts and tighten to 8–10 ft. lbs. (10–14 Nm).

32. Install the lower intake manifold. Install the CMP sensor.

33. Lubricate the pushrods and rocker arms with engine assembly lubricant or equivalent. Install the pushrods into their original positions. Position each rocker arm onto its related pushrod.

34. Rotate the crankshaft to set each lifter on its base circle, then tighten the rocker arm bolts in 2 steps. Tighten the rocker arm bolts first to 5–11 ft. lbs. (7–15 Nm) and then to 19–28 ft. lbs. (26–38 Nm).

35. If new valve lifters are being installed, check the collapsed valve lifter gaps. Clearance should be 0.085–0.185 inch (2.15–4.69mm) with the valve lifter installed and the camshaft lobe on its base circle.

36. Install the engine front cover and the engine oil pan.

37. Install the crankshaft damper and pulley. Tighten the crankshaft damper bolt to 93–121 ft. lbs.

(125–165 Nm) and the 4 pulley bolts to 30–40 ft. lbs. (40–55 Nm).

38. Install the alternator and brackets.

39. Install the drive belt tensioner and the accessory drive belt.

40. Install the cylinder head covers.

41. Install the fuel injector harness to each fuel injector. Secure the harness to the cylinder head cover stud bolts.

42. Install the ignition coil to the left cylinder head. Tighten the retaining bolts to 29–41 ft. lbs. (40–55 Nm).

43. Install the upper intake manifold.

44. Install the ignition wire harness retainers to the cylinder head cover stud bolts and connect the ignition wires to the spark plugs and ignition coil.

45. Install the engine assembly into the vehicle.

46. Fill the engine cooling system.

47. Fill the crankcase with the correct amount and type of engine oil.

48. Connect the negative battery cable.

49. Start the engine and check for leaks and proper engine operation.

50. Check and adjust the ignition timing, if needed.

3.8L Engine

1. Disconnect the negative battery cable. Remove the engine from the vehicle and place on a suitable engine stand.

2. Remove the upper plenum and the intake manifold. Remove the valve covers, rocker arms, pushrods, guide plates and lifters.

NOTE: If the hydraulic lifters are to be reused, mark their locations for reassembly.

3. Remove the engine oil pan.

4. Remove the timing chain front cover. Remove the timing chain and sprockets.

5. Remove the camshaft thrust plate.

6. Remove the camshaft through the front of the engine, being careful not to damage the camshaft bearing surfaces.

To install:

NOTE: Inspect the camshaft rear bearing cover. If damaged or leaking, replace the camshaft rear bearing cover. Inspect the camshaft and camshaft bearings for signs of wear or damage and replace as necessary. If the camshaft is replaced, new lifters should also be installed.

7. Lubricate the cam lobes and journals with engine assembly lubricant.

8. Install the camshaft, being careful not to damage the lobes or bearing surfaces while sliding it into position.

9. Install the camshaft thrust plate. Install and tighten the 2 retaining bolts to 72–120 inch lbs. (8–14 Nm).

10. Check the camshaft end-play as follows:

   a. Temporarily install the camshaft sprocket retaining bolt.

   b. Push the camshaft toward the rear of the engine.

   c. Install a dial indicator to the front of the cylinder block and position the dial indicator to rest on the face of the camshaft sprocket retaining bolt.

   d. Zero the dial indicator.

   e. Pull the camshaft forward.

   f. Note the reading on the dial indicator. The end-play should measure between 0.001–0.006 inch (0.025–0.150mm).

   g. If the reading is excessive, replace the camshaft thrust plate and recheck. If the camshaft end-play is still excessive, check for a worn camshaft or cylinder block.

   h. Remove the camshaft sprocket retaining bolt.

11. Install the engine oil pan.

12. Install the timing chain and sprockets making sure the camshaft and crankshaft are properly timed.

13. Install the engine front cover.

14. Install the lifters, guide plates, pushrods, rocker arms and the valve covers.

15. Install the intake manifold and the upper plenum.

16. Remove the engine from the engine stand and install the engine in vehicle.

17. Restore all fluid levels.

18. Connect the negative battery cable.

19. Run the engine and check for leaks and proper operation.

4.6L Engine

1. Disconnect the negative battery cable. Remove the engine from the vehicle and position on a suitable workstand.

2. Remove the cylinder head covers and the engine front cover.

3. Remove the timing chains.

NOTE: If the rocker arms are to be reused, label them as they are removed so they can be reinstalled in their original locations.

4. Remove the rocker arms from the cylinder head and the camshaft that is being serviced. If both bank camshafts are to be serviced, all the rocker arms must be removed.

5. Remove 13 bolts retaining the exhaust camshaft cap cluster assemblies to the cylinder head to remove the exhaust camshaft.

6. Remove 12 bolts retaining the intake camshaft cap cluster assemblies to the cylinder head to remove the intake camshaft.

7. Tap upward lightly on the camshaft cap clusters and gradually lift the camshaft cap clusters from the cylinder head.

8. Lift the camshaft straight upward to avoid bearing surface damage.

NOTE: The previous Steps will remove only one bank of camshafts. If both banks are being serviced, repeat the removal procedure for the remaining camshafts.

9. Inspect the camshaft lobes and journals for wear and/or damage. Replace as necessary.

To install:

NOTE: When cleaning the sealing surfaces of the engine front cover and the oil pan–to–cylinder block joints, use extreme caution not to damage the rubber bead of the oil pan gasket. If damaged, the oil pan gasket must be replaced.

10. Clean and inspect the cylinder head cover, engine front cover, and cylinder head sealing surfaces.

11. If removed, install the lash adjusters (valve tappets).

12. Apply clean engine oil to the journals and lobes of the camshaft. Install the camshaft to the cylinder head.

13. Install and seat the camshaft cap cluster assemblies. Hand-start 12 (52mm long) bolts into the intake camshaft caps, 7 (52mm long) bolts into the inboard side of the exhaust camshaft and 6 (42mm long) bolts into the outboard side of the exhaust camshaft caps.

NOTE: Each camshaft cap cluster is tightened individually.

14. Tighten the camshaft cap cluster retaining bolts, in sequence, to 71–106 inch lbs. (8–12 Nm).

15. Loosen the camshaft cap cluster retaining bolts approximately 2 turns or until the head of the bolt is free.

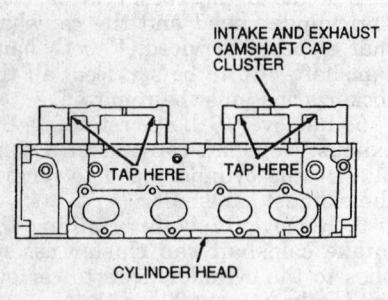

Camshaft cap clusters — 4.6L engine

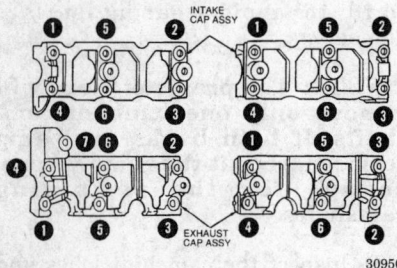

Camshaft cap cluster torque sequence — 4.6L engine

16. Tighten all bolts again, in sequence, to 71–106 inch lbs. (8–12 Nm).

NOTE: Once installed, the camshaft should turn freely but with a slight drag.

17. Check the camshaft end-play using a dial indicator.
18. Install the timing chains.
19. Install Valve Spring Compressor T91P-6565-A or equivalent, under the exhaust camshaft and on top of the exhaust valve spring retainer. Compress the valve spring far enough to install the rocker arm.
20. Install Valve Spring Compressor T93P-6565-A or equivalent, under the intake camshaft and on top of the primary intake valve spring retainer. Compress the valve spring far enough to install the rocker arm.
21. Install Valve Spring Compressor T93P-6565-A or equivalent, under the intake camshaft and on top of the secondary intake valve spring retainer. Compress the valve spring far enough to install the rocker arm.

NOTE: The previous installation Steps must be repeated if both banks are being serviced to install the remaining camshafts.

22. Inspect and replace the crankshaft front seal and timing chain front cover gasket.
23. Install the engine front cover and the cylinder head covers.
24. Remove the engine from the workstand and install in the vehicle.
25. Start the engine and check for leaks.
26. Road test the vehicle and check for proper engine operation.

Piston and Connecting Rod

POSITIONING

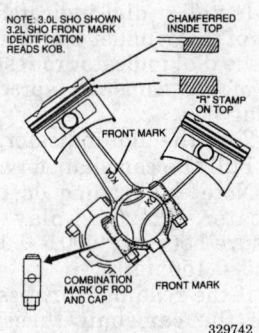

Piston to cylinder block positioning — 3.0L (VIN Y) and 3.2L SHO Engines

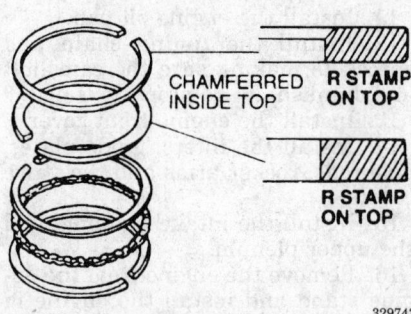

Piston ring installation — 3.0L (VIN Y) and 3.2L Engines

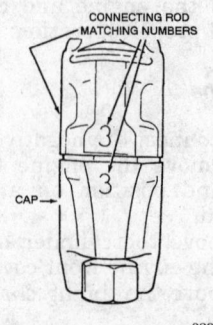

Connecting rod and cap positioning — 3.0L (VIN 1, U and S) and 3.8L Engines

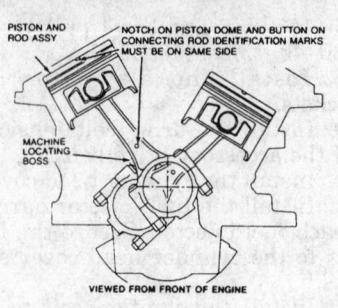

Piston to cylinder block positioning (notch on piston must face front of engine) — 3.0L (VIN 1 and U) and 3.8L engines

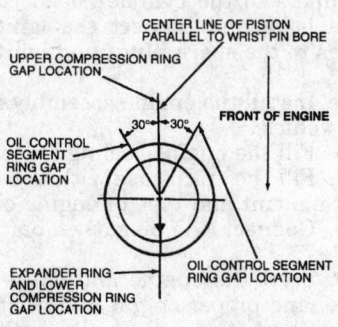

Piston ring positioning — 3.0L (VIN 1 and U), 3.8L and 4.6L engines

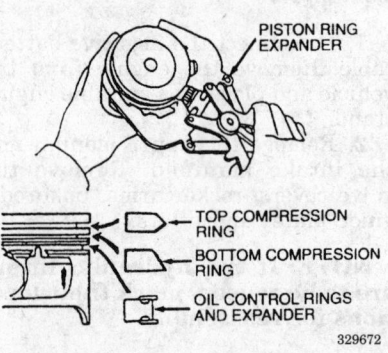

Piston ring installation — 3.0L (VIN S) and 4.6L engines

ENGINE LUBRICATION

Oil Pan

REMOVAL AND INSTALLATION

3.0L (VIN Y) and 3.2L SHO Engines

1. Disconnect the negative battery cable.

2. Remove the oil level dipstick.

3. Raise the vehicle and support it safely. Drain the engine oil.

4. If equipped with a low oil level sensor, remove the retainer clip and the electrical connector from the sensor.

5. Remove the starter motor.

6. Disconnect the oxygen sensors.

7. Remove the catalyst and pipe assembly.

8. Remove the lower flywheel dust cover from the bell housing.

9. Remove the oil pan attaching bolts and the oil pan.

To install:

10. Clean the gasket surfaces of the cylinder block and the oil pan.

11. Position the oil pan gasket on the oil pan and secure with silicone sealer.

12. Position the oil pan and tighten the retaining bolts in sequence to 11–16 ft. lbs. (15–23 Nm).

13. Install the remaining components in the reverse order of removal.

14. Connect the negative battery cable. Fill the crankcase with the proper type and quantity of oil, then start the vehicle and check for leaks.

3.0L (VIN 1 and U) Engines

1. Disconnect the negative battery cable. Remove the engine oil level dipstick.

2. Raise and safely support the vehicle. Drain the engine oil.

3. If equipped with a low oil level sensor, remove the retainer clip at the sensor. Disconnect the electrical connector from the sensor.

4. Remove the starter motor and brace.

5. Detach the connector from the O2 sensor(s).

6. Remove the catalytic converter and exhaust Y-pipe assembly.

7. Remove the engine rear plate from the torque converter housing.

8. Remove the 16 oil pan retaining bolts and carefully remove the engine oil pan from the cylinder block. Remove the oil pan gasket.

To install:

9. Clean the gasket sealing surfaces on the cylinder block and the engine oil pan. Apply a ¼ inch (6mm) bead of silicone sealer to the junction of the front cover assembly and the cylinder block and to the junction of the rear main bearing cap and cylinder block.

NOTE: When using a silicone sealer, the assembly process should occur within 5 minutes after the sealer has been applied.

Make sure the sealer does not fall into the engine and form plugs that could block oil passages.

10. Place the oil pan gasket on the engine oil pan and secure with a suitable contact adhesive.

11. Place the engine oil pan into position on the cylinder block. Install the 16 engine oil pan retaining bolts. Tighten the retaining bolts to 8–10 ft. lbs. (10–14 Nm). Back off all bolts and retighten.

12. Install the remaining components in the reverse order of removal.

13. Connect the negative battery cable. Fill the crankcase with the proper type and quantity of engine oil, then start the engine and check for leaks and proper operation.

3.0L (VIN S) Engine

1. Disconnect the negative battery cable.

2. Raise and safely support the vehicle. Drain the engine oil.

3. Remove the dual catalytic converter Y-pipe retaining nuts from the exhaust manifolds.

4. Remove the bolt and nut retainers from the transaxle.

5. Remove the 2 remaining nuts and bolts from the dual converter Y-pipe connection. Remove the Y-pipe from the vehicle.

6. Reinstall the oil pan drain plug using a new gasket and torque to 16–22 ft. lbs. (22–30 Nm).

7. Remove the oil pan retaining bolts from the transaxle housing.

8. Remove the access plug from the engine rear plate.

9. Remove the support bracket from the oil pan and transaxle.

10. Remove the oil pan retaining bolts and studs from the lower cylinder block following the correct bolt removal sequence.

11. Remove the oil pan and the oil pan gasket from the vehicle.

12. If required, remove the oil pump screen and tube assembly from the oil pump.

To install:

13. Clean the oil pan to lower cylinder block gasket sealing surfaces.

14. Thoroughly clean the oil pan and mating surfaces with soap and water and dry completely with compressed air.

15. Clean the mating surfaces with Metal Surface Cleaner F4AZ9A536-RA or equivalent to remove all the residues that may cause oil leakage.

16. If removed, install the oil pump screen and tube assembly to the oil pump using a new O-ring. Tighten the retaining bolts to 71–106 inch lbs.

(8–12 Nm). Install a new self-locking nut and tighten to 71–106 inch lbs. (8–12 Nm).

17. Install a new oil pan gasket into the groove of the oil pan.

18. Apply a 0.31 inch (8mm) bead of silicone sealer to the gasket area where the pan meets the parting lines of the lower cylinder block and the front engine cover.

19. Carefully install the oil pan with gasket to the lower cylinder block. Install the bolts and studs but do not tighten.

20. Push the oil pan against the transaxle case and tighten the oil pan bolts and studs, finger-tight.

21. Tighten the oil pan retaining bolts and studs in the proper sequence to 15–22 ft. lbs. (20–30 Nm).

22. Tighten the oil pan to transaxle case bolts to 25–34 ft. lbs. (34–46 Nm).

23. Install the transaxle support bracket to the oil pan retaining stud bolts and transaxle. Tighten the retaining nuts to 71–106 inch lbs. (8–12 Nm). Tighten the retaining bolts to 15–22 ft. lbs. (20–30 Nm).

24. Install the access plug into the engine rear plate.

25. Position the Y-pipe assembly using a new flange gasket and install all the retaining nuts and bolts loosely.

26. Starting at the front of the system tighten the Y-pipe to exhaust manifold nuts to 26–34 ft. lbs. (34–46 Nm). Tighten the converter to transaxle nut and bolt to 30 ft. lbs. (40.3 Nm). Tighten the converter outlet bolts to 26–34 ft. lbs. (34–46 Nm).

27. Replace the engine oil filter.

28. Lower the vehicle.

29. Fill the crankcase with the correct amount and grade of engine oil.

30. Connect the negative battery cable. Run the engine and check for leaks and proper operation.

3.8L Engine

1. Disconnect the negative battery cable. Raise the vehicle and support safely.

2. Drain the engine oil into a suitable container and remove the oil filter element. Install the drain plug and tighten it to 15–25 ft. lbs. (20–34 Nm) and move the drain pan aside.

3. Remove the catalytic converter Y-pipe assembly.

4. Remove the starter motor and the rear engine plate.

5. Remove the oil pan retaining bolts and remove the oil pan.

6. To remove the oil pump screen and tube, remove the 2 retaining bolts and the support bracket nut

and remove the oil pump screen and tube and gasket.

To install:

7. If removed, clean the oil pump screen and tube mounting gasket surfaces and install a new oil pump screen and tube gasket.

8. Position the oil pump screen and tube to the mounting gasket and install the 2 bolts and the support bracket nut. Tighten the 2 retaining bolts to 15–22 ft. lbs. (20–30 Nm) and the nut to 30–40 ft. lbs. (40–55 Nm).

9. Clean the gasket surfaces on the cylinder block and the oil pan.

10. Trial fit the oil pan to cylinder block. Ensure that enough clearance has been provided to allow the oil pan to be installed without sealant being accidentally scraped off when the oil pan is positioned under the engine.

11. Make sure there is no engine oil on the gasket mating surfaces. Apply a bead of silicone sealer to the oil pan flange. Also apply a bead of sealer to the front cover/cylinder block joint and fill the grooves on both sides of the rear main seal cap.

NOTE: When using silicone rubber sealer, assembly must occur within 15 minutes after sealer application. After this time, the sealer may start to harden and its sealing effectiveness may be reduced.

12. Install the oil pan and secure to the block with the retaining bolts. Tighten the 18 retaining bolts to 84–108 inch lbs. (9–12 Nm).

13. Install a new oil filter element.

14. Install the rear engine cover and the starter motor.

15. Install the catalytic converter Y-pipe assembly.

16. Lower the vehicle.

17. Fill the crankcase with the correct grade and amount of engine oil.

18. Connect the negative battery cable. Run the engine and check for oil leaks.

4.6L Engine

------ **CAUTION** ------

The air suspension switch, located in the left side of the luggage compartment, must be turned OFF before raising the vehicle. Failure to do so may result in unexpected inflation or deflation of the air springs which may result in shifting of the vehicle during service.

1. Turn the air suspension switch, located in the left side of the luggage compartment, to the **OFF** position.

2. Disconnect the negative battery cable.

3. Remove the oil level dipstick.

4. Raise and safely support the vehicle. Drain the engine oil.

5. Remove the dual converter Y-pipe from the exhaust manifolds.

6. Disconnect the wiring from the low oil sensor connector.

7. Remove the power steering pressure hose retainer brackets from the engine front cover studs in 2 locations. Position the power steering pressure hose aside.

8. Remove 16 bolts retaining the oil pan to the cylinder block and remove the oil pan and gasket.

To install:

9. Clean the gasket surfaces on the cylinder block, oil pan and front cover with a clean cloth. If scraping is required, only use plastic-tipped scrapers to prevent damage to aluminum surfaces. Thoroughly clean the inside of the oil pan.

10. Apply a bead of silicone sealer to the oil pan flange. Also apply a bead of sealer to the front cover/cylinder block joint and fill the grooves on both sides of the rear main seal cap.

NOTE: When using silicone rubber sealer, assembly must occur within 15 minutes after sealer application. Make sure the sealer does not fall into the engine where it may plug oil passages.

11. Install a new gasket on the oil pan.

12. Carefully install the oil pan using 16 retaining bolts. Tighten the bolts, in sequence, to 15–22 ft. lbs. (20–30 Nm).

13. Install the power steering pressure hose retainer bracket onto the engine front cover studs. Install and tighten the retaining nuts to 71–106 inch lbs. (8–12 Nm).

14. Install the low oil sensor wiring on the sensor connector.

15. Install the dual converter Y-pipe to the exhaust manifolds.

16. Lower the vehicle. Install the oil level dipstick.

17. Fill the crankcase with the correct type and quantity of engine oil.

18. Connect the negative battery cable. Start the engine and check for leaks.

19. Turn the air suspension switch to the **ON** position.

Oil Pump

REMOVAL AND INSTALLATION

3.0L (VIN Y) and 3.2L SHO Engines

1. Disconnect the negative battery cable. Raise and safely support the vehicle.

2. Drain the crankcase into a suitable container and remove the oil pan. Remove accessory drive belt.

3. Remove the timing belt from the engine.

4. Remove the crankshaft sprocket.

5. Remove the oil pump pickup tube and screen retaining bolts. Remove the cover and tube.

6. Remove the oil pump to block bolts and remove the pump.

To install:

7. Align the oil pump on the crankshaft and install the oil pump retaining bolts. Tighten the bolts to 11–17 ft. lbs. (15–23 Nm).

8. Install the oil pump pickup tube and screen and tighten the retaining bolts to 72–96 inch lbs. (7–11 Nm).

9. Install the retaining components in the reverse order of removal.

10. Fill the crankcase with the proper type and quantity of oil. Connect the negative battery cable, then start the engine and check for leaks and proper oil pressure.

3.0L (VIN 1 and U) Engines

1. Disconnect the negative battery cable.

2. Raise and safely support the vehicle. Drain the engine oil.

3. Remove the engine oil pan.

4. Remove the oil pump retaining bolt and remove the oil pump and the oil pump intermediate shaft from the engine.

5. If replacing the engine oil pump, separate the intermediate shaft from the oil pump.

To install:

6. If replacing the engine oil pump, insert the oil pump intermediate shaft into the new oil pump assembly until the intermediate shaft retaining ring clicks into place.

7. Prime the new oil pump by filling either the inlet or the outlet port with engine oil. Rotate the pump shaft to distribute the oil within the oil pump body cavity.

8. Insert the oil pump intermediate shaft assembly through the hole in the rear main bearing cap and place the oil pump onto the locating pins.

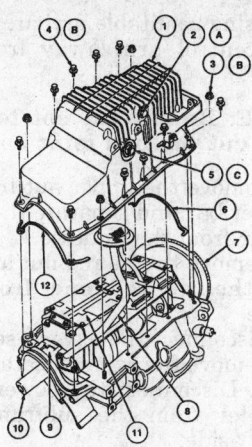

Item	Part Number	Description
1	6675	Oil Pan
2A	6730	Oil Pan Drain Plug
3B	9S702-08500	Nut (4 Req'd)
4B	97522-08525	Bolt (10 Req'd)
5C	6C824	Low Oil Level Sensor
6	6723	Oil Pan Rear Seal
7	6375	Flywheel
8	6622	Oil Pump Screen Cover and Tube
9	6600	Oil Pump
10	6303	Crankshaft
11	6687	Oil Pan Baffle
12	6722	Oil Pan Seal—Front
A		Tighten to 20-33 N·m (15-24 Lb-Ft)
B		Tighten to 15-23 N·m (11-17 Lb-Ft)
C		Tighten to 21-33 N·m (15-24 Lb-Ft)

236843

Oil pan and oil pump pickup tube and screen — SHO engines

9. Install the oil pump retaining bolt and torque the retaining bolt to 30–40 ft. lbs. (40–55 Nm).

10. Install the engine oil pan.

11. Lower the vehicle. Fill the crankcase with the proper type and quantity of engine oil.

12. Connect the negative battery cable. Start engine and check for leaks, proper oil pressure and proper engine operation.

3.0L (VIN S) Engine

1. Disconnect the negative battery cable. Remove the engine from the vehicle.

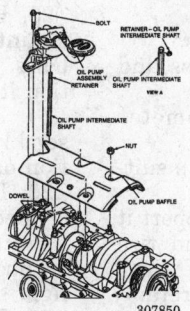

307850

Oil pump installation — 3.0L (VIN 1 and U)

2. Remove the engine oil pan and the engine front cover.

3. Remove the timing chains and the crankshaft sprockets.

4. Remove the oil pump screen cover/tube nut/bolts and the tube from the engine.

5. Remove the 4 oil pump retaining bolts in sequence. Remove the oil pump from the vehicle.

To install:

6. Rotate the inner rotor of the oil pump to align with the flats on the crankshaft. Install the oil pump flush to the cylinder block.

7. Install the oil pump retaining bolts and tighten in sequence to 71–106 inch lbs. (8–12 Nm).

8. Inspect the oil pump screen and tube O–ring and replace if needed.

9. Position the oil pump screen and tube with the O-ring to the oil pump. Tighten the retaining bolts to 71–106 inch lbs. (8–12 Nm).

10. Install a new self-locking tube support nut to the lower cylinder block stud. Torque the nut to 15–22 ft. lbs. (20–30 Nm).

11. Install the crankshaft sprockets and the timing chains.

12. Install the oil pan and the engine front cover.

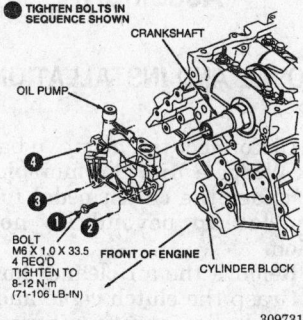

309731

Oil pump retaining bolt torque sequence — 3.0L (VIN S) engine

13. Install the engine in the vehicle. Fill the crankcase with the correct amount of engine oil.

14. Connect the negative battery cable. Run the engine and check for leaks and proper operation.

3.8L Engine

1. Disconnect the negative battery cable.

2. Raise and safely support the vehicle. Drain the oil, then remove the filter.

3. Remove the oil pump cover-to-timing chain front cover bolts and remove the oil pump cover.

4. Remove the oil pump gears.

5. Inspect the gears, oil pump cover and timing chain front cover for wear and/or damage.

To install:

6. If reusing the oil pump cover, clean the gasket contact surface. Place a straight-edge across the oil pump cover mounting surface and check for wear or warpage using a feeler gauge. If the surface is out of flat by more than 0.0016 in. (0.04mm), replace the cover.

7. Lightly pack the gear pocket with petroleum jelly or coat all pump gear surfaces with oil conditioner.

8. Install the gears in the pocket. Make certain the petroleum jelly fills the gap between the gears and the pocket.

—— **WARNING** ——

Failure to properly coat the oil pump gears may result in failure of the pump to prime when the engine is started.

9. Position the oil pump cover gasket and install the oil pump cover. Tighten the oil pump cover retaining bolts to 18–22 ft. lbs. (25–30 Nm).

10. Connect the negative battery cable. Fill the crankcase with the proper type and quantity of engine oil.

11. Start the engine and check for leaks and proper oil pressure.

12. Check the ignition timing and curb idle speed, adjust as required.

13. Install the air cleaner assembly and air intake duct.

4.6L Engine

1. Disconnect the negative battery cable. Remove the engine from vehicle and position on a suitable workstand.

2. Remove the engine front cover. Remove the engine oil pan.

3. Remove the oil pump screen cover/tube bolts and the tube; discard the O-ring.

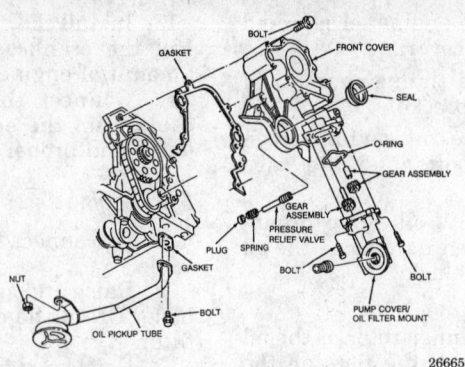

GASKET · FRONT COVER
BOLT
SEAL
O-RING
GEAR ASSEMBLY
GEAR ASSEMBLY · PRESSURE RELIEF VALVE
PLUG · SPRING
NUT
GASKET
BOLT
OIL PICKUP TUBE · BOLT
PUMP COVER/ OIL FILTER MOUNT · BOLT

266653

View of the oil pump and timing chain cover — 3.8L engine

4. Remove the timing chains. Remove the crankshaft sprockets.

5. Remove the oil pump-to-cylinder block bolts and the oil pump.

To install:

6. Thoroughly clean the oil pump and cylinder block mounting surfaces and the oil pump screen cover and tube.

7. Turn the inner rotor of the oil pump to align with the flats on the crankshaft and install the oil pump flush with the cylinder block. Install 4 retaining bolts and tighten to 71–106 inch lbs. (8–12 Nm).

8. Replace the engine oil filter. Install the crankshaft sprockets. Install both timing chains.

9. Position the oil pump screen cover and tube on the oil pump with a new O-ring and hand-start 2 retaining bolts.

10. Install the bolt retaining the oil pump screen cover and tube to the main bearing stud spacer finger-tight.

11. Tighten the oil pump screen cover and tube-to-oil pump bolts to 71–106 inch lbs. (8–12 Nm). Tighten the oil pump screen cover and tube-to-main bearing stud spacer bolt to 15–22 ft. lbs. (20–30 Nm).

12. Install the engine oil pan. Install the engine front cover.

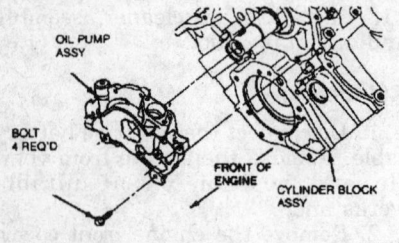

OIL PUMP ASSY
BOLT 4 REQ'D
FRONT OF ENGINE
CYLINDER BLOCK ASSY

307966

Oil pump installation — 4.6L engine

13. Remove the engine from the workstand and install in the vehicle.

14. Restore all fluid levels. Connect the negative battery cable.

15. Run the engine and check for leaks and proper oil pressure. Road test the vehicle and check for proper operation.

TRANSAXLE

Manual Transaxle Assembly

REMOVAL AND INSTALLATION

1. Disconnect the negative battery cable. Wedge a 7 in. (178mm) block of wood under the clutch pedal to hold the pedal up beyond its normal position.

2. Remove the air cleaner hose.

3. Grasp the clutch cable and pull it forward, disconnecting it from the clutch release shaft assembly.

4. Disconnect the clutch cable casing from the rib on top of the transaxle case.

5. Install engine lifting eyes.

6. Tie up the wiring harness and power steering cooler hoses.

7. Disconnect the speedometer cable and speed sensor wire.

8. Support the engine using a suitable engine support fixture.

9. Raise the vehicle and support it safely. Remove the wheel and tire assemblies.

10. Remove the nut and bolt retaining the lower control arm ball joint to the steering knuckle assembly. Discard the removed nut and bolt. Repeat on the opposite side.

11. Using a suitable prybar, pry the lower control arm away from the knuckle.

NOTE: Be careful not to damage or cut the ball joint boot.

12. Remove the upper nut from the stabilizer bar and separate the stabilizer bar from the knuckle.

13. Remove the tie rod nut and separate the tie rod end from the knuckle.

14. Disconnect the oxygen sensor.

15. Remove the exhaust catalyst assembly. Disconnect the power steering cooler from the subframe and place it aside.

16. Disconnect the battery cable bracket from the subframe.

17. Using a suitable prybar, pry the left inboard CV-joint assembly from the transaxle. Install a plug into the seal to prevent fluid leakage. Remove the CV-joint from the transaxle by grasping the left steering knuckle and swinging the knuckle and half-shaft outward from the transaxle. Repeat the procedure on the right side.

NOTE: If the CV-joint assembly cannot be pried from the transaxle, insert a suitable tool through the left side and tap the joint out. The tool can be used from either side of the transaxle.

18. Support the halfshaft assembly with wire in a near level position to prevent damage to the assembly during the remaining operations. Repeat the procedure on the opposite side.

19. Remove the retaining bolts from the center support bearing and remove the right halfshaft from the transaxle.

20. Remove the 2 steering gear retaining nuts from the subframe. Support the steering gear by wiring up the tie rod ends to the coil springs.

21. Remove the transaxle to engine retaining bolts. Disconnect the 2 shift rods from the transaxle.

22. Remove the engine mount bolts.

23. Position jacks under the body mount positions and remove the 4 bolts, lower the subframe and position it aside.

24. Remove the starter motor assembly. Remove the left engine vibration dampener lower bracket.

25. Remove the backup light switch connector from the transaxle backup light switch, located on top of the transaxle and remove the backup light switch.

26. Position a suitable support jack under the transaxle. Lower the transaxle, remove it from the engine and lower it from the vehicle.

To install:

27. Raise the transaxle into position. Engage the input shaft spline into the clutch disc and work the transaxle onto the dowel sleeves. Make sure the transaxle assembly is flush with the rear face of the engine before installation of the retaining bolts.

28. Install the engine to transaxle retaining bolts. Tighten to 28–31 ft. lbs. (38–42 Nm).

29. Install the remaining components in the reverse order of removal. Torque the following:

Starter bolts to 30–40 ft. lbs. (41–54 Nm)

Subframe bolts to 65–85 ft. lbs. (90–115 Nm)

Engine mount bolts to 40–55 ft. lbs. (54–75 Nm)

Stabilizer bolt to 35–46 ft. lbs. (47–63 Nm)

Shift rod clamp bolt and nut to 80–106 inch lbs. (9–12 Nm)

Engine-to-transaxle bolts to 28–31 ft. lbs. (38–42 Nm)

Steering gear nuts to 85–100 ft. lbs. (115–135 Nm)

Center support bearing bolts to 85–100 ft. lbs. (115–135 Nm)

Exhaust catalyst retaining bolts to 25–34 ft. lbs. (34–47 Nm)

Tie rod retaining nut to 35–47 ft. lbs. (47–64 Nm)

Lower control arm ball joint-to-steering knuckle nut and bolt to 37–44 ft. lbs. (50–60 Nm)

30. Check the transaxle fluid level.

31. Lower the vehicle.

32. Remove the engine support tool.

33. Install the speedometer cable. Connect the speedometer cable and speed sensor wire.

34. Remove the engine lifting eyes.

35. Connect the clutch cable to the transaxle. Install the air cleaner hose and remove the wood block from the clutch pedal.

36. Connect the negative battery cable. Road test and check transaxle operation. Check the transaxle for fluid leaks.

Clutch Assembly

REMOVAL AND INSTALLATION

1. Disconnect the negative battery cable. Raise the vehicle and support it safely.

2. Remove the transaxle assembly.

3. If the pressure plate is to be reused, mark its location on the flywheel so it can be reinstalled in the same position.

4. Loosen the pressure plate bolts one turn at a time in a criss-cross pattern, until spring tension is relieved, to prevent pressure plate distortion. Support the pressure plate and remove the bolts. Remove the pressure plate and clutch disc from the flywheel.

5. Inspect the flywheel, clutch disc, pressure plate, throwout bearing and the clutch fork for wear. Replace parts as required. If the flywheel shows any signs of overheating (blue discoloration) or if it is badly grooved or scored, it should be refaced or replaced.

To install:

6. Install the flywheel, if removed. Tighten the attaching bolts to 54–64 ft. lbs. (73–87 Nm) on all except the 3.0L SHO and 3.2L SHO engine. On the 3.0L SHO and 3.2L SHO engine, tighten the bolts to 51–58 ft. lbs. (69–78 Nm).

7. Clean the pressure plate and flywheel surfaces thoroughly. Place the clutch disc and pressure plate into the installed position. Align the marks made during the removal procedure if components are being reused. Support the clutch disc and pressure plate with a suitable dummy shaft or clutch aligning tool.

8. Install the pressure plate-to-flywheel bolts. Tighten them gradually in a criss-cross pattern to 12–24 ft. lbs. (17–32 Nm). Remove the alignment tool.

9. Lubricate the release bearing and install it in the fork.

10. Install the transaxle and lower the vehicle.

11. Connect the negative battery cable. Road test the vehicle and check for proper clutch and transaxle operation.

Clutch Cable

ADJUSTMENT

The clutch pedal mechanism is self-adjusting. No adjustment is necessary.

Automatic Transaxle Assembly

REMOVAL AND INSTALLATION

1993–95 Taurus and Sable

1. Disconnect the battery cables, negative cable first.

2. Remove the battery and battery tray. Remove the engine air cleaner assembly.

3. Disconnect the electrical connectors from the engine and remove the bolt retaining the main wiring harness bracket.

4. Remove the shift lever retaining nut and the shift lever from the manual control lever shaft on the Transmission Range (TR) sensor.

5. Remove the EGR bracket and throttle body bracket retaining bolts and install engine lifting eyes.

6. Remove radiator sight shield.

7. Secure the wiring harness aside and remove the radiator sight shield.

8. Secure the engine with the necessary lifting brackets to Three Bar Engine Support D88L-6000-A, or equivalent.

9. Remove the dipstick and disconnect the power steering line bracket.

10. Remove the upper transaxle-to-engine retaining bolts from the top of the transaxle.

11. Raise and safely support the vehicle. Remove the front wheel and tire assemblies.

12. Loosen the transaxle oil pan and drain the transaxle fluid into a suitable container.

13. Remove the cotter pins and nuts from both outer tie rod ends. Separate the tie rod ends from the steering knuckles. Discard the cotter pins and nuts.

14. Remove the brake lines from the support brackets.

15. Remove the retaining bolts from the front stabilizer bar assembly. Disconnect both link assemblies.

16. Remove 2 steering gear retaining nuts from the subframe.

17. Disconnect the Heated Oxygen Sensor (HO$_2$S) electrical connectors.

18. Remove the exhaust pipe, converter assembly and mounting bracket.

19. Remove 2 bolts from the transaxle mount and 4 bolts from the left engine support. Remove the engine support.

20. Position Sub-frame Removal tool 014-00751, or equivalent with a suitable transaxle jack.

21. Raise the steering gear from the subframe and secure to the rear of the engine compartment with wire.

22. Remove 4 subframe-to-body retaining bolts and lower the subframe.

23. Remove the dust cover.

24. Remove 2 starter motor retaining bolts and position the starter motor aside.

25. Rotate the engine at the crankshaft pulley to align the torque converter bolts with the starter drive

hole. Remove 4 torque converter-to-flywheel retaining nuts.

26. Remove the transaxle cooler line retaining clips. Disconnect and plug the transaxle cooler lines.

27. Remove the engine-to-transaxle retaining bolts.

28. Remove the Vehicle Speed Sensor (VSS) heat shield. Remove the VSS from the transaxle.

NOTE: Vehicles with electronic instrument clusters do not use a speedometer cable.

29. Position a suitable transaxle jack to the transaxle and secure with a strap or chain.

30. Remove the halfshafts from transaxle and install 2 transaxle shipping plugs.

31. Remove the remaining transaxle housing bolts. Carefully separate the transaxle from the engine and lower from the vehicle.

To install:

NOTE: Verify the transaxle cooler lines are thoroughly cleaned before installing the transaxle assembly.

32. If removed, place the transaxle on a suitable transaxle jack and secure with a strap or chain.

33. Slowly raise and position the transaxle assembly while aligning the torque converter studs to the appropriate holes in the flywheel and the transaxle housing to the dowel pins on the cylinder block.

34. Install 2 lower cylinder block-to-transaxle bolts, alternately tighten the bolts while making sure transaxle pulls in flush with cylinder block. Tighten the bolts to 41–50 ft. lbs. (55–68 Nm).

35. Install 4 flywheel-to-torque converter nuts. Tighten to 23–39 ft. lbs. (31–53 Nm).

36. Remove the transaxle jack.

37. Install the remaining components in the reverse order of removal. Torque the following:

VSS retaining bolt to 31–39 inch lbs. (3.4–4.5 Nm)

Starter motor bolts to 30–40 ft. lbs. (41–54 Nm)

Sub-frame-to-body bolts (make sure the insulators are properly installed) to 55–75 ft. lbs. (75–102 Nm)

Left engine support bolts to 40–55 ft. lbs. (54–75 Nm)

Engine mount-to-subframe retaining bolts to 60–85 ft. lbs. (81–116 Nm)

Engine mount-to-engine support nut to 55–75 ft. lbs. (74–102 Nm)

Steering gear-to-subframe nuts to 85–100 ft. lbs. (115–135 Nm)

Lower control arm-to-steering knuckle to 40–53 ft. lbs. (53–72 Nm), using a new pinch bolt and nut

Stabilizer bar links-to-stabilizer bar nuts to 30–40 ft. lbs. (40–55 Nm)

Tie rod end-to-steering knuckle nuts to 23–35 ft. lbs. (31–37 Nm), using new castellated nuts and new cotter pins

Remaining transaxle housing bolts, if not already installed, to 41–50 ft. lbs. (55–68 Nm)

Shift lever-to-manual control lever shaft at the TR sensor nut to 12–16 ft. lbs. (16–22 Nm)

38. Connect both battery cables, negative cable last.

NOTE: Whenever the vehicles subframe is removed or lowered, the wheel alignment should be checked.

39. Check the front end alignment.

40. Start the engine and check for leaks. Add transaxle fluid as needed.

41. Road test the vehicle and check for proper operation.

1996–97 Taurus and Sable

1. Disconnect both battery cables, negative cable first.

2. Remove the battery and battery tray. Remove the engine air cleaner assembly.

3. Detach the transaxle harness and the Transmission Range (TR) sensor connectors.

4. Remove the shift cable actuator fitting (cable retaining clip) and 1 retaining nut and disconnect the shift cable from the shift cable bracket on the transaxle.

5. Disconnect the transaxle cooler lines.

6. For the 3.0L OHV engine, remove 4 upper transaxle-to-engine retaining bolts and 1 transaxle-to-engine stud.

7. For the 3.0L DOHC engine, remove 5 transaxle-to-engine retaining bolts.

8. Install 2 engine lifting brackets on the engine assembly.

9. Install and secure the engine using Three Bar Engine Support D88L-6000-A, or equivalent.

10. Raise and safely support the vehicle.

11. Loosen the transaxle oil pan retaining bolts and drain the transaxle fluid into a suitable container.

12. Remove both front wheel and tire assemblies.

13. Remove both halfshafts.

14. Detach 4 Heated Oxygen Sensor (HO$_2$S) electrical connectors.

15. Remove 3 bolts and 7 nuts securing the converter Y-pipe assembly and remove from the vehicle.

16. Detach the 2 starter motor connectors, then remove the starter.

17. Remove 1 bolt and 1 stud securing the starter motor and remove the starter motor from the transaxle.

18. For the OHV engine, remove 1 bolt and the transaxle housing cover.

19. Support the rack and pinion assembly using wire attached to the strut and spring assembly to hold it in position. Remove 2 rack and pinion assembly retaining nuts from the subframe.

20. Remove 2 lower control arm-to-ball joint retaining nuts and separate the lower control arms from the steering knuckles and ball joints.

21. Remove the retaining nuts from the front engine support insulators (mounts) at the subframe.

22. Remove the sway bar (stabilizer bar) link retaining nuts at each end of the sway bar and separate the links from the sway bar.

23. Remove the engine and transaxle support insulator through-bolts from the subframe.

24. Place High Lift Transmission Jack 014-00210 or equivalent, using a suitable subframe adapter under the subframe and support the subframe.

25. Remove 4 subframe-to-body retaining bolts. Carefully lower the subframe and set aside.

26. Place High Lift Transmission Jack 014-00210 or equivalent, using Adapter 014-00461 or equivalent, under the transaxle and support the transaxle assembly. Secure the transaxle to the transaxle adapter using a strap or chain.

27. For the OHV engine, remove 1 lower engine-to-transaxle bolt.

28. For the DOHC engine, remove 4 lower engine-to-transaxle bolts.

29. Remove 4 flywheel-to-torque converter nuts.

30. Remove 3 bolts and 2 nuts securing the rear engine support to the transaxle and remove the rear engine support.

31. Remove 1 bolt from the right engine mount brace, then slowly lower the transaxle from the vehicle.

To install:

NOTE: Flush the transaxle cooler lines thoroughly before installing the transaxle assembly.

32. If removed, place the transaxle assembly on High Lift Transmission Jack 014-00210 or equivalent, using Adapter 014-00461 or equivalent. Secure the transaxle to the transaxle adapter using a strap or chain.

33. Slowly raise the transaxle assembly into place. Align the torque converter studs with the appropriate holes in the flywheel and engage the transaxle housing to the engine dowel pins.

34. Install 1 bolt in the right engine mount brace and tighten to 39–53 ft. lbs. (53–72 Nm).

35. Install the rear engine support to the transaxle and install 3 bolts and 2 nuts. Tighten the bolts and nuts to 39–53 ft. lbs. (53–72 Nm).

36. Install 4 flywheel-to-torque converter nuts and tighten to 20–34 ft. lbs. (27–46 Nm).

37. For the OHV engine, install 1 lower transaxle-to-engine bolt and tighten to 39–53 ft. lbs. (53–72 Nm).

38. For the DOHC engine, install 4 lower transaxle-to-engine bolts and tighten to 39–53 ft. lbs. (53–72 Nm).

39. Place the subframe on High Lift Transmission Jack 014-00210 or equivalent, using a suitable subframe adapter and raise the subframe into position.

40. Install the subframe insulators, if removed and loosely install 4 subframe-to-body bolts.

41. Install a ¾ inch (19mm) outside diameter pipe or similar tool into the front left subframe and body alignment holes and align the holes. Slightly tighten the front left subframe-to-body bolt.

42. Repeat the subframe alignment procedure on the front right subframe and body alignment holes. Slightly tighten the right subframe-to-body bolt.

43. Check the left alignment holes again and adjust if necessary.

44. After the subframe alignment is complete, tighten all 4 subframe-to-body bolts to 57–76 ft. lbs. (77–103 Nm).

45. Install the remaining components in the reverse order of removal. Torque the following:

Engine and transaxle support insulator-to-subframe bolts to 65–87 ft. lbs. (88–118 Nm).

Sway bar link-to-sway bushings and nuts to 35–46 ft. lbs. (47–63 Nm).

Rack and pinion-to-subframe nuts to 84–113 ft. lbs. (113–133 Nm)

Front engine support insulator-to-subframe nuts to 57–76 ft. lbs. (77–103 Nm)

Ball joint-to-lower control arm nuts to 51–67 ft. lbs. (68–92 Nm), using new nuts

For the OHV engine: transaxle housing cover bolt to 80–106 inch lbs. (9–12 Nm)

Starter motor bolt and stud to 15–21 ft. lbs. (21–29 Nm)

Converter Y-pipe 3 bolts and 7 nuts to 26–34 ft. lbs. (34–46 Nm)

For the OHV engine: 4 upper transaxle-to-engine bolts and 1 upper transaxle-to-engine stud to 39–53 ft. lbs. (53–72 Nm)

For the DOHC engine: 5 transaxle-to-engine bolts to 39–53 ft. lbs. (53–72 Nm)

Shift actuator cable fitting nut to 14–19 ft. lbs. (19–26 Nm)

46. Connect both battery cables, negative cable last.

47. If the transaxle is empty of transaxle fluid, add several quarts of MERCON or equivalent transaxle fluid to the transaxle.

48. Start the engine and continue to fill the transaxle until the correct level is reached. Check for leaks and proper operation.

NOTE: Whenever the vehicles subframe is removed or lowered, the wheel alignment should be checked.

49. Check the front end alignment.
50. Road test the vehicle and check the transaxle for proper operation.

1993–94 Continental (AX4S Transaxle)

1. Disconnect the battery cables and remove the battery and battery tray.

2. Remove the air cleaner assembly, hoses and tubes.

3. Disconnect the electrical connectors from the engine and remove the bolt retaining the main wiring harness bracket.

4. Remove the shift lever. Remove the EGR bracket and throttle body bracket retaining bolts and install engine lifting eyes.

5. Secure the wiring harness aside and remove the radiator sight shield. Position a suitable engine support fixture.

6. Turn the air suspension switch located in the luggage compartment to the **OFF** position.

7. Remove the dipstick and disconnect the power steering line bracket. Remove the 4 torque converter housing bolts from the top of the transaxle.

8. Raise and safely support the vehicle. Remove the front wheel and tire assemblies.

9. Disconnect the left outer tie rod end. Remove the suspension height sensor, if equipped. Disconnect the brake line support brackets.

10. Remove the retaining bolts from the front stabilizer bar assembly. Disconnect both lower arm assemblies.

11. Remove the steering gear retaining nuts from the subframe. Remove the front oxygen sensor, exhaust pipe, converter assembly and mounting bracket.

12. Remove 2 bolts from the transaxle mount and the 4 bolts from the left engine support. Remove the engine support.

13. Position a suitable subframe removal tool. Remove the steering gear from the subframe and secure to the rear of the engine compartment. Remove the subframe-to-body bolts and lower the subframe.

14. Remove the starter and the dust cover.

15. Rotate the engine at the crankshaft pulley to align the torque converter bolts with the starter drive hole. Remove the 4 torque converter-to-flywheel retaining nuts.

16. Disconnect the transaxle cooler lines. Remove the engine-to-transaxle retaining bolts.

17. Remove the speedometer sensor heat shield. Remove the vehicle speed sensor from the transaxle.

NOTE: Vehicles with electronic instrument clusters do not use a speedometer cable.

18. Position a suitable transaxle jack. Remove the halfshafts.

19. Remove the last 2 torque converter housing bolts, carefully separate the transaxle from the engine and lower it out of the vehicle.

To install:

20. Installation is the reverse of the removal procedure. During installation be sure to observe the following:

   a. Clean the transaxle oil cooler lines.

   b. Install new circlips on the CV-joint seals.

   c. Carefully install the halfshafts in the transaxle by aligning the splines of the CV-joint with the splines of the differential.

   d. Attach the lower ball joint to the steering knuckle with a new nut and bolt. Tighten the nut to 37–44 ft. lbs.

   e. When installing the transaxle to the engine, verify that the converter-to-transaxle engagement is maintained. Prevent the converter from moving forward and disengaging during installation.

   f. Adjust the TV and manual linkages. Check the transaxle fluid level.

21. Tighten the following bolts to the torque specifications listed:

   a. Transaxle-to-engine bolts: 41–50 ft. lbs. (55–68 Nm)

   b. Control arm-to-knuckle bolts: 36–44 ft. lbs. (50–60 Nm)

c. Stabilizer U-clamp-to-bracket bolts: 60–70 ft. lbs. (81–95 Nm)

d. Tie rod-to-knuckle nut: 23–35 ft. lbs. (31–47 Nm)

e. Starter-to-transaxle bolts: 30–40 ft. lbs. (41–54 Nm)

f. Converter-to-flywheel bolts: 23–39 ft. lbs. (31–53 Nm)

g. Insulator-to-bracket bolts: 55–70 ft. lbs. (75–90 Nm)

1995–97 Continental

CAUTION

Do not remove an air spring under any circumstances when pressurized. Do not remove any components supporting an air spring without exhausting the air. Failure to follow these instructions may result in unexpected inflation or deflation of the air springs which may result in shifting of the vehicle during service.

NOTE: Make sure to have available new halfshaft circlips and wheel hub retainer nuts.

1. Properly deflate the air springs.
2. Disconnect both battery cables, negative cable first. Remove the battery and battery tray.
3. Remove the hoses and electrical connectors from the engine air cleaner, and remove the air cleaner assembly.
4. Remove 2 screws and the accelerator control splash shield from the throttle body.
5. Disconnect the transaxle harness electrical connector and the Transmission Range (TR) sensor electrical connectors.
6. Remove the nut from shift lever, remove the clip from the cable bracket and lift the cable from the bracket.
7. Disconnect the transaxle oil cooler lines from the transaxle.
8. Remove 4 transaxle housing-to-engine bolts from the top of the transaxle.
9. Install 2 engine lifting eyes to the front and rear locations of the engine. Install 3 Bar Engine Support Bracket Set D88L-6000-A or equivalent, and suitably support the engine.
10. Raise and safely support the vehicle. Remove the front wheel and tire assemblies.
11. Loosen the transaxle oil pan and drain the fluid into a suitable container. When most of the fluid has drained, install the pan bolts.
12. Disconnect the height sensors.

13. Remove the cotter pins and castellated nuts from both tie rod ends. Separate the tie rod ends from the steering knuckles. Discard the cotter pins.
14. Disconnect the sway bar links from the sway bar.
15. Separate the lower control arms from the steering knuckles.
16. Remove the wheel hub retainer nuts and remove both halfshafts from the vehicle. Discard the nuts.
17. Disconnect the Heated Oxygen Sensor (HO$_2$S) electrical connectors.
18. Remove the dual converter Y-pipe assembly from the vehicle.
19. Remove the starter motor and the dust cover.
20. Support the subframe using Powertrain Lift 014-00765 and Removal Bracket 014-00766, or equivalents.
21. Remove 4 subframe-to-body bolts and lower the subframe.
22. Remove the transaxle housing cover from the transaxle.
23. Remove 4 torque converter-to-flywheel retaining nuts.
24. Remove 5 bolts retaining the rear engine support and remove the support from the transaxle.
25. Support the transaxle with High Lift Transmission Jack 014-00210 and Adapter 014-00461, or equivalents.
26. Remove 2 lower engine-to-transaxle bolts.
27. Remove 2 bolts and 1 nut from the right side of engine mount to transaxle case.
28. Carefully separate the transaxle from the engine and slowly lower the transaxle out of the vehicle.

To install:

NOTE: Verify the transaxle cooler lines are thoroughly cleaned before installing the transaxle assembly.

29. If removed, place the transaxle on High Lift Transmission Jack 014-00210 and Adapter 014-00461, or equivalents.
30. Slowly raise the transaxle assembly into position until the transaxle housing engages the engine dowel pins. Align the torque converter studs to the flywheel.
31. Install 2 lower engine-to-transaxle bolts. Alternately tighten the bolts while making sure the transaxle pulls in flush with the engine. Tighten the bolts to 80–106 inch lbs. (9–12 Nm).

NOTE: When installing the transaxle to the engine, verify that the converter-to-transaxle engagement is maintained. Pre-

vent the converter from moving forward and disengaging during installation.

32. Install 4 flywheel-to-torque converter nuts. Tighten to 20–34 ft. lbs. (27–46 Nm).
33. Mount the rear engine/transaxle support to the transaxle. Tighten 5 retaining bolts to 44–60 ft. lbs. (60–80 Nm).
34. Install the transaxle housing cover and securely tighten the bolts.
35. Install 2 bolts and 1 nut to the right side of the engine mount to transaxle case.
36. Install the starter motor and tighten the bolt and stud to 15–21 ft. lbs. (21–29 Nm).
37. Connect the electrical connectors to the starter motor.
38. If removed, place the subframe on Powertrain Lift 014-00765 and Removal Bracket 014-00766, or equivalents.
39. Raise the subframe into position.
40. Loosely install 4 subframe to body bolts.
41. Align the subframe using a 0.75 inch (19mm) outside diameter pipe or similar tool installed through both alignment holes. Tighten the bolts to 55–75 ft. lbs. (75–102 Nm).
42. Remove the powertrain lift.
43. Install the remaining components in the reverse order of removal. Torque the following:

Ball joint stud retaining nuts to 50–68 ft. lbs. (68–92 Nm)

Sway bar link-to-sway bar nuts to 30–40 ft. lbs. (40–55 Nm)

Tie rod end-to-steering knuckle nuts to 35–46 ft. lbs. (47–63 Nm)

Manual control lever nut to 12–16 ft. lbs. (16–22 Nm)

44. Install the battery tray and the battery. Connect both battery cables, negative cable last.
45. Refill the air springs. Install the engine appearance cover. Tighten the retaining nuts to 61–79 inch lbs. (7–9 Nm).
46. Fill the transaxle with the proper type and quantity of fluid. Check the front end alignment.
47. Check the transaxle for leaks and proper shift selector functioning. Road test the vehicle and check for proper operation.

Throttle Valve Cable

ADJUSTMENT

1. Position the gear selector lever in the **OD** position. If equipped with floor shift, the shift lever must be

held in the rearward position using a constant force of 3 lbs. (1.4 Kg) while the linkage is being adjusted. If equipped with a column shift, a 3 lb. (1.4 Kg) weight should be hung on the shift lever to make sure the lever is firmly located in the overdrive detent.

2. Loosen the manual lever-to-control cable retaining nut.

3. Move the transaxle manual lever to the **OD** position, second detent from the most rearward position.

4. Tighten the retaining nut to 12–19 ft. lbs. (16–27 Nm).

5. Check the operation of the transaxle in each selector lever position. Make sure the park and neutral start switch are functioning properly.

DRIVE AXLE

Halfshaft

REMOVAL AND INSTALLATION

NOTE: Due to the 3-speed automatic transaxle case configuration, the right halfshaft assembly must be removed first. Differential rotator tool T81P-4026-A or equivalent, is then inserted into the transaxle to drive the left inboard CV-joint assembly from the transaxle. If only the left halfshaft assembly is to be removed for service, remove the right halfshaft assembly from the transaxle first. After removal, support it with a length of wire, then drive the left halfshaft assembly from the transaxle.

NOTE: Do NOT begin this procedure unless the following parts are available:

- New hub retainer nut
- New lower control arm-to-steering knuckle bolt and nut
- New stub shaft/link shaft circlip

Once removed, these parts must not be reused during assembly. Their torque holding ability or retention capability is diminished during removal.

Except 1995–97 Continental

───────── WARNING ─────────
When removing both halfshafts on vehicles equipped with manual transaxle or 3-speed automatic transaxle, install transaxle plug tools T81P-1177-B or equivalent, to prevent disloca-

tion of the differential side gears. Should the gears become misaligned, the differential will have to be removed from the transaxle to re-align the side gears.

1. Remove the wheel cover/hub cover from the wheel and tire assembly. Loosen the hub retainer nut and the lug nuts.

2. Raise the vehicle and support safely.

3. Remove the wheel and tire assembly. Remove the hub retainer nut and washer and discard the hub nut.

4. Remove the nut from the ball joint-to-steering knuckle attaching bolts. Drive the bolt out of the steering knuckle using a punch and hammer. Discard the bolt and nut.

5. If equipped with ABS, remove the anti-lock brake sensor and position aside. If equipped with air suspension, remove the height sensor bracket retaining bolt and wire sensor bracket to inner fender. Position the sensor link aside.

6. Separate the ball joint from the steering knuckle using a suitable prybar. Position the end of the prybar outside of the bushing pocket to avoid damage to the bushing. Use care to prevent damage to the ball joint boot. Remove the stabilizer bar link at the stabilizer bar.

NOTE: The remaining removal procedures differ according to transaxle application: manual transaxle, 4-speed automatic overdrive (AXOD or AXODE) transaxle or 3-speed automatic (ATX/FLC) transaxle.

7. If equipped with AXOD or AXODE transaxle and removing the right or left halfshaft, or if equipped with manual transaxle and removing the left halfshaft, proceed as follows:

a. Install CV-joint puller tool T86P-3514-A1 or equivalent, between the CV-joint and transaxle case. Turn the steering hub and/or wire the strut assembly aside.

b. Screw extension tool T86P-3514-A2 or equivalent, into the CV-joint puller and hand tighten. Screw an impact slide hammer onto the extension and remove the CV-joint from the transaxle.

c. Support the end of the shaft by suspending it from a convenient underbody component with a piece of wire. Do not allow the shaft to hang unsupported; damage to the outboard CV-joint may occur.

───────── WARNING ─────────
Never use a hammer to separate the outer CV-joint stub shaft from the hub. Damage to the CV-joint threads and internal components may result.

d. Separate the outboard CV-joint from the hub using front hub remover tool T81P-1104-C or equivalent, metric adapter tools T83-P-1104-BH, T86P-1104-Al and front hub installer T81P-1104-A or equivalent.

e. Remove the halfshaft assembly from the vehicle.

8. If equipped with ATX/FLC or manual transaxle and removing the right halfshaft, proceed as follows:

a. Remove the bolts attaching the bearing support to the bracket. Slide the link shaft out of the transaxle. Support the end of the shaft by suspending it from a convenient underbody component with a piece of wire. Do not allow the shaft to hang unsupported, damage to the outboard CV-joint may occur.

b. Separate the outboard CV-joint from the hub using front hub remover tool T81P-1104-C or equivalent, metric adapter tools T83-P-1104-BH, T86P-1104-Al and front hub installer T81P-1104-A or equivalent.

───────── WARNING ─────────
Never use a hammer to separate the outboard CV-joint stub shaft from the hub. Damage to the CV-joint threads and internal components may result. The right side link shaft and halfshaft assembly is removed as a complete unit.

9. If equipped with ATX/FLC transaxle and removing the left halfshaft, proceed as follows:

a. Support the end of the shaft by suspending it from a convenient underbody component with a piece of wire. Do not allow the shaft to hang unsupported as damage to the outboard CV-joint may occur.

b. Separate the outboard CV-joint from the hub front hub remover tool T81P-1104-C or equivalent, metric- adapter tools T83-P-1104-BH, T86P-1104-Al and front hub installer T81P-1104-A or equivalent.

c. Remove the halfshaft assembly from the vehicle.

To install:

10. Install a new circlip on the inboard CV-joint stub shaft and/or link shaft. The outboard CV-joint does not have a circlip. When installing the

circlip, start one end in the groove and work the circlip over the stub shaft end into the groove. This will avoid over expanding the circlip.

NOTE: The circlip must not be re-used. A new circlip must be installed each time the inboard CV-joint is installed into the transaxle differential.

11. Carefully align the splines of the inboard CV-joint stub shaft with the splines in the differential. Exerting some force, push the CV-joint into the differential until the circlip is felt to seat in the differential side gear. Use care to prevent damage to the differential oil seal. If equipped, torque the link shaft bearing retaining bolts to 16–23 ft. lbs. (21–32 Nm).

NOTE: A non-metallic mallet may be used to aid in seating the circlip into the differential side gear groove. If a mallet is necessary, tap only on the outboard CV-joint stub shaft.

12. Carefully align the splines of the outboard CV-joint stub shaft with the splines in the hub and push the shaft into the hub as far as possible.

13. Temporarily fasten the rotor to the hub with washers and 2 wheel lug nuts. Insert a steel rod into the rotor and rotate clockwise to contact the knuckle to prevent the rotor from turning during the CV-joint installation.

14. Install the hub nut washer and a new hub nut. Manually thread the retainer onto the CV-joint as far as possible.

15. Install the remaining components in the reverse order of removal. Torque the following:

Control arm-to-steering knuckle nut and bolt: 40–55 ft. lbs. (54–74 Nm) for 1993–95 vehicles or to 50–68 ft. lbs. (68–92 Nm) for 1996–97 vehicles, using a new nut and bolt

Stabilizer link-to-stabilizer bar/front strut to 35–48 ft. lbs. (47–65 Nm) for 1993–95 vehicles or to 57–75 ft. lbs. (77–103 Nm) for 1996–97 vehicles

Hub retainer nut to 180–200 ft. lbs. (245–270 Nm)

Lug nuts to 85–105 ft. lbs. (115–142 Nm)

16. Fill the transaxle to the proper level with the specified fluid.

1995–97 Continental

NOTE: Make sure to have available a New Generation Star (NGS) tester or equivalent scan tool with the proper software to

properly deflate the air springs before servicing the suspension system.

---——— CAUTION ————---
Do not remove an air spring under any circumstances when pressurized. Do not remove any components supporting an air spring without exhausting the air. Failure to follow these instructions may result in unexpected inflation or deflation of the air springs which may result in shifting of the vehicle during service.

1. Place the vehicle over a frame contact hoist.
2. Properly deflate the air springs.
3. Turn the ignition switch to the **OFF** position. Position the steering wheel in the UNLOCKED position.
4. Disconnect the negative battery cable.
5. Loosen and remove the wheel hub retainer nut. Discard the nut.
6. Raise and safely support the vehicle. Remove the wheel and tire assembly.
7. Remove the brake anti-lock sensor mounting bolt and the anti-lock sensor from the steering knuckle. Move the sensor aside.
8. Remove the air suspension height sensor at the lower control arm ball stud attachment. Move the sensor aside.
9. Remove the nuts from the sway bar link at the strut. Remove the sway bar link from the strut and move aside.
10. Using Hub Remover/Replacer T81P-1104-C, Stud Adapter T86P-1104-A1, Hub Remover Adapters T83P-1104-BH and Hub Replacer T81P-1104-A or equivalents, push the CV-joint only to the point that the splines are free in the hub.
11. Remove the lower ball joint-to-lower control arm nut. Using Puller T64P-3590-F or equivalent, separate the lower ball joint from the lower control arm. Discard the nut.
12. Pull down on the lower control arm until the lower ball joint clears the lower control arm.

NOTE: Do not allow the half-shaft to hang unsupported. Damage to the CV-joint or boots may result.

13. Push the outboard CV-joint stub shaft in while placing a wooden dowel between the frame pocket and steering knuckle. Remove the outer CV-joint from the hub.

14. Assemble CV-Joint Puller Adapter T89P-3415-B, Puller Extension T86P-3514-A2 or equivalents, and a slide hammer. Insert the puller adapter behind the inner CV-joint housing at the 4 o'clock position on the left side or at the 5 o'clock position on the right side.
15. Use the slide hammer to release the halfshaft bearing retainer circlip.
16. Carefully remove halfshaft and tools from the vehicle. Discard the circlip.

To install:
17. Install a new circlip on the inner CV-joint stub shaft. When installing the circlip, start one end in the groove and work the circlip over the stub shaft end into the groove. This will avoid over expanding the circlip.
18. Carefully align the splines of the inner CV-joint stub shaft with the splines in the transaxle. Exerting some force, push the CV-joint into the transaxle until the circlip is felt to seat in the differential side gear. Use care to prevent damage to the oil seal.

NOTE: A non-metallic (plastic or rawhide) mallet may be used to aid in seating the halfshaft bearing retainer circlip into the differential side gear groove. If a mallet is necessary, tap only on the outer CV-joint.

19. Remove the wooden dowel and carefully align the splines of the outer CV-joint with the splines in the wheel hub.
20. Carefully push the shaft into the wheel hub as far as possible.
21. Install a new wheel hub retainer nut. Manually thread the hub retainer nut onto the outer CV-joint as far as possible.
22. Install the remaining components in the reverse order of removal:

Lower ball joint-to-lower control arm, using a new retaining nut, to 39–53 ft. lbs. (53–72 Nm)

Sway bar link-to-strut nut to 57–75 ft. lbs. (77–107 Nm)

Anti-lock sensor bolt to 40–60 inch lbs. (4.5–6.8 Nm)

Hub retainer to 180–200 ft. lbs. (245–270 Nm)

Lug nuts to 80–105 ft. lbs. (108–144 Nm)

23. Lower the vehicle. Connect the negative battery cable. Refill the air spring.
24. Check the transaxle fluid level and add, as necessary. Road test the vehicle and check for proper operation.

CV-Joint Boot

REPLACEMENT

Outboard CV-Joint Boot

Except 1995-97 Continental

1. Disconnect the negative battery cable. Raise and safely support the vehicle, then remove the halfshaft assembly from the vehicle.
2. Clamp the halfshaft in a vise that is equipped with soft jaw covers. Do not allow the vise jaws to contact the boot or boot clamp.
3. Cut the large boot clamp with a pair of side cutters and peel the clamp away from the boot. Roll the boot back over the shaft after the clamp has been removed.
4. Clamp the interconnecting shaft in a soft jawed vise and angle the CV-joint so the inner bearing race is exposed.
5. Using a brass drift and hammer, give a sharp tap to the inner bearing race to dislodge the internal snapring and separate the CV-joint from the interconnecting shaft. Take care to secure the CV-joint so it does not drop after separation. Remove the clamp and boot from the shaft.
6. Remove and discard the circlip at the end of the interconnecting shaft. Remove the stop ring, located just below the circlip. The stop ring should be replaced only if damaged or worn.

NOTE: The left and right interconnecting shafts are different. The outboard end of the shaft is shorter from the end of the shaft to the end of the boot groove than the inboard end. Take a measurement to insure correct installation.

7. Install the new boot. Make sure the boot is seated in the mounting groove and secure it in position with a new clamp. Tighten the clamp securely using suitable boot clamp pliers, but not to the point where the clamp bridge is cut or the boot is damaged.
8. Clean the interconnecting shaft splines and install a new circlip and stop ring. To install the circlip correctly, start one end in the groove and work the circlip over the shaft end and into the groove.
9. Pack the CV-joint with grease. Any grease remaining in the tube is to be spread evenly inside the CV-joint boot. Total amount of grease required is 6 oz. (170g) for all except 3.0L without ABS. For 3.0L without

ABS, the total amount of grease required is 5.1 oz. (145g).

10. With the boot peeled back, position the CV-joint on the shaft and tap into position using a plastic tipped hammer. The CV-joint is fully seated when the circlip locks into the groove cut into the CV-joint inner bearing race. Check for seating by attempting to pull the joint away from the shaft.
11. Remove all excess grease form the CV-joint external surface and position the boot over the joint.
12. Before installing the boot clamp, make sure all air pressure that may have built up in the boot is removed. Pry up on the boot lip to allow the air to escape.
13. The large end clamp should be installed after making sure the boot is seated in its groove. Tighten the clamp securely, but not to the point where the clamp bridge is cut or the boot is damaged.
14. Install the halfshaft assembly and lower the vehicle. Connect the negative battery cable.

1995-97 Continental

NOTE: The inner CV-joint is permanently retained to the shaft. The outer CV-joint must be removed before removing the inner CV-joint boot.

1. Turn the air suspension switch to the **OFF** position.
2. Raise and safely support the vehicle, then remove the halfshaft from the vehicle.
3. Clamp the halfshaft in a vise that is equipped with soft jaw covers. Do not allow the vise jaws to contact the boot or boot clamp.
4. Cut and remove both CV-joint boot clamps. Slide the boot back on the shaft.
5. Remove the outboard CV-joint.
6. Remove the halfshaft stop ring, and remove the boot.
To install:
7. Slide the boot and clamps on the interconnecting shaft.
8. Pack the CV-joint with CV-joint grease and fill the boot with grease. Combined fill should be 16.75 oz. (475 grams).
9. Install the boot on the CV-joint boot groove. Make sure all excess grease is cleaned from the groove.
10. Before installing the boot clamp, make sure all the air pressure that may have built up in the boot is relieved. Pry up on the boot lip to allow the air to escape.
11. Tighten the boot clamps using a suitable tool. Install the outer CV-joint and boot.

12. Install the halfshaft assembly. Lower the vehicle.
13. Turn the air suspension switch to the **ON** position. Road test the vehicle and check for proper operation.

NOTE: Make sure to have available, new halfshaft circlips.

Inboard CV-Joint Boot

For Taurus and Sable three different design inboard CV-joints are used:
- **1993-94 3.0L with ABS, 3.8L and 3.0L SHO with Manual Transaxle:** The CV-joint tripod can be removed from the halfshaft to allow replacement of the CV-joint boot.
- **1993-94 3.0L without ABS/1995-97 All Except 3.2L SHO with Automatic Transaxle:** The CV-joint tripod cannot be removed from the halfshaft. In order to replace the inboard CV-joint boot, the outboard CV-joint and boot must first be removed from the halfshaft.
- **3.2L SHO with Automatic Transaxle:** The CV-joint is removed from the halfshaft with the tripod, as an assembly, to allow replacement of the CV-joint boot.

1993-94 Taurus and Sable — 3.0L with ABS, 3.8L and 3.0L SHO with Manual Transaxle

1. Disconnect the negative battery cable. Raise and safely support the vehicle. Remove the halfshaft assembly from the vehicle.
2. Clamp the halfshaft in a vise that is equipped with soft jaw covers. Do not allow the vise jaws to contact the boot or boot clamp.
3. Cut and remove both boot clamps and slide the boot back on the shaft. Remove the clamp by engaging the pincer jaws of boot clamp pliers D87P-1090-A or equivalent, in the closing hooks on the clamp and draw together. Disengage the windows and locking hooks and remove the clamp.
4. Mark the position of the outer race in relation to the shaft and remove the outer race.
5. Move the stop ring back on the shaft using snapring pliers. Move the tripod assembly back on the shaft to allow access to the circlip.
6. Remove the circlip from the shaft. Mark the position of the tripod on the shaft and remove the tripod assembly. Remove the boot.
7. Check the CV-joint grease for contamination. If the CV-joints are operating properly and the grease is not contaminated, add grease and replace the boot. If the grease appears contaminated, clean the CV-joint

Halfshaft Assembled Lengths

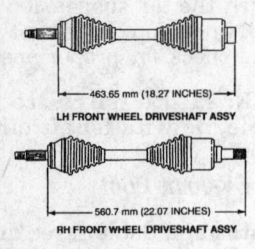

463.65 mm (18.27 INCHES)
LH FRONT WHEEL DRIVESHAFT ASSY

560.7 mm (22.07 INCHES)
RH FRONT WHEEL DRIVESHAFT ASSY

301410

**Halfshaft assembled lengths —
1995–97 Continental**

components and inspect for damage. Replace the CV-joint, if necessary.

To install:

8. Install the CV-joint boot. Make sure the boot is seated in the boot groove on the shaft. Tighten the clamp using crimping pliers, but do not tighten to the point where the clamp bridge is cut or the boot is damaged.

9. Install the tripod assembly with chamfered side toward the stop ring. Be sure to align the marks that were made during the removal procedure.

10. Install a new circlip. Compress the circlip and slide the tripod assembly forward over the circlip to expose the stop ring groove.

11. Move the stop ring into the groove using snapring pliers, making sure it is fully seated in the groove.

12. Fill the CV-joint outer race with grease and spread the remaining grease evenly inside the boot. Total combined fill is 9 oz. (250g).

13. Install the outer race over the tripod assembly, aligning the marks made during the removal procedure.

14. Remove all excess grease from the CV-joint external surfaces and mating boot surface. Position the boot over the CV-joint making sure the boot is seated in the groove. Move the CV-joint in and out, as necessary, to adjust the length to the following specifications:

3.0L and 3.8L left halfshaft — 18.7 in. (475mm)

3.0L and 3.8L right halfshaft — 23.74 in. (603mm)

3.0L SHO left halfshaft — 21.4 in. (544mm)

3.0L SHO right halfshaft — 21.81 in. (554mm)

15. Before installing the boot clamp, make sure any air pressure that may have built up in the boot is relieved. Insert a small prybar between the boot and outer race to allow the trapped air to escape. Release

the air only after adjusting the length dimension.

16. Seat the boot in the groove and clamp in position using crimping pliers D87P-1098-A or equivalent. Install the clamp as follows:

a. With the boot seated in the groove, place the clamp over the boot.

b. Engage hook C in the window.

c. Place the pincer jaws of the crimping pliers in closing hooks A and B.

d. Secure the clamp by drawing the closing hooks together. When windows 1 and 2 are above locking hooks D and E, the spring tab will press the windows over the locking hooks and engage the clamp.

17. Install the halfshaft and lower the vehicle. Connect the negative battery cable.

Taurus and Sable: 1993–94 3.0L without ABS/1995–97 Except 3.2L SHO with Automatic Transaxle

1. Disconnect the negative battery cable. Raise and safely support the vehicle. Remove the halfshaft assembly from the vehicle.

2. Clamp the halfshaft in a vise that is equipped with soft jaw covers. Do not allow the vise jaws to contact the boot or boot clamp.

3. Cut and remove both boot clamps and slide the boot back on the shaft. Remove the clamp by engaging the pincer jaws of boot clamp pliers D87P-1090-A or equivalent, in the closing hooks on the clamp and draw together. Disengage the windows and locking hooks and remove the clamp.

4. Mark the position of the outer race in relation to the shaft and remove the outer race.

5. Check the CV-joint grease for contamination by rubbing it between 2 fingers. A gritty feeling indicates a contaminated CV-joint, which should be replaced.

6. Remove the outboard CV-joint and boot. Remove the outboard CV-joint stop ring and circlip.

7. Remove the inboard CV-joint boot from the halfshaft.

8. Remove the trilobe insert from the CV-joint outer race. Remove the grease from the outer race and inspect the outer race and tripod assembly.

To install:

9. Install the inboard CV-joint boot on the halfshaft. Position the boot to allow CV-joint outer race installation.

10. Install the trilobe insert on the CV-joint outer race, positioning it in the groove on the outer race.

11. Fill the CV-joint outer race with grease and spread the remaining grease inside the boot. The total fill is 9 oz. (250g).

12. Install the outer race on the tripod assembly, aligning the race and shaft with the marks made during removal.

13. Remove all excess grease from the CV-joint external surfaces and mating boot surface. Position the boot over the CV-joint making sure the boot is seated in the groove. Move the CV-joint in and out, as necessary, to adjust the length to the following specifications:

AX4N automatic transaxle left halfshaft — 18.7 in. (475mm)

AX4N automatic transaxle right halfshaft — 21.4 in. (544mm)

AX4S automatic transaxle left halfshaft — 18.7 in. (475mm)

AX4S automatic transaxle right halfshaft — 23.74 in. (603mm) Manual transaxle left halfshaft — 21.4 in. (544mm)

Manual transaxle right halfshaft — 21.81 in. (554mm)

14. Before installing the boot clamp, make sure any air pressure that may have built up in the boot is relieved. Insert a small prybar between the boot and outer race to allow the trapped air to escape. Release the air only after adjusting the length dimension.

15. Seat the boot in the groove and clamp in position using crimping pliers D87P-1098-A or equivalent. Install the clamp as follows:

a. With the boot seated in the groove, place the clamp over the boot.

b. Engage hook C in the window.

c. Place the pincer jaws of the crimping pliers in closing hooks A and B.

d. Secure the clamp by drawing the closing hooks together. When windows 1 and 2 are above locking hooks D and E, the spring tab will press the windows over the locking hooks and engage the clamp.

16. Install the halfshaft and lower the vehicle. Connect the negative battery cable.

3.2L SHO with Automatic Transaxle/1993–94 Continental

1. Disconnect the negative battery cable.

2. Raise and safely support the vehicle. Remove the halfshaft assembly from the vehicle.

3. Clamp the halfshaft in a vise that is equipped with soft jaw covers. Do not allow the vise jaws to contact the boot or boot clamp.

4. Cut and remove both boot clamps and slide the boot back on the shaft.

5. Wipe the grease away from the CV-joint to expose the CV-joint retaining snapring. Check the CV-joint grease for contamination by rubbing it between 2 fingers. A gritty feeling indicates a contaminated joint which should be replaced.

6. After the grease has been removed, expand the snapring using suitable pliers and pull the CV-joint from the halfshaft.

7. Remove the boot from the halfshaft.

To install:

8. Install the CV-joint boot on the halfshaft, sliding it down far enough to allow installation of the inboard CV-joint.

9. Position the inboard CV-joint to the halfshaft, making sure the splines are engaged.

10. Tap the CV-joint into position using a rawhide mallet. The joint is fully seated when the retaining snapring locks into the groove on the shaft. Check by attempting to pull the joint from the shaft.

11. Fill the CV-joint outer race with grease, then spread grease evenly inside the boot for a total combined fill of 17.3 oz. (490g).

12. Remove all excess grease from the CV-joint external surfaces and mating boot surface. Position the boot over the CV-joint making sure the boot is seated in the groove. Move the CV-joint in and out, as necessary, to adjust the length to 18.7 in. (475mm) for left halfshaft or 23.85 in. (606mm) for right halfshaft.

13. Before installing the boot clamp, make sure any air pressure that may have built up in the boot is relieved. Insert a small prybar between the boot and outer race to allow the trapped air to escape. Release the air only after adjusting the length dimension.

14. Seat the boot in the groove and clamp in position using crimping pliers D87P-1098-A or equivalent.

15. Install the halfshaft and lower the vehicle. Connect the negative battery cable.

1995–97 Continental

CAUTION
The air suspension switch, located in the left side of the luggage compartment, must be turned OFF before raising the vehicle. Failure to do so may result in unexpected inflation or defla-

tion of the air springs which may result in shifting of the vehicle during service.

1. Turn the air suspension switch, located in the left side of the luggage compartment, to the **OFF** position.

2. Raise and safely support the vehicle. Remove the halfshaft from the vehicle.

3. Clamp the halfshaft in a vise that is equipped with soft jaw covers. Do not allow the vise jaws to contact the boot or boot clamp.

4. Cut and remove both boot clamps and slide the boot back on the shaft. Remove the clamp using side cutters and pull away from the shaft joint boot.

5. Angle the CV-joint to expose the inner bearing race.

6. Using a brass drift and hammer, give a sharp tap to the inner bearing race to dislodge the internal snapring and separate the CV-joint from the interconnecting shaft. Take care to secure the CV-joint so it does not fall after separation. Remove the clamp and boot from the shaft.

7. Remove and discard the circlip at the end of the interconnecting shaft. The stop ring, located just below the circlip should be removed and replaced only if damaged or worn.

8. Check the lubricate for contamination by rubbing between 2 fingers, any gritty feelings indicates a contaminated joint.

9. Clamp the joint in a vise with the outer face facing up. Be careful not to damage the front wheel dust shield.

10. Press down on the inner race until it tilts enough to allow removal of the ball.

NOTE: If the cage and inner race assembly is tight, tap the inner race with a wooden dowel and hammer.

11. Continue to tilt the cage and inner race assembly until all the balls are removed.

NOTE: If the balls are tight in the cage, a screwdriver can be used to pry the balls from the cage. Be careful not to scratch the inner race.

12. With the cage pivoted and aligned, lift the cage from the outer race.

13. Clean and inspect all CV-joint components for wear and/or damage. Replace the joint, if necessary.

To install:

14. Apply a light coating of CV-joint Grease to the inner and outer ball races.

15. Install the assembly vertically and pivot into position.

16. Tilt the inner race and cage and install a ball. Repeat this step until all 6 balls are installed.

17. Install the boot and clamps and the stop ring.

18. Make sure the boot is seated in the shaft groove and tighten the boot clamp with a suitable tool.

19. Install a new halfshaft circlip, do not reuse the old circlip.

NOTE: Be careful not to over–expand or twist the circlip. When installing the circlip start one end in the groove and carefully work the clip around the shaft.

20. Pack the CV-joint with 6.3 oz. (180 grams) of suitable high temperature CV-joint grease. Any grease remaining from the tube supplied with the boot kit should be spread evenly inside the boot. Position the joint on the halfshaft assembly.

21. With the boot held back, tap the joint on the shaft using a plastic tipped hammer.

22. Remove all remaining grease from the joint boot groove, and position the boot into the boot groove.

23. Make sure the boot is seated in the groove, then position and tighten the boot clamp with a suitable tool.

24. Install the halfshaft in the vehicle. Lower the vehicle.

25. Turn the air suspension switch to the **ON** position.

26. Road test the vehicle and check for proper operation.

STEERING

Air Bag

CAUTION
Some vehicles are equipped with an air bag system, also known as the Supplemental Inflatable Restraint (SIR) system. The system must be disabled before performing service on or around system components, steering column, instrument panel components, wiring and sensors. Failure to follow safety and disabling procedures could result in accidental air bag deployment, possible personal injury and unnecessary system repairs.

PRECAUTIONS

Several precautions must be observed when handling the inflator module to avoid accidental deployment and possible personal injury.

- Never carry the inflator module by the wires or connector on the underside of the module.
- When carrying a live inflator module, hold securely with both hands, and ensure that the bag and trim cover are pointed away.
- Place the inflator module on a bench or other surface with the bag and trim cover facing up.
- With the inflator module on the bench, never place anything on or close to the module which may be thrown in the event of an accidental deployment.

DISARMING

— CAUTION —

The Supplemental Inflatable Restraint (SIR) system must be disarmed before performing service around SIR system components or SIR system wiring. Failure to do so may cause accidental deployment of the air bag, resulting in unnecessary SIR system repairs and/or personal injury.

1. Disconnect both battery cables from the battery, negative cable first.
2. Wait one minute before proceeding with the service procedure. This is the time required for the backup power supply in the air bag diagnostic monitor to deplete its stored energy.
3. After service is completed, reconnect the battery cables, negative cable last.
4. Turn the ignition switch to the **RUN** position. The air bag indicator should light continuously for approximately 6 seconds and then turn OFF. If the indicator fails to light, flashes or remains lit continuously, there is a fault in the air bag system.

Steering Wheel

REMOVAL AND INSTALLATION

— CAUTION —

The Supplemental Inflatable Restraint (SIR) system must be disarmed before performing service around SIR system components or SIR system wiring. Failure to do so may cause accidental deploy-

ment of the air bag, resulting in unnecessary SIR system repairs and/or personal injury.

NOTE: Before proceeding, make sure a new steering wheel retaining bolt is available as the bolt removed must not be reused.

1. Make sure the front wheels are in the straight-ahead position.
2. If equipped, propely disarm the SIR (air bag) system.
3. Remove the steering wheel spoke covers, if equipped.
4. Remove the air bag module retaining nuts or bolts. Lift the air bag module from the steering wheel and detach the air bag wiring connector from the air bag module. Remove the air bag module from the steering wheel and properly place aside.

— CAUTION —

When carrying a live air bag, make sure the air bag and trim cover are pointed away from the body. In the unlikely event of an accidental deployment, the bag will then deploy with minimal chance if injury. When placing a live air bag on a bench or other surface, always face the bag and trim cover up, away from the surface. This will reduce the motion of the module if it is accidently deployed.

5. Detach the cruise control wiring connector from the steering wheel. Remove and discard the steering wheel retaining bolt.
6. Install Steering Wheel Puller T67L-3600-A, or equivalent and remove the steering wheel. Route the contact assembly wiring harness through the steering wheel as the wheel is lifted off the shaft.

To install:

7. Make sure the vehicle's front wheels are in the straight-ahead position. Route the contact assembly wire harness through the steering

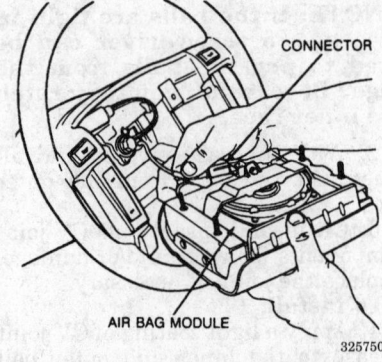

AIR BAG MODULE

CONNECTOR

325750

Air bag module removal (typical)

wheel opening at the 3 o'clock position and install the steering wheel on the shaft. The steering wheel and shaft alignment marks should be aligned.

8. Install a new steering wheel retaining bolt and tighten to 25–34 ft. lbs. (34–46 Nm).
9. Connect the cruise control wiring harness to the steering wheel and snap the connector assembly into the steering wheel clip.
10. Attach the air bag wiring connector to the air bag module, install the air bag module to the steering wheel. Install and tighten the 4 air bag module retaining nuts to 36–47 inch lbs. (4–5 Nm) for 1993 vehicles, except SHO models, and the 2 bolts to 89–124 inch lbs. (10–14 Nm) for 1994–97 vehicles including SHO models.
11. Install the steering wheel spoke covers, if equipped. Connect the battery cables, negative cable last.
12. Prove out the air bag system by turning the ignition key to the **RUN** position and visually monitoring the air bag indicator lamp in the instrument cluster. The indicator lamp should illuminate for approximately 6 seconds and then turn OFF. If the indicator lamp does not illuminate, stays ON, or flashes at any time, a fault has been detected by the air bag diagnostic monitor.
13. Check the steering wheel, horn and steering column for proper operation. Road test the vehicle and check the speed control for proper operation.

Tie Rod Ends

REMOVAL AND INSTALLATION

NOTE: Be sure to have new cotter pins and castellated nuts for the tie rod ends.

1. Hold the tie rod end with a suitable wrench and loosen the tie rod jam nut. Remove and discard the cotter pin and castellated nut from the tie rod end.
2. Separate the tie rod end from the steering knuckle, using Tie Rod Remover tool 3290-D or equivalent.
3. Note the depth to which the tie rod end was located using the jam nut or paint as a marker, then grip the tie rod end with a pair of suitable pliers and remove the tie rod end from the front wheel spindle (inner tie rod).

To install:

4. Clean the tie rod threads. Thread the tie rod end onto the inner

tie rod to the same depth as the one removed.

5. Place the tie rod end stud into the steering knuckle.

NOTE: Make sure the front wheels are pointed straight-ahead before fitting the tie rod end stud to the steering knuckle.

6. Install a new castellated nut on the tie rod end stud. Tighten the nut to 35–46 ft. lbs. (47–63 Nm) and continue tightening until the next castellation on the nut is aligned with the cotter pin hole in the stud. Install a new cotter pin.

7. Set the toe to specification. Tighten the jam nuts to 35–46 ft. lbs. (47–63 Nm).

8. Road test the vehicle and check for proper operation.

Power Rack and Pinion

REMOVAL AND INSTALLATION

1993–95 Taurus and Sable with Integral Power Rack and Pinion

1. Disconnect the negative battery cable. Remove the nuts retaining the steering shaft weather boot to the dash panel.

2. Remove the bolts retaining the intermediate shaft to the steering column shaft. Set the weather boot aside.

3. Remove the pinch bolt at the steering gear input shaft and remove the intermediate shaft. Raise the vehicle and support safely.

4. Remove the left front wheel and tire assembly. Remove the heat shield. Cut the bundling strap retaining the lines to the gear.

5. Remove the tie rod ends from the spindles. Place a drain pan under the vehicle and remove the hydraulic pressure and return lines from the steering gear.

NOTE: The pressure and return lines are on the front of the housing. Do not confuse them with the transfer lines on the side of the valve.

6. Remove the nuts from the gear housing bolts.

7. Push the weather boot end into the vehicle and lift the gear out of the mounting holes. Rotate the gear so the input shaft will pass between the brake booster and the floor pan.

8. Rotate the input shaft so it clears the left fender apron opening and complete the removal of the steering gear.

To install:

9. Install new plastic seals on the hydraulic line fittings. Insert the steering gear through the left fender apron.

10. Install the remaining components in the reverse order of removal.

11. Align the steering gear bolts, install the nuts and torque to 85–100 ft. lbs. (115–135 Nm). Lower the vehicle.

12. Tighten the hydraulic pressure line and return line to 15–25 ft. lbs. (20–35 Nm).

13. Torque the tie rod end castle nuts to 35 ft. lbs. (48 Nm). If necessary, torque the nuts a little bit more to align the slot in the nut for the cotter pin. Install the cotter pin.

14. Fill and bleed the power steering system. Check the system for leaks and proper operation. Adjust the toe setting as necessary.

Continental
1993–97 Taurus and Sable with Variable Assist Power Steering (VAPS)

----- **CAUTION** -----
On 1995–97 Continental, the air suspension switch, located in the left side of the luggage compartment, must be turned OFF before raising the vehicle. Failure to do so may result in unexpected inflation or deflation of the air springs which may result in shifting of the vehicle during service.

1. On 1995–97 Continental, turn the air suspension switch, located in the left side of the luggage compartment, to the **OFF** position.

2. Disconnect the negative battery cable.

3. On 1995–97 Continental, perform the following procedures:

  a. Disconnect the engine control sensor wiring from the Powertrain Control Module (PCM) and position the wiring aside.

  b. Directly above the engine control sensor wiring, remove the ground straps from the dash panel.

4. Working from inside the vehicle, remove the nuts retaining the steering shaft weather boot to the dash panel.

5. Remove the 2 bolts retaining the intermediate shaft to the steering column shaft. Set the weather boot aside.

6. Remove the pinch bolt at the steering gear input shaft and remove the intermediate shaft. Raise the vehicle and support safely.

7. On 1995–97 Continental, support the subframe with Powertrain

Lift 014-00765 and Bracket 014-00766 or equivalents.

8. Remove the front wheel and tire assemblies. Support the vehicle under the rear edge of the subframe.

9. On 1996–97 Taurus and Sable, disconnect the heated oxygen sensor wiring harness and remove the Y-pipe from the vehicle.

10. Remove the tie rod cotter pins and nuts. Remove the tie rod ends from the spindle. Remove the tie rod ends from the shaft. Mark the position of the jam nut to maintain the alignment.

11. On 1995–97 Continental, remove both air suspension and height sensor attachments.

12. Remove the nuts from the gear-to-subframe attaching bolts. On 1993–94 Continental, remove both height sensor attachments.

13. On 1993–95 Taurus and Sable install jackstands under the vehicle to steady the vehicle. Position a Rotunda Powertrain Lift 014-00765 or equivalent under the subframe.

14. On 1995–97 Continental, remove the dual converter Y-pipe.

15. Remove the rear subframe-to-body attaching bolts. On 1993–94 Continental and 1993–95 Taurus and Sable, remove the exhaust pipe-to-catalytic converter attachment.

16. Lower the vehicle carefully until the subframe separates from the body; approximately 4 in. Remove the heat shield band and fold the shield down.

17. On 1996–97 Taurus and Sable, perform the following procedures:

  a. Remove the heat shield push-pin retainers from the power steering hose bracket. Remove the steering shaft U-joint shield.

  b. Remove the retaining screw and the power steering left turn pressure hose from the power steering hose bracket.

  c. Remove the power steering hose bracket screws.

18. Disconnect the VAPS electrical connector from the actuator assembly.

19. On 1995–97 Continental, disconnect the electrical connectors from the auxiliary actuator and the Power Steering Pressure (PSP) switch.

20. On 1996–97 Taurus and Sable, disconnect the Power Steering Pressure (PSP) switch electrical connector.

21. Rotate the gear to clear the bolts from the subframe and pull to the left to facilitate line fitting removal.

22. Position a drain pan under the vehicle and remove the line fittings.

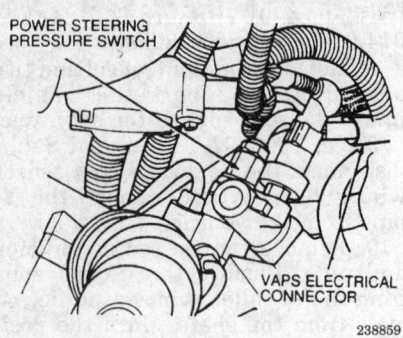

VAPS electrical connector — 1993–95 Taurus and Sable

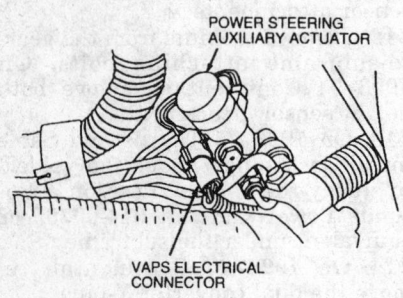

VAPS electrical connector — 1996–97 Taurus and Sable

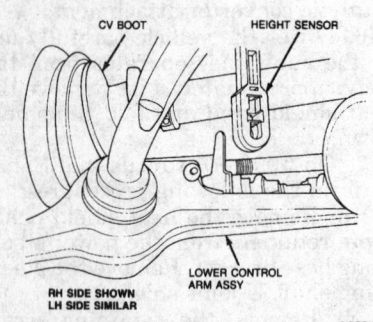

Disconnecting the height sensor — 1993–94 Continental

Remove the O-rings from the fitting connections and replace with new.

23. On 1993–94 Continentals and 1993–95 Taurus and Sable, remove the left sway bar link. On 1996–97 Taurus and Sable, remove the left and right stabilizer bar links.

24. Remove the steering gear assembly through the left wheel well.

To install:

25. Install new O-rings into the line fittings.

26. Place the gear attachment bolts in the gear housing.

27. Install the steering gear assembly through the left wheel well.

28. Connect and tighten the line fittings to the steering gear assembly.

29. On 1996–97 Taurus and Sable, perform the following procedures:

a. Connect the power steering pressure and return hoses to the fittings on the rack and pinion assembly and tighten to 24–30 ft. lbs. (33–41 Nm).

b. Install the power steering left turn pressure hose and screw. Tighten to 80–106 inch lbs. (9–12 Nm).

30. Connect the VAPS and the PSP (1996–97 Taurus and Sable) for electrical connector.

31. On 1995–97 Continental, connect the PSP switch and auxiliary actuator electrical connectors.

32. Position the steering gear into the subframe. Install the tie rod ends onto the shaft. Install the heat shield band.

33. Attach the tie rod ends onto the spindle. Install the nuts and secure with new cotter pins. Attach the sway bar link.

34. Raise the vehicle until the subframe contacts the body. Install the subframe attaching bolts.

35. Install the gear-to-subframe nuts and torque to 85–100 ft. lbs. (115–135 Nm) for 1993–94 Continental and 1993–95 Taurus and Sable or to 72–97 ft. lbs. (97–132 Nm) for 1996–97 Taurus and Sable.

36. On 1995–97 Continental, perform the following procedures:

a. Install the dual converter Y-pipe.

b. Connect the air suspension height sensors.

37. Attach the exhaust pipe to the catalytic converter.

38. For 1996–97 Taurus and Sable, torque the following items to:

a. Power steering hose bracket screws: 80–106 inch lbs. (9–12 Nm).

b. Tie rod end-to-steering knuckle nuts: 35–46 ft. lbs. (47–63 Nm).

c. Left stabilizer bar link nuts: 56–76 ft. lbs. (77–104 Nm).

d. Rear subframe bolts: 100–144 ft. lbs. (135–195 Nm).

e. Steering column intermediate shaft coupling bolt: 30–38 ft. lbs. (41–51 Nm)

f. Steering column intermediate shaft coupling-to-steering column lower yoke: 16–21 ft. lbs. (21–28 Nm).

39. On 1995–97 Continental, torque the following items to:

a. Power steering line fittings-to-steering gear: 24–30 ft. lbs. (33–41 Nm)

b. Tie rod ends-to-steering knuckles: 35–46 ft. lbs. (47–63 Nm)

c. Subframe bolts: 100–144 ft. lbs. (135–195 Nm)

d. Steering gear-to-subframe nuts: 83–113 ft. lbs. (113–135 Nm)

e. Intermediate shaft-to-steering gear input shaft bolt: 30–38 ft. lbs. (41–51 Nm)

f. Intermediate shaft-to-steering column shaft nuts: 15–24 ft. lbs. (21–33 Nm)

40. On 1993–94 Continental, attach the height sensors.

41. Install the wheel and tire assemblies and lower the vehicle. Fill the power steering system.

42. Working from inside the vehicle, pull the weather boot end out of the vehicle and install it over the valve housing. Install the intermediate shaft to the steering gear input shaft. Install the inner weather boot to the floor pan.

43. On 1995–97 Continental, perform the following procedures:

a. Connect engine control sensor wiring to the PCM.

b. Install the screw retaining the ground straps to the dash panel directly above the engine control sensor wiring.

44. Install the intermediate shaft to the steering column shaft. Fill the power steering system.

45. On 1995–97 Continental, turn the air suspension switch to the **ON** position.

46. Check the system for leaks and proper operation. Adjust the toe setting as necessary.

Power Steering Pump

BLEEDING

1. If equipped with a distributor, disconnect the ignition coil. If equipped with distributorless ignition, disable the ignition by disconnecting the ignition module or disconnecting the camshaft position sensor.

2. Raise and safely support the vehicle so the front wheels are off the floor.

3. Place jackstands under the front of the vehicle.

4. Fill the power steering fluid reservoir.

5. Crank the engine with the starter motor and add fluid until the level remains constant.

6. While cranking the engine, rotate the steering wheel from lock-to-lock.

NOTE: The front wheels must be off the floor during lock-to-lock rotation of the steering wheel. Do not hold the steering wheel on the stops.

7. Check the fluid level and add fluid, if necessary.
8. Enable the ignition system. Start the engine and allow it to run for several minutes.
9. Rotate the steering wheel from lock-to-lock.
10. Stop the engine and check the fluid level. Add fluid, if necessary.
11. If air is still present in the system, purge the system of air using Power Steering Air Evacuator 021–00014 or equivalent, as follows:

a. Make sure the power steering pump reservoir is full to the COLD FULL mark on the dipstick or to just above the minimum indication on the reservoir.

b. Tightly insert the rubber stopper of the air evacuator assembly into the pump reservoir fill neck.

c. Apply 20 inch Hg maximum vacuum on the pump reservoir for a minimum of 3 minutes with the engine idling. As air purges from the system, vacuum will fall off. Maintain adequate vacuum with the vacuum source.

d. Release the vacuum and remove the vacuum source. Fill the reservoir to the COLD FULL mark or to just above the minimum indication on the reservoir. Install the evacuator tool again.

e. With the engine idling, apply 20 inch Hg vacuum to the pump reservoir. Slowly cycle the steering wheel from lock-to-lock every 30 seconds for approximately 5 minutes. Do not hold the steering wheel on the stops while cycling. Maintain adequate vacuum with the vacuum source as the air purges.

f. Release the vacuum and remove the vacuum source. Add fluid, if necessary.

g. Start the engine and cycle the steering wheel. Check for oil leaks at all connections. In severe cases of aeration, it may be necessary to repeat the procedure.
12. Remove the jackstands.
13. Lower the vehicle.
14. Check the power steering system for proper operation.

REMOVAL AND INSTALLATION

1993–95 Vehicles, Except 1995 Continental

3.0L Engine, Except SHO

1. Disconnect the negative battery cable. Loosen the idler pulley and remove the power steering belt.
2. Remove the radiator overflow bottle in order to gain access to the 3 screws attaching the pulleys to the pulley hub.
3. Matchmark both pulley to hub positions. Remove the pulleys from the pulley hub.
4. Remove the return line from the pump. Be prepared to catch any spilled fluid in a suitable container.
5. Back off the pressure line attaching nut completely. The line will separate from the pump connection when the pump is removed.
6. Remove the pump mounting bolts and remove the pump.
7. Installation is the reverse of the removal procedure. Fill the pump with fluid, bleed the air from the system and check for proper operation.

3.0L and 3.2L SHO Engines

1. Disconnect the negative battery cable.
2. Remove the engine damper strut. Remove the power steering belt.
3. Raise and support the vehicle safely.
4. Remove the right front wheel and tire assembly.
5. Position a jack under the engine and remove the right rear engine mount.
6. Remove the pump pulley. Place a drain pan under the pump and remove the pressure and return lines from the pump.
7. Remove the 4 pump retaining bolts and remove the pump.
8. Installation is the reverse of the removal procedure. Tighten the pump retaining bolts to 15–24 ft. lbs. (20–33 Nm). Fill and bleed the system and check for proper operation.

3.8L Engines

1. Disconnect the negative battery cable. Loosen the tensioner pulley attaching bolts and using the ½ in. drive hole provided in the tensioner pulley, rotate the tensioner pulley clockwise and remove the belt from the alternator and power steering pulley.
2. Position a drain pan under the power steering pump from underneath the vehicle. Disconnect the hydraulic pressure and return lines.

3. Remove the pulley from the pump shaft using hub puller T69L-10300-B or equivalent. Remove the bolts retaining pump to bracket and remove the power steering pump.
4. Installation is the reverse of the removal procedure. Fill the pump with fluid, bleed the air from the system and check the system for proper operation.

NOTE: To install the power steering pump pulley, use steering pump pulley replacer T65P-3A733-C or equivalent. When using this tool, the small diameter threads must be fully engaged in the pump shaft before pressing on the pulley. Hold the head screw and turn the nut to install the pulley. Install the pulley face flush with the pump shaft within ±0.100 in. (0.25mm).

1996–97 Taurus and Sable

3.0L OHV Engine With CII Pump

1. Disconnect the negative battery cable. Remove the accessory drive belt.
2. Remove the alternator. Drain and remove the radiator coolant recovery reservoir.
3. Position a drain pan under the power steering pump underneath the vehicle. Disconnect the hydraulic pressure and return lines and allow to drain.
4. Remove the idler pulley from the power steering pump support.
5. Remove the bracket mounting bolt located under the belt tensioner mounting.
6. Remove 2 retaining nuts from the bracket mounting studs. Remove both mounting studs, and pull off the power steering pump support with the pump attached.
7. Clamp the pump support bracket in a suitable vise.
8. Remove the power steering pump pulley from the pump shaft using Pump Pulley Remover T69L-10300-B, or equivalent. Remove the 3 bolts retaining the power steering pump to the power steering pump support and remove the power steering pump.
   To install:
9. Install the power steering pump to the power steering pump support.
10. Install the power steering pump pulley using Steering Pump Pulley Replacer T65P-3A733-C, or equivalent; the pulley face must be flush within 0.010 inch (0.25mm) of the pump shaft.

11. Install the power steering pump/pump support and torque nuts/bolt to 17–24 ft. lbs. (23–32 Nm).

12. Complete the installation by reversing the removal procedure.

13. Fill the power steering reservoir with power steering fluid. Connect the negative battery cable.

14. Run the engine and check for leaks and proper operation. Bleed the power steering system if needed.

3.0L DOHC Engine With CIII Pump

1. Disconnect the negative battery cable. Remove the accessory drive belt.

2. Drain and remove the radiator coolant recovery reservoir.

3. Position a drain pan under the power steering pump. Disconnect the power steering pump reservoir hose and allow it to drain.

4. Using Pulley Remover T69L-10300-B, or equivalent, remove the pulley from the power steering pump shaft .

5. Disconnect the left turn pressure hose from the power steering pump and allow it to drain.

6. Remove the power steering pump retaining bolts and the pump.

To install:

7. Install the power steering pump and tighten bolts to 15–22 ft. lbs. (20–30 Nm).

8. Connect the left turn pressure hose and secure clamp.

9. Using a Steering Pump Pulley Replacer T65P-3A733-C, or equivalent, install the power steering pump pulley: the pulley face must be flush with the pump shaft or within 0.010 inch (0.25mm).

10. Fill the power steering reservoir with power steering fluid. Connect the negative battery cable.

11. Start the engine and check for leaks and proper operation. Bleed the power steering system if needed.

1995–97 Continental

—————— CAUTION ——————
The air suspension switch, located in the left side of the luggage compartment, must be turned OFF before raising the vehicle. Failure to do so may result in unexpected inflation or deflation of the air springs which may result in shifting of the vehicle during service.

1. Turn the air suspension switch, located in the left side of the luggage compartment, to the **OFF** position.

2. Disconnect the negative battery cable. Remove the radiator upper sight shield. Remove the accessory drive belt.

3. Properly recover the refrigerant from the air conditioning system using approved equipment.

4. Raise and safely support the vehicle.

5. Remove the lower radiator air deflector push pins/screws and the air deflector.

6. Position a drain pan and remove the return hose from the combination power steering and transaxle fluid cooler. Drain the power steering system.

7. After the system has drained, install the power steering return hose to the power steering and transaxle fluid cooler.

8. Remove the A/C manifold and tube. Detach the A/C clutch connector at the A/C compressor.

9. Remove A/C compressor bolts and remove the compressor. Cap all openings in the A/C system to prevent the entrance of dirt and moisture.

10. Disconnect the pressure and reservoir hoses from the power steering pump.

11. Remove the power steering pump bolts and the pump.

To install:

12. Install the power steering pump and tighten the bolts to 15–22 ft. lbs. (20–30 Nm).

13. Install the remaining components in the reverse order of removal. Torque the following:

Power steering pressure hose tube nut to 42–54 ft. lbs. (57–73 Nm)

A/C compressor bolts to 15–22 ft. lbs. (20–30 Nm)

NOTE: Swivel or endplay of the power steering pressure hose tube end is normal and does not indicate a loose fitting.

14. Connect the negative battery cable. Turn the air suspension switch to the **ON** position.

15. Refill the power steering pump reservoir. Bleed the power steering system and check for leaks.

16. Evacuate and recharge the air conditioning system. Road test the vehicle and check for proper operation.

BRAKES

Anti-Lock Brake System Service

PRECAUTIONS

• Certain components within the Anti-Lock Brake System (ABS) are not intended to be serviced or repaired individually. Only those components with removal and installation procedures should be serviced.

• Do not use rubber hoses or other parts not specifically specified for and ABS system. When using repair kits, replace all parts included in the kit. Partial or incorrect repair may lead to functional problems and require the replacement of components.

• Lubricate rubber parts with clean, fresh brake fluid to ease assembly. Do not use lubricated shop air to clean parts; damage to rubber components may result.

• Use only specified brake fluid from an unopened container.

• If any hydraulic component or line is removed or replaced, it may be necessary to bleed the entire system.

• A clean repair area is essential. Always clean the reservoir and cap thoroughly before removing the cap. The slightest amount of dirt in the fluid may plug an orifice and impair the system function. Perform repairs after components have been thoroughly cleaned; use only denatured alcohol to clean components. Do not allow ABS components to come into contact with any substance containing mineral oil; this includes used shop rags.

• The Anti-Lock control unit is a microprocessor similar to other computer units in the vehicle. Ensure that the ignition switch is **OFF** before removing or installing controller harnesses. Avoid static electricity discharge at or near the controller.

• If any arc welding is to be done on the vehicle, the control unit should be unplugged before welding operations begin.

DEPRESSURIZING

Before servicing any components which contain high pressure, it is mandatory that the hydraulic pressure in the system be discharged. To discharge the system, turn the ignition **OFF** and pump the brake pedal a minimum of 20 times until an increase in pedal force is clearly felt.

Master Cylinder

REMOVAL AND INSTALLATION

1. Disconnect the negative battery cable.

2. Apply the brake pedal several times to exhaust all vacuum in the brake booster.

3. Disconnect the brake lines from the primary and secondary outlet ports of the master cylinder and the pressure control valve on station wagons. Plug the brake lines.

4. Disconnect the brake warning indicator wire connector.

5. If equipped with ABS, disconnect the HCU supply hose at the brake master cylinder and secure in a position to prevent loss of brake fluid.

6. Remove 2 nuts retaining the master cylinder to the brake booster. Slide the master cylinder forward and upward from the vehicle.

To install:

7. Install a new seal in the groove in the brake master cylinder mounting face.

8. Bench bleed the new master cylinder.

9. Mount the master cylinder on the booster. Tighten the nuts to 19–28 ft. lbs. (26–39 Nm).

10. Unplug and connect the brake lines to the master cylinder and tighten to 12–15 ft. lbs. (16–20 Nm) except for 1995–97 Continental or 16–21 ft. lbs. (21–29 Nm) for 1995–97 Continental. Connect brake pressure control valve (wagon) outlet ports.

11. If equipped with ABS, connect the HCU supply hose to the brake master cylinder fitting and secure with a hose clamp.

12. Connect the brake warning light or fluid level indicator switch wire connector.

13. On 1996–97 Taurus and Sable, position the cowl top inner panel tube on the brake master cylinder mounting stud and install the retaining nut.

14. Connect the negative battery cable. Bleed the system and check for leaks.

15. Road test the vehicle and check the brake system for proper operation.

Brake Caliper

REMOVAL AND INSTALLATION

―――― **CAUTION** ――――
On Continentals, the air suspension switch, located in the left side of the luggage compartment, must be turned OFF before raising the vehicle. Failure to do so may result in unexpected inflation or deflation of the air springs which may result in shifting of the vehicle during service.

Front

1. Remove brake fluid from the brake master cylinder reservoir until the reservoir is ½ full.

2. For Continentals, turn the air suspension switch, located in the left side of the luggage compartment, to the **OFF** position.

3. Raise and safely support the vehicle. Remove the wheel and tire assembly.

4. Mark the disc brake caliper to ensure that it is reinstalled in the correct location.

5. Remove the hollow bolt connecting the brake hose to the disc brake caliper and plug the brake hose. Discard the 2 copper sealing washers.

―――― **WARNING** ――――
Do not pry directly against the piston or damage to the piston will result.

6. Remove the caliper locating pins and lift the caliper off the rotor using a rotating motion.

To install:

7. Retract the disc brake caliper piston fully in the piston bore, using an old brake pad or block of wood and a C-clamp or equivalent.

NOTE: Be sure to clean all dirt from the mating surfaces of the caliper locating pins and housing ears. Also, make sure the clip-on insulators are attached to the brake pads.

8. Install the disc brake pads to the caliper. Make sure the brake pad insulators are correctly attached to the brake pad plate.

9. Position the disc brake caliper and pad assembly above the rotor and install it with a rotating motion. Make sure the inner and outer pads are properly positioned and the outer anti-rattle spring is properly positioned.

10. Lubricate the locating pins and the inside of the insulators with sili-

cone grease. Tighten the locating pins to 25 ft. lbs. (34 Nm).

11. Remove the plug and install the brake hose to the disc brake caliper. Use 2 new copper washers and torque the hollow bolt to 30–40 ft. lbs. (41–54 Nm).

12. Bleed the brake system, filling the master cylinder as required.

13. Install the wheel and tire assembly; tighten the nuts to 85–104 ft. lbs. (115–142 Nm).

14. Lower the vehicle.

15. Pump the brake pedal several times to position the brake pads prior to moving the vehicle.

16. For Continentals, turn the air suspension service switch to the **ON** position.

17. Road test the vehicle and check for proper brake system operation.

Rear

1. Remove brake fluid from the brake master cylinder reservoir until the reservoir is ½ full.

2. For Continentals, turn the air suspension switch, located in the left side of the luggage compartment, to the **OFF** position.

3. Raise and safely support the vehicle. Remove the wheel and tire assembly.

4. Remove the retaining bolt and disconnect the brake hose from the caliper assembly. Discard the copper sealing washers.

5. Remove the retaining clip from the parking brake at the caliper. Disengage the parking brake cable end from the lever arm.

6. Lift the rear disc brake caliper away from the rear disc support bracket.

7. Remove the disc brake caliper locating pins and boots from the rear disc support bracket.

To install:

8. Using rear caliper piston adjuster tool T87P-2588-A or equivalent, rotate the rear disc brake piston and adjuster clockwise until fully seated.

NOTE: Make sure one of the 2 slots in the rear disc brake piston and adjuster face is positioned so it will engage the nib on the disc brake pad.

9. Apply silicone dielectric compound or equivalent, to the inside of the slider pin boots and the slider pins.

10. Position the slider pins and boots in the support bracket. Position the caliper assembly on the support bracket. Make sure the brake pads are installed correctly.

11. Remove the residue from the pin retainer threads and apply 1 drop of threadlock and sealer. Install the pin retainers and tighten to 23–26 ft. lbs. (31–35 Nm).

12. Attach the cable end to the parking brake lever. Install the cable retaining clip on the caliper assembly.

13. Using new washers, connect the brake flex hose to the caliper. Tighten the retaining bolt to 40 ft. lbs. (54 Nm).

14. Bleed the brake system, filling the master cylinder as required.

15. Install the wheel and tire assembly; tighten the nuts to 85–104 ft. lbs. (115–142 Nm).

16. Lower the vehicle.

17. Pump the brake pedal several times to position the brake pads prior to moving the vehicle.

18. For Continentals, turn the air suspension service switch to the **ON** position.

19. Road test the vehicle and check for proper brake system operation.

Disc Brake Pads

REMOVAL AND INSTALLATION

— CAUTION —

For Continentals, the air suspension switch, located in the left side of the luggage compartment, must be turned OFF before raising the vehicle. Failure to do so may result in unexpected inflation or deflation of the air springs which may result in shifting of the vehicle during service.

Front

1. Remove the master cylinder reservoir cap and check the fluid level in the reservoir. Remove brake fluid until the reservoir is ½ full. Discard the removed fluid.

2. On Continentals, turn the air suspension switch, located in the left side of the luggage compartment, to the **OFF** position.

3. Raise and safely support the vehicle. Remove the wheel and tire assembly.

4. Remove the disc brake caliper locating pins. Lift the caliper assembly from the anchor plate and rotor using a rotating motion.

— WARNING —

Do not pry directly against the metal caliper piston or damage will occur.

5. Suspend the caliper inside the fender housing with wire. Do not allow the caliper to hang from the brake hose.

6. Remove the inner and outer brake pads. Inspect the rotor braking surfaces for scoring and machine as necessary.

To install:

7. Use a C-clamp and an old brake pad or block of wood to seat the caliper piston in its bore.

8. Remove any rust buildup from the inside of the caliper in the brake pad contact area.

9. Install the inner pad in the caliper piston. Install the outer pad. Make sure the clips are properly seated.

NOTE: Make sure the insulators are installed on the brake pads.

10. Install the disc brake caliper over the rotor and install the wheel. Lower the vehicle.

11. Pump the brake pedal several times prior to moving the vehicle to position the brake pads to the rotor.

12. Refill the master cylinder reservoir as necessary, using only clean DOT 3 or equivalent brake fluid from a closed container.

13. On Continentals, turn the air suspension service switch to the **ON** position.

14. Road test the vehicle and check the brake system for proper operation.

Rear

1. Remove the master cylinder reservoir cap and check the fluid level in the reservoir. Remove brake fluid until the reservoir is ½ full. Discard the removed fluid.

2. On Continentals, turn the air suspension switch, located in the left side of the luggage compartment, to the **OFF** position.

3. Raise and safely support the vehicle. Remove the wheel and tire assembly.

4. Remove the screw retaining the brake hose bracket to the frame side rail.

5. Remove the retaining clip from the parking brake cable at the disc brake caliper. Remove the cable end from the parking brake lever.

6. Remove the upper disc brake caliper locating pin at the support bracket. Rotate the caliper away from the rotor.

7. Remove the disc brake pads.

8. Inspect the rotor braking surfaces for scoring and machine as necessary.

To install:

9. Using Rear Caliper Piston Adjuster T87P-2588-A or equivalent, rotate the piston clockwise until it is fully seated. Make sure one of the slots in the piston face is positioned so it will engage the nib on the brake pad.

10. Install the brake pads in the support bracket. Rotate the caliper assembly over the rotor into position on the support bracket. Make sure the brake pads are installed correctly.

11. Remove the residue from the rear brake pin retainer bolt threads and apply one drop of a suitable threadlock sealer. Install and tighten the disc brake caliper locating pin to 23–26 ft. lbs. (31–35 Nm).

12. Attach the cable end to the parking brake lever. Install the cable retaining clip on the caliper assembly. Position the brake flex hose and bracket assembly to the side rail, and install the retaining screw. Tighten to 8–11 ft. lbs. (11–16 Nm).

13. Install the wheel and tire, then lower the vehicle. Pump the brake pedal several times prior to moving the vehicle, to position the brake pads to the rotor.

14. Refill the master cylinder reservoir if necessary, using only clean

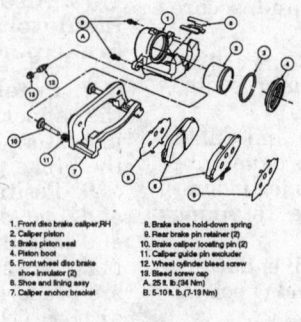

1. Front disc brake caliper, RH
2. Caliper piston
3. Brake piston seal
4. Piston boot
5. Front wheel disc brake shoe insulator (2)
6. Shoe and lining assy
7. Caliper anchor bracket
8. Brake shoe hold-down spring
9. Rear brake pin retainer (2)
10. Brake caliper locating pin (2)
11. Caliper guide pin excluder
12. Wheel cylinder bleed screw
13. Bleed screw cap
A. 25 ft. lb (34 Nm)
B. 5-10 ft. lb (7-19 Nm)

300308

Front disc brake caliper, pads and related components

DOT 3 or equivalent brake fluid from a closed container.

15. On Continentals, turn the air suspension service switch to the **ON** position.

16. Road test the vehicle and check the brake system for proper operation.

Brake Rotor

REMOVAL AND INSTALLATION

─────── **CAUTION** ───────
On Continentals, the air suspension switch, located in the left side of the luggage compartment, must be turned OFF before raising the vehicle. Failure to do so may result in unexpected inflation or deflation of the air springs which may result in shifting of the vehicle during service.
────────────────────────

NOTE: Make sure to have available, 2 new caliper anchor bracket bolts.

Front

1. On Continentals, turn the air suspension switch, located in the left side of the luggage compartment, to the **OFF** position.

2. Raise and safely support the vehicle. Remove the wheel and tire assembly.

3. Remove the disc brake caliper and the caliper anchor bracket as an assembly. Discard the bracket retaining bolts. Position the caliper aside and support it with a length of wire. Do not allow the caliper to hang by the brake hose.

NOTE: To prevent contamination of the disc brake caliper guide pin journals, do not separate the disc brake caliper from the caliper anchor bracket.

4. Separate the disc brake rotor from the hub assembly by pulling it off the hub studs. If additional force is required to remove the rotor, apply rust penetrate on the front and rear rotor/hub mating surfaces and then strike the rotor between the studs with a plastic hammer. If this does not work, attach a suitable 3-jaw puller and remove the rotor.

NOTE: If excessive force must be used to remove the rotor, it should be checked for lateral runout before installation.

5. Check the rotor for scoring and/or other wear. Machine or replace, as necessary.

To install:
6. If the disc brake rotor is being replaced, remove the protective coating from the new rotor. If the original rotor is being installed, make sure the rotor braking surfaces are clean.

7. Apply a small amount of silicone dielectric compound or equivalent, to the pilot diameter of the disc brake rotor. Install the rotor on the hub assembly.

8. Install remaining components in the reverse order of removal. Torque the following:
Caliper anchor bracket bolts to 84 ft. lbs. (115 Nm)
Lug nuts to 85–105 ft. lbs. (115–142 Nm)

9. Pump the brake pedal several times prior to moving the vehicle to seat the disc brake pads to the rotor.

10. On Continentals, turn the air suspension switch to the **ON** position.

11. Road test the vehicle and check the brake system for proper operation.

Rear

1. On Continentals, turn the air suspension switch, located in the left side of the luggage compartment, to the **OFF** position.

2. Raise and safely support the vehicle. Remove the wheel and tire assembly.

3. Remove the caliper assembly from the disc brake rotor and support it with a length of wire. Do not let the caliper hang from the brake line.

4. Remove the upper and lower support bracket retaining bolts. Remove the support bracket.

5. Remove 2 retainer nuts and remove the disc brake rotor from the hub. Check the rotor for scoring and/or other wear. Machine or replace, as necessary.

To install:
6. If the disc brake rotor is being replaced, remove the protective coating from the new rotor. If the original rotor is being installed, make sure the rotor braking surfaces are clean.

7. Install the disc brake rotor on the hub assembly.

NOTE: Lubricate the hub pilot with a suitable caliper slide grease to ease future rotor removal.

8. Install 2 retainer nuts to hold the disc brake rotor in position.

9. Install the disc brake caliper support bracket and tighten the support bracket retaining bolts to 64–88 ft. lbs. (87–119 Nm).

10. Install the brake pads and disc brake caliper assembly.

11. Install the wheel and tire assembly. Install the lug nuts and torque to 85–105 ft. lbs. (115–142 Nm).

12. Pump the brake pedal several times prior to moving the vehicle to seat the brake pads to the rotor.

13. On Continentals, turn the air suspension service switch to the **ON** position.

14. Road test the vehicle and check the brake system for proper operation.

Brake Drums

REMOVAL AND INSTALLATION

Taurus and Sable

1. Raise and safely support the vehicle. Remove the wheel and tire assembly.

2. Remove the brake drum.

NOTE: If the drum will not come off, pry the rubber plug from the backing plate inspection hole. Remove the brake line-to-axle retention bracket. This will allow sufficient room to insert suitable brake tools through the inspection hole to disengage the adjusting lever and back off the adjusting screw.

3. Inspect the drum for scoring and/or other wear. Machine or replace, as necessary.

To install:
4. Measure the brake drum inside diameter using D81L-1103-A or equivalent brake adjustment gauge.

5. Using the brake adjustment gauge, adjust the brake shoes to the same dimensions as the brake drum.

6. Position the brake drum over the brake shoes on the axle hub.

7. Install the wheel and tire assembly. Tighten the lug nuts to 85–104 ft. lbs. (115–141 Nm).

8. Lower the vehicle.

9. Pump the brake pedal several times to position the brake shoes and complete the adjustment.

10. Road test the vehicle and check for proper brake system operation.

Brake Shoes

REMOVAL AND INSTALLATION

Taurus and Sable

1. Raise and safely support the vehicle. Remove the wheel and tire assembly.

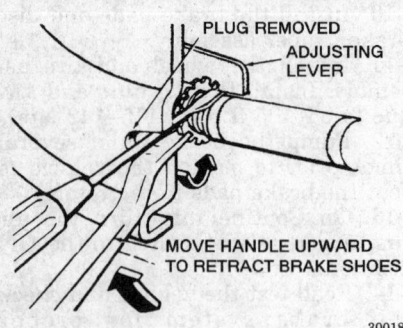

PLUG REMOVED
ADJUSTING
LEVER

MOVE HANDLE UPWARD
TO RETRACT BRAKE SHOES

300188

Retracting the brake shoes to allow drum removal — Taurus and Sable

2. Remove the brake drum.
3. Remove the parking brake cable from the parking brake lever.
4. Remove the 2 brake shoe hold-down springs and pins.
5. Lift the brake shoes, springs and adjuster assembly off the backing plate and wheel cylinder assembly. When removing the assembly, be careful not to bend the adjusting lever.
6. Remove the retracting springs from the lower brake attachments and upper shoe-to-adjusting lever attachment points.

7. Remove the horse shoe retaining clip and spring washer and slide the lever off the parking brake lever pin on the trailing shoe. Discard the horseshoe clip.

To install:

8. Apply a light coating of disc brake caliper slide grease at the points where the brake shoes contact the backing plate.
9. Apply a thin coat of lubricant to the adjuster screw threads and socket end of the adjusting screw. Install the stainless steel washer over the socket end of the adjusting screw and install the socket. Turn the adjusting screw into the adjusting pivot nut to the limit of the threads and then back off ½ turn.
10. Assemble the parking brake lever to the trailing shoe by installing the spring washer and a new horse shoe retaining clip. Crimp the clip until it retains the lever to the shoe securely.
11. Position the trailing shoe on the backing plate and attach the rear parking brake cable.
12. Position the leading shoe on the backing plate and attach the lower brake shoe adjusting spring to the brake shoes.

13. Install the adjuster assembly in the slots on the brake shoes. The wide slot on the dual slotted end must fit into the leading shoe. The narrow slot on the dual slotted end fits into the shoe adjusting lever. The single slotted side of the adjuster assembly must fit into the slots on the trailing shoe and the rear parking brake cable bracket.

NOTE: The adjuster socket blade is marked R for the right or L for the left brake assemblies. The adjuster blade must be installed with the letter R or L in the upright position, facing the wheel cylinder. Make sure the adjuster socket fits into the parking brake lever.

14. Complete the installation by reversing the removal procedures.
15. Pump the brake pedal several times to position the brake shoes and finish the brake shoe adjustment.
16. Road test the vehicle and check the brake system for proper operation.

Wheel Cylinder

REMOVAL AND INSTALLATION

Taurus and Sable

1. Raise and safely support the vehicle. Remove the wheel and tire assembly.
2. Remove the brake drum.
3. Remove the brake shoes, retainers and springs from the backing plate.
4. Disconnect and plug the brake line at the rear of the wheel cylinder.
5. Remove the 2 bolts securing the wheel cylinder to the backing plate and remove the wheel cylinder.

To install:

NOTE: Wipe the end of the brake line to remove any foreign matter before connecting to the wheel cylinder.

6. Install the wheel cylinder and tighten the bolts to 8–10 ft. lbs. (10–14 Nm).
7. Tighten the brake line fitting, using a tube nut wrench, to 12–14 ft. lbs. (16–20 Nm).
8. Complete the installation by reversing the removal procedures.
9. Bleed the rear brakes.
10. Road test the vehicle and check for leaks and proper brake system operation.

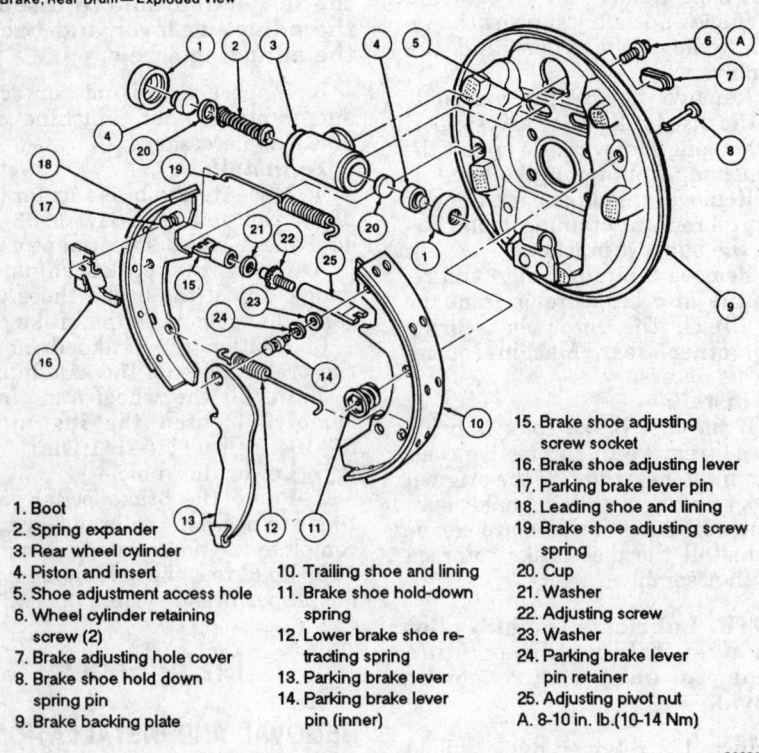

Brake, Rear Drum — Exploded View

1. Boot
2. Spring expander
3. Rear wheel cylinder
4. Piston and insert
5. Shoe adjustment access hole
6. Wheel cylinder retaining screw (2)
7. Brake adjusting hole cover
8. Brake shoe hold down spring pin
9. Brake backing plate
10. Trailing shoe and lining
11. Brake shoe hold-down spring
12. Lower brake shoe retracting spring
13. Parking brake lever
14. Parking brake lever pin (inner)
15. Brake shoe adjusting screw socket
16. Brake shoe adjusting lever
17. Parking brake lever pin
18. Leading shoe and lining
19. Brake shoe adjusting screw spring
20. Cup
21. Washer
22. Adjusting screw
23. Washer
24. Parking brake lever pin retainer
25. Adjusting pivot nut
A. 8-10 in. lb. (10-14 Nm)

300212

Brake shoes and related components — Taurus and Sable

Parking Brake Cable

ADJUSTMENT

1993–95 Vehicles, Except 1995 Continental

1. Make sure the parking brake control is fully released.
2. Place the transaxle in **N**.
3. Raise and safely support the vehicle.
4. For drum brakes, tighten the adjusting nut against the cable equalizer, causing a rear wheel brake drag. Loosen the adjusting nut until the rear brakes are fully released. There should be no brake drag.
5. For disc brakes, tighten the adjusting nut against the rear parking brake cable adjuster until there is a slight movement, less than 1/16 inch (1.59mm), of either parking brake lever at the rear disc brake calipers.
6. Apply the parking brake control with approximately 100 lbs. (445 N) of pedal effort, then release the parking brake control. Repeat this step a second time.
7. Lower the vehicle.
8. Check the parking brake system for proper operation.

1996–97 Taurus and Sable

1. Make sure the parking brake is fully released.
2. Place the transaxle in **N**.
3. Partially raise and safely support the vehicle.
4. Place jackstands under the rear suspension.
5. Using Cable Tension Gauge 021-00018 or equivalent, tighten the adjusting nut against the rear parking brake cable adjuster until the cable tension is 18–26 lbs.
6. Apply the parking brake control fully and then release.
7. Verify the cable tension is still 18–26 lbs. and no drag is present on the rear brakes.
8. Remove the jackstands.
9. Lower the vehicle.
10. Check the parking brake for proper operation.

1995–97 Continental

———— CAUTION ————
The air suspension switch, located in the left side of the luggage compartment, must be turned OFF before raising the vehicle. Failure to do so may result in unexpected inflation or defla- *tion of the air springs which may result in shifting of the vehicle during service.*

1. Turn the air suspension switch, located in the left side of the luggage compartment, to the **OFF** position.
2. Make sure the parking brake is fully released.
3. Raise and safely support the vehicle.
4. Remove the clip from the parking brake adjuster. Install the clip after the parking brake cable adjuster has removed slack from the cables.
5. Apply 157 lbs. (700 N) of force to the parking brake control. Apply the force for 20 minutes.
6. Repeat Step 4.
7. Lower the vehicle.
8. Check the operation of the parking brake.
9. Turn air suspension switch to the **ON** position.

REMOVAL AND INSTALLATION

1993–95 Vehicles, Except 1995 Continental

———— CAUTION ————
On Continentals, the air suspension switch, located in the left side of the luggage compartment, must be turned OFF before raising the vehicle. Failure to do so may result in unexpected inflation of deflation of the air springs which may result in shifting of the vehicle during service.

Front Cable

1. If equipped, turn **OFF** the air suspension switch, located in the left side of the luggage compartment.
2. Raise the vehicle and support safely.
3. Loosen the adjuster nut at the cable adjuster bracket.
4. Lower the vehicle.
5. Disconnect the cable from the control assembly at the clevis.
6. Raise the vehicle and support safely. At the cable connector, disconnect the front cable from the rear cable.
7. Remove the cable and push-in prong retainer from the cable bracket, using a 13mm box end wrench to depress the retaining prongs. Allow the cable to hang.
8. Push the grommet up through the floor pan and lower the vehicle.
9. Remove the left cowl side panel. Pull the carpet away from the cowl panel.
10. From inside the vehicle, remove the cable end from the clevis and re- move the conduit retainer from the control assembly.
11. Pull the cable assembly through the floorpan hole.
To install:
12. Position the cable assembly through the floorpan hole.
13. From inside the vehicle, install the cable end from the clevis and install the conduit retainer to the control assembly.
14. Adjust the parking brake and check for proper operation. Install the remaining components in the reverse order of removal.
15. If equipped, turn the air suspension switch to the **ON** position.

Rear Cable — Left Side

1. If equipped, turn **OFF** the air suspension switch, located in the left side of the luggage compartment.
2. Raise the vehicle and support safely. Remove the parking brake cable adjusting nut.
3. Remove the rear cable end fitting from the front cable connector.
4. Remove the wheel and drum assembly if equipped with drum brakes.
5. Disconnect the brake cable from the parking brake actuating lever.
• On drum brake vehicles, use a 13mm box end wrench to depress the conduit retaining prongs and remove the cable end pronged fitting from the backing plate.
• On disc brake vehicles, remove the E-clip from the conduit end of the fitting at the caliper.
6. Push the plastic snap-in grommet rearward to disconnect it from the side rail bracket.
7. Remove the pronged connector from the parking park adjuster bracket. Remove the cable assembly.
To install:
8. Install the pronged connector to the parking park adjuster bracket. Install the cable assembly.
9. Push the plastic snap-in grommet forward to connect it from the side rail bracket.
10. Install the remaining components in the reverse order of removal.
11. Adjust the parking brake. Lower the vehicle. Check for proper operation.
12. If equipped, turn the air suspension switch to the **ON** position.

Rear Cable — Right Side

1. If equipped, turn **OFF** the air suspension switch, located in the left side of the luggage compartment.
2. Raise the vehicle and support it safely. Remove the parking brake cable adjuster nut.

3. Use a 13mm box wrench to remove the conduit retainer prongs and remove the cable from the frame side rail bracket.

4. Remove the rear wheel and drum assembly if equipped with drum brakes.

5. Disconnect the brake cable from the parking brake actuating lever.

• On drum brake vehicles, use a 13mm box end wrench to depress the conduit retaining prongs and remove the cable end pronged fitting from the backing plate.

• On disc brake vehicles, remove the E-clip from the conduit end of the fitting at the caliper.

6. On sedan:

a. Remove the brake pressure control valve bracket at the control arm.

b. Remove the cable retaining screw and clip from the lower suspension arm.

c. Remove one screw from the cable bracket at the crossmember.

d. Remove the entire right cable assembly.

7. On station wagon:

a. Remove the cable retaining clip and screw from each lower suspension arm.

b. Remove the cable clip retaining screw from lower suspension arm inner mounting bracket.

To install:

8. On sedan:

a. Install the brake pressure control valve bracket at the control arm.

b. Install the cable retaining screw and clip to the lower suspension arm.

c. Install one screw to the cable bracket at the crossmember.

d. Install the right cable assembly.

9. On station wagon:

a. Install the cable retaining clip and screw to each lower suspension arm.

b. Install the cable clip retaining screw to the lower suspension arm inner mounting bracket.

10. Connect the brake cable to the parking brake actuating lever.

11. Install the rear drum and tire and wheel assembly if equipped with drum brakes. Install the parking brake cable adjuster nut.

12. Lower the vehicle and support it safely. Ensure the pronged fitting is securely locked in place. Adjust the parking brake.

13. If equipped, turn the air suspension switch to the **ON** position.

1996–97 Taurus and Sable

Front Cable

1. Raise and safely support the vehicle. Loosen the adjuster nut at the cable adjuster bracket.

2. Remove the grommet from the floor. Remove the cable and push-in prong retainer from the cable bracket, using a ½ inch or 13mm box end wrench to depress the retaining prongs.

3. Lower the vehicle.

4. Remove the floor scuff plate (cowl trim panel) and position the carpeting aside.

5. From inside the vehicle, remove the cable end from the clevis using a ½ inch or 13mm box end wrench to press the retaining prongs and remove the conduit from the parking brake control.

6. Raise and safely support the vehicle. Pull the cable assembly through the floorpan hole and remove.

To install:

7. Position the cable assembly through the floorpan hole and secure the rubber grommet.

8. Position the cable through the bracket at the inner floor side member. Push the prong into the bracket.

9. Install the cable adjusting nut and adjuster.

10. Complete the installation by reversing the removal procedures.

11. Adjust the parking brake. Check the parking brake for proper operation.

Rear Cable — Left Side

1. Raise and safely support the vehicle. Loosen the parking brake cable adjusting nut.

2. Remove the rear cable end fitting from the front cable connector.

3. Remove the wheel and tire assembly. Remove the brake drum, if equipped with drum brakes.

4. Disconnect the brake cable from the parking brake actuating lever:

• On drum brake vehicles, use a ½ inch or 13mm box end wrench to depress the conduit retaining prongs and remove the cable end pronged fitting from the backing plate.

• On disc brake vehicles, remove the E-clip from the conduit end of the fitting at the caliper.

5. Remove the bolt retaining the cable to the side rail bracket.

6. Remove the pronged connector from the parking brake adjuster bracket. Remove the left side parking brake cable assembly.

To install:

7. Install the pronged connector to the parking brake adjuster bracket. Install the cable assembly.

8. Position the cable along the LH rear frame rail and secure into position with the retaining bolt. Tighten to 97–124 inch lbs. (11–14 Nm).

9. Connect the brake cable from the parking brake actuating lever:

• On drum brake vehicles, insert the cable and conduit end into the backing plate hole. Make sure the prongs are locked into place.

• On disc brake vehicles, insert cable and conduit into the caliper and install the E-clip.

10. Complete the installation by reversing the removal procedures.

11. Adjust the parking brake. Check the parking brake for proper operation.

Rear Cable — Right Side

1. Raise and safely support the vehicle. Loosen the parking brake cable adjuster nut.

2. Disconnect the LH rear cable from the equalizer.

3. Use a ½ inch or 13mm box wrench to remove the conduit retainer prongs and remove the RH cable from the equalizer.

4. Push the plastic snap-in grommet rearward to disconnect from inner floor side member brackets.

5. Remove the rear wheel and tire assembly. Remove the brake drum if equipped with drum brakes.

6. Disconnect the brake cable from the parking brake actuating lever:

• On drum brake vehicles, use a ½ inch or 13mm box end wrench to depress the conduit retaining prongs and remove the cable end pronged fitting from the backing plate.

• On disc brake vehicles, remove the E-clip from the conduit end of the fitting at the caliper.

7. Remove the cable retaining screw and clip from the lower suspension arm and 1 screw from the cable bracket at the rear floor crossmember.

8. Remove the right side parking brake cable assembly.

To install:

9. Install the cable into the opening in the inner floor side member bracket and into the equalizer. Make sure the plastic grommet is pressed into the inner floor side member bracket and securely locked in place and that the pronged fitting is fully seated in the cable equalizer.

10. Route the RH cable along the rear suspension arm.

11. Connect the brake cable from the parking brake actuating lever:

• On drum brake vehicles, insert the cable and conduit end into the backing plate hole. Make sure the prongs are locked into place.

• On disc brake vehicles, insert the cable and conduit into the caliper and install the E-clip.

12. Install the cable clip and retaining bolts to the lower suspension arm. Tighten the bolts to 71–97 inch lbs. (8–11 Nm).

13. Install the brake drum, if equipped with drum brakes.

14. Complete the installation by reversing the removal procedures.

15. Adjust the parking brake. Check the parking brake for proper operation.

1995–97 Continental

---CAUTION---

The air suspension switch, located in the left side of the luggage compartment, must be turned OFF before raising the vehicle. Failure to do so may result in unexpected inflation or deflation of the air springs which may result in shifting of the vehicle during service.

Front Cable

1. Turn **OFF** the air suspension switch, located in the left side of the luggage compartment.

2. Raise and safely support the vehicle. Remove the clip at the parking brake cable adjuster.

3. Lower the vehicle.

4. Disconnect the front parking brake cable and conduit from the parking brake control assembly at the clevis, using a 13mm box-end wrench to depress the conduit retaining prongs, and remove the cable end pronged fitting from the parking brake control.

5. Remove the left cowl side trim panel and pull the carpet back to expose the cable.

6. Raise and safely support the vehicle. Disconnect the parking brake cable and conduit from the parking brake rear cable and conduit at the cable connector.

7. Remove the front parking brake cable and conduit and push–in prong retainer from the cable bracket. Use a 13mm box–end wrench to depress the retaining prongs.

8. Pull the parking brake cable assembly down from the floorpan.

To install:

9. Position the parking brake cable assembly through the hole in the floor pan.

NOTE: Make sure the cable insulator is in place. Prongs must be securely locked in place.

10. Complete the installation by reversing the removal procedures.

11. Check the parking brake for proper operation. Adjust the parking brake cable if necessary.

12. Turn the air suspension switch to the **ON** position.

Rear Cable — Left Side

1. Turn **OFF** the air suspension switch, located in the left side of the luggage compartment.

2. Raise and safely support the vehicle. Remove the wire hook from frame and parking brake cable adjuster.

3. Separate the rear cable end fitting from the connector at the parking brake cable adjuster.

4. Using a 13mm box-end wrench, depress the conduit retaining prongs and remove the rear cable from the retaining bracket.

5. Remove the clip from the rear disc brake caliper. Remove the parking brake rear cable from the rear disc brake caliper and parking brake lever.

To install:

6. Install the parking brake rear cable into the rear disc brake caliper and parking brake lever. Install the clip.

7. Insert the parking brake rear cable into the retaining bracket. Make sure the pronged cable end is securely attached to the retaining bracket.

8. Connect the rear cable end fitting to the parking brake cable adjuster. Install the wire hook on the parking brake cable adjuster and frame.

9. Check the parking brake for proper operation. Adjust the parking brake cable if necessary.

10. Turn the air suspension switch to the **ON** position.

Right Side

1. Turn **OFF** the air suspension switch, located in the left side of the luggage compartment.

2. Raise and safely support the vehicle.

3. Separate the parking brake rear cable from the front parking brake cable at the connector. Disengage the parking brake rear cable from the rear parking brake guide.

4. Use a 13mm box-end wrench to depress the conduit retainer prongs and remove the parking brake rear cable and conduit from the parking brake cable adjuster.

5. Remove the parking brake rear cable from the retaining bracket. Remove the parking brake rear cable and conduit from the routing clips.

6. Remove the clip from the rear disc brake caliper. Remove the parking brake rear cable and conduit from the parking brake lever.

To install:

7. Install the parking brake rear cable and conduit into the rear disc brake caliper and parking brake lever. Install the clip.

8. Install the parking brake rear cable into the routing clips.

9. Insert the parking brake rear cable through the mounting bracket. Insert the cable end into the parking brake cable adjuster. Make sure the pronged cable end is securely attached to the parking brake cable adjuster.

10. Connect the parking brake rear cable and the front parking brake cable. Lower the vehicle.

11. Check the parking brake for proper operation. Adjust the parking brake cable if necessary.

12. Turn the air suspension switch to the **ON** position.

Brake System Bleeding

PROCEDURE

Conventional Procedure

The brake fluid level in the master cylinder reservoir should be no more than 0.16 in. (4mm) below the MAX line on the side of the reservoir. If the brake fluid is low, the red BRAKE lamp will illuminate. To add brake fluid, clean and remove the cap and pour clean brake fluid into the top of the reservoir. If brake fluid has to be added often, check all hydraulic connections for leaks.

NOTE: Never let the master cylinder reservoir run empty of brake fluid when bleeding the brake system. Only use clean DOT 3 or equivalent, brake fluid from a sealed container.

1. Clean all dirt from the master cylinder filler cap.

2. If the master cylinder is known or suspected to have air in the bore, it must be bled before any of the calipers. To bleed the master cylinder, position a shop towel under the front brake line fitting and loosen the fit-

ting approximately ¾ turn. Have an assistant depress the brake pedal slowly through its full travel. Close the fitting and let the pedal return slowly to the fully released position. Wait 5 seconds and then repeat the operation until all air bubbles disappear.

3. Repeat Step 2 with the rear brake line fitting.

4. Continue to bleed the brake system by removing the rubber dust cap from the caliper bleeder fitting at the right rear of the vehicle. Place a suitable box wrench on the bleeder fitting and attach a rubber drain tube to the fitting. The end of the tube should fit snugly around the bleeder fitting. Submerge the other end of the tube in a container partially filled with clean brake fluid and loosen the fitting ¾ turn.

5. Have an assistant push the brake pedal down slowly through its full travel. Close the bleeder fitting and allow the pedal to slowly return to its full release position. Wait 5 seconds and repeat the procedure until no bubbles appear at the submerged end of the bleeder tube. Secure the bleeder fitting and remove the bleeder tube. Install the rubber dust cap on the bleeder fitting.

6. Repeat the procedure in Steps 4 and 5 in the following sequence: left rear, right front, left front. Refill the master cylinder reservoir after each caliper has been bled and install the master cylinder filler cap. When brake bleeding is completed, the fluid level should be filled to the maximum level indicated on the reservoir.

NOTE: Never reuse brake fluid that has been drained from the hydraulic system or has been allowed to stand in an open container for an extended period of time. Never allow the master cylinder reservoir to run dry while bleeding the brake system.

7. Always make sure the disc brake pistons are returned to their normal positions by depressing the brake pedal several times until normal pedal travel is established. If the pedal feels spongy, repeat the bleeding procedure.

Anti-Lock Procedure

If a spongy brake pedal is present and air in the hydraulic control unit is suspected, use the following procedure:

1. Bleed the brake system.
2. Connect a New Generation Star (NGS) tester or equivalent scan tool, to the serial data link connector be-

low the instrument panel as though retrieving codes.

3. Make sure the ignition switch is in the **RUN** position.

4. Follow the instructions on the NGS screen. Verify correct vehicle and model year go to the "Diagnostic Data Link" menu item, choose ABS Module, choose "Function Tests", and choose "Service Bleed".

5. Bleed the right front wheel as follows:

   a. Open the caliper bleed screw and pump the brake pedal for 3 seconds. Repeat the procedure again.

   b. When the fluid runs clear, begin the program and continue to pump the brake pedal.

   c. Continue bleeding for approximately 1–2 minutes after the program ends and then tighten the bleed screw.

6. Repeat the bleeding procedure to the left front, left rear and finally the right rear wheel.

7. Remove the pressure bleeding device and adjust the brake fluid level.

Wheel Speed Sensor

REMOVAL AND INSTALLATION

1993–95 Vehicles, Except 1995 Continental

──────── **CAUTION** ────────
On Continentals, the air suspension switch, located in the left side of the luggage compartment, must be turned OFF before raising the vehicle. Failure to do so may result in unexpected inflation or deflation of the air springs which may result in shifting of the vehicle during service.

Front Sensor

1. If equipped, turn the air suspension switch located in the luggage compartment, to the **OFF** position.
2. Disconnect the negative battery cable.
3. Detach the anti-lock sensor connector located in the engine compartment.
4. For the right front sensor, remove 2 plastic push studs to loosen the front section of the splash shield in the wheel well. For the left front sensor, remove 2 plastic push studs to loosen the rear section of the splash shield.
5. Thread the sensor wires through the holes in the fender apron. For the right front sensor, re-

move 2 retaining clips behind the splash shield.

6. Raise and safely support the vehicle. Remove the wheel and tire assembly.

7. Disengage the sensor wire grommets at the height sensor bracket and from the retainer clip on the strut just above the steering knuckle.

8. Loosen the sensor retaining screw and remove the sensor assembly from the steering knuckle.

To install:

9. Position the anti-lock sensor with its mounting holes on the steering knuckle. Tighten the retaining screw 40–60 inch lbs. (4.6–6.8 Nm).

10. Install the grommets at the height sensor bracket and the retainer clip at the strut.

11. Thread the wire through the holes in the front fender apron. Install the retainer clips (for the right brake sensor only). Secure the front fender splash shield with plastic push studs.

12. Install the wheel and tire assembly. Torque the lug nuts to 85–105 ft. lbs. (115–142 Nm).

13. Lower the vehicle.

14. Attach the sensor connector to the wiring harness in the engine compartment.

15. Connect the negative battery cable.

16. If equipped, turn the air suspension switch to the **ON** position.

17. Turn the ignition key to the **RUN** position while watching the amber anti-lock brake warning light in the instrument cluster. The light should illuminate for approximately 2 seconds proving out the anti-lock brake system. If the light does not illuminate or stays illuminated, a problem with the anti-lock brake system exist and must be corrected.

18. Road test the vehicle and check for proper operation.

Rear Sensor — Taurus and Sable Sedan

1. Disconnect the negative battery cable. Remove the rear seat and seat back insulation.

2. Disconnect the anti-lock sensor from the electrical harness and tie the sensor connector to the rear seat sheetmetal bracket with wire.

3. Push the sensor wire grommet and connector through the floorpan pulling the wire from the sensor from underneath the vehicle.

4. Disconnect the wire from the sensor from underneath the vehicle.

5. Raise and safely support the vehicle.

6. Disconnect the routing clips from the suspension arms and remove the sensor retaining bolt from the rear brake adapter.

7. Remove the sensor assembly from the brake adapter.

To install:

8. Install the sensor assembly into the brake adapter hole. Install the retaining bolt and tighten to 40–60 inch lbs. (4.6–6.8 Nm).

9. Install the sensor routing clips to the suspension arms.

10. Attach a length of wire to the new sensor connector, then pull the connector through the hole in the floorpan. Install the grommet.

11. Lower the vehicle.

12. Remove the wire, the plug the sensor electrical connector into the harness.

13. Install the rear seat back and cushion. Connect the negative battery cable.

14. Turn the ignition key to the **RUN** position while watching the amber anti-lock brake warning light in the instrument cluster. The light should illuminate for approximately 2 seconds and extinguish proving out the anti-lock brake system. If the light does not illuminate or stays illuminated, a problem with the anti-lock brake system exist and must be corrected.

15. Road test the vehicle and check for proper operation.

Rear Sensor — Taurus and Sable Wagon

1. Disconnect the negative battery cable. Raise and safely support the vehicle.

2. Remove the sensor wire with the attached grommet from the hole in the floorpan.

3. Detach the sensor from the harness. Remove the routing clips, then remove the sensor attaching bolt and sensor.

To install:

4. Install the sensor assembly into the brake adapter hole. Install the retaining bolt and tighten to 40–60 inch lbs. (4.6–6.8 Nm).

5. Route the harness and install the sensor clips. Attach the sensor into the harness. Push through the hole in the floorpan and install the grommet.

6. Lower the vehicle. Connect the negative battery cable.

7. Turn the ignition key to the **RUN** position while watching the amber anti-lock brake warning light in the instrument cluster. The light should illuminate for approximately

2 seconds and extinguish proving out the anti-lock brake system. If the light does not illuminate or stays illuminated, a problem with the anti-lock brake system exist and must be corrected.

8. Road test the vehicle and check for proper operation.

Rear Sensor — Continental

1. Disconnect the negative battery cable. Turn the air suspension switch in the luggage compartment to the **OFF** position.

2. Disconnect the speed sensor connector in the luggage compartment.

3. Push the rubber grommet through the sheetmetal floorpan.

4. Raise and safely support the vehicle.

5. Remove the retainer clips for the sensor wire and remove the wire.

6. Loosen the speed sensor retaining screw at the caliper anchor plate and remove the sensor.

To install:

7. Install the speed sensor and tighten the retaining bolt 40–60 inch lbs. (4.5–6.8 Nm).

8. Position the wire in its normal routing position and install the retaining clips.

9. Push the speed sensor wire connector up through the hole in the floor and seat the large round grommet into the hole.

10. Lower the vehicle. Attach the speed sensor connector.

11. Connect the negative battery cable. Turn the air suspension switch to the **ON** position.

12. Turn the ignition key to the **RUN** position. The amber anti-lock brake indicator in the instrument cluster will illuminate for approximately 2 seconds and then extinguish indicating the anti-lock brake system is functional. If the amber indicator does not illuminate during the prove-out test or, at any time the anti-lock brake indicator illuminates after the initial prove-out or stays illuminated, a problem has been found in the anti-lock system and required immediate attention.

13. Road test the vehicle and check for proper operation.

1996–97 Taurus and Sable

Front Sensor

1. Disconnect the negative battery cable. Detach the anti-lock speed sensor wire located in the engine compartment.

2. Raise and safely support the vehicle.

3. Remove the plastic studs to loosen the front fender splash shield.

4. Remove the anti-lock speed sensor wire grommets at the rail bracket and from the retainer on the strut housing just above the steering knuckle.

5. Remove the anti-lock speed sensor retaining bolt and the sensor from the steering knuckle.

To install:

6. Install the anti-lock speed sensor and the retaining bolt. Tighten the retaining bolt to 90–120 inch lbs. (10.2–13.8 Nm).

7. Install the grommets at the rail bracket and the retainer at the strut housing.

8. Install the plastic studs in the front fender splash shield.

9. Lower the vehicle.

10. Connect the sensor wire to the wire harness in the wheel opening at the frame rail.

11. Connect the negative battery cable. Road test the vehicle and check for proper operation.

Rear Sensor

1. Disconnect the negative battery cable. Raise and safely support the vehicle.

2. Disconnect the anti-lock speed sensor from the mating body connector located in the center of the crossmember.

3. Disconnect the clips from the suspension arm and crossmember.

4. On station wagons, remove the sensor retaining bolt from the brake adapter and on sedans remove the bolt from the spindle.

5. Remove the anti-lock speed sensor assembly.

To install:

6. Install the anti-lock speed sensor and retaining bolt. Tighten the retaining bolt 7 ft. lbs. (10 Nm).

7. Install the sensor wiring clips to the suspension arm and crossmember.

8. Connect the anti-lock speed sensor to the mating body connector in the center of the crossmember.

9. Lower the vehicle.

10. Connect the negative battery.

11. Road test the vehicle and check for proper operation.

12. If the amber or red brake indicator lamps illuminate at any time during the road test, check the brake system for possible faults.

1995–97 Continental

CAUTION

The air suspension switch, located in the left side of the luggage compartment, must be turned OFF before raising the vehicle. Failure to do so may result in unexpected inflation or deflation of the air springs which may result in shifting of the vehicle during service.

Front Sensor

1. Turn the air suspension switch to the **OFF** position.
2. Disconnect the negative battery cable. Raise and safely support the vehicle.
3. Disconnect the speed sensor wire from the wiring harness.
4. Remove the wire from the retaining clips on the wire harness shield, front fender splash shield and the strut bracket. A prytool or similar tool may be used to open the clips on the front fender splash shield.
5. Remove the speed sensor retaining bolt and sensor from the steering knuckle.
   To install:
6. Install the speed sensor and retaining bolt. Tighten the retaining bolt to 7 ft. lbs. (10 Nm).
7. Insert the sensor routing grommets into the strut bracket. Route the wire into the fender apron clips and harness shield retainers.
8. Connect the speed sensor to the vehicle harness. Install the splash shield.
9. Turn the air suspension switch to the **ON** position.
10. Lower the vehicle. Connect the negative battery cable.
11. Turn the ignition key to the **RUN** position. The amber anti-lock brake indicator in the instrument cluster will illuminate for approximately 2 seconds and then extinguish indicating the anti-lock brake system is functional. If the amber indicator does not illuminate during the prove-out test or, at any time the anti-lock brake indicator illuminates after the initial prove-out or stays illuminated, a problem has been found in the anti-lock system and required immediate attention.
12. Road test the vehicle and check for proper operation.

Rear Sensor

1. Turn the air suspension switch to the **OFF** position.
2. Disconnect the negative battery cable. Remove the rear seat cushion by pulling upward on the front of the cushion.
3. Disconnect the speed sensor connector and feed the wire through the floorpan.
4. Raise and safely support the vehicle. Disconnect the wire from the crossmember and the lower suspension arms.
5. Remove the speed sensor retaining bolt and remove the sensor.
   To install:
6. Install the speed sensor and retaining bolt. Tighten the retaining bolt to 7 ft. lbs. (10 Nm).
7. Connect the wire to the lower control arm and the crossmember.
8. Push the speed sensor wire connector up through the hole in the floor and seat the large round grommet into the hole.
9. Lower the vehicle.
10. Connect the wiring connector and install the rear seat cushion.
11. Turn the air suspension switch to the **ON** position. Connect the negative battery cable.
12. Turn the ignition key to the **RUN** position. The amber anti-lock brake indicator in the instrument cluster will illuminate for approximately 2 seconds and then extinguish indicating the anti-lock brake system is functional. If the amber indicator does not illuminate during the prove-out test, or at any time the anti-lock brake indicator illuminates after the initial prove-out test or stays illuminated, a problem has been found in the anti-lock system and required immediate attention.
13. Road test the vehicle and check for proper operation.

FRONT SUSPENSION

Air Spring

INFLATION AND DEFLATION

The air spring must be deflated before component replacement. To properly inflate or deflate the air spring a New Generation Star (NGS) Tester or equivalent scan tool, must be used as follows:
1. When inflating or deflating air springs, make sure the vehicle is raised off the ground and safely supported on the subframe rails and rear fender-to-floorpan pinch point notches.
2. Make sure the suspension switch is in the **ON** position.
3. Turn the ignition switch to the **RUN** position.
4. Install a battery charger to reduce the battery drain.
5. Connect NGS Tester 007–00500 or equivalent, to the data link connector inside the passenger compartment.
6. Configure the Tester for the model and year vehicle.
7. Enter ACTIVE COMMAND MODES on the tester.
8. Enter AIR SUSPENSION DIAGNOSTIC CONTROL.
9. At this time, the tester will present a list of component choices.
10. Using a combination of choices on the tester will result in the inflation or deflation of individual air springs.
11. When the air spring has reached its desired fill or vent level, turn the air suspension switch to the **OFF** position. Disconnect the tester and turn the ignition switch **OFF**.

Strut

REMOVAL AND INSTALLATION

1993–95 Taurus and Sable

NOTE: Make sure to have available for each strut a new hub retainer nut, tie rod end castellated nut and cotter pin, lower control arm-to-steering knuckle bolt and nut and a strut-to-steering knuckle pinch bolt.

1. Place the ignition switch in the **OFF** position and the steering column in the UNLOCKED position.
2. With all wheels on the ground, remove the hub retainer nut. Discard the nut.
3. Loosen 3 top mount-to-strut tower retaining nuts. Do not remove the nuts at this time.
4. Raise and safely support the vehicle. Remove the wheel and tire assembly.

NOTE: When raising the vehicle, do not lift by using the lower control arms.

5. Remove the brake caliper and hang it out of the work area with wire. Do not disconnect the brake hose from the brake caliper.
6. Remove the brake rotor.
7. Remove the cotter pin and castellated nut from the tie rod end. Discard the cotter pin and nut.
8. Using Tie Rod End Remover 3290-D and Adapter T81P-3504-W or

equivalents, separate the tie rod from the steering knuckle.

9. Remove the stabilizer bar link nut and the link from the strut.

10. Remove the lower control arm-to-steering knuckle pinch bolt and nut. It may be necessary to use a drift punch to remove the bolt. Discard the pinch bolt and nut.

11. Spread the pinch joint if needed and carefully pry the lower control arm down and away from the steering knuckle.

12. If necessary, use a wheel puller to press the halfshaft out of the hub. Support the halfshaft with wire so it is not hanging on the inner CV-joint.

NOTE: Do not let the halfshaft hang by the inner CV-joint or move too far outward. The internal parts of the tripod CV-joint could be pulled apart.

13. Remove the strut-to-steering knuckle pinch bolt. Using a small prybar, spread the pinch joint and separate the strut from the steering knuckle. Remove the steering knuckle/hub assembly from the strut. Discard the pinch bolt.

14. Support the strut and remove 3 top mount-to-strut tower nuts. Lower the strut assembly from the vehicle.

To install:

15. Raise and position the strut assembly in the strut tower. Hand start 3 top mount-to-strut tower retaining nuts.

16. Install the steering knuckle and hub assembly to the strut. Install a new pinch bolt and nut and tighten to 73–97 ft. lbs. (98–132 Nm).

17. Carefully align the splines and install the halfshaft into the hub. Loosely install a new hub retainer nut.

18. Install the lower control arm to the steering knuckle, making sure the ball joint stud groove is properly positioned. Install a new bolt and nut. Tighten the nut to 40–53 ft. lbs. (53–72 Nm).

19. Install the stabilizer link to the strut and install a new nut. Tighten to 55–75 ft. lbs. (75–101 Nm).

20. Install the tie rod end to the steering knuckle using a new castellated nut. Tighten the nut to 23–34 ft. lbs. (31–47 Nm) on 1993–94 vehicles or 35–46 ft. lbs. (47–63 Nm) on 1995 vehicles and install a new cotter pin.

21. Install the disc brake rotor and caliper.

22. Install the wheel and tire assembly. Lower the vehicle.

23. Tighten 3 top mount-to-strut tower nuts to 23–30 ft. lbs. (30–40

Nm). Tighten the hub retainer nut to 180–200 ft. lbs. (244–271 Nm).

24. Apply the brake pedal several times before moving the vehicle to position the brake pads.

25. Check the front wheel alignment. Road test the vehicle and check for proper operation.

1996–97 Taurus and Sable

NOTE: Make sure new wheel hub retainer nuts, tie rod end castellated nuts, stabilizer bar link nuts and knuckle-to-strut pinch bolt/nuts are available. These parts lose their torque holding/retention capabilities during removal and must not be reused.

1. Turn the ignition switch to the **OFF** position. Leave the steering column in the UNLOCKED position.

2. With all 4 wheels on the ground, remove the wheel hub retainer nut. Discard the nut.

3. Loosen the 3 top mount-to-shock tower nuts, but do not remove the nuts at this time.

4. Raise and safely support the vehicle. Remove the wheel and tire assembly.

NOTE: When raising the vehicle, do not lift by using the lower control arms.

5. Remove the disc brake caliper. Hang the caliper out of the work area with wire to prevent damage to the brake hose.

6. Remove the disc brake rotor.

7. Remove the anti-lock brake sensor wiring harness clip and the mounting screw from the brake hose bracket on the strut assembly. Move the anti-lock brake sensor aside.

8. Remove the cotter pin and the castellated nut from the tie rod end. Discard the cotter pin and nut.

9. Using Removal tool 3290-D, or equivalent and Adapter tool T81P-3504-W or equivalent, separate the tie rod end from the steering knuckle.

10. Remove the sway bar (stabilizer bar) link nut and link from the strut. Discard the link nut.

11. Remove and discard the lower ball joint retaining nut. Using Ball Joint Remover T96P-3010-A or equivalent, separate the ball joint from the lower control arm.

12. Using Spring Compressor 164-R-3571 or equivalent, compress the coil spring until the ball joint clears the lower arm.

13. Remove the pinch bolt and nut from the bottom of the steering knuckle. It may be necessary to use a

drift punch to remove the bolt. Discard the pinch bolt and nut.

14. Separate the halfshaft from the wheel hub using Front Hub Remover/Replacer T81P-1104-C or equivalent and the required adapters.

15. Support the halfshaft with wire in a level position to prevent it from hanging by the inner CV-joint.

NOTE: Do not let the halfshaft hang by the inner CV-joint or move too far outward. The internal parts of the tripod CV-joint could be pulled apart.

16. Remove the 3 top mount-to-strut tower nuts while supporting the strut. Lower the strut assembly from the vehicle.

To install:

17. Install the strut assembly with the spring compressor installed to the vehicle and install the 3 top mount-to-strut tower nuts loosely.

NOTE: Further compress the coil spring if added clearance is required for installation.

18. Install the steering knuckle and hub assembly to the strut. Install a new pinch bolt and nut. Tighten to 73–97 ft. lbs. (98–132 Nm).

19. Install the halfshaft into the hub using care to align the splines.

20. Install the sway bar link to the strut and install a new sway bar link nut. Tighten to 55–75 ft. lbs. (75–101 Nm).

21. Install the tie rod end onto the steering knuckle using a new castellated nut. Tighten the nut to 35–46 ft. lbs. (47–63 Nm). Continue to tighten the nut until a slot lines up with the opening in the tie rod end stud and install a new cotter pin.

22. Install the anti-lock brake sensor wiring routing clip and the brake hose bracket mounting screw and tighten to 11 ft. lbs. (15 Nm).

23. Install the disc brake rotor and caliper. Tighten the caliper anchor bracket bolts to 65–87 ft. lbs. (88–118 Nm).

24. Install the wheel and tire assembly. Torque the lug nuts to 85–105 ft. lbs. (115–142 Nm).

25. Tighten the 3 top mount-to-shock tower nuts to 22–29 ft. lbs. (30–40 Nm).

26. Lower the vehicle.

27. Install a new wheel hub retainer nut. Tighten the nut to 170–202 ft. lbs. (230–275 Nm).

28. Pump the brake pedal several times prior to moving the vehicle, to position the brake pads.

29. Road test the vehicle and check for proper operation.

1993–94 Continental

1. Turn OFF the air suspension switch, located in the left side of the luggage compartment.

2. Place the ignition switch in the **OFF** position and the steering column in the **UNLOCKED** position.

3. Remove the plastic cover from the shock tower to gain access to the upper mounting nuts and dual damping actuator.

4. Remove the actuator retaining screws. Remove the actuator and place it aside.

5. Remove the hub nut.

6. Loosen the 3 top mount-to-shock tower nuts but do not remove them at this time.

7. Raise the vehicle and support it safely. Remove the wheel and tire assembly.

─────── CAUTION ───────

Do not raise the vehicle by the lower control arms.

8. Remove the brake line bracket from the strut assembly.

9. Disconnect the height sensor link from the ball stud pin at the lower control arm.

10. Disconnect the air line from the solenoid valve.

11. Disconnect the electrical connector at the solenoid valve.

12. Remove the brake caliper and the disc brake rotor. Support the caliper with wire; do not let the caliper hang by the brake hose.

13. Remove the cotter pin and castle nut from the tie rod end. Discard the cotter pin and castle nut.

14. Using tie rod end remover TOOL–3290–D and tie rod end remover adapter T81P–3504–W or equivalent, remove the tie rod from the knuckle.

15. Remove the stabilizer bar link nut and the link from the strut.

16. Remove and discard the lower arm-to-steering knuckle pinch bolt and nut. A suitable drift punch may be used to remove the bolt. Using a small prybar, slightly spread the knuckle-to-lower arm pinch joint and remove the lower arm from the steering knuckle.

17. Remove the halfshaft from the hub.

─────── CAUTION ───────

When removing the halfshaft, do not allow the halfshaft to move outward. This could result in separation of the internal parts of the tripod CV-joint, causing failure of the joint.

18. Remove the strut-to-steering knuckle pinch bolt. Using a small prybar, slightly spread the knuckle-to-strut pinch joint to remove the strut from the steering knuckle.

19. Remove the 3 top mount-to-shock tower nuts and remove the strut from the vehicle.

To install:

20. Install the strut with the 3 top mount-to-shock tower nuts and leave the nuts loose.

21. Install the steering knuckle and hub assembly to the strut. Install a new strut-to-steering knuckle pinch bolt. Tighten the bolt to 73–97 ft. lbs. (98–132 Nm).

22. Install the halfshaft into the hub. Install the lower arm to the steering knuckle and install a new pinch bolt and nut. Tighten to 40–53 ft. lbs. (54–72 Nm).

23. Install the stabilizer bar link to the strut and install a new stabilizer bar link nut. Tighten to 57–75 ft. lbs. (77–101 Nm).

24. Install the tie rod end onto the knuckle using a new castle nut. Before tightening the nut, make sure the steering wheel and wheels are in the straight-ahead position. Tighten the castle nut to 23–35 ft. lbs. (31–47 Nm). Install a new cotter pin in the castle nut.

25. Install the brake caliper and rotor.

26. Connect the electrical connector and the air line to the solenoid valve and position them properly.

27. Install the height sensor link on the ball stud pin on the control arm.

28. Install the brake line bracket to the strut assembly.

29. Install the wheel and tire assembly.

30. Tighten the 3 top mount-to-shock tower nuts to 20–30 ft. lbs. (27–40 Nm).

31. Install the dual damping actuator and the plastic shock tower cover. Correctly position the actuator wiring.

32. Refill the air spring prior to fully lowering the vehicle.

33. Lower the vehicle and tighten the hub nut to 170–202 lbs. (230–275 Nm).

34. Turn ON the air suspension.

35. Depress the brake pedal several times before moving the vehicle. Check the front end alignment.

1995–97 Continental

─────── CAUTION ───────

Do not remove an air spring under any circumstances when pressurized. Do not remove any components supporting an air spring without exhausting the air. Failure to follow these instructions may result in unexpected inflation or deflation of the air springs which may result in shifting of the vehicle during service.

1. Properly deflate the air springs.

2. Turn the ignition switch to the **OFF** position.

3. Make sure to leave the steering column in the UNLOCKED position.

4. Loosen, but do not remove, 3 top mount-to-strut tower retaining nuts.

5. Raise and safely support the vehicle. Remove the wheel and tire assembly.

6. Remove the brake line bracket from the strut assembly.

7. Disconnect the air line from the solenoid valve.

8. Disconnect the electrical connector at the air spring solenoid valve. Disconnect the front spring and strut actuator electrical connector at the body harness connector.

9. Remove the stabilizer bar link nut and the link from the strut.

10. Disconnect the air suspension height sensor from the lower height sensor ball stud and position aside.

─────── CAUTION ───────

Do not fully release the solenoid until the air is completely bled from the air spring.

11. Remove the solenoid valve as follows:

   a. Remove the air spring solenoid retainer.

   b. Rotate the solenoid valve counterclockwise to the first stop.

   c. Pull the solenoid valve straight out slowly to the second stop to bleed the air from the system.

   d. After the air is fully bled from the system, rotate the solenoid valve counterclockwise to the third stop and remove the solenoid valve from the housing.

12. Remove and discard the front suspension arm-to-ball joint nut. Using Tie Rod End Remover tool 3290-D or equivalent, separate the ball joint from the front suspension lower arm.

─────── CAUTION ───────

Do not allow the halfshaft to move outward. This could result in separation of the internal parts of the tripod CV-joint, causing failure of the joint.

13. Remove the strut-to-steering knuckle pinch bolt and nut. Using a large prybar, slightly spread the

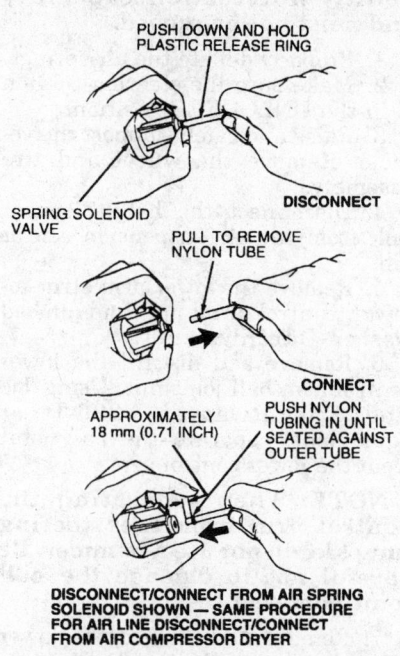

PUSH DOWN AND HOLD
PLASTIC RELEASE RING

SPRING SOLENOID
VALVE

DISCONNECT

PULL TO REMOVE
NYLON TUBE

CONNECT

APPROXIMATELY
18 mm (0.71 INCH)

PUSH NYLON
TUBING IN UNTIL
SEATED AGAINST
OUTER TUBE

**DISCONNECT/CONNECT FROM AIR SPRING
SOLENOID SHOWN — SAME PROCEDURE
FOR AIR LINE DISCONNECT/CONNECT
FROM AIR COMPRESSOR DRYER**

322291

Air line disconnect/connect procedure —
1995–97 Continental

strut-to steering knuckle pinch joint, if required, for removal.

— **WARNING** —

Do not remove the front spring and strut large center nut while the front spring and strut is in the vehicle as this may result in a permanent air leak through the top of the air spring.

---

14. Remove 3 top mount-to-strut tower retaining nuts and remove the strut/air spring assembly from the vehicle.

To install:

15. Install the strut with 3 top mount-to-shock tower nuts, leaving the nuts loose.

16. Install the steering knuckle and hub assembly to the strut. Install a new strut-to-steering knuckle pinch bolt and nut. Tighten the bolt to 73–97 ft. lbs. (98–132 Nm). Install a new control arm-to-ball joint nut. Tighten to 50–68 ft. lbs. (68–92 Nm).

17. Install the stabilizer bar link to the strut and install a new stabilizer bar link nut. Tighten to 57–75 ft. lbs. (77–101 Nm).

18. Install the height sensor link on the ball stud pin on the control arm.

19. Install the solenoid valve as follows:

   a. Check the solenoid O-ring for abrasions or cuts and replace, as necessary.

   b. Lightly grease the solenoid O-ring prior to installation using a suitable silicone dielectric compound.

   c. Insert the solenoid O-ring into the air spring end cap and rotate clockwise to the third stop. Push in to the second stop, then rotate clockwise to the first stop.

   d. Install the air solenoid valve retainer.

20. Connect the electrical connector to the solenoid valve.

21. Connect the air line to the solenoid valve.

22. Connect the front spring and shock actuator electrical connector.

23. Install the screw and brake hose bracket to the front spring and shock. Tighten the screw to 11 ft. lbs. (15 Nm).

24. Install the wheel and tire assembly. Install and tighten the lug nuts to 85–105 ft. lbs. (115–142 Nm).

25. Lower the vehicle enough to gain access to 3 top mount-to-strut tower nuts.

26. Tighten 3 top mount–to–strut tower nuts to 23–29 ft. lbs. (30–40 Nm).

27. Fill the air spring. Lower the vehicle. Turn the air suspension switch to the **ON** position. Road test the vehicle and check for proper operation.

Lower Ball Joints

REMOVAL AND INSTALLATION

1993–95 Taurus and Sable
1993–94 Continental

The lower ball joint is an integral component of the lower control arm. If the ball joint is defective, the entire lower control arm must be replaced.

1996–97 Taurus and Sable
1995–97 Continental

The lower ball joint is an integral part of the steering knuckle. If the lower ball joint is found to be defective, the entire steering knuckle must be replaced.

Lower Control Arms

REMOVAL AND INSTALLATION

1993–95 Taurus and Sable
1993–94 Continental

NOTE: Make sure to have available a new lower control arm-to-steering knuckle pinch bolt and nut, tension strut-to-control arm nut and lower control arm-to-body bolt and nut.

1. On 1993–94 Continental, turn **OFF** the air suspension switch, located in the left side of the luggage compartment.

2. Disconnect the negative battery cable. Leave the steering column UNLOCKED.

3. Raise and safely support the vehicle. Remove the wheel and tire assembly.

4. On 1993–94 Continental, disconnect the height sensor link from the ball stud pin.

5. Remove the tension strut-to-control arm nut and the dished washer. Discard the nut.

6. Remove and discard the lower control arm-to-steering knuckle pinch bolt and nut at the ball joint. Using a small prybar, spread the pinch joint and pry the control arm from the steering knuckle. Do not use a hammer. Use care not to damage the ball joint boot.

7. Remove the lower control arm-to-body nut and bolt, then remove the control arm from the body and tension strut. Discard the nut and bolt.

NOTE: Do not allow the half-shaft to move outward or the tri-pod CV-joint internal parts could separate, causing failure of the joint.

To install:

8. Insert the tension strut arm into the lower control arm inner bushing.

9. Position the lower control arm into the body bracket. Install a new bolt and nut. Tighten to 73–97 ft. lbs. (98–132 Nm).

10. Install the lower control arm ball joint stud to the steering knuckle.

11. Install a new pinch bolt and nut. Tighten to 40–53 ft. lbs. (53–72 Nm).

12. Clean the lower control arm tension strut threads. Install the dished washer and a new nut. Make sure the dished side is facing away from lower control arm bushing. Tighten the nut to 73–97 ft. lbs. (98–132 Nm).

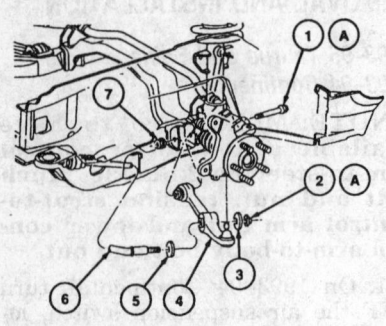

1 Bolt (2 Req'd)
2 Nut (2 Req'd)
3 Washer (2 Req'd)
4 Front Suspension Lower Arm
5 Washer (2 Req'd)
6 Front Suspension Lower Arm
  Strut
7 Nut (2 Req'd)
A Tighten to 53-72 N·m (40-53
  Lb-Ft)

323021

Lower control arm and related components —
1993–95 Taurus and Sable

13. Install the wheel and tire assembly. Lower the vehicle.
14. Check the front wheel alignment. Road test the vehicle and check for proper operation.

1996–97 Taurus and Sable

1. Leave the steering column in the UNLOCKED position.
2. Raise and safely support the vehicle. Remove the wheel and tire assembly.
3. Remove and discard the lower ball joint nut. Separate the lower ball joint from the lower control arm using Ball Joint tool T96P-3010-A or equivalent and Tie Rod End Adapter T881P-3504-W or equivalent.
4. Using Rotunda Spring Compressor 164-R-3571 or equivalent, compress the coil spring until the ball joint clears the lower arm.
5. Remove the front and rear lower control arm mounting nuts and bolts. Remove the lower control arm from the vehicle.
To install:
6. Place the lower control arm into position and install the front and rear mounting nuts and bolts. Tighten the rear nut and bolt to 72–97 ft. lbs. (98–132 Nm). Tighten the front nut and bolt to 57–75 ft. lbs. (77–103 Nm).

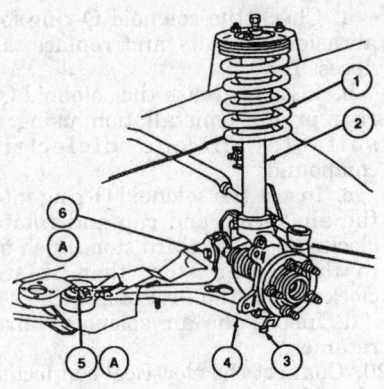

1. Front coil spring
2. Front shock absorber
3. Lower ball joint
4. Front suspension lower arm
5. Forward lower suspension
   arm mounting bolt
6. Rear lower suspension arm
   mounting bolt
A. 72-97 ft. lb.(98-132 Nm)

323037

Lower control arm and strut assembly —
1996–97 Taurus and Sable

7. Slowly release the spring compressor while guiding the ball joint into the lower control arm. Remove the spring compressor.
8. Install a new ball joint retaining nut and tighten to 50–67 ft. lbs. (68–92 Nm).
9. Install the wheel and tire assembly. Tighten the lug nuts to 85–104 ft. lbs. (115–142 Nm).
10. Lower the vehicle.
11. Road test the vehicle and check for proper operation.

1995–97 Continental

— **CAUTION** —
Do not remove an air spring under any circumstances when pressurized. Do not remove any components supporting an air spring without exhausting the air. Failure to follow these instructions may result in unexpected inflation or deflation of the air springs which may result in shifting of the vehicle during service.

NOTE: Make sure to have available, a new lower control arm-to-sub frame bracket bolt and nut, ball joint stud nut and a lower arm strut nut, per side. Once these parts are removed,

they loose their torque holding ability or retention capability and must not be reused.

1. Properly deflate the air springs.
2. Make sure the steering column is in the UNLOCKED position.
3. Raise and safely support the vehicle. Remove the wheel and tire assembly.
4. Disconnect the height sensor link from the air suspension sensor pin.
5. Remove the lower arm strut-to-lower control arm nut and dished washer. Discard the nut.
6. Remove and discard the lower control arm ball joint nut. Using Tie Rod End Remover 3290-D or equivalent, separate the ball joint from the lower control arm.

NOTE: When separating the control arm from the steering knuckle, do not use a hammer. Be careful not to damage the ball joint boot seal.

7. Remove and discard the lower control arm pivot bolt and nut.

— **WARNING** —
Do not allow the halfshaft to move outward or the tripod CV-joint internal parts could separate, causing failure of the joint.

8. Separate the lower control arm from the lower arm strut and remove from the vehicle.
To install:
9. Make sure the front washer is positioned on the lower arm strut.
10. Install the lower control arm to the lower arm strut.
11. Position the lower control arm to the subframe bracket. Install a new bolt and nut and tighten to 73–97 ft. lbs. (98–132 Nm).
12. Install the lower control arm to the ball joint stud at the steering knuckle. Be extremely careful not to damage the ball joint boot seal.
13. Install a new nut to the lower ball joint stud and tighten to 50–68 ft. lbs. (68–92 Nm).
14. Clean the lower arm strut threads to remove any dirt and contamination.
15. Install the dished washer, dished side away from the lower arm rear strut bushing, and install a new nut on the lower arm strut. Tighten to 73–97 ft. lbs. (98–132 Nm).
16. Attach the height sensor link to the front air suspension sensor pin.
17. Install the wheel and tire assembly. Install the lug nuts and tighten to 85–105 ft. lbs. (115–142 Nm).
18. Refill the air spring.

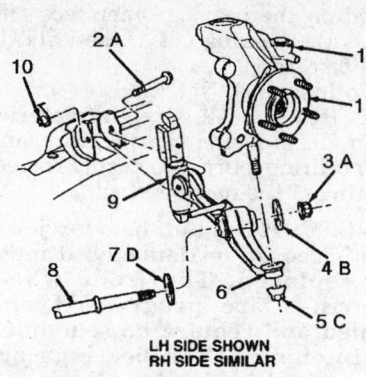

1. Front wheel knuckle
2. Bolt (2 req'd)
3. Nut (2 req'd)
4. Washer
5. Nut (2 req'd)
6. Front suspension lower arm
7. Washer (2 req'd)
8. Front suspension lower arm strut
9. Air suspension height sensor
10. Nut (2 req'd)
11. Wheel hub

A - Tighten to 98-132 Nm (72-98 lb. ft.)
B - Install with "this side out" toward rear of vehicle
C - Tighten to 68-92 Nm (50-68 lb. ft.)
D - Install with "this side out" toward front of vehicle

323017

Lower control arm and related components — 1995–97 Continental

19. Lower the vehicle.
20. Road test the vehicle and check for proper operation.

Sway Bar

REMOVAL AND INSTALLATION

Except 1995–97 Continental

NOTE: Make sure 4 new sway bar link retaining nuts, 2 sway bar insulators, 4 sway bar U-bracket bolts and 2 subframe to body retaining bolts are available. These parts lose their torque holding/retention capabilities during removal and must not be reused.

1. Raise and safely support the vehicle.

NOTE: Do not raise the vehicle on the lower control arms or subframe.

2. Remove the sway bar (stabilizer bar) link-to-sway bar nuts at each strut assembly by holding the link stud with an 8mm box wrench while removing the retaining nut with an 18mm open end wrench. Discard the sway bar link nuts.

3. Remove the sway bar link-to-strut nuts and the links; discard the nuts.

NOTE: Use care not to damage the boot seals on the sway bar links. Do not use power tools for removal or installation.

4. Remove the rack and pinion-to-subframe nuts and move the rack and pinion assembly off the subframe to allow for removal of the sway bar.
5. Place a set of jackstands under the rear of the subframe and remove the rear subframe-to-body bolts. Lower the rear of the subframe enough to gain access to the sway bar brackets.
6. Remove the sway bar U-bracket bolts, 2 per side and remove the U-brackets. Discard the 4 U-bracket bolts.
7. Remove the sway bar from the vehicle. Remove the 2 sway bar insulators and discard.
To install:
8. Clean the sway bar of contamination in the areas that the sway bar insulators are positioned.
9. Lubricate the inside diameter of the new sway bar insulators with a lubricant designed for rubber suspension insulators.
10. Install the new sway bar insulators onto the sway bar and position the insulators in their approximate locations. Make sure the slits are positioned towards the front of the vehicle.
11. Install the sway bar onto the vehicle.
12. Install the U-brackets and 4 new retaining bolts securing the sway bar to the subframe. Tighten the bolts to 22–29 ft. lbs. (30–40 Nm).
13. Raise the subframe and install 2 new subframe to body retaining bolts. Tighten the bolts to 57–76 ft. lbs. (77–103 Nm).
14. Place the rack and pinion assembly in position on the subframe and install 2 retaining nuts. Tighten the retaining nuts to 85–99 ft. lbs. (115–135 Nm).
15. Install the sway bar links to the sway bar and struts. Note the letters TOP LH and TOP RH on each link for correct positioning. Install new nuts. Hold each link stud with an 8mm box wrench while installing the new retaining nut with an 18mm open end wrench. Tighten the nuts at the strut to 57–75 ft. lbs. (77–103

Nm) and the nuts at the sway bar to 35–46 (47–63 Nm).

NOTE: Use care not to damage the boot seals on the sway bar links. Do not use power tools for removal or installation.

16. Remove the jackstands.
17. Lower the vehicle. Check the front wheel alignment.

NOTE: Whenever the vehicles subframe is removed or lowered, the wheel alignment should be checked.

18. Road test the vehicle and check for proper operation.

1995–97 Continental

------- **CAUTION** -------
The air suspension switch, located in the left side of the luggage compartment, must be turned OFF before raising the vehicle. Failure to do so may result in unexpected inflation or deflation of the air springs which may result in shifting of the vehicle during service.

NOTE: Make sure to have available 2 new sway bar (stabilizer bar) insulators, 4 sway bar bracket retaining nuts, 2 subframe to body retaining bolts and 2 sway bar link-to-sway bar retaining nut.

1. Turn the air suspension switch, located in the left side of the luggage compartment, to the **OFF** position.
2. Raise and safely support the vehicle. Support the vehicle with jackstands behind the subframe.
3. Remove and discard the nuts retaining each sway bar (stabilizer bar) link to the sway bar using an 8mm box-end wrench to hold the sway bar link stud stationary while removing the nut using an 18mm open-end wrench.

NOTE: Use care not to damage the sway bar link boot seals. Do not use power tools for removal or installation.

4. Position another set of jackstands under the rear of the subframe and remove 2 rear subframe to body retaining bolts. Lower the rear of the subframe just enough to gain access to the sway bar brackets.
5. Remove the sway bar bracket bolts and remove the sway bar from the vehicle. Discard the sway bars insulators and retaining bolts.
6. Inspect the sway bar and brackets and replace as necessary.

To install:

7. Clean the sway bar to remove dirt and contamination in the insulator mounting area.

8. Lubricate the inside diameter of 2 new insulators with a suitable rubber lubricant. Do not use petroleum or mineral based lubricants, as they will deteriorate the rubber.

9. Install the sway bar/insulators onto the sway bar and position the insulators in their approximate location.

10. Install the sway bar to the vehicle. Install new retaining bolts and tighten to 23–29 ft. lbs. (30–40 Nm).

11. Raise the rear of the subframe and install 2 new subframe-to-body retaining bolts. Tighten to 50–68 ft. lbs. (68–92 Nm).

12. Install the sway bar links to the sway bar. Install new nuts. Hold the sway bar link stud stationary and tighten to 35–46 ft. lbs. (47–63 Nm).

13. Remove the safety stands.

14. Lower the vehicle.

15. Turn the air suspension switch to the **ON** position.

NOTE: Whenever the vehicles subframe is removed or lowered, the wheel alignment should be checked.

16. Check the front wheel alignment. Road test the vehicle and check for proper operation.

Front Wheel Bearings

ADJUSTMENT

There is no adjustment for the front wheel bearings due to the cartridge design which contains both inner and outer bearings. The cartridge bearings are sealed and permanently lubricated.

REMOVAL AND INSTALLATION

1993–95 Vehicles, Except 1995 Continental

NOTE: Before beginning this procedure, make sure to have available a new hub retainer nut, tie rod end castellated nut and a lower control arm-to-steering knuckle pinch bolt and nut.

1. Turn the ignition switch to the **OFF** position. Position the steering wheel in the UNLOCKED position.

2. With all wheels on the ground, loosen and remove the hub retaining nut. Discard the nut.

3. Raise and safely support the vehicle. Remove the wheel and tire assembly.

4. Remove the cotter pin from the tie rod end stud and remove the castellated nut. Discard the cotter pin and nut.

5. Using a tie rod end removal tool, separate the tie rod end from the steering knuckle.

6. Remove the stabilizer bar link assembly from the strut.

7. Remove the brake caliper and wire it aside in order to gain working clearance.

8. Loosen but do not remove the 3 upper strut retaining nuts from the top of the strut tower.

9. Remove and discard the lower control arm-to-steering knuckle pinch bolt and nut. Using a prybar or similar tool, spread the pinch joint apart and carefully pry the lower control arm from the steering knuckle.

NOTE: Be sure the steering column is in the UNLOCKED position. Do not use a hammer to perform this operation. Use extreme care not to damage the boot seal.

10. Remove the strut-to-steering knuckle pinch bolt.

——— WARNING ———
Do not allow the halfshaft to move outboard or to hang by the inner CV-joint. Over extension of the CV-joint could result in separation of internal parts, causing failure of the joint.

11. Press the halfshaft from the hub with a wheel puller. Wire the halfshaft to the body to maintain a level position. If equipped, remove the rotor splash shield.

12. Remove the steering knuckle and hub assembly from the strut.

13. Install Front Hub Puller D80L-1002-L, or equivalent and Shaft Protector D80L-625-1 or equivalent, with the jaws of the puller on the knuckle bosses. Make sure the shaft protector is centered, clears the bearing inside diameter and rests on the end face of the hub journal. Remove the hub.

14. Remove the snapring that retains the bearing in the knuckle assembly and discard.

15. Using a suitable hydraulic press, place Front Bearing Spacer T86P-1104-A2 or equivalent, on the press plate with the step side facing up and position the knuckle with the outboard side up on the spacer. Install Front Bearing Remover T83P-1104-AH2 or equivalent, cen-

tered on the bearing inner race and press the bearing out of the knuckle and discard.

To install:

16. Remove all foreign material from the knuckle bearing bore and hub bearing journal to ensure correct seating of the new bearing.

NOTE: If the hub bearing journal is scored or damaged it must be replaced. The front wheel bearings are pregreased and sealed and require no scheduled maintenance. The bearings are preset and cannot be adjusted. If a bearing is disassembled for any reason, it must be replaced as a unit, as individual service seals, rollers and races are not available.

17. Place Front Bearing Spacer T86P-1104-A2 or equivalent, with the step side down on the hydraulic press plate and position the knuckle with the outboard side down on the spacer. Position a new bearing in the inboard side of the knuckle. Install Bearing Installer T86P-1104-A3 or equivalent, with the undercut side facing the bearing, on the bearing outer race and press the bearing into the knuckle. Make sure the bearing seats completely against the shoulder of the knuckle bore.

NOTE: Bearing Installer T86P-1104-A3 or equivalent, must be positioned as indicated above to prevent bearing damage during installation.

18. Install a new snapring (part of the bearing kit) in the knuckle groove.

19. Place Front Bearing Spacer T86P-1104-A2 or equivalent, on the press plate and position the hub on the tool with the lugs facing downward. Position the knuckle assembly with the outboard side down on the hub barrel. Place Bearing Remover T83P-1104-AH2 or equivalent, flat side down, centered on the inner race of the bearing and press down on the tool until the bearing is fully seated onto the hub. Make sure the hub rotates freely in the knuckle after installation.

20. Prior to hub/bearing/knuckle installation, replace the bearing dust seal on the outboard CV-joint with a new seal from the bearing kit. Make sure the seal flange faces outboard toward the bearing. Use Drive Tube T83T-3132-A1 and front bearing Dust Seal Installer T86P-1104-A4, or equivalent.

21. Install the rotor splash shield using new rivets, if equipped.

22. Install the steering knuckle onto the strut. Loosely install a new pinch bolt in the knuckle to retain the strut.

23. Install the steering knuckle and hub onto the halfshaft. Loosely install a new hub retaining nut.

24. Install the lower control arm to the knuckle. Be sure the ball stud groove is properly positioned. Install a new nut and bolt. Torque to 40–53 ft. lbs. (53–72 Nm).

25. Tighten the strut-to-knuckle pinch bolt to 70–95 ft. lbs. (98–132 Nm).

26. Install the disc brake rotor and brake caliper.

27. Position the tie rod to the steering knuckle, install a new castellated nut and tighten to 35 ft. lbs. (47 Nm). If necessary advance the nut to align the slot and install a new cotter pin.

28. Install the stabilizer bar link to the strut. Torque the nut to 57–75 ft. lbs. (77–103 Nm).

29. Install the wheel and tire assembly. Torque the lug nuts to 85–105 ft. lbs. (115–142 Nm).

30. Lower the vehicle.

31. Tighten 3 upper strut retaining nuts to 23–29 ft. lbs. (30–40 Nm).

32. With all wheels on the ground, tighten the hub retaining nut to 180–200 ft. lbs. (245–275 Nm).

33. Pump the brake pedal several times prior to moving the vehicle, in order to position the brake pads.

34. Road test the vehicle and check for proper operation.

1996–97 Taurus and Sable

NOTE: Make sure new wheel hub retainer nuts, tie rod end castellated nuts, hub-to-knuckle retaining bolts, knuckle-to-strut pinch bolt/nut and inboard half-shaft circlips are available. These parts lose their torque holding/retention capabilities during removal and must not be reused.

1. Turn the ignition switch to the **OFF** position. Place the steering column in the UNLOCKED position.

2. Remove the wheel hub retainer nut before raising the vehicle off the ground. Discard the wheel hub retainer nut.

3. Raise and safely support the vehicle. Remove the wheel and tire assembly.

NOTE: When raising the vehicle, do not lift by using the lower control arms.

4. Remove the wheel and tire assembly.

5. Remove the cotter pin and the castellated nut from the tie rod end. Discard the cotter pin and nut.

6. Separate the tie rod end from the steering knuckle using Remover tool 3290-D and adapter T81P-3504-W, or equivalents.

7. Remove the stabilizer link from the strut. Remove the disc brake caliper and hang it aside.

8. Remove the disc brake rotor.

9. Remove the anti-lock sensor and move it aside.

10. Remove and discard the lower ball joint retaining nut. Using Ball Joint Remover T96P-3010-A or equivalent, separate the ball joint from the lower control arm.

11. Using Rotunda Spring Compressor 164-R-3571 or equivalent, compress the coil spring until the ball joint clears the lower control arm.

12. Remove and discard the steering knuckle-to-strut pinch bolt and nut.

13. Separate the halfshaft from the wheel hub using Front Hub Remover/Replacer T81P-1104-C or equivalent and adapters.

14. Support the halfshaft with wire in a level position to prevent it from hanging by the inner CV-joint.

NOTE: Do not let the halfshaft hang by the inner CV-joint or move too far outward. The internal parts of the tripod CV-joint could be pulled apart.

15. Separate the steering knuckle from the strut assembly and place on a suitable workbench.

16. Remove the 3 hub and bearing retainer bolts from the back of the steering knuckle while using a prybar to steady the assembly. Discard the 3 hub and bearing retainer bolts.

——— WARNING ———
The wheel hub is not pressed into the front wheel knuckle. DO NOT USE a slide hammer to remove a stuck wheel hub. Do not strike the back of the inner bearing race.

17. Remove the wheel hub from the steering knuckle using a suitable prybar.

18. Inspect all components and replace as necessary. The wheel bearings are not serviceable and must be replaced with a new wheel hub assembly.

To install:

19. Install the disc brake rotor shield using new rivets, if removed.

NOTE: If the hub bearing journal is scored or damaged, replace the steering knuckle. If the wheel hub is damaged or any endplay is detectable, replace the wheel hub.

20. Remove all foreign material from the knuckle bearing bore and hub bearing journal to ensure correct seating of the new hub.

NOTE: The knuckle must be clean enough to allow the wheel hub to be completely seated by hand. Do not press or draw the wheel hub into place.

21. Place the wheel hub to the steering knuckle using light oil. Push the wheel hub assembly into the steering knuckle. Install bolts and tighten to 61–78 ft. lbs. (83–107 Nm).

22. Position the steering knuckle assembly to the vehicle.

23. Install the steering knuckle to the strut and loosely install a new pinch bolt.

24. Install the steering knuckle and hub assembly onto the halfshaft. Make sure the splines are properly aligned.

25. Slowly release Rotunda Spring Compressor 164-R-3571 or equivalent, while guiding the lower ball joint into the lower control arm.

26. Remove the spring compressor.

27. Install a new nut on the lower ball joint stud and tighten to 50–67 ft. lbs. (68–92 Nm).

28. Install a new nut on the steering knuckle-to-strut pinch bolt. Tighten the pinch bolt nut to 72–97 ft. lbs. (98–132 Nm)

29. Position the tie rod end to the steering knuckle. Install a new castellated nut and tighten to 35–46 ft. lbs. (47–63 Nm). Install a new cotter pin.

30. Install the sway bar link and tighten the nut to 57–75 ft. lbs. (77–103 Nm).

NOTE: Use care not to damage the sway bar link boot seals. Do not use power tools to tighten the nuts or seal damage will result.

31. Install the disc brake rotor and disc brake caliper. Tighten the caliper anchor bracket bolts to 65–87 ft. lbs. (88–118 Nm).

32. Install the wheel and tire assembly.

33. Lower the vehicle.

34. Install a new wheel hub retainer nut. Tighten the nut to 170–202 ft. lbs. (230–275 Nm).

35. Pump the brake pedal several times prior to moving the vehicle, to position the brake pads.

36. Road test the vehicle and check for proper operation.

1995–97 Continental

— **CAUTION** —

Do not remove an air spring under any circumstances when pressurized. Do not remove any components supporting an air spring without exhausting the air. Failure to follow these instructions may result in unexpected inflation or deflation of the air springs which may result in shifting of the vehicle during service.

NOTE: Make sure to have available a new wheel hub retainer nut, tie rod end castellated nut and cotter pin, lower control arm-to-ball joint nut, steering knuckle-to-strut pinch bolt and nut, stabilizer bar link nut and 3 hub and bearing retaining bolts, per side.

1. Properly deflate the air springs.
2. Loosen, but do not remove nuts from the top of the strut tower.
3. Remove and discard the wheel hub retainer nut.
4. Make sure the steering column is in the UNLOCKED position.
5. Raise and safely support the vehicle. Remove the wheel and tire assembly.
6. Remove the air suspension height sensor and position aside. Remove the brake hose bracket mounting screw and bracket from the strut.
7. Remove and discard the cotter pin and nut from the tie rod end.
8. Using Tie Rod End Remover 3290-D or equivalent, separate the tie rod end from the steering knuckle.
9. Remove the stabilizer bar link from the strut assembly.
10. Remove the brake caliper and rotor. Wire the caliper aside.

NOTE: Do not allow the brake caliper to hang by the brake hose or damage to the hose may result.

11. Remove and discard the lower control arm-to-ball joint nut. Using Tie Rod End Remover 3290-D or equivalent, separate the ball joint from the lower control arm.

NOTE: Do not allow the halfshaft to move outboard. Over extension of the CV-joint could result in separation of internal parts, causing failure of the joint.

12. Using Front Hub Remover/Replacer T81P-1104-C or equivalent, press the outer CV-joint from the hub.
13. Support the halfshaft with wire to maintain a level position.
14. Remove the strut-to-steering knuckle pinch bolt and nut. Using a small prybar, slightly spread the strut-to-knuckle pinch joint and remove the knuckle and hub from the strut.
15. Remove the brake rotor splash shield, if damaged.
16. Remove and discard 3 hub and bearing retainer bolts from the steering knuckle.

— **WARNING** —

The wheel hub and bearing assembly is not pressed into the steering knuckle. DO NOT use a slide hammer to remove a stuck wheel hub. Do not strike back of inner bearing race.

17. Remove the wheel hub from the steering knuckle.

To install:

NOTE: If the hub bearing journal is scored or damaged, replace the knuckle. If the wheel hub is damaged or any endplay is detectable, replace the wheel hub and bearing assembly.

18. Remove all foreign material from the knuckle bearing bore and hub bearing journal to ensure correct seating of the new hub.

NOTE: The steering knuckle must be clean enough to allow the wheel hub to be completely seated by hand. Do not press or draw the wheel hub into place.

19. Install the wheel hub and tighten bolts to 61–79 ft. lbs. (83–107 Nm).
20. Install a new brake rotor splash shield using new rivets, if necessary.
21. Install the steering knuckle onto the strut and loosely install a new pinch bolt and nut.
22. Install the steering knuckle and hub to the halfshaft.
23. Install the lower control arm to the steering knuckle.
24. Install a new nut to the lower ball joint stud and tighten to 50–68 ft. lbs. (68–92 Nm). Tighten the strut-to-knuckle pinch bolt to 73–97 ft. lbs. (98–132 Nm)
25. Install the brake rotor and brake caliper.
26. Position the tie rod to the steering knuckle. Install a new castellated nut and tighten to 23–34 ft. lbs. (31–47 Nm).

27. Install a new wheel hub retainer nut and tighten to 170–202 ft. lbs. (230–275 Nm).
28. Install the stabilizer bar link and tighten nut to 56–77 ft. lbs. (76–104 Nm). Install the sway bar link cover.
29. Install the brake hose bracket. Install the air suspension height sensor. Install the wheel and tire assembly.
30. Lower the vehicle enough to gain access to 3 top mount-to-strut tower retaining bolts. Torque the nuts to 23–29 ft. lbs. (30–40 Nm).
31. Fill the air springs. Lower the vehicle.
32. Pump the brake pedal several times prior to moving the vehicle, to position the brake pads to the rotor. Road test the vehicle and check for proper operation.

REAR SUSPENSION

Air Spring

INFLATION AND DEFLATION

The air spring must be deflated before component replacement. To properly inflate or deflate the air spring a New Generation Star (NGS) Tester or equivalent scan tool, must be used as follows:

1. When inflating or deflating air springs, make sure the vehicle is raised off the ground and safely supported on the subframe rails and rear fender-to-floorpan pinch point notches.
2. Make sure the suspension switch is in the **ON** position.
3. Turn the ignition switch to the **RUN** position.
4. Install a battery charger to reduce the battery drain.
5. Connect NGS Tester 007–00500 or equivalent, to the data link connector inside the passenger compartment.
6. Configure the Tester for the model and year vehicle.
7. Enter ACTIVE COMMAND MODES on the tester.
8. Enter AIR SUSPENSION DIAGNOSTIC CONTROL.
9. At this time, the tester will present a list of component choices.
10. Using a combination of choices on the tester will result in the inflation or deflation of individual air springs.

11. When the air spring has reached its desired fill or vent level, turn the air suspension switch to the **OFF** position. Disconnect the tester and turn the ignition switch **OFF**.

Strut

REMOVAL AND INSTALLATION

Taurus and Sable Sedans

NOTE: Before continuing, make sure to have available, 3 new upper strut-to-body retaining nuts and 1 new strut-to-spindle pinch bolt; per strut.

1. Raise the luggage compartment lid. Loosen but do not remove the upper strut-to-body nuts.
2. Raise and support the rear of the vehicle safely. Remove the wheel and tire assembly.
3. Remove the brake differential control valve-to-control arm bolt. Using a wire, secure the differential valve to the body to ensure there is enough clearance for the strut removal.
4. Remove the brake hose-to-strut bracket clip and move the hose aside.
5. If equipped, disconnect the electronic strut actuator.
6. If equipped, remove the stabilizer bar U-bracket from the vehicle.
7. If equipped, remove the stabilizer bar-to-stabilizer link nut, washer and insulator, then separate the stabilizer bar from the link.

NOTE: When removing the strut, be sure the rear brake flex hose is not stretched or the steel brake tube is not bent.

8. Remove the tension strut-to-spindle nut, washer and insulator. Move the spindle rearward to separate it from the tension strut.
9. Remove the strut-to-spindle pinch bolt. If necessary, use a medium prybar and spread the strut-to-spindle pinch joint to remove the strut. Discard the bolt.
10. Lower the jackstand and separate the strut from the spindle.
11. Remove the nut, washer and insulator attaching link to strut and remove link.
12. Support the strut, then remove the top strut-to-body nuts and the strut.

To install:
13. Install the stabilizer link in the strut bracket, the insulator, washer and nut. Tighten to 5–7 ft. lbs. (7–9.5 Nm).

14. Insert the 3 upper mount studs into the strut tower in the apron and hand start 3 new nuts.
15. Partially raise and safely support the vehicle.
16. Complete the installation by reversing the removal procedures and torquing the following components to:
   a. Strut-to-spindle pinch bolt — 50–68 ft. lbs. (68–92 Nm).
   b. Tension strut-to-spindle nut — 35–46 ft. lbs. (47–63 Nm).
   c. Link-to-stabilizer bar nut — 5–7 ft. lbs. (7–9.5 Nm).
   d. Stabilizer bar U-bracket-to-body bolt — 25–33 ft. lbs. (34–46 Nm).
   e. Strut top mount-to-body nuts — 19–25 ft. lbs. (26–34 Nm).
17. Install the wheel and tire assembly and lower the vehicle.

1993–94 Continental

1. Turn **OFF** the air suspension switch located in the luggage compartment.
2. From inside the luggage compartment, disconnect the electrical connector from the dual dampening actuator.
3. Loosen the strut-to-upper body nuts.
4. Raise and support the vehicle safely. Remove the wheel and tire assembly.

— **CAUTION** —
Do not raise the vehicle by the tension strut.

5. Disconnect the air line and electrical connector from the solenoid valve.
6. Remove the brake hose retainer at the strut bracket.
7. Disconnect the parking brake cable from the brake caliper. Remove all the wire retainers and parking brake cable retainers from the lower suspension arm.
8. Disconnect the height sensor link from the ball stud pin on the lower arm.
9. Remove the caliper and position it aside with a piece of wire. Do not kink or place a load on the brake hose.
10. Bleed the air spring by performing the following:
   a. Remove the solenoid clip.
   b. Rotate the solenoid counterclockwise to the first stop.
   c. Slowly pull the solenoid straight out to the second stop and bleed the air from the system.

— **CAUTION** —
Do not fully release the solenoid until the air is fully bled from the spring or personal injury may result.

   d. After the air is fully bled from the system, rotate the solenoid to the 3rd stop and remove the solenoid from the housing.
11. Mark the position of the notch on the toe adjustment cam.
12. Remove the torsion spring clamp from the spindle-to-strut bolt.
13. Remove the suspension arm nut from the inboard bushing.
14. Install torsion spring remover tool T88P–5310–A or equivalent, on the suspension arm. Pry up on the tool and arm using a ¾ in. drive ratchet to relieve the pressure on the pivot bolt. An assistant may be required to pull outboard on the spindle simultaneously to fully relieve the tension on the bolt. Remove the bolt and lower arm. Repeat this procedure for the opposite arm.
15. Remove the torsion spring from the arms. Remove the stabilizer U-bracket from the body.
16. Remove the stabilizer bar-to-link nut, washer and insulator. Separate the stabilizer bar from the link.
17. Remove the tension strut-to-spindle nut, washer and insulator. Move the spindle rearward enough to separate it from the tension strut.
18. Remove and discard the strut-to-spindle pinch bolt. With a suitable prybar, spread the strut-to-spindle pinch joint as required to assist in removing the bolt.
19. Separate the spindle from the strut. Remove the spindle as an assembly with the arms attached.
20. From inside the luggage compartment area, support the shock strut by hand and remove and discard the upper mount-to-body nuts. Care should be taken not to drop the strut when removing the upper nuts. Guide the electric actuator wire through the opening to prevent snagging and damage while removing the strut assembly.

To install:
21. Install the solenoid valve on the air spring.
22. Guide the electric actuator wire through the opening and install the strut assembly. Install new upper mount nuts.
23. Install the spindle and arms to the strut. Install a new strut-to-spindle pinch bolt. Do not tighten the bolt until the control arms are attached to the body and the cams are centered.

24. Complete the installation by reversing the removal procedures and torque the following items to:

a. Tension strut-to-spindle nut — 35–50 ft. lbs. (48–68 Nm).

b. Stabilizer bar-to-link nut — 5–7 ft. lbs. (7–9.5 Nm).

c. Stabilizer U-bracket-to-body bolt — 25–37 ft. lbs. (34–50 Nm).

d. Spindle-to-strut bolt — 51–70 ft. lbs. (68–95 Nm).

25. Set the toe adjustment cam to the alignment mark.

26. Tighten the 3 nuts retaining the strut to the upper body to 19–26 ft. lbs. (26–35 Nm).

27. From inside the luggage compartment, connect the electrical connector for the dual dampening actuator.

28. Refill the air springs. Lower the vehicle all of the way.

29. Check the toe setting and adjust if necessary. Tighten the inboard bushing nut to 45–65 ft. lbs. (61–88 Nm).

Shock Absorber

REMOVAL AND INSTALLATION

Taurus and Sable Wagons

NOTE: Before continuing, make sure new mounting bolts and nuts and shock absorber insulator bushings are available.

1. Raise the vehicle enough to allow wheel and tire removal. Safely support the vehicle.

2. Remove the wheel and tire assembly. Position a jackstand under the lower control arm.

3. On 1993–95 models, remove the 2 nuts retaining the shock absorber to the lower suspension arm.

4. From inside the vehicle, remove the rear compartment access panel.

── **WARNING** ──
The lower control arm must be supported before removal of the upper or lower shock absorber attachments to prevent injury or damage to attached components.

5. Remove the top shock absorber retaining nut using a crow foot wrench and ratchet while holding the shock absorber shaft stationary with an open-end wrench. Do not grip the shaft of the shock absorber if it is to be reused. Discard the retaining nut.

6. Remove the upper washer and insulator from the shock absorber.

NOTE: The shock absorbers are gas filled. It will require an effort to collapse the shock to remove it from the lower control arm.

7. On 1996–97 models, remove the lower shock absorber mounting nut and bolt.

8. Remove the shock absorber from the vehicle. Discard the nut and bolt.

To install:

9. Install a new washer and insulator on the upper shock absorber rod.

10. Maneuver the upper part of the shock absorber into the shock tower opening in the body. Push slowly on the lower part of the shock absorber until the lower bracket is aligned with the mounting holes in the lower control arm.

11. On 1996–97 models, install a new retaining bolt and nut; then, tighten to 50–68 ft. lbs. (68–92 Nm).

12. From inside the vehicle, install a new insulator, washer and nut on top of the shock absorber shaft. Torque the nut to 19–25 ft. lbs. (26–34 Nm.).

13. Install the rear compartment access panel.

14. On 1993–95 models, torque the 2 lower attaching nuts to 15–19 ft. lbs. (19–26 Nm).

15. Install the wheel and tire assembly. Remove the jackstand.

16. Lower the vehicle. Road test the vehicle and check for proper operation.

Coil Spring

REMOVAL AND INSTALLATION

Taurus and Sable Wagons

1. Raise and safely support the vehicle. Remove the wheel and tire assembly

2. Position a floor jack under the lower suspension control arm.

── **CAUTION** ──
The lower control arm must be supported before removal of the upper or lower shock absorber mounts to prevent injury or damage to the related components due to tension applied by the coil spring.

3. On 1996–97 models, perform the following procedures:

a. Remove the bolt retaining the rear brake hose bracket to the body.

b. Remove the stabilizer bar and bracket from the lower control arm.

c. Using the floor jack, slowly raise the lower control arm to normal curb height.

d. From inside the vehicle, remove the rear compartment access panel.

4. Remove and discard the top shock absorber retaining nut using a crows foot wrench while holding the shaft with and open end wrench.

── **CAUTION** ──
The shock absorbers are gas-filled. It will require an effort to collapse the shock in order to remove the shock from the lower arm.

5. Remove the lower shock mounting nut and bolt and remove the shock absorber.

6. On 1993–95 models, perform the following procedures:

a. Disconnect and remove the parking brake cable and clip from the lower suspension arm.

b. If equipped with rear disc brakes, remove the ABS cable from the clips on the lower suspension arm.

c. Remove and discard the bolt and nut attaching the tension strut to the lower suspension arm.

d. Suspend the spindle and upper suspension arms from the body with a piece of wire to prevent them from dropping.

e. Remove the nut, bolt, washer and adjusting cam that retain the lower suspension arm to the spindle. Discard the nut, bolt and washer and replace with new. Set the cam aside.

7. On 1996–97 models, perform the following procedures:

a. Install Spring Cage 164-R3555 or equivalent, on the coil spring.

b. Remove and discard the upper ball joint nut. Separate the upper ball joint from the wheel spindle.

8. Slowly lower the lower control arm using the floor jack until the tension is relaxed on the coil spring. Remove the coil spring and the upper and lower spring insulators.

To install:

9. Place the lower spring insulator on the lower control arm. Press the insulator downward into place, making certain that the insulator is properly seated.

10. Position the upper insulator on top of the coil spring. Install the coil spring on the lower control arm. Make certain the spring is properly seated.

11. Using the floor jack, slowly raise the lower control arm. Guide the upper spring insulator onto the upper spring seat on the underbody.

12. Position the upper ball joint into the upper control arm. Install a new nut and tighten to 50–68 ft. lbs. (68–92 Nm).

13. On 1993–95 models, perform the following procedures:

   a. Position the spindle in the lower suspension arm with a new bolt, nut, washer, and the existing cam. Install the bolt with the head of the bolt toward the front of the vehicle. Do not tighten the bolt at this time.

   b. Remove the wire supporting the spindle and suspension arms.

   c. Install the tension strut in the lower suspension arm using a new nut and bolt; do not tighten at this time.

   d. Attach the parking brake cable and clip to the lower suspension arm.

   e. If equipped with rear disc brakes, install the ABS cable into the clips on the lower suspension arm.

14. Position the shock absorber into the tower opening with a new washer and insulator installed. Push on the lower end of the shock until the lower bracket is lined up with the mounting holes in the lower control arm. Install a new lower retaining bolt and nut. Tighten to 50–68 ft. lbs. (68–92 Nm).

15. From inside the vehicle, install a new upper shock absorber insulator and washer. Tighten the nut to 19–25 ft. lbs. (25–34 Nm).

16. On 1993–95 models, perform the following procedures:

   a. Attach the sway bar U-bracket to the lower suspension arm using a new bolt. Torque the bolt to 23–30 ft. lbs. (30–40 Nm).

   b. Attach the flexible brake hose to the body and tighten the bolt to 8–12 ft. lbs. (11–16 Nm).

   c. With the floor jack, raise the rear suspension arm and bushing to normal position when at curb height. Tighten the rear suspension arm and bushing to rear wheel spindle nut to 40–52 ft. lbs. (54–71 Nm). Tighten the rear suspension tension strut and bushing to body bracket bolt to 40–52 ft. lbs. (54–71 Nm).

17. On 1996–97 models, perform the following procedures:

   a. Install the rear compartment access panel.

   b. Install the stabilizer bar and bracket to the lower control arm. Tighten to 15–19 ft. lbs. (19–26 Nm).

18. Install the brake hose support bracket to the body and install the retaining bolt. Tighten the bolt to 10 ft. lbs. (12 Nm).

19. Install the wheel and tire assembly. Torque the lug nuts to 85–105 ft. lbs. (115–142 Nm).

20. Remove the floor jack.

21. Lower the vehicle.

22. Check the rear wheel alignment and adjust if necessary.

23. Road test the vehicle and check for proper operation.

Air Spring

REMOVAL AND INSTALLATION

1995–97 Continental

CAUTION

Do not remove an air spring under any circumstances when pressurized. Do not remove any components supporting an air spring without exhausting the air. Failure to follow these instructions may result in unexpected inflation or deflation of the air springs which may result in shifting of the vehicle during service.

---

1. Properly deflate the air springs.

2. Raise and safely support the vehicle. Remove the wheel and tire assembly.

3. Disconnect the air spring solenoid electrical connector and air line.

4. Depress 2 plastic tabs on the air spring cap and rotate downward to

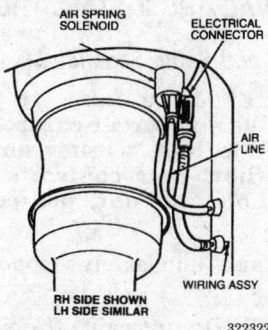

Air spring and solenoid assembly — 1995–97 Continental

disengage the air spring cap from the body bracket.

5. Remove 3 air spring-to-lower suspension bolts. Remove the air spring from the vehicle.

To install:

6. Position the air spring in the vehicle.

7. Install 3 air spring-to-lower suspension arm bolts. Tighten to 50–59 ft. lbs. (68–82 Nm).

8. Engage the air spring cap to the body bracket.

9. Connect the solenoid air line and the electrical connector.

10. Fill the air spring.

11. Install the wheel and tire assembly. Install and tighten the lug nuts to 80–105 ft. lbs. (108–144 Nm).

12. Lower the vehicle.

13. Turn the air suspension switch to the **ON** position.

14. Road test the vehicle and check for proper operation.

Upper Control Arms

REMOVAL AND INSTALLATION

Taurus and Sable Station Wagons

1993–95 Models

1. Raise the vehicle and support it safely with wood blocks on jackstands so the suspension is at normal curb height.

2. Remove the wheel and tire assembly.

3. Remove the brake line flexible hose bracket from the body.

4. Loosen, but do not remove the nuts attaching the spindle to the upper and lower suspension arms.

5. Remove and discard the nuts and bolts attaching the front and rear upper suspension arms to the body brackets. Make sure the spindle does not fall outward.

6. Tilt the top of the spindle outward, letting it pivot on the lower suspension arm attaching bolt until the ends of the upper suspension arms are clear of the body bracket. Support the spindle with wire in this position.

7. Remove and discard the nut attaching the upper suspension arms to the spindle and remove the arms from the vehicle.

To install:

8. Install the upper suspension arms on the spindle and install a new nut but do not tighten the nut at this time.

9. Position the upper suspension arm ends to the body bracket and install new nuts and bolts. Tighten to

70–95 ft. lbs. (95–129 Nm). Remove the wire from the spindle.

10. Tighten the nut attaching the upper suspension arms to the spindle to 150–190 ft. lbs. (204–257 Nm). Tighten the nut attaching the lower suspension arm to the spindle to 40–52 ft. lbs. (54–71 Nm).

11. Install the brake line bracket to the body. Tighten to 8–12 ft. lbs. (11–16 Nm).

12. Install the wheel and tire assembly, remove the jackstand and wood block and lower the vehicle.

13. Check the rear wheel alignment.

1996–97 Models

NOTE: Make sure a new upper ball joint nut is available.

1. Raise and safely support the vehicle.

2. Place a jackstand under the lower control arm and raise the lower control arm to normal curb height.

3. Remove the wheel and tire assembly. Remove the brake hose bracket retaining bolt and separate the bracket from the body.

4. Remove and discard the upper ball joint nut. Separate the ball joint from the wheel spindle.

5. Remove the upper control arm-to-body nuts and bolts. Remove the upper control arm.

6. Inspect the upper control arm and bushings. The bushings are not serviceable.

To install:

7. Place the upper control arm in position and install the upper control arm-to-body nuts and bolts. Do not tighten at this time.

8. Install the ball joint stud into the upper control arm and install a new nut. Tighten the nut to 50–68 ft. lbs. (68–92 Nm).

9. Tighten the upper control arm-to-body nuts and bolts to 73–97 ft. lbs. (98–132 Nm).

10. Install the brake line bracket and retaining bolt to the body. Tighten the bolt to 10 ft. lbs. (12 Nm).

11. Install the wheel and tire assembly. Torque the lug nuts to 85–105 ft. lbs. (115–142 Nm).

12. Remove the jackstand. Lower the vehicle.

13. Check the rear wheel alignment. Road test the vehicle and check for proper operation.

1995–97 Continental

————— **CAUTION** —————

Do not remove an air spring under any circumstances when there is pressure in the air spring.

Do not remove any components supporting an air spring without exhausting the air. Failure to follow these instructions may result in unexpected inflation or deflation of the air springs which may result in shifting of the vehicle during service.

NOTE: Make sure to have available a new ball joint stud-to-upper control arm retaining nut.

1. Properly deflate the air springs.

2. Raise and safely support the vehicle. Remove the wheel and tire assembly.

3. Disconnect the air suspension height sensor from the ball stud pin.

4. Loosen but do not remove the upper ball joint-to-upper control arm nut. Using Tie Rod End Remover 3290-D or equivalent, separate the ball joint from the upper control arm. Remove the nut and discard.

5. Remove the upper control arm-to-body nuts and bolts. Remove the upper control arm from the vehicle.

To install:

6. Place the upper control arm in the vehicle.

7. Install the upper control arm-to-body bolts and nuts. Tighten to 72–98 ft. lbs. (97–132 Nm).

8. Install the ball joint stud into the upper control arm. Install a new nut and tighten to 50–68 ft. lbs. (68–92 Nm).

9. Install the air suspension height sensor onto the ball stud pin.

10. Install the wheel and tire assembly. Install and tighten the lug nuts to 85–105 ft. lbs. (115–142 Nm).

11. Fill the air springs. Lower the vehicle.

12. Turn the air suspension switch to the **ON** position. Road test the vehicle and check for proper operation.

Lower Control Arms

REMOVAL AND INSTALLATION

Taurus and Sable Sedans

NOTE: Make sure to have available a new rear control arm-to-spindle bolt, washer and nut and a new rear control arm-to-body bolt and nut, per control arm.

1. Raise and safely support vehicle.

NOTE: Do not raise the vehicle by the tension strut.

2. Disconnect the brake load sensor proportioning valve from the left front control arm, if that control arm is being replaced.

3. Disconnect the parking brake rear cable from the control arm being removed.

4. Remove and discard the control arm-to-spindle bolt, washer and nut.

5. Remove and discard the control arm-to-body bolt and nut. Remove the control arm from the vehicle.

6. Inspect the control arm bushings. If worn or damaged, the control arm must be replaced as an assembly.

To install:

NOTE: When installing new control arms, the offset on all arms must face up. The arms are stamped BOTTOM on the lower edge. The flange edge of the right rear arm stamping must face the front of the vehicle. The other 3 control arms must face the rear of the vehicle. The rear control arms have 2 adjustment cams that fit inside the bushings at the control arm-to-body attachment points.

7. Position the control arm and if equipped, the adjustment cam. Insert a new bolt and install a new nut, but do not tighten at this time.

8. Move the control arm end up to the spindle and insert a new bolt, washer and nut. Tighten the nut to 44–59 ft. lbs. (59–81 Nm).

9. Tighten the control arm-to-body nut to 50–68 ft. lbs. (68–92 Nm) while preventing the bolt from turning.

10. If a front control arm was removed, attach the parking brake cable.

11. If the left front control arm was removed, install the brake load sensor proportioning valve. Lower the vehicle.

12. Check the rear wheel alignment and adjust as necessary. Road test the vehicle and check for proper operation.

Taurus and Sable Station Wagons

1993–95 Models

1. Raise and support the vehicle safely on the lifting pads on the underbody forward of the tension strut body bracket.

2. Remove the wheel and tire assembly. Place a floor jack under the lower suspension arm.

3. Remove the bracket retaining the flexible brake hose to the body. Remove the stabilizer bar U-bracket from the lower suspension arm.

4. Remove and discard the nuts attaching the shock absorber to the lower suspension arm. Remove the

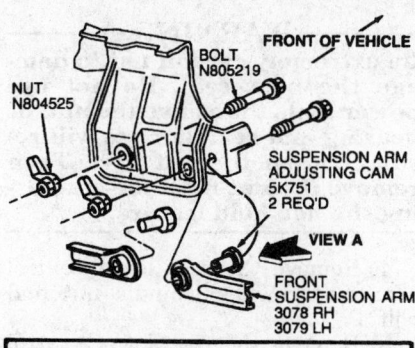

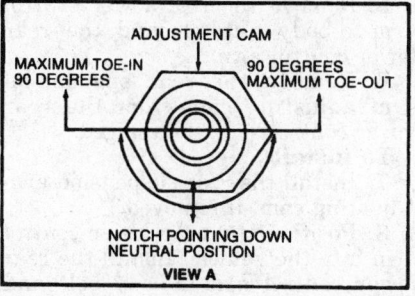

Lower control arm installation — 1993–95
Taurus and Sable sedan

239415

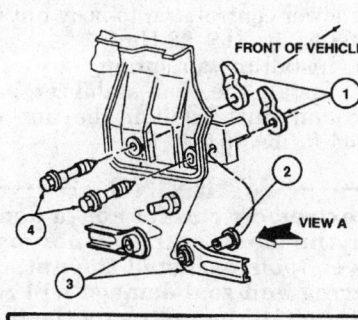

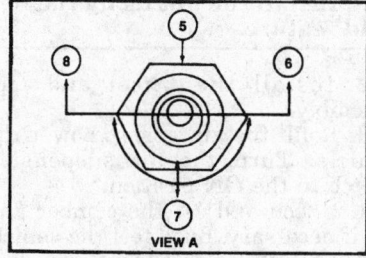

1. Nut
2. Rear suspension arm adjusting
   Cam kit (2 req'd)
3. Rear suspension arm and bushing
4. Bolt
5. Adjustment Cam
6. 90 degrees - maximum toe-out
7. Notch pointing down - neutral position
8. 90 degrees - maximum toe-out

323076

Lower control arm installation — 1996–97
Taurus and Sable sedan

parking brake cable and clip from the lower suspension arm.

5. Remove and discard the bolt and nut attaching the tension strut to the lower suspension arm.

6. Support the spindle and upper suspension arms by wiring them to the body, to prevent them from dropping down.

7. Remove the nut, bolt, washer and adjusting cam retaining the lower suspension arm to the spindle. Discard the nut, bolt and washer.

8. Lower the suspension arm with the floor jack until the spring can be removed. Remove and discard the bolt and nut attaching the lower suspension arm to the center body bracket and remove the arm.

To install:

9. Position the lower suspension arm-to-center body bracket and install but do not tighten a new bolt and nut with the bolt head toward the front of the vehicle.

10. Position the lower insulator on the lower suspension arm and press the insulator downward into place. Make sure the insulator is properly seated.

11. Position the upper insulator on top of the spring. Install the spring on the lower suspension arm, making sure the spring is properly seated.

12. Raise the suspension arm with the floor jack and guide the upper spring insulator onto the upper spring seat on the underbody.

13. Position the spindle in the lower suspension arm and install, but do not tighten, a new bolt, nut, washer and the existing cam, with the bolt head toward the front of the vehicle.

14. Remove the wire from the spindle and suspension arms.

15. Install the tension strut in the lower suspension arm using a new bolt and nut but do not tighten at this time.

16. Install the parking brake cable and clip to the lower suspension arm.

17. Position the shock absorber on the lower suspension arm and install 2 new nuts. Tighten the nuts to 13–20 ft. lbs. (17–27 Nm).

18. Install the stabilizer bar and U-bracket to the lower suspension arm using a new bolt. Tighten the bolt to 20–30 ft. lbs. (27–40 Nm).

19. Install the flexible brake hose bracket to the body. Tighten the bolt to 8–12 ft. lbs. (11–16 Nm).

20. Using the floor jack, raise the lower suspension arm to normal curb

height and tighten components to the following specifications:

   a. Lower suspension arm-to-body bracket nut: 40–55 ft. lbs. (54–74 Nm).

   b. Lower suspension arm-to-spindle nut: 40–55 ft. lbs. (54–74 Nm)

   c. Tension strut-to-body bracket bolt: 40–55 ft. lbs. (54–74 Nm)

21. Install the wheel and tire assembly and lower the vehicle. Check the rear wheel alignment.

1996–97 Models

NOTE: Make sure to have available a new lower control arm-to-body bolt and nut.

1. Raise and safely support the vehicle.

NOTE: If using a twin-post lift, floor jacks must be placed under the lifting pads on the underbody forward of the tension strut body bracket and the rear lift post lowered aside.

2. Remove the wheel and tire assembly. Properly remove the shock absorber and coil spring.

3. Remove the control arm-to-spindle bolt and nut. Discard the bolt and nut. Remove the wheel spindle.

4. Remove and discard the lower control arm-to-body bolt and nut. Remove the lower control arm from the vehicle.

5. Inspect the lower control arm and bushings for damage or wear. The bushings are not serviceable and if worn, the control arm must be replaced as an assembly.

To install:

6. Place the lower control arm to the body bracket and install a new retaining bolt and nut with the bolt head toward the front of the vehicle. Do not tighten at this time.

7. Properly install the coil spring and shock absorber. Install the wheel spindle. Install a new lower control arm-to-spindle bolt and nut. Do not tighten at this time.

8. Using a floor jack or equivalent, raise the lower control arm to normal curb height.

9. Tighten the lower control arm-to-body bolt/nut to 40–52 ft. lbs. (54–71 Nm) and the lower control arm-to-spindle bolt/nut to 50–68 ft. lbs. (68–92 Nm).

10. Remove the floor jack from underneath the vehicle. Lower the vehicle.

11. Check the rear wheel alignment and adjust as necessary. Road test the vehicle and check for proper operation.

1993–94 Continental

1. Turn **OFF** the air suspension switch located in the luggage compartment. Raise and support the vehicle safely.

2. Remove all wire retainers and parking brake cable retainers from the lower suspension arm. Disconnect the height sensor link from the ball stud pin on the lower arm.

3. Mark the position of the notch on the toe adjustment cam. Remove the torsion spring retaining clamp at the spindle.

4. Remove the nut from the inboard bushing on the suspension arm.

5. Install torsion spring remover T88P–5310–A or equivalent, on the arm. Using a 3/4 in. ratchet, pry up on the tool and arm to relieve the pressure on the pivot bolt. Remove the bolt and lower the arm.

6. Remove the torsion spring-to-arm nut and separate the spring from the arm.

7. Remove the outboard attaching bolt at the spindle. Repeat the removal procedure for the other arm.

To install:

NOTE: When installing new control arms, the offset must face up. The arms are stamped bottom on the lower edge. The rear control arms have adjustment cams that fit inside the bushings at the arm-to-body attachment. The cams are installed from the front of both arms.

8. Loosely attach the arm(s) at the spindle. Attach the torsion spring(s) to the arm(s).

9. Position the inboard bushing using torsion spring remover T88P–5310–A or equivalent, and install the bolt. Repeat this step for the opposite side.

10. Set the toe adjustment cam to the alignment mark for rear arm only. Connect the height sensor link to the ball stud pin on the lower arm for right front only.

11. Install all wire retainers and parking brake cable retainers to the lower suspension arm. Lower the vehicle and turn **ON** the air suspension switch.

12. With the vehicle suspension at curb height, tighten the control arm-to-spindle bolt to 44–59 ft. lbs. (60–80 Nm) and the control arm-to-body bolt to 50–68 ft. lbs. (68–92 Nm).

13. Check the rear toe setting.

1995–97 Continental

CAUTION

Do not remove an air spring under any circumstances when pressurized. Do not remove any components supporting an air spring without exhausting the air. Failure to follow these instructions may result in unexpected inflation or deflation of the air springs which may result in shifting of the vehicle during service.

Lower Front

1. Properly deflate the air springs.

2. Raise and safely support the vehicle. Remove the wheel and tire assembly.

3. Remove the air spring assembly.

4. Remove the brake anti-lock sensor wiring and rear parking brake cable and clips from the front lower control arm.

5. Remove the front lower control arm-to-wheel spindle nut, washer and bolt. Remove the front lower control arm-to-body nut/bolt and the front lower control arm.

6. Remove the rear suspension arm adjusting cam from the front lower control arm, if damaged.

To install:

7. Install the rear suspension arm adjusting cam, if removed.

8. Position the front lower control arm into the vehicle. Install the front lower control arm-to-body bolt and nut. Hand-tighten the nut.

9. Install the front lower control arm-to-wheel spindle bolt, washer and nut. Tighten the front lower control arm-to-spindle nut to 50–68 ft. lbs. (68–92 Nm) and the front lower control arm-to-body nut to 50–68 ft. lbs. (68–92 Nm).

10. Position the rear parking brake cable and rear brake anti-lock sensor wiring routing clips as necessary.

11. Install the air spring. Install the wheel and tire assembly.

12. Refill the air springs. Lower the vehicle. Turn the air suspension switch to the **ON** position.

13. Check and set the camber and toe if necessary. Road test the vehicle and check for proper operation.

Lower Rear

1. Properly deflate the air springs. Raise and safely support the vehicle.

2. Remove the wheel and tire assembly. Remove the air spring.

3. Remove the stabilizer bar link-to-lower control arm nut. Separate the stabilizer bar link and bushing from the rear lower control arm.

WARNING

Be extremely careful not to damage the boot seal. Do not use power tools to remove the nut, or bearing and seal damage will result. Loosen the nut first. Then remove the nut from the stud using the hex hold feature.

---

4. Remove the rear lower control arm-to-rear wheel spindle nut and bolt.

5. Remove the rear lower control arm-to-body nut/bolt and the rear lower control arm.

6. Remove the rear suspension arm adjusting cam from the rear lower control arm, if damaged.

To install:

7. Install the rear suspension arm adjusting cam, if removed.

8. Position the rear lower control arm into the vehicle. Install the rear lower control arm-to-body bolt and nut. Hand-tighten the nut.

9. Install the rear lower control arm-to-rear wheel spindle bolt, washer and nut. Tighten the rear lower control arm-to-spindle nut to 50–68 ft. lbs. (68–92 Nm) and the rear lower control arm-to-body nut to 50–68 ft. lbs. (68–92 Nm).

10. Install the air spring.

11. Install the rear stabilizer bar link and nut. Tighten the nut to 25–34 ft. lbs. (34–46 Nm).

WARNING

Be extremely careful not to damage the boot seal. Do not use power tools to install the nut, or bearing and seal damage will result. Install the nut using the hex hold feature.

---

12. Install the wheel and tire assembly.

13. Refill the air spring. Lower the vehicle. Turn the air suspension switch to the **ON** position.

14. Check and set the camber and toe if necessary. Road test the vehicle and check for proper operation.

Sway Bar

REMOVAL AND INSTALLATION

Taurus and Sable

NOTE: Make sure new sway bar (stabilizer bar) link nuts and sway bar bracket bolts and nuts are available.

Sedan

1. Raise and safely support the vehicle. Remove the sway bar-to-link nuts, washers and insulators; discard the nuts.

2. Remove the 2 bolts attaching the sway bar brackets to the body and remove the sway bar. Discard the bolts.

3. Inspect the bracket insulators and replace if damaged or worn.

4. If required, remove the nut, washer and insulator retaining each sway bar link to the strut brackets and remove the sway bar links. Check the link insulators and replace if damaged or worn.

To install:

5. Apply a rubber suspension lubricant to the inside of the insulators. Install the insulators to the sway bar.

6. If removed, position each sway bar link to the strut brackets and install the insulator, washer and a new nut. Tighten the nuts to 5–7 ft. lbs. (7–10 Nm).

7. Place the sway bar with the insulators and brackets into position on the body. Install 2 new bolts and tighten to 25–33 ft. lbs. (34–46 Nm).

8. Position the sway bar links on the sway bar ends. Install the insulators, washers and new nuts. Tighten the nuts to 5–7 ft. lbs. (7–10 Nm).

9. Lower the vehicle. Check for proper operation.

Station Wagon

1. Raise and safely support the vehicle.

2. Support the vehicle with jackstands under the lower control arm to unload the sway bar link insulators.

3. Remove the 2 bolts and nuts retaining the sway bar brackets and insulators to the lower control arms. Discard the bolts and nuts.

4. Clean the sway bar of any contamination and slide the sway bar bracket and insulator off the sway bar end.

5. Inspect the brackets and insulators. Replace if damaged or worn.

6. Remove the 2 bolts and nuts attaching the sway bar link assemblies to the body brackets. Discard the bolts and nuts.

7. Remove the sway bar and link assemblies from the vehicle.

8. Inspect the sway bar link assemblies and replace if damaged or worn.

To install:

9. Lubricate the sway bar link bushings with a suitable rubber suspension lubricant. Install the sway bar and link assemblies to the body

brackets using 2 new nuts and bolts. Tighten to 40–52 ft. lbs. (54–71 Nm) on 1993–95 or 44–59 ft. lbs. (60–80 Nm) on 1996–97.

10. Clean the sway bar insulator inside diameter and apply a suitable rubber suspension lubricant. Slide the insulators on both ends of the sway bar into their approximate positions and install the brackets.

11. Position the sway bar brackets on the lower control arms and install 2 new bolts and nuts. Tighten to 23–30 ft. lbs. (30–40 Nm) on 1993–95 or 14–19 ft. lbs. (19–26 Nm) on 1996–97.

12. Lower the vehicle. Check for proper operation.

1993–94 Continental

1. Raise vehicle on hoist.

2. Remove nuts, washers and insulators attaching stabilizer bar to right and left side links.

3. Remove bolts attaching brackets and stabilizer bar to body and remove stabilizer bar.

4. Inspect bracket insulators and replace if damaged or worn.

5. Remove nut, washer and insulator retaining link to strut bracket. Check link insulators and replace if damaged or worn.

To install:

6. Position link into strut bracket and install the insulator, washer and a new nut. Tighten to 6–12 ft. lbs. (8–16 Nm).

7. Position stabilizer bar, brackets and insulators on body. Install new bolt and tighten to 15–25 ft. lbs. (20–34 Nm).

8. Position stabilizer bar onto links. Install insulators, washers and new nuts. Tighten to 6–12 ft. lbs. (8–16 Nm).

1995–97 Continental

> ─────── **CAUTION** ───────
> *The air suspension switch, located in the left side of the luggage compartment, must be turned OFF before raising the vehicle. Failure to do so may result in unexpected inflation or deflation of the air springs which may result in shifting of the vehicle during service.*

1. Turn the air suspension switch located in the luggage compartment, to the **OFF** position. Raise and safely support the vehicle.

2. Remove both rear wheel and tire assemblies. Remove both sway bar link-to-sway bar retaining nuts.

> ─────── **WARNING** ───────
> **Be extremely careful not to damage the boot seal. Do not use power tools to remove the nut, or bearing and seal damage will result. Loosen the nut first. Then remove the nut from the stud using the hex hold feature.**

3. Mark the position of the rear stabilizer bar to insure proper installation. Remove the stabilizer bar insulator bracket mounting bolts and the stabilizer brackets.

4. Remove the rear stabilizer bar from the vehicle. Remove the insulators from the stabilizer bar.

To install:

5. Install the insulators on the stabilizer bar with the opening facing the rear of the vehicle.

NOTE: Green color code to the right side of the vehicle.

6. Position the sway bar brackets and sway bar bracket mounting bolts. Tighten to 19–25 ft. lbs. (26–34 Nm).

7. Install the sway bar links to the sway bar. Install the nuts and tighten to 25–34 ft. lbs. (34–46 Nm).

8. Install the wheel and tire assemblies. Install and tighten the lug nuts to 85–105 ft. lbs. (115–142 Nm).

9. Lower the vehicle. Turn the air suspension switch to the **ON** position. Road test the vehicle and check for proper operation.

Wheel Bearings

ADJUSTMENT

The rear wheel bearings are pregreased, sealed and require no scheduled maintenance. The wheel bearings cannot be adjusted. If a bearing is disassembled for any reason the hub and bearing must be replaced. The rear axle hub retainer nut must be replaced whenever it is backed off or removed. Never use power tools to install the retainer nut.

REMOVAL AND INSTALLATION

Taurus and Sable

Sedan

1. Raise and safely support the vehicle. Remove the wheel and tire assembly.

2. Remove the bolt retaining the brake hose to the strut bracket. Remove the rear disc/drum brake assembly.

3. Remove the grease cap from the bearing and hub assembly and discard the grease cap.

4. Remove the bearing and hub assembly retaining nut and discard. Remove the bearing and hub assembly from the spindle.

5. Remove and discard the suspension arm to spindle bolts, washers and nuts.

6. Remove the tension strut nut and discard. Remove the washer and bushing.

7. Remove and discard the pinch bolt retaining the spindle to the strut and remove the spindle.

To install:

8. Loosely, assemble a new bolt through the spindle boss holes. Position the spindle onto the tension strut and then onto the strut.

9. Install a new spindle pinch bolt; do not tighten at this time. Install the tension strut bushing, washer and new nut; do not tighten at this time.

10. Install new suspension arm to spindle washers and nuts. Position the suspension at curb height using a jack before tightening the fasteners.

11. Tighten the suspension arm to spindle bolt to 35–46 ft. lbs. (47–63 Nm).

12. Tighten the tension strut to body nut to 35–46 ft. lbs. (47–63 Nm).

13. Tighten the strut pinch bolt to 50–68 ft. lbs. (68–92 Nm).

14. Install the disc/drum brake assembly. Position the hub on the wheel spindle.

15. Install a new wheel hub retainer nut and tighten to 188–254 ft. lbs. (255–345 Nm).

16. Install a new hub cap grease seal using Shaft Protector T89P-19623-FH, or equivalent. Tap on the tool until the hub cap grease seal is fully seated.

17. Install the brake hose onto the strut with the retaining bolt and tighten to 8–12 ft. lbs. (11–16 Nm).

18. Install the wheel and tire assembly. Lower the vehicle.

19. Check the front end alignment. Road test the vehicle and check for proper operation.

Station Wagon

1. Raise and safely support the vehicle.

NOTE: If a frame contact lift is used, a jackstand must be placed under the lower suspension arm to raise it to normal curb height.

2. Remove the wheel and tire assembly. Remove the bolt retaining the brake hose to the shock absorber bracket.

3. Remove the disc/drum brake assembly. Remove the grease cap from the bearing and hub assembly and discard the grease cap seal.

4. Remove the bearing and hub assembly retaining nut and discard the nut.

5. Remove the bearing and hub assembly from the spindle.

6. Remove and discard the bolt and nut attaching the upper front suspension arm to the body crossmember.

7. Remove the bolt, washer and adjusting cam kit and nut attaching the lower arm to the spindle. Discard the bolt, washer and nut.

8. Remove the upper front suspension arm nut from the rear wheel spindle and discard the nut. Remove the upper front arm and the spindle from the vehicle.

To install:

9. Position the spindle and the upper front control using a new retaining nut. Do not tighten at this time.

10. Position the spindle on the lower control arm with a new bolt, washer and existing cam kit. Do not tighten at this time.

11. Position the upper front control arm to the body bracket and install a new bolt and nut. Do not tighten at this time.

12. Make sure the suspension is supported under the control arm and is at curb height before tightening retaining bolts.

13. Tighten the upper arm to body bolts to 73–97 ft. lbs. (98–132 Nm).

14. Tighten the nut attaching the upper front arm to the rear arm to 150–190 ft. lbs. (203–258 Nm).

15. Tighten the nut attaching the spindle to the lower suspension arm to 40–52 ft. lbs. (54–71 Nm).

16. Install the disc/drum brake assembly. Position the bearing and hub assembly on the wheel spindle.

17. Install a new wheel hub retaining nut and tighten to 188–254 ft. lbs. (255–345 Nm).

18. Install a new hub cap grease seal using Shaft Protector T89P-19623-FH, or equivalent. Make sure the hub cap grease seal is fully seated.

19. Install the brake hose support bracket and tighten to 11 ft. lbs. (15 Nm).

20. Install the wheel and tire assembly. Lower the vehicle.

21. Check the front end alignment. Road test the vehicle and check for proper operation.

1993–94 Continental

CAUTION

The air suspension switch, located in the left side of the luggage compartment, must be turned OFF before raising the vehicle. Failure to do so may result in unexpected inflation or deflation of the air springs which may result in shifting of the vehicle during service.

NOTE: Make sure to have available, a new wheel hub retainer nut, strut-to-spindle pinch bolt and grease cap, per side.

1. Turn the air suspension switch, located in the luggage compartment, to the **OFF** position.

2. From inside the luggage compartment, loosen, but do not remove 3 nuts retaining the upper strut to the body.

3. Raise and safely support the vehicle. Remove the wheel and tire assembly.

4. Remove the brake hose retainer at the strut bracket. Disconnect the parking brake cable from the brake caliper.

5. Remove all the wire retainers and parking brake cable retainers from the lower control arm.

6. Disconnect the height sensor link from the ball stud pin on the lower arm.

7. Remove the brake caliper assembly from the brake adapter. Support the caliper assembly with a length of wire to prevent damaging the brake hose.

8. If equipped, remove the push on nuts that retain the brake rotor to the hub and remove the rotor.

9. Remove the grease cap from the bearing and hub assembly and discard the grease cap.

10. Remove the bearing and hub assembly retaining nut and remove the bearing and hub assembly from the spindle.

11. Remove the torsion spring bracket from the spindle.

12. Mark the position of the notch on the toe adjustment cam. Remove the nut from the inboard bushing on the suspension arm.

13. Install Torsion Spring Remover T88P-5310-A or equivalent on the arm. Pry up on the tool and arm using a 3/4 inch drive ratchet to relieve pressure on the pivot bolt. Remove the bolt and lower arm.

14. Remove and discard shock strut-to-spindle pinch bolt. Using a suitable prybar spread the pinch joint for removal.

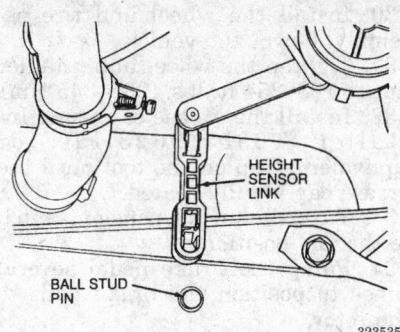

Disconnecting height sensor at ball stud pin — 1993–94 Continental

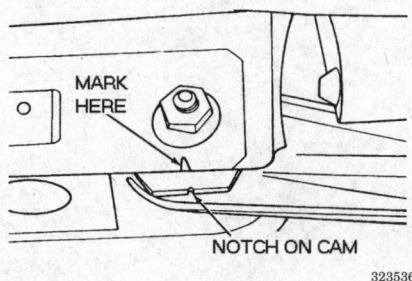

Marking position for adjustment cam — 1993–94 Continental

15. Remove the nut, washer and insulator, retaining the tension strut to the spindle. Move the spindle rearward enough to separate it from the tension strut.

16. Separate the spindle from the strut and remove the spindle.

To install:

17. Install the spindle to the strut. Install a new strut-to-spindle pinch bolt; do not tighten at this time.

18. Position the tension strut to the spindle. Install the insulator, washer, and nut retaining the tension strut to the spindle; do not tighten at this time.

19. Position the spindle to the arms; do not tighten at this time. Install the torsion spring to arms.

20. Position the inboard bushing using Torsion Spring Remover T88P-5310-A or equivalent and install the bolt; do not tighten at this time.

21. Center the toe cam and tighten the spindle-to-strut bolt. Set the toe adjustment to the alignment mark.

22. Install the torsion bar clamp to the spindle.

23. Tighten the strut-to-spindle pinch bolt to 50–68 ft. lbs. (68–92

Nm). Tighten the tension strut nut to 34–47 ft. lbs. (47–63 Nm). Tighten the control arm bolts to 59 ft. lbs. (80 Nm).

24. Position the bearing and hub assembly on the spindle and install the hub retaining nut. Do not tighten at this time.

25. Install a new grease cap using tool T89P-19623-FH or equivalent. Tap on the tool until the grease cap is fully seated.

26. Install the brake adapter plate, splash shield, anchor plate and rotor hub to the spindle.

27. Remove the wire from the caliper and install the caliper to the spindle. Connect the height sensor link to the ball stud pin on the lower arm.

28. Install all the wire retainers and parking brake cable retainers to the lower suspension arm.

29. Connect the parking brake cable to the brake caliper. Install the brake hose retainer at the strut bracket.

30. Install the wheel and tire assembly. Lower the vehicle.

31. Tighten the wheel hub retainer nut to 188–254 ft. lbs. (255–345 Nm). Turn the air suspension switch to the **ON** position.

32. Pump the brake pedal several times to position the brake pads to the rotor. Road test the vehicle and check for proper operation.

1995–97 Continental

——— CAUTION ———
The air suspension switch, located in the left side of the luggage compartment, must be turned OFF before raising the vehicle. Failure to do so may result in unexpected inflation or deflation of the air springs which may result in shifting of the vehicle during service.

NOTE: Make sure to have available 1 new wheel hub retainer nut, 1 upper ball joint nut and 2 tension strut nuts and 1 tension strut bolt, per side.

1. Turn the air suspension switch, located in the luggage compartment, to the **OFF** position.

2. Raise and safely support the vehicle. Remove the wheel and tire assembly.

3. Remove the brake caliper assembly from the brake adapter. Support the caliper assembly with a length of wire to prevent damage to the brake hose.

4. If equipped, remove the push on nuts that retain the rotor to the hub and remove the rotor.

5. Remove the grease cap from the bearing and hub assembly.

6. Remove the wheel hub retainer nut and remove the bearing and hub assembly from the spindle. Discard the retainer nut.

NOTE: The wheel bearings are permanently sealed and lubricated. If wheel bearing replacement is required, the bearing and hub must be replaced as an assembly.

7. Remove the air suspension height sensor.

8. Remove the lower shock mounting bolt and nut and move the shock aside.

NOTE: Note the position of the washers on the tension strut before removing the tension strut.

9. Remove and discard the nut retaining the rear tension strut to the spindle.

10. Remove the mastic patch covering the tension strut nut access hole. Remove the nut and bolt retaining the tension strut to the body and remove the strut.

11. Remove the stabilizer bar link nut and separate from the lower arm.

12. Remove the suspension damper.

13. Loosen, but do not remove the upper ball joint nut. Separate the upper ball joint from the upper control arm.

14. Remove the bolts and nuts attaching the front lower control arm and rear lower control arm to the spindle.

15. Remove the ball joint nut and remove the spindle. Discard the ball joint nut.

To install:

16. Install the spindle to the upper control arm. Install a new ball joint nut but do not tighten at this time.

17. Position the front and rear lower control arms to the spindle and install the retaining bolts, washers, and nuts. Tighten the bolts to 50–68 ft. lbs. (68–92 Nm).

18. Tighten the ball joint nut to 50–68 ft. lbs. (68–92 Nm).

19. Install the suspension damper.

20. Install the stabilizer bar link to the lower arm and tighten the nut to 25–34 ft. lbs. (34–46 Nm).

21. Install the bushings and washers on the tension strut noting their correct position and position the tension strut into the spindle. Install a

new nut and only hand-tighten the nut at this time.

NOTE: Make sure the rear suspension plate contacts the body before tightening the bolt.

22. Position the tension strut into the body and install a new nut and bolt. Tighten to 50–68 ft. lbs. (68–92 Nm).

23. Tighten the spindle nut to 35–46 ft. lbs. (47–63 Nm).

24. Install a new mastic patch covering the tension strut access hole.

25. Install the lower shock mounting bolt and nut. Tighten to 50–68 ft. lbs. (68–92 Nm).

26. Position the bearing and hub assembly on the spindle and install a new wheel hub retainer nut. Do not tighten at this time.

27. Install the brake adapter plate, splash shield, anchor plate and rotor hub to the spindle.

28. Remove the wire from the brake caliper and install the brake caliper to the spindle.

29. Connect the height sensor link to the ball stud pin on the lower arm.

30. Install the wheel and tire assembly. Lower the vehicle.

31. Tighten the wheel hub retainer nut to 188–254 ft. lbs. (255–345 Nm).

32. Install the grease cap using Installer T89P-19623-FH, or equivalent. Tap on the tool until the grease cap is fully seated.

33. Turn the air suspension switch to the **ON** position.

34. Pump the brake pedal several times to position the brake pads to the rotor.

35. Road test the vehicle and check for proper operation.

FORD MOTOR CO.

Front Wheel Drive

FORD-Escort MERCURY-Tracer

14

FIRING ORDERS

NOTE: To avoid confusion, always replace spark plug wires one at a time.

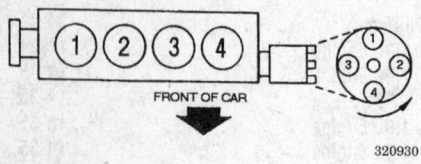

320930

1.8L Engine
Engine Firing Order: 1–3–4–2
Distributor Rotation: Counterclockwise

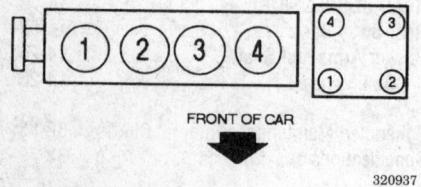

320937

1.9L Engine
Engine Firing Order: 1–3–4–2
Distributorless Ignition System

ENGINE ELECTRICAL

NOTE: Disconnecting the negative battery cable on some vehicles may interfere with the functions of the on board computer systems and may require the computer to undergo a relearning process, once the negative battery cable is reconnected.

Distributor

REMOVAL AND INSTALLATION

1.8L Engine

1. Disconnect the negative battery cable.

2. Disconnect the ignition coil wire at the distributor cap.
3. Remove 2 distributor cap retaining screws and position the distributor cap aside leaving the ignition wires attached.
4. Disconnect the distributor electrical connector.
5. If the distributor unit is not being replaced, scribe a reference mark across the distributor base flange and the cylinder head.
6. Remove 2 distributor retaining bolts and remove the distributor from the engine.

NOTE: Do not rotate the engine while the distributor assembly is removed. Do not attempt to disassemble and service the distributor. The distributor should be replaced as an assembly.

To install:

7. Fit a new O-ring to the distributor unit, if required, and lubricate with engine oil.

NOTE: The distributor can only be installed in one direction. Make sure the drive tangs are engaged in the camshaft slots.

8. Install the distributor to the engine. Make sure the drive tangs fully engage with the camshaft slots.
9. If the same distributor is being installed, align the reference mark. Install and snug 2 retaining bolts.
10. If a new distributor is being installed, position the engine at Top Dead Center (TDC) for No. 1 cylinder on its compression stroke and install the distributor so the tip of the rotor inside the distributor is pointing to the No. 1 cylinder ignition wire on the distributor cap. Install and snug 2 retaining bolts.
11. Install the distributor cap and tighten 2 retaining screws.
12. Connect the distributor electrical connector.

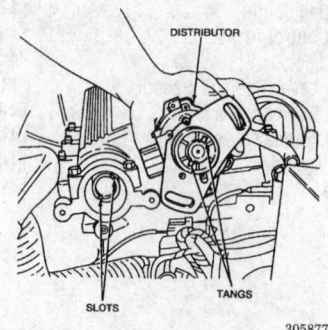

305877

Distributor installation showing tang-to-slot alignment — 1.8L engine

13. Connect the ignition coil wire to the distributor cap.
14. Connect the negative battery cable.
15. Start the engine and set the ignition timing to 10 degrees BTDC. Tighten the distributor retaining bolts to 14–19 ft. lbs. (19–25 Nm).
16. Road test the vehicle and check for proper operation.

Ignition Timing

ADJUSTMENT

1.8L Engine

1. Place the gear selector lever in **P** or **N**. Apply the parking brake.
2. Turn all accessories **OFF**.
3. Connect a timing light to the engine (Rotunda 059–00005, or equivalent).
4. Start the engine and allow it to reach operating temperature.
5. Using a jumper wire, connect the **GROUND** terminal to the **TEN** terminal of the Data Link Connector (DLC).
6. Connect the positive lead of a tachometer (Rotunda 059–00010, or equivalent) to the **IG** terminal of the DLC and the negative lead to the negative battery post.
7. Check the ignition timing. Timing should be 10 degrees BTDC at 700–800 rpm. The mark on the pulley should be aligned with the corresponding mark on the timing belt cover.
8. If necessary, loosen the distributor mounting bolt and turn the distributor until the marks are aligned. Tighten the distributor mounting bolt to 14–19 ft. lbs. (19–25 Nm).
9. Verify proper ignition timing.
10. Remove the jumper wire from the DLC, timing light and tachometer.

1.9L Engine

The base ignition timing is set from the factory at 10 ± 2 degrees Before Top Dead Center (BTDC) and is checked with the spout plug removed. With the spout plug installed, ignition timing is computer controlled and is not adjustable.

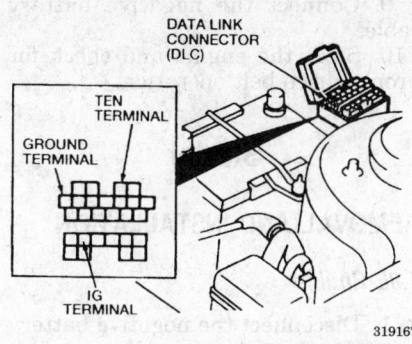

DATA LINK CONNECTOR (DLC)

TEN TERMINAL

GROUND TERMINAL

IG TERMINAL

319167

Data link connector — 1.8L engine

Alternator

PRECAUTIONS

Several precautions must be observed with alternator equipped vehicles to avoid damage to the unit.

• If the battery is removed for any reason, make sure it is reconnected with the correct polarity. Reversing the battery connections may result in damage to the 1-way rectifiers.

• When utilizing a booster battery as a starting aid, always connect the positive to positive terminals and the negative terminal from the booster battery to a good engine ground on the vehicle being started.

• Never use a fast charger as a booster to start vehicles.

• Disconnect the battery cables when charging the battery with a fast charger.

• Never attempt to polarize the alternator.

• Do not use test lights of more than 12 volts when checking diode continuity.

• Do not short across or ground any of the alternator terminals.

• The polarity of the battery, alternator and regulator must be matched and considered before making any electrical connections within the system.

• Never separate the alternator on an open circuit. Make sure all connections within the circuit are clean and tight.

• Disconnect the battery ground terminal when performing any service on electrical components.

• Disconnect the battery if arc welding is to be done on the vehicle.

REMOVAL AND INSTALLATION

1.8L Engine

1. Disconnect the negative battery cable.

2. Remove the retaining nut securing the battery wiring connector to the alternator and remove the wire.

3. Remove the field terminal wiring connector.

4. Remove the upper retaining bolt securing the alternator to the alternator bracket.

5. Loosen the lower alternator retaining bolt and pivot the alternator to relieve tension on the drive belt.

6. Remove the alternator drive belt from the pulley and position the drive belt aside.

7. Raise and safely support the vehicle.

8. Remove the lower splash shield located under the accessory drive belts.

9. Remove the alternator lower retaining bolt and remove the alternator from the vehicle.

To install:

10. Position the alternator to the engine and install the lower retaining bolt.

11. Install the lower splash shield.

12. Lower the vehicle.

13. Place the alternator drive belt onto the alternator pulley.

14. Install the upper retaining bolt and adjust the drive belt tension. Tighten the upper retaining bolt to 14–19 ft. lbs. (19–25 Nm) and the lower retaining bolt to 27–38 ft. lbs. (37–52 Nm).

15. Install the field terminal wiring connector.

16. Place the battery wiring connector to the alternator and secure it with the retaining nut.

17. Connect the negative battery cable.

18. Start the engine and check for proper charging system operation.

1.9L Engine

1. Disconnect the negative battery cable.

2. Insert a ⅜ inch drive ratchet or breaker bar in the automatic tensioner. Release the belt tension by pulling the tool toward the front of the vehicle.

3. Remove the drive belt from the tensioner pulley and slip it off the alternator pulley.

4. Remove the nut securing the wiring connector to the alternator.

5. Disconnect 2 snap-in type wiring connectors from the alternator.

6. If equipped with A/C, remove the A/C accumulator tube support clip from the alternator bracket and move aside.

7. Remove the alternator lower retaining bolt and then remove the alternator upper retaining bolt.

8. Remove the bolts securing the power steering pump auxiliary reservoir and move the reservoir aside.

9. Remove the alternator from the bracket and remove from the vehicle.

To install:

10. Place the alternator to the bracket.

11. Place the power steering pump auxiliary reservoir in position and secure.

12. Install the alternator upper and lower retaining bolts. Torque the upper bolt to 14–22 ft. lbs. (20–30 Nm) and the lower bolt to 29–40 ft. lbs. (40–55 Nm).

13. If equipped, install the A/C accumulator tube support clip.

14. Install 2 snap-in wiring connectors to the alternator.

15. Install the battery terminal wire and nut. Secure the nut but do not overtighten.

16. Insert the ⅜ inch drive ratchet or breaker bar in the automatic tensioner, and pull the tool toward the front of the vehicle. While holding the tool in this position, slip the drive belt behind the tensioner pulley and over the alternator pulley. Release and remove the tool.

17. Check that all V-grooves of the drive belt are properly installed in the pulleys.

18. Connect the negative battery cable.

19. Start the engine and check for proper charging system operation.

Drive Belt

REMOVAL AND INSTALLATION

1.8L Engine

NOTE: The procedure for the power steering pump drive belt remains the same with or without air conditioning.

Power Steering Pump With Or Without A/C

1. Disconnect the negative battery cable.

2. Raise and safely support the vehicle.

3. Loosen the power steering pump retaining bolt and nuts.

4. Rotate the steering pump as necessary and remove the drive belt.

To install:

5. Install the new drive belt to the power steering, crankshaft and if equipped, the A/C compressor pulleys making sure of a proper fit on all pulleys.

6. Adjust the belt tension by turning the pump adjusting bolt. Proper belt deflection of a new belt should be

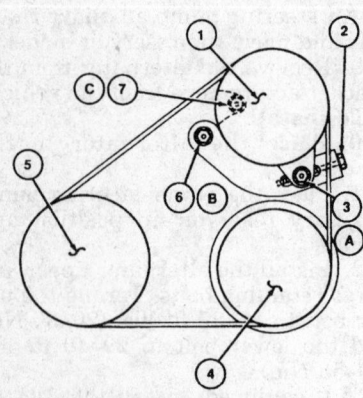

1. Power steering pump pulley
2. Power steering pump adjusting bolt
3. Power steering pump nut (near adjusting bolt)
4. A/C clutch pulley
5. Crankshaft pulley
6. Power steering pump nut
7. Power steering pump bolt
A. 28-38 ft. lb.(37-52 Nm)
B. 23-34 ft. lb.(31-46 Nm)
C. 27-39 ft. lb.(36-55 Nm)

306582

Drive belt installation and adjustment with A/C — 1.8L engine

0.31–0.35 inch (8–9mm). Deflection of a used belt with more than 10 minutes of running time should be 0.35–0.39 inch (9–10mm).

7. Tighten the power steering pump retaining nut, located near the adjusting bolt, to 27–38 ft. lbs. (37–52 Nm).

8. Tighten the pump retaining bolt behind the pulley to 27–40 ft. lbs. (36–54 Nm), and the remaining pump retaining nut to 23–34 ft. lbs. (31–46 Nm).

9. Lower the vehicle.

10. Connect the negative battery cable.

11. Start the engine and check for proper drive belt operation.

Alternator and Water Pump

1. Disconnect the negative battery cable.

2. Remove the power steering pump drive belt.

3. Loosen the alternator (generator) adjusting bolt.

4. Raise and safely support the vehicle.

5. Loosen the alternator pivot bolt. Rotate the alternator as necessary and remove the belt.

NOTE: Do not pry against the alternator housing. Position the prybar against a stronger point, such as the area around a case bolt.

To install:

6. Install the new drive belt to the alternator, crankshaft and water pump pulleys.

7. Position a suitable belt tension gauge on the longest accessible span of the belt. Adjust the tension to 85.8–103.4 lbs. for a new belt or 68.2–85.8 lbs. for a used belt.

8. If a belt tension gauge is not available, adjust the tension to 0.31–0.35 inch (8–9mm) deflection for a new belt or 0.35–0.39 inch (9–10mm) deflection for a used belt.

9. Tighten the alternator adjusting bolt to 14–19 ft. lbs. (19–25 Nm).

10. Tighten the alternator pivot bolt to 27–38 ft. lbs. (37–52 Nm).

11. Lower the vehicle.

12. Replace or reinstall the power steering pump drive belt.

13. Connect the negative battery cable.

14. Start the engine and check for proper drive belt operation.

1.9L Engine

NOTE: Movement of the automatic tensioner assembly during engine operation is not a sign of a malfunctioning tensioner. The movement is required to maintain constant belt tension.

1. Disconnect the negative battery cable.

2. Install a ⅜ inch drive ratchet or breaker bar inserted in the automatic tensioner. Pull the tool toward the front of the vehicle.

3. While releasing belt tension, remove the drive belt from the tensioner pulley and slip it off the remaining accessory pulleys.

To install:

4. Position the drive belt over the accessory pulleys.

5. Install the ⅜ in. drive ratchet or breaker bar inserted in the automatic tensioner. Pull the tool toward the front of the vehicle.

6. While holding the tool in this position, slip the drive belt behind the tensioner pulley and slowly release the tool.

7. Remove the tool from the automatic tensioner.

8. Check that all V-grooves make proper contact with the pulley.

9. Connect the negative battery cable.

10. Start the engine and check for proper drive belt operation.

Starter

REMOVAL AND INSTALLATION

1.8L Engine

1. Disconnect the negative battery cable.

2. Remove the engine air cleaner to intake manifold tube.

3. Remove the starter motor upper retaining bolts.

4. Raise and safely support the vehicle.

5. Remove the intake manifold support bracket retaining bolts and remove the bracket.

6. Disconnect the S terminal connector from the starter solenoid.

NOTE: When disconnecting the plastic hard shell connector at the solenoid S terminal, grasp the plastic connector, depress the plastic tab and pull off the lead assembly. Do not pull on the lead wire or damage may result.

7. Remove the B (battery) terminal retaining nut and disconnect the cable from the terminal.

8. Remove the starter motor lower retaining bolt and remove the starter motor.

To install:

9. Place the starter motor into its mounting position and loosely install the lower retaining bolt.

10. Lower the vehicle and install the upper starter motor retaining bolts. Tighten the upper starter motor retaining bolts to 15–20 ft. lbs. (20–27 Nm).

11. Raise and safely support the vehicle.

12. Tighten the lower starter motor retaining bolt to 15–20 ft. lbs. (20–27 Nm).

13. Connect the cable to the starter solenoid B terminal and install the retaining nut to the terminal. Tighten the nut to 80–120 inch lbs. (9–14 Nm).

14. Connect the electrical connector to the starter solenoid S terminal.

15. Install the intake manifold support bracket. Tighten the retaining bolts to 27–38 ft. lbs. (37–52 Nm) and the retaining nut to 14–19 ft. lbs. (19–25 Nm).

16. Lower the vehicle.

17. Install the engine air cleaner to intake manifold tube.

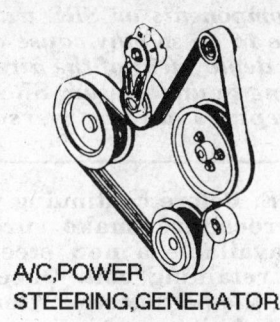

**A/C,POWER
STEERING,GENERATOR**

GENERATOR ONLY

**POWER
STEERING,GENERATOR**

306588

Drive belt positioning — 1.9L engine

18. Connect the negative battery cable.

19. Check the starter motor for proper operation.

1.9L Engine

1. Disconnect the negative battery cable.

2. If equipped with an automatic transaxle, remove the accelerator cable bracket from the cylinder block.

3. Disconnect the wire from the starter solenoid **S** terminal.

NOTE: When disconnecting the plastic hard shell connector at the solenoid S terminal, grasp the plastic connector, depress the plastic tab and pull off the lead assembly. Do not pull on the lead wire or damage may result.

4. Remove the retaining nut from the starter solenoid **B** terminal and disconnect the cable from the terminal.

5. Remove the starter motor retaining bolts and remove the starter motor.

To install:

6. Place the starter motor to its mounting position and install the retaining bolts.

NOTE: Make sure the starter drive housing is fully engaged and flush to the cylinder block.

7. Tighten the starter motor retaining bolts to 15–20 ft. lbs. (20–27 Nm).

8. Connect the cable to the starter solenoid **B** terminal and install the retaining nut. Tighten the nut to 80–120 inch lbs. (9–13 Nm).

9. Connect the wire to the starter solenoid **S** terminal.

10. If equipped with an automatic transaxle, install the accelerator cable bracket to the cylinder block.

11. Connect the negative battery cable.

12. Check for proper starter motor operation.

CHASSIS ELECTRICAL

Blower Motor

REMOVAL AND INSTALLATION

1. Disconnect the negative battery cable.

2. Remove the trim panel below the glove compartment.

3. Remove the wiring bracket and retaining bolt.

4. Disconnect the blower motor electrical connector.

5. Remove 3 blower motor retaining bolts and remove the blower motor assembly from the evaporator housing.

6. Remove the blower wheel retaining clip and remove the blower wheel from the blower motor.

To install:

7. If removed, assemble the blower wheel to the motor shaft and install the retaining clip.

8. Install the blower motor and wheel assembly to the evaporator housing and secure with 3 retaining bolts.

9. Connect the blower motor electrical connector.

10. Install the wiring bracket and retaining bolt.

11. Install the trim panel below the glove compartment.

12. Connect the negative battery cable.

13. Check for proper blower motor operation.

Windshield Wiper Motor

REMOVAL AND INSTALLATION

Front Wiper Motor

1. Make sure the wiper motor switch is in the **OFF** position.

2. Disconnect the negative battery cable.

3. Remove the windshield wiper arms.

4. With the hood closed, remove 7 screw covers. Remove 7 cowl grille retaining screws and remove the cowl grille.

5. Pry up 4 baffle retaining clips and remove the baffle trim piece.

6. Make sure the wiper motor is in the park position before disconnecting the linkage.

7. Remove the wiper linkage retaining clip and disconnect the wiper linkage from the motor.

8. Disconnect 2 wiper motor electrical connectors.

9. Remove 3 wiper motor retaining bolts until they are loose from the sheetmetal mounting surface.

10. Remove the wiper motor from the vehicle.

To install:

11. Place the wiper motor into position and install 3 retaining bolts. Tighten the bolts to 61–79 inch lbs. (7–9 Nm).

12. Connect the wiring connectors to the wiper motor.

13. Temporarily install the negative battery cable.

14. Turn the wiper switch to the **ON** and then **OFF** positions and allow the wiper motor to stop in the **PARK** position.

15. Remove the negative battery cable.

16. Connect the wiper linkage and the retaining clip.

17. Install the baffle trim piece and 4 retaining clips.

18. Install the cowl grille and 7 retaining screws. Install the screw covers.

19. Properly position the wiper arms. Tighten the wiper arm pivot nuts to 10–14 ft. lbs. (14–19 Nm).

20. Connect the negative battery cable.

21. Run the wipers at all speeds and make sure the wiper arms park in their correct positions when the wiper switch is turned **OFF**. Adjust the wiper arms if necessary.

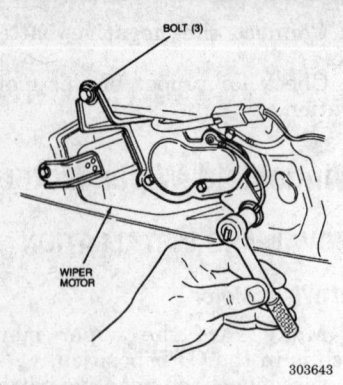

Front wiper motor assembly

Rear Wiper Motor

1. Disconnect the negative battery cable.

2. Remove the wiper arm by lifting the wiper arm attaching nut cover, removing the attaching nut and pulling the wiper arm from the pivot shaft.

3. Remove the shaft seal from the outer bushing retaining nut.

4. Remove the outer bushing attaching nut and remove the outer bushing.

5. Remove the liftgate trim panel.

6. Disconnect the wiper motor electrical connector.

7. Remove 3 wiper motor retaining bolts and washers and remove the wiper motor.

To install:

8. Place the rear wiper motor into position and install 3 retaining bolts. Tighten the wiper motor bolts to 61–79 inch lbs. (7–9 Nm).

9. Install the outer bushing and tighten the nut to 35–52 inch lbs. (4–6 Nm). Install the shaft seal.

10. Connect the rear wiper motor wiring connector.

11. Install the negative battery cable.

12. Turn the wiper switch to the **ON** and then **OFF** positions and allow the wiper motor to stop in the **PARK** position.

13. Install the wiper arm. Tighten the wiper arm nut to 61–87 inch lbs. (7–9 Nm).

14. Run the rear wiper motor at all speeds and make sure the wiper arm parks in its correct position when the wiper switch is turned **OFF**. Adjust the wiper arm if necessary.

15. Install the liftgate trim panel.

Combination Switch

REMOVAL AND INSTALLATION

1993 Models

1. Disconnect the negative battery cable.

2. Remove the steering wheel cover retaining screws from the back side of the steering wheel and remove the cover.

3. Disconnect the horn electrical connector and the speed control electrical connector, if equipped.

4. Remove the steering wheel mounting nut or bolt.

—————— CAUTION ——————

Do not attempt to remove the steering wheel by hitting the column shaft with a hammer. The column may collapse.

5. Remove the steering wheel using a suitable puller.

6. Remove the 4 retaining screws from the steering column lower cover.

7. Disconnect the ignition switch lock cylinder illumination lamp socket if equipped.

8. Remove the steering column lower and upper covers.

9. Disconnect the 3 combination switch electrical connectors.

10. Remove the 2 combination switch retaining screws.

11. Remove the electrical connectors from the retaining brackets and remove the switch.

To install:

12. Position the combination switch to the steering column.

13. Install the switch retaining screws and tighten.

14. Position the electrical connector harnesses to the retaining brackets and install the connectors to the switch.

15. Install the steering column upper cover.

16. Install the ignition switch lock cylinder illumination lamp socket to the lower cover, if equipped.

17. Install the steering column lower cover and the 4 retaining screws.

18. Install the steering wheel and retaining nut or bolt.

19. Torque the steering wheel mounting bolt to 34–46 ft. lbs. (46–63 Nm).

1994–97 Models

—————— CAUTION ——————

The Supplemental Restraint System (SRS) must be disarmed before performing service around SRS components or SRS wiring. Failure to do so may cause accidental deployment of the air bag, resulting in unnecessary SRS system repairs and/or personal injury.

NOTE: Before continuing with this procedure, make sure to have available a new steering wheel retaining bolt. Once removed, the bolt looses its torque holding ability or retention capability and must not be reused.

1. Make sure the front wheels are in the straight-ahead position.

2. Disarm the air bag system.

3. Remove 2 air bag module retaining bolts from the back side of the steering wheel.

4. Raise the air bag module and disconnect the air bag module electrical connector.

5. Remove the drivers side air bag module and place on a work bench with the soft side face up.

—————— CAUTION ——————

When carrying a live air bag, make sure the bag and trim cover are pointed away from the body. In the unlikely event of an accidental deployment, the bag will then deploy with minimal chance of injury. When placing a live air bag on a bench or other surface, always face the bag and trim cover up, away from the surface. This will reduce the motion of the module if it is accidently deployed.

6. Remove the steering wheel retaining bolt and discard.

7. Remove the steering wheel using a suitable puller.

NOTE: Do not attempt to remove the steering wheel by hitting the column shaft with a hammer. The column may collapse.

8. Place tape on the air bag sliding contact to prevent it from moving.

9. Remove 4 retaining screws from the steering column lower cover.

10. Disconnect the ignition switch lock cylinder illumination lamp socket, if equipped.

11. Remove the steering column lower and upper covers.

12. Disconnect 3 combination switch electrical connectors.

13. Remove the combination switch retaining screws.

14. Remove the electrical connectors from the retaining brackets and remove the switch.

To install:

15. Place the combination switch in position to the steering column.

16. Install the switch retaining screws and tighten securely.

17. Position the electrical connector harnesses to the retaining brackets and install the connectors to the switch.

18. Install the steering column upper cover.

19. Install the ignition switch lock cylinder illumination lamp socket to the lower cover, if equipped.

20. Install the steering column lower cover and 4 retaining screws.

21. Remove the tape from the air bag sliding contact.

22. Install the steering wheel and a new retaining bolt. Torque the retaining bolt to 34–46 ft. lbs. (46–63 Nm).

23. Position the drivers side air bag to the steering wheel and install the air bag electrical connector.

24. Install 2 air bag module retaining bolts at the back of the steering wheel. Tighten the retaining bolts to 35–53 inch lbs. (4–6 Nm).

25. Connect the negative battery cable.

26. Prove out the air bag system by turning the ignition key to the **RUN** position and visually monitoring the air bag indicator lamp in the instrument cluster. The indicator lamp should illuminate for approximately 6 seconds and then turn OFF. If the indicator lamp does not illuminate, stays ON, or flashes at any time, a fault has been detected by the air bag diagnostic monitor.

27. Check all combination switch functions for proper operation.

Ignition Lock Cylinder

REMOVAL AND INSTALLATION

— **CAUTION** —

The Supplemental Restraint System (SRS) must be disarmed before performing service around SRS components or SRS wiring. Failure to do so may cause accidental deployment of the air bag, resulting in unnecessary SRS system repairs and/or personal injury.

NOTE: Before continuing with this procedure, make sure to have available a new steering wheel retaining bolt and 2 lock cylinder retaining bolts. Once removed, these parts loose their

torque holding ability or retention capability and must not be reused.

1. If equipped, disarm the air bag system and disconnect the negative battery cable.

2. Remove the combination switch.

3. Disconnect the wiring connector to the ignition switch.

4. Remove the ignition and shifter inter-lock cable mounting bracket and cable and move it aside.

5. Remove 4 steering column bracket bolts and lower the steering column.

6. Cut a groove in the ignition cylinder bracket bolts with a chisel and hammer or similar method.

7. Using a screwdriver, remove and discard the bracket bolts.

8. Remove the ignition switch lock cylinder.

To install:

9. Install the new ignition switch and the bracket to the steering column tube. Install 2 new bolts and tighten the bolts enough to hold the switch in position.

10. Check for proper switch operation by rotating the key to each position.

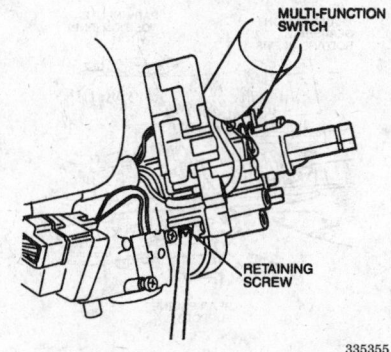

Combination (Multi-function) switch

11. If positioned properly, tighten the retaining bolts until the bolt head shears off.

12. Raise and position the steering column. Tighten 4 bracket bolts 80–124 inch lbs. (9–14 Nm).

13. If equipped with tilt column, remove the upper mounting bracket retaining pin.

14. Position the inter-lock cable bracket. Install and tighten the retaining bolt 35–53 inch lbs. (4–6 Nm).

15. Connect the wiring connector to the ignition switch.

16. Install the combination switch.

17. Connect the negative battery cable.

18. Prove out the air bag system by turning the ignition key to the **RUN** position and visually monitoring the air bag indicator lamp in the instrument cluster. The indicator lamp should illuminate for approximately 6 seconds and then turn OFF. If the indicator lamp does not illuminate, stays ON, or flashes at any time, a fault has been detected by the air bag diagnostic monitor.

19. Check the steering wheel, horn and combination switch functions for proper operation.

Ignition Switch

REMOVAL AND INSTALLATION

— **CAUTION** —

If equipped with an air bag, the Supplemental Restraint System (SRS) must be disarmed before performing service around SRS components or SRS wiring. Failure to do so may cause accidental deployment of the air bag, resulting in unnecessary SRS system repairs and/or personal injury.

1. If equipped, disarm the air bag system.

2. Disconnect the negative battery cable.

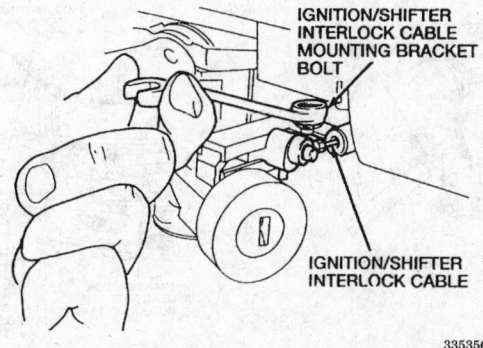

Ignition shifter interlock cable connection

3. Remove the combination switch.

4. Disconnect the ignition switch electrical connector.

NOTE: The ignition key reminder switch is part of the ignition switch and is only serviced with the ignition switch.

5. Remove the ignition switch and ignition key reminder switch retaining screws and remove the ignition switch.

To install:

6. Position the ignition switch and ignition key reminder switch to the steering column.

7. Install the retaining screws and tighten securely.

8. Connect the ignition switch electrical connector.

9. Install the combination switch and steering column trim.

10. Connect the negative battery cable.

11. Check the ignition switch and combination switch for proper operation.

12. If equipped with an air bag, prove out the air bag system by turning the ignition key to the **RUN** position and visually monitoring the air bag indicator lamp in the instrument cluster. The indicator lamp should illuminate for approximately 6 seconds and then turn OFF. If the indicator lamp does not illuminate, stays ON, or flashes at any time, a fault has been detected by the air bag diagnostic monitor.

Park/Neutral Safety Switch

REMOVAL AND INSTALLATION

NOTE: The Park/Neutral Position (PNP) switch is only used with automatic transaxles. The PNP switch informs the Powertrain Control Module (PCM)

when the transaxle is in the N position. The PCM uses this information to control engine idle speed.

1. Disconnect the negative battery cable.

2. Raise and safely support the vehicle.

3. Drain the transaxle fluid into a suitable container for recycling.

4. Install the drain plug.

5. Disconnect the PNP switch electrical connector.

6. Remove the PNP switch and gasket.

To install:

7. Install the PNP switch using a new gasket.

8. If equipped with 1.8L engine, torque the PNP switch to 14–22 ft. lbs. (20–29 Nm). If equipped with 1.9L engine, torque the PNP switch to 14–18 ft. lbs. (20–25 Nm).

9. Connect the PNP electrical connector.

10. Lower the vehicle.

11. Fill the transaxle with the proper type and amount of transaxle fluid.

12. Connect the negative battery cable.

13. Check the PNP switch for proper operation.

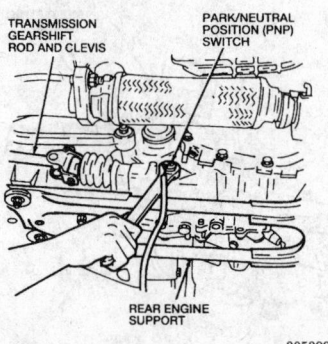

Park/neutral switch — 1.8L engine

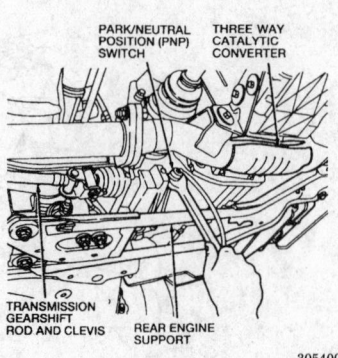

Park/neutral switch — 1.9L engine

Manual Lever Position Sensor

ADJUSTMENT

1. Disconnect the negative battery cable.

2. Remove the clip securing the shift cable and bracket from the manual control shift outer lever.

3. Remove the shift cable and bracket from the manual control shift outer lever.

4. Turn the manual control shift outer lever to the **N** position.

5. Loosen the MLP sensor bolts.

6. Remove the screw from the access hole in the MLP sensor.

7. Using a ⁵⁄₁₆ inch (2.0mm) drill bit or pin, align the indentation inside the MLP sensor with the access hole.

8. Tighten the MLP sensor bolts to 69–95 inch lbs. (8–11 Nm).

9. Remove the drill bit or pin and install the screw in the access hole on the MLP sensor.

10. Install the shift cable and bracket to the manual control shift outer lever and install the clip.

11. Connect the negative battery cable.

12. Check the MLP sensor for proper operation. The engine must only be able to crank when the transaxle range selector is in the **P** and **N** positions.

REMOVAL AND INSTALLATION

NOTE: The Manual Lever Position (MLP) sensor, also known as the Transaxle Range (TR) sensor, is used only on vehicles equipped with automatic transaxles and allows the engine to crank only in the P and N positions.

1. Disconnect the negative battery cable.

2. Remove the engine air cleaner outlet tube.

3. Remove the manual control lever nut securing the manual control shift outer lever to the manual control lever at the transaxle.

4. Remove the manual control shift outer lever from the manual control lever.

5. Disconnect 3 electrical connectors located on top of the transaxle.

6. Disconnect the electrical connector located on the front side of the transaxle.

7. Remove 2 bolts securing the MLP sensor.

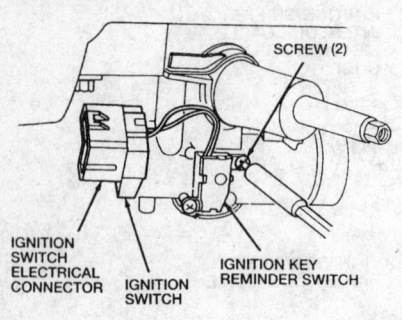

Ignition switch removal

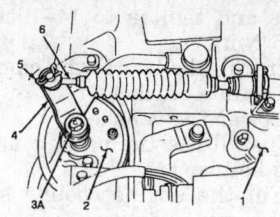

1. Transaxle case
2. Transmission range (TR) switch
3. Manual control lever nut
4. Manual control shift outer lever
5. Clip
6. Shift cable and bracket
A. Tighten to 44-64 Nm (33-47 lb-ft)

305372

Manual lever position (transaxle range) sensor

8. Remove the bolts securing the electrical connector brackets to the top of the transaxle case.

9. Remove the MLP sensor from the manual control lever.

To install:

10. Position the MLP sensor to the manual control lever.

11. Install the bolts securing the electrical connector brackets to the transaxle case.

12. Install the bolts securing the MLP sensor but do not tighten.

13. Install the manual control shift outer lever and nut. Torque the nut to 33–47 ft. lbs. (44–64 Nm).

14. Connect the electrical connectors located on the top and sides of the transaxle case.

15. Adjust the MLP sensor.

16. Install the engine air cleaner outlet tube.

17. Connect the negative battery cable.

18. Check the MLP sensor for proper operation. The engine must only be able to crank when the transaxle range selector is in the **P** and **N** positions.

Powertrain Control Module

REMOVAL AND INSTALLATION

1.8L Engine

— CAUTION —
Electronic modules are sensitive to static electrical charges. Wear a static discharge strap to prevent damage to electronic components.

1. Disconnect the negative battery cable.

2. Remove the trim panels on the sides of the console panel under the instrument panel.

3. Remove the heater and ventilation intake duct located in front of the Powertrain Control Module (PCM).

4. Wear a static discharge strap grounded to the vehicle.

5. Remove the PCM retaining bolts and nut.

6. Disconnect the PCM electrical connectors.

7. Remove the nut from the relay mounting bracket and separate from the PCM.

8. Remove the PCM.

To install:

9. Wear a static discharge strap grounded to the vehicle.

10. Position the PCM in the vehicle.

11. Install the relay mounting bracket and nut.

12. Connect the PCM electrical connectors.

13. Install the retaining bolts and nut to secure the PCM.

14. Install the heater and ventilation intake duct.

15. Install the trim panels to the sides of the console panel.

16. Connect the negative battery cable.

17. Road test the vehicle and check for proper operation.

1.9L Engine

— CAUTION —
Electronic modules are sensitive to static electrical charges. Wear a static discharge strap to prevent damage to electronic components.

1. Disconnect the negative battery cable.

2. The Powertrain Control Module (PCM) is located on the floor tunnel ahead of the shift console. Remove the shift console box side covers.

3. Wear a static discharge strap and ground to the vehicle chassis.

4. Disconnect the harness connector from the PCM.

5. Remove the PCM relay and fuel pump relay from the PCM bracket.

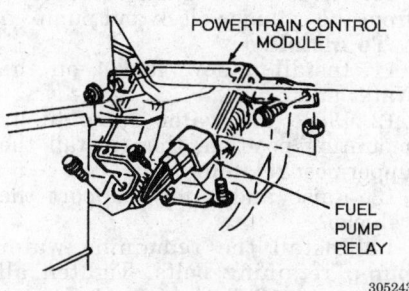

305243

Powertrain control module mounting — 1.8L engine

6. Remove the rear seat heater duct adapter and pad from the top of the PCM bracket.

7. Remove 3 PCM bracket retaining bolts and loosen 1 retaining nut.

8. Remove the PCM and bracket from the vehicle floor tunnel.

9. Remove the PCM from the bracket by sliding it straight out.

To install:

10. Position the PCM to the bracket and install by sliding it straight in.

11. Install the PCM and bracket to the floor tunnel and tighten the retaining nut.

12. Install 3 PCM and bracket to floor tunnel retaining bolts and tighten to 72 inch lbs. (8 Nm).

13. Install the rear seat heater duct adapter and pad.

14. Install the PCM relay and fuel pump relay to the PCM bracket.

15. Carefully connect the harness connector to the PCM. Tighten the harness connector bolt to 55 inch lbs. (6.2 Nm).

16. Install the shift console box side covers.

17. Connect the negative battery cable.

18. Road test the vehicle and check for proper operation.

ENGINE COOLING

Radiator

REMOVAL AND INSTALLATION

1. Disconnect the negative battery cable.

2. Place protective covers on the fender aprons.

3. Raise and safely support the vehicle.

4. Drain the cooling system into a suitable container for recycling or reuse.

5. Remove the left-hand and right-hand front splash shields.

6. Remove the lower radiator hose from the radiator.

7. If equipped with an automatic transaxle, remove the lower oil cooler line from the radiator. Remove the oil cooler line brackets from the bottom of the radiator.

8. Lower the vehicle.

9. If equipped with an automatic transaxle and A/C, remove the seal located between the radiator and fan shroud.

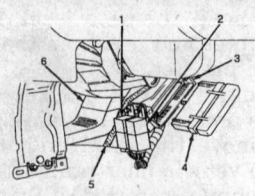

1. Powertrain control module relay (PCM relay)
2. Fuel pump relay
3. Powertrain control module (PCM)
4. EEC-IV harness connector
5. Pad
6. Rear seat heater outlet duct adapter

305252

Powertrain control module mounting — 1.9L engine

10. If equipped with automatic transaxle, remove the upper oil cooler line from the radiator.

11. If equipped with a 1.8L engine, remove the resonance duct from the radiator isomounts (upper support brackets).

12. Disconnect the cooling fan motor electrical connector and the cooling fan thermoswitch electrical connector.

13. Remove 3 fan shroud retaining bolts and remove the shroud assembly by pulling it straight up.

14. Remove the upper radiator hose from the radiator.

15. Remove 2 upper radiator isomounts.

16. Remove the oil cooler tube bracket bolts from the radiator, if equipped.

17. Remove the radiator coolant overflow hose from the radiator.

18. Carefully remove the radiator from the vehicle by lifting it straight up.

To install:

19. Make sure the radiator lower isomounts (rubber insulators), are installed over the bolts on the radiator support.

20. Position the radiator on the radiator support, making sure the radiator lower brackets are positioned properly on the lower isomounts.

21. Install the radiator upper isomounts, making sure the radiator locating pegs are positioned correctly.

22. Install the oil cooler tube bracket bolts to secure the oil cooler tubes.

23. Install the upper radiator hose.

24. Lower the cooling fan shroud assembly into place and install 3 fan shroud retaining bolts.

25. Connect the cooling fan motor electrical connector and thermoswitch electrical connector.

26. If equipped with a 1.8L engine, install the resonance duct on the radiator isomounts.

27. If equipped with an automatic transaxle, install the upper oil cooler line to the radiator.

28. If equipped with an automatic transaxle and A/C, install the air deflector seal between the radiator and fan shroud.

29. Raise and safely support the vehicle.

30. If equipped with an automatic transaxle, install the lower oil cooler line to the radiator.

31. Install the lower radiator hose.

32. Install the left-hand and right-hand front splash shields.

33. Lower the vehicle.

34. Install the radiator coolant overflow hose to the radiator.

35. Fill and bleed the cooling system with the proper coolant.

36. Connect the negative battery cable.

37. Start the engine and allow to reach normal operating temperature.

38. Check for coolant leaks and proper cooling system operation.

Water Pump

REMOVAL AND INSTALLATION

1.8L Engine

1. Disconnect the negative battery cable.

2. Drain the engine cooling system into a suitable container for recycling or reuse.

3. Remove the timing belt covers and the timing belt.

4. Raise and safely support the vehicle.

5. Remove the engine oil dipstick tube bracket bolt from the water pump.

6. Remove 2 bolts and the gasket from the water inlet pipe.

7. Remove all but the uppermost water pump retaining bolt.

8. Lower the vehicle.

9. Remove the remaining bolt and the water pump assembly.

10. Remove all gasket material from the cylinder block and pump.

To install:

11. Install a new gasket on the water pump.

12. Place the water pump to its mounting position, then install the uppermost retaining bolt.

13. Raise and safely support the vehicle.

14. Install the remaining water pump retaining bolts. Tighten all bolts to 14–19 ft. lbs. (19–25 Nm).

15. Install a new gasket onto the water inlet pipe and install the water inlet pipe to the water pump. Install

2 bolts and tighten to 14–19 ft. lbs. (19–25 Nm).

16. Install the bolt to the engine oil dipstick tube bracket.

17. Lower the vehicle.

18. Install the timing belt and the timing belt covers.

19. Fill the engine cooling system with the proper coolant.

20. Connect the negative battery cable.

21. Run the engine and allow it to reach normal operating temperature.

22. Check for coolant leaks and proper engine operation.

23. Check the coolant level and add as necessary.

1.9L Engine

1. Disconnect the negative battery cable.

2. Drain the engine cooling system into a suitable container for recycling or reuse.

3. Remove the accessory drive belt and its automatic tensioner.

4. Remove the timing belt cover and the timing belt.

5. Raise and safely support the vehicle.

6. Remove the lower radiator hose from the water pump.

7. Remove the heater hose from the water pump.

8. Lower the vehicle.

9. Support the engine with a suitable floor jack.

10. Remove the right-hand engine mount retaining bolts and roll the engine mount aside.

11. Remove 4 water pump retaining bolts.

12. Using the floor jack, raise the engine enough to provide clearance for removing the water pump.

13. Remove the water pump and gasket from the engine through the top of the engine compartment.

To install:

14. Make sure the mating surfaces of the cylinder block and water pump are clean and free of gasket material.

15. If the water pump is to be replaced, transfer the timing belt tensioner components to the new water pump.

16. With the engine supported and raised with a suitable floor jack, place the water pump and the gasket on the cylinder block and install 4 retaining bolts. Tighten the bolts to 15–22 ft. lbs. (20–30 Nm).

17. Rotate the right-hand engine mount into position and install the mounting bolts.

18. Remove the floor jack.

19. Raise and safely support the vehicle.

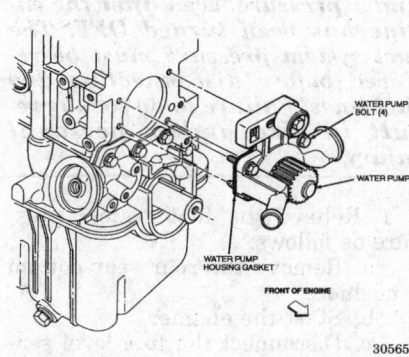

Water pump to cylinder block mounting — 1.9L engine

20. Install the lower radiator hose to the water pump.
21. Install the heater hose to the water pump.
22. Lower the vehicle.
23. Install the timing belt and cover.
24. Install the accessory drive belt automatic tensioner and the accessory drive belt.
25. Connect the negative battery cable.
26. Refill the cooling system.
27. Start the engine and allow it to reach normal operating temperature.
28. Check for coolant leaks and proper engine operation.
29. Check the coolant level and add as necessary.

Thermostat

REMOVAL AND INSTALLATION

1.8L Engine

1. Disconnect the negative battery cable.
2. Drain the engine cooling system into a suitable container for recycling or reuse.
3. Remove the engine air cleaner to intake manifold tube.

4. Disconnect the low speed fan control switch electrical connector, engine wiring harness ground strap from the connector above the housing and the Heated Oxygen Sensor (HO$_2$S) electrical connector.
5. Remove the upper radiator hose from the housing.
6. Remove the thermostat housing retaining bolt and nut and remove the housing. Remove the gasket and the thermostat.
   To install:
7. Clean the thermostat housing and cylinder head gasket surfaces.
8. Position the thermostat, gasket and housing on the cylinder head.
9. Install the retaining bolt and nut and tighten to 14–19 ft. lbs. (19–26 Nm).
10. Install the upper radiator hose.
11. Connect the HO$_2$S electrical connector, engine wiring harness ground strap and the low speed fan control switch electrical connector.
12. Install the engine air cleaner to intake manifold tube.
13. Fill and bleed the cooling system.
14. Connect the negative battery cable.
15. Start the engine and allow to reach normal operating temperature.
16. Check for coolant leaks and proper cooling system operation.

1.9L Engine

1. Disconnect the negative battery cable.
2. Drain the engine cooling system into a suitable container for recycling or reuse.
3. Remove the engine air cleaner outlet tube.
4. Remove the crankcase ventilation hose.
5. Remove the ignition coil and bracket.
6. Remove the upper radiator hose from the thermostat housing.
7. Remove the heater hose from the thermostat housing.
8. Remove 3 thermostat housing retaining bolts.
9. Carefully remove the thermostat housing, gasket and thermostat from the vehicle. Do not pry on the housing.
   To install:
10. Make sure the thermostat housing and the cylinder head mating surfaces are clean and free of gasket material.
11. Place the thermostat to the thermostat housing with the tabs on the thermostat properly positioned. Press the thermostat fully into the housing to compress the rubber seal.

12. Make sure the retaining bolt threads are clean.
13. Position the thermostat housing with a new gasket to the cylinder head. Install and tighten the retaining bolts to 6–8 ft. lbs. (8–11 Nm).
14. Install the heater hose.
15. Install the top radiator hose.
16. Install the ignition coil and bracket.
17. Connect the crankcase ventilation hose.
18. Connect the engine air cleaner outlet tube.
19. Fill the engine cooling system.
20. Connect the negative battery cable.
21. Start the engine and allow to reach normal operating temperature.
22. Check for coolant leaks and proper cooling system operation.

Electric Cooling Fan

REMOVAL AND INSTALLATION

1.8L and 1.9L Engines

1. Disconnect the negative battery cable.
2. If equipped with a 1.8L engine, remove the engine air intake resonator from the radiator support upper bracket.
3. Disconnect the wiring connector from the cooling fan motor by pushing down on 2 lock fingers and pulling the electrical connector from the motor end. Disconnect the wire loom from the clip on the shroud.
4. Remove 3 fan shroud bolts retaining the cooling fan motor and shroud assembly to the radiator.
5. Remove the cooling fan and shroud assembly from the vehicle.
6. Remove the retaining clip from the cooling fan motor shaft and remove the fan blade.

NOTE: A metal burr may be present on the cooling fan motor shaft after the retaining clip has been removed. If necessary, remove the burr to facilitate fan removal.

7. Remove 3 cooling fan motor retaining screws and remove the cooling fan motor from the shroud.
   To install:
8. Install the cooling fan motor in position in the fan shroud. Install 3 retaining screws and tighten to 44–66 inch lbs. (5–7.5 Nm).
9. Position the cooling fan blade on the motor shaft and install the retaining clip.
10. Position the fan, motor and shroud assembly into the clips on the

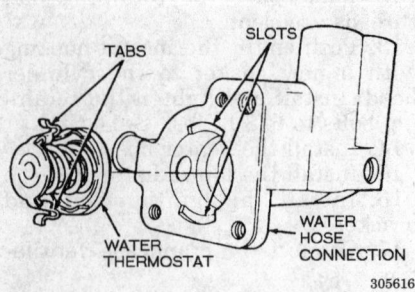

Install thermostat with locating tabs properly positioned — 1.9L engine

bottom of the radiator. Install and tighten the retaining bolts.

11. Install the cooling fan motor wire loom in the clip provided on the fan shroud. Connect the wiring connector to the cooling fan motor ensuring that the connector locks in position.

12. If equipped with a 1.8L engine, install the engine air intake resonator to the radiator support upper bracket.

13. Connect the negative battery cable.

14. Start the engine and allow to reach normal operating temperature.

15. Check the cooling fan for proper operation.

Cooling System

BLEEDING

When the entire cooling system is drained, the following procedure should be used to ensure a complete fill.

1. Install the block drain plug, if removed, and close the radiator draincock. With the engine OFF, add a 50/50 mixture of coolant to the radiator until it reaches the radiator filler neck seat.

2. Install the radiator cap to the first notch to keep spillage to a minimum.

3. Start the engine and let it idle until the upper radiator hose is warm. This indicates that the thermostat is open and coolant is flowing through the entire system.

4. Carefully remove the radiator cap and top off the radiator with water. Install the cap on the radiator securely.

5. Fill the coolant recovery reservoir with a 50/50 mixture of coolant.

6. Check for leaks at the radiator draincock and the block drain plug.

FUEL SYSTEM

Fuel System Service Precautions

Safety is the most important factor when performing not only fuel system maintenance but any type of maintenance. Failure to conduct maintenance and repairs in a safe manner may result in serious personal injury or death. Maintenance and testing of the vehicle's fuel system components can be accomplished safely and effectively by adhering to the following rules and guidelines.

• To avoid the possibility of fire and personal injury, always disconnect the negative battery cable unless the repair or test procedure requires that battery voltage be applied.

• Always relieve the fuel system pressure prior to disconnecting any fuel system component (injector, fuel rail, pressure regulator, etc.), fitting or fuel line connection. Exercise extreme caution whenever relieving fuel system pressure to avoid exposing skin, face and eyes to fuel spray. Please be advised that fuel under pressure may penetrate the skin or any part of the body that it contacts.

• Always place a shop towel or cloth around the fitting or connection prior to loosening to absorb any excess fuel due to spillage. Ensure that all fuel spillage (should it occur) is quickly removed from engine surfaces. Ensure that all fuel soaked cloths or towels are deposited into a suitable waste container.

• Always keep a dry chemical (Class B) fire extinguisher near the work area.

• Do not allow fuel spray or fuel vapors to come into contact with a spark or open flame.

• Always use a backup wrench when loosening and tightening fuel line connection fittings. This will prevent unnecessary stress and torsion to fuel line piping. Always follow the proper torque specifications.

• Always replace worn fuel fitting O-rings with new. Do not substitute fuel hose or equivalent, where fuel pipe is installed.

Fuel System Pressure

RELIEVING

— CAUTION —

Fuel injection systems remain under pressure, even after the engine has been turned OFF. The fuel system pressure must be relieved before disconnecting any fuel lines. Failure to do so may result in fire and/or personal injury.

1. Relieve the fuel system pressure as follows:
   a. Remove the rear seat bottom cushion.
   b. Start the engine.
   c. Disconnect the fuel level sensor and fuel pump electrical connectors.
   d. Wait for the engine to stall, then turn OFF the ignition switch.
   e. Disconnect the negative battery cable.
   f. Reconnect the fuel level sensor and fuel pump electrical connectors.
   g. Install the rear seat cushion.

2. If the engine is unable to run, the fuel pressure may be relieved using a fuel pressure gauge connected to the Schrader valve fitting at the fuel supply manifold on the engine.

Idle Speed

ADJUSTMENT

1.8L Engine

Idle Speed Adjustment

1. Apply the parking brake and make sure the transaxle is in P or N.

2. Start the engine and allow it to warm up to normal operating temperature.

3. Turn OFF all electrical loads and accessories.

4. Using a jumper wire, connect the GROUND terminal to the TEN terminal on the Data Link Connector (DLC) located next to the battery in the engine compartment.

5. Connect a tachometer positive lead (+) to the IG terminal on the DLC and the negative lead (-) to the negative battery terminal.

6. Check the vehicles idle speed using the tachometer. Idle speed should be 700–800 rpm. For Canadian vehicles with automatic transaxles, the idle speed should be ap-

proximately 800 rpm when the parking brake is not applied.

NOTE: Do not check or adjust the engine idle speed if the electric cooling fan comes ON during the procedure. Wait for the engine cooling fan to turn OFF before continuing.

7. If the idle speed is not within specifications, turn the idle speed adjusting screw on the throttle body until the idle speed is within specifications.

8. Remove the jumper wire from the DLC.

9. Remove the tachometer.

10. Road test the vehicle and check for proper operation.

Dashpot Adjustment

1. Apply the parking brake and make sure the transaxle is in **P** or **N**.

2. Start the engine and allow it to warm up to normal operating temperature.

3. Turn OFF all electrical loads and accessories.

4. Connect a tachometer positive lead (+) to the **IG** terminal on the Data Link Connector (DLC) located next to the battery in the engine compartment.

5. Increase the engine speed to 4000 rpm.

6. Slowly decrease the engine speed and verify that the bolt on the throttle lever touches the dashpot rod at approximately 3500 rpm.

7. If the bolt does not touch the dashpot rod at approximately 3500 rpm, loosen the bolt and adjust so it will touch the throttle lever at approximately 3500 rpm by repeating the procedure until correct.

8. Remove the tachometer.

9. Road test the vehicle and check for proper operation.

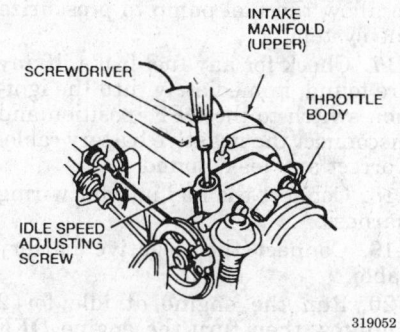

Idle speed adjusting screw location — 1.8L engine

1.9L Engine

The idle speed adjustment is set from the factory and controlled by the Powertrain Control Module (PCM) and is not adjustable. The idle speed should be 780 ± 50 rpm with the transaxle in **N** and the accessories turned OFF. If the idle speed is not within specifications, there may be a problem(s) associated with fuel quality, cooling system, exhaust system, vacuum leaks, charging system or contamination with the throttle bore or the idle air bypass valve.

Mixture

ADJUSTMENT

The air/fuel mixture is electronically controlled by the Powertrain Control Module (PCM) and the Idle Air Control (IAC) valve, and cannot be adjusted. The PCM continuously adjust the air/fuel ratio in response to signals received from operator controls, sensor and switch signals monitoring the engines running condition.

Fuel Filter

REMOVAL AND INSTALLATION

———— **CAUTION** ————
Fuel injection systems remain under pressure even after the engine has been turned OFF. The fuel system pressure must be relieved before disconnecting any fuel lines. Failure to do so may result in fire and/or personal injury.

1. Relieve the fuel system pressure.

2. Position a suitable container below the fuel filter to collect any excess fuel that may leak from the filter and lines.

3. Remove the retaining clip from the fuel filter upper hose.

4. Disconnect the upper hose from the fuel filter and drain any excess fuel into the container. Plug the hose.

5. Loosen the fuel filter mounting clamp.

6. Raise and safely support the vehicle.

7. Remove the retaining clip from the fuel filter lower hose.

8. Disconnect the lower hose from the fuel filter and drain any excess fuel into the container. Plug the hose.

9. Lower the vehicle.

10. Remove the fuel filter.

To install:

11. Position the fuel filter and tighten the filter mounting clamp. Be sure the filter is installed in the proper fuel flow direction.

12. Connect the upper hose to the fuel filter and install the upper hose retaining clip. Replace the retaining clip if damaged.

13. Raise and safely support the vehicle.

14. Connect the lower hose to the fuel filter and install the lower hose retaining clip. Replace the retaining clip if damaged.

15. Lower the vehicle.

16. Connect the negative battery cable.

17. Start the engine and check for fuel leaks.

Fuel Pump

REMOVAL AND INSTALLATION

———— **CAUTION** ————
Fuel injection systems remain under pressure, even after the engine has been turned OFF. The fuel system pressure must be relieved before disconnecting any fuel lines. Failure to do so may result in fire and/or personal injury.

1. Relieve the fuel system pressure.

2. Remove 4 screws securing the fuel level sensor and pump cover and remove the cover with the ground strap.

3. Remove any dirt that has accumulated around the fuel pump retainer ring flange to prevent dirt from entering the fuel tank.

4. Remove the fuel tube clips and disconnect the fuel supply and return lines using the proper spring lock coupling tools.

5. Using Fuel Tank Sender Wrench D84P-9275-A or equivalent, remove the fuel pump locking retainer ring by turning the ring in a counterclockwise direction.

6. Remove the fuel level sensor and fuel pump assembly from the fuel tank.

7. Remove and discard the fuel pump mounting gasket.

To install:

NOTE: A new fuel filter should be installed whenever the fuel pump is replaced.

8. Clean the fuel pump mounting flange, fuel tank mounting surface and mounting gasket groove.

9. Apply a light coating of grease on a new mounting gasket to hold it in place during assembly and install the mounting gasket in the fuel tank groove.

10. Install the fuel pump assembly carefully to ensure the filter and fuel level sensor are not damaged. Make sure the locating keys are in the keyways and the mounting gasket remains in the groove.

11. Hold the pump assembly in place and install the locking ring finger-tight. Make sure all the locking tabs are under the tank lock ring tabs.

12. Rotate the locking ring clockwise until the ring is against the stops.

13. Connect the fuel supply and return lines to the fuel pump and ensure that the spring lock couplings are fully engaged. Properly install the fuel tube clips.

14. Position the fuel level sensor and fuel pump cover with the ground strap. Secure with 4 retaining screws.

15. Connect the fuel level sensor and fuel pump electrical connector.

16. Install the rear seat cushion.

17. Connect the negative battery cable.

18. Run the engine and check for leaks and proper operation.

Fuel Injector

REMOVAL AND INSTALLATION

1.8L Engine

——— CAUTION ———
Fuel injection systems remain under pressure, even after the engine has been turned OFF. The fuel system pressure must be relieved before disconnecting any fuel lines. Failure to do so may result in fire and/or personal injury.

1. Relieve the fuel system pressure.

2. Disconnect fuel supply and return lines using the proper tool for spring lock coupling removal.

3. Disconnect the crankcase ventilation hose from the valve cover and the upper intake manifold.

4. Disconnect the vacuum line from the fuel pressure regulator.

5. Disconnect the fuel injector wiring harness connectors from each fuel injector.

6. Remove 3 fuel injection supply manifold (fuel rail) retaining bolts and remove the fuel injection supply manifold with the fuel injectors attached.

7. Carefully remove the fuel injectors, grommets and insulators from the fuel injection supply manifold as required.

8. Once the fuel injector is removed from the fuel injection supply manifold, remove and discard the fuel injector O-rings.

To install:

9. Inspect the injector end caps (covering the injector pintle) for signs of deterioration. Replace as required. If a cap is missing, look for it in intake manifold.

10. Install 2 new O-rings onto each injector and apply a small amount of clean engine oil to the O-rings.

11. Install the fuel injectors with grommets and insulators to the fuel injection supply manifold using a light twisting motion while pushing the injector into position.

12. Carefully seat the fuel injection supply manifold and injector assembly onto the intake manifold making sure the fuel injectors are fully seated. Install 3 retaining bolts. Torque the bolts to 14–19 ft. lbs. (19–25 Nm).

13. Connect the fuel injector wiring harness connectors to each fuel injector.

14. Connect the vacuum line to the fuel pressure regulator.

15. Connect the crankcase ventilation hose.

16. Connect the fuel supply and fuel return lines. Make sure the spring lock couplings are properly attached and that the fuel tube clips are properly installed.

17. Connect the negative battery cable.

18. Run the engine and check for fuel leaks.

19. Check the entire assembly for proper alignment and seating.

1.9L Engine

——— CAUTION ———
Fuel injection systems remain under pressure, even after the engine has been turned OFF. The fuel system pressure must be relieved before disconnecting any fuel lines. Failure to do so may result in fire and/or personal injury.

1. Relieve the fuel system pressure.

2. Disconnect fuel supply and return lines using the proper tool for spring lock coupling removal.

3. Disconnect the fuel injector electrical connectors.

4. Remove 2 fuel injection supply manifold (fuel rail) retaining bolts.

5. Carefully disengage the fuel injection supply manifold from the fuel injectors.

6. Disconnect the vacuum line from the fuel pressure regulator and set the fuel injection supply manifold aside.

7. Grasp the fuel injectors body and pull up while gently rocking the fuel injector from side to side.

8. Once removed, inspect the fuel injector end caps and washers for signs of deterioration. Replace as required.

9. Remove the O-rings and discard. If an O-ring or end cap is missing, look in the intake manifold for the missing part.

To install:

10. Install 2 new O-rings onto each injector and apply a small amount of clean engine oil to the O-rings.

11. Install the fuel injectors using a light twisting, pushing motion into the intake manifold.

12. Carefully position the fuel injection supply manifold on top of the fuel injectors. Push the fuel injection supply manifold down onto the injectors to fully seat the O-rings.

13. Install the 2 fuel injection supply manifold retaining bolts and torque to 15–22 ft. lbs. (20–30 Nm).

14. Connect the fuel supply and return lines making sure the spring lock couplings are fully engaged and that the fuel tube clips are properly installed.

15. Connect the vacuum line to the fuel pressure regulator.

16. Before connecting the fuel injector electrical harness connectors, connect the negative battery cable and turn the ignition to the **ON** position to allow the fuel pump to pressurize the system.

17. Check for any fuel leaks. If any are found, immediately turn the ignition switch to the **OFF** position and disconnect the negative battery cable. Correct any leaks found.

18. Connect the fuel injector wiring harness.

19. Connect the negative battery cable.

20. Run the engine at idle for 2 minutes, then turn the engine **OFF** and check for fuel leaks and proper operation.

EMISSION CONTROLS

Emission Warning Lamps

RESETTING

These vehicles have a CHECK ENGINE or SERVICE ENGINE SOON light that will light when there is a fault in the engine control system. Depending upon the system or sensor involved, the light may go out if the fault is intermittent. However, the fault code will remain stored in the PCM until the system is serviced and the PCM memory cleared. When a fault is detected in certain systems or sensors, the light will remain lit until the system is serviced. When the system is being diagnosed, the problem corrected and the PCM memory cleared, the light will go out.

ENGINE MECHANICAL

Engine Assembly

REMOVAL AND INSTALLATION

1.8L Engine

CAUTION

Fuel injection systems remain under pressure, even after the engine has been turned OFF. The fuel system pressure must be relieved before disconnecting any fuel lines. Failure to do so may result in fire and/or personal injury.

With Automatic Transaxle

The engine can be separated from the transaxle and removed from the engine compartment.

1. Relieve the fuel system pressure.
2. Mark the position of the hood on the hinges and remove the hood.
3. If equipped with A/C, properly discharge the system using suitable recovery equipment. Do not dis-charge the air conditioner into the atmosphere.
4. Drain the engine cooling system into a suitable container.
5. Drain the engine oil into a suitable container and reinstall the drain plug.
6. Remove the engine air cleaner outlet tube from the throttle body and the engine air intake resonator.
7. Disconnect the power brake vacuum supply hose from the power brake booster.
8. If equipped with speed control, disconnect the necessary vacuum hoses from the upper intake manifold.
9. Disconnect the electrical connectors from the power steering pressure switch, water temperature sending unit, engine coolant temperature sensor, oil pressure sensor, fuel injector wiring harness, heated oxygen sensor, throttle position sensor and distributor.

NOTE: Tag the connectors prior to removal to ease installation.

10. Disconnect all engine ground straps.
11. Disconnect the ignition coil high tension lead from the distributor.
12. Disconnect the accelerator and kickdown cables from the throttle cam.
13. Remove the accelerator and kickdown cable bracket from the upper intake manifold and set the assembly aside.
14. Disconnect the heater core inlet and outlet hoses at the bulkhead.
15. Remove the necessary fuel line clips and disconnect the fuel supply and return lines.
16. Remove the upper radiator hose.
17. Disconnect the electrical connectors from the cooling fan and the radiator thermoswitch.
18. Remove the starter motor.
19. Raise and safely support the vehicle.
20. Remove the right upper and both left and right lower splash shields.
21. Remove the radiator lower hose.
22. Disconnect 2 transaxle cooling lines from the radiator and plug the lines.
23. If equipped, remove the A/C line routing bracket from the radiator and position the line aside.
24. Remove the halfshaft bearing support.
25. Remove the inspection plate from the oil pan, place a wrench on the crankshaft pulley, and rotate the crankshaft to gain access to 4 torque converter nuts. Remove the nuts.
26. Remove the power steering and A/C drive belt.
27. Remove the crankshaft pulley.
28. Remove the exhaust inlet pipe and gasket from the exhaust manifold.
29. If equipped with A/C, remove the compressor. Cap all openings to prevent the entrance of dirt and/or moisture.
30. Remove the power steering pump and bracket assembly with the hoses still connected. Secure the pump with wire away from the work area.
31. Remove all accessible transaxle-to-engine retaining bolts from the cylinder block.
32. Lower the vehicle.
33. Remove the radiator mounting brackets and the engine air intake resonator.
34. Remove the radiator, fan and shroud assembly from the vehicle.
35. Remove the vacuum reservoir located next to the upper intake manifold.
36. Remove the pressure regulator and bracket assembly and set it aside.
37. Remove the High Speed Inlet Air (HSIA) shutter valve actuator and bracket assembly and set it aside.
38. Remove the alternator and water pump drive belt and remove the alternator.
39. Install a suitable engine removal sling onto the engine lifting brackets. Place a suitable engine hoist into position and support the engine.
40. Remove the oil pan-to-transaxle retaining bolts and the remaining transaxle-to-engine bolts from the cylinder block.
41. Remove the front engine support damper.
42. Remove the right-hand engine support insulator.
43. Remove the transaxle-to-engine upper right-hand bolt.
44. Carefully separate the engine from the transaxle, then remove the engine from the vehicle.
45. If the engine is to be placed on an engine stand, first remove the flexplate.

To install:
46. Install a suitable engine removal sling onto the engine lifting brackets.
47. Place a suitable engine hoist into position and install the engine sling. Remove the engine from the engine stand.

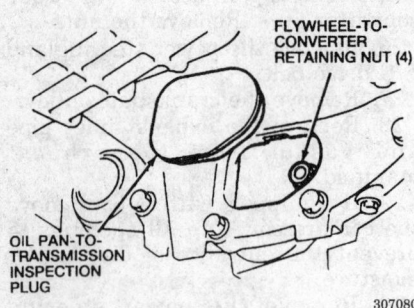

Transaxle inspection plug — 1.8L engine

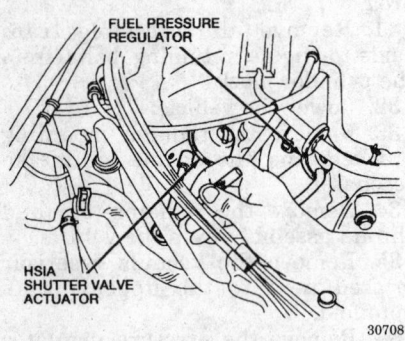

HSIA shutter valve actuator location — 1.8L engine

48. If removed, install the flexplate. Torque the retaining bolts to 71–76 ft. lbs. (96–103 Nm).

49. Carefully lower the engine assembly into the engine compartment and position to the transaxle.

NOTE: Make sure the torque converter studs are properly seated in the flexplate mounting holes.

50. Install the transaxle-to-engine upper right retaining bolt and tighten to 41–59 ft. lbs. (55–80 Nm).

51. Install the right-hand engine support insulator. Tighten the bolt and nuts to 49–69 ft. lbs. (67–93 Nm).

52. Install the front engine support damper. Tighten the bolt and nuts to 41–50 ft. lbs. (55–80 Nm).

53. Remove the engine sling from the lifting brackets and remove the engine hoist.

54. Install the remaining transaxle-engine retaining bolts and tighten to 41–59 ft. lbs. (55–80 Nm).

55. Install the alternator and the alternator and water pump drive belt.

56. Install the HSIA shutter valve actuator and bracket assembly.

57. Install the pressure regulator and bracket assembly.

58. Install the vacuum reservoir located next to the upper intake manifold.

59. Place the power steering pump and bracket assembly to its mounting position.

60. Place the radiator, fan and shroud assembly to its mounting position.

61. Install the radiator mounting brackets along with the engine air intake resonator. Tighten the retaining bolts to 69–95 inch lbs. (7.8–11.0 Nm).

62. Connect the engine cooling fan and radiator thermoswitch electrical connectors.

63. Raise and safely support the vehicle.

64. Install the oil pan-to-transaxle retaining bolts and tighten to 27–38 ft. lbs. (37–52 Nm).

65. Install the power steering pump and bracket assembly. Tighten the bolts to 27–38 ft. lbs. (37–52 Nm).

66. Install the lower radiator hose and clamps.

67. Unplug and connect 2 transaxle cooling lines to the radiator.

68. If equipped, install the A/C compressor.

69. If equipped, install the A/C hose routing bracket to the radiator. Tighten the bracket retaining nuts to 56–82 inch lbs. (6.4–9.3 Nm).

70. Install the crankshaft pulley and tighten the bolts to 109–152 inch lbs. (12–17 Nm).

71. Place a wrench on the crankshaft pulley and rotate the crankshaft to gain access to the torque converter studs. Install 4 torque converter retaining nuts and tighten to 25–36 ft. lbs. (34–49 Nm). Install the transaxle inspection plate.

72. Install the power steering drive belt.

73. Install the halfshaft bearing support and tighten the bolts to 31–46 ft. lbs. (42–62 Nm).

74. Install the starter motor.

75. Connect the heater core inlet and outlet hoses at the bulkhead.

76. Install the exhaust inlet pipe, with a new gasket, to the exhaust manifold. Tighten the pipe-to-converter retaining nuts to 23–34 ft. lbs. (31–46 Nm).

77. Install the right and left lower splash shields and the right upper splash shield. Tighten the bolts to 69–95 inch lbs. (7.8–11.0 Nm).

78. Lower the vehicle.

79. Install the upper radiator hose and clamps.

80. Unplug the fuel supply and return lines and connect them to the fuel rail. Install the necessary fuel line clips.

81. Install the accelerator and kickdown cable bracket onto the upper intake manifold. Tighten the bolts to 69–95 inch lbs. (7.8–11.0 Nm). Install the accelerator and kickdown cables onto the throttle cam.

82. Connect the power brake vacuum supply hose to the vacuum booster.

83. If equipped, connect the vehicle speed control vacuum hoses to the upper intake manifold.

84. Connect all engine ground straps.

85. Connect all remaining electrical connectors to their original locations, as marked during the removal procedure.

86. Connect the ignition coil high-tension lead into the distributor.

87. Install the engine air cleaner outlet tube, engine air intake resonator assembly and the throttle body.

88. Fill the engine cooling system.

89. Fill the crankcase with the correct grade and amount of engine oil.

90. If equipped, evacuate and recharge the A/C system.

91. Install the hood, aligning the marks that were made during the removal procedure.

92. Install the rear seat cushion.

93. Connect the negative battery cable.

94. Run the engine and check for leaks.

95. Road test the vehicle and check for proper operation.

96. Recheck all fluid levels.

With Manual Transaxle

The engine is removed from the engine compartment with the transaxle, as an assembly.

1. Relieve the fuel system pressure.

2. Mark the position of the hood on the hinges and remove the hood.

3. If equipped with A/C, properly discharge the system using suitable recovery equipment. Do not discharge the air conditioner into the atmosphere.

4. Drain the engine cooling system into a suitable container.

5. Drain the engine oil into a suitable container and reinstall the drain plug.

6. Remove the engine air cleaner outlet tube and the engine air intake resonator.

7. Remove the battery and the battery tray.

8. Disconnect the accelerator cable from the throttle cam and remove the

accelerator cable bracket from the upper intake manifold.

9. Remove the upper radiator hose and disconnect the radiator overflow hose from the radiator filler neck.

10. Disconnect the radiator thermoswitch and engine cooling fan electrical connectors.

11. Remove the retaining nuts to the radiator mounting brackets and remove the brackets.

12. Disconnect the alternator, oil pressure sensor, throttle position sensor, idle air control valve, park/neutral position switch, fuel injector wiring harness, back-up light switch, water temperature sender unit, heated oxygen sensor, power steering pressure switch and distributor electrical connectors.

NOTE: Mark the position of the connectors prior to removal to ease installation.

13. Disconnect all engine ground straps.

14. Remove the ignition coil high-tension lead from the distributor.

15. Disconnect the fuel supply and return lines and plug the lines.

16. Disconnect the heater core inlet and outlet hoses.

17. Remove the power brake vacuum supply, purge control vacuum and, if equipped, speed control vacuum hoses.

NOTE: Mark the position of the hoses prior to removal to ease installation.

18. Raise and safely support the vehicle.

19. Remove the right upper and lower splash shields.

20. Remove the clutch slave cylinder pipe bracket from the transaxle with the hose still connected. Position the slave cylinder aside.

NOTE: Be careful not to damage the pipe or the hose.

21. Disconnect the shift control rod and the extension bar from the transaxle.

22. Remove the battery cover air duct.

23. Remove the radiator lower hose.

24. Remove the power steering drive belt.

25. Remove the power steering pump and bracket assembly with the hoses still connected. Suspend the pump with wire away from the work area.

26. Remove the A/C hose routing bracket, if equipped, from the rear engine support and position the A/C hose aside.

27. If equipped, remove the A/C compressor with the hoses still connected. Suspend the compressor with wire aside of the work area.

28. Disconnect the speedometer cable from the transaxle.

29. Remove the exhaust inlet pipe and muffler bracket and gasket from the exhaust manifold.

30. Mark the location and disconnect the wires from the starter motor.

31. Remove the front stabilizer bar.

32. Remove the cotter pins and castellated nuts from the tie rod ends. Separate the tie rod ends from the steering knuckles. Discard the cotter pins.

33. Remove the halfshafts from the transaxle.

34. Remove the front and rear transaxle support insulator nuts from the rear engine support.

35. Lower the vehicle.

36. Remove the radiator, fan and shroud assembly from the vehicle.

37. Install a suitable engine removal sling onto the engine lifting brackets.

38. Place a suitable engine hoist into position and support the engine.

39. Remove the crankshaft pulley.

40. Remove the front engine support damper nut and bolt. Remove the engine insulator mounting bolt and nut. Remove the right-hand engine support insulator-to-engine retaining nuts.

41. Remove the left-hand transaxle support insulator and bracket.

42. Carefully raise and remove the engine and transaxle assembly.

43. To separate the engine from the transaxle, remove the upper intake manifold support bracket.

44. Remove the starter motor.

45. Remove the front transaxle support insulator from the transaxle.

46. Remove all oil pan-to-transaxle bolts and transaxle-to-engine retaining bolts from the cylinder block and separate the transaxle from the engine.

47. Remove the flywheel and clutch assembly from the engine.

48. Install the engine to a suitable engine stand.

To install:

49. Install a suitable engine removal sling onto the engine lifting brackets. Place a suitable engine hoist into position and install the engine sling.

50. Remove the engine from the engine stand and lower the engine with the hoist still supporting it.

51. Install the flywheel and clutch assembly.

52. Install the transaxle onto the engine.

53. Install the transaxle-to-engine bolts and tighten to 47–66 ft. lbs. (64–89 Nm).

54. Install the oil pan-to-transaxle retaining bolts and tighten to 27–38 ft. lbs. (37–52 Nm).

55. Position the front transaxle support insulator onto the transaxle and install the retaining bolts. Tighten the bolts to 27–38 ft. lbs. (37–52 Nm).

56. Position the starter motor into the transaxle housing and install the retaining bolts. Tighten the bolts to 27–38 ft. lbs. (37–52 Nm).

57. Install the upper intake manifold support bracket. Tighten the bolts to 27–38 ft. lbs. (37–52 Nm) and the nut to 14–19 ft. lbs. (19–25 Nm).

58. Using the engine hoist, carefully position the engine and transaxle assembly into the engine compartment and align the engine mounting points with the engine mount and the mounting holes in the transaxle crossmember.

59. Install the retaining nuts to the front transaxle support insulator and the rear engine support.

60. Position the right-hand engine support insulator into the vehicle.

61. Install the engine insulator retaining bolt and nut. Tighten them to 49–69 ft. lbs. (67–93 Nm).

62. Install the right-hand engine support insulator-to-engine retaining nuts. Tighten the nuts to 54–76 ft. lbs. (74–103 Nm).

63. Install the front engine support damper and retaining bolt and nut. Tighten the bolt and nut to 41–59 ft. lbs. (55–80 Nm).

64. Place the clutch slave cylinder assembly into its proper mounting position.

65. Install the left-hand transaxle support bracket and retaining bolts. Tighten the bolts to 41–59 ft. lbs. (55–80 Nm).

66. Install the left-hand transaxle support insulator and install the retaining bolts. Tighten the bolts to 32–45 ft. lbs. (43–61 Nm).

67. Install the left-hand transaxle support insulator retaining nuts. Tighten the nuts to 49–69 ft. lbs. (67–93 Nm).

68. Place the radiator, fan and shroud assembly to its mounting position.

69. Install the radiator mounting brackets and tighten the nuts to 69–95 inch lbs. (7.8–11.0 Nm).

70. Install the upper radiator hose and connect the expansion reservoir

overflow tube to the radiator filler neck.

71. Connect the engine cooling fan and radiator thermoswitch electrical connectors.

72. Raise and safely support the vehicle.

73. Install the lower radiator hose.

74. Install the halfshafts.

75. Install the tie rod ends to the steering knuckles. Install the castellated nuts and tighten to 25–33 ft. lbs. (34–46 Nm). Install new cotter pins.

76. Install the stabilizer bar.

77. Connect the wires to the starter motor according to their positions as marked during the removal procedure.

78. Install the exhaust inlet pipe to the exhaust manifold while making sure to install a new gasket. Tighten the flange-to-manifold retaining nuts to 23–34 ft. lbs. (31–46 Nm).

79. Install the exhaust pipe support bracket. Tighten the bracket retaining bolts to 27–38 ft. lbs. (37–52 Nm).

80. Install the speedometer cable to the transaxle.

81. If equipped, install the A/C compressor. Tighten the mounting bolts to 15–22 ft. lbs. (20–30 Nm).

82. Install the A/C routing bracket, if equipped, to the transaxle crossmember. Tighten the bolt to 56–82 inch lbs. (6.4–9.3 Nm).

83. Install the power steering pump and bracket assembly. Tighten the pump retaining bolts to 27–38 ft. lbs. (37–52 Nm).

84. Install the power steering drive belt.

85. Install the battery cover air duct and tighten the retaining bolts to 69–95 inch lbs. (7.8–11.0 Nm).

86. Install the extension bar to the transaxle and tighten the retaining nut to 23–34 ft. lbs. (31–46 Nm).

87. Connect the shift control rod to the transaxle and tighten the retaining nut to 12–17 ft. lbs. (16–23 Nm).

88. Install the clutch slave cylinder retaining bolts and tighten to 12–17 ft. lbs. (16–23 Nm).

89. Position the slave cylinder pipe and install the routing bracket and retaining bolt. Tighten the bolt to 12–17 ft. lbs. (16–23 Nm).

90. Install the right upper and lower splash shields. Tighten the bolts to 69–95 inch lbs. (7.8–11.0 Nm).

91. Lower the vehicle.

92. Connect the heater core and vacuum hoses according to their original positions as marked during the removal procedure.

93. Unplug and connect the fuel supply and return lines and clips.

94. Connect the ignition coil high tension lead into the distributor.

95. Connect all engine ground straps.

96. Connect all remaining electrical connectors according to the locations marked during the removal procedure.

97. Install the accelerator cable bracket to the upper intake manifold and connect the accelerator cable to the throttle cam.

98. Install the battery tray and the battery.

99. Install the engine air cleaner inlet tube and the engine air intake resonator.

100. Fill the engine cooling system.

101. Fill the crankcase with the proper amount and type of engine oil.

102. If equipped, evacuate and recharge the A/C system.

103. Install the hood, aligning the marks that were made during the removal procedure.

104. Install the rear seat cushion.

105. Connect both battery cables, negative cable last.

106. Run the engine and check for leaks.

107. Road test the vehicle and check for proper engine and transaxle operation.

108. Recheck all fluid levels.

1.9L Engine

——— CAUTION ———
Fuel injection systems remain under pressure, even after the engine has been turned OFF. The fuel system pressure must be relieved before disconnecting any fuel lines. Failure to do so may result in fire and/or personal injury.

With Automatic Transaxle

The engine can be separated from the transaxle and removed from the engine compartment.

1. Relieve the fuel system pressure.

2. Mark the position of the hood on the hinges and remove the hood.

3. Drain the engine cooling system into a suitable container.

4. Drain the engine oil into a suitable container and install the drain plug.

5. Remove the engine air cleaner intake tube.

6. Remove the crankcase ventilation hose from the cylinder head cover and the vacuum hose from the bottom side of the throttle body.

7. Disconnect the power brake booster supply hose.

8. Disconnect the following electrical connectors:

   a. Fuel charging harness, located at the right strut tower.

   b. Alternator harness, from the back side of the alternator.

   c. Heated oxygen sensor.

   d. Ignition coil.

   e. Radio suppressor, mounted on the coil bracket.

   f. Engine coolant temperature sensor, cooling fan sensor and temperature gauge sending unit, mounted on a common water tube near the thermostat housing.

   g. Radiator cooling fan.

NOTE: Mark the position of the electrical connectors to aid installation.

9. Remove the Idle Air Control (IAC) valve.

10. Remove the ground strap from the stud on the left side of the cylinder head near the ignition coil.

11. Disconnect the accelerator cable and the transaxle kickdown cable from the throttle lever. Remove the cable bracket from the intake manifold and position aside.

12. Disconnect both heater hoses at the engine compartment bulkhead.

13. Disconnect the fuel supply and return hoses at the fuel supply manifold and plug the fuel lines.

14. Remove the upper radiator hose.

15. Raise and safely support the vehicle.

16. Remove the right side and the right and left front splash shields.

17. Remove the lower radiator hose from the radiator.

18. Position a drain pan under the radiator and remove the lower transaxle oil cooler line.

19. Remove 2 oil cooler line retaining bracket bolts from the bottom of the radiator.

20. Remove the radiator fan shroud lower retaining bolts.

21. Lower the vehicle.

22. Remove the radiator fan shroud upper retaining bolts and remove the fan and shroud assembly from the vehicle.

23. Remove the upper transaxle oil cooler line from the radiator and remove the radiator from the vehicle.

24. If equipped with A/C, properly discharge the system using suitable recovery equipment. Do not discharge the A/C into the atmosphere.

25. Disconnect the A/C suction line at the suction accumulator/drier. Plug or cap the openings to prevent the entrance of dirt and moisture.

26. Remove the accessory drive belt.

27. Remove the power steering return hose from the pump reservoir and the high-pressure hose from the power steering pump.

28. Remove the power steering and air conditioner line retainer bracket bolts from the alternator bracket. Position the hoses aside.

29. Remove the accessory drive belt automatic tensioner assembly.

30. Raise and safely support the vehicle.

31. Remove the drive belt idler pulley.

32. If equipped, remove 4 A/C compressor retaining bolts. Remove the compressor assembly with the lines attached and position aside. Safety wire the compressor to the vehicle sub-frame.

33. If required, remove the catalytic converter inlet pipe.

34. Remove the transaxle kickdown cable support bracket from the back side of the cylinder block. Position the cable and the bracket aside.

35. Disconnect the oil pressure sensor electrical connector.

36. Disconnect the starter solenoid wire and the positive battery cable from the starter.

37. Remove the flexplate inspection shield.

38. Remove 4 torque converter retaining nuts.

39. Remove the crankshaft pulley.

40. Remove 5 engine-to-transaxle retaining bolts.

41. Lower the vehicle.

42. Remove 3 starter motor retaining bolts and remove the starter motor out of the top of the engine compartment.

43. Remove 2 transaxle-to-engine retaining bolts.

44. Connect an engine removal sling to suitable engine lifting brackets. Position a suitable engine hoist and support the engine.

45. Remove the front engine support damper and the right-hand engine support insulator.

46. With the engine assembly supported by the engine hoist, carefully separate the engine from the transaxle. Make sure the torque converter does not fall out.

47. Carefully lift the engine out of the vehicle.

48. Remove the flexplate.

49. Install the engine onto a suitable engine stand.

To install:

50. Attach the engine removal sling to the engine lifting brackets and remove the engine from the stand with the engine hoist.

51. Install the flexplate, if removed. Torque the retaining bolts to 54–67 ft. lbs. (73–91 Nm).

52. Carefully lower the engine into the vehicle and join the engine to the transaxle. Make sure the torque converter studs correctly engage the flexplate and the alignment dowels engage the transaxle housing.

53. Install 2 upper transaxle-to-engine bolts, but do not fully tighten them at this time.

54. Install the right-hand engine support insulator and the front engine support damper.

55. Position the engine hoist aside and remove the sling from the engine lifting brackets.

56. Raise and safely support the vehicle.

57. Install 5 engine-to-transaxle bolts, but do not fully tighten them at this time.

58. Install the crankshaft pulley and tighten the retaining bolt to 81–96 ft. lbs. (110–130 Nm).

59. Install 4 torque converter retaining nuts and tighten to 25–36 ft. lbs. (34–49 Nm). Install the transaxle inspection plate.

60. Connect the oil pressure sensor electrical connector.

61. Install the kickdown cable support bracket.

62. If equipped, position the A/C compressor on the bracket and install 4 retaining bolts. Tighten the bolts to 15–22 ft. lbs. (20–30 Nm).

63. Install the catalytic converter inlet pipe, if removed.

64. Lower the vehicle.

65. From above, position the starter motor and install 3 retaining bolts. Torque the bolts to 15–20 ft. lbs. (20–27 Nm).

66. Connect the positive battery cable and the starter relay cable to the starter motor.

67. Tighten 2 upper transaxle-to-engine bolts to 40–59 ft. lbs. (55–80 Nm).

68. Install the power steering high-pressure hose on the pump.

69. Install the accessory drive belt idler pulley and automatic tensioner.

70. Install the power steering return hose on the pump reservoir.

71. Install the power steering hose retainer bracket on the alternator bracket.

72. Install the accessory drive belt.

73. If equipped, connect the A/C suction line to the accumulator.

74. Install the radiator assembly.

75. Connect the upper transaxle oil cooler line at the radiator.

76. Position the cooling fan and shroud assembly and install the upper retaining bolts.

77. Raise and safely support the vehicle.

78. Install the lower shroud retaining bolts and connect the lower transaxle oil cooler line.

79. Install the oil cooler line retaining bracket bolts.

80. Install the lower radiator hose.

81. Tighten 5 engine-to-transaxle bolts to 27–38 ft. lbs. (37–52 Nm).

82. Install the left and right front splash shields and the right side splash shield.

83. Lower the vehicle.

84. Install the upper radiator hose.

85. Connect both heater hoses at the engine compartment bulkhead.

86. Install the accelerator cable bracket and attach the accelerator and kickdown cables to the throttle lever.

87. Install the IAC valve.

88. Install the ground strap on the stud at the front left side of the cylinder head, near the ignition coil.

89. Connect the remaining electrical connectors according to the positions marked during the removal procedure.

90. Unplug and connect the fuel supply and return lines. Be sure to install the fuel line safety clips.

91. Connect the power brake supply hose, the vacuum hose on the bottom side of the throttle body, and the crankcase ventilation hose to the cylinder head cover.

92. Install the engine air cleaner intake tube.

93. Install the rear seat cushion.

94. Connect the negative battery cable.

95. Fill the engine cooling system.

96. Fill the crankcase with the proper grade and amount of engine oil.

97. Install the hood, aligning the marks that were made during the removal procedure.

98. Run the engine and check for leaks.

99. Road test the vehicle and check for proper engine and transaxle operation.

100. Recheck all fluid levels.

101. If equipped, evacuate and recharge the A/C system.

With Manual Transaxle

The engine is removed from the engine compartment with the transaxle, as an assembly.

1. Relieve the fuel system pressure.

2. Disconnect the positive battery cable and remove the battery and the battery tray.

3. Drain the engine cooling system into a suitable container.

4. Drain the engine oil into a suitable container and reinstall the drain plug.

5. Remove the engine air cleaner.

6. Disconnect the crankcase ventilation hose from the cylinder head cover and the vacuum hose from the bottom side of the throttle body.

7. Remove the power brake supply hose.

8. Disconnect the following electrical connectors:

   a. Fuel charging harness, located at the right shock tower.

   b. Alternator harness, from the back side of the alternator.

   c. Heated oxygen sensor.

   d. Ignition coil.

   e. Radio suppressor, mounted on the coil bracket.

   f. Engine coolant temperature sensor, cooling fan sensor and temperature gauge sending unit, mounted on a common water tube near the thermostat housing.

   g. Radiator cooling fan.

NOTE: Mark the position of the electrical connectors to aid installation.

9. Remove the Idle Air Control (IAC) valve.

10. Remove the ground strap from the stud on the left side of the cylinder head near the ignition coil.

11. Disconnect the accelerator cable from the throttle lever. Remove the cable bracket from the intake manifold and position aside.

12. Disconnect both heater hoses at the engine compartment bulkhead.

13. Disconnect the fuel supply and return hoses at the fuel supply manifold and plug the lines.

14. Remove the upper radiator hose.

15. If equipped, properly discharge the A/C system using suitable recovery equipment. Do not discharge the A/C into the atmosphere.

16. Disconnect the suction line at the accumulator.

17. Remove the accessory drive belt and the automatic tensioner and idler pulley.

18. Disconnect the power steering return hose from the pump reservoir and the high pressure hose from the pump.

19. Remove the power steering hose and air conditioning line retainer brackets from the alternator bracket.

20. Raise and safely support the vehicle.

21. Remove the right-hand, left-hand and front splash shields.

22. Disconnect the lower radiator hose from the radiator and remove the radiator fan shroud lower retaining bolts.

23. If equipped, remove 4 A/C compressor retaining bolts. Remove the compressor assembly with the lines attached and position aside. Safety wire the compressor to the vehicle sub-frame.

24. Remove the catalytic converter from the exhaust manifold. If a 2-piece unit a used, also remove the inlet pipe.

25. Disconnect the oil pressure sensor electrical connector.

26. Disconnect the starter motor solenoid wire and the positive battery cable at the starter.

27. Remove the transaxle extension bar and shift control rod.

28. Remove the crankshaft pulley.

29. Remove both halfshaft assemblies.

30. Install suitable transaxle plugs into the differential side gears.

NOTE: Failure to install the transaxle plugs may allow the differential side gears to move out of position.

31. Disconnect the speedometer cable at the transaxle.

32. Remove the clutch slave cylinder and line as an assembly from the transaxle and set it aside.

33. Remove the transaxle front, rear and left-hand support insulator through-bolts.

34. Lower the vehicle.

35. Remove the radiator fan shroud upper retaining bolts and remove the fan shroud assembly from the vehicle.

36. Connect a suitable engine removal sling to the engine lifting brackets. Connect the sling to a suitable engine hoist, position the hoist and support the engine.

37. Remove the right-hand engine support bracket and the right-hand engine support insulator through-bolt.

38. Remove the right-hand engine support insulator.

39. Carefully lift the engine and transaxle assembly out of the vehicle.

40. Separate the engine from the transaxle as required.

To install:

41. If separated, install the engine to the transaxle.

42. Carefully lower the engine and transaxle assembly into the vehicle with the engine hoist.

43. Position the transaxle onto its mounts and install the right-hand engine support insulator.

44. Install the right-hand engine support insulator through-bolt and the right-hand engine support bracket. Torque the support bracket bolts to 69–86 ft. lbs. (93–117 Nm) and the through-bolt to 49–69 ft. lbs. (67–93 Nm).

45. Remove the engine removal sling and the hoist.

46. Position the fan shroud assembly and install the upper retaining bolts.

47. Raise and safely support the vehicle.

48. Install the front, rear and left-hand transaxle support insulator through-bolts.

49. Install the clutch slave cylinder and line assembly.

50. Connect the speedometer cable.

51. Remove the transaxle plugs and install the halfshaft assemblies.

52. Install the crankshaft pulley and tighten the bolt to 81–96 ft. lbs. (110–130 Nm).

53. Install the transaxle extension bar and shift control rod.

54. Connect the starter solenoid wire and the positive battery cable to the starter motor.

55. Connect the oil pressure sensor electrical connector.

56. Install the catalytic converter/inlet pipe.

57. If equipped, position the A/C compressor on its bracket and install 4 retaining bolts. Torque the bolts to 15–22 ft. lbs. (20–30 Nm).

58. Install the radiator fan shroud lower retaining bolts and install the lower radiator hose.

59. Install the left and right side and front splash shields.

60. Lower the vehicle.

61. Install the power steering hoses and install the power steering hose and A/C line retainer brackets.

62. Install the accessory drive belt idler pulley and automatic tensioner and install the accessory drive belt.

63. If equipped, connect the A/C suction line.

64. Install the upper radiator hose.

65. Unplug and connect the fuel supply and return hoses to the fuel supply manifold. Install the fuel line clips.

66. Connect both heater hoses at the engine compartment bulkhead.

67. Install the accelerator cable bracket on the intake manifold and connect the cable to the throttle lever.

68. Install the ground strap on the stud at the front left side of the cylinder head.

69. Install the IAC valve.

70. Connect the remaining electrical connectors according to the positions marked during the removal procedure.

71. Connect the power brake vacuum hose, the crankcase ventilation hose and the vacuum line at the bottom of the throttle body.

72. Install the engine air cleaner.

73. Install the rear seat cushion.

74. Install the battery tray and the battery. Connect the battery cables, negative cable last.

75. Fill the engine cooling system.

76. Fill the crankcase with the proper grade and amount of engine oil.

77. Install the hood, aligning the marks that were made during the removal procedure.

78. Run the engine and check for leaks.

79. Road test the vehicle and check for proper engine and transaxle operation.

80. Recheck all fluid levels.

81. If equipped, evacuate and recharge the A/C system.

Engine Mounts

REMOVAL AND INSTALLATION

1.8L and 1.9L Engines with 4EAT or MX5 Transaxle

Rear Engine Support

1. Disconnect the negative battery cable.

2. Support the engine/transaxle assembly using Three Bar Engine Support D88L-6000-A, or equivalent.

3. Raise and safely support the vehicle.

4. Remove 3 splash shield retaining screws and the wiring harness clip from the rear engine support.

5. Remove 4 transaxle support insulator nuts from the rear engine support.

6. Remove 2 rear engine support retaining nuts.

7. While supporting the rear engine support, remove 2 rear engine support rebound insulator bolts.

8. Remove the rear engine support.

To install:

9. Position the rear engine support and install 2 rear engine support rebound insulator bolts.

10. Install 2 rear engine support retaining nuts. Tighten the rear engine

support bolts and nuts to 47–66 ft. lbs. (64–89 Nm).

11. Install 4 transaxle support insulator retaining nuts to the rear engine support. Torque the nuts to 27–38 ft. lbs. (37–52 Nm).

12. Install the splash shield screws and the wiring harness clip.

13. Lower the vehicle.

14. Remove the engine support tool.

15. Connect the negative battery cable.

Left-Hand Engine Support Insulator

1. Disconnect both battery cables, negative cable first.

2. Remove the battery and the battery tray.

3. Remove 3 left-hand engine support insulator mounting bracket retaining nuts.

4. Remove 4 left-hand engine support insulator retaining bolts.

5. Remove the left-hand support insulator and bracket assembly.

6. To disassemble the bracket from the insulator, remove the nut securing the assembly.

To install:

7. If disassembled, install the bracket to the insulator.

8. Position the left-hand support insulator and install 4 insulator bolts. Torque the bolts to 32–45 ft. lbs. (43–61 Nm).

9. Install 3 left-hand engine support insulator mounting bracket nuts. Torque the bracket nuts to 49–69 ft. lbs. (67–93 Nm).

10. Install the battery tray and battery. Properly secure the battery.

11. Install the battery cables, negative cable last.

Front Transaxle Support Insulator

1. Disconnect the negative battery cable.

2. Remove the rear engine support.

3. Remove the front transaxle support insulator through-bolt and retaining nut.

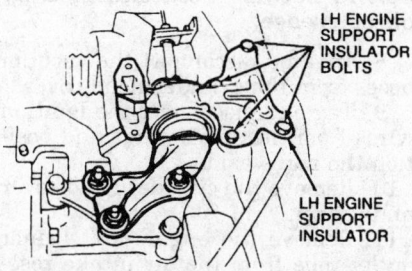

307163

Left-hand engine support insulator — 1.8L and 1.9L engines with 4EAT or MX5 transaxle

4. Remove the front transaxle support insulator.

5. If required, remove 3 front transaxle support bracket bolts and the bracket.

To install:

6. If removed, install the front transaxle support bracket and bolts. Torque the bolts to 27–38 ft. lbs. (37–52 Nm).

7. Position the front transaxle support insulator and install the through-bolt and retaining nut. Torque the through-bolt and nut to 12–17 ft. lbs. (16–23 Nm).

8. Install the rear engine mount.

9. Connect the negative battery cable.

Rear Transaxle Support Insulator

1. Disconnect the negative battery cable.

2. Support the engine/transaxle assembly using Three Bar Engine Support D88L-6000-A, or equivalent.

3. Remove the left-hand engine support insulator and bracket.

4. Raise and safely support the vehicle.

5. Remove the rear engine support.

NOTE: It may be necessary to raise or lower the engine slightly using the engine support for removal of the rear transaxle support insulator.

6. Remove 2 rear transaxle support bracket-to-transaxle retaining bolts.

7. Remove the rear transaxle support insulator and bracket assembly.

8. If required, separate the bracket from the support insulator by removing through-bolt and retaining nut.

To install:

9. If separated, assemble the rear transaxle support insulator and bracket with the through-bolt and retaining nut. Torque the through-bolt and nut to 49–69 ft. lbs. (67–93 Nm).

10. Position the rear transaxle support insulator and bracket and install 2 retaining bolts.

11. Install the rear engine support.

12. Lower the vehicle.

13. Install the left-hand engine support insulator and bracket.

14. Remove the engine support tool.

15. Connect the negative battery cable.

Front Engine Support Damper

1. Disconnect the negative battery cable.

2. Remove the engine insulator bolt and the engine damper nut.

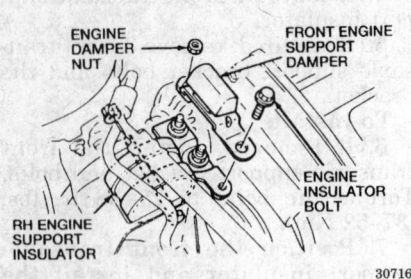

Front engine support damper — 1.8L and 1.9L engines with 4EAT or MX5 transaxle

3. Remove the front engine damper.

To install:

4. Position the front engine support damper to the right-hand engine support insulator.

5. Install the engine insulator bolt and the engine damper nut and torque to 41–59 ft. lbs. (55–80 Nm).

6. Connect the negative battery cable.

Right-Hand Engine Support Insulator

1. Disconnect the negative battery cable.

2. Remove the front engine support damper.

3. If equipped, remove the low side A/C hose bracket retaining nut and bolt, and pivot the A/C hose aside.

4. Remove the power steering hose bracket retaining nuts and move the hose aside.

5. Support the engine with a floor jack.

6. Remove 2 engine insulator nuts and washers.

7. Remove the right-hand engine support insulator through-bolt and retaining nut.

8. Remove the right-hand engine support insulator.

9. If required, remove 3 mounting bracket bolts and remove the right-hand engine support insulator mounting bracket.

To install:

10. If removed, install the right-hand engine support insulator mounting bracket and 3 retaining bolts. Torque the bracket bolts to 69–86 ft. lbs. (93–117 Nm).

11. Position the right-hand engine support insulator and install the through-bolt and retaining nut. Torque the through-bolt and nut to 49–69 ft. lbs. (67–93 Nm).

12. Install 2 engine insulator nuts and washers. Torque the nuts to 54–76 ft. lbs. (74–103 Nm).

13. Remove the floor jack.

14. Position the power steering hose and install the bracket retaining nuts.

15. If equipped, position the A/C hose and install the bracket nut and bolt.

16. Install the front engine support damper.

17. Connect the negative battery cable.

Cylinder Head

REMOVAL AND INSTALLATION

1.8L Engine

——— **CAUTION** ———

Fuel injection systems remain under pressure, even after the engine has been turned OFF. The fuel system pressure must be relieved before disconnecting any fuel lines. Failure to do so may result in fire and/or personal injury.

1. Relieve the fuel system pressure.

2. Disconnect the negative battery cable if not already done.

3. Drain the engine cooling system into a suitable container.

4. Remove the retaining bolts from the timing belt upper and middle covers. Remove the covers and gaskets.

5. Rotate the crankshaft in the direction of normal engine rotation and align the timing marks located on the camshaft sprockets and the seal plate.

6. Loosen the timing belt tensioner lock bolt and temporarily secure the tensioner spring in the fully extended position.

7. Remove the timing belt from the camshaft sprockets and secure it aside to prevent damage during the removal and installation of the cylinder head.

NOTE: Do not allow the timing belt to become contaminated by oil or grease.

8. Tag and disconnect the vacuum hoses from the cylinder head cover.

9. Tag and disconnect the ignition wires from the spark plugs and position the wires aside.

10. Remove the cylinder head cover and gasket.

11. Remove the engine air cleaner outlet tube from the air intake resonator and throttle body.

12. Disconnect the accelerator cable and, if equipped with an automatic transaxle, the kickdown cable from the throttle cam. Remove the cable bracket from the intake manifold.

13. Tag and disconnect all vacuum lines from the intake manifold.

14. Tag and disconnect all necessary electrical connectors from the cylinder head, exhaust manifold, intake manifold and throttle body. Disconnect the ground strap.

15. Remove the upper radiator hose.

16. Remove the transaxle-to-engine block upper-right bolt.

17. Disconnect the fuel pressure and return lines and plug the lines.

18. Disconnect the ignition coil high-tension lead from the distributor.

19. Tag and disconnect the necessary hoses connected to the cylinder head and intake plenum.

20. Remove 2 bolts from the transaxle vent tube routing brackets.

21. Raise and safely support the vehicle.

22. Remove the bolt from the water pump-to-cylinder head hose bracket.

23. Remove the exhaust front mounting flange and exhaust pipe support bracket from the exhaust manifold.

24. Remove the intake manifold support bracket.

25. Lower the vehicle.

26. Loosen and remove the cylinder head bolts in reverse of the torque sequence.

27. Remove the cylinder head assembly, with the intake manifold and exhaust manifold attached, from the vehicle.

28. Remove the intake manifold and exhaust manifolds.

29. Inspect the cylinder head for damage, cracks, and leakage of water and oil.

To install:

30. Remove all dirt, oil and old gasket material from all gasket contact surfaces.

31. Install the intake manifold and exhaust manifolds using new gaskets.

32. Install a new cylinder head gasket to the top of the cylinder block, using the dowel pins for reference.

33. Carefully place the cylinder head into its mounting position on top of the cylinder block.

34. Lubricate the cylinder head bolts with clean engine oil and install them finger-tight. Tighten the bolts in the proper sequence to 56–60 ft. lbs. (76–81 Nm).

35. Install 2 bolts to the transaxle vent tube routing brackets.

36. Connect the heater hoses to the cylinder head and install the clamps.

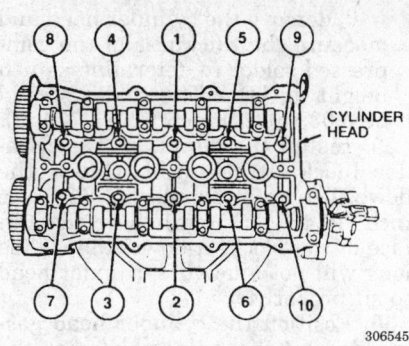

Cylinder head bolt tightening sequence — 1.8L engine

37. Connect the ignition coil high-tension lead to the distributor.

38. Connect the fuel pressure and return lines to the fuel supply manifold and install the safety clips.

39. Install the transaxle-to-engine block upper-right bolt; tighten the bolt to 47–66 ft. lbs. (64–89 Nm) for a manual transaxle or to 41–59 ft. lbs. (55–80 Nm) for an automatic transaxle.

40. Install the upper radiator hose and clamps.

41. Connect the ground strap and the electrical connectors that were disconnected at the cylinder head, exhaust manifold, intake manifold, and throttle body.

42. Connect the vacuum lines to the intake manifold.

43. Install the accelerator and kickdown cable bracket onto the intake manifold and tighten the bolts to 69–95 inch lbs. (7.8–11.0 Nm). Connect the cable(s) to the throttle cam.

44. Install the cylinder head cover and gasket, then connect the hose running from the plenum to the cylinder head cover. Tighten the cover bolts to 43–78 inch lbs. (4.9–8.8 Nm).

45. Install the air cleaner outlet tube to the engine air intake resonator and throttle body and tighten the clamps. Connect the hose going from the air duct to the cylinder head cover.

46. Install and connect the ignition wires.

47. Raise and safely support the vehicle.

48. Install the intake manifold support bracket. Tighten the bolts to 27–38 ft. lbs. (37–52 Nm) and the nut to 14–19 ft. lbs. (19–25 Nm).

49. Install the bolt to the water pump-to-cylinder head hose bracket.

50. Install the exhaust front mounting flange with a new gasket to the exhaust manifold. Tighten the flange-to-manifold retaining nuts to 23–34 ft. lbs. (31–46 Nm).

51. Install the exhaust pipe support bracket. Tighten the bracket retaining bolts to 27–38 ft. lbs. (37–52 Nm).

52. Iinstall and adjust the timing belt.

53. Turn the crankshaft 1⁵⁄₆ turns clockwise and align the 4th tooth to the right of the **I** and **E** timing marks on the camshaft sprockets with the seal plate alignment marks.

54. Loosen the timing belt tensioner lock bolt and apply tension to the timing belt. Tighten the tensioner lock bolt to 27–38 ft. lbs. (37–52 Nm).

55. Turn the crankshaft 2⅙ turns (780 degrees) clockwise and verify that the timing marks on the camshaft sprockets and the seal plate are aligned.

56. Install new gaskets onto the timing belt upper and middle covers and install the covers. Tighten the retaining bolts to 69–95 inch lbs. (8–11 Nm).

57. Fill the engine cooling system.

58. Check for contamination in the engine oil. Replace the engine oil and filter as required.

59. Connect the negative battery cable.

60. Run the engine and check for leaks.

61. Road test the vehicle and check for proper engine operation.

1.9L Engine

—— CAUTION ——
Fuel injection systems remain under pressure, even after the engine has been turned OFF. The fuel system pressure must be relieved before disconnecting any fuel lines. Failure to do so may result in fire and/or personal injury.

NOTE: The cylinder head bolts are a torque-to-yield design and cannot be reused. Before beginning this procedure, make sure new cylinder head bolts are available.

1. Relieve the fuel system pressure.

2. Disconnect the negative battery cable if not already done.

3. Drain the engine cooling system into a suitable container.

4. Remove the engine air cleaner intake tube.

5. Remove the crankcase breather hose from the cylinder head cover and the vacuum hose from the bottom of the throttle body.

6. Remove the power brake supply hose.

7. Label and disconnect the following:
   a. Fuel charging harness.
   b. Alternator harness.
   c. Crankshaft Position (CKP) sensor.
   d. Heated Oxygen Sensor (HO₂S).
   e. Ignition coil.
   f. Radio noise suppressor.
   g. Engine Coolant Temperature (ECT) sensor, cooling fan sensor and temperature sending unit.

8. Remove the ground strap from the stud on the left side of the cylinder head.

9. Disconnect the accelerator and the transaxle kickdown cables from the throttle lever and remove the cable bracket from the intake manifold.

10. Disconnect the heater hose containing the coolant temperature switches at the bulkhead.

11. Remove the upper radiator hose.

12. Disconnect the fuel supply and return lines.

13. Remove the oil dipstick tube retaining nut from the cylinder head stud.

14. Remove the power steering hose and the air conditioner line retainer bracket bolts from the alternator bracket.

15. Remove the accessory drive belt and the automatic tensioner.

16. Remove the alternator.

17. Raise and safely support the vehicle.

18. Remove the right side splash shield and remove the crankshaft dampener.

19. Remove the catalytic converter inlet pipe.

20. Remove the starter wiring harness from the retaining clip below the intake manifold.

21. Rotate the crankshaft to TDC for No. 1 cylinder.

22. Lower the vehicle.

NOTE: Removal of the engine mount dampener and right engine mount for removal of the timing belt, may not be necessary on all vehicles.

23. Support the engine with a suitable floor jack.

24. Remove the right engine mount dampener and the right engine mount retaining bolts from the mount bracket on the engine. Loosen the right engine mount through-bolt and roll the mount back.

25. Remove the timing belt cover.

26. Loosen the belt tensioner retaining bolt and pry the tensioner as far toward the rear of the engine as

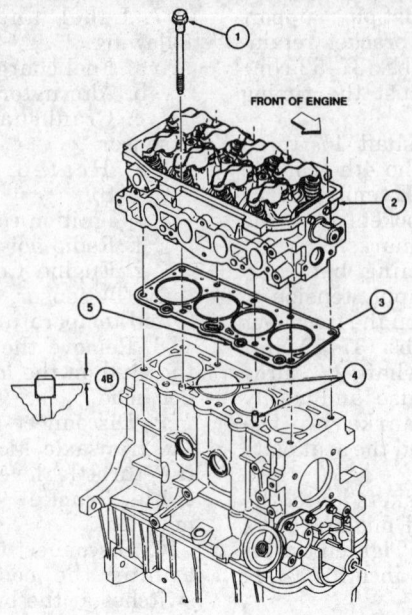

FRONT OF ENGINE

1. Cylinder Head Bolts (10 Req'd)
2. Cylinder Head
3. Head Gasket
4. Cylinder Head to Block Dowel (2 Req'd)

4B. 10.40 - 11.75mm Dowel Height
5. Gasket Identification Tab to be in Forward Position With Lettering on Top (Shown)

306560

Cylinder head installation — 1.9L engine

possible. Tighten the retaining bolt in this position.

27. Remove the timing belt.
28. Roll the right engine mount back into position and install the retaining bolts. Lower the floor jack.
29. Remove the heater hose support bracket retaining bolt and the alternator bracket-to-cylinder head retaining bolt.
30. Remove the cylinder head cover.
31. Remove and discard the cylinder head bolts.
32. Remove the cylinder head with the exhaust and intake manifolds attached. Discard the cylinder head gasket.

NOTE: Do not lay the cylinder head flat. Damage to the spark plugs, valves or gasket surfaces may result.

To install:

33. Clean all gasket material from the mating surfaces on the cylinder head and cylinder block. Clean out the head bolt holes in the cylinder block.

NOTE: Before final installation of the cylinder head to the engine, check the piston squish height. The old cylinder head bolts and cylinder head gasket may be used to check squish height before discarding.

Squish height is the clearance of the piston dome to the combustion chamber at piston TDC. Cylinder block deck machining or use of replacement crankshaft, pistons or connecting rods all effect squish height. If only the head gasket is being replaced, the squish height should be within specification. If the cylinder block or cylinder head are machined or if other parts are replaced, check the squish height before installing the cylinder head. If the squish height is out of specification, the connecting rods and/or pistons must be replaced or modified as required. The correct squish height is critical for correct compression and valve-to-piston clearance.

34. To check piston squish height:
    a. Place a small amount of soft lead solder or shot of an appropriate thickness on the top flat area of the piston.
    b. Rotate the crankshaft to lower the piston in the bore and install the old (compressed) head gasket and the cylinder head.
    c. Install the old head bolts and tighten them to 30–44 ft. lbs. (40–60 Nm) following proper torque sequence.
    d. Rotate the crankshaft to move the piston through its TDC position.

e. Remove the cylinder head and measure the thickness of the compressed solder to determine squish height at TDC. The solder should be 0.039–0.070 inch (1.0–1.77mm).

35. Install the dowels in the cylinder block, if removed. Check the dowel height, it should be 0.41–0.46 inch (10.40–11.75mm) above the surface of the block. A dowel that is too long will not allow the cylinder head to sit properly.

36. Position the cylinder head gasket and set the cylinder head to the cylinder block.

NOTE: The cylinder head bolts cannot be tightened to the specified torque more than once. Always use new bolts when installing the cylinder head.

37. Apply a light coat of engine oil to the threads of 10 new cylinder head bolts and install the new bolts finger-tight.
38. To properly torque the cylinder head bolts:
    a. Torque the new cylinder head bolts in sequence to 44 ft. lbs. (60 Nm).
    b. Loosen all cylinder head bolts approximately 2 turns and then torque again to 44 ft. lbs. (60 Nm) in sequence.
    c. Turn each bolt in sequence an additional 90 degrees.
    d. Turn each bolt in sequence again an additional 90 degrees.
39. Install the cylinder head cover and the alternator bracket-to-cylinder head bolt.

NOTE: If removal of the right engine mount and dampener was necessary for removal of the timing belt, continue with the following steps, otherwise go to the installation of the timing belt.

40. Support the engine with a suitable floor jack.
41. Remove the right engine mount-to-bracket bolts. Roll the mount aside.
42. Make sure No. 1 cylinder is at TDC.
43. Install the timing belt. Make sure the timing marks align properly and install the timing belt cover.
44. Roll the right engine mount into place and install 2 retaining bolts and the mount dampener. Remove the floor jack.
45. Raise and safely support the vehicle.
46. Install the crankshaft dampener.
47. Install the starter wiring harness on the retaining clip below the intake manifold.

```
9   3   1   5   7
O   O   O   O   O    INTAKE

O   O   O   O   O    EXHAUST
8   6   2   4   10
```
306561

Cylinder head bolt tightening sequence — 1.9L engine

48. Install the catalytic converter inlet pipe and the right side splash shield.
49. Lower the vehicle.
50. Install the alternator and the accessory drive belt tensioner. Install the accessory drive belt.
51. Install both the power steering hose and air conditioner line retainer bracket bolts. Install the oil dipstick tube retainer bolt.
52. Connect the fuel supply and return lines.
53. Install the upper radiator hose and connect the heater hose at the engine compartment bulkhead. Install the heater hose support bracket retaining bolt.
54. Install the accelerator cable bracket on the intake manifold and connect the accelerator and kickdown cables to the throttle lever.
55. Install the ground strap at the left side of the cylinder head.
56. Connect all remaining electrical connectors according to their positions marked during the removal procedure.
57. Connect the power brake supply hose, crankcase breather hose and the vacuum line at the bottom of the throttle body.
58. Install the engine air cleaner intake tube.
59. Connect the negative battery cable.
60. Fill and bleed the engine cooling system.
61. Check the condition of the engine oil, and if contaminated, change the engine oil and filter.
62. Run the engine and check for leaks and proper fluid levels.
63. Road test the vehicle and check for proper engine operation.

Valve Lifters

REMOVAL AND INSTALLATION

1.8L Engine

1. Disconnect the negative battery cable.
2. Remove the camshafts.
3. Mark the hydraulic lash adjusters so they can be installed in their original position and location.
4. Remove the hydraulic valve lifters from the cylinder head.

To install:
5. Apply clean engine oil to the hydraulic valve lifter friction surfaces.
6. If the hydraulic valve lifters are being reused, install them in the positions from which they were removed.
7. Make sure the hydraulic valve lifters move smoothly in their bores.
8. Install the camshafts.
9. Connect the negative battery cable.

1.9L Engine

1. Disconnect the negative battery cable.
2. Remove air cleaner assembly. Remove valve cover and gasket.
3. Remove rocker arms, lifter guides, lifter retainers and lifters.

NOTE: Always return lifters to the original bores unless they are being replaced.

To install:
4. Rotate the crankshaft so the camshaft keyway is at the 12 o'clock position with cylinder No. 1 at TDC.
5. Lubricate each lifter bore with engine oil.
6. If equipped with flat bottom lifters, install with oil hole in plunger upward.
7. If equipped with roller lifters, install with plunger upward and position guide flats of lifters to be parallel with centerline of camshaft. Color

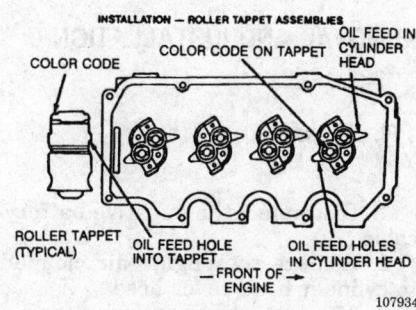

Roller tappet assembly — 1.9L engine

orientation dots on lifters should be opposite the oil feed holes in cylinder head.
8. With roller lifters, install lifter guide plates over tappet guide flats with tab toward exhaust side. For flat lifters, no guide plate is required.
9. Lubricate lifter plunger cap and valve tip with engine oil.
10. Install lifter guide plate retainers into rocker arm fulcrum slots. Notch should be toward the exhaust lifter.
11. Install the intake rocker arms on cylinders 3 and 4; then, the exhaust rocker arms on cylinders 2 and 4.
12. Lubricate rocker arms and install the fulcrums. Fulcrums must be fully seated in slots of cylinder head.
13. Install the fulcrum bolts and tighten to 17–22 ft. lbs. (23–30 Nm).
14. Rotate the crankshaft until the camshaft sprocket keyway is in the 6 o'clock position.
15. Install the remaining rocker arms, fulcrums and fulcrum bolts.
16. Install valve cover and gasket. Install air cleaner assembly.
17. Connect negative battery cable.

BLEEDING

1.8L and 1.9L Engines

Bleeding the valve lifters is not required for removal purposes. On the 1.9L engine, a Tappet Collapser T81P-6500-A or equivalent, is used to bleed down the valve lifters for checking valve clearance only.

New valve lifters should always be bled of air (pumped up) before installing into the cylinder head. This is normally accomplished by placing the valve lifters into a bucket of clean engine oil and physically operating the valve lifter between the thumb and index finger until it is fully pumped up or no more air is expelled from the valve lifter.

ADJUSTMENT

1.8L Engine

The valve lifters, are hydraulic and are not adjustable. It is important that all valve components are in good condition and installed and torqued properly.

1.9L Engine

The valve lifters are hydraulic and adjustment is not possible, however the collapsed lifter clearance can be checked. The clearance may be checked to help isolate a problem by

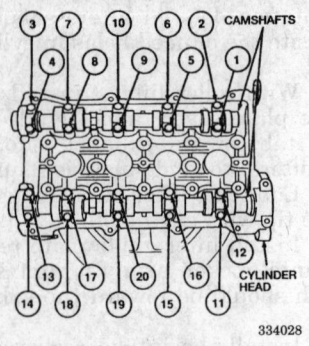

Camshaft cap bolt removal sequence — 1.8L engine

334028

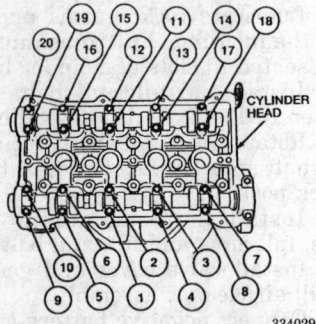

Camshaft cap bolt tightening sequence — 1.8L engine

334029

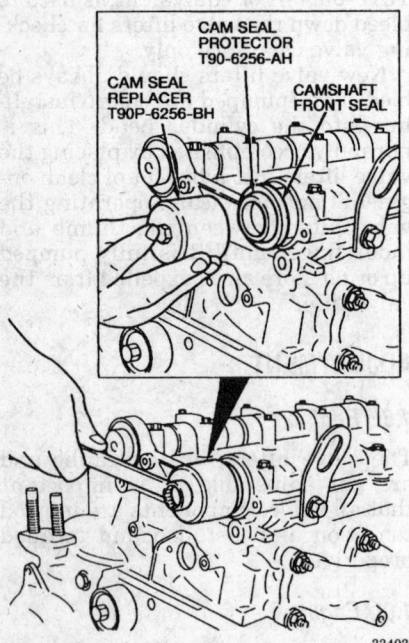

CAM SEAL PROTECTOR T90-6256-AH

CAM SEAL REPLACER T90P-6256-BH

CAMSHAFT FRONT SEAL

Camshaft oil seal installation — 1.8L engine

334030

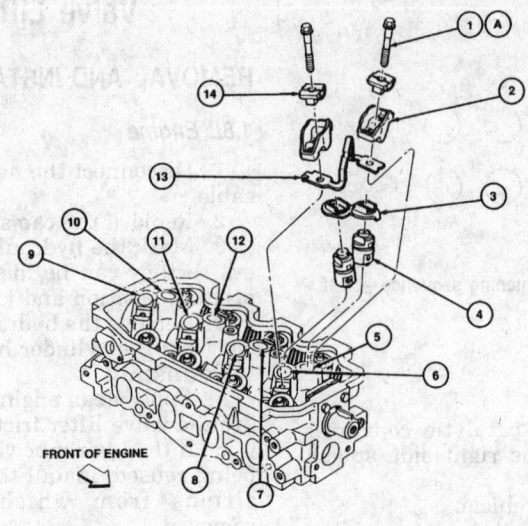

FRONT OF ENGINE

1. Rocker Arm Bolt (8 Req'd)
2. Rocker Arm (8 Req'd)
3. Valve Tappet Guide Plate (Tab To Front) (4 Req'd)
4. Valve Tappet
5. Valve Tappet # 1 Position
6. Valve Tappet # 2 Position
7. Valve Tappet # 3 Position
8. Valve Tappet # 4 Position
9. Valve Tappet # 8 Position
10. Valve Tappet # 7 Position
11. Valve Tappet # 6 Position
12. Valve Tappet # 5 Position
13. Valve Tappet Guide Plate Retainer (Tab To Rear) (4 Req'd)
14. Rocker Arm Seat
A. Tighten to 23–30 N·m (17–22 Lb-Ft)

309549

Valve lifters and related valve train components — 1.9L engine

first collapsing the valve lifter and placing the valve lifter on the base circle of the camshaft lobe. Measurements are taken with a feeler gauge between the valve stem tip and the rocker arm. Nominal clearance should be approximately 0.087 inch (2.2mm), with a minimum clearance of 0.000 inch (0.0mm) and a maximum clearance of 0.177 inch (4.5mm). Possible problems may be a worn rocker arm seat, camshaft lobe. valve tip or a collapsed or worn valve lifter.

Rocker Arms

REMOVAL AND INSTALLATION

1.9L Engine

1. Disconnect the negative battery cable.
2. Remove the engine air cleaner to cylinder head cover hose.
3. Remove the ignition wire separators and ignition wires from the cylinder head cover.

4. Remove the cylinder head cover bolts and remove the cover and gasket.

NOTE: If the components are to be reused, mark or lay out the components so they can be reinstalled in their original positions.

5. Examine the rocker arms and related parts for wear and/or damage and replace as necessary.

To install:

6. Coat the valve stem tips, rocker arms and the fulcrum contact areas with an appropriate engine assembly lubricant.

7. Rotate the engine until the lash adjuster (valve tappet) is on the base circle of the cam (valve closed) for the rocker arm being installed.

NOTE: Be sure to turn the engine only in a clockwise rotation. A counterclockwise rotation will cause the timing belt to possibly slip or lose teeth, altering the valve timing and causing serious engine damage.

8. Install each rocker arm, rocker arm seat and bolt. Torque the rocker arm bolts to 17–22 ft. lbs. (23–30 Nm). Be sure the lash adjuster is on the base circle of the cam for each rocker arm as it is installed.

9. Install a new cylinder head cover gasket and the cylinder head cover. Install the retaining bolts and torque to 4–9 ft. lbs. (5–12 Nm).

NOTE: Do not use any type of sealer with the cylinder head cover silicone gasket.

10. Install the ignition wire retainers, ignition wires and the engine air cleaner hose.
11. Connect the negative battery cable.
12. Run the engine and check for leaks.
13. Road test the vehicle and check for proper engine operation.

Intake Manifold

REMOVAL AND INSTALLATION

1.8L Engine

———— CAUTION ————
Fuel injection systems remain under pressure, even after the engine has been turned OFF. The fuel system pressure must be relieved before disconnecting any fuel lines. Failure to do so may result in fire and/or personal injury.

Upper Intake Manifold Only

1. Disconnect the negative battery cable.
2. Partially drain the engine cooling system.
3. Remove the engine air cleaner outlet tube at the throttle body.
4. Disconnect the accelerator cable and if equipped with an automatic transaxle, the TV cable from the throttle body control lever and the bracket from the upper intake manifold.
5. Disconnect the throttle body electrical connectors and vacuum hoses.
6. Disconnect the Idle Air Control (IAC) valve and the Bypass Air (BPA) valve coolant hoses from the upper intake manifold.
7. Remove the idle air control hose from the upper intake manifold.
8. Remove the upper intake manifold retaining bolts and nuts.
9. Raise and safely support the vehicle.
10. Remove the upper intake manifold lower retaining bolts.
11. Lower the vehicle.
12. Remove the upper intake manifold from the lower intake manifold.
13. If needed, separate the throttle body from the upper intake manifold.

14. Remove the upper intake manifold gasket.

To install:
15. Clean the gasket sealing surfaces of both intake manifolds.
16. If removed, install the throttle body to the upper intake manifold.
17. Install a new upper intake manifold gasket to the lower intake manifold.
18. Place the upper intake manifold in position and install the upper retaining bolts and nuts. Tighten the bolts and nuts to 14–19 ft. lbs. (19–25 Nm).
19. Raise and safely support the vehicle.
20. Install the upper intake manifold lower retaining bolts. Tighten the bolts to 14–19 ft. lbs. (19–25 Nm).
21. Lower the vehicle.
22. Connect the idle air control coolant hose to the upper intake manifold.
23. Connect the throttle body electrical and vacuum connectors to the upper intake manifold.
24. Install the accelerator cable bracket to the upper intake manifold. Tighten the retaining bolt to 69–95 inch lbs. (7.8–11 Nm).
25. Connect the accelerator cable and TV cable if equipped, to the throttle body control lever.
26. Connect the coolant hoses to the IAC valve and the BPA valve.
27. Install the engine air cleaner outlet tube to the throttle body.
28. Fill the cooling system.
29. Connect the negative battery cable.
30. Start the engine and check for leaks.
31. Road test the vehicle and check for proper operation.

Intake Manifold Assembly

1. Relieve the fuel system pressure.
2. Disconnect the negative battery cable, if not already done.
3. Tag and disconnect the necessary vacuum hoses from the upper intake manifold and throttle body.
4. Drain the engine cooling system into a suitable container below the level of the intake manifold.
5. Remove the vacuum reservoir from the upper intake manifold.
6. Disconnect the Idle Air Control (IAC) valve and the Bypass Air (BPA) valve coolant hoses from the upper intake manifold.
7. Disconnect the accelerator cable and if equipped with an automatic transaxle, the TV cable from the throttle body control lever and the bracket from the upper intake manifold.

8. Disconnect the throttle body electrical connectors and vacuum hoses.
9. Disconnect the fuel supply and return line spring lock couplings.
10. Disconnect the fuel supply and return lines.
11. Disconnect the PCV hose from the upper intake manifold and cylinder head cover.
12. Disconnect the fuel pressure regulator vacuum hose and the fuel injector wiring harness electrical connectors.
13. Remove the fuel injection supply manifold (fuel rail) retaining bolts and remove the fuel injection supply manifold.
14. Remove 2 bolts from the transaxle breather tube and remove the tube.
15. Remove 5 upper intake manifold-to-cylinder head retaining nuts.
16. Raise and safely support the vehicle.
17. Remove 4 bolts securing the intake manifold support and remove the support.
18. Remove 4 lower intake manifold-to-cylinder block retaining nuts.
19. Lower the vehicle.
20. Remove the intake manifold, upper intake manifold and throttle body as an assembly from the vehicle.
21. Remove the intake manifold gasket.
22. If necessary, separate the upper intake manifold and throttle body from the intake manifold.
23. Clean all gasket mating surfaces.

To install:
24. If necessary, install the throttle body and upper intake manifold to the intake manifold.
25. Install the intake manifold gasket.
26. Install the intake manifold, upper intake manifold and throttle body assembly onto the lower intake manifold studs.
27. Install the intake manifold-to-cylinder head retaining nuts and tighten in the proper sequence to 14–19 ft. lbs. (19–25 Nm).
28. Raise and safely support the vehicle.
29. Install the intake manifold support bracket. Tighten 3 larger bolts to 27–38 ft. lbs. (37–52 Nm) and the smaller bolt to 14–19 ft. lbs. (19–25 Nm).
30. Lower the vehicle.
31. Place the fuel injection supply manifold in position and install the mounting bolts. Tighten the bolts to 14–19 ft. lbs. (19–25 Nm).

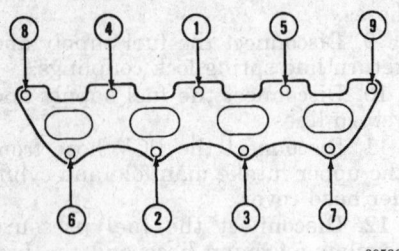

Intake manifold torque sequence — 1.8L engine

305967

32. Connect the fuel supply and return lines.

33. Connect the fuel supply and return line spring lock couplings.

34. Connect the fuel injector wiring harness electrical connectors and connect the vacuum hose to the pressure regulator.

35. Install the transaxle breather tube and 2 retaining bolts.

36. Install the vacuum reservoir to the upper intake manifold.

37. Connect the electrical connectors to the throttle body and the necessary vacuum hoses to the intake plenum and throttle body.

38. Connect the coolant hoses to the IAC valve and the BPA valve.

39. Connect the PCV hose to the intake plenum and cylinder head cover.

40. Install the cable bracket onto the upper intake manifold and connect the accelerator and, if equipped, the TV cable to the throttle lever.

41. Install the engine air cleaner outlet tube to the throttle body.

42. Fill the engine cooling system.

43. Connect the negative battery cable.

44. Start the engine and check for leaks.

45. Road test the vehicle and check for proper operation.

1.9L Engine

> **CAUTION**
>
> *Fuel injection systems remain under pressure, even after the engine has been turned OFF. The fuel system pressure must be relieved before disconnecting any fuel lines. Failure to do so may result in fire and/or personal injury.*

1. Relieve the fuel system pressure.

2. Partially drain the engine cooling system into a suitable container.

3. Remove the engine air intake tube.

4. Disconnect the fuel injector harness from the engine control harness at the right-hand strut tower.

5. Disconnect the Crankshaft Position (CKP) sensor.

6. Disconnect the fuel supply and return lines using the proper spring lock coupling removal tool.

7. If equipped, remove the Camshaft Position (CMP) sensor electrical connector.

8. Remove the accelerator cable and, if equipped with an automatic transaxle, the kickdown cable from the throttle lever. Remove the cable bracket from the intake manifold and position the cables aside.

9. Remove the power brake vacuum supply hose, PCV line and the vacuum line from the bottom of the throttle body.

10. If equipped, loosen and separate the nut connecting the EGR tube to the EGR valve.

11. Remove 7 retaining nuts from the intake manifold studs and slide the manifold assembly off the studs and remove it from the cylinder head.

12. Remove and discard the intake manifold gasket.

To install:

13. Clean and inspect the gasket mounting surfaces of the intake manifold and cylinder head. Both surfaces must be clean and flat.

14. Clean and lightly oil the manifold studs and position a new intake manifold gasket.

15. Install the intake manifold and 7 retaining nuts. Tighten the nuts to 12–15 ft. lbs. (16–20 Nm).

16. Install the vacuum lines to the bottom of the throttle body, the power brake vacuum supply hose and the PCV line.

17. If equipped, position the EGR tube to the EGR valve and install the EGR tube retaining nut. Torque the nut to 18–25 ft. lbs. (25–35 Nm).

18. Install the accelerator cable bracket and connect the accelerator

cable and, if equipped, kickdown cable on the throttle lever.

19. Connect the CMP sensor electrical connector.

20. Connect the fuel supply and return lines. Properly install the fuel line retaining clips.

21. Connect 2 fuel injector harness connectors to the engine control harness at the right-hand strut tower.

22. Connect the CKP sensor electrical connector.

23. Install the engine air cleaner intake tube.

24. Refill the engine cooling system.

25. Connect the negative battery cable.

26. Start the engine and allow it to reach normal operating temperature.

27. Check for fuel, coolant and vacuum leaks.

28. Road test the vehicle and check for proper operation.

Exhaust Manifold

REMOVAL AND INSTALLATION

1.8L Engine

1. Disconnect the negative battery cable.

2. Remove the engine air intake resonator.

3. Partially drain the cooling system into a suitable container.

4. Remove the upper radiator hose.

5. Remove the engine cooling fan motor and shroud assembly.

6. Raise and safely support the vehicle.

7. Remove the engine and transaxle splash shield.

8. Remove the exhaust inlet pipe from the exhaust manifold and remove the gasket.

9. Remove 2 bolts from the exhaust pipe support bracket.

10. Lower the vehicle.

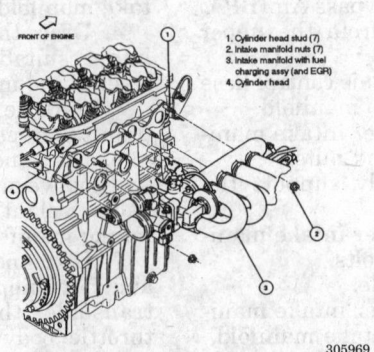

1. Cylinder head stud (7)
2. Intake manifold nuts (7)
3. Intake manifold with fuel charging assy (and EGR)
4. Cylinder head

305969

Intake manifold assembly — 1.9L engine

11. Disconnect the Heated Oxygen Sensor (HO$_2$S) electrical connector.

12. Remove the exhaust manifold heat shield retaining bolts and remove the shield.

13. Remove 9 exhaust manifold retaining nuts and remove the exhaust manifold.

14. Remove all gasket material from the cylinder head and exhaust manifold.

To install:

15. Install a new gasket onto the exhaust manifold studs.

16. Place the exhaust manifold onto the manifold studs and install 9 manifold retaining nuts. Torque the exhaust manifold retaining nuts to 28–34 ft. lbs. (38–46 Nm).

17. Install the exhaust manifold heat shield to its mounting position and install 4 shield retaining bolts. Tighten the bolts to 69–95 inch lbs. (7.8–11.0 Nm).

18. Connect the HO$_2$S electrical connector.

19. Install the engine cooling fan motor and shroud assembly.

20. Connect the upper radiator hose.

21. Install the engine air intake resonator.

22. Raise and safely support the vehicle.

23. Install the exhaust pipe support bracket.

24. Install a new gasket and install the exhaust pipe to the exhaust manifold. Tighten the retaining nuts to 23–34 ft. lbs. (31–46 Nm).

25. Install the engine and transaxle splash shield. Tighten the bolts to 69–95 inch lbs. (7.8–11.0 Nm).

26. Lower the vehicle.

27. Refill the cooling system.

28. Connect the negative battery cable.

29. Run the engine and check for leaks and proper operation.

1.9L Engine

1. Disconnect the negative battery cable.

2. Remove the accessory drive belt.

3. Remove the alternator.

4. Remove the radiator cooling fan motor and shroud assembly.

5. If equipped with an Exhaust Gas Recirculation (EGR) valve tube, perform the following:

   a. Loosen the nut on each end of the EGR tube, while holding the tube.

   b. Remove 2 retaining bolts at the ignition coil bracket securing the EGR tube and remove the EGR tube.

6. Remove 3 exhaust manifold heat shield retaining nuts and remove the exhaust manifold heat shield.

7. Raise and safely support the vehicle.

8. Remove 2 catalytic converter-to-exhaust manifold retaining nuts.

9. Lower the vehicle.

10. Remove 8 exhaust manifold retaining nuts.

11. Remove the exhaust manifold and gasket from the vehicle.

To install:

12. Clean the cylinder head and exhaust manifold gasket surfaces. Use care not to damage the aluminum cylinder head gasket surface.

13. Position the new gasket onto the exhaust manifold studs.

14. Position the exhaust manifold on the cylinder head and install 8 retaining nuts. Torque the nuts to 16–19 ft. lbs. (21–26 Nm).

15. Raise and safely support the vehicle.

16. Install the catalytic converter-to-exhaust manifold retaining nuts. Torque the retaining nuts to 25–33 ft. lbs. (34–47 Nm).

17. Lower the vehicle.

18. Install the exhaust manifold heat shield. Tighten 3 retaining nuts to 3–5 ft. lbs. (5–7 Nm).

19. If equipped with an EGR valve tube, proceed as follows:

   a. Install the EGR valve tube and hand-start both tube nuts.

   b. Torque the tube nuts to 18–25 ft. lbs. (25–35 Nm).

   c. Install 2 EGR valve to exhaust manifold tube retaining bolts at the ignition coil bracket and tighten.

20. Install the engine cooling fan motor and shroud assembly.

21. Install the alternator.

22. Install the accessory drive belt.

23. Connect the negative battery cable.

24. Run the engine and check for leaks and proper operation.

Front Crankshaft Seal

REMOVAL AND INSTALLATION

1.8L Engine

1. Disconnect the negative battery cable.

2. Remove the accessory drive belts.

3. Raise and safely support the vehicle.

4. Remove the right front wheel and tire assembly.

5. Remove the right upper and lower splash shields.

6. Remove the middle and lower timing belt covers.

7. Remove 4 crankshaft pulley retaining bolts and the crankshaft pulley.

8. If equipped with a crankshaft sprocket and no hub, remove the inner and outer guide plates, and remove the sprocket bolt.

9. If equipped with a crankshaft hub and a crankshaft sprocket, remove the crankshaft sprocket bolt and the hub.

10. Position the engine to TDC for No. 1 cylinder and remove the timing belt.

NOTE: If the timing belt is to be reused, mark an arrow on the belt to indicate its direction of rotation for reference during installation.

11. Remove the crankshaft sprocket and key.

12. If necessary, cut the lip of the front oil seal to ease removal.

13. Use Seal Remover T92C-6700-CH or equivalent prying tool to remove the oil seal.

To install:

14. Lubricate the lip of the new oil seal with clean engine oil.

15. Using Crankshaft Front Seal Replacer T88C-6701-AH or equivalent, install the seal evenly until it is flush with the edge of the oil pump body.

16. Install the crankshaft timing belt sprocket onto the crankshaft while making sure to match the alignment grooves.

17. Install the key with the tapered end facing the oil pump.

18. If equipped with a crankshaft sprocket and hub, Install the sprocket and hub and install the retaining bolt. Tighten the bolt to 80–87 ft. lbs. (108–118 Nm).

19. If equipped with a sprocket and no hub, install the sprocket and tighten the retaining bolt to 80–87 ft. lbs. (108–118 Nm).

20. Temporarily secure the timing belt tensioner in the far left position with the spring fully extended, then tighten the lock bolt if not already done.

21. Verify that the timing marks on the timing belt crankshaft sprocket and the cylinder block are aligned.

22. Verify that the timing marks on the camshaft sprockets and the seal plate are aligned.

23. Install the timing belt.

24. Loosen the tensioner lock bolt. Using a suitable prying tool, position the timing belt tensioner so the tim-

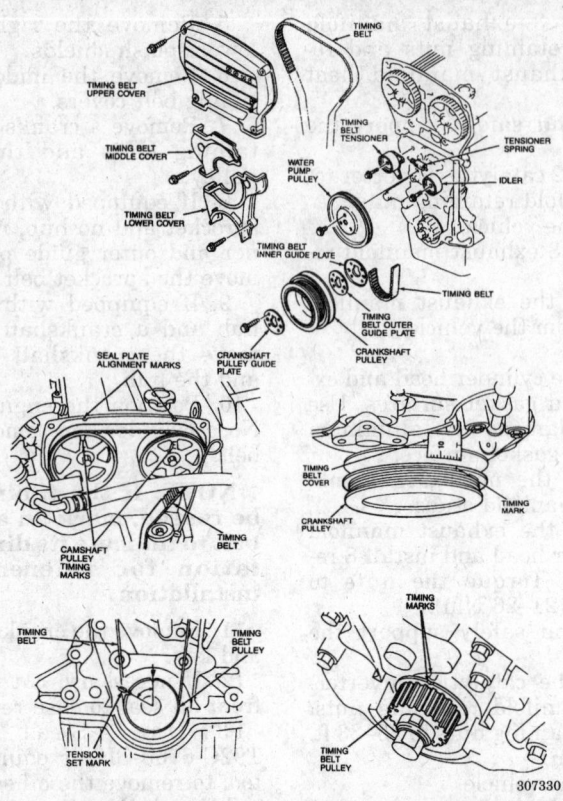

Timing belt, sprockets and timing marks — 1.8L engine

ing belt is taut, then tighten the tensioner lock bolt.

25. Rotate the crankshaft 2 full turns clockwise and align the timing belt sprocket mark with the mark on the cylinder block. Verify that the camshaft sprocket marks are aligned with the seal plate marks. If the marks are not aligned, remove the timing belt and repeat the procedure.

26. Turn the crankshaft 1⅚ turns clockwise and align the timing belt pulley mark with the tension set mark. This is approximately the 10 o'clock position.

27. Apply tension to the timing belt tensioner and install the tensioner lock bolt. Tighten the bolt to 27–38 ft. lbs. (37–52 Nm).

28. Turn the crankshaft 2⅙ turns (780 degrees) clockwise and verify that the timing marks are aligned.

29. Measure the timing belt deflection by applying approximately 22 pounds (98 N) of pressure on the belt between the camshaft pulleys. The timing belt deflection should be 0.35–0.45 inch. If necessary, adjust the timing belt deflection.

30. Rotate the crankshaft 2 turns clockwise and verify that the timing marks are aligned. If not, repeat the timing belt alignment procedure.

31. Install the inner and outer guide plates to the crankshaft sprocket.

32. Install the timing belt covers. Tighten the retaining bolts to 110–150 inch lbs. (12–17 Nm).

33. Install the pulleys, accessory drive belts, and splash shields.

34. Install the right front wheel and tire assembly. Torque the lug nuts to 65–84 ft. lbs. (88–118 Nm).

35. Lower the vehicle.

36. Connect the negative battery cable.

37. Run the engine and check for leaks.

38. Road test the vehicle and check for proper operation.

1.9L Engine

1. Disconnect the negative battery cable.

2. Remove the accessory drive belt.

3. Remove 2 timing belt cover retaining nuts and remove the cover.

4. Align the timing mark on the camshaft sprocket with the timing mark on the cylinder head. Confirm that the timing mark on the crankshaft sprocket is aligned with the timing mark on the oil pump housing.

5. Loosen the belt tensioner retaining bolt and pry the tensioner away from the timing belt. Tighten the tensioner retaining bolt in this position.

6. Tag and separate the ignition wires from the spark plugs. Remove the spark plugs.

7. Remove the front engine support insulator (mount).

8. Raise and safely support the vehicle.

9. Remove the right side splash shield.

10. Remove the flywheel inspection shield.

11. Use a suitable tool to hold the flywheel in place.

12. Remove the crankshaft retaining bolt and washer and remove the crankshaft damper.

13. Remove the timing belt. If the belt is to be reused, mark the direction of rotation on the belt for proper installation.

14. Remove the crankshaft sprocket and belt guide.

15. Using Seal Remover T92C-6700-CH or equivalent, remove the front crankshaft seal from the oil pump body.

To install:

16. Lubricate the lip of the new front crankshaft seal with clean engine oil.

17. Install the new seal using Oil pump Seal Replacer T81P-6700-A or equivalent seal installation tool.

18. Install the timing belt guide and crankshaft sprocket.

19. Install the timing belt over the sprockets in a counterclockwise direction starting at the crankshaft. Keep the belt span from the crankshaft to the camshaft tight while installing over the remaining sprocket. Loosen the tensioner retaining bolt allowing the tensioner to snap against the belt.

20. Rotate the crankshaft 2 complete revolutions stopping at TDC. This will allow the tensioner spring to load the timing belt. Do not rotate the crankshaft with the spark plugs installed. After rotation, recheck camshaft and crankshaft alignment. This will verify that the belt has not skipped a tooth during rotation. Tighten the belt tensioner bolt to 17–22 ft. lbs. (23–30 Nm).

21. Place the crankshaft damper on the crankshaft. Install the retaining bolt and washer and tighten to 81–96 ft. lbs. (110–130 Nm).

22. Remove the flywheel holding tool and install the inspection shield.

23. Install the right splash shield.

24. Lower the vehicle.

CAUTION: With the timing belt removed and the No. 1 piston at TDC, DO NOT rotate the camshaft. If the camshaft must be rotated, align the crankshaft damper 90 degrees BTC.

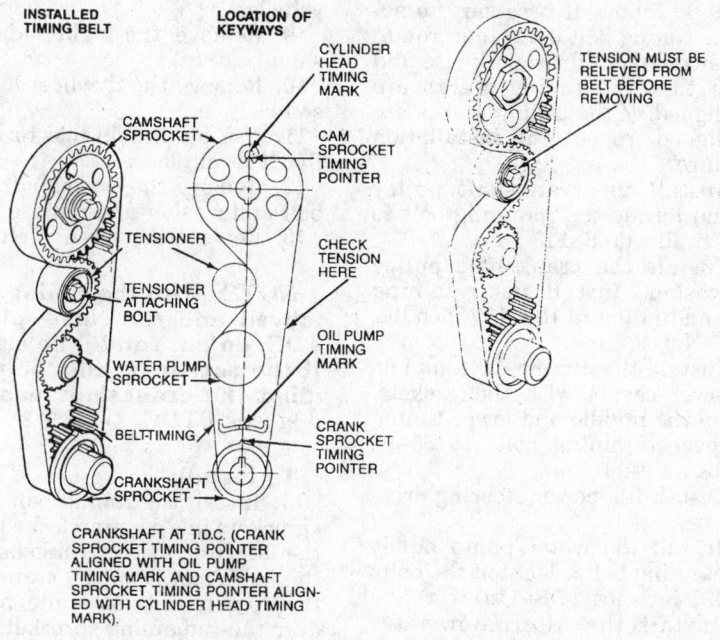

Timing marks and sprocket alignment — 1.9L engine

25. Install the front engine support insulator.
26. Install the spark plugs and attach the ignition wires.
27. Install the timing belt cover and 2 retaining nuts.
28. Install the accessory drive belt.
29. Connect the negative battery cable.
30. Run the engine and check for leaks.
31. Road test the vehicle and check for proper operation.

Timing Belt, Sprockets, Tensioner and Front Cover

ADJUSTMENT

1.8L Engine

1. Loosen the tensioner lockbolt.
2. With No. 1 piston at TDC on its compression stroke, turn the crankshaft 1⅚ turns clockwise to align the timing belt sprocket mark with the tension set mark, at approximately the 10 o'clock position.
3. Apply tension to the timing belt tensioner and tighten the tensioner lockbolt to 27–38 ft. lbs. (37–52 Nm).
4. Turn the crankshaft 2⅙ turns clockwise making sure the timing marks are still aligned.

5. Measure the timing belt deflection by applying 22 pounds (98 N) of pressure on the timing belt between the camshaft sprockets. The timing belt deflection should be 0.35–0.45 inch (9.0–11.5mm). If necessary, repeat the procedure until the timing belt deflection is correct.
6. Turn the crankshaft 2 turns clockwise and verify that the timing marks are aligned.

1.9L Engine

The timing belt adjustment is set by the timing belt tensioner at the time of timing belt installation and should not require adjustment during the life of the timing belt. If timing belt adjustment is required, loosen the timing belt tensioner bolt and allow the tensioner spring to adjust the belt tension. Tighten the tensioner bolt to 17–22 ft. lbs. (23–30 Nm).

REMOVAL AND INSTALLATION

1.8L Engine

1. Disconnect the negative battery cable.
2. Remove the timing belt upper cover and gasket.
3. Remove the accessory drive belts.

4. Remove the water pump pulley bolts and remove the pulley.
5. Raise and safely support the vehicle.
6. Remove the right front wheel and tire assembly.
7. Remove the right upper and lower splash shields.
8. Remove the timing belt middle and lower covers along with the gaskets.
9. Remove the crankshaft pulley hub bolt and hub.
10. Rotate the crankshaft and align the timing marks located on the camshaft sprockets and seal plate.
11. Check that the crankshaft sprocket and the oil pump are aligned.

NOTE: If the timing belt is to be reused, mark an arrow on the belt to indicate it's rotational direction for installation reference.

12. Loosen the timing belt tensioner bolt.
13. Turn the timing belt tensioner counterclockwise and hand-tighten the tensioner bolt to relieve the tension on the timing belt.
14. Remove the timing belt.
15. If the camshaft sprockets are to be removed, continue as follows:
   a. Disconnect and tag the ignition wires wires and vacuum lines blocking the removal of the cylinder head cover.
   b. Remove the cylinder head cover retaining bolts and remove the cover and gasket.
   c. While holding the camshaft with a wrench, remove the camshaft sprocket retaining bolt.
   d. Remove the camshaft sprocket.
   e. If removing both camshaft sprockets, tag the sprockets for identification at reassembly.
16. If removing the crankshaft sprocket, remove the crankshaft pulley bolt and hub, if not already done. Slide the crankshaft sprocket off the crankshaft.
17. Inspect the timing belt tensioner and spring and replaced, if necessary.
To install:
18. If the crankshaft sprocket was removed, install the crankshaft key with the tapered end facing the oil pump. Install the crankshaft sprocket onto the crankshaft while making sure to match the alignment grooves.
19. If removed, install the camshaft sprockets as follows
   a. Turn the camshaft until the dowel pins face straight up.

b. Install the camshaft sprocket with the **I** mark straight up for the intake camshaft or with the **E** mark straight up for the exhaust camshaft.

c. Align the camshaft sprockets with the timing marks on the seal plate.

d. While holding each camshaft with a wrench, install the camshaft sprocket retaining bolts. Tighten the bolts to 36–45 ft. lbs. (49–61 Nm).

e. Install a new cylinder head cover gasket onto the cylinder head.

f. Place the cylinder head cover into its mounting position and install the retaining bolts. Tighten the cylinder head cover bolts to 43–78 inch lbs. (4.9–8.8 Nm).

g. Install the ignition wires to the spark plugs and connect the vacuum hoses to the cylinder head cover.

20. Temporarily secure the timing belt tensioner in the far left position.

21. Verify that the timing marks on the crankshaft sprocket and the oil pump are aligned.

22. Verify that the timing marks on the camshaft sprockets and the seal plate are aligned.

23. Install the timing belt in a counterclockwise motion. Make sure there is no looseness on the idler side of the timing belt or between the camshaft sprockets.

NOTE: If using the old timing belt, make sure to install the belt as it was removed.

24. Loosen the timing belt tensioner bolt. Allow the tensioner spring to apply tension to the timing belt.

25. Rotate the crankshaft 1⁵⁄₆ turns clockwise and align the timing belt pulley mark with the tension set mark which is located at approximately the 10 o'clock position.

26. Turn the crankshaft 2 turns clockwise and align the crankshaft sprocket with the tension set mark on the oil pump.

27. Verify that all timing marks are aligned. If not, remove the timing belt and repeat the installation procedures.

28. Apply tension to the timing belt tensioner and tighten the tensioner lock bolt to 27–38 ft. lbs. (37–52 Nm).

29. Rotate the crankshaft 2¹⁄₆ (780 degrees) turns clockwise and verify that the camshaft and crankshaft timing marks are aligned.

30. Measure the timing belt deflection by applying 22 lbs. (98 N) of pressure on the timing belt between the camshaft sprockets. The timing belt deflection should be 0.35–0.45 inch (9–11.5mm). If necessary to adjust the timing belt deflection, rotate the crankshaft 2 turns clockwise and ensure that the timing marks are still aligned. If the timing marks are not aligned, repeat the installation procedure.

31. Install the crankshaft pulley hub and torque the retaining bolt to 80–87 ft. lbs. (108–118 Nm).

32. Install the crankshaft pulley and washer. Install the retaining bolts and tighten to 109–152 inch lbs. (12–17 Nm).

33. Install the timing belt middle and lower covers with the gaskets. Tighten the middle and lower timing belt cover retaining bolts to 65–95 inch lbs. (7.8–11 Nm,).

34. Install the power steering drive belt.

35. Install the water pump pulley and retaining bolts. Tighten the bolts to 69–95 inch lbs. (7.8–11.0 Nm).

36. Install the alternator/water pump drive belt.

37. Install the splash shields. Tighten the bolts to 69–95 inch lbs. (7.8–11.0 Nm).

38. Install the right wheel and tire assembly. Torque the lug nuts to 65–87 ft. lbs. (88–118 Nm).

39. Lower the vehicle.

40. Install the timing belt upper cover and gasket. Tighten the bolts to 69–95 inch lbs. (7.8–11.0 Nm).

41. Connect the negative battery cable.

42. Run the engine and check for leaks.

43. Road test the vehicle and check for proper engine operation.

1.9L Engine

1. Disconnect the negative battery cable.

2. Remove the accessory drive belt automatic tensioner and the accessory drive belt.

3. Remove the timing belt cover.

4. Align the timing mark on the camshaft sprocket with the timing mark on the cylinder head.

5. Confirm that the timing mark on the crankshaft sprocket is aligned with the timing mark on the oil pump housing.

6. Loosen the belt tensioner attaching bolt, pry the tensioner away from the timing belt and retighten the bolt.

7. Remove the spark plugs. Remove the right engine mount.

8. Raise and safely support the vehicle.

9. Remove the right side splash shield.

10. Remove the flywheel inspection shield.

11. Use a suitable tool to hold the flywheel in place.

12. Remove the crankshaft damper bolt and washer and remove the bolt.

13. Remove the timing belt.

NOTE: With the timing belt removed and the No. 1 piston at TDC, do not rotate the camshaft. If the camshaft must be rotated, align the crankshaft damper 90 degrees BTDC.

To install:

14. Install the timing belt over the sprockets in a counterclockwise direction starting at the crankshaft. Keep the belt span from the crankshaft to the camshaft tight while installing over the remaining sprocket.

15. Loosen the belt tensioner attaching bolt, allowing the tensioner to snap against the belt.

16. Rotate the crankshaft clockwise 2 complete revolutions, stopping at TDC. This will allow the tensioner spring to load the timing belt.

NOTE: Do not turn the engine counterclockwise to align the timing marks. Do not rotate the crankshaft with the spark plugs installed.

17. Recheck the camshaft and crankshaft timing marks for alignment, to make sure the timing belt has not skipped a tooth during rotation. Repeat the procedure if the timing marks are not aligned.

18. Tighten the tensioner attaching bolt to 17–22 ft. lbs. (23–30 Nm).

19. Install the crankshaft dampener and the bolt and washer. Tighten the bolt to 81–96 ft. lbs. (110–130 Nm).

20. Install the flywheel inspection shield.

21. Install the splash shield and lower the vehicle.

22. Install the right engine mount. Install the spark plugs.

23. Install the timing belt cover.

24. Install the accessory drive belt automatic tensioner and the accessory drive belt.

25. Connect the negative battery cable.

Camshaft

REMOVAL AND INSTALLATION

1.8L Engine

1. Disconnect the negative battery cable.
2. Remove the distributor assembly and ignition wires.
3. Remove 2 uppermost engine front cover bolts from the upper engine front cover.
4. Disconnect the crankcase breather hose from the cylinder head cover.
5. Remove the cylinder head cover retaining bolts and the cylinder head cover and gasket.
6. Remove 2 remaining bolts securing the upper engine front cover and remove the cover.
7. Remove the timing belt.
8. Remove both camshaft sprockets.
9. Remove the seal plate retaining bolts and remove the seal plate.
10. Loosen the camshaft cap retaining bolts in sequence.
11. Remove the camshaft caps and note their mounting locations for installation reference.

NOTE: The camshaft caps are numbered and have arrow marks for installation and direction reference.

12. Remove the camshafts and camshaft oil seals.
13. If the lash adjusters (valve tappets) are to be removed from the cylinder head, note their positions as they are removed for installation to their original locations.

NOTE: If new camshafts are installed, the lash adjusters should also be replaced.

14. Inspect the camshafts and lash adjusters for wear and/or damage and replace, as necessary.

To install:

15. Coat the lash adjusters with clean engine oil if removed, and install into the lifter bores in the cylinder head.

NOTE: If the original lash adjusters are being installed, make sure they are returned to the bores they were removed.

16. Verify that the lash adjusters move smoothly in their bores.
17. Apply clean engine oil to the camshaft journals and bearings.
18. Place the camshafts into their mounting positions.

NOTE: The exhaust camshaft has a groove which must be installed into the distributor drive gear.

19. Apply silicone sealer to the bottom of both front bearing caps (pulley end).
20. Install the camshaft caps according to the cap numbers and arrow marks.
21. Install the camshaft cap bolts and tighten them in sequence to 100–126 inch lbs. (11.3–14.2 Nm).
22. Apply a small amount of clean engine oil to the lips of 2 new camshaft oil seals. Using Cam Seal Replacer T90P-6256-BH and Seal Protector T90P-6256-AH or equivalents, install the new seals.
23. Place the seal plate to its mounting position and install 6 retaining bolts. Tighten the bolts to 69–95 inch lbs. (7.8–11.0 Nm).
24. Install the camshaft sprockets.
25. Install the timing belt.
26. Clean the gasket mating surfaces for the cylinder head cover and install the cylinder head cover. If needed, install a new gasket and tighten the retaining bolts to 43–78 inch lbs. (4.9–8.8 Nm).
27. Install the crankcase breather hose.

28. Install the upper engine front cover and tighten 4 retaining bolts to 69–95 inch lbs. (7.8–11 Nm).
29. Install the distributor and ignition wires.
30. Connect the negative battery cable.
31. Run the engine and check for leaks.
32. Road test the vehicle and check for proper engine operation.

1.9L Engine

1. Disconnect the negative battery cable.
2. Remove the engine air cleaner intake tube.
3. Remove the engine air cleaner hose from the cylinder head cover.
4. Move the ignition wires from the cylinder head cover.
5. Remove the cylinder head cover retaining bolts and remove the cover and gasket.

NOTE: Note the position of the rocker arm assemblies and lash adjusters (valve tappets) as they are removed, so they can be reinstalled in their original locations.

6. Remove the rocker arm retaining bolts, fulcrums and rocker arms.
7. Remove the lash adjuster guide plates and retainers. Remove the lash adjusters.
8. Remove the ignition coil pack.
9. Set the engine to TDC for No. 1 cylinder prior to removing the timing belt.
10. Remove the timing belt cover.

NOTE: Make sure the crankshaft is positioned at TDC No. 1 cylinder. Do not turn the crankshaft until the timing belt is reinstalled.

11. Remove the timing belt.
12. Remove the camshaft sprocket retaining bolt and the camshaft sprocket and key.
13. Remove the camshaft thrust plate.
14. Remove the cup plug from the back of the cylinder head.
15. Carefully remove the camshaft through the back of the cylinder head toward the transaxle.
16. Remove the camshaft front oil seal.

NOTE: If the camshaft is replaced, new lash adjusters should also be installed.

17. Inspect the camshaft for wear and/or damage and replace, as necessary.

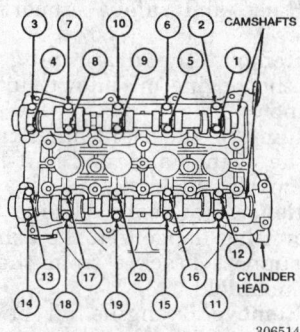

Camshaft cap bolt removal
sequence — 1.8L engine

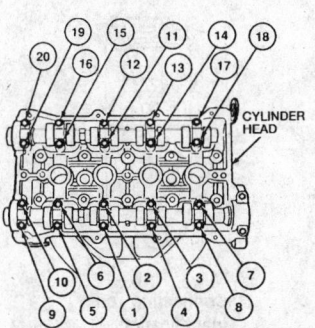

Camshaft cap bolt tightening
sequence — 1.8L engine

To install:

18. Apply clean engine oil to the lip of a new camshaft front oil seal. Install the camshaft front oil seal using Seal Replacer T81P-6292-A or equivalent. The seal depth should be 0.002–0.04 inch (0.05–1.0mm) below flush with the front face of the cylinder head.

19. Thoroughly coat the camshaft bearing journals, cam lobe surfaces and thrust plate groove with a suitable engine assembly lubricant.

NOTE: If the camshaft is being replaced, verify that the camshaft contains a plastic camshaft oil flow control rod and the oil gallery plug in the rear of the camshaft or low oil pressure will result.

20. Carefully install the camshaft through the rear of the cylinder head. Rotate the camshaft during installation.

21. Install the camshaft thrust plate. Tighten the retaining bolts to 6–9 ft. lbs. (8–13 Nm).

22. Position the camshaft key onto the front of the camshaft. Align and install the camshaft sprocket over the camshaft key.

23. Apply pipe sealant to the cleaned threads of the camshaft sprocket retaining bolt and install the bolt and washer. While holding the camshaft stationary, tighten the bolt to 71–84 ft. lbs. (95–115 Nm).

24. Install a new cup plug into the back of the cylinder head using a small amount of sealer.

25. Install the timing belt.

26. Install the timing belt cover.

27. Lubricate the lash adjusters with an appropriate engine assembly lubricant and install in their bores.

28. Install the lash adjuster guides and retainers.

29. Lubricate and install the rocker arms and fulcrums. Install the rocker arm retaining bolts. Tighten to 17–22 ft. lbs. (23–30 Nm).

30. Install the ignition coil pack.

31. Install the cylinder head cover using a new gasket if needed. Tighten the retaining bolts to 4–9 ft. lbs. (5–12 Nm).

32. Install the ignition wires to the spark plugs and fit the wires to the retainers on the valve cover.

33. Install the engine air cleaner hose to the valve cover.

34. Install the engine air cleaner intake tube.

35. Connect the negative battery cable.

36. Run the engine and check for leaks.

37. Road test the vehicle and check for proper engine operation.

Piston and Connecting Rod

POSITIONING

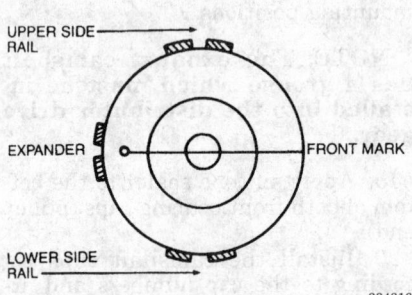

Oil ring end gap positions — 1.8L engine

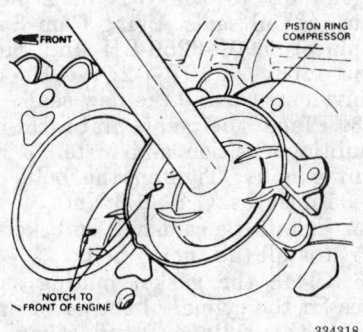

Piston and ring installation; notch on piston faces front of engine — 1.8L and 1.9L engine

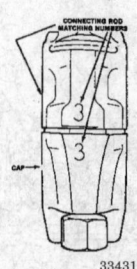

Connecting rod identification marks — 1.8L and 1.9L engines

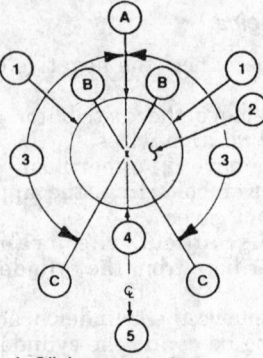

PISTON RING SPACING

1. Oil ring segment gap
2. Piston
3. 150 degrees (spacing)
4. Pin bore
5. Front of vehicle
A. Oil ring spacer gap
B. Compression ring gap
C. Compression ring(opposite placement)gap

Piston ring position — 1.9L engine

ENGINE LUBRICATION

Oil Pan

REMOVAL AND INSTALLATION

1.8L Engine

1. Disconnect the negative battery cable.

2. Remove the engine oil filler cap.

3. Raise and safely support the vehicle.

4. Remove the drain plug and drain the engine oil into a suitable container.

5. Remove the right-hand upper and the right-hand and left-hand lower splash shields.

6. Remove the exhaust pipe front its mounting flange and the exhaust pipe support bracket from the exhaust manifold.

7. Remove 6 engine oil pan-to-transaxle retaining bolts.

8. Support the engine oil pan with a suitable jack stand.

9. Remove 18 oil pan-to-cylinder block retaining bolts.

NOTE: Do not force a prying tool between the cylinder block and the oil pan contact surface when trying to remove the oil pan. This may damage the oil pan contact surface and cause oil leakage.

10. Only at the most rearward points of the oil pan, next to the transaxle, use a suitable tool to carefully pry the oil pan away from the cylinder block and remove the oil pan.

11. Use a suitable tool to pry the crankcase stiffeners away from the cylinder block and/or oil pan.

NOTE: When removing the crankcase stiffeners and sealant material from the oil pan and cylinder block, be careful not to damage the oil pan and cylinder block contact surfaces.

12. Remove the front and rear oil pan gaskets and end seals. Remove all sealant material from the cylinder block and oil pan.

To install:

13. Apply a bead of silicone sealant to the crankcase stiffeners along the inside of the bolt holes.

14. Install the crankcase stiffeners onto the engine oil pan.

15. Apply sealant to the proper areas of the end seals. Be sure to install the end seals with the projections in the notches.

16. Install the front and rear end seals to the oil pan.

17. Apply a continuous bead of silicone sealant to the oil pan along the inside of the bolt holes. Overlap the sealant ends.

18. Place the engine oil pan in position and install the oil pan-to-cylinder block retaining bolts. Tighten the bolts to 69–95 inch lbs. (7.8–11.0 Nm).

NOTE: If the oil pan retaining bolts are to be reused, the old sealant must be removed from the bolt threads. Tightening the old retaining bolts with old sealant still on them may cause cracking inside the bolt holes.

19. Install the engine oil pan-to-transaxle retaining bolts and torque to 27–38 ft. lbs. (37–52 Nm).

20. Install the engine oil drain plug and torque to 22–30 ft. lbs. (29–41 Nm).

21. Install the exhaust front mounting flange to the exhaust manifold using a new gasket. Torque the mounting flange-to-exhaust manifold

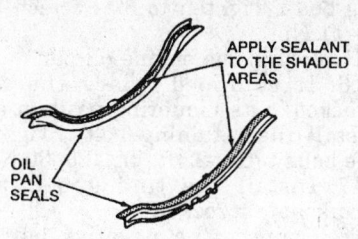

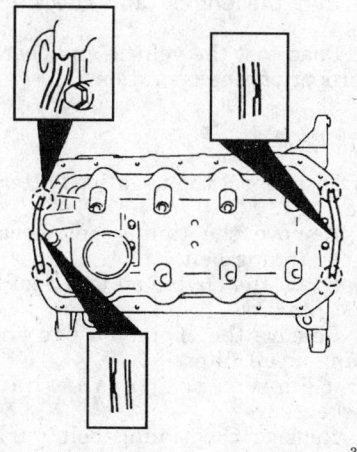

Applying sealant to the oil pan seals — 1.8L engine

retaining nuts to 23–34 ft. lbs. (31–46 Nm).

22. Install the exhaust pipe support bracket and torque the bolts to 27–38 ft. lbs. (37–52 Nm).

23. Install the splash shields. Tighten the bolts to 69–95 inch lbs. (7.8–11.0 Nm).

24. Lower the vehicle.

25. Fill the crankcase with the proper type and quantity of engine oil.

26. Install the engine oil filler cap.

27. Connect the negative battery cable.

28. Run the engine and check for oil leaks.

29. Road test the vehicle and check for proper engine operation.

1.9L Engine

1. Disconnect the negative battery cable.

2. Raise and safely support the vehicle.

3. Drain the engine oil into a suitable container.

4. Remove the catalytic converter.

5. Remove 2 engine oil pan to transaxle retaining bolts.

6. Remove 10 oil pan to cylinder block retaining bolts.

7. Gently pry the oil pan away from the cylinder block and remove.

8. Remove the oil pan gasket material and discard.

To install:

9. Clean the oil pan gasket surface and the mating surface on the cylinder block. Wipe the oil pan rail with a solvent-soaked cloth to remove all traces of engine oil.

10. Remove and clean the oil pump pick up tube and screen assembly. Install the tube and screen assembly using a new gasket. Tighten 2 retaining bolts to 7–9 ft. lbs. (10–13 Nm).

11. Apply a bead of suitable silicone rubber sealer at the corner of the oil pan front and rear seals and at the seating point of the oil pump to the cylinder block retainer joint.

12. Install the gasket in the oil pan ensuring the press fit tabs are fully engaged in the oil pan gasket channel.

13. Install the oil pan and 10 retaining bolts. Tighten the bolts lightly until 2 oil pan to transaxle bolts can be installed.

NOTE: If the oil pan is installed on the engine outside of the vehicle, a transaxle case or equivalent assembly fixture must be bolted to the cylinder block to align the oil pan flush with the rear face of the block.

14. Tighten 2 oil pan to transaxle retaining bolts to 30–40 ft. lbs. (40–54 Nm), then loosen ½ turn.

15. Tighten the oil pan to cylinder block retaining bolts in sequence to 15–22 ft. lbs. (20–30 Nm).

16. Tighten 2 oil pan to transaxle bolts to 30–40 ft. lbs. (40–55 Nm).

17. Install the transaxle inspection plate.

18. Install the catalytic converter.

19. Lower the vehicle.

20. Fill the crankcase with the proper grade and amount of engine oil.

21. Connect the negative battery cable.

22. Run the engine and check for oil leaks.

23. Road test the vehicle and check for proper engine operation.

Oil Pump

REMOVAL AND INSTALLATION

1.8L Engine

1. Disconnect the negative battery cable.

2. Remove the timing belt and crankshaft sprocket.

3. Remove the engine oil pan.

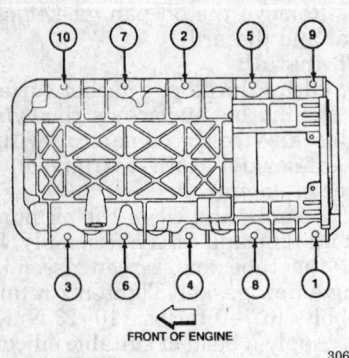

Oil pan tightening sequence — 1.9L engine

4. Remove 2 oil pump pickup tube retaining bolts and remove the oil pump pickup tube and gasket.

5. If equipped, remove the A/C compressor retaining bolts and position the compressor out of the way. Do not disconnect any A/C hoses.

6. Remove the A/C compressor mounting bracket.

7. Remove the retaining bolt from the engine oil dipstick tube bracket and remove the alternator lower retaining bolt.

8. Remove 6 oil pump retaining bolts and remove the oil pump.

9. Remove all gasket material from the oil pump and cylinder block mating surfaces.

To install:

10. Install a new gasket to the oil pump.

11. Place the oil pump in position and install 6 oil pump retaining bolts. Torque the bolts to 14–19 ft. lbs. (19–25 Nm).

12. Place the dipstick tube bracket bolt into its mounting position and install the retaining bolt. Tighten the bolt to 68–95 inch lbs. (8–11 Nm).

13. Install the alternator lower retaining bolt and torque to 27–38 ft. lbs. (37–52 Nm).

14. Install a new gasket to the oil pump pickup tube. Place the pickup tube in position and install 2 retaining bolts. Tighten to 69–95 inch lbs. (8–11 Nm).

15. Install the engine oil pan.

16. If equipped, place the A/C bracket to its mounting position and install the retaining bolts. Tighten the bolts to 15–22 ft. lbs. (20–30 Nm).

17. Install the timing belt and crankshaft sprocket.

18. Connect the negative battery cable.

19. Run the engine and check for leaks.

20. Road test the vehicle and check for proper engine operation.

1.9L Engine

1. Disconnect the negative battery cable.

2. Remove the timing belt cover and the timing belt.

3. Drain the engine oil into a suitable container.

4. Remove the engine oil pan and the engine oil filter.

5. Remove the crankshaft sprocket.

6. Remove the timing belt guide from the crankshaft.

7. Disconnect the Crankshaft Position (CKP) sensor.

8. Remove 6 oil pump-to-cylinder block retaining bolts and remove the oil pump assembly from the engine. Remove and discard the gasket.

9. Remove 2 oil pump pickup tube retaining bolts and remove the pickup tube and gasket from the oil pump assembly.

10. Remove the crankshaft seal from the oil pump and discard if the oil pump is to be reinstalled.

To install:

11. Make sure the oil pump mating surfaces on the cylinder block and oil pump are clean and free of gasket material.

12. Lubricate the outside diameter of a new crankshaft seal with clean engine oil and install the seal using Seal Replacer T81P-6700-A, or equivalent. Lubricate the seal lip with clean engine oil.

13. Position the oil pump gasket to the cylinder block.

14. Prime the oil pump with engine oil and place the pump in position. Using a small prybar or suitable tool, index the oil pump drive gear to pilot over the crankshaft and seat firmly on the cylinder block.

NOTE: The pump drive gear can be accessed through the oil pickup hole in the body of the pump. Do not install the oil pump pickup tube and screen until the pump has been correctly installed on the cylinder block.

15. Install 6 oil pump retaining bolts. Check the oil pump-to-cylinder block positioning and tighten the retaining bolts to 8–12 ft. lbs. (11–16 Nm).

NOTE: When the oil pump bolts are tightened, the gasket must not be below the cylinder block sealing surface.

16. Install the pickup tube on the oil pump using a new gasket. Tighten 2 retaining bolts to 7–9 ft. lbs. (10–13 Nm).

17. Install the timing belt guide over the end of the crankshaft.

18. Install the crankshaft sprocket.

19. Connect the CKP sensor.

20. Install the engine oil pan.

21. Install a new engine oil filter.

22. Install the timing belt and the timing belt cover.

23. Fill the crankcase with the proper grade and amount of engine oil.

24. Connect the negative battery cable.

25. Run the engine and check for leaks.

26. Road test the vehicle and check for proper engine operation.

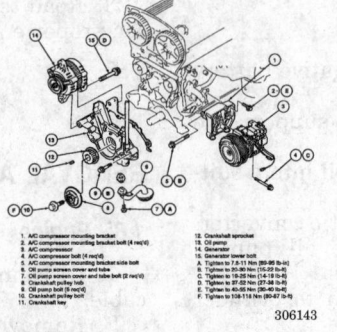

Oil pump and related components — 1.8L engine

TRANSAXLE

Manual Transaxle Assembly

REMOVAL AND INSTALLATION

MX5 Transaxle

1. Disconnect both battery cables, negative cable first.

2. Remove the battery and battery tray.

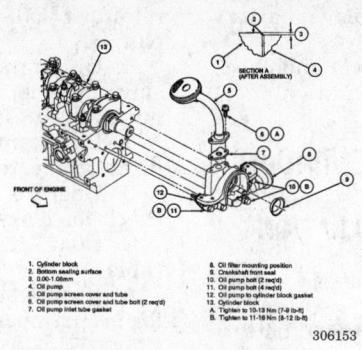

1. Cylinder block
2. Bottom sealing surface
3. 0.00-1.08mm
4. Oil pump
5. Oil pump screen cover and tube
6. Oil pump screen cover and tube bolt (2 req'd)
7. Oil pump inlet tube gasket
8. Oil filter mounting position
9. Crankshaft front seal
10. Oil pump bolt (2 req'd)
11. Oil pump bolt (4 req'd)
12. Oil pump-to-cylinder block gasket
13. Cylinder block
A. Tighten to 10-13 Nm (7-9 lb-ft)
B. Tighten to 11-16 Nm (8-12 lb-ft)

306153

**Oil pump and related components —
1.9L engine**

3. Remove the engine air cleaner tube and the air intake resonator.

4. Disconnect the speedometer cable at the transaxle.

5. Remove the retaining clip, then disconnect the slave cylinder line from the slave cylinder hose and plug the hose.

6. Disconnect the ground strap from the transaxle.

7. Remove the tie wrap and disconnect 3 electrical connectors located above the transaxle. Remove the electrical connector support bracket.

8. Install Engine Support Bar D88L-6000-A or equivalent, and attach it to the engine lifting eyes with suitable chains or cables.

9. Remove 3 nuts from the left-hand engine support bracket.

10. Loosen the mount pivot nut and rotate the mount out of the way.

11. Remove 3 bolts and the left-hand engine support mount.

12. Remove 2 upper transaxle-to-engine bolts.

13. Raise and safely support the vehicle.

14. Remove the front wheel and tire assemblies.

15. Remove the inner fender splash shields.

16. Drain the transaxle fluid and install the drain plug.

17. Remove the halfshafts.

18. Install Transaxle Plugs T88C-7025-AH or equivalent, between the differential side gears.

**———— WARNING ————
Failure to install the transaxle plugs may cause the differential side gears to become improperly positioned. If the gears become misaligned, the differential will have to be removed from the transaxle to align them.**

19. If equipped with the 1.8L engine, remove 4 intake manifold support bolts and the support.

20. Remove the starter motor.

21. Remove the gearshift stabilizer bar nut and washers. Remove the stabilizer bar and bracket from the transaxle.

22. Remove the bolt and nut and remove the shift control rod from the transaxle.

23. Remove both lower splash shields.

24. Remove the transaxle mount-to-crossmember bolts and nuts and remove the lower crossmember (rear engine support).

25. Position and secure a suitable transaxle jack under the transaxle.

26. Remove the front transaxle mount and bracket.

27. Remove 5 lower engine-to-transaxle bolts and slowly lower the transaxle out of the vehicle.

To install:

28. Apply a thin coating of suitable grease to the spline of the input shaft.

29. Place the transaxle onto a suitable transaxle jack. Make sure the transaxle is secure.

30. Raise the transaxle into position.

31. Install 5 lower engine-to-transaxle bolts and tighten to 27–38 ft. lbs. (37–52 Nm).

32. Install the front transaxle mount and bracket. Tighten the bolts to 12–17 ft. lbs. (16–23 Nm).

33. Remove the transaxle jack.

34. Install the lower crossmember. Tighten the support bolts to 47–66 ft. lbs. (64–89 Nm) and 4 transaxle support insulator-to-rear engine support nuts to 27–38 ft. lbs. (37–52 Nm).

35. Install both lower splash shields.

36. Install the shift control rod bolt and nut and tighten to 12–17 ft. lbs. (16–23 Nm).

37. Install the stabilizer bracket and stabilizer bar with the nut and washers and tighten to 23–34 ft. lbs. (31–46 Nm).

38. Install the starter motor.

39. If equipped with the 1.8L engine, install the intake manifold support and torque the retaining bolts to 27–38 ft. lbs. (37–52 Nm).

40. Remove the transaxle plugs and install the halfshafts.

41. Install the inner fender splash shields.

42. Install the wheel and tire assemblies. Torque the lug nuts to 65–87 ft. lbs. (88–118 Nm).

43. Lower the vehicle.

44. Install 2 upper engine-to-transaxle bolts and tighten to 47–66 ft. lbs. (64–89 Nm).

45. Install the left-hand engine support mount and tighten 3 bolts to 32–45 ft. lbs. (43–61 Nm) for 1.8L engine or 41–59 ft. lbs. (55–80 Nm) for 1.9L engine.

46. Rotate the left-hand engine support bracket into position and tighten the pivot nut.

47. Install and tighten 3 left-hand engine support bracket nuts and tighten to 47–66 ft. lbs. (64–89 Nm).

48. Remove the engine support bar.

49. Install the electrical connector support bracket. Connect 3 electrical connectors and secure with a new tie wrap.

50. Connect the ground strap to the transaxle.

51. Connect the slave cylinder line to the slave cylinder hose and install the retaining clip.

52. Add the proper type and amount of fluid to the transaxle.

53. Connect the speedometer cable.

54. Install the engine air cleaner tube and the air intake resonator.

55. Install the battery tray and the battery.

56. Connect the battery cables, negative cable last.

57. Check for fluid leaks and proper clutch operation.

58. Road test the vehicle and check for proper transaxle operation.

Clutch Assembly

REMOVAL AND INSTALLATION

1. Disconnect the negative battery cable.

2. Raise and safely support the vehicle.

3. Remove the transaxle assembly.

4. If the clutch assembly is to be reused, matchmark the pressure plate and the flywheel so they can be assembled in the same position.

5. Loosen the pressure plate-to-flywheel retaining bolts one turn at a time, in a criss-cross pattern, until

the spring tension is relieved, to prevent pressure plate cover distortion.

6. Support the pressure plate and remove the retaining bolts. Remove the pressure plate and clutch disc from the flywheel.

NOTE: If the flywheel shows any signs of overheating (blue discoloration) or if it is badly grooved or scored, it should be refaced or replaced.

7. Inspect the flywheel, clutch disc, pressure plate, release bearing, pilot bearing and the clutch fork for wear. Replace parts as needed.

To install:

8. If removed, install a new pilot bearing using a suitable installation tool.

9. If removed, install the flywheel. Make sure the flywheel and crankshaft flange mating surfaces are clean. Tighten the flywheel retaining bolts to 71–76 ft. lbs. (96–103 Nm) on the 1.8L engine or 54–67 ft. lbs. (73–91 Nm) on the 1.9L engine.

10. Clean the pressure plate and flywheel surfaces thoroughly. Position the clutch disc and pressure plate into the installed position and support them with a dummy shaft or clutch aligning tool. If the clutch assembly is being reused, align the matchmarks that were made during the removal procedure.

11. Install the pressure plate-to-flywheel retaining bolts. Tighten the bolts in the correct sequence to 13–20 ft. lbs. (18–26 Nm). Remove the alignment tool.

12. If the release bearing was removed, lubricate the release fork where it contacts the bearing and install the bearing in the fork.

13. Install the transaxle assembly.

14. Lower the vehicle.

15. Bleed the hydraulic clutch system, if needed.

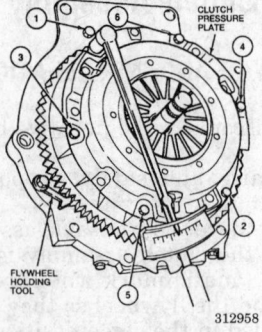

Pressure plate retaining bolt
torque sequence — MX5
transaxle

16. Connect the negative battery cable.

17. Road test the vehicle and check the clutch for proper operation.

Clutch Master Cylinder

REMOVAL AND INSTALLATION

1. Disconnect both battery cables, negative cable first.

2. Remove the battery and battery tray.

3. Disconnect the clutch tube from the clutch master cylinder using a suitable line wrench.

4. Disengage the clamp and remove the master cylinder reservoir hose from the clutch master cylinder. Prevent excess fluid loss by plugging the hose.

5. Remove the external retaining nut.

6. Remove the internal retaining nut and remove the clutch master cylinder.

To install:

7. Align the clutch pedal push rod and install the clutch master cylinder.

8. Install the external and internal retaining nuts and tighten to 14–19 ft. lbs. (19–25 Nm).

9. Connect the clutch tube and tighten the tube nut to 10–16 ft. lbs. (13–22 Nm).

10. Install the master cylinder reservoir hose and the clamp to the clutch master cylinder.

11. Install the battery and battery tray.

12. Bleed the air from the hydraulic clutch system.

13. Test the clutch system and make sure there is no leakage.

14. Connect both battery cables, negative cable last.

15. Road test the vehicle and check for proper operation.

Clutch Slave Cylinder

REMOVAL AND INSTALLATION

1. Disconnect the negative battery cable.

2. Disconnect the clutch slave cylinder tube from the clutch slave cylinder. Plug the line to prevent leakage.

3. Remove 2 clutch slave cylinder retaining bolts.

4. Remove the slave cylinder.

To install:

5. Position the slave cylinder and install 2 retaining bolts. Torque the

retaining bolts to 12–17 ft. lbs. (16–23 Nm).

6. Unplug and connect the clutch slave cylinder tube. Torque the tube nut to 10–16 ft. lbs. (13–22 Nm).

7. Bleed any air from the hydraulic clutch system.

8. Operate the clutch pedal and check for leaks.

9. Connect the negative battery cable.

10. Road test the vehicle and check for proper operation.

Hydraulic Clutch System

BLEEDING

1. Check that the brake master cylinder is at least ¾ full during the entire bleeding process.

2. Remove the bleeder screw cap from the clutch slave cylinder and attach a hose to the bleeder screw. Place the other end of the hose into a container to catch the fluid.

3. Have an assistant slowly pump the clutch pedal several times and then hold the clutch pedal down.

4. Loosen the bleeder screw to release the fluid and air. Tighten the bleeder screw.

5. Repeat the bleeding procedure until no more air bubbles are seen in the fluid.

6. Tighten the bleeder screw to 52–78 inch lbs. (6–9 Nm).

7. Top off the brake master cylinder to the full line.

8. Check for proper clutch system operation.

ADJUSTMENT

To determine if the clutch pedal free-play needs adjusting, depress the clutch pedal by hand until resistance is felt. Using a ruler, measure the distance between the upper clutch pedal height and where the resistance is felt.

Free-play should be 0.20–0.51 inch (5–13mm). If not, proceed as follows:

1. Loosen the clutch pedal to clutch master cylinder rod locknut.

2. Turn the clutch pedal to clutch master cylinder rod until the free-play is within specification.

3. Check that the disengagement height is within specification. Minimum disengagement height is 1.6 inches (41mm).

4. Tighten the clutch pedal to clutch master cylinder rod locknut to 9–12 ft. lbs. (12–17 Nm).

5. Check clutch pedal free-play to verify proper adjustment.

Automatic Transaxle Assembly

REMOVAL AND INSTALLATION

4EAT Transaxle

1. Disconnect both battery cables, negative cable first.

2. Remove the battery and battery tray.

3. Disconnect the wiring harness retaining clip from the battery tray.

4. Remove the engine air cleaner assembly.

5. Disconnect the shift control cable from the manual shift lever on the transaxle.

6. Disconnect the speedometer cable from the transaxle by unsnapping the cable at the Vehicle Speed Sensor (VSS).

7. Disconnect the transaxle electronic control electrical connectors and separate the harness from the transaxle clips.

8. Remove the Manual Lever Position (MLP) sensor wiring brackets and disconnect the ground cables from the top of the transaxle.

9. Remove the starter motor.

10. Disconnect the MLP sensor wiring connectors.

11. Install Engine Support D88L-6000-A or equivalent, to the engine. The engine must be properly supported for transaxle removal.

12. Disconnect the kickdown cable at the throttle cam on the throttle body.

13. Place a suitable drain pan under the transaxle and disconnect the transaxle cooler lines at the transaxle.

14. Remove the left–hand engine mount bolts and the mount.

15. Remove the upper transaxle housing bolts.

16. Disconnect the Heated Oxygen Sensor (HO₂S) electrical connector,

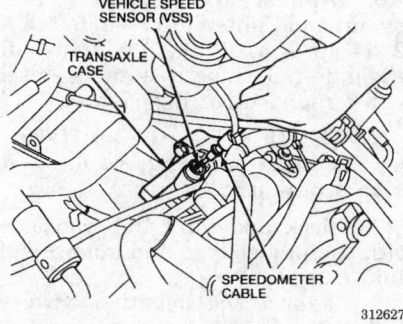

VEHICLE SPEED SENSOR (VSS)
TRANSAXLE CASE
SPEEDOMETER CABLE
312627

Speedometer cable and Vehicle Speed Sensor (VSS) location — 4EAT transaxle

the transaxle vent hose and the electrical connector at the VSS.

17. Raise and safely support the vehicle.

18. Remove the front wheel and tire assemblies.

19. Raise the staked portion of the wheel hub retainer nuts securing the halfshafts to the steering knuckles. Remove the wheel hub retainer nuts and discard.

20. Remove the lower ball joint-to-steering knuckle pinch bolt and nuts. Separate the lower ball joints from the steering knuckles using a prybar or similar tool.

21. Remove the cotter pins and castellated nuts securing the tie rod ends to the steering knuckles. Separate the tie rod ends from the steering knuckles using a suitable removal tool. Discard the cotter pins.

22. Remove 3 retaining bolts connecting the axle bearing bracket to the back of the engine.

23. Separate the halfshafts from both steering knuckles.

24. Remove 3 engine/transaxle lower splash shield bolts and the torque converter inspection plate.

25. Remove 4 nuts securing the torque converter to the flexplate.

26. Remove the bolts securing the lower transaxle to the engine oil pan.

27. Remove the rear engine support-to-vehicle chassis bolt and remove the transaxle support crossmember from the vehicle chassis.

28. Remove the rear engine support-to-transaxle mount nuts and remove the transaxle support crossmember from the transaxle mounts.

29. Remove both halfshafts from the transaxle. Install 2 Transaxle Plugs T88C-7025-AH or equivalent, into the differential side gears.

--- **WARNING** ---

Failure to install the transaxle plugs may cause the differential side gears to become improperly positioned. If the gears become misaligned, the differential will have to be removed from the transaxle to align them.

30. Position a drain pan and remove the drain plug from the transaxle. Drain the fluid from the differential cavity. Remove the transaxle pan and drain the transaxle fluid, then install the pan and drain plug.

31. Position a suitable transaxle jack under the transaxle. Secure the transaxle to the jack.

32. Remove the lower transaxle-to-engine bolts.

33. Slowly lower the transaxle out of the vehicle.

34. Inspect all components including mounts and brackets.

To install:

NOTE: A pin is used for securing the throttle control lever in a fixed position on new and rebuilt transaxles. This pin must be removed to allow proper transaxle operation. If the pin is not removed, the throttle lever will remain in a fixed position. After removing the pin, apply sealant to the bolt from the previous transaxle. Install the bolt and tighten to 69–95 inch lbs. (8–11 Nm).

35. If rebuilding or replacing the transaxle, make sure to completely flush the transaxle cooler and cooler lines before installing the transaxle.

36. Remove the pin securing the throttle control lever, if installed.

37. Secure the transaxle on the transaxle jack, if removed.

38. Carefully raise the transaxle into position and install the lower transaxle-to-engine bolts. Tighten the bolts to 41–59 ft. lbs. (55–80 Nm).

39. Position the torque converter to the flexplate and install 4 retaining nuts. Tighten the nuts to 25–36 ft. lbs. (34–49 Nm). Install the torque converter inspection plate.

40. Remove 2 transaxle plugs and install the halfshafts.

41. Connect the crossmember to the transaxle mounts and the chassis. Tighten the crossmember-to-transaxle mount nuts to 27–38 ft. lbs. (37–52 Nm). Tighten the crossmember-to-chassis nuts and bolts to 47–66 ft. lbs. (64–89 Nm).

42. Install the lower transaxle-to-engine oil pan bolts and tighten to 27–38 ft. lbs. (37–52 Nm). Install the engine/transaxle splash shields.

43. Install the starter motor.

44. Place the lower ball joints to the steering knuckles and secure with the pinch bolts and nuts. Tighten the nuts to 32–43 ft. lbs. (43–59 Nm).

45. Position the tie rod ends to the steering knuckles and install the castellated nuts. Tighten to 31–42 ft. lbs. (42–57 Nm). Install new cotter pins.

46. Install the wheel and tire assemblies and tighten the lug nuts to 65–88 ft. lbs. (88–118 Nm).

47. Lower the vehicle.

48. Install the upper transaxle-to-engine bolts and tighten to 41–59 ft. lbs. (55–80 Nm).

49. Install the left–hand engine mount and tighten the nuts to 49–69 ft. lbs. (67–93 Nm).

50. Connect the transaxle vent hose, the electrical connector at the VSS, the speedometer cable and the HO₂S connector.

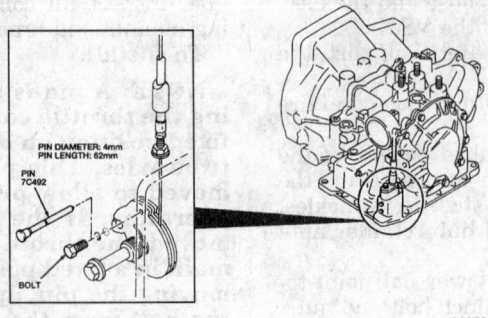

PIN DIAMETER: 4mm
PIN LENGTH: 62mm

PIN
7C492

BOLT

312630

Pin used for securing the throttle control lever — 4EAT transaxle

51. Connect the transaxle cooler lines and connect the kickdown cable at the throttle body.

52. Remove the engine support.

53. Connect the ground wires to the transaxle and connect the MLP sensor bracket and wiring connectors.

54. Connect the shift control cable to the manual shift lever on the transaxle. Tighten the selector lever attaching locknut to 33–47 ft. lbs. (44–64 Nm).

NOTE: Do not use any type of power wrench to tighten the locknut. Damage to the transaxle may result.

55. Install the battery tray and battery. Connect the wiring harness retaining clip to the battery tray.

56. Install the engine air cleaner assembly.

57. Connect both battery cables, negative cable last.

58. Add the proper type and quantity of transaxle fluid.

59. Check the transaxle for leaks and for proper operation.

60. Check the manual lever position sensor for proper adjustment.

61. Road test the vehicle and check for proper operation.

DRIVE AXLE

Halfshaft

REMOVAL AND INSTALLATION

1.9L Engine and
Left Side — 1.8L Engine

NOTE: Before continuing with any halfshaft procedure, make sure to have available, new halfshaft retaining nuts and circlips. Once removed, these parts loose

their torque holding ability or retention capability and must not be reused.

1. Disconnect the negative battery cable.

2. With the vehicle sitting on the ground, carefully raise the staked portion of the halfshaft retaining nut using a suitable small chisel. Loosen the nut.

3. Raise and safely support the vehicle.

4. Remove the wheel and tire assembly.

5. Remove the splash shield.

6. Remove and discard the halfshaft retaining nut.

7. Remove the cotter pin and castellated nut from the tie rod end and separate the tie rod end from the steering knuckle using a suitable removal tool. Discard the cotter pin.

8. Remove the lower ball joint pinch bolt. Carefully pry down on the lower control arm to separate the ball joint stud from the steering knuckle.

9. Pull outward on the steering knuckle/brake assembly. Carefully pull the halfshaft from the hub and position it aside.

10. Removal of the left side halfshaft requires removal of the crossmember to allow access with a prybar:

 a. Support the transaxle with a suitable transaxle jack.

 b. Remove 4 transaxle mount-to-crossmember retaining nuts.

 c. Remove 2 crossmember retaining nuts at the rear of the crossmember.

 d. While supporting the rear of the crossmember, remove 2 front mounting bolts. Remove the crossmember.

11. Position a drain pan under the transaxle.

12. Insert a prybar between the halfshaft and the transaxle case. Gently pry outward to release the halfshaft from the differential side

gear. Be careful not to damage the transaxle case, oil seal, CV-joint or CV-joint boot.

13. Remove the halfshaft.

NOTE: Install suitable plugs after removing the halfshafts to prevent the differential side gears from moving out of place. Should the gears become misaligned, the differential will have to be removed from the transaxle to align the gears.

To install:

14. Position a new circlip on the inner CV-joint spline so the circlip gap is at the top. Lubricate the splines lightly with a suitable grease.

15. Remove the plugs that were installed in the differential side gears.

16. Position the halfshaft so the CV-joint splines are aligned with the differential side gear splines. Push the halfshaft into the differential. When seated properly, the circlip can be felt snapping into the differential side gear groove.

17. Pull outward on the steering knuckle/brake assembly and insert the halfshaft into the hub.

18. Pry downward on the lower control arm and position the lower ball joint stud in the steering knuckle.

19. Install the crossmember and tighten the crossmember-to-frame nuts and bolts. Torque the nuts to 27–38 ft. lbs. (37–52 Nm) and the bolts to 47–66 ft. lbs. (64–89 Nm).

20. Install 4 transaxle mount-to-crossmember nuts and tighten to 27–38 ft. lbs. (37–52 Nm).

21. Remove the transaxle jack.

22. Install the lower ball joint pinch bolt and tighten to 32–43 ft. lbs. (43–59 Nm).

23. Install the tie rod end to the steering knuckle. Install the castellated nut and tighten to 31–42 ft. lbs. (42–57 Nm). Install a new cotter pin.

24. Install the splash shield.

25. Install the wheel and tire assembly. Torque the lug nuts to 65–87 ft. lbs. (88–118 Nm).

26. Install a new halfshaft retaining nut and tighten to 174–235 ft. lbs. (235–319 Nm). Stake the halfshaft retaining nut using a suitable chisel with a rounded cutting edge.

NOTE: If the nut splits or cracks after staking, replace it with a new nut.

27. Check and refill the transaxle with the proper type and quantity of fluid.

28. Connect the negative battery cable.

29. Road test the vehicle and check for proper operation.

Right Side — 1.8L Engine

NOTE: The right side halfshaft assembly is a 2 piece shaft with a bearing support bracket positioned between the 2 halves. The bearing support bracket is mounted on the cylinder block and must be unbolted if the entire halfshaft assembly is to be removed. If only the CV-joints/boots are to be serviced, the outboard shaft assembly may be removed, leaving the bearing support bracket mounted on the engine cylinder block.

1. Disconnect the negative battery cable.
2. With the vehicle sitting on the ground, carefully raise the staked portion of the halfshaft retaining nut using a suitable small chisel. Loosen the nut.
3. Raise and safely support the vehicle.
4. Remove the right front wheel and tire assembly.
5. Remove the splash shield.
6. Remove and discard the halfshaft retaining nut.
7. Remove the cotter pin and castellated nut from the tie rod end and separate the tie rod end from the steering knuckle using a suitable removal tool. Discard the cotter pin.
8. Remove the lower ball joint pinch bolt. Carefully pry down on the lower control arm to separate the ball joint stud from the steering knuckle.
9. Pull outward on the steering knuckle/brake assembly. Carefully pull the halfshaft from the hub and position it aside.
10. Position a drain pan under the transaxle.
11. Remove 3 bearing support bracket mounting bolts.
12. Insert a prybar between the bearing support bracket and the starter bracket. Gently pry outward on the damper until the halfshaft disengages from the differential side gear.
13. Remove the halfshaft assembly. Install an appropriate differential plug in the differential side gear.
To install:
14. Position a new circlip on the inner CV-joint spline so the circlip gap is at the top. Lubricate the splines lightly with a suitable grease.
15. Remove the differential plug from the side gear. Position the halfshaft assembly so the shaft splines are aligned with the differential side gear splines. Push the halfshaft into the differential. When seated prop-

erly, the circlip can be felt snapping into the differential side gear groove.
16. Pull outward on the steering knuckle/brake assembly and insert the halfshaft into the hub.
17. Pry downward on the lower control arm and position the lower ball joint stud into the steering knuckle. Install the lower ball joint pinch bolt and tighten to 32–43 ft. lbs. (43–59 Nm).
18. Install the tie rod end to the steering knuckle. Install the castellated nut and tighten to 31–42 ft. lbs. (42–57 Nm). Install a new cotter pin.
19. Position the bearing support bracket and install 3 retaining bolts. Tighten the outer bolt first, then the top inner, then the bottom inner. Tighten the bolts to 31–46 ft. lbs. (42–62 Nm).
20. Install the splash shield.
21. Install the wheel and tire assembly. Torque the lug nuts to 65–87 ft. lbs. (88–118 Nm).
22. Lower the vehicle.
23. Install a new halfshaft retaining nut and tighten to 174–235 ft. lbs. (235–319 Nm). Stake the retaining nut using a suitable chisel with the cutting edge rounded off.

NOTE: If the nut splits or cracks after staking, it must be replaced with a new nut.

24. Check and refill the transaxle with the proper type and quantity of fluid.
25. Connect the negative battery cable.
26. Road test the vehicle and check for proper operation.

CV-Joint Boot

REPLACEMENT

NOTE: The outboard CV-joint is permanently fitted onto the halfshaft and cannot be removed. To replace the outboard CV-joint boot, the inner CV-joint must be removed. If a boot has failed due to age or wear, all boots should be replaced at the same time.

1. Raise and safely support the vehicle.
2. Remove the halfshaft from the vehicle. Support the assembly in a vise with soft jaws.
3. Remove the large boot clamp from the inboard CV-joint and roll the boot back over the shaft.
4. Matchmark the outer race, axle shaft and tripot bearing for reassembly. Remove the wire ring bearing retainer from inside the outer

race/housing and remove the outer race.
5. Matchmark the tripot bearing and the shaft. Remove the tripot bearing snapring and remove the tripot bearing from the shaft. It may be necessary to drive the tripot off the shaft with a brass drift.
6. Remove the small clamp and the CV-joint boot from the halfshaft.
7. To remove the outboard CV-joint boot, remove the clamps and slide the boot off the shaft from the inboard side. If equipped with the 1.9L engine, remove the dynamic damper and mark its location on the halfshaft for installation.
To install:
8. The inner and outer CV-joint boots are different. To make sure the boot is being installed in the correct location, measure the outer diameter of the large end of the boot:
1.9L engine
    Oouter boot — 3.50 inch (89.0mm)
    Right side inner boot — 3.31 inch (84.0mm)
    Left side inner boot — 3.54 inch (90.0mm)
1.8L engine
    Outer boot — 3.35 inch (85.2mm)
    Inner boot — 3.54 inch (89.9mm)
9. Wrap smooth electrical tape around the halfshaft spline to ease installation of the boot. Slide the clamps and the outboard boot onto the shaft.
10. Before positioning the boot over the CV-joint, pack the CV-joint and the boot with grease. Be sure to use all of the grease in the pouch supplied with the boot kit.
11. Fit the boot into place on the CV-joint, making sure it is fully seated in the grooves in the shaft and outer race. Insert a suitable tool between the boot and the outer bearing race to allow trapped air to escape from the boot.
12. Install the boot clamps, wrapping them around the boots in the opposite direction of halfshaft rotation. Pull the clamps tight with a suitable tool and bend the locking tabs to secure in position.
13. On 1.9L engine vehicles, after installing the outboard CV-joint boot, install the dynamic damper onto the halfshaft at its original location. Make sure the band is tightened properly and the locking clip is secured.
14. Fit the inboard CV-joint boot and clamps onto the halfshaft and remove the electrical tape.
15. Install the tripot assembly on the halfshaft with the matchmarks

aligned. Install the tripot retaining ring.

16. Fill the CV-joint outer race with CV-joint grease. Make sure the boot is also greased to prevent wear from internal contact of the ribs.

17. Install the outer race over the tripot joint with the matchmarks aligned and install the wire ring bearing retainer.

18. Fit the boot into place on the CV-joint, making sure it is fully seated in the grooves in the shaft and outer race. Insert a suitable tool between the boot and the outer bearing race to allow trapped air to escape from the boot.

19. Install the boot clamps, wrapping them around the boots in the opposite direction of halfshaft rotation. Pull the clamps tight with a suitable tool and bend the locking tabs to secure in position.

20. Work the CV-joint through its full range of travel at various angles. The joint should flex, extend, and compress smoothly. Wipe away any excess grease.

21. Install the halfshaft in the vehicle using new circlips.

22. Lower the vehicle.

23. Road test the vehicle and check for proper operation.

STEERING

Air Bag

—— CAUTION ——

Some vehicles are equipped with an air bag system, also known as the Supplemental Inflatable Restraint (SIR) or Supplemental Restraint System (SRS). The system must be disabled before performing service on or around system components, steering column, instrument panel components, wiring and sensors. Failure to follow safety and disabling procedures could result in accidental air bag deployment, possible personal injury and unnecessary system repairs.

PRECAUTIONS

Several precautions must be observed when handling the inflator module to avoid accidental deployment and possible personal injury.

• Never carry the inflator module by the wires or connector on the underside of the module.

• When carrying a live inflator module, hold securely with both hands, and ensure that the bag and trim cover are pointed away.

• Place the inflator module on a bench or other surface with the bag and trim cover facing up.

• With the inflator module on the bench, never place anything on or close to the module which may be thrown in the event of an accidental deployment.

DISARMING

1994–97 Models

—— CAUTION ——

The Supplemental Restraint System (SRS) must be disarmed before performing service around SRS components or SRS wiring. Failure to do so may cause accidental deployment of the air bag, resulting in unnecessary SRS repairs and/or personal injury.

1. Disconnect both battery cables, negative cable first.

2. Wait at least 1 minute. This allows time for the backup power supply to to deplete its stored energy.

3. Remove the drivers side air bag module and then the passenger side if required.

4. Use caution when carrying live air bags. Always place the air bag with the cover up.

—— CAUTION ——

When carrying a live air bag, make sure the bag and trim cover are pointed away from the body. In the unlikely event of an accidental deployment, the bag will then deploy with minimal chance of injury. When placing a live air bag on a bench or other surface, always face the bag and trim cover up, away from the surface. This will reduce the motion of the module if it is accidently deployed.

5. If the battery needs to be reconnected while one or both of the air bags are removed from the system, install Air Bag Simulator 105-00010 or equivalent, to the drivers side and/or passenger side air bag harness connectors as required. Before removing either air bag simulator, disconnect both battery cables and wait at least 1 minute before continuing.

6. Once service is completed and the air bag modules are back in place, connect the negative battery cable and prove out the air bag system by turning the ignition key to the **RUN** position and visually monitoring the air bag indicator lamp in the instrument cluster. The indicator lamp should illuminate for approximately 6 seconds and then turn OFF. if the indicator lamp does not illuminate, stays ON, or flashes at any time, a fault has been detected by the air bag diagnostic monitor requiring immediate attention.

Steering Wheel

REMOVAL AND INSTALLATION

1993 Models

1. Disconnect the negative battery cable.

2. Remove the steering wheel cover retaining screws from the back side of the steering wheel and remove the cover.

NOTE: On 2-spoke steering wheel, there are 2 retaining screws; on 4-spoke steering wheel, there are 4 retaining screws.

3. Disconnect the horn electrical connector and the speed control electrical connector, if equipped.

4. Remove the steering wheel mounting nut and remove the steering wheel with a suitable puller.

—— WARNING ——

Do not attempt to remove the steering wheel by hitting the column shaft with a hammer; the column may collapse.

To install:

5. Position the steering wheel and install the mounting nut. Tighten the nut to 29–36 ft. lbs. (39–49 Nm).

6. Connect the horn electrical connector and the speed control electrical connector, if equipped.

7. Position the steering wheel cover and install the retaining screws.

8. Connect the negative battery cable.

9. Check the horn and speed control for proper operation.

1994–97 Models

―――――― **CAUTION** ――――――
The Supplemental Restraint System (SRS) must be disarmed before performing service around SRS components or SRS wiring. Failure to do so may cause accidental deployment of the air bag, resulting in unnecessary SRS repairs and/or personal injury.

NOTE: Before continuing with this procedure, make sure to have available a new steering wheel retaining bolt. Once removed, the bolt looses its torque holding ability or retention capability and must not be reused.

1. Make sure the front wheels are in the straight-ahead position.
2. Disarm the air bag system.
3. Remove 2 air bag module retaining bolts from the rear of the steering wheel. Pull the air bag module up and away from the steering wheel and disconnect the module electrical connector from the air bag sliding contact.

―――――― **CAUTION** ――――――
When carrying a live air bag, make sure the bag and trim cover are pointed away from the body. In the unlikely event of an accidental deployment, the bag will then deploy with minimal chance of injury. When placing a live air bag on a bench or other surface, always face the bag and trim cover up, away from the surface. This will reduce the motion of the module, if accidently deployed.

4. Remove the air bag module from the steering wheel.
5. Remove the steering wheel retaining bolt and the steering wheel, using a suitable puller. Discard the steering wheel retaining bolt.
6. Tape the air bag sliding contact to prevent disturbing the air bag sliding contact alignment.
   To install:
7. Remove the tape from the air bag sliding contact.
8. Install the steering wheel with a new retaining bolt. Tighten to 34–46 ft. lbs. (46–63 Nm).
9. Connect the air bag module electrical connector and install the module to the steering wheel. Install 2 module retaining bolts and tighten to 35–53 inch lbs. (4–6 Nm).
10. Connect the negative battery cable.

11. Prove out the air bag system by turning the ignition key to the **RUN** position and visually monitoring the air bag indicator lamp in the instrument cluster. The indicator lamp should illuminate for approximately 6 seconds and then turn OFF. If the indicator lamp does not illuminate, stays ON, or flashes at any time, a fault has been detected by the air bag diagnostic monitor.
12. Check the steering wheel, horn and if equipped cruise control for proper operation.

Tie Rod Ends

REMOVAL AND INSTALLATION

1. Disconnect the negative battery cable.
2. Raise and safely support the vehicle.
3. Remove the wheel and tire assembly.
4. Remove and discard the cotter pin and remove the castellated nut from the tie rod end.
5. Loosen the tie rod end jam nut.
6. Separate the tie rod end from the steering knuckle, using Removal Tool 3290-D and Adapter T81P-3504-W, or equivalents.
7. Count and record the number of turns to unscrew the tie rod end from the inner tie rod or use the jam nut as a marker for reassembly.
   To install:
8. Clean the inner tie rod threads. Apply a light coating of suitable grease to the threads.
9. If removed, install the jam nut.
10. Thread the new tie rod end onto the inner tie rod to the same depth as the one removed. Snug the jam nut.
11. Place the tie rod end stud to the steering knuckle. Replace the dust boot if damaged.
12. Install the castellated nut onto the tie rod end stud. Tighten the nut to 25–34 ft. lbs. (34–46 Nm) and align the next slot with the cotter pin hole in the tie rod end stud. Install a new cotter pin.
13. Install the wheel and tire assembly. Torque the lug nuts to 65–87 ft. lbs. (88–118 Nm).
14. Lower the vehicle.
15. Check the front wheel alignment and adjust as necessary.
16. Tighten the tie rod jam nut to 25–29 ft. lbs. (34–49 Nm).
17. Road test the vehicle and check for proper steering system operation.

Manual Rack and Pinion

REMOVAL AND INSTALLATION

1. Disconnect the negative battery cable.
2. Working from inside the vehicle, remove 5 nuts securing the steering column tube boot and remove the boot.
3. Remove the intermediate shaft-to-pinion shaft coupling bolt from inside the vehicle.
4. Raise and safely support the vehicle.
5. Remove the front wheel and tire assemblies.
6. Remove the cotter pins and castellated nuts securing the tie rod ends to the steering knuckles. Separate the tie rod ends from the steering knuckles. Discard the cotter pins.
7. If equipped with a manual transaxle, disconnect the gearshift lever stabilizer bar.
8. Remove the nuts securing the rack and pinion (steering gear) mounting brackets to the bulkhead. Remove the mounting brackets.
9. Remove the rack and pinion assembly from the vehicle.
   To install:
10. Position the rack and pinion assembly to its mounting position and install the mounting brackets and retaining nuts. Tighten the nuts to 27–38 ft. lbs. (37–52 Nm).
11. If equipped with a manual transaxle, connect the gearshift lever stabilizer bar. Tighten the nut to 23–34 ft. lbs. (31–46 Nm).
12. Install the tie rod ends to the steering knuckles. Install the castellated nuts and tighten to 31–42 ft. lbs. (42–57 Nm) for 1993 vehicles or 25–34 ft. lbs. (34–46 Nm) for 1994–97 vehicles. Install new cotter pins.
13. Install the front wheel and tire assemblies. Torque the lug nuts to 65–87 ft. lbs. (88–118 Nm).
14. Lower the vehicle.
15. Install the intermediate shaft-to-pinion shaft bolt and tighten to 13–20 ft. lbs. (18–27 Nm) for 1993–94 vehicles or 33 ft. lbs. (45 Nm) for 1995–97 vehicles.
16. Install the steering column tube boot and 5 retaining nuts. Tighten to 35 inch lbs. (4 Nm).
17. Connect the negative battery cable.
18. Check the alignment and adjust the toe as required.
19. Road test the vehicle and check the steering system for proper operation.

Power Rack and Pinion

REMOVAL AND INSTALLATION

1. Disconnect the negative battery cable.

2. From inside the passenger compartment, remove 5 steering column tube boot nuts and remove the boot plate.

3. Remove the intermediate shaft-to-pinion shaft bolt.

4. Raise and safely support the vehicle.

5. Remove the front wheel and tire assemblies.

6. Remove the left-hand splash shield.

7. Remove the cotter pins and castellated nuts from the tie rod ends. Using a suitable tool, separate the tie rod ends from the steering knuckles. Discard the cotter pins.

8. Remove the strap that holds the power steering lines to the rack and pinion (steering gear) housing and discard the strap.

9. Disconnect the pressure and return lines from the rack and pinion assembly and plug the lines.

10. If equipped with a manual transaxle, disconnect the gearshift stabilizer bar and shift control rod from the transaxle.

11. Remove the retaining nuts from the rack and pinion mounting brackets.

12. Remove the rack and pinion assembly from the left-hand side of the vehicle.

To install:

13. Place the rack and pinion assembly in its mounting location.

14. Install 2 rack and pinion mounting brackets. Install the retaining nuts and tighten to 27–38 ft. lbs. (37–52 Nm).

15. If equipped with a manual transaxle, connect the gearshift stabilizer bar and shift control rod. Tighten the gearshift stabilizer bar nut to 23–34 ft. lbs. (31–46 Nm) and the shift control rod nut to 12–17 ft. lbs. (16–23 Nm).

16. Remove the plugs and connect the pressure and return lines to the rack and pinion assembly. Tighten the flare nuts to 22–28 ft. lbs. (29–39 Nm).

17. Install a new strap to hold the power steering lines to the rack and pinion housing.

18. Install the tie rod ends to the steering knuckles and install the castellated nuts. Tighten the nuts to 31–42 ft. lbs. (42–57 Nm) on 1993 vehicles or 25–33 ft. lbs. (34–46 Nm) on

1994–97 vehicles. Install new cotter pins.

19. Install the left-hand splash shield.

20. Install the wheel and tire assemblies. Torque the lug nuts to 65–87 ft. lbs. (88–118 Nm).

21. Lower the vehicle.

22. From inside the vehicle, install the intermediate shaft-to-pinion shaft bolt. Tighten the bolt to 13–20 ft. lbs. (18–27 Nm) for 1993–94 vehicles or 30–36 ft. lbs. (40–50 Nm) for 1995–97 vehicles.

23. Install the steering column tube boot and 5 retaining nuts. Tighten the retaining nuts to 35 inch lbs. (4 Nm).

24. Connect the negative battery cable.

25. Fill and bleed the power steering system.

26. Check the alignment and adjust as required.

27. Start the engine and check for leaks.

28. Road test the vehicle and check for proper steering system operation.

Power Steering Pump

BLEEDING

Normally the power steering system will expel any trapped air during normal operation. If the power steering system is noisy due to trapped air, turn the steering wheel left to right several times with the front wheels off the ground and the vehicle properly secured, to release the trapped air.

REMOVAL AND INSTALLATION

1.8L Engine

1. Disconnect the negative battery cable.

2. Loosen the power steering fluid reservoir-to-pump hose clamp and pull the hose from the reservoir. Plug the hose.

3. Remove 2 reservoir retaining bolts and lift the reservoir from its mounting position.

4. Loosen the return hose clamp and pull the return hose from the reservoir. Plug the hose and remove the reservoir.

5. Disconnect the electrical connector from the Power Steering Pressure (PSP) switch.

6. Loosen the pressure line fitting and disconnect the line from the pump. Plug the line.

7. Raise and safely support the vehicle.

8. Remove 5 right front splash shield retaining bolts and remove the splash shield.

9. Remove the belt tensioner adjustment bolt and remove the accessory drive belt from the pulley.

10. Lower the vehicle.

11. Remove 3 power steering pump mounting bracket retaining bolts and remove the steering pump and the bracket.

12. Remove the bolt attaching the pump to the mounting bracket.

13. Remove the nut and bolt attaching the tensioner to the pump mounting bracket and remove the nut and bolt attaching the tensioner to the pump.

To install:

14. Position the tensioner to the pump and install the bolt and nut. Tighten the nut to 14–19 ft. lbs. (19–25 Nm).

15. Position the tensioner to the pump mounting bracket and install the bolt and nut. Tighten the nut to 23–34 ft. lbs. (31–46 Nm).

16. Install the bolt that attaches the pump to the mounting bracket and tighten to 27–40 ft. lbs. (36–54 Nm).

17. Position the pump and bracket and install 3 pump mounting bracket retaining bolts. Tighten the bolts to 27–38 ft. lbs. (37–54 Nm).

18. Raise and safely support the vehicle.

19. Position the accessory drive belt on the pulley and install the belt tensioner adjustment bolt.

20. Position the right front splash shield and install 5 retaining bolts.

21. Lower the vehicle.

22. Unplug and connect the pressure line to the power steering pump. Tighten the line fitting to 12–17 ft. lbs. (16–24 Nm).

23. Connect the PSP switch electrical connector.

24. Unplug and connect the return hose to the reservoir. Tighten the clamp.

25. Position the reservoir and install 2 retaining bolts.

26. Unplug and connect the pump hose to the reservoir. Tighten the clamp.

27. Adjust the accessory drive belt tension.

28. Fill the system with power steering fluid.

29. Connect the negative battery cable.

30. Bleed air from the power steering system.

31. Start the engine and check for leaks.

32. Road test the vehicle and check for proper steering system operation.

1.9L Engine

1. Disconnect the negative battery cable.

2. Drain the engine cooling system into a suitable container.

3. Loosen the drive belt tensioner and remove the drive belt from the pulley. Remove the belt tensioner bolt and remove the tensioner.

4. Support the engine with a suitable floor jack.

5. If equipped, remove the speed control servo and bracket and move aside.

6. Remove the right-hand engine support insulator nut and through-bolt. Remove the support insulator.

7. Remove 2 front engine mount nuts. Loosen the engine mount pivot bolt and nut and position the engine mount aside.

8. Raise the engine using the floor jack, to gain access to the power steering pump pulley.

9. Hold the pulley in position with a suitable tool and remove 3 pulley retaining bolts. Remove the pulley and lower the engine.

10. Position the engine mount and install 2 retaining nuts.

11. Loosen the clamp and disconnect the return line from the pump. Loosen the line fitting from the pressure line and disconnect the line from the pump. Plug both power steering lines.

12. If required, remove the A/C tube bracket bolt from the power steering pump support.

13. Raise and safely support the vehicle.

14. Remove 2 passenger side splash shields.

15. If equipped with A/C, remove 4 A/C compressor retaining bolts and position the A/C compressor aside.

16. Remove the lower radiator hose.

17. Remove 3 power steering pump retaining bolts and remove the power steering pump.

To install:

18. Place the power steering pump in position and install 3 retaining bolts. Tighten the bolts to 30–45 ft. lbs. (40–62 Nm).

19. Install the lower radiator hose.

20. If equipped with A/C, position the A/C compressor and install 4 retaining bolts. Tighten the bolts to 30–40 ft. lbs. (40–55 Nm).

21. Install 2 passenger side splash shields.

22. Lower the vehicle.

23. Connect the pressure line to the power steering pump and tighten the fitting. Connect the return line to the pump and position the clamp.

24. Support the engine with a suitable floor jack.

25. Remove 2 front engine mount retaining nuts and raise the engine to gain access to the pulley using the floor jack.

26. Place the pulley in position and loosely install 3 retaining bolts. Hold the pulley in place with a suitable tool and tighten the bolts to 15–22 ft. lbs. (20–30 Nm).

27. Lower the engine.

28. Position the engine mount and install 2 retaining nuts. Tighten the engine mount pivot bolt and nut.

29. Position the engine support insulator and install the bolt and nut.

30. Remove the floor jack.

31. Position the belt tensioner and loosely install the bolt. Position the accessory drive belt on the pulley and tighten the tensioner mounting bolt to 30–41 ft. lbs. (40–55 Nm).

32. Fill the engine cooling system.

33. Add the proper type and quantity of power steering fluid to the power steering fluid reservoir.

34. Connect the negative battery cable.

35. Bleed the air from the power steering system.

36. Start the engine and check for leaks.

37. Road test the vehicle and check the power steering system for proper operation.

BRAKES

Anti-Lock Brake System Service

PRECAUTIONS

• Certain components within the Anti-Lock Brake System (ABS) are not intended to be serviced or repaired individually. Only those components with removal and installation procedures should be serviced.

• Do not use rubber hoses or other parts not specifically specified for and ABS system. When using repair kits, replace all parts included in the kit. Partial or incorrect repair may lead to functional problems and require the replacement of components.

• Lubricate rubber parts with clean, fresh brake fluid to ease assembly. Do not use lubricated shop air to clean parts; damage to rubber components may result.

• Use only specified brake fluid from an unopened container.

• If any hydraulic component or line is removed or replaced, it may be necessary to bleed the entire system.

• A clean repair area is essential. Always clean the reservoir and cap thoroughly before removing the cap. The slightest amount of dirt in the fluid may plug an orifice and impair the system function. Perform repairs after components have been thoroughly cleaned; use only denatured alcohol to clean components. Do not allow ABS components to come into contact with any substance containing mineral oil; this includes used shop rags.

• The Anti-Lock control unit is a microprocessor similar to other computer units in the vehicle. Ensure that the ignition switch is **OFF** before removing or installing controller harnesses. Avoid static electricity discharge at or near the controller.

• If any arc welding is to be done on the vehicle, the control unit should be unplugged before welding operations begin.

Master Cylinder

REMOVAL AND INSTALLATION

1. Disconnect both battery cables, negative cable first.

2. Remove the battery.

3. Remove the speed control cable from its bracket, if equipped.

4. Disconnect the low fluid level sensor electrical connector.

5. Pump the brake pedal to exhaust the brake booster vacuum.

6. Loosen the brake line fittings and disconnect the brake lines from the master cylinder.

7. Cap the brake lines and the master cylinder ports.

8. If equipped with a manual transaxle, remove the clamp and pull the clutch hose from the brake/clutch fluid reservoir.

9. Cap the clutch master cylinder hose and reservoir port.

10. Remove 2 retaining nuts and remove the master cylinder assembly.

To install:

11. Adjust the master cylinder piston to brake booster pushrod clearance as follows:

a. Insert a pencil in the pushrod socket of the master cylinder. Mark the point on the pencil that is even with the end of the master cylinder with a hacksaw blade.

b. Measure the length of the pencil to the saw mark with a ruler.

c. Using the ruler, measure how far the master cylinder pushrod protrudes out of the booster assembly.

d. Measure the length of the master cylinder boss with the ruler. Subtract the length of the boss from the length of the pencil. The difference in length between the master cylinder pushrod and the corrected pencil length is equal to the pushrod clearance.

e. Adjust the pushrod length to get the correct clearance. It should be 0.025 inch (1mm) shorter than the pushrod socket.

12. Before installation, bench bleed the new master cylinder.

13. Position the master cylinder over the booster pushrod and booster mounting studs. Install 2 nuts and tighten to 8–12 ft. lbs. (10–16 Nm).

14. Install the speed control cable to its bracket, if removed.

15. If equipped with manual transaxle, connect the clutch hose onto the brake/clutch fluid reservoir and install the clamp.

16. Remove the caps and connect the brake lines. Tighten the brake line fittings.

17. Make sure the master cylinder reservoir is full and bleed it.

18. Connect the low fluid level sensor electrical connector.

19. Install the battery.

20. Connect both battery cables, negative cable last.

21. Make sure the master cylinder reservoir is full. Bleed the brakes at each wheel, if necessary.

22. Bleed the clutch hydraulic system, if required.

23. Check for brake fluid leaks and brake pedal operation.

24. Road test the vehicle and check the brake system for proper operation.

Brake Caliper

REMOVAL AND INSTALLATION

Front Caliper

1. Raise and safely support the vehicle.

2. Remove the wheel and tire assembly.

3. Remove the disc brake pads.

4. Remove the banjo bolt securing the brake hose to the brake caliper. Discard 2 copper sealing washers.

5. Remove 2 brake caliper retaining bolts and remove the brake caliper.

To install:

6. Place the brake caliper in position and install 2 brake caliper retaining bolts. Torque the bolts to 29–36 ft. lbs. (39–49 Nm).

7. Install the brake hose to the brake caliper and install the banjo bolt with 2 new copper sealing washers. Torque the banjo bolt to 16–22 ft. lbs. (22–29 Nm).

8. Install the disc brake pads.

9. Bleed the brake system.

10. Install the wheel and tire assembly. Torque the lug nuts to 65–87 ft. lbs. (88–118 Nm).

11. Lower the vehicle.

12. Pump the brake pedal several times to position the brake pads to the brake rotor.

13. Road test the vehicle and check the brake system for proper operation.

Rear Caliper

1. Raise and safely support the vehicle.

2. Remove the wheel and tire assembly.

3. Remove the disc brake pads.

4. Remove the parking brake cable bracket bolt and position the bracket aside.

5. Remove the parking brake cable from the operating lever.

6. Remove the banjo bolt securing the brake hose to the brake caliper. Discard 2 copper sealing washers.

7. Remove the upper brake caliper retaining bolt.

8. Slide the brake caliper off the mounting bracket and remove from the vehicle.

To install:

9. Place the brake caliper on the mounting bracket and slide into position.

10. Install the upper brake caliper retaining bolt. Torque the bolt to 33–43 ft. lbs. (45–59 Nm).

11. Install the brake hose to the brake caliper and secure with the banjo bolt using 2 new copper sealing washers. Tighten the banjo bolt to 16–22 ft. lbs. (22–29 Nm).

12. Connect the parking brake cable to the operating lever. Position the bracket and install the bracket bolt.

13. Install the disc brake pads.

14. Bleed the brake system.

15. Install the wheel and tire assembly. Torque the lug nuts to 65–87 ft. lbs. (88–118 Nm).

16. Lower the vehicle.

17. Pump the brake pedal several times to position the brake pads to the brake rotor.

18. Road test the vehicle and check the brake system for proper operation.

Disc Brake Pads

REMOVAL AND INSTALLATION

Front Disc Brake Pads

1. Remove the master cylinder reservoir cap and check the fluid level in the reservoir. Remove brake fluid until the reservoir is ½ full. Discard the removed fluid.

2. Raise and safely support the vehicle.

3. Remove the wheel and tire assembly.

4. Remove the W-spring.

5. Remove 2 disc brake pad locating pins and remove the M-spring.

6. Remove the brake pads and shims from the brake caliper.

7. Inspect the brake rotor. Resurface or replace if needed.

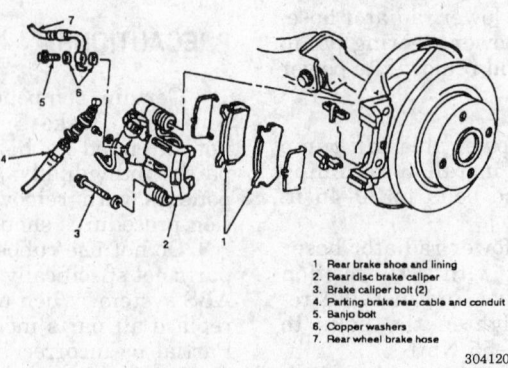

1. Rear brake shoe and lining
2. Rear disc brake caliper
3. Brake caliper bolt (2)
4. Parking brake rear cable and conduit
5. Banjo bolt
6. Copper washers
7. Rear wheel brake hose

304120

Rear disc brake assembly

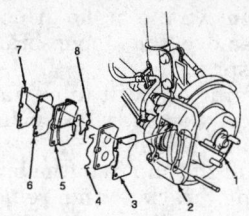

1. Disc brake caliper locating pin
2. Front disc brake caliper
3. Outer shim
4. W-spring
5. Brake shoe and lining
6. Inner shim
7. Anti-rattle shim
8. M-spring

304247

Front disc brake pads

To install:

8. Use a suitable tool to push the piston into the brake caliper bore.

9. Apply a suitable grease between the shims and the disc brake pad guide plates and position the brake pads and shims to the brake caliper.

10. Install the W-spring and 2 brake pad locating pins.

11. Install the M-spring.

12. Install the wheel and tire assembly. Torque the lug nuts to 65–87 ft. lbs. (88–118 Nm).

13. Lower the vehicle.

14. Pump the brake pedal several times prior to moving the vehicle to position the brake pads.

15. Check the fluid level in the master cylinder reservoir and fill as needed.

16. Road test the vehicle and check for proper brake system operation.

Rear Disc Brake Pads

1. Remove the master cylinder reservoir cap and check the fluid level in the reservoir. Remove brake fluid until the reservoir is ½ full. Discard the removed fluid.

2. Raise and safely support the vehicle.

3. Remove the wheel and tire assembly.

4. If necessary, remove the screw plug and turn the adjustment gear counterclockwise with an Allen® wrench to pull the piston fully inward.

5. Remove the lower brake caliper bolt.

6. Using a small prybar, pivot the caliper on its mounting bracket to access the brake pads. If the upper lock bolt requires lubrication or service, remove it and suspend the caliper with mechanics wire.

7. Remove the brake pads, shims, spring and guides.

8. Inspect the brake rotor. Resurface or replace if needed.

To install:

9. Apply an appropriate brake pad grease between the shims and the brake pads.

10. Pivot the caliper on its mounting bracket and position the brake pads, shims, spring and guides to the brake rotor.

11. Lubricate and install the lower caliper bolt. Tighten the bolt to 33–43 ft. lbs. (45–59 Nm).

12. Turn the adjustment gear clockwise with an Allen wrench until the brake pads just touch the rotor, then loosen the gear ⅓ of a turn. Install the screw plug and tighten to 9–12 ft. lbs. (12–16 Nm).

13. Install the wheel and tire assembly. Torque the lug nuts to 65–87 ft. lbs. (88–118 Nm).

14. Lower the vehicle.

15. Pump the brake pedal several times prior to moving the vehicle to position the brake pads.

16. Check the brake fluid level in the master cylinder reservoir and fill if needed.

17. Road test the vehicle and check for proper brake system operation.

Brake Rotor

REMOVAL AND INSTALLATION

Front Brake Rotor

1. Raise and safely support the vehicle.

2. Remove the wheel and tire assembly.

3. Remove 2 disc brake caliper retaining bolts.

4. Remove the disc brake caliper and secure the caliper aside with mechanics wire. Do not disconnect the brake hose.

5. Pull the brake rotor from the hub.

6. Inspect the rotor and refinish or replace, as necessary. If refinishing, the minimum thickness rotor thickness is stamped on the brake rotor.

To install:

7. Place the brake rotor to the hub. Make sure the mating surfaces are free of rust which may prevent the brake rotor from seating properly.

8. Remove the mechanics wire and position the brake caliper.

9. Install 2 brake caliper retaining bolts and torque to 29–36 ft. lbs. (39–49 Nm).

10. Install the wheel and tire assembly. Torque the lug nuts to 65–87 ft. lbs. (88–118 Nm).

11. Lower the vehicle.

12. Pump the brake pedal several times to position the brake pads before moving the vehicle.

13. Road test the vehicle and check the brake system for proper operation.

Rear Brake Rotor

1. Raise and safely support the vehicle.

2. Remove the wheel and tire assembly.

3. Remove 2 disc brake rotor retaining screws.

4. Remove the lower caliper retaining bolt.

5. Pivot the caliper and remove the disc brake pads.

6. Pivot the caliper and remove the disc brake rotor from the hub.

7. Inspect the rotor and refinish or replace, as necessary. If refinishing, the minimum thickness is stamped on the brake rotor.

To install:

8. Pivot the brake caliper and position the disc brake rotor onto the hub.

9. Install the disc brake pads and shims.

10. Install the lower brake caliper retaining bolt and torque to 33–43 ft. lbs. (45–59 Nm).

11. Install 2 brake rotor retaining screws.

12. Install the wheel and tire assembly. Torque the lug nuts to 65–87 ft. lbs. (88–118 Nm).

13. Lower the vehicle.

14. Pump the brake pedal several times to position the brake pads before moving the vehicle.

15. Road test the vehicle and check the brake system for proper operation.

Brake Drums

REMOVAL AND INSTALLATION

1. Raise and safely support the vehicle.

2. Remove the wheel and tire assembly.

3. Remove 2 brake drum retaining screws.

4. Pull the brake drum from the hub. Inspect the drum and refinish or replace, as necessary. If refinishing, check the maximum inside diameter specification.

To install:

5. If needed, adjust the brake shoes to fit the brake drum.

6. Place the brake drum on the hub.

7. Install 2 brake drum retaining screws. Tighten to 89–123 inch lbs. (10–14 Nm).

8. Install the wheel and tire assembly. Torque the lug nuts to 65–87 ft. lbs. (88–118 Nm).

9. Lower the vehicle.

10. Check the brake system for proper operation.

Brake Shoes

REMOVAL AND INSTALLATION

1. Raise and safely support the vehicle.

2. Remove the wheel and tire assembly.

3. Remove 2 brake drum retaining screws and remove the brake drum.

4. Remove 2 brake shoe return springs.

5. Remove the right-hand anti-rattle spring.

6. Push and turn to release 2 brake shoe hold-down springs and remove the springs.

7. Remove the leading and trailing shoes from the brake backing plate.

To install:

8. Use a suitable high temperature grease to lightly lubricate the brake shoe contact points on the backing plate.

NOTE: Make sure the wheel cylinder retaining bolts are tight.

9. Position the trailing brake shoe on the backing plate and install one of the brake shoe hold-down springs.

10. Position the leading brake shoe on the backing plate and install the other brake shoe hold-down spring.

11. Install the right-hand anti-rattle spring.

12. Install 2 brake shoe return springs.

13. Using a brake adjusting gauge, measure the inside diameter of the brake drum.

14. Compare the brake drum measurement to the brake shoes.

15. Adjust the brake shoes by inserting a screwdriver into the knurled quadrant of the rear quad operating lever and adjust the shoes to the same measurement as the brake drum.

16. Install the brake drum.

17. Install 2 brake drum retaining screws.

18. Install the wheel and tire assembly. Torque the lug nuts to 65–87 ft. lbs. (88–118 Nm).

19. Lower the vehicle.

20. Complete the brake shoe adjustment by sharply applying the brakes several times while driving the vehicle alternating between forward and reverse gears.

21. Check the brake system operation by making several stops while driving forward.

Wheel Cylinder

REMOVAL AND INSTALLATION

1. Raise and safely support the vehicle.

2. Remove the wheel and tire assembly.

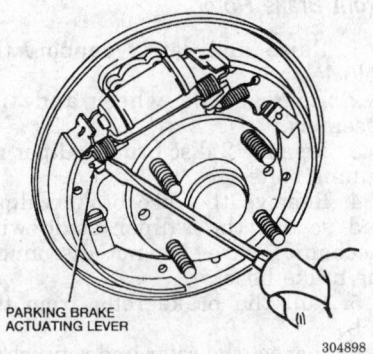

PARKING BRAKE
ACTUATING LEVER

304898

Adjusting the brake shoes

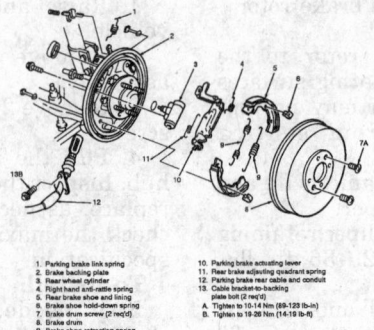

1. Parking brake link spring
2. Brake backing plate
3. Rear wheel cylinder
4. Right-hand anti-rattle spring
5. Rear brake shoe and lining
6. Brake shoe hold-down spring
7. Brake drum screw (2 req'd)
8. Brake drum
9. Brake shoe retracting spring
10. Parking brake actuating lever
11. Rear brake adjusting quadrant spring
12. Parking brake rear cable and conduit
13. Cable bracket-to-backing
    plate bolt (2 req'd)
A. Tighten to 10-14 Nm (89-123 lb-in)
B. Tighten to 19-28 Nm (14-19 lb-ft)

304895

Brake shoes and related components

3. Remove the brake drum.

4. Remove the upper brake shoe return spring.

5. Using a suitable flare nut wrench, loosen the wheel cylinder-to-brake line flare nut.

6. Pull the clip from the brake line retaining bracket and remove the brake line from the retaining bracket.

7. Remove the brake line from the wheel cylinder and cap the brake line fitting.

8. Remove 2 wheel cylinder retaining bolts and remove the wheel cylinder from the backing plate.

9. Remove and discard the wheel cylinder gasket.

To install:

10. Install a new wheel cylinder gasket onto the backing plate.

11. Position the wheel cylinder onto the backing plate and loosely install 2 retaining bolts.

12. Install the brake line to the wheel cylinder and tighten the flare nut to 12–16 ft. lbs. (16–22 Nm).

13. Tighten the wheel cylinder retaining bolts to 89–115 inch lbs. (10–13 Nm).

14. Position the brake hose to the retaining bracket and install the clip.

15. Install the upper brake shoe return spring.

16. Have and assistant gently press the brake pedal to verify the operation of the automatic brake adjuster.

17. Manually adjust the brakes if required.

18. Install the brake drum.

19. Install the wheel and tire assembly. Torque the lug nuts to 65–87 ft. lbs. (88–118 Nm).

20. Bleed the brake system.

21. Lower the vehicle.

22. Fill the brake master cylinder as required.

23. Road test the vehicle and check the brake system for proper operation.

Parking Brake Cable

ADJUSTMENT

——————— **WARNING** ———————
Use caution when performing a parking brake adjustment which requires operating the vehicle in reverse.

1. Start the engine and place the transaxle in **R**.

2. With the vehicle slowly moving in reverse, apply the parking brake lever several times.

3. Stop the vehicle and move the gearshift selector to the **N** or **P** position.

4. Shut the engine **OFF**.

5. Turn the parking brake adjusting nut until the parking brake control clicks 5–7 notches when pulled up with a force of 22 pounds (98 N).

6. Check the parking brake system for proper operation.

REMOVAL AND INSTALLATION

Front Parking Brake Cable

1. Disconnect the negative battery cable.

2. If required, remove the parking brake console as follows:

   a. Remove the rear seat ash tray.

   b. Position both front seats to the rear-most position.

   c. Remove 2 front retaining screws from the parking brake console.

   d. Recline both front seats.

   e. Remove 2 rear retaining screws and with the parking brake engaged, remove the parking brake console.

   f. Release the parking brake lever.

3. Disconnect the parking brake signal switch electrical connector.

4. Remove the cable adjusting nut.

5. Remove 2 parking brake control retaining bolts and remove the parking brake control.

6. Remove the front parking brake cable and conduit from the cable guide.

7. Raise and safely support the vehicle.

8. Remove the rear exhaust pipe and resonator heat shields.

9. Remove the equalizer return spring and disconnect both rear parking brake cables from the equalizer.

10. Remove the front parking brake cable, conduit and equalizer from the vehicle.

To install:

11. Position the front parking brake cable, conduit and equalizer to the vehicle.

12. Connect the rear parking brake cables to the equalizer and install the equalizer return spring.

13. Install the exhaust heat shields.

14. Lower the vehicle.

15. Position the front parking brake cable and conduit to the cable guide.

16. Install the parking brake control and 2 retaining bolts. Torque the bolts to 14–19 ft. lbs. (19–26 Nm).

17. Fit the cable to the parking brake control and install the adjusting nut.

18. Connect the parking brake signal switch electrical connector.

19. If removed, install the parking brake console as follows:

   a. Raise the parking brake lever and position the console.

   b. With both seats reclined, install 2 rear retaining screws.

   c. With both seats at the rear-most position, install 2 front retaining screws.

   d. Install the rear seat ash tray.

20. Connect the negative battery cable.

CAUTION

Use care when performing a parking brake adjustment which requires running the vehicle in R.

---

21. Adjust the parking brake.

Rear Parking Brake Cable

1. Disconnect the negative battery cable.

2. If required, remove the parking brake console as follows:

   a. Remove the rear seat ash tray.

   b. Position both front seats to the rear-most position.

   c. Remove 2 front retaining screws from the parking brake console.

   d. Recline both front seats.

   e. Remove 2 rear retaining screws and with the parking brake engaged, remove the parking brake console.

   f. Release the parking brake lever.

3. Disconnect the parking brake signal switch electrical connector.

4. Remove the cable adjusting nut.

5. Raise and safely support the vehicle.

6. Remove the rear exhaust pipe and resonator heat shields.

7. Remove the equalizer return spring and disconnect both rear parking brake cables from the equalizer.

8. Remove the clip that attaches the cable to the retaining bracket located near the equalizer. Remove the cable from the bracket.

9. Remove the cable routing bracket bolt from the floorpan and remove the bracket.

10. Remove 2 cable routing bracket nuts from the trailing link and remove the bracket.

11. If equipped with rear drum brakes, remove 2 cable retaining bracket bolts from the backing plate and remove the bracket.

12. If equipped with rear disc brakes, remove the parking brake bracket bolt and position the parking brake bracket aside.

13. Remove the rear cable and conduit from the parking brake actuating lever.

14. Remove the rear cable from the vehicle.

To install:

15. Position the cable onto the parking brake actuating lever.

16. If equipped with rear drum brakes, position the parking brake cable bracket onto the backing plate and install 2 bolts. Tighten the bolts to 14–19 ft. lbs. (19–25 Nm).

17. If equipped with rear disc brakes, install the parking brake bracket and retaining bolt.

18. Position the cable routing bracket onto the trailing link and install 2 nuts. Tighten the nuts to 12–17 ft. lbs. (16–23 Nm).

19. Position the cable routing bracket to the floor pan and install the mounting bolt. Tighten the bolt to 14–19 ft. lbs. (19–25 Nm).

20. Position the cable into the retaining bracket near the equalizer and install the clip.

21. Install the cables into the equalizer and install the equalizer return spring.

22. Install the rear exhaust pipe and resonator heat shields.

23. Lower the vehicle.

24. Install the adjusting nut.

25. If removed, install the parking brake console as follows:

   a. Raise the parking brake lever and position the console.

   b. With both seats reclined, install 2 rear retaining screws.

   c. With both seats at the rear-most position, install 2 front retaining screws.

   d. Install the rear seat ash tray.

26. Connect the negative battery cable.

CAUTION

Use care when performing a parking brake adjustment which requires running the vehicle in R.

---

27. Adjust the parking brake.

Brake System

BLEEDING

The hydraulic brake system is bled the same for ABS brake systems as well as non–ABS brake systems. The only exception is that the ABS brake system must be bled at least twice to ensure that no more air exist in the hydraulic system.

1. Clean the area around the master cylinder filler cap and remove the cap.

2. Fill the reservoir with clean brake fluid and install the cap.

3. If the master cylinder is known or suspected to have air in the bore, it must be bled before any of the wheel cylinders or calipers. To bleed the master cylinder, position a shop towel under the rear master cylinder outlet fitting and loosen the fitting approximately ¾ turn. Have an assistant depress the brake pedal slowly through its full travel. Close the outlet fitting and let the pedal return slowly to the fully released position. Wait 5 seconds, then repeat the operation until all air bubbles disappear.

——————— WARNING ———————
Be careful not to spill brake fluid on painted surfaces, as it can destroy the finish. If any brake fluid is spilled, rinse the area immediately with water.

4. Repeat the master cylinder bleeding procedure at the front master cylinder outlet fitting.

5. Position the correct box–end wrench on the brake bleeder on the right–rear wheel.

NOTE: Always bleed the longest line first, which is the right–rear.

6. Place a drain tube to the brake bleeder screw and the free end into a container half full of clean brake fluid.

7. Loosen the bleeder screw while an assistant apply the brake pedal slowly through its entire travel.

8. Close the brake bleeder and then have the assistant release the brake pedal.

9. Continue bleeding the right–rear wheel until the fluid is free of air bubbles.

NOTE: DO NOT allow the master cylinder to run out of brake fluid. Only use fresh DOT 3 or equivalent brake fluid from a closed container.

10. Repeat the procedure on the left–front wheel, then the left–rear and finally the right–front wheel.

11. Throughout the brake bleeding process, continually check the brake master cylinder and top off the fluid as required.

12. If the brake pedal is spongy, repeat the brake bleeding procedure, or if available bleed the brake system

using a pressure bleeder following the manufacturers recommendations.

Wheel Speed Sensor

REMOVAL AND INSTALLATION

1994–97 Models

Front Wheel

1. Disconnect both battery cables, negative cable first.

2. Remove the battery to gain access to the left speed sensor connector.

3. Disconnect the left speed sensor connector. Remove the grommet and feed the wire through the hole in the strut tower.

4. Raise and safely support the vehicle.

5. Remove the left-front wheel and tire assembly.

6. Remove the retaining clips holding the wiring harness to the wheel well.

7. Remove the wiring harness bracket from the wheel well.

8. Remove 2 retaining clips holding the wiring harness to the top and bottom of the strut tower.

9. Remove 2 speed sensor bracket retaining bolts and remove the speed sensor.

To install:

10. Install 2 speed sensor bracket retaining bolts and tighten to 12–17 ft. lbs. (16–23 Nm).

11. Position the wiring and install 2 retaining clips holding the wiring harness to the top and bottom of the strut tower. Feed the wiring connector through the hole in the strut tower. Connect the plug and position the grommet.

12. Install the wiring harness bracket to the wheel well.

13. Install the retaining clips holding the wiring harness to the wheel well.

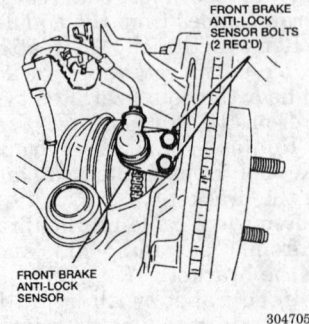

Front wheel speed speed sensor and bracket — 1994–97 models

14. Install the wheel and tire assembly. Tighten the lug nuts to 65–87 ft. lbs. (88–118 Nm).

15. Install the battery. Connect both battery cables, negative cable last.

16. Turn the ignition key to the **RUN** position. In the key ON, engine OFF position, the amber anti-lock brake indicator in the instrument cluster will illuminate continuously. Once the engine has started and for up to 60 seconds after engine start-up while the system performs a self-diagnostic check, the indicator will extinguish indicating that the anti-lock brake system is functional. If the amber indicator does not illuminate during the prove-out test, or at any time the anti-lock brake indicator illuminates after the initial prove-out or stays illuminated, a problem has been found in the anti-lock system and required immediate attention.

17. Road test the vehicle and check for proper operation.

Rear Wheel

1. Disconnect the negative battery cable.

2. Remove the interior trim panel.

3. Disconnect the speed sensor electrical connector. Remove the grommet and feed the wire connection through the hole in the panel.

4. Raise and safely support the vehicle.

5. Remove the wheel and tire assembly.

6. Remove the upper and lower retaining clips holding the wiring harness to the wheel well.

7. Remove the speed sensor retaining bolt and remove the sensor.

To install:

8. Install the speed sensor and the retaining bolt. Tighten to 12–17 ft. lbs. (16–23 Nm).

9. Position the wire and install the upper and lower retaining clips holding the wiring harness to the wheel well.

10. Feed the wire connection through the hole in the panel.

11. Install the interior trim panel.

12. Install the wheel and tire assembly. Tighten the lug nuts to 65–87 ft. lbs. (88–118 Nm).

13. Lower the vehicle.

14. Connect the negative battery cable.

15. Turn the ignition key to the **RUN** position. In the key ON, engine OFF position, the amber anti-lock brake indicator in the instrument cluster will illuminate continuously. Once the engine has started and for up to 60 seconds after engine start-up while the system performs a self-di-

agnostic check, the indicator will extinguish indicating that the anti-lock brake system is functional. If the amber indicator does not illuminate during the prove-out test, or at any time the anti-lock brake indicator illuminates after the initial prove-out or stays illuminated, a problem has been found in the anti-lock system and required immediate attention.

16. Road test the vehicle and check for proper operation.

FRONT SUSPENSION

Strut

REMOVAL AND INSTALLATION

1. Disconnect the negative battery cable.
2. Raise and safely support the vehicle.
3. Remove the front wheel and tire assembly.
4. Remove the clip securing the brake hose to the strut (spring and shock) assembly. If equipped with anti-lock brakes, remove the anti-lock brake harness cable and clip.
5. Remove 2 nuts and 2 bolts securing the strut assembly to the steering knuckle.
6. Partially lower the vehicle.
7. Remove 4 upper strut retaining nuts and remove the strut assembly from the vehicle.

——— **WARNING** ———
Never remove the strut piston rod nut unless the coil spring is compressed. Always wear safety glasses when using a spring compressor.

8. Inspect all components and replace as needed.
To install:
9. Position the strut assembly into the wheel housing. Make sure the direction indicator on the upper mounting bracket faces inboard.
10. Secure the upper mounting bracket to the strut tower with 4 retaining nuts. Tighten the nuts to 22–30 ft. lbs. (29–40 Nm).
11. Attach the strut assembly to the steering knuckle and install the retaining bolts and nuts. Tighten to 69–93 ft. lbs. (93–127 Nm).

12. Place the brake hose to the strut assembly and secure it with the brake hose clip. If equipped with anti-lock brakes, install the anti-lock harness cable and clip.
13. Install the front wheel and tire assembly. Torque the lug nuts to 65–87 ft. lbs. (88–118 Nm).
14. Lower the vehicle.
15. Connect the negative battery cable.
16. Check the front wheel alignment.
17. Road test the vehicle and check for proper operation.

Lower Control Arm

REMOVAL AND INSTALLATION

1. Raise and safely support the vehicle.
2. Remove the front wheel and tire assembly.
3. Remove the stabilizer bar link nuts, retainers, bushings, bolts and sleeves.
4. Remove the pinch bolt and nut securing the ball joint to the steering knuckle.
5. Separate the lower ball joint from the steering knuckle by prying the lower control arm down with a prybar.
6. Remove the bolt and washer securing the front bushing of the lower control arm.
7. Remove the bolts securing the lower control arm rear bushing retaining strap.
8. Remove the lower control arm.
9. Inspect the lower control arm, lower control arm bushings and the lower ball joint. The ball joint and bushings can be replaced individually.
To install:
10. If required, install new bushings to the lower control arm.
11. If replacing the lower ball joint, position the ball joint to the steering knuckle. Install the ball joint retaining nuts and bolt and tighten to 32–43 ft. lbs. (43–59 Nm). Apply a suitable thread lock sealer to the nut and bolt threads prior to installation.
12. Install the lower control arm rear bushing retaining strap to the lower frame. Install the bolts and tighten to 69–86 ft. lbs. (93–117 Nm).
13. Install the lower control arm front pivot bolt and washer. Tighten the nut to 69–93 ft. lbs. (93–127 Nm).
14. Install the stabilizer bolts, washers, bushings, sleeves and nuts.

Tighten the stabilizer nuts so 0.67–0.75 inches (17–19mm) of thread is exposed at the end of the bolt.
15. Install the wheel and tire assembly. Torque the lug nuts to 65–87 ft. lbs. (88–118 Nm).
16. Lower the vehicle.
17. Check the front wheel alignment.
18. Road test the vehicle and check for proper operation.

Lower Ball Joints

REMOVAL AND INSTALLATION

1. Raise and safely support the vehicle.
2. Remove the wheel and tire assembly.
3. Remove the pinch bolt and nut securing the lower ball joint to the steering knuckle.
4. Separate the lower ball joint from the steering knuckle by prying on the lower control arm with a prybar or similar tool.
5. Remove 2 nuts and 1 bolt securing the lower ball joint to the lower control arm.
6. Remove the lower ball joint from the lower control arm.
7. If the lower ball joint dust boot is to be reused, place a suitable chisel between the ball joint and the dust boot. Lightly tap on the chisel to separate the dust boot from the ball joint.
To install:
8. Position the dust boot over the new lower ball joint and press it down to secure the dust boot to the ball joint, using a suitable tool.
9. Install the lower ball joint into the lower control arm and install 2 retaining nuts and 1 retaining bolt. Apply a suitable thread locking compound to the threads and tighten the fasteners to 69–86 ft. lbs. (93–117 Nm).
10. Apply a suitable thread locking compound to the ball joint pinch bolt threads. Install the lower ball joint stud into the steering knuckle. Install the pinch bolt and nut. Tighten the nut to 32–43 ft. lbs. (43–59 Nm).
11. Install the wheel and tire assembly. Torque the lug nuts to 65–87 ft. lbs. (88–118 Nm).
12. Lower the vehicle.
13. If required, check the front end alignment.
14. Road test the vehicle and check for proper operation.

Sway Bar

REMOVAL AND INSTALLATION

1. Disconnect the negative battery cable.

2. Support the engine and transaxle assembly with Engine Support D88L-6000-A, or equivalent.

3. Raise and safely support the vehicle.

4. Remove the front wheel and tire assemblies.

5. Remove 4 retaining nuts securing the rack and pinion (steering gear) mounting brackets and position the rack and pinion slightly forward.

6. Remove the sway bar (stabilizer bar) nuts, washers, bushings, sleeves and bolts from the lower control arm.

7. Remove the rear crossmember retaining nuts from the rear transaxle mount and the vehicle frame.

8. Loosen the front crossmember retaining bolts and nuts from the front transaxle mount and the vehicle frame. Lower the rear end of the crossmember.

9. Remove the nuts and bolts securing the sub-frame to the vehicles frame. Lower the sub-frame enough for removal of the sway bar.

NOTE: The engine and transaxle mounts will support the sub-frame from falling when it is unbolted from the vehicle frame.

10. Unbolt the sway bar from the sub-frame and remove.

To install:

11. Position the sway bar to the sub-frame.

12. Secure the sway bar to the sub-frame with the retaining bolts. Tighten the bolts to 32–43 ft. lbs. (43–59 Nm).

13. Install the sub-frame to the vehicle frame with the bolts and nuts. Tighten the bolts and nuts to 69–93 ft. lbs. (93–127 Nm).

14. Position the crossmember to the vehicle frame and the transaxle mounts. Tighten the retaining nuts and/or bolts to insulator mounts to 27–38 ft. lbs. (37–52 Nm) and the crossmember to frame retaining nuts and/or bolts to 47–66 ft. lbs. (64–89 Nm).

15. Install the sway bar bolts, sleeves, bushings, washers and nuts. Tighten the sway bar bolts so 0.67–0.75 inches (17–19mm) of thread is exposed at the end of the bolt.

16. Position the rack and pinion (steering gear) and secure it with the brackets and nuts. Tighten the nuts to 28–38 ft. lbs. (37–52 Nm).

17. Install the front wheel and tire assemblies. Tighten the lug nuts to 65–87 ft. lbs. (88–118 Nm).

18. Lower the vehicle.

19. Remove the engine support.

20. Connect the negative battery cable.

NOTE: Whenever the vehicles sub-frame is removed or lowered, the wheel alignment should be checked.

21. Check the wheel alignment.

22. Road test the vehicle and check for proper operation.

Front Wheel Bearings

ADJUSTMENT

The bearings on the front and rear wheels are a one piece cartridge design and cannot be adjusted. Wheel bearing play can be checked with a dial indicator. If wheel bearing play exceeds 0.002 inch (0.05mm) check the wheel hub retainer nut for proper torque. If the torque is correct, replacement of the wheel bearing is required.

REMOVAL AND INSTALLATION

1. Raise and safely support the vehicle.

2. Remove the front wheel and tire assembly.

3. Remove the brake caliper and support it out of the way with mechanics wire. Do not disconnect the brake hose or allow the caliper to hang by the brake hose.

4. Remove the disc brake rotor from the hub.

5. Carefully raise the staked portion of the hub retainer nut using a small cape chisel. Remove and discard the nut.

6. Remove the cotter pin and castellated nut from the tie rod end. Separate the tie rod end from the steering knuckle, using Separator Tool T85M-3395-A, or equivalent.

7. Remove the strut-to-steering knuckle retaining nuts and bolts and separate the strut assembly from the steering knuckle.

8. Remove the nut and bolt securing the lower ball joint to the steering knuckle.

9. Separate the lower ball joint from the steering knuckle by prying down on the lower control arm with a prybar or similar tool.

10. Remove the hub/knuckle assembly from the halfshaft.

11. Place the hub/knuckle assembly on a suitable workbench and remove the oil seal from the rear of the steering knuckle.

12. Position the hub/knuckle assembly on a hydraulic press and press the hub out of the steering knuckle using a suitable removal tool.

—— **CAUTION** ——
Always wear safety glasses whenever working with a grinder, chisel or a hydraulic press.

13. If the bearing inner race remains on the hub, use a grinder to grind a section of the bearing inner race until only 0.020 inch (0.5mm) remains. Remove the inner race with a suitable chisel.

14. Remove the retainer ring from the steering knuckle.

15. Place the steering knuckle on a hydraulic press and press the wheel bearing out of the steering knuckle, using an appropriate bearing removal tool.

NOTE: The wheel bearings are of a cartridge design and are permanently lubricated.

16. If the dust shield is damaged and must be removed, scribe an alignment mark between the dust shield and the steering knuckle. Using a suitable chisel, remove the dust shield.

NOTE: If the dust shield is removed, it must be replaced.

To install:

17. Scribe a mark on the new dust shield in the same position as on the old dust shield. Align the mark on the steering knuckle to the mark on the dust shield and using a suitable tool, press the dust shield onto the steering knuckle.

18. Apply a suitable threadlocking compound to the wheel bearing outer race and press the wheel bearing assembly into the steering knuckle. Make sure the press tool contacts only the outer bearing race or the bearing will be damaged.

19. Install the retainer ring.

20. Place the wheel hub to the steering knuckle and press the hub into the bearing. Make sure the inner bearing race is supported on the press table or the bearing will be destroyed.

21. Using an appropriate seal installer, install a new oil seal onto the inboard side of the steering knuckle. Make sure the oil seal mounts flush with the steering knuckle.

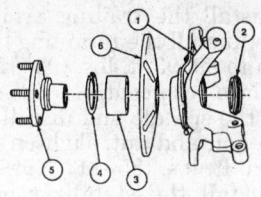

1. Front Wheel Knuckle
2. Inner Wheel Bearing Oil Seal
3. Front Wheel Bearing
4. Retainer Ring
5. Wheel Hub, Front
6. Front Disc Brake Rotor Shield

311991

Steering knuckle and hub assembly

22. Install the hub/steering knuckle assembly onto the ball joint and at the same time, align the splines of the halfshaft to the hub/knuckle assembly and push the assembly onto the halfshaft as far as possible.

23. Apply a suitable threadlocking compound to the nut and bolt securing the ball joint to the steering knuckle and install. Tighten to 32–43 ft. lbs. (43–59 Nm).

24. Install the tie rod end to the steering knuckle and install the castellated nut. Torque the nut to 25–33 ft. lbs. (34–46 Nm) and install a new cotter pin.

25. Install the steering knuckle to the strut assembly. Tighten the nuts and bolts to 69–93 ft. lbs. (93–127 Nm).

26. Install a new hub retainer nut securing the halfshaft to the front hub. Tighten the hub retainer nut to 174–235 ft. lbs. (235–319 Nm). Stake the locknut to prevent it from loosening.

27. Install the front disc brake rotor and caliper. Torque the caliper bolts to 29–36 ft. lbs. (39–49 Nm).

28. Install the wheel and tire assembly. Torque the lug nuts to 65–87 ft. lbs. (88–118 Nm).

29. Lower the vehicle.

30. Pump the brake pedal several times to position the brake pads.

31. Road test the vehicle and check for proper operation.

REAR SUSPENSION

Strut

REMOVAL AND INSTALLATION

1. Disconnect the negative battery cable.

2. Raise and safely support the vehicle.

3. Remove the wheel and tire assembly.

4. Remove the clip securing the brake hose to the rear strut assembly.

5. Remove the nuts and bolts securing the rear strut assembly to the wheel spindle.

6. If equipped with anti-lock brakes, remove the ABS harness from the clip on the strut assembly.

7. Partially lower the vehicle.

8. On hatchback and wagon models, remove the quarter trim panel.

9. Remove 2 upper strut insulator retaining nuts and remove the rear strut assembly from the vehicle.

10. Position the strut assembly in a vise and secure the assembly at the upper insulator bracket.

11. Remove the cap and loosen the piston rod nut 1 turn. Do not remove the piston rod nut at this time.

CAUTION

Attempting to remove the spring from the strut without first compressing the spring with a tool designed for that purpose could cause bodily injury.

12. Install an appropriate coil spring compressor and compress the coil spring.

13. Remove the piston rod nut, washer, retainer, anti-rattle plate and insulator.

14. Remove the coil spring.

15. Remove the stopper seat and dust boot from the strut piston.

To install:

16. Position the strut assembly in a vise and secure.

17. Install the dust boot and the stopper seat onto the strut piston rod.

18. Compress the coil spring and install on the strut assembly.

19. Install the upper strut insulator, then align the upper strut insulator mounting studs and the lower bracket of the strut assembly.

20. Install the retainer, washer and piston rod nut. Tighten the nut to 41–50 ft. lbs. (55–68 Nm).

21. Make sure the spring is properly aligned and carefully release the spring into the seats of the strut.

22. Remove the spring compressor from the coil spring and install the cap.

23. Position the strut assembly into the vehicle wheel housing.

24. Install 2 upper strut insulator retaining nuts and tighten to 22–27 ft. lbs. (29–40 Nm).

25. On hatchback and wagon models, install the lower trim panel.

26. Install the nuts and bolts securing the strut assembly to the rear wheel spindle assembly. Tighten the lower strut bolts and nuts to 69–93 ft. lbs. (93–127 Nm).

27. Install the clip securing the flexible brake hose to the rear strut assembly.

28. If equipped with anti-lock brakes, install the ABS harness to the clip.

29. Install the wheel and tire assembly. Torque the lug nuts to 65–87 ft. lbs. (88–118 Nm).

30. Lower the vehicle.

31. Connect the negative battery cable.

32. Check the rear wheel alignment.

33. Road test the vehicle and check for proper operation.

Sway Bar

REMOVAL AND INSTALLATION

1. Raise and safely support the vehicle.

2. Remove the sway bar (stabilizer bar) link nuts, retainers, washers, bushings, sleeves and bolts.

3. Remove 2 sway bar bracket retaining bolts and brackets.

4. Remove the sway bar from the vehicle.

5. Inspect the sway bar bracket insulators and link bushings Replace as needed.

To install:

6. If removed, install the insulators on the sway bar and align the insulators to the positions painted on the sway bar.

7. Place the sway bar to the floor crossmember and secure it in place with 2 sway bar brackets and bolts. Tighten the bolts to 32–43 ft. lbs. (43–59 Nm).

8. Install the sway bar link bolts, retainers, bushings, sleeves and nuts. Tighten the sway bar link nuts so 0.64–0.72 inch (16.2–17.0mm) of thread is exposed at the end of the bolt.

9. Install the wheel and tire assembly. Torque the lug nuts to 65–87 ft. lbs. (88–118 Nm).

10. Lower the vehicle.

11. Road test the vehicle and check for proper operation.

Wheel Bearings

ADJUSTMENT

The bearings on the front and rear wheels are alone piece cartridge design and cannot be adjusted. Wheel bearing play can be checked with a dial indicator. If wheel bearing play exceeds 0.002 inch (0.05mm) check the wheel hub retainer nut for proper torque. If the torque is correct, replacement of the wheel bearing is required.

REMOVAL AND INSTALLATION

NOTE: The wheel bearings are a cartridge design and are not serviceable. If bearing replacement is required, the bearings and hub must be replaced as an assembly. Do not continue with this procedure without having available a new wheel hub retainer nut. Once removed, the nut looses its torque holding ability or retention capability and must not be reused.

1. Raise and safely support the vehicle.

2. Remove the wheel and tire assembly.

3. Remove the hub grease cap.

4. Remove the brake drum or disc brake caliper and rotor, as necessary.

5. Unstake and remove the wheel hub retainer nut securing the hub to the spindle and remove the hub and bearing assembly. Discard the hub retainer nut.

6. If equipped with disc brakes, remove 4 disc brake shield retaining bolts and the shield.

7. If equipped with drum brakes, remove the backing plate.

8. Remove 2 strut-to-spindle retaining bolts and nuts.

9. Remove the trailing arm bolt securing the trailing arm to the spindle.

10. Remove the stabilizer bar link nuts, retainers, bushings, sleeves and bolts.

11. Remove the control arm nut and bolt securing both control arms to the wheel spindle and remove the spindle from the vehicle.

12. Inspect all components. If the wheel spindle is damaged, replace it. Wheel bearings are sealed and must be replaced if damaged with the wheel hub.

To install:

13. Place the wheel spindle in position to the strut and install 2 retaining bolts and nuts. Tighten to 69–93 ft. lbs. (93–127 Nm).

14. Install the trailing arm to the spindle. Install the retaining bolt and tighten to 69–93 ft. lbs. (93–127 Nm).

15. Place both control arms in position to the spindle and install the retaining bolt and nut. Tighten the nut to 63–86 ft. lbs. (85–117 Nm).

16. Install the stabilizer bar link nuts, retainers, bushings, sleeves and bolts.

17. If equipped with drum brakes, install the brake backing plate.

18. If equipped with disc brakes, install the brake shield and 4 disc brake shield retaining bolts. Tighten securely.

19. Install the wheel hub and bearing assembly to the wheel spindle.

20. Install a new wheel hub retainer nut and tighten to 130–174 ft. lbs. (177–235 Nm).

21. Stake the wheel hub retainer nut using a cape or round end chisel. Do not use a sharp chisel to stake the hub nut.

22. Install the brake drum or the disc brake rotor and caliper, as equipped.

23. Install the hub grease cap.

24. Install the wheel and tire assembly. Tighten the lug nuts to 65–87 ft. lbs. (88–118 Nm).

25. Lower the vehicle.

26. Pump the brake pedal several times to position the brake lining before attempting to move the vehicle.

27. Road test the vehicle and check for proper operation.

FORD MOTOR CO.

Rear Wheel Drive

FORD-Mustang • Thunderbird **LINCOLN**-Mark VIII
MERCURY-Cougar

15

FIRING ORDERS

NOTE: To avoid confusion, always replace spark plug wires one at a time.

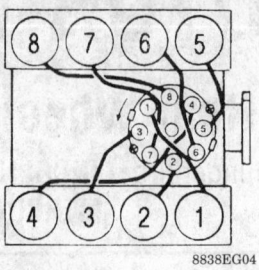
8838EG04

5.0L Engine
Engine Firing Order:
1–3–7–2–6–5–4–8
Distributor Rotation:
Counterclockwise

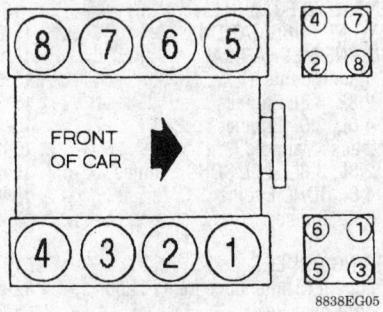
8838EG05

4.6L Engine
Engine Firing Order: 1–3–7–2–6–5–4–8
Distributorless Ignition System

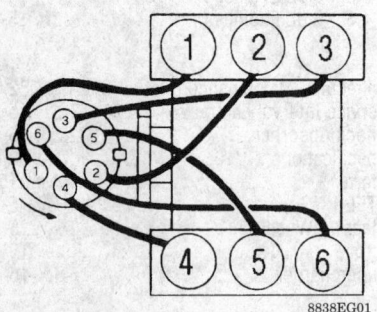
8838EG01

3.8L Engine (except SC)
Engine Firing Order: 1–4–2–5–3–6
Distributor Rotation: Counterclockwise

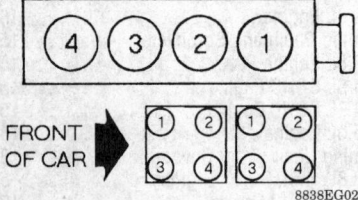
8838EG02

2.3L Engine
Engine Firing Order: 1–3–4–2
Distributorless Ignition System

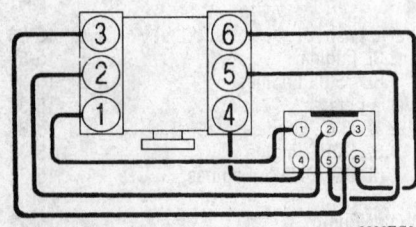
8838EG03

3.8L SC Engine
Engine Firing Order: 1–4–2–5–3–6
Distributorless Ignition System

ENGINE ELECTRICAL

NOTE: Disconnecting the negative battery cable on some vehicles may interfere with the functions of the on board computer systems and may require the computer to undergo a relearning process, once the negative battery cable is reconnected.

Distributor

REMOVAL

1. Disconnect the negative battery cable.
2. Mark the position of the No. 1 cylinder wire tower on the distributor base.

NOTE: This reference is necessary in case the engine is disturbed while the distributor is removed.

3. Remove the distributor cap and position the cap and ignition wires to the side. Disconnect the wiring harness plug from the distributor connector.
4. Scribe a mark on the distributor body to indicate the position of the rotor tip. Scribe a mark on the distributor housing and engine block or timing cover to indicate the position of the distributor in the engine.
5. Remove the hold-down bolt and clamp located at the base of the distributor. Remove the distributor from the engine. Note the direction the rotor tip points if it moves from the No. 1 position when the drive gear disengages. For reinstallation purposes, the rotor should be at this point to insure proper gear mesh and timing.
6. Cover the distributor opening in the engine to prevent the entry of dirt or foreign material.
7. Avoid turning the crankshaft, if possible, while the distributor is removed. If the engine is disturbed, the No. 1 cylinder piston will have to be brought to TDC on the compression stroke before the distributor is installed.

INSTALLATION

NOTE: Before installing, visually inspect the distributor. The drive gear should be free of nicks, cracks and excessive wear. The distributor driveshaft should move freely, without binding. The O-ring should fit tightly and be free of cuts.

Timing Not Disturbed

1. Position the distributor in the engine, aligning the rotor and distributor housing with the marks that were made during removal. If the distributor does not fully seat in the engine block or timing cover, it may be because the distributor is not engaging properly with the oil pump intermediate shaft. Remove the distributor and, using a suitable tool, turn the intermediate shaft until the distributor will seat properly.
2. Install the hold-down clamp and bolt. Snug the mounting bolt so the distributor can be turned for ignition timing purposes.
3. Install the distributor cap and connect the distributor to the wiring harness.
4. Connect the negative battery cable. Check and, if necessary, adjust the ignition timing.

5. After the timing has been set, tighten the distributor hold-down clamp bolt to:

3.8L engine 15–22 ft. lbs. (20–30 Nm)

5.0L engine 18–26 ft. lbs. (24–35 Nm)

6. Recheck the ignition timing after tightening the bolt.

Timing Disturbed

1. Disconnect the No. 1 cylinder spark plug wire and remove the No. 1 cylinder spark plug.

2. Place a finger over the spark plug hole and crank the engine slowly until compression is felt.

3. Align the TDC mark on the crankshaft pulley with the pointer on the timing cover. This places the piston in No. 1 cylinder at TDC on the compression stroke.

4. Turn the distributor shaft until the rotor points to the distributor cap No. 1 spark plug tower, as marked during the removal procedure.

5. Install the distributor in the engine, aligning the rotor and distributor housing with the marks that were made during removal. If the distributor does not fully seat in the engine block or timing cover, it may be because the distributor is not engaging properly with the oil pump intermediate shaft. Remove the distributor and, using a suitable tool, turn the intermediate shaft until the distributor will seat properly.

6. Install the hold-down clamp and bolt. Snug the mounting bolt so the distributor can be turned for ignition timing purposes.

7. Install the No. 1 cylinder spark plug and connect the spark plug wire. Install the distributor cap and connect the distributor to the wiring harness.

8. Connect the negative battery cable and set the ignition timing.

9. After the timing has been set, tighten the distributor hold-down clamp bolt to:

3.8L engine 15–22 ft. lbs. (20–30 Nm)

5.0L engine 18–26 ft. lbs. (24–35 Nm)

10. Recheck the ignition timing after tightening the bolt.

Distributorless Ignition System (DIS)

The 3.8L SC, 2.3L and 4.6L engines are equipped with Distributorless Ignition Systems. The DIS consists of the following components: crankshaft sensor, ignition module, ignition coil pack, the spark angle portion of the ECU and the related wiring. The system used on the 3.8L SC engine includes a camshaft sensor.

REMOVAL AND INSTALLATION

Crankshaft Sensor

2.3L Engine

1. Disconnect the negative battery cable.

2. Disconnect the sensor electrical connectors from the engine harness.

3. Remove the large electrical connector from the crankshaft position sensor assembly.

4. Remove the accessory drive belt. Remove the 4 screws retaining the crankshaft pulley hub assembly and remove the pulley.

5. Remove the timing belt outer cover. Rotate the crankshaft by hand till the keyway is at the 10 o'clock position.

6. Remove the 2 sensor retaining bolts and the plastic wire harness retainer that secures the sensor harness to it's mounting bracket. Remove the sensor, sliding the electrical wires out from behind the inner timing belt cover.

To install:

7. Remove the large electrical connector from the new crankshaft timing sensor.

8. Slide the electrical wires behind the inner timing belt cover and position the sensor. Hold the sensor loosely in place with the retaining bolts, but do not tighten at this time.

9. Install the large electrical connector onto the sensor.

NOTE: Make sure the 4 wires to the large electrical connector are installed in the proper locations as indicated. The sensor will not function properly if the wires are installed in the wrong locations.

10. Reconnect both sensor electrical connectors to the engine harness.

11. Rotate the crankshaft so the outer vane on the crankshaft pulley hub assembly engages both sides of the crankshaft Hall effect sensor positioner T89P-6316-A or equivalent, and tighten the sensor retaining bolts.

12. Rotate the crankshaft so the vane on the crankshaft pulley hub is no longer engaged in the positioning tool. Remove the tool.

13. Install the new plastic wire harness retainer to secure the sensor

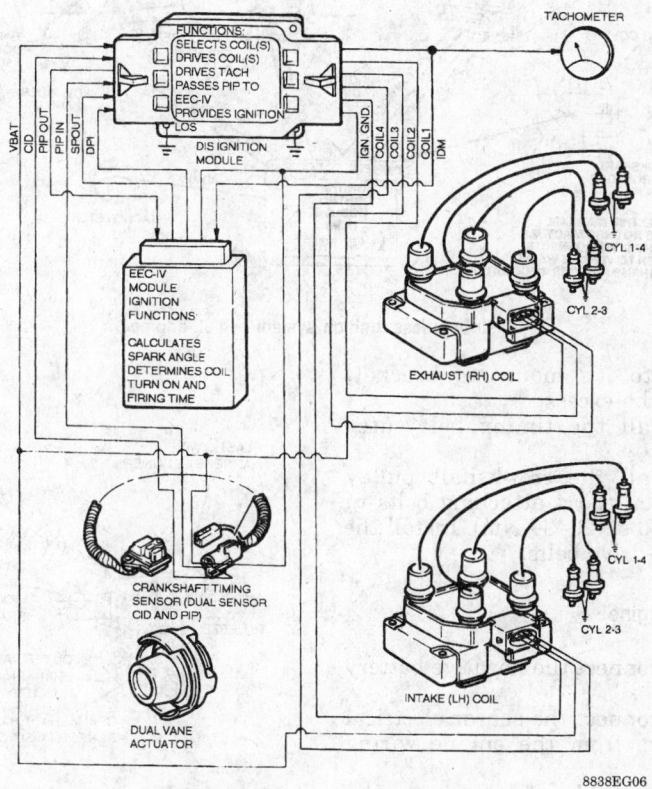

Distributorless ignition system — 2.3L engine

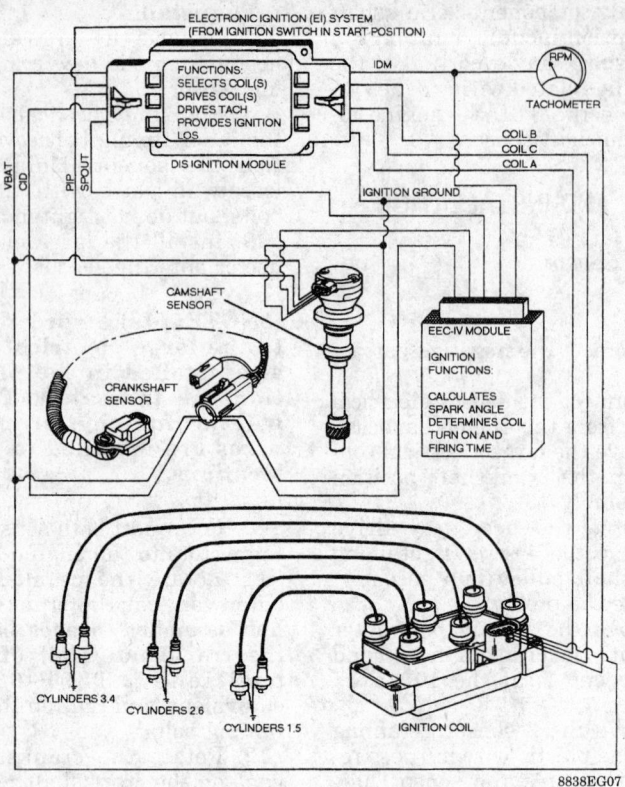

Distributorless ignition system — 3.8L SC engine

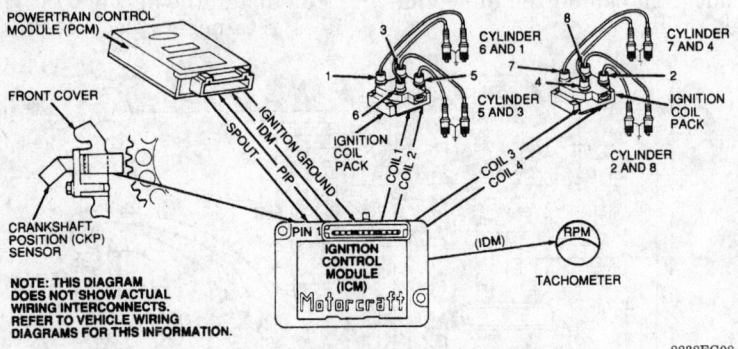

Distributorless ignition system — 4.6L engine

harness to it's mounting bracket. Trim off the excess.

14. Install the timing belt outer cover.

15. Install the crankshaft pulley and tighten the 4 attaching bolts to 15–22 ft. lbs. (20–30 Nm). Install the accessory drive belts.

3.8L SC Engine

1. Disconnect the negative battery cable.

2. Disconnect the sensor electrical connectors from the engine wiring harness.

3. Raise and safely support the vehicle.

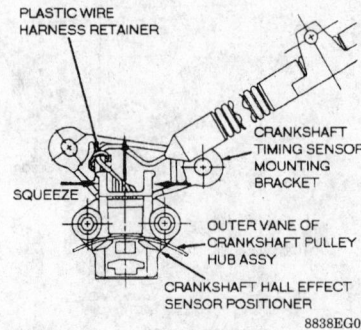

Crankshaft Hall effect sensor positioning — 2.3L engine

4. Remove the upper and lower damper shield assemblies.

5. Rotate the crankshaft by hand to position the metal vane of the shutter, attached to the rear of the damper, outside of the sensor air gap.

6. Remove the crankshaft sensor retaining screws and remove the sensor.

To install:

7. Position the crankshaft sensor assembly on the bracket.

8. Install 2 sensor retaining screws but do not tighten at this time.

9. Install crankshaft sensor gauge T89P-6316-AH or equivalent, to the outside surface of 1 vane of the shutter.

NOTE: The gauge is magnetic and will conform to the shape of the vane.

10. Rotate the crankshaft by hand to align the shutter vane with the gauge into the sensor air gap.

11. Push the sensor housing inward to contact the gauge and tighten the screws to 22–31 inch lbs. (2.5–3.5 Nm).

—— WARNING ——
This is a critical torque. Overtightening can cause damage to the timing sensor.

12. Rotate the crankshaft by hand to position the shutter vane with the gauge outside of the air gap. Remove the magnetic gauge.

13. Install the upper and lower damper shields and tighten the nuts to 9–11 ft. lbs. (12–15 Nm) and the bolts to 6–8.5 ft. lbs. (8–11.5 Nm).

14. Lower the vehicle.

15. Route the sensor wiring harness and connect both electrical connectors.

16. Connect the negative battery cable.

4.6L Engine

1. Disconnect the negative battery cable.

2. Remove the accessory drive belt.

3. Raise and safely support the vehicle.

4. Disconnect the crankshaft position sensor and air conditioning compressor electrical connectors from the harness.

5. Properly discharge the air conditioning system and remove the compressor.

6. Remove the crankshaft sensor retaining screws and remove the sensor.

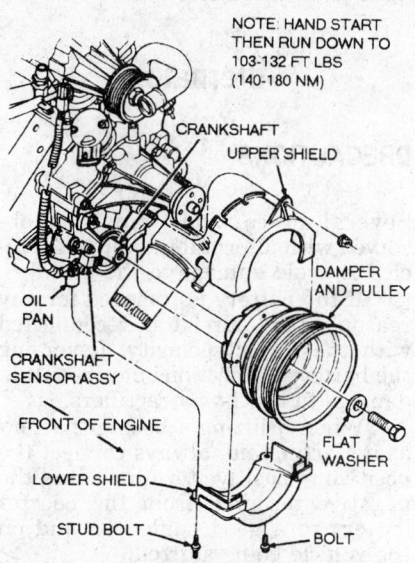

NOTE: HAND START THEN RUN DOWN TO 103-132 FT LBS (140-180 NM)

CRANKSHAFT UPPER SHIELD

DAMPER AND PULLEY

OIL PAN

CRANKSHAFT SENSOR ASSY

FRONT OF ENGINE

FLAT WASHER

LOWER SHIELD

STUD BOLT

BOLT

NOTE: LUBRICATE SEAL SURFACE (O.D.) AND I.D. OF DAMPER HUB WITH ESE-M2C39-F OIL OR ESE-M1C104-A GREASE PRIOR TO ASSEMBLY. SEAL SURFACE MUST BE FREE OF DIRT OR GRIT.

8838EG10

Crankshaft position sensor removal and installation — 3.8L engine

To install:

7. Make sure the sensor mounting surface is clean and O-ring is in the proper location. Position the crankshaft sensor assembly on the bracket.

8. Install 2 sensor retaining screws and tighten to 71–106 inch lbs. (8–12 Nm).

WARNING

This is a critical torque. Overtightening can cause damage to the timing sensor.

9. Install the A/C compressor. Evacuate and recharge the A/C system.

10. Connect the crankshaft position sensor and air conditioning compressor electrical connectors to the harness.

11. Lower the vehicle.

12. Install the accessory drive belt.

13. Connect the negative battery cable.

Camshaft Sensor

3.8L SC Engine

1. Disconnect the negative battery cable.

2. Disconnect the camshaft sensor electrical connector.

3. Remove the camshaft sensor retaining screws and remove the sensor.

4. Installation is the reverse of the removal procedure. Tighten the retaining screws to 22–31 inch lbs. (2.5–3.5 Nm).

Synchronizer Assembly

3.8L SC Engine

The synchronizer assembly mounts in place of the distributor. It provides the mechanical link between the camshaft sensor and the camshaft.

NOTE: Before starting this procedure, set the No. 1 cylinder to 26 degrees after TDC on the compression stroke. Then note the position of the camshaft sensor electrical connector. The installation procedure requires that the connector be located in the same position.

1. Disconnect the negative battery cable.

2. Remove the camshaft sensor assembly.

3. Remove the synchronizer clamp, bolt and washer.

4. Remove the synchronizer from the front engine cover, by pulling it out. The oil pump intermediate shaft will come out with the assembly.

To install:

WARNING

If the replacement synchronizer does not contain a plastic locator cover tool, a special service tool such as synchro positioner tool T89P-12200-A or equivalent, must be used to install the synchronizer. Failure to use this special tool will cause the synchronizer timing to be out of adjustment, and could lead to engine damage.

5. If the plastic locator cover tool is not attached to the replacement synchronizer, attach synchro positioner tool T89P-12200-A or equivalent, as follows:

a. Engage the synchronizer vane into the tool's radial slot.

b. Rotate the tool on the synchronizer base until the tool boss engages the base notch. The cover tool should be square and in contact with the entire top surface of the synchronizer base.

6. Install the intermediate oil pump shaft onto the replacement synchronizer.

7. Position the synchronizer so gear engagement occurs when the arrow on the locator tool is pointed 30

degrees counterclockwise from the front face of the engine block. This will locate the camshaft sensor electrical connector to the position it was in before removal.

8. Install the synchronizer base clamp and tighten the mounting bolt to 15–22 ft. lbs. (20–30 Nm).

9. Remove the positioner tool and install the camshaft sensor. Connect the sensor electrical lead and connect the negative battery cable.

WARNING

If the camshaft sensor electrical connector is not positioned properly — contacting the A/C bracket or forward of the supercharger drive belt, do not reposition the connector by rotating the synchronizer base. This will result in the ignition and fuel systems being out of time with the engine, possibly causing engine damage. If the sensor electrical connector is not properly positioned, remove the synchronizer and repeat the installation procedure.

Ignition Module

1. Disconnect the negative battery cable.

2. Disconnect the electrical connectors at the module.

3. Remove the module retaining screws and remove the module.

To install:

4. Apply an even coating of silicone dielectric compound WA-IO, D7AZ-19A331-A or equivalent to the mounting surface of the module.

5. Install the module and the retaining screws. Tighten the screws to 22–31 inch lbs. (2.5–3.5 Nm).

6. Connect the electrical connectors to the module and connect the negative battery cable.

Ignition Coil Pack

2.3L Engine

1. Disconnect the negative battery cable.

2. Squeeze the locking tabs of the coil wire retainer by hand and remove the spark plug wires with a twisting and pulling motion. Do not pull on the wire.

3. Disconnect the engine harness electrical connector from the ignition coil assembly.

4. Remove the 4 retaining screws and remove the ignition coil.

5. Installation is the reverse of the removal procedure.

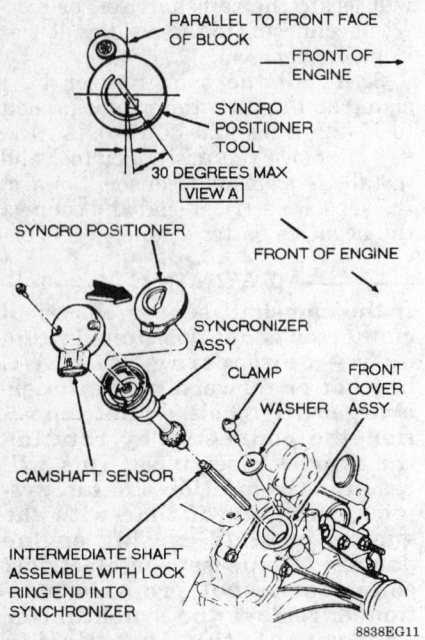

Synchronizer assembly installation — 3.8L SC engine

3.8L SC and 4.6L Engines

1. Disconnect the negative battery cable.

2. Disconnect the electrical harness connector from the coil pack. On 4.6L engine, disconnect the capacitor.

3. Remove the spark plug wires by squeezing the locking tabs to release the coil boot retainers.

4. Remove the coil pack retaining screws and remove the coil pack. On 4.6L engines, save the capacitor for installation with the new coil pack.

5. Installation is the reverse of the removal procedure. Tighten the screws to 40–62 inch lbs. (4.5–7.0 Nm).

Ignition Timing

ADJUSTMENT

NOTE: Always refer to the Vehicle Emission Information Label to verify the timing adjustment procedure.

Distributorless Ignition Systems

Base timing for distributorless engines is set from the factory at 10 degrees BTDC and is not adjustable.

Distributor Ignition System

1. Locate the timing marks and pointer on the crankshaft pulley and the timing cover. Clean the marks so they will be visible with a timing light. Apply chalk or bright-colored paint, if necessary.

2. Place the transaxle in **P** or **N**. The air conditioning and heater controls should be in the **OFF** position.

3. Connect a suitable tachometer and inductive timing light according to the manufacturer's instructions.

NOTE: The tachometer can be connected to the ignition coil without removing the coil connector. Insert an alligator clip into the back of the connector, onto the dark green/yellow dotted wire. Do not let the clip accidently ground to a metal surface as it may permanently damage the coil.

4. Disconnect the single wire inline SPOUT connector or remove the shorting bar from the double wire SPOUT connector.

5. Start the engine and allow it to warm up to operating temperature.

NOTE: To set timing correctly, a remote starter should not be used. Use the ignition key only to start the vehicle. Disconnecting the start wire at the starter relay will cause the TFI module to revert to start mode timing after the vehicle is started. Reconnecting the start wire after the vehicle is running will not correct the timing.

6. With the engine at the timing rpm specified, check the initial timing by aiming the timing light at the timing marks and pointer. Refer to the underhood Vehicle Emission Information Label for specifications.

7. If the marks align, shut **OFF** the engine and proceed to Step 8. If the marks do not align, shut **OFF** the engine and loosen the distributor hold-down clamp bolt. Start the engine, aim the timing light and turn the distributor until the timing marks align. Shut **OFF** the engine and tighten the distributor hold-down clamp bolt. Recheck the timing after the bolt has been tightened.

8. Reconnect the single wire inline SPOUT connector or reinstall the shorting bar on the double wire SPOUT connector. Check the timing advance to verify the distributor is advancing beyond the initial setting.

9. Remove the inductive timing light and tachometer.

Alternator

PRECAUTIONS

Several precautions must be observed with alternator equipped vehicles to avoid damage to the unit.

• If the battery is removed for any reason, make sure it is reconnected with the correct polarity. Reversing the battery connections may result in damage to the 1-way rectifiers.

• When utilizing a booster battery as a starting aid, always connect the positive to positive terminals and the negative terminal from the booster battery to a good engine ground on the vehicle being started.

• Never use a fast charger as a booster to start vehicles.

• Disconnect the battery cables when charging the battery with a fast charger.

• Never attempt to polarize the alternator.

• Do not use test lights of more than 12 volts when checking diode continuity.

• Do not short across or ground any of the alternator terminals.

• The polarity of the battery, alternator and regulator must be matched and considered before making any electrical connections within the system.

• Never separate the alternator on an open circuit. Make sure all connections within the circuit are clean and tight.

• Disconnect the battery ground terminal when performing any service on electrical components.

• Disconnect the battery if arc welding is to be done on the vehicle.

REMOVAL AND INSTALLATION

1. Disconnect the negative battery cable.

2. Tag and disconnect the wiring connectors from the rear of the alternator.

3. Loosen the alternator pivot bolt and remove the adjusting bolt. Disengage the drive belt from the alternator pulley.

4. Remove the alternator pivot bolt and the alternator.

5. Installation is the reverse of the removal procedure.

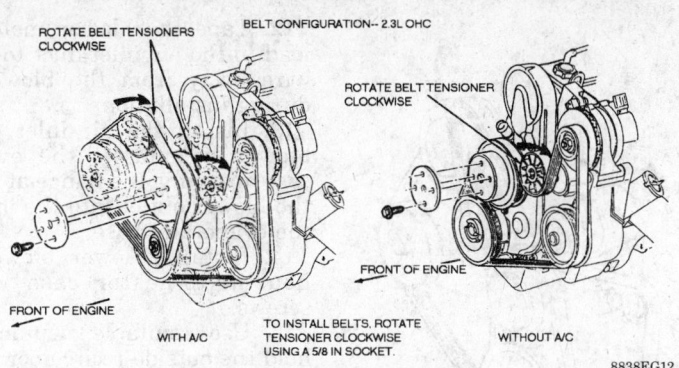

Belt configuration — 2.3L engine

8838EG12

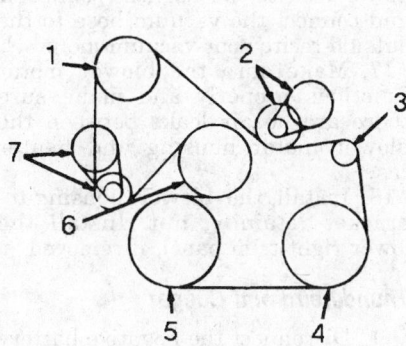

1. A/C compressor
2. Idler Assembly
3. Generator
4. Power steering pump
5. Crankshaft
6. Water pump
7. Tensioner assembly

8838EG13

Belt configuration — 3.8L engine

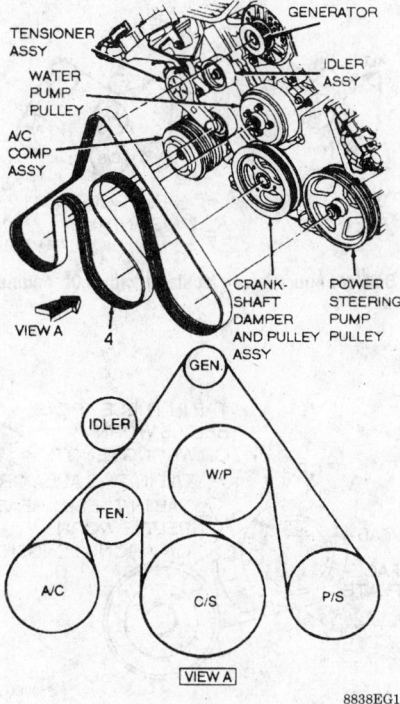

Belt configuration — 4.6L (DOHC) engine

8838EG15

Drive Belt

REMOVAL AND INSTALLATION

Alternator

All vehicles are equipped with an automatic belt tensioner. No adjustment is necessary or possible. The belt tensioner is equipped with a belt wear indicator; when 1 percent belt stretch is indicated, the drive belt must be replaced. If the wear indicator is difficult to see on the 3.8L or 5.0L engines, locate the tab on the tensioner face plate. The tab should be approximately between the stops.

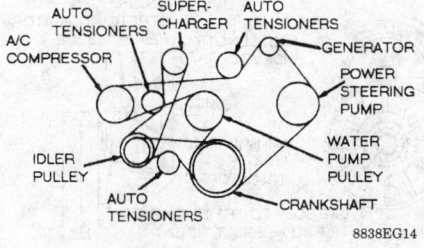

Belt configuration — 3.8L SC engine

8838EG14

Starter

REMOVAL AND INSTALLATION

1. Disconnect the negative battery cable.
2. Raise the vehicle and support it safely.
3. Disconnect the starter cable from the starter. If equipped with starter mounted solenoid, disconnect the push-on connector from the solenoid.

NOTE: To disconnect the hard-shell connector from the solenoid S terminal, grasp the plastic shell and pull off; do not pull on the wire. Pull straight off to prevent damage to the connector and S terminal.

4. Remove the starter bolts and the starter.
5. Position the starter to the engine and tighten the mounting bolts to 15–20 ft. lbs. (20–27 Nm).
6. Reconnect the starter cable and, if equipped, solenoid wire. Connect the negative battery cable.

CHASSIS ELECTRICAL

Blower Motor

REMOVAL AND INSTALLATION

Mustang

1. Disconnect the negative battery cable.
2. Squeeze the sides of the glove compartment together to disengage the retaining tabs. Let the glove compartment hang down in front of the instrument panel.
3. Disconnect the blower motor electrical connector and the vacuum hose from the outside-recirc door vacuum motor.
4. Remove the housing assembly-to-bracket case retaining nut. Close the glove compartment door and remove the lower screws from the blower motor housing.
5. Lift the blower motor housing and air inlet duct/recirc door assemblies away from the heater case. Removing the lower right trim panel will allow for easier removal.
6. Disconnect the cooling tube from the blower motor.

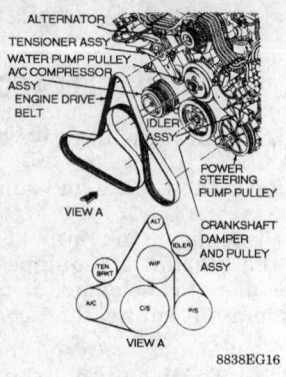

Belt configuration — 4.6L (SOHC) engine

8838EG16

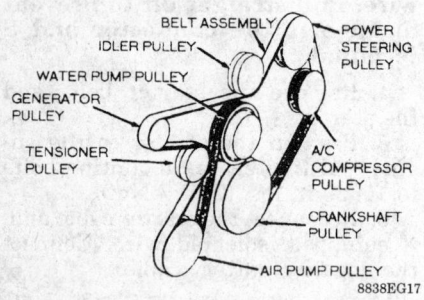

Belt configuration — Thunderbird and Cougar with 5.0L engine

8838EG17

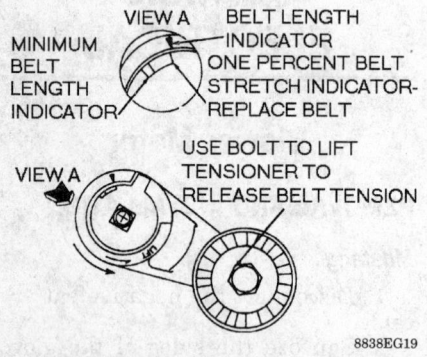

Belt tensioner — 3.8L and 5.0L engines

8838EG19

7. Remove the retaining screws and pull the blower motor and wheel from the housing.

8. Remove the pushnut from the blower motor shaft and remove the wheel.

To install:

9. Install gasket material, jumper wire harness, wheel and pushnut onto a new blower motor.

10. Install the blower motor into the housing and secure with the screws.

11. Connect the blower motor cooling tube.

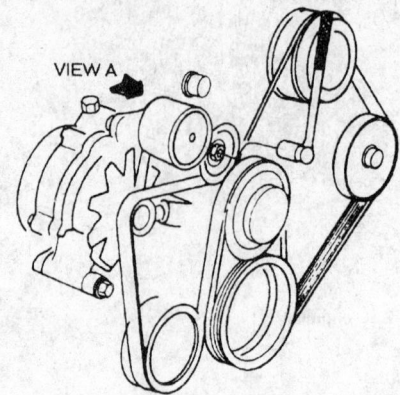

VIEW A

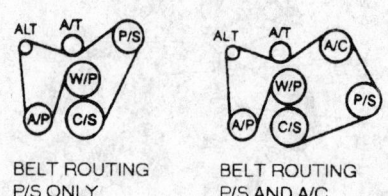

BELT ROUTING P/S ONLY

BELT ROUTING P/S AND A/C

8838EG18

Belt configuration — Mustang with 5.0L engine

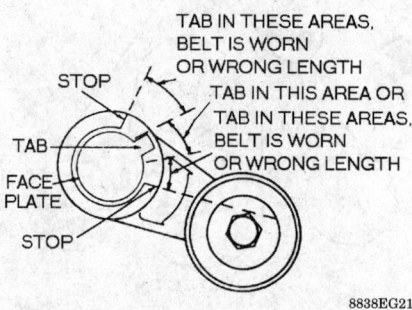

Belt replacement checking — 3.8L 4.6L and 5.0L engines

8838EG21

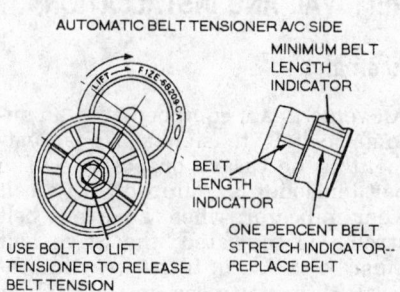

AUTOMATIC BELT TENSIONER A/C SIDE

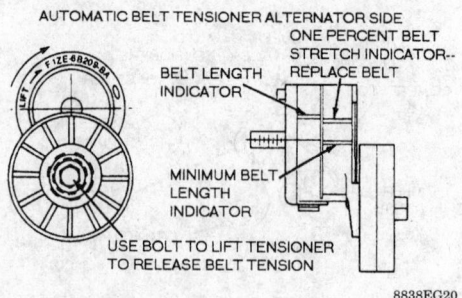

AUTOMATIC BELT TENSIONER ALTERNATOR SIDE

8838EG20

Belt wear indicator marks — 2.3L engine

12. Tape the blower motor power lead to the air inlet duct to keep the wire away from the blower outlet during installation.

13. Install the air inlet duct and blower housing to the evaporator case, inserting the flange at the top of the blower outlet into the opening in the evaporator case.

14. Install 2 lower blower motor housing-to-heater case retaining screws.

15. Use a suitable vacuum pump to hold the outside-recirc door open and rotate the blower wheel to make sure it rotates freely. If there is interference, remove the blower motor and wheel and correct the problem.

16. Connect the blower motor power lead to the harness connector and connect the vacuum hose to the outside-recirc door vacuum motor.

17. Make sure the blower motor functions properly and make sure there are no air leaks between the blower motor housing and heater case.

18. Install the blower housing-to-bracket retaining nut. Install the lower right trim panel, if removed.

Thunderbird and Cougar

1. Disconnect the negative battery cable.

2. Remove the glove compartment liner to gain access to the blower motor mounting screws.

3. Remove the 4 retaining screws and remove the blower motor and wheel assembly from the blower housing.

4. Remove the pushnut from the blower motor shaft and remove the blower wheel from the shaft.

5. Installation is the reverse of removal. Connect the negative battery cable.

Mark VIII

1. Disconnect the negative battery cable.

2. Lower the glove compartment door to gain access to the rear of the A/C evaporator housing.

3. Remove the 2 retaining screws and remove the muffler and automatic temperature control sensor hose and elbow.

4. Remove the screw and pull the A/C blower motor out of the housing. Pull the blower motor wheel retainer off of the shaft.

5. Installation is the reverse of the removal. Connect the negative battery cable.

Windshield Wiper Motor

REMOVAL AND INSTALLATION

Mustang

1. Disconnect the negative battery cable.

2. Remove the right hand wiper arm assembly as follows:

   a. Raise the wiper blade off the windshield.

   b. Move the slide latch away from the pivot shaft and slowly lower the arm onto the latch. This unlocks the arm from the pivot shaft and holds the blade off the glass.

   c. Pull the arm from the pivot shaft. The use of tools is unnecessary.

3. Remove the cowl top grille retaining screws and grille.

4. Disconnect the linkage drive arm from the motor crankpin after removing the clip.

5. Disconnect the electrical connector from the wiper motor, remove the 3 retaining bolts and remove the motor from the vehicle.

   To install:

6. Install the motor and tighten the bolts to 60–80 inch lbs. (7–9 Nm). Connect the electrical connector.

7. Connect the linkage drive arm to the motor crankpin and install the clip.

8. Install the cowl top grille and secure with the screws.

9. Make sure the motor is in the PARK position. Install the wiper arm so the blade is 2.3–3.5 in. (58.42–88.90mm) from the bottom windshield moulding. Install the arm as follows:

   a. Install the arm head over the pivot shaft.

   b. While applying downward pressure on the arm head, raise the other end of the arm enough to let the latch slide under the pivot shaft to the latched position, using

finger pressure only to slide the latch.

   c. Lower the blade. If the blade does not touch the windshield, the slide latch is not completely in place.

Thunderbird and Cougar

1. Disconnect the negative battery cable.

2. With the wipers in the PARK position, remove the arm and blade assemblies as follows:

   a. Raise the wiper blade off the windshield.

   b. Move the slide latch away from the pivot shaft and slowly lower the arm onto the latch. This unlocks the arm from the pivot shaft and holds the blade off the glass.

   c. Pull the arm from the pivot shaft. The use of tools is unnecessary.

3. Remove the left-hand cowl vent screen.

4. Remove the vacuum manifolds from the wiper module and disconnect the electrical connectors.

5. Remove the 5 screws and 1 nut from the wiper module and remove the wiper module.

6. Disconnect the linkage drive arm from the motor crankpin after removing the clip.

7. Remove the 3 wiper motor retaining screws and pull the motor from the opening.

   To install:

8. Install the motor and secure with the retaining screws.

9. Connect the linkage arm to the motor crankpin and install the clip.

10. Install the wiper module and secure with the screws and nut.

11. Connect the electrical connectors and install the vacuum manifolds. Install the left-hand cowl vent screen.

12. Make sure the motor is in the PARK position. Install the wiper arms as follows:

   a. Align the keyway on the pivot shaft with the wiper arm and install the arm head over the pivot shaft.

   b. While applying downward pressure on the arm head, raise the other end of the arm enough to let the latch slide under the pivot shaft to the latched position, using finger pressure only to slide the latch.

   c. Lower the blade. If the blade does not touch the windshield, the slide latch is not completely in place.

Mark VIII

1. Turn the wipers ON until they reach midway (straight up) on the windshield then turn the key **OFF**.

2. Disconnect the negative battery cable and remove the right and left wiper arms

3. Remove the cowl top to hood seal and the right and left cowl vent screens.

4. Remove the 4 screws and the cowl top extension.

5. Disconnect the electrical connector from the wiper motor, remove the 5 retaining bolts and 1 nut.

6. Lift the module to disengage the support bracket from the mounting stud, move the module about 2 inches toward the passenger side and remove the module.

7. Disconnect the linkage drive arm from the motor crankpin after removing the clip. Check the location of the support bracket on the module.

8. Remove the 3 screws and separate the motor from the module.

   To install:

9. Connect the motor to the module assembly and tighten the retaining bolts to 7.5–10 ft. lbs. (10–14 Nm).

10. Connect the linkage arm to the motor crankpin and install the clip.

11. Install the 5 retaining bolts and 1 nut and connect the electrical connector. Install the cowl top extension.

12. Install the left and right cowl vent screens and cowl top to hood seal.

NOTE: Before installing the wiper arms to the pivot shafts, cycle the motor to make sure the linkage is in the park position.

13. Connect the wiper arms and the negative battery cable.

Headlight Switch

REMOVAL AND INSTALLATION

Mustang

1. Disconnect the negative battery cable.

2. Disengage the 2 locking tabs on the left side of the switch, under the paddles, by pushing the tabs in with a small prybar and pulling on the paddles.

3. Using a small prybar, pry the right side of the switch out of the instrument panel.

4. Pull the switch out of the opening and disconnect the 2 connectors.

5. To install, assemble the connectors, insert the switch into the panel opening and push until the locking

tabs on both sides of the switch snap into place.

Thunderbird and Cougar

1. Disconnect the negative battery cable.
2. Remove the 2 cluster finish panel retaining screws.
3. Pull the headlight switch knob off.
4. Unsnap the cluster finish panel.
5. Disconnect the electrical connector to the headlight dimmer sensor assembly.
6. Through the opening in the instrument panel, depress the shaft release button on the switch and remove the shaft. The switch must be in the full **ON** position to release the shaft.
7. Remove the headlight switch retaining nut and pull the switch through the opening to disconnect the wiring connector.
8. Installation is the reverse of removal.

Mark VIII

1. Disconnect the negative battery cable.
2. Pull off the headlight switch knob.
3. Remove the lamp switch knob applique by pulling at the left end to unsnap it from the finish panel and twist out the bulb socket.
4. Remove the 2 screws retaining the left end of the finish panel to the instrument panel.
5. Pull up on the forward edge to unsnap the 4 snap in tabs from the upper steering column cover.
6. Pull back on the left finish panel far enough to disconnect the 2 electrical connectors.
7. Remove the 2 screws retaining the switch to the center finish panel.
8. Installation is the reverse of removal procedure.

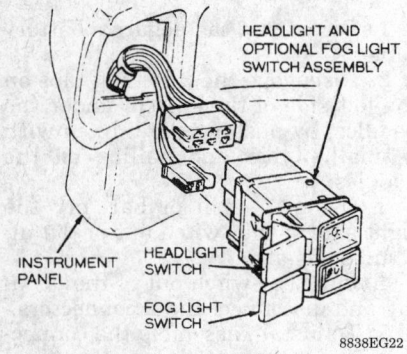

Headlight switch installation — Mustang

Combination Switch

The combination switch incorporates the turn signal, dimmer and wiper switch functions.

REMOVAL AND INSTALLATION

Mustang, Thunderbird and Cougar

1. Disconnect the negative battery cable.
2. Remove the shroud retaining screws and remove the upper and lower shrouds.
3. Remove the switch retaining screws and lift the switch assembly.
4. With the wiring connectors exposed, carefully lift the connector retainer tabs and disconnect the connectors.
5. Installation is the reverse of the removal procedure.

Mark VIII

1. Disconnect the negative battery cable.
2. If equipped with tilt column, move to the lowest position and remove the tilt lever.
3. Remove the ignition lock cylinder.
4. Remove the shroud screws and remove the upper and lower shrouds.
5. Remove the 2 self-tapping screws attaching the combination switch to the steering column casting and remove the switch.
6. Disconnect the 2 electrical connectors.
7. Installation is the reverse of the removal procedure.

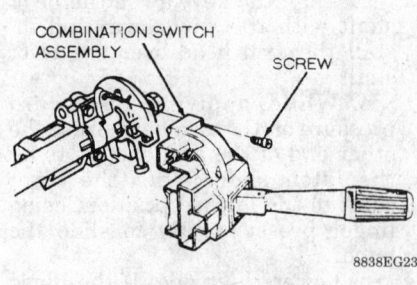

Combination switch — Mustang

Ignition Lock Cylinder

REMOVAL AND INSTALLATION

Functional Lock

The following procedure is for vehicles with functioning lock cylinders. Ignition keys are available for these vehicles, or the ignition key numbers are known and the proper key can be made.

1. Disconnect the negative battery cable. If equipped, properly disarm the air bag system.
2. On Thunderbird, Cougar and Mustang equipped with tilt column, remove the upper extension shroud by unsnapping the shroud retaining clip at the 9 o'clock position.
3. On all except Mark VIII, remove the trim shroud halves by removing the attaching screws. Remove the electrical connector from the key warning switch.
4. Place the gear shift lever in **P**, for column shift only, and turn the ignition to the **RUN** position.
5. Place a 1/8 in. diameter wire pin or small drift punch in the hole in the casting surrounding the lock cylinder and depress the retaining pin while pulling out on the lock cylinder to remove it from the column housing.
   To install:
6. To install the lock cylinder, turn it to the **RUN** position and depress the retaining pin. Insert the lock cylinder into its housing in the lock cylinder casting.
7. Make sure the cylinder is fully seated and aligned in the interlocking washer before turning the key to the **OFF** position. This action will permit the cylinder retaining pin to extend into the hole in the lock cylinder housing.
8. Using the ignition key, rotate the cylinder to ensure the correct mechanical operation in all positions.
9. Check for proper start in **P** or **N**. Also make sure the start circuit cannot be actuated in **D** or **R** positions and that the column is locked in the **LOCK** position.
10. Connect the key warning buzzer electrical connector and install the trim shrouds, if required.

Non-Functional Lock

The following procedure is for vehicles with non-functioning locks. On these vehicles, the lock cylinder cannot be rotated due to a lost or broken key, the key number is not known, or the lock cylinder cap is damaged

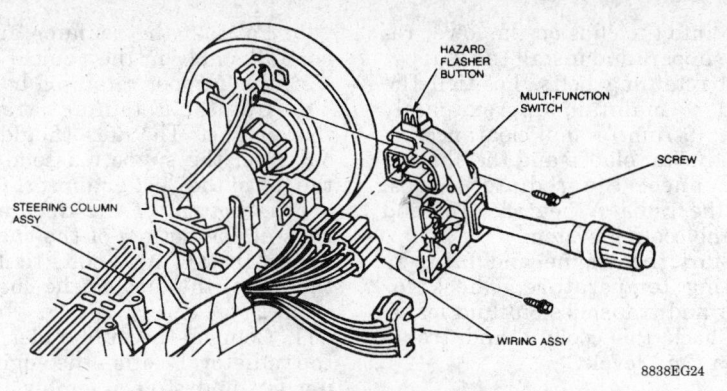

Combination switch — Mark VIII

8838EG24

and/or broken, preventing the lock cylinder from rotating.

1. Disconnect the negative battery cable. If equipped, properly disarm the air bag system.

2. Remove the steering wheel.

3. On Thunderbird, Cougar and and Mustang equipped with tilt column, remove the upper extension shroud by unsnapping the shroud retaining clip at the 9 o'clock position.

4. On all except Mark VIII:

a. Remove the trim shroud halves by removing the attaching screws. Remove the electrical connector from the key warning switch.

b. Drill out the retaining pin using a 1/8 in. diameter drill, being careful not to drill deeper than 1/2 in.

c. Position a chisel at the base of the ignition lock cylinder. Strike the chisel with sharp blows, using a hammer, to break the cap away from the lock cylinder.

5. On Mark VIII, use suitable pliers to to twist lock cylinder clip until it separates from lock cylinder.

6. Drill approximately 1 3/4 in. down the middle of the ignition key slot, using a 3/8 in. diameter drill bit, until the lock cylinder breaks loose from the breakaway base of the lock cylinder. Remove the lock cylinder and drill shavings from the lock cylinder housing.

7. Remove the snapring or retainer, washer and steering column lock gear. Thoroughly clean all drill shavings and other foreign materials from the casting.

8. Inspect the lock cylinder housing for damage and replace, as necessary.

To install:

9. Install the ignition lock drive gear, washer and retainer.

10. Install the ignition lock cylinder and check for smooth operation.

11. Connect the electrical connector to the key warning switch.

12. Install the trim shrouds, if necessary.

13. Install the new lock cylinder housing assembly.

14. Install the steering wheel and connect the negative battery cable.

Ignition Switch

REMOVAL AND INSTALLATION

1. Disconnect the negative battery cable.

2. Remove the steering column shroud. On Mark VIII, remove the steering column opening trim cover and remove the cover.

3. Disconnect the switch electrical connector.

4. Turn the ignition lock cylinder to **RUN**.

5. Remove the screws attaching the switch and disengage the switch from the actuator.

To install:

6. Adjust the new ignition switch by sliding the carrier to the **RUN** position.

7. Make sure the ignition key lock cylinder is in the **RUN** position. The **RUN** position is achieved by rotating the key lock cylinder approximately 90 degrees from the **LOCK** position.

8. Install the ignition switch onto the actuator pin or install the switch pin in the column hole, as required.

9. Align the switch mounting holes and install the attaching screws. Tighten the screws to 50–69 inch lbs. (5.6–7.9 Nm).

10. Connect the electrical connector to the ignition switch.

11. Connect the negative battery cable. Check the ignition switch for proper function in **START** and **ACC** positions. Make sure the column is locked in the **LOCK** position.

12. Install the remaining components in the reverse order of removal.

Neutral Safety Switch

REMOVAL AND INSTALLATION

Mustang with 2.3L Engine

1. Disconnect the negative battery cable.

2. Disconnect the switch wiring harness connector.

3. Remove the neutral safety switch and O-ring using socket tool T74P-77247-A or equivalent.

NOTE: Use of different tools could crush or puncture the walls of the switch.

To install:

4. Install the neutral safety switch and new O-ring using socket tool T74P-77247-A or equivalent.

5. Tighten the switch to 7–10 ft. lbs. (10–14 Nm).

6. Connect the neutral safety switch to the wiring harness.

7. Connect the negative battery cable.

8. Check that the vehicle starts only in the **N** or **P** position.

Mustang with 5.0L Engine, Thunderbird, Cougar

1. Place the selector lever in the manual **L** position.

2. Disconnect the negative battery cable.

3. Raise and support the vehicle safely.

4. Disconnect the neutral safety switch electrical harness from the switch by pushing the harness straight up off the switch using a small long-bladed prybar under the rubber plug section of the harness.

5. Install socket tool T74P-77247-A or equivalent, and rachet on the neutral safety switch. Once the ratchet and socket tool are over the switch, reach from the rear of the transmission over the extension housing area and remove the neutral safety switch and O-ring.

NOTE: Use of different tools could crush or puncture the walls of the switch.

To install:

6. Install the neutral safety switch and new O-ring using socket tool T74P-77247-A or equivalent.

7. Tighten the switch to 8–11 ft. lbs. (11–15 Nm).

8. Connect the neutral safety switch to the wiring harness.

9. Lower the vehicle and connect the negative battery cable.

10. Check that the vehicle starts only in the **N** or **P** position.

ENGINE COOLING

Radiator

REMOVAL AND INSTALLATION

Except 3.8L SC Engine

1. Disconnect the negative battery cable.

2. Remove the radiator cap. Place a drain pan under the radiator, open the draincock and drain the coolant.

---- CAUTION ----

Never remove the radiator cap while the engine is running or personal injury from scalding hot coolant or steam may result. If possible, wait until the engine has cooled to remove the radiator cap. If this is not possible, wrap a thick cloth around the radiator cap and turn it slowly to the first stop. Step back while the pressure is released from the cooling system. When it is certain all the pressure has been released, press down on the cap, still with the cloth, and turn and remove it.

---

3. Disconnect the upper, lower and overflow hoses at the radiator.

4. If equipped with an automatic transmission, disconnect the fluid cooler lines at the radiator.

5. On Mustang with 2.3L engine and Mark VIII, remove the electric cooling fan/shroud assembly. On all other vehicles, remove the 2 upper fan shroud retaining bolts at the radiator support, lift the fan shroud sufficiently to disengage the lower retaining clips and lay the shroud back over the fan.

6. Remove the radiator upper support retaining bolts and remove the supports. Lift the radiator from the vehicle.

To install:

7. If a new radiator is to be installed, transfer the petcock from the old radiator to the new one. If equipped with automatic transmission, transfer the fluid cooler line fittings from the old radiator.

8. Position the radiator assembly into the vehicle. Install the upper supports and the retaining bolts. If equipped with automatic transmission, connect the fluid cooler lines.

9. On Mustang with the 2.3L engine and Mark VIII, install the electric cooling fan/shroud assembly. On all other vehicles, place the fan

shroud into the clips on the lower radiator support and install the 2 upper shroud retaining bolts. Position the shroud to maintain approximately 0.38 in. (9.7mm) radial clearance between the fan blades and the shroud.

10. Connect the radiator hoses. Close the radiator petcock. Fill and bleed the cooling system.

11. Start the engine and bring to operating temperature. Check for coolant and transmission fluid leaks.

12. Check the coolant and transmission fluid levels.

3.8L SC Engine

1. Disconnect the negative battery cable.

2. Remove the charge air cooler.

3. Remove the radiator cap. Place a drain pan under the radiator, open the draincock and drain the coolant.

---- CAUTION ----

Never remove the radiator cap while the engine is running or personal injury from scalding hot coolant or steam may result. If possible, wait until the engine has cooled to remove the radiator cap. If this is not possible, wrap a thick cloth around the radiator cap and turn it slowly to the first stop. Step back while the pressure is released from the cooling system. When it is certain all the pressure has been released, press down on the cap, still with the cloth, and turn and remove it.

---

4. Disconnect the upper and lower radiator hoses and the overflow hose at the radiator.

5. If equipped with an automatic transmission, disconnect the fluid cooler lines at the radiator.

6. Remove the overflow hose from the clip on the fan shroud. Remove the 2 shroud upper retaining bolts at the radiator support and remove the wiring harness retaining clip from the fan shroud. Lift the electric cooling fan/shroud assembly from the radiator, disengaging the shroud from the lower retaining clips.

7. Remove the 2 bolts retaining the top of the air duct to the charge air cooler and remove the upper 2 radiator retaining bolts. Tilt the radiator and support assembly toward the engine and lift the radiator from the vehicle.

To install:

8. If a new radiator is to be installed and the vehicle is equipped with automatic transmission, transfer the fluid cooler line fittings from the old radiator.

9. Position the radiator and support assembly in the vehicle and install the 2 upper retaining bolts.

10. Cut the retaining strap from the air duct. The duct should spring out from the support assembly. Lift the top of the duct and insert the tabs on the bottom of the duct into the clips at the bottom of the charge air cooler. Install the 2 bolts that retain the top of the duct to the charge air cooler.

11. Connect the fluid cooler lines to the radiator. Position the engine cooling fan and stud assembly into the radiator lower clips. Attach the top of the radiator to the top of the support with the 2 bolts.

12. Connect the radiator and overflow hoses to the radiator. Route the overflow hose through the retaining clip. Make sure the draincock is closed and fill the cooling system.

13. Install the charge air cooler. Connect the cooling fan electrical connector and install the harness clip to the fan shroud.

14. Start the engine and bring to operating temperature. Check for coolant and transmission fluid leaks.

15. Check the coolant and transmission fluid levels.

Water Pump

REMOVAL AND INSTALLATION

2.3L Engine

1. Disconnect the negative battery cable and drain the cooling system.

2. Remove the 4 bolts retaining the pulley to the water pump shaft. Remove the fan and shroud.

3. Remove the air conditioning and power steering belts. Remove the water pump pulley.

4. Remove the heater hose to the water pump and the lower radiator hose.

5. Remove the timing belt outer cover bolt, release the interlocking tabs and remove the cover.

6. Remove the water pump retaining bolts and remove the water pump.

7. Installation is the reverse of the removal procedure. Clean all gasket mating surfaces prior to installation. Apply pipe sealant to the water pump bolts and tighten to 14–21 ft. lbs. (20–30 Nm). Tighten the pulley retaining bolts to 15–22 ft. lbs. (20–30 Nm).

8. Fill and bleed the cooling system. Operate the engine until normal operating temperatures have been reached and check for leaks.

3.8L Engine

1. Disconnect the negative battery cable and drain the cooling system.
2. On all except supercharged engine, remove the fan/clutch assembly and shroud.
3. Rotate the main accessory drive belt tensioner. Remove the main drive belt and water pump pulley.
4. Remove the power steering pump pulley and remove the water pump to power steering pump brace.

NOTE: On supercharged engines, it may be necessary to remove the charge air cooler to gain access to the power steering pump pulley.

5. On all except supercharged engine, disconnect the coolant bypass hose(s) and the heater hose at the water pump. On supercharged engine, disconnect the oil cooler tube and bypass hose and remove the upper crankshaft sensor cover.
6. Disconnect the lower radiator hose. Remove the water pump retaining bolts and the pump. If a prybar is used to assist removal, be careful not to damage the mating surfaces.
7. Installation is the reverse of the removal procedure. Clean all gasket mating surfaces prior to installation. Tighten the water pump retaining bolts to 15–22 ft. lbs. (20–30 Nm). Fill and bleed the cooling system. Operate the engine until normal operating temperatures have been reached and check for leaks.

NOTE: The threads of the No. 1 water pump retaining bolt must be coated with pipe sealant before installing.

4.6L (DOHC) Engine

1. Disconnect the negative battery cable and drain the cooling system.
2. Remove the 4 bolts retaining the pulley to the water pump shaft.
3. Release belt tensioner and remove the drive belts. Remove 4 bolts retaining water pump pulley and remove the pulley.
4. Remove the 4 water pump retaining bolts and remove the water pump.
5. Installation is the reverse of the removal procedure. Clean all gasket mating surfaces prior to installation. Tighten the water pump bolts and pulley retaining bolts to 15–22 ft. lbs. (20–30 Nm).
6. Fill and bleed the cooling system. Operate the engine until normal operating temperatures have been reached and check for leaks.

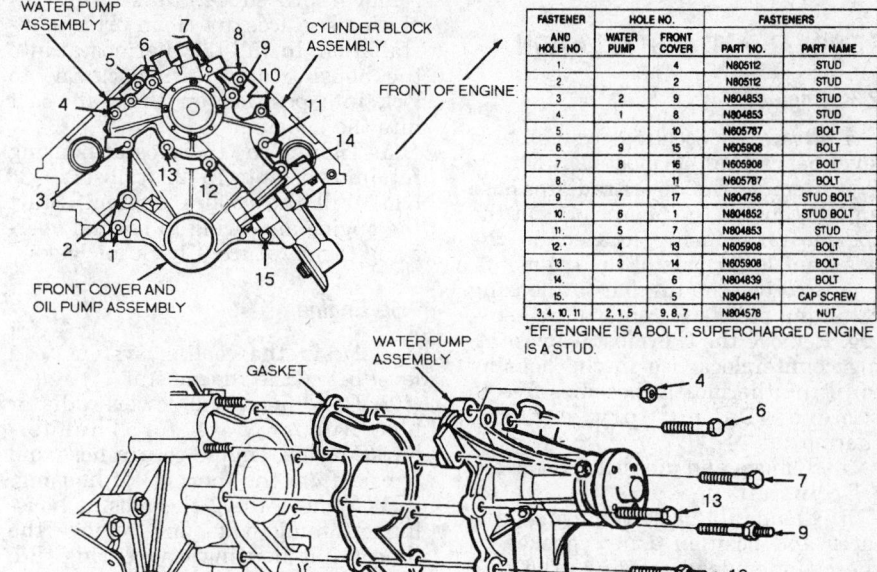

FASTENER AND HOLE NO.	HOLE NO.		FASTENERS	
	WATER PUMP	FRONT COVER	PART NO.	PART NAME
1.		4	N805112	STUD
2.		2	N805112	STUD
3.	2	9	N804853	STUD
4.	1	8	N804853	STUD
5.		10	N805787	BOLT
6.	9	15	N805908	BOLT
7.	8	16	N805908	BOLT
8.		11	N805787	BOLT
9.	7	17	N804756	STUD BOLT
10.	6	1	N804852	STUD BOLT
11.	5	7	N804853	STUD
12.*	4	13	N805908	BOLT
13.	3	14	N805908	BOLT
14.		6	N804839	BOLT
15.			N804841	CAP SCREW
3, 4, 10, 11	2, 1, 5	9, 8, 7	N804578	NUT

*EFI ENGINE IS A BOLT, SUPERCHARGED ENGINE IS A STUD.

8838EG25

Water pump fastener and hole location — 3.8L engine

4.6L (SOHC) Engine

1. Disconnect the negative battery cable and drain the cooling system.
2. Release the drive belt tensioner and remove the belt.
3. Remove the water pump pulley bolts and pulley.
4. Remove the pump-to-block bolts and pump.
5. Installation is the reverse of the removal procedure. Clean all gasket mating surfaces prior to installation. Tighten the water pump bolts and pulley retaining bolts to 15–22 ft. lbs. (20–30 Nm).
6. Fill and bleed the cooling system. Operate the engine until normal operating temperatures have been reached and check for leaks.

5.0L Engine

1. Disconnect the negative battery cable.
2. Drain the cooling system. Remove the air inlet tube, if equipped.
3. On Thunderbird and Cougar, disconnect the upper radiator hose at the engine.
4. On all except Thunderbird and Cougar, remove the fan shroud attaching bolts and position the shroud over the fan. Remove the fan and clutch assembly from the water pump shaft and remove the shroud.

5. On Thunderbird and Cougar, remove the fan and clutch assembly from the water pump shaft using fan clutch holding tool T84T-6312-C or equivalent, and fan clutch nut wrench T84T-6312-D or equivalent, and position the fan and clutch assembly in the fan shroud. The nut is turned counterclockwise. Remove the fan shroud and fan/clutch as an assembly.
6. Loosen the water pump pulley bolts. Rotate the tensioner away from the accessory drive belt and remove the belt. Remove the water pump pulley.
7. Remove all accessory brackets that attach to the water pump.
8. Disconnect the lower radiator hose, heater hose and water pump bypass hose at the water pump.
9. Remove the water pump attaching bolts and remove the water pump. Discard the gasket.
10. Installation is the reverse of the removal procedure. Clean all gasket mating surfaces prior to installation. Tighten the water pump attaching bolts to 12–18 ft. lbs. (16–24 Nm).
11. Fill and bleed the cooling system. Operate the engine until normal operating temperatures have been reached and check for leaks.

Thermostat

REMOVAL AND INSTALLATION

2.3L Engine

1. Drain the cooling system to a level below the thermostat.

2. Remove the upper radiator hose and disconnect the heater hose at the thermostat housing located on the left front lower side of the engine.

3. Remove the thermostat housing retaining bolts and remove the housing. Remove the thermostat by rotating counterclockwise in the housing until the thermostat becomes free to remove. Do not pry out the thermostat.

4. Remove and discard the gasket.

To install:

5. Clean all gasket mating surfaces and position a new gasket on the cylinder head opening. The gasket must be positioned on the cylinder head, before the thermostat is installed.

6. Install the thermostat into the housing with the bridge section in the housing. Turn the thermostat clockwise to lock it into position on the flats cast into the housing.

NOTE: It is important that the rubber thermostat gasket be pressed and the correct thermostat installation alignment be made to provide coolant flow to the heater. Insert and rotate the thermostat to the left or right until it stops in the thermostat housing. Visually check for full width of heater outlet tube opening to be visible within the thermostat port in assembly. This port alignment at assembly is required to provide maximum coolant flow to the heater.

7. Position the thermostat housing against the gasket on the cylinder head. Install and tighten the retaining bolts to 14–21 ft. lbs. (19–29 Nm).

8. Connect the upper radiator hose and the heater hose to the thermostat housing. Fill the cooling system. Start the engine and bring to normal operating temperature. Check for leaks.

3.8L Engine

1. Drain the cooling system to a level below the thermostat.

2. Disconnect the upper radiator hose at the thermostat housing.

3. Remove the 2 thermostat housing retaining bolts and remove the thermostat housing and gasket. Remove the thermostat.

4. Installation is the reverse of the removal procedure. Make sure all mating surfaces are clean prior to installation. Install the thermostat into the housing and turn clockwise to lock into position on the flats cast into the housing.

5. Tighten the thermostat housing retaining bolts to 15–22 ft. lbs. (20–30 Nm). Fill the cooling system. Start the engine and bring to normal operating temperature. Check for leaks.

4.6L Engine

1. Drain the cooling system to a level below the thermostat.

2. Disconnect the lower radiator hose (upper hose for Thunderbird/Cougar), engine return hose and bypass from the thermostat housing.

3. Remove the 2 thermostat housing retaining bolts and remove the thermostat housing and O-ring. Remove the thermostat.

4. Installation is the reverse of the removal procedure. Make sure all mating surfaces are clean prior to installation.

5. Tighten the thermostat housing retaining bolts to 15–22 ft. lbs. (20–30 Nm). Fill the cooling system. Start the engine and bring to normal operating temperature. Check for leaks.

5.0L Engine

1. Drain the cooling system to a level below the thermostat.

2. Disconnect the upper radiator hose and the bypass hose at the thermostat housing.

3. To gain access to the thermostat housing, either mark the location of the distributor, loosen the hold-down clamp and rotate the distributor, or remove the distributor cap and rotor.

4. Remove the thermostat housing retaining bolts and the housing and gasket. Remove the thermostat.

To install:

5. Clean the gasket mating surfaces. Position a new gasket on the intake manifold.

6. Install the thermostat in the housing, rotating slightly to lock the thermostat in place on the flats cast into the housing. Install the housing on the manifold and tighten the bolts to 12–18 ft. lbs. (16–24 Nm).

7. Install the distributor cap and rotor, or reposition the distributor for correct ignition timing, as necessary. Tighten the hold-down bolt to 17–25 ft. lbs. (23–34 Nm).

8. Connect the bypass hose and the upper radiator hose to the thermostat housing. Fill the cooling system.

9. Start the engine and bring to normal operating temperature. Check for leaks.

Electric Cooling Fan

REMOVAL AND INSTALLATION

Mustang with 2.3L Engine

1. Disconnect the negative battery cable.

2. Remove the fan wiring harness from the routing clip. Disconnect the wiring harness from the fan motor connector by pulling up on the single lock finger to separate the connectors.

3. Remove the 4 mounting bracket attaching screws and remove the fan assembly from the vehicle.

4. Remove the retaining clip from the end of the motor shaft and remove the fan.

NOTE: A metal burr may be present on the motor after the retaining clip is removed. Deburring of the shaft may be required to remove the fan.

5. Remove the nuts attaching the fan motor to the mounting bracket.

6. Installation is the reverse of the removal procedure.

Thunderbird and Cougar with 3.8L SC Engine

1. Disconnect the negative battery cable.

2. Disconnect the fan motor wiring connector at the side of the fan shroud. Remove the male terminal connector retaining clip from the shroud mounting tab.

3. Remove the overflow hose from the fan shroud retaining clip and remove the 2 shroud upper retaining bolts at the radiator support.

4. Lift the cooling fan module past the radiator, disengaging the shroud from the 2 lower retaining clips.

5. Installation is the reverse of the removal procedure. Tighten the shroud retaining bolts to 36 inch lbs. (4 Nm).

Thunderbird, Cougar and Mark VIII with 4.6L engine

1. Disconnect the negative battery cable.

2. Remove the fan shroud bolts and fan shroud from the mounts. Remove the lower radiator hose from the fan shroud.

3. Lift the radiator electric fan/shroud assembly from the vehicle.

4. Disconnect the radiator electric motor wiring harness at the electric motor.

5. Installation is the reverse of the removal procedure.

Cooling System Bleeding

PROCEDURE

When the entire cooling system is drained, the following procedure should be used to ensure a complete fill.

1. Install the block drain plug, if removed and close the draincock. On 3.8L engine, remove the vent plug on the intake manifold behind the thermostat housing. With the engine OFF, add a 50/50 mixture of water and anti-freeze to the radiator until it reaches the radiator filler neck seat.

NOTE: On Mustang equipped with the 2.3L engine, disconnect the heater hose at the connection on the thermostat housing. Fill the radiator until coolant is visible at the connection in the thermostat housing or the coolant level in the radiator reaches the radiator filler neck seat. Install the heater hose and tighten the hose clamps.

2. Install the radiator cap to the first notch to keep spillage to a minimum. On 3.8L engine, install the vent plug.

3. Start the engine and let it idle until the upper radiator hose is warm. This indicates that the thermostat is open and coolant is flowing through the entire system.

4. Carefully remove the radiator cap and top off the radiator with the water/anti-freeze mixture. Install the cap on the radiator securely.

5. Fill the coolant recovery reservoir to the FULL HOT mark with the water/anti-freeze mixture. Install the reservoir cap.

FUEL SYSTEM

Fuel System Service Precautions

Safety is the most important factor when performing not only fuel system maintenance but any type of maintenance. Failure to conduct maintenance and repairs in a safe manner may result in serious personal injury or death. Maintenance and testing of the vehicle's fuel system components can be accomplished safely and effectively by adhering to the following rules and guidelines.

• To avoid the possibility of fire and personal injury, always disconnect the negative battery cable unless the repair or test procedure requires that battery voltage be applied.

• Always relieve the fuel system pressure prior to disconnecting any fuel system component (injector, fuel rail, pressure regulator, etc.), fitting or fuel line connection. Exercise extreme caution whenever relieving fuel system pressure to avoid exposing skin, face and eyes to fuel spray. Please be advised that fuel under pressure may penetrate the skin or any part of the body that it contacts.

• Always place a shop towel or cloth around the fitting or connection prior to loosening to absorb any excess fuel due to spillage. Ensure that all fuel spillage (should it occur) is quickly removed from engine surfaces. Ensure that all fuel soaked cloths or towels are deposited into a suitable waste container.

• Always keep a dry chemical (Class B) fire extinguisher near the work area.

• Do not allow fuel spray or fuel vapors to come into contact with a spark or open flame.

• Always use a backup wrench when loosening and tightening fuel line connection fittings. This will prevent unnecessary stress and torsion to fuel line piping. Always follow the proper torque specifications.

• Always replace worn fuel fitting O-rings with new. Do not substitute fuel hose or equivalent, where fuel pipe is installed.

Fuel System Pressure

RELIEVING

Fuel supply lines on all fuel injected engines will remain pressurized for some period of time after the engine is shut OFF. This pressure must be relieved before servicing the fuel system. Pressure is relieved through the fuel pressure relief valve. To relieve the fuel system pressure, first remove the fuel tank cap to relieve pressure in the tank, then remove the cap on the fuel pressure relief valve, located on the fuel rail. Attach fuel pressure gauge T80L-9974-A or equivalent, and drain the system through the drain tube into a suitable container. Remove the fuel pressure gauge and replace the cap on the relief valve.

Idle Speed

ADJUSTMENT

2.3L Engine

1. Connect the SUPER STAR II tester, tool number 007-00028 or other suitable scan tool to the Self-Test connector. Activate the Key On Engine Running (KOER) Self-Test.

2. After Code 1 or 111 has been displayed, unlatch and within 4 seconds, latch the STI button.

3. A single pulse code indicates the entry mode, then observe the Self-Test Output (STO) on the tester for the following:

a. A constant tone, solid light or **STO LO** readout means the base idle speed is within the correct range. To exit the test, unlatch the STI button, then wait 4 seconds for reinitialization. After 10 minutes, the tool will exit by itself.

b. A beeping tone, flashing light or **STO LO** readout at 8 Hz indicates the Throttle Position Sensor (TPS) is out of range due to over adjustment. Adjustment may be required.

c. A beeping tone, flashing light or **STO LO** readout at 4 Hz indicates the base idle speed is too fast and adjustment is required. Proceed to Step 5.

d. A beeping tone, flashing light or **STO LO** readout at 1 Hz indicates the base idle speed is too low and adjustment is required. Proceed to Step 4.

4. If the idle speed is too low, check for the presence of a throttle plate orifice plug. If there is no plug, turn the throttle screw clockwise until the conditions in Step 3a exist. If there is a plug from previous service, remove the plug and then adjust the screw in either direction, as required. The screw must be in contact with the lever pad after adjustment.

5. If the idle speed is too high, proceed as follows:

a. Turn the engine **OFF**.

b. Block off the orifice in the throttle plate temporarily with tape. If the orifice already has a plug, proceed to Step d.

c. Reattach the air intake hose. Restart the engine and check the idle speed using the Self-Test. If the engine stalled, crack open the plate with the throttle return screw.

d. If the idle speed continues to be fast, run the Key On Engine Off (KOEO) Self-Test and check for a TPS output code.

e. If the output code is within range, remove the tape and check for vacuum leaks, throttle linkage binding, or other causes for excessive high idle.

f. If the output code is out of range, adjust the throttle screw to obtain the proper code. The lever pad must be in contact with the screw after adjustment.

g. If the idle speed drops to or below the desired level, as indicated by the Self-Test Output tone, turn the engine OFF, disconnect the air cleaner hose and remove the tape.

h. Install the proper plug in the throttle plate orifice.

i. Reconnect the air cleaner hose. Start the engine and turn the throttle plate stop screw clockwise until the conditions in Step 3a exist. Do not turn the screw counterclockwise as this may cause the throttle plate to stick at idle.

6. Run the KOEO Self-Test for proper TPS output code.

7. Make sure the throttle is not stuck in the bore and the linkage is not preventing the throttle from closing.

3.8L Engine, Except SC

1. Place the transaxle in P and apply the parking brake.

2. Start the engine and bring to normal operating temperature. Make sure the heater, air conditioning and all other accessories are OFF.

3. Check and if necessary, adjust the ignition timing.

4. Make sure the fuel pressure is correct. Any indicated vehicle malfunction service codes should be resolved before proceeding further.

5. Connect the SUPER STAR II tester, tool number 007-00028 or other suitable scan tool to the Self-Test connector. Activate the Key On Engine Running (KOER) Self-Test.

6. After Code 1 or 111 has been displayed, unlatch and within 4 seconds, latch the STI button.

7. A single pulse code indicates the entry mode, then observe the Self-Test Output (STO) on the tester for the following:

a. A constant tone, solid light or STO LO readout means the base idle speed is within the correct range. To exit the test, unlatch the STI button, then wait 4 seconds for

reinitialization. After 10 minutes, the tool will exit by itself.

b. A beeping tone, flashing light or STO LO readout at 8 Hz indicates the Throttle Position Sensor (TPS) is out of range due to over adjustment. Adjustment may be required.

c. A beeping tone, flashing light or STO LO readout at 4 Hz indicates the base idle speed is too fast and adjustment is required. Proceed to Step 9.

d. A beeping tone, flashing light or STO LO readout at 1 Hz indicates the base idle speed is too low and adjustment is required. Proceed to Step 8.

8. If the idle speed is too low, check for the presence of a throttle plate orifice plug. If there is no plug, turn the throttle screw clockwise until the conditions in Step 7a exist. If there is a plug from previous service, remove the plug and then adjust the screw in either direction, as required. The screw must be in contact with the lever pad after adjustment.

9. If the idle speed is too high, proceed as follows:

a. Turn the engine OFF.

b. Block off the orifice in the throttle plate temporarily with tape. If the orifice already has a plug, proceed to Step d.

c. Reattach the air intake hose. Restart the engine and check the idle speed using the Self-Test. If the engine stalled, crack open the plate with the throttle return screw.

d. If the idle speed continues to be fast, run the Key On Engine Off (KOEO) Self-Test and check for a TPS output code.

e. If the output code is within range, remove the tape and check for vacuum leaks, throttle linkage binding, or other causes for excessive high idle.

f. If the output code is out of range, adjust the throttle screw to obtain the proper code. The lever pad must be in contact with the screw after adjustment.

g. If the idle speed drops to or below the desired level, as indicated by the Self-Test Output tone, turn the engine OFF, disconnect the air cleaner hose and remove the tape.

h. Install the proper plug in the throttle plate orifice.

i. Reconnect the air cleaner hose. Start the engine and turn the throttle plate stop screw clockwise until the conditions in Step 7a exist. Do not turn the screw counter-

clockwise as this may cause the throttle plate to stick at idle.

10. Run the KOEO Self-Test for proper TPS output code.

11. Make sure the throttle is not stuck in the bore and the linkage is not preventing the throttle from closing.

12. Check the Throttle Valve (TV) pressure adjustment.

3.8L SC Engine

1. Place the transaxle in P and apply the parking brake.

2. Start the engine and bring to normal operating temperature. Make sure the heater, air conditioning and all other accessories are OFF.

3. Check and if necessary, adjust the ignition timing.

4. Make sure the fuel pressure is correct. Any indicated vehicle malfunction service codes should be resolved before proceeding further.

5. Connect the SUPER STAR II tester, tool number 007-00028 or other suitable scan tool to the Self-Test connector. Activate the Key On Engine Running (KOER) Self-Test.

6. After Code 1 or 111 has been displayed, unlatch and within 4 seconds, latch the STI button.

7. A single pulse code indicates the entry mode, then observe the Self-Test Output (STO) on the tester for the following:

a. A constant tone, solid light or STO LO readout means the base idle speed is within the correct range. To exit the test, unlatch the STI button, then wait 4 seconds for reinitialization. After 10 minutes, the tool will exit by itself.

b. A beeping tone, flashing light or STO LO readout at 8 Hz indicates the Throttle Position Sensor (TPS) is out of range due to over adjustment. Adjustment may be required.

c. A beeping tone, flashing light or STO LO readout at 4 Hz indicates the base idle speed is too fast and adjustment is required. Proceed to Step 9.

d. A beeping tone, flashing light or STO LO readout at 1 Hz indicates the base idle speed is too low and adjustment is required. Proceed to Step 8.

8. If the idle speed is too low, do not clean the throttle body. Turn the air trim screw counterclockwise until the conditions in Step 7a are satisfied.

9. If the idle speed is too high, do not clean the throttle body. Turn the air trim screw clockwise until the conditions in Step 7a are satisfied.

4.6L Engine

The idle speed is not adjustable.

5.0L Engine

1993 Thunderbird and Cougar

1. Place the transaxle in **P** and apply the parking brake.
2. Start the engine and bring to normal operating temperature. Make sure the heater, air conditioning and all other accessories are **OFF**.
3. Check and if necessary, adjust the ignition timing.
4. Make sure the fuel pressure is correct. Any indicated vehicle malfunction service codes should be resolved before proceeding further.
5. Connect the SUPER STAR II tester tool 007-00028 or other suitable scan tool, to the Self-Test connector. Activate the Key On Engine Running (KOER) Self-Test.
6. After Code 1 or 111 has been displayed, unlatch and within 4 seconds, latch the STI button.
7. A single pulse code indicates the entry mode, then observe the Self-Test Output (STO) on the tester for the following:

   a. A constant tone, solid light or **STO LO** readout means the base idle speed is within the correct range. To exit the test, unlatch the STI button, then wait 4 seconds for reinitialization. After 10 minutes, the tool will exit by itself.

   b. A beeping tone, flashing light or **STO LO** readout at 8 Hz indicates the Throttle Position Sensor (TPS) is out of range due to over adjustment. Adjustment may be required.

   c. A beeping tone, flashing light or turn the air trim screw counterclockwise until the conditions in Step 7a are satisfied.

8. If the idle speed is too high, do not clean the throttle body. Turn the air trim screw clockwise until the conditions in Step 7a are satisfied.

1993 Mustang

1. Place the transaxle in **P** and apply the parking brake.
2. Start the engine and bring to normal operating temperature. Make sure the heater, air conditioning and all other accessories are **OFF**.
3. Check and if necessary, adjust the ignition timing.
4. Make sure the fuel pressure is correct. Any indicated vehicle malfunction service codes should be resolved before proceeding further.
5. Disconnect the negative battery terminal for 5 minutes, then reconnect. Start the engine and stabilize for 2 minutes, then goose the engine and let it return to idle. Lightly depress and release the accelerator and let the engine idle. Check the engine idle.
6. If the engine does not idle properly, shut the engine **OFF** and place a 0.025 in. feeler gauge between the throttle plate stop screw and throttle lever.
7. Start the engine and let it idle. Check the idle speed; it should be 675 ± 50 rpm.
8. If the idle speed is too low, proceed as follows:

   a. Shut the engine **OFF**. Do not clean the throttle body, but check the throttle plate for an orifice plug.

   b. If there is no plug, start the engine and let it idle for 2 minutes, then adjust the idle to the desired speed, ± 25 rpm.

   c. If there is a plug, remove it, then start the engine and let it idle for 2 minutes. Adjust the idle to the desired speed, ± 25 rpm.

   d. The screw must be in contact with the lever pad after adjustment.

9. If the idle speed is too high, proceed as follows:

   a. Shut the engine **OFF** and disconnect the air cleaner hose.

   b. Block off the orifice in the throttle plate temporarily with tape. If the orifice already has a plug, proceed to Step d.

   c. Reattach the air intake hose. Restart the engine and check the idle speed. If the engine stalled, crack open the plate with the throttle return screw.

   d. If the idle speed continues to be fast, connect a suitable scan tool and run the Key On Engine Off (KOEO) Self-Test and check for a TPS output code.

   e. If the output code is within range, remove the tape and check for vacuum leaks, throttle linkage binding, or other causes for excessive high idle.

   f. If the output code is out of range, adjust the throttle screw to obtain the proper code. The lever pad must be in contact with the screw after adjustment.

   g. If the idle speed drops to or below the desired level, turn the engine **OFF**, disconnect the air cleaner hose and remove the tape.

   h. Install the proper plug in the throttle plate orifice.

   i. Reconnect the air cleaner hose. Start the engine and turn the throttle plate stop screw clockwise to the nominal idle speed, ± 25 rpm. Do not turn the screw counterclockwise as this may cause the throttle plate to stick at idle.

10. Remove the feeler gauge from between the throttle plate stop screw and throttle lever.
11. Shut the engine **OFF** and disconnect the battery for 10 minutes minimum.
12. Run the KOEO Self-Test for proper TPS output code.
13. Start the engine and let the idle stabilize for 2 minutes. Rev the engine and let it return to idle. Lightly depress and release the accelerator; let the engine idle.
14. If equipped with automatic overdrive transmission, check the throttle valve pressure adjustment.

Mixture

ADJUSTMENT

The idle mixture is controlled by the electronic control unit and cannot be adjusted.

Fuel Filter

REMOVAL AND INSTALLATION

1. Disconnect the negative battery cable and relieve the fuel system pressure.
2. Raise and safely support the vehicle. On Mark VIII, remove the right front fender liner.
3. Remove the push connect fittings at both ends of the filter. Install new retainer clips in each push connect fitting.
4. Remove the fuel filter from the bracket by loosening the worm gear clamp on all vehicles except the Mark VIII. Note the direction of the flow arrow as installed in the bracket to ensure proper direction of fuel flow through the replacement filter.
5. On Mark VIII, remove the screw from the apron and remove retainer tab from sheetmetal.

To install:

6. Install the fuel filter into the bracket, ensuring the proper direction of flow. Tighten the worm gear clamp to 15–25 inch lbs. (1.7–2.8 Nm) on all vehicles except the Mark VIII.
7. On Mark VIII, insert tab into sheetmetal and install screw.
8. Install the push connect fittings onto the filter ends. Start the engine and check for leaks. On Mark VIII, install the fender liner.
9. Lower the vehicle.

Fuel Pump

REMOVAL AND INSTALLATION

1. Disconnect the negative battery cable and properly relieve the fuel system pressure.
2. Remove the fuel tank and place it on a bench.
3. Remove any dirt that has accumulated around the fuel pump retaining flange so it will not enter the tank during pump removal and installation.
4. Turn the fuel pump locking ring counterclockwise and remove the locking ring.
5. Remove the fuel pump and bracket assembly. Remove and discard the seal ring.
   To install:
6. Clean the fuel pump mounting flange, fuel tank mounting surface and seal ring groove.
7. Apply a light coating of grease on a new seal ring to hold it in place during assembly and install in the seal ring groove.
8. Install the fuel pump and bracket assembly carefully to ensure the filter is not damaged. Make sure the locating keys are in the keyways and the seal ring remains in the groove.
9. Hold the pump assembly in place and install the locking ring finger-tight. Make sure all the locking tabs are under the tank lock ring tabs.
10. Rotate the locking ring clockwise until the ring is against the stops.
11. Install the fuel tank in the vehicle. Add a minimum of 10 gallons of fuel to the tank and check for leaks.
12. Install a suitable fuel pressure gauge to the valve on the fuel rail.
13. Turn the ignition switch from **OFF** to **ON** for 3 seconds. Repeat this procedure 5–10 times until the pressure gauge shows at least 35 psi. Check for fuel leaks.
14. Remove the pressure gauge, start the engine and check for leaks.

Fuel Injector

REMOVAL AND INSTALLATION

2.3L Engine

1. Disconnect the negative battery cable.
2. Remove the fuel tank cap and relieve the fuel system pressure.
3. Disconnect the air intake, electrical connectors, throttle linkage,

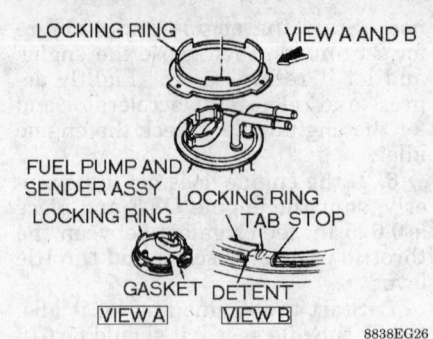

Electric fuel pump installation — Except Mark VIII

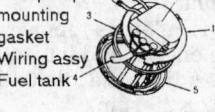
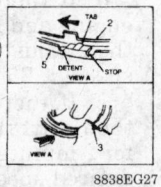

1. Fuel tank sending unit and pump
2. Fuel pump locking retainer ring
3. Fuel pump mounting gasket
4. Wiring assy
5. Fuel tank

8838EG27

Electric fuel pump installation — Mark VIII

vacuum lines and EGR tube from the upper intake manifold and throttle body. Tag the electrical connectors and vacuum lines prior to removal for installation reference.
4. Remove the upper intake manifold retaining bolts and remove the upper intake manifold and throttle body assembly.
5. Disconnect the electrical connectors from the injectors.
6. Disconnect the fuel lines from the fuel supply manifold.
7. Remove the fuel supply manifold retaining bolts, carefully disengage the manifold and fuel injectors from the engine and remove the manifold and injectors.
8. Remove the fuel injectors from the manifold.
   To install:
9. Lubricate new O-rings with clean light grade oil and install 2 on each injector.

NOTE: Never use silicone grease as it will clog the injectors.

10. Install the fuel supply manifold and injectors into the intake manifold. Push the fuel rail down to make sure all the fuel injector O-rings are

fully seated in the fuel rail cups and intake manifold.
11. Install the fuel manifold assembly retaining bolts and tighten to 15–22 ft. lbs. (20–30 Nm) while holding the assembly down.
12. Connect the fuel lines to the manifold assembly.
13. After the fuel rail assembly has been installed and before the fuel injector wire connectors have been connected, connect the negative battery cable and turn the key to the **ON** position. This will cause the fuel pump to run for 2–3 seconds and pressurize the system.
14. Check for fuel leaks, especially where the fuel injector is installed into the fuel rail.
15. Disconnect the negative battery cable.
16. Install the upper intake manifold in the reverse order of removal. Tighten the retaining bolts, in sequence, to 15–22 ft. lbs. (20–30 Nm).
17. Connect the fuel injector wire connectors.
18. Connect the negative battery cable. Start the engine and let it idle.
19. Turn the engine **OFF** and check for fuel leaks.

3.8L Engine

Except Supercharged Engine

1. Disconnect the negative battery cable.
2. Remove the fuel tank cap and relieve the fuel system pressure.
3. Disconnect the electrical connectors at the idle air bypass valve, TP sensor and EGR position sensor.
4. Disconnect the throttle linkage at the throttle ball and the transmission linkage from the throttle body.
5. Remove the 2 bolts securing the bracket to the intake manifold and position the bracket with the cables aside.
6. Disconnect all the vacuum lines and PCV from the upper intake manifold and throttle assembly. Tag all lines prior to removal for ease of reinstallation.
7. Remove the upper intake manifold retaining bolts and remove the upper intake manifold and throttle body assembly.
8. Disconnect the fuel lines from the fuel rail assembly.
9. Remove the fuel pressure regulator.
10. Disconnect the electrical connectors from the fuel injectors. Remove the injector retaining clips, as required.
11. Remove the fuel rail retaining bolts. Carefully disengage the fuel

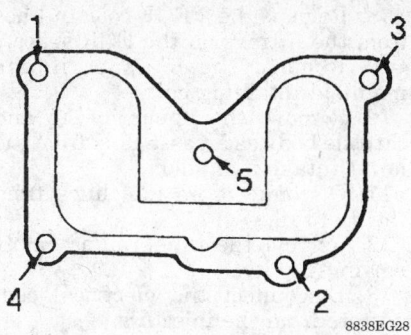

8838EG28

Upper intake manifold bolt torque — 2.3L engine

rail from the fuel injectors and remove the fuel rail.

NOTE: It may be easier to remove the injectors with the fuel rail as an assembly.

12. Grasping the injector body, pull while gently rocking the injector from side-to-side to remove the injector from the fuel rail or the intake manifold.

13. Inspect the pintle protection cap (plastic hat) and washer for signs of deterioration. Replace the complete injector, as required. If the cap is missing, look for it in the intake manifold.

NOTE: The pintle protection cap is not available as a separate part.

To install:

14. Lubricate new O-rings with light grade oil and install 2 on each injector.

NOTE: Never use silicone grease as it will clog the injectors.

15. Install the injectors in the intake manifold using a light, twisting pushing motion.

16. Install the fuel rail, pushing it down to ensure all injector O-rings are fully seated in the fuel rail cups and intake manifold.

17. Install the retaining bolts while holding the fuel rail down and tighten to 87 inch lbs. (10 Nm). Reinstall the injector retaining clips, as required.

18. Install the fuel pressure regulator retaining bolt and tighten to 15–22 ft. lbs. (20–30 Nm).

19. Connect the fuel lines to the fuel rail.

20. With the injector wiring disconnected, connect the negative battery cable and turn the ignition to the **RUN** position to allow the fuel pump to pressurize the system. Check for fuel leaks.

21. Disconnect the negative battery cable.

22. Connect the fuel injector wiring harness.

23. Install the upper intake manifold and throttle body assembly by reversing the removal procedure. Tighten the upper intake manifold retaining bolts to 24 ft. lbs. (32 Nm).

24. Connect the negative battery cable, start the engine and check for fuel leaks.

Supercharged Engine

1. Disconnect the negative battery cable.

2. Remove the fuel tank cap and relieve the fuel system pressure.

3. Remove the supercharger assembly.

4. Disconnect the fuel lines from the fuel rail assembly.

5. Remove the 4 fuel rail assembly retaining bolts and remove the fuel pressure regulator bracket retaining bolt.

6. Disconnect the electrical connectors from the injectors.

7. Carefully disengage the fuel rail from the fuel injectors and remove the fuel rail.

NOTE: It may be easier to remove the injectors with the fuel rail as an assembly.

8. Grasping the injector body, remove the injector from the fuel rail or intake manifold by pulling while gently rocking the injector from side-to-side.

9. Inspect the pintle protection cap (plastic hat) and washer for signs of deterioration. Replace the complete injector, as required. If the cap is missing, look for it in the intake manifold.

NOTE: The pintle protection cap is not available as a separate part.

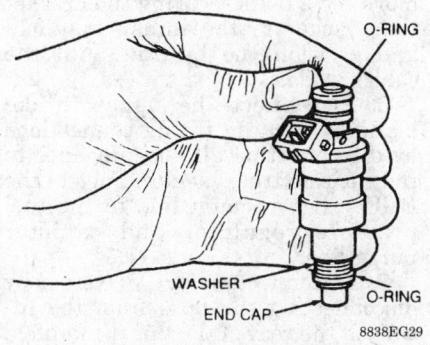

8838EG29

Fuel injector — typical

To install:

10. Lubricate new O-rings with light grade oil and install 2 on each injector.

NOTE: Never use silicone grease as it will clog the injectors.

11. Install the injectors, using a light, twisting, pushing motion.

12. Place the fuel rail assembly over each of the injectors and seat the injectors into the fuel rail.

NOTE: It may be easier to seat the injectors in the fuel rail and then seat the entire assembly in the lower intake manifold.

13. Install the fuel rail assembly retaining bolts and tighten to 70–97 inch lbs. (8–11 Nm). Install the fuel pressure regulator bracket retaining bolt and tighten to 15–22 ft. lbs. (20–30 Nm).

14. Install the supercharger assembly.

15. Connect the negative battery cable. Turn the ignition from **OFF** to **ON** several times without starting the engine to check for fuel leaks. Check all connections at the fuel rail and injectors.

16. Start the engine and warm to operating temperature. Check for fuel or coolant leaks.

4.6L (SOHC) Engine

1. Disconnect the negative battery cable.

2. Remove the fuel tank cap and relieve the fuel system pressure.

3. Disconnect the vacuum line at the pressure regulator.

4. Disconnect the fuel lines from the fuel rail.

5. Disconnect the electrical connectors from the injectors.

6. Remove the fuel rail assembly retaining bolts.

7. Carefully disengage the fuel rail from the fuel injectors and remove the fuel rail.

NOTE: It may be easier to remove the injectors with the fuel rail as an assembly.

8. Grasping the injector body, pull while gently rocking the injector from side-to-side to remove the injector from the fuel rail or intake manifold.

9. Inspect the pintle protection cap and washer for signs of deterioration. Replace the complete injector, as required. If the cap is missing, look for it in the intake manifold.

NOTE: The pintle protection cap is not available as a separate part.

To install:

10. Lubricate new O-rings with light grade oil and install 2 on each injector.

NOTE: Never use silicone grease as it will clog the injectors.

11. Install the injectors using a light, twisting, pushing motion.
12. Install the fuel rail, pushing it down to ensure all injector O-rings are fully seated in the fuel rail cups and intake manifold.
13. Install the retaining bolts while holding the fuel rail down and tighten to 71–106 inch lbs. (8–12 Nm).
14. Connect the fuel lines to the fuel rail and the vacuum line to the pressure regulator.
15. With the injector wiring disconnected, connect the negative battery cable and turn the ignition switch to the **RUN** position to allow the fuel pump to pressurize the system.
16. Check for fuel leaks.
17. Disconnect the negative battery cable.
18. Connect the electrical connectors to the fuel injectors.
19. Connect the negative battery cable and start the engine. Let it idle for 2 minutes.
20. Turn the engine **OFF** and check for leaks.

4.6L (DOHC) Engine

1. Disconnect the negative battery cable.
2. Remove the fuel tank cap and relieve the fuel system pressure.
3. Remove engine cover and air inlet tube.
4. Disconnect the vacuum line at the fuel pressure regulator and brake booster.
5. Disconnect the fuel lines from the fuel rail assembly.
6. Remove the 10 fuel injection supply manifold assembly bolts.

NOTE: It may be easier to remove the injectors with the fuel rail as an assembly.

7. Disconnect the electrical connectors from the fuel injectors.
8. Carefully disengage the fuel rail from the fuel injectors and remove the fuel rail.
9. Grasping the injector body, pull while gently rocking the injector from side-to-side to remove the injector from the fuel rail or the intake manifold.

To install:

10. Lubricate new O-rings with light grade oil and install 2 on each injector.

NOTE: Never use silicone grease as it will clog the injectors.

11. Install the injectors in the intake manifold using a light, twisting pushing motion.
12. Install the fuel rail, pushing it down to ensure all injector O-rings are fully seated in the fuel rail cups and intake manifold.
13. Install the retaining bolts while holding the fuel rail down and tighten to 71–106 inch lbs. (8–12 Nm).
14. Connect the fuel lines to the fuel rail.
15. Connect the vacuum line to the fuel pressure regulator and brake booster.
16. With the injector wiring disconnected, connect the negative battery cable and turn the ignition to the **RUN** position to allow the fuel pump to pressurize the system. Check for fuel leaks.
17. Disconnect the negative battery cable.
18. Connect the fuel injector wiring harness. Install air inlet tube.
19. Connect the negative battery cable, start the engine and check for fuel leaks. Turn engine **OFF** and install the engine cover.

5.0L Engine

1. Disconnect the negative battery cable.
2. Remove the fuel tank cap and relieve the fuel system pressure.
3. Partially drain the cooling system into a suitable container.
4. Disconnect the electrical connectors at the idle air bypass valve, TP sensor and EGR sensor.
5. Disconnect the throttle linkage at the throttle ball and transmission linkage from the throttle body. Remove the 2 bolts securing the bracket the bracket to the intake manifold and position the bracket with the cables aside.
6. Disconnect the upper intake manifold vacuum fitting connections by disconnecting all vacuum lines to the vacuum tree, vacuum lines to the EGR valve, vacuum line to the fuel pressure regulator and canister purge line.
7. Disconnect the PCV system by disconnecting the hose from the fitting on the rear of the upper manifold and disconnect the PCV vent closure tube at the throttle body.

8. Remove the 2 EGR coolant lines from the fittings on the EGR spacer.
9. Remove the 6 upper intake manifold retaining bolts.
10. Remove the upper intake and throttle body as an assembly from the lower intake manifold.
11. Disconnect the fuel lines from the fuel rail.
12. Remove the 4 fuel rail assembly retaining bolts.
13. Disconnect the electrical connectors from the injectors.
14. Carefully disengage the fuel rail from the fuel injectors.

NOTE: It may be easier to remove the injectors with the fuel rail as an assembly.

15. Grasping the injector body, pull up while gently rocking the injector from side-to-side to remove the injector from the fuel rail or intake manifold.
16. Inspect the pintle protection cap (plastic hat) and washer for signs of deterioration. Replace the complete injector, as required. If the cap is missing, look for it in the intake manifold.

NOTE: The pintle protection cap is not available as a separate part.

To install:

17. Lubricate new O-rings with light grade oil and install 2 on each injector.

NOTE: Never use silicone grease as it will clog the injectors.

18. Install the injectors using a light, twisting, pushing motion.
19. Install the fuel rail, pushing it down to ensure all the injector O-rings are fully seated in the fuel rail cups and intake manifold.
20. Install the retaining bolts while holding the fuel rail down and tighten to 70–105 inch lbs. (8–12 Nm).
21. Connect the fuel lines to the fuel rail.
22. With the injector wiring disconnected, connect the negative battery cable and turn the ignition switch to the **RUN** position to allow the fuel pump to pressurize the system.
23. Check for fuel leaks.
24. Disconnect the negative battery cable.
25. Connect the electrical connectors to the injectors.
26. Install the upper intake manifold and throttle body assembly by reversing the removal procedure. Tighten the retaining bolts to 12–17 ft. lbs. (16–24 Nm).

27. Refill the cooling system and connect the negative battery cable.

28. Start the engine and let it idle for 2 minutes. Turn the engine **OFF** and check for leaks.

EMISSION CONTROLS

Emission Warning Lamps

RESETTING

These vehicles have a CHECK ENGINE lamp that will light when there is a fault in the engine control system. This light cannot be reset without diagnosing the fault in the system. When the system has been diagnosed and the problem corrected, the light will go out.

Service Interval Lamp

RESETTING

Thunderbird and Cougar

The optional Vehicle Maintenance Monitor (VMM) alerts the vehicle operator to when engine oil needs to be changed and when fuel, oil, coolant and washer fluids are low. To reset the VMM after an oil change, proceed as follows:

1. Turn the ignition key **OFF**, then turn it **ON**, but do not start the engine.

2. Within 16 seconds of turning the key to **ON**, stick a straightened paperclip into the reset switch hole and firmly push in the switch. The left side of the display will now flash.

NOTE: The reset switch is very small and is located to the left of the word OK on the VMM panel.

3. Keep pushing down on the reset switch with the paperclip until the left side of the display stops flashing. The VMM is now reset. Do not stop pushing in the switch until the display stops flashing, or the VMM will not be reset.

ENGINE MECHANICAL

Engine Assembly

REMOVAL AND INSTALLATION

2.3L Engine

1. Disconnect the negative battery cable and relieve the fuel system pressure.

2. Drain the cooling system and the crankcase.

3. Mark the position of the hood on the hinges and remove the hood.

4. Remove the air cleaner outlet hose.

5. Remove the radiator upper and lower hoses. Disconnect the electrical connector to the cooling fan and remove the fan and shroud. If equipped with automatic transmission, disconnect the oil cooler lines from the radiator. Remove the radiator.

6. Disconnect the heater hose from the heater core. Tag and disconnect the wires from the alternator and starter, Disconnect the accelerator cable from the throttle body.

7. If equipped with air conditioning, remove the compressor from the mounting bracket and position it aside, leaving the refrigerant lines attached.

8. If equipped with power steering, remove the pump and position aside, leaving the hoses attached.

9. Disconnect the flexible fuel line at the fuel rail and plug the fuel line.

10. Disconnect the coil primary wire, the water temperature sending unit connector and the injector wiring harness connectors from the main wiring harness.

11. Remove the starter and remove the engine mount bolts.

12. Raise and safely support the vehicle. Remove the flywheel or converter housing upper retaining bolts.

13. Disconnect the muffler inlet pipe at the exhaust manifold. Disconnect the engine right and left mounts at the No. 2 crossmember pedestals. Remove the flywheel or converter housing cover.

14. If equipped with a manual transmission, remove the flywheel housing lower retaining bolts. If equipped with an automatic transmission, disconnect the converter from the flywheel and disconnect the transmission oil cooler lines, if attached to the engine at the pan rail.

Remove the converter housing lower retaining bolts.

15. Lower the vehicle. Support the transmission and flywheel or converter housing with a jack.

16. Attach suitable engine lifting equipment to the engine lifting brackets. Carefully lift the engine out of the engine compartment and install on a workstand.

To install:

17. Install the clutch, if removed.

18. Carefully lower the engine into the engine compartment. Make sure the studs on the exhaust manifold are aligned with the holes in the muffler inlet pipe.

19. If equipped with an automatic transmission, start the converter pilot into the crankshaft. If equipped with a manual transmission, start the transmission input shaft into the clutch disc. It may be necessary to adjust the position of the transmission in relation to the engine if the input shaft will not enter the clutch disc.

NOTE: If the engine hangs up after the shaft enters, turn the crankshaft slowly in a clockwise direction, with the transmission in gear, until the shaft splines mesh with the clutch disc splines.

20. Install the flywheel or converter housing upper retaining bolts. Remove the engine lifting equipment.

21. Remove the jack from the transmission. Raise and safely support the vehicle.

22. Install the flywheel or converter housing lower retaining bolts. If equipped with an automatic transmission, attach the converter to the flywheel and tighten the retaining nuts to 20–34 ft. lbs. (27–46 Nm).

23. Install the flywheel or converter housing dust cover. Install the left and right engine mounts to the No. 2 crossmember pedestal. Tighten the nuts and bolts to 80–106 ft. lbs. (108–144 Nm).

24. Connect the muffler inlet pipe to the manifold. Connect the fuel line to the fuel rail.

25. Install the starter and connect the starter cable.

26. Lower the vehicle. Connect the oil pressure and water temperature sending unit connectors. Connect the coil and alternator wires. Connect the accelerator cable and the heater hoses.

27. If equipped with air conditioning, install the compressor in the mounting bracket. If equipped with power steering, install the pump. Install the drive belt.

28. Install the radiator, cooling fan and shroud. Connect the fan electrical connector. If equipped with automatic transmission, connect the oil cooler lines to the radiator. Install the upper and lower radiator hoses.

29. Install the air cleaner outlet hose.

30. Fill the crankcase with the proper type and quantity of oil. Fill and bleed the cooling system.

31. Connect the negative battery cable, start the engine and bring to normal operating temperature. Check for leaks. Check all fluid levels.

32. Align the hood on the hinges with the marks that were made during removal. Secure with the mounting bolts.

3.8L Engine

1. Disconnect the negative battery cable. Drain the crankcase and the cooling system.

2. Relieve the fuel system pressure and discharge the air conditioning system.

3. Disconnect the electrical connector to the underhood lamp. Mark the position of the hood on the hinges and remove the hood.

4. Remove the left cowl vent screen and wiper module. On non-supercharged engines, disconnect the alternator to voltage regulator wiring assembly.

5. On supercharged engines, remove the upper charge air cooler tube at the supercharger and cooler assemblies. Remove the bolt retaining the cooler tube to the alternator bracket and remove the tube.

6. Remove the radiator upper sight shield. Release the belt tension and remove the drive and accessory/supercharger belts. Remove the air cleaner-to-throttle body tube.

7. On supercharged engines, disconnect the cooling fan electrical connector and remove the cooling fan/shroud assembly. On non-supercharged engines, remove the fan and shroud.

8. Remove the upper radiator hose and disconnect the heater hoses. If equipped with an automatic transmission, disconnect the oil cooler lines from the radiator.

9. Disconnect the lower radiator hose at the water pump. Remove the radiator. On supercharged engines it will also be necessary to remove the 2 push pins retaining the charge air cooler to the radiator assembly.

10. Disconnect the power steering pressure hose assembly. On non-supercharged engines, remove the power steering pump and bracket assembly and position aside.

11. Disconnect the air conditioner compressor clutch wire. Disconnect and plug the refrigerant lines. Remove the compressor.

12. Remove the coolant recovery reservoir and remove the wiring shield. Remove the accelerator cable mounting bracket and position aside.

13. Disconnect the fuel lines from the fuel rail. Tag and disconnect the engine control module (PCM) wiring, engine feed harnesses and vacuum hoses.

14. On non-supercharged engines, disconnect the ground and coil wires. On supercharged engines, disconnect the DIS module wiring, remove the coil pack retaining bolts and position the coil pack aside.

15. On supercharged engines, remove the nuts retaining the lower charge air cooler tube to the supercharger elbow and lower charge air cooler tube bracket and remove the charge air cooler tube retaining bolt and nut at the alternator bracket.

16. On supercharged engines, remove the alternator bracket bolts, disconnect the alternator wiring and remove the alternator. Remove the power steering pump bracket assembly and position aside.

17. Disconnect the canister purge line and disconnect 1 end of the throttle control valve cable.

18. Raise and safely support the vehicle. Remove the oil filter element.

19. On supercharged engines, remove the 2 nuts retaining the lower charge air cooler tube to the charge air cooler and remove the charge air cooler and charge air cooler tube.

20. Remove the exhaust pipe-to-manifold nuts and remove the left exhaust shield. Disconnect the oxygen sensors.

21. If equipped with an automatic transmission, remove the inspection plug and remove the torque converter bolts.

22. Remove the engine-to-transmission bolts and remove the engine mount through bolts. On supercharged engines, remove the left mount retaining strap bolt.

23. Remove the crankshaft pulley assembly.

NOTE: If the crankshaft pulley and vibration damper have to be separated, mark the damper and pulley so they may be reassembled in the same relative position. This is important as the damper and pulley are initially balanced as a unit. If the crankshaft damper is being replaced, check if the original damper has balance pins installed. If so, new balance pins must be installed on the new damper in the same position as the original damper.

24. Remove the starter. Remove the ground cable and remove the left and right starter harness retainers.

25. Disconnect the oil level indicator sensor and partially lower the vehicle. Disconnect the oil pressure sending unit gauge assembly.

26. Position a floor jack under the transmission and position suitable engine lifting equipment.

27. Remove the engine from the vehicle and position on a workstand.

To install:

28. Remove the engine assembly from the workstand and install engine lifting equipment.

29. Position the engine in the vehicle and install 2 engine-to-transmission bolts. Lower the engine onto the mounting seats, left side first, and remove the lifting equipment. Remove the jacks.

30. Tighten the 2 engine-to-transmission bolts to 40–50 ft. lbs. (55–68 Nm) and connect the oil pressure sending unit gauge assembly. Raise and safely support the vehicle.

31. Install the remaining engine-to-transmission bolts and tighten to 40–50 ft. lbs. (55–68 Nm).

32. Install the torque converter bolts and tighten to 20–34 ft. lbs. (27–46 Nm). Install the inspection plug.

33. Install and tighten the engine mount through bolts to 35–50 ft. lbs. (47–68 Nm). On supercharged engines, install the left mount retaining strap bolt and tighten to 33–45 ft. lbs. (45–61 Nm).

34. Install the starter. Install the starter harness retainer, ground cable and transmission oil cooler line bracket. Install the exhaust pipe-to-manifold nuts.

35. Install the crankshaft pulley assembly and tighten the bolts to 20–28 ft. lbs. (26–30 Nm).

36. Connect the oxygen sensors and the oil level indicator sensor. Install a new oil filter and lower the vehicle.

37. Connect the throttle control valve cable and the canister purge line.

38. On supercharged engines, perform the following:

   a. Install the lower charge air cooler tube, charge air cooler and power steering pump bracket assembly.

   b. Install the alternator, connect the wiring and install the alternator bracket bolts.

c. Install the charge air cooler tube bolts at the power steering bracket and install the nuts retaining the lower charge air cooler tube to the lower charge air cooler tube bracket and supercharger elbow.

d. Install the coil pack and retaining bolts.

39. Install the coolant recovery reservoir.

40. Connect the alternator-to-voltage regulator wiring, the engine control module (PCM) wiring assembly and engine feed harnesses. Connect the vacuum hoses.

41. On non-supercharged engines, connect the wiring assembly ground and coil wire. On supercharged engines, connect the DIS module wiring.

42. Connect the fuel lines to the fuel rail. Install the accelerator cable mounting bracket and the wiring shield.

43. Install the air conditioning compressor and retaining bolts. Tighten the bolts to 30–45 ft. lbs. (41–61 Nm).

44. Remove the plugs from the air conditioner compressor lines and connect the lines to the compressor. Connect the compressor clutch wire.

45. On non-supercharged engines, install the power steering pump bracket assembly. Connect the power steering hoses.

46. Install the radiator. On supercharged engines, install the charge air cooler to the radiator and install the retaining push pins.

47. Connect the lower radiator hose to the water pump and install the heater hoses. If equipped with an automatic transmission, install the oil cooler lines to the radiator.

48. Install the upper radiator hose and the fan and fan shroud. On supercharged engines, connect the cooling fan electrical connector.

49. Position the drive belts and the accessory/supercharger belts. Install the radiator sight shield.

50. On supercharged engines, install the charge air cooler tube and bolts retaining the tube to the power steering bracket. Install the upper charge air cooler tube to the supercharger and cooler assemblies.

51. Install the cowl vent screen and wiper module. Install the hood, aligning the marks that were made during removal. Connect the underhood lamp wiring.

52. Fill the crankcase with the proper type and quantity of engine oil. Fill and bleed the cooling system.

53. Connect the negative battery cable, start the engine and bring to normal operating temperature. Check for leaks. Check all fluid levels.

54. Leak test, evacuate and charge the air conditioning system. Observe all safety precautions.

4.6L (SOHC) Engine

1. Disconnect the negative battery cable, drain the cooling system and discharge/recycle the A/C refrigerant.

2. Remove the air cleaner outlet tube and air cleaner assembly.

3. Remove the fan blade and shroud.

4. Relieve the fuel system pressure.

5. Disconnect the 42-pin and 8-pin connectors and position out of the way.

6. Label all components for reassembly. Disconnect the accelerator cable, speed control actuator, throttle valve control cable, canister purge electrical connector and vacuum lines.

7. Disconnect the power supply from the power distribution box and starter relay.

8. Disconnect the transmission oil cooler tubes from the transmission, upper radiator hose and heater hoses.

9. Disconnect the engine-to-frame ground straps. Partially raise the vehicle and support safely. Remove the front wheels.

10. Disconnect the right and left front anti-lock sensor and brackets.

11. Remove the right and left disc brake caliper bolts. Remove the calipers and support with wire to a frame member.

12. Disconnect the right and left front suspension upper arms from the spindles.

13. Disconnect the front springs and shocks from the lower arms.

14. Raise the vehicle and support safely.

15. Drain the engine oil.

16. Disconnect the dual converter Y-pipe from the manifolds. Disconnect the transmission shift cable and bracket.

17. Index the driveshaft centering socket yoke to the rear axle universal joint flange.

18. Remove the 4 bolts connecting the driveshaft centering socket yoke to the rear axle universal joint flange. Support the rear axle assembly with a jackstand.

19. Remove the rear axle assembly to rear sub-frame bolts. Loosen the rear differential bracket-to-body bolts and lower.

20. Slide the driveshaft rearward until it is free of the extension housing.

21. Remove the lower radiator hose, power steering lines and steering oil cooler.

22. Disconnect the wiring connector at the bulkhead. Support the front sub-frame using Rotunda Powertrain Lift 014-00765 and Adapter 014-00341 or equivalent.

23. Remove the rear engine support insulator bolts and disconnect the steering coupling at the pinch bolt joint.

24. Remove the 8 sub-frame bolts and lower the engine/transmission assembly.

25. Label and remove all needed components. Separate the engine from the transmission and place the engine on a suitable workstand.

To install:

26. Install engine lifting brackets 014-00334. Install the transmission to the engine. Make sure the torque converter studs align with the holes in the flywheel. Torque the bellhousing retaining bolts to 30–44 ft. lbs. (40–60 Nm). Torque the torque converter bolts to 25 ft. lbs. (35 Nm).

27. Install the transmission housing cover, starter motor and transmission-to-engine block brackets.

28. Position the transmission oil cooler tube bracket to the transmission case.

29. Raise the engine/transmission assembly and carefully lower onto the front sub-frame.

30. Install the right and left front engine support insulator through bolts. Torque the bolts to 22 ft. lbs. (30 Nm).

31. Install the power steering lines and remove the engine lift brackets from the cylinder heads.

32. Raise the engine/transmission/sub-frame assembly using the Powertrain Lift 014-00765 and Adapter 014-00341 into the vehicle.

33. Align the sub-frame-to-body and install the bolts. Torque bolts to 100 ft. lbs. (130 Nm).

34. Connect the steering pinch joint and driveshaft. Raise the rear axle assembly into the vehicle and torque the rear sub-frame bolts to 89 ft. lbs. (120 Nm). Torque the 2 rear axle differential insulator nuts to 94 ft. lbs. (127 Nm).

35. Align the driveshaft centering socket yoke to the rear axle universal joint flange and install the 4 retaining bolts and torque to 162 ft. lbs. (220 Nm).

36. Connect the transmission shift linkage, front suspension arms and lower radiator hose.

37. Connect the power steering hoses to steering cooler, ground straps and dual converter Y-pipe.

38. Connect the front suspension upper arms to the spindles and torque to 98 ft. lbs. (92 Nm).

39. Connect the anti-lock sensor and install the calipers. Install the front wheels.

40. Lower the vehicle and connect the A/C lines, transmission cooler lines and radiator hoses.

41. Connect all remaining hoses, lines, electrical connectors and cables.

42. Install the fan/ shroud and air cleaner.

43. Refill the cooling system. Evacuate, recharge and leak test the A/C system.

44. Connect the negative battery cable, fill the engine with oil, start the engine and check for leaks.

4.6L (DOHC) Engine

1. Disconnect the negative battery cable. Drain the crankcase and the cooling system.

2. Properly relieve the fuel system pressure and discharge the air conditioning system.

3. Disconnect IAT sensor connector and crankcase vent tube from air cleaner outlet tube. Loosen air cleaner outlet to throttle body bolt and disconnect the tube.

4. Remove remaining bolt from support bracket clamp on right valve cover and remove air cleaner outlet tube assembly and resonator assembly.

5. Remove the hush panel to expose the windshield wiper module, remove screw, lower module and bracket assembly and disconnect electrical connector. Remove the wiper module.

6. Remove the cooling fan and shroud.

7. Disconnect fuel lines. Remove 42 pin connector from retaining bracket on left fender well. Disconnect connector and position aside.

8. Remove power distribution box and disconnect alternator **B+** connector from inside of box.

9. Disconnect engine harness connector from canister purge solenoid assembly. Disconnect accelerator and speed control cables from throttle body and from accelerator cable bracket.

10. Disconnect canister purge vacuum line at throttle body and chassis vacuum supply hose at connection on cowl support.

11. Disconnect heater supply and return hose at rear of right cylinder head and upper radiator hose at coolant crossover tube.

12. Remove power steering return and supply hoses from reservoir and drain fluid into appropriate container. Remove reservoir retaining bolt and stud from left coil bracket and remove reservoir.

13. Remove upper transmission cooler line from radiator.

14. Install suitable engine lifting eyes. Disconnect and plug air conditioning compressor lines and disconnect retaining clips from pump.

15. Raise and safely support the vehicle. Remove the wheel and tire assemblies.

16. Disconnect right and left ride height sensor electrical connectors.

17. Remove right and left caliper bolts and remove calipers from rotors. Support calipers with mechanics wire.

18. Disconnect right and left upper control arms from spindles.

19. Remove Y-pipe from both exhaust manifolds and resonator. Disconnect ground strap from right fender apron.

20. Disconnect power steering pressure line from steering rack at power steering pump connection.

21. Remove lower radiator hose. Remove right and left lower strut to control arm bolts and nuts.

22. Disconnect transmission wiring harness and shift linkage. Disconnect lower transmission cooler line from radiator.

23. Index driveshaft to rear axle companion flange.

24. Remove 4 bolts connecting driveshaft to rear axle companion flange. Support rear axle housing with jackstand.

25. Remove 2 nut and bolt assemblies retaining the front of the differential to the undercarriage. Loosen rear differential bracket to body bolts and lower.

26. Slide driveshaft rearward until it is free of transmission extension housing.

27. Remove 2 nuts from bottom front cover studs and remove starter cable.

28. Disconnect low oil level sensor connector and wiring harness from oil pan. Remove starter.

29. Support subframe with Rotunda Powertrain Lift 014-00765 and adapter 014-00341, or equivalent. Remove rear transmission mount bolts.

30. Disconnect steering shaft coupler at rag joint. Remove 8 subframe bolts.

31. Lower engine and transmission assembly from vehicle.

32. Remove motor mount through bolts.

33. Install suitable engine lifting equipment on the engine lifting eyes and lift engine and transmission from subframe.

34. Lower engine and transmission. Support the transmission on a level, stationary surface for transmission storage.

35. Remove transmission to cylinder block mounting bolts and separate engine from transmission/torque converter assembly. Place engine on suitable workstand. Remove engine lifting equipment.

36. Remove right and left motor mounts from block. Remove right and left water jacket pipe plugs in cylinder block and drain coolant. Reinstall plugs.

37. Disconnect EGR to exhaust manifold tube from left exhaust manifold connector, differential pressure feedback EGR hose connections and loosen EGR tube connector at EGR control valve. Remove EGR valve to exhaust manifold tube.

38. Remove exhaust manifolds and discard gaskets.

To install:

39. Install exhaust manifolds with new gaskets. Tighten nuts in sequence to 15–22 ft. lbs. (20–30 Nm).

40. Install EGR to exhaust manifold tube to EGR control valve, left exhaust manifold and differential pressure feedback EGR hose.

41. Install left and right motor mounts to block. Install engine lifting bracket.

42. Remove engine from stand using suitable lifting equipment.

43. Connect transmission to cylinder block. Tighten bolts to 30–44 ft. lbs. (40–60 Nm).

44. Position engine and transmission on subframe assembly. Install mount through bolts and tighten to 15–22 ft. lbs. (20–30 Nm).

45. Raise engine, transmission and subframe assembly into position.

46. Align subframe to body and install bolts. Tighten to 73–100 ft. lbs. (95–130 Nm).

47. Connect steering shaft coupler at rag joint. Install rear transmission mount bolt and tighten to 15–22 ft. lbs. (20–30 Nm). Remove lifting equipment.

48. Install starter, low oil level sensor and lower engine wiring harness.

49. Install starter cable and 2 retaining nuts to lower front cover studs.

50. Install driveshaft to transmission. Lift rear differential into position. Install 2 nut and bolt assemblies retaining the front of the differential to the rear subframe and tighten to 72–89 ft. lbs. (98–120 Nm).

51. Tighten 2 differential mounts to 75–89 ft. lbs. (102–127 Nm). Align driveshaft to companion flange and install 4 bolts. Tighten to 70–96 ft. lbs. (95–130 Nm).

52. Connect lower transmission cooler line to the radiator. Connect the transmission wiring harness and shift linkage.

53. Position lower control arms to strut and install bolt. Tighten to 118–162 ft. lbs. (160–220 Nm).

54. Install lower radiator hose and power steering pressure line. Connect ground strap to right fender apron.

55. Install exhaust Y-pipe.

56. Connect upper control arms to spindles. Tighten to 50–68 ft. lbs. (68–92 Nm). Connect right and left ride height sensor electrical connectors.

57. Install brake calipers. Install front tire and wheel assemblies. Lower the vehicle.

58. Unplug and connect the air conditioning lines. Remove the engine lifting brackets and eyelets.

59. Connect upper transmission cooler line.

60. Install power steering reservoir, power steering return line and power steering supply hose.

61. Install coolant supply hose to reservoir. Connect upper radiator hose at crossover pipe and connect heater supply and return hoses.

62. Connect chassis vacuum supply and canister purge vacuum hoses.

63. Connect accelerator and speed control cables. Connect engine wiring harness to canister purge solenoid assembly.

64. Connect alternator **B+** connector and install power distribution box. Connect 42 pin connector and attach to retaining bracket.

65. Connect the fuel lines. Install cooling fan and shroud.

66. Install windshield wiper module. Install engine air cleaner outlet tube assembly and resonator assembly.

67. Connect IAT sensor and crankcase vent tube.

68. Fill the engine with the proper type and quantity of engine oil. Fill and bleed the cooling system.

69. Fill the power steering system with the proper type and quantity of fluid. Connect the negative battery cable.

70. Start the engine and bring to normal operating temperature. Check for leaks. Check all fluid levels.

71. Leak test, evacuate and charge the air conditioning system. Observe all safety precautions.

5.0L Engine

Mustang

1. Disconnect the negative battery cable. Drain the crankcase and the cooling system.

2. Properly relieve the fuel system pressure and discharge the air conditioning system.

3. Mark the position of the hood on the hinges and remove the hood. Disconnect the battery ground cables from the cylinder block.

4. Remove the air intake duct.

5. Disconnect the upper radiator hose from the thermostat housing and the lower hose from the water pump. If equipped with an automatic transmission, disconnect the oil cooler lines from the radiator.

6. Remove the bolts attaching the radiator fan shroud to the radiator. Remove the radiator. Remove the fan, belt, pulley and shroud.

7. Remove the alternator bolts and position the alternator aside.

8. Disconnect the oil pressure sending unit wire from the sending unit and, if equipped, the low oil level sensor wire from the left side of the oil pan. Disconnect the flexible fuel line at the fuel tank line. Plug the fuel tank line.

9. Disconnect the accelerator cable from the throttle body. Disconnect the TV rod if equipped with an automatic transmission. Disconnect the cruise control cable, if equipped.

10. Disconnect the transmission filler tube bracket from the cylinder block.

11. If equipped with air conditioning, disconnect the lines and electrical connectors at the compressor and remove the compressor. Plug the lines and the compressor fittings to prevent the entrance of dirt and moisture.

12. Disconnect the power steering pump bracket from the cylinder head. Position the power steering pump aside in a position that will prevent the fluid from leaking.

13. Disconnect the power brake vacuum line from the intake manifold.

14. Disconnect the heater hoses from the heater tubes. Disconnect the electrical connector from the coolant temperature sending unit.

15. Remove the flywheel or converter housing-to-engine upper bolts.

16. Disconnect the wiring harness at the two 10-pin connectors.

17. Raise and safely support the vehicle. Disconnect the starter cable from the starter and remove the starter.

18. Disconnect the muffler inlet pipes from the exhaust manifolds. Disconnect the engine mounts from the chassis. Disconnect the downstream thermactor tubing and check valve from the right exhaust manifold stud, if equipped.

19. If equipped with automatic transmission, disconnect the transmission cooler lines from the retainer and remove the converter housing inspection cover. Disconnect the flywheel from the converter and secure the converter assembly in the housing.

20. Remove the remaining converter or flywheel housing-to-engine bolts.

21. Lower the vehicle and then support the transmission. Attach engine lifting equipment and hoist the engine.

22. Raise the engine slightly and carefully pull it from the transmission. Carefully lift the engine out of the engine compartment. Avoid bending or damaging the rear cover plate or other components. Install the engine on a workstand.

To install:

23. Attach the engine lifting equipment and remove the engine from the workstand.

24. Lower the engine carefully into the engine compartment. Make sure the exhaust manifolds are properly aligned with the muffler inlet pipes.

25. Start the converter pilot, or manual transmission input shaft, into the crankshaft. Align the paint mark on the flywheel to the paint mark on the torque converter.

26. Install the flywheel or converter housing upper bolts, making sure the dowels in the cylinder block engage the housing.

27. Install the engine mount-to-chassis attaching fasteners and remove the engine lifting equipment.

28. Raise and safely support the vehicle. Connect both muffler inlet pipes to the exhaust manifolds. Install the starter and connect the starter cable.

29. If equipped with automatic transmission, remove the retainer

holding the converter in the housing. Attach the converter to the flywheel. Install the converter housing inspection cover.

30. Install the remaining flywheel or converter housing attaching bolts. Remove the support from the transmission and lower the vehicle.

31. Connect the wiring harness at the two 10-pin connectors.

32. Connect the coolant temperature sending unit wire and connect the heater hoses. Connect the wiring to the sensors.

33. Connect the transmission filler tube bracket, if equipped with automatic transmission.

34. Connect the accelerator cable and TV cable. Connect the cruise control cable, if equipped.

35. Remove the plug from the fuel tank line and connect the fuel line and the oil pressure sending unit wire.

36. Install the pulley, water pump belt and fan/clutch assembly.

37. Position the alternator bracket and install the alternator bolts. Connect the alternator and ground cables.

38. Install the air conditioning compressor. Unplug and connect the refrigerant lines and connect the electrical connector to the compressor.

39. Install the power steering pump bracket and the accessory drive belt. Connect the power brake vacuum line.

40. Place the shroud over the fan and install the radiator. Connect the radiator hoses and the transmission oil cooler lines. Position the shroud and install the bolts.

41. Connect the heater hoses to the heater tubes. Fill and bleed the cooling system. Fill the crankcase with the proper type and quantity of engine oil. Adjust the transmission throttle linkage, if equipped with automatic transmission.

42. Connect the negative battery cable. Start the engine and bring to normal operating temperature. Check for leaks. Check all fluid levels.

43. Install the air intake duct assembly. Install the hood, aligning the marks that were made during removal.

44. Leak test, evacuate and charge the air conditioning system. Observe all safety precautions.

Thunderbird and Cougar

1. Disconnect the negative battery cable. Drain the crankcase and the cooling system.

2. Properly relieve the fuel system pressure and discharge the air conditioning system.

3. Disconnect the electrical connector for the underhood lamp. Mark the position of the hood on the hinges and remove the hood.

4. Remove the oil dipstick. Disconnect and plug the refrigerant lines at the air conditioning compressor.

5. Disconnect the compressor clutch and power steering pressure switch electrical connectors. Disconnect the alternator wiring harness from the alternator and position the harness aside.

6. Remove the fan shroud and the fan. Remove the upper radiator hose.

7. Remove the air cleaner-to-throttle body tube. Disconnect and plug the transmission oil cooler lines at the radiator.

8. Disconnect the throttle and kickdown cables from the throttle body and remove the cable bracket retaining bolts. Position the cable and bracket assembly aside.

9. Tag and disconnect the vacuum lines at the upper intake manifold vacuum tree, air conditioning control panel vacuum supply hose, thermactor valve and EGR valve. Disconnect the electrical connector at the EGR valve.

10. Remove the upper intake manifold as follows:

    a. Disconnect the electrical connectors at the idle air bypass valve, throttle position sensor and EGR position sensor.

    b. Disconnect the vacuum line from the fuel pressure regulator and the PCV hose from the fitting on the rear of the upper manifold.

    c. Remove the upper intake manifold retaining bolts and remove the manifold.

11. Disconnect the main engine wiring harness connectors at the right side of the dash panel. Position the engine wiring harness so it can be removed with the engine.

12. Disconnect the heater hoses at the engine. Disconnect the wiring harness from the coil and distributor and position the harness aside.

13. Disconnect and plug the fuel lines at the fuel supply manifold.

14. Disconnect the lower radiator hose from the water pump. Remove the radiator retaining bolts and remove the radiator.

15. Raise and safely support the vehicle. Remove the oil filter.

16. Remove the starter. Disconnect the oxygen sensors for the right and left catalytic converters. Disconnect the negative battery cable from the left side of the engine.

17. On the right side of the engine, disconnect the brackets for the transmission cooler lines, engine-to-body ground straps and the starter wiring harness.

18. Remove the torque converter inspection cover and mark 1 of the converter studs to the flywheel for alignment during reassembly. Remove the torque converter attaching nuts.

19. Remove the exhaust manifold heat shield at the left manifold flange and disconnect the exhaust pipe from the flange. Disconnect the right exhaust manifold flange.

20. Loosen the transmission mount retaining nut. Remove the converter housing to engine bolts and the motor mount through bolts.

21. Lower the vehicle and disconnect the power steering lines. Cap the lines to prevent contamination.

22. Support the transmission with a floor jack. Install suitable engine lifting equipment on the engine lifting eyes.

23. Lift the engine assembly clear of the engine mounts and remove the engine from the vehicle. Place the engine on a workstand.

To install:

24. Install suitable engine lifting equipment on the engine lifting eyes and lift the engine from the workstand.

25. Carefully lower the engine into the engine compartment. Make sure the exhaust manifolds are properly aligned with the muffler inlet pipes.

26. Start the converter pilot into the crankshaft. Align the mark on the flywheel to the mark on the torque converter.

27. Position the retaining clip for the left oxygen sensor wiring near the left upper transmission-to-engine bolt. Install the converter housing upper bolts. Make sure the dowels in the cylinder block engage the converter housing.

28. Raise and safely support the vehicle. Install the remaining converter housing bolts and install the motor mount through bolts. Tighten the transmission mount retaining nut to 65–85 ft. lbs. (88–115 Nm).

29. Connect the right exhaust manifold flange. Connect the exhaust pipe to the left exhaust manifold flange and install the heat shield.

30. Install the torque converter retaining nuts and the inspection cover.

31. On the right side of the engine, install the brackets for the transmission cooler lines, engine-to-body

ground strap and the starter wiring harness.

32. Connect the negative battery cable to the left side of the engine. Connect the oxygen sensors for the catalytic converters.

33. Install the starter. Install a new oil filter and the oil pan drain plug. Lower the vehicle and connect the power steering lines.

34. Install the radiator. Connect the coolant overflow hose and the lower radiator hose.

35. Connect the fuel lines to the fuel supply manifold.

36. Position and connect the wiring harness for the coil and distributor. Connect the heater hoses at the engine. Connect the main engine wiring harness connectors at the right side of the dash panel.

37. Install the upper intake manifold in the reverse order of removal. Be sure to use a new gasket.

38. Connect the vacuum lines at the upper intake manifold vacuum tee, air conditioning control panel vacuum supply hose, thermactor valve and EGR valve. Connect the electrical connector to the EGR valve.

39. Connect the throttle and kickdown cables to the throttle body and install the cable bracket retaining bolts.

40. Connect the transmission oil cooler lines and the upper radiator hose. Install the fan shroud and the fan. Install the air cleaner-to-throttle body tube assembly.

41. Position and connect the wiring harness for the alternator. Connect the compressor clutch electrical connector and connect the refrigerant lines to the compressor.

42. Install the hood, aligning the marks that were made during removal. Connect the wiring connector for the underhood lamp.

43. Fill the engine with the proper type and quantity of engine oil. Fill and bleed the cooling system. Install the dipstick.

44. Fill the power steering system with the proper type and quantity of fluid. Connect the negative battery cable.

45. Start the engine and bring to normal operating temperature. Check for leaks. Check all fluid levels.

46. Leak test, evacuate and charge the air conditioning system. Observe all safety precautions.

Engine Mounts

REMOVAL AND INSTALLATION

2.3L Engine

Front

1. Disconnect the negative battery cable. Raise and safely support the vehicle. Support the engine using a wood block and jack placed under the engine.

2. Remove the through bolts attaching both mounts to the No. 2 crossmember pedestal bracket. On convertible, remove the nuts.

3. Disconnect shift linkage.

4. Raise the engine sufficiently to disengage the mount from the crossmember pedestal bracket.

5. Remove the bolts attaching the mount to the engine and remove the mount.

To install:

6. Position the mount on the engine and install the attaching bolts. Tighten to 35–46 ft. lbs. (47–63 Nm).

7. Lower the engine into position making sure the mounts are seated flat on the No. 2 crossmember. Hand start the bolts, lower the engine completely, then tighten the through bolts to 72–98 ft. lbs. (97–133 Nm).

8. On convertible, tighten the flange nut to 73–106 ft. lbs. (98–144 Nm).

9. Install shift linkage. Lower the vehicle and connect the negative battery cable.

Rear

1. Disconnect the negative battery cable. Raise and safely support the vehicle.

2. Support the transmission with a jack and a wood block. Remove the nut(s) retaining the rear mount to the crossmember.

3. Remove the 2 bolts and nuts retaining the crossmember to the body brackets. Remove the crossmember by raising the transmission slightly with the jack.

4. Remove the 2 bolts retaining the rear mount to the transmission and remove the mount and retainer. If equipped with automatic transmission, remove the 2 bolts retaining rear mount to the intermediate bracket.

To install:

5. Position the rear mount and retainer on the transmission. Install the 2 retaining bolts and tighten to 51–70 ft. lbs. (68–96 Nm).

6. Install the crossmember to the body brackets. Tighten the retaining

nuts and bolts to 35–50 ft. lbs. (47–68 Nm).

7. Lower the transmission and install the mount to crossmember retaining nuts. Tighten to 26–35 ft. lbs. (34–48 Nm). If equipped with automatic transmission, tighten the nut to 65–87 ft. lbs. (88–119 Nm).

8. Lower the vehicle. Connect the negative battery cable.

3.8L Engine

Front

1. Disconnect the negative battery cable.

2. Remove fan shroud retaining screws. Remove the air tube to the remote air cleaner.

3. Raise and safely support the vehicle. Support engine using a jack and wood block placed under the engine.

4. Remove the engine mount through bolt. On supercharged engines, remove the retaining strap bolt from the left side.

5. Remove shift linkage.

6. Raise engine high enough to clear clevis brackets.

NOTE: Raise the engine carefully so as not to damage the lines and hoses at the rear of the engine.

7. Remove any accessory and oil cooler line retaining clips from the engine support brackets.

8. Remove bolts retaining the engine mount and bracket assembly to engine. Remove the mount and bracket assembly.

NOTE: The left hand front engine mount removal on the supercharged engine may require lowering the front sub frame.

To install:

9. Position the engine mount and bracket assembly to the engine, install the retaining bolts and tighten to 26–34 ft. lbs. (34–47 Nm).

10. Install the accessories to the lower front engine mount support bracket stud. Tighten to 26–34 ft. lbs. (34–47 Nm).

11. Lower the engine into position and make sure the engine mounts are seated flat on the front sub frame; the left mount must seat first. Install the through-bolt and tighten to 35–50 ft. lbs. (47–68 Nm). On supercharged engines, install the retaining strap bolt and tighten to 34–44 ft. lbs. (45–61 Nm).

12. Lower the vehicle and install the air tube and the fan shroud retaining screws. Connect the negative battery cable.

Rear

1. Disconnect the negative battery cable. Raise and safely support the vehicle.

2. Support the transmission with a jack and a wood block. Remove the rear nut attaching the mount-to-crossmember. Keep transmission weight on the mount during nut removal.

3. Remove the 2 bolts retaining the crossmember-to-body brackets. Remove the crossmember by raising the transmission slightly with the jack.

4. Remove the bolts retaining the rear engine mount to the transmission. Remove the mount.

To install:

5. Position the engine mount and retainer on the transmission. Install the 2 retaining bolts and tighten to 35–50 ft. lbs. (47–68 Nm).

6. Install the crossmember-to-body brackets. Tighten the retaining bolts or nuts to 34–47 ft. lbs. (45–65 Nm).

7. Lower the transmission. Install the mount-to-crossmember nut. Tighten to 65–84 ft. lbs. (88–115 Nm).

8. Lower the vehicle and connect the negative battery cable.

4.6L Engine

Front

1. Disconnect the negative battery cable.

2. Remove the engine cover. Remove the air tube to the remote air cleaner.

3. Install engine lifting bracket. Install Three Bar Engine Support D88L-6000-A or equivalent, to engine lifting bracket and support engine.

4. Raise and safely support the vehicle.

5. Remove the engine mount through bolt. Remove any accessory and oil cooler clips from front subframe assembly.

6. Remove front anti-lock brake wire harness from body and front subframe assembly retainers.

7. Remove steering column pinch bolt and separate steering column from intermediate steering shaft.

8. Support front lower control arms and remove front strut through bolts. Separate studs from control arms.

9. Slowly lower control arms until they hang freely.

10. Support the front subframe with jackstands. Remove 8 front subframe retaining bolts.

11. Slowly lower the front subframe to gain access to the engine mounts. Remove engine mount and bracket assembly bolts.

12. Remove bolts retaining the engine mount and bracket assembly to engine. Remove the mount and bracket assembly.

To install:

13. Installation is the reverse of the removal procedure.

Engine mounts and bracket assembly to engine bolts and studs to 40–53 ft. lbs. (53–71 Nm).

Subframe assembly bolts to 72–97 ft. lbs. (97–132 Nm).

Front strut to control arm bolts to 90–118 ft. lbs. (120–160 Nm).

Steering coupling pinch bolt to 20–29 ft. lbs. (28–40 Nm).

Engine mount through bolt to 35–50 ft. lbs. (47–68 Nm).

Rear

1. Disconnect the negative battery cable. Raise and safely support the vehicle.

2. Support the transmission with a jack and a wood block. Remove the rear nuts attaching the mount-to-crossmember.

3. Remove the 2 bolts retaining the crossmember-to-body brackets. Remove the crossmember.

4. Remove the 2 nuts retaining the rear engine mount to the transmission bracket. Remove the mount.

To install:

5. Position the engine mount and retainer on the transmission. Install the 2 retaining bolts and tighten to 50–72 ft. lbs. (68–96 Nm).

6. Install the crossmember-to-body brackets. Tighten the retaining bolts or nuts to 35–50 ft. lbs. (47–68 Nm).

7. Lower the transmission. Install the mount-to-crossmember nuts. Tighten to 25–35 ft. lbs. (34–48 Nm).

8. Lower the vehicle and connect the negative battery cable.

5.0L Engine

Front

1. Disconnect the negative battery cable. Remove fan shroud attaching screws.

2. Raise and safely support the vehicle. Support the engine using a jack and wood block placed under the engine.

3. Remove the nuts or bolts attaching the mounts to the No. 2 crossmember. On Thunderbird and Cougar, remove the through bolts.

4. Disconnect shift linkage on all except Thunderbird and Cougar.

5. Raise the engine sufficiently with the jack to disengage the mount from the crossmember. If equipped, remove the transmission brace attached at the left or right engine mount bracket.

6. Remove the engine mount and bracket assembly to the cylinder block attaching bolts. Remove the engine mount.

To install:

7. Position the mount on the engine and install the attaching bolts. Tighten the bolts to 45–59 ft. lbs. (61–81 Nm).

8. Attach the transmission brace to the right or left engine mount, if equipped. Tighten the nut to 45–59 ft. lbs. (60–81 Nm).

9. Lower the engine into position making sure the mounts are seated flat on the No. 2 crossmember and the insulator studs are at the bottom of the slots.

NOTE: On Thunderbird and Cougar, the left mount, with the locating pin, must seat before the right mount.

10. Install and tighten the mount nuts to 73–106 ft. lbs. (98–144 Nm). On Thunderbird and Cougar, install the through bolts and tighten to 35–45 ft. lbs. (45–61 Nm).

11. Lower the vehicle and install the fan shroud attaching screws. Connect the negative battery cable.

Rear — Mustang

1. Disconnect the negative battery cable. Raise and safely support the vehicle.

2. Support the transmission with a jack and wood block. Remove the 2 nuts attaching the mount to the crossmember.

3. Remove the 2 bolts and nuts attaching the crossmember to the body brackets and remove the crossmember by raising the transmission slightly with the jack.

4. Remove the 2 bolts attaching the rear mount to the transmission and remove the mount and retainer.

To install:

5. Position the rear mount and retainer on the transmission. Install the 2 attaching bolts and tighten to 51–70 ft. lbs. (68–96 Nm).

6. Install the crossmember to the body brackets. Tighten the attaching nuts to 35–50 ft. lbs. (47–68 Nm).

7. Lower the transmission and install the mount-to-crossmember attaching nuts. Tighten to 26–35 ft. lbs. (34–48 Nm).

8. Lower the vehicle and connect the negative battery cable.

Rear — Thunderbird and Cougar

1. Disconnect the negative battery cable. Raise and safely support the vehicle.

2. Remove the nut attaching the rear mount-to-crossmember.

NOTE: This must be done while the transmission weight is still on the mount.

3. Support the transmission with a jack and a wood block. Remove the bolts that attach the crossmember to the body brackets and remove the crossmember.

4. Remove the bolts attaching the mount to the transmission bracket and remove the mount.

To install:

5. Position the mount on the transmission bracket. Install the bolts and tighten to 35–50 ft. lbs. (47–68 Nm).

6. Install the crossmember to the body brackets and tighten the bolts to 34–47 ft. lbs. (45–65 Nm).

7. Lower the transmission and install the nut. Tighten the nut to 65–85 ft. lbs. (88–115 Nm).

8. Lower the vehicle and connect the negative battery cable.

Cylinder Head

REMOVAL AND INSTALLATION

2.3L Engine

1. Disconnect the negative battery cable. Drain the cooling system and relieve the fuel system pressure.

2. Remove the air cleaner assembly.

3. Remove the engine and alternator wiring harnesses. Remove the heater hose retaining screw from the rocker arm cover, if equipped.

4. Tag and disconnect the spark plug wires from the spark plugs. Remove the spark plug wires and, if equipped, the distributor cap. Remove the spark plugs.

5. Tag and disconnect the required vacuum hoses. Remove the dipstick and disconnect the dipstick tube from the bracket.

6. Remove the upper intake manifold and throttle body as follows:

   a. Tag and disconnect the electrical connectors and vacuum hoses.

   b. Disconnect the throttle linkage, cruise control and kickdown cable. Unbolt the accelerator cable from the bracket and position the cable aside.

   c. Disconnect the crankcase vent hose. Disconnect the PCV hose from the fitting on the underside of the upper intake manifold.

   d. Disconnect the EGR tube from the EGR valve. Remove the upper intake manifold mounting bolts and the manifold.

7. Remove the rocker cover retaining bolts and remove the cover. Remove the intake manifold retaining bolts.

8. Remove the accessory drive belt, loosen the retaining bolt and swing the alternator aside.

9. Remove the upper radiator hose. Remove the timing belt cover retaining bolts and remove the cover.

10. Loosen the timing belt idler retaining bolts. Position the idler in the unloaded position and tighten the retaining bolts.

11. Remove the timing belt from the camshaft sprocket and the auxiliary sprocket.

12. Remove the exhaust manifold retaining bolts. Remove the timing belt idler and 2 bracket bolts. Remove the timing belt idler spring stop from the cylinder head.

13. Disconnect the oil sending unit wire, if necessary.

14. Remove the cylinder head bolts and the cylinder head. Clean all gasket mating surfaces and blow the oil out of the cylinder head bolt block holes.

15. Check the cylinder head for flatness using a straight-edge and a feeler gauge. If the head gasket surface is warped greater than 0.006 in., it must be resurfaced. Do not grind more than 0.010 in. from the cylinder head.

To install:

16. Position the head gasket on the block. Position the camshaft with the pin approximately 30 degrees to the right of the 6 o' clock position when facing the front of the cylinder head. The camshaft must be positioned this way to protect protruding valves.

17. Position the cylinder head on the block and install new cylinder head bolts. Tighten the bolts, in sequence, in 2 steps, 1st to 50–60 ft. lbs. (60–81 Nm) and then to 80–90 ft. lbs. (108–122 Nm).

18. Connect the oil sending unit wire, if necessary. Install the timing belt tensioner spring stop to the cylinder head.

19. Position the timing belt tensioner and tensioner spring to the cylinder head and install the retaining bolts. Rotate the tensioner against the spring with belt tensioner

tool T74P-6254-A or equivalent, and temporarily tighten.

20. Install the exhaust manifold retaining bolts. Tighten the bolts, in sequence, in 2 steps, 1st to 178–204 inch lbs. (20–23 Nm) and then to 20–30 ft. lbs. (27–40 Nm).

21. If equipped with a distributor, align the distributor rotor with the No. 1 plug location on the distributor cap. Align the camshaft sprocket with the pointer and align the crankshaft pulley with the pointer on the timing belt cover.

22. Install the timing belt over the sprockets. Loosen the tensioner retaining bolts, rotate the engine by hand 1 complete revolution and check the timing alignment.

23. Tighten the 10mm tensioner bolt to 28–40 ft. lbs. (38–54 Nm) and the 8mm bolt to 14–21 ft. lbs. (19–29 Nm).

24. Install the timing belt cover and tighten the retaining bolts to 6–9 ft. lbs. (8–12 Nm).

25. Install the rocker arm cover and tighten the retaining bolts to 62–97 inch lbs. (7–11 Nm).

26. Install the intake manifold. Tighten the bolts, in sequence, to 19–28 ft. lbs. (26–38 Nm).

27. Install the upper intake manifold and throttle body in the reverse order of removal. Tighten the upper intake-to-lower intake bolts, in sequence, to 15–22 ft. lbs. (20–30 Nm).

28. Position the alternator and install the drive belt. Install the upper radiator hose.

29. Install the dipstick and connect the necessary vacuum hoses. Install the spark plugs, spark plug wires and distributor cap, if equipped.

30. Position and connect the engine and alternator wiring harnesses. Install the hose from the air cleaner to the throttle body. If equipped, install the retaining heater hose screw to the rocker cover.

31. Fill and bleed the cooling system. Connect the negative battery cable, start the engine and bring to normal operating temperature. Check for leaks. If equipped with distributor ignition, check the ignition timing.

3.8L Engine

1. Disconnect the negative battery cable.

2. Relieve the fuel system pressure. Drain the cooling system.

3. Remove air cleaner assembly including the air intake duct and heat tube.

4. Loosen accessory drive belt idler. Remove drive belt.

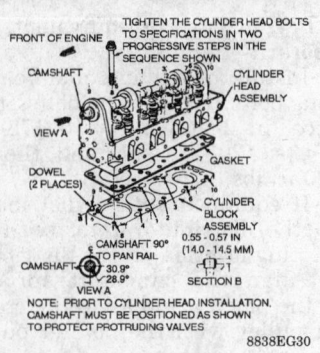

TIGHTEN THE CYLINDER HEAD BOLTS
TO SPECIFICATIONS IN TWO
PROGRESSIVE STEPS IN THE
SEQUENCE SHOWN

FRONT OF ENGINE

CAMSHAFT

VIEW A

CYLINDER
HEAD
ASSEMBLY

GASKET

DOWEL
(2 PLACES)

CYLINDER
BLOCK
ASSEMBLY

CAMSHAFT 90°
TO PAN RAIL

0.55 - 0.57 IN
(14.0 - 14.5 MM)

CAMSHAFT

30.9°
28.9°

VIEW A

SECTION B

NOTE: PRIOR TO CYLINDER HEAD INSTALLATION,
CAMSHAFT MUST BE POSITIONED AS SHOWN
TO PROTECT PROTRUDING VALVES

8838EG30

**Cylinder head bolt torque
sequence — 2.3L engine**

5. If the left cylinder head is being removed, perform the following:

a. On supercharged engines, remove the charge air cooler and charge air cooler tubes.

b. Remove oil fill cap.

c. Remove the power steering pump front mounting bracket attaching bolts.

d. Remove the alternator assembly and accessory drive belt main idler.

e. Remove the power steering/pump alternator bracket retaining bolts.

f. Leaving the hoses connected, place the power steering pump/alternator bracket assembly aside in a position to prevent the fluid from leaking out.

6. If the right cylinder head is being removed, perform the following:

a. If equipped, disconnect the thermactor tube support bracket from the rear of the cylinder head. Remove the thermactor pump pulley and remove the pump.

b. If equipped, remove the air conditioner compressor belt and main drive belt.

c. If equipped, remove the compressor mounting bracket retaining bolts. Leave the hoses connected and position the compressor aside.

d. Remove the PCV valve.

7. On supercharged engines, remove the supercharger. On non-supercharged engines, remove the upper intake manifold as follows:

a. Disconnect the electrical connectors at the idle air bypass valve, throttle position sensor and EGR position sensor.

b. Disconnect the throttle and transmission linkage from the throttle body. Remove the cable bracket from the manifold and position the bracket and cables aside.

c. Tag and disconnect the vacuum lines at the upper manifold

vacuum tree, EGR valve and fuel pressure regulator.

d. Disconnect the PCV hose from the fitting at the rear of the upper manifold.

e. Remove the retaining bolts and remove the upper intake manifold.

8. Remove valve rocker arm cover attaching screws. Remove the fuel rail and the lower intake manifold.

9. Remove the exhaust manifold(s).

10. Loosen rocker arm fulcrum attaching bolts enough to allow the rocker arm to be lifted off the pushrod and rotated to 1 side.

11. Remove the pushrods. Identify the position of each rod. The rods should be installed in their original position during assembly.

12. Remove the cylinder head attaching bolts and discard.

13. Remove the cylinder head(s). Clean all gasket mating surfaces.

14. Check the flatness of the cylinder head gasket surface using a straight-edge and a feeler gauge. The allowable warpage is 0.003 in. for every 6.0 inches. Do not machine more than 0.010 in.

To install:

NOTE: Lightly oil all bolt and stud bolt threads before installation except those specifying special sealant.

15. Position new head gasket(s) on the cylinder block using the dowels for alignment.

16. Position the cylinder head(s) on the block.

17. Install new cylinder head bolts.

NOTE: Always use new cylinder head bolts to assure a leak-tight assembly. Torque retention with used bolts can vary, which may result in coolant or compression leakage at the cylinder head mating surface area.

18. Tighten the new cylinder head attaching bolts in numerical sequence as follows:

a. 37 ft. lbs. (50 Nm)

b. 45 ft. lbs. (60 Nm)

c. 52 ft. lbs. (70 Nm)

d. 59 ft. lbs. (80 Nm)

e. Back-off the attaching bolts 2–3 turns.

f. On supercharged engines, tighten each long and short bolt to 48–55 ft. lbs. (65–75 Nm), rotate an additional 90–110 degrees, then go to the next bolt in sequence.

g. On non-supercharged engines, tighten each long bolt to 11–18 ft. lbs. (15–25 Nm), rotate an additional 85–96 degrees, then go to the

next bolt in sequence. Do the same for each short bolt except only rotate the short bolts 85–96 degrees.

NOTE: When cylinder head attaching bolts have been tightened using multi-step torque procedure, it is not necessary to retighten the bolts after extended engine operation.

19. Lubricate each pushrod with heavy engine oil and install, in their original positions.

20. For each valve, rotate the crankshaft until the lifter rests on the base circle of the camshaft lobe, before tightening the fulcrum attaching bolts to 43 inch lbs. (5 Nm).

21. Lubricate the rocker arm assemblies with heavy engine oil and final tighten the fulcrum bolts to 19–25 ft. lbs. (25–35 Nm). Fulcrums must be fully seated in cylinder head and pushrods must be seated in rocker arm sockets prior to final tightening. Final tightening can be done with the camshaft in any position.

NOTE: If the original valve train components are being installed, a valve clearance check is not required. If a component has been replaced, perform a valve clearance check.

22. Install the exhaust manifold(s).

23. Install the lower intake manifold and the fuel rail.

24. Position cover and new gasket on the cylinder head and install attaching bolts. Note the location of spark plug wire routing clip stud bolts. Tighten attaching bolts to 80–106 inch lbs. (9–12 Nm).

25. Install the upper intake manifold. On supercharged engines, install the supercharger.

26. Install the spark plugs, if removed.

27. Connect the spark plug wires to the spark plugs.

28. If the left cylinder head is being installed, perform the following:

a. Install the oil filler cap.

b. Install the alternator/power steering pump mounting bracket.

c. Install the alternator assembly.

d. Install the main accessory drive belt tensioner assembly.

e. Install the power steering pump assembly.

f. Install the power steering pump support bracket.

g. On supercharged engines, install the charge air cooler tubes.

29. If the right cylinder head is being installed, perform the following:

a. Install PCV valve.

b. If equipped with air conditioning, install the compressor mounting and support brackets and install the compressor.

c. If equipped, install the thermactor pump and pump pulley.

d. If equipped, install the accessory drive belt idler pulley.

e. If equipped, install the thermactor air control valve or idle air bypass valve hose. Tighten the clamps securely to the air pump assembly.

30. Install the accessory drive belt. If equipped, attach the thermactor tube(s) support bracket to the rear of the cylinder head. Tighten attaching bolts to 30–40 ft. lbs. (40–55 Nm).

31. Connect the negative battery cable.

32. Fill and bleed the cooling system.

33. Start engine and check for coolant, fuel and oil leaks.

34. Check and, if necessary, adjust the curb idle speed.

35. Install the air cleaner assembly including the air intake duct and heat tube.

4.6L (SOHC) Engine

1. Disconnect the negative battery cable.

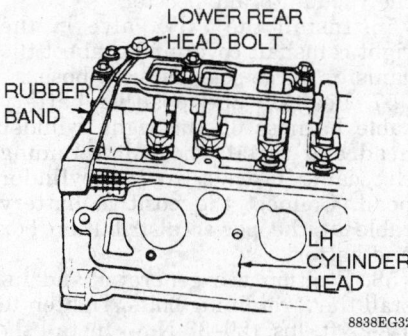

Supporting left lower head bolt with rubberband — 4.6L SOHC engine

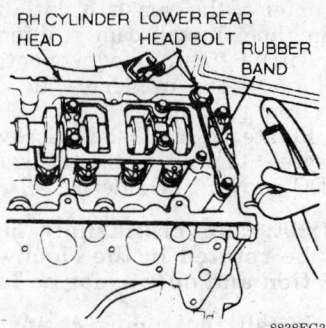

Supporting right lower head bolt with rubberband — 4.6L SOHC engine

2. Drain the cooling system and remove the cooling fan and shroud.

3. Relieve the fuel system pressure and disconnect the fuel lines.

4. Remove the air inlet tube and the wiper module. Release the belt tensioner and remove the accessory drive belt.

5. Tag and disconnect the ignition wires from the spark plugs. Disconnect the ignition wire brackets from the camshaft cover studs and remove the 2 bolts retaining the ignition wire tray to the coil brackets.

6. Remove the bolt retaining the air conditioner high pressure line to the right coil bracket. Disconnect both ignition coils and CID sensor.

7. Remove the nuts retaining the coil brackets to the front cover. Slide the ignition coil brackets and ignition wire assembly off the mounting studs and remove from the vehicle.

8. Remove the water pump pulley. Disconnect the generator wiring harness from the junction block, fender apron and generator. Disconnect the bolts retaining the generator to the intake manifold and engine block and remove the generator.

9. Disconnect the positive battery cable at the power distribution box. Remove the retaining bolt from the positive battery cable bracket located on the side of the right cylinder head.

10. Disconnect the vent hose from the canister purge solenoid and position the positive battery cable out of the way. Disconnect the canister purge solenoid vent hose from the PCV valve and remove the PCV valve from the camshaft cover.

11. Remove the 42-pin engine harness connector from the retaining bracket on the brake vacuum booster, disconnect and position out of the way.

12. Disconnect the HDR sensor, air conditioning compressor clutch and canister purge solenoid connectors.

13. Raise and safely support the vehicle.

14. Remove the bolts retaining the power steering pump to the engine block and front cover. The front lower bolt on the power steering pump will not come all the way out. Wire the power steering pump out of the way.

15. Remove the 4 bolts retaining the oil pan to the front cover. Remove the crankshaft damper retaining bolt and remove the damper, using a suitable puller.

16. Disconnect the EVO sensor and oil sending unit. Position the EVO sensor and oil pressure sending unit harness out of the way.

17. Disconnect the EGR tube from the right exhaust manifold. Disconnect the exhaust pipes from the exhaust manifolds. Lower the exhaust pipes and hang with wire from the crossmember.

18. Remove the bolt retaining the starter wiring harness to the rear of the right cylinder head. Lower the vehicle.

19. Remove the bolts and stud bolts retaining the camshaft covers to the cylinder heads and remove the covers.

20. Disconnect the accelerator, cruise control and throttle valve cables. Remove the accelerator cable bracket from the intake manifold and position out of the way.

21. Disconnect the vacuum hose from the throttle body elbow vacuum port, both oxygen sensors and the heater supply hose.

22. Remove the 2 bolts retaining the thermostat housing to the intake manifold and position the upper hose and thermostat housing out of the way.

NOTE: Two thermostat housing bolts also retain the intake manifold.

23. Remove the 9 bolts retaining the intake manifold to the cylinder heads and remove the intake manifold and gaskets.

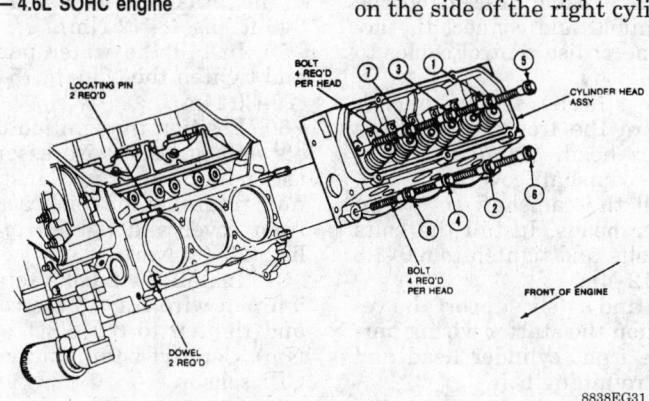

Cylinder head bolt torque sequence — 3.8L engine

24. Remove the 7 stud bolts and 4 bolts retaining the front cover to the engine and remove the front cover.

25. Remove the timing chains.

26. Remove the 10 bolts retaining the left cylinder head to the engine block and remove the head.

NOTE: The lower rear bolt cannot be removed due to interference with the brake vacuum booster. Use a rubber band to hold the bolt away from the engine block.

27. Remove the ground strap, 1 stud and 1 bolt retaining the heater return line to the right cylinder head.

28. Remove the 10 bolts retaining the right cylinder head to the engine block and remove the head.

NOTE: The lower rear bolt cannot be removed due to interference with the evaporator housing. Use a rubber band to hold the bolt away from the engine block.

29. Clean all gasket mating surfaces. Check the cylinder head and engine block for flatness. Check the cylinder head for scratches near the coolant passage and combustion chamber that could provide leak paths. Machine as necessary.

To install:

30. Rotate the crankshaft counterclockwise 45 degrees. The crankshaft keyway should be at the 9 o'clock position viewed from the front of the engine. This ensures that all pistons are below the top of the engine block deck face.

31. Rotate the camshaft to a stable position where the valves do not extend below the head face.

32. Position new head gaskets on the engine block. Install the lower rear bolts on both cylinder heads and retain with rubber bands as explained during the removal procedure.

33. Position the cylinder heads on the engine block dowels, being careful not to score the surface of the head face. Apply clean oil to the head bolts, remove the rubber band from the lower rear bolt and install all bolts hand-tight.

34. Tighten the head bolts as follows:

   a. Tighten the bolts, in sequence, to 15–22 ft. lbs. (20–30 Nm).

   b. Rotate each bolt, in sequence, 85–96 degrees.

   c. Rotate each bolt, in sequence, an additional 85–96 degrees.

35. Position the heater return hose and install the 2 bolts. Rotate the camshafts using the flats matched at the center of the camshaft until both are in time. Install cam positioning tools T91P-6256-A or equivalent, on the flats of the camshafts to keep them from rotating.

36. Rotate the crankshaft clockwise 45 degrees to position the crankshaft at TDC on No. 1 cylinder.

NOTE: The crankshaft must only be rotated in the clockwise direction and only as far as TDC.

37. Install the timing chains according to the proper procedure.

38. Install a new front cover seal and gasket. Apply silicone sealer to the lower corners of the cover where it meets the junction of the oil pan and cylinder block and to the points where the cover contacts the junction of the cylinder block and cylinder head.

39. Install the front cover and the stud bolts and bolts. Tighten to 15–22 ft. lbs. (20–30 Nm).

40. Position new intake manifold gaskets on the cylinder heads. Make sure the alignment tabs on the gaskets are aligned with the holes in the cylinder heads.

NOTE: Before installing the intake manifold, inspect it for nicks and cuts that could provide leak paths.

41. Position the intake manifold on the cylinder heads and install the retaining bolts. Tighten the bolts, in sequence, to 15–22 ft. lbs. (20–30 Nm).

42. Install the thermostat and O-ring, then position the thermostat housing and upper hose and install the 2 bolts. Tighten to 15–22 ft. lbs. (20–30 Nm).

43. Connect the heater supply hose and both oxygen sensors. Connect the vacuum hose to the throttle body adapter vacuum port.

44. Connect and, if necessary, adjust the throttle valve cable. Install the accelerator cable bracket on the intake manifold and connect the accelerator and cruise control cables to the throttle body.

45. Apply silicone sealer to both places where the front cover meets the cylinder head. Install new gaskets on the camshaft covers.

46. Install the camshaft covers on the cylinder heads. Install the bolts and stud bolts and tighten to 6.0–8.8 ft. lbs. (8–12 Nm).

47. Raise and safely support the vehicle. Position the starter wiring harness to the right cylinder head and install the retaining bolt.

48. Cut the wire and position the exhaust pipes to the exhaust manifolds. Tighten the 4 nuts to 20–30 ft. lbs. (27–41 Nm).

NOTE: Make sure the exhaust system clears the No. 3 crossmember. Adjust as necessary.

49. Connect the EGR tube to the right exhaust manifold and tighten the line nut to 26–33 ft. lbs. (35–45 Nm). Connect the EVO sensor and oil sending unit.

50. Apply a small amount of silicone sealer in the rear of the keyway on the damper. Position the damper on the crankshaft, making sure the crankshaft key and keyway are aligned.

51. Using damper installer T74P-6316-B or equivalent, install the crankshaft damper. Install the damper bolt and washer and tighten to 114–121 ft. lbs. (155–165 Nm).

52. Install the 4 bolts retaining the oil pan to the front cover and tighten to 15–22 ft. lbs. (20–30 Nm).

53. Position the power steering pump on the engine and install the 4 retaining bolts. Tighten to 15–22 ft. lbs. (20–30 Nm). Lower the vehicle.

54. Connect the air conditioning compressor, HDR sensor and canister purge solenoid.

55. Connect the 42-pin engine harness connector and transmission harness connector. Install the 42-pin connector on the retaining bracket on the vacuum brake booster.

56. Install the PCV valve in the right camshaft cover and connect the canister purge solenoid vent hose.

57. Position the positive battery cable harness on the right cylinder head and install the bolt retaining the cable bracket to the cylinder head. Connect the positive battery cable at the power distribution box and battery.

58. Position the generator and install the 2 retaining bolts. Tighten to 15–22 ft. lbs. (20–30 Nm). Install the 2 bolts retaining the generator brace to the intake manifold and tighten to 6–8 ft. lbs. (8–12 Nm).

59. Install the water pump pulley and tighten the bolts to 15–22 ft. lbs. (20–30 Nm).

60. Position the ignition coil brackets and ignition wire assembly onto the mounting studs. Install the 7 nuts retaining the coil brackets to the front cover and tighten to 15–22 ft. lbs. (20–30 Nm).

61. Install the 2 bolts retaining the ignition wire tray to the coil bracket and tighten to 6.0–8.8 ft. lbs. (8–12 Nm). Connect both ignition coils and CID sensor.

62. Position the air conditioner high pressure line on the right coil

bracket and install the bolt. Connect the ignition wires to the spark plugs and install the bracket onto the camshaft cover studs.

63. Install the accessory drive belt and the wiper module. Connect the fuel lines and install the cooling fan and shroud. Fill and bleed the cooling system.

64. Install the air inlet tube and connect the negative battery cable. Start the engine and bring to normal operating temperature. Check for leaks. Check all fluid levels.

4.6L (DOHC) Engine

1. Disconnect the negative battery cable.
2. Relieve the fuel system pressure. Drain the cooling system.
3. Remove the engine front cover and valve covers.
4. Raise and safely support the vehicle. Disconnect the exhaust Y-pipe and support aside with mechanics wire. Lower the vehicle.
5. Disconnect the EGR tube from left catalyst.
6. Remove the intake manifold assembly.
7. Remove the crankshaft position sensor tooth wheel and rotate the engine by hand to No. 1 cylinder TDC.
8. Install camshaft positioning tool T93P-6256-A, or equivalent, in rear D-slots of camshafts.
9. Remove 2 right tensioner bolts and remove tensioner. Remove 2 left tensioner bolts and remove tensioner.
10. Remove right and left primary timing chains.
11. Install crankshaft holding tool T93P-6303-A, or equivalent. Remove the the roller followers from intake and exhaust valves on cylinder head being removed.
12. Loosen the 10 cylinder head bolts using the reverse order of the tightening sequence. Discard the bolts.

NOTE: The lower rear bolt cannot be removed because of interference with the brake vacuum booster. Use a rubber band or suitable device to hold the away from the block.

13. Remove the cylinder head and discard the gasket.
14. Clean all gasket mating surfaces. Check the flatness of the cylinder head using a straight-edge and a feeler gauge. If there is any warpage, machine cylinder as necessary.

   To install:
15. Rotate the crankshaft 90 degrees to ensure all pistons are below top of engine block deck face.

16. Install new exhaust manifold gaskets and install exhaust manifolds.
17. Position new cylinder gasket on engine block. Install lower rear bolt on right and left heads and hold in place with rubber band or suitable device.
18. Position left head on dowels. Oil all cylinder head bolts with clean engine oil prior to installation.
19. Remove rubber band and install the remaining head bolts and hand-tighten.
20. Tighten cylinder head bolts in 3 steps:
   Tighten bolts in sequence to 27–32 ft. lbs. (37–43 Nm).
   Rotate bolts, in sequence 85–96 degrees.
   Rotate bolts, in sequence an additional 85–96 degrees.
21. Position right head on dowels. Oil all cylinder head bolts with clean engine oil prior to installation.
22. Remove rubber band and install the remaining head bolts and hand-tighten.
23. Tighten cylinder head bolts in 3 steps:
   Tighten bolts in sequence to 27–32 ft. lbs. (37–43 Nm).
   Rotate bolts, in sequence 85–96 degrees.
   Rotate bolts, in sequence an additional 85–96 degrees.
24. Install heater return hose. Install primary timing chains and set valve timing. Make sure the copper link on the timing chain aligns the timing on the camshaft sprocket.
25. Install the rocker followers. Install the crankshaft position sensor tooth wheel.
26. Install the intake manifold.
27. Install front engine cover. Connect the EGR tube to the left catalyst.
28. Raise and safely support the vehicle. Connect the exhaust Y-pipe. Lower the vehicle.
29. Install valve covers and connect the negative battery cable.
30. Fill and bleed the cooling system.
31. Start engine and check for coolant, fuel and oil leaks.

5.0L Engine

1. Disconnect the negative battery cable.
2. Drain the cooling system and relieve the fuel system pressure.
3. Remove the upper and lower intake manifold and throttle body assembly.
4. If the left cylinder head is to be removed and the vehicle is equipped

with air conditioning, proceed as follows:
   a. Discharge the air conditioning system.
   b. Disconnect and plug the refrigerant lines at the compressor. Cap the openings on the compressor.
   c. Disconnect the electrical connector to the compressor.
   d. Remove the compressor and the necessary mounting brackets.
5. If the left cylinder head is to be removed, disconnect the power steering pump bracket from the cylinder head. Position the pump aside in a position that will prevent the oil from draining out.
6. Disconnect the oil level indicator tube bracket from the exhaust manifold stud.
7. If the right cylinder head is to be removed, remove the alternator mounting bracket from the cylinder head.
8. Remove the thermactor crossover tube from the rear of the cylinder heads. Remove the fuel line from the clip at the front of the right cylinder head.
9. Raise and safely support the vehicle. Disconnect the exhaust manifolds from the muffler inlet pipes. Lower the vehicle.
10. Loosen the rocker arm fulcrum bolts so the rocker arms can be rotated to the side. Remove the pushrods in sequence so they may be installed in their original positions.
11. Remove the cylinder head attaching bolts and the cylinder heads. If necessary, remove the exhaust manifolds to gain access to the lower bolts. Remove and discard the head gaskets.
12. Clean all gasket mating surfaces. Check the flatness of the cylinder head using a straight-edge and a feeler gauge. The cylinder head must not be warped any more than 0.003 in. in any 6.0 in.; 0.006 in. overall. Machine as necessary.

   To install:
13. Position the new cylinder head gasket over the dowels on the block. Position the cylinder heads on the block and install the attaching bolts.
14. There are 2 types of bolts, flanged hex head and non-flanged hex head. If equipped with non-flanged hex head bolts, tighten the bolts, in sequence, in 2 steps, 1st to 55–65 ft. lbs. (75–88 Nm), then to 65–72 ft. lbs. (88–97 Nm). If equipped with flanged hex head bolts, tighten the bolts, in sequence, in 3 steps, 1st to 25–35 ft. lbs. (34–47 Nm), then to 45–55 ft. lbs. (61–75 Nm), then

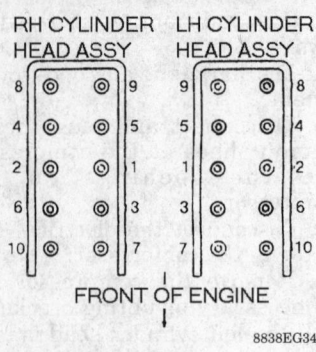

Cylinder head bolt torque sequence — 4.6L engine

tighten an additional ¼ turn (85–96 degrees).

NOTE: When the cylinder head bolts have been tightened following this procedure, it is not necessary to retighten the bolts after extended operation.

15. If removed, install the exhaust manifolds. Tighten the retaining bolts to 18–24 ft. lbs. (24–32 Nm).

16. Clean the pushrods, making sure the oil passages are clean. Check the ends of the pushrods for wear. Visually check the pushrods for straightness or check for runout using a dial indicator. Replace pushrods, as necessary.

17. Apply a suitable grease to the ends of the pushrods and install them in their original positions. Position the rocker arms over the pushrods and the valves.

18. Before tightening each fulcrum bolt, bring the lifter for the fulcrum bolt to be tightened onto the base circle of the camshaft by rotating the engine. When the lifter is on the base circle of the camshaft, tighten the fulcrum bolt to 18–25 ft. lbs. (24–34 Nm).

NOTE: If all the original valve train parts are reinstalled, a valve clearance check is not necessary. If any valve train components are replaced, a valve clearance check must be performed.

19. Install new rocker arm cover gaskets on the rocker arm covers and install the covers on the cylinder heads. Tighten the retaining bolts to 10–13 ft. lbs. (14–18 Nm), wait 2 minutes, then retighten to the same specification.

20. Raise and safely support the vehicle. Connect the exhaust manifolds to the muffler inlet pipes. Lower the vehicle.

21. If necessary, install the air conditioning compressor and brackets. Connect the refrigerant lines and

electrical connector to the compressor.

22. If necessary, install the alternator bracket.

23. If the left cylinder head was removed, install the power steering pump.

24. Install the drive belt. Install the thermactor tube at the rear of the cylinder heads.

25. Install the intake manifold. Fill and bleed the cooling system.

26. Connect the negative battery cable, start the engine and bring to normal operating temperature. Check for leaks. Check all fluid levels.

27. If necessary, leak test, evacuate and charge the air conditioning system. Observe all safety precautions.

Valve Lifters

REMOVAL AND INSTALLATION

2.3L Engine

The engine is equipped with hydraulic lash adjusters which, while not being exactly the same as a conventional hydraulic lifter, perform the same function — maintain proper valve train clearance.

1. Disconnect the negative battery cable.

2. If equipped with distributor ignition, tag and disconnect the spark plug wires from the spark plugs. Move the wires aside.

3. Remove the hose and the retaining bolts from the rocker arm cover and remove the cover.

4. Rotate the camshaft so the base circle of the cam is facing the cam follower to be removed.

5. Using valve spring compressor tool T88T-6565-BH or equivalent, compress the lash adjuster as required and/or depress the valve spring if necessary and slide the cam

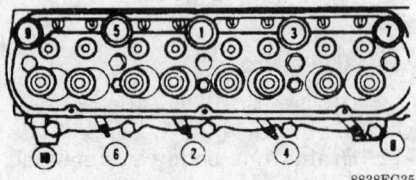

Cylinder head bolt torque sequence — 5.0L engine

follower over the lash adjuster and out.

6. Lift out the hydraulic lash adjuster.

To install:

7. Rotate the camshaft so the base circle of the camshaft is facing the lash adjuster and cam follower to be installed. Lubricate the hydraulic lash adjuster with clean engine oil and position it in the bore.

8. Using valve spring compressor tool T88T-6565-BH or equivalent, compress the lash adjuster, as necessary, to position the cam follower over the lash adjuster and the valve stem.

9. Before rotating the camshaft to the next position, make sure the lash adjuster just installed is fully compressed and released.

10. Clean the gasket mating surface of the rocker arm cover and cylinder head. Install a new gasket and the rocker arm cover. Install the mounting screws and tighten to 62–97 inch lbs. (7–11 Nm).

11. Install the remaining components in the reverse order of removal. Start the engine and check for oil leaks.

3.8L Engine

NOTE: Before replacing a lifter for noisy operation, be sure the noise is not caused by improper valve to rocker arm clearance or by worn rocker arms or pushrods.

1. Disconnect the negative battery cable. Tag and disconnect the spark plug wires at the spark plugs.

2. Remove plug wire routing clips from the studs on the rocker arm cover attaching bolts. Lay the plug wires, with the routing clips toward the front of the engine.

3. Remove the upper intake manifold. On supercharged engine, remove the supercharger.

4. Remove the rocker arm covers. Remove the lower intake manifold.

5. Sufficiently loosen each rocker arm fulcrum attaching bolt to allow the rocker arm to be lifted off the pushrod and rotated to 1 side.

6. Remove the pushrods. The location of each pushrod should be identified. When the engine is assembled each rod should be installed in its original position.

7. Remove the 4 bolts holding the 2 guide plate retainers in place; the bolts are held captive in the retainers. Remove the 6 guide plates from the adjacent lifters.

8. Remove the lifters using a magnet. The location of each lifter should

be identified. When the engine is assembled, each lifter should be installed in its original position.

NOTE: If the lifters are stuck in the bores due to excessive varnish or gum deposits, it may be necessary to use a claw-type tool to aid removal. When using a remover tool, rotate the lifter back and forth to loosen it from gum or varnish that may have formed on the lifter.

To install:

9. Clean the rocker arm cover and cylinder head mating surfaces.

10. Install each lifter in the bore from which it was removed. If new lifters are being installed, check the new lifters for free fit in the bores.

11. Align the flats on the side of the lifters and install the 6 guide plates between the adjacent lifters. Make sure the word **UP** is showing. Install the 2 guide plate retainers and tighten the 4 captive bolts to 7–10 ft. lbs. (9–14 Nm).

12. Dip each pushrod in heavy engine oil and install in its original position.

13. For each valve, rotate the crankshaft until the lifter rests on the base circle of the camshaft lobe. Position the rocker arms over the pushrods. Install the fulcrums and tighten the bolts to 5–11 ft. lbs. (7–15 Nm).

14. Lubricate all rocker arm assemblies with heavy engine oil. Final tighten the fulcrum bolts to 19–25 ft. lbs. (25–35 Nm). For final tightening the camshaft may be in any position.

NOTE: The fulcrums must be fully seated in the cylinder head and the pushrods must be seated in the rocker arm sockets prior to final tightening.

15. Install the lower intake manifold and the rocker arm covers. On non-supercharged engines, install the upper intake manifold. On supercharged engines, install the supercharger.

16. Install the spark plug wire routing clips and connect the wires to the spark plugs. Connect the negative battery cable, start the engine and check for oil and coolant leaks.

4.6L (SOHC) Engine

The 4.6L engine is equipped with hydraulic lash adjusters which, while not being exactly the same as a conventional hydraulic lifter, perform

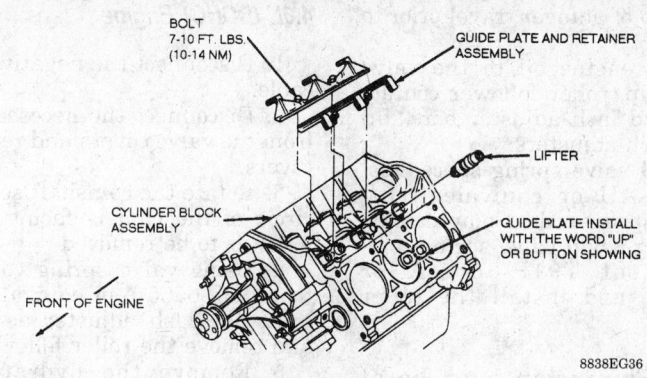

BOLT
7-10 FT. LBS.
(10-14 NM)

GUIDE PLATE AND RETAINER
ASSEMBLY

LIFTER

CYLINDER BLOCK
ASSEMBLY

GUIDE PLATE INSTALL
WITH THE WORD "UP"
OR BUTTON SHOWING

FRONT OF ENGINE

8838EG36

Valve lifter installation — 3.8L engine

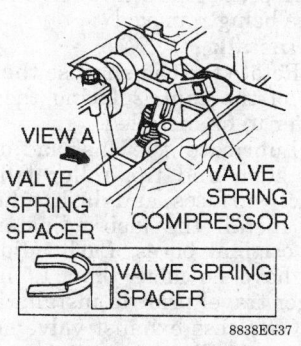

VIEW A
VALVE
SPRING
SPACER

VALVE
SPRING
COMPRESSOR

VALVE SPRING
SPACER

8838EG37

Valve spring compressor and spacer 4.6L SOHC engine

the same function — maintain proper valve train clearance.

1. Disconnect the negative battery cable.

2. Remove the right camshaft cover as follows:

a. Disconnect the positive battery cable at the battery and at the power distribution box. Remove the retaining bolt from the positive battery cable bracket located on the side of the right cylinder head.

b. Disconnect the High Data Rate (HDR) sensor, air conditioning compressor clutch and canister purge solenoid connectors. Position the harness out of the way.

c. Disconnect the vent hose from the purge solenoid and position the positive battery cable out of the way.

d. Disconnect the ignition wires from the spark plugs. Remove the ignition wire brackets from the camshaft cover studs and position the wires out of the way.

e. Remove the PCV valve from the camshaft cover grommet and position out of the way.

f. Remove the bolts and stud bolts and remove the camshaft cover.

3. Remove the left camshaft cover as follows:

a. Remove the air inlet tube. Relieve the fuel system pressure and disconnect the fuel lines.

b. Raise and safely support the vehicle.

c. Disconnect the EVO sensor and oil pressure sending unit and position the harness out of the way. Lower the vehicle.

d. Remove the 42-pin engine harness connector from the retaining bracket on the brake vacuum booster. Disconnect and position out of the way.

e. Remove the windshield wiper module.

f. Disconnect the ignition wires from the spark plugs. Remove the ignition wire brackets from the studs and position the wires out of the way.

g. Remove the bolts and stud bolts and remove the camshaft cover.

4. Position the piston of the cylinder being serviced at the bottom of its stroke and position the camshaft lobe on the base circle.

5. Install valve spring spacer tool T91P-6565-AH or equivalent, between the spring coils to prevent valve seal damage.

NOTE: If the valve spring spacer tool is not used, the retainer will hit the valve stem seal and damage the seal.

6. Install valve spring compressor tool T91P-6565-A or equivalent, under the camshaft and on top of the valve spring retainer.

7. Compress the valve spring and remove the roller follower. Remove the valve spring compressor and spacer.

8. Remove the hydraulic lash adjuster.

To install:

9. Check the hydraulic lash adjusters. They must have no more

than 1.5mm of plunger travel prior to installation.

10. Apply engine oil to the valve stem and tip, roller follower contact surfaces and lash adjuster bore. Install the lash adjusters.

11. Install valve spring spacer tool T91P-6565-AH or equivalent, between the spring coils. Compress the valve spring using valve spring compressor tool T91P-6565-A or equivalent, and install the roller follower.

NOTE: The piston must be at the bottom of its stroke and the camshaft at the base circle.

12. Remove the valve spring compressor and spacer.

13. Clean the sealing surfaces of the camshaft covers and cylinder heads. Apply silicone sealer to the places where the front cover meets the cylinder head.

14. Position new gaskets onto the camshaft covers and install the covers. Install the bolts and stud bolts and tighten to 6.0–8.8 ft. lbs. (8–12 Nm).

15. When installing the right camshaft cover, proceed as follows:

a. Install the PCV valve into the camshaft cover grommet.

b. Install the ignition wire brackets on the studs and connect the wires to the spark plugs.

c. Position the harness and connect the canister purge solenoid, air conditioning compressor clutch and HDR sensor.

d. Position the positive battery cable harness on the right cylinder head. Install the bolt retaining the cable bracket to the cylinder head.

e. Connect the positive battery cable at the power distribution box and the battery.

16. When installing the left camshaft cover, proceed as follows:

a. Install the ignition wire brackets on the studs and connect the wires to the spark plugs.

b. Install the windshield wiper module.

c. Connect the 42-pin connector and transmission harness connector. Install the connector on the retaining bracket.

d. Raise and safely support the vehicle. Position and connect the EVO sensor and oil pressure sending unit harness.

e. Lower the vehicle. Connect the fuel lines.

17. Connect the negative battery cable. Start the engine and check for leaks.

4.6L (DOHC) Engine

1. Disconnect the negative battery cable.

2. Disconnect the necessary hoses from the valve covers and remove the covers.

3. Rotate the camshaft so the base circle of the cam is facing the cam follower to be removed.

4. Using valve spring compressor tool T93P-6565-A or equivalent, compress the lash adjuster as required and remove the roller follower.

5. Remove the hydraulic lash adjuster.

6. Repeat steps 3 through 5 for adjusters being removed.

To install:

7. Rotate the camshaft so the base circle of the cam is facing the cam follower to be installed.

8. Lubricate the lash adjuster, valve stem and tip, roller follower contact surfaces and lash adjuster bore. Install the lash adjusters in their original bores. Lash adjusters must have no more than 1.5mm of plunger travel prior to installation.

9. Compress exhaust valve spring and install roller follower.

10. Compress primary intake valve spring and install roller follower. Compress secondary intake valve spring and install roller follower.

11. Remove valve spring compressor. Repeat steps 7 through 11 until all lash adjusters are installed.

12. Install the rocker arm covers and necessary hoses. Connect the negative battery cable, start the engine and check for leaks.

5.0L Engine

1. Disconnect the negative battery cable. Remove the intake manifold and related parts.

2. Disconnect the necessary hoses from the rocker arm covers. Tag and disconnect the spark plug wires, then remove the wires and brackets from the rocker arm cover attaching studs. Remove the upper intake manifold.

3. Remove the rocker arm covers. Loosen the rocker arm fulcrum bolts and rotate the rocker arms to the side.

4. Remove the valve pushrods and identify them so they can be installed in their original position.

5. Remove the lifter guide retainer bolts. Remove the retainer and lifter guide plates. Identify the guide plates so they may be reinstalled in their original positions.

6. Using a magnet, remove the lifters and place them in a rack so they can be installed in their original bores.

NOTE: If the lifters are stuck in the bores due to excessive varnish or gum deposits, it may be necessary to use a claw-type tool to aid removal. When using a remover tool, rotate the lifter back and forth to loosen it from gum or varnish that may have formed on the lifter.

To install:

7. Lubricate the lifters and install them in their original bores. If new lifters are being installed, check them for free fit in their respective bores.

8. Install the lifter guide plates in their original positions, then install the guide plate retainer.

9. Install the pushrods in their original positions. Apply grease to the ends prior to installation.

10. Lubricate the rocker arms and fulcrum seats with heavy engine oil. Position the rocker arms over the pushrods and install the fulcrum bolts.

11. Before tightening each fulcrum bolt, rotate the crankshaft until the lifter is on the base circle of the cam. Tighten the fulcrum bolt to 18–25 ft. lbs. (24–34 Nm). Check the valve clearance.

12. Install the rocker arm covers and the intake manifold. Connect the negative battery cable, start the engine and check for leaks.

Valve Lash

ADJUSTMENT

2.3L Engine

1. Disconnect the negative battery cable.

2. Remove the valve cover assembly.

3. Position the camshaft so the base circle of the lobe is facing the cam follower of the valve to be checked.

4. Using valve spring compressor tool T88T-6565-BH or equivalent, slowly apply pressure to the cam follower until the the lash adjuster is completely collapsed.

5. With follower collapsed, insert a feeler gauge between the base circle of the camshaft and follower. The clearance should not be more than 0.035–0.055 in.

6. If the clearance is excessive, remove the cam follower and inspect for damage.

7. If the cam follower appears to be intact and not excessively worn,

measure the valve spring assembled height to make sure the valve is not sticking.

8. If the valve spring assembled height is correct, check the camshaft for wear. If the camshaft dimensions are correct, replace the lash adjuster.

9. Install the valve cover and all remaining components.

3.8L Engine

The valve lash is not adjustable. If the collapsed lifter clearance is found to be incorrect, there are replacement pushrods available to compensate for excessive or insufficient clearance.

1. Disconnect the negative battery cable.

2. Remove the valve cover assembly on the side to be checked.

3. Turn the engine until the No. 1 piston is at TDC on the compression stroke.

4. The following valves can be checked with the engine in this position:

   a. No. 1 intake — No. 1 exhaust
   b. No. 3 intake — No. 2 exhaust
   c. No. 6 intake — No. 4 exhaust

5. Rotate the engine 360 degrees and check the following valves:

   a. No. 2 intake — No. 3 exhaust
   b. No. 4 intake — No. 5 exhaust
   c. No. 5 intake — No. 6 exhaust

6. Check each of the lifters by placing hydraulic lifter compressor tool T71P-6513-B or equivalent, on the rocker arm and slowly applying pressure to the lifter, until the lifter is collapsed.

7. Hold the lifter in this position and check the clearance between the rocker arm and the valve stem tip. The clearance should be 0.09–0.19 in. (2.25–4.79mm).

8. Repeat this operation for each valve to be checked.

9. If the clearance is greater than specification, replace the pushrod with a longer one. If the clearance is less than specified, replace the pushrod with a shorter one.

5.0L Engine

The valve lash is not adjustable. If the collapsed lifter clearance is found to be incorrect, there are replacement pushrods available to compensate for excessive or insufficient clearance.

1. Install an auxiliary starter switch. Crank the engine with the ignition switch **OFF** until the No. 1 piston is at TDC on the compression stroke.

2. With the crankshaft in the positions designated in Steps 4, 5 and 6, position lifter bleed down wrench tool T71P-6513-B or equivalent, on the rocker arm. Slowly apply pressure to bleed down the lifter until the plunger is completely bottomed. Hold the lifter in this position and check the available clearance between the rocker arm and the valve stem tip with a feeler gauge.

3. The clearance should be 0.123–0.146 in. If the clearance is less than specification, install a shorter pushrod. If the clearance is greater than specification, install a longer pushrod.

4. The following valves can be checked with the engine in position 1, No. 1 piston at TDC on the compression stroke.

   a. No. 1 intake — No. 1 exhaust
   b. No. 4 intake — No. 3 exhaust
   c. No. 8 intake — No. 7 exhaust

5. Rotate the engine 360 degrees (1 revolution) from the 1st position and check the following valves:

   a. No. 3 intake — No. 2 exhaust
   b. No. 7 intake — No. 6 exhaust

6. Rotate the engine 90 degrees (¼ revolution) from the 2nd position and check the following valves:

   a. No. 2 intake — No. 4 exhaust
   b. No. 5 intake — No. 5 exhaust
   c. No. 6 intake — No. 8 exhaust

Rocker Arms

REMOVAL AND INSTALLATION

2.3L Engine

1. Disconnect the negative battery cable.

2. If equipped with distributor ignition, tag and disconnect the spark plug wires from the spark plugs. Move the wires aside.

3. Remove the hose and the retaining bolts from the rocker arm cover and remove the cover.

4. Rotate the camshaft so the base circle of the cam is facing the cam follower to be removed.

5. Using valve spring compressor tool T88T-6565-BH or equivalent, compress the lash adjuster as required and/or depress the valve spring if necessary and slide the cam follower over the lash adjuster and out.

To install:

6. Using valve spring compressor tool T88T-6565-BH or equivalent, compress the lash adjuster, as necessary, to position the cam follower over the lash adjuster and the valve stem.

7. Before rotating the camshaft to the next position, make sure the lash adjuster just installed is fully compressed and released.

8. Clean the gasket mating surface of the rocker arm cover and cylinder head. Install a new gasket and the rocker arm cover. Install the mounting screws and tighten to 62–97 inch lbs. (7–11 Nm).

9. Install the remaining components in the reverse order of removal. Start the engine and check for oil leaks.

3.8L Engine

1. Disconnect the negative battery cable.

2. Disconnect the spark plug wires from the spark plugs. Remove the spark plug wire routing clips from the rocker arm cover attaching bolt studs.

3. To remove the left rocker arm cover, proceed as follows:

   a. Remove the oil fill cap.
   b. Remove the crankcase vent tube.
   c. On supercharged engines, remove the charge air cooler tubes and the oil cooler inlet tube.

4. To remove the right rocker arm cover, proceed as follows:

   a. Remove the PCV valve.
   b. Position the air cleaner assembly aside, if necessary.

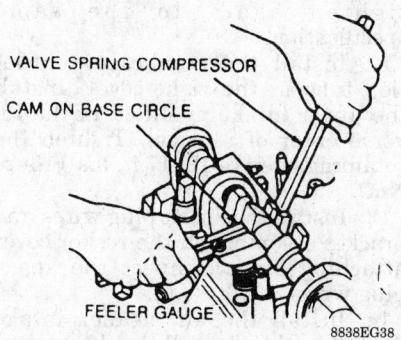

VALVE SPRING COMPRESSOR

CAM ON BASE CIRCLE

FEELER GAUGE

8838EG38

Checking collapsed lifter valve clearance — 2.3L engine

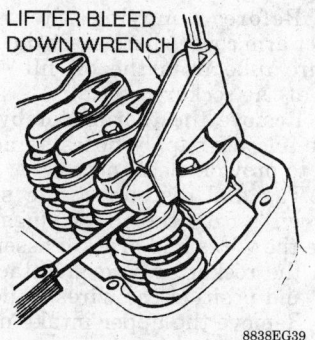

LIFTER BLEED DOWN WRENCH

8838EG39

Checking collapsed lifter valve clearance — 3.8L and 5.0L engine

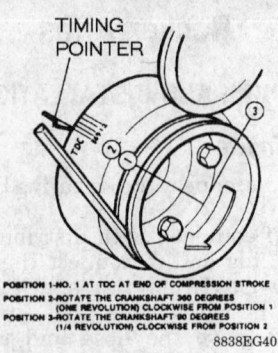

TIMING
POINTER

POSITION 1-NO. 1 AT TDC AT END OF COMPRESSION STROKE
POSITION 2-ROTATE THE CRANKSHAFT 360 DEGREES
(ONE REVOLUTION) CLOCKWISE FROM POSITION 1
POSITION 3-ROTATE THE CRANKSHAFT 90 DEGREES
(1/4 REVOLUTION) CLOCKWISE FROM POSITION 2

8838EG40

Engine valve adjusting
positions — 5.0L engine

c. On supercharged engines, remove the air inlet tube and remove the throttle body assembly.

5. Remove the rocker arm cover attaching screws and remove the rocker arm covers.

6. Remove the rocker arm, fulcrum and bolt assemblies. Keep each assembly together and identify the assemblies so they may be reinstalled in their original positions.

To install:

7. Clean all gasket mating surfaces on the rocker arm covers and cylinder heads. Clean the rocker arms and fulcrums and inspect for wear or damage. Replace as necessary.

8. Apply grease to the pushrod tips and valve stem tips. Lubricate the fulcrums and rocker arms with heavy engine oil and install them over the pushrods and valve stems.

9. For each valve, rotate the crankshaft until the lifter is on the base circle of the camshaft. Install the fulcrum bolt and tighten to 5–11 ft. lbs. (7–15 Nm). Make sure the pushrod and fulcrum are fully seated prior to tightening.

10. Lubricate all rocker arm assemblies with engine oil. Final tighten the fulcrum bolts to 19–25 ft. lbs. (25–35 Nm). When final tightening, the camshaft may be in any position. Make sure the pushrod and fulcrum are fully seated prior to tightening.

11. Position new gaskets on the cylinder heads and install the rocker arm covers. Tighten the attaching bolts to 80–106 inch lbs. (9–12 Nm). Note the location of the spark plug wire routing clip stud bolts prior to installation.

12. After installing the left rocker arm cover, proceed as follows:

a. Install the oil fill cap.

b. Install the crankcase vent tube.

c. On supercharged engines, install the charge air cooler tubes and the oil cooler inlet tube.

13. After installing the right valve cover, proceed as follows:

a. Install the PCV valve.

b. Install the air cleaner assembly, if necessary.

c. On supercharged engines, install the air inlet tube and the throttle body assembly.

14. Install the spark plug wire routing clips and connect the wires to the spark plugs.

15. Connect the negative battery cable, start the engine and check for leaks.

4.6L Engine

1. Disconnect the negative battery cable.

2. Disconnect the necessary hoses from the valve covers and remove the covers.

3. Rotate the camshaft so the base circle of the cam is facing the cam follower to be removed.

4. Using valve spring compressor tool T93P-6565-A or equivalent, compress the lash adjuster as required and remove the roller follower.

5. Repeat steps 3 and 4 for followers being removed.

To install:

6. Rotate the camshaft so the base circle of the cam is facing the cam follower to be installed.

7. Compress exhaust valve spring and install roller follower.

8. Compress primary intake valve spring and install roller follower. Compress secondary intake valve spring and install roller follower.

9. Remove valve spring compressor. Repeat steps 6 through 10 until all followers are installed.

10. Install the rocker arm covers and necessary hoses. Connect the negative battery cable, start the engine and check for leaks.

5.0L Engine

1. Disconnect the negative battery cable.

2. Before removing the right rocker arm cover, disconnect the PCV closure tube from the oil fill stand pipe at the rocker cover.

3. Remove the thermactor bypass valve and air supply hoses as necessary to provide clearance.

4. Tag and disconnect the spark plug wires from the spark plugs. Remove the wires and bracket assembly from the rocker arm cover attaching stud and position the wires aside.

5. Remove the upper intake manifold as follows:

a. Disconnect the electrical connectors at the idle air bypass valve,

throttle position sensor and EGR position sensor.

b. Disconnect the throttle and transmission linkage from the throttle body. Remove the cable bracket from the manifold and position the cables and bracket aside.

c. Tag and disconnect the vacuum lines at the upper intake manifold vacuum tree, EGR valve, fuel pressure regulator and evaporative canister.

d. Disconnect the PCV hose from the fitting on the rear of the upper manifold and disconnect the PCV vent closure tube at the throttle body.

e. Partially drain the cooling system and remove the 2 EGR coolant lines from the fittings on the EGR spacer.

f. Remove the retaining bolts and remove the upper intake manifold and throttle body assembly.

6. Remove the attaching bolts and remove the covers.

7. Remove the rocker arm fulcrum bolt, fulcrum seat and rocker arm. Keep all rocker arm assemblies together. Identify each assembly so it may be reinstalled in its original position.

To install:

8. Clean all gasket mating surfaces of the rocker arm covers and cylinder heads. Clean and inspect the rocker arm assemblies for wear and/or damage. Replace as necessary.

9. Apply grease to the pushrod and valve stem tips and the underside of the fulcrum seats.

10. Rotate the crankshaft until the lifter is on the camshaft base circle and install the rocker, fulcrum seat and fulcrum bolt. Tighten the bolts to 18–25 ft. lbs. (24–34 Nm). Make sure the pushrod and fulcrum are fully seated prior to tightening.

11. Position new rocker arm cover gaskets and install the rocker arm covers. Tighten the bolts to 10–13 ft. lbs. (14–18 Nm), wait 2 minutes and tighten again to the same specification.

12. Install the crankcase ventilation tube in the right cover. Install the upper intake manifold in the reverse order of removal. Tighten the retaining bolts to 12–17 ft. lbs. (16–24 Nm).

13. Install the spark plug wires and bracket assembly on the rocker cover attaching stud. Connect the spark plug wires.

14. Install the air cleaner intake duct assembly. Install the thermactor bypass valve and air supply hoses, if required.

15. Connect the negative battery cable, start the engine and check for leaks.

Intake Manifold

REMOVAL AND INSTALLATION

2.3L Engine

1. Disconnect the negative battery cable.
2. Relieve the fuel system pressure and drain the cooling system.
3. Disconnect and label the electrical connectors at the following:
   a. Idle air control valve
   b. Throttle positioning sensor
   c. Injector wiring harness
   d. Air charge temperature sensor
   e. Engine coolant temperature sensor
   f. EGR valve, if necessary
   g. Fan switch, if necessary
   h. Ignition control assembly, if equipped
4. Tag and disconnect the necessary vacuum lines.
5. Remove the throttle linkage shield. Disconnect the throttle linkage and if equipped, the cruise control and kickdown cables. Unbolt the accelerator cable from the bracket and position the cable aside.
6. Disconnect the air intake hose and crankcase vent hose.
7. Disconnect the PCV system hose from the fitting on the underside of the upper intake manifold.
8. Disconnect the water bypass hose at the lower intake manifold.
9. Loosen the EGR flange nut and disconnect the EGR tube.
10. Remove the engine oil dipstick bracket retaining bolt.
11. Remove the upper intake manifold retaining bolts and/or studs and remove the upper intake manifold assembly.

12. Disconnect the fuel lines from the fuel supply manifold.
13. Disconnect the electrical connectors from the fuel injectors and move the harness aside.
14. Remove the fuel supply manifold retaining bolts and remove the manifold carefully. Injectors can be removed at this time by exerting a slight twisting/pulling motion.
15. Remove the lower intake manifold retaining bolts and remove the lower intake manifold. The front 2 bolts also secure an engine lifting bracket.

   To install:
16. Clean all gasket mating surfaces. Clean and oil the manifold bolt threads. Install a new intake manifold gasket.
17. Position the lower intake manifold to the head with the engine lift bracket. Install the manifold retaining bolts finger-tight.
18. Tighten the manifold retaining bolts, in sequence, to 15–22 ft. lbs. (20–30 Nm).
19. Install the fuel supply manifold and injectors. Connect the electrical connectors to the injectors.
20. Install a new gasket and the upper intake manifold. Tighten the bolts to 15–22 ft. lbs. (20–30 Nm) in the proper sequence. Connect the fuel lines to the fuel supply manifold.
21. Install the engine oil dipstick and retaining bolt. Connect the EGR tube, water bypass line and PCV hose.
22. Connect the electrical connectors and vacuum lines to their original locations. Connect the throttle linkage.
23. Fill and bleed the cooling system. Connect the negative battery cable, start the engine and check for leaks.

3.8L Engine

1. Disconnect the negative battery cable.

2. Drain the cooling system and relieve the fuel system pressure.
3. Remove the air cleaner assembly or air inlet tube.
4. Disconnect the accelerator cable at the throttle body. Disconnect the cruise control cable, if equipped.
5. If equipped with an automatic transmission, disconnect the transmission linkage at the upper intake manifold. Remove the retaining bolts from the accelerator cable mounting bracket and position the cables aside.
6. If equipped, disconnect the thermactor air supply hose at the check valve. The valve is located in the Y-pipe assembly.
7. Disconnect the fuel lines. If equipped, remove the supercharger.
8. Disconnect the radiator hose at the thermostat housing and the coolant bypass hose at the manifold.
9. Disconnect the heater tube at the intake manifold and remove the tube support bracket retaining nut. Remove the heater hose at the rear of the heater tube. Loosen the hose clamp at the heater elbow and remove the heater tube with the hose attached. Remove the heater tube with the lines attached and set the assembly aside.
10. Tag and disconnect the vacuum lines at the fuel rail assembly and intake manifold. Tag and disconnect the necessary electrical connectors.
11. If equipped with air conditioning, remove the compressor support bracket. Disconnect the 1 PCV line at the upper intake manifold and at the valve. Remove the 2nd PCV line from the left rocker arm cover.
12. Remove the throttle body assembly. Remove the EGR valve assembly from the upper manifold.
13. Remove the retaining nut and remove the wiring retainer bracket located at the left front of the intake manifold and set aside with the spark plug wires.
14. Remove the upper intake manifold retaining bolts/studs and remove the upper intake manifold.
15. Remove the injectors and fuel rail assembly. Remove the heater water outlet hose.
16. Remove the lower intake manifold retaining bolts/studs and remove the lower intake manifold.

NOTE: The manifold is sealed at each end with RTV-type sealer. To break the seal, it may be necessary to pry on the front of the manifold with a small prybar. If it is necessary to pry on the manifold, use care to prevent damage to the machined surfaces.

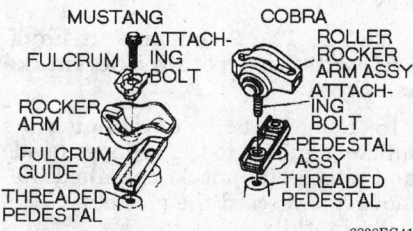

Rocker arm assembly — 5.0L engine

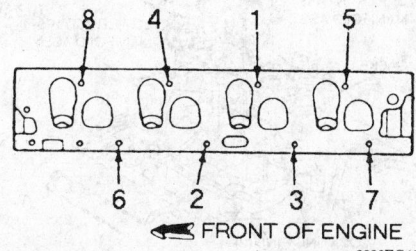

Intake manifold to cylinder head torque sequence — 2.3L engine

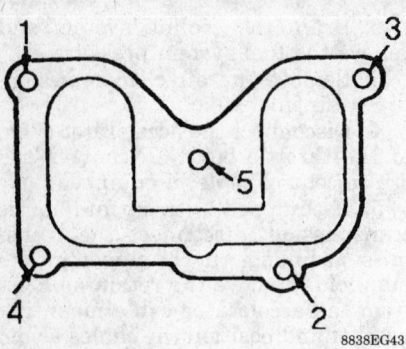

Upper intake manifold bolt torque sequence — 2.3L engine

To install:

17. Clean all gasket mating surfaces. Lightly oil all retaining bolt and stud threads.

18. Apply a dab of gasket adhesive to each cylinder head mating surface. Press new intake manifold gaskets in place, using location pins as necessary to aid in installation.

19. Apply a ⅛ in. bead of silicone sealer at each corner where the cylinder head joins the cylinder block. Install the front and rear intake manifold end seals.

20. Carefully lower the intake manifold into place on the cylinder heads and cylinder block. Use locating pins as necessary to guide the manifold.

21. Install the bolts and stud bolts in their original locations.

22. On all supercharged engines and 1993 non-supercharged engines, tighten the bolts, in sequence, in 2 steps, 1st to 8 ft. lbs. (11 Nm) and then to 11 ft. lbs. (15 Nm).

23. Connect the rear PCV line to the upper intake tube. Install the front PCV tube so the mounting bracket sits over the lower intake manifold stud. Tighten the nut on the stud to 15–22 ft. lbs. (20–30 Nm).

24. Install the injectors and the fuel rail. On non-supercharged engines, install the upper intake manifold assembly. Install the bolts and stud bolts in their original locations. Tighten the 4 center bolts and then the end bolts in 3 steps, 1st to 8 ft. lbs. (10 Nm), then to 15 ft. lbs. (20 Nm), and finally to 24 ft. lbs. (32 Nm).

25. On supercharged engines, install the supercharger.

26. Install the EGR valve. Install the throttle body and cross-tighten the retaining nuts to 15–22 ft. lbs. (20–30 Nm).

27. Connect the rear PCV line at the PCV valve on the upper intake manifold. If equipped with air conditioning, install the compressor support bracket.

28. Connect the necessary electrical connectors and vacuum hoses. Connect the heater tube hose to the heater elbow and position the heater tube support bracket. Tighten the retaining nut to 15–22 ft. lbs. (20–30 Nm).

29. Connect the heater hose to the heater tube and connect the coolant bypass hose and radiator upper hose.

30. Connect the fuel lines. Position the accelerator cable mounting bracket and tighten the mounting bolts to 15–22 ft. lbs. (20–30 Nm).

31. Connect the transmission linkage at the upper intake manifold. If equipped, connect the cruise control cable.

32. Fill and bleed the cooling system. Connect the negative battery cable, start the engine and check for leaks.

33. Check and if necessary, adjust the engine idle speed, transmission throttle linkage and cruise control.

4.6L (SOHC) Engine

1993–95

1. Disconnect the negative battery cable.

2. Drain the cooling system. Relieve the fuel system pressure and disconnect the fuel lines.

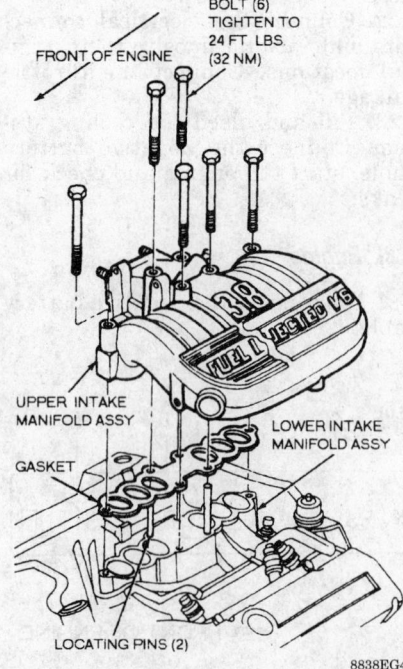

Upper intake manifold installation — 3.8L engine

3. Remove the wiper module and the air inlet tube. Release the belt tensioner and remove the accessory drive belt.

4. Tag and disconnect the ignition wires from the spark plugs. Disconnect the ignition wire brackets from the camshaft cover studs.

5. Disconnect both ignition coils and CID sensor. Tag and disconnect all ignition wires from both ignition coils. Remove the 2 bolts retaining the ignition wire tray to the coil brackets and remove the ignition wire assembly.

6. Disconnect the generator wiring harness from the junction block at the fender apron and generator. Remove the bolts retaining the generator brace to the intake manifold and the generator to the engine block and remove the generator.

7. Raise and safely support the vehicle. Disconnect the oil sending unit and EVO harness sensor and position the wiring harness out of the way.

8. Disconnect the EGR tube from the right exhaust manifold and lower the vehicle.

9. Remove the 42-pin engine harness connector from the retaining bracket on the vacuum brake booster and disconnect the connector.

10. Disconnect the air conditioning compressor, HDR sensor and canister purge solenoid.

11. Remove the PCV valve from the camshaft cover and disconnect the canister purge vent hose from the PCV valve.

12. Disconnect the accelerator and cruise control cables from the throttle body. Remove the accelerator cable bracket from the intake manifold and position out of the way.

13. Disconnect the throttle valve cable from the throttle body and the vacuum hose from the throttle body adapter port.

14. Disconnect both oxygen sensors and the heater supply hose.

15. Remove the 2 bolts retaining the thermostat housing to the intake manifold and position the upper hose and thermostat housing out of the way.

NOTE: The 2 thermostat housing bolts also retain the intake manifold.

16. Remove the bolts retaining the intake manifold to the cylinder heads and remove the intake manifold. Remove and discard the gaskets.

To install:

17. Clean all gasket mating surfaces. Position new intake manifold gaskets on the cylinder heads. Make sure the alignment tabs on the gas-

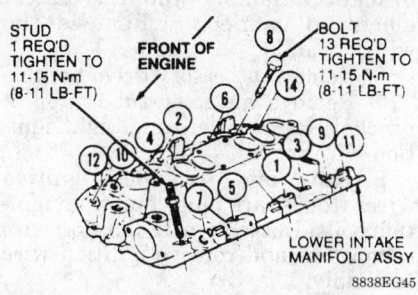

Lower intake manifold bolt torque sequence — 1993–95 3.8L engine

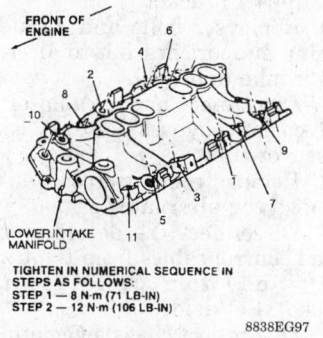

Lower intake manifold bolt torque sequence — 1996–97 3.8L engine

kets are aligned with the holes in the cylinder heads.

18. Install the intake manifold and the retaining bolts. Tighten the bolts, in sequence, to 15–22 ft. lbs. (20–30 Nm).

19. Inspect and if necessary, replace the O-ring seal on the thermostat housing. Position the housing and upper hose and install the 2 bolts. Tighten to 15–22 ft. lbs. (20–30 Nm).

20. Connect the heater supply hose and connect both oxygen sensors.

21. Connect the vacuum hose to the throttle body adapter vacuum port.

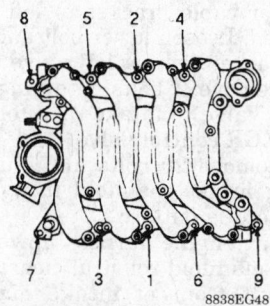

Intake manifold torque sequence — 1993–95 with 4.6L (SOHC) engine

Connect and, if necessary, adjust the throttle valve cable.

22. Install the accelerator cable bracket on the intake manifold and connect the accelerator and cruise control cables to the throttle body.

23. Install the PCV valve in the camshaft cover and connect the canister purge solenoid vent hose. Connect the air conditioning compressor, HDR sensor and canister purge solenoid.

24. Connect the 42-pin engine harness connector. Install the connector on the retaining bracket on the vacuum brake booster.

25. Raise and safely support the vehicle. Connect the EGR tube to the right exhaust manifold and tighten the line nut to 26–33 ft. lbs. (35–45 Nm).

26. Connect the EVO sensor and oil sending unit. Lower the vehicle.

27. Position the generator and install the retaining bolts. Tighten to 15–22 ft. lbs. (20–30 Nm). Install the 2 bolts retaining the generator brace to the intake manifold and tighten to 6.0–8.8 ft. lbs. (8–12 Nm).

28. Connect the generator wiring harness to the generator, right-hand fender apron and junction block.

29. Position the ignition wire assembly on the engine and install the 2 bolts retaining the ignition wire tray to the coil brackets. Tighten the bolts to 6.0–8.8 ft. lbs. (8–12 Nm).

30. Connect the ignition wires to the ignition coils in their proper positions. Connect the ignition wires to the spark plugs.

31. Connect the ignition wire brackets on the camshaft cover studs. Connect both ignition coils and CID sensor.

32. Install the accessory drive belt and the air inlet tube. Install the wiper module and connect the fuel lines.

33. Fill and bleed the cooling system. Connect the negative battery cable, start the engine and check for leaks.

1996–97

— CAUTION —
Fuel injection systems remain under pressure, even after the engine has been turned OFF. The fuel system pressure must be relieved before disconnecting any fuel lines. Failure to do so may result in fire and/or personal injury.

1. If equipped with air suspension, the air suspension switch, located on the right-hand side of the luggage compartment, must be turned to the **OFF** position before raising the vehicle.

2. Disconnect the negative battery cable.

3. Drain the engine cooling system.

4. Relieve the fuel system pressure.

5. Disconnect the fuel supply and return lines.

6. Remove the windshield wiper governor (module).

7. Remove the engine air cleaner outlet tube.

8. Release the drive belt tensioner and remove the accessory drive belt.

9. Tag and disconnect the ignition wires from the spark plugs. Disconnect the ignition wire brackets from the cylinder head cover studs.

10. Disconnect the wiring from both ignition coils and the Camshaft Position (CMP) sensor. Tag and disconnect all ignition wires from both ignition coils. Remove 2 bolts retaining the ignition wire bracket to the ignition coil brackets and remove the ignition wire assembly.

11. Disconnect the alternator wiring harness from the junction block at the fender apron and alternator. Remove the bolts retaining the alternator brace to the intake manifold and the alternator to the cylinder block and remove the alternator.

12. Raise and safely support the vehicle.

13. Disconnect the oil pressure sensor and power steering control valve actuator wiring and position the wiring harness out of the way.

14. Disconnect the EGR valve to exhaust manifold tube from the right-hand exhaust manifold.

15. Lower the vehicle.

16. Remove and disconnect the engine/transmission harness connector from the retaining bracket on the power brake booster.

17. Disconnect the A/C compressor clutch, Crankshaft Position (CKP) sensor and the canister purge solenoid wiring connectors.

18. Remove the PCV valve from the cylinder head cover and disconnect the canister purge vent hose from the PCV valve.

19. Disconnect the accelerator and cruise control cables from the throttle body. Remove the accelerator cable bracket from the intake manifold and position out of the way.

20. Disconnect the vacuum hose from the throttle body adapter port.

21. Disconnect both Heated Oxygen Sensors (HO_2S) and the heater water hose.

22. Remove 2 bolts retaining the thermostat housing to the intake manifold and position the upper hose and thermostat housing out of the way.

NOTE: The 2 thermostat housing bolts are also used to retain the intake manifold.

23. Remove 9 bolts retaining the intake manifold to the cylinder heads and remove the intake manifold. Remove and discard the gaskets.

24. If replacing the intake manifold, swap over the necessary parts.

To install:

25. Clean all gasket mating surfaces.

26. Position new intake manifold gaskets on the cylinder heads. Make sure the alignment tabs on the gaskets are aligned with the holes in the cylinder heads.

27. Install the intake manifold and 9 retaining bolts. Hand-tighten the right-rear bolt (viewed from the front of the engine) before final tightening, then torque the bolts, in sequence, to 15–22 ft. lbs. (20–30 Nm).

28. Inspect and if necessary, replace the O-ring seal on the thermostat housing. Position the housing and upper hose and install 2 retaining bolts. Tighten to 15–22 ft. lbs. (20–30 Nm).

29. Connect the heater water hose.

30. Connect both HO2S wiring connectors.

31. Connect the vacuum hose to the throttle body adapter vacuum port.

32. Install the accelerator cable bracket on the intake manifold and connect the accelerator and cruise control cables to the throttle body.

33. Install the PCV valve in the cylinder head cover and connect the canister purge solenoid vent hose. Connect the A/C compressor clutch, CKP sensor and canister purge solenoid wiring connectors.

34. Connect the engine/transmission harness connector.

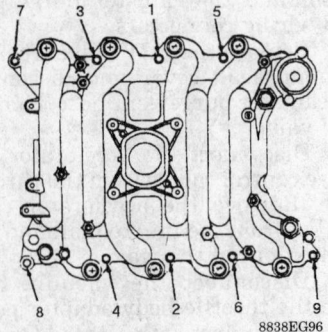

Intake manifold torque sequence — 1996-97 with 4.6L (SOHC) engine

8838EG96

Install the connector on the retaining bracket on the power brake booster.

35. Raise and safely support the vehicle.

36. Connect the EGR valve to exhaust manifold tube to the right-hand exhaust manifold. Tighten the tube nut to 26–33 ft. lbs. (35–45 Nm).

37. Connect the power steering control valve actuator and the oil pressure sensor wiring connectors.

38. Lower the vehicle.

39. Position the alternator and install 2 retaining bolts. Tighten to 15–22 ft. lbs. (20–30 Nm). Install 2 bolts retaining the alternator brace to the intake manifold and tighten to 71–106 inch lbs. (8–12 Nm).

40. Connect the alternator wiring harness to the alternator, right-hand fender apron and junction block.

41. Position the ignition wire assembly on the engine and install 2 bolts retaining the ignition wire bracket to the ignition coil brackets. Tighten the bolts to 71–106 inch lbs. (8–12 Nm).

42. Connect the ignition wires to the ignition coils in their proper positions. Connect the ignition wires to the spark plugs.

43. Connect the ignition wire brackets on the cylinder head cover studs. Connect the wiring connectors to both ignition coils and the CMP sensor.

44. Install the accessory drive belt.

45. Install the engine air cleaner outlet tube.

46. Install the windshield wiper governor.

47. Connect the fuel supply and return lines.

48. Fill and bleed the engine cooling system.

49. Connect the negative battery cable.

50. If equipped with air suspension, turn the air suspension switch to the **ON** position.

51. Start the engine and check for leaks.

52. Road test the vehicle and check for proper operation.

4.6L (DOHC) Engine

All Except 1996–97 Mustang

1. Disconnect the negative battery cable.

2. Drain the cooling system and relieve the fuel system pressure.

3. Remove engine cover. Remove air cleaner outlet tube and resonator assembly.

4. Disconnect fuel lines.

5. Remove the hush panel to expose the windshield wiper module, remove screw, lower module and

bracket assembly and disconnect electrical connector. Remove the wiper module.

6. Remove accessory drive belt.

7. Remove air inlet tube support bracket. Remove left and right ignition wire covers.

8. Label and disconnect ignition wires from spark plugs and from ignition coils. Detach center ignition wire separators and remove ignition wire assembly.

9. Disconnect alternator wiring harness at **B+** terminal and stator connector plug. Disconnect wiring harness retaining clip from alternator support bracket.

10. Remove 2 bolts and 2 studs retaining support bracket to alternator and intake manifold.

11. Disconnect upper radiator hose and coolant bypass hose at coolant crossover tube.

12. Remove coolant crossover tube. Remove the alternator.

13. Disconnect the accelerator and speed control cables from the throttle body and from cable mounting bracket. Set aside.

14. Disconnect chassis vacuum supply hose from intake manifold and position aside.

15. Tag and disconnect the following vacuum lines:
Fuel pressure regulator
Right and left IMRC vacuum controls
EGR control valve
EGR vacuum regulator control
IMRC vacuum control solenoids
Intake manifold connections.

16. Disconnect PCV hose connector from throttle body. Remove PCV valve from left valve cover and position PCV tube assembly aside.

17. Disconnect 4 engine harness retaining clips from intake manifold studs and 1 clip from fuel regulator bracket.

18. Disconnect fuel injector electrical connectors.

19. Remove upper bolt retaining accelerator cable bracket to left cylinder head. Loosen lower bolt and position bracket against left valve cover.

20. Remove 2 bolts retaining EGR control valve to intake manifold. Remove EGR control valve.

21. Remove 20 bolts and studs retaining intake manifold using a reverse torque pattern.

22. Lift engine harness upward for intake manifold removal clearance.

23. Lift front of intake manifold and IMRC assembly and move assembly forward to obtain access to the rear of the intake manifold and disconnect connector at idle air con-

trol, throttle position switch and harness clip at idle air control retaining stud.

24. Remove intake manifold assembly. Remove fuel injector supply manifold from fuel injectors.

25. Remove right and left IMRC units. Remove gaskets and load limiting spacers. Discard the gaskets.

26. Remove and discard EGR and lower intake gaskets.

To install:

27. Clean and inspect all sealing surfaces. Remove paper from adhesive backing on new upper intake manifold gasket. Install onto top of right IMRC unit. Use tapered pins at end bolt hole locations for proper alignment. Repeat for left IMRC.

28. Assemble IMRC units onto intake manifold and hand-tighten 4 bolts.

29. Install fuel injection supply manifold onto fuel injectors and tighten 4 retaining screws to 26–45 inch lbs. (3–5 Nm). Install 2 fuel pressure regulator retaining bracket screws and tighten to 6–8.8 ft. lbs. (8–12 Nm).

30. Position new IMRC gaskets onto cylinder heads. Use gaskets integral location pins to align gasket.

31. Place intake manifold under engine harness and connect wiring harness plugs to idle air control valve, throttle position sensor and connect wiring clip to idle air control valve retaining clip.

32. Move intake manifold into position and install new EGR valve gasket. Hand-tighten 2 retaining bolts.

33. Install 10 long bolts and studs in outer holes of intake manifold. Do not tighten.

34. Install 10 short bolts and studs in inner holes of intake manifold. Hand-tighten all 20 fasteners.

35. Tighten bolts and studs in sequence as follows:

Numbers 5, 7, 9 and 11 to 8.8–11 ft. lbs. (12–15 Nm)

All others to 13–16 ft. lbs. (18–22 Nm)

Rotate, in sequence, 85–96 degrees.

36. Tighten 4 IMRC retaining bolts to 6–7.5 ft. lbs. (8–10 Nm). Rotate, in sequence, 85–96 degrees.

37. Tighten 2 EGR retaining bolts to 15–22 ft. lbs. (20–30 Nm).

38. Install upper retaining bolt into accelerator cable bracket and tighten both upper and lower bolts to 15–22 ft. lbs. (20–30 Nm).

39. Connect engine harness to fuel injectors.

40. Connect engine harness clips to intake manifold studs and connect harness clip to fuel pressure regulator bracket. Connect PCV system.

41. Connect vacuum harness to fuel pressure regulator, right and left IMRC vacuum controls, EGR vacuum regulator control, IMRC vacuum control solenoids and intake manifold vacuum at rear connection.

42. Connect chassis vacuum supply hose and spring clamp to intake manifold vacuum at forward connection.

43. Connect the accelerator and speed control cables to the throttle body and to the cable mounting bracket.

44. Replace O-rings on coolant crossover tubes if necessary. Insert crossover tube support tube support braces over intake manifold inner studs and press crossover tube into place. Install nuts and tighten to 6–8.8 ft. lbs. (8–12 Nm).

45. Connect ECT and temperature gauge sensor connections. Connect upper radiator hose and coolant bypass hose at coolant crossover tube.

46. Install alternator and tighten 2 bolts to 15–22 ft. lbs. (20–30 Nm). Install alternator support bracket and tighten 2 bolts and 2 studs to 6–8.8 ft. lbs. (8–12 Nm).

47. Connect alternator wiring harness connector plug and **B+** terminal wire to the alternator. Connect wiring harness retaining clip to support bracket.

48. Attach center ignition wire separators and install ignition wire assembly. Connect ignition wires to spark plugs and to ignition coils.

49. Install accessory drive belt.

50. Install air inlet tube support bracket. Tighten nuts to 6–8.8 ft. lbs. (8–12 Nm).

51. Connect the fuel lines. Install air cleaner outlet tube and resonator assembly. Install engine cover.

52. Install windshield wiper module.

53. Fill and bleed the cooling system. Connect the negative battery cable, start the engine and check for leaks.

1996–97 Mustang (Upper)

——————— **CAUTION** ———————

Fuel injection systems remain under pressure, even after the engine has been turned OFF. The fuel system pressure must be relieved before disconnecting any fuel lines. Failure to do so may result in fire and/or personal injury.

1. Turn the ignition switch to the **ON** position. Operate the windshield wipers until they are in the straight up position, then turn the ignition switch to the **OFF** position.

2. Disconnect the negative battery cable.

3. Drain the engine cooling system.

4. Remove the engine air cleaner outlet tube.

5. Relieve fuel system pressure as follows:

a. Remove the fuel tank fill cap to relieve the pressure in the fuel tank.

b. Remove the cap from the Schrader valve located on the fuel supply manifold (fuel rail).

c. Attach Fuel Pressure Gauge T80L-9974-B or equivalent, to the Schrader valve and drain the fuel through the drain tube into a suitable container.

d. After the fuel system pressure is relieved, remove the fuel pressure gauge and install the cap on the Schrader valve. Install the fuel tank fill cap.

6. Remove the windshield wiper module assembly.

7. Release the belt tensioner and remove the accessory drive belt.

8. Remove 8 bolts retaining the left-hand and right-hand ignition wire covers and remove.

9. Tag and disconnect the ignition wires from the spark plugs. Disconnect the ignition wires from both ignition coils and remove the wires.

10. Disconnect the alternator wiring harness. Remove 2 bolts and nuts retaining the alternator mounting bracket to the alternator and remove the bracket.

11. Disconnect the upper radiator hose and the water bypass tube-to-thermostat hose at the water bypass tube.

12. Disconnect the engine control wiring from the Throttle Position (TP) sensor, Idle Air Control (IAC) valve, Engine Coolant Temperature (ECT) sensors and the water temperature sensor.

13. Remove 2 retaining nuts supporting the water bypass tube to the intake manifold. Remove the bypass tube.

14. Remove 2 bolts retaining the alternator to the cylinder block and remove the alternator.

15. Disconnect the accelerator and cruise control cables from the throttle body. Remove the accelerator cable bracket from the intake manifold and position out of the way.

16. Disconnect the main vacuum source connector from the intake manifold, fuel pressure regulator,

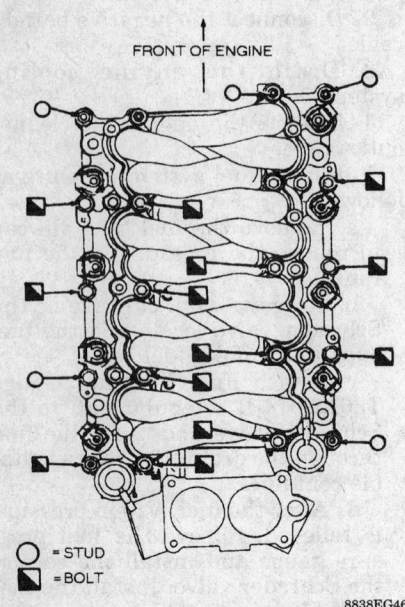

FRONT OF ENGINE

○ = STUD
◼ = BOLT

8838EG46

Intake manifold bolt and stud locations — 4.6L (DOHC) engine except 1996–97 Mustang

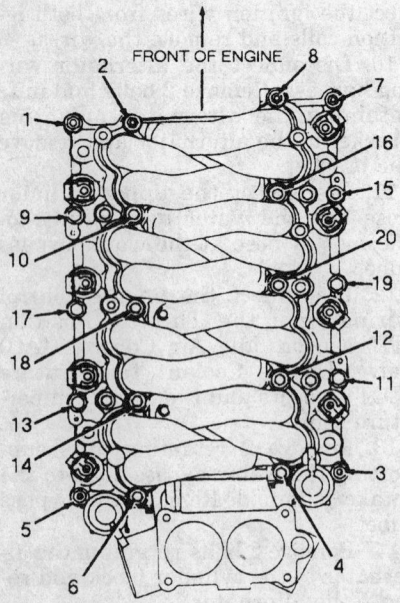

FRONT OF ENGINE

8838EG47

Intake manifold bolt torque sequence — 4.6L (DOHC) engine except 1996–97 Mustang

EGR valve and the EGR solenoid and move aside.

17. Disconnect the PCV valve and hose from the left-hand cylinder head cover.

18. Remove the EGR valve and gasket from the upper intake manifold.

19. If required, remove the throttle body and IAC valve from the upper intake manifold.

20. Remove the upper manifold and gasket from the intake manifold.

21. Proceed to lower intake manifold removal if necessary.

To install:

22. Clean and inspect all gasket sealing surfaces. Only clean the mating surfaces with a plastic scraper to prevent gouging the aluminum housings.

23. Place a new upper intake manifold gasket in position and install the upper intake manifold. Install 7 retaining bolts and tighten in sequence, to 71–106 inch lbs. (8–12 Nm).

24. Install the EGR valve using a new gasket. Tighten 2 retaining bolts to 15–22 ft. lbs. (20–30 Nm).

25. If removed, install the throttle body and the IAC valve.

26. Connect the PCV valve and hose and the IAC valve inlet tube to the intake manifold.

27. Connect the main vacuum connector to the fuel pressure regulator, upper intake manifold connector, EGR valve and the EGR valve solenoid.

28. Install the accelerator cable bracket, if removed and tighten to 71–106 inch lbs. (8–12 Nm). Connect the accelerator and cruise control cables to the throttle body.

29. Replace the O-rings on the water bypass tube and coat with a suitable rubber lubricant. Install the water bypass tube with 2 retaining nuts. Tighten the nuts to 71–106 inch lbs. (8–12 Nm).

30. Connect the engine control wiring to the Throttle Position (TP) sen-

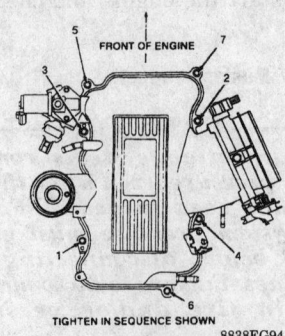

FRONT OF ENGINE

TIGHTEN IN SEQUENCE SHOWN

8838EG94

Upper intake manifold torque sequence — 1996–97 Mustang with 4.6L (DOHC) engine

sor, Idle Air Control (IAC) valve, Engine Coolant Temperature (ECT) sensors and the water temperature sensor.

31. Verify proper operation of the water temperature indicator sending unit. Using a suitable multi-meter set on the ohms scale, touch 1 meter probe to the sending unit body and the other probe end to ground. If resistance is greater than 1 ohm, remove the water bypass tube and remove any paint from the bottom of the mounting tabs and reinstall to improve the ground of the water bypass tube.

32. Connect the upper radiator hose and the water bypass tube-to-water thermostat housing hose.

33. Place the alternator in position on the cylinder block and install 2 retaining bolts. Tighten the bolts to 15–22 ft. lbs. (20–30 Nm). Install the alternator mounting bracket and secure with 2 bolts and 2 nuts. Tighten to 71–106 inch lbs. (8–12 Nm).

34. Connect the alternator wiring harness connectors.

35. Place the ignition wires in position and and connect the ignition wire separators to the alternator mounting bracket.

36. Connect the ignition wires to the ignition coils in their proper positions. Connect the ignition wires to the spark plugs.

37. Connect the ignition wire separators on the cylinder head cover slots. Install 8 retaining nuts and tighten to 45–60 inch lbs. (5–7 Nm).

38. Install the accessory drive belt.

39. Connect the fuel supply and return lines.

40. Install the engine air cleaner outlet tube.

41. Install the windshield wiper module assembly.

42. Fill and bleed the engine cooling system.

43. Connect the negative battery cable.

44. Start the engine and bring to normal operating temperature while checking for leaks.

45. Road test the vehicle and check for proper engine operation.

1996–97 Mustang (Lower)

1. Disconnect the negative battery cable.

2. Remove the upper intake manifold.

3. Disconnect 4 wiring harness retaining clips from the intake manifold studs.

4. Disconnect 8 wiring connectors to the fuel injectors. Move the wiring out of the way.

5. Using the proper sequence, loosen and then remove 10 intake manifold retaining bolts and stud bolts.

6. Remove the intake manifold and the lower intake manifolds, or Intake Manifold Runner Controls (IMRC's), as an assembly.

7. If further disassembly is required, disconnect the engine control wiring from the IMRC actuator.

8. Disconnect the IMRC actuator cables from the IMRC levers and retaining brackets.

9. Remove 4 bolts retaining the fuel injection supply manifold (fuel rail) to the lower intake manifolds (IMRC's). Remove the fuel injection supply manifold and the fuel injectors as an assembly.

10. Remove 2 bolts retaining the left-hand and right-hand lower intake manifolds (IMRC's) to the intake manifold. Remove the lower intake manifolds.

11. Remove the intake manifold upper gaskets and load limiting spacers from both lower intake manifolds and discard. Remove the lower intake manifold gaskets from the cylinder heads and discard.

To install:

12. Clean all gasket mating surfaces. Ensure that all load limiting spacers are removed.

13. Remove the protective paper from the adhesive backing on the new intake manifold upper gaskets. Position the gaskets on the top of the left-hand and right-hand lower intake manifold (IMRC's) surfaces using the tapered pins at each end of the manifolds for alignment.

14. Assemble the lower intake manifolds (IMRC's) to the intake manifold and install 4 retaining bolts, hand-tight.

15. Install the fuel injectors and the fuel injection supply manifold. Install 4 retaining bolts and tighten to 71–106 inch lbs. (8–12 Nm).

16. Place new lower intake manifold gaskets on the cylinder heads using the gaskets integral locating pins to align the gaskets.

17. Place the intake manifold under the engine control wiring and connect the wiring to the IMRC actuator.

18. Place the intake manifold assembly into position on the engine.

19. Install the intake manifold retaining bolts and stud bolts.

20. Tighten the intake manifold bolts, in sequence, to 15–22 ft. lbs. (20–30 Nm).

21. Install the 2 bolts retaining the intake manifold-to-lower intake

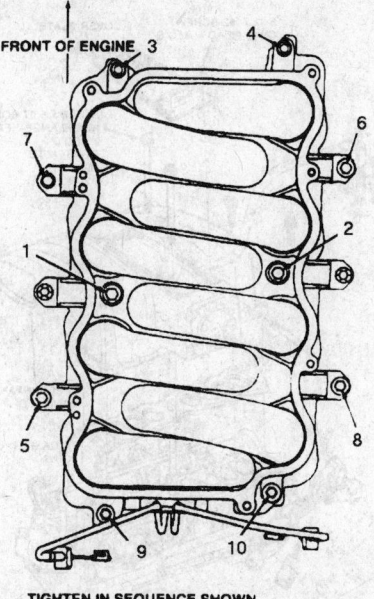

TIGHTEN IN SEQUENCE SHOWN

8838EG95

Lower intake manifold torque sequence — 1996–97 Mustang with 4.6L (DOHC) engine

manifolds. Tighten the bolts to 71–106 inch lbs. (8–12 Nm).

22. Connect the IMRC actuator cables to the retaining brackets and the IMRC levers.

23. Connect the engine control wiring to the fuel injectors.

24. Connect the engine control wiring harness retaining clips to the intake manifold studs.

25. Install the upper intake manifold.

26. Connect the negative battery cable.

27. Start the engine and bring to normal operating temperature while checking for leaks.

28. Road test the vehicle and check for proper engine operation.

5.0L Engine

1. Disconnect the negative battery cable.

2. Drain the cooling system and relieve the fuel system pressure.

3. Disconnect the accelerator cable and cruise control linkage, if equipped, from the throttle body. Disconnect the TV cable, if equipped. Tag and disconnect the vacuum lines at the intake manifold fitting.

4. Tag and disconnect the spark plug wires from the spark plugs. Re-

move the wires and bracket assembly from the rocker arm cover attaching stud. Remove the distributor cap and wires assembly.

5. Disconnect the fuel lines and the distributor wiring connector. Mark the position of the rotor on the distributor housing and the distributor housing in the block. Remove the hold-down bolt and remove the distributor.

6. Disconnect the upper radiator hose at the thermostat housing and the water temperature sending unit wire at the sending unit. Disconnect the heater hose from the intake manifold and disconnect the 2 throttle body cooler hoses, except on Cobra.

7. Disconnect the water pump bypass hose from the thermostat housing. Tag and disconnect the connectors from the engine coolant temperature, air charge temperature, throttle position and EGR sensors and the idle speed control solenoid. Disconnect the injector wire connections and the fuel charging assembly wiring.

8. Remove the PCV valve from the grommet at the rear of the lower intake manifold. Disconnect the fuel evaporative purge hose from the plastic connector at the front of the upper intake manifold.

9. Remove the upper intake manifold cover plate, except on Cobra, and upper intake bolts. Remove the upper intake manifold.

10. Remove the heater tube assembly from the lower intake manifold.

11. Remove the lower intake manifold retaining bolts and remove the lower intake manifold.

NOTE: If it is necessary to pry the intake manifold away from the cylinder heads, be careful to avoid damaging the gasket sealing surfaces.

To install:

12. Clean all gasket mating surfaces. Apply a ⅛ in. bead of silicone sealer to the points where the cylinder block rails meet the cylinder heads.

13. Position new seals on the cylinder block and new gaskets on the cylinder heads with the gaskets interlocked with the seal tabs. Make sure the holes in the gaskets are aligned with the holes in the cylinder heads.

14. Apply a ¹⁄₁₆ in. bead of sealer to the outer end of each intake manifold seal for the full width of the seal.

15. Using guide pins to ease installation, carefully lower the intake

manifold into position on the cylinder block and cylinder heads.

NOTE: After the intake manifold is in place, run a finger around the seal area to make sure the seals are in place. If the seals are not in place, remove the intake manifold and position the seals.

16. Make sure the holes in the manifold gaskets and the manifold are in alignment. Remove the guide pins. Install the intake manifold attaching bolts and tighten, in sequence, in sequence, to 8 ft. lbs. (11 Nm), then to 16 ft. lbs. (22 Nm), finally to 24 ft. lbs. (32 Nm).

17. Install the heater tube assembly to the lower intake manifold.

18. Install the water pump bypass hose on the thermostat housing. Install the hoses to the heater tubes.

19. Connect the upper radiator hose and connect the heater hose at the intake manifold. Connect the fuel lines.

20. Install the distributor, aligning the housing and rotor with the marks that were made during removal. Install the distributor cap. Position the spark plug wires in the harness brackets on the rocker arm cover attaching stud and connect the wires to the spark plugs.

21. Install a new gasket and the upper intake manifold. Tighten the bolts to 12–18 ft. lbs. (16–24 Nm). Install the cover plate and connect the crankcase vent tube.

22. Connect the TV cable and cruise control cable, if equipped, to the throttle body. Connect the electrical connectors and vacuum lines.

23. Connect the coolant hoses to the EGR spacer. Fill and bleed the cooling system.

24. Connect the negative battery cable, start the engine and check for leaks. Check the ignition timing.

25. Operate the engine at fast idle. When engine temperatures have stabilized, tighten the intake manifold bolts to 23–25 ft. lbs. (31–34 Nm).

26. Connect the air intake duct and the crankcase vent hose.

Exhaust Manifold

REMOVAL AND INSTALLATION

2.3L Engine

1. Disconnect the negative battery cable.

2. Remove the air cleaner and duct assembly.

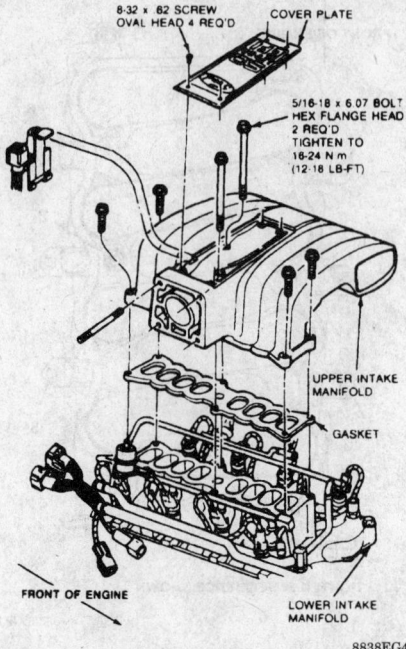

Upper intake manifold installation — 5.0L engine

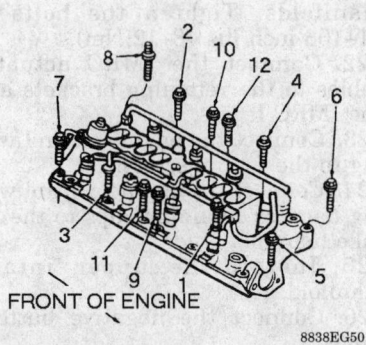

Lower intake manifold bolt torque sequence — 5.0L engine

3. Remove the EGR tube at the exhaust manifold and loosen at the EGR valve.

4. Disconnect and, if necessary, remove the oxygen sensor from the exhaust manifold.

5. Raise and safely support the vehicle. Remove the 2 exhaust pipe bolts and lower the vehicle.

6. Remove the 8 exhaust manifold bolts and remove the exhaust manifold.

7. Installation is the reverse of the removal procedure. Tighten the manifold bolts, in sequence, in 2 steps, 1st to 15–17 ft. lbs. (20–30 Nm) and then

to 20–30 ft. lbs. (27–41 Nm). Tighten the exhaust pipe bolts to 25–34 ft. lbs. (36–46 Nm).

3.8L Engine

Left Side

1. Disconnect the negative battery cable. Remove oil level dipstick tube support bracket.

2. Disconnect the oxygen sensor at the wiring connector.

3. Tag and disconnect the wires from the spark plugs.

4. Raise and safely support the vehicle.

5. Remove the manifold to exhaust pipe attaching nuts.

6. Lower the vehicle.

7. On supercharged engines, remove the charge air cooler tubes and remove the oil cooler tube and dipstick tube support brackets from the studs.

8. Remove exhaust manifold attaching bolts and manifold.

9. Installation is the reverse of the removal procedure. Tighten the manifold retaining bolts to 15–22 ft. lbs. (20–30 Nm).

Right Side

1. Disconnect the negative battery cable. On supercharged engines, remove the air cleaner inlet tube.

2. On non-supercharged engines, disconnect the coil secondary wire from the coil. Tag and disconnect the wires from the spark plugs.

3. On non-supercharged engines, remove the spark plugs and the outer heat shield.

4. Raise and safely support the vehicle. Disconnect the EGR tube.

5. If equipped with automatic transmission, remove the dipstick tube.

6. Remove the manifold-to-exhaust pipe retaining nuts and lower the vehicle.

7. Remove the exhaust manifold retaining bolts and remove the manifold.

8. Installation is the reverse of the removal procedure. Tighten the exhaust manifold retaining bolts to 15–22 ft. lbs. (20–30 Nm).

4.6L (SOHC) Engine

1. Disconnect the battery cables. Remove the air inlet tube.

2. Drain the cooling system and remove the cooling fan and shroud. Relieve the fuel system pressure and disconnect the fuel lines.

3. Remove the upper radiator hose. Remove the wiper module and support bracket.

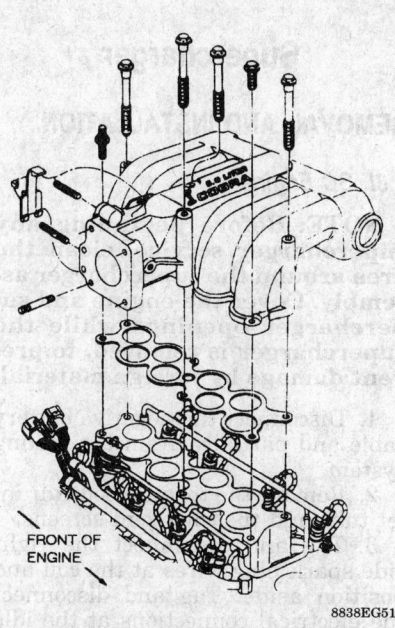

Upper intake manifold installation — 5.0L Cobra engine

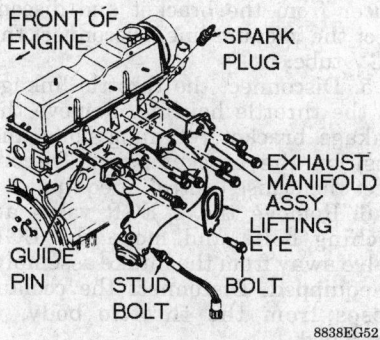

Exhaust manifold installation and torque sequence — 2.3L engine

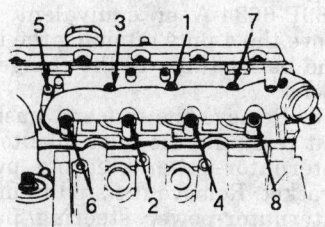

NOTE: ENGINE SHOWN REMOVED FOR CLARITY

NOTE: LH EXHAUST MANIFOLD SHOWN RH EXHAUST MANIFOLD TYPICAL

8838EG53

Exhaust manifold torque sequence — 4.6L (SOHC) engine

4. Discharge the air conditioning system. Disconnect and plug the compressor outlet hose at the compressor and remove the bolt retaining the hose assembly to the right coil bracket. Cap the compressor opening.

5. Remove the 42-pin engine harness connector from the retaining bracket on the brake vacuum booster. Disconnect the connector.

6. Disconnect the throttle valve cable from the throttle body. Disconnect the heater outlet hose.

7. Remove the nut retaining the ground strap to the right cylinder head. Remove the upper stud and lower bolt retaining the heater outlet hose to the right cylinder head and position out of the way.

8. Remove the blower motor resistor and remove the bolt retaining the right engine insulator to the lower engine bracket. Disconnect both oxygen sensors.

9. Raise and safely support the vehicle. Remove the engine mount through bolts.

10. Remove the EGR tube line nut from the right exhaust manifold.

11. Disconnect the exhaust pipes from the manifolds. Lower the exhaust system and hang it from the crossmember with wire.

12. To remove the left exhaust manifold, remove the engine mount from the engine block and remove the 8 bolts retaining the exhaust manifold.

13. Position a jack and a block of wood under the oil pan, rearward of the oil drain hole. Raise the engine approximately 4 in. (100mm).

14. Remove the 8 bolts retaining the right exhaust manifold and remove the manifold.

To install:

15. If the exhaust manifolds are being replaced, transfer the oxygen sensors and tighten to 27–33 ft. lbs. (37–45 Nm). On the right manifold, transfer the EGR tube connector and tighten to 33–48 ft. lbs. (45–65 Nm).

16. Clean the mating surfaces of the exhaust manifolds and cylinder heads.

17. Position the exhaust manifolds to the cylinder heads and install the retaining bolts. Tighten, in sequence, to 15–22 ft. lbs. (20–30 Nm).

18. Position and connect the EGR valve and tube assembly to the exhaust manifold. Tighten the line nut to 26–33 ft. lbs. (35–45 Nm).

19. Install the left engine mount and tighten the bolts to 15–22 ft. lbs. (20–30 Nm). Lower the engine onto the mounts and remove the jack. Install the engine mount through bolts

and tighten to 15–22 ft. lbs. (20–30 Nm).

20. Cut the wire and position the exhaust system. Tighten the nuts to 20–30 ft. lbs. (27–41 Nm).

NOTE: Make sure the exhaust system clears the No. 3 crossmember. Adjust as necessary.

21. Lower the vehicle. Connect both oxygen sensors and install the bolt retaining the right engine mount to the frame. Tighten to 15–22 ft. lbs. (20–30 Nm).

22. Install the blower motor resistor. Position the heater outlet hoses. Install the upper stud and lower bolt and tighten to 15–22 ft. lbs. (20–30 Nm). Install the ground strap onto the stud and tighten the nut to 15–22 ft. lbs. (20–30 Nm).

23. Connect the heater outlet hose. Connect and if necessary, adjust the throttle valve cable.

24. Connect the 42-pin connector and transmission harness connector. Install the connector to the retaining bracket on the brake vacuum booster.

25. Connect the air conditioning compressor outlet hose to the compressor and install the bolt retaining the hose assembly to the right coil bracket.

26. Install the upper radiator hose and connect the fuel lines. Install the wiper module and retaining bracket.

27. Install the cooling fan and shroud. Install the air inlet tube. Connect the battery cables, start the engine and check for leaks.

28. Leak test, evacuate and charge the air conditioning system according to the proper procedure. Observe all safety precautions.

4.6L (DOHC) Engine

Right Side

1. Disconnect the negative battery cable.

2. Remove engine appearance cover. Disconnect intake air temperature sensor connector and crankcase vent tube from air cleaner outlet tube.

3. Completely loosen air cleaner outlet to throttle body bolt. Disconnect air cleaner outlet tube.

4. Remove remaining bolt from support bracket clamp on right valve cover to air cleaner outlet tube assembly and remove assembly and resonator.

5. Disconnect the oxygen sensor connector.

6. Remove 4 manifold to cylinder head bolts from upper side of exhaust manifold.

7. Disconnect exhaust pipe from catalyst and remove catalyst. Remove 4 lower exhaust manifold retaining bolts and remove manifold. Remove exhaust manifold gasket and discard.

8. Installation is the reverse of the removal procedure. Tighten the manifold retaining bolts in sequence to 15–22 ft. lbs. (20–30 Nm).

Left Side

1. Turn **OFF** air suspension switch located in luggage compartment.

2. Disconnect the negative and positive battery cables.

3. Disconnect the oxygen sensor. Install engine support tripod onto engine compartment and attach to engine.

4. Raise and safely support the vehicle.

5. Remove the front wheel and tire assemblies.

6. Disconnect EGR tube at left catalyst. Disconnect the left exhaust pipe at the catalyst and right catalyst at the exhaust manifold.

7. Disconnect left and right ride height sensor wiring connectors and remove right and left brake caliper bolts. Remove brake calipers and support with mechanics wire.

8. Disconnect left and right upper control arms from spindles. Disconnect steering shaft coupler at rag joint.

9. Remove power steering line clamp from subframe.

10. Support subframe with engine jack.

11. Remove 2 motor mount through bolts. Remove left and right lower strut control arm nuts and bolts.

12. Remove 8 subframe bolts. Lower the subframe.

13. Remove the 8 manifold attaching nuts and remove left exhaust manifold. Remove and discard gaskets.

14. Installation is the reverse of the removal procedure. Tighten the manifold retaining bolts in sequence to 15–22 ft. lbs. (20–30 Nm).

5.0L Engine

Thunderbird and Cougar

1. Disconnect the negative battery cable.

2. If removing the left manifold, remove the oil dipstick tube nut and pull the bracket from the manifold stud.

3. Raise and safely support the vehicle.

4. Carefully tap upward on the dipstick tube and remove it from the vehicle.

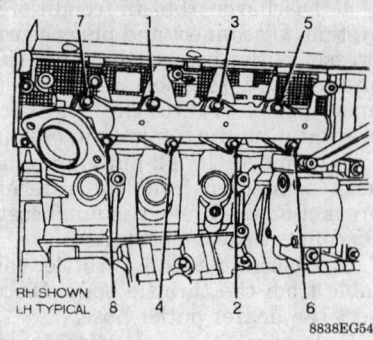

RH SHOWN
LH TYPICAL   8838EG54

Exhaust manifold bolt torque sequence — 4.6L (DOHC) engine

5. If removing the left manifold, disconnect the oxygen sensor connector.

6. Disconnect the exhaust manifold(s) from the exhaust pipe(s). Lower the vehicle.

7. If removing the right exhaust manifold, disconnect the electrical connector from the mass air flow sensor, located on the air cleaner assembly. Remove the air cleaner and inlet duct assembly.

8. Tag and disconnect the spark plug wires.

9. If removing the right exhaust manifold, remove the alternator rear brace and the thermactor hose assembly and EGR tube.

10. Remove the retaining bolts and remove the exhaust manifold(s) through the top of the engine compartment.

11. Installation is the reverse of the removal procedure. Clean the manifold and cylinder head mating surfaces prior to installation. Working from the center to the ends, tighten the exhaust manifold-to-cylinder head bolts to 18–24 ft. lbs. (24–32 Nm).

Mustang

1. Disconnect the negative battery cable.

2. If removing the right exhaust manifold, remove the thermactor hardware.

3. Tag and disconnect the spark plug wires. Remove the spark plugs.

4. Raise and safely support the vehicle.

5. Disconnect the exhaust pipe(s) from the manifold(s). Lower the vehicle.

6. Remove the retaining bolts and remove the exhaust manifold(s).

7. Installation is the reverse of the removal procedure. Clean the manifold and cylinder head mating surfaces prior to installation. Working from the center to the ends, tighten

the exhaust manifold attaching bolts to 26–32 ft. lbs. (32–43 Nm).

Supercharger

REMOVAL AND INSTALLATION

3.8L SC Engine

NOTE: Before beginning any supercharger service, clean the area around the supercharger assembly. Cover the engine and supercharger openings while the supercharger is removed, to prevent damage by foreign material.

1. Disconnect the negative battery cable and partially drain the cooling system.

2. Remove the throttle body air inlet tube and the cowl vent screens.

3. Tag and disconnect the right side spark plug wires at the coil and position aside. Tag and disconnect the electrical connections at the idle air bypass valve, throttle position sensor and air charge temperature sensors.

4. Tag and disconnect the vacuum lines from the inlet/plenum assembly. If equipped, remove the EGR transducer from the bracket and disconnect the vacuum line. Disconnect the PCV tube.

5. Disconnect the throttle linkage at the throttle housing. Remove the linkage bracket retaining bolts and position the bracket aside. Disconnect the cruise control, if equipped.

6. Remove the 2 EGR valve attaching bolts and move the EGR valve away from the intake assembly, if equipped. Disconnect the coolant hoses from the throttle body, if equipped.

7. Remove the supercharger drive belt. Remove the charge air cooler inlet and outlet tubes as follows:

a. Disconnect the inlet tube from the supercharger outlet adapter using spanner nut wrench tool T89P-6634-A or equivalent. Remove the 4 nuts retaining the inlet and outlet tubes to the charge air cooler.

b. Remove the nut and push-on nut retaining the inlet tube to the alternator-power steering pump bracket. Remove the stud from the alternator-power steering pump bracket. Remove the inlet tube.

NOTE: Use extreme care during removal and installation of the charge air cooler tubes so as not to scratch, nick or contaminate the sealing surfaces.

c. Remove the 2 nuts retaining the outlet tube to the intake elbow assembly. Raise and safely support the vehicle.

d. Remove the bolt retaining the outlet tube to the cylinder block front upper support bracket. Loosen, but do not remove the support bracket.

NOTE: The bracket must be close to the front face of the cylinder block to allow the bracket to pivot during outlet tube reinstallation.

e. Remove the nut and push-on nut retaining the outlet tube to the alternator-power steering pump bracket. Remove the power steering pump drive belt.

f. Tag and disconnect the spark plug wires from the coil. Remove the power steering pump bracket brace to water pump retaining stud nuts. Remove 2 power steering pump bracket to cylinder head retaining bolts and 1 stud nut.

g. Install a 10 x 1.5mm x 170mm bolt, 6½ in. long into the top hole in the power steering pump bracket. Thread the bolt into the cylinder head approximately 5 turns. This will aid in holding the power steering pump bracket in position.

h. Remove the power steering pump filler cap. Slide the power steering pump bracket assembly forward on the stud and bolt that was installed in the previous step.

i. Remove the outlet tube by pulling underneath the power steering pump bracket assembly and up through the engine compartment. It may be necessary to pivot the outlet tube clamping connector to gain clearance during removal.

8. Remove the 3 intake elbow retaining bolts and the 3 supercharger retaining bolts. Lift the supercharger and intake elbow assembly from the vehicle as a unit.

To install:

9. Clean and inspect all gasket surfaces. Position a new gasket on the intake manifold using guide pins, if available.

10. Install the supercharger, throttle body and intake elbow as an assembly. Tighten the two 8mm bolts to 15–22 ft. lbs. (20–30 Nm). Tighten the 12mm bolt to 52–70 ft. lbs. (70–96 Nm).

11. Install the 3 intake elbow retaining bolts and tighten to 20–28 ft. lbs. (26–38 Nm). Install the charge air cooler tubes as follows:

a. Clean and inspect the sealing surfaces of the supercharger outlet adapter intake elbow, charge air cooler and tubes.

NOTE: Make sure there are no foreign particles on the sealing surfaces of the tubes. It is important that the charge air cooler tubes seal completely. Any air leak will cause poor operation and performance.

b. Install gasket sealant tape ESE-M4G168-B or equivalent, to the spherical seat surfaces of the charge air cooler tubes. Install the tape approximately ⅛ in. (3mm) from the inner diameter of the tubes. Overlap the tape ends approximately ¼ in. (6mm). Do not stretch the tape during installation or the seal may leak. During proper installation, a slight wrinkling will occur on the tape edge at the inner diameter.

NOTE: The system must be torqued in sequence and to the specification for that step. This is required for proper alignment of the system to ensure sealing of the charge air cooler tubes.

c. Guide the outlet tube down through the engine compartment and underneath the power steering pump bracket assembly. It may be necessary to rotate the lower outlet tube clamping connector to gain clearance.

NOTE: Use extreme care during installation of the charge air cooler tubes so as not to scratch, nick or contaminate the sealing surfaces.

d. Slide the power steering pump bracket assembly into position. Install the power steering pump bracket retaining stud nut and tighten to 30–40 ft. lbs. (40–55 Nm).

e. Remove the bolt installed in Step 7g of the removal procedure. Install the power steering pump bracket to cylinder head bolts and tighten to 30–40 ft. lbs. (40–55 Nm).

f. Install the power steering pump bracket brace to water pump retaining stud nuts and tighten to 15–22 ft. lbs. (20–30 Nm). Install the outlet tube over the lower stud on the alternator-power steering pump bracket.

g. Install the push-on nut onto the stud, tight enough to retain the tube against the alternator-power steering pump bracket surface but free enough to allow tube movement to ensure seating of the

spherical seat on the outlet tube to intake elbow assembly.

h. Install the outlet tube clamping connector over the studs on the intake elbow assembly and secure with the 2 nuts. Tighten both nuts to 15–22 ft. lbs. (20–30 Nm). The clamping connector should be installed so it is visually parallel to the stud mounting face of the intake elbow assembly.

i. Install the nut to the stud on the alternator-power steering pump bracket and tighten to 30–40 ft. lbs. (40–55 Nm). Install the bolt to secure the outlet tube to the cylinder block support bracket and tighten to 30–40 ft. lbs. (40–55 Nm). Tighten the support bracket to front of cylinder block retaining nut to 15–22 ft. lbs. (20–30 Nm) and bolt to 52–70 ft. lbs. (70–96 Nm).

j. Apply anti-seize compound to the inner backside spherical seat surface and threads of the supercharger outlet adapter collar. Position the inlet tube, then install the upper stud into the alternator-power steering pump bracket.

k. Install the push-on nut onto the stud, tight enough to retain the tube against the alternator-power steering pump bracket surface but free enough to allow tube movement to ensure seating of the spherical seat on the inlet tube to supercharger outlet adapter.

l. Fully hand-tighten the supercharger outlet adapter collar onto the threaded tube end of the inlet tube assembly. Install the charge air cooler assembly to the inlet and outlet tubes. Install the nuts to the studs tight enough to retain the charge air cooler and tubes together but free enough to allow movement on the spherical seats. Do not tighten at this time.

m. Tighten the supercharger outlet adapter collar to inlet tube to 148 ft. lbs. (200 Nm).

n. Wait 10 minutes minimum and retighten the supercharger outlet collar to 148 ft. lbs. (200 Nm).

NOTE: When 1st compressed, the sealant tape flows and forms to the sealing surface. If the collar is not retightened, the torque of the collar will drop causing a leak at this joint.

o. Tighten the inlet and outlet tube to charge air cooler nuts to 15–22 ft. lbs. (20–30 Nm). The clamping connectors should be installed so they are visually parallel to the stud mounting face of the charge air cooler assembly.

p. Install the nut retaining the inlet tube to the alternator-power steering pump support bracket and tighten to 30–40 ft. lbs. (40–55 Nm).

12. Install the supercharger drive belt. Connect the coolant hoses to the throttle body, if equipped.

13. Connect the EGR valve with a new gasket to the intake manifold, if equipped. Tighten the retaining bolts to 14–22 ft. lbs. (20–30 Nm).

14. Install the throttle linkage bracket and connect the throttle linkage. Tighten to 10–15 ft. lbs. (14–20 Nm).

15. Connect the vacuum lines to the inlet assembly and connect the PCV tube. If equipped, connect the vacuum line to the EGR transducer and install the transducer in the bracket.

16. Install the right side spark plug wires. Connect the electrical connectors at the idle air bypass valve, throttle position sensor and air charge temperature sensor.

17. Install the cowl covers and the throttle body air inlet tube.

18. Fill and bleed the cooling system. Connect the negative battery cable. Start the engine and check for leaks and proper operation.

Front Cover Seal

REMOVAL AND INSTALLATION

3.8L Engine

1. Disconnect the negative battery cable.

2. On non-supercharged engines, remove the fan shroud and position it back over the fan. Remove the fan/clutch assembly and shroud.

3. On supercharged engines, disconnect the electric cooling fan connector and remove the fan assembly.

4. Loosen the accessory drive belt idlers. Raise and safely support the vehicle.

5. Disengage the drive belts and remove the crankshaft pulley. On supercharged engines, remove the upper and lower crankshaft shields.

6. Remove the crankshaft damper retaining bolt and remove the damper using a puller.

7. Using a small prybar, remove the seal from the front cover. Use care to prevent damage to the cover and crankshaft.

To install:

8. Inspect the front cover and crankshaft damper for damage, nicks, burrs or other roughness which may cause the seal to fail. Ser-

vice or replace components as necessary.

9. Lubricate the seal lip using clean engine oil. Install the seal using a suitable seal installer.

10. Lubricate the seal surface on the damper with clean engine oil. Install the damper using a suitable installation tool.

11. Install the damper retaining bolt and tighten to 103–132 ft. lbs. (140–180 Nm). Install the crankshaft pulley and tighten the retaining bolts to 20–28 ft. lbs. (26–38 Nm).

12. Install the remaining components in the reverse order of their removal. Connect the negative battery cable, start the engine and check for leaks.

4.6L Engine

1. Disconnect the negative battery cable.

2. Remove accessory drive belt.

3. Raise and safely support the vehicle.

4. Remove crankshaft damper bolt and washer. Using a suitable puller, remove the damper.

5. Using a suitable seal remover, remove the front cover oil seal.

To install:

6. Lubricate oil seal lip and bore with clean engine oil. Install seal with a suitable seal installer. Install the remaining components in the reverse order of their removal.

5.0L Engine

1. Disconnect the negative battery cable.

2. Remove the fan shroud and position it back over the fan. Remove the fan/clutch assembly and shroud.

3. Remove the accessory drive belts.

4. Remove the crankshaft pulley from the damper and remove the damper retaining bolt. Remove the damper using a puller.

5. Remove the seal using a seal removal tool.

To install:

6. Lubricate the seal lip with clean engine oil and install using a seal installer.

7. Apply clean engine oil to the sealing surface of the vibration damper. Line up the crankshaft damper keyway with the crankshaft key and install the damper using a damper installation tool.

8. Install the damper retaining bolt and tighten to 70–90 ft. lbs. (95–122 Nm).

9. Install the remaining components in the reverse order of their removal.

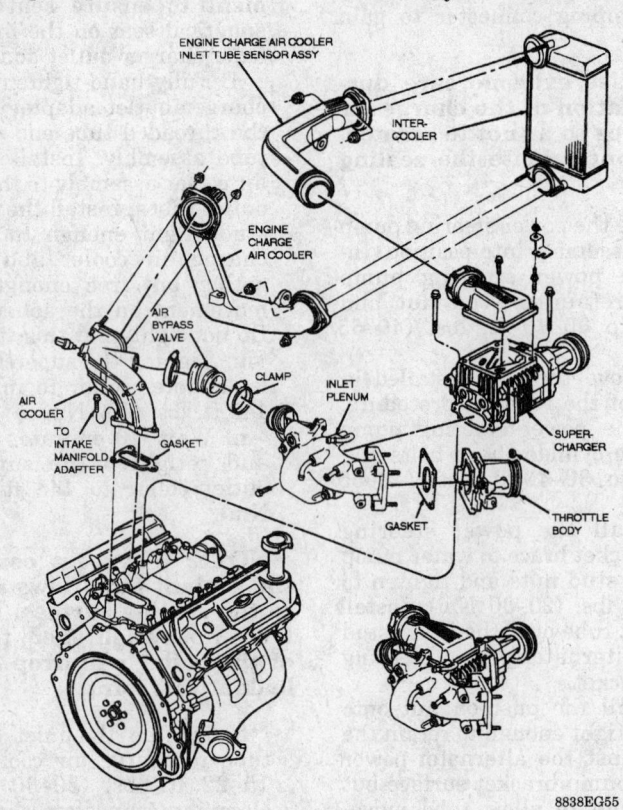

ENGINE CHARGE AIR COOLER INLET TUBE SENSOR ASSY

INTER-COOLER

ENGINE CHARGE AIR COOLER

AIR BYPASS VALVE

CLAMP

AIR COOLER TO INTAKE MANIFOLD ADAPTER

GASKET

INLET PLENUM

SUPER-CHARGER

GASKET

THROTTLE BODY

8838EG55

Supercharger system components — 3.8L SC engine

Timing Chain Front Cover

REMOVAL AND INSTALLATION

3.8L Engine

1. Disconnect the negative battery cable and drain the cooling system.
2. Remove the air cleaner assembly and air intake duct.
3. On non-supercharged engines, remove the fan/clutch assembly and shroud. On supercharged engines, remove the electric cooling fan assembly.
4. Remove the accessory drive belt idlers, drive belts and the water pump pulley.
5. Remove the power steering pump bracket retaining bolts. Leaving the hoses connected, place the pump/bracket assembly aside in a position to prevent fluid from leaking out.
6. If equipped with air conditioning, remove the compressor front support bracket but leave the compressor in place.
7. Disconnect the coolant bypass hose and heater hose at the water pump. Disconnect the upper radiator hose at the thermostat housing.
8. On non-supercharged engines, disconnect the coil wire from the distributor cap and remove the cap with the secondary wires attached.
9. On non-supercharged engines, mark the position of the rotor in relation to the distributor housing and mark the position of the distributor housing on the front cover. Remove the distributor hold-down clamp and lift the distributor out of the front cover.
10. On supercharged engines, remove the hold-down clamp and lift the camshaft synchronizer from the front cover.
11. Raise and safely support the vehicle. Remove the crankshaft damper and pulley using a puller.

NOTE: If the crankshaft pulley and vibration damper have to be separated, mark the damper and pulley so they may be reassembled in the same relative position. This is important as the damper and pulley are initially balanced as a unit. If the crankshaft damper is being replaced, check if the original damper has balance pins installed. If so, new balance pins must be installed on the new damper in the same position as the original damper. The crankshaft pulley, new or origi-

nal, must also be installed in the same relative position as originally installed.

12. Remove the oil filter. On supercharged engines, remove the oil cooler.
13. Disconnect the lower radiator hose at the water pump. Remove the oil pan.

NOTE: The front cover cannot be removed without lowering the oil pan.

14. Lower the vehicle. Remove the front cover retaining bolts. It is not necessary to remove the water pump.

NOTE: Do not overlook the cover retaining bolt located behind the oil filter adapter. The front cover will break if pried on and all retaining bolts are not removed.

15. Remove the front cover and water pump as an assembly. Remove and discard the cover gasket.

NOTE: The front cover contains the oil pump and water pump. If a new front cover is to be installed, remove the water pump and oil pump from the old front cover.

To install:
16. Clean all gasket mating surfaces. If reusing the front cover, replace the front cover oil seal.
17. Position a new gasket on the cylinder block and install the front cover using dowels for proper alignment. Install the front cover retaining bolts and tighten to 15–22 ft. lbs. (20–30 Nm).
18. Raise and safely support the vehicle. Install the oil pan. Connect the lower radiator hose and install the oil filter.
19. Coat the crankshaft damper sealing surface with clean engine oil. Apply a small amount of silicone sealer to the crankshaft keyway.
20. Position the crankshaft pulley key in the crankshaft keyway and install the damper, using a suitable installation tool.
21. Install the damper washer and retaining bolt and tighten to 103–132 ft. lbs. (140–180 Nm). Install the crankshaft pulley and tighten the retaining bolts to 20–28 ft. lbs. (26–38 Nm).
22. Lower the vehicle. Connect the coolant bypass hose.
23. On non-supercharged engines, install the distributor, aligning the marks that were made during the removal procedure. Install the distributor cap and coil wire. On super-

charged engines, install the camshaft synchronizer.

24. Connect the upper radiator hose at the thermostat housing. Connect the heater hose.
25. If equipped with air conditioning, install the compressor and mounting brackets. Tighten retaining bolts to 30–45 ft. lbs. (41–61 Nm).
26. Install the power steering pump and mounting bracket. Tighten the retaining bolts to 30–45 ft. lbs. (41–61 Nm).
27. Install the water pump pulley. Position the accessory drive belts over the pulleys.
28. On non-supercharged engines, install the fan/clutch assembly and fan shroud. Cross-tighten the fan/clutch assembly retaining bolts to 12–18 ft. lbs. (16–24 Nm).
29. On supercharged engines, install the electric cooling fan assembly and connect the harness connector to the fan motor.
30. Fill the crankcase with the proper type and quantity of engine oil. Fill and bleed the cooling system. Connect the negative battery cable.
31. Start the engine and check for leaks. Check the ignition timing and curb idle speed and adjust, as necessary.

4.6L (SOHC) Engine

1. Disconnect the negative battery cable.
2. Remove the cooling fan and shroud. Loosen the water pump pulley bolts, remove the accessory drive belt and remove the water pump pulley.
3. Raise and safely support the vehicle.
4. Remove the bolts retaining the power steering pump to the engine block and cylinder front cover. The lower front bolt on the power steering pump will not come all the way out. Wire the power steering pump out of the way.
5. Remove the 4 bolts retaining the oil pan to the front cover. Remove the crankshaft damper retaining bolt and washer. Remove the damper using a puller.
6. Lower the vehicle. Remove the bolt retaining the air conditioner high pressure line to the right coil bracket.
7. Remove the front bolts and loosen the remaining bolts on the camshaft covers. Using plastic wedges or similar tools, prop up both camshaft covers. Disconnect both ignition coils and CID sensor.
8. Remove the 3 nuts retaining the right coil bracket to the front cover.

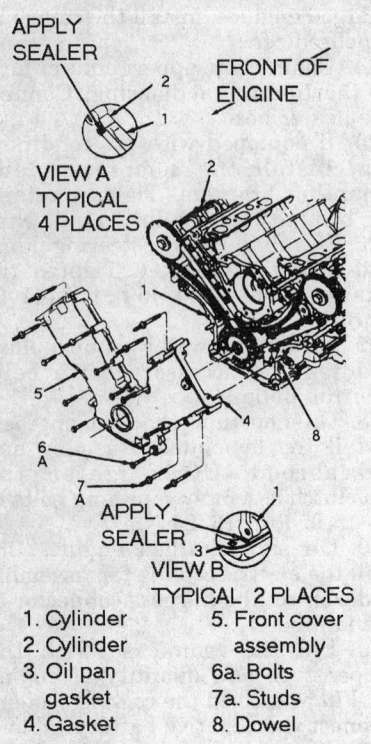

APPLY SEALER

FRONT OF ENGINE

VIEW A
TYPICAL 4 PLACES

APPLY SEALER

VIEW B
TYPICAL 2 PLACES

1. Cylinder
2. Cylinder
3. Oil pan gasket
4. Gasket
5. Front cover assembly
6a. Bolts
7a. Studs
8. Dowel

8838EG56

Timing chain front cover installation — 4.6L (SOHC) engine

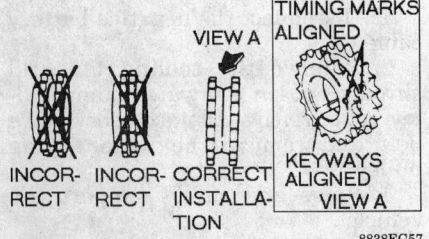

VIEW A

TIMING MARKS ALIGNED

INCORRECT   INCORRECT   CORRECT INSTALLATION

KEYWAYS ALIGNED
VIEW A

8838EG57

Crankshaft sprocket positioning — 4.6L (SOHC) engine

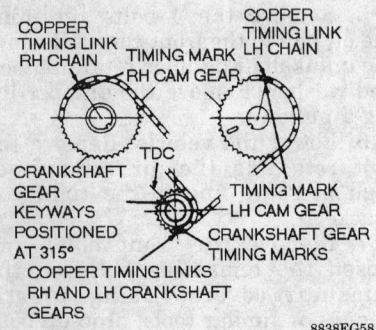

COPPER TIMING LINK RH CHAIN
COPPER TIMING LINK LH CHAIN
TIMING MARK RH CAM GEAR
TDC
CRANKSHAFT GEAR KEYWAYS POSITIONED AT 315°
TIMING MARK LH CAM GEAR
CRANKSHAFT GEAR TIMING MARKS
COPPER TIMING LINKS RH AND LH CRANKSHAFT GEARS

8838EG58

Timing chain and sprocket alignment — 4.6L (SOHC) engine

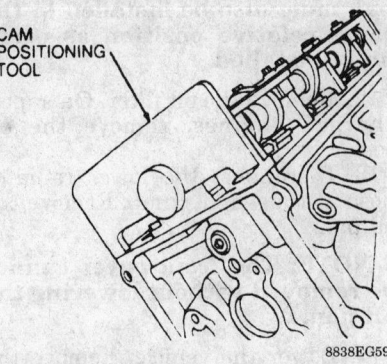

CAM POSITIONING TOOL

8838EG59

Cam positioning tool — 4.6L (SOHC) engine

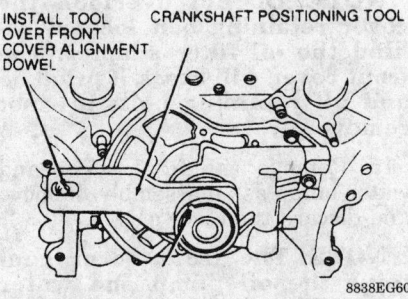

INSTALL TOOL OVER FRONT COVER ALIGNMENT DOWEL
CRANKSHAFT POSITIONING TOOL

8838EG60

Crankshaft positioning tool — 4.6L (SOHC) engine

Position the power steering hose out of the way.

9. Remove the 4 nuts retaining the left coil bracket to the front cover. Slide both coil brackets and ignition wires off the mounting studs and lay the assembly on top of the engine.

10. Disconnect the High Data Rate (HDR) sensor. Remove the 7 stud bolts and 4 bolts retaining the front cover to the engine and remove the front cover.

To install:

11. Inspect and replace the front cover seal as necessary and clean the sealing surfaces of the cylinder block. Apply silicone sealer to the oil pan where it meets the cylinder block and to the points where the cylinder head meets the cylinder block.

12. Install the front cover and the attaching studs and bolts. Tighten to 15–22 ft. lbs. (20–30 Nm). Connect the HDR sensor.

13. Position the coil brackets and ignition wires as an assembly onto the mounting studs. Position the power steering hose and install the 7 nuts retaining the coil brackets to the front cover. Tighten the nuts to 15–22 ft. lbs. (20–30 Nm). Connect both ignition coils and CID sensor.

14. Remove the plastic wedges holding up the camshaft covers. Ap-

ply silicone sealer where the front cover meets the cylinder head and make sure the camshaft cover gaskets are properly positioned. Install the front retaining bolts into the camshaft cover and tighten the bolts to 6.0–8.8 ft. lbs. (8–12 Nm).

15. Position the air conditioner high pressure line on the right coil bracket and install the bolt. Raise and safely support the vehicle.

16. Apply a small amount of silicone sealer in the rear of the keyway in the damper. Position the damper on the crankshaft and install, using a suitable installation tool. Install the damper bolt and washer and tighten to 114–121 ft. lbs. (155–165 Nm).

17. Install the 4 bolts retaining the oil pan to the front cover. Tighten to 15–22 ft. lbs. (20–30 Nm).

18. Position the power steering pump on the engine and install the 4 retaining bolts. Tighten to 15–22 ft. lbs. (20–30 Nm). Lower the vehicle.

19. Install the water pump pulley with the 4 bolts. Tighten to 15–22 ft. lbs. (20–30 Nm). Install the accessory drive belt and the cooling fan and shroud.

20. Connect the negative battery cable, start the engine and check for leaks.

4.6L (DOHC) Engine

1. Disconnect the negative battery cable and drain the cooling system.

2. Remove the windshield wiper module assembly.

3. Remove the engine appearance cover.

4. Remove coolant bypass and upper radiator hoses at coolant crossover tube. Loosen water pump pulley bolts.

5. Remove accessory drive belt. Remove water pump pulley and lower water pump to cylinder block bolt for front cover removal clearance.

6. Remove 2 bolts retaining power steering pump reservoir to left coil bracket.

7. Raise and safely support the vehicle. Using a suitable puller, remove the power steering pump pulley.

8. Remove power steering pump to engine block and cylinder front cover bolts. The front lower bolt on the power steering pump will not come all the way out. Position pump and reservoir aside.

9. Remove 4 bolts retaining oil pan to front cover. Remove the crankshaft damper bolt and washer. Using a suitable puller, remove the damper.

10. Lower the vehicle.

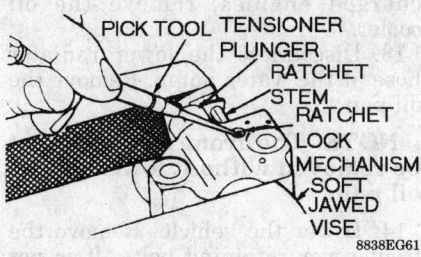

**Timing chain tensioner bleeding procedure —
4.6L (SOHC) engine**

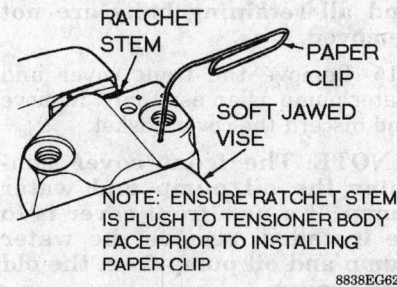

NOTE: ENSURE RATCHET STEM
IS FLUSH TO TENSIONER BODY
FACE PRIOR TO INSTALLING
PAPER CLIP

**Timing chain tensioner locking procedure —
4.6L (SOHC) engine**

11. Remove front bolts and loosen valve cover bolts. Prop up valve covers with plastic wedges or equivalent devices.

12. Disconnect both ignition coils and camshaft position sensor.

13. Remove 3 right coil bracket to front cover bolts. Remove 3 nuts and 1 bolt retaining left coil bracket to front cover. Slide both coil brackets and ignition wires off mounting studs and position aside.

14. Remove accessory drive belt idler pulley. Disconnect crankshaft position sensor.

15. Remove 7 studs and 8 bolts retaining front cover to engine block and remove the front cover.

To install:

16. Clean all mating surfaces. Replace front cover gasket and front cover oil seal.

17. Apply silicone sealer to six locations. Place bottom of front cover on oil pan and roll cover into place. Do not slide cover into place.

18. Attach front cover to engine. Install 7 studs and 8 bolts attaching front cover to engine block. Tighten in sequence to 15–22 ft. lbs. (20–30 Nm).

19. Connect crankshaft position sensor.

20. Install right and left coil brackets. Tighten to 15–22 ft. lbs. (20–30 Nm). Connect ignition coils and camshaft position sensor.

21. Apply silicone sealer to point where cylinder head meets front cover. Make sure camshaft cover gaskets are in proper position.

22. Remove plastic wedges and install valve covers. Tighten to 6–8.8 ft. lbs. (8–12 Nm).

23. Raise and safely support the vehicle.

24. Apply silicone sealer in keyway of damper. Install crankshaft damper and tighten bolt to 114–121 ft. lbs. (155–165 Nm).

25. Install 4 oil pan to front cover bolts. Tighten to 15–22 ft. lbs. (20–30 Nm).

26. Install power steering pump and tighten retaining bolts to 15–22 ft. lbs. (20–30 Nm). Install power steering pump pulley and lower the vehicle.

27. Install water pump lower bolt. Tighten to 15–22 ft. lbs. (20–30 Nm). Install water pump pulley and accessory drive belt.

28. Install air cleaner outlet tube and resonator. Install engine appearance cover.

29. Fill the crankcase with the proper type and quantity of engine oil. Fill and bleed the cooling system. Connect the negative battery cable.

30. Start the engine and check for leaks.

5.0L Engine

1. Disconnect the negative battery cable.

2. Drain the cooling system. Remove the air inlet tube, if equipped.

3. On Thunderbird and Cougar, disconnect the upper radiator hose at the engine.

4. On all except Thunderbird and Cougar, remove the fan shroud attaching bolts and position the shroud over the fan. Remove the fan and

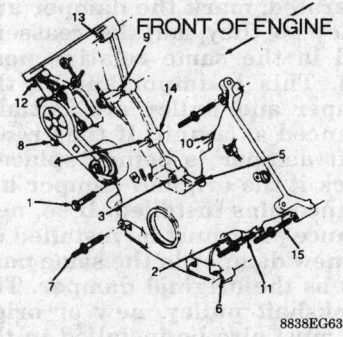

**Front cover bolt torque sequence —
4.6L (DOHC) engine**

clutch assembly from the water pump shaft and remove the shroud.

5. On Thunderbird and Cougar, remove the fan and clutch assembly from the water pump shaft using fan clutch holding tool T84T-6312-C or equivalent, and fan clutch nut wrench T84T-6312-D or equivalent, and position the fan and clutch assembly in the fan shroud. The nut is turned counterclockwise. Remove the fan shroud and fan/clutch as an assembly.

6. Loosen the water pump pulley bolts. Rotate the tensioner away from the accessory drive belt and remove the belt. Remove the water pump pulley.

7. Remove all accessory brackets that attach to the water pump.

8. Disconnect the lower radiator hose, heater hose and water pump bypass hose at the water pump.

9. Remove the crankshaft pulley from the crankshaft vibration damper. Remove the damper attaching bolt and washer and remove the damper using a puller.

10. Remove the oil pan-to-front cover attaching bolts. Use a thin blade knife to cut the oil pan gasket flush with the cylinder block face prior to separating the cover from the cylinder block.

11. Remove the cylinder front cover and water pump as an assembly.

NOTE: Cover the front oil pan opening while the cover assembly is off to prevent foreign material from entering the pan.

To install:

12. If a new front cover is to be installed, remove the water pump from the old front cover and install it on the new front cover.

13. Clean all gasket mating surfaces. Pry the old oil seal from the front cover and install a new 1, using a seal installer.

14. Coat the gasket surface of the oil pan with sealer, cut and position the required sections of a new gasket on the oil pan and apply silicone sealer at the corners. Apply sealer to a new front cover gasket and install on the block.

15. Position the front cover on the cylinder block. Use care to avoid seal damage or gasket mislocation. It may be necessary to force the cover downward to slightly compress the pan gasket. Use front cover aligner tool T61P-6019-B or equivalent to assist the operation.

16. Coat the threads of the front cover attaching screws with pipe sealant and install. While pushing in on the alignment tool, tighten the oil

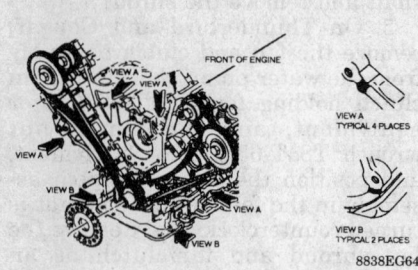

Front cover sealer points — 4.6L (DOHC) engine

8838EG64

pan to cover attaching screws to 9–12 ft. lbs. (12–16 Nm).

17. Tighten the front cover to cylinder block attaching bolts to 12–18 ft. lbs. (16–24 Nm). Remove the alignment tool.

18. Apply multi-purpose grease to the sealing surface of the vibration damper. Apply silicone sealer to the keyway of the vibration damper.

19. Line up the vibration damper keyway with the crankshaft key and install the damper using a suitable installation tool. Tighten the retaining bolt to 70–90 ft. lbs. (95–122 Nm). Install the crankshaft pulley.

20. Install the remaining components in the reverse order of their removal.

21. Fill the crankcase with the proper type and quantity of engine oil. Fill and bleed the cooling system.

22. Connect the negative battery cable, start the engine and check for leaks.

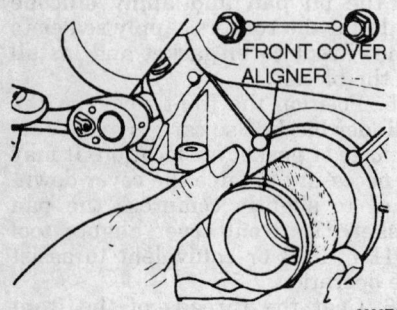

FRONT COVER ALIGNER

8838EG65

Timing chain front cover alignment — 5.0L engine

Timing Chain and Sprockets

REMOVAL AND INSTALLATION

3.8L Engine

1. Disconnect the negative battery cable and drain the cooling system.
2. Remove the air cleaner assembly and air intake duct.
3. On non-supercharged engines, remove the fan/clutch assembly and shroud. On supercharged engines, remove the electric cooling fan assembly.
4. Remove the accessory drive belt idlers, drive belts and the water pump pulley.
5. Remove the power steering pump bracket retaining bolts. Leaving the hoses connected, place the pump/bracket assembly aside in a position to prevent fluid from leaking out.
6. If equipped with air conditioning, remove the compressor front support bracket but leave the compressor in place.
7. Disconnect the coolant bypass hose and heater hose at the water pump. Disconnect the upper radiator hose at the thermostat housing.
8. On non-supercharged engines, disconnect the coil wire from the distributor cap and remove the cap with the secondary wires attached.
9. On non-supercharged engines, remove the distributor hold-down clamp and lift the distributor out of the front cover.
10. On supercharged engines, remove the hold-down clamp and lift the camshaft synchronizer from the front cover.
11. Raise and safely support the vehicle. Remove the crankshaft damper and pulley using a puller.

NOTE: If the crankshaft pulley and vibration damper have to be separated, mark the damper and pulley so they may be reassembled in the same relative position. This is important as the damper and pulley are initially balanced as a unit. If the crankshaft damper is being replaced, check if the original damper has balance pins installed. If so, new balance pins must be installed on the new damper in the same position as the original damper. The crankshaft pulley, new or original, must also be installed in the same relative position as originally installed.

12. Remove the oil filter. On supercharged engines, remove the oil cooler.
13. Disconnect the lower radiator hose at the water pump. Remove the oil pan.

NOTE: The front cover cannot be removed without lowering the oil pan.

14. Lower the vehicle. Remove the front cover retaining bolts. It is not necessary to remove the water pump.

NOTE: Do not overlook the cover retaining bolt located behind the oil filter adapter. The front cover will break if pried on and all retaining bolts are not removed.

15. Remove the front cover and water pump as an assembly. Remove and discard the cover gasket.

NOTE: The front cover contains the oil pump and water pump. If a new front cover is to be installed, remove the water pump and oil pump from the old front cover.

16. Remove the camshaft bolt and washer from the end of the camshaft.
17. Remove the distributor drive gear, camshaft sprocket, crankshaft sprocket and timing chain.

NOTE: If the crankshaft sprocket is difficult to remove, pry the sprocket off the shaft using a pair of large prybars positioned on both sides of the sprocket.

To install:

18. Clean all gasket mating surfaces. If reusing the front cover, replace the front cover oil seal.
19. Rotate the crankshaft to position the No. 1 piston at TDC and the crankshaft keyway at the 12 o' clock position.
20. Lubricate the timing chain with engine oil.
21. Install the camshaft sprocket, crankshaft sprocket and timing chain. Make sure the timing marks align.
22. Install the distributor drive gear. Install the bolt and washer assembly on the end of the camshaft and tighten to 30–37 ft. lbs. (40–50 Nm).
23. Position a new gasket on the cylinder block and install the front cover using dowels for proper alignment. Install the front cover retaining bolts and tighten to 15–22 ft. lbs. (20–30 Nm).
24. Raise and safely support the vehicle. Install the oil pan. Connect the

lower radiator hose and install the oil filter.

25. Coat the crankshaft damper sealing surface with clean engine oil. Apply a small amount of silicone sealer to the crankshaft keyway.

26. Position the crankshaft pulley key in the crankshaft keyway and install the damper, using a suitable installation tool.

27. Install the damper washer and retaining bolt and tighten to 103–132 ft. lbs. (140–180 Nm). Install the crankshaft pulley and tighten the retaining bolts to 20–28 ft. lbs. (26–38 Nm).

28. Lower the vehicle. Connect the coolant bypass hose.

29. On non-supercharged engines, install the distributor with the rotor pointing at the No. 1 distributor cap tower. Install the distributor cap and coil wire. On supercharged engines, install the camshaft synchronizer.

30. Connect the upper radiator hose at the thermostat housing. Connect the heater hose.

31. If equipped with air conditioning, install the compressor and mounting brackets. Tighten retaining bolts to 30–45 ft. lbs. (41–61 Nm).

32. Install the power steering pump and mounting bracket. Tighten the retaining bolts to 30–45 ft. lbs. (41–61 Nm).

33. Install the water pump pulley. Position the accessory drive belts over the pulleys.

34. On non-supercharged engines, install the fan/clutch assembly and fan shroud. Cross-tighten the fan/clutch assembly retaining bolts to 12–18 ft. lbs. (16–24 Nm).

35. On supercharged engines, install the electric cooling fan assembly and connect the harness connector to the fan motor.

36. Fill the crankcase with the proper type and quantity of engine oil. Fill and bleed the cooling system. Connect the negative battery cable.

37. Start the engine and check for leaks. Check the ignition timing and curb idle speed and adjust, as necessary.

4.6L (SOHC) Engine

NOTE: This is not a free wheeling engine. If it has "jumped time" there will be damage to the valves and/or pistons and will require the removal of the cylinder heads.

1. Disconnect the negative battery cable.

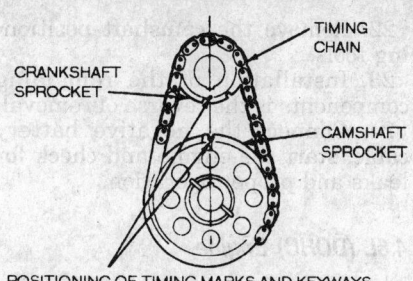

POSITIONING OF TIMING MARKS AND KEYWAYS IN CAMSHAFT AND CRANKSHAFT SPROCKETS MUST BE IN LINE AS SHOWN WITH NO. 1 PISTON AT TOP DEAD CENTER FIRING.

8838EG66

Timing chain sprocket alignment — 3.8L engine

2. Remove the camshaft covers and the timing chain front cover.

3. Remove the High Data Rate (HDR) wheel.

4. Rotate the engine to set the No. 1 piston at TDC on the compression stroke.

5. Install cam positioning tools T91P-6256-A or equivalent, on the flats of the camshaft. This will prevent accidental rotation of the camshafts.

6. Remove the 2 bolts retaining the right tensioner to the cylinder head and remove the tensioner. Remove the right tensioner arm.

7. Remove the 2 bolts retaining the right chain guide to the cylinder head and remove the chain guide. Remove the right chain and right crankshaft sprocket. Remove the right camshaft sprocket retaining bolt, washer, sprocket and spacer.

NOTE: Cam positioning tools T91P-6256-A or equivalent, must be installed on the camshaft to prevent the camshaft from rotating.

8. Remove the 2 bolts retaining the left tensioner to the cylinder head and remove the tensioner. Remove the left tensioner arm.

9. Remove the 2 bolts retaining the left chain guide to the cylinder head and remove the chain guide. Remove the left chain and left crankshaft sprocket. Remove the left camshaft sprocket retaining bolt, washer, sprocket and spacer.

NOTE: Cam positioning tools T91P-6256-A or equivalent, must be installed on the camshaft to prevent the camshaft from rotating.

10. Inspect the friction material on the tensioner arms and chain guides. If worn or damaged, remove and clean the oil pan and replace the oil pickup tube.

NOTE: At no time, when the timing chains are removed and the cylinder heads are installed, may the crankshaft and/or camshafts be rotated. Failure to follow these directions will result in valve and/or piston damage.

To install:

11. Make sure cam positioning tools T91P-6256-A or equivalent, are installed on the camshafts to prevent them from rotating.

12. Position the camshaft spacers and sprockets on the camshafts and install the washers and retaining bolts. Tighten the retaining bolts to 81–96 ft. lbs. (110–130 Nm).

13. Install the left crankshaft sprocket with the tapered part of the sprocket facing away from the engine block.

NOTE: The crankshaft sprockets are identical. They may only be installed one way, with the tapered part of the sprocket facing each other.

14. Install the left timing chain on the camshaft and crankshaft sprockets. Make sure the copper links of the

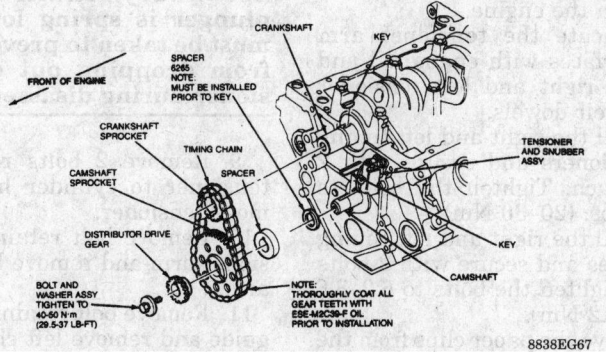

Timing chain and sprockets installation — 3.8L engine

8838EG67

chain line up with the timing marks of the sprockets.

NOTE: If the copper links of the timing chain are not visible, pull the chain taught until the opposite sides of the chain contact one another and lay it on a flat surface. Mark the links at each end of the chain and use them in place of the copper links.

15. Install the right crankshaft sprocket with the tapered part of the sprocket facing the left crankshaft sprocket.

16. Install the right timing chain on the camshaft and crankshaft sprockets. Make sure the copper links of the chain line up with the timing marks of the sprockets.

17. It is necessary to bleed the timing chain tensioners before installation. Proceed as follows:

   a. Position the timing chain tensioner in a soft-jawed vice.

   b. Using a small pick or similar tool, hold the ratchet lock mechanism away from the ratchet stem and slowly compress the tensioner plunger by rotating the vise handle.

NOTE: The tensioner must be compressed slowly or damage to the internal seals will result.

   c. Once the tensioner plunger bottoms in the tensioner bore, continue to hold the ratchet lock mechanism and push down on the ratchet stem until flush with the tensioner face.

   d. While holding the ratchet stem flush to the tensioner face, release the ratchet lock mechanism and install a paper clip or similar tool in the tensioner body to lock the tensioner in the collapsed position.

   e. The paper clip must not be removed until the timing chain, tensioner, tensioner arm and timing chain guide are completely installed on the engine.

18. Lubricate the tensioner arm contact surfaces with engine oil and install the right and left tensioner arms on their dowels.

19. Install the right and left timing chain tensioners and secure with 2 bolts on each. Tighten the bolts to 15–22 ft. lbs. (20–30 Nm).

20. Install the right and left timing chain guides and secure with 2 bolts on each. Tighten the bolts to 6.0–8.8 ft. lbs. (8–12 Nm).

21. Remove the paper clips from the timing chain tensioners and make sure all timing marks are aligned.

22. Remove the camshaft positioning tools.

23. Installation of the remaining components is the reverse of removal.

24. Connect the negative battery cable, start the engine and check for leaks and proper operation.

4.6L (DOHC) Engine

NOTE: Because the engine is not free wheeling, damage to the valves and pistons may occur if the timing chain is out of alignment.

1. Disconnect the negative battery cable and drain the cooling system.

2. Remove the timing chain front cover.

3. Remove the crankshaft position sensor tooth wheel. Rotate the engine to No. 1 TDC.

4. Install camshaft positioning tool T93P-6256-A, or equivalent, in rear D-slots of camshafts. Remove 2 bolts retaining right tensioner to cylinder head and remove tensioner.

5. Remove bolt retaining right tensioner arm and remove right tensioner arm.

6. Remove bolt retaining right chain guide and remove right chain guide. Note the location of the longer bolts which retain the chain guide to the cylinder head.

7. Remove right chain from camshaft and crankshaft gears. Install camshaft torquing tool at center area of camshafts.

NOTE: Camshaft positioning tool must be installed to prevent the camshafts from rotating.

8. Remove cam sprocket retaining bolts. Compress secondary chain tensioners and remove secondary timing chain and sprockets.

WARNING

Secondary chain tensioner plunger is spring loaded. Care must be taken to prevent plunger from dropping out of the tensioner during disassembly.

---

9. Remove 2 bolts retaining left tensioner to cylinder head and remove tensioner.

10. Remove bolt retaining left tensioner arm and remove left tensioner arm.

11. Remove bolt retaining left chain guide and remove left chain guide.

12. Remove left chain from camshaft and crankshaft gears. In-

stall camshaft torquing tool at center area of camshafts.

NOTE: Camshaft positioning tool must be installed to prevent the camshafts from rotating.

13. Remove cam sprocket retaining bolts. Compress secondary chain tensioners and remove secondary timing chain and sprockets.

WARNING

Secondary chain tensioner plunger is spring loaded. Care must be taken to prevent plunger from dropping out of the tensioner during disassembly.

---

14. Inspect friction material on tensioner arms and chain guides. If worn or damaged, remove and clean oil pan and oil pickup tube.

To install:

15. Install left secondary tensioner. Tighten bolts to 6–9 ft. lbs. (8–12 Nm). Install secondary sprockets and chain as an assembly. Note direction of hubs.

16. Install left primary camshaft sprocket and spacer on left camshaft.

17. Install washers and camshaft sprocket retaining bolts. Hand-tighten bolts.

NOTE: Secondary sprockets must be free to turn.

18. Install secondary chain tensioning tool on left secondary tensioner.

19. If removed, install camshaft positioning tool T93P-6256-A, or equivalent, into D-slots. Install camshaft torquing tool at center area of camshafts. Tighten camshaft sprocket bolts to 81–96 ft. lbs. (110–130 Nm). Do not remove camshaft positioning tool.

20. Repeat steps 14 through 18 for the right side.

21. Install left crankshaft sprocket. Make sure tapered part of sprocket faces away from engine block.

22. Install primary timing chain on left primary camshaft sprocket. Make sure copper link of chain aligns with the timing mark on the camshaft sprocket.

23. Install primary timing chain on crankshaft sprocket. Make sure copper link of chain aligns with the timing mark on the crankshaft sprocket. Make sure tapered part of sprocket faces toward engine block.

24. Repeat steps 20 through 22 for right primary timing chain.

25. Install right and left chain guides. Tighten bolts to 15–22 ft. lbs. (20–30 Nm).

26. Lubricate tensioner arm contact surfaces with clean engine oil and in-

stall right and left tensioner arms and tighten bolts to 7–11 ft. lbs. (10–15 Nm). Do not remove tensioner locking pins until timing chain guides are installed.

27. Install right and left timing chain tensioners. Tighten 2 bolts on each tensioner to 15–22 ft. lbs. (20–30 Nm).

28. Remove locking pins from timing chain tensioners and make sure all timing marks are aligned. Remove camshaft positioning tool.

29. Install the remaining parts and fill the crankcase with the proper type and quantity of engine oil. Fill and bleed the cooling system. Connect the negative battery cable.

30. Start the engine and check for leaks. Check the ignition timing and curb idle speed and adjust, as necessary.

5.0L Engine

1. Disconnect the negative battery cable and drain the cooling system.
2. Remove the timing chain front cover.
3. Rotate the crankshaft until the timing marks on the sprockets are aligned.

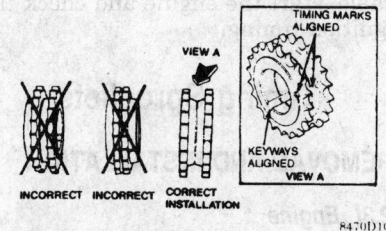

Crankshaft sprocket alignment — 4.6L engine

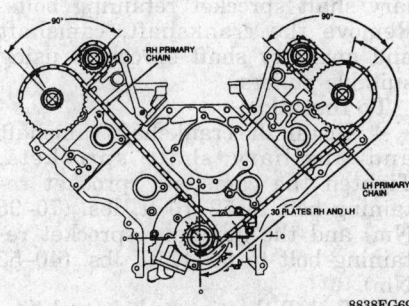

Timing chain alignment to TDC — 4.6L (DOHC) engine

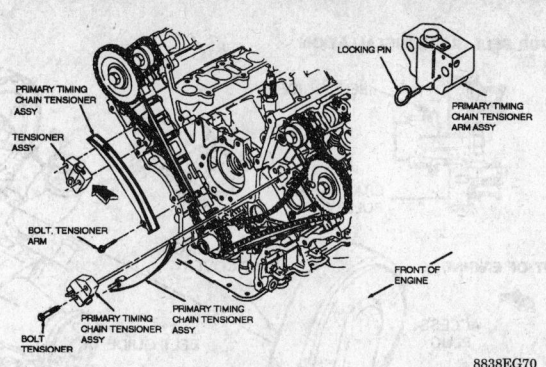

Timing chain tensioner installation — 4.6L (DOHC) engine

4. Remove the camshaft retaining bolt, washer and eccentric, if equipped. Slide both sprockets and the timing chain forward and remove them as an assembly.

To install:

5. Position the sprockets and timing chain on the camshaft and crankshaft simultaneously. Make sure the timing marks on the sprockets are aligned.
6. Install the washer, eccentric if equipped, and camshaft sprocket retaining bolt. Tighten the bolt to 40–45 ft. lbs. (54–61 Nm).
7. Install the timing chain front cover and remaining components.
8. Fill and bleed the cooling system. Connect the negative battery cable, start the engine and check for leaks.
9. Check and adjust the ignition timing and idle speed, as necessary.

Timing Belt Front Cover

REMOVAL AND INSTALLATION

2.3L Engine

1. Disconnect the negative battery cable and drain the cooling system. Remove the 4 water pump pulley bolts.

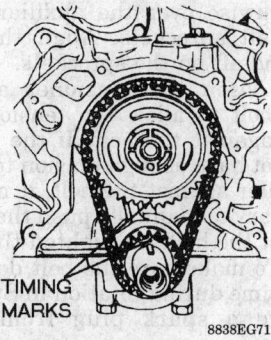

Timing chain sprocket alignment — 5.0L engine

2. Remove the automatic belt tensioner and accessory drive belt. Remove the upper radiator hose.
3. Remove the crankshaft pulley bolt and pulley. Remove the thermostat housing and gasket.
4. Remove the timing belt outer cover retaining bolt(s). Release the cover interlocking tabs, if equipped, and remove the cover.

To install:

5. Position the timing belt front cover. Snap the interlocking tabs into place, if necessary. Install the timing belt outer cover retaining bolt(s) and tighten to 71–106 inch lbs. (8–12 Nm).
6. Install the thermostat housing and a new gasket. Install the upper radiator hose.
7. Install the crankshaft pulley and retaining bolt. Tighten to 114–151 ft. lbs. (155–205 Nm) on 1993 vehicles.
8. Install the water pump pulley and the automatic belt tensioner. Install the accessory drive belt.
9. Connect the negative battery cable, start the engine and check for leaks.

OIL SEAL REPLACEMENT

2.3L Engine

1. Disconnect the negative battery cable.
2. Remove the timing belt front cover and timing belt.
3. Use a suitable puller to remove the crankshaft, camshaft and auxiliary shaft sprockets, as necessary.
4. Use seal remover tool T74P-6700-B or equivalent, to remove the crankshaft, camshaft and auxiliary shaft seals, as necessary. Position the tool so the jaws are gripping the thin edge of the seal. Operate the jackscrew on the tool to remove the seal.

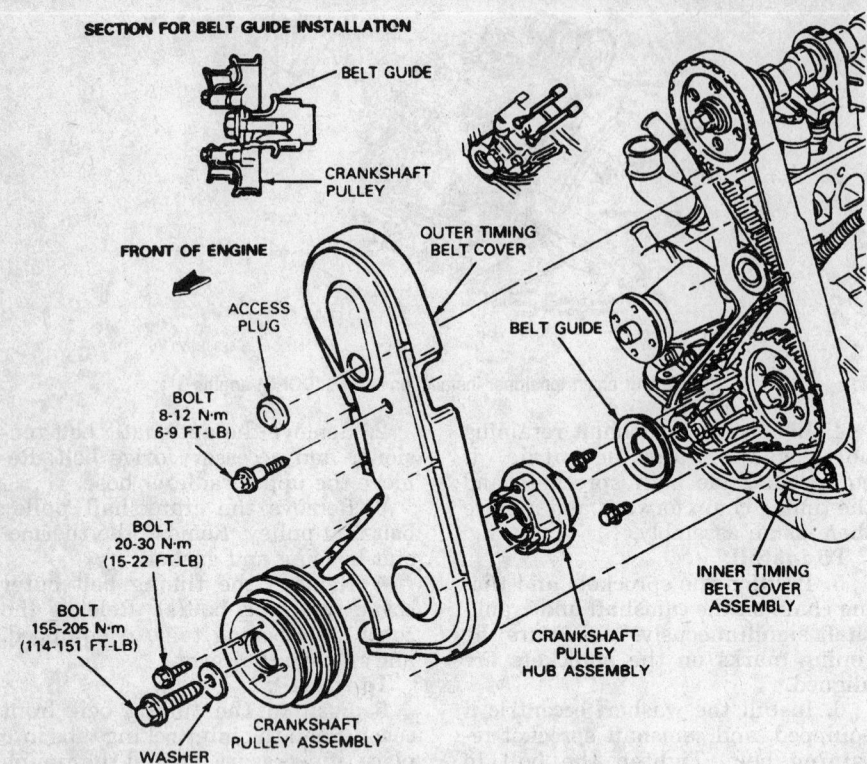

SECTION FOR BELT GUIDE INSTALLATION

BELT GUIDE

CRANKSHAFT PULLEY

FRONT OF ENGINE

OUTER TIMING BELT COVER

ACCESS PLUG

BELT GUIDE

BOLT 8-12 N·m 6-9 FT-LB)

BOLT 20-30 N·m (15-22 FT-LB)

BOLT 155-205 N·m (114-151 FT-LB)

INNER TIMING BELT COVER ASSEMBLY

CRANKSHAFT PULLEY HUB ASSEMBLY

WASHER

CRANKSHAFT PULLEY ASSEMBLY

8838EG72

Timing belt front cover installation — 2.3L engine

To install:

5. Lubricate the lips of the new seal(s) with clean engine oil.

6. Use seal replacer tool T74P-6150-A or equivalent, to install the seal(s).

7. Install the crankshaft, camshaft and auxiliary shaft sprockets, as necessary. Tighten the camshaft sprocket retaining bolt to 52–70 ft. lbs. (70–96 Nm) and the auxiliary sprocket retaining bolt to 30–41 ft. lbs. (40–55 Nm).

8. Install the timing belt and timing belt front cover.

9. Connect the negative battery cable, start the engine and check for leaks.

Timing Belt and Tensioner

REMOVAL AND INSTALLATION

2.3L Engine

1. Disconnect the negative battery cable.

2. Remove the timing belt front cover.

3. Loosen the belt tensioner adjustment screw, position belt tensioner tool T74P-6254-A or equivalent, on the tension spring roll pin and release the belt tensioner.

Tighten the adjustment screw to hold the tensioner in the released position.

4. On 1993 vehicles, remove the bolts holding the timing sensor in place and pull the sensor assembly free of the dowel pin.

5. Remove the crankshaft pulley, hub and belt guide. Remove the timing belt. If the belt is to be reused, mark the direction of rotation so it may be reinstalled in the same direction.

To install:

6. Position the crankshaft sprocket to align with the TDC mark and the camshaft sprocket to align with the camshaft timing pointer.

7. Install the timing belt over the crankshaft sprocket and then counterclockwise over the auxiliary and camshaft sprockets. Align the belt fore-and-aft on the sprockets.

8. Loosen the tensioner adjustment bolt to allow the tensioner to move against the belt. If the spring does not have enough tension to move the roller against the belt, it may be necessary to manually push the roller against the belt and tighten the bolt.

9. To make sure the belt does not jump time during rotation in Step 10, remove a spark plug from each cylinder.

10. Rotate the crankshaft 2 complete turns in the direction of normal

rotation to remove the slack from the belt. Tighten the tensioner adjustment to 29–40 ft. lbs. (40–55 Nm) and pivot bolts to 14–22 ft. lbs. (20–30 Nm). Check the alignment of the timing marks.

11. Install the crankshaft belt guide.

12. On 1993 vehicles, proceed as follows:

   a. Install the timing sensor onto the dowel pin and tighten the 2 longer bolts to 14–22 ft. lbs. (20–30 Nm).

   b. Rotate the crankshaft 45 degrees counterclockwise and install the crankshaft pulley and hub assembly. Tighten the bolt to 114–151 ft. lbs. (155–205 Nm).

   c. Rotate the crankshaft 90 degrees clockwise so the vane of the crankshaft pulley engages with timing sensor positioner tool T89P-6316-A or equivalent. Tighten the 2 shorter sensor bolts to 14–22 ft. lbs. (20–30 Nm).

   d. Rotate the crankshaft 90 degrees counterclockwise and remove the sensor positioner tool.

   e. Rotate the crankshaft 90 degrees clockwise and measure the outer vane to sensor air gap. The air gap must be 0.018–0.039 in. (0.458–0.996mm).

13. Install the timing belt front cover, spark plugs and remaining components.

14. Connect the negative battery cable, start the engine and check the ignition timing.

Timing Sprockets

REMOVAL AND INSTALLATION

2.3L Engine

1. Disconnect the negative battery cable.

2. Remove the timing belt front cover and the timing belt.

3. Remove the camshaft and auxiliary shaft sprocket retaining bolts. Remove the crankshaft, camshaft and auxiliary shaft sprockets using suitable pullers.

To install:

4. Install the crankshaft, camshaft and auxiliary shaft sprockets. Tighten the camshaft sprocket retaining bolt to 52–70 ft. lbs. (70–96 Nm) and the auxiliary sprocket retaining bolt to 30–41 ft. lbs. (40–55 Nm).

5. Install the timing belt and timing belt front cover.

6. Connect the negative battery cable.

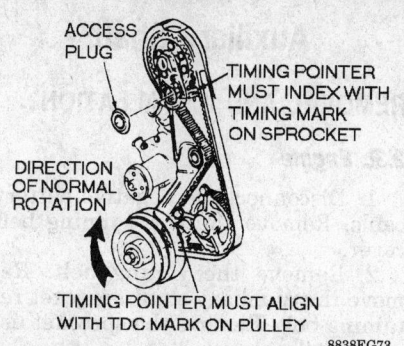

ACCESS PLUG

TIMING POINTER MUST INDEX WITH TIMING MARK ON SPROCKET

DIRECTION OF NORMAL ROTATION

TIMING POINTER MUST ALIGN WITH TDC MARK ON PULLEY

8838EG73

Timing belt and sprockets positioning — 2.3L engine

Camshaft

REMOVAL AND INSTALLATION

2.3L Engine

1. Disconnect the negative battery cable and drain the cooling system.
2. Remove the air intake and the throttle body.
3. Disconnect the radiator hoses. Remove the cooling fan, shroud and radiator assembly.
4. Tag and disconnect the spark plug wires and position aside.
5. Tag and disconnect the necessary electrical connectors and vacuum lines and position aside.
6. Remove the rocker cover retaining bolts and the rocker cover.
7. Remove the timing belt front cover and the timing belt.
8. Compress the valve springs using valve spring compressor lever T88T-6565-BH or equivalent and remove the cam followers.
9. Remove the camshaft sprocket retaining bolt. Remove the camshaft sprocket using a suitable puller. Remove the camshaft seal using a seal removal tool.
10. Remove the 2 screws and the camshaft rear retainer.
11. Raise and safely support the vehicle. Remove the right and left engine support bolts and nuts.
12. Position a block of wood and a jack under the engine. Raise the engine as high as it will go. Place blocks of wood between the engine mounts and chassis brackets and remove the jack.
13. Lower the vehicle and remove the camshaft.
To install:
14. Make sure the threaded plug is in the rear of the camshaft. If not, remove the plug from the old camshaft and install.
15. Coat the camshaft lobes with multi-purpose grease and lubricate

the journals with heavy engine oil before installation. Carefully slide the camshaft through the bearings.
16. Install the camshaft rear retainer and tighten the 2 screws to 6–9 ft. lbs. (8–12 Nm). Install a new camshaft seal using a suitable seal installer.
17. Install the camshaft sprocket and tighten the retaining bolt to 52–70 ft. lbs. (70–96 Nm).
18. Install the timing belt and timing belt front cover.
19. Raise and safely support the vehicle. Position a block of wood and a jack and raise the engine. Remove the blocks of wood, lower the engine and remove the jack.
20. Install the engine support bolts and nuts and lower the vehicle.
21. Install the remaining components in the reverse order of removal.
22. Connect the negative battery cable, start the engine and check for leaks. Check the ignition timing, if necessary.

3.8L and 5.0L Engines

1. Disconnect the negative battery cable and drain the cooling system.
2. Relieve the fuel system pressure and discharge the air conditioning system.
3. Remove the radiator. If equipped with air conditioning, remove the condenser.
4. Remove the grille.
5. Remove the intake manifolds and the lifters. On the 3.8L engine, remove the oil pan.
6. Remove the timing chain front cover, the timing chain and spacer.
7. Remove the thrust plate. Remove the camshaft, being careful not to damage the bearing surfaces.
To install:
8. Lubricate the cam lobes and journals with heavy engine oil. Install the camshaft, being careful not to damage the bearing surfaces while sliding into position.
9. Install the thrust plate. Tighten the bolts to 6–10 ft. lbs. (8–14 Nm) on the 3.8L engine or 9–12 ft. lbs. (12–16 Nm) on the 5.0L engine.
10. Install the timing chain and sprockets. Install the engine front cover.
11. Install the lifters and the intake manifolds. On 3.8L engine, install the oil pan.
12. Install the grille. If equipped with air conditioning, install the condenser.
13. Install the radiator. Fill and bleed the cooling system.

14. Connect the negative battery cable. Start the engine and check for leaks.

4.6L (SOHC) Engine

1. Disconnect the negative battery cable and drain the cooling system. Relieve the fuel system pressure.
2. Remove the right and left camshaft covers.
3. Remove the timing chain front cover. Remove the timing chains.
4. Rotate the crankshaft counterclockwise 45 degrees from TDC to make sure all pistons are below the top of the engine block deck face.

NOTE: The crankshaft must be in this position prior to rotating the camshafts or damage to the pistons and/or valve train will result.

5. Install valve spring compressor tool T91P-6565-A or equivalent, under the camshaft and on top of the valve spring retainer.

NOTE: Valve spring spacer tool T91P-6565-AH or equivalent, must be installed between the spring coils and the camshaft must be at the base circle before compressing the valve spring.

6. Compress the valve spring far enough to remove the roller follower. Repeat Steps 5 and 6 until all roller followers are removed.
7. Remove the bolts retaining the camshaft cap cluster assemblies to the cylinder heads. Tap upward on the camshaft caps at points near the upper bearing halves and gradually lift the camshaft clusters from the cylinder heads.
8. Remove the camshafts straight upward to avoid bearing damage.
To install:
9. Apply heavy engine oil to the camshaft journals and lobes. Position the camshafts on the cylinder heads.
10. Install and seat the camshaft cap cluster assemblies. Hand start the bolts.
11. Tighten the camshaft cluster retaining bolts in sequence to 6.0–8.8 ft. lbs. (8–12 Nm).

NOTE: Each camshaft cap cluster assembly is tightened individually.

12. Loosen the camshaft cap cluster retaining bolts approximately 2 turns or until the heads of the bolts are free. Retighten all bolts, in sequence, to 6.0–8.8 ft. lbs. (8–12 Nm).

NOTE: The camshafts should turn freely with a slight drag.

13. Install cam positioning tools T91P-6256-A or equivalent, on the flats of the camshafts and install the spacers and camshaft sprockets. Install the bolts and washers and tighten to 81–96 ft. lbs. (110–130 Nm).

14. Install valve spring compressor T91P-6565-A or equivalent, under the camshaft and on top of the valve spring retainer.

NOTE: Valve spring spacer tool T91P-6565-AH or equivalent, must be installed between the spring coils and the camshaft must be at the base circle before compressing the valve spring.

15. Compress the valve spring far enough to install the roller followers.

16. Repeat Steps 14 and 15 until all roller followers are installed.

17. Rotate the crankshaft clockwise 45 degrees to position the crankshaft at TDC.

NOTE: The crankshaft must only be rotated in the clockwise direction and only as far as TDC.

18. Install the timing chains and install the timing chain front cover. Install the camshaft covers.

19. Install the remaining components in the reverse order of removal.

20. Connect the negative battery cable. Start the engine and check for leaks.

4.6L (DOHC) Engine

1. Disconnect the negative battery cable and drain the cooling system.

2. Remove the windshield wiper module.

3. Remove the valve covers.

4. Remove the the front cover.

5. Remove the timing chains.

6. Rotate the crankshaft counter clockwise 45 degrees to ensure all pistons are below the top of the cylinder block deck face.

7. Install valve spring compressor T91P-6565-A, or equivalent, under the exhaust camshaft and on top of the exhaust valve spring retainer.

8. Install valve spring compressor T91P-6565-A, or equivalent, under the intake camshaft and on top of the primary intake valve spring retainer.

9. Install valve spring compressor T91P-6565-A, or equivalent, under the intake camshaft and on top of the secondary valve spring retainer.

10. Compress valve spring far enough to remove roller follower. Repeat steps 7 through 10 until all followers are removed.

11. Remove 13 bolts retaining exhaust camshaft cap cluster assemblies to cylinder head to remove exhaust camshaft.

12. Remove 12 bolts retaining intake camshaft cap cluster assemblies to cylinder head to remove intake camshaft.

13. Lightly tap upward on camshaft cap with rubber or plastic hammer and gradually lift camshaft cap cluster from cylinder head.

14. Remove camshaft straight up to avoid bearing damage. Repeat steps 7 through 14 to remove the other camshafts.

To install:

15. Clean and inspect valve cover, front cover and cylinder head mating surfaces.

16. Apply clean engine oil to all journals and lobes of the camshaft. Position camshaft on cylinder head.

17. Install and seat camshaft cap cluster assemblies. Hand-tighten all 25 cap cluster retaining bolts. Tighten in sequence the cap cluster retaining bolts to 6–8.8 ft. lbs. (8–12 Nm). Tighten each camshaft cluster cap assembly individually.

18. Loosen all 25 cap cluster retaining bolts 2 turns or until head of bolt is free. Retighten in sequence the cap cluster retaining bolts to 6–8.8 ft. lbs. (8–12 Nm). The camshaft should turn freely with a slight drag.

19. Check the camshaft endplay. The play should be between 0.001–0.009 inches.

20. Install timing chains and set valve timing.

21. Install valve spring compressor T91P-6565-A, or equivalent, under the exhaust camshaft and on top of the exhaust valve spring retainer.

22. Install valve spring compressor T91P-6565-A, or equivalent, under the intake camshaft and on top of the primary intake valve spring retainer.

23. Install valve spring compressor T91P-6565-A, or equivalent, under the intake camshaft and on top of the secondary valve spring retainer.

24. Compress valve spring far enough to install the roller follower. Repeat steps 21 through 24 until all followers are installed.

25. Inspect and replace if necessary front cover seal. Replace front cover gasket.

26. Install front cover. Install valve covers.

27. Install windshield wiper module.

28. Fill and bleed the cooling system. Connect the negative battery cable, start the engine and check for leaks.

Auxiliary Shaft

REMOVAL AND INSTALLATION

2.3L Engine

1. Disconnect the negative battery cable. Remove the front timing belt cover.

2. Remove the timing belt. Remove the auxiliary shaft sprocket retaining bolt. Remove the sprocket using a puller.

3. Remove the auxiliary shaft cover and thrust plate.

4. Withdraw the auxiliary shaft from the block being careful not to damage the bearings.

To install:

5. Dip the auxiliary shaft in engine oil before installing. Slide the auxiliary shaft into the cylinder block, being careful not to damage the bearings.

6. Install the thrust plate. Tighten the thrust plate screws to 6–9 ft. lbs. (8–12 Nm).

7. Install a new gasket and auxiliary shaft cover. Tighten the cover screws to 6–9 ft. lbs. (8–12 Nm).

NOTE: The auxiliary shaft cover and cylinder front cover share a common gasket. Cut off the old gasket around the cylinder cover and use half of the new gasket on the auxiliary shaft cover.

8. Insert the distributor, aligning the housing-to-engine block marks, and install the auxiliary shaft sprocket.

9. Align the timing marks and install the timing belt.

10. Install the timing belt cover.

11. Check the ignition timing.

Piston and Connecting Rod

POSITIONING

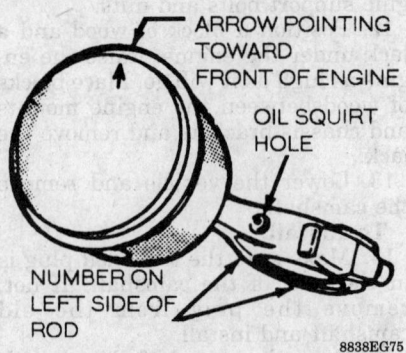

ARROW POINTING TOWARD FRONT OF ENGINE

OIL SQUIRT HOLE

NUMBER ON LEFT SIDE OF ROD

8838EG75

Piston and rod assembly — 2.3L engine

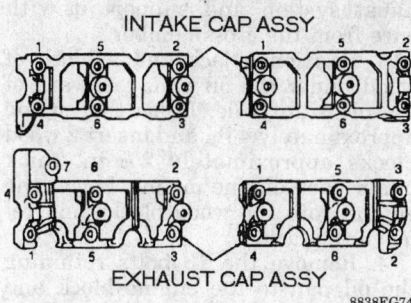

Bolt torque sequence for camshaft cap cluster assemblies — 4.6L (DOHC) engine

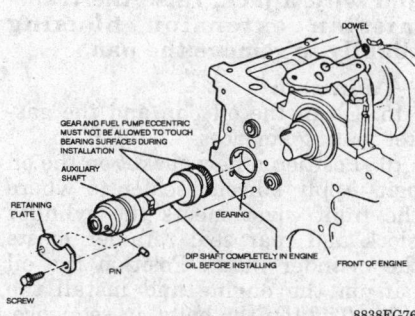

Auxiliary shaft installation — 2.3L engine

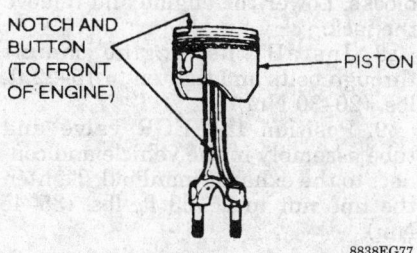

Piston and rod assembly — 3.8L engine

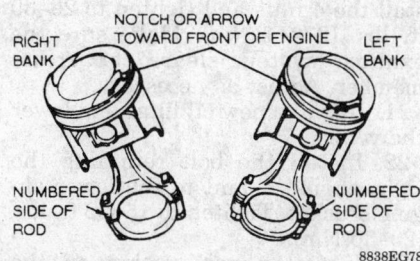

Piston and rod assembly — 4.6L and 5.0L engines

ENGINE LUBRICATION

Oil Pan

REMOVAL AND INSTALLATION

2.3L Engine

1. Disconnect the negative battery cable. Remove the air cleaner outlet tube at the throttle body.
2. Remove the engine oil dipstick.
3. Install engine support fixture D88L-6000-A or equivalent.
4. Raise and safely support the vehicle.
5. Remove the engine mount through bolts.
6. Drain the engine oil.
7. Disconnect the cable from the starter and remove the starter.
8. Disconnect the exhaust manifold tube to the inlet pipe bracket and disconnect the catalytic converter at the inlet pipe.
9. Remove the transmission. If equipped with manual transmission, remove the clutch pressure plate and disc.
10. Remove the flywheel retaining bolts and remove the flywheel.
11. If equipped with automatic transmission, remove the oil cooler lines from the retainer at the block.
12. Lower the vehicle. Raise the engine using the engine support fixture, then raise and safely support the vehicle.
13. Remove the oil pan attaching bolts and lower the oil pan to the chassis. Remove the oil pump and pickup tube and lay the assembly in the oil pan.
14. Remove the pan and pump from the vehicle.
15. Clean the oil pan and all gasket mating surfaces. Clean the oil pump exterior and pickup tube screen.

To install:

16. Install the oil pan gasket in the groove in the oil pan.
17. Lay the oil pump and pickup tube assembly in the oil pan and position the pan on the crossmember.
18. Install the oil pump and pickup tube assembly. Tighten the oil pump mounting bolts to 14–21 ft. lbs. (19–29 Nm) and the oil pump strap nut to 30–41 ft. lbs. (40–55 Nm).
19. Apply silicone sealer to the points where the rear main bearing cap meets the cylinder block, to the corners of the engine front cover and to where the front cover meets the cylinder block.
20. Install the oil pan assembly. Install the oil pan flange bolts tight enough to compress the oil pan gasket to the point that the 2 transmission holes are aligned with the 2 tapped holes in the oil pan, but loose enough to allow movement of the pan, relative to the block.
21. Install the 2 oil pan/transmission bolts and tighten to 30–36 ft. lbs. (40–50 Nm) to align the oil pan with the transmission, then loosen the bolts ½ turn.
22. Tighten all oil pan flange bolts to 90–120 inch lbs. (10–13 Nm). Tighten the 2 oil pan/transmission bolts to 30–39 ft. lbs. (40–54 Nm).
23. Install a new oil filter.
24. Lower the vehicle. Lower the engine onto the engine mounts, then raise and safely support the vehicle.
25. Install the flywheel and tighten the attaching bolts to 54–64 ft. lbs. (73–87 Nm). If equipped with manual transmission, install the clutch pressure plate and disc assembly.
26. Install the transmission.
27. Install the engine mount through bolts and tighten to 65–85 ft. lbs. (88–115 Nm).
28. If equipped with automatic transmission, connect the oil cooler line retainer clip to the engine.
29. Connect the exhaust pipe and the inlet pipe.
30. Install the starter and connect the starter cable.
31. Lower the vehicle. Remove the engine support fixture.
32. Connect the air cleaner outlet tube to the throttle body.
33. Install the dipstick and fill the crankcase with the proper type and quantity of engine oil.
34. Connect the negative battery cable, start the engine and check for leaks.

3.8L Engine

1. Disconnect the negative battery cable and remove the air inlet tube.
2. Remove the 2 bolts retaining the sight shield and position aside. Remove the hood weather seal.
3. Remove the wipers. Remove the left cowl vent screen and the wiper module. On supercharged engines, remove the charge air cooler tubes.
4. Install engine support fixture tool D88L-6000-A or equivalent. Raise and safely support the vehicle.
5. Remove the engine mount through bolts. On supercharged engine, remove the left mount retaining strap bolt.

6. Partially lower the vehicle and raise the engine with the support fixture.

7. Raise and safely support the vehicle. Remove the starter.

8. Drain the crankcase and remove the oil filter.

9. Remove the wire loom, ground strap and automatic transmission oil cooler lines, if equipped.

10. Remove the oil pan-to-bellhousing bolts and the bolts at the crankshaft position sensor shield, if equipped. Remove the remaining oil pan retaining bolts.

11. Remove the steering shaft pinch bolts and separate the steering shaft. Position a jack under the front of the sub-frame.

12. Remove the 6 rearward bolts on the front of the sub-frame. Loosen the 2 front sub-frame bolts.

13. Remove the lower strut-to-control arm bolts and nuts and lower the sub-frame. Remove the oil pan.

To install:

14. Clean the gasket mating surfaces and the oil pan. Apply silicone sealer to the oil pan.

15. Fit the oil pan to the cylinder block. Make sure enough clearance has been provided to allow the oil pan to be installed without sealer being scraped off under the cylinder block.

16. Install the oil pan retaining bolts at the cylinder block and bellhousing and install the lower crankshaft sensor shield, if equipped. Tighten the bolts to 80–106 inch lbs. (9–12 Nm).

17. Raise the sub-frame into position and install the lower strut mount-to-control arm bolts. Tighten to 103–144 ft. lbs. (140–195 Nm).

18. Install the 2 front sub-frame bolts and the 6 bolts at the rear of the front sub-frame member. Install a ¾ in. outside diameter pipe or equivalent, into both front left and right sub-frame and body alignment holes. Tighten 1 bolt at each corner. Remove the alignment tools and tighten the bolts to 70–96 ft. lbs. (95–130 Nm).

19. Connect the steering shaft and install the pinch bolt. Tighten to 30–42 ft. lbs. (41–57 Nm).

20. Install the transmission cooler lines, wire loom and ground strap. Install a new oil filter.

21. Install the starter and partially lower the vehicle.

22. Lower the engine with the support fixture. Seat the left side locating pin before the right. Partially raise the vehicle and support safely.

23. Install the engine mount through bolts and tighten to 35–50 ft.

lbs. (47–68 Nm). On supercharged engine, install the left mount retaining strap bolt and tighten to 33–45 ft. lbs. (45–61 Nm). Lower the vehicle.

24. Remove the engine support fixture. On supercharged engine, install the charge air cooler tubes.

25. Install the wiper module and the left cowl vent screen. Install the wipers and the hood weather seal.

26. Install the sight shield and the 2 retaining bolts. Install the air duct assembly. Fill the crankcase with the proper type and quantity of engine oil.

27. Connect the negative battery cable, start the engine and check for leaks.

4.6L (SOHC) Engine

1. Disconnect the battery cables and remove the air inlet tube.

2. Drain the cooling system and remove the cooling fan and shroud. Relieve the fuel system pressure and disconnect the fuel lines.

3. Remove the upper radiator hose. Remove the wiper module and support bracket.

4. Discharge the air conditioning system. Disconnect and plug the compressor outlet hose at the compressor and remove the bolt retaining the hose assembly to the right coil bracket. Cap the compressor outlet.

5. Remove the 42-pin engine harness connector from the retaining bracket on the brake vacuum booster and disconnect the connector and transmission harness connector.

6. Disconnect the throttle valve cable from the throttle body and disconnect the heater outlet hose.

7. Remove the nut retaining the ground strap to the right cylinder head. Remove the upper stud and loosen the lower bolt retaining the heater outlet hose to the right cylinder head and position out of the way.

8. Remove the blower motor resistor. Remove the bolt retaining the right engine mount to the lower engine bracket.

9. Disconnect the vacuum hoses from the EGR valve and tube. Remove the 2 bolts retaining the EGR valve to the intake manifold.

10. Raise and safely support the vehicle. Drain the crankcase and remove the engine mount through bolts.

11. Remove the EGR tube line nut from the right exhaust manifold and remove the EGR valve and tube assembly.

12. Disconnect the exhaust from the exhaust manifolds. Lower the ex-

haust system and support it with wire from the crossmember.

13. Position a jack and a block of wood under the oil pan, rearward of the oil drain hole. Raise the engine approximately 4 in. and insert 2 wood blocks approximately 2½ in. thick under each engine mount. Lower the engine onto the wood blocks and remove the jack.

14. Remove the 16 bolts retaining the oil pan to the engine block and remove the oil pan.

NOTE: It may be necessary to loosen, but not remove, the 2 nuts on the rear transmission mount and with a jack, raise the transmission extension housing slightly to remove the pan.

To install:

15. Clean the oil pan and the gasket mating surfaces.

16. Position a new gasket on the oil pan. Apply silicone sealer to where the front cover meets the cylinder block and rear seal retainer meets the cylinder block. Position the oil pan on the engine and install the bolts. Tighten the bolts, in sequence, to 15–22 ft. lbs. (20–30 Nm).

17. Position the jack and wood block under the oil pan, rearward of the oil drain hole and raise the engine enough to remove the wood blocks. Lower the engine and remove the jack.

18. Install the engine mount through bolts and tighten to 15–22 ft. lbs. (20–30 Nm).

19. Position the EGR valve and tube assembly in the vehicle and connect to the exhaust manifold. Tighten the line nut to 26–33 ft. lbs. (35–45 Nm).

NOTE: Loosen the line nut at the EGR valve prior to installing the assembly into the vehicle. This will allow enough movement to align the EGR valve retaining bolts.

20. Cut the wire and position the exhaust system to the manifolds. Install the 4 nuts and tighten to 20–30 ft. lbs. (27–41 Nm). Make sure the exhaust system clears the crossmember. Adjust as necessary.

21. Install a new oil filter and lower the vehicle.

22. Install the bolt retaining the right engine mount to the lower engine bracket. Tighten to 15–22 ft. lbs. (20–30 Nm).

23. Install a new gasket on the EGR valve and position on the intake manifold. Install the 2 bolts retaining the EGR valve to the intake manifold and tighten to 15–22 ft. lbs. (20–30

Nm). Tighten the EGR tube line nut at the EGR valve to 26–33 ft. lbs. (35–45 Nm). Connect the vacuum hoses to the EGR valve and tube.

24. Install the blower motor resistor. Position the heater outlet hose, install the upper stud and tighten the upper and lower bolts to 15–22 ft. lbs. (20–30 Nm). Install the ground strap on the stud and tighten to 15–22 ft. lbs. (20–30 Nm).

25. Connect the heater outlet hose and the throttle valve cable. If necessary, adjust the throttle valve cable.

26. Connect the 42-pin connector and transmission harness connector. Install the harness connector on the brake vacuum booster.

27. Connect the air conditioning compressor outlet hose to the compressor and install the bolt retaining the hose to the right coil bracket.

28. Install the upper radiator hose and connect the fuel lines. Install the wiper module and retaining bracket.

29. Install the cooling fan and shroud and fill the cooling system. Fill the crankcase with the proper type and quantity of engine oil.

30. Connect the negative battery cable and install the air inlet tube. Start the engine and check for leaks.

31. Evacuate and recharge the air conditioning system.

4.6L (DOHC) Engine

1. Turn **OFF** air suspension switch located in luggage compartment, if equipped.

2. Disconnect the negative and positive battery cables.

3. Remove oil dipstick. Install engine support tripod onto engine compartment and attach to engine.

4. Raise and safely support the vehicle.

5. Remove the front wheel and tire assemblies.

6. Drain the engine oil.

7. Disconnect left and right ride height sensor wiring connectors and remove right and left brake caliper bolts. Remove brake calipers and support with mechanics wire.

8. Disconnect left and right upper control arms from spindles. Disconnect steering shaft coupler at rag joint.

9. Remove power steering line clamp from subframe.

10. Support subframe with engine jack.

11. Remove 2 motor mount through bolts. Remove left and right lower strut control arm nuts and bolts.

12. Remove 8 subframe bolts. Lower the subframe.

13. Disconnect low oil level sensor connector and sensor wiring clips from oil pan rail in 2 places.

14. Remove 16 oil pan bolts and remove the pan. Clean the oil pan and mating surfaces and inspect for damage.

To install:

15. Position new gasket on oil pan. Apply silicone sealer where front cover meets cylinder block and rear seal retainer meets cylinder block. Position oil pan on cylinder block and install 16 oil pan bolts. Tighten the bolts to 15–22 ft. lbs. (20–30 Nm).

16. Connect low oil level sensor connector and sensor wiring clips to oil pan rail.

17. Raise the subframe into position. Install lower strut to control arm bolts.

18. Install the motor mount through bolts. Install the 8 subframe bolts. Install power steering line clamps.

19. Connect steering shaft at rag joint. Partially lower the vehicle.

20. Install the left and right ride height sensor wiring connectors. Install right and left brake calipers and caliper retaining bolts.

21. Connect left and right upper control arms to spindles. Install wheel and tire assemblies. Lower the vehicle.

22. Remove engine support. Install dipstick.

23. Fill the crankcase with the proper type and quantity of engine oil.

24. Connect the negative battery cable, start the engine and check for leaks.

25. Turn **ON** air suspension switch located in luggage compartment.

5.0L Engine

Mustang

1. Disconnect the negative battery cable and remove the air cleaner tube.

2. Remove the oil level indicator from the left side of the cylinder block. Remove the fan shroud and position the shroud over the fan.

3. Raise and safely support the vehicle. Drain the crankcase and remove the oil level sensor wiring from the oil pan.

4. Disconnect the electrical connectors from the starter and remove the starter. Remove the catalytic converter and muffler inlet pipes.

5. Remove the engine mount-to-No. 2 crossmember attaching bolts or nuts. Support the transmission and remove the No. 3 crossmember and rear mount support assemblies.

6. Remove the steering gear attaching bolts and position the steering gear forward aside.

7. Position a jack and wood block under the oil pan. Raise the engine and install wood blocks between the engine mounts and frame. Lower the engine onto the wood blocks and remove the jack.

8. Remove the oil pan attaching bolts and lower the pan to the crossmember. Remove the oil pump and pickup tube assembly and allow to drop into the pan. Remove the pan.

To install:

9. Clean the oil pan and the gasket mating surfaces. Clean the oil pump exterior and pickup tube screen. Apply gasket sealer to the gasket mating surfaces and install new oil pan gaskets.

10. With the oil pump and pickup tube assembly positioned in the oil pan, raise the pan onto the crossmember. Install the oil pump and then the pan. Tighten the oil pan bolts to 9 ft. lbs. (12 Nm).

11. Position the oil pan and the wood block under the oil pan. Raise the engine and remove the wood blocks. Lower the engine and remove the jack. Install the engine mount-to-No. 2 crossmember attaching nuts or bolts. Tighten to 80–106 ft. lbs. (108–144 Nm).

12. Position the steering gear and install the retaining bolts. Install the starter and connect the electrical connectors. Connect the oil level sensor wire to the oil pan.

13. Install the rear mount and the No. 3 crossmember. Tighten the attaching bolts to 80–106 ft. lbs. (108–144 Nm). Install the catalytic converter and muffler inlet pipes. Lower the vehicle.

14. Install the fan shroud and install the oil level indicator to the side of the cylinder block. Install the air cleaner assembly.

15. Fill the crankcase with the proper type and quantity of engine oil. Connect the negative battery cable, start the engine and check for leaks.

Thunderbird and Cougar

1. Disconnect the negative battery cable and remove the oil level dipstick. Disconnect the air cleaner cover retaining clips to allow free movement when the engine is raised.

2. Remove the 2 bolts retaining the radiator shroud to the radiator and pull the shroud loose from the lower retaining clips.

3. Install engine support fixture tool D88L-6000-A or equivalent. Raise and safely support the vehicle.

4. Drain the crankcase and remove the engine mount through bolts. Loosen the transmission mount nut to allow the mount to move when the engine is raised. Partially lower the vehicle.

5. Raise the engine approximately 2 in. using the support fixture. Raise and safely support the vehicle.

6. Remove the power steering cooler line retaining clips. Remove the bolt securing the transmission lines to the right side of the engine block.

7. Disconnect the electrical connector from the low oil level sensor located in the oil pan, if equipped. Remove the oil pan retaining bolts.

8. Remove the steering shaft pinch bolt and separate the steering shaft from the power steering rack assembly.

9. Position 2 jackstands under the engine support sub-frame. Remove the lower strut-to-control arm bolts and nuts from both sides.

10. While supporting the engine support sub-frame on jackstands, remove the 6 rearward bolts on the subframe. Loosen the 2 froward bolts on the sub-frame. Lower the sub-frame.

11. Remove the oil pump/pickup tube assembly and place it in the oil pan. Remove the pan.

To install:

12. Clean the oil pan and the gasket mating surfaces. Clean the oil pump exterior and the pickup tube screen.

13. Apply a thin coat of silicone sealer to the engine block and to the engine block side of a new oil pan gasket. Allow the adhesive to set-up for approximately 5 minutes, before positioning the gasket to the engine.

14. Place the oil pump and pickup tube assembly in the oil pan and position the pan on the sub-frame. Install the oil pump/pickup tube and oil pump drive to the engine. Tighten the oil pump retaining bolts to 22–32 ft. lbs. (30–43 Nm).

15. Position the oil pan to the engine and install all the pan bolts hand-tight, then tighten the bolts evenly to 9 ft. lbs. (12 Nm). Connect the electrical connector to the low oil level sensor, if equipped.

16. Raise the sub-frame into position while supporting the sub-frame on the jackstands. Install a ¾ in. outside diameter pipe or equivalent, into both front left and right sub-frame and body alignment holes. Tighten 1 bolt at each corner. Remove the alignment tools and tighten the bolts to 70–96 ft. lbs. (95–130 Nm).

17. Install the lower strut-to-control arm bolts and nuts and tighten to 103–144 ft. lbs. (140–195 Nm). Remove the 2 jackstands used for installing the sub-frame.

18. Connect the steering shaft and install the steering shaft pinch bolt. Tighten the pinch bolt to 30–42 ft. lbs. (41–57 Nm). Install the bolt securing the transmission lines to the right side of the engine block.

19. Secure the power steering cooler line retaining clips and partially lower the vehicle. Lower the engine onto the engine mounts and remove the engine support fixture.

20. Raise and safely support the vehicle. Tighten the transmission mount nut to 65–85 ft. lbs. (88–115 Nm). Install the engine mount through bolts.

21. Install a new oil filter and lower the vehicle. Position the fan shroud into the lower retaining clips and install the 2 bolts. Connect the air filter cover retaining clips.

22. Fill the crankcase with the proper type and quantity of engine oil. Install the dipstick and connect the negative battery cable. Start the engine and check for leaks.

Oil Pump

REMOVAL AND INSTALLATION

Except 3.8L and 4.6L Engines

1. Disconnect the negative battery cable. Remove the oil pan.

2. Remove the oil pump inlet tube and screen assembly.

3. Remove the oil pump attaching bolts and gasket. Remove the oil pump intermediate shaft.

To install:

4. Prime the oil pump by filling either the inlet or outlet ports with engine oil and rotating the pump shaft to distribute the oil within the pump body.

5. Position the intermediate driveshaft into the distributor socket. With the shaft firmly seated in the distributor socket, the stop on the shaft should touch the roof of the crankcase. Remove the shaft and position the stop, as necessary.

6. Position a new gasket on the pump body, insert the intermediate shaft into the oil pump and install the pump and shaft as an assembly.

NOTE: Do not attempt to force the pump into position if it will not seat readily. The driveshaft hex may be misaligned with the distributor shaft. To align, rotate the intermediate shaft into a new position.

7. Tighten the oil pump attaching screws to 14–21 ft. lbs. (19–29 Nm) on the 2.3L engine or 22–32 ft. lbs. (30–43 Nm) on the 5.0L engine.

8. Clean and install the oil pump inlet tube and screen assembly.

9. Install the oil pan and the remaining components.

3.8L Engine

NOTE: The timing chain front cover houses the oil pump on the 3.8L engine. If the oil pump housing is scored, worn or grooved, the entire front cover will have to be replaced.

1. Disconnect the negative battery cable. Raise and safely support the vehicle.

2. Remove the oil filter.

3. Remove the cover/filter mount assembly. On supercharged engines, remove the oil cooler assembly.

4. Lift the pump gears from their mounting pocket in the front cover.

5. Clean all gasket mounting surfaces.

6. Inspect the mounting pocket for wear. If excessive wear is present, complete timing cover assembly replacement is necessary.

7. Inspect the cover/filter mount gasket to timing cover surface for flatness. Place a straight-edge across the flat and check clearance with a feeler gauge. If the measured clearance exceeds 0.0016 in. (0.04mm), replace the cover/filter mount.

8. Replace the pump gears if wear is excessive.

9. Remove the plug from the end of the pressure relief valve passage using a small drill and slide hammer. Use caution when drilling.

10. Remove the spring and valve from the bore. Clean all dirt and metal chips from the bore and valve. Inspect all parts for wear. Replace as necessary.

To install:

11. Install the valve and spring after lubricating them with engine oil. The end with the smaller diameter goes in first.

12. Install a new plug. The plug can be tapped into the bore using a plastic tipped hammer. Make sure the plug is 0–0.010 in. (0–0.25mm) below the machined surface.

13. Lightly pack the gear pocket with petroleum jelly. Install the gears in the cover pocket, making

sure petroleum jelly fills all the voids between the gears and pockets.

NOTE: Failure to properly coat the oil pump gears may result in failure of the pump to prime when the engine is started.

14. Position the pump body O-ring seal and install the pump body to the front cover using alignment dowels on the front cover.

15. Tighten the pump body retaining bolts to 18–22 ft. lbs. (25–30 Nm) for M8 bolts and 30–40 ft. lbs. (40–55 Nm) for M10 bolts.

16. Install the oil cooler on supercharged engine. Install a new oil filter. Fill the crankcase with the proper type and quantity of engine oil.

17. Connect the negative battery cable, start the engine and check for leaks and proper oil pressure.

4.6L Engine

1. Disconnect the negative battery cable. Remove the front cover.

2. Remove the oil pan and pickup tube.

3. Remove primary timing chains. Remove crankshaft timing sprockets.

4. Remove 4 bolts retaining the oil pump to cylinder block. Remove the oil pump. Clean mating surfaces and inspect for damage.

To install:

5. Rotate the inner rotor of oil pump to align with flats on crankshaft and install oil pump flush with cylinder block.

6. Install 4 retaining bolts and tighten to 6–8.8 ft. lbs. (8–12 Nm).

7. Replace oil filter. Install timing chains.

8. Install pickup tube and oil pan.

9. Install front cover. Fill the crankcase with the proper type and quantity of engine oil.

10. Connect the negative battery cable, start the engine and check for leaks and proper oil pressure.

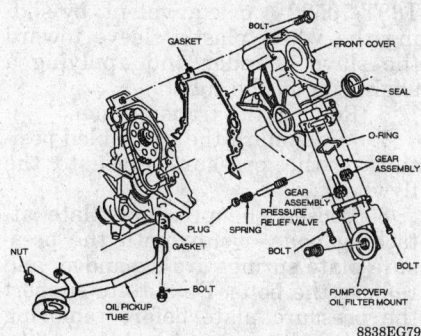

Oil pump and timing chain front cover exploded view — 3.8L engine

---

TRANSMISSION

Manual Transmission Assembly

REMOVAL AND INSTALLATION

Mustang

1. Disconnect the negative battery cable.

2. Raise and support the vehicle safely.

3. Mark the position of the driveshaft on the axle flange so it can be reinstalled in the same position. Disconnect the driveshaft from the flange. Slide the driveshaft off the transmission output shaft and install a suitable plug to prevent lubricant from leaking.

4. Remove the catalytic converter.

5. Remove the 2 nuts attaching the rear transmission support to the crossmember. Remove the bolts.

6. Support the engine and transmission with a suitable jack.

7. Remove the 2 nuts from the crossmember bolts. Remove the bolts, raise the jack slightly and remove the crossmember.

8. Lower the transmission to expose the 2 bolts securing the shift handle to the shift tower. Remove the 2 nuts and bolts and remove the shift handle.

9. Disconnect the wiring harness from the backup lamp switch. On the 5.0L engine, disconnect the neutral sensing switch.

10. Remove the bolt from the speedometer cable retainer and remove the speedometer driven gear from the transmission.

11. Remove the 4 bolts that secure the transmission to the flywheel housing.

12. Move the transmission and jack rearward until the transmission input shaft clears the flywheel housing. If necessary lower the engine enough to obtain clearance for removing the transmission.

NOTE: Do not depress the clutch while the transmission is removed.

To install:

13. Make sure the mounting surface of the transmission and flywheel housing are clean and free of dirt, paint and burrs.

14. Install 2 guide pins in the flywheel housing lower mounting bolt holes. Raise the transmission and move forward on the guide pins until the input shaft splines enter the clutch hub splines and the case is positioned against the flywheel housing.

15. Install the 2 upper transmission-to-flywheel housing mounting bolts snug and remove the 2 guide pins. Install the 2 lower mounting bolts and tighten all the bolts to 45–65 ft. lbs. (61–88 Nm).

16. Raise the transmission with a jack until the shift handle can be secured to the shift tower. Install and tighten the attaching bolts and washers to 23–32 ft. lbs. (31–43 Nm).

17. Connect the speedometer cable to the extension housing and tighten the attaching screw to 54–115 inch lbs. (6–13 Nm).

18. Raise the rear of the transmission with the jack and install the transmission support. Install and tighten the attaching bolts to 36–50 ft. lbs. (48–68 Nm).

19. With the transmission extension housing resting on the engine rear support, install the attaching bolts and tighten to 35 ft. lbs. (48 Nm).

20. Connect the backup lamp switch wiring harness. On 5.0L engine, connect the neutral sensing switch to the wiring harness.

21. Install the catalytic converter. Tighten the attaching bolts to 20–30 ft. lbs. (27–41 Nm).

22. Remove the extension housing installation tool and slide the forward end of the driveshaft over the transmission output shaft. Connect the driveshaft to the axle flange. Make sure the marks align that were made during removal. Tighten the U-bolt nuts to 42–57 ft. lbs. (56–77 Nm).

23. Fill the transmission with the proper type and quantity of fluid.

24. Lower the vehicle. Check the shift and crossover motion for full shift engagement and smooth crossover operation.

Thunderbird and Cougar

1. Disconnect the negative battery cable.

2. Shift the transmission into the N position.

3. Remove the shift knob and the console top cover.

4. Remove the 2 shifter retaining bolts and remove the shifter.

5. Raise and support the vehicle safely.

6. Remove the drain plug and drain the oil from the transmission.

7. Remove the body reinforcement in front of the axle.

8. Disconnect the rear exhaust assembly from the resonator.

9. Remove the 4 bolts retaining the driveshaft to the companion flange. If the rear driveshaft yoke and companion flange are not marked, mark the position for reassembly.

10. Position an axle stand under the front axle housing and remove the forward retaining nuts and bushings. Loosen the rear retaining nuts to allow the axle to tilt for driveshaft removal.

11. Pull the vent tube from the hole in the sub-frame.

12. Lower the front of the axle housing with the axle stand and slide the driveshaft out of the transmission above the axle housing. Let the driveshaft rest on the front driveshaft support and axle assembly.

13. Remove the catalytic converter.

14. Disconnect the hydraulic clutch line.

15. Disconnect the electrical connectors and remove the starter.

16. Position a transmission jack under the transmission. Remove the crossmember and the bellhousing to engine bolts.

17. Move the transmission to the rear until the input shaft clears the flywheel and lower the transmission from the vehicle.

To install:

18. Install guide studs in the engine block and raise the transmission until the input shaft splines are aligned with the clutch disc splines.

19. Slide the transmission forward on the guide studs until it is against the bellhousing. Install the bellhousing-to-engine retaining bolts and tighten to 28–38 ft. lbs. (38–51 Nm).

20. Install the crossmember and tighten the bolts to 35–50 ft. lbs. (47–68 Nm). Remove the transmission jack.

21. Install the starter and connect the electrical connectors. Connect the hydraulic clutch line.

22. Install the catalytic converter assembly.

23. Lubricate the splines with grease and slide the driveshaft into the transmission.

24. Raise the axle housing with the axle stand and install the bushings and retaining nuts. Tighten the retaining nuts to 68–100 ft. lbs. (92–136 Nm) and remove the axle stand.

25. Position the vent tube in the hole of the sub-frame.

26. Align the driveshaft yoke and companion flange and install the retaining bolts. Tighten to 71–96 ft. lbs. (95–129 Nm).

27. Connect the exhaust pipe muffler assembly to the resonator. Lower the vehicle.

28. Position the shifter and install the retaining bolts. Tighten to 18–24 ft. lbs. (24–33 Nm). Install the console top cover and the shifter knob.

29. Connect the negative battery cable. Check transmission operation.

Clutch Assembly

REMOVAL AND INSTALLATION

Mustang

1. Disconnect the negative battery cable. Lift the clutch pedal to its uppermost position to disengage the pawl and quadrant. Push quadrant forward, unhook cable from quadrant and allow quadrant to slowly swing rearward.

2. Raise and safely support the vehicle. Remove the dust shield, if equipped.

3. Disconnect cable from the release lever. Remove the retaining clip and remove the clutch cable from the flywheel housing.

4. Remove the starter. If equipped with 5.0L engine, remove the bolts that secure engine rear plate to front lower part of flywheel housing. If equipped with 2.3L engine, remove the flywheel housing-to-oil pan bolts.

5. Remove the transmission, then the flywheel housing.

6. Remove the clutch release lever boot. Remove clutch release lever from housing by pulling it through the window in housing until retainer spring is disengaged from pivot. Remove release bearing from release lever.

7. Loosen the pressure plate cover attaching bolts evenly to release spring tension gradually and avoid distorting cover. If same pressure plate and cover are to be installed, mark cover and flywheel so pressure plate can be installed in its original position.

8. Inspect the flywheel for scoring, cracks or other damage and machine or replace, as necessary. Inspect the pilot bearing for damage and free movement. Replace, as necessary.

To install:

9. If removed, install the flywheel. Make sure the mating surfaces of the flywheel and the crankshaft flange are clean prior to installation. Tighten the flywheel bolts to 56–64 ft. lbs. (73–87 Nm) on 2.3L engines or 75–85 ft. lbs. (102–115 Nm) on 5.0L engine.

10. Position the clutch disc and pressure plate assembly on the flywheel. The 3 dowel pins on the flywheel must be properly aligned with the pressure plate. Bent, damaged or missing dowels must be replaced. Start the pressure plate bolts but do not tighten them.

11. Align the clutch disc using a suitable alignment tool inserted in the pilot bearing. Alternately tighten the bolts a few turns at a time, until they are all tight. Final torque the bolts to 12–24 ft. lbs. (17–32 Nm). Remove the alignment tool.

12. Apply a light coating of multi-purpose long-life grease to the release bearing contact surface of the transmission bearing retainer, the pressure plate fingers contact surface of the release bearing, the release lever pivot pocket, release lever fork and flywheel housing pivot ball. Fill the grease groove of the release bearing hub with the same grease. Clean all excess grease from the inside bore of the bearing hub.

13. Install the release bearing on the release lever and install the lever in the flywheel housing. Install the boot.

14. Install the flywheel housing. Tighten the bolts to 29–38 ft. lbs. (38–52 Nm) on the 2.3L engine or 38–55 ft. lbs. (52–74 Nm) on the 5.0L engine.

15. Install the remaining components in the reverse order of removal.

Thunderbird and Cougar

1. Disconnect the negative battery cable.

2. Disconnect the clutch hydraulic system master cylinder from the clutch pedal.

3. Raise and support the vehicle safely.

4. Remove the starter.

5. Disconnect the hydraulic coupling at the transmission with tool T88T-70522-A or equivalent, by sliding the white plastic sleeve toward the slave cylinder and applying a slight tug on the tube.

6. Remove the transmission.

7. Matchmark the assembled position of the pressure plate to the flywheel.

8. Loosen the pressure plate attaching bolts evenly until the pressure plate springs are expanded, and remove the bolts. Be sure to support the pressure plate before removing the last bolt.

9. Remove the pressure plate and clutch disc from the flywheel.

10. Inspect the flywheel for scoring, cracks or other damage and machine or replace, as necessary. Inspect the pilot bearing for damage and free movement. Replace, as necessary.

To install:

11. If removed, install the flywheel. Make sure the mating surfaces of the flywheel and the crankshaft flange are clean prior to installation. Tighten the flywheel bolts to 54–64 ft. lbs. (73–87 Nm).

12. Position the clutch disc on the flywheel so a suitable alignment tool can enter the clutch pilot bearing and align the disc.

13. If reinstalling the original pressure plate, align the matchmarks. Position the pressure plate on the flywheel and install the retaining bolts hand-tight. Tighten the bolts, in sequence, to 20–28 ft. lbs. (27–39 Nm). Remove the alignment tool.

14. Install the remaining components in the reverse order of removal. Tighten the flywheel housing-to-engine bolts to 40–49 ft. lbs. (54–67 Nm).

NOTE: Reuse the aluminum washers under the attaching bolts to prevent galvanic corrosion.

Clutch Cable

ADJUSTMENT

Mustang

NOTE: Whenever the clutch cable is disconnected, it is mandatory that the proper method for installing the clutch cable be followed.

1. Lift the clutch pedal to its upward most position to disengage the pawl and quadrant. Push the quadrant forward, unhook the cable from the quadrant and allow it to slowly swing rearward.

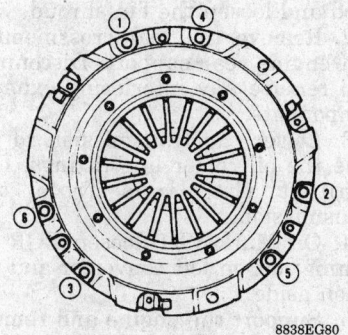

8838EG80

Pressure plate bolt torque sequence — Thunderbird and Cougar

2. Remove the screw that holds the cable insulator to the dash panel and pull the cable through the dash panel and into the engine compartment.

3. Remove the cable bracket screw from the fender apron.

4. Raise and support the vehicle safely.

5. On 5.0L engine, remove the dust cover from the bellhousing.

6. Remove the clip retainer holding the cable to the bellhousing.

7. On the 5.0L engine, slide the ball on the end of the cable through the hole in the clutch release lever and remove the cable.

8. On the 2.3L engine, remove the hairpin clip, clevis pin and clevis from the end of the cable.

To install:

NOTE: The clutch pedal must be lifted to disengage the adjusting mechanism during cable installation. Failure to do so will cause damage the self-adjuster mechanism. A prying instrument should never be used to install the cable into the quadrant.

9. Slide the cable through the hole in the bellhousing and through the hole in the the release lever. On the 5.0L engine, slide the ball on the end of the cable assembly into the cable ball pocket on the clutch release lever. On the 2.3L engine, place the cable ball into the clevis. Install the clevis and clevis pin onto the clutch release lever and into the clevis pin.

10. Install the clutch cable retaining clip on the bellhousing.

11. On the 5.0L engine, install the dust shield on the bellhousing.

12. Push the cable assembly into the engine compartment and lower the vehicle. Install the cable bracket screw in the fender apron.

13. Push the cable into the hole in the dash panel and secure the insulator with a screw.

14. Install the cable assembly by lifting the clutch pedal to disengage the pawl and quadrant. Then, pushing the quadrant forward, hook the end of the cable over the rear of the quadrant.

15. Depress the clutch pedal several times to adjust the cable.

Clutch Master Cylinder

REMOVAL AND INSTALLATION

Thunderbird and Cougar

1. Disconnect the negative battery cable.

2. Disconnect the clutch pedal from the pushrod.

3. Disconnect the hydraulic line from the slave cylinder by depressing the white retainer bushing with tool T88T-70522-A or equivalent, while pulling slightly on the line.

4. Remove the 2 push pins retaining the clutch master cylinder reservoir to the left shock tower.

5. Rotate the master cylinder 45 degrees counterclockwise, then carefully pull the master cylinder through the dash panel, noting the routing of the hydraulic line to the slave cylinder.

6. If the master cylinder is to be replaced, position the master cylinder in a vise and drive out the roll pin using a drift. Remove the O-ring from the tube connection of the master cylinder.

To install:

7. Install a new O-ring onto the clutch tube and install the tube into the master cylinder. Install the roll pin.

8. Position the clutch master cylinder in the engine compartment and route the hydraulic line to the slave cylinder.

9. Install the master cylinder to the dash panel and install the clutch master cylinder fluid reservoir.

10. Push the hydraulic line male connector onto the slave cylinder female connector. Connect the pushrod to the clutch pedal.

11. Fill the reservoir and bleed the system.

Clutch Slave Cylinder

REMOVAL AND INSTALLATION

Thunderbird and Cougar

1. Disconnect the negative battery cable.

2. Disconnect the master cylinder pushrod from the clutch pedal.

3. Raise and support the vehicle safely.

4. Disconnect the hydraulic line from the slave cylinder by depressing the white retainer bushing with tool T88T-70522-A or equivalent, while pulling slightly on the line.

5. Remove the transmission.

6. Remove the clutch release bearing by rotating the assembly against the spring tension until the spring pushes the bearing off the slave cylinder.

7. Remove the clutch slave cylinder retaining bolts and remove the slave cylinder.

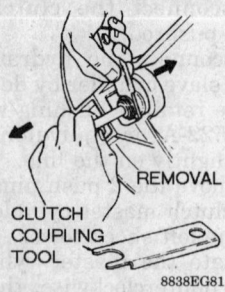

REMOVAL

CLUTCH
COUPLING
TOOL

8838EG81

Disconnecting the clutch
hydraulic line from the
slave cylinder —
Thunderbird and Cougar

To install:

8. Position the slave cylinder over
the input shaft aligning the bleeder
screw and line coupling with holes in
the transmission housing.

9. Install the slave cylinder retaining
bolts and tighten to 15–19 ft. lbs.
(20–27 Nm).

10. Install the release bearing and
transmission.

11. Push the hydraulic line male
connector onto the slave cylinder female
connector.

12. Connect the master cylinder
pushrod to the clutch pedal. Bleed
the system.

Hydraulic Clutch System Bleeding

PROCEDURE

Thunderbird and Cougar

**NOTE: Be sure to pump the
clutch at least 30 times to make
sure air is in the system. If the
slave cylinder is pushed off the
clutch plate, a similar pedal feel
may occur. Pumping the clutch
pushes fluid from the clutch reservoir
into the slave cylinder,
pushing it out to meet the clutch
plate.**

1. Clean all dirt and grease from
the cap to make sure no foreign substances
enter the system.

2. Remove the cap and diaphragm
and fill the reservoir to the top with
the proper fluid.

3. Raise and support the vehicle
safely.

4. Attach a hose to the bleeder
valve at the slave cylinder.

**NOTE: Keep the clutch fluid
reservoir full at all times to prevent
air from being pulled into
the system.**

5. While the clutch pedal is being
depressed, slightly open the bleeder
valve and observe air bubbles in the
clutch fluid at the end of the hose.

6. Close the bleeder valve before
releasing the clutch pedal.

7. Repeat Steps 5 and 6, as necessary,
until no air bubbles are
observed.

8. Lower the vehicle and fill the
reservoir. Road test the vehicle.

Automatic Transmission Assembly

REMOVAL AND INSTALLATION

1. Disconnect the negative battery
cable. Raise the vehicle and support
safely.

2. Drain the fluid from the transmission
by removing all oil pan bolts
except the 2 at the front. Loosen the 2
at the front and drop the oil pan at
the rear to allow the fluid to drain
into a container. When drained, reinstall
a few of the bolts to hold the pan
in place.

3. Remove the access cover and remove
the converter drain plug, if
equipped, to allow the converter to
drain. After the converter has
drained, reinstall the drain plug and
tighten. Remove the converter to flywheel
nuts by turning the converter
to expose the bolts.

**NOTE: Crank the engine over
with a wrench on the crankshaft
pulley attaching bolt. On belt
driven OHC engines, never rotate
the pulley in a counterclockwise
direction as viewed from the
front.**

4. On Mustang, mark the position
of the driveshaft on the axle flange so
it can be reinstalled in the same position.
Disconnect the driveshaft from
the flange and slide the driveshaft
from the transmission. Install a suitable
plug in the extension housing to
prevent fluid leakage.

5. On Thunderbird and Cougar,
proceed as follows:

a. Remove the catalytic
converter.

b. Remove the body
reinforcement.

c. Remove the exhaust pipe and
muffler assembly.

d. Mark the position of the
driveshaft on the axle flange so it
can be reinstalled in the same position.
Disconnect the driveshaft
from the flange.

e. Loosen the differential housing
assembly rear mounting nuts
approximately ¼ in.

f. Position an axle stand under
the front of the differential housing
and remove the forward mounting
nuts and bushings. Pull the vent
tube from the hole in the subframe.

g. Lower the front of the differential
housing with the axle stand
and slide the driveshaft out of the
transmission above the axle housing.
Let the driveshaft rest on the
front driveshaft support and axle
assembly.

6. On Mark VIII, proceed as
follows:

a. Mark the position of the
driveshaft to rear axle so it can be
reinstalled in the same position.
Disconnect the driveshaft from the
rear axle.

b. Remove the catalytic converter
inlet pipe.

c. Lower the exhaust pipe and
muffler assembly.

d. Loosen the rear axle housing
bolts.

e. Lower the housing enough for
driveshaft clearance. Slide the
driveshaft out of the transmission
and position over rear axle
housing.

f. Remove transmission oil inlet
tube.

7. Remove the speedometer cable
or sensor from the extension housing.

8. Disconnect the manual control
shift rod or cable and the downshift
rod or cable from the transmission
control levers.

9. Remove the starter cable and
remove the starter.

10. Remove the electrical wires and
vacuum lines, as required from the
transmission assembly. Remove the
bellcrank bracket, if equipped, from
the converter housing.

11. Place a support under the
transmission and slightly raise it. It
may be necessary to raise the engine
hood and loosen the fan shroud.

12. Remove the rear crossmember
and engine rear support. Disconnect
and remove any interfering exhaust
components.

13. Lower the transmission to expose
the oil cooler line fittings. Disconnect
the lines from the
transmission.

14. On Mark VIII, loosen EGR retaining
nut on left converter and position
aside.

15. Support the engine and remove
the dipstick tube and all the bellhousing
retaining bolts except for
the top 2.

16. Chain the transmission to the jack or support unit for safety.

17. Remove the 2 top bolts from the converter housing and move the transmission rearward and down from under the vehicle. Hold the converter in place to avoid having it drop from the transmission.

18. On Mark VIII, remove the transmission pad from the left side of the converter housing using a hacksaw. Wear proper eye protection.

To install:

19. Tighten the converter drain plug to 8–28 ft. lbs. (11–38 Nm) or to 21–23 ft. lbs. (28–30 Nm) on Mark VIII.

20. Position the converter to the transmission and rotate into position to make sure the drive flats are fully engaged in the pump gear.

NOTE: Lubricate the pilot with chassis grease.

21. Raise the converter and transmission assembly into position. Rotate the converter until the studs and drain plug are in alignment with the holes in the flywheel. Align the orange balancing marks on the converter stud and flywheel bolt hole if balancing marks are present.

NOTE: The converter face must rest squarely against the flywheel. This indicates that the converter pilot is not binding in the engine crankshaft. To ensure the converter is properly seated, grasp a converter stud. It should move freely back and forth in the flywheel hole. If the converter will not move, the transmission must be removed and the converter repositioned so the impeller hub is properly engaged in the pump gear.

22. Install the transmission-to-engine attaching bolts. Tighten the bolts to 40–50 ft. lbs. (55–68 Nm) on all except 2.3L engine. On 2.3L engine, tighten the bolts to 28–38 ft. lbs. (38–51 Nm).

23. Remove the safety chain from around the transmission.

24. Install a new O-ring on the lower end of the transmission filler tube, if equipped. Install the tube to the transmission case and secure with the retaining bolt.

25. Connect the speedometer cable to the transmission case, if equipped.

26. Connect the oil cooler lines to the right side of the transmission case.

27. Position the crossmember on the side supports. Position the rear mount on the crossmember and install the attaching bolts and/or nuts.

28. Secure the engine rear support to the extension housing.

29. Install any exhaust system components, if removed.

30. Lower the transmission and remove the jack.

31. Secure the crossmember to the side supports with the attaching bolts.

32. Connect the TV linkage rod or cable and the manual linkage rod.

33. Install the converter to flywheel attaching nuts and tighten to 20–34 ft. lbs. (27–46 Nm). Install the converter housing cover.

34. Secure the starter motor in place and connect all electrical connections.

35. Install the driveshaft, making sure the index marks are aligned. On Thunderbird and Cougar, proceed as follows:

   a. Raise the differential housing with the axle stand and install the bushings and retaining nuts. Tighten to 68–100 ft. lbs. (92–136 Nm). Remove the axle stand.

   b. Tighten the differential rear retaining nuts to 122–156 ft. lbs. (165–211 Nm).

   c. Position the vent tube in the hole of the sub-frame.

   d. Align the driveshaft yoke and companion flange and install the retaining bolts. Tighten to 70–96 ft. lbs. (95–129 Nm).

   e. Install the catalytic converter, exhaust pipe and muffler.

   f. Install the body reinforcement.

36. On Mark VIII, proceed as follows:

   a. Install transmission oil inlet tube.

   b. Slide the driveshaft into position.

   c. Align marks made during removal and connect the driveshaft to the rear axle.

   d. Tighten the rear axle housing bolts.

   e. Install exhaust system.

37. Lower the vehicle. Fill the transmission with the proper type and quantity of fluid, start the engine and check the transmission for leakage. Adjust the linkage as required.

Throttle Valve Cable

ADJUSTMENT

Automatic Overdrive Transmission

1. Set the parking brake and place the shift selector in **N**.

2. Remove the air cleaner cover and inlet tube from the throttle body inlet to access the throttle lever and cable.

3. Using a small prybar, pry the grooved pin on the cable assembly out of the grommet on the throttle body. Then push out the white locking tab.

4. Check the plastic block with pin and tab; it should slide freely on the notched rod. If not, the white tab may not be pushed out far enough.

5. While holding the throttle lever firmly against the idle stop, push the grooved pin into the grommet on the throttle lever as far as it will go.

6. Make sure the throttle lever does not move while pushing the pin into the grommet.

7. Install the air cleaner cover and inlet tube.

DRIVE AXLE

Driveshaft and U-Joints

REMOVAL AND INSTALLATION

Except Mark VIII, Thunderbird and Cougar

1. Raise and safely support the vehicle. Matchmark the rear driveshaft yoke and the companion flange so they can be reassembled in the same position to maintain balance.

2. Remove the flange bolts and disconnect the driveshaft from the axle companion flange.

3. Allow the rear of the driveshaft to drop down slightly. Pull the driveshaft and slip yoke out of the transmission extension housing.

4. Plug the transmission to prevent fluid leakage.

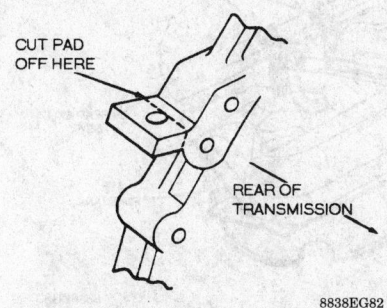

CUT PAD OFF HERE

REAR OF TRANSMISSION

8838EG82

Transmission pad removal — Mark VIII

To install:

5. Lubricate the yoke splines and install the yoke into the transmission extension housing, aligning the splines. Be careful not to bottom the slip yoke hard against the transmission seal.

6. Rotate the pinion flange, as necessary, to align the matchmarks made during removal. Install the driveshaft yoke to the pinion flange. Install the bolts and tighten to 71–96 ft. lbs. (95–130 Nm).

Thunderbird and Cougar

1. Drain the fuel tank.
2. Raise and safely support the vehicle by the frame.
3. Remove the crossmember on the forward side of the fuel tank.
4. Remove the exhaust pipe at the muffler. Lower the pipe and support with a wire.
5. Remove the exhaust pipe rear insulator from the exhaust pipe hanger stud.
6. Remove the muffler insulator from the hanger stud. Remove the exhaust system from the vehicle.
7. Remove the driveshaft hoop on the rear side of the tank.
8. Remove the fuel tank filler tube retaining bolt from the right frame rail.
9. Carefully place a transmission jack under the fuel tank and remove the front heat shield.
10. Remove the support on the forward side of the fuel tank.
11. Remove the fuel tank support straps and lower the tank approximately 6 in.
12. Locate the original paint mark on the axle companion flange and mark the driveshaft flange in the same location. If the original mark is not visible matchmark both flanges.
13. Remove the driveshaft retaining bolts and separate the driveshaft from the axle companion flange. Pull the driveshaft rearward to remove. Install a plug in the extension housing to prevent fluid loss.

To install:

14. Lubricate the slip yoke splines and remove the plug from the transmission extension. Install the driveshaft assembly. Do not allow the slip yoke to bottom on the output shaft with excessive force.
15. Align the marks on the driveshaft with the axle companion flange. Install and tighten the bolts to 71–96 ft. lbs. (95–130 Nm).
16. Raise the fuel tank and install the support straps. Tighten the re-

taining bolts to 21–29 ft. lbs. (28–40 Nm).

17. Install the fuel tank filler tube retaining bolt. Tighten to 36–48 inch lbs. (4.0–5.5 Nm).
18. Install the driveshaft hoop and tighten the retaining bolts to 30–44 ft. lbs. (40–61 Nm).
19. Install the support on the forward side of the fuel tank and tighten the bolts to 30–44 ft. lbs. (40–61 Nm).
20. Raise the exhaust pipe and support with wire. Install the muffler and exhaust pipe insulators on the hanger studs.
21. Install the exhaust pipe to the muffler and tighten the bolts to 21–29 ft. lbs. (28–40 Nm).
22. Install the crossmember on the forward side of the fuel tank and tighten the bolts to 12–17 ft. lbs. (16–24 Nm).
23. Lower the vehicle.

Mark VIII

1. Drain the fuel tank.
2. Raise and safely support the vehicle by the frame.
3. Remove the crossmember on the forward side of the fuel tank.
4. Remove the exhaust pipe at the muffler. Lower the pipe and support with a wire.
5. Remove the exhaust pipe rear insulator from the exhaust pipe hanger stud.
6. Support mufflers and remove muffler rear hangers from rear frame rails. Remove the exhaust system from the vehicle.
7. Remove the fuel tank filler tube retaining bolt from the right frame rail.
8. Carefully place a transmission jack under the fuel tank.
9. Remove the support on the forward side of the fuel tank.
10. Remove the driveshaft hoop on the rear side of the tank.

11. Remove the fuel tank support straps and lower the tank approximately 6 inches.
12. Locate the original paint mark on the axle companion flange and mark the driveshaft flange in the same location. If the original mark is not visible matchmark both flanges.
13. Remove the driveshaft retaining bolts and separate the driveshaft from the axle companion flange. Pull the driveshaft rearward to remove. Install a plug in the extension housing to prevent fluid loss.

To install:

14. Lubricate the slip yoke splines and remove the plug from the transmission extension. Install the driveshaft assembly. Do not allow the slip yoke to bottom on the output shaft with excessive force.
15. Align the marks on the driveshaft with the axle companion flange. Install and tighten the bolts to 71–96 ft. lbs. (95–130 Nm).
16. Raise the fuel tank and install the support straps. Tighten the retaining bolts to 22–30 ft. lbs. (29–41 Nm).
17. Install the fuel tank filler tube retaining bolt. Tighten to 16–23 inch lbs. (2.7–3.7 Nm).
18. Install the driveshaft hoop and tighten the retaining bolts to 20–30 ft. lbs. (14–22 Nm).
19. Install the support on the forward side of the fuel tank and tighten the bolts to 14–22 ft. lbs. (20–30 Nm).
20. Raise the exhaust pipe and support with wire. Install the muffler and exhaust pipe insulators on the hanger studs.
21. Install the exhaust pipe to the muffler and tighten the bolts to 21–29 ft. lbs. (28–40 Nm).
22. Install the crossmember on the forward side of the fuel tank and tighten the bolts to 12–17 ft. lbs. (16–24 Nm).
23. Lower the vehicle.

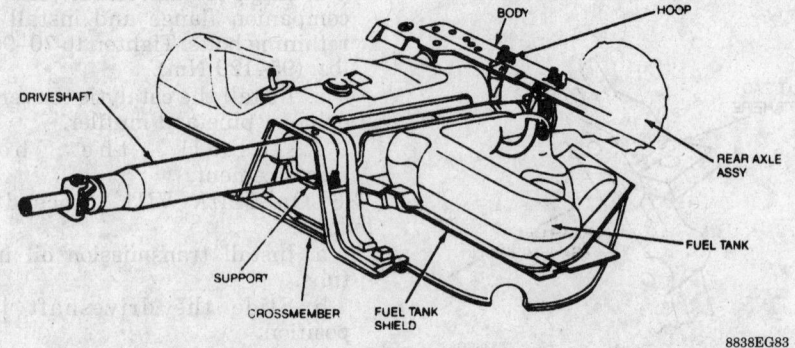

Driveshaft installation — Mark VIII, Thunderbird and Cougar

8838EG83

Rear Halfshaft

REMOVAL AND INSTALLATION

Mark VIII, Thunderbird and Cougar

NOTE: Before removing the rear halfshafts, new inboard CV-joint stub shaft circlips, new differential oil seals and new hub retainer nuts must be available for assembly.

1. Remove the wheelcover/hub cover and remove the hub retainer nut. Loosen the wheel nuts.
2. Raise and support the vehicle safely by the frame only. Remove the wheel nuts and remove the wheel and tire assembly.
3. If equipped with drum brakes, remove the brake drum.
4. If equipped with disc brakes, perform the following:
   a. Remove the anti-lock brake sensors, if equipped.
   b. Use needle-nose pliers to slide the parking brake cable adjusting clip downward until the cable is free.
   c. Remove the parking brake cable from the brake caliper.
   d. Remove the upper and lower caliper retaining bolts and remove the caliper. Support the caliper aside with a wire, do not allow it to hang from the brake hose.
   e. Remove the brake rotor, except Mark VIII.
5. Remove the upper control arm nuts and bolt. Wire the upper control arm to the top of the shock absorber.
6. Using a paint marker, mark the position of the lower control arm in relation to the knuckle with the lower bushings in the relaxed position.

NOTE: Failure to mark this relationship will result in bushing wind-up on assembly and incorrect ride height, causing misalignment and premature tire wear.

7. Use a suitable puller to free the halfshaft from the hub.
8. On Thunderbird and Cougar, proceed as follows:
   a. Remove the lower control arm to knuckle attaching bolts. Remove the knuckle assembly while supporting the outboard CV-joint and boot. Carefully rest the halfshaft on the lower control arm.
   b. If equipped with drum brakes, wire the knuckle assembly to the top of the shock. Do not allow the knuckle assembly to hang from the brake hose.
   c. Remove the halfshaft from the differential using CV-joint remover tool T89P-3514-A or equivalent. Push the tool outward until the CV-joint is freed from the differential side gear.

NOTE: Be careful not to damage the differential oil seal, differential housing and/or CV-joint boot.

   d. Remove the halfshaft from the vehicle. Insert plugs into the differential housing to prevent fluid loss.
9. On Mark VIII, proceed as follows:
   a. Remove the lower control arm to knuckle attaching nuts. Push the halfshaft through the hub while positioning CV-joint and knuckle to allow the front lower bolt to clear the CV-joint and remove the bolt. Remove and save the washers.
   b. Remove rear bolt, washer, and knuckle assembly.

NOTE: Be careful not to damage the differential oil seal, differential housing and/or CV-joint boot.

   c. Remove the halfshaft from the vehicle. Insert plugs into the differential housing to prevent fluid loss.
To install:
10. Remove the differential plugs and install new differential oil seals.
11. Install a new circlip on the halfshaft. Start the ends in the groove and push the circlip into the groove, to prevent over expanding the circlip.
12. Lightly lubricate the stub shaft splines and carefully align the splines on the shaft with the splines in the differential.
13. Push the halfshaft inward to seat the circlip in the differential side gear groove. Use care not to damage the seal.
14. Engage the hub splines with the outboard CV-joint splines.
15. Install the lower control arm bolts and nuts. Align the paint marks and tighten the bolts to 119–147 ft. lbs. (160–200 Nm) on Thunderbird and Cougar or to 94–131 ft. lbs. (128–178 Nm) on Mark VIII.
16. Install a new hub retaining nut and pull the CV-joint into the hub as far as possible by hand.

17. Install the upper arm retaining bolt and nut and tighten to 119–147 ft. lbs. (160–200 Nm) on Thunderbird and Cougar or to 117–142 ft. lbs. (158–193 Nm) or Mark VIII.
18. If equipped with drum brakes, install the brake drum. If equipped with disc brakes, proceed as follows:
   a. Install the brake rotor, except Mark VIII.
   b. Install the brake caliper assembly to the rotor with the outer brake shoe against the rotor's braking surface. This prevents pinching the piston boot between the inner brake shoe and the piston.
   c. Install the upper and lower caliper retaining bolts and tighten to 80–99 ft. lbs. (108–135 Nm) on Thunderbird and Cougar or to 64–88 ft. lbs. (87–119 Nm) on Mark VIII.
   d. Install the parking brake cable to the brake caliper. Install the cable adjustment clip.
   e. Install the anti-lock brake sensor, if equipped. Tighten the retaining bolts to 15–19 ft. lbs. (19–27 Nm).
19. Check inboard CV-joint circlip engagement by attempting to pull the inboard CV-joint from the axle. If the CV-joint circlip is not seated, push the CV-joint in until the circlip is fully engaged in the side gear.
20. Check the axle lube level and fill, as necessary.
21. Install the wheel and tire assembly and tighten the wheel nuts to 80–106 ft. lbs. (108–144 Nm). Lower the vehicle.
22. Tighten the hub nut to 250 ft. lbs. (340 Nm). Install the wheelcover/hub cover.

CV-Joint Boot

REMOVAL AND INSTALLATION

Mark VIII, Thunderbird and Cougar

1. Remove the halfshaft from the vehicle and clamp in a vise. Do not allow the vise jaws to contact the boot or its clamp.

NOTE: The vise should be equipped with jaw caps to prevent damage to any machined surfaces.

2. Cut and remove both boot clamps and slide the boot back on the shaft.

3. Slide the outer race off the tripod.

NOTE: When replacing damaged CV-joint boots, the grease should be checked for contamination or gritty feeling. If the CV-joints are operating satisfactory and the grease does not feel contaminated, add grease and replace the boot. If the grease appears contaminated, the CV-joint should be disassembled and inspected.

4. Remove trilobe insert from outer race.
5. Move the stopring back on the shaft using snapring pliers.
6. Move the tripod back on the shaft to allow access to the circlip.
7. Remove the circlip and the tripod from the shaft.
8. Remove the stopring and remove the inboard CV-joint boot.
9. Reposition the halfshaft in the vise and remove the outboard CV-joint boot.

NOTE: The outboard CV-joint is permanently retained to the inter-connecting shaft and cannot be disassembled. Outboard CV-joints are serviced as an assembly, including the inter-connecting shaft, boot, clamps grease and circlips.

To install:
10. Slide the outboard boot on the shaft. Before positioning the boot over the CV-joint, pack the CV-joint and boot with grease.
The total amount of grease required is 5.28 ounces (150 grams) with 3.8L engine
The total amount of grease required is 7.92 ounces (225 grams) with 3.8L SC, 4.6L or 5.0L engine.
11. Position the boot on the CV-joint and install the boot clamps.
12. Slide the inboard CV-joint boot on the shaft.
13. With the stopring installed past the splines, install the tripod assembly with the chamfered side toward the stopring.
14. Start 1 end of a new circlip in the groove of the halfshaft and work the circlip over the stub shaft end and into the groove. This will avoid over-expanding the circlip.
15. Compress the circlip and slide the tripod assembly forward over the circlip to expose the stopring groove.
16. Move the stopring into the groove using snapring pliers and make sure it is fully seated in the groove.

17. Fill the CV-joint outer race and boot with grease.
The total amount of grease required is 5 ounces (140 grams) with 3.8L engine
The total amount of grease required is 8.8 ounces (250 grams) with 3.8L SC, 4.6L or 5.0L engine.
18. Install the outer race on the tripod assembly.
19. Position the boot over the CV-joint. Move the CV-joint in and out, as necessary, to adjust to the proper length.
20. Release any air pressure by inserting a small prybar with a dulled blade between the boot and the outer bearing race.
21. Seat the boot in the groove and clamp in position without cutting the boot.

STEERING

Air Bag

CAUTION

Some vehicles are equipped with an air bag system, also known as the Supplemental Inflatable Restraint (SIR) or Supplemental Restraint System (SRS). The system must be disabled before performing service on or around system components, steering column, instrument panel components, wiring and sensors. Failure to follow safety and disabling procedures could result in accidental air bag deployment, possible personal injury and unnecessary system repairs.

PRECAUTIONS

Several precautions must be observed when handling the inflator module to avoid accidental deployment and possible personal injury.

• Never carry the inflator module by the wires or connector on the underside of the module.
• When carrying a live inflator module, hold securely with both hands, and ensure that the bag and trim cover are pointed away.
• Place the inflator module on a bench or other surface with the bag and trim cover facing up.
• With the inflator module on the bench, never place anything on or close to the module which may be thrown in the event of an accidental deployment.

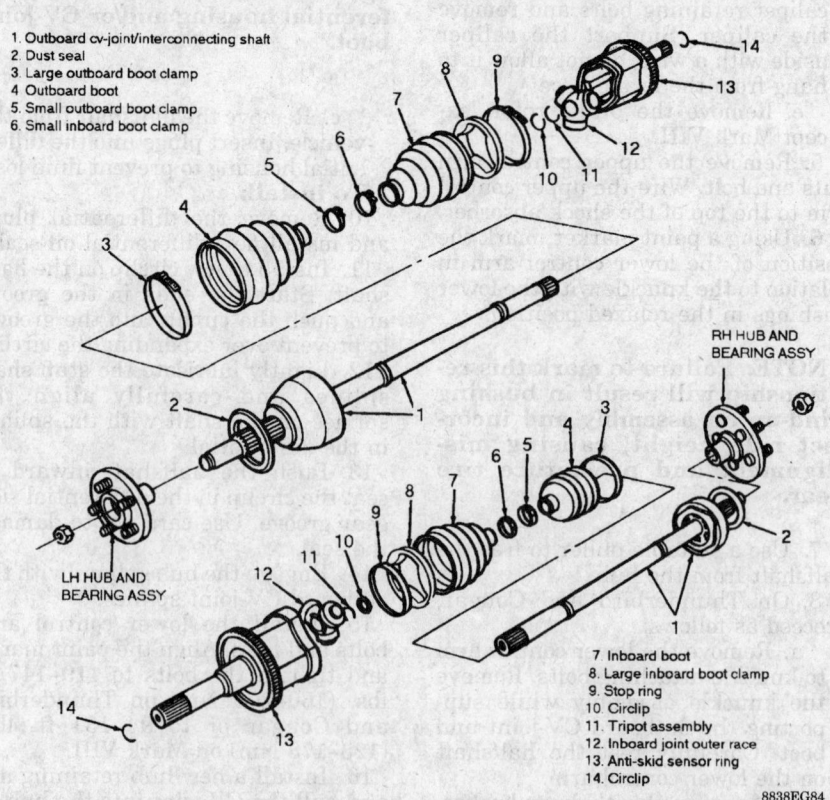

1. Outboard cv-joint/interconnecting shaft
2. Dust seal
3. Large outboard boot clamp
4. Outboard boot
5. Small outboard boot clamp
6. Small inboard boot clamp
7. Inboard boot
8. Large inboard boot clamp
9. Stop ring
10. Circlip
11. Tripot assembly
12. Inboard joint outer race
13. Anti-skid sensor ring
14. Circlip

LH HUB AND BEARING ASSY

RH HUB AND BEARING ASSY

8838EG84

Disassembled view of the halfshafts — Mark VIII, Thunderbird and Cougar

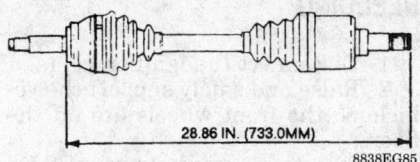

28.86 IN. (733.0MM)

8838EG85

Halfshaft assembled length — Mark VIII, Thunderbird and Cougar

DISARMING

1. Disconnect the positive battery cable. Wait 1 minute for the backup power supply in the diagnostic monitor to deplete its stored energy.
2. Remove the 4 nut and washer assemblies (2 screws on Mark VIII) retaining the driver air bag module to the steering wheel.

CAUTION

When carrying a live air bag, make sure the bag and trim cover are pointed away from the body. In the unlikely event of an accidental deployment, the bag will then deploy with minimal chance of injury. When placing a live air bag on a bench or other surface, always face the bag and trim cover up, away from the surface. This will reduce the motion of the module if it is accidently deployed.

---

3. Disconnect the driver air bag connector. Connect air bag simulator tool 105-00008 or equivalent, to the vehicle harness at the top of the steering wheel.

Steering Wheel

REMOVAL AND INSTALLATION

Mustang, Mark VIII With Air Bag

1. Center the front wheels in the straight-ahead position.
2. Disarm the air bag system.
3. Disconnect the cruise control wire harness from the steering wheel, if equipped.
4. Remove and discard the steering wheel bolt. Remove the steering wheel using a suitable puller. Route

the contact assembly wire harness through the steering wheel as the wheel is lifted off the shaft.

NOTE: Do not use a knock-off type steering wheel puller or strike the retaining bolt with a hammer. This could cause damage to the steering shaft bearing.

To install:
5. Make sure the front wheels are in the straight-ahead position.
6. Route the contact assembly wire harness through the steering wheel opening at the 3 o'clock position and install the steering wheel on the steering shaft. The steering wheel and shaft alignment marks should be aligned. Make sure the air bag contact wire is not pinched.
7. Install a new steering wheel retaining bolt and tighten to 23–33 ft. lbs. (31–48 Nm).
8. If equipped, connect the cruise control wire harness to the wheel and snap the connector assembly into the steering wheel clip. Make sure the wiring does not get trapped between the steering wheel and contact assembly.
9. Connect the air bag wire harness to the air bag module and install the module to the steering wheel. Tighten the module retaining nuts to 3–4 ft. lbs. (4–6 Nm) on Mustang or tighten the screws to 8–10 ft. lbs. (10–14 Nm) on Mark VIII.
10. Connect the air bag backup power supply and battery cable. Verify the air bag warning indicator.

Except Mustang With Air Bag

1. Disconnect the negative battery cable.
2. Remove the horn pad and cover assembly. Disconnect the horn electrical connector.

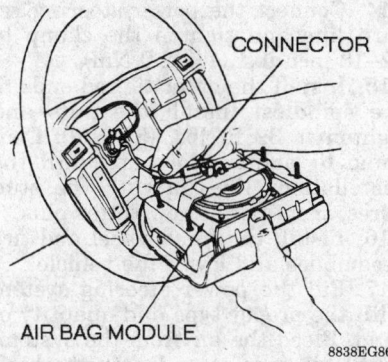

CONNECTOR

AIR BAG MODULE

8838EG86

Air bag module removal — Mustang shown

3. Disconnect the cruise control switch electrical connector, if equipped.
4. Remove and discard the steering wheel bolt. Remove the steering wheel using a suitable puller.

NOTE: Do not use a knock-off type steering wheel puller or strike the retaining bolt with a hammer. This could cause damage to the steering shaft bearing.

To install:
5. Align the index marks on the steering wheel and shaft and install the steering wheel.
6. Install a new steering wheel retaining bolt and tighten to 23–33 ft. lbs. (31–45 Nm).
7. Connect the cruise control electrical connector, if equipped.
8. Connect the horn electrical connector and install the horn pad and cover.
9. Connect the negative battery cable.

Tie Rod Ends

REMOVAL AND INSTALLATION

1. Raise and safely support the vehicle.
2. Remove the cotter pin and nut from the tie rod end ball stud. Disconnect the tie rod end from the spindle using ball stud remover tool 3290-D or equivalent.
3. Holding the tie rod end with a wrench, loosen the tie rod jam nut. Grip the tie rod end with pliers and remove the assembly from the tie rod, but 1st note the depth to which the tie rod was located by using the jam nut as a marker.
To install:
4. Clean the tie rod threads.
5. Thread the new tie rod end onto the tie rod to the same depth as the removed tie rod end.
6. Place the tie rod end ball stud into the spindle and install the nut. Make sure the front wheels are in the straight-ahead position.
7. Tighten the nut to 35 ft. lbs. (48 Nm) and continue tightening the nut to align the next castellation of the nut with the cotter pin hole in the stud. Install a new cotter pin.
8. Set the toe to specification. Tighten the jam nut to 35–50 ft. lbs. (48–68 Nm).

Power Rack and Pinion

REMOVAL AND INSTALLATION

Mustang

1. Disconnect the negative battery cable. Turn the ignition switch to the **RUN** position.
2. Raise and safely support the vehicle. Position a drain pan to catch the fluid from the power steering lines.
3. Remove the 1 bolt retaining the flexible coupling to the input shaft.
4. Remove the cotter pins and nuts from the tie rod ends and separate the tie rod studs from the spindles.
5. Remove the 2 nuts, insulator washers and bolts retaining the steering gear to the crossmember. Remove the front rubber insulators.
6. Position the gear to allow access to the hydraulic lines and disconnect the lines.
7. Remove the steering gear.

To install:

8. Install new plastic seals on the hydraulic line fittings.
9. Install the gear on the mounting spikes and install the hydraulic lines. Tighten the fittings to 20–25 ft. lbs. (27–33 Nm).

NOTE: The hoses are designed to swivel when properly tightened. Do not attempt to eliminate looseness by over-tightening the fittings.

10. Install the front rubber insulators. Make sure all rubber insulators are pushed completely inside the gear housing before installing the mounting bolts.
11. Insert the input shaft into the flexible coupling. Install the mounting bolts, insulator washers and nuts. Tighten the nuts to 30–40 ft. lbs. (41–54 Nm) while holding the bolts. Install and tighten the flexible coupling bolt to 20–30 ft. lbs. (28–40 Nm).
12. Connect the tie rod ends to the spindle arms and install the retaining nuts. Tighten to 35–47 ft. lbs. (48–63 Nm). After tightening, tighten the nuts to their nearest cotter pin castellation and install 2 new cotter pins.
13. Lower the vehicle. Turn the ignition switch to **OFF** and connect the negative battery cable.
14. Fill the power steering system with the proper type and quantity of fluid. Bleed the air from the system. If the tie rod ends were loosened, check and adjust the front end alignment.

Mark VIII, Thunderbird and Cougar

1. Disconnect the negative battery cable. Raise and safely support the vehicle.
2. Remove the front wheel and tire assemblies.
3. Remove the cotter pins and nuts from the tie rod ends. Separate the tie rods from the spindles using a suitable tool.
4. Place a drain pan under the vehicle. Disconnect and plug the power steering return line hose. Disconnect the power steering pressure line at the intermediate fitting and position aside.
5. Remove the steering shaft retaining bolt. Remove the rack-to-subframe bolts and nuts. The nuts are accessed through the hole in the front crossmember.
6. Lower the rack as necessary to remove the pressure line inlet tube. Remove and discard the plastic seal on the inlet tube. Cut the tie strap securing the pressure line to each tube.
7. Remove the steering rack from the vehicle.

To install:

8. Install a new seal on the pressure line inlet tube.
9. Install the insulators from the rear side of the rack housing making sure they are fully seated. Use a suitable rubber lubricant to aid in installation.
10. Install and position the rack to the front crossmember. Install the pressure line inlet tube to the rack.
11. Align the steering shaft to allow the rack to completely seat on the crossmember. Install the steering rack retaining bolts and nuts. Tighten the bolts to 100–144 ft. lbs. (135–195 Nm).
12. Install the steering shaft retaining bolt and tighten to 20–30 ft. lbs. (28–40 Nm).
13. Secure the pressure line to the rack tube with a new tie strap. Connect the power steering pressure line.
14. Connect the power steering return line and tighten the clamp to 12–18 inch lbs. (1.4–2.0 Nm).
15. Install the outer tie rod ends to the spindles. Install the nuts and tighten to 39 ft. lbs. (53 Nm). Continue to tighten the nuts until the castellations line up with the stud bores, then install new cotter pins.
16. Install the front wheel and tire assemblies and lower the vehicle.
17. Fill the power steering system with the proper type and quantity of fluid. Bleed the air from the system. If the tie rods were loosened, check and adjust the front end alignment.

Power Steering Pump

BLEEDING

1. Disconnect the ignition coil(s).
2. Raise and safely support the vehicle so the front wheels are off the floor.
3. Fill the power steering fluid reservoir.
4. Crank the engine with the starter and add fluid until the level remains constant.
5. While cranking the engine, rotate the steering wheel from lock-to-lock.

NOTE: The front wheels must be off the floor during lock-to-lock rotation of the steering wheel.

6. Check the fluid level and add fluid, if necessary.
7. Connect the ignition coil wire. Start the engine and allow it to run for several minutes.
8. Rotate the steering wheel from lock-to-lock.
9. Shut off the engine and check the fluid level. Add fluid, if necessary.
10. If air is still present in the system, purge the system of air using power steering pump air evacuator tool 021-00014 or equivalent, as follows:

  a. Make sure the power steering pump reservoir is full to the COLD FULL mark on the dipstick.

  b. Tightly insert the rubber stopper of the air evacuator assembly into the pump reservoir fill neck.

  c. Apply 15 in. Hg maximum vacuum on the pump reservoir for a minimum of 3 minutes with the engine idling. As air purges from the system, vacuum will fall off. Maintain adequate vacuum with the vacuum source.

  d. Release the vacuum and remove the vacuum source. Fill the reservoir to the COLD FULL mark.

  e. With the engine idling, apply 15 in. Hg vacuum to the pump reservoir. Slowly cycle the steering wheel from lock-to-lock every 30 seconds for approximately 5 minutes. Do not hold the steering wheel on the stops while cycling. Maintain adequate vacuum with the vacuum source as the air purges.

  f. Release the vacuum and remove the vacuum source. Fill the reservoir to the COLD FULL mark.

  g. Start the engine and cycle the steering wheel. Check for oil leaks at all connections. In severe cases of aeration, it may be necessary to repeat Steps 10b–10f.

REMOVAL AND INSTALLATION

NOTE: On the 3.8L SC engine, the charge air cooler and charge air cooler tubes must be removed to gain access to the power steering pump.

1. Disconnect the negative battery cable.
2. Disconnect the fluid return hose at the reservoir and drain the fluid into a container.
3. Remove the pressure hose from the pump fitting, but do remove the fitting from the pump.
4. Remove the pump mounting bracket. Disconnect the belt from the pulley and remove the pump.
5. On engines with the fixed pump system, remove the pulley before removing the pump.

To install:
6. On non-fixed pump systems, install the pulley on the pump, if removed.
7. Place the pump on the mounting bracket and install the bolts at the front of the pump. Tighten to 30–45 ft. lbs. (40–62 Nm) except Mark VIII or to 15–22 ft. lbs. (20–30 Nm) on Mark VIII.
8. On fixed pump systems, install the pulley.
9. Place the belt on the pump pulley and adjust the tension, if necessary.
10. Install the pressure hose to the pump fitting. Tighten the tube nut with a tube nut wrench rather than with an open-end wrench. Tighten to 20–25 ft. lbs. (27–34 Nm) except Mark VIII or to 34–45 ft. lbs. (47–60 Nm) on Mark VIII.

NOTE: Do not overtighten this fitting. Swivel and/or end-play of the fitting is normal and does not indicate a loose fitting. Overtightening the tube nut can collapse the tube nut wall, resulting in a leak and requiring replacement of the entire pressure hose assembly. Use of an open-end wrench to tighten the nut can deform the tube nut hex which may result in improper torque and may make further servicing of the system difficult.

11. Connect the return hose to the pump and tighten the clamp. Fill the reservoir with the proper type and quantity of fluid. Bleed the air from the system.

BRAKES

Anti-Lock Brake System Service

PRECAUTIONS

• Certain components within the Anti-Lock Brake System (ABS) are not intended to be serviced or repaired individually. Only those components with removal and installation procedures should be serviced.
• Do not use rubber hoses or other parts not specifically specified for and ABS system. When using repair kits, replace all parts included in the kit. Partial or incorrect repair may lead to functional problems and require the replacement of components.
• Lubricate rubber parts with clean, fresh brake fluid to ease assembly. Do not use lubricated shop air to clean parts; damage to rubber components may result.
• Use only specified brake fluid from an unopened container.
• If any hydraulic component or line is removed or replaced, it may be necessary to bleed the entire system.
• A clean repair area is essential. Always clean the reservoir and cap thoroughly before removing the cap. The slightest amount of dirt in the fluid may plug an orifice and impair the system function. Perform repairs after components have been thoroughly cleaned; use only denatured alcohol to clean components. Do not allow ABS components to come into contact with any substance containing mineral oil; this includes used shop rags.
• The Anti-Lock control unit is a microprocessor similar to other computer units in the vehicle. Ensure that the ignition switch is **OFF** before removing or installing controller harnesses. Avoid static electricity discharge at or near the controller.
• If any arc welding is to be done on the vehicle, the control unit should be unplugged before welding operations begin.

DEPRESSURIZING

———— **CAUTION** ————
Before servicing any component which contains high pressure, it is mandatory that the hydraulic pressure in the system be discharged or personal injury could result.

To discharge the system, turn the ignition **OFF** and pump the brake pedal a minimum of 20 times until an increase in pedal force is clearly felt.

Master Cylinder

REMOVAL AND INSTALLATION

Except Thunderbird and Cougar With Anti-Lock Brakes

NOTE: The master cylinder on Thunderbird and Cougar with anti-lock brakes is part of the hydraulic actuation assembly and cannot be removed separately.

1. Disconnect the negative battery cable.
2. Remove the brake lines from the primary and secondary outlet ports of the master cylinder, except on Mark VIII.
3. On Mark VIII, depress brake pedal several times to exhaust vacuum in the system. Disconnect the hydraulic control unit supply hose at the master cylinder and plug the hose.
4. Disconnect the brake warning indicator connector.
5. Remove the nuts attaching master cylinder to the brake booster assembly.
6. Slide the master cylinder forward and upward from the vehicle.

To install:
7. Position the master cylinder over the booster pushrod and onto the 2 studs on the booster. Install the retaining nuts and tighten to 14–25 ft. lbs. (18–34 Nm) or to 16–21 ft. lbs. (21–29 Nm) on Mark VIII.
8. Install short brake lines in the master cylinder outlet ports and position them so they point back into the reservoir and the ends of the lines are submerged in brake fluid, except Mark VIII.
9. Fill the reservoir with brake fluid and cover the reservoir with a shop towel.
10. Pump the brakes until clear, bubble-free fluid comes out of both brake lines. If any brake fluid spills on the paint, wash it off immediately with water.
11. Remove the short brake lines and connect the vehicle brake lines to the master cylinder. Bleed each brake line at the master cylinder using the following procedure, except on Mark VIII:
    a. Have an assistant pump the brake pedal 10 times and then hold firm pressure on the pedal.

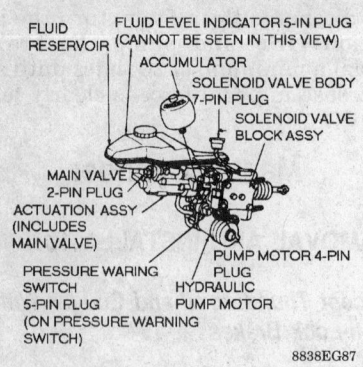

FLUID RESERVOIR
FLUID LEVEL INDICATOR 5-IN PLUG (CANNOT BE SEEN IN THIS VIEW)
ACCUMULATOR
SOLENOID VALVE BODY 7-PIN PLUG
SOLENOID VALVE BLOCK ASSY
MAIN VALVE 2-PIN PLUG
ACTUATION ASSY (INCLUDES MAIN VALVE)
PRESSURE WARING SWITCH 5-PIN PLUG (ON PRESSURE WARNING SWITCH)
PUMP MOTOR 4-PIN PLUG
HYDRAULIC PUMP MOTOR

8838EG87

Anti-Lock Brake Hydraulic Control Unit (HCU)

b. Crack the rear most brake line fitting with a tubing wrench until a stream of brake fluid comes out. Have the assistant maintain pressure on the brake pedal until the brake line fitting is tightened again.

c. Repeat this operation until clear, bubble free fluid comes out from around the brake line fitting.

d. Repeat this bleeding operation at the front brake line fitting.

12. On Mark VIII, unplug the hose and connect the hydraulic control unit supply hose to the master cylinder.

13. Connect the brake warning indicator switch connector.

14. Bleed the system. Operate the brakes several times, then check for external hydraulic leaks.

Brake Caliper

REMOVAL AND INSTALLATION

Front

1. Raise and safely support the vehicle. Remove the front wheel and tire assembly.

2. On all except Mustang with 2.3L engine, remove the hollow brake hose retaining bolt and plug the brake hose.

3. On Mustang with 2.3L engine, loosen the brake line fitting that connects the brake hose to the brake line at the frame bracket. Remove the retaining clip from the hose and bracket and disengage the hose from the bracket. Unscrew the hose from the caliper.

4. Remove the caliper locating pins and remove the caliper. If removing both calipers, mark the right and left sides so they may be reinstalled correctly.

To install:

5. Install the caliper over the rotor with the outer brake shoe against the rotor's braking surface. On Thunderbird and Cougar, make sure the ant-rattle spring is under the arm of the knuckle.

6. Lubricate the inside of the locating pin insulators with silicone dielectric grease. Install the caliper locating pins and start the threads by hand. Tighten to 45–65 ft. lbs. (61–88 Nm) on Mustang. On Thunderbird and Cougar, tighten the locating pins to 19–25 ft. lbs. (25–34 Nm). On Mark VIII, tighten caliper side pin to 16–24 ft. lbs. (22–32 Nm)

NOTE: On Mustang with 2.3L engine, new caliper locating pins must be used.

7. On all except Mustang with 2.3L engine, install new copper washers on each side of the brake hose fitting outlet and install the bolt, through the hose fitting and into the caliper. Tighten the bolt to 30–40 ft. lbs. (40–55 Nm) on Mark VIII, Thunderbird and Cougar. On Mustang, tighten the bolt to 17–25 ft. lbs. (23–34 Nm).

8. On Mustang with 2.3L engine, thread the brake hose into the caliper and tighten to 20–30 ft. lbs. (28–41 Nm).

NOTE: This is a special self-sealing fitting that does not require a gasket. When the hose is correctly tightened, there should be 1 or 2 threads of the fitting still showing at the caliper. It is not necessary for the hose fitting to be flush with the caliper for sealing, so do not over-tighten.

9. On Mustang with 2.3L engine, position the brake hose in its bracket and install the retaining clip. Connect the brake line to the hose and tighten the line fitting nut.

10. Bleed the brake system, install the wheel and tire assembly and lower the vehicle.

11. To position the brake pads, apply the brake pedal several times before moving the vehicle.

Rear

Mark VIII, Thunderbird, Cougar

1. Raise and safely support the vehicle. Remove the rear wheel and tire assembly.

2. Remove the brake hose from the caliper.

3. Release the parking brake cable tension, if necessary. Remove the cable retaining clip and disconnect the cable end from the lever.

4. Hold the slider pin hex heads with an open end wrench and remove the pinch bolts.

5. Lift the caliper assembly away from the anchor plate. Remove the slider pins and boots from the anchor plate.

To install:

6. Apply silicone dielectric compound to the inside of the slider pin boots and to the slider pins.

7. Position the slider pins and boots in the anchor plate. Position the caliper assembly on the anchor plate. Make sure the brake shoes and anti-rattle springs are installed correctly.

8. Remove the residue from the pinch bolt threads and apply locking compound. Install the pinch bolts and tighten to 23–26 ft. lbs. (31–35 Nm) while holding the slider pins with an open end wrench.

9. Attach the cable end to the parking brake lever and install the cable retaining clip. Adjust the parking brake.

10. Using new washers, connect the brake flex hose to the caliper. Tighten the retaining bolt to 30–45 ft. lbs. (40–60 Nm).

11. Bleed the brake system, install the wheel and tire assembly and lower the vehicle.

12. Pump the brake pedal prior to moving the vehicle to position the linings.

Disc Brake Pads

REMOVAL AND INSTALLATION

Front

1. Remove and discard half the brake fluid from the master cylinder reservoir.

2. Raise and safely support vehicle. Remove the front wheel and tire assemblies.

3. Remove the caliper locating pins and remove the caliper from the anchor plate and rotor, but do not disconnect the brake hose.

4. Lift the caliper assembly from the knuckle or spindle.

5. Remove the outer brake pad from the caliper assembly and remove the inner brake pad from the caliper piston.

6. Inspect the disc brake rotor for scoring and wear. Replace or machine, as necessary. If machining, observe the minimum thickness specification.

7. Suspend the caliper inside the fender housing with a length of wire.

Do not let the caliper hang by the brake hose.

To install:

8. Use a large C-clamp and wood block to push the caliper piston back into its bore.

9. Install the inner brake pad, then the outer brake pad, making sure the clips are properly seated.

10. Install the caliper and the wheel and tire assembly. Lower the vehicle.

11. Pump the brake pedal prior to moving the vehicle to seat the brake pads. Refill the master cylinder.

Rear

Mark VIII, Thunderbird, Cougar

1. Remove and discard half the brake fluid from the master cylinder.

2. Raise and safely support vehicle. Remove the rear wheel and tire assembly.

3. Remove the caliper from the anchor plate and rotor, but do not disconnect the brake hose. Suspend the caliper inside the fender housing with a length of wire. Do not let the caliper hang by the brake hose.

4. Remove the brake pads from the anchor plate.

5. Inspect the disc brake rotor for scoring and wear. Replace or machine, as necessary. If machining, observe the minimum thickness specification.

To install:

6. Using brake piston turning tool T87P-2588-A or equivalent, rotate the caliper piston clockwise until it is fully seated. Make sure 1 of the 2 slots in the piston face is positioned so it will engage the nib on the brake pad.

7. Install the brake pads on the anchor plate. Install the caliper and wheel and tire assembly and lower the vehicle.

8. Pump the brake pedal prior to moving the vehicle to seat the brake pads. Refill the master cylinder.

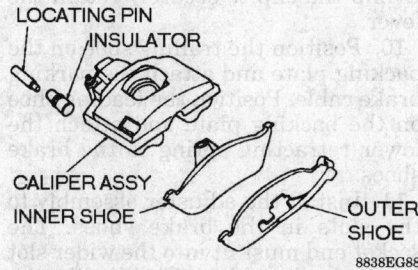

Front disc brake assembly — Thunderbird and Cougar

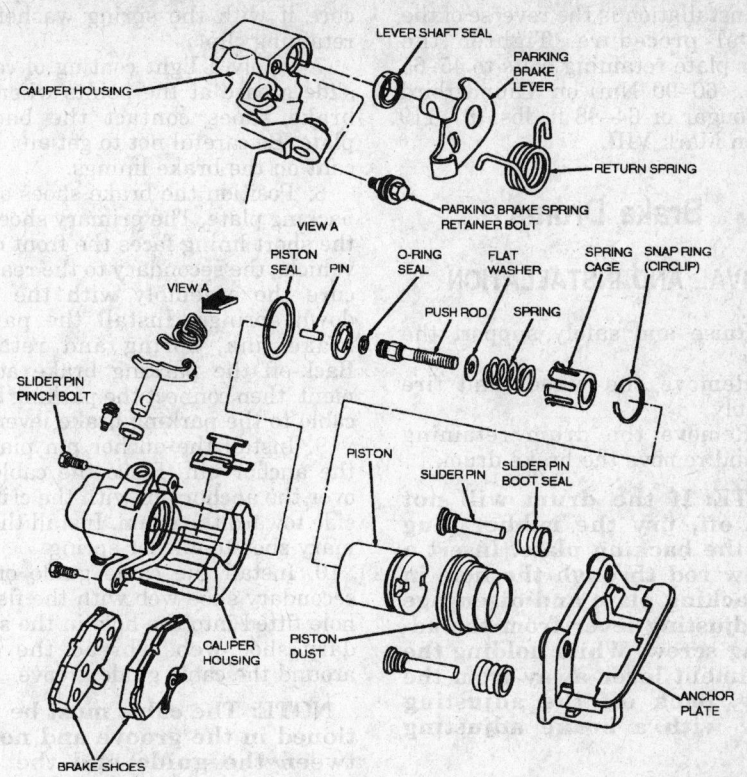

Exploded view of the rear disc brakes — Mark VIII, Thunderbird, Cougar

Brake Rotor

REMOVAL AND INSTALLATION

Front

Mustang

1. Raise and safely support the vehicle. Remove the wheel and tire assembly.

2. Remove the caliper, but do not disconnect the brake hose. Suspend the caliper inside the fender housing with a length of wire. Do not let the caliper hang by the brake hose.

3. Remove the grease cap from the hub and remove the cotter pin, nut lock, adjusting nut and flat washer.

4. Remove the outer roller bearing assembly and remove the hub and rotor assembly.

5. Inspect the rotor for scoring and wear. Replace or machine as necessary. If machining, observe the minimum thickness specification.

6. Installation is the reverse of removal. Make sure the grease in the rotor is clean and adequate. Adjust the wheel bearings.

Mark VIII, Thunderbird and Cougar

1. Raise and safely support the vehicle. Remove the wheel and tire assembly.

2. Remove the caliper, but do not disconnect the brake hose. Suspend the caliper inside the fender housing with a length of wire. Do not let the caliper hang by the brake hose.

3. Remove the rotor retaining push nuts, if equipped, and remove the rotor from the hub.

4. Inspect the rotor for scoring and wear. Replace or machine as necessary. If machining, observe the minimum thickness specification.

5. Installation is the reverse of the removal procedure.

Rear

1. Raise and safely support the vehicle. Remove the wheel and tire assembly.

2. Remove the caliper, but do not disconnect the brake hose. Suspend the caliper inside the fender housing with a length of wire. Do not let the caliper hang by the brake hose.

3. Remove the caliper anchor plate.

4. Remove the rotor retaining push nuts, if equipped, and remove the rotor from the hub.

5. Inspect the rotor for scoring and wear. Replace or machine as necessary. If machining, observe the minimum thickness specification.

6. Installation is the reverse of the removal procedure. Tighten the anchor plate retaining bolts to 45–65 ft. lbs. (60–90 Nm) on Thunderbird and Cougar or 64–88 ft. lbs. (87–119 Nm) on Mark VIII.

Brake Drums

REMOVAL AND INSTALLATION

1. Raise and safely support the vehicle.
2. Remove the wheel and tire assembly.
3. Remove the drum retaining nuts and remove the brake drum.

NOTE: If the drum will not come off, pry the rubber plug from the backing plate. Insert a narrow rod through the hole in the backing plate and disengage the adjusting lever from the adjusting screw. While holding the adjustment lever away from the screw, back off the adjusting screw with a brake adjusting tool.

4. Inspect the brake drum for scoring and wear. Replace or machine as necessary. If machining, observe the maximum diameter specification.
5. Installation is the reverse of removal.

Brake Shoes

REMOVAL AND INSTALLATION

Mustang

1. Raise and safely support the vehicle. Remove the rear wheel and tire assembly. Remove the brake drum.
2. Remove the shoe-to-anchor springs and unhook the cable eye from the anchor pin. Remove the anchor pin plate.
3. Remove the shoe hold-down springs, shoes, adjusting screw, pivot nut, socket and automatic adjustment parts.
4. Remove the parking brake link, spring and retainer. Disconnect the parking brake cable from the parking brake lever.
5. After removing the rear brake secondary shoe, disassemble the parking brake lever from the shoe by removing the retaining clip and spring washer.
To install:
6. Before installing the rear brake shoes, assemble the parking brake lever to the secondary shoe and se-

cure it with the spring washer and retaining clip.
7. Apply a light coating of caliper slide grease at the points where the brake shoes contact the backing plate. Be careful not to get any lubricant on the brake linings.
8. Position the brake shoes on the backing plate. The primary shoe with the short lining faces the front of the vehicle, the secondary to the rear. Secure the assembly with the hold-down springs. Install the parking brake link, spring and retainer. Back-off the parking brake adjustment, then connect the parking brake cable to the parking brake lever.
9. Install the anchor pin plate on the anchor pin. Place the cable eye over the anchor pin with the crimped side toward the drum. Install the primary shoe-to-anchor spring.
10. Install the cable guide on the secondary shoe web with the flanged hole fitted into the hole in the secondary shoe web. Thread the cable around the cable guide groove.

NOTE: The cable must be positioned in the groove and not between the guide and the shoe web.

11. Install the secondary shoe-to-anchor spring. Make sure the cable eye is not cocked or binding on the anchor pin when installed. All parts should be flat on the anchor pin.
12. Apply a thin coat of lubricant to the threads and the socket end of the adjusting screw. Turn the adjusting screw into the adjusting pivot nut to the limit of the threads, then back-off ½ turn.

NOTE: Make sure the socket end of the adjusting screw is stamped with an R or L, indicating the right or left side of the vehicle. The adjusting screw assemblies must be installed on the correct side for proper brake shoe adjustment.

13. Place the adjusting socket on the screw and install the assembly between the shoe ends with the adjusting screw toothed wheel nearest the secondary shoe.
14. Hook the cable hook into the hole in the adjusting lever. The adjusting levers are stamped with an **R** or **L** to indicate their installation on the right or left side.
15. Position the hooked end of the adjuster spring completely into the large hole in the primary shoe web. Connect the loop end of the spring to the adjuster lever hole.
16. Pull the adjuster lever, cable and automatic adjuster spring down

and toward the rear, engaging the pivot hook in the large hole of the secondary shoe web.
17. Make sure the upper ends of the brake shoes are seated against the anchor pin and the shoes are centered on the backing plate.
18. Adjust the brakes using brake adjustment gauge D81L-1103-A or equivalent.
19. Install the brake drum, wheel and tire assemblies and lower the vehicle.
20. Apply the brakes several times while backing up the vehicle. After each stop, the vehicle must be moved forward.

Thunderbird and Cougar

1. Raise and safely support the vehicle. Remove the rear wheel and tire assembly. Remove the brake drum.
2. Disconnect the parking brake cable from the parking brake lever.
3. Remove the 2 brake shoe hold-down retainers, springs and pins.
4. Spread the brake shoes over the piston shoe guide slots. Lift the brake shoes, springs and adjuster off the backing plate as an assembly. Be careful not to bend the adjusting lever.
5. Remove the adjuster spring. To separate the shoes, remove the retracting springs.
6. Remove the parking brake lever retaining clip and spring washer. Remove the lever from the pin.
To install:
7. Apply a light coating of caliper slide grease to the backing plate brake shoe contact areas.
8. Apply a light coat of lubricant to the threaded areas of the adjuster screw and socket. Assemble the brake adjuster with the stainless steel washer. Turn the socket all the way down on the screw, then back off ½ turn.
9. Install the parking brake lever to the trailing shoe with the spring washer and new retaining clip. Crimp the clip to securely retain the lever.
10. Position the trailing shoe on the backing plate and attach the parking brake cable. Position the leading shoe on the backing plate and attach the lower retracting spring to the brake shoes.
11. Install the adjuster assembly to the slots in the brake shoes. The socket end must fit into the wider slot in the leading shoe. The slot in the adjuster nut must fit into the slots in the trailing shoe and parking brake lever.

12. Install the adjuster lever on the pin on the leading shoe and to the slot in the adjuster socket.

13. Install the upper retracting spring in the slot on the trailing shoe and the slot in the adjuster lever. The adjuster lever should contact the star and adjuster assembly.

14. Install the brake shoe anchor pins, springs and retainers.

15. Adjust the brake shoes using brake adjusting gauge D81L-1103-A or equivalent.

16. Install the brake drum, wheel and tire assemblies and lower the vehicle.

17. Apply the brakes several times while backing up the vehicle. After each stop, the vehicle must be moved forward.

Wheel Cylinder

REMOVAL AND INSTALLATION

1. Remove the wheel and tire assembly and the brake drum.

2. Remove the brake shoe assembly.

3. Disconnect the brake line from the wheel cylinder at the backing plate.

4. Remove the wheel cylinder attaching bolts and remove the wheel cylinder.

5. Installation is the reverse of the removal procedure. Tighten the wheel cylinder attaching bolts to 10–20 ft. lbs. (14–28 Nm) on Mustang. On Thunderbird and Cougar, tighten the attaching bolts to 106–160 inch lbs. (12–18 Nm).

6. Bleed the brake system.

Parking Brake Cable

ADJUSTMENT

Mustang

1. Make sure the parking brake is fully released.

2. Place the transmission in **N**. Raise and safely support the vehicle.

3. Tighten the adjusting nut against the cable equalizer, causing a rear wheel brake drag. Loosen the adjusting nut until the rear brakes are fully released. There should be no brake drag.

4. Lower the vehicle and check the operation of the parking brake.

Mark VIII, Thunderbird and Cougar

1. Apply the parking brake control fully on. Release the parking brake

control. Repeat the application and release.

2. Place the transmission in **N**. Raise and safely support the vehicle by the axles.

3. On 1993 vehicles, proceed as follows:

a. With the parking brake control in the **OFF** position, grasp the tensioner around the housing, then using a hook tool, hook the end into the rounded end of the clip between the clip and the housing.

b. Unlock the clip by pulling downward with the tool and support tensioner; the tensioner spring will take up cable slack and preload the cables.

c. While holding the tensioner, lock the clip by pushing up on the bottom of the clip. If the clip does not slide up, move the assembly slightly to align the closest groove on the adjuster rod to the clip.

4. Examine the tensioner for remaining cable take up capability. If none is present, check all cables, parking brake control and brackets for possible damage or deflection.

REMOVAL AND INSTALLATION

Front Cable

Mark VIII, Thunderbird and Cougar, Except Super Coupe

1. Make sure the parking brake is fully released. Raise and safely support the vehicle on the axles.

2. Remove the cable tension as follows:

a. Unlock the tensioner by pulling downward on the clip.

b. While the clip is disengaged, have an assistant apply the parking brake control fully to the last notch position. The tensioner spring will compress allowing cable slack to return.

c. Lock the tensioner by pushing up on the clip.

d. Make sure the locking lever is secure by rotating it toward the threaded rod. Wrap tape or wire around the locking lever and threaded rod to prevent any accidental release.

3. Disconnect the front cable from the intermediate cable at the connector.

4. Remove the cable snap-in retainer from the cable bracket and allow the cable to hang. Lower the vehicle.

5. Remove the left cowl trim panel.

6. Disconnect the cable from the control assembly at the clevis. Re-

move the cable snap-in retainer and pull the cable and grommet up through the floor pan.

7. Installation is the reverse of the removal procedure. Adjust the parking brake.

Thunderbird Super Coupe

1. Make sure the parking brake is fully released. Raise and safely support the vehicle on the axles.

2. Remove the cable tension as follows:

a. Unlock the tensioner by pulling downward on the clip.

b. While the clip is disengaged, have an assistant apply the parking brake control fully to the last notch position. The tensioner spring will compress allowing cable slack to return.

c. Lock the tensioner by pushing up on the clip.

d. Make sure the locking lever is secure by rotating it toward the threaded rod. Wrap tape or wire around the locking lever and threaded rod to prevent any accidental release.

3. Disconnect the front cable from the right rear cable at the connector.

4. Remove the cable snap-in retainer from the tensioner housing. Remove the cable routing clip from the body rear crossmember by squeezing the clip together between the cable and crossmember. Allow the cable to hang.

5. Lower the vehicle. Remove the rear seat and the console.

6. Disconnect the cable from the control at the clevis hook. Remove the cable snap-in retainer from the hand control assembly.

7. Pull the cable and grommet up through the rear floor. Pull the cable out from under the carpet.

8. Installation is the reverse of the removal procedure. Adjust the parking brake.

Intermediate Cable

Mark VIII, Thunderbird and Cougar, Except Super Coupe

1. Make sure the parking brake is fully released. Raise and safely support the vehicle on the axles.

2. Remove the cable tension as follows:

a. Unlock the tensioner by pulling downward on the clip.

b. While the clip is disengaged, have an assistant apply the parking brake control fully to the last notch position. The tensioner spring will compress allowing cable slack to return.

c. Lock the tensioner by pushing up on the clip.

d. Make sure the locking lever is secure by rotating it toward the threaded rod. Wrap tape or wire around the locking lever and threaded rod to prevent any accidental release.

3. Disconnect the intermediate cable from the right rear cable and front cable at the connector.

4. Remove the cable snap-in retainer from the tensioner housing and body bracket.

5. Remove the cable routing clips for body side rails and rear crossmember by squeezing the clip together between the cable and the crossmembers. Remove the cable.

6. Installation is the reverse of the removal procedure. Adjust the parking brake.

Rear Cables

Mustang, Except Cobra

1. Place the parking brake control in the released position. Release the cable tension as follows:

a. Remove the floor console.

b. With an assistant inside the vehicle, raise and safely support the vehicle.

c. Have another assistant pull the equalizer rearward approximately 1–2½ in. to rotate the self-adjuster reel backward.

d. Insert a steel lockpin through the holes in the lever and control assembly. This locks the ratchet wheel in the cable released position.

NOTE: Do not remove the steel lockpin until the cables are connected to the equalizer. Pin removal releases the tension in the ratchet wheel causing the spring to unwind and release tension. If the pin is removed without the cables attached, the entire assembly must be removed to reset the spring tension.

2. Raise and safely support the vehicle. Remove the rear cables from the equalizer.

3. Remove the cable snap fitting from the body. Remove the retaining clip that attaches the cable to the underbody.

4. Remove the wheel and tire assemblies and the brake drums.

5. Remove the self adjuster springs and remove the cable retainers from the backing plates.

6. Disconnect the cable ends from the parking brake levers, compress the cable retainer prongs and pull the cable ends from the backing plates.

7. Installation is the reverse of the removal procedure. Adjust the parking brake.

Mustang Cobra

1. Place the parking brake control in the released position. Release the cable tension as follows:

a. Remove the floor console.

b. With an assistant inside the vehicle, raise and safely support the vehicle.

c. Have another assistant pull the equalizer rearward approximately 1–2½ in. to rotate the self-adjuster reel backward.

d. Insert a steel lockpin through the holes in the lever and control assembly. This locks the ratchet wheel in the cable released position.

NOTE: Do not remove the steel lockpin until the cables are connected to the equalizer. Pin removal releases the tension in the ratchet wheel causing the spring to unwind and release tension. If the pin is removed without the cables attached, the entire assembly must be removed to reset the spring tension.

2. Raise and safely support the vehicle. Remove the rear cables from the rear calipers.

3. Remove trailing arm brackets. Remove the rear cables from the equalizer.

4. Remove cables from tunnel brackets and remove cables.

5. Installation is the reverse of the removal procedure. Adjust the parking brake.

Thunderbird and Cougar With Drum Brakes

1. Make sure the parking brake is fully released. Raise and safely support the vehicle on the axles.

2. Remove the cable tension as follows:

a. Unlock the tensioner by pulling downward on the clip.

b. While the clip is disengaged, have an assistant apply the parking brake control fully to the last notch position. The tensioner spring will compress allowing cable slack to return.

c. Lock the tensioner by pushing up on the clip.

d. Make sure the locking lever is secure by rotating it toward the threaded rod. Wrap tape or wire around the locking lever and threaded rod to prevent any accidental release.

3. Remove the wheel and tire assemblies and the brake drums.

4. Disconnect the parking brake cable end from the parking brake actuating lever. Depress the conduit retaining prongs and remove the cable and pronged fitting from the backing plate.

5. Remove the cable snap-in retainer from the frame bracket.

6. Disconnect the cable end from the tensioner or intermediate cable at the connector. Remove the rear cable by sliding the cable through the clip on the lower control arm.

7. Installation is the reverse of the removal procedure. Adjust the parking brake.

Mark VIII, Thunderbird and Cougar With Disc Brakes

1. Make sure the parking brake is fully released. Raise and safely support the vehicle on the axles.

2. Remove the cable tension as follows:

a. Unlock the tensioner by pulling downward on the clip.

b. While the clip is disengaged, have an assistant apply the parking brake control fully to the last notch position. The tensioner spring will compress allowing cable slack to return.

c. Lock the tensioner by pushing up on the clip.

d. Make sure the locking lever is secure by rotating it toward the threaded rod. Wrap tape or wire around the locking lever and threaded rod to prevent any accidental release.

3. Disconnect the rear cable end from the tensioner or intermediate/front cable at the connector.

4. Remove the cable snap-in retainer from the frame bracket. Disconnect the rear cable end from the caliper housing and remove the cable from the parking brake lever arm on the caliper.

5. Remove the cable retainer from the rear stabilizer bar.

6. Installation is the reverse of the removal procedure. Adjust the parking brake.

Brake System Bleeding

PROCEDURE

Without Anti-Lock Brakes

1. Clean all dirt from the master cylinder filler cap.

2. If the master cylinder is known or suspected to have air in the bore, it must be bled before any of the wheel cylinders or calipers. To bleed the master cylinder, loosen the upper

secondary left front outlet fitting approximately ¾ turn. Have an assistant depress the brake pedal slowly through it's full travel. Close the outlet fitting and let the pedal return slowly to the fully released position. Wait 5 seconds and then repeat the operation until all air bubbles disappear.

3. Repeat Step 2 with the right-hand front outlet fitting.

4. Continue to bleed the brake system by removing the rubber dust cap from the wheel cylinder bleeder fitting or caliper fitting at the right-hand rear of the vehicle. Place a suitable box wrench on the bleeder fitting and attach a rubber drain tube to the fitting. The end of the tube should fit snugly around the bleeder fitting. Submerge the other end of the tube in a container partially filled with clean brake fluid and loosen the fitting ¾ turn.

5. Have an assistant push the brake pedal down slowly through it's full travel. Close the bleeder fitting and allow the pedal to slowly return to it's full release position. Wait 5 seconds and repeat the procedure until no bubbles appear at the submerged end of the bleeder tube. Secure the bleeder fitting and remove the bleeder tube. Install the rubber dust cap on the bleeder fitting.

6. Repeat the procedure in Steps 4 and 5 in the following sequence: left front, left rear and right front. Refill the master cylinder reservoir after each wheel cylinder or caliper has been bled and install the master cylinder cover and gasket. When brake bleeding is completed, the fluid level should be filled to the maximum level indicated on the reservoir.

7. Always make sure the disc brake pistons are returned to their normal positions by depressing the brake pedal several times until normal pedal travel is established. If the pedal feels spongy, repeat the bleeding procedure.

With Anti-Lock Brakes

The front brakes can be bled in the same manner as a vehicle without anti-lock brakes or they can be bled with a pressure bleeder. The rear brakes must be bled with a pressure bleeder or with a fully charged accumulator.

Pressure Bleeding

1. Clean all dirt from the reservoir filler cap area. Attach a suitable pressure bleeder to the reservoir cap opening.

2. Maintain 35 psi pressure on the system through the pressure bleeder.

3. Remove the dust cap from the right front caliper bleeder fitting. Attach a rubber drain tube to the fitting, making sure the tube fits snugly.

4. With the ignition switch in the **OFF** position and the brake pedal in the fully released position, open the bleeder fitting for 10 seconds at a time until an air-free stream of brake fluid flow is observed.

5. Repeat the procedure at the left front, right rear and left rear calipers, in that order.

6. Place the ignition switch in the **RUN** position and pump the brake pedal several times to complete the bleeding procedure and to fully charge the accumulator.

7. Turn the ignition switch to the **OFF** position and remove the pressure bleeder. Siphon off the excess fluid in the reservoir to adjust the level to the **MAX** mark with a fully charged accumulator.

Rear Brake Bleeding With a Fully Charged Accumulator

1. Remove the dust cap from the right rear caliper bleeder fitting. Attach a rubber drain tube to the fitting, making sure the tube fits snugly.

2. Turn the ignition switch to the **RUN** position. This will turn on the electric pump to charge the accumulator, as required.

3. Have an assistant hold the brake pedal in the applied position. Open the bleeder fitting for 10 seconds at a time until an air-free stream of brake fluid flow is observed.

─────── CAUTION ───────
To prevent possible injury, care must be used when opening the bleeder screws due to the high pressures available from a fully charged accumulator.
─────────────────────

4. Repeat the procedure at the left rear caliper.

5. Pump the brake pedal several times to complete the bleeding procedure.

6. Adjust the fluid level in the reservoir to the MAX mark with a fully charged accumulator.

NOTE: If the pump motor is allowed to run continuously for approximately 20 minutes, a thermal safety switch inside the motor may shut the motor off to prevent it from overheating. If that happens, a 2–10 minute cool

down period is typically required before normal operation can resume.

Front Wheel Speed Sensor

REMOVAL AND INSTALLATION

Thunderbird and Cougar

1. Disconnect the negative battery cable. Raise and safely support the vehicle.

2. From under the vehicle, near the front radiator support, disconnect the sensor electrical connector for right or left front sensor.

3. Remove routing clips along wiring harness.

4. Remove Torx® head screws securing sensor to front spindle.

NOTE: If the toothed speed indicator ring is damaged, replace it.

To install:

5. Install sensor into hole in spindle. No adjustment is necessary. Install Torx® head screw and tighten to 40–60 inch lbs. (4.5–6.8 Nm).

6. Route wiring using clips previously removed. Ensure wiring is routed properly.

7. Connect sensor wiring connector to harness connector.

Rear Wheel Speed Sensor

REMOVAL AND INSTALLATION

Thunderbird and Cougar

1. Disconnect the negative battery cable.

2. From inside luggage compartment, disconnect wheel sensor electrical connector located rearward of wheel well, behind carpeting on sides of luggage compartment.

3. Lift luggage compartment carpet and push sensor wire grommet through hole in luggage compartment floor.

4. Raise and safely support the vehicle.

5. Remove the plastic clip holding sensor wire to axle carrier housing. Do not bend the clip open more than the amount necessary to remove the clip from the axle housing.

6. Remove wheel sensor retaining bolt using a ½ in. socket.

To install:

7. Align sensor locating tab and bolt hole with axle housing and push into position.

8. Install sensor retaining bolt and tighten to 14–20 ft. lbs. (19–27 Nm).

9. Install plastic clip retaining sensor wire to axle carrier housing and push electrical connector through hole in floor into luggage compartment. Ensure that rubber grommet is properly seated in hole in floor.

10. Lower the vehicle. Connect sensor electrical connector to connector on harness.

FRONT SUSPENSION

NOTE: If equipped with the level ride air suspension, power to the air system must be shut OFF before servicing the suspension. The switch is located in the luggage compartment, on the drivers side rear fender well.

Strut

REMOVAL AND INSTALLATION

Mustang

1. Disconnect the negative battery cable.

2. Place the ignition switch in the UNLOCKED position to permit free movement of the front wheels.

3. Raise the vehicle by the lower control arms until the wheels are just off the ground. From the engine compartment, remove and discard the 3 upper mount retaining nuts. Do not remove the pop-rivet holding the camber plate position.

4. Continue to raise the front of the vehicle by the lower control arms and position safety stands under the frame jacking pads, rearward of the wheels.

5. Remove the wheel and tire assembly and remove the brake caliper. Support the caliper with a length of wire; do not let the caliper hang by the brake hose.

6. Remove the 2 lower nuts that attach the strut to the spindle, leaving the bolts in place. Carefully remove both spindle-to-strut bolts, push the bracket free of the spindle and remove the strut.

7. Compress the strut to clear the upper mount of the body mounting pad. Remove the upper mount and jounce bumper, if necessary.

To install:

8. Install the upper mount and jounce bumper, if removed.

9. Position the 3 upper mount studs into the body mounting pad and camber plate and start 3 new nuts.

10. Compress the strut and position into the spindle. Install 2 new lower retaining bolts and hand start the nuts. Remove the suspension load from the control arms by lowering the vehicle. Tighten the lower retaining nuts to 140–200 ft. lbs. (190–271 Nm).

11. Raise the suspension control arms and tighten the 3 new upper mount retaining nuts to 45–59 ft. lbs. (60–81 Nm).

12. Install the brake caliper and the wheel and tire assembly.

13. Lower the vehicle to the ground and check the front end alignment.

Thunderbird and Cougar

1. Remove the plastic cover at the upper strut mount, if equipped. If equipped with automatic ride control, remove the actuator assembly as follows:

   a. Make sure the vehicle is level. Turn the ignition switch OFF.

   b. Disconnect the actuator connector from the wiring harness connector. Remove the actuator cover by snapping off.

   c. Slide the actuator connector off the cover by inserting a small prybar tip between the connector and track to separate the 2 parts prior to sliding the connector off.

   d. Squeeze the 2 actuator retaining tabs firmly inward with 1 hand and lift the actuator off the mounting bracket with the other hand.

   e. Grasp the piston rod end at the 9mm hex with a socket wrench.

   f. Loosen the nut retaining the actuator mounting bracket to the strut with a 19mm box wrench while holding the socket wrench.

   g. Remove the nut and mounting bracket.

2. Remove the 3 upper strut retaining nuts and collar plate from the mounting studs in the engine compartment.

3. Raise and safely support the vehicle. Remove the wheel and tire assembly.

4. Remove the lower strut mounting bolt and nut and remove the nut at the stabilizer link upper mounting stud. Separate the link from the spindle using a suitable joint separator tool.

5. Support the lower control arm with a jack. Raise the control arm and spindle with the jack until the stabilizer link can be completely sep-

arated from the spindle. Position the link aside.

6. Remove and discard the spindle to upper control arm retaining nut and bolt. Lower the jack to separate the spindle from the upper control arm. Support the spindle with a length of wire; do not let it hang free.

7. Lower the support for the lower control arm and remove the strut assembly from the vehicle.

To install:

8. Position the strut over the lower arm. Insert the lower strut bolt into the control arm.

9. Using a jack, raise the control arm and strut into position. Align the upper strut mounting studs with the holes.

10. Remove the wire supporting the spindle and position the spindle to the upper control arm. Raise the lower control arm using the jack and attach the spindle to the upper control arm.

11. Install a new spindle retaining bolt from the front of the vehicle and install the nut. Tighten to 59–66 ft. lbs. (80–90 Nm).

12. Position the stabilizer bar link and lower the spindle assembly until the link can be installed. Install the nut on the link stud and tighten to 39–53 ft. lbs. (53–77 Nm).

13. Remove the jack from the lower arm. Install the lower strut nut, but do not tighten at this time.

14. Install the wheel and tire assembly and lower the vehicle. Make sure the upper strut mounting studs are aligned with the holes.

15. Install the collar plate and 3 nuts to the upper mounting studs. Tighten to 17–22 ft. lbs. (22–31 Nm).

16. Install the washer, nut and automatic ride actuator, if equipped. Install the plastic cover, if equipped.

17. Neutralize the front suspension bushings by pushing down and releasing on the front of the vehicle. Then tighten the lower strut nut to 140–162 ft. lbs. (190–220 Nm).

NOTE: The lower strut nut must be tightened with the vehicle weight on the wheels.

Coil Spring

REMOVAL AND INSTALLATION

Mustang

1. Raise and safely support the vehicle, allowing the control arms to hang free.

2. Remove the wheel and tire assembly and the brake caliper. Sus-

pend the caliper with a length of wire; do not let the caliper hang by the brake hose.

3. Disconnect the tie rod end from the steering spindle and disconnect the stabilizer link from the lower arm.

4. Remove the steering gear bolts, if necessary and position the gear so the suspension arm bolt can be removed.

5. If equipped with 2.3L engine, use spring compressor tool T82P-5310-A or equivalent to place the upper plate in position into the spring pocket cavity on the crossmember. The hooks on the plate should be facing the center of the vehicle.

6. If equipped with 5.0L engine, use spring compressor tool D78P-5310-A or equivalent, to install a plate between the coils near the toe of the spring. Mark the location of the upper plate on the coils for installation.

7. Install the compression rod into the lower arm spring pocket hole, through the coil spring and into the upper plate.

8. Install the lower plate, lower ball nut, thrust washer and bearing and forcing nut onto the compression rod. Tighten the forcing nut until a drag on the nut is felt.

9. Remove the suspension arm-to-crossmember nuts and bolts. The compressor tool forcing nut may have to be tightened or loosened for easy bolt removal.

10. Loosen the compression rod forcing nut until spring tension is relieved and remove the forcing nut. Remove the compression rod and coil spring.

To install:

11. Place the insulator on top of the spring. Position the spring into the lower arm pocket. Make sure the spring pigtail is positioned between the 2 holes in the lower arm spring pocket.

12. Position the spring into the upper spring seat in the crossmember.

13. If equipped with 2.3L engine, insert the compression rod through the control arm and spring, then hook it to the upper plate. The upper plate is installed with the hooks facing the center of the vehicle.

14. If equipped with 5.0L engine, install the upper plate between the coils in the location marked during removal.

15. Install the lower plate, ball nut, thrust washer and bearing and forcing nut onto the compression rod.

16. Tighten the forcing nut, position the lower arm into the crossmember and install new lower arm-to-crossmember bolts and nuts. Do not tighten at this time.

17. Remove the spring compressor tool from the vehicle. Raise the suspension arm to a normal attitude position with a jack. Tighten the lower arm-to-crossmember attaching nuts to 110–150 ft. lbs. (149–203 Nm). Remove the jack.

18. Install the steering gear-to-crossmember bolts and nuts, if removed. Hold the bolts and tighten the nuts to 90–100 ft. lbs. (122–135 Nm).

19. Connect the stabilizer bar link to the lower suspension arm. Tighten the attaching nut to 6–17 ft. lbs. (8–24 Nm).

20. Position the tie rod into the steering spindle and install the retaining nut. Tighten the nut to 35 ft. lbs. (47 Nm) and continue tightening the nut to align the next castellation with the hole in the stud. Install a new cotter pin.

21. Install the brake caliper and the wheel and tire assembly. Lower the vehicle.

Thunderbird and Cougar

1. Remove the strut assembly from the vehicle.

NOTE: The upper strut mount cannot be rotated when the strut and spring are assembled. Mark the position of the upper mount to the coil spring with chalk or paint, prior to disassembly. If the upper mount is not properly positioned during assembly, it will not install in the vehicle.

2. Position the strut assembly in spring compressor tool 086-00029 or equivalent.

3. Compress the spring. Remove the strut nut and washer and remove the upper mount.

4. Release the spring compressor to remove the coil spring.

To install:

5. If installing a new spring or upper mount, transfer the reference marks from the removed part to the new part.

6. Position the strut and the spring in the spring compressor tool and compress the spring to install the upper mount.

7. Install the upper mount, aligning the reference marks. Install the washer and nut and tighten to 37–52 ft. lbs. (50–71 Nm).

8. Release the spring compressor, making sure the spring is properly seated at top and bottom.

9. Install the strut assembly in the vehicle.

Air Springs

REMOVAL AND INSTALLATION

Mark VIII

1. Turn the air suspension switch **OFF**.

2. Raise and safely support the vehicle. Make sure the suspension is in full rebound. Remove the wheel and tire assembly.

3. Remove the forward section of the front wheel splash shield far enough to disconnect the connector. Pull out the height sensor clips. Remove the sensor.

4. Disconnect the air spring solenoid connector and air line. Remove the solenoid clip.

5. Rotate the solenoid counterclockwise to the 1st stop. Pull the solenoid straight out slowly to the 2nd stop to bleed air from the system. When air is fully bled from the system, rotate the solenoid to the 3rd stop and remove the solenoid.

6. From inside the engine compartment, remove the appearance cover from over the air spring attachments.

7. Remove the 3 nuts and and the collar retaining the 3 air spring studs to the shock tower. Remove the lower nut and bolt to detach shock from the lower suspension arm. Remove the shock.

NOTE: Do not remove the front shock's large center nut. It may cause a permanent air leak.

To install:

8. Insert the solenoid into the air spring end cap and rotate clockwise to the 3rd stop, push into the 2nd stop, then rotate clockwise to the 1st stop.

9. Install the solenoid clip. Position and fasten the 3 upper air spring studs through the shock tower and hand-tighten the 3 nuts and the collar. The solenoid socket for the left or right spring should be oriented forward of the spring.

10. Place the lower end of the shock over the lower suspension arm and hand-tighten the nut and bolt.

11. Tighten the 3 upper attachment bolts to 17–23 ft. lbs. (23–32 Nm). Attach the appearance cover over the top of the shock tower.

12. Connect the air line and electrical connector to the solenoid. Install the height sensor and connect the connector.

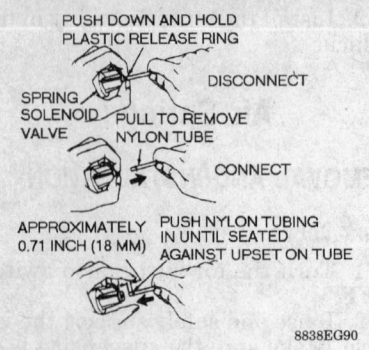

PUSH DOWN AND HOLD
PLASTIC RELEASE RING

SPRING
SOLENOID
VALVE

DISCONNECT

PULL TO REMOVE
NYLON TUBE

CONNECT

APPROXIMATELY
0.71 INCH (18 MM)

PUSH NYLON TUBING
IN UNTIL SEATED
AGAINST UPSET ON TUBE

8838EG90

Air line disconnect/connect procedure — Mark VIII

13. Refill the air spring as follows:

a. Make sure the air suspension switch is turned **OFF**.

b. Connect SUPER STAR II diagnostic tool, or equivalent, to the data link connector located on the front side of the right shock tower.

c. Connect a battery charger to reduce battery drain.

d. Enter function test to pump appropriate air springs. Left front spring = 212. Right front spring = 214. Any further leveling will be done when the vehicle is in normal operation.

14. Install the tire and wheel assembly. Lower the vehicle.

15. Neutralize the front suspension bushings by pushing down then releasing the front end.

16. Tighten the lower shock nut to 199–243 ft. lbs. (170–230 Nm).

17. Turn the air suspension switch **ON**. Check the front end alignment.

Upper Ball Joints

REMOVAL AND INSTALLATION

Mark VIII, Thunderbird and Cougar

The ball joint is an integral part of the upper control arm. If the ball joint is defective, the entire upper control arm must be replaced.

Lower Ball Joints

REMOVAL AND INSTALLATION

Mustang

The ball joint is an integral part of the lower control arm. If the ball joint is defective, the entire lower control arm must be replaced.

Mark VIII, Thunderbird and Cougar

1. Remove the lower control arm.

2. Remove and discard the joint boot seal.

3. Press out the ball joint using ball joint remover tool D89P-3010-A and cup tool D84P-3395-A4 or equivalent, and a suitable press.

To install:

4. When installing a new ball joint, leave the protective cover in place during installation to protect the ball joint seal. It may be necessary to cut off the end of the cover to allow it to pass through the receiving cup.

5. Install the ball joint with ball joint replacer tool D89P-3010-B, cup tool D84P-3395-A4 or equivalent, and a suitable press.

6. Make sure the ball joint is fully seated in the control arm and the ball joint seal is free of cuts or tears.

7. Install the lower control arm. Check the front end alignment.

Upper Control Arms

REMOVAL AND INSTALLATION

Mark VIII, Thunderbird and Cougar

1. Raise and safely support the vehicle. Remove the wheel and tire assembly.

2. On Mark VIII, disconnect the height sensor.

3. Remove and discard the upper spindle-to-ball joint bolt and nut. Slightly spread the spindle at the slot and remove the ball joint.

4. Lower the vehicle. Break off the flags on the upper control arm pivot bolt heads.

5. Remove the upper control arm bolts and the control arm.

To install:

6. Position the upper control arm and install new bolts without the flags and nuts.

7. Hold the upper control arm at a horizontal position and tighten the nuts to 72–88 ft. lbs. (98–120 Nm).

NOTE: If it is necessary to tighten the bolts, due to nut access, tighten the bolts to 82–88 ft. lbs. (110–120 Nm).

8. Raise the vehicle. Attach the spindle to the upper control arm. Install a new bolt and nut from the front of the vehicle and tighten to 59–66 ft. lbs. (80–90 Nm).

9. On Mark VIII, connect the height sensor.

10. Install the wheel and tire assembly and lower the vehicle. Check the front end alignment.

Lower Control Arms

REMOVAL AND INSTALLATION

Mustang

1. Raise and safely support the vehicle. Allow the control arms to hang free. Remove the wheel and tire assembly.

2. If necessary, remove the brake caliper and suspend with a length of wire; do not let the caliper hang by the brake hose. Remove the brake rotor and dust shield.

3. Disconnect the tie rod end from the steering spindle. Disconnect the stabilizer bar link from the lower arm.

4. Remove the steering gear bolts and lower the gear aside to provide clearance, if necessary, for suspension arm bolt removal.

5. Remove the cotter pin and loosen the lower ball joint stud nut 1–2 turns. Do not remove the nut at this time. Tap the spindle boss sharply to relieve the stud pressure.

6. On Mustang, install a suitable spring compressor and compress the spring so it is free in the seat.

7. Remove and discard the ball joint nut and raise the entire strut and spindle assembly. Wire aside to obtain working room.

8. Remove and discard the control arm-to-crossmember nuts and bolts. Remove the lower control arm and, on Mustang, remove the coil spring.

To install:

9. On Mustang, position the coil spring into the lower arm pocket. Make sure the spring pigtail is positioned between the 2 holes in the pocket.

10. Position the lower arm to the crossmember and install new arm-to-crossmember bolts and nuts. Do not tighten at this time.

11. Remove the wire from the strut and spindle assembly and attach the spindle to the ball joint stud. Install a new ball joint stud nut, but do not tighten at this time.

12. On Mustang, raise the control arm with a jack to a normal attitude position and remove the spring compressor.

13. With the jack in place, tighten the lower arm-to-crossmember attaching nuts to 110–150 ft. lbs. (149–203 Nm).

14. Tighten the ball joint stud nut to 100–120 ft. lbs. (136–163 Nm) and install a new cotter pin. Remove the jack.

15. Install the dust shield, rotor and brake caliper, if removed. Install

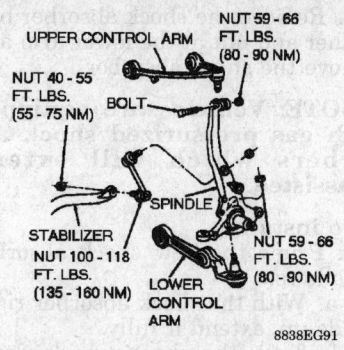

UPPER CONTROL ARM
NUT 59 - 66 FT. LBS. (80 - 90 NM)
NUT 40 - 55 FT. LBS. (55 - 75 NM)
BOLT
SPINDLE
STABILIZER
NUT 100 - 118 FT. LBS. (135 - 160 NM)
LOWER CONTROL ARM
NUT 59 - 66 FT. LBS. (80 - 90 NM)
8838EG91

Front suspension assembly — Mark VIII, Thunderbird and Cougar

the steering gear-to-crossmember bolts and nuts, if removed. Hold the bolts and tighten the nuts to 90–100 ft. lbs. (122–136 Nm).

16. Position the tie rod into the steering spindle and install the retaining nut. Tighten the nut to 35 ft. lbs. (47 Nm) and continue tightening the nut to align the next castellation with the hole in the stud. Install a new cotter pin.

17. Connect the stabilizer bar link to the lower control arm. Tighten the retaining nut to 6–17 ft. lbs. (8–24 Nm).

18. On Mustang, install the wheel and tire assembly and lower the vehicle. Check the front end alignment.

NOTE: Any further vehicle leveling will be done when the vehicle is in normal operation on the ground.

Mark VIII, Thunderbird and Cougar

1. Raise and safely support the vehicle. Remove the wheel and tire assembly.

2. Loosen the ball joint nut 3–4 turns. Rap the spindle to separate the ball joint. Leave the nut attached.

3. Support the spindle with a wire. Mark the position of the camber adjusting cam. Remove and discard the nut attaching the tension strut to the control arm.

4. Remove the lower strut bolt and remove the pivot bolt.

5. Remove the ball joint nut and remove the control arm.

To install:

6. Position the control arm in the vehicle and loosely install the pivot bolt and new nut.

7. Install the tension strut washer and insulators and loosely install the strut to control arm attaching nut.

8. Loosely install a new ball joint nut. Install a new lower strut bolt and nut, but do not tighten at this time.

9. Tighten the ball joint nut to 82–118 ft. lbs. (110–160 Nm). Tighten the tension strut to control arm nut to 103–118 ft. lbs. (140–160 Nm).

10. Remove the wire holding the spindle. Install the wheel and tire assembly and lower the vehicle.

11. Push down on the front of the vehicle and release to neutralize the suspension. Tighten the lower strut nut to 140–162 ft. lbs. (190–220 Nm).

NOTE: The lower strut nut must be tightened with the vehicle weight on the wheels.

12. Align the camber marks at the pivot bolt and tighten the nut to 98–114 ft. lbs. (135–155 Nm).

13. Check the front end alignment.

Sway Bar

REMOVAL AND INSTALLATION

Mustang

1. Raise the front of the vehicle and place jackstands under the lower control arms.

2. Disconnect the stabilizer bar from the links and the insulator mounting clamps. Remove the stabilizer bar.

3. Cut the worn insulators from the stabilizer bar.

4. Installation is the reverse of the removal procedure. Coat the necessary parts of the stabilizer bar with rubber lubricant prior to installation.

Mark VIII, Thunderbird and Cougar

1. Disconnect the negative battery cable.

2. Remove the air inlet tube. Remove the stabilizer bar retaining bracket bolts and brackets.

3. Remove the serpentine drive belt. Raise and safely support the vehicle.

4. Remove the wheel and tire assemblies. Remove the crankshaft vibration damper.

5. Remove the cotter pins and nuts from the tie rod ends. Separate the tie rod ends from the spindles.

6. Remove the transmission oil cooler line bracket. Remove the stabilizer bar to lower link retaining nuts.

7. Remove the stabilizer bar link from the stabilizer bar using joint separator tool D88L-3006-A or equivalent. Be careful not to damage the ball joint seal.

8. Remove the stabilizer bar through the right wheel opening. Remove the bushings from the stabilizer bar.

To install:

9. Install the bushings onto the stabilizer bar and position the bar in the vehicle.

10. Attach the stabilizer links to the bar and tighten the retaining nuts to 48–55 ft. lbs. (65–75 Nm).

11. Install the stabilizer bar bracket and retaining bolts. Tighten to 48–55 ft. lbs. (65–75 Nm).

12. Install the transmission oil cooler lines. Install the tie rod ends to the spindles. Tighten the nuts to 39–54 ft. lbs. (53–73 Nm) and install new cotter pins.

13. Install the crankshaft damper. Install the wheel and tire assemblies and lower the vehicle.

14. Install the serpentine drive belt and the air inlet tube. Connect the negative battery cable.

Front Wheel Bearings

ADJUSTMENT

Mustang

1. Raise and safely support the front of the vehicle.

2. Remove the wheel cover and grease cap.

3. Remove the cotter pin and nut retainer.

4. Loosen the adjusting nut 3 turns and rock the wheel back and forth a few times to release the brake pads from the rotor.

5. While rotating the wheel and hub assembly in a counterclockwise direction, tighten the adjusting nut to 17–25 ft. lbs. (23–34 Nm).

6. Back off the adjusting nut ½ turn, then retighten to 10–28 inch lbs. (1.1–3.2 Nm).

7. Install the nut retainer and a new cotter pin. Check the wheel rotation. If it is noisy or rough, the bearings either need to be cleaned and repacked or replaced. After adjustment is completed, replace the grease cap.

8. Lower the vehicle. Before driving the vehicle, pump the brake pedal several times to restore normal brake pedal travel.

Mark VIII, Thunderbird and Cougar

The front wheel bearings are of a hub unit design and are pregreased, sealed and require no maintenance. The bearings are preset and cannot be adjusted.

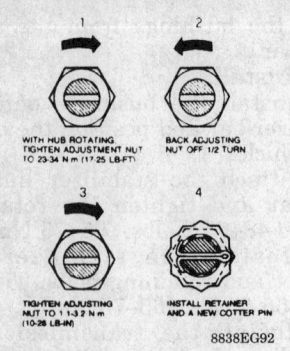

Wheel bearing adjustment procedure — Mustang

REMOVAL AND INSTALLATION

Mustang

1. Raise and support the vehicle safely. Remove the wheel and tire assembly and the caliper. Suspend the caliper with a length of wire; do not let it hang from the brake hose.
2. Pry off the dust cap. Remove the cotter pin, nut retainer, adjusting nut and flat washer. Remove the outer roller bearing assembly.
3. Pull off the brake disc and wheel hub assembly.
4. Remove the inner grease seal using a prybar. Remove the inner roller bearing assembly.
5. Clean the bearings with solvent and inspect them for pits, scratches and excessive wear. Wipe all the old grease from the hub and inspect the bearing races. If either bearings or races are damaged, the bearing races must be removed and the bearings and races replaced as an assembly.
6. If the bearings are to be replaced, drive out the races from the hub using a brass drift.
7. Make sure the spindle, hub and bearing assemblies are clean prior to installation.

To install:

8. If the bearing races were removed, install new ones using a suitable bearing race installer. Pack the bearings with a bearing packer. If a packer is not available, work as much grease as possible between the rollers and cages.
9. Coat the inner surface of the hub and bearing races with grease.
10. Install the inner bearing in the hub. Lubricate the lips of a new seal with grease and install the seal in the hub, using a seal installer.
11. Install the hub/disc assembly on the spindle, being careful not to damage the oil seal.
12. Install the outer bearing, washer and spindle nut. Install the

caliper and the wheel and tire assembly and adjust the bearings.

Mark VIII, Thunderbird and Cougar

1. Raise and safely support the vehicle. Remove the wheel and tire assembly.
2. Remove and discard the grease cap from the hub.
3. Remove the brake caliper. Suspend the caliper with a length of wire; do not let it hang from the brake hose.
4. Remove the rotor. Remove and discard the wheel hub nut.
5. Remove the hub and bearing assembly.

To install:

6. Install the hub and bearing assembly. Install a new wheel hub nut and tighten to 238 ft. lbs. (322 Nm).
7. Install the rotor and a new grease cap. Install the brake caliper.
8. Install the wheel and tire assembly and lower the vehicle.

REAR SUSPENSION

Shock Absorber

REMOVAL AND INSTALLATION

1. On Mark VIII, Thunderbird and Cougar, proceed as follows to remove the actuation assembly, if equipped with automatic ride control:
    a. Make sure the vehicle is on a flat surface and the ignition is in the **OFF** position.
    b. Remove the luggage compartment side trim panel and disconnect the actuator wiring connector.
    c. Squeeze the 2 actuator retaining tabs firmly inward with 1 hand and lift the actuator off the mounting bracket with the other hand.
    d. Grasp the actuator mounting bracket with water pump pliers and hold firmly. While holding the bracket, loosen the bracket retaining nut. Remove the bracket.
2. Raise the vehicle and support it by the rear axle housing. Open the luggage compartment. On Mustang 3-door, open the hatch back door.
3. Remove the trim panels, as necessary, to gain access to the shock absorber. Remove the shock absorber retaining nut washer and insulator.

4. Remove the shock absorber bolt washer and nut at the lower arm and remove the shock absorber.

NOTE: Vehicles are equipped with gas pressurized shock absorbers which will extend unassisted.

To install:

5. Prime the new shock absorber as follows:
    a. With the shock absorber right side up, extend it fully.
    b. Turn the shock upside down and fully compress it.
    c. Repeat the previous 2 steps at least 3 times to make sure any trapped air has been expelled.
6. Place the inner washer and insulator on the upper retaining stud and position the stud through the shock tower mounting hole.
7. Attach the lower end of the shock absorber with the retaining bolt and nut. Tighten the bolt to 45–60 ft. lbs. (61–81 Nm) on Mustang with handling package or 57–70 ft. lbs. (76–96 Nm) on Mustang without handling package. Tighten the nut to 72–97 ft. lbs. (97–132 Nm) on Thunderbird and Cougar or to 82–113 ft. lbs. (113–153 Nm) on Mark VIII.
8. Install the upper insulator, washer and retaining nut and tighten to 20–25 ft. lbs. (26–35 Nm) on Mustang or 27–35 ft. lbs. (37–47 Nm) on Thunderbird and Cougar or to 17–23 ft. lbs. (23–31 Nm) on Mark VIII.
9. On Mark VIII, Thunderbird and Cougar, install the shock actuator, if necessary.
10. Lower the vehicle.

Coil Spring

REMOVAL AND INSTALLATION

Mustang

1. Raise and safely support the vehicle. Support the body at the rear body crossmember.
2. Remove the stabilizer bar, if equipped.
3. Support the axle with a suitable jack or jackstands.
4. Place another jack under the lower arm axle pivot bolt. Remove and discard the bolt and nut. Lower the jack slowly until the coil spring load is relieved.
5. Remove the coil spring and insulator from the vehicle.

To install:

6. Place the upper spring insulator on top of the spring. Place the lower spring insulator on the lower arm.

7. Position the coil spring on the lower arm spring seat, so the pigtail on the lower arm is at the rear of the vehicle and pointing toward the left side of the vehicle.

8. Slowly raise the jack until the arm is in position. Insert a new rear pivot bolt and nut with the nut facing outward. Do not tighten at this time.

9. Raise the axle to curb height. Tighten the lower arm-to-axle pivot bolt to 70–100 ft. lbs. (95–135 Nm).

10. Install the stabilizer bar, if equipped. Remove the crossmember supports and lower the vehicle.

Thunderbird and Cougar

1. Raise and safely support the vehicle. Remove the rear wheel and tire assembly.

2. Remove the rear stabilizer bar link nuts at both ends of the bar. Rotate the bar up and out of the way.

3. Disconnect the parking brake cable at the brake caliper.

4. Install 3 spring cage tools 086-00031 or equivalent to the rear spring as follows:

   a. Install 1 spring cage without an adjuster link to the inboard side, the innermost "bend" of the spring.

   b. Install 2 more spring cages, with adjusters, at 120 degree angles to the previously installed cage.

5. Place a jack under the lower rear control arm as far outboard as possible.

6. Support the rear knuckle and caliper assembly by wiring the upper control arm to the body.

7. Remove the lower shock absorber mounting bolt and nut. Mark the toe adjustment cam-to-subframe position and loosen both inboard pivot bolts on the lower control arm.

NOTE: The control arm must not be lowered until the pivot bolts are loose. Do not attempt to remove the plastic cap on the front pivot nut.

8. Remove the 2 bolts and nuts attaching the lower control arm to the knuckle. Lower the control arm by lowering the jack. Make sure the spring cages properly seat on the spring as the control arm is dropped.

9. Remove the jack, pull the control arm down fully by hand and remove the rear spring with the cages in place. Remove the spring insulators, if necessary.

10. If the springs are to be replaced, use a suitable coil spring compressor to compress the spring and remove the spring cages.

To install:

11. If a new spring is to be installed, it 1st must be compressed and caged. Compress the spring to the length of the original spring. If replacing a broken spring, compress the spring to approximately 10½ in. (267mm).

12. Install the spring insulators, if removed. Install the spring, with the cages in place, onto the upper and lower control arm seats.

NOTE: The short cage, without the adjuster, must be inboard. the spring pigtails may be in any position.

13. Position 2 jackstands under the front bumper reinforcement to prevent the rear of the vehicle lifting off the hoist.

14. Position a jack under the lower control arm and raise the lower control arm up to the knuckle bores. Make sure the spring seats properly. Install the bolts and nuts attaching the lower control arm to the knuckle and tighten the bolts to 118–148 ft. lbs. (160–200 Nm).

15. Remove the wire supporting the knuckle, caliper and upper control arm. Install the lower shock absorber mount bolt and nut and tighten the nut to 110–120 ft. lbs. (150–162 Nm).

16. Remove the jack and the jackstands. Remove the spring cages.

17. Connect the parking brake cable to the caliper. Install the rear stabilizer bar links and retaining nuts.

18. Install the wheel and tire assembly and lower the vehicle.

19. Set the toe adjustment cam to the mark made at the time of removal. Tighten the front lower control arm-to-sub-frame nut to 185–228 ft. lbs. (250–310 Nm). Tighten the rear lower control arm-to-sub-frame nut to 126–169 ft. lbs. (170–230 Nm).

20. Check the rear wheel toe setting and adjust as necessary.

Air Springs

REMOVAL AND INSTALLATION

Mark VIII

1. Turn the air suspension switch **OFF**.

2. Raise and safely support the vehicle on the frame. The suspension must be at full rebound.

3. Remove the wheel and tire assembly. Remove the air spring solenoid as follows:

   a. Disconnect the electrical connector and then disconnect the air line.

   b. Remove the solenoid clip.

   c. Rotate the solenoid counterclockwise to the 1st stop.

   d. Pull the solenoid straight out slowly to the 2nd stop to bleed air from the system.

NOTE: Do not fully release the solenoid until the air is completely bled from the air spring.

   e. After the air is fully bled from the system, rotate counterclockwise to the 3rd stop and remove the solenoid from the solenoid housing. Remove the large O-ring from the solenoid housing.

4. Press the 4 plastic locking fingers in the bottom of the air spring's piston and remove the piston from the lower suspension arm.

5. Remove the air spring.

To install:

6. Install the air spring solenoid as follows:

   a. Check the solenoid O-rings for cuts or abrasion. Replace the O-rings as required. Lightly grease the O-ring area of the solenoid and the larger solenoid housing O-ring with silicone dielectric compound.

   b. Insert the solenoid into the air spring end cap and rotate clockwise to the 3rd stop, push in to the 2nd stop, then rotate clockwise to the 1st stop.

   c. Install the solenoid clip. Inspect the wire harness connector and ensure the rubber gasket is in place at the bottom of the connector cavity.

7. For left side installations, position the notch on the collar to be in-line with the centerline of the solenoid. For right side installations, the flat on the collar is to be in-line with the centerline of the solenoid.

8. Insert the air spring piston's 4 plastic locking fingers into the lower suspension arm until they lock in place. Make sure all 4 fingers are locked in place.

9. Connect the air line and electrical connector to the solenoid.

10. Align and secure the lower arm-to-spring attachment with the suspension at full rebound and supported by the shock absorbers.

NOTE: The air springs may be damaged if the suspension is allowed to compress before the spring is inflated.

11. Refill the air spring as follows:

a. Make sure the air suspension switch is turned **OFF**.

b. Connect SUPER STAR II diagnostic tool, or equivalent, to the data link connector located on the front side of the right shock tower.

c. Connect a battery charger to reduce battery drain.

d. Enter function test to pump appropriate air springs. Left rear spring = 216. Right rear spring = 218. Any further leveling will be done when the vehicle is in normal operation.

12. Install the wheel and tire assembly and lower the vehicle. Turn the air suspension switch **ON**.

NOTE: Any further vehicle leveling will be done when the vehicle is in normal operation on the ground.

Rear Control Arms

REMOVAL AND INSTALLATION

Mustang

Upper Arm

NOTE: If 1 arm needs to be replaced, replace the other arm also.

1. Raise and safely support the vehicle at the rear crossmember.
2. Remove and discard the upper arm pivot bolts and nuts and remove the control arm.

To install:

3. Place the upper arm into the bracket of the body side rail. Install a new pivot bolt and nut with the nut facing outboard. Do not tighten at this time.
4. Using a jack, raise the suspension until the upper arm-to-axle pivot hole is in position with the hole in the axle bushing. Install a new pivot bolt and nut with the nut facing inboard. Do not tighten at this time.
5. Raise the suspension to curb height. Tighten the front upper arm bolt to 77–105 ft. lbs. (104–142 Nm) and the rear upper arm bolt to 70–100 ft. lbs. (95–135 Nm).
6. Remove the supports and lower the vehicle.

Lower Arm

NOTE: If 1 arm needs to be replaced, replace the other arm also.

1. Raise and safely support the vehicle at the rear crossmember.

2. Remove the stabilizer bar, if equipped.
3. Place a jack under the lower arm-to-axle pivot bolt. Remove and discard the bolt and nut. Lower the jack slowly until the coil spring can be removed.
4. Remove and discard the lower arm-to-frame pivot bolt and nut. Remove the lower arm.

To install:

5. Position the lower arm assembly into the front arm bracket. Install a new pivot bolt and nut with the nut facing outwards. Do not tighten at this time.
6. Position the coil spring on the lower arm spring seat, so the pigtail on the lower arm is at the rear of the vehicle and pointing toward the left side of the vehicle.
7. Slowly raise the jack until the arm is in position. Insert a new rear pivot bolt and nut with the nut facing outward. Do not tighten at this time.
8. Raise the axle to curb height. Tighten the lower arm front bolt to 77–105 ft. lbs. (104–142 Nm) and the rear bolt to 70–100 ft. lbs. (95–135 Nm).
9. Install the stabilizer bar, if equipped. Remove the crossmember supports and lower the vehicle.

Thunderbird and Cougar

Upper Arm

1. Raise and safely support the vehicle. Remove the rear wheel and tire assembly.
2. Support the knuckle and hub assembly so it cannot swing outward.
3. Remove the inner and outer pivot bolts and nuts at the upper control arm and remove the arm.

To install:

4. Install the upper arm. Loosely install the bolts and nuts.

NOTE: The inner pivot bolt used for camber adjustment has a specially shaped washer under the bolt head. Make sure fasteners are used in the correct locations.

5. Install the wheel and tire assembly and lower the vehicle.
6. Tighten the outboard nut to 118–148 ft. lbs. (160–200 Nm).
7. Set the camber and tighten the inner pivot nut to 50–68 ft. lbs. (68–92 Nm).

Lower Arm

1. Remove the coil spring.

2. Remove the inner control arm pivot bolts and nuts and remove the arm.

NOTE: Do not attempt to remove the plastic cap on the front pivot nut.

3. Remove the toe compensating link from the control arm.

To install:

4. Inspect the large nut used at the inner front arm attachment for condition of plastic cap. Use a new nut if the cap is cracked, loose or missing.
5. Install the toe compensating link on the arm.
6. Install the lower control arm to the sub-frame and loosely install the pivot bolts and nuts.
7. Tighten the toe compensating link nut to 118–148 ft. lbs. (160–200 Nm).
8. Install the spring and reattach the control arm at the knuckle.
9. Check and adjust the rear toe.

Mark VIII

Upper Arm

1. Raise and safely support the vehicle. Remove the rear wheel and tire assembly.
2. Support the knuckle and hub assembly so it cannot swing outward.
3. Remove the inner and outer pivot bolts and nuts at the upper control arm and remove the arm.

To install:

4. Install the upper arm. Loosely install the bolts and nuts.

NOTE: The inner pivot bolt used for camber adjustment has a specially shaped washer under the bolt head. Make sure fasteners are used in the correct locations.

5. Install the wheel and tire assembly and lower the vehicle.
6. Tighten the outboard nut to 116–142 ft. lbs. (157–192 Nm).
7. Set the camber and tighten the inner pivot nut to 50–68 ft. lbs. (68–92 Nm).

Lower Arm

1. Remove the rear knuckle/hub assembly.
2. Deflate the rear air springs. On left side only, disengage the height sensor from the lower arm ballstud attachment. Make sure the integral lower air spring clip is disengaged from lower arm.
3. Disengage the rear stabilizer bar straps from the emergency brake cables and remove the stabilizer bar link nuts at the lower arm attach-

ment ends. Remove the link bushings, push the link ends up and out of the lower arm and rotate the bar aside.

4. Remove the inner control arm pivot bolts and nuts and remove the arm.

NOTE: Do not attempt to remove the plastic cap on the front pivot nut.

5. Mark the toe adjustment cam-to-subframe position. Loosen the inboard pivot nuts at the lower arm to subframe position. Do not remove the bolt.

6. Remove the shock to lower arm bolt, nut and washer. Note the location of the washers.

7. Remove the inner pivot bolts and nuts. Remove the control arm from the vehicle.

To install:

8. Inspect the large nut used at the inner front arm attachment for condition of plastic cap. Use a new nut if the cap is cracked, loose or missing.

9. Install the toe compensating link on the arm.

10. On left side only, install height sensor bracket assembly to lower control arm.

11. Install the lower control arm to the sub-frame and loosely install the pivot bolts and nuts.

12. Attach the shock and tighten to 83–113 ft. lbs. (113–153 Nm).

13. Install air spring assembly into the lower arm spring pocket opening and make sure the base of the air spring is fully seated and the air spring is not kinked.

14. Position the control arm to curb/design position and tighten lower arm to subframe to 166–203 ft. lbs. (225–275 Nm). Align the inner pivot cam bolt to the marks made during removal. Tighten to 141–191 ft. lbs. (191–258 Nm).

15. Tighten the toe compensating link nut to 83–113 ft. lbs. (113–153 Nm).

16. Install the stabilizer bar. tighten nuts to 7.5–10.2 ft. lbs. (10.2–13.8 Nm).

17. On left side only, install height sensor to lower arm bracket.

18. Install the knuckle/hub assembly. Make sure inflated air spring is properly seated and centered in the lower arm pocket. Inflate the air springs.

19. Check and adjust the rear toe.

Wheel Bearings

REMOVAL AND INSTALLATION

Except Mark VIII, Thunderbird and Cougar

1. Raise and safely support the vehicle. Remove wheel and tire assembly and remove brake drum or brake rotor.

2. If equipped, remove the anti-lock brake speed sensor.

3. Clean all dirt from the area of the carrier cover. Drain the axle lubricant by removing the housing cover.

4. Remove differential pinion shaft lock bolt and pinion shaft.

5. Push flanged end of axle shafts toward the center of the vehicle and remove the C-lock from button end of the axle shaft. Remove the axle shaft from the housing, being careful not to damage the oil seal.

6. Insert wheel bearing and seal replacer tool T85L-1225-AH or equivalent, in the bore and position it behind the bearing so the tangs on the tool engage the bearing outer race. Remove bearing and seal as a unit using an impact slide hammer.

To install:

7. Lubricate the new bearing with rear axle lubricant. Install the bearing into the housing bore using a suitable bearing installer.

8. Install a new axle seal using a seal installer.

NOTE: On 8.8 in. axle, check for the presence of an axle shaft O-ring on the spline end of the shaft and install, if not present.

9. Carefully slide the axle shaft into the axle housing, without damaging the bearing or seal assembly. Start the splines into the side gear and push firmly until the button end of the axle shaft can be seen in the differential case.

10. Install the C-lock on the button end of the axle shaft splines, then push the shaft outboard until the shaft splines engage and the C-lock seats in the counterbore of the differential side gear.

11. Insert the differential pinion shaft through the case and pinion gears, aligning the hole in the shaft with the lock bolt hole. Apply a suitable locking compound to the lock bolt and install in the case and pinion shaft. Tighten to 15–30 ft. lbs. (20–41 Nm).

12. Cover the inside of the differential case with a shop rag and clean the machined surface of the carrier and cover. Remove the shop rag.

13. Apply a bead of silicone sealer to the cover and install on the carrier. Tighten the bolts in a criss-cross pattern. Final torque the cover retaining bolts to 25–34 ft. lbs. (34–47 Nm) if the cover is metal or 15–19 ft. lbs. (20–27 Nm) if the cover is plastic.

14. Add rear axle lubricant to the carrier to a level $1/4$–$9/16$ in. below the bottom of the fill hole. Install the filler plug and tighten to 15–30 ft. lbs. (20–41 Nm).

15. Install the anti-lock speed sensor, if equipped. Tighten the retaining bolt to 40–60 inch lbs. (4.5–6.8 Nm).

16. Install the brake calipers and rotors or the brake drums, as required. Install the wheel and tire assembly and lower the vehicle.

Mark VIII, Thunderbird and Cougar

NOTE: A new hub retainer nut must be used in this procedure.

1. Remove the wheelcover/hub cover from the wheel and tire assembly and loosen the lug nuts.

2. Remove and discard the hub nut and washer.

3. Raise and safely support the vehicle. Remove the wheel and tire assembly.

4. Use needle-nose pliers to slide the parking brake cable adjusting clip downward, until the cable is free.

5. If equipped with disc brakes, remove the parking brake cable from the caliper.

6. Remove the caliper from the disc brake rotor, leaving the brake hose connected. Wire the caliper to the brake line junction bracket; do not let it hang by the brake hose.

7. Remove the brake rotor or brake drum except on Mark VIII.

8. If equipped with disc brakes, remove the splash shield.

9. If equipped with drum brakes, disconnect the parking brake cable and disconnect the brake line from the wheel cylinder.

10. Remove the upper control arm nut and bolt. Wire the upper control arm to the body to prevent damage to the CV-joint boots when the knuckle and hub assembly is removed.

11. Attach hub removal tool T81P-1104-C or other equivalent puller tool, to the hub studs and turn the tool shaft until the halfshaft is free in the hub.

12. Mark the position of the control arm in relation to the knuckle with the bushings in the relaxed position. When the upper control arm bolt is removed from the knuckle, the lower

arm bushings will return to the relaxed position.

NOTE: Failure to mark the position will cause bushing wind up at assembly resulting in improper ride height. This can cause incorrect alignment and tire wear.

13. If the knuckle is being replaced, note the approximate angle of the knuckle in the relaxed position by measuring the distance from the upper bushing to a point on the vehicle body.

14. Remove the lower control arm-to-knuckle retaining bolts and nuts and remove the knuckle assembly from the halfshaft.

15. Position the knuckle and hub assembly in a vise.

16. If equipped with drum brakes, remove the brake shoes, springs and adjuster from the backing plate. Remove the screws retaining the backing plate to the knuckle.

17. Position a suitable 3-jaw puller on the knuckle and press the hub out of the knuckle.

18. Remove the backing plate and remove the bearing retainer snapring.

19. Position the knuckle and bearing assembly on a press and, using suitable tools, press the bearing from the knuckle.

To install:

20. Place the knuckle in the press and position the bearing in the knuckle bore. Press the bearing into the knuckle using suitable tools.

21. Install the bearing retainer snapring.

22. Position the backing plate on the knuckle with the retaining bolts. Tighten the bolts to 45–59 ft. lbs. (61–81 Nm).

23. Support the knuckle on a suitable fixture. Position the hub on the bearing and press into place using suitable tools.

24. If equipped with drum brakes, install the brake shoes, springs and adjuster.

25. Using a hammer and chisel, drive the bearing dust seal from the outer CV-joint. Using a suitable installation tool, install a new seal on the CV-joint, making sure the seal flange faces out toward the bearing.

26. Place the knuckle and hub assembly on the halfshaft splines and install the lower control arm-to-knuckle bolts and nuts. Position the knuckle so the marks made during the removal procedure align with the marks on the control arm. If a new knuckle is being installed, set the knuckle at the approximate angle noted during removal before tightening the bolts.

27. Push the knuckle and hub assembly firmly onto the halfshaft

splines. Install the upper control arm bolt and nut and tighten the nut to 118–148 ft. lbs. (160–200 Nm).

28. Install a new hub nut and washer and tighten by hand.

29. If equipped with disc brakes, install the splash shield to the knuckle and tighten the retaining bolts to 45–59 ft. lbs. (61–81 Nm).

30. If equipped with drum brakes, connect the brake line to the wheel cylinder and connect the parking brake cable.

31. Install the brake rotor or drum.

32. If equipped with disc brakes, install the caliper over the rotor with the outer brake pad against the rotor, to prevent pinching the piston boot between the inner pad and piston. Install the caliper-to-knuckle bolts and tighten to 44–60 ft. lbs. (59–81 Nm) except Mark VIII or to 64–88 ft. lbs. (87–119 Nm) on Mark VIII.

33. If equipped with disc brakes, connect the parking brake cable to the caliper and install the adjustment clip.

34. Bleed the brake system, if equipped with drum brakes.

35. Install the wheel and tire assembly, lower the vehicle and apply the parking brake.

36. Tighten the hub retainer nut to 250 ft. lbs. (340 Nm).

37. Install the wheelcover/hub cover.

FORD MOTOR CO. 16

Rear Wheel Drive

FORD-Crown Victoria **LINCOLN**-Town Car
MERCURY-Grand Marquis

FIRING ORDERS

NOTE: To avoid confusion, always replace spark plug wires one at a time.

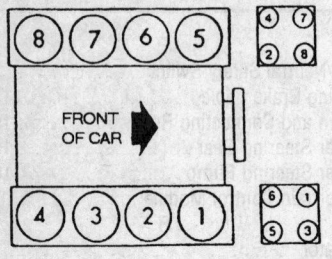

8838FG01

4.6L Engine
Engine Firing Order: 1–3–7–2–6–5–4–8
Distributorless Ignition System

ENGINE ELECTRICAL

NOTE: Disconnecting the negative battery cable on some vehicles may interfere with the functions of the on board computer systems and may require the computer to undergo a relearning process, once the negative battery cable is reconnected.

Distributorless Ignition System (DIS)

The engine is equipped with a distributorless ignition system. The DIS consists of the following components: crankshaft sensor, ignition module, ignition coil packs, the spark angle portion of the ECU and the related wiring.

The DIS eliminates the need for a distributor by using multiple ignition coils. Each coil fires 2 spark plugs at the same time. The plugs are paired so as 1 fires during the compression cycle, the other fires during the exhaust stroke. The next time the coil is fired, the plug that was on exhaust will be on compression and the 1 that was on compression will be on exhaust. The spark in the exhaust cylinder is wasted but little of the coil energy is lost. The ignition coils are mounted together in coil packs. There

are 2 coil packs used, each containing 2 ignition coils.

The crankshaft sensor is a variable reluctance-type sensor triggered by a 36-minus-1 tooth trigger wheel located inside the front cover. The signal generated by this sensor is called a Variable Reluctance Sensor (VRS) signal. The VRS signal provides engine position and rpm information to the ignition module.

The ignition module is a microprocessor that receives input from the crankshaft sensor in regards to engine position, base timing and engine speed and input from the ECU pertaining to spark advance. The ignition module uses this information to direct which coil to fire and to calculate the turn ON and turn OFF times of the coils required to achieve the correct dwell and spark advance.

Base ignition timing is referenced to the position of the crankshaft sensor, and is set at 10 ± 2 degrees BTDC and is not adjustable.

REMOVAL AND INSTALLATION

Crankshaft Sensor

1. Disconnect the negative battery cable.
2. Remove the serpentine belt.
3. Raise and safely support the vehicle.
4. Disconnect the crankshaft sensor and air conditioning compressor electrical connectors from the engine wiring harness.
5. Properly discharge the air conditioning system and remove the air conditioning compressor.
6. Remove the crankshaft position sensor retaining screw and remove the sensor.
To install:
7. Make sure the sensor mounting surface is clean and the sensor O-ring is in the proper location on the sensor assembly.
8. Position the sensor assembly and install the retaining screw. Tighten to 71–106 inch lbs. (8–12 Nm).

NOTE: Do not overtighten the screw.

9. Install the air conditioning compressor. Evacuate and recharge the system according to the proper procedure.
10. Properly route the engine wiring harness and connect the electrical connectors to the air conditioning compressor and crankshaft sensor.

11. Lower the vehicle.
12. Install the serpentine belt and connect the negative battery cable.

Ignition Module

1. Disconnect the negative battery cable.
2. Disconnect the electrical connectors at the module by pushing in on the connector finger ends while grasping the connector body and pulling away from the module.
3. Remove 2 module retaining screws and remove the module.
To install:
4. Position the module to the inner fender and install retaining screws. Tighten the screws to 24–35 inch lbs. (3–4 Nm).
5. Connect the electrical connectors to the module by pushing until the connector fingers are locked over the locking wedge feature on the module.

NOTE: Locking the connector is important to ensure sealing of the connector/module interface.

6. Connect the negative battery cable.

Ignition Coil Pack

1. Disconnect the negative battery cable.
2. Disconnect the electrical connectors from the coil pack and capacitor.
3. Disconnect the spark plug wires by squeezing the locking tabs and twisting while pulling upward.
4. Remove the 4 coil pack retaining bolts and remove the coil pack and capacitor. Save the capacitor for installation with the new coil pack.

NOTE: Apply silicone dielectric compound D7AZ–19A331–A or equivalent, to all spark plug wire boots prior to installation.

5. Installation is the reverse of the removal procedure. Tighten the retaining bolts to 40–61 inch lbs. (5–7 Nm).

Ignition Timing

ADJUSTMENT

NOTE: Always refer to the Vehicle Emission Information Label to verify the timing adjustment procedure.

Base timing for the distributorless engine is set by the factory at 10 ± 2 degrees BTDC and is not adjustable.

Generator

NOTE: To conform with J1930 standardized terminology, Ford Motor Co. will refer to an alternator as a generator.

PRECAUTIONS

Several precautions must be observed with generator equipped vehicles to avoid damage to the unit.

• If the battery is removed for any reason, make sure it is reconnected with the correct polarity. Reversing the battery connections may result in damage to the 1-way rectifiers.

• When utilizing a booster battery as a starting aid, always connect the positive to positive terminals and the negative terminal from the booster battery to a good engine ground on the vehicle being started.

• Never use a fast charger as a booster to start vehicles.

• Disconnect the battery cables when charging the battery with a fast charger.

• Never attempt to polarize the generator.

• Do not use test lights of more than 12 volts when checking diode continuity.

• Do not short across or ground any of the generator terminals.

• The polarity of the battery, generator and regulator must be matched and considered before making any electrical connections within the system.

• Never separate the generator on an open circuit. Make sure all connections within the circuit are clean and tight.

• Disconnect the battery ground terminal when performing any service on electrical components.

• Disconnect the battery if arc welding is to be done on the vehicle.

BELT TENSION ADJUSTMENT

The engine is equipped with an automatic belt tensioner. To remove the drive belt, rotate the tensioner away from the belt using a ½ in. breaker bar and remove old belt. To install, position new belt over pulleys, ensure that all V-grooves make proper contact and relax pressure on the tensioner. No adjustment is necessary or possible.

REMOVAL AND INSTALLATION

1. Disconnect negative battery cable.

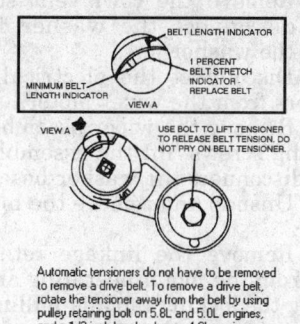

Drive belt tensioner

2. Disconnect the wiring harness attachments to the integral generator/regulator assembly.

3. Disengage the generator drive belt from the generator pulley.

4. Remove generator brace and mounting bolts and lift out generator.

5. Installation is reverse of the removal procedure.

Drive Belt

REMOVAL AND INSTALLATION

1. Disconnect the negative battery cable.

2. Install a ½ inch breaker bar or equivalent, in the square slot in the automatic tensioner arm.

3. Rotate the tensioner away from the drive belt with the breaker bar or similar tool.

4. Lift the drive belt over the alternator pulley flange and remove it.

5. Relax the drive belt tensioner.

6. Inspect the drive belt and replace as needed.

To install:

7. Install the drive belt over all pulleys except the generator pulley flange.

8. Rotate the tensioner away from the drive belt with the breaker bar or similar tool.

9. Install the belt over the generator pulley flange.

10. Relax the drive belt tensioner and remove the tool.

11. Check that the drive belt is routed properly. Make sure the ribs on the belt properly contact the grooves on the pulleys.

12. Connect the negative battery cable.

13. Start the engine and check for proper drive belt operation.

Starter

REMOVAL AND INSTALLATION

────── **WARNING** ──────

When servicing starter or performing any maintenance in the area of the starter, note the heavy gauge input lead connected to the starter solenoid is hot at all times. Make sure the protective cap is installed over the terminal and is replaced after service.

REMOVAL AND INSTALLATION

1. Disconnect the negative battery cable.

2. Raise the vehicle and support it safely.

3. Disconnect the starter cable from the starter. If equipped with starter mounted solenoid, disconnect the push-on connector from the solenoid.

NOTE: To disconnect the hardshell connector from the solenoid S terminal, grasp the plastic shell and pull off; do not pull on the wire. Pull straight off to prevent damage to the connector and S terminal.

4. Remove the upper and lower starter bolts and remove starter.

To install:

5. Position the starter to the engine and tighten the mounting bolts to 15–20 ft. lbs. (20–27 Nm).

6. Reconnect the electrical leads. Be careful to push straight on and make sure connector locks in position with a notable click or detent. Install starter cable nut to starter terminal, tighten to 80–124 inch lbs. (9–14 Nm).

7. Replace red solenoid safety cap. Lower vehicle to the floor.

8. Connect the negative battery cable.

CHASSIS ELECTRICAL

Blower Motor

REMOVAL AND INSTALLATION

1. Disconnect the negative battery cable.

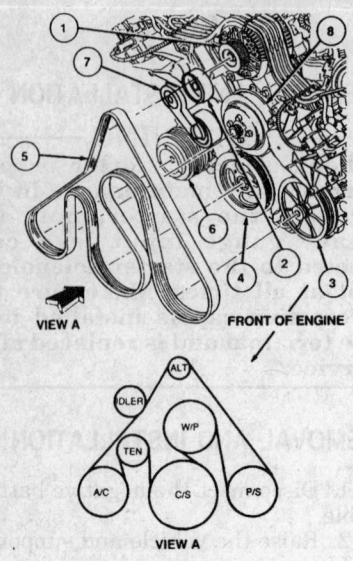

VIEW A   FRONT OF ENGINE

VIEW A

1	Generator
2	Idler
3	Power Steering Pump Pulley
4	Crankshaft Pulley
5	Engine Drive Belt
6	A/C Compressor Assy
7	Tensioner Assy
8	Water Pump Pulley

328758

Drive belt routing

2. Disconnect the blower motor lead connector from the wiring harness connector.

3. Remove the blower motor cooling tube from the blower motor.

4. Remove the 4 retaining screws.

5. Turn the motor and wheel assembly slightly to the right so the bottom edge of the mounting plate follows the contour of the wheel well splash panel. Lift up on the blower and remove it from the blower housing.

6. Installation is the reverse of the removal. Connect the negative battery cable.

Windshield Wiper Motor

REMOVAL AND INSTALLATION

1. Disconnect the negative battery cable.

2. Remove the rear hood seal. Remove the wiper arm assemblies by raising the wiper blade off the windshield. Move the slide latch away from the pivot shaft and slowly lower the arm onto the latch. This unlocks the arm from the pivot shaft and holds the blade off the glass. Pull the arm from the pivot shaft.

3. Remove the cowl vent screws and disconnect the washer hoses from the washer jets.

4. Disconnect the electrical connectors from the wiper motor.

5. Remove the wiper assembly attaching screws, lift the assembly out and disconnect the washer hose.

6. Unsnap and remove the linkage cover.

7. Remove the linkage retaining clip from the motor operating arm by lifting the locking tab and pulling the clip away from the pin.

8. Remove the motor retaining screws and remove the motor from the vehicle.

To install:

9. Installation is the reverse of removal. Install the wiper arms by aligning the wiper arm key with the pivot shaft keyway.

10. Install the arm head over the pivot shaft. While applying downward pressure on the arm head, raise the other end of the arm enough to let the latch slide under the pivot shaft to the latched position, using finger pressure only to slide the latch.

11. Lower the blade. If the blade does not touch the windshield, the slide latch is not completely in place.

Headlight Switch

REMOVAL AND INSTALLATION

Town Car

1. Disconnect the negative battery cable.

2. Remove the headlight switch knob and auto dimmer knob, if equipped.

3. Remove the right and left mouldings from the instrument panel by pulling away from the instrument panel and snapping out of the retainers.

4. Remove 12 screws retaining the finish panel and remove the panel.

5. Remove the 2 headlight switch bracket retaining screws and pull the bracket and switch from the instrument panel.

6. Remove the nut retaining the switch to the bracket, disconnect the connector and remove the switch.

7. Installation is the reverse of the removal procedure.

Crown Victoria and Grand Marquis

1. Disconnect the negative battery cable.

2. Remove the right and left mouldings from the instrument panel by pulling up and snapping out of the retainers.

3. Remove the screws retaining the finish panel to the instrument panel.

4. Remove the headlight switch knob from the shaft by depressing the spring, in the knob slot, with a hooked tool. Remove the finish panel.

5. Remove the 2 headlight bracket retaining screws and pull the bracket and switch from from the instrument panel.

6. Remove the nut retaining the switch to the bracket.

7. Disconnect the electrical connector and remove the switch.

8. Installation is the reverse of removal.

Combination Switch

The combination switch incorporates the turn signal, dimmer and wiper switch functions on the turn signal lever.

REMOVAL AND INSTALLATION

1. Disconnect the negative battery cable.

2. If equipped with tilt column, move to the lowest position and remove the tilt lever.

3. Remove the ignition lock cylinder.

4. Remove the shroud screws and remove the upper and lower shrouds.

5. Remove the 2 self-tapping screws attaching the combination switch to the steering column casting and remove the switch.

6. Remove the wiring harness retainer and disconnect the 2 electrical connectors.

7. Installation is the reverse of the removal procedure.

Ignition Lock Cylinder

REMOVAL AND INSTALLATION

Functional Lock

The following procedure is for vehicles with functioning lock cylinders. Ignition keys are available for these vehicles or the ignition key numbers are known and the proper key can be made.

1. Disconnect the negative battery cable. If equipped, properly disarm the air bag system.

2. Remove the steering column shroud by removing the 4 or 5 self tapping screws. Remove tilt lever if equipped.

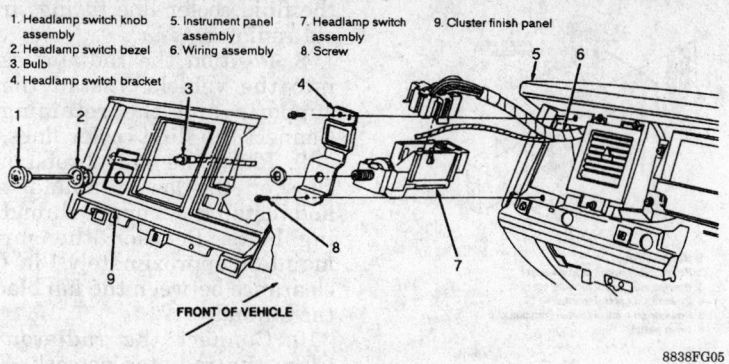

1. Headlamp switch knob assembly
2. Headlamp switch bezel
3. Bulb
4. Headlamp switch bracket
5. Instrument panel assembly
6. Wiring assembly
7. Headlamp switch assembly
8. Screw
9. Cluster finish panel

FRONT OF VEHICLE

8838FG05

Headlight switch installation — Town Car

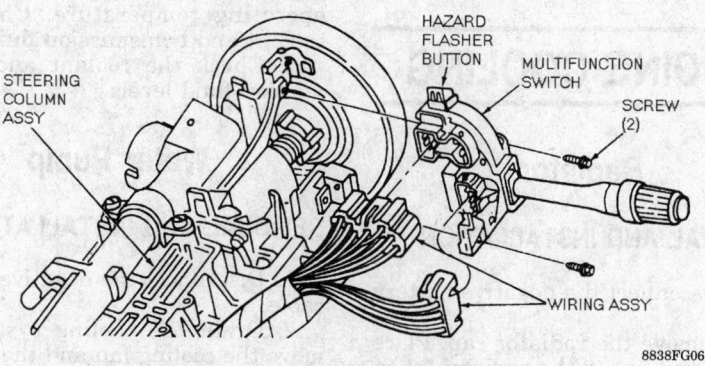

STEERING COLUMN ASSY

HAZARD FLASHER BUTTON

MULTIFUNCTION SWITCH

SCREW (2)

WIRING ASSY

8838FG06

Combination switch installation

3. Turn the ignition to the **RUN** position.

4. Place a ¹⁄₈ in. diameter wire pin or small drift punch in the hole in the casting surrounding the lock cylinder and depress the retaining pin while pulling out on the lock cylinder to remove it from the column housing.

To install:

5. To install the lock cylinder, turn it to the **RUN** position and depress the retaining pin. Insert the lock cylinder into its housing in the lock cylinder casting.

6. Make sure the cylinder is fully seated and aligned in the interlocking washer before turning the key to the **OFF** position. This action will permit the cylinder retaining pin to extend into the hole in the lock cylinder housing.

7. Using the ignition key, rotate the cylinder to ensure the correct mechanical operation in all positions.

8. Check for proper start in **P** or **N**. Also make sure the start circuit cannot be actuated in **D** or **R** positions and that the column is locked in the **LOCK** position.

9. Connect the key warning buzzer electrical connector and install the trim shrouds, if required.

Non-functional Lock

The following procedure is for vehicles with non-functioning locks. On these vehicles, the lock cylinder cannot be rotated due to a lost or broken key, the key number is not known, or the lock cylinder cap is damaged and/or broken, preventing the lock cylinder from rotating.

1. Disconnect the negative battery cable. If equipped, properly disarm the air bag system.

2. Remove the steering wheel.

3. Use channel lock or vise grip type pliers to twist the lock cylinder cap until it separates from the lock cylinder.

4. Drill approximately 1³⁄₄ in. down the middle of the ignition key slot, using a ³⁄₈ in. diameter drill bit, until the lock cylinder breaks loose from the breakaway base of the lock cylinder. Remove the lock cylinder and drill shavings from the lock cylinder housing.

5. Remove the snapring or retainer, washer and steering column lock gear. Thoroughly clean all drill shavings and other foreign materials from the casting.

6. Inspect the lock cylinder housing for damage and replace, as necessary.

To install:

7. Install the ignition lock cylinder and check for smooth operation.

8. Connect the electrical connector to the key warning switch and install the trim shrouds, if necessary.

9. Install the steering wheel and connect the negative battery cable.

Ignition Switch

REMOVAL AND INSTALLATION

1. Disconnect the negative battery cable.

2. Remove the steering column shroud.

3. Remove the instrument panel lower steering column cover.

4. Disconnect the electrical connector from the ignition switch.

5. Rotate the ignition key lock cylinder to the **RUN** position.

6. Remove the 2 screws attaching the ignition switch.

7. Disengage the ignition switch from the actuator pin and remove the switch.

To install:

8. Adjust the new ignition switch by sliding the carrier to the **RUN** position.

9. Check to ensure that the ignition key lock cylinder is in the **RUN** position. The **RUN** position is achieved by rotating the key lock cylinder approximately 90 degrees from the **LOCK** position.

10. Install the ignition switch onto the actuator pin.

11. Align the switch mounting holes and install the attaching screws. Tighten the screws to 50–69 inch lbs. (5.6–7.9 Nm).

12. Connect the electrical connector to the ignition switch.

13. Connect the negative battery cable. Check the ignition switch for proper function in **START** and **ACC** positions. Make sure the column is locked in the **LOCK** position.

14. Install the remaining components in the reverse order of removal.

Park/Neutral Safety Switch

REMOVAL AND INSTALLATION

1. Set the parking brake.

2. Place the selector lever in the manual **L** position.

3. Remove the air cleaner assembly.

4. Disconnect the negative battery cable.

5. Disconnect the neutral safety switch electrical harness from the switch by lifting the harness straight up off the switch without side-to-side motion.

6. Reach in the area of the left hand dash panel, using a 24 inch extension, universal adapter and socket tool T74P–77247–A or equivalent, and remove the neutral safety switch and O-ring.

NOTE: Use of different tools could crush or puncture the walls of the switch.

To install:

7. Install the neutral safety switch and new O-ring using socket tool T74P–77247–A or equivalent.

8. Tighten the switch to 8–11 ft. lbs. (11–15 Nm).

9. Connect the neutral safety switch to the wiring harness.

10. Connect the negative battery cable.

11. Check that the vehicle starts in the **N** or **P** position.

Powertrain Control Module

It is located in the engine compartment, attached to the firewall on the driver's side, near the master cylinder.

REMOVAL AND INSTALLATION

WARNING

Never disconnect a Powertrain Control Module (PCM) with the battery connected. Be sure to wear a grounding device when removing or installing a PCM to prevent damage to the unit due to static electricity.

1. Disconnect the negative battery cable.

2. Wearing a grounding device, loosen the engine control sensor wiring harness connector retaining bolt and carefully disconnect the harness connector from the PCM.

3. Remove the PCM bracket clip from the module bracket and remove the PCM.

To install:

4. Wearing a grounding device, install the PCM to the mounting bracket and secure the bracket clip.

5. Carefully fit the engine control sensor wiring harness connector to the PCM and secure with the retaining bolt. Tighten the retaining bolt to 32 inch lbs. (3.7 Nm).

6. Connect the negative battery cable.

7. Check for proper operation.

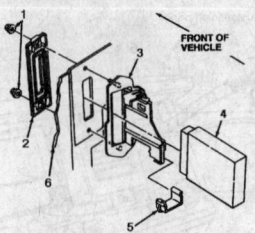

1 Nut (2 req'd)
2 Powertrain control module seal
3 Powertrain control module bracket
4 Powertrain control module
5 Powertrain control module bracket clip
6 Dash panel

327470

Powertrain control module mounting

ENGINE COOLING

Radiator

REMOVAL AND INSTALLATION

1. Disconnect the negative battery cable.

2. Remove the radiator cap. Place a drain pan under the radiator, open the draincock and drain the coolant.

CAUTION

Never remove the radiator cap while the engine is running or personal injury from scalding hot coolant or steam may result. If possible, wait until the engine has cooled to remove the radiator cap. If this is not possible, wrap a thick cloth around the radiator cap and turn it slowly to the first stop. Step back while the pressure is released from the cooling system. When it is certain all the pressure has been released, press down on the cap, still with the cloth, and turn and remove it.

3. Disconnect the upper, lower and coolant reservoir hoses at the radiator.

4. Disconnect the fluid cooler lines at the radiator.

5. Remove the 2 upper fan shroud retaining bolts at the radiator support, lift the fan shroud sufficiently to disengage the lower retaining clips and lay the shroud back over the fan.

6. Remove the radiator upper support retaining bolts and remove the supports. Lift the radiator from the vehicle.

To install:

7. If a new radiator is to be installed, transfer the petcock from the old radiator to the new one. Transfer the fluid cooler line fittings from the old radiator.

8. Position the radiator assembly into the vehicle. Install the upper supports and the retaining bolts. Connect the fluid cooler lines.

9. Place the fan shroud into the clips on the lower radiator support and install the 2 upper shroud retaining bolts. Position the shroud to maintain approximately 1 in. (25mm) clearance between the fan blades and the shroud.

10. Connect the radiator hoses. Close the radiator petcock. Fill and bleed the cooling system.

11. Start the engine and bring to operating temperature. Check for coolant and transmission fluid leaks.

12. Check the coolant and transmission fluid levels.

Water Pump

REMOVAL AND INSTALLATION

1. Disconnect the negative battery cable.

2. Drain the cooling system, remove the cooling fan and the shroud.

3. Release the belt tensioner and remove the accessory drive belt.

4. Remove the 4 bolts retaining the water pump pulley to the water pump and remove the pulley.

5. Remove the 4 bolts retaining the water pump to the engine assembly and remove the water pump.

To install:

6. Installation is the reverse of the removal procedure. Be sure to clean the sealing surfaces of the water pump and block and use a new O-ring. Lubricate the O-ring with clean anti-freeze prior to installation.

7. Tighten the water pump-to-engine bolts and the pulley-to-water pump bolts to 15–22 ft. lbs. (20–30 Nm). Fill and bleed the cooling system. Operate the engine until normal operating temperatures have been reached and check for leaks.

Thermostat

REMOVAL AND INSTALLATION

1. Drain the cooling system to a level below the thermostat.

2. Disconnect the upper radiator hose at the thermostat housing.

3. Remove the 2 thermostat housing retaining bolts and remove the thermostat housing.

4. Remove the thermostat and O-ring seal.

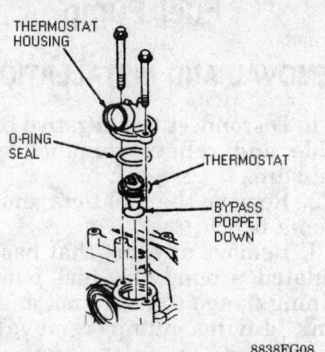

THERMOSTAT
HOUSING

O-RING
SEAL

THERMOSTAT

BYPASS
POPPET
DOWN

8838FG08

Thermostat installation

To install:

5. Installation is the reverse of the removal procedure. Make sure all mating surfaces are clean prior to installation. Use a new O-ring seal.

6. Tighten the thermostat housing retaining bolts to 15–22 ft. lbs. (20–30 Nm). Fill the cooling system. Start the engine and bring to normal operating temperature. Check for leaks.

Electric Cooling Fan

REMOVAL AND INSTALLATION

1. Disconnect the negative battery cable.

2. Unplug the electric cooling fan wiring connector at the side of the fan shroud.

3. Turn the lower fan shroud in the upper fan shroud to allow clearance for shroud removal.

4. Remove the radiator upper sight shield.

5. Loosen the fan shroud from its radiator mounting and remove the lower radiator hose from the supports on the fan shroud.

6. Lift the fan shroud from the vehicle.

7. Remove the retaining screws and remove the fan blade and motor assembly.

To install:

8. Mount the fan blade and motor assembly. Tighten the retaining screws to 27–53 inch lbs. (3–6 Nm).

9. Install the shroud and tighten the retaining screws to 36 inch lbs. (5 Nm).

10. Install the lower radiator hose to the supports on the fan shroud.

11. Plug in the electric cooling fan wiring connector and secure the wire to the shroud.

12. Turn the lower fan shroud in the upper fan shroud to the closed position.

13. Install the radiator upper sight shield.

14. Connect the negative battery cable.

15. Start the engine and operate the A/C system to check cooling fan operation.

Cooling System Bleeding

PROCEDURE

When the entire cooling system is drained, the following procedure should be used to ensure a complete fill.

1. Install the block drain plug, if removed and close the draincock. With the engine OFF, add a 50/50 mixture of water and anti-freeze to the reservoir filler neck seat.

2. Install the radiator cap to the first notch to keep spillage to a minimum.

3. Place the heater temperature selector in the maximum heat position.

4. Start the engine and let it idle until the upper radiator hose is warm. This indicates that the thermostat is open and coolant is flowing through the entire system.

5. Stop the engine and carefully remove the radiator cap. Fill the reservoir to the minimum level with the coolant mixture. Install the pressure cap securely.

FUEL SYSTEM

Fuel System Service Precautions

Safety is the most important factor when performing not only fuel system maintenance but any type of maintenance. Failure to conduct maintenance and repairs in a safe manner may result in serious personal injury or death. Maintenance and testing of the vehicle's fuel system components can be accomplished safely and effectively by adhering to the following rules and guidelines.

• To avoid the possibility of fire and personal injury, always disconnect the negative battery cable unless the repair or test procedure requires that battery voltage be applied.

• Always relieve the fuel system pressure prior to disconnecting any fuel system component (injector, fuel rail, pressure regulator, etc.), fitting or fuel line connection. Exercise extreme caution whenever relieving fuel system pressure to avoid exposing skin, face and eyes to fuel spray. Please be advised that fuel under pressure may penetrate the skin or any part of the body that it contacts.

• Always place a shop towel or cloth around the fitting or connection prior to loosening to absorb any excess fuel due to spillage. Ensure that all fuel spillage (should it occur) is quickly removed from engine surfaces. Ensure that all fuel soaked cloths or towels are deposited into a suitable waste container.

• Always keep a dry chemical (Class B) fire extinguisher near the work area.

• Do not allow fuel spray or fuel vapors to come into contact with a spark or open flame.

• Always use a backup wrench when loosening and tightening fuel line connection fittings. This will prevent unnecessary stress and torsion to fuel line piping. Always follow the proper torque specifications.

• Always replace worn fuel fitting O-rings with new. Do not substitute fuel hose or equivalent, where fuel pipe is installed.

Fuel System Pressure

RELIEVING

Fuel supply lines on all fuel injected engines will remain pressurized for some period of time after the engine is shut OFF. This pressure must be relieved before servicing the fuel system. Pressure is relieved through the fuel pressure relief valve, located on the fuel rail.

To relieve the fuel system pressure, first remove the fuel tank cap to relieve pressure in the tank, then remove the cap on the fuel pressure relief valve. Attach fuel pressure gauge T80L–9974–A or equivalent, and drain the system through the drain tube into a suitable container. Remove the fuel pressure gauge and replace the cap on the relief valve.

Idle Speed

ADJUSTMENT

1. Place the transmission in **P** and apply the parking brake.

2. Start the engine and bring to normal operating temperature. Make sure the heater, air conditioning and all other accessories are OFF.

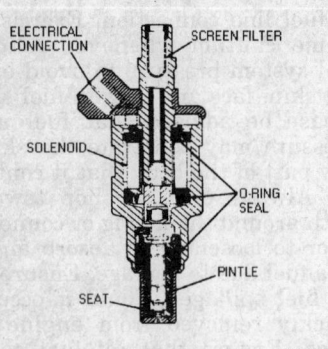

ELECTRICAL CONNECTION — SCREEN FILTER

SOLENOID

O-RING SEAL

PINTLE

SEAT

8838FG09

MFI fuel injector

3. Check and if necessary, adjust the ignition timing.

4. Make sure the fuel pressure is correct. Any indicated vehicle malfunction service codes should be resolved before proceeding further.

5. Connect the SUPER STAR II tester, tool number 007–00028 or other suitable scan tool to the Self-Test connector. Activate the Key On Engine Running (KOER) Self-Test.

6. After Code 1 or 111 has been displayed, unlatch and within 4 seconds, latch the STI button.

7. A single pulse code indicates the entry mode, then observe the Self-Test Output (STO) on the tester for the following:

a. A constant tone, solid light or STO LO readout means the base idle speed is within the correct range. To exit the test, unlatch the STI button, then wait 4 seconds for reinitialization. After 10 minutes, the tool will exit by itself.

b. A beeping tone, flashing light or STO LO readout at 8 Hz indicates the Throttle Position Sensor (TPS) is out of range due to over adjustment. Adjustment may be required.

c. A beeping tone, flashing light or STO LO readout at 4 Hz indicates the base idle speed is too fast and adjustment is required. Proceed to Step 9.

d. A beeping tone, flashing light or STO LO readout at 1 Hz indicates the base idle speed is too low and adjustment is required. Proceed to Step 8.

8. If the idle speed is too low, do not clean the throttle body. Turn the air trim screw counterclockwise until the conditions in Step 7a are satisfied.

9. If the idle speed is too high, do not clean the throttle body. Turn the air trim screw clockwise until the conditions in Step 7a are satisfied.

Mixture

ADJUSTMENT

The idle mixture is controlled by the electronic control unit and cannot be adjusted.

Fuel Filter

REMOVAL AND INSTALLATION

In-line Fuel Filter

1. Disconnect the negative battery cable and relieve the fuel system pressure.

2. Raise and safely support the vehicle.

3. Remove the push connect fittings at both ends of the filter. Install new retainer clips in each push connect fitting.

4. Remove the fuel filter and retainer from the metal bracket. Remove the filter from the retainer. Note that the direction of the flow arrow points to the open end of the retainer. Remove the rubber insulator rings.

To install:

5. Install the rubber insulator rings, place the filter into the retainer with the flow arrow pointing out of the retainer open end, and install the retainer on the metal bracket. Tighten the retaining bolts to 27–44 inch lbs. (3–5 Nm).

6. Install the push connect fittings onto the filter ends. Start the engine and check for leaks.

7. Lower the vehicle.

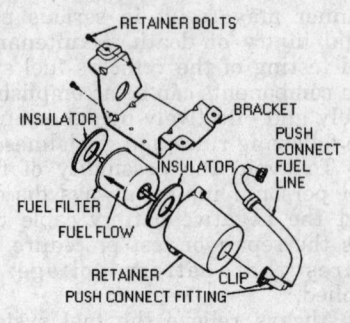

RETAINER BOLTS

INSULATOR

BRACKET

PUSH CONNECT FUEL LINE

INSULATOR

FUEL FILTER

FUEL FLOW

RETAINER

CLIP

PUSH CONNECT FITTING

8838FG10

Inline fuel filter

Fuel Pump

REMOVAL AND INSTALLATION

1. Disconnect the negative battery cable and relieve the fuel system pressure.

2. Remove the fuel tank and place it on a bench.

3. Remove any dirt that has accumulated around the fuel pump retaining flange so it will not enter the tank during pump removal and installation.

4. Turn the fuel pump locking ring counterclockwise and remove the locking ring.

5. Remove the fuel pump and bracket assembly. Remove and discard the seal ring.

To install:

6. Clean the fuel pump mounting flange, fuel tank mounting surface and seal ring groove.

7. Apply a light coating of grease on a new seal ring to hold it in place during assembly and install in the seal ring groove.

8. Install the fuel pump and bracket assembly carefully to ensure the filter is not damaged. Make sure the locating keys are in the keyways and the seal ring remains in the groove.

9. Hold the pump assembly in place and install the locking ring finger-tight. Make sure all the locking tabs are under the tank lock ring tabs.

10. Rotate the locking ring clockwise until the ring is against the stops.

11. Install the fuel tank in the vehicle. Add a minimum of 10 gallons of fuel to the tank and check for leaks.

12. Install a suitable fuel pressure gauge to the valve on the fuel rail.

13. Turn the ignition switch from OFF to ON for 3 seconds. Repeat this procedure 5–10 times until the pressure gauge shows at least 35 psi. Check for fuel leaks.

14. Remove the pressure gauge, start the engine and check for leaks.

Fuel Injector

REMOVAL AND INSTALLATION

1. Disconnect the negative battery cable.

2. Remove the fuel tank cap and relieve the fuel system pressure.

3. Disconnect the vacuum line at the pressure regulator.

4. Disconnect the fuel lines from the fuel rail.

5. Disconnect the electrical connectors from the injectors.

6. Remove the fuel rail assembly retaining bolts.

7. Carefully disengage the fuel rail from the fuel injectors and remove the fuel rail.

NOTE: It may be easier to remove the injectors with the fuel rail as an assembly.

8. Grasping the injector body, pull while gently rocking the injector from side-to-side to remove the injector from the fuel rail or intake manifold.

9. Inspect the pintle protection cap and washer for signs of deterioration. Replace the complete injector, as required. If the cap is missing, look for it in the intake manifold.

NOTE: The pintle protection cap is not available as a separate part.

To install:

10. Lubricate new O-rings with light grade oil and install 2 on each injector.

NOTE: Never use silicone grease as it will clog the injectors.

11. Install the injectors using a light, twisting, pushing motion.

12. Install the fuel rail, pushing it down to ensure all injector O-rings are fully seated in the fuel rail cups and intake manifold.

13. Install the retaining bolts while holding the fuel rail down and tighten to 71–106 inch lbs. (8–12 Nm).

14. Connect the fuel lines to the fuel rail and the vacuum line to the pressure regulator.

15. With the injector wiring disconnected, connect the negative battery cable and turn the ignition switch to the **RUN** position to allow the fuel pump to pressurize the system.

16. Check for fuel leaks.

17. Disconnect the negative battery cable.

18. Connect the electrical connectors to the fuel injectors.

19. Connect the negative battery cable and start the engine. Let it idle for 2 minutes.

20. Turn the engine **OFF** and check for leaks.

EMISSION CONTROLS

Emission Warning Lamps

RESETTING

These vehicles have an CHECK ENGINE lamp that will light when there is a fault in the engine control system. This light cannot be reset without diagnosing the fault in the system. When the system has been diagnosed and the problem corrected, the light will go out.

ENGINE MECHANICAL

NOTE: Disconnecting the negative battery cable on some vehicles may interfere with the functions of the on board computer systems and may require the computer to undergo a relearning process, once the negative battery cable is reconnected.

Engine Assembly

REMOVAL AND INSTALLATION

1. Disconnect the battery cables. Drain the crankcase and the cooling system.

2. Relieve the fuel system pressure and discharge the air conditioning system.

3. Mark the position of the hood on the hinges and remove the hood.

4. Remove the cooling fan, shroud and radiator.

5. Remove the wiper module and support bracket. Remove the air inlet tube.

6. Remove the 42-pin connector from the retaining bracket on the brake vacuum booster. Disconnect the 42-pin connector and transmission harness connector and position aside.

7. Disconnect the accelerator and cruise control cables. Disconnect the throttle valve cable.

8. Disconnect the electrical connector and vacuum hose from the purge solenoid. Disconnect the power supply from the power distribution box and starter relay.

9. Disconnect the vacuum supply hose from the throttle body adapter vacuum port. Disconnect the heater hoses.

10. Disconnect the generator harness from the fender apron and junction block. Disconnect the air conditioning hoses from the compressor.

11. Disconnect the EVO sensor connector from the power steering pump and disconnect the body ground strap from the dash panel.

12. Raise and safely support the vehicle.

13. Disconnect the exhaust system from the exhaust manifolds and support with wire hung from the crossmember.

14. Remove the retaining nut from the transmission line bracket and remove the 3 bolts and stud retaining the engine to the transmission knee braces.

15. Remove the starter. Remove the 4 bolts retaining the power steering pump to the engine block and position aside.

16. Remove the plug from the engine block to access the torque converter retaining nuts. Rotate the crankshaft until each of the 4 nuts is accessible and remove the nuts.

17. Remove the 6 transmission-to-engine retaining bolts. Remove the engine mount through bolts, 2 on the left mount and 1 on the right mount.

18. Lower the vehicle. Support the transmission with a floor jack and remove the bolt retaining the right engine mount to the lower engine bracket.

19. Install an engine lifting bracket to the left cylinder head on the front and the right cylinder head on the rear. Connect engine lifting equipment to the lifting brackets.

20. Raise the engine slightly and carefully separate the engine from the transmission.

21. Carefully lift the engine out of the engine compartment and position on a workstand. Remove the engine lifting equipment.

To install:

22. Install engine lifting brackets as in Step 19. Connect engine lifting equipment to the brackets and remove the engine from the workstand.

23. Carefully lower the engine into the engine compartment. Start the converter pilot into the flexplate and align the paint marks on the flexplate and torque converter. Make sure the studs on the torque converter align with the holes in the flexplate.

24. Fully engage the engine to the transmission and lower onto the mounts. Remove the engine lifting equipment and brackets. Install the bolt retaining the right engine mount to the frame.

25. Raise and safely support the vehicle. Install the 6 engine-to-transmission bolts and tighten to 30–44 ft. lbs. (40–60 Nm).

26. Install the engine mount through bolts and tighten to 15–22 ft. lbs. (20–30 Nm). Install the 4 torque converter retaining nuts and tighten to 22–25 ft. lbs. (20–30 Nm). Install the plug into the access hole in the engine block.

27. Position the power steering pump on the engine block and install the 4 retaining nuts. Tighten to 15–22 ft. lbs. (20–30 Nm). Install the starter.

28. Position the engine to transmission braces and install the 3 bolts and 1 stud. Tighten the bolts and stud to 18–31 ft. lbs. (25–43 Nm).

29. Position the transmission line bracket to the knee brace stud and install the retaining nut. Tighten to 15–22 ft. lbs. (20–30 Nm).

30. Cut the wire and position the exhaust system to the manifolds. Install the 4 nuts and tighten to 20–30 ft. lbs. (27–41 Nm).

NOTE: Make sure the exhaust system clears the No. 3 crossmember. Adjust as necessary.

31. Lower the vehicle and connect the EVO sensor.

32. Connect the air conditioner lines to the compressor and connect the generator harness from the fender apron and junction block.

33. Connect the heater hoses and connect the vacuum supply hose to the throttle body adapter vacuum port.

34. Connect the power supply to the power distribution box and starter relay. Connect the electrical connector and vacuum hose to the purge solenoid.

35. Connect and if necessary, adjust the throttle valve cable. Connect the accelerator and cruise control cables.

36. Connect the 42-pin engine harness connector and transmission harness connector. Install the 42-pin connector to the retaining bracket on the brake vacuum booster.

37. Install the wiper module and support bracket. Connect the fuel lines.

38. Install the radiator, cooling fan and shroud. Install the air inlet tube.

39. Fill the crankcase with the proper type and quantity of engine oil. Fill and bleed the cooling system.

40. Install the hood, aligning the marks that were made during removal. Connect the battery cables.

41. Start the engine and bring to operating temperature. Check for leaks. Check all fluid levels. Leak test, evacuate and charge the air conditioning system according to the proper procedure. Observe all safety precautions.

Engine Mounts

REMOVAL AND INSTALLATION

Front

1. Disconnect the battery cables. Drain the cooling system, relieve the fuel system pressure and discharge the air conditioning system.

2. Remove the air inlet tube and the cooling fan and shroud. Remove the upper radiator hose.

3. Disconnect the fuel lines from the fuel rail. Remove the wiper module and support bracket.

4. Disconnect the air conditioning compressor outlet hose at the compressor and remove the bolt retaining the hose assembly to the right coil bracket.

5. Remove the 42-pin engine harness connector from the retaining bracket on the brake vacuum booster. Disconnect the 42-pin connector and transmission harness connector.

6. Disconnect the throttle valve cable from the throttle body. Disconnect the heater outlet hose.

7. Remove the upper stud and loosen the lower bolt retaining the heater outlet hose to the right cylinder head and position aside.

8. Remove the blower motor resistor. Remove the bolt retaining the right engine mount to the lower engine bracket.

9. Disconnect the vacuum hoses from the EGR valve and EGR tube. Remove the 2 bolts retaining the EGR valve to the intake manifold. Disconnect both oxygen sensors.

10. Raise and safely support the vehicle. Remove the engine mount through bolts, 2 from the left side and 1 from the right.

11. Remove the EGR tube line nut from the right exhaust manifold and remove the EGR valve and tube assembly.

12. Disconnect the exhaust pipes from the manifolds. Lower the exhaust and hang the pipes with wire from the crossmember.

13. Position a jack and a block of wood under the oil pan, rearward of the oil drain hole. Raise the engine approximately 4 in. (100mm).

14. Install a block of wood under the oil pan and lower the engine onto the wood block. Remove 3 retaining bolts each from the right and left engine mounts and remove the mounts.

To install:

15. Position the mounts on the engine block, install 3 retaining bolts and tighten to 45–60 ft. lbs. (60–81 Nm). Raise the engine and remove the wood block.

16. Lower the engine onto the mounts. Position and connect the EGR valve and tube assembly to the exhaust manifold. Tighten the line nut to 26–33 ft. lbs.

NOTE: Loosen the line nut at the EGR valve prior to installing the assembly onto the vehicle. This will allow enough movement to align the EGR valve retaining bolts.

17. Install the engine mount through bolts and tighten to 15–22 ft. lbs. (20–30 Nm).

18. Cut the wire and position the exhaust manifolds. Install the 4 nuts and tighten to 20–30 ft. lbs. (27–41 Nm). Make sure the exhaust system clears the No. 3 crossmember; adjust as necessary.

19. Lower the vehicle and connect the oxygen sensors. Install the bolt retaining the right engine mount to the frame. Tighten to 45–60 ft. lbs. (60–81 Nm).

20. Install a new gasket on the EGR valve and position to the intake manifold. Install the 2 EGR valve retaining bolts and tighten to 15–22 ft. lbs. (20–30 Nm).

21. Tighten the EGR tube line nut at the EGR valve to 26–33 ft. lbs. (35–45 Nm). Connect the vacuum hoses to the EGR valve and tube.

22. Install the blower motor resistor. Position the heater outlet hose. Install the upper stud and tighten the upper stud and lower bolt to 15–22 ft. lbs. (20–30 Nm). Install the ground strap onto the stud and tighten the nut to 15–22 ft. lbs. (20–30 Nm). Connect the heater outlet hose.

23. Connect and if necessary, adjust the throttle valve cable. Connect the 42-pin connector and transmission harness connector. Install the 42-pin engine harness connector to the retaining bracket on the brake vacuum booster.

24. Connect the air conditioning compressor outlet hose to the compressor and install the bolt retaining

the hose assembly to the right coil bracket.

25. Install the upper radiator hose and connect the fuel lines. Install the wiper module and retaining bracket.

26. Install the cooling fan and shroud. Install the air inlet tube.

27. Fill and bleed the cooling system. Connect the battery cables, start the engine and check for leaks. Leak test, evacuate and charge the air conditioning system according to the proper procedure. Observe all safety precautions.

Rear

1. Disconnect the negative battery cable. Raise and safely support the vehicle.

2. Support the transmission with a jack and wood block. Remove the 2 nuts attaching the rear mount to the crossmember.

3. Remove the 2 bolts attaching the mount to the transmission.

4. Raise the transmission with the jack and remove the mount.

To install:

5. Position the mount on the transmission. Install the 2 retaining bolts and tighten to 50–70 ft. lbs. (68–95 Nm).

6. Lower the transmission. Install the rear mount-to-crossmember retaining nuts and tighten to 35–50 ft. lbs. (48–68 Nm).

7. Lower the vehicle and connect the negative battery cable.

Cylinder Head

REMOVAL AND INSTALLATION

1. Disconnect the negative battery cable.

2. Drain the cooling system and remove the cooling fan and shroud.

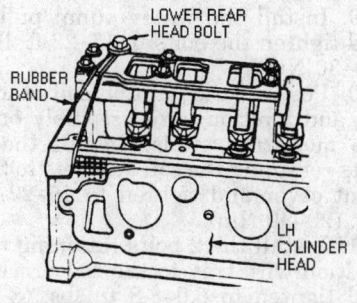

Supporting left lower head bolt with rubberband

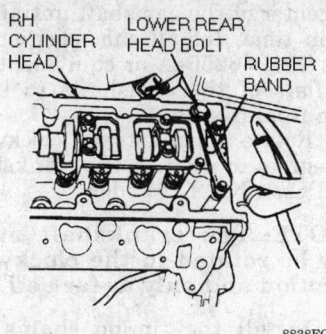

Supporting right lower head bolt with rubberband

3. Relieve the fuel system pressure and disconnect the fuel lines.

4. Remove the air inlet tube and the wiper module. Release the belt tensioner and remove the accessory drive belt.

5. Tag and disconnect the ignition wires from the spark plugs. Disconnect the ignition wire brackets from the camshaft cover studs and remove the 2 bolts retaining the ignition wire tray to the coil brackets.

6. Remove the bolt retaining the air conditioner high pressure line to the right coil bracket. Disconnect both ignition coils and CID sensor.

7. Remove the nuts retaining the coil brackets to the front cover. Slide the ignition coil brackets and ignition wire assembly off the mounting studs and remove from the vehicle.

8. Remove the water pump pulley. Disconnect the generator wiring harness from the junction block, fender apron and generator. Disconnect the bolts retaining the generator to the intake manifold and engine block and remove the generator.

9. Disconnect the positive battery cable at the power distribution box. Remove the retaining bolt from the positive battery cable bracket located on the side of the right cylinder head.

10. Disconnect the vent hose from the canister purge solenoid and position the positive battery cable out of the way. Disconnect the canister purge solenoid vent hose from the PCV valve and remove the PCV valve from the camshaft cover.

11. Remove the 42-pin engine harness connector from the retaining bracket on the brake vacuum booster, disconnect and position out of the way.

12. Disconnect the HDR sensor, air conditioning compressor clutch and canister purge solenoid connectors.

13. Raise and safely support the vehicle.

14. Remove the bolts retaining the power steering pump to the engine block and front cover. The front lower bolt on the power steering pump will not come all the way out. Wire the power steering pump out of the way.

15. Remove the 4 bolts retaining the oil pan to the front cover. Remove the crankshaft damper retaining bolt and remove the damper, using a suitable puller.

16. Disconnect the EVO sensor and oil sending unit. Position the EVO sensor and oil pressure sending unit harness out of the way.

17. Disconnect the EGR tube from the right exhaust manifold. Disconnect the exhaust pipes from the exhaust manifolds. Lower the exhaust pipes and hang with wire from the crossmember.

18. Remove the bolt retaining the starter wiring harness to the rear of the right cylinder head. Lower the vehicle.

19. Remove the bolts and stud bolts retaining the camshaft covers to the cylinder heads and remove the covers.

20. Disconnect the accelerator, cruise control and throttle valve cables. Remove the accelerator cable bracket from the intake manifold and position out of the way.

21. Disconnect the vacuum hose from the throttle body elbow vacuum port, both oxygen sensors and the heater supply hose.

22. Remove the 2 bolts retaining the thermostat housing to the intake manifold and position the upper hose and thermostat housing out of the way.

NOTE: Two thermostat housing bolts also retain the intake manifold.

23. Remove the 9 bolts retaining the intake manifold to the cylinder heads and remove the intake manifold and gaskets.

24. Remove the 7 stud bolts and 4 bolts retaining the front cover to the engine and remove the front cover.

25. Remove the timing chains.

26. Remove the 10 bolts retaining the left cylinder head to the engine block and remove the head.

NOTE: The lower rear bolt cannot be removed due to interference with the brake vacuum booster. Use a rubber band to hold the bolt away from the engine block.

27. Remove the ground strap, 1 stud and 1 bolt retaining the heater return line to the right cylinder head.

28. Remove the 10 bolts retaining the right cylinder head to the engine block and remove the head.

NOTE: The lower rear bolt cannot be removed due to interference with the evaporator housing. Use a rubber band to hold the bolt away from the engine block.

29. Clean all gasket mating surfaces. Check the cylinder head and engine block for flatness. Check the cylinder head for scratches near the coolant passage and combustion chamber that could provide leak paths. Machine as necessary.

To install:

30. Rotate the crankshaft counterclockwise 45 degrees. The crankshaft keyway should be at the 9 o'clock position viewed from the front of the engine. This ensures that all pistons are below the top of the engine block deck face.

31. Rotate the camshaft to a stable position where the valves do not extend below the head face.

32. Position new head gaskets on the engine block. Install the lower rear bolts on both cylinder heads and retain with rubber bands as explained during the removal procedure.

33. Position the cylinder heads on the engine block dowels, being careful not to score the surface of the head face. Apply clean oil to the head bolts, remove the rubber band from the lower rear bolt and install all bolts hand-tight.

34. Tighten the head bolts as follows:
   a. Tighten the bolts, in sequence, to 15–22 ft. lbs. (20–30 Nm).
   b. Rotate each bolt, in sequence, 85–95 degrees.
   c. Rotate each bolt, in sequence, an additional 85–95 degrees.

35. Position the heater return hose and install the 2 bolts. Rotate the camshafts using the flats matched at the center of the camshaft until both are in time. Install cam positioning tools T91P–6256–A or equivalent, on the flats of the camshafts to keep them from rotating.

36. Rotate the crankshaft clockwise 45 degrees to position the crankshaft at TDC on No. 1 cylinder.

NOTE: The crankshaft must only be rotated in the clockwise direction and only as far as TDC.

37. Install the timing chains according to the proper procedure.

38. Install a new front cover seal and gasket. Apply silicone sealer to the lower corners of the cover where it meets the junction of the oil pan and cylinder block and to the points where the cover contacts the junction of the cylinder block and cylinder head.

39. Install the front cover and the stud bolts and bolts. Tighten to 15–22 ft. lbs. (20–30 Nm).

40. Position new intake manifold gaskets on the cylinder heads. Make sure the alignment tabs on the gaskets are aligned with the holes in the cylinder heads.

NOTE: Before installing the intake manifold, inspect it for nicks and cuts that could provide leak paths.

41. Position the intake manifold on the cylinder heads and install the retaining bolts. Tighten the bolts, in sequence, to 15–22 ft. lbs. (20–30 Nm).

42. Install the thermostat and O-ring, then position the thermostat housing and upper hose and install the 2 bolts. Tighten to 15–22 ft. lbs. (20–30 Nm).

43. Connect the heater supply hose and both oxygen sensors. Connect the vacuum hose to the throttle body adapter vacuum port.

44. Connect and, if necessary, adjust the throttle valve cable. Install the accelerator cable bracket on the intake manifold and connect the accelerator and cruise control cables to the throttle body.

45. Apply silicone sealer to both places where the front cover meets the cylinder head. Install new gaskets on the camshaft covers.

46. Install the camshaft covers on the cylinder heads. Install the bolts and stud bolts and tighten to 6.0–8.8 ft. lbs. (8–12 Nm).

47. Raise and safely support the vehicle. Position the starter wiring harness to the right cylinder head and install the retaining bolt.

48. Cut the wire and position the exhaust pipes to the exhaust manifolds. Tighten the 4 nuts to 20–30 ft. lbs. (27–41 Nm).

NOTE: Make sure the exhaust system clears the No. 3 crossmember. Adjust as necessary.

49. Connect the EGR tube to the right exhaust manifold and tighten the line nut to 26–33 ft. lbs. (35–45 Nm). Connect the EVO sensor and oil sending unit.

50. Apply a small amount of silicone sealer in the rear of the keyway on the damper. Position the damper on the crankshaft, making sure the crankshaft key and keyway are aligned.

51. Using damper installer T74P–6316–B or equivalent, install the crankshaft damper. Install the damper bolt and washer and tighten to 114–121 ft. lbs. (155–165 Nm).

52. Install the 4 bolts retaining the oil pan to the front cover and tighten to 15–22 ft. lbs. (20–30 Nm).

53. Position the power steering pump on the engine and install the 4 retaining bolts. Tighten to 15–22 ft. lbs. (20–30 Nm). Lower the vehicle.

54. Connect the air conditioning compressor, HDR sensor and canister purge solenoid.

55. Connect the 42-pin engine harness connector and transmission harness connector. Install the 42-pin connector on the retaining bracket on the vacuum brake booster.

56. Install the PCV valve in the right camshaft cover and connect the canister purge solenoid vent hose.

57. Position the positive battery cable harness on the right cylinder head and install the bolt retaining the cable bracket to the cylinder head. Connect the positive battery cable at the power distribution box and battery.

58. Position the generator and install the 2 retaining bolts. Tighten to 15–22 ft. lbs. (20–30 Nm). Install the 2 bolts retaining the generator brace to the intake manifold and tighten to 6–8 ft. lbs. (8–12 Nm).

59. Install the water pump pulley and tighten the bolts to 15–22 ft. lbs. (20–30 Nm).

60. Position the ignition coil brackets and ignition wire assembly onto the mounting studs. Install the 7 nuts retaining the coil brackets to the front cover and tighten to 15–22 ft. lbs. (20–30 Nm).

61. Install the 2 bolts retaining the ignition wire tray to the coil bracket and tighten to 6.0–8.8 ft. lbs. (8–12 Nm). Connect both ignition coils and CID sensor.

62. Position the air conditioner high pressure line on the right coil

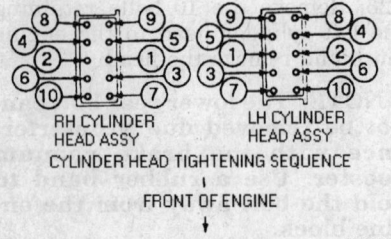

RH CYLINDER HEAD ASSY LH CYLINDER HEAD ASSY
CYLINDER HEAD TIGHTENING SEQUENCE
FRONT OF ENGINE

8838FG13

Cylinder head torque sequence

bracket and install the bolt. Connect the ignition wires to the spark plugs and install the bracket onto the camshaft cover studs.

63. Install the accessory drive belt and the wiper module. Connect the fuel lines and install the cooling fan and shroud. Fill and bleed the cooling system.

64. Install the air inlet tube and connect the negative battery cable. Start the engine and bring to normal operating temperature. Check for leaks. Check all fluid levels.

Valve Lifters

BLEEDING

Lash adjusters (valve tappets), can be bled of air by placing in a container of clean engine oil, then pumping the lash adjuster plunger with the lash adjuster positioned between 2 fingers until no more air bubbles are seen. Lash adjusters must always be pumped up before installing into the cylinder head.

REMOVAL AND INSTALLATION

The engine is equipped with hydraulic lash adjusters which, while not being exactly the same as a conventional hydraulic lifter, perform the same function — maintain proper valve train clearance.

1. Disconnect the negative battery cable.

2. Remove the right camshaft cover as follows:

    a. Disconnect the positive battery cable at the battery and at the power distribution box. Remove the retaining bolt from the positive battery cable bracket located on the side of the right cylinder head.

    b. Disconnect the High Data Rate (HDR) sensor, air conditioning compressor clutch and canister

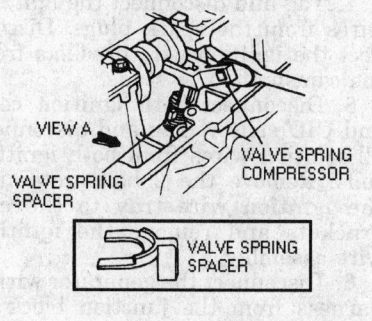

Valve spring compressor and spacer

purge solenoid connectors. Position the harness out of the way.

    c. Disconnect the vent hose from the purge solenoid and position the positive battery cable out of the way.

    d. Disconnect the ignition wires from the spark plugs. Remove the ignition wire brackets from the camshaft cover studs and position the wires out of the way.

    e. Remove the PCV valve from the camshaft cover grommet and position out of the way.

    f. Remove the bolts and stud bolts and remove the camshaft cover.

3. Remove the left camshaft cover as follows:

    a. Remove the air inlet tube. Relieve the fuel system pressure and disconnect the fuel lines.

    b. Raise and safely support the vehicle.

    c. Disconnect the EVO sensor and oil pressure sending unit and position the harness out of the way. Lower the vehicle.

    d. Remove the 42-pin engine harness connector from the retaining bracket on the brake vacuum booster. Disconnect and position out of the way.

    e. Remove the windshield wiper module.

    f. Disconnect the ignition wires from the spark plugs. Remove the ignition wire brackets from the studs and position the wires out of the way.

    g. Remove the bolts and stud bolts and remove the camshaft cover.

4. Position the piston of the cylinder being serviced at the bottom of its stroke and position the camshaft lobe on the base circle.

5. Install valve spring spacer tool T91P–6565–AH or equivalent, between the spring coils to prevent valve seal damage.

NOTE: If the valve spring spacer tool is not used, the retainer will hit the valve stem seal and damage the seal.

6. Install valve spring compressor tool T91P–6565–A or equivalent, under the camshaft and on top of the valve spring retainer.

7. Compress the valve spring and remove the roller follower. Remove the valve spring compressor and spacer.

8. Remove the hydraulic lash adjuster.

To install:

9. Check the hydraulic lash adjusters. They must have no more

than 1.5mm of plunger travel prior to installation.

10. Apply engine oil to the valve stem and tip, roller follower contact surfaces and lash adjuster bore. Install the lash adjusters.

11. Install valve spring spacer tool T91P–6565–AH or equivalent, between the spring coils. Compress the valve spring using valve spring compressor tool T91P–6565–A or equivalent, and install the roller follower.

NOTE: The piston must be at the bottom of its stroke and the camshaft at the base circle.

12. Remove the valve spring compressor and spacer.

13. Clean the sealing surfaces of the camshaft covers and cylinder heads. Apply silicone sealer to the places where the front cover meets the cylinder head.

14. Position new gaskets onto the camshaft covers and install the covers. Install the bolts and stud bolts and tighten to 6.0–8.8 ft. lbs. (8–12 Nm).

15. When installing the right camshaft cover, proceed as follows:

    a. Install the PCV valve into the camshaft cover grommet.

    b. Install the ignition wire brackets on the studs and connect the wires to the spark plugs.

    c. Position the harness and connect the canister purge solenoid, air conditioning compressor clutch and HDR sensor.

    d. Position the positive battery cable harness on the right cylinder head. Install the bolt retaining the cable bracket to the cylinder head.

    e. Connect the positive battery cable at the power distribution box and the battery.

16. When installing the left camshaft cover, proceed as follows:

    a. Install the ignition wire brackets on the studs and connect the wires to the spark plugs.

    b. Install the windshield wiper module.

    c. Connect the 42-pin connector and transmission harness connector. Install the connector on the retaining bracket.

    d. Raise and safely support the vehicle. Position and connect the EVO sensor and oil pressure sending unit harness.

    e. Lower the vehicle. Connect the fuel lines.

17. Connect the negative battery cable. Start the engine and check for leaks.

Valve Lash

ADJUSTMENT

The valve lash is not adjustable. If the collapsed lash adjuster clearance is incorrect, check the camshaft, roller follower and valve for wear or damage.

1. Disconnect the negative battery cable.
2. Remove the camshaft covers.
3. Rotate the crankshaft until the camshaft base circle is contacting the roller follower.
4. Use a suitable tool to bleed down the lash adjuster. Slowly compress the lash adjuster until the plunger is bottomed.
5. Use a feeler gauge to check the clearance between the camshaft and the roller follower. The clearance should be 0.018–0.033 in. (0.45–0.85mm).

Rocker Arms

REMOVAL AND INSTALLATION

1. Disconnect the negative battery cable.
2. Remove the right camshaft cover as follows:
    a. Disconnect the positive battery cable at the battery and at the power distribution box. Remove the retaining bolt from the positive battery cable bracket located on the side of the right cylinder head.
    b. Disconnect the High Data Rate (HDR) sensor, air conditioning compressor clutch and canister purge solenoid connectors. Position the harness out of the way.
    c. Disconnect the vent hose from the purge solenoid and position the positive battery cable out of the way.
    d. Disconnect the ignition wires from the spark plugs. Remove the ignition wire brackets from the camshaft cover studs and position the wires out of the way.
    e. Remove the PCV valve from the camshaft cover grommet and position out of the way.
    f. Remove the bolts and stud bolts and remove the camshaft cover.
3. Remove the left camshaft cover as follows:
    a. Remove the air inlet tube. Relieve the fuel system pressure and disconnect the fuel lines.
    b. Raise and safely support the vehicle.

c. Disconnect the EVO sensor and oil pressure sending unit and position the harness out of the way. Lower the vehicle.
    d. Remove the 42-pin engine harness connector from the retaining bracket on the brake vacuum booster. Disconnect and position out of the way.
    e. Remove the windshield wiper module.
    f. Disconnect the ignition wires from the spark plugs. Remove the ignition wire brackets from the studs and position the wires out of the way.
    g. Remove the bolts and stud bolts and remove the camshaft cover.
4. Position the piston of the cylinder being serviced at the bottom of its stroke and position the camshaft lobe on the base circle.
5. Install valve spring spacer tool T91P–6565–AH or equivalent, between the spring coils to prevent valve seal damage.

NOTE: If the valve spring spacer tool is not used, the retainer will hit the valve stem seal and damage the seal.

6. Install valve spring compressor tool T91P–6565–A or equivalent, under the camshaft and on top of the valve spring retainer.
7. Compress the valve spring and remove the roller follower. Remove the valve spring compressor and spacer.
To install:
8. Apply engine oil to the valve stem and tip and roller follower contact surfaces.
9. Install valve spring spacer tool T91P–6565–AH or equivalent, between the spring coils. Compress the valve spring using valve spring compressor tool T91P–6565–A or equivalent, and install the roller follower.

NOTE: The piston must be at the bottom of its stroke and the camshaft at the base circle.

10. Remove the valve spring compressor and spacer.
11. Clean the sealing surfaces of the camshaft covers and cylinder heads. Apply silicone sealer to the places where the front cover meets the cylinder head.
12. Position new gaskets onto the camshaft covers and install the covers. Install the bolts and stud bolts and tighten to 6.0–8.8 ft. lbs. (8–12 Nm).

13. When installing the right camshaft cover, proceed as follows:
    a. Install the PCV into the camshaft cover grommet.
    b. Install the ignition wire brackets on the studs and connect the wires to the spark plugs.
    c. Position the harness and connect the canister purge solenoid, air conditioning compressor clutch and HDR sensor.
    d. Position the positive battery cable harness on the right cylinder head. Install the bolt retaining the cable bracket to the cylinder head.
    e. Connect the positive battery cable at the power distribution box and the battery.
14. When installing the left camshaft cover, proceed as follows:
    a. Install the ignition wire brackets on the studs and connect the wires to the spark plugs.
    b. Install the windshield wiper module.
    c. Connect the 42-pin connector and transmission harness connector. Install the connector on the retaining bracket.
    d. Raise and safely support the vehicle. Position and connect the EVO sensor and oil pressure sending unit harness.
    e. Lower the vehicle. Connect the fuel lines.
15. Connect the negative battery cable. Start the engine and check for leaks.

Intake Manifold

REMOVAL AND INSTALLATION

1. Disconnect the negative battery cable.
2. Drain the cooling system. Relieve the fuel system pressure and disconnect the fuel lines.
3. Remove the wiper module and the air inlet tube. Release the belt tensioner and remove the accessory drive belt.
4. Tag and disconnect the ignition wires from the spark plugs. Disconnect the ignition wire brackets from the camshaft cover studs.
5. Disconnect both ignition coils and CID sensor. Tag and disconnect all ignition wires from both ignition coils. Remove the 2 bolts retaining the ignition wire tray to the coil brackets and remove the ignition wire assembly.
6. Disconnect the generator wiring harness from the junction block at the fender apron and generator. Remove the bolts retaining the genera-

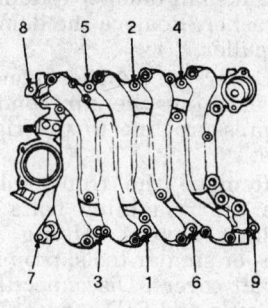

Intake manifold torque sequence

tor brace to the intake manifold and the generator to the engine block and remove the generator.

7. Raise and safely support the vehicle. Disconnect the oil sending unit and EVO harness sensor and position the wiring harness out of the way.

8. Disconnect the EGR tube from the right exhaust manifold and lower the vehicle.

9. Remove the 42-pin engine harness connector from the retaining bracket on the vacuum brake booster and disconnect the connector.

10. Disconnect the air conditioning compressor, HDR sensor and canister purge solenoid.

11. Remove the PCV valve from the camshaft cover and disconnect the canister purge vent hose from the PCV valve.

12. Disconnect the accelerator and cruise control cables from the throttle body. Remove the accelerator cable bracket from the intake manifold and position out of the way.

13. Disconnect the throttle valve cable from the throttle body and the vacuum hose from the throttle body adapter port.

14. Disconnect both oxygen sensors and the heater supply hose.

15. Remove the 2 bolts retaining the thermostat housing to the intake manifold and position the upper hose and thermostat housing out of the way.

NOTE: The 2 thermostat housing bolts also retain the intake manifold.

16. Remove the bolts retaining the intake manifold to the cylinder heads and remove the intake manifold. Remove and discard the gaskets.

To install:

17. Clean all gasket mating surfaces. Position new intake manifold gaskets on the cylinder heads. Make sure the alignment tabs on the gaskets are aligned with the holes in the cylinder heads.

18. Install the intake manifold and the retaining bolts. Tighten the bolts, in sequence, to 15–22 ft. lbs. (20–30 Nm).

19. Inspect and if necessary, replace the O-ring seal on the thermostat housing. Position the housing and upper hose and install the 2 bolts. Tighten to 15–22 ft. lbs. (20–30 Nm).

20. Connect the heater supply hose and connect both oxygen sensors.

21. Connect the vacuum hose to the throttle body adapter vacuum port. Connect and, if necessary, adjust the throttle valve cable.

22. Install the accelerator cable bracket on the intake manifold and connect the accelerator and cruise control cables to the throttle body.

23. Install the PCV valve in the camshaft cover and connect the canister purge solenoid vent hose. Connect the air conditioning compressor, HDR sensor and canister purge solenoid.

24. Connect the 42-pin engine harness connector. Install the connector on the retaining bracket on the vacuum brake booster.

25. Raise and safely support the vehicle. Connect the EGR tube to the right exhaust manifold and tighten the line nut to 26–33 ft. lbs. (35–45 Nm).

26. Connect the EVO sensor and oil sending unit. Lower the vehicle.

27. Position the generator and install the retaining bolts. Tighten to 15–22 ft. lbs. (20–30 Nm). Install the 2 bolts retaining the generator brace to the intake manifold and tighten to 6.0–8.8 ft. lbs. (8–12 Nm).

28. Connect the generator wiring harness to the generator, right-hand fender apron and junction block.

29. Position the ignition wire assembly on the engine and install the 2 bolts retaining the ignition wire tray to the coil brackets. Tighten the bolts to 6.0–8.8 ft. lbs. (8–12 Nm).

30. Connect the ignition wires to the ignition coils in their proper positions. Connect the ignition wires to the spark plugs.

31. Connect the ignition wire brackets on the camshaft cover studs. Connect both ignition coils and CID sensor.

32. Install the accessory drive belt and the air inlet tube. Install the wiper module and connect the fuel lines.

33. Fill and bleed the cooling system. Connect the negative battery cable, start the engine and check for leaks.

Exhaust Manifold

REMOVAL AND INSTALLATION

1. Disconnect the battery cables. Remove the air inlet tube.

2. Drain the cooling system and remove the cooling fan and shroud. Relieve the fuel system pressure and disconnect the fuel lines.

3. Remove the upper radiator hose. Remove the wiper module and support bracket.

4. Discharge the air conditioning system. Disconnect and plug the compressor outlet hose at the compressor and remove the bolt retaining the hose assembly to the right coil bracket. Cap the compressor opening.

5. Remove the 42-pin engine harness connector from the retaining bracket on the brake vacuum booster. Disconnect the connector.

6. Disconnect the throttle valve cable from the throttle body. Disconnect the heater outlet hose.

7. Remove the nut retaining the ground strap to the right cylinder head. Remove the upper stud and lower bolt retaining the heater outlet hose to the right cylinder head and position out of the way.

8. Remove the blower motor resistor and remove the bolt retaining the right engine insulator to the lower engine bracket. Disconnect both oxygen sensors.

9. Raise and safely support the vehicle. Remove the engine mount through bolts.

10. Remove the EGR tube line nut from the right exhaust manifold.

11. Disconnect the exhaust pipes from the manifolds. Lower the exhaust system and hang it from the crossmember with wire.

12. To remove the left exhaust manifold, remove the engine mount from the engine block and remove the 8 bolts retaining the exhaust manifold.

13. Position a jack and a block of wood under the oil pan, rearward of the oil drain hole. Raise the engine approximately 4 in. (100mm).

14. Remove the 8 bolts retaining the right exhaust manifold and remove the manifold.

To install:

15. If the exhaust manifolds are being replaced, transfer the oxygen sensors and tighten to 27–33 ft. lbs. (37–45 Nm). On the right manifold, transfer the EGR tube connector and tighten to 33–48 ft. lbs. (45–65 Nm).

16. Clean the mating surfaces of the exhaust manifolds and cylinder heads.

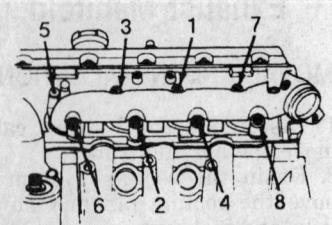

NOTE: ENGINE SHOWN REMOVED FOR CLARITY
NOTE: LH EXHAUST MANIFOLD SHOWN RH EXHAUST MANIFOLD TYPICAL

8838FG16

Exhaust manifold torque sequence

17. Position the exhaust manifolds to the cylinder heads and install the retaining bolts. Tighten, in sequence, to 15–22 ft. lbs. (20–30 Nm).

18. Position and connect the EGR valve and tube assembly to the exhaust manifold. Tighten the line nut to 26–33 ft. lbs. (35–45 Nm).

19. Install the left engine mount and tighten the bolts to 15–22 ft. lbs. (20–30 Nm). Lower the engine onto the mounts and remove the jack. Install the engine mount through bolts and tighten to 15–22 ft. lbs. (20–30 Nm).

20. Cut the wire and position the exhaust system. Tighten the nuts to 20–30 ft. lbs. (27–41 Nm).

NOTE: Make sure the exhaust system clears the No. 3 crossmember. Adjust as necessary.

21. Lower the vehicle. Connect both oxygen sensors and install the bolt retaining the right engine mount to the frame. Tighten to 15–22 ft. lbs. (20–30 Nm).

22. Install the blower motor resistor. Position the heater outlet hoses. Install the upper stud and lower bolt and tighten to 15–22 ft. lbs. (20–30 Nm). Install the ground strap onto the stud and tighten the nut to 15–22 ft. lbs. (20–30 Nm).

23. Connect the heater outlet hose. Connect and if necessary, adjust the throttle valve cable.

24. Connect the 42-pin connector and transmission harness connector. Install the connector to the retaining bracket on the brake vacuum booster.

25. Connect the air conditioning compressor outlet hose to the compressor and install the bolt retaining the hose assembly to the right coil bracket.

26. Install the upper radiator hose and connect the fuel lines. Install the wiper module and retaining bracket.

27. Install the cooling fan and shroud. Install the air inlet tube.

Connect the battery cables, start the engine and check for leaks.

28. Leak test, evacuate and charge the air conditioning system according to the proper procedure. Observe all safety precautions.

Front Cover Seal

REMOVAL AND INSTALLATION

1. Disconnect the negative battery cable.

2. Release the belt tensioner and remove the accessory drive belt.

3. Raise and safely support the vehicle.

4. Remove the crankshaft damper retaining bolt and washer. Remove the damper using a puller.

5. Using a small prybar, remove the front cover seal.

To install:

6. Lubricate the seal bore in the front cover and seal lip with clean engine oil. Install the seal, using a seal installer.

7. Apply a small amount of silicone sealer to the rear of the damper keyway. Using a damper installer, install the crankshaft damper. Be sure the key on the crankshaft aligns with the keyway in the damper.

8. Install the crankshaft damper retaining bolt and washer and tighten to 114–121 ft. lbs. (155–165 Nm).

9. Lower the vehicle and install the accessory drive belt.

10. Connect the negative battery cable, start the engine and check for leaks.

Timing Chain Front Cover

REMOVAL AND INSTALLATION

1. Disconnect the negative battery cable.

2. Remove the cooling fan and shroud. Loosen the water pump pulley bolts, remove the accessory drive belt and remove the water pump pulley.

3. Raise and safely support the vehicle.

4. Remove the bolts retaining the power steering pump to the engine block and cylinder front cover. The lower front bolt on the power steering pump will not come all the way out. Wire the power steering pump out of the way.

5. Remove the 4 bolts retaining the oil pan to the front cover. Remove

the crankshaft damper retaining bolt and washer. Remove the damper using a puller.

6. Lower the vehicle. Remove the bolt retaining the air conditioner high pressure line to the right coil bracket.

7. Remove the front bolts and loosen the remaining bolts on the camshaft covers. Using plastic wedges or similar tools, prop up both camshaft covers. Disconnect both ignition coils and CID sensor.

8. Remove the 3 nuts retaining the right coil bracket to the front cover. Position the power steering hose out of the way.

9. Remove the 4 nuts retaining the left coil bracket to the front cover. Slide both coil brackets and ignition wires off the mounting studs and lay the assembly on top of the engine.

10. Disconnect the High Data Rate (HDR) sensor. Remove the 7 stud bolts and 4 bolts retaining the front cover to the engine and remove the front cover.

To install:

11. Inspect and replace the front cover seal as necessary and clean the sealing surfaces of the cylinder block. Apply silicone sealer to the oil pan where it meets the cylinder block and to the points where the cylinder head meets the cylinder block.

12. Install the front cover and the attaching studs and bolts. Tighten to 15–22 ft. lbs. (20–30 Nm). Connect the HDR sensor.

13. Position the coil brackets and ignition wires as an assembly onto the mounting studs. Position the power steering hose and install the 7 nuts retaining the coil brackets to the front cover. Tighten the nuts to 15–22 ft. lbs. (20–30 Nm). Connect both ignition coils and CID sensor.

14. Remove the plastic wedges holding up the camshaft covers. Apply silicone sealer where the front cover meets the cylinder head and make sure the camshaft cover gaskets are properly positioned. Install the front retaining bolts into the camshaft cover and tighten the bolts to 6.0–8.8 ft. lbs. (8–12 Nm).

15. Position the air conditioner high pressure line on the right coil bracket and install the bolt. Raise and safely support the vehicle.

16. Apply a small amount of silicone sealer in the rear of the keyway in the damper. Position the damper on the crankshaft and install, using a suitable installation tool. Install the damper bolt and washer and tighten to 114–121 ft. lbs. (155–165 Nm).

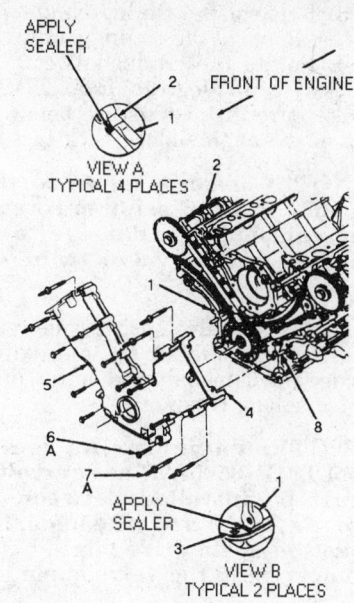

1. Cylinder block
2. Cylinder head
3. Oil pan gasket
4. Gasket
5. Front cover assy
6A. Bolts
7A. Studs
8. Dowel

8838FG17

Timing chain front cover installation

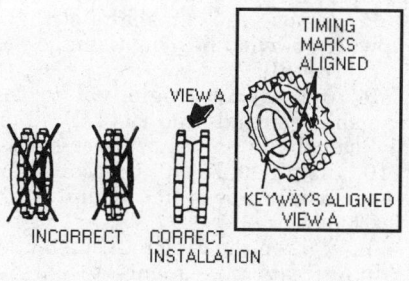

8838FG18

Crankshaft sprocket positioning

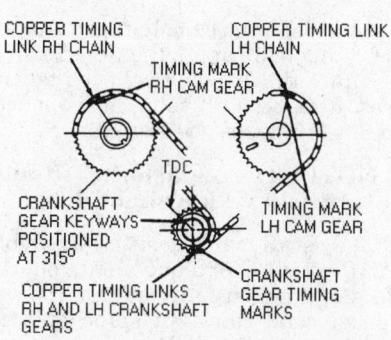

8838FG19

Timing chain and sprocket alignment

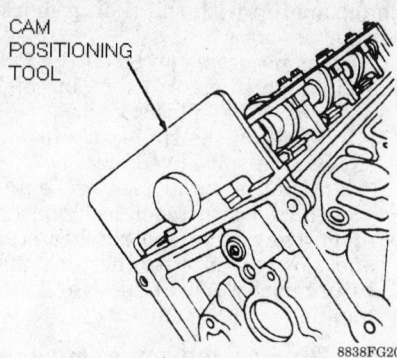

8838FG20

Cam positioning tool

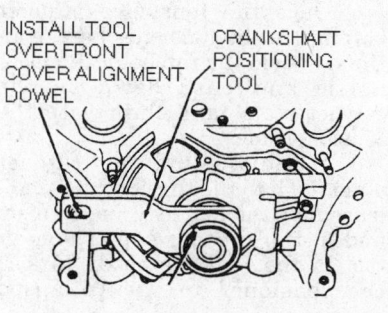

8838FG21

Crankshaft positioning tool

17. Install the 4 bolts retaining the oil pan to the front cover. Tighten to 15–22 ft. lbs. (20–30 Nm).

18. Position the power steering pump on the engine and install the 4 retaining bolts. Tighten to 15–22 ft. lbs. (20–30 Nm). Lower the vehicle.

19. Install the water pump pulley with the 4 bolts. Tighten to 15–22 ft. lbs. (20–30 Nm). Install the accessory drive belt and the cooling fan and shroud.

20. Connect the negative battery cable, start the engine and check for leaks.

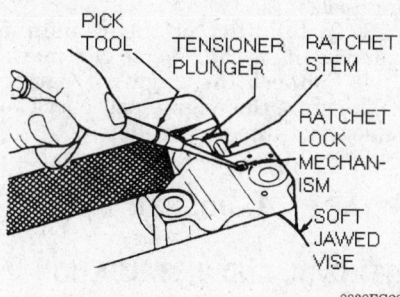

8838FG22

Timing chain tensioner bleeding procedure

Timing Chain and Sprockets

REMOVAL AND INSTALLATION

NOTE: This is not a free wheeling engine. If it has "jumped time" there will be damage to the valves and/or pistons and will require the removal of the cylinder heads.

1. Disconnect the negative battery cable.

2. Remove the camshaft covers and the timing chain front cover.

3. Remove the High Data Rate (HDR) wheel.

4. Rotate the engine to set the No. 1 piston at TDC on the compression stroke.

5. Install cam positioning tools T91P–6256–A or equivalent, on the flats of the camshaft. This will prevent accidental rotation of the camshafts.

6. Remove the 2 bolts retaining the right tensioner to the cylinder head and remove the tensioner. Remove the right tensioner arm.

7. Remove the 2 bolts retaining the right chain guide to the cylinder head and remove the chain guide. Remove the right chain and right crankshaft sprocket. Remove the right camshaft sprocket retaining bolt, washer, sprocket and spacer.

NOTE: Cam positioning tools T91P–6256–A or equivalent, must be installed on the camshaft to prevent the camshaft from rotating.

8. Remove the 2 bolts retaining the left tensioner to the cylinder head and remove the tensioner. Remove the left tensioner arm.

9. Remove the 2 bolts retaining the left chain guide to the cylinder head and remove the chain guide. Remove the left chain and left crankshaft sprocket. Remove the left camshaft sprocket retaining bolt, washer, sprocket and spacer.

NOTE: Cam positioning tools T91P–6256–A or equivalent, must be installed on the camshaft to prevent the camshaft from rotating.

10. Inspect the friction material on the tensioner arms and chain guides. If worn or damaged, remove and

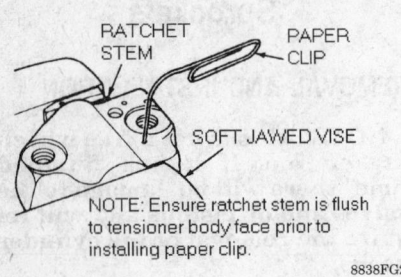

RATCHET STEM PAPER CLIP

SOFT JAWED VISE

NOTE: Ensure ratchet stem is flush to tensioner body face prior to installing paper clip.

8838FG23

Timing chain tensioner locking procedure

clean the oil pan and replace the oil pickup tube.

NOTE: At no time, when the timing chains are removed and the cylinder heads are installed, may the crankshaft and/or camshafts be rotated. Failure to follow these directions will result in valve and/or piston damage.

To install:

11. Make sure cam positioning tools T91P–6256–A or equivalent, are installed on the camshafts to prevent them from rotating.

12. Position the camshaft spacers and sprockets on the camshafts and install the washers and retaining bolts. Tighten the retaining bolts to 81–95 ft. lbs. (110–130 Nm).

13. Install the left crankshaft sprocket with the tapered part of the sprocket facing away from the engine block.

NOTE: The crankshaft sprockets are identical. They may only be installed one way, with the tapered part of the sprocket facing each other.

14. Install the left timing chain on the camshaft and crankshaft sprockets. Make sure the copper links of the chain line up with the timing marks of the sprockets.

NOTE: If the copper links of the timing chain are not visible, pull the chain taught until the opposite sides of the chain contact one another and lay it on a flat surface. Mark the links at each end of the chain and use them in place of the copper links.

15. Install the right crankshaft sprocket with the tapered part of the sprocket facing the left crankshaft sprocket.

16. Install the right timing chain on the camshaft and crankshaft sprockets. Make sure the copper links of the

chain line up with the timing marks of the sprockets.

17. It is necessary to bleed the timing chain tensioners before installation. Proceed as follows:

   a. Position the timing chain tensioner in a soft-jawed vice.

   b. Using a small pick or similar tool, hold the ratchet lock mechanism away from the ratchet stem and slowly compress the tensioner plunger by rotating the vise handle.

NOTE: The tensioner must be compressed slowly or damage to the internal seals will result.

   c. Once the tensioner plunger bottoms in the tensioner bore, continue to hold the ratchet lock mechanism and push down on the ratchet stem until flush with the tensioner face.

   d. While holding the ratchet stem flush to the tensioner face, release the ratchet lock mechanism and install a paper clip or similar tool in the tensioner body to lock the tensioner in the collapsed position.

   e. The paper clip must not be removed until the timing chain, tensioner, tensioner arm and timing chain guide are completely installed on the engine.

18. Lubricate the tensioner arm contact surfaces with engine oil and install the right and left tensioner arms on their dowels.

19. Install the right and left timing chain tensioners and secure with 2 bolts on each. Tighten the bolts to 15–22 ft. lbs. (20–30 Nm).

20. Install the right and left timing chain guides and secure with 2 bolts on each. Tighten the bolts to 6.0–8.8 ft. lbs. (8–12 Nm).

21. Remove the paper clips from the timing chain tensioners and make sure all timing marks are aligned.

22. Remove the camshaft positioning tools.

23. Installation of the remaining components is the reverse of removal.

24. Connect the negative battery cable, start the engine and check for leaks and proper operation.

Camshaft

REMOVAL AND INSTALLATION

1. Disconnect the negative battery cable and drain the cooling system. Relieve the fuel system pressure.

2. Remove the right and left camshaft covers.

3. Remove the timing chain front cover. Remove the timing chains.

4. Rotate the crankshaft counterclockwise 45 degrees from TDC to make sure all pistons are below the top of the engine block deck face.

NOTE: The crankshaft must be in this position prior to rotating the camshafts or damage to the pistons and/or valve train will result.

5. Install valve spring compressor tool T91P–6565–A or equivalent, under the camshaft and on top of the valve spring retainer.

NOTE: Valve spring spacer tool T91P–6565–AH or equivalent, must be installed between the spring coils and the camshaft must be at the base circle before compressing the valve spring.

6. Compress the valve spring far enough to remove the roller follower. Repeat Steps 5 and 6 until all roller followers are removed.

7. Remove the bolts retaining the camshaft cap cluster assemblies to the cylinder heads. Tap upward on the camshaft caps at points near the upper bearing halves and gradually lift the camshaft clusters from the cylinder heads.

8. Remove the camshafts straight upward to avoid bearing damage.

To install:

9. Apply heavy engine oil to the camshaft journals and lobes. Position the camshafts on the cylinder heads.

10. Install and seat the camshaft cap cluster assemblies. Hand start the bolts.

11. Tighten the camshaft cluster retaining bolts in sequence to 6.0–8.8 ft. lbs. (8–12 Nm).

NOTE: Each camshaft cap cluster assembly is tightened individually.

12. Loosen the camshaft cap cluster retaining bolts approximately 2 turns or until the heads of the bolts are free. Retighten all bolts, in sequence, to 6.0–8.8 ft. lbs. (8–12 Nm).

NOTE: The camshafts should turn freely with a slight drag.

13. Install cam positioning tools T91P–6256–A or equivalent, on the flats of the camshafts and install the spacers and camshaft sprockets. Install the bolts and washers and tighten to 81–95 ft. lbs. (110–130 Nm).

14. Install valve spring compressor T91P–6565–A or equivalent, under

the camshaft and on top of the valve spring retainer.

NOTE: Valve spring spacer tool T91P-6565-AH or equivalent, must be installed between the spring coils and the camshaft must be at the base circle before compressing the valve spring.

15. Compress the valve spring far enough to install the roller followers.

16. Repeat Steps 14 and 15 until all roller followers are installed.

17. Rotate the crankshaft clockwise 45 degrees to position the crankshaft at TDC.

NOTE: The crankshaft must only be rotated in the clockwise direction and only as far as TDC.

18. Install the timing chains and install the timing chain front cover. Install the camshaft covers.

19. Install the remaining components in the reverse order of removal.

20. Connect the negative battery cable. Start the engine and check for leaks.

Piston and Connecting Rod

POSITIONING

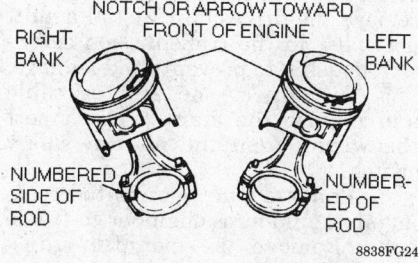

Piston position — 8 cylinder engine.

ENGINE LUBRICATION

Oil Pan

REMOVAL AND INSTALLATION

1. Disconnect the battery cables and remove the air inlet tube.

2. Drain the cooling system and remove the cooling fan and shroud.

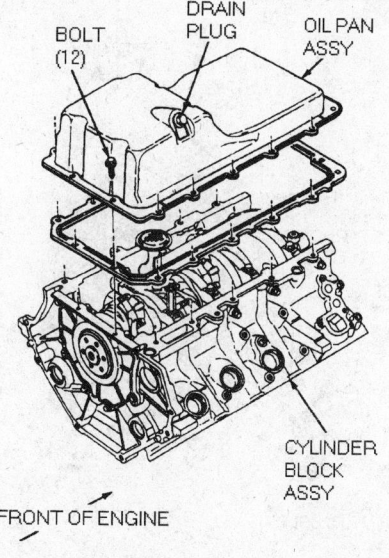

Oil pan replacement

Relieve the fuel system pressure and disconnect the fuel lines.

3. Remove the upper radiator hose. Remove the wiper module and support bracket.

4. Discharge the air conditioning system. Disconnect and plug the compressor outlet hose at the compressor and remove the bolt retaining the hose assembly to the right coil bracket. Cap the compressor outlet.

5. Remove the 42-pin engine harness connector from the retaining bracket on the brake vacuum booster and disconnect the connector and transmission harness connector.

6. Disconnect the throttle valve cable from the throttle body and disconnect the heater outlet hose.

7. Remove the nut retaining the ground strap to the right cylinder head. Remove the upper stud and loosen the lower bolt retaining the heater outlet hose to the right cylinder head and position out of the way.

8. Remove the blower motor resistor. Remove the bolt retaining the right engine mount to the lower engine bracket.

9. Disconnect the vacuum hoses from the EGR valve and tube. Remove the 2 bolts retaining the EGR valve to the intake manifold.

10. Raise and safely support the vehicle. Drain the crankcase and re-

move the engine mount through bolts.

11. Remove the EGR tube line nut from the right exhaust manifold and remove the EGR valve and tube assembly.

12. Disconnect the exhaust from the exhaust manifolds. Lower the exhaust system and support it with wire from the crossmember.

13. Position a jack and a block of wood under the oil pan, rearward of the oil drain hole. Raise the engine approximately 4 in. and insert 2 wood blocks approximately 2½ in. thick under each engine mount. Lower the engine onto the wood blocks and remove the jack.

14. Remove the 16 bolts retaining the oil pan to the engine block and remove the oil pan.

NOTE: It may be necessary to loosen, but not remove, the 2 nuts on the rear transmission mount and with a jack, raise the transmission extension housing slightly to remove the pan.

To install:

15. Clean the oil pan and the gasket mating surfaces.

16. Position a new gasket on the oil pan. Apply silicone sealer to where the front cover meets the cylinder block and rear seal retainer meets the cylinder block. Position the oil pan on the engine and install the bolts. Tighten the bolts, in sequence, to 15-22 ft. lbs. (20-30 Nm).

17. Position the jack and wood block under the oil pan, rearward of the oil drain hole and raise the engine enough to remove the wood blocks. Lower the engine and remove the jack.

18. Install the engine mount through bolts and tighten to 15-22 ft. lbs. (20-30 Nm).

19. Position the EGR valve and tube assembly in the vehicle and connect to the exhaust manifold. Tighten the line nut to 26-33 ft. lbs. (35-45 Nm).

NOTE: Loosen the line nut at the EGR valve prior to installing the assembly into the vehicle. This will allow enough movement to align the EGR valve retaining bolts.

20. Cut the wire and position the exhaust system to the manifolds. Install the 4 nuts and tighten to 20-30 ft. lbs. (27-41 Nm). Make sure the exhaust system clears the crossmember. Adjust as necessary.

21. Install a new oil filter and lower the vehicle.

22. Install the bolt retaining the right engine mount to the lower engine bracket. Tighten to 15–22 ft. lbs. (20–30 Nm).

23. Install a new gasket on the EGR valve and position on the intake manifold. Install the 2 bolts retaining the EGR valve to the intake manifold and tighten to 15–22 ft. lbs. (20–30 Nm). Tighten the EGR tube line nut at the EGR valve to 26–33 ft. lbs. (35–45 Nm). Connect the vacuum hoses to the EGR valve and tube.

24. Install the blower motor resistor. Position the heater outlet hose, install the upper stud and tighten the upper and lower bolts to 15–22 ft. lbs. (20–30 Nm). Install the ground strap on the stud and tighten to 15–22 ft. lbs. (20–30 Nm).

25. Connect the heater outlet hose and the throttle valve cable. If necessary, adjust the throttle valve cable.

26. Connect the 42-pin connector and transmission harness connector. Install the harness connector on the brake vacuum booster.

27. Connect the air conditioning compressor outlet hose to the compressor and install the bolt retaining the hose to the right coil bracket.

28. Install the upper radiator hose and connect the fuel lines. Install the wiper module and retaining bracket.

29. Install the cooling fan and shroud and fill the cooling system. Fill the crankcase with the proper type and quantity of engine oil.

30. Connect the negative battery cable and install the air inlet tube. Start the engine and check for leaks.

31. Evacuate and recharge the air conditioning system.

Oil Pump

REMOVAL AND INSTALLATION

1. Disconnect the negative battery cable.

2. Remove the camshaft covers, front cover, and oil pan.

3. Remove the timing chains.

4. Remove the 4 bolts retaining the oil pump to the cylinder block and remove the pump.

5. Remove the 2 bolts retaining the oil pickup tube to the oil pump and remove the bolt retaining the oil pickup tube to the main bearing stud spacer. Remove the pickup tube.

To install:

6. Clean the oil pickup tube and replace the O-ring.

7. Position the tube on the oil pump and hand-start the 2 bolts. In-

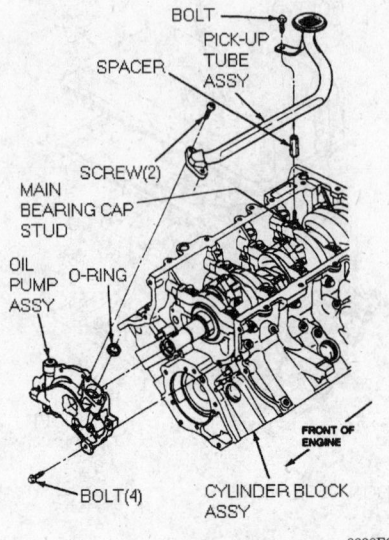

Oil pump

stall the bolt retaining the pickup tube to the main bearing stud spacer hand-tight.

8. Tighten the pickup tube-to-oil pump bolts to 6.0–8.8 ft. lbs. (8–12 Nm). Tighten the pickup tube to main bearing stud spacer bolt to 15–22 ft. lbs. (20–30 Nm).

9. Rotate the inner rotor of the oil pump to align with the flats on the crankshaft and install the oil pump flush with the cylinder block. Install the 4 retaining bolts and tighten to 6.0–8.8 ft. lbs. (8–12 Nm).

10. Install a new oil filter. Install the timing chains.

11. Install the oil pan, front cover and camshaft covers.

12. Fill the crankcase with the proper type and quantity of engine oil. Connect the negative battery cable, start the engine and check for leaks.

TRANSMISSION

Automatic Transmission Assembly

REMOVAL AND INSTALLATION

1. Disconnect the negative battery cable. Raise the vehicle and support safely.

2. Drain the fluid from the transmission by removing all oil pan bolts except the 2 at the front. Loosen the 2 at the front and drop the oil pan at the rear to allow the fluid to drain into a container. When drained, reinstall a few of the bolts to hold the pan in place.

3. Remove the converter bottom cover and remove the converter drain plug, to allow the converter to drain. After the converter has drained, reinstall the drain plug and tighten. Remove the converter to flywheel nuts by turning the converter to expose the nuts.

NOTE: Crank the engine over with a wrench on the crankshaft pulley attaching bolt.

4. Mark the position of the driveshaft on the rear axle flange and remove the driveshaft. Install a suitable plug in the transmission extension housing to prevent fluid leakage.

5. Disconnect the starter cable and remove the starter. Disconnect the wiring from the neutral safety switch.

6. Remove the mount-to-crossmember and crossmember-to-frame bolts. Remove the mount-to-transmission bolts.

7. Disconnect the shift and throttle valve cables from the transmission.

8. Remove the bellcrank bracket from the converter housing.

9. Position a suitable jack and raise the transmission. Remove the transmission mount and crossmember.

NOTE: It may be necessary to disconnect or remove interfering exhaust system components.

10. Lower the transmission to gain access to the oil cooler lines. Disconnect the oil cooler lines from the transmission.

11. Disconnect the speedometer cable from the extension housing.

12. Remove the transmission dipstick tube-to-engine block retaining

bolt and remove the tube and dipstick from the transmission.

13. Secure the transmission to the jack with a chain and remove the transmission-to-engine bolts.

14. Carefully pull the transmission and converter assembly rearward and lower them from the vehicle.

To install:

15. Tighten the converter drain plug to 8–28 ft. lbs. (11–38 Nm).

16. If removed, position the converter on the transmission and rotate into position to make sure the drive flats are fully engaged in the pump gear.

NOTE: Lubricate the pilot with chassis grease.

17. Raise the converter and transmission assembly into position. Rotate the converter until the studs and drain plug are in alignment with the holes in the flywheel. Align the orange balancing marks on the converter stud and flywheel bolt hole if balancing marks are present.

NOTE: The converter face must rest squarely against the flywheel. This indicates that the converter pilot is not binding in the engine crankshaft. To ensure the converter is properly seated, grasp a converter stud. It should move freely back and forth in the flywheel hole. If the converter will not move, the transmission must be removed and the converter repositioned so the impeller hub is properly engaged in the pump gear.

18. Install the transmission-to-engine attaching bolts. Tighten the bolts to 40–50 ft. lbs. (55–68 Nm).

19. Remove the safety chain from around the transmission.

20. Install a new O-ring on the lower end of the transmission dipstick tube and install the tube to the transmission case.

21. Connect the speedometer cable to the transmission case.

22. Connect the oil cooler lines to the right side of the transmission case.

23. Position the crossmember on the side supports. Position the rear mount on the crossmember and install the attaching bolt/nut.

24. Secure the engine rear support to the transmission extension housing.

25. Install any exhaust system components, if removed.

26. Lower the transmission and remove the jack.

27. Secure the crossmember to the side supports with the attaching bolts.

28. Connect the TV linkage and the manual linkage rod. Connect the shift cable.

29. Install the converter to flywheel attaching nuts and tighten to 20–34 ft. lbs. (27–46 Nm). Install the converter housing cover.

30. Secure the starter motor in place and connect all electrical connections.

31. Install the driveshaft, aligning the marks that were made during removal.

32. Lower the vehicle. Fill the transmission with the proper type and quantity of fluid, start the engine and check the transmission for leakage. Adjust the linkage as required.

Manual Linkage

ADJUSTMENT

1. From the passenger compartment, place the steering column selector lever in **OVERDRIVE** and hold the selector lever in position by placing a 3 lb. weight on the lever.

2. Place a prybar in the slot of the slide adjuster to open the adjuster.

3. Move transmission manual shift lever to **OVERDRIVE** position, 2nd detent from most rearward position.

4. Push slide adjuster closed.

5. Check the shift lever for proper operation. Ensure that park/neutral start switch is functioning properly.

Throttle Valve Cable

ADJUSTMENT

Automatic Overdrive Transmission

1. Set the parking brake and place the shift selector in **N**.

2. Remove the air cleaner cover and inlet tube from the throttle body inlet to access the throttle lever and cable.

3. Using a small prybar, pry the grooved pin on the cable assembly out of the grommet on the throttle body lever. Push out the white locking tab.

4. Check the plastic block with pin and tab; it should slide freely on the notched rod. If not, the white tab may not be pushed out far enough.

5. While holding the throttle lever firmly against the idle stop, push the grooved pin into the grommet on the throttle lever as far as it will go.

6. Make sure the throttle lever does not move while pushing the pin into the grommet.

7. Install the air cleaner cover and inlet tube.

DRIVE AXLE

Driveshaft and U-Joints

REMOVAL AND INSTALLATION

1. Raise and safely support the vehicle. Mark the position of the driveshaft yoke on the axle companion flange so they can be reassembled in the same way to maintain balance.

2. Remove the flange bolts and disconnect the driveshaft from the axle companion flange.

3. Allow the rear of the driveshaft to drop slightly. Pull the driveshaft and slip yoke out of the transmission extension housing.

4. Plug the transmission to prevent fluid leakage.

5. Place driveshaft on a suitable workbench with a vise.

6. Prior to disassembly, mark the positions of the driveshaft components relative to the driveshaft tube. All components must be reassembled in the same relationship to maintain proper balance.

7. Using a U-joint C-clamp mounted in the vise, press out one side bearing cup. Rotate the driveshaft 180 degrees and press out bearing cap.

8. Remove slip yoke from the U-joint. Remove remaining bearing cups in the same manner.

9. Clean all foreign matter from the yoke areas of the driveshaft.

To install:

10. Start a new bearing cup into the yoke of the driveshaft

11. Position the new U-joint in the driveshaft yoke and press the bearing cup ¼ in. (6.3mm) below the yoke surface with the C-clamp and install a new snapring.

12. Start a new bearing cup into the opposite side of the yoke. Check needles for proper position.

13. Position driveshaft in C-clamp and press the cup until the opposite cup contacts the snapring. Install snapring. Tap yoke with plastic hammer to seat the snaprings.

14. Lubricate the yoke splines and install the yoke into the transmission extension housing, aligning the

splines. Be careful not to bottom the slip yoke hard against the transmission seal.

15. Rotate the axle flange, as necessary, to align the marks made during removal. Install the driveshaft yoke to the axle flange. Install the bolts and tighten to 71–95 ft. lbs. (95–130 Nm).

Rear Axle Shaft, Bearing and Seal

REMOVAL AND INSTALLATION

1. Raise and safely support the vehicle. Remove wheel and tire assembly and remove brake drum or brake rotor.

2. If equipped, remove the anti-lock brake speed sensor.

3. Clean all dirt from the area of the carrier cover. Drain the axle lubricant by removing the housing cover.

4. Remove differential pinion shaft lock bolt and pinion shaft.

5. Push flanged end of axle shafts toward the center of the vehicle and

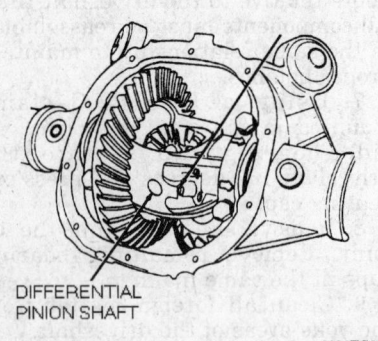

DIFFERENTIAL
PINION SHAFT

8838FG29

Removal of differential pinion shaft

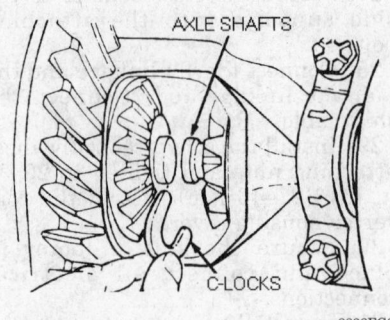

AXLE SHAFTS

C-LOCKS

8838FG30

Removing axle shaft C-lock clips

remove the C-lock from button end of the axle shaft. Remove the axle shaft from the housing, being careful not to damage the oil seal.

6. Insert wheel bearing and seal replacer tool T85L–1225–AH or equivalent, in the bore and position it behind the bearing so the tangs on the tool engage the bearing outer race. Remove bearing and seal as a unit using an impact slide hammer.

To install:

7. Lubricate the new bearing with rear axle lubricant. Install the bearing into the housing bore using a suitable bearing installer.

8. Install a new axle seal using a seal installer.

NOTE: Check for the presence of an axle shaft O-ring on the spline end of the shaft and install, if not present.

9. Carefully slide the axle shaft into the axle housing, without damaging the bearing or seal assembly. Start the splines into the side gear and push firmly until the button end of the axle shaft can be seen in the differential case.

10. Install the C-lock on the button end of the axle shaft splines, then push the shaft outboard until the

shaft splines engage and the C-lock seats in the counterbore of the differential side gear.

11. Insert the differential pinion shaft through the case and pinion gears, aligning the hole in the shaft with the lock bolt hole. Apply locking compound to the lock bolt and install in the case and pinion shaft. Tighten to 15–30 ft. lbs. (20–41 Nm).

12. Cover the inside of the differential case with a shop rag and clean the machined surface of the carrier and cover. Remove the shop rag.

13. Apply a bead of silicone sealer to the cover and install on the carrier. Tighten the bolts in a criss-cross pattern. Final torque the cover retaining bolts to 25–35 ft. lbs. (34–47 Nm).

14. Add rear axle lubricant to the carrier to a level $\frac{1}{4}$–$\frac{9}{16}$ in. below the bottom of the fill hole. If equipped with limited slip, add friction modifier C8AZ–19B564–A or equivalent. Install the filler plug and tighten to 15–30 ft. lbs. (20–41 Nm).

15. Install the anti-lock speed sensor, if equipped. Tighten the retaining bolt to 40–60 inch lbs. (4.5–6.8 Nm).

16. Install the brake calipers and rotors or the brake drums, as required. Install the wheel and tire assembly and lower the vehicle.

STEERING

Air Bag

— CAUTION —

Some vehicles are equipped with an air bag system, also known as the Supplemental Inflatable Restraint (SIR) or Supplemental Restraint System (SRS). The system must be disabled before performing service on or around system components, steering column, instrument panel components, wiring and sensors. Failure to follow safety and disabling procedures could result in accidental air bag deployment, possible personal injury and unnecessary system repairs.

PRECAUTIONS

Several precautions must be observed when handling the inflator

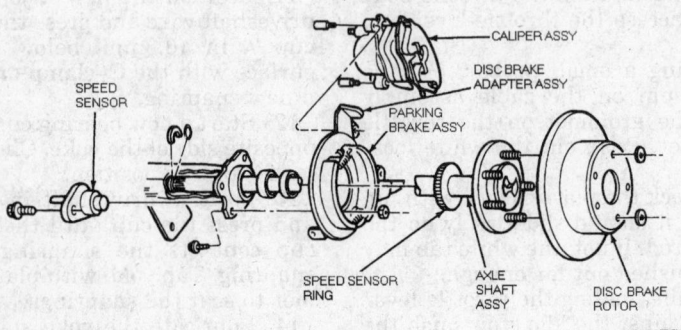

CALIPER ASSY

DISC BRAKE
ADAPTER ASSY

PARKING
BRAKE ASSY

SPEED
SENSOR

SPEED SENSOR
RING

AXLE
SHAFT
ASSY

DISC BRAKE
ROTOR

8838FG28

Axle shaft exploded view

module to avoid accidental deployment and possible personal injury.

• Never carry the inflator module by the wires or connector on the underside of the module.

• When carrying a live inflator module, hold securely with both hands, and ensure that the bag and trim cover are pointed away.

• Place the inflator module on a bench or other surface with the bag and trim cover facing up.

• With the inflator module on the bench, never place anything on or close to the module which may be thrown in the event of an accidental deployment.

DISARMING

Driver Side

1. Disconnect the positive battery cable. Wait 1 minute for the backup power supply in the diagnostic monitor to deplete its stored energy.

2. Remove the 4 nut and washer assemblies retaining the driver air bag module to the steering wheel.

─────── CAUTION ───────
When carrying a live air bag, make sure the bag and trim cover are pointed away from the body. In the unlikely event of an accidental deployment, the bag will then deploy with minimal chance of injury. When placing a live air bag on a bench or other surface, always face the bag and trim cover up, away from the surface. This will reduce the motion of the module if it is accidently deployed.

3. Disconnect the driver air bag connector.

Passenger Side

1. Remove the right-hand instrument panel lower moulding.

2. Remove the cluster finish panel retaining screws and remove the panel.

3. Open the glove compartment, press the sides inward and lower the glove compartment to the floor.

4. Remove the air bag module retaining bolts. Disconnect the electrical connector and remove the module.

Steering Wheel

─────── CAUTION ───────
If equipped with an air bag, the air bag system must be disarmed, before working on the system.

Failure to do so may result in deployment of the air bag and possible personal injury.

REMOVAL AND INSTALLATION

With Air Bag

1. Center the front wheels in the straight-ahead position.

2. Disarm the air bag system as follows:

   a. Disconnect the positive battery cable. Wait 1 minute for the backup power supply in the diagnostic monitor to deplete its stored energy.

   b. Remove the 4 nut and washer assemblies retaining the driver air bag module to the steering wheel and remove the air bag module.

─────── CAUTION ───────
When carrying a live air bag, make sure the bag and trim cover are pointed away from the body. In the unlikely event of an accidental deployment, the bag will then deploy with minimal chance of injury. When placing a live air bag on a bench or other surface, always face the bag and trim cover up, away from the surface. This will reduce the motion of the module if it is accidently deployed.

   c. Disconnect the driver air bag connector. Connect air bag simulator tool 105–00008 or equivalent, to the vehicle harness at the top of the steering wheel.

3. Disconnect the cruise control wire harness from the steering wheel, if equipped.

4. Remove and discard the steering wheel bolt. Remove the steering wheel using a suitable puller. Route the contact assembly wire harness

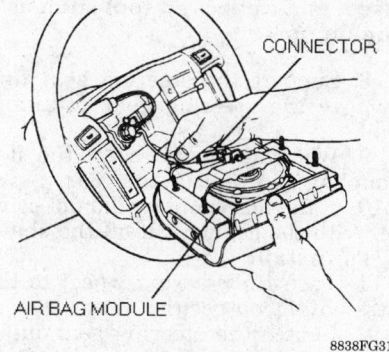

CONNECTOR

AIR BAG MODULE

8838FG31

Air bag module removal

through the steering wheel as the wheel is lifted off the shaft.

NOTE: Do not use a knock-off type steering wheel puller or strike the retaining bolt with a hammer. This could cause damage to the steering shaft bearing.

To install:

5. Make sure the front wheels are in the straight-ahead position.

6. Route the contact assembly wire harness through the steering wheel opening at the 3 o'clock position and install the steering wheel on the steering shaft. The steering wheel and shaft alignment marks should be aligned. Make sure the air bag contact wire is not pinched.

7. Install a new steering wheel retaining bolt and tighten to 23–33 ft. lbs. (31–48 Nm).

8. If equipped, connect the cruise control wire harness to the wheel and snap the connector assembly into the steering wheel clip. Make sure the wiring does not get trapped between the steering wheel and contact assembly.

9. Connect the air bag wire harness to the air bag module and install the module to the steering wheel. Tighten the module retaining nuts to 3–4 ft. lbs. (4–6 Nm).

10. Connect the air bag backup power supply and negative battery cable. Verify the air bag warning indicator.

Without Air Bag

1. Disconnect the negative battery cable.

2. Remove the horn pad and cover assembly. Disconnect the horn electrical connector.

3. Disconnect the cruise control switch electrical connector, if equipped.

4. Remove and discard the steering wheel bolt. Remove the steering wheel using a suitable puller.

NOTE: Do not use a knock-off type steering wheel puller or strike the retaining bolt with a hammer. This could cause damage to the steering shaft bearing.

To install:

5. Align the index marks on the steering wheel and shaft and install the steering wheel.

6. Install a new steering wheel retaining bolt and tighten to 30 ft. lbs. (41 Nm).

7. Connect the cruise control electrical connector, if equipped.

8. Connect the horn electrical connector and install the horn pad and cover.

9. Connect the negative battery cable.

Tie Rod Ends

REMOVAL AND INSTALLATION

1. Raise and support the vehicle safely.

2. Remove the cotter pin and nut from the tie rod end ball stud.

3. Loosen the tie rod adjusting sleeve clamp bolts and remove the rod end from the spindle arm or center link, using ball stud remover tool 3290–D or equivalent.

4. Remove the tie rod end from the sleeve, counting the exact number of turns required to do so.

To install:

5. Install the new tie rod end into the sleeve, using the exact number of turns it took to remove the old one. Install the tie rod end ball studs into the spindle arm or center link.

6. Install the stud and stud nut. Tighten to 43–47 ft. lbs. (59–63 Nm), then continue tightening the nut to align its next castellation with the cotter pin hole in the stud. Install a new cotter pin.

NOTE: Never loosen the nut to align the nut castellation and cotter pin hole.

7. Check the toe and adjust if necessary. Loosen the clamps from the sleeve and oil the sleeve, clamps, bolts and nuts. Position the adjusting sleeve clamps so the bolts are horizontal, with the threaded end pointing toward the front of the vehicle, and tighten the clamp nuts to 20–22 ft. lbs. (27–29 Nm).

Power Steering Gear

ADJUSTMENT

Adjust the total-over-center position load to eliminate excessive lash between the sector and rack teeth as follows:

1. Disconnect the pitman arm from the sector shaft.

2. Disconnect the fluid return line at the reservoir. Cap the reservoir return line pipe.

3. Place the end of the return line in a clean container and turn the steering wheel from left stop to right stop several times to discharge the fluid from the gear.

4. Turn the steering wheel to 45 degrees from the left stop.

5. Using a ft. lb. torque wrench on the steering wheel nut, determine the torque required to rotate the shaft slowly approximately 1/4 turn from the 45 degree position. If equipped with tilt column, place the steering wheel in the center tilt position.

6. Turn the steering wheel back to center and determine the torque required to rotate the shaft back and forth across the center position. If the reading is not to specification, loosen the nut and turn the adjuster screw until the reading is to specification. Tighten the wheel nut while holding the screw in place.

7. Check the readings and replace the pitman arm and steering wheel hub cover.

8. Connect the fluid return line to the reservoir and fill the reservoir. Check the belt tension and adjust, if necessary.

REMOVAL AND INSTALLATION

1. Disconnect the negative battery cable.

2. Remove the stone shield.

3. Tag the pressure and return lines so they may be reassembled in their original positions.

4. Disconnect the pressure and return lines from the steering gear. Plug the lines and ports in the gear to prevent the entry of dirt.

5. Remove the clamp bolts retaining the flexible coupling to the steering gear.

6. Raise and safely support the vehicle. Remove the nut from the sector shaft.

7. Remove the pitman arm from the sector shaft with pitman arm remover tool T64P–3590–F or equivalent. Remove the tool from the pitman arm.

NOTE: Do not damage the seals and/or gear housing. Do not use a non-approved tool such as a pickle fork.

8. Support the steering gear and remove the steering gear retaining bolts.

9. Work the gear free of the flex coupling and remove the gear.

10. If the flex coupling did not come off with the gear, lift it off the shaft.

To install:

11. Turn the steering wheel to the straight-ahead position.

12. Center the steering gear input shaft with the centerline of the 2 indexing flats at 4 o'clock.

13. Slide the steering gear input shaft into the flex coupling and into place on the frame side rail. Install the retaining bolts and tighten to 50–65 ft. lbs. (68–88 Nm).

14. Make sure the wheels are in the straight-ahead position. Install the pitman arm on the sector shaft and install the lockwasher and nut. Tighten the nut to 200–250 ft. lbs. (271–339 Nm). Install and tighten the sector shaft and retaining bolts.

15. Move the flex coupling into place on the steering gear input shaft. Install the retaining bolt and tighten to 20–30 ft. lbs. (27–41 Nm).

16. Connect the pressure and return lines to the steering gear and tighten the lines. Fill the reservoir and turn the steering wheel from stop-to-stop to distribute the fluid. Check the fluid level and add fluid, if necessary.

17. Start the engine and turn the steering wheel from left to right. Check for leaks. Install the stone shield.

Power Steering Pump

BLEEDING

1. Disconnect the ignition coil. Raise and safely support the vehicle so the front wheels are off the floor.

2. Fill the power steering fluid reservoir.

3. Crank the engine with the starter and add fluid until the level remains constant.

4. While cranking the engine, rotate the steering wheel from lock-to-lock.

NOTE: The front wheels must be off the floor during lock-to-lock rotation of the steering wheel.

5. Check the fluid level and add fluid, if necessary.

6. Connect the ignition coil wire. Start the engine and allow it to run for several minutes.

7. Rotate the steering wheel from lock-to-lock.

8. Shut **OFF** the engine and check the fluid level. Add fluid, if necessary.

9. If air is still present in the system, purge the system of air using power steering pump air evacuator tool 021–00014 or equivalent, as follows:

   a. Make sure the power steering pump reservoir is full to the COLD FULL mark on the dipstick or to just above the minimum indication on the reservoir.

b. Tightly insert the rubber stopper of the air evacuator assembly into the pump reservoir fill neck.

c. Apply 15 in. Hg maximum vacuum on the pump reservoir for a minimum of 3 minutes with the engine idling. As air purges from the system, vacuum will fall off. Maintain adequate vacuum with the vacuum source.

d. Release the vacuum and remove the vacuum source. Fill the reservoir to the COLD FULL mark or to just above the minimum indication on the reservoir.

e. With the engine idling, apply 15 in. Hg vacuum to the pump reservoir. Slowly cycle the steering wheel from lock-to-lock every 30 seconds for approximately 5 minutes. Do not hold the steering wheel on the stops while cycling. Maintain adequate vacuum with the vacuum source as the air purges.

f. Release the vacuum and remove the vacuum source. Add fluid, if necessary.

g. Start the engine and cycle the steering wheel. Check for oil leaks at all connections. In severe cases of aeration, it may be necessary to repeat Steps 9b–9f.

REMOVAL AND INSTALLATION

1. Disconnect the negative battery cable.

2. Disconnect the fluid return hose at the pump and drain the fluid into a container.

3. Remove the pressure hose from the pump and, if necessary, drain the fluid into a container. Do not remove the fitting from the pump.

4. Disconnect the belt from the pulley.

5. Remove the mounting bolts and remove the pump.

To install:

6. Place the pump on the mounting bosses of the engine block and install the bolts at the side of the pump. Tighten to 15–22 ft. lbs. (20–30 Nm).

7. Place the belt on the pump pulley and adjust the tension, if necessary.

8. Install the pressure hose to the pump fitting. Tighten the tube nut with a tube nut wrench rather than with an open-end wrench. Tighten to 20–25 ft. lbs. (27–34 Nm).

NOTE: Do not overtighten this fitting. Swivel and/or endplay of the fitting is normal and does not indicate a loose fitting. Overtightening the tube nut can col- lapse the tube nut wall, resulting in a leak and requiring replacement of the entire pressure hose assembly. Use of an open-end wrench to tighten the nut can deform the tube nut hex which may result in improper torque and may make further servicing of the system difficult.

9. Connect the return hose to the pump and tighten the clamp. Fill the reservoir with the proper type and quantity of fluid. Bleed the air from the system.

BRAKES

Anti-Lock Brake System Service

PRECAUTIONS

• Certain components within the Anti-Lock Brake System (ABS) are not intended to be serviced or repaired individually. Only those components with removal and installation procedures should be serviced.

• Do not use rubber hoses or other parts not specifically specified for and ABS system. When using repair kits, replace all parts included in the kit. Partial or incorrect repair may lead to functional problems and require the replacement of components.

• Lubricate rubber parts with clean, fresh brake fluid to ease assembly. Do not use lubricated shop air to clean parts; damage to rubber components may result.

• Use only specified brake fluid from an unopened container.

• If any hydraulic component or line is removed or replaced, it may be necessary to bleed the entire system.

• A clean repair area is essential. Always clean the reservoir and cap thoroughly before removing the cap. The slightest amount of dirt in the fluid may plug an orifice and impair the system function. Perform repairs after components have been thoroughly cleaned; use only denatured alcohol to clean components. Do not allow ABS components to come into contact with any substance containing mineral oil; this includes used shop rags.

• The Anti-Lock control unit is a microprocessor similar to other computer units in the vehicle. Ensure that the ignition switch is **OFF** before removing or installing controller harnesses. Avoid static electricity discharge at or near the controller.

• If any arc welding is to be done on the vehicle, the control unit should be unplugged before welding operations begin.

DEPRESSURIZING

— **CAUTION** —

Before servicing any component which contains high pressure, it is mandatory that the hydraulic pressure in the system be discharged or personal injury could result.

To discharge the system, turn the ignition **OFF** and pump the brake pedal a minimum of 20 times until an increase in pedal force is clearly felt.

Master Cylinder

REMOVAL AND INSTALLATION

1. Disconnect the negative battery cable.

2. If equipped with anti-lock brakes, depress the brake pedal several times to exhaust all vacuum in the system.

3. Remove the brake lines from the primary and secondary outlet ports of the master cylinder.

4. Disconnect the brake warning indicator connector.

5. If equipped with anti-lock brakes, disconnect the Hydraulic Control Unit (HCU) supply hose at the master cylinder and secure in a position to prevent loss of brake fluid.

6. Remove the nuts attaching master cylinder to the brake booster assembly.

7. Slide the master cylinder forward and upward from the vehicle.

To install:

8. If equipped with anti-lock brakes, install a new seal in the groove in the master cylinder mounting face.

9. Install the master cylinder on the booster studs and install the mounting nuts. Tighten the nuts to 13–25 ft. lbs. (18–34 Nm) on all except vehicles with anti-lock brakes. If equipped with anti-lock brakes, tighten the nuts to 16–21 ft. lbs. (21–29 Nm).

10. Install short brake lines in the master cylinder outlet ports and position them so they point back into the reservoir and the ends of the lines are submerged in brake fluid.

11. Fill the reservoir with brake fluid and cover the reservoir with a shop towel.

12. Pump the brakes until clear, bubble-free fluid comes out of both brake lines. If any brake fluid spills on the paint, wash it off immediately with water.

13. Remove the short brake lines and connect the vehicle brake lines to the master cylinder. Bleed each brake line at the master cylinder using the following procedure:

a. Have an assistant pump the brake pedal 10 times and then hold firm pressure on the pedal.

b. Crack the rear most brake line fitting with a tubing wrench until a stream of brake fluid comes out. Have the assistant maintain pressure on the brake pedal until the brake line fitting is tightened again.

c. Repeat this operation until clear, bubble free fluid comes out from around the brake line fitting.

d. Repeat this bleeding operation at the front brake line fitting.

14. Attach the HCU supply hose to the master cylinder.

15. Connect the brake warning indicator switch connector.

16. Bleed the system. Operate the brakes several times, then check for external hydraulic leaks.

Brake Caliper

REMOVAL AND INSTALLATION

Front

1. Raise and safely support the vehicle. Remove the front wheel and tire assembly.

2. Loosen the brake line fitting that connects the brake hose to the brake line at the frame bracket. Remove the retaining clip from the hose and bracket and disengage the hose from the bracket. Remove the hose from the caliper.

3. Remove the caliper locating pins and remove the caliper. If removing both calipers, mark the right and left sides so they may be reinstalled correctly.

To install:

4. Install the caliper over the rotor with the outer brake shoe against the rotor's braking surface.

5. Lubricate the inside of the locating pin insulators with silicone dielectric grease. Install the caliper locating pins and tighten to 45–60 ft. lbs. (61–81 Nm).

6. Install new sealing washers on each side of the brake hose fitting outlet and install the bolt, through the hose fitting and into the caliper. Tighten the bolt to 30 ft. lbs. (41 Nm).

7. Position the other end of the brake hose in the bracket and install the retaining clip. Make sure the hose is not twisted.

8. Connect the brake line to the brake hose and tighten the fitting nut.

9. Bleed the brake system, install the wheel and tire assembly and lower the vehicle.

10. Apply the brake pedal several times before moving the vehicle, to position the brake pads.

Rear

1. Raise and safely support the vehicle. Remove the rear wheel and tire assembly.

2. Remove the brake fitting retaining bolt from the caliper and disconnect the flexible brake hose from the caliper. Plug the hose and the caliper fitting.

3. Remove the caliper locating pins. Lift the caliper off the rotor and anchor plate using a rotating motion.

NOTE: Do not pry directly against the plastic piston or damage to the piston will occur.

To install:

4. Position the caliper assembly above the rotor with the anti-rattle spring located on the lower adapter support arm. Install the caliper over the rotor with a rotating motion. Make sure the inner pad is properly positioned.

5. Install the caliper locating pins and start them in the threads by hand. Tighten them to 19–25 ft. lbs. (26–34 Nm).

6. Install the brake hose on the caliper with a new gasket on each side of the fitting outlet. Insert the retaining bolt and tighten to 30–40 ft. lbs. (40–54 Nm).

7. Bleed the brake system, install the wheel and tire assembly and lower the vehicle.

8. Pump the brake pedal prior to moving the vehicle to position the linings.

Disc Brake Pads

REMOVAL AND INSTALLATION

Front

1. Remove and discard half the brake fluid from the master cylinder.

2. Raise and safely support vehicle. Remove the front wheel and tire assemblies.

3. Remove the caliper locating pins and remove the caliper from the anchor plate and rotor, but do not disconnect the brake hose.

4. Remove the outer brake pad from the caliper assembly and remove the inner brake pad from the caliper piston.

5. Inspect the disc brake rotor for scoring and wear. Replace or machine, as necessary.

6. Suspend the caliper inside the fender housing with a length of wire. Do not let the caliper hang by the brake hose.

To install:

7. Use a large C-clamp and wood block to push the caliper piston back into its bore.

8. Install the inner brake pad, then the outer brake pad, making sure the clips are properly seated.

9. Install the caliper and the wheel and tire assembly. Lower the vehicle.

10. Pump the brake pedal prior to moving the vehicle to seat the brake pads. Refill the master cylinder.

Rear

1. Remove and discard half the brake fluid from the master cylinder.

2. Raise and safely support vehicle. Remove the rear wheel and tire assemblies.

3. Remove the caliper locating pins and remove the caliper from the anchor plate and rotor, but do not disconnect the brake hose.

4. Remove the inner and outer brake pads.

5. Inspect the disc brake rotor for scoring and wear. Replace or machine, as necessary.

6. Suspend the caliper inside the fender housing with a length of wire. Do not let the caliper hang by the brake hose.

To install:

7. Use a large C-clamp and wood block to push the caliper piston back into its bore.

8. Install the inner brake pad, then the outer brake pad, making sure the clips are properly seated.

9. Install the caliper and the wheel and tire assembly. Lower the vehicle.

10. Pump the brake pedal prior to moving the vehicle to seat the brake pads. Refill the master cylinder.

Brake Rotor

REMOVAL AND INSTALLATION

Front

1. Raise and safely support the vehicle. Remove the wheel and tire assembly.
2. Remove the caliper, but do not disconnect the brake hose. Suspend the caliper inside the fender housing with a length of wire. Do not let the caliper hang by the brake hose.
3. Remove the rotor retaining push nuts, if equipped, and remove the rotor from the hub.
4. Inspect the rotor for scoring and wear. Replace or machine as necessary. If machining, observe the minimum thickness specification.
5. Installation is the reverse of the removal procedure.

Rear

1. Raise and safely support the vehicle. Remove the wheel and tire assembly.
2. Remove the caliper, but do not disconnect the brake hose. Suspend the caliper inside the fender housing with a length of wire. Do not let the caliper hang by the brake hose.
3. Remove the rotor retaining push nuts and remove the rotor from the hub.
4. Inspect the rotor for scoring and wear. Replace or machine as necessary. If machining, observe the minimum thickness specification.
5. Installation is the reverse of the removal procedure.

Brake Drums

REMOVAL AND INSTALLATION

1. Raise and safely support the vehicle.
2. Remove the wheel and tire assembly.
3. Remove the drum retaining clips and discard. Remove the brake drum.

NOTE: If the drum will not come off, pry the rubber plug from the backing plate. Insert a narrow rod through the hole in the backing plate and disengage the adjusting lever from the adjusting screw. While holding the adjustment lever away from the screw, back off the adjusting screw with a brake adjusting tool.

4. Inspect the brake drum for scoring and wear. Replace or machine as necessary. If machining, observe the maximum diameter specification.
5. Installation is the reverse of removal.

Brake Shoes

REMOVAL AND INSTALLATION

1. Raise and safely support the vehicle. Remove the rear wheel and tire assemblies. Remove the brake drum.
2. Remove the shoe-to-anchor springs and unhook the cable eye from the anchor pin. Remove the anchor pin plate.
3. Remove the shoe hold-down springs, shoes, adjusting screw, pivot nut, socket and automatic adjustment parts.
4. Remove the parking brake link, spring and retainer. Disconnect the parking brake cable from the parking brake lever.
5. After removing the rear brake secondary shoe, disassemble the parking brake lever from the shoe by removing the retaining clip and spring washer.

To install:

6. Before installing the rear brake shoes, assemble the parking brake lever to the secondary shoe and secure it with the spring washer and retaining clip.
7. Apply a light coating of caliper slide grease at the points where the brake shoes contact the backing plate. Be careful not to get any lubricant on the brake linings.
8. Position the brake shoes on the backing plate. The primary shoe with the short lining faces the front of the vehicle, the secondary shoe with the long lining, to the rear. Secure the assembly with the hold-down springs. Install the parking brake link, spring and retainer. Back-off the parking brake adjustment, then connect the parking brake cable to the parking brake lever.
9. Install the anchor pin plate on the anchor pin. Place the cable eye over the anchor pin with the crimped side toward the drum. Install the primary shoe to the anchor pin.
10. Install the cable guide on the secondary shoe web with the flanged hole fitted into the hole in the secondary shoe web. Thread the cable around the cable guide groove.

NOTE: The cable must be positioned in the groove and not between the guide and the shoe web.

11. Install the secondary shoe-to-anchor spring. Make sure the cable eye is not cocked or binding on the anchor pin when installed. All parts should be flat on the anchor pin.
12. Apply a thin coat of lubricant to the threads and the socket end of the adjusting screw. Turn the adjusting screw into the adjusting pivot nut to the limit of the threads, then back-off ½ turn.

NOTE: Make sure the socket end of the adjusting screw is stamped with an R or L, indicating the right or left side of the vehicle. The adjusting screw assemblies must be installed on the correct side for proper brake shoe adjustment.

13. Place the adjusting socket on the screw and install the assembly between the shoe ends with the adjusting screw toothed wheel nearest the secondary shoe.
14. Hook the cable hook into the hole in the adjusting lever. The adjusting levers are stamped with an **R** or **L** to indicate their installation on the right or left side.
15. Position the hooked end of the adjuster spring completely into the large hole in the primary shoe web. Connect the loop end of the spring to the adjuster lever hole.
16. Pull the adjuster lever, cable and automatic adjuster spring down and toward the rear, engaging the pivot hook in the large hole of the secondary shoe web.
17. Make sure the upper ends of the brake shoes are seated against the anchor pin and the shoes are centered on the backing plate.
18. Adjust the brakes using brake adjustment gauge D81L–1103–A or equivalent.
19. Install the brake drum, wheel and tire assemblies and lower the vehicle.
20. Apply the brakes several times while backing up the vehicle. After each stop, the vehicle must be moved forward.

Wheel Cylinder

REMOVAL AND INSTALLATION

1. Remove the wheel and tire assembly and the brake drum.
2. Remove the brake shoe assembly.
3. Disconnect the brake line from the wheel cylinder at the backing plate.

4. Remove the wheel cylinder attaching bolts and remove the wheel cylinder.

5. Installation is the reverse of the removal procedure. Tighten the wheel cylinder attaching bolts to 10–20 ft. lbs. (14–28 Nm).

6. Bleed the brake system.

Parking Brake Cable

ADJUSTMENT

NOTE: The following procedure is to be used only if a new parking brake control assembly is installed. All components of the parking brake system must be installed prior to the adjustment procedure. The parking brake control with automatic tensioning is preset by means of a shipping clip. The following procedure must be followed in sequence and must be done with the vehicle weight on the axle.

1. Verify removal of the shipping clip. The take up reel will apply tension to the system.

2. Depress the parking brake control to the 8th notch.

3. Push the parking brake control pedal to release.

4. Check function as follows:

    a. Apply the parking brake with a full stroke, to the 9th or 10th notch.

    b. Release the parking brake by shifting the vehicle into a forward gear with the engine running. The control must release.

    c. Apply the parking brake with a full stroke, to the 9th or 10th notch.

    d. Manually release the parking brake with the push to release feature.

NOTE: With the control in the OFF position, the rear brakes must not drag. Check for movement of the rear cables from their conduits when the intermediate cable is deflected with a force of 10–15 lbs.

REMOVAL AND INSTALLATION

Front Cable

1. Raise and safely support the vehicle.

2. Disconnect the cable from the rear of the cable connector located along the left frame side rail.

3. Use a 13mm box end wrench to depress the retaining tabs and remove the conduit retainer from the

frame. Remove screw holding the plastic inner fender apron to the frame, at the rear of the fender panel.

4. Pull back the fender apron. If equipped, remove the spring clip retainer that holds the parking brake cable to the frame.

5. Pull the cable through the frame and let it hang in the wheel housing. Lower the vehicle.

6. Inside the passenger compartment, remove the sound deadener cover from the cable at the dash panel.

7. Pull the cable until the parking brake control take up spring tang is at full clockwise position. Use a fabricated tool to retain the reel spring and disconnect the cable from the take up reel.

8. Using a 13mm box end wrench, depress the retaining tabs and remove the conduit from the cable assembly. Push the cable down through the dash panel and remove cable from inside the wheel housing.

9. Installation is the reverse of the removal procedure. Check the parking brake adjustment.

Rear Cables

With Drum Brakes

1. Raise and safely support the vehicle.

2. Remove the wheel and tire assemblies and the brake drums.

3. Working on the wheel side of the rear brake, remove the brake automatic adjuster spring. Compress the prongs on the parking brake cable so they can pass through the hole in the backing plate. Pull the cable retainer through the hole.

4. With the tension off the cable spring at the parking brake lever, lift the cable end out of the slot in the lever. Remove the cable through the backing plate hole.

5. Installation is the reverse of the removal procedure. Adjust the parking brake, if necessary.

With Disc Brakes

1. Raise and safely support the vehicle. Disconnect the control cable from the rear cable at the connector.

2. Disconnect the parking brake cable retainer spring at the frame, if equipped with dual exhaust.

3. Disconnect the left cable from the right cable at the adjuster bracket. Release the right cable tabbed conduit retainer from the frame, using a 13mm box end wrench.

4. Remove the cable retainer from the left shock bracket, the wire retainer on the left axle bracket and

disconnect the cable from the retainer on the right axle tube by removing the bolt and retainer.

5. Remove the cable retaining E-clip and cable eyelet from the brake lever. Pull the cable out of the disc brake adapter boss. Remove the cables.

6. Installation is the reverse of the removal procedure. Check parking brake operation.

Brake System Bleeding

PROCEDURE

Without Anti-Lock Brakes

1. Clean all dirt from the master cylinder filler cap.

2. If the master cylinder is known or suspected to have air in the bore, it must be bled before any of the wheel cylinders or calipers. To bleed the master cylinder, loosen the upper secondary left front outlet fitting approximately ¾ turn. Have an assistant depress the brake pedal slowly through it's full travel. Close the outlet fitting and let the pedal return slowly to the fully released position. Wait 5 seconds and then repeat the operation until all air bubbles disappear.

3. Repeat Step 2 with the right-hand front outlet fitting.

4. Continue to bleed the brake system by removing the rubber dust cap from the wheel cylinder bleeder fitting or caliper fitting at the right-hand rear of the vehicle. Place a suitable box wrench on the bleeder fitting and attach a rubber drain tube to the fitting. The end of the tube should fit snugly around the bleeder fitting. Submerge the other end of the tube in a container partially filled with clean brake fluid and loosen the fitting ¾ turn.

5. Have an assistant push the brake pedal down slowly through it's full travel. Close the bleeder fitting and allow the pedal to slowly return to it's full release position. Wait 5 seconds and repeat the procedure until no bubbles appear at the submerged end of the bleeder tube. Secure the bleeder fitting and remove the bleeder tube. Install the rubber dust cap on the bleeder fitting.

6. Repeat the procedure in Steps 4 and 5 in the following sequence: left rear, right front, left front. Refill the master cylinder reservoir after each wheel cylinder or caliper has been bled and install the master cylinder cover and gasket. When brake bleeding is completed, the fluid level

should be filled to the maximum level indicated on the reservoir.

7. Always make sure the disc brake pistons are returned to their normal positions by depressing the brake pedal several times until normal pedal travel is established. If the pedal feels spongy, repeat the bleeding procedure.

With Anti-Lock Brakes

NOTE: The anti-lock brake system must be bled in 2 steps.

The master cylinder and hydraulic control unit must be bled using the Rotunda Anti-Lock Brake Breakout Box/Bleeding Adapter tool T90P–50–ALA or equivalent. If this procedure is not followed, air will be trapped in the hydraulic control unit which will eventually lead to a spongy brake pedal. To bleed the master cylinder and the hydraulic control unit, disconnect the 55-pin plug from the electronic control unit and install the Anti-Lock Brake Breakout Box/Bleeding Adapter to the wire harness 55-pin plug.

1. Place the Bleed/Harness switch in the **BLEED** position.

2. Turn the ignition to the ON position. At this point the red off light should come ON.

3. Push the motor button on the adapter down to start the pump motor. The red light will turn OFF and the green light will turn ON. The pump motor will run for 60 seconds after the motor button is pushed. If the pump motor is to be turned off for any reason before the 60 seconds has elapsed, push the ABORT button to turn the pump motor OFF.

   a. After 20 seconds of pump motor operation, push and hold the VALVE button down. Hold the VALVE button down for 20 seconds and then release it.

   b. The pump motor will continue to run for an additional 20 seconds after the valve button is released.

4. The brake lines can now be bled in the normal fashion. Bleed the brake system by removing the rubber dust cap from the caliper fitting at the right-hand rear of the vehicle. Place a suitable box wrench on the bleeder fitting and attach a rubber drain tube to the fitting. The end of the tube should fit snugly around the bleeder fitting. Submerge the other end of the tube in a container partially filled with clean brake fluid and loosen the fitting ¾ turn.

5. Have an assistant push the brake pedal down slowly through it's full travel. Close the bleeder fitting and allow the pedal to slowly return

to it's full release position. Wait 5 seconds and repeat the procedure until no bubbles appear at the submerged end of the bleeder tube. Secure the bleeder fitting and remove the bleeder tube. Install the rubber dust cap on the bleeder fitting.

6. Repeat the bleeding procedure at the left front, left rear and right front in that order. Refill the master cylinder reservoir after each caliper has been bled and install the master cylinder and gasket. When brake bleeding is completed, the fluid level should be filled to the maximum level indicated on the reservoir.

7. Always make sure the disc brake pistons are returned to their normal positions by depressing the brake pedal several times until normal pedal travel is established. If the pedal feels spongy, repeat the bleeding procedure.

Wheel Speed Sensor

REMOVAL AND INSTALLATION

Front

1. Disconnect the negative battery cable.

2. From inside engine compartment, disconnect sensor assembly 2-pin connector from the wiring harness.

3. Remove the steel routing clip attaching the sensor wire to the tube bundle on the left sensor or remove the plastic routing clip attaching the sensor wire to the frame on the right sensor.

4. Remove the rubber coated spring steel clip holding the sensor wire to the frame.

5. Remove the sensor wire from the steel routing clip on the frame and from the dust shield.

6. Remove the sensor attaching bolt from the front spindle and slide the sensor out of the mounting hole.

To install:

7. Install the sensor into the mounting hole in the front spindle and attach with the mounting bolt. Tighten to 40–60 inch lbs. (4.5–6.8 Nm).

8. Insert the sensor routing grommets into the dust shield and steel bracket on the frame. Route the wire into the engine compartment.

9. Install the rubber coated steel clip that holds the sensor wire to frame into the hole in the frame.

10. Install the steel clip that holds sensor wire to tube bundle on left side or plastic clip that holds sensor to frame on right side.

11. Connect the 2-pin connector to wire harness. Connect the negative battery cable.

Rear

1. Disconnect the negative battery cable.

2. From inside luggage compartment, disconnect 2-pin sensor connector from wiring harness and push sensor wire through hole in floor.

3. From below vehicle, remove sensor wire from routing bracket located on top of rear axle carrier housing and remove steel clip holding sensor wire and brake tube against axle housing.

4. Remove screw from clip holding sensor wire and brake tube to bracket on axle.

5. Remove sensor to rear adapter retaining bolt and remove sensor.

To install:

6. Insert sensor adapter and install retaining bolt. Tighten to 40–60 inch lbs. (4.5–6.8 Nm).

7. Attach clip holding sensor and brake tube to bracket on axle housing and secure with screw. Tighten to 40–60 inch lbs. (4.5–6.8 Nm).

8. Install steel clip around axle tube that holds sensor wire and brake tube against axle tube and push spool-shaped grommet into clip located on top of axle carrier housing.

9. Push sensor wire connector up through hole in floor and seat large round grommet into hole.

10. Push sensor wire connector up through hole in floor and seat large round grommet into hole.

11. Connect sensor 2-pin connector to wiring harness inside luggage compartment.

FRONT SUSPENSION

Shock Absorber

REMOVAL AND INSTALLATION

NOTE: Purge a new shock of air by repeatedly extending it in its normal position and compressing it while inverted.

1. Remove the nut, washer and bushing from the upper end of the shock absorber.

2. Raise and safely support the vehicle by the frame rails allowing the front wheels to hang.

3. Remove the 2 bolts securing the shock absorber to the lower control arm and remove the shock absorber.

To install:

4. Install a new bushing and washer on the top of the shock absorber and position the unit inside the front spring. Install the 2 lower attaching bolts and torque them to 12–18 ft. lbs. (16–24 Nm).

5. Lower the vehicle.

6. Place a new bushing and washer on the shock absorber top stud and install a new attaching nut. Tighten to 22–26 ft. lbs. (30–41 Nm).

Coil Spring

REMOVAL AND INSTALLATION

1. Raise and safely support the vehicle. Remove the wheel and tire assembly.

2. Remove the shock absorber. Remove the steering link from the pitman arm.

3. Using spring compressor tool D78P–5310–A or equivalent, install 1 plate with the pivot ball seat facing downward into the coils of the spring. Rotate the plate, so it is flush with the upper surface of the lower arm.

4. Install the other plate with the pivot ball seat facing upward into the coils of the spring. Insert the upper ball nut through the coils of the spring, so the nut rests in the upper plate.

5. Insert the compression rod into the opening in the lower arm, through the upper and lower plate and upper ball nut. Insert the securing pin through the upper ball nut and compression rod.

NOTE: This pin can only be inserted 1 way into the upper ball nut because of a stepped hole design.

6. With the upper ball nut secured, turn the upper plate so it walks up the coil until it contacts the upper spring seat. Then back off ½ turn.

7. Install the lower ball nut and thrust washer on the compression rod and screw on the forcing nut. Tighten the forcing nut until the spring is compressed enough so it is free in its seat.

8. Remove the 2 lower arm pivot bolts, disengage the lower arm from the frame crossmember and remove the spring.

9. If a new spring is to be installed, perform the following:

   a. Mark the position of the upper and lower plates on the spring with chalk.

   b. With an assistant, compress a new spring for installation and measure the compressed length and the amount of curvature of the old spring.

10. Loosen the forcing nut to relieve the spring tension and remove the tools from the spring.

To install:

11. Assemble the spring compressor and locate in the same position as indicated in Step 10a.

12. Before compressing the coil spring, make sure the upper ball nut securing the pin is inserted properly.

13. Compress the coil spring until the spring height reaches the dimension obtained in Step 10b.

14. Position the coil spring assembly into the lower arm and reverse the removal procedure.

Upper Ball Joints

REMOVAL AND INSTALLATION

1. Raise and safely support the vehicle with safety stands under the frame behind the lower arm. Remove the wheel and tire assembly.

2. Position a floor jack under the lower arm at the lower ball joint area. The floor jack will support the spring load on the lower arm.

3. Remove the retaining nut and pinch bolt from the upper ball joint stud.

4. Mark the position of the alignment cams. When replacing the ball joint this will approximate the current alignment.

5. Remove the 2 nuts retaining the ball joint to the upper arm. Remove the ball joint and spread the slot with a suitable prybar to separate the ball joint stud from the spindle.

To install:

NOTE: The upper ball joints differ from side to side. Be sure to use the proper ball joint on each side.

6. Position the ball joint on the upper arm and insert the ball stud into the spindle.

7. Install the pinch bolt and retaining nut. Tighten to 51–67 ft. lbs. (68–92 Nm).

8. Install the alignment cams to the approximate position at removal.

If not marked, install in neutral position.

9. Install the 2 nuts attaching the ball joint to the arm. Hold the cams and tighten the nuts to 90–109 ft. lbs. (122–149 Nm).

10. Remove the floor jack from the lower arm and install the wheel and tire assembly. Remove the safety stands and lower the vehicle.

11. Check and adjust the front end alignment.

Lower Ball Joints

REMOVAL AND INSTALLATION

The ball joint is an integral part of the lower control arm. If the ball joint is defective, the entire lower control arm must be replaced.

Upper Control Arms

REMOVAL AND INSTALLATION

1. Raise and safely support the vehicle on safety stands positioned on the frame just behind the lower arm.

2. Remove the wheel and tire assembly and position a floor jack under the lower arm.

3. Remove the retaining nut from the upper ball joint stud to spindle pinch bolt. Tap the pinch bolt to remove from the spindle.

4. Using a suitable prybar, spread the slot to allow the ball joint stud to release out of the spindle.

5. Remove the upper arm retaining bolts and the upper arm.

To install:

6. Transfer the rebound bumper from the old arm to the new arm, or replace the bumper if worn or damaged.

7. Use reference marks from the camber and caster cams as initial settings.

8. Position the upper arm shaft to the frame bracket. Install the 2 retaining bolts and washers. Position the arm in the center of the slot adjustment range and tighten to 100 ft. lbs. (136 Nm).

9. Connect the upper ball joint stud to the spindle and install the retaining pinch bolt and nut. Tighten the nut to 52–66 ft. lbs. (70–90 Nm).

10. Install the wheel and tire assembly and lower the vehicle. Check the front end alignment.

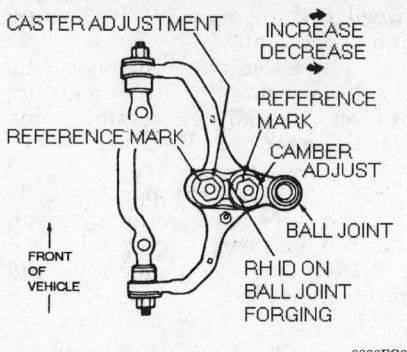

CASTER ADJUSTMENT

INCREASE
DECREASE

REFERENCE
MARK

REFERENCE MARK

CAMBER
ADJUST

FRONT
OF
VEHICLE

BALL JOINT

RH ID ON
BALL JOINT
FORGING

8838FG32

Upper control arm

Lower Control Arms

REMOVAL AND INSTALLATION

1. Raise the front of the vehicle and position safety stands on the frame behind the lower control arms. Remove the wheel and tire assembly.
2. Remove the brake caliper and suspend with a length of wire; do not let the caliper hang by the brake hose. Remove the brake rotor and dust shield. Remove the anti-lock brake sensor, if equipped.
3. Remove the jounce bumper; inspect and save for installation if in good condition. Remove the shock absorber.
4. Disconnect the steering center link from the pitman arm.
5. Remove the cotter pin and loosen the lower ball joint stud nut 1–2 turns.

NOTE: Do not remove the nut at this time.

6. Install a suitable ball joint press tool to place the ball joint stud under compression. With the stud under compression, tap the spindle

BALL JOINT COVER

NEW
OK

WORN IF BELOW
SURFACE OF COVER

CHECKING SURFACE

8838FG33

Lower ball joint checking cover surface

sharply with a hammer to loosen the stud in the spindle. Remove the ball joint press tool.
7. Place a floor jack under the lower arm and install a suitable spring compression tool.
8. Remove the coil spring, the ball joint nut and remove the lower control arm.
To install:
9. Position the arm assembly ball joint stud into the spindle and install the nut. Tighten to 80–120 ft. lbs. (108–163 Nm). Continue to tighten until the slot for the cotter pin is aligned. Install a new cotter pin.
10. Position the coil spring into the upper spring pocket and raise the lower arm, aligning the holes in the arm with the holes in the crossmember. Install the bolts and nuts with the washer installed on the front bushing. Do not tighten at this time.

NOTE: Make sure the pigtail of the lower coil of the spring is in the proper location of the seat on the lower arm, between the 2 holes.

11. Remove the spring compressor tool.
12. Connect the steering center link at the pitman arm and install the nut. Tighten to 43–47 ft. lbs. (59–63 Nm). Continue to tighten until the slot for the cotter pin is aligned. Install a new cotter pin.
13. Install the shock absorber and the jounce bumper.
14. Install the dust shield, rotor and caliper. Install the anti-lock brake sensor, if equipped.
15. Install the wheel and tire assembly and lower the vehicle. With the vehicle supported on the wheels and tires at normal curb height, tighten the lower control arm-to-crossmember bolts to 100–140 ft. lbs. (136–190 Nm).
16. Check the front end alignment.

Sway Bar

REMOVAL AND INSTALLATION

1. Raise the front of the vehicle and place jackstands under the lower control arms.
2. Remove the retaining nuts from the pinch bolts at the spindles. Spread the slots in the spindles with a prybar to free the ball studs.

3. Remove the stabilizer bar brackets from the frame and remove the stabilizer bar. If worn, cut the insulators from the stabilizer bar.
4. Remove the retaining nuts from the ball joint studs at the end of the bar. Use removal tool 3290–D or equivalent to separate the links from the ends of the stabilizer bar.
To install:
5. Coat the necessary parts of the stabilizer bar with rubber lubricant. Slide new insulators onto the stabilizer bar.
6. Install the ball joint links into the ends of the bar with the retaining nuts. Tighten to 30–40 ft. lbs. (40–55 Nm).
7. Position the bar under the vehicle and engage the upper ball joint links to the spindles. Install the insulator brackets with the retaining nuts. Tighten the pinch bolts and nuts at the spindles to 30–40 ft. lbs. (40–55 Nm). Tighten the bracket-to-frame nuts to 44–59 ft. lbs. (59–81 Nm).

Front Wheel Bearings

ADJUSTMENT

The front wheel bearings are of a hub unit design and are pregreased, sealed and require no maintenance. The bearings are preset and cannot be adjusted.

REMOVAL AND INSTALLATION

1. Raise and safely support the vehicle. Remove the wheel and tire assembly.
2. Remove and discard the grease cap from the hub.
3. Remove the brake caliper. Suspend the caliper with a length of wire; do not let it hang from the brake hose.
4. Remove the rotor. Remove and discard the wheel hub nut.
5. Remove the hub and bearing assembly.
To install:
6. Install the hub and bearing assembly. Install a new wheel hub nut and tighten to 238 ft. lbs. (322 Nm).
7. Install the rotor and a new grease cap. Install the brake caliper.
8. Install the wheel and tire assembly and lower the vehicle.

REAR SUSPENSION

Shock Absorber

REMOVAL AND INSTALLATION

Without Automatic Leveling

1. If equipped with air suspension, turn the air suspension switch **OFF**.
2. Raise and safely support the vehicle. Make sure the rear axle is supported.
3. Remove the shock absorber retaining nut, washer and insulator from the stud on the upper side of the frame. Discard the nut. Compress the shock to clear the hole in the frame and remove the inner insulator and washer from the upper retaining stud.

NOTE: All vehicles, except police applications, are equipped with gas pressurized shock absorbers which will extend unassisted.

4. Remove the self-locking retaining nut and disconnect the shock absorber lower stud from the mounting bracket on the rear axle.

To install:

5. Prime the new shock absorber as follows:
   a. With the shock absorber right side up, extend it fully.
   b. Turn the shock upside down and fully compress it.
   c. Repeat the previous 2 steps at least 3 times to make sure any trapped air has been expelled.
6. Place the inner washer and insulator on the upper retaining stud and position the shock absorber with the stud through the hole in the frame.
7. While holding the shock absorber in position, install the outer insulator, washer and a new stud nut on the upper side of the frame. Tighten the nut to 21 ft. lbs. (29 Nm).
8. Extend the shock absorber and place the lower stud in the mounting bracket hole on the rear axle housing. Install a new self-locking nut and tighten to 52–85 ft. lbs. (70–115 Nm).
9. Lower the vehicle and, if equipped, turn the air suspension switch **ON**.

With Automatic Leveling

NOTE: Disconnect the height sensor connector link before allowing the rear axle to hang free.

Then, raise the vehicle on a hoist so the suspension arms hang free with the ignition switch in the OFF position. The rear shock absorbers will vent air through the compressor and a hissing noise will be heard. When the noise stops, the air lines can be disconnected. A residual pressure of 8–24 psi will remain in the air lines.

1. Disconnect the air line by pushing in on the retainer ring and pulling the line out.
2. Remove the top retaining nut, washer and bushing.
3. Remove the bottom retaining nut and washer. Remove the shock absorber.

To install:

4. Position the shock absorber and install the bottom retaining washer and nut. Tighten to 52–85 ft. lbs. (70–115 Nm).
5. Install the top bushing, washer and retaining nut. Tighten to 14–26 ft. lbs. (19–35 Nm).

NOTE: Check the rubber sleeve on the shock absorber to be sure it is not wrapped up. To assist in identifying wrap-up during installation, a white stripe is on the rubber sleeve and on the shock absorber body. The stripes should align. To correct a wrap-up condition, loosen the upper shock retaining nut and turn the shock to align the stripes. Retighten the retaining nut.

6. Connect the air line to the shock absorber by pushing in on the retainer ring and installing the air line.
7. Connect the height sensor connecting link and lower the vehicle.

Coil Spring

REMOVAL AND INSTALLATION

1. Raise the vehicle and support the rear axle housing. Place jackstands under the frame side rails.
2. Remove the rear stabilizer bar, if equipped.
3. Disconnect the lower studs of both rear shock absorbers from the mounting brackets on the axle tube.
4. Unsnap the right parking brake cable from the right upper arm retainer before lowering the axle.
5. Lower the axle housing until the coil springs are released. Remove the springs and insulators.

To install:

6. Position the spring in the upper and lower seats with an insulator between the upper end of the spring and frame seat.
7. Raise the axle and connect the shock absorbers to the mounting brackets. Install new retaining nuts and tighten to 52–85 ft. lbs. (70–115 Nm).
8. Snap the right parking cable into the upper arm retainer. Install the stabilizer bar, if equipped.
9. Remove the jackstands and lower the vehicle.

Air Springs

REMOVAL AND INSTALLATION

1. Turn the air suspension switch **OFF**.
2. Raise and safely support the vehicle on the frame. The suspension must be at full rebound.
3. Remove the heat shield, as required. Remove the spring retainer clip. Remove the air spring solenoid as follows:
   a. Disconnect the electrical connector and then disconnect the air line.
   b. Remove the solenoid clip.
   c. Rotate the solenoid counterclockwise to the first stop.
   d. Pull the solenoid straight out slowly to the 2nd stop to bleed air from the system.

— CAUTION —
Do not fully release the solenoid until the air is completely bled from the air spring or personal injury may result.

   e. After the air is fully bled from the system, rotate counterclockwise to the 3rd stop and remove the solenoid from the solenoid housing. Remove the large O-ring from the solenoid housing.
4. Remove the spring piston-to-axle spring seat as follows:
   a. Insert air spring removal tool T90P–5310–A or equivalent, between the axle tube and the spring seat on the forward side of the axle.
   b. Position the tool so its flat end rests on the piston knob. Push downward, forcing the piston and retainer clip off the axle spring seat.
5. Remove the air spring.

To install:

6. Install the air spring solenoid as follows:
   a. Check the solenoid O-rings for cuts or abrasion. Replace the O-rings as required. Lightly grease the O-ring area of the solenoid and

the larger solenoid housing O-ring with silicone dielectric compound.

b. Insert the solenoid into the air spring end cap and rotate clockwise to the 3rd stop, push into the 2nd stop, then rotate clockwise to the 1st stop.

c. Install the solenoid clip. Inspect the wire harness connector and ensure the rubber gasket is in place at the bottom of the connector cavity.

7. Install the air spring into the frame spring seat, taking care to keep the solenoid air and electrical connections clean and free of damage.

8. Connect the push on spring retainer clip to the knob of the spring cap from the top side of the frame spring seat.

9. Connect the air line and electrical connector to the solenoid. Install the heat shield to frame spring seat, if required.

10. Align the air spring piston to axle seats. Squeeze to increase pressure and push downward on the piston, snapping the piston to axle seat at rebound and supported by the shock absorber.

NOTE: The air springs may be damaged if the suspension is allowed to compress before the spring is inflated.

11. Refill the air spring as follows:

a. Turn the air suspension switch **ON**. The ignition switch must be **ON** and the engine running or a battery charger must be connected to the battery to reduce battery drain.

b. Remove the right luggage compartment trim panel and connect Super Star II tester 007–0041–A or equivalent to the air suspension diagnostic connector.

c. Set the tester to EEC-IV/MCU mode. Also set the tester to FAST mode. Release the tester button to the HOLD (up) position and turn the tester **ON**.

d. Depress the tester button to TEST (down) position. A Code **10** will be displayed. Within 2 minutes a Code **13** will be displayed. After Code **13** is displayed, release the tester button to HOLD (up) position, wait 5 seconds and depress the tester button to TEST (down) position. Ignore any codes displayed.

e. Release the tester button to the HOLD (up) position. Wait at least 20 seconds, then depress the tester button to TEST (down) posi-

tion. Within 10 seconds, the following codes will be displayed in the order shown.

f. Within 4 seconds after Code **26** is displayed, release the tester button to the HOLD (up) position. Waiting longer than 4 seconds may result in Functional Test **31** being entered. The compressor will fill the air springs with air as long as the tester button is in the HOLD (up) position. To stop filling the air springs, depress the tester button to the TEST (down) position.

NOTE: It is possible to overheat the compressor during this operation. If the compressor overheats, the self-resetting circuit breaker in the compressor will open and remain open for about 15 minutes. This allows the compressor to cool down.

g. To exit Functional Test **26**, disconnect the tester and turn the ignition switch **OFF**.

Upper Control Arms

REMOVAL AND INSTALLATION

NOTE: If both arms are to be replaced, remove and install 1 at a time to prevent the axle from rolling or slipping sideways.

1. If equipped, turn the air suspension switch **OFF**.
2. Raise the vehicle and support the frame side rails with jackstands.
3. Support the rear axle under the differential pinion nose as well as under the axle.
4. Unsnap the parking brake cable from the upper arm retainer. If equipped, disconnect the height sensor from the ball stud on the left upper control arm.
5. Remove and discard the nut and bolt retaining the upper arm to the axle housing. Disconnect the arm from the housing.
6. Remove and discard the nut and bolt retaining the upper arm to the frame bracket and remove the arm.
To install:
7. Hold the upper arm in place on the front arm bracket and install a new retaining bolt and self-locking nut. Do not tighten at this time.
8. Secure the upper arm to the axle housing with new retaining bolts and nuts. The bolts must be pointed toward the front of the vehicle.
9. Raise the suspension with a jack until the upper arm rear pivot

hole is in position with the hole in the axle bushing. Install a new pivot bolt and nut with the nut facing inboard.
10. Tighten the upper arm-to-axle pivot bolts to 103–133 ft. lbs. (140–180 Nm) and upper arm-to-frame pivot bolts to 120–150 ft. lbs. (162–203 Nm).
11. Snap the parking brake cable into the upper arm retainer. Connect the height sensor to the ball stud on the left upper arm, if equipped.
12. Remove the supports from the frame and axle and lower the vehicle. If equipped, turn the air suspension switch **ON**.

Lower Control Arms

REMOVAL AND INSTALLATION

1. If equipped, turn the air suspension switch **OFF**.
2. Raise the vehicle and support the frame side rails with jackstands.
3. Remove the stabilizer bar, if equipped.
4. Support the axle with jackstands under the differential pinion nose as well as under the axle.
5. Remove and discard the lower arm pivot bolts and nuts and remove the lower arm.
To install:
6. Position the lower arm to the frame bracket and axle. Install new bolts and nuts.
7. Raise the axle. Tighten the lower arm-to-axle pivot bolt to 103–133 ft. lbs. (140–180 Nm) and lower arm-to-frame pivot bolt to 120–150 ft. lbs. (162–203 Nm).
8. Install the stabilizer bar, if equipped.
9. Remove the jackstands and lower the vehicle. If equipped, turn the air suspension switch **ON**.

Sway Bar

REMOVAL AND INSTALLATION

1. If equipped with air suspension, the air suspension switch, located on the right-hand side of the luggage compartment, must be turned to the **OFF** position before raising the vehicle.
2. Raise and safely support the vehicle.
3. Place suitable jack stands under the frame side rails.
4. The rear axle housing must hang unsupported allowing the shock absorbers to fully extend.

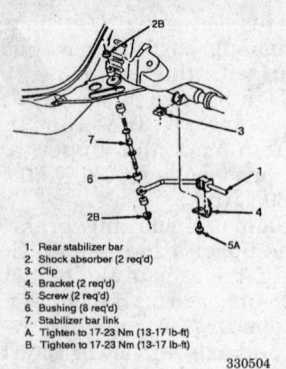

1. Rear stabilizer bar
2. Shock absorber (2 req'd)
3. Clip
4. Bracket (2 req'd)
5. Screw (2 req'd)
6. Bushing (8 req'd)
7. Stabilizer bar link
A. Tighten to 17-23 Nm (13-17 lb-ft)
B. Tighten to 17-23 Nm (13-17 lb-ft)

330504

Sway bar and link assembly

5. Remove the sway bar link retaining nuts, washers and insulators from the sway bar ends.

6. Remove 2 bolts retaining both sway bar brackets to the rear axle and remove the sway bar brackets and the sway bar.

7. Inspect the sway bar rubber insulators and replace, if necessary.

To install:

8. Install new insulators on the sway bar, if removed.

9. Apply a suitable rubber insulator lubricant to the sway bar insulators.

10. Install 2 sway bar brackets onto the sway bar insulators and hook both brackets into the T-slot of the rear axle bracket. Install 2 retaining bolts and tighten to 16–21 ft. lbs. (21–29 Nm).

11. Install the sway bar links to the sway bar ends using insulators, nuts and washers. Tighten to 13–17 ft. lbs. (17–23 Nm).

12. Remove the jack stands.

13. Lower the vehicle.

14. If equipped with air suspension, turn the air suspension switch to the **ON** position.

15. Road test the vehicle and check for proper operation.

Wheel Bearings

REMOVAL AND INSTALLATION

1. Raise and safely support the vehicle. Remove wheel and tire assembly and remove brake drum or brake rotor.

2. If equipped, remove the anti-lock brake speed sensor.

3. Clean all dirt from the area of the carrier cover. Drain the axle lubricant by removing the housing cover.

4. Remove differential pinion shaft lock bolt and pinion shaft.

5. Push flanged end of axle shafts toward the center of the vehicle and remove the C-lock from button end of the axle shaft. Remove the axle shaft from the housing, being careful not to damage the oil seal.

6. Insert wheel bearing and seal replacer tool T85L–1225–AH or equivalent, in the bore and position it behind the bearing so the tangs on the tool engage the bearing outer race. Remove bearing and seal as a unit using an impact slide hammer.

To install:

7. Lubricate the new bearing with rear axle lubricant. Install the bearing into the housing bore using a suitable bearing installer.

8. Install a new axle seal using a seal installer.

NOTE: Check for the presence of an axle shaft O-ring on the spline end of the shaft and install, if not present.

9. Carefully slide the axle shaft into the axle housing, without damaging the bearing or seal assembly. Start the splines into the side gear and push firmly until the button end of the axle shaft can be seen in the differential case.

10. Install the C-lock on the button end of the axle shaft splines, then push the shaft outboard until the shaft splines engage and the C-lock seats in the counterbore of the differential side gear.

11. Insert the differential pinion shaft through the case and pinion gears, aligning the hole in the shaft with the lock bolt hole. Apply locking compound to the lock bolt and install in the case and pinion shaft. Tighten to 15–30 ft. lbs. (20–41 Nm).

12. Cover the inside of the differential case with a shop rag and clean the machined surface of the carrier and cover. Remove the shop rag.

13. Apply a bead of silicone sealer to the cover and install on the carrier. Tighten the bolts in a criss-cross pattern. Final torque the cover retaining bolts to 25–35 ft. lbs. (34–47 Nm).

14. Add rear axle lubricant to the carrier to a level $1/4–9/16$ in. below the bottom of the fill hole. If equipped with limited slip, add friction modifier C8AZ–19B564–A or equivalent. Install the filler plug and tighten to 15–30 ft. lbs. (20–41 Nm).

15. Install the anti-lock speed sensor, if equipped. Tighten the retaining bolt to 40–60 inch lbs. (4.5–6.8 Nm).

16. Install the brake calipers and rotors or the brake drums, as required. Install the wheel and tire assembly and lower the vehicle.

GM "A" BODY

Front Wheel Drive

BUICK-Century OLDSMOBILE-Cutlass Ciera • Cutlass Cruiser

FIRING ORDERS

NOTE: To avoid confusion, always replace spark plug wires one at a time.

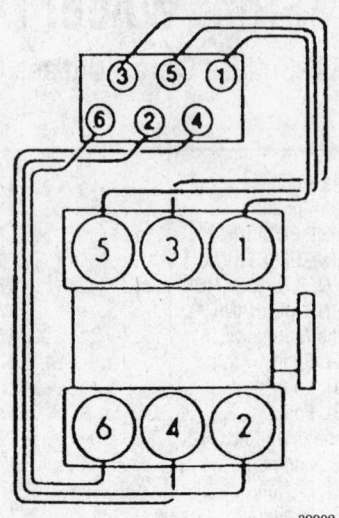

3.3L (VIN N) Engine
Firing Order: 1–6–5–4–3–2
Distributorless Ignition System

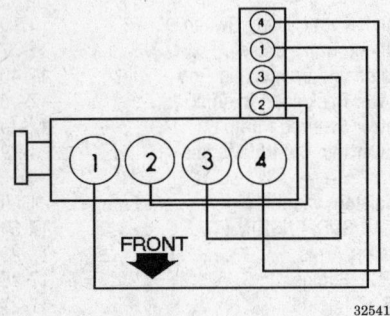

1993 2.2L (VIN 4) Engine
Firing Order: 1–3–4–2
Distributorless Ignition System

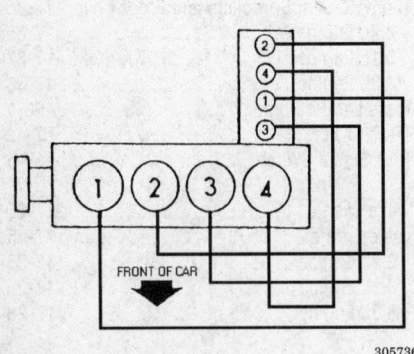

1994–95 2.2L (VIN 4) Engine
Firing Order: 1–3–4–2
Distributorless Ignition System

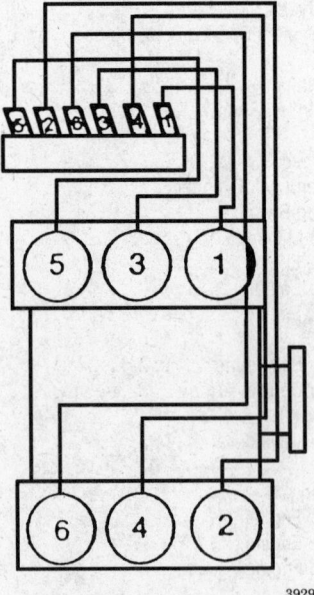

3.1 (VIN M) Engine
Firing Order: 1–2–3–4–5–6
Distributorless ignition system

ENGINE ELECTRICAL

NOTE: Disconnecting the negative battery cable on some vehicles may interfere with the functions of the on board computer systems and may require the computer to undergo a relearning process, once the negative battery cable is reconnected.

Alternator

PRECAUTIONS

Several precautions must be observed with alternator equipped vehicles to avoid damage to the unit.

• If the battery is removed for any reason, make sure it is reconnected with the correct polarity. Reversing the battery connections may result in damage to the 1–way rectifiers.

• When utilizing a booster battery as a starting aid, always connect the positive to positive terminals and the negative terminal from the booster battery to a good engine ground on the vehicle being started.

• Never use a fast charger as a booster to start vehicles.

• Disconnect the battery cables when charging the battery with a fast charger.

• Never attempt to polarize the alternator.

• Do not use test lights of more than 12 volts when checking diode continuity.

• Do not short across or ground any of the alternator terminals.

• The polarity of the battery, alternator and regulator must be matched and considered before making any electrical connections within the system.

• Never separate the alternator on an open circuit. Make sure all connections within the circuit are clean and tight.

• Disconnect the battery ground terminal when performing any service on electrical components.

• Disconnect the battery if arc welding is to be done on the vehicle.

REMOVAL AND INSTALLATION

2.2L (VIN 4) and 3.3L (VIN N) Engines

1. Disconnect the negative battery cable.

2. Disconnect the electrical connector from the alternator case. Remove the protective cover from the rear of the alternator and remove the nut securing the battery wire to the alternator post.

3. Remove the serpentine belt.

4. Remove the 3 (4-cyl.) or 2 (6-cyl.) alternator mounting bolts.

5. Remove the alternator.

To install:

6. Position the alternator in the mounting bracket and install the mounting bolts.

7. Tighten the front mounting bolt(s) to 37 ft. lbs. (50 Nm) and the rear mounting bolt to 18 ft. lbs. (25 Nm).

8. Install the serpentine belt.

9. Connect the battery wire to the alternator post and tighten the securing nut to 71 inch lbs. (8 Nm). Install the cover over the battery wire and post.

10. Connect the electrical connector to the alternator.

11. Connect the negative battery cable.

3.1L (VIN M) Engine

1. Disconnect the the negative battery cable.

2. Disconnect the electrical connector from the alternator.

3. Remove the serpentine belt.

4. Remove bolt, stud and nuts from the alternator braces.

5. Remove the lower alternator bolt.

6. Remove the plastic cover from the post on the back of the alternator. Remove the nut and disconnect battery positive lead from alternator output terminal.

7. Remove the alternator.

8. Remove the alternator bracket bolts.

9. Remove the alternator bracket from the alternator.

To install:

10. Install the alternator bracket on the alternator and tighten the bracket bolt to 37 ft. lbs. (50 Nm).

11. Install the alternator.

12. Connect the battery positive lead and nut to alternator output terminal. Tighten the mounting nut to 71 inch lbs. (8 Nm). Install the protective cap.

13. Install the lower alternator bolt and tighten to 37 ft. lbs. (50 Nm).

14. Install the brace stud and bolts and nuts to alternator. Tighten the brace nuts and stud to 18 ft. lbs. (25 Nm).

15. Install the serpentine belt.

16. Connect the electrical connector to the alternator.

17. Connect the negative battery cable.

Drive Belt

REMOVAL AND INSTALLATION

2.2L (VIN 4) Engine

1. Disconnect negative battery cable.

2. Using a 15mm wrench, pivot tensioner and remove belt from alternator.

3. Remove the serpentine belt.

To Install:

4. Route the serpentine belt as follows:

   a. Under the crankshaft and A/C compressor, if equipped.

   b. Over the tensioner with the back of the belt driving the pulley.

   c. Around the water pump with the back of the belt driving the pulley.

   d. Around the power steering pump.

5. Rotate the tensioner and loop the belt around the alternator.

6. Release the tensioner.

7. Verify the belt is properly seated in all the pulleys.

8. Connect the negative battery cable.

3.3L (VIN N) Engine

1. Disconnect negative battery cable.

2. Rotate the serpentine belt tensioner counterclockwise using a 18mm breaker bar.

3. With the tension removed from the belt, slip the belt off of the alternator pulley.

4. Release the tensioner slowly.

5. Remove the serpentine belt from the remainder of the pulleys.

To Install:

6. Install the serpentine belt as follows:

   a. Loop the belt under the crankshaft.

   b. Bring the belt around the water pump pulley so the back of the belt drives the pulley.

   c. Loop the belt under A/C compressor pulley.

   d. Route the belt over the power steering pulley.

   e. Bring the belt under the tensioner assembly.

7. Rotate the belt tensioner counterclockwise using a 18mm breaker bar.

8. Install the serpentine belt over the alternator pulley.

9. Release the tensioner and verify the belt is properly seated on the drive pulleys.

10. Connect the negative battery cable.

3.1L (VIN M) Engine

The route many serpentine belts use can be very complicated. A sketch of the belt routing before removal may save much time at installation. An incorrectly installed belt could cause the water pump to turn backwards from its normal rotation which could lead to engine overheating and possible engine damage.

1. Disconnect the negative battery cable.

2. Remove belt shield bolt and belt shield from above the water pump pulley.

3. Rotate the belt tensioner counterclockwise using a 3/8 inch breaker bar.

4. With the tension removed from the belt, remove the serpentine belt from around the alternator pulley.

5. Slowly release the tensioner.

6. Remove the belt from the remaining pulleys.

To install:

7. Loop the serpentine belt under the crankshaft and A/C compressor pulleys.

8. Route the belt around the water pump pulley so the back of the belt drives the pulley.

9. Route the belt around the tensioner pulley so the back of the belt drives the pulley.

10. Route the belt over the power steering pump pulley.

11. Rotate the belt tensioner counterclockwise using a 3/8 inch breaker bar.

12. Loop the serpentine around the alternator pulley.

13. Release the tensioner and verify the belt is properly seated in the the drive pulleys.

14. Install the belt shield and mounting bolt and tighten to 89 inch lbs. (10 Nm).

15. Connect the negative battery cable.

Starter

REMOVAL AND INSTALLATION

2.2L (VIN 4) Engine

1. Disconnect the negative battery cable.

2. Remove the torque strut from the vehicle.

3. Raise and safely support the vehicle.

4. Remove the mounting bolts from the flywheel inspection cover and remove the cover.

5. Remove the bolt securing the starter front bracket to the engine block.

6. Tag and disconnect the starter electrical connections.

7. Remove the two starter mounting bolts and remove the starter.

8. If the starter is being replaced, remove the two bracket mounting nuts and remove the bracket.

To install:

9. Install the front bracket on the starter and tighten the nuts to 80 inch lbs. (9 Nm).

10. Connect the electrical connectors to the starter. Tighten the solenoid wire nut to 22 inch lbs. (2.5 Nm) and the battery terminal wire nut to 12 ft. lbs. (16 Nm).

11. Install the starter in the vehicle and tighten the two starter mounting bolts to 32 ft. lbs. (43 Nm).

12. Install the starter bracket to engine bolt and tighten to 26 ft. lbs. (32 Nm).

13. Install the flywheel cover and tighten the mounting bolts to 89 inch lbs. (10 Nm).

14. Lower the vehicle.

15. Install the torque strut.

16. Connect the negative battery cable.

3.3L (VIN N) Engine

1. Disconnect the negative battery cable.

2. Raise and safely support the vehicle.

3. Disconnect the battery cable and solenoid wire from the starter.

4. Remove the flywheel cover bolts and flywheel cover.

5. Remove the two starter mounting bolts.

6. Remove the starter by rotating the solenoid side toward the front of the vehicle.

To Install:

7. Install the starter into the vehicle.

8. Install the two starter mounting bolts and tighten to 32 ft. lbs. (43 Nm).

9. Install the flywheel cover and cover mounting bolts.

10. Connect the battery cable and solenoid wire to the starter. Tighten the battery terminal bolt to 12 ft. lbs. (16 Nm) and the solenoid wire nut to 27 inch lbs. (3 Nm).

11. Lower the vehicle.

12. Connect the negative battery cable.

3.1L (VIN M) Engine

1. Disconnect the negative battery cable.

2. Raise and safely support the vehicle.

3. Disconnect the wiring clip from the front frame rail.

4. Remove the bolts from the flywheel cover and remove the cover.

5. Disconnect the electrical connections from the starter.

6. Remove the starter mounting bolts and remove the starter.

To install:

7. Position the starter in the vehicle and tighten the starter mounting bolts to 32 ft. lbs. (43 Nm).

8. Connect the electrical wiring to the starter. Tighten the battery cable nut to 12 ft. lbs. (16 Nm) and the solenoid wire nut to 27 inch lbs. (3 Nm).

9. Install the flywheel cover and cover bolts.

10. Lower the vehicle.

11. Connect the negative battery cable.

CHASSIS ELECTRICAL

Blower Motor

REMOVAL AND INSTALLATION

All Vehicles with 2.2L (VIN 4) Engine

1. Disconnect the negative battery cable.

2. Remove the air inlet resonator.

3. Remove the torque strut mounting bolt from the engine side bracket and rotate the engine forward.

4. Disconnect the blower motor electrical connector.

5. Disconnect the blower motor vent tube.

6. Remove the blower motor retaining screws and remove the blower motor.

To install:

7. Position the blower motor in the blower case and install the mounting screws.

8. Connect the blower motor vent tube and electrical connector.

9. Rotate the engine rearward and connect the torque strut to the engine

side mounting bracket. Tighten the through bolt to 41 ft. lbs. (55 Nm).

10. Install the air inlet resonator.

11. Connect the negative battery cable.

All Vehicles with 3.1L (VIN M) or 3.3L (VIN N) Engines

1. Disconnect the negative battery cable.

2. Remove the serpentine belt.

3. Remove the alternator from the mounting bracket and position it out of the way.

4. Disconnect the blower motor electrical connector.

5. Disconnect the blower motor vent tube.

6. Remove the blower motor retaining screws and remove the blower motor.

To install:

7. Position the blower motor in the blower case and install the mounting screws.

8. Connect the blower motor vent tube and electrical connector.

9. Install the alternator in the mounting bracket and tighten the mounting bolts.

10. Install the serpentine belt.

11. Connect the negative battery cable.

Windshield Wiper Motor

REMOVAL AND INSTALLATION

Front Wiper Motor

1. Disconnect negative battery cable.

2. Raise hood and remove cowl screen or front cowl panel.

3. Loosen transmission drive link to motor crank arm attaching nuts.

4. Disconnect wiring and washer hoses.

5. Remove the three motor attaching screws.

6. Remove motor while guiding crank arm through hole.

To install:

7. Install motor while guiding crank arm through hole.

8. Install the three motor attaching screws.

9. Connect wiring and washer hoses.

10. Tighten motor crank arm attaching nuts to transmission drive link. Make sure motor is in park position before assembling crank arm to transmission dive link.

11. Install cowl screen or cowl panel.

12. Connect negative battery cable.

Rear Wiper Motor

1. Disconnect the negative battery cable.
2. Remove the wiper arm assembly.
3. Remove the nut and washer from the wiper motor drive shaft.
4. Remove the liftgate glass opening upper finishing molding.
5. Disconnect the electrical connector.
6. Remove the two mounting screws.
7. Guide the wiper motor assembly from the tailgate recess.
   To install :
8. Position the wiper motor assembly into tailgate recess.
9. Install the two mounting screws and tighten to 53 inch lbs. (6 Nm).
10. Install the washer and nut and tighten the nut to 62 inch lbs. (7 Nm).
11. Connect the electrical connector.
12. Install the liftgate glass opening upper finishing molding.
13. Install the wiper arm assembly.
14. Connect the negative battery cable.
15. Verify proper operation.

Headlight Switch

REMOVAL AND INSTALLATION

Century

1. Disconnect the negative battery cable.
2. Remove the left lower sound insulator.
3. Remove the ashtray, disconnect the electrical connector from the ashtray track assembly. Remove the mounting screws from the track assembly.
4. Remove the DLC and position it aside.
5. Remove the four screws and remove the knee bolster.
6. Remove the seven screws securing the left side trim plate to the instrument panel.
7. Apply the parking brake and shift the transaxle selector lever into the **1** detent.
8. Pull the trim plate away from the instrument panel to disengage the two clips and remove the trim plate.
9. Remove the two switch mounting screws and remove the switch from the instrument panel.
   To install:
10. Position the switch in the instrument cluster, making sure the

electrical connector is fully seated. Install the mounting screws.
11. Install the instrument cluster trim plate, carefully engaging the two center mounted clips in the instrument panel.
12. Place the transaxle range selector in the **P** detent.
13. Install the seven instrument panel trim plate screws.
14. Install the drivers side knee bolster and mounting screws.
15. Position the DLC and tighten the mounting screws.
16. Install the ashtray assembly mounting screws, connect the electrical connector and install the ashtray.
17. Install the left lower sound insulator.
18. Connect the negative battery cable.

1993–95 Cutlass Ciera

1. Disconnect the negative battery cable.
2. Remove the steering column collar by carefully prying it rearward to release the five mounting clips.
3. Remove the left and right side outside air deflector outlets.
4. Working through the deflector outlet openings remove the mounting bolt and screw securing the trim plate to the dashboard.
5. Remove the bolt in the steering column collar opening.
6. Open the ashtray and remove the two trim plate screws.
7. Place the transaxle range selector into the **1** detent.
8. Pull the trim plate rearward to disengage the mounting clips. Remove the trim plate.
9. Remove the cluster trim plate mounting screws and remove the plate.
10. Remove the headlight switch mounting screws and remove the switch from the instrument panel.
    To install:
11. Install the headlight switch into the dashboard, making sure the electrical connector is fully seated. Install the mounting screws.
12. Install the instrument cluster trim plate and install the mounting screws.
13. Line up the clips on the back of the accessory trim plate and firmly press the plate into place.
14. Place the range selector into the **P** detent.
15. Install the trim plate mounting screws.
16. Install the air deflector outlet vents.

17. Install the steering column filler panel making sure all five clips engage.
18. Connect the negative battery cable.

1996 Cutlass Ciera and Cutlass Cruiser

1. Disconnect the negative battery cable.
2. Remove the accessory trim plate.
3. Remove the mounting screws from the switch.
4. Remove switch from its contact.
   To install:
5. Install the switch, making sure to push it solidly into the connector.
6. Install the switch mounting screws. Tighten the mounting screws to 18 inch lbs. (2 Nm).
7. Install the accessory trim plate.
8. Connect the negative battery cable.

Turn Signal Switch

REMOVAL AND INSTALLATION

───── CAUTION ─────
The Supplemental Inflatable Restraint (SIR) system must be disarmed before removing the steering wheel. Failure to do so may cause accidental deployment of the air bag, resulting in unnecessary SIR system repairs and/or personal injury.

1. Properly disarm the SIR system using the recommended procedure.
2. Ensure the wheels are in the straight ahead position and remain that way through the entire procedure.
3. Disconnect the negative battery cable.
4. Remove the coil assembly retaining ring.
5. Remove the inflatable restraint coil assembly. Let the switch hang freely if removal is not needed.
6. Remove the wave washer.
7. Remove the shaft lock retaining ring using special tool J-23653-C to push down shaft lock. Dispose of the ring.
8. Remove the shaft lock.
9. Remove the turn signal cancelling cam assembly.
10. Remove the upper bearing spring.
11. Remove the upper bearing inner race seat.
12. Remove the inner race.
13. Position the turn signal to **RIGHT TURN** position.

14. Remove the multi-Function lever and hazard knob assembly.

15. Remove the screw and signal switch arm.

16. Remove the turn signal switch screws.

17. Disconnect the turn signal switch connector from the bulkhead protector and pull wire harness through the column. Remove the turn signal switch assembly.

To install:

18. Install the turn signal switch and connect switch connector to bulkhead protector.

19. Install the turn signal switch screws.

20. Install the screw and signal switch arm.

21. Install the multi-Function lever and hazard knob assembly.

22. Install the inner race assembly.

23. Install the upper bearing inner race seat.

24. Install the upper bearing spring.

25. Install the turn signal cancelling cam assembly.

26. Install the new shaft lock retaining ring.

27. Install the wave washer.

28. Install the inflatable restraint coil assembly, ensuring it is centered and was not allowed to rotate while removed from steering column.

29. Install the coil assembly retaining ring.

30. Connect the negative battery cable.

31. Enable SIR system.

Combination Switch

REMOVAL AND INSTALLATION

───── CAUTION ─────

The Supplemental Inflatable Restraint (SIR) system must be disarmed before working on or near

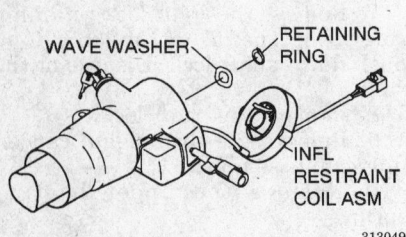

WAVE WASHER · RETAINING RING · INFL RESTRAINT COIL ASM

313049

Removing coil assembly off shaft

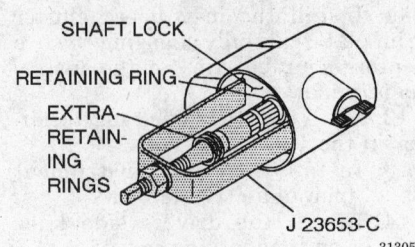

SHAFT LOCK · RETAINING RING · EXTRA RETAINING RINGS · J 23653-C

313050

Removing shaft lock retaining ring

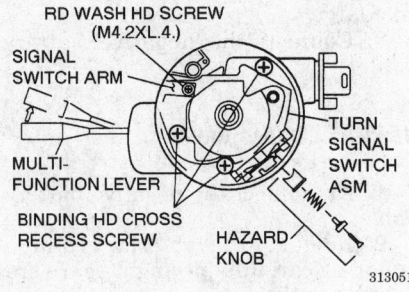

RD WASH HD SCREW (M4.2XL.4.) · SIGNAL SWITCH ARM · MULTI-FUNCTION LEVER · BINDING HD CROSS RECESS SCREW · TURN SIGNAL SWITCH ASM · HAZARD KNOB

313051

Turn signal switch removal

the steering column. Failure to do so may cause accidental deployment of the air bag, resulting in unnecessary SIR system repairs and/or personal injury.

1. Properly disarm the SIR system.

2. Disconnect the negative battery cable.

3. Remove the steering wheel.

4. Remove the self tapping screws securing the upper cover to the lower cover and remove the upper cover from the steering column.

5. Remove the two straps from the steering column wiring harness and separate the combination switch harness from the remainder of the column wiring.

6. Disconnect the interlock solenoid electrical connector.

7. Disconnect the pass key wiring harness from the grey combination switch connector.

8. Remove the two screws securing the combination switch to the steering column.

9. Remove the combination switch.

To Install:

10. Position the combination switch on the steering column and using a small blade screwdriver compress the electrical contact and move the multifunction switch into position. The electrical contact must rest on the cancelling cam assembly.

11. Tighten the mounting screws to 46 inch lbs. (5 Nm).

12. Connect the pass key wiring harness to the grey turn signal switch connector.

13. Connect the interlock solenoid electrical connector to the combination switch harness.

14. Route the combination switch wiring in with the remainder of the steering column wiring and install the wiring straps.

15. Install the upper cover on the steering column and tighten the three self tapping screws.

16. Install the steering column in the vehicle.

17. Install the steering wheel.

18. Connect the negative battery cable.

19. Enable the SIR system.

20. Verify proper operation of the steering column and all column mounted switches.

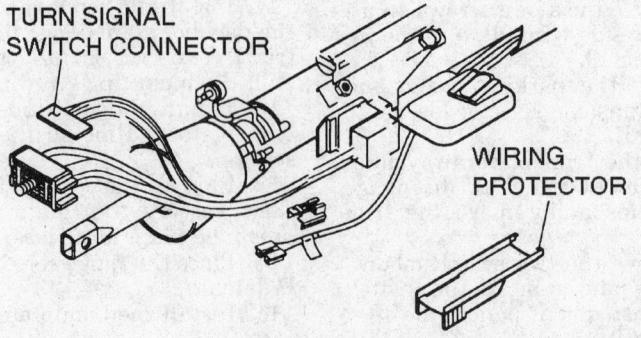

TURN SIGNAL SWITCH CONNECTOR · WIRING PROTECTOR

313052

Removing turn signal switch

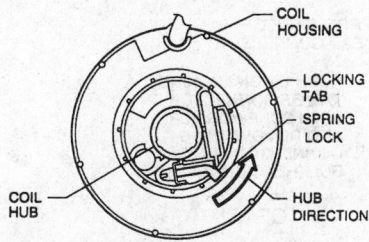

PERFORM THE FOLLOWING STEPS TO CENTER COIL ASSEMBLY

A. WHEELS STRAIGHT AHEAD.
B. REMOVE COIL ASSEMBLY.
C. HOLD COIL ASSEMBLY WITH BOTTOM UP.
D. WHILE HOLDING COIL ASSEMBLY, DEPRESS SPRING LOCK TO ROTATE HUB IN DIRECTION OF ARROW UNTIL IT STOPS.
E. THE COIL RIBBON SHOULD BE WOUND UP SNUG AGAINST CENTER HUB.
F. ROTATE COIL HUB IN OPPOSITE DIRECTION APPROXIMATELY TWO AND A HALF (2-1/2) TURNS. RELEASE SPRING LOCK BETWEEN LOCKING TABS.

313053

Centering coil assembly

Ignition Lock Cylinder

REMOVAL AND INSTALLATION

With Air Bag

———— **CAUTION** ————
The Supplemental Inflatable Restraint System (SIR) must be disarmed before performing service around SIR components or wiring. Failure to do so may cause accidental deployment of the air bag, resulting in unnecessary SIR repairs and/or personal injury.

NOTE: When replacing the lock cylinder, the key code must be read with interrogator tool J 35628-A or equivalent. A new key must be ordered with the new lock cylinder and must be cut to dummy key in the new lock cylinder.

1. Disconnect the negative battery cable.
2. Disable the air bag system by removing the fuse from the fuse block.
3. Remove the 2 screws in back of the steering wheel
4. Position the inflator module to gain access to the wire connectors. Note the wiring routing for reinstal-

lation. Push in and rotate the horn contact lead counterclockwise and remove the steering column horn tower. Remove the horn contact lead.
5. Disconnect the air bag connector retainer and then disconnect the connector. Remove the air bag inflator module and store trim side face up for safety reasons.

———— **CAUTION** ————
When carrying a live air bag, make sure the air bag and trim cover are pointed away from the body. In an unlikely event of an accidental deployment, the bag will then deploy with minimal chance of injury. When placing a live air bag on a bench or other surface, always face the bag and trim cover up, away from the surface. This will reduce the motion of the module if it is accidently deployed.

6. Remove the steering wheel control switch connector, if equipped.
7. Remove the steering wheel fastener. Install steering wheel removal tools J 1859-03 and J 28720 or equivalent, and remove the steering wheel while guiding the wires through the steering wheel hub. Mark the steering wheel to shaft position, prior to removal for re-installation purposes.
8. Remove the SIR coil retaining ring.
9. Remove the left side sound insulators from under the dash.
10. Remove the wire connectors from the retainers and disconnect the bulkhead connector to remove the turn signal switch connector.
11. Gently pull up on the SIR coil assembly while pushing up on the SIR coil harness on the bottom side of the steering column and let the coil hang freely.
12. Remove the turn signal canceling cam assembly and remove the screws to the turn signal switch arm and turn signal switch-to-steering column, then remove the hazard switch button.
13. Remove the upper bearing spring. Pull up gently on the turn signal switch assembly and let it hang, then remove the key buzzer switch.
14. Remove the 13-way secondary lock on the turn signal switch connector at the lower end of the steering column and remove terminals 12 and 13 which run to the ignition lock cylinder.
15. Attach a piece of wire to terminals for guiding the harness through the steering column.

16. Remove the retaining bolt to the ignition lock cylinder and insert the key in the lock cylinder to remove.
   To install:

———— **CAUTION** ————
Route wires from the pass key lock cylinder as shown and the retaining clip into hole in housing. Failure to do so may result in component damage or the malfunctioning of the steering column may occur.

17. Install the pass key lock cylinder. Route the wires correctly and insert the terminals in the turn signal connector.
18. Install the lock cylinder fastener and torque to 22 inch lbs. (2.5 Nm)
19. Install the buzzer switch and the turn signal switch assembly. Torque the bolts to 30 inch lbs. (3.4 Nm). When installing the switch be sure to remove all slack out of the wiring harness by pulling gently on the harness.

———— **CAUTION** ————
SIR coil assembly wires must be kept tight with no slack while installing the SIR coil assembly. Failure to do so may cause the wires to be kinked near the lock plate, causing the wires to be cut when the steering wheel is turned.

20. Install the turn signal switch arm and torque to 20 inch lbs. (2.3 Nm). Install the hazard switch button
21. Install the upper bearing spring.
22. Install the turn signal canceling cam assembly and lock plate.
23. Using compressor tool J 23653-SIR or equivalent, install the retaining ring after compressing the upper bearing spring.
24. Install the wave washer and SIR coil assembly, pulling gently on the harness on the bottom side of the steering column to remove slack from the wiring harness. Be sure to align the SIR coil assembly properly.
25. Install the snaring on top of the SIR coil.
26. Install the steering wheel, guiding the wires through for the air bag and steering wheel controls, if equipped.
27. Connect the negative battery cable.
28. Enable the SIR system by installing the fuse. Count the flashes on the SIR warning light. It should flash 7 times and turn OFF.

29. Test the turn signals, high beams and wipers.

30. Install the lower sound insulator.

Ignition Switch

REMOVAL AND INSTALLATION

──────── CAUTION ────────

The Supplemental Inflatable Restraint (SIR) system must be disarmed before working around the steering column to replace the dimmer switch. Failure to do so may cause accidental deployment of the air bag, resulting in unnecessary SIR system repairs and/or personal injury.

1. Disable the SIR system, if equipped.

2. Disconnect the negative battery cable.

3. Remove the lower dash panel on the left side.

4. Remove the nut and bolt mounting the dimmer switch and disconnect the dimmer switch from the actuator rod.

5. Remove the ignition and dimmer switch mounting stud.

6. Disconnect the ignition and dimmer switch assemblies from the ignition switch actuator.

7. Disconnect the electrical connector from the bulkhead connector.

To install:

8. Connect the electrical connector to the bulkhead and tighten the mounting screw to 22 inch lbs. (2 Nm).

9. Make sure the ignition lock cylinder is in the **OFF–LOCK** position.

10. Install the ignition switch assembly and dimmer switch assembly onto the steering column jacket and connect the actuators. The new ignition switch will be pinned in the **OFF–LOCK** position. DO NOT remove the pin until after the switch is installed on the column.

11. Adjust the ignition switch as follows:

   a. Move the switch slider to the far right position.

   b. Move the switch to the left one position. This is the **OFF–LOCK** position.

   c. Install a 3/32 inch drill bit into the switch alignment hole to limit switch movement.

12. Position the ignition switch on the column and install the mounting stud and tighten to 35 inch lbs. (4 Nm).

13. Remove the drill bit from the switch.

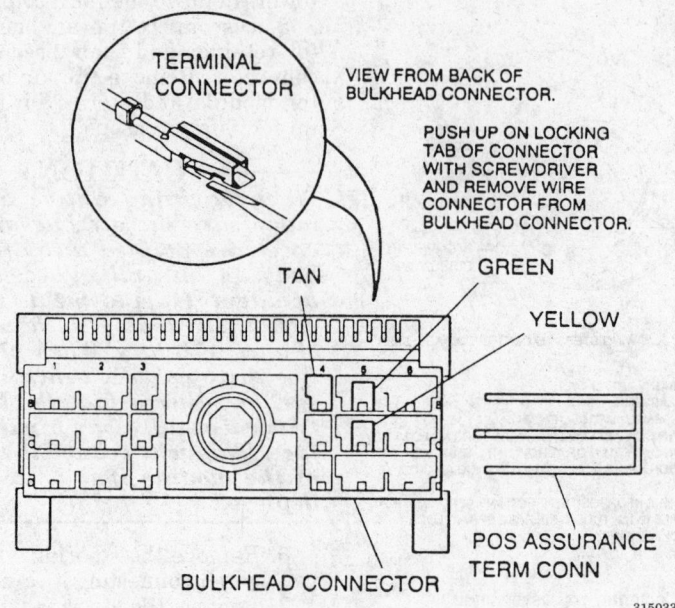

Ignition switch wire locations

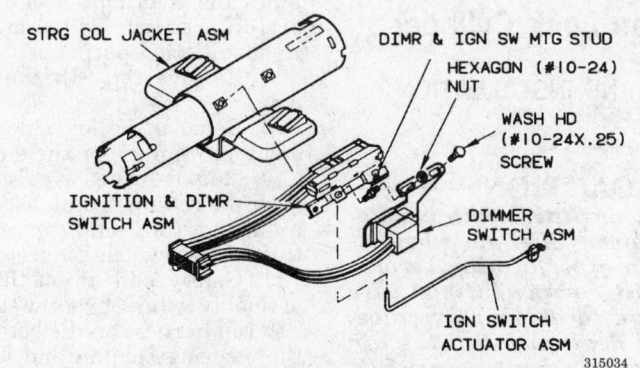

Dimmer switch assembly

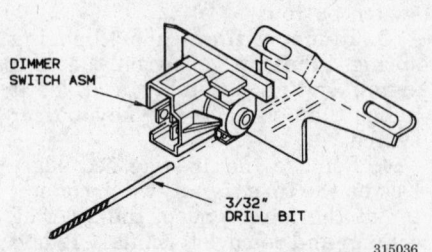

Use a drill bit to limit switch travel and remove all actuator rod free-play during installation and adjustment

14. Install the dimmer switch and adjust as follows:

   a. Install a 3/32 inch drill bit into the switch alignment hole to limit switch movement.

   b. Position the switch on the column so all the lash in the actuator rod is taken up.

   c. Remove the drill bit.

15. Tighten the mounting nut and bolt to 35 inch lbs. (4 Nm).

16. Install the under dash panel.

17. Connect the negative battery cable.

18. Enable the SIR system, if equipped.

Park/Neutral Safety Switch

REMOVAL AND INSTALLATION

1. Disconnect negative battery cable.
2. Place gear selector in neutral position.
3. Disconnect electrical connector at switch.
4. Press mounting tabs together on switch while pulling forward from shift tube.

To Install:

5. Align actuator on switch with hole in shift tube
6. Position rearward portion of switch (connector side) to fit into cut-out in lower jacket.
7. Push down on front of switch. The two tangs on the housing back will snap into place in the rectangular holes in the jacket.
8. Connect electrical connector at switch.
9. Adjust switch by moving gear selector to park and moving the main housing forward or rearward for proper adjustment.
10. Connect negative battery cable.

Powertrain Control Module

REMOVAL AND INSTALLATION

1. Disconnect the negative battery cable.
2. Remove the passenger side interior access panel.
3. Disconnect the electrical connectors from the PCM.
4. Release the clips holding the PCM to the mounting bracket and remove the PCM.

To install:

5. Install the PCM into the mounting bracket and seat the mounting clips.
6. Connect the electrical connectors to the PCM.
7. Install the passenger side interior access panel.
8. Connect the negative battery cable.

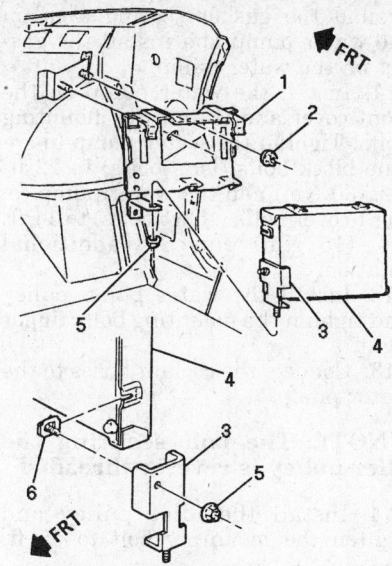

1. Bracket asm
2. Nut (2) - 19-30 in. lb.(2-3 Nm)
3. Bracket asm(PCM)
4. Module asm(PCM)
5. Nut (2) - 70-97 in. lb.(8-11 Nm)
6. Bolt/screw (L27)3800 engine

301314

PCM assembly and bracket

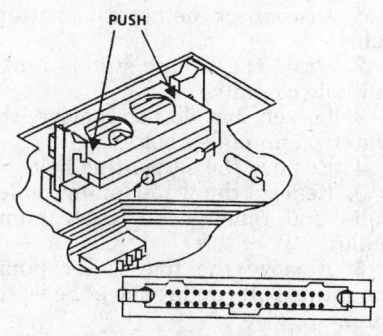

301315

MEM CAL unit installation

ENGINE COOLING

Radiator

REMOVAL AND INSTALLATION

1. Disconnect the negative battery cable.
2. Drain the cooling system into a suitable container.

3. Remove the air cleaner assembly, mounting stud and all duct work.
4. Remove the front torque strut mounting bolt and nut. Loosen but do not remove the rear nut and bolt and swing the strut onto the rear bracket.
5. On the 4-cyl., remove the air intake resonator mounting nut and remove the resonator.
6. Disconnect the fan electrical connector from the fan motor.
7. Remove the fan motor mounting bolts and remove the fan assembly.
8. Disconnect the upper and lower radiator hoses from the radiator.
9. Disconnect the coolant overflow hose from the radiator neck.
10. Disconnect and cap the transaxle cooler lines from the radiator.
11. Remove the upper radiator mounting panel attaching bolts.
12. Remove the upper radiator mounting panel and insulators.
13. Remove the radiator from the vehicle.

To install:

14. Install the radiator in the vehicle, making sure the bottom of the radiator is in the lower mounting pads.
15. Install the upper radiator mounting panel and panel mounting bolts. Tighten the mounting panel bolts to 89 inch lbs. (10 Nm).
16. Connect the transaxle cooler lines and tighten the fittings to 20 ft. lbs. (27 Nm).
17. Connect the upper and lower radiator hoses to the radiator.
18. Connect the coolant overflow hose to the radiator neck and position the clamp in place.
19. Install the cooling fan assembly and tighten the mounting bolts to 89 inch lbs. (10 Nm).
20. Connect the fan electrical connector.
21. On the 4-cyl. install the air intake resonator and mounting nut.
22. Swing the torque strut and bracket until the bracket contacts the radiator support. Install the bracket-to-radiator support attaching bolts and tighten to 17 ft. lbs. (23 Nm). Make sure the engine ground strap is connected to the torque strut brace, if equipped.
23. Install the air cleaner duct work, mounting stud and air cleaner assembly.
24. Refill the cooling system.
25. Connect the negative battery cable.
26. Start the engine and verify no leaks.
27. Check and, if necessary, add transmission fluid.

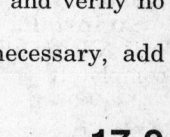

Water Pump

REMOVAL AND INSTALLATION

2.2L (VIN 4) Engine

1. Disconnect the negative battery cable.
2. Drain the cooling system into a suitable container.
3. Loosen, but do not remove, the water pump pulley bolts.
4. Remove the serpentine belt.
5. Remove the alternator mounting bolts and set the alternator aside.
6. Remove the water pump pulley bolts and remove the water pump pulley.
7. Remove the four water pump mounting bolts and remove the water pump.

To install:

8. Clean all the gasket surfaces completely.
9. Apply a thin bead of sealer around the outer edge of the water pump gasket seating area and place he gasket on the pump.
10. Install the water pump on the engine and tighten the four mounting bolts to 18 ft. lbs. (25 Nm).
11. Install the water pump pulley and tighten the mounting bolts finger tight.
12. Install the alternator in the mounting bracket.
13. Install the serpentine belt.
14. Tighten the water pump pulley mounting bolts to 22 ft. lbs. (30 Nm).
15. Connect the negative battery cable.
16. Refill and bleed the cooling system.

3.3L (VIN N) Engine

1. Disconnect the negative battery cable.
2. Drain the engine coolant into a suitable container.
3. Loosen, but do not remove, the water pump pulley bolts.
4. Remove the serpentine belt.

NOTE: The bolt securing the idler pulley is reverse threaded.

5. Remove the idler pulley mounting bolt and remove the idler pulley.
6. Disconnect the coolant hoses from the water pump.
7. Remove the water pump pulley bolts and remove the water pump pulley.
8. Remove the water pump mounting bolts and remove the water pump.

To install:

9. Clean all the gasket surfaces completely.

10. Apply a thin bead of sealer around the gasket seating area on the water pump and install the gasket on the water pump.
11. Install the water pump on the front cover and install the mounting bolts. Tighten the water pump-to-engine block bolts (long bolts) to 22 ft. lbs. (30 Nm) and the water pump-to-front cover bolts (short bolts) to 11 ft. lbs. (15 Nm) plus 80° additional rotation.
12. Install the water pump pulley and tighten the mounting bolts finger tight.
13. Connect the coolant hoses to the water pump.

NOTE: The bolt securing the idler pulley is reverse threaded.

14. Install the idler pulley and tighten the mounting bolt to 55 ft. lbs. (75 Nm).
15. Install the serpentine belt.
16. Tighten the water pump pulley bolts 22 ft. lbs. (30 Nm).
17. Connect the negative battery cable.
18. Refill and bleed the cooling system.

3.1L (VIN M) Engine

1. Disconnect the negative battery cable.
2. Drain the cooling system into a suitable container.
3. Loosen, but do not remove, the water pump pulley bolts.
4. Remove the serpentine belt.
5. Remove the water pump pulley bolts and remove the water pump pulley.
6. Remove the five water pump mounting bolts and remove the water pump.

To install:

7. Clean all the gasket surfaces completely.
8. Apply a thin bead of sealer around the outside edge of the water pump along the gasket sealing area and install the gasket onto the water pump.
9. Install the water pump on the engine and tighten the water pump mounting bolts to 89 inch lbs. (10 Nm).
10. Install the water pump pulley and tighten the pulley bolts finger tight.
11. Install the serpentine belt.
12. Tighten the water pump pulley bolts to 18 ft. lbs. (25 Nm).
13. Connect the negative battery cable.
14. Refill and bleed the cooling system.

Thermostat

REMOVAL AND INSTALLATION

2.2L (VIN 4) Engine

1. Disconnect the negative battery cable.
2. Drain the cooling system to a level below the thermostat housing.
3. Disconnect the upper radiator hose from the thermostat housing.
4. Remove the thermostat housing mounting nuts and remove the thermostat housing.
5. Remove the thermostat.

To install:

6. Clean all the gasket surfaces completely.
7. Install the thermostat in the thermostat housing.
8. Apply a thin bead of sealer around the gasket contact surface of the housing and install the gasket.
9. Install the thermostat housing on the engine and tighten the mounting nuts to 89 inch lbs. (10 Nm).
10. Connect the upper radiator hose to the thermostat housing.
11. Refill the cooling system.
12. Connect negative battery cable.
13. Start engine and check for leaks.

3.3L (VIN N) Engine

1. Disconnect the negative battery cable.
2. Drain the cooling system to a level below the thermostat housing.
3. Disconnect the air inlet duct from the throttle body.
4. Disconnect the upper radiator hose from the thermostat housing outlet.
5. Remove the thermostat housing mounting bolts and remove the coolant pipe mounting bracket.
6. Remove the thermostat housing and thermostat.

To install:

7. Clean all gasket surfaces completely.
8. Apply a thin bead of sealer around the gasket sealing area on the thermostat housing and install the gasket on the housing.
9. Install the thermostat, thermostat housing and coolant pipe mounting bracket and tighten the mounting bolts to 20 ft. lbs. (27 Nm).
10. Connect the upper radiator hose to the thermostat housing.
11. Connect the air inlet duct to the throttle body.
12. Refill the cooling system.
13. Connect the negative battery cable.

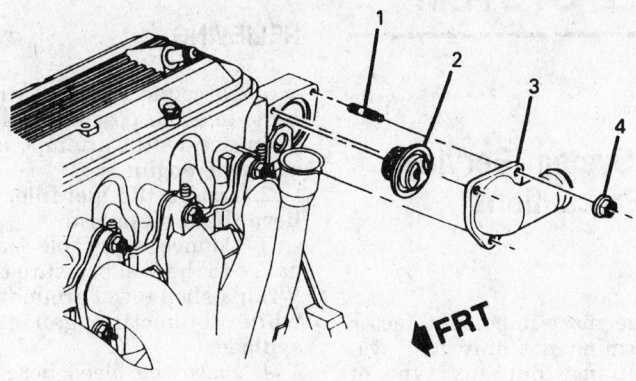

1. Water outlet stud
2. Engine coolant thermostat assembly
3. Water outlet
4. Water outlet nut

296090

Thermostat components — 2.2L (VIN 4) engine

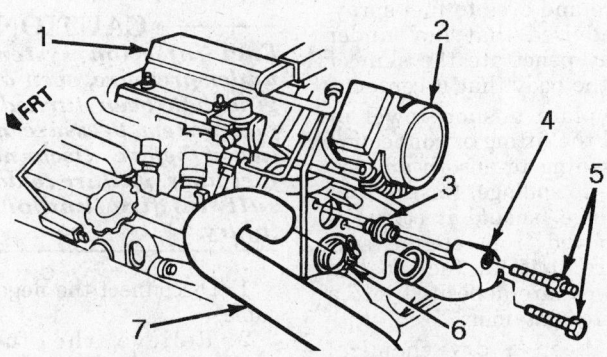

1. Intake manifold
2. Throttle body
3. Thermostat
4. Outlet asm
5. 20 ft. lb.(27 Nm)
6. Clamp 27 in. lb. (3 Nm)
7. Outlet hose

177614

Thermostat components — 3.3L (VIN N) engine

3.1L (VIN M) Engine

1. Disconnect the negative battery cable.
2. Drain the cooling system to a level below the thermostat housing.
3. Disconnect the air inlet from the throttle body.
4. Disconnect the upper radiator hose from the thermostat housing.
5. Remove the thermostat housing mounting bolts and remove the thermostat housing.
6. Remove the thermostat.
To install:
7. Clean all the gasket surfaces completely.
8. Install the thermostat in the thermostat housing.
9. Apply a thin bead of sealer around the gasket contact surface of the housing and install the gasket.
10. Install the thermostat housing on the engine and tighten the mounting bolts to 18 ft. lbs. (25 Nm).
11. Connect the upper radiator hose to the thermostat housing.
12. Connect the air inlet duct to the throttle body.
13. Refill the cooling system.
14. Connect negative battery cable.
15. Start engine and check for leaks.

Electric Cooling Fan

REMOVAL AND INSTALLATION

NOTE: Depending on vehicle equipment the cooling fan design may vary slightly. The following procedure is for both the three and four bolt mounted fans.

1. Disconnect the negative battery cable.
2. Loosen the torque strut mounting bolt at the engine–side bracket. Remove the three bolts securing the torque strut mounting bracket to the radiator panel and lay the torque strut on the engine.
3. Remove the air cleaner and resonator assembly, if necessary for clearance.
4. Disconnect the cooling fan electrical connector.
5. Remove the cooling fan assembly mounting bolts.
6. Remove the cooling fan assembly.
7. If the fan motor is to be replaced, remove the fan blade from the fan motor and remove the fan motor to fan frame mounting bolts and remove the motor from the frame.

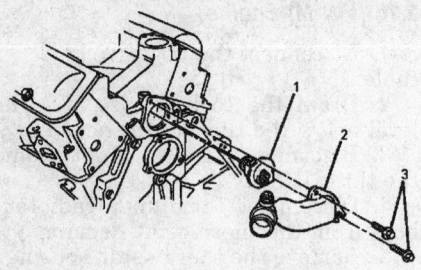

1 THERMOSTAT ASSEMBLY, ENGINE COOLANT
2 OUTLET ASSEMBLY, WATER
3 BOLT/SCREW, WATER OUTLET

296167

Thermostat components — 3.1L (VIN M) engine

To install:

8. Install the fan motor on the fan frame and tighten the mounting bolts to 89 inch lbs. (10 Nm).

9. Install the fan blade on the fan motor and tighten the nut to 29 inch lbs. (3.3 Nm).

10. Install the fan assembly in the vehicle and tighten the mounting bolts to 89 ft. lbs. (10 Nm).

11. Connect the cooling fan motor electrical connector.

12. Install the air cleaner and resonator assembly.

13. Install the torque strut mounting bracket on the radiator panel and tighten the bolts at the radiator panel and the torque strut engine-side mount.

14. Connect the negative battery cable.

Cooling System Bleeding

1. Fill the radiator with the correct mix of antifreeze and water.

2. Fill the coolant overflow tank to the **FULL** mark and re-install the overflow tank cap.

3. Run the engine with the radiator cap removed until the thermostat opens.

4. Top off the radiator until the coolant is at the base of the fill neck.

5. Install the radiator cap.

6. Allow the engine to cool.

7. Top off the radiator and overflow tank as necessary.

FUEL SYSTEM

Fuel System Service Precautions

Safety is the most important factor when performing not only fuel system maintenance but any type of maintenance. Failure to conduct maintenance and repairs in a safe manner may result in serious personal injury or death. Maintenance and testing of the vehicle's fuel system components can be accomplished safely and effectively by adhering to the following rules and guidelines.

• To avoid the possibility of fire and personal injury, always disconnect the negative battery cable unless the repair or test procedure requires that battery voltage be applied.

• Always relieve the fuel system pressure prior to disconnecting any fuel system component (injector, fuel rail, pressure regulator, etc.), fitting or fuel line connection. Exercise extreme caution whenever relieving fuel system pressure to avoid exposing skin, face and eyes to fuel spray. Please be advised that fuel under pressure may penetrate the skin or any part of the body that it contacts.

• Always place a shop towel or cloth around the fitting or connection prior to loosening to absorb any excess fuel due to spillage. Ensure that all fuel spillage (should it occur) is quickly removed from engine surfaces. Ensure that all fuel soaked cloths or towels are deposited into a suitable waste container.

• Always keep a dry chemical (Class B) fire extinguisher near the work area.

• Do not allow fuel spray or fuel vapors to come into contact with a spark or open flame.

• Always use a backup wrench when loosening and tightening fuel line connection fittings. This will prevent unnecessary stress and torsion to fuel line piping. Always follow the proper torque specifications.

• Always replace worn fuel fitting O-rings with new. Do not substitute fuel hose or equivalent, where fuel pipe is installed.

Fuel System Pressure

RELIEVING

1. Disconnect the negative battery cable to avoid possible fuel discharge if an accidental attempt is made to start the engine.

2. Loosen the fuel filler cap to relieve the tank pressure.

3. Connect a suitable fuel pressure gauge to the fuel pressure test fitting. Wrap a shop towel around the fitting while connecting gauge to avoid spillage.

4. Place the bleed hose in an approved container and open the valve on the pressure gauge to relieve system pressure.

5. Dispose of the discharged liquid fuel promptly.

Idle Speed

ADJUSTMENT

Idle speed and mixture are controlled by the engine control module and cannot be adjusted.

Fuel Filter

REMOVAL AND INSTALLATION

— CAUTION —
Fuel injection systems remain under pressure, even after the engine has been turned OFF. The fuel system pressure must be relieved before disconnecting any fuel lines. Failure to do so may result in fire and/or personal injury.

1. Disconnect the negative battery cable.

2. Relieve the fuel system pressure.

3. Raise and safely support the vehicle.

4. Place a pan under the fuel filter.

5. Rotate each of the fuel filter quick-Connects 1/4 turn to loosen any dirt trapped under the fittings.

6. Squeeze the plastic tabs on the male ends of the quick-Connects and disconnect the fuel lines from the filter.

7. Remove the filter bracket mounting bolt.

8. Remove the filter from the vehicle.

To install:

9. Install the filter assembly in position and install the filter bracket mounting bolt.

10. Apply a few drops of clean engine oil to the male ends of the fuel filter assembly and fuel line.

11. Snap the fuel lines onto the fuel filter. The lock tabs on the quick-Connects will lock in place. Give each line a light tug to make sure the fittings are tight.

12. Lower the vehicle.

13. Connect the negative battery cable.

14. Pressurize the fuel system and verify no leaks.

Fuel Pump

REMOVAL AND INSTALLATION

The fuel pump is located in the fuel tank. The fuel tank must be removed from the vehicle for this procedure. Note that the fuel pump has a strainer attached to reduce sediment and debris that might enter the pump. These strainers may become clogged on high–mileage vehicles or where the vehicles has had contamination in the tank. A clogged strainer can keep the pump from drawing suf-

ficient fuel to maintain enough fuel pressure to keep the engine running smoothly. Keep this in mind when troubleshooting a driveability and/or low fuel pressure complaint. This strainer should be replaced whenever the tank is removed for any kind of service.

----- **CAUTION** -----

Fuel injection systems remain under pressure, even after the engine has been turned OFF. The fuel system pressure must be relieved before disconnecting any fuel lines. Failure to do so may result in fire and/or personal injury.

1. Properly relieve the fuel system pressure.

2. Disconnect the negative battery cable.

----- **CAUTION** -----

Observe all applicable safety precautions when working around fuel. Do not allow fuel spray or fuel vapors to come into contact with a spark or open flame. Keep a dry chemical (Class B) fire extinguisher near the work area. Never drain or store fuel in an open container due to the possibility of fire or explosion.

3. Remove the fuel tank from the vehicle using the recommended procedure and place in a suitable work area.

4. Remove the fuel sender assembly locking cam using a spanner wrench, J-24187, or an equivalent.

5. Remove the fuel sender assembly from the fuel tank.

6. Place the sender assembly on a work bench with the fuel strainer facing up.

7. Note the direction the fuel strainer is facing and remove the strainer from the fuel pump by pulling it off the pump.

8. Disconnect the electrical connectors from the fuel pump.

9. Push the fuel pump toward the fuel outlet hose to unseat the pump from the lower mounting bracket. Once the pump is clear of the mounting bracket, tilt the bottom of the pump outward away from the mounting bracket.

10. Remove the plastic clamp on the outlet line and disconnect the fuel pump from the outlet hose.

To install:

11. Connect the new pump to the outlet hose and install the plastic clamp.

12. Install the rubber insulator on the fuel pump.

13. Push the pump toward the outlet hose until the bottom of the pump can be seated in the mounting bracket.

14. Connect the electrical connectors to the fuel pump.

15. Install a new fuel strainer on the pump making sure the direction the strainer is facing is the same as on the old pump.

16. Install the fuel sender into the fuel tank using a new O-ring on top of the tank. It may help to lubricate the O-ring with a silicone grease.

17. Install the locking cam and rotate into the locked position with a spanner wrench, J-24187, or an equivalent.

18. Install the fuel tank in the vehicle. Refill with gasoline. Never attempt to test run a fuel pump dry, even for a short period of time. It is made to be immersed in fuel for both lubrication and cooling.

19. Connect the negative battery cable.

20. Pressurize the fuel system and verify no fuel leaks. This can be done by rotating the ignition switch to the **ON** position without starting the vehicle.

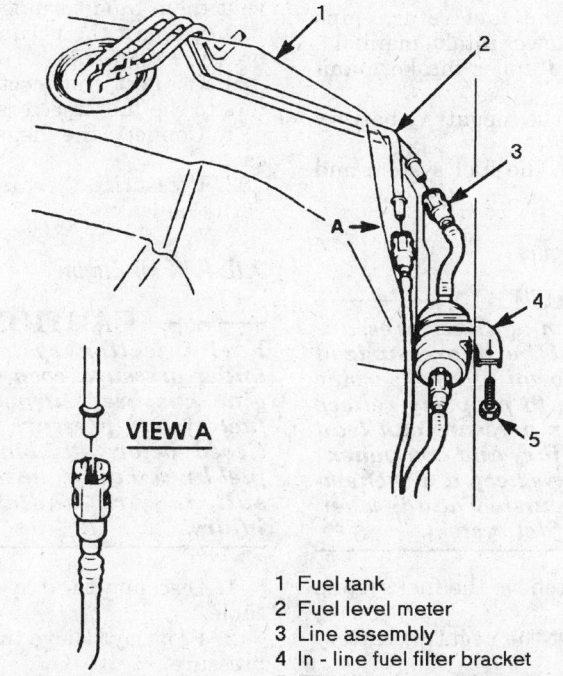

VIEW A

TYPICAL HOSE ASM.

1 Fuel tank
2 Fuel level meter
3 Line assembly
4 In - line fuel filter bracket
5 Filter bracket attaching screw

291582

Fuel filter mounting location and quick-connect diagram

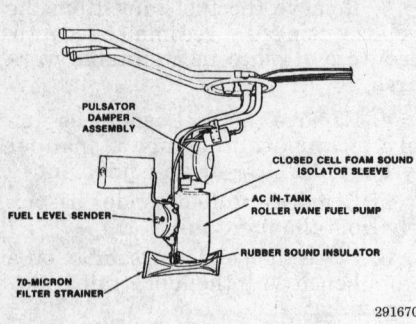

PULSATOR
DAMPER
ASSEMBLY

CLOSED CELL FOAM SOUND
ISOLATOR SLEEVE

FUEL LEVEL SENDER

AC IN-TANK
ROLLER VANE FUEL PUMP

RUBBER SOUND INSULATOR

70-MICRON
FILTER STRAINER

291670

Fuel pump and sender assembly

Fuel Injector

REMOVAL AND INSTALLATION

2.2L (VIN 4) Engine

---------- CAUTION ----------
Fuel injection systems remain under pressure, even after the engine has been turned OFF. The fuel system pressure must be relieved before disconnecting any fuel lines. Failure to do so may result in fire and/or personal injury.

---

1. Disconnect the negative battery cable.
2. Relieve the fuel system pressure.
3. Remove the upper intake manifold assembly.
4. Remove the fuel return line bracket mounting bolt and remove the bracket.
5. Position the return line away from the pressure regulator.
6. Remove the pressure regulator as follows:
    a. Disconnect the vacuum hose from the regulator.
    b. Remove the fuel return pipe clamp.
    c. Disconnect the fuel return pipe from the pressure regulator and discard the O-ring.
    d. Remove the pressure regulator bracket mounting bolt.
    e. Remove the regulator assembly and O-ring and discard the O-ring.
7. Remove the fuel injector retaining bracket mounting bolts.
8. Remove the fuel injector retaining bracket by carefully sliding it off to clear the fuel injector slots.
9. Disconnect the fuel injector electrical connector(s).
10. Pull the injector(s) out.

11. Remove the old O-rings and discard.
    To install:

NOTE: **Each fuel injector is calibrated for a specific flow rate. Be sure to use the correct part number when ordering replacement fuel injectors.**

12. Coat the new O-rings with clean engine oil and install the O-rings on the injector.
13. Install the injector into the cavity in the lower intake manifold with the electrical connector facing inward.
14. Install the injector retaining bracket an mounting screws.
15. Connect the fuel injector electrical connector.
16. Install the pressure regulator as follows:
    a. Lubricate the regulator assembly and O-ring and install the O-ring on the regulator.
    b. Install the regulator onto the intake manifold.
    c. Install the pressure regulator bracket mounting bolt after coating the threads with thread locking compound. Tighten the mounting bolt to 76 inch lbs. (8.5 Nm).
    d. Connect the vacuum hose to the regulator.
    e. Connect the fuel return pipe to the pressure regulator and tighten the fitting to 13 ft. lbs. (17 Nm).
    f. Install the fuel return pipe clamp to the lower intake manifold.
17. Install the upper intake manifold assembly.
18. Connect the negative battery cable.
19. Pressurize the fuel system and verify no leaks.

3.3L (VIN N) Engine

---------- CAUTION ----------
The fuel system is under pressure and the must be depressurized prior to performing any service work. Failure to properly relieve the fuel system pressure can lead to personal injury and component damage. Always keep a dry chemical fire extinguisher handy when servicing the fuel system.

---

1. Properly relieve the fuel system pressure.
2. Disconnect the negative battery cable.
3. Disconnect the fuel injector electrical connectors.
4. Disconnect the vacuum line from the pressure regulator.

5. Disconnect and cap the fuel lines from the fuel rail. Use a backup wrench on the fuel rail fittings to prevent them from turning.
6. Remove the O-rings from the lines and discard.
7. Remove the fuel rail mounting nuts.
8. Remove the fuel rail assembly.
9. Remove the injector retaining clip and discard.
10. Remove the injector assembly from the fuel rail.
11. Remove the old O-rings and discard.
    To Install:
12. Coat the new O-rings with clean engine oil and install the O-rings on the injector.
13. Install a new retaining clip on the fuel injector. Position the open end of the clip toward the injector electrical connector.
14. Install the injector onto the fuel rail assembly with the injector electrical connector facing outward.
15. Install the fuel rail onto the intake manifold so all the injector fit into the injector cavities.
16. Install the fuel rail mounting nuts and tighten to 20 ft. lbs. (27 Nm).
17. Install new O-rings coated with engine oil on the fuel lines.
18. Connect the fuel feed and return lines and tighten the fittings to 20 ft. lbs. (26 Nm). Use a backup wrench on the fuel rail fittings to prevent them from turning.
19. Connect the vacuum line to the pressure regulator.
20. Connect the electrical connectors to the fuel injectors.
21. Connect the negative battery cable.
22. Pressurize the fuel system and verify no leaks.

3.1L (VIN M) Engine

---------- CAUTION ----------
Fuel injection systems remain under pressure, even after the engine has been turned OFF. The fuel system pressure must be relieved before disconnecting any fuel lines. Failure to do so may result in fire and/or personal injury.

---

1. Disconnect the negative battery cable.
2. Properly relieve the fuel system pressure.
3. Remove the upper intake manifold assembly.
4. Disconnect and cap the fuel feed line from the fuel rail.

1. Upper manifold asm
2. Upper inlet manifold stud
3. EGR valve and fuel pressure regulator vacuum harness asm
4. Map sensor seal
5. Manifold absolute pressure (map) sensor
6. Map sensor attaching bolt
7. Fitting and washer asm
8. Idle air control (IAC) valve asm
9. IAC valve attaching screw
10. IAC valve o-ring
11. EGR transport tube asm
12. Upper inlet manifold bolt
13. Upper inlet manifold gasket
14. Power brake fitting
15. Fuel injector o-ring

16. (Bottom feed) MFI fuel injector asm
17. Lower o-ring
18. Fuel pressure regulator asm
19. Fuel pressure regulator attaching screw
20. Fuel return line o-ring
21. Fuel injector fuel return pipe asm
22. Filter (if so equipped) screen
23. Fuel inlet fitting o-ring
24. Lower manifold asm
25. Injector retainer attaching screw
26. TP sensor attaching screw
27. Throttle position (TP) sensor
28. Injector retainer
29. Fuel feed line o-ring
30. Fuel injection fuel feed pipe asm
31. Fuel feed nut

291595

Fuel injector and intake manifold components — 2.2L (VIN 4) engine

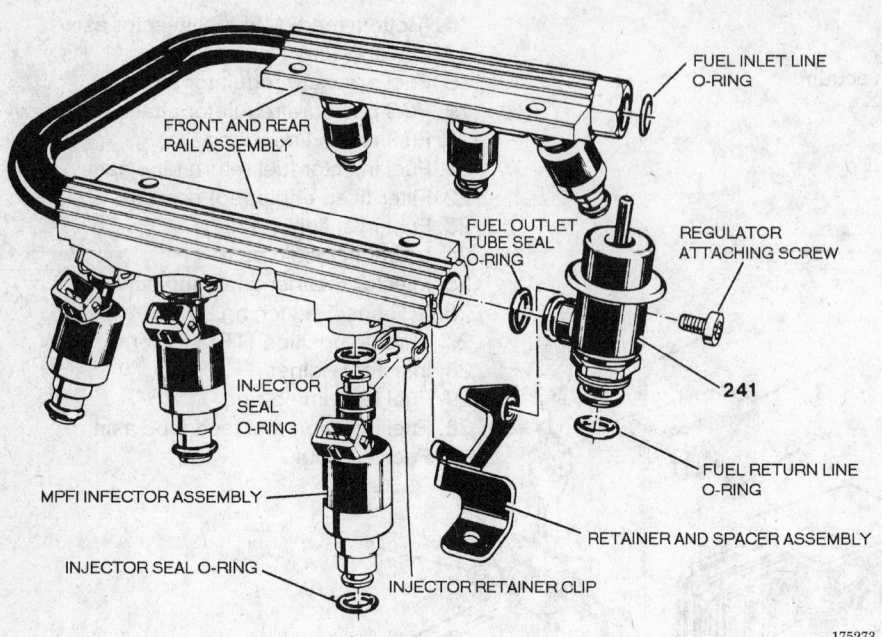

FRONT AND REAR RAIL ASSEMBLY

FUEL INLET LINE O-RING

FUEL OUTLET TUBE SEAL O-RING

REGULATOR ATTACHING SCREW

INJECTOR SEAL O-RING

241

MPFI INFECTOR ASSEMBLY

FUEL RETURN LINE O-RING

INJECTOR SEAL O-RING

RETAINER AND SPACER ASSEMBLY

INJECTOR RETAINER CLIP

175273

Fuel rail and injector components — 3.3L (VIN N) engine

e. Install the pressure regulator mounting bolts and tighten to 76 inch lbs. (9 Nm).

25. Install the upper intake manifold assembly.

26. Connect the negative battery cable.

27. Pressurize the fuel system and verify no leaks.

EMISSION CONTROLS

Service Interval Lamp

RESETTING

1993 Vehicles

Vehicles equipped with an **ENGINE OIL LIFE INDEX** display as a part of the **DRIVER INFORMATION SYSTEM (DIS)**, have a display that will show when to change the engine oil.

The oil change interval is determined by the driver information system and will usually fall at or between the 2 recommended alternative intervals of 3000 miles and 7500 miles, but it could be shorter than 3000 miles under some severe driving conditions. The driver information system will also signal the need for an oil change at 7500 miles or one year passed since the last oil change. If the driver information system does not indicate the need for an oil change after 7500 miles or one year or if the **ENGINE OIL LIFE INDEX** display fails to appear, the oil should be changed and the driver information system serviced.

When the **ENGINE OIL LIFE INDEX** reaches 10 percent or less, the **CHANGE OIL** light display will function as a reserve trip odometer (indicating the distance to an oil change). Until the **ENGINE OIL LIFE INDEX** reset is performed, the driver information system will display the distance to the oil change and sound a beep when the ignition switch is turned to the **ACCESSORY** or **RUN** position the first time each day.

When the distance to the next oil change reaches 0, the driver information system will display the **CHANGE OIL NOW** light. Until an **ENGINE OIL LIFE INDEX** reset is

5. Remove the fuel pressure regulator as follows:

  a. Remove the fuel pressure regulator retaining screw.

  b. Place a shop towel under the pressure regulator and remove the regulator from the fuel rail.

  c. Remove the retainer and spacer bracket from the fuel rail.

  d. Disconnect the regulator from the fuel return line.

6. Disconnect the fuel injector main harness.

7. Disconnect the coolant temperature sensor electrical connector.

8. Remove the fuel rail retaining bolts and remove the fuel rail.

9. Disconnect the individual injector electrical connector.

10. Remove the injector retaining clip and discard.

11. Remove the injector assembly from the fuel rail.

12. Remove the old O-rings and discard. Save the O-ring backups for installation on the new injector.

To install:

13. Install the O-ring backups on the fuel injector.

14. Coat the new O-rings with clean engine oil and install the O-rings on the injector.

15. Install a new retaining clip on the fuel injector. Position the open end of the clip toward the injector electrical connector.

16. Install the injector onto the fuel rail assembly with the injector electrical connector facing outward.

17. Connect the electrical connector to the fuel injector.

18. Install the fuel rail onto the intake manifold so all the injector fit into the injector cavities.

19. Install the fuel rail mounting bolts and tighten to 89 inch lbs. (10 Nm).

20. Connect the electrical connector to the coolant temperature switch.

21. Connect the main fuel injector harness to the fuel rail assembly.

22. Install new O-rings on the pressure regulator and fuel inlet line.

23. Connect the fuel feed line and tighten the fitting to 22 ft. lbs. (30 Nm).

24. Install the fuel pressure regulator as follows:

  a. Connect the fuel regulator to the fuel return line.

  b. Install a new retainer and spacer bracket into the slot on the fuel rail.

  c. Install the pressure regulator onto the fuel rail.

  d. Tighten the return line fitting to 13 ft. lbs. (17 Nm).

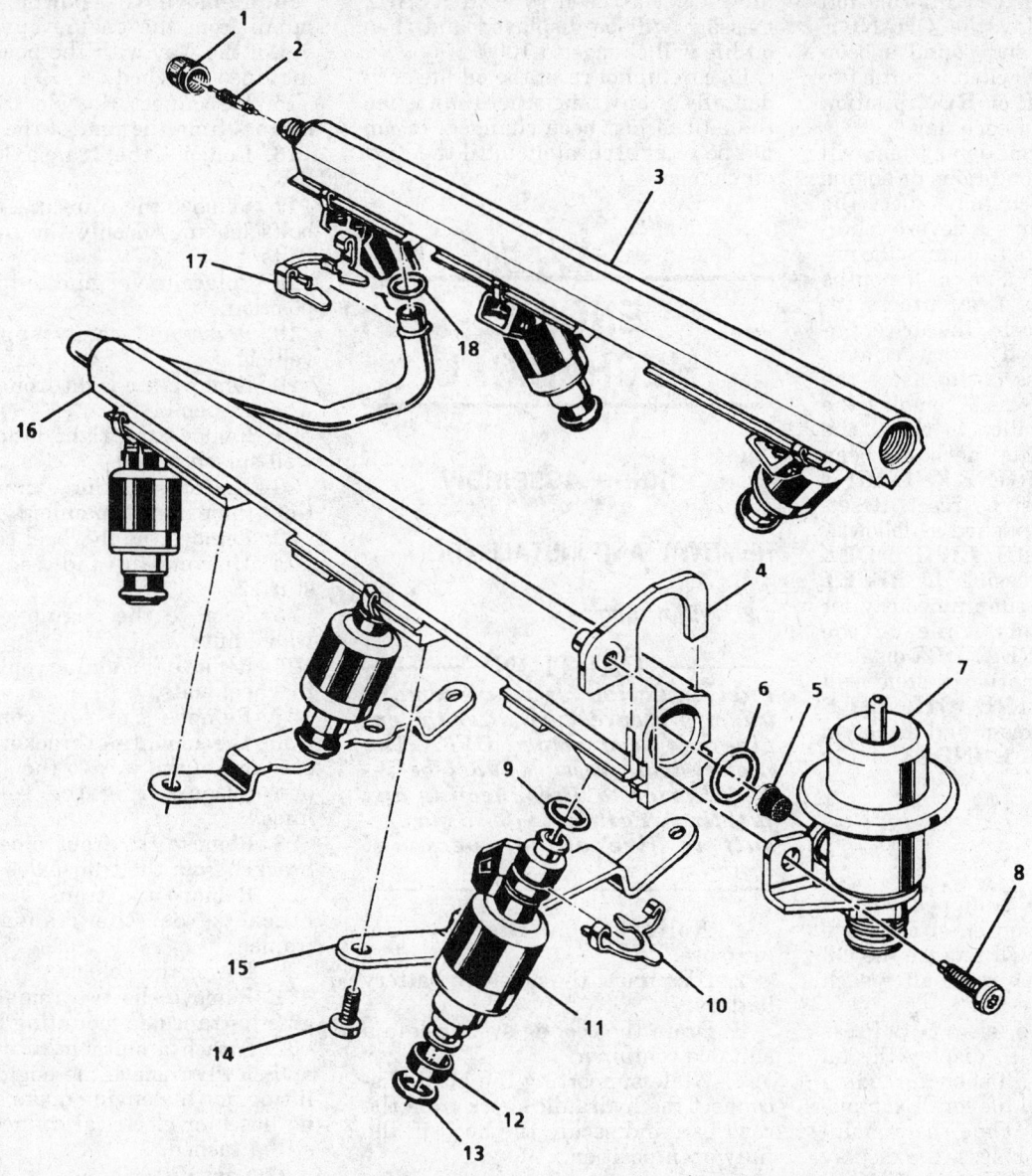

1   CAP - FUEL PRESSURE CONNECTION

2   VALVE ASSEMBLY - FUEL PRESSURE CONNECTION

3   RAIL ASSEMBLY - RH FUEL

4   BRACKET - RETAINER AND SPACER

5   SCREEN - FILTER

6   O-RING - PRESSURE REGULATOR INLET

7   REGULATOR ASSEMBLY - FUEL PRESSURE

8   SCREW - PRESSURE REGULATOR ATTACHING

9   O-RING - UPPER INJECTOR (BLACK)

10   CLIP - INJECTOR RETAINER

11   INJECTOR ASSEMBLY - FUEL

12   BACKUP - O-RING

13   O-RING - LOWER INJECTOR (BROWN)

14   SCREW - RAIL MOUNTING BRACKET ATTACHING

15   BRACKET - RAIL MOUNTING (2 EA.)

16   RAIL ASSEMBLY - LH FUEL (WITH CROSSOVER TUBE)

17   CLIP - CROSSOVER TUBE RETAINER

18   O-RING - CROSSOVER TUBE

291789

Fuel rail and pressure regulator components — 3.1L (VIN M) engine

performed, the driver information system will display the **CHANGE OIL NOW** light and sound a beep when the ignition switch is turned to the **ACCESSORY** or **RUN** position at the beginning of each day.

The driver information system will not detect dusty conditions or engine malfunctions which may affect the engine oil. If driving in severe conditions exists, be sure to change the engine oil every 3000 miles or 3 months which ever comes first, unless instructed otherwise by the driver information system. The driver information center does not measure the engine oil level. It still remains the owner's responsibility to check the engine oil level. After the oil has been changed, the **ENGINE OIL LIFE INDEX** light must be reset. Resetting can be accomplished as follows:

The **ENGINE OIL LIFE INDEX** can be reset by pressing the **RESET** and **OIL** buttons simultaneously for at least 5 seconds while on the **ENGINE OIL LIFE INDEX** display. The driver information system will reset the **ENGINE OIL LIFE INDEX** to 100 percent and display a **ENGINE OIL LIFE INDEX** of 100 percent.

1994–96 Vehicles

Vehicles that are equipped with a Drivers Information Center and have an Oil life index, will require the Oil Life indicator to be reset after each oil change.

Press the **SEL** to select **OIL**. Press **SEL** if necessary to display the oil life. The display will show a reading of the estimated oil life left. Example: **OIL LIFE 85%**. When the remaining oil life is 9% or less, the display will show **CHANGE OIL SOON**. Then the vehicle is started tone will sound an the **CHANGE OIL SOON** message will display each time the vehicle is started.

When the oil life is zero, a tone will sound and the display will show, **CHANGE OIL NOW**. Then when the vehicle is started a tone will sound and the **CHANGE OIL NOW** message will display each time the vehicle is started. Reset the Oil Life Display as follows:

1. Acknowledge all diagnostic messages in the Drivers Information Center by pressing **RESET**.
2. Press the **SEL** button on the left to select **OIL**. Press **SEL** button on the right if necessary to display oil life.
3. Press and hold the **RESET** button for about 5 seconds. Once the oil life index has been reset, a **RESET** message will be displayed and then oil life will change to 100%.

Be careful not reset the oil life accidentally at any time other than when the oil has just been changed. It can not be reset accurately until the next oil change.

ENGINE MECHANICAL

Engine Assembly

REMOVAL AND INSTALLATION

2.2L (VIN 4) Engine

————— **CAUTION** —————
Fuel injection systems remain under pressure, even after the engine has been turned OFF. The fuel system pressure must be relieved prior to disconnecting any fuel lines. Failure to do so may result in fire and/or personal injury.

1. Relieve the fuel system pressure.
2. Disconnect the negative battery battery.
3. Drain the cooling system into a suitable container.
4. While supporting the hood, disconnect the hydraulic shock from the cowl pan and secure the hood in the fully open position.
5. Remove the air cleaner and air duct assembly.
6. Disconnect the control cables from the throttle body and remove the control cable bracket from the intake manifold and rocker arm cover. Set the assembly out of the way.
7. Disconnect and cap the fuel lines from the throttle body and manifold mounting bracket.
8. Tag and disconnect any vacuum hoses that will interfere with engine removal.
9. Remove the upper and lower radiator hoses.
10. Disconnect the heater hoses from the intake manifold and water pump and secure out of the way.
11. Remove the torque strut.
12. Disconnect the engine harness connector.
13. Rotate the engine forward.

14. Remove the power steering pump from the engine and support out of the way with the power steering lines attached.
15. Disconnect the electrical connectors from the rear of the engine.
16. Remove the transaxle oil fill tube.
17. Remove the transaxle-to-engine bolts leaving in only the upper two bolts.
18. Rotate the engine to its normal position.
19. Raise and safely support the vehicle.
20. Remove the right front tire and wheel assembly.
21. Remove the right inner fender well splash shield.
22. Disconnect the exhaust pipe from the exhaust manifold.
23. Remove the flywheel cover.
24. Disconnect and remove the starter.
25. Remove the engine mount-to-frame nuts.
26. Remove the torque converter-to-flywheel bolts.
27. Remove the A/C compressor from the mounting bracket and secure out of the way to the side without disconnecting the refrigerant lines.
28. Remove the front pipe support bracket from the transaxle.
29. Remove the transaxle support bracket from the transaxle and frame.
30. Lower the vehicle.
31. Remove the two remaining engine-to-transaxle mounting bolts.
32. Attach a suitable lifting device and slowly remove the engine. While lifting out the engine make sure that no hoses or electrical connectors are still attached.

To install:
33. Lower the engine into the vehicle and connect the engine to the transaxle.
34. Install the two upper engine-to-transaxle bolts. Tighten the bolts only until snug.
35. Raise and safely support the vehicle.
36. Install the transaxle support bracket and tighten the mounting bolts to 38 ft. lbs. (52 Nm).
37. Install the front exhaust pipe-to-transaxle mounting bolt.
38. Install the A/C compressor in the mounting bracket.
39. Install the torque converter-to-flywheel bolts and tighten to 46 ft. lbs. (62 Nm).
40. Install the engine mount-to-frame nuts and tighten to 33 ft. lbs. (45 Nm).

41. Connect and install the starter and tighten the mounting bolts to 33 ft. lbs. (45 Nm).
42. Install the flywheel cover.
43. Connect the exhaust pipe to the exhaust manifold and tighten the mounting bolts to 22 ft. lbs. (30 Nm).
44. Install the right inner fender well splash shield.
45. Install the tire and wheel assembly and tighten to specification.
46. Lower the vehicle.
47. Install the remainder of the engine-to-transaxle bolts and tighten all the bolts to 37 ft. lbs. (50 Nm).
48. Install the transaxle fill tube.
49. Connect the electrical connectors to the rear of the engine.
50. Install the power steering pump on the mounting bracket.
51. Connect the engine harness.
52. Install the torque strut and tighten the mounting bolts to 41 ft. lbs. (56 Nm).
53. Connect the heater hoses to the intake manifold and water pump.
54. Install the upper and lower radiator hoses.
55. Connect any vacuum hoses disconnected for engine removal.
56. Connect the fuel lines to the throttle body and fuel line brackets.
57. Install the throttle cable bracket on the intake manifold and rocker arm cover and connect the control cables to the throttle body.
58. Install the air cleaner assembly and air inlet ductwork.
59. Connect the hood hydraulic shock to the cowl panel.
60. Refill the cooling system. An oil and filter change is recommended.
61. Connect the negative battery cable.
62. Start the vehicle and verify no leaks.

3.3L (VIN N) Engine

CAUTION

The fuel system is under pressure and the pressure must be relieved prior to disconnecting ay fuel lines. Failure to properly relieve the fuel pressure can result in personal injury.

---

1. Relieve the fuel system pressure.
2. Disconnect the negative battery cable.
3. Drain the cooling system.
4. Remove the air cleaner and air inlet duct assembly.
5. Disconnect the heater hoses from the front cover and intake manifold.
6. Remove the exhaust crossover pipe.

7. Disconnect the fuel line quick-Connects from the fuel lines.
8. Remove the torque strut.
9. Remove the serpentine belt.
10. Place a catch pan under the power steering pump and disconnect and cap the power steering lines from the power steering pump.
11. Disconnect the vacuum line from the power brake booster.
12. Disconnect the control cables from the throttle body lever and mounting bracket at the intake manifold.
13. Remove the alternator.
14. Disconnect the engine electrical connectors and position the harness out of the way.
15. Tag and disconnect any vacuum lines interfering with engine removal.
16. Remove the engine ground wires from the transaxle mounting bolts.
17. Remove the wiring harness clips from the right side of the engine compartment.
18. Raise and safely support the vehicle.
19. Drain the engine oil.
20. Remove the right front tire and wheel assembly.
21. Disconnect the exhaust pipe from the exhaust manifold.
22. Remove the right inner fender splash shield.
23. Remove the lower radiator hose.
24. Remove the A/C compressor from the mounting bracket and set aside without disconnecting the refrigerant lines.
25. Remove the flywheel cover.
26. Remove the starter.
27. Remove the torque converter-to-flywheel bolts.
28. Remove the engine mount-to-frame nuts.
29. Disconnect the oil pressure sender, knock sensor and ground wire near the power steering pump bracket.
30. Remove the transaxle support bracket bolts from the transaxle.
31. Remove the lower transaxle-to-engine bolt.
32. Lower the vehicle.
33. Remove the transaxle-to-engine bolts.
34. Install a suitable engine lifting device.
35. Support the transaxle with a floor jack.
36. Remove the engine from the vehicle.

To install:
37. Install the engine carefully into the engine compartment and mate the engine to the transaxle.

38. Install the transaxle-to-engine bolts finger tight.
39. Remove the lifting device and floor jack.
40. Tighten the mounting bolts to 55 ft. lbs. (75 Nm).
41. Raise and safely support the vehicle.
42. Install the lower engine-to-transaxle mounting bolt and tighten to 55 ft. lbs. (75 Nm).
43. Connect the transaxle support to the transaxle and tighten the mounting bolt to 32 ft. lbs. (43 Nm).
44. Connect the oil pressure sender, knock sensor and ground near from the power steering pump bracket.
45. Install the engine mount-to-frame nuts and tighten to 34 ft. lbs. (46 Nm).
46. Line up the torque converter with the flywheel and install the mounting bolts and tighten to 46 ft. lbs. (62 Nm).
47. Install the starter and tighten the mounting bolts to 33 ft. lbs. (45 Nm).
48. Install the flywheel cover and tighten the mounting bolts to 48 inch lbs. (6 Nm).
49. Install the A/C compressor in the mounting bracket.
50. Install the lower radiator hose.
51. Install the right inner fender splash shield.
52. Connect the exhaust pipe to the exhaust manifold and tighten the mounting bolts to 22 ft. lbs. (30 Nm).
53. Install the right front tire and wheel assembly and tighten to specification.
54. Lower the vehicle.
55. Connect the wiring harness clips at the right side of the engine compartment.
56. Connect the ground wires to the transaxle mounting bolts.
57. Connect any vacuum hoses disconnected.
58. Connect any electrical connectors disconnected for engine removal.
59. Install the alternator.
60. Connect the control cables to the mounting bracket and throttle lever.
61. Connect the vacuum line to the power brake booster.
62. Connect the power steering lines to the pump and tighten the fittings to 21 ft. lbs. (28 Nm).
63. Install the serpentine belt.
64. Install the torque strut and tighten the mounting bolts to 41 ft. lbs. (56 Nm).
65. Connect the fuel lines at the quick-Connects.

66. Install the exhaust crossover pipe and tighten the mounting nuts to 19 ft. lbs. (26 Nm).
67. Install the upper radiator hose.
68. Connect the heater hoses to the water pump and intake manifold.
69. Install the air cleaner and air inlet duct assembly.
70. Refill the crankcase with oil.
71. Connect the negative battery cable.
72. Fill the cooling system.

3.1L (VIN M) Engine

——————— CAUTION ———————
Fuel injection systems remain under pressure, even after the engine has been turned OFF. The fuel system pressure must be relieved before disconnecting any fuel lines. Failure to do so may result in fire and/or personal injury.

1. Relieve the fuel system pressure.
2. Disconnect the negative battery cable.
3. Scribe reference marks at the hood supports and remove the hood. Install covers on both fenders.
4. Remove the air cleaner assembly.
5. Drain the cooling system.
6. Remove the radiator hoses from the engine.
7. Remove the engine torque strut.
8. Remove the serpentine belt.
9. Remove the heater hoses.
10. Remove the throttle body bracket and cable.
11. Disconnect the fuel lines.
12. Tag and disconnect electrical connections, from the engine wiring harness and position the harness aside.
13. Tag and disconnect the vacuum lines, from the engine, to non-engine mounted components.
14. Disconnect the power steering lines.
15. Remove the power steering brace.
16. Remove the coolant reservoir.
17. Remove the two A/C compressor upper bolts.
18. Remove the electrical connections from transaxle.
19. Remove the electrical grounds from transaxle mounting bolts.
20. Remove the four transaxle top bolts.
21. Safely raise and support the vehicle.
22. Disconnect the front exhaust pipe from manifold.
23. Remove the oil drip shield bolts and shield.

24. Remove the flywheel cover bolts and cover.
25. Disconnect and remove the starter.
26. Disconnect the electrical connection to the lower engine.
27. Remove the flywheel to converter bolts.
28. Remove the A/C compressor, position aside.

NOTE: Do not disconnect the A/C compressor lines when removing.

29. Remove transaxle support bracket to transaxle bolts.
30. Remove engine mounts.
31. Remove two remaining transaxle bolts.
32. Lower the vehicle.
33. Attach a lifting device to the engine.
34. Carefully remove the engine assembly.
To install:
35. Install engine assembly into vehicle with lifting device.
36. Start two transaxle to engine bolts.
37. Remove lifting device and transaxle support.
38. Safely raise and support the vehicle.
39. Install remaining transaxle bolts and tighten to 37 ft. lbs. (50 Nm).
40. Install the engine mounts.
41. Install transaxle support bracket to transaxle bolts.
42. Properly position A/C compressor.
43. Install flywheel to converter bolts and tighten to 47 ft. lbs. (63 Nm).
44. Install flywheel cover and bolts. Torque bolts to 89 inch lbs. (10 Nm).
45. Install the electrical grounds to the transaxle mounting bolts.
46. Connect the lower engine wiring to the engine.
47. Install the oil drip shield and bolts.
48. Connect the front exhaust pipe to manifold.
49. Safely lower the vehicle.
50. Install the four transaxle top bolts and tighten to 37 ft. lbs. (50 Nm).
51. Connect the electrical grounds to transaxle mounting bolts.
52. Connect the upper engine electrical connections, to the engine.
53. Install the A/C compressor upper bolts.
54. Install the coolant reservoir.
55. Install the power steering brace.
56. Connect the power steering lines.

57. Connect all previously removed vacuum lines to the engine.
58. Connect the fuel lines.
59. Install the throttle body bracket and cable.
60. Install the heater hoses and secure with clamps.
61. Install the serpentine belt.
62. Connect the radiator hoses and secure with clamps.
63. Align the hood-to-hinge mating marks and install the hood.
64. Connect the negative battery cable.
65. Install air cleaner assembly.
66. Fill the coolant and engine oil.
67. Start the engine and allow to reach normal operating temperature. Check for leaks and refill the cooling system.

Engine Mounts

REMOVAL AND INSTALLATION

2.2L (VIN 4) Engine

Front Engine Mount

1. Disconnect the negative battery cable.
2. Install an engine support tool J-28467-A and J-35953, or their equivalents.
3. Remove the torque strut.
4. Raise and safely support the vehicle.
5. Remove the right front tire and wheel assembly.
6. Remove inner fender splash shield.
7. Remove the engine mount-to-frame nuts.
8. Lower the vehicle.
9. Raise the engine using the support fixture.
10. Raise and safely support the vehicle.
11. Remove engine mount-to-engine bracket nuts.
12. Remove the engine mount.
To install:
13. Install the engine mount.
14. Install the engine mount-to-bracket nuts and tighten to 39 ft. lbs. (53 Nm).
15. Lower the vehicle.
16. Lower the engine into position. Make sure when lowering the engine the engine mount studs line up with the holes in the support frame.
17. Raise and safely support the vehicle.
18. Install the engine mount-to-frame nuts and tighten to 32 ft. lbs. (43 Nm).
19. Install inner fender splash shield.

20. Install the right front tire and wheel assembly and tighten to specification.

21. Lower the vehicle.

22. Install the torque strut and tighten the mounting nuts to 44 ft. lbs. (60 Nm) or the through bolts to 40 ft. lbs. (55 Nm).

23. Remove the engine support fixture.

24. Connect negative battery cable.

Torque Strut

1. Remove the torque strut nut and through bolt from the radiator–side mount.

2. Remove the torque strut nut and through bolt from the engine–side mount and remove the torque strut.

To install:

3. Install the torque strut in the engine–side mount and install the through bolt and nut.

4. Install the radiator side through bolt and nut.

5. Tighten the mounting nuts to 44 ft. lbs. (60 Nm) or the through bolts to 40 ft. lbs. (55 Nm).

3.3L (VIN N) Engine

Front Engine Mount

1. Disconnect the negative battery cable.

2. Support the engine using tool J-28467 or an equivalent engine support fixture.

3. Raise and safely support the vehicle.

4. Remove the engine mount bracket nuts.

5. Lower the vehicle.

6. Remove the torque strut.

7. Using the engine support fixture raise the engine off of the engine mount.

8. Remove the engine mount-to-frame mounting nuts.

9. Remove the engine mount.

To Install:

10. Install the engine mount.

11. Install the engine mount-to-bracket nuts and tighten to 34 ft. lbs. (46 Nm).

12. Lower the vehicle.

13. Lower the engine. When lowering the engine make sure the studs on the engine mount line up with the holes in the support frame.

14. Raise and safely support the vehicle.

15. Install engine mount-to-support frame nuts and tighten to 34 ft. lbs. (46 Nm).

16. Lower the vehicle.

17. Install the torque strut.

18. Remove the engine support fixture.

19. Connect the negative battery cable.

Torque Strut

1. Remove the radiator–side torque strut nut and through bolt.

2. Remove the engine–side torque strut nut and through bolt and remove the torque strut.

To Install:

3. Install the torque strut into the engine–side mount and install the through bolt and nut.

4. Line up the torque strut with the radiator side mount and install the through bolt and nut.

5. Tighten the mounting nuts to 42 ft. lbs. (57 Nm).

3.1L (VIN M) Engine

Front Engine Mount

1. Disconnect the negative battery cable.

2. Remove the torque strut assembly.

3. Raise and safely support the vehicle.

4. Remove the nuts securing the engine mount to the frame.

5. Remove the transaxle mount-to-frame nuts.

6. Lower the vehicle.

7. Raise engine to provide clearance, using tool J-28467-A and J-36462 or their equivalents.

8. Raise and safely support the vehicle.

9. Remove mount to engine mount bracket nuts.

10. Remove engine mount.

To install:

11. Install the engine mount in the vehicle and install the engine mount-to-bracket nuts and tighten to 39 ft. lbs. (53 Nm).

12. Lower the vehicle.

13. Lower the engine.

14. Raise and safely support the vehicle.

15. Install the engine mount to frame nuts and tighten to 39 ft. lbs. (53 Nm).

16. Install the transaxle mount nuts and tighten to 32 ft. lbs. (43 Nm).

17. Lower the vehicle.

18. Install the engine torque strut.

19. Remove the engine support fixture.

20. Connect the negative battery cable.

Torque Strut

1. Remove the radiator–side torque strut mounting nut and through bolt.

2. Remove the engine–side torque strut mounting nut and through bolt and remove the torque strut.

To install:

3. Install the torque strut in the engine side bracket and install the through bolt and nut.

4. Install the radiator side through bolt and bracket and tighten the mounting nuts to 44 ft. lbs. (60 Nm).

Cylinder Head

REMOVAL AND INSTALLATION

2.2L (VIN 4) Engine

――――――― CAUTION ―――――――

Fuel injection systems remain under pressure, even after the engine has been turned OFF. The fuel system pressure must be relieved before disconnecting any fuel lines. Failure to do so may result in fire and/or personal injury.

―――――――――――――――――――――

1. Disconnect the negative battery cable.

2. Relieve the fuel system pressure.

3. Drain the cooling system into a suitable container.

4. Remove the air cleaner and air duct assembly.

5. Remove the air inlet resonator upper tie bar.

6. Remove the lower air inlet.

7. Remove the serpentine belt.

8. Remove the alternator.

9. Remove the power steering pump and position it aside without disconnect the power steering lines.

10. Disconnect the spark plug wires and lay them out of the way.

11. Disconnect the control cables from the throttle body and remove the cable bracket at the throttle body and rocker arm cover.

NOTE: Use care when removing valve train components. Parts to be reused must be returned to their original locations.

12. Remove the rocker arm cover, rocker arm nuts, rocker arms and pushrods.

13. Disconnect the electrical connectors from the intake manifold, throttle body and cylinder head.

14. Disconnect the oxygen (O_2) sensor connector.

15. Remove the power steering pump bracket from the intake manifold brace, located under the intake manifold.

16. Remove the torque strut and engine–side torque strut mounting bracket.

17. Remove the alternator rear bracket.

18. Tag and disconnect the vacuum lines at the intake manifold and cylinder head.

19. Disconnect the upper radiator hose from the engine.

20. Raise and safely support the vehicle.

21. Disconnect the exhaust pipe from the exhaust manifold.

22. Lower the vehicle.

23. Disconnect and cap the fuel lines at the quick disconnects.

24. Remove the transaxle fill tube.

25. Remove the cylinder head bolts.

26. Remove the cylinder head with both manifolds. Remove the intake and exhaust manifolds from the cylinder head.

To install:

27. Clean all the gasket surfaces completely. Clean the threads on the cylinder head bolts and the block threads.

28. Install the intake and exhaust manifolds on the cylinder head.

29. Place a new cylinder head gasket in position over the dowel pins on the block. Carefully guide the cylinder head into position.

30. New head bolts are recommended. Install all the cylinder head bolts finger tight. The long bolts go in bolt positions 1, 4, 5, 8 and 9. The short bolt are in positions 2, 3, 6 and 7. The stud is in position 10.

31. Tighten the bolts in sequence. The long bolts to 46 ft. lbs. (63 Nm) and the short bolts and stud to 43 ft. lbs. (58 Nm). Make second pass tightening the long bolts to 46 ft. lbs. (63 Nm) and the short bolts to 43 ft. lbs. (58 Nm). Make a final pass over all bolts tightening each an additional 90 degrees (¼ turn).

32. Install the transaxle fill tube.

33. Connect the fuel lines to the throttle body.

34. Raise and safely support the vehicle.

35. Connect the exhaust pipe to the exhaust manifold. Tighten the mounting bolts to 22 ft. lbs. (30 Nm).

36. Lower the vehicle.

37. Connect the upper radiator hose.

38. Connect the vacuum lines to the intake manifold.

39. Install the engine–side torque strut bracket and torque strut.

40. Install the alternator rear bracket.

41. Install the power steering pump bracket to the intake manifold brace, located under the intake manifold.

42. Connect the electrical connections at the intake manifold, throttle body and cylinder head.

43. Connect the oxygen (O₂) sensor connector.

44. Install the pushrods, rocker arms and rocker arm nuts and tighten the nuts to 22 ft. lbs. (30 Nm).

45. Install the rocker arm cover.

46. Connect the control cables to the throttle body and install the cable brackets at the throttle body and rocker arm cover.

47. Connect the spark plug wires.

48. Install the power steering pump in the mounting bracket.

49. Install the alternator.

50. Install the serpentine belt.

51. Install the lower air inlet duct.

52. Install the air inlet resonator tie bar.

53. Install the air cleaner and duct assembly.

54. Refill the cooling system.

55. An oil and filter change is recommended since coolant can enter the oil system when the head is removed.

56. Connect the negative battery cable.

57. Start the vehicle and verify no leaks.

3.3L (VIN N) Engine

Right Side (Rear)

— **CAUTION** —

Care should be used when working around the fuel system. DO NOT smoke or expose the fuel system to any open flames or sparks. The fuel system is under pressure and must be properly bled prior to disconnecting any fuel line connections. Failure to properly bleed the fuel system can lead to personal injury or component damage. Always keep a suitable fire extinguisher handy when servicing the fuel system.

1. Relieve the fuel system pressure.

2. Disconnect the negative battery cable.

3. Drain the cooling system into a suitable container.

4. Remove the intake manifold.

5. Remove the crossover pipe.

6. Remove the nut from the heater pipe and slide the pipe out of the cover housing.

7. Remove the power steering pump mounting bolts from the mounting bracket and pull the pump forward slightly.

8. Remove the engine lift bracket and coolant pipe.

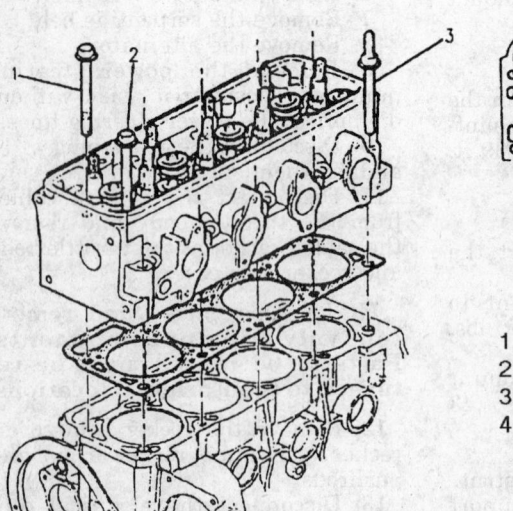

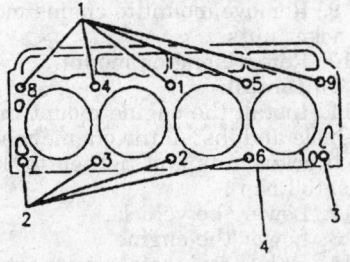

1. Long bolts
2. Short bolts
3. Stud
4. Numbers on gasket indicate torque sequence

291090

Cylinder head bolt tightening sequence — 2.2L (VIN 4) engine

9. Install the exhaust manifold heat shield.

10. Remove the transaxle fill tube.

11. Disconnect the spark plug wires from the spark plugs and wire looms.

12. Remove the spark plugs.

13. Raise and safely support the vehicle.

14. Remove the exhaust pipe-to-exhaust manifold bolts.

15. Lower the vehicle.

16. Remove the exhaust manifold bolts and remove the exhaust manifold.

17. Remove the rocker arm cover, rocker arm bolts, rocker arms, pedestals and pushrods.

18. Remove the cylinder head bolts.

19. Remove the cylinder head and gasket from the engine.

To install:

20. Thoroughly clean and dry all bolts, bolt holes and mating surfaces.

21. Install the head gasket to the block and carefully position the cylinder head in place.

22. Coat the cylinder head bolts with thread lock compound and start all the cylinder head bolts finger tight.

23. Tighten the cylinder head bolts in sequence to the following:

    a. Step 1: Tighten to 35 ft. lbs. (47 Nm).

    b. Step 2: Tighten each bolt an additional 130°.

    c. Step 3: Tighten the four center bolts an additional 30°.

24. Install the pushrods, rocker arms and pedestals and tighten the mounting bolts to 28 ft. lbs. (38 Nm).

25. Install the rocker arm cover.

26. Install the exhaust manifold and tighten the mounting bolts and studs to 38 ft. lbs. (52 Nm).

27. Raise and safely support the vehicle.

28. Connect the exhaust pipe to the exhaust manifold and tighten the bolts to 22 ft. lbs. (30 Nm).

29. Lower the vehicle.

30. Install the spark plugs and tighten to 11 ft. lbs. (15 Nm).

31. Connect spark plugs wires to the wire looms and spark plugs.

32. Install the transaxle fill tube.

33. Install the exhaust manifold heat shield.

34. Install the coolant pipe and engine lift bracket.

35. Install the power steering pump mounting bolts.

36. Slide the heater pipe into the front cover housing and install the mounting nut.

37. Install the crossover pipe.

38. Install the intake manifold.

39. Connect the negative battery cable.

40. Start the vehicle and verify no leaks.

Left Side (Front)

CAUTION

Care should be used when working around the fuel system. DO NOT smoke or expose the fuel system to any open flames or sparks. The fuel system is under pressure and must be properly bled prior to disconnecting any fuel line connections. Failure to properly bleed the fuel system can lead to personal injury or component damage. Always keep a suitable fire extinguisher handy when servicing the fuel system.

1. Relieve the fuel system pressure.

2. Disconnect the negative battery cable.

3. Drain the cooling system into a suitable container.

4. Remove the intake manifold.

5. Disconnect the spark plug wires from the spark plugs and wire looms.

6. Remove the engine lift bracket.

7. Remove the rocker arm cover.

8. Remove the spark plugs.

9. Remove the exhaust crossover.

10. Remove the oil level indicator tube.

11. Remove the cooling fan assembly.

12. Install the exhaust manifold heat shield.

13. Remove the exhaust manifold support bracket.

14. Remove the exhaust manifold bolts and remove the exhaust manifold.

15. Remove the rocker arm bolts, rocker arms, pedestals and pushrods.

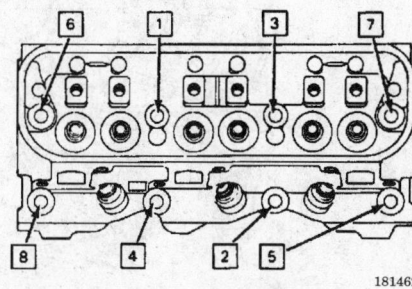

Cylinder head bolt tightening sequence — 3.3L (VIN N) engine

16. Remove the A/C bracket bolt from the cylinder head.

17. Remove the bolts securing the alternator and ignition coil mounting bracket and position the bracket aside.

18. Disconnect the vacuum line at the rear of the cylinder head.

19. Remove the cylinder head bolts.

20. Remove the cylinder head and gasket from the engine.

To install:

21. Thoroughly clean and dry all bolts, bolt holes and mating surfaces.

22. Install the head gasket to the block and carefully position the cylinder head in place.

23. Coat the cylinder head bolts with thread lock compound and start all the cylinder head bolts finger tight.

24. Tighten the cylinder head bolts in sequence to the following:

    a. Step 1: Tighten to 35 ft. lbs. (47 Nm).

    b. Step 2: Tighten each bolt an additional 130°.

    c. Step 3: Tighten the four center bolts an additional 30°.

25. Connect the vacuum line to the rear of the cylinder head.

26. Install the A/C bracket bolt to the head.

27. Install the ignition assembly and alternator mounting bracket.

28. Install the pushrods, rocker arms and pedestals and tighten the mounting bolts to 28 ft. lbs. (38 Nm).

29. Install the exhaust manifold and tighten the mounting bolts and studs to 38 ft. lbs. (52 Nm).

30. Install the exhaust manifold support bracket.

31. Install the exhaust manifold heat shield and tighten the nuts to 20 ft. lbs. (27 Nm).

32. Install the cooling fan assembly.

33. Install the oil level indicator and tighten the mounting nut to 20 ft. lbs. (27 Nm).

34. Install the exhaust crossover pipe.

35. Install the spark plugs and tighten to 11 ft. lbs. (15 Nm).

36. Install the rocker arm cover.

37. Install the engine lift bracket.

38. Connect spark plugs wires to the wire looms and spark plugs.

39. Install the intake manifold.

40. Connect the negative battery cable.

41. Start the vehicle and verify no leaks.

3.1L (VIN M) Engine

Left Side (Front)

— **CAUTION** —

Fuel injection systems remain under pressure, even after the engine has been turned OFF. The fuel system pressure must be relieved before disconnecting any fuel lines. Failure to do so may result in fire and/or personal injury.

1. Disconnect the negative battery cable.
2. Relieve the fuel system pressure.
3. Drain the cooling system into a suitable container.

— **CAUTION** —

When servicing the A/C system the correct tools and procedures must be used. Protective eye covering must be worn. DO NOT smoke or expose the refrigerant to any open flames or sparks.

4. Recover the A/C refrigerant using a refrigerant recovery and recycling station.
5. Remove the rocker arm covers.
6. Remove upper intake plenum and lower intake manifold.
7. Remove the exhaust crossover pipe.
8. Disconnect the spark plug wires from spark plugs and wire looms and route the wires out of the way.

NOTE: When removing the valve train components use care to identify any components that will be reused. Valve train components must be kept in order for installation in the same locations from which they were removed.

9. Remove rocker arms nut, rocker arms, balls and pushrods.
10. Remove oil level indicator tube.
11. Remove any A/C compressor bolts accessible from the top.
12. Raise and safely support the vehicle.
13. Remove the lower A/C compressor mounting bolts.
14. Remove the refrigerant lines from the rear of the compressor.
15. Disconnect the A/C compressor electrical connections and remove the A/C compressor.
16. Remove the A/C compressor lower bracket bolts.
17. Lower the vehicle.
18. Remove the A/C compressor upper bracket bolts.

19. Remove the compressor brackets.
20. Remove the cylinder head bolts evenly.
21. Remove the cylinder head.
To install:
22. Clean all the gasket surfaces completely. Clean the threads on the cylinder head bolts and block threads.
23. Place the cylinder head gasket in position over the dowel pins on the cylinder block so the words **THIS SIDE UP** or other gasket identification are showing.
24. Coat the bolt threads with sealer and install finger tight.
25. Tighten the cylinder head bolts in sequence to 33 ft. lbs. (45 Nm). With all the bolts tightened make a second pass tightening all the bolts an additional 90 degrees (¼ turn).
26. Install compressor bracket and tighten the upper bracket bolts to 35 ft. lbs. (47 Nm).
27. Raise vehicle and safely support.
28. Install the compressor lower bracket bolts and tighten to 35 ft. lbs. (47 Nm).
29. Connect the electrical connection to the rear of compressor.
30. Install the A/C compressor in the mounting bracket.

31. Connect the A/C lines to the rear of the compressor with NEW seals.
32. Install the compressor lower mounting bolts and tighten to 18 ft. lbs. (25 Nm).
33. Lower the vehicle.
34. Tighten the compressor upper mounting bolts to 18 ft. lbs. (25 Nm).
35. Install the oil level indicator tube.
36. Install the pushrods, rocker arms, balls and rocker arm nuts. Tighten the rocker arm nuts to 20 ft. lbs. (27 Nm).
37. Connect the spark plug wires to spark plugs and wire looms.
38. Install the exhaust crossover pipe.
39. Install the lower intake manifold and upper intake plenum.
40. Install the rocker arm covers.
41. Refill the cooling system.
42. Evacuate and charge A/C system.
43. Connect negative battery cable.
44. An oil and filter change are recommended since coolant can enter the oil system when the head is being removed.
45. Start vehicle and verify no leaks.

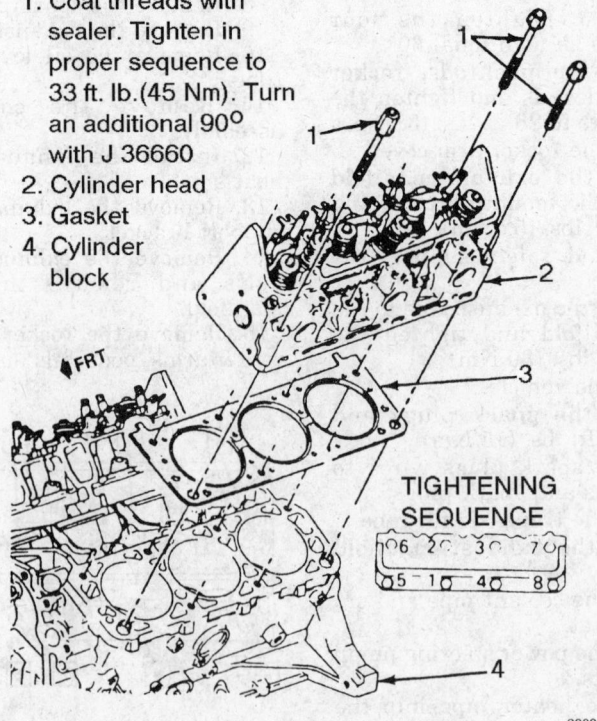

1. Coat threads with sealer. Tighten in proper sequence to 33 ft. lb. (45 Nm). Turn an additional 90° with J 36660
2. Cylinder head
3. Gasket
4. Cylinder block

TIGHTENING SEQUENCE

290998

Cylinder head components and bolt tightening sequence — 3.1L (VIN M) engine

Right Side (Rear)

CAUTION

Fuel injection systems remain under pressure, even after the engine has been turned OFF. The fuel system pressure must be relieved before disconnecting any fuel lines. Failure to do so may result in fire and/or personal injury.

1. Disconnect the negative battery cable.
2. Relieve the fuel system pressure.
3. Drain the cooling system into a suitable container.
4. Remove the rocker arm covers.
5. Remove upper intake plenum and lower intake manifold.
6. Disconnect the electrical connector from the ignition assembly.
7. Remove the alternator.
8. Remove the exhaust crossover pipe.
9. Disconnect the oxygen (O_2) sensor connector.
10. Raise and safely support the vehicle.
11. Disconnect the exhaust pipe from the exhaust manifold.
12. Lower the vehicle.
13. Remove the exhaust manifold.
14. Disconnect the spark plug wires from spark plugs and wire looms and route the wires out of the way.

NOTE: When removing the valve train components use care to identify any components that will be reused. Valve train components must be kept in order for installation in the same locations from which they were removed.

15. Remove rocker arms nut, rocker arms, balls and pushrods.
16. Remove the cylinder head bolts evenly.
17. Remove the cylinder head.

To install:

18. Clean all the gasket surfaces completely. Clean the threads on the cylinder head bolts and block threads.
19. Place the cylinder head gasket in position over the dowel pins on the cylinder block so the words **THIS SIDE UP** or other gasket identification is showing.
20. Coat the bolt threads with sealer and install finger tight.
21. Tighten the cylinder head bolts in sequence to 33 ft. lbs. (45 Nm). With all the bolts tightened make a second pass tightening all the bolts an additional 90 degrees (¼ turn).
22. Install the pushrods, rocker arms, balls and rocker arm nuts.

Tighten the rocker arm nuts to 20 ft. lbs. (27 Nm).
23. Install the exhaust manifold.
24. Raise vehicle and safely support.
25. Connect the exhaust pipe to the exhaust manifold.
26. Lower the vehicle.
27. Connect the oxygen (O_2) sensor connector.
28. Connect the spark plug wires to spark plugs and wire looms.
29. Install the exhaust crossover pipe.
30. Install the alternator.
31. Connect the electrical connector to the ignition assembly.
32. Install the lower intake manifold and upper intake plenum.
33. Install the rocker arm covers.
34. Refill the cooling system.
35. An oil and filter change are recommended since coolant can enter the oil system when the head is being removed.
36. Connect negative battery cable.
37. Start vehicle and verify no leaks.

Valve Lash

ADJUSTMENT

NOTE: These engines come from the factory with non-Adjustable valve lash, BUT if the valves and the valve seats are reconditioned adjustable rocker arm studs must be installed. The adjustment procedure is ONLY for adjustable rocker arm studs.

1. The engine should be in the #1 firing position. This may be determined by placing fingers on the #1 rocker arms as the engine assembly alignment marks on the front face of the torsional damper aligns with the arrow on the front cover. If the valves are not moving, the engine is in the #1 firing position. If the valves move the engine is in the #4 firing position and should be rotated one revolution to reach the #1 position.
2. With the engine in the #1 firing position, the following valves should be adjusted:
   - Exhaust — 1, 2, 3
   - Intake– 1, 5, 6
3. Tighten the adjusting nut until there is zero lash + 1½ turns.
4. Crank the engine one revolution. This is the #4 firing position. With the engine in this position, the following valves should be adjusted to zero lash + 1½ turns:
   - Exhaust — 4, 5, 6
   - Intake — 2, 3, 4

Valve Lifters

REMOVAL AND INSTALLATION

2.2L (VIN 4) Engine

CAUTION

Fuel injection systems remain under pressure, even after the engine has been turned OFF. The fuel system pressure must be relieved before disconnecting any fuel lines. Failure to do so may result in fire and/or personal injury.

The cylinder head must be removed from the engine for this procedure.
1. Disconnect the negative battery cable.
2. Relieve the fuel system pressure.
3. Drain the engine coolant into a suitable container.
4. Remove the rocker arm cover and gasket.

NOTE: If any valvetrain components are to be reused, keep them in order so they can be reinstalled in the same locations from which they were removed.

5. Remove the rocker arm nuts. Remove the rocker arms and balls.
6. Remove the pushrods.
7. Remove the engine lift bracket from the rear of the engine.
8. Disconnect the spark plug wires and route them under the intake manifold.
9. Remove the cylinder head with the intake and exhaust manifold as an assembly.
10. Remove the bolt securing the anti–rotation brackets and remove the brackets.
11. Remove the lifters.

NOTE: Whenever new valve lifters are being installed, coat the valve lifters with camshaft assembly lube GM No. 1052365, or equivalent.

To install:
12. Coat the base of the lifters with camshaft lube or engine oil supplement prior to installation in the lifter bore.
13. If any lifters are being reused, install the lifters in the same bores from which they were removed.
14. Install the anti–rotation brackets and tighten the mounting bolts to 97 inch lbs. (11 Nm).
15. Install the cylinder head and manifolds.

16. Install the rear engine lift bracket and tighten the mounting nut to 37 ft. lbs. (50 Nm).

17. Connect the spark plug wires to the spark plugs.

18. Install the pushrods making sure they seat in the lifters correctly. Install the rocker arms, balls and mounting nuts and tighten the nuts to 22 ft. lbs. (30 Nm).

19. Install the rocker cover gasket and the rocker cover. Torque the bolts to 89 inch lbs. (10 Nm).

20. Refill the engine with coolant. An oil and filter change is recommended.

21. Connect the negative battery cable.

3.3L (VIN N) Engine

1. Disconnect the negative battery cable.

2. Drain the cooling system.

3. Remove the valve cover and the intake manifold.

NOTE: Be sure to keep all valve train parts in order so they may be reinstalled in their original locations and with the same mating surfaces as when removed.

4. Remove the rocker arm mounting bolts making note of the rocker arm mounting studs.

5. Remove the rocker arms and pedestals.

6. Remove the pushrods.

7. Remove the lifter guide retainer bolts and retainer.

8. Remove the lifter guides.

9. Remove the valve lifters. **To Install:**

10. Lubricate the valve lifter surfaces with Molykote® or equivalent.

11. Install the lifters in their original locations.

12. Install the lifter guides.

13. Install the lifter guide retainer and tighten the mounting bolts to 22 ft. lbs. (30 Nm).

14. Install the pushrods.

15. Install rocker arms, pedestals and bolts. Apply thread lock compound to bolt threads before assembly. Tighten the mounting bolts to 11 ft. lbs. (15 Nm) plus 110° additional rotation.

16. Install the intake manifold and rocker arm covers.

17. Refill the cooling system.

18. Connect the negative battery cable.

3.1L (VIN M) Engine

------ CAUTION ------

Fuel injection systems remain under pressure, even after the engine has been turned OFF. The fuel system pressure must be relieved before disconnecting any fuel lines. Failure to do so may result in fire and/or personal injury.

1. Disconnect the negative battery cable.

2. Relieve the fuel system pressure.

3. Drain the cooling system.

4. Remove the rocker arm covers and the intake manifold.

NOTE: Be sure to keep all valve train parts in order so they may be reinstalled in their original locations and with the same mating surfaces as when removed.

5. Remove the rocker arm retaining nuts, rocker arm balls, rocker arms and pushrods.

6. Remove the 2 guide bolts from the right or left side lifter guide and remove the guide.

7. Remove the valve lifter(s) from the lifter bores.

To install:

8. Lubricate the bearing surfaces with Molykote® or equivalent.

NOTE: Installation of a new camshaft or a wear pattern on the old valve lifter will require the replacement of the camshaft and lifters together. If camshaft replacement is not necessary, make sure to install the used valve lifters in their original position upon reinstallation.

9. Install the lifters in their original locations.

10. Install the lifter guide and lifter guide bolts and tighten the guide bolts to 89 inch lbs. (10 Nm).

11. Install the pushrods, rocker arms, rocker balls and rocker arm nuts. Tighten the rocker arm nuts to 18 ft. lbs. (25 Nm).

12. Install the intake manifold and rocker arm covers.

13. Refill the cooling system. An oil and filter change is recommended.

14. Connect the negative battery cable.

15. Start the vehicle and verify no leaks.

Rocker Arms

REMOVAL AND INSTALLATION

2.2L (VIN 4) Engine

1. Disconnect the negative battery cable.

2. Remove the air cleaner and air duct assembly.

3. Tag and disconnect the spark plug wires from the spark plugs and disconnect the spark plug wire clips from the rocker arm cover and position aside.

4. Disconnect the throttle cables from the throttle body and remove the throttle cable bracket from the intake plenum and move aside.

5. Remove the rocker arm cover bolts.

6. Remove the rocker arm cover.

NOTE: If any valvetrain components are to be reused, keep in order. They should be installed in the same locations from which they were removed.

7. Remove the rocker arm nut(s), ball(s) and rocker arms.

To install:

8. Clean the gasket surfaces completely.

9. Coat the bearing surfaces of the rocker arm(s) and the rocker arm ball(s) with Molykote® or its equivalent.

10. Seat the push rods in the lifters.

11. Install the rocker arm(s), ball(s) and nut(s) in the same positions they were removed from and tighten the rocker arm nut(s) to 22 ft. lbs. (30 Nm).

12. Install a new gasket in the cut out in the rocker arm cover.

13. Install the rocker arm cover on the cylinder head and tighten the rocker arm cover bolts to 89 inch lbs. (10 Nm).

14. Install the throttle cable bracket on the intake plenum and connect the throttle control cables to the throttle body.

15. Connect the spark plug wire clips to the rocker arm cover and connect the spark plug wires to the spark plugs.

16. Install the air cleaner and air duct assembly.

17. Connect the negative battery cable.

18. Start the engine and verify no oil leaks.

3.3L (VIN N) Engine

Left Side (Front)

1. Disconnect the negative battery cable.
2. Remove the serpentine belt.
3. Remove the bolt from the alternator brace.
4. Remove the alternator brace to intake manifold and remove the brace.
5. Tag and disconnect the spark plug wires from the plugs and move the wires away from the rocker arm cover.
6. Remove the six rocker arm cover bolts.
7. Remove the rocker arm cover.
8. Remove the rocker arm mounting bolt(s).
9. Remove the pedestal(s) and rocker arm(s).
10. If all six rocker arms are being removed from one head, remove the pushrods and pushrod guide from the cylinder head.

   To Install:

11. Clean all gasket surfaces completely.
12. Coat all the valve train components with engine oil prior to installation. Clean all thread locking compound from the rocker arm bolts and apply new thread locking compound.
13. Install the pushrod guide and pushrods.
14. Install the rocker arm(s) and pedestal(s) and tighten the mounting bolt(s) to 28 ft. lbs. (38 Nm).
15. Install the rocker arm cover with a new gasket and tighten the rocker arm cover mounting bolts to 88 ft. lbs. (10 Nm).
16. Connect the spark plug wires to the plugs.
17. Install the rear alternator brace and tighten the brace bolts to 18 ft. lbs. (25 Nm).
18. Install the serpentine belt.
19. Connect the negative battery cable.

Right Side (Rear)

1. Disconnect the negative battery cable.
2. Remove the serpentine belt.
3. Loosen the power steering pump bolts and slide the pump forward.
4. Remove the power steering pump braces.
5. Tag and disconnect the spark plug wires from the plugs and move the wires away from the rocker arm cover.
6. Remove the six rocker arm cover bolts.
7. Remove the rocker arm cover.
8. Remove the rocker arm mounting bolt(s).
9. Remove the pedestal(s) and rocker arm(s).
10. If all six rocker arms are being removed from one head, remove the pushrods and pushrod guide from the cylinder head.

   To Install:

11. Clean all gasket surfaces completely.
12. Coat all the valve train components with engine oil prior to installation. Clean all thread locking compound from the rocker arm bolts and apply new thread locking compound.
13. Install the pushrod guide and pushrods.
14. Install the rocker arm(s) and pedestal(s) and tighten the mounting bolt(s) to 28 ft. lbs. (38 Nm).
15. Install the rocker arm cover with a new gasket and tighten the rocker arm cover mounting bolts to 88 ft. lbs. (10 Nm).
16. Connect the spark plug wires to the plugs.
17. Install the power steering pump braces and tighten the brace bolts to 19 ft. lbs. (26 Nm).
18. Install the serpentine belt.
19. Connect the negative battery cable.

3.1L (VIN M) Engine

Left Side (Front)

1. Disconnect the negative battery cable.
2. Drain the cooling system to a level below the coolant pipe on the front of the engine.
3. Remove the coolant bypass hose clamp at the coolant tube.
4. Remove the two bolts and nut securing the coolant tube to the cylinder head and position the tube out of the way.

NOTE: If any valvetrain components are to be reused, keep in order. They should be installed in the same locations from which they were removed.

5. Remove the four rocker arm cover bolts and remove the rocker arm cover.
6. Remove the rocker arm nut(s), ball(s) and rocker arm(s).

   To install:

7. Clean all the gasket surfaces completely.
8. Coat all the valve train components with engine oil prior to installation.
9. Install the rocker arm(s) on the stud(s). Install the rocker arm ball(s) and mounting nuts. Make sure the pushrods are properly seated in the lifter and rocker arm.
10. Tighten the rocker arm nuts to 89 inch lbs. (10 Nm) and then an additional 30 degrees.
11. Install the rocker arm cover using a new gasket and tighten the rocker cover bolts to 90 inch lbs. (10 Nm).
12. Position the coolant tube and connect the thermostat bypass hose.
13. Install the coolant tube mounting nut and bolts. Tighten the screw at the water pump to 106 inch lbs. (12 Nm), the bolt at the corner of the cylinder head to 18 ft. lbs. (25 Nm) and the nut to 18 ft. lbs. (25 Nm).
14. Refill the cooling system.
15. Connect the negative battery cable.
16. Start the vehicle and verify no oil or coolant leaks.

Right Side (Rear)

1. Disconnect the negative battery cable.
2. Drain the cooling system into a suitable container.
3. Remove the serpentine belt.
4. Remove the alternator.
5. Remove the alternator mounting bracket.
6. Remove the four rocker arm cover bolts and remove the rocker arm cover.

NOTE: If any valvetrain components are to be reused, keep in order. They should be installed in the same locations from which they were removed.

7. Remove the rocker arm nut(s), ball(s) and rocker arm(s).

   To install:

8. Clean all the gasket surfaces completely.
9. Coat all the valve train components with engine oil prior to installation.
10. Install the rocker arm(s) on the stud(s). Install the rocker arm ball(s) and mounting nuts. Make sure the pushrods are properly seated in the lifter and rocker arm.
11. Tighten the rocker arm nuts to 89 inch lbs. and then an additional 30 degrees.
12. Install the rocker arm cover using a new gasket and tighten the rocker cover bolts to 90 inch lbs. (10 Nm).
13. Install the alternator mounting bracket.
14. Install the alternator.
15. Install the serpentine belt.
16. Refill the cooling system.
17. Connect the negative battery cable.

18. Start the vehicle and verify no oil or coolant leaks.

Intake Manifold

REMOVAL AND INSTALLATION

2.2L (VIN 4) Engine

These vehicles use a two–piece intake manifold. The upper half, sometimes called a plenum, contains the throttle body and control cable connections. The lower half has individual port runners to each intake port on the cylinder head. The lower half of the manifold bolts to the cylinder head and houses the fuel injectors. Note that these pieces are cast aluminum. Care should be exercised when working with any light alloy component.

———— CAUTION ————

Fuel injection systems remain under pressure, even after the engine has been turned OFF. The fuel system pressure must be relieved before disconnecting any fuel lines. Failure to do so may result in fire and/or personal injury.

1. Relieve the fuel system pressure.
2. Disconnect the negative battery cable.
3. Remove the throttle body air intake duct.
4. Drain the cooling system.
5. Identify, tag and disconnect all necessary vacuum lines.
6. Disconnect the control cables from the throttle body lever and remove the control cable bracket from the intake manifold.
7. Remove the serpentine belt.
8. Remove the power steering pump and lay it aside, without disconnecting the fluid lines.
9. Remove the transaxle fluid fill tube.
10. Identify, tag and disconnect the electrical connectors from the MAP sensor, EGR solenoid valve, Idle Air Control (IAC) valve, Throttle Position Sensor (TPS) and fuel injectors.
11. Remove the MAP sensor.
12. Remove the upper intake manifold mounting bolts and the upper intake manifold.
13. Disconnect the fuel lines from the fuel rail.
14. Remove the EGR valve injector.
15. Remove the EGR valve.
16. Remove the fuel injector retainer bracket, regulator and injectors.

17. Remove the control cable bracket.
18. Remove the six intake manifold nuts.
19. Remove the intake manifold.
   To install:
20. Clean the gasket mounting surfaces.
21. Install a new gasket and position the lower intake manifold. Tighten the lower intake manifold nuts in the proper sequence to 24 ft. lbs. (33 Nm).
22. Connect the control cables and cable bracket.
23. Install the EGR valve.
24. Connect the fuel lines to the fuel rail.
25. Install the fuel injectors, regulator and injector retainer bracket and tighten the retaining bolts to 22 inch lbs. (3.5 Nm).
26. Install the EGR valve injector so that the port is facing directly towards the throttle body.
27. Install the upper intake manifold assembly. Tighten the upper intake manifold nuts in the proper sequence to 22 ft. lbs. (30 Nm).
28. Install the MAP sensor.
29. Connect the electrical connectors to the MAP sensor, EGR solenoid valve, Idle Air Control (IAC) valve, Throttle Position Sensor (TPS), and the fuel injectors.

30. Install the transaxle fill tube.
31. Install the power steering pump.
32. Install the serpentine belt.
33. Connect the vacuum lines.
34. Install the air intake duct.
35. Connect the negative battery cable.
36. Refill the coolant system.
37. Start the vehicle and verify no coolant or vacuum leaks.

3.3L (VIN N) Engine

———— CAUTION ————

Prior to performing any service work on the fuel system or related components the fuel system pressure must be bled off. Failure to bleed the fuel system down prior to component service can lead to personal injury and component damage. DO NOT smoke or expose the fuel system to any open flames or sparks. Always keep a suitable fire extinguisher handy when servicing the fuel system.

1. Relieve the fuel system pressure.
2. Disconnect the negative battery cable.
3. Drain the cooling system to a level below the intake manifold.

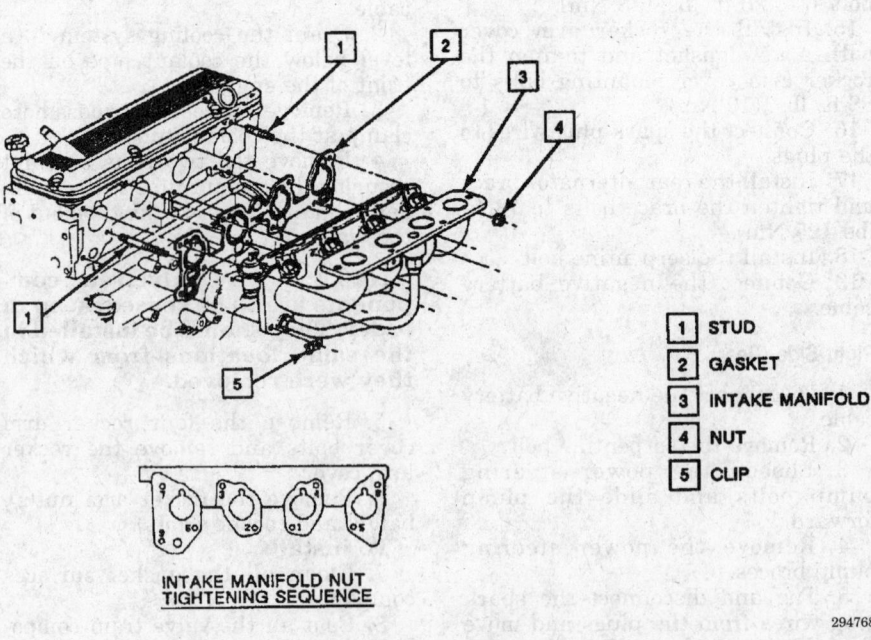

1	STUD
2	GASKET
3	INTAKE MANIFOLD
4	NUT
5	CLIP

INTAKE MANIFOLD NUT TIGHTENING SEQUENCE

294768

Lower intake manifold torque sequence — 2.2L (VIN 4) engine

29. Install the remainder of the intake manifold bolts after coating the threads with thread lock compound.

30. Tighten the bolts **TWICE** to 89 inch lbs. (10 Nm) in the proper sequence.

31. Install the throttle cable bracket to the intake manifold.

32. Install the TV cable bracket and connect the TV cable to the throttle body lever.

33. Connect the throttle cable to the throttle body lever.

34. Install the fuel line bracket.

35. Connect the fuel lines quick connects.

36. Connect the following electrical connectors to their respective components:
- Throttle Position Sensor (TPS)
- Idle Air Control (IAC)
- Fuel injector
- Coolant Temperature Sensor
- Mass Air Flow Sensor (MAF)

37. Connect the rocker arm cover breather hose.

38. Route the rear spark plug wires and connect to the ignition coil terminals.

39. Connect any vacuum lines disconnected.

40. Connect the bypass and heater hoses to the intake manifold.

41. Connect the upper radiator hose to the thermostat housing.

42. Refill the cooling system.

43. Install the torque strut bracket to the intake manifold and install the torque strut. Tighten the bracket mounting bolts to 38 ft. lbs. (52 Nm) and the mounting nuts to 22 ft. lbs. (30 Nm). Tighten the torque strut mounting bolts and nuts to 42 ft. lbs. (57 Nm).

44. Connect and install the alternator.

45. Install the serpentine belt.

46. Install the air cleaner assembly.

47. Connect the negative battery cable.

48. Pressurize the fuel system and verify no leaks.

49. Start the vehicle and check for oil and coolant leaks.

3.1L (VIN M) Engine

1994–95 Vehicles

CAUTION

The fuel system is under pressure and must be properly relieved before disconnecting the fuel lines. Failure to properly relieve the fuel system pressure can lead to personal injury and component damage.

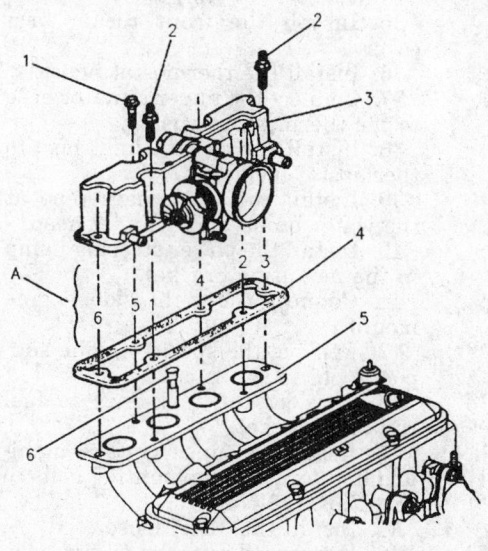

A UPPER INTAKE MANIFOLD ASSEMBLY TIGHTENING SEQUENCE
1 BOLT
2 STUD
3 UPPER INTAKE MANIFOLD ASSEMBLY
4 GASKET
5 LOWER INTAKE MANIFOLD
6 EGR VALVE INJECTOR

294769

Upper intake manifold torque sequence — 2.2L (VIN 4) engine

4. Remove the air cleaner assembly.

5. Remove the serpentine belt.

6. Disconnect and remove the alternator.

7. Remove the torque strut and remove the torque strut bracket from the intake manifold.

8. Remove the rear power steering pump brace.

9. Disconnect the rear spark plug wires from the ignition coil and lay aside.

10. Tag and disconnect the vacuum lines from the intake manifold.

11. Disconnect the upper radiator hose from the thermostat housing.

12. Disconnect the bypass and heater hoses from the intake manifold.

13. Disconnect the rocker arm breather hose from the rocker arm.

14. Disconnect the following electrical connectors from their respective components:
- Throttle Position Sensor (TPS)
- Idle Air Control (IAC)
- Fuel injector
- Coolant Temperature Sensor
- Mass Air Flow Sensor (MAF)

15. Disconnect the fuel line quick disconnects from the fuel rail.

16. Remove the fuel line bracket from the intake manifold.

17. Disconnect the throttle cable from the throttle body lever.

18. Disconnect the TV cable from the throttle body lever and remove the TV cable bracket from the intake manifold.

19. Remove the throttle cable bracket from the intake manifold.

20. Remove the intake manifold bolt securing the power steering pump rear brace to the intake manifold.

21. Remove the manifold vacuum line clamp bolt at the alternator bracket.

22. Remove the intake manifold bolts and remove the intake manifold.

To Install:

23. Clean all the gasket surfaces completely.

24. Apply gasket sealer to the intake manifold end gaskets.

25. Install the intake manifold side gaskets and intake manifold.

26. Slide the vacuum line between the fuel rail and intake manifold.

27. Install the vacuum line clamp bolt at the alternator mounting bracket.

28. Install the power steering pump rear brace and install the intake manifold bolt securing it.

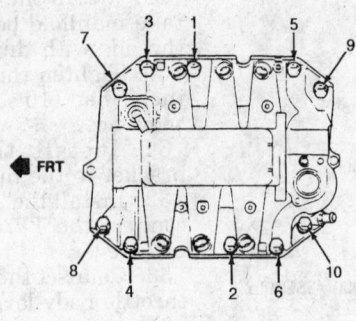

Intake manifold bolt tightening sequence — 3.3L
(VIN N) engine

1. Relieve the fuel system pressure.

2. Disconnect the negative battery cable.

3. Remove top half of the air cleaner assembly and throttle body duct.

4. Drain and recover the cooling system.

5. Remove the EGR pipe from exhaust manifold.

6. Remove the serpentine belt.

7. Remove the brake vacuum pipe at the intake plenum.

8. Disconnect the control cables from the throttle body and intake plenum mounting bracket.

9. Remove the power steering lines at the alternator bracket.

10. Remove the alternator.

11. Disconnect the spark plug wires from the spark plugs and wire retainers on the intake plenum.

12. Remove the Ignition assembly and the EVAP canister purge solenoid together.

13. Disconnect the upper engine wiring harness connectors at the following components:
- Throttle Position Sensor (TPS)
- Idle Air Control (IAC)
- Fuel Injectors
- Coolant temperature sensor
- MAP sensor
- Camshaft Position (CMP) sensor

14. Disconnect the vacuum lines from the following components:
- Vacuum modulator
- Fuel pressure regulator
- PCV valve

15. Remove the MAP sensor from upper intake manifold.

16. Remove the upper intake plenum mounting bolts and remove the plenum.

17. Disconnect the fuel lines from the fuel rail and fuel line bracket.

18. Install engine support fixture special tool J 28467-A or an equivalent.

19. Remove the right side engine mount.

20. Remove the power steering mounting bolts and support the pump out of the way without disconnecting the power steering lines.

21. Disconnect the coolant inlet pipe from coolant outlet housing.

22. Remove the coolant bypass hose from the water pump and the cylinder head.

23. Disconnect the upper radiator hose at thermostat housing.

24. Remove the thermostat housing.

25. Remove both rocker arm covers.

26. Remove the lower intake manifold bolts. Make sure the washers on the four center bolts are installed in their original locations.

NOTE: When removing the valve train components they should be kept in order for installation the original locations.

27. Remove the rocker arm retaining nuts and remove the rocker arms and pushrods.

28. Remove the intake manifold from the engine.

To Install:

29. Clean gasket material from all mating surfaces. Remove all excess RTV sealant from front and rear ridges of cylinder block.

30. Place a 3mm bead of RTV, on each ridge, where the front and rear of the intake manifold contact the block.

31. Using a new gasket, install the intake manifold on the engine. Torque the bolts evenly and gradually, working from the center towards the ends.

32. Install the pushrods, rocker arms and mounting nuts. Make sure the pushrods are properly seated in the valve lifters and rocker arms.

33. Install rocker arm nuts and tighten the rocker arm nuts to 18 ft. lbs. (24 Nm).

34. Install lower the intake manifold attaching bolts. Apply sealant PN 12345739 or equivalent to the threads of bolts, and torque bolts to 115 inch lbs. (13 Nm).

35. Install the front rocker arm cover.

36. Install the thermostat housing.

37. Connect the upper radiator hose to the thermostat housing.

38. Install the coolant inlet pipe to thermostat housing.

39. Install coolant bypass pipe at the water pump and cylinder head.

40. Install the power steering pump in the mounting bracket.

41. Connect the right side engine mount.

42. Remove the special engine support tool.

43. Connect the fuel lines to fuel rail and bracket.

44. Install the upper intake manifold and torque the mounting bolts to 18 ft. lbs. (25 Nm).

45. Install the MAP sensor.

46. Connect the upper engine wiring harness connectors to the following components:
- Throttle Position Sensor (TPS)
- Idle Air Control (IAC)
- Fuel Injectors
- Coolant temperature sensor
- MAP sensor
- Camshaft Position (CMP) sensor

47. Connect the vacuum lines to the following components:
- Vacuum modulator
- Fuel pressure regulator
- PCV valve

48. Install the EVAP canister purge solenoid and ignition assembly.

49. Install the alternator assembly.

50. Connect the power steering line to the alternator bracket.

51. Install the serpentine belt.

52. Connect the spark plug wires to the spark plugs and intake plenum wire retainer.

53. Install the EGR pipe to the exhaust manifold.

54. Connect the control cables to the throttle body lever and upper intake plenum mounting bracket.

55. Install air intake assembly and top half of the air cleaner assembly.

56. Install the brake vacuum pipe.

57. Fill the cooling system.

58. Connect the negative battery cable.

59. Start the vehicle and verify no leaks.

3.1L (VIN M) Engine

1996 Vehicles

These vehicles use a two-piece intake manifold. These pieces are aluminum and care should be exercised when working with these components.

CAUTION
Fuel injection systems remain under pressure, even after the engine has been turned OFF. The fuel system pressure must be relieved before disconnecting any fuel lines. Failure to do so may result in fire and/or personal injury.

1. Disconnect the negative battery cable.
2. Relieve the fuel system pressure.
3. Remove the air cleaner assembly.
4. Remove the throttle cables from the throttle body and bracket.
5. Remove the fuel line retaining clamp from control cable bracket.
6. Remove the control cable bracket from the manifold.
7. Tag and disconnect the vacuum lines from the upper intake manifold.
8. Remove the EGR valve.
9. Position the heater inlet pipe hose clamps out of the way.
10. Remove the rear ignition coil nuts.

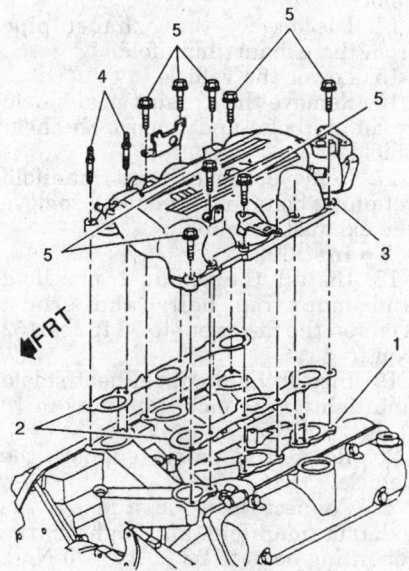

1. Manifold, lower intake
2. Gasket, upper intake manifold
3. Manifold, upper intake
4. Stud, upper intake manifold
5. Bolt, upper intake manifold

294885

Upper intake manifold view — 1996 3.1L (VIN M) engine

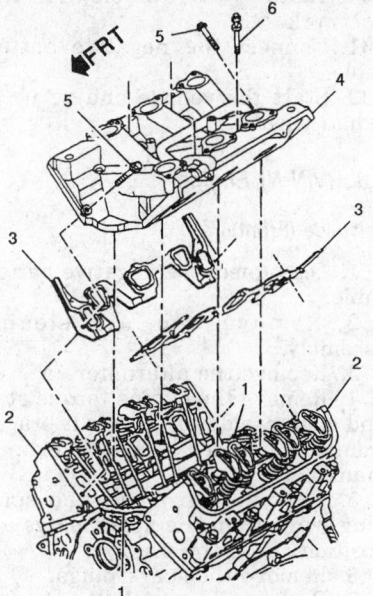

1. Apply sealant
2. Head, cylinder
3. Gasket, lower intake manifold
4. Manifold, lower intake
5. Bolt, lower intake manifold
6. Bolt, lower intake manifold

294886

Lower intake manifold view — 1996 3.1L (VIN M) engine

11. Remove the front ignition coil bolts.
12. Remove the power steering line clip from the alternator brace.
13. Remove the alternator and brackets.
14. Disconnect the electrical connections from the throttle body.
15. Remove the upper intake manifold bolts and remove upper intake manifold with throttle body.
16. Remove the heater hoses from the upper intake manifold manifold pipes.
17. Remove the fuel rail.
18. Disconnect the coolant hose.
19. Remove the power steering pump.
20. Remove the heater hose from the thermostat housing.
21. Remove the ignition coil.
22. Remove the engine mount strut.
23. Remove the valve covers.
24. Remove the lower intake manifold bolts.
25. Remove the lower intake manifold.

To install :
26. Clean all gasket material from mating surfaces. Then clean all sealing surfaces with a degreaser product and blow dry with compressed air.

27. Apply a small bead (0.08–0.11 inch) of RTV sealer on each ridge where the intake manifold contacts the engine block.
28. Install the gasket and position the lower intake manifold into place.
29. Apply sealant, GM P/N 12345382 or equivalent to threads of the manifold bolts.

NOTE: When installing the manifold bolts, you must tighten the vertical bolts before the diagonal bolts. Failure to follow this step may result in an oil leak.

30. Hand tighten the vertical bolts.
31. Hand tighten the diagonal bolts.
32. Tighten the vertical bolts to 115 inch lbs. (13 Nm).
33. Tighten the diagonal bolts to 115 inch lbs. (13 Nm).
34. Install the valve covers.
35. Install the engine mount strut.
36. Install the ignition coil.
37. Install the heater hose to the thermostat housing.
38. Install the upper radiator hose.
39. Install the power steering pump.
40. Install the alternator brackets.
41. Install the fuel rail and lines.
42. Install the gasket on top of the lower intake manifold and place upper intake manifold into place.
43. Connect the coolant hoses to the manifold pipes.
44. Install the upper intake manifold bolts and tighten them to 18 ft. lbs. (25 Nm).
45. Connect the electrical connections to the throttle body.
46. Install the upper alternator brace and alternator.
47. Install the power steering line clip to the alternator brace.
48. Install the mounting nuts for the coil.
49. Properly position the heater inlet pipe hose clamps.
50. Install the EGR valve.
51. Connect the vacuum lines to the upper intake manifold.
52. Install the control cable bracket and return spring.
53. Install the fuel line retaining clamp to the control cable bracket.
54. Install the throttle control cables to the throttle body and bracket.
55. Install the air cleaner assembly.
56. Refill the coolant system.
57. Connect the negative battery cable.
58. Start the vehicle and verify no leaks.

Exhaust Manifold

REMOVAL AND INSTALLATION

2.2L (VIN 4) Engine

1. Disconnect the negative battery cable.
2. Remove the air cleaner and air duct assembly.
3. Remove the lower air inlet duct.
4. Remove the engine drive belt.
5. Remove the alternator.
6. Remove the engine torque strut from the engine bracket and radiator support bracket.
7. Remove the torque strut bracket from the cylinder head.
8. Remove the alternator rear support bracket.
9. Raise and safely support the vehicle.
10. Remove the two bolts securing the exhaust pipe to the exhaust manifold.
11. Lower the vehicle.
12. Remove the dipstick tube mounting bolt and dipstick tube and dipstick.
13. Disconnect the electrical connector from the oxygen sensor (O_2).
14. Remove the exhaust manifold mounting bolts.
15. Remove the exhaust manifold.

To install:

16. Clean all gasket surfaces completely.
17. Install the exhaust manifold and tighten the mounting bolts to 116 inch lbs. (13 Nm).
18. Connect the O_2 sensor electrical connector.
19. Install the dipstick tube and mounting bolt.
20. Raise and safely support the vehicle.
21. Connect the front exhaust pipe to the exhaust manifold and tighten the mounting bolts to 18 ft. lbs. (25 Nm).
22. Lower the vehicle.
23. Install the alternator rear support bracket and tighten the three mounting bolts to 74 ft. lbs. (100 Nm).
24. Install the engine torque strut bracket to the engine and tighten the nuts to 41 ft. lbs. (56 Nm) and the bolt to 40 ft. lbs. (55 Nm).
25. Install the torque strut and tighten the through bolts to 40 ft. lbs. (55 Nm).
26. Install the alternator.
27. Install the drive belt.
28. Install the lower air inlet duct.
29. Install the air inlet resonator to the upper tie bar.
30. Install the air cleaner and ductwork.
31. Connect the negative battery cable.
32. Start the vehicle and verify no exhaust leaks.

3.3L (VIN N) Engine

Left Side (Front)

1. Disconnect the negative battery cable.
2. Remove the air cleaner assembly.
3. Remove the alternator.
4. Remove the engine torque strut and engine–side torque strut bracket from the cylinder head and exhaust manifold.
5. Tag and disconnect the spark plug wires from the spark plugs and position out of the way.
6. Remove the spark plugs.
7. Remove the two bolts connecting the exhaust crossover pipe to the manifold.
8. Remove the oil level indicator and tube.
9. Remove the cooling fan assembly.
10. Remove the exhaust manifold heat shield.
11. Remove the exhaust manifold bracket nut and bolt and remove the bracket.
12. Remove the exhaust manifold mounting bolts and studs and remove the exhaust manifold.

To install:

13. Install the exhaust manifold and mounting studs and bolts and tighten to 38 ft. lbs. (52 Nm).
14. Install the exhaust manifold bracket and mounting nut and bolt.
15. Install the exhaust manifold heat shield.
16. Install the cooling fan assembly.
17. Install the oil level indicator tube and tighten the mounting nut to 20 ft. lbs. (27 Nm).
18. Install the two bolts connecting the exhaust crossover pipe to the manifold and tighten to 19 ft. lbs. (26 Nm).
19. Install the spark plugs and tighten to 11 ft. lbs. (15 Nm).
20. Connect the spark plug wires.
21. Install the torque strut bracket on the cylinder head and tighten the fasteners to 18 ft. lbs. (24 Nm).
22. Install the torque strut and tighten the mounting nut to 42 ft. lbs. (57 Nm).
23. Install the alternator.
24. Install the air cleaner assembly.
25. Connect the negative battery cable.
26. Start the vehicle and verify no exhaust leaks.

Right Side (Rear)

1. Disconnect the negative battery cable.
2. Remove the air cleaner assembly.
3. Remove the serpentine belt.
4. Drain the cooling system into a suitable container.
5. Remove the power steering pump brace and remove the power steering pump mounting bolts and pull the pump forward.
6. Tag and disconnect the spark plug wires from the rear plugs and disconnect the wires from the wire loom on the rocker arm cover and position the wires out of the way.
7. Disconnect the oxygen sensor (O_2) connector.
8. Remove the engine lift bracket nuts from the exhaust manifold.
9. Disconnect the coolant tube. Remove the engine lift bracket and fuel line clamp screw from the coolant tube and position the tube out of the way.
10. Remove the rear plugs.
11. Remove the two bolts connecting the exhaust crossover pipe to the exhaust manifold.
12. Remove the transaxle fill tube bolt and reposition the tube for access.
13. Raise and safely support the vehicle.
14. Disconnect the exhaust pipe from the exhaust manifold.
15. Lower the vehicle.
16. Remove the exhaust heat shield retaining bolts and remove the heat shield.
17. Remove the exhaust manifold retaining bolts and studs and remove the exhaust manifold.

To install:

18. Install the exhaust manifold and mounting bolts and studs. Tighten the fasteners to 38 ft. lbs. (52 Nm).
19. Install the manifold heat shield and tighten the mounting nuts to 18 ft. lbs. (26 Nm).
20. Raise and safely support the vehicle.
21. Connect the exhaust pipe to the exhaust manifold and tighten the mounting bolts to 22 ft. lbs. (30 Nm).
22. Lower the vehicle.
23. Position the transaxle fill tube and tighten the mounting bolt to 18 ft. lbs. (25 Nm).
24. Connect the crossover pipe to the exhaust manifold and tighten the mounting bolts to 19 ft. lbs. (26 Nm).
25. Install the spark plugs and tighten to 11 ft. lbs. (15 Nm).

26. Install the coolant tube. Install the engine lift bracket and install the fuel line bracket screw.

27. Install the engine lift bracket nuts and tighten to 22 ft. lbs. (30 Nm).

28. Connect the O₂ sensor connector.

29. Route the spark plug wires through the loom on the rocker arm cover and connect the wires to the spark plugs.

30. Position the power steering pump and install the pump brace. Tighten the power steering pump mounting bolts to 21 ft. lbs. (29 Nm) and the brace bolt to 19 ft. lbs. (26 Nm).

31. Refill the cooling system.

32. Install the serpentine belt.

33. Install the air cleaner assembly.

34. Connect the negative battery cable.

35. Start the vehicle and verify no exhaust leaks.

3.1L (VIN M) Engine

----- **CAUTION** -----
Fuel injection systems remain under pressure, even after the engine has been turned OFF. The fuel system pressure must be relieved before disconnecting any fuel lines. Failure to do so may result in fire and/or personal injury.

Left Side (Front)

1. Relieve the fuel system pressure.

2. Disconnect the negative battery cable.

3. Drain the engine coolant.

4. Disconnect the electrical connectors and the heater hoses from the throttle body.

5. Remove the throttle body from the intake.

6. Disconnect the fuel lines and position aside.

7. Disconnect the coolant hose from the thermostat housing.

8. Remove the crossover pipe heat shield, then remove the crossover pipe.

9. Remove the engine torque strut.

10. If equipped with air conditioning, perform the following steps:

   a. Recover the air conditioning refrigerant, using an approved recovery system.

   b. Remove the serpentine belt cover and belt.

   c. Remove the compressor front mounting bolts.

d. Remove the two bolts at the top of the compressor mounting bracket.

   e. Raise and properly support the vehicle.

   f. Disconnect the A/C lines from the rear of the compressor.

   g. Disconnect the electrical connector from the compressor.

   h. Remove the compressor rear mounting bolts.

   i. Remove the compressor assembly from the vehicle.

   j. Remove the mounting bracket lower bolts.

   k. Remove the compressor mounting bracket.

   l. Lower the vehicle.

11. Remove the exhaust manifold retaining nuts and manifold.

To install:

12. Thoroughly clean all gasket material from the manifold and cylinder head mating surfaces.

13. Install the exhaust manifold and retaining nuts. Tighten the retaining nuts to 18 ft. lbs. (25 Nm).

14. If equipped with air conditioning, perform the following steps:

   a. Raise and properly support the vehicle.

   b. Install the mounting bracket lower bolts and tighten to 35 ft. lbs. (47 Nm).

   c. Position the compressor assembly in the mounting bracket.

   d. Install the rear compressor mounting bolts and tighten to 18 ft. lbs. (25 Nm).

   e. Connect the electrical connector to the compressor.

   f. Using new O-rings, connect the A/C lines to the rear of the compressor. Tighten the mounting bolt to 24 ft. lbs. (35 Nm).

   g. Lower the vehicle.

   h. Install the front compressor mounting bolts and tighten to 37 ft. lbs. (50 Nm).

   i. Install the two bolts at the top of the compressor mounting bracket and tighten to 35 ft. lbs. (47 Nm).

   j. Install the serpentine belt and belt cover.

15. Install the torque strut and tighten to 39 ft. lbs. (53 Nm).

16. Connect the exhaust crossover pipe to the manifold and tighten the retaining nuts to 18 ft. lbs. (25 Nm).

17. Install the heat shield and tighten the bolts to 89 inch lbs. (10 Nm).

18. Connect the coolant hose to the thermostat housing and secure with retaining clamp.

19. Connect the fuel lines.

20. Install the throttle body to the intake manifold.

21. Connect the coolant hoses and electrical wiring to the throttle body.

22. Refill the engine coolant and bleed the cooling system as required.

23. Following proper procedures, evacuate and recharge the A/C system.

24. Connect the negative battery cable.

25. Start the engine and check for coolant, fuel, refrigerant, and exhaust leaks.

Right Side (Rear)

1. Relieve the fuel system pressure.

2. Disconnect the negative battery cable.

3. Drain the engine coolant.

4. Disconnect the electrical connectors and the heater hoses from the throttle body.

5. Remove the throttle body from the intake.

6. Disconnect the fuel lines and position aside.

7. Disconnect the upper radiator hose from the thermostat housing.

8. Disconnect the EGR tube from the manifold.

9. Remove the crossover pipe heat shield, then remove the crossover pipe.

10. Disconnect the oxygen sensor electrical connector and remove the sensor.

11. Remove the exhaust manifold upper heat shield.

12. Raise and properly support the vehicle.

13. Disconnect the front (converter) exhaust pipe at the manifold.

14. Remove the automatic transaxle dipstick and fill tube from the transaxle.

15. Remove the exhaust manifold lower heat shield.

16. Remove the exhaust manifold retaining nuts and manifold.

To install:

17. Thoroughly clean all gasket material from the manifold and cylinder head mating surfaces.

18. Install the exhaust manifold and retaining nuts. Tighten the retaining nuts to 18 ft. lbs. (25 Nm).

19. Install the manifold lower heat shield and tighten the bolts to 89 inch lbs. (10 Nm).

20. Install the transaxle fill tube and dipstick.

21. Connect the front exhaust pipe to the manifold assembly and tighten the bolts to 22 ft. lbs (30 Nm).

22. Lower the vehicle.

23. Install the upper manifold heat shield and tighten the bolts to 89 inch lbs. (10 Nm).

24. Install the oxygen sensor and connect the wiring.

25. Connect the exhaust crossover pipe to the manifold and tighten the retaining nuts to 18 ft. lbs. (25 Nm).

26. Install the crossover pipe heat shield and tighten the bolts to 89 inch lbs. (10 Nm).

27. Connect the upper radiator hose to the thermostat housing and secure with retaining clamp.

28. Connect the fuel lines.

29. Connect the EGR tube to the manifold.

30. Install the throttle body to the intake manifold.

31. Connect the coolant hoses and electrical wiring to the throttle body.

32. Refill the engine coolant and bleed the cooling system as required.

33. Connect the negative battery cable.

34. Start the engine and check for coolant, fuel and exhaust leaks.

3.3L (VIN N) Engine

Front

1. Disconnect the negative battery cable.

2. Disconnect the air cleaner inlet duct and disconnect the spark plug wires.

3. Remove the exhaust crossover pipe to manifold bolts.

4. Remove the engine lift hook and manifold heat shield.

5. Remove the oil level indicator tube and the indicator.

6. Remove the manifold studs and remove the manifold.

To Install:

7. Clean all manifold the cylinder head mating surfaces.

8. Install the exhaust manifold and manifold studs.

9. Install the oil indicator tube and the indicator.

10. Install the manifold heat shield and engine lift hook.

11. Install exhaust crossover pipe to manifold bolts.

12. Connect the air cleaner inlet duct and spark plug wires.

13. Connect the negative battery cable.

Rear

1. Disconnect the negative battery cable.

2. Disconnect the spark plug wires.

3. Disconnect the wire from the oxygen sensor.

4. Remove the throttle cable bracket and remove the cables from the throttle body.

5. Remove the brake booster hose from the manifold.

6. Remove the 2 exhaust crossover pipe to manifold bolts.

7. Remove the exhaust pipe to manifold bolts.

8. Remove the engine lift hook.

9. Remove the transaxle oil level indicator tube.

10. Remove the manifold heat shield.

11. Remove the manifold studs and bolt and remove the manifold.

To Install:

12. Install the manifold and manifold studs.

13. Install manifold heat shield.

14. Install transaxle oil level indicator tube.

15. Install engine lift hook.

16. Install the exhaust pipe to manifold bolts.

17. Install the 2 exhaust crossover pipe to manifold bolts.

18. Install the brake booster hose from the manifold.

19. Install the throttle cable bracket and install cables to the throttle body.

20. Connect wire to the oxygen sensor.

21. Connect the spark plug wires.

22. Connect the negative battery cable.

Front Cover Seal

REMOVAL AND INSTALLATION

2.2L (VIN 4) Engine

1. Disconnect the negative battery cable.

2. Remove the serpentine belt.

3. Raise and properly support the vehicle.

4. Remove the right front tire.

5. Remove the inner fender splash shield.

6. Separate the crankshaft pulley, from the pulley hub, by removing the three mounting bolts.

7. Install special tool J-24420-B or equivalent damper puller on crankshaft hub, turn puller screw and remove hub. Use care not to loose the small key on the crankshaft nose.

8. Pry out oil seal with a large screwdriver.

NOTE: Take care not to damage the seal seat in the front cover or the crankshaft surface.

To install:

9. Clean out the seal recess in the front cover.

10. Press the oil seal into the front cover, using special tool J-35468 or equivalent driver tool.

11. If removed, install the key in the slot in the crankshaft nose. A small amount of silicone sealer can be used to hold it in place. Align the pulley hub keyway (slot) with the key, and install the hub to the crankshaft. Make sure the key was not displaced from its groove.

12. Install the pulley to the hub, and tighten the mounting bolts to 37 ft. lbs. (50 Nm).

13. Install the pulley/hub center bolt and tighten to 77 ft. lbs. (105 Nm).

14. Install the inner fender splash shield.

15. Install the front tire and wheel assembly and tighten to specification.

16. Lower the vehicle.

17. Install the serpentine belt.

18. Connect negative battery cable.

19. Start the engine and check for leaks.

3.3L (VIN N) Engine

1. Disconnect the negative battery cable.

2. Remove the engine drive belt.

3. Raise and safely support the vehicle.

4. Remove the tire and wheel assembly.

5. Remove the right inner fender well splash shield.

6. Remove the crankshaft balancer as follows:

   a. Remove the crankshaft balancer bolt.

   b. Using tool J-38197 or an equivalent balancer puller, remove the balancer from the crankshaft.

7. Remove the four crankshaft sensor shield mounting bolts and remove the shield.

8. Pry the front cover seal out of the cover using a suitable prying tool. Use care not to nick the crankshaft or damage the front cover.

To Install:

9. Clean out the recess in the front cover where the front cover oil seal sits.

10. Install a new front cover oil seal in the front cover and seat in place using J-35354, or an equivalent seal installer.

11. Install the crankshaft sensor shield and four mounting bolts.

12. Install the crankshaft balancer as follows:

   a. Coat the seal contact area of the balancer with clean engine oil.

b. Line up the notch in the balancer with the crankshaft key. Slide the balancer on until the key is inside the notch.

c. Install the balancer bolt and tighten to 110 ft. lbs. (150 Nm) plus 76° additional rotation. The balancer bolt being installed will fully seat the balancer on the crankshaft.

13. Install the right inner fender well splash shield.

14. Install the tire and wheel assembly and tighten to specification.

15. Lower the vehicle.

16. Install the engine drive belt.

17. Connect the negative battery cable.

18. Warm the engine to operating temperature and verify no leaks.

3.1L (VIN M) Engine

1. Disconnect the negative battery cable.

2. Remove the serpentine belt.

3. Raise and safely support the vehicle.

4. Remove the right front tire and wheel assembly.

5. Remove the right inner fender well splash shield.

6. Remove the flywheel cover at the transaxle.

7. Remove the torsional dampener as follows:

a. Remove the torsional damper mounting bolt while holding the crankshaft from turning.

b. Install tool J 24420-B or an equivalent puller and remove the damper from the crankshaft. Use care not to lose the key from the crankshaft nose.

8. Pry the seal out of the front cover using a suitable prying tool. Use care not to damage the front cover or the crankshaft surface.

To install:

9. Clean out the oil seal recess in the front cover.

10. Install the new seal in the front cover using J-34995, or an equivalent installation tool, to fully seat the seal in the cover.

11. Install the torsional damper as follows:

a. Coat the seal contact area on the damper with clean engine oil.

b. A small amount of silicone sealer may help hold the key in position in the crankshaft nose.

c. Line up the notch in the damper with the crankshaft key and slide the damper onto the crankshaft until the key is in the notch.

d. Using tool J-29113, pull the damper onto the crankshaft.

e. Verify that the key has not been displaced from its slot in the crankshaft nose.

f. Install the damper mounting bolt and tighten to 76 ft. lbs. (102 Nm).

12. Install the flywheel cover at transaxle.

13. Install the right inner fender splash shield.

14. Install the tire and wheel assembly and tighten to specification.

15. Lower the vehicle.

16. Install the serpentine belt.

17. Refill the cooling system.

18. Connect the negative battery cable.

19. Start the vehicle and verify no coolant leaks or oil leaks.

Timing Chain, Sprockets and Front Cover

REMOVAL AND INSTALLATION

2.2L (VIN 4) Engine

1. Disconnect the negative battery cable.

2. Drain the engine coolant into a suitable container.

3. Remove the serpentine belt.

4. Remove the coolant reservoir.

5. Remove the front alternator mounting bolts.

6. Remove the three mounting bolts from the power steering pump. These bolts can be reached by going through the holes in the drive pulley on the pump. Lay the pump aside without disconnecting the hoses.

7. Remove the four tensioner mounting bolts and remove the tensioner.

8. Raise and safely support the vehicle.

9. Drain the engine oil into a suitable container.

10. Remove the engine oil pan.

11. Remove the right front tire and wheel assembly.

12. Remove the right inner fender well splash shield.

13. Remove the crankshaft pulley.

14. Remove the crankshaft balancer.

15. Remove the front cover bolts and remove the front cover. If the cover is difficult to remove, use a soft faced mallet to lightly tap the cover to loosen it.

16. Rotate the crankshaft until the piston in No. 1 cylinder is at TDC on the compression stroke (firing position). The marks on the camshaft and crankshaft sprockets should be in alignment.

17. Loosen, but do not remove, the timing chain tensioner nut.

18. Remove the camshaft sprocket bolt and remove the sprocket and chain together. If the sprocket does not slide from the camshaft easily, a light blow with a soft mallet at the lower edge of the sprocket will dislodge it.

19. Use puller tool J-22888 or equivalent, and remove the crankshaft sprocket.

To install:

20. Install the crankshaft sprocket, using installation tool J 5590 or equivalent.

21. Install the timing chain over the camshaft sprocket and then around the crankshaft sprocket. Make sure that the marks on the two sprockets are in alignment. Lubricate the thrust surface with Molykote® or its equivalent.

22. Align the dowel in the camshaft with the dowel hole in the sprocket and then install the sprocket onto the camshaft. Use the mounting bolt to draw the sprocket onto the camshaft and then tighten to 77 ft. lbs. (105 Nm).

23. Lubricate the timing chain with clean engine oil. Torque the bolts on the chain tensioner to 18 ft. lbs. (24 Nm).

24. Clean all gasket surfaces completely.

25. Apply a thin layer of sealer to the front cover and install a new gasket onto the cover.

26. Install the cover to the engine making sure the dowel pins line up with the holes in the front cover.

27. Install the cover mounting bolts and tighten them to 97 inch lbs. (11 Nm).

28. Install the crankshaft and pulley and tighten the pulley bolts to 37 ft. lbs. (50 Nm) and the center balancer bolt to 77 ft. lbs. (105 Nm).

29. Install the right side splash shield.

30. Install the tire and wheel assembly.

31. Install the oil pan.

32. Lower the vehicle.

33. Install the belt adjuster and tighten the bolts to 37 ft. lbs. (50 Nm).

34. Install the power steering pump and tighten the mounting bolts to 25 ft. lbs. (34 Nm).

35. Install the alternator mounting bolts and tighten the upper bolt to 22 ft. lbs. (30 Nm) and tighten the lower to 37 ft. lbs. (50 Nm).

36. Install the coolant reservoir.

37. Install the serpentine belt.

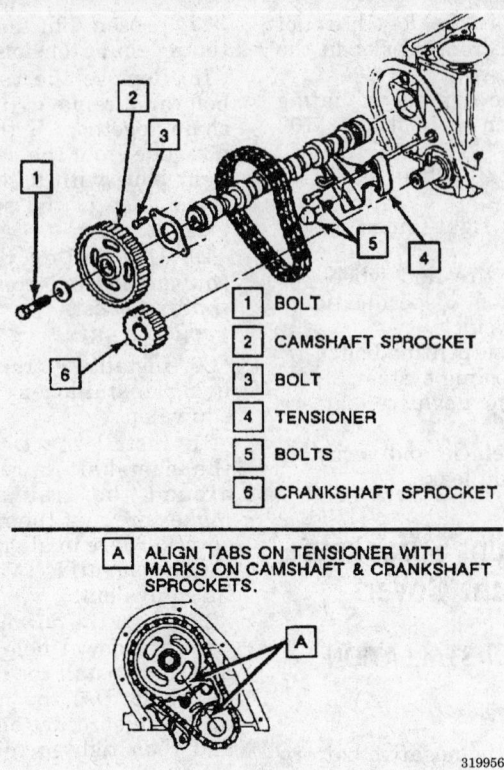

1	BOLT
2	CAMSHAFT SPROCKET
3	BOLT
4	TENSIONER
5	BOLTS
6	CRANKSHAFT SPROCKET

A ALIGN TABS ON TENSIONER WITH MARKS ON CAMSHAFT & CRANKSHAFT SPROCKETS.

319956

Timing cover, chain and sprocket exploded view and timing mark alignment — 2.2L (VIN 4) engine

38. Fill the engine crankcase with clean oil.

39. Refill the coolant system.

40. Connect the negative battery cable.

41. Start the vehicle and verify no leaks.

42. Road test the vehicle and ensure proper operation.

3.3L (VIN N) Engine

1. Disconnect the negative battery cable.

2. Remove the timing chain front cover.

3. Rotate the crankshaft until the timing marks are in alignment. The crankshaft sprocket mark should point straight up at the camshaft mark which is pointing straight down. The marks should be on the closest approach.

4. Remove the timing chain damper assembly.

5. Remove the camshaft sprocket mounting bolt.

6. Remove the camshaft sprocket from the camshaft and remove the sprocket and chain assembly from the vehicle.

7. Remove the crankshaft sprocket from the crankshaft using a suitable puller.

To Install:

8. Install the crankshaft sprocket onto the crankshaft. It may be necessary to use a sprocket installer to fully seat the sprocket on the crankshaft. Make sure the crankshaft timing mark is pointing straight up.

9. Temporarily install the camshaft sprocket and rotate the camshaft as necessary to align the timing mark with the crankshaft timing mark.

10. Assemble timing chain on the camshaft sprocket.

11. Hold the sprocket so the mark is pointing downward with the chain hanging off the sprocket.

12. Loop the timing chain under the crankshaft sprocket and with the marks in alignment install the camshaft sprocket on the camshaft.

13. Verify all timing marks are in alignment. If the marks are out of alignment repeat steps 9 to 12.

14. With the marks aligned, tighten the camshaft sprocket mounting bolts to 27 ft. lbs. (37 Nm).

15. Install the timing chain dampener assembly and tighten the mounting bolt to 14 ft. lbs. (19 Nm).

16. Install the front timing chain cover.

17. Connect the negative battery cable.

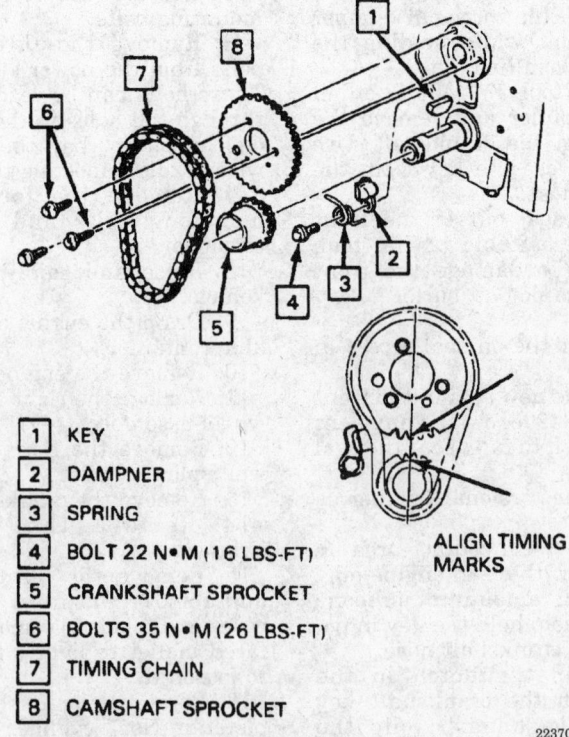

1	KEY
2	DAMPNER
3	SPRING
4	BOLT 22 N•M (16 LBS-FT)
5	CRANKSHAFT SPROCKET
6	BOLTS 35 N•M (26 LBS-FT)
7	TIMING CHAIN
8	CAMSHAFT SPROCKET

ALIGN TIMING MARKS

223703

Timing chain and sprocket components and timing mark alignment — 3.3L (VIN N) engine

3.1L (VIN M) Engine

1. Disconnect the negative battery cable.

2. Drain the cooling system into a suitable container.

3. Remove the right engine mount bracket.

4. Remove the serpentine belt.

5. Remove the crankshaft balancer as follows:

   a. Raise and safely support the vehicle.

   b. Remove the right front tire and wheel assembly.

   c. Remove the right inner fender well splash shield.

   d. Remove the flywheel cover and install a flywheel holding tool.

   e. Remove the balancer mounting bolt and washer.

   f. Using a suitable puller, J-24420-B or the equivalent remove the balancer from the crankshaft.

6. Remove the serpentine belt tensioner mounting bolt and tensioner.

7. Remove the oil pan following the recommended procedure.

8. Remove coolant bypass pipe from the water pump and the intake manifold.

9. Disconnect the lower radiator hose to from the front cover outlet.

10. Remove the front cover mounting bolts and remove the front cover.

11. Rotate the crankshaft until the timing marks on the camshaft and crankshaft sprockets are in alignment at their closest approach.

12. Remove the camshaft sprocket mounting bolt and remove the camshaft sprocket and timing chain.

13. Remove the crankshaft sprocket, using a gear puller J-5825-A or equivalent.

14. Remove the two mounting bolts from the timing chain damper and remove the damper.

To install:

15. Install the timing chain damper and tighten the mounting bolts to 15 ft. lbs. (21 Nm).

16. Install the crankshaft sprocket onto the crankshaft making sure the notch in the sprocket fits over the crankshaft key. Fully seat the sprocket on the crankshaft using J-38612, or an equivalent gear installer.

17. Make sure the timing mark on the crankshaft sprocket is pointing straight up.

18. Install the timing chain over the camshaft sprocket and hold the sprocket in such a way, that the timing mark is pointing down, and the timing chain is hanging down off the sprocket.

19. Loop the timing chain under the crankshaft sprocket and install the camshaft sprocket on the camshaft. The sprocket will only fit on the camshaft if, the dowel on the camshaft lines up with the hole in the sprocket.

20. Verify that the marks are aligned (the camshaft sprocket will be at the 6 o'clock position and the crankshaft sprocket will be in the 12 o'clock position).

21. On 1994–1995 vehicles, tighten the camshaft sprocket mounting bolt to 74 ft. lbs. (100 Nm). On 1996 vehicles tighten the bolt to 81 ft. lbs. (110 Nm).

22. Lubricate the timing chain components with engine oil.

23. Clean all gasket surfaces completely.

24. Apply a thin bead of sealer around the gasket sealing area of the front cover. Install a new front cover seal on the front cover.

25. Install the front cover on the engine and install the mounting bolts. Tighten the small bolts to 18 ft. lbs. (24 Nm). Tighten the large bolts to 41 ft. lbs. (55 Nm). On 1996 vehicles, tighten the small bolts to 15 ft. lbs. (21 Nm) and tighten the large bolts to 35 ft. lbs. (47 Nm).

26. Connect the radiator hose to the coolant outlet.

27. Install coolant by pass pipe to the water pump and the intake manifold.

28. Install the oil pan following the recommended procedure.

29. Install crankshaft balancer as follows:

   a. Coat the seal contact surface of the crankshaft balancer with clean engine oil.

   b. Line up the notch in the balancer with the crankshaft key and slide the balancer on until the key is in the balancer.

   c. Using J-29113 or an equivalent puller, seat the balancer on the crankshaft.

   d. Install the balancer mounting bolt and washer and tighten to 76 ft. lbs. (103 Nm).

   e. Install the flywheel cover.

   f. Install the right inner fender well splash shield.

   g. Install the tire and wheel assembly and tighten to specification.

30. Install the serpentine belt tensioner and tighten the mounting bolt to 40 ft. lbs. (54 Nm).

31. Install the serpentine belt.

32. Install the right engine mount bracket and tighten the bracket-to-mount bolts to 96 ft. lbs. (130 Nm).

33. Refill the cooling system.

34. Check the engine oil level and top off as necessary.

35. Connect the negative battery cable.

36. Start the vehicle and verify no oil leaks.

Camshaft

REMOVAL AND INSTALLATION

2.2L (VIN 4) Engine

Please note that the engine must be removed from the vehicle for this procedure. Use care when disassembling valvetrain components. Any part that is to be reused must be returned to its original location. In addition, if the camshaft is being replaced, all new lifters must also be installed. Installing used lifters on a new camshaft will fail the new camshaft.

— **CAUTION** —

Fuel injection systems remain under pressure, even after the engine has been turned OFF. The fuel system pressure must be relieved before disconnecting any fuel lines. Failure to do so may result in fire and/or personal injury.

1. Disconnect the negative battery cable.

2. Relieve the fuel system pressure.

3. Remove the engine assembly from the vehicle using the recommended procedure. Mount engine assembly on a suitable engine stand.

4. Remove the serpentine belt.

5. Remove the serpentine belt tensioner assembly with the alternator attached.

6. Remove the strut bracket and the rear alternator bracket.

7. Remove the front engine mount bracket.

8. Remove the oil level indicator tube.

9. Drain the engine oil into a suitable container.

10. Remove the oil pan.

11. Remove the crankshaft balancer and front cover.

12. Remove the timing chain and camshaft sprocket.

13. Disconnect the spark plug wires.

14. Remove the rocker arm cover.

NOTE: When removing the valve train components they must be kept in order for installation in the same locations from which they were removed.

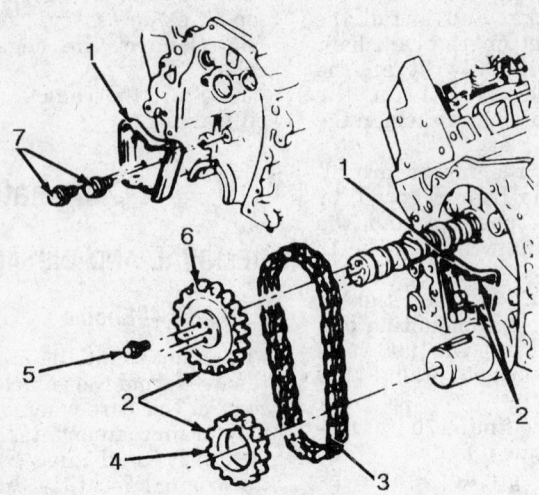

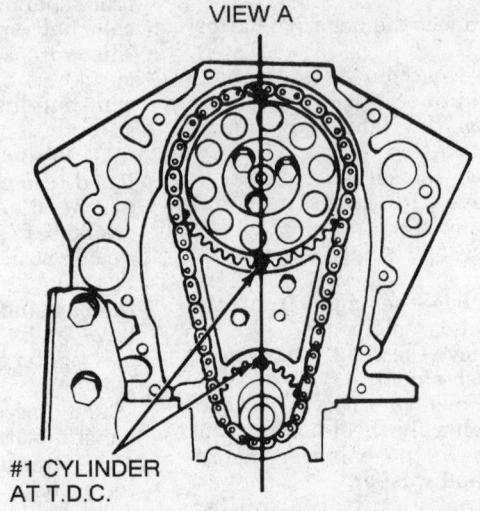

VIEW A

#1 CYLINDER
AT T.D.C.

NOTE - ALIGN TIMING MARKS ON CAM
& CRANK SPROCKETS USING ALIGNMENT
MARKS ON DAMPER STAMPING OR CAST
ALIGNMENT MARKS ON CYL & CASE

NOTE - CAMSHAFT SPROCKET
MARK AT 6 O'CLOCK
CRANKSHAFT SPROCKET
MARK AT 12 O'CLOCK

1. Damper
2. Alignment marks
3. Timing chain
4. Crank sprocket
5. 28 Nm (21 lb. ft.)
6. Camshaft sprocket
7. 21 Nm (15 lb. ft.)

297628

Timing chain and sprocket timing mark alignment — 3.1L (VIN M) engine

15. Remove the rocker arm nuts, rocker arms, balls and pushrods.

16. Remove the power steering pump pencil brace.

17. Remove the cylinder head with the intake and exhaust manifolds attached.

18. Remove the valve lifters.

19. Remove the camshaft thrust plate mounting bolts and remove the thrust plate.

20. Remove the oil pump drive assembly.

21. Remove the camshaft carefully from the engine.

To install:

22. Coat the camshaft lobes with and bearings with GM Engine Oil Supplement (E.O.S.) part number 1051396 or equivalent camshaft break–in lube and insert the camshaft carefully into the engine.

23. Install the oil pump drive assembly.

24. Install the thrust plate and tighten the mounting bolts to 106 inch lbs. (12 Nm).

25. Install the valve lifters.

26. Install the cylinder head and manifold assemblies.

27. Install the power steering pump pencil brace.

28. Install the pushrods, rocker arms, balls and rocker arm nuts. Tighten the nuts to 22 ft. lbs. (30 Nm).

29. Install the rocker arm cover.

30. Connect the spark plug wires.

31. Install the timing chain and camshaft sprocket. Verify that the camshaft and crankshaft sprocket timing marks are correctly aligned.

32. Install the timing chain front cover and crankshaft balancer.

33. Install the oil pan.

34. Install the oil level indicator tube.

35. Install the front engine bracket, rear alternator bracket and strut bracket.

36. Install the drive belt tensioner and alternator assembly and install the serpentine belt.

37. Install the engine assembly in the vehicle.

38. Install the oil filter and refill with oil.

39. Connect the negative battery cable.

40. Refill the crankcase with the recommended engine oil and the cooling system with 50–50 water anti-freeze mix.

41. Start the vehicle and verify no leaks.

3.3L (VIN N) Engine

— **CAUTION** —

Care should be used when working around the fuel system. DO NOT smoke or expose the fuel system to any open flames or sparks. The fuel system is under pressure and must be bled prior to performing any service on fuel system components. Failure to properly relieve the fuel system pressure can lead to personal injury and component damage. Always keep a suitable fire extinguisher handy when servicing the fuel system.

1. Relieve the fuel system pressure.

2. Disconnect the negative battery cable.

3. Remove the intake manifold.

4. Remove the rocker arm covers.

NOTE: When removing the valve train components they must be kept in order for installation in the same locations they were removed from.

5. Remove the rocker arm bolts, pedestals, rocker arms, pushrods and pedestal guides.

6. Remove the lifter guide retainer bolts, lifter guide retainer, lifter guides and lifters.

7. Remove the water pump pulley.

8. Disconnect the lower radiator hose from the water pump.

9. Raise and safely support the vehicle.

10. Remove the tire and wheel assemblies.

11. Remove the right inner fender well splash shield.

12. Remove the flywheel covers.

13. Remove the balancer bolt and using tool J-37096, or an equivalent, to hold the crankshaft from turning.

14. Remove the balancer with a puller, J-38197, or the equivalent.

15. Remove the crankshaft sensor shield.

16. Disconnect the crankshaft sensor and oil pressure sender connections.

17. Remove the front cover mounting bolts.

18. Lower the vehicle.

19. Remove the heater pipe mounting nut and disconnect the heater pipe.

20. Remove the power steering pump pencil brace and remove the power steering pump and position aside. It is not necessary to disconnect the power steering hoses from the pump.

21. Disconnect the ground wires from the tensioner bracket.

22. Remove the upper front cover bolts and remove the front cover.

23. Remove the timing chain and sprocket assembly.

24. Install an engine support fixture, J-28467-A and J-36462 or their equivalents.

25. Raise and safely support the vehicle.

26. Remove the steering gear pinch bolt.

27. Remove the right ball joint pinch bolt and separate the ball joint from the steering knuckle.

28. Disconnect the wiring harness at the right front frame rail.

29. Disconnect the front exhaust pipe from the exhaust manifold.

30. Remove the cotter pin and castle nut from the tie rod end and separate the tie rod end from the right steering knuckle.

31. Install the jackstands to support the right side frame.

32. Disconnect the knock and VSS electrical connectors.

33. Remove the right side subframe mounting bolts.

34. Loosen the left side frame bolts.

35. Remove the right side air deflector screws and remove the deflector.

36. Lower the right side of the frame 6 inches.

37. Remove the camshaft thrust plate.

38. Lower the right side of the engine and remove the camshaft from the engine carefully.

To Install:

39. Coat the camshaft with pre-lube, 1234551, or the equivalent.

40. Carefully install the camshaft into the engine.

41. Install the thrust plate and tighten the mounting bolts to 11 ft. lbs. (15 Nm).

42. Raise the right side of the engine and frame into position.

43. Install the right side frame bolt and tighten all four frame bolts to 103 ft. lbs. (150 Nm).

44. Connect the VSS and knock sensor electrical connectors.

45. Remove the jackstands.

46. Connect the right tie rod end to the steering knuckle and tighten the castle nut to 52 ft. lbs. (70 Nm) and install a new cotter pin.

47. Connect the exhaust pipe to the exhaust manifold and tighten the mounting bolts to 22 ft. lbs. (30 Nm).

48. Connect the wiring harness at the right front frame rail.

49. Connect the ball joint to the steering knuckle and tighten the pinch bolt to 32 ft. lbs. (43 Nm).

50. Connect the steering shaft to the rack and pinion and tighten the pinch bolt to 35 ft. lbs. (48 Nm).

51. Lower the vehicle.

52. Remove the engine support fixture.

53. Install the timing chain and sprockets.

54. Install the front timing cover and tighten the front cover bolts to 22 ft. lbs. (30 Nm).

55. Connect the ground wire to the tensioner bracket.

56. Install the power steering pump and tighten the mounting bolts to 21 ft. lbs. (29 Nm).

57. Install the power steering pump pencil brace.

58. Connect the heater pipe and install the mounting nut.

59. Raise and safely support the vehicle.

60. Install the oil pan-to-front cover bolts and tighten to 124 inch lbs. (14 Nm).

61. Connect the crankshaft sensor and oil sender electrical connectors.

62. Install the crankshaft sensor shield.

63. Install the crankshaft balancer and tighten the mounting bolt to 111 ft. lbs. (150 Nm) plus 76° addition rotation, while holding the engine from turning with J-37096.

64. Install the flywheel covers.

65. Install the right inner fender well splash shield.

66. Install the front tire and wheel assemblies and tighten to specification.

67. Lower the vehicle.

68. Connect the lower radiator hose to the water pump.

69. Install the water pump pulley.

NOTE: If the camshaft was replaced the valve lifters should be replaced also. The lifters have already worn to the old camshaft and if re-used can damage the new camshaft.

70. Install the valve lifters.

71. Install the lifter guides and lifter guide retainer. Tighten the retainer mounting bolts to 22 ft. lbs. (30 Nm).

72. Install the pushrods, pedestal guide, rocker arms, pedestals and mounting bolts. Tighten the mounting bolts to 11 ft. lbs. (15 Nm) plus 115° additional rotation.

73. Install the intake manifold.

74. Install the rocker arm covers.

75. Connect the negative battery cable.

76. Refill the engine with coolant and crankcase with new oil.

77. Start the vehicle and verify no leaks.

3.1L (VIN M) Engine

Please note that the engine must be removed from the vehicle to perform this procedure. When removing valve train components, any parts that are to be reused must be returned to their original locations. Lay all parts out in an orderly fashion and mark them for identification. In addition, if the camshaft is being replaced, all of the valve lifters must also be replaced with new parts. Installing used lifters on a new camshaft will quickly wear the camshaft.

CAUTION

Fuel injection systems remain under pressure, even after the engine has been turned OFF. The fuel system pressure must be relieved before disconnecting any fuel lines. Failure to do so may result in fire and/or personal injury.

1. Relieve the fuel system pressure following the recommended procedure.

2. Disconnect the negative battery cable.

3. Remove the engine assembly.

4. Remove the intake manifold, valve cover, rocker arms, pushrods and valve lifters.

5. Remove the crankshaft balancer and front cover.

6. Remove the timing chain and sprockets.

7. Remove the oil pump driven gear mounting bolt and remove the oil pump driven gear.

8. Remove the two bolts and remove the camshaft thrust plate.

9. Carefully remove the camshaft. Avoid marring the camshaft bearing surfaces.

To install:

10. Coat the camshaft with lubricant GM part number 1052365 or equivalent camshaft break–in lubricant or quality engine oil supplement, and install the camshaft.

11. Install the camshaft thrust plate and tighten the mounting bolts to 89 inch lbs. (10 Nm).

12. Install the oil pump driven gear and tighten the mounting bolt to 27 ft. lbs. (36 Nm).

13. Install the timing chain and sprocket.

14. Install the camshaft thrust button and front cover.

15. Install the crankshaft balancer and tighten the bolt to 76 ft. lbs. (103 Nm).

16. Install the intake manifold, valve cover, rocker arms, pushrods and valve lifters.

17. Install the engine assembly.

18. Connect the negative battery cable.

19. Fill the crankcase with fresh oil.

20. Adjust the valves, as required.

21. Start the engine and verify no oil leaks.

Piston and Connecting Rod

POSITIONING

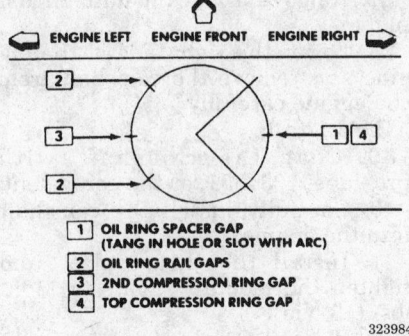

ENGINE LEFT   ENGINE FRONT   ENGINE RIGHT

1 OIL RING SPACER GAP
  (TANG IN HOLE OR SLOT WITH ARC)
2 OIL RING RAIL GAPS
3 2ND COMPRESSION RING GAP
4 TOP COMPRESSION RING GAP

323984

Piston ring gap location — all engines

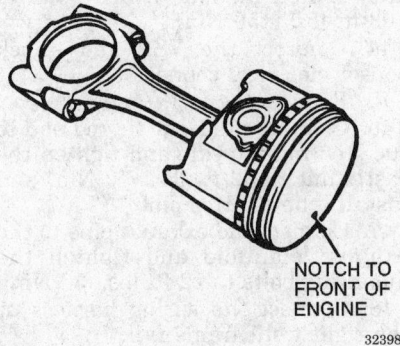

NOTCH TO FRONT OF ENGINE

323986

Piston notch — all engines

ENGINE LUBRICATION

Oil Pan

REMOVAL AND INSTALLATION

2.2L (VIN 4) Engine

1. Disconnect the negative terminal from the battery.

2. Remove the air cleaner and air duct assembly.

3. Remove the serpentine belt.

4. Remove the engine torque strut.

5. Install engine support fixture J-28467-A.

6. Raise and safely support the vehicle.

7. Remove the right front tire and wheel assembly.

8. Remove the right inner fender well splash shield.

9. Remove the flywheel cover bolts.

10. Disconnect and remove the starter and starter bracket.

11. Remove the flywheel cover.

12. Remove the exhaust pipe and converter.

13. Remove the A/C compressor mounting bolts and set the compressor aside without disconnecting the refrigerant lines.

14. Remove the front engine mount nuts from the frame.

15. Remove the front engine mount bolts from the engine.

16. Lower the vehicle.

17. Raise the engine about three inches using the support fixture.

18. Raise and safely support the vehicle.

19. Drain the engine oil.

20. Remove the front engine mount bracket.

21. Remove the oil pan mounting nuts and bolts.

22. Remove the oil pan.

To install:

23. Clean all the gasket surfaces completely.

24. Apply a thin bead of sealer around the outside edge of the oil pan and install the oil pan gasket onto the sealer.

25. Install the oil pan onto the engine and loosely install all the fasteners.

26. Tighten the nuts and bolts to 89 inch lbs. (10 Nm).

27. Install the front engine mount bracket and loosely install the mount-to-engine bolts.

28. Lower the vehicle.

29. Lower the engine into place.

30. Raise and safely support the vehicle.

31. Tighten the front engine mount bolts.

32. Install the engine mount nuts to 33 ft. lbs. (45 Nm).

33. Install the A/C compressor in the mounting bracket and tighten to 37 ft. lbs. (50 Nm).

34. Install the exhaust pipe and converter.

35. Connect the starter and install the starter and support bracket.

36. Install the flywheel cover and cover mounting bolts.

37. Install the right fender well splash shield.

38. Install the right front tire and wheel assembly and tighten to specification.

39. Lower the vehicle.

40. Remove the engine support fixture.

41. Install the engine torque strut.

42. Refill the crankcase with oil.

43. Install the serpentine belt.
44. Install the air cleaner and air duct assembly.
45. Connect the negative battery cable.
46. Start the vehicle and verify no leaks.

3.3L (VIN N) Engine

1. Disconnect the negative battery cable.
2. Raise and support the vehicle safely.
3. Drain the engine oil into a suitable container.
4. Remove the flywheel cover.
5. Remove the right front tire and wheel assembly.
6. Remove the right fender well splash shield.
7. Place a drain pan under the oil filter and remove the oil filter.
8. Remove the oil pan mounting bolts.
9. Remove the oil pan.

To install:
10. Clean all the gasket surfaces completely.
11. Apply a thin bead of sealer around the edge of the oil pan and install the gasket onto the oil pan.
12. Install the oil pan and loosely install the oil pan fasteners.
13. Tighten the oil pan-to-engine bolts to 12 ft. lbs. (16 Nm) and the oil pan to front cover bolts to 124 inch lbs. (14 Nm).
14. Coat the seal on the oil filter with clean engine oil and install.
15. Install the right side splash shield.
16. Install the right front tire and wheel assembly and tighten to specification.
17. Install the flywheel cover.
18. Lower the vehicle.
19. Refill the crankcase with engine oil.
20. Connect the negative battery cable.
21. Start the vehicle and verify no leaks.

3.1L (VIN M) Engine

1994–95 Vehicles

1. Disconnect the negative battery cable.
2. Remove the serpentine belt and the tensioner.
3. Support the engine with tool J-28467 or equivalent.
4. Raise and safely support the vehicle. Drain the engine oil.
5. Remove the right tire and wheel assembly. Remove the right inner fender splash shield.

6. Remove the steering gear pinch bolt. Remove the transaxle mount retaining bolts. Failure to disconnect intermediate shaft from rack and pinion stub shaft can result in damage to the steering gear and/or intermediate shaft. This could cause a loss of steering control which could result in personal injury.
7. Remove the engine-to-cradle mounting nuts. Remove the front engine collar bracket from the block.
8. Remove the starter shield and the flywheel cover. Remove the starter.
9. Loosen, but do not remove the rear engine cradle bolts. Remove electrical connector at DIS sensor, if equipped.
10. Remove the front cradle bolts and lower front of frame. Remove the oil pan retaining bolts and nuts. Remove the oil pan.

To install:
11. Clean the gasket mating surfaces.
12. Install a new gasket on the oil pan. Apply silicon sealer to the portion of the pan that contacts the rear of the block.
13. Install the oil pan, nuts and retaining bolts. Tighten rear bolts to 18 ft. lbs. (18–25 Nm), and remaining nuts and bolts to 89 inch lbs. (10 Nm).
14. Install the front cradle bolts and tighten the rear cradle bolts. Install DIS connector, if equipped. Install the starter and splash shield. Install the flywheel shield.
15. Attach the collar bracket to the block, install the engine-to-cradle nuts. Install the transaxle mount nuts.
16. Install the steering pinch bolt. Install the right inner fender splash shield and tire assembly. Lower the vehicle.
17. Remove the engine support tool. Install the serpentine belt and tensioner.
18. Fill the crankcase to the correct level. Connect the negative battery cable. Run the engine to normal operating temperature and check for leaks.

3.1L (VIN M) Engine

1996 Vehicles

1. Disconnect the negative battery cable.
2. Install engine support fixture J-28467-A or equivalent.
3. Remove the engine mount strut bolt.
4. Raise and safely support the vehicle.

5. Drain the engine oil into a suitable container.
6. Remove the oil drip shield bolts and shield.
7. Remove the engine mount.
8. Remove the transaxle mount nuts.
9. Disconnect the exhaust pipe from the exhaust manifold.
10. Raise the engine using the special tools J-28467-A and J-36462 or equivalents.
11. Lower the vehicle.
12. Place the jackstands under the frame at the front and rear center crossmembers.
13. Loosen the rear frame bolts, but do not remove.
14. Remove the front frame bolts and lower front of frame.
15. Remove the front engine mount bracket bolts, bracket and mount.
16. Disconnect the electrical leads from the starter and remove the starter.
17. Remove the brackets from the oil pan.
18. Remove the oil pan retaining bolts and remove the oil pan.
19. Remove the oil deflector.

NOTE: Clean all gasket mating surfaces.

To Install:

NOTE: Apply a small amount of sealer on either side of the rear main bearing cap, where the seal surface on the cap meets the cylinder block.

20. Install the oil deflector and tighten the mounting nuts to 18 ft. lbs. (25 Nm).
21. Install the gasket and oil pan and hand tighten retaining bolts.
22. After all bolts are hand tight, tighten to 18 ft. lbs. (25 Nm). Tighten the side bolts to 37 ft. lbs. (50 Nm).
23. Install the electrical brackets to the pan.
24. Install the starter and connect the electrical leads.
25. Install the engine mount bracket and mount.
26. Install the front engine mount bracket bolts.
27. Raise the frame to proper position. Using new bolts tighten to 76 ft. lbs. (103 Nm).
28. Remove the jackstands.
29. Raise and safely support the vehicle.
30. Lower the engine to correct position.
31. Install the exhaust pipe to the manifold.
32. Install the transaxle mount nuts.

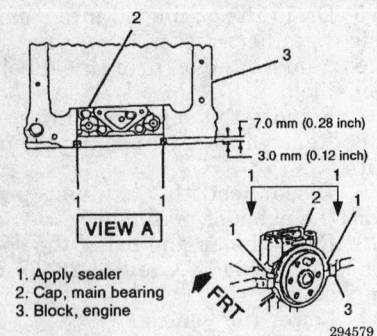

VIEW A

1. Apply sealer
2. Cap, main bearing
3. Block, engine

7.0 mm (0.28 inch)
3.0 mm (0.12 inch)

FRT

294579

Apply sealer as indicated for oil pan sealing — 1996 3.1L (VIN M) engine

33. Install the engine mount.
34. Install the oil drip shield.
35. Lower the vehicle.
36. Fill the crankcase with new engine oil.
37. Install the engine mount strut bolt.
38. Remove the engine support fixture.
39. Connect the negative battery cable.
40. Start the vehicle and verify no leaks.

Oil Pump

REMOVAL AND INSTALLATION

2.2L (VIN 4) Engine

1. Disconnect the negative battery cable.
2. Raise and safely support the vehicle.
3. Drain the engine oil into a suitable container.
4. Remove the oil pan-to-engine bolts and the oil pan.
5. Remove the oil pump-to-rear main bearing cap bolt, the oil pump and extension shaft.
To install:

— CAUTION —

Heat the extension shaft retainer in hot water prior to assembly. Be sure the retainer does not crack upon installation.

6. Install the extension shaft, oil pump and pump-to-rear main cap bolt. Tighten the oil pump-to-bearing cap bolt to 32 ft. lbs. (43 Nm) and the upper oil pump drive bolt to 18 ft. lbs. (25 Nm).

— CAUTION —

To avoid engine damage, all oil pump cavities must be filled with petroleum jelly before installing the gears into the pump body. This seals the gears, acts like a

prime and allows the pump to start drawing oil as soon as the engine begins to crank the first time after pump service. Also use only original equipment gaskets. Gasket thickness is critical to proper oil pump operation.

7. Install the oil pan and attaching bolts.
8. Lower the vehicle.
9. Fill the crankcase with clean engine oil.
10. Connect the negative battery cable.
11. Start the engine and check oil pressure and check for leaks.
12. Turn the engine OFF and allow to stand. Check oil level, add as necessary.

3.3L (VIN N) Engine

1. Disconnect the negative battery cable.
2. Raise and safely support the vehicle.
3. Drain the engine oil into a suitable container.
4. Remove the front cover assembly.
5. Remove the oil filter adapter, pressure regulator valve and spring.
6. Remove the oil pump cover attaching screws and remove the cover.
7. Remove the oil pump
To Check Components:
8. Measure the following components and replace any components that does not fall with the specification:
• Measure inner gear tip clearance: 0.006 in. (0.152mm)
• Outer gear diameter clearance: 0.008–0.015 in. (0.203–0.381mm)
• Gear end clearance (Gear drop in housing): 0.001–0.0035 in. (0.025–0.089mm)
• Gear pocket depth: 0.4610–0.4625 in. (11.71–11.75mm)
• Gear pocket diameter: 3.508–3.512 in. (89.10–89.20mm)
To Install:
9. Lubricate the gears with petroleum jelly and install the gears into the housing.
10. Pack the gear cavity with petroleum jelly after the gears have been installed in the housing.
11. Install the oil pump cover and screws and tighten to 97 inch lbs. (10 Nm).
12. Install the oil filter adapter with new gasket, pressure regulator valve and spring.
13. Install the front cover assembly.
14. Lower the vehicle.
15. Refill the crankcase with clean engine oil.

16. Start the engine and verify the oil pressure is correct and their are no leaks.

3.1L (VIN M) Engine

1. Disconnect the negative battery cable.
2. Raise and safely support the vehicle.
3. Drain the engine oil into a suitable container.
4. Remove the oil pan.
5. Remove the crankshaft oil deflector bolts.
6. Remove the crankshaft oil deflector.
7. Remove the oil pump retaining bolts and remove the oil pump and pump driveshaft.
To install:
8. Install the oil pump and pump driveshaft. Tighten the oil pump mounting bolts to 30 ft. lbs. (41 Nm).
9. Install the crankshaft oil deflector and mounting bolts. Tighten the mounting bolts to 18 ft. lbs. (25 Nm).
10. Install the oil pan.
11. Lower the vehicle.
12. Fill the crankcase to the correct level with oil.
13. Start the engine, check the oil pressure and check for leaks.

TRANSAXLE

Automatic Transaxle Assembly

REMOVAL AND INSTALLATION

1993–94 Vehicles

3T40 Transaxle

1. Disconnect the negative battery cable from the transaxle. Tape the wire to the upper radiator hose out of the way.
2. Remove the air cleaner and disconnect the detent cable. Slide the detent cable in the opposite direction of the cable to remove it from the carburetor or throttle body, as equipped.
3. Unbolt the detent cable attaching bracket at the transaxle.
4. Pull up on the detent cable cover at the transaxle until the cable is exposed. Disconnect the cable from the rod.
5. Remove the two transaxle strut bracket bolts at the transaxle, if equipped.

6. Remove all the engine-to-transaxle bolts except the one near the starter. The one nearest the firewall is installed from the engine side requiring a short handled box wrench or ratchet.

7. Loosen, but do not remove the engine-to-transaxle bolt near the starter.

8. Disconnect the speedometer cable at the upper and lower coupling. On cars with cruise control, remove the speedometer cable at the transducer.

9. Remove the retaining clip and washer from the shift linkage at the transaxle. Remove the two shift linkages at the transaxle. Remove the two shift linkage bracket bolts.

10. Disconnect and plug the two fluid cooler lines at the transaxle. Use a back–up wrench to avoid twisting the lines.

11. Install an engine holding chain or hoist. Raise the engine enough to take its weight off the mounts.

12. Unlock the steering column, raise and safely support the vehicle.

13. Remove the two nuts holding the anti–sway (stabilizer) bar to the left lower control arm (driver's side).

14. Remove the four bolts attaching the covering plate over the stabilizer bar to the engine cradle on the left side (driver's side).

15. Loosen but do not remove the four bolts holding the stabilizer bar bracket to the right side (passenger's side) of the engine cradle. Pull the bar down on the driver's side.

16. Disconnect the front and rear transaxle mounts at the engine cradle.

17. Remove the two rear center crossmember bolts.

18. Remove the three right (passenger) side front engine cradle attaching bolts. The nuts are accessible under the splash shield next to the frame rail.

19. Remove the top bolt from the lower front transaxle shock absorber, if equipped (V6 engine only).

20. Remove the left (driver) side front and rear cradle-to-body bolts.

21. Remove the left front wheel. Attach an axle shaft removing tool (G.M. part no. J-28468 or the equivalent) to a slide hammer. Place the tool behind the axle shaft cones and pull the cones out away from the transaxle. Remove the right shaft in the same manner. Set the shafts out of the way. Plug the openings in the transaxle to prevent fluid leakage and the entry of dirt.

22. Swing the partial engine cradle to the left (driver) side and wire it out

of the way outboard of the fender well.

23. Remove the four torque converter and starter shield bolts. Remove the two transaxle extension bolts from the engine-to-transaxle bracket.

24. Attach a transaxle jack to the case.

25. Matchmark the torque converter and flywheel so they can be reinstalled in the same relationship. Remove the three torque converter-to-flywheel bolts.

26. Remove the transaxle-to-engine bolt near the starter. Remove the transaxle by sliding it to the left, away from the engine.

To install:

27. Place the transaxle on a jack and raise it into the vehicle. As the transaxle is installed, slide the right axle shaft into the case.

28. Align the matchmarks made earlier and connect the torque converter to the flywheel. Install the transaxle-to-engine bolt near the starter.

29. Install the engine-to-transaxle bracket extension bolts. Install the torque converter and starter shield bolts.

30. Install the partial engine cradle.

31. Install the left axle shaft.

32. Install the drivers side front and rear cradle to body bolts.

33. Install the top bolt to the lower front transaxle shock absorber, if equipped (V6 engine only).

34. Install the three right (passenger) side front engine cradle attaching bolts.

35. Install the two rear center crossmember bolts.

36. Connect the front and rear transaxle mounts at the engine cradle.

37. Install the stabilizer bar. Tighten the four bolts holding the stabilizer bar bracket to the right side (passenger's side) of the engine cradle.

NOTE: To aid in stabilizer bar installation, a pry hole has been provided in the engine cradle.

38. Install the four bolts attaching the covering plate over the stabilizer bar to the engine cradle on the left side (driver's side).

39. Install the two nuts holding the anti–sway (stabilizer) bar to the left lower control arm (driver's side).

40. Lower the vehicle and remove the engine support device.

41. Connect the two fluid cooler lines at the transaxle.

42. Install the two shift linkage bracket bolts.

43. Connect the two shift linkages at the transaxle and install the retaining clips and washers.

44. Connect the speedometer cable at the upper and lower coupling. On cars with cruise control, connect the speedometer cable at the transducer.

45. Tighten the engine-to-transaxle bolt near the starter.

46. Install all remaining engine-to-transaxle bolts. The one nearest the firewall is installed from the engine side.

47. Install the two transaxle strut bracket bolts at the transaxle, if equipped.

48. Install the detent cable and air cleaner.

49. Refill the transaxle with the proper quantity of DEXRON®II or equivalent automatic transmission fluid. Do not overfill.

50. Connect the negative battery cable to the transaxle.

440–T4 Transaxle

1. Disconnect the negative battery cable.

2. Remove the air cleaner and disconnect the TV cable at the throttle body.

3. Disconnect the shift linkage at the transaxle.

4. Install engine support fixture tool J-28467 or equivalent.

5. Disconnect all electrical connectors.

6. Remove the 3 bolts from the transaxle to the engine.

7. Disconnect the vacuum line at the modulator.

8. Raise and safely support the vehicle.

9. Remove the left front wheel and tire assembly.

10. Remove the left side ball joint from the steering knuckle.

11. Disconnect the brake line bracket at the strut.

NOTE: A halfshaft seal protector tool J-34754 or equivalent should be modified and installed on any halfshaft prior to service procedures on or near the halfshaft. Failure to do so could result in seal damage or joint failure.

12. Remove the halfshafts from the transaxle.

13. Disconnect the pinch bolt at the intermediate steering shaft. Failure to do so could cause damage to the steering gear.

14. Remove the frame to stabilizer bolts.

15. Remove the stabilizer bolts at the control arm.

16. Remove the left front frame assembly.

17. Disconnect the speedometer cable or wire connector from the transaxle.

18. Remove the extension housing to engine block support bracket.

19. Disconnect the cooler pipes.

20. Remove the converter cover and converter-to-flywheel bolts.

21. Remove all of the remaining transaxle-to-engine bolts except one.

22. Position a jack under the transaxle.

23. Remove the remaining transaxle-to-engine bolt and remove the transaxle.

To install:

24. Install the transaxle in the vehicle. Install the engine-to-transaxle bolt accessible from under the vehicle. Tighten to 55 ft. lbs. (75 Nm).

25. Install all of the remaining transaxle-to-engine bolts. Tighten to 55 ft. lbs. (75 Nm).

26. Remove the jack.

27. Install the converter-to-flywheel bolts and the converter cover.

28. Connect the cooler pipes.

29. Install the extension housing to engine block support bracket.

30. Connect the speedometer cable or wire connector to the transaxle.

31. Install the left front frame assembly.

32. Install the stabilizer bolts at the control arm.

33. Install the frame-to-stabilizer bolts.

34. Connect the pinch bolt at the intermediate steering shaft.

35. Install the halfshafts to the transaxle.

36. Connect the brake line bracket at the strut.

37. Install the left side ball joint to the steering knuckle.

38. Install the left front wheel and tire assembly.

39. Lower the vehicle.

40. Connect the vacuum line at the modulator.

41. Install the 3 bolts from the transaxle to the engine.

42. Connect all electrical connectors.

43. Remove the engine support tool.

44. Connect the shift linkage to the transaxle.

45. Connect the TV cable at the throttle body and adjust as necessary. Install the air cleaner.

46. Connect the negative battery cable.

1995–96 Vehicles

1. Disconnect the negative battery cable.

2. Remove the air cleaner assembly.

3. Remove the engine torque strut from the engine.

4. Remove the shift control cable bracket from the transaxle case and lever.

5. Disconnect the electrical connector from the torque converter clutch solenoid.

6. Disconnect the vacuum hose from the modulator.

7. Remove the upper transaxle bolts, including the grounds.

8. Install engine support fixture J-28467-A or equivalent.

9. Raise and safely support the vehicle.

10. Remove the front tire and wheel assemblies.

11. Remove the engine splash shields.

12. Remove the pinch bolts from the control arms.

13. Remove the pinch bolt from the intermediate steering shaft.

14. Remove the stabilizer shaft bolts and reinforcement plates from the frame.

15. Remove the stabilizer shaft nuts and bracket from the control arm and separate the stabilizer shaft from the control arm.

16. Using a $7/16$ inch drill bit, drill through two spot welds located between the front and rear holes of the left front stabilizer shaft mounting.

17. Remove the front and rear transaxle mounting nuts.

18. Disconnect the engine wiring harness from the frame.

19. Remove the power steering cooler line bolts.

20. Remove the right frame to left frame retaining bolt. Position a jackstand under the frame for support.

21. Loosen the two right frame mounts and discard the bolts.

22. Remove the two left frame bolts from the frame.

23. Remove the left frame with the aid of an assistant.

24. Disconnect the right lower ball joint from the steering knuckle.

25. Remove the transaxle support bracket bolts from the transaxle.

26. Remove the power steering cooler line support from the transaxle.

27. Remove the torque converter cover.

28. Remove the starter.

29. Remove the torque converter bolts.

30. Remove the transaxle mount bolts from the transaxle case and remove the mount.

31. Disconnect the transaxle range switch electrical connectors.

32. Remove both drive axles from the transaxle.

33. Disconnect the vehicle speed sensor from the transaxle.

34. Position transaxle jack under the transaxle.

35. Remove the oil cooler lines and plug openings.

36. Remove the remaining transaxle bolts.

37. Remove the transaxle.

To install:

38. Install the transaxle into the vehicle and install the transaxle bolts. Torque to 55 ft. lbs. (75 Nm).

39. Install the oil cooler lines.

40. Remove the transaxle jack.

41. Connect the vehicle speed sensor connector to the transaxle.

42. Install the drive axles.

43. Connect the transaxle range switch electrical connector.

44. Install the transaxle mount and mounting bolts to the transaxle case.

45. Install the torque converter bolts and tighten them to 47 ft. lbs. (63 Nm).

46. Install the starter motor.

47. Install the torque converter cover.

48. Install the power steering cooler line support to the transaxle.

49. Install the transaxle brace and the bolts to the transaxle.

50. Install the pinch bolts to the control arms.

51. Install the left frame (with the aid of an assistant).

52. Tighten the frame bolts to 40 ft. lbs. (54 Nm).

53. Install the right frame bolts to the body and tighten them to 40 ft.lbs. (54 Nm).

54. Remove the jackstand.

55. Install the right frame to the left frame retaining bolts, and tighten to 40 ft. lbs. (54 Nm).

56. Install the engine wiring harness to the frame.

57. Install the power steering cooler lines.

58. Install the transaxle rear mount nuts to frame and tighten to 39 ft. lbs. (53 Nm).

59. Install the left stabilizer shaft to the control arm.

60. Install the stabilizer shaft bolts and reinforcement plates to frame using support.

61. Install the steering shaft pinch bolt and tighten to 35 ft. lbs. (48 Nm).

62. Install the engine splash shields.

63. Install the front tire and wheel assemblies.

64. Lower the vehicle.

65. Remove the engine support fixture.

66. Install the three upper transaxle bolts and the ground wires to the engine and tighten to 55 ft. lbs. (75 Nm).

67. Connect the vacuum hose to the modulator.

68. Connect the torque converter clutch switch electrical connector.

69. Install the transaxle shift cable, mounting bracket and lever.

70. Install the engine torque strut.

71. Install the air cleaner assembly.

72. Connect the negative battery cable.

73. Start the vehicle and check transaxle oil level.

NOTE: Whenever the vehicle subframe is removed or lowered, the wheel alignment should be checked.

74. Check the front end alignment and adjust as required.

75. Road test the vehicle and verify proper operation.

DRIVELINE

Halfshaft

REMOVAL AND INSTALLATION

1. Raise and safely support the vehicle.

2. Remove the tire and wheel assembly.

3. Because the hub nut is torqued so tightly, the halfshaft must be kept from turning while the hub nut is being loosened. Insert a drift punch through the opening in the top of the caliper and down into the brake rotor cooling fins to keep the assembly from turning.

4. Remove the halfshaft nut and washer.

5. Remove the drift punch.

6. Remove the caliper mounting bolts and remove the caliper from the steering knuckle and support out of the way. Hang the caliper on a stiff piece of wire. DO NOT allow the brake hose to support the caliper.

7. Remove the brake rotor.

8. Remove the lower ball joint pinch bolt and separate the ball joint from the steering knuckle.

9. Separate the halfshaft from the hub bearing using J-28733-B, or an equivalent puller.

10. Pull the steering knuckle outward to slip the halfshaft out of the hub bearing.

11. Disconnect the halfshaft from the transaxle using J-33008, J-29794 and J-2619–01 or their equivalents.

To Install:

12. If installing the right side halfshaft install J-37292-B, or an equivalent halfshaft seal protector over the end of the CV-joint. The handle of the tool should be between 5 and 7 o'clock for ease of removal.

13. Seat the halfshaft into the transaxle until the inner lock ring locks in the transaxle. This can be checked by pulling lightly on the inner CV-joint body, not the halfshaft, to make sure it stays seated in the transaxle. DO NOT pull on the halfshaft or the inboard joint could come apart damaging internal components.

14. Remove the halfshaft seal protector by cutting it off and removing all pieces.

15. Seat the halfshaft into the hub bearing assembly. Make sure the splines engage smoothly. If the halfshaft will not engage smoothly, coat the internal splines on the hub bearing with grease.

16. Connect the lower ball joint to the steering knuckle and tighten the pinch bolt to 38 ft. lbs. (52 Nm).

17. Install the washer and halfshaft nut loosely.

18. Install the brake rotor.

19. Install the caliper over the steering knuckle and tighten the caliper mounting bolts to 38 ft. lbs. (52 Nm).

20. Insert a drift punch through the caliper opening and into the brake rotor cooling fins.

21. Tighten the halfshaft nut to 103 ft. lbs. (140 Nm) plus 20 degrees additional rotation.

22. Remove the drift.

23. Install the tire and wheel assembly and tighten the wheel nuts to 100 ft. lbs. (140 Nm).

24. Lower the vehicle.

CV-Joint Boot

REPLACEMENT

Outer CV-Joint Boot

1. Raise and safely support the vehicle.

2. Remove the front wheel assembly. Mark the wheel to the wheel stud for re–installation purposes.

3. Remove the front halfshaft using the recommended procedure and carefully place it in a vise using a protective covering on the vise jaws.

4. Cut the large and small clamps and discard them.

5. Slide the boot down the shaft uncovering the outer joint.

6. Clean the grease away from the joint to locate the snap ring.

7. Using J 8059 or equivalent, open the snapring and slide the outer joint off the shaft.

8. Remove the outer boot from the shaft.

9. Clean the outer joint completely and inspect for damage to the cage or balls. Wipe any grease off the axle shaft.

To install:

10. Slide the small clamp onto the halfshaft and push the boot down several inches past the seal mounting area.

11. Check the snapring in the outer joint and replace as necessary. Pack the outer joint with half of the grease supplied in the boot kit and install it on the end of the halfshaft.

12. Gently pull down on the joint until the splines engage. With a brass drift lightly tap the joint down until the snapring engages.

13. Pack the remaining grease from the boot kit into the boot and then pull the boot up over the end of the joint. Seat the small end on the seal mounting area.

14. Slide the small clamp into position and using J 35910 or equivalent, crimp the clamp to 130 ft. lbs. (176 Nm).

15. Install the large clamp to the proper position and crimp to 130 ft. lbs. (176 Nm).

— **CAUTION** —

The boot must not be dimpled, stretched or out of shape in any way. If boot is not shaped correctly, carefully insert a thin flat blunt tool at the large end of the boot to equalize pressure. Shape the boot properly by hand and then remove the tool.

16. Install the halfshaft using the recommended procedure. Install the wheel and tire assembly and torque the wheel lug nuts to 100 ft. lbs. (130 Nm). Road test the vehicle to check for abnormal noise or vibration.

Inner CV-Joint Boot

1. Raise and safely support the vehicle.

2. Remove the front wheel assembly.

3. Remove the front halfshaft using the recommended procedure and carefully place it in a vise using a protective covering on the vise jaws.

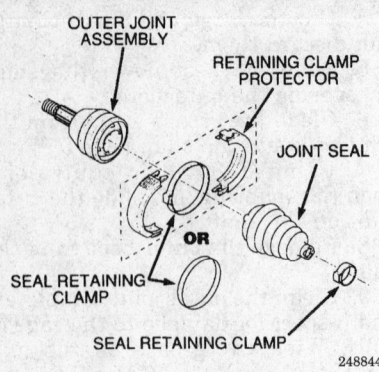

Outer CV-Joint disassembly

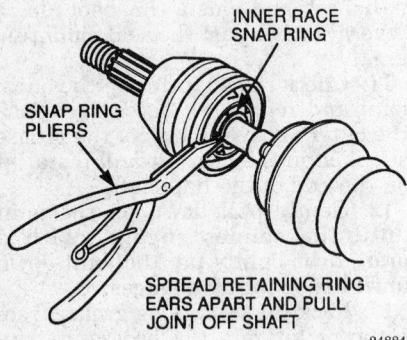

Outer CV-Joint snapring removal

4. Remove the outer CV-joint and boot.

5. Cut the large and small boot clamps and discard.

— CAUTION —

Do not cut through the boot and damage the sealing surface of the the tripot outer housing.

6. Separate the boot from the tripot bushing at the large diameter end and slide the boot away from the joint along the axle.

7. Remove the tripot housing from the spider and axle. Clean thoroughly and set aside.

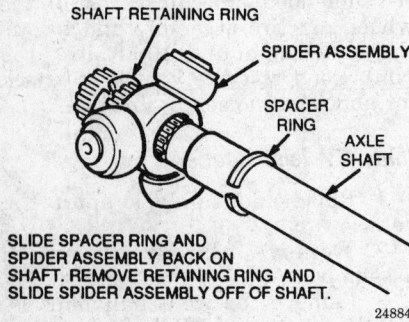

Outer CV-Joint snapring and spider assembly removal

8. Clean the grease from the spider assembly to expose the spacer ring located on the outboard spider assembly. Slide the spacer ring down the halfshaft.

9. Push the spider assembly down the shaft to uncover the snap ring on the end of the shaft. Using J 8059 or equivalent snapring pliers, remove the snapring. Slide the spider assembly off the end of the shaft. Clean completely, using care not to knock the caps off of the bearings. Set aside.

10. Remove the tripot bushing from the tripot housing.

11. Remove the spacer ring from the halfshaft and slide the boot off. Clean any grease off the shaft.

To install:

12. Slide the small clamp onto the halfshaft.

13. Slide the boot onto the shaft and position the neck of the boot in the groove on the axle shaft.

14. Crimp the seal retaining clamp with J-35910 or equivalent crimping tool to 100 ft. lbs. (136 Nm).

15. Put the spacer ring on the shaft, several inches below the second spacer ring groove.

16. Install the spider assembly far enough down the shaft to expose the top snapring groove. Make sure that the counterbored face of the tripot spider faces the end of the shaft.

17. Install the top snapring and pull the spider assembly back up into position.

18. Using J 8059 or equivalent snapring pliers lock the spacer ring in the spacer ring groove.

19. Pack the housing with half the grease supplied and put the rest in the boot.

20. Install the tripot bushing into the tripot housing.

21. Slide the larger clamp over the boot.

22. Push the tripot housing over the spider assembly.

23. Slide the larger diameter of the boot, into position and clamp in place over the outside of the bushing and locate lip of the boot in the bushing groove.

24. Position the tripot assembly to the dimension shown in the illustration. Install the large clamp in position and using J-35910 or equivalent crimping tool and tighten clamp.

— CAUTION —

The boot must not be dimpled, stretched or out of shape in any way. If boot is not shaped correctly, carefully insert a thin flat blunt tool at the large end of the

boot to equalize pressure. Shape the boot properly by hand and then remove the tool.

25. Install the outer CV-Joint and boot.

26. Install halfshaft using the recommended procedure.

27. Install the wheel and tire assembly and torque the wheel lug nuts to 100 ft. lbs. (140 Nm). Road test for proper operation. Check for abnormal noise or vibration.

STEERING

Air Bag

— CAUTION —

Some vehicles are equipped with the Supplemental Inflatable Restraint (SIR) or air bag system. The SIR system must be disabled before performing service on or around SIR system components, steering column, instrument panel components, wiring and sensors. Failure to follow safety and disabling procedures could result in accidental air bag deployment, possible personal injury and unnecessary SIR system repairs.

PRECAUTIONS

Several precautions must be observed when handling the inflator module to avoid accidental deployment and possible personal injury.

• Never carry the inflator module by the wires or connector on the underside of the module.

• When carrying a live inflator module, hold securely with both hands, and ensure that the bag and trim cover are pointed away.

• Place the inflator module on a bench or other surface with the bag and trim cover facing up.

• With the inflator module on the bench, never place anything on or close to the module which may be thrown in the event of an accidental deployment.

DISARMING

1993–94 Vehicles

1. Disconnect the negative battery cable.

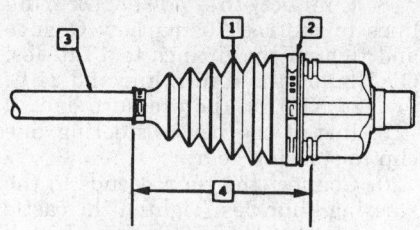

1. INBOARD THERMOPLASTIC SEAL
2. TRILOBAL TRIPOT JOINT
3. AXLE SHAFT
4. 130 MM (5 1/16") – JOINT AND SEAL ARE TO BE COMPRESSED TO THIS DIMENSION BEFORE CRIMPING CLAMPS.

248847

Inner boot collapsed dimension

2. Turn the steering wheel so that the vehicle's wheels are pointing straight ahead.

3. Turn the ignition switch to the **LOCK** position and remove the key.

4. Remove the **AIR BAG** fuse from the fuse block.

5. Remove the left sound insulator.

6. Remove the Connector Position Assurance (CPA) clip from the yellow 2–way connector at the base of the steering column and, if equipped, disconnect the lead going to the passenger side module.

7. Disconnect the yellow 2–way connector.

1995–96 Vehicles

1. Turn the steering wheel so that the vehicle's wheels are pointing straight ahead.

2. Turn the ignition switch to the LOCK position and remove the key.

3. Remove the AIR BAG fuse from the instrument panel fuse block.

4. Remove the left hand sound insulator.

5. Disconnect the Connector Position Assurance (CPA) and yellow 2–way connector at the base of the steering column.

6. Remove the right hand sound insulator.

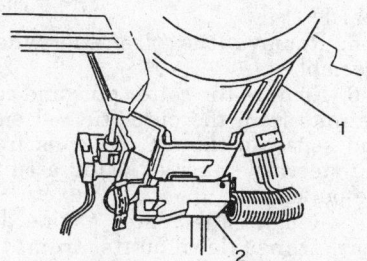

1. Steering column
2. Connector, Sir (yellow)

256779

Disconnect the SIR yellow 2–way connector

7. Disconnect the CPA and yellow 2–way connector from the passenger inflator module pigtail.

ARMING

1993–94 Vehicles

1. Turn the ignition switch to the **LOCK** position and remove the key.

2. Connect the yellow 2–way connector at the base of steering column and secure it with the Connector Position Assurance (CPA) clip.

3. Install the left sound insulator.

4. Install the **AIR BAG** fuse in the fuse block.

5. Turn the ignition switch to the **RUN** position and verify that the **AIR BAG** warning lamp flashes 7 times and then turns off.

6. Connect the negative battery cable.

1995–96 Vehicles

1. Turn the ignition switch to the LOCK position and remove the key.

2. Connect the yellow 2–way connector and CPA to the passenger inflator module pigtail.

3. Install the right hand sound insulator.

4. Connect the yellow 2–way connector and CPA at the base of the steering column. After installing the CPA, clip the connector to flange on the steering column support.

5. Install the left hand insulator.

6. Install the AIR BAG fuse into the instrument panel fuse block.

7. Turn the ignition switch to the RUN position and verify that the AIR BAG warning lamp flashes seven times and then turns to the OFF position.

Steering Wheel

REMOVAL AND INSTALLATION

With Air Bag

———— **CAUTION** ————
The Supplemental Inflatable Restraint (SIR) system must be disarmed before removing the steering wheel. Failure to do so may cause accidental deployment of the air bag, resulting in unnecessary SIR system repairs and/or personal injury.

1. Disable SIR system using the recommended procedure.

2. Remove the four inflator module mounting screws from the back of the steering wheel.

3. Pull the inflator module from the steering wheel. Remove the CPA and disconnect the coil connector from the inflator module.

4. Disconnect the horn lead from the steering column.

———— **CAUTION** ————
When carrying a live air bag, make sure the bag and trim cover are pointed away from the body. In the unlikely event of an accidental deployment, the bag will then deploy with minimal chance of injury. When placing a live air bag on a bench or other surface, always face the bag and trim cover up, away from the surface. The will reduce the motion of the module if it is accidently deployed.

5. Set the inflator module aside.

6. Remove the retainer and steering wheel mounting nut.

———— **CAUTION** ————
While attaching J-1859-03 to the steering wheel, use care to prevent threading the bolts all the way through the steering wheel hub into the coil assembly and damaging the coil assembly.

7. Install a steering wheel puller, J-1859–03, or the equivalent. Remove the steering wheel from the steering shaft.

To install:

8. Align the mark on steering wheel with the mark on the steering shaft and install the steering wheel.

9. Install the steering wheel mounting nut and tighten the nut to 30 ft. lbs. (41 Nm).

10. Install the nut retainer.

11. Connect the horn lead from the inflator module to the steering column.

12. Connect the coil assembly connector to the inflator module and install the CPA.

13. Route the coil lead around the mounting post and secure it under the clip.

14. Position the inflator module on the steering wheel and tighten the four mounting screws to 25 inch lbs. (3 Nm).

15. Enable the SIR system.

Without Air Bag

1. Using a thin blade screwdriver at the top of the horn pad, gently pry it away from the steering wheel.

2. Disconnect the horn lead from the turn signal switch by pushing it down and turning it counterclockwise.

3. Remove the retainer and steering wheel mounting nut.

— CAUTION —

While attaching J-1859–03 to the steering wheel, use care to prevent threading the bolts all the way through the steering wheel hub and into the electrical components below.

---

4. Install a steering wheel puller, J-1859–03, or the equivalent. Remove the steering wheel from the steering shaft.

To install:

5. Align the mark on steering wheel with the mark on the steering shaft and install the steering wheel.

6. Install the steering wheel mounting nut and tighten the nut to 30 ft. lbs. (41 Nm).

7. Install the nut retainer.

8. Connect the horn lead to the turn signal switch and lock in place by pushing down and rotating clockwise.

9. Snap the horn pad in place.

Tie Rod Ends

REMOVAL AND INSTALLATION

1. Raise and safely support the vehicle.

2. Remove the tire and wheel assembly.

3. Remove the cotter pin and castle nut from the outer tie rod end.

4. Loosen the inner tie rod jam nut ½ turn.

5. Separate the outer tie rod end from the steering knuckle with a suitable puller.

6. Unthread the outer tie rod end from the rack and pinion assembly counting the number of turns required for removal.

To install:

7. Thread the tie rod end onto the rack the same number of turns required for removal of the old one.

8. Connect the tie rod end to the steering knuckle and tighten the castle nut to 31 ft. lbs. (42 Nm). If necessary to align the cotter pin holes tighten the nut up to 60 degrees additional. Install a new cotter pin.

9. Temporarily tighten the jam nut to 45 ft. lbs. (60 Nm).

10. Install the tire and wheel assembly and tighten the wheel nuts to 100 ft. lbs. (140 Nm).

11. Lower the vehicle.

12. Check the front end alignment and adjust as necessary.

Power Rack and Pinion

REMOVAL AND INSTALLATION

With 2.2L (VIN 4) Engine

1. Disconnect the negative battery cable.

2. Install an engine support fixture J-28467-A, or the equivalent.

3. Raise and safely support the vehicle.

4. Remove the tire and wheel assemblies.

5. Remove the intermediate shaft lower pinch bolt.

6. Disconnect the intermediate shaft from stub shaft.

— CAUTION —

Failure to disconnect the intermediate shaft from the rack and pinion stub shaft can result in damage to the steering gear and/or intermediate shaft. This damage can cause loss of steering control which could result in personal injury.

---

7. Remove the exhaust pipe hanger bracket near rear of frame including the brake line retainer and rubber exhaust pipe hangers.

8. Remove the engine and transaxle mount nuts at the subframe.

9. Support the rear of the subframe with jack stands.

10. Remove the rear subframe bolts and discard.

11. Lower the subframe for access to rack and pinion.

NOTE: Do not lower rear of cradle too far as damage to engine components nearest to the cowl may result.

12. Remove the cotter pins and castle nuts from the outer tie rod ends and separate the tie rod ends from the steering knuckles using a suitable puller.

13. Disconnect the power steering lines from the steering gear.

14. Remove the rack and pinion mounting bolts and nuts.

15. Remove the rack and pinion through left wheel opening.

To install:

16. Install the rack and pinion through left wheel opening.

17. Install the rack and pinion mounting bolts and nuts and tighten to 66 ft. lbs. (90 Nm).

18. Connect the power steering lines to the rack using new O-rings and tighten the fittings to 13 ft. lbs. (17 Nm) on the return line and 21 ft. lbs. (28 Nm). on the pressure line.

19. Install the power steering line clip to steering gear.

20. Connect the tie rod ends to the steering knuckles. Tighten the castle nuts to 31 ft. lbs. (42 Nm). Install new cotter pins.

21. Raise the subframe into position and install new frame mounting bolts. Tighten the bolts to 103 ft. lbs. (140 Nm).

22. Remove the jack.

23. Install the engine mount and transaxle mount nuts and tighten to 39 ft. lbs. (53 Nm).

24. Install the exhaust pipe hanger bracket near the rear of frame including brake line retainer and rubber exhaust pipe hanger.

25. Connect the intermediate shaft to the stub shaft and tighten the pinch bolt to 29 ft. lbs. (40 Nm).

26. Install the tire and wheel assemblies.

27. Lower the vehicle.

28. Remove the support fixture.

29. Connect the negative battery cable.

30. Refill the power steering reservoir and bleed the system.

NOTE: Whenever the vehicle subframe is removed or lowered, the wheel alignment should be checked.

31. Check the front end alignment and adjust as necessary.

With 3.1L (VIN M) or 3.3L (VIN N) Engines

1. Disconnect the negative battery cable.

2. Remove engine torque strut from engine.

3. Install an engine support fixture J-28467-A, or the equivalent.

4. Raise and safely support the vehicle.

5. Remove the tire and wheel assemblies.

6. Remove the cotter pins and castle nuts from the outer tie rod ends and separate the tie rod ends from the steering knuckles using a suitable puller.

7. Remove the center engine and rear transaxle mounts from the frame.

8. Remove the intermediate shaft lower pinch bolt.

9. Disconnect the intermediate shaft from stub shaft.

CAUTION

Failure to disconnect the intermediate shaft from the rack and pinion stub shaft can result in damage to the steering gear and/or intermediate shaft. This damage can cause loss of steering control which could result in personal injury.

10. Remove the brace bolts and brace, including brake line brace.

11. Support the rear of the subframe with jack stands.

12. Remove the rear subframe bolts and discard.

13. Lower the subframe for access to rack and pinion.

14. Remove the steering gear heat shield.

15. Remove the clip holding lines at rack assembly.

16. Disconnect the power steering lines and O-rings.

17. Remove the rack and pinion mounting bolts and nuts.

18. Remove the rack and pinion unit through the left wheel well.

To install:

19. Install the rack and pinion through left wheel opening.

20. Install the rack and pinion mounting bolts and nuts and tighten the mounting bolts to 66 ft. lbs. (90 Nm).

21. Connect the power steering lines to the rack using new O-rings and tighten the fittings to 13 ft. lbs. (17 Nm) on the return line and 21 ft. lbs. (28 Nm). on the pressure line.

22. Install the power steering line clip to steering gear.

23. Install the steering gear heat shield.

24. Raise the subframe into position and install new frame mounting bolts. Tighten the bolts to 103 ft. lbs. (140 Nm).

25. Remove the jackstands.

26. Install the engine mount and transaxle mount nuts and tighten to 39 ft. lbs. (53 Nm).

27. Install the brace and bolts including brake line brace.

28. Connect the intermediate shaft to the stub shaft and tighten the pinch bolt to 29 ft. lbs. (40 Nm).

29. Connect the tie rod ends to the steering knuckles. Tighten the castle nuts to 31 ft. lbs. (42 Nm). Install new cotter pins.

30. Install the tire and wheel assemblies.

31. Lower the vehicle.

32. Remove the support fixture.

33. Connect the negative battery cable.

34. Refill the power steering reservoir and bleed the system.

NOTE: Whenever the vehicle subframe is removed or lowered, the wheel alignment should be checked.

35. Check the front end alignment and adjust as necessary.

Power Steering Pump

REMOVAL AND INSTALLATION

2.2L (VIN 4) and 3.1L (VIN M) Engines

1. Remove the serpentine belt.

2. Disconnect the power steering lines from the pump.

3. On the 3.1L, remove the spark plug wire clip from the power steering pump.

4. Remove the power steering mounting bolts.

5. Remove the power steering pump from the mounting bracket.

6. Remove the pulley from the pump using a suitable puller, J-25034-B, or the equivalent.

To install:

7. Install the pulley onto the pump shaft and seat in place using J-25033-B, or an equivalent pulley installer.

8. Position the pump in the mounting bracket and install the mounting bolts. Tighten the front mounting bolts to 25 ft. lbs. (34 Nm).

9. On the 3.1L, connect the spark plug wire clip to the pump.

10. Connect the power steering lines to the pump and tighten the fitting on the pressure line to 21 ft. lbs. (28 Nm).

11. Install the serpentine belt.

12. Refill the power steering reservoir and bleed the power steering system.

3.3L (VIN N) Engine

1. Remove the serpentine belt from the power steering pulley. It is not necessary to remove the drive belt from the remainder of the pulleys.

2. Remove the bolt securing the rear brace to the power steering pump.

3. Disconnect the power steering lines from the pump.

4. Remove the power steering pump mounting bolts working through the holes in the pulley. Remove the pump from the mounting bracket.

5. If pump replacement is necessary, remove the reservoir as follows:

   a. Release the lock tabs on the reservoir mounting clips and drive the clips off the reservoir.

   b. Remove the reservoir from the pump body.

   c. Remove the O-rings from the pump and discard.

6. Remove the pulley from the pump using a suitable puller, J-25034-B, or the equivalent.

To install:

7. Install a new O-ring onto the reservoir and coat it with clean power steering fluid.

8. Install the reservoir onto the pump and install the mounting clips until they lock into place.

9. Install the pulley onto the pump shaft and seat in place using J-25033-B, or an equivalent pulley installer.

10. Position the pump on the mounting bracket and tighten the mounting bolts to 21 ft. lbs. (29 Nm).

11. Connect the power steering lines to the pump.

12. Install the bolt at the rear brace where it connects to the pump.

13. Install the serpentine belt.

14. Refill the power steering reservoir and bleed the power steering system.

POWER STEERING SYSTEM BLEEDING

1. Fill the power steering reservoir to the **FULL** mark.

2. Let fluid stand undisturbed for 2 minutes, then crank engine for about 2 seconds. Refill reservoir if necessary.

3. Repeat Steps 1 and 2 until fluid level remains constant after cranking the engine.

4. Raise the front of the vehicle until both wheels are off the ground, then start the engine. Increase engine speed to 1500 rpm.

5. Turn the wheels lightly against the stop to the left and right, checking the fluid level and refilling, as necessary.

BRAKES

Anti-Lock Brake System Service

PRECAUTIONS

• Certain components within the ABS system are not intended to be serviced or repaired individually. Only those components with removal and installation procedures should be serviced.

• Do not use rubber hoses or other parts not specified for an ABS system. When using repair kits, replace all parts included in the kit. Partial or incorrect repair may lead to functional problems and require the replacement of components.

• Lubricate rubber parts with clean, fresh brake fluid to ease assembly. Do not use lubricated shop air to clean parts; damage to rubber components may result.

• Use only DOT 3 brake fluid from an unopened container.

• If any hydraulic component or line is removed or replaced, it may be necessary to bleed the entire system.

• A clean repair area is essential. Always clean the reservoir and cap thoroughly before removing the cap. The slightest amount of dirt in the fluid may plug an orifice and impair the system function. Perform repairs after components have been thoroughly cleaned; use only denatured alcohol to clean components. Do not allow ABS components to come into contact with any substance containing mineral oil; this includes used shop rags.

• The Anti-Lock control unit is a microprocessor similar to other computer units in the vehicle. Ensure that the ignition switch is **OFF** before removing or installing controller harnesses. Avoid static electricity discharge at or near the controller.

• If any arc welding is to be done on the vehicle, the control unit should be unplugged before welding operations begin.

Master Cylinder

REMOVAL AND INSTALLATION

1993 Vehicles

1. Disconnect the electrical connector from the fluid level sensor.

NOTE: **When disconnecting the brake lines and removing the master cylinder, do not allow brake fluid to contact any painted surfaces or electrical connectors.**

2. Disconnect and cap the brake lines from the master cylinder.
3. Remove the two mounting nuts securing the master cylinder to the power booster.
4. Remove the master cylinder from the vehicle.
   To Install:
5. Bench bleed the master cylinder prior to installation.
6. Position the master cylinder on the power booster and tighten the mounting nuts to 20 ft. lbs. (27 Nm).
7. Connect the brake lines to the master cylinder and tighten the fittings to 24 ft. lbs. (32 Nm).
8. Connect the electrical connector to the brake fluid level sensor.
9. Bleed the brake system.

1994–96 Vehicles

——— **CAUTION** ———
When servicing the master cylinder on this vehicle the procedure below must be followed in the correct sequence. DO NOT perform any services to the master cylinder without first performing the gear tension release procedure.

When the ABS modulator cylinder pistons are in their uppermost position, each motor has been prevailing torque due to the force necessary to ensure each piston is held firmly at the top of its travel. This torque re-

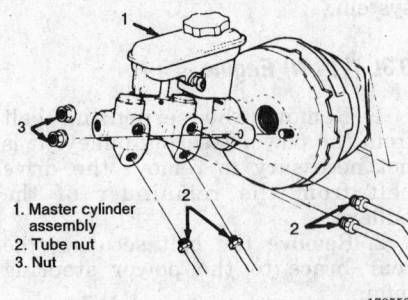

1. Master cylinder assembly
2. Tube nut
3. Nut

170553

Removing the master cylinder — 1993 vehicles

sults in "gear tension," or force on each gear that makes motor pack separation difficult. To avoid injury, or damage to the gears, the "Gear Tension Relief Sequence" briefly reverses each motor to eliminate the prevailing torque. This procedure is one of the many functions of GM's Tech 1 scan tool. Use care when using a substitute. In general, make sure the ignition switch is in the OFF position. Install the Tech 1 or equivalent with the correct chassis cartridge. Turn the ignition switch to the ON position, leaving the engine OFF. Select the proper function. The "Gear Tension Relief Sequence" is F5 on the Tech 1 scan tool. Note that the same scan tool is needed to bleed the system after repairs to the hydraulic system. Always perform the "Gear Tension Relief Sequence" prior to removing the hydraulic modulator/master cylinder assembly from the vehicle. Each hydraulic modulator gear (large gears) should be able to be turned in one direction and then in the opposite direction when the motor pack is removed. If any gear will not move, replace the hydraulic modulator.

1. Perform the gear tension release procedure.
2. Disconnect the electrical connector from the brake fluid level switch.
3. Disconnect the electrical connector from the ABS system solenoids.
4. Disconnect the motor pack electrical connectors.

NOTE: **When disconnecting the brake lines from the master cylinder and removing the unit, use care not to spill brake fluid on any painted surfaces or electrical connectors.**

5. Disconnect and cap the brake lines from the master cylinder assembly. Plug the master cylinder ports to prevent excessive fluid loss.
6. Remove the master cylinder mounting nuts.
7. Remove the master cylinder and modulator assembly.
   To Install:
8. Install the master cylinder and modulator assembly on the power booster and tighten the mounting nuts to 20 ft. lbs. (27 Nm).
9. Connect the brake lines to the master cylinder and tighten the fittings to 18 ft. lbs. (25 Nm).
10. Connect the electrical connector to the brake fluid level sensor.
11. Connect the electrical connectors to the ABS system solenoids.

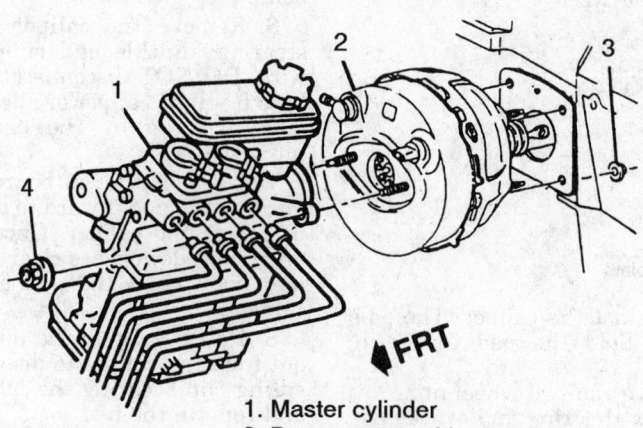

FRT

1. Master cylinder
2. Booster assembly, vacuum
3. Nut, vacuum booster assembly
4. Nut, master cylinder

295544

Removing the master cylinder — vehicles with ABS

12. Connect the motor pack electrical connectors.

13. Fill the master cylinder to the proper level. The proper level is to the **MAX** level indicator on the reservoir.

14. Bleed the brake system following the recommended procedure.

15. Road test the vehicle and verify proper operation.

Brake Caliper

REMOVAL AND INSTALLATION

Front Caliper

1. Raise and safely support the vehicle.

2. Remove the wheel assembly.

3. If the caliper is to be removed from the vehicle for bench unit repair, disconnect the brake hose from the caliper. Cap the brake hose to avoid leakage and/or contamination. If removal is to service other components, leave the brake hose connected and after caliper removal, hang the caliper out of the way on a stiff piece of wire. Do not allow the caliper to hang by the brake hose.

4. Remove the caliper mounting bolts.

5. Install two wheel nuts to hold the rotor in place when the caliper is removed.

6. Remove the caliper from the steering knuckle.

To install:

7. Install the caliper on the steering knuckle and tighten the mounting bolts to 38 ft. lbs. (51 Nm).

8. If removed, connect the brake hose to the caliper and tighten the mounting bolt to 33 ft. lbs. (45 Nm).

9. Measure the caliper clearance at the top and bottom of the caliper. The clearance should be 0.005–0.0012 inch (0.13–0.30mm).

10. Remove the nuts securing the rotor.

11. Install the tire and wheel assembly and tighten the wheel nuts to specification.

12. Lower the vehicle.

13. If the brake hose had been disconnected, bleed the brake system following the recommended procedure.

Rear Caliper

1. Raise and safely support the vehicle.

2. Remove the wheel assembly.

3. If the caliper is to be removed from the vehicle for bench unit repair, disconnect the brake hose from the caliper. Cap the brake hose to avoid leakage and/or contamination. If removal is to service other components, leave the brake hose connected and after caliper removal, hang the caliper out of the way on a stiff piece of wire, using care not to kink the parking brake cable. Do not allow the caliper to hang by the brake hose.

4. Remove the brake pads following the recommended procedure.

5. f the caliper is to be removed from the vehicle for bench unit repair, disconnect the parking brake cable from the parking brake lever and remove the return spring.

6. Remove the parking brake cable from the parking brake cable bracket. This can be done easily by inserting a 13mm box wrench over the cable end lock tabs and pulling the cable out.

7. Remove the two caliper mounting bolts and remove the caliper from the vehicle.

To install:

8. Install the caliper over the rotor and install the caliper mounting bolts.

9. Tighten the mounting bolts to 74 ft. lbs. (100 Nm).

10. Install the parking brake cable into the cable mounting bracket.

11. Install the return spring and connect the cable end to the parking brake lever.

12. Install the brake pads following the recommended procedure.

13. If removed, connect the brake hose to the caliper and tighten the mounting bolt to 33 ft. lbs. (45 Nm).

14. Install the tire and wheel assembly and tighten to specification.

15. If the brake hose was disconnected, bleed the brake system following the recommended procedure.

16. Lower the vehicle.

Disc Brake Pads

REMOVAL AND INSTALLATION

Front Brake Pads

Light Duty Brakes

1. Siphon ⅔ of the brake fluid out of the master cylinder reservoir.

2. Raise and safely support the vehicle.

3. Remove the tire and wheel assembly.

4. Install two wheel nuts to secure the rotor in place when the caliper is removed.

5. Unstake the outboard pads using a hammer and chisel.

6. Remove the caliper mounting bolts.

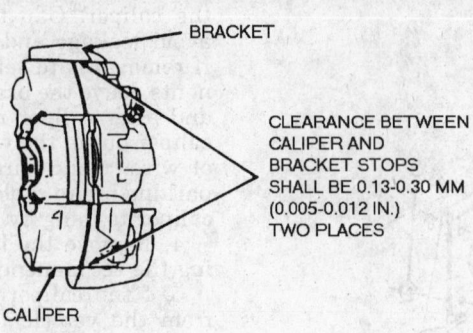

Front caliper clearance measurement points

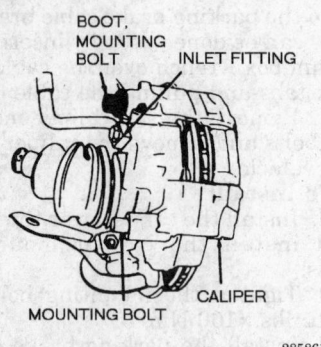

Front caliper mounting bolts

7. Remove the caliper from the steering knuckle and support on a wire. DO NOT disconnect the brake hose or allow the brake hose to support the weight of the caliper.

8. Remove the outboard pad from the caliper.

9. Pull the top of the inboard pad out from the caliper to disengage the spring clip securing the inboard pad and remove the pad.

10. Using a suitable C-Clamp compress the caliper piston fully into the caliper bore.

To install:

11. Install the inboard pad with a new spring clip into the caliper piston. This can be done by locking the bottom edge of the spring clip into the piston and push the top of the pad back.

12. Install the outboard pad into the caliper.

13. Install the caliper on the steering knuckle and tighten the mounting bolts to 38 ft. lbs. (51 Nm).

14. Pump up the brake pedal until the pads seat against the rotor. Using a suitable tool lock the brake pedal down keeping the front wheels locked.

15. Using a hammer and punch, stake the pad ears down until they

are flat against the caliper. The pad ears must hold the pad firmly in place.

16. Remove the two wheel nuts.

17. Install the tire and wheel assembly and tighten to specification.

18. Lower the vehicle.

19. Check and top off the master cylinder.

Heavy Duty Brakes

1. Siphon ⅔ of the brake fluid out of the master cylinder reservoir.

2. Raise and safely support vehicle.

3. Remove the tire and wheel assembly.

4. Install two wheel nuts to secure the rotor in place when the caliper is removed.

5. Remove the caliper mounting bolts.

6. Remove the caliper from the steering knuckle and support on a wire. DO NOT disconnect the brake hose from the caliper or allow the caliper to hang by the brake hose unsupported.

7. Remove the outboard pad by pushing the pad inward to unseat the pad from the caliper. Once the pad alignment dowels are clear of the caliper body push the pad out of the caliper.

8. Pull the top of the inboard pad out from the caliper to disengage the spring clip securing the inboard pad and remove the pad.

9. Using a suitable C-Clamp or large pliers, compress the caliper piston fully into the caliper bore.

To Install:

10. Install the inboard pad with a new spring clip into the caliper piston. This can be done by locking the bottom edge of the spring clip into the piston and push the top of the pad back.

11. Install the outboard pad into the caliper making sure the dowels in

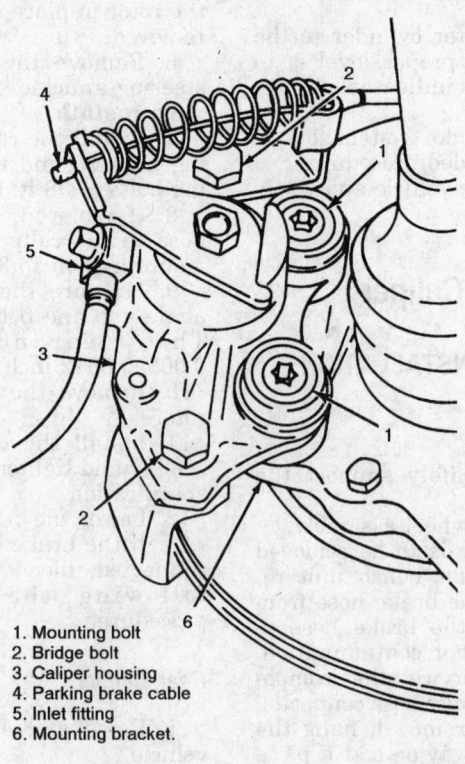

1. Mounting bolt
2. Bridge bolt
3. Caliper housing
4. Parking brake cable
5. Inlet fitting
6. Mounting bracket.

Rear caliper mounting components and parking brake assembly

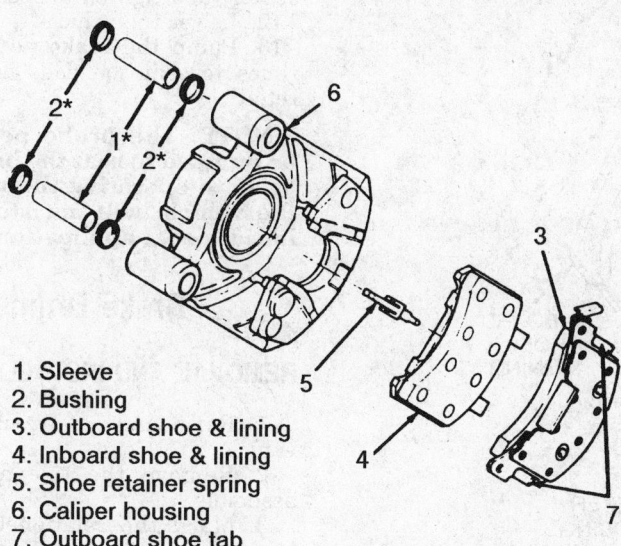

1. Sleeve
2. Bushing
3. Outboard shoe & lining
4. Inboard shoe & lining
5. Shoe retainer spring
6. Caliper housing
7. Outboard shoe tab

285944

Brake components — light duty front brakes

1. Inboard shoe & lining
2. Shoe retainer spring
3. Retention lug

285945

Installing the spring clip on the inboard pad

the brake pad seat in the holes in the caliper.

12. Install the caliper on the steering knuckle and install the mounting bolts. Tighten the mounting bolts to 38 ft. lbs. (51 Nm).
13. Remove the two wheel nuts.
14. Install the tire and wheel assembly and tighten the nuts to specification.
15. Lower the vehicle.
16. Pump the brake pedal several times to seat the pads against the rotor.
17. Check and top off the master cylinder.

Rear Brake Pads

1. Siphon ⅔ of the brake fluid out of the master cylinder reservoir.
2. Raise and safely support the vehicle.
3. Remove the tire and wheel assembly.
4. Remove the spring pins using spring pin removal tool, J 6125–1B and J 36620. The tool will thread into the spring pin. Using the tool remove the spring pins from the caliper.
5. Remove the outboard brake pad from the caliper. Loosen the parking brake cable adjustment if necessary to remove the pad.
6. Remove the inboard pad from the caliper. It may be necessary to slide the caliper in slightly to remove the pad.
7. Remove the two check valve from the end of the piston assembly, using a small screwdriver.
8. Bottom the piston in the caliper bore using a caliper spin back tool. Make sure the two notches in the caliper piston are at the 6 and 12 o'clock positions.
To install:
9. Install a new lubricated two way check valve into the piston assembly.

10. Install the inboard pad so the dowels on the back of the pad seat into the piston notches.
11. Install the outboard pad.
12. Install one of the spring pins using a hammer and soft brass drift.
13. Install the pad spring over the outboard pad with the spring center section over the top of the pad. Slide the second spring pin into the caliper until the pad spring can be hooked under it.
14. Install the pad spring over the inboard pad with the spring center section over the top of the pad. Slide the second spring pin into the caliper until the pad spring can be hooked under it.
15. With both pad springs in place drive the spring pin to the fully seated position in the caliper.
16. Install the tire and wheel assembly and tighten to specification.
17. Lower the vehicle.
18. Pump the brake several times to seat the pads against the rotor.

Brake Rotor

REMOVAL AND INSTALLATION

Front Brake Rotor

1. Raise and safely support the vehicle.
2. Remove the tire and wheel assembly.
3. Install two wheel nuts to hold the rotor on the vehicle.
4. Remove the two caliper mounting bolts and remove the caliper from the steering knuckle.
5. Using a piece of wire support the caliper out of the way. DO NOT disconnect the brake hose or allow the caliper to hang from the brake hose.
6. Remove the two wheel nuts and remove the rotor.
To Install:
7. Install the rotor on the vehicle and install two wheel nuts finger tight to hold it in place while the caliper is being installed.
8. Install the caliper onto the steering knuckle. Make sure the brake hose is not twisted.
9. Tighten the caliper mounting bolts to 38 ft. lbs. (51 Nm).
10. Remove the two wheel nuts.
11. Install the tire and wheel assembly and torque the wheel nuts to specification.
12. Lower the vehicle.

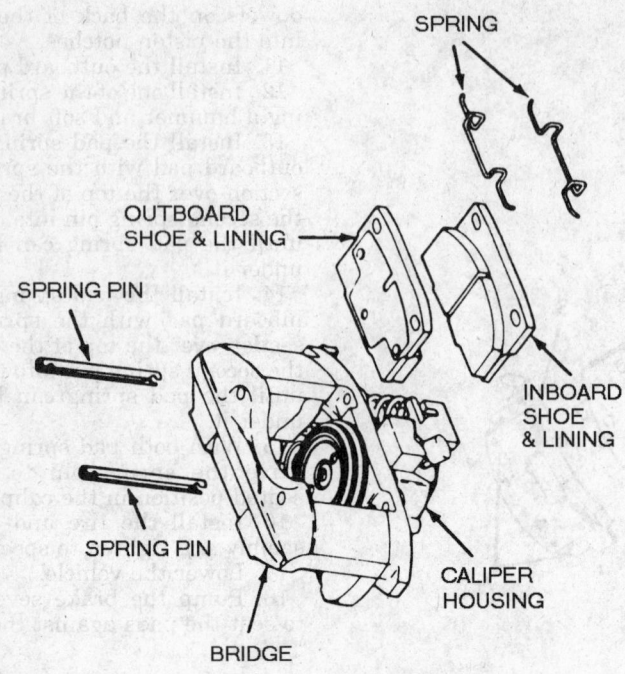

Rear brake components

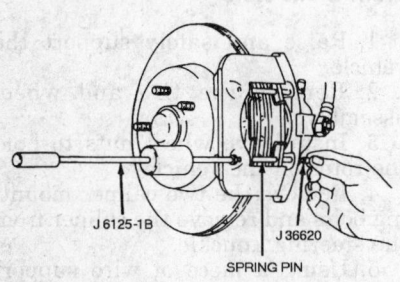

Installing the spring pin removal tool — rear brakes

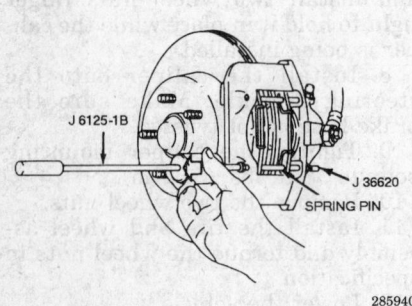

Removing the spring pin — rear brakes

13. Pump the brake pedal several times to seat the pads against the rotor.

NOTE: The brake pedal must be pumped to seat the brake pads prior to operating the vehicle or the vehicle will not stop on the initial pedal application.

Rear Brake Rotor

1. Raise and safely support the vehicle.
2. Remove the tire and wheel assembly.
3. Remove the brake pads following the recommended procedure.
4. Remove the two caliper mounting bolts and remove the caliper from the knuckle.
5. Using a piece of wire support the caliper out of the way. DO NOT disconnect the brake hose or allow the caliper to hang from the brake hose.
6. Remove the brake rotor from the vehicle.
To Install:
7. Install the rotor on the vehicle.
8. Install the caliper onto the steering knuckle. Make sure the brake hose is not twisted.
9. Tighten the mounting bolts to 74 ft. lbs. (100 Nm).

10. Install the brake pads following the recommended procedure.
11. Install the tire and wheel assembly and tighten to specification.
12. Lower the vehicle.
13. Pump the brake pedal several times to seat the pads against the rotor.

NOTE: The brake pedal must be pumped to seat the brake pads prior to operating the vehicle or the vehicle will not stop on the initial pedal application.

Brake Drums

REMOVAL AND INSTALLATION

1. Raise and safely support the vehicle.
2. Remove the tire and wheel assembly.
3. Mark the relationship of the brake drum to the axle flange so that, it the drum is reused, it can be installed in its original location. Remove the brake drum from the vehicle. Do not pry against the splash shield that surrounds the backing plate in an attempt to free the drum. This will bend the splash shield. If difficulty is encountered in removing the the brake drum, try the following.
   a. Make sure the parking brake is released.
   b. Back off the parking brake cable adjustment.
   c. Remove the access hole plug from the backing plate.
   d. Use a small brake spoon to reach in and back off the adjusting screw (sometimes called a star wheel).
   e. After loosening the adjustment, install the access hole plug to keep out dirt and debris.
   f. Use a small amount of penetrating oil applied around the drum pilot hole (on the face of the drum where the end of the axle shaft hub comes through the center of the brake drum).
To install:
4. Clean all parts well. Inspect the drum for cracks or excessive wear. If in doubt, replace the drum.
5. If it is suspected that the self-Adjuster(s) are not working, disassemble the brakes far enough to remove the adjuster. Soak the adjuster in penetrating oil and disassemble the adjuster, Clean the threads and lubricate with GM lubricant 1052196 or equivalent anti–seize compound to the adjuster screw threads, the inside diameter of the adjuster socket and the socket face.

6. Pre-Adjust the brake shoes and install the brake drum(s).

7. Properly adjust the brakes.

8. Install the tire and wheel assembly and tighten to specification.

9. Lower the vehicle.

10. Road test the vehicle and verify proper brake operation.

Brake Shoes

REMOVAL AND INSTALLATION

1993 Vehicles

1. Raise and safely support the vehicle.

2. Remove the tire and wheel assembly.

3. Remove the brake drum.

4. Remove the return springs using tool J 8049, or the equivalent.

5. Remove the hold down springs and pins using a suitable pair of pliers or hold down spring removal tool.

6. Remove the parking brake lever pivot.

7. Remove the actuator link from the top pad stop.

8. Remove the actuator lever.

9. Remove the parking brake strut and spring.

10. Remove the primary shoe, adjuster assembly and lower spring.

11. Remove the C–lock clip from the parking brake lever pivot and pull out the pivot pin.

12. Remove the secondary shoe.

To install:

13. Clean the backing plate completely and lubricate the pad contact points on the backing plate with white grease.

14. Disassemble the self adjuster and clean all components thoroughly. Lubricate the adjuster threads and re-Assemble.

15. Connect the secondary shoe to the parking brake lever and install the mounting pin and C–lock clip.

16. Install the primary shoe, adjuster assembly and lower spring onto the secondary shoe.

17. Position the assembly on the backing plate and install the primary shoe hold down pin and spring.

18. Install the parking brake strut and spring between the brake shoes so all the notches in the strut match up with the shoes.

19. Install the actuator lever and actuator link. Once the actuator lever is in place install the actuator pivot and hold down pin and spring.

20. Install the actuator lever return spring.

21. Install the brake return springs.

22. Install the brake drum.

23. Install the tire and wheel assembly and tighten to specification.

24. Adjust the rear brakes.

25. Lower the vehicle.

1994–96 Vehicles

1. Raise and safely support vehicle. Remove the tire and wheel assembly.

2. Remove the brake drum. If the drum is difficult to remove, remove the access plug from the backing plate and back off the adjusting screw.

3. Remove the return springs from the anchor using appropriate brake spring pliers.

4. Remove the hold-Down springs and retaining pins. Remove the lever pivot, actuator link, actuator lever, actuator pivot and lever return spring, parking brake strut and strut spring.

5. Remove the brake shoes, then disconnect the parking brake cable.

6. Remove the adjusting screw assembly and spring. Note position of adjusting spring.

NOTE: Do not interchange the adjusting screws or adjusting screw springs from right to left brake assembly

7. Remove the retaining ring, pin and parking brake lever from the secondary shoe.

To install:

8. Lubricate the shoe contact surfaces on the backing plate and adjusting screw assembly.

9. Install the parking brake lever on the secondary shoe with the pin and retaining ring.

10. Install the adjusting screw assembly and spring.

11. Install the shoe and lining assemblies after attaching the parking brake cable.

12. Install the parking brake strut and spring by spreading the shoes apart. Ensure the strut is properly positioned. The end with the spring engages the primary shoe and the end without the spring engages the parking brake lever.

13. Install the actuator pivot, actuator lever and return spring.

14. Install the actuator link in the shoe retainer.

15. Install the link into the lever while holding up on the lever.

16. Install the hold-Down pins, lever pivot and hold-Down spring.

17. Install the shoe return springs.

18. Install the brake drum, wheel and tire assembly, then lower vehicle. Apply the brake pedal several times to seat the brake shoes. Check and adjust the parking brake, as required.

19. Check the master cylinder reservoir and add fluid if required.

Wheel Cylinder

REMOVAL AND INSTALLATION

1993–95 Vehicles

With Spring Clip Mounting

1. Raise and safely support the vehicle.

2. Remove the tire and wheel assembly.

3. Remove the brake drum.

4. Remove the brake shoes following the recommended procedure.

5. Disconnect and cap the brake line from the wheel cylinder.

6. Using two awls, release the spring clip securing the wheel cylinder to the backing plate.

7. Remove the wheel cylinder from the vehicle.

8. On some vehicles it may be necessary to remove the bleeder screw from the wheel cylinder to remove it from the vehicle.

To Install:

9. Remove the bleeder screw from the wheel cylinder and position it in the backing plate.

10. Hold the wheel cylinder in place with a small prybar and using a 1⅛ inch socket on the end of an extension push the spring clip into place. Make sure both spring clip ears are seated correctly.

11. Connect the brake line to the wheel cylinder and tighten the fitting to 11 ft. lbs. (16 Nm).

12. Install the bleeder screw and temporarily tighten.

13. Install the rear brake shoes following the recommended procedure.

14. Install the brake drum.

15. Install the tire and wheel assembly and tighten to specification.

16. Adjust the brakes.

17. Bleed the brakes following the recommended procedure.

18. Lower the vehicle.

With Mounting Bolts

1. Raise and safely support vehicle.

2. Remove the tire and wheel assembly.

3. Remove the brake drum.

4. Remove the brake shoes following the recommended procedure.

5. Disconnect and cap the brake line from the wheel cylinder.

1. Return spring
2. Return spring
3. Hold-down spring
4. Lever pivot
5. Hold-down pin
6. Actuator link
7. Actuator lever
8. Lever return spring
9. Parking brake strut
10. Strut spring
11. Primary shoe & lining
12. Secondary shoe & lining
13. Adjusting screw spring
14. Socket
15. Pivot nut
16. Adjusting screw
17. Retaining ring
18. Pin
19. Parking brake lever
20. Bleeder valve
21. Bolt
22. Boot
23. Piston
24. Seal
25. Spring assy
26. Wheel cylinder
27. Backing plate

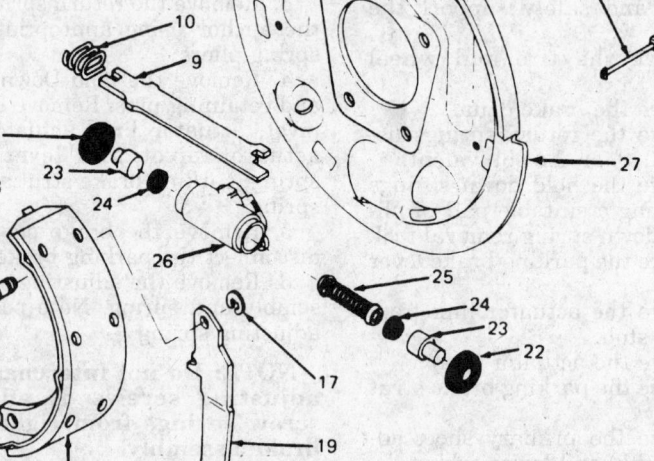

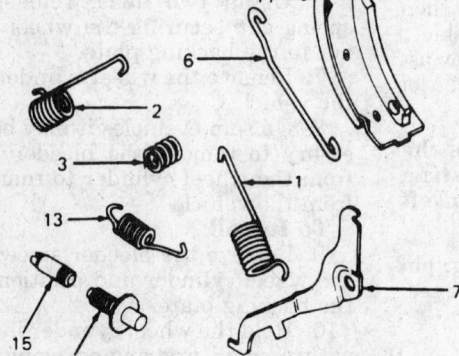

286048

Rear drum brake components — 1993 vehicles

6. Remove the two wheel cylinder mounting bolts.
7. Remove the wheel cylinder from the backing plate.
   To install:
8. Install the wheel cylinder in the backing plate and tighten the mounting bolts to 15 ft. lbs. (20 Nm).
9. Connect the brake line to the wheel cylinder and tighten the fitting to 12 ft. lbs. (17 Nm).
10. Install the rear brake shoes.
11. Install the brake drum.
12. Install the tire and wheel assembly and tighten to specification.
13. Adjust the rear brakes.
14. Bleed the brake system.
15. Lower the vehicle.

1996 Vehicles

1. Raise and safely support the vehicle.
2. Remove the wheel and tire assembly.
3. Remove the brake drum.
4. Remove the brake components.
5. Remove the bleeder valve.
6. Disconnect the brake line from the back of the wheel cylinder, and plug opening in line to prevent excess fluid loss.
7. Remove the mounting bolts from the wheel cylinder and remove the wheel cylinder.
   To install :
8. Apply Loctite® gasket maker or equivalent sealer to the wheel cylin-

der shoulder face that contacts the backing plate.
9. Position the wheel cylinder and hold in place, and install the mounting bolts.
10. Tighten the bolts to 110 inch lbs. (12 Nm).
11. Connect the brake line to the wheel cylinder and tighten to 13 ft. lbs. (17 Nm).
12. Install the bleeder valve.
13. Install the brake components.
14. Install the brake drum and wheel and tire assembly.
15. Bleed the wheel cylinder.
16. Lower the vehicle.
17. Check and refill the master cylinder as needed.

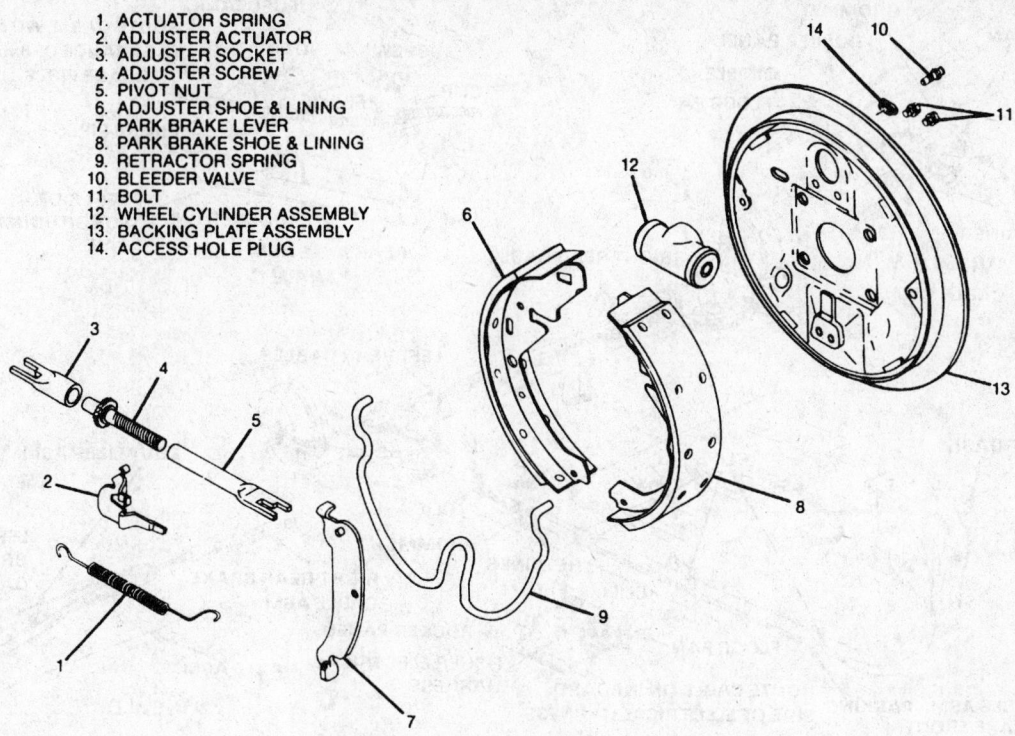

1. ACTUATOR SPRING
2. ADJUSTER ACTUATOR
3. ADJUSTER SOCKET
4. ADJUSTER SCREW
5. PIVOT NUT
6. ADJUSTER SHOE & LINING
7. PARK BRAKE LEVER
8. PARK BRAKE SHOE & LINING
9. RETRACTOR SPRING
10. BLEEDER VALVE
11. BOLT
12. WHEEL CYLINDER ASSEMBLY
13. BACKING PLATE ASSEMBLY
14. ACCESS HOLE PLUG

290775

Rear drum brake components — 1994–96 vehicles

18. Road test the vehicle and verify proper operation.

Parking Brake Cable

ADJUSTMENT

1993–94 Vehicles

NOTE: The rear brake shoes must be adjusted properly before parking brake adjustment can be performed. Failure to properly adjust the brakes can lead to overtightening the parking brake cable adjuster resulting in rear brake drag.

1. Raise and support the car with both rear wheels off the ground.
2. Depress the parking brake pedal exactly three ratchet clicks.
3. Loosen the equalizer locknut, then tighten the adjusting nut until the left rear wheel can just be turned backward using two hands, but is locked in forward rotation.
4. Tighten the locknut.
5. Release the parking brake. Rotate the rear wheels; there should be no drag.
6. Lower the car.

1995–96 Vehicles

NOTE: The rear brake shoes must be adjusted properly before parking brake adjustment can be performed. Failure to properly adjust the brakes can lead to overtightening the parking brake cable adjuster resulting in rear brake drag.

1. Raise and support the car with both rear wheels off the ground.
2. Depress the parking brake pedal exactly three ratchet clicks.
3. Loosen the equalizer locknut, then tighten the adjusting nut until the left rear wheel can just be turned backward using two hands, but is locked in forward rotation.
4. Tighten the locknut.
5. Release the parking brake. Rotate the rear wheels. There should be no drag.
6. Lower the car.

REMOVAL AND INSTALLATION

Front Parking Brake Cable

1. Raise and safely support the vehicle.
2. On the rear suspension member, loosen the nut on the parking brake cable equalizer.
3. Disconnect the front parking brake cable from the equalizer assembly.
4. Disconnect the parking brake cable from the frame clip at the left rear of the vehicle.
5. Pull the parking brake cable through the cable hanger in front of the clip mounting point.
6. Lower the vehicle.
7. Remove the driver's side sound insulator panel.
8. Remove the carpet finish moulding.
9. Pull the carpet back to uncover the parking brake cable.
10. Remove the cable retaining clip at the parking brake lever assembly.
11. Disengage the parking brake cable housing locking fingers from the parking brake assembly.
12. Disconnect the cable end from the parking brake lever assembly.
13. Remove the grommet retainer from the floor pan.
14. Pull the cable through the floor pan and remove.
To install:
15. Pull the cable through the floor pan until the grommet can be seated in the hole in the floor.
16. Install the grommet retainer on the floor pan.
17. Connect the cable end to the parking brake assembly lever.

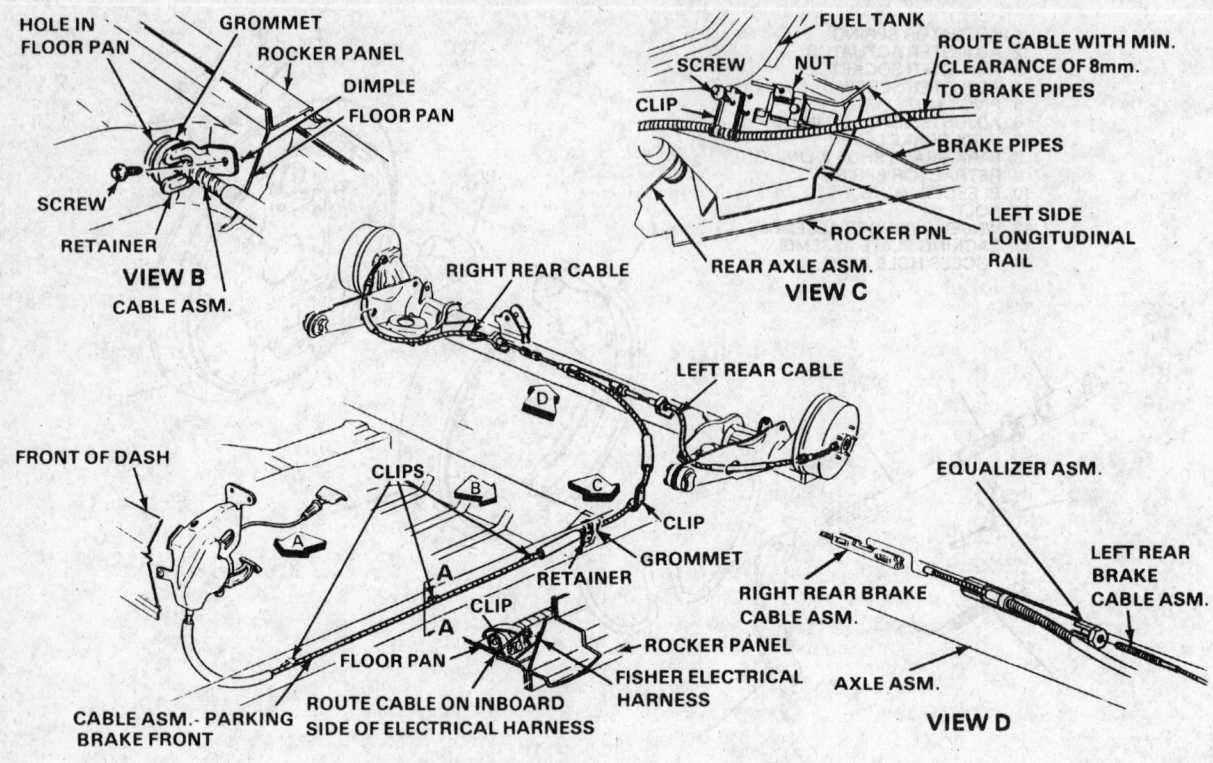

Parking brake cable adjustment — with rear drum brakes

324920

18. Seat the parking brake cable housing into the parking brake assembly until the locking fingers engage.

19. Install the cable retaining clip at the parking brake assembly.

20. Place the carpet back in place and install the carpet finish moulding.

21. Install the drivers side sound insulator cable.

22. Raise and safely support the vehicle.

23. Route the cable through the cable hanger at the rear of the vehicle.

24. Install the cable mounting clip at the left rear of the vehicle.

25. Connect the cable to the parking brake equalizer on the rear suspension member.

26. Adjust the parking brake following the recommended procedure.

27. Lower the vehicle.

Rear Parking Brake Cable

Drum Brakes

1. Raise and safely support the vehicle.

2. On the rear suspension member, loosen the nut on the parking brake cable equalizer.

3. Disconnect the front parking brake cable from the equalizer assembly.

4. Remove the tire and wheel assembly.

5. Remove the brake drum.

6. Disconnect the parking brake cable from the parking brake lever.

7. Depress the locking fingers on the parking brake cable and pull the cable out of the backing plate.

To Install:

8. Install the cable through the hole in the backing plate and seat the locking fingers in the backing plate.

9. Connect the cable end to the parking brake lever.

10. Install the brake drum.

11. Install the tire and wheel assembly and tighten to specification.

12. Connect the cable to the equalizer assembly.

13. Adjust the parking brake following the recommended procedure.

14. Lower the vehicle.

Disc Brakes

1. Raise and safely support the vehicle.

2. On the rear suspension member, loosen the nut on the parking brake cable equalizer.

3. Disconnect the front parking brake cable from the equalizer assembly.

4. Disconnect the cable from the mounting bracket on the rear axle assembly.

5. Remove the tire and wheel assembly.

6. Disconnect the parking brake cable end from the parking brake lever on the caliper and remove the return spring.

7. Disconnect the cable guide from the strut mounting bracket.

To Install:

8. Connect the cable to the mounting bracket on the rear axle assembly.

9. Connect the cable end to the equalizer assembly.

10. Connect the cable to the cable guide on the strut.

11. Connect the cable to the bracket at the caliper, seat the locking fingers completely.

12. Install the return spring on the cable and connect the cable end to the caliper lever.

13. Install the tire and wheel assembly.

14. Adjust the parking brake cable.

15. Lower the vehicle.

1. Front cable assembly
2. Right rear cable
3. Left rear cable
4. Equalizer assembly
5. Guide
6. Clip

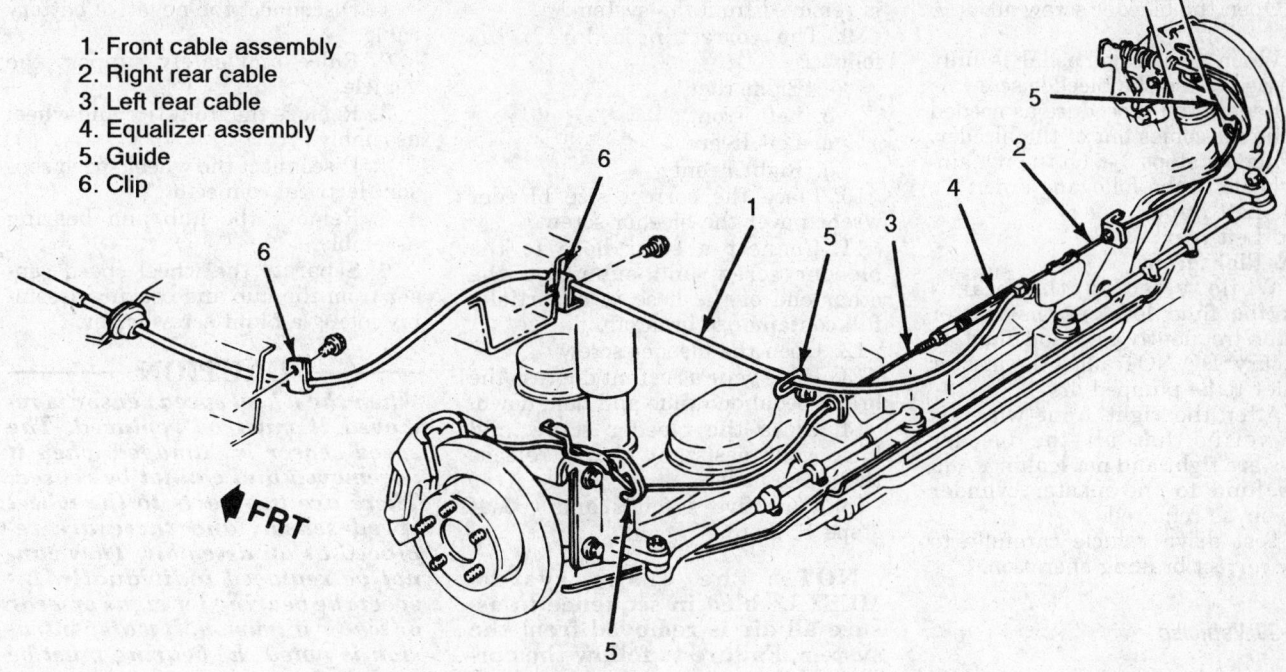

286214

Rear parking brake cable routing — with rear disc brakes

Brake Hydraulic System

BLEEDING

1993 Century

NOTE: When bleeding the brake system use care not to spill brake fluid on any painted surfaces or electrical connectors. If brake fluid is spilled on a painted surface flush the area immediately with water.

1. Remove the master cylinder cap and fill the master cylinder to the FULL mark. During the bleeding procedure the master cylinder level should be kept at least half full.

NOTE: DO NOT bleed the brake system with the master cylinder cap removed.

2. Raise and safely support the vehicle.
3. Have an assistant in the vehicle.
4. The wheels must be bled in proper sequence to assure that all air is removed from the system.
5. The correct procedure is as follows:
   a. Right rear
   b. Left Front
   c. Left Rear

d. Right Front
6. Place the correct size bleeder wrench over the bleeder screw.
7. Connect a clear hose to the bleeder screw and submerge the other end of the hose in a partially full container of brake fluid.
8. Have your assistant depress the brake pedal one time and hold down. Loosen the bleeder screw to allow the air to exit the system.
9. Tighten the bleeder screw and have your assistant slowly release the brake pedal.
10. Wait fifteen seconds and repeat steps 8 and 9.

NOTE: The brake system MUST be bled in sequence to assure all air is removed from the system. Failure to follow the correct bleeding sequence can lead to a increased stopping distances and increased pedal effort.

11. When all of the air is out of the component proceed to the next one in sequence. Check the master cylinder frequently to prevent sucking the master cylinder dry.
12. After bleeding all four wheels refill the master cylinder to the proper level.
13. Verify a firm brake pedal and no fluid leaks.

14. Road test the vehicle for brake feel and stopping effort.

1993 Cutlass Ciera and Cutlass Cruiser

NOTE: When bleeding this vehicle it is important that the correct bleeding sequence is followed. Failure to follow the correct sequence may result in air becoming trapped in the brake system. If air remains in the system decreased braking performance and increased pedal travel will result.

1. Clean off the master cylinder cap and remove the cap. Fill the master cylinder to the proper level and install the cap.
2. Have an assistant in the vehicle pump the brake pedal several times using smooth even strokes, pausing 10 seconds between each stroke. After the last stroke hold the pedal down.

NOTE: When pumping the brakes DO NOT force or stab the pedal down or damage to the master cylinder could result.

3. Starting with the right rear wheel, connect a clear rubber hose to the bleeder screw and submerge the

other end of the hose in a container to collect the brake fluid.

4. Open the bleeder screw about ½ turn.

5. When the brake pedal is fully depressed, close the bleeder screw.

6. Repeat the procedure as needed until no air comes out of the bleeder.

7. Repeat steps 3–6 on the remaining wheels in the following order:
   a. Left front
   b. Left rear
   c. Right front

8. While bleeding the brakes check the fluid level in the master cylinder frequently and add fluid as necessary. DO NOT allow the master cylinder to be pumped dry.

9. After the right front wheel is done verify that all the bleeder screws are tight and not leaking. Add brake fluid to the master cylinder reservoir as required.

10. Test drive vehicle carefully to verify correct braking operation.

1994–96 Vehicles

WARNING

DO NOT place your foot on the brake pedal through this entire procedure unless specifically directed to do so.

NOTE: When bleeding the brake system use care not to spill brake fluid on any painted surfaces or electrical connectors. If brake fluid is spilled on a painted surface flush the area immediately with water.

1. Start the engine and allow to run for ten seconds and verify the ABS warning lamp is not ON.

2. If the lamp is ON, the ABS problem must be fixed before bleeding the system.

3. If the ABS warning light came ON for three seconds the turned OFF and remained OFF, turn off the vehicle.

4. Repeat steps 1 to 3.

5. Remove the master cylinder cap and fill the master cylinder to the FULL mark. During the bleeding procedure the master cylinder level should be kept at least half full.

NOTE: DO NOT bleed the brake system with the master cylinder cap removed.

6. Raise and safely support the vehicle.

7. Have an assistant in the vehicle.

8. The wheels must be bled in proper sequence to assure that all air is removed from the system.

9. The correct procedure is as follows:
   a. Right rear
   b. Left Front
   c. Left Rear
   d. Right Front

10. Place the correct size bleeder wrench over the bleeder screw.

11. Connect a clear hose to the bleeder screw and submerge the other end of the hose in a partially full container of brake fluid.

12. Open the bleeder screw.

13. Have your assistant depress the brake pedal one time and hold down.

14. Close the bleeder screw and have your assistant slowly release the brake pedal.

15. Wait five seconds and repeat steps 11 to 13.

NOTE: The brake system MUST be bled in sequence to assure all air is removed from the system. Failure to follow the correct bleeding sequence can lead to a increased stopping distances and increased pedal effort.

16. When all of the air is out of the component proceed to the next one in sequence. Check the master cylinder frequently to prevent sucking the master cylinder dry.

17. After bleeding all four wheels refill the master cylinder to the proper level.

18. Verify a firm brake pedal and no fluid leaks.

19. Road test the vehicle for brake feel and stopping effort.

Wheel Speed Sensor

REMOVAL AND INSTALLATION

The front wheel speed sensors are of a variable reluctance type. Each sensor is attached to the knuckle assembly in close proximity to a toothed ring. This results in (as the teeth pass by the sensor) an A/C voltage with a frequency proportional to the speed of the wheel. The magnitude of the voltage and frequency increase with speed. The sensor is not repairable, or adjustable. The rear wheel speed sensors operate in the same manner as the front wheel speed sensors. If a rear wheel speed sensor fails, the entire integral hub/bearing and speed sensor assembly must be replaced.

Front Wheel Speed Sensor

1. Disconnect the negative battery cable.

2. Raise and safely support the vehicle.

3. Remove the front tire and wheel assembly.

4. Disconnect the wheel speed sensor electrical connector.

5. Remove the hub and bearing assembly.

6. Separate the wheel speed sensor from the hub and bearing assembly, using a blunt screwdriver.

CAUTION

When the wheel speed sensor is removed it must be replaced. The speed sensor is damaged when it is removed and cannot be reused. There are two parts to the wheel speed sensor, and these are replaced as an assembly. They cannot be replaced individually. Inspect the bearing for signs of wear or water intrusion. If water intrusion is noted the bearing must be replaced.

To install:

7. Apply Loctite®620 or equivalent to the mating surfaces of the wheel speed sensor that contacts the hub and bearing assembly.

8. Using special tool J-38764 or equivalent, press the wheel speed sensor on to the hub and bearing assembly.

9. Install the hub and bearing assembly to the vehicle.

10. Connect the wheel speed sensor electrical connector.

11. Install the tire and wheel assembly.

12. Lower the vehicle.

13. Connect the negative battery cable.

14. Road test the vehicle and verify proper operation.

Rear Wheel Speed Sensor

1. Disconnect the negative battery cable.

2. Raise and safely support the vehicle.

3. Remove the tire and wheel assembly.

4. Remove the brake drum.

5. Remove the bolts and nuts attaching the rear wheel bearing and speed sensor assembly.

6. Remove the wheel bearing and speed sensor assembly and disconnect the electrical connector from the speed sensor.

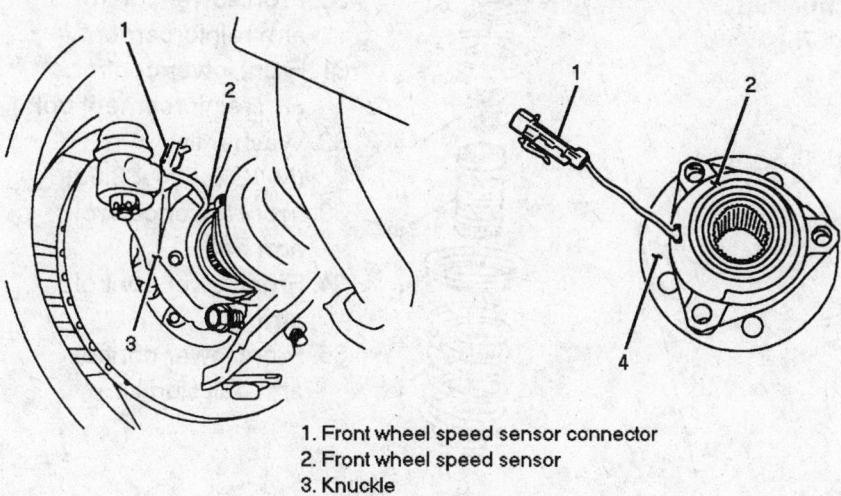

1. Front wheel speed sensor connector
2. Front wheel speed sensor
3. Knuckle
4. Hub and bearing assy

326497

Front wheel speed sensor and hub/bearing assembly

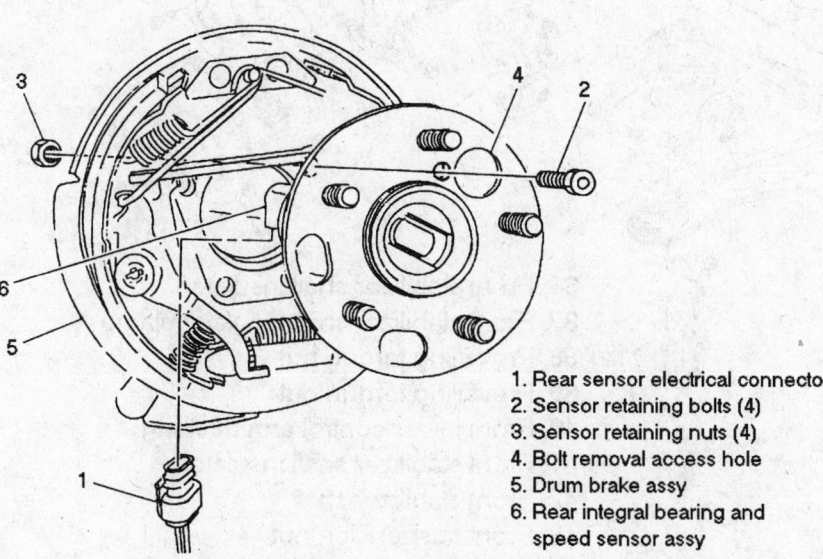

1. Rear sensor electrical connector
2. Sensor retaining bolts (4)
3. Sensor retaining nuts (4)
4. Bolt removal access hole
5. Drum brake assy
6. Rear integral bearing and speed sensor assy

326499

Rear wheel speed sensor and hub/bearing assembly

CAUTION

The brake assembly will be held in place by the brake pipe connection at this point. Use care not to bump or exert force on the brake assembly, or damage can occur to the brake pipe.

To install :

7. Connect the electrical connector to the wheel speed sensor.
8. Install the wheel bearing and speed sensor assembly into place.
9. Install the mounting bolts and nuts and tighten to 46 ft. lbs. (63 Nm).
10. Install the brake drum.
11. Install the tire and wheel assembly.
12. Lower the vehicle.
13. Connect the negative battery cable.
14. Road test the vehicle and verify proper operation.

FRONT SUSPENSION

Strut and Spring

REMOVAL AND INSTALLATION

1. Raise and safely support the vehicle.
2. Remove the tire and wheel assembly.
3. Remove the bolt securing the brake hose bracket to the strut and position the hose out of the way.
4. Matchmark the strut to the steering knuckle.
5. Remove the lower strut mounting nuts and through bolts.
6. Lower the vehicle enough to access the upper strut plate mounting bolts.
7. Remove the upper mounting nuts and remove the strut from the vehicle.

To install:

8. Install the strut into the vehicle so the upper plate studs pass through the body. Install the mounting nuts loosely.
9. Raise the vehicle.
10. Install the lower strut bracket through bolts. Install the mounting nuts and with the matchmarks in alignment tighten the nuts to 122 ft. lbs. (165 Nm).
11. Connect the brake line bracket to the strut and tighten the mounting bolt to 15 ft. lbs. (21 Nm).

1. Front wheel drive shaft kit
2. Front suspension strut mount washer
3. Front suspension strut mount nut
4. Prevailing torque nut (M 12 x 1.75)
5. Front suspension strut bolt
6. Front suspension strut nut
7. Front suspension strut mount
8. Front spring seat
9. Front suspension strut bumper
10. Front suspension strut shield
11. Front spring upper insulator
12. Front spring

29. Front wheel drive shaft kit
30. Front lower control arm reinforcement
31. Front lower control arm reinforcement bolt
32. Washer flat (M 12.2 x 24 x 3.38)
33. Front lower control arm bolt
34. Front lower control arm
35. Front lower control arm ball stud kit

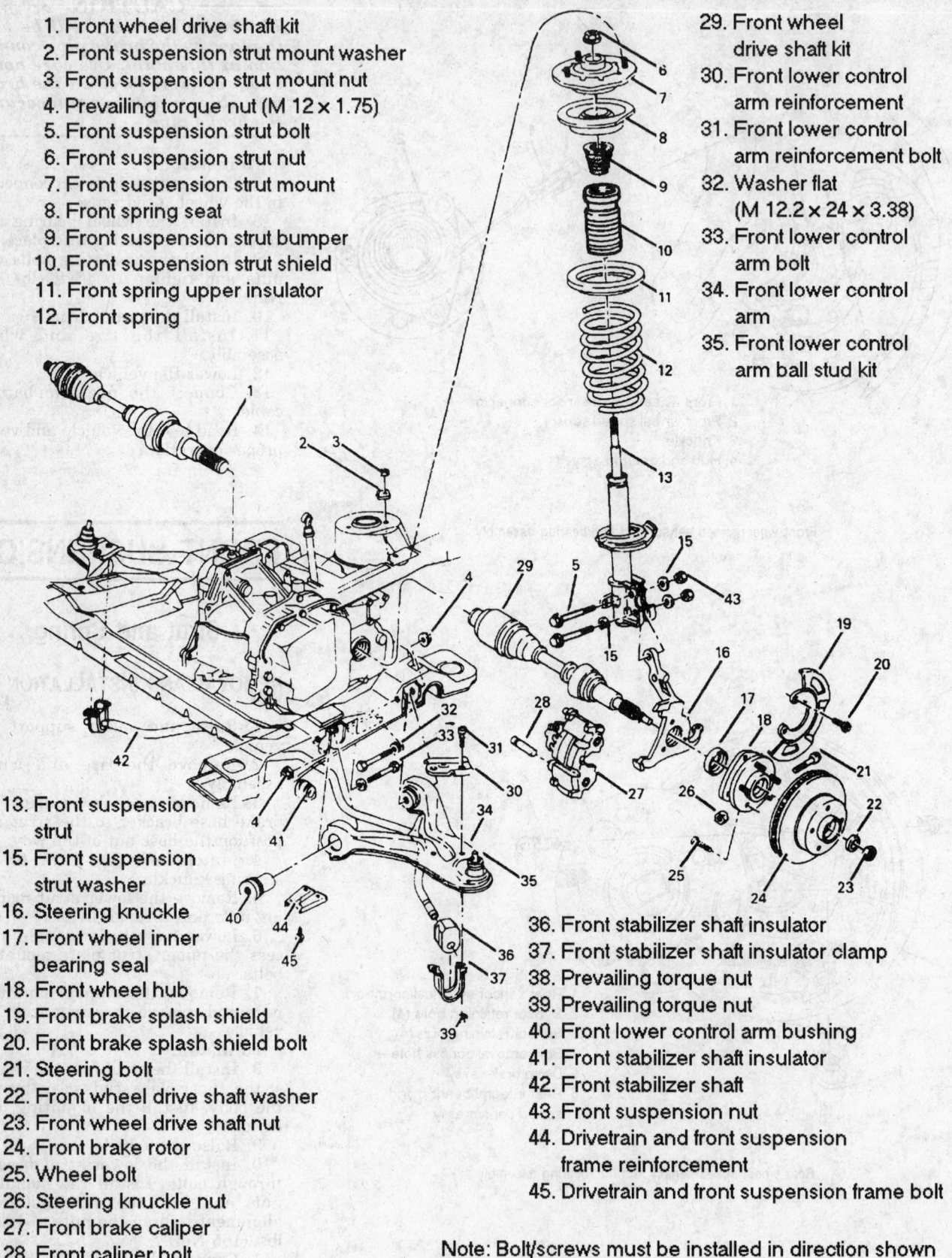

13. Front suspension strut
15. Front suspension strut washer
16. Steering knuckle
17. Front wheel inner bearing seal
18. Front wheel hub
19. Front brake splash shield
20. Front brake splash shield bolt
21. Steering bolt
22. Front wheel drive shaft washer
23. Front wheel drive shaft nut
24. Front brake rotor
25. Wheel bolt
26. Steering knuckle nut
27. Front brake caliper
28. Front caliper bolt

36. Front stabilizer shaft insulator
37. Front stabilizer shaft insulator clamp
38. Prevailing torque nut
39. Prevailing torque nut
40. Front lower control arm bushing
41. Front stabilizer shaft insulator
42. Front stabilizer shaft
43. Front suspension nut
44. Drivetrain and front suspension frame reinforcement
45. Drivetrain and front suspension frame bolt

Note: Bolt/screws must be installed in direction shown

230700

Front suspension components

12. Install the tire and wheel assembly.

13. Lower the vehicle.

14. Tighten the upper strut plate mounting bolts to 16 ft. lbs. (22 Nm).

15. Check the front end alignment and adjust as necessary.

Lower Ball Joints

REMOVAL AND INSTALLATION

1. Raise and safely support the vehicle.

2. Remove the tire and wheel assembly.

3. Remove the nut from the ball joint pinch bolt nut and remove the pinch bolt.

4. Drill a pilot hole into each of the three ball joint rivets using a 1/8 inch drill bit.

5. Using a 1/2 inch drill bit, drill off the rivet heads.

6. With a hammer and punch tap the rivet bodies out of the ball joint and control arm.

7. Pull the control arm down to disengage the ball joint stud from the steering knuckle.

8. Remove the ball joint from the control arm.

To Install:

9. Install the ball joint in the control arm and install the mounting nuts and bolts provided with the service kit. Tighten the nuts to the specifications in the service kit.

10. Connect the ball joint to the steering knuckle and install a new pinch bolt and nut. Tighten the nut to 38 ft. lbs. (52 Nm). If the replacement joint is supplied with a grease fitting, lubricate the ball joint with quality chassis grease.

11. Install the tire and wheel assembly and tighten the wheel nuts to 100 ft. lbs. (140 Nm).

12. Lower the vehicle.

Lower Control Arms

REMOVAL AND INSTALLATION

1. Raise and safely support the vehicle.

2. Remove the tire and wheel assembly.

3. Remove the outer sway bar clamp nuts from the control arm studs and remove the clamp.

4. Remove the four bolts and remove the sway bar plate at the suspension crossmember.

5. Remove the nut from the ball joint pinch bolt nut and remove the pinch bolt.

6. Pull the control arm down to disengage the ball joint stud from the steering knuckle.

7. Remove the control arm through bolts and nuts and remove the control arm from the vehicle.

To Install:

8. Position the control arm and loosely install the mounting bolts and nuts.

9. Connect the ball joint stud to the steering knuckle. Verify that the slot in the ball joint stud is aligned with the pinch bolt hole. Install the pinch bolt and a new nut and tighten to 38 ft. lbs. (52 Nm).

10. Position the sway bar and install the outboard sway bar mounting clamp and nuts. Tighten the nuts to 32 ft. lbs. (43 Nm).

11. Install the sway bar plate and tighten the four mounting bolts to 30 ft. lbs. (40 Nm).

12. Install the tire and wheel assembly and tighten the wheel nuts to 100 ft. lbs. (140 Nm).

13. Lower the vehicle.

14. With the weight on the control arm tighten the inboard control arm mounting bolts to 61 ft. lbs. (83 Nm).

Sway Bar

REMOVAL AND INSTALLATION

Before removing the sway bar place a piece of tape or a paint mark on the sway bar where the bar contacts the right side frame bushing. This mark will aid in centering the sway bar during installation.

1. Raise and support the vehicle.

2. Remove the nuts securing the sway bar clamps to the lower control arms and remove the clamps. DO NOT remove the studs from the lower control arms.

3. Remove the four bolts from the sway bar plates at the subframe and remove the plates.

4. Remove the sway bar. If the sway bar is being replaced remove the bushings. **To install:**

5. Install the the bushings on the sway bar with the slits toward the front of the vehicle. Make sure the right side bushing is against the matchmark on the bar.

6. Install the sway bar in the vehicle and install the mounting plates with the bolts finger tight.

7. Install the sway bar-to-control arm clamps and tighten the mounting nuts to 32 ft. lbs. (43 Nm).

8. Tighten the mounting plate bolts to 40 ft. lbs. (55 Nm).

9. Lower the vehicle.

Front Wheel Bearings

REMOVAL AND INSTALLATION

The front wheel bearings on these vehicles are not serviced separately. The wheel bearings are pressed into the hub. If the wheel bearings are noisy or defective, the hub and bearing assembly must be replaced.

1. Disconnect the negative battery cable.

2. Raise and safely support the vehicle.

3. Remove the front tire and wheel assembly.

4. Matchmark the lower strut bracket to the steering knuckle.

5. To keep the hub and rotor from turning when removing the axle nut, iInsert a drift punch through the caliper and into the brake rotor cooling fins to lock the assembly.

6. Remove the axle nut and washer. Remove the drift.

7. Remove the caliper mounting bolts and remove the caliper from the steering knuckle and support it out of the way. DO NOT allow the caliper to hang unsupported from the brake hose.

8. Remove the brake rotor.

9. Remove the three hub and bearing assembly mounting bolts and remove the backing plate.

10. Disconnect the ABS speed sensor, if equipped.

11. Using a puller, separate the axle from the hub bearing assembly. Remove the hub bearing assembly.

12. If the steering knuckle must also be removed, continue this procedure. Remove the cotter pin and castle nut from the outer tie rod end and using a steering linkage puller, separate the tie rod from the steering knuckle.

13. Remove the strut mounting nuts and through bolts.

14. Remove the ball joint pinch bolt and lift the steering knuckle off of the ball joint.

To install:

15. If removed, install the steering knuckle onto the lower ball joint and install a new pinch bolt and nut and tighten to 38 ft. lbs. (52 Nm).

16. Connect the steering knuckle to the lower strut bracket and install the through bolts and nuts. With the matchmarks in alignment, tighten the nuts to 122 ft. lbs. (165 Nm).

17. Connect the tie rod end to the steering knuckle and tighten the castle nut to 30 ft. lbs. (41 Nm). If necessary to align the cotter pin holes tighten the castle nut up to 60° additional. Install a new cotter pin.

18. Install the hub bearing onto the steering knuckle and position the backing plate. Make sure the ABS sensor is properly routed.

19. Install the hub bearing mounting bolts and tighten to 63 ft. lbs. (85 Nm).

20. Install the brake rotor.

21. Install the caliper onto the steering knuckle and tighten the mounting bolts to 38 ft. lbs. (52 Nm).

22. Insert a drift punch into the rotor cooling fins. Install the washer and axle nut onto the axle and tighten to 103 ft. lbs. (140 Nm) plus 20 degrees additional rotation.

23. Install the tire and wheel assembly.

24. Lower the vehicle.

25. Connect the negative battery cable.

26. Check the front end alignment and adjust as necessary.

REAR SUSPENSION

Shock Absorber

REMOVAL AND INSTALLATION

NOTE: If the vehicle is equipped with Super Lift® shocks, bleed the air out of the system before disconnecting the air lines.

1. Raise and safely support the vehicle at a height where the upper shock plate mounting nuts in the trunk can be reached.

2. Support the rear axle with a jack.

3. Pull the trunk side trim away from the rear shock tower.

4. Remove the cap from the top of the shock and remove the two shock plate mounting nuts.

5. Disconnect the air line from the shock absorber, if equipped with Super Lift® shocks.

6. Remove the lower shock through bolt and nut and remove the shock assembly from the vehicle.

7. Remove the center shock nut and washer and remove the shock plate from the shock.

To install:

8. Install the shock plate on the new shock absorber and install the washer and mounting nut. Tighten the nut to 21 ft. lbs. (28 Nm).

9. Position the shock absorber in the vehicle so the upper plate studs line up with the holes in the shock tower. Loosely install the mounting nuts.

10. Install the lower shock through bolt and nut and tighten to 50 ft. lbs. (72 Nm).

11. Remove the jack under the suspension and lower the vehicle.

12. Tighten the shock plate mounting nuts to 18 ft. lbs. (25 Nm).

13. Install the cap on the top of the shock and position the trunk side trim in place.

Coil Spring

REMOVAL AND INSTALLATION

1. Raise and safely support the vehicle.

2. Support the rear axle with a suitable jack.

3. Remove the left and right side brake line bracket attaching screws and allow the brake lines to hang freely.

4. Remove the lower track arm bolt from the suspension and position the bar out of the way.

5. Remove the lower shock absorber mounting bolts.

6. Lower the rear axle.

7. Remove the coil springs and insulators.

To install:

POSITION RIGHT HAND AND LEFT HAND REAR SPRINGS SUCH THAT THE LOWER PIGTAIL (END OF SPRING) IS POINTING TOWARD THE REAR OF VEHICLE.

231164

Proper alignment of the rear springs

8. Install the springs and insulators in the vehicle and orient the springs so the lower spring ends point toward the rear of the vehicle.

9. Raise the rear axle into place and install the lower shock mounting bolts and nuts. Tighten the nuts to 53 ft. lbs. (72 Nm).

10. Connect the track arm to the suspension and install the mounting nut and bolt and tighten to 53 ft. lbs. (72 Nm). Position the brake line mounting brackets and tighten the mounting bolts.

11. Position the brake line mounting brackets and tighten the mounting bolts.

12. Remove the jack from under the suspension.

13. Lower the vehicle.

Wheel Bearings

REMOVAL AND INSTALLATION

The rear wheel bearings on this vehicle are not serviced separately. The wheel bearings are pressed into the rear hub. If the wheel bearings are noisy or defective, the hub and bearing assembly must be replaced.

1. Disconnect the negative battery cable.

2. Raise and safely support the vehicle.

3. Remove the tire and wheel assembly.

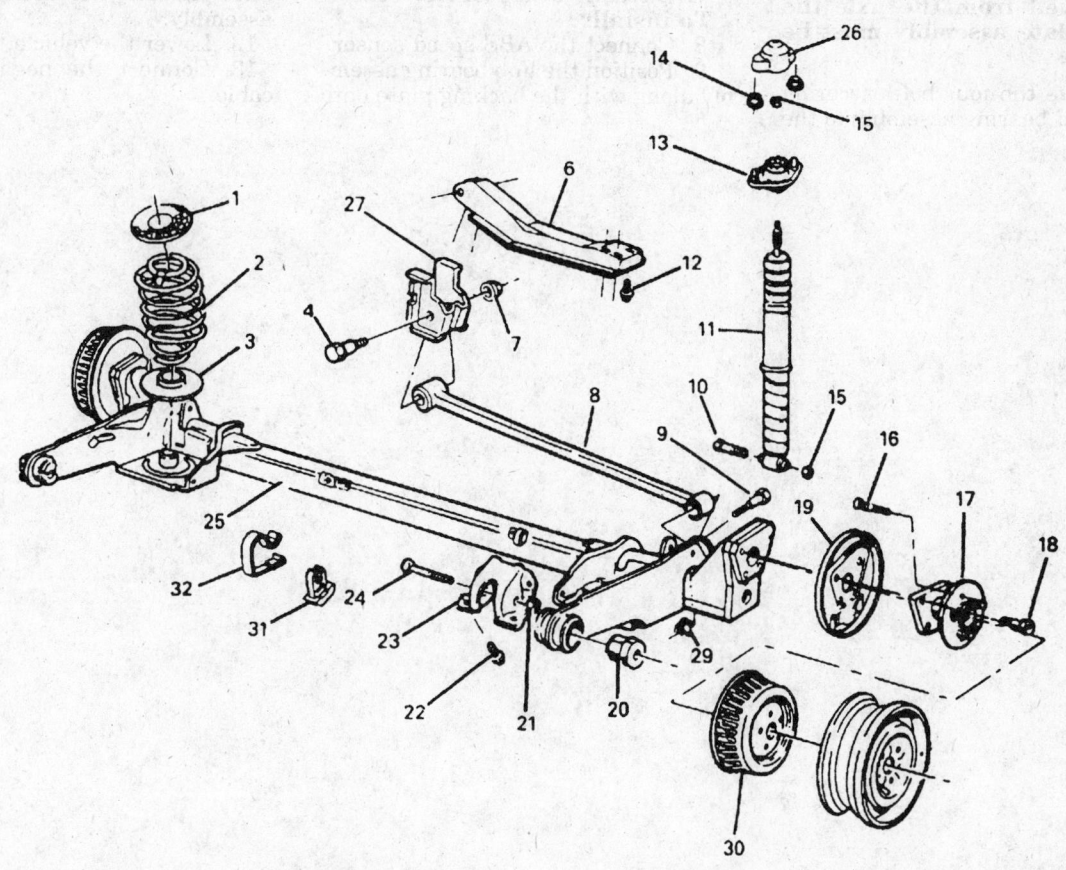

1 REAR SPRING UPPER INSULATOR	**17** HUB AND BEARING ASSEMBLY, REAR WHEEL
2 REAR SPRING	**18** BOLT/SCREW REAR WHEEL HUB AND BEARING
3 REAR SPRING LOWER INSULATOR	**19** PLATE, REAR BRAKE BACKING
4 BOLT/SCREW, REAR AXLE TIE ROD BRACKET BRACE	**20** BUSHING, REAR SUSPENSION CONTROL ARM
6 BRACE, REAR AXLE TIE ROD BRACKET (WAGON)	**21** NUT, REAR SUSPENSION CONTROL ARM BRACKET
7 NUT, REAR AXLE TIE ROD BRACKET BRACE	**22** BOLT/SCREW, REAR SUSPENSION CONTROL ARM BRACKET
8 ROD, REAR AXLE TIE (TRACK BAR)	**23** BRACKET, REAR SUSPENSION CONTROL ARM
9 BOLT/SCREW REAR AXLE TIE ROD	**24** BOLT/SCREW, REAR SUSPENSION CONTROL ARM
10 BOLT/SCREW, REAR SHOCK ABSORBER	**25** AXLE, REAR
11 ABSORBER, REAR SHOCK	**26** COVER, REAR SHOCK ABSORBER UPPER
12 BOLT/SCREW, REAR AXLE TIE ROD BRACKET BRACE	**27** BRACKET, REAR AXLE TIE ROD
13 MOUNT, SHOCK ABSORBER UPPER	**29** NUT, REAR AXLE TIE ROD
14 NUT, REAR SHOCK ABSORBER UPPER MOUNT	**30** DRUM, REAR BRAKE
15 NUT, REAR SHOCK ABSORBER	**31** INSULATOR, REAR STABILIZER SHAFT
16 STUD, WHEEL	**32** CLAMP, REAR STABILIZER SHAFT INSULATOR

292713

Rear axle, suspension, wheel bearing and hub assembly and related components

4. Remove the brake drum.

NOTE: When the hub bearing mounting bolts and hub bearing are removed from the axle the backing plate assembly must be supported.

5. Remove the four bolts securing the hub and bearing assembly to the axle.

6. Remove the hub and bearing assembly and support the backing plate.

7. Disconnect the rear ABS sensor.

To install:

8. Connect the ABS speed sensor.

9. Position the hub bearing assembly along with the backing plate onto the axle. Install and tighten the mounting bolts to 60 ft. lbs. (82 Nm).

10. Install the brake drum.

11. Install the tire and wheel assembly.

12. Lower the vehicle.

13. Connect the negative battery cable.

GM "C AND H" BODY

Front Wheel Drive

BUICK-LeSabre • Park Avenue **CADILLAC**-DeVille • Sixty Special
OLDSMOBILE-Eighty-Eight Royale • Ninety-Eight **PONTIAC**-Bonneville

18

FIRING ORDERS

NOTE: To avoid confusion, always replace spark plug wires one at a time.

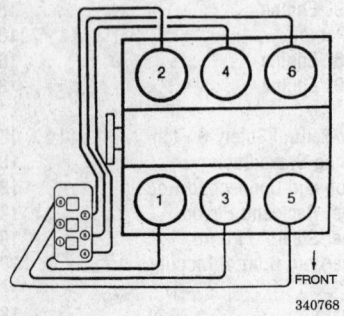

340768

**3.8L (VIN L, VIN K and VIN 1) Engines
Engine Firing Order: 1–6–5–4–3–2
Distributorless Ignition System**

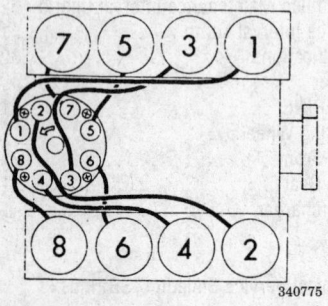

340775

**4.9L (VIN B) Engine
Engine Firing Order:
1–8–4–3–6–5–7–2
Distributor Rotation:
Counterclockwise**

ENGINE ELECTRICAL

NOTE: Disconnecting the negative battery cable on some vehicles may interfere with the functions of the on board computer systems and may require the computer to undergo a relearning process, once the negative battery cable is reconnected.

Distributor

REMOVAL AND INSTALLATION

4.9L (VIN B) Engine

1. Disconnect the negative battery cable.
2. Disconnect the battery feed wire from the distributor cap.
3. Detach the coil connections from the distributor cap.
4. Remove the 4 distributor cap mounting bolts and move the cap out of the way. It is not necessary to disconnect the ignition wires.
5. Detach the 6-terminal ECM connector from the distributor.
6. Mark the position of the rotor with respect to the distributor housing and mark the position of the distributor housing with respect to the engine.
7. Remove the distributor hold-down nut and hold-down clamp.
8. Remove the distributor.

NOTE: Do not crank the engine after the distributor is removed. If the engine is accidentally cranked, the engine will have to be positioned at No. 1 cylinder TDC.

To install:
9. Install the distributor, aligning the marks made during removal.
10. If the engine was accidentally cranked after the distributor was removed, proceed as follows:
   a. Remove the No. 1 cylinder spark plug.
   b. Place a finger over the No. 1 cylinder spark plug hole and crank the engine slowly until compression is felt.
   c. Align the timing mark on the crankshaft pulley to **0** on the engine timing indicator.
   d. Position the distributor rotor to point towards the No. 1 spark

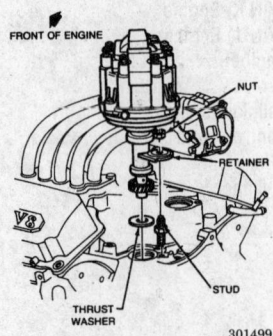

FRONT OF ENGINE

NUT

RETAINER

STUD

THRUST WASHER

301499

Distributor and mounting components — 4.9L (VIN B) engines

plug wire terminal on the distributor cap.
   e. Install the distributor, aligning the mark on the distributor housing with the mark on the engine.
   f. Install the spark plug in the No. 1 cylinder.
11. Install the distributor hold-down clamp and nut. DO NOT tighten the nut at this time.
12. Attach the 6-terminal connector. Install the distributor cap and mounting screws.
13. Connect the coil leads to the cap and the battery feed wire to the distributor cap.
14. Connect the negative battery cable.
15. Start the engine, bring the engine up to operating temperature and jumper pins A to B at the diagnostic connector under the dash board.
16. Set the ignition timing.
17. Tighten the distributor hold-down nut to 20 ft. lbs. (27 Nm).
18. Remove the jumper from the diagnostic connector and road test for proper operation.

Ignition Timing

ADJUSTMENT

3.8L (VIN L, VIN K, and VIN 1) Engines

The ignition timing is not adjustable, and is set according to engine demand electronically. The Powertrain Control Module (PCM) controls the ignition timing for all driving conditions.

4.9L (VIN B) Engine

NOTE: Check the Vehicle Emission Control Information (VECI) label, located in the engine compartment. If the procedure below differs from that described on the label, follow the instructions on the label. The information on the VECI label often reflects changes made during production.

1. With the ignition **OFF**, connect the inductive pickup from the timing light to the number one spark plug wire. DO NOT pierce any ignition wires or install adapters between the spark plug and ignition wire. Connect the power leads for the timing light according to the manufacturers directions.
2. Jumper pins A and B together at the Assembly Line Diagnostic Link

(ALDL) connector while not in diagnostic display.

3. Start the engine and aim the timing light at the timing indicator and crankshaft pulley. Refer to the VECI label for ignition timing specifications.

4. If adjustment is required, loosen the distributor hold-down bolt and rotate the distributor until the timing marks are in alignment.

5. Tighten the hold-down bolt and check the timing to make sure the distributor didn't move.

6. Turn OFF the engine and disconnect the timing light.

Alternator

PRECAUTIONS

Several precautions must be observed with alternator equipped vehicles to avoid damage to the unit.

- If the battery is removed for any reason, make sure it is reconnected with the correct polarity. Reversing the battery connections may result in damage to the 1-way rectifiers.
- When utilizing a booster battery as a starting aid, always connect the positive to positive terminals and the negative terminal from the booster battery to a good engine ground on the vehicle being started.
- Never use a fast charger as a booster to start vehicles.
- Disconnect the battery cables when charging the battery with a fast charger.
- Never attempt to polarize the alternator.
- Do not use test lights of more than 12 volts when checking diode continuity.
- Do not short across or ground any of the alternator terminals.
- The polarity of the battery, alternator and regulator must be matched and considered before making any electrical connections within the system.
- Never separate the alternator on an open circuit. Make sure all connections within the circuit are clean and tight.
- Disconnect the battery ground terminal when performing any service on electrical components.
- Disconnect the battery if arc welding is to be done on the vehicle.

REMOVAL AND INSTALLATION

3.8L (VIN L) Engine

1. Disconnect the negative battery cable.
2. Use a suitable wrench to move the automatic belt tensioner and release the tension on the drive belt. Remove the serpentine drive belt from the alternator pulley.
3. Detach the alternator connector and remove the nut securing the battery wire to the alternator stud.
4. Remove the 2 alternator mounting bolts.
5. Remove the alternator from the mounting bracket.

To install:

6. Install the alternator in the mounting bracket.
7. Install the remaining components in the reverse order of removal. Tighten the mounting bolts to 20 ft. lbs. (27 Nm) and the battery wire nut to 71 inch lbs. (8 Nm).
8. Connect the negative battery cable. Run the engine and check for proper charging system operation.

1993–95 3.8L (VIN 1) Engine

1. Disconnect the negative battery cable.
2. Use a suitable wrench to move the automatic belt tensioner and release the drive belt tension. Remove the serpentine drive belt from the alternator pulley.
3. Disconnect the electrical connector from the alternator and remove the nut securing the battery wire to the alternator stud.
4. Remove the nut and bolt, and the through bolt from the alternator.
5. Remove the alternator from the mounting bracket.

To install:

6. Install the alternator in the mounting bracket.
7. Install the remaining components in the reverse order of removal. Tighten the alternator bolts and through bolt to to 21 ft. lbs. (29 Nm). Tighten the battery wire nut to 66 inch lbs. (7.5 Nm).
8. Connect the negative battery cable. Run the engine and check for proper charging system operation.

3.8L (VIN K) and 1996–97 3.8L (VIN 1) Engines

1. Disconnect the negative battery cable.

2. Using a 15mm box end wrench, push down on the automatic belt tensioner to release the tension on the drive belt. Remove the belt from the alternator pulley.
3. If equipped, remove the engine dress-up cover.
4. Disconnect the battery cable from output terminal nut and the electrical connections from the alternator.
5. Remove the 2 nuts and the bracket from the front of the alternator.
6. Remove the alternator mounting bolts.
7. Remove the rear pencil brace.
8. Remove the alternator.

To install:

9. Install the alternator to the vehicle.
10. Install the remaining components in the reverse order of removal. Tighten the pencil brace bolts to 22 ft. lbs. (30 Nm), the front mounting bolts and brace nuts to 37 ft. lbs. (50 Nm), and tighten the battery cable nut to 15 ft. lbs. (20 Nm).
11. Connect the negative battery cable.
12. Start the engine and check for proper charging system operation.

4.9L (VIN B) Engine

1. Disconnect the negative battery cable.
2. Disconnect the 3 terminal connector and battery charging wire from the rear of the alternator.
3. Relieve the tension on the drive belt tensioner and slip the belt off the alternator pulley. It is not necessary to remove the belt from any other accessory pulleys.
4. Remove the 3 alternator mounting bolts and remove the alternator from the mounting bracket.

To install:

5. Install the alternator in the alternator mounting bracket and install the bolts. Tighten the front bolts to 32 ft. lbs. (44 Nm) and the rear bolt to 20 ft. lbs. (27 Nm).
6. While raising the automatic belt tensioner, slip the drive belt over the alternator pulley. Release the tensioner.
7. Connect the battery feed wire and tighten to 15 ft. lbs. (20 Nm). Install the 3 terminal connector to the rear of the alternator.
8. Connect the negative battery cable and test the charging system for proper operation.

Drive Belt

REMOVAL AND INSTALLATION

3.8L (VIN L and VIN K) Engines

1. Disconnect the negative battery cable.
2. Rotate the tensioner assembly counterclockwise using an 18mm (1993–95 vehicles) or 15mm (1996–97 vehicles) box end wrench on the tensioner pulley nut.
3. Slip the belt off of the alternator pulley and slowly release the tensioner.
4. Remove the wrench from the pulley and remove the belt from the remainder of the pulleys.
5. Inspect the belt for wear and/or damage and replace as necessary.

To install:
6. For 1993–95 vehicles, install and route the serpentine belt as follows:
   a. Loop the belt under the crankshaft.
   b. Bring the front of the belt up and around the water pump pulley so the back of the belt drives the pulley.
   c. Loop the belt under the A/C compressor.

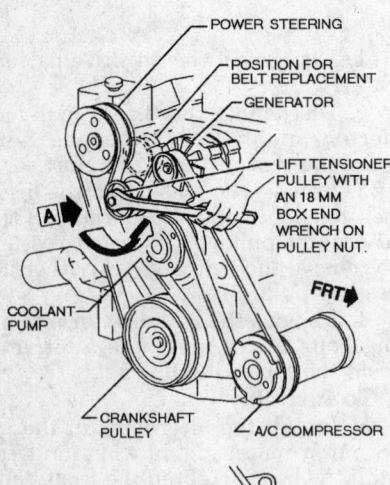
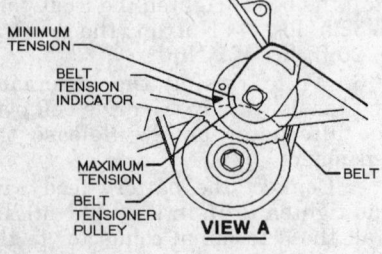

Serpentine drive belt routing — 1993–95 3.8L (VIN L and VIN K) engines

250197

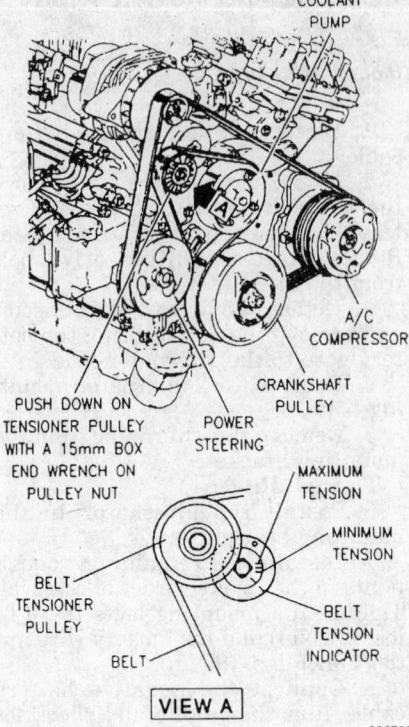

VIEW A

336708

Drive belt routing — 1996–97 3.8L (VIN K) engines

   d. Bring the rear of the belt over the power steering pump then under the tensioner assembly so the back of the belt drives the tensioner pulley.
7. For 1996–97 vehicles, install the belt and route as follows:
   a. Loop the belt under the crankshaft pulley.
   b. Bring the rear of the belt over the tensioner so the back of the belt drives the pulley.
   c. Loop the belt around the power steering pump.
   d. Take the front of the belt around the idler pulley so the back of the belt drives the pulley.
8. Install the wrench over on the tensioner nut and rotate the tensioner counterclockwise.
9. Route the belt over the alternator pulley.
10. Slowly release the tensioner assembly and verify the belt is properly seated in all the pulleys.
11. Connect the negative battery cable.
12. Run the engine and check operation.

3.8L (VIN 1) Engine

1. Disconnect the negative battery cable.

2. For 1993–95 vehicles, use an 18mm wrench on the rear tensioner pulley mounting bolt to rotate the pulley clockwise.
3. For 1996–97 vehicles, use a 15mm wrench on the rear tensioner pulley mounting bolt to rotate the pulley counterclockwise.
4. Slip the belt off of the alternator pulley and slowly release the tensioner.
5. Remove the wrench, then remove the belt from the remainder of the pulleys.
6. Install an 18mm (1993–95 vehicles) or 15mm (1996–97 vehicles) wrench on the front tensioner pulley mounting bolt and rotate the pulley clockwise.
7. For 1996–97 vehicles, install a 15mm wrench on the front tensioner pulley mounting bolt and rotate the pulley clockwise.
8. Slip the belt off the supercharger pulley and slowly release the tensioner.
9. Remove the belt from the remainder of the pulleys.
10. Inspect the belt for wear and/or damage and replace as necessary.

To install:
11. Install the back belt and route the belt as follows:
   a. Loop the belt under the crankshaft pulley.
   b. Bring the front of the belt around the water pump so the back of the belt drives the pump.
   c. Loop the belt under the A/C compressor.
   d. Bring the rear of the belt over the idler and under the tensioner.
12. Rotate the tensioner clockwise and loop the belt over the supercharger pulley.
13. Install the front belt and route as follows:
   a. Loop the belt under the crankshaft pulley.
   b. Bring the rear of the belt over the tensioner so the back of the belt drives the pulley.
   c. Loop the belt around the power steering pump.
   d. Take the front of the belt around the idler pulley so the back of the belt drives the pulley.
14. Rotate the tensioner clockwise for 1993–95 vehicles or counterclockwise for 1996–97 vehicles, and loop the belt over the alternator pulley.
15. Connect the negative battery cable. Run the engine and check operation.

4.9L (VIN B) Engine

1. Disconnect the negative battery cable.

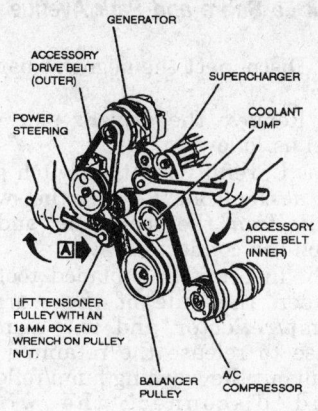

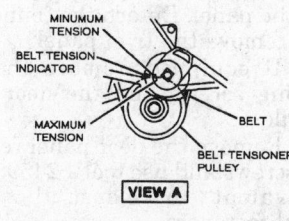

VIEW A

250195

Serpentine belt routing and tensioner locations — 1993-95 3.8L (VIN 1) engines

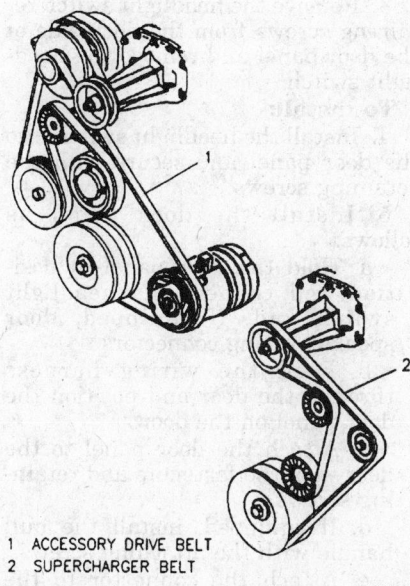

1 ACCESSORY DRIVE BELT
2 SUPERCHARGER BELT

291375

Drive belt routings — 1996-97 3.8L (VIN 1) engines

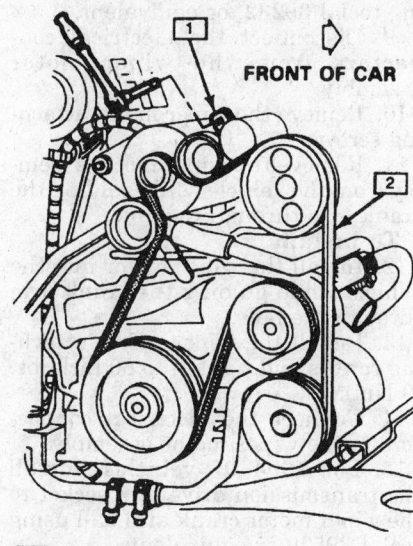

⇨ **FRONT OF CAR**

1 DRIVE BELT TENSIONER
2 SERPENTINE DRIVE BELT

241860

Drive belt routing — 4.9L (VIN B) engines

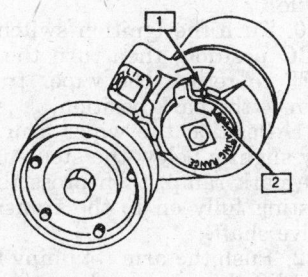

1 CAST ARROW
2 RANGE INDICATOR

241861

Automatic belt tensioner — 4.9L (VIN B) engines

2. Using a ½ inch breaker bar, rotate the tensioner mechanism upward off the drive belt.

3. Remove the drive belt.

4. Inspect the belt for wear and/or damage; replace as necessary.

To install:

5. Install the drive belt loosely making sure it is aligned properly with all the pulley grooves.

6. Rotate the tensioner mechanism upward and slide the belt under the tensioner.

7. Release the tensioner.

8. Connect the negative battery cable. Run the engine and check operation.

Starter

REMOVAL AND INSTALLATION

3.8L (VIN L, VIN K, and VIN 1) Engines

1. Disconnect the negative battery cable. Raise and safely support the vehicle.

2. Remove the 4 flywheel inspection cover retaining screws and remove the cover.

3. If equipped, remove the starter motor shielding.

4. Remove the starter motor bolts and carefully move the starter motor away from the block.

5. Disconnect the wiring from the starter and solenoid assembly.

To install:

6. With the starter motor and flywheel cover removed, inspect the teeth on the engine flywheel and the electrical connections.

7. Connect the starter wiring and tighten the solenoid battery terminal nut to 12 ft. lbs. (16 Nm) and the solenoid S terminal nut to 22 inch lbs. (2.5 Nm).

8. Install the starter to the engine block and tighten the starter mounting bolts to 32 ft. lbs. (43 Nm).

9. If equipped, install the starter motor shielding.

10. Install the flywheel inspection cover and tighten the mounting bolts to 48–62 inch lbs. (6–8 Nm).

11. Lower the vehicle.

12. Connect the negative battery cable and check the starter for proper operation.

4.9L (VIN B) Engine

1. Disconnect the negative battery cable. Raise and safely support the vehicle.

2. Remove the flywheel covers.

3. Disconnect the solenoid wires and battery feed cable.

4. Remove the starter-to-engine bolts, then remove the starter.

To install:

5. Install the starter and starter-to-engine mounting bolts.

6. Tighten the starter mounting bolts to 33 ft. lbs. (45 Nm).

7. Connect the solenoid wires to the starter.

8. Connect the battery feed cable to the starter and tighten the mounting nut to 13 ft. lbs. (17 Nm).

9. Install the flywheel covers.

10. Lower the vehicle.
11. Connect the negative battery cable. Check starter operation.

CHASSIS ELECTRICAL

Blower Motor

REMOVAL AND INSTALLATION

1. Disconnect the negative battery cable.
2. Remove the nuts and cross brace bar, if equipped.
3. Detach the blower motor connector. Remove the cooling tube from the blower motor.
4. Remove the screws attaching the blower motor to the heater and A/C module assembly.
5. Remove the blower motor and fan from the vehicle.
To install:
6. Install the blower motor into the heater and A/C module assembly. Position the blower motor and install the attaching screws. Tighten the attaching screws to 5 inch lbs. (0.6 Nm).
7. Install the remaining components in the reverse order of removal. Tighten the cross brace bar nuts to 18 ft. lbs. (25 Nm).
8. Connect the negative battery cable.

Windshield Wiper Motor

REMOVAL AND INSTALLATION

1. Turn the ignition switch to the **ACC** position.
2. Turn the windshield wipers ON. When the wiper arm assembly is at mid-wipe position, turn the ignition switch **OFF**.
3. Disconnect the negative battery cable.
4. Lift the wiper blade from the windshield and disengage the wiper arm retaining latch using a small prying tool. Remove the arm from the wiper transmission drive shaft.
5. Disconnect the washer hoses.
6. Remove the air inlet screen assembly.
7. For 1993 vehicles, remove the link from the wiper motor crank arm by loosening the 2 drive link adjustment nuts.

8. For 1994–97 vehicles, remove the transmission drive link socket from the wiper motor crank arm using tool J 39232, or equivalent.
9. Disconnect the electrical connectors from the wiper motor assembly.
10. Remove the wiper motor attaching screws.
11. Remove the wiper motor assembly from the vehicle while guiding the crank arm through the hole.
To install:
12. Install the wiper motor into the vehicle while guiding the crank arm through the hole.
13. Install the wiper motor attaching screws and tighten to 80 inch lbs. (9 Nm).
14. Connect the electrical connectors to the wiper motor assembly.
15. For 1994–97 vehicles, install the transmission drive link socket to the wiper motor crank arm ball using tool J 39529 or equivalent.
16. For 1993 vehicles, install the drive link to the wiper motor crank arm. Tighten the nuts to 71 inch lbs. (8 Nm).
17. Install the air inlet screen assembly.
18. Connect the washer hoses.
19. Connect the negative battery cable.
20. Turn the ignition switch to the **ACC** position, then turn the wipers OFF to return the wiper transmission to the park position.
21. Position the wiper arm assembly slightly below the stop surface of the park ramp, then press the head casting fully on to the transmission drive shaft.
22. Push the arm retaining latch in and lift the wiper arm assembly over the park ramp.
23. Operate the wipers and check the wipe pattern. When the wipers are turned OFF, the wiper arms must be against the park ramps.

Headlight Switch

REMOVAL AND INSTALLATION

— **CAUTION** —
The Supplemental Restraint System (SRS) or Supplemental Inflatable Restraint (SIR) system must be disarmed before working around the air bag or SRS/SIR wiring. Failure to do so may cause accidental deployment of the air bag, resulting in unnecessary SRS/SIR repairs and/or personal injury.

Buick Le Sabre and Park Avenue

1. Disconnect the negative battery cable.
2. Remove the driver's side door panel as follows:
   a. Carefully pry the switch plate up at the front. Remove the switch plate from the door panel and detach the connectors.
   b. Insert a thin bladed tool between the side of the warning lamp/reflector and the armrest base to release the retaining tab. Pull out the warning lamp/reflector and disconnect the wiring connector.
   c. Remove the lock control lever trim panel retaining screw and pull out the panel. Detach the connector and remove the trim panel.
   d. If equipped, remove the retaining screws and the door pull handle.
   e. Remove the door panel retaining screws and use tool J 24595C or equivalent, to release the door panel fasteners.
   f. Feed the wiring harness through the door and pull the door panel away from the door slightly. Disconnect the headlight switch and, if equipped, door speaker wiring connectors and remove the door panel.
3. Remove the headlight switch retaining screws from the back side of the door panel and remove the headlight switch.
To install:
4. Install the headlight switch into the door panel and secure with the retaining screws.
5. Install the door panel as follows:
   a. Hold the door panel in position and connect the headlight switch and, if equipped, door speaker wiring connectors.
   b. Feed the wiring harness through the door and position the door panel on the door.
   c. Attach the door panel to the door with the fasteners and retaining screws.
   d. If equipped, install the pull handle with the attaching screws.
   e. Attach the connector to the lock control lever trim panel and secure the panel with the retaining screw.
   f. Attach the connector to the warning lamp/reflector and install the warning lamp/reflector to the door panel.
   g. Attach the connector and install the switch plate.

6. Connect the negative battery cable. Check headlight switch operation.

Oldsmobile Eighty Eight and Ninety Eight

1993–94 Models

1. Properly disarm the SIR air bag system, if equipped. Disconnect the negative battery cable.

2. Remove the steering column lower cover or the instrument panel trim plate covering the headlight switch, if equipped with a rocker-type headlight switch.

3. Disconnect the electrical harness retainer below headlight switch assembly. The switch connector is integral to the instrument panel. Pull the switch outward to disconnect it.

4. Remove screw with ground wire at bottom of switch housing and all other mounting screws.

5. Pull assembly down and rearward, disconnect wiring harness connectors, bulb(s) and remove assembly.

To install:

6. Connect the wiring connectors, bulb(s) and install the assembly.

7. Install the screw with ground wire at the bottom of switch housing and all other mounting screws.

8. Attach the electrical harness retainer below headlight switch assembly. Push the switch inward to connect it.

9. Install the steering column lower cover or the instrument panel trim plate covering the headlight switch, if equipped with a rocker-type headlight switch.

10. Connect the negative battery cable and enable the SIR system.

1995–97 Models

1. Disconnect the negative battery cable. Disable the air bag system.

2. Open the glove compartment door to access the instrument panel moulding fasteners. Remove the fasteners, release the clips and carefully pull the instrument panel moulding rearward.

3. Remove the cluster trim plate fasteners, tilt the top of the trim plate rearward, then pull the bottom of the trim plate rearward.

4. Detach the switch connector.

5. Remove the headlight switch by carefully pushing one side outward.

To install:

6. Route the headlight switch connector through the hole in the cluster trim plate. Install the cluster trim plate and tighten the fasteners to 17 inch lbs. (1.9 Nm).

7. Connect the headlight switch electrical connector and install the switch in the cluster trim plate.

8. Install the instrument panel moulding by carefully pushing forward to secure the clips. Install the moulding fasteners that were accessed through the glove compartment.

9. Connect the negative battery cable. Enable the air bag system.

10. Check headlight switch operation.

Pontiac Bonneville

1. Disconnect the negative battery cable. Disable the air bag system.

2. Carefully pry out the instrument panel trim plate. Pull the trim plate up, then rearward.

3. If equipped, disconnect the subwoofer gain control switch electrical connector and remove the instrument panel trim plate from the vehicle.

4. Remove the instrument cluster trim plate fasteners.

5. Remove the interior light dimmer and twilight sentinel control knobs from the headlight switch. Be careful not to lose the dimmer knob retainer.

6. Disconnect the cigar lighter electrical connector and remove the instrument cluster trim plate from the vehicle.

7. Remove the instrument cluster fasteners and pull the cluster rearward, to gain access to the headlight switch.

8. Remove the headlight switch fasteners and pull the switch rearward. Disconnect the switch wiring connector and remove the headlight switch from the vehicle.

To install:

9. Connect the headlight switch wiring connector and install the switch. Tighten the fasteners to 17 inch lbs. (2 Nm).

10. Push the instrument cluster into position and install the fasteners. Tighten to 14 inch lbs. (1.6 Nm).

11. Connect the cigar lighter connector and install the instrument cluster trim plate, carefully inserting the tab.

12. Install the interior light dimmer and twilight sentinel control knobs to the headlight switch. Properly align the retainer in the dimmer knob before installing to insure proper retention.

13. Install the instrument cluster trim plate fasteners and tighten to 17 inch lbs. (1.9 Nm).

14. If equipped, connect the subwoofer gain control switch electrical connector.

15. Install the instrument panel trim plate and press into place.

16. Connect the negative battery cable. Enable the air bag system.

17. Check headlight switch operation.

Cadillac DeVille and Sixty Special

1. Disconnect the negative battery cable. Disarm the air bag system.

2. Remove the lower steering column filler panel by carefully prying out on the lower edge.

3. Remove the upper screws and 2 lower screws at the steering column opening, pull back gently and lift up to disengage the clips on the lower end and remove.

4. Remove the screws from the top of the left side trim plate and remove the plate.

5. Remove the 4 headlight switch mounting panel screws.

6. Pull the assembly out of the instrument panel and disconnect the electrical connectors.

7. Remove the assembly from the vehicle.

8. Remove the knob from the switch by pressing the release button.

9. Remove the nut and remove the switch from the trim plate.

To install:

10. Install the switch on the trim plate and install the mounting nut.

11. Insert the control knob onto the switch until it is fully seated.

12. Connect the electrical connectors to the switch assembly.

13. Position the switch assembly in the dash board and install the 4 mounting screws.

14. Install the left side trim panel and trim panel mounting screws.

15. Install the instrument panel trim panel by engaging the lower clips, seating the upper half of the panel and installing the screws.

16. Snap in the lower filler panel.

17. Connect the negative battery cable. Enable the air bag system, if equipped.

18. Check headlight switch operation.

Turn Signal Switch

REMOVAL AND INSTALLATION

Except Cadillac DeVille and Sixty Special

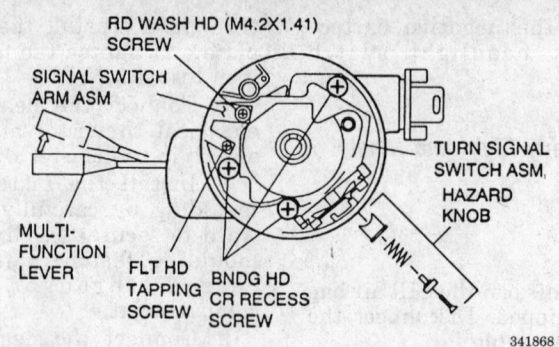

RD WASH HD (M4.2X1.41) SCREW

SIGNAL SWITCH ARM ASM

TURN SIGNAL SWITCH ASM, HAZARD KNOB

MULTI-FUNCTION LEVER

FLT HD TAPPING SCREW

BNDG HD CR RECESS SCREW

341868

Turn signal switch removal preparation — Except Cadillac DeVille and Sixty Special models

CAUTION

The Supplemental Restraint System (SRS) must be disarmed before working around the air bag or SRS wiring. Failure to do so may cause accidental deployment of the air bag, resulting in unnecessary SRS repairs and/or personal injury.

1. Place the vehicle's front wheels in the straight-ahead position.
2. Turn the ignition switch to the **LOCK** position and remove the key.
3. Properly disable the SRS system. Disconnect the negative battery cable.
4. Remove the steering wheel and inflator module.

CAUTION

When carrying a live air bag, make sure the bag and trim cover are pointed away from the body. In the unlikely event of an accidental deployment, the bag will then deploy with minimal chance of injury. When placing a live inflator module on a bench or other surface, always place the bag and trim cover up, away from the surface. This will reduce the motion of the module if it is accidently deployed.

5. Remove the SRS coil retaining rings and remove the coil assembly. Let the coil hang freely.
6. Remove the wave washer.
7. Remove the shaft lock retaining ring using tool J 23653 or equivalent, to push down on the shaft lock. Dispose of the ring.
8. Remove the shaft lock.
9. Remove the turn signal cam assembly. Remove the upper bearing spring.
10. Place the turn signal switch to the RIGHT TURN position.
11. Remove the the multifunction lever and hazard knob assembly.
12. Remove the screws attaching the signal switch arm and remove the arm from the vehicle.

13. Remove the turn signal connector from the bulkhead connector and wire harness strap.
14. Disconnect the electrical connections at the turn signal switch.
15. Remove the wiring protector and wire harness strap.
16. Remove the turn signal attaching nut and disconnect the ground wire from the stud.
17. Gently pull the wire harness through the column and remove the turn signal switch from the vehicle.

To install:
18. Route the turn signal switch assembly wire harness through the steering column and let the switch hang freely.
19. Install the turn signal assembly connector to the bulkhead connector.
20. Install the ground wire to the stud and tighten the turn signal stud nut to 35 inch lbs. (4 Nm).
21. Install the coil assembly wire harness through the steering column. Connect the wiring connections to the turn signal switch.
22. Install and tighten the turn signal switch screws to 30 inch lbs. (3 Nm).
23. Install and tighten the turn signal switch arm screws to 20 inch lbs. (2 Nm).
24. Install the hazard knob assembly with the multifunction lever.
25. Install the upper bearing spring.
26. Lubricate the turn signal cancelling cam with synthetic grease and install the turn signal cancelling cam assembly.
27. Install the shaft lock.
28. Install the new shaft lock retaining ring using tool J 23653 or equivalent, to push down the shaft

lock. Firmly seat the ring in the groove on the shaft.

NOTE: Set the steering shaft so that the block tooth on the upper steering shaft is at the 12 o'clock position. The vehicle's front wheels should be straight ahead. Set the ignition switch to the LOCK position to ensure no damage to the coil assembly.

29. Install the wave washer.
30. Check for proper centering of the SRS coil assembly.

NOTE: SRS coil assembly wires must be kept tight with no slack while installing the SRS coil assembly. Failure to do so may cause the wires to be kinked near the shaft lock area and cut when the steering wheel is turned.

31. Pull the SRS coil wire snug while positioning the coil to the steering column.
32. Align the opening in the coil with the horn tower and locating bump between the tabs on the housing cover.
33. Seat the coil assembly into the steering wheel.
34. Install and firmly seat the the coil assembly retaining ring into the groove on the shaft.

NOTE: Gently pull the lower coil assembly, turn signal, and pass key wires to remove any wire kinks that may be inside the steering column assembly. Failure to do so may cause damage to the wire harness.

35. Install the wiring protector and strap. Install the steering wheel and inflator module.
36. Connect the negative battery cable. Enable the SRS system.

Combination Switch

REMOVAL AND INSTALLATION

Cadillac DeVille and Sixty Special

---— CAUTION ——---
The Supplemental Inflatable Restraint (SIR) system must be disarmed before working around the steering column. Failure to do so may cause accidental deployment of the air bag resulting in unnecessary SIR system repairs and/or personal injury.

1. Disable the SIR system. Disconnect the negative battery cable.

---— CAUTION ——---
When carrying a live air bag, make sure the bag and trim cover are pointed away from the body. In the unlikely event of accidental deployment, the bag will then deploy with minimal chance of injury. When placing a live air bag on a bench or other surface, always face the bag and trim cover up, away from the surface. This will reduce the motion of the module if it is accidently deployed.

2. Remove the SIR inflator module (air bag), if equipped.
3. Remove the steering wheel.
4. Remove the steering shaft bumper.
5. Remove the carrier snap ring retainer.
6. Remove the lock plate by installing a lock plate compressor and removing the snap ring from the steering shaft. Slowly back off the compressor until spring tension is relieved. Remove the lock plate.
7. Remove the turn signal lever.
8. Remove the turn signal switch mounting screws.
9. At the base of the column, disconnect the turn signal switch electrical connector.
10. Remove the lower steering column trim plates and steering column wire guide.
11. Remove the turn signal switch while guiding the wiring harness through the column.
   To install:
12. Feed the turn signal harness down into the column.
13. Position the turn signal switch and install the mounting screws.
14. Install the turn signal arm.
15. Install the lock plate and lock plate retainer.

16. Install the carrier snap ring retainer. Install the steering shaft bumper.
17. Install the steering wheel.
18. Install the SIR inflator module, if equipped.
19. Connect the negative battery cable. Enable the SIR system.

Ignition Lock Cylinder

REMOVAL AND INSTALLATION

---— CAUTION ——---
The Supplemental Restraint System (SRS) must be disarmed before working around the air bag or SRS wiring. Failure to do so may cause accidental deployment of the air bag, resulting in unnecessary SRS repairs and/or personal injury.

NOTE: When replacing the lock cylinder, the key code must be read with interrogator tool J 35628-A or equivalent. A new key must be ordered with the new lock cylinder and must be cut to the dummy key in the new lock cylinder.

1. Disconnect the negative battery cable. Disable the SIR/SRS system.
2. Remove the 2 screws in back of the steering wheel
3. Position the inflator module to gain access to the wire connectors. Note the wiring routing for reinstallation. Push in and rotate the horn contact lead counterclockwise and remove the steering column horn tower. Remove the horn contact lead.
4. Disconnect the air bag connector retainer and then disconnect the connector. Remove the air bag inflator module and store trim side face up for safety reasons.

---— CAUTION ——---
When carrying a live air bag, make sure the air bag and trim cover are pointed away from the body. In an unlikely event of an accidental deployment, the bag will then deploy with minimal chance of injury. When placing a live air bag on a bench or other surface, always face the bag and trim cover up, away from the surface. This will reduce the motion of the module if it is accidentally deployed.

5. Remove the steering wheel control switch connector, if equipped.
6. Remove the steering wheel fastener. Install steering wheel removal

tools J 1859-03 and J 28720 or equivalent, and remove the steering wheel while guiding the wires through the steering wheel hub. Mark the steering wheel to shaft position, prior to removal for re-installation purposes.
7. Remove the SIR coil retaining ring.
8. Remove the left side sound insulators from under the dash.
9. Remove the wire connectors from the retainers and disconnect the bulkhead connector to remove the turn signal switch connector.
10. Gently pull up on the SIR coil assembly while pushing up on the coil harness on the bottom side of steering column, and let the coil hang freely.
11. Remove the wave washer. Install compressor tool J 23653-SIR or equivalent, and compress the lock plate to remove the shaft lock retaining ring.
12. Remove the turn signal canceling cam assembly and remove the screws to the turn signal switch arm and turn signal switch-to-steering column, then remove the hazard switch button.
13. Remove the upper bearing spring. Pull up gently on the turn signal switch assembly and let it hang, then remove the key buzzer switch.
14. Remove the 13-way secondary lock on the turn signal switch connector at the lower end of the steering column and remove terminals **12** and **13** which are routed to the ignition lock cylinder.
15. Attach a piece of wire to the terminals for guiding the harness through the steering column.
16. Remove the retaining bolt to the ignition lock cylinder and insert the key in the lock cylinder to remove.
   To install:

---— CAUTION ——---
Route wires from the pass key lock cylinder as shown and the retaining clip into hole in housing. Failure to do so may result in component damage or the malfunctioning of the steering column may occur.

17. Install the pass key lock cylinder. Route the wires correctly and insert the terminals in the turn signal connector.
18. Install the lock cylinder fastener and torque to 22 inch lbs. (2.5 Nm)
19. Install the buzzer switch and the turn signal switch assembly. Torque the bolts to 30 inch lbs. (3.4 Nm). When installing the switch be sure to

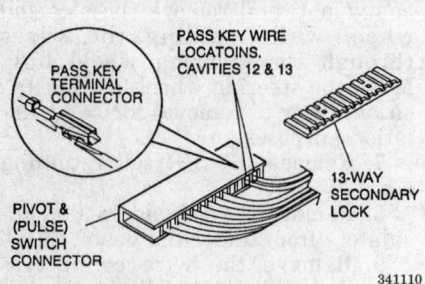

Turn signal switch wire locations

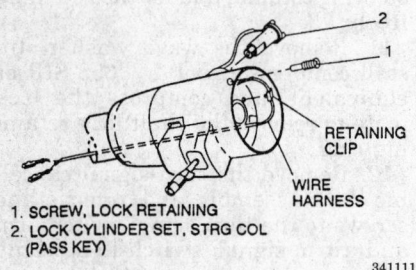

1. SCREW, LOCK RETAINING
2. LOCK CYLINDER SET, STRG COL (PASS KEY)

341111

Installing the pass key lock cylinder

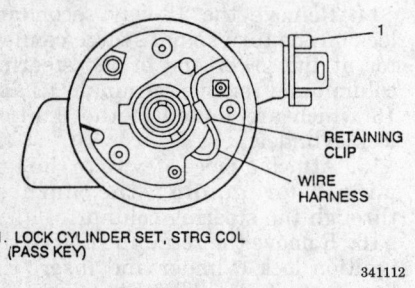

RETAINING CLIP

WIRE HARNESS

1. LOCK CYLINDER SET, STRG COL (PASS KEY)

341112

Routing the pass key wire harness

remove all slack out of the wiring harness by pulling gently on the harness.

CAUTION

SIR coil assembly wires must be kept tight with no slack while installing the coil assembly. Failure to do so may cause the wires to be kinked near the lock plate, causing the wires to be cut when the steering wheel is turned.

20. Install the turn signal switch arm and torque to 20 inch lbs. (2.3

Nm). Install the hazard switch button

21. Install the upper bearing spring.

22. Install the turn signal cancelling cam assembly and lock plate.

23. Using compressor tool J 23653-SIR or equivalent, install the retaining ring after compressing the upper bearing spring.

24. Install the wave washer and SIR coil assembly, pulling gently on the harness on the bottom side of the steering column to remove slack from the wiring harness. Be sure to align the SIR coil assembly properly.

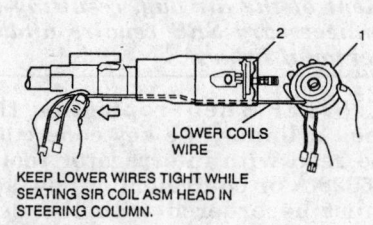

LOWER COILS WIRE

KEEP LOWER WIRES TIGHT WHILE SEATING SIR COIL ASM HEAD IN STEERING COLUMN.

1. COIL ASM, SIR
2. LOCK, SHAFT

341113

Routing SRS coil wires

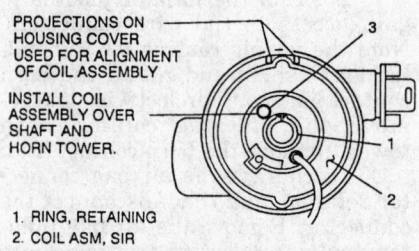

1. RING, RETAINING
2. COIL ASM, SIR
3. CAM ASM, TURN SIG CANCEL

341114

Coil assembly installed

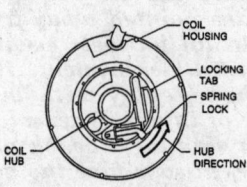

341115

Centering the coil assembly

25. Install the snapring on top of the coil.

26. Install the steering wheel, guiding the wires through for the air bag and steering wheel controls, if equipped.

27. Connect the negative battery cable. Enable the air bag system.

28. Test the turn signals, high beams and wipers.

29. Install the lower sound insulator.

Ignition Switch

REMOVAL AND INSTALLATION

CAUTION

The Supplemental Restraint System (SRS) must be disarmed before working around the air bag or SRS wiring. Failure to do so may cause accidental deployment of the air bag, resulting in unnecessary SRS repairs and/or personal injury.

1. Disconnect the negative battery cable. Disable the SRS system.
2. Remove the left side sound insulator.
3. Remove the lower steering column trim plate and lower the steering column.
4. Position the ignition switch in the **OFF** position.
5. Remove the hexagon bolt and nut holding the ignition switch and dimmer switch in place.
6. Remove the ground wire with ring terminal from the stud.
7. Remove the dimmer switch assembly from the dimmer switch rod.
8. Remove the wire harness strap.
9. Remove the turn signal switch assembly connector from the main steering harness connector.
10. Remove the pivot and pulse switch assembly connector from the main steering harness connector.
11. Remove the axial positive assurance connector from the Brake Transmission Shift Interlock (BTSI) actuator, if equipped.
12. Remove the ignition switch fastener and the ignition switch from the vehicle.

WARNING

If installing a new ignition switch, the new ignition switch will pinned in the OFF LOCK position. Remove the plastic pin after the switch is assembled to the steering column. Failure to do so may cause ignition switch damage.

To install:

13. Install the switch to the steering column and install the bolt loosely.

14. Adjust the ignition switch using the following procedure:

a. Move the switch slider to the extreme left and then move the switch slider one detent to the left of the **OFF LOCK** position.

b. Install a ³⁄₃₂ in. drill bit in the hole on the ignition switch to limit travel.

15. Torque the ignition switch to 35 inch lbs. (4 Nm).

16. Install the dimmer switch: Adjust the dimmer switch using the following procedure:

a. Place a ³⁄₃₂ in. drill bit in hole on the switch to limit the travel.

b. Position the switch on the steering column and push against the dimmer switch rod to remove all slack.

c. Remove the drill bit and tighten the hexagon nut to 35 inch lbs. (4 Nm).

17. Install the turn signal switch assembly connector to the main steering harness connector. Seat the connector body until securely locked in.

18. Install the pivot and pulse switch assembly connector to the main steering harness connector.

19. Install the wire harness strap.

20. Install the connector to the electrical BTSI actuator, if equipped.

21. Position the steering column cover in place and install the column mounting bolts. Tighten to 20 ft. lbs. (27 Nm).

22. Install the column trim plate and replace the sound insulator.

23. Connect the negative battery cable and properly enable the air bag system.

24. Verify that all steering wheel switches and controls operate correctly.

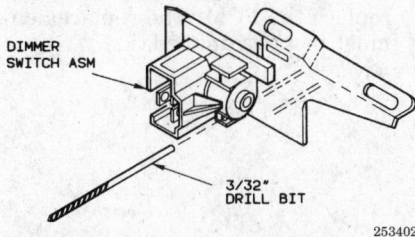

DIMMER SWITCH ASM

3/32" DRILL BIT

253402

Use a drill bit to limit switch travel and remove all actuator rod free-play during installation and adjustment

Park/Neutral Safety Switch

REMOVAL AND INSTALLATION

1993–95 Models

1. Disconnect the negative battery cable. Apply the parking brake and place the shift lever in **N**.

2. Remove the retaining nut and disconnect the linkage cable and bracket from the transaxle shaft.

3. Remove the park/neutral switch retaining bolts.

4. Detach the switch connector. If equipped, remove the nut and disconnect the switch-to-starter lead from the starter.

5. Remove the switch from the vehicle.

To install:

6. If equipped, install the switch lead to the starter and install the nut. Tighten to 35 inch lbs. (4 Nm).

7. Connect the switch electrical connectors to the vehicle harness.

8. Make sure that the transaxle manual shaft is still in the NEUTRAL position.

9. Align the flats in the park/neutral switch with the transaxle manual shaft flats. Press the switch onto the shaft and fully seat against the transaxle.

10. If installing the original switch, insert a ³⁄₃₂ in. drill bit into the service slots. A new replacement switch should already be pinned in the NEUTRAL position.

11. Tighten the park/neutral switch bolts to 20 ft. lbs. (28 Nm).

12. Remove the drill bit.

13. Install the linkage cable and bracket and secure with the nut.

14. Connect the negative battery cable.

15. After switch installation, make sure that the engine starts only in the **P** or **N** position.

1996–97 Models

1. Disconnect the negative battery cable.

2. Set the parking brake and place the transaxle selector lever to the **N** position.

3. Remove the linkage retaining nut, cable and lever from the transaxle manual shaft.

4. Detach the connectors.

5. Remove the park/neutral switch from the vehicle.

To install:

6. Make sure that the transaxle shaft is still in the **N** position.

7. Install the park/neutral switch onto the vehicle. A new park/neutral switch is already pinned in the NEUTRAL position. If installing a switch with a sheared pin, proceed as follows:

a. Insert alignment tool J 41545 or equivalent, into the 2 slots on the switch in the area of the transaxle shaft. Rotate the tool until the rear leg of the tool falls into the slot on the switch near the hose. Verify the tool is properly seated in all 3 slots and remove the tool.

b. Check for a cracked carrier if the pin is sheared or if installing the original switch.

8. Align the flats in the park/neutral switch with the transaxle manual shaft flats. Press the switch onto the shaft and fully seat against the transaxle.

9. Install the attaching bolts. If using the alignment tool, make sure the switch is properly aligned with the tool before tightening the bolts. Tighten the bolts to 18 ft. lbs. (25 Nm).

10. Remove the alignment tool.

11. Connect the cable and lever and install the linkage retaining nut. Tighten the nut to 15 ft. lbs. (20 Nm).

12. Attach the connectors.

13. Connect the negative battery cable.

14. After installing the park/neutral switch, verify that the engine only starts in **P** or **N** positions.

Powertrain Control Module

REMOVAL AND INSTALLATION

1993–95 Models

Except Cadillac DeVille and Sixty Special

1. Disconnect the negative battery cable. Turn the ignition to the **OFF** position.

2. Remove the right side hush panel from the passenger compartment.

3. Disconnect the connectors from the PCM.

4. Remove the PCM mounting hardware, then remove it from the passenger compartment.

5. Remove the PCM access cover.

6. Remove the MEM CAL. Using 2 fingers, push both retaining clips back away from the the MEM CAL. At the same time, grasp it at both ends and lift it up out of the socket.

NOTE: Replacing the PCM requires the removal of the MEM CAL. Care should be used when removing the MEM CAL from the old PCM because it will be used in the new PCM.

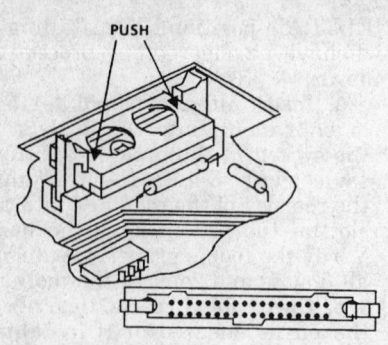

333621

MEM CAL unit installation — 1993–95 vehicles

To install:

7. Install the MEM CAL in the MEM CAL socket. Align small notches in the MEM CAL socket. Gently press down on the ends on the MEM CAL until the clips are against the sides of the MEM CAL. Press inward on the clips until they snap into place.

8. Install the access cover on the PCM.

9. Install the PCM into the passenger compartment. Attach the connectors to the PCM.

10. Install the right side hush panel. Connect the negative battery cable.

Cadillac DeVille and Sixty Special

1. Disconnect the negative battery cable. Remove the right side hush panel.

2. Disconnect the electrical harnesses from the PCM.

3. Remove the PCM retaining nut.

4. Pull the PCM down.

5. Lift the unit up to get the front mounting pin to clear the mounting bracket.

6. Remove the PCM.

7. If replacing the PCM remove the MEM-CAL cover and remove the MEM-CAL (sometimes called the PROM chip).

To install:

8. If replacing the PCM install the MEM-CAL into the new PCM.

9. Install the PCM so the front mounting pin goes through the opening in the bracket.

10. Install the PCM mounting nut.

11. Connect the electrical harnesses to the PCM.

12. Install the right side hush panel.

NOTE: When connecting the negative battery cable make sure the ignition is in the LOCK or OFF position. Connecting the cable with the key in the ON or RUN position can cause damage to the PCM.

13. Connect the negative battery cable.

1996–97 Models

NOTE: To prevent possible electrostatic discharge damage to the PCM, do not touch the connector pins or soldered components on the circuit board. Service of the PCM consists of either

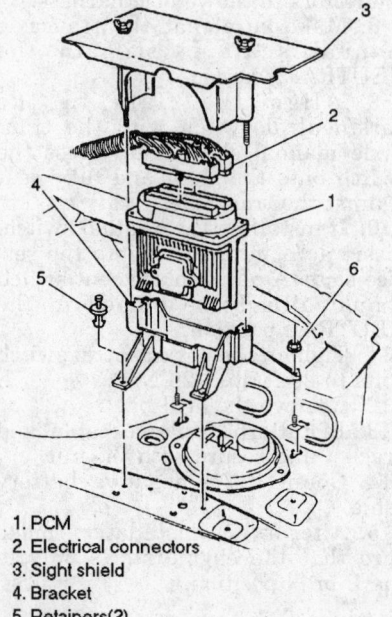

1. PCM
2. Electrical connectors
3. Sight shield
4. Bracket
5. Retainers(2)
6. Nuts(2)

333623

Powertrain control module mounting — 1996–97 models

replacement of the PCM or EEPROM programming. If the diagnostic procedures call for the PCM to be replaced, the PCM should be checked first to see if it is the correct part. If it is, remove the faulty PCM and install the new service PCM. THE SERVICE PCM EEPROM WILL NOT BE PROGRAMMED, DTC PO602 indicates the EEPROM is not programmed or has malfunctioned.

WARNING

To prevent internal PCM damage, the ignition must be OFF when disconnecting or reconnecting power to the PCM.

1. Disconnect the negative battery cable. Remove the sight shield.

2. Remove the harness connectors from the PCM.

3. Remove the PCM from the vehicle.

To install:

4. If replacing the PCM, remove the knock sensor from the PCM and install it into the new PCM.

5. Install the PCM into the mounting brackets.

6. Connect the wiring harness to the PCM.

7. Install the sight shield and connect the negative battery cable.

8. If a new PCM is being installed, the new PCM EEPROM must be programmed as follows:

a. Make sure that the battery is fully charged, the ignition is ON and the vehicle interface module cable connection at the DLC is secure.

b. Program the PCM using the latest software matching the vehicle. Refer to up-to-date Techline equipment users instructions.

c. If the PCM fails to program, make sure that all PCM connections are OK. Check the techline equipment for the latest software version and attempt to program the PCM again. If the PCM still cannot be programmed properly, replace the PCM. The replacement must be programmed.

ENGINE COOLING

Radiator

REMOVAL AND INSTALLATION

—— CAUTION ——

Never remove the radiator cap under any conditions while the engine is operating. Failure to follow these instructions could result in personal injury and/or damage to the cooling system or engine. To avoid having scalding hot coolant or steam blow out of the radiator, use extreme care when removing the radiator cap from a hot radiator. Wait until the engine has cooled, then wrap a thick cloth around the radiator cap and turn it slowly to the first stop. Step back while the pressure is released from the cooling system. When it is certain that all pressure has been released, press down on the radiator cap, with the cloth, turn and remove.

1. Disconnect the negative battery cable. Drain the cooling system.
2. Remove the air cleaner duct, as necessary, for clearance.
3. Remove the cooling fan or fans, as applicable.
4. If necessary, remove the upper radiator panel.
5. Disconnect the upper and lower radiator hoses and coolant overflow hose from the radiator.
6. Disconnect and cap the transaxle oil cooler lines.
7. Remove the radiator from the vehicle.

To install:

8. Install the radiator in the vehicle. Tighten the mounting bolts 88 inch lbs. (10 Nm).
9. Install the remaining components in the reverse order of removal.
10. Connect the negative battery cable. Fill and bleed the cooling system.
11. Run the engine and check for leaks. Recheck the coolant level when the engine has cooled.

Water Pump

REMOVAL AND INSTALLATION

3.8L (VIN L and K) Engines

1. Disconnect the negative battery cable. Drain the cooling system.

2. Remove the accessory drive belt. Disconnect the coolant hoses from the water pump.
3. Remove the water pump pulley bolts. The long bolt can be removed by lining the head of the bolt up with the hole in the frame rail. Remove the pulley.
4. Support the engine using engine support fixture tool J 28467-A or equivalent, and remove the front engine mount.
5. Remove the water pump mounting bolts and remove the water pump.

To install:

6. Clean all the gasket mating surfaces.
7. Apply a thin bead of sealer around the outside edge of the water pump and install the gasket on the pump.
8. Install the water pump on the engine. Tighten the water pump-to-engine block bolts to 22 ft. lbs. (30 Nm) and the water pump-to-front cover bolts to 11 ft. lbs. (15 Nm) plus an additional 80 degree turn.
9. Install the remaining components in the reverse order of removal. Tighten the water pump pulley bolts to 115 inch lbs. (13 Nm), after installing the drive belt.
10. Connect the negative battery cable. Refill and bleed the cooling system.
11. Run the engine and check for leaks.

1995 3.8L (VIN 1) Engine

1. Disconnect the negative battery cable. Drain the cooling system.
2. Remove the outer and inner drive belts. Remove the alternator and brace.
3. Disconnect the hoses and pipes from the water pump.
4. Remove the pulley bracket assembly.
5. Raise and safely support the vehicle. Remove the power steering pump and lines.

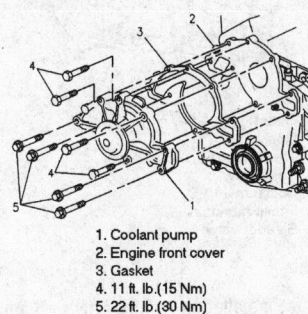

1. Coolant pump
2. Engine front cover
3. Gasket
4. 11 ft. lb.(15 Nm)
5. 22 ft. lb.(30 Nm)

334206

Water pump installation — 1996–97 3.8L (VIN L and K) engines shown

6. Lower the vehicle.
7. Support the engine using engine support fixture J 28467-A or equivalent, and remove the front engine mount.
8. Remove the water pump pulley.
9. Remove the water pump mounting bolts and remove the water pump from the vehicle.

To install:

10. Clean all gasket mating surfaces.
11. Apply a thin bead of sealant to the gasket mating surface of the water pump and install a new gasket to the pump.
12. Install the water pump on the engine. Tighten the pump-to-block bolts to 22 ft. lbs. (30 Nm) and the pump-to-front cover bolts to 11 ft. lbs. (15 Nm) plus an additional 80 degree turn.
13. Install the water pump pulley and tighten the bolts to 115 inch lbs. (13 Nm).
14. Install the remaining components in the reverse order of removal.
15. Connect the negative battery cable. Fill and bleed the cooling system.
16. Run the engine and check for leaks.

1996–97 3.8L (VIN 1) Engines

1. Disconnect the negative battery cable. Drain the cooling system.
2. Remove the A/C compressor splash shield.
3. Remove the supercharger and accessory drive belts.
4. Remove the coil pack and position out of the way.
5. Remove the supercharger belt tensioner.
6. Support the engine using engine support fixture J 28467-A or equivalent, and remove the front engine mount.
7. Remove the power steering pump.
8. Remove the engine mount bracket and the idler pulley.
9. Remove the water pump pulley.
10. Remove the water pump mounting bolts and remove the water pump.

To install:

11. Clean all gasket mating surfaces.
12. Apply a thin bead of sealer around the outside edge of the water pump and install a new gasket on the pump.
13. Install the water pump on the engine and tighten the bolts to specification.
14. Install the remaining components in the reverse order of removal.

15. Connect the negative battery cable. Fill and bleed the cooling system. Run the engine and check for leaks.

4.9L (VIN B) Engine

1. Disconnect the battery cables from the battery, negative cable first. Remove the battery from the vehicle.

2. Force the accessory drive belt tensioner into the drive belt with a breaker bar, to apply tension to the water pump pulley and keep it from rotating. With the pulley held in place, remove the pulley mounting bolts.

3. Remove the water pump pulley and the accessory drive belt.

4. Raise and safely support the vehicle.

5. Remove the wheelhouse engine splash shield.

6. Remove the lower water pump mounting fasteners.

7. Lower the vehicle.

8. Remove the remaining water pump fasteners and remove the water pump. Remove and discard the water pump gasket.

To install:

9. Clean all gasket mating surfaces.

10. Install a new water pump gasket on the front cover studs.

11. Install the water pump on the front cover and install the upper fasteners hand-tight.

12. Raise and safely support the vehicle.

13. Install the lower water pump fasteners and torque to specification.

14. Install the wheelhouse engine splash shield.

15. Lower the vehicle.

16. Torque the upper water pump fasteners to specification.

17. Install the accessory drive belt and water pump pulley. Tighten the water pump pulley bolts to 22 ft. lbs. (30 Nm), using the accessory drive belt drag to hold the pulley in place. If the pulley turns while torquing, apply additional drag using a breaker bar on the belt tensioner to load the belt.

18. Connect the negative battery cable. Fill and bleed the cooling system. Run the engine and check for leaks.

Thermostat

REMOVAL AND INSTALLATION

3.8L (VIN L, K and 1) Engines

1. Disconnect the negative battery cable. Remove the engine cover if necessary.

2. Drain the cooling system to a level below the intake manifold.

3. Disconnect the upper radiator hose from the thermostat housing.

4. Remove the thermostat housing mounting bolts.

5. Remove the housing and gasket.

6. Remove the thermostat from the intake manifold.

To install:

7. Install the thermostat into the recess in the intake manifold.

8. Install the remaining components in the reverse order of removal. Use a new gasket and tighten the mounting bolt to 20 ft. lbs. (27 Nm).

9. Connect the negative battery cable. Fill and bleed the cooling system. Run the engine and check for leaks.

4.9L (VIN B) Engine

1. Disconnect the negative battery cable.

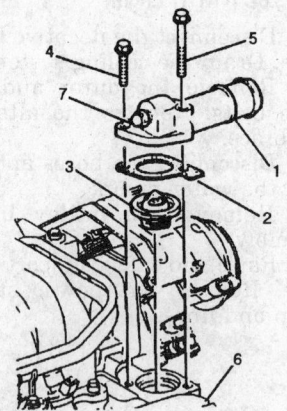

1 Outlet
2 Gasket
3 Thermostat
4 Bolt/screw
5 Bolt/screw
6 Intake manifold
7 Bleeder valve

249008

Thermostat installation — 3.8L (VIN L, K and 1) engines

2. Drain the cooling system to a level below the intake manifold.

3. Remove the upper air filter assembly, if necessary.

4. Disconnect the upper radiator hose from the thermostat housing.

5. Remove the 2 bolts fastening the upper and lower thermostat housings together.

6. Remove the upper thermostat housing.

7. Remove the thermostat and gasket from the lower housing.

To install:

8. Clean all gasket surfaces completely.

9. Install the thermostat and gasket on the lower thermostat housing.

10. Install the remaining components in the reverse order of removal.

11. Connect the negative battery cable. Refill and bleed the cooling system. Run the engine and check for leaks.

Electric Cooling Fan

REMOVAL AND INSTALLATION

3.8L (VIN L, K and 1) Engines

1. Disconnect the negative battery cable.

2. Detach the cooling fan motor harness from the motor and fan frame.

3. Remove the upper fan guard and hose support.

4. Remove the fan assembly from the radiator support.

To install:

5. Install the fan assembly on the radiator support. Tighten the attaching bolts to 88 inch lbs. (10 Nm).

6. Install the hose support and upper fan guard.

7. Connect the cooling fan harness to the fan frame and fan motor.

8. Connect the negative battery cable and test the cooling fan for proper operation.

4.9L (VIN B) Engine

NOTE: The vehicle is equipped with twin electric cooling fans. The procedure that follows covers the removal of either fan.

1. Disconnect the negative battery cable. Raise and safely support the vehicle.

2. Detach the fan connector.

3. Remove the cooling fan-to-lower radiator support mounting bolts.

4. Lower the vehicle.

5. Remove the air cleaner intake duct.

1 UPPER HOUSING
2 GASKET
3 THERMOSTAT ASSEMBLY
4 LOWER HOUSING
5 GASKET

334033

Thermostat installation — 4.9L (VIN B) engines

6. Remove the cooling fan-to-upper radiator support.

7. Remove the upper radiator mounting panel.

8. Remove the cooling fan.

To install:

9. Install the cooling fan.

10. Install the upper radiator mounting panel.

11. Install the cooling fan upper mounting bolts finger tight.

12. Raise and safely support the vehicle. Attach the fan connector.

13. Install the and tighten the lower mounting screws to 84 inch lbs. (9.5 Nm).

14. Lower the vehicle.

15. Tighten the upper mounting screws to 84 inch lbs. (9.5 Nm).

16. Install the air cleaner inlet duct. Connect the negative battery cable.

17. Start the engine and bring to operating temperature. Make sure the cooling fans are working correctly.

Cooling System Bleeding

PROCEDURE

CAUTION

Never remove the radiator cap under any conditions while the engine is operating. Failure to follow these instructions could result in personal injury and/or damage to the cooling system or engine. To avoid having scalding hot coolant or steam blow out of the radiator, use extreme care when removing the radiator cap from a hot radiator. Wait until the engine has cooled, then wrap a thick cloth around the radiator cap and turn it slowly to the first stop. Step back while the pressure is released from the cooling system. When it is certain that all pressure has been released, press down on the radiator cap, with the cloth, turn and remove.

1993–95 Vehicles

1. Park the vehicle on a level surface.

2. Remove the radiator cap when the engine is cool.

3. Open the bleeder valve on the thermostat housing.

4. Completely drain the cooling system by opening the drain cock at the bottom of the radiator and the plugs in the engine block. If the coolant is dirty, or if there are deposits in the radiator, flush the cooling system before refilling.

5. Disconnect all hoses from the coolant reservoir. Remove the reservoir and pour out any coolant. Clean the inside of the reservoir. Flush it well with clean water, then drain it. Install the reservoir and hoses.

6. Close the radiator drain cock and install any plugs that were removed from the block.

7. Refill the cooling system with a 50/50 mixture of antifreeze and clean water.

8. Close the bleeder valve on the thermostat housing.

9. Place the heater and A/C control in any A/C mode except MAX and the temperature to the highest setting.

10. Allow the engine to continue idling until the lower radiator hose to the water pump is hot.

11. Cycle the engine speed up to about 3000 rpm and back to idle five times. Slowly open the bleed valve on the thermostat housing for 15 seconds to expel any trapped air in the cooling system.

12. After the air has been expelled, top off the radiator. Install the radiator cap, making sure the arrows on the cap line up with the coolant recovery tube.

13. Allow the engine to cool to outside temperature, then check the coolant level in the reservoir. If not at the ADD mark, add coolant to the FULL mark.

14. Road test the vehicle to ensure the cooling system is operating properly.

1996–97 Vehicles

CAUTION

These vehicles have a newly developed engine coolant. DEX-COOL was developed to last for 100,000 miles or 5 years whichever occurs first. Make sure only DEX-COOL is used when coolant is added or changed.

A 50/50 mixture of DEX-COOL and water will provide the following protection:

• Giving protection down to -34 degrees F (-37 degrees C).

• Give a boiling point of 265 degrees F (129 degrees C).

• Protect against rust and corrosion.

• Helps keep the engine the proper temperature.

• Lets the warning lights and gauges work as they should.

DEX-COOL anti-freeze can be added to raise the boiling point of the coolant, but too much will affect the freezing point. DO NOT use a solution stronger than 70 percent anti-freeze, as the freeze level rises rapidly after this point. Pure anti-freeze will freeze at -8°F (-22°C).

1. Refill the cooling system to its proper level. Making sure not to use a mixture stronger than 70 percent antifreeze.

2. Place the heater-AC control in any AC mode except MAX and the temperature to the highest setting.

3. Allow the engine to idle until the lower radiator to coolant pump hose is hot.

4. Cycle the engine speed up to 3000 rpm and back to idle 5 times. Locate the cooling system bleeder valve. It is a plug on the back of the thermostat housing. Slowly open the bleed valve as the engine is running to release any trapped air in the cooling system.

5. Shut the engine OFF. Allow the engine to cool to outside temperature, then check the coolant level in the coolant reservoir. If it is below the

FULL COLD mark than add some coolant until it is at that level.

6. Make sure the radiator pressure cap is properly installed. Line up the arrows on the cap with the overflow tube.

FUEL SYSTEM

Fuel System Service Precautions

Safety is the most important factor when performing not only fuel system maintenance but any type of maintenance. Failure to conduct maintenance and repairs in a safe manner may result in serious personal injury or death. Maintenance and testing of the vehicle's fuel system components can be accomplished safely and effectively by adhering to the following rules and guidelines.

• To avoid the possibility of fire and personal injury, always disconnect the negative battery cable unless the repair or test procedure requires that battery voltage be applied.

• Always relieve the fuel system pressure prior to disconnecting any fuel system component (injector, fuel rail, pressure regulator, etc.), fitting or fuel line connection. Exercise extreme caution whenever relieving fuel system pressure to avoid exposing skin, face and eyes to fuel spray. Please be advised that fuel under pressure may penetrate the skin or any part of the body that it contacts.

• Always place a shop towel or cloth around the fitting or connection prior to loosening to absorb any excess fuel due to spillage. Ensure that all fuel spillage (should it occur) is quickly removed from engine surfaces. Ensure that all fuel soaked cloths or towels are deposited into a suitable waste container.

• Always keep a dry chemical (Class B) fire extinguisher near the work area.

• Do not allow fuel spray or fuel vapors to come into contact with a spark or open flame.

• Always use a backup wrench when loosening and tightening fuel line connection fittings. This will prevent unnecessary stress and torsion to fuel line piping. Always follow the proper torque specifications.

• Always replace worn fuel fitting O-rings with new. Do not substitute fuel hose or equivalent, where fuel pipe is installed.

Fuel System Pressure

RELIEVING

—— CAUTION ——
Fuel injection systems remain under pressure, even when the engine has been turned OFF. The fuel system pressure must be relieved before disconnecting any fuel lines. Failure to do so may result in fire and/or personal injury.

1. Disconnect the negative battery cable to avoid possible fuel discharge is an accidental attempt is made to start the engine.

2. Remove the fuel tank cap to relieve tank pressure. Do not tighten until service procedure has been completed.

3. Connect a suitable fuel pressure gauge with bleed valve such as J-34730–1 or equivalent, to the fuel pressure test port. Wrap a shop towel around the fitting while connecting the gauge to catch any spilled fuel.

4. Install the bleed hose into an approved container and open the valve to bleed off the fuel system pressure.

5. Drain any fuel remaining in the gauge into an approved container.

—— CAUTION ——
There may still be residual fuel in the system, and a small amount of fuel may be released when servicing fuel lines or connections. In order to reduce the chance of personal injury, cover the fuel line fittings with a shop towel before disconnecting to catch any fuel that may leak out.

Idle Speed and Mixture

ADJUSTMENT

These adjustments are controlled by the Powertrain Control Module (PCM). No adjustments are necessary or possible.

Fuel Filter

REMOVAL AND INSTALLATION

—— CAUTION ——
Fuel injection systems remain under pressure, even after the engine has been turned OFF. The fuel system pressure must be relieved before disconnecting any fuel lines. Failure to do so may result in fire and/or personal injury.

3.8L (VIN L, K and 1) Engines

1. Disconnect the negative battery cable. Properly relieve the fuel system pressure.

2. Raise and safely support the vehicle.

3. Disconnect the inlet and outlet fuel lines from the filter by detaching the fuel line quick-connects.

4. Remove the filter bracket mounting bolt and remove the fuel filter from the vehicle.

To install:

5. Position the fuel filter in the mounting bracket, making sure it is facing in the proper direction.

6. Connect the inlet and outlet lines to the fuel filter. After connecting each line make sure the line is correctly attached to the filter by giving it a quick tug.

7. Install the filter bracket mounting bolt and tighten to 10 ft. lbs. (14 Nm).

8. Lower the vehicle. Connect the negative battery cable.

9. Pressurize the fuel system by turning the ignition switch to the **ON** position, then check for leaks.

4.9L (VIN B) Engine

1. Disconnect the negative battery cable. Properly relieve the fuel system pressure.

2. Raise and safely support the vehicle.

3. Release the fuel filter bracket tabs or remove the fuel filter bracket mounting bolt, as required.

4. Disconnect the quick release clip at one end of the filter by depressing the 2 plastic tabs.

5. Release the clip at the other end and remove the filter.

To install:

6. Make sure the fuel line fittings are clean.

7. Apply a few drops of clean engine oil to the pipe ends of the filter.

8. Install the filter with the flow arrow facing the proper direction. Snap the filter quick connects together.

9. Install the fuel filter bracket mounting bolt or install the filter into the retainer and engage the bracket tabs, as required.

10. Lower the vehicle. Connect the negative battery cable.

11. Turn the ignition switch **ON** for 2 seconds, then turn **OFF** for 5 seconds. Again turn to **ON** position (to pressurize the fuel system) and check for leaks.

Fuel Pump

REMOVAL AND INSTALLATION

——————— CAUTION ———————

Fuel injection systems remain under pressure, even after the engine has been turned OFF. The fuel system pressure must be relieved before disconnecting any fuel lines. Failure to do so may result in fire and/or personal injury.

3.8L (VIN L, K and 1) Engines

1. Disconnect the negative battery cable. Properly relieve the fuel system pressure.

2. Raise and safely support the vehicle. Remove the fuel tank from the vehicle.

3. Clean the fuel tank in the area of the fuel sender assembly, to prevent dirt and debris from entering the tank when the fuel sender is removed.

4. Rotate the lock ring on top of the tank counterclockwise and remove the fuel sender assembly from the tank.

5. Note the direction the strainer is pointing and remove the strainer from the pump by pulling it down and twisting.

6. Disconnect the pump electrical wires and hoses.

7. Pull the pump assembly out of the rubber connectors.

To install:

8. Transfer any insulators and grommets from the old pump to the new one.

9. Connect the pump to the fuel hose and tilt the bottom of the pump into the mounting bracket.

10. Install a new strainer on the pump so it points in the same direction as the old one.

11. Connect the electrical connectors and fuel lines to the pump.

12. Install a new O-ring on top of the fuel tank and install the fuel sender assembly carefully into the tank.

13. Install the lock ring and rotate clockwise until the tabs are against the stops.

14. Install the fuel tank in the vehicle. Lower the vehicle.

15. Refill the fuel tank. Connect the negative battery cable.

16. Turn the ignition switch to the **ON** position, to pressurize the fuel system, and check for leaks.

17. Run the engine and check for leaks and proper engine operation.

4.9L (VIN B) Engine

1. Disconnect the negative battery cable. Properly relieve the fuel system pressure.

2. Remove the fuel tank from the vehicle.

3. Clean the fuel tank in the area of the fuel pump/gauge sending unit.

4. Remove the fuel pump/gauge sending unit cam ring.

5. Remove the fuel pump/gauge sending unit assembly along with the rubber O-ring.

6. Remove the fuel pump from the gauge sending unit.

To install:

7. Install the fuel pump to the gauge sending unit.

8. Using a new O-ring, install the fuel pump/gauge sending unit assembly in the tank.

9. Install the fuel pump/gauge sending unit locking cam ring.

10. Install the fuel tank in the vehicle. Connect the negative battery cable.

11. Turn the ignition **ON**, to pressurize the fuel system, and check for leaks.

12. Run the engine and check for leaks and proper engine operation.

Fuel Injector

REMOVAL AND INSTALLATION

——————— CAUTION ———————

Fuel injection systems remain under pressure, even after the engine has been turned OFF. The fuel system pressure must be relieved before disconnecting any fuel lines. Failure to do so may result in fire and/or personal injury.

3.8L (VIN L) Engine

1. Disconnect the negative battery cable. Properly relieve the fuel system pressure.

2. Remove the plastic cover from the fuel rail.

3. Disconnect and cap the fuel lines at the fuel rail.

4. Disconnect the vacuum line from the pressure regulator and the connectors from the injectors.

5. Remove the 4 fuel rail mounting screws and remove the fuel rail assembly.

6. Release the fuel injector retaining clip and remove the injector from the fuel rail.

7. Remove the O-rings from the injector and discard.

To install:

8. Coat new injector O-rings with clean engine oil and install them on the injectors.

9. Line up the fuel injectors with the six cavities in the intake manifold and carefully push the fuel rail down.

10. Install the fuel rail mounting bolts and tighten to 10 ft. lbs. (14 Nm).

11. Connect the vacuum line to the pressure regulator and the connectors to the injectors.

12. Connect the fuel lines to the fuel rail. Connect the negative battery cable.

13. Turn the ignition **ON**, to pressurize the fuel system, and check for leaks.

14. Install the plastic cover over the fuel rail.

15. Run the engine and check for leaks and proper engine operation.

3.8L (VIN K) and 3.8L (VIN 1) Engines

1. Disconnect the negative battery cable. Properly relieve the fuel system pressure.

2. Disconnect and cap the fuel feed and return lines at the fuel rail.

3. Disconnect the vacuum line from the fuel pressure regulator.

4. If equipped, disconnect the vacuum line from the supercharger.

5. For the VIN K engine, disconnect the vacuum line from the throttle body.

6. For the VIN 1 engine, disconnect the vacuum line from the supercharger.

7. Label and disconnect the wires from the ignition coils and the retainer clips on top of the supercharger, if equipped.

8. For 1996–97 3.8L (VIN 1) engines, remove the alternator and the rear alternator mounting bracket.

9. Disconnect the fuel injector electrical connectors.

10. Remove the fuel rail mounting nuts and remove the fuel rail from the intake manifold. When removing, lift both sides of the fuel rail with equal force.

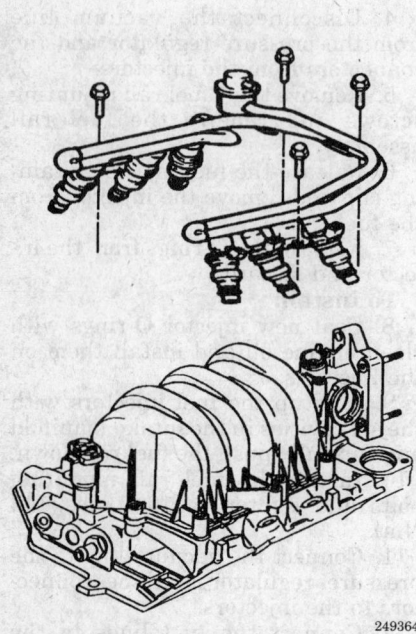

Fuel rail and injectors assembly — 3.8L (VIN L) engines

249364

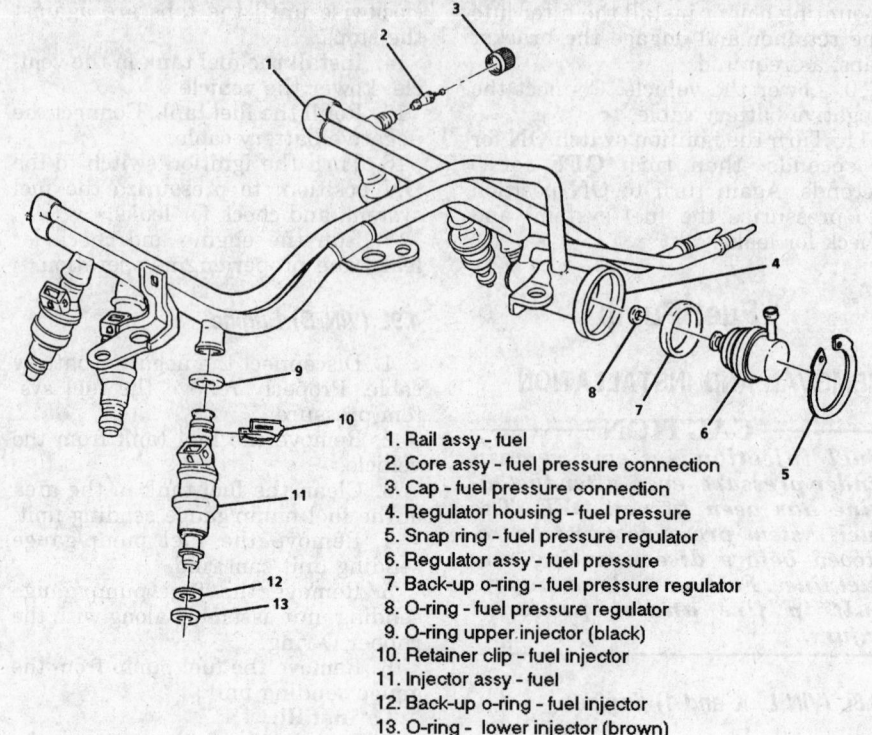

1. Rail assy - fuel
2. Core assy - fuel pressure connection
3. Cap - fuel pressure connection
4. Regulator housing - fuel pressure
5. Snap ring - fuel pressure regulator
6. Regulator assy - fuel pressure
7. Back-up o-ring - fuel pressure regulator
8. O-ring - fuel pressure regulator
9. O-ring upper injector (black)
10. Retainer clip - fuel injector
11. Injector assy - fuel
12. Back-up o-ring - fuel injector
13. O-ring - lower injector (brown)

297396

Fuel rail assembly — 3.8L (VIN 1) engines

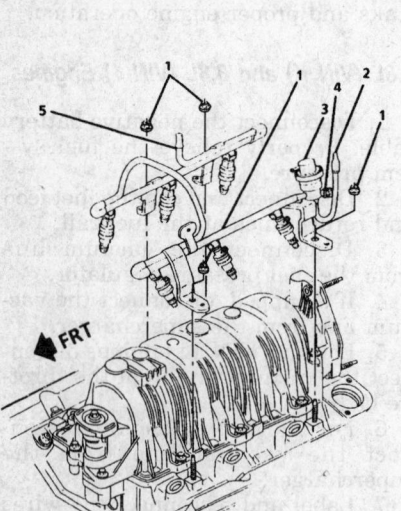

1. NUTS 10 N·m (7 lb. ft.)
2. FUEL FEED
3. FUEL PRESSURE REGULATOR
4. FUEL PRESSURE CONNECTION
5. FUEL RETURN

334541

Fuel rail mounting nuts — 3.8L (VIN K) engines

11. Remove the fuel injector retaining clip and remove the injector from the rail.

12. Remove the O-rings from the injector and discard the retaining clip and O-rings.

—————— CAUTION ——————
In order to reduce the risk of fire and personal injury that may result from a fuel leak, always install the proper fuel injector O-rings in the proper position. If the upper and lower O-rings are different colors (black or brown), be sure to install the black in the upper position and the brown in the lower position on the fuel injector. The O-rings are the same size but are made from different materials.

To install:

13. Coat the new O-rings with clean engine oil and install them on the fuel injector, with the brown O-ring in the lower position.

14. Install the injector on the fuel rail and install the retaining clip. The electrical connector on the injector must point outward.

15. Install the fuel rail onto the intake manifold. Be sure to seat all the injectors in their cavities.

16. With the rail seated by hand, tighten the mounting nuts to 7 ft. lbs. (10 Nm).

17. Connect the fuel pressure and return lines.

18. Connect the electrical connectors to the fuel rail.

19. If necessary, install the alternator and rear alternator mountings.

20. Connect the ignition wires to the coils.

21. Connect the vacuum line to the throttle body or supercharger, as applicable and to the pressure regulator.

22. Connect the negative battery cable.

23. Turn the ignition **ON** to pressurize the fuel system, then check for leaks.

24. Run the engine and check for leaks and proper engine operation.

4.9L (VIN B) Engine

1. Disconnect the negative battery cable.

2. Loosen the power steering pump bracket bolts and maneuver the bracket out of the way.

3. Properly relieve the fuel system pressure.

4. Remove the upper intake manifold, if necessary.

5. Disconnect the vacuum line from the pressure regulator.

6. Remove the inlet fitting screw assemblies and bracket from the rear of the fuel rail and pressure regulator.

7. Disconnect the fuel feed line and O-ring from the rear of the fuel rail assembly.

8. Disconnect the fuel return line from the return fitting.

9. Remove the 4 fuel rail attaching bolts.

10. Disconnect the electrical connectors at the front and rear wiring harnesses.

11. Remove the fuel rail assembly from the intake manifold.

12. Label and disconnect the electrical connector from the injector(s) to be removed.

13. Remove the injector retainer clip from the fuel injector and discard.

14. Remove the injector assembly.

To install:

15. Lubricate new injector O-rings with petroleum based grease and install on the injector.

16. Install the injector on the fuel rail and install the retaining clip.

17. Connect the electrical connector to the injector. The electrical connector should face toward the center of the assembly except on cylinder Nos.

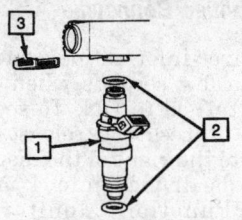

1	INJECTOR ASSEMBLY
2	SEAL-O-RING-INJECTOR
3	CLIP-INJECTOR RETAINER

241179

Injector seals and retainer — 4.9L (VIN B) engines

1 and 8. Injectors 1 and 8 should face away from the ends of the fuel rail.

18. Install the fuel rail assembly on the intake manifold and install the fuel rail mounting bolts. The bolts should be tightened to 18 ft. lbs. (25 Nm).

19. Connect the vacuum line to the fuel pressure regulator.

20. Lubricate a new inlet fitting O-ring and connect the fuel feed line to the fuel rail.

21. Install the inlet fitting bracket and screw assemblies to the rear of the fuel rail and pressure regulators. Tighten the screws to 44 inch lbs. (5 Nm).

22. Connect the fuel return line to the return line fitting. Tighten the fitting to 22 ft. lbs. (30 Nm).

23. Connect the electrical connectors to the front and rear wiring harnesses.

24. Install the upper intake manifold, if removed.

25. Install the power steering pump brackets. Connect the negative battery cable.

26. Turn the ignition **ON**, to energize the fuel pump, and check for leaks. Correct as required.

27. Run the engine and check for leaks and proper engine operation.

EMISSION CONTROLS

Service Interval Lamp

RESETTING

Buick Park Avenue

After the engine oil has been changed, the Engine Oil Monitor must be reset. The reset button is located in the glove box. Reset as follows:

With the ignition key in the **RUN** position, push the reset button, hold it in for at least 5 seconds but not more than 60 seconds. The CHANGE OIL SOON light will flash 4 times and then go off. This indicates that the Oil Life Monitor System has been reset.

Cadillac DeVille and Sixty Special

1. Press **RANGE** and **FUEL USED** buttons at the same time to display oil life index.

2. Press **RANGE** and **RESET** buttons at the same time until the **CHANGE OIL SOON** light flashes and hold for 5 seconds. The oil life index will not remain displayed.

Oldsmobile Eighty Eight and Ninety Eight

1993 Models

Vehicles equipped with an **ENGINE OIL LIFE INDEX** display as a part of the **DRIVER INFORMATION SYSTEM (DIS)**, have a display that will show when to change the engine oil.

After the oil has been changed, the **ENGINE OIL LIFE INDEX** light

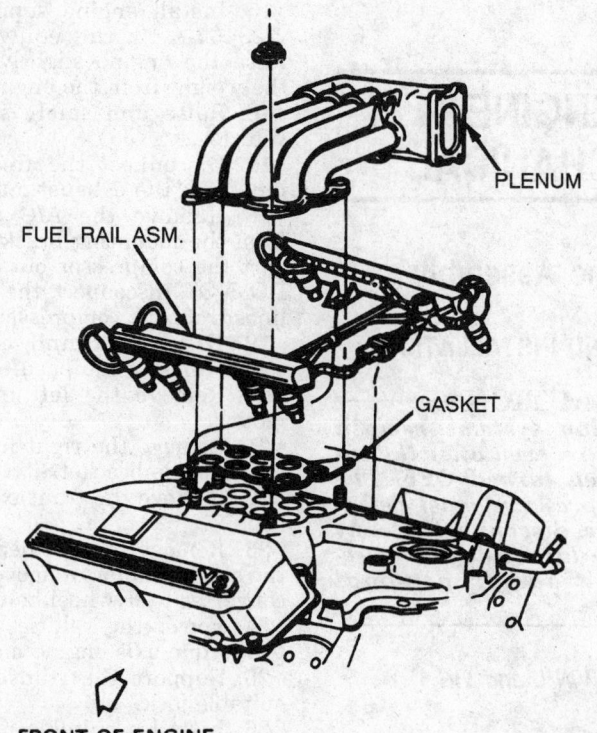

FRONT OF ENGINE

241177

Removing the fuel rail — 4.9L (VIN B) engines

269719

The reset button is located inside the hole under the passenger side of the instrument panel, near the door — 1993 Park Avenue shown

must be reset. Resetting can be accomplished as follows:

1. Eighty Eight — the **ENGINE OIL LIFE INDEX** can be reset by pressing the **RESET** and **OIL** buttons simultaneously for at least 5 seconds while on the **ENGINE OIL LIFE INDEX** display. The driver information system will reset the **ENGINE OIL LIFE INDEX** to 100 percent and display a **ENGINE OIL LIFE INDEX** of 100 percent.

2. Ninety Eight — Press the **OIL** button to display the oil message and then press the **RESET** button for at least 5 seconds. The oil life monitor will be reset to 100 percent.

NOTE: The Engine Oil Life Index is stored on a non-volatile memory chip and will not require resetting by disconnecting the battery cables and/or fuse.

1994–95 Models

Vehicles that are equipped with a Drivers Information Center and have an Oil life index, will require the Oil Life indicator to be reset after each oil change.

Reset the Oil Life Display as follows:

1. Acknowledge all diagnostic messages in the Drivers Information Center by pressing **RESET**.

2. Press the **SEL** button on the left to select **OIL**. Press **SEL** button on the right if necessary to display oil life.

3. Press and hold the **RESET** button for about 5 seconds. Once the oil life index has been reset, a **RESET** message will be displayed and then oil life will change to 100%.

Be careful not reset the oil life accidentally at any time other than when the oil has just been changed. It can not be reset accurately until the next oil change.

1993 Pontiac Bonneville

The Driver Information Center lights up for a few seconds when the ignition is turned to **ON**. To see the entire DIC, press and release the DIC button to the right of the display. The system is divided into 4 main systems: Function Monitor, Lamp Check, Security and Service Reminder.

To use the service reminder, the vehicle owner must first decide which service item to check. For example, if the owner wants to know when to next rotate the tires, push and hold the DIC button until the **SERVICE REMINDER** light comes ON. Then push and release the DIC button until **ROTATE TIRES** appears. The number at the bottom then shows how many miles there are to go before the tires should be rotated.

To reset the service reminder, push the DIC until the service item desired appears and do not release the button. In 5–10 seconds the display will start to count down, 500 miles at a time. When the display reads the distance the vehicle owner wishes to set, release the button. Note that sometimes a service reminder display will stay ON even though it has been reset, but the reminder should go OFF after the vehicle has been driven about 10 miles.

ENGINE MECHANICAL

Engine Assembly

REMOVAL AND INSTALLATION

—— CAUTION ——
Fuel injection systems remain under pressure, even after the engine has been turned OFF. The fuel system pressure must be relieved before disconnecting any fuel lines. Failure to do so may result in fire and/or personal injury.

1993–95 3.8L (VIN L and VIN 1) Engines

1. Disconnect the negative battery cable. Properly relieve the fuel system pressure.

2. Mark the position of the hood hinges for re-installation reference and remove the hood.

3. Drain the cooling system and engine crankcase.

4. Remove the strut tower cross brace, if necessary.

5. Disconnect the heater hoses from the engine. Remove the upper and lower radiator hoses.

6. Disconnect the starter wiring.

7. Disconnect the main wiring harness to the engine and the battery harness connectors located near the relay center.

8. Remove the serpentine belt(s).

9. Remove the power steering pump. Remove the air cleaner and air inlet duct.

10. Disconnect the control cables from the throttle body and cable mounting bracket.

11. Disconnect the Throttle Position Sensor (TPS), Idle Air Control (IAC) valve, MAT Sensor and Oxygen Sensor (O_2) connectors, oil pressure switch connectors, power steering cut off switch connectors, and low oil level sensor.

12. Disconnect the ignition coil assembly ground wire from the inner fender panel.

13. Disconnect and cap the fuel lines from the fuel rail.

14. Disconnect and tag all necessary vacuum lines from the engine.

15. Install engine support fixture J-28467-A or the equivalent, and raise the engine slightly to remove the weight from the engine mounts.

16. Raise and safely support the vehicle.

17. Disconnect the front exhaust pipe from the exhaust manifold.

18. Remove the A/C compressor from the mounting bracket and support the compressor out of the way. DO NOT disconnect the refrigerant lines from the compressor.

19. Disconnect and cap the oil cooler lines, if equipped.

20. Remove the left front engine mount.

21. Remove the right front engine-to-transaxle bracket.

22. Remove the transaxle-to-engine bolts.

23. Remove the flywheel cover and the starter motor. Remove the torque converter-to-flywheel bolts.

24. Lower the vehicle and remove the torque axis engine mount.

25. Support the transaxle with a suitable jack.

26. Install a suitable lifting device and remove the support fixture.

27. Remove the engine from the vehicle.

To install:

28. Carefully install the engine in the vehicle. Install the transaxle-to-engine bolts and tighten to 55 ft. lbs. (75 Nm).

29. Install the engine support fixture and remove the lifting device.

30. Raise and safely support the vehicle.

31. Install the torque axis engine mount. Install the right front engine-to-transaxle bracket.

32. Install the left front engine mount.

33. Install the torque converter-to-flywheel bolts and tighten to 46 ft. lbs. (62 Nm).

34. Install the flywheel cover.

35. Connect the engine oil cooler lines. Install the A/C compressor.

36. Connect the exhaust pipe to the exhaust manifold.

37. Lower the vehicle.

38. Connect the Throttle Position Sensor, (TPS), Idle Air Control (IAC) valve, MAT Sensor and Oxygen (O$_2$) Sensor connector, oil pressure switch, low oil level sensor and power steering cut off switch connector.

39. Connect the control cables to the throttle body and cable bracket.

40. Install the air cleaner and air duct assembly.

41. Install the power steering pump.

42. Connect the main engine harness and battery harness connectors.

43. Connect the wiring to the starter.

44. Connect the fuel lines to the fuel rail. Connect the engine vacuum lines.

45. Connect the heater hoses and install the upper and lower radiator hoses.

46. Install the strut tower cross brace, if necessary.

47. Refill the cooling system and engine crankcase. Connect the negative battery cable.

48. Install the hood, aligning the marks made during removal.

49. Start the engine and check for leaks. Test all systems for proper operation.

3.8L (VIN K) and 1996–97 3.8L (VIN 1) Engines

1. Disconnect the negative battery cable.

2. Mark the position of the hood hinges, for installation reference, and remove the hood from the vehicle.

3. Properly relieve the fuel system pressure. Drain the engine coolant and engine oil.

4. Remove the radiator and heater supply hoses.

5. Remove the negative battery cable from the engine block.

6. Disconnect the engine harness at the bulkhead.

7. Remove the drive belts.

8. Remove the power steering pump and set aside.

9. Remove the air flow duct.

10. Remove the throttle cable from the throttle mounting bracket and other applicable cables.

11. Disconnect wiring connectors from the MAT sensor, TP sensor, Idle Air Control valve, Oxygen sensor, A/C compressor, oil pressure switch, power steering cutoff switch, vehicle speed sensor, and low oil level sensor.

12. Remove the ignition assembly ground strap from the fender inner panel attaching screws.

13. Remove the fuel feed and return lines from the fuel rail and fuel pressure regulator.

14. Remove the emission control canister hoses from the throttle body connections.

15. Remove the brake booster and heater control hoses from the engine vacuum connections.

16. Disconnect the vacuum hoses at the cruise control and servo assembly.

17. Raise and safely support vehicle.

18. Remove the exhaust pipe from the right manifold.

19. Remove the A/C compressor and tie back the compressor away from the engine. Do not disconnect the refrigerant lines.

20. Remove the right front engine-to-transaxle bracket.

21. Remove the flywheel cover.

22. Remove the starter motor.

23. Remove the torque converter-to-flywheel bolts. Use a scribe to mark the flywheel to torque converter relationship for reassembly.

24. Lower the vehicle.

25. Attach a suitable lifting hook and chain to the engine lifting brackets. Raise the engine slightly to take the weight off the engine mounts.

26. Remove the engine torque axis engine mount.

27. Support the transaxle and remove the engine-to-transaxle bolts.

28. Separate the engine from the transaxle, raise the engine and remove from the vehicle.

To install:

29. Lower the engine assembly into the vehicle.

30. Install and tighten the engine-to-transaxle bolts to 55 ft. lbs. (75 Nm).

31. Install and tighten the torque axis engine mount bolts to 52 ft. lbs. (87 Nm).

32. Raise and safely support the vehicle.

33. Align the torque converter and flywheel with the marks made during removal and tighten the torque converter to flywheel bolts to 46 ft. lbs. (62 Nm).

34. Install the remaining components in the reverse order of removal.

35. Fill the cooling system and fill the engine with the proper type and quantity of engine oil.

36. Install the hood, aligning the marks made during removal.

37. Connect the negative battery cable. Start the engine and check for leaks.

38. Road test the vehicle and check operation.

4.9L (VIN B) Engine

1. Disconnect the negative battery cable. Properly relieve the fuel system pressure.

2. Properly recover the refrigerant from the A/C system.

3. Drain the cooling system into a suitable container.

4. Remove the air cleaner.

5. Mark the position of the hood on the hinges, for assembly reference, then remove the hood.

6. Remove the cooling fan.

7. Remove the drive belt.

8. Disconnect the upper radiator hose and heater hoses.

9. Label and disconnect the wiring connectors for the oil pressure switch, distributor, EGR solenoid, engine temperature switch, Idle Speed Control (ISC) motor, Throttle Position Sensor (TPS), coolant temperature sensor, fuel injectors at rail connector, Manifold Air Temperature (MAT) sensor, oxygen sensor and alternator. Disconnect the ground wire at the alternator bracket.

10. Disconnect the throttle cable from the throttle lever.

11. Remove the cruise control diaphragm with bracket and move out of the way.

12. Remove the exhaust crossover pipe.

13. Disconnect the oil and transmission cooler lines from the radiator. Remove the radiator.

14. Disconnect the oil cooler lines from the oil filter adapter and remove the lines. Remove the oil cooler line bracket at the transmission.

15. Remove the air cleaner mounting bracket. Remove the oil filter housing adapter.

16. Remove the strut tower cross car brace. If equipped with heated windshield, disconnect from the alternator.

17. Disconnect the right front heater hose.

18. Remove the power steering line brace from the right cylinder head. Remove the power steering pump and tensioner assembly and position out of the way, toward the front of the engine.

19. Disconnect the refrigerant lines from the accumulator and condenser. Cap all openings to prevent the entrance of dirt and moisture.

20. Disconnect the fuel lines at the fuel rail. Remove the fuel line bracket from the transmission and move the fuel lines out of the way.

21. Disconnect the vacuum modulator line and power brake vacuum line and reposition.

22. Raise and safely support the vehicle.

23. Remove the starter heat shield.

24. Disconnect the wiring from the starter and disconnect the ground wires from the block.

25. Disconnect the exhaust pipe from the manifold.

26. Remove the flexplate covers and remove the starter.

27. Mark the position of the torque converter on the flexplate, then remove the flexplate-to-converter bolts.

28. Remove the A/C compressor lower dust shield.

29. Remove the right front wheel and the outer wheelhouse plastic shield.

30. Remove the right rear transaxle-to-engine support and bolt.

31. Remove the lower engine damper nut.

32. Remove the front engine mount nuts and right rear transmission mount bolts.

33. Disconnect the lower radiator hose.

34. Remove the oxygen sensor wires and heater bypass bracket from the right side of the vehicle.

35. Lower the vehicle.

36. Attach suitable lifting equipment and support the engine.

WARNING

Be careful when attaching the right rear lift hook to the bracket to assure clearance to the A/C accumulator line.

37. Remove the 5 upper transmission-to-engine bolts.

38. Remove the engine from the vehicle and position on a suitable workstand.

To install:

WARNING

During engine installation, be careful not to damage any items in the engine compartment (transmission dipstick tube, throttle linkage, etc.). Some twisting and maneuvering of the engine is sometimes required.

39. Raise the transaxle with the jack.

40. Lower the engine into the vehicle on an angle matching the transaxle. Engage the dowels on the block with the holes in the transaxle case.

NOTE: **Make sure that the torque converter is properly positioned to the flexplate and engaged in the front pump of the transaxle.**

41. Install the 5 upper transmission-to-engine bolts and tighten to 55 ft. lbs. (75 Nm).

42. Lower and remove the floor jack from the transaxle.

43. Lower the engine, making sure it is properly seated on the mount. Remove the engine lifting equipment.

44. Raise and safely support the vehicle. Support the engine.

45. Install the lower right-hand transaxle-to-engine bolt and tighten to 55 ft. lbs. (75 Nm).

46. Install the right side engine brace and remove the engine support.

47. Install the oxygen sensor wires and the heater bypass bracket from the right side of the vehicle.

48. Install the front engine mount nuts and right rear transmission mount bolts.

49. Install the lower engine damper nut.

50. Align the flexplate and torque converter with the marks made during removal. Install the flexplate-to-converter bolts and tighten to 46 ft. lbs. (63 Nm).

51. Install the flexplate covers and the starter.

52. Connect the ground wires to the block and connect the starter wiring.

53. Install the starter heat shield.

54. Connect the exhaust pipe to the manifold.

55. Connect the lower radiator hose. Install the A/C compressor lower dust shield.

56. Install the outer wheelhouse plastic shield and the right front wheel. Lower the vehicle.

57. Install the remaining components in the reverse order of removal.

58. Install the air cleaner and connect the negative battery cable.

59. Refill the cooling system. Check all fluid levels.

60. Evacuate, recharge and leak test the A/C system.

61. Run the engine and check for leaks and proper engine operation.

Engine Mounts

REMOVAL AND INSTALLATION

3.8L (VIN L and 1) Engines

1. Disconnect the negative battery cable.

2. Install engine support fixture J-28467-A or equivalent. Using the fixture, raise the engine slightly to remove the engine weight from the engine mount.

3. Remove the engine drive belt(s).

4. Remove the engine mount through bolt and nut.

5. Raise the engine with the support fixture until the power steering reservoir is touching the strut tower cross brace.

6. Loosen the 2 lower mounting bolts. It is not necessary to remove the bolts to remove the engine mount.

7. Remove the A/C line clip from the engine mount.

8. Remove the 2 engine mount top bolts. Remove the engine mount.

To install:

9. Position the engine mount on the frame so the cut outs in the front of the mount go around the lower mounting bolts.

10. Install the upper mounting bolts and tighten the 4 engine mount bolts to 62 ft. lbs. (87 Nm).

11. Install the A/C line clip.

12. Lower the engine with the fixture until the through bolt can be installed.

13. Tighten the through bolt nut to 52 ft. lbs. (70 Nm).

14. Install the serpentine belt(s).

15. Remove the engine support fixture. Connect the negative battery cable.

3.8L (VIN K) Engines

1. Disconnect the negative battery cable.

2. Remove the vacuum reservoir.

3. Install engine support fixture J-28467-A , lifting bracket J 38854, core support wedges J 28467 350 and engine core support arm J 28467 330 or equivalents. Using the fixture, raise the engine slightly to remove the engine weight from the engine mount.

4. Remove the 2 bottom bolts on the torque axis engine mount.

5. Remove the 2 through bolts from the front torque axis engine mount.

6. Remove the torque axis engine mount. Remove the drive belts.

7. Remove the power steering pump.

8. Raise the engine to gain clearance and remove the engine mount bracket.

To install:

9. Install the engine mount bracket to the front of the engine.

10. Install the power steering pump. Install the drive belts.

11. Install the torque axis engine mount to the rail and tighten the attaching bolts to 52 ft. lbs. (87 Nm).

12. Safely lower the engine support fixture to align the engine mount bracket wit the torque axis engine mount.

13. Install and tighten the torque axis through bolt to 52 ft. lbs. (70 Nm).

14. Remove the engine support fixture.

15. Install the vacuum reservoir.

16. Connect the negative battery cable.

4.9L (VIN B) Engine

Right Side Mount

1. Disconnect the negative battery cable. Raise and safely support the vehicle.

2. Remove the 2 heat shield screws and remove the heat shield.

3. Remove the bolt from the engine mount brace at the engine mount bracket.

4. Loosen the nut at the top of the brace from the exhaust manifold and position the brace out of the way.

5. Support the engine with a transmission jack.

6. Remove the 2 bolts holding the engine mount to the transaxle.

7. Remove the 4 nuts at the top and bottom of the mount.

8. Raise the engine enough to remove the engine mount.

To install:

9. Install the engine mount in the vehicle. Lower the engine.

10. Install the mount to the transaxle and tighten the bolts to 50 ft. lbs. (68 Nm).

11. Install the 4 nuts at the top and bottom of the mount and tighten to 30 ft. lbs. (41 Nm).

12. Install the brace from the bracket to the engine and tighten the nut and bolt to 25 ft. lbs. (34 Nm).

13. Install the heat shield and heat shield mounting screws.

14. Remove the transmission jack and lower the vehicle.

15. Connect the negative battery cable. Start the engine and check for proper operation.

Left Side Mount

1. Disconnect the negative battery cable. Remove the air cleaner assembly.

2. Remove the serpentine belt.

---------- **CAUTION** ----------
When servicing the A/C system all of the recommended procedures must be followed and the correct tools and safety equipment must be used. Failure to use the proper safety equipment or using the incorrect tool for the job can cause serious personal injury and/or damage to the vehicle's A/C system.

3. Properly recover the refrigerant from the A/C system.

4. Remove the lower center exhaust manifold nuts.

5. Raise and safely support the vehicle.

6. Remove the right side engine compartment splash shield.

7. Remove the A/C splash shield.

8. Remove the A/C compressor brackets, then remove the compressor.

9. Remove the engine mount bracket bolts from the engine block and cradle.

10. Raise the engine with a transmission jack enough to remove the mount.

11. Remove the engine mount from the bracket.

To install:

12. Place the mount in a vise and position the mount bracket onto the mount.

13. Install the bracket-to-mount nuts and tighten to 30 ft. lbs. (41 Nm).

14. Install the bracket assembly in the vehicle.

15. Install the mount assembly to engine block bolts and tighten to 50 ft. lbs. (68 Nm).

16. Lower the engine, install the mount to frame bolts and tighten to 30 ft. lbs. (41 Nm).

17. Remove the transmission jack.

18. Install the remaining components in the reverse order of removal.

19. Connect the negative battery cable.

20. Install the air cleaner assembly.

21. Evacuate and recharge the A/C system.

22. Start the engine and check for proper operation.

Cylinder Head

REMOVAL AND INSTALLATION

---------- **CAUTION** ----------
Fuel injection systems remain under pressure, even after the engine has been turned OFF. The fuel system pressure must be relieved before disconnecting any fuel lines. Failure to do so may result in fire and/or personal injury.

3.8L (VIN L, K and 1) Engines

Left Side (Front)

1. Disconnect the negative battery cable. Properly relieve the fuel system pressure.

2. Drain the cooling system and remove the supercharger assembly.

3. Remove the intake manifold and the left side exhaust manifold.

4. Remove the valve covers and remove the rocker arm assemblies and pushrods, keeping everything in order for reinstallation.

5. Tag and disconnect the ignition wires and remove the spark plugs.

6. Remove the alternator front mounting bracket and the ignition module with bracket.

7. Remove the one bolt securing the A/C bracket to the cylinder head.

8. Remove the cylinder head bolts in the reverse order of the torque sequence, and remove the cylinder head.

9. Remove the power steering pump and set to the side. Complete removal of the steering pump is not needed.

10. Clean all gasket mating surfaces and the cylinder head bolt holes in the block.

To install:

11. Place the cylinder head gasket on the engine block dowels with the note **THIS SIDE UP** facing the cylinder head and the arrow facing the front of the engine. Position the cylinder head on the engine block.

---------- **WARNING** ----------
In order to prevent damage to the gasket, when installing the cylinder head, do not slide the cylinder head on the gasket. Head gasket are not interchangeable. Failure to install with arrow pointing to the front will cause gasket failure and possible engine damage. Gaskets are identified by either an L or an R stamped in it next to the arrow.

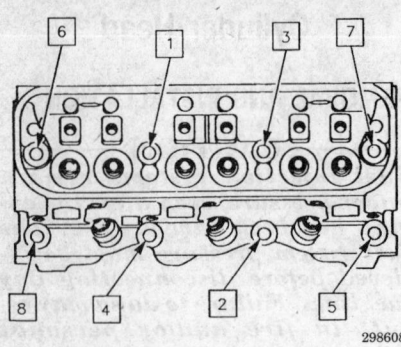

Cylinder head bolt torque sequence — 3.8L (VIN L, K and 1) engines

298608

12. Install the cylinder head bolts and tighten as follows:

NOTE: This engine uses special torque to yield head bolts. The procedure must be followed carefully and new bolts must be used whenever the head is removed. Total bolt torque should not exceed 60 ft. lbs. (81 Nm).

    a. Tighten the cylinder head bolts, in sequence, to 37 ft. lbs. (50 Nm).

    b. Rotate each bolt 130 degrees, in sequence.

    c. Rotate the center 4 bolts an additional 30 degrees, in sequence.

13. Install the pushrods, rocker arm assemblies and valve covers. (Apply approved thread lock compound to the rocker arm pedestal bolts before assembly.)

14. Install the intake and supercharger assembly.

15. Install the exhaust manifold.

16. Install the alternator front mount bracket and ignition module with bracket.

17. Install the spark plugs and wires.

18. Install the A/C compressor bracket bolt, and torque to 52 ft. lbs. (70 Nm).

19. Fill the cooling system and connect the negative battery cable.

20. Start the engine and check for leaks and proper operation.

Right Side (Rear)

1. Disconnect the negative battery cable. Properly relieve the fuel system pressure.

2. Drain the cooling system and disconnect the exhaust crossover pipe.

3. Remove the supercharger assembly and the intake manifold.

4. Remove the right side exhaust manifold.

5. Remove the valve covers.

6. Remove the serpentine drive belt and the belt tensioner pulley.

7. Remove the power steering pump mounting bracket and lay the pump aside.

8. Remove the fuel line heat shield.

9. Tag and disconnect the ignition wires and remove the spark plugs.

10. Remove the rocker arm assemblies and pushrods, keeping everything in order for reinstallation.

11. Remove the cylinder head bolts in reverse order of installation and remove the cylinder head.

12. Clean all gasket mating surfaces and the cylinder head bolt holes in the block.

To install:

13. Place the cylinder head gasket on the engine block dowels with the note **THIS SIDE UP** facing the cylinder head and the arrow facing the front of the engine. Position the cylinder head on the engine block.

---------- **WARNING** ----------
In order to prevent damage to the gasket, when installing the cylinder head, do not slide the cylinder head on the gasket. Head gasket are not interchangeable. Failure to install with arrow pointing to the front will cause gasket failure and possible engine damage. Gaskets are identified by either an L or an R stamped in it next to the arrow.

14. Install the cylinder head bolts and tighten as follows:

NOTE: This engine uses special torque to yield head bolts. The procedure must be followed carefully and new bolts must be used whenever the head is removed. Total bolt torque should not exceed 60 ft. lbs. (81 Nm).

    a. Tighten the cylinder head bolts, in sequence, to 37 ft. lbs. (50 Nm).

    b. Rotate each bolt 130 degrees, in sequence.

    c. Rotate the center 4 bolts an additional 30 degrees, in sequence.

15. Install the intake manifold and the supercharger assembly.

16. Install the right side exhaust manifold.

17. Install the pushrods and rocker arm assemblies.

18. Install the valve cover(s).

19. Install the spark plugs and wires.

20. Install the power steering pump bracket and torque the bolts to 35 ft. lbs. (47 Nm).

21. Install the belt tensioner pulley and serpentine belt.

22. Connect the exhaust crossover pipe.

23. Fill the cooling system and connect the negative battery cable.

24. Start the engine end check for leaks and proper operation.

4.9L (VIN B) Engine

1. Disconnect the negative battery cable. Properly relieve the fuel system pressure.

2. Drain the cooling system.

3. Disconnect and label the spark plug wires, harnesses and vacuum hoses (if necessary).

4. Remove the rocker arm covers.

5. Remove the belt driven components.

NOTE: Label all valve train components as they are removed, so they can be reinstalled in their original locations.

6. Remove the rocker arm support retaining nuts/bolts and remove the rocker arm support.

---------- **WARNING** ----------
The rocker arm support must be removed with the rocker arms and pivots attached. The pivot assemblies or support bar threads may be damaged if pivot bolt torque is removed against valve spring pressure. If necessary, secure the support in a vise and remove the rocker arms and pivots.

7. Remove the pushrods.

8. Remove the intake manifold.

9. For the right cylinder head, disconnect the exhaust pipe from the exhaust manifold and disconnect the oxygen sensor connector.

10. Remove the right or left side exhaust manifold.

11. For the left cylinder head, remove the engine left bracket and dipstick tube.

12. Remove the cylinder head bolts, loosening them a little at a time in the reverse order of the torque sequence. Note the locations of the stud headed bolts, so they can be reinstalled in their proper locations.

13. Remove the cylinder head.

14. Clean all gasket mating surfaces. Clean the head bolt threads and the corresponding bolt holes in the cylinder block.

To install:

15. Install a new cylinder head gasket on the engine block. Make sure the gasket aligns with the dowel pins on the block.

16. Install the cylinder head.

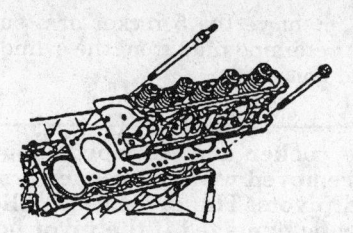

| 8 | 4 | 1 | 3 | 7 | INBOARD STUDS |
| 10 | 6 | 2 | 5 | 9 | OUTBOARD BOLTS |

348417

Cylinder head bolt torque sequence — 4.9L (VIN B) engines

17. Lubricate the cylinder head bolts with GM lubricant 1052356 or equivalent; do not use engine oil.

18. If equipped with the "short" head bolts, as determined during the removal procedure, install the cylinder head bolts finger-tight, then tighten, in sequence, as follows:

  a. Tighten all bolts and studs, in sequence, to 29 ft. lbs. (40 Nm).

  b. Tighten all bolts and studs, in sequence, to 51 ft. lbs. (70 Nm).

  c. Tighten all bolts and studs, in sequence, to 85 ft. lbs. (115 Nm).

  d. Tighten the 3 center inboard studs, 1, 3 and 4 to 88 ft. lbs. (120 Nm).

19. If equipped with the "long" head bolts, as determined during the removal procedure, install the cylinder head bolts finger-tight, then tighten, in sequence, as follows:

  a. Tighten all bolts and studs, in sequence, to 29 ft. lbs. (40 Nm).

  b. Tighten all bolts and studs, in sequence, to 51 ft. lbs. (70 Nm).

  c. Tighten all bolts and studs, in sequence, to 81 ft. lbs. (110 Nm).

  d. Tighten all bolts and studs, in sequence, to 88 ft. lbs. (120 Nm).

  e. Tighten the 3 center inboard studs, 1, 3 and 4 to 96 ft. lbs. (130 Nm).

20. Install the engine lift bracket. For the left cylinder head, install dipstick tube.

21. Install the exhaust manifold and the oxygen sensor connector (for the right cylinder head).

22. For the right cylinder head, connect the exhaust system pipe to the manifold and the cross-over pipe.

23. Install the intake manifold.

24. Install the pushrods, making sure they are seated in the lifters.

25. If removed, install the rocker arms and pivots to the rocker arm support. Loosely install the pivot bolts, then tighten the pivot bolts to 22 ft. lbs. (30 Nm).

26. Install the rocker arm support, with the rocker arms and pivots installed, over the 5 studded head bolts. Make sure each pushrod is seated in a rocker arm.

27. Loosely install the rocker arm support retaining nuts and bolts.

28. Tighten the 5 rocker arm support retaining nuts alternately and evenly until snug.

NOTE: Check the pushrods from time to time during the tightening procedure to make sure that they are correctly positioned.

29. Tighten the 4 rocker arm retaining bolts until snug.

30. Tighten the 5 rocker arm support retaining nuts to 37 ft. lbs. (50 Nm).

31. Tighten the 4 rocker arm retaining bolts to 7 ft. lbs. (10 Nm).

32. For the right cylinder head, install the wedges and coat them with RTV for improved sealing.

33. Install the rocker arm covers.

34. Install the belt driven components.

35. Refill the cooling system.

36. Connect the negative battery cable.

37. Run the engine and check for leaks and proper engine operation.

Valve Lifters

BLEEDING

There is no valve lifter bleeding to be performed on these engines.

REMOVAL AND INSTALLATION

———— CAUTION ————

Fuel injection systems remain under pressure, even after the engine has been turned OFF. The fuel system pressure must be relieved before disconnecting any fuel lines. Failure to do so may result in fire and/or personal injury.

NOTE: When removing valve train components, it is very important that they are marked for installation reference, so that they can be reinstalled in the same location from which they were removed.

3.8L (VIN L, K and 1) Engines

1. Disconnect the negative battery cable. Properly relieve the fuel system pressure.

2. Drain the cooling system.

3. Remove the intake manifold.

4. Remove the rocker arm covers.

5. Remove the rocker arm mounting bolts, rocker arms, pedestals and pushrods. Keep all parts in order so they can be reinstalled in their original locations.

6. Remove the guide retainer mounting bolts and retainer.

7. Remove the lifter guides and remove the lifters. Keep all parts in order so they can be reinstalled in their original locations.

8. Inspect the lifters for wear and/or damage; replace as necessary.

9. Inspect the pushrod ends for wear and/or damage. Roll the pushrods on a flat surface to check for a bent condition. Replace any pushrods that are bent and/or damaged.

10. Clean all gasket mating surfaces. Make sure all thread adhesive is removed from the rocker arm pedestal bolts and rocker arm cover bolts.

To install:

11. Coat the lifters with assembly lubricant and install in their original locations.

12. Install the lifter guides and guide retainer. Tighten the retainer mounting bolts to 22 ft. lbs. (30 Nm).

13. Lubricate the pushrod tips with assembly lubricant and install them in their proper locations.

14. Lubricate the rocker arms and pedestals with assembly lubricant and install. Be sure the pushrod tips are properly seated in the rocker arms.

15. Apply suitable thread locking compound to the rocker arm pedestal bolt threads and install. Tighten the bolts to 19 ft. lbs. (25 Nm) plus an additional 70 degree turn on 1993–94 VIN 1 and 1993–95 VIN L engines or to 11 ft. lbs. (15 Nm) plus an additional 90 degree turn on VIN K and 1995–97 VIN 1 engines.

16. Install the rocker arm covers, using new gaskets, and tighten the mounting bolts to 89 inch lbs. (10 Nm).

17. Install the intake manifold.

18. Connect the negative battery cable. Fill and bleed the cooling system.

19. Run the engine and check for leaks and proper engine operation.

4.9L (VIN B) Engine

1. Disconnect the negative battery cable. Relieve the fuel system pressure.

2. Remove the rocker arm covers.

3. Remove the 4 bolts from the cylinder head bosses which retain the rocker arm support.

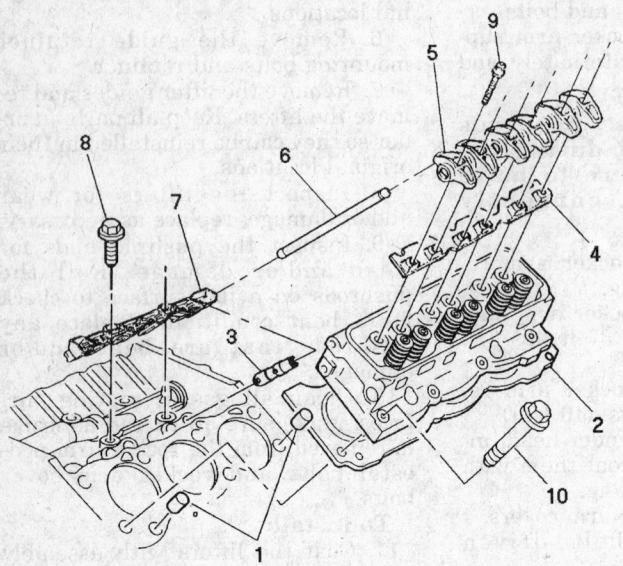

1. Dowel pin
2. Head gasket
3. Valve lifter
4. Pivot retainer
5. Rocker arm
6. Pushrod
7. Lifter guide
8. Bolt
9. Bolt
10. Head bolt

251043

Valve train components — 1995–96 3.8L (VIN K and VIN 1) engines shown, others are similar

ROCKER ARM PIVOT

PUSHROD

ROCKER ARM SUPPORT STUD (HEAD BOLT)

337919

Rocker arm support components — 4.9L (VIN B) engines

4. Remove the 5 rocker arm support retaining nuts from the cylinder head bolt studs.

WARNING
The rocker arm support should be removed with the rocker arms and pivots. The pivot assemblies may be damaged if the pivot bolt torque is not released evenly against valve spring pressure.

5. Remove the pushrods, keeping them in order so they can be reinstalled in their original locations.
6. Remove the intake manifold.
7. Remove the valve lifter retainer bolts and remove the valve lifter retainer.
8. Remove the valve lifters and lifter guides. Keep all parts in order so they can be reinstalled in their original locations.
9. Inspect the valve lifters for wear and/or damage; replace as necessary.
10. Inspect the pushrod ends for wear and/or damage. Roll the pushrods on a flat surface to check for a bent condition. Replace any pushrods that are bent and/or damaged.

To install:
11. Thoroughly clean all gasket mating surfaces.
12. Lubricate the valve lifters with assembly lubricant and install them in the lifter bores.
13. Install the valve lifter guides.
14. Install the valve lifter retainer and tighten the mounting bolts to 15 ft. lbs. (20 Nm).
15. Install the intake manifold.
16. Lubricate the pushrod ends with assembly lubricant and install the pushrods.
17. Install the rocker arm support over the five mounting studs.
18. Guide each pushrod into the rocker arm.
19. Install the 5 rocker arm support nuts and 4 rocker arm support bolts loosely.
20. Snug down the 5 rocker arm support nuts evenly until snug. While tightening, check the pushrods to make sure they are still aligned.
21. Snug down the 4 rocker arm support bolts evenly.
22. Tighten the rocker arm support nuts 37 ft. lbs. (50 Nm).
23. Tighten the rocker arm support bolts to 7 ft. lbs. (9 Nm).
24. Install the rocker arm covers.
25. Connect the negative battery cable.
26. Run the engine and check for leaks and proper engine operation.

Valve Lash

ADJUSTMENT

The valve clearance cannot be adjusted on these engines. The hydraulic lifters function to maintain a zero clearance when the valves are opening and closing. Any clearance is instantaneously taken up by the hydraulic action. "Valve lifter noise" complaints may require cleaning of sludge coated or varnished valve lifters, or replacement.

Rocker Arms

REMOVAL AND INSTALLATION

NOTE: When removing valve train components, it is very important that they are marked for installation reference, so that they can be reinstalled in the same location from which they were removed.

3.8L (VIN L, K and 1) Engines

1. Disconnect the negative battery cable.
2. If removing the left (front) rocker arm cover, proceed as follows:
   a. Remove the accessory drive belt(s).
   b. Remove the alternator-to-brace bolt and remove the alternator brace.
   c. Disconnect the spark plug wires from the spark plugs and position the wires aside, out of the way.
3. If removing the right (rear) rocker arm cover, proceed as follows:
   a. Remove the accessory drive belt(s).
   b. Loosen the power steering pump bolts and slide the pump forward.
   c. Disconnect the spark plug wires from the spark plugs and position the wires aside, out of the way.
4. Remove the rocker arm cover bolts and the rocker arm cover.
5. Remove the rocker arm pedestal bolts and remove the rocker arm and pedestal assembly. Keep all parts in order so they can be reinstalled in their original locations.
6. Inspect the rocker arms and pedestals for wear and/or damage; replace as necessary.
7. Remove the pushrods. Keep them in order so they can be reinstalled in their original locations.
8. Inspect the pushrod tips for wear and/or damage. Roll the pushrods on a flat surface to check for a bent condition. Replace any pushrod that is bent and/or damaged.
9. Clean all parts and all gasket mating surfaces. Make sure all thread adhesive is removed from the rocker arm pedestal bolts and rocker arm cover bolts.

To install:

10. Lubricate the pushrod tips with assembly lubricant and install them in their proper locations.
11. Lubricate the rocker arms and pedestals with assembly lubricant and install. Be sure the pushrod tips are properly seated in the rocker arms.
12. Apply suitable thread locking compound to the rocker arm pedestal bolt threads and install. Tighten the bolts to 19 ft. lbs. (25 Nm) plus an additional 70 degree turn for 1993–95 3.8L (VIN L and 1) engines. For 3.8L (VIN K) and 1996–97 3.8L (VIN 1) engines tighten to 11 ft. lbs. (15 Nm) plus an additional 90 degree turn for .
13. Install the rocker arm cover using a new gasket. Apply suitable thread locking compound to the rocker arm cover bolts and install. Tighten to 89 inch lbs. (10 Nm).
14. Reposition the spark plug wires and connect them to the spark plugs.
15. Reinstall the power steering pump and/or alternator brace, as required.
16. Install the accessory drive belt(s). Connect the negative battery cable.
17. Run the engine and check for leaks and proper engine operation.

4.9L (VIN B) Engine

1. Disconnect the negative battery cable.
2. Remove the rocker arm cover.
3. Remove the 4 bolts from the cylinder head bosses which retain the rocker arm support.
4. Remove the 5 rocker arm support retaining nuts from the cylinder head bolt studs.

— WARNING —
The rocker arm support should be removed with the rocker arms and pivots. The pivot assemblies may be damaged if the pivot bolt torque is not released evenly against valve spring pressure.

5. Support the rocker arm support in a vise and remove the rocker arms and pivots.

To install:

6. Install the rocker arm and pivots to the rocker arm support.
7. Install the pivot bolts loosely.

NOTE: The pivot bolts are self tapping. The new rocker arm support does not require tapping prior to installation.

8. Tighten the pivot bolts to 22 ft. lbs. (29 Nm).
9. Install the rocker arm support, with all components in place, over the five mounting studs.
10. Guide each pushrod into the rocker arm.
11. Install the 5 rocker arm support nuts and 4 bolts loosely.
12. Snug down the 5 rocker arm support nuts evenly until snug. While tightening, check the pushrods to make sure they are still aligned.
13. Snug down the 4 rocker arm support bolts evenly.
14. Tighten the rocker arm support nuts 37 ft. lbs. (50 Nm).
15. Tighten the rocker arm support bolts to 7 ft. lbs. (9 Nm).
16. Install the rocker arm cover.
17. Connect the negative battery cable.
18. Run the engine and check operation.

Intake Manifold

REMOVAL AND INSTALLATION

— CAUTION —
Fuel injection systems remain under pressure, even after the engine has been turned OFF. The fuel system pressure must be relieved before disconnecting any fuel lines. Failure to do so may result in fire and/or personal injury.

3.8L (VIN L) Engine

1. Disconnect the negative battery cable. Properly relieve the fuel system pressure.
2. Drain the cooling system.
3. Remove the plastic engine cover clipped to the fuel rail.
4. Disconnect the air inlet duct from the throttle body.
5. Label and disconnect the spark plug wires from the rear of the engine and position the wires out of the way.
6. Disconnect the electrical connectors from the fuel injectors.
7. Disconnect the vacuum line from the pressure regulator.
8. Disconnect and cap the fuel lines from the fuel rail.
9. Remove the fuel rail mounting bolts and remove the fuel rail from the intake.

10. Remove the heat shield from the exhaust crossover pipe.

11. Remove the control cable bracket.

12. Disconnect the control cables, vacuum lines and electrical connectors from the throttle body and intake manifold.

13. Remove the power steering pump support bracket.

14. Remove the serpentine belt.

15. Remove the alternator.

16. Remove the alternator mounting bracket.

17. Disconnect the upper radiator hose from the thermostat housing.

18. Disconnect the heater and bypass hoses from the intake manifold.

19. Remove the intake manifold bolts and remove the manifold and gasket.

To install:

20. Thoroughly clean all gasket mating surfaces.

21. Install new intake manifold gaskets and position the intake manifold on the engine.

22. Install the intake manifold bolts and tighten, in sequence, to 89 inch lbs. (10 Nm). After all the bolts have been tightened, make a second pass making sure all bolts are still torqued to 89 inch lbs. (10 Nm).

23. Install the remaining components in the reverse order of removal. Tighten the fuel rail mounting bolts to 10 ft. lbs. (14 Nm).

24. Connect the negative battery cable.

25. Turn the ignition **ON**, to pressurize the fuel system, and check for leaks.

26. Install the plastic engine cover over the fuel rail.

27. Fill and bleed the cooling system.

28. Run the engine and check for leaks and proper engine operation.

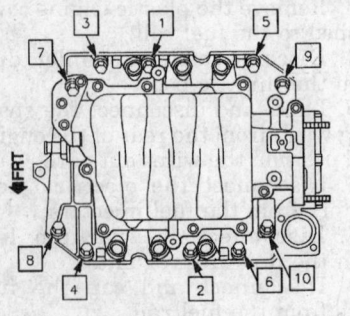

Intake manifold bolt torque sequence — 3.8L (VIN L) engines

3.8L (VIN K) Engine

1. Disconnect the negative battery cable. Properly relieve the fuel system pressure.

2. Remove the fuel injector sight shield and air inlet duct.

3. Disconnect the spark plug wires from the right side spark plugs and route the wires out of the way.

4. Label and disconnect the vacuum lines from the intake manifold.

5. Disconnect the fuel lines, vacuum lines and electrical connectors from the fuel rail. Remove the fuel rail mounting bolts and remove the fuel rail from the intake manifold.

6. Remove the EGR heat shield.

7. Remove the throttle cable bracket from the cylinder head mounting bracket and disconnect the cables from the throttle body lever.

8. Remove the throttle body support bracket.

9. Remove the upper intake plenum mounting bolts and remove the plenum and gasket.

10. Drain the cooling system.

11. Disconnect the upper radiator hose from the thermostat housing.

12. Remove the serpentine belt and the alternator.

13. Remove the 4 bolts and remove the drive belt tensioner assembly.

14. Disconnect the EGR valve outlet pipe, if necessary.

15. Remove the lower intake manifold mounting bolts and remove the intake manifold and gaskets.

To install:

16. Thoroughly clean all gasket mating surfaces.

17. Install the intake manifold using new manifold gaskets and install the intake manifold bolts finger-tight. With all the bolts in place, tighten the bolts in sequence to 11 ft. lbs. (15 Nm). With all bolts tightened, make a second pass, again tightening each bolt to 11 ft. lbs. (15 Nm).

18. Install the remaining components in the reverse order of removal. Torque the following:

Drive belt tensioner bolts to 37 ft. lbs. (50 Nm)

Intake plenum bolts to 11 ft. lbs. (15 Nm)

Fuel rail bolts to 7 ft. lbs. (10 Nm)

19. Connect the negative battery cable.

20. Turn the ignition **ON**, to pressurize the fuel system, and check for leaks.

21. Refill and bleed the cooling system.

22. Run the engine and check for leaks and proper engine operation.

3.8L (VIN 1) Engine

1993–95 Vehicles

1. Disconnect the negative battery cable.

2. Remove the supercharger dress up cover and air intake duct.

3. Properly relieve the fuel system pressure. Drain the cooling system.

4. Label and disconnect the spark plug wires on the right side of the engine and set aside.

5. Remove the manifold vacuum source harness.

6. Disconnect the fuel supply and return lines from the fuel rail.

7. Disconnect the vacuum hoses. Identify and tag as they are removed for installation reference.

8. Disconnect the upper radiator hose and bypass hose.

9. Disconnect the electrical connectors from the EGR valve, Throttle Position (TP) sensor, Idle Air Control (IAC) solenoid, fuel injectors and Manifold Absolute Pressure (MAP) sensor. Identify and tag as they are removed for installation reference.

10. Remove the EGR outlet pipe.

11. Disconnect the throttle and cruise control cables.

12. Remove the throttle bracket with the steering reservoir and set aside.

13. Remove the inner accessory drive belt.

14. Disconnect the heater hose pipe from the intake.

15. Remove the tensioner bracket-to-supercharger retaining stud using the double-nut method.

16. Remove the manifold bolts and remove the supercharger and intake manifold as an assembly. Discard the gasket and seals.

17. If the intake manifold is damaged or is to be replaced with another, the supercharger must be removed.

To install:

18. Clean the cylinder block, heads and intake manifold sealing surface of all old gasket material and oil.

19. Clean the intake manifold bolts and bolt holes of adhesive compound. Apply thread lock compound GM No. 12345493 or equivalent to the intake manifold bolt threads before assembly.

20. Install new gaskets and seals. Apply sealant to the ends of the manifold seals.

21. Install the intake manifold/supercharger assembly. Tighten the bolts, in sequence, to 11 ft. lbs. (15 Nm). Repeat the torque sequence and make sure all bolts are torqued to 11 ft. lbs. (15 Nm).

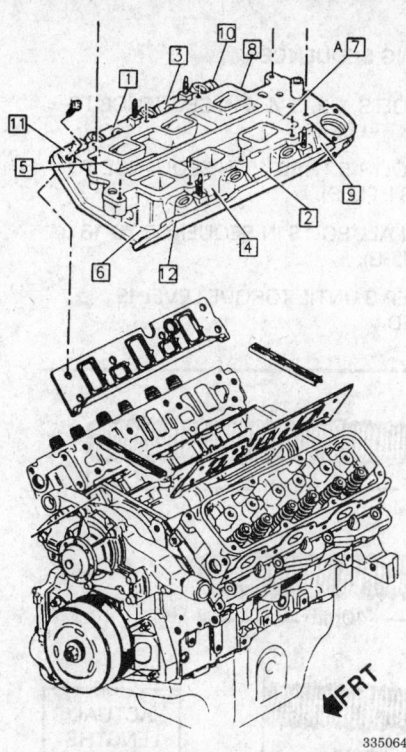

Lower intake manifold mounting bolt torque sequence — 1996–97 3.8L (VIN K and 1) engines

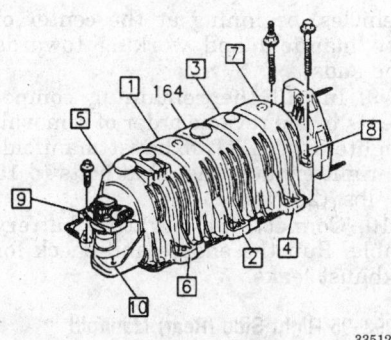

Upper intake manifold mounting bolt torque sequence — 3.8L (VIN K) engines

22. Install the remaining components in the reverse order of removal.

23. Connect the negative battery cable. Fill and bleed the cooling system.

24. Run the engine and check for leaks and proper engine operation.

1996–97 Vehicles

1. Disconnect the negative battery cable. Properly relieve the fuel system pressure.

2. Drain the radiator coolant into a suitable container.

3. Remove the supercharger assembly. Remove the thermostat housing.

4. Disconnect the EGR tube at the intake manifold.

5. Disconnect the electrical connection at the temperature sensor.

6. Remove the intake manifold.

To install:

7. Thoroughly clean all gasket mating surfaces.

8. Clean all old sealant from the intake manifold bolts and bolt holes.

9. Install new gaskets and manifold seals. Apply sealant to the ends of the manifold seals.

10. Install the intake manifold.

11. Install the intake manifold bolts and torque, in sequence, to 11 ft. lbs. (15 Nm).

12. Install the electrical connector to the temperature sensor.

13. Install the EGR tube to the intake manifold.

14. Install the thermostat housing.

15. Install the supercharger assembly.

16. Connect the negative battery cable. Fill and bleed the cooling system.

17. Run the engine and check for leaks and proper engine operation.

4.9L (VIN B) Engine

1. Disconnect the negative battery cable. Properly relieve the fuel system pressure.

2. Remove the air cleaner assembly. Drain the cooling system.

3. Remove the cross car brace.

4. Remove the accessory drive belt.

5. Remove the power steering and tensioner bracket assembly and reposition toward the front of the engine.

6. Remove the alternator and bracket. Disconnect the cable at the throttle body.

7. Label and disconnect the wiring connectors at the distributor, oil pressure switch, coolant sensor, EGR solenoid, Idle Speed Control (ISC) motor, Throttle Position Sensor (TPS), fuel injectors and Manifold Air Temperature (MAT) sensor.

8. Disconnect the Manifold Absolute Pressure (MAP) sensor hoses.

9. Disconnect the upper radiator hose and heater hoses at the thermostat housing.

10. Remove the A/C hose bracket.

11. Remove the distributor cap, spark plug wire protectors and reposition the cap.

12. Mark the position of the distributor rotor in relation to the distributor housing and mark the position of the distributor housing in relation to the engine. Remove the distributor.

13. Disconnect the fuel lines at the transmission bracket.

14. Label and disconnect the vacuum and fuel lines at the throttle body.

15. Loosen the vacuum line clip at the lift bracket.

16. Remove the vacuum line bracket at the left rear engine lift bracket and disconnect the vacuum supply line at the throttle body.

17. Disconnect the transmission modulator vacuum line.

18. Remove the EGR solenoid and bracket assembly.

19. Remove the rocker arm covers.

NOTE: **Label all valve train components as they are removed, so they can be reinstalled in their original locations.**

20. Remove the rocker arm support retaining nuts/bolts and remove the rocker arm support.

— WARNING —

The rocker arm support must be removed with the rocker arms and pivots attached. The pivot assemblies or support bar threads may be damaged if pivot bolt torque is removed against valve spring pressure. If necessary, secure the support in a vise and remove the rocker arms and pivots.

21. Remove the pushrods.

22. Remove the power steering bracket. Remove the right front engine lift bracket.

23. Remove the intake manifold bolts and remove the intake manifold.

24. Remove and discard the gaskets and seals. Clean all gasket mating surfaces.

To install:

25. Install new intake manifold end seals.

26. Apply RTV sealant to the four corners where the end seals meet the side gaskets.

27. Install new intake manifold gaskets.

28. Install the intake manifold bolts and tighten, in sequence, to specification.

29. Install the pushrods, making sure they are seated in the lifters.

30. If removed, install the rocker arms and pivots to the rocker arm support. Loosely install the pivot bolts, then tighten the pivot bolts to 22 ft. lbs. (30 Nm).

31. Install the rocker arm support, with the rocker arms and pivots installed, over the 5 stud headed head bolts. Make sure each pushrod is seated in a rocker arm.

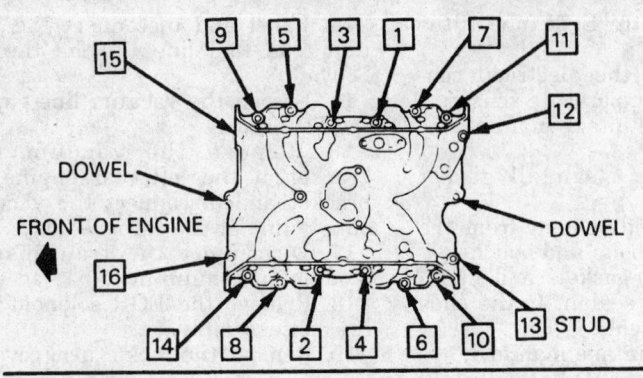

DOWEL

FRONT OF ENGINE

DOWEL

STUD

BOLT TIGHTENING SEQUENCE

1. TIGHTEN BOLTS 1, 2, 3, & 4 IN SEQUENCE TO 12.0 N·m (8 FT-LBS).

2. TIGHTEN BOLTS 5 THRU 16 IN SEQUENCE TO 12.0 N·m (8 FT-LBS).

3. RETIGHTEN ALL BOLTS IN SEQUENCE TO 16.0 N·m (12 FT-LBS).

4. REPEAT STEP 3 UNTIL TORQUE LEVEL IS MAINTAINED.

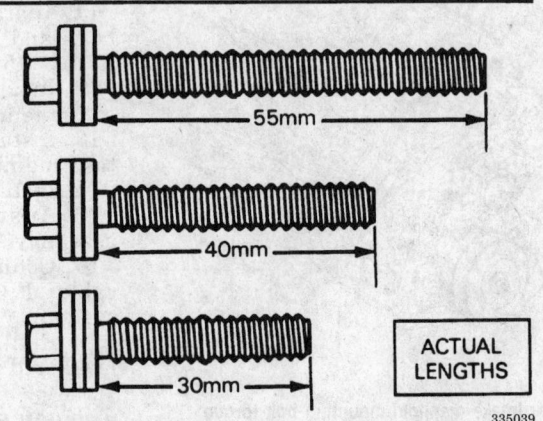

BOLT POSITION	BOLT LENGTH (MM)	BOLT POSITION	BOLT LENGTH (MM)
1	55	9	40
2	55	10	40
3	55	11	40
4	55	12	55
5	30	13	40 W/Studhead
6	30	14	40
7	30	15	55
8	30	16	40

55mm

40mm

30mm

ACTUAL LENGTHS

335039

Intake manifold bolt torque sequence — 4.9L (VIN B) engines

32. Loosely install the rocker arm support retaining nuts and bolts.

33. Tighten the 5 rocker arm support retaining nuts and bolts alternately and evenly until snug.

NOTE: Check the pushrods from time to time during the tightening procedure to make sure they are correctly positioned.

34. Tighten the 4 rocker arm retaining bolts until snug.

35. Tighten the 5 rocker arm support retaining nuts to 37 ft. lbs. (50 Nm).

36. Tighten the 4 rocker arm retaining bolts to 7 ft. lbs. (10 Nm).

37. Install the remaining components in the reverse order of removal.

38. Connect the negative battery cable. Refill the cooling system.

39. Start the engine and set the ignition timing. Run the engine and check for leaks and proper engine operation.

Exhaust Manifold

REMOVAL AND INSTALLATION

1993–95 3.8L (VIN L, K and 1) Engines

Left Side (Front) Manifold

1. Disconnect the negative battery cable.

2. Remove the 2 bolts attaching the left exhaust manifold to the right exhaust manifold.

3. Label and disconnect the spark plug wires from the spark plugs and position out of the way.

4. Remove the engine oil dipstick and tube.

5. Remove the manifold mounting bolts and studs and remove the exhaust manifold.

To install:

6. Make sure that the manifold and cylinder head mating surfaces are clean and free of any debris that might cause an exhaust leak.

7. Install the manifold to the cylinder head and right exhaust manifold.

8. Install the manifold mounting studs and bolts. Tighten the studs and bolts, gradually and evenly, to 38 ft. lbs. (52 Nm) for 1993–95 vehicles or to 22 ft. lbs. (30 Nm) for 1996–97

vehicles, beginning at the center of the manifold and working towards the ends.

9. Install the remaining components in the reverse order of removal. Tighten the 2 left exhaust manifold-to-right exhaust manifold bolts to 15 ft. lbs. (20 Nm).

10. Connect the negative battery cable. Run the engine and check for exhaust leaks.

1993–95 Right Side (Rear) Manifold

1. Disconnect the negative battery cable.

2. Label and disconnect the spark plug wires from the spark plugs and position out of the way.

3. Remove the throttle cable bracket. Remove the crossover pipe heat shield.

4. Remove the transaxle fluid dipstick and tube.

5. Disconnect the oxygen sensor electrical connector.

6. Remove the 2 bolts attaching the right exhaust manifold to the left exhaust manifold.

7. Remove the vacuum reservoir from the cowl.

8. Raise and safely support the vehicle.

9. Remove the catalytic converter heat shield and pipe hanger.

10. Remove the exhaust pipe-to-manifold mounting nuts and disconnect the exhaust pipe from the manifold.

11. Lower the vehicle.

12. Remove the engine lift bracket.

13. Remove the exhaust manifold bolts and studs and remove the exhaust manifold.

To install:

14. Make sure that the manifold and cylinder head mating surfaces are clean and free of any debris that might cause an exhaust leak.

15. Install the manifold to the cylinder head and left exhaust manifold.

16. Install the manifold mounting studs and bolts. Tighten the studs and bolts, gradually and evenly, to 38 ft. lbs. (52 Nm), beginning at the center of the manifold and working towards the ends.

17. Install the remaining components in the reverse order of removal. Torque the following:

Exhaust pipe-to-manifold nuts to 18 ft. lbs. (25 Nm)

Right exhaust manifold-to-left exhaust manifold bolts to 15 ft. lbs. (20 Nm)

18. Connect the negative battery cable. Run the engine and check for leaks.

1996–97 Right Side (Rear) Manifold

1. Disconnect the negative battery cable.

2. Label and disconnect the spark plug wires from the spark plugs.

3. Remove the transaxle fluid dipstick and tube.

4. Disconnect the oxygen sensor electrical connector.

5. Remove the 2 bolts attaching the right exhaust manifold to the crossover pipe.

6. Raise and safely support the vehicle.

7. Remove the front exhaust pipe-to-exhaust manifold nuts and disconnect the exhaust pipe from the manifold.

8. Lower the vehicle.

9. Remove the engine lift bracket.

10. Remove the exhaust manifold mounting studs and bolts and remove the exhaust manifold. Remove and discard the manifold-to-crossover pipe gasket and manifold-to-front exhaust pipe gasket.

To install:

11. Make sure that the manifold, cylinder head and crossover pipe mating surfaces are clean and free of any debris that might cause an exhaust leak.

12. Install the manifold to the cylinder head and crossover pipe using a new manifold-to-crossover pipe gasket.

13. Install the manifold mounting studs and bolts. Tighten the studs and bolts, gradually and evenly, to 22 ft. lbs. (30 Nm), beginning at the center of the manifold and working towards the ends.

14. Install the remaining components in the reverse order of removal. Tighten the front exhaust pipe-to-manifold nuts to 18 ft. lbs. (25 Nm) and the manifold-to-crossover pipe bolts and tighten to 15 ft. lbs. (20 Nm).

15. Connect the negative battery cable. Run the engine and check for exhaust leaks.

4.9L (VIN B) Engine

Left Side (Front) Manifold

1. Disconnect the negative battery cable. Remove the air cleaner assembly.

2. Disconnect the AIR pipe from the AIR pump and reposition it out of the way.

3. Detach the O^2 sensor wire.

4. Remove the drive belt.

5. Remove the power steering and tensioner bracket from the front of the manifold.

6. Remove the A/C hose bracket.

7. Remove the cooling fans.

8. Disconnect and tag the spark plug wires from the front spark plugs.

9. Raise and safely support the vehicle. Disconnect the Y-pipe from the left exhaust manifold.

10. Remove the engine and A/C brace from the front of the manifold.

11. Remove the exhaust manifold bolts, then remove the manifold.

To install:

12. Clean all gasket surfaces completely.

13. Install the exhaust manifold using a new gasket. Install the bolts and tighten to 16 ft. lbs. (22 Nm).

14. Install the remaining components in the reverse order of removal.

15. Connect the negative battery cable. Start the vehicle and verify no exhaust leaks.

Right Side (Rear) Manifold

1. Disconnect the negative battery cable. Remove the air cleaner assembly.

2. Disconnect the EGR pipe from the exhaust manifold.

3. Raise and safely support the vehicle.

4. Disconnect the exhaust Y-pipe from the right side manifold.

5. Disconnect the engine mount brace from the front of the exhaust manifold.

6. Detach the O^2 sensor wire.

7. Remove the exhaust manifold bolts. Remove the exhaust manifold.

To install:

8. Clean all gasket surfaces completely.

9. Install the exhaust manifold using a new gasket and install the bolts finger tight to hold the assembly in place.

10. Tighten the bolts to 16 ft. lbs. (22 Nm).

11. Install the remaining components in the reverse order of removal.

12. Connect the negative battery cable. Start the vehicle and verify no exhaust leaks.

Supercharger

REMOVAL AND INSTALLATION

NOTE: A small amount of oil seepage through the front seal, behind the pulley, of the supercharger is normal. This seepage is caused by minute traces of oil escaping around the seal due to normal pressure build up in the oil cavity within the supercharger. A build up of dust can stick to the thin oil film, which causes the oil seepage to appear worse than it really is. The supercharger should not be replaced for this seepage, however, if supercharger oil is visually dripping or puddling from the supercharger front seal, the supercharger will need to be replaced. The supercharger oil level should be checked every 30,000 miles or every 36 months.

―――――― **CAUTION** ――――――

Fuel injection systems remain under pressure, even after the engine has been turned OFF. The fuel system pressure must be relieved before disconnecting any fuel lines. Failure to do so may result in fire and/or personal injury.

1993–95 Engines

1. Disconnect the negative battery cable. Properly relieve the fuel system pressure.

2. Drain the cooling system.

3. Remove the drive belt from the supercharger pulley.

4. Remove the engine dress up cover.

5. Disconnect and cap the fuel lines from the fuel rail.

6. Label and disconnect the vacuum lines from the supercharger.

7. Disconnect the electrical connectors from the fuel injector.

8. Remove the electrical harness shield from the front of the supercharger and disconnect the electrical connector from the supercharger.

9. Remove the fuel rail mounting bolts and remove the fuel rail.

10. Tag and detach the connectors from the IAC valve, Throttle Position Sensor (TPS), MAP sensor, MAF sensor, EGR valve, boost control solenoid and the coolant temperature sensor.

11. Disconnect the air intake duct from the throttle body.

12. Disconnect the EGR pipe from the supercharger.

13. Disconnect the control cables from the throttle body.

14. Remove the boost pressure manifold and vacuum block.

15. Remove the control cable bracket.

16. Remove the tensioner bracket-to-supercharger mounting stud using the double-nut method.

NOTE: The stud must be removed or the supercharger can not be lifted off the lower intake manifold mounting dowels.

17. Remove the throttle body from the supercharger.

18. Remove the supercharger-to-intake manifold bolts.

19. Remove the supercharger, gasket and coolant passage O-rings.

To install:

20. Thoroughly clean the intake manifold and supercharger mating surfaces. Make sure the locator pins are in their proper location on the intake manifold.

21. Install new coolant passage O-rings and new supercharger-to-intake manifold gasket. Do not use any sealant on the gasket.

22. Install the supercharger and the mounting bolts. Only tighten the bolts finger-tight at this time.

23. Install the tensioner bracket-to-supercharger stud and tighten to 88 inch lbs. (10 Nm).

24. Tighten the supercharger mounting bolts, gradually and evenly, to 19 ft. lbs. (26 Nm).

25. Install the tensioner bracket nut and tighten to 37 ft. lbs. (50 Nm).

26. Install the throttle body, using a new gasket, and tighten the mounting nuts to 11 ft. lbs. (15 Nm).

27. Install the boost pressure manifold.

28. Install the vacuum block with a new gasket and tighten the mounting bolt to 62 inch lbs. (7 Nm).

29. Install the control cable bracket.

30. Install the remaining components in the reverse of removal.

31. Connect the negative battery cable. Fill and bleed the cooling system.

32. Run the engine and check for leaks and proper engine operation.

1996–97 Engines

1. Disconnect the negative battery cable. Properly relieve the fuel system pressure.

2. Drain the cooling system.

3. Remove the drive belt from the supercharger pulley.

4. Remove the engine dress up cover.

5. Disconnect and cap the fuel lines from the fuel rail.

6. Label and disconnect the right side spark plug wires from the ignition module and set aside.

7. Remove the alternator brace.

8. Disconnect the electrical connectors from the fuel injectors.

9. Remove the MAP sensor bracket.

10. Remove the fuel rail mounting bolts and remove the fuel rail.

11. Remove the boost control solenoid.

12. Remove the throttle body nuts and remove the throttle body.

13. Remove the supercharger assembly.

To install:

14. Thoroughly clean the supercharger and intake manifold mating surfaces.

15. Install a new supercharger-to-intake manifold gasket. Do not use any sealer on this gasket.

16. Install the supercharger with the mounting bolts. Torque the bolts, gradually and evenly, to 17 ft. lbs. (23 Nm).

17. Install the MAP sensor bracket.

18. Install the throttle body to the supercharger. Torque the nuts to 89 inch lbs. (10 Nm).

19. Install the boost control solenoid. Torque the nut to 72 inch lbs. (8 Nm).

20. Install the fuel rail and connect the fuel lines.

21. Connect the electrical connectors to the fuel injectors.

22. Install the alternator support brace.

23. Connect the right side spark plug wires to the ignition module.

24. Install the supercharger belt.

25. Install the engine dress up cover.

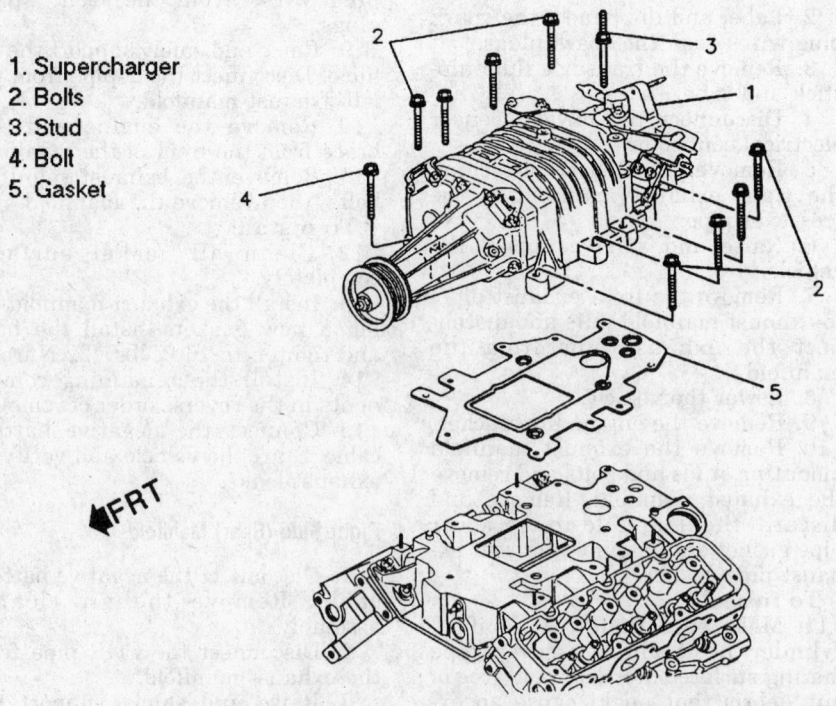

1. Supercharger
2. Bolts
3. Stud
4. Bolt
5. Gasket

FRT

Supercharger installation — 1996–97 engines shown, 1993–95 similar

297631

26. Connect the negative battery cable. Fill and bleed the cooling system.

27. Run the engine and check for leaks and proper engine operation.

Front Cover Seal

REMOVAL AND INSTALLATION

3.8L (VIN L, K and 1) Engines

1. Disconnect the negative battery cable.

2. Remove the serpentine drive belt(s). Raise and safely support the vehicle.

3. Remove the right front wheel.

4. Remove the right inner fender well splash shield.

5. Remove the crankshaft balancer bolt and washer.

6. While preventing the engine from turning, remove the crankshaft balancer using J-38197 or an equivalent puller.

7. Remove the seal from the front cover using a suitable prying tool.

To install:

8. Install the seal in the front cover using seal installation tool J-35354 or the equivalent.

9. Coat the seal contact area on the balancer with clean oil and install the balancer onto the crankshaft.

10. Using J-38197 or equivalent, seat the balancer on the crankshaft.

11. Install the washer and mounting bolt and tighten to 110 ft. lbs. (150 Nm) plus an additional 76 degree turn.

12. Install the remaining components in the reverse order of removal.

13. Check the engine oil level.

14. Connect the negative battery cable. Start the engine and check for oil leaks.

4.9L (VIN B) Engine

1. Disconnect the negative battery cable. Remove the drive belt.

2. Raise and safely support the vehicle.

3. Remove the right front wheel.

4. Remove the right front air deflector.

5. Remove the crankshaft damper-to-crankshaft bolt.

NOTE: It may be necessary to lower the engine cradle on the right side to remove the damper.

6. Using J-24420-B crankshaft damper puller or equivalent, remove the crankshaft damper.

7. Using tools J-1859–03 and J-23129 or their equivalents, remove the seal from the front cover.

To install:

8. Lubricate the new seal with clean engine oil and install with the garter spring towards the engine.

9. Install the seal in the bore using tool J-29662 and a hammer. Make sure the seal is fully installed in the bore.

10. Install the crankshaft damper using installer tool J 29774 or equivalent.

11. Install the damper-to-crankshaft bolt and tighten to 70 ft. lbs. (95 Nm).

12. Install the remaining components in the reverse order of removal.

13. Connect the negative battery cable.

14. Check the engine oil level. Run the engine and check for leaks.

Timing Chain, Sprockets and Front Cover

REMOVAL AND INSTALLATION

3.8L (VIN L, K and 1) Engines

1. Disconnect the negative battery cable. Drain the cooling system.

2. Disconnect the coolant hoses from the timing chain front cover.

3. Support the engine using support fixture J 28467-A or equivalent, and remove the engine mount.

4. Remove the drive belt(s).

5. Raise and safely support the vehicle.

6. Remove the right front wheel.

7. Remove the right inner fender access panel.

8. Detach the camshaft position sensor, crankshaft position sensor and oil pressure sensor connectors.

9. Keep the flywheel from turning using holder tool J 37096 or equivalent. Remove the crankshaft

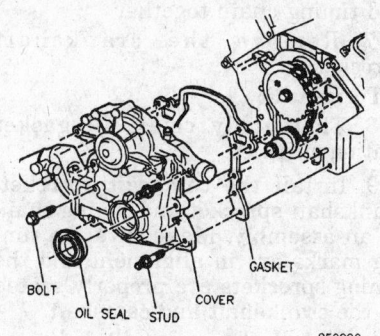

Timing chain front cover — 3.8L (VIN L, K and 1) engines

250990

balancer retaining bolt, then use puller tool J 38197 or equivalent, to remove the balancer from the crankshaft.

10. Remove the crankshaft position sensor shield and the crankshaft position sensor.

11. Remove the oil pan-to-front cover bolts.

12. Remove the front cover attaching bolts and remove the timing chain front cover.

13. Align the timing marks on the camshaft and crankshaft sprockets so they are as close together as possible.

14. Remove the timing chain damper.

15. Remove the camshaft sprocket retaining bolt and remove the camshaft sprocket and timing chain.

16. Remove the crankshaft sprocket.

NOTE: Do not rotate the camshaft or crankshaft while the timing chain and sprockets are removed.

17. Thoroughly clean all gasket mating surfaces.

To install:

18. Assemble the timing chain and sprockets with the timing marks aligned. Install the timing chain and sprockets to the camshaft and crankshaft.

19. Install the camshaft sprocket bolt and tighten to 74 ft. lbs. (100 Nm) plus an additional 90 degree turn. Recheck the camshaft and crankshaft sprocket timing marks to make sure they are still aligned.

20. Install the timing chain damper and tighten the bolt to 14 ft. lbs. (19 Nm).

WARNING

The oil pump is built into the front cover. When the cover is removed, oil drains from the pump. Since the pump "looses its prime" it may not establish oil pressure as soon as the engine starts. Therefore, it is important to remove the oil pump cover from the back of the timing chain front cover and pack the space around the oil pump gears completely full of petroleum jelly. If this is not done, the oil pump may not pump engine oil when the engine is started, resulting in severe engine damage.

21. Remove the screws and the oil pump cover from the back of the timing chain front cover. Pack the space around the oil pump gears completely full of petroleum jelly. There must be no air space left inside the pump. Re-

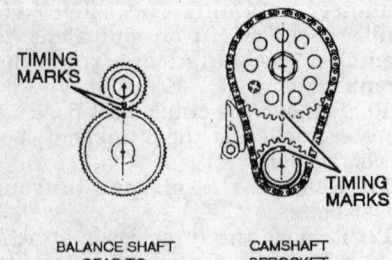

TIMING MARKS

TIMING MARKS

BALANCE SHAFT GEAR TO BALANCE SHAFT DRIVE GEAR

CAMSHAFT SPROCKET TO CRANKSHAFT SPROCKET

337879

Timing chain sprocket and balance shaft gear alignment — 3.8L (VIN L, K and 1) engines

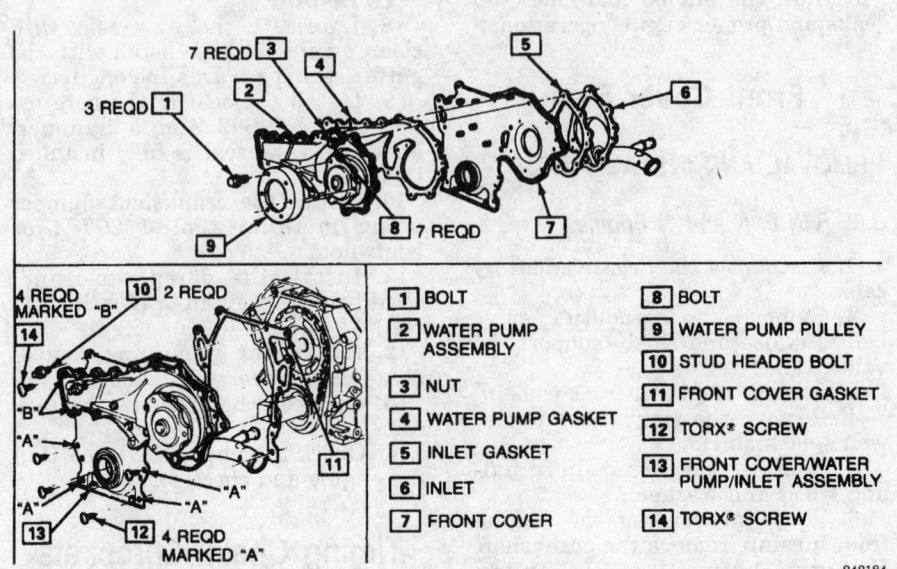

1	BOLT	8	BOLT
2	WATER PUMP ASSEMBLY	9	WATER PUMP PULLEY
3	NUT	10	STUD HEADED BOLT
4	WATER PUMP GASKET	11	FRONT COVER GASKET
5	INLET GASKET	12	TORX® SCREW
6	INLET	13	FRONT COVER/WATER PUMP/INLET ASSEMBLY
7	FRONT COVER	14	TORX® SCREW

242184

Timing chain front cover and related components — 4.9L (VIN B) engines

install the pump cover and tighten the screws to 97 inch lbs. (11 Nm).

22. Using new gaskets, install the timing chain front cover to the block. Tighten the front cover-to-engine block bolts to 22 ft. lbs. (30 Nm) and the oil pan-to-front cover bolts to 125 inch lbs. (14 Nm).

23. Install the crankshaft position sensor and tighten the bolts to 14–28 ft. lbs. (20–40 Nm). Install the crankshaft position sensor shield.

24. Keep the flywheel from turning using holder tool J 37096 or equivalent. Install the crankshaft balancer and tighten the bolt to 111 ft. lbs. (150 Nm) plus an additional 76 degree turn.

25. Attach the camshaft position sensor, crankshaft position sensor and oil pressure sensor connectors.

26. Install the right inner fender access panel and the right front wheel.

27. Lower the vehicle.

28. Install the drive belt(s).

29. Install the engine mount and remove the engine support fixture.

30. Connect the coolant hoses.

31. Connect the negative battery cable. Fill and bleed the cooling system.

32. Start the vehicle and check for leaks and proper engine operation.

4.9L (VIN B) Engine

1. Disconnect the negative battery cable. Drain the cooling system.

2. Remove the cross car brace.

3. Remove the coolant reservoir.

4. Remove the drive belt.

5. Remove the water pump pulley.

6. Remove the water pump.

7. Raise and safely support the vehicle.

8. Remove the right front wheel and the right front air deflector.

9. Support the body of the vehicle and the right side of the engine cradle.

10. Remove the right side cradle body bolts and lower the right side of the cradle for balancer clearance.

11. Remove the bolt from the end of the crankshaft. Using J 24420-B crankshaft damper puller or equivalent, remove the damper.

12. Remove the timing chain front cover bolts and remove the front cover.

13. Remove the oil slinger from the crankshaft.

14. Rotate the engine until the timing marks are in proper alignment.

15. Remove the thrust button and the camshaft sprocket mounting bolt. Discard the thrust button.

16. Remove the camshaft sprocket and timing chain together.

17. Remove the crankshaft sprocket.

To install:

18. Thoroughly clean all gasket mating surfaces.

19. Install the camshaft sprocket, crankshaft sprocket and timing chain as an assembly. Make sure the timing marks are in alignment and the timing sprockets are properly seated on the crankshaft and camshaft.

20. Install the camshaft bolt and tighten to 37 ft. lbs. (50 Nm).

21. Install a new thrust button.

22. Install the crankshaft oil slinger against the crankshaft sprocket.

23. Apply a bead of RTV sealant to the front cover lip on the oil pan sealing surface. The bead must be placed along the front cover lip behind the 2 oil pan-to-front cover bolts, in order to prevent leakage through the bolt threads.

24. Apply a ¼ inch bead of RTV sealant on the oil pan where the oil pan, block and front cover join together. The front corner has a rounder corner, instead of square; if the RTV sealant is not applied an oil leak may occur.

25. Install the front cover using a new gasket.

26. Install the front cover bolts and tighten to 17 ft. lbs. (22 Nm).

27. Lubricate the crankshaft damper hub bore and the front cover seal with EP lubricant. Position the damper on the crankshaft, aligning the crankshaft key with the hub keyway.

28. Install the crankshaft damper using installer tool J 29774 or equivalent, until the damper hub bottoms on the crankshaft. Remove the installer tool and install the bolt. Torque the bolt to 70 ft. lbs. (95 Nm).

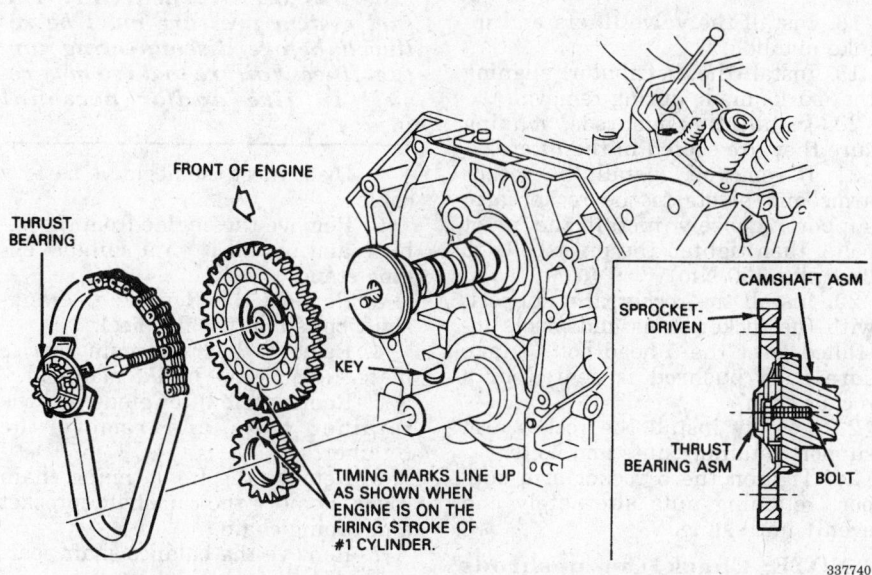

FRONT OF ENGINE

THRUST BEARING

KEY

TIMING MARKS LINE UP AS SHOWN WHEN ENGINE IS ON THE FIRING STROKE OF #1 CYLINDER.

CAMSHAFT ASM

SPROCKET-DRIVEN

THRUST BEARING ASM

BOLT

337740

Timing chain and sprocket installation — 4.9L (VIN B) engines

29. Raise the right side of the cradle and install the cradle-to-body bolts. Tighten to 66 ft. lbs. (90 Nm).

30. Install the remaining components in the reverse order of removal.

31. Connect the negative battery cable. Refill and bleed the cooling system.

32. Start the engine and check for leaks and proper engine operation.

Camshaft

REMOVAL AND INSTALLATION

─────── CAUTION ───────
Fuel injection systems remain under pressure, even after the engine has been turned OFF. The fuel system pressure must be relieved before disconnecting any fuel lines. Failure to do so may result in fire and/or personal injury.

3.8L (VIN L, K and 1) Engines

1. Disconnect the negative battery cable. Properly relieve the fuel system pressure.

2. Remove the engine assembly from the vehicle and mount on a suitable engine stand.

3. If equipped, remove the supercharger.

4. Remove the intake manifold.

5. Remove the rocker arm covers.

6. Remove the rocker arm assemblies, pushrods and lifters. Identify all parts as they are removed, so they can be reinstalled in their original locations.

7. Remove the crankshaft balancer center bolt and, using a suitable puller, remove the balancer from the crankshaft.

8. Remove the timing chain front cover.

9. Set the engine to Top Dead Center (TDC) No. 1 cylinder (firing position) to align the timing marks, before disassembling the timing chain and sprockets.

NOTE: Align the timing marks of the camshaft and crankshaft sprockets to avoid burring the camshaft journals by the crankshaft.

10. Remove the camshaft sprocket and timing chain.

11. Remove the camshaft thrust plate bolts and remove the thrust plate.

12. Carefully remove the camshaft from the engine block.

13. Inspect the camshaft lobes and journals for wear and/or damage; replace as necessary.

NOTE: If the camshaft was replaced the lifters must also be replaced. The old lifters have developed a wear pattern and will cause the new camshaft to wear prematurely.

To install:

14. Coat the camshaft lobes and bearings with lubricant GM part number 1052365 or equivalent camshaft break-in prelube prior to installation.

15. Carefully install the camshaft into the engine.

16. Install the camshaft thrust plate and tighten the mounting bolts to 10 ft. lbs. (14 Nm).

17. Install the camshaft sprocket and timing chain. Be sure the timing marks are aligned. Tighten the camshaft sprocket retaining bolt to 74 ft. lbs. (100 Nm) plus an additional 90 degree (¼) turn.

18. Install the timing chain front cover.

19. Install the crankshaft balancer and tighten the mounting bolt to 111 ft. lbs. (150 Nm) plus an additional 76 degree turn.

20. Coat the valve lifters with camshaft prelube and install the lifters in the lifter bores.

21. Install the lifter guides and lifter guide retainer. Tighten the retainer mounting bolts to 22 ft. lbs. (30 Nm).

22. Install the pushrods and rocker arms and tighten the rocker arm bolts to 11 ft. lbs. (15 Nm) plus an additional 90 degree turn.

23. Install the rocker arm covers.

24. Install the intake manifold.

25. If equipped, install the supercharger.

26. Install the engine in the vehicle.

27. Connect the negative battery cable.

28. Verify that all fluid levels are full and correct.

29. Start the engine and verify no leaks.

4.9L (VIN B) Engine

1. Disconnect the negative battery cable. Properly relieve the fuel system pressure.

2. Remove the engine from the vehicle and mount on a suitable workstand.

3. Remove the rocker arm covers.

NOTE: Label all valve train components as they are removed, so they can be reinstalled in their original locations. If the camshaft is being replaced, the valve lifters should also be replaced.

4. Remove the rocker arm support retaining nuts/bolts and remove the rocker arm support.

—————— WARNING ——————

The rocker arm support must be removed with the rocker arms and pivots attached. The pivot assemblies or support bar threads may be damaged if pivot bolt torque is removed against valve spring pressure. If necessary, secure the support in a vise and remove the rocker arms and pivots.

5. Remove the pushrods.
6. Remove the timing chain front cover.
7. Rotate the crankshaft until the timing chain sprocket timing marks are aligned.
8. Mark the position of the distributor rotor in relation to the distributor body and mark the relation of the distributor body to the intake manifold. Remove the distributor.
9. Remove the intake manifold.
10. Remove the valve lifters.
11. Remove the timing chain and camshaft sprocket.
12. Temporarily reinstall the camshaft sprocket to the camshaft (to provide a suitable hand hold). Remove the camshaft, being careful not to damage the lobes, journals or camshaft bearings.
13. Inspect the camshaft journals and lobes for wear and/or damage. Replace as necessary.

To install:

NOTE: If a new camshaft is required, new lifters and a new distributor gear must also be installed.

14. Temporarily reinstall the camshaft sprocket to the camshaft (to provide a suitable hand hold). Lubricate the camshaft journals, lobes and distributor drive gear with prelube 1052365 or equivalent.
15. Install the camshaft, being careful not to damage the lobes, journals or camshaft bearings. Remove the camshaft sprocket.
16. Install the timing chain and sprocket, making sure the sprocket timing marks are aligned. Install the camshaft sprocket retaining bolt and

tighten to 37 ft. lbs. (50 Nm). Install a new thrust button, if equipped.

17. Install the timing chain front cover.
18. Install the valve lifters and intake manifold.
19. Install the distributor, aligning the marks made during removal.
20. Install the pushrods, making sure they are seated in the lifters.
21. If removed, install the rocker arms and pivots to the rocker arm support. Loosely install the pivot bolts, then tighten the pivot bolts to 22 ft. lbs. (30 Nm).
22. Install the rocker arm support, with the rocker arms and pivots installed, over the 5 head bolts. Make sure each pushrod is seated in a rocker arm.
23. Loosely install the rocker arm support retaining nuts and bolts.
24. Tighten the 5 rocker arm support retaining nuts alternately and evenly until snug.

NOTE: Check the pushrods from time to time during the tightening procedure to make sure they are correctly positioned.

25. Tighten the 4 rocker arm retaining bolts until snug.
26. Tighten the 5 rocker arm support retaining nuts to 37 ft. lbs. (50 Nm).
27. Tighten the 4 rocker arm retaining bolts to 7 ft. lbs. (10 Nm).
28. Install the rocker arm covers.
29. Install the engine in the vehicle.
30. Connect the negative battery cable.
31. Restore all fluid levels and make all necessary adjustments.
32. Run the engine and check for leaks and proper engine performance. Check the ignition timing.

Balance Shaft

REMOVAL AND INSTALLATION

3.8L (VIN L, K and 1) Engines

NOTE: The balance shaft bearing retainer bolts are designed to permanently stretch when tightened. The correct part number must be used to replace this type of fastener. DO NOT use a bolt that is stronger in this application. If the correct bolt is not used, the parts will not be tightened correctly. Part or system damage may occur. Use care to obtain correct factory replacement parts before beginning this service.

—————— CAUTION ——————

Fuel injection systems remain under pressure, even after the engine has been turned OFF. The fuel system pressure must be relieved before disconnecting any fuel lines. Failure to do so may result in fire and/or personal injury.

1. Disconnect the negative battery cable.
2. Remove the engine from the vehicle and install it on a suitable engine stand.
3. Remove the flywheel-to-crankshaft bolts and the flywheel.
4. Remove the rear main carrier plate. Remove the intake manifold.
5. Remove the lifter guide and the retainer bolts and remove the retainer.
6. Remove the front timing chain cover. Remove the camshaft sprocket and timing chain.
7. Remove the balance shaft gear-to-shaft bolt and the gear.
8. Remove the balance shaft retainer-to the-engine bolts and the retainer.
9. Using J 6125-B or equivalent slide hammer tool, pull the balance shaft from the front of the engine.
10. Remove the balance shaft rear plug.
11. Examine the balance shaft bearings. If in good condition, they may remain in the block. If the are worn or damaged, special tools are recommended for their removal and installation. If using substitutes, first examine the bearings for their factory installed positions, depth and orientation to the oil feed holes. Remove the balance shaft rear bearing using the special tool J 36995–5.

NOTE: It may take a considerable amount of force to loosen the bushing from the block bore.

To install:
12. Make sure the rear main oil seal housing gasket is removed and there is no remaining debris.
13. Dip the rear bearing in clean engine oil before the installation.
14. Using J 36995–5, or an equivalent balance shaft installer, install the balance shaft rear bearing into the engine and remove the installation tool. Install the bearing with the rolled side facing into the engine and the manufacturers marking facing the flywheel side.
15. Remove the tool J 36995 or equivalent.
16. Dip the front balance shaft bearings in clean engine oil before the installation.

17. Install the balance shaft into the block using J 36996 or equivalent.

18. Temporarily install the balance shaft bearing retainer and bolts.

NOTE: This bolt is designed to permanently stretch when tightened. The correct part number must be used to replace this type of fastener. DO NOT use a bolt that is stronger in this application. If the correct bolt is not used, the parts will not be tightened correctly. Part or system damage may occur.

19. Install and tighten the balance shaft drive gear and the bolt with thread locking compound on the threads to 16 ft. lbs. + 70 degrees (22 Nm + 70 degrees) using the special tool J 36660 or equivalent. Install the rear balance shaft plugs, if equipped.

20. Install the rear main seal carrier plate.

21. Measure the end play for the following applications:

a. End play should be 0.008 inch (0.023 mm).

b. Front radial play should be 0.0011 inch (0.028 mm).

c. Rear radial play should be 0.0005–0.0047 inch (0.0127–0.119 mm).

22. Turn the camshaft sprocket with the camshaft gear temporarily installed to set the timing mark pointing straight down.

23. With the camshaft sprocket and the camshaft gear removed, turn the balance shaft so that the timing mark on the gear points straight down.

24. Install the camshaft gear.

25. Align the marks on the balance shaft gear and the camshaft gear by turning the balance shaft.

26. Turn the crankshaft so that the number one piston is at Top Dead Center compression stroke (firing position).

27. Install the timing chain and camshaft sprocket. Verify that ALL timing marks are correctly aligned.

28. Measure the gear lash at four different places taking four measurements total with lash not exceeding 0.002 to 0.005 inch or (0.050mm to 0.127mm).

29. Install and tighten the balance shaft front bearing retainer and bolts to 22 ft. lbs. (30 Nm).

30. Verify one last time that ALL timing marks are correctly aligned.

When satisfied that all timing marks are correct, install the engine front cover.

31. Install the lifter guide retainer.

32. Install the intake manifold.

33. Install and tighten the flywheel bolts to 46 ft. lbs. (62 Nm).

34. Install the engine into the vehicle using the recommended procedure. Refill and check all fluid levels.

35. Connect the negative battery cable.

36. Start the engine and check for leaks and proper engine operation.

Piston and Connecting Rod

POSITIONING

For 3.8L (VIN K, VIN L and VIN 1) engines, the piston may be installed on the connecting rod in either direction. Inspect the underside of the piston for identification ridges which should face the front of the engine when installed.

The 4.9L (VIN B) pistons are installed with the arrow on the crown facing forward. The 3.8L (VIN K, VIN L and VIN 1) pistons are installed with the ridges on the skirt facing forward.

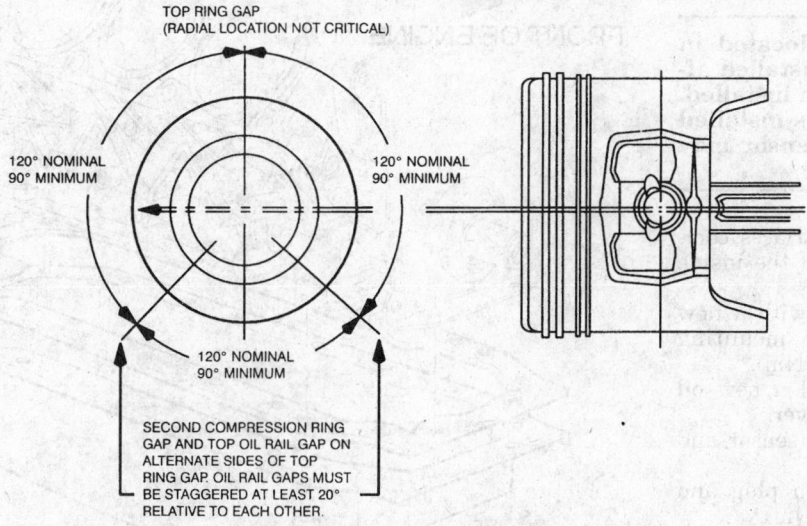

TOP RING GAP
(RADIAL LOCATION NOT CRITICAL)

120° NOMINAL
90° MINIMUM

120° NOMINAL
90° MINIMUM

120° NOMINAL
90° MINIMUM

SECOND COMPRESSION RING GAP AND TOP OIL RAIL GAP ON ALTERNATE SIDES OF TOP RING GAP. OIL RAIL GAPS MUST BE STAGGERED AT LEAST 20° RELATIVE TO EACH OTHER.

340904

Piston ring orientation — all (VIN B) engines

ENGINE LUBRICATION

Oil Pan

REMOVAL AND INSTALLATION

3.8L (VIN L, K and 1) Engines

— **WARNING** —
The oil level sensor, located in the oil pan, must be removed prior to removal of the oil pan. If the oil pan is removed first, damage to the oil level sensor may occur.

1. Disconnect the negative battery cable. Raise and safely support the vehicle.
2. Remove the oil drain plug and drain the engine oil into a suitable container.
3. For 1996–97 vehicles, remove the flywheel cover.
4. Disconnect the oil level sensor connector and remove the sensor.
5. For 1996–97 vehicles, remove the oil filter.
6. Remove the oil pan mounting bolts and remove the oil pan.
To install:

— **WARNING** —
The oil level sensor, located in the oil pan, must be installed after the oil pan has been installed. If the oil level sensor is installed first, damage to the sensor may occur.

7. Clean all gasket surfaces completely. Thoroughly clean the inside of the oil pan.
8. Install the oil pan with a new gasket and tighten the mounting bolts to 125 inch lbs. (14 Nm).
9. If necessary, install a new oil filter and the flywheel cover.
10. Install the oil level sensor and attach the connector.
11. Install the oil drain plug and tighten to 30 ft. lbs. (40 Nm).
12. Lower the vehicle.
13. Refill the crankcase with the proper type and quantity of engine oil.
14. Connect the negative battery cable. Run the engine and check for leaks.

4.9L (VIN B) Engine

1. Disconnect the negative battery cable. Raise and safely support the vehicle.
2. Remove the oil pan drain plug and drain the engine oil into a suitable container.
3. Remove the flywheel covers.
4. Remove the exhaust Y-pipe.
5. Remove the oil pan bolts and nuts and loosen the oil pan from the cylinder block.
6. Drop the oil pan down about 2 inches, then remove the oil pump housing mounting bolts and nut. Drop the oil pump down into the oil pan, then remove the oil pan from the vehicle.

— **WARNING** —
Attempting to remove the oil pan without first dropping the oil pump will result in breaking the oil pump housing.

To install:
7. Before installing the oil pan, the excess material on the rear edge must be ground off ¹/₁₆ in. (1.5mm) on both sides for oil pump clearance.

— **WARNING** —
Failure to grind off the excess material will result in breaking the oil pump housing upon installation.

8. Clean all gasket surfaces completely. Thoroughly clean the inside of the oil pan.
9. Install a new O-ring on the oil pump outlet pipe. Install the oil pump, making sure the pump driveshaft engages the distributor gear. Tighten the nut to 22 ft. lbs. (30 Nm) and the bolts to 15 ft. lbs. (20 Nm).
10. Apply a ¹/₄ in. (6mm) bead of sealant at the rear main bearing cap and at the front cover-to-block joints.
11. Install the oil pan using a new gasket.
12. Install the oil pan bolts and nuts and tighten to 14 ft. lbs. (18 Nm).
13. Install the flywheel cover.
14. Install the oil pan drain plug and tighten to 22 ft. lbs. (30 Nm).
15. Install the exhaust Y-pipe.
16. Lower the vehicle.
17. Refill the crankcase with the proper type and quantity of engine oil.
18. Connect the negative battery cable. Run the engine and check for leaks.

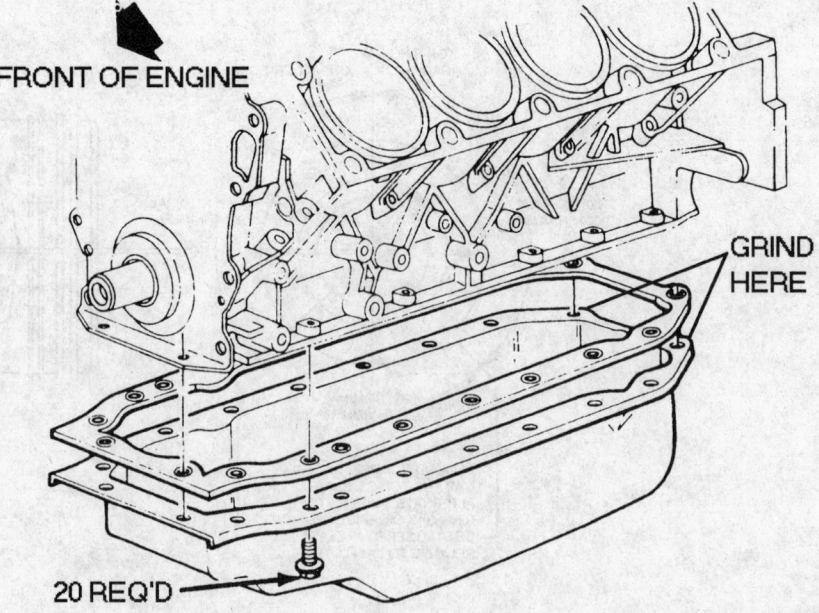

FRONT OF ENGINE

GRIND HERE

20 REQ'D

335446

Grind the oil pan at these locations — 4.9L (VIN B) engines

Oil Pump

REMOVAL AND INSTALLATION

3.8L (VIN L, K and 1) Engines

1. Disconnect the negative battery cable.
2. Remove the engine drive belts and tensioner assembly.
3. Remove the drive belt idler pulley and bracket, if necessary.
4. Raise and safely support the vehicle.
5. Support the engine using engine support fixture J 28467 or equivalent, and remove the torque axis mount and bracket assembly.
6. Remove the engine front cover assembly.
7. Remove the four bolts securing the oil filter adapter to the front cover assembly and oil filter adapter, pressure regulator valve and spring.
8. Remove the four oil pump cover attaching screws and remove the cover.
9. Remove the inner and outer pump gears and inspect.
   To install:
10. Lubricate the oil pump gears with petroleum jelly and install the gears into the housing.
11. Pack the gear cavity with petroleum jelly after the gears have been installed in the housing. This seals the pump and acts like a "prime" so oil will begin to drawn from the oil pan as soon as the engine begins to turn. Do not neglect this step. DO NOT use any type of grease. Petroleum jelly has a low melting point and will correctly dissipate when oil begins to flow and it is no longer needed.
12. Install the oil pump cover and screws and tighten to 97 inch lbs. (11 Nm).
13. Install the oil filter adapter with new gasket, pressure regulator

1. 97 inch lbs. (11 Nm)
2. Oil pump cover
3. Pump outer gear
4. Pump inner gear
5. Front cover

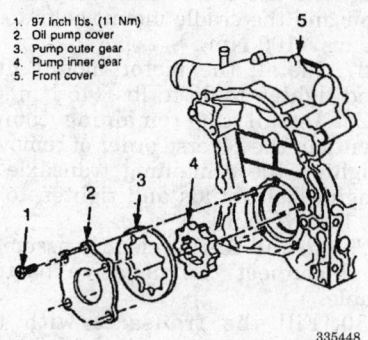

335448

Oil pump assembly — 3.8L (VIN L, K and 1) engines

valve and spring. Tighten the mounting bolts to 24 ft. lbs. (33 Nm). Apply sealant to the bolt threads.
14. Install the front cover assembly.
15. Install the tensioner assembly.
16. Install the drive belt idler pulley and bracket, if removed.
17. Install the torque axis mount assembly and remove the engine support fixture.
18. Verify the correct engine oil level. A new oil filter is recommended.
19. Connect the negative battery cable. Start the vehicle and verify no leaks and proper oil pressure.

4.9L (VIN B) Engine

1. Disconnect the negative battery cable. Raise and safely support the vehicle.
2. Remove the oil pan drain plug and drain the engine oil into a suitable container.
3. Remove the flywheel covers.
4. Remove the oil pan bolts and nuts and loosen the oil pan from the cylinder block.
5. Drop the oil pan down about 2 inches, then remove the oil pump housing mounting bolts and nut. Drop the oil pump down into the oil pan, then remove the oil pan from the vehicle.

WARNING
Attempting to remove the oil pan without first dropping the oil pump will result in breaking the oil pump housing.
---

6. Remove the oil pump from the pan and discard the O-ring.
7. If necessary, disassemble and inspect the oil pump for wear and/or damage. Replace as necessary.
   To install:
8. Reassemble the oil pump, as required, lubricating all internal parts with clean engine oil during assembly. Tighten the pump cover-to-housing screws to 96 inch lbs. (10 Nm) and the pipe-to-housing screws to 120 inch lbs. (12 Nm). Be sure to use a new O-ring when installing the outlet pipe.
9. Temporarily install the driveshaft to the pump. Prime the pump by pouring clean engine oil into the pump pickup screen and rotating the driveshaft until oil emerges from the passage in the pump.
10. Install the oil pump and driveshaft to the block, engaging the driveshaft to the distributor gear. Use a new O-ring at the pump outlet pipe.

11. Install the mounting nut and tighten to 22 ft. lbs. (30 Nm). Install the mounting bolts and tighten to 15 ft. lbs. (20 Nm).
12. Instal the oil pan. Tighten the bolts and nuts to 14 ft. lbs. (18 Nm).
13. Install the oil pan drain plug and tighten to 22 ft. lbs. (30 Nm).
14. Lower the vehicle.
15. Refill the crankcase with the proper type and quantity of engine oil.
16. Connect the negative battery cable.
17. Run the engine; check for proper oil pressure and check for leaks.

TRANSAXLE

Automatic Transaxle Assembly

REMOVAL AND INSTALLATION

Except Cadillac DeVille and Sixty Special

1. Raise the hood and place protective covers over the fenders.
2. Disconnect the negative battery cable.
3. Loosen the cross brace assembly through bolts and remove the inboard strut nuts.
4. Reinstall the the inboard strut nuts. Remove the air intake duct.
5. Remove the cruise control servo assembly.
6. Remove the shift control linkage and mounting bracket at the transaxle.
7. Disconnect the wiring connectors at the transaxle park/neutral position switch, backup switch, transaxle electrical connector and vehicle speed sensor.
8. Remove the fuel pipe retainers.
9. Remove the vacuum modulator hose at the modulator.
10. Remove the 3 top transaxle-to-engine bolts.
11. Install special tool J 28467 A or equivalent engine support fixture.
12. Load the engine support fixture by tightening the wing nuts several turns to relieve tension on the frame and mounts.
13. Turn the steering wheel to the full left position.

CAUTION

To help avoid personal injury when a vehicle is on a hoist, provide additional support to the rear of the vehicle while removing the transaxle.

---

14. Raise and safely support the vehicle.

15. Remove both front wheels.

16. Remove the right and left front ball joint nuts.

17. Separate the right and left control arms from the steering knuckle.

18. Remove the right halfshaft from the transaxle only. Do not remove the halfshaft from the steering knuckle assembly.

19. Remove the left halfshaft from both the transaxle and steering knuckle assembly.

20. Support the transaxle with a suitable jack.

21. Remove the left front transaxle mount.

22. Remove the torque strut bracket from the transaxle.

23. Remove the left rear transaxle mount-to-transaxle bolts.

24. Remove the transaxle brace from the engine bracket.

25. Remove the stabilizer shaft link-to-control arm bolt.

26. Remove the flywheel cover bolts and the flywheel cover.

27. Mark the flywheel to the converter with a reference line and remove the flywheel-to-converter bolts.

28. Remove the bolts attaching the rear frame member to the front dog leg.

29. Remove the front left frame-to-body attaching bolts.

30. Remove the frame assembly by swinging it aside and supporting with a suitable stand.

31. Remove the oil cooler lines from the transaxle.

32. Remove the remaining lower transaxle-to-engine block bolts.

NOTE: One transaxle bolt is located between the transaxle case and the engine block and is installed in the opposite direction.

33. Remove the transaxle assembly and lower it away from the vehicle.

To install:

34. Install the transaxle into the vehicle.

35. Install the lower transaxle-to-engine block bolts and tighten to 55 ft. lbs. (76 Nm).

36. Connect the cooler lines at the transaxle.

37. Raise the frame assembly into place and tighten the frame-to-body bolts to 83 ft. lbs. (112 Nm).

38. Install and tighten the front frame dog leg-to-right frame member bolts.

39. Install and tighten the left frame-to-body attaching bolts to 83 ft. lbs. (112 Nm).

40. Install and tighten the bolts attaching the rear frame member to the front frame dog leg.

41. Install the flywheel-to-converter bolts, aligning the marks made during removal, and tighten to 46 ft. lbs. (62 Nm).

42. Install the remaining components in the reverse order of removal. Tighten the transaxle brace-to-engine bracket bolts to 70 ft. lbs. (95 Nm).

43. Connect the negative battery cable and start the engine.

44. Check the transaxle fluid level and check for leaks.

45. Adjust the transaxle shifter cable, as necessary.

46. Road test the vehicle.

Cadillac DeVille and Sixty Special

1. Disconnect the negative battery cable. Remove the air cleaner assembly.

2. Label and disconnect the necessary electrical connectors.

3. Remove the transaxle harness and shift cable brackets.

4. Remove the engine oil cooler line, vacuum hose and fuel line bracket.

5. Disconnect the vacuum modulator. Remove the transaxle filler tube and transaxle mounting bracket.

6. Remove the 4 transaxle-to-engine bolts/studs from positions 2, 3, 4 and 5.

7. Support the engine with engine support tool J–28467 or equivalent.

NOTE: Be careful when attaching the right rear lift hook to the bracket to make sure there is clearance to the A/C accumulator line.

8. Raise and safely support the vehicle. Remove the front wheels.

9. Remove the left stabilizer link bolts and the ball joint cotter pins and nuts. Separate the ball joints from the steering knuckles.

10. Disconnect the left outer tie rod end from the spindle.

11. Disconnect the halfshafts from the transaxle.

12. Remove the A/C splash shield and the right and left wheel well splash shields.

13. Remove the power steering return line bracket and the ABS pump from the bracket.

14. Remove the flexplate splash shields, then remove the torque converter-to-flexplate bolts.

15. Remove the power steering line and the transaxle oil cooler lines.

16. Remove the cradle mount bolts and motor mount nuts on the right side, then remove the cradle mount bolts on the left side.

17. Remove the No. 1 cradle mount insulator bolt and the left cradle member, separating the right front corner first.

18. Support the transaxle with a suitable jack.

19. Remove the bracket assembly from the transaxle mount bracket and remove the engine-to-transaxle bracket.

20. Remove the transaxle-to-engine bolts from positions 1 and 6.

NOTE: To remove the bolt at position No. 6, use a 3 foot extension and socket to access the bolt through the right wheel well.

21. Remove the transaxle from the vehicle.

To install:

22. Position the transaxle in the vehicle and install the lower transaxle-to-engine bolts/studs. Tighten the bolts/studs to 55 ft. lbs. (75 Nm).

NOTE: Studs must be installed in positions 2 and 3. Bolts must be installed in positions 1, 4, 5 and 6.

23. Install the torque converter-to-flexplate bolts and tighten to 46 ft. lbs. (63 Nm). Install the flexplate splash shields.

24. Connect the oil cooler lines at the transaxle. Tighten the oil cooler line fitting nuts to 15 ft. lbs. (21 Nm).

25. Install the engine-to-transaxle bracket and tighten the bolts to 37 ft. lbs. (50 Nm).

26. Install the cradle member. Tighten the No. 1 insulator mount bolt and the cradle mount bolts to 74 ft. lbs. (100 Nm).

27. Install the motor mount nuts and tighten to 35 ft. lbs. (45 Nm).

28. Install the remaining components in the reverse order of removal. Tighten the remaining transaxle-to-engine bolts/studs and tighten to 55 ft. lbs. (75 Nm).

29. Install the air cleaner assembly and connect the negative battery cable.

30. Fill the transaxle with the proper type and quantity of fluid. Adjust the shift linkage.

31. Road test the vehicle.

DRIVE AXLE

Halfshaft

REMOVAL AND INSTALLATION

—————— **WARNING** ——————
Use care when removing the half-shaft to prevent the inner CV-joint from becoming over-extended. Over-extension of the joint could result in separation of internal components and possible joint failure.

1. Raise and safely support the vehicle.

2. Remove the front wheel. Mark the position of the wheel to the wheel stud, prior to removal, for installation reference.

3. Install J-34754 boot protector, or equivalent on the outer CV-joint boot.

4. If necessary, loosen and remove the stabilizer shaft link assembly bolt.

5. Remove the ball joint cotter pin and nut. Loosen the joint using tool J-36226 or equivalent. The grease fitting may have to be removed from ball joint for tool access. If removing the right halfshaft, turn the wheel to the left. If removing the left half-shaft, turn the wheel to the right. Use a suitable prybar between the suspension support and the lower control arm to separate the joint.

6. Remove the hub nut. It requires a lot of torque to break free and loosen the hub nut. Insert a drift pin or punch through the opening in the caliper into the ventilation openings in the brake rotor to keep the rotor from turning as the nut is loosened. Discard the hub nut; it must not be reused.

7. Partially install the hub nut to protect the threads, then remove the halfshaft from the hub using tool J-28733-B or equivalent.

8. Move the strut and knuckle rearward.

9. For all vehicles except DeVilles and Sixty Specials, remove the half-shaft from the transaxle using special tools J-33008 and J-2619–01 or their equivalents.

10. For DeVilles and Sixty Specials, remove the halfshaft from the transaxle using a suitable prying tool and a wood block fulcrum to protect the transaxle case.

NOTE: If equipped with anti-lock brakes, care must be used to prevent damage to the toothed sensor ring on the halfshaft and the wheel speed sensor on the steering knuckle.

To install:

11. If installing the right side half-shaft, install J-37292-B seal protector tool or equivalent, so that it can be pulled out after the halfshaft is installed.

12. Install the halfshaft into the transaxle by placing a drift pin or punch into the groove on the joint housing and tapping lightly until seated. Verify that the halfshaft is seated by grasping the inner joint housing and pulling. DO NOT pull on the halfshaft.

13. Install the halfshaft into the hub and bearing assembly. Loosely install a new hub nut.

14. Install the ball joint into the steering knuckle.

15. Install the castle nut and torque the nut to 88 inch lbs. (10 Nm), then tighten an additional 120 degrees (1/3 turn) during which a torque of 41 ft. lbs. (55 Nm) must be obtained. If necessary to install the cotter pin, the nut can be tightened up to 20 degrees additional. NEVER loosen the castle nut to install the cotter pin.

16. Insert a drift in the brake rotor cooling fins to keep the rotor from turning and tighten the nut to 110 ft. lbs. (149 Nm), unless equipped with the J55 brake option, in which case the torque is 130 ft. lbs. (177 Nm).

17. If necessary, install the stabilizer shaft link assembly. Torque the nut to 14 ft. lbs. (17 Nm).

18. If J-37292-B seal protector was installed, remove it by pulling in line with the handle.

19. Install the wheel, aligning the marks made during removal.

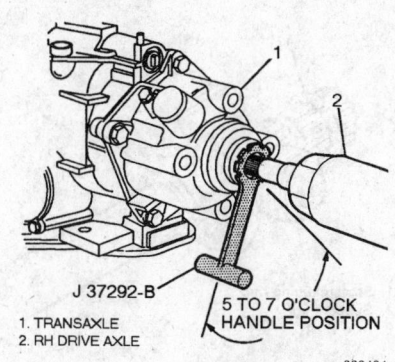

J 37292-B

1. TRANSAXLE
2. RH DRIVE AXLE

5 TO 7 O'CLOCK HANDLE POSITION

333434

Installing axle seal protector tool

20. Lower the vehicle and road test for proper operation.

CV-Joint Boot

REPLACEMENT

NOTE: Two types of inner CV-joint are used: the Tripot Ball type and the Bearing Block type. The following procedure contains information for both types. Also, depending upon application, the small end of the CV-joint boot may be retained with either a conventional type boot clamp or with a swage clamp.

Outer CV-Joint Boot

1. Remove the halfshaft from the vehicle. Mount the shaft in a soft-jawed vise.

2. Cut the large boot retaining clamp from the CV-joint with side cutters and discard.

3. If equipped with a conventional small boot clamp, cut the small boot retaining clamp with side cutters and discard. If equipped with a swage ring, use a hand grinder to cut through the ring, being careful not to damage the axle shaft. Remove and discard the swage ring.

4. Separate the boot from the CV-joint and slide it away from the joint along the shaft.

5. Wipe the grease from the face of the CV-joint inner race. Spread the retaining snapring with snapring pliers and remove the CV-joint from the shaft.

6. Remove the boot from the shaft.

7. Using a brass drift and hammer, gently tap on the CV-joint cage until it is tilted enough to remove the first chrome ball. Tilt the cage in the opposite direction to remove the opposing ball. Repeat until all 6 balls are removed.

8. Position the cage and inner race 90 degrees to the centerline of the outer race and align the cage windows with the lands of the outer race. Remove the cage and inner race from the outer race.

9. Rotate the inner race 90 degrees to the centerline of the cage with the lands of the inner race aligned with the windows of the cage. Pivot the inner race into the cage window and remove the inner race.

10. Thoroughly clean the inner and outer races and the cage and balls with clean solvent and allow to dry. Inspect all parts for damage and replace the CV-joint, if necessary.

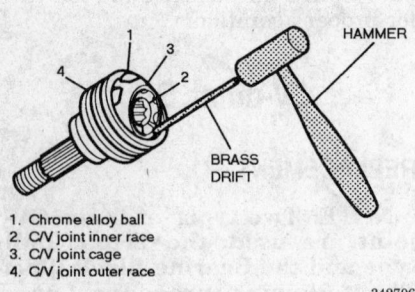

HAMMER

BRASS
DRIFT

1. Chrome alloy ball
2. C/V joint inner race
3. C/V joint cage
4. C/V joint outer race

342706

Removing the CV-joint balls

To install:

11. As required, install a new small boot clamp on the neck of a new boot, but do not crimp yet, or install a new swage ring on the neck of a new boot, but do not swage yet.

12. Slide the boot onto the shaft and position the boot neck in the groove on the shaft. As required, crimp the boot clamp with crimping tool J–35910 or equivalent, or swage the swage ring using tool J 41048 or equivalent.

13. Put a light coat of grease from the CV-joint service kit on the ball grooves of the inner and outer races.

14. Hold the inner race 90 degrees to the centerline of the cage with the lands of the inner race aligned with the windows of the cage. Insert the inner race into the cage.

15. Hold the cage and inner race 90 degrees to the centerline of the outer race and align the cage windows with the lands of the outer race. Install the cage and inner race into the outer race, making sure the snapring side of the inner race faces the shaft.

16. Insert the first chrome ball, then tilt the cage in the opposite direction to insert the opposing ball. Repeat until all 6 balls are in place.

17. Place approximately half of the grease from the service kit inside the boot and pack the CV-joint with the remaining grease.

18. Push the CV-joint onto the shaft until the snapring is seated in the groove on the shaft.

19. Slide the boot over the CV-joint until the boot lip is in the groove on the outer race. The boot must not be dimpled, stretched or out of shape in any way. If it is not shaped correctly, equalize the pressure in the boot and shape the boot properly by hand.

20. Install the large boot clamp and crimp, using crimping tool J–35910 or equivalent.

21. Install the halfshaft in the vehicle.

Inner CV-Joint Boot — Tripot Ball Type

1. Remove the halfshaft from the vehicle. Mount the shaft in a soft-jawed vise.

2. Cut the large boot retaining clamp from the CV-joint with side cutters and discard.

3. If equipped with a conventional small boot clamp, cut the small boot retaining clamp with side cutters and discard. If equipped with a swage ring, use a hand grinder to cut through the ring, being careful not to damage the axle shaft. Remove and discard the swage ring.

4. Separate the boot from the CV-joint and slide it away from the joint along the shaft.

5. Remove the CV-joint housing from the spider and shaft. Remove the trilobal tripot bushing from the housing.

6. Spread the spacer ring and slide the spacer ring and spider back on the axle shaft.

7. Remove the shaft retaining ring from the groove on the axle shaft and slide the spider assembly off of the shaft.

8. Remove the spacer ring and boot from the axle shaft.

9. Thoroughly clean the spider assembly and housing with clean solvent and allow to dry. Inspect all parts for damage and replace the CV-joint, if necessary.

To install:

10. As required, install a new small boot clamp on the neck of a new boot, but do not crimp yet, or install a new swage ring on the neck of a new boot, but do not swage yet.

11. Slide the boot onto the shaft and position the boot neck in the groove on the shaft. As required, crimp the boot clamp with crimping tool J–35910 or equivalent, or swage the swage ring using tool J 41048 or equivalent.

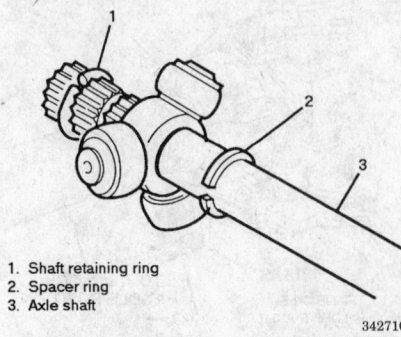

1. Shaft retaining ring
2. Spacer ring
3. Axle shaft

342710

CV-joint spider assembly — Tripot ball type

12. Slide the spacer ring on the axle shaft beyond the second groove.

13. Slide the spider assembly towards the spacer ring as far as it will go on the axle shaft. Make sure that the counterbored face of the spider faces the end of the shaft.

14. Install the shaft retaining ring in the groove of the axle shaft.

15. Slide the spider towards the end of the shaft and reseat the spacer ring in the groove on the shaft.

16. Place approximately ½ of the grease from the service kit in the boot and use the remainder to repack the CV-joint housing.

17. Install the trilobal tripot bushing to the housing, making sure the bushing is flush with the face of the housing.

18. Position the large boot clamp on the boot.

19. Slide the housing over the spider assembly on the shaft. Slide the boot, with the clamp in place, over the outside of the trilobal tripot bushing and locate the lip of the boot in the groove.

20. Position the CV-joint at the proper dimension. The boot must not be dimpled, stretched or out of shape in any way. If it is not shaped correctly, equalize the pressure in the boot and shape the boot properly by hand.

21. Install the large boot clamp and crimp, using crimping tool J–35910 or equivalent.

22. Install the halfshaft in the vehicle.

Inner CV-joint Boot — Bearing Block Type

1. Remove the halfshaft from the vehicle. Mount the shaft in a soft-jawed vise.

2. Cut the large boot retaining clamp from the CV-joint with side cutters and discard.

3. If equipped with a conventional small boot clamp, cut the small boot retaining clamp with side cutters and discard. If equipped with a swage ring, use a hand grinder to cut through the ring, being careful not to damage the axle shaft. Remove and discard the swage ring.

4. Separate the boot from the CV-joint and slide it away from the joint along the shaft.

5. Remove the outer CV-joint housing from the spider and shaft.

6. Spread the spacer ring with snapring pliers and slide the spacer ring and spider back on the shaft.

7. Remove the shaft retaining ring from the groove on the shaft and slide the spider assembly off the shaft.

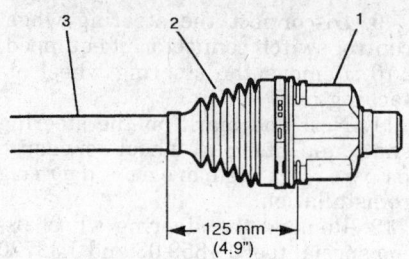

1. Retainer & housing asm.
2. Inboard boot
3. Axle shaft

342716

Boot installation measurement

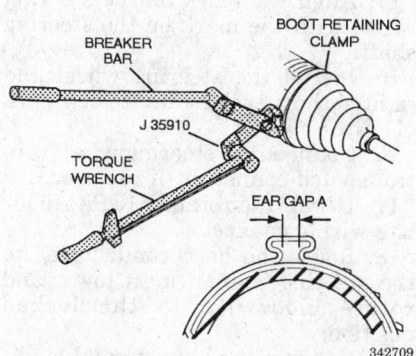

Installing the large boot retaining clamp

8. Thoroughly clean the bearing blocks, spider and housing with clean solvent and allow to dry.

9. Remove the tripot bushing from the housing.

10. Remove the spacer ring and boot from the shaft.

11. Inspect the spider, housing, bearing blocks and tripot bushing for damage. Replace the CV-joint, if necessary.

To install:

12. As required, install a new small boot clamp on the neck of a new boot, but do not crimp yet, or install a new swage ring on the neck of a new boot, but do not swage yet.

13. Slide the boot onto the shaft and position the boot neck in the groove on the shaft. As required, crimp the boot clamp with crimping tool J–35910 or equivalent, or swage the swage ring using tool J 41048 or equivalent.

14. Install the spacer ring on the shaft beyond the second groove.

15. Apply a small amount of grease to the inside of the bearing blocks. Align the flats on the opening in the bearing block with the flats on the spider trunnion. Attach the bearing

block to the spider trunnion, then rotate 90 degrees to secure the block to the spider.

16. Slide the spider assembly against the spacer ring on the shaft, making sure that the counterbored face of the spider faces the end of the shaft.

17. Install the shaft retaining ring in the groove of the shaft.

18. Slide the spider towards the end of the shaft and reseat the spacer ring in the groove on the shaft.

19. Place approximately half the grease from the CV-joint service kit in the boot and use the remainder to repack the housing.

20. Place a slotted 6 in. square piece of sheet metal between the boot and bearing blocks, to maintain proper bearing block alignment during reassembly.

21. Install the tripot bushing to the housing.

22. Position the large boot clamp on the boot.

23. Slide the housing over the spider assembly on the shaft and remove the slotted sheet metal plate.

24. Slide the boot, with the large boot clamp in place, over the outside of the tripot bushing and locate the lip of the boot in the groove.

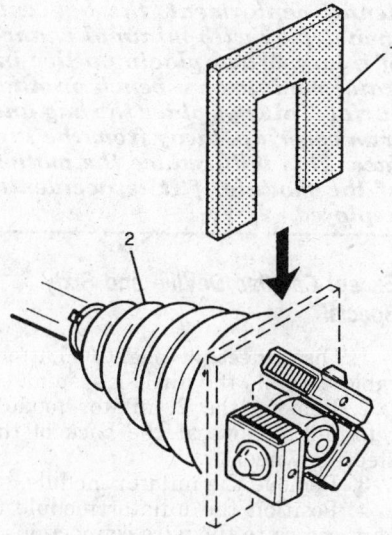

1. Slotted 6" square sheet metal
2. Inboard boot

342715

Proper bearing block alignment

25. Position the CV-joint to the proper dimension. The boot must not be dimpled, stretched or out of shape in any way. If the boot is not shaped correctly, carefully insert a thin, flat, blunt tool (no sharp edges) between the boot and the tripot bushing to equalize pressure. Shape the boot properly by hand and remove the tool.

26. Install the large boot clamp and crimp, using crimping tool J–35910 or equivalent. Make sure that the boot, housing and large clamp all remain in alignment while crimping.

27. Install the halfshaft in the vehicle.

STEERING

Air Bag

— **CAUTION** —

Some vehicles are equipped with an air bag system, also known as the Supplemental Inflatable Restraint (SIR) or Supplemental Restraint System (SRS). The system must be disabled before performing service on or around system components, steering column, instrument panel components, wiring and sensors. Failure to follow safety and disabling procedures could result in accidental air bag deployment, possible personal injury and unnecessary system repairs.

PRECAUTIONS

Several precautions must be observed when handling the inflator module to avoid accidental deployment and possible personal injury.

• Never carry the inflator module by the wires or connector on the underside of the module.

• When carrying a live inflator module, hold securely with both hands, and ensure that the bag and trim cover are pointed away.

• Place the inflator module on a bench or other surface with the bag and trim cover facing up.

• With the inflator module on the bench, never place anything on or close to the module which may be thrown in the event of an accidental deployment.

DISARMING

—— **CAUTION** ——

The Supplemental Restraint System (SRS) must be disarmed before performing service around the air bag or SRS wiring. Failure to do so may cause accidental deployment of the air bag, resulting in unnecessary SRS repairs and/or personal injury.

Disarming

1. Turn the steering wheel so the front wheels are in the straight ahead position.
2. Turn the ignition switch to the **LOCK** position.
3. Disconnect the negative battery cable.
4. Remove the Air Bag fuse from the fuse panel. The position of the fuse on the panel varies according to model and year. Consult the vehicle owner's manual for fuse location.
5. Remove the left-hand sound insulator.
6. Remove the Connector Position Assurance (CPA) clip and disconnect the yellow 2-way connector at the base of the steering column.
7. Remove the CPA and disconnect the passenger side yellow 2-way connector, if equipped. Positions for the connector vary from behind the glove box to removing the right side sound insulator and finding the yellow connector.

Enabling

1. Turn the steering wheel so the front wheels are in the straight ahead position.
2. Turn the ignition switch to the **LOCK** position.
3. Disconnect the negative battery cable.
4. Connect the yellow 2-way connector at the base of the steering column and install the CPA.

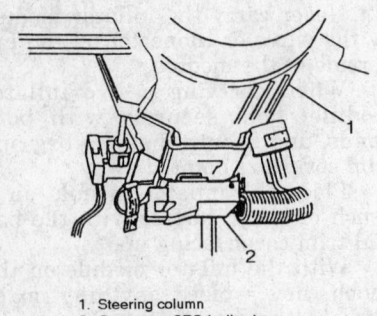

1. Steering column
2. Connector, SRS (yellow)

341964

Disconnect the SRS yellow 2-way connector

5. Install the left-hand sound insulator.
6. Connect the yellow 2-way connector on the right side and install the CPA, if equipped. Install the sound insulator and/or glove box.
7. Install the air bag fuse.
8. Connect the negative battery cable.
9. Turn the ignition switch to the **RUN** position. Verify that the INFLATABLE RESTRAINT indicator lamp flashes 7–9 times and then remains OFF. If the lamp does not function as specified, there is a malfunction in the air bag system.

Steering Wheel

REMOVAL AND INSTALLATION

—— **CAUTION** ——

The Supplemental Restraint System (SRS) must be disarmed before working around the air bag or SRS wiring. Failure to do so may cause accidental deployment of the air bag, resulting in unnecessary SRS repairs and/or personal injury.

—— **CAUTION** ——

When carrying a live air bag, make sure the bag and trim cover are pointed away from the body. In the unlikely event of an accidental deployment, the bag will then deploy with minimal chance of injury. When placing a live inflator module on a bench or other surface, always place the bag and trim cover up, away from the surface. This will reduce the motion of the module if it is accidently deployed.

Except Cadillac DeVille and Sixty Special

1. Disconnect the negative battery cable. Disable the air bag system.
2. Remove the 2 inflator module retaining screws at the back of the steering wheel.
3. Remove the inflator module.
4. Position the inflator module to gain access to the wire connections.
5. Push in and rotate the horn contact lead counterclockwise and remove from the steering column horn tower.
6. Remove the horn contact lead.
7. Remove the retainer (CPA) from the air bag wiring connector.
8. Remove the air bag wiring connector from the inflator module.

9. Disconnect the steering wheel control switch connector, if equipped.
10. Remove the steering wheel attaching nut.
11. Note the mark on the steering shaft and steering wheel to ensure proper alignment during reinstallation.
12. Remove the steering wheel using special tool J 1859 03 and J 38720 or equivalent. Be sure to use the correct size bolts for pulling the steering wheel.

To install:

13. Route the wiring through the steering wheel.
14. Align the mark on the steering wheel with the mark on the steering shaft.
15. Install the steering wheel and tighten the attaching nut to 30 ft. lbs. (41 Nm).
16. Connect the steering wheel control switch connector, if removed.
17. Install the retainer (CPA) to air bag wiring connector.
18. Route the horn contact lead to the steering column horn tower and rotate clockwise to the locked position.
19. Connect the horn ground lead.
20. Connect the air bag connector and retainer (CPA).
21. Install the inflator module to the steering wheel. Install the top surface of the module first to ease assembly and fit.
22. Install the 2 screws through the back of the steering wheel to the inflator module and tighten to 27 inch lbs. (3 Nm).
23. Enable the air bag system.
24. Connect the negative battery cable.

Cadillac DeVille and Sixty Special

1. Disconnect the negative battery cable. Disable the air bag system.
2. Turn the steering wheel so the spoke arms are at 6 and 12 o'clock.
3. Remove the 4 inflator module retaining screws.
4. Disconnect the horn contact wire.
5. Disconnect the Connector Position Assurance (CPA) and coil assembly connector from the inflator module.
6. Remove the inflator module.
7. Center the steering wheel. Mark the position of the steering wheel on the steering shaft, for installation reference.
8. Remove the steering column shaft nut.
9. Remove the steering wheel using J 1859–03 or the equivalent

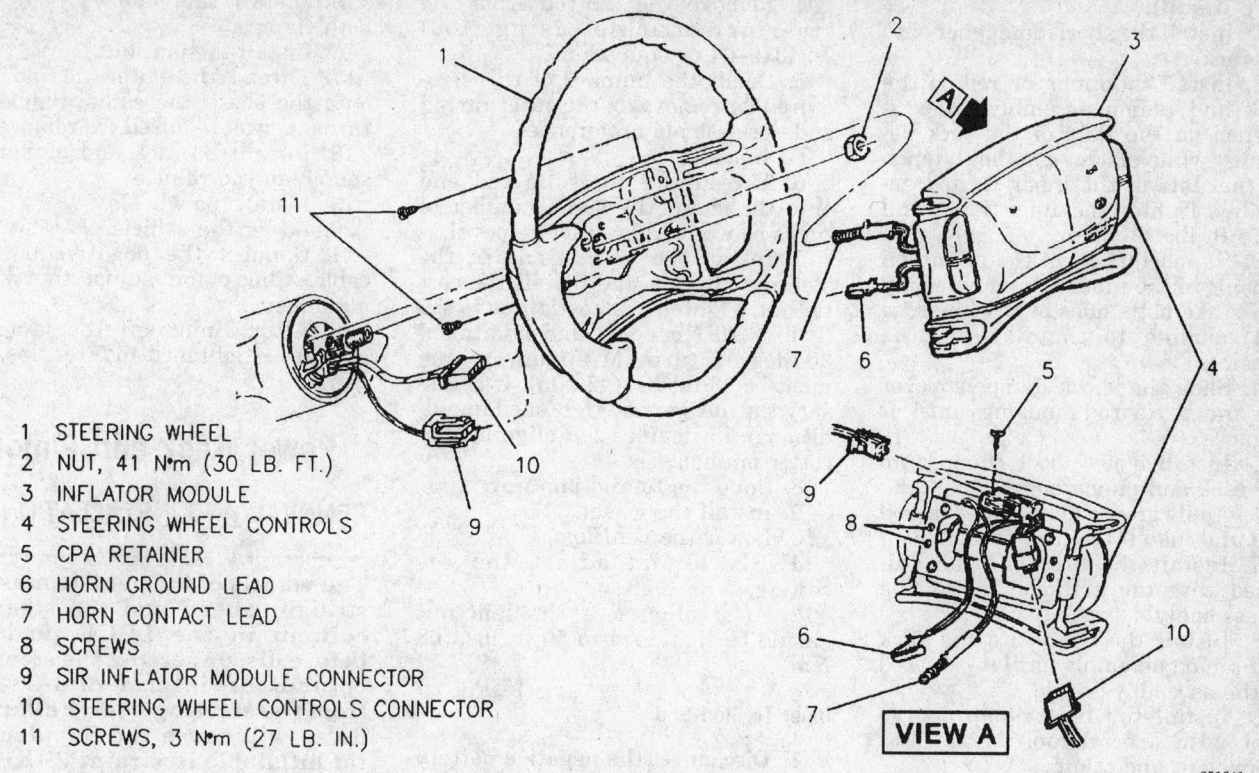

1   STEERING WHEEL
2   NUT, 41 N·m (30 LB. FT.)
3   INFLATOR MODULE
4   STEERING WHEEL CONTROLS
5   CPA RETAINER
6   HORN GROUND LEAD
7   HORN CONTACT LEAD
8   SCREWS
9   SIR INFLATOR MODULE CONNECTOR
10 STEERING WHEEL CONTROLS CONNECTOR
11 SCREWS, 3 N·m (27 LB. IN.)

VIEW A

251641

Steering wheel assembly — except Cadillac DeVille and Sixty Special models

puller. Remove the puller from the wheel.

To install:

10. Feed the coil assembly lead through the slot in the steering wheel.

11. Install the steering wheel on the column shaft, aligning the marks made during removal.

12. Install the column shaft nut and tighten to 30 ft. lbs. (41 Nm).

13. Connect the horn contact to the steering column.

14. Connect the CPA and coil assembly connector to the inflator module.

15. Install the inflator module on the steering wheel and install the 4 mounting screws.

16. Enable the air bag system.

17. Connect the negative battery cable.

Tie Rod Ends

REMOVAL AND INSTALLATION

Except Cadillac DeVille and Sixty Special

Outer Tie Rod End

1. Raise and safely support the vehicle.

2. Remove the front wheel.

3. Remove the cotter pin and hex slotted nut from the outer tie rod assembly.

4. Loosen the jam nut.

5. Remove the outer tie rod end from the steering knuckle with special tool J 24319 01 or equivalent.

6. Remove the outer tie rod from the inner tie rod, counting the number of turns required for removal.

To install:

7. Thread the outer tie rod end onto the tie rod, the same number of turns as was required for removal. Do not tighten the jam nut at this time.

8. Install the outer tie rod to the steering knuckle and tighten the hex slotted nut at the outer tie rod end to 35 ft. lbs. (47 Nm). If required, the nut may be tightened as much as 52 ft. lbs. (70 Nm) to align for cotter pin insertion. Do not back off the nut for cotter pin insertion.

9. Insert the cotter pin into the hole in the tie rod stud and peen the ends of the cotter pin over the tie rod stud.

10. Install the wheel and lower the vehicle.

11. Adjust the toe to specifications.

12. Tighten the jam nut against the outer tie rod to 50 ft. lbs. (68 Nm).

Inner Tie Rod End

1. Raise and safely support the vehicle. Remove the front wheels.

2. Remove the rack and pinion assembly from the vehicle.

3. Loosen the jam nut.

4. Remove the outer tie rod from the inner tie rod, counting the number of turns required for removal.

5. Remove the hex jam nut from the inner tie rod assembly.

6. Remove the tie rod end clamp.

7. Remove the boot clamp with side cutter and discard.

NOTE: Mark the location of the breather tube on the rack and pinion assembly before removing the tube or rack and pinion boot.

8. Remove the rack and pinion boot and breather tube.

9. Remove the shock dampener from the inner tie rod assembly and slide back on the rack and pinion assembly.

10. Remove the inner tie rod assembly while holding the rack and pinion assembly. Place a wrench on the flats of the rack and pinion assembly while placing a wrench on the flats of the inner tie rod housing. Rotate the inner tie rod housing counterclockwise until the inner tie rod separates from the rack and pinion assembly.

To install:

11. Install the shock dampener onto the rack.

12. Install the inner tie rod on the rack and pinion assembly. Place a wrench on the flats of the rack assembly while placing another wrench on the flats of the inner tie rod assembly. Tighten the inner tie rod end to 74 ft. lbs. (100 Nm).

13. Support the rack and pinion housing of the inner tie rod assembly and stake both sides of the inner tie rod housing to the flats on the housing.

14. Slide the shock dampener over the inner tie rod housing until it engages.

15. Install a new boot clamp onto the rack and pinion boot.

16. Apply grease to the inner tie rod end and install the boot assembly.

17. Install the breather tube aligned with the marks made during disassembly.

18. Install the boot onto the rack and pinion assembly until it is seated in the assembly groove.

19. Install the boot clamp on the boot with special tool J 22610 or equivalent and crimp.

20. Install the tie rod end clamp with pliers on the boot.

21. Install the hex nut to the inner tie rod assembly.

22. Install the rack and pinion assembly into the vehicle.

23. Install the outer tie rod assembly to the inner tie rod end, threading it on the same number of turns as was required to remove. Do not tighten the jam nut at this time.

24. Install the outer tie rod to the steering knuckle and tighten the hex slotted nut at the outer tie rod end to 35 ft. lbs. (47 Nm). If required, the nut may be tightened as much as 52 ft. lbs. (70 Nm) to align for cotter pin insertion. Do not back off the nut for cotter pin insertion.

25. Insert the cotter pin into the hole in the tie rod stud and peen the ends of the cotter pin over the tie rod stud.

26. Install the wheels and lower the vehicle.

27. Adjust the toe to specifications.

28. Tighten the jam nut against the outer tie rod to 50 ft. lbs. (68 Nm).

Cadillac DeVille and Sixty Special

Outer Tie Rod End

1. Raise and safely support the vehicle. Remove the front wheel.

2. Loosen the tie rod jam nut.

3. Remove the cotter pin and castle nut from the outer tie rod end.

4. Remove the tie rod from the steering knuckle using tool J-24319–01 or equivalent.

5. Count the number of turns required for removal of the outer tie rod end, for assembly reference.

To install:

6. Thread the outer tie rod end onto the shaft, the same number of turns as was required for removal.

7. Install the tie rod end in the steering knuckle and install the castle nut. Tighten the castle nut to 7.5 ft. lbs. (10 Nm) plus and additional 30 degree turn. Minimum torque must be 33 ft. lbs. (45 Nm). If necessary the nut may be tightened an additional 15 degrees for aligning the cotter pin holes.

8. Snug the tie rod jam nut.

9. Install the wheel.

10. Lower the vehicle.

11. Check and adjust the toe setting.

12. After alignment the jam nut should be tightened to 50 ft. lbs. (68 Nm).

Inner Tie Rod End

1. Disconnect the negative battery cable. Raise and safely support the vehicle.

2. Remove the front wheels.

3. Remove the rack and pinion assembly from the vehicle.

4. With the rack and pinion unit supported in a vise, loosen the tie rod jam nut and remove the outer tie rod end noting the number of turns required for removal.

5. Thread the jam nut off the end of the rack.

6. Remove the clamp on the small end of the bellows boot.

7. Remove the clamp on the large end of the bellows boot and remove the bellows boot.

8. Slide the shock damper off the inner tie rod end and back onto the rack.

9. Loosen the inner tie rod socket.

10. Put a wrench on the rack flat to prevent damage to the unit during tie rod removal.

11. Place a second wrench on the inner tie rod end and unscrew the inner tie rod socket from the end of the rack.

To install:

12. Install the inner tie rod assembly on the rack and tighten to 70 ft. lbs. (95 Nm).

13. Stake the inner tie rod housing to the rack and pinion unit. Both sides of the inner tie rod socket must be staked.

14. Slide the shock damper over the end of the inner tie rod socket.

15. Install the bellows boot and both clamps.

16. Install the jam nut.

17. Thread the outer tie rod end onto the shaft, the same number of turns as was required for removal.

18. Install the rack and pinion assembly in the vehicle.

19. Install the wheels.

20. Lower the vehicle.

21. Connect the negative battery cable. Check and adjust the wheel alignment.

22. After alignment the jam nut should be tightened to 50 ft. lbs. (68 Nm).

Power Rack and Pinion

REMOVAL AND INSTALLATION

——————— **WARNING** ———————
The wheels of the vehicle must be straight ahead and the steering column in the LOCK position before disconnecting the steering column or intermediate shaft from the steering gear. Failure to do so will cause the Supplemental Inflatable Restraint (SIR) coil assembly in the steering column to become uncentered which will cause damage to the coil assembly.

Except Cadillac DeVille and Sixty Special

1. Disconnect the negative battery cable. Raise and safely support the vehicle.

2. Remove both front wheels.

3. Remove the pinch bolt from the intermediate shaft at the rack and pinion assembly and separate the shaft from the rack.

——————— **CAUTION** ———————
Failure to disconnect the intermediate shaft from the rack and pinion stub shaft can result in damage to the steering gear and/or intermediate shaft. This damage can cause loss of steering control which could result in personal injury.

4. Remove the cotter pins and castle nuts from the tie rod ends and separate the tie rods from the steering knuckles using special tool J 24319–01 or equivalent.

5. Remove the power steering line retainers and retaining clips.

6. Disconnect the power steering lines from the rack and pinion unit.

7. Support the rear of the subframe with a suitable jack. Loosen

the front frame bolts and remove the rear bolts, then lower the frame about 3 inches for clearance purposes.

WARNING

Do not lower rear of frame too far as damage to engine components nearest to the cowl may result. Lower the rear of the frame no more than 3 inches (76mm).

---

8. Remove the rack and pinion mounting bolts.

9. Remove the rack and pinion through the left wheel opening.

To install:

10. Install the rack and pinion through the left wheel opening.

11. Raise the rear of the frame.

12. Apply Loctite™ or equivalent, to the threads of the rack and pinion mounting bolts. Install the rack and pinion mounting bolts and tighten, in sequence, to 50 ft. lbs. (68 Nm).

13. Raise the subframe into position and tighten the mounting bolts to 76 ft. lbs. (103 Nm).

14. Remove the jack.

15. Connect the power steering hoses and tighten the fittings to 20 ft. lbs. (27 Nm).

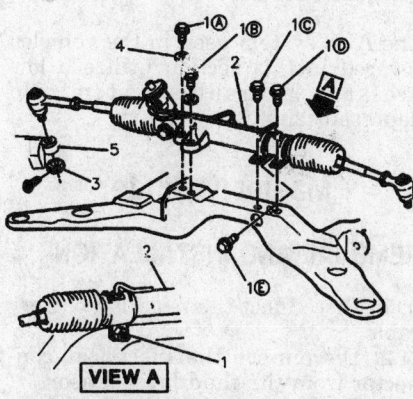

VIEW A

1 BOLT; 68 N·m (50 LB. FT.). TIGHTEN IN SEQUENCE A THRU E.
2 STEERING GEAR
3 NUT; 47 N·m (35 LB. FT.). MAXIMUM PERMISSIBLE TORQUE TO ALIGN COTTER PIN SLOT IS 70 N·m (52 LB. FT.).
4 WASHER
5 STEERING KNUCKLE

338988

Rack and pinion mounting bolt torque sequence — except Cadillac DeVille and Sixty Special models

16. Connect the power steering lines to the retainers.

17. Connect the tie rod ends to the steering knuckles and tighten the castle nuts to 35 ft. lbs. (47 Nm). If necessary to align the holes for the cotter pin tighten the nuts up to 60 degrees additional. NEVER loosen the castle nuts to align the holes.

18. Connect the intermediate shaft to the rack and pinion and tighten the pinch bolt to 35 ft. lbs. (47 Nm).

19. Install the wheels and torque the lug nuts to 100 ft. lbs. (140 Nm).

20. Lower the vehicle.

21. Connect the negative battery cable.

22. Fill the reservoir with fluid and bleed the air from the system.

23. Start the engine, check for leaks and proper steering operation.

24. Check the wheel alignment.

Cadillac DeVille and Sixty Special

1. Disconnect the negative battery cable. Raise and safely support the vehicle.

2. Remove the front wheels.

3. Remove the bolt holding the intermediate shaft to the steering shaft. Disconnect the intermediate shaft lower coupling.

4. Remove the cotter pin and castle nuts from both outer tie rods. Using tool J-24319 or equivalent, separate the outer tie rod ends from the steering knuckles.

5. Support the rear of the subframe with a screw jack.

6. Remove the rear subframe bolts. Slowly lower the subframe to gain access.

7. Remove the heat shield.

8. Remove the power steering line retainer.

9. Place a catch pan under the power steering rack and the line fittings.

10. Disconnect the power steering pressure and return lines from the rack and pinion unit.

11. Disconnect the Speed Sensitive Steering solenoid valve connector, if equipped.

12. Remove the 5 power steering rack-to-subframe bolts.

13. Slide the rack and pinion unit out the left wheel well.

To install:

14. Install the rack and pinion through the left wheel well and into position on the subframe.

15. Install the 5 mounting bolts and tighten to 50 ft. lbs. (68 Nm).

16. Connect the Speed Sensitive Steering solenoid valve connector, if equipped.

17. Connect the power steering pressure and return hoses and tighten the fittings to 20 ft. lbs. (27 Nm).

18. Install the power steering line retainer. Install the heat shield.

19. Raise the subframe into position and install the subframe mounting bolts.

20. Connect the outer tie rod ends to the steering knuckle. Install the castle nuts and torque to 7.5 ft. lbs. plus an additional 30 degree turn. A minimum torque of 33 ft. lbs. (45 Nm) must be attained. Install the cotter pins. Up to 15 degrees additional tightening is acceptable for lining up the cotter pin holes. NEVER loosen the castle nut to install the cotter pin.

21. Connect the intermediate shaft and install the intermediate shaft-to-steering shaft bolt. Tighten the bolt to 35 ft. lbs. (47 Nm).

22. Install the wheels.

23. Lower the vehicle.

24. Connect the negative battery cable.

25. Refill the power steering reservoir and bleed the power steering system.

26. Check the wheel alignment.

27. Road test for proper operation.

Power Steering Pump

BLEEDING

1. With the engine **OFF** and the front wheels off the ground, turn wheels all the way to the left and add power steering fluid to the **FULL COLD** mark on the level indicator.

2. Bleed the system by turning the wheels from side to side several times (as many as 20 times could be required), without hitting stops. Be sure to keep the level to the **FULL COLD** mark.

3. Start the engine and run at fast idle momentarily, then recheck the fluid level with the engine idling. If necessary add fluid to bring the level back to the **FULL COLD** mark.

4. Return the wheels to the center position and continue running the engine for a few minutes. Road test to check the operation of the steering.

5. Recheck the fluid level it should now be stabilized at the **FULL HOT** level on the indicator.

REMOVAL AND INSTALLATION

Except Cadillac DeVille and Sixty Special

NOTE: On TFE (Two Flow Electronic) systems, if the pump is to be replaced, be sure to replace the pump with one that is specified for the TFE system. The TFE pump looks identical to non-TFE pumps, but the two are not interchangeable.

1. Disconnect the negative battery cable.
2. Remove the strut housing upper tie bar, as needed for access.
3. Remove the power steering pump drive belt.
4. On the TFE pump, disconnect the electrical connector.
5. Remove the lower bolt to the mounting bracket.
6. Remove the power steering gear outlet and inlet hoses from the pump. On supercharged engines, also disconnect the remote reservoir hose from the power steering pump reservoir.
7. Remove the power steering pump. On supercharged engines, access the pump area from beneath the engine. Transfer the power steering pulley, as needed.

To install:

8. Hand start the power steering gear inlet hose connection to the power steering pump.
9. Install the power steering pump with the pulley. On non supercharged engines, install and tighten the lower bolt first and check that clearance of 0.080–0.200 in. (2–5mm) exists between the hose and heater pipes as contact may cause noise after installation.
10. Install the power steering gear outlet hose to the power steering pump. The hose must be routed outboard and under the engine harness and heater hoses.
11. On supercharged engines, reconnect the hose from the remote reservoir to the power steering pump reservoir.
12. Tighten the power steering gear inlet hose to the power steering pump to 20 ft. lbs. (27 Nm). Tighten the power steering pump mounting bolts to 20 ft. lbs. (27 Nm).
13. Install the remaining components in the reverse order of removal. Tighten the strut housing upper tie bar to 18 ft. lbs. (25 Nm).
14. Connect the negative battery cable. Fill and bleed the power steering pump.

15. Run the engine and test the steering system for proper operation.

Cadillac DeVille and Sixty Special

1. Disconnect the negative battery cable. Remove the serpentine belt.
2. Drain the power steering fluid from the reservoir.
3. Remove the bolt retaining the power steering return line to the pump.
4. Remove the power steering pump mounting bolts.
5. Reposition the pump to disconnect the adapter from the pump inlet.
6. Disconnect the return pipe adapter clip from the adaptor, using a suitable prying tool.
7. Disconnect the return pipe from the adapter.
8. Remove the adapter from the reservoir tubes.
9. Remove the power steering pump from the pump bracket.

To install:

10. Install the adapter on the return pipe and install the clip.
11. Install the adapter on the reservoir tubes.
12. Position the pump so the adapter pump inlet tube is positioned in the pump inlet.
13. Install the power steering pump mounting bolts and tighten to 18 ft. lbs. (24 Nm).
14. Install the bolt retaining the return pipe and tighten to 18 ft. lbs. (24 Nm).
15. Install the serpentine belt.
16. Connect the negative battery cable. Refill the power steering pump and bleed the system.
17. Road test for proper operation.

BRAKES

Anti-Lock Brake System Service

PRECAUTIONS

• Certain components within the Anti-Lock Brake System (ABS) are not intended to be serviced or repaired individually. Only those components with removal and installation procedures should be serviced.
• Do not use rubber hoses or other parts not specifically specified for and ABS system. When using repair kits, replace all parts included in the kit.

Partial or incorrect repair may lead to functional problems and require the replacement of components.
• Lubricate rubber parts with clean, fresh brake fluid to ease assembly. Do not use lubricated shop air to clean parts; damage to rubber components may result.
• Use only specified brake fluid from an unopened container.
• If any hydraulic component or line is removed or replaced, it may be necessary to bleed the entire system.
• A clean repair area is essential. Always clean the reservoir and cap thoroughly before removing the cap. The slightest amount of dirt in the fluid may plug an orifice and impair the system function. Perform repairs after components have been thoroughly cleaned; use only denatured alcohol to clean components. Do not allow ABS components to come into contact with any substance containing mineral oil; this includes used shop rags.
• The Anti-Lock control unit is a microprocessor similar to other computer units in the vehicle. Ensure that the ignition switch is **OFF** before removing or installing controller harnesses. Avoid static electricity discharge at or near the controller.
• If any arc welding is to be done on the vehicle, the control unit should be unplugged before welding operations begin.

DEPRESSURIZING

The ABS system used in the vehicles covered in this section utilize a low pressure hydraulic system. No depressurizing is necessary.

Master Cylinder

REMOVAL AND INSTALLATION

1. Disconnect the negative battery cable.
2. Disconnect the electrical connector from the fluid level sensor.
3. Drain as much brake fluid from the master cylinder reservoir as possible.
4. Remove the brake lines from the master cylinder. Plug the open brake lines to prevent fluid loss and to prevent fluid contamination.
5. Remove the 2 nuts retaining the master cylinder to the power brake booster.
6. Remove the master cylinder.

To install:

7. Bench bleed the new master cylinder.

8. Install the master cylinder. Torque the retaining nuts to 20 ft. lbs. (27 Nm).

9. Install the brake lines and snug the fittings.

10. Bleed the master cylinder and the rear brake line fitting.

11. Final tighten the brake line fittings to 11 ft. lbs. (15 Nm).

12. Attach the fluid level sensor connector.

13. Bleed the brake system and top off the brake fluid level in the master cylinder.

14. Road test and check for proper brake operation.

Brake Caliper

REMOVAL AND INSTALLATION

NOTE: **The Bosch 2U ABS system cannot increase brake pressure above master cylinder pressure applied by during braking. There is no need to depressurize the system prior to service.**

1. Disconnect the negative battery cable.

2. Remove brake fluid from the master cylinder reservoir until the reservoir is approximately 1/3 full. Discard the removed fluid.

3. Raise and safely support the vehicle.

4. Remove the front wheel. Mark the position of the wheel to the wheel studs, prior to removal, for installation reference.

5. Install 2 lug nuts to retain the rotor once the caliper is removed.

6. Using a large C-clamp, bottom the piston in the caliper bore by positioning the C-clamp on the outboard pad and on the round portion of the brake caliper where the piston is housed.

7. Remove the banjo bolt that fastens the brake hose to the brake caliper. Discard the gaskets.

8. Cap the brake line to avoid excessive fluid loss or fluid contamination.

9. Remove the rubber dust boots from the caliper mounting bolt heads.

10. Remove the caliper mounting bolts and the sleeve assemblies.

11. Remove the caliper from the vehicle.

12. Remove the brake pads from the caliper, if the caliper is being replaced.

To install:

13. Install the brake pads in the caliper. Lubricate the slides where the caliper mounts on the steering knuckle with high-temperature

grease (disc-brake rated wheel bearing grease).

14. Install the caliper over the rotor.

15. Install the caliper mounting bolts and sleeve assemblies. Tighten the mounting bolts to 38 ft. lbs. (51 Nm).

16. Install the rubber dust boots over the caliper mounting bolt heads.

17. Check the clearance between the brake caliper and caliper bracket stops. If the clearance is too tight, check the caliper leading and trailing edges for build up. File down as necessary.

18. Connect the brake hose to the caliper. Install the brake hose banjo bolt, using new gaskets, and tighten to 33 ft. lbs. (45 Nm).

19. Remove the lug nuts used to secure the rotor.

20. Refill the master cylinder and bleed the brake system.

21. Install the wheel, aligning the reference marks made during removal, and torque the lug nuts to 100 ft. lbs. (140 Nm).

22. Lower the vehicle.

23. Road test the vehicle and check for proper braking performance.

Disc Brake Pads

REMOVAL AND INSTALLATION

1. Disconnect the negative battery cable.

2. Remove brake fluid from the master cylinder reservoir until the reservoir is approximately 1/3 full. Discard the removed fluid.

3. Raise and safely support the vehicle.

4. Remove the front wheel. Mark the position of the wheel to the wheel studs, prior to removal, for installation reference.

5. Install 2 lug nuts to retain the rotor once the caliper is removed.

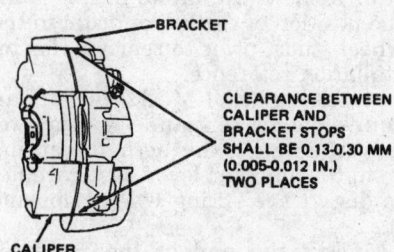

BRACKET

CLEARANCE BETWEEN CALIPER AND BRACKET STOPS SHALL BE 0.13–0.30 MM (0.005–0.012 IN.) TWO PLACES

CALIPER

332769

Measuring the caliper clearance

6. Remove the caliper and support it out of the way on a wire. DO NOT disconnect the brake hose or allow the caliper to hang from the brake hose.

7. Remove the outboard pad by pushing it in toward the piston until the mounting tabs clear the holes in the caliper body. With the tabs clear of the holes, push the pad out the bottom of the caliper.

8. Remove the inboard pad from the piston by pulling the top of the pad out to disengage the retainer spring.

To install:

9. Install the inboard pad in the caliper by inserting the top pad ears in first, then sliding the bottom of the pad into place until the spring clip snaps into place. Make sure the inboard pad seats flush against the caliper piston.

10. Install the outboard pad by lining up the tabs on the rear of the pad with the mounting holes in the caliper body. Press the pad firmly down into the caliper until the tabs snap into the mounting holes.

11. Remove the caliper bolts from the caliper. Clean and lubricate them and reinstall them into the caliper. Lubricate the caliper slides and mountings.

12. Install the caliper over the rotor and tighten the mounting bolts to 38 ft. lbs. (51 Nm).

13. Remove the 2 lug nuts used to secure the rotor in place.

14. Install the wheel, aligning the marks made during removal, and torque the lug nuts to 100 ft. lbs. (140 Nm).

15. Lower the vehicle.

16. Pump the brake pedal several times to seat the pads against the rotor.

17. Refill the master cylinder using DOT 3 brake fluid only and road test to verify proper brake operation.

Brake Rotor

REMOVAL AND INSTALLATION

1. Disconnect the negative battery cable.

2. Remove brake fluid from the master cylinder reservoir until the reservoir is approximately 1/3 full. Discard the removed fluid.

3. Raise and safely support the vehicle.

4. Remove the front wheel. Mark the position of the wheel to the wheel studs, prior to removal, for installation reference.

5. Remove the caliper and support out of the way on a wire. DO NOT disconnect the brake hose from the caliper or allow the caliper to hang from the brake hose.

6. Remove the disc brake rotor. Mark the position of the rotor to the wheel studs, prior to removal, for installation reference.

7. Inspect the brake rotor for scoring, cracks or other wear; machine or replace as necessary. If machining, observe the rotor minimum thickness specification.

To install:

8. Install the rotor on the hub, aligning the marks made during removal, and loosely install 2 lug nuts to secure the rotor until the caliper is installed. Make sure there is no rust built-up on the hub that may cause a wobble condition.

9. Install the caliper over the rotor and tighten the mounting bolts to 38 ft. lbs. (51 Nm).

10. Remove the 2 lug nuts securing the rotor.

11. Install the wheel, aligning the marks made during removal.

12. Lower the vehicle.

13. Pump the brake pedal several times to seat the pads against the rotor.

14. Check the brake fluid and top off as necessary.

15. Road test the vehicle and check for proper brake performance.

Brake Drums

REMOVAL AND INSTALLATION

1. Raise and safely support the vehicle.

2. Remove the rear wheel. Mark the position of the wheel on the wheel studs, prior to removal, for installation reference.

3. Remove the brake drum from the hub. Mark the position of the drum on the wheel studs, prior to removal, for installation reference.

4. If the brake drum is difficult to remove, check the following:

   a. Make sure that the parking brake is released.

   b. Back off on the parking brake cable adjustment.

   c. Use a hammer and a small metal punch to bend in the backing plate knockout to provide access to the park brake lever.

   d. Insert a screwdriver through the hole and press in to push the park brake lever off its stop. This

lets the brake shoes retract slightly.

   e. Apply a small amount of penetrating oil around the drum pilot hole.

5. Remove the brake drum from the vehicle.

NOTE: After the brake drum has been removed from the vehicle, be sure to remove the knockout slug using pliers. Insert a rubber access hole plug into the hole to prevent dirt or water contamination.

6. Inspect the wheel cylinder for leakage and inspect the brake shoes for wear; replace as necessary.

7. Inspect the brake drum for scoring, cracks or other wear; machine or replace as necessary. If machining, observe the maximum diameter specification.

To install:

8. Measure the inside diameter of the brake drum using clearance gauge J 21177-A or equivalent.

9. Turn the star wheel on the adjusting screw assembly until the brake shoe diameter is 0.050 in. (1.27mm) less than the drum inside diameter.

10. Install the brake drum onto the wheel hub, aligning the marks made during removal.

11. Install the wheel, aligning the marks made during removal. Tighten the lug nuts to 100 ft. lbs. (140 Nm).

12. Lower the vehicle.

13. Road test the vehicle and check brake operation.

Brake Shoes

REMOVAL AND INSTALLATION

1. Raise and safely support the vehicle.

2. Remove the rear wheel. Mark the position of the wheel to the wheel studs, prior to removal, for installation reference.

3. Remove the brake drum. Mark the position of the brake drum to the wheel studs, prior to removal, for installation reference.

4. Using tool J 38400 or an equivalent brake spanner tool and remover, remove the actuator spring from the adjuster lever. Use care not to distort the spring when removing it.

5. Lift the end of the retractor spring from the adjuster shoe assembly. Insert the hook end of the J-38400 between the retractor spring and the shoe. Pry slightly to remove

the spring end from the hole in the shoe.

6. Pry the end of the retractor spring toward the axle with the flat end of the tool until the spring snaps down off the shoe web onto the backing plate.

7. Remove the one brake shoe and remove the adjuster assembly.

8. Disconnect the parking brake lever from the shoe. DO NOT remove the parking brake lever from the cable end unless it is being replaced.

9. Using J 38400 or equivalent, lift the end of the retractor spring from the adjuster shoe assembly. Insert the hook end of the tool between the retractor spring and the shoe. Pry slightly to remove the spring end from the hole in the shoe. Pry the end of the retractor spring toward the axle with the flat end of the tool, until the spring snaps down off the shoe web onto the backing plate. Leave the spring there, do not remove.

10. Remove the brake shoe.

To install:

11. Check the backing plate attaching bolts to make sure that they are tight. Use fine emery cloth to clean all rust and dirt from the shoe contact surfaces on the plate and lubricate with high temperature grease. Check the wheel cylinder for signs of leakage.

12. Clean all parts completely in brake solvent and air dry.

13. Clean the backing plate shoe contact points.

14. Inspect the inside of the brake drum. If worn, heavily grooved or if the opening is distorted, the drum should be machined or replaced. If machining, observe the maximum drum diameter specification.

15. Disassemble, clean and lubricate the adjuster screw.

16. Position the brake shoe that connects to the parking brake lever, on the backing plate. Using J-38400 or equivalent, pull the end of the retractor spring up to rest on the web of the shoe. Pull the end of the retractor spring up until it snaps into the slot in the brake shoe.

17. Connect the parking brake lever.

18. Install the remaining shoe and the adjuster screw assembly.

19. Position the brake shoe using J-38400 or equivalent, and pull the end of the retractor spring up to rest on the web of the shoe. Pull the end of the retractor spring up until it snaps into the slot in the brake shoe.

20. Using J-38400 or equivalent, spread the brake shoes and work the adjuster screw into position.

11. Bleed the brake system. Road test and check the brake system for proper operation.

1 ADJUSTER SOCKET
2 ADJUSTER SCREW
3 PIVOT NUT
4 RETRACTOR SPRING
5 ADJUSTER SHOE AND LINING
6 WHEEL CYLINDER
7 BLEEDER VALVE
8 BOLT
9 BACKING PLATE
10 PARK BRAKE SHOE AND LINING
11 PARK BRAKE LEVER
12 ACTUATOR SPRING
13 ADJUSTER ACTUATOR

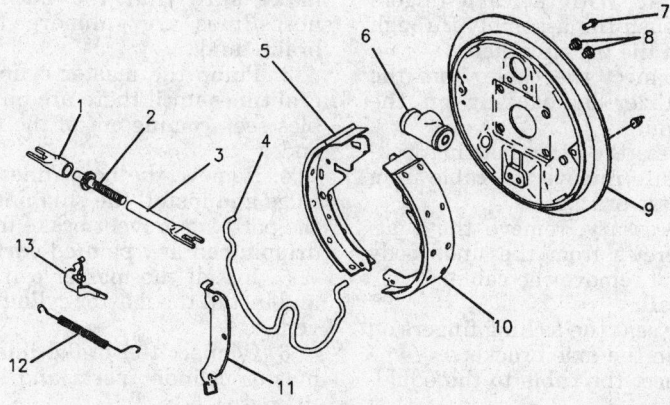

247128

Exploded view of the drum brake components

21. Install the actuator spring with the U-shaped end going through the web.

22. Measure the inside diameter of the brake drum using clearance gauge J 21177-A or equivalent.

23. Turn the star wheel on the adjusting screw assembly until the brake shoe diameter is 0.050 in. (1.27mm) less than the drum inside diameter.

24. Install the brake drum onto the wheel hub, aligning the marks made during removal.

25. Install the wheel, aligning the marks made during removal. Tighten the lug nuts to 100 ft. lbs. (140 Nm).

26. Repeat the procedure for the brake shoe assembly on the opposite side of the vehicle.

27. Lower the vehicle.

28. Apply and release the brake pedal 30–35 times with normal pedal force. Pause about 1 second between applications.

29. Road test the vehicle and check brake operation.

Wheel Cylinder

REMOVAL AND INSTALLATION

1. Raise and safely support the vehicle.

2. Remove the rear wheel. Mark the position of the wheel to the wheel studs, prior to removal, for installation reference.

3. Remove the brake drum. Mark the position of the drum to the wheel studs, prior to removal, for installation reference.

4. Remove the brake shoes.

5. From the rear of the backing plate, disconnect the steel brake line and cap the line opening to prevent excessive fluid loss and possible fluid contamination.

6. Remove the two wheel cylinder mounting screws.

7. Remove the wheel cylinder from the backing plate. Due to tight clearances it may be necessary to remove the bleeder screw from the wheel cylinder prior to removing the wheel cylinder from the vehicle.

To install:

8. Install the wheel cylinder on the backing plate. Apply gasket sealant to the wheel cylinder shoulder that contacts the backing plate.

9. Install the mounting bolts and tighten to 18 ft. lbs. (24 Nm).

10. Install the remaining components in the reverse order of removal. Tighten the brake line to to 11 ft. lbs. (15 Nm).

Parking Brake Cable

ADJUSTMENT

NOTE: The rear brake shoes must be adjusted properly before parking brake adjustment can be performed. Failure to properly adjust the brakes can lead to overtightening the parking brake cable adjuster resulting in rear brake drag.

1. Apply the parking brake 6 to 10 clicks.

2. Release the parking brake pedal.

3. Check the parking brake pedal assembly for full release by turning the ignition to **ON** and checking the **BRAKE** warning light. The light should be OFF. If the **BRAKE** light is ON and the brake appears to be released, operate the pedal release lever and pull downward on the front parking brake cable to remove slack from the assembly.

4. Raise and safely support the vehicle.

5. Remove the access hole plug in the brake backing plate.

6. Adjust the parking brake cable until a ⅛ inch drill can be inserted through the access hole into the space between the shoe web and parking brake lever. Satisfactory cable adjustment is achieved when a ⅛ inch drill bit will fit into the space but a ¼ inch drill bit will not fit.

7. Check for free wheel rotation.

8. Replace the access hole plug.

9. Safely lower the vehicle.

REMOVAL AND INSTALLATION

Front Cable

1. Raise and safely support the vehicle.

2. Loosen the equalizer assembly at the front parking brake cable.

3. Disconnect the front cable from the equalizer assembly.

4. Remove the cable casing retaining nut at the underbody.

5. From under the dash, disconnect the cable casing and cable from the control assembly.

To install:

6. Install the cable casing through the mounting hole in the control assembly, making sure the locking fingers are fully engaged.

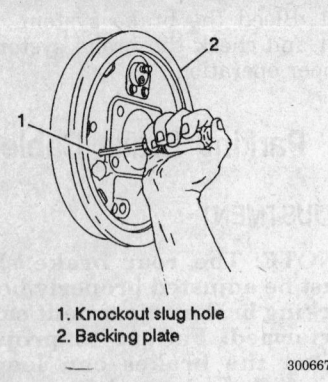

1. Knockout slug hole
2. Backing plate

300667

Moving the parking brake lever off the stop

7. Connect the cable end to the control assembly connector.

8. From under the vehicle, install the cable casing retaining nut and tighten to 22 ft. lbs. (30 Nm).

9. Connect the front cable to the equalizer assembly.

10. Adjust the parking brake.

11. Lower the vehicle.

Intermediate Cable

1. Raise and safely support the vehicle.

2. Loosen the equalizer assembly and disconnect the intermediate cable from the equalizer assembly.

3. Disengage the locking fingers on the intermediate cable housing from the front bracket.

4. Remove the intermediate cable housing clip and guide from the underbody.

5. Disconnect the cable from the rear equalizer assembly.

To install:

6. Connect the cable to the rear equalizer assembly.

7. Install the housing clip and guide on the underbody.

8. Install the cable housing to the front bracket. Make sure the locking fingers are fully engaged in the bracket.

9. Connect the cable to the front equalizer assembly.

10. Adjust the parking brake.

11. Lower the vehicle.

Left and Right Rear Cables

1. Raise and safely support the vehicle.

2. Loosen the adjuster nut at the front equalizer until the cable is slack.

3. Remove the rear wheel. Mark the position of the wheel to the wheel studs, prior to removal, for installation reference.

4. Remove the brake drum. Mark the position of the drum to the wheel studs, prior to removal, for installation reference.

5. Remove the brake shoes.

6. Remove the parking brake lever by twisting the lever outward.

7. Depress the cable retaining fingers to remove the cable from the backing plate. With the cable fingers released, push the assembly through the hole in the backing plate.

8. Disconnect the cable from the rear equalizer by backing off the equalizer nut.

9. Depress the cable housing locking fingers to remove the cable from the equalizer bracket.

10. If necessary, remove the 2 attaching screws from the underbody bracket and remove the cable.

To install:

11. Fully seat the locking fingers on the cable in the axle bracket.

12. Connect the cable to the equalizer bracket.

13. If necessary, position the cable on the underbody bracket and install the attaching screws. Tighten the screws to 84 inch lbs. (10 Nm).

14. Seat the cable locking fingers in the backing plate hole.

15. Install the remaining components in the reverse order of removal.

16. Adjust the parking brake.

17. Lower the vehicle.

Brake System Bleeding

PROCEDURE

Use only fresh Delco supreme 11 brake fluid GM part # 1052535 or an equivalent DOT 3 brake fluid from a sealed container. Under NO circumstances should DOT 4 or 5 be used; brake system contamination will result and every brake system component will have to be replaced.

NOTE: When bleeding the brake system it is important that the correct bleeding sequence be followed. Failure to follow the correct sequence may result in air becoming trapped in the brake system. If air remains in the system decreased braking performance and increased pedal travel will result.

Master Cylinder Bench Bleeding

When the master cylinder is replaced, the new master cylinder should be bench bled as follows:

1. Install the master cylinder in a vise, clamping at the flange. DO NOT clamp the master cylinder by the body, it may distort the body and cause a pressure loss.

2. Install short lengths of hoses/lines and fittings to the master cylinder ports, aiming the ends of the hoses/lines into the master cylinder reservoir. Fill the master cylinder with fresh DOT 3 brake fluid and make sure that the ends of the hoses/lines are submerged in the brake fluid.

3. Pump the master cylinder several times until there are no air bubbles seen coming out of the hose/line ends.

4. Remove the hoses/lines and fittings and install the shipping caps to the ports, to prevent brake fluid from dripping on any painted surface.

5. Install the master cylinder cap and install the master cylinder on the vehicle.

6. Connect the brake lines to the master cylinder ports and snug the fittings.

7. Have an assistant depress the brake pedal slowly one time and hold. Loosen the forward brake line fitting to purge air from the master cylinder bore. Tighten the brake line fitting and have the assistant slowly release the brake pedal. Wait 15 seconds and repeat the sequence, including the 15 second wait, until all air is purged from the master cylinder bore.

8. Repeat the bleeding procedure at the rear brake line fitting.

9. Finally tighten the brake line fittings to 11 ft. lbs. (15 Nm).

10. Check the master cylinder fluid level and bleed the system. Road test and check for proper brake operation.

Master Cylinder Bleeding On-Vehicle

If the master cylinder is known or suspected to have air in the bore, it must be bled first, before bleeding the rest of the system. Bleed the master cylinder as follows:

1. Loosen the brake line fittings at the master cylinder.

2. Pour fresh brake fluid into the master cylinder reservoir until fluid flows from the master cylinder ports, then tighten the fittings.

3. Have an assistant depress the brake pedal slowly one time and hold. Loosen the forward brake line fitting to purge air from the master cylinder bore. Tighten the brake line fitting and have the assistant slowly release the brake pedal. Wait 15 seconds and repeat the sequence, including the 15 second wait, until all air is purged from the master cylinder bore.

4. Repeat the bleeding procedure at the rear brake line fitting.

5. Final tighten the brake line fittings to 11 ft. lbs. (15 Nm).

6. Connect the electrical connector to the fluid level sensor.

7. Check the master cylinder fluid level and bleed the system. Road test and check for proper brake operation.

Brake System Bleeding

1. Have an assistant pump the brake pedal several times using smooth even strokes, pausing 10 seconds between each stroke. After the last stroke hold the pedal down.

--- **WARNING** ---
When pumping the brakes DO NOT force or stab the pedal down or damage to the master cylinder could result.

2. Starting with the right rear wheel, connect a clear rubber hose to the bleeder screw and submerge the other end of the hose in a container partially filled with brake fluid.

3. Open the bleeder screw about ½ turn.

4. When the brake pedal is fully depressed, close the bleeder screw.

5. Repeat the procedure as needed until no air comes out of the bleeder.

6. Repeat Steps 2–5 on the remaining wheels in the following order:
   a. Left rear
   b. Right front
   c. Left front

7. While bleeding the brakes check the fluid level in the master cylinder frequently and add fluid as necessary. DO NOT allow the master cylinder to be pumped dry.

8. After the right front wheel is done verify that all the bleeder screws are tight and not leaking. Add brake fluid to the master cylinder reservoir as required.

9. Test drive the vehicle carefully to verify correct braking operation.

Wheel Speed Sensor

REMOVAL AND INSTALLATION

Front Wheel Speed Sensor

1. Raise and safely support the vehicle.

2. Remove the front wheel. Mark the position of the wheel to wheel studs, prior to removal, for installation reference.

3. Insert a drift punch through the caliper and into the rotor cooling fins to prevent the rotor from turning.

4. Remove the halfshaft nut and washer.

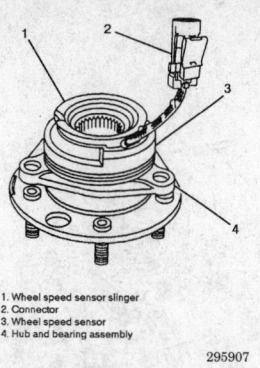

1. Wheel speed sensor slinger
2. Connector
3. Wheel speed sensor
4. Hub and bearing assembly

295907

Front wheel speed sensor assembly

5. Remove the caliper mounting bolts, remove the caliper from the steering knuckle and support out of the way. DO NOT allow the brake hose to support the weight of the caliper.

6. Remove the brake rotor. Mark the position of the rotor to the wheel studs, prior to removal, for installation reference.

7. Disconnect the speed sensor connector and unclip from the dust shield.

8. Remove the 3 hub and bearing bolts and remove the backing plate.

9. Place the transaxle selector lever in **P**.

10. Install J-28733 or an equivalent puller and separate the hub and bearing from the halfshaft.

11. Remove the hub and bearing assembly from the steering knuckle.

12. Gently pry off the wheel speed sensor slinger. Discard the used slinger.

13. Remove the wheel speed sensor by gently prying off the hub and bearing assembly.

NOTE: Do not allow debris to enter the bearing when sensor is removed. Do not add lubricant to the bearing through the sensor housing opening; the bearing is lubricated for the life of the vehicle. Do not clean grease from the toothed sensor ring; grease does not effect wheel speed sensor operation. Inspect the wheel speed sensor cap for any indication of water or debris entry. If indication of water or debris exists, replace the entire hub and bearing assembly.

To install:

14. Apply Loctite® 620 or equivalent, to the outer diameter of the bearing hub per instructions in the service kit.

15. Place the hub and bearing assembly on J 38764–4 or equivalent, to

prevent possible damage to the wheel studs.

16. Install the wheel speed sensor using tool J 38764–1 or equivalent, to press the wheel speed sensor into place.

17. Install the hub and bearing over the halfshaft splines. Make sure the splines engage smoothly.

18. Apply a light coating of grease to the steering knuckle bore.

19. Slide the hub assembly onto the halfshaft as far as possible. If the hub will not bottom out on the halfshaft, install the hub mounting bolts and use the hub nut to draw the hub onto the halfshaft.

20. Once the hub is flush with the steering knuckle remove the mounting bolts and install the dust shield and mounting bolts. Tighten the mounting bolts to 70 ft. lbs. (95 Nm).

21. Place the transaxle in **N**.

22. Install the remaining components in the reverse order of removal.

23. Install the wheel, aligning the marks made during removal. Torque the lug nuts to 100 ft. lbs. (140 Nm).

24. Lower the vehicle. Road test and check operation.

Rear Wheel Speed Sensor

1. Raise and safely support the vehicle.

2. Remove the wheel. Mark the position of the wheel to the wheel studs, prior to removal, for installation reference.

3. Disconnect the wheel speed sensor electrical connector.

4. Remove the lower rear strut retainer bolt and nut only. Do not loosen the upper bolt and nut or a wheel alignment must be performed.

5. Clean all the dirt and debris away from sensor interface area.

6. Remove the 3 mounting screws and discard.

7. Remove the wheel speed sensor.

8. Inspect the the sensor ring and the sensor plate for contact. If contact is found, the hub and bearing must be replaced as an assembly.

To install:

9. Lubricate the new O-ring and install the sensor to the housing.

10. Install the mounting screws. Torque the screws to 25 inch lbs. (2.8 Nm).

11. Install the lower rear strut retaining bolt and nut and torque to 140 ft. lbs. (190 Nm).

12. Connect the electrical connector to the wheel speed sensor.

13. Install the wheel, aligning the marks made during removal, and torque the lug nuts to 100 ft. lbs. (140 Nm).

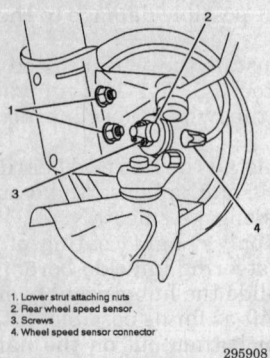

1. Lower strut attaching nuts
2. Rear wheel speed sensor
3. Screws
4. Wheel speed sensor connector

295908

**Rear wheel speed sensor
assembly**

14. Lower the vehicle. Road test and check operation.

FRONT SUSPENSION

Strut

REMOVAL AND INSTALLATION

Except Cadillac DeVille and Sixty Special

——— **WARNING** ———
The steering knuckle must be retained after the strut to steering knuckle bolts have been removed. Failure to observe this may cause ball joint and/or half-shaft damage.

1. Loosen the strut housing tie bar through bolts on both ends of the tie bar.
2. If equipped with electronic ride control option, disconnect the electrical connection.
3. Remove the 3 strut-to-body nuts at the strut mount.
4. Raise and safely support the vehicle, allowing the control arms to hang free.
5. Remove the front wheel.
6. Disconnect the ABS front wheel speed sensor connector.
7. Remove the wheel speed sensor bracket from the strut.
8. Remove the brake line bracket from the strut.
9. Remove the strut-to-steering knuckle bolts and remove the strut from the vehicle.
10. Install the strut assembly into compressor tool J 34013-B or equivalent, to compress the coil spring.

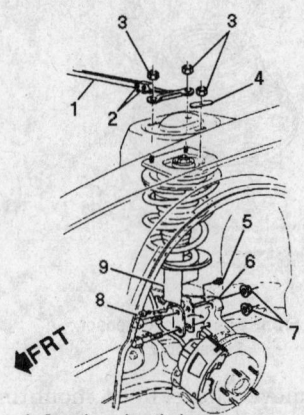

1. Strut housing tie bar
2. Through-bolts, 27 ft. lb. (37 Nm)
3. Strut mount nut, 18 ft. lb. (24 Nm)
4. Washer
5. Front wheel speed sensor bracket bolt 13 ft. lb. (17 Nm)
6. Retain knuckle once strut is removed
7. Strut to knuckle nut, 140 ft. lb. (190 Nm)
8. Brake line bracket bolt, 13 ft. lb. (17 Nm)
9. Strut

252488

Strut mounting and related components

11. Hold the strut shaft from turning using a number 50 Torx® socket and remove the 24mm nut on the top end of the strut.
12. Install rod J 34013–38 to help guide the strut shaft out of the upper mount assembly.
13. Loosen the spring compressor tool until the coil spring and mount can be removed as an assembly. Remove the lower spring insulator, if equipped.

To install:

14. Install the strut into spring compressor tool J 34013-B or equivalent. Install the lower insulator, coil spring and upper mount and compress the coil spring while guiding the strut shaft through the upper mount, using tool J 34013–38 or equivalent.
15. Install and torque the upper nut to 55 ft. lbs. (75 Nm) while holding the strut shaft with a socket.
16. Remove the strut assembly from the spring compressor by loosening the spring compressor nut.
17. Install the strut in the vehicle.
18. Install and tighten the 3 strut mount nuts to 18 ft. lbs. (24 Nm).
19. Connect the electronic ride control option electrical connector, if removed.

20. Install the strut-to-knuckle bolts and tighten to 140 ft. lbs. (190 Nm).
21. Install the remaining components in the reverse order of removal.
22. Lower the vehicle. Check and adjust the wheel alignment.

Cadillac DeVille and Sixty Special

——— **WARNING** ———
When working near the half-shafts, be careful not to overextend the inner CV-joints. If the joint is overextended, the internal components may separate resulting in joint failure.

1. Disconnect the negative battery cable.
2. Raise and safely support the vehicle. The vehicle must be supported by the frame with the control arms hanging free.
3. Remove the front wheel.
4. If equipped, loosen the strut housing tie bar through bolts at each end of the bar.
5. Remove the nuts attaching the top of the strut to the body.
6. If equipped with anti-lock brakes, disconnect the wheel speed sensor and remove the sensor bracket from the strut.
7. Remove the brake line bracket from the strut.
8. Remove the strut-to-knuckle bolts and remove the strut. Support the knuckle with wire. The knuckle must be supported to prevent damage to the ball joint and/or halfshaft.

——— **WARNING** ———
Be careful not to scratch or crack the coating on the coil spring, as damage may cause premature spring failure.

9. Mount the strut assembly in strut compressor tool J–34013 or equivalent.
10. Turn the compressor forcing screw until the spring compresses slightly.
11. Use a T–50 Torx® bit to keep the strut shaft from turning, then remove the nut on the top of the strut shaft.
12. Loosen the compressor screw while guiding the strut shaft out of the assembly. Continue loosening the compressor screw until the strut and spring can be removed.

To install:

13. Mount the strut in the strut compressor tool. Use clamping tool J–34013–20 or equivalent, to hold the strut shaft in place.

14. Install the spring over the strut. The flat on the upper spring seat must face out from the centerline of the vehicle, or when mounted in the strut compressor, the spring seat must face the same direction as the steering knuckle mounting flange.

NOTE: If the bearing was removed from the upper spring seat, it must be reinstalled in the spring seat in the same position before attaching to the strut mount.

15. Turn the compressor screw to compress the spring, while guiding the strut shaft through the top of the strut assembly. If necessary, use tool J–34013–38 or equivalent, to guide the shaft.

16. When the strut shaft threads are visible through the top of the strut assembly, install the washer and nut.

17. Remove clamping tool J–34013–20 from the strut shaft.

18. Tighten the strut shaft nut to 55 ft. lbs. (75 Nm) while holding the strut shaft with a T–50 Torx® bit.

19. Remove the strut assembly from the compressor tool.

20. Attach the strut to the knuckle with the bolts and nuts.

21. Attach the strut to the body with the nuts. If equipped, position the washer and tie bar prior to installing the strut-to-body nuts.

22. Install the brake line bracket to the strut.

23. If equipped with anti-lock brakes, install the wheel speed sensor bracket and connect the sensor.

24. Tighten the strut-to-knuckle bolts to 140 ft. lbs. (190 Nm), the strut-to-body nuts to 18 ft. lbs. (24 Nm) and if equipped, the tie bar through bolts to 20 ft. lbs. (27 Nm).

25. Install the wheel and lower the vehicle. Check and adjust the front wheel alignment.

Lower Ball Joints

REMOVAL AND INSTALLATION

——— WARNING ———
If the ball joint is separated for related suspension/driveline service, the ball joint seal should be inspected for damage. A damaged seal will cause ball joint failure. The ball joint should should be replaced if seal damage is found.

1. Raise and safely support the vehicle, allowing the control arms to hang free.

2. Remove the front wheel.

3. Remove the cotter pin from the lower ball joint and loosen the nut.

4. Using separator tool J 36226 or equivalent, remove the ball joint from the steering knuckle.

5. Remove the stabilizer shaft link assembly.

6. Drill out the 3 rivets retaining the lower ball joint to the lower control arm.

7. Remove the lower ball joint from the steering knuckle and control arm.

To install:

8. Install the lower ball joint to the steering knuckle and align the holes with the holes in the lower control arm.

9. Attach the ball joint to the lower control arm with bolts and nuts, making sure the bolts face down. Tighten the nuts to 50 ft. lbs. (68 Nm).

10. Tighten the stabilizer shaft link nut to 13 ft. lbs. (17 Nm).

11. Tighten the lower ball joint nut to 88 inch lbs. (10 Nm). Then tighten an additional 120 degree turn, during which a torque of 41 ft. lbs. (55 Nm) must be obtained.

12. Install the cotter pin to the lower ball joint. To align the slot in the nut with the hole in the lower ball joint stud, tighten the ball joint nut up to one more flat. Never loosen the nut to align the slot.

13. Install the wheel and torque the lug nuts to 100 ft. lbs. (140 Nm).

14. Lower the vehicle.

15. Check and adjust the wheel alignment.

Lower Control Arms

REMOVAL AND INSTALLATION

——— WARNING ———
If the ball joint is separated for related suspension/driveline service, the ball joint seal should be inspected for damage. A damaged seal will cause ball joint failure. The ball joint should should be replaced if seal damage is found.

1. Raise and safely support the vehicle, allowing the control arms to hang free.

2. Remove the front wheel.

3. For DeVille and Sixty Special, note the position of all stabilizer bar link kit washers and insulators and remove the link kit.

4. Remove the cotter pin from the lower ball joint and loosen or remove the nut, as necessary.

5. Using separator tool J 36226 or equivalent, remove the ball joint from the steering knuckle.

6. For all vehicles, except DeVille and Sixty Special, loosen the stabilizer shaft link nut.

7. Remove the lower control arm mounting bolts, then remove the lower control arm from the frame.

To install:

8. Install the lower control arm to the frame with the attaching bolts, washers and nuts.

NOTE: Do not tighten the lower control arm nuts at this time. The weight of the vehicle must be supported by the control arms such that the vehicle design trim heights are obtained before tightening the lower control arm mounting nuts.

9. Install the stabilizer shaft link nut or the link kit and tighten to 13 ft. lbs. (17 Nm).

10. Connect the ball joint stud to the steering knuckle and tighten the lower ball joint nut to 88 inch lbs. (10 Nm). Then tighten an additional 120 degree turn, during which a torque of 41 ft. lbs. (55 Nm) must be obtained.

11. Install the cotter pin to the lower ball joint. To align the slot in the nut with the hole in the lower ball joint stud, tighten the ball joint nut up to one more flat. Never loosen the nut to align the slot.

12. Install the wheel and lower the vehicle. Check trim height specifications.

13. Tighten the front lower control arm nut to 140 ft. lbs. (190 Nm), then tighten the rear lower control arm nut to 91 ft. lbs. (123 Nm).

14. Check and adjust wheel alignment.

Sway Bar

REMOVAL AND INSTALLATION

1. Raise and safely support the vehicle, allowing the control arms to hang.

2. Remove both front wheels.

3. Remove the left and right sway bar link kits. Note the position of the washers and bushings.

4. Remove the sway bar bracket bolts, then remove the brackets.

5. Remove the cotter pins and castle nuts from the outer tie rod ends and separate the tie rod ends from the steering knuckles using J-24319-B or equivalent.

6. Disconnect the front exhaust pipe from the exhaust manifold.

7. For DeVille and Sixty Special, remove the bolt connecting the AIR pipe to the exhaust pipe.

8. Turn the passenger side strut completely to the right. Slide the sway bar over the steering knuckle and pull down until the sway bar clears the frame.

9. Remove the sway bar from the vehicle.

To install:

10. Position the sway bar in the vehicle. Loosely install the sway bar bushings and bracket.

11. Loosely install the sway bar link kits.

12. Connect the tie rod ends to the steering knuckles. Tighten the castle nuts to 35 ft. lbs. (48 Nm). Install new cotter pins. If the cotter pin holes do not line up tighten the nut as necessary. NEVER loosen the nut to align the cotter pin holes.

13. Tighten the sway bar bracket bolts to 35 ft. lbs (47 Nm) and tighten the sway bar link kits to 13 ft. lbs. (17 Nm).

14. Connect the exhaust pipe to the manifold and tighten the bolts to 18 ft. lbs. (25 Nm).

15. Connect the AIR pipe bolt to the pipe and tighten to 7 ft. lbs. (9 Nm).

16. Install the front wheels and torque the lug nuts to 100 ft. lbs. (140 Nm). Lower the vehicle.

Front Wheel Bearings

ADJUSTMENT

The wheel bearings are not adjustable. If a wheel bearing is out of specifications, it must be replaced. Using a dial indicator, check for looseness. If play exceeds 0.005 inch (0.127mm) the bearing wear is excessive and the hub and bearing should be replaced.

REMOVAL AND INSTALLATION

1. Raise and safely support the vehicle. Remove the front wheel.

2. Insert a drift punch through the caliper and into the rotor cooling fins to prevent the rotor from turning.

3. Remove the axle nut and washer.

4. Remove the caliper mounting bolts. Remove the caliper from the steering knuckle and support out of the way. DO NOT allow the brake hose to support the weight of the caliper.

5. Remove the brake rotor.

6. Disconnect the ABS speed sensor connector and unclip from the dust shield.

7. Remove the 3 hub and bearing bolts and remove the dust shield.

8. Place the transaxle selector in the **P** detent.

9. Install J-28733 or an equivalent puller, and separate the hub and bearing from the drive axle.

10. Remove the hub and bearing assembly from the steering knuckle.

To install:

11. Install the hub and bearing over the half shaft splines. Make sure the splines engage smoothly.

12. Apply a light coating of grease to the steering knuckle bore.

13. Slide the hub assembly onto the axle as far as possible. If the hub will not bottom out on the axle, install the hub mounting bolts and use the axle nut to draw the hub onto the axle.

14. Once the hub is flush with the steering knuckle, remove the mounting bolts and install the dust shield. Reinstall the mounting bolts and tighten to 70 ft. lbs. (95 Nm).

15. Place the transaxle in **N**.

16. Connect the ABS front wheel speed sensor connector and clip to the dust shield.

17. Install the brake rotor.

18. Install the caliper and tighten the mounting bolts to 38 ft. lbs. (51 Nm).

19. Insert a drift punch through the rotor to prevent the axle from turning.

20. Tighten the drive axle shaft nut to 107 ft. lbs. (145 Nm).

21. Remove the drift punch.

22. Install the wheel and lower the vehicle. Road test the vehicle.

REAR SUSPENSION

Strut

REMOVAL AND INSTALLATION

1. Remove the trunk side cover or the rear seat cushion and seat back, as required, to gain access to the upper strut mounting nuts.

2. Raise and safely support the vehicle so the rear wheels are just off the ground.

3. Remove the rear wheel.

4. Using a suitable floor jack and a block of wood, support the lower control arm.

5. Remove the two upper strut mounting nuts.

6. If equipped with Electronic Leveling Control (ELC), disconnect the air tube from the strut.

7. Remove the lower strut bracket-to-knuckle nuts and bolts. Remove the sway bar bracket from the lower strut bracket.

8. If equipped with Computer Control Ride (CCR), lower the strut and disconnect the electrical connector.

9. Remove the strut from the vehicle.

To install:

10. If equipped with CCR, connect the electrical connector. As the strut is positioned, make sure the CCR wiring is routed properly.

11. Position the strut in the vehicle and install the upper strut mounting nuts. Install the sway bar bracket and the lower strut bracket-to-knuckle bolts.

12. Tighten the upper strut mounting nuts to 30 ft. lbs. (41 Nm) on all except Cadillac. On Cadillac, tighten the upper strut mounting nuts to 19 ft. lbs. (25 Nm). Tighten the strut bracket-to-knuckle bolts to 140 ft. lbs. (190 Nm).

13. If equipped with ELC, connect the air tube.

14. For ELC equipped vehicles, lightly pressurize the ELC system by momentarily grounding terminal B at the height sensor connector before lowering the vehicle.

15. Install the wheel and torque the lug nuts to 100 ft. lbs. (140 Nm).

16. Remove the jack from under the control arm.

17. Lower the vehicle. Check and, if necessary, adjust the wheel alignment.

Coil Spring

REMOVAL AND INSTALLATION

———— **CAUTION** ————

The coil springs are under a considerable amount of tension. Be very careful when removing or installing them; they can exert enough force to cause very serious injury.

1. Raise and safely support the vehicle. Remove the rear wheel.

2. If removing the right side coil spring, disconnect the height sensor link from the right control arm, if equipped with Electronic Leveling Control (ELC).

3. If removing the left side coil spring, disconnect the parking brake cable retaining clip from the left control arm.

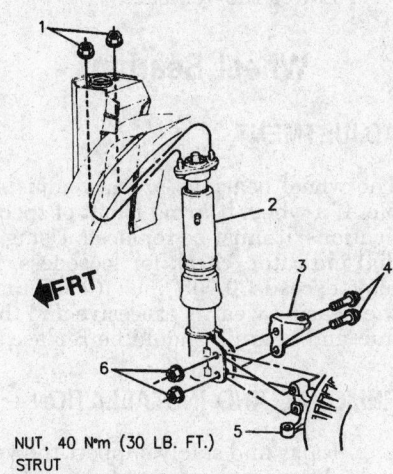

1 NUT, 40 N·m (30 LB. FT.)
2 STRUT
3 STABILIZER SHAFT BRACKET
4 BOLT; INSTALL IN DIRECTION SHOWN
5 KNUCKLE
6 NUT, 190 N·m (140 LB. FT.)

340354

Strut mounting components

4. Remove the rear sway bar link from the bracket on the knuckle.
5. Mount tool J-23028–01 or an equivalent control arm support adapter, on a transmission jack and position to cradle the control arm bushings.

— **CAUTION** —
Tool J-23028–01 or equivalent, must be secured to the jack or personal injury could result.

6. Place a chain around the spring and through the control arm as a safety measure.
7. Raise the jack to remove tension from the control arm pivot bolts.
8. Remove the rear nut and through bolt.
9. Slowly maneuver the jack to relieve tension from the front control arm bolt.
10. Remove the front nut and through bolt.

NOTE: Do not apply force to the control arm and/or ball joint to remove the spring. Proper maneuvering of the spring will allow for easy removal.

11. Lower the jack to pivot the control arm downward. When all compression is removed from the spring,

remove the safety chain, spring and insulators.
12. Inspect the spring insulators and replace them if they are cut or torn. If the vehicle has been driven more than 50,000 miles, replace them regardless of condition.

To install:
13. Snap the upper insulator onto the spring.
14. Position the lower insulator and spring in the vehicle.
15. Using the jack and J-23028–01 or equivalent, raise the control arm into place.
16. Slowly maneuver the jack to permit installation of the control arm bolts and nuts. DO NOT tighten the nuts until the weight of the vehicle is on the suspension.
17. Connect the rear sway bar to the knuckle bracket with the link assembly. Do not tighten the link bolt yet.
18. Connect the height sensor link to the right control arm or connect the parking brake cable retaining clip to the left control arm, as required.
19. Install the rear wheel and torque the lug nuts to 100 ft. lbs. (140 Nm).
20. Lower the vehicle.
21. With the vehicle resting on its wheels, tighten the control arm through bolt nuts to 85 ft. lbs. (115 Nm) and the sway bar link bolt to 13 ft. lbs. (17 Nm).
22. Check and, if necessary, adjust the rear wheel alignment.
23. Road test the vehicle for proper operation.

— **CAUTION** —
The coil springs are under a considerable amount of tension. Be very careful when removing or installing them; they can exert enough force to cause very serious injury.

24. Raise and safely support the vehicle. Remove the rear wheel.
25. If removing the right side coil spring, disconnect the height sensor link from the right control arm, if equipped with Electronic Leveling Control (ELC).
26. If removing the left side coil spring, disconnect the parking brake cable retaining clip from the left control arm.
27. Remove the rear sway bar link from the bracket on the knuckle.
28. Mount tool J-23028–01 or an equivalent control arm support adapter, on a transmission jack and position to cradle the control arm bushings.

— **CAUTION** —
Tool J-23028–01 or equivalent, must be secured to the jack or personal injury could result.

29. Place a chain around the spring and through the control arm as a safety measure.
30. Raise the jack to remove tension from the control arm pivot bolts.
31. Remove the rear nut and through bolt.
32. Slowly maneuver the jack to relieve tension from the front control arm bolt.
33. Remove the front nut and through bolt.

NOTE: Do not apply force to the control arm and/or ball joint to remove the spring. Proper maneuvering of the spring will allow for easy removal.

34. Lower the jack to pivot the control arm downward. When all compression is removed from the spring, remove the safety chain, spring and insulators.
35. Inspect the spring insulators and replace them if they are cut or torn. If the vehicle has been driven more than 50,000 miles, replace them regardless of condition.

To install:
36. Snap the upper insulator onto the spring.
37. Position the lower insulator and spring in the vehicle.
38. Using the jack and J-23028–01 or equivalent, raise the control arm into place.
39. Slowly maneuver the jack to permit installation of the control arm bolts and nuts. DO NOT tighten the nuts until the weight of the vehicle is on the suspension.
40. Connect the rear sway bar to the knuckle bracket with the link assembly. Do not tighten the link bolt yet.
41. Connect the height sensor link to the right control arm or connect the parking brake cable retaining clip to the left control arm, as required.
42. Install the rear wheel and torque the lug nuts to 100 ft. lbs. (140 Nm).
43. Lower the vehicle.
44. With the vehicle resting on its wheels, tighten the control arm through bolt nuts to 85 ft. lbs. (115 Nm) and the sway bar link bolt to 13 ft. lbs. (17 Nm).
45. Check and, if necessary, adjust the rear wheel alignment.
46. Road test the vehicle for proper operation.

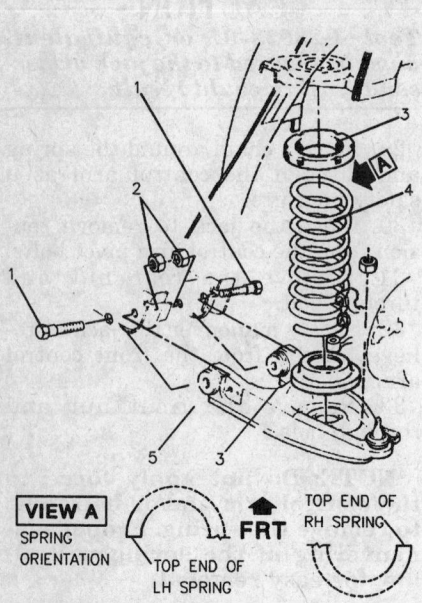

VIEW A
SPRING
ORIENTATION

TOP END OF
LH SPRING

FRT

TOP END OF
RH SPRING

1 BOLT, 160 N•m (118 LB. FT.)
  (OPTIONAL TORQUE)
2 NUT, 119 N•m (88 LB. FT.)
3 INSULATOR
4 SPRING
5 CONTROL ARM

340352

Coil spring installation

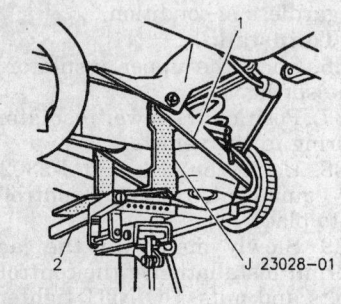

1 CONTROL ARM
2 TRANSMISSION JACK

J 23028-01

340353

Special tool J-23028-01 in position on the jack

Sway Bar

REMOVAL AND INSTALLATION

Except Cadillac DeVille and Sixty Special

1. Raise and safely support the rear of the vehicle.
2. Remove the rear wheels.
3. Remove the sway bar link assembly bolt, nut, retainer and insulators from the sway bar bracket.
4. Remove the sway bar insulators.

To install:

5. Install the sway bar and insulators, installing the insulator into the link with the slit rearward.
6. Bend the link upward to close around the insulator and install the link bolt. Tighten the link bolt to 17 ft. lbs. (23 Nm).
7. Install the link assembly insulators, retainers and tighten the support bolt and nut to 13 ft. lbs. (17 Nm).
8. Install the wheels and torque the lug nuts to 100 ft. lbs. (140 Nm).
9. Lower the vehicle and road test.

Cadillac DeVille and Sixty Special

1. Raise and safely support the vehicle.
2. Remove the rear wheels.
3. Note the position of all washers and insulators in the rear link kits and remove the link kit upper nut. Remove the link kit components from the knuckle bracket.
4. Remove the bushing clip bolts.
5. Bend open the sway bar bushing clamps.
6. Remove the sway bar and bushings.

To install:

7. Install the sway bar in the bushing clamps.
8. Bend the bushing clamps closed, install the clamp bolts and tighten to 17 ft. lbs. (23 Nm).

9. Install the link kits and tighten the nuts to 13 ft. lbs. (17 Nm).
10. Install the wheels.
11. Lower the vehicle.

Wheel Bearings

ADJUSTMENT

The wheel bearings are not adjustable. If a wheel bearing is out of specifications, it must be replaced. Using a dial indicator, check for looseness. If play exceeds 0.005 inch (0.127mm) the bearing wear is excessive and the hub and bearing should be replaced.

REMOVAL AND INSTALLATION

1. Raise and safely support the vehicle. Remove the rear wheel.
2. Remove the brake drum.
3. Disconnect the ABS sensor wire connector, if equipped.

— **WARNING** —
The hub assembly mounting bolts also secure the backing plate assembly. Once the bolts are removed, the backing plate must be supported with wire or other means. Do not let the brake line or ABS electrical wire support the brake assembly.

4. Remove the 4 hub and bearing mounting bolts and remove the hub assembly.

To install:

5. Install hub and bearing assembly onto the rear knuckle. Install the 4 mounting bolts and tighten to 52 ft. lbs. (70 Nm).
6. Connect the ABS sensor connector, if equipped.
7. Install the brake drum.
8. Install the wheel and torque the nuts to 100 ft. lbs. (140 Nm).
9. Lower the vehicle and road test for proper operation.

FIRING ORDERS

NOTE: To avoid confusion, always replace spark plug wires one at a time.

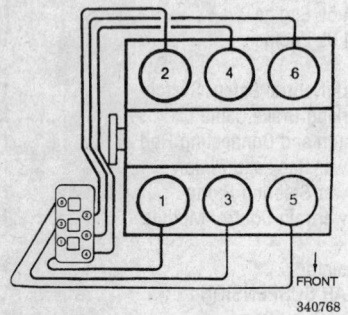

3.8L Engine
Engine Firing Order: 1–6–5–4–3–2
Distributorless Ignition System

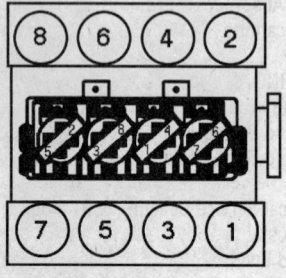

4.6L Engine
Engine Firing Order:
1–2–7–3–4–5–6–8
Distributorless Ignition System

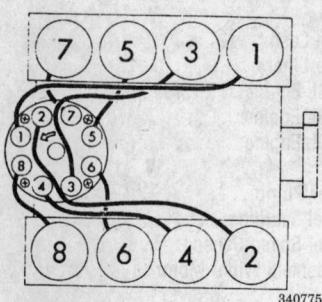

4.9L Engine
Engine Firing Order:
1–8–4–3–6–5–7–2
Distributor Rotation:
Counterclockwise

ENGINE ELECTRICAL

NOTE: Disconnecting the negative battery cable on some vehicles may interfere with the functions of the on board computer systems and may require the computer to undergo a relearning process, once the negative battery cable is reconnected.

Distributor

REMOVAL AND INSTALLATION

4.9L Engine

1. Disconnect the negative battery cable.
2. Disconnect the feed wire and coil connection from the distributor cap.
3. Remove the distributor cap mounting bolts, then move the cap out of the way. It is not necessary to disconnect the ignition wires.
4. Disconnect the PCM harness from the distributor.
5. Mark the position of the rotor with respect to the distributor housing and scribe the position of the distributor housing with respect to the engine.
6. Remove the distributor hold-down nut and clamp.
7. Remove the distributor.

NOTE: Do not crank the engine after the distributor is removed. If the engine is accidentally cranked, the engine will have to be positioned at No. 1 cylinder TDC.

To install:

8. Install the distributor, aligning the marks made during removal.

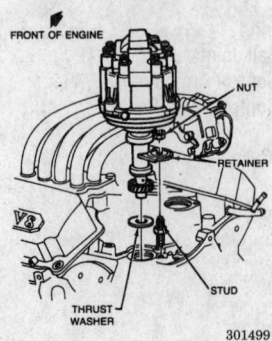

FRONT OF ENGINE
NUT
RETAINER
V8
STUD
THRUST WASHER

301499

Distributor and mounting components — 4.9L engine

9. If the engine was accidentally cranked after the distributor was removed, proceed as follows:
   a. Remove the No. 1 cylinder spark plug.
   b. Place a finger over the No. 1 cylinder spark plug hole and crank the engine slowly until compression is felt.
   c. Align the timing mark on the crankshaft pulley to **0** on the engine timing indicator.
   d. Position the distributor rotor to point towards the No. 1 spark plug wire terminal on the distributor cap.
   e. Install the distributor, aligning the mark on the distributor housing with the mark on the engine.
   f. Install the spark plug in the No. 1 cylinder.
10. Install the distributor hold-down clamp and nut. DO NOT tighten the nut at this time.
11. Connect the PCM connector.
12. Install the distributor cap and mounting screws.
13. Connect the coil leads and feed wire to the cap.
14. Connect the negative battery cable.
15. Set the ignition timing.
16. Tighten the distributor hold-down nut to 20 ft. lbs. (26 Nm).

Ignition Timing

ADJUSTMENT

3.8L Engine

The ignition timing is not adjustable, and is set according to engine demand electronically. The Powertrain Control Module (PCM) controls the ignition timing under all driving conditions. The PCM monitors input signals from the engine coolant sensor, intake air temperature, mass air flow sensor, park/neutral switch, throttle position sensor and vehicle speed sensor, and in turn adjusts the timing based on these inputs. No adjustments are possible or should be attempted.

4.6L Engines

The 4.6L Northstar engine is equipped with a Distributorless Ignition System (DIS). The system consists of 2 crankshaft position sensors, crankshaft reluctor ring, camshaft position sensor, ignition control module, 4 ignition coils, 8 plug wires and spark plugs, knock sensor and the Powertrain Control Module (PCM).

The base ignition timing is determined by the relationship of the crankshaft position sensors to the crankshaft reluctor ring. This relationship is not adjustable. Base ignition timing is 10 degrees Before Top Dead Center (BTDC).

The PCM controls spark advance under all driving conditions. The PCM incorporates a permanent spark control override, which electronically lowers the base timing if spark knock (detonation) is encountered during normal operation due to the use of low octane fuel.

4.9L Engine

NOTE: Check the Vehicle Emission Control Information (VECI) label, located in the engine compartment. If the procedure below differs from that described on the label, follow the instructions on the label. The information on the VECI label often reflects changes made during production.

1. With the ignition OFF, connect a timing light to the No. 1 spark plug wire. Connect the power leads for the timing light according to the manufacturers directions.

2. Jumper pins A and B together at the Assembly Line Diagnostic Link (ALDL) connector while not in diagnostic display.

3. Start the engine and allow it to reach normal operating temperature. Aim the timing light at the timing indicator and crankshaft pulley. Refer to the VECI label for ignition timing specifications.

4. If adjustment is required, loosen the distributor hold-down bolt and rotate the distributor until the timing marks are in alignment.

5. Tighten the hold-down bolt and check the timing to make sure the distributor didn't move.

6. Turn OFF the engine and disconnect the timing light.

Alternator

PRECAUTIONS

Several precautions must be observed with alternator equipped vehicles to avoid damage to the unit.

• If the battery is removed for any reason, make sure it is reconnected with the correct polarity. Reversing the battery connections may result in damage to the 1-way rectifiers.

• When utilizing a booster battery as a starting aid, always connect the positive to positive terminals and the negative terminal from the booster battery to a good engine ground on the vehicle being started.

• Never use a fast charger as a booster to start vehicles.

• Disconnect the battery cables when charging the battery with a fast charger.

• Never attempt to polarize the alternator.

• Do not use test lights of more than 12 volts when checking diode continuity.

• Do not short across or ground any of the alternator terminals.

• The polarity of the battery, alternator and regulator must be matched and considered before making any electrical connections within the system.

• Never separate the alternator on an open circuit. Make sure all connections within the circuit are clean and tight.

• Disconnect the battery ground terminal when performing any service on electrical components.

• Disconnect the battery if arc welding is to be done on the vehicle.

REMOVAL AND INSTALLATION

3.8L Engine

1. Disconnect the negative battery cable.

2. Turn the automatic tensioner to relieve the belt tension, then remove the drive belt from the alternator.

3. Disconnect the harness from the alternator. Remove the nut securing the battery wire to the alternator stud.

4. Remove the alternator mounting bolts, then remove the alternator from the bracket.

To install:

5. Install the alternator in the bracket.

6. Install the mounting bolts and tighten to 20 ft. lbs. (27 Nm).

7. Connect the battery wire to the alternator stud and tighten the nut to 71 inch lbs. (8 Nm). Connect the harness to the alternator.

8. Install the drive belt.

9. Connect the negative battery cable.

10. Start the engine and check for proper charging system operation.

4.6L Engines

1. Disconnect the negative battery cable.

2. Remove the cover from the headlights and radiator shroud.

3. Remove the air cleaner assembly.

4. Disconnect the left engine torque strut.

5. Disconnect the upper transaxle oil cooler line.

6. Remove the right and left cooling fans.

7. Remove the serpentine belt from the alternator pulley.

8. Remove the bolt from the front top and the lower front of the alternator.

9. Disconnect the harness and output cable from the alternator.

10. Remove the bolts from the alternator rear bracket.

NOTE: The bolt nearest to the exhaust manifold cannot be removed from the bracket but can be backed out enough to allow for alternator removal.

11. Remove the A/C splash shield.

12. Remove the access panel from the bottom side of the radiator support and disconnect the harness clip from the cradle.

13. Move the alternator away from the engine, and out from the bottom side of the engine compartment.

To install:

14. Install the alternator to the bracket with the bolts to the rear of the alternator. Hand-tighten the bolts.

15. Connect the harness clip to the cradle.

16. Install the access panel and A/C splash panel.

17. Connect the harness and output cable to the alternator. Torque the output cable to 15 ft. lbs. (20 Nm).

18. Tighten the rear mounting bolts to 36 ft. lbs. (47 Nm).

19. Install the bolts to the lower front and front top of the alternator and tighten to 36 ft. lbs. (47 Nm).

20. Rotate the drive belt tensioner and install the serpentine belt.

21. Install the remaining components.

22. Connect the negative battery cable.

23. Run the engine and check charging system operation.

4.9L Engine

1. Disconnect the negative battery cable.

2. Disconnect the terminal connector and battery charging wire from the rear of the alternator.

3. Remove the drive belt.

4. Remove the alternator mounting bolts, then remove the alternator from the bracket.

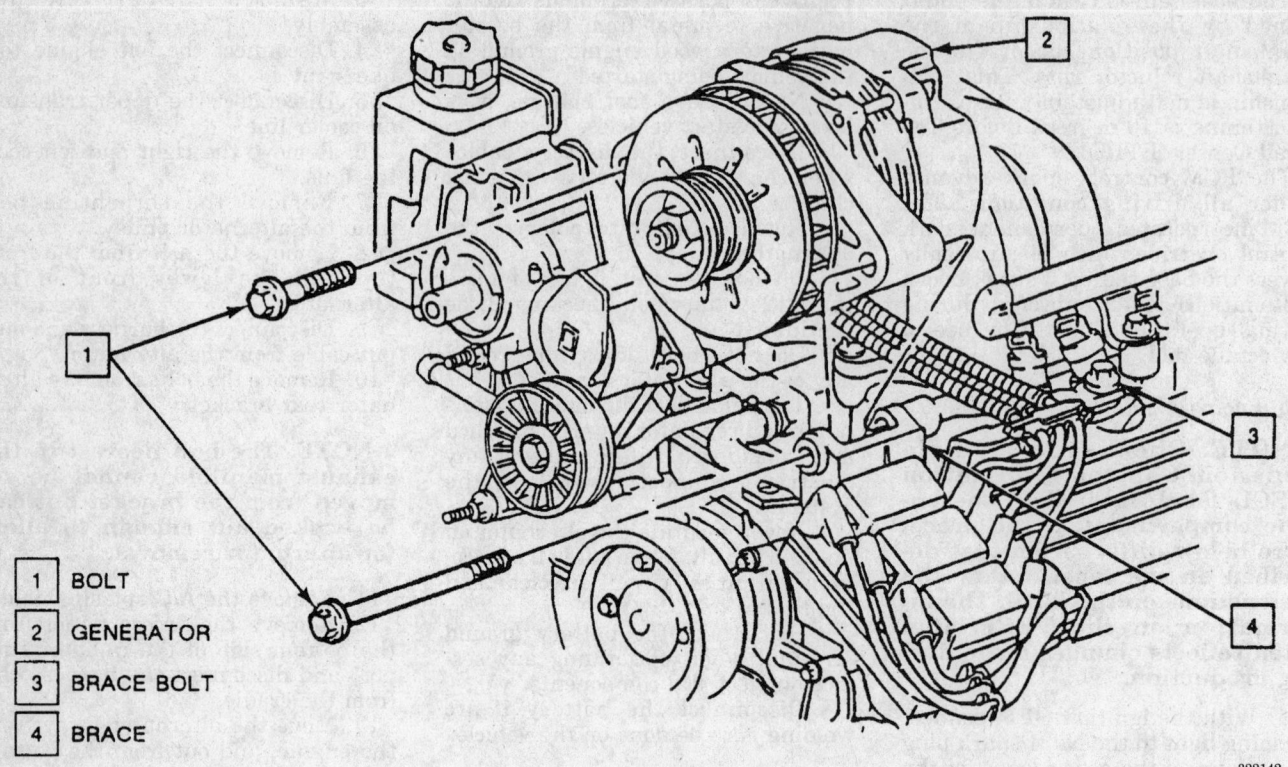

1 BOLT

2 GENERATOR

3 BRACE BOLT

4 BRACE

322142

Alternator mounting — 3.8L engine

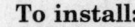
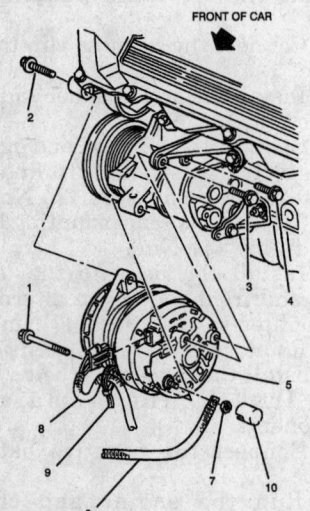

FRONT OF CAR

1 BOLT – FRONT LOWER
2 BOLT – FRONT UPPER
3 BOLT – REAR
4 BOLT – REAR
5 GENERATOR
6 BATTERY CHARGING CABLE
7 OUTPUT STUD NUT
8 HARNESS CONNECTOR
10 CAP

NOTE: BOLTS 1, 2, 3 AND 4 TORQUE IN ORDER SHOWN TO AVOID BREAKAGE

233714

Alternator mounting components — 4.6L engine

To install:

5. Install the alternator in the bracket and install the bolts. Tighten the front bolts to 32 ft. lbs. (44 Nm) and the rear bolt to 20 ft. lbs. (27 Nm).

6. Install the drive belt.

7. Connect the feed wire and tighten to 15 ft. lbs. (20 Nm). Install the harness to the rear of the alternator.

8. Connect the negative battery cable and test the charging system for proper operation .

Drive Belt

REMOVAL AND INSTALLATION

3.8L Engine

1. Disconnect the negative battery cable.

2. Rotate the tensioner assembly counterclockwise using an 18mm box end wrench on the tensioner pulley nut.

3. Slip the belt off of the alternator pulley and slowly release the tensioner.

4. Remove the wrench from the pulley, then remove the belt from the remainder of the pulleys.

5. Inspect the belt for wear and/or damage and replace as necessary.

To install:

6. Route the serpentine belt as follows:

a. Loop the belt under the crankshaft.

b. Bring the front of the belt up and around the water pump pulley so the back of the belt drives the pulley.

c. Loop the belt under the A/C compressor.

d. Bring the rear of the belt over the power steering pump then under the tensioner assembly so the back of the belt drives the tensioner pulley.

7. Rotate the tensioner counterclockwise.

8. Route the belt over the alternator pulley.

9. Slowly release the tensioner assembly and verify the belt is properly seated.

10. Connect the negative battery cable.

11. Start the engine and check operation.

4.6L and 4.9L Engines

1. Disconnect the negative battery cable.

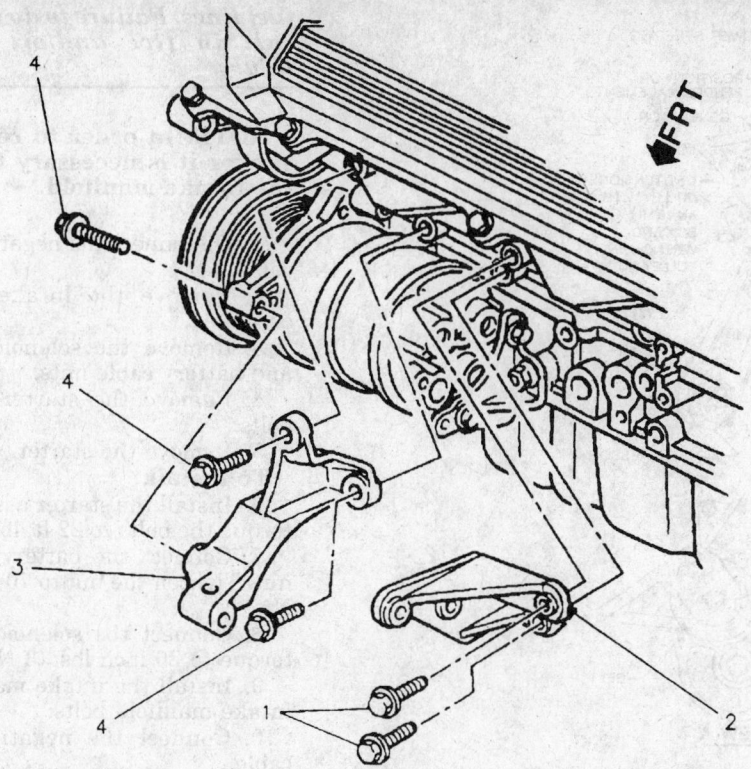

1. Bolt
2. Rear bracket
3. A/C compressor-
   generator bracket
4. Bolt

303232

Alternator mounting brackets — 4.6L engine

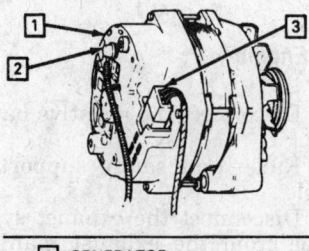

1 GENERATOR
2 POSITIVE BATTERY CABLE
3 P, L, F & S TERMINAL
  CONNECTOR

241990

**Alternator wiring connectors — 4.9L
engine**

2. Using a ½ inch drive breaker
bar, rotate the belt tensioner clockwise to release tension on the drive
belt.
3. Slip the belt off the tensioner
pulley and slowly release the
tensioner.
4. Remove the belt from the drive
pulleys.
5. Inspect the belt for wear, cracks
or other damage; replace as
necessary.
To install:
6. Install the belt over the drive
pulleys, except the tensioner.

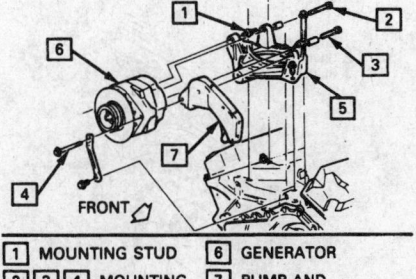

1 MOUNTING STUD   6 GENERATOR
2 3 4 MOUNTING   7 PUMP AND
    BOLTS   TENSIONER BRACKET
5 MOUNTING BRACKET

241991

Alternator mounting components — 4.9L engine

7. Move the tensioner into the
fully released position and slip the
belt under the tensioner.
8. Release the tensioner.
9. Connect the negative battery
cable.
10. Start the engine and check
operation.

Starter

REMOVAL AND INSTALLATION

3.8 Engine

1. Disconnect the negative battery
cable.
2. Raise and safely support the
vehicle.
3. Remove the flywheel inspection
cover retaining screws, then remove
the cover.
4. If equipped, remove the starter
motor shielding.
5. Remove the starter motor bolts,
then move the starter away from the
block.
6. Disconnect the wiring from the
starter and solenoid assembly.
To install:
7. Inspect the teeth on the engine
flywheel. Damaged teeth can cause
starter problems. Inspect the electrical connections for corrosion and
clean as required.
8. Connect the starter wiring and
tighten the battery terminal nut to
12 ft. lbs. (16 Nm) and the solenoid S
terminal nut to 22 inch lbs. (2 Nm).
9. Install the starter to the engine
block and tighten the starter bolts to
32 ft. lbs. (43 Nm).
10. If equipped, install the starter
motor shielding.

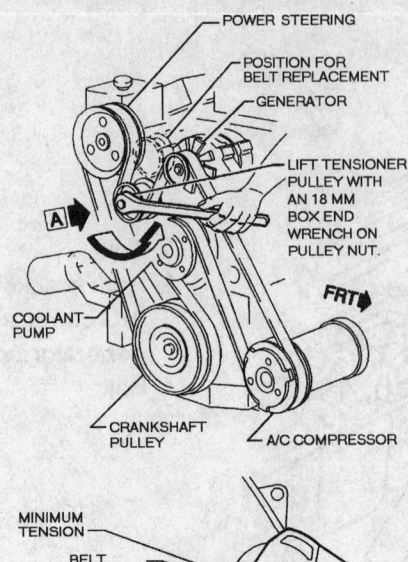

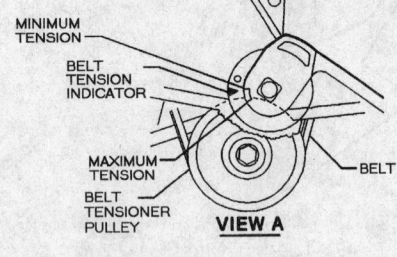

Serpentine drive belt routing — 3.8L engine

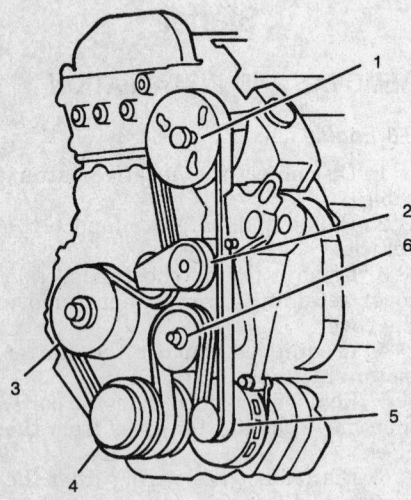

1	POWER STEERING PUMP PULLEY
2	TENSIONER PULLEY
3	CRANKSHAFT PULLEY
4	A/C COMPRESSOR PULLEY
5	GENERATOR PULLEY
6	IDLER PULLEY

348073

Serpentine drive belt routing — 4.6L and 4.9L engines

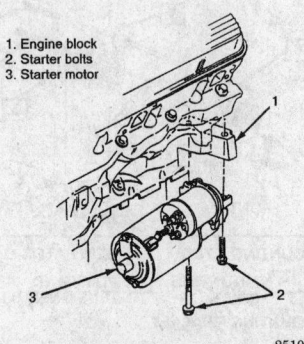

1. Engine block
2. Starter bolts
3. Starter motor

251099

Starter motor mounting — 3.8L engine

11. Install the flywheel inspection cover and tighten the bolts to 48–62 inch lbs. (6–8 Nm).

12. Lower the vehicle.

13. Connect the negative battery cable and check the starter for proper operation.

4.6L Engine

─────── CAUTION ───────

Fuel injection systems remain under pressure, even after the engine has been turned OFF. The fuel system pressure must be relieved before disconnecting any fuel lines. Failure to do so may result in fire and/or personal injury.

NOTE: In order to remove the starter it is necessary to remove the intake manifold.

1. Disconnect the negative battery cable.

2. Remove the intake manifold assembly.

3. Remove the solenoid terminal and battery cable nuts.

4. Remove the starter mounting bolts.

5. Remove the starter.

To install:

6. Install the starter assembly and torque the bolts to 22 ft. lbs. (30 Nm).

7. Connect the battery cable and nut. Tighten the nut to 70 inch lbs. (8 Nm).

8. Connect the solenoid wire and torque to 26 inch lbs. (3 Nm).

9. Install the intake manifold and intake manifold bolts.

10. Connect the negative battery cable.

11. Start the engine several times to ensure proper starter engagement into the flywheel.

4.9L Engine

1. Disconnect the negative battery cable.

2. Raise and safely support the vehicle.

3. Disconnect the exhaust system Y-pipe from the exhaust manifolds and let the pipe hang.

4. Remove the starter heat shield.

5. Disconnect the solenoid wires and battery feed cable.

6. Remove the starter-to-engine bolts.

7. Remove the starter.

To install:

8. Install the starter using the mounting bolts. Tighten the bolts to 33 ft. lbs. (48 Nm).

9. Connect the solenoid wires to the starter.

10. Connect the battery feed cable to the starter, and tighten the mounting nut to 13 ft. lbs. (17 Nm).

11. Install the starter heat shield.

12. Connect the exhaust system.

13. Lower the vehicle.

14. Connect the negative battery cable.

15. Check starter operation.

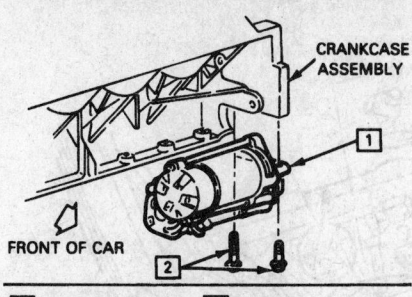

Starter mounting — 4.9L engine

① STARTER MOTOR ② MOUNTING BOLTS

348334

CHASSIS ELECTRICAL

Blower Motor

REMOVAL AND INSTALLATION

Riviera

1. Disconnect the negative battery cable.
2. Remove the cross-tower brace for easier removal of the blower motor.
3. Remove the relay center, if necessary.
4. On some 1993–95 models, it may be necessary to cut the rubber insulator at the guide lines. Once cut, remove the screws to the blower motor.
5. Lift up on the rubber insulator to expose the single screw and remove.
6. On 1996–97 models, do the following:
   a. Disconnect the 4 electrical connectors from the Ignition Control Module (ICM).
   b. Remove spark plug wires No. 5 and No. 7 and label for reinstalla-

tion purposes. Position the spark plug harness out of the way.
   c. Remove the solenoid purge valve.
   d. Remove the inertial plate from the blower motor. Remove the three 8mm screws.
7. Remove the harness and cooling tube.
8. Remove the screws to the blower motor, then remove the motor.
9. Check inside the case for any debris; clean out if needed.

To install:
10. Install the blower motor and tighten the screws to 7 inch lbs. (1 Nm).
11. On 1996–97 models, do the following:
   a. Install the inertial plate.
   b. Install the solenoid purge valve.
   c. Install the spark plug wires.
   d. Connect the electrical connectors to the ICM.
12. If the rubber insulator seal was removed, install a new one
13. Install the cooling tube and harness.
14. Install the relay center, if removed.
15. Install the cross-tower brace.
16. Connect the negative battery cable.
17. Start the engine and test the blower motor for proper operation.

Windshield Wiper Motor

REMOVAL AND INSTALLATION

1. Disconnect the negative battery cable.
2. Remove the A/C pipe shroud.
3. Remove wiper arms.
4. Remove the windshield reveal molding.
5. Remove the cowl cover.
6. Remove the wiper arm drive link from the wiper motor.

7. Disconnect the harness from the motor and module.
8. Remove the wiper motor attaching bolts.
9. On Allante models, remove the suspension accelerometer and position it out of the way.
10. Remove the wiper motor.

To install:
11. Install the wiper motor. Tighten the bolts to to 80 inch lbs. (9 Nm).
12. Connect the electrical harnesses.
13. Install the wiper arm drive link to the motor.
14. On Allante models, install the suspension accelerometer.
15. Install cowl cover and reveal molding.
16. Install the wiper arms.
17. Install the A/C pipe shroud.
18. Connect the negative battery cable.
19. Check wiper operation.

Headlight Switch

REMOVAL AND INSTALLATION

Riviera

1. Disconnect the negative battery cable.
2. Remove the instrument cluster trim plate.
3. Remove the retaining screws, then the switch assembly.
4. Pull the switch assembly from the harness.

To install:
5. Install the switch to the harness.
6. Install the retaining screws, then the trim plate.
7. Connect the negative battery cable.
8. Test headlight operation.

Allante

NOTE: The headlight switch is located in the right side switch pod.

1. Disconnect the negative battery cable.
2. Remove the switch pod trim plate mounting screws, then remove the switch pod mounting screws.
3. Pull the switch pod out of the dashboard and disconnect the electrical harness.

To install:
4. Connect the electrical harness to the switch pod.
5. Position the switch pod in the dashboard and install the mounting screws.

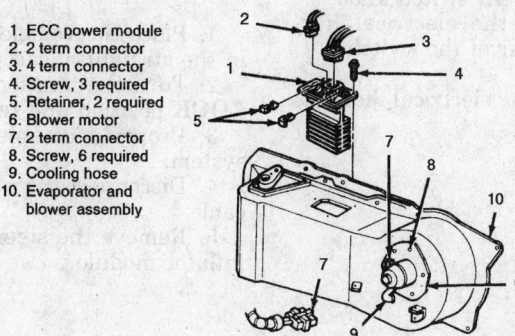

1. ECC power module
2. 2 term connector
3. 4 term connector
4. Screw, 3 required
5. Retainer, 2 required
6. Blower motor
7. 2 term connector
8. Screw, 6 required
9. Cooling hose
10. Evaporator and blower assembly

312337

Blower motor housing — Riviera models

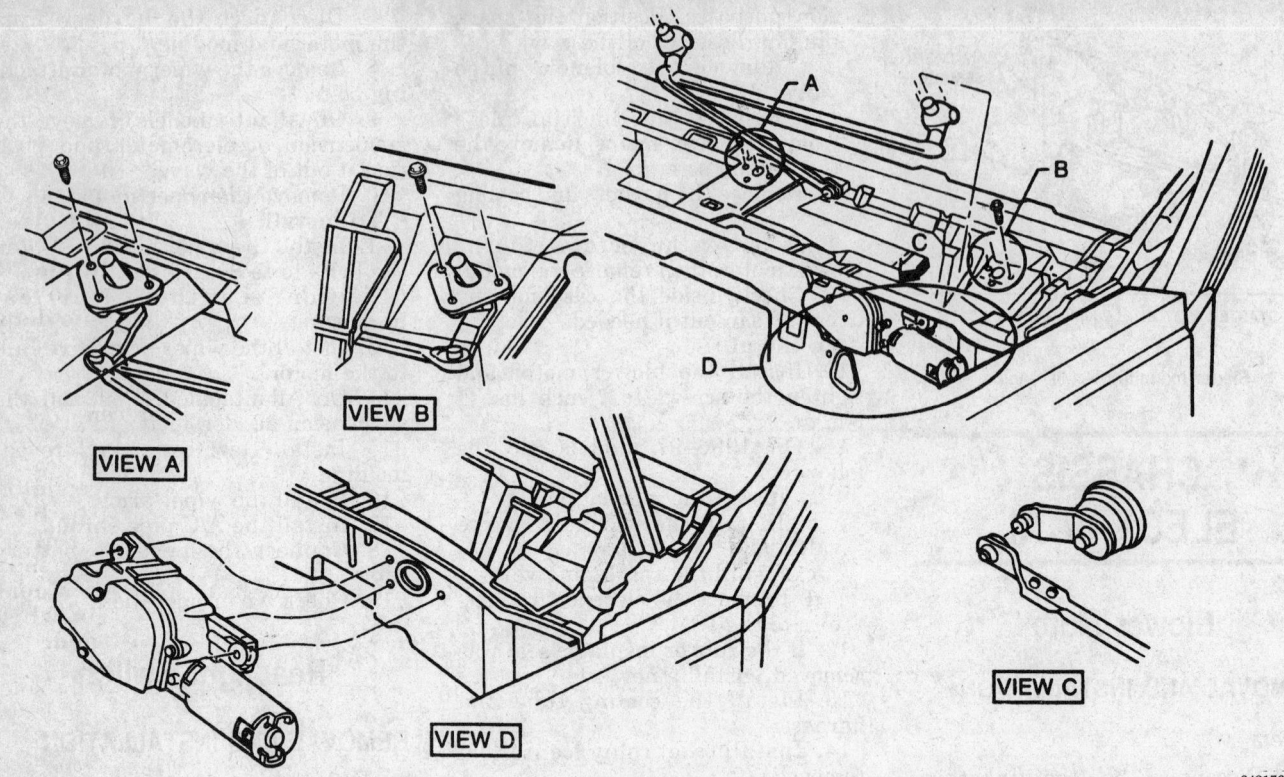

VIEW B

VIEW A

VIEW D

VIEW C

342176

Wiper transmission and motor

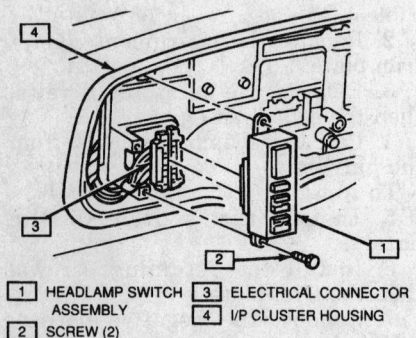

| 1 | HEADLAMP SWITCH ASSEMBLY | 3 | ELECTRICAL CONNECTOR |
| 2 | SCREW (2) | 4 | I/P CLUSTER HOUSING |

320521

Headlight switch assembly — Riviera

6. Install the switch pod trim plate and trim plate mounting screws.
7. Connect the negative battery cable.
8. Test headlight operation.

DeVille, Eldorado and Seville

1. Disconnect the negative battery cable.

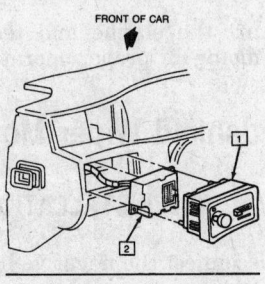

FRONT OF CAR

| 1 | HEADLAMP SWITCH |
| 2 | TWILIGHT SENTINEL/DRL MODULE |

306827

Headlight switch assembly — DeVille, Eldorado and Seville

2. Remove the headlight switch from the dashboard by firmly pulling out on the headlight switch knob.
3. Disconnect the electrical harness from the rear of the switch.
To install:
4. Connect the electrical harness to the switch.

5. Push the headlight switch in until it locks in place.
6. Connect the negative battery cable.
7. Test headlight operation.

Turn Signal Switch

REMOVAL AND INSTALLATION

— **CAUTION** —
The Air Bag System must be disarmed before working around the Air Bag or wiring. Failure to do so may cause accidental deployment and/or personal injury.

1. Place the vehicle's front wheels in the straight-ahead position.
2. Turn the ignition switch to the **LOCK** position and remove the key.
3. Properly disable the Air Bag system.
4. Disconnect the negative battery cable.
5. Remove the steering wheel and inflator module.

CAUTION

When carrying a live Air Bag, make sure the bag and trim cover are pointed away from the body. In the unlikely event of an accidental deployment, the bag will then deploy with minimal chance of injury. When placing a live inflator module on a bench or other surface, always place the bag and trim cover up, away from the surface. This will reduce the motion of the module if it is accidently deployed.

6. Remove the coil retaining rings and coil assembly. Let the coil hang freely.

7. Remove the wave washer.

8. Remove the shaft lock retaining ring using tool J-23653 or equivalent, to push down on the shaft lock. Discard the ring.

9. Remove the shaft lock.

10. Remove the turn signal cam assembly, then the upper bearing spring.

11. Place the turn signal switch to the RIGHT position.

12. Remove the the multifunction lever and hazard knob assembly.

13. Remove the screws attaching the signal switch arm, then remove the arm.

14. Remove the turn signal connector from the bulkhead harness.

15. Disconnect the harness at the turn signal switch.

16. Remove the turn signal attaching nut and unfasten the ground wire from the stud.

17. Remove the turn signal switch and harness.

To install:

18. Route the turn signal switch harness through the steering column.

19. Install the turn signal harness to the bulkhead connector.

20. Install the ground wire to the stud and tighten the nut to 35 inch lbs. (4 Nm).

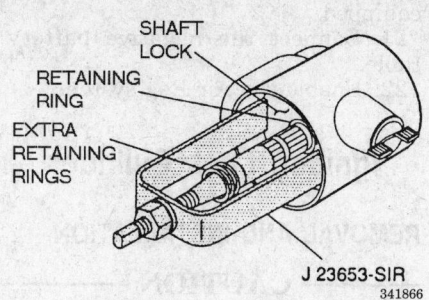

SHAFT LOCK
RETAINING RING
EXTRA RETAINING RINGS

J 23653-SIR
341866

Removing shaft lock retaining ring

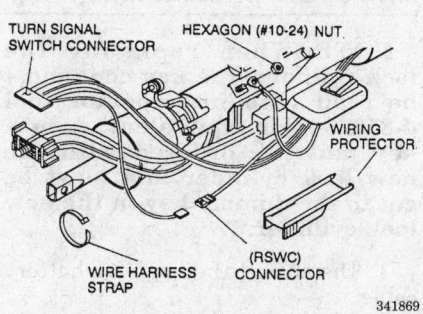

TURN SIGNAL SWITCH CONNECTOR
HEXAGON (#10-24) NUT
WIRING PROTECTOR
(RSWC) CONNECTOR
WIRE HARNESS STRAP

341869

Turn signal switch wiring

21. Install the coil assembly wire harness through the steering column. Connect the wiring harnesses to the turn signal switch.

22. Install and tighten the turn signal switch screws to 30 inch lbs. (3.4 Nm).

23. Install and tighten the turn signal switch arm screws to 20 inch lbs. (2.3 Nm).

24. Install the hazard knob assembly with the multifunction lever.

25. Install the upper bearing spring.

26. Lubricate the turn signal cancelling cam with synthetic grease and install.

27. Install the shaft lock and retainer ring using tool J-23653 or equivalent, to push down the shaft lock. Firmly seat the ring in the groove on the shaft.

NOTE: Set the steering shaft so that the block tooth on the upper steering shaft is at the 12 o'clock position. The vehicle's front wheels should be straight ahead. Set the ignition switch to the LOCK position to ensure no damage to the coil assembly.

28. Install the wave washer.

29. Check for proper centering of the coil assembly.

NOTE: Air Bag coil assembly wires must be kept tight with no slack while installing the coil assembly. Failure to do so may cause the wires to be kinked near the shaft lock area and cut when the steering wheel is turned.

30. Pull the Air Bag coil wire snug while positioning the coil to the steering column.

31. Align the opening in the coil with the horn tower and locating the bump between the tabs on the housing cover.

32. Seat the coil assembly into the steering wheel.

33. Install and firmly seat the the coil assembly retaining ring into the groove on the shaft.

NOTE: Gently pull the lower coil assembly, turn signal, and pass key wires to remove any wire kinks that may be inside the steering column assembly. Failure to do so may cause damage to the wire harness.

34. Install the steering wheel and inflator module.

35. Connect the negative battery cable.

36. Enable the Air Bag system.

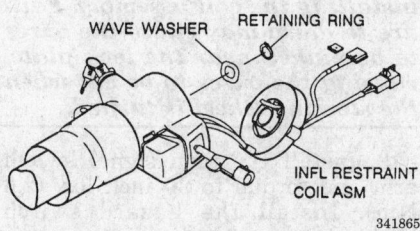

WAVE WASHER
RETAINING RING
INFL RESTRAINT COIL ASM

341865

Removing SRS coil assembly from the steering shaft

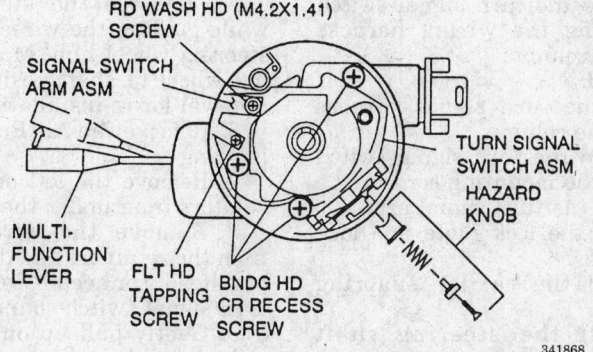

RD WASH HD (M4.2X1.41) SCREW
SIGNAL SWITCH ARM ASM
TURN SIGNAL SWITCH ASM, HAZARD KNOB
MULTI-FUNCTION LEVER
FLT HD TAPPING SCREW
BNDG HD CR RECESS SCREW

341868

Turn signal switch removal preparation

Combination Switch

REMOVAL AND INSTALLATION

------- **CAUTION** -------
The Air Bag system must be disarmed before working around the steering column. Failure to do so may cause accidental deployment and/or personal injury.

1. Disable the Air Bag system following the recommended procedure.
2. Disconnect the negative battery cable.

------- **CAUTION** -------
When carrying a live Air Bag, make sure the bag and trim cover are pointed away from the body. In the unlikely event of accidental deployment, the bag will then deploy with minimal chance of injury. When placing a live Air Bag on a bench or other surface, always face the bag and trim cover up, away from the surface. This will reduce the motion of the module if it is accidently deployed.

3. Remove the inflator module, if equipped.
4. Remove the steering wheel.
5. Remove the steering shaft bumper.
6. Remove the carrier snapring retainer.
7. Remove the lock plate by installing a lock plate compressor and removing the snapring from the steering shaft. Slowly back off the compressor until spring tension is relieved. Remove the lock plate.
8. Remove the turn signal lever.
9. Remove the turn signal switch mounting screws.
10. At the base of the column, disconnect the turn signal switch harness.
11. Remove the lower steering column trim plates and steering column wire guide.
12. Remove the turn signal switch while guiding the wiring harness through the column.
To install:
13. Feed the turn signal harness down into the column.
14. Position the turn signal switch and install the mounting screws.
15. Install the turn signal arm.
16. Install the lock plate and lock plate retainer.
17. Install the carrier snapring retainer.
18. Install the steering shaft bumper.
19. Install the steering wheel.

20. Install the inflator module, if equipped.
21. Connect the negative battery cable.
22. Enable the Air Bag system.

Ignition Lock Cylinder

REMOVAL AND INSTALLATION

------- **CAUTION** -------
The Air Bag system must be disarmed before working around the Air Bag or wiring. Failure to do so may cause accidental deployment and/or personal injury.

NOTE: When replacing the lock cylinder, the key code must be read with interrogator tool J-35628-A or equivalent. A new key must be ordered with the new lock cylinder and must be cut to the dummy key in the new lock cylinder.

1. Disconnect the negative battery cable.
2. Disable the Air Bag system.
3. Remove the Air Bag inflator module and store trim side face up for safety reasons.

------- **CAUTION** -------
When carrying a live Air Bag, make sure the Air Bag and trim cover are pointed away from the body. In an unlikely event of an accidental deployment, the bag will then deploy with minimal chance of injury. When placing a live Air Bag on a bench or other surface, always face the bag and trim cover up, away from the surface. This will reduce the motion of the module if it is accidentally deployed.

4. Remove the steering wheel control switch, if equipped.
5. Remove the steering wheel fastener. Remove the steering wheel while guiding the wires through the steering wheel hub. Mark the steering wheel to shaft position, prior to removal for re-installation purposes.
6. Remove the Air Bag coil retaining ring.
7. Remove the left side sound insulators from under the dash.
8. Remove the wire harnesses from the retainers and disconnect the bulkhead connector to remove the turn signal switch connector.
9. Gently pull up on the Air Bag coil assembly while pushing up on the coil harness on the bottom side of

the steering column, and let the coil hang freely.
10. Remove the wave washer. Install compressor tool J-23653-SIR or equivalent, and compress the lock plate to remove the shaft lock retaining ring.
11. Remove the turn signal canceling cam assembly, then the screws to the turn signal switch arm and turn signal switch-to-steering column. Remove the hazard switch button.
12. Remove the upper bearing spring. Pull up gently on the turn signal switch assembly and let it hang, then remove the key buzzer switch.
13. Remove the 13-way secondary lock on the turn signal switch harness at the lower end of the steering column, and remove terminals **12** and **13** which are routed to the ignition lock cylinder.
14. Attach a piece of wire to the terminals for guiding the harness through the steering column.
15. Remove the retaining bolt to the ignition lock cylinder and insert the key in the lock cylinder to remove.
To install:

------- **CAUTION** -------
Route wires from the pass key lock cylinder into the hole in the housing. Failure to do so may result in component damage or the malfunctioning of the steering column may occur.

16. Install the pass key lock cylinder. Route the wires correctly and insert the terminals in the turn signal connector.
17. Install the lock cylinder fastener and torque to 22 inch lbs. (2.5 Nm)
18. Install the buzzer and turn signal switch assembly. Torque the bolts to 30 inch lbs. (3.4 Nm). When installing the switch be sure to remove all slack out of the wiring harness by pulling gently on the harness.

------- **CAUTION** -------
Air Bag coil assembly wires must be kept tight with no slack while installing the coil assembly. Failure to do so may cause the wires to be kinked near the lock plate, causing the wires to be cut when the steering wheel is turned.

19. Install the turn signal switch arm and torque to 20 inch lbs. (2.3 Nm). Install the hazard switch button
20. Install the upper bearing spring.
21. Install the turn signal cancelling cam assembly and lock plate.

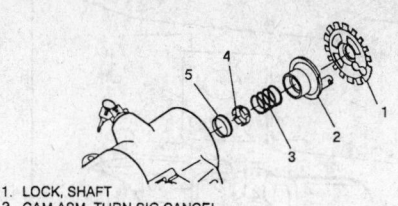

1. LOCK, SHAFT
2. CAM ASM, TURN SIG CANCEL
3. SPRING, UPPER BEARING
4. SEAT, UPPER BEARING INNER RACE
5. RACE, INNER

341108

Removing components from the upper shaft

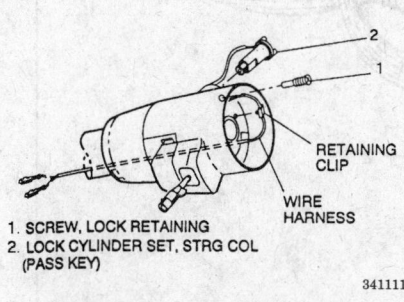

RETAINING CLIP

WIRE HARNESS

1. SCREW, LOCK RETAINING
2. LOCK CYLINDER SET, STRG COL (PASS KEY)

341111

Installing the pass key lock cylinder

22. Using compressor tool J-23653-SIR or equivalent, install the retaining ring after compressing the upper bearing spring.

23. Install the wave washer and Air Bag coil assembly, pulling gently on the harness on the bottom side of the steering column to remove slack from the wiring harness. Be sure to align the Air Bag coil assembly properly.

24. Install the snapring on top of the coil.

25. Install the steering wheel, guiding the wires through for the Air Bag and steering wheel controls, if equipped.

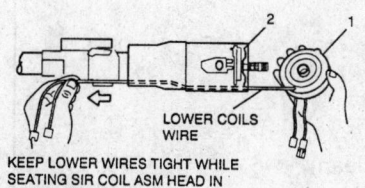

LOWER COILS WIRE

KEEP LOWER WIRES TIGHT WHILE SEATING SIR COIL ASM HEAD IN STEERING COLUMN.

1. COIL ASM, SIR
2. LOCK, SHAFT

341113

Routing SRS coil wires

26. Connect the negative battery cable.

27. Enable the Air Bag system.

28. Test the turn signals, high beams and wipers.

29. Install the lower sound insulator.

Ignition Switch

REMOVAL AND INSTALLATION

1993–94 Models

Except Riviera

——— **CAUTION** ———
The Air Bag system must be disarmed before servicing steering column components. Failure to do so may result in accidental deployment, causing personal injury and/or system repairs.

NOTE: The ignition switch is hardwired. The wiring harness with the column harness connector must be replaced with the ignition switch. Do not splice the new switch to the existing column wiring harness.

1. Disconnect the negative battery cable.

2. Disarm the Air Bag system.

3. Remove the knee bolster as follows:

a. Remove the left side sound insulator mounting screws and nuts. Remove the courtesy lamp from the sound insulator, then remove the sound insulator from the vehicle.

b. Remove the left side console floor trim panel and the left side shroud panel. Each is secured with a screw.

c. Remove the 2 hood release lever retaining screws and position the hood release lever aside.

d. Remove the knee bolster retaining bolts/screws and remove the knee bolster.

4. Remove the ignition switch wire protector and remove the switch mounting screws.

5. Disconnect the ignition switch and turn signal switch column harnesses from the dash connector.

6. Disconnect the turn signal switch harness from the column harness.

7. Remove the ignition switch and harness as an assembly.

To install:

8. Install the ignition switch and secure with mounting screws.

9. Connect the turn signal switch harness to the ignition switch.

10. Connect the ignition and turn signal switch harness to the dash connector.

11. Install the ignition switch wire protector.

12. Install the knee bolster.

13. Connect the negative battery cable.

14. Enable the Air Bag system.

15. Check the ignition switch for proper operation.

Riviera

——— **CAUTION** ———
The Air Bag system must be disarmed before working around the Air Bag or wiring. Failure to do so may cause accidental deployment and/or personal injury.

1. Disconnect the negative battery cable.

2. Disable the Air Bag system.

3. Remove the left side sound insulator.

4. Remove the lower steering column trim plate and lower the steering column.

5. Position the ignition switch in the **OFF** position.

6. Remove the hexagon bolt and nut holding the ignition and dimmer switch in place.

7. Remove the ground wire with ring terminal from the stud.

8. Remove the dimmer switch assembly from the dimmer switch rod.

9. Remove the wire harness strap.

10. Remove the turn signal switch assembly harness from the main steering harness.

11. Remove the pivot and pulse switch assembly connector from the main steering harness.

12. Remove the axial positive assurance connector from the Brake Transmission Shift Interlock (BTSI) actuator, if equipped.

13. Remove the ignition switch fastener and the ignition switch from the vehicle.

——— **WARNING** ———
If installing a new ignition switch, the new ignition switch will be pinned in the OFF LOCK position. Remove the plastic pin after the switch is assembled to the steering column. Failure to do so may cause ignition switch damage.

To install:

14. Install the switch to the steering column and install the bolt loosely.

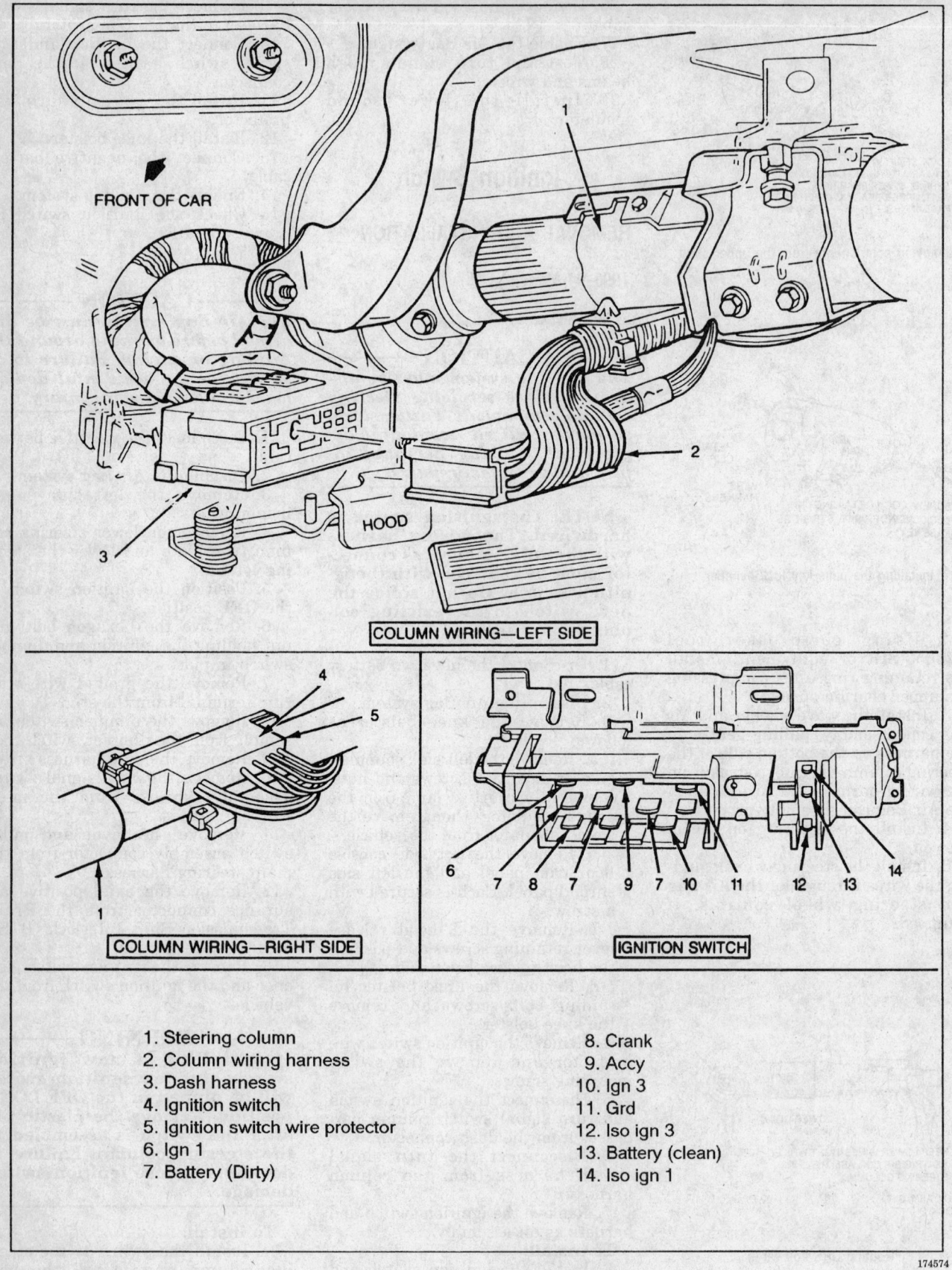

FRONT OF CAR

COLUMN WIRING—LEFT SIDE

HOOD

COLUMN WIRING—RIGHT SIDE

IGNITION SWITCH

1. Steering column
2. Column wiring harness
3. Dash harness
4. Ignition switch
5. Ignition switch wire protector
6. Ign 1
7. Battery (Dirty)
8. Crank
9. Accy
10. Ign 3
11. Grd
12. Iso ign 3
13. Battery (clean)
14. Iso ign 1

174574

Ignition switch location — except 1993–94 Riviera

MOVE SWITCH SLIDER TO EXTREME RIGHT POSITION
AND THEN MOVE SLIDER ONE DETENT TO THE LEFT
"OFF-LOCK"

IGNITION & DIMR
SWITCH ASM

253429

Adjusting the ignition switch — 1993–94 Riviera

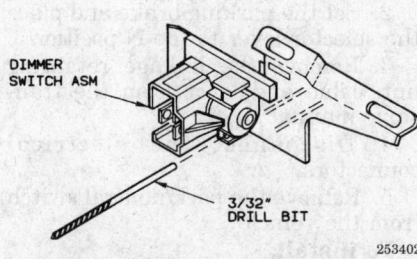

DIMMER
SWITCH ASM

3/32"
DRILL BIT

253402

Use a drill bit to limit switch travel and remove all actuator rod free-play during installation and adjustment — 1993–94 Riviera

15. Adjust the ignition switch using the following procedure:

    a. Move the switch slider to the extreme left, then move the switch slider one detent to the left of the **OFF LOCK** position.

    b. Install a $3/32$ in. drill bit in the hole on the ignition switch to limit travel.

16. Torque the ignition switch to 35 inch lbs. (4 Nm).

17. Install the dimmer switch: Adjust the dimmer switch using the following procedure:

    a. Place a $3/32$ in. drill bit in hole on the switch to limit the travel.

    b. Position the switch on the steering column and push against the dimmer switch rod to remove all slack.

    c. Remove the drill bit and tighten the hexagon nut to 35 inch lbs. (4 Nm).

18. Install the turn signal switch connector to the main steering harness. Seat the connector body until securely locked in.

19. Install the pivot and pulse switch assembly harness to the main steering harness.

20. Install the wire harness strap.

21. Install the harness to the electrical BTSI actuator, if equipped.

22. Position the steering column cover in place and install the column mounting bolts. Tighten to 20 ft. lbs. (27 Nm).

23. Install the column trim plate and replace the sound insulator.

24. Connect the negative battery cable and enable the Air Bag system.

25. Verify that all steering wheel switches and controls operate correctly.

1995–97 Models

CAUTION

The Air Bag system must be disarmed before working around the Air Bag or wiring. Failure to do so may cause accidental deployment and/or personal injury.

1. Disconnect the negative battery cable.

2. Disable the Air Bag system.

3. Lower the steering column to gain access to the ignition switch.

NOTE: Depending on the type of column, it may be necessary to remove the gear selector indicator for the steering column to be lowered.

4. Remove the hex nut and screw and remove the PRNDL adjuster bracket.

5. Remove the hex nut and screw and remove the dimmer switch from the rod.

6. Remove the dimmer and ignition switch mounting stud.

7. Remove the ignition switch from the ignition switch actuator.

8. Disconnect the turn signal switch connector from the bulkhead.

9. Disconnect the dimmer connector from the dimmer switch.

To install:

10. Set the ignition lock cylinder to the **OFF LOCK** position.

WARNING

The ignition switch must be installed with the switch in the OFF LOCK position. A new ignition switch will be pinned in the OFF LOCK position. Remove the plastic pin after the switch is assembled to the column. Failure to do so may cause switch damage.

11. Move the ignition switch slider to the extreme right position, then move the slider one detent to the left, to the **OFF LOCK** position.

12. Install the ignition switch to the actuator assembly.

13. Install the dimmer and ignition switch mounting stud and tighten the stud to 35 inch lbs. (4 Nm).

14. Connect the dimmer switch to the rod and install the dimmer switch hex nut and pan head screw, finger-tight.

15. Place a $3/32$ in. drill bit in the hole on the dimmer switch to limit travel. Position the switch on the column and push against the dimmer switch rod to remove all lash. Remove the drill bit and tighten the nut and screw to 35 inch lbs. (4 Nm).

16. Install the selector adjustment adjuster bracket with the hex nut and pan head screw. Tighten the nut and screw to 35 inch lbs. (4 Nm).

17. Connect the dimmer switch connector to the dimmer switch.

18. Connect the pivot and pulse switch connector to the bulkhead. Seat the connector body until it is securely locked in.

19. Connect the turn signal switch to the bulkhead. Seat the connector body until it is securely locked in.

20. Raise the steering column in the vehicle and tighten to 20 ft. lbs. (27 Nm).

21. Connect the negative battery cable.

22. Adjust the interlock solenoid assembly as follows:

    a. Lock the cylinder to the **RUN** position.

    b. Shift the column out of the **P** position.

    c. Unlock the solenoid cable. Disengage the locking button to allow cable movement.

    d. Install the ball joint spring and ball joint socket to the solenoid and cable assembly with the locking button disengaged.

    e. Pull the cable away from the solenoid until tight.

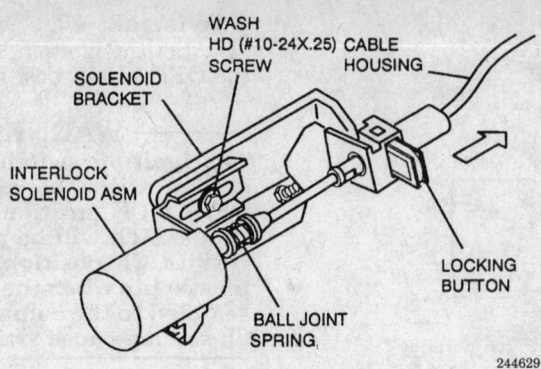

WASH
HD (#10-24X.25) CABLE
SCREW HOUSING

SOLENOID
BRACKET

INTERLOCK
SOLENOID ASM

LOCKING
BUTTON

BALL JOINT
SPRING

244629

Interlock solenoid adjustment — 1995–97 models

f. Let the cable move back (1–2mm), then engage the locking button.

NOTE: During inspection and adjustment of the interlock and solenoid assembly, the solenoid will become hot.

23. Function check the interlock solenoid as follows:

a. The solenoid must lock the gearshift lever bowl assembly whenever the steering column is in **P**, and when trying to shift from **P** without pressing the brake pedal (solenoid is energized).

b. The solenoid must release the gearshift bowl assembly when pressure is applied to the brake pedal (solenoid is de-energized).

c. Readjust the interlock solenoid, if necessary.

24. Enable the Air Bag system.

25. Check the ignition switch and dimmer switch for proper operation.

Park/Neutral Safety Switch

ADJUSTMENT

1. Disconnect the negative battery cable.

2. Engage the parking brake and set the transaxle lever to the **N** position.

3. Remove the nut and linkage cable bracket from the transaxle shaft.

4. Loosen the attaching bolts to the switch assembly.

5. Verify the transaxle manual shaft is still in the **N** position. If it had been moved, rotate the shaft clockwise from **P** through **R** into **N** being careful not to damage the shaft flats, corners or threads.

6. Install tool J-41545 or the equivalent and adjust the switch.

7. Tighten the attaching bolts to 20 ft. lbs. (28 Nm).

8. Install and tighten the linkage cable bracket on the manual shaft, torquing the nut to 18 ft. lbs. (24 Nm).

9. Connect the negative battery cable.

10. Verify that the engine starts in **P** or **N** positions only.

REMOVAL AND INSTALLATION

1993–95 Models

1. Disconnect the negative battery cable.

2. Apply the parking brake and place the shift lever in **N**.

3. Remove the retaining nut and disconnect the linkage cable and bracket from the transaxle shaft.

4. Remove the park/neutral switch retaining bolts.

5. Disconnect the switch connectors from the harness. If equipped, remove the nut and disconnect the switch-to-starter lead from the starter.

6. Remove the switch from the vehicle.

To install:

7. If equipped, install the switch lead to the starter and install the nut. Tighten to 35 inch lbs. (4 Nm).

8. Connect the switch connectors to the vehicle harness.

9. Make sure that the transaxle manual shaft is still in the neutral position.

10. Align the flats in the park/neutral switch with the transaxle manual shaft flats. Press the switch onto the shaft and fully seat against the transaxle.

11. If installing the original switch, insert a $^{3}/_{32}$ in. drill bit into the service slots. A new replacement switch

should already be pinned in the neutral position.

12. Tighten the park/neutral switch bolts to 20 ft. lbs. (28 Nm).

13. Remove the drill bit.

14. Install the linkage cable and bracket and secure with nut.

15. Connect the negative battery cable.

16. Make sure that the engine starts only in the **P** or **N** position.

1996–97 Models

1. Disconnect the negative battery cable.

2. Set the parking brake and place the selector lever to the **N** position.

3. Remove the linkage retaining nut, cable and lever from the transaxle manual shaft.

4. Disconnect the electrical connectors.

5. Remove the park/neutral switch from the vehicle.

To install:

6. Make sure that the transaxle shaft is still in the **N** position.

7. Install the park/neutral switch onto the vehicle. A new park/neutral switch is already pinned in the neutral position. If installing a switch with a sheared pin, proceed as follows:

a. Insert alignment tool J-41545 or equivalent, into the 2 slots on the switch in the area of the transaxle shaft. Rotate the tool until the rear leg of the tool falls into the slot on the switch near the hose. Verify the tool is properly seated in all 3 slots and remove the tool.

b. Check for a cracked carrier if the pin is sheared or if installing the original switch.

8. Align the flats in the park/neutral switch with the transaxle manual shaft flats. Press the switch onto the shaft and fully seat against the transaxle.

9. Install the attaching bolts. If using the alignment tool, make sure the switch is properly aligned with the tool before tightening the bolts. Tighten the bolts to 18 ft. lbs. (25 Nm).

10. Remove the alignment tool.

11. Connect the cable and lever and install the linkage retaining nut. Tighten the nut to 15 ft. lbs. (20 Nm).

12. Connect the electrical connectors.

13. Connect the negative battery cable.

14. Verify that the engine only starts in **P** or **N** positions.

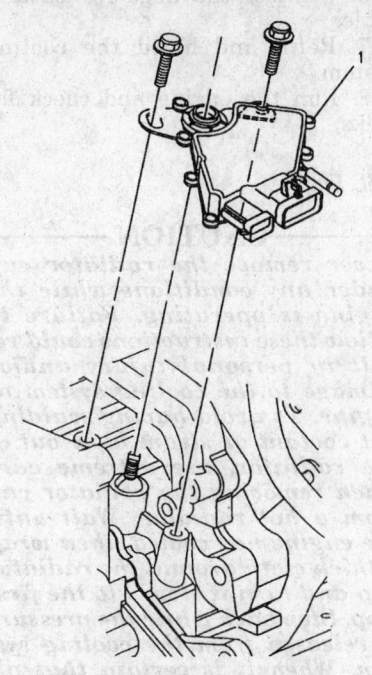

1. Park/neutral position switch

300937

Park/Neutral switch assembly

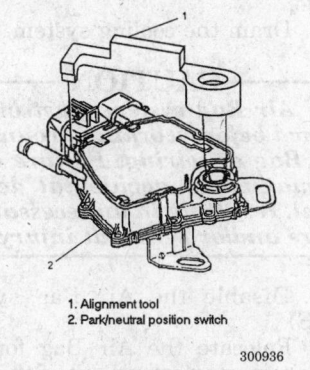

1. Alignment tool
2. Park/neutral position switch

300936

Park/Neutral switch adjustments

Powertrain Control Module

REMOVAL AND INSTALLATION

1993–94 Models

Except Riviera

1. Disconnect the negative battery cable.
2. Remove the right hand hush panel.
3. Disconnect the electrical connectors from the Powertrain Control Module (PCM).

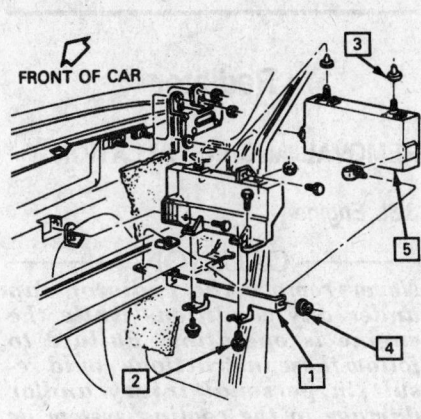

FRONT OF CAR

1	BRACKET
2	SCREW
3	RETAINER
4	NUT
5	ECM

312369

PCM mounting location — 1993–94 models

4. Remove the 2 PCM-to-mounting bracket screws.
5. Remove the PCM.
To install:
6. Install the PCM in the bracket and install the mounting screws.
7. Connect the electrical connectors to the PCM.
8. Install the right side hush panel.
9. Connect the negative battery cable.
10. Start the engine and verify proper engine operation.

Riviera

NOTE: Replacement Powertrain Control Modules (PCM) are supplied without a PROM, so care should be taken when removing the PROM from the defective unit because it will be reused in the new PCM. Using two fingers, push both retaining clips back away from the PROM. At the same time grasp it at both ends and lift it up and out of its socket.

1. Disconnect the negative battery cable.
2. Remove the right hand sound insulator.

3. Disconnect the electrical connectors from the Powertrain Control Module (PCM).
4. Remove the retaining fasteners from the heater case and remove the PCM.
5. If replacing the PCM, remove the PROM access cover.
6. Remove the PROM from its socket.
7. If replacing the PCM, be sure to transfer the mounting clips to their correct locations.
To install:
8. Install the PROM into its socket and clip the retaining clips in until a snap is heard.
9. Install the PROM access cover.
10. Install the retainers to the heater case.
11. Connect the electrical connectors to the PCM.
12. Install the right hand sound insulator.
13. Connect the negative battery cable.
14. Verify proper system operation and no service lights are ON.

1995–97 Models

1. Turn the ignition key **ON**. Connect a suitable diagnostic scan tool and record the oil life and the transaxle oil life index.
2. Turn the ignition key **OFF**, remove the scan tool and disconnect the negative battery cable.
3. Remove the air cleaner housing.
4. Lift the air inlet housing onto the front end sheet metal and pull up firmly.
5. Open the air inlet housing to gain access to the Powertrain Control Module (PCM).
6. Unbolt and disconnect the PCM connectors.
7. Remove the knock sensor module.
8. Remove the PCM insulators.
To install:
9. Install the PCM insulators and the knock sensor module.
10. Install the PCM connectors and torque the bolt to 7 ft. lbs. (10 Nm).
11. Install the PCM with the mounting blocks and connectors attached into the air inlet housing.
12. Close and install the air inlet housing (press firmly on the air inlet housing to engage the locating pins).
13. Install the air cleaner housing.

NOTE: Ensure the air cleaner is installed without any kinking or bending of the intake inlet hose.

14. Connect the negative battery cable.

1. PCM housing
2. PCM harness

304212

PCM housing — 1995-97 models

15. Start the engine; the engine should start and run. If does not, verify that the PCM contains the correct program, the PCM is properly connected to the harness, and that fuses A1, C7, D7 are present and not blown.

16. Connect the diagnostic tool and verify that Diagnostic Trouble Code (DTC) DO603 and P1623 appear. Clear the PCM DTC's.

17. Enter the transaxle oil life recorded prior to the PCM replacement.

18. Display the Oil Life Left on the IPC and press the RESET button to reset the oil life index.

19. Perform a TP sensor learning procedure.

ENGINE COOLING

Radiator

REMOVAL AND INSTALLATION

3.8L Engine

— CAUTION —
Never remove the radiator cap under any conditions while the engine is operating. Failure to follow these instructions could result in personal injury and/or damage to the cooling system or engine. To avoid having scalding hot coolant or steam blow out of the radiator, use extreme care when removing the radiator cap from a hot radiator. Wait until the engine has cooled, then wrap a thick cloth around the radiator cap and turn it slowly to the first stop. Step back while the pressure is released from the cooling system. When it is certain that all pressure has been released, press down on the radiator cap, with the cloth, turn and remove.

1. Disconnect the negative battery cable.

2. Drain the cooling system.

3. Remove the cooling fan assembly and torque strut, if necessary.

4. Remove the air cleaner duct, if necessary.

5. Remove the upper radiator panel.

6. Disconnect the upper and lower radiator hoses and coolant overflow hose from the radiator.

7. Disconnect and cap the transaxle oil cooler lines.

8. Remove the radiator from the vehicle.

To install:

9. Install the radiator in the vehicle.

10. Install the radiator mounting bolts and tighten to 88 inch lbs. (10 Nm).

11. Connect the transaxle oil cooler lines and tighten to 20 ft. lbs. (27 Nm).

12. Connect the upper and lower radiator hoses and overflow hose to the radiator.

13. Install the upper radiator panel.

14. Install the air inlet duct, if removed.

15. Install the cooling fan assembly.

16. Connect the negative battery cable.

17. Refill and bleed the cooling system.

18. Run the engine and check for leaks.

4.6L Engine

— CAUTION —
Never remove the radiator cap under any conditions while the engine is operating. Failure to follow these instructions could result in personal injury and/or damage to the cooling system or engine. To avoid having scalding hot coolant or steam blow out of the radiator, use extreme care when removing the radiator cap from a hot radiator. Wait until the engine has cooled, then wrap a thick cloth around the radiator cap and turn it slowly to the first stop. Step back while the pressure is released from the cooling system. When it is certain that all pressure has been released, press down on the radiator cap, with the cloth, turn and remove.

1. Disconnect the negative battery cable.

2. Drain the cooling system.

— CAUTION —
The Air Bag system must be disarmed before working around the Air Bag or wiring. Failure to do so may cause accidental deployment, resulting in unnecessary repairs and/or personal injury.

3. Disable the Air Bag system (SRS).

4. Relocate the Air Bag forward discriminating sensor out of the way.

5. Remove the air cleaner assembly.

6. Remove the cooling fans.

7. Disconnect the upper and lower radiator hoses from the radiator.

8. Disconnect the engine oil cooler lines from the radiator.

9. Disconnect the transaxle oil cooler lines from the radiator.

10. Remove the left and right engine support torque struts.

11. Remove the radiator upper support bolts and upper support.

12. Lift the radiator up and out of the vehicle.

To install:

13. Install the radiator.

14. Install the upper support and upper support mounting bolts and tighten to 53 inch lbs. (6 Nm).

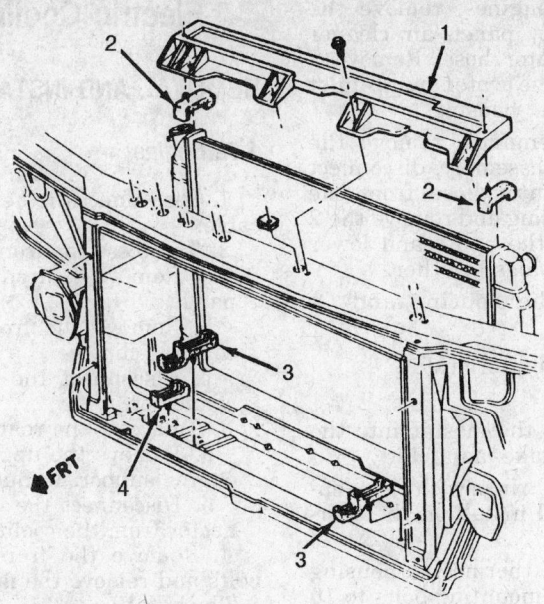

1 Upper radiator panel
2 Upper insulator
3 Lower insulator
4 Lower insulator
(LN3 with HD cooling)

321495

Radiator and mounting components — Riviera

15. Connect the transaxle cooler lines and tighten the fittings to 20 ft. lbs. (27 Nm).

16. Connect the engine oil cooler lines and tighten the fittings to 13 ft. lbs. (17 Nm).

17. Install the engine support torque struts.

18. Connect the upper and lower radiator hoses to the radiator.

19. Install the cooling fans.

20. Install the air cleaner assembly.

21. Install the forward discriminating sensor.

22. Enable the Air Bag system.

23. Connect the negative battery cable.

24. Refill and bleed the cooling system.

25. Run the engine and check for leaks.

4.9L Engine

CAUTION

Never remove the radiator cap under any conditions while the engine is operating. Failure to follow these instructions could result in personal injury and/or damage to the cooling system or engine. To avoid having scalding hot coolant or steam blow out of the radiator, use extreme care when removing the radiator cap from a hot radiator. Wait until the engine has cooled, then wrap a thick cloth around the radiator cap and turn it slowly to the first stop. Step back while the pressure is released from the cooling system. When it is certain that all pressure has been released, press down on the radiator cap, with the cloth, turn and remove.

1. Disconnect the negative battery cable.

2. Drain the cooling system into a suitable container.

3. Remove the upper plastic radiator support cover.

4. Remove the rear cooling fan.

5. Disconnect the coolant reservoir hose from the radiator neck.

6. Disconnect the upper and lower radiator hoses.

7. Disconnect the engine oil cooler lines from the radiator.

8. Disconnect the transaxle oil cooler lines from the radiator.

9. Remove the radiator top support.

10. Lift the radiator up and out of the vehicle.

To install:

11. Install the radiator in the vehicle.

12. Install the radiator support and tighten the retaining bolts to 18 ft. lbs. (25 Nm).

13. Connect the transaxle oil cooler lines and tighten the fittings to 20 ft. lbs. (27 Nm).

14. Connect the engine oil cooler lines and tighten the fittings to 13 ft. lbs. (18 Nm).

15. Connect the upper and lower radiator hoses.

16. Connect the coolant reservoir hose to the radiator neck.

17. Install the rear cooling fan.

18. Install the plastic radiator support cover.

19. Connect the negative battery cable.

20. Refill and bleed the cooling system.

21. Run the engine and check for leaks and proper cooling system operation.

Water Pump

REMOVAL AND INSTALLATION

NOTE: On 4.6L engines, there was a design change to the water pump inlet housing. If it is a black plastic housing, the housing must be replaced. There are no seals available for the plastic housings. The new housings are made of aluminum. When ordering parts, be sure to ask for a water pump, water pump seal and an inlet housing seal and housing if needed.

1. Disconnect the negative battery cable.

2. Drain the coolant into a suitable container.

3. Remove the accessory drive belt. Remove the water pump pulley.

4. On 4.6L engines, remove the following:

 a. Remove the air cleaner.

 b. Remove the lower radiator hose and remove the thermostat by-pass hose from the coolant inlet housing.

5. On 4.9L engines, remove the coolant overflow tank.

6. Remove the water pump mounting bolts.

7. On 4.6L engines, remove the water pump from the vehicle by rotating clockwise with tool J-38816-A or equivalent. Remove the O-ring and clean out the groove.

To install:

8. Clean all gasket mating surfaces.

9. On 3.8L engines, apply a thin bead of sealer around the outside

edge of the water pump and install the gasket on the pump.

10. On 4.6L engines, install the O-ring seal into the groove, then install the water pump, turning it counterclockwise until it stops, using tool J-38816-A or equivalent.

11. Install the water pump housing and tighten the bolts to 88 inch lbs. (10 Nm).

12. On 4.9L engines, install the new water pump gasket on the front cover studs with the raised sealing surface facing outward.

13. Install the water pump on the engine and tighten the mounting bolts.

14. Install the water pump pulley. Install the pulley bolts finger-tight.

15. Install the accessory drive belt.

16. Tighten the water pump pulley bolts to 115 inch lbs. (13 Nm).

17. On 4.6L engines, install the air cleaner.

18. On 4.9L engines, install and connect the coolant reservoir.

19. Connect the negative battery cable.

20. Refill and bleed the cooling system.

21. Run the engine and check for leaks.

Thermostat

REMOVAL AND INSTALLATION

1. Disconnect the negative battery cable.

2. Drain the cooling system to a level below the thermostat housing.

3. On 3.8L Engines, remove the fuel rail cover from the engine to ease access to the thermostat housing retaining bolt. Disconnect the upper radiator hose from the thermostat housing.

4. On 4.6L engines, remove the front end beauty panel, air cleaner and lower radiator hose. Remove 2 bolts securing the water pump inlet to the thermostat housing.

5. On 4.9L engines, remove the upper air filter assembly, disconnect the upper radiator hose from the thermostat housing and remove the 2 bolts fastening the upper and lower thermostat housings together.

6. Remove the housing and O-rings.

7. Remove the thermostat.

To install:

8. Install the thermostat into the recess in the intake manifold.

9. Coat new O-rings with clean engine coolant and install on the thermostat housing.

10. Install the thermostat housing and tighten the mounting bolts to 10 ft. lbs. (14 Nm).

11. Install or connect the remaining components.

12. Connect the negative battery cable.

13. Refill and bleed the cooling system.

14. Run the engine and check for leaks and proper cooling system operation.

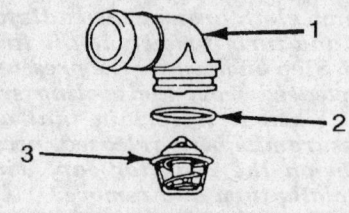

1. Thermostat housing
2. O-ring
3. Thermostat

343321

Thermostat and housing

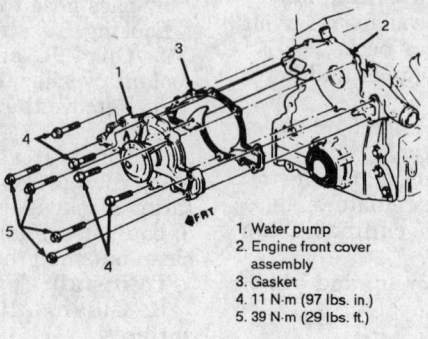

1. Water pump
2. Engine front cover assembly
3. Gasket
4. 11 N·m (97 lbs. in.)
5. 39 N·m (29 lbs. ft.)

343326

Water pump installation

Electric Cooling Fan

REMOVAL AND INSTALLATION

3.8L Engine

1. Disconnect the negative battery cable.

2. To access the front fan:
   a. Remove the front center finish panel.
   b. Remove the front fan guard cover (4 clips).
   c. Disconnect the fan electrical connector.

3. To access the rear fan:
   a. Remove the upper radiator to engine support torque strut.
   b. Disconnect the electrical connector from the cooling fan.

4. Remove the front cooling fan bolts and remove the fan assembly.

To install:

5. Install the fan and the bolts, tighten to 89 inch lbs. (10 Nm).

6. Connect the cooling fan electrical connector.

7. Install the fan guard cover (4 clips).

8. If installing a front fan, install the front end center finish panel.

9. If installing a rear fan, install the upper torque strut and tighten the bolts to 18 ft. lbs. (25 Nm).

10. Connect the negative battery cable.

11. Start the engine, bring up to operating temperature and check for proper cooling fan operation.

4.6L and 4.9L Engines

1. Disconnect the negative battery cable.

2. Remove the air cleaner assembly, if needed.

3. Drain the engine coolant.

4. On 4.9L models with a front fan, remove the plastic radiator cover.

5. Remove the front headlamp panel, if needed.

6. Relocate the upper radiator hose for removal of the fan.

7. Remove the left side torque strut.

8. On 4.9L models with a rear fan, remove the engine-to-upper radiator support torque strut.

9. Disconnect the upper transaxle oil cooler hose and position it out of the way.

10. Disconnect the electric cooling fan connector.

11. Remove the bolts from the cooling fan.

12. Remove the fan assembly.

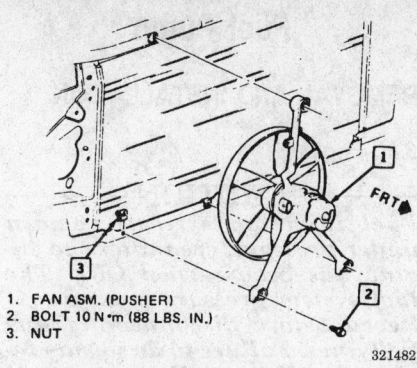

1. FAN ASM. (PUSHER)
2. BOLT 10 N·m (88 LBS. IN.)
3. NUT

321482

Front cooling fan

To install:

13. Install the fan assembly so the tab on the bottom of the fan assembly bracket sits in the slot at the bottom of the radiator support.

14. Install the cooling fan to body bolts and tighten to 88 inch lbs. (10 Nm).

15. Connect the wiring harness to the fan motor.

16. Install the upper transaxle cooler hose. Torque to 20 ft. lbs. (27 Nm).

17. Install the remaining components.

18. Fill the system with coolant, if drained. Connect the negative battery cable.

19. Run the engine and check cooling fan operation.

Cooling System Bleeding

PROCEDURE

NOTE: Use only DEX-COOL type coolant in 1996-97 4.6L engines.

A newly developed engine coolant, DEX-COOL, must be used on 1996-97 4.6L engines. Only DEX-COOL should be used when coolant is added or changed.

DEX-COOL was developed to last for 100,000 miles (161,000 km) or 5 years whichever occurs first. A 50/50 mixture of DEX-COOL and water will provide the following:

• Give protection down to -34°F (-37°C).

• Give a boiling point of 265°F (129°C).

• Protect against rust and corrosion.

• Helps keep the engine the proper temperature.

• Lets the warning lights and gauges work as they should.

DEX-COOL anti-freeze can be added to raise the boiling point of the coolant, but too much will affect the freezing point. DO NOT use a solution stronger than 70 percent anti-freeze, as the freeze level rises rapidly after this point. Pure anti-freeze will freeze at -8°F (-22°C).

When the cooling system has been drained, the following procedure should be used to ensure a complete refill.

CAUTION

Never remove the radiator cap under any conditions while the engine is operating. Failure to follow these instructions could result in personal injury and/or damage to the cooling system or engine. To avoid having scalding hot coolant or steam blow out of the radiator, use extreme care when removing the radiator cap from a hot radiator. Wait until the engine has cooled, then wrap a thick cloth around the radiator cap and turn it slowly to the first stop. Step back while the pressure is released from the cooling system. When it is certain that all pressure has been released, press down on the radiator cap, with the cloth, turn and remove.

When the cooling system has been drained, the following procedure should be used to ensure a complete refill.

1. With the cooling system completely drained, fill the system with at least a 50/50 mixture of ethylene glycol antifreeze and water but no more than a 70/30 mixture of water to antifreeze.

2. Fill the radiator to just below the filler neck. Fill the coolant recovery reservoir to the **COLD FULL** mark.

3. Run the engine with the radiator cap removed until normal operating temperature is reached, with the radiator inlet hose hot.

4. With the engine idling, add coolant to the radiator until it reaches the bottom of the filler neck.

5. Position the heating system controls on maximum; allowing coolant to circulate through the heater core.

6. Check the coolant level again and add, as necessary.

7. Install the radiator cap.

FUEL SYSTEM

Fuel System Service Precautions

Safety is the most important factor when performing not only fuel system maintenance but any type of maintenance. Failure to conduct maintenance and repairs in a safe manner may result in serious personal injury or death. Maintenance and testing of the vehicle's fuel system components can be accomplished safely and effectively by adhering to the following rules and guidelines.

• To avoid the possibility of fire and personal injury, always disconnect the negative battery cable unless the repair or test procedure requires that battery voltage be applied.

• Always relieve the fuel system pressure prior to disconnecting any fuel system component (injector, fuel rail, pressure regulator, etc.), fitting or fuel line connection. Exercise extreme caution whenever relieving fuel system pressure to avoid exposing skin, face and eyes to fuel spray. Please be advised that fuel under pressure may penetrate the skin or any part of the body that it contacts.

• Always place a shop towel or cloth around the fitting or connection prior to loosening to absorb any excess fuel due to spillage. Ensure that all fuel spillage (should it occur) is quickly removed from engine surfaces. Ensure that all fuel soaked cloths or towels are deposited into a suitable waste container.

• Always keep a dry chemical (Class B) fire extinguisher near the work area.

• Do not allow fuel spray or fuel vapors to come into contact with a spark or open flame.

• Always use a backup wrench when loosening and tightening fuel line connection fittings. This will prevent unnecessary stress and torsion to fuel line piping. Always follow the proper torque specifications.

• Always replace worn fuel fitting O-rings with new. Do not substitute fuel hose or equivalent, where fuel pipe is installed.

Fuel System Pressure

RELIEVING

—————— CAUTION ——————

Fuel injection systems remain under pressure, even when the engine has been turned OFF. The fuel system pressure must be relieved before disconnecting any fuel lines. Failure to do so may result in fire and/or personal injury.

1. Loosen the fuel filler cap to relieve tank vapor pressure.
2. Make sure the ignition switch is in the **OFF** position.
3. Disconnect the negative battery cable.
4. Remove the engine dress-up cover.

—————— CAUTION ——————

There may still be residual fuel in the system, and a small amount of fuel may be released when servicing fuel lines or connections. In order to reduce the chance of personal injury, cover the fuel line fittings with a shop towel before disconnecting to catch any fuel that may leak out.

5. Install a fuel pressure gauge with a drain hose attached, J-34730-1 or equivalent. Wrap a shop towel around the fitting while connecting the gauge to avoid spillage.
6. Install the drain hose into an approved container and open the valve to drain the system pressure. Fuel connections are now safe for servicing.
7. Drain any remaining fuel from inside the gauge into the approved container.

Idle Speed

ADJUSTMENT

The engine idle speeds are controlled through the Powertrain Control Module (PCM) and the Idle Air Control (IAC) valve. During idle, the proper position of the IAC valve is calculated based on inputs from the battery voltage, coolant temperature, engine load, and engine RPM. There are no external service adjustments that can be made.

Mixture

ADJUSTMENT

The air/fuel mixture is controlled by the Powertrain Control Module (PCM). External adjustment is not possible. The Heated Oxygen Sensor (HO_2S), located in the exhaust manifold or exhaust pipe, tells the PCM how much oxygen is in the exhaust gas. The PCM controls the fuel injector's opening duration to control the air/fuel mixture. The air/fuel mixture is constantly adjusted to suit vehicle operating conditions and to ensure efficient operation of the catalytic converter. No air/fuel mixture adjustment is possible nor should any adjustment be attempted.

Fuel Filter

REMOVAL AND INSTALLATION

—————— CAUTION ——————

Fuel Injection systems remain under pressure, even after the engine has been turned OFF. The fuel system pressure must be relieved before disconnecting any fuel lines. Failure to do so may result in fire and/or personal injury.

1. Properly relieve the fuel system pressure.
2. Disconnect the negative battery cable.
3. Release the locking tabs on the fuel filter retainer.
4. Disconnect the fuel lines from the fuel filter by releasing the locking tabs on the fuel filter quick-connects. Wrap a shop towel around the fitting when removing to catch excess fuel leaking out of the filter.
5. Remove the fuel filter.
To install:
6. Apply a few drops of engine oil to the tips of the fuel filter and install the fuel filter in the mounting bracket.
7. Connect the fuel lines to the fuel filter and snap the quick-connects into place. Make sure the tabs on the quick-connects lock into place.
8. Engage the locking tabs on the fuel filter retainer.
9. Connect the negative battery cable.
10. Turn the ignition key **ON** for 2 seconds, then **OFF** for 5 seconds. Again turn the ignition key **ON** and check for fuel leaks.

Fuel Pump

REMOVAL AND INSTALLATION

3.8L Engine

—————— CAUTION ——————

Fuel injection systems remain under pressure, even after the engine has been turned OFF. The fuel system pressure must be relieved before disconnecting any fuel lines. Failure to do so may result in fire and/or personal injury.

1. Disconnect the negative battery cable.
2. Relieve the fuel system pressure.
3. Drain the fuel tank into an approved container using a hand operated pump.

—————— CAUTION ——————

The modular fuel sender assembly may spring up from its position. When removing the assembly, be aware that the reservoir bucket is full of fuel. Tip the assembly slightly during removal to avoid damaging the float. Have a shop towel ready to absorb any leakage.

4. Remove the fuel tank from the vehicle.
5. Remove the fuel sender assembly using tool J-24187 or the equivalent.
6. Note the position of the fuel pump strainer on the fuel pump.
7. Support the pump with one hand and grasp the strainer with the other hand. Rotate the strainer in one direction, pull it off the pump and discard the strainer.
8. Remove the fuel pump electrical connector.
9. Loosen the fuel pump dampener.
10. Place the fuel pump assembly upside down on a bench. Pull the fuel pump downward to remove from the mounting bracket, then tilt the pump outward and remove pump and rubber bumper from the dampener.
To install:
11. Install a new rubber bumper on the fuel pump.
12. Position the sender assembly upside down. Install the fuel pump between the dampener and the mounting bracket.
13. Center the dampener between the fuel pump and the fuel feed pipe.
14. Install the electrical connector to the fuel pump.

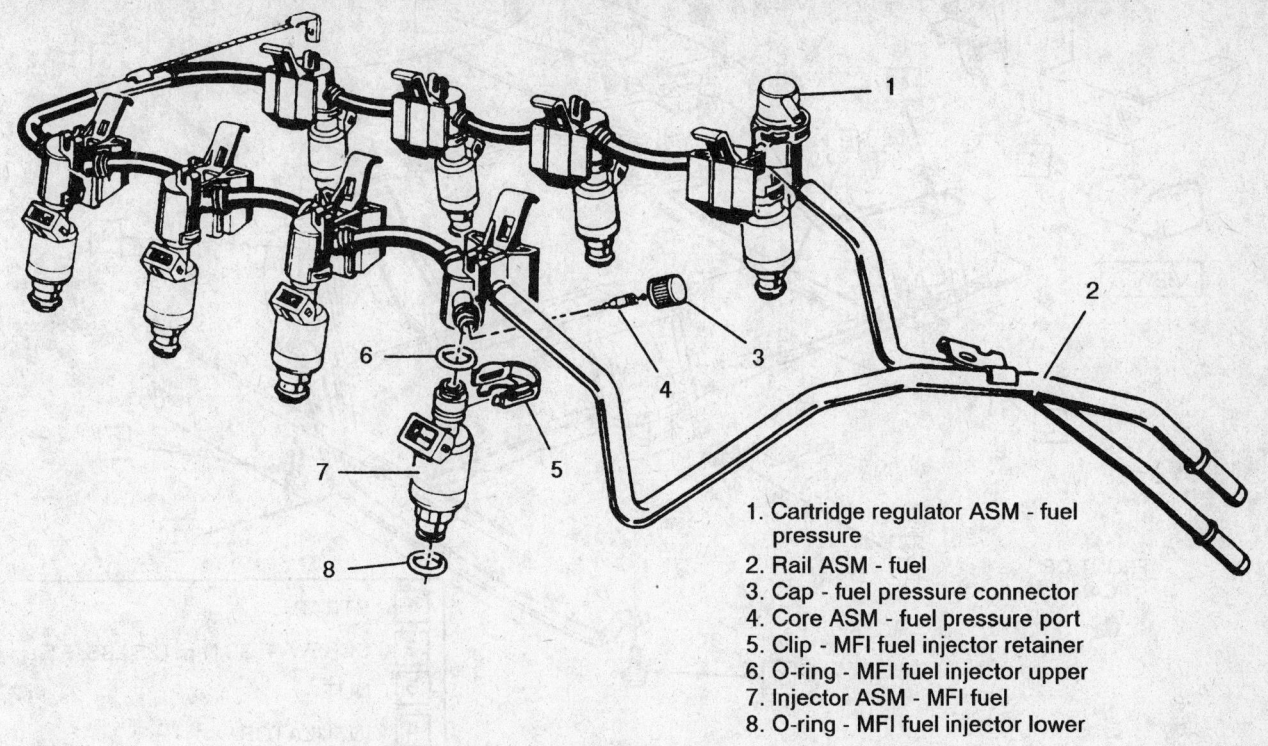

1. Cartridge regulator ASM - fuel pressure
2. Rail ASM - fuel
3. Cap - fuel pressure connector
4. Core ASM - fuel pressure port
5. Clip - MFI fuel injector retainer
6. O-ring - MFI fuel injector upper
7. Injector ASM - MFI fuel
8. O-ring - MFI fuel injector lower

343708

View of the fuel rail assembly showing fuel system service port location

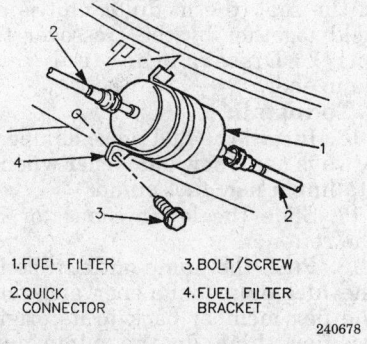

1. FUEL FILTER
2. QUICK CONNECTOR
3. BOLT/SCREW
4. FUEL FILTER BRACKET

240678

Fuel filter installation — DeVille, Eldorado and Seville

15. Support the pump with one hand and position a new fuel pump strainer on the pump in the same position as noted during removal. Push on the outer edge of the ferrule until it is fully seated.
16. Install the sender assembly into the fuel tank using a new O-ring.
17. Install the fuel tank into the vehicle and connect the fuel feed, return and the vapor lines. Connect the electrical connectors.
18. Fill the fuel tank. Turn the ignition **ON** to pressurize the fuel system and check for any fuel leaks before starting the vehicle.

19. Start the engine and check for proper operation.

4.6L and 4.9L Engines

> **CAUTION**
>
> *Fuel injection systems remain under pressure, even after the engine has been turned OFF. The fuel system pressure must be relieved before disconnecting any fuel lines. Failure to do so may result in fire and/or personal injury.*

NOTE: The modular fuel sender assembly must be disassembled in the exact order described.

1. Disconnect the negative battery cable.
2. Relieve the fuel system pressure.
3. Remove the fuel tank from the vehicle and position on a suitable workbench.
4. Clean the fuel tank in the area of the modular fuel sender assembly.
5. Remove the locking nut by turning counterclockwise using tool J-39348 or equivalent.
6. Remove the modular fuel sender assembly from the fuel tank.

> **CAUTION**
>
> *The modular fuel sender assembly may spring up from its position. When removing the assembly, be aware that the reservoir bucket is full of fuel. Tip the assembly slightly during removal to avoid damaging the float. Have a shop towel ready to absorb any leakage.*

7. Slide the fuel sender seal downward, past the reservoir and carefully over the float arm assembly. Discard the seal.

> **CAUTION**
>
> *Observe all applicable safety precautions when working around gasoline. Do not allow fuel spray or fuel vapors to come in contact with a spark or open flame. Keep a dry chemical (Class B) fire extinguisher near the work area. Never drain or store fuel in an open container due to the possibility of fire or explosion.*

8. Remove the Connector Position Assurance (CPA) clip from the wire harness under the modular fuel sender assembly cover. Depress the black connector tabs to remove the electrical connector from the cover.

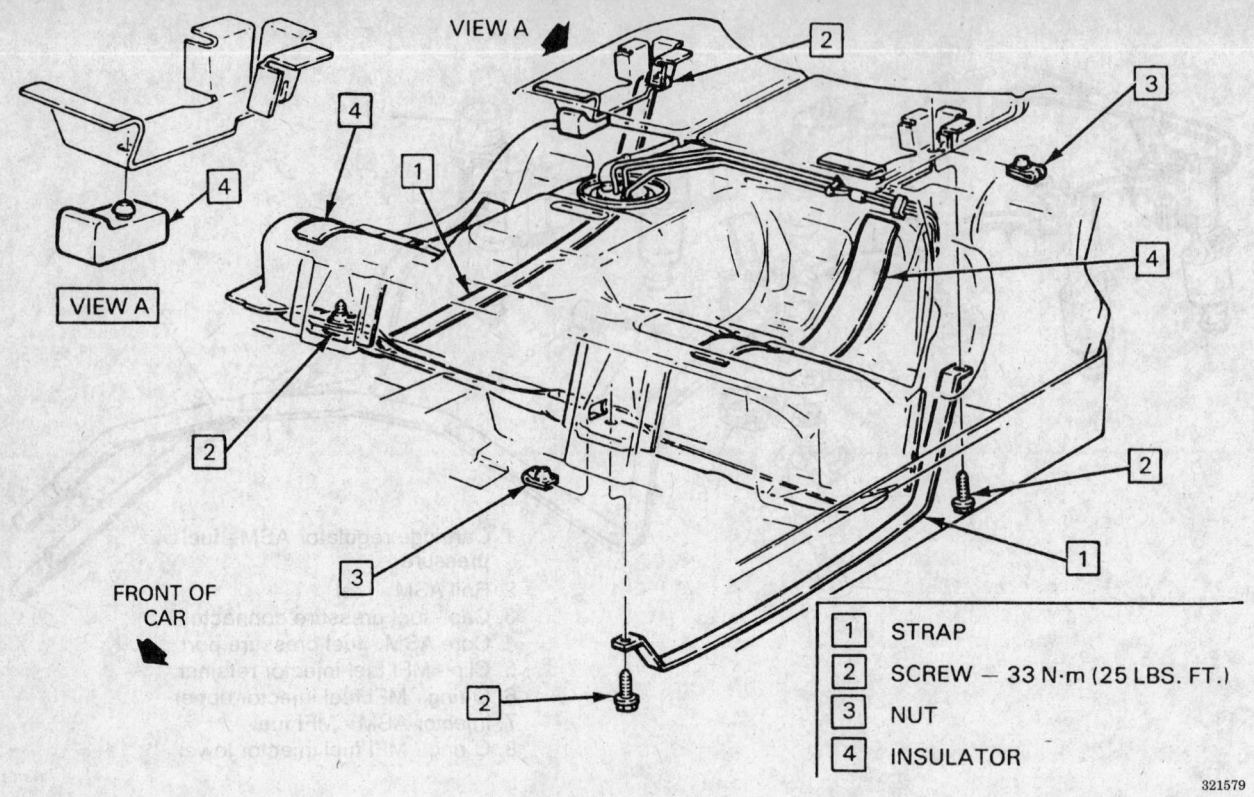

VIEW A

FRONT OF
CAR

1	STRAP
2	SCREW — 33 N·m (25 LBS. FT.)
3	NUT
4	INSULATOR

321579

Fuel tank mountings

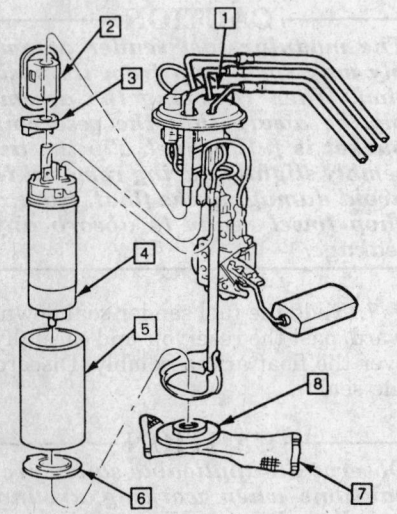

1	FUEL SENDER ASSY.
2	FUEL PULSE DAMPENER
3	BUMPER
4	FUEL PUMP
5	SOUND ISOLATOR SLEEVE
6	SOUND INSULATOR
7	FILTER STRAINER
8	DEFLECTOR

321582

Fuel sender assembly

9. Locate the curved side of the modular unit's reservoir. Beginning at locking tab 1, squeeze the reservoir to release the first locking tab.

10. Moving clockwise to locking tab 2, apply gentle pressure to the guide rod to release the second locking tab.

11. At locking tab 3, gently twist and squeeze to release the reservoir from the retainer.

12. Remove the cover and the retainer from the reservoir. Be careful not to damage the crossover tube; the unit will still be attached by the convoluted fuel pipe and the crossover tube.

13. Remove the external strainer by carefully prying the strainer ferrule off the reservoir. Excessive force may dislodge the jet pump. Note the position of the strainer for installation reference. Discard the strainer.

14. Remove the rubber bumper pad and discard. The fuel pump and the sleeve assembly are attached to the retainer when pulled from the reservoir. Depress the flex member on the pump sleeve and rotate the sleeve counterclockwise to remove the fuel pump from the retainer. Note the orientation of the pump to the retainer.

15. Slide the lower connector assembly out of the retainer to remove the fuel pulse dampener from the lower connector. Note the orientation

of the seal (the modular unit is now held together by the crossover tube only). Discard the fuel pulse dampener.

To install:

16. Install the fuel pulse dampener. Always use a new dampener when installing a new fuel pump.

17. Slide the lower connector into the retainer.

18. Push the pump outlet tube into the fuel pulse dampener and rotate the flex member back to its original position. Line up the pump outlet tube into the retainer opening. All 3 sleeve tabs should protrude through the retainer before rotating. Rotate the pump clockwise until a click is heard, be sure fit is snug before rotating. Place the fuel pump back into its reservoir. The crossover tube must be placed in its proper slot.

19. Install a new rubber bumper pad. Insert the drain tube into the proper retainer and bumper pad slots.

20. Install a new strainer, being careful not to dislodge the jet pump.

21. Install the fuel pump wire connector, the undercover wire harness connector and the CPA clip.

22. Install a new lip seal on the modular fuel sender assembly. Always use a new seal when servicing the modular fuel sender assembly.

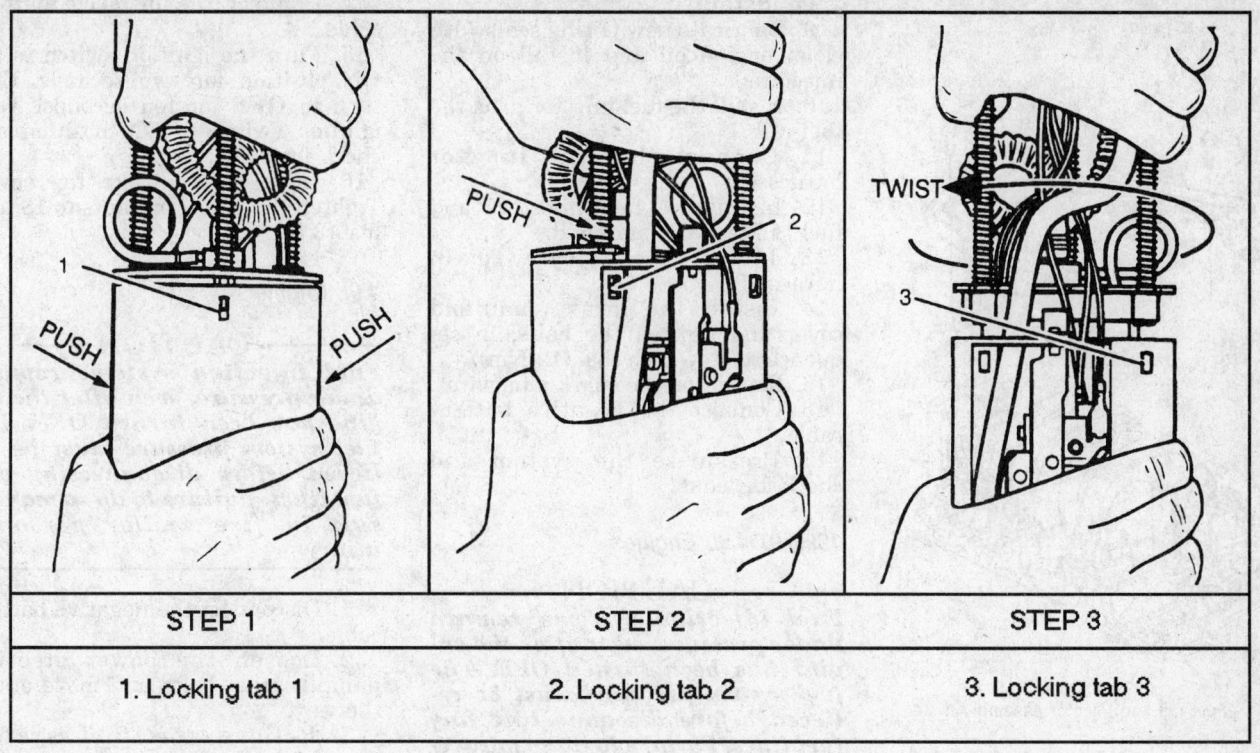

STEP 1	STEP 2	STEP 3
1. Locking tab 1	2. Locking tab 2	3. Locking tab 3

301947

Modular fuel sender disassembly

Lightly lubricate the inside diameter of the lip seal with clean engine oil. The lip seal should be positioned over the float arm assembly, moved up over the reservoir and half-way up the guide posts.

23. Install the modular fuel assembly into the tank. Seat the lip seal into the tank opening.

24. Align the arrows on top of the fuel tank to the arrow on the modular assembly.

25. Slowly apply pressure to the top of the spring loaded sender until the lip seal is flush between the fuel tank and the modular cover.

26. Install the locking nut, torquing it to 37 ft. lbs. (50 Nm).

27. Install the fuel tank in the vehicle.

28. Replace the fuel that was drained before removal.

29. Connect the negative battery cable.

30. Pressurize the fuel system and verify there are no fuel leaks.

Fuel Injector

REMOVAL AND INSTALLATION

3.8L Engine

———— **CAUTION** ————

Fuel Injection systems remain under pressure, even after the engine has been turned OFF. The fuel system pressure must be relieved before disconnecting any fuel lines. Failure to do so may result in fire and/or personal injury.

1. Disconnect the negative battery cable.

2. Relieve the fuel system pressure.

3. Remove the plastic cover from the fuel rail.

4. Disconnect and cap the fuel lines at the rail.

5. Disconnect the vacuum line from the pressure regulator and the harnesses from the injectors.

6. Remove the fuel rail mounting screws and remove the rail assembly.

7. Release the injector retaining clip and remove the injector from the fuel rail.

8. Remove and discard the O-ring from the injector.

To install:

9. Coat new injector O-rings with clean engine oil and install on the injectors.

10. Line up the fuel injectors with the cavities in the manifold and carefully push down.

11. Install the fuel rail mounting bolts and tighten to 10 ft. lbs. (14 Nm).

12. Connect the vacuum line to the pressure regulator and harnesses to the injectors.

13. Connect the lines to the fuel rail.

14. Connect the negative battery cable.

15. Turn the ignition **ON**, to pressurize the fuel system, and check for leaks.

16. Install the plastic cover over the fuel rail.

17. Run the engine and check for leaks and proper engine operation.

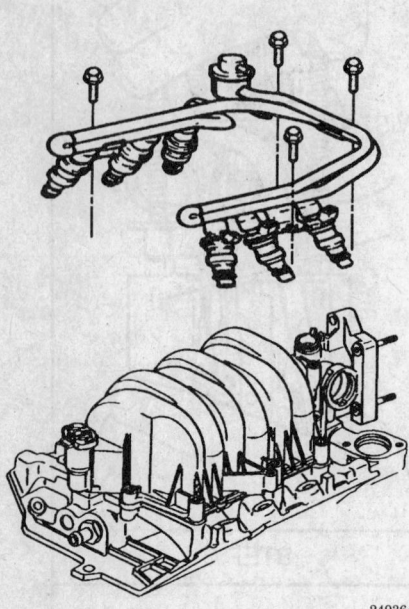

Fuel rail and injector assembly

249364

1993–95 4.6L Engines

CAUTION

Fuel injection systems remain under pressure, even after the engine has been turned OFF. The fuel system pressure must be relieved before disconnecting any fuel lines. Failure to do so may result in fire and/or personal injury.

1. Disconnect the negative battery cable.
2. Relieve the fuel system pressure.
3. Reposition the spark plug wires and remove the intake manifold top cover.
4. Remove the injector main harness and harness at the Idle Speed Control (ISC) actuator and cruise control bracket.
5. Lift the fuel rail with the injector assemblies out of the intake manifold housing enough to access the fuel injectors.
6. Label and disconnect the fuel injector harnesses.
7. Remove the fuel injector(s) from the rail.
8. Remove the O-ring seals from the injector(s) and discard.

To install:

9. Lubricate new O-ring seals with clean engine oil and install on the injector(s).
10. Install the fuel injector(s) on the fuel rail.
11. Connect the fuel injector harnesses.
12. Install the fuel injectors and fuel rail into the manifold.
13. Install the injector main harness.
14. Install the intake manifold cover and tighten the bolts, in sequence, to 106 inch lbs. (12 Nm).
15. Reposition the spark plug wires.
16. Connect the negative battery cable.
17. Pressurize the system and check for leaks.

1996–97 4.6L Engines

CAUTION

Fuel injection systems remain under pressure, even after the engine has been turned OFF. The fuel system pressure must be relieved before disconnecting any fuel lines. Failure to do so may result in fire and/or personal injury.

1. Disconnect the negative battery cable.
2. Relieve the fuel system pressure.
3. Remove the intake manifold top cover.
4. Disconnect the harness from the injector.
5. Release the fuel rail to manifold locking tab by pushing toward the center of the manifold.
6. Pry the injector up out of the manifold using tool J-41081 or equivalent.
7. Spread the injector retainer clip to release the injector from the fuel rail.
8. Remove the fuel injector.

NOTE: It may be necessary to release the adjacent injector(s) from the intake manifold to allow increased fuel rail movement for injector access.

To install:

9. Lubricate the upper and lower injector O-rings with clean engine oil and install them on the injector.
10. Install a new retainer clip onto the injector.
11. Push the fuel rail down onto the intake manifold until the locks latch.
12. Connect the fuel injector harness.
13. Tighten the fuel filler cap.

14. Connect the negative battery cable.
15. Turn the ignition switch to the **ON** position for two seconds, then turn to **OFF** for ten seconds. Turn ignition switch to **ON** position and check for leaks.
16. Install the intake top cover. Tighten the mounting nuts to 18 inch lbs. (2 Nm).

4.9L Engine

CAUTION

Fuel injection systems remain under pressure, even after the engine has been turned OFF. The fuel system pressure must be relieved before disconnecting any fuel lines. Failure to do so may result in fire and/or personal injury.

1. Disconnect the negative battery cable.
2. Loosen the power steering pump bracket bolts and move out of the way.
3. Relieve the fuel system pressure.
4. Remove the upper intake manifold, if necessary.
5. Disconnect the vacuum line from the pressure regulator.
6. Remove the inlet fitting screw assemblies and bracket from the rear of the fuel rail and pressure regulator.
7. Disconnect the fuel feed line and O-ring from the rear of the fuel rail assembly.
8. Disconnect the fuel return line from the return fitting.
9. Remove the 4 fuel rail attaching bolts.
10. Remove the fuel rail assembly from the intake manifold.
11. Label and disconnect the harnesses from the injector(s).
12. Remove the injector retainer clip from the fuel injector and discard.
13. Remove the injector assembly.

To install:

14. Lubricate new injector O-rings with petroleum based grease and install on the injector.
15. Install the injector on the fuel rail and install the retaining clip.
16. Connect the harnesses to the injectors. The electrical connector should face toward the center of the assembly except on cylinder Nos. 1 and 8. Injectors 1 and 8 should face away from the ends of the fuel rail.
17. Install the fuel rail assembly on the intake manifold and bolts. Tightened to 18 ft. lbs. (25 Nm).

1. Injector assembly - fuel
A. Part number identification
B. Build date code
C. Month 1-9(Jan-Sept)
   O,N,D(Oct-Dec)
D. Day
E. Year

241098

Fuel injector

1 INJECTOR ASSEMBLY
2 CLIP-INJECTOR RETAINER

241180

Installing the injector on the fuel rail

18. Connect the vacuum line to the fuel pressure regulator.

19. Lubricate a new inlet fitting O-ring and connect the fuel feed line to the fuel rail.

20. Install the inlet fitting bracket and screw assemblies to the rear of the fuel rail and pressure regulators. Tighten the screws to 44 inch lbs. (5 Nm).

21. Connect the fuel return line to the fitting. Tighten to 22 ft. lbs. (30 Nm).

22. Install the upper intake manifold, if removed.

23. Install the power steering pump brackets.

24. Connect the negative battery cable.

25. Turn the ignition **ON**, to energize the fuel pump, and check for leaks. Correct as required.

26. Run the engine and check for leaks and proper engine operation.

EMISSION CONTROLS

Emission Warning Lamps

RESETTING

1993 Models

1. On Allante models, press the **RANGE** button until the oil life index is displayed.

2. On 1993 Eldorado and Seville models, press the **INFORMATION** button until the oil life index is displayed.

3. Press **AVE SPEED** and **RANGE** buttons at the same time until the index resets to 100 and hold for 5 seconds.

1994–95 Models

Vehicles are equipped with an Engine Oil Life Index (EOLI) feature as part of the Driver Information Center display (DIC). Engine oil life is displayed through engine data as the "OIL LIFE INDEX" and as a "CHANGE ENGINE OIL" message. The "OIL LIFE INDEX" is displayed following a number between 0 and 100. This is the percentage of oil life REMAINING based on driving conditions, engine oil temperature, and mileage driven since the last time the oil life indicator was reset. When the oil life index reaches 1035106370 or less, a "CHANGE OIL SOON" message will appear as a reminder to schedule an oil change. When the oil life index reaches 0, the "CHANGE ENGINE OIL" message will appear indicating that the oil should be changed within the next 200 miles (320 km). After the oil has been changed, display the "OIL LIFE INDEX" message by pressing the "INFORMATION" button several times. Press and hold the "RESET" button until the display shows "100". This will reset the oil life index. The "CHANGE ENGINE OIL" message will remain off until the next oil change is needed. The percentage of oil life remaining may be checked at any time by pressing the "INFORMATION" button several times until the "OIL LIFE INDEX" appears.

Service Interval Lamp

RESETTING

1994–95 Eldorado and Seville

All Cadillacs with 4.6L Northstar engines and 4T80-E transaxles are equipped with a transaxle fluid change indicator. A "CHANGE TRANS FLUID" message will display on the information center when the powertrain control module monitors actual operating conditions and displays a change trans fluid message based on calculations based on those conditions or 100,000 miles (160,000 km.) Change fluid by removing lower pan and side cover drain plug.

When the "CHANGE TRANS FLUID" message appears, change the fluid in both the pan and side cover. Reset the indicator as follows:

Turn the key **ON** with the engine stopped. Press and hold the "OFF" and "REAR DEFOG" buttons on the climate control simultaneously until the "TRANS FLUID RESET" message appears in the Information Center (between 5 and 20 seconds) The system is now reset.

ENGINE MECHANICAL

Engine Assembly

REMOVAL AND INSTALLATION

3.8L Engine

―――― **CAUTION** ――――
Fuel Injection systems remain under pressure, even after the engine has been turned OFF. The fuel system pressure must be relieved before disconnecting any fuel lines. Failure to do so may result in fire and/or personal injury.

1. Disconnect the negative battery cable.

2. Relieve the fuel system pressure.

3. Mark the position of the hood hinges and remove the hood.

4. Drain the cooling system and engine oil into suitable containers.

5. Disconnect the heater and both radiator hoses from the engine.

6. Remove the starter.

7. Disconnect the harnesses to the engine and the battery near the relay center.

8. Remove the serpentine belt(s).

9. Remove the power steering pump and set aside.

10. Remove the air cleaner and air inlet duct.

11. Disconnect the control cables from the throttle body and cable mounting bracket.

12. Disconnect the Throttle Position Sensor (TPS), Idle Air Control (IAC) valve, MAT Sensor and Oxygen Sensor (O_2) connectors.

13. Disconnect the ignition coil assembly ground wire from the inner fender panel.

14. Disconnect and cap the fuel lines from the fuel rail.

15. Install engine support fixture tool J-28467-A or the equivalent. Raise the engine slightly to remove the weight from the engine mounts.

16. Raise and safely support the vehicle.

17. Disconnect the front exhaust pipe from the exhaust manifold.

18. Remove the A/C compressor from the bracket and support out of the way. Do not disconnect the refrigerant lines from the compressor.

19. Remove the left front engine mount.

20. Remove the right front engine-to-transaxle bracket.

21. Remove the lower transaxle-to-engine bolts.

22. Remove the flywheel cover.

23. Remove the torque converter-to-flywheel bolts.

24. Lower the vehicle.

25. Remove the torque strut.

26. Remove the vibration absorber.

27. Remove the transaxle-to-engine bolts.

28. Support the transaxle with a suitable jack.

29. Install a suitable engine lifting device and remove the support fixture.

30. Remove the engine from the vehicle.

To install:

31. Install the engine in the vehicle, install the transaxle-to-engine bolts and tighten to 55 ft. lbs. (75 Nm).

32. Install the engine support fixture and remove the engine lifting device.

33. Raise and safely support the vehicle.

34. Install the right front engine-to-transaxle bracket.

35. Install the left front engine mount.

36. Install the torque strut. Install the vibration absorber.

37. Install the torque converter-to-flywheel bolts and tighten to 46 ft. lbs. (62 Nm).

38. Install the remaining components.

39. Refill the cooling system and engine crankcase.

40. Connect the negative battery cable.

41. Install the hood.

42. Turn the ignition key **ON** and inspect for any fuel leaks.

43. Run the engine and check for leaks and proper vehicle performance.

1993 4.6L and 4.9L Engines

CAUTION

Fuel injection systems remain under pressure, even after the engine has been turned OFF. The fuel system pressure must be relieved before disconnecting any fuel lines. Failure to do so may result in fire and/or personal injury.

---

1. Disconnect the negative battery cable.

2. Relieve the fuel system pressure.

3. Properly remove the refrigerant from the A/C system.

4. Drain the cooling system and crankcase oil into suitable containers.

5. Remove the air cleaner assembly.

6. Mark the position of the hood on the hinges for assembly reference. Remove the hood.

7. Remove the cooling fan. Remove the accessory drive belt.

8. Disconnect the upper radiator and heater hose from the thermostat housing.

9. Label and disconnect the wiring connectors at the following:

  a. Oil pressure switch
  b. Coolant temperature sensor
  c. Distributor
  d. EGR solenoid
  e. Engine temperature switch
  f. Idle Speed Control (ISC) motor
  g. Throttle Position Sensor (TPS)
  h. Fuel injectors at rail connector
  i. Manifold Air Temperature (MAT) sensor
  j. Oxygen sensor
  k. Alternator
  l. Ground wires from the alternator bracket.

10. Disconnect the cable from the throttle lever.

11. Remove the cruise control diaphragm and bracket and move aside.

12. Disconnect the transaxle oil cooler and engine oil cooler lines from the radiator.

13. Remove the radiator.

14. Remove the oil filter housing adapter.

15. Remove the crossbrace. Remove the right front heater hose.

16. Remove the power steering line brace from the right side cylinder head.

17. Remove the power steering pump and tensioner bracket and set aside.

18. Disconnect the A/C lines to the accumulator and condenser. Cap all openings to prevent the entrance of dirt and moisture.

19. Disconnect the fuel lines at the fuel rail.

20. Remove the fuel line bracket at the transaxle.

21. Remove the EGR lines and brackets.

22. Disconnect the vacuum modulator line and power brake vacuum line and reposition.

23. Raise and safely support the vehicle.

24. Remove the flexplate covers.

25. Mark the position of the torque converter on the flexplate. Remove the 3 flexplate-to-torque converter bolts.

26. Remove the A/C compressor lower shield.

27. Remove the right front wheel. Remove the right front wheel well plastic shield.

28. Remove the right rear transaxle-to-engine support.

29. Remove the right rear transaxle-to-engine bolt.

30. Remove the lower engine damper nut.

31. Remove the front engine mount nuts and right rear transaxle mount bolts.

32. Remove the oxygen sensor wires and heater bypass bracket from the right side of the vehicle.

33. Remove the right side engine brace.

34. Lower the vehicle.

35. Remove the 5 top engine-to-transaxle mounting bolts.

WARNING

Be careful when attaching the right rear lift hook to the bracket to assure clearance to the A/C accumulator line.

---

36. Install a suitable engine hoist. Remove the engine from the vehicle.

To install:

37. Lower the engine into the vehicle on an angle matching the transaxle. Engage the dowels on the block with the holes in the transaxle case.

NOTE: Make sure that the torque converter is properly positioned to the flexplate and engaged in the front pump of the transaxle.

38. Install the top engine-to-transaxle mounting bolts and tighten to 55 ft. lbs. (75 Nm).
39. Lower the engine. When the engine is lowered make sure it seats on the mount correctly.
40. Remove the hoist.
41. Raise and safely support the vehicle. Support the engine.
42. Install the right rear transaxle-to-engine mounting bolt and tighten to 55 ft. lbs. (75 Nm).
43. Install the right side engine brace.
44. Install the oxygen sensor wires and heater bypass bracket to the right side of the vehicle.
45. Install the front engine mount nuts and right rear transaxle mount bolt.
46. Install the lower engine damper nut.
47. Install the wheel well plastic shield.
48. Install the right front wheel.
49. Install the A/C compressor lower dust shield.
50. Align the flexplate and converter with the marks made during removal. Install the 3 flexplate-to-torque converter bolts and tighten to 46 ft. lbs. (63 Nm).
51. Install the flexplate covers.
52. Install the starter.
53. Install the exhaust crossover pipe.
54. Lower the vehicle.
55. Connect the vacuum modulator and power brake vacuum lines.
56. Install the EGR lines and bracket.
57. Install the fuel line bracket at the transaxle.
58. Connect the fuel lines to the fuel rail.
59. Connect the A/C lines to the accumulator and condenser.
60. Install the power steering pump and tensioner assembly.
61. Connect the power steering line brace to the right side cylinder head.
62. Connect the right front heater hose.
63. Install the cross car brace.
64. Install the oil filter adapter.
65. Install the oil cooler line bracket at the transaxle.

66. Connect the oil cooler lines to the oil filter adapter.
67. Install or connect the remaining components.
68. Install the hood, aligning the marks made during removal.
69. Install the air cleaner assembly.
70. Refill the cooling system. Check all fluid levels.
71. Connect the negative battery cable.
72. Evacuate, recharge and leak test the A/C system.
73. Run the engine and check for leaks and proper operation.

1994–97 4.6L Engines

—————— **CAUTION** ——————
Fuel injection systems remain under pressure, even after the engine has been turned OFF. The fuel system pressure must be relieved before disconnecting any fuel lines. Failure to do so may result in fire and/or personal injury.

1. Disconnect the negative battery cable.
2. Relieve the fuel system pressure.
3. Remove the air cleaner assembly.
4. Remove the left and right torque struts. Install the left front strut bolt back into the bracket.
5. Disconnect the radiator hoses at the water crossover.
6. Remove both cooling fans from the engine.
7. Properly remove the refrigerant from the A/C system.
8. Disconnect the cruise control servo connections.
9. Disconnect the ISC motor connectors.
10. Disconnect the throttle cable from the throttle body cam. Disconnect the shift cable from the park/neutral switch. Remove the cable bracket at the transaxle.
11. Remove the park/neutral switch connector and disconnect the power brake vacuum hose.
12. Disconnect the fuel inlet and fuel return lines using special tool J37088 or equivalent.
13. Remove the fuel line retainer at the transaxle case.
14. Disconnect the hoses from the coolant reservoir. Remove the reservoir.
15. Disconnect the heater hoses from the front of the right cylinder head. Disconnect the temperature switch.
16. Remove the starter.

17. Disconnect the power steering pump pressure and return lines at the cooler. Remove power steering line retainer from the right front of the crankcase.
18. Disconnect the engine harness connectors at the PCM.
19. From under the hood, remove the wiring harness retainer screws at the cowl and pull the engine harness through.
20. Disconnect the refrigerant high temperature switch.
21. To remove the engine harness on the left wheelhouse, remove 1 screw holding the connector halves together and separate. The engine portion of the harness will be removed with the engine.
22. Remove the serpentine drive belt.
23. Raise and safely support the vehicle.
24. Remove both front wheels.
25. Disconnect the oil cooler lines at the oil filter adapter.
26. Disconnect the exhaust Y-pipe.
27. Disconnect the coupling between the steering rack and the column.
28. Disconnect the speed sensitive steering solenoid and the knock sensor.
29. Disconnect the power steering switch.
30. Separate the lower ball joints and stabilizer links (the struts will stay in the vehicle).
31. Disconnect the A/C hoses from the accumulator and the condenser.
32. Move powertrain dolly J-36295 or equivalent, into position and lower the vehicle onto the table.
33. Remove the 6 engine cradle mounting bolts and remove the powertrain assembly by lifting the vehicle or lowering the table.
34. Remove the torque converter splash shield and remove the 4 converter-to-flywheel bolts.
35. Separate the engine and transaxle assemblies.

To install:

36. Install the engine to the transaxle and torque the bolts to 55 ft. lbs. (75 Nm).
37. Install the exhaust manifolds and torque to 18 ft. lbs. (25 Nm).
38. Install the exhaust system Y-pipe and the transaxle to oil pan brace.
39. With the powertrain on the dolly, move the assembly into approximate position under the vehicle.
40. Lower the vehicle onto the powertrain.
41. Align the engine cradle to the body and install the 6 engine cradle

mounting bolts. Torque the bolts to 75 ft. lbs. (100 Nm).

42. Raise the vehicle and remove the dolly.

43. Connect the oil cooler lines at the oil filter adapter.

44. Install the exhaust Y-pipe.

45. Connect the coupling between the steering rack and the column.

46. Connect the speed sensitive steering solenoid and the knock sensor.

47. Connect the power steering switch.

48. Install the A/C hoses to the accumulator and the condenser.

49. Connect the lower ball joint and the stabilizer shaft link.

50. Position the ABS/TCS assembly to the engine cradle.

51. Lower the vehicle.

52. Connect the engine harness connectors at the PCM.

53. Connect the refrigerant high temperature switch.

54. Position the engine harness connector on the left wheelhouse and install the one screw holding the connector halves together.

55. Connect or install the remaining components.

56. Install the right and left torque struts and tighten the retainer bolts as follows:

NOTE: It is important during installation that the engine torque struts are not preloaded in their installed position. Adjustment is provided at the point the strut fastens to the core support bracket. Make sure this bolt is loose during assembly. Tighten to 45 ft. lbs. (60 Nm) as the final step of assembly.

a. Torque strut bracket to cylinder head (M10) bolt: 35 ft. lbs. (50 Nm).

b. Torque strut bracket to cylinder head (M10) stud: 35 ft. lbs. (50 Nm).

c. Torque strut bracket to water manifold (M8) bolts: 20 ft. lbs. (25 Nm).

d. Torque strut bracket to cylinder head (M10) bolt: 35 ft. lbs. (50 Nm).

e. Torque strut to core support bracket bolt: 45 ft. lbs. (60 Nm).

57. Connect the negative battery cable.

58. Evacuate and recharge the A/C system.

59. Install the air cleaner.

60. Run the engine and check for leaks and proper vehicle operation.

NOTE: A wheel alignment is recommended after the removal of the sub-frame assembly.

Engine Mounts

REMOVAL AND INSTALLATION

3.8L Engine

1. Disconnect the negative battery cable.

2. Install engine support fixture tool J-28467-A, or the equivalent.

3. Raise and safely support the vehicle.

4. Remove the front engine mount-to-frame nuts.

5. Lower the vehicle.

6. Raise the engine off of the mount using the support fixture.

7. Remove the engine mount-to-engine bracket nuts and remove the engine mount.

To install:

8. Position the engine mount in the vehicle and tighten the engine mount-to-engine bracket nuts to 30 ft. lbs. (41 Nm).

9. Lower the engine into place and remove the support fixture.

10. Raise and safely support the vehicle.

11. Install the engine mount-to-frame nuts and tighten to 30 ft. lbs. (41 Nm).

12. Lower the vehicle.

13. Connect the negative battery cable.

4.6L Engines

1. Disconnect the negative battery cable.

2. Remove the 2 engine cooling fans.

3. Remove the right and left torque struts.

4. Install engine support fixture tool J-28467-A or equivalent, using only one support at the left rear engine bracket.

5. Raise and safely support the vehicle.

6. Remove the 2 nuts securing the engine mount to the cradle assembly.

7. Remove the 2 bolts securing the engine mount to the engine block.

8. Lower the vehicle.

9. Remove the 2 bolts from the engine mount to the cylinder head.

10. Remove the 2 nuts securing the engine mount to the bracket.

11. Raise the engine by tightening the support chain until the mount and bracket can be removed.

To install:

12. Install the engine mount and bracket in position in front of the engine.

13. Loosely install the 2 engine mount-to-bracket nuts.

14. Loosely install the 2 bolts securing the mount assembly to the cylinder head.

15. Lower the engine and guide the engine mount studs into the cradle.

16. Raise and safely support the vehicle.

17. Install the 2 bolts securing the mount to the block and tighten to 22 ft. lbs. (30 Nm).

18. Install the nuts securing the mount assembly to the cradle and tighten to 22 ft. lbs. (30 Nm).

19. Lower the vehicle.

20. Remove the support fixture.

21. Tighten the mount assembly to cylinder head bolts to 22 ft. lbs. (30 Nm).

22. Tighten the mount-to-bracket nuts to 22 ft. lbs. (30 Nm).

23. Install the torque struts.

NOTE: When installing the torque struts, it is important that they are not preloaded in their installed position. Adjustment is provided at the point the strut fastens to the core support bracket; be sure this bolt is loose during assembly. Tighten this bolt to 44 ft. lbs. (60 Nm) as the final step of assembly.

24. Install the cooling fans.

25. Connect the negative battery cable.

1993 4.9L Engine

Right Side Mount

1. Disconnect the negative battery cable.

2. Raise and safely support the vehicle.

3. Remove the heat shield screws, then remove the heat shield.

4. Remove the bolt from the engine mount brace at the bracket.

5. Loosen the nut at the top of the brace from the exhaust manifold and position the brace out of the way.

6. Support the engine with a transmission jack.

7. Remove the 2 bolts holding the engine mount to the transaxle.

8. Remove the 4 nuts at the top and bottom of the mount.

9. Raise the engine enough to remove the engine mount.

To install:

10. Install the engine mount in the vehicle.

11. Lower the engine.

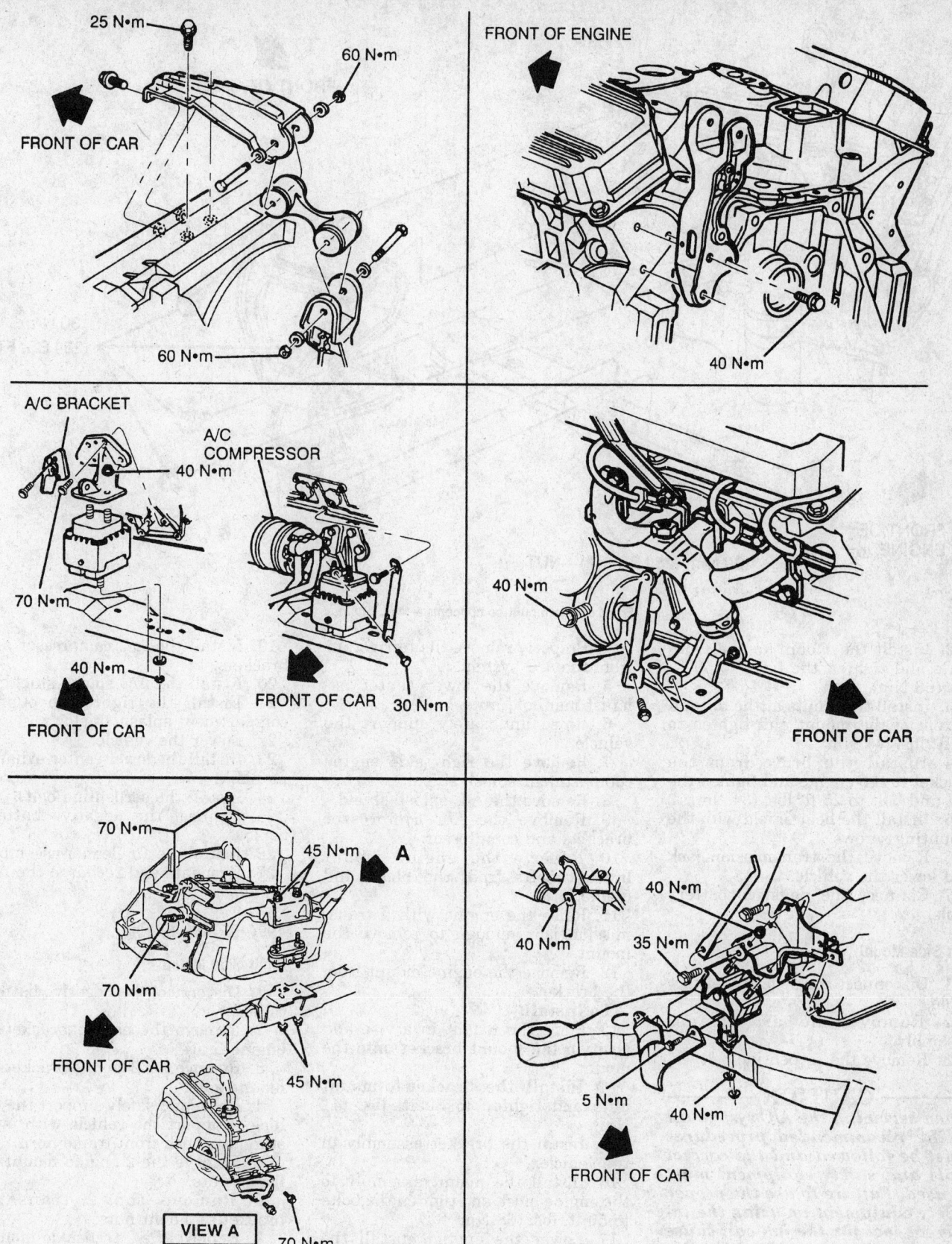

Engine and transaxle mounts — 4.6L engines

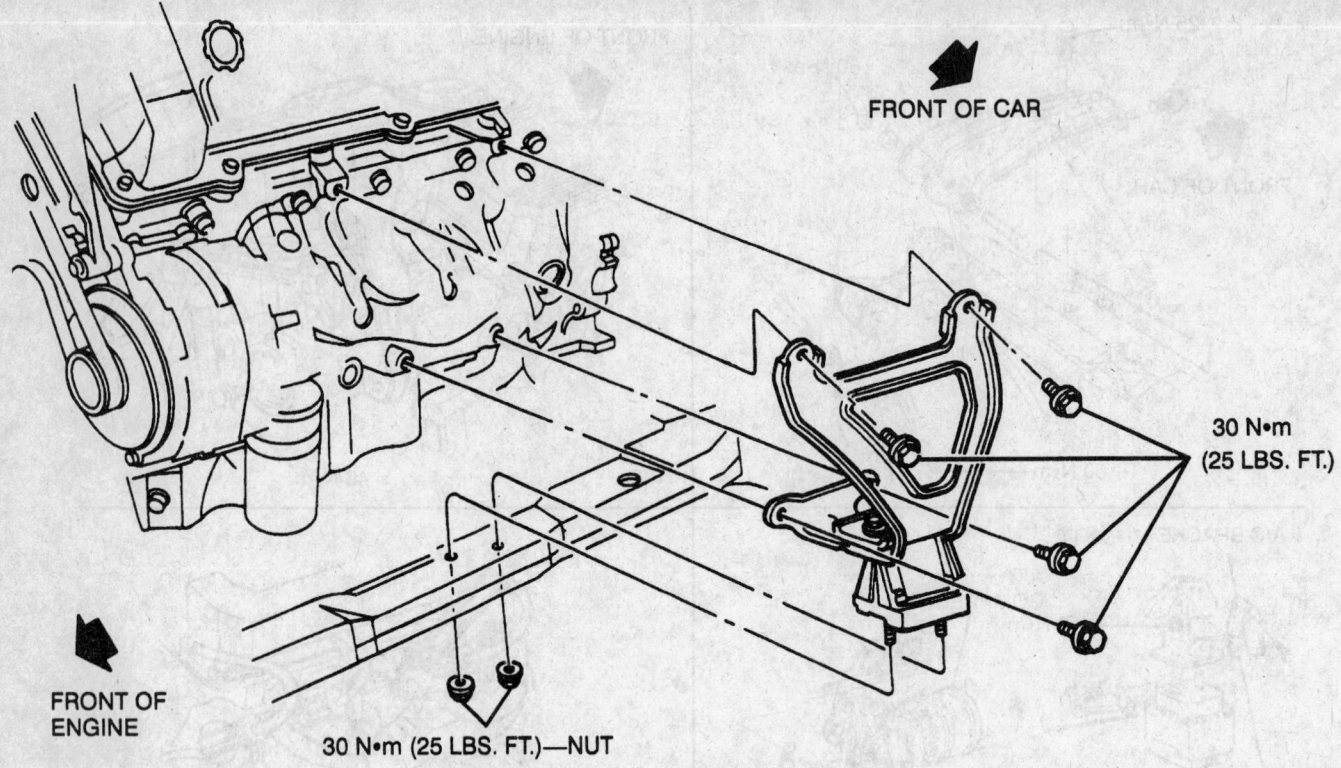

FRONT OF CAR

30 N•m
(25 LBS. FT.)

FRONT OF
ENGINE

30 N•m (25 LBS. FT.)—NUT

234035

Front engine mount components — 4.6L engines

12. Install the mount to the transaxle and tighten the bolts to 50 ft. lbs. (68 Nm).

13. Install the 4 nuts at the top and bottom of the mount and tighten to 30 ft. lbs. (41 Nm).

14. Install the brace from the bracket to the engine and tighten the nut and bolt to 25 ft. lbs. (34 Nm).

15. Install the heat shield with the mounting screws.

16. Remove the transmission jack and lower the vehicle.

17. Connect the negative battery cable.

Left Side Mount

1. Disconnect the negative battery cable.

2. Remove the air cleaner assembly.

3. Remove the serpentine belt.

CAUTION
When servicing the A/C system all of the recommended procedures must be followed and the correct tools and safety equipment must be used. Failure to use the proper safety equipment or using the incorrect tool for the job can cause serious personal injury and/or damage to the vehicle's A/C system.

4. Properly remove the refrigerant from the A/C system.

5. Remove the lower center exhaust manifold nuts.

6. Raise and safely support the vehicle.

7. Remove the right side engine compartment splash shield.

8. Remove the A/C splash shield.

9. Remove the A/C compressor brackets and compressor.

10. Remove the engine mount bracket bolts from the block and cradle.

11. Raise the engine with a transmission jack enough to remove the mount.

12. Remove the engine mount from the bracket.

To install:

13. Place the mount in a vise and position the mount bracket onto the mount.

14. Install the bracket-to-mount nuts and tighten to 30 ft. lbs. (41 Nm).

15. Instal the bracket assembly in the vehicle.

16. Install the mount assembly to the engine block and tighten the bolts to 50 ft. lbs. (68 Nm).

17. Lower the engine, install the mount to frame bolts and tighten to 30 ft. lbs. (41 Nm).

18. Remove the transmission jack.

19. Install the A/C compressor and brackets.

20. Install the A/C splash shield.

21. Install the right side engine compartment splash shield.

22. Lower the vehicle.

23. Install the lower center exhaust manifold nuts.

24. Install the serpentine belt.

25. Connect the negative battery cable.

26. Install the air cleaner assembly.

27. Evacuate and recharge the A/C system.

1994–95 4.9L Engine

Right Side Mount

1. Disconnect the negative battery cable.

2. Remove the engine bracket-to-engine brace.

3. Remove the 2 engine bracket-to-mount nuts.

4. Raise and safely support the vehicle. Support the vehicle with jackstands at each front frame horn.

5. Remove the 2 engine mount-to-frame nuts.

6. Remove the 2 transaxle bracket-to-mount nuts.

7. Remove the 2 transaxle mount-to-frame bracket nuts.

8. Raise the engine using support tool J-28467 or equivalent, until the

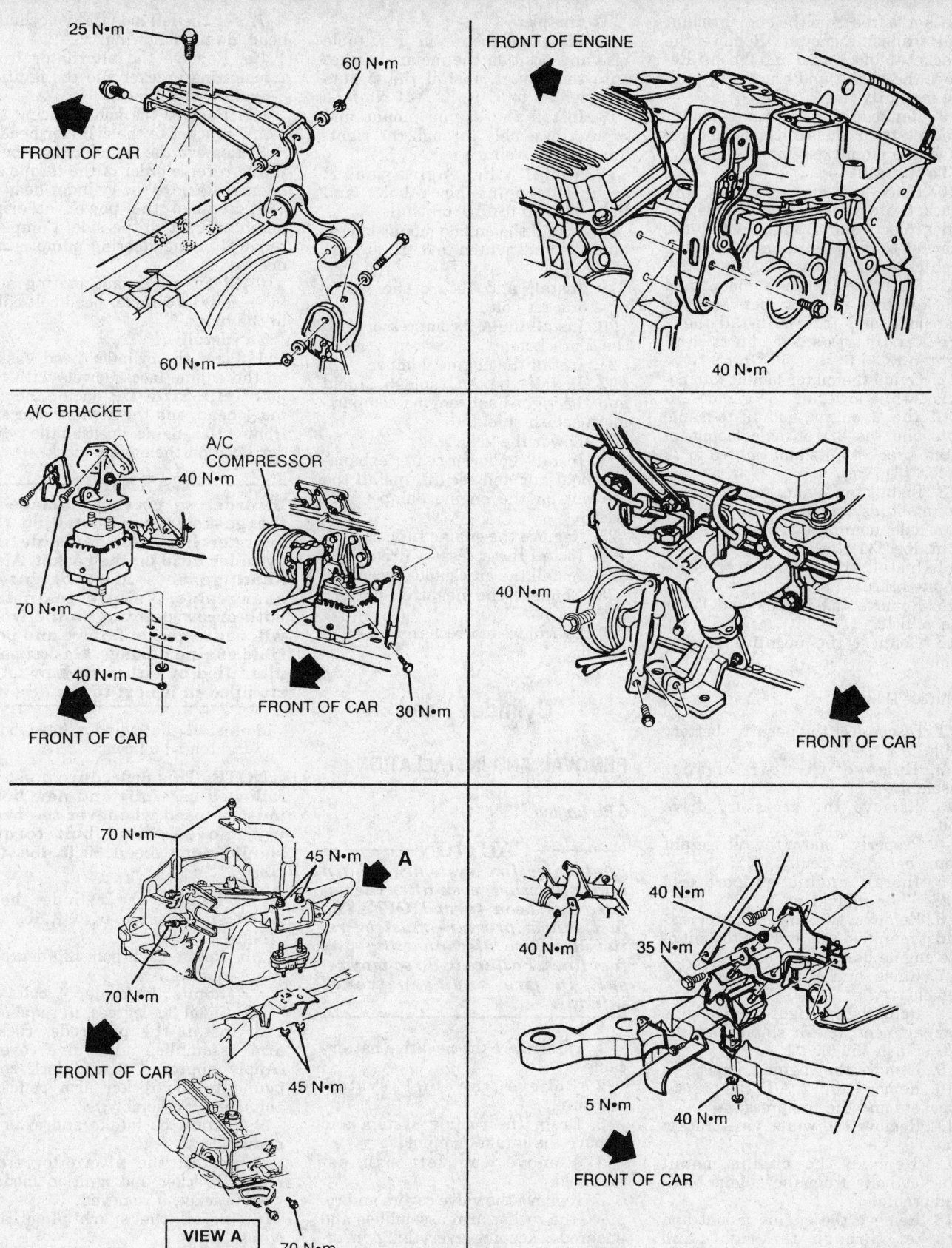

Engine and transaxle mounts — 1993 4.9L engine

bracket is free from the engine mount and transaxle mount. Remove the bracket-to-block stud and 2 bolts. Remove the mount and bracket by pulling forward.

9. Remove the transaxle mount bracket from the transaxle and remove the mount assembly.

To install:

10. Place the engine mount and bracket into position between the cylinder block and frame. Install the bracket-to-block stud and 2 bolts. Tighten to 34 ft. lbs. (46 Nm).

11. Place the transaxle mount and bracket into position between the transaxle and frame. Install the 2 bracket-to-transaxle bolts and tighten to 34 ft. lbs. (46 Nm).

12. Guide the motor mount into position while lowering the engine. Install the 2 engine mount-to-frame nuts and the 2 transaxle mount-to-frame bracket nuts and tighten to 22 ft. lbs. (31 Nm).

13. Install the 2 nuts to the engine mount studs and the 2 nuts to the transaxle mount studs and tighten to 22 ft. lbs. (31 Nm).

14. Install the engine bracket-to-engine brace.

15. Remove the stands and lower the vehicle.

16. Connect the negative battery cable.

Left Side Mount

1. Disconnect the negative battery cable.

2. Remove the air cleaner assembly.

3. Remove the accessory drive belt.

4. Properly remove the refrigerant from the A/C system.

5. Install engine support tool J-28467 or equivalent.

6. Remove the lower center exhaust manifold nut and the top nut of the engine damper.

7. Raise and safely support the vehicle.

8. Remove the right-hand engine compartment splash shield and the A/C splash shield.

9. Remove the engine damper.

10. Remove the 2 A/C compressor brackets and the compressor.

11. Remove the water pipe bracket bolt.

12. Remove the engine mount bracket bolts from the engine block and cradle.

13. Remove the engine mount and bracket through the right-hand wheel well.

14. Remove the engine mount from the bracket.

To install:

15. Place the mount in a suitable vise and position the mount bracket onto the mount. Install the 2 nuts and tighten to 31 ft. lbs. (41 Nm).

16. Install the engine mount and bracket assembly through the right-hand wheel well.

17. Install the engine mount bracket-to-engine block bolts and tighten to 50 ft. lbs. (68 Nm).

18. Install the engine mount-to-cradle bolts and tighten to 31 ft. lbs. (41 Nm).

19. Install and secure the water pipe bracket bolt.

20. Install the A/C compressor with the 2 brackets.

21. Install the engine damper.

22. Install the A/C splash shield and the right-hand engine compartment splash shield.

23. Lower the vehicle.

24. Install the lower center exhaust manifold nut and secure. Install the top nut on the engine damper and secure.

25. Remove the engine support tool.

26. Install the accessory drive belt.

27. Install the air cleaner assembly.

28. Connect the negative battery cable.

29. Evacuate and recharge the A/C system.

Cylinder Head

REMOVAL AND INSTALLATION

3.8L Engine

---- **CAUTION** ----
Fuel injection systems remain under pressure, even after the engine has been turned OFF. The fuel system pressure must be relieved before disconnecting any fuel lines. Failure to do so may result in fire and/or personal injury.

---

1. Disconnect the negative battery cable.

2. Relieve the fuel system pressure.

3. Drain the cooling system and remove the intake manifold.

4. Remove the left exhaust manifold.

5. Remove the valve covers and remove the rocker arm assemblies and pushrods, keeping everything in order for reinstallation.

6. Tag and disconnect the ignition wires and spark plugs.

7. For the left side (front) cylinder head, do the following:

    a. Remove the alternator front mounting bracket and the ignition module with bracket.

    b. Remove the bolt securing the A/C bracket to the cylinder head.

8. Remove the cylinder head bolts in the reverse order of the torque sequence. Remove the cylinder head.

9. Remove the power steering pump and set to the side. Complete removal of the steering pump is not needed.

10. Clean all gasket mating surfaces and the cylinder head bolt holes in the block.

To install:

11. Place the cylinder head gasket on the engine block dowels with the note **THIS SIDE UP** facing the cylinder head, and the arrow facing the front of the engine. Position the cylinder head on the engine block.

---- **WARNING** ----
In order to prevent damage to the gasket, when installing the cylinder head, do not slide the cylinder head on the gasket. Also, head gaskets are not interchangeable. Failure to install with arrow pointing to the front will cause gasket failure and possible engine damage. Gaskets are identified by either an L or an R stamped in it next to the arrow.

---

12. Install the cylinder head bolts and tighten as follows:

NOTE: This procedure must be followed carefully and new bolts must be used whenever the head is removed. Total bolt torque should not exceed 60 ft. lbs. (81 Nm).

    a. Tighten the cylinder head bolts, in sequence, to 37 ft. lbs. (50 Nm).

    b. Rotate each bolt 130 degrees, in sequence.

    c. Rotate the center 4 bolts an additional 30 degrees, in sequence.

13. Install the pushrods, rocker arm assemblies and valve covers. (Apply approved thread lock compound to the rocker arm pedestal bolts before assembly.)

14. Install the intake and exhaust manifolds.

15. Install the alternator front mount bracket and ignition module with bracket, if removed.

16. Install the spark plugs and wires.

17. Install the A/C compressor bracket bolt, and torque to 52 ft. lbs. (70 Nm), if removed.

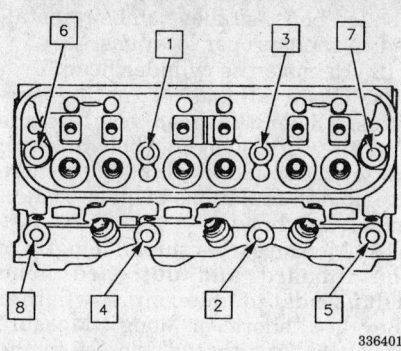

Cylinder head bolt torque sequence — 3.8L engine

336401

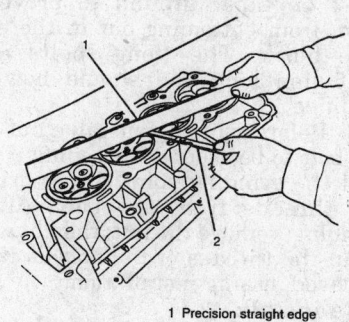

1 Precision straight edge
2 Feeler gage

347777

Measuring cylinder head flatness

18. Fill the cooling system and connect the negative battery cable.
19. Start the engine and check for leaks and proper operation.

4.6L Engines

NOTE: The manufacturer recommends that the entire powertrain be removed from the vehicle before removing the cylinder heads.

—— **CAUTION** ——

Fuel injection systems remain under pressure, even after the engine has been turned OFF. The fuel system pressure must be relieved before disconnecting any fuel lines. Failure to do so may result in fire and/or personal injury.

1. Disconnect the negative battery cable.
2. Properly relieve the fuel system pressure.
3. Drain the cooling system into a suitable container.
4. Remove the powertrain assembly.
5. Remove the intake manifold, cam covers, harmonic balancer, timing chain front cover and oil pump.

—— **WARNING** ——

Align all timing marks before performing the next Step.

6. Remove the chain tensioner from the timing chain.
7. Remove the cam sprockets. The timing chain remains in the chain case.
8. Removing the timing chain guides. Access for the retaining screws is through the plugs at the front of the cylinder head.
9. Remove the water crossover.
10. Remove the exhaust manifold.
11. Remove the cylinder head bolts, a little at a time, in the reverse order of the tightening sequence.
12. Remove the cylinder head and gasket.

—— **WARNING** ——

With the camshafts remaining in the cylinder head, some valves will be open at all times. Do not rest the cylinder head on a flat service with the cylinder face down, or valve damage will result.

13. Clean all gasket mating surfaces. Clean the head bolt holes in the crankcase with compressed air.

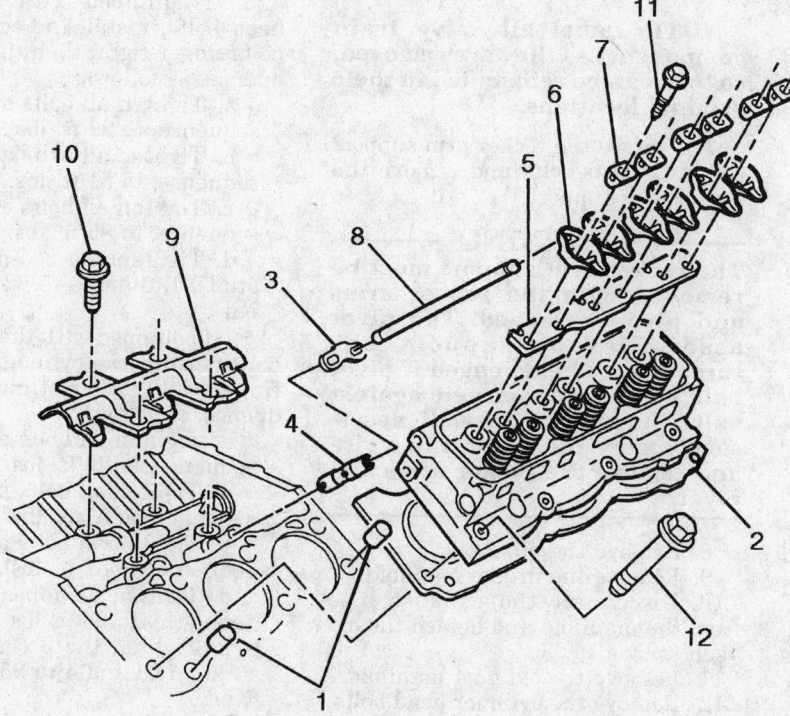

1. Dowel pin
2. Head gasket
3. Lifter guide
4. Valve lifter
5. Pushrod guide
6. Rocker arm
7. Rocker arm pivot
8. Pushrod
9. Lifter guide retainer
10. Bolt
11. Bolt
12. Head bolt

336400

Cylinder head and components — 3.8L engine

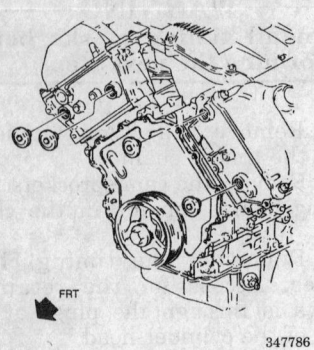

Chain guide access plugs — 4.6L engine

347786

---------- **WARNING** ----------

Be careful when cleaning aluminum gasket surfaces to prevent damage to the sealing surfaces. Use only plastic, wood or "dull" gasket scrapers. Chemical agents to dissolve gasket materials can also be used by following the manufacturers directions.

14. Check the cylinder head for warpage using a straightedge and feeler gauge. Measure along each edge, at the center and across both ends.

15. If warpage is less than 0.002 in. (0.05mm), the cylinder head surface is usable. If warpage is 0.002–0.008 in. (0.05–0.2mm), the cylinder head must be resurfaced. After resurfacing, the dimension between the combustion chamber gauge pad and the deck surface must be at least 10.5mm.

To install:

16. Using a new cylinder head gasket, install the cylinder head and the M11 and M6 head bolts. Lube the washer and the underside of the bolt head with engine oil prior to installation.

17. Tighten the M11 bolts, in sequence, to 22 ft. lbs. (30 Nm) plus 90 degrees. Repeat the sequence, turning each bolt an additional 90 degrees (total 180 degrees). Tighten the M6 bolts to 10 ft. lbs. (12 Nm).

18. Install the camshafts and set the camshaft timing.

19. Install the camshaft guide bolt access hole plugs in the cylinder heads. The plugs should be seated and snug.

20. Install the intake cam covers, oil pump, timing chain front cover, harmonic balancer, intake manifold and water crossover.

21. Install the exhaust manifold. Tighten the nuts to 22 ft. lbs. (30 Nm) or the bolts to 18 ft. lbs. (25 Nm).

22. Install the powertrain assembly.

23. Connect the negative battery cable.

24. Fill the cooling system and check all fluid levels.

25. Properly charge the A/C system.

26. Run the engine and check for leaks and proper engine performance.

NOTE: A wheel alignment is recommended after removal of the sub-frame assembly.

4.9L Engine

---------- **CAUTION** ----------

Fuel injection systems remain under pressure, even after the engine has been turned OFF. The fuel system pressure must be relieved before disconnecting any fuel lines. Failure to do so may result in fire and/or personal injury.

1. Disconnect the negative battery cable.

2. Relieve the fuel system pressure.

3. Drain the cooling system into a suitable container.

4. Disconnect and label the spark plug wires and harnesses.

5. Remove the rocker arm covers.

6. Remove the drive belts and components.

NOTE: Label all valve train components as they are removed, so they can be reinstalled in their original locations.

7. Remove the rocker arm support retaining nuts/bolts and remove the rocker arm support.

---------- **WARNING** ----------

The rocker arm support must be removed with the rocker arms and pivots attached. The pivot assemblies or support bar threads may be damaged if pivot bolt torque is removed against valve spring pressure. If necessary, secure the support in a vise and remove the rocker arms and pivots.

8. Remove the pushrods.

9. Remove the intake manifold.

10. Disconnect the exhaust pipe from the manifold and detach the oxygen sensor.

11. Remove the exhaust manifold.

12. Remove the cylinder head bolts, loosening them a little at a time in the reverse order of the torque sequence. Note the locations of the stud headed bolts, so they can be reinstalled in their proper locations.

13. Remove the cylinder head.

14. Clean all gasket mating surfaces. Clean the head bolt threads and the corresponding bolt holes in the cylinder block. Make sure there are no accumulations of oil or coolant in the bolt holes.

15. Measure the thread length of the inboard and outboard head bolts/studs to determine whether they are "short" or "long". Measure from the first thread (closest to the bolt head) to the end of the bolt. The "short" bolts are 39–41mm long and must be taper ground to prevent them from bottoming out in the engine block. The "long" bolts are 47–50mm long and should not be modified.

16. Before grinding, install a nut on the bolt to be ground, threading it on past the grinding point. Taper grind 1½ threads from the bolt. After grinding, remove the nut; the nut will clean the threads of the bolt as it is removed, easing installation.

To install:

17. Install a new cylinder head gasket on the engine block. Make sure the gasket aligns with the dowel pins on the block.

18. Install the cylinder head.

19. Lubricate the cylinder head bolts with GM lubricant 1052356 or equivalent. Do not use engine oil.

20. If equipped with the "short" head bolts, install the cylinder head bolts finger-tight, then tighten, in sequence, as follows:

   a. Tighten all bolts and studs, in sequence, to 29 ft. lbs. (40 Nm).

   b. Tighten all bolts and studs, in sequence, to 51 ft. lbs. (70 Nm).

   c. Tighten all bolts and studs, in sequence, to 85 ft. lbs. (115 Nm).

   d. Tighten the 3 center inboard studs, 1, 3 and 4 to 88 ft. lbs. (120 Nm).

21. If equipped with the "long" head bolts, install the cylinder head bolts finger-tight, then tighten, in sequence, as follows:

   a. Tighten all bolts and studs, in sequence, to 29 ft. lbs. (40 Nm).

   b. Tighten all bolts and studs, in sequence, to 51 ft. lbs. (70 Nm).

   c. Tighten all bolts and studs, in sequence, to 81 ft. lbs. (110 Nm).

   d. Tighten all inboard studs, in sequence, to 88 ft. lbs. (120 Nm).

   e. Tighten the 3 center inboard studs, 1, 3 and 4 to 96 ft. lbs. (130 Nm).

22. Install the engine lift bracket.

23. Install the exhaust manifold and the oxygen sensor connector.

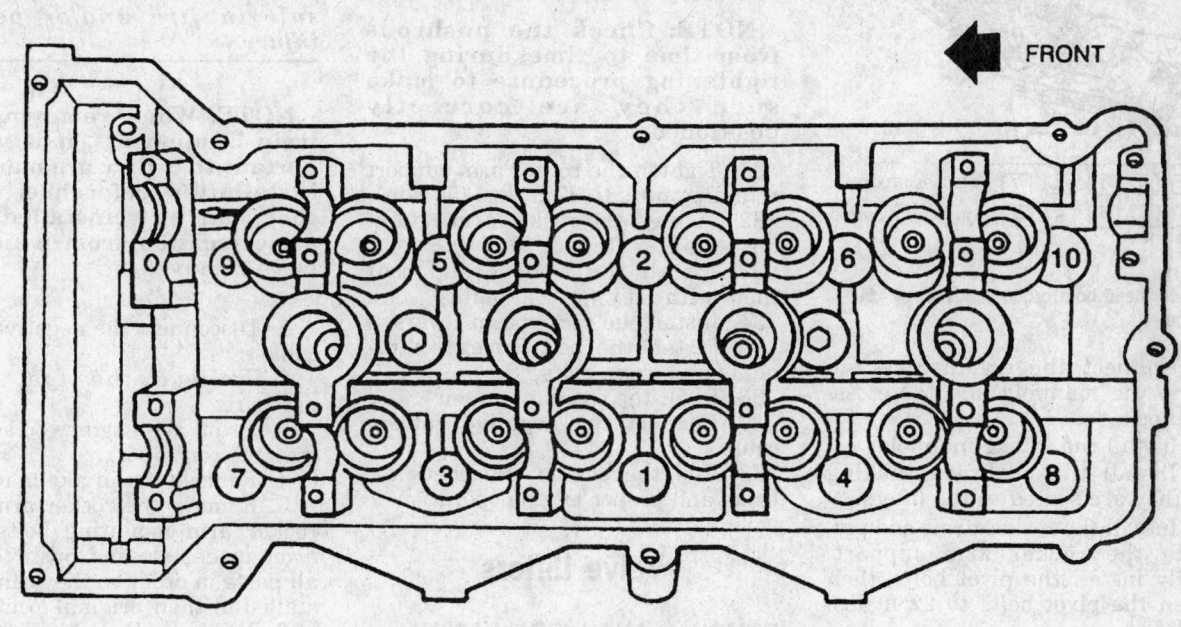

FRONT

Cylinder head bolt tightening sequence — 4.6L engine

347780

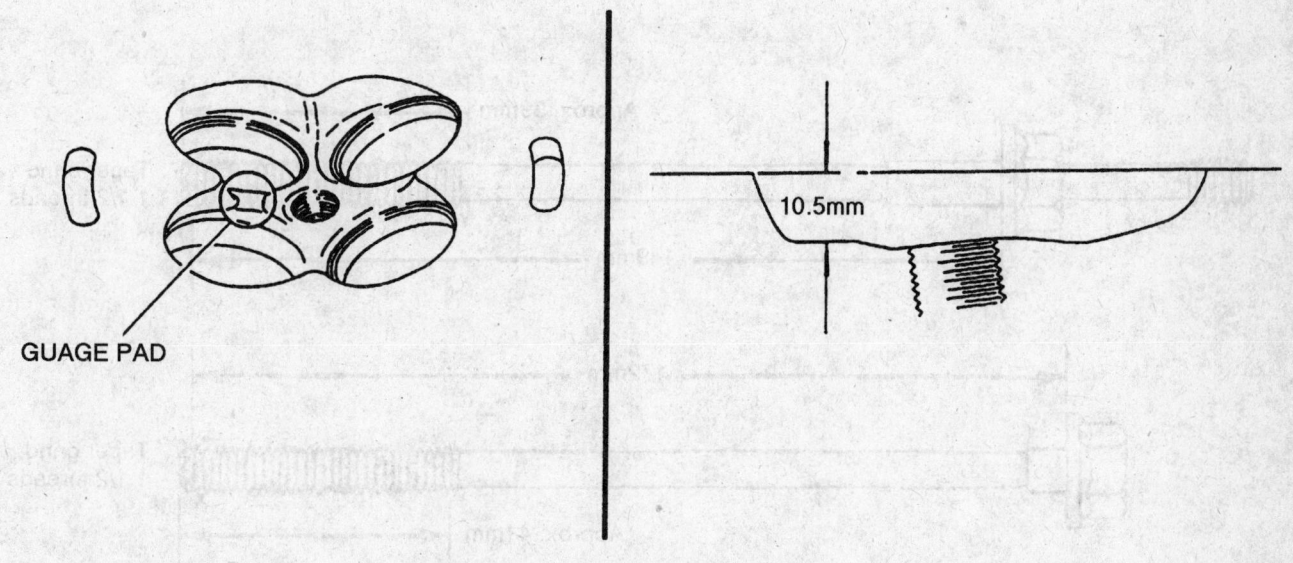

GUAGE PAD

10.5mm

Minimum head resurface dimension — 4.6L engine

347788

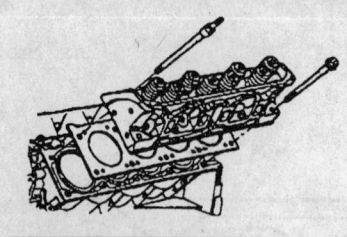

```
 8  4  1  3  7     INBOARD STUDS
10  6  2  5  9     OUTBOARD BOLTS
```

348417

Cylinder head bolt torque sequence — 4.9L engine

24. Connect the exhaust system pipe to the manifold and the cross-over pipe.

25. Install the intake manifold.

26. Install the pushrods, making sure they are seated in the lifters.

27. Install the rocker arms and pivots to the rocker arm support. Loosely install the pivot bolts, then tighten the pivot bolts to 22 ft. lbs. (30 Nm).

28. Install the rocker arm support, with the rocker arms and pivots over the 5 studded head bolts. Make sure each pushrod is seated in a rocker arm.

29. Loosely install the rocker arm support retaining nuts and bolts.

30. Tighten the rocker arm support retaining nuts alternately and evenly until snug.

NOTE: Check the pushrods from time to time during the tightening procedure to make sure they are correctly positioned.

31. Tighten the rocker arm support retaining nuts to 37 ft. lbs. (50 Nm).

32. Tighten the rocker arm retaining bolts to 7 ft. lbs. (10 Nm).

33. Install the wedges and coat them with RTV for seal ability.

34. Install the rocker arm covers.

35. Install the belt driven components and drive belts.

36. Refill the cooling system.

37. Connect the negative battery cable.

38. Run the engine and check for leaks and proper engine operation.

Valve Lifters

REMOVAL AND INSTALLATION

3.8L Engine

--- CAUTION ---

Fuel injection systems remain under pressure, even after the engine has been turned OFF. The fuel system pressure must be relieved before disconnecting any fuel lines. Failure to do so may result in fire and/or personal injury.

NOTE: When removing valve train components, it is very important that they are marked for installation reference, so that they can be reinstalled in the same location from which they were removed.

1. Disconnect the negative battery cable.

2. Relieve the fuel system pressure.

3. Drain the engine coolant into a suitable container.

4. Remove the intake manifold.

5. Remove the rocker arm covers, rocker arm mounting bolts, rocker arms, pedestals and pushrods. Keep all parts in order so they can be reinstalled in their original locations.

6. Remove the guide retainer mounting bolts and retainer.

7. Remove the lifter guides, then remove the lifters. Keep all parts in order so they can be reinstalled in their original locations.

8. Inspect the lifters for wear and/or damage; replace as necessary.

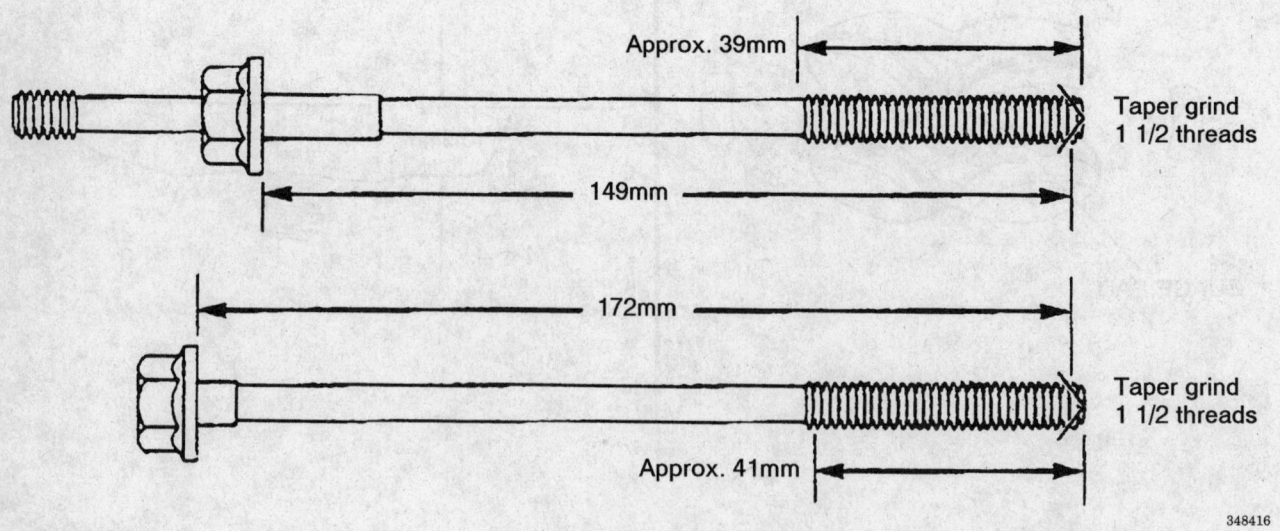

Taper grind the "short" cylinder head bolts

348416

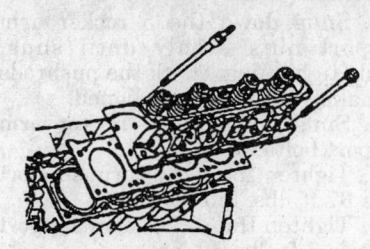

INBOARD STUDS
OUTBOARD BOLTS

348417

Cylinder head bolt torque sequence — 4.9L

9. Inspect the pushrod ends for wear and/or damage. Roll the pushrods on a flat surface to check for a bent condition. Replace any pushrods that are bent and/or damaged.

10. Clean all gasket mating surfaces. Make sure all thread adhesive is removed from the rocker arm pedestal bolts and rocker arm cover bolts.

To install:

11. Coat the lifters with assembly lubricant and install.

12. Install the lifter guides and guide retainer. Tighten the retainer mounting bolts to 22 ft. lbs. (30 Nm).

13. Lubricate the pushrod tips with assembly lubricant and install.

14. Lubricate the rocker arms and pedestals with assembly lubricant and install. Be sure the pushrod tips are properly seated in the rocker arms.

15. Apply thread locking compound to the rocker arm pedestal bolt threads and install. Tighten the bolts to 19 ft. lbs. (25 Nm) plus an additional 70 degree turn on 1993–95 engines.

16. Install the rocker arm covers, using new gaskets, and tighten the mounting bolts to 89 inch lbs. (10 Nm).

17. Install the intake manifold.

18. Connect the negative battery cable.

19. Fill and bleed the cooling system.

20. Run the engine and check for leaks and proper engine operation.

4.6L Engines

1. Disconnect the negative battery cable.

2. Remove the camshafts.

3. Remove the valve lifters. Store the lifters on their camshaft face so that the residual oil is retained.

NOTE: Retain the valve lifters in order so that they can be installed in the same bores.

4. Inspect the lifters for wear and/or damage; replace as necessary.

To install:

5. Install the valve lifters in the same bore from which they were removed.

6. Install the camshafts.

7. Run the engine and check for leaks and proper engine operation.

4.9L Engine

── CAUTION ──
Fuel injection systems remain under pressure, even after the engine has been turned OFF. The fuel system pressure must be relieved before disconnecting any fuel lines. Failure to do so may result in fire and/or personal injury.

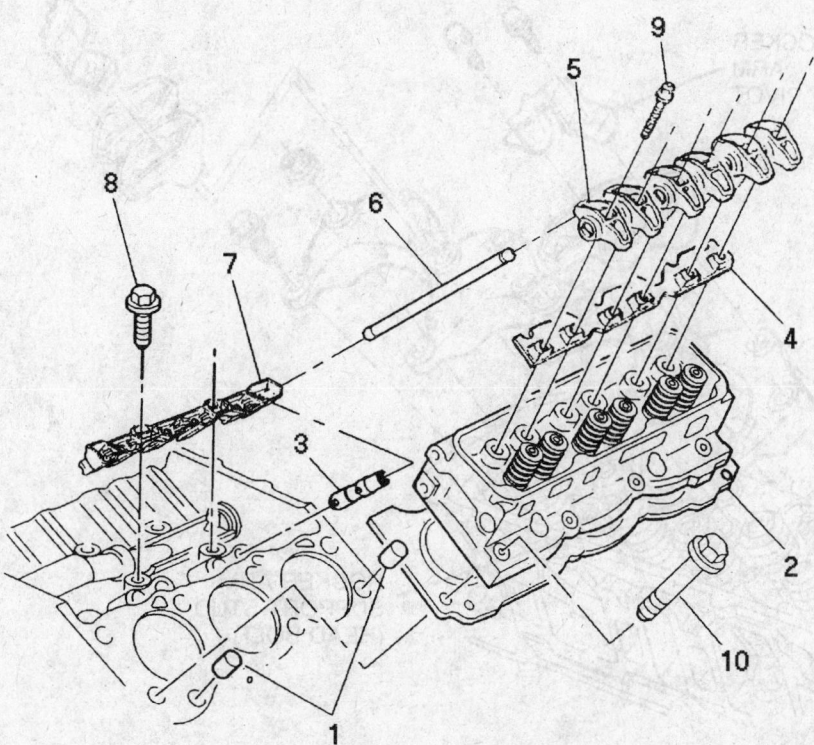

1. Dowel pin
2. Head gasket
3. Valve lifter
4. Pivot retainer
5. Rocker arm
6. Pushrod
7. Lifter guide
8. Bolt
9. Bolt
10. Head bolt

251043

Valve train components — 3.8L engine

NOTE: When removing valve train components, it is very important that they are marked for installation reference, so that they can be reinstalled in the same location from which they were removed.

1. Disconnect the negative battery cable.
2. Relieve the fuel system pressure.
3. Remove the rocker arm covers.
4. Remove the 4 bolts from the cylinder head bosses which retain the rocker arm support.
5. Remove the 5 rocker arm support retaining nuts from the cylinder head bolt studs.

------- **WARNING** -------
The rocker arm support should be removed with the rocker arms and pivots. The pivot assemblies may be damaged if the pivot bolt torque is not released evenly against valve spring pressure.

6. Remove the pushrods, keeping them in order so they can be reinstalled in their original locations.
7. Remove the intake manifold.
8. Remove the valve lifter retainer bolts and remove the valve lifter retainer.

9. Remove the valve lifters and lifter guides. Keep all parts in order so they can be reinstalled in their original locations.
10. Inspect the valve lifters for wear and/or damage; replace as necessary.
11. Inspect the pushrod ends for wear and/or damage. Roll the pushrods on a flat surface to check for a bent condition. Replace any pushrods that are bent and/or damaged.

To install:
12. Thoroughly clean all gasket mating surfaces.
13. Lubricate the valve lifters with assembly lubricant and install them in the lifter bores.
14. Install the valve lifter guides.
15. Install the valve lifter retainer and tighten the mounting bolts to 15 ft. lbs. (20 Nm).
16. Install the intake manifold.
17. Lubricate the pushrod ends with assembly lubricant and install the pushrods.
18. Install the rocker arm support over the five mounting studs.
19. Guide each pushrod into the rocker arm.
20. Install the 5 rocker arm support nuts loosely.
21. Install the 4 rocker arm support bolts loosely.

22. Snug down the 5 rocker arm support nuts evenly until snug. While tightening, check the pushrods to make sure they are aligned.
23. Snug down the 4 rocker arm support bolts evenly.
24. Tighten the rocker arm support nuts 37 ft. lbs. (50 Nm).
25. Tighten the rocker arm support bolts to 7 ft. lbs. (9 Nm).
26. Install the rocker arm covers.
27. Connect the negative battery cable.
28. Run the engine and check for leaks and proper engine operation.

Valve Lash

ADJUSTMENT

The valve clearance cannot be adjusted on these engines. The hydraulic lifters function to maintain a zero clearance when the valves are opening and closing. Any clearance is instantaneously taken up by the hydraulic action. The hydraulic lifter consists of 3 main parts: the body, plunger and valve. Oil under pressure from the engine lubricating system passes through the check valve and forces the plunger upwards in the body of the lifter and takes up any excess clearance. "Valve lifter

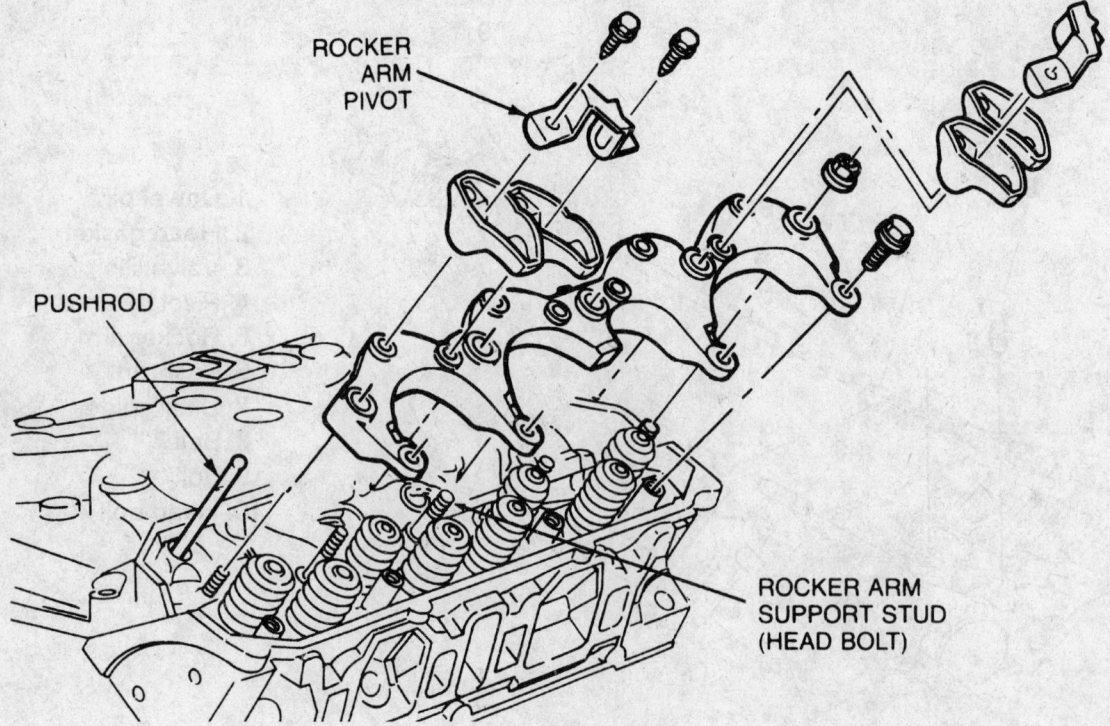

ROCKER ARM PIVOT

PUSHROD

ROCKER ARM SUPPORT STUD (HEAD BOLT)

337919

Valve lifters and lifter retainer — 4.9L engine

noise" complaints may require cleaning of sludged or varnished valve lifters, or replacement.

Rocker Arms

REMOVAL AND INSTALLATION

3.8L Engine

Left Side (Front) Rocker Arms

1. Disconnect the negative battery cable.
2. Remove the serpentine belt.
3. On the left cylinder head, remove the alternator-to-brace bolt.
4. Remove the alternator brace from the intake manifold.
5. On the right cylinder head, Loosen the power steering pump mounting bolts and pull the pump forward for clearance.
6. Remove the rear power steering pump braces from the intake manifold.
7. Disconnect the spark plug wires from the plugs and route the wires out of the way.
8. Remove the rocker arm cover bolts and cover.

NOTE: Identify all valve train components as they are removed, so they can be reinstalled in their original location.

9. Remove the rocker arm mounting bolts and remove the pedestals and rocker arms.
10. Remove the pushrods.
11. Inspect the rocker arms and pedestals for wear and/or damage; replace as necessary. Inspect the pushrod ends for wear and/or damage. Roll the pushrods on a flat surface to check for a bent condition. Replace any pushrods that are worn, damaged and/or bent.

To install:

12. Lubricate the rocker arms and pushrod ends with clean engine oil or suitable engine assembly lubricant.
13. Install the pushrods, making sure they are seated properly in the lifters.
14. Install the rocker arms and pedestals. Tighten the mounting bolts to 19 ft. lbs. (25 Nm) plus an additional 70 degree turn.
15. Install the rocker arm cover with a new gasket and tighten the mounting bolts to 88 inch lbs. (10 Nm).
16. Connect the spark plug wires to the spark plugs.
17. If working on the left cylinder head, install the alternator brace loosely on the intake manifold and tighten the alternator-to-brace bolt to 20 ft. lbs. (27 Nm).
18. If working on the right cylinder head, install the power steering pump braces on the intake manifold and position the pump. Tighten the mounting bolts to 20 ft. lbs. (27 Nm).
19. Tighten the brace mounting nut to 20 ft. lbs. (27 Nm).
20. Install the serpentine belt.
21. Connect the negative battery cable.
22. Run the engine and check for leaks and proper engine operation.

4.9L Engine

NOTE: When removing valve train components, it is very important that they are marked for installation reference, so that they can be reinstalled in the same location from which they were removed.

1. Disconnect the negative battery cable.
2. Remove the rocker arm cover.
3. Remove the 4 bolts from the cylinder head bosses which retain the rocker arm support.
4. Remove the 5 rocker arm support retaining nuts from the cylinder head bolt studs.

------ WARNING ------

The rocker arm support should be removed with the rocker arms and pivots. The pivot assemblies may be damaged if the pivot bolt torque is not released evenly against valve spring pressure.

5. Support the rocker arm support in a vise and remove the rocker arms and pivots.

To install:

6. Install the rocker arm and pivots to the rocker arm support.
7. Install the pivot bolts loosely.

NOTE: The pivot bolts are self tapping. The new rocker arm support does not require tapping prior to installation.

8. Tighten the pivot bolts to 22 ft. lbs. (29 Nm).
9. Install the rocker arm support, with all components in place, over the five mounting studs.
10. Guide each pushrod into the rocker arm.
11. Install the 5 rocker arm support nuts loosely.
12. Install the 4 rocker arm support bolts loosely.
13. Snug down the 5 rocker arm support nuts evenly until snug. While tightening, check the pushrods to make sure they are still aligned.
14. Snug down the 4 rocker arm support bolts evenly.
15. Tighten the rocker arm support nuts 37 ft. lbs. (50 Nm).
16. Tighten the rocker arm support bolts to 7 ft. lbs. (9 Nm).
17. Install the rocker arm cover.
18. Connect the negative battery cable.
19. Run the engine and check operation.

Intake Manifold

REMOVAL AND INSTALLATION

3.8L Engine

------ CAUTION ------

Fuel Injection systems remain under pressure, even after the engine has been turned OFF. The fuel system pressure must be relieved before disconnecting any fuel lines. Failure to do so may result in fire and/or personal injury.

1. Disconnect the negative battery cable.
2. Relieve the fuel system pressure.
3. Drain the cooling system into a suitable container.
4. Remove the plastic engine cover clipped to the fuel rail.
5. Disconnect the air inlet duct from the throttle body.
6. Disconnect and the spark plug wires from the rear of the engine. Position the wires out of the way.
7. Disconnect the harnesses from the fuel injectors.
8. Disconnect the vacuum line from the pressure regulator.
9. Disconnect and cap the fuel lines from the fuel rail.
10. Remove the fuel rail from the intake.
11. Remove the heat shield from the exhaust crossover pipe.
12. Remove the control cable bracket.
13. Disconnect the control cables, vacuum lines and electrical connectors from the throttle body and intake manifold.
14. Remove the power steering pump support bracket.
15. Remove the serpentine belt.
16. Remove the alternator and mounting bracket.
17. Disconnect the upper radiator hose from the thermostat housing.
18. Disconnect the heater and bypass hoses from the intake manifold.

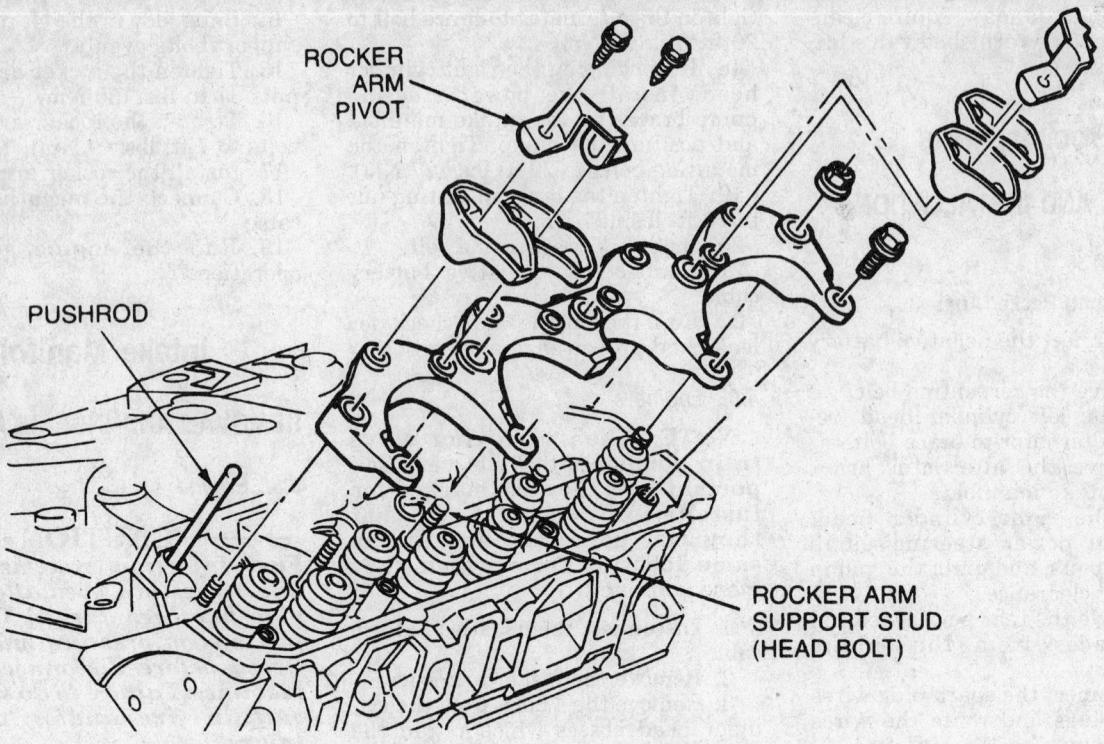

ROCKER ARM PIVOT

PUSHROD

ROCKER ARM SUPPORT STUD (HEAD BOLT)

337919

Rocker arm support components — 4.9L engine

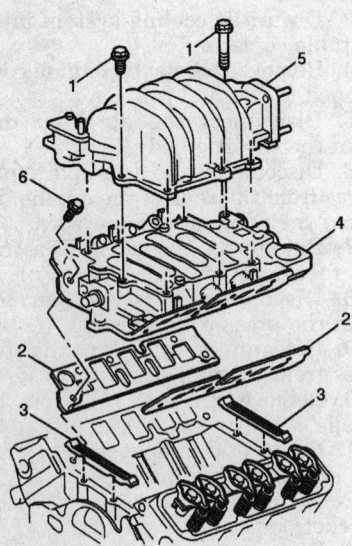

1 Intake manifold upper bolt
2 Intake manifold gasket
3 Intake manifold seal
4 Intake manifold lower
5 Intake manifold upper
6 Intake manifold lower bolt

321820

Intake manifold components — 3.8L engine

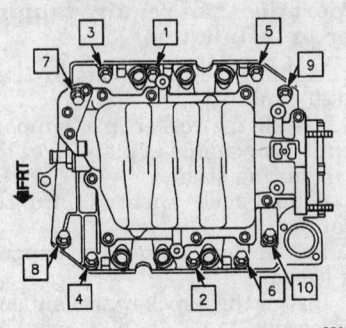

321821

Intake manifold bolt torque sequence — 3.8L engine

19. Remove the intake manifold bolts and remove the manifold and gasket.

To install:

20. Thoroughly clean all gasket mating surfaces.

21. Install new intake manifold gaskets and position the intake manifold on the engine.

22. Install the intake manifold bolts and tighten, in sequence, to 88 inch lbs. (10 Nm). After all the bolts have been tightened.

23. Connect the heater hoses and bypass hose to the intake manifold.

24. Connect the upper radiator hose to the thermostat housing.

25. Install the alternator mounting bracket and alternator.

26. Install the serpentine belt.

27. Install the power steering pump support bracket.

28. Connect the electrical connectors, vacuum hoses and control cables to the intake manifold and throttle body.

29. Install the throttle body cable bracket.

30. Install the heat shield on the crossover pipe.

31. Install the fuel rail, and connect the fuel lines to the fuel rail.

32. Connect the vacuum line to the pressure regulator and harnesses to the fuel injectors.

33. Connect the rear spark plug wires.

34. Connect the air inlet duct to the throttle body.

35. Connect the negative battery cable.

36. Pressurize the fuel system and verify no leaks.

37. Install the plastic engine cover over the fuel rail.

38. Fill and bleed the cooling system.

39. Start the vehicle and check for oil or coolant leaks.

1993-94 4.6L Engine

—————— CAUTION ——————

Fuel Injection systems remain under pressure, even after the engine has been turned OFF. The fuel system pressure must be relieved before disconnecting any fuel lines. Failure to do so may result in fire and/or personal injury.

1. Disconnect the negative battery cable.
2. Relieve the fuel system pressure.
3. Drain the cooling system.
4. Remove the intake duct from the throttle body.
5. Disconnect the coolant hoses at the throttle body.
6. Label and disconnect the necessary harnesses at the intake manifold. This includes the throttle position sensor, idle speed control motor, cruise control servo and EVAP solenoid, if equipped.
7. Label and disconnect the vacuum hoses at the brake vacuum booster, fuel pipe bundle, throttle body, cruise control servo and to body, as necessary.
8. Disconnect the PCV valve at the intake manifold.
9. Disconnect the accelerator cable and position out of the way.
10. Remove the cruise control servo and bracket, if necessary.
11. Disconnect the fuel pipe quick-connects at the fuel pipe bundle in the engine compartment.
12. Remove the EVAP solenoid bracket at the rear cam cover, if equipped.
13. Label and disconnect the spark plug wires at the front bank and lay aside.
14. Reposition the transaxle range control cable away from the cruise control servo or ISC actuator bracket, as necessary.
15. Remove the intake manifold bolts and lift the manifold with the throttle body out of the engine compartment.
16. Remove the intake manifold seals and spacers at the heads.
17. Remove the fuel pipe retainer at the ISC actuator bracket.
   To install:
18. Thoroughly clean all intake manifold seal mating surfaces.
19. Position a new O-ring seal on the throttle body.

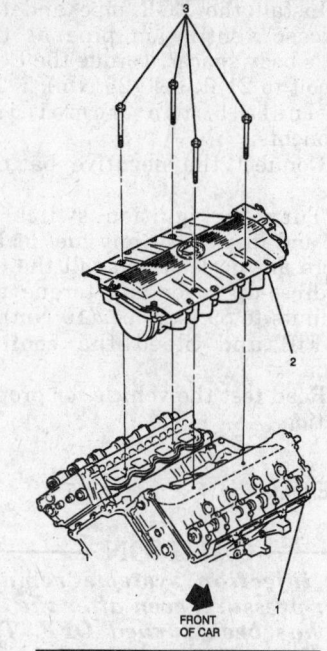

1	ENGINE
2	INTAKE MANIFOLD WITH THROTTLE BODY
3	BOLT

241437

Intake manifold and mounting bolts — 4.6L engine

—————— WARNING ——————

Do not reuse the old throttle body seal because it may not seat properly. Do not use any type of silicone lubricant on the seal or damage could result.

20. Install the throttle body to the intake manifold and tighten the bolts to 106 inch lbs. (12 Nm).
21. Install the fuel pipe retainer at the ISC bracket.
22. Install the intake manifold seals and spacers at the cylinder heads.
23. Install the intake manifold and throttle body. Tighten the bolts, in sequence, to 53 inch lbs. (6 Nm) and then, in sequence, an additional $1/3$ turn.
24. Reposition the transaxle range control cable.
25. Install the EVAP solenoid bracket at the rear cam cover, if equipped.
26. Connect or install the remaining components.
27. Connect the negative battery cable.
28. Fill and bleed the cooling system.
29. Start the engine and check for leaks and proper operation.

1995-97 4.6L Engines

—————— CAUTION ——————

Fuel injection systems remain under pressure, even after the engine has been turned OFF. The fuel system pressure must be relieved before disconnecting any fuel lines. Failure to do so may result in fire and/or personal injury.

1. Disconnect the negative battery cable.
2. Remove the dress-up cover by removing the 4 nuts.
3. Relieve the fuel system pressure.
4. Drain the cooling system into a suitable container.
5. Remove the intake duct from the throttle body.
6. Remove the transaxle vent hose and the transaxle shift cable at the bracket.
7. Remove the Throttle Position (TP) sensor and the Idle Air Control (IAC) valve connectors.
8. Disconnect the throttle cable and the cruise control cable at the throttle body.
9. Remove the throttle body coolant hoses and the surge tank pipe.
10. Remove the EGR pipe and the crankcase ventilation pipe at the throttle body spacer.
11. Remove the brake booster vacuum hose at the intake manifold.
12. Remove the fuel rail ground wire at the rear cylinder head.
13. Remove the quick-disconnect fuel rail fittings using tool J-37088-A or equivalent, insert the tool into the female connector and push inward to release the locking tabs and pull the connection apart.
14. Disconnect the fuel rail bracket at the EGR valve.
15. Disconnect the PCV hose at the intake manifold.
16. Disconnect the main injector harness connector.
17. Remove the 6 bolts and the 4 studs in the intake manifold.

NOTE: The intake manifold carrier gaskets are attached to the intake manifold through a snap-lock feature. When removing the intake manifold, the carrier gaskets will remain attached to the intake manifold. DO NOT replace the intake manifold gaskets after the intake manifold removal. The gaskets are reusable. The gaskets should only be replaced if the plastic housing or the rubber seals are damaged.

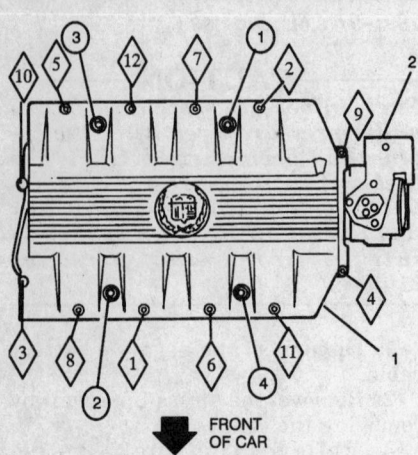

FRONT
OF CAR

○ TORQUE (4) INTAKE MANIFOLD BOLTS IN
SEQUENCE ABOVE TO 8 N•m (71 LB. IN.)
THEN AN ADDITIONAL 120○ (1/3 TURN).

◇ TORQUE (12) INTAKE MANIFOLD COVER
BOLTS IN SEQUENCE ABOVE TO 12 N•m
(106 LB. IN.).

1	INTAKE MANIFOLD
2	THROTTLE BODY
(#)	INTAKE MANIFOLD BOLTS
(#)	INTAKE MANIFOLD COVER BOLTS

241439

Intake manifold bolt torque sequence — 4.6L
engine

To install:

18. Install the intake manifold to
the engine assembly.
19. Install the 6 bolts and 4 studs in
their proper locations. Torque the
bolts and studs to 89 inch lbs. (10
Nm). When tightening the intake
manifold bolts and studs, start at the
center of the manifold and work outward in a circular pattern. DO NOT
tighten the intake manifold bolts
when the engine is HOT or at operating temperature. This repair should
only be done on a cool engine.
20. Connect the injector main wiring harness.
21. Install the PCV hose to the intake manifold.
22. Install the fuel bracket at the
EGR valve.
23. Install the fuel lines to the intake manifold. Apply a few drops of
clean engine oil to the male ends of
the fuel rail inlet and outlet tubes
prior to assembly. When installed,
pull outward to ensure a good
connection.
24. Install the removed ground wire
to the rear cylinder head.
25. Install the brake booster vacuum hose to the intake manifold vacuum fitting.

26. Install the EGR pipe and the
crankcase ventilation pipe at the
throttle body spacer. Torque the EGR
pipe bolt to 21 ft. lbs. (28 Nm).
27. Install the remaining
components.
28. Connect the negative battery
cable.
29. Turn the ignition switch to
RUN and inspect for any fuel leaks.
If there are no leaks, install the engine dress-up cover and torque the
cover nuts to 89 inch lbs. (10 Nm).
30. Fill and bleed the cooling
system.
31. Road test the vehicle for proper
operation.

4.9L Engine

CAUTION

*Fuel injection systems remain
under pressure, even after the engine has been turned OFF. The
fuel system pressure must be relieved before disconnecting any
fuel lines. Failure to do so may result in fire and/or personal
injury.*

1. Disconnect the negative battery
cable.
2. Relieve the fuel system
pressure.
3. Remove the air cleaner
assembly.
4. Drain the engine coolant into a
suitable container.
5. Remove the cross car brace.
6. Remove the drive belt.
7. Remove the power steering and
tensioner bracket assembly and position the pump toward the front of the
engine.
8. Remove the alternator and
bracket.
9. Disconnect the cable at the
throttle body.
10. Label and disconnect the wiring
at the distributor, oil pressure
switch, coolant sensor, EGR solenoid,
Idle Speed Control (ISC) motor,
Throttle Position Sensor (TPS), fuel
injectors and Manifold Air Temperature (MAT) sensor and Manifold Absolute Pressure (MAP) sensor hoses.
11. Disconnect the upper radiator
and heater hoses at the thermostat
housing.
12. Remove the A/C hose bracket.
13. Remove the distributor.
14. Disconnect the fuel lines at the
transaxle bracket.
15. Label and disconnect the vacuum and fuel lines at the throttle
body.

16. Loosen the vacuum line clip at
the lift bracket.
17. Remove the vacuum line
bracket at the left rear engine lift
bracket, and disconnect the vacuum
supply line at the throttle body.
18. Disconnect the transaxle modulator vacuum line.
19. Remove the EGR solenoid and
bracket assembly.
20. Remove the rocker arm covers.

NOTE: Label all valve train
components as they are removed,
so they can be reinstalled in their
original locations.

21. Remove the rocker support retaining nuts/bolts and remove the
rocker arm support.

WARNING

The rocker arm support must be
removed with the rocker arms
and pivots attached. The pivot
assemblies or support bar
threads may be damaged if pivot
bolt torque is removed against
valve spring pressure. If necessary, secure the support in a vise
and remove the rocker arms and
pivots.

22. Remove the pushrods.
23. Remove the power steering
bracket.
24. Remove the right front engine
lift bracket.
25. Remove the intake manifold
bolts and remove the intake
manifold.
26. Remove and discard the gaskets
and seals. Clean all gasket mating
surfaces.

To install:

27. Install new intake manifold end
seals.
28. Apply RTV sealant to the four
corners where the end seals meet the
side gaskets.
29. Install new intake manifold
gaskets.
30. Install the intake manifold
bolts and tighten, in sequence.
31. Install the pushrods, making
sure they are seated in the lifters.
32. Install the rocker arms and pivots to the rocker arm support.
Loosely install the pivot bolts, then
tighten the pivot bolts to 22 ft. lbs.
(30 Nm).
33. Install the rocker arm support,
with the rocker arms and pivots installed, over the 5 stud headed head
bolts. Make sure each pushrod is
seated in a rocker arm.
34. Loosely install the rocker arm
support retaining nuts and bolts.

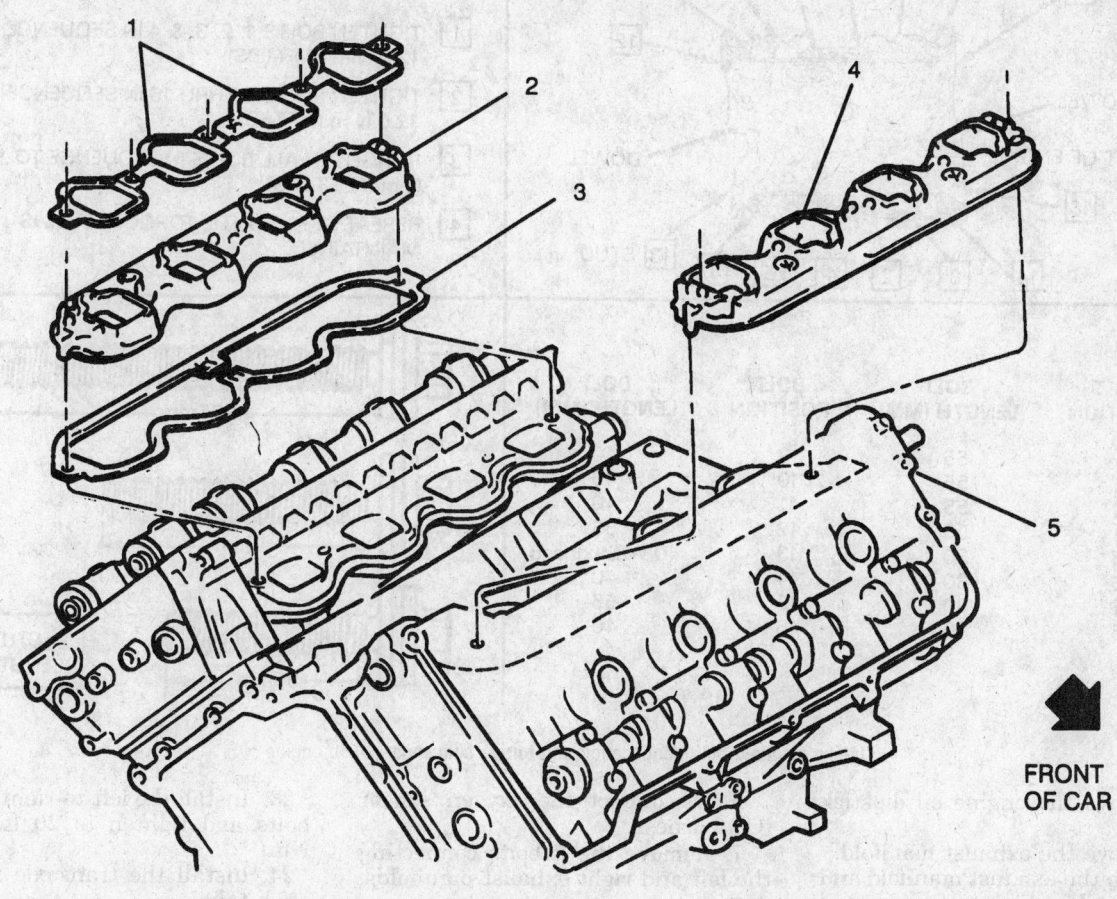

FRONT
OF CAR

1	UPPER INTAKE MANIFOLD SEAL	4	SPACER WITH SEALS ATTACHED
2	SPACER	5	ENGINE
3	LOWER INTAKE MANIFOLD SEAL		

241438

Intake manifold seals and spacers — 4.6L engine

35. Tighten the 5 rocker arm support retaining nuts alternately and evenly until snug.

NOTE: Check the pushrods from time to time during the tightening procedure to make sure they are correctly positioned.

36. Tighten the 4 rocker arm retaining bolts until snug.
37. Tighten the 5 rocker arm support retaining nuts to 37 ft. lbs. (50 Nm).
38. Tighten the 4 rocker arm retaining bolts to 7 ft. lbs. (10 Nm).
39. Install the rocker arm covers.

40. Install the remaining components.
41. Install the air cleaner.
42. Refill the cooling system.
43. Connect the negative battery cable.
44. Start the engine and set the ignition timing. Run the engine and check for leaks and proper engine operation.

Exhaust Manifold

REMOVAL AND INSTALLATION

3.8L Engine

Left Side (Front)

1. Disconnect the negative battery cable.
2. Remove the bolts attaching the left exhaust manifold to the right manifold.
3. Remove the spark plug wires from the plugs and position out of the way.
4. Remove the exhaust manifold bolts.

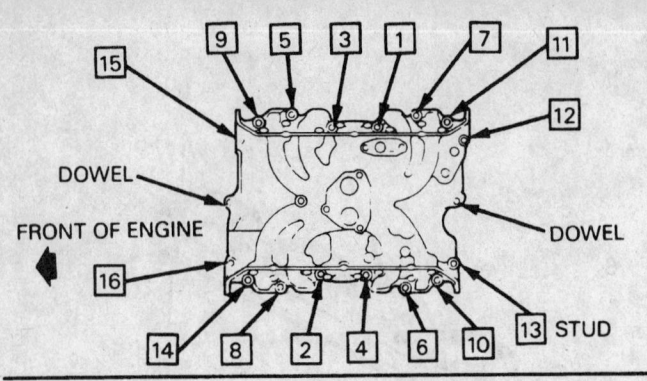

BOLT TIGHTENING SEQUENCE

| 1 | TIGHTEN BOLTS 1, 2, 3, & 4 IN SEQUENCE TO 12.0 N·m (8 FT-LBS). |

| 2 | TIGHTEN BOLTS 5 THRU 16 IN SEQUENCE TO 12.0 N·m (8 FT-LBS). |

| 3 | RETIGHTEN ALL BOLTS IN SEQUENCE TO 16.0 N·m (12 FT-LBS). |

| 4 | REPEAT STEP 3 UNTIL TORQUE LEVEL IS MAINTAINED. |

BOLT POSITION	BOLT LENGTH (MM)	BOLT POSITION	BOLT LENGTH (MM)
1	55	9	40
2	55	10	40
3	55	11	40
4	55	12	55
5	30	13	40 W/Studhead
6	30	14	40
7	30	15	55
8	30	16	40

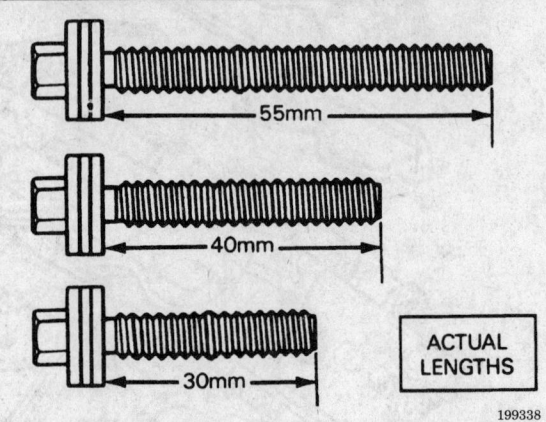

ACTUAL LENGTHS

199338

Intake manifold bolt identification and torque sequence — 4.9L engine

5. Remove the engine oil dipstick tube.

6. Remove the exhaust manifold.

7. Clean the exhaust manifold and cylinder head mating surfaces.

To install:

8. Install the exhaust manifold and loosely install the mounting bolts.

9. Install the engine oil dipstick tube.

10. Tighten the exhaust manifold mounting bolts to 41 ft. lbs. (55 Nm).

11. Connect the plug wires to the spark plugs.

12. Install the bolts connecting the right and left exhaust manifolds and tighten to 20 ft. lbs. (27 Nm).

13. Connect the negative battery cable.

14. Run the engine and check for exhaust leaks.

Right Side Manifold (Rear)

1. Disconnect the negative battery cable.

2. Label and disconnect the spark plug wires from the plugs and position out of the way.

3. Remove the throttle cable bracket.

4. Remove the crossover pipe heat shield.

5. Remove the transaxle fluid dipstick tube.

6. Disconnect the oxygen sensor (O₂) harness.

7. Remove the 2 bolts connecting the left and right exhaust manifolds.

8. Remove the plastic vacuum tank mounted on the cowl.

9. Raise and safely support the vehicle.

10. Remove the converter heat shield and pipe hanger.

11. Disconnect the exhaust pipe from the manifold.

12. Lower the vehicle.

13. Remove the rear engine lift bracket.

14. Remove the exhaust manifold bolts and remove the manifold.

15. Clean the mating surfaces.

To install:

16. Install the exhaust manifold and tighten the mounting bolts to 41 ft. lbs. (55 Nm).

17. Install the rear engine lift bracket.

18. Raise and safely support the vehicle.

19. Connect the front pipe to the manifold and tighten the mounting bolts to 18 ft. lbs. (25 Nm).

20. Install the exhaust hanger and converter heat shield.

21. Lower the vehicle.

22. Install the vacuum tank on the cowl.

23. Install the left-to-right manifold bolts and tighten to 20 ft. lbs. (27 Nm).

24. Install the transaxle fluid dipstick tube.

25. Connect the oxygen sensor (O₂) harness.

26. Install the throttle cable bracket.

27. Connect the plug wires to the spark plugs.

28. Connect the negative battery cable.

29. Run the engine and check for exhaust leaks.

4.6L Engine

Left Side Exhaust Manifold

1. Disconnect the negative battery cable.

2. Remove the radiator cover panel.

3. Remove the air cleaner assembly.

4. Disconnect the left and right engine torque struts and position out of the way.

5. Remove the engine cooling fans.

6. Support the engine using engine support fixture J-28467-A or equivalent.

7. Raise and safely support the vehicle.

8. Remove the nuts securing the engine mount to the engine cradle.

9. Remove the bolts securing the engine mount bracket to the crankcase.

10. Remove the bolts securing the engine mount bracket to the cylinder head.

11. Remove the nuts securing the engine mount to the mount bracket.

12. Disconnect the Y-pipe from the front of the catalytic converter and position the converter out of the way.

13. Lower the vehicle.

14. Raise the engine by adjusting the engine support fixture.

15. Remove the engine mount and bracket.

16. Raise and safely support the vehicle.

17. Remove the rear alternator bracket.

18. Remove the bolts at the manifold outlet flange.

19. Disconnect the oxygen sensor.

20. Remove the exhaust manifold from the cylinder head, then remove the manifold from the vehicle.

21. Remove the gasket. Thoroughly clean the cylinder head and exhaust manifold mating surfaces.

To install:

22. Position the exhaust manifold by inserting the outlet pipe partially into the exhaust crossover pipe and moving the manifold into position.

23. Install a new gasket to the manifold. Insert 2 bolts to hold the gasket in place.

24. Install the remainder of the bolts and tighten all the bolts to 18 ft. lbs. (25 Nm).

25. Coat the oxygen sensor threads with Hi temperature anti-seize compound and install. Tighten the sensor nut to 30 ft. lbs. (40 Nm). Connect the oxygen sensor harness.

26. Install the rear alternator bracket. Tighten the crankcase bolts to 40 ft. lbs. (60 Nm) and the alternator bolts to 25 ft. lbs. (30 Nm).

27. Install 2 new bolts at the manifold outlet flange and tighten to 25 ft. lbs. (30 Nm).

28. Position the engine mount and bracket.

29. Loosely install the 2 nuts securing the mount to the bracket.

30. Lower the vehicle.

31. Loosely install the 2 bolts securing the mount bracket at the cylinder head.

32. Lower the engine into position guiding the engine mount studs in the cradle holes.

33. Raise and safely support the vehicle.

34. Loosely install the 2 nuts to the bottom of the engine mount.

35. Loosely install the 2 bolts securing the mount bracket to the crankcase.

36. Tighten the 2 nuts securing the mount to the bracket, the 2 nuts at the bottom of the engine mount, and the 2 bolts securing the mount bracket to the crankcase to 22 ft. lbs. (30 Nm).

37. Install the 2 bolts at the converter to exhaust Y-pipe and tighten to 20 ft. lbs. (25 Nm).

38. Lower the vehicle.

39. Remove the engine support fixture.

40. Tighten the 2 bolts securing the mount bracket to the cylinder head to 22 ft. lbs. (30 Nm).

41. Install the engine cooling fans.

42. Install the air cleaner assembly.

43. Connect the left and right engine torque struts. Be sure to tighten the bolts that attach the struts to the core support bracket last, and tighten them to 44 ft. lbs. (60 Nm).

44. Install the radiator trim panel.

45. Connect the negative battery cable.

46. Run the engine and check for exhaust leaks.

Right Side Exhaust Manifold

1. Disconnect the negative battery cable.

2. Disconnect the oxygen sensor at the rear of the right cam cover. Disconnect the harness clip.

3. Raise and safely support the vehicle.

4. Disconnect the Y-pipe from the front of the catalytic converter.

5. Disconnect the suspension position sensor at the lower control arm from both sides.

6. Disconnect the intermediate shaft from the steering gear.

7. Place a support below the rear crossmember of the engine cradle and remove the 4 cradle to body bolts.

8. Lower the rear of the engine cradle and disconnect the Y-pipe from the exhaust crossover and from the manifold.

9. Remove the manifold nuts and manifold.

10. Remove the gasket from the manifold.

11. Thoroughly clean all cylinder head and exhaust manifold contact surfaces.

To install:

12. Coat the oxygen sensor threads with hi-temperature anti-seize compound. Tighten the sensor to 30 ft. lbs. (40 Nm).

13. Install the gasket, manifold and nuts. Tighten the nuts to 25 ft. lbs. (30 Nm).

14. Install the exhaust Y-pipe and install 4 new bolts. Tighten the M10 bolts to 35 ft. lbs. (50 Nm) and the M8 bolts to 25 ft. lbs. (30 Nm).

15. Raise the engine cradle into position and tighten the bolts to 75 ft. lbs. (100 Nm).

16. Connect the intermediate shaft to the steering gear and tighten to 35 ft. lbs. (50 Nm).

17. Connect the exhaust Y-pipe to the catalytic converter and tighten the bolts to 35 ft. lbs. (50 Nm).

18. Connect the suspension position sensors to the lower control arms.

19. Lower the vehicle and connect the oxygen sensor. Install the harness retainer.

20. Connect the negative battery cable.

21. Run the engine and check for exhaust leaks.

4.9L Engine

Right Side Manifold

1. Disconnect the negative battery cable.

2. Remove the air cleaner assembly.

3. Remove the 2 heat shield screws.

4. Raise and safely support the vehicle.

5. Disconnect the exhaust Y-pipe from the right side manifold.

6. Disconnect the engine mount brace from the right side of the exhaust manifold.

7. Disconnect the oxygen sensor wire.

8. Remove the heat shield.

9. Support the engine cradle with screw jacks and remove the rear cradle bolts on both sides. Loosen the front cradle bolts to act as a pivot.

10. Slightly lower the engine cradle. Be careful not to lower the cradle more than 2 inches, as the rack and pinion intermediate shaft may become disconnected.

11. Remove the exhaust manifold bolts and manifold.

To install:

12. Thoroughly clean all mounting surfaces.

13. Install the exhaust manifold and torque the bolts to 16 ft. lbs. (20 Nm).

14. Raise the cradle back into position, install the frame bolts and torque to 75 ft. lbs. (100 Nm).

15. Connect the oxygen sensor wire.

16. Install the heat shield and screws.

17. Connect the engine mount brace to the front of the exhaust manifold.

18. Connect the exhaust Y-pipe to the right side manifold.

19. Lower the vehicle.

20. Install the air cleaner assembly.

21. Connect the negative battery cable.

22. Run the engine and check for exhaust leaks.

Left Side Manifold

1. Disconnect the negative battery cable.

2. Remove the air cleaner assembly.

3. Disconnect the oxygen sensor wire.

4. Remove the power steering and tensioner bracket from the front of the manifold.

5. Remove the A/C hose bracket.

6. Remove the cooling fans.

7. Label and disconnect the spark plug wires from the left side spark plugs.

8. Raise and safely support the vehicle.

9. Disconnect the Y-pipe from the left exhaust manifold.

10. Remove the engine and A/C brace from the front of the manifold.

11. Remove the exhaust manifold bolts and remove the manifold.

To install:

12. Thoroughly clean all mounting surfaces.

13. Install the exhaust manifold and tighten the bolts to 16 ft. lbs. (20 Nm).

14. Install the engine and A/C brace to the front of the manifold.

15. Connect the Y-pipe to the left side exhaust manifold.

16. Lower the vehicle.

17. Connect the spark plug wires to the plugs.

18. Install the cooling fans.

19. Install the A/C hose bracket.

20. Install the power steering and tensioner bracket.

21. Connect the oxygen sensor wire.

22. Install the air cleaner assembly.

23. Connect the negative battery cable.

24. Run the engine and check for exhaust leaks.

Front Cover Seal

REMOVAL AND INSTALLATION

3.8L Engine

1. Disconnect the negative battery cable.

2. Remove the serpentine belt.

3. Raise and safely support the vehicle.

4. Remove the right front wheel.

5. Remove the right inner fender well splash shield.

6. Remove the crankshaft balancer bolt and washer.

7. Remove the crankshaft balancer using J-38197 or an equivalent puller.

8. Remove the seal from the front cover using a suitable prying tool. Be careful not to damage the crankshaft sealing surface or the front cover seal bore.

To install:

9. Install the seal in the front cover using seal installation tool J-35354 or the equivalent.

10. Coat the seal contact area on the balancer with clean engine oil and install the balancer onto the crankshaft.

11. Using tool J-38197 or equivalent, seat the balancer on the crankshaft.

12. Install the washer and mounting bolt and tighten to 105 ft. lbs. (140 Nm) plus an additional 76 degree turn.

13. Install the right inner fender well splash shield.

14. Install the right front wheel.

15. Lower the vehicle.

16. Install the serpentine belt.

17. Connect the negative battery cable.

18. Check the engine oil level; add oil as necessary.

19. Run the engine and check for leaks.

4.6L Engines

1. Disconnect the negative battery cable.

2. Remove the accessory drive belt.

3. Raise and safely support the vehicle.

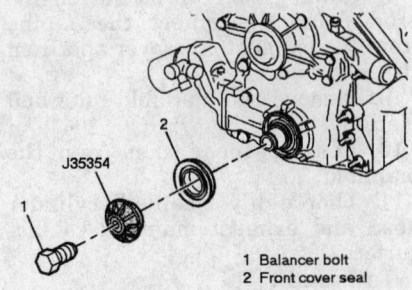

1 Balancer bolt
2 Front cover seal

322046

Front cover oil seal and installation tool — 3.8L engine

4. Remove the right front wheel.

5. Remove the 2 splash shields from the wheel well.

6. Remove the flywheel cover and install a suitable flywheel holder, to keep the engine from turning.

7. Remove the crankshaft balancer bolt.

8. Support the engine cradle with a screw type jack and remove the 3 right side engine cradle bolts.

9. Disconnect the RSS sensor from the right control arm.

10. Lower the cradle to gain access for the crankshaft balancer puller.

11. Using puller J-38416 or the equivalent, remove the crankshaft balancer.

12. Using a suitable prying tool, remove the seal from the front cover. Use care not to damage the crankshaft sealing surface or the front cover seal bore.

To install:

13. Lubricate the lip of the new oil seal with clean engine oil.

14. Install the the seal using tools J-38818 and J-39444 or their equivalents.

15. Install the balancer using tool J-39344 or the equivalent.

16. Apply engine oil to the balancer bolt threads and tighten the bolt to 44 ft. lbs. (60 Nm) plus an additional 120 degrees.

17. Raise the screw jack until the three cradle bolts can be installed. Tighten the bolts to 75 ft. lbs.

18. Connect the RSS sensor.

19. Remove the flywheel holding tool and install the flywheel cover.

20. Install the wheel well splash shields.

21. Lower the vehicle.

22. Install the drive belt.

23. Connect the negative battery cable.

24. Check the engine oil level; add oil as necessary.

25. Run the engine and check for leaks.

4.9L Engine

1. Disconnect the negative battery cable.

2. Remove the drive belt.

3. Raise and safely support the vehicle.

4. Remove the right front wheel.

5. Remove the right front air deflector.

6. Remove the crankshaft damper-to-crankshaft bolt.

NOTE: It may be necessary to lower the engine cradle on the right side to remove the damper.

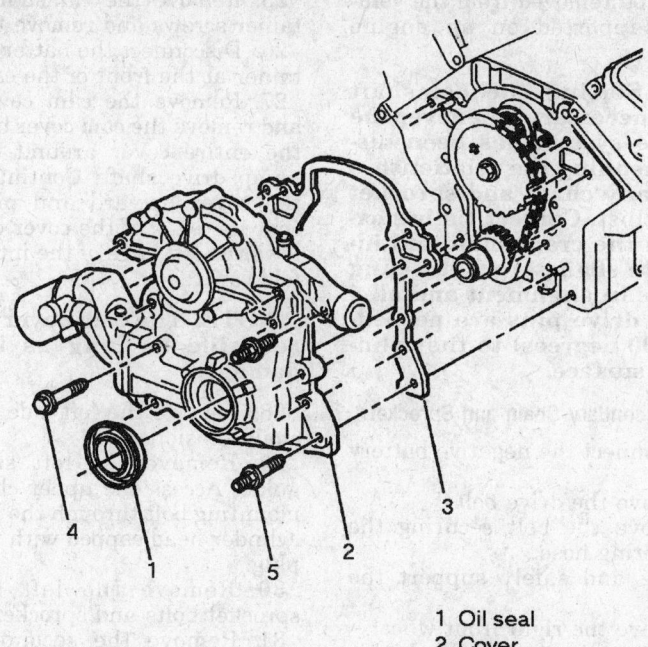

1 Oil seal
2 Cover
3 Gasket
4 Bolt
5 Stud

322045

Front cover and seal — 4.6L engine

7. Using J-24420-B crankshaft damper puller or equivalent, remove the crankshaft damper.

8. Using tools J-1859-03 and J-23129 or their equivalents, remove the seal from the front cover.

To install:

9. Lubricate the new seal with clean engine oil and install with the garter spring towards the engine.

10. Install the seal in the bore using tool J-29662 and a hammer. Make sure the seal is fully installed in the bore.

11. Install the crankshaft damper using installer tool J-29774 or equivalent.

12. Install the damper-to-crankshaft bolt and tighten to 70 ft. lbs. (95 Nm) on 1993–94 vehicles or 118 ft. lbs. (160 Nm) on 1995 and later vehicles.

13. Install the right front air deflector.

14. Install the wheel.

15. Lower the vehicle.

16. Install the drive belt.

17. Connect the negative battery cable.

18. Check the engine oil level.

19. Run the engine and check for leaks.

Timing Chain, Sprockets and Front Cover

REMOVAL AND INSTALLATION

3.8L Engine

1. Disconnect the negative battery cable.

2. Drain the engine coolant.

3. Disconnect the coolant hoses from the timing chain front cover.

4. Support the engine using support fixture J-28467-A or equivalent, and remove the engine mount.

5. Remove the drive belts.

6. Raise and safely support the vehicle.

7. Remove the right front wheel. Remove the right inner fender access panel.

8. Disconnect the harnesses at the camshaft and crankshaft position sensors and oil pressure sensor.

9. Keep the flywheel from turning using holder tool J-37096 or equivalent. Remove the crankshaft balancer retaining bolt, then use puller tool J-38197 or equivalent, to remove the balancer from the crankshaft.

10. Remove the crankshaft position sensor shield and the crankshaft position sensor.

11. Remove the oil pan-to-front cover bolts.

12. Remove the front cover attaching bolts and remove the timing chain front cover.

13. Align the timing marks on the camshaft and crankshaft sprockets so they are as close together as possible.

14. Remove the timing chain damper.

15. Remove the camshaft sprocket retaining bolt, then remove the camshaft sprocket and timing chain.

16. Remove the crankshaft sprocket.

NOTE: Do not rotate the camshaft or crankshaft while the timing chain and sprockets are removed.

17. Thoroughly clean all gasket mating surfaces.

To install:

18. Assemble the timing chain and sprockets with the timing marks aligned. Install the timing chain and sprockets to the camshaft and crankshaft.

19. Install the camshaft sprocket bolt and tighten to 74 ft. lbs. (100 Nm) plus an additional 90 degree turn. Recheck the camshaft and crankshaft sprocket timing marks to make sure they are still aligned.

20. Install the timing chain damper and tighten the bolt to 14 ft. lbs. (19 Nm).

— **WARNING** —
The oil pump is built into the front cover. When the cover is removed, oil drains from the pump. Since the pump "looses its prime" it may not establish oil pressure as soon as the engine starts. Therefore, it is important to pack the space around the oil pump gears completely full of petroleum jelly. If this is not done, the oil pump may not pump engine oil when the engine is started, resulting in severe engine damage.

21. Remove the screws and the oil pump cover from the back of the timing chain front cover. Pack the space around the oil pump gears completely full of petroleum jelly. There must be no air space left inside the pump. Reinstall the pump cover and tighten the screws to 97 inch lbs. (11 Nm).

22. Using new gaskets, install the timing chain front cover to the block. Tighten the front cover-to-engine block bolts to 22 ft. lbs. (30 Nm) and the oil pan-to-front cover bolts to 125 inch lbs. (14 Nm).

23. Install the crankshaft position sensor and tighten the bolts to 14–28

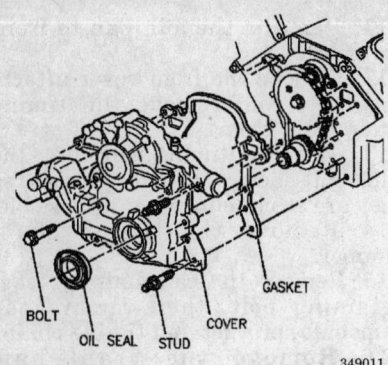

Timing chain front cover — 3.8L engine

349011

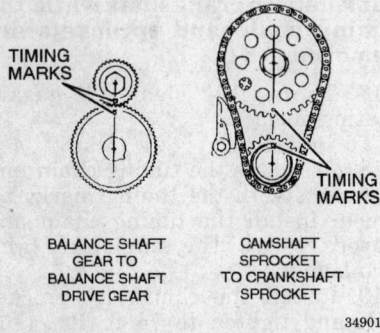

349018

Timing chain sprocket and balance shaft gear alignment — 3.8L engine

ft. lbs. (20–40 Nm). Install the crankshaft position sensor shield.

24. Keep the flywheel from turning using holder tool J-37096 or equivalent. Install the crankshaft balancer and tighten the bolt to 111 ft. lbs. (150 Nm) plus an additional 76 degree turn.

25. Connect the harnesses to the camshaft position sensor, crankshaft position sensor and oil pressure sensor.

26. Install the right inner fender access panel and the right front wheel.

27. Lower the vehicle.

28. Install the drive belt(s).

29. Install the engine mount and remove the engine support fixture.

30. Connect the coolant hoses.

31. Connect the negative battery cable.

32. Fill and bleed the cooling system.

33. Start the vehicle and check for leaks and proper engine operation.

4.6L Engines

The left and right side secondary timing chains can be removed with the engine in the vehicle. If the primary timing chain or intermediate shaft sprocket need to be replaced, the en-gine must be removed from the vehicle and supported on an engine stand.

NOTE: Setting the camshaft timing is necessary whenever the cam drive system has been disturbed, meaning the relationship between any chain and sprocket has been lost. Correct timing exists when the crankshaft and intermediate shaft sprocket timing marks are in alignment and all 4 camshaft drive pins are perpendicular (90 degrees) to the cylinder head surface.

Left Side Secondary Chain and Sprocket

1. Disconnect the negative battery cable.

2. Remove the drive belt.

3. Remove the bolt securing the power steering hose.

4. Raise and safely support the vehicle.

5. Remove the right front wheel.

6. Remove the 2 splash shields from the wheel well.

7. Remove the flywheel cover and install a suitable flywheel holder.

8. Remove the crankshaft balancer bolt.

9. Support the engine cradle with a screw type jack and remove the 3 right side engine cradle bolts.

10. Disconnect the RSS sensor from the right control arm.

11. Lower the cradle to gain access for the crankshaft damper puller.

12. Using puller J-38416 or equivalent, remove the crankshaft damper.

13. Remove the drive belt tensioner.

14. Remove the drive belt idler pulley.

15. Remove the front cover bolts.

16. Remove the front cover and gasket.

NOTE: The front cover gasket is reusable as long as it is not damaged.

17. Partially drain the coolant from the radiator.

18. Disconnect the upper radiator hose at the water crossover.

19. Label and disconnect the spark plug wires.

20. Remove the right side fan.

21. Disconnect the battery cable at the alternator and disconnect the cable harness at the cam cover.

22. Disconnect the PCV fresh air tube from the cam cover.

23. Remove the right and left torque struts.

24. Remove the water pump pulley with tool J-38825 or equivalent.

25. Remove the camshaft seal retainer screws and remove the seal.

26. Disconnect the battery cable retainer at the front of the cam cover.

27. Remove the cam cover screws and remove the cam cover by pivoting the entire cover around the water pump drive shaft. Continue moving the cover upward and pivoting so that the edge of the cover closely follows the left edge of the intake manifold cover.

NOTE: The cam cover gasket is reusable as long as it is not damaged.

28. Remove the left side secondary chain tensioner.

29. Remove the left side chain guide. Access the upper chain guide mounting bolt through the hole in the cylinder head capped with the plastic plug.

30. Remove the left side cam sprocket bolts and sprockets.

31. Remove the secondary drive chain.

Right Side Secondary Chain and Sprocket

1. Raise and safely support the vehicle.

2. Disconnect the exhaust Y-pipe at the converter.

3. Lower the vehicle.

4. Remove the tower-to-tower brace.

5. Disconnect the ICM wiring connectors and mounting bolts.

6. Remove the ICM and the plug wires on the right bank.

7. Disconnect the PCV valve.

8. Remove the purge canister solenoid from the rear of the cover.

9. Remove the cam cover screws.

10. Safely support the front of the engine cradle and remove the 2 mounting bolts at the front of the cradle.

11. Remove the right and left torque struts.

12. Lower the engine cradle (or raise the vehicle) to provide clearance at the rear of the engine compartment.

13. Remove the cam cover.

NOTE: The cam cover gasket is reusable as long as it is not damaged.

14. Remove the right side secondary chain tensioner.

15. Remove the right side chain guide. Access the upper chain guide mounting bolt through the hole in the cylinder head capped with the plastic plug.

16. Remove the right side cam sprocket bolts and cam sprockets.

17. Remove the secondary drive chain.

Primary Chain and Sprocket

1. Remove the intermediate shaft sprocket-to-intermediate shaft bolt, then remove the sprocket.

2. Slide the primary timing sprockets and primary chain off the engine.

To install:

NOTE: The following procedure must be followed to set the camshaft timing on the vehicle.

3. Install the primary and secondary chain guide.

4. Rotate the crankshaft until the sprocket drive key is at the 1 o'clock position. Use tool J-39946 or the equivalent, to rotate the crankshaft.

5. Install the crankshaft sprocket and intermediate shaft sprocket in the primary timing chain so the timing marks are aligned. Install the assembly in position on the engine. The crankshaft sprocket key way will have to slide over the key on the crankshaft. If it is necessary to turn the crankshaft sprocket, the intermediate shaft sprocket will also have to be turned so the timing mark still lines up with the crankshaft sprocket.

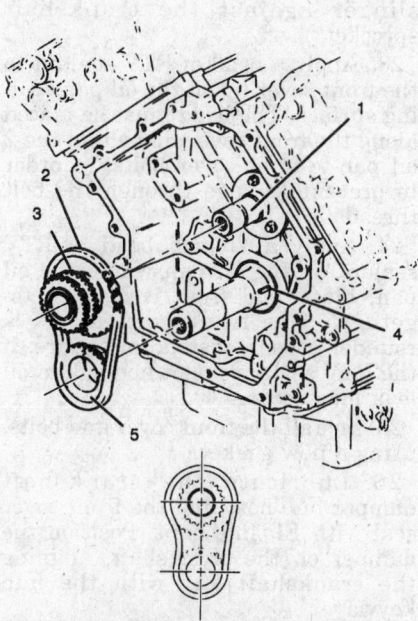

1 INTERMEDIATE SHAFT
2 PRIMARY CHAIN
3 INTERMEDIATE SHAFT SPROCKET
4 CRANKSHAFT SPROCKET KEY
5 SPROCKET

242506

Primary drive chain components — 4.6L engine

6. Install the intermediate shaft sprocket-to-intermediate shaft bolt and tighten to 45 ft. lbs. (61 Nm).

7. Install the primary timing chain tensioner. Tighten the tensioner mounting bolts to 20 ft. lbs. (27 Nm).

8. Install a suitable flywheel holder to lock the crankshaft in position.

9. Install the secondary timing chain over the inner row of teeth on the intermediate shaft sprocket. Route the chain over the chain guide and install the exhaust cam sprocket so the **RE** (Right Head Exhaust) pin engages the sprocket notch. There should be no slack in the lower section of the timing chain and the cam drive pin **must** be perpendicular to the cylinder head face.

10. Install the intake cam sprocket into the chain so the sprocket notch **RI** (Right Head Intake) engages the cam and the camshaft drive pin remains perpendicular to the cylinder head face. A hex is cast into the camshafts behind the lobes for cylinder No. 1, so an open end wrench may be used to provide minor repositioning of the cams.

11. Loosely install the exhaust cam sprocket bolt and intake sprocket bolt.

12. Install the chain tensioner and tighten the mounting bolts to 20 ft. lbs. (27 Nm).

13. Tighten the camshaft sprocket bolts to 90 ft. lbs. (120 Nm).

14. Route the secondary timing chain for the left side over the outer row of intermediate sprocket teeth.

15. Install the secondary timing chain over the inner row of teeth on the intermediate shaft sprocket. Route the chain over the chain guide and install the exhaust cam sprocket so the **LE** (Left Head Exhaust) pin engages the sprocket notch. There should be no slack in the lower section of the timing chain and the cam

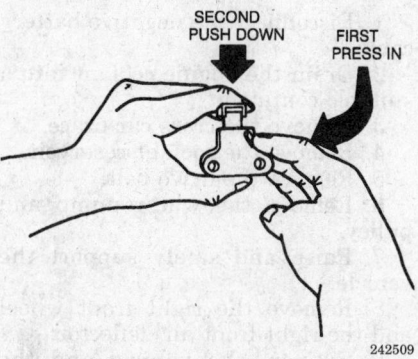

SECOND PUSH DOWN   FIRST PRESS IN

242509

Rotating tensioner release lever — 4.6L engine

drive pin **must** be perpendicular to the cylinder head face.

16. Install the intake cam sprocket into the chain so the sprocket notch **LI** (Left Head Intake) engages the cam and the camshaft drive pin remains perpendicular to the cylinder head face. A hex is cast into the camshafts behind the lobes for cylinder No. 2, so an open end wrench may be used to provide minor repositioning of the cams.

17. Loosely install the exhaust cam sprocket bolt and intake sprocket bolt.

18. Install the chain tensioner and tighten the mounting bolts to 20 ft. lbs. (27 Nm).

19. Tighten the camshaft sprocket bolts to 90 ft. lbs. (120 Nm).

NOTE: The RE cam sprocket must contain the cam position sensor pick-up.

20. Install the front cover gasket on the dowel pins on the block.

21. Install the front cover on the dowel pins and install the front cover mounting bolts. Tighten the bolts to 89 inch lbs. (10 Nm). Apply a dab of RTV to the split line between the upper and lower crankcase assemblies.

22. Install the drive belt idler pulley and tighten the mounting bolt to 35 ft. lbs. (47 Nm).

23. Install the drive belt tensioner and tighten the tensioner mounting nut to 35 ft. lbs. (47 Nm).

24. Install the crankshaft balancer using tool J-39344 or the equivalent.

25. Apply engine oil to the balancer bolt threads and tighten the bolt to 44 ft. lbs (60 Nm) plus an additional 120 degree turn.

26. Raise the screw jack until the three cradle bolts can be installed. Tighten the bolts to 75 ft. lbs. (102 Nm).

27. Connect the RSS sensor.

28. Remove the flywheel holding tool and install the flywheel cover.

29. Install the wheel well splash shields.

Left Side Cam Cover

1. Install the spark plugs and cam cover seals.

2. Insert the intake cam through the hole in the cam cover and using fingers, guide the cam cover up over the edge of the cylinder head.

— **WARNING** —
Use care to prevent the exposed section of the cam cover seal from being damaged by the edge of the cylinder head casting.

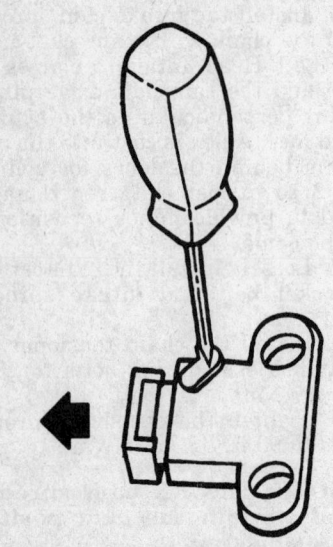

1 RELEASE TO FIRST CLICK
2 INSTALL LOCK PIN

242511

Locking the tensioner in the collapsed position — 4.6L engine

3. Work the cover into position by allowing the top edge of the cover to follow the left side edge of the intake manifold.

4. Install the cam cover screws and tighten to 7 ft. lbs. (10 Nm).

5. Connect the battery cable retainer to the front of the cam cover.

6. Connect the battery cable at the alternator.

7. Lubricate the seal lips and install the camshaft seal to the end of the intake cam. Seal the screw threads with sealer.

8. Install the water pump pulley with tool J-38825 or equivalent.

9. Connect the PCV fresh air tube to the cam cover.

10. Install the right side fan.

11. Connect the spark plug wires.

12. Connect the upper radiator hose to the water crossover.

13. Refill the cooling system.

Right Side Cam Cover

1. Install spark plug and cam cover seals.

2. Install the cam cover and tighten the screws to 7 ft. lbs. (10 Nm).

3. Raise the engine cradle into position and install and tighten the 2 mounting bolts to 75 ft. lbs. (100 Nm).

4. Install the right and left torque struts and torque the retaining bolts as follows:

NOTE: It is important during installation that the engine torque struts are not preloaded in their installed position. Adjustment is provided at the point the strut fastens to the core support bracket. Make sure this bolt is loose during assembly. Tighten to 45 ft. lbs. (60 Nm) as the final step of assembly.

 a. Torque strut bracket to cylinder head (M10) bolt: 35 ft. lbs. (50 Nm).

 b. Torque strut bracket to cylinder head (M10) stud: 35 ft. lbs. (50 Nm).

 c. Torque strut bracket to water manifold (M8) bolts: 20 ft. lbs. (25 Nm).

 d. Torque strut bracket to cylinder head (M10) bolt: 35 ft. lbs. (50 Nm).

 e. Torque strut to core support bracket bolt: 45 ft. lbs. (60 Nm) (see note above).

5. Install the screws retaining wiring harness to the cover.

6. Install the purge canister solenoid to the rear of the cover.

7. Connect the PCV valve.

8. Install the ICM and the spark plug wires on the right bank.

9. Connect the ICM wiring connectors.

10. Install the tower-to-tower brace.

11. Raise and support the vehicle safely. Connect the exhaust Y-pipe to the converter and tighten the bolts to 20 ft. lbs. (25 Nm).

12. Connect the negative battery cable.

13. Start the engine. Check for leaks and inspect for proper operation.

4.9L Engine

1. Disconnect the negative battery cable.

2. Drain the engine coolant into a suitable container.

3. Remove the cross car brace.

4. Remove the coolant reservoir.

5. Remove the drive belt.

6. Remove the water pump and pulley.

7. Raise and safely support the vehicle.

8. Remove the right front wheel and the right front air deflector.

9. Support the vehicle and the right side of the engine cradle.

10. Remove the right side cradle body bolts and lower the right side of the cradle for balancer clearance.

11. Remove the bolt from the end of the crankshaft. Using J-24420-B crankshaft damper puller or equivalent, remove the crankshaft damper.

12. Remove the timing chain front cover bolts and cover.

13. Remove the oil slinger from the crankshaft.

14. Rotate the engine until the timing marks are in proper alignment.

15. Remove the thrust button and the camshaft sprocket mounting bolt. Discard the thrust button.

16. Remove the camshaft sprocket and timing chain together.

17. Remove the crankshaft sprocket.

 To install:

18. Thoroughly clean all gasket mating surfaces.

19. Install the camshaft and crankshaft sprockets and timing chain as an assembly. Make sure the timing marks are in alignment and the sprockets are properly seated on the crankshaft and camshaft.

20. Install the camshaft bolt and tighten to 37 ft. lbs. (50 Nm).

21. Install a new thrust button.

22. Install the crankshaft oil slinger against the crankshaft sprocket.

23. Apply a bead of RTV sealant to the front cover lip on the oil pan sealing surface. The bead must be placed along the front cover lip behind the 2 oil pan-to-front cover bolts, in order to prevent leakage through the bolt threads.

24. Apply a ¼ inch bead of RTV sealant on the oil pan where the oil pan, block and front cover join together. The front corner has a rounder corner, instead of square; if the RTV sealant is not applied an oil leak may occur.

25. Install the front cover and bolts, using a new gasket.

26. Lubricate the crankshaft damper hub bore and the front cover seal with EP lubricant. Position the damper on the crankshaft, aligning the crankshaft key with the hub keyway.

27. Install the crankshaft damper using installer tool J-29774 or equivalent, until the damper hub bottoms on the crankshaft. Remove the installer tool and attach the bolt. Torque the bolt to 70 ft. lbs. (95 Nm) on 1993–94 vehicles or 118 ft. lbs. (160 Nm) on 1995 vehicles.

28. Raise the right side of the cradle and install the cradle-to-body bolts. Tighten to 66 ft. lbs. (90 Nm).

29. Install the remaining components.

30. Connect the negative battery cable.

31. Refill and bleed the cooling system.

32. Start the engine and check for leaks and proper engine operation.

Camshaft

REMOVAL AND INSTALLATION

3.8L Engine

CAUTION

Fuel Injection systems remain under pressure, even after the engine has been turned OFF. The fuel system pressure must be relieved before disconnecting any fuel lines. Failure to do so may result in fire and/or personal injury.
---

1. Disconnect the negative battery cable.

2. Relieve the fuel system pressure.

3. Remove the engine from the vehicle and mount on a suitable engine stand.

4. Remove the intake manifold.

5. Remove the rocker arm covers.

6. Remove the rocker arm assemblies, pushrods and lifters. Identify all parts as they are removed, so they can be reinstalled in their original locations.

7. Remove the crankshaft balancer center bolt and, balancer from the crankshaft.

8. Remove the timing chain front cover.

9. Set the engine to Top Dead Center (TDC) No. 1 cylinder (firing position), and remove the camshaft sprocket and timing chain.

NOTE: Align the timing marks of the camshaft and crankshaft sprockets to avoid burring the camshaft journals by the crankshaft.

10. Remove the camshaft thrust plate bolts and remove the thrust plate.

11. Carefully remove the camshaft from the engine block.

12. Inspect the camshaft lobes and journals for wear and/or damage; replace as necessary.

NOTE: If the camshaft is to be replaced, the lifters must also be replaced. The old lifters have developed a wear pattern and will cause the new camshaft to wear prematurely.

To install:

13. Coat the camshaft lobes and bearings with lubricant GM part number 1052365 or equivalent camshaft break-in prelube prior to installation.

14. Carefully install the camshaft into the engine.

15. Install the camshaft thrust plate and tighten the mounting bolts to 10 ft. lbs. (14 Nm).

16. Install the camshaft sprocket and timing chain. Be sure the timing marks are aligned. Tighten the camshaft sprocket retaining bolt to 74 ft. lbs. (100 Nm) plus an additional 90 degree ($\frac{1}{4}$) turn.

17. Install the timing chain front cover.

18. Install the crankshaft balancer and tighten the mounting bolt to 111 ft. lbs. (150 Nm) plus an additional 76 degree turn.

19. Coat the valve lifters with camshaft prelube and install the lifters in the lifter bores.

20. Install the lifter guides and lifter guide retainer. Tighten the retainer mounting bolts to 22 ft. lbs. (30 Nm).

21. Install the pushrods and rocker arms and tighten the rocker arm bolts to 11 ft. lbs. (15 Nm) plus an additional 90 degree turn.

22. Install the rocker arm covers.

23. Install the intake manifold.

24. Install the engine in the vehicle.

25. Connect the negative battery cable.

26. Verify that all fluid levels (coolant, engine oil, etc.) are full and correct.

27. Start the engine and verify no leaks.

4.6L Engines

Left Side

1. Disconnect the negative battery cable.

2. Drain the coolant from the radiator.

3. Disconnect the upper radiator hose at the water crossover.

4. Label and disconnect the spark plug wires.

5. Remove the right side fan.

6. Disconnect the battery cable at the alternator. Disconnect the cable harness at the cam cover and move out of the way.

7. Disconnect the PCV fresh air tube from the cam cover.

8. Remove the right and left torque struts.

9. Disconnect the water pump drive belt and pulley.

10. Remove the camshaft seal retainer screws and remove the seal.

11. Remove the cam cover screws and remove the cam cover by moving the cam drive end of the cover up and then pivot the entire cover around the water pump drive shaft. Continue moving the cover upward and pivoting so that the edge of the cover closely follows the left edge of the intake manifold cover.

12. Discard the cam cover seal if it is damaged. The spark plug seals may be reused if undamaged.

13. Secure the cam sprocket to the timing chain by installing tie-wraps through the cam sprocket holes. Use 4 tie-wraps per sprocket.

NOTE: The sprocket/chain relationship must be maintained throughout this procedure or camshaft timing will be lost and require further engine disassembly to retime.

14. Working from behind the sprockets, install cam chain holder J-38222 so that it is positioned between the chain tensioner and chain guide. Apply tension to the tool by tightening the tension adjusting screw.

15. Remove both cam sprocket bolts. Note the relative location of the cam drive pins in the end of the camshafts.

16. Work the sprockets off the cams using play in the chain.

17. Alternately loosen the cam bearing cap screws a few turns at a time until all valve spring pressure has been released. Remove the bolts and caps.

18. Remove the camshaft.

19. Inspect the camshaft for excessive lobe wear. Check the bearing journals, making sure they are not scored or burned. Replace the camshaft, as necessary.

To install:

20. Lubricate the camshaft lobes with camshaft prelube 1052365 or equivalent. Lubricate the camshaft journals with clean engine oil.

21. Install the camshaft.

22. Position the cam bearing caps to the cylinder head.

NOTE: Each cap is identified for position and direction. The arrow points towards the front of the engine. An "E" indicates a cap for the exhaust cam. An "I" indicates a cap for the intake cam. Position No. 1 is towards the front of the engine.

23. Loosely install the cam bearing cap bolts.

24. Alternately tighten the cam bearing cap bolts a few turns at a time against valve spring pressure until all the bolts are snug. Tighten the bolts to 106 inch lbs. (12 Nm).

25. Using the hex cast into the camshaft, rotate the cams until the drive pins are in position to engage the cam sprockets over the cams, and install the retaining bolts.

26. Attach the cam sprockets and install the retaining bolts. Tighten the bolts to 90 ft. lbs. (120 Nm).

27. Remove the chain holder J-38222.

28. Remove the tie-wraps from the cam sprockets.

29. Install the remaining components.

30. Install the cam cover screws and tighten to 89 inch lbs. (10 Nm).

31. Connect the battery cable retainer to the front of the cam cover.

32. Connect the battery cable at the alternator.

33. Lubricate the seal lips and install the camshaft seal to the end of the intake cam. Seal the screw threads with sealer.

34. Install the water pump pulley with tool J-38825 or equivalent, the install the drive belt.

35. Install the right and left torque struts and torque the retaining bolts as follows:

NOTE: It is important during installation that the engine torque struts are not preloaded in their installed position. Adjustment is provided at the point the strut fastens to the core support bracket. Make sure this bolt is loose during assembly. Tighten to 45 ft. lbs. (60 Nm) as the final step of assembly.

a. Torque strut bracket to cylinder head (M10) bolt: 35 ft. lbs. (50 Nm).

b. Torque strut bracket to cylinder head (M10) stud: 35 ft. lbs. (50 Nm).

c. Torque strut bracket to water manifold (M8) bolts: 20 ft. lbs. (25 Nm).

d. Torque strut bracket to cylinder head (M10) bolt: 35 ft. lbs. (50 Nm).

e. Torque strut to core support bracket bolt: 45 ft. lbs. (60 Nm) (see note above).

36. Connect the PCV fresh air tube to the cam cover.

37. Install the right side fan.

38. Connect the spark plug wires.

39. Connect the upper radiator hose to the water crossover.

40. Fill the cooling system.

41. Connect the negative battery cable.

42. Run the engine and check for leaks and proper engine operation.

Right Side

1. Disconnect the negative battery cable.

2. Raise and support the vehicle safely.

3. Disconnect the exhaust Y-pipe at the converter. Position the converter out of the way.

4. Lower the vehicle.

5. Remove the tower-to-tower brace.

6. Disconnect the ICM wiring harnesses and mounting bolts.

7. Remove the ICM and the spark plug wires on the right bank. Tag wires for installation reference.

8. Disconnect the PCV valve.

9. Remove the purge canister solenoid from the rear of the cover.

10. Remove the 3 screws retaining the wiring harness to the cover.

11. Remove the cam cover screws.

12. Support the front of the engine cradle and remove the mounting screws at the front of the cradle.

13. Remove the right and left torque struts.

14. Lower the engine cradle (or raise the vehicle) to provide clearance at the rear of the engine compartment.

15. Remove the cam cover.

16. Discard the cover seal if damaged. The spark plug seals may be reused if undamaged.

17. Secure the cam sprocket to the timing chain by installing tie-wraps through the cam sprocket holes. Use 4 tie-wraps per sprocket.

NOTE: The sprocket/chain relationship must be maintained throughout this procedure or camshaft timing will be lost and require further engine disassembly to retime.

18. Working from behind the sprockets, install cam chain holder J-38222 so that it is positioned between the chain tensioner and chain guide. Apply tension to the tool by tightening the tension adjusting screw.

19. Remove both cam sprocket bolts. Note the relative location of the cam drive pins in the end of the camshafts.

20. Work the sprockets off the cams.

21. Alternately loosen the cam bearing cap screws a few turns at a time until all valve spring pressure has been released. Remove the bolts and caps.

22. Remove the camshaft.

23. Inspect the camshaft for excessive lobe wear. Check the bearing journals, making sure they are not scored or burned. Replace the camshaft, as necessary.

To install:

24. Lubricate the camshaft lobes with camshaft prelube 1052365 or equivalent. Lubricate the camshaft journals with clean engine oil.

25. Install the camshaft.

26. Position the cam bearing caps to the cylinder head.

NOTE: Each cap is identified for position and direction. The arrow points towards the front of the engine. An "E" indicates a cap for the exhaust cam. An "I" indicates a cap for the intake cam. Position No. 1 is towards the front of the engine.

27. Loosely install the cam bearing cap bolts.

28. Alternately tighten the cam bearing cap bolts a few turns at a time against valve spring pressure until all the bolts are snug. Tighten the bolts to 106 inch lbs. (12 Nm).

29. Using the hex cast into the camshaft, rotate the cams until the drive pins are in position to engage the cam sprockets over the cams and install the retaining bolts.

30. Attach the cam sprockets and install the retaining bolts. Tighten the bolts to 90 ft. lbs. (120 Nm).

31. Remove the chain holder J-38222.

32. Remove the tie-wraps from the cam sprockets.

33. Install spark plug and cam cover seals, as required.

34. Install the cam cover and tighten the screws to 7 ft. lbs. (10 Nm).

35. Raise the engine cradle into position. Install and tighten the 2 mounting bolts to 75 ft. lbs. (100 Nm).

36. Install the right and left torque struts and torque the retaining bolts as follows:

NOTE: It is important during installation that the engine torque struts are not preloaded in their installed position. Adjustment is provided at the point the strut fastens to the core support bracket. Make sure this bolt is loose during assembly. Tighten to 45 ft. lbs. (60 Nm) as the final step of assembly.

    a. Torque strut bracket to cylinder head (M10) bolt: 35 ft. lbs. (50 Nm).

    b. Torque strut bracket to cylinder head (M10) stud: 35 ft. lbs. (50 Nm).

    c. Torque strut to core support bracket bolt: 45 ft. lbs. (60 Nm) (see note above).

37. Install the remaining components.

38. Connect the negative battery cable.

39. Run the engine and check for leaks and proper engine operation.

4.9L Engine

— **CAUTION** —

Fuel injection systems remain under pressure, even after the engine has been turned OFF. The fuel system pressure must be relieved before disconnecting any fuel lines. Failure to do so may result in fire and/or personal injury.

1. Disconnect the negative battery cable.

2. Relieve the fuel system pressure.

3. Remove the engine from the vehicle and mount on a suitable workstand.

4. Remove the rocker arm covers.

NOTE: Label all valve train components as they are removed, so they can be reinstalled in their original locations. If the camshaft is being replaced, the valve lifters should also be replaced.

5. Remove the rocker arm support retaining nuts/bolts and remove the rocker arm support.

— **WARNING** —

The rocker arm support must be removed with the rocker arms and pivots attached. The pivot assemblies or support bar threads may be damaged if the pivot bolt torque is removed

against valve spring pressure. If necessary, secure the support in a vise and remove the rocker arms and pivots.

6. Remove the pushrods.

7. Remove the timing chain front cover.

8. Rotate the crankshaft until the timing chain sprocket timing marks are aligned.

9. Remove the distributor.

10. Remove the intake manifold.

11. Remove the valve lifters.

12. Remove the timing chain and camshaft sprocket.

13. Temporarily reinstall the camshaft sprocket to the camshaft (to provide a suitable hand hold). Remove the camshaft, being careful not to damage the lobes, journals or camshaft bearings.

14. Inspect the camshaft journals and lobes for wear and/or damage. Replace as necessary.

To install:

NOTE: If a new camshaft is required, new lifters and a new distributor gear must also be installed.

15. Temporarily reinstall the camshaft sprocket to the camshaft (to provide a suitable hand hold). Lubricate the camshaft journals, lobes and distributor gear with prelube 1052365 or equivalent.

16. Install the camshaft, being careful not to damage the lobes, journals or camshaft bearings. Remove the camshaft sprocket.

17. Install the timing chain and sprocket. Install the camshaft sprocket retaining bolt and tighten to 37 ft. lbs.

18. Install the timing chain front cover.

19. Install the valve lifters and intake manifold.

20. Install the distributor.

21. Install the pushrods and rocker arms. Loosely install the pivot bolts, then tighten the pivot bolts to 22 ft. lbs. (30 Nm). Install the rocker arm support. Make sure each pushrod is seated in a rocker arm.

22. Loosely install the rocker arm support retaining nuts and bolts.

23. Tighten the rocker arm support retaining nuts alternately and evenly until snug.

NOTE: Check the pushrods from time to time during the tightening procedure to make sure they are correctly positioned.

24. Tighten the rocker arm retaining bolts until snug.

25. Tighten the rocker arm support retaining nuts to 37 ft. lbs. (50 Nm).

26. Tighten the rocker arm retaining bolts to 7 ft. lbs. (10 Nm).

27. Install the rocker arm covers.

28. Install the engine in the vehicle.

29. Connect the negative battery cable.

30. Restore all fluid levels and make all necessary adjustments.

31. Run the engine and check for leaks and proper engine performance. Check the ignition timing.

Balance Shaft

REMOVAL AND INSTALLATION

3.8L Engine

1. Disconnect the negative battery cable.

2. Remove the engine from the vehicle and install on a suitable engine stand.

3. Remove the flywheel-to-crankshaft bolts and the flywheel.

4. Remove the intake manifold.

5. Remove the lifter guide retainer bolts and remove the retainer.

6. Remove the timing chain front cover.

7. Remove the camshaft sprocket and timing chain.

8. Remove the balance shaft gear-to-shaft bolt and the gear.

9. Remove the balance shaft retainer-to-engine bolts and the retainer.

10. Using the J-6125-B slide hammer tool, pull the balance shaft from the front of the engine.

To install:

11. Using J-36995-5 or equivalent balance shaft installer, install the balance shaft into the engine.

12. Install the balance shaft retainer and tighten the shaft retainer-to-engine bolts to 22 ft. lbs (30 Nm).

13. Align the balance shaft gear with the camshaft gear timing marks. Install the balance shaft gear onto the balance shaft and tighten the balance shaft gear-to-balance shaft bolt to 15 ft. lbs. (20 Nm) plus an additional 35 degree turn.

14. Turn the crankshaft so the No. 1 piston is at TDC.

15. Install the timing chain and sprocket.

16. Replace the balance shaft front bearing retainer and bolts. Tighten the bolts to 26 ft. lbs. (34 Nm).

17. Install the front timing chain cover.

18. Install the lifter guide retainer and tighten the mounting bolts to 27 ft. lbs. (37 Nm).

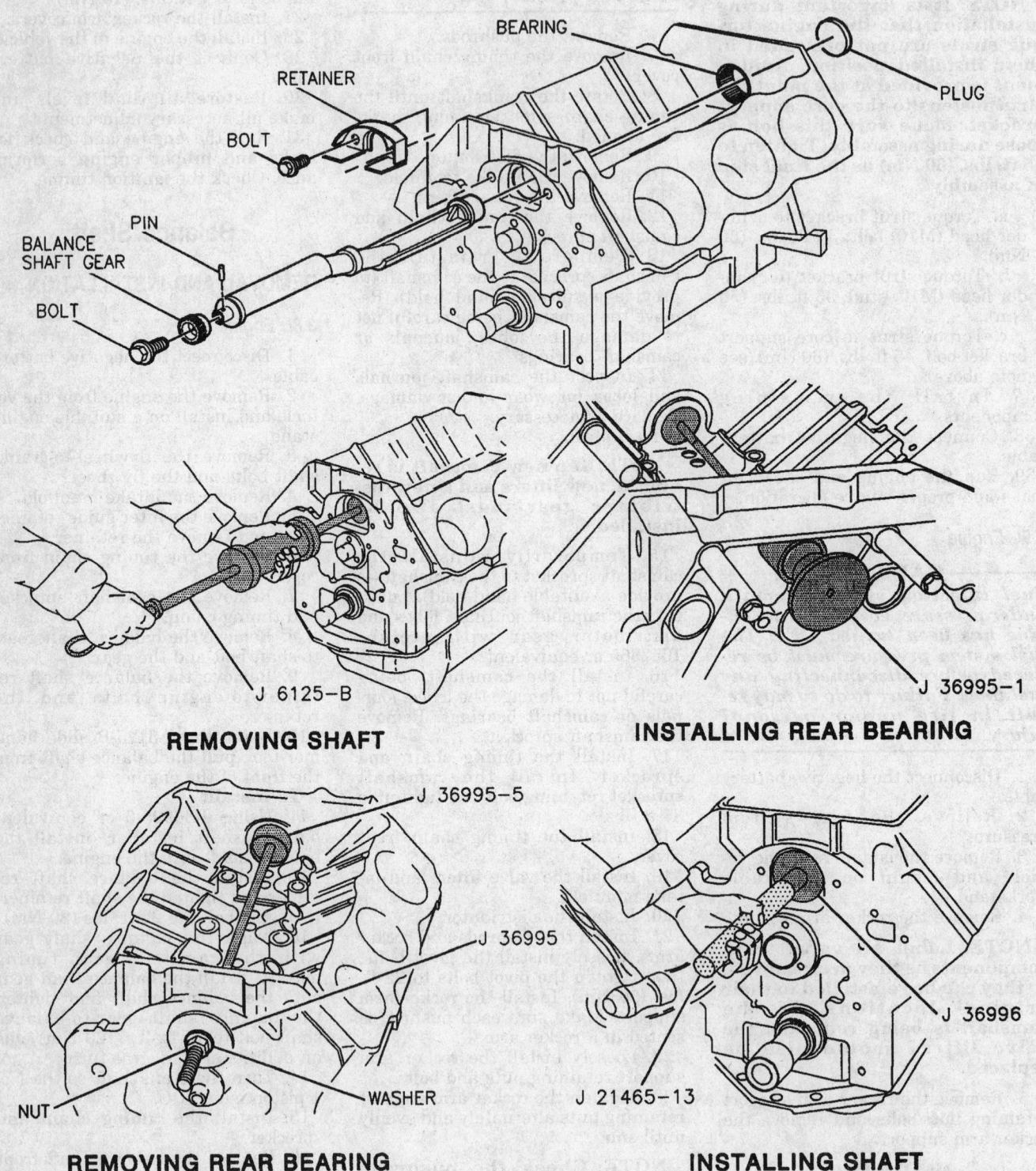

Balance shaft removal and installation — 3.8L engine

19. Install the intake manifold.

20. Install the flywheel and tighten the flywheel bolts to 61 ft. lbs. (82 Nm) plus an additional 90 degree turn.

21. Install the engine assembly in the vehicle.

22. Connect the negative battery cable.

23. Run the engine and check for leaks and proper engine operation.

Piston and Connecting Rod

POSITIONING

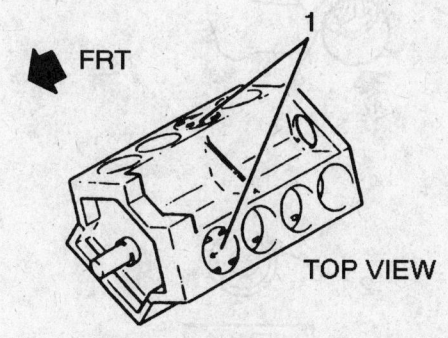

TOP VIEW

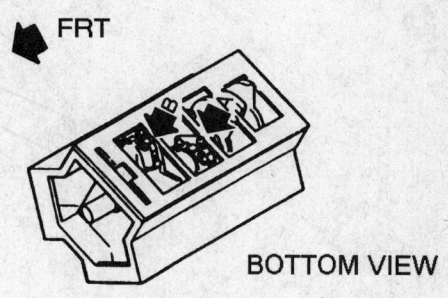

BOTTOM VIEW

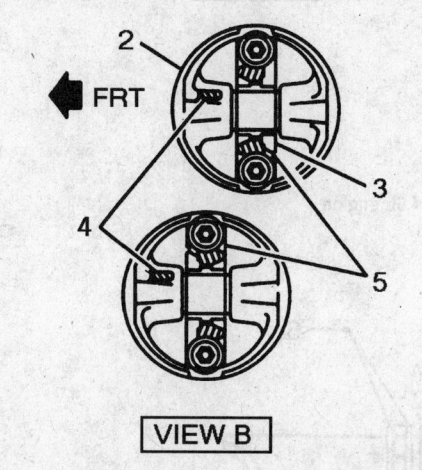

VIEW B

BOTTOM VIEW OF PISTON/ROD ASM

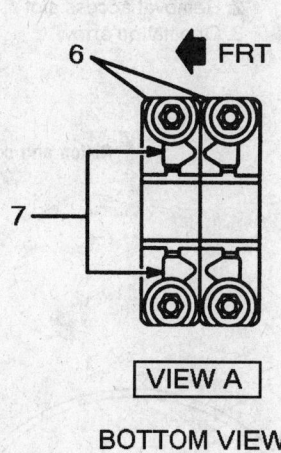

VIEW A

BOTTOM VIEW

1. Piston arrow toward chain case on both sides
2. Piston
3. Rod cap
4. Locater lugs indicate piston front towards engine front
5. Bearing cap arrows point toward each other on paired rods
6. Rod caps
7. Bearing cap arrows point toward each other on paired rods

311647

Piston and connecting rod orientation — 1993 4.6L engine

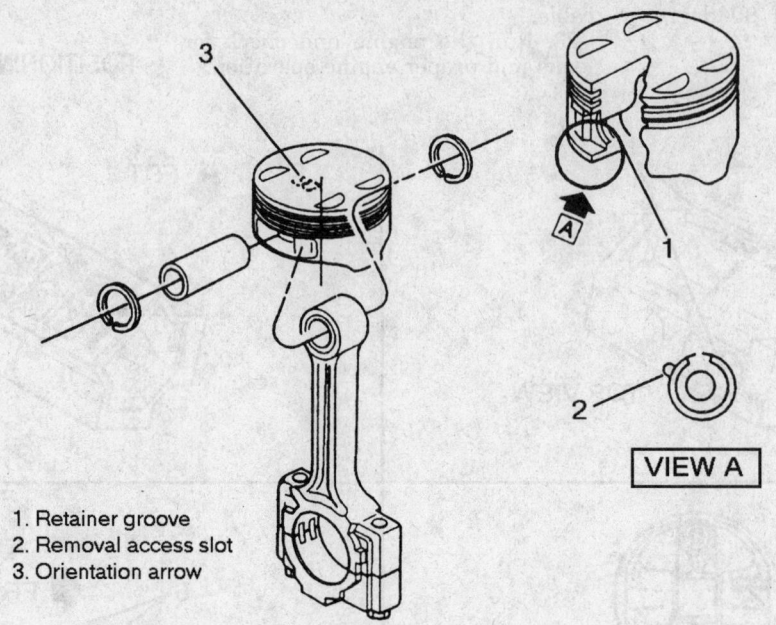

VIEW A

1. Retainer groove
2. Removal access slot
3. Orientation arrow

311649

Piston and connecting rod assembly — 1993 4.6L engine

30°
30°

1. Oil ring segment gap
2. Upper compression ring gap
3. Oil ring segment gap
4. Expander & lower compression ring gaps

5. Expander ring
6. Oil segment rings
7. Lower compression ring
8. Upper compression ring

311648

Piston ring positioning — 1993 4.6L engine

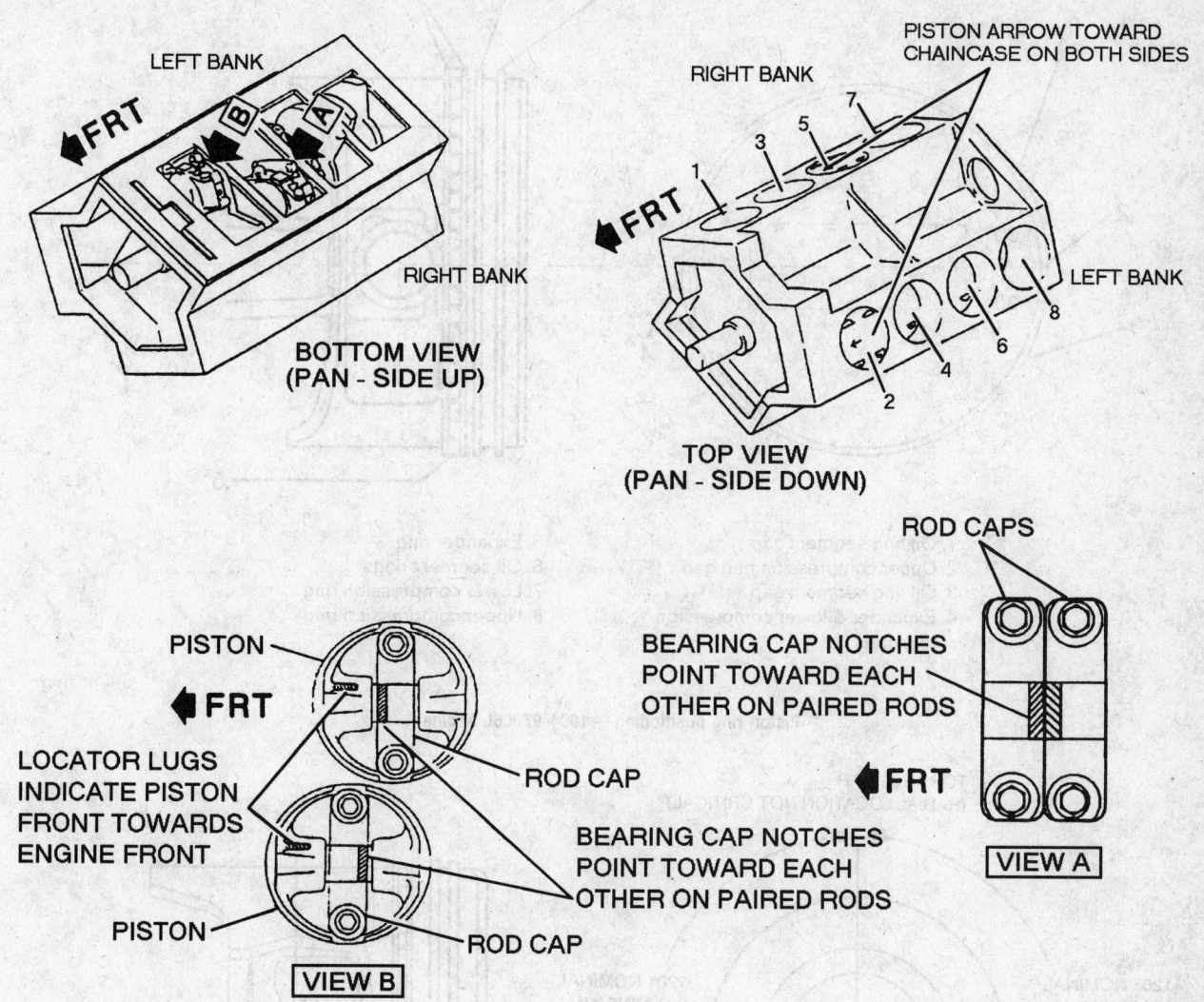

LEFT BANK

FRT

B   A

RIGHT BANK

BOTTOM VIEW (PAN - SIDE UP)

RIGHT BANK

PISTON ARROW TOWARD CHAINCASE ON BOTH SIDES

FRT

1 3 5 7

2 4 6 8

LEFT BANK

TOP VIEW (PAN - SIDE DOWN)

ROD CAPS

BEARING CAP NOTCHES POINT TOWARD EACH OTHER ON PAIRED RODS

FRT

VIEW A

PISTON

FRT

LOCATOR LUGS INDICATE PISTON FRONT TOWARDS ENGINE FRONT

ROD CAP

BEARING CAP NOTCHES POINT TOWARD EACH OTHER ON PAIRED RODS

PISTON

ROD CAP

VIEW B

311635

Piston and connecting rod orientation — 1994–97 4.6L engine

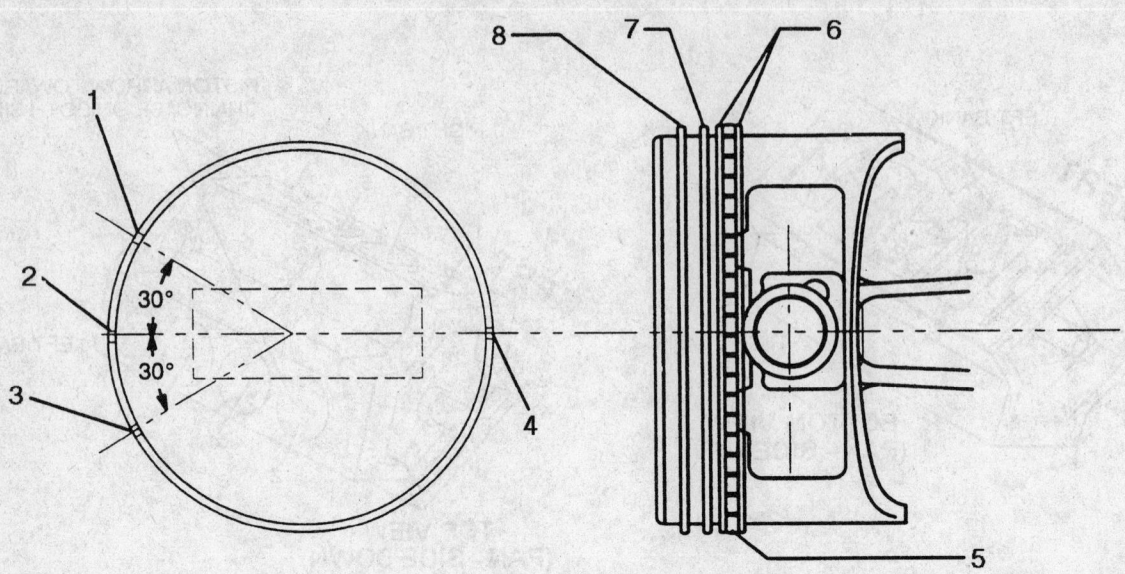

1. Oil ring segment gap
2. Upper compression ring gap
3. Oil ring segment gap
4. Expander & lower compression ring gaps
5. Expander ring
6. Oil segment rings
7. Lower compression ring
8. Upper compression ring

311637

Piston ring positioning — 1994–97 4.6L engine

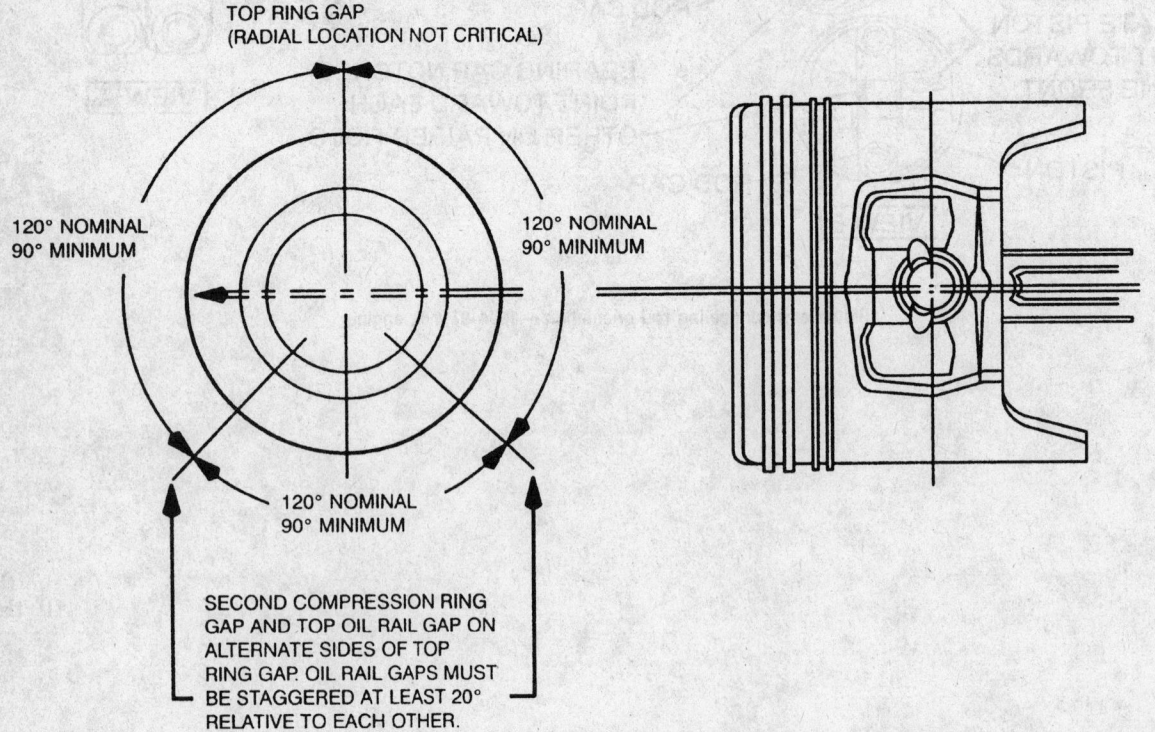

TOP RING GAP
(RADIAL LOCATION NOT CRITICAL)

120° NOMINAL
90° MINIMUM

120° NOMINAL
90° MINIMUM

120° NOMINAL
90° MINIMUM

SECOND COMPRESSION RING
GAP AND TOP OIL RAIL GAP ON
ALTERNATE SIDES OF TOP
RING GAP. OIL RAIL GAPS MUST
BE STAGGERED AT LEAST 20°
RELATIVE TO EACH OTHER.

340904

Piston ring orientation — 4.6L engine

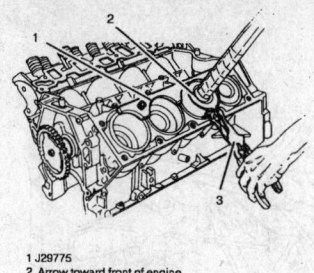

1 J29775
2 Arrow toward front of engine
3 Ring compressor J29789

340905

Install the piston with the arrow toward the front of the engine — 4.6L engine

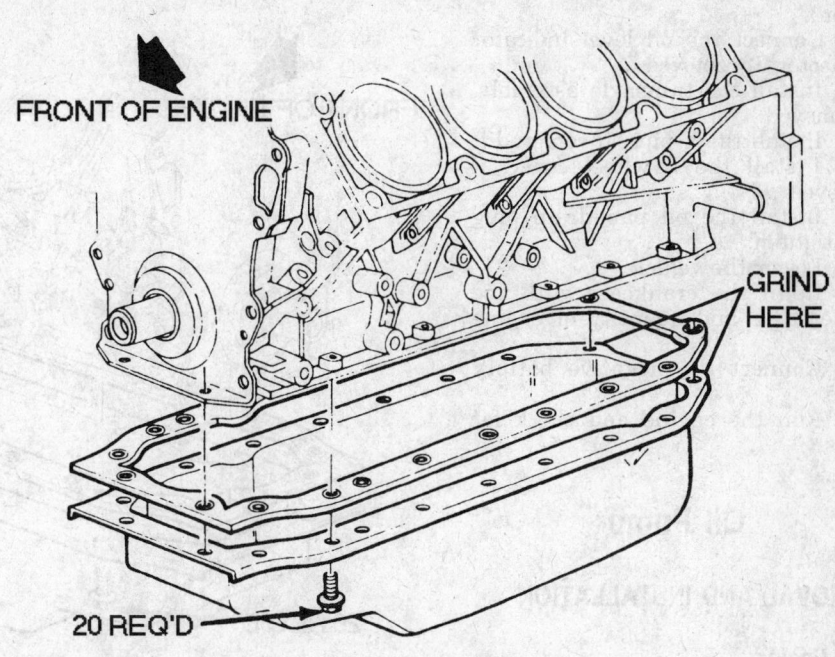

FRONT OF ENGINE

GRIND HERE

20 REQ'D

303722

Oil pan grinding location — 4.9L engine

ENGINE LUBRICATION

Oil Pan

REMOVAL AND INSTALLATION

1. Disconnect the negative battery cable.
2. Raise and safely support the vehicle.
3. Remove the oil pan drain plug and drain the engine oil into a suitable container.
4. Disconnect the oil level indicator harness, if equipped.
5. Remove the exhaust Y-pipe, if needed.
6. On 4.6L engines, remove the transaxle assembly from the vehicle.
7. On 4.9L engines, remove the flywheel cover.

WARNING
Attempting to remove the oil pan without first dropping the oil pump will result in breaking the oil pump housing.

8. On 3.8L and 4.6L engines, remove the oil pan mounting bolts and remove the oil pan.
9. On 4.9L engines, remove the oil pan mounting bolts, then drop the oil pan down about 2 inches, then remove the oil pump housing mounting bolts and nut. Drop the oil pump down into the oil pan, then remove the oil pan from the vehicle.

WARNING
Attempting to remove the oil pan without first dropping the oil pump will result in breaking the oil pump housing.

To install:
10. On 3.8L engines, do the following:
   a. Thoroughly clean the inside of the oil pan and all gasket mating surfaces.
   b. Install the oil pan with a new gasket and tighten the mounting bolts to 124 inch lbs. (14 Nm).
11. On 4.6L engines, do the following:
   a. The oil pan gasket is reusable unless it is damaged. Do not remove the gasket from the oil pan groove unless gasket replacement is required.
   b. Thoroughly clean the inside of the oil pan and the cylinder block contact surface. If the oil pan gasket is being reused, be careful not to damage it. Do not expose the gasket to cleaning solvents.
   c. If a new gasket is being installed, start the gasket into the oil pan groove and work the gasket into the groove in both directions. Once the gasket is exposed to oil, it will expand and no longer stay in the groove without wrinkles. If this

condition exists, replace the gasket.
   d. Install the oil pan and secure using the mounting bolts. Tighten, in sequence, to 89 inch lbs. (10 Nm).
12. On 4.9L engines, do the following:
   a. Before installing the oil pan, the excess material on the rear edge must be ground off 1/16 in. (1.5mm) on both sides for oil pump clearance.

WARNING
Failure to grind off the excess material will result in breaking the oil pump housing upon installation.

   b. Clean all gasket surfaces completely. Thoroughly clean the inside of the oil pan.
   c. Install a new O-ring on the oil pump outlet pipe. Install the oil pump, making sure the pump driveshaft engages the distributor gear. Tighten the nut to 22 ft. lbs. (30 Nm) and the bolts to 15 ft. lbs. (20 Nm).
   d. Apply a 1/4 in. bead of sealant at the rear main bearing cap and at the front cover-to-block joints.
   e. Install the oil pan using a new gasket.

f. Install the oil pan bolts and nuts and tighten to 14 ft. lbs. (18 Nm).

13. Connect the oil level indicator connector, if removed.

14. Install the transaxle assembly, if removed.

15. Install the Y-pipe, if removed.

16. Install the flywheel cover, if removed.

17. Install the oil pan drain plug and tighten.

18. Lower the vehicle.

19. Refill the crankcase with the proper type and quantity of engine oil.

20. Connect the negative battery cable.

21. Run the engine and check for leaks.

Oil Pump

REMOVAL AND INSTALLATION

3.8L Engine

1. Disconnect the negative battery cable.

2. Remove the drive belts and tensioner assembly.

3. Raise and safely support the vehicle.

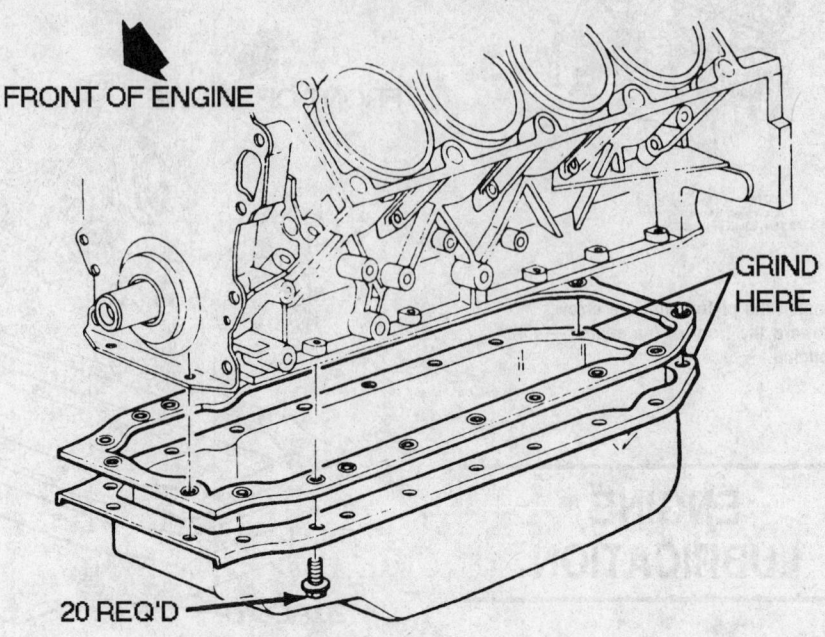

FRONT OF ENGINE

GRIND HERE

20 REQ'D

Oil pan grinding location — 4.9L engine

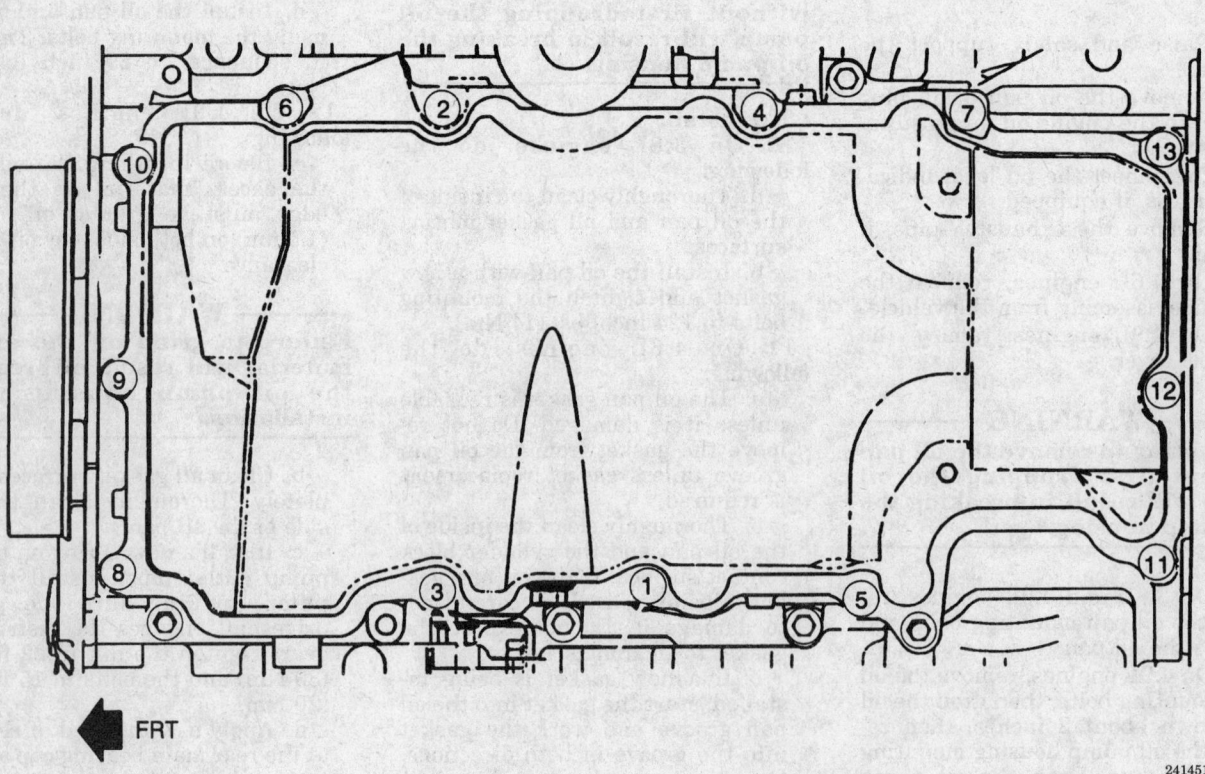

FRT

Oil pan bolt torque sequence — 4.6L engine

4. Support the engine using engine support fixture J-28467 or equivalent, and remove the torque axis mount and bracket assembly.

5. Remove the engine front cover assembly.

6. Remove the bolts securing the oil filter adapter to the front cover assembly, pressure regulator valve and spring.

7. Remove the oil pump cover attaching screws, then remove the cover.

8. Remove the inner and outer pump gears and inspect.

To install:

9. Lubricate the oil pump gears with petroleum jelly and install the gears into the housing.

10. Pack the gear cavity with petroleum jelly after the gears have been installed in the housing. This seals the pump and acts like a "prime" so oil will begin to drawn from the oil pan as soon as the engine begins to turn. Do not neglect this step. DO NOT use any type of grease. Petroleum jelly has a low melting point and will correctly dissipate when oil begins to flow and it is no longer needed.

11. Install the oil pump cover and screws and tighten to 97 inch lbs. (11 Nm).

12. Install the oil filter adapter with a new gasket, pressure regulator valve and spring. Tighten the mounting bolts to 24 ft. lbs. (33 Nm). Apply sealant to the bolt threads.

13. Install the front cover assembly.

14. Install the tensioner assembly and drive belts.

15. Install the torque axis mount assembly and remove the engine support fixture.

16. Verify the correct engine oil level. A new oil filter is recommended.

17. Connect the negative battery cable.

18. Start the vehicle and verify no leaks and proper oil pressure.

4.6L Engines

1. Disconnect the negative battery cable.

2. Remove the drive belt.

3. Remove the bolt securing the power steering hose.

4. Raise and safely support the vehicle.

5. Remove the right front wheel.

6. Remove the 2 splash shields from the wheel well.

7. Remove the brace between the engine oil pan and transaxle case.

8. Install flywheel holder tool J-39411 or equivalent, to keep the crankshaft from turning.

9. Remove the crankshaft balancer bolt.

10. Support the engine cradle with a screw type jack and remove the 3 right side engine cradle bolts.

11. Disconnect the RSS sensor from the right lower control arm.

12. Lower the cradle to gain access for the crankshaft balancer puller.

13. Using puller tool J-38416 or equivalent, remove the crankshaft balancer.

14. Remove the accessory drive belt tensioner and idler pulley.

15. Remove the front cover bolts and remove the front cover and gasket.

NOTE: The front cover gasket is reusable as long as it is not damaged.

16. Remove the 3 oil pump mounting bolts and remove the oil pump and drive spacer.

17. If necessary, disassemble and inspect the pump as follows:

a. Remove the drive spacer from the pump housing.

b. Remove the 2 screws holding the pump housing halves together.

c. Remove the inner (drive) and outer (driven) rotors from the housing. Indicate the mating surfaces (dimples).

d. Remove the pressure relief valve.

e. Inspect the pump housing for nicks, burrs, chips or debris that might cause a leak or binding condition in the rotor pocket.

f. Inspect the drive and driven rotors for nicks or burrs.

g. Check the pump cover and interior surface for excessive wear or score marks. Check for flatness.

h. If any components show signs of excessive wear or damage, replace the pump assembly.

To install:

18. If the pump was disassembled, reassemble as follows:

a. Install the inner and outer rotors to the pump cover in the same orientation as removed.

b. Install the pressure relief valve seat, spring and pilot in the pump housing.

c. Pack the pump housing halves with Amojell™ or white petroleum grease to ensure pump priming.

d. Assemble the housing and cover over the locating dowel.

e. Insert a 9mm drill in the pump mounting hole on the opposite side to aid alignment of the housing and cover. Install the 2

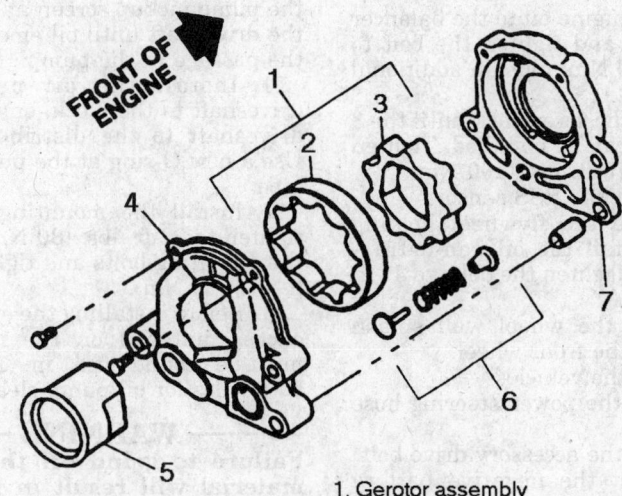

1. Gerotor assembly
2. Outer gear
3. Inner gear
4. Housing
5. Drive spacer
6. Relief valve
7. Cover

244635

Exploded view of the oil pump — 4.6L engine

screws and tighten to 108 inch lbs. (12 Nm).

19. Install the oil pump drive spacer into the oil pump from the rear so the drive flat engages the pump rotor.

20. Install the oil pump over the crankshaft and loosely install the mounting bolts.

21. On 1993 vehicles, hold the pump in its furthest up position and tighten the mounting bolts to 7 ft. lbs. (10 Nm) plus an additional 35 degree turn.

22. On 1994–97 vehicles, hold the pump in its furthest up position and tighten the mounting bolts (1, 2 and 3) in sequence, in 2 Steps. First tighten in sequence to 89 inch lbs. (10 Nm), then tighten in sequence to 20 ft. lbs. (26 Nm).

23. Place a small amount of RTV sealant at the split line of the upper and lower crankcases.

24. Install the front cover gasket on the dowel pins on the block.

25. Install the front cover on the dowel pins and install the front cover mounting bolts. Tighten the bolts to 89 inch lbs. (10 Nm).

26. Install the drive belt idler pulley and tighten the mounting bolt to 37 ft. lbs. (50 Nm).

27. Install the drive belt tensioner and tighten the tensioner mounting bolt to 37 ft. lbs. (50 Nm).

28. Install the crankshaft balancer using tool J-39344 or equivalent.

29. Apply engine oil to the balancer bolt threads and tighten the bolt to 37 ft. lbs. (50 Nm) plus an additional 120 degrees

30. Raise the screw jack until the 3 cradle bolts can be installed. Tighten the bolts to 75 ft. lbs. (100 Nm).

31. Connect the RSS sensor.

32. Remove the flywheel holding tool and install the oil pan-to-transaxle brace. Tighten the bolts to 37 ft. lbs. (50 Nm).

33. Install the wheel well splash shields and the front wheel.

34. Lower the vehicle.

35. Install the power steering hose retainer.

36. Install the accessory drive belt.

37. Connect the negative battery cable.

38. Run the engine and check for proper engine oil pressure. Check for leaks.

4.9L Engine

1. Disconnect the negative battery cable.

2. Raise and safely support the vehicle.

3. Remove the oil pan drain plug and drain the engine oil into a suitable container.

4. Remove the flywheel covers.

5. Remove the oil pan bolts and nuts and loosen the oil pan from the cylinder block.

6. Drop the oil pan down about 2 inches, then remove the oil pump housing mounting bolts and nut. Drop the oil pump down into the oil pan, then remove the oil pan from the vehicle.

—— WARNING ——
Attempting to remove the oil pan without first dropping the oil pump will result in breaking the oil pump housing.

7. Remove the oil pump from the pan and discard the O-ring.

8. If necessary, disassemble and inspect the oil pump for wear and/or damage. Replace as necessary.

To install:

9. Reassemble the oil pump, as required, lubricating all internal parts with clean engine oil during assembly. Tighten the pump cover-to-housing screws to 96 inch lbs. (10 Nm) and the pipe-to-housing screws to 120 inch lbs. (12 Nm). Be sure to use a new O-ring when installing the outlet pipe.

10. Temporarily install the driveshaft to the pump. Prime the pump by pouring clean engine oil into the pump pickup screen and rotating the driveshaft until oil emerges from the passage in the pump.

11. Install the oil pump and driveshaft to the block, engaging the driveshaft to the distributor gear. Use a new O-ring at the pump outlet pipe.

12. Install the mounting nut and tighten to 22 ft. lbs. (30 Nm). Install the mounting bolts and tighten to 15 ft. lbs. (20 Nm).

13. Before installing the oil pan, the excess material on the rear edge must be ground off $^{1}/_{16}$ in. (1.5mm) on both sides for oil pump clearance.

—— WARNING ——
Failure to grind off the excess material will result in breaking the oil pump housing upon installation.

14. Clean all gasket surfaces completely. Thoroughly clean the inside of the oil pan.

15. Apply a $^{1}/_{4}$ in. bead of sealant at the rear main bearing cap and at the front cover-to-block joints.

16. Install the oil pan using a new gasket.

17. Install the oil pan bolts and nuts and tighten to 14 ft. lbs. (18 Nm).

18. Install the flywheel cover.

19. Install the oil pan drain plug and tighten to 22 ft. lbs. (30 Nm).

20. Lower the vehicle.

21. Refill the crankcase with the proper type and quantity of engine oil.

22. Connect the negative battery cable.

23. Run the engine: check for proper oil pressure and check for leaks.

TRANSAXLE

Automatic Transaxle Assembly

REMOVAL AND INSTALLATION

All Models Except Riviera

1. Disconnect the negative battery cable.

2. Remove the headlight housing upper filler panel and diagonal brace.

3. Remove the air cleaner assembly.

4. Disconnect the shift control cable and bracket at the transaxle.

5. Remove the torque struts.

6. Disconnect the oil cooler lines at the cooler and the oil sending line at the transaxle.

7. Remove the 2 upper transaxle-to-engine bolts.

8. Disconnect the power steering return hose at the auxiliary cooler. Plug the cooler and return hose to prevent leakage.

9. Support the engine with engine support fixture J-28467 or equivalent. Tighten the wing nuts several turns to take the weight of the powertrain off of the mounts and frame.

10. Raise and safely support the vehicle. Remove the front wheels.

11. Remove the splash shields from both front wheel wells.

12. Disconnect both front suspension position sensors from the lower control arms and position aside.

13. Remove both stabilizer links from the struts.

14. Separate the tie rod ends from the steering knuckles.

15. Separate the lower ball joints from the steering knuckles.

16. Remove the halfshafts.

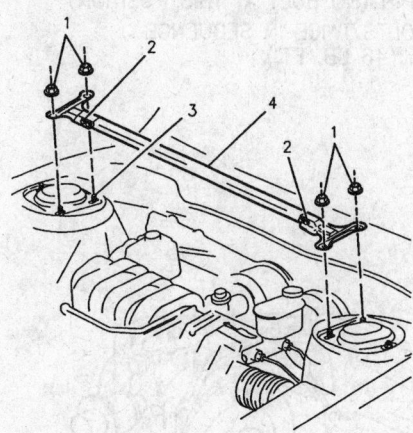

1 NUT
2 THROUGH–BOLTS
3 STUD, STRUT INBOARD
4 BAR, CROSS BRACE

323901

Cross brace to strut towers

17. Remove the power steering filter at the cradle and the A/C splash shield from the frame.

18. Remove the ABS modulator from the bracket and support. Remove the engine oil pan-to-transaxle bracket.

19. Remove the torque converter cover, then remove the torque converter-to-flexplate bolts. Prior to bolt removal, mark the torque converter position in relation to the flexplate so they can be reassembled in the same position.

20. Remove the powertrain mount nuts from the cradle.

21. Rotate the intermediate steering shaft until the steering gear stub shaft clamp bolt is accessible from the left wheel well. Remove the clamp bolt and disconnect the intermediate steering shaft from the steering gear.

CAUTION

If the intermediate steering shaft is not disconnected from the steering gear stub shaft, damage to the steering gear and/or intermediate shaft may result. This damage can cause loss of steering control which could result in personal injury.

WARNING

Do not turn the steering wheel or move the position of the steering gear once the intermediate steering shaft is disconnected as this will uncenter the Air Bag coil in the steering column. If the Air Bag coil becomes uncentered, it may be damaged during vehicle operation.

22. Remove the electrical harness and connector from the front of the cradle.

23. Support the rear of the cradle with a suitable jack, then remove the 4 rear cradle bolts.

24. Lower the jack a few inches to gain access to the power steering gear heat shield and return line fitting. Remove the heat shield and disconnect the return line. Plug the line and the opening in the gear to prevent fluid leakage.

25. Disconnect the power steering electrical connector.

26. Raise the jack and reinstall one rear cradle bolt on each side finger-tight to support the cradle. Remove the jack.

27. Support the frame with a suitable jack and remove the 6 frame mount bolts. Lower the frame and/or raise the vehicle with the steering gear attached.

28. Label and disconnect the electrical connectors to the transaxle, vehicle speed sensor and ground. Remove the transaxle harness from the transaxle clip.

29. Remove the fuel line bundle from the transaxle.

30. Remove the left and right transaxle mount and bracket from the transaxle.

31. Support the transaxle with a suitable jack.

32. Remove the engine-to-transaxle heat shield and bracket and the remaining transaxle-to-engine bolts. Lower the transaxle.

33. Remove the manual shaft linkage and neutral safety switch. Remove the vehicle speed sensor and oil return line.

To install:

34. Install the oil return line and the vehicle speed sensor.

35. Install the neutral safety switch and tighten the bolts to 106 inch lbs. (12 Nm).

36. Install the manual shaft linkage and tighten the manual shaft nut to 15 ft. lbs. (20 Nm).

37. Raise the transaxle into position and install the 2 lower transaxle-to-engine bolts. Tighten to 35 ft. lbs. (47 Nm).

38. Install the engine-to-transaxle bracket and heat shield. Tighten the bolts to 35 ft. lbs. (47 Nm).

39. Remove the transaxle jack.

40. Install the right and left transaxle bracket and mount to the transaxle. Tighten the bolts and nuts to 35 ft. lbs. (47 Nm).

41. Install the fuel line bundle to the transaxle.

42. Connect the electrical connectors to the transaxle, vehicle speed sensor and ground. Install the transaxle harness to the transaxle clip.

43. Raise the frame and/or lower the vehicle while locating the engine and transaxle mount studs into the frame, harnesses at the cradle, and frame mount bolt holes to the underbody.

44. Install 2 front and 2 rear cradle bolts finger-tight to support the cradle, then remove the cradle support.

45. Support the rear of the cradle with a suitable jack and remove the 2 rear cradle bolts.

46. Lower the jack a few inches to gain access to the power steering gear. Connect the hose at the steering gear and tighten the fitting to 20 ft. lbs. (27 Nm). Connect the power steering gear electrical connector and install the steering gear heat shield.

47. Raise the cradle with the jack. Install the 6 frame mount bolts beginning with the No. 2 mount bolt into the body, followed by the No. 1 mount bolt into the body, followed by the remaining frame mount bolts. Tighten the bolts to 74 ft. lbs (100 Nm).

48. Install the electrical harness to the front of the cradle.

49. Connect the intermediate steering shaft to the steering gear and install the clamp bolt. Tighten the bolt to 35 ft. lbs. (47 Nm).

WARNING

Do not turn the steering wheel or move the position of the steering gear while the intermediate steering shaft is disconnected as this will uncenter the Air Bag coil in the steering column. If the Air Bag coil becomes uncentered, it may be damaged during vehicle operation.

50. Install the left and right transaxle mount nuts and right engine mount nuts at the frame. Tighten the nuts to 35 ft. lbs. (47 Nm).

51. Align the flexplate and torque converter using the marks made during the removal procedure. Install the flexplate-to-converter bolts and tighten to 35 ft. lbs. (47 Nm).

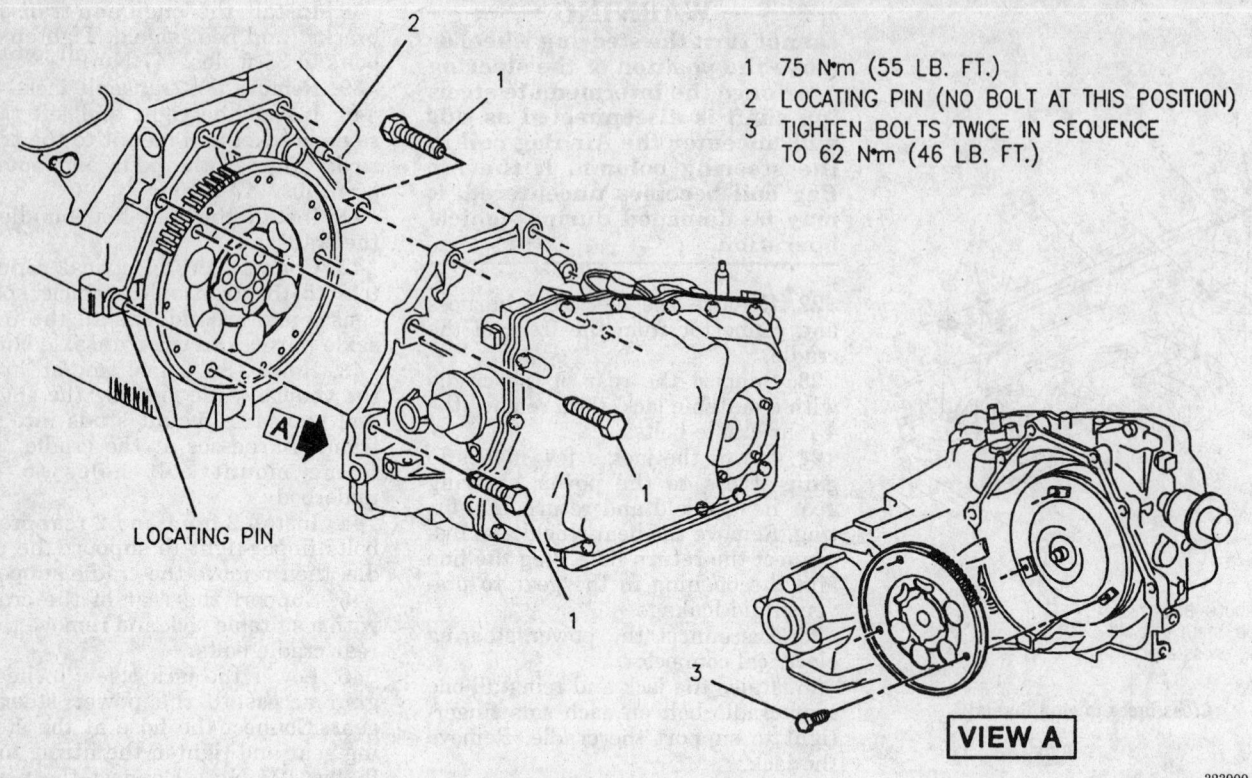

1 75 N·m (55 LB. FT.)
2 LOCATING PIN (NO BOLT AT THIS POSITION)
3 TIGHTEN BOLTS TWICE IN SEQUENCE
   TO 62 N·m (46 LB. FT.)

LOCATING PIN

VIEW A

323900

Engine to transaxle attachments

52. Install the torque converter cover and tighten the bolts to 106 inch lbs. (12 Nm).

53. Install the engine oil pan-to-transaxle bracket and tighten the bolts to 35 ft. lbs. (47 Nm).

54. Install the ABS modulator to the bracket and install the A/C splash shield at the frame.

55. Install the halfshafts. Tighten the halfshaft nuts to 110 ft. lbs. (145 Nm).

56. Install the lower ball joints into the steering knuckles and install the nuts.

57. Install new cotter pins.

58. Install the tie rod ends into the steering knuckles and install the nuts.

59. Install new cotter pins.

60. Connect the stabilizer links to the struts and tighten the nuts to 49 ft. lbs. (65 Nm).

61. Install both front suspension position sensors to the lower control arms.

62. Install the power steering filter to the cradle and install the splash shields in the wheel wells.

63. Install the wheels and lower the vehicle. Remove the engine support fixture.

64. Connect the power steering hose at the auxiliary cooler.

65. Install the remaining transaxle-to-engine bolts and tighten to 35 ft. lbs. (47 Nm).

66. Flush the transaxle oil cooler using flushing tool J-35944 and flushing solution J-35944-20. The transaxle oil cooler and lines should be flushed before the oil cooler lines are connected to the transaxle.

67. Connect the transaxle oil cooler lines to the transaxle. Start the fittings by hand and tighten them finger-tight, then torque the fittings to 16 ft. lbs. (22 Nm).

68. Install the torque struts.

69. Adjust the neutral safety switch.

70. Install the shift control cable and bracket to the transaxle and tighten the bracket bolts to 106 inch lbs. (12 Nm). Adjust the shift control cable.

71. Install the air cleaner assembly. Install the headlight housing upper filler panel and diagonal brace.

72. Connect the negative battery cable.

73. Fill the transaxle with the proper type and quantity of transaxle fluid. Bleed the power steering system.

74. Check the front suspension alignment and adjust as necessary.

75. The Powertrain Control Module (PCM) maintains 3 types of transaxle adapt parameters which are used to modify transaxle line pressure. The line pressure is modified to maintain shift quality regardless of wear or tolerance variations within the transaxle. Whenever the transaxle is replaced, the transaxle adapts must be reset as follows:

a. Turn the ignition key **ON**. Enter the self-diagnostic system.

b. Select Powertrain Control Module (PCM) override PS13 (TP SENSOR LEARN).

c. Press the WARMER button. The Driver Information Center (DIC) should display 09, indicating that the Garage Shift Adapt value has been reset.

d. Select PCM override PS14 (TRAN ADAPT).

e. Press the COOLER button. The DIC should display 90, indicating the Upshift Adapt (UA) value has been reset.

f. Press the WARMER button. The DIC should display 09, indicating the Steady State Adapt (SSA) value has been reset.

76. The PCM maintains a value for transaxle oil life. This value indicates the percentage of oil life remaining and is calculated based on transaxle temperature and speed. When the vehicle is new, the transaxle oil life value is 100 As the vehicle operates,

the percentage will decrease. Whenever the transaxle is replaced, the transaxle oil life indicator should be reset to 100 as follows:

a. Turn the ignition key **ON**, but leave the engine OFF.

b. Press and hold the OFF and REAR DEFOG buttons on the DIC until the message TRANSAXLE OIL LIFE RESET is displayed on the DIC.

Riviera

1. Position the front wheels in the straight ahead position and lock the steering column with the ignition lock cylinder. This will prevent the Air Bag coil from becoming uncentered and accidently deploying the Air Bag.
2. Raise the hood and cover the fenders.
3. Disconnect the negative battery cable.
4. Remove the cross brace.
5. Remove the air intake duct.
6. Remove the cruise servo assembly.
7. Remove the shift control linkage and mounting bracket at the transaxle.
8. Disconnect the wiring connectors at the transaxle park/neutral position switch, backup switch, transaxle electrical connector and vehicle speed sensor.
9. Remove the fuel pipe retainers.
10. Remove the vacuum modulator hose at the modulator.
11. Disconnect the power steering pressure lines.
12. Install J-28467 A or equivalent engine support fixture.
13. Load the engine support fixture by tightening the wing nuts several turns to relieve tension on the frame and mounts.

CAUTION

To help avoid personal injury when a vehicle is on a hoist, provide additional support to the rear of the vehicle while removing the transaxle.

---

14. Raise and safely support the vehicle.
15. Remove both front wheels.
16. Remove both wheel opening splash shields.
17. Remove the intermediate shaft bolt from the left wheel well opening and carefully pry up on the shaft to disengage from the rack and pinion.
18. Remove both stabilizer links from the strut assemblies.
19. Separate both tie rods from the steering knuckle.

20. Separate the right and left ball joints from the steering knuckles.
21. Remove the right and left halfshafts from the steering knuckles.
22. Remove the A/C splash shield.
23. Remove the ABS brake modulator from the sub-frame.
24. Remove the transaxle cooler lines.
25. Remove the flywheel cover and the bolts. Prior to removal, mark the relationship between the flywheel and the converter for installation reference.
26. Remove the front exhaust pipe.
27. Remove the left and right transaxle mount nuts at the frame.
28. Support the frame using a suitable tool and remove the 6 frame mounting bolts.
29. Lower the frame with the steering gear attached.
30. Remove the left transaxle mount and the bracket assembly.
31. Remove the right rear transaxle mount and bracket from the transaxle.
32. Remove the engine to transaxle bracket.
33. Install a suitable transaxle jack to remove the transaxle assembly.
34. Remove the engine to transaxle bolts. One of the transaxle bolts is located between the transaxle case and the engine block in the opposite direction.
35. Remove the transaxle assembly.

NOTE: If replacing the transaxle, proper flushing of the transaxle oil cooler is recommended.

To install:
36. Position the transaxle into the vehicle.
37. Install and tighten the lower transaxle to engine block bolts to 55 ft. lbs. (75 Nm).
38. Install the engine to transaxle bracket.
39. Install the right rear transaxle mount and bracket to the transaxle.
40. Install the left transaxle mount and the bracket assembly.
41. Install the sub-frame assembly and install the frame mounting bolts, torquing them to 83 ft. lbs. (112 Nm).
42. Install the left and the right transaxle mount nuts at the frame.
43. Install the front exhaust pipe assembly.
44. Line up and install the torque converter bolts. Tighten the bolts, in 2 passes, to 46 ft. lbs. (62 Nm).
45. Install the transaxle cooler lines.
46. Install the ABS modulator into the mounting bracket.
47. Install the A/C splash shield.

48. Install the right and left halfshafts.
49. Connect the lower ball joints to the steering knuckle.
50. Install the tie rods to the steering knuckle.
51. Install the stabilizer links to the strut assembly.
52. Connect the intermediate shaft to the rack and pinion and torque the bolt to 35 ft. lbs. (47 Nm).
53. Install the wheel opening splash shields.
54. Install the front wheels and lower the vehicle.
55. Relieve the tension on the engine support fixture and remove.
56. Connect the power steering pressure lines.
57. Connect the vacuum line to the modulator.
58. Install the fuel line retainers.
59. Connect the electrical connectors to the transaxle.
60. Install the shift control linkage and mounting bracket.
61. Install the cruise control servo and the vacuum harness.
62. Install the air intake duct.
63. Install the cross car brace.
64. Connect the negative battery cable.
65. Fill and bleed the power steering system.
66. Fill the transaxle with the proper type and quantity of fluid.
67. Start the engine and check for proper operation and fluid leaks.
68. Adjust the transaxle shifter cable, as necessary.

NOTE: A 4-wheel alignment is recommended after the removal of the sub-frame assembly.

DRIVE AXLE

Halfshaft

REMOVAL AND INSTALLATION

WARNING

Use care when removing the halfshaft to prevent the inner CV-joint from becoming over-extended. Over-extension of the joint could result in separation of internal components and possible joint failure.

---

1. Raise and safely support the vehicle.
2. Remove the front wheel. Mark the position of the wheel to the wheel

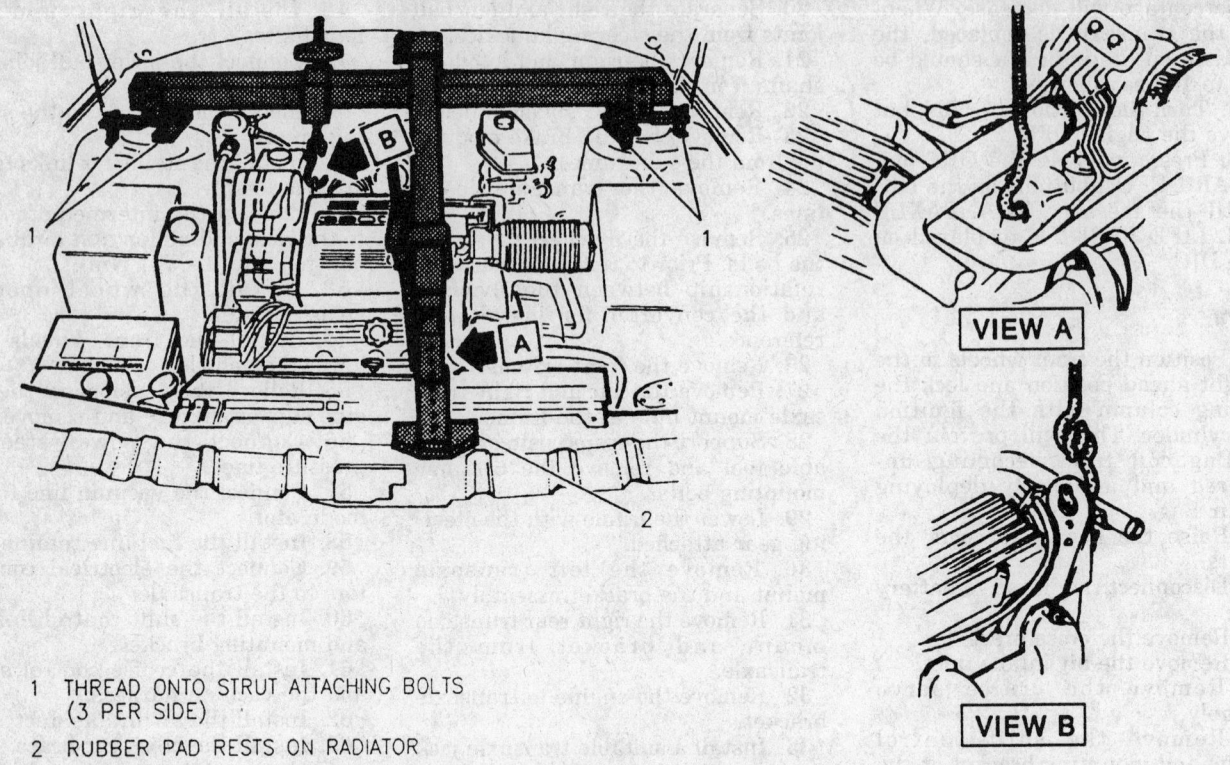

1 THREAD ONTO STRUT ATTACHING BOLTS (3 PER SIDE)
2 RUBBER PAD RESTS ON RADIATOR

VIEW A

VIEW B

323899

Engine support fixture

studs, prior to removal, for installation reference.

3. Install J-34754 boot protector, or the equivalent, on the outer CV-joint boot.

4. Remove the hub nut. To prevent the rotor from turning insert a drift punch in the rotor cooling fins. Discard the hub nut; it must not be reused.

5. Disconnect the stabilizer link, if necessary.

6. Remove the ball joint cotter pin and nut. Loosen the ball joint in the knuckle using tool J-36226 or equivalent, being careful not to damage the ball joint and grease seal. If removing the right halfshaft, turn the wheel to the left. If removing the left halfshaft, turn the wheel to the right.

7. Using a suitable prybar between the suspension support and the lower control arm, separate the ball joint from the steering knuckle.

8. Partially install the hub nut to protect the threads, then remove the halfshaft from the hub using tool J-28733-B or equivalent.

9. Move the strut and knuckle rearward.

10. Remove the halfshaft from the transaxle using a suitable prying tool and a wood block fulcrum to protect the transaxle case.

NOTE: If equipped with anti-lock brakes, care must be used to prevent damage to the toothed sensor ring on the halfshaft and the wheel speed sensor on the steering knuckle.

To install:

11. If installing the right side halfshaft, install tool J-37292-B or equivalent, so it can be pulled out after the halfshaft is installed.

12. Install the halfshaft into the transaxle. To verify the halfshaft is properly seated, grasp the inner CV-joint housing and pull it outward. DO NOT pull on the halfshaft. If the CV-joint is properly seated, the halfshaft will not pull back out.

13. Install the halfshaft into the hub and bearing assembly. Loosely install a new hub nut.

14. Install the ball joint into the steering knuckle.

15. Install the castle nut and tighten to 84 inch lbs. plus 120 degrees (1/3 turn). A minimum of 37 ft. lbs. (51 Nm) of torque must be attained. If necessary to install the cotter pin, the nut can be tightened up to 20 degrees additional. NEVER loosen the castle nut to install the cotter pin.

16. With a drift in the brake rotor cooling fins to keep the rotor from turning, tighten the nut to 110 ft. lbs. (149 Nm) on all models, except DeVille with J55 brake option, in which case the torque is 130 ft. lbs. (177 Nm).

17. Connect the stabilizer link, if removed.

18. Remove the boot protector.

19. If tool J-37292-B was installed, remove it by pulling in line with the handle.

20. Install the wheel, aligning the marks made during removal.

21. Lower the vehicle.

22. Road test and check vehicle operation.

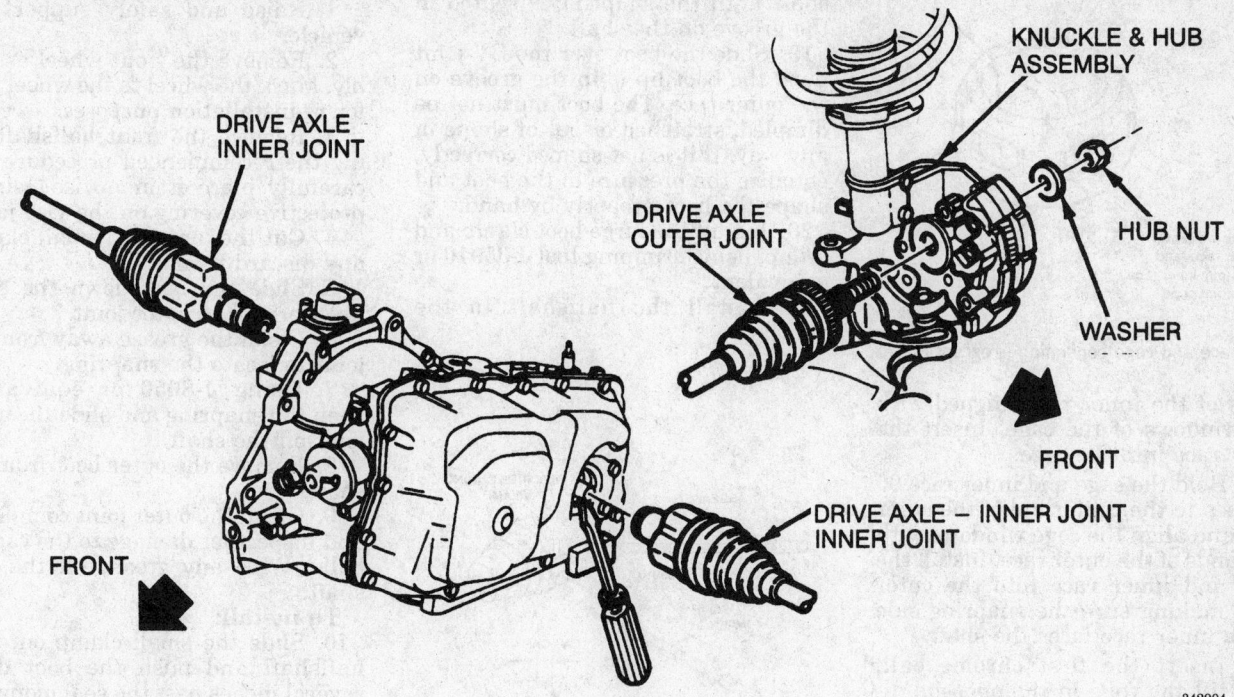

DRIVE AXLE INNER JOINT

DRIVE AXLE OUTER JOINT

KNUCKLE & HUB ASSEMBLY

HUB NUT

WASHER

FRONT

FRONT

DRIVE AXLE – INNER JOINT INNER JOINT

342904

Removing the halfshaft

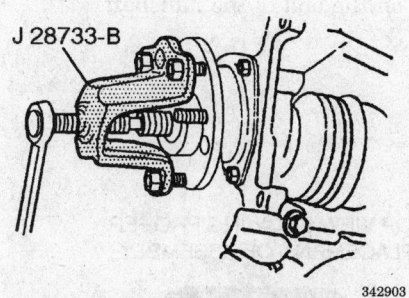

J 28733-B

342903

Removing the halfshaft from the hub

CV-Joint Boot

REPLACEMENT

Outer

All Models Except Riviera

   NOTE: Two types of inner CV-joint are used: the Tripot Ball type and the Bearing Block type. The following procedure contains information for both types. Also, depending upon application, the small end of the CV-joint

boot may be retained with either a conventional type boot clamp or with a swage clamp.

   1. Remove the halfshaft from the vehicle. Mount the shaft in a soft-jawed vise.
   2. Cut the large boot retaining clamp from the CV-joint with side cutters and discard.
   3. If equipped with a conventional small boot clamp, cut the small boot retaining clamp with side cutters and discard. If equipped with a swage ring, use a hand grinder to cut through the ring, being careful not to damage the axle shaft. Remove and discard the swage ring.
   4. Separate the boot from the CV-joint and slide it away from the joint along the shaft.
   5. Wipe the grease from the face of the CV-joint inner race. Spread the retaining snapring with snapring pliers and remove the CV-joint from the shaft.
   6. Remove the boot from the shaft.
   7. Using a brass drift and hammer, gently tap on the CV-joint cage until it is tilted enough to remove the first chrome ball. Tilt the cage in the opposite direction to remove the opposing ball. Repeat until all 6 balls are removed.

   8. Position the cage and inner race 90 degrees to the centerline of the outer race and align the cage windows with the lands of the outer race. Remove the cage and inner race from the outer race.
   9. Rotate the inner race 90 degrees to the centerline of the cage with the lands of the inner race aligned with the windows of the cage. Pivot the inner race into the cage window and remove the inner race.
   10. Thoroughly clean the inner and outer races and the cage and balls with clean solvent and allow to dry. Inspect all parts for damage and replace the CV-joint, if necessary.
   To install:
   11. As required, install a new small boot clamp on the neck of a new boot, but do not crimp yet, or install a new swage ring on the neck of a new boot, but do not swage yet.
   12. Slide the boot onto the shaft and position the boot neck in the groove on the shaft. As required, crimp the boot clamp with crimping tool J-35910 or equivalent, or swage the swage ring using tool J-41048 or equivalent.
   13. Put a light coat of grease from the CV-joint service kit on the ball grooves of the inner and outer races.
   14. Hold the inner race 90 degrees to the centerline of the cage with the

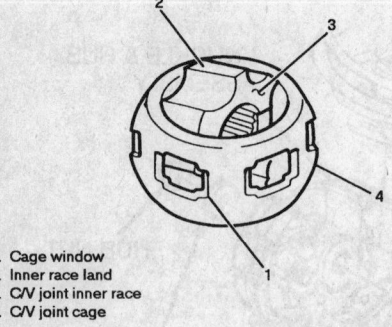

1. Cage window
2. Inner race land
3. C/V joint inner race
4. C/V joint cage

342708

Inner race and cage separation — except Rivera

lands of the inner race aligned with the windows of the cage. Insert the inner race into the cage.

15. Hold the cage and inner race 90 degrees to the centerline of the outer race and align the cage windows with the lands of the outer race. Install the cage and inner race into the outer race, making sure the snapring side of the inner race faces the shaft.

16. Insert the first chrome ball, then tilt the cage in the opposite direction to insert the opposing ball. Repeat until all 6 balls are in place.

17. Place approximately half of the grease from the service kit inside the boot and pack the CV-joint with the remaining grease.

18. Push the CV-joint onto the shaft until the snapring is seated in the groove on the shaft.

19. Slide the boot over the CV-joint until the boot lip is in the groove on the outer race. The boot must not be dimpled, stretched or out of shape in any way. If it is not shaped correctly, equalize the pressure in the boot and shape the boot properly by hand.

20. Install the large boot clamp and crimp, using crimping tool J-35910 or equivalent.

21. Install the halfshaft in the vehicle.

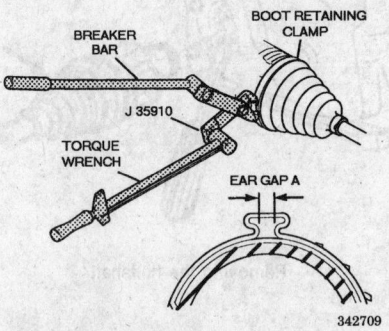

Installing the large boot retaining clamp — except Rivera

Riviera

1. Raise and safely support the vehicle.

2. Remove the front wheel assembly. Mark the wheel to the wheel stud for re-installation purposes.

3. Remove the front halfshaft using the recommended procedure and carefully place it in a vise using a protective covering on the vise jaws.

4. Cut the large and small clamps and discard them.

5. Slide the boot down the shaft uncovering the outer joint.

6. Clean the grease away from the joint to locate the snapring.

7. Using J-8059 or equivalent, open the snapring and slide the outer joint off the shaft.

8. Remove the outer boot from the shaft.

9. Clean the outer joint completely and inspect for damage to the cage or balls. Wipe any grease off the axle shaft.

To install:

10. Slide the small clamp onto the halfshaft and push the boot down several inches past the seal mounting area.

11. Check the snapring in the outer joint and replace as necessary. Pack the outer joint with half of the grease supplied in the boot kit and install it on the end of the halfshaft.

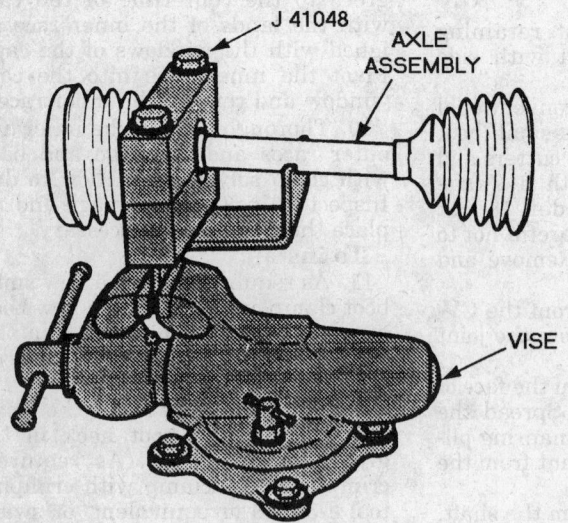

Swage clamp installation — except Rivera

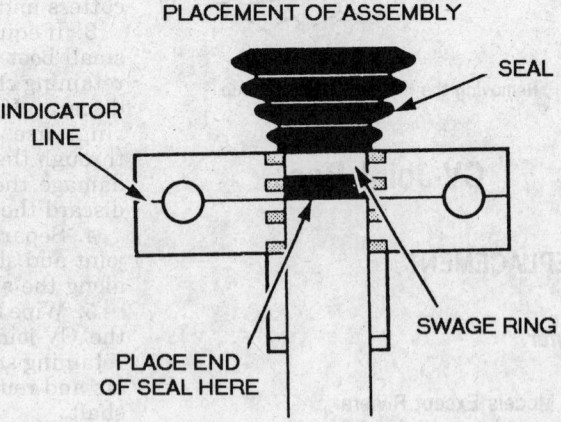

342718

12. Gently pull down on the joint until the splines engage. With a brass drift lightly tap the joint down until the snapring engages.

13. Pack the remaining grease from the boot kit into the boot and then pull the boot up over the end of the joint. Seat the small end on the seal mounting area.

14. Slide the small clamp into position and using J-35910 or equivalent, crimp the clamp to 130 ft. lbs. (176 Nm)

15. Install the large clamp to the proper position and crimp to 130 ft. lbs. (176 Nm).

— CAUTION —

The boot must not be dimpled, stretched or out of shape in any way. If boot is not shaped correctly, carefully insert a thin flat blunt tool at the large end of the boot to equalize pressure. Shape the boot properly by hand and then remove the tool.

16. Install the halfshaft using the recommended procedure. Install the wheel and tire assembly and torque the wheel lug nuts to 100 ft. lbs. (130 Nm). Road test the vehicle to check for abnormal noise or vibration.

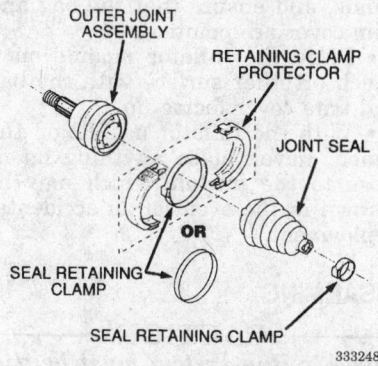

Outer CV-Joint disassembly — Rivera

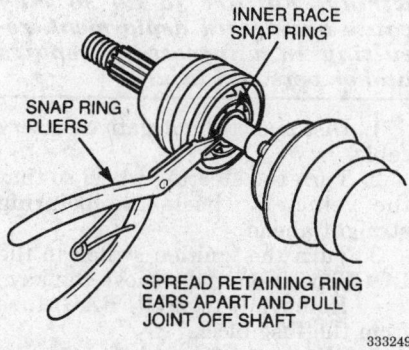

Outer CV-Joint snapring removal — Rivera

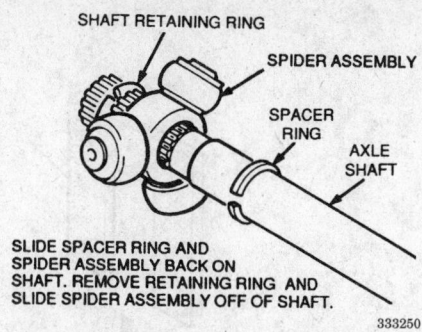

Outer CV-Joint snapring and spider assembly removal — Rivera

Inner

Tripot Ball Type

1. Remove the halfshaft from the vehicle. Mount the shaft in a soft-jawed vise.

2. Cut the large boot retaining clamp from the CV-joint with side cutters and discard.

3. If equipped with a conventional small boot clamp, cut the small boot retaining clamp with side cutters and discard. If equipped with a swage ring, use a hand grinder to cut through the ring, being careful not to damage the axle shaft. Remove and discard the swage ring.

4. Separate the boot from the CV-joint and slide it away from the joint along the shaft.

5. Remove the CV-joint housing from the spider and shaft. Remove the tripot bushing from the housing.

6. Spread the spacer ring and slide the spacer ring and spider back on the axle shaft.

7. Remove the shaft retaining ring from the groove on the axle shaft and slide the spider assembly off of the shaft.

8. Remove the spacer ring and boot from the axle shaft.

9. Thoroughly clean the spider assembly and housing with clean solvent and allow to dry. Inspect all parts for damage and replace the CV-joint, if necessary.

To install:

10. As required, install a new small clamp on the neck of a new boot, but do not crimp yet, or install a new swage ring on the neck of a new boot, but do not swage yet.

11. Slide the boot onto the shaft and position the boot neck in the groove on the shaft. As required, crimp the boot clamp with crimping tool J-35910 or equivalent, or swage the swage ring using tool J-41048 or equivalent.

12. Slide the spacer ring on the axle shaft beyond the second groove.

13. Slide the spider assembly towards the spacer ring as far as it will go on the axle shaft. Make sure that the counterbored face of the spider faces the end of the shaft.

14. Install the shaft retaining ring in the groove of the axle shaft.

15. Slide the spider towards the end of the shaft and reseat the spacer ring in the groove on the shaft.

16. Place approximately ½ of the grease from the service kit in the boot and use the remainder to repack the CV-joint housing.

17. Install the tripot bushing to the housing, making sure the bushing is flush with the face of the housing.

18. Position the large boot clamp on the boot.

19. Slide the housing over the spider assembly on the shaft. Slide the boot, with the clamp in place, over the outside of the tripot bushing and locate the lip of the boot in the groove.

20. Position the CV-joint at the proper dimension. The boot must not be dimpled, stretched or out of shape in any way. If it is not shaped correctly, equalize the pressure in the boot and shape the boot properly by hand.

21. Install the large boot clamp and crimp, using crimping tool J-35910 or equivalent.

22. Install the halfshaft in the vehicle.

Bearing Block Type

1. Remove the halfshaft from the vehicle. Mount the shaft in a soft-jawed vise.

2. Cut the large boot retaining clamp from the CV-joint with side cutters and discard.

3. If equipped with a conventional small boot clamp, cut the small boot retaining clamp with side cutters and discard. If equipped with a swage ring, use a hand grinder to cut through the ring, being careful not to damage the axle shaft. Remove and discard the swage ring.

4. Separate the boot from the CV-joint and slide it away from the joint along the shaft.

5. Remove the outer CV-joint housing from the spider and shaft.

6. Spread the spacer ring with snapring pliers and slide the spacer ring and spider back on the shaft.

7. Remove the shaft retaining ring from the groove on the shaft and slide the spider assembly off the shaft.

8. Thoroughly clean the bearing blocks, spider and housing with clean solvent and allow to dry.

9. Remove the tripot bushing from the housing.

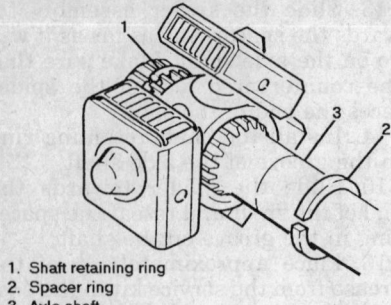

1. Shaft retaining ring
2. Spacer ring
3. Axle shaft

342712

Removing the spider assembly — Bearing block type

10. Remove the spacer ring and boot from the shaft.

11. Inspect the spider, housing, bearing blocks and tripot bushing for damage. Replace the CV-joint, if necessary.

To install:

12. As required, install a new small boot clamp on the neck of a new boot, but do not crimp yet, or install a new swage ring on the neck of a new boot, but do not swage yet.

13. Slide the boot onto the shaft and position the boot neck in the groove on the shaft. As required, crimp the boot clamp with crimping tool J-35910 or equivalent, or swage the swage ring using tool J-41048 or equivalent.

14. Install the spacer ring on the shaft beyond the second groove.

15. Apply a small amount of grease to the inside of the bearing blocks. Align the flats on the opening in the bearing block with the flats on the spider trunnion. Attach the bearing block to the spider trunnion, then rotate 90 degrees to secure the block to the spider.

16. Slide the spider assembly against the spacer ring on the shaft, making sure that the counterbored face of the spider faces the end of the shaft.

17. Install the shaft retaining ring in the groove of the shaft.

18. Slide the spider towards the end of the shaft and reseat the spacer ring in the groove on the shaft.

19. Place approximately half the grease from the CV-joint service kit in the boot and use the remainder to repack the housing.

20. Place a slotted 6 in. square piece of sheet metal between the boot and bearing blocks, to maintain proper bearing block alignment during reassembly.

21. Install the tripot bushing to the housing.

22. Position the large boot clamp on the boot.

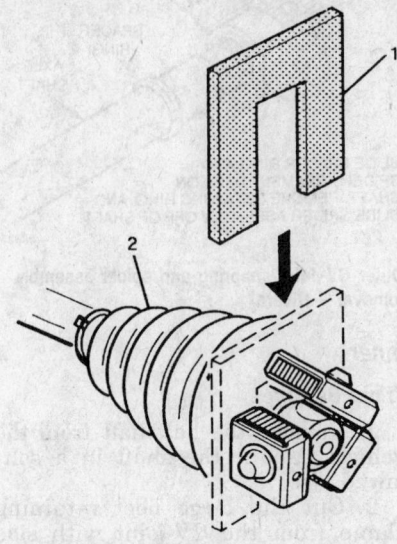

1. Slotted 6" square sheet metal
2. Inboard boot

342715

Proper bearing block alignment

23. Slide the housing over the spider assembly on the shaft and remove the slotted sheet metal plate.

24. Slide the boot, with the large boot clamp in place, over the outside of the tripot bushing and locate the lip of the boot in the groove.

25. Position the CV-joint to the proper dimension. The boot must not be dimpled, stretched or out of shape in any way. If the boot is not shaped correctly, carefully insert a thin, flat, blunt tool (no sharp edges) between the boot and the tripot bushing to equalize pressure. Shape the boot properly by hand and remove the tool.

26. Install the large boot clamp and crimp, using crimping tool J-35910 or equivalent. Make sure that the boot, housing and large clamp all remain in alignment while crimping.

27. Install the halfshaft in the vehicle.

STEERING

Air Bag

---CAUTION---

Some vehicles are equipped with an Air Bag system, also known as the Supplemental Inflatable Restraint (SIR) or Supplemental Restraint System (SRS). The system must be disabled before performing service on or round system components, steering column, instrument panel components, wiring and sensors. Failure to follow safety and disabling procedures could result in accidental Air Bag deployment, possible personal injury and unnecessary system repairs.

PRECAUTIONS

Several precautions must be observed when handling the inflator module to avoid accidental deployment and possible personal injury.

• Never carry the inflator module by the wires or connector on the underside of the module.

• When carrying a live inflator module, hold securely with both hands, and ensure that the bag and trim cover are pointed away.

• Place the inflator module on a bench or other surface with the bag and trim cover facing up.

• With the inflator module on the bench, never place anything on or close to the module which may be thrown in the event of an accidental deployment.

DISARMING

---CAUTION---

The Air Bag system must be disarmed before performing service procedures around the Air Bag or wiring. Failure to do so may cause accidental deployment, resulting in unnecessary repairs and/or personal injury.

1. Disconnect the negative battery cable.

2. Turn the steering wheel so that the vehicle's wheels are pointing straight ahead.

3. Turn the ignition switch to the **LOCK** position and remove the key.

4. Remove the **AIR BAG** fuse from the fuse block.

5. Remove the left sound insulator.

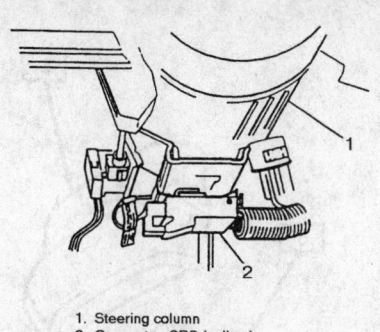

1. Steering column
2. Connector, SRS (yellow)

333823

Disconnect the SRS yellow 2-way connector

6. Remove the Connector Position Assurance (CPA) clip from the yellow 2-way connector at the base of the steering column, and disconnect the harness. If equipped with a passenger's side Air Bag, remove the CPA and disconnect the yellow 2-way connector from the passenger Air Bag lead.

ARMING

1. Turn the ignition switch to the **LOCK** position and remove the key.
2. Connect the yellow 2-way connector at the base of steering column and secure it with the CPA clip. If equipped with a passenger's side Air Bag, connect the yellow 2-way connector at the passenger Air Bag lead and secure it with the CPA clip.
3. Install the left sound insulator.
4. Install the **AIR BAG** fuse in the fuse block.
5. Turn the ignition switch to the **RUN** position and verify that the **AIR BAG** warning lamp flashes 7 times and then turns OFF.
6. Connect the negative battery cable.

Steering Wheel

REMOVAL AND INSTALLATION

─────── **CAUTION** ───────

The Air Bag system must be disarmed before working. Failure to do so may cause accidental deployment, resulting in unnecessary repairs and/or personal injury.

1. Disconnect the negative battery cable.
2. Properly disarm the Air Bag system.
3. Remove the 4 screws from the back of the inflator module. Note that

rotating the steering wheel so that the access holes on the back side of the steering wheel are at the 12 and 6 o'clock positions will allow tool access and reduce the potential of marring the steering column. Remove the inflator module from the steering wheel. Disconnect the horn contact by pushing down slightly and twisting counterclockwise.

4. Remove the CPA and coil assembly connector from the inflation module.

─────── **CAUTION** ───────

When carrying a live inflator module, make sure the bag opening is pointed away from you. In case of accidental deployment, the bag will then deploy with a minimal chance of personal injury. Never carry the inflator module by the wires or connectors on the underside of the module. When placing a live inflator module on a bench or other surface, always face the air bag and trim cover up, away from the surface. This is necessary so that a free space is provided to allow the Air Bag to expand in the unlikely event of accidental deployment.

5. Mark the relationship of the steering wheel to the steering shaft.

Remove the steering column shaft nut.

6. Using J-1859-03 Steering Wheel Puller or equivalent, remove the steering wheel. Use care when handling the steering wheel. The steering wheel is a leather wrapped design and special care must be taken in handling it. It should be stored on a soft cloth so that the leather wrapping is not torn or damaged.

NOTE: Do not hammer on the puller or damage could result to the steering column.

To install:

7. Feed the Air Bag coil assembly lead through the slot in the steering wheel. Make sure the turn signal lever is in the neutral position. Install the steering wheel. Torque the center nut to 30 ft. lbs. (41 Nm).
8. Install the horn contact and coil assembly connector. Make sure the CPA is in place.
9. Align the inflator module to the steering wheel taking precautions so that the wires at the back of the inflator module do not get pinched during assembly. Insert the 4 screws through the back of the steering wheel to the inflator module and torque to 27 inch lbs. (3 Nm).

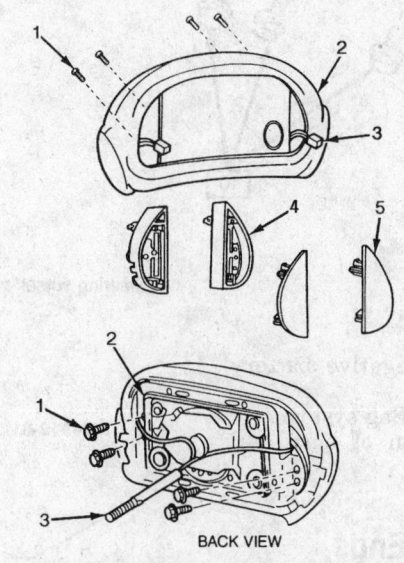

1. Horn switch mounting screw(4)
2. Inflator module
3. Horn lead(2)
4. Horn switch housing(2)
5. Horn switch button(2)

323025

Horn switch assemblies

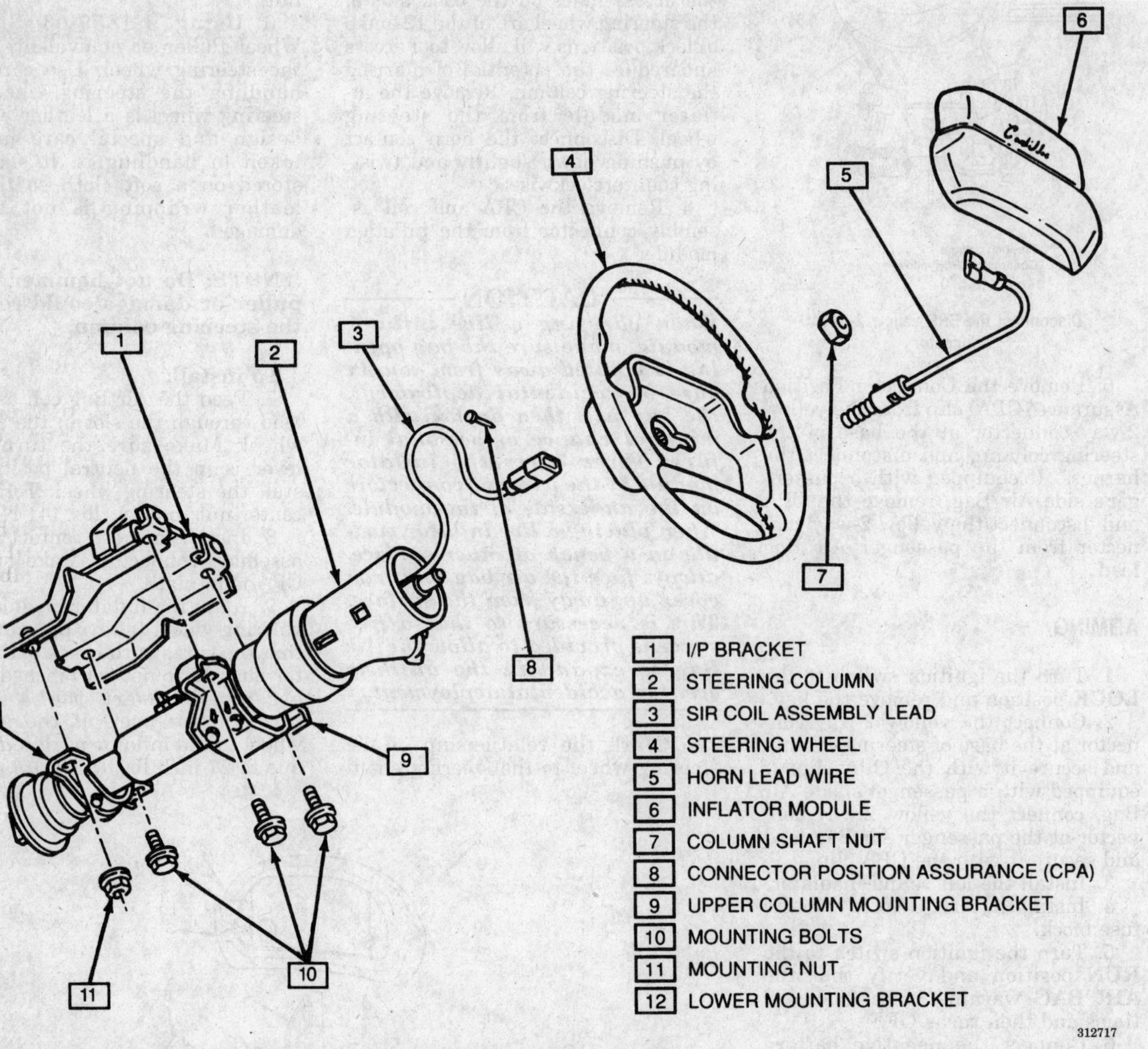

1	I/P BRACKET
2	STEERING COLUMN
3	SIR COIL ASSEMBLY LEAD
4	STEERING WHEEL
5	HORN LEAD WIRE
6	INFLATOR MODULE
7	COLUMN SHAFT NUT
8	CONNECTOR POSITION ASSURANCE (CPA)
9	UPPER COLUMN MOUNTING BRACKET
10	MOUNTING BOLTS
11	MOUNTING NUT
12	LOWER MOUNTING BRACKET

312717

Steering wheel and related components

10. Connect the negative battery cable.

11. Enable the Air Bag system.

12. Check operation of horn and turn signals.

Tie Rod Ends

REMOVAL AND INSTALLATION

Outer

1. Raise and safely support the vehicle.

2. Remove the front wheel.

3. Loosen the tie rod jam nut.

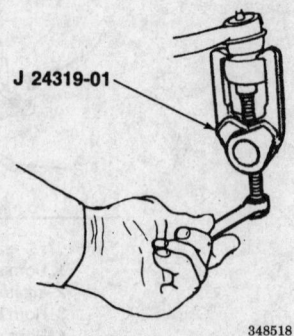

J 24319-01

348518

Separating the tie rod end from the steering knuckle

4. Remove the cotter pin and castle nut from the outer tie rod end.

5. Remove the tie rod end from the steering knuckle using tool J-24319-01 or equivalent.

6. Count the number of turns required for removal of the outer tie rod end, for assembly reference.

To install:

7. Thread the outer tie rod end onto the shaft, the same number of turns required for removal.

8. Install the tie rod end in the steering knuckle and install the castle nut. Tighten the castle nut to 7.5 ft. lbs. plus and additional 30 degree turn. Minimum torque must be 33 ft. lbs. (45 Nm). If necessary the nut

may be tightened an additional 15 degrees for aligning the cotter pin holes.

9. Snug the tie rod end jam nut.
10. Install the wheel.
11. Lower the vehicle.
12. Check and adjust the wheel alignment.
13. After alignment the jam nut should be tightened to 50 ft. lbs. (68 Nm).

Inner

1. Disconnect the negative battery cable.
2. Raise and safely support the vehicle.
3. Remove the front wheels.
4. Remove the rack and pinion assembly from the vehicle.
5. With the rack and pinion unit supported in a vise, loosen the tie rod end jam nut and remove the outer tie rod end, noting the number of turns required for removal.
6. Thread the jam nut off the end of the inner tie rod.
7. Remove the clamp on the small end of the bellows boot.
8. Remove the clamp on the large end of the bellows boot and remove the bellows boot.
9. Slide the shock damper off the inner tie rod end and back onto the rack.
10. Remove the inner tie rod end assembly.
11. Put a wrench on the rack flat to prevent damage to the unit during inner tie rod end removal.
12. Place a second wrench on the inner tie rod end and unscrew the inner tie rod socket from the end of the rack.

To install:

13. Install the inner tie rod assembly on the rack and tighten to 70 ft. lbs. (95 Nm).
14. Stake the inner tie rod end housing to the rack and pinion unit.

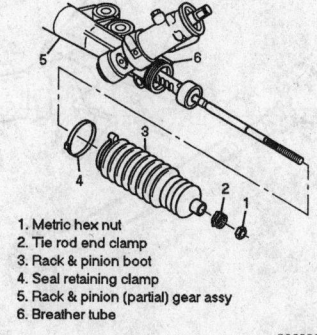

1. Metric hex nut
2. Tie rod end clamp
3. Rack & pinion boot
4. Seal retaining clamp
5. Rack & pinion (partial) gear assy
6. Breather tube

306328

Inner tie rod end assembly

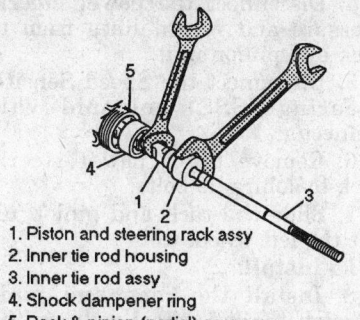

1. Piston and steering rack assy
2. Inner tie rod housing
3. Inner tie rod assy
4. Shock dampener ring
5. Rack & pinion (partial) gear assy

306330

Inner tie rod end removal

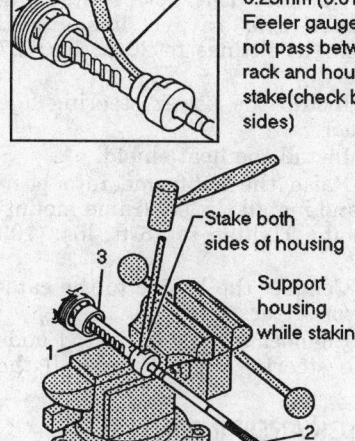

0.25mm (0.010 in.) Feeler gauge must not pass between rack and housing stake(check both sides)

Stake both sides of housing

Support housing while staking

1. Piston and steering rack assy
2. Inner tie rod assy
3. Shock dampener ring

306334

Inner tie rod end staking

Both sides of the inner tie rod socket must be staked so a 0.010 in. (25mm) feeler gauge does not fit between rack and the housing stake.

15. Slide the shock damper over the end of the inner tie rod socket.
16. Install the bellows boot and both clamps.
17. Install the jam nut.
18. Thread the outer tie rod end on the same number of turns as was required for removal.
19. Install the rack and pinion assembly in the vehicle.
20. Install the wheels.
21. Lower the vehicle.

22. Connect the negative battery cable.
23. Check and adjust the wheel alignment.
24. After the alignment the jam nut should be tightened to 50 ft. lbs. (68 Nm).

Power Rack and Pinion

REMOVAL AND INSTALLATION

All Models Except Riviera

—————— **WARNING** ——————
The wheels of the vehicle must be straight ahead and the steering column in the LOCK position before disconnecting the steering column or intermediate shaft from the steering gear. Failure to do so will cause the coil assembly in the steering column to become uncentered which will cause damage to the coil assembly.

—————— **CAUTION** ——————
Failure to disconnect the intermediate shaft from the rack and pinion stub shaft can result in damage to the steering gear and/or intermediate shaft. This damage can cause loss of steering control which could result in personal injury.

1. Disconnect the negative battery cable.
2. Raise and safely support the vehicle.
3. Remove the front wheels.
4. Remove the bolt holding the intermediate shaft to the steering shaft. Disconnect the intermediate shaft lower coupling.
5. Disconnect the Road Sensing Suspension (RSS) sensor from the lower control arm.
6. Remove the cotter pin and castle nuts from both outer tie rods and using tool J-24319 or the equivalent, separate the outer tie rod ends from the steering knuckles.
7. Disconnect the Y-pipe from the catalytic converter.
8. Support the rear of the subframe with a screw jack.
9. Remove the rear subframe bolts.
10. Slowly lower the subframe to gain access.
11. Remove the heat shield.
12. Remove the power steering line retainer.
13. Place a catch pan under the power steering rack and the line fittings.

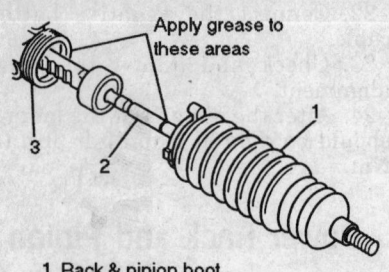

1. Rack & pinion boot
2. Inner tie rod assy
3. Rack & pinion (partial) gear assy

306336

Boot seal grease application

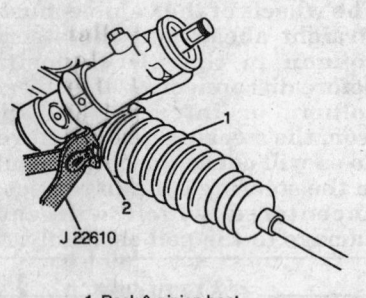

1. Rack & pinion boot
2. Seal retaining clamp

J 22610

306337

Boot clamp installation

14. Disconnect the power steering pressure and return lines from the rack and pinion unit.

15. Disconnect the Speed Sensitive Steering (SSS) solenoid valve connector.

16. Remove the 5 power steering rack-to-subframe bolts.

17. Slide the rack and pinion unit out the left wheel well.

To install:

18. Install the rack and pinion through the left wheel well and into position on the subframe.

19. Install the 5 mounting bolts and tighten to 50 ft. lbs. (68 Nm).

20. Connect the Speed Sensitive Steering (SSS) solenoid valve connector.

21. Connect the power steering pressure and return hoses and tighten the fittings to 20 ft. lbs. (27 Nm).

22. Install the power steering line retainer.

23. Install the heat shield.

24. Raise the subframe into position and install the subframe mounting bolts. Tighten to 76 ft. lbs. (103 Nm).

25. Connect the Y-pipe to the catalytic converter.

26. Connect the outer tie rod ends to the steering knuckle. Install the

castle nuts and torque. Install the cotter pins.

27. Connect the RSS sensor to the lower control arm.

28. Connect the intermediate shaft and install the intermediate shaft-to-steering shaft bolt. Tighten the bolt to 30 ft. lbs. (41 Nm).

29. Install the front wheels.

30. Lower the vehicle.

31. Connect the negative battery cable.

32. Refill the power steering reservoir and bleed the power steering system.

33. Check the wheel alignment. Road test the vehicle.

Riviera

─── **CAUTION** ───

The Air Bag system must be disarmed before working. Failure to do so may cause accidental deployment, resulting in unnecessary repairs and/or personal injury.

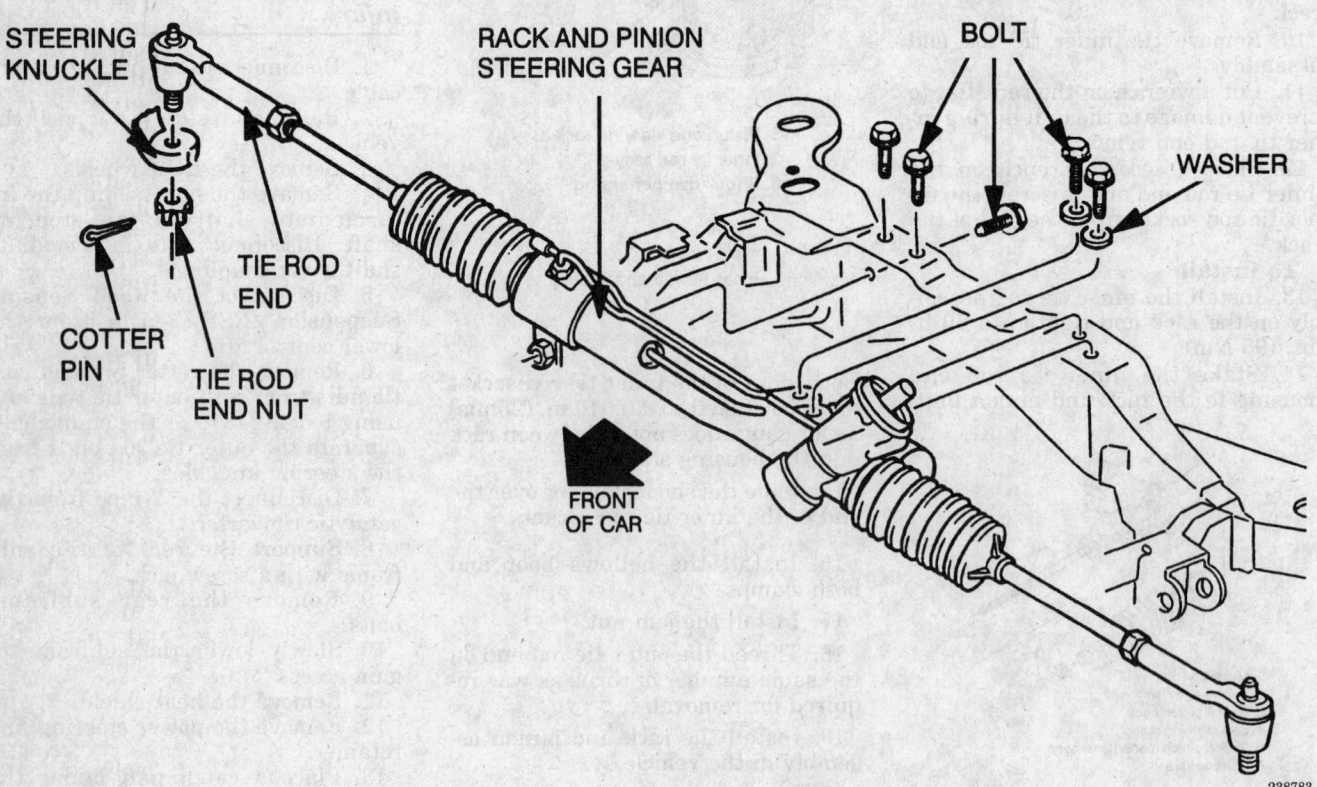

Power steering rack components

238783

--- WARNING ---

If equipped with an Air Bag, the front wheels of the vehicle must be pointing in the straight ahead position, the steering column must be in the LOCK position and the ignition key removed before disconnecting the intermediate shaft. Failure to do so may result in damage to the SRS coil and the possibility of the Air Bag deploying causing unnecessary Air Bag system repairs.

1. Disconnect the negative battery cable.
2. If equipped, properly disable the Air Bag system.
3. Raise and safely support the vehicle.
4. Remove the front wheels.
5. Remove the bolt retaining the intermediate shaft to the steering shaft. Disconnect the intermediate shaft lower coupling.
6. Remove the cotter pin and castle nuts from both outer tie rods and using tool J-24319 or the equivalent, separate the outer tie rod ends from the steering knuckles.
7. Remove the power steering line retainer.
8. Place a catch pan under the power steering rack and the line fittings.
9. Disconnect the power steering pressure and return lines from the rack and pinion unit.
10. Disconnect and remove the power steering pressure switch.
11. Safely support the subframe with a jackstand and remove the 4 rear sub-frame retaining bolts. Carefully lower the subframe about 4 inches.
12. Remove the 5 power steering rack-to-subframe bolts.
13. Slide the rack and pinion unit out the left wheel well.
To install:
14. Install the rack and pinion through the left wheel well and into position on the subframe.
15. Install the 5 mounting bolts and tighten to 50 ft. lbs. (68 Nm).
16. Raise the subframe and install the retaining bolts. Torque the bolts to 76 ft. lbs. (103 Nm).
17. Connect the power steering pressure and return hoses and tighten the fittings to 20 ft. lbs. (27 Nm).
18. Install the power steering line retainer.
19. Connect the tie rod ends to the steering knuckle. Install the castle nuts and torque. Install the cotter pins.

20. Connect the intermediate shaft and install the intermediate shaft-to-steering shaft bolt. Tighten the bolt to 35 ft. lbs. (47 Nm).
21. Install and connect the power steering pressure switch.
22. Install the wheels.
23. Lower the vehicle.
24. Connect the negative battery cable.
25. Refill the power steering reservoir and bleed the power steering system.
26. If equipped, enable the Air Bag system.
27. Check the wheel alignment. A 4-wheel alignment is recommended after any steering/suspension repairs are performed.
28. Road test the vehicle to ensure proper steering assist and proper operation.

Power Steering Pump

BLEEDING

1. With the engine **OFF** and the front wheels off the ground, turn wheels all the way to the left and add power steering fluid to the FULL COLD mark on the level indicator.
2. Bleed the system by turning the wheels from side to side several times (as many as 20 times could be required), without hitting stops. Be sure to keep the level to the FULL COLD mark.
3. Start the engine and run at fast idle momentarily, then recheck the fluid level with the engine idling. If necessary add fluid to bring the level back to the FULL COLD mark.
4. Return the wheels to the center position and continue running the engine for a few minutes. Road test to check the operation of the steering.
5. Recheck the fluid level it should now be stabilized at the FULL HOT level on the indicator.

REMOVAL AND INSTALLATION

1. Disconnect the negative battery cable.
2. Remove the accessory drive belt.
3. Drain the power steering fluid from the reservoir.
4. Remove the power steering return line from the pump.
5. Disconnect the power steering pressure line from the pump.
6. Remove the power steering pump mounting bolt.
7. Remove the power steering pump, reservoir, pulley and bracket.

8. Remove the pump and reservoir from the bracket.
9. Remove the retaining clips that secure the reservoir to the pump and remove the reservoir.
10. Remove the pulley from the pump using puller J-25034-B or equivalent.
To install:
11. Install the power steering pulley using tool J-25033-B or equivalent.
12. Install the reservoir and reservoir-to-power steering pump clips.
13. Install the power steering pump bracket to the pump and reservoir assembly. Tighten the bracket-to-pump bolts to 18 ft. lbs. (24 Nm).
14. Install the power steering pump assembly on the engine. Tighten the pump mounting bolts to 35 ft. lbs. (47 Nm).
15. Connect the power steering pressure line and tighten the fitting to 20 ft. lbs.
16. Connect the return hose to the reservoir.
17. Install the accessory drive belt.
18. Connect the negative battery cable.
19. Refill the power steering pump and bleed the system.
20. Check steering operation and check for leaks.

BRAKES

Anti-Lock Brake System Service

PRECAUTIONS

• Certain components within the Anti-Lock Brake System (ABS) are not intended to be serviced or repaired individually. Only those components with removal and installation procedures should be serviced.
• Do not use rubber hoses or other parts not specifically specified for and ABS system. When using repair kits, replace all parts included in the kit. Partial or incorrect repair may lead to functional problems and require the replacement of components.
• Lubricate rubber parts with clean, fresh brake fluid to ease assembly. Do not use lubricated shop air to clean parts; damage to rubber components may result.
• Use only specified brake fluid from an unopened container.

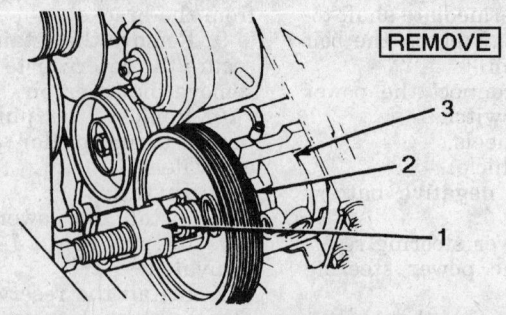

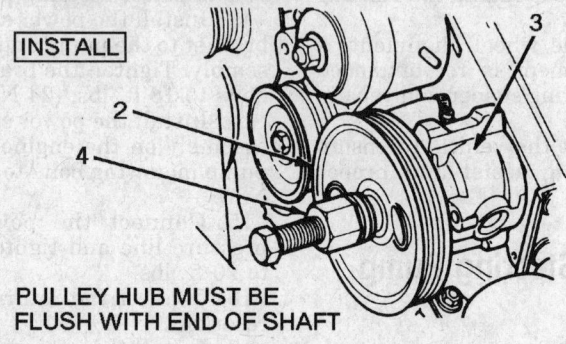

PULLEY HUB MUST BE FLUSH WITH END OF SHAFT

| 1. J-29785-A | 3. Pump |
| 2. Pulley | 4. J-25033-B |

304622

Special tools required to install or remove the power steering pump pulley

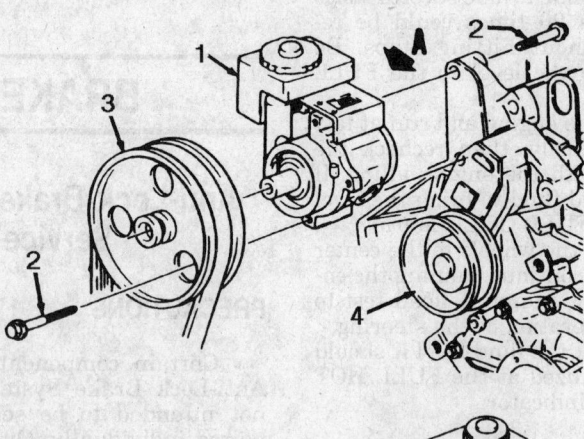

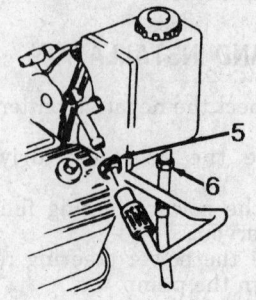

1. Power steering pump
2. Bolt - 27 Nm (20 lb. ft.)
3. Pulley
4. Belt tensioner
5. Clamp
6. 27 Nm (20 lb. ft.)

304618

Power steering pump installation — Riviera models

• If any hydraulic component or line is removed or replaced, it may be necessary to bleed the entire system.

• A clean repair area is essential. Always clean the reservoir and cap thoroughly before removing the cap. The slightest amount of dirt in the fluid may plug an orifice and impair the system function. Perform repairs after components have been thoroughly cleaned; use only denatured alcohol to clean components. Do not allow ABS components to come into contact with any substance containing mineral oil; this includes used shop rags.

• The Anti-Lock control unit is a microprocessor similar to other computer units in the vehicle. Ensure that the ignition switch is **OFF** before removing or installing controller harnesses. Avoid static electricity discharge at or near the controller.

• If any arc welding is to be done on the vehicle, the control unit should be unplugged before welding operations begin.

Master Cylinder

REMOVAL AND INSTALLATION

1. Disconnect the negative battery cable.
2. Use a syringe or similar tool to remove as much fluid as possible from the master cylinder reservoir.
3. Disconnect the electrical connector from the low fluid level sensor.
4. Disconnect the brake lines from the master cylinder. The lines should be capped and the master cylinder ports plugged immediately to prevent excessive fluid loss or contamination.
5. Remove the 2 master cylinder-to-power booster mounting nuts.
6. Pull the master cylinder forward enough to gain access to the rubber hose attached to the master cylinder reservoir.
7. Disconnect the rubber hose from the fluid reservoir. Cap the nozzle on the reservoir and plug the hose.
8. Remove the master cylinder from the vehicle.
To install:
9. Bench bleed the new master cylinder as follows:
   a. Install the master cylinder in a vise, clamping at the flange. DO NOT clamp the master cylinder by the body, it may distort the body and cause a pressure loss.
   b. Make sure the reservoir nozzle is capped to prevent fluid loss.

c. Install short lengths of hoses/lines and fittings to the master cylinder ports, aiming the ends of the hoses/lines into the master cylinder reservoir. Fill the master cylinder with fresh DOT 3 brake fluid and make sure that the ends of the hoses/lines are submerged in the brake fluid.

d. Pump the master cylinder several times until there are no air bubbles seen coming out of the hose/line ends.

e. Remove the hoses/lines and fittings and install the shipping caps to the ports, to prevent brake fluid from dripping on any painted surface.

f. Install the master cylinder cap.

10. Position the master cylinder on the vehicle, remove the reservoir nozzle cap and connect the rubber hose. Secure the hose to the nozzle using a new clamp.

11. Install the master cylinder on the booster studs and torque the retaining nuts to 20 ft. lbs. (27 Nm).

12. Install the brake lines and snug the fittings.

13. Final tighten the brake line fittings to 24 ft. lbs. (33 Nm).

14. Connect the electrical connector to the fluid level sensor.

15. Bleed the brake system and top off the brake fluid level in the master cylinder.

16. Connect the negative battery cable.

17. Road test and check for proper brake operation.

Brake Caliper

REMOVAL AND INSTALLATION

Front

1. Disconnect the negative battery cable.

2. Remove ⅔ of the brake fluid from the master cylinder.

3. Raise and safely support the vehicle.

4. Remove the front wheel. Mark the relationship between the wheel and the wheel stud for re-installation purposes.

5. Install 2 wheel nuts to keep the rotor in place.

6. Using a large C-clamp against the inboard pad, compress the caliper piston into the caliper to provide clearance during removal.

7. Place a catch pan under the caliper.

8. Disconnect the brake hose from the caliper. Cap the line to prevent excessive fluid loss or contamination.

9. Remove the caliper mounting bolts and remove the caliper from the vehicle.

10. If the caliper is being replaced, remove the brake pads from the caliper.

11. Inspect the mounting bolts; sleeves and boots for wear and/or damage. Replace parts as necessary.

To install:

12. Before installing the caliper, make sure the piston is fully seated in the bore and the brake pads are properly seated.

13. Lubricate the mounting bolt shafts and inner diameter of the sleeves with silicone grease.

14. Install the caliper in the caliper mounting bracket and install the mounting bolts. Tighten the mounting bolts to 38 ft. lbs. (51 Nm).

15. Connect the brake hose with the bolt and new gaskets. Tighten the brake hose bolt to 33 ft. lbs. (45 Nm).

16. Refill the master cylinder and bleed the brake system.

17. Remove the 2 wheel nuts securing the rotor.

18. Install the wheel and lower the vehicle.

19. Connect the negative battery cable.

20. Road test the vehicle for proper brake system operation.

Rear

1. Disconnect the negative battery cable.

2. Remove ⅔ of the brake fluid from the master cylinder.

3. Raise and safely support the vehicle.

4. Remove the rear wheel.

5. Install 2 wheel nuts to keep the rotor in place.

6. Place a catch pan under the caliper.

7. Disconnect the brake hose from the caliper. Cap the line to prevent fluid loss or contamination.

8. Loosen the tension on the parking brake at the equalizer.

9. Remove the parking brake cable mounting lever, and remove the cable end by lifting up and disengaging the end.

10. Remove the caliper sleeve bolt.

11. Lift the caliper up and slide the caliper inboard off of the pin sleeve to remove the caliper from the vehicle.

12. Use a suitable tool in the caliper piston slots to turn the piston and thread it into the caliper. After bottoming the piston, lift the inner edge of the boot next to the piston and press out any trapped air the boot must lay flat.

To install:

13. Inspect the pin boot, bolt boot and sleeve boot for cuts, tears or deterioration and replace as necessary.

14. Inspect the bolt sleeve and pin sleeve for corrosion or damage. Pull the boots to gain access to the sleeves for inspection or replacement. Replace corroded or damaged sleeves; do not try to polish away corrosion.

15. If not replaced, remove the pin boot from the caliper and install the small end over the pin sleeve (installed on caliper support) until the boot seats in the pin groove. This prevents cutting the pin boot when sliding the caliper onto the pin sleeve.

16. Hold the caliper in the position as removed and start it over the end of the pin sleeve. As the caliper approaches the pin boot, work the large end of the pin boot in the caliper groove, then push the caliper fully onto the pin.

17. Pivot the caliper down, being careful not to damage the piston boot on the inboard disc brake pad. Compress the sleeve boot by hand as the caliper moves into position to prevent boot damage.

18. After the caliper is in position, recheck the position of the pad clips. If necessary, use a small prybar to reseat or enter the pad clips on the bracket abutments.

19. Install the sleeve bolt and tighten to 20 ft. lbs. (27 Nm).

20. Install the parking brake cable bracket, with the cable attached, and torque the bolt to 32 ft. lbs. (43 Nm).

21. Install the parking brake cable onto the parking brake lever and the retaining clip onto the parking brake cable.

22. Connect the brake hose with the bolt and new gaskets and tighten the bolt to 32 ft. lbs. (43 Nm).

23. Adjust the parking brake cable.

24. Refill the master cylinder and bleed the brake system.

25. Remove the wheel nuts retaining the rotor and install the wheel. Lower the vehicle.

26. Connect the negative battery cable.

27. Road test the vehicle for proper brake system operation.

Disc Brake Pads

REMOVAL AND INSTALLATION

Front

1. Remove ⅔ of the brake fluid from the master cylinder reservoir.
2. Raise and safely support the vehicle.
3. Remove the front wheel.
4. Remove the caliper mounting bolts.
5. Remove the caliper from the steering knuckle without disconnecting the brake hose. Suspend the caliper from the coil spring with wire. Do not let the caliper hang from the brake hose.
6. Remove the outboard pad by pushing in enough to unseat the back of the pad from the caliper housing. Once the button on the back of the pad is clear of the caliper, push the pad out the bottom of the caliper.
7. Remove the inboard pad by unsnapping the shoe spring from the piston.
   To install:
8. Lubricate all the brake caliper mounting surfaces.
9. Before installing the pads in the caliper, the piston must be fully seated in the bore. A large C-clamp can be used to compress the piston.
10. Install the inboard pad by snapping the pad retainer spring into the caliper piston. The pad must lay flat against the piston.
11. Install the outer pad by pushing the pad straight in from the bottom of the caliper. A click will be heard when the button on the back of the pad seats in the recess in the caliper. Be careful when installing the new pad, not to distort the spring clip on the outside of the pad. Make sure the wear indicator is at the trailing edge of the pad during forward wheel rotation.

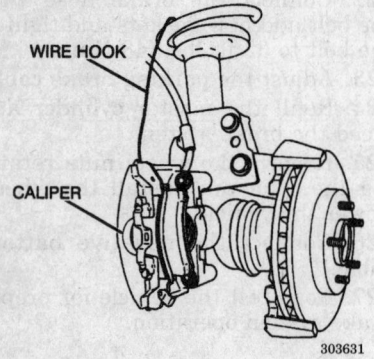

Supporting the front caliper

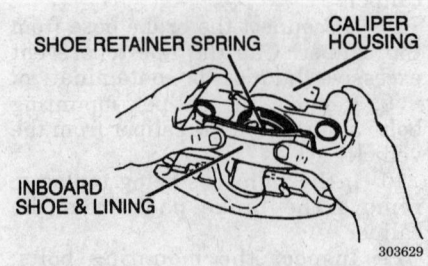

Installing the inboard disc brake pad in the front caliper

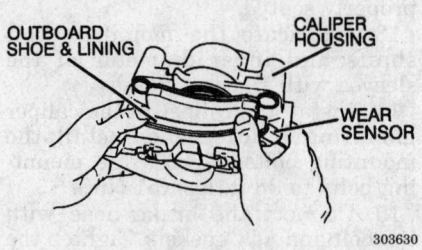

Installing the outboard disc brake pad in the front caliper

12. Install the caliper on the steering knuckle and install the bolts. Tighten the mounting bolts to 38 ft. lbs. (51 Nm).
13. Install the wheel and lower the vehicle.
14. Fill the master cylinder reservoir to the FULL mark and pump the brake pedal several times to seat the brake pads.
15. Check the fluid level in the master cylinder again and top off, as necessary.
16. Road test the vehicle for proper brake operation.

Rear

1. Remove ⅔ of the brake fluid from the master cylinder reservoir.
2. Raise and safely support the vehicle.
3. Remove the rear wheel.
4. Remove the caliper mounting bolt. Remove the parking brake cable bracket.
5. Remove the caliper from the caliper mounting bracket without disconnecting the brake hose. Suspend the caliper from the strut with a wire. Do not let the caliper hang from the brake hose.
6. Remove the inboard and the outboard brake pads from the caliper mounting bracket.

To install:
7. Lubricate mounting surfaces and install new brake pad clips.
8. The caliper piston must be fully seated in the bore. A spanner type tool can be used to bottom the piston into its bore by turning it in. Make sure the slots in the piston are straight across from each other so the notches on the brake pad will seat.
9. Install the inboard pad by inserting the pad into the straight tabs on the retainer, then pressing down and snapping the pad under the S-shaped tabs. The pad must lay flat against the rotor. Make sure the D-shaped notches are in line with the buttons on the back of the pad lining. If they are not in alignment, rotate the piston until the D-shaped notches face the caliper mounting bolt holes.
10. Install the outer pad into the brake pad clips. Make sure the wear indicator is at the leading edge of the pad during forward wheel rotation.
11. Install the caliper on the mounting bracket and install the bolt. Tighten the mounting bolt to 20 ft. lbs. (27 Nm). Install the bolt to the parking brake cable bracket.
12. Install the wheel and lower the vehicle.
13. Fill the master cylinder reservoir to the FULL mark and pump the brake pedal several times to seat the brake pads.
14. Check the fluid level in the master cylinder again and top off, as necessary.
15. Road test the vehicle for proper brake operation.

Brake Rotor

REMOVAL AND INSTALLATION

1. Raise and safely support the vehicle.
2. Remove the front wheel.
3. Disconnect the parking brake cable bracket from the rear caliper if removing a rear rotor.
4. Remove the caliper mounting bolts.
5. Remove the caliper and suspend it with wire from the coil spring. Do not let the caliper hang from the brake hose.
6. If equipped, remove the retainers from the wheel studs; they do not have to be reinstalled.
7. Remove the rotor from the hub.
8. Inspect the rotor for discoloration due to overheating, excessive glazing or scoring. Machine or replace the rotor, as necessary. If

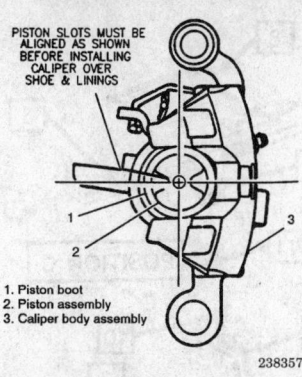

PISTON SLOTS MUST BE ALIGNED AS SHOWN BEFORE INSTALLING CALIPER OVER SHOE & LININGS

1. Piston boot
2. Piston assembly
3. Caliper body assembly

238357

Aligning the slots on the rear caliper

machining, observe the minimum rotor thickness specification.

To install:

9. Install the rotor over the wheel studs onto the hub.

10. Install the caliper with the mounting bolts. Tighten the bolts to 38 ft. lbs. (51 Nm).

11. Install the parking brake cable bracket and tighten to 32 ft. lbs. (43 Nm).

12. Install the wheel and lower the vehicle.

13. Pump the brake pedal several times to seat the pads against the rotor. Check the brake fluid level in the master cylinder reservoir.

14. Road test for proper brake operation.

Parking Brake Cable

ADJUSTMENT

1. Apply the brake pedal 3 times with a pedal force of approximately 175 lbs. (778 N). Apply and release the parking brake 3 times.

2. Raise and safely support the vehicle.

3. Check the parking brake lever for full release:

  a. Turn the ignition **ON**.

  b. The brake warning light should be **OFF**. If the brake warning light is still **ON** and the parking brake lever is completely released, pull downward on the front parking brake cable to remove slack from the lever assembly.

  c. Turn the ignition switch **OFF**.

4. Remove the rear wheels. Reinstall 2 wheel nuts on each side to retain the brake rotors.

5. Pull the parking lever 4 clicks. The parking brake levers on both calipers should be against the lever stops on the caliper housings. If the levers are not against the stops,

check for binding in the rear cables and/or loosen the cables at the equalizer nut until both left and right levers are against their stops.

6. Adjust the equalizer adjusting nut until the parking brake levers on both calipers just begin to move off their stops.

7. Back off the adjuster nut until the levers move back, barely touching their stops.

8. Operate the parking brake lever several times to check the adjustment. After cable adjustment, the parking brake lever should travel no more than 14 ratchet clicks. The rear wheels should not turn forward when the parking brake lever is applied 8–16 ratchet clicks.

9. Release the parking brake lever. Both rear wheels must turn freely in both directions. The parking brake levers on both calipers should be resting on their stops.

10. Remove the wheel nuts retaining the rotors. Install the wheels.

11. Lower the vehicle.

REMOVAL AND INSTALLATION

Front Cable

1. Raise and safely support the vehicle.

2. Disconnect the front cable from the intermediate cable at the adjuster.

3. Disengage the cable housing retainer at the point where the cable passes through the underbody.

4. Lower the vehicle.

5. Remove the left side lower dash panel.

6. Disconnect the cable from the parking brake pedal assembly.

7. Remove the cable from the vehicle.

To install:

8. Position the cable in the vehicle.

9. Connect the cable end to the pedal assembly.

10. Install the left side lower dash panel.

11. Raise and safely support the vehicle.

12. Connect the front cable to the intermediate cable.

13. Adjust the parking brake.

14. Lower the vehicle.

Intermediate Cable

1. Raise and safely support the vehicle.

2. Disconnect the front cable from the intermediate cable at the adjuster.

3. Remove the adjuster from the intermediate cable.

4. Disconnect the left and right rear cables from the equalizer.

5. Disconnect the intermediate cable from the equalizer.

6. Remove the intermediate cable from the vehicle.

To install:

7. Position the intermediate cable in the vehicle, through the cable retainers.

8. Connect the intermediate cable, right and left rear cables to the equalizer.

9. Install the adjuster on the intermediate cable forward end.

10. Connect the front cable to the intermediate cable at the adjuster.

11. Adjust the parking brake.

12. Lower the vehicle.

Rear Cable

1. Raise and safely support the vehicle.

2. Remove the rear wheel at the side of the vehicle where the cable is to be replaced.

3. Loosen the cable adjuster.

4. Disconnect the left and right rear cables from the equalizer.

5. Disconnect the intermediate cable from the equalizer.

6. Disconnect the cable housing retainer from the rear suspension crossmember assembly.

7. Remove the retainer clip from the caliper actuator lever.

8. Remove the cable from the lever slot.

9. Disconnect the cable housing retainer from the caliper slot.

10. Remove the cable from the vehicle.

To install:

11. Position the cable in the retainers.

12. Install the cable and housing retainer in the caliper bracket.

13. Connect the cable to the actuator lever.

14. Install the retainer clip on the actuator lever.

15. Connect the left rear, right rear and intermediate cables at the equalizer.

16. Connect the cable housing retainer to the rear suspension crossmember assembly.

17. Adjust the parking brake assembly.

18. Install the wheel.

19. Lower the vehicle.

Brake System Bleeding

Whenever air gets into the brake hydraulic system, the system must be bled to remove it. If air has entered

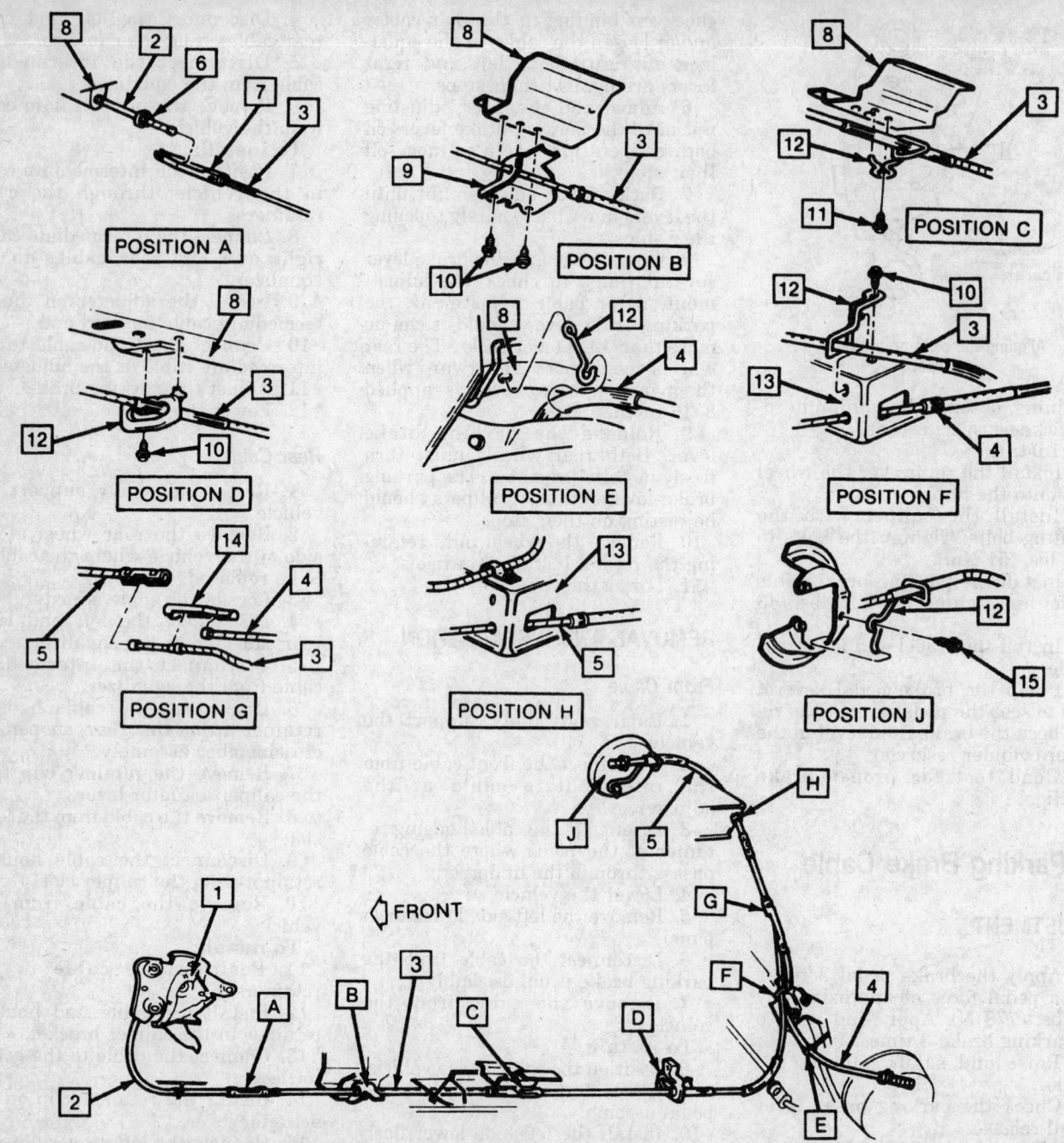

POSITION A

POSITION B

POSITION C

POSITION D

POSITION E

POSITION F

POSITION G

POSITION H

POSITION J

FRONT

1	PEDAL ASSEMBLY	9	BRACKET
2	FRONT CABLE	10	SCREWS
3	INTERMEDIATE CABLE	11	SCREW
4	LEFT REAR CABLE	12	CABLE GUIDE
5	RIGHT REAR CABLE	13	REAR SUSPENSION CROSSMEMBER
6	ADJUSTER NUT	14	EQUALIZER
7	ADJUSTER	15	BOLT
8	PART OF UNDERBODY		

321339

Parking brake cable mounting

the system due to low fluid level, or from brake lines having been disconnected at the master cylinder, the system should be bled at all 4 wheels. If a brake line is disconnected at one wheel, only that wheel cylinder or caliper needs to be bled. If lines are disconnected at any fitting located between the master cylinder and the wheels, then the brake system served by the disconnected line must be bled.

The proper fluid level in the master cylinder reservoir must be maintained during bleeding. Periodically check the fluid level in the reservoir throughout the bleeding procedure, however, be sure the reservoir cap is installed before the brake pedal is pressed, to prevent brake fluid from spraying from the reservoir.

Have an assistant press the brake pedal, when needed, during bleeding.

NOTE: If the Brake Pressure Modulator (BPM) valve, located in the left front of the engine compartment, is equipped with a bleed screw and the entire brake system must be bled, the master cylinder prime pipe should be bled first. The prime pipe is bled at the BPM valve. The BPM valve prime pipe must be bled any time the master cylinder, reservoir, prime pipe or BPM valve is replaced; also if the reservoir fluid level becomes too low due to a fluid leak or brake system service.

1. If necessary, bleed the BPM valve as follows:
   a. Make sure the master cylinder is filled to the proper level. Leave the cap off the reservoir while bleeding.
   b. Remove the engine compartment close-out panel above the radiator.
   c. Remove the air cleaner box.
   d. Remove the bleed screw cap and place a wrench over the BPM valve bleed screw.
   e. Install a short piece of clear tube over the bleed screw and put the other end of the tube in a clear container.
   f. Open the bleed screw and allow fluid to flow into the container until all air is removed.
   g. Tighten the BPM valve bleed screw to 106 inch lbs. (12 Nm), making sure it seals. Install the bleed screw cap.
   h. Install the air cleaner box and the close-out panel.
2. Deplete the power brake booster vacuum reserve by applying the brakes several times with the ignition **OFF**.
3. Fill the master cylinder reservoir with brake fluid.
4. If the master cylinder is known or suspected to have air in the bore, bleed it as follows before bleeding any wheel cylinder or caliper.
   a. Disconnect the front brake line connection at the master cylinder.
   b. Allow brake fluid to flow from the front line connector port.
   c. Connect the front brake line to the master cylinder and tighten.
   d. Depress the brake pedal slowly one time and hold. Loosen the front brake line connection at the master cylinder to purge air from the cylinder. Tighten the connection and then release the brake pedal slowly. Wait 15 seconds, then repeat the sequence, until all air is removed from the bore.
   e. After all air has been removed at the forward connection, bleed the master cylinder at the rear connection in the same manner.
5. Bleed the calipers only after all air is removed from the master cylinder. Bleed the wheel circuits in the following sequence: Left front, Right front, Left rear, Right rear.
6. Raise and safely support the vehicle.
7. Remove the bleeder valve cap and place the proper size box-end wrench over the bleeder valve.
8. Attach a transparent tube to the bleeder valve. Submerge the other end of the tube in a clear container, partially filled with clean brake fluid.
9. Depress the brake pedal slowly one time and hold. Loosen the bleeder valve to purge the air from the caliper. Tighten the bleeder valve and slowly release the brake pedal. Wait 15 seconds, then repeat the sequence, until all air is removed. It may be necessary to repeat the sequence several times to remove all the air.

NOTE: Depress the brake pedal slowly. Rapid pedal pumping pushes the master cylinder secondary piston down the bore in a way that makes it difficult to bleed the system.

10. Install the bleeder valve caps and lower the vehicle.
11. Check the master cylinder fluid level and add fluid as necessary.
12. Check the brake pedal for "sponginess". Repeat the entire bleeding procedure to correct "sponginess".

Wheel Speed Sensor

REMOVAL AND INSTALLATION

Front

All wheel speed sensors consist of a permanent magnet and coil, which generate a small amount of AC voltage due to the magnetic induction caused by a toothed sensor ring passing near the sensor face at a specified gap. The front and rear sensor gap is not adjustable. Left and right rear sensors are not interchangeable.

1. Raise and safely support the vehicle.
2. Remove the front wheel.
3. Insert a drift punch through the caliper and into the rotor cooling fins to prevent the rotor from turning.
4. Remove the axle nut and washer.
5. Remove the caliper mounting bolts. Remove the caliper from the steering knuckle and suspend from the coil spring with wire. Do not allow the brake hose to support the weight of the caliper.
6. Remove the brake rotor.
7. Disconnect the ABS speed sensor connector and unclip from the dust shield.
8. Remove the 3 hub and bearing bolts and remove the backing plate.
9. Place the transaxle selector in the **P** detent.
10. Install J-28733 or an equivalent puller, and separate the hub and bearing from the drive axle.
11. Remove the hub and bearing assembly from the steering knuckle.
12. Gently pry off the wheel speed sensor slinger. Discard the used slinger.
13. Remove the wheel speed sensor by gently prying off the hub and bearing assembly.

NOTE: Observe the following when servicing the wheel speed sensor: Do not allow debris to enter the bearing when the sensor is removed. Do not add lubricant to the bearing through the sensor housing opening; the bearing is lubricated for the life of the vehicle. Do not clean grease from the toothed sensor ring; grease does not effect wheel speed sensor operation. Inspect the wheel speed sensor cap for any indication of water or debris entry. If indication of water or debris exists, replace the entire hub and bearing assembly.

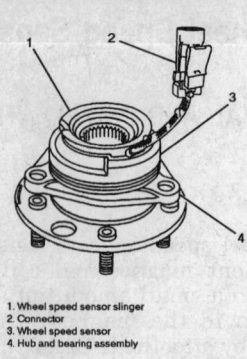

1. Wheel speed sensor slinger
2. Connector
3. Wheel speed sensor
4. Hub and bearing assembly

307217

Front wheel speed sensor assembly

To install:

14. Apply Loctite® 620 or equivalent to the outer diameter of the bearing hub.

15. Place the hub and bearing assembly on J-38764-4 or equivalent, to prevent possible damage to the wheel studs.

16. Install the wheel speed sensor using tool J-38764-1 or equivalent, to press the wheel speed sensor into place.

17. Install the hub and bearing over the half shaft splines. Make sure the splines engage smoothly.

18. Apply a light coating of grease to the steering knuckle bore.

19. Slide the hub assembly onto the axle as far as possible. If the hub will not bottom out on the axle, install the hub mounting bolts and use the axle nut to draw the hub onto the axle.

20. Once the hub is flush with the steering knuckle remove the mounting bolts, install the dust shield and mounting bolts. Tighten the mounting bolts to 70 ft. lbs. (95 Nm).

21. Place the transaxle in **N**.

22. Connect the ABS front wheel speed sensor connector and clip to the dust shield.

23. Install the brake rotor.

24. Install the caliper and tighten the mounting bolts to 38 ft. lbs. (51 Nm).

25. Insert a drift punch through the rotor to prevent the axle from turning.

26. Tighten the drive axle shaft nut to 107 ft. lbs. (145 Nm).

27. Remove the drift punch.

28. Install the wheel and torque the wheel nuts to 100 ft. lbs. (140 Nm).

29. Lower the vehicle and road test for proper operation.

Rear

The rear wheel speed sensor and the rear toothed sensor ring are integral parts of the rear hub and bearing assembly. They cannot be serviced separately and no attempt should be made to remove them from the hub and bearing assembly.

FRONT SUSPENSION

Strut and Spring

REMOVAL AND INSTALLATION

———— WARNING ————
When working near the half-shafts, use care to prevent the inner tripot CV-joint from being overextended. If the joint is overextended, the internal joint components could separate, resulting in CV-joint failure.

1. Disconnect the negative battery cable.

2. Raise and safely support the vehicle.

3. Remove the front wheel.

4. If equipped, disconnect the electrical connector from the top of the strut.

5. Remove the upper strut mounting nuts.

6. Remove the Road Sensing Suspension (RSS) position sensor from the lower control arm, if equipped.

7. Disconnect the ABS wheel speed sensor, if equipped.

8. Remove the wheel speed sensor from the bracket on the strut, if equipped.

9. Remove the brake line bracket from the strut.

10. Remove the stabilizer link from the strut.

11. Scribe a mark on the strut referencing the lower strut bracket to the steering knuckle.

12. Remove the nuts and through bolts from the lower strut bracket and remove the strut from the vehicle.

———— WARNING ————
Be careful not to scratch or crack the protective coating on the coil spring, as damage may cause premature spring failure.

13. Mount the strut assembly in strut compressor tool J-34013 or equivalent strut compressor.

14. Turn the compressor forcing screw until the spring compresses slightly.

15. Use a T-50 Torx® bit to keep the strut shaft from turning, then remove the nut on the top of the strut shaft.

16. Loosen the compressor screw while guiding the strut shaft out of the assembly. Continue loosening the compressor screw until the strut and spring can be removed.

To install:

17. Mount the strut in the compressor tool. Use clamping tool J-34013-20 or the equivalent, to hold the strut shaft in place.

18. Install the spring over the strut. The flat on the upper spring seat must face out from the centerline of the vehicle, or when mounted in the strut compressor, the spring seat must face the same direction as the steering knuckle mounting flange.

NOTE: If the bearing was removed from the upper spring seat, it must be reinstalled in the spring seat in the same position before attaching to the strut mount.

19. Turn the compressor screw to compress the spring, while guiding the strut shaft through the top of the strut assembly. If necessary, use tool J-34013-38 or equivalent, to guide the shaft.

20. When the strut shaft threads are visible through the top of the strut assembly, install the nut.

21. Remove clamping tool J-34013-20 from the strut shaft.

22. Tighten the strut shaft nut to 55 ft. lbs. (75 Nm) while holding the strut shaft with a T-50 Torx® bit.

23. Remove the strut assembly from the compressor tool.

24. Install the strut assembly into the vehicle. Install the strut-to-knuckle bolts and nuts, but do not tighten yet.

25. Connect the stabilizer link to the strut assembly, but do not tighten the nuts yet.

26. Connect the brake line bracket to the strut.

27. Install the speed sensor bracket on the strut, if equipped.

28. Connect the ABS sensor, if equipped.

29. Install the RSS sensor to the lower control arm, if equipped.

30. Install the upper strut mounting nuts. If equipped, connect the electrical connector to the top of the strut.

31. Tighten the upper strut mounting nuts to 18 ft. lbs. (24 Nm). Tighten the stabilizer link nuts to 48 ft. lbs. (65 Nm). Align the marks made during removal and tighten the strut-to-knuckle bolt nuts to 140 ft. lbs. (190 Nm).

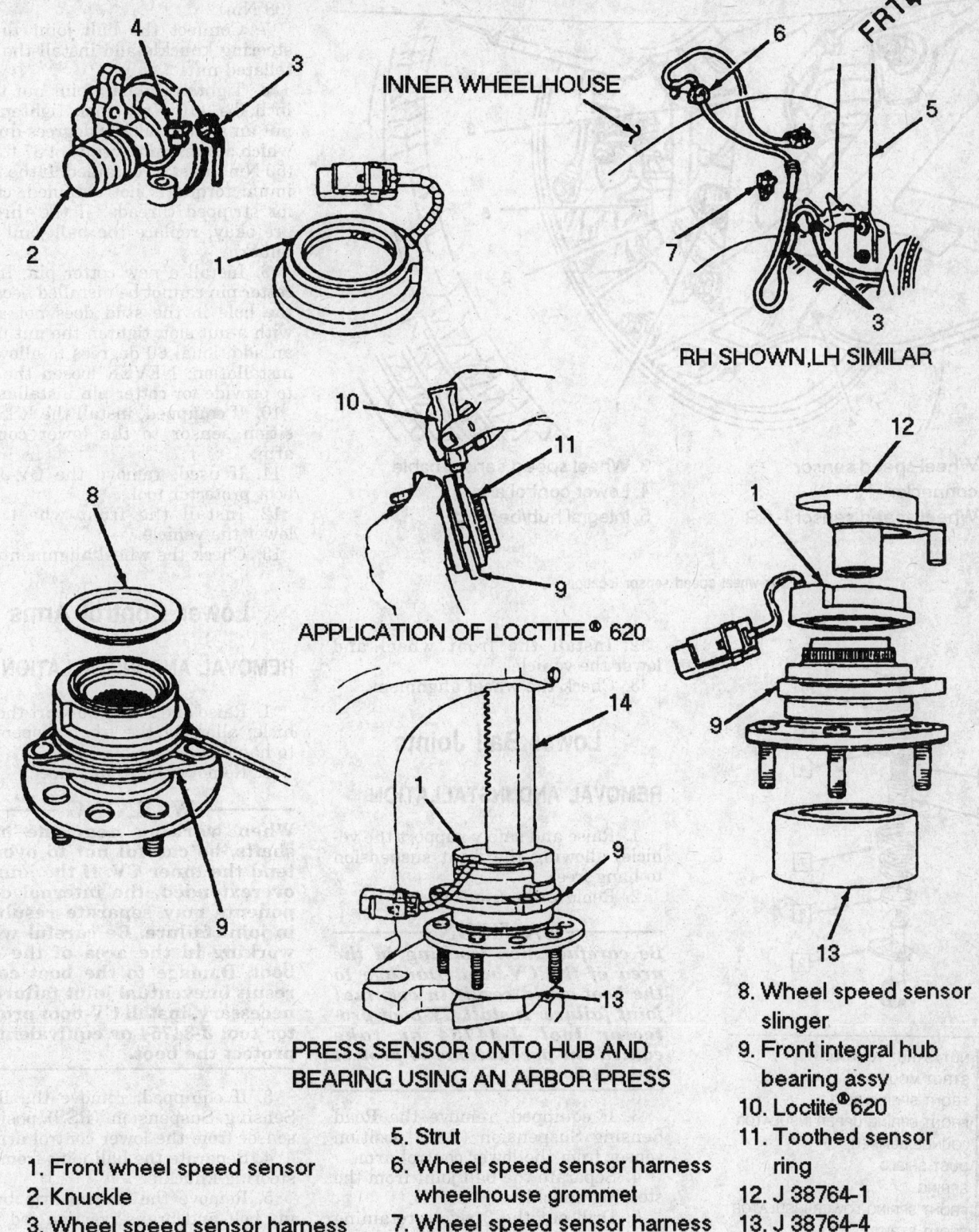

INNER WHEELHOUSE

RH SHOWN,LH SIMILAR

APPLICATION OF LOCTITE ® 620

PRESS SENSOR ONTO HUB AND
BEARING USING AN ARBOR PRESS

1. Front wheel speed sensor
2. Knuckle
3. Wheel speed sensor harness
4. Wheel speed sensor connector

5. Strut
6. Wheel speed sensor harness
   wheelhouse grommet
7. Wheel speed sensor harness
   bracket

8. Wheel speed sensor
   slinger
9. Front integral hub
   bearing assy
10. Loctite® 620
11. Toothed sensor
    ring
12. J 38764-1
13. J 38764-4
14. Arbor press

307273

Front sensor removal

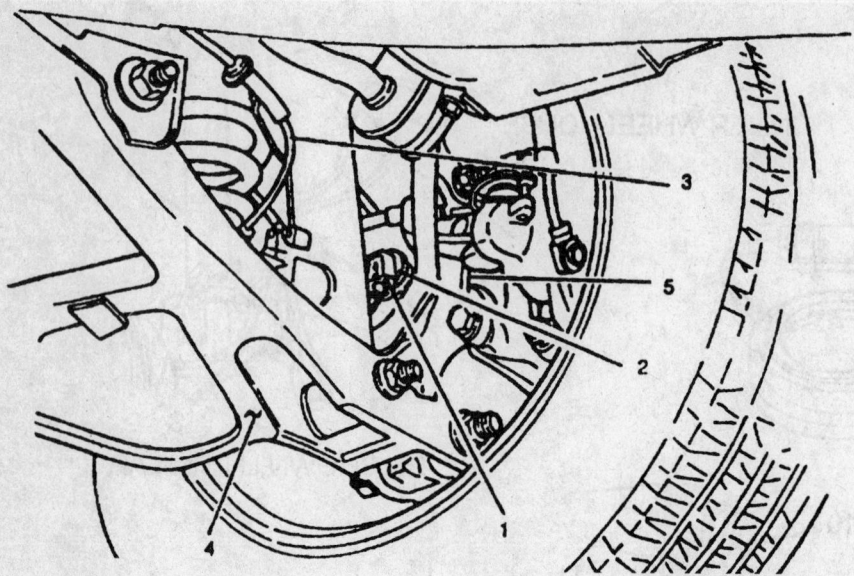

1. Wheel speed sensor connector - RR
2. Wheel speed sensor - RR
3. Wheel speed sensor cable
4. Lower control arm
5. Integral hub/bearing

307272

Rear wheel speed sensor location

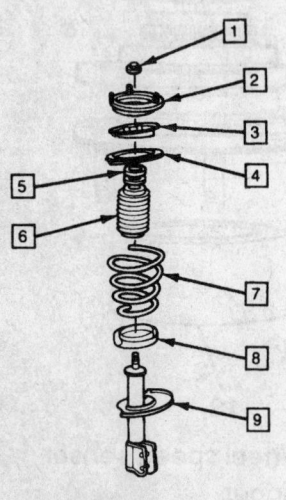

1	NUT, STRUT TO MOUNT
2	STRUT MOUNT
3	FRONT SPRING SEAT
4	FRONT SPRING UPPER INSULATOR
5	JOUNCE BUMPER
6	DUST SHIELD
7	SPRING
8	FRONT SPRING LOWER INSULATOR
9	FRONT STRUT

351169

Disassembled view of strut

32. Install the front wheel and lower the vehicle.
33. Check the wheel alignment.

Lower Ball Joints

REMOVAL AND INSTALLATION

1. Raise and safely support the vehicle, allowing the front suspension to hang free.
2. Remove the front wheel.

— **CAUTION** —
Be careful when working in the area of the CV-boot. Damage to the boot could result in eventual joint failure. Install CV-boot protector tool J-34754 or take equivalent precautions to protect the soft boot.

3. If equipped, remove the Road Sensing Suspension (RSS) position sensor from the lower control arm.
4. Separate the ball joint from the steering knuckle.
5. Drill out the 3 rivets retaining the ball joint to the lower control arm and remove the ball joint.
To install:
6. Attach the new ball joint to the lower control arm with 3 mounting bolts and nuts. The bolts must be in-

stalled from the bottom of the control arm. Tighten the nuts to 50 ft. lbs. (68 Nm).
7. Connect the ball joint to the steering knuckle and install the castellated nut.
8. Tighten the ball joint nut to 84 inch lbs. (10 Nm), then tighten the nut an additional 120 degrees during which a minimum torque of 37 ft. lbs. (50 Nm) must be obtained. If the minimum torque is not obtained, check for stripped threads. If the threads are okay, replace the ball joint and knuckle.
9. Install a new cotter pin. If the cotter pin cannot be installed because the hole in the stud does not align with a nut slot, tighten the nut up to an additional 60 degrees to allow for installation. NEVER loosen the nut to provide for cotter pin installation.
10. If equipped, install the RSS position sensor to the lower control arm.
11. If used, remove the CV-Joint boot protector tool.
12. Install the front wheel and lower the vehicle.
13. Check the wheel alignment.

Lower Control Arms

REMOVAL AND INSTALLATION

1. Raise and safely support the vehicle, allowing the front suspension to hang free.
2. Remove the front wheel.

— **WARNING** —
When working near the half-shafts, be careful not to overextend the inner CV. If the joint is overextended, the internal components may separate resulting in joint failure. Be careful when working in the area of the CV-boot. Damage to the boot could result in eventual joint failure. If necessary, install CV-boot protector tool J-34754 or equivalent, to protect the boot.

3. If equipped, remove the Road Sensing Suspension (RSS) position sensor from the lower control arm.
4. Separate the ball joint from the steering knuckle.
5. Remove the control arm bushing bolt and brake reaction rod nut, retainer and insulator. Remove the control arm.
To install:
6. Position the control arm to the frame and install the bushing bolt and brake reaction rod nut, retainer

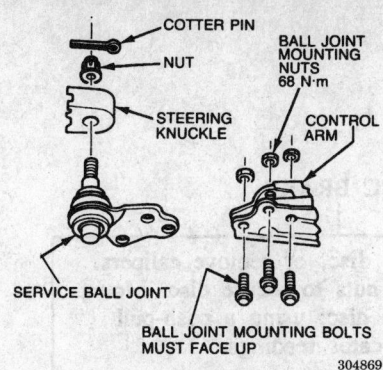

Lower ball joint mounting

and insulator. Do not tighten at this time.

7. Connect the ball joint to the steering knuckle and install the castellated nut. Tighten the ball joint nut.

8. Install a new cotter pin in the ball joint stud. If the cotter pin cannot be installed because the hole in the stud does not align with a nut opening, tighten the nut up to an additional 60 degrees to allow for installation. NEVER loosen the nut to provide for cotter pin installation.

9. If equipped, install the RSS position sensor to the lower control arm.

10. Install the wheel and lower the vehicle.

11. When the weight of the vehicle is supported by the lower control arms, tighten the control arm bushing bolt to 105 ft. lbs. (140 Nm) or nut to 90 ft. lbs. (123 Nm). Tighten the brake reaction rod nut to 60 ft. lbs. (78 Nm).

12. Check the wheel alignment.

Sway Bar

REMOVAL AND INSTALLATION

1. Raise and safely support the vehicle, allowing the front suspension to hang free.

2. Remove the right front wheel.

3. If equipped, remove the Road Sensing Suspension (RSS) position sensor from the lower control arm.

4. Remove the stabilizer links. Use pliers or a Torx® bit, as required, to keep the ball stud from turning while loosening the nut.

5. Remove the bracket bolts, brackets and bushings.

6. Disconnect the exhaust pipe from the manifold.

7. Remove the stabilizer bar.

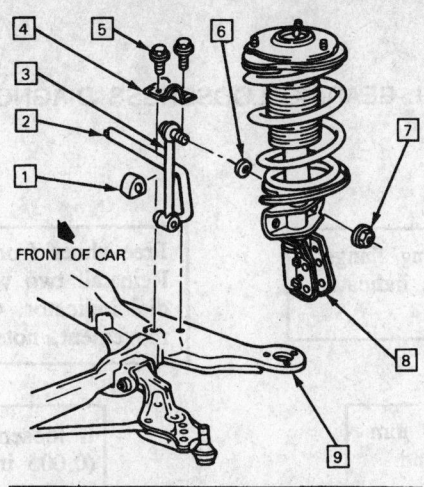

FRONT OF CAR

1	INSULATOR – INSTALL WITH SLIT TO REAR OF CAR
2	STABILIZER SHAFT
3	STABILIZER LINK
4	BRACKET
5	BOLT
6	WASHER
7	NUT
8	SPRING & STRUT ASSEMBLY
9	FRAME

239627

Sway bar components

To install:

8. Position the stabilizer bar in the vehicle.

9. Connect the exhaust pipe to the manifold.

10. Install the bushings with the slits to the rear of the vehicle. Install the brackets and loosely install the mounting bolts.

11. Install the stabilizer links.

12. Tighten the bracket bolts to 18 ft. lbs. (24 Nm). Tighten the stabilizer link nuts to 48 ft. lbs. (65 Nm), using pliers or a Torx® bit to keep the ball stud from turning.

13. If equipped, install the RSS position sensor to the lower control arm.

14. Install the wheel and lower the vehicle.

Front Wheel Bearings

ADJUSTMENT

The front wheel bearings are not adjustable. If a wheel bearing is out of specifications, it must be replaced. Using a dial indicator, check for looseness. If play exceeds 0.005 inch (0.127mm) the bearing wear is excessive and the hub and bearing should be replaced.

REMOVAL AND INSTALLATION

1993–94 Models

1. Raise and safely support the vehicle, allowing the front suspension to hang free.

2. Remove the wheel and tire assembly.

— **WARNING** —

Be careful when working in the area of the CV-joint boot. Damage to the boot could result in eventual CV-joint failure. If necessary, install CV-joint boot protector tool J-34754 or equivalent, to protect the boot.

3. Clean the halfshaft threads of all dirt and lubricant. Insert a drift punch through the caliper and into the rotor to keep the rotor from turning, then remove the hub nut and washer.

4. Remove the caliper without disconnecting the brake hose. Suspend the caliper from the coil spring with wire; do not let it hang by the brake hose.

5. Remove the disc brake rotor.

6. Disconnect the anti-lock brake wheel speed sensor connector.

7. Remove the hub and bearing assembly retaining bolts and the dust shield.

8. Use tool J-28733-A or equivalent, to separate the hub and bearing assembly from the halfshaft.

To install:

9. Apply a light coating of grease to the knuckle bore. Install the new hub and bearing assembly. Draw the hub and bearing assembly onto the halfshaft with a new hub nut.

10. Install the dust shield and the hub and bearing assembly retaining bolts. Tighten to 70 ft. lbs. (95 Nm).

11. Connect the anti-lock brake wheel speed sensor connector.

12. Install the disc brake rotor and caliper.

13. Insert a drift punch into the caliper and rotor to prevent the rotor from turning. Tighten the hub nut to 110 ft. lbs. (145 Nm).

14. Install the wheel and tire assembly and lower the vehicle.

1995–97 Models

1. Raise and safely support the vehicle.

2. Remove the front wheel.

3. Insert a drift punch through the caliper and into the rotor cooling fins to prevent the rotor from turning.

4. Remove the axle nut and washer.

WHEEL BEARING LOOSENESS DIAGNOSIS

DRUM BRAKE

Mount dial indicator. Grasp bearing flange; using a push-pull movement, note indicator readings.

If looseness exceeds 0.1270 mm (0.005 inch), replace hub and bearing assembly.

DISC BRAKE

Free shoes from the disc, or remove calipers. Reinstall two wheel nuts to secure disc. Mount dial indicator. Grasp disc; using a push-pull movement, note indicator readings.

If looseness exceeds 0.1270 mm (0.005 inch), replace hub and bearing assembly.

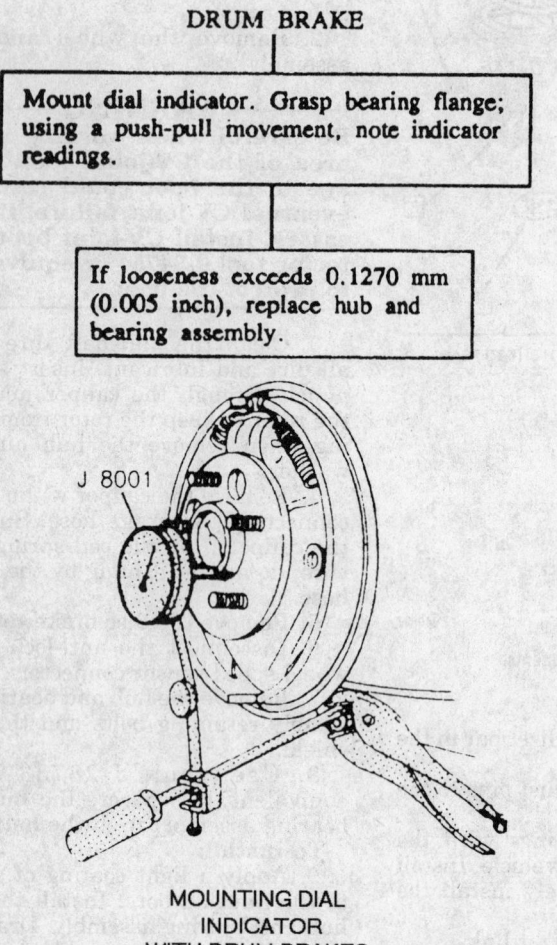

J 8001

MOUNTING DIAL
INDICATOR
WITH DRUM BRAKES

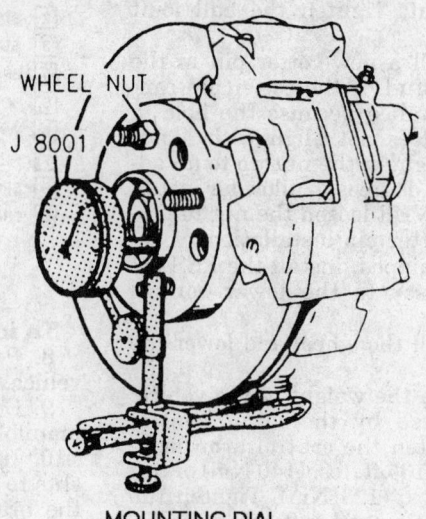

WHEEL NUT

J 8001

MOUNTING DIAL
INDICATOR
WITH DISK BRAKES

339655

Wheel bearing check

5. Remove the caliper mounting bolts. Remove the caliper from the steering knuckle and support out of the way. DO NOT allow the brake hose to support the weight of the caliper.

6. Remove the brake rotor.

7. Disconnect the ABS speed sensor connector and unclip from the dust shield.

8. Remove the 3 hub and bearing bolts and remove the dust shield.

9. Place the transaxle selector in the **P** detent.

10. Install J-28733 or an equivalent puller, and separate the hub and bearing from the drive axle.

11. Remove the hub and bearing assembly from the steering knuckle.
To install:
12. Install the hub and bearing over the half shaft splines. Make sure the splines engage smoothly.

13. Apply a light coating of grease to the steering knuckle bore.

14. Slide the hub assembly onto the axle as far as possible. If the hub will not bottom out on the axle, install the hub mounting bolts and use the axle nut to draw the hub onto the axle.

15. Once the hub is flush with the steering knuckle, remove the mounting bolts and install the dust shield. Reinstall the mounting bolts and tighten to 70 ft. lbs. (95 Nm).

16. Place the transaxle in **N**.

17. Connect the ABS front wheel speed sensor connector and clip to the dust shield.

18. Install the brake rotor.

19. Install the caliper and tighten the mounting bolts to 38 ft. lbs. (51 Nm).

20. Insert a drift punch through the rotor to prevent the axle from turning.

21. Tighten the drive axle shaft nut to 107 ft. lbs. (145 Nm).

22. Remove the drift punch.

23. Install the wheel and lower the vehicle.

24. Road test for proper operation.

REAR SUSPENSION

Strut and Spring

REMOVAL AND INSTALLATION

Riviera

1. Raise and safely support the vehicle.
2. Remove the rear wheel.
3. Install 2 wheel nuts to hold the rotor on the hub and bearing assembly.
4. Remove the stabilizer link mounting, if equipped.
5. Remove the caliper mounting bolts and remove the caliper without disconnecting the brake hose. Support the caliper out of the way on a wire.
6. Compress the rear spring with a suitable jack or equivalent to remove the strut assembly.
7. Loosen the knuckle pivot bolt on the outboard end of the control arm. Do not remove the pivot bolt at this time.
8. Remove the strut cap, nut, retainer and insulator.
9. Compress the strut and remove the lower insulator from the top of the strut.
10. Rotate the strut and knuckle assembly outward by pivoting the assembly on the knuckle pivot bolt.
11. Remove the knuckle pinch bolt.

NOTE: It may be necessary to carefully pry open the area at the pinch bolt to remove the strut without too much resistance.

12. Remove the strut from the knuckle.
   To install:
13. Position the strut in the steering knuckle. The strut must be fully seated in the steering knuckle with

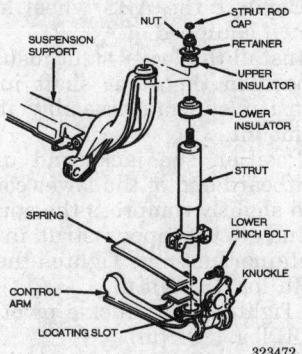

Strut and related components — Riviera

the tang on the strut bottomed in the knuckle slot. There is a cut out on the strut the bolt must pass through.
14. Install the knuckle pinch bolt and tighten to 40 ft. lbs. (54 Nm).
15. Install the lower strut insulator on the top of the strut.
16. Pivot the assembly back into position.
17. Position the strut shaft in the suspension support upper strut mount.
18. Install the upper insulator, retainer and nut. Tighten the nut to 65 ft. lbs. (88 Nm).
19. Tighten the knuckle pivot bolt to 59 ft. lbs. (80 Nm).
20. Install the strut cap.
21. Relieve the pressure from the jack and remove.
22. Remove the 2 wheel nuts used to hold the rotor in place.
23. Install the caliper and new caliper bracket mounting bolts. Tighten the bolts to 83 ft. lbs. (112 Nm).
24. Connect the stabilizer link mounting and torque to 43 ft. lbs. (58 Nm), if equipped.
25. Install the wheel and lower the vehicle.
26. Check the wheel alignment and adjust, as necessary.

Shock Absorber

REMOVAL AND INSTALLATION

All Models Except Riviera

1. Disconnect the negative battery cable.
2. Raise and safely support the vehicle.
3. Remove the rear wheel.
4. If equipped, disconnect the shock absorber electrical connector from the rear suspension support.
5. Support the lower control arm with a jack to relieve the tension on the shock absorber.
6. Remove the lower shock absorber mounting bolt and nut.
7. Remove the upper mounting nut, retainer, and insulator.
8. Compress the shock absorber by hand and remove through the upper control arm.
   To install:
9. Position the the top of the shock absorber with the insulator attached into the suspension support.
10. Install the upper shock insulator, retainer and nut.
11. Install the shock absorber lower mounting nut and bolt.
12. Tighten the upper nut to 55 ft. lbs. (74 Nm) and the lower nut to 75 ft. lbs. (102 Nm).

13. If equipped, connect the shock absorber electrical connector to the rear suspension support.
14. Install the rear wheel and lower the vehicle.
15. Connect the negative battery cable.

Coil Spring

REMOVAL AND INSTALLATION

All Models Except Riviera

1. Raise and safely support the vehicle.
2. Remove the rear wheel.
3. Support the inboard end of the lower control arm with a transmission jack.
4. Remove the sway bar link lower mounting bolt.
5. Remove the shock absorber lower mounting bolt and push the shock up and out of the way.
6. Remove the lower control arm inner mounting nuts and bolts.
7. Slowly lower the transmission jack until the spring tension has been released.
8. Remove the coil spring.
   To install:
9. Install the coil spring and spring insulators.
10. Raise the control arm with the transmission jack until the lower control arm bolts can be installed.
11. Install the mounting nuts. Do not tighten at this time.
12. Remove the transmission jack.
13. Extend the shock absorber and install the mounting nuts. Tighten the nut to 75 ft. lbs. (102 Nm).
14. Install the link kit and tighten the nut to 44 ft. lbs. (60 Nm).
15. Install the rear wheel and lower the vehicle.
16. Tighten the lower control arm inner bolts to 75 ft. lbs. (102 Nm).
17. Check the alignment and adjust, as necessary.

Leaf Spring

REMOVAL AND INSTALLATION

Riviera

NOTE: Removal and installation of the transverse mounted rear spring requires the disassembly of either the left or the right suspension while leaving the other side in contact. The spring may be removed from either side of the vehicle.

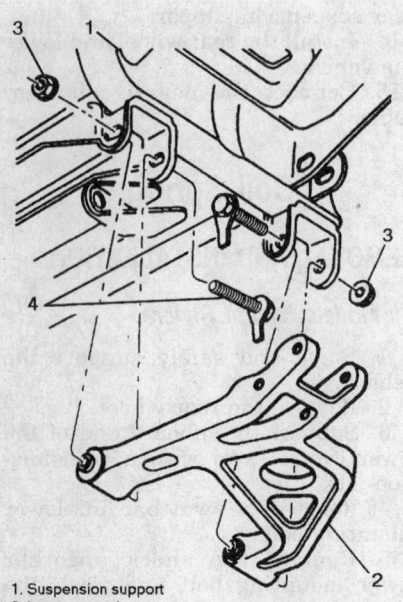

1. Suspension support
2. Lower control arm
3. Lower control arm inner nut - 102 Nm (75 lb. ft.)
4. Lower control arm inner bolt

305876

Lower control arm mounting

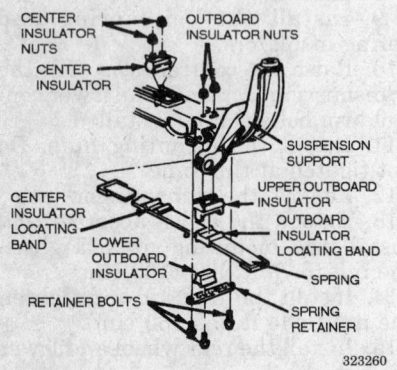

Leaf spring and related components — Riviera

323260

1. Disconnect the negative battery cable.
2. Raise and safely support the vehicle.
3. Remove the rear wheels.
4. Disconnect the height sensor link, if removing the left side control arm.
5. Disconnect the stabilizer link, if equipped.
6. Install 2 wheel lugs on each side to hold the rotor on the hub and bearing assembly.
7. Remove the caliper mounting bolts and remove the caliper without disconnecting the brake line. Support the caliper on a wire out of the way.
8. Loosen the knuckle pivot bolt on the outboard end of the lower control arm. Do not remove the pivot bolt.
9. Support the outboard end of the control arm with a jackstand to slightly compress the spring.

CAUTION

The jackstand must be able to support the vehicle weight. The jackstand must be securely positioned or personal injury may result.

---

10. Remove the strut cap, mounting nut, retainer and upper insulator.
11. Slowly remove the jackstand to relieve spring pressure.
12. Compress the strut by hand and remove the lower insulator from the top of the strut.
13. Disconnect the ABS wheel speed sensor connector, if equipped.
14. Remove the lower control arm inner nuts.
15. While supporting the knuckle and control arm, remove the inner control arm bolts. Remove the control arm, knuckle, strut, rotor and hub and bearing as an assembly.
16. Raise the vehicle and place a jackstand under the end of the leaf spring.

CAUTION

The jackstand must be able to support the vehicle weight. The jackstand must be securely positioned or personal injury may result.

---

17. Lower the vehicle so the weight of the vehicle loads the spring downward on the safety stand.
18. Remove the 3 spring retainer bolts, retainer and lower insulator from the end nearest the supported end of the spring.

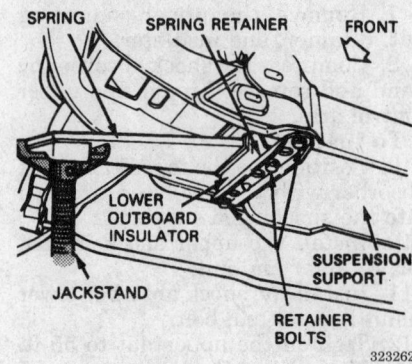

323262

Supporting the leaf spring — Riviera

19. Slowly raise the vehicle until the spring is no longer pressing against the jackstand.
20. Remove the 3 spring retainer bolts, retainer and lower insulator from the opposite side of the vehicle.
21. Remove the spring from the vehicle through the side of the suspension that was disassembled.
22. Remove the spring insulators. Inspect all the spring insulators and control arm contact points; replace worn parts as needed.

To install:

WARNING

Improper positioning of the leaf spring will result in reduced vehicle handling.

---

23. Install the spring insulators. When installing the upper outboard insulators, the molded arrow on the insulator must point toward the centerline of the vehicle.
24. Install the spring in the vehicle through the side of the suspension that was disassembled.
25. With the spring in position, install the lower insulator, retainer and mounting bolts on the side of the vehicle that was not disassembled. Tighten the bolts to 21 ft. lbs. (28 Nm).
26. Place the jackstand under the opposite end of the spring.
27. Lower the vehicle slowly. Allow the weight of the vehicle to load the spring and deflect the free end of the spring into position in the suspension support.
28. Install the lower insulator, retainer and retainer bolts. Tighten the bolts to 21 ft. lbs. (28 Nm).
29. Raise the vehicle and remove the jackstand.
30. Position the assembled control arm, knuckle, strut, rotor and hub bearing in the suspension support and install the inner control arm bolts and nuts. Do not tighten the bolts at this time.
31. Connect the ABS wheel speed sensor, if equipped.
32. Install the lower strut insulator and position the strut shaft in the suspension support assembly upper strut mount.
33. Position the jackstand under the outboard end of the lower control arm to slightly compress the spring.
34. Install the upper strut insulator, retainer and nut. Tighten the nut to 65 ft. lbs. (88 Nm).
35. Tighten the knuckle pivot bolt to 59 ft. lbs. (80 Nm).
36. Tighten the inner control arm bolts to 66 ft. lbs. (90 Nm).
37. Remove the jackstand.

38. Install the strut cap.
39. Install the stabilizer shaft mounting bolt and tighten to 43 ft. lbs. (58 Nm).
40. Remove the 2 wheel nuts used to hold the rotor in position.
41. Install the caliper and new caliper mounting bracket bolts.
42. Install the wheels and lower the vehicle.
43. Connect the negative battery cable.
44. Check the wheel alignment and adjust, as necessary.

Upper Control Arms

REMOVAL AND INSTALLATION

All Models Except Riviera

1. Raise and safely support the vehicle.
2. Remove the rear wheel.
3. If equipped, disconnect the Road Sensing Suspension (RSS) position sensor and bracket from the shock tower.
4. Remove the inner and outer control arm bolts and nuts.
5. Lift the control arm up and over the shock tower to remove from the vehicle.

To install:

6. Install the upper control arm.
7. Install the inner and outer control arm nuts and bolts. Do not tighten the bolts at this time.
8. Place a jack under the lower control arm and raise the suspension to its loaded position.
9. Tighten the control arm mounting nuts to 42 ft. lbs. (57 Nm).
10. If equipped, install the RSS sensor and bracket to the shock tower.
11. Install the wheel and lower the vehicle.
12. Check the wheel alignment; adjust as necessary.

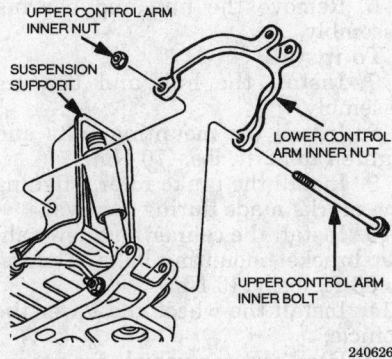

240626

Rear upper control arm components

Lower Control Arms

REMOVAL AND INSTALLATION

All models Except Riviera

1. Raise and safely support the vehicle.
2. Remove the rear wheel.
3. Support the inboard end of the lower control arm with a transmission jack.
4. Remove the sway bar link lower mounting bolt.
5. Remove the shock absorber lower mounting bolt and push the shock up and out of the way.
6. Remove the lower control arm inner mounting nuts and bolts.
7. Slowly lower the transmission jack until the spring tension has been released.
8. Remove the coil spring.
9. Remove the outboard lower control arm nut and bolt, and remove the control arm.

To install:

10. Install the lower control arm and lower control arm outer mounting bolt and nut. Tighten the nut to 75 ft. lbs. (102 Nm).
11. Install the coil spring and spring insulators.
12. Raise the control arm with the transmission jack until the lower control arm bolts can be installed.
13. Install the mounting nuts. Do not tighten at this time.
14. Remove the transmission jack.
15. Extend the shock absorber and install the mounting bolt. Tighten to 75 ft. lbs. (102 Nm).
16. Install the link kit and tighten the nut to 44 ft. lbs. (60 Nm).
17. Install the wheel and lower the vehicle.
18. Tighten the lower control arm inner bolts to 80 ft. lbs. (108 Nm).
19. Check the wheel alignment; adjust as necessary.

Riviera

1. Disconnect the negative battery cable.
2. Raise and safely support the vehicle.
3. Remove the rear wheels.
4. Disconnect the height sensor link, if removing the left side control arm.
5. Remove the stabilizer shaft mounting bolt.
6. If equipped, disconnect the speed sensor from the knuckle.
7. Install 2 wheel nuts to hold the rotor on the hub and bearing assembly.

8. Remove the caliper mounting bolts and remove the caliper without disconnecting the brake hose. Support the caliper out of the way on a wire.
9. Loosen the knuckle pivot bolt on the outboard end of the control arm. Do not remove the pivot bolt at this time.
10. Remove the strut cap, nut, retainer and insulator.
11. Compress the strut.
12. While supporting the knuckle, remove the knuckle pivot bolt and remove the knuckle, strut, rotor and hub and bearing assembly as an unit.
13. Remove both inner control arm bolts and nuts.
14. Remove the control arm from the vehicle.

To install:

15. Position the control arm in the vehicle and insert the inner control arm bolts and nuts. Do not tighten the nuts and bolts at this time.
16. Position a jackstand under the outboard end of the lower control arm to slightly compress the spring.
17. Position the knuckle, strut, rotor and hub and bearing assembly on the control arm and install the pivot bolt. Do not tighten the bolt at this time.
18. Position the strut shaft in the suspension support upper strut mount.
19. Install the upper insulator, retainer and nut. Tighten the nut to 65 ft. lbs. (88 Nm).
20. Tighten the knuckle pivot bolt to 60 ft. lbs. (81 Nm).
21. Tighten the inner control arm bolts to 65 ft. lbs. (88 Nm).
22. Install the strut cap.
23. Remove the 2 wheel nuts used to hold the rotor in place.
24. Install the caliper and new caliper bracket mounting bolts. Tighten the bolts to 83 ft. lbs. (112 Nm).
25. If equipped, connect the speed sensor to the knuckle.
26. Install the wheel and lower the vehicle.
27. Connect the negative battery cable.
28. Check the wheel alignment; adjust as necessary.

Sway Bar

REMOVAL AND INSTALLATION

All Models Except Riviera

1. Raise and safely support the vehicle.
2. Remove the rear wheels.

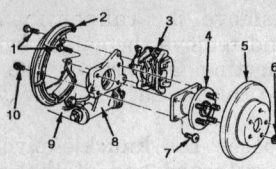

1. Caliper mounting bolts - 113 Nm (83 ft. lb.)
2. Splash shield
3. Caliper assembly
4. Hub and bearing assembly
5. Rotor
6. Rotor retainer
7. Hub mounting bolts - 70 Nm (52 ft. lb.)
8. Knuckle
9. Control arm
10. Bolt - 50 Nm (37 ft. lb.)

323442

Hub and bearing assembly and related components — Riviera

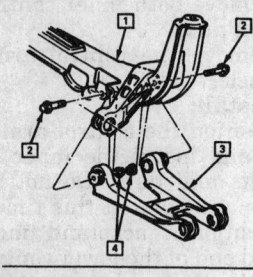

1 SUSPENSION SUPPORT
2 INNER CONTROL ARM BOLTS
3 CONTROL ARM
4 NUTS

323412

Lower control arm mounting — Riviera

3. Remove the lower link bolts on both sides.
4. Remove the upper link nuts on both sides.
5. Remove the sway bar bracket mounting bolts and sway bar brackets.
6. Remove the insulators from the sway bar.

7. Disconnect the rear exhaust hangers. Lower and support the exhaust system.
8. Remove the sway bar from the vehicle.

To install:

9. Install the sway bar in the vehicle.
10. Raise the exhaust into position and connect the hangers.
11. Install the bushings on the sway bar.
12. Install the sway bar bushing brackets and mounting bolts. Tighten the mounting bolts to 44 ft. lbs. (60 Nm).
13. Install the upper link nuts on both sides and tighten to 40 ft. lbs. (54 Nm).
14. Install the lower link bolts on both sides and tighten to 45 ft. lbs. (61 Nm).
15. Install the wheels and lower the vehicle.

Riviera

NOTE: Removal of the sway bar is easier if the vehicle is at curb height with the wheels on the ground or safely supported in such a way that the rear suspension is not hanging.

1. Remove the sway bar mounting bolts and nuts from the struts.
2. Remove the sway bar link bolt and nut at the suspension support.
3. Remove the sway bar from the vehicle.
4. Remove the sway bar link from the sway bar.
5. Remove the insulators as necessary.

To install:

6. Install the insulators on the sway bar.

7. Install the sway bar links on the sway bar.
8. Install the sway bar assembly in the vehicle.
9. Install the nuts and bolts to the suspension support.
10. Install the nuts and bolts to the struts.
11. With all fasteners in place, tighten the nuts to 45 ft. lbs. (58 Nm).

Wheel Bearings

ADJUSTMENT

The wheel bearings are not adjustable. If a wheel bearing is out of specifications, it must be replaced. Using a dial indicator, check for looseness. If play exceeds 0.005 inch (0.127mm) the bearing wear is excessive and the hub and bearing should be replaced.

REMOVAL AND INSTALLATION

NOTE: The wheel bearing and hub are serviced as an assembly. The individual components are not serviceable separately.

1. Raise and safely support the vehicle.
2. Remove the rear wheel.
3. Remove the caliper bracket mounting bolts and remove the caliper assembly from the knuckle. Support the caliper out of the way on a wire. Do not disconnect the brake hose from the caliper.
4. Remove the brake rotor. Mark the relationship between the wheel stud and the brake rotor for installation reference.
5. Remove the 4 hub and bearing assembly mounting bolts.
6. Remove the hub and bearing assembly.

To install:

7. Install the hub and bearing assembly.
8. Install the mounting bolts and tighten to 52 ft. lbs. (70 Nm).
9. Install the brake rotor, aligning the marks made during removal.
10. Install the caliper and new caliper bracket mounting bolts. Tighten to 83 ft. lbs. (113 Nm).
11. Install the wheel and lower the vehicle.
12. Road test the vehicle for proper operation.

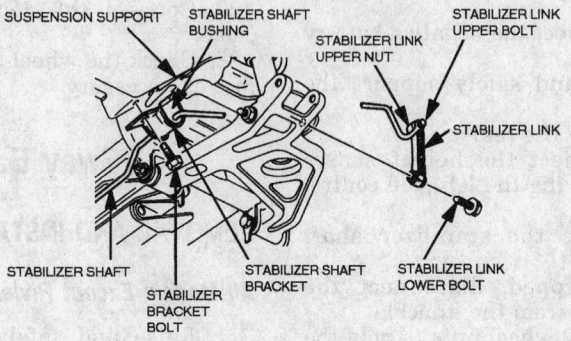

SUSPENSION SUPPORT
STABILIZER SHAFT BUSHING
STABILIZER LINK UPPER NUT
STABILIZER LINK UPPER BOLT
STABILIZER LINK
STABILIZER SHAFT
STABILIZER BRACKET BOLT
STABILIZER SHAFT BRACKET
STABILIZER LINK LOWER BOLT

240471

Sway bar mounting components

FIRING ORDERS

NOTE: To avoid confusion, always replace spark plug wires one at a time.

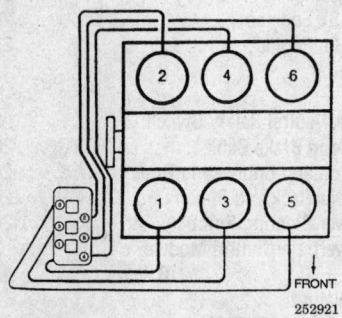

3.8L (VIN 1) and (VIN K) Engines
Firing Order: 1–6–5–4–3–2
Distributorless Ignition System

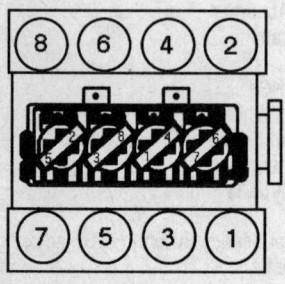

4.0L (VIN C) Engine
Firing Order: 1–2–7–3–4–5–6–8
Distributorless Ignition System

ENGINE ELECTRICAL

NOTE: Disconnecting the negative battery cable on some vehicles may interfere with the functions of the on board computer systems and may require the computer to undergo a relearning process, once the negative battery cable is reconnected.

Alternator

PRECAUTIONS

Several precautions must be observed with alternator equipped vehicles to avoid damage to the unit.

• If the battery is removed for any reason, make sure it is reconnected with the correct polarity. Reversing the battery connections may result in damage to the 1–way rectifiers.

• When utilizing a booster battery as a starting aid, always connect the positive to positive terminals and the negative terminal from the booster battery to a good engine ground on the vehicle being started.

• Never use a fast charger as a booster to start vehicles.

• Disconnect the battery cables when charging the battery with a fast charger.

• Never attempt to polarize the alternator.

• Do not use test lights of more than 12 volts when checking diode continuity.

• Do not short across or ground any of the alternator terminals.

• The polarity of the battery, alternator and regulator must be matched and considered before making any electrical connections within the system.

• Never separate the alternator on an open circuit. Make sure all connections within the circuit are clean and tight.

• Disconnect the battery ground terminal when performing any service on electrical components.

• Disconnect the battery if arc welding is to be done on the vehicle.

REMOVAL AND INSTALLATION

3.8L (VIN 1) Engine

1. Disconnect the negative battery cable.
2. Remove serpentine drive belt from alternator pulley. It is not necessary to remove the belt from the remainder of the pulleys.
3. Disconnect the electrical connector from the alternator and remove the nut securing the battery wire to the alternator stud.
4. Remove the alternator-to-mounting bracket bolt.
5. Remove the lower alternator mounting nut and through bolt.
6. Remove the alternator from the mounting bracket.
   To install:
7. Install the alternator in the mounting bracket.

8. Install the mounting bolts and nut and tighten to 21 ft. lbs. (29 Nm).
9. Connect the battery wire to the alternator stud and tighten the mounting nut to 71 inch lbs. (8 Nm). Connect the electrical connector the alternator.
10. Install the serpentine belt.
11. Connect the negative battery cable.

3.8L (VIN K) Engine

1. Disconnect the negative battery cable.
2. Remove serpentine drive belt from alternator pulley. It is not necessary to remove the belt from the remainder of the pulleys.
3. Disconnect the electrical connector from the alternator and remove the nut securing the battery wire to the alternator stud.
4. Remove the 3 alternator mounting bolts.
5. Remove the front generator mounting nut.
6. Remove the alternator from the mounting bracket.
   To install:
7. Install the alternator in the mounting bracket.
8. Install the front alternator mounting nut and tighten to 30 ft. lbs. (40 Nm).
9. Install the 3 alternator mounting bolts and tighten to 30 ft. lbs. (40 Nm).
10. Connect the battery wire to the alternator stud and tighten the mounting nut to 71 inch lbs. (8 Nm). Connect the electrical connector the alternator.
11. Install the serpentine belt.
12. Connect the negative battery cable.

4.0L (VIN C) Engine

1. Disconnect the negative battery cable.
2. Drain the cooling system.
3. Remove the upper radiator shroud.
4. Disconnect and remove the cooling fans.
5. Remove the serpentine belt from the power steering pump pulley then the alternator pulley. The belt can remain on the rest of the pulleys.
6. Remove the top alternator bolts and disconnect the electrical connections from the alternator.
7. Disconnect the engine ground strap and power steering line bracket to provide clearance for removal of the alternator.
8. Raise and safely support the vehicle and remove the lower alternator mounting bolts.

9. Remove the lower radiator hose.

10. Lower the vehicle.

11. Remove the air inlet duct.

12. Remove the alternator out the top left side of the engine compartment.

To install:

13. Position the alternator in the mounting bracket.

14. Loosely install the upper alternator mounting bolts.

15. Install the air inlet duct.

16. Raise and safely support the vehicle.

17. Install the lower radiator hose.

18. Install the lower alternator mounting bolts and tighten to 28 ft. lbs. (38 Nm).

19. Lower the vehicle.

20. Connect the ground strap and power steering line bracket.

21. Tighten the upper alternator mounting bolts to 35 ft. lbs. (47 Nm).

22. Connect the electrical connectors to the alternator.

23. Install the serpentine belt around the alternator and power steering pulleys.

24. Install the cooling fan assemblies.

25. Install the upper radiator shroud.

26. Connect the negative battery cable.

27. Refill the cooling system and bleed.

Drive Belt

REMOVAL AND INSTALLATION

3.8L (VIN 1) Engine

1. Disconnect the negative battery cable.

2. Using an 18mm wrench on the rear tensioner pulley mounting bolt, and rotate the pulley clockwise.

3. Slip the belt off of the alternator pulley and slowly release the tensioner.

4. Remove the wrench then remove the belt from the remainder of the pulleys.

5. Install an 18mm wrench on the front tensioner pulley mounting bolt and rotate the pulley clockwise.

6. Slip the belt off the supercharger pulley and slowly release the tensioner.

7. Remove the belt from the remainder of the pulleys.

To install:

8. Install the back belt and route the belt as follows:

a. Loop the belt under the crankshaft pulley.

b. Bring the front of the belt around the water pump so the back of the belt drives the pump.

c. Loop the belt under the A/C compressor.

d. Bring the rear of the belt over the idler and under the tensioner.

9. Rotate the tensioner clockwise and loop the belt over the supercharger pulley.

10. Install the front belt and route as follows:

a. Loop the belt under the crankshaft pulley.

b. Bring the rear of the belt over the tensioner so the back of the belt drives the pulley.

c. Loop the belt around the power steering pump.

d. Take the front of the belt around the idler pulley so the back of the belt drives the pulley.

11. Rotate the tensioner clockwise and loop the belt over the alternator pulley.

12. Connect the negative battery cable.

3.8L (VIN K) Engine

1. Disconnect the negative battery cable.

2. Rotate the tensioner downward with a 15mm wrench and slip the belt off of the alternator pulley.

3. Slowly release the tensioner and remove the wrench.

4. Remove the belt from the remainder of the pulleys.

To install:

5. Loop the serpentine belt under the crankshaft pulley and A/C compressor pulley.

6. Bring the back of the belt under and then around the water pump pulley so the back of the belt drives the pulley.

7. Continue the belt upward over the tensioner so the back of the belt drives the tensioner.

8. Loop the belt under the power steering pump pulley.

9. Install a 15mm wrench on the tensioner pulley bolt and rotate the pulley downward and loop the belt over the alternator pulley.

10. Slowly release the tensioner and remove the wrench. Verify the belt is correctly seated on all the drive pulleys.

11. Connect the negative battery cable.

4.0L (VIN C) Engine

1. Disconnect the negative battery cable.

2. Release the serpentine belt tensioner using a ⅜ inch ratchet.

3. Slip the belt off of the power steering pump pulley and release the belt tensioner slowly.

4. Remove the belt from the alternator pulley.

5. Twist the belt for clearance to remove the belt from the tensioner pulley and to slip the belt between the power steering pulley and engine mount bracket.

6. Raise and safely support the vehicle.

7. Turn the wheels to the right and remove the right inner fender well splash shield.

8. Remove the belt from the A/C compressor pulley and crankshaft pulley. Twist the belt for clearance to remove the belt from the idler pulley.

9. Remove the belt from the vehicle.

To install:

10. Install the belt around the power steering pulley, it will be necessary to twist the belt for clearance between the power steering pulley and front cover bolts.

11. Twist the belt to slip it around the alternator pulley.

12. Remove the belt from the power steering pump and set aside. This will provide slack in the belt for installation around the remainder of the pulleys.

13. Raise and safely support the vehicle.

14. Loop the belt under the crankshaft and A/C compressor pulleys.

15. Install the right side splash shield.

16. Lower the vehicle.

17. Compress the belt tensioner and loop the belt around the power steering pulley.

18. Slowly release the tensioner and remove the ratchet.

19. Make sure the belt is properly seated in all the drive pulleys.

Starter

REMOVAL AND INSTALLATION

3.8L (VIN 1) and (VIN K) Engines

1. Disconnect the negative battery cable.

2. Raise and safely support the vehicle.

3. Remove the 4 flywheel inspection cover retaining screws and remove the cover.

4. Remove the starter mounting bolts and carefully move starter motor away from the block.

5. Remove the wiring from the starter and solenoid assembly.

To install:

6. Connect the starter wiring and tighten the solenoid battery terminal nut to 12 ft. lbs. (16 Nm) and the solenoid S terminal nut to 22 inch lbs. (2.5 Nm).

7. Install the starter to the engine block and tighten the starter mounting bolts to 32 ft. lbs. (43 Nm).

8. Install the flywheel inspection cover and tighten the mounting bolts to 55 inch lbs. (7 Nm).

9. Lower the vehicle.

10. Connect the negative battery cable and check the starter for proper operation.

4.0L (VIN C) Engine

On the 4.0L (VIN C) engine, the starter motor is mounted inside the engine, in the valley between the cylinder banks, driving the flywheel ring gear from the top. This means the intake manifold assembly must be removed to access the starter motor.

─────── **CAUTION** ───────
Fuel injection systems remain under pressure, even after the engine has been turned OFF. The fuel system pressure must be relieved before disconnecting any fuel lines. Failure to do so may result in fire and/or personal injury.

1. Disconnect the negative battery cable.

2. Relieve the fuel system pressure using the recommended procedure.

3. Disconnect the positive battery cable.

4. Remove the intake manifold assembly using the recommended procedure.

5. Remove the solenoid S terminal nut and battery cable nut and disconnect the electrical connections from the starter.

6. Remove the 2 starter mounting nuts.

7. Remove the starter from the lifter valley.

To install:

8. Before installing the new starter make sure the inner nuts on the solenoid are tight.

9. Position the starter in the lifter valley and install the mounting bolts. Tighten the bolts alternately and evenly to 32 ft. lbs. (43 Nm).

10. Connect the battery cable to the solenoid and tighten the mounting nut to 88 inch lbs. (10 Nm).

11. Connect the solenoid wire to the terminal and tighten the mounting nut to 26 inch lbs. (3 Nm).

12. Install the intake manifold using the recommended procedure.

13. Connect the positive battery cable.

14. Connect the negative battery cable.

15. Start the vehicle and verify proper starter operation and no leaks around the intake manifold.

─────────────────────
CHASSIS ELECTRICAL
─────────────────────

Blower Motor

REMOVAL AND INSTALLATION

1. Disconnect the negative battery cable.

2. Remove the right side sound insulator panel.

3. Disconnect the electrical connector from the blower motor.

4. Remove the blower motor cooling tube.

5. Remove the blower motor mounting screws and remove the blower motor and fan assembly from the heater module.

To install:

6. Install the blower motor into the heater module and install the blower motor mounting screws.

7. Install the blower motor cooling tube.

8. Connect the electrical connector to the blower motor.

9. Install the right side sound insulator panel.

10. Connect the negative battery cable.

Windshield Wiper Motor

REMOVAL AND INSTALLATION

1. Turn the ignition switch to the **ACC** position.

2. Turn the wiper switch to the pulse position.

3. Turn the ignition switch off when the wipers are at the bottom of the wipe cycle.

4. Lift the wiper blade up from the window.

5. Pull out the retaining latch with a flat blade screwdriver.

6. Remove the wiper arm from the vehicle.

7. Disconnect the negative battery cable.

8. Disconnect the washer hose.

9. Remove the air inlet panel from the vehicle by releasing the push button retainers.

10. Disconnect the transmission drive link socket from the wiper motor crank arm using J-39232, or an equivalent.

11. Remove the electrical connector retainers from the connectors on the wiper motor and disconnect the wiring harnesses.

12. Remove the 2 mounting screws and mounting stud and remove the wiper motor from the firewall.

To install:

13. Position the wiper motor on the firewall and install the mounting bolts and stud. Tighten the fasteners alternately and evenly to 11 ft. lbs. (15 Nm).

14. Connect the electrical harnesses to the wiper motor and install the connector retainers.

15. Connect the transmission drive link to the wiper motor crank arm using J-39529, or the equivalent.

16. Install the air inlet panel and the push button retainers.

17. Connect the washer hose.

18. Connect the negative battery cable.

19. Return the new wiper motor to the park position as follows:

   a. Turn the ignition switch to the **ACC** position.

   b. Turn the wiper switch **ON** then **OFF**.

   c. Turn the ignition switch to the **OFF** position.

20. Position the wiper arm assembly on the shaft so the arm is slightly below the stop surface on the park ramp.

21. Push the wiper arm down onto the transmission drive until it bottoms.

22. Pivot the wiper upward and engage the retaining latch.

23. Verify proper wiper operation.

Combination Switch

REMOVAL AND INSTALLATION

────── **CAUTION** ──────

The Supplemental Inflatable Restraint (SIR) system must be disarmed before removing the combination switch. Failure to do so may cause accidental deployment of the air bag, resulting in unnecessary SIR system repairs and/or personal injury.

1. Properly disarm the SIR system.
2. Disconnect the negative battery cable.
3. Remove the steering wheel.
4. Remove the steering column from the vehicle.
5. Remove the self tapping screws securing the upper cover to the lower cover and remove the upper cover from the steering column.
6. Remove the 2 straps from the steering column wiring harness and separate the combination switch harness from the remainder of the column wiring.
7. Disconnect the interlock solenoid electrical connector.
8. Disconnect the pass key wiring harness from the grey combination switch connector.
9. Remove the 2 screws securing the combination switch to the steering column.
10. Remove the combination switch.

To install:

11. Position the combination switch on the steering column and using a small blade screwdriver compress the electrical contact and move the multifunction switch into position. The electrical contact must rest on the cancelling cam assembly.
12. Tighten the mounting screws to 46 inch lbs. (5 Nm).
13. Connect the pass key wiring harness to the grey turn signal switch connector.
14. Connect the interlock solenoid electrical connector to the combination switch harness.
15. Route the combination switch wiring in with the remainder of the steering column wiring and install the wiring straps.
16. Install the upper cover on the steering column and tighten the 3 self tapping screws.
17. Install the steering column in the vehicle.
18. Install the steering wheel.
19. Connect the negative battery cable.

20. Enable the SIR system.
21. Verify proper operation of the steering column and all column mounted switches.

Ignition Lock Cylinder and Switch

REMOVAL AND INSTALLATION

────── **CAUTION** ──────

The Supplemental Inflatable Restraint (SIR) system must be disarmed before working around the steering column. Failure to do so may cause accidental deployment of the air bag, resulting in unnecessary SIR system repairs and/or personal injury.

1. Properly disarm the SIR system.
2. Disconnect the negative battery cable.
3. Remove the steering wheel.
4. Remove the steering column from the vehicle.
5. Remove the self tapping screws securing the upper cover to the lower cover and remove the upper cover from the steering column.
6. Remove the screw securing the tilt lever knob and remove the knob from the column.
7. If equipped with a column shift, remove the seal from around the shifter lever.
8. Remove the close out knob from around the lock cylinder by gently prying out the lock tangs.
9. Remove the 3 self tapping screws securing the lower cover to the column and remove the lower cover.
10. Remove the 2 straps from the steering column wiring harness and separate the ignition switch harness from the remainder of the column wiring.
11. Remove the 2 screws securing the ignition switch to the steering column.
12. Remove the ignition switch assembly.

To install:

13. Position the ignition switch on the steering column and tighten the mounting screws to 12 inch lbs. (1.5 Nm).
14. Route the ignition switch wiring in with the remainder of the steering column wiring and install the wiring straps.
15. Position the lower cover on the steering column and install the 3 self tapping screws.

16. Align the tabs on the close out knob with the slots on the lock cylinder housing and push the knob in until the tabs lock in place.
17. Install the shifter lever seal, if equipped.
18. Install the tilt lever and mounting screw.
19. Install the steering column in the vehicle.
20. Install the steering wheel.
21. Connect the negative battery cable.
22. Enable the SIR system.
23. Verify proper operation of the steering column and all column mounted switches.

Park/Neutral Safety Switch

REMOVAL AND INSTALLATION

4T60-E Transaxle

1. Disconnect the negative battery cable.
2. Engage the parking brake and place the gear selector in the **N** detent.
3. Raise and safely support the vehicle.
4. Remove the front air deflector assembly.
5. Remove the nut securing the park/neutral switch wire to the starter solenoid.
6. Lower the vehicle.
7. Remove the nut securing the shift lever to the transaxle and disconnect the lever from the transaxle stud.
8. Remove the bolts securing the shift cable bracket and position the shift cable and bracket out of the way.
9. Disconnect the 2 electrical connectors.
10. Remove the 2 bolts and remove the park/neutral switch from the vehicle.

To install:

11. Make sure the transaxle is still in the **N** detent.
12. Align the flats on the park neutral switch with the flats on the shift shaft on the transaxle. Install the switch over the shaft and seat it fully on the transaxle.
13. If installing the old switch install a gage pin into the service slots to properly align the switch.
14. Install the mounting bolts and tighten to 20 ft. lbs. (28 Nm).
15. Remove the gage pin, if installed.
16. Install the shifter cable bracket and mounting bolts.

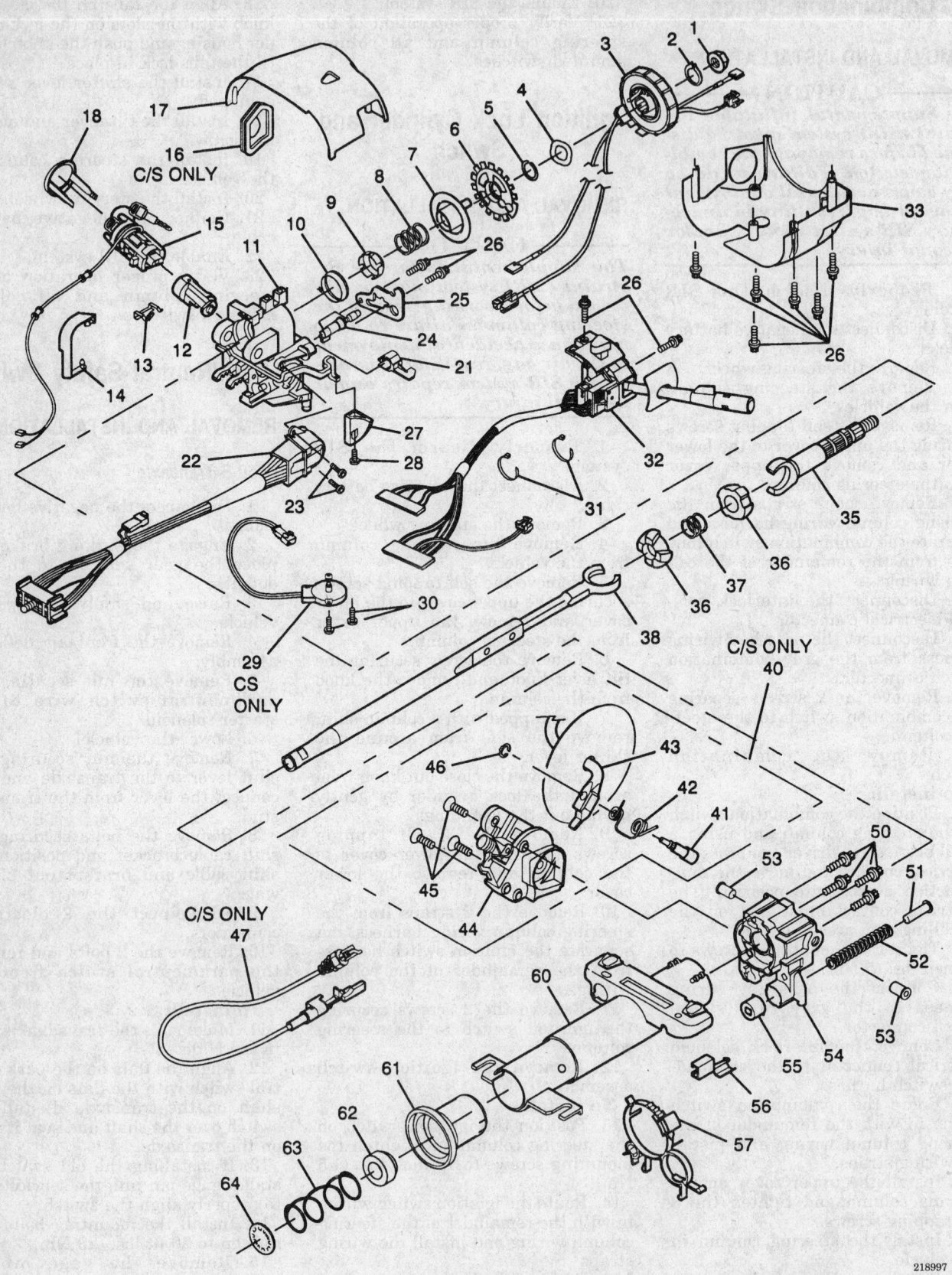

Exploded view of the steering column assembly

218997

1- NUT, HEXAGON LOCKING (M14x1.5)
2- RING, RETAINING
3- COIL ASM, SIR
4- WASHER, WAVE
5- RING, RETAINING
6- LOCK, SHAFT
7- CAM ASM, T/SIG CANCEL
8- SPRING, UPPER BEARING
9- SEAT, UPPER BEARING INNER RACE
10- RACE, INNER
11- HOUSING ASM, BRG &
12- ACTUATOR ASM, IGNITION LOCK
13- SPRING, LOCK PRE-LOAD
14- PROTECTOR, WIRING
15- LOCK CYL SET, STRG COLUMN
16- SEAL ASM, SHIFT LEVER
17- COVER, UPPER
18- KNOB, CLOSE OUT
21- BRACKET, TILT LEVER
22- SWITCH ASM, IGN & KEY ALARM
23- SCREW, TAPPING
24- BOLT ASM, LOCK
25- BRACKET, LOCK BOLT SUPPORT
26- SCREW, TAPPING
27- KNOB, TILT LEVER
28- SCREW, PAN HEAD
29- SOLENOID ASM, INTERLOCK
30- SCREW, TAPPING
31- STRAP, WIRE HARNESS
32- SWITCH ASM, T/S & MULTIFUNC
33- COVER, LOWER
35- SHAFT ASM, UPPER
36- SPHERE, CENTERING
37- SPRING, SHAFT PRELOAD
38- SHAFT ASM, LOWER STRG
40- SHIFT ASM, LINEAR
41- PIN, SHIFT LEVER
42- SPRING, SHIFT LEVER
43- LEVER ASM, LINEAR SHIFT
44- ADAPTER, LINEAR SHIFT BASE (NC)
45- SCREW, FLT HD TAPPING
46- CLIP, SHIFT LEVER
47- CABLE ASM, PARK LOCK
50- SCREW, SUPPORT

51- GUIDE, SPRING
52- SPRING, TILT WHEEL
53- PIN, PIVOT
54- SUPPORT ASM, STRG COL JACKET
55- RETAINER, SPRING
56- CLIP, WIRE RESTRAINT
57- STRAP, WIRE HARNESS
60- JACKET ASM, STRG COL
61- BEARING ASM, ADAPTER &
62- SEAT, LOWER BEARING
63- SPRING, LOWER BEARING
64- RETAINER, LOWER SPRING

Service Kits

201- SPRING SERV KIT, TILT COLUMN
-INCLUDES: 9,10,51,52,53,55
202- SPHERE SERV KIT, TILT COLUMN
-INCLUDES: 36,37
203- GREASE SERV KIT, (SYNTHETIC)

221447

Component list for steering column exploded view

17. Connect the 2 electrical connectors to the main engine harness. Route the solenoid wire down toward the starter.

18. Connect the shift linkage to the shift shaft and install the mounting nut.

19. Raise and safely support the vehicle.

20. Connect the switch wire to the solenoid and tighten the mounting nut to 71 inch lbs, (8 Nm).

21. Install the front air deflector.

22. Lower the vehicle.

23. Connect the negative battery cable.

24. Verify the vehicle will only start in the **P** and **N** detents.

4T80-E Transaxle

1. Disconnect the negative battery cable.

2. Engage the parking brake and place the gear selector in the **N** detent.

3. Remove the nut securing the shift lever to the transaxle and disconnect the lever from the transaxle stud.

4. Disconnect the electrical connector.

5. Remove the 2 bolts and remove the park/neutral switch from the vehicle.

To install:

6. Make sure the transaxle is still in the **N** detent.

7. Align the flats on the park neutral switch with the flats on the shift shaft on the transaxle. Install the switch over the shaft and seat it fully on the transaxle.

8. Install the mounting bolts finger tight.

9. Rotate the switch so a gage pin into the service slots to properly align the switch.

10. Tighten the mounting bolts to 106 inch lbs. (12 Nm).

11. Remove the gage pin.

12. Connect the electrical connector to the main engine harness.

13. Connect the shift linkage to the shift shaft and install the mounting nut and tighten to 15 ft. lbs. (20 Nm).

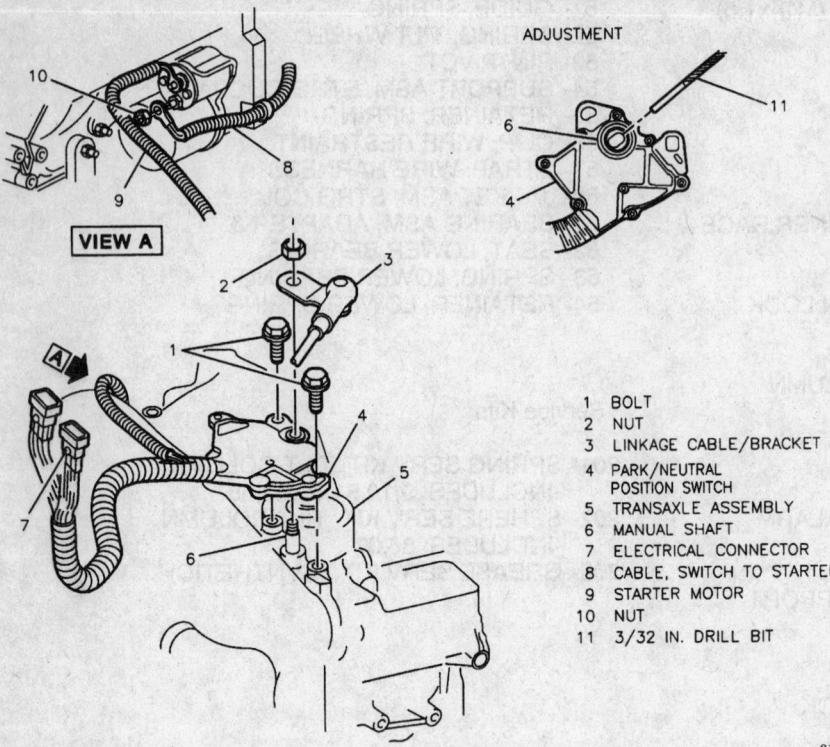

ADJUSTMENT

VIEW A

1 BOLT
2 NUT
3 LINKAGE CABLE/BRACKET
4 PARK/NEUTRAL
   POSITION SWITCH
5 TRANSAXLE ASSEMBLY
6 MANUAL SHAFT
7 ELECTRICAL CONNECTOR
8 CABLE, SWITCH TO STARTER
9 STARTER MOTOR
10 NUT
11 3/32 IN. DRILL BIT

311599

Park/neutral switch arrangement — 4T60E transaxle

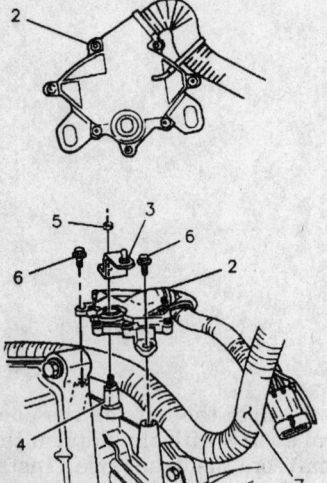

1 TRANSAXLE
2 SWITCH, TRANSAXLE RANGE
3 LEVER, RANGE SELECTION
4 SHAFT, TRANSAXLE MANUAL
5 NUT, 20 N•m (15 LB. FT.)
6 BOLT, 28 N•m (21 LB. FT.)
7 ENGINE WIRING HARNESS

311600

Park/neutral switch arrangement — 4T80E transaxle

14. Connect the negative battery cable.

15. Verify the vehicle will only start in the **P** and **N** detents.

Powertrain Control Module

REMOVAL AND INSTALLATION

1. Disconnect the negative battery cable.

2. Remove the right side hush panel under the dash.

3. Release the 2 lock tangs on the bottom of the PCM mounting bracket and tilt the bottom of the PCM out of the mounting bracket.

4. Pull the PCM down and out of the mounting bracket.

5. Disconnect the harness connectors from the PCM and remove the PCM from the vehicle.

To install:

6. Connect the electrical connectors to the PCM.

7. Install the top of the PCM under the mounting clip on top of the mounting bracket and tilt the bottom in until the locking tangs snap in place.

8. Install the right side hush panel.

9. Connect the negative battery cable.

ENGINE COOLING

Radiator

REMOVAL AND INSTALLATION

3.8L (VIN 1) and (VIN K) Engines

1. Disconnect the negative battery cable.

2. Raise and safely support the vehicle.

3. Remove the lower air dam.

4. Drain the cooling system.

5. Install the lower air dam and lower the vehicle.

6. Remove the upper radiator panel mounting bolts and upper radiator panel.

7. Disconnect the electrical connectors from the cooling fans.

8. Remove the left and right cooling fans.

9. Disconnect the electrical connector from the coolant level sensor.

10. Disconnect the reservoir hose from the radiator neck.

11. Disconnect the upper and lower radiator hoses.

12. Remove the radiator-to-condenser bolts.

13. Disconnect and cap the transaxle cooler lines from the radiator.

14. Remove the radiator from the vehicle.

To install:

15. Position the radiator in the radiator cradle.

16. Connect the transaxle cooler lines to the radiator and tighten the line fittings to 20 ft. lbs. (27 Nm).

17. Install the radiator-to-condenser bolts.

18. Connect the upper and lower radiator hoses.

19. Connect the recovery reservoir hose to the radiator neck.

20. Connect the electrical connector to the coolant level sensor.

21. Install the cooling fan assemblies and connect the electrical wiring.

22. Install the upper radiator panel and mounting bolts and tighten to 89 inch lbs. (10 Nm).

23. Connect the negative battery cable.

24. Refill the cooling system.

25. Start the vehicle and verify no leaks. With the engine at operating temperature check the transaxle fluid level and add fluid as necessary.

4.0L (VIN C) Engine

1. Disconnect the negative battery cable.
2. Raise and safely support the vehicle.
3. Remove the lower air dam.
4. Drain the cooling system.
5. Install the lower air dam and lower the vehicle.
6. Remove the upper radiator panel mounting bolts and upper radiator panel.
7. Disconnect the electrical connectors from the cooling fans.
8. Remove the left and right cooling fans.
9. Disconnect the electrical connector from the coolant level sensor.
10. Disconnect the reservoir hose from the radiator neck.
11. Disconnect the upper and lower radiator hoses.
12. Remove the radiator-to-condenser bolts.
13. Disconnect and cap the transaxle cooler lines from the radiator.
14. Disconnect and cap the engine oil cooler lines from the radiator.
15. Remove the radiator from the vehicle.

To install:
16. Position the radiator in the radiator cradle.
17. Connect the transaxle cooler lines to the radiator and tighten the line fittings to 13 ft. lbs. (18 Nm).
18. Connect the engine oil cooler lines and tighten the fittings to 10 ft. lbs. (14 Nm).
19. Install the radiator-to-condenser bolts.
20. Connect the upper and lower radiator hoses.
21. Connect the recovery reservoir hose to the radiator neck.
22. Connect the electrical connector to the coolant level sensor.
23. Install the cooling fan assemblies and connect the electrical wiring.
24. Install the upper radiator panel and mounting bolts and tighten to 89 inch lbs. (10 Nm).
25. Connect the negative battery cable.
26. Refill the cooling system.
27. Start the vehicle and verify no leaks. With the engine at operating temperature check the transaxle fluid level and add fluid as necessary.

Water Pump

REMOVAL AND INSTALLATION

3.8L (VIN 1) and (VIN K) Engines

1. Disconnect the negative battery cable.
2. Drain the cooling system into a suitable container.
3. Remove the serpentine belt.
4. Disconnect the heater and by-pass hoses from the water pump.
5. Remove the water pump pulley bolts. The long bolt can be removed by lining the head of the bolt up with the hole in the frame rail. Remove the pulley.
6. Install an engine support fixture and remove the torque axis mount.
7. Remove the water pump mounting bolts and remove the water pump.

To install:
8. Clean all the gasket surfaces completely.
9. Apply a thin bead of sealer around the outside edge of the water pump and install the gasket on the pump.
10. Install the water pump on the engine and tighten the water pump bolts on the sides of the water inlet and outlet to 29 ft. lbs. (39 Nm) and the remainder of the bolts to 97 inch lbs. (11 Nm).
11. Install the torque axis mount and remove the support fixture.
12. Install the water pump pulley and tighten the bolts finger tight.
13. Connect the hoses to the water pump.
14. Install the serpentine belt.
15. Tighten the water pump pulley bolts to 115 inch lbs. (13 Nm).
16. Refill the cooling system.
17. Connect the negative battery cable.
18. Start the vehicle and verify no leaks.

4.0L (VIN C) Engine

1. Disconnect the negative battery cable.
2. Drain the cooling system.
3. Remove the air inlet duct.
4. Remove the water pump drive belt cover.
5. Remove the water pump drive belt and set aside.
6. Disconnect the lower radiator hose and bypass hose.
7. Remove the thermostat housing from the water pump housing.
8. Remove the water pump from the water pump housing by rotating turning the locking ring with

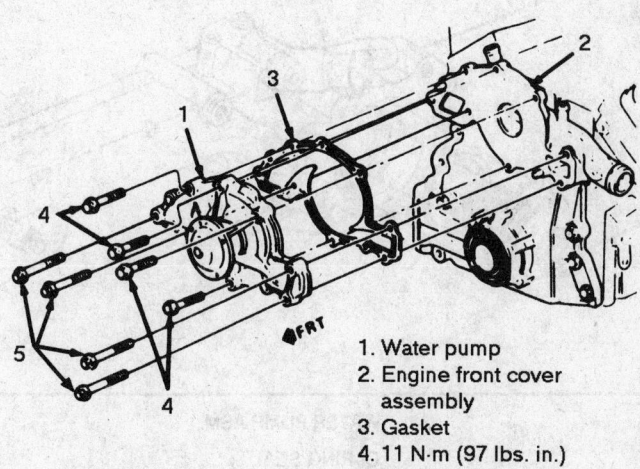

1. Water pump
2. Engine front cover assembly
3. Gasket
4. 11 N·m (97 lbs. in.)
5. 39 N·m (29 lbs. ft.)

249275

Water pump and mounting bolts — 3.8L engine

J-38816, or an equivalent water pump remover/installer.

To install:

9. Install the water pump and seat the locking ring using J-38816.

10. Install the thermostat housing to the water pump housing.

11. Connect the lower radiator hose and coolant bypass hose.

12. Install the drive belt and drive belt cover.

13. Install the air inlet duct.

14. Connect the negative battery cable.

15. Refill and bleed the cooling system.

Thermostat

REMOVAL AND INSTALLATION

3.8L (VIN K) Engine

1. Disconnect the negative battery cable.

2. Drain the cooling system to a level below the intake manifold.

3. Remove the air inlet duct.

4. Disconnect the upper radiator hose from the thermostat housing.

5. Remove the 2 bolts and remove the thermostat housing from the intake manifold.

6. Remove the thermostat from the intake.

To install:

7. Clean all the gasket surfaces completely.

8. Install the thermostat in the recess in the intake manifold.

9. Apply a thin bead of sealer around the thermostat housing and install the thermostat housing on the intake manifold with a new gasket.

10. Install the mounting bolts and tighten to 15 ft. lbs. (21 Nm).

11. Connect the upper radiator hose.

12. Install the air inlet duct.

13. Connect the negative battery cable.

14. Refill and bleed the cooling system.

4.0L (VIN C) Engine

1. Disconnect the negative battery cable.

2. Drain coolant to a level below the thermostat housing.

3. Remove the air inlet duct.

4. Disconnect the radiator hose.

5. Remove 2 bolts securing the water inlet to the thermostat housing.

6. Remove thermostat and O-ring from the housing.

To install:

7. Clean all sealing surfaces.

8. Install the thermostat and a new O-ring.

9. Attach the water inlet to the thermostat housing and tighten the thermostat housing bolts to 84 inch lbs. (9.5 Nm).

10. Connect the radiator hose and tighten the clamp.

11. Install the air inlet duct.

12. Connect the negative battery cable.

13. Fill and bleed the cooling system.

14. Start engine and check for coolant leaks. Allow engine to come to normal operating temperature. Recheck for coolant leaks.

Electric Cooling Fan

REMOVAL AND INSTALLATION

This vehicle is equipped with twin electrical cooling fans. The removal and installation procedures are the same for either fan.

1. Disconnect the negative battery cable.

2. Disconnect the cooling fan wiring from the fan frame and cooling fan motor.

3. Remove the fan guard and hose support.

4. Remove the cooling fan mounting bolts and lift the fan out of the lower clips.

To install:

5. Install the fan into the vehicle and engage the lower fan tangs into the radiator clips.

6. Install the fan mounting bolts and tighten the outboard mounting bolt to 53 inch lbs. (6 Nm) and the inboard mounting bolt to 35 ft. lbs. (4 Nm).

7. Connect the electrical connector to the fan motor and fan frame.

8. Connect the negative battery cable.

Cooling System

BLEEDING

3.8L (VIN 1) and (VIN K) Engines

1. Fill the cooling system with a suitable mix of coolant and water.

2. Start the vehicle and allow to idle.

3. Place the heater-A/C control in any A/C setting except MAX and the temperature control to the highest setting.

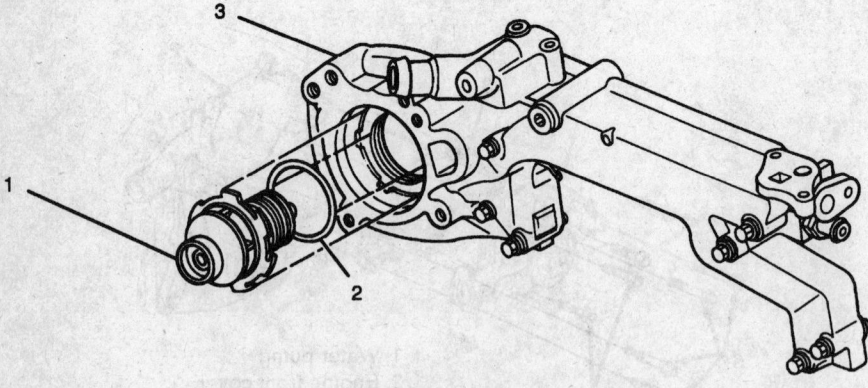

1	WATER PUMP ASM.
2	O–RING SEAL
3	WATER PUMP HOUSING ASM.

311781

Water pump housing and water pump — 4.0L (VIN C) engine

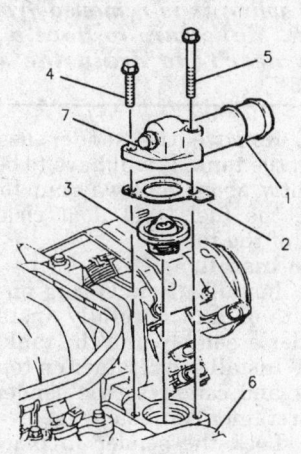

1	OUTLET	5	BOLT/SCREW
2	GASKET	6	INTAKE MANIFOLD
3	THERMOSTAT	7	BLEEDER VALVE
4	BOLT/SCREW		

304396

Thermostat housing mounting components — 3.8L (VIN 1) supercharged engine

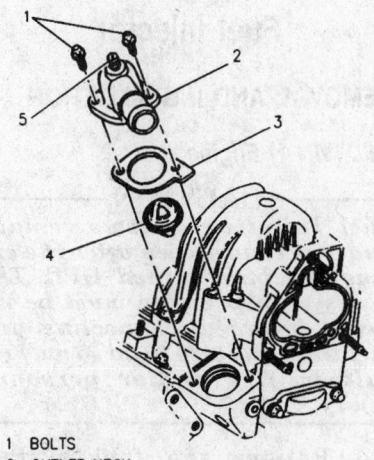

1	BOLTS
2	OUTLET NECK
3	GASKET
4	THERMOSTAT
5	BLEEDER VALVE

304397

Thermostat housing mounting components — 3.8L (VIN K) engine

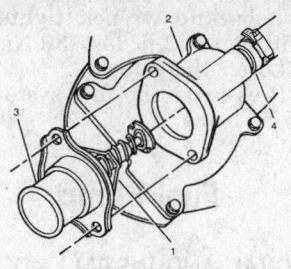

1	THERMOSTAT
2	THERMOSTAT HOUSING
3	COOLANT PUMP INLET
4	THERMOSTAT BY-PASS HOSE

311741

Thermostat and housing — 4.0L (VIN C) engine

4. Run the engine until the lower hose to the water pump is hot.

5. With the engine hot run the engine speed up to 3000 rpm and allow it return to idle. Repeat this 5 times.

6. Slowly open the bleed valve on the back of the thermostat housing for fifteen seconds to expel any trapped air.

7. Top off the coolant as necessary and install the radiator cap.

8. Fill the reservoir to the full cold mark and shut off the vehicle.

9. After the vehicle has cooled, check the reservoir level and top off as required.

4.0L (VIN C) Engine

1. Fill the cooling system with a suitable mix of coolant and water.

NOTE: This vehicle uses a coolant solution and a coolant supplement, part number 3634621, specially designed for aluminum engines. Failure to use the coolant supplement and the correct antifreeze can lead to major engine damage. When refilling the engine coolant add 3 pellets of engine coolant supplement sealant to the lower radiator hose.

2. Start the vehicle and allow to idle.

3. Place the heater-A/C control in any A/C setting except MAX and the temperature control to the highest setting.

4. Run the engine until the lower hose to the water pump is hot.

5. With the engine hot, top off the coolant as necessary and install the radiator cap.

6. Fill the reservoir to the full cold mark and shut off the vehicle.

7. After the vehicle has cooled, check the reservoir level and top off as required.

FUEL SYSTEM

Fuel System Service Precautions

Safety is the most important factor when performing not only fuel system maintenance but any type of maintenance. Failure to conduct maintenance and repairs in a safe manner may result in serious personal injury or death. Maintenance and testing of the vehicle's fuel system components can be accomplished safely and effectively by adhering to the following rules and guidelines.

• To avoid the possibility of fire and personal injury, always disconnect the negative battery cable unless the repair or test procedure requires that battery voltage be applied.

• Always relieve the fuel system pressure prior to disconnecting any fuel system component (injector, fuel rail, pressure regulator, etc.), fitting or fuel line connection. Exercise extreme caution whenever relieving fuel system pressure to avoid exposing skin, face and eyes to fuel spray. Please be advised that fuel under pressure may penetrate the skin or any part of the body that it contacts.

• Always place a shop towel or cloth around the fitting or connection prior to loosening to absorb any excess fuel due to spillage. Ensure that all fuel spillage (should it occur) is quickly removed from engine surfaces. Ensure that all fuel soaked cloths or towels are deposited into a suitable waste container.

• Always keep a dry chemical (Class B) fire extinguisher near the work area.

• Do not allow fuel spray or fuel vapors to come into contact with a spark or open flame.

• Always use a backup wrench when loosening and tightening fuel line connection fittings. This will prevent unnecessary stress and torsion to fuel line piping. Always follow the proper torque specifications.

• Always replace worn fuel fitting O-rings with new. Do not substitute fuel hose or equivalent, where fuel pipe is installed.

Fuel System Pressure

RELIEVING

1. Disconnect the negative battery cable.
2. Remove the fuel filler cap from the filler neck.
3. Connect J-34730-1, or an equivalent fuel pressure gauge to the fuel pressure test port. Wrap a shop towel around the fitting while connecting the gauge to prevent fuel spillage.
4. Install the gauge bleed hose into a suitable container and open the valve to bleed the system.
5. Drain any remaining fuel from the gauge into the container and remove the gauge from the test port.

Idle Speed

ADJUSTMENT

All fuel system operations are computer controlled. No adjustments are possible.

Fuel Filter

REMOVAL AND INSTALLATION

——— CAUTION ———
Fuel Injection systems remain under pressure, even after the engine has been turned OFF. The fuel system pressure must be relieved before disconnecting any fuel lines. Failure to do so may result in fire and/or personal injury.

1. Relieve the fuel system pressure.
2. Disconnect the negative battery cable.
3. Raise and safely support the vehicle.
4. Disconnect the quick connect fitting at the fuel filter inlet.
5. Disconnect the fuel filter outlet fitting from the fuel filter while holding the filter fitting with a backup wrench.
6. Remove the fuel filter from the vehicle.
   To install:
7. Install a new plastic retainer on the fuel inlet line.
8. Apply a drop of oil on the fuel filter inlet fitting and snap the fitting onto the fuel filter.
9. Connect the fuel outlet line to the filter and while holding the filter

with a backup wrench tighten the line fitting to 22 ft. lbs. (30 Nm).
10. Lower the vehicle.
11. Pressurize the fuel system and verify no leaks.

Fuel Pump

REMOVAL AND INSTALLATION

——— CAUTION ———
Fuel Injection systems remain under pressure, even after the engine has been turned OFF. The fuel system pressure must be relieved before disconnecting any fuel lines. Failure to do so may result in fire and/or personal injury.

1. Relieve the fuel system pressure.
2. Disconnect the negative battery cable.

——— CAUTION ———
Observe all applicable safety precautions when working around fuel. Do not allow fuel spray or fuel vapors to come into contact with a spark or open flame. Keep a dry chemical (Class B) fire extinguisher near the work area. Never drain or store fuel in an open container due to the possibility of fire or explosion.

3. Drain the fuel tank with a hand held siphon until the level is less than 1/4 full.
4. Working in the trunk, remove the spare tire cover, jack and spare tire.
5. Pull back the floor trunk liner.
6. Remove the cover from the fuel sender access cover.
7. Disconnect the quick connect fittings from the fuel sender assembly.
8. Disconnect the electrical connector from the fuel sender assembly.

——— CAUTION ———
When the lock ring is removed from the fuel sender the sender assembly will spring up. Downward pressure should be kept on the assembly and slowly released to ensure the sender assembly does not get damaged.

9. Remove lock ring from the fuel sender using J-39765, or an equivalent spanner wrench.
10. Slowly release the spring pressure on the sender.

——— CAUTION ———
The reservoir bucket on the fuel sender assembly will be full of fuel when it is removed from the tank. Make sure to have a catch pan nearby to drain the sender into.

11. Remove the sender assembly from the tank. It will have to be tilted slightly about half way out to make sure the fuel level float clears the side of the tank.
   To install:
12. Install a new O-ring on top of the tank and carefully install the sender assembly into the tank.
13. Install the retainer on top of the tank and compress the sender until the retainer can be engaged.
14. Lock the sender in place with J-39765, or the equivalent.
15. Connect the quick connect fittings to the fuel sender assembly.
16. Connect the electrical connector to the sender assembly.
17. Connect the negative battery cable.
18. Pressurize the fuel system and verify no leaks.
19. Install the fuel sender access panel.
20. Reposition the trunk liner.
21. Install the spare tire, jack and spare tire cover.
22. Refill the fuel tank.

Fuel Injector

REMOVAL AND INSTALLATION

3.8L (VIN 1) Engine

——— CAUTION ———
Fuel Injection systems remain under pressure, even after the engine has been turned OFF. The fuel system pressure must be relieved before disconnecting any fuel lines. Failure to do so may result in fire and/or personal injury.

1. Relieve the fuel system pressure.
2. Disconnect the negative battery cable.
3. Disconnect and cap the fuel feed and return lines from the fuel rail.
4. Disconnect the vacuum line from the fuel pressure regulator.
5. Disconnect the vacuum line from the supercharger.
6. Disconnect the fuel injector electrical connectors.
7. Remove the fuel rail mounting bolts and remove the fuel rail from

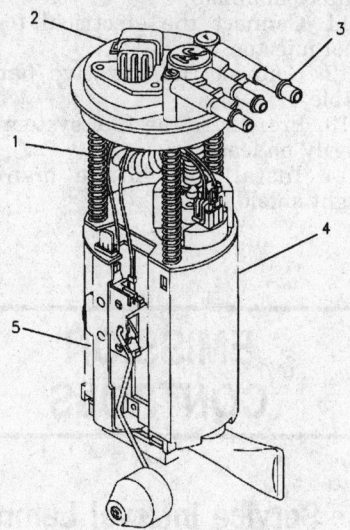

1 SUPPORT ASSEMBLY – FUEL SENDER
2 COVER ASSEMBLY – FUEL SENDER
3 FUEL PIPES (ABOVE COVER)
4 RESERVOIR – FUEL PUMP FUEL
5 SENSOR ASSEMBLY – FUEL LEVEL

320677

Fuel pump and sending unit module assembly

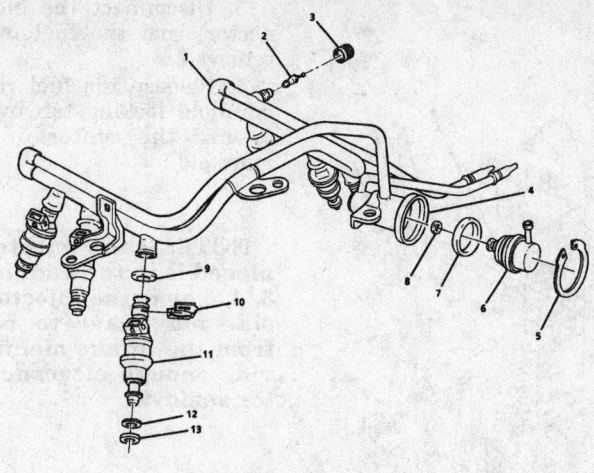

1 RAIL ASSEMBLY - FUEL
2 CORE ASSEMBLY - FUEL PRESSURE CONNECTION
3 CAP - FUEL PRESSURE CONNECTION
4 REGULATOR HOUSING - FUEL PRESSURE
5 SNAP RING - FUEL PRESSURE REGULATOR
6 REGULATOR ASSEMBLY - FUEL PRESSURE
7 BACK-UP O-RING - FUEL PRESSURE REGULATOR
8 O-RING - FUEL PRESSURE REGULATOR
9 O-RING UPPER INJECTOR (BLACK)
10 RETAINER CLIP - FUEL INJECTOR
11 INJECTOR ASSEMBLY - FUEL
12 BACK-UP O-RING - FUEL INJECTOR
13 O-RING - LOWER INJECTOR (BROWN)

304765

Fuel rail components — 3.8L (VIN 1) engine

the intake manifold. Make sure to lift both side of the fuel rail with equal force.

8. Remove the fuel injector retaining clip and remove the injector from the rail.

9. Remove the O-rings from the injector and discard the retaining clip and O-rings.

To install:

10. Coat the new O-rings with clean engine oil and install the O-rings on the fuel injector with the brown O-ring in the lower position.

11. Install the injector on the fuel rail and install the retaining clip. The electrical connector on the injector must point outward.

12. Install the fuel rail onto the intake manifold and make sure to seat all the injectors in their cavities.

13. With the rail seated by hand, tighten the mounting bolts to 8 ft. lbs. (11 Nm).

14. Connect the fuel pressure and return lines.

15. Connect the electrical connectors to the fuel rail.

16. Connect the vacuum line to the supercharger and the pressure regulator.

17. Connect the negative battery cable.

18. Pressurize the fuel system and verify no leaks.

3.8L (VIN K) Engine

CAUTION

Fuel Injection systems remain under pressure, even after the engine has been turned OFF. The fuel system pressure must be relieved before disconnecting any fuel lines. Failure to do so may result in fire and/or personal injury.

1. Relieve the fuel system pressure.

2. Disconnect the negative battery cable.

3. Disconnect and cap the fuel feed and return lines from the fuel rail.

4. Disconnect the vacuum line from the fuel pressure regulator.

5. Disconnect the vacuum line from the throttle body.

6. Disconnect and tag the wires from the ignition coils.

7. Disconnect the fuel injector electrical connectors.

8. Remove the fuel rail mounting nuts and remove the fuel rail from the intake manifold. Make sure to lift both side of the fuel rail with equal force.

9. Remove the fuel injector retaining clip and remove the injector from the rail.

10. Remove the O-rings from the injector and discard the retaining clip and O-rings.

CAUTION

In order to reduce the risk of fire and personal injury that may result from a fuel leak, always install the proper fuel injector O-rings in the proper position. If the upper and lower O-rings are different colors (black or brown), be sure to install the black in the upper position and the brown in the lower position on the fuel injector. The O-rings are the same size but are made from different materials.

To install:

11. Coat the new O-rings with clean engine oil and install the O-rings on the fuel injector with the brown O-ring in the lower position.

12. Install the injector on the fuel rail and install the retaining clip. The electrical connector on the injector must point outward.

13. Install the fuel rail onto the intake manifold and make sure to seat all the injectors in their cavities.

14. With the rail seated by hand, tighten the mounting nuts to 7 ft. lbs. (10 Nm).

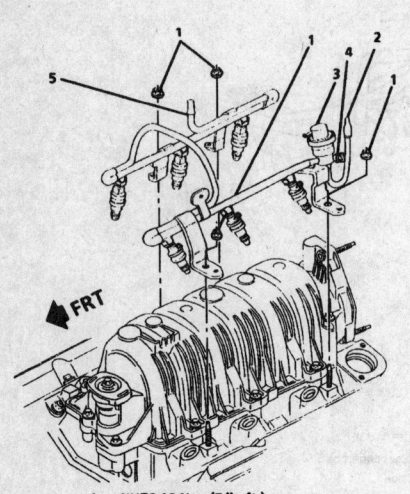

1 NUTS 10 N·m (7 lb. ft.)
2 FUEL FEED
3 FUEL PRESSURE REGULATOR
4 FUEL PRESSURE CONNECTION
5 FUEL RETURN

249372

Fuel rail mounting nuts — 3.8L (VIN K) engine

15. Connect the fuel pressure and return lines.

16. Connect the electrical connectors to the fuel rail.

17. Connect the ignition wires to the coils.

18. Connect the vacuum line to the throttle body and the pressure regulator.

19. Connect the negative battery cable.

20. Pressurize the fuel system and verify no leaks.

4.0L (VIN C) Engine

> **CAUTION**
> *Fuel Injection systems remain under pressure, even after the engine has been turned OFF. The fuel system pressure must be relieved before disconnecting any fuel lines. Failure to do so may result in fire and/or personal injury.*

1. Relieve the fuel system pressure.

2. Disconnect the negative battery cable.

3. Disconnect the electrical connector from the fuel injector being removed.

4. Release the fuel rail to intake manifold locking tab by pushing in toward the center of the intake manifold.

NOTE: If the injector being replaced is in the middle, cylinders 3, 4, 5 or 6, the injector on either side may have to be removed from the intake manifold to provide enough clearance for injector removal.

5. Pry the injector out of the intake manifold using J-41081 or an equivalent and prying between the injector and intake manifold.

6. Spread the injector retaining clip and remove the injector from the fuel rail.

7. Discard the clip and the injector O-rings.

To install:

8. Coat the O-rings with clean engine oil and install the O-rings on the injector.

9. Install the fuel injector onto the fuel rail and install a new retaining clip.

10. Snap the fuel rail down onto the intake manifold.

11. Connect the electrical to the fuel injector.

12. Connect the negative battery cable.

13. Pressurize the fuel system and verify no leaks.

14. Install the intake manifold sight shield.

EMISSION CONTROLS

Service Interval Lamp

RESETTING

After the engine oil has been changed, display the oil life index on the Driver's Information Center (DIC), then hold the **RESET** button for 5 seconds. When a DIC message of **RESET** is displayed and the oil life index equals 100 percent, the reset is complete.

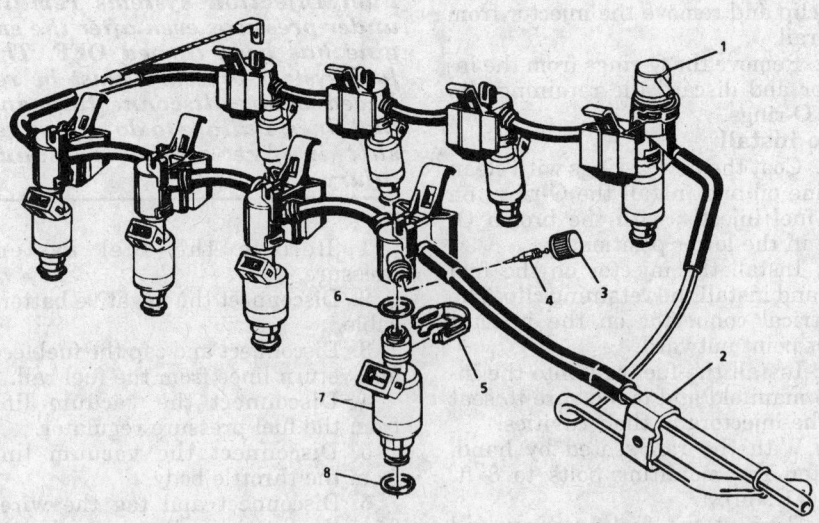

1 CARTRIDGE REGULATOR ASSEMBLY - FUEL PRESSURE
2 RAIL ASSEMBLY - FUEL
3 CAP - FUEL PRESSURE CONNECTION
4 CORE ASSEMBLY - FUEL PRESSURE
5 CLIP - MFI FUEL INJECTOR RETAINER
6 O-RING - MFI FUEL INJECTOR UPPER
7 INJECTOR ASSEMBLY - MFI FUEL
8 O-RING - MFI FUEL INJECTOR LOWER

310218

Fuel injector and fuel rail mounting components — 4.0L (VIN C) engine

ENGINE MECHANICAL

Engine Assembly

REMOVAL AND INSTALLATION

3.8L (VIN 1) and (VIN K) Engines

— CAUTION —

Fuel Injection systems remain under pressure, even after the engine has been turned OFF. The fuel system pressure must be relieved before disconnecting any fuel lines. Failure to do so may result in fire and/or personal injury.

1. Relieve the fuel system pressure.
2. Disconnect the negative battery cable.
3. Matchmark the hood to the hood hinges and remove the hood.
4. Drain the cooling system.
5. Remove the radiator hoses and disconnect the heater hoses from the heater core.
6. Disconnect the negative battery cable from the engine block.
7. Disconnect the engine harness connector at the bulkhead.
8. Remove the serpentine belt(s).
9. Remove the power steering pump from the mounting bracket and position the pump aside. DO NOT disconnect the power steering lines from the pump.
10. Remove the air inlet duct.
11. Disconnect the throttle cables from the linkage mounting bracket and disconnect all the cables from the throttle body lever.
12. Disconnect the following wiring connectors from their related components:
- MAT sensor
- Throttle Position Sensor
- Idle Air Control valve
- Oxygen Sensor
- A/C compressor
- Oil Pressure Switch
- Power steering cutout switch
- Vehicle Speed Sensor
- Low oil level sensor
13. Remove the screws securing the ignition assembly ground strap to the inner fender panel.
14. Disconnect and cap the fuel feed and sender lines from the fuel rail and pressure regulator.
15. Disconnect the EVAP cannister hoses from the throttle body.
16. Disconnect the vacuum lines from the power brake booster and heater control hoses.
17. Disconnect and remove the cruise control servo.
18. Raise and safely support the vehicle.
19. Drain the engine oil.
20. Disconnect the exhaust pipe from the right side exhaust manifold.
21. Remove the A/C compressor from the mounting bracket and support the compressor out of the way. DO NOT disconnect the refrigerant lines from the compressor.
22. Remove the right front engine-to-transaxle brace.
23. Remove the flywheel cover mounting screws and remove the cover.
24. Disconnect and remove the starter.
25. Matchmark the flywheel to the converter. Remove the torque converter-to-flywheel bolts.
26. Lower the vehicle.
27. Install a suitable engine lifting device and slightly raise the engine to take the weight off the front engine mount.
28. Remove the torque axis mount.
29. Support the transaxle with a floor jack. Remove the engine-to-transaxle bolts.
30. Carefully remove the engine from the vehicle. Make sure as the engine is being raised that there are no electrical or vacuum connections still attached.

To install:

31. Lower the engine into the vehicle. Guide the engine into position against the transaxle and make sure the alignment dowels are correctly seated in the transaxle.
32. Install the engine-to-transaxle bolts and tighten to 55 ft. lbs. (75 Nm).
33. Install the torque axis mount. Tighten the mount-to-frame bolts to 64 ft. lbs. (87 Nm), the engine side through bolt to 65 ft. lbs. (87 Nm) and the frame side through bolt to 52 ft. lbs. (70 Nm).
34. Remove the engine lifting assembly and floor jack.
35. Raise and safely support the vehicle.
36. Line up the matchmarks on the torque converter and flywheel and install the mounting bolts. Tighten the mounting bolts to 46 ft. lbs. (62 Nm).
37. Connect and install the starter.
38. Install the flywheel cover and tighten the mounting screws to 88 inch lbs. (10 Nm).
39. Install the right side engine-to-transaxle bracket.
40. Position the A/C compressor in the mounting bracket and tighten the front mounting bolts to 44 ft. lbs. (60 Nm), the rear mounting bolts to 18 ft. lbs. (25 Nm) and the rear mounting nut to 18 ft. lbs. (25 Nm).
41. Connect the exhaust pipe to the rear exhaust manifold and tighten the mounting nuts to 18 ft. lbs. (25 Nm).
42. Lower the vehicle.
43. Install the cruise control servo and connect the vacuum lines.
44. Connect the vacuum lines to the power brake booster and heater control hoses.
45. Connect the EVAP cannister hoses to the throttle body.
46. Connect the fuel feed and return lines to the fuel rail and pressure regulator.
47. Connect the ignition assembly ground strap to the inner fender well mounting screw.
48. Connect the following wiring connectors to their related components:
- MAT sensor
- Throttle Position Sensor
- Idle Air Control valve
- Oxygen Sensor
- A/C compressor
- Oil Pressure Switch
- Power steering cutout switch
- Vehicle Speed Sensor
- Low oil level sensor
49. Connect the control cables to the throttle body lever and connect the throttle cable to the mounting bracket.
50. Install the air inlet duct.
51. Install the power steering pump in the mounting bracket.
52. Install the serpentine belt(s).
53. Connect the main engine harness at the bulkhead connector.
54. Connect the negative battery cable to the engine block.
55. Connect the heater hoses to the heater core and install the upper and lower radiator hoses.
56. Refill the crankcase with engine oil.
57. Connect the negative battery cable.
58. Install the hood on the vehicle.
59. Pressurize the fuel system and verify no leaks.
60. Refill and bleed the cooling system.

4.0L (VIN C) Engine

On this vehicle, some engine-related service procedures require that the entire Powertrain Assembly be removed from the vehicle. The requires specialized lifts and supports to lower the powertrain as the vehicle body is lifted up and off of the powertrain support cradle. Use care during this operation.

CAUTION

Fuel Injection systems remain under pressure, even after the engine has been turned OFF. The fuel system pressure must be relieved before disconnecting any fuel lines. Failure to do so may result in fire and/or personal injury.

1. Relieve the fuel system pressure.
2. Disconnect the negative battery cable.
3. Remove the air inlet duct.
4. Drain the cooling system.
5. Recover the refrigerant from the A/C system.
6. Disconnect and cap the fuel feed and return lines from the fuel rail.
7. Disconnect the vacuum harness from the rear of the intake manifold.
8. Disconnect the throttle cables from the throttle body lever and intake manifold bracket and position out of the way.
9. Disconnect the engine harness connector at the bulkhead.
10. Disconnect the coolant hoses from the radiator surge tank.
11. Disconnect the upper radiator hose from the coolant crossover.
12. Disconnect the lower radiator hose from the thermostat housing.
13. Disconnect and cap the upper transaxle cooler line from the radiator and the lower cooler line from the transaxle.
14. Disconnect the heater hoses from the engine compartment heater pipes.
15. Disconnect the vacuum supply line from the power booster.
16. Disconnect the shift lever from the transaxle manual shift lever and remove the transaxle cable bracket and position the cable assembly out of the way.
17. Remove the vacuum reservoir.
18. Disconnect the electrical connectors and vacuum lines from the cruise control servo.
19. Install an engine support fixture, J-28467-A, or the equivalent.
20. Disconnect the positive battery cable at the junction block.
21. Remove the through bolt from the torque axis mount.
22. Remove the front engine mount bolts from the torque axis mount bracket and set the mount aside.
23. Disconnect the negative battery cable from the engine block.
24. Raise and safely support the vehicle.

25. Remove both front tire and wheel assemblies.
26. Remove the left and right inner fender well splash shields.
27. Disconnect the electrical connector from the power steering rack.
28. Remove the cotter pins and castle nuts from the lower ball joints and separate the ball joints from the steering knuckles using J-36226, or an equivalent puller.
29. Remove the cotter pins and castle nuts from the outer tie rod ends and separate the ends from the steering knuckles using J-36226, or the equivalent.
30. Remove the axle nuts and separate the halfshafts from the wheel bearing and hub assemblies.
31. Remove the pinch bolt from the power steering rack at the intermediate shaft.
32. Remove the complete exhaust system.
33. Disconnect the engine oil cooler lines at the adapter.
34. Remove the engine-to-transaxle brace located at the oil pan, right cylinder bank in the rear of the engine and the left bank on the front of the engine.
35. Remove the flywheel cover.
36. Matchmark the converter to the flywheel and remove the mounting bolts.
37. Remove the oil cooler adapter from the engine.
38. Disconnect the A/C hose and muffler from the rear of the A/C compressor.
39. Position a frame support table under the powertrain assembly.
40. Remove the left transaxle mount.
41. Remove the subframe mounting bolts.
42. Release the engine support fixture.
43. Raise the vehicle off of the engine table slowly and verify no electrical or vacuum connections are still hooked to the engine.
44. Remove the transaxle-to-engine.
45. Using a suitable engine lift, separate the engine from the transaxle and subframe assembly.

To install:
46. Position the engine assembly onto the subframe. Line up the engine dowels with the holes in the transaxle and install the mounting bolts. Tighten the bolts to 55 ft. lbs. (75 Nm).
47. Remove the engine lift and position the powertrain assembly under the vehicle.

48. Lower the vehicle and line up the subframe with the vehicle and install the mounting bolts and tighten to 142 ft. lbs. (192 Nm).
49. Install the engine support fixture.
50. Raise the vehicle and remove the engine dolly.
51. Install the left side transaxle mount.
52. Connect the A/C hose and muffler to the A/C compressor.
53. Install the oil filter adapter to the engine block and tighten the mounting bolts to 12 ft. lbs. (16 Nm).
54. With the converter-to-flywheel matchmarks in alignment, tighten the mounting bolts to 35 ft. lbs. (47 Nm).
55. Install the rear engine-to-transaxle brace and tighten the mounting bolts to 44 ft. lbs. (60 Nm).
56. Install the front engine-to-transaxle brace and tighten the mounting bolts to 44 ft. lbs. (60 Nm).
57. Install the oil pan brace and tighten the mounting bolts to 37 ft. lbs. (50 Nm).
58. Install the flywheel cover.
59. Connect the oil cooler hoses to the adapter and tighten the fittings to 12 ft. lbs. (18 Nm).
60. Install the exhaust system.
61. Connect the halfshafts to the hub bearing assemblies and tighten the axle nuts to 107 ft. lbs. (148 Nm).
62. Connect the tie rod ends to the steering knuckle and tighten the castle nuts to 41 ft. lbs. (55 Nm). If necessary tighten the castle nuts up to 60° additional to align the cotter pin holes. NEVER loosen the castle nut and DO NOT exceed 52 ft. lbs. (70 Nm) to make the holes align. Install new cotter pins.
63. Connect the ball joints to the steering knuckles and tighten the castle nuts to 88 inch lbs. (10 Nm) plus 120° additional. If necessary tighten the castle nuts up to 60° additional to align the cotter pin holes. NEVER loosen the castle nuts to make the holes align. Install new cotter pins.
64. Connect the intermediate shaft to the rack and pinion and tighten the pinch bolt to 35 ft. lbs. (47 Nm).
65. Connect the electrical connector to the rack and pinion assembly.
66. Install the left and right inner fender splash shields.
67. Install the front tire and wheel assemblies and tighten the wheel nuts to 100 ft. lbs. (140 Nm).
68. Lower the vehicle.
69. Connect the negative battery cable to the engine block.

70. Install the front engine mount bracket-to-torque axis mount bolts.

71. Install the torque axis mount through bolt.

72. Connect the positive battery cable to the junction block.

73. Remove the engine support fixture.

74. Connect the electrical and vacuum connectors to the cruise control servo.

75. Install the vacuum reservoir.

76. Connect the shift cable to the transaxle manual shift lever and install the cable mounting bracket.

77. Connect the vacuum hose to the power booster.

78. Connect the heater hoses at the rear of the engine.

79. Connect the transaxle cooler lines.

80. Connect the lower radiator hose to the thermostat housing.

81. Connect the upper radiator hose to the coolant crossover.

82. Connect the hoses to the coolant surge tank.

83. Connect the main engine harness to the bulkhead connector.

84. Connect the throttle cables to the throttle body lever and install the intake manifold bracket.

85. Connect the vacuum hoses to the rear of the intake manifold.

86. Connect the fuel feed and return lines.

87. Fill the cooling system, engine crankcase and recharge the A/C system.

88. Install the air inlet duct.

89. Connect the negative battery cable.

90. Start the vehicle and verify no leaks.

Engine Mounts

REMOVAL AND INSTALLATION

3.8L (VIN 1) and (VIN K) Engines

1. Disconnect the negative battery cable.

2. Remove the vacuum reservoir.

3. Install an engine support fixture, J-28467-A or the equivalent and raise the engine so the weight is removed from the torque axis mount.

4. Loosen the 2 lower torque axis-to-frame bolts.

5. Remove the 2 torque axis through bolt nuts and remove the through bolts.

6. Remove the torque axis mount.

To install:

7. Position the torque axis mount in the vehicle so the lower mounting

bracket slips around the 2 mounting bolts in the frame.

8. Install the 2 torque axis mount through bolts and nuts.

9. Tighten the torque axis mount-to-frame bolts to 52 ft. lbs. (70 Nm) and tighten the through bolts to 65 ft. lbs. (87 Nm).

10. Remove the engine support fixture.

11. Install the vacuum reservoir.

12. Connect the negative battery cable.

4.0L (VIN C) Engine

1. Disconnect the negative battery cable.

2. Install an engine support fixture, J-28467-A or the equivalent.

3. Remove the engine mount-to-body through bolt nut and through bolt.

4. Remove the through bolt nut and bolt from the engine side of the mount.

5. Raise and safely support the vehicle.

6. Remove the front tire and wheel assembly.

7. Remove the right inner fender well splash shield.

8. Remove the lower center air deflector.

9. Remove the left transaxle mount-to-frame through bolt.

10. Remove the lower retaining bolt and nut from the engine mount bracket.

11. Remove the bolt and disconnect the power steering line retainer from the bracket.

12. Lower the vehicle.

13. Remove the upper retaining nut and bolt from the engine mount bracket.

14. Remove the serpentine belt from the power steering pump pulley.

15. Remove the fuel plastic sight shield on the intake manifold.

16. Remove the power steering pump from the mounting bracket and set aside. DO NOT disconnect the power steering lines from the pump.

17. Raise the engine using the support fixture.

18. Remove the engine mount bracket from the engine.

19. Remove the torque axis mount from the frame rail.

To install:

20. Install the torque axis mount on the body.

21. Install the engine mount bracket on the engine and loosely install the mounting nuts and bolts. With all the fasteners in place tighten the nuts to 30 ft. lbs. (40 Nm) and the bolts to 41 ft. lbs. (55 Nm).

22. Lower the engine support fixture until the engine is at its normal height.

23. Install the power steering pump in the mounting bracket.

24. Install the serpentine belt.

25. Connect the power steering line retainer to the engine mount bracket and install the mounting bolt.

26. Raise and safely support the vehicle.

27. Install the left transaxle mount-to-frame through bolt and tighten to 63 ft. lbs. (85 Nm).

28. Install the lower center air deflector.

29. Install the right inner fender well splash shield.

30. Install the tire and wheel assembly and tighten the wheel nuts to 100 ft. lbs. (140 Nm).

31. Lower the vehicle.

32. Install the torque axis mount-to-engine mount bracket through bolt and nut and tighten the bolt to 70 ft. lbs. (95 Nm).

33. Install the torque axis mount-to-frame through bolt and nut and tighten the through bolt to 37 ft. lbs. (50 Nm).

34. Remove the engine support fixture.

35. Connect the negative battery cable.

Cylinder Head

REMOVAL AND INSTALLATION

3.8L (VIN 1) and (VIN K) Engines

NOTE: Head gaskets are not interchangeable. Failure to install them with the arrow pointing to the front, will cause gasket failure and possible engine failure. The left–hand gasket can be identified with the letter "L" punched in it (next to the arrow).

Left Side (Front)

1. Disconnect the negative battery cable.

2. Following proper procedures, relieve the fuel system pressure.

3. Drain the cooling system and remove the intake manifold.

4. Remove the left exhaust manifold.

5. Remove the valve covers and remove the rocker arm assemblies and pushrods, keeping everything in order for reinstallation.

6. Tag and disconnect the ignition wires and remove the spark plugs.

7. Remove the alternator front mounting bracket and the ignition module with bracket.

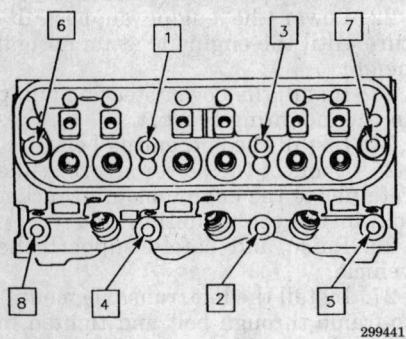

Cylinder head tightening sequence — 3.8L (VIN 1) and (VIN K) engine

8. Remove the one bolt securing the A/C bracket to the cylinder head.

9. Remove the cylinder head bolts in the reverse order of installation, and remove the cylinder head.

10. Clean all gasket mating surfaces and the cylinder head bolt holes in the block.

To install:

11. Place the cylinder head gasket on the engine block dowels with the note **THIS SIDE UP** facing the cylinder head and the arrow facing the front of the engine. Position the cylinder head on the engine block.

——— **CAUTION** ———

In order to prevent damage to the gasket, when installing the cylinder head, do not slide the cylinder head on the gasket.

12. Install the cylinder head bolts and tighten as follows:

NOTE: This engine uses special torque to yield head bolts. The procedure must be followed carefully and NEW BOLTS MUST BE USED WHENEVER THE HEAD IS REMOVED. Total bolt torque should not exceed 60 ft. lbs. (81 Nm).

   a. Tighten the cylinder head bolts, in sequence, to 35 ft. lbs. (47 Nm).

   b. Rotate each bolt 130 degrees, in sequence.

   c. Rotate the center 4 bolts an additional 30 degrees, in sequence.

13. Install the pushrods, rocker arm assemblies and valve covers.

14. Install the intake and exhaust manifolds.

15. Install the alternator front mount bracket and ignition module with bracket.

16. Install the spark plugs and wires.

17. Install the A/C compressor bracket bolt, and torque to 52 ft. lbs. (70 Nm).

18. Fill the cooling system and connect the negative battery cable.

Right Side (Rear)

1. Disconnect the negative battery cable.

2. Following proper procedures, relieve the fuel system pressure.

3. Drain the cooling system and disconnect the exhaust crossover pipe.

4. Remove the intake manifold.

5. Remove the right side exhaust manifold.

6. Remove the valve covers.

7. Remove the serpentine drive belt.

8. Remove the belt tensioner pulley.

9. Remove the power steering pump mounting bracket and lay the pump aside.

10. Remove the fuel line heat shield.

11. Tag and disconnect the ignition wires and remove the spark plugs.

12. Remove the rocker arm assemblies and pushrods, keeping everything in order for reinstallation.

13. Remove the cylinder head bolts in reverse order of installation and remove the cylinder head.

14. Clean all gasket mating surfaces and the cylinder head bolt holes in the block.

To install:

15. Place the cylinder head gasket on the engine block dowels with the note **THIS SIDE UP** facing the cylinder head and the arrow facing the front of the engine. Position the cylinder head on the engine block.

——— **CAUTION** ———

In order to prevent damage to the gasket, when installing the cylinder head, do not slide the cylinder head on the gasket.

16. Install the cylinder head bolts and tighten as follows:

NOTE: This engine uses special torque to yield head bolts. The procedure must be followed carefully and NEW BOLTS MUST BE USED WHENEVER THE HEAD IS REMOVED. Total bolt torque should not exceed 60 ft. lbs. (81 Nm).

   a. Tighten the cylinder head bolts, in sequence, to 35 ft. lbs. (47 Nm).

   b. Rotate each bolt 130 degrees, in sequence.

   c. Rotate the center 4 bolts an additional 30 degrees, in sequence.

17. Install the exhaust manifold and intake manifold.

18. Install the pushrods and rocker arm assemblies.

19. Install the valve cover(s).

20. Install the spark plugs and wires.

21. Install the power steering pump bracket and torque the bolts to 35 ft. lbs. (47 Nm).

22. Install the belt tensioner pulley and serpentine belt.

23. Install the exhaust crossover pipe.

24. Fill the cooling system and connect the negative battery cable.

4.0L (VIN C) Engine

NOTE: The manufacturer recommends that the entire powertrain be removed from the vehicle before removing the cylinder heads.

——— **CAUTION** ———

Fuel injection systems remain under pressure, even after the engine has been turned OFF. The fuel system pressure must be relieved before disconnecting any fuel lines. Failure to do so may result in fire and/or personal injury.

1. Disconnect the negative battery cable.

2. Properly relieve the fuel system pressure.

3. Drain the cooling system into a suitable container.

4. Remove the powertrain assembly.

5. Remove the intake manifold, cam covers, harmonic balancer, timing chain front cover and oil pump.

——— **WARNING** ———

Align all timing marks before performing the next Step.

6. Remove the chain tensioner from the timing chain for the cylinder head being removed.

7. Remove the cam sprockets from the head being removed. The timing chain remains in the chain case.

8. Removing the timing chain guides. Access for the retaining screws is through the plugs at the front of the cylinder head.

9. Remove the water crossover.

10. Remove the exhaust manifold.

11. Remove the cylinder head bolts, a little at a time, in the reverse order of the tightening sequence.

12. Remove the cylinder head and gasket.

13. Clean all gasket mating surfaces. Clean the head bolt holes in the crankcase with compressed air and clean the head bolt bosses in the cylinder head.

14. Check the cylinder head for warpage using a straightedge and feeler gauge. Measure along each edge, at the center and across both ends.

15. If warpage is less than 0.002 inch (0.05mm), the cylinder head surface is usable. If warpage is 0.002–0.008 inch (0.05–0.2mm), the cylinder head must be resurfaced. After resurfacing, the dimension between the combustion chamber gauge pad and the deck surface must be at least 10.5mm.

To install:

16. Using a new cylinder head gasket, install the cylinder head and the ten M11 and three M6 head bolts. Lube the washer and the underside of the bolt head with engine oil prior to installation. New replacement head bolts are recommended.

17. Tighten the ten M11 bolts, in sequence, to 22 ft. lbs. (30 Nm) plus 90 degrees (¼ turn). Repeat the sequence, turning each bolt an additional 75 degrees (total 165 degrees). Tighten the three M6 bolts to 10 ft. lbs. (12 Nm).

18. Install the camshafts and set the camshaft timing.

19. Install the camshaft guide bolt access hole plugs in the cylinder heads. The plugs should be seated and snug.

20. Install the intake cam covers, oil pump, timing chain front cover, harmonic balancer, intake manifold and water crossover.

21. Install the exhaust manifold. Tighten the nuts to 22 ft. lbs. (30 Nm) or the bolts to 18 ft. lbs. (25 Nm).

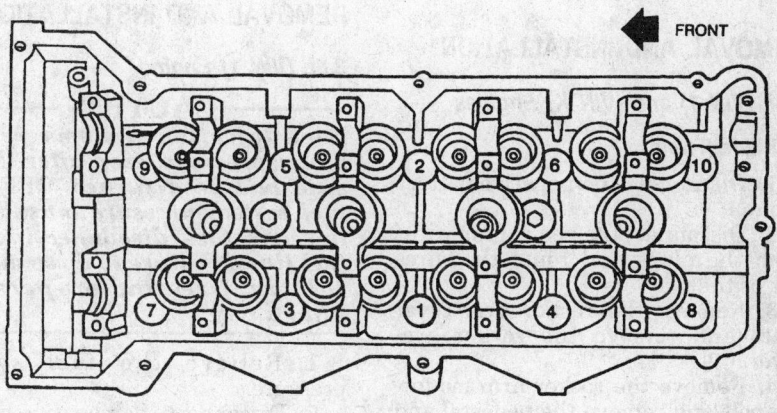

Cylinder head bolt tightening sequence — 4.0L (VIN C) engine

310510

22. Install the powertrain assembly into the vehicle.

23. Connect the negative battery cable.

24. Fill the cooling system and check all fluid levels.

25. Properly charge the A/C system.

26. Run the engine and check for leaks and proper engine performance.

Valve Lifters

REMOVAL AND INSTALLATION

3.8L (VIN 1) and (VIN K) Engines

1. Disconnect the negative battery cable.

2. Remove the intake manifold.

3. Remove the rocker arm covers.

NOTE: When removing the valve train components they must be kept in order for installation in the same locations they were removed from.

4. Remove the rocker arm mounting bolts, rocker arms, pedestals and pushrods.

5. Remove the lifter guide retainer mounting bolts and retainer.

6. Remove the lifter guides.

7. Remove the lifters from their bores.

To install:

8. Clean all parts well. Note that a new camshaft requires new lifters.

9. Coat the lifters with engine lube or engine oil supplement. Any lifters being reused MUST be installed in their original locations.

10. Install the lifter guides and lifter guide retainer. Tighten the retainer mounting bolts to 22 ft. lbs. (30 Nm).

11. Install the pushrods, rocker arms, pedestals, and mounting bolts to 11 ft. lbs. (15 Nm) plus 90 degrees (¼ turn) additional rotation.

12. Install the rocker arm covers and tighten the mounting bolts to 88 inch lbs. (10 Nm).

13. Install the intake manifold.

14. Connect the negative battery cable.

15. Start the vehicle and verify no leaks.

Rocker Arms

REMOVAL AND INSTALLATION

3.8L (VIN 1) and (VIN K) Engines

Left Side (Front)

1. Disconnect the negative battery cable.

2. Disconnect the spark plug wires from the plugs and route the wires out of the way.

3. Remove the rocker arm cover bolts and remove the rocker arm cover.

4. Remove the rocker arm mounting bolt and remove the pedestal and rocker arm.

5. Remove the pushrod.

To install:

6. Install the pushrod into the lifter and install the rocker arm and pedestal. Tighten the mounting bolt to 19 ft. lbs. (25 Nm) plus 70 degrees additional rotation.

7. Install the rocker arm cover with a new gasket and tighten the mounting bolt to 88 ft. lbs. (10 Nm).

8. Connect the spark plug wires to the front plugs.

9. Connect the negative battery cable.

Right Side (Rear)

1. Disconnect the negative battery cable.

2. Remove the serpentine belt.

3. Remove the alternator rear brace.

4. Disconnect the spark plug wires from the plugs and route the wires out of the way.

5. Remove the rocker arm cover bolts and remove the rocker arm cover.

6. Remove the rocker arm mounting bolt and remove the pedestal and rocker arm.

7. Remove the pushrod.

To install:

8. Install the pushrod into the lifter and install the rocker arm and pedestal. Tighten the mounting bolt to 19 ft. lbs. (25 Nm) plus 70 degrees additional rotation.

9. Install the rocker arm cover with a new gasket and tighten the mounting bolt to 88 ft. lbs. (10 Nm).

10. Connect the spark plug wires to the front plugs.

11. Install the alternator rear brace.

12. Install the serpentine belt.

13. Connect the negative battery cable.

Intake Manifold

REMOVAL AND INSTALLATION

3.8L (VIN 1) Engine

———— CAUTION ————

Fuel Injection systems remain under pressure, even after the engine has been turned OFF. The fuel system pressure must be relieved before disconnecting any fuel lines. Failure to do so may result in fire and/or personal injury.

1. Relieve the fuel system pressure.

2. Disconnect the negative battery cable.

3. Remove the fuel injector sight shield and air inlet duct.

4. Disconnect the spark plug wires from the right side spark plugs and route the wires out of the way.

5. Disconnect and tag the vacuum lines from the intake manifold.

6. Drain the cooling system.

7. Disconnect the fuel feed and return lines from the fuel rail.

8. Disconnect the upper radiator hose from the thermostat housing and disconnect the bypass hose from the coolant pipe.

9. Disconnect the electrical connectors from the following components:
- EGR valve
- Throttle Position sensor
- Idle Air Control solenoid
- Fuel injectors
- MAP sensor

10. Disconnect the EGR outlet pipe.

11. Remove the throttle cable bracket from the cylinder head mounting bracket and disconnect the cables from the throttle body lever.

12. Remove the throttle body support bracket with the steering reservoir attached.

13. Remove the inner serpentine belt from the supercharger drive pulley.

14. Disconnect the heater hose pipe from the intake manifold.

15. Remove the tensioner bracket retaining stud from the supercharger. The best way to remove the stud is to install 2 nuts on the stud and back the stud out by turning the inner of the 2 nuts.

16. Remove the intake manifold mounting bolts and remove the intake manifold and gaskets.

To install:

17. Clean all the gasket surfaces completely.

18. Install the intake manifold using new manifold gaskets and install the intake manifold bolts finger tight. With all the bolts in place tighten the bolts in sequence to 11 ft. lbs. (15 Nm). With all the bolts tightened, make a second pass tightening each bolt to 11 ft. lbs. (15 Nm).

19. Install the tensioner bracket-to-supercharger mounting stud using 2 nuts and tightening the stud in using the outer nut.

20. Connect the heater hose pipe to the intake manifold.

21. Install the serpentine belt around the supercharger pulley.

22. Install the throttle bracket and steering reservoir.

23. Connect the throttle cables to the throttle body lever.

24. Connect the EGR valve outlet pipe.

25. Connect the electrical connectors to the following components:
- EGR valve
- Throttle Position sensor
- Idle Air Control solenoid
- Fuel injectors
- MAP sensor

26. Connect the upper radiator hose to the thermostat housing and the bypass hose to the coolant pipe.

27. Connect the fuel lines to the fuel rail.

28. Connect the vacuum lines to the intake manifold.

29. Connect the spark plug wires to the rear bank of plugs.

30. Install the air inlet duct and fuel injector sight shield.

31. Connect the negative battery cable.

32. Pressurize the fuel system and verify no leaks.

33. Refill and bleed the cooling system.

3.8L (VIN K) Engine

———— CAUTION ————

Fuel Injection systems remain under pressure, even after the engine has been turned OFF. The fuel system pressure must be relieved before disconnecting any fuel lines. Failure to do so may result in fire and/or personal injury.

1. Relieve the fuel system pressure.

2. Disconnect the negative battery cable.

3. Remove the fuel injector sight shield and air inlet duct.

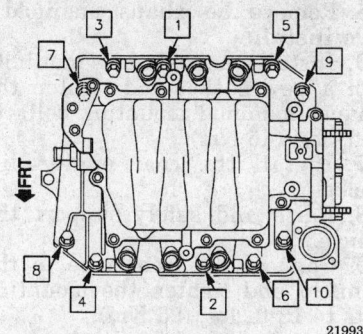

Intake manifold bolt tightening sequence — 3.8L (VIN 1) engine

219938

4. Disconnect the spark plug wires from the right side spark plugs and route the wires out of the way.

5. Disconnect and tag the vacuum lines from the intake manifold.

6. Disconnect the fuel lines, vacuum lines and electrical connectors from the fuel rail. Remove the fuel rail mounting bolts and remove the fuel rail from the intake manifold.

7. Remove the EGR heat shield.

8. Remove the throttle cable bracket from the cylinder head mounting bracket and disconnect the cables from the throttle body lever.

9. Remove the throttle body support bracket.

10. Remove the upper intake plenum mounting bolts and remove the plenum and gasket.

11. Drain the cooling system.

12. Disconnect the upper radiator hose from the thermostat housing.

13. Remove the alternator.

14. Remove the 4 bolts and remove the drive belt tensioner assembly.

15. Disconnect the EGR valve outlet pipe.

16. Remove the lower intake manifold mounting bolts and remove the intake manifold and gaskets.

To install:

17. Clean all the gasket surfaces completely.

18. Install the intake manifold using new manifold gaskets and install the intake manifold bolts finger tight. With all the bolts in place tighten the bolts in sequence to 11 ft. lbs. (15 Nm). With all the bolts tightened, make a second pass tightening each bolt to 11 ft. lbs. (15 Nm).

19. Connect the EGR valve outlet pipe.

20. Install the drive belt tensioner and tighten the mounting bolts to 37 ft. lbs. (50 Nm).

21. Install the alternator and serpentine belt.

22. Connect the upper radiator hose to the thermostat housing.

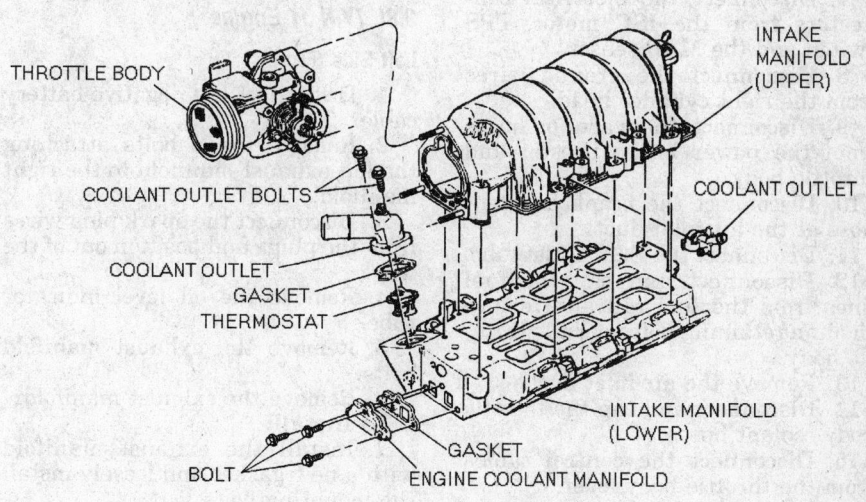

Upper intake plenum components — 3.8L (VIN K) engine

219899

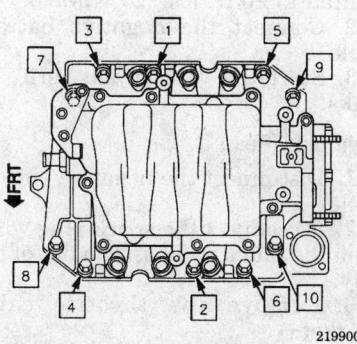

Lower intake manifold mounting bolt tightening sequence — 3.8L (VIN K) engine

219900

23. Install the intake plenum on the lower intake manifold using a new gasket. Install the mounting bolts and with all the bolts in place, tighten in sequence to 89 inch lbs. (10 Nm).

24. Install the throttle body support bracket and connect the throttle cables to the throttle body lever.

25. Install the EGR heat shield.

26. Install the fuel rail assembly and tighten the mounting bolts to 7 ft. lbs. (10 Nm). Connect the vacuum lines, fuel lines and electrical connectors.

27. Connect the vacuum lines to the intake manifold.

28. Connect the spark plug wires to the rear bank of plugs.

29. Install the air inlet duct and fuel injector sight shield.

30. Connect the negative battery cable.

31. Pressurize the fuel system and verify no leaks.

32. Refill and bleed the cooling system.

4.0L (VIN C) Engine

——— CAUTION ———
Fuel Injection systems remain under pressure, even after the engine has been turned OFF. The fuel system pressure must be relieved before disconnecting any fuel lines. Failure to do so may result in fire and/or personal injury.

1. Relieve the fuel system pressure.

2. Disconnect the negative battery cable.

3. Remove the 4 cap nut and washers and remove the intake manifold sight shield.

4. Disconnect the PCV hose from the intake manifold.

5. Disconnect the front spark plug wires and route them out of the way.

6. Disconnect the main fuel injector harness.

7. Disconnect the electrical connectors from the ISC motor, TPS switch and the MAP sensor.

8. Disconnect the ground wires from the right cylinder head.

9. Disconnect the vacuum hoses from the power brake booster and throttle body.

10. Disconnect the crankcase vent hose at the air inlet duct.

11. Disconnect the EGR outlet tube.

12. Disconnect and cap the fuel lines from the fuel rail and remove the line retaining bolt near the throttle body.

13. Remove the air inlet duct.

14. Disconnect and cap the throttle body coolant hoses.

15. Disconnect the control cables from the throttle body lever.

16. Take note of the positions of the 4 studs and remove the 6 intake manifold mounting bolts and 4 mounting studs.

17. Remove the intake manifold.

To install:

18. Install a new intake manifold gasket and position the manifold in place and install the 6 mounting bolts and 4 studs. With all the bolts in place tighten the bolts to 89 inch lbs. (10 Nm) starting in the center and working in a circular pattern.

19. Connect the control cables to the throttle body lever.

20. Connect the throttle body coolant hoses and add coolant as necessary.

21. Install the air inlet duct.

22. Connect the fuel lines to the fuel rail and install the line retaining bolt near the throttle body.

23. Connect the EGR outlet pipe.

24. Connect the crankcase vent pipe to the air inlet duct.

25. Connect the vacuum hoses to the power booster and throttle body.

26. Connect the ground wires to the right cylinder head.

27. Connect the electrical connectors to the ISC motor, the TPS switch and the MAP sensor.

28. Connect the main fuel injector harness.

29. Connect the spark plug wires to the front plugs.

30. Connect the PCV hose to the intake manifold.

31. Connect the negative battery cable.

32. Pressurize the fuel system and verify no leaks.

33. Install the intake manifold sight shield.

Exhaust Manifold

REMOVAL AND INSTALLATION

3.8L (VIN 1) Engine

Left Side (Front)

1. Disconnect the negative battery cable.

2. Remove the 2 bolts attaching the left exhaust manifold to the right manifold.

3. Disconnect the spark plug wires from the plugs and position out of the way.

4. Remove the oil level indicator tube.

5. Remove the exhaust manifold bolts.

6. Remove the exhaust manifold.

To install:

7. Install the exhaust manifold with a new gasket and loosely install the mounting bolts.

8. Install the oil level indicator tube.

9. Tighten the exhaust manifold mounting bolts to 33 ft. lbs. (45 Nm).

10. Connect the spark plug wires to the plugs.

11. Install the bolts connecting the right and left exhaust manifolds and tighten to 20 ft. lbs. (27 Nm).

12. Connect the negative battery cable.

13. Start the vehicle and verify no leaks.

Right Side (Rear)

1. Disconnect the negative battery cable.

2. Disconnect the spark plug wires from the plugs and position out of the way.

3. Remove the throttle cable bracket.

4. Remove the crossover pipe heat shield.

5. Remove the transaxle level indicator tube.

6. Disconnect the oxygen sensor (O_2) electrical connector.

7. Remove the 2 bolts connecting the left and right and left exhaust manifolds.

8. Remove the plastic vacuum tank mounted on the cowl.

9. Raise and safely support the vehicle.

10. Remove the converter heat shield and pipe hanger.

11. Disconnect the exhaust pipe from the manifold.

12. Lower the vehicle.

13. Remove the rear engine lift bracket.

14. Remove the exhaust manifold bolts.

15. Remove the exhaust manifold.

To install:

16. Install the exhaust manifold with a new gasket and tighten the exhaust manifold mounting bolts to 33 ft. lbs. (45 Nm).

17. Install the rear engine lift bracket.

18. Raise and safely support the vehicle.

19. Connect the front pipe to the manifold and tighten the mounting bolts to 18 ft. lbs. (25 Nm).

20. Install the exhaust hanger and converter heat shield.

21. Lower the vehicle.

22. Install the vacuum tank on the cowl.

23. Install the left-to-right manifold bolts and tighten to 20 ft. lbs. (27 Nm).

24. Install the transaxle level indicator tube.

25. Connect the oxygen sensor (O_2) electrical connector.

26. Install the throttle cable bracket.

27. Connect the spark plug wires to the plugs.

28. Connect the negative battery cable.

29. Start the vehicle and verify no leaks.

3.8L (VIN K) Engine

Left Side (Front)

1. Disconnect the negative battery cable.

2. Remove the 2 bolts attaching the left exhaust manifold to the crossover pipe.

3. Disconnect the spark plug wires from the plugs and position out of the way.

4. Remove the oil level indicator tube.

5. Remove the exhaust manifold bolts.

6. Remove the exhaust manifold.

To install:

7. Install the exhaust manifold with a new gasket and loosely install the mounting bolts.

8. Install the oil level indicator tube.

9. Tighten the exhaust manifold mounting bolts to 33 ft. lbs. (45 Nm).

10. Connect the spark plug wires to the plugs.

11. Install the bolts connecting the left exhaust manifold to the crossover pipe and tighten to 20 ft. lbs. (27 Nm).

12. Connect the negative battery cable.

13. Start the vehicle and verify no leaks.

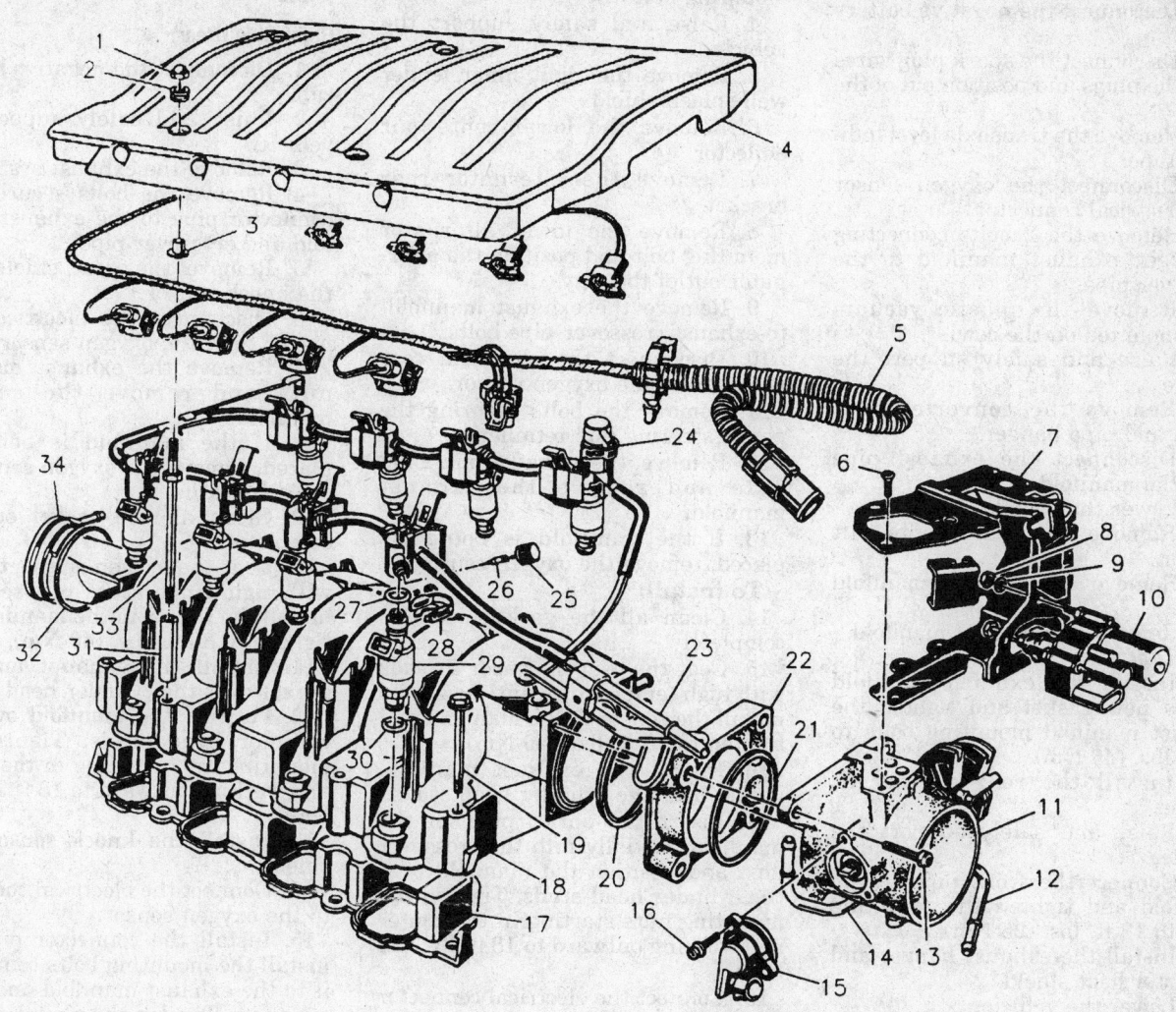

1. NUT, ATTACHING COVER
2. GROMMET, COVER
3. WASHER, COVER GROMMET
4. COVER, INTAKE MANIFOLD
5. WIRING HARNESS ASM
6. SCREW, ISC BRACKET ASM ATTACHING
7. BRACKET ASM, IDLE SPEED CONTROL (ISC)
8. NUT, HEX
9. WASHER, LOCK
10. ACTUATOR ASM, IDLE SPEED CONTROL (ISC)
11. BODY ASM, THROTTLE
12. TUBE, COOLANT OUTLET
13. BOLT, THROTTLE BODY ATTACHING
14. TUBE, COOLANT INLET
15. SENSOR, THROTTLE POSITION (TP)
16. SCREW ASM, TP SENSOR ATTACHING
17. GASKET, INTAKE MANIFOLD

18. HOUSING ASM, INTAKE MANIFOLD
19. BOLT, INTAKE MANIFOLD ATTACHING
20. SEAL, EGR TRANSFER SPACER TO INTAKE MANIFOLD
21. SEAL, THROTTLE BODY TO EGR TRANSFER SPACER
22. SPACER, EGR TRANSFER
23. RAIL ASM, FUEL
24. CARTRIDGE REGULATOR ASM, FUEL PRESSURE
25. CAP, FUEL PRESSURE CONNECTION
26. CORE ASM, FUEL PRESSURE CONNECTION VALVE
27. O-RING, MFI FUEL INJECTOR UPPER
28. CLIP, MFI FUEL INJECTOR RETAINER
29. INJECTOR ASM, MFI FUEL
30. O-RING, MFI FUEL INJECTOR LOWER
31. SPACER, INTAKE MANIFOLD
32. O-RING, PRESSURE RELIEF VALVE
33. STUD, INTAKE MANIFOLD/COVER ATTACHING
34. VALVE ASM, PRESSURE RELIEF

311489

Intake manifold and related components — 4.0L (VIN C) engine

Right Side (Rear)

1. Disconnect the negative battery cable.
2. Disconnect the spark plug wires from the plugs and position out of the way.
3. Remove the transaxle level indicator tube.
4. Disconnect the oxygen sensor (O₂) electrical connector.
5. Remove the 2 bolts connecting the right exhaust manifold to the crossover pipe.
6. Remove the plastic vacuum tank mounted on the cowl.
7. Raise and safely support the vehicle.
8. Remove the converter heat shield and pipe hanger.
9. Disconnect the exhaust pipe from the manifold.
10. Lower the vehicle.
11. Remove the rear engine lift bracket.
12. Remove the exhaust manifold bolts.
13. Remove the exhaust manifold.

To install:

14. Install the exhaust manifold with a new gasket and tighten the exhaust manifold mounting bolts to 33 ft. lbs. (45 Nm).
15. Install the rear engine lift bracket.
16. Raise and safely support the vehicle.
17. Connect the front pipe to the manifold and tighten the mounting bolts to 18 ft. lbs. (25 Nm).
18. Install the exhaust hanger and converter heat shield.
19. Lower the vehicle.
20. Install the vacuum tank on the cowl.
21. Install the bolts connecting the right exhaust manifold to the crossover pipe and tighten the bolts to 20 ft. lbs. (27 Nm).
22. Install the transaxle level indicator tube.
23. Connect the oxygen sensor (O₂) electrical connector.
24. Connect the spark plug wires to the plugs.
25. Connect the negative battery cable.
26. Start the vehicle and verify no leaks.

4.0L (VIN C) Engine

Left Side (Front)

1. Disconnect the negative battery cable.
2. Release the tension on the serpentine belt and remove the belt from the power steering and alternator pulleys.

3. Remove the alternator upper mounting bolt.
4. Raise and safely support the vehicle.
5. Remove the right inner fender well splash shield.
6. Remove the lower center air deflector.
7. Remove the alternator rear bracket.
8. Remove the lower alternator mounting bolt and position the alternator out of the way.
9. Remove the exhaust manifold-to-exhaust crossover pipe bolts.
10. Disconnect the electrical connector from the oxygen sensor.
11. Remove the bolts securing the power steering line retainers.
12. Remove the exhaust manifold nuts and remove the exhaust manifold.
13. If the manifold is being replaced, remove the oxygen sensor.

To install:

14. Clean all the gasket surfaces completely.
15. Coat the oxygen sensor threads with high temperature anti-seize and install the sensor in the manifold and tighten to 30 ft. lbs. (40 Nm).
16. Install the exhaust manifold gasket over the cylinder head studs.
17. Insert the outlet pipe on the manifold partially into the crossover pipe and position the manifold over the cylinder head studs. Tighten the mounting nuts starting in the center and working outward to 18 ft. lbs. (24 Nm).
18. Connect the electrical connector to the oxygen sensor.
19. Install the power steering line retainers and tighten the mounting bolts to 10 ft. lbs. (14 Nm).
20. Position the alternator and loosely install the lower retaining bolt and the rear mounting bracket. Tighten the bolts to 35 ft. lbs. (47 Nm) and the nuts to 28 ft. lbs. (38 Nm).
21. Install the crossover pipe mounting bolts and tighten to 37 ft. lbs. (50 Nm).
22. Install the lower center air deflector.
23. Install the right inner fender well splash shield.
24. Lower the vehicle.
25. Install the upper alternator retaining bolt and tighten to 35 ft. lbs. (47 Nm).
26. Install the serpentine belt around the alternator and power steering pulleys.
27. Connect the negative battery cable.

28. Start the vehicle and verify no leaks.

Right Side (Rear)

1. Disconnect the negative battery cable.
2. Raise and safely support the vehicle.
3. Remove the exhaust system.
4. Remove the bolts securing the connector pipe to the exhaust manifold and crossover pipe.
5. Remove the heat shield from the knock sensor.
6. Disconnect the electrical connector from the oxygen sensor.
7. Remove the exhaust manifold nuts and remove the exhaust manifold.
8. If the manifold is being replaced, remove the oxygen sensor.

To install:

9. Clean all the gasket surfaces completely.
10. Coat the oxygen sensor threads with high temperature anti-seize and install the sensor in the manifold and tighten to 30 ft. lbs. (40 Nm).
11. Install the exhaust manifold gasket over the cylinder head studs.
12. Position the manifold over the cylinder head studs. Tighten the mounting nuts starting in the center and working outward to 18 ft. lbs. (24 Nm).
13. Install the knock sensor heat shield.
14. Connect the electrical connector to the oxygen sensor.
15. Install the connector pipe and install the mounting bolts connecting it to the exhaust manifold and crossover pipe. Tighten the mounting bolts to 30 ft. lbs. (40 Nm).
16. Install the exhaust system and tighten the spring loaded nuts to 18 ft. lbs. (25 Nm).
17. Lower the vehicle.
18. Connect the negative battery cable.
19. Start the vehicle and verify no leaks.

Supercharger

REMOVAL AND INSTALLATION

3.8L (VIN 1) Engine

The supercharger oil level should be checked every 30,000 miles or 36 months. Remove the oil only when the engine is cold. The oil level should be maintained to the bottom of the threads in the inspection hole. Do not use petroleum based oil. Use only GM 12345982 synthetic oil or equivalent 5W-30 synthetic oil. Use of the wrong oil can cause the supercharger to fail.

CAUTION

Fuel Injection systems remain under pressure, even after the engine has been turned OFF. The fuel system pressure must be relieved before disconnecting any fuel lines. Failure to do so may result in fire and/or personal injury.

1. Relieve the fuel system pressure.
2. Disconnect the negative battery cable.
3. Remove the drive belt from the supercharger pulley. It is not necessary to remove the drive belt from the remainder of the pulleys.
4. Remove the plastic fuel injector shield.
5. Disconnect and cap the fuel lines from the fuel rail.
6. Disconnect the vacuum lines from the supercharger.
7. Disconnect the electrical connectors from the fuel injector.
8. Remove the electrical harness shield from the front of the supercharger and disconnect the electrical connector from the supercharger.
9. Remove the fuel rail mounting bolts and remove the fuel rail.
10. Disconnect the following electrical connectors and lay the harness aside:
   • Idle Air Control (IAC) valve
   • Throttle Position Sensor (TPS)
   • MAP sensor
   • MAF sensor
   • EGR valve
   • Boost control solenoid
   • Coolant temperature sensor
11. Disconnect the air intake duct from the throttle body.
12. Disconnect the EGR pipe from the supercharger.
13. Disconnect the control cables from the throttle body.
14. Remove the boost pressure manifold and vacuum block.
15. Remove the control cable bracket.
16. Remove the tensioner bracket-to-supercharger mounting stud.

NOTE: The stud must be removed or the supercharger can not be lifted off the lower intake manifold mounting dowels.

17. Disconnect the throttle body from the supercharger.
18. Remove the supercharger-to-intake manifold bolts.
19. Remove the supercharger, gasket and coolant passage O-rings.

To install:

20. Install new coolant passage O-rings and new supercharger-to-intake manifold gasket.
21. Install the supercharger and the mounting bolts. Only tighten the bolts finger tight.
22. Install the tensioner bracket-to-supercharger stud and tighten to 88 inch lbs. (10 Nm).
23. Tighten the supercharger mounting bolts to 19 ft. lbs. (26 Nm).
24. Install the tensioner bracket nut and tighten to 37 ft. lbs. (50 Nm).
25. Install the throttle body and tighten the mounting nuts to 11 ft. lbs. (15 Nm).
26. Install the boost pressure manifold.
27. Install the vacuum block with a new gasket and tighten the mounting bolt to 62 inch lbs. (7 Nm).
28. Install the control cable bracket.
29. Connect the control cables to the throttle body.
30. Connect the EGR pipe to the supercharger.
31. Install the air intake duct.
32. Connect the electrical connectors to the following components:
   • Idle Air Control (IAC) valve
   • Throttle Position Sensor (TPS)
   • MAP sensor
   • MAF sensor
   • EGR valve
   • Boost control solenoid
   • Coolant temperature sensor
33. Install the fuel rail and mounting bolts.
34. Connect the electrical connectors to the fuel injectors.
35. Connect the vacuum hoses to the supercharger.
36. Connect the harness to the front of the supercharger and install the harness shield.
37. Install the drive belt.
38. Connect the negative battery cable.
39. Start the vehicle and verify no leaks.
40. Install the fuel injector sight shield.

Front Cover Seal

REMOVAL AND INSTALLATION

3.8L (VIN 1) and (VIN K) Engines

1. Disconnect the negative battery cable.
2. Remove the serpentine belt(s).
3. Raise and safely support the vehicle.
4. Remove the right front tire and wheel assembly.
5. Remove the right inner fender well splash shield.
6. Remove the crankshaft balancer bolt and washer.
7. While preventing the engine from turning, remove the crankshaft balancer using J-38197, or an equivalent puller.
8. Remove the seal from the front cover using a suitable prying tool.

To install:

9. Install the seal in the front cover using a seal installation tool, J-35354, or the equivalent.
10. Coat the seal contact area on the balancer with clean oil and install the balancer onto the crankshaft.
11. Using J-38197, seat the balancer on the crankshaft.
12. Install the washer and mounting bolt and tighten to 110 ft. lbs. (150 Nm) plus 76° additional rotation.
13. Install the right inner fender well splash shield.
14. Install the right front tire and wheel assembly.
15. Lower the vehicle.
16. Install the serpentine belt(s).
17. Connect the negative battery cable.

4.0L (VIN C) Engine

1. Disconnect the negative battery cable.
2. Install an engine support fixture, J-28467-A or the equivalent.
3. Remove the engine mount-to-body through bolt nut and through bolt.
4. Remove the through bolt nut and bolt from the engine side of the mount.
5. Raise and safely support the vehicle.
6. Remove the front tire and wheel assembly.
7. Remove the right inner fender well splash shield.
8. Remove the lower center air deflector.
9. Remove the left transaxle mount-to-frame through bolt.
10. Remove the lower retaining bolt and nut from the engine mount bracket.
11. Remove the bolt and disconnect the power steering line retainer from the bracket.
12. Lower the vehicle.
13. Remove the upper retaining nut and bolt from the engine mount bracket.
14. Remove the serpentine belt from the power steering pump pulley.
15. Remove the fuel plastic sight shield on the intake manifold.

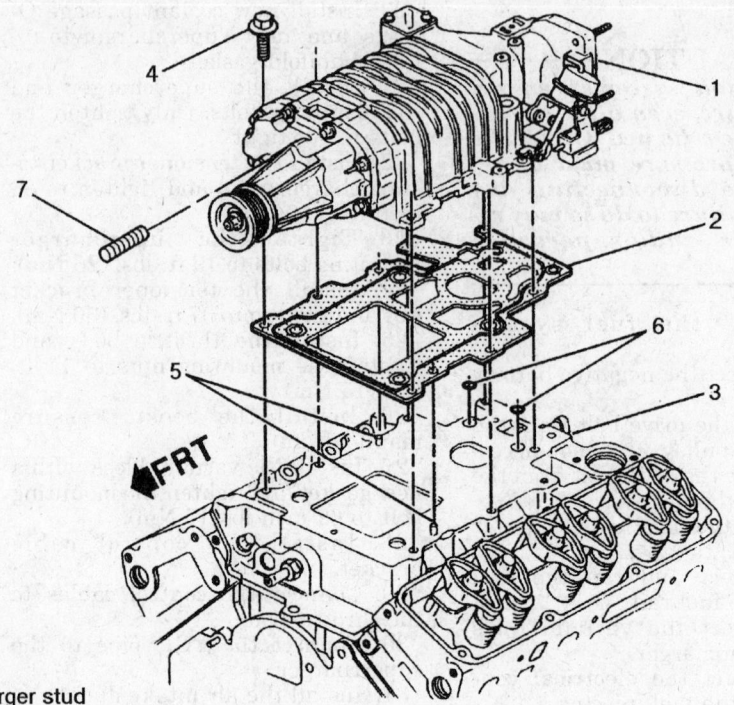

1 Supercharger
2 Supercharger gasket
3 Lower intake manifold
4 Supercharger bolts (8)
5 Locator pins
6 Coolant passage o-rings (2)
7 Tensioner bracket to supercharger stud

304385

Supercharger components

16. Remove the power steering pump from the mounting bracket and set aside. DO NOT disconnect the power steering lines from the pump.

17. Raise the engine using the support fixture.

18. Remove the engine mount bracket from the engine.

19. Remove the torque axis mount from the frame rail.

20. Lower the engine using the support fixture until clearance for J-38416-B, or an equivalent puller is attained.

21. Raise and safely support the vehicle.

22. Remove the crankshaft balancer retaining bolt.

23. Remove the crankshaft balancer using J-38416-B.

24. Carefully pry the seal out of the front cover.

To install:

25. Lubricate the seal lips with clean engine oil and position the front seal in the cover. Using seal installer J-38818 and balancer installer J-39344, or their equivalents seat the front seal in the cover.

26. Coat the seal contact area on the balancer with engine oil and install the balancer onto the end of the crankshaft with the notch lined up with the crankshaft key. Using

J-39344 or the equivalent push the balancer into place.

27. Install the balancer center bolt and tighten to 44 ft. lbs. (60 Nm) plus 120 degrees (1/3 turn) additional rotation.

28. Lower the vehicle.

29. Raise the engine with the support fixture.

30. Install the torque axis mount on the body.

31. Install the engine mount bracket on the engine and loosely install the mounting nuts and bolts. With all the fasteners in place tighten the nuts to 30 ft. lbs. (40 Nm) and the bolts to 41 ft. lbs. (55 Nm).

32. Lower the engine support fixture until the engine is at its normal height.

33. Install the power steering pump in the mounting bracket.

34. Install the serpentine belt.

35. Connect the power steering line retainer to the engine mount bracket and install the mounting bolt.

36. Raise and safely support the vehicle.

37. Install the left transaxle mount-to-frame through bolt and tighten to 63 ft. lbs. (85 Nm).

38. Install the lower center air deflector.

39. Install the right inner fender well splash shield.

40. Install the tire and wheel assembly and tighten the wheel nuts to 100 ft. lbs. (140 Nm).

41. Lower the vehicle.

42. Install the torque axis mount-to-engine mount bracket through bolt and nut and tighten the bolt to 70 ft. lbs. (95 Nm).

43. Install the torque axis mount-to-frame through bolt and nut and tighten the through bolt to 37 ft. lbs. (50 Nm).

44. Remove the engine support fixture.

45. Connect the negative battery cable.

Timing Chain, Sprockets and Front Cover

REMOVAL AND INSTALLATION

3.8L (VIN 1) and (VIN K) Engines

1. Disconnect the negative battery cable.

2. Drain the coolant into a suitable container.

3. Remove the vacuum reservoir.

4. Install an engine support fixture, J-28467-A or equivalent.

5. Raise the engine so that the weight is removed from the torque axis mount.

6. Remove the 2 lower torque axis-to-frame bolts.

7. Remove the torque axis mount.

8. Remove the serpentine belt(s).

9. Remove the water pump on vehicles equipped with supercharger.

10. Raise the engine with the support fixture and remove the engine mount bracket bolts and bracket.

11. Remove the alternator and the 3 bolts securing the drive belt tensioner and remove the tensioner.

12. Remove the crankshaft balancer.

13. Remove the crankshaft sensor shield.

14. Disconnect the electrical connector from the crankshaft sensor.

15. Remove the front oil pan-to-front cover bolts.

16. Remove the front cover bolts and remove the front cover.

17. Rotate the crankshaft until the timing mark on the camshaft sprocket is lined up with the crankshaft sprocket timing mark.

18. Remove timing chain damper mounting bolt and remove the chain damper assembly.

19. Remove camshaft sprocket bolt.

20. Remove camshaft sprocket along with the timing chain.

21. Using a suitable gear puller remove the crankshaft sprocket from the crankshaft.

To install:

22. Line up the notch in the crankshaft sprocket with the crankshaft key and slide the gear on the crankshaft. It may be necessary to use a gear installer to fully seat the gear. Make sure the timing mark on the crankshaft gear is pointing straight up.

23. Insert the camshaft gear into the timing chain. Hold the sprocket with the timing mark straight down and with the chain hanging down off of the sprocket.

24. Loop the chain under the crankshaft sprocket.

25. Install the camshaft sprocket onto the camshaft so the notch in the sprocket fits over the camshaft key.

26. The camshaft and crankshaft timing marks should be in line.

27. If the marks are not in alignment perform the following:

    a. Remove the camshaft sprocket and timing chain.

    b. Install the camshaft sprocket onto the camshaft and rotate the camshaft until the camshaft and crankshaft marks are aligned.

    c. Remove the camshaft sprocket.

    d. Repeat Steps 9 to 11.

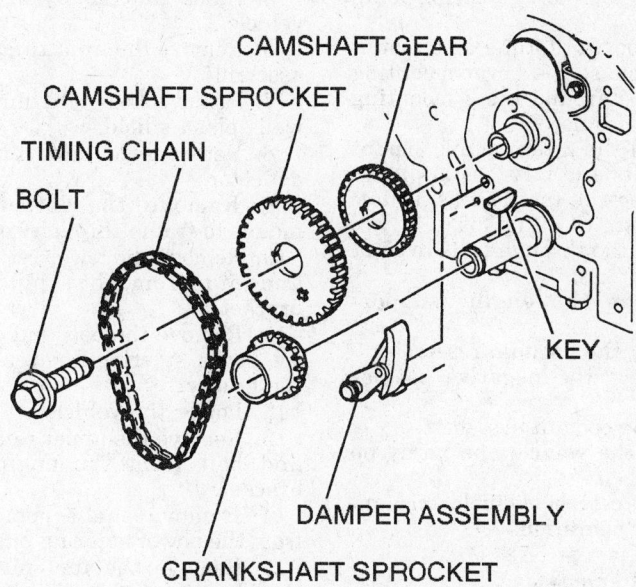

Timing chain and sprockets — 3.8L (VIN 1) and (VIN K) engines

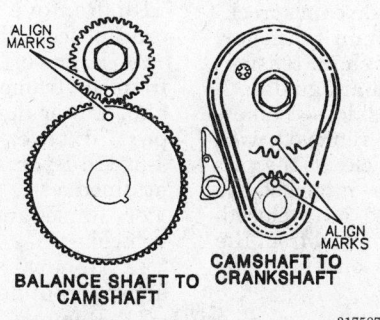

Balance shaft-to-camshaft and camshaft-to-crankshaft timing mark alignment — 3.8L (VIN 1) and (VIN K) engines

28. Install the camshaft sprocket bolt and tighten the bolt to 74 ft. lbs. (100 Nm) plus 90 degrees (¼ turn) additional rotation.

29. Install the timing chain damper and tighten the mounting bolts to 16 ft. lbs. (22 Nm).

30. Clean all gasket surfaces completely.

31. Install the front cover on the engine using a new gasket. Install the mounting bolts and tighten them to 22 ft. lbs. (30 Nm).

32. Install the oil pan-to-front cover bolts and tighten them to 125 inch lbs. (14 Nm).

33. Install the belt tensioner and tighten the bolts to 37 ft. lbs. (50 Nm).

34. Install the alternator.

35. Connect the crankshaft sensor electrical connector.

36. Install the crankshaft sensor shield.

37. Install the crankshaft balancer onto the crankshaft.

38. Install the crankshaft balancer bolt and draw the balancer into position. Tighten the bolt to 111 ft. lbs. (150 Nm) plus an additional 75 degrees.

39. Install the engine mount bracket and tighten the mounting bolts to 65 ft. lbs. (87 Nm).

40. Install the power steering pump and belt(s).

41. Position the torque axis mount in the vehicle so the lower mounting bracket slips around the 2 mounting bolts in the frame.

42. Install the 2 torque axis mount through bolts and nuts.

43. Tighten the torque axis mount-to-frame bolts to 52 ft. lbs. (70 Nm) and tighten the through bolts to 65 ft. lbs. (87 Nm).

44. Remove the engine support fixture.

45. Install the vacuum reservoir.

46. Connect the negative battery cable.

47. Fill the coolant system.

48. Start the vehicle and verify no leaks.

49. Road test the vehicle and ensure proper operation.

4.0L (VIN C) Engine

The following procedure covers removal and installation of the crankshaft sprocket, intermediate shaft sprocket, primary timing chain, left side secondary timing chain, right side secondary timing chain, left side cam sprockets, right side cam sprockets, primary timing chain tensioner, both secondary timing chain tensioners, and both timing chain guides.

The left and right side secondary timing chains can be removed with the engine in the vehicle. If the primary timing chain or intermediate shaft sprocket need to be replaced, the engine must be removed from the vehicle and supported on an engine stand.

NOTE: Setting the camshaft timing is necessary whenever the cam drive system has been disturbed, meaning the relationship between any chain and sprocket has been lost. Correct timing exists when the crankshaft and intermediate shaft sprocket timing marks are in alignment and all 4 camshaft drive pins are perpendicular (90 degrees) to the cylinder head surface.

Removal — Left Side Secondary Chain and Sprocket

1. Disconnect the negative battery cable.

2. Install an engine support fixture, J-28467-A or the equivalent.

3. Remove the engine mount-to-body through bolt nut and through bolt.

4. Remove the through bolt nut and bolt from the engine side of the mount.

5. Raise and safely support the vehicle.

6. Remove the front tire and wheel assembly.

7. Remove the right inner fender well splash shield.

8. Remove the lower center air deflector.

9. Remove the left transaxle mount-to-frame through bolt.

10. Remove the lower retaining bolt and nut from the engine mount bracket.

11. Remove the bolt and disconnect the power steering line retainer from the bracket.

12. Lower the vehicle.

13. Remove the upper retaining nut and bolt from the engine mount bracket.

14. Remove the serpentine belt from the power steering pump pulley.

15. Remove the fuel plastic sight shield on the intake manifold.

16. Remove the power steering pump from the mounting bracket and set aside. DO NOT disconnect the power steering lines from the pump.

17. Raise the engine using the support fixture.

18. Remove the engine mount bracket from the engine.

19. Remove the torque axis mount from the frame rail.

20. Lower the engine using the support fixture until clearance for J-38416-B, or an equivalent puller is attained.

21. Raise and safely support the vehicle.

22. Remove the crankshaft balancer retaining bolt.

23. Remove the crankshaft balancer using J-38416-B.

24. Remove the serpentine belt tensioner.

25. Remove the serpentine belt idler pulley.

26. Remove the front cover bolts and remove the cover and gasket. DO NOT discard the gasket, if it is undamaged it can be re-used.

NOTE: The front cover gasket is reusable as long as it is not damaged.

27. Partially drain the cooling system to a level below the water pump assembly.

28. Remove the oil level indicator tube.

29. Lower the vehicle.

30. Disconnect the upper radiator hose from the thermostat housing.

31. Disconnect the spark plug wires from the front plugs and route the wires out of the way.

32. Remove the upper radiator support assembly.

33. Disconnect the PCV fresh air tube from the left side camshaft cover.

34. Remove the air inlet duct.

35. Disconnect the EGR outlet pipe.

36. Remove the water pump drive belt shield.

37. Remove the water pump drive belt and drive belt tensioner.

38. Remove the coolant pump pulley using J-38825, or an equivalent pulley remover.

39. Remove the camshaft seal retainer screws and remove the camshaft seal.

40. Remove the camshaft cover bolts.

41. Remove the camshaft by pivoting up the intake manifold side of the cover 10 inches. Lift up the exhaust manifold side of the cover 2 inches. Swing the oil fill cap end of the cover up over the intake manifold and slide the cover over the camshafts.

42. Remove the left side secondary chain tensioner.

43. Remove the left side chain guide. Access the upper chain guide mounting bolt through the hole in the cylinder head capped with the plastic plug.

44. Remove the left side cam sprocket bolts and cam sprockets.

45. Remove the secondary drive chain.

Removal — Right Side Secondary Timing Chain and Sprocket

1. Remove the vacuum reservoir.

2. Remove the cruise control servo from its mount and position out of the way.

3. Disconnect the ignition assembly electrical connectors.

4. Disconnect the spark plug wires from the right side plugs and remove the ignition assembly from its mount and position out of the way.

5. Remove the PCV valve from the camshaft cover.

6. Remove the EVAP cannister solenoid from the right camshaft cover.

7. Raise and safely support the vehicle.

8. Disconnect the knock sensor, vehicle speed sensor and power steering pressure switch.

9. Lower the vehicle.

10. Disconnect the wiring harness retainers from the cover.

11. Remove the camshaft cover bolts and remove the camshaft cover.

12. Remove the right side secondary chain tensioner.

13. Remove the right side chain guide. Access the upper chain guide mounting bolt through the hole in the cylinder head capped with the plastic plug.

14. Remove the right side cam sprocket bolts and cam sprockets.

15. Remove the secondary drive chain.

Removal — Primary Timing Chain and Sprocket

1. Remove the intermediate shaft sprocket-to-intermediate shaft bolt and remove the intermediate shaft sprocket.

2. Slide the primary timing sprockets and primary chain off the engine.

Installation — Primary and Secondry Chains and Sprockets

NOTE: The following procedure must be followed to set the camshaft timing on the vehicle.

1. Install the primary and secondary chain guide.

2. Rotate the crankshaft until the sprocket drive key is at the 1 o'clock position. Use tool J-39946, or the equivalent to rotate the crankshaft.

3. Install the crankshaft sprocket and intermediate shaft sprocket in the primary timing chain so the timing marks are aligned. Install the assembly in position on the engine. The crankshaft sprocket key way will have to slide over the key on the crankshaft. If it is necessary to turn the crankshaft sprocket the intermediate shaft sprocket will also have to be turned so the timing mark still lines up with the crankshaft sprocket.

4. Install the intermediate shaft sprocket-to-intermediate shaft bolt and tighten to 45 ft. lbs. (61 Nm).

5. Install the primary timing chain tensioner. Tighten the tensioner mounting bolts to 20 ft. lbs. (27 Nm).

6. Install a flywheel holder to lock the crankshaft in position.

7. Install the secondary timing chain over the inner row of teeth on the intermediate shaft sprocket. Route the chain over the chain guide and install the exhaust cam sprocket so the **RE** (Right Head Exhaust) pin engages the sprocket notch. There should be no slack in the lower section of the timing chain and the cam drive pin **must** be perpendicular to the cylinder head face.

8. Install the intake cam sprocket into the chain so the sprocket notch

RI (Right Head Intake) engages the cam and the camshaft drive pin remains perpendicular to the cylinder head face. A hex is cast into the camshafts behind the lobes for cylinder No. 1, so an open end wrench may be used to provide minor repositioning of the cams.

9. Loosely install the exhaust cam sprocket bolt and intake sprocket bolt.

10. Install the chain tensioner and tighten the mounting bolts to 20 ft. lbs. (27 Nm).

11. Tighten the camshaft sprocket bolts to 90 ft. lbs. (122 Nm).

12. Route the secondary timing chain for the left side over the outer row of intermediate sprocket teeth.

13. Install the secondary timing chain over the inner row of teeth on the intermediate shaft sprocket. Route the chain over the chain guide and install the exhaust cam sprocket so the **LE** (Left Head Exhaust) pin engages the sprocket notch. There should be no slack in the lower section of the timing chain and the cam drive pin **must** be perpendicular to the cylinder head face.

14. Install the intake cam sprocket into the chain so the sprocket notch **LI** (Left Head Intake) engages the cam and the camshaft drive pin remains perpendicular to the cylinder head face. A hex is cast into the camshafts behind the lobes for cylinder No. 2, so an open end wrench may be used to provide minor repositioning of the cams.

15. Loosely install the exhaust cam sprocket bolt and intake sprocket bolt.

16. Install the chain tensioner and tighten the mounting bolts to 20 ft. lbs. (27 Nm).

17. Tighten the camshaft sprocket bolts to 90 ft. lbs. (122 Nm).

NOTE: The RE cam sprocket must contain the cam position sensor pick-up.

18. Install the front cover gasket on the dowel pins on the block.

19. Install the front cover on the dowel pins and install the front cover mounting bolts. Tighten the bolts to 7 ft. lbs. (9.5 Nm).

20. Install the drive belt idler pulley and tighten the mounting bolt to 35 ft. lbs. (48 Nm).

21. Install the drive belt tensioner and tighten the tensioner mounting nut to 35 ft. lbs. (48 Nm).

22. Install the crankshaft balancer using tool J-39344 or the equivalent.

23. Apply engine oil to the balancer bolt threads and tighten the bolt to

105 ft. lbs. (143 Nm) plus and additional 120 degrees (1/3 turn).

24. Raise the screw jack until the 3 cradle bolts can be installed. Tighten the bolts to 75 ft. lbs. (102 Nm).

25. Connect the RSS sensor.

26. Remove the flywheel holding tool and install the flywheel cover.

27. Install the wheel well splash shields.

Installation — Left Side Cam Cover

1. Install a new camshaft cover gasket and spark plug seals in the camshaft cover.

2. Install the back end of the camshaft cover first while keeping the front end of the cover high. Pivot the oil fill end of the cover into place once the back edge of the cover is clear of the tensioner assembly. Hold the seal in place and position the cover so it ids square with the cylinder head. Slide the cover left and down onto the cylinder head.

3. Install the camshaft cover bolts and tighten to 89 inch lbs. (10 Nm).

4. Lubricate the seal lip with petroleum jelly and install the camshaft seal retainer over the end of the intake camshaft. Install the mounting bolts and tighten to 10 inch lbs. (1.1 Nm).

5. Install the water pump pulley using J-38823, or an equivalent pulley installer.

6. Install the drive belt tensioner and drive belt.

7. Install the water pump belt shield.

8. Connect the EGR outlet pipe.

9. Connect the upper radiator hose to the thermostat housing.

10. Connect the PCV hose to the camshaft cover.

11. Install the air inlet duct.

12. Connect the spark plug wires to the front plugs.

13. Install the upper radiator support.

14. Raise and safely support the vehicle.

15. Install the oil level indicator tube.

16. Lower the vehicle.

17. Refill the cooling system.

Installation — Right Side Cam Cover

1. Install a new camshaft cover gasket and spark plug seals in the camshaft cover.

2. Install the camshaft cover on the cylinder head and tighten the mounting bolts to 89 inch lbs. (10 Nm).

3. Connect the wiring harness retainers to the camshaft cover.

4. Raise and safely support the vehicle.

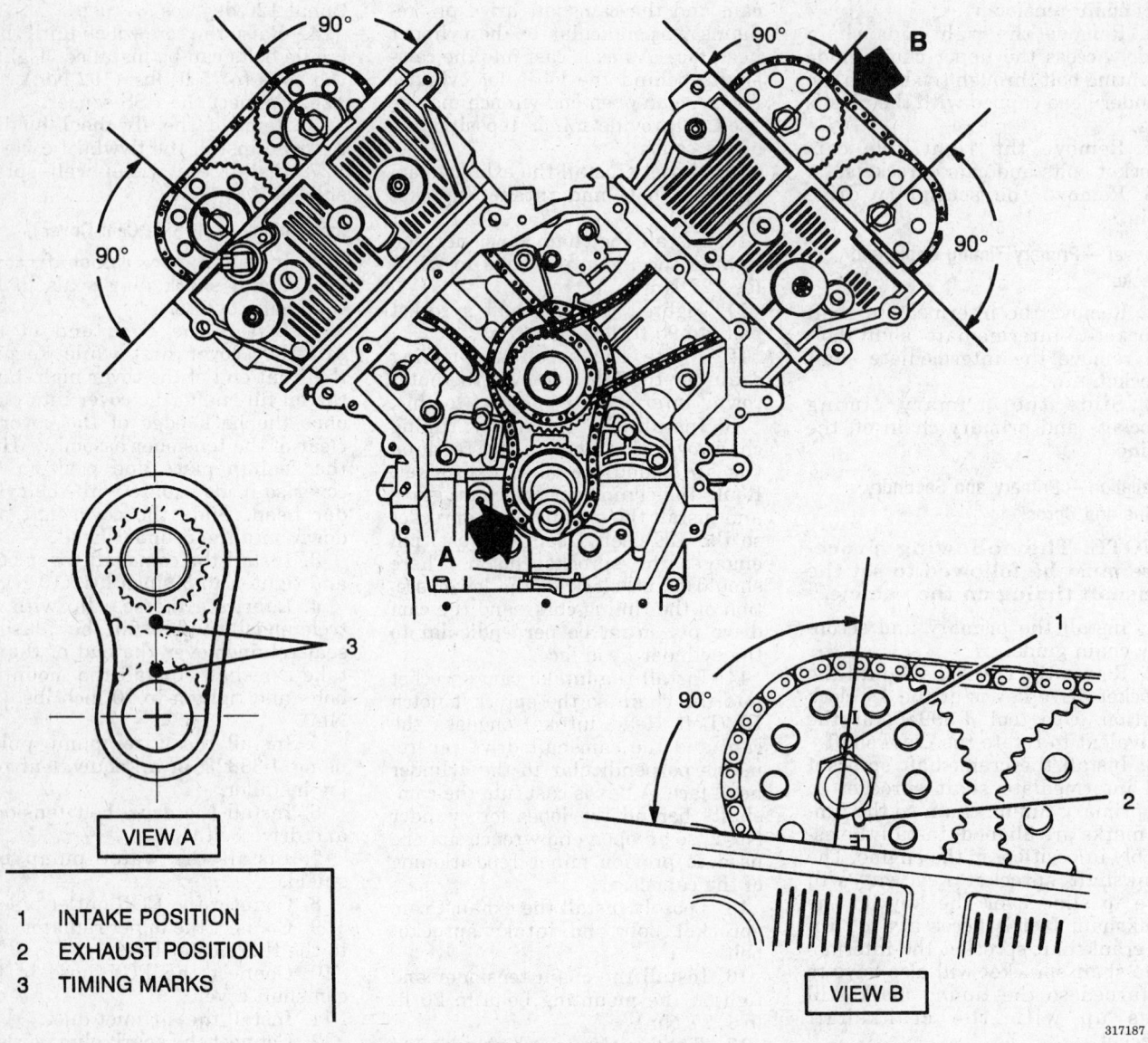

1 INTAKE POSITION
2 EXHAUST POSITION
3 TIMING MARKS

VIEW A

VIEW B

317187

Primary and secondary timing mark alignment — 4.0L (VIN C) engine

5. Connect the knock sensor, vehicle speed sensor and power steering pressure switch.

6. Lower the vehicle.

7. Install the EVAP cannister solenoid on the right side camshaft cover.

8. Install the PCV valve.

9. Install the ignition assembly on the right side cover and connect the spark plugs wires to the plugs and connect the electrical connector to the ignition assembly.

10. Install the cruise control servo.

11. Install the vacuum reservoir.

12. Install the front cover gasket over the dowel pins on the block.

13. Install the front cover on the engine and tighten the mounting bolts to 89 inch lbs. (10 Nm).

14. Install the serpentine belt idler pulley and tighten the mounting bolt to 37 ft. lbs. (50 Nm).

15. Install the serpentine belt tensioner and tighten the mounting bolt to 37 ft. lbs. (50 Nm).

16. Coat the seal contact area on the balancer with engine oil and install the balancer onto the end of the crankshaft with the notch lined up with the crankshaft key. Using J-39344 or the equivalent push the balancer into place.

17. Install the balancer center bolt and tighten to 44 ft. lbs. (60 Nm) plus

120 degrees (⅓ turn additional rotation.

18. Lower the vehicle.

19. Raise the engine with the support fixture.

20. Install the torque axis mount on the body.

21. Install the engine mount bracket on the engine and loosely install the mounting nuts and bolts. With all the fasteners in place tighten the nuts to 30 ft. lbs. (40 Nm) and the bolts to 41 ft. lbs. (55 Nm).

22. Lower the engine support fixture until the engine is at its normal height.

23. Install the power steering pump in the mounting bracket.

24. Install the serpentine belt.

25. Connect the power steering line retainer to the engine mount bracket and install the mounting bolt.

26. Raise and safely support the vehicle.

27. Install the left transaxle mount-to-frame through bolt and tighten to 63 ft. lbs. (85 Nm).

28. Install the lower center air deflector.

29. Install the right inner fender well splash shield.

30. Install the tire and wheel assembly and tighten the wheel nuts to 100 ft. lbs. (140 Nm).

31. Lower the vehicle.

32. Install the torque axis mount-to-engine mount bracket through bolt and nut and tighten the bolt to 70 ft. lbs. (95 Nm).

33. Install the torque axis mount-to-frame through bolt and nut and tighten the through bolt to 37 ft. lbs. (50 Nm).

34. Remove the engine support fixture.

35. Check all engine fluid levels and refill as required.

36. Connect the negative battery cable.

37. Test run engine to verify correct engine performance.

Camshaft

REMOVAL AND INSTALLATION

3.8L (VIN 1) and (VIN K) Engines

— CAUTION —

Fuel Injection systems remain under pressure, even after the engine has been turned OFF. The fuel system pressure must be relieved before disconnecting any fuel lines. Failure to do so may result in fire and/or personal injury.

1. Relieve the fuel system pressure.

2. Disconnect the negative battery cable.

3. Remove the engine from the vehicle and mount the engine on a suitable engine stand.

4. Remove the intake manifold.

5. Remove the rocker arm covers.

6. Remove the rocker arm assemblies, pushrods and lifters.

7. Remove the crankshaft balancer center bolt and using a puller remove the balancer from the crankshaft.

8. Remove the crankshaft sensor cover and disconnect the crankshaft sensor.

9. Remove the timing chain front cover.

NOTE: Align the timing marks of the camshaft and crankshaft sprockets to avoid burring the camshaft journals by the crankshaft.

10. Remove the camshaft sprocket and timing chain.

11. Remove the camshaft thrust plate bolts and remove the thrust plate.

12. Carefully remove the camshaft.

To install:

13. Coat the camshaft lobes and bearings with prelube prior to installation.

14. Install the camshaft into the engine.

15. Install the camshaft thrust plate and tighten the mounting bolts to 10 ft. lbs. (14 Nm).

16. Install the camshaft sprocket and timing chain.

17. Install the front timing chain cover.

18. Connect the crankshaft sensor and install the crankshaft sensor cover.

19. Install the crankshaft balancer and tighten the mounting bolt to:
   a. 1995 3.8L (VIN 1) and (VIN K) enginess: 103 ft. lbs. (140 Nm) plus 56 degrees additional rotation.
   b. 1996–97 3.8L (VIN 1) and (VIN K) enginess: 111 ft. lbs. (150 Nm) plus 76 degrees additional rotation.

NOTE: If the camshaft was replaced the lifters should also be replaced. The old lifters have developed a wear pattern and will cause the new camshaft to wear prematurely. New lifters MUST be installed with a new camshaft.

20. Coat the valve lifters with prelube and install the lifters in the lifter bores.

21. Install the lifter guides and lifter guide retainer. Tighten the retainer mounting bolts to 27 ft. lbs. (37 Nm).

22. Install the pushrods and rocker arms and tighten the rocker arm bolts to 28 ft. lbs. (38 Nm).

23. Install the rocker arm covers.

24. Install the intake manifold.

25. Install the engine in the vehicle.

26. Connect the negative battery cable.

27. Start the vehicle and verify no leaks.

4.0L (VIN C) Engine

Left Side

1. Disconnect the negative battery cable.

2. Remove the intake manifold sight shield.

3. Raise and safely support the vehicle.

4. Partially drain the cooling system to a level below the water pump assembly.

5. Remove the oil level indicator tube.

6. Lower the vehicle.

7. Disconnect the upper radiator hose from the thermostat housing.

8. Disconnect the spark plug wires from the front plugs and route the wires out of the way.

9. Remove the upper radiator support assembly.

10. Disconnect the PCV fresh air tube from the left side camshaft cover.

11. Remove the air inlet duct.

12. Disconnect the EGR outlet pipe.

13. Remove the water pump drive belt shield.

14. Remove the water pump drive belt and drive belt tensioner.

15. Remove the coolant pump pulley using J-38825, or an equivalent pulley remover.

16. Remove the camshaft seal retainer screws and remove the camshaft seal.

17. Remove the camshaft cover bolts.

18. Remove the camshaft by pivoting up the intake manifold side of the cover 10 inches. Lift up the exhaust manifold side of the cover 2 inches. Swing the oil fill cap end of the cover up over the intake manifold and slide the cover over the camshafts.

19. Secure the cam sprocket to the timing chain by installing tie–wraps through the cam sprocket holes. Use 4 tie–wraps per sprocket.

— CAUTION —

The sprocket/chain relationship must be maintained throughout this procedure or camshaft timing will be lost and require further engine disassembly to retime.

20. Working from behind the sprockets, install cam chain holder J 38815 so that it is positioned between the chain tensioner and chain guide. Apply tension to the tool by tightening the tension adjusting screw.

21. Remove both cam sprocket bolts. Note the relative location of the cam drive pins in the end of the camshafts.

22. Work the sprockets off the cams using play in the chain.

23. Alternately loosen the cam bearing cap screws a few turns at a time until all valve spring pressure has been released. Remove the bolts and caps.

24. Remove the camshaft.

25. Inspect the camshaft for excessive lobe wear such as the evidence of grooves, scoring or flaking. Check the bearing journals, making sure they are not scored or burned. Replace the camshaft, as necessary.

To install:

26. Lubricate the camshaft lobes with camshaft prelube 1052365 or equivalent. Lubricate the camshaft journals with clean engine oil.

27. Install the camshaft.

28. Position the cam bearing caps to the cylinder head.

NOTE: Each cap is identified for position and direction. The arrow points towards the front of the engine. An "E" indicates a cap for the exhaust cam. An "I" indicates a cap for the intake cam. Position No. 1 is towards the front of the engine.

29. Loosely install the cam bearing cap bolts.

30. Alternately tighten the cam bearing cap bolts a few turns at a time against valve spring pressure until all the bolts are snug. Tighten the bolts to 9 ft. lbs. (12 Nm).

31. Using the hex cast into the camshaft, rotate the cams until the drive pins are in position to engage the cam sprockets over the cams and install the retaining bolts.

32. Work the cam sprockets over cams and install the retaining bolts. Tighten the bolts to 90 ft. lbs. (120 Nm).

33. Remove the chain holder J 38815.

34. Remove the tie-wraps from the cam sprockets.

35. Install a new camshaft cover gasket and spark plug seals in the camshaft cover.

36. Install the back end of the camshaft cover first while keeping the front end of the cover high. Pivot the oil fill end of the cover into place once the back edge of the cover is clear of the tensioner assembly. Hold the seal in place and position the cover so it is square with the cylinder head. Slide the cover left and down onto the cylinder head.

37. Install the camshaft cover bolts and tighten to 89 inch lbs. (10 Nm).

38. Lubricate the seal lip with petroleum jelly and install the camshaft seal retainer over the end of the intake camshaft. Install the mounting bolts and tighten to 10 inch lbs. (1.1 Nm).

39. Install the water pump pulley using J-38823, or an equivalent pulley installer.

40. Install the drive belt tensioner and drive belt.

41. Install the water pump belt shield.

42. Connect the EGR outlet pipe.

43. Connect the upper radiator hose to the thermostat housing.

44. Connect the PCV hose to the camshaft cover.

45. Install the air inlet duct.

46. Connect the spark plug wires to the front plugs.

47. Install the upper radiator support.

48. Raise and safely support the vehicle.

49. Install the oil level indicator tube.

50. Lower the vehicle.

51. Refill the cooling system.

52. Install the intake manifold sight shield.

53. Connect the negative battery cable.

54. Run the engine and check for leaks and proper engine operation.

Right Side

1. Disconnect the negative battery cable.

2. Remove the intake manifold sight shield.

3. Remove the vacuum reservoir.

4. Remove the cruise control servo from its mount and position out of the way.

5. Disconnect the ignition assembly electrical connectors.

6. Disconnect the spark plug wires from the right side plugs and remove the ignition assembly from its mount and position out of the way.

7. Remove the PCV valve from the camshaft cover.

8. Remove the EVAP cannister solenoid from the right camshaft cover.

9. Raise and safely support the vehicle.

10. Disconnect the knock sensor, vehicle speed sensor and power steering pressure switch.

11. Lower the vehicle.

12. Disconnect the wiring harness retainers from the cover.

13. Remove the camshaft cover bolts and remove the camshaft cover.

14. Secure the cam sprocket to the timing chain by installing tie-wraps through the cam sprocket holes. Use 4 tie-wraps per sprocket.

CAUTION

The sprocket/chain relationship must be maintained throughout this procedure or camshaft timing will be lost and require further engine disassembly to retime.

---

15. Working from behind the sprockets, install cam chain holder J 38815 so that it is positioned between the chain tensioner and chain guide. Apply tension to the tool by tightening the tension adjusting screw.

16. Remove both cam sprocket bolts. Note the relative location of the cam drive pins in the end of the camshafts.

17. Work the sprockets off the cams using play in the chain.

18. Alternately loosen the cam bearing cap screws a few turns at a time until all valve spring pressure has been released. Remove the bolts and caps.

19. Remove the camshaft.

20. Inspect the camshaft for excessive lobe wear such as the evidence of grooves, scoring or flaking. Check the bearing journals, making sure they are not scored or burned. Replace the camshaft, as necessary.

To install:

21. Lubricate the camshaft lobes with camshaft prelube 1052365 or equivalent. Lubricate the camshaft journals with clean engine oil.

22. Install the camshaft.

23. Position the cam bearing caps to the cylinder head.

NOTE: Each cap is identified for position and direction. The arrow points towards the front of the engine. An "E" indicates a cap for the exhaust cam. An "I" indicates a cap for the intake cam. Position No. 1 is towards the front of the engine.

24. Loosely install the cam bearing cap bolts.

25. Alternately tighten the cam bearing cap bolts a few turns at a time against valve spring pressure until all the bolts are snug. Tighten the bolts to 9 ft. lbs. (12 Nm).

26. Using the hex cast into the camshaft, rotate the cams until the drive pins are in position to engage the cam sprockets over the cams and install the retaining bolts.

27. Work the cam sprockets over cams and install the retaining bolts. Tighten the bolts to 90 ft. lbs. (120 Nm).

28. Remove the chain holder special tool J 38815.

29. Remove the tie-wraps from the cam sprockets.

30. Install a new camshaft cover gasket and spark plug seals in the camshaft cover.

31. Install the camshaft cover on the cylinder head and tighten the mounting bolts to 89 inch lbs. (10 Nm).

32. Connect the wiring harness retainers to the camshaft cover.

33. Raise and safely support the vehicle.

34. Connect the knock sensor, vehicle speed sensor and power steering pressure switch.

35. Lower the vehicle.

36. Install the EVAP cannister solenoid on the right side camshaft cover.

37. Install the PCV valve.

38. Install the ignition assembly on the right side cover and connect the spark plugs wires to the plugs and connect the electrical connector to the ignition assembly.

39. Install the cruise control servo.

40. Install the vacuum reservoir.

41. Install the intake manifold sight shield.

42. Connect the negative battery cable.

43. Run the engine and check for leaks and proper engine operation.

Balance Shaft

REMOVAL AND INSTALLATION

3.8L (VIN 1) and (VIN K) Engines

The engine must be removed from the vehicle for this procedure.

——— **CAUTION** ———
Fuel injection systems remain under pressure, even after the engine has been turned OFF. The fuel system pressure must be relieved before disconnecting any fuel lines. Failure to do so may result in fire and/or personal injury.

1. Disconnect the negative battery cable.

2. Remove the engine from the vehicle and install it on a suitable engine stand.

3. Remove the flywheel-to-crankshaft bolts and the flywheel.

4. Remove the rear main carrier plate.

5. Remove the intake manifold.

6. Remove the lifter guide and the retainer bolts and remove the retainer.

7. Remove the front timing chain cover.

8. Remove the camshaft sprocket and timing chain.

9. Remove the balance shaft gear-to-shaft bolt and the gear.

10. Remove the balance shaft retainer–to the–engine bolts and the retainer.

11. Using the tool J 6125–B or equivalent slide hammer tool, pull the balance shaft from the front of the engine.

12. Remove the balance shaft rear plug.

13. Remove the balance shaft rear bearing using the special tool J 36995–5 or equivalent.

NOTE: It may take a considerable amount of force to loosen the bushing from the block bore.

To install:

14. Make sure the rear main oil seal housing gasket is removed and there is no remaining debris.

15. Dip the rear bushing in clean engine oil before the installation.

——— **CAUTION** ———
Engine VIN codes K and 1 use different style rear balance shaft bearings. DO NOT interchange bearing for these engines or damage may result.

16. Using J 36995–5, or an equivalent balance shaft installer, install the balance shaft rear bearing into the engine and remove the installation tool. Install the bearing with the rolled side facing into the engine and the manufactures marking facing the flywheel side.

17. Remove the tool J 36995 or equivalent.

18. Dip the front balance shaft bearings in clean engine oil before the installation.

19. Install the balance shaft into the block using J 36996.

20. Temporarily install the balance shaft bearing retainer and bolts.

NOTE: This bolt is designed to permanently stretch when tightened. The correct part number must be used to replace this type of fastener. Do not use a bolt that is stronger in this application. If the correct bolt is not used, the parts will not be tightened correctly. Part or system damage may occur.

21. Install and tighten the balance shaft drive gear and the bolt with thread locking compound on the threads to 16 ft. lbs. + 70 degrees (22 Nm + 70 degrees). Using the special tool J 36660 or equivalent, install the rear balance shaft plugs.

22. Install the rear main seal carrier plate.

23. Measure the end play for the following applications:

   a. End play should be 0.008 inch (0.023mm).

   b. Front radial play should be 0.0011 inch (0.028mm).

   c. Rear radial play should be 0.0005–0.0047 inch (0.0127–0.119mm).

24. Turn the camshaft sprocket with the camshaft gear temporarily installed to set the timing mark pointing straight down.

25. With the camshaft sprocket and the camshaft gear removed, turn the balance shaft so that the timing mark on the gear points straight down.

26. Install the camshaft gear.

27. Align the marks on the balance shaft gear and the camshaft gear by turning the balance shaft.

28. Turn the crankshaft so that the number one piston is at top dead center.

29. Install the timing chain and camshaft sprocket.

30. Measure the gear lash at 4 different places taking 4 measurements total with lash not exceeding 0.002 to 0.005 inch or (0.050mm to 0.127mm).

31. Install and tighten the balance shaft front bearing retainer and bolts to 22 ft. lbs. (30 Nm).

32. Install the engine front cover.

33. Install the lifter guide retainer.

34. Install the intake manifold.

35. Install and tighten the flywheel bolts to 46 ft. lbs. (62 Nm).

36. Install the engine into the vehicle. Refill and check all fluid levels.

37. Connect the negative battery cable.

38. Start the engine and check for leaks and proper engine operation.

Piston and Connecting Rod

POSITIONING

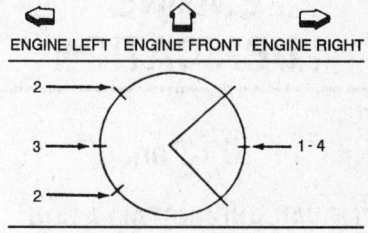

1. Oil ring spacer gap (tang in hole or slot with ARC)
2. Oil ring rail gaps
3. 2nd compression ring gap
4. Top compression ring gap

250084

Ring gap positioning — 3.8L engine

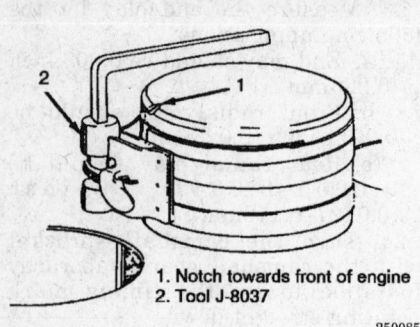

1. Notch towards front of engine
2. Tool J-8037

250085

Piston installation — 3.8L engine

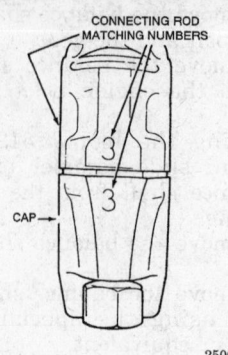

CONNECTING ROD MATCHING NUMBERS

CAP

250076

Cylinder identification marks —
all engines

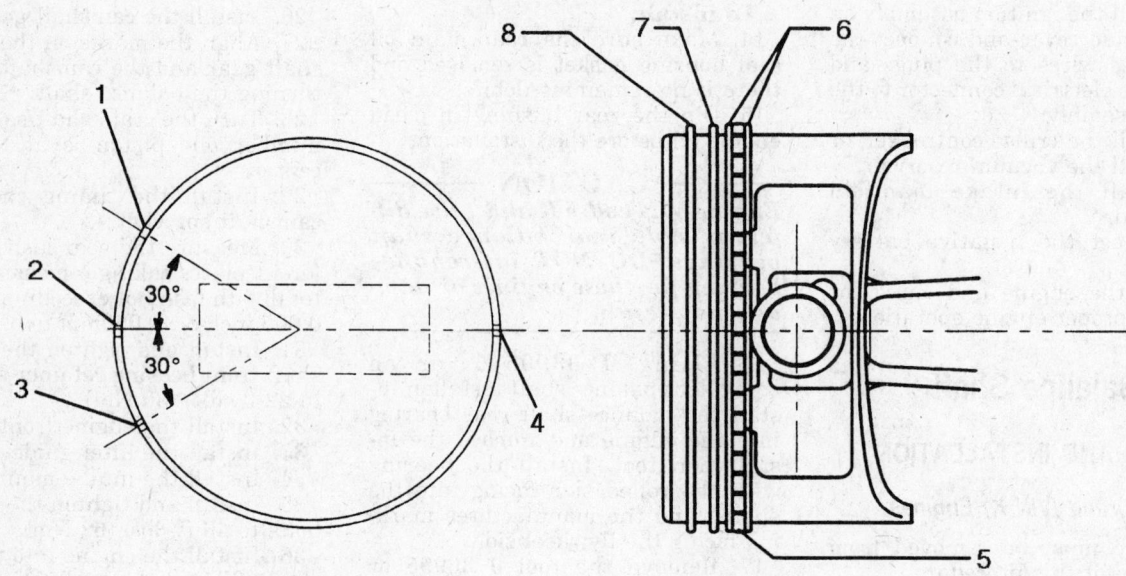

1. Oil ring segment gap
2. Upper compression ring gap
3. Oil ring segment gap
4. Expander & lower compression ring gaps
5. Expander ring
6. Oil segment rings
7. Lower compression ring
8. Upper compression ring

314750

Ring gap positioning — 4.0L (VIN C) engine

ENGINE LUBRICATION

Oil Pan

REMOVAL AND INSTALLATION

3.8L (VIN 1) and (VIN K) Engines

1. Disconnect the negative battery cable.
2. Raise and safely support the vehicle.
3. Drain the engine oil.
4. Disconnect the oil level indicator connector.
5. Remove the oil pan mounting bolts and remove the oil pan.

To install:
6. Clean all the gasket surfaces completely.
7. Install the oil pan with a new gasket and tighten the mounting bolts to 10 ft. lbs. (14 Nm).
8. Connect the oil level indicator connector.
9. Lower the vehicle.
10. Refill the crankcase with engine oil.
11. Connect the negative battery cable.
12. Start the vehicle and verify no oil leaks.

4.0L (VIN C) Engine

On the 4.0L (VIN C) Engine Oldsmobile Aurora, remove the transaxle assembly, not the engine, for engine oil pan removal.

1. Disconnect the negative battery cable. The battery is located under the passenger side of the rear seat cushion. Raise the rear seat cushion to access the battery.
2. Raise and safely support the vehicle.

3. Remove the oil pan drain plug and drain the engine oil into a suitable container.

4. Remove the transaxle assembly.

5. Remove the oil pan bolts and remove the oil pan from the vehicle.

NOTE: The oil pan gasket is reusable unless it is damaged. Do not remove the gasket from the oil pan groove unless gasket replacement is required.

To install:

6. Install the oil pan and seal in the vehicle.

7. Install the oil pan mounting bolts and tighten in sequence to 9 ft. lbs. (13 Nm).

8. Install the transaxle assembly.

9. Install the oil pan drain plug and tighten to 15 ft. lbs. (20 Nm).

10. Lower the vehicle.

11. Refill the crankcase with the proper type and quantity of engine oil.

12. Connect the negative battery cable.

Oil Pump

REMOVAL AND INSTALLATION

3.8L (VIN 1) and (VIN K) Engines

1. Disconnect the negative battery cable.

2. Raise and safely support the vehicle.

3. Drain the engine oil.

4. Remove the front cover assembly.

5. Remove the 4 bolts securing the oil filter adapter to the front cover assembly and oil filter adapter, pressure regulator valve and spring.

6. Remove the 4 oil pump cover attaching screws and remove the cover.

7. Remove the inner and outer pump gears.

8. Make the following measurements and replace any components not within specification:

   a. Gear pocket depth: 0.461–0.4265 in. (11.71–11.75mm)

   b. Gear pocket diameter: 3.508–3.512 in. (89.10–89.20mm)

   c. Inner gear tip clearance: 0.006 in. (0.152mm)

   d. Outer gear diameter clearance: 0.008–0.015 in. (0.203–0.381mm)

   e. Gear end clearance: 0.001–0.0035 in. (0.025–0.089mm)

9. If measurement **A** or **B** is out of specification the front cover assembly must be replaced. If measurement **C**, **D** or **E** is out of specification replace the gears.

To install:

10. Lubricate the gears with petroleum jelly and install the gears into the housing.

11. Pack the gear cavity with petroleum jelly after the gears have been installed in the housing. This seals the gears and acts as a prime to allow the oil pump to draw oil as soon as it starts to turn as the engine is cranked. Do not overlook this step. DO NOT use chassis grease. Use only petroleum jelly since its low–temperature nature means it will dissolve as the engine warms up.

12. Install the oil pump cover and screws and tighten to 97 inch lbs. (11 Nm).

13. Install the oil filter adapter with new gasket, pressure regulator valve and spring. Tighten the mounting bolts to 24 ft. lbs. (33 Nm).

14. Install the front cover assembly.

15. Refill the crankcase with clean engine oil.

16. Connect the negative battery cable.

17. Start the vehicle and verify no leaks and proper oil pressure.

4.0L (VIN C) Engine

1. Disconnect the negative battery cable.

2. Install an engine support fixture, J-28467-A or the equivalent.

3. Remove the engine mount-to-body through bolt nut and through bolt.

4. Remove the through bolt nut and bolt from the engine side of the mount.

5. Raise and safely support the vehicle.

6. Remove the front tire and wheel assembly.

7. Remove the right inner fender well splash shield.

8. Remove the lower center air deflector.

9. Remove the left transaxle mount-to-frame through bolt.

10. Remove the lower retaining bolt and nut from the engine mount bracket.

11. Remove the bolt and disconnect the power steering line retainer from the bracket.

12. Lower the vehicle.

13. Remove the upper retaining nut and bolt from the engine mount bracket.

14. Remove the serpentine belt from the power steering pump pulley.

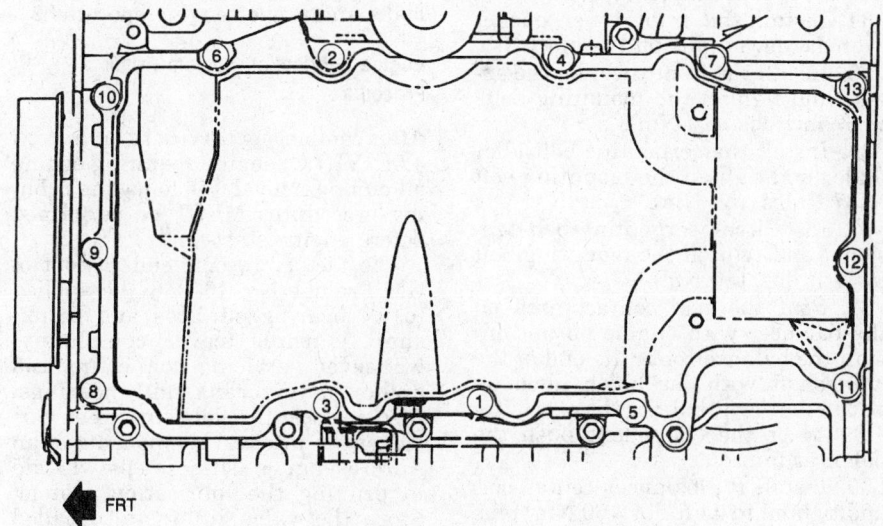

FRT

321926

Oil pan bolt tightening sequence — 4.0L (VIN C) engine

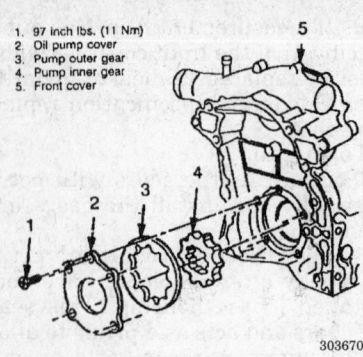

1. 97 inch lbs. (11 Nm)
2. Oil pump cover
3. Pump outer gear
4. Pump inner gear
5. Front cover

303670

Oil pump assembly — 3.8L (VIN 1) and (VIN K) engines

15. Remove the fuel plastic sight shield on the intake manifold.

16. Remove the power steering pump from the mounting bracket and set aside. DO NOT disconnect the power steering lines from the pump.

17. Raise the engine using the support fixture.

18. Remove the engine mount bracket from the engine.

19. Remove the torque axis mount from the frame rail.

20. Lower the engine using the support fixture until clearance for J-38416-B, or an equivalent puller is attained.

21. Raise and safely support the vehicle.

22. Remove the crankshaft balancer retaining bolt.

23. Remove the crankshaft balancer using J-38416-B.

24. Remove the serpentine belt tensioner.

25. Remove the serpentine belt idler pulley.

26. Remove the front cover bolts and remove the cover and gasket. DO NOT discard the gasket, if it is undamaged it can be re-used.

27. Remove the 3 oil pump mounting bolts and remove the oil pump and drive spacer.

28. If necessary, disassemble and inspect the pump as follows:

　a. Remove the drive spacer from the pump housing.

　b. Remove the 2 screws holding the pump housing halves together.

　c. Remove the inner (drive) and outer (driven) rotors from the housing. Indicate the mating surfaces (dimples).

　d. Remove the pressure relief valve.

　e. Inspect the pump housing for nicks, burrs, chips or debris that might cause a leak or binding condition in the rotor pocket.

　f. Inspect the drive and driven rotors for nicks or burrs.

　g. Check the pump cover and interior surface for excessive wear or score marks. Check for flatness.

　h. If any components show signs of excessive wear or damage, replace the pump assembly.

To install:

29. If the pump was disassembled, reassemble as follows:

　a. Install the inner and outer rotors to the pump cover in the same orientation as removed.

　b. Install the pressure relief valve seat, spring and pilot in the pump housing.

　c. Pack the pump housing halves with Amojell™ or white petroleum grease to ensure pump priming.

　d. Assemble the housing and cover over the locating dowel.

　e. Insert a 9mm drill in the pump mounting hole on the opposite side to aid alignment of the housing and cover. Install the 2 screws and tighten to 108 inch lbs. (12 Nm).

30. Install the oil pump drive spacer into the oil pump from the rear so the drive flat engages the pump rotor.

31. Install the oil pump over the crankshaft and loosely install the mounting bolts.

32. Hold the pump in its furthest up position and tighten the mounting bolts to 89 inch lbs. (10 Nm) plus 35 degrees additional rotation.

33. Place a small amount of RTV sealant at the split line of the upper and lower crankcases.

34. Install the front cover gasket over the dowel pins on the block.

35. Install the front cover on the engine and tighten the mounting bolts to 89 inch lbs. (10 Nm).

36. Install the serpentine belt idler pulley and tighten the mounting bolt to 37 ft. lbs. (50 Nm).

37. Install the serpentine belt tensioner and tighten the mounting bolt to 37 ft. lbs. (50 Nm).

38. Coat the seal contact area on the balancer with engine oil and install the balancer onto the end of the crankshaft with the notch lined up with the crankshaft key. Using J-39344 or the equivalent push the balancer into place.

39. Install the balancer center bolt and tighten to 44 ft. lbs. (60 Nm) plus 120 degrees (²/₃ turn) additional rotation.

40. Lower the vehicle.

41. Raise the engine with the support fixture.

42. Install the torque axis mount on the body.

43. Install the engine mount bracket on the engine and loosely install the mounting nuts and bolts. With all the fasteners in place tighten the nuts to 30 ft. lbs. (40 Nm) and the bolts to 41 ft. lbs. (55 Nm).

44. Lower the engine support fixture until the engine is at its normal height.

45. Install the power steering pump in the mounting bracket.

46. Install the serpentine belt.

47. Connect the power steering line retainer to the engine mount bracket and install the mounting bolt.

48. Raise and safely support the vehicle.

49. Install the left transaxle mount-to-frame through bolt and tighten to 63 ft. lbs. (85 Nm).

50. Install the lower center air deflector.

51. Install the right inner fender well splash shield.

52. Install the tire and wheel assembly and tighten the wheel nuts to 100 ft. lbs. (140 Nm).

53. Lower the vehicle.

54. Install the torque axis mount-to-engine mount bracket through bolt and nut and tighten the bolt to 70 ft. lbs. (95 Nm).

55. Install the torque axis mount-to-frame through bolt and nut and tighten the through bolt to 37 ft. lbs. (50 Nm).

56. Remove the engine support fixture.

57. Connect the negative battery cable.

58. Run the engine and check for leaks and proper engine operation.

Engine Lubrication System Priming Procedure

After completing service to an Aurora 4.0L (VIN C) engine requiring engine oil pump removal, the following priming procedures MUST be performed before engine start–up.

The factory recommends a coat of GM Prelube No. 1052367 be applied to all bearing surfaces and crankshaft journals (cover completely) whenever servicing connecting rod and/or main crankshaft bearings. Also perform the following when servicing the internal components of an Aurora engine. These steps will aid in priming the lubrication system: Store the valve lifters (also called tappets or lash adjusters) with the camshaft contact surface down, so engine oil will not drain from the lifters; Apply liberal amounts of GM Camshaft Prelube, No. 1052365 to the camshaft lobes, bearing caps and lifter surfaces; Piston rings and pis-

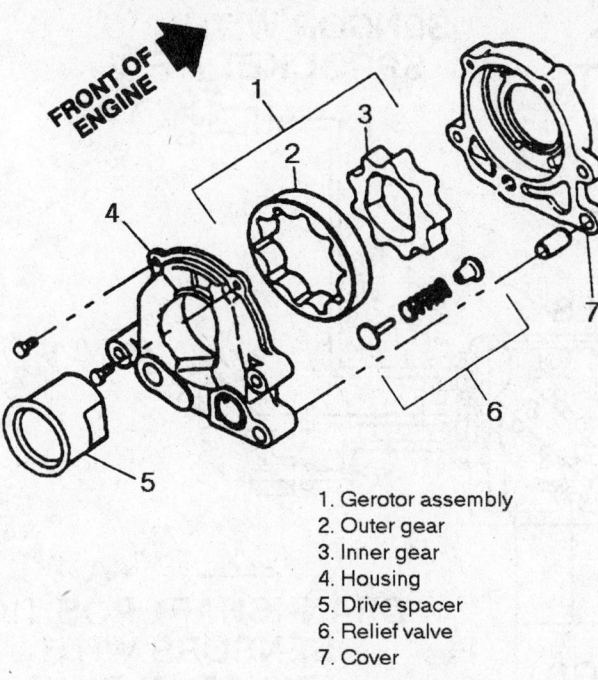

1. Gerotor assembly
2. Outer gear
3. Inner gear
4. Housing
5. Drive spacer
6. Relief valve
7. Cover

305701

Oil pump components — 4.0L (VIN C) engine

TRANSAXLE

Transaxle Assembly

REMOVAL AND INSTALLATION

Riviera

1. Disconnect the negative battery cable.
2. Remove the air intake duct.
3. Disconnect the cruise control cable from the throttle body lever, disconnect the vacuum hose at the control servo and remove the servo from the engine.
4. Remove the nut from the shift linkage at the manual shaft on the transaxle. Remove the 2 linkage bracket bolts and remove the cable and bracket from the transaxle and position out of the way.
5. Disconnect the vacuum line from the modulator.
6. Disconnect the electrical connectors from the park neutral switch, backup light switch, transaxle harness and vehicle speed sensor.
7. Remove the left-to-right exhaust manifold bolts.
8. Remove the vacuum reservoir.
9. Remove the heater hose retainer bracket.
10. Disconnect the oxygen sensor connector.
11. Install an engine support fixture, J-28467-A, or the equivalent. Make sure the support fixture is tight and the weight of the powertrain is removed from the mounts.
12. Remove the upper transaxle-to-engine bolts.
13. Raise and safely support the vehicle.
14. Remove the front tire and wheel assemblies.
15. Remove the left side power steering rack mounting bolts.
16. Remove the left front splash shield.
17. Remove the cotter pins and castle nuts from the lower ball joints and using a suitable ball joint separator, disconnect the ball joints from the steering knuckles.
18. Remove the front lower air deflector.
19. Remove the power steering line retaining clamp from the frame.
20. Remove the remaining rack mounting bolt and raise the rack off its frame mount and support.
21. Remove the left and right transaxle mount through bolts.

tons should be completely covered with proper specification motor oil during installation.

To perform the factory–required Engine Lubrication System Priming Procedure, use the following procedures.

1. Thoroughly pack the oil pump with Amogell™ or white petroleum jelly during reassembly and fill the oil filter with correct specification engine oil before installation.
2. Verify engine oil is at the proper level. System capacity is 7 quarts with the oil filter full. Fill to proper level if necessary.
3. Disconnect the right front connector (power lead) from the Distributorless Ignition System (DIS) module and crank the engine for 30 seconds.
4. Reconnect the power lead to the DIS and start the engine. Check the Driver Information Center (DIC) for **Low Oil Pressure** message and listen for any audible noise such as lifters "ticking".
5. If engine noises persist, stop the engine and remove the Oil Pressure Switch from the Oil Filter Adapter and install a mechanical oil pressure gauge. **Be careful not to damage the threads in the adapter.** If is

made of soft magnesium and is easily damaged. Also note that the ports are sealed with O-rings which must be in place and in good condition to seal properly. The bypass valves in the adapter are non–serviceable.

6. If oil is indicated on the gauge and no unusual sounds are heard, oil pressure is present and both the oil pump and engine lubrication system are primed. If no oil pressure is recorded, repeat Step 3 of this procedure and then proceed to Step 7.
7. If no oil pressure is recorded after repeating the process given above, remove the Oil Filter Adapter and force engine oil under pressure (using shop air) in the engine block outlet port (the port closest to the front of the engine). Reinstall the Oil Filter Adapter with the mechanical oil pressure gauge installed in the sender port and start the engine.
8. If oil pressure is obtained, stop the engine and reinstall the Oil Pressure Switch in the Oil Filter Adapter. Once connected, check the instrument display for no oil pressure or **Low Oil Pressure** message. If that message is present, check the switch connection or begin low oil pressure complaint troubleshooting.

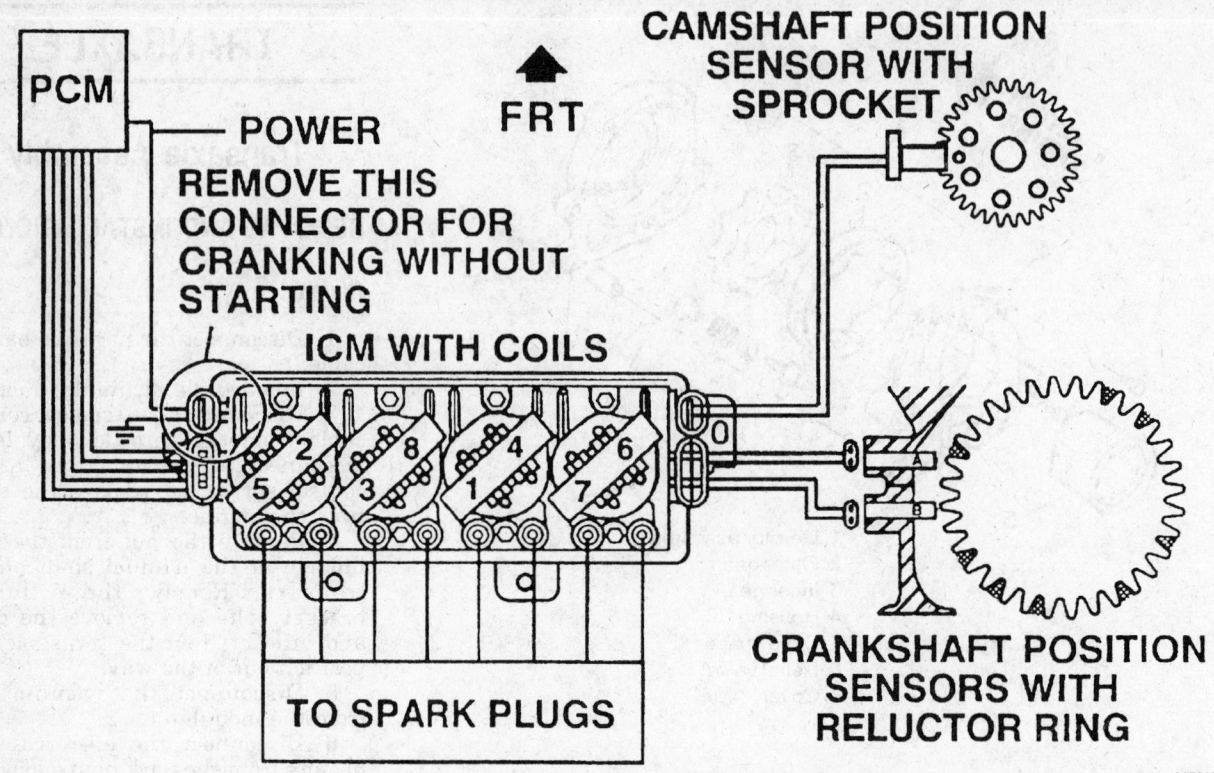

Connector removal for engine priming — 4.0L (VIN C) engine

22. Remove the complete exhaust system.

23. Support the subframe assembly.

24. Remove the subframe mounting bolts and remove the subframe assembly.

25. Remove the torque converter cover mounting screws and remove the cover.

26. Matchmark the converter to the flywheel and remove the 3 converter mounting bolts.

27. Disconnect the electrical connectors from the starter.

28. Remove the starter.

29. Remove the halfshafts.

30. Support the transaxle with a suitable jack.

31. Remove the rear transaxle mount-to-bracket retaining nuts and remove the rear transaxle mount bracket from the vehicle.

32. Disconnect the rear spark plug wires from the plugs and route them out of the way.

33. Remove the transaxle filler tube.

34. Remove the right side exhaust manifold.

35. Remove the lower transaxle-to-engine mounting bolts.

36. Disconnect and cap the transaxle oil cooler lines from the transaxle.

37. Lower the transaxle from the vehicle.

To install:

38. Raise the transaxle assembly into the vehicle and guide the transaxle onto the engine alignment dowels.

39. Connect the upper and lower transaxle oil cooler lines.

40. Install the transaxle-to-engine bolts and with all the bolts in place tighten them alternately an evenly to 55 ft. lbs. (75 Nm).

41. Install the right side exhaust manifold and tighten the mounting bolts to 38 ft. lbs. (52 Nm).

42. Install the transaxle filler tube.

43. Connect the spark plug wires to the rear plugs.

44. Install the rear transaxle mount to the body and tighten the mounting bolts to 37 ft. lbs. (50 Nm), the transaxle mount-to-transaxle bolts to 55 ft. lbs. (75 Nm) and the mount-to-transaxle bracket nuts to 29 ft. lbs. (40 Nm).

45. Remove the transaxle jack.

46. Install the halfshafts.

47. Install the starter and connect the electrical connectors.

48. Make sure the matchmarks on the converter and flywheel are in alignment. Install the converter mounting bolts and tighten to 44 ft. lbs. (60 Nm).

49. Install the flywheel cover and cover mounting bolts.

50. Raise the subframe into position and tighten the mounting bolts to 142 ft. lbs. (192 Nm).

51. Install the exhaust system.

52. Install the left transaxle mount-to-frame bolt and tighten to 63 ft. lbs. (85 Nm).

53. Install the right transaxle mount-to-transaxle bolt and tighten to 75 ft. lbs. (102 Nm).

54. Lower the steering rack into position and install the through bolt on the right side of the rack.

55. Install the power steering line retaining bracket to the subframe.

56. Connect the ball joints to the steering knuckles and tighten the castle nuts to 41 ft. lbs. (55 Nm). If necessary tighten the nuts up to 60 degrees ($^{1}/_{6}$ turn) additional to align the cotter pin holes. NEVER loosen the nuts to make the holes align.

57. Install the left side splash shield.

58. Install the left side rack and pinion mounting bolts.

59. Install the front tire and wheel assemblies and tighten the wheel nuts to 100 ft. lbs. (140 Nm).

60. Lower the vehicle.

61. Remove the engine support fixture.

62. Connect the oxygen sensor electrical connector.

63. Install the heater hose retainer bracket.

64. Install the vacuum reservoir.

65. Install the right-to-left exhaust manifold mounting bolts.

66. Connect the electrical connectors to the park neutral switch, backup light switch, transaxle harness and vehicle speed sensor.

67. Connect the vacuum line to the modulator.

68. Connect the shift cable to the manual shaft and install the cable mounting bracket. Tighten the manual shaft nut to 15 ft. lbs. (20 Nm).

69. Install the cruise control servo, connect the vacuum line and linkage.

70. Install the air inlet duct.

71. Connect the negative battery cable.

72. Refill the transaxle with fluid. Start the vehicle and check the fluid level again and top off as necessary.

73. Road test the vehicle and verify proper transaxle operation and no fluid leaks.

Aurora

1. Disconnect the negative battery cable.

2. Remove the nut from the shift linkage at the manual shaft on the transaxle. Remove the 2 linkage bracket bolts and remove the cable and bracket from the transaxle and position out of the way.

3. Disconnect the electrical connectors from the park neutral switch.

4. Install an engine support fixture, J-28467-A, or the equivalent. Make sure the support fixture is tight and the weight of the powertrain is removed from the mounts.

5. Drain the cooling system.

6. Disconnect the vacuum line at the brake booster.

7. Disconnect the transaxle vent hose.

8. Disconnect the speed sensor connector and power steering gear connector.

9. Disconnect and cap the upper transaxle oil cooler line from the radiator.

10. Disconnect and cap the lower transaxle oil cooler line from the transaxle.

11. Remove the retaining bracket nut for the cooler lines at the transaxle.

12. Disconnect the coolant bypass pipe from the thermostat housing and position out of the way.

13. Remove the left and right transaxle mount bolts.

14. Raise and safely support the vehicle.

15. Remove the left front tire and wheel assembly.

16. Remove the left front splash shield.

17. Remove the cotter pins and castle nuts from the outer tie rod end and using a suitable ball joint separator, disconnect the tie rod ends from the steering knuckles.

18. Remove the cotter pins and castle nuts from the lower ball joints and using a suitable ball joint separator, disconnect the ball joints from the steering knuckles.

19. Remove the halfshafts.

20. Remove the engine oil pan-to-transaxle bracket.

21. Remove the torque converter cover mounting screws and remove the cover.

22. Matchmark the converter to the flywheel and remove the 3 converter mounting bolts.

23. Remove the complete exhaust system.

24. Remove the exhaust manifold rear pipe.

25. Remove the steering rack-to-right transaxle mount bolts.

26. Remove the bolt from the right transaxle mount.

27. Remove the frame-to-right transaxle mount and remove the right side mount.

28. Remove the power steering line retaining clamp from the frame.

29. Remove the remaining rack mounting bolt and raise the rack off its frame mount and support.

30. Support the subframe assembly.

31. Remove the knock sensor shield.

32. Remove the engine-to-transaxle brace.

33. Remove the rear transaxle mount-to-bracket retaining nuts and remove the rear transaxle mount bracket from the vehicle.

34. Remove the right and left lower transaxle-to-engine bolts.

35. Remove the subframe mounting bolts and remove the subframe assembly.

36. Support the transaxle with a suitable jack.

37. Remove the remainder of the transaxle-to-engine bolts.

38. Lower the transaxle from the vehicle.

To install:

39. Raise the transaxle assembly into the vehicle and guide the transaxle onto the engine alignment dowels.

40. Install the transaxle-to-engine bolts and with all the bolts in place tighten them alternately an evenly to 55 ft. lbs. (75 Nm).

41. Install the rear transaxle mount to the body and tighten the mounting bolts to 37 ft. lbs. (50 Nm).

42. Install the rear transaxle mount bracket to the transaxle and tighten the mounting bolts to 43 ft. lbs. (58 Nm).

43. Install the engine-to-transaxle brace and tighten the brace mounting bolts to 35 ft. lbs. (47 Nm).

44. Install the knock sensor shield.

45. Connect the electrical connectors to the speed and knock sensors.

46. Make sure the matchmarks on the converter and flywheel are in alignment. Install the converter mounting bolts and tighten to 44 ft. lbs. (60 Nm).

47. Install the flywheel cover and cover mounting bolts.

48. Install the transaxle-to-oil pan brace.

49. Raise the subframe into position and tighten the mounting bolts to 142 ft. lbs. (192 Nm).

50. Install the left transaxle mount-to-frame bolt and tighten to 63 ft. lbs. (85 Nm).

51. Lower the steering rack into position and install the through bolt on the right side of the rack.

52. Install the power steering line retaining bracket to the subframe.

53. Install the right transaxle mount to the frame and tighten the mounting bolts to 54 ft. lbs. (73 Nm).

54. Install the right transaxle mount-to-transaxle bolt and tighten to 81 ft. lbs. (110 Nm).

55. Install the rear exhaust manifold pipe.

56. Install the exhaust system.

57. Install the halfshafts.

58. Connect the ball joints to the steering knuckles and tighten the castle nuts to 41 ft. lbs. (55 Nm). If necessary tighten the nuts up to 60 degrees (1/6 turn) additional to align the cotter pin holes. NEVER loosen the nuts to make the holes align.

59. Connect the tie rod ends to the steering knuckles and tighten the castle nuts to 52 ft. lbs. (70 Nm). If necessary tighten the nuts up to 60 degrees (1/6 turn) additional to align the cotter pin holes. NEVER loosen the nuts to make the holes align.

60. Install the left side splash shield.

61. Install the front tire and wheel assemblies and tighten the wheel nuts to 100 ft. lbs. (140 Nm).

62. Lower the vehicle.

63. Remove the engine support fixture.

64. Connect the coolant bypass pipe to the thermostat housing.

65. Install the transaxle oil cooler line bracket and mounting bolt.

66. Connect the upper and lower transaxle oil cooler lines.

67. Connect the shift cable to the manual shaft and install the cable mounting bracket. Tighten the manual shaft nut to 15 ft. lbs. (20 Nm).

68. Connect the vacuum line to the brake booster.

69. Connect the transaxle vent.

70. Refill the cooling system.

71. Connect the negative battery cable.

72. Refill the transaxle with fluid. Start the vehicle and check the fluid level again and top off as necessary. Service this vehicle with DEXRON®II or DEXRON®IIE automatic transmission fluid.

73. Road test the vehicle and verify proper transaxle operation and no fluid leaks.

DRIVELINE

Halfshaft

REMOVAL AND INSTALLATION

1. Raise and safely support the vehicle.

2. Remove the tire and wheel assembly.

3. Remove the sway bar link kit.

4. Remove the cotter pin and castle nut from the lower ball joint. Using a suitable ball joint separator, J-36226 or the equivalent, disconnect the ball stud from the steering knuckle.

5. If removing the right halfshaft turn the wheel to the left and if removing the left halfshaft turn the wheel to the right.

6. Using a suitable prying tool, pry the lower control arm down away from the steering knuckle.

7. Insert a drift punch through the caliper and into the rotor cooling fins. Remove the hub nut from the end of the halfshaft.

8. Separate the halfshaft from hub using J-28733-B, or an equivalent puller. Once the halfshaft is clear of the knuckle assembly,

swing the strut assembly toward the rear of the vehicle.

9. Separate the halfshaft from the transaxle using J-33008 and J-2619-01, or an equivalent slide hammer and adapter.

10. Remove the halfshaft.

To install:

11. If installing the right side halfshaft, install J-37292-B, or an equivalent tear away axle seal protector over the seal.

12. Install the halfshaft into the transaxle and seat in place by inserting a suitable pry bar in the groove in the inboard joint and tapping the joint into place. Make sure the joint is properly seated by grasping the inboard joint and making sure it wont pull out of the transaxle. DO NOT pull on the shaft itself or the inboard joint can become damaged.

13. Insert the halfshaft through the hub and install the washer and mounting nut. Insert a drift through the caliper and into the rotor cooling fins and tighten the mounting nut to 107 ft. lbs. (145 Nm).

14. Connect the ball joint to the steering knuckle and tighten the castle nut to 41 ft. lbs. (55 Nm). If necessary to align the cotter pin holes rotate the nut up to 60 degrees (⅙ turn) additional. NEVER loosen the nut to align the holes.

15. Install a new cotter pin.

16. Install the sway bar link kit and tighten the bolt to 13 ft. lbs. (17 Nm).

17. Remove the tear away seal protector. Make sure no pieces of the tool remain in the transaxle.

18. Install the tire and wheel assembly and tighten the wheel nuts to 100 ft. lbs. (140 Nm).

19. Lower the vehicle.

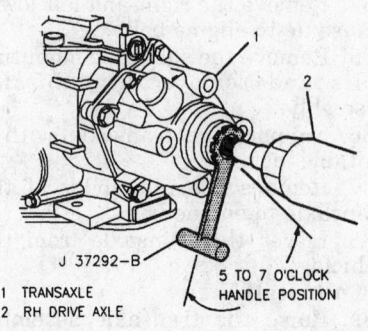

1 TRANSAXLE
2 RH DRIVE AXLE

5 TO 7 O'CLOCK HANDLE POSITION

J 37292-B

303390

Installing the right side axle with the tear away seal protector

CV-Joint Boot

REPLACEMENT

Outer CV-Joint

1. Remove the half shaft from the vehicle and support in a suitable vise.

2. Remove the large clamp from the outboard boot using a pair of side cutters and discard the clamp.

3. Cut the small clamp off the inboard end of the CV-joint boot and discard.

4. Separate the large end of the boot from the joint and push the boot down the axle to provide clearance to work on the CV-joint.

5. Clean the grease off the face of the CV-joint until the retaining ring is visible in the cutout.

6. Spread the ears on the retaining ring and slide the CV-joint off of the halfshaft.

7. Remove the CV-joint boot from the shaft.

To install:

8. Install a new swage ring on the inboard end of the CV-joint boot and install the boot onto the axle shaft until the seal lip is in the largest groove near the bottom site groove.

9. The inner swage clamp can be compressed as follows:

   a. Position the inboard end of the halfshaft in the compression tool, J-41048 or the equivalent.

   b. Align the swage ring with the outboard edge of the ring even with the front edge of the tool where the taper ends.

   c. Install the top half of the tool.

   d. Insert the bolts and tighten by hand until snug.

   e. Check that the tool and ring are still properly aligned.

   f. Tighten the bolts 180 degrees (½ turn) at a time, alternating from one to the other until both bolts are bottomed out.

   g. Remove the tool and reset the halfshaft in the vise.

10. Install a new retaining ring in the CV-joint.

11. Pack the joint with grease and put the remainder of the grease in the CV-joint boot.

12. Install the large clamp onto the outboard end of the CV-joint boot.

13. With the splines aligned, slide the CV-joint onto the shaft until the retainer ring engages in the groove in the halfshaft.

14. Slide the CV-joint boot over the end of the joint and seat the boot in the groove on the joint.

15. Crimp the seal retaining clamp.

16. Install the halfshaft in the vehicle.

Inner CV-Joint

Except Aurora Left Side

1. Remove the half shaft from the vehicle and support in a suitable vise.
2. Remove the large clamp from the boot using a pair of side cutters and discard the clamp.
3. Cut the small clamp off the inboard end of the CV-joint boot and discard.
4. Separate the large end of the boot from the joint and push the boot

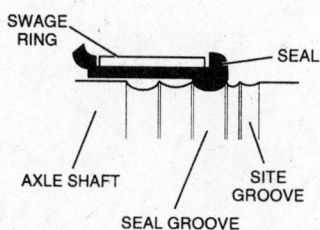

Aligning the outboard CV-joint boot on the halfshaft grooves

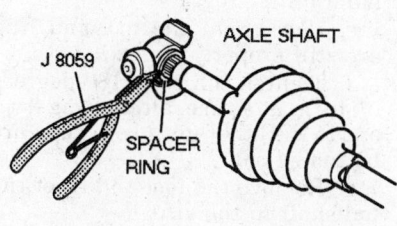

Removing the inboard spacer ring below the spider assembly

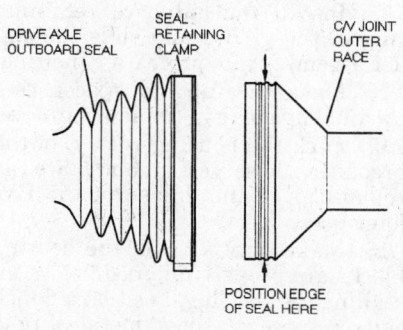

Aligning the boot on the outboard joint large end

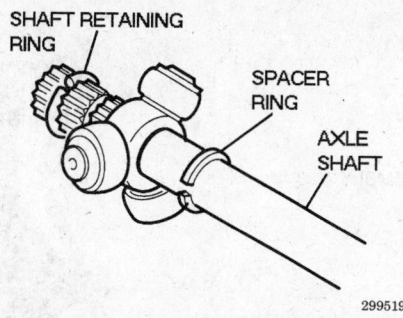

Spider assembly with upper and lower retainers on the halfshaft

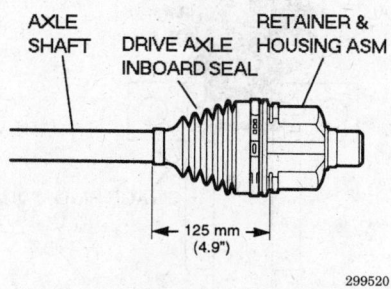

Inboard boot installation measurement

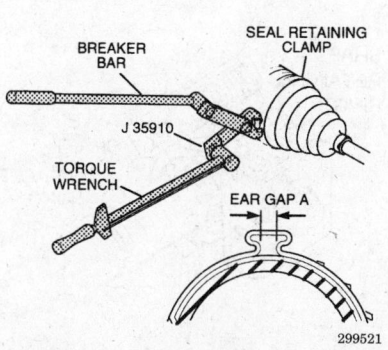

Compressing the large clamp on the inboard and outboard CV-joint boots

down the axle to provide clearance to work on the CV-joint.
5. Remove the housing from the spider and shaft by sliding it carefully off the end of the shaft.
6. Remove the tripot bushing from the housing.
7. Spread the spacer ring on the inboard side of the spider assembly and slide the ring down onto the shaft.
8. Push the spider down onto the axle to uncover the outboard retainer ring.

9. Remove the outboard retainer ring and remove the spider and inboard spacer from the shaft.
10. Remove the CV-joint boot from the shaft.

To install:

11. Install a new swage ring on the inboard end of the CV-joint boot and install the boot onto the axle shaft until the seal lip is in the largest groove near the bottom site groove.
12. The inner swage clamp can be compressed as follows:
   a. Position the inboard end of the halfshaft in the compression tool, J-41048 or the equivalent.
   b. Align the swage ring with the outboard edge of the ring even with the front edge of the tool where the taper ends.
   c. Install the top half of the tool.
   d. Insert the bolts and tighten by hand until snug.
   e. Check that the tool and ring are still properly aligned.
   f. Tighten the bolts 180 degrees (½ turn) at a time, alternating from one to the other until both bolts are bottomed out.
   g. Remove the tool and reset the halfshaft in the vise.
13. Install the inboard spacer ring onto the shaft and situate it below the second groove.
14. Slide the spider assembly onto the shaft so the top groove is visible.
15. Install the outboard retaining ring in the groove and slide the spider assembly into place over the ring.
16. Move the inboard spacer ring up and engage it in the lower groove.
17. Pack the tripot housing with ½ of the grease in the service kit and place the remainder of the grease in the CV-joint boot.
18. Install the tripot bushing on the housing making sure it is seated flush.
19. Install the large clamp on the CV-joint boot.
20. Slide the tripot hosing over the spider assembly until the CV-joint boot can be slid over the end of the joint.
21. With the boot seated in the grooves on the tripot bushing, compress the retainer ring.
22. Install the halfshaft in the vehicle.

Aurora Left Side

1. Remove the half shaft from the vehicle and support in a suitable vise.
2. Remove the large clamp from the boot using a pair of side cutters and discard the clamp.
3. Cut the small clamp off the inboard end of the CV-joint boot and discard.

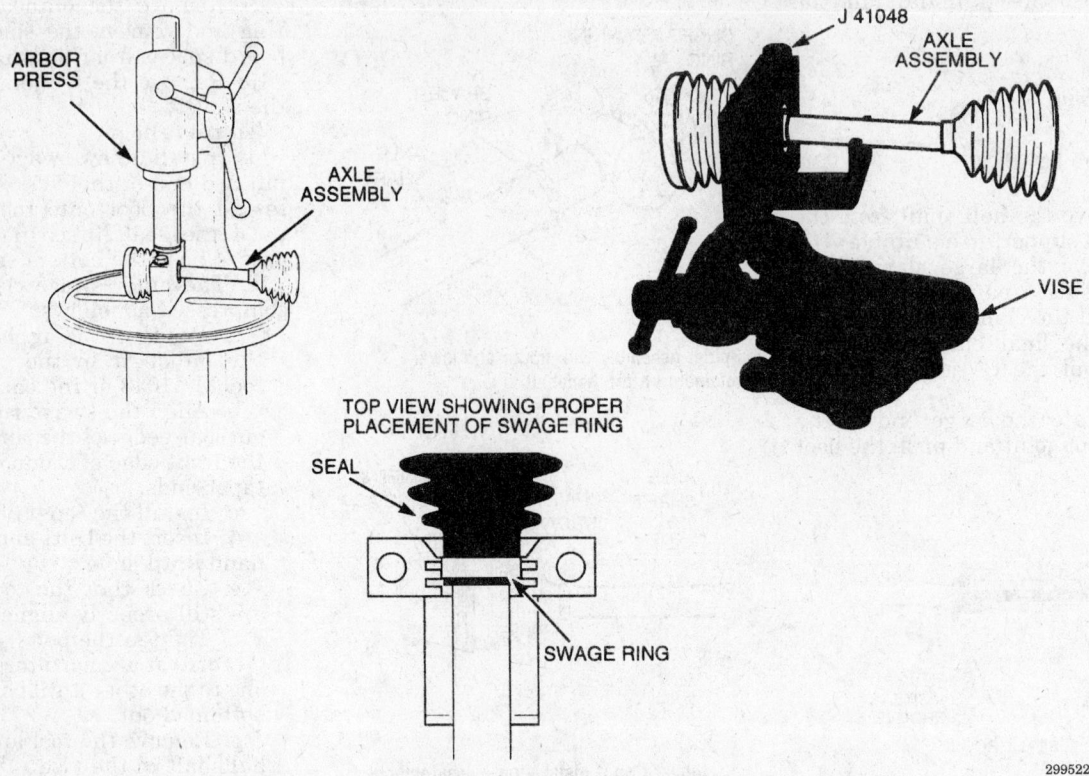

Compressing the swage ring on the inboard and outboard CV-joint boots

4. Separate the large end of the boot from the joint and push the boot down the axle to provide clearance to work on the CV-joint.

5. Remove the housing from the spider and shaft by sliding it carefully off the end of the shaft.

6. Remove the bushing from the housing.

7. Spread the spacer ring on the inboard side of the spider assembly and slide the ring down onto the shaft.

8. Push the spider down onto the axle to uncover the outboard retainer ring.

9. Remove the outboard retainer ring and remove the spider and inboard spacer from the shaft.

10. Remove the CV-joint boot from the shaft.

To install:

11. Install a new swage ring on the inboard end of the CV-joint boot and install the boot onto the axle shaft until the seal lip is in the largest groove near the bottom site groove.

12. The inner swage clamp can be compressed as follows:

   a. Position the inboard end of the halfshaft in the compression tool, J-41048 or the equivalent.

   b. Align the swage ring with the outboard edge of the ring even with

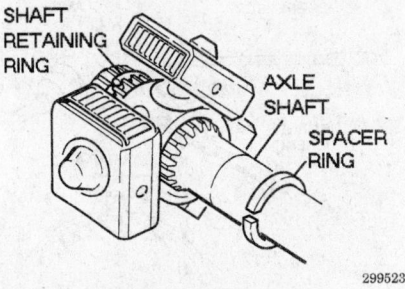

Positioning the inboard retaining ring for access to the outboard retainer

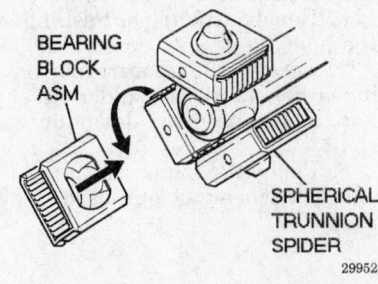

Removing the bearing blocks from the spider

the front edge of the tool where the taper ends.

   c. Install the top half of the tool.

   d. Insert the bolts and tighten by hand until snug.

   e. Check that the tool and ring are still properly aligned.

   f. Tighten the bolts 180 degrees (½ turn) at a time, alternating from one to the other until both bolts are bottomed out.

   g. Remove the tool and reset the halfshaft in the vise.

13. Install the inboard spacer ring onto the shaft and situate it below the second groove.

14. Slide the spider assembly onto the shaft so the top groove is visible.

15. Install the outboard retaining ring in the groove and slide the spider assembly into place over the ring.

16. Move the inboard spacer ring up and engage it in the lower groove.

17. Pack the housing with ½ of the grease in the service kit and place the remainder of the grease in the CV-joint boot.

18. Make sure all 3 of the bearing blocks are properly aligned before installing the housing. Position a 6 inch square piece of sheet metal with a notch cut out over the axle and against the bearing blocks to keep them from rotating.

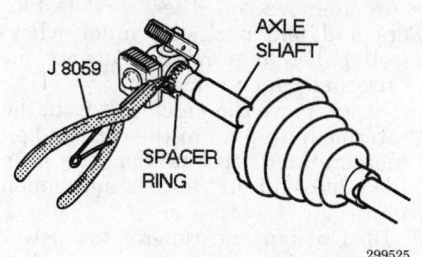

Installation of the inboard retaining ring

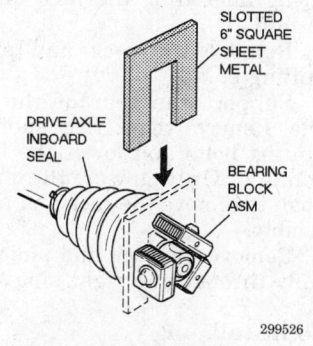

Six inch sheet metal square used
for bearing block alignment

19. Install the bushing on the housing making sure it is seated flush.

20. Install the large clamp on the CV-joint boot.

21. Slide the hosing over the spider assembly until the CV-joint boot can be slid over the end of the joint.

22. Remove the 6 inch piece of sheet metal.

23. With the boot seated in the grooves on the tripot bushing, compress the retainer ring.

24. Install the halfshaft in the vehicle.

STEERING

Air Bag

— CAUTION —

Some vehicles are equipped with the Supplemental Inflatable Restraint (SIR) or air bag system. The SIR system must be disabled before performing service on or around SIR system components, steering column, instrument panel components, wiring and sensors. Failure to follow safety and disabling procedures could result in accidental air bag deployment, possible personal injury and unnecessary SIR system repairs.

PRECAUTIONS

Several precautions must be observed when handling the inflator module to avoid accidental deployment and possible personal injury.

• Never carry the inflator module by the wires or connector on the underside of the module.

• When carrying a live inflator module, hold securely with both hands, and ensure that the bag and trim cover are pointed away.

• Place the inflator module on a bench or other surface with the bag and trim cover facing up.

• With the inflator module on the bench, never place anything on or close to the module which may be thrown in the event of an accidental deployment.

DISARMING

1. Turn the steering wheel so the vehicle wheels are pointing straight ahead.

2. Turn the ignition key to the **LOCK** position and remove the key.

3. Remove the AIR BAG fuse from the fuse block.

4. Remove the left side sound insulator.

5. Disconnect the Connector Position Assurance (CPA) and yellow 2–way SIR connector at the multi–use bracket near the base of the steering column. The drivers side air bag is disabled.

6. Remove the right side sound insulator.

7. Disconnect the Connector Position Assurance (CPA) and yellow 2–way SIR connector at the DERM mounting bracket. The passenger side air bag is disabled.

ARMING

1. Make sure the ignition is locked and the key is removed.

2. Connect the yellow 2–way SIR connector and Connector Position Assurance (CPA) at the DERM mounting bracket.

3. Install the right side sound insulator.

4. Connect the yellow 2–way SIR connector and Connector Position Assurance (CPA) at the multi–use bracket at the base of the column.

5. Install the left side sound insulator.

6. Install the AIR BAG fuse.

7. Turn the ignition switch to the **RUN** position and verify the AIR BAG light flashes 7 times and then shuts off.

Steering Wheel

REMOVAL AND INSTALLATION

— CAUTION —

The Supplemental Inflatable Restraint (SIR) system must be disarmed before removing the steering wheel. Failure to do so may cause accidental deployment of the air bag, resulting in unnecessary SIR system repairs and/or personal injury.

1. Properly disarm the SIR system.

2. If equipped with inserts over the back of the inflator module, remove the inserts to access the bolts.

3. Remove the inflator module mounting screws.

4. Push down and twist the horn contact lead counterclockwise and disconnect the horn lead from the cam tower.

5. Disconnect the horn ground lead.

6. Disconnect the CPA retainer from the inflator module.

7. Disconnect the SIR connector from the inflator module.

8. Disconnect the steering wheel control switch connector, if equipped.

9. Remove the inflator module.

— CAUTION —

When carrying a live air bag, make sure the bag and trim cover are pointed away from the body. In the unlikely event of an accidental deployment, the bag will then deploy with minimal chance of injury. When placing a live air bag on a bench or other surface, always face the bag and trim cover up, away from the surface. The will reduce the motion of the module if it is accidently deployed.

10. Remove the steering wheel mounting nut.

11. Take note of the alignment marks on the steering wheel and shaft and using a suitable puller, J-1859–03 and J-38720 or their equivalents, remove the steering wheel from the shaft.

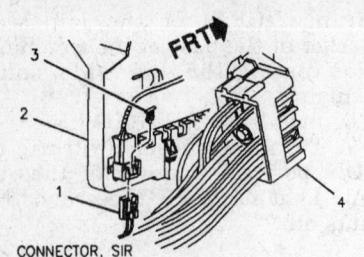

1 CONNECTOR, SIR
2 BRACKET, MULTIUSE MODULE
3 CONNECTOR POSITION ASSURANCE (CPA)
4 CONNECTOR, STEERING COLUMN
   WIRING HARNESS

220718

2–way connector — driver's side

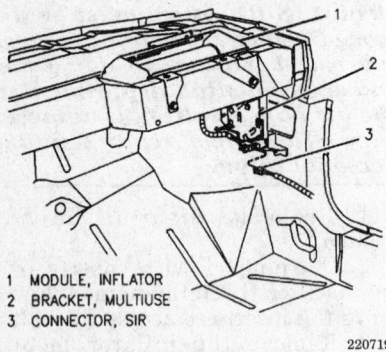

1 MODULE, INFLATOR
2 BRACKET, MULTIUSE
3 CONNECTOR, SIR

220719

2–way connector — passenger's side

To install:

12. Route the inflator module wiring through the steering wheel.

13. Align the mark on the wheel with the mark on the shaft and install the wheel on the shaft.

14. Install the steering wheel mounting nut and tighten to 30 ft. lbs. (41 Nm).

15. Connect the SIR connector to the inflator module.

16. Install the CPA retainer to the inflator module.

17. Connect the horn contact lead into the cam tower and rotate clockwise to lock in position.

18. Connect the horn ground lead.

19. Connect the steering wheel control switch connector.

20. Install the inflator module on the steering wheel. For ease of assembly install the top of the assembly in first.

21. Install the inflator module mounting screws and tighten to 27 inch lbs. (3 Nm).

22. Install the inserts covering the mounting screws, if equipped.

23. Enable the SIR system.

Tie Rod Ends

REMOVAL AND INSTALLATION

1. Raise and safely support the vehicle.

2. Remove the tire and wheel assembly.

3. Remove the cotter pin and castle nut from the outer tie rod end ball stud.

4. Back off the inner tie rod jam nut 1/2 a turn.

5. Using a suitable puller, J-24319–01 or the equivalent, separate the outer tie rod end from the steering knuckle.

6. Count the number of rotations required to remove the outer tie rod end from the inner tie rod.

To install:

7. Thread the outer tie rod end onto the inner tie rod the same number of turns required to remove the old tie rod end. The tie rod should be within 1/2 a turn of the jam nut.

8. Connect the tie rod end to the steering knuckle and tighten the castle nut to 35 ft. lbs. (47 Nm). If necessary to align the cotter pin holes, tighten the nut up to 60 degrees additional rotation. NEVER loosen the castle nut to align the cotter pin holes.

9. Install a new cotter pin.

10. Install the tire and wheel assembly and tighten the wheel nuts to 100 ft. lbs. (140 Nm).

11. Lower the vehicle.

12. Check the front end alignment and adjust as necessary.

13. Tighten the jam nut to 50 ft. lbs. (68 Nm).

Power Rack and Pinion

REMOVAL AND INSTALLATION

1. Lock the steering wheel in the straight ahead position.

2. Raise and safely support the vehicle.

3. Remove the tire and wheel assembly.

4. Remove the cotter pin and castle nut from the outer tie rod end ball stud.

5. Back off the inner tie rod jam nut 1/2 turn.

6. Using a suitable puller, J-24319–01 or the equivalent, separate the outer tie rod end from the steering knuckle.

7. Disconnect the exhaust pipe from the rear manifold and remove the intermediate pipe hangers. Lower the exhaust system to provide clearance.

8. Remove the wheel well fasteners and fold back the inner wheel well panel to provide clearance for rack and pinion removal.

9. Remove the pinch bolt from the intermediate shaft at the rack and pinion unit and separate the shaft from the stub shaft on the rack and pinion unit.

10. Unsnap and remove the power steering unit heat shield.

11. Disconnect the Magnasteer® electrical connector.

12. Disconnect and cap the power steering lines from the rack and pinion unit.

13. Remove the rack and pinion mounting bolts.

14. Support the rear of the subframe. Remove the 2 rear subframe mounting bolts and lower the frame assembly. Only lower the frame enough to remove the rack and pinion assembly.

15. Remove the rack and pinion assembly through the right side wheel well.

To install:

16. Install the rack and pinion unit through the right side wheel well.

17. Loosely install the 3 mounting bolts.

18. Raise the subframe into position and tighten the frame mounting bolts to 142 ft. lbs. (192 Nm). Remove the support.

19. Tighten the rack and pinion mounting bolts in sequence to 48 ft. lbs. (65 Nm). Start with the vertically installed bolt closest to the pinion housing and work toward the passenger side of the vehicle.

20. Connect the power steering hoses to the rack and pinion unit and tighten the fittings to 20 ft. lbs. (27 Nm).

21. Connect the Magnasteer® electrical connector.

22. Install the rack and pinion heat shield.

23. Connect the intermediate shaft to the rack and pinion stub shaft and tighten the pinch bolt to 35 ft. lbs. (47 Nm).

24. Connect the tie rod end to the steering knuckle and tighten the castle nut to 35 ft. lbs. (47 Nm). If necessary to align the cotter pin holes, tighten the nut up to 60° additional rotation. NEVER loosen the castle nut to align the cotter pin holes.

25. Install a new cotter pin.

26. Connect the exhaust pipe to the exhaust manifold and connect the intermediate pipe hangers.

27. Install the wheel well moulding.

28. Install the tire and wheel assembly and tighten the wheel nuts to 100 ft. lbs. (140 Nm).

29. Lower the vehicle.

NOTE: Whenever the vehicle sub–frame is removed or lowered, the front wheel alignment should be checked.

30. Refill and bleed the power steering system. Verify no leaks.

31. Check the front end alignment and adjust as necessary.

Power Steering Pump

BLEEDING

1. Raise the front wheels off the ground.

2. Turn the steering wheel full left.

3. Fill the power steering reservoir to the **FULL COLD** mark and leave the cap off.

4. The engine must be off.

5. Turn the steering wheel lock to lock a minimum of 20 times while periodically checking the fluid level.

6. Top off the fluid as necessary and install the reservoir cap.

7. Lower the front wheels and start the vehicle.

8. Center the steering wheel and allow the engine to run for 2 minutes.

9. Turn the steering wheel in both directions and verify smooth quiet operation.

REMOVAL AND INSTALLATION

Riviera

1. Remove the serpentine belt from the power steering pump pulley. It is not necessary to remove the serpentine belt from the remainder of the pulleys.

2. Disconnect the power steering pressure and return lines from the pump and plug the lines and pump ports.

3. If equipped with a supercharger, disconnect and cap the reservoir hose from the pump.

4. Working through the holes in the power steering pump pulley remove the pump mounting bolts.

5. Raise and safely support the vehicle.

6. Remove the power steering pump from the vehicle.

7. If the pump is being replaced, remove the pulley using J-25034-B, or an equivalent puller.

To install:

8. Install the power steering pump pulley on the pump and seat the pulley on the pump using J-25033-B, or the equivalent.

9. Install the power steering pump assembly on the engine and loosely install the mounting bolts.

10. Lower the vehicle.

11. Tighten the mounting bolts to 20 ft. lbs. (27 Nm).

12. Connect the reservoir fluid line to the power steering pump.

13. Connect the power steering lines to the pump and tighten the fitting on the pressure line to 20 ft. lbs. (27 Nm).

14. Install the serpentine drive belt over the power steering pump pulley.

15. Refill and bleed the power steering system. Verify no leaks.

Aurora

Service this vehicle with Power Steering Fluid GM 1050017, 1052884 or equivalent. In cold-Climate conditions, use GM 12345866, 12345867 or equivalent.

1. Remove the serpentine belt from the power steering pump drive pulley. It is not necessary to remove the belt from the remainder of the drive pulleys.

2. Disconnect the power steering lines from the pump and cap the lines and pump ports.

3. Remove the power steering pump mounting bolt.

4. Remove the power steering pump and mounting bracket from the vehicle.

5. If the power steering pump is being replaced, remove the pulley from the pump using a suitable puller, J-38825 or the equivalent.

6. Remove the 2 bolts and remove the mounting bracket from the pump.

To install:

7. Install the mounting bracket on the power steering pump and tighten the bracket mounting bolts to 20 ft. lbs. (27 Nm).

8. Install the power steering pump pulley using a suitable puller, J-25033-B or the equivalent.

9. Install the power steering pump assembly on the vehicle and tighten the mounting bolt to 20 ft. lbs. (27 Nm).

10. Connect the power steering lines to the pump and tighten the power steering pressure hose fitting to 20 ft. lbs. (27 Nm).

11. Install the serpentine belt around the power steering pulley.

12. Refill and bleed the power steering system. Verify no leaks.

BRAKES

Anti-Lock Brake System Service

PRECAUTIONS

• Certain components within the ABS system are not intended to be serviced or repaired individually. Only those components with removal and installation procedures should be serviced.

• Do not use rubber hoses or other parts not specifically specified for and ABS system. When using repair kits, replace all parts included in the kit. Partial or incorrect repair may lead to functional problems and require the replacement of components.

• Lubricate rubber parts with clean, fresh brake fluid to ease assembly. Do not use lubricated shop air to clean parts; damage to rubber components may result.

• Use only DOT 3 brake fluid from an unopened container.

• If any hydraulic component or line is removed or replaced, it may be necessary to bleed the entire system.

• A clean repair area is essential. Always clean the reservoir and cap thoroughly before removing the cap. The slightest amount of dirt in the fluid may plug an orifice and impair the system function. Perform repairs after components have been thoroughly cleaned; use only denatured alcohol to clean components. Do not allow ABS components to come into contact with any substance containing mineral oil; this includes used shop rags.

• The Anti–Lock control unit is a microprocessor similar to other computer units in the vehicle. Ensure that the ignition switch is **OFF** before removing or installing controller harnesses. Avoid static electricity discharge at or near the controller.

• If any arc welding is to be done on the vehicle, the control unit should be unplugged before welding operations begin.

Master Cylinder

REMOVAL AND INSTALLATION

1. Remove the retainer from the fluid level sensor connector and dis-

connect the sensor lead from the reservoir.

NOTE: Be careful when working on the brake hydraulic system not to spill any brake fluid on painted surfaces of electrical connectors.

2. Drain the brake fluid from the master cylinder reservoir.

3. Disconnect the reservoir hose from the master cylinder and cap the hose to prevent excessive fluid loss and or contamination.

4. Disconnect the brake pipes from the master cylinder and cap the pipes and ports to fluid loss and possible contamination.

5. Remove the master cylinder attaching nuts and remove the master cylinder from the power booster.

To install:

6. Bench bleed the master cylinder prior to installation on the power booster.

7. Install the master cylinder onto the power booster and tighten the mounting nuts to 20 ft. lbs. (27 Nm).

8. Connect the brake pipes to the master cylinder and tighten the fittings to 11 ft. lbs. (15 Nm).

9. Connect the reservoir hose to the master cylinder.

10. Refill the master cylinder to the FULL mark.

11. Connect the fluid level sensor connector and install the retainer.

12. Bleed the brake system and top off the fluid as necessary.

13. Verify no hydraulic leaks and proper brake pedal feel.

Brake Caliper

REMOVAL AND INSTALLATION

Front Caliper

1. Siphon ⅔ of the brake fluid out of the master cylinder reservoir.

2. Raise and safely support the vehicle.

3. Remove the tire and wheel assembly.

4. Install 2 wheel nuts loosely to secure the rotor when the caliper is removed.

5. Remove the bolts securing the brake hose to the caliper and disconnect the brake hose from the caliper. Plug the hose to prevent excessive fluid loss and possible fluid contamination.

6. Compress the piston into the caliper bore to provide clearance for removal.

7. Remove the caliper mounting bolts and sleeves.

8. Remove the caliper from the steering knuckle.

9. If the caliper is being replaced remove the brake pads from the caliper.

To install:

10. Coat the inside of the caliper bushings with silicone grease.

11. Seat the caliper piston fully in its bore.

12. Install the brake pads in the caliper.

13. Install the caliper on the steering knuckle and install the mounting bolts and sleeves. Tighten the caliper mounting bolts to 38 ft. lbs. (51 Nm).

14. Connect the brake hose to the caliper using new washers. Tighten the fitting bolt to 33 ft. lbs. (45 Nm).

15. Remove the 2 lug nuts securing the brake rotor.

16. Install the wheel and tire assembly and tighten the wheel nuts to 100 ft. lbs. (140 Nm).

17. Refill the master cylinder with fluid and bleed the brake system.

18. Pump the brake pedal several times to seat the brake pads against the rotor. Verify no hydraulic leaks and a good firm brake pedal.

Rear Caliper

1. Siphon ⅔ of the brake fluid out of the master cylinder reservoir.

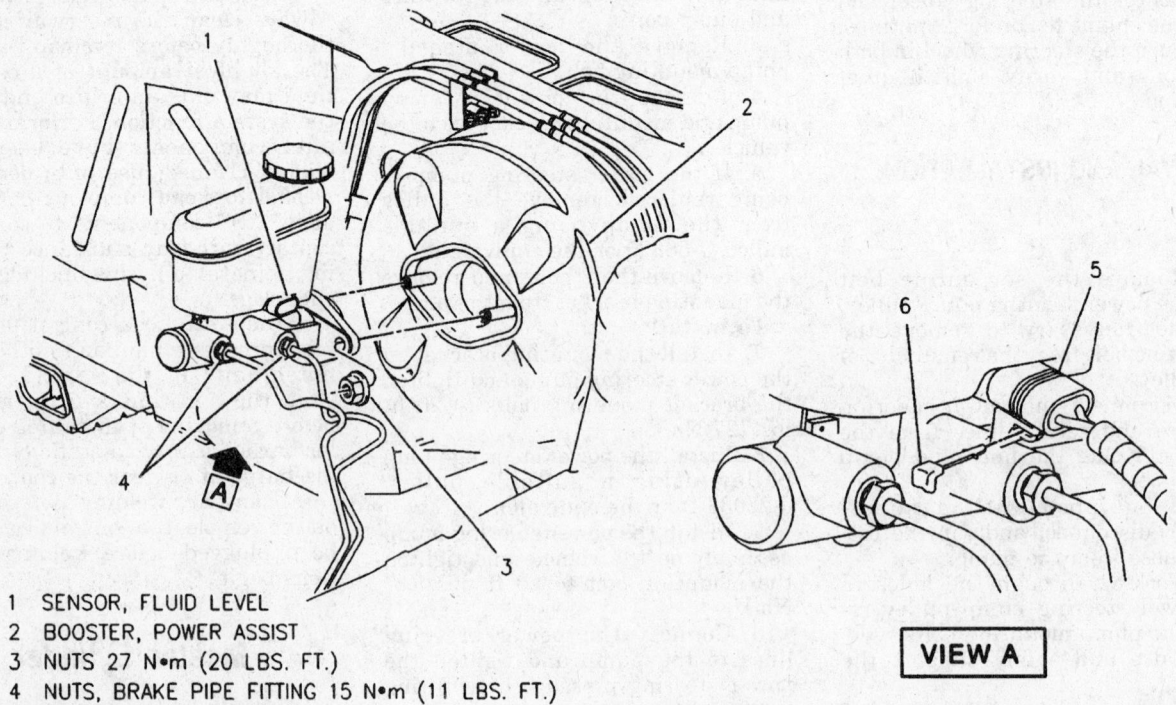

1 SENSOR, FLUID LEVEL
2 BOOSTER, POWER ASSIST
3 NUTS 27 N•m (20 LBS. FT.)
4 NUTS, BRAKE PIPE FITTING 15 N•m (11 LBS. FT.)
5 CONNECTOR, LEVEL SENSOR
6 RETAINER, CONNECTOR

VIEW A

325764

Master cylinder components and fluid level sensor with retainer

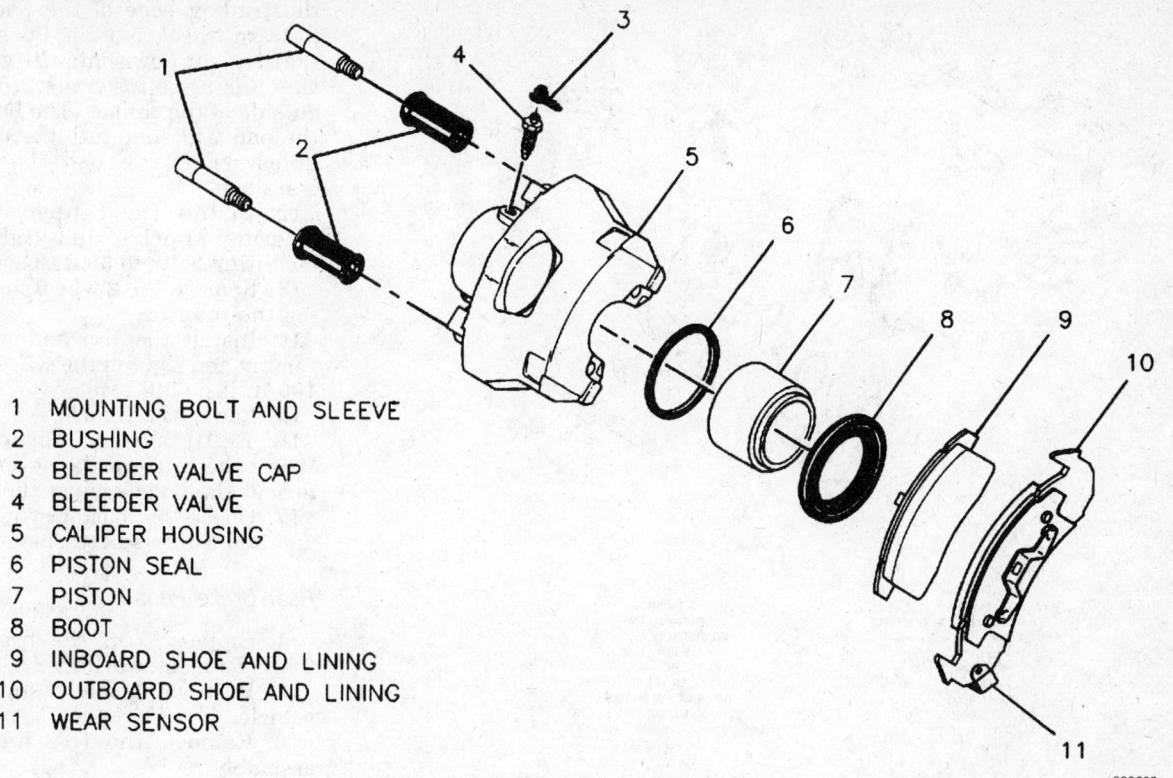

1 MOUNTING BOLT AND SLEEVE
2 BUSHING
3 BLEEDER VALVE CAP
4 BLEEDER VALVE
5 CALIPER HOUSING
6 PISTON SEAL
7 PISTON
8 BOOT
9 INBOARD SHOE AND LINING
10 OUTBOARD SHOE AND LINING
11 WEAR SENSOR

299602

Front caliper and brake pad components

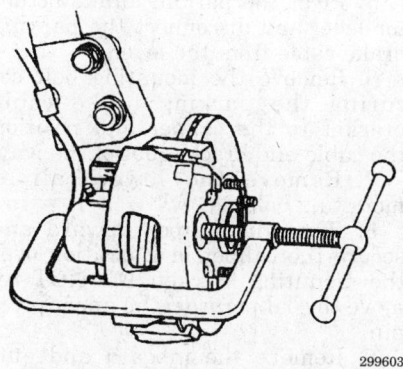

Compressing the front caliper piston

299603

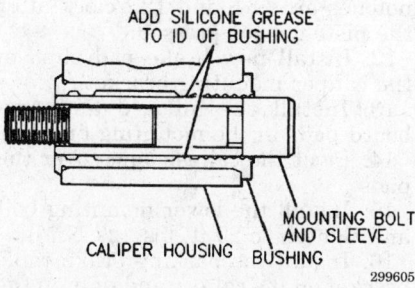

ADD SILICONE GREASE
TO ID OF BUSHING

MOUNTING BOLT
AND SLEEVE
BUSHING

CALIPER HOUSING

299605

Caliper mounting bolt and sleeve lubrication points

2. Raise and safely support the vehicle.

3. Remove the tire and wheel assembly.

4. Install 2 wheel nuts loosely to secure the rotor when the caliper is removed.

5. Remove the bolts securing the brake hose to the caliper and disconnect the brake hose from the caliper. Plug the hose to prevent excessive fluid loss and possible fluid contamination.

6. Compress the parking brake actuator lever on the caliper and disconnect the parking brake cable.

7. Remove the bolt and washer securing the parking brake cable bracket to the caliper.

8. Remove the lower caliper mounting bolt.

9. Pivot the caliper upwards until it clears the brake rotor. Push the caliper inward to remove it from the upper mounting pin.

To install:

10. Coat the inside of the caliper bushings with silicone grease.

11. Seat the caliper piston fully in its bore.

12. Make sure the notches in the caliper piston are at 6 and 12 o'clock.

13. Install the caliper onto the upper pivot pin. Turn the caliper downward until it is seated over the rotor.

14. Install the sleeve bolt and tighten to 20 ft. lbs. (27 Nm).

15. Install the parking brake cable bracket and tighten the mounting bolt to 32 ft. lbs. (43 Nm).

16. Pivot the parking brake actuator lever and connect the cable end to the actuator arm.

17. Connect the brake hose to the caliper using new washers. Tighten the mounting bolt to 33 ft. lbs. (45 Nm).

18. Remove the wheel nuts securing the rotor.

19. Install the tire and wheel assembly and tighten the wheel nuts to 100 ft. lbs. (140 Nm).

20. Refill the master cylinder with fluid and bleed the brake system.

21. Pump the brake pedal several times to seat the brake pads against the rotor. Verify no hydraulic leaks and a good firm brake pedal.

Disc Brake Pads

REMOVAL AND INSTALLATION

Front Brake Pads

1. Siphon ⅔ of the brake fluid from the master cylinder.

2. Raise and safely support the vehicle.

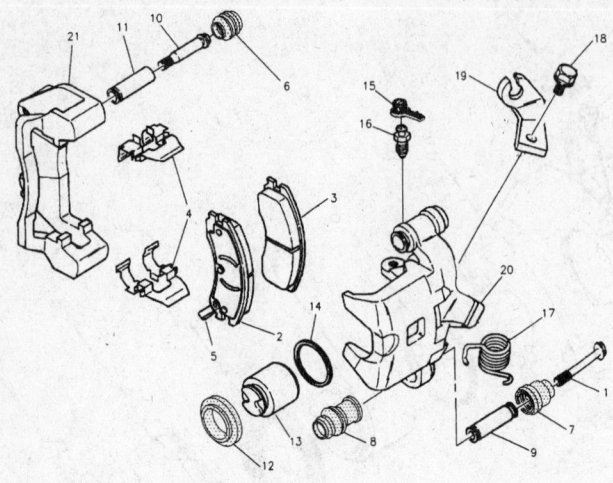

1	SLEEVE BOLT	12	PISTON BOOT
2	OUTBOARD SHOE & LINING	13	PISTON ASSEMBLY
3	INBOARD SHOE & LINING	14	PISTON SEAL
4	PAD CLIP	15	BLEEDER VALVE CAP
5	WEAR SENSOR	16	BLEEDER VALVE
6	PIN BOOT	17	LEVER RETURN SPRING
7	BOLT BOOT	18	BOLT AND WASHER
8	SLEEVE BOLT	19	CABLE SUPPORT BRACKET
9	BOLT SLEEVE	20	CALIPER BODY ASSEMBLY
10	PIN BOLT	21	CALIPER SUPPORT
11	PIN SLEEVE		

299606

Rear caliper and brake pad components

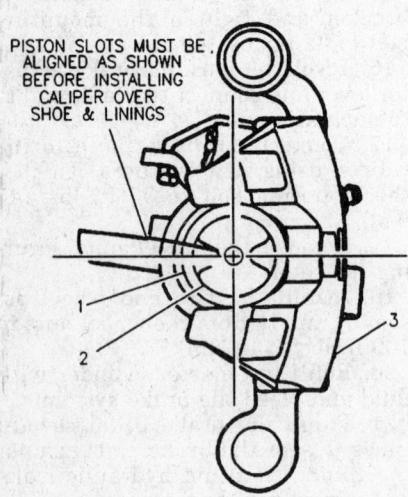

PISTON SLOTS MUST BE
ALIGNED AS SHOWN
BEFORE INSTALLING
CALIPER OVER
SHOE & LININGS

1 PISTON BOOT
2 PISTON ASSEMBLY
3 CALIPER BODY ASSEMBLY

299609

Aligning the notches in the piston

3. Remove the tire and wheel assembly.

4. Install 2 wheel nuts to secure the rotor on the hub.

5. Remove the caliper mounting bolts and sleeves.

6. Remove the caliper from the mounting bracket and support the caliper. DO NOT allow the caliper to hang unsupported from the brake hose.

7. Remove the outboard brake pad from the caliper by pushing the pad inward toward the piston to unseat the buttons on the back of pad from the holes in the caliper. Once the buttons are unseated push the pad out of the caliper.

8. Remove the inboard pad from the caliper by pulling the top of the pad away from the piston and disengaging the the spring clip from the caliper.

9. Compress the piston back into the bore using a C-Clamp.

To install:

10. Install the inboard pad into the caliper so the spring clip seats in the piston. Make sure lower edge of the spring clip is engaged in the piston and the pad is even against the base of the piston. Push the pad flat against caliper piston.

11. Install the outboard pad into the caliper so the wear indicator is at the trailing edge of the pad during forward wheel rotation. Push the pad the straight down into the caliper so the spring clips rode along the outside of the caliper. The buttons on the pad will snap into the mounting holes when the pad is correctly installed.

12. Install the caliper onto the steering knuckle and tighten the mounting bolts to 38 ft. lbs. (51 Nm).

13. Remove the 2 wheel nuts securing the rotor.

14. Install the tire and wheel assembly and tighten the wheel nuts to 100 ft. lbs. (140 Nm).

15. Lower the vehicle.

16. Refill the master cylinder. Pump the brake pedal several times to seat the pads against the rotor.

17. Check the master cylinder level and add fluid as necessary.

Rear Brake Pads

1. Siphon 2/3 of the brake fluid from the master cylinder.

2. Raise and safely support the vehicle.

3. Remove the tire and wheel assembly.

4. Install 2 wheel nuts to secure the rotor on the hub.

5. Pivot the parking brake actuator level and disconnect the parking brake cable from the lever.

6. Remove the mounting bolt securing the parking brake cable bracket to the caliper and position the cable and bracket out of the way.

7. Remove the lower caliper mounting bolt.

8. Pivot the caliper upward and secure the caliper in a position over the mounting bracket. DO NOT remove the caliper from the upper pivot pin.

9. Remove the inboard and outboard pads from the caliper bracket.

10. Remove the brake pad clips from the mounting bracket.

To install:

11. Spin the caliper piston back into the bore. Make sure the piston notches are at 6 and 12 o'clock after the piston is compressed.

12. Install new brake pad clips in the caliper mounting bracket.

13. Install the inboard and outboard pads in the mounting bracket.

14. Pivot the caliper down over the pads.

15. Install the lower mounting bolt and tighten to 20 ft. lbs. (27 Nm).

16. Install the parking brake cable bracket on the caliper and tighten the mounting bolt to 32 ft. lbs. (43 Nm).

17. Pivot the actuator lever and connect the cable end to the actuator.

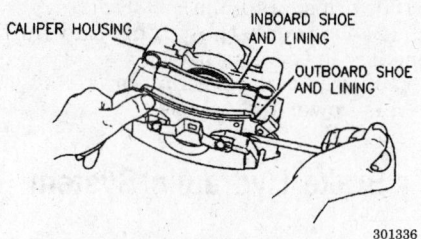

CALIPER HOUSING
INBOARD SHOE AND LINING
OUTBOARD SHOE AND LINING

301336

Removing the outboard front pad

18. Remove the 2 wheel nuts securing the rotor.
19. Install the tire and wheel assembly and tighten the wheel nuts to 100 ft. lbs. (140 Nm).
20. Lower the vehicle.
21. Refill the master cylinder. Pump the brake pedal several times to seat the pads against the rotor.
22. Check the master cylinder level and add fluid as necessary.

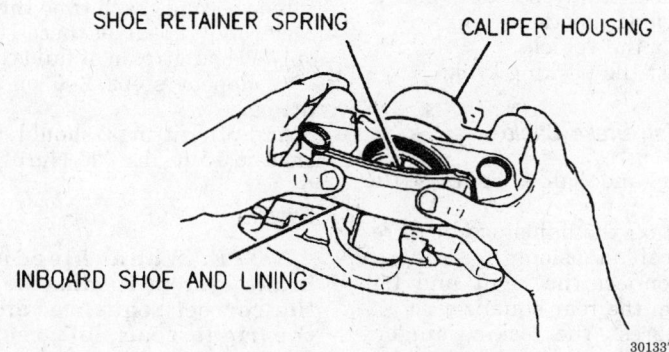

SHOE RETAINER SPRING   CALIPER HOUSING

INBOARD SHOE AND LINING

301337

Installing the inboard front pad

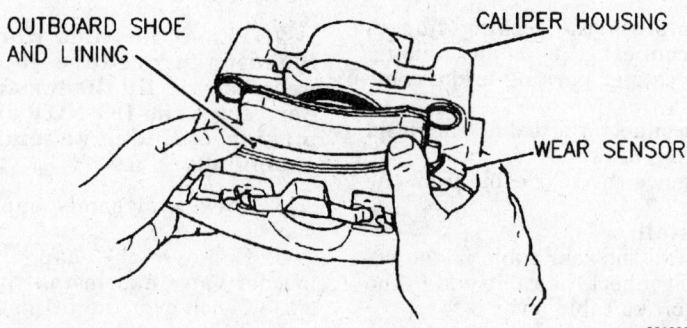

OUTBOARD SHOE AND LINING   CALIPER HOUSING

WEAR SENSOR

301338

Installing the outboard front pad

Brake Rotor

REMOVAL AND INSTALLATION

Front Rotor

1. Siphon ⅔ of the brake fluid out of the master cylinder reservoir.
2. Raise and safely support the vehicle.
3. Remove the tire and wheel assembly.
4. Compress the caliper piston back into the caliper bore.
5. Remove the caliper mounting bolts.
6. Remove the caliper from the steering knuckle and support out of the way. DO NOT disconnect the brake hose from the caliper or allow the hose to support the weight of the caliper.
7. Remove the rotor from the hub.
To install:
8. Install the rotor onto the hub.
9. Install the caliper onto the steering knuckle and tighten the caliper mounting bolts and tighten to 38 ft. lbs. (51 Nm).
10. Install the tire and wheel assembly and tighten the wheel nuts to 100 ft. lbs. (140 Nm).

11. Lower the vehicle.
12. Refill the master cylinder.
13. Pump the brake pedal several times to seat the pads against the brake rotor.
14. Check and add brake fluid to the master cylinder as necessary.

Rear Rotor

1. Siphon ⅔ of the brake fluid out of the master cylinder reservoir.
2. Raise and safely support the vehicle.
3. Remove the tire and wheel assembly.
4. Compress the parking brake actuator lever on the caliper and disconnect the parking brake cable.
5. Remove the bolt and washer securing the parking brake cable bracket to the caliper.
6. Remove the lower caliper mounting bolt.
7. Pivot the caliper upwards until it clears the brake rotor. Push the caliper inward to remove it from the upper mounting pin.
8. Support the caliper out of the way. DO NOT disconnect the brake hose from the caliper or allow the caliper to hang unsupported from the brake hose.
9. Remove the 2 bolts and remove the caliper mounting bracket.
10. Remove the rotor from the hub.
To install:
11. Compress the caliper piston slightly to provide the clearance necessary to install the caliper over the new rotor.
12. Make sure the notches in the caliper piston are at 6 and 12 o'clock.
13. Install the rotor on the hub.
14. Install the caliper mounting bracket and tighten the mounting bolts to 41 ft. lbs. (55 Nm).
15. Install the caliper onto the upper pivot pin. Turn the caliper downward until it is seated over the rotor. Install the sleeve bolt and tighten to 20 ft. lbs. (27 Nm).
16. Install the sleeve bolt and tighten to 20 ft. lbs. (27 Nm).
17. Install the parking brake cable bracket and tighten the mounting bolt to 32 ft. lbs. (43 Nm).
18. Pivot the parking brake actuator lever and connect the cable end to the actuator arm.
19. Install the tire and wheel assembly and tighten the wheel nuts to 100 ft. lbs. (140 Nm).
20. Refill the master cylinder with fluid and bleed the brake system.
21. Pump the brake pedal several times to seat the brake pads against the rotor. Verify no hydraulic leaks and a good firm brake pedal.

Parking Brake Cable

ADJUSTMENT

1. Cycle the brake system as follows:

a. Apply the service brake with a force of 150 lbs. (660 N).

b. Fully apply the parking brake using 125 lbs. (550 N) pedal force and then release. Repeat this 2 more times.

c. Make sure the parking brake light illuminates after 2 clicks.

d. Apply the parking brake and drive the vehicle about fifty feet and verify the parking brake chime sounds.

2. Release the parking brake and verify the light is off and the chime is not sounding.

3. Raise and safely support the vehicle.

4. Check the cables at the rear calipers and make sure the levers are seated against the caliper stops.

5. Tighten the parking brake cable adjuster until either of the levers comes off the stop.

6. Loosen the adjuster until the levers are both seated against the stops.

7. Apply the parking brake several times to ensure the pedal engages before completion of a full stroke.

8. Lower the vehicle.

REMOVAL AND INSTALLATION

Front Parking Brake Cable

1. Pull back the edge of the driver's side carpet enough to access the grommet where it passes through the floor.

2. Unseat the grommet from the floor pan.

3. Raise and safely support the vehicle.

4. Disconnect the front cable housing from the intermediate cable adjuster.

5. Lower the vehicle.

6. Pull the cable slack and disconnect the cable end from the parking brake lever assembly.

7. Compress the locking fingers on the cable end and disconnect the cable housing from the parking brake lever.

8. Remove the cable from the vehicle.

To install:

9. Position the cable in the vehicle and seat the grommet in the floor pan.

10. Connect the cable housing to the parking brake lever assembly

and make sure the locking fingers are seated.

11. Connect the cable end to the pivot on the lever assembly.

12. Raise and safely support the vehicle.

13. Connect the cable housing to the intermediate cable adjuster.

14. Lower the vehicle.

15. Install the carpet back in place.

16. Adjust the parking brake.

Intermediate Parking Brake Cable

1. Raise and safely support the vehicle.

2. Disconnect the intermediate cable from the adjuster assembly.

3. Disconnect the intermediate cable housing from the front cable housing retaining bracket.

4. Disconnect the cable from the rear equalizer.

5. Disconnect the intermediate cable housing from the rear housing retaining bracket.

6. Remove the intermediate cable from the vehicle.

To install:

7. Install the intermediate cable in the vehicle and seat the locking fingers in the front and rear cable housing brackets.

8. Connect the cable end to the rear equalizer.

9. Connect the front cable end to the front cable junction.

10. Lower the vehicle.

11. Adjust the parking brake.

Rear Parking Brake Cable

1. Raise and safely support the vehicle.

2. Back off the adjuster nut to release the cable tension.

3. Disconnect the right and left cables from the rear equalizer.

4. Compress the locking fingers and disconnect the cable housing from the backing plate.

5. Remove the tire and wheel assembly.

6. Compress the locking fingers and disconnect the cable housing from the caliper parking brake cable bracket.

7. Disconnect the cable end from the parking brake lever.

8. Remove the rear cable from the vehicle.

To install:

9. Install the rear cable in the vehicle and connect the cable end to the parking brake cable lever.

10. Engage the locking fingers in the caliper parking brake cable bracket and backing plate.

11. Install the tire and wheel assembly and tighten the wheel nuts to 100 ft. lbs. (140 Nm).

12. Connect the right and left cables to the equalizer.

13. Adjust the parking brake.

14. Lower the vehicle.

Brake Hydraulic System

BLEEDING

Master Cylinder

1. Fill the master cylinder to the **FULL** mark.

NOTE: At anytime during the bleeding procedure it may be necessary to fill the master cylinder reservoir. DO NOT allow the master cylinder to empty out completely.

2. Loosen the forward brake pipe and allow brake fluid to flow from the connection.

3. Tighten the fitting to 15 ft. lbs. (21 Nm).

4. Have an assistant depress the brake pedal one time and hold it down.

5. Loosen the fitting allowing the air to escape. Once the pedal is all the way down tighten the fitting. Wait fifteen seconds each time through the cycle and repeat as necessary until only a clear stream of fluid comes out.

6. Repeat steps 2–5 on the rear fitting.

7. Both fittings should be tightened to 15 ft. lbs. (20 Nm).

System

NOTE: When bleeding the brake system it must be bled in the correct sequence. Start with the right rear, left rear, right front and then the left front.

1. Fill the master cylinder to the **FULL** mark.

NOTE: At anytime during the bleeding procedure it may be necessary to fill the master cylinder reservoir. DO NOT allow the master cylinder to empty out completely.

2. Raise and safely support the vehicle.

3. Remove the cap from the bleeder valve and install the correct size wrench over the fitting.

4. Connect a clear hose to the fitting and place the other end of the hose in a container.

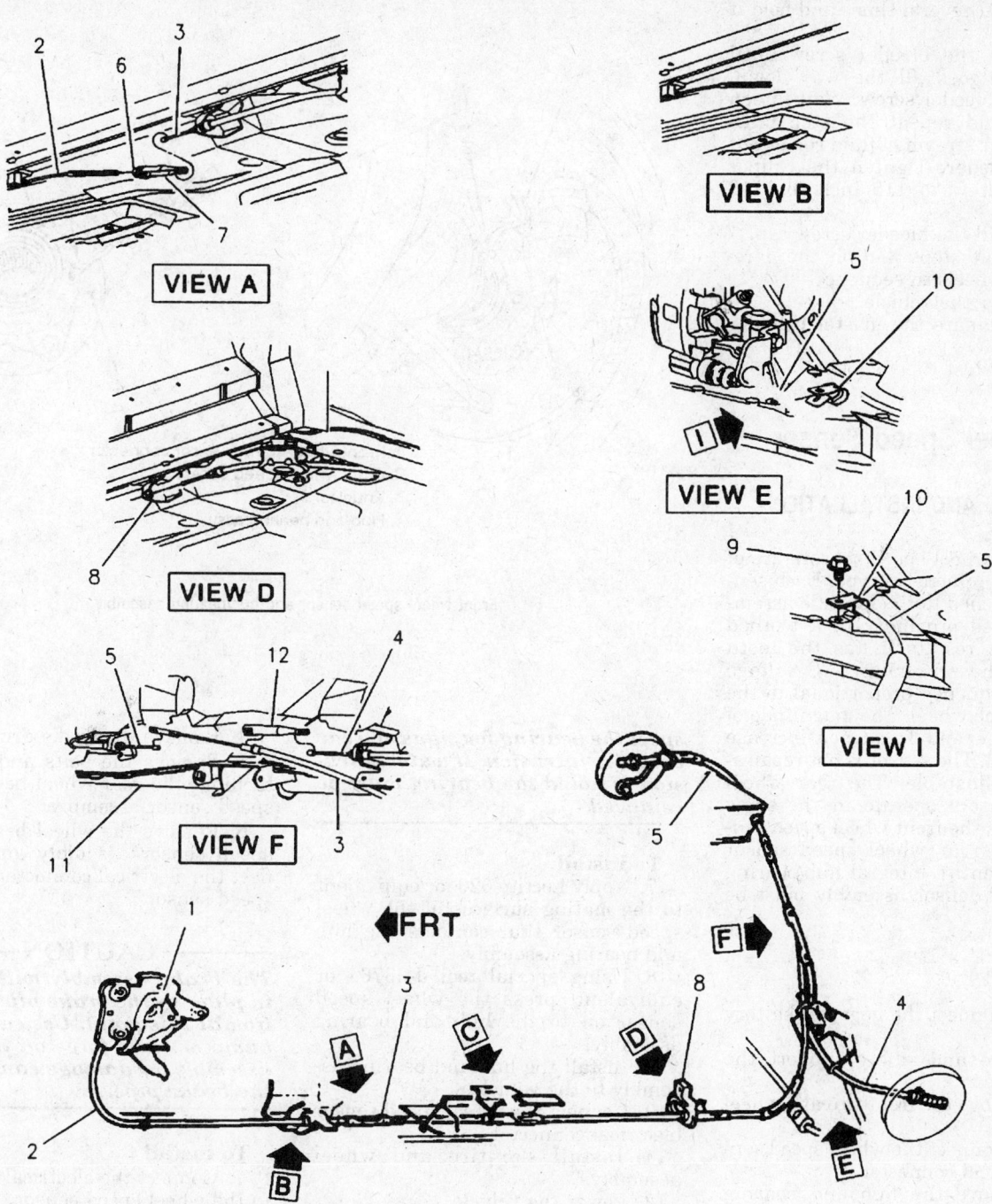

VIEW A

VIEW B

VIEW D

VIEW E

VIEW I

VIEW F

FRT

1 PEDAL ASSEMBLY	8 BRACKET
2 FRONT CABLE	9 SCREW, 18 N•m (13 LB. FT.)
3 INTERMEDIATE CABLE	10 CABLE GUIDE
4 LEFT REAR CABLE	11 REAR SUSPENSION SUPPORT
5 RIGHT REAR CABLE	12 EQUALIZER
6 ADJUSTER NUT	
7 ADJUSTER	

311502

Parking brake cable routing and junction points

5. Have an assistant pump the brake pedal several times and hold it down.

6. Open the bleeder screw until the pedal goes all the way down. Close the bleeder screw. Wait fifteen seconds and repeat the step until only a clear stream of fluid comes out of the bleeder. Tighten the caliper bleeder screw to 115 inch lbs. (13 Nm).

7. Install the bleeder screw cap.

8. Repeat steps 4–8 on the 3 remaining wheels in sequence.

9. Lower the vehicle.

10. Make sure the master cylinder is full.

11. Road test the vehicle.

Wheel Speed Sensor

REMOVAL AND INSTALLATION

The front wheel speed sensors are of a variable reluctance type. Each sensor is attached to the knuckle assembly in close proximity to a toothed ring. This results in (as the teeth pass by the sensor) an A/C voltage with a frequency proportional to the speed of the wheel. The magnitude of the voltage and frequency increase with speed. The sensor is not repairable, or adjustable. The rear wheel speed sensors operate in the same manner as the front wheel speed sensors. If a rear wheel speed sensor fails, the entire integral hub/bearing and speed sensor assembly must be replaced.

Front Wheel

1. Disconnect the negative battery cable.

2. Raise and safely support the vehicle.

3. Remove the front tire and wheel assembly.

4. Disconnect the wheel speed sensor electrical connector.

5. Remove the hub and bearing assembly.

6. Separate the wheel speed sensor from the hub and bearing assembly, using a blunt screwdriver.

—— **CAUTION** ——

When the wheel speed sensor is removed it must be replaced. The speed sensor is damaged when it is removed and cannot be reused. There are 2 parts to the wheel speed sensor, and these are replaced as an assembly. They cannot be replaced individually. In-

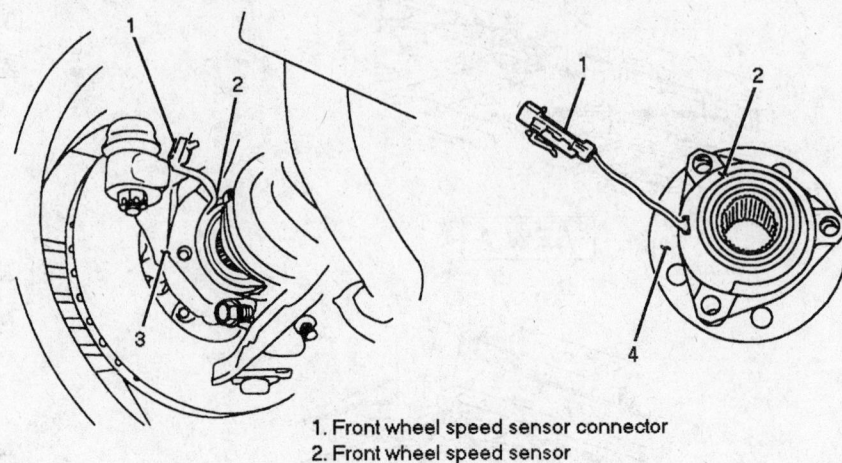

1. Front wheel speed sensor connector
2. Front wheel speed sensor
3. Knuckle
4. Hub and bearing assy

326497

Front wheel speed sensor and hub/bearing assembly

spect the bearing for signs of wear or water intrusion. If water intrusion is noted the bearing must be replaced.

To install:

7. Apply Loctite 620 or equivalent to the mating surfaces of the wheel speed sensor that contacts the hub and bearing assembly.

8. Using special tool J-38764 or equivalent, press the wheel speed sensor on to the hub and bearing assembly.

9. Install the hub and bearing assembly to the vehicle.

10. Connect the wheel speed sensor electrical connector.

11. Install the tire and wheel assembly.

12. Lower the vehicle.

13. Connect the negative battery cable.

14. Road test the vehicle and verify proper operation.

Rear Wheel

1. Disconnect the negative battery cable.

2. Raise and safely support the vehicle.

3. Remove the tire and wheel assembly.

4. Remove the brake drum.

5. Remove the bolts and nuts attaching the rear wheel bearing and speed sensor assembly.

6. Remove the wheel bearing and speed sensor assembly and disconnect the electrical connector from the speed sensor.

—— **CAUTION** ——

The brake assembly will be held in place by the brake pipe connection at this point. Use care not to bump or exert force on the brake assembly, or damage can occur to the brake pipe.

To install :

7. Connect the electrical connector to the wheel speed sensor.

8. Install the wheel bearing and speed sensor assembly into place.

9. Install the mounting bolts and nuts and tighten to 46 ft. lbs. (63 Nm).

10. Install the brake drum.

11. Install the tire and wheel assembly.

12. Lower the vehicle.

13. Connect the negative battery cable.

14. Road test the vehicle and verify proper operation.

23. Check the front end alignment and adjust as necessary.

24. Connect the negative battery cable.

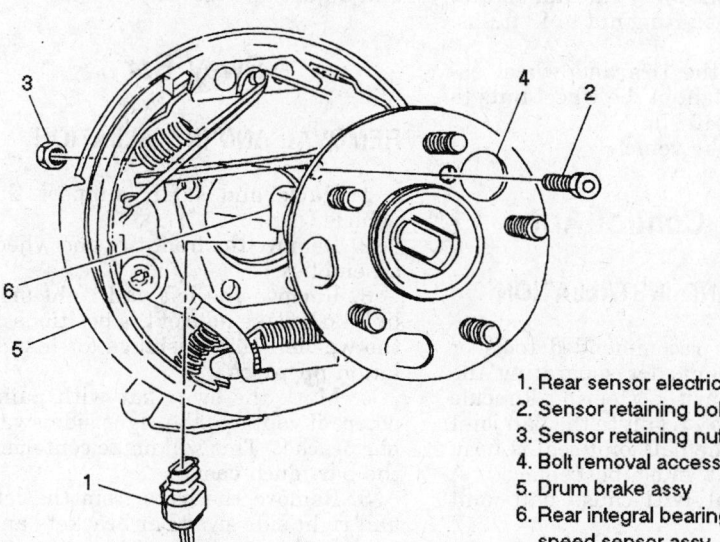

1. Rear sensor electrical connector
2. Sensor retaining bolts (4)
3. Sensor retaining nuts (4)
4. Bolt removal access hole
5. Drum brake assy
6. Rear integral bearing and speed sensor assy

326499

Rear wheel speed sensor and hub/bearing assembly

FRONT SUSPENSION

Strut and Spring

REMOVAL AND INSTALLATION

1. Disconnect the negative battery cable.

2. Raise and safely support the vehicle.

3. Remove the tire and wheel assembly.

4. Disconnect the ABS wheel speed sensor.

5. Remove the ABS speed sensor bracket from the strut.

6. If removing the left side strut, disconnect the brake line bracket from the strut.

7. Support the steering knuckle so when the strut is disconnected the brake line does not get stretched.

8. Scribe a mark on the strut referencing the lower strut bracket to the steering knuckle.

9. Remove the nut and through bolts from the lower strut bracket.

10. From under the hood, remove the 3 upper strut plate mounting nuts and washers,

11. Remove the strut from the vehicle.

12. Place the strut in an approved fixture to disassemble the coil spring from the strut. With the spring compressed, hold the strut shaft from turning using a socket and remove the 24mm strut shaft nut. Relieve the pressure on the spring and separate the front coil spring from the strut.

To install:

13. If the spring and strut were disassembled, assemble using an approved fixture.

14. Position the strut in the vehicle guiding the upper strut plate studs into the body and loosely install the nuts.

15. Install the lower strut through bolts. With the matchmarks in alignment, tighten the nuts to 136 ft. lbs. (185 Nm).

16. Remove the support from the steering knuckle.

17. Connect the brake line to the strut, if removed.

18. Install the speed sensor bracket on the strut.

19. Connect the ABS sensor.

20. Install the tire and wheel assembly and tighten the wheel nuts to 100 ft. lbs. (140 Nm).

21. Lower the vehicle.

22. Tighten the upper strut plate mounting nuts to 35 ft. lbs. (47 Nm).

Lower Ball Joints

REMOVAL AND INSTALLATION

Ball joints must be replaced if any looseness is detected in the joint or if the ball joint seal is cut. To inspect the ball joints, raise the front of the vehicle allowing the suspension to hang free. Grasp the tire at the top and bottom and move the top of the tire in an in-And-out motion. Check for any horizontal movement of the knuckle relative to the control arm. If movement is in the wheel bearing, the bearing and hub must be replaced. If the ball joint stud is disconnected from the knuckle and looseness can be detected or if the ball stud can be twisted in its socket using finger pressure, replace the ball joint.

Ball joint tightness in the knuckle boss should also be checked. This may be done by shaking the wheel and feeling for movement of the stud end or nut at the knuckle boss. Worn or damaged ball joints and knuckles must be replaced.

1. Raise and safely support the vehicle.

2. Remove the front tire and wheel assembly.

3. Remove the sway bar link kit. Take note of the positions of the washers and insulators for installation purposes.

4. Remove the cotter pin and castle nut from the lower ball joint and using a suitable ball joint separator, J-36226 or the equivalent separate the ball joint from the steering knuckle.

5. Using a ½ inch drill bit, drill the heads off of the 3 rivets securing the lower ball joint to the control arm. With the rivet heads removed, drive the bodies of the rivets out of the control arm using a suitable punch.

6. Remove the ball joint.

To install:

7. Position the lower ball joint on the control arm and install the 3 mounting bolts so the bolt threads point down. Install the mounting nuts and tighten to 50 ft. lbs. (68 Nm).

8. Connect the lower ball joint to the steering knuckle and tighten the castle nut to 41 ft. lbs. (55 Nm). If necessary tighten the castle nut up to 60 degrees (⅙ turn) additional rota-

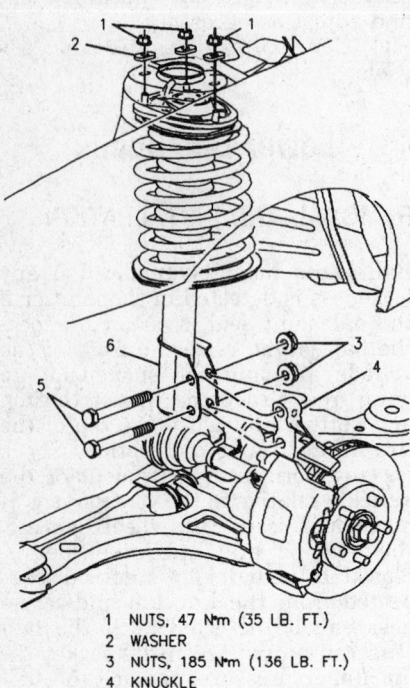

1 NUTS, 47 N•m (35 LB. FT.)
2 WASHER
3 NUTS, 185 N•m (136 LB. FT.)
4 KNUCKLE
5 BOLT
6 STRUT

304352

Upper and lower strut mounting components

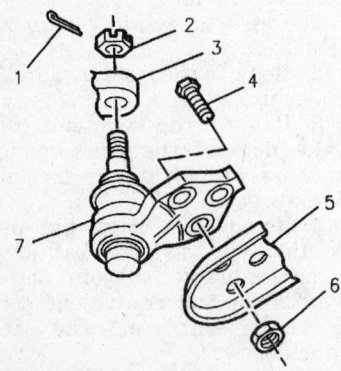

1 PIN
2 NUT, BALL JOINT TO KNUCKLE;
  TIGHTEN TO 10 N•m (88 LB. IN.)
  THEN TIGHTEN 2 FLATS TO
  55 N•m (41 LB. FT.), MIN.
3 KNUCKLE
4 BALL JOINT MOUNTING BOLTS MUST
  FACE DOWN
5 CONTROL ARM
6 BALL JOINT MOUNTING NUTS
  68 N•m (50 LB. FT.)
7 SERVICE BALL JOINT

305250

Lower ball joint and replacement mounting bolts and nuts

tion to align the cotter pin holes. NEVER loosen the nut to make the alignment. Install a new cotter pin.

9. Install the sway bar link kit and tighten the mounting nut to 13 ft. lbs. (17 Nm).

10. Install the tire and wheel assembly and tighten the wheel nuts to 100 ft. lbs. (140 Nm).

11. Lower the vehicle.

Lower Control Arms

REMOVAL AND INSTALLATION

Use only the recommended tools or their equivalents for separating the ball joint from the steering knuckle or damage may result to the ball joint and seal. If the ball joint seal is torn, the ball joint must be replaced. A damages seal **will** cause ball joint failure.

1. Raise and safely support the vehicle.

2. Remove the front tire and wheel assembly.

3. Remove the sway bar link kit. Take note of the positions of the washers and insulators for installation purposes.

4. Remove the cotter pin and castle nut from the lower ball joint and using a suitable ball joint separator, J-36226 or the equivalent separate the ball joint from the steering knuckle.

5. Working at the inboard end of the control arm remove the 2 control arm mounting nuts and through bolts and remove the lower control arm.

To install:

6. Position the control arm on the frame and loosely install the mounting nut and bolts. DO NOT tighten the nuts and bolts at this time, the weight of the vehicle must be on the suspension to ensure proper control arm alignment.

7. Connect the lower ball joint to the steering knuckle and tighten the castle nut to 41 ft. lbs. (55 Nm). If necessary tighten the castle nut up to 60 degrees (1/6 turn) additional rotation to align the cotter pin holes. NEVER loosen the nut to make the alignment. Install a new cotter pin.

8. Install the sway bar link kit and tighten the mounting nut to 13 ft. lbs. (17 Nm).

9. Install the tire and wheel assembly and tighten the wheel nuts to 100 ft. lbs. (140 Nm).

10. Lower the vehicle.

11. Tighten the vertical mounting nut (front) to 93 ft. lbs. (126 Nm) and

the horizontal mounting nut (rear) to 117 ft. lbs. (158 Nm).

12. Check the front end alignment and adjust as necessary.

Sway Bar

REMOVAL AND INSTALLATION

1. Raise and safely support the vehicle.

2. Remove the front tire and wheel assemblies.

3. Remove the left and right side link kits. Take note of the positions of the washers and bushings for installation purposes.

4. Mark the sway bar with paint where it contacts the right side sway bar bracket. This will make centering the bar much easier.

5. Remove the bolts from the left and right side sway bar brackets and remove the brackets.

6. Remove the cotter pin and castle nut from the left outer tie rod end and using J-24319-B, separate the tie rod end from the steering knuckle.

7. Disconnect the exhaust pipe from the exhaust manifold and disconnect the intermediate pipe hangers. Lower the exhaust.

8. Turn the left strut completely to the right and slide the sway bar out the opening and over the left steering knuckle until the right end clears the right side suspension member.

9. Remove the sway bar out the bottom of the vehicle, between the suspension rails.

To install:

10. Install the sway bar up from the bottom on the vehicle. The left side of the bar should protrude from the left wheel opening until the right side of the bar clears the suspension member.

11. Loosely install the sway bar link kits.

12. Loosely install the sway bar bushings and brackets to the frame. Make sure the mark on the bar lines up with the right side frame bracket.

13. Connect the left outer tie rod end to the steering knuckle and tighten the castle nut to 35 ft. lbs. (48 Nm). If necessary to align the cotter pin holes tighten the castle nut up to 60 degrees (1/6 turn) additional rotation. NEVER loosen the nut to make the alignment. Install a new cotter pin.

14. Connect the exhaust pipe to the exhaust manifold and tighten the mounting nuts to 18 ft. lbs. (25 Nm).

15. Install the intermediate pipe exhaust hangers.

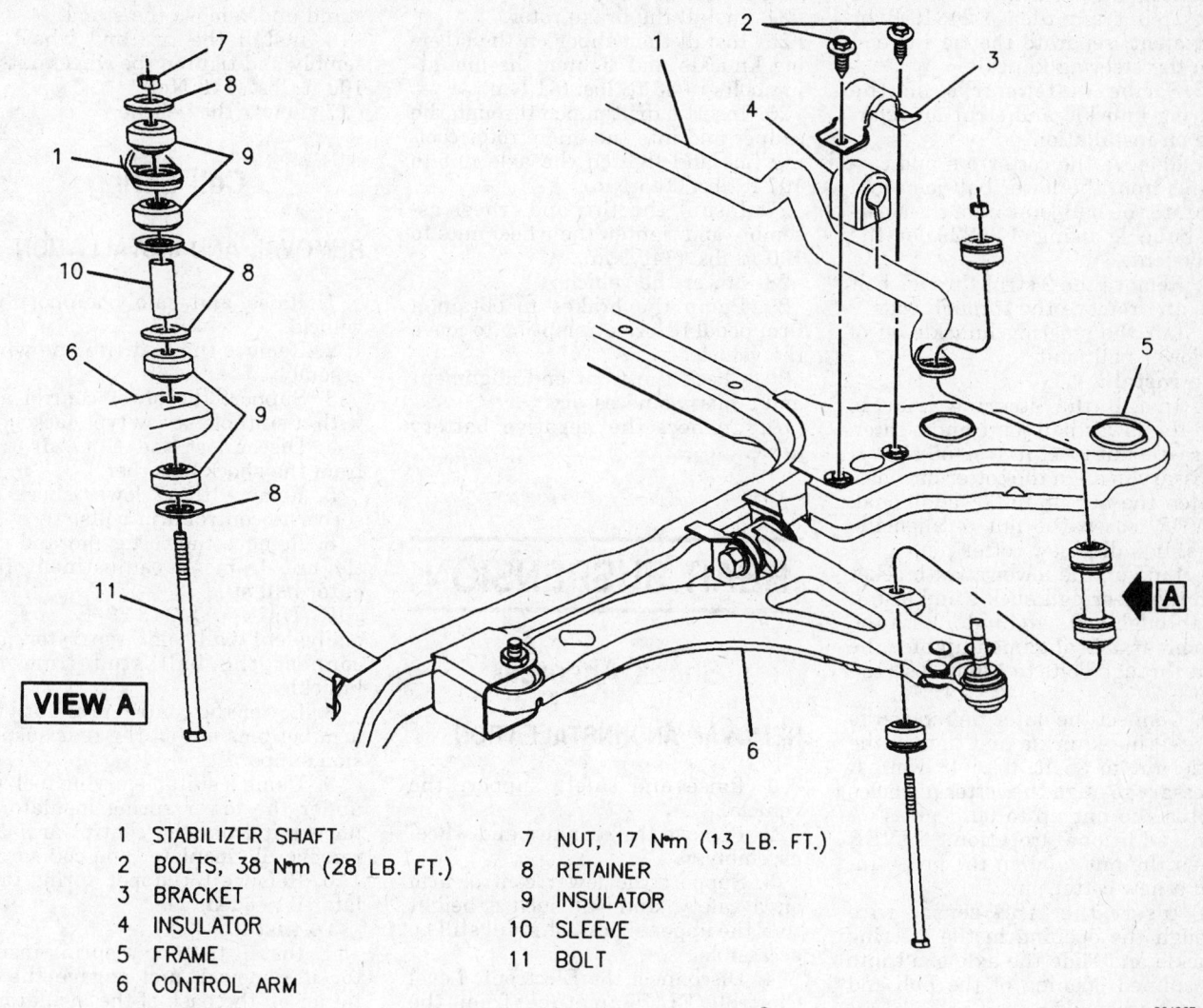

1   STABILIZER SHAFT
2   BOLTS, 38 N·m (28 LB. FT.)
3   BRACKET
4   INSULATOR
5   FRAME
6   CONTROL ARM

7   NUT, 17 N·m (13 LB. FT.)
8   RETAINER
9   INSULATOR
10 SLEEVE
11 BOLT

304967

Front sway bar mounting components

16. Tighten the sway bar frame bracket bolts to 28 ft. lbs. (38 Nm) and the link kit nuts to 13 ft. lbs. (17 Nm).

17. Install the front tire and wheel assemblies and tighten the wheel nuts to 100 ft. lbs. (140 Nm).

18. Lower the vehicle.

Front Wheel Bearings

REMOVAL AND INSTALLATION

The front wheel bearings are not serviced separately. If the front wheel bearings are defective, the hub and bearing assembly must be replaced.

1. Disconnect the negative battery cable.

2. Raise and safely support the vehicle.

3. Remove the tire and wheel assembly.

4. Lubricate the threads on the drive axle with clean engine oil.

5. Install a drift punch through the caliper and into the brake rotor cooling fins. This keeps the hub from turning when removing the hub nut. Remove the hub nut from the drive axle.

6. Remove the caliper mounting bolts and remove the caliper from the steering knuckle. Support the caliper out of the way. DO NOT allow the brake hose to support the weight of the caliper.

7. Remove the brake rotor.

8. Disconnect the ABS sensor and unclip the sensor from the backing plate.

9. Remove the 3 hub and bearing mounting bolts and remove the backing plate.

10. Using axle puller J-28733 or equivalent hub puller, separate the hub and bearing assembly from the halfshaft.

11. Remove the hub and bearing assembly.

12. If the steering knuckle is also to be removed, remove the cotter pin and castle nut from the outer tie rod end. Using a puller J-24319-B or equivalent, separate the tie rod end from the steering knuckle.

13. Scribe matchmarks on the steering knuckle and strut for reference on installation.

14. Remove the cotter pin and castle nut from the lower ball joint and separate the ball joint from the steering knuckle using J-36226, or the equivalent.

15. Remove the 2 strut through bolt nuts and remove the through bolts.

16. Lift the steering knuckle off of the lower ball joint.

To install:

17. Install the steering knuckle onto the lower ball joint and tighten the castle nut to 41 ft. lbs. (55 Nm). If necessary to align the cotter pin holes tighten the nut up to 60° additional. NEVER loosen the nut to align the holes. Install a new cotter pin.

18. Line up the lower strut bracket with the steering knuckle and install the through bolts and nuts. With the matchmarks in alignment tighten the strut through bolts to 136 ft. lbs. (185 Nm).

19. Connect the outer tie rod end to the steering knuckle and tighten the castle nut to 35 ft. lbs. (48 Nm). If necessary to align the cotter pin holes tighten the nut up to 60 degrees (¹⁄₆ turn) additional rotation. NEVER loosen the nut to align the holes. Install a new cotter pin.

20. Insert the ABS sensor wire through the opening in the steering knuckle and slide the axle shaft into the splined opening of the hub and bearing assembly.

21. Install a new axle nut and draw the hub and bearing assembly in to place.

22. Install the backing plate and install the 3 mounting bolts. Tighten the bolts alternately and evenly to 70 ft. lbs. (95 Nm).

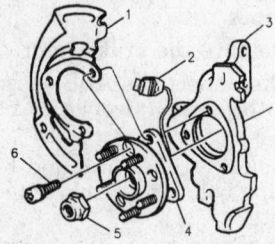

1 DUST SHIELD
2 WHEEL SPEED
  SENSOR CONNECTOR
3 STEERING KNUCKLE
4 HUB AND BEARING
5 NUT, DRIVE AXLE,
  145 Nm (107 LB. FT.)
6 RETAINING BOLT,
  95 Nm (75 LB. FT.)

304847

Front hub and bearing components

23. Connect the ABS sensor to the backing plate and connect the wiring harness connector.

24. Install the brake rotor.

25. Install the caliper on the steering knuckle and tighten the mounting bolts to 38 ft. lbs. (51 Nm).

26. Insert a drift punch through the caliper and into the brake rotor cooling fins and tighten the axle nut to 107 ft. lbs. (145 Nm).

27. Install the tire and wheel assembly and tighten the wheel nuts to 100 ft. lbs. (140 Nm).

28. Lower the vehicle.

29. Pump the brakes to obtain a firm pedal before attempting to move the vehicle.

30. Check the front end alignment and adjust as necessary.

31. Connect the negative battery cable.

REAR SUSPENSION

Shock Absorber

REMOVAL AND INSTALLATION

1. Raise and safely support the vehicle.

2. Remove the rear tire and wheel assembly.

3. Support the lower control arm on a safety stand at such a height that the upper shock bolts will still be accessible.

4. Disconnect the Electronic Level Control (ELC) air tube from the shock.

5. Remove the 2 bolts from under the control arm securing the lower shock mount.

6. From inside the trunk, remove the trunk trim panel to access the upper shock mount bolts.

7. Remove the upper shock cap.

8. Remove the 2 nuts from the upper shock mount and remove the reinforcement.

9. Remove the shock from the vehicle.

To install:

10. Install the shock in the vehicle.

11. Install the reinforcement and upper shock mounting nuts. Tighten the mounting nuts to 15 ft. lbs. (20 Nm).

12. Install the upper shock cap and reposition the inner trunk trim.

13. Extend the shock and install the lower shock mounting bolts and tighten to 18 ft. lbs. (24 Nm).

14. Connect the ELC air tube to the shock.

15. Raise the vehicle off the safety stand and remove the stand.

16. Install the tire and wheel assembly and tighten the wheel nuts to 100 ft. lbs. (140 Nm).

17. Lower the vehicle.

Coil Spring

REMOVAL AND INSTALLATION

1. Raise and safely support the vehicle.

2. Remove the rear tire and wheel assembly.

3. Support the lower control arm with a suitable screw type jack.

4. Disconnect the ELC air tube from the shock absorber.

5. Remove the 2 lower shock absorber-to-control arm bolts.

6. Remove the cotter pin and castle nut from the adjustment link outer ball stud.

7. Using J-24319-B, or an equivalent ball joint separator, disconnect the ball stud from the knuckle.

8. Lower the control arm until the arm bottoms out on the rear suspension support.

9. Using a suitable prying tool, pry under the lower spring insulator to unseat it from the control arm and remove the insulator and coil spring.

10. Remove the upper spring insulator if needed.

To install:

11. Install the upper spring insulator if removed, and engage the retainer on the back of the insulator in the upper mount hole.

12. Install the coil spring and lower insulator in the vehicle and seat the lower insulator in the control arm hole.

13. Raise the control arm until the shock bolts can be installed. Tighten the 2 bolts to 18 ft. lbs. (24 Nm).

14. Remove the jack.

15. Connect the adjustment link ball stud to the knuckle and tighten the castle nut to 88 inch lbs. (10 Nm) plus 180 degrees (¹⁄₂ turn) additional rotation. If necessary to align the cotter pin holes tighten the nut up to 60 degrees (¹⁄₆ turn) more. NEVER loosen the castle nut to align the cotter pin holes.

16. Install a new cotter pin.

17. Connect the ELC air tube to the shock absorber.

18. Install the tire and wheel assembly and tighten the wheel nuts to 100 ft. lbs. (140 Nm).

19. Lower the vehicle.

Lower Control Arms

REMOVAL AND INSTALLATION

1. Raise and safely support the vehicle.
2. Remove the rear tire and wheel assemblies.
3. Remove the rear section of the exhaust system.
4. Support the lower control arm with a suitable screw type jack.
5. Disconnect the ELC air tube from the shock absorber.
6. Remove the 2 lower shock absorber-to-control arm bolts.
7. Remove the cotter pin and castle nut from the adjustment link outer ball stud.
8. Using J-24319-B, or an equivalent ball joint separator, disconnect the ball stud from the knuckle.
9. Lower the control arm until the arm bottoms out on the rear suspension support.
10. Using a suitable prying tool, pry under the lower spring insulator to unseat it from the control arm and remove the insulator and coil spring.
11. Remove the brake caliper mounting bracket bolts and remove the caliper and bracket assembly and support out of the way. DO NOT allow the brake hose to support the caliper and bracket assembly.
12. Remove the brake rotor.
13. Disconnect the parking brake cables from the rear calipers.
14. Loosen the parking brake cable then disconnect the cable at the rear suspension support assembly.
15. Disconnect the electrical connector attached to the suspension support.
16. Disconnect the ELC electrical connector and vent hose.
17. Disconnect the ELC air tube from the ELC compressor.
18. Support the rear suspension assembly with a transmission jack.
19. Remove the 3 bolts per side securing the support assembly to the vehicle.
20. Remove the 2 front and 2 rear bolts and lower the suspension assembly.
21. Remove the sway bar link kit.
22. If the left control arm is being removed, disconnect the ELC height sensor link.
23. Disconnect the ABS sensor connector.

24. Remove the 4 bolts and remove the hub and bearing assembly along with the backing plate.
25. Remove the control arm mounting nuts and through bolts and remove the control arm from the suspension support.

To install:

26. Install the control arm into the suspension support and install the through bolts. The bolts must be installed pointing in toward each other. Loosely install the mounting nuts.
27. Install the hub and bearing assembly and backing plate and tighten the mounting bolts to 52 ft. lbs. (70 Nm).
28. Connect the ABS sensor electrical connector.
29. Install the sway bar link kit and tighten the mounting bolt to 13 ft. lbs. (17 Nm).
30. Connect the ELC height sensor link to the left control arm.
31. Raise the suspension assembly on the transmission jack into position on the underside of the vehicle.
32. Install all of the mounting bolts loosely. Tighten the front mounting bolts to 141 ft. lbs. (191 Nm), the rear mounting bolts to 122 ft. lbs. (162 Nm) and the 6 support bracket bolts to 63 ft. lbs. (86 Nm).
33. Connect the ELC air tube to the ELC compressor.
34. Connect the ELC electrical connector and vent tube.
35. Connect the electrical harness connectors at the support assembly.
36. Connect the parking brake cable to the support assembly.
37. Connect the parking brake cables to the calipers.
38. Install the brake rotors.
39. Install the brake calipers and mounting brackets and tighten the bracket mounting bolts to 35 ft. lbs. (47 Nm).
40. Adjust the parking brake.
41. Install the coil spring and lower insulator in the vehicle and seat the lower insulator in the control arm hole.
42. Raise the control arm until the shock bolts can be installed. Tighten the 2 bolts to 18 ft. lbs. (24 Nm).
43. Remove the jack.
44. Connect the adjustment link ball stud to the knuckle and tighten the castle nut to 88 inch lbs. (10 Nm) plus 180 degrees (½ turn) additional rotation. If necessary to align the cotter pin holes tighten the nut up to 60 degrees (⅙ turn) more. NEVER loosen the castle nut to align the cotter pin holes.

45. Install a new cotter pin.
46. Connect the ELC air tube to the shock absorber.
47. Install the rear section of the exhaust.
48. Install the tire and wheel assemblies and tighten the wheel nuts to 100 ft. lbs. (140 Nm).
49. Lower the vehicle.
50. With the vehicle at normal ride height tighten the rear control arm mounting nuts to 78 ft. lbs. (106 Nm).
51. Check the rear alignment and adjust as necessary.

Sway Bar

REMOVAL AND INSTALLATION

1. Disconnect the negative battery cable.
2. Raise and safely support the vehicle.
3. Remove the rear tire and wheel assemblies.
4. Disconnect the ELC (Electronic Level Control) height sensor link from the lower control arm.
5. Remove the 2 bolts securing the ELC height sensor, and position the sensor aside for clearance.
6. Remove the sway bar link kits. Take note of the positions of the washers and insulators for installation.
7. Remove the sway bar clamp bolts.
8. Bend open the sway bar clamps and remove the sway bar and bushings.

To install:

9. Install the sway bar bushings on the sway bar with the slits in the bushing facing forward.
10. Install the sway bar assembly in the vehicle and bend the sway bar clamps closed. Loosely install the clamp bolts.
11. Install the link kits and tighten the bolts to 115 inch lbs. (13 Nm).
12. Make sure the sway bar is centered in the vehicle and tighten the clamp bolts to 24 ft. lbs. (33 Nm).
13. Position the ELC height sensor and tighten the 2 mounting bolts to 62 inch lbs. (7 Nm).
14. Connect the ELC height sensor link to the control arm.
15. Install the wheel and tire assemblies and tighten the wheel nuts to 100 ft. lbs. (140 Nm).
16. Lower the vehicle.
17. Connect the negative battery cable.

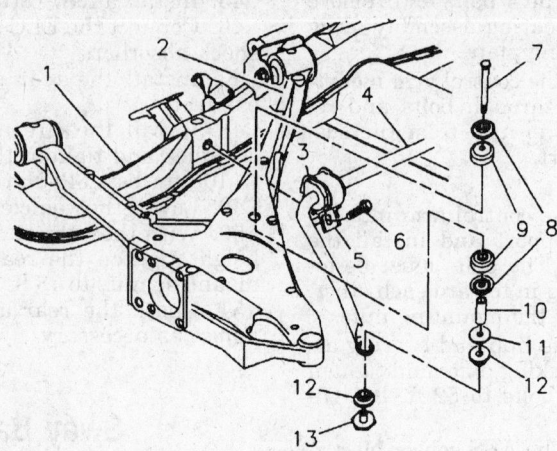

1 REAR SUSPENSION
  SUPPORT ASSEMBLY
2 CONTROL ARM
3 INSULATOR,
  STABILIZER SHAFT
4 SHAFT, STABILIZER
5 CLAMP, STABILIZER
  SHAFT
6 BOLT,
  33 N•m (24 LB. IN.)

7 BOLT,
  13 N•m (115 LB. IN.)
8 RETAINER, UPPER
9 INSULATOR, UPPER
10 SLEEVE
11 RETAINER, LOWER
12 INSULATOR, LOWER
13 NUT

305014

Rear sway bar mounting components

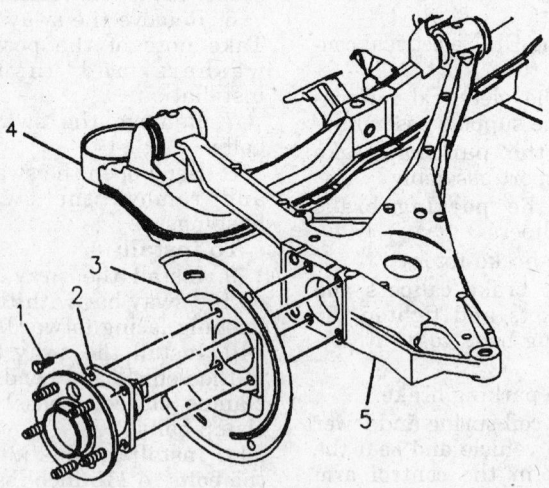

1 BOLT
2 HUB & BEARING
3 BRAKE SHIELD
4 REAR SUSPENSION SUPPORT
  ASSEMBLY
5 CONTROL ARM

303482

Rear hub and bearing assembly

Wheel Bearings

REMOVAL AND INSTALLATION

The rear wheel bearings are not serviced separately. If the rear wheel bearings are defective, the hub and bearing assembly must be replaced.

  1. Disconnect the negative battery cable.

  2. Raise and safely support the vehicle.

  3. Remove the tire and wheel assembly.

  4. Remove the caliper mounting bracket bolts and remove the caliper and bracket assembly from the brake rotor. Support the assembly out of the way. DO NOT allow the brake hose to support the weight of the caliper and bracket.

  5. Remove the brake rotor.

  6. Disconnect the ABS sensor electrical connector.

  7. Remove the 4 hub assembly mounting bolts and remove the hub and bearing assembly along with the backing plate from the rear control arm.

To install:

  8. Install the hub and bearing assembly along with the backing plate onto the control arm and loosely install the mounting bolts.

  9. Make sure the ABS sensor is properly routed and connect the ABS sensor harness.

  10. Tighten the hub and bearing mounting bolts alternately and evenly to 52 ft. lbs. (70 Nm).

  11. Install the brake rotor.

  12. Install the caliper and bracket assembly and tighten the bracket bolts to 35 ft. lbs. (48 Nm).

  13. Install the tire and wheel assembly and tighten the wheel nuts to 100 ft. lbs. (140 Nm).

  14. Lower the vehicle.

  15. Connect the negative battery cable.

FIRING ORDERS

NOTE: To avoid confusion, always replace spark plug wires one at a time.

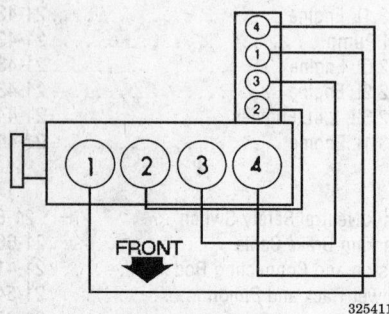

1993 2.0L and 2.2L Engines
Engine Firing Order: 1–3–4–2
Distributorless Ignition System

325411

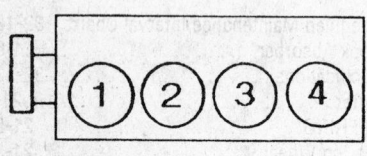

FRONT OF CAR

8838LG01

2.3L and 2.4L Engines
Engine Firing Order: 1–3–4–2
Distributorless Ignition System

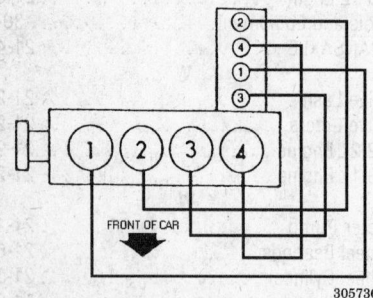

FRONT OF CAR

305736

1994–97 2.2L Engine
Engine Firing Order: 1–3–4–2
Distributorless Ignition System

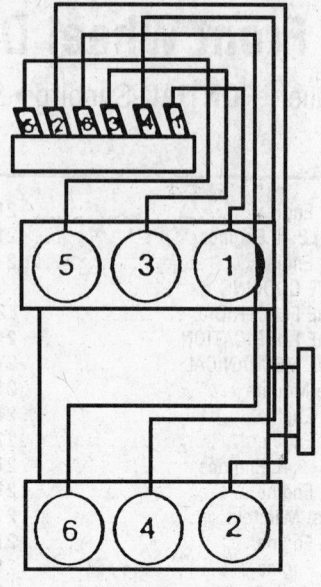

1993–94 3.1L Engine
Engine Firing Order: 1–2–3–4–5–6
Distributorless Ignition System

39299

ENGINE ELECTRICAL

NOTE: Disconnecting the negative battery cable on some vehicles may interfere with the functions of the on board computer systems and may require the computer to undergo a relearning process, once the negative battery cable is reconnected.

Ignition Timing

ADJUSTMENT

Ignition timing is controlled the PCM. No adjustment is necessary or possible.

Alternator

PRECAUTIONS

Several precautions must be observed with alternator equipped vehicles to avoid damage to the unit.

• If the battery is removed for any reason, make sure it is reconnected with the correct polarity. Reversing the battery connections may result in damage to the 1-way rectifiers.

• When utilizing a booster battery as a starting aid, always connect the positive to positive terminals and the negative terminal from the booster battery to a good engine ground on the vehicle being started.

• Never use a fast charger as a booster to start vehicles.

• Disconnect the battery cables when charging the battery with a fast charger.

• Never attempt to polarize the alternator.

• Do not use test lights of more than 12 volts when checking diode continuity.

• Do not short across or ground any of the alternator terminals.

• The polarity of the battery, alternator and regulator must be matched and considered before making any electrical connections within the system.

• Never separate the alternator on an open circuit. Make sure all connections within the circuit are clean and tight.

• Disconnect the battery ground terminal when performing any service on electrical components.

• Disconnect the battery if arc welding is to be done on the vehicle.

REMOVAL AND INSTALLATION

2.0L Engine

1. Disconnect the negative battery cable.

2. Remove the serpentine belt from the alternator pulley. It is not necessary to remove the serpentine belt from the remainder of the pulleys.

3. Detach the electrical connectors from the alternator.

4. Remove the 3 alternator mounting bolts and remove the alternator.

To install:

5. Install the alternator in the mounting bracket and install the 3 mounting bolts.

6. Tighten the long mounting bolt to 37 ft. lbs. (50 Nm), the front mounting bolt to 18 ft. lbs. (25 Nm)

and the rear mounting bolt to 18 ft. lbs. (25 Nm).

7. Install the serpentine belt.

8. Attach the electrical connectors to the alternator.

9. Connect the negative battery cable.

2.2L Engine

1. Disconnect the negative battery cable.

2. Disconnect the electrical connector from the alternator case. Remove the protective cover from the rear of the alternator and remove the nut securing the battery wire to the alternator post.

3. Remove the serpentine belt.

4. Remove the 3 alternator mounting bolts.

5. Remove the alternator.

To install:

6. Position the alternator in the mounting bracket and install the mounting bolts.

7. On 1993–95 vehicles, tighten the front mounting bolts to 37 ft. lbs. (50 Nm) and the rear mounting bolt to 18 ft. lbs. (25 Nm).

8. On 1996–97 vehicles, perform the following:

  a. Tighten the front upper mounting bolt to 22 ft. lbs. (30 Nm).

  b. Tighten the front lower mounting bolt to 37 ft. lbs. (50 Nm).

9. Install the serpentine belt.

10. On 1993–95 vehicles, connect the battery wire to the alternator post and tighten the securing nut to 71 inch lbs. (8 Nm). Install the cover over the battery wire and post.

11. Connect the electrical connector to the alternator.

12. Connect the negative battery cable.

2.3L and 2.4L Engines

1. Disconnect the negative battery cable.

2. Remove the serpentine belt.

―――――― **CAUTION** ――――――

To avoid personal injury when rotating the serpentine belt tensioner, use a tight-fitting 13mm wrench that is at least 24 inches (61cm) long. This operation can be done using tool J-37059 or equivalent.

3. Raise and safely support the vehicle.

4. Remove the lower alternator mounting bolts.

5. Lower the vehicle.

6. Remove the upper alternator bolt.

7. Detach the electrical connections at the alternator.

8. Remove the alternator assembly.

To install:

9. Install the alternator assembly.

10. Connect the **B+** (battery) connection on the alternator and tighten the nut to 65 inch lbs. (7.5 Nm).

11. Plug in the alternator connection.

12. Install the upper alternator bolt and tighten it to 22 ft. lbs. (30 Nm).

13. Raise the vehicle and safely support it.

14. Install the lower alternator mounting bolts and tighten them to 37 ft. lbs. (50 Nm).

15. Lower the vehicle.

16. Install the serpentine belt.

17. Connect the negative battery cable.

3.1L Engine

1. Disconnect the negative battery cable.

2. Detach the alternator electrical connector.

3. Remove the serpentine belt.

4. Remove the front alternator bolt, rear rear bolt and alternator rear brace bolt.

5. Remove the alternator from the mounting bracket and tilt forward.

6. Remove the cap from the alternator post and remove the nut securing the battery wire to the post and disconnect the wire.

7. Remove the alternator from the vehicle.

8. Remove the rear alternator bracket bolt and bracket from the alternator.

To install:

9. Install the alternator rear bracket on the alternator, then loosely install the mounting bolt.

10. Install the alternator assembly in the vehicle.

11. Connect the battery wire to the alternator and tighten the mounting nut to 71 inch lbs. (8 Nm). Install the protective cap over the post.

12. Install all of the mounting bolts and tighten as follows:

- Long bolt: 37 ft. lbs. (50 Nm)
- Short bolts: 18 ft. lbs. (25 Nm)
- Brace bolt: 37 ft. lbs. (50 Nm)

13. Install the serpentine belt.

14. Attach the electrical connector to the alternator.

15. Connect the negative battery cable.

Drive Belt

REMOVAL AND INSTALLATION

2.0L Engine

1. Disconnect the negative battery cable.

2. Remove the coolant recovery tank.

3. Using a 19mm wrench, pivot the tensioner, then remove the belt from the alternator.

4. Slowly release the tensioner and remove the wrench.

5. Raise and safely support the vehicle.

6. Remover the right inner fender splash shield.

7. Rotate the tensioner and remove the A/C belt, if equipped.

8. Lower the vehicle.

9. Remove the serpentine belt from the remainder of the pulleys.

To install:

10. Install the serpentine belt and route as follows:

  a. Loop the belt under the crankshaft pulley.

  b. Bring the rear of the belt up and around the tensioner so the back of the belt drives the pulley.

  c. Route the belt around the power steering pulley.

11. Rotate the tensioner and loop the belt over the alternator pulley.

12. Slowly release the tensioner and verify the belt is properly seated in all the drive belt pulleys.

13. Raise and safely support the vehicle.

14. Install the serpentine belt around the A/C compressor and tensioner.

15. Compress the tensioner and loop the belt around the crankshaft pulley.

16. Verify the belt is properly seated in all the drive belt pulleys.

17. Install the inner fender splash shield.

18. Lower the vehicle.

19. Install the coolant recovery tank.

20. Connect the negative battery cable.

2.2L, 2.3L and 2.4L Engines

1. Disconnect the negative battery cable.

2. For 1993–94 vehicles, remove the coolant reservoir.

3. Using a 15mm wrench, pivot tensioner and remove belt from alternator.

4. Remove the serpentine belt.

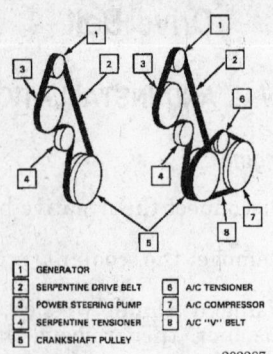

1. GENERATOR
2. SERPENTINE DRIVE BELT
3. POWER STEERING PUMP
4. SERPENTINE TENSIONER
5. CRANKSHAFT PULLEY
6. A/C TENSIONER
7. A/C COMPRESSOR
8. A/C "V" BELT

202287

Serpentine belt routing — 2.0L engine

To install:

5. Route the serpentine belt as follows:

    a. Under the crankshaft and A/C compressor, if equipped.

    b. Over the tensioner with the back of the belt driving the pulley.

    c. Around the water pump with the back of the belt driving the pulley.

    d. Around the power steering pump.

6. Rotate the tensioner and loop the belt around the alternator.

7. Release the tensioner.

─────── **WARNING** ───────

Do not exceed torque of 30 ft. lbs. (40 Nm) on the tensioner center bolt when installing or removing the belt. If these conditions are not followed, parts or system damage could result.

8. Verify the belt is properly seated in all the pulleys.

9. For 1993–94 vehicles, install the coolant reservoir.

10. Connect the negative battery cable.

11. Start the engine and check the belt alignment.

3.1L Engine

1. Disconnect the negative battery cable.

2. Rotate the belt tensioner counterclockwise using a ½ in. breaker bar.

3. Remove the serpentine belt from the alternator pulley.

4. Slowly release the belt tensioner.

5. Remove the serpentine belt from the remainder of the drive belt pulleys.

To install:

6. Install the serpentine belt as follows:

    a. Loop the serpentine belt under the crankshaft and A/C compressor.

    b. Bring the belt up the front side of the engine around the water pump pulley so the back of the belt drives the pulley.

    c. Loop the belt over the power steering pump pulley.

    d. Bring the belt up the back of the engine around the tensioner pulley so the back of the belt drives the pulley.

7. Rotate the belt tensioner counterclockwise using a ½ breaker bar.

8. With the tensioner compressed loop the belt around the alternator pulley.

9. Release the belt tensioner and verify the belt is properly seated in all the drive pulleys.

10. Connect the negative battery cable.

Starter

REMOVAL AND INSTALLATION

2.0L Engine

1. Disconnect the negative battery cable.

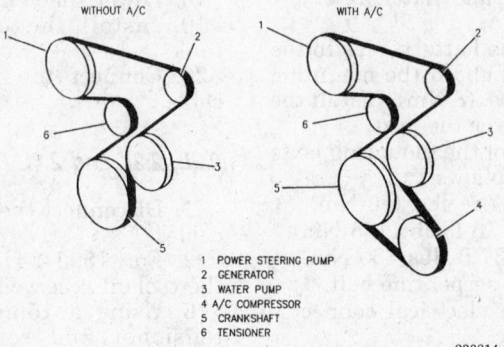

WITHOUT A/C — WITH A/C

1. POWER STEERING PUMP
2. GENERATOR
3. WATER PUMP
4. A/C COMPRESSOR
5. CRANKSHAFT
6. TENSIONER

230814

Drive belt routing — 2.2L engine

2. Disconnect the wire loom strap from the upper starter bolt.

3. Detach the shift and selector level cables at the external selector lever.

4. Remove the upper and lower transaxle control lever cable brackets and disconnect the cables.

5. Remove the drive axle support brace.

6. Detach the starter electrical connectors.

7. Remove the starter mounting bolts and remove the starter.

To install:

8. Install the starter in position and tighten the mounting bolts to 32 ft. lbs. (43 Nm).

9. Attach the starter electrical connectors.

10. Install the drive axle support brace.

11. Install the upper and lower transaxle control lever cable brackets and connect the cables.

12. Install the wire loom strap to the upper starter bolt.

13. Connect the negative battery cable.

2.2L Engine

1. Disconnect the negative battery cable.

2. Raise and safely support the vehicle.

3. Tag and detach the starter electrical connections.

4. Remove the wiring clamp at the support bracket.

5. Remove the bolt securing the starter front bracket to the engine block.

6. Remove the 2 starter mounting bolts, then remove the starter.

7. If the starter is being replaced, remove the 2 bracket mounting nuts and remove the bracket.

To install:

8. Install the front bracket on the starter and tighten the nuts to 80 inch lbs. (9 Nm).

9. Install the starter in the vehicle and tighten the 2 starter mounting bolts to 32 ft. lbs. (43 Nm).

10. Install the starter bracket-to-engine bolt and tighten to 26 ft. lbs. (32 Nm).

11. Connect the wiring clamp at the support bracket.

12. Attach the electrical connectors to the starter. Tighten the solenoid wire nut to 27 inch lbs. (3 Nm) and the battery terminal wire nut to 12 ft. lbs. (16 Nm).

13. Lower the vehicle.

14. Connect the negative battery cable.

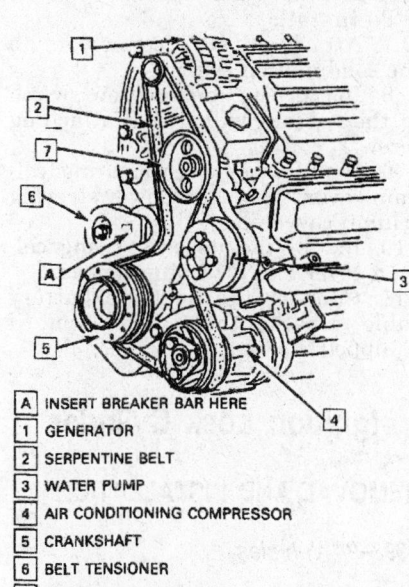

A INSERT BREAKER BAR HERE
1 GENERATOR
2 SERPENTINE BELT
3 WATER PUMP
4 AIR CONDITIONING COMPRESSOR
5 CRANKSHAFT
6 BELT TENSIONER
7 POWER STEERING PUMP

202354

Serpentine belt routing — 3.1L engine

2.3L and 2.4L Engines

1. Disconnect negative battery cable.
2. Remove air inlet duct to throttle body.
3. Remove top starter bolt.
4. Raise vehicle and safely support vehicle.
5. Remove lower starter bolt.
6. Position the starter to access solenoid wiring, and detach the electrical connections.
7. Remove the starter.
   To install:
8. Attach the electrical connections to the starter solenoid.
9. Position the starter to the engine, then install the lower bolt. Tighten the lower bolt to 66 ft. lbs. (90 Nm).
10. Lower the vehicle.
11. Install the top starter bolt and tighten to 66 ft. lbs. (90 Nm).
12. Fasten the air inlet duct to the throttle body.
13. Connect negative battery cable.

3.1L Engine

1. Disconnect the negative battery cable.
2. Raise and safely support the vehicle.
3. Remove the 2 starter mounting bolts.

4. Lower the starter to access the wiring.
5. Disconnect the battery terminal and solenoid wire.
6. Remove the starter from the vehicle.
   To install:
7. Connect the starter wiring. Tighten the battery terminal nut to 71 inch lbs. (8 Nm) and the solenoid wire nut to 27 inch lbs. (3 Nm).
8. Position the starter on the engine and tighten the mounting nuts to 33 ft. lbs. (45 Nm).
9. Lower the vehicle.
10. Connect the negative battery cable.

CHASSIS ELECTRICAL

Blower Motor

REMOVAL AND INSTALLATION

The blower motor and fan are serviced as an assembly only.

Except 3.1L Engine

1. Disconnect the negative battery cable.
2. For 1995–97 vehicles, remove the right side sound insulator.
3. Detach the electrical connections from the blower motor.
4. For 1993–94 vehicles, remove the blower motor cooling tube.
5. Remove the blower motor retaining screws, then remove blower motor and fan assembly.
   To install:
6. Install blower motor and fan assembly, then secure using the retaining screws.
7. If equipped, connect the blower motor cooling tube.
8. Attach the electrical connections to the blower motor.
9. For 1995–97 vehicles, install the right side sound insulator.
10. Connect the negative battery cable.

3.1L Engine

1. Disconnect the negative battery cable.
2. Remove the tower-to-tower brace.
3. Tag and detach the electrical connections at the blower motor.

4. Disconnect the blower motor cooling tube.
5. Remove the alternator assembly.
6. Unfasten the blower motor retaining screws, then remove the blower motor and fan assembly.
   To install:
7. Place the blower motor and fan in the correct position.
8. Install the and evenly tighten the fan assembly retaining screws.
9. Install the alternator assembly. Adjust the drive belt.
10. Connect the blower motor cooling tube.
11. Attach the electrical connections, then install the tower-to-tower brace.
12. Connect the negative battery cable.

Windshield Wiper Motor

REMOVAL AND INSTALLATION

1993–94 Vehicles

1. Disconnect the negative battery cable.
2. Remove the wiper arm and blade assemblies.
3. Remove the shroud top vent grille.
4. Detach the electrical connector.
5. Remove the transmission drive link from the wiper motor crank arm using special tool J-39232.
6. Remove the 3 retaining screws.
7. Remove the wiper assembly guiding the crank arm through the hole.
8. Remove the seal.
   To install:
9. Install seal on the wiper motor assembly.
10. Install the wiper motor assembly guiding crank arm through the hole.
11. Install the 3 attaching screws and torque the screws to 80 inch lbs. (9 Nm).
12. Attach the electrical connector to the motor.
13. Install the shroud top vent grille.
14. Connect the negative battery cable.
15. Install the wiper arm and blade assemblies.
16. Check the operation of wiper system.

1995–97 Vehicles

1. Disconnect the negative battery cable.

2. Remove the wiper arm assemblies the wiper transmission assembly drive shaft.

3. Remove the wiper cowl assembly from the vehicle.

4. Disconnect the electrical connector from the wiper motor.

5. Remove the 3 retaining screws from the wiper drive module and remove from the vehicle.

6. Disconnect the wiper transmission assembly from the wiper motor crank arm assembly, using tool J-39232 or equivalent.

7. Disconnect the wiper crank arm assembly from the wiper motor assembly.

   a. Loosen the wiper motor crank screw.

   b. Tap on the wiper motor crank screw with a plastic mallet while holding up on the wiper motor crank arm assembly, until the crank arm assembly is loose.

   c. Remove the wiper motor screw and wiper motor crank arm assembly.

   d. Remove the 3 mounting screws from the bracket and remove the wiper motor.

To install:

8. Install the wiper motor to the mounting bracket and install the 3 mounting screws. Tighten the screws to 62 inch lbs. (7 Nm).

9. Attach the wiper motor crank arm assembly to the wiper motor assembly.

10. Attach the electrical connector to the wiper motor.

11. Connect the negative battery cable.

12. Turn the ignition switch to the **ACCY** position and set the wiper switch to **PULSE**. Verify wiper motor operation, turn the ignition switch **OFF** while the motor is still operating. Wiper motor assembly will then return to PARK.

13. Disconnect the negative battery cable and then disconnect the electrical connector from the wiper motor.

NOTE: Do not rotate the wiper motor assembly shaft during installation of wiper motor crank arm assembly.

14. Install the wiper motor crank arm assembly onto the wiper motor assembly, while maintaining a 4–8mm gap between the wiper motor crank arm and the bracket tab.

15. Install the wiper motor crank arm screw and tighten to 144 inch lbs. (16 Nm). Make sure the gap between wiper motor crank arm and bracket tab is still 0.16–0.32 in. (4–8mm).

16. Install the wiper transmission assembly onto the wiper motor crank arm assembly using tool J-39529 or equivalent.

17. Install the wiper drive system module and tighten the 3 screws to 88 inch lbs. (10 Nm).

18. Install the electrical connector to the wiper motor assembly.

19. Install the wiper cowl assembly.

20. Install the wiper arm assemblies onto the vehicle.

21. Connect the negative battery cable.

22. Operate the wipers and check for proper operation.

Headlight Switch

REMOVAL AND INSTALLATION

1993–94 Sunbird

1. Disconnect the negative battery cable.

2. Pull out the left pad trim plate.

3. Remove the screws securing the switch to the trim plate.

4. Detach the switch electrical connection, then remove the switch.

To install:

5. Attach the switch electrical connector.

6. Install the screws securing the switch to the trim plate.

7. Fasten the left pad trim plate.

8. Connect the negative battery cable.

Combination Switch

REMOVAL AND INSTALLATION

——————— CAUTION ———————
The Supplemental Inflatable Restraint (SIR) system must be disarmed before working around the steering column. Failure to do so may cause accidental deployment of the air bag, resulting in unnecessary SIR system repairs and/or personal injury.

1. If equipped, disable the SIR system. Disconnect the negative battery cable.

2. Remove the steering column cover and cover screws.

3. Remove the upper steering column cover.

4. Remove the lower steering column cover screws and lower the steering column cover.

5. Remove the 2 switch mounting screws and remove the switch.

6. Detach all wire connectors from the combination switch. Push locking tabs to remove the wires.

To install:

7. Attach all wire connectors to the combination switch.

8. Install the combination switch to the column with the 2 attaching screws.

9. Install the lower steering column cover screws and lower steering column cover.

10. Install the upper steering column cover and attaching screws.

11. Connect the negative battery cable. Enable the SIR system, if equipped.

Ignition Lock Cylinder

REMOVAL AND INSTALLATION

1993–94 Vehicles

Cavalier

1. Disconnect the negative battery cable.

2. Remove the steering column from the vehicle.

3. Turn the ignition key to the **RUN** position and, if applicable, place the shift lock cable lever in the **PARK** position.

4. Drill off the heads of the lock cylinder assembly using a ¼ in. (6.5mm) drill bit.

5. Remove the lock cylinder housing assembly from the steering column.

6. Using a pair of locking pliers, remove the threaded ends of the shear bolts.

To install:

7. Thoroughly clean all metal shavings from the steering column assembly.

8. Place the lock cylinder key in the **RUN** position, then install the cylinder housing to the steering column assembly. Tighten the shear bolts until the head separates from the rest of the body, this should occur at approximately 97 inch lbs. (11 Nm) of torque.

9. Install the steering column in the vehicle.

10. Connect the negative battery cable.

Sunbird

1. Disconnect the negative battery cable.

2. Remove the steering wheel.

3. Remove the turn signal switch from the steering column and allow it to hang from the wiring harness.

4. Make sure the ignition is in the **LOCK** position, then remove the key from the lock cylinder. Remove the buzzer switch assembly, reinsert the key, remove the retaining screw, then remove the lock cylinder assembly from the vehicle.

NOTE: Use extreme caution when removing the lock cylinder retaining screw. If the screw is dropped during removal, it could fall into the column, requiring complete column disassembly to retrieve the screw.

To install:

5. Align the lock cylinder set with the steering column housing.

6. Push the lock all the way in.

7. Install the cylinder retaining screw and tighten to 22 inch lbs. (2.5 Nm).

8. With the key still in the same position as during removal, install the buzzer switch.

9. Install the turn signal switch.

10. Install the steering wheel.

11. Connect the negative battery cable.

1995–97 Vehicles

— **CAUTION** —
The Supplemental Inflatable Restraint (SIR) system must be disarmed before performing service around SIR components or wiring. Failure to do so may cause accidental deployment of the air bag, resulting in unnecessary SIR repairs and/or personal injury.

NOTE: When replacing the lock cylinder, the key code must be read with interrogator tool J-35628-A or equivalent. A new key must be ordered with the new lock cylinder and must be cut to the dummy key in the new lock cylinder.

1. Disconnect the negative battery cable.

2. If equipped, disable the SIR (air bag) system.

3. Remove the 2 screws in back of the steering wheel.

4. Position the inflator module to gain access to the wire connectors. Note the wiring routing for reinstallation. Push in and rotate the horn contact lead counterclockwise and remove the steering column horn tower. Remove the horn contact lead.

5. Disconnect the air bag connector retainer, then detach the connector. Remove the air bag inflator module and store trim side face up for safety reasons.

— **CAUTION** —
When carrying a live air bag, make sure the air bag and trim cover are pointed away from the body. In the unlikely event of an accidental deployment, the bag will then deploy with minimal chance of injury. When placing a live air bag on a bench or other surface, always face the bag and trim cover up, away from the surface. This will reduce the motion of the module if it is accidently deployed.

6. Remove the steering wheel control switch connector, if equipped.

7. Remove the steering wheel fastener. Install steering wheel removal tools J-1859-03 and J-28720 or equivalent, and remove the steering wheel while guiding the wires through the steering wheel hub. Mark the steering wheel-to-shaft position, prior to removal for re-installation purposes.

8. Remove the SIR coil retaining ring.

9. Remove the left side sound insulators from under the dash.

10. Remove the wire connectors from the retainers and detach the bulkhead connector to remove the turn signal switch connector.

11. Gently pull up on the SIR coil assembly while pushing up on the SIR coil harness on the bottom side of the steering column and let the coil hang freely.

12. Remove the wave washer. Install compressor tool J-23653-SIR or equivalent, and compress the lock plate to remove the shaft lock retaining ring.

13. Remove the turn signal canceling cam assembly and remove the screws to the turn signal switch arm and turn signal switch-to-steering column, then remove the hazard switch button.

14. Remove the upper bearing spring. Pull up gently on the turn sig-

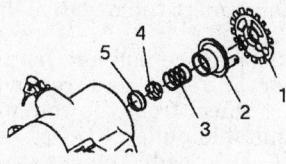

1. Shaft lock
2. Turn sig cancel cam asm
3. Upper bearing spring
4. Upper bearing inner race seat
5. Inner race

8838LG11

Removing components from the upper shaft — 1996–97 vehicles shown

nal switch assembly and let it hang, then remove the key buzzer switch.

15. Remove the 13-way secondary lock on the turn signal switch connector at the lower end of the steering column and remove terminals **12** and **13** which run to the ignition lock cylinder.

16. Attach a piece of wire to terminals for guiding the harness through the steering column.

17. Remove the retaining bolt to the ignition lock cylinder and insert the key in the lock cylinder to remove.

To install:

— **WARNING** —
Route wires from the pass key lock cylinder as shown and the retaining clip into hole in housing. Failure to do so may result in component damage or the malfunctioning of the steering column may occur.

18. Install the pass key lock cylinder. Route the wires correctly and insert the terminals in the turn signal connector.

19. Install the lock cylinder fastener and torque to 22 inch lbs. (2.5 Nm)

20. Install the buzzer switch and the turn signal switch assembly. Torque the bolts to 30 inch lbs. (3.4 Nm). When installing the switch be sure to remove all slack from the wiring harness by pulling gently on the harness.

— **WARNING** —
SIR coil assembly wires must be kept tight with no slack while installing the SIR coil assembly. Failure to do so may cause the wires to be kinked near the lock plate, causing the wires to be cut when the steering wheel is turned.

21. Install the turn signal switch arm and torque to 20 inch lbs. (2.3 Nm). Install the hazard switch button

22. Install the upper bearing spring.

23. Install the turn signal cancelling cam assembly and lock plate.

24. Using compressor tool J-23653-SIR or equivalent, install the retaining ring after compressing the upper bearing spring.

25. Install the wave washer and SIR coil assembly, pulling gently on the harness on the bottom side of the steering column to remove slack from the wiring harness. Be sure to align the SIR coil assembly properly.

26. Install the snapring on top of the SIR coil.

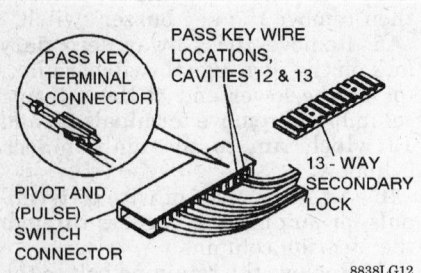

Turn signal switch wire locations — 1996–97 vehicles shown

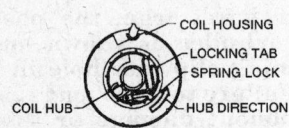

Perform the following steps to center coil assembly
A. Wheels straight ahead.
B. Remove coil assembly.
C. Hold coil assembly with bottom up.
D. While holding coil assembly, depress spring lock to rotate hub in direction of arrow until it stops.
E. The coil ribbon should be wound up snug against center hue.
F. Rotate coil hub in opposite direction approximately two and a half (2 -1/2) turns. Release spring lock between looking tabs.

8838LG13

Centering the SIR coil assembly — 1996–97 vehicles shown

27. Install the steering wheel, guiding the wires through for the air bag and steering wheel controls, if equipped.
28. Connect the negative battery cable.
29. If equipped, enable the SIR system.
30. Test the turn signals, high beams and wipers.
31. Install the lower sound insulator.

Ignition Switch

REMOVAL AND INSTALLATION

Except 1993–94 Sunbird

━━ CAUTION ━━
The Supplemental Inflatable Restraint (SIR) system must be disarmed before working around the steering column. Failure to do may cause accidental deployment of the air bag, resulting in unnecessary SIR system repairs and/or personal injury.

1. Disconnect the negative battery cable.
2. If equipped, disable the SIR system.

━━ CAUTION ━━
The DERM can maintain sufficient voltage to cause an air bag deployment for up to 2 minutes after the ignition has been turned to OFF, or the battery is disconnected.

3. Remove the steering column.
4. The ignition switch is located at the base of the steering column. Disconnect the vehicle wire harness from the ignition switch.
5. Unfasten the mounting screws, then remove the ignition switch assembly.

To install:

━━ WARNING ━━
When fasteners are removed, always reinstall them at the same location from which they were removed. If a fastener needs to be replaced, use the correct part number fastener for that application. If the correct part number is not available, a fastener of equal size and strength (or stronger) may be used. Fasteners that are to be reused, and those requiring thread locking compound will be called out. The correct torque value must be used when installing fasteners that require it. If the above conditions are not followed, parts or system damage could result.

6. Position the ignition switch assembly.
7. Install the 2 mounting screws, then tighten them to 12 inch lbs. (1.4 Nm).
8. Connect the vehicle wire harness to the switch assembly.
9. Install the steering column.
10. Enable the SIR system.
11. Connect the negative battery cable.

1993–94 Sunbird

1. Disconnect the negative battery cable.
2. Remove the pad and horn lead, retainer and nut and remove the steering wheel from the column, using a suitable puller.
3. If equipped, remove the tilt lever.
4. Unfasten the 2 lower column cover retaining screws.
5. Reove the 3 bottom retaining screws from the lower column cover.
6. Unfasten the washer head screw and hex nut.
7. Remove the dimmer switch assembly from the rod. Remove the

dimmer switch and ignition switch mounting stud.
8. Disconnect the ignition switch from the actuator assembly, by unfastening the two cross recess screws. Detach the electrical connector, then remove the switch.

To install:

NOTE: **Install the ignition switch to the jacket with the switch in the ""OFF-LOCK" position. a new switch will be pinned in the ""OFF-LOCK" position. Don't forget to remove the plastic pin after the switch is attached to the column.**

9. Connect the ignition switch to the switch actuator assembly and adjust as follows:
   a. Move the switch slider to the extreme left position.
   b. Move the switch slider one detent to the right ""OFF-LOCK" position.
10. Install the ignition switch assembly and stud. Tighten the retaining screws to 30 inch lbs. (3.4 Nm).
11. Fasten the dimmer switch assembly to the rod.
12. Install the hex nut and washer head screw, then tighten finger-tight.
13. Adjust the dimmer switch as follows:
   a. Place a 3/32 in. (2.4mm) drill bit in the hole in the switch to limit travel.
   b. Position the switch on the column and push against the dimmer switch rod to remove all lash.
   c. Remove the drill bit.
14. Attach the switch electrical connector.
15. Install the steering column.
16. Connect the negative battery cable.

Park/Neutral Safety Switch

REMOVAL AND INSTALLATION

1. Disconnect the negative battery cable.
2. Remove the shift linkage.
3. Detach the electrical connector from the switch.
4. Unfasten the switch-to-transaxle screws/bolts, then remove the switch.

To install:

NOTE: **After switch engagement, make sure the engine will only start in P or N.**

5. If installing the old switch, proceed as follows:
   a. Place the shift shaft/control lever in N.

1 PARK/NEUTRAL POSITION SWITCH
2 BOLT – 24 N•m (18 LBS. FT.) TIGHTEN FIRST
3 BOLT – 24 N•m (18 LBS. FT.) TIGHTEN SECOND

8838LG02

Park/Neutral position switch — 1995 4T40-E automatic transaxle shown

b. Align the flats of the shift shaft with the switch, then loosely install the mounting bolts.

c. Insert a gauge pin in the service adjustment hole and rotate the switch until the pin drops to a depth of 9/64 in. (9mm).

d. Tighten the bolts to 18 ft. lbs. (24 Nm), then remove the gauge pin.

6. If installing a new switch, proceed as follows:

a. Place the shift shaft/control lever in **N**.

b. Align the flats of the shift shaft to the flats in the switch, then install the switch assembly. Tighten the bolts to 18 ft. lbs. (24 Nm).

c. If the bolt holes do not align with the mounting boss on the transaxle, verify the transaxle is in the **N** position, do not rotate the switch.

d. If the shift has been rotated and the pin broken, the switch can be adjusted by following the procedure in Step 5.

7. Attach the electrical connector to the switch.

8. Install the shift linkage.

9. Connect the negative battery cable, then start the engine and check the switch operation.

Powertrain Control Module

REMOVAL AND INSTALLATION

1. Turn the ignition switch **OFF**.

2. Disconnect the negative battery cable.

3. For 1993–94 vehicles, remove the right side hush panel, glove box or interior access panel, as required for access to the control module.

4. For 1995–97 vehicles, remove the right hand engine splash shield. Unfasten the horn attaching bolt, then detach the electrical connector and remove the horn.

5. Detach the electrical harness connectors from the computer control module.

6. Unfasten the module-to-bracket retaining screws, then remove the PCM.

7. If replacement of the calibration unit is required, unfasten the access cover retaining screws, then remove the cover from the computer control module. Carefully remove the calibration unit from the PCM, as follows:

a. If the PCM contains a PROM carrier, use the rocker type PROM removal tool.

b. If the PCM contains a CAL-PAK, grasp the CAL-PAK carrier (at the narrow end only), using the removal tool. Remove the CAL-PAK carrier.

c. If the PCM contains a MEM-CAL, EPROM or KS Module, push both retaining clips back away from the MEM-CAL/EPROM. At the same time, grasp it at both ends and lift it up out of the socket. Do not remove the cover of the MEM-CAL/EPROM/KS module.

NOTE: Before replacement of a defective computer control module, first check the resistance of each PCM controlled solenoid.

To install:

8. Fit the replacement calibration unit into the socket.

NOTE: The small notch of the carrier should be aligned with the small notch in the socket. Press on only the ends of the carrier until it is firmly seated in the socket.

9. Install the access cover, then secure using the retaining screws.

10. Position the computer control module in the vehicle, then install the module-to-bracket retaining screws.

11. Attach the module electrical harness connectors.

12. For 1993–94 vehicles, install the right side hush panel, glove box or interior access panel as required.

13. For 1995–97 vehicles, install the horn, attach the electrical connector, then fasten with the horn attaching bolt. Tighten the bolt to 6–9 ft. lbs. (8–12 Nm). Install the right hand splash shield.

14. Check that the ignition switch is **OFF**, then connect the negative battery cable.

15. If equipped with a 3.1L engine which is equipped with a PCM with EPROM, the EPROM must be reprogrammed using a scan tool and the latest available software. In all likelihood, the vehicle must be towed to a dealer or repair shop containing the suitable equipment for this service.

16. Enter the self-diagnostic system and check for trouble codes to be sure the module and calibration unit are properly installed. For details, refer to the procedure for checking trouble codes.

ENGINE COOLING

Radiator

REMOVAL AND INSTALLATION

1993–94 Vehicles

1. Disconnect the negative battery cable.

2. Drain the cooling system into a suitable container.

3. Disconnect the forward lamp harness from the fan frame.

4. Detach the electrical connector from the cooling fan motor.

5. Remove the mounting bolts, then remove the cooling fan assembly.

6. Scribe a mark around the hood latch assembly on the upper radiator panel. Remove the bolts securing the hood latch, then move the latch assembly out of the way.

7. Remove the radiator air inlet ducts.

8. Disconnect the upper and lower radiator hoses from the radiator.

9. Detach the coolant overflow hose from the radiator neck.

10. Disconnect and cap the transaxle cooler lines from the radiator.

11. If equipped with A/C, remove the 4 radiator-to-condenser assembly mounting bolts, then remove the ra-

diator tank-to-refrigerant line clamp bolt.

12. Remove the radiator-to-radiator support mounting bolts.

13. Remove the radiator from the vehicle. If equipped with A/C, it may be necessary to raise the right hand side of the radiator first, so the radiator neck will clear the A/C compressor.

To install:

14. Install the radiator in the vehicle, placing the bottom of the radiator in the lower mounting pads.

15. Install the radiator-to-radiator support and mounting bolts and tighten to 89 inch lbs. (10 Nm).

16. If equipped with A/C, install the refrigerant line clamp-to-radiator tank mounting bolt and install the 4 radiator-to-condenser mounting bolts.

17. Connect the transaxle cooler lines and tighten the fittings to 20 ft. lbs. (27 Nm.).

18. Attach the upper and lower radiator hoses to the radiator.

19. Connect the coolant overflow hose to the radiator neck and position the clamp.

20. Install the radiator air inlet ducts.

21. Line up the previously made marks and install the hood latch to the radiator support. Tighten the mounting bolts to 18 ft. lbs. (25 Nm).

22. Install the cooling fan assembly and mounting bolts and tighten to 7 ft. lbs. (10 Nm).

23. Attach the cooling fan electrical connector.

24. Connect the forward lamp harness to the fan frame.

25. Refill the cooling system.

26. Connect the negative battery cable.

27. Start the engine and check for leaks.

1995–97 Vehicles

An aluminum cross-flow radiator core with plastic tanks is used.

Starting in 1996, a new type of antifreeze/coolant is used called GM Goodwrench DEX-COOL™. **Propylene glycol is not recommended for use in GM vehicles. DO NOT use DEX-COOL™ in pre-1996 vehicles. DO NOT mix DEX-COOL™ with any other type of antifreeze.**

— CAUTION —
The Supplemental Inflatable Restraint (SIR) system must be disarmed before performing this pro-

cedure. Failure to do so may cause accidental deployment of the air bag, resulting in unnecessary SIR system repairs and/or personal injury.

1. Disconnect the negative battery cable.

2. Discharge and recover the A/C refrigerant.

3. Disable the SIR system.

4. Drain the engine coolant into a suitable container.

5. Remove the hood latch from the mounting plate.

6. Remove the headlight assemblies.

7. Remove the radiator mounts.

8. Raise and safely support the vehicle.

9. Disconnect the forward SIR sensor harness.

10. Remove the cooling fan assembly.

11. Remove the lower radiator hose from the radiator.

12. Disconnect the lower transmission oil cooler line from the radiator.

13. Lower the vehicle.

14. Remove the hood latch support bracket and the forward sensor with harness.

15. Disconnect the upper oil cooler line from the radiator.

16. Remove the upper radiator hose from the radiator.

17. Disconnect the compressor and accumulator hoses from the condenser.

18. Disconnect the overflow hose from the radiator.

19. Remove the radiator/condenser assembly from the vehicle.

20. Remove the condenser from the radiator, if equipped.

To install:

21. Attach the condenser to the radiator.

22. Install the radiator assembly into the vehicle.

23. Connect the overflow hose to the radiator.

24. Attach the hood latch bracket and route the forward sensor harness.

25. Raise and safely support the vehicle.

26. Install the cooling fan assembly.

27. Connect the lower transmission oil cooler line to the radiator.

28. Install the lower radiator hose.

29. Connect the forward SIR sensor harness connector.

30. Lower the vehicle.

31. Install the upper radiator hose.

32. Connect the upper transmission oil cooler line to the radiator.

33. Install the radiator mounts.

34. Attach the hood latch support.

35. Connect the compressor and accumulator hoses to condenser, using new O-rings.

36. Install the headlight assemblies.

37. Attach the hood latch assembly and adjust.

38. Refill the cooling system.

39. Recharge the A/C system.

40. Enable the SIR system.

41. Connect the negative battery cable.

42. Start the vehicle and verify no leaks.

Water Pump

REMOVAL AND INSTALLATION

2.0L Engine

1. Disconnect the negative battery cable.

2. Remove the timing belt cover.

3. Drain the cooling system.

4. Rotate the engine until the crankshaft and camshaft marks are lined up with the marks on the camshaft carrier and engine block.

5. Loosen the water pump bolts and rotate the water pump to release tension on the timing belt.

6. Remove the timing belt.

7. Remove the rear timing belt cover.

8. Disconnect the hose from the water pump.

9. Remove the water pump mounting bolts, then remove the water pump. Discard the O-ring.

To install:

10. Coat a new O-ring with clean anti-freeze, then install it on the water pump.

11. Install the water pump and loosely install the mounting bolts.

12. Connect the hose to the water pump.

13. Install the rear timing belt cover.

14. Install the timing belt and with all the marks in alignment adjust the belt to specification. Tighten the water pump mounting bolts to 21 ft. lbs. (28 Nm).

15. Refill the cooling system.

16. Install the timing belt cover.

17. Connect the negative battery cable.

18. Start the vehicle and check for leaks.

2.2L Engine

WARNING

When adding coolant to 1996–97 vehicles, it is important that you use GM Goodwrench DEX-COOL coolant meeting GM Specification 6277M.

---

1. Disconnect the negative battery cable.
2. Drain the cooling system into a suitable container.
3. Loosen, but do not remove, the water pump pulley bolts.
4. Remove the serpentine belt.
5. Remove the alternator mounting bolts and set the alternator aside.
6. Remove the water pump pulley bolts, then remove the pulley.
7. Remove the 4 water pump mounting bolts, then remove the water pump.

To install:

8. Clean all the gasket surfaces completely.
9. Apply a thin bead of sealer around the outer edge of the water pump gasket seating area and place he gasket on the pump.
10. Install the water pump on the engine and tighten the 4 mounting bolts to 18 ft. lbs. (25 Nm).
11. Install the water pump pulley and tighten the mounting bolts finger-tight.
12. Install the alternator in the mounting bracket.
13. Install the serpentine belt.
14. Tighten the water pump pulley mounting bolts to 22 ft. lbs. (30 Nm).
15. Connect the negative battery cable.
16. Refill and bleed the cooling system.

2.3L and 2.4L Engines

1. Disconnect the negative battery cable
2. Detach the oxygen sensor connector.
3. Properly drain the engine coolant into a suitable container. Remove the heater hose from the thermostat housing for more complete coolant drain.
4. Remove upper exhaust manifold heat shield.
5. Remove the bolt that attaches the exhaust manifold brace to the manifold.
6. Remove the lower exhaust manifold heat shield.
7. Break loose the manifold to exhaust pipe spring loaded bolts using a 13mm box wrench.

8. Raise and safely support the vehicle.

NOTE: It is necessary to relieve the spring pressure from 1 bolt prior to removing the second bolt. If the spring pressure is not relieved, it will cause the exhaust pipe to twist and bind up the bolt as it is removed.

9. Unfasten the 2 radiator outlet pipe-to-water pump cover bolts.
10. Remove the manifold to exhaust pipe bolts from the exhaust pipe flange as follows:
    a. Unscrew either bolt clockwise 4 turns.
    b. Remove the other bolt.
    c. Remove the first bolt.

WARNING

On the 2.4L engines, DO NOT rotate the flex coupling more than 4 degrees or damage may occur.

---

11. Pull down and back on the exhaust pipe to disengage it from the exhaust manifold bolts.
12. Remove the radiator outlet pipe from the oil pan and transaxle. If equipped with a manual transaxle, remove the exhaust manifold brace. Leave the lower radiator hose attached and pull down on the outlet pipe to remove it from the water pump. Leave the radiator outlet pipe hang.
13. Carefully lower the vehicle.
14. Unfasten the exhaust manifold-to-cylinder head retaining nuts, then remove the exhaust manifold, seals and gaskets.
15. For the 2.4L engine, remove the front timing chain cover and the chain tensioner.
16. Unfasten the water pump-to-cylinder block bolts. Remove the water pump-to-timing chain housing nuts. Remove the water pump and cover mounting bolts and nuts. Remove the water pump and cover as an assembly, then separate the 2 pieces.

To install:

17. Thoroughly clean and dry all mounting surfaces, bolts and bolt holes. Using a new gasket, install the water pump to the cover and tighten the bolts finger-tight.
18. Lubricate the splines of the water pump with clean grease and install the assembly to the engine using new gaskets. Install the mounting bolts and nuts finger-tight.
19. Lubricate the radiator outlet pipe O-ring with antifreeze and slid the pipe onto the water pump cover. Instal the bolts finger-tight.

20. With all gaps closed, tighten the bolts, in the following sequence, to the proper values:
    a. Pump assembly-to-chain housing nuts — 19 ft. lbs. (26 Nm).
    b. Pump cover-to-pump assembly — 106 inch lbs. (12 Nm).
    c. Cover-to-block, bottom bolt first — 19 ft. lbs. (26 Nm).
    d. Radiator outlet pipe assembly-to-pump cover — 125 inch lbs. (14 Nm).
21. Using new gaskets, install the exhaust manifold.
22. Raise and safely support the vehicle.
23. Index the exhaust manifold bolts into the exhaust pipe flange.
24. Connect the exhaust pipe to the manifold. Install the exhaust pipe flange bolts evenly and gradually to avoid binding. Turn the bolts in until fully seated.
25. Connect the radiator outlet pipe to the transaxle and oil pan. Install the exhaust manifold brace, if removed.
26. On the 2.4L engine, install the timing chain tensioner and front cover.
27. Install the lower heat shield.
28. Carefully lower the vehicle.
29. Fasten the bolt that attaches the exhaust manifold brace to the manifold.
30. Tighten the manifold-to-exhaust pipe nuts to specification.
31. Install the upper heat shield.
32. Attach the oxygen sensor connector.
33. Fill the radiator with coolant until it comes out the heater hose outlet at the thermostat housing. Then connect the heater hose. Leave the radiator cap off.
34. Connect the negative battery cable, then start the engine. Run the vehicle until the thermostat opens, fill the radiator and recovery tank to their proper levels, then turn the engine **OFF**.
35. Once the vehicle has cooled, recheck the coolant level.

3.1L Engine

1. Disconnect the negative battery cable.
2. Drain the cooling system into a suitable container.
3. Loosen, but do not remove, the water pump pulley mounting bolts.
4. Remove the serpentine belt.
5. Remove the pulley mounting bolts and remove the pulley.
6. Unfasten the 5 water pump attaching bolts, then remove the water pump.

To install:

7. Clean all the gasket surfaces completely.

8. Apply a thin bead and of sealer around the gasket seating area on the water pump and install the gasket on the pump. The gasket is marked **TOP** and must be installed so the indicator is up.

9. Install the water pump and gasket on the engine and tighten the water pump mounting bolts to 7 ft. lbs. (10 Nm).

10. Install the water pump pulley and tighten the bolts finger-tight.

11. Install the serpentine belt.

12. Tighten the water pump pulley bolts to 18 ft. lbs. (24 Nm).

13. Connect the negative battery cable.

14. Refill and bleed the cooling system.

Thermostat

REMOVAL AND INSTALLATION

2.0L Engine

1. Disconnect the negative battery cable.

2. Remove the cap from the thermostat housing.

3. Use a suitable pry tool under the wire handle of the thermostat to carefully unseat the thermostat.

4. Remove the thermostat from the housing.

To install:

5. Install the thermostat into the housing push down until it is fully seated.

6. Install the cap on the thermostat housing and lock in place so the arrows on the cap line up with the outlet hoses.

7. Connect the negative battery cable.

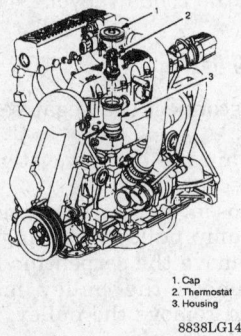

1. Cap
2. Thermostat
3. Housing

8838LG14

Thermostat and housing — 2.0L engine

2.2L Engine

Starting with 1996, the vehicles are filled with GM Goodwrench DEX-COOL® antifreeze. **DO NOT mix DEX-COOL™ with any other type of antifreeze.**

1. Disconnect the negative battery cable.

2. Drain the cooling system to a level below the thermostat housing.

3. Disconnect the upper radiator hose from the thermostat housing.

4. Remove the thermostat housing mounting nuts and remove the thermostat housing.

5. Remove the thermostat.

To install:

6. Clean all the gasket surfaces completely.

7. Install the thermostat in the thermostat housing.

8. Apply a thin bead of sealer around the gasket contact surface of the housing and install the gasket.

9. Install the thermostat housing on the engine and tighten the mounting nuts to 89 inch lbs. (10 Nm).

10. Connect the upper radiator hose to the thermostat housing.

11. Refill the cooling system.

12. Connect negative battery cable.

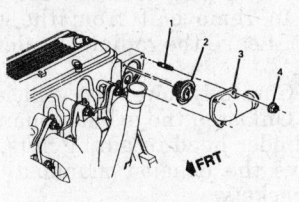

1. Water outlet stud
2. Engine coolant thermostat assembly
3. Water outlet
4. Water outlet nut

323397

Thermostat components — 1993–94 2.2L engine

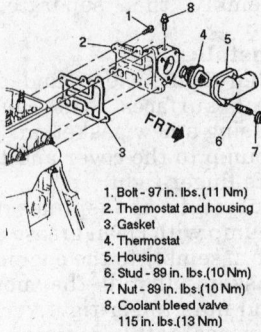

1. Bolt - 97 in. lbs.(11 Nm)
2. Thermostat and housing
3. Gasket
4. Thermostat
5. Housing
6. Stud - 89 in. lbs.(10 Nm)
7. Nut - 89 in. lbs.(10 Nm)
8. Coolant bleed valve 115 in. lbs.(13 Nm)

326701

Thermostat installation — 1996–97 2.2L engine

13. Refill the cooling system with the proper coolant. 1996–97 vehicles use DEX-COOL™. Bleed the cooling system and check for leaks with the engine running at idle.

2.3L and 2.4L Engines

1. Disconnect the negative battery cable.

2. Drain the engine coolant into a suitable container.

3. Remove the cover to outlet pipe bolt through exhaust manifold runners.

4. Raise and safely support the vehicle.

5. Disconnect the radiator and heater hoses from the outlet pipe.

6. Remove the outlet pipe to oil pan bolt.

7. Remove the cover to outlet pipe bolt, then remove the thermostat.

To install:

8. Clean all sealing surfaces.

9. Install the thermostat.

10. Install the cover to outlet pipe bolt, tighten the bolt to 124 inch lbs. (14 Nm).

11. Install the outlet pipe to oil pan bolt and tighten to 19 ft. lbs. (26 Nm).

12. Install the pipe to transaxle and tighten to 40 ft. lbs. (54 Nm).

13. Connect the radiator and heater hoses to the outlet pipe.

14. Lower the vehicle.

15. Install the cover to outlet pipe bolt through the exhaust manifold runner.

16. Fill the cooling system. Use the proper coolant. 1996–97 vehicles must use DEX-COOL™ (orange-color) coolant.

17. Connect the negative battery cable.

18. Start the vehicle and verify no leaks.

3.1L Engine

1. Disconnect the negative battery cable.

2. Drain the cooling system to a level below the thermostat housing.

3. Remove the air cleaner assembly.

4. Disconnect the upper radiator hose from the thermostat housing.

5. Remove the thermostat housing mounting bolts.

6. Remove the thermostat housing, then remove the thermostat.

To install:

7. Clean all the gasket surfaces completely.

8. Install the thermostat in the recess in the intake manifold.

9. Apply a thin bead of sealer around the gasket seating area on

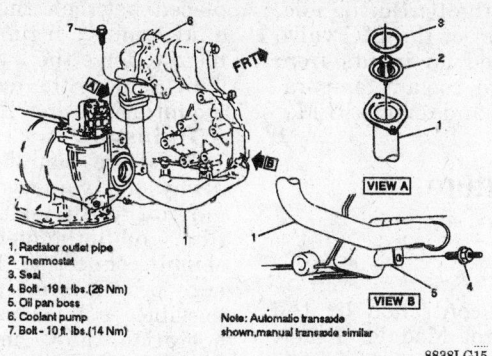

1. Radiator outlet pipe
2. Thermostat
3. Seal
4. Bolt - 19 ft. lbs.(26 Nm)
5. Oil pan boss
6. Coolant pump
7. Bolt - 10 ft. lbs.(14 Nm)

Note: Automatic transaxle
shown, manual transaxle similar

8838LG15

Thermostat assembly — 2.3L and 2.4L engines

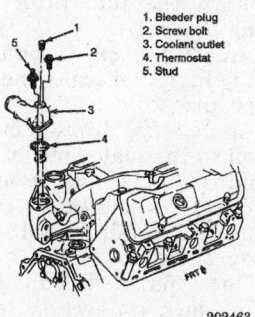

1. Bleeder plug
2. Screw bolt
3. Coolant outlet
4. Thermostat
5. Stud

202463

Exploded view of thermostat components — 3.1L engine

the intake manifold and install the gasket on the manifold.

10. Install the thermostat housing and tighten the mounting bolts to 18 ft. lbs. (25 Nm).

11. Connect the upper radiator hose to the thermostat housing.

12. Install the air cleaner assembly.

13. Refill the cooling system.

14. Connect the negative battery cable.

15. Start the vehicle and verify no leaks.

Electric Cooling Fan

REMOVAL AND INSTALLATION

───── **CAUTION** ─────
If a fan blade is damaged, replace with a new fan assembly. A fan assembly that is not in proper balance could fly apart creating an dangerous situation.

1. Disconnect negative battery cable.

2. Disconnect the wiring harness from the motor and the fan frame.

3. Remove the fan assembly from the radiator support. On vehicles with a 2.3L or 2.4L engine, remove the fan from the bottom.

To install:

4. Secure the fan assembly to the radiator support. Tighten the mounting bolts to 7 ft. lbs. (10 Nm).

5. Connect the wiring harness.

6. Connect the negative battery cable.

7. Start vehicle and verify proper cooling fan operation.

───── **CAUTION** ─────
Keep hands, tools and clothing away from the engine coolant fan. The fan can turn ON whether or not the engine is running; can start automatically in response to a heat sensor.

Cooling System Bleeding

PROCEDURE

1. Make sure the radiator drain cock is closed and, if removed, that the block drain plugs are installed.

2. Fill the cooling system through the reservoir tank or the radiator until the FULL COLD mark is reached. Be sure to use a solution which is at least 50 percent coolant but no more than 70 percent ethylene glycol anti-freeze and the balance water.

3. For the 2.0L engine, remove the thermostat cap and the thermostat and fill the coolant to a level just below the housing cap seat and install the thermostat.

4. Install the radiator cap, surge and/or thermostat housing caps.

5. Start the engine and allow it to run until normal operating temperature is reached and the upper radiator hose becomes hot. Stop the engine and observe the coolant level in the surge tank. The coolant should now be at the FULL or FULL HOT line. Check for any leaks at this time.

6. Allow the engine to cool until the ambient temperature is reached, then check the coolant level in the

reservoir, it should be at or above the FULL COLD line. If necessary, add coolant to top-off the system.

FUEL SYSTEM

Fuel System Service Precautions

Safety is the most important factor when performing not only fuel system maintenance but any type of maintenance. Failure to conduct maintenance and repairs in a safe manner may result in serious personal injury or death. Maintenance and testing of the vehicle's fuel system components can be accomplished safely and effectively by adhering to the following rules and guidelines.

• To avoid the possibility of fire and personal injury, always disconnect the negative battery cable unless the repair or test procedure requires that battery voltage be applied.

• Always relieve the fuel system pressure prior to disconnecting any fuel system component (injector, fuel rail, pressure regulator, etc.), fitting or fuel line connection. Exercise extreme caution whenever relieving fuel system pressure to avoid exposing skin, face and eyes to fuel spray. Please be advised that fuel under pressure may penetrate the skin or any part of the body that it contacts.

• Always place a shop towel or cloth around the fitting or connection prior to loosening to absorb any excess fuel due to spillage. Ensure that all fuel spillage (should it occur) is quickly removed from engine surfaces. Ensure that all fuel soaked cloths or towels are deposited into a suitable waste container.

• Always keep a dry chemical (Class B) fire extinguisher near the work area.

• Do not allow fuel spray or fuel vapors to come into contact with a spark or open flame.

• Always use a backup wrench when loosening and tightening fuel line connection fittings. This will prevent unnecessary stress and torsion to fuel line piping. Always follow the proper torque specifications.

• Always replace worn fuel fitting O-rings with new. Do not substitute fuel hose or equivalent, where fuel pipe is installed.

Fuel System Pressure

RELIEVING

---- CAUTION ----

Fuel injection systems remain under pressure, even after the engine has been turned OFF. The fuel system pressure must be relieved before disconnecting any fuel lines. Failure to do so may result in fire and/or personal injury.

2.0L and 2.2L (1993–94) Engines

1. Loosen the fuel filler cap to relieve tank vapor pressure.
2. Remove the fuel pump fuse.
3. Start the engine and run until the fuel supply remaining in the fuel pipes is consumed. Engage the starter for 3 seconds to assure relief of any remaining pressure.
4. Disconnect the negative battery cable to avoid possible fuel discharge if an accidental attempt is made to start the engine.

2.2L (1995–97), 2.3L, 2.4L and 3.1L Engines

1. Disconnect the negative battery cable.
2. Release the fuel vapor pressure in the fuel tank by momentarily removing the tank filler cap.
3. Connect J-34730-1 or equivalent fuel pressure gauge, to the fuel pressure connection located on the end of the fuel rail assembly. Wrap a cloth around the fitting to absorb any fuel leakage.
4. Install the bleed hose into an approved container and open the valve to bleed system pressure. The fuel pipe connections are now safe for servicing.
5. Drain any fuel remaining in the fuel pressure gauge into an approved container.
6. Once the tests or repairs are completed, prime the fuel system by cycling the ignition switch **ON** for 2 seconds, **OFF** for 10 seconds and then **ON** again. Repeat, if necessary to build system pressure.

Idle Speed

ADJUSTMENT

The engine idle speed is controlled by the Powertrain Control Module (PCM) and the Idle Air Control (IAC) valve. There are no adjustments that can be made externally. During idle, the proper position of the IAC valve is calculated based on inputs from the battery voltage, coolant temperature, engine load, and engine RPM.

Mixture

ADJUSTMENT

Fuel mixture is controlled by the Powertrain Control Module (PCM). No adjustments are necessary or possible.

Fuel Filter

REMOVAL AND INSTALLATION

The fuel filter is located under the rear of the vehicle, rearward of the fuel tank. Note that there is an additional filter/strainer inside the fuel tank attached to the fuel pump/sending unit assembly.

---- CAUTION ----

Fuel injection systems remain under pressure, even after the engine has been turned OFF. The fuel system pressure must be relieved before disconnecting any fuel lines. Failure to do so may result in fire and/or personal injury.

1. Relieve the fuel system pressure using the recommended procedure.
2. Disconnect the negative battery cable.
3. Raise and safely support the vehicle.
4. Using a backup wrench on the fuel filter, disconnect the fuel line from the filter.

---- CAUTION ----

If a nylon fuel line becomes kinked and cannot be straightened, replace it. Some technicians use compressed air to blow dirt from the filter's quick-connect fittings; be sure to wear safety glasses.

5. Grasp the filter and nylon connection line fitting. Twist the quick-connect fitting ¼ turn in each direction to loosen any dirt within the fitting. Disconnect the quick-connect fitting from the fuel filter by compressing the tabs while pulling outward on the line. GM also has available special tool J-38778, that is placed between the fuel filter and quick-connect fitting release mechanism to force the 2 apart.
6. Remove the fuel filter from the mounting bracket.
To install:
7. Before installing a new filter, always apply a few drops of clean engine oil to the male tube end of the filter and to the fuel sending unit assembly connection. This will help ensure proper connection and prevent possible fuel leaks. During normal operation, the O-rings located in the female connector will swell and may prevent proper connection if not lubricated.
8. Install the fuel filter in the mounting bracket.
9. Connect the quick-connect fitting to the fuel filter using the following procedure:
   a. Apply a few drops of clean engine oil to the male ends of the filter and the fuel sender assembly.
   b. Push the connectors together to cause the retaining tabs/fingers to snap into place.
   c. Once installed, pull on both ends of each connection to make sure they are secure.
10. Using a new O-ring, tighten the fuel line using a backup wrench on the fuel filter. Tighten the fuel line fitting to 20 ft. lbs. (27 Nm).
11. Lower the vehicle.
12. Connect the negative battery cable.
13. Pressurize the fuel system and verify no leaks.

Fuel Pump

REMOVAL AND INSTALLATION

---- CAUTION ----

Fuel injection systems remain under pressure, even after the engine has been turned OFF. The fuel system pressure must be relieved before disconnecting any fuel lines. Failure to do so may result in fire and/or personal injury.

1. Relieve the fuel system pressure using the recommended procedure.
2. Disconnect the negative battery cable.
3. Drain the fuel tank, then remove the fuel tank from the vehicle.
4. While holding the modular fuel sender assembly down, remove the snapring from the designated slots located on the retainer.

WARNING

The modular fuel sender assembly may spring up from its position. When removing the modular fuel sender from the tank, be aware that the reservoir bucket is full of fuel. It must be tipped slightly during removal to avoid damage to the float.

5. Remove the external fuel strainer.

6. Detach the Connector Position Assurance (CPA) piece from the electrical connector and detach the fuel pump electrical connector.

7. Gently release the tabs on the sides of the fuel sender at the cover assembly. Begin by squeezing the sides of the reservoir and releasing the tab opposite the fuel level sensor. Move clockwise to release the second and third tabs in the same manner.

8. Lift the cover assembly out far enough to detach the fuel pump electrical connection.

9. Rotate the fuel pump baffle counterclockwise and remove the baffle and pump assembly from the retainer.

10. Slide the fuel pump outlet out of slot, then remove the fuel pump outlet seal.

To install:

11. Install the fuel pump outlet seal, then slide the fuel pump outlet in the slots of reservoir cover.

12. Install the fuel pump and baffle assembly onto the reservoir retainer and rotate clockwise until seated.

13. Install the lower retainer assembly partially into the reservoir. Line up all 3 sleeve tabs. Press the retainer onto the reservoir making sure all 3 tabs are firmly seated.

NOTE: Gently pull on the fuel pump reservoir from retainer to assure a secure fastening. If not secure, replace the entire fuel sender.

14. Attach the fuel pump connector.

15. Fasten the CPA connector to the fuel sender cover.

16. Install a new external fuel strainer.

17. Install the modular fuel sender.

18. Install the fuel tank in the vehicle.

19. Connect the negative battery cable.

20. Pressurize the fuel system and verify no leaks.

Fuel Injector

REMOVAL AND INSTALLATION

CAUTION

Fuel injection systems remain under pressure, even after the engine has been turned OFF. The fuel system pressure must be relieved before disconnecting any fuel lines. Failure to do so may result in fire and/or personal injury.

2.0L Engine

1. Relieve the fuel system pressure.

2. Disconnect the negative battery cable.

3. Detach and cap the fuel lines from the fuel rail.

4. Disconnect the fuel line from the pressure regulator and remove the regulator mounting bolts and the regulator.

5. Detach the electrical connectors from the fuel rail.

6. Remove the 3 fuel rail mounting bolts and remove the fuel rail from the intake manifold.

7. Release the clip on the injector and remove the injector from the fuel rail.

To install:

8. Install new O-rings on the fuel injector and a new clip on the fuel rail.

9. Snap the injector onto the fuel rail with the electrical connector pointing up.

10. Install the fuel rail onto the intake manifold so all the injectors seat in their respective cavities.

11. Tighten the 3 mounting bolts to 20 ft. lbs. (27 Nm).

12. Attach the electrical connectors to the fuel injectors.

13. Install the pressure regulator and tighten the mounting bolts to 20 ft. lbs. (27 Nm). Connect the fuel line to the regulator.

14. Connect the fuel lines to the fuel rail.

15. Connect the negative battery cable.

16. Pressurize the fuel system and verify that there are no leaks.

2.2L Engine

The fuel injectors used in this engine are called Bottom Feed Multiport

Fuel Injectors. These injectors are solenoid-operated devices controlled by the Powertrain Control Module (PCM).

1. Relieve the fuel system pressure using the recommended procedure.

2. Disconnect the negative battery cable.

3. Remove the upper intake manifold assembly.

4. Remove the fuel return line bracket mounting bolt and remove the bracket.

5. Position the return line away from the pressure regulator.

6. Remove the pressure regulator as follows:

   a. Disconnect the vacuum hose from the regulator.

   b. Remove the fuel return pipe clamp.

   c. Disconnect the fuel return pipe from the pressure regulator and discard the O-ring.

   d. Remove the pressure regulator bracket mounting bolt.

   e. Remove the regulator assembly and O-ring and discard the O-ring.

WARNING

To prevent damage to the injector retaining bracket and/or the injectors, do not try to remove the injectors by lifting up the injector retaining bracket while the injectors are still installed in the bracket slots. Do not attempt to remove the bracket without first removing the fuel pressure regulator.

7. Remove the fuel injector retaining bracket mounting bolts.

8. Remove the fuel injector retaining bracket by carefully sliding it off to clear the fuel injector slots.

9. Disconnect the fuel injector electrical connector(s), then pull the injector(s) out.

CAUTION

To reduce the risk of fire and personal injury, verify that the lower (small) O-ring of each injector does not remain in the lower manifold. If the O-ring is not removed with the injector, the replacement injector with new O-rings installed, will not seat properly in the injector socket and could cause a fuel leak.

10. Remove and discard the old O-rings.

NOTE: Each injector is calibrated for a specific flow rate. When replacing fuel injectors, be sure to order the correct injector for the application being serviced. Cover injector sockets to prevent dirt and other contaminates from entering into open fuel passages.

To install:

11. Coat new O-rings with clean engine oil, then install them in the injector(s).

12. Install the injector into the cavity in the lower intake manifold with the electrical connector facing inward.

13. Install the injector retaining bracket and mounting screws. Carefully install the bracket so the injector retaining slots and regulator are aligned with the bracket slots.

14. Connect the fuel injector electrical connector.

15. Install the pressure regulator as follows:

    a. Lubricate the regulator assembly and O-ring and install the O-ring on the regulator.

    b. Install the regulator onto the intake manifold.

    c. Install the pressure regulator bracket mounting bolt after coating the threads with thread locking compound. Tighten the mounting bolt to 31 inch lbs. (3.5 Nm).

    d. Connect the vacuum hose to the regulator.

    e. Connect the fuel return pipe to the pressure regulator and tighten the fitting to 22 ft. lbs. (30 Nm).

    f. Install the fuel return pipe clamp to the lower intake manifold.

16. Install the upper intake manifold assembly.

17. Connect the negative battery cable.

18. Pressurize the fuel system and verify that there are no leaks.

2.3L and 2.4L Engines

1. Relieve the fuel system pressure.

2. Disconnect the negative battery cable.

3. Detach the electrical connector from the fuel injector(s).

4. Remove the fuel rail assembly from the cylinder head.

NOTE: It is not necessary to separate the rail from the fuel pipes.

5. Spread the injector retainer clip to release injector from the rail extrusion flange.

6. Remove the fuel injector assembly.

NOTE: Each fuel injector is calibrated for a specific application, therefore it is important to order the identical part number that is inscribed on the old injector.

To install:

7. Lubricate the new injector O-ring seals with clean engine oil and install on injector assembly.

8. Position a new retainer clip on the injector (on the right side of the electrical connector).

9. Attach the fuel injector to the fuel rail with the electrical connector facing outward.

10. Install the fuel rail assembly.

11. Attach the electrical connector.

12. Connect the negative battery cable.

13. Turn the ignition key to the **ON** position to pressurize the fuel rail and verify no leaks.

3.1L Engine

1. Relieve the fuel system pressure.

2. Disconnect the negative battery cable.

3. Remove the upper intake manifold assembly.

4. Remove the fuel line bracket bolt.

5. Disconnect and cap the fuel lines from the fuel rail. Use a backup wrench on the fuel rail fittings to prevent them from turning.

6. Remove the O-rings from the lines and discard.

7. Disconnect the vacuum line from the pressure regulator.

8. Remove the fuel rail mounting bolts.

9. Disconnect the fuel injector electrical connectors.

10. Remove the fuel rail assembly.

11. Remove the injector retaining clip and discard.

12. Remove the injector assembly from the fuel rail, then remove and discard the old O-rings.

To install:

13. Coat new O-rings with clean engine oil, then install them on the injector.

14. Install a new retaining clip on the fuel injector. Position the open end of the clip toward the injector electrical connector.

15. Install the injector onto the fuel rail assembly with the injector electrical connector facing outward.

16. Install the fuel rail onto the intake manifold so all the injectors fit into their cavities.

17. Install the fuel rail mounting bolts and tighten to 89 ft. lbs. (10 Nm).

18. Attach the electrical connectors to the fuel injectors.

19. Connect the vacuum line to the pressure regulator.

20. Install new O-rings coated with engine oil on the fuel lines, then connect the fuel feed and return lines and tighten the fittings to 17 ft. lbs. (23 Nm). Use a backup wrench on the fuel rail fittings to prevent them from turning.

21. Install the fuel line bracket bolt.

22. Install the upper intake manifold assembly.

23. Connect the negative battery cable.

24. Pressurize the fuel system and verify that there are no leaks.

ENGINE MECHANICAL

Engine Assembly

REMOVAL AND INSTALLATION

2.0L Engine

CAUTION

Fuel injection systems remain under pressure, even after the engine has been turned OFF. The fuel system pressure must be relieved before disconnecting any fuel lines. Failure to do so may result in fire and/or personal injury.

---

1. Relieve the fuel system pressure.

2. Disconnect the negative, then the positive battery cables.

3. Remove the battery.

4. Drain the cooling system.

5. If equipped, properly discharge the A/C system and remove the compressor.

6. Remove the air cleaner assembly.

7. Disconnect the upper and lower radiator hoses.

8. Detach the following electrical connectors:
- Engine harness at the bulkhead
- Master cylinder
- A/C cluster relay switches
- Wiper motor
- Cooling fan relay and ground
- ECM and pull the harness through the bulkhead
- Coolant switch at thermostat housing
- A/C pressure sensor
- Blower motor
- ABS wiring
- Ground wires

9. Disconnect the vacuum hoses at the throttle body and intake manifold.

10. Detach the heater hoses and the return hose from the water pump.

11. Disconnect the control cables from the throttle body and intake bracket.

12. Disconnect the shift control cable from the transaxle and mounting bracket.

13. Install an engine support fixture, J-28467-A or the equivalent.

14. Remove the transaxle mount and bracket.

15. Disconnect and cap the fuel lines.

16. Disconnect and plug the power steering pump pressure and return hoses.

17. Raise and safely support the vehicle.

18. Detach the VSS connector at the PM generator.

19. Disconnect the exhaust pipe from the exhaust manifold and hangers.

20. If equipped with an automatic transaxle, disconnect the transaxle cooler lines.

21. Disconnect the refrigerant lines from the A/C compressor.

22. Remove the front tire and wheel assemblies.

23. Remove the front air deflector.

24. Remove the right and left side splash shields.

25. Disconnect the right and left ABS wheel sensors connectors.

26. Remove the lower ball joint cotter pins and castle nuts, then separate the ball joints from the steering knuckles.

27. Remove the transaxle mount strut.

28. Disconnect all wiring from the transaxle.

29. Remove the right and left suspension supports.

30. Disconnect the drive axles from the transaxle.

31. Remove the starter.

32. Remove the flywheel cover.

33. Remove the rear engine mount.

34. Position an engine dolly under the engine and transaxle assembly.

35. Lower the vehicles weight onto the dolly lightly.

36. With the engine and transaxle supported on the dolly, remove the engine support fixture.

37. Raise the vehicle off the dolly, leaving the engine and transaxle on the dolly.

38. Separate the engine and transaxle.

To install:

39. Install the transaxle on the engine and tighten the mounting bolts to 55 ft. lbs. (75 Nm).

40. Position the assembly on the engine dolly and locate the dolly under the engine bay.

41. Lower the vehicle until the powertrain assembly is in the engine bay.

42. Install the engine support fixture.

43. Raise and safely support the vehicle and remove the dolly.

44. Install the rear engine mount.

45. Install the front engine mount.

46. Install the flywheel housing cover.

47. Install the starter.

48. Connect the transaxle cooler lines.

49. Attach the heater hoses to the heater core.

50. Connect the refrigerant lines to the A/C compressor.

51. Attach the cooling fan connector.

52. Connect the driveaxles to the transaxle.

53. Install the right and left suspension supports.

54. Install the transaxle strut.

55. Connect the ball joints to the steering knuckles and tighten the castle nuts to 55 ft. lbs. (75 Nm). Install new cotter pins.

56. Attach the ABS sensor connectors.

57. Install the left and right splash shields.

58. Install the front air deflector.

59. Install the front tire and wheel assemblies and tighten to specification.

60. Uncap and connect the fuel lines.

61. Connect the exhaust pipe to the exhaust manifold.

62. Attach the VSS connector.

63. Lower the vehicle.

64. Connect the control cables to the transaxle and throttle body lever.

65. Connect the vacuum lines the engine and transaxle.

66. Attach the following electrical connectors:
- Engine harness at the bulkhead
- Master cylinder
- A/C cluster relay switches
- Wiper motor
- Cooling fan relay and ground
- ECM and push the harness through the bulkhead
- Coolant switch at thermostat housing
- A/C pressure sensor
- Blower motor
- ABS wiring
- Ground wires

67. Install the engine-to-transaxle bracket.

68. Remove the support fixture.

69. Connect the upper and lower radiator hoses.

70. Refill the cooling system and recharge the A/C system, if equipped.

71. Install the battery.

72. Connect the positive, then the negative battery cables.

73. Start the vehicle and verify that there are no leaks.

2.2L Engine

CAUTION

Fuel Injection systems remain under pressure, even after the engine has been turned OFF. The fuel system pressure must be relieved before disconnecting any fuel lines. Failure to do so may result in fire and/or personal injury.

---

1. Relieve the fuel system pressure.

2. Disconnect the negative, then the positive battery cables.

3. Drain the cooling system into a suitable container.

4. Remove the throttle body air inlet duct.

5. Remove the battery from the vehicle.

6. Remove the air cleaner assembly.

7. Remove the upper radiator.

8. Disconnect the vacuum hose from power brake booster.

9. Remove the alternator upper brace and disconnect the wiring from the alternator.

10. Tag and disconnect the electrical wiring from the following components:
- Oxygen (O_2) sensor
- Fuel injector harness
- Idle Air Control (IAC)
- Throttle Position Sensor (TPS)
- Coolant Temperature Sender
- Park/neutral switch
- TCC solenoid
- Transaxle shift solenoid

- Transaxle ground
- MAP sensor
- EGR
- Cooling fan

11. If equipped, properly discharge and recover the refrigerant from the A/C system.

12. Disconnect the compressor-to-condenser and accumulator lines.

13. Remove the slave cylinder from the transaxle, if equipped with a manual transaxle.

14. Disconnect and cap the fuel lines.

15. Disconnect the shift linkage from the transaxle bracket.

16. Disconnect the control cables from the throttle body lever and remove the cable bracket from the rocker arm cover and intake manifold.

17. Place a pan under the power steering pump and disconnect and cap the power steering lines.

18. Install an engine support fixture, J-28467, or the equivalent.

19. Safely raise and support the vehicle.

20. Remove the front tire and wheel assemblies.

21. Remove the right and left inner fender splash shields.

22. Disconnect the front exhaust pipe from manifold and remove the front exhaust pipe.

23. Tag and disconnect the following electrical connectors from the lower engine components:
- Ignition assembly
- Starter
- Vehicle Speed Sensor (VSS)
- Transaxle ground wire

24. Remove the flywheel cover.

25. Remove the lower radiator hose.

26. Disconnect the heater hoses from the heater core pipes.

27. Disconnect the transaxle cooler lines at the radiator.

28. Remove the mounting bolt from the lower engine strut at the suspension crossmember.

29. Disconnect the both front ABS wheel sensor connectors and unfasten the harnesses from the suspension crossmember.

30. Remove the cotter pins and castle nuts from the lower ball joints, then separate the joints from the steering knuckles.

31. Remove the suspension crossmembers.

32. Disconnect the halfshafts from the transaxle and support out of the way.

33. Position a suitable table under the engine and transaxle assembly.

34. Lower the vehicle until the powertrain assembly is on the table.

35. Remove the transaxle-to-frame mount bolts.

36. Remove the intermediate bracket from the right engine mount support bracket.

37. Remove the engine support fixture.

38. Raise the vehicle, leaving the powertrain assembly on the table. When raising the vehicle, take it up slowly and verify that no lines are still connected to the powertrain assembly.

39. Remove the torque converter-to-flywheel bolts and transaxle-to-engine mounting bolts and remove the transaxle from the engine.

To install:

40. Install the transaxle on the engine and tighten the mounting bolts to 68 ft. lbs. (93 Nm).

41. Install flywheel-to-converter bolts and tighten to 46 ft. lbs. (62 Nm).

42. Position the engine on the table and lower the vehicle until the engine is in the engine bay.

43. Install an engine support fixture, J-28467, or the equivalent.

44. Raise and safely support the vehicle and remove the engine table.

45. Connect the transaxle cooler lines to the radiator.

46. Attach the heater hoses to the heater core outlet pipes.

47. Install the lower radiator hose.

48. Connect the engine mount support bracket to the intermediate bracket and tighten the bolts to 96 ft. lbs. (103 Nm).

49. Install the transaxle mount-to-body bolts and tighten to 40 ft. lbs. (54 Nm).

50. Connect the halfshafts to the transaxle.

51. Install the suspension crossmembers and control arms in the vehicle and tighten the mounting bolts to 89 ft. lbs. (120 Nm).

52. Connect the lower ball joints to the steering knuckles and tighten the mounting bolts to 48 ft. lbs. (65 Nm) and install 2 new cotter pins.

53. Attach the front ABS wheel speed sensors to the harness connector and the subframe clips.

54. Connect the engine strut to the suspension support and tighten the mounting bolt to 89 ft. lbs. (120 Nm).

55. Install the transaxle cover.

56. Connect the following electrical connectors to the lower engine components:
- Ignition assembly
- Stater
- Vehicle Speed Sensor (VSS)
- Transaxle ground wire

57. Install the front exhaust pipe and tighten the pipe-to-manifold bolts to 22 ft. lbs. (30 Nm).

58. Install the left and right inner fender splash shields.

59. Install the front tire and wheel assemblies and tighten to specification.

60. Lower the vehicle.

61. Attach the power steering lines to the pump.

62. Connect the control cables to the throttle body lever and mounting bracket.

63. Attach the shift linkage to the transaxle and mounting bracket.

64. Connect the fuel lines.

65. Install the clutch slave cylinder, if equipped.

66. Connect the A/C compressor-to-accumulator and condenser lines.

67. Recharge the A/C system.

68. Connect the electrical wiring to the following components:
- Oxygen sensor (O_2)
- Fuel injector harness
- Idle Air Control (IAC)
- Throttle Position Sensor (TPS)
- Coolant Temperature Sender
- Park/neutral switch
- TCC solenoid
- Transaxle shift solenoid
- Transaxle ground
- MAP sensor
- EGR
- Cooling fan

69. Install the alternator upper mounting bracket and connect the alternator wiring.

70. Connect the vacuum hose to the power brake booster.

71. Install the upper radiator hoses.

72. Install the air cleaner.

73. Install the battery cable and connect the positive cable.

74. Install the throttle body inlet duct.

75. Connect the negative battery cable.

76. Fill the engine with the correct amounts and types of coolant and engine oil.

77. Refill and bleed the power steering system.

78. Start the vehicle and verify no leaks.

NOTE: Whenever the vehicle sub-frame is removed or lowered, the wheel alignment should be checked.

2.3L and 2.4L Engines

1. If equipped with A/C, have a repair shop recover the refrigerant using the proper equipment.

2. Disconnect the negative battery.

3. Properly drain the cooling system into an approved container.

4. Relieve the fuel system pressure.

5. Remove the left sound insulator, then disconnect the clutch pushrod from the pedal assembly.

6. Disconnect the heater hose at the thermostat assembly, then detach the radiator inlet (upper) hose.

7. Remove the air cleaner assembly and the coolant fan.

8. If equipped with A/C, disconnect the compressor/condenser hose assembly at the compressor, then discard the O-rings.

9. Disconnect the 2 vacuum hoses from the front of the engine.

10. Tag and detach the following electrical connectors:
- Alternator
- A/C compressor (if equipped)
- Fuel injector harness
- Idle Air Control (IAC) and TP sensor at the throttle body
- Manifold Absolute Pressure (MAP) sensor
- Intake Air Temperature (IAT) sensor
- EVAP canister purge solenoid
- Starter solenoid
- Ground connections
- Negative battery cable from the transaxle
- Electronic ignition coil and module assembly
- Engine Coolant Temperature (ECT) sensor(s)
- Oil pressure sensor/switch
- Oxygen (O₂) sensor
- Crankshaft Position (CKP) sensor
- Back-up lamp switch, then position the harness aside

11. Disconnect the power brake vacuum hose from the throttle body. Detach the power brake vacuum tube-to-check valve hose from the tube.

12. Remove the throttle cable and bracket.

13. Unfasten the power steering pump rear bracket, then remove the bracket and vacuum tube as an assembly.

14. Unfasten the power steering pivot bolt, then remove the pump and drive belt. Position the pump aside, with the lines still attached.

15. Carefully disconnect the fuel lines.

16. Disconnect the shift cables. Detach the clutch actuator line.

17. Remove the exhaust manifold and heat shield.

18. Disconnect the radiator outlet (lower) hose from the radiator.

19. Install J-28467-A or equivalent engine support fixture.

20. Unfasten the bolt attaching the coolant recovery/surge tank, then position the tank aside with the hoses still connected.

21. Remove the engine mount assembly.

22. Raise and safely support the vehicle, then remove the front wheel and tire assemblies. Remove the right splash shield.

23. Remove the radiator air deflector.

24. Tag and detach the following electrical connections:
- Vehicle Speed Sensor (VSS)
- Knock sensor
- Starter solenoid
- If equipped, both front ABS wheel speed sensors

25. Remove the engine mount strut and the transaxle mount.

26. Separate the ball joints from the steering knuckles.

27. Remove the suspension supports, crossmember, and stabilizer shaft as an assembly.

28. Disconnect the heater outlet hose from the radiator outlet pipe.

29. Remove the axle shaft from the transaxle and intermediate shaft, then position aside.

30. If equipped, disconnect the A/C lines from the oil pan.

31. Remove the flywheel housing cover.

32. Position a suitable support below the engine, then carefully lower the car onto the support.

33. Matchmark the threads on the support fixture hooks so the setting can be duplicated when reinstalling the engine. Remove the engine support fixture J-hooks.

34. Raise the vehicle slowly off the engine and transaxle assembly. If may be necessary to move the engine/transaxle assembly rearward to clear the intake manifold.

35. Noting the position of the bolts, separate the engine from the transaxle.

To install:

------------ **WARNING** ------------
Be sure the retaining bolts are in their correct locations. If not, engine damage may occur.

36. Assemble the engine to the transaxle.

37. Position the engine and transaxle assembly under the engine compartment, then slowly lower the vehicle over the assembly until the transaxle mount is indexed, then install the retaining bolt.

38. Install engine support fixture J-28467-A or equivalent, making sure to adjust it to the previous setting.

39. Install the engine mount assembly and transaxle mount.

40. Carefully raise the vehicle off the support.

41. Attach the axle shafts to the transaxle.

42. Connect the heater outlet hose to the radiator outlet pipe.

43. Install the suspension supports, crossmember and stabilizer shaft assembly.

44. Attach the ball joints to the steering knuckles, then secure with the nuts.

45. Install the engine strut mount.

46. If equipped, connect the A/C line to the oil pan.

47. Attach the following electrical connectors, as tagged during removal:
- Vehicle Speed Sensor (VSS)
- Knock sensor
- Starter solenoid
- If equipped, both front ABS wheel speed sensors

48. Install the flywheel housing cover.

49. Fasten the radiator air deflector.

50. Connect the lower radiator hose.

51. Install the right splash shield, then the front wheel and tire assemblies.

52. Carefully lower the vehicle, then remove the engine support fixture.

53. Install the coolant recovery/surge tank, then secure using the retaining bolt.

54. Attach the following electrical connections, as tagged during removal:
- Alternator
- A/C compressor (if equipped)
- Fuel injector harness
- Idle Air Control (IAC) and TP sensor at the throttle body
- Manifold Absolute Pressure (MAP) sensor
- Intake Air Temperature (IAT) sensor
- EVAP canister purge solenoid
- Starter solenoid
- Ground connections
- Negative battery cable to the transaxle
- Electronic ignition coil and module assembly
- Engine Coolant Temperature (ECT) sensor(s)
- Oil pressure sensor/switch

- Oxygen (O$_2$) sensor
- Crankshaft Position (CKP) sensor
- Back-up lamp switch

55. Attach the vacuum hoses.

56. If equipped with A/C, attach the compressor/condenser hose assembly to the compressor.

57. Fasten the clutch actuator line.

58. Install the exhaust manifold and heat shield.

59. Connect the fuel lines.

60. Connect the positive battery cable.

61. Fasten the power steering pump pivot-to-block bolt. Install the power steering pump rear bracket and tension belt.

62. Connect the vacuum hoses to the intake manifold and to the tube from the brake booster.

63. Install the throttle cable and bracket.

64. Install the coolant fan and air cleaner assembly.

65. Attach the radiator outlet (upper) hose. Fill the cooling system with the proper type and quantity of coolant.

66. Connect the clutch pushrod to the pedal assembly, then install the left sound insulator.

67. Attach the heater hose at the thermostat housing.

68. Fill the transaxle with fluid, then fill the crankcase with oil.

69. Connect the negative battery cable.

70. If equipped with A/C, evacuate, charge and leak test the system.

71. Start the engine and check the fluid levels, proper operation of the engine and/or fluid leakage.

3.1L Engine

This engine is removed from the top of the vehicle.

1. Remove the air cleaner and duct assembly.

───── CAUTION ─────
Fuel injected vehicles maintain high fuel pressure within the fuel system, even after the engine has been shut OFF. Fuel system residual pressure must be properly relieved, prior to attempting this procedure; failure to do so can result in serious personal injury and/or property damage.

2. Relieve fuel system pressure, then disconnect the negative battery cable and the engine ground wire. Disconnect the positive battery cable,

then remove the battery from the vehicle.

3. Drain the cooling system.

4. If equipped, discharge the A/C system, using a suitable recovery system.

5. Remove the exhaust manifold crossover assembly.

6. Remove the serpentine drive belt, then remove the tensioner from the front of the engine. If equipped, remove the idler from the engine.

7. Disconnect the radiator hoses from the engine.

8. Detach the cables from the plenum cable bracket.

9. Remove the alternator from the engine.

10. Tag and disconnect the wiring harness, at the engine.

11. Disconnect and cap the fuel lines, then remove the coolant bypass and overflow hoses from the engine.

12. Install J-28467-A, or an equivalent engine support fixture, to the engine assembly.

13. Raise the vehicle and support it safely.

14. Remove the right inner fender splash shield, then remove the flywheel cover.

15. Remove the starter assembly from the engine.

16. If equipped, make sure the refrigerant has been properly recovered, then remove the A/C compressor from the engine.

17. Disconnect the exhaust pipe from the rear of the exhaust manifold.

18. Remove the flywheel-to-torque converter bolts.

19. Remove the engine mounts from the vehicle.

20. If equipped with a manual transaxle, disconnect the intermediate shaft bracket from the engine.

21. Disconnect the shift cable bracket at the transaxle, then remove the lower bellhousing-to-engine bolts.

22. Lower the vehicle.

23. Disconnect the heater hoses.

24. Install an engine lifting device, then remove the engine support fixture.

25. Support the transaxle, then remove the remaining transaxle-to-engine bolts.

NOTE: Although it is not necessary to remove the hood from the engine compartment, it may be easier to remove the engine if done. Using an awl, scribe marks around the hood hinges to help aid correct hood alignment upon installation.

26. Remove the engine assembly from the vehicle.

To install:

27. Lower the engine assembly into position in the vehicle.

28. Install the engine support fixture, then remove the lifting device.

29. Install the upper transaxle-to-engine bolts and tighten to 55 ft. lbs. (75 Nm).

30. Install the heater hoses, then raise and support the vehicle safely.

31. Install and tighten the lower transaxle bolts to 55 ft. lbs. (75 Nm).

32. Connect the shift cable bracket to the transaxle.

33. If equipped with a manual transaxle, install the intermediate shaft bracket to the engine.

34. Install the engine mounts to the vehicle.

35. Install the flywheel-to-converter bolts and tighten to 52 ft. lbs. (70 Nm).

36. Reconnect the exhaust pipe to the rear of the manifold, then install the air conditioning compressor.

37. Install the starter assembly to the engine, then install the flywheel cover.

38. Install the right inner fender splash shield, then lower the vehicle and remove the engine support fixture.

39. Attach the coolant bypass and overflow hoses to the engine.

40. Uncap and connect the fuel lines.

41. Attach the wiring harness to the engine, then install the alternator assembly.

42. Install the cables to the bracket on the plenum, then connect the radiator hoses.

43. If equipped, install the idler.

44. Install the drive belt tensioner assembly, then install the serpentine belt to the front of the engine.

45. Install the exhaust manifold crossover assembly.

46. Install the battery to the vehicle, then connect the positive battery cable followed by the negative battery cable.

47. Install the air cleaner and duct assembly.

48. Fill cooling system and check for leaks. Start the engine and allow it to warm until normal operating temperature is reached, then top-off the coolant.

49. If equipped, evacuate, recharge and leak test the air conditioning system.

50. Recheck for coolant, fuel, oil and transaxle fluid for proper levels or leaks.

Engine Mounts

REMOVAL AND INSTALLATION

2.0L Engine

Front Engine Mount

1. Disconnect the negative battery cable.
2. Install an engine support fixture, J-28467-A, or the equivalent.
3. Remove the 2 engine mount-to-engine bracket bolts.
4. Remove the 2 top engine mount-to-frame bolts.
5. Raise and safely support the vehicle.
6. Remove the right side splash shield.
7. Remove the lower engine mount-to-frame bolt.
8. Remove the engine mount from the vehicle.

To install:

9. Install the engine mount and install the lower mounting bolt and tighten to 40 ft. lbs. (54 Nm).
10. Install the right side splash shield.
11. Lower the vehicle.
12. Install the 2 upper engine mount bolts and tighten to 40 ft. lbs. (54 Nm).
13. Install the engine mount-to-bracket bolts and tighten to 40 ft. lbs. (54 Nm).
14. Remove the engine support fixture.
15. Connect the negative battery cable.

Rear Engine Mount

1. Disconnect the negative battery cable.
2. Install an engine support fixture, J-28467-A, or the equivalent.
3. Remove the 2 engine mount-to-engine bracket bolts.
4. Raise and safely support the vehicle.
5. Remove the lower engine mount nuts and reinforcement.
6. Remove the engine mount from the vehicle.

To install:

7. Install the engine mount and install the lower mounting nuts and reinforcement. Tighten the mounting nuts to 18 ft. lbs. (24 Nm).
8. Lower the vehicle.
9. Install the engine mount-to-bracket bolts and tighten to 40 ft. lbs. (54 Nm).
10. Remove the engine support fixture.
11. Connect the negative battery cable.

1993–94 2.2L Engine

Front Engine Mount

1. Disconnect the negative battery cable.
2. Install an engine support fixture, J-28467-A, or the equivalent.
3. Raise the engine with the support fixture until the weight is removed from the engine mount.
4. Remove the upper engine mount-to-body bolts.
5. Remove the engine mount-to-engine bracket bolts.
6. Raise and safely support the vehicle.
7. Remove the front air deflector.
8. Remove the right front splash shield.
9. Remove the lower engine mount-to-body bolt.
10. Remove the engine mount.

To install:

11. Install the engine mount in position and tighten the lower mounting engine mount-to-frame bolt to 58 ft. lbs. (78 Nm).
12. Install the right front splash shield.
13. Install the front air deflector.
14. Lower the vehicle.
15. Lower the engine until the engine mount-to-bracket bolts can be installed. Tighten the bolts to 50 ft. lbs. (68 Nm).
16. Install the upper engine mount-to-body bolts and tighten to 58 ft. lbs. (78 Nm).
17. Remove the support fixture.
18. Connect the negative battery cable.

Rear Engine Mount

1. Disconnect the negative battery cable.
2. Install an engine support fixture, J-28467-A, or the equivalent.
3. Raise the engine with the support fixture until the weight is removed from the engine mount.
4. Remove the engine mount-to-frame nuts.
5. Remove the engine mount-to-bracket bolts.
6. Remove the engine mount.

To install:

7. Position the engine mount in the vehicle and install the engine mount-to-bracket bolts and tighten to 38 ft. lbs. (52 Nm).
8. Install the engine mount-to-frame nuts and tighten to 38 ft. lbs. (52 Nm).
9. Lower the vehicle.
10. Remove the support fixture.
11. Connect the negative battery cable.

1995–97 2.2L Engine

Torque axis mounting systems allow more engine movement than others; this movement may seem excessive but that is the design intent of the mounting system. Do **NOT** replace any torque axis mounts for excessive engine movement unless there is damage to the mount or the rubber.

1. Disconnect negative battery cable.
2. Install engine support fixture J-28467-A or equivalent.
3. Raise the engine at the front lift hook with a support fixture enough to remove the engine weight from the engine mount assembly.
4. Remove the coolant recovery tank attaching bolt, and position the tank aside, leaving the hoses connected.
5. Remove the engine mount assembly-to-serpentine drive belt tensioner bracket.
6. Remove the 2 nuts on top of the engine mount assembly.
7. Remove the 2 body attaching nuts from the engine mount assembly.
8. Remove engine mount assembly bracket, then remove the engine mount assembly.

To install:

9. Install the engine mount assembly, then install the bracket.
10. Install the engine mount assembly-to-body attaching nuts, then tighten to 49 ft. lbs. (66 Nm).
11. Install the 2 nuts on top of engine mount assembly and tighten to 33 ft. lbs. (45 Nm).
12. Install the engine mount assembly-to-serpentine drive belt tensioner bracket bolts, and tighten to 96 ft. lbs. (130 Nm).
13. Install the coolant recovery tank.
14. Remove the engine support fixture.
15. Connect the negative battery cable.

2.3L and 2.4L Engines

The mounts are a design called torque axis mounting; they allow more engine movement than other mounting systems. The movement may seem excessive but that is the design intent of the mounting system. Do **NOT** replace any torque axis mounts for excessive engine movement unless there is damage to the mount or the rubber.

Engine Mount Assembly

1. Disconnect negative battery cable.

2. Remove the coolant recovery tank attaching bolt and position tank aside with hoses attached.

3. Install engine support fixture J-28467-A or equivalent to hold the weight of the engine.

4. Remove the nuts holding mount to body.

5. Remove the bolts holding engine bracket to mount, then remove the mount assembly.

To install:

6. Install engine mount assembly, then install the nuts holding the mount to the body, but do not tighten them at this time.

7. Install new replacement bolts holding engine bracket to mount.

8. Tighten the bolts and nuts to specifications.

9. Remove the engine support fixture.

10. Install coolant recovery tank.

11. Connect negative battery cable.

Engine Mount Strut

1. Disconnect the negative battery cable.

2. Raise and safely support the vehicle.

3. Remove the right front splash shield.

4. Remove the engine mount strut bolts.

5. Remove the engine mount struts.

To install:

6. Install the engine mount strut and bolts.

7. Tighten the bolts to 88 ft. lbs. (120 Nm).

8. Install the right front splash shield.

9. Lower the vehicle.

10. Connect the negative battery cable.

3.1L Engine

Front Engine Mount

1. Disconnect the negative battery cable.

2. Install an engine support fixture, J-28467-A, or the equivalent.

3. Remove the engine mount-to-body bracket bolts.

4. Remove the engine mount-to-engine bracket bolts.

5. Raise and safely support the vehicle.

6. Remove the right side splash shield.

7. Remove the lower engine mount-to-body bracket bolt.

8. Remove the lower engine mount-to-engine bracket bolt and remove the engine mount.

To install:

9. Position the engine mount in the vehicle and install the lower engine mount-to-engine bracket bolt and tighten to 50 ft. lbs. (68 Nm).

10. Install the both engine mount-to-body bracket bolt and tighten to 58 ft. lbs. (78 Nm).

11. Install the right side splash shield.

12. Lower the vehicle.

13. Install the 2 remaining engine mount-to-body bolts and tighten to 54 ft. lbs. (73 Nm).

14. Remove the engine support fixture.

15. Connect the negative battery cable.

Rear Engine Mount

1. Disconnect the negative battery cable.

2. Install an engine support fixture, J-28467-A, or the equivalent.

3. Raise and safely support the vehicle.

4. Remove the engine mount-to-engine bolts.

5. Remove the engine mount-to-frame nuts and remove the engine mount.

To install:

6. Position the engine mount and install the engine mount-to-frame nuts and tighten to 17 ft. lbs. (23 Nm).

7. Install the engine mount-to-engine bolts and tighten to 38 ft. lbs. (52 Nm).

8. Lower the vehicle.

9. Remove the engine support fixture.

10. Connect the negative battery cable.

Cylinder Head

REMOVAL AND INSTALLATION

2.0L Engine

———— WARNING ————
The engine must be overnight cold before removing the cylinder head and camshaft carrier. Working on an engine that has not sufficiently cooled will result in damage to the aluminum engine components.

———— CAUTION ————
Fuel injection systems remain under pressure, even after the engine has been turned OFF. The fuel system pressure must be relieved before disconnecting any fuel lines. Failure to do so may result in fire and/or personal injury.

1. Relieve the fuel system pressure.

2. Disconnect the negative battery cable.

3. Remove the air cleaner assembly.

4. Drain the cooling system.

5. Remove the coolant recovery tank.

6. Remove the fuel vapor pipe assembly.

7. Remove the serpentine belt.

8. Remove the timing belt cover upper mounting bolts and nuts.

9. Remove the serpentine belt tensioner assembly.

10. Raise and safely support the vehicle.

11. Remove the right inner fender splash shield and lower splash shield.

12. If equipped, remove the A/C belt.

13. Remove the crankshaft pulley.

14. Remove the flywheel inspection cover.

15. Unfasten the lower timing cover bolts, then remove the timing belt cover.

16. Rotate the crankshaft until all the timing marks are in alignment.

17. Loosen the water pump mounting bolts and rotate the water pump to release tension on the timing belt. Remove the timing belt from the camshaft sprocket.

18. Lower the vehicle.

19. Disconnect the fuel vapor pipe.

20. Remove the rear timing belt cover.

21. Detach the PCV hose.

22. Disconnect and tag all electrical connections from the throttle body and intake manifold.

23. Disconnect the power steering pressure and return hoses.

24. Remove the exhaust manifold mounting nuts and remove the manifold.

25. Remove the alternator and mounting bracket from the camshaft carrier.

26. Remove the power steering pump.

27. Remove the front and rear engine lift brackets.

28. Remove the breather tube bracket and breather tube.

29. Disconnect the control cables from the throttle body and intake manifold bracket.

30. Disconnect and cap the fuel lines.

31. Tag and disconnect all vacuum lines.

32. Disconnect the heater hoses from the intake manifold.

33. Disconnect the spark plug wires from the spark plugs.

34. Label and disconnect wiring at engine harness and thermostat housing.

35. Remove the camshaft carrier/cylinder head bolts in sequence.

36. Remove the camshaft carrier assembly.

37. Remove the rocker arms, lash adjusters and thrust pieces and keep in order for installation.

38. Remove the cylinder head with the intake manifold.

39. Remove the old cylinder head gasket.

40. Remove the intake manifold from the cylinder head.

To install:

41. Thoroughly clean and dry the mating surfaces and bolt holes. Apply a $\frac{1}{8}$ in. (3mm) continuous bead of RTV sealant to the sealing surface of camshaft carrier.

42. Install the intake and exhaust manifolds on the cylinder head.

43. Install a new head gasket on the cylinder block and install the cylinder head.

44. Install the rocker arms, lash compensators and thrust pieces.

45. Install the camshaft carrier on the cylinder head.

WARNING

The cylinder head bolt tightening sequence requires the engine to be warmed to operating temperature before the final rotation is made on the cylinder head/camshaft carrier mounting bolts. This step must be performed to ensure proper seating of the head gasket.

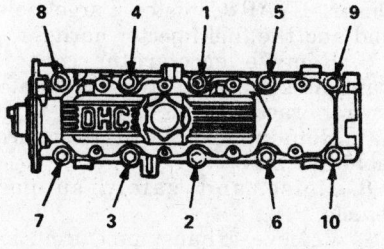

Cylinder head/camshaft carrier bolt tightening sequence — 2.0L engine

201873

46. Install the mounting bolts finger-tight. With all the mounting bolts in place tighten the bolts as follows:
- Step 1: Tighten all bolts in sequence to 18 ft. lbs. (25 Nm).
- Step 2: Tighten each bolt in sequence an additional 60 degrees
- Step 3: Tighten each bolt in sequence another 60 degrees
- Step 4: Tighten each bolt a third time in sequence 60 degrees
- Step 5: Tighten all bolts in sequence one final time 30–50 degrees after engine warm up.

47. Connect the breather to the camshaft carrier.

48. Attach the heater hoses.

49. Connect all wiring to the engine harness and thermostat housing.

50. Attach all vacuum and fuel lines.

51. Connect all electrical connectors to the throttle body and intake manifold.

52. Attach the control cables to the throttle body.

53. Connect the spark plug wires.

54. Install the power steering pump and mounting bracket. Connect the pressure and return lines to the power steering pump.

55. Install the alternator and mounting bracket.

56. Install the camshaft cover.

57. Connect the PCV hose.

58. Install the rear timing belt cover.

59. Connect the fuel vapor pipe.

60. Raise and safely support the vehicle.

61. Connect the front exhaust pipe to the exhaust manifold.

62. Make sure the timing marks are in alignment and install the timing belt.

63. Install the timing belt cover and lower mounting bolts.

64. Install the crankshaft pulley.

65. Install the serpentine belt around the crankshaft pulley.

66. Install the A/C belt.

67. Install the right side splash shields.

68. Install the flywheel cover.

69. Lower the vehicle.

70. Install the serpentine belt tensioner.

71. Install the coolant reservoir tank.

72. Fill all fluids to their proper levels.

73. Connect the negative battery cable.

74. Start the engine and check for leaks.

75. Warm the engine to operating temperature and complete the cylinder head torque sequence.

2.2L Engine

1993–94 Vehicles

CAUTION

Fuel injection systems remain under pressure, even after the engine has been turned OFF. The fuel system pressure must be relieved before disconnecting any fuel lines. Failure to do so may result in fire and/or personal injury.

---

1. Disconnect the negative battery cable.

2. Relieve the fuel system pressure.

3. Drain the cooling system into a suitable container.

4. Remove the air cleaner and air duct assembly.

5. Remove the air inlet resonator upper tie bar.

6. Remove the lower air inlet.

7. Remove the serpentine belt.

8. Remove the alternator.

9. Remove the power steering pump and position it aside without disconnect the power steering lines.

10. Disconnect the spark plug wires and lay them aside.

11. Disconnect the control cables from the throttle body and remove the cable bracket at the throttle body and rocker arm cover.

NOTE: Use care when removing valve train components. Parts to be reused must be returned to their original locations.

12. Remove the rocker arm cover, rocker arm nuts, rocker arms and pushrods.

13. Disconnect the electrical connectors from the intake manifold, throttle body and cylinder head.

14. Detach the oxygen (O_2) sensor connector.

15. Remove the power steering pump bracket from the intake manifold brace, located under the intake manifold.

16. Remove the torque strut and engine-side torque strut mounting bracket.

17. Remove the alternator rear bracket.

18. Tag and disconnect the vacuum lines at the intake manifold and cylinder head.

19. Disconnect the upper radiator hose from the engine.

20. Raise and safely support the vehicle.

21. Disconnect the exhaust pipe from the exhaust manifold.

22. Lower the vehicle.

23. Disconnect and cap the fuel lines at the quick disconnects.

24. Remove the transaxle fill tube.

25. Remove the cylinder head bolts.

26. Remove the cylinder head with both manifolds. Remove the intake and exhaust manifolds from the cylinder head.

To install:

27. Clean all the gasket surfaces completely. Clean the threads on the cylinder head bolts and the block threads.

28. Install the intake and exhaust manifolds on the cylinder head.

29. Place a new cylinder head gasket in position over the dowel pins on the block. Carefully guide the cylinder head into position.

30. New head bolts are recommended. Install all the cylinder head bolts finger-tight. The long bolts go in bolt positions 1, 4, 5, 8 and 9. The short bolt are in positions 2, 3, 6 and 7. The stud is in position 10.

31. Tighten the bolts in sequence. The long bolts to 46 ft. lbs. (63 Nm) and the short bolts and stud to 43 ft. lbs. (58 Nm). Make second pass tightening the long bolts to 46 ft. lbs. (63 Nm) and the short bolts to 43 ft. lbs. (58 Nm). Make a final pass over all bolts tightening each an additional 90 degrees (¼ turn).

32. Install the transaxle fill tube.

33. Connect the fuel lines to the throttle body.

34. Raise and safely support the vehicle.

35. Connect the exhaust pipe to the exhaust manifold. Tighten the mounting bolts to 22 ft. lbs. (30 Nm).

36. Lower the vehicle.

37. Connect the upper radiator hose.

38. Connect the vacuum lines to the intake manifold.

39. Install the engine-side torque strut bracket and torque strut.

40. Install the alternator rear bracket.

41. Install the power steering pump bracket to the intake manifold brace, located under the intake manifold.

42. Attach the electrical connections at the intake manifold, throttle body and cylinder head.

43. Connect the oxygen (O_2) sensor connector.

44. Install the pushrods, rocker arms and rocker arm nuts and tighten the nuts to 22 ft. lbs. (30 Nm).

45. Install the rocker arm cover.

46. Connect the control cables to the throttle body and install the cable brackets at the throttle body and rocker arm cover.

47. Connect the spark plug wires.

48. Install the power steering pump in the mounting bracket.

49. Install the alternator.

50. Install the serpentine belt.

51. Install the lower air inlet duct.

52. Install the air inlet resonator tie bar.

53. Install the air cleaner and duct assembly.

54. Refill the cooling system.

55. An oil and filter change is recommended since coolant can enter the oil system when the head is removed.

56. Connect the negative battery cable.

57. Start the vehicle and verify that there are no leaks.

1995–97 Vehicles

— **CAUTION** —

Fuel injection systems remain under pressure, even after the engine has been turned OFF. The fuel system pressure must be relieved before disconnecting any fuel lines. Failure to do so may result in fire and/or personal injury.

1. Relieve fuel system pressure using the recommended procedure.

— **CAUTION** —

After relieving system pressure a small amount of fuel may be released when servicing fuel pipes or connections. In order to reduce the chance of personal injury, cover fuel pipes fittings with a shop towel before disconnecting, to catch any fuel that may leak out. Place the towel in an approved container when disconnect is complete.

2. Disconnect negative battery cable.

3. Remove air cleaner outlet duct assembly.

4. Label and disconnect vacuum lines.

5. Disconnect and tag for identification the electrical connections on the Engine Coolant Temperature (ECT) sensor, Oxygen Sensor (O_2S), IAC, Throttle Position Sensor, MAP sensor, EVAP Canister Purge Solenoid and the fuel injector harness.

6. Remove accelerator control, cruise and TV cables from accelerator control bracket.

7. Remove accelerator control cable bracket.

8. Raise and safely support vehicle.

9. Remove exhaust pipe from exhaust manifold.

10. Drain and recover coolant into a suitable container.

11. Lower vehicle.

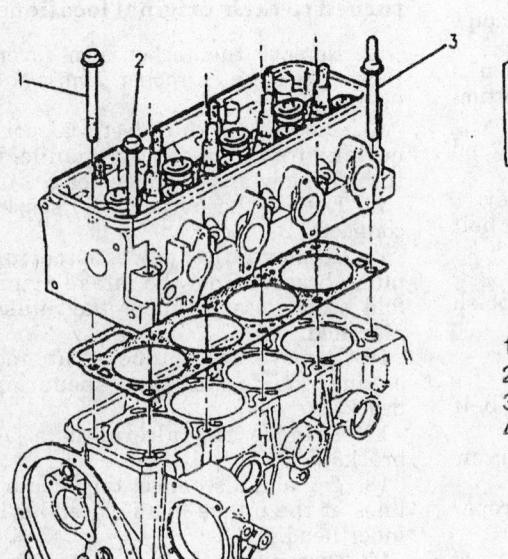

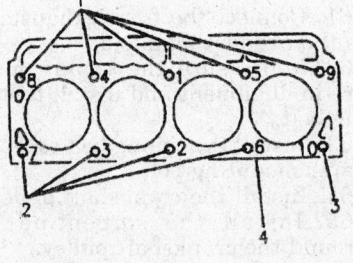

1. Long bolts
2. Short bolts
3. Stud
4. Numbers on gasket indicate torque sequence

291090

Cylinder head bolt tightening sequence — 2.2L engine

12. Remove serpentine drive belt.

13. Remove the alternator.

14. Remove the power steering pump and position aside with lines attached.

15. Remove power steering pump bracket.

16. Install engine support fixture J-28467-A or equivalent.

17. Remove serpentine drive belt tensioner bracket.

18. Tag and disconnect the spark plug wires.

19. Disconnect EVAP canister purge line, from under manifold.

20. Remove upper hose from coolant outlet.

21. Remove heater hose from coolant outlet.

22. Remove attaching nut holding automatic transaxle filler tube from intake manifold, if equipped.

23. Disconnect fuel lines.

24. Remove valve cover.

NOTE: Whenever valve train components are removed for service, they should be kept in order. They should be installed in the same locations and with the same mating surfaces as when removed.

25. Remove rocker arms and pushrods.

26. Remove cylinder head bolts. Two sizes of bolts are used. Note the location of each. These bolts are called "torque-to-yield." This means that, at assembly, after the bolts are tightened to a specific torque, they are tightened another quarter turn. This stretches the bolts slightly. Therefore, new cylinder head bolts are recommended.

27. Remove cylinder head with both manifolds attached.

28. Remove the intake and exhaust manifolds from the cylinder head.

To install:

29. Clean all the gasket surfaces completely. Clean the threads on cylinder head bolts and make sure all bolt holes are clean and free of foreign material. It is good practice to clean all internally threaded openings with the proper size thread cutting tap. This removes rust, dirt and old sealer build-up that can prevent getting a proper torque reading when tightening bolts.

30. Inspect cylinder head and block surface for cracks, nicks, heavy scratches and flatness.

31. Install exhaust and intake manifolds on cylinder head prior to installing cylinder head.

32. Place a new cylinder head gasket in position over the dowel pins on the engine block. Carefully guide the cylinder head into position.

33. Install cylinder head bolts finger-tight. New cylinder head bolts are recommended.

34. Tighten bolts in sequence, tighten the long bolts to 46 ft. lbs. (63 Nm) plus 90 degrees. Tighten the short bolts to 43 ft. lbs. (58 Nm) plus 90 degrees.

35. Install pushrods and rocker arms and rocker arm nuts. Tighten nuts to 22 ft. lbs. (30 Nm).

36. Install valve cover and tighten bolts to 89 inch lbs. (10 Nm).

37. Connect fuel lines.

38. Install transaxle filler tube nut and tighten to 20 ft. lbs. (27 Nm).

39. Install heater hose to coolant outlet.

40. Install upper radiator hose to coolant.

41. Connect EVAP canister purge line.

42. Attach the ignition wires to spark plugs, as tagged during removal.

43. Install serpentine drive belt tensioner bracket and bolts. Tighten bolts to 37 ft. lbs. (27 Nm).

44. Remove engine support fixture.

45. Install power steering pump bracket and power steering pump.

46. Install alternator and brace.

47. Install serpentine drive belt.

48. Raise and safely support the vehicle.

49. Connect the exhaust pipe to the exhaust manifold.

50. Lower the vehicle.

51. Install the accelerator control cable bracket and bolts. Tighten bolts to 18 inch lbs. (25 Nm).

52. Connect the accelerator control, cruise and TV cables to control bracket.

53. Attach all of the electrical connections for the sensors.

54. Connect the vacuum lines.

55. Install air cleaner outlet duct assembly.

56. Refill the coolant system.

57. Connect negative battery cable.

58. Start vehicle and inspect for leaks.

59. Bleed air from coolant system as follows:

a. Loosen the engine coolant air bleed screw, (located on the top side of the engine coolant outlet) and add coolant until all of the air is evacuated through the air bleed. Tighten the air bleed screw.

b. When filling the coolant system, use coolant meeting GM specifications.

2.3L and 2.4L Engines

CAUTION

Fuel injection systems remain under pressure, even after the engine has been turned OFF. The fuel system pressure must be relieved before disconnecting any fuel lines. Failure to do so may result in fire and/or personal injury.

---

1. Relieve the fuel system pressure.

CAUTION

After relieving system pressure, a small amount of fuel may be released when servicing fuel pipes or connections. In order to reduce the chance of personal injury, cover fuel pipe fittings with a shop towel before disconnecting, to catch any fuel that may leak out. Place the towel in an approved container when disconnect is complete.

---

2. Disconnect the negative battery cable.

3. Drain and recover the coolant into a suitable container.

4. Disconnect the heater inlet and throttle body heater hoses from water outlet.

5. Remove the exhaust manifold.

6. Remove the intake camshaft housing and lifters, then remove the exhaust camshaft housing and lifters.

7. Remove the oil fill tube.

8. Remove the throttle body-to-air cleaner duct.

9. Disconnect the power brake vacuum hose from throttle body.

10. Remove the throttle cable bracket.

11. Remove the throttle body from intake manifold, with electrical harness and throttle cable attached. Position it aside.

12. Disconnect the MAP sensor vacuum hose from intake manifold.

13. Remove the intake manifold brace.

14. Disconnect electrical connections from the following sensors: MAP sensor, intake air temperature sensor and evap canister purge solenoid.

15. Disconnect the upper radiator hose from water outlet.

16. Detach the coolant temperature sensors connectors.

17. Unfasten the cylinder head bolt, then remove cylinder head and gasket.

To install:

18. This is an aluminum cylinder head and must be treated with care.

Do not use abrasive pads to clean the cylinder head or block surfaces. An abrasive pad may damage the cylinder head and block. GM says that abrasive pads should not be used for the following reasons:

a. Abrasive pads will produce a fine grit that the oil filter will not be able to remove from the oil. This grit is abrasive and has been known to cause internal engine damage.

b. Abrasive pads can easily remove enough metal to round cylinder head edges. This has been known to affect the gasket's ability to seal, especially in the narrow areas between the combustion chambers and coolant jackets. The cylinder head gasket is likely to leak if these edges are rounded.

c. Abrasive pads can also remove enough metal to affect cylinder head flatness. It takes only about 15 seconds to remove 0.008 inch (0.20mm) of metal from the cylinder head with an abrasive pad. If the cylinder head flatness is out of specification, the gasket will not be able to seal and the gasket will leak.

19. Use a razor blade gasket scraper to clean the cylinder head and cylinder block gasket surfaces. Be careful not to gouge or scratch the gasket surfaces. Do not gouge or scrape the combustion chamber surfaces. Use a new razor blade for each cylinder head. Hold the scraper so the razor blade is as parallel to the gasket surface as possible. Do not use any other method or technique to clean these gasket surfaces. In addition, GM warns not to use a tap to clean cylinder head bolt holes.

20. When working on an aluminum head, do not remove spark plugs from an aluminum cylinder head until the cylinder head has cooled. Always clean all dirt and debris from the spark plug recess area. If the spark plug opening threads are damaged and NOT restorable with a Thread Chaser, replace the cylinder head. GM **DOES NOT** approve of the installation of thread inserts into the spark plug openings on this engine. If threads are installed into the spark plug openings, severe engine damage will occur.

21. Clean all the gasket surfaces completely. Clean the threads on cylinder head bolts and make sure all bolt holes are clean and free of debris. New bolts are recommended.

22. Inspect the cylinder head and block surface for cracks, nicks, heavy scratches and flatness.

23. Place a new cylinder head gasket on the block. Do not use any sealing material.

24. Carefully place the cylinder head on dowel pins, being careful not to disturb the gasket.

25. Apply a small amount of clean engine oil to the threads of the cylinder head bolts, and install fingertight.

26. Torque head bolts in sequence. Tighten bolts 1 through 8 to 40 ft. lbs. (65 Nm); then, tighten bolts 9 and 10 to 30 ft. lbs. (40 Nm). Turn all 10 bolts an additional 90 degrees (1/4 turn) in sequence.

27. Attach the coolant temperature sensor connections.

28. Connect upper radiator hose to coolant outlet.

29. Install manifold brace and tighten to 19 ft. lbs. (26 Nm).

30. Attach all sensor connections.

31. Connect the MAP sensor vacuum hose to intake manifold.

32. Install throttle body to intake manifold, using a new gasket.

33. Install accelerator control cable bracket to the throttle body, and tighten the bolts to 106 inch lbs. (12 Nm). Tighten the nut to 19 ft. lbs. (26 Nm).

34. Install the throttle body-to-air cleaner duct.

35. Install oil fill tube, tighten attaching bolt to 71 inch lbs. (8 Nm).

36. Install the lifters and camshaft housing.

37. Install the exhaust manifold, then tighten the exhaust nuts to 26 ft. lbs. (35 Nm).

38. Connect negative battery.

39. Fill coolant system and bleed off air from system. An oil and filter change is recommended.

40. Check and verify that vehicle has no coolant or vacuum leaks.

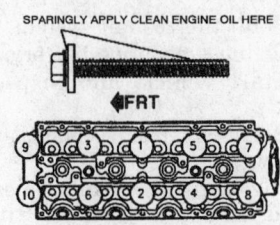

SPARINGLY APPLY CLEAN ENGINE OIL HERE

◄FRT

A. Tighten the bolts to the following N.m (lb. ft.) specification in sequence:
bolts 1 through 8: 65 N·m (40 ft. lb.)
bolts 9 and 10: 40 N·m (30 ft. lb.)
B. Then turn all 10 bolts an additional 90 degrees in sequence

329614

Head bolt torque sequence — 2.3L and 2.4L engines

3.1L Engine

Left Cylinder Head (Front)

— **CAUTION** —

Fuel injection systems remain under pressure, even after the engine has been turned OFF. The fuel system pressure must be relieved before disconnecting any fuel lines. Failure to do so may result in fire and/or personal injury.

1. Relieve the fuel system pressure.
2. Disconnect the negative, then the positive battery cables.
3. Remove the battery.
4. Remove the air cleaner assembly.
5. Drain the cooling system into a suitable container.
6. Remove the exhaust crossover pipe.
7. Remove the rocker arm covers.
8. Remove the upper intake plenum and lower intake manifold.
9. Remove the front exhaust manifold.
10. Remove the oil level indicator tube.
11. Tag and disconnect the spark plug wires from the spark plugs and remove the wires from the wire looms. Position the wires out of the way.

NOTE: When removing the valve train components they must be kept in order for installation in the same locations they were removed from.

12. Remove the rocker arm nuts, rocker arms, rocker arm balls and pushrods.
13. Remove the cylinder head bolts.
14. Remove the cylinder head.
To install:
15. Clean all the gasket surfaces completely. Clean the cylinder head bolt threads and cylinder block threads.
16. Install a new cylinder head gasket on the block. The gasket should fit over the dowel pins and the words **THIS SIDE UP** must be visible.
17. Install the cylinder head.
18. Coat the cylinder head bolts with sealer and install the bolts finger-tight. With all the bolts installed tighten the bolts in sequence to 33 ft. lbs. (45 Nm). Make a second pass tightening each bolt an additional 90 degrees.
19. Install the pushrods, rocker arms, balls and nuts.
20. Tighten the rocker arm nuts to 18 ft. lbs. (25 Nm).

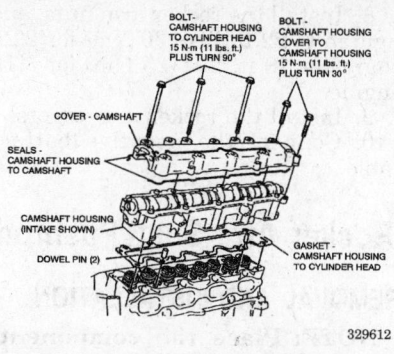

Camshaft housing cover — 2.3L and 2.4L engines

21. Install the lower intake manifold and upper intake plenum.
22. Install the rocker arm covers.
23. Install the oil level indicator tube.
24. Connect the spark plug wires to the spark plugs and the wire looms.
25. Install the front exhaust manifold.
26. Install the crossover pipe.
27. Refill the cooling system.
28. Install the battery.
29. Install the air cleaner assembly.
30. Connect the positive, then the negative battery cables.
31. Start the vehicle and verify that there are no leaks.

Right Cylinder Head (Rear)

──────── CAUTION ────────

Fuel injection systems remain under pressure, even after the engine has been turned OFF. The fuel system pressure must be relieved before disconnecting any fuel lines. Failure to do so may result in fire and/or personal injury.

1. Relieve the fuel system pressure.
2. Disconnect the negative, then the positive battery cables.
3. Remove the battery.

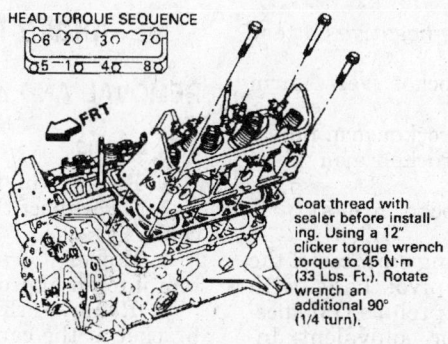

Coat thread with sealer before installing. Using a 12" clicker torque wrench torque to 45 N·m (33 Lbs. Ft.). Rotate wrench an additional 90° (1/4 turn).

202471

Cylinder head components and bolt tightening sequence — 3.1L engine

4. Remove the air cleaner assembly.
5. Drain the cooling system into a suitable container.
6. Detach the oxygen sensor connector.
7. Disconnect the crossover pipe from the right manifold.
8. Remove the right exhaust manifold.
9. Tag and disconnect the spark plug wires from the spark plugs and disconnect the wires from the looms.
10. Remove the rocker arm covers.
11. Remove the upper intake plenum and lower intake manifold.

NOTE: When removing the valve train components they must be kept in order for installation in the same locations they were removed from.

12. Remove the rocker arm nuts, rocker arms, rocker arm balls and pushrods.
13. Remove the cylinder head bolts.
14. Remove the cylinder head.
To install:
15. Clean all the gasket surfaces completely. Clean the cylinder head bolt threads and cylinder block threads.
16. Install a new cylinder head gasket on the block. The gasket should fit over the dowel pins and the words **THIS SIDE UP** must be visible.
17. Install the cylinder head.
18. Coat the cylinder head bolts with sealer and install the bolts finger-tight. With all the bolts installed tighten the bolts in sequence to 33 ft. lbs. (45 Nm). Make a second pass tightening each bolt an additional 90 degrees.
19. Install the pushrods, rocker arms, balls and nuts.
20. Tighten the rocker arm nuts to 18 ft. lbs. (25 Nm).
21. Install the lower intake manifold and upper intake plenum.
22. Install the rocker arm covers.

23. Connect the spark plug wires to the spark plugs and the wire looms.
24. Install the rear exhaust manifold.
25. Connect the crossover pipe to the right exhaust manifold.
26. Refill the cooling system.
27. Install the battery.
28. Install the air cleaner assembly.
29. Connect the positive, then the negative battery cables.
30. Start the vehicle and verify no leaks.

Lash Adjusters

BLEEDING

2.0L Engine

1. Disconnect the negative battery cable.
2. Remove the air cleaner.
3. Disconnect the breather hoses from the camshaft cover.
4. Remove the cover bolts and remove the cover.
5. Rotate the engine to Top Dead Center (TDC) with the timing pointer at 0 degrees.
6. Compress the valve springs using J-33302, or an equivalent valve spring compressor.
7. Remove the rocker arms and valve lash compensators. All valve train components removed must be kept in order for installation in their original locations.
To install:
8. Coat the rocker arms and valve lash compensators with clean engine oil.
9. Compress the valve springs using J-33302, or an equivalent valve spring compressor and install the rocker arms and valve lash compensators in their original locations.
10. Install the camshaft cover and tighten the mounting bolts to 5 ft. lbs. (7 Nm).
11. Connect the breather hoses to the cover.
12. Install the air cleaner.
13. Connect the negative battery cable.
14. Start the vehicle and verify no leaks.

Valve Lifters

REMOVAL AND INSTALLATION

NOTE: Hydraulic valve lifters do not require adjustment. Valve lifters must be kept in order so they may be reinstalled in their original position.

2.2L Engine

1. Disconnect the negative battery cable.
2. Remove the rocker arm (valve) cover.
3. Loosen the rocker arm nut, then position the rocker arm aside.
4. Remove the pushrod/rocker arm shaft.
5. Remove the engine lift bracket located at the rear of the engine.
6. Route the spark plug wires below the lower intake manifold.
7. Remove the cylinder head and gasket.
8. Remove the anti-rotation brackets.
9. Remove the lifters.
To install:

WARNING
Improper installation of the lifters or anti-rotation brackets could result in damage to the components and/or engine.

10. Coat the foot of the valve lifters with camshaft assembly lube GM No. 1052365 or equivalent whenever new lifters are being installed.
11. Install the lifters, being careful to position the flat sides of the lifters to allow installation of the anti-rotation bracket.
12. Install the anti-rotation bracket and bolts. Tighten the bolts to 97 inch lbs. (11 Nm).
13. Install a new cylinder head gasket, then install the cylinder head.
14. Install the engine lift bracket and nut. Tighten the nut to 37 ft. lbs. (50 Nm).
15. Route the spark plug wires up through the lower intake manifold.
16. Install the pushrods, making sure they seat in the lifters.
17. Position the rocker arms, then tighten the nuts to 22 ft. lbs. (30 Nm).
18. Install the valve rocker arm cover.
19. Connect the negative battery cable.

3.1L Engine

NOTE: If the lifters need to be replaced, use lifters with a narrow flat ground along the lower ¾ of the lifter. These flats provide additional oil to the cam lobe and lifter surfaces.

1. Disconnect the negative battery cable.
2. Properly drain the cooling system into a suitable container.
3. Remove the intake manifold.
4. Remove the rocker arms.
5. Remove the valve lifter.

To install:
6. Whenever new lifters are being installed, coat the foot of the valve lifters with camshaft assembly lube GM No. 1052365, or equivalent.
7. Install the valve lifters.
8. Install the rocker arms.
9. Install the intake manifold.
10. Fill the engine cooling system, then connect the negative battery cable.

NOTE: On the 3.1L engine, a Tech 1® scan tool is needed to perform the "Idle Learn" procedure.

11. Perform the following "Idle Learn" procedure;
   a. Install a Tech 1® scan tool.
   b. Turn the ignition to the **ON** position, engine not running.
   c. Select **IAC SYSTEM**, then **IDLE LEARN** in the **MISC TEST** mode.
   d. Place the transaxle in **P** or **N**, as applicable.
   e. Proceed with idle learn as directed by the scan tool.

Valve Lash

ADJUSTMENT

All of the engines are equipped with hydraulic valve lifters that do not require periodic valve lash adjustment. Adjustment to zero lash is maintained automatically by hydraulic pressure in the lifters.

Rocker Arms

REMOVAL AND INSTALLATION

NOTE: Place the components in a rack on order to be sure they are installed at the same location and with the same mating surface as when removed.

1. Disconnect the negative battery cable.
2. Remove the rocker (valve) arm cover(s).
3. Unfasten the rocker arm nuts.
4. Remove the rocker arm pivot ball(s).
5. Remove the rocker arm(s).
To install:
6. Coat the bearing surfaces of the rocker arms and pivot balls with camshaft and lifter prelube GM specification 1052365 or equivalent. Install the rocker arm(s).
7. Install the rocker arm pivot ball(s).

8. Install the rocker arm nuts, and tighten to 22 ft. lbs. (30 Nm) for 2.2L engine or 18 ft. lbs. (25 Nm) for 3.1L engine.
9. Install the rocker arm covers.
10. Connect the negative battery cable.

Rocker Arm Shaft/Pushrod

REMOVAL AND INSTALLATION

NOTE: Place the components in a rack in order to be sure they are installed at the same location and with the same mating surface as when removed.

1. Disconnect the negative battery cable.
2. Remove the rocker (valve) arm cover(s).
3. Unfasten the rocker arm nuts.
4. Remove the rocker arm pivot ball(s).
5. Remove the rocker arm(s).
6. Remove the pushrods. For the 3.1L engine, the intake pushrods are marked orange and are 6 in. (15.2cm) long and the exhaust pushrods are marked blue and are 6⅜ in. (16.2cm) long.
To install:
7. Install the pushrods. Make sure to install the pushrods in the correct positions, and be sure they seat properly in the lifters.
8. Coat the bearing surfaces of the rocker arms and pivot balls with camshaft and lifter prelube GM specification 1052365 or equivalent. Install the rocker arm(s).
9. Install the rocker arm pivot ball(s).
10. Install the rocker arm nuts, and tighten to 22 ft. lbs. (30 Nm) for 2.2L engine or 18 ft. lbs. (25 Nm) for 3.1L engine.
11. Install the rocker arm covers.
12. Connect the negative battery cable.

Intake Manifold

REMOVAL AND INSTALLATION

2.0L Engine

1. Release the fuel pressure.
2. Disconnect the negative battery cable.
3. Properly drain the cooling system into an approved container.
4. Remove the alternator and bracket at the camshaft carrier.
5. Remove the power steering pump, then position it aside with the lines still attached.

6. Remove the ignition coil.

7. Disconnect the throttle cable from the intake manifold bracket.

8. Detach the throttle and TV cables from the throttle body.

9. Disconnect the wiring from the throttle body.

10. Detach the vacuum brake hose at the filter.

11. Disconnect the coolant hoses at the water pump and intake manifold and the inlet tube-to-water pump.

12. For access to the lower retaining nuts, detach the ECM electrical harness.

13. Unfasten the retaining nuts and washers, then remove the intake manifold. Clean the mating surfaces of the cylinder head and intake manifold.

To install:

14. Place a new gasket on the cylinder head, then position the intake manifold. Tighten intake manifold retaining nuts and washers evenly in sequence to 16 ft. lbs. (22 Nm). Start with the inner nuts and work outward.

15. Attach the ECM/PCM wiring harness.

16. Connect the coolant hoses at the water pump and intake manifold and the inlet tube-to-water pump.

17. Fasten the inlet and return fuel lines.

18. Connect the vacuum brake hose at the filter.

19. Attach the wiring at the throttle body.

20. Connect the throttle and TV cables to the throttle body. Attach the throttle cable to the intake manifold bracket.

21. Install the ignition coil.

22. Attach the power steering bracket to the intake manifold.

23. Install the power steering pump.

24. Attach the alternator and bracket to the camshaft carrier.

25. Fill the engine cooling system.

26. Install the air cleaner.

27. Connect the negative battery cable, then start the engine and check for leaks.

2.2L Engine

These vehicles use a two-piece intake manifold. The upper half, sometimes called a plenum, contains the throttle body and the control cable connections. The lower half has individual port runners to each intake port on the cylinder head. The lower half of the manifold bolts to the cylinder head and houses the fuel injectors. Note that these pieces are cast aluminum. Care should be exercised when working with any light alloy component.

1. Properly relieve the fuel system pressure.

2. Disconnect the negative battery cable.

3. Remove the throttle body air intake duct.

4. Drain the cooling system into an approved container.

5. Identify, tag and disconnect all necessary vacuum lines.

6. Disconnect the control cables from the throttle body lever and remove the control cable bracket from the intake manifold.

7. Remove the serpentine belt.

8. Remove the power steering pump and lay it aside with the fluid lines attached.

9. Remove the transaxle fill tube.

10. Tag and disconnect the following electrical wires:

• Idle Air Control (IAC) valve
• Throttle Position (TP) sensor
• Manifold Absolute Pressure (MAP) sensor
• EVAP Emission solenoid
• Fuel injector harness
• Exhaust Gas Recirculation (EGR) valve

11. Remove the MAP sensor.

12. Unfasten the upper intake manifold mounting bolts, then remove the upper intake manifold.

13. Disconnect the fuel lines from the fuel rail.

14. Remove the EGR valve injector, then remove the EGR valve.

15. Remove the fuel injector retainer bracket, regulator and injectors.

16. Unfasten and remove the control cable bracket.

17. If necessary for access, raise and safely support the vehicle.

18. Unfasten the 6 intake manifold nuts, then remove the manifold.

19. Clean the gasket mounting surfaces.

To install:

20. Install a new gasket, then position the lower intake manifold. Tighten the lower intake manifold nuts in the proper sequence to 24 ft. lbs. (33 Nm).

21. Connect the control cables and cable bracket.

22. Install the EGR valve.

23. Attach the fuel lines to the fuel rail.

24. Install the fuel injectors, regulator and injector retainer bracket, then tighten the retaining bolts to 22 inch lbs. (3.5 Nm).

25. Install the EGR valve injector that the port is facing directly towards the throttle body.

26. Install the upper intake manifold assembly. Tighten the upper intake manifold nuts in the proper sequence to 22 ft. lbs. (30 Nm).

27. Install the MAP sensor.

28. Attach the electrical connectors to the MAP sensor, EGR solenoid valve, Idle Air Control (IAC) valve, Throttle Position (TP) sensor, and the fuel injectors.

29. Install the transaxle fill tube.

30. Install the power steering pump, then install the serpentine belt.

31. Connect the vacuum lines, as tagged during removal.

32. Install the air intake duct.

33. Refill the coolant system.

34. Connect the negative battery cable, then start the engine and check for leaks.

2.3L and 2.4L Engines

1. Properly relieve the fuel system pressure.

2. Disconnect the negative battery cable, then properly drain the cooling system.

3. Tag and detach the following electrical connectors:

• Manifold Absolute Pressure (MAP) sensor
• Intake Air Temperature (IAT) sensor
• EVAP canister purge solenoid
• Fuel injector harness

4. Label and disconnect the vacuum hoses from the fuel regulator and EVAP canister purge solenoid to canister.

5. Unfasten the air cleaner duct.

6. Remove the accelerator control cable bracket.

7. For the 2.4L engine, remove the stud-ended alternator mount bolt, then detach the EGR pipe from the EGR adapter.

8. For the 2.3L engine, perform the following procedures:

a. Remove the oil air separator (crankcase ventilation system) as an assembly. Leave the hoses attached to the separator. Disconnect the hoses from the oil fill, chain cover, intake duct and the intake manifold.

b. Detach the oil/air separator from the oil fill tube.

c. Remove the oil fill cap and oil level indicator assembly.

d. Unfasten the oil fill tube bolt/screw, then pull the tube upward to remove.

9. Remove the fill tube out the top, rotating as necessary to gain clearance for the oil/air separator nipple between the intake tubes and fuel rail electrical harness.

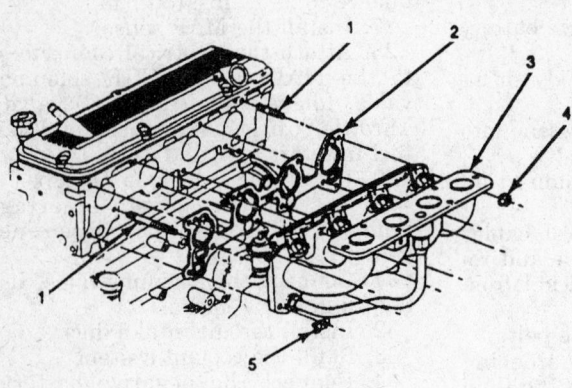

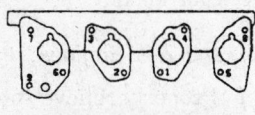

INTAKE MANIFOLD NUT
TIGHTENING SEQUENCE

1	STUD
2	GASKET
3	INTAKE MANIFOLD
4	NUT
5	CLIP

8838LG03

Lower intake manifold tightening sequence — 2.2L engine

10. For the 2.4L engine, raise and safely support the vehicle.

11. Remove the intake manifold support brace.

12. If raised, carefully lower the vehicle.

13. Unfasten the manifold retaining nuts and bolts, then remove the intake manifold from the engine.

NOTE: If installing a new intake manifold, transfer all necessary parts from the old manifold to the new one.

14. Using a suitable scraping tool, clean the old gasket material from the intake manifold mating surfaces. Do not let any debris fall into the engine!

To install:

15. Install the manifold with a new gasket.

NOTE: Make sure the numbers stamped on the gasket are facing towards the manifold surface.

16. Follow the tightening sequence in the accompanying figure, then tighten the bolts/nuts to 19 ft. lbs. (26 Nm) for 2.3L engine or 18 ft. lbs. (24 Nm) for 2.4L engine.

17. For the 2.4L engine, raise and safely support the vehicle.

18. Install the intake manifold brace and retainers.

19. If raised, carefully lower the vehicle.

20. For the 2.3L engine, install the oil/air separator assembly.

21. For the 2.3L engine, lubricate a new oil fill tube O-ring seal with clean engine oil, then install the tube down between intake manifold. Rotate as needed to gain clearance for the oil/air separator nipple on the fill tube.

22. If removed, position the oil fill tube in its cylinder block opening. Align the fill tube so it is in about its proper position. Place the palm of your hand over the oil fill opening and press straight down to seat the fill tube and O-ring into the cylinder block.

23. If necessary, connect the oil/air separator hose to the oil fill tube. You can lubricate the hose as necessary to ease installation. Install the oil fill tube bolt/screw. Fasten the cap.

24. For the 2.4L engine, attach the EGR pipe to the adapter; tighten the fasteners to 19 ft. lbs. (26 Nm). Install the stud-ended alternator bolt.

25. Install the accelerator control cable bracket.

26. Connect the vacuum hoses to the fuel regulator and EVAP canister purge solenoid.

27. Attach all electrical connectors, as tagged during removal.

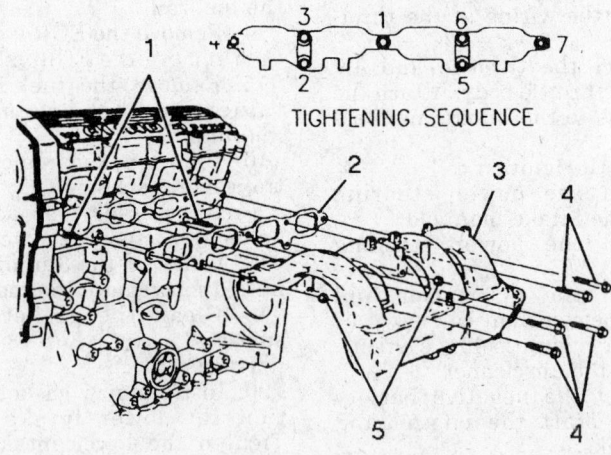

TIGHTENING SEQUENCE

1. STUD — 11 N·m (96 LBS. IN.)
2. INTAKE MANIFOLD GASKET
3. INTAKE MANIFOLD
4. BOLT — 26 N·m (19 LBS. FT.)
5. NUT — 26 N·m (19 LBS. FT.)

8838LG04

View of the intake manifold and torque sequence — 2.3L engine

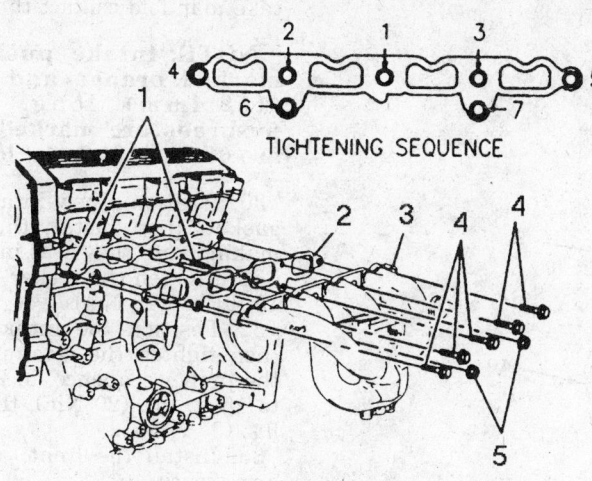

TIGHTENING SEQUENCE

1. STUD — 12 N•M (100 LB. IN.)
2. INTAKE MANIFOLD GASKET
3. INTAKE MANIFOLD
4. BOLT — 24 N•M (17 LB. FT.)
5. NUT — 24 N•M (17 LB. FT.)

8838LG05

View of the intake manifold and torque sequence — 2.4L engine

28. Install the air cleaner duct.
29. Refill the coolant to it's proper level.
30. Connect the negative battery cable, then start the engine and inspect for leaks.

3.1L Engine

1993 Vehicles

1. Properly relieve the fuel system pressure.
2. Disconnect the negative battery cable.
3. Detach the cables at the plenum.
4. Unfasten the brake vacuum pipe at the plenum.
5. Disconnect the throttle body and the EGR pipe from the EGR valve.
6. Remove the plenum assembly.
7. Disconnect the fuel line along the fuel rail.
8. Remove the serpentine drive belt.
9. Unfasten the alternator, then move it to one side. Loosen the alternator bracket.
10. Remove the power steering pump and move to one side.
11. Disconnect the idle air vacuum hose at the throttle body.

12. Detach the the wires at the fuel injectors.
13. Remove the fuel rail, breather tube and the fuel runners from the engine.
14. Tag and disconnect the plug wires at the manifold.
15. Remove the rocker arm covers; refer to the necessary service procedure located in this section.
16. Drain the cooling system, the disconnect the radiator hose at the thermostat housing. Disconnect the heater hose from the thermostat housing and tag and disconnect the thermostat wiring.
17. Unfasten the intake manifold bolts and remove the intake manifold from the engine.

NOTE: Retain the Belleville washer in the same orientation on the 4 center bolts.

18. Loosen the rocker arms and remove the pushrods.
19. Remove and discard the intake manifold gasket.
  To install:
20. Clean the gasket material and grease from the mating surfaces.
21. Place a 0.08–0.12 in. (2–3mm) bead of RTV sealant on each ridge where the front and rear of the intake manifold contact the block.

22. Install the new intake manifold gasket.
23. Install the pushrods and make sure they are seated properly in the lifter.

NOTE: Intake pushrods are marked orange and are 6 in. (152.4mm) long. Exhaust pushrods are marked blue and are 6⅜ in. (161.9mm) long.

24. Install the rocker arm nuts and tighten to 18 ft. lbs. (24 Nm).
25. Install the intake manifold and retaining bolts. Tighten the bolts to 15 ft. lbs. (20 Nm) in the proper sequence shown, then retighten in sequence to 24 ft. lbs. (33 Nm).
26. Install the heater pipe to the manifold.
27. Connect the coolant sensor and any other necessary wiring.
28. Fasten the upper radiator hose.
29. Install the rocker arm covers.
30. Connect the breather tube.
31. Install the fuel rail and attach the wires to the injectors.
32. Connect the idle air vacuum hose at the throttle body.
33. Install the alternator and bracket.
34. Install the power steering gear and pump.
35. Connect the power steering line at the alternator bracket.
36. Install the serpentine drive belt.
37. Connect the fuel lines to the fuel rail and to the bracket.
38. Install the plenum, plug harness and EGR valve.
39. Install the throttle body.
40. Attach the cables at the plenum.
41. Connect the negative battery cable and fill the cooling system.

1994 Vehicles

NOTE: An "Idle Learn" procedure must be performed which requires the use of a scan tool.

1. Properly relieve the fuel system pressure.
2. Disconnect the negative battery cable.
3. Remove the battery and the air cleaner assembly.
4. Remove the serpentine belt.
5. Unfasten and remove the exhaust crossover pipe.
6. Disconnect the EGR transfer tube from the exhaust manifold.
7. Properly drain and recover the engine cooling system.
8. Remove the radiator surge or overflow tank.
9. Disconnect the brake vacuum pipe at the plenum.

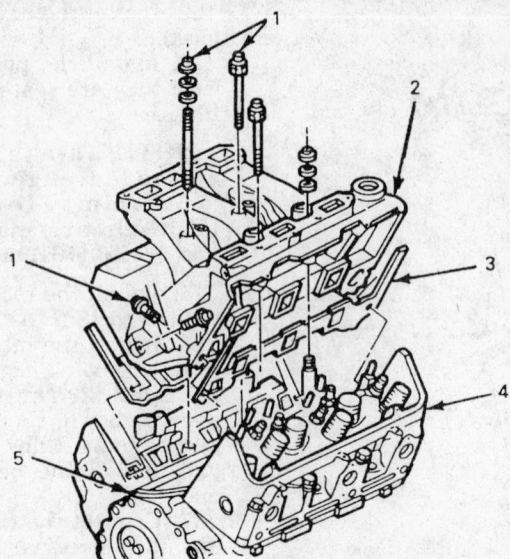

1 TIGHTEN IN
PROPER SEQUENCE TO
20 N·m (15 LB. FT.), THEN
RETIGHTEN TO 33 N·m
(24 LB. FT.)

2 INTAKE MANIFOLD
3 GASKET
4 CYLINDER HEAD
5 SEALER

⑦ ④ ③ ⑥
⑧ ① ② ⑤

8838LG06

View of the intake manifold and torque sequence — 1993 3.1L engine — 1994 similar

10. Remove the throttle cable and the vacuum line bracket at the plenum.

11. Disconnect the power steering lines at the alternator bracket.

12. Unfasten the rear alternator brace, then remove the alternator.

13. Tag and disconnect the spark plug wires from the plug, then unroute them.

14. Remove the rear valve (rocker arm) cover.

15. Tag and disconnect, then unroute the following electrical connections:
- Throttle Position (TP) sensor
- Idle Air Control (IAC) sensor
- Exhaust Gas Recirculation (EGR) valve
- Engine Coolant Temperature (ECT) sensor
- Fuel injector wiring harness connector

16. Tag and disconnect the PCV hose and any other necessary vacuum lines.

17. Detach the cables at the throttle body.

18. Unfasten the throttle body heater hoses.

19. Remove the plenum.

20. Disconnect the fuel lines from the fuel rail and the fuel lines at the bracket.

21. Unfasten the power steering pump mounting bolts, then position the pump aside with the lines attached.

22. Disconnect the coolant bleed pipe from the thermostat housing.

23. Detach the heater pipe from the cylinder heads, thermostat housing and the water pump. Disconnect the upper radiator hose at the thermostat housing.

24. Remove the thermostat housing.

25. Remove the front valve (rocker arm) cover.

NOTE: When removing the manifold, make sure to retain the washers in the same orientation on the 4 center bolts.

26. Unfasten the retaining bolts and/or nuts, then remove the intake manifold.

27. Loosen the rocker arms, then remove the pushrods.

28. Remove and discard the intake gasket, then clean the mating surfaces free of gasket material and old RTV sealant.

To install:

29. Place a 0.08–0.12 in. (2–3mm) bead of RTV sealant on each ridge where the front and rear of the intake manifold contact the block.

NOTE: Intake pushrods are marked orange and are 6 in. (152.4mm) long. Exhaust pushrods are marked blue and are 6³⁄₈ in. (161.9mm) long.

30. Position a new intake manifold gasket, then install the pushrods, making sure they seat in the lifters.

31. Tighten the rocker arm nuts to 18 ft. lbs. (25 Nm).

32. Position the intake manifold, then tighten the retaining bolts, in the proper sequence, in 2 steps; first to 15 ft. lbs. (20 Nm), then to 24 ft. lbs. (33 Nm).

33. Install the front valve (rocker arm) cover.

34. Attach the heater pipe to the manifold.

35. Install the thermostat housing, then attach the upper radiator hose to the housing.

36. Connect the heater pipe to the cylinder heads, water pump and thermostat housing.

37. Attach the coolant bleed pipe to the thermostat housing.

38. Install the power steering pump.

39. Attach the fuel lines to the fuel rail, the fasten fuel lines at the bracket.

40. Install the plenum.

41. Attach the throttle body heater hoses, then connect the cables to the throttle body.

42. Connect the PCV hose, and any other vacuum lines disconnected during removal.

43. Route and attach the following electrical connections, as tagged during removal:
- TP sensor
- IAC valve
- EGR valve
- ECT sensor
- Fuel injector wiring harness connector

44. Install the rear valve (rocker arm) cover.

45. Route and connect all spark plug wires, as tagged during removal.

46. Install the alternator and the alternator brace, then attach the power steering line to the brace.

47. Attach the throttle cable and vacuum lines bracket to the plenum.

48. Connect the brake vacuum pipe to the plenum.

49. Install the radiator surge or overflow tank.

50. Attach the EGR transfer tube to the exhaust manifold.

51. Connect the exhaust crossover to the exhaust manifolds.

52. Install the air cleaner assembly and the battery, then connect the negative battery cable.

NOTE: A Tech 1® scan tool is needed to perform the "Idle Learn" procedure.

53. Perform the following "Idle Learn" procedure;
    a. Install a Tech 1® scan tool.
    b. Turn the ignition to the **ON** position, engine not running.
    c. Select **IAC SYSTEM**, then **IDLE LEARN** in the **MISC TEST** mode.
    d. Place the transaxle in **P** or **N**, as applicable.
    e. Proceed with idle learn as directed by the scan tool.
54. Fill the cooling system, then start the engine and check for leaks.

Exhaust Manifold

REMOVAL AND INSTALLATION

2.0L Engine

1. Disconnect the negative battery cable.
2. Remove the air cleaner assembly.
3. Tag and disconnect the spark plug wires from the spark plugs, then remove the retainers.
4. Remove the oil dipstick tube and the breather.
5. Detach the Oxygen (O_2) sensor electrical connector.
6. Raise and safely support the vehicle.
7. Remove the exhaust pipe from the manifold, then support the pipe.
8. Unfasten the retaining nuts, then remove the exhaust manifold by lowering it from the bottom of the vehicle. Remove and discard the gasket. Clean the exhaust manifold and cylinder head gasket mating surfaces.

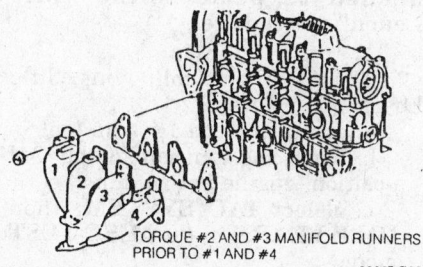

TORQUE #2 AND #3 MANIFOLD RUNNERS PRIOR TO #1 AND #4

8838LG08

View of the exhaust manifold and torque sequence — 2.0L engine except turbo

To install:

9. Use a new gasket, position the exhaust manifold, then secure using the retaining nuts. Tighten the nuts to 16 ft. lbs. (22 Nm), in the sequence.
10. Connect the exhaust pipe to the manifold. Tighten the retaining nuts to 16–19 ft. lbs. (22–26 Nm).
11. Carefully lower the vehicle.
12. Attach the (O_2) sensor electrical connector.
13. Install the breather and the oil dipstick tube.
14. Connect the spark plugs as tagged during remove, then install the retainers.
15. Install the air cleaner assembly.
16. Connect the negative battery cable.

2.2L Engine

1. Disconnect the negative battery cable.
2. Detach the oxygen sensor wire.
3. Remove the serpentine belt or alternator drive belt.
4. Remove the alternator-to-bracket bolts, then support the alternator (with the wires attached) out of the way.
5. Raise and support the vehicle safely.
6. Unfasten the exhaust pipe-to-exhaust manifold bolts, then carefully lower the vehicle.
7. If necessary for access to remove the manifold, remove the oil fill tube and disconnect the heater outlet hose assembly nut from the exhaust manifold.
8. Remove the exhaust manifold-to-cylinder head bolts.
9. Detach the exhaust manifold from the exhaust pipe flange.
10. Unfasten the retaining nuts, then remove the exhaust manifold from the vehicle. Remove and discard the gasket(s).

To install:

11. Using a gasket scraper, carefully clean the gasket mounting surfaces.
12. To install, use new gaskets and reverse the removal procedures. Tighten the exhaust manifold-to-cylinder head nuts to 3–12 ft. lbs. (4–16 Nm) and the bolts to 6–13 ft. lbs. (8–18 Nm).
13. Start the engine and check for exhaust leaks.

2.3L and 2.4L Engines

1. Disconnect the negative battery cable
2. Detach the Oxygen (O_2) sensor connector.
3. Raise and safely support the vehicle.

4. Unfasten the exhaust manifold brace-to-manifold bolt and the oil pan nuts, if necessary.
5. For 2.3L engine, remove the manifold-to-exhaust pipe spring loaded nuts.

NOTE: Do not bend the exhaust flex decoupler more than necessary to remove it. Excessive movement will damage the flex decoupler.

6. For 2.4L engine, remove the manifold-to-exhaust flex decoupler fasteners.
7. Pull down and back on the exhaust pipe to disengage it from the exhaust manifold bolts.
8. Carefully lower the vehicle.
9. Unfasten the exhaust manifold-to-cylinder head retaining nuts/bolts, then remove the manifold. Remove and discard the gaskets and/or seals. Clean the mating surfaces.

To install:

10. Use gaskets, then position the exhaust manifold. Tighten the retaining nuts to 31 ft. lbs. (42 Nm) for 2.3L engine or to 110 inch lbs. (12.5 Nm) for 2.4L engine, in the sequence.
11. Raise and safely support the vehicle.
12. Install the heat shield. Tighten the bolts to 124 inch lbs. (14 Nm).
13. Fasten the exhaust manifold brace-to-manifold bolt and the oil pan nuts. Tighten the bolts to 41 ft. lbs. (56 Nm) and the nuts to 19 ft. lbs. (26 Nm).
14. For 2.3L engine, install the manifold-to-exhaust pipe nuts. Be sure to turn both nuts in evenly to avoid cocking the exhaust pipe and binding the nuts.
15. For 2.4L engine, install the manifold-to-flex decoupler fasteners. Tighten the bolts to 26 ft. lbs. (35 Nm).
16. Carefully lower the vehicle.
17. Attach the O_2 connector. Coat the threads of the sensor with anti-seize compound 5613695 or equivalent.
18. Connect the negative battery cable and check for leaks.

3.1L Engine

1993 Vehicles (Left Side)

1. Disconnect the negative battery cable.
2. Remove the air cleaner assembly.
3. Remove the air cleaner inlet hose and the mass air flow sensor, if equipped.
4. Properly drain the coolant, then remove the coolant by-pass pipe
5. Remove the heat shield.

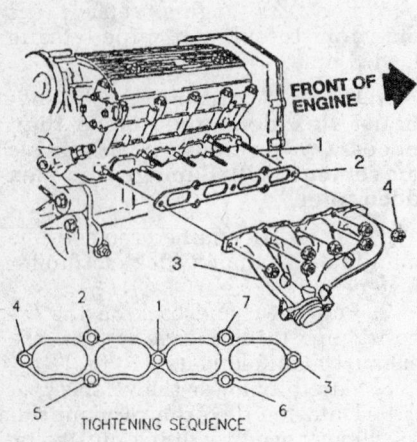

View of the exhaust manifold and torque sequence — 2.3L engine

1. STUD, EXHAUST MANIFOLD
2. GASKET, EXHAUST MANIFOLD
3. MANIFOLD, EXHAUST
4. NUT, EXHAUST MANIFOLD
   42 N•m (31 LB. FT.)

TIGHTENING SEQUENCE

8838LG09

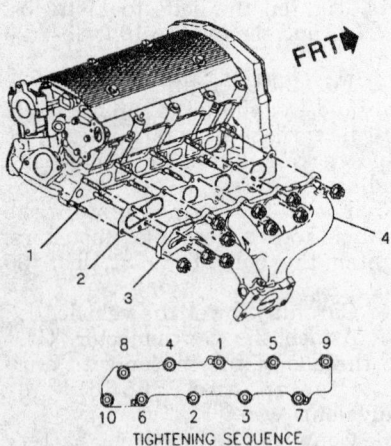

View of the exhaust manifold and torque sequence — 2.4L engine

1. STUD, EXHAUST MANIFOLD
2. GASKET, EXHAUST MANIFOLD
3. MANIFOLD, EXHAUST
4. NUT, EXHAUST MANIFOLD, MUST BE TIGHTENED IN SEQUENCE SHOWN TO 12.5 N•m (110 LB. IN.)

TIGHTENING SEQUENCE

8838LG10

6. Disconnect the exhaust crossover pipe at the manifold.

7. Unfasten the exhaust manifold bolts and remove the exhaust manifold or manifold/crossover assembly.

To install:

8. Clean the surfaces of the cylinder head and manifold.

9. Using a new gasket, place the manifold into position and loosely install the manifold bolts.

10. Attach the exhaust crossover pipe to the manifold.

11. Tighten the exhaust manifold bolts evenly to 18–21 ft. lbs. (24–28 Nm).

12. Install the heat shield. Tighten the nuts to 89 inch lbs. (10 Nm).

13. Connect the coolant pipe.

14. Install the air cleaner inlet hose and the air cleaner assembly

15. Fill the cooling system.

16. Connect the negative battery cable.

1993 Vehicles (Right Side)

1. Disconnect the negative battery cable.

2. Remove the air cleaner assembly.

3. Raise and safely support the vehicle.

4. Remove the heat shield.

5. Disconnect the exhaust pipe at the crossover.

6. Carefully lower the vehicle.

7. Remove the EGR valve or pipe.

8. Disconnect the oxygen sensor wire.

9. Unfasten the exhaust manifold bolts, then remove the manifold.

10. Clean the mating surfaces of the cylinder head and manifold. Discard the gasket.

To install:

11. Use a new manifold gasket, place the manifold into position and loosely install the manifold bolts.

12. Install the crossover at the manifold.

13. Tighten the manifold bolts evenly to 21–25 ft. lbs. (28–34 Nm).

14. Install the EGR valve or pipe and connect the oxygen sensor wire.

15. Raise and support the vehicle safely.

16. Connect the exhaust pipe to the crossover.

17. Install the engine heat shield.

18. Carefully lower the vehicle.

19. Connect the negative battery cable.

1994 Vehicles (Left Side)

NOTE: An Idle Learn procedure must be performed which requires the use of a scan tool.

1. Disconnect the negative, then the positive battery cables.

2. Remove the battery.

3. Remove the air cleaner assembly.

4. Raise and safely support the vehicle.

5. Unfasten the lower cooling fan bolt and detach the electrical connector.

6. Carefully lower the vehicle.

7. Remove the cooling fan assembly.

8. Disconnect the exhaust crossover pipe.

9. Tag and disconnect the left side spark plug wires from the spark plugs and position aside.

10. Remove the exhaust manifold heat shield.

11. Unfasten the retaining bolts, then remove the exhaust manifold.

12. Clean the mating surfaces of the cylinder head and manifold. Discard the gasket.

To install:

13. Use a new manifold gasket, place the manifold into position and loosely install the manifold bolts.

14. Tighten the exhaust manifold bolts to 21 ft. lbs. (28 Nm).

15. Install the exhaust manifold heat shield. Tighten the retaining nuts to 89 inch lbs. (10 Nm).

16. Attach the spark plug wires to the plugs, as tagged during removal.

17. Fasten the exhaust crossover pipe to the manifolds.

18. Install the cooling fan, and secure with the upper retaining bolts.

19. Raise and safely support the vehicle.

20. Install the lower cooling fan assembly bolt and attach the electrical connector.

21. Carefully lower the vehicle.

22. Install the air cleaner assembly.

23. Install the battery, then connect the positive, then the negative battery cables.

NOTE: A Tech 1® scan tool is needed to perform the "Idle Learn" procedure.

24. Perform the following "Idle Learn" procedure;

a. Install a Tech 1® scan tool.

b. Turn the ignition to the **ON** position, engine not running.

c. Select **IAC SYSTEM**, then **IDLE LEARN** in the **MISC TEST** mode.

d. Place the transaxle in **P** or **N**, as applicable.

e. Proceed with idle learn as directed by the scan tool.

1994 Vehicles (Right Side)

NOTE: An Idle Learn procedure must be performed which requires the use of a scan tool.

1. Disconnect the negative, then the positive battery cables.
2. Remove the battery.
3. Remove the air cleaner assembly.
4. Disconnect the exhaust crossover pipe from the manifolds.
5. Detach the Oxygen sensor electrical connector.
6. Disconnect the EGR pipe from the exhaust manifold.
7. If equipped with an automatic transaxle, remove the transaxle fluid level indicator and fill tube.
8. Raise and safely support the vehicle.
9. If equipped with an automatic transaxle, remove the suspension support crossmember brace.
10. If equipped with a manual transaxle, remove the heat shield. If equipped with an automatic transaxle, remove the nut but do not remove the heat shield from the vehicle at this time.
11. Disconnect the exhaust pipe from the exhaust manifold.
12. If equipped with an automatic transaxle, remove the exhaust manifold extension pipe.
13. Unfasten the retaining bolts, then remove the exhaust manifold from the vehicle. If equipped with an automatic transaxle, remove the heat shield with the exhaust manifold.
14. Clean the mating surfaces of the cylinder head and manifold. Discard the gasket.

To install:

15. Use a new manifold gasket, place the manifold (and heat shield, if automatic) into position and loosely install the manifold bolts.
16. Tighten the exhaust manifold bolts to 21 ft. lbs. (28 Nm).
17. If equipped with an automatic transaxle, install the exhaust manifold extension pipe.
18. If applicable, install the exhaust manifold heat shield, then tighten the nut to 89 inch lbs. (10 Nm).
19. Attach the exhaust pipe to the manifold.
20. If equipped with an automatic transaxle, install the suspension support crossmember brace.
21. Carefully lower the vehicle.
22. If equipped with an automatic transaxle, install the transaxle fluid level indicator and fill tube.
23. Attach the EGR pipe to the exhaust manifold.

24. Engage the Oxygen sensor connector.
25. Fasten the exhaust crossover to the manifolds.
26. Install the air cleaner assembly.
27. Install the battery, then connect the positive, then negative battery cables.

NOTE: A Tech 1® scan tool is needed to perform the "Idle Learn" procedure.

28. Perform the following "Idle Learn" procedure;
    a. Install a Tech 1® scan tool.
    b. Turn the ignition to the **ON** position, engine not running.
    c. Select **IAC SYSTEM**, then **IDLE LEARN** in the **MISC TEST** mode.
    d. Place the transaxle in **P** or **N**, as applicable.
    e. Proceed with idle learn as directed by the scan tool.

Front Cover Seal

REMOVAL AND INSTALLATION

2.2L, 2.3L, 2.4L and 3.1L Engines

The oil seal can be replaced with the cover either on or off the engine. If the cover is on the engine, remove the crankshaft pulley and hub first. Pry out the seal using a suitable prying tool, being careful not to distort the seal mating surface. Install the new seal that the open side or lip side is towards the engine. Press it into place with a seal driver made for the purpose. General Motors recommends a tool, J-35468 Seal Centering tool. Install the hub, if removed.

Timing Chain, Sprockets and Front Cover

REMOVAL AND INSTALLATION

2.2L Engine

NOTE: The following procedure requires the use of a special tool.

1. Disconnect the negative battery cable.
2. Remove the serpentine belt and tensioner.

NOTE: Although not absolutely necessary, removal of the right front inner fender splash shield will facilitate access to the front cover.

3. Install engine support fixture J-28467-A or equivalent.
4. Remove the engine mount assembly.
5. Remove the alternator rear brace, then remove the alternator.
6. Remove the power steering pump, then position it aside with the lines still attached.
7. Raise and safely support the vehicle.
8. Remove the oil pan.
9. Unscrew the center bolt from the crankshaft pulley and slide the pulley and hub from the crankshaft.
10. Unfasten the front cover-to-block bolts and then remove the front cover. If the cover is difficult to remove, use a plastic mallet to carefully loosen the cover.
11. Place the No. 1 piston at TDC of the compression stroke that the marks on the camshaft and crankshaft sprockets are in alignment.
12. Loosen the timing chain tensioner nut as far as possible without actually removing it.
13. Remove the camshaft sprocket bolts and remove the sprocket and chain together. If the sprocket does not slide from the camshaft easily, a light blow with a soft mallet at the lower edge of the sprocket will dislodge it.
14. Use J-2288-8-20 or equivalent gear puller, and remove the crankshaft sprocket.

To install:

15. Press the crankshaft sprocket back onto the crankshaft.
16. Install the timing chain over the camshaft sprocket and then around the crankshaft sprocket. Make sure the marks on the 2 sprockets are in alignment. Lubricate the thrust surface with Molykote® or equivalent.
17. Align the dowel in the camshaft with the dowel hole in the sprocket and then install the sprocket onto the camshaft. Use the mounting bolts draw the sprocket onto the camshaft and then tighten to 66–68 ft. lbs. (89–92 Nm).
18. Lubricate the timing chain with clean engine oil. Tighten the chain tensioner.
19. The surfaces of the block and front cover must be clean and free of oil. Install a new gasket, then position the front cover on the block using a centering tool (J-23042). Tighten the retaining bolts to 6–9 ft. lbs. (8–12 Nm).
20. Installation of the remaining components is the reverse of the removal procedure.

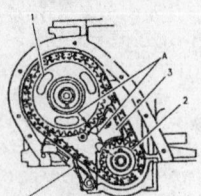

1. Camshaft sprocket
2. Crankshaft sprocket
3. Timing chain tensioner
A. Line up timing marks on sprockets with tabs on timing chain tensioner
B. Remove pin after timing chain is installed

8838LG17

Align sprocket timing marks with the alignment tabs on the tensioner — 2.2L engine

21. Connect the negative battery cable.

2.3L and 2.4L Engines

——— **WARNING** ———

The timing chain on the 1996–97 2.4L DOHC engine is NOT to be replaced with a timing chain from any other model year. The timing sprockets are different on these engines and the shape of the links matches the sprockets. Engine damage may result if the wrong timing chain is used.

1. Disconnect the negative battery cable.
2. Remove the coolant recovery reservoir.
3. Remove the serpentine belt, using a 13mm wrench that is at least 24 in. (61cm) long.
4. For the 2.3L engine, remove the alternator, then position it aside. Install engine support J-28467-A or equivalent. Reinstall the alternator through-bolt, then attach the engine support fixture.
5. For the 2.4L engine, install tool J-28467-400 onto the alternator stud-ended bolt, and attach the fixture.
6. Remove the upper cover fasteners.
7. Detach the cover vent hose.
8. Remove the right engine mount and the engine mount bracket or bracket adapter. Whenever the engine mounting bracket adapter is removed, the bolts must be replaced.
9. Raise and safely support the vehicle.
10. Remove the right front wheel and tire assembly and the splash shield.
11. Remove the crankshaft balancer assembly.

NOTE: Do not install an automatic transaxle equipped engine balancer on a manual transaxle equipped engine or vice-versa.

12. Remove the lower cover fasteners.
13. Carefully lower the vehicle.
14. Remove the front cover and gasket. Inspect the gasket for damage and replace if necessary.
15. Rotate the crankshaft clockwise, as viewed from the front of engine/normal rotation, until the camshaft sprocket timing dowel pin holes line up with the holes in the timing chain housing. The crankshaft sprocket keyway should point upwards and line up with the centerline of the cylinder bores; this is the ""timed" position.
16. Remove the timing chain guides.
17. Raise and safely support the vehicle.
18. Make sure all of the slack in the timing chain is above the tensioner assembly, then remove the tensioner. The timing chain must be disengaged from any wear grooves in the tensioner shoe in order to remove the shoe. Slide a suitable prytool under the timing chain while pulling the shoe outward.

——— **WARNING** ———

Do NOT attempt to pry the socket off the camshaft or damage to the sprocket or chain housing could occur.

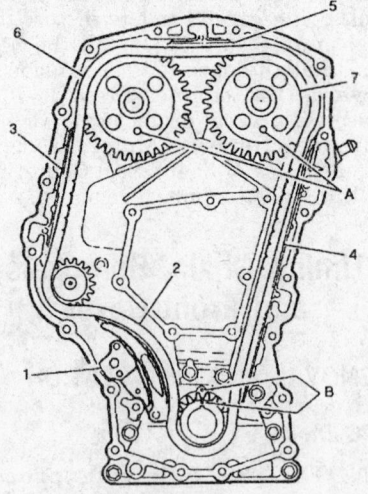

A Camshaft timing alignment pin location
B Crankshaft gear timing mark
1 Shoe assembly timing chain tensioner
2 Timing chain
3 R.H. timing chain guide
4 L.H. timing chain guide
5 Upper timing chain guide
6 Exhaust camshaft sprocket
7 Intake camshaft sprocket

8838LG18

The chain must be in the "timed" position — 2.3L and 2.4L engines

19. If difficulty is encountered in removing the chain tensioner shoe, remove the intake camshaft sprocket, as follows:
   a. Carefully lower the vehicle.
   b. Hold the intake camshaft sprocket with tool J-39579, or equivalent, and remove the sprocket bolt and washer.
   c. Remove the washer from the bolts and rethread the bolt back into the camshaft by hand. The bolt provides a surface to push against.
   d. Remove the camshaft sprocket using a three-jaw puller in the 3 relief holes in the sprocket.
20. Unfasten the tensioner assembly retaining bolts, then remove the tensioner.

NOTE: The timing chain and crankshaft sprocket MUST be marked before removal. If the chain or sprocket is installed with the wear pattern in the opposite direction, noise and increased wear may occur.

21. Mark the crankshaft sprocket and timing chain outer surface for reassembly, then remove the chain.
22. Clean the old sealant off the bolt with a wire brush. Clean the threaded hole in the camshaft with a round nylon brush. Inspect the parts for wear and replace as necessary. Note that some scoring of the chain shoe and guides is normal.

To install:

——— **WARNING** ———

Failure to follow this procedure may result in severe engine damage.

23. Position the intake camshaft sprocket onto the camshaft with the surface marked during removal showing.
24. Install the intake camshaft sprocket retaining bolt and washer, tighten to 52 ft. lbs. (70 Nm) while holding the sprocket with tool J-39579, if removed. Use sealant 12345493 or equivalent on the camshaft sprocket bolt.
25. Place tool J-36008, or equivalent camshaft aligning pins, through the holes in the camshaft sprockets into the holes in the timing chain housing. This positions the cams for correct timing.
26. If the camshafts are out of position and must be rotated more than $\frac{1}{8}$ turn in order to install the alignment dowel pins, proceed as follows:
   a. The crankshaft MUST be rotated 90 degrees clockwise off TDC in order to give the valves adequate clearance to open.

b. Once the camshafts are in position and the dowels installed, rotate the crankshaft counterclockwise back to TDC.

WARNING

Do not rotate the crankshaft clockwise to TDC; valve or piston damage could result.

27. Place the timing chain over the exhaust camshaft sprockets, around the idler sprocket and around the camshaft sprocket.

28. Set the camshafts at the timed position and install the timing chain. Remove the alignment dowel pin from the intake camshaft. Using tool J-39579, rotate the intake camshaft sprocket counterclockwise enough to slide the timing chain over the intake camshaft sprocket. Release the camshaft sprocket wrench J-39579. The length of the chain between the 2 camshaft sprockets will tighten. If properly timed, the intake camshaft alignment dowel pin should slide in easily. If the dowel pin does not fully index, the camshafts are NOT timed correctly and the procedure must be repeated.

29. Leave the alignment dowel pins installed. Raise and safely support the vehicle.

30. With the slack removed from the chain between the intake camshaft sprocket and the crankshaft sprocket, the timing marks on the crankshaft and cylinder block should be aligned. If the marks are not aligned, move the chain one tooth forward or rearward, remove the slack and recheck the marks.

31. Reload the timing chain tensioner assembly to it "zero" position as follows:

a. Form a keeper from a piece of heavy gauge wire.

b. Apply slight force on the tensioner blade to compress the plunger.

c. Insert a small prytool into the reset access hole, and pry the ratchet pawl away from the ratchet teeth while forcing the plunger completely in the hole.

d. Install the keeper between the access hole and the blade.

32. Install the tensioner assembly to the timing chain housing. Recheck the plunger assembly installation, it is correctly installed when the long end is toward the crankshaft. Install the tensioner retaining bolts; tighten to 89 inch lbs. (10 Nm).

33. Carefully lower the vehicle enough to reach and remove the alignment dowel pins.

WARNING

Severe engine damage could result if the engine is not properly timed.

34. Rotate the crankshaft clockwise (normal rotation) 2 full rotations. Align the crankshaft keyway with the mark on the cylinder block and reinstall the alignment dowel pins. The pins will slide in easily if the engine of correctly timed.

35. Install the timing chain guides, then install the front (timing chain) cover. Tighten the timing chain cover fasteners to 106 inch lbs. (12 Nm). Tighten the balancer attaching bolt to 74 ft. lbs. (100 Nm).

36. Installation is the reverse of the removal procedure.

37. Connect the negative battery cable.

3.1L Engine

1. Disconnect the negative battery cable.

2. Drain the cooling system.

3. Remove the serpentine belt adjustment pulley.

4. Remove the alternator and disconnect the electrical wires.

5. Remove the power steering pump bracket.

6. Disconnect the heater pipe at the power steering bracket.

7. Raise and safely support the vehicle.

8. Remove the wheel and tire assembly, then remove the inner splash shield.

9. If equipped, remove the A/C compressor belt.

10. Remove the flywheel cover at the transaxle.

11. Remove the harmonic balancer with tool J-23523-1 or equivalent.

NOTE: The outer ring (weight) of the harmonic balancer is bonded to the hub with rubber. Breakage may occur if the balancer is hammered back onto the crankshaft. A press or special installation tool is necessary.

12. Remove the serpentine belt idler pulley.

13. Unfasten the pan-to-front cover bolts, and remove the lower cover bolts.

14. Carefully lower the vehicle.

15. Disconnect the radiator hose at the water pump.

16. Remove the heater pipe at the goose neck.

17. Detach the bypass and overflow hoses.

18. Disconnect the canister purge hose.

19. Unfasten the upper front cover bolts and remove the front cover.

20. Place the No. 1 piston at TDC and the stamped timing marks on both sprockets are closest to one another and in line between the shaft centers (No. 4 firing position).

21. Take out the 3 bolts that hold the camshaft sprocket to the camshaft. This sprocket is a light press fit on the camshaft and should come off readily. If the sprocket does not come off easily, a light blow on the lower edge of the sprocket with a plastic mallet should dislodge the sprocket. The chain comes off with the camshaft sprocket. A gear puller will be required to remove the crankshaft sprocket.

To install:

22. Without disturbing the position of the engine, mount the new crank sprocket on the shaft, then mount the chain over the camshaft sprocket. Arrange the camshaft sprocket in such a way that the timing marks will line up between the shaft centers and the camshaft locating dowel will enter the dowel hole in the cam sprocket.

23. Place the cam sprocket, with the chain mounted over it, in position on the front of the camshaft and pull up with the 3 bolts that hold it to the camshaft. Tighten the bolts to 18 ft. lbs. (25 Nm).

24. Lubricate the timing chain with clean engine oil.

25. After the sprockets are in place, turn the engine 2 full revolutions to make certain that the timing marks are in correct alignment between the shaft centers.

26. Clean all the gasket mounting surfaces on the front cover and block and place a new gasket to the front cover sealing surface. Apply a continuous 1/8 in. (3mm) wise bead of sealer (1052357 or equivalent) to the oil pan sealing surface of the front cover.

27. Place the front cover on the engine and install the upper front cover bolts.

28. Raise and safely support the vehicle.

29. Install the lower cover bolts.

30. Install the oil pan-to-cover screws.

31. Install the serpentine belt idler pulley.

32. Install the harmonic balancer.

33. Fasten the flywheel cover to the transaxle.

34. If equipped, install the A/C compressor belt.

35. Install the inner splash shield and the tire and wheel assembly.

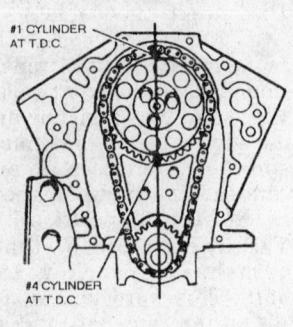

#1 CYLINDER
AT T.D.C.

#4 CYLINDER
AT T.D.C.

8838LG19

View of the alignment marks with the
No. 1 and 4 cylinders at TDC

36. Lower the vehicle, and install the remainder of the parts in the reverse order of removal.

Timing Belt, Sprockets, Tensioner and Front Cover

REMOVAL AND INSTALLATION

2.0L Engine

1. Disconnect the negative battery cable.
2. Remove the coolant recovery reservoir.
3. If equipped, remove the A/C belt.
4. Remove the serpentine belt. Unfasten the belt tensioner bolt, the tensioner arm will swing down, then remove the tensioner.
5. Disconnect the evaporative emission pipe assembly.
6. Remove the crankshaft pulley assembly.
7. Either unsnap (upper first) or unfasten the cover bolts and nuts, then remove the cover.
8. Align the marks on the crankshaft sprocket with the marks on the rear timing belt cover by rotating the crankshaft.
9. Loosen the water pump retaining bolts, and release the tensioner with tool J-33039-A or equivalent.
10. Carefully lower the vehicle, then remove the timing belt.
To install:
11. Turn the crankshaft and camshaft gears clockwise to align the timing marks on the gears with the timing marks on the rear cover.
12. Install the timing belt, making sure the portion between the camshaft gear and crankshaft gear is in tension.
13. Using tool J-33039-A or equivalent adjustment tool, turn the coolant pump eccentric clockwise until the tensioner arm contacts the

high torque stop. Tighten the water pump screws slightly.
14. Turn the engine by the crankshaft gear bolt 720 degrees clockwise to fully seat the belt into the gear teeth.
15. Turn the coolant pump eccentric counterclockwise until the hole in the tensioner arm is aligned with the hole in the base. This must be done with the engine at room temperature, approximately 68°F (20°C).
16. Tighten the water pump screws while checking that the tensioner holes remain as adjusted as in Step 15.
17. Install the timing belt cover by snapping it into place or secure with the retaining bolts and nut. Tighten to 89 inch lbs. (10 Nm).
18. Install the crankshaft pulley.
19. Connect the evaporative emission pipe assembly.
20. Install the serpentine drive belt and the A/C belt, if equipped.
21. Install the coolant recovery reservoir.
22. Properly fill the cooling system, then connect the negative battery cable.

Camshaft

REMOVAL AND INSTALLATION

2.0L Engine

NOTE: The following procedure requires the use of a special tool.

1. Disconnect the negative battery cable.
2. Remove camshaft carrier cover.
3. Using valve train compressing fixture J-33302, which holds the valves in place, compress valve springs and remove rocker arms.
4. Remove the timing belt front cover and the timing belt.
5. Remove camshaft sprocket.
6. If equipped, remove the distributor.
7. Remove camshaft thrust plate from rear of camshaft carrier.
8. Slide camshaft rearward, then carefully remove it from the carrier.
To install:
9. Install a new camshaft carrier front oil seal using tool J-33085.
10. Place camshaft in the carrier.

WARNING
Take care not to damage the carrier front oil seal when installing the camshaft.

11. Install camshaft thrust plate retaining bolts. Tighten the bolts to 70 inch lbs. (7.8 Nm).
12. Check camshaft end-play, which should be within 0.0006–0.0025 in. (0.016–0.064mm).
13. Install distributor.
14. Install camshaft sprocket.
15. Install timing belt.
16. Install timing belt front cover.
17. Using valve train compressing fixture J-33302, compress valve springs and replace rocker arms.
18. Install camshaft carrier cover.

2.2L Engine

1. Remove the engine and place it on a suitable engine stand.
2. Remove the cylinder head cover, pivot the rocker arms to the sides, and remove the pushrods, keeping them in order. Remove the valve lifters, keeping them in order. There are special tools which make lifter removal easier.
3. Remove the front cover.
4. If necessary, remove the oil pump drive.
5. Remove the fuel pump and it's pushrod.
6. Remove the timing chain and sprocket .
7. Carefully pull the camshaft from the block, being sure the camshaft lobes do not contact the bearings.
To install:
8. To install, lubricate the camshaft journals with clean engine oil. Lubricate the lobes with Molykote® or the equivalent.
9. Install the camshaft into the engine, being extremely careful not to contact the bearings with the cam lobes.
10. Install the timing chain and sprocket. Install the fuel pump and pushrod. Install the timing cover.
11. Install the valve lifters. If a new camshaft has been installed, new lifters should be used to ensure durability of the cam lobes.
12. Install the pushrods and rocker arms and the intake manifold. Adjust the valve lash after installing the engine. Install the cylinder head cover.

2.3L and 2.4L Engines

Intake Camshaft

NOTE: Any time the camshaft housing to cylinder head bolts are loosened or removed, the camshaft housing to cylinder head gasket must be replaced.

1. Relieve the fuel system pressure. Disconnect the negative battery cable.

2. Label and detach the ignition coil and module assembly electrical connections.

3. Unfasten the ignition coil and module assembly to camshaft housing bolts, then remove the assembly by pulling straight up. Use a special spark plug boot wire remover tool to remove connector assemblies, if they have stuck to the spark plugs.

4. If equipped, remove the idle speed power steering pressure switch connector.

5. Loosen the 3 power steering pump pivot bolts and remove drive belt.

6. Disconnect the 2 rear power steering pump bracket-to-transaxle bolts.

7. Remove the front power steering pump bracket to cylinder block bolt.

8. Disconnect the power steering pump assembly, then position it aside.

9. Using the special tool, remove the power steering pump drive pulley from the intake camshaft.

10. Remove oil/air separator bolts and hoses. Leave the hoses attached to the separator, disconnect from the oil fill, chain housing and intake manifold. Remove as an assembly.

11. Remove vacuum line from fuel pressure regulator and detach the fuel injector harness connector.

12. Disconnect fuel line attaching clamp from bracket on top of intake camshaft housing.

13. Unfasten the fuel rail-to-camshaft housing attaching bolts, then remove the fuel rail from the cylinder head. Cover or plug injector openings in cylinder head and the injector nozzles. Leave the fuel lines attached, then position fuel rail aside.

14. Disconnect the timing chain and housing, but do NOT remove from the engine.

15. Remove the intake camshaft housing cover-to-camshaft housing attaching bolts.

16. Unfasten the intake camshaft housing-to-cylinder head attaching bolts. Use the reverse of the tightening sequence when loosening the bolts. Leave 2 of the bolts loosely in place to hold the camshaft housing while separating the camshaft cover from housing.

17. Push the cover off the housing by threading 4 of the housing-to-head attaching bolts into the tapped holes in the cam housing cover. Tighten the bolts evenly so the cover does not bind on the dowel pins.

18. Remove the 2 loosely installed camshaft housing to head bolts and remove the cover. Discard the gaskets.

19. Note the position of the chain sprocket dowel pin for reassembly.

20. Remove intake camshaft oil seal from camshaft and discard seal. This seal must be replaced any time the housing and cover are separated.

21. Remove the camshaft carrier from the cylinder head and remove the gasket. Discard the gasket.

To install:

22. Thoroughly clean the mating surfaces of the camshaft carrier and the cylinder head, bolts and bolt holes. Install a new gasket and place the housing on the head. Install one bolt loosely to hold it in place.

23. Install the lifters into their bores. If the camshaft is being replaced, the lifters must also be replaced. Lubricate camshaft lobes, journals and lifters with camshaft and lifter prelube. The camshaft lobes and journals must be adequately lubricated or engine damage could occur upon start up.

24. Install the camshaft in the same position as when removed. The timing chain sprocket dowel pin should be straight up and align with the centerline of the lifter bores.

25. Install new camshaft housing to camshaft housing cover seals into cover; do not use sealer. Make sure the correct color seal is placed in each groove. Install the cover to the housing.

26. Apply thread locking compound to the camshaft housing and cover attaching bolt threads.

27. Install the bolts, then tighten to 11 ft. lbs. (15 Nm). Rotate the bolts (except the 2 rear bolts that hold the fuel pipe to the camshaft housing) an additional 75 degrees, in sequence. Tighten the 2 rear bolts to 16 ft. lbs. (15 Nm), then rotate an additional 25 degrees.

28. Install the timing chain housing and the timing chain.

29. Uncover fuel injectors, then install new fuel injector O-ring seals lubricated with oil. Install the fuel rail.

30. Fasten the fuel line attaching clamp and retainer to bracket on top of the intake camshaft housing.

31. Connect the vacuum line to the fuel pressure regulator.

32. Attach the fuel injectors harness connector.

33. Install the oil/air separator assembly.

34. Lubricate the inner sealing surface of the intake camshaft seal with oil and install the seal to the housing.

35. Install the power steering pump pulley onto the intake camshaft.

36. Install the power steering pump assembly and drive belt.

37. Connect the idle speed power steering pressure switch connector.

38. Clean any loose lubricant that is present on the ignition coil and module assembly to camshaft housing bolts. Apply Loctite® 592 or equivalent, onto the ignition coil and module assembly to camshaft housing bolts. Install the bolts and tighten to 13 ft. lbs. (18 Nm).

39. Attach the electrical connectors to ignition coil and module assembly.

40. Connect the negative battery cable, then start the engine and check for leaks.

Exhaust Camshaft

NOTE: Any time the camshaft housing-to-cylinder head bolts are loosened or removed, the camshaft housing to cylinder head gasket must be replaced.

1. Relieve the fuel system pressure. Disconnect the negative battery cable.

2. Label and disconnect the ignition coil and module assembly electrical connections.

3. Unfasten the ignition coil and module assembly-to-camshaft housing bolts, then remove the assembly by pulling straight up. Use a special tool to remove connector assemblies if they have stuck to the spark plugs.

4. If equipped, remove the idle speed power steering pressure switch connector.

5. Remove the transaxle fluid level indicator tube assembly from exhaust camshaft cover and position aside.

6. Remove exhaust camshaft cover and gasket.

7. Disconnect the timing chain and housing but do not remove from the engine.

8. Remove exhaust camshaft housing to cylinder head bolts. Use the reverse of the tightening procedure when loosening camshaft housing while separating camshaft cover from housing.

9. Push the cover off the housing by threading 4 of the housing to head attaching bolts into the tapped holes in the camshaft cover. When threading the bolt, tighten them evenly so the cover does not bind on the dowel pins.

10. Remove the 2 loosely installed camshaft housing to cylinder head bolts and remove cover, discard gaskets.

11. Loosely reinstall one camshaft housing to cylinder head bolt to retain the housing during camshaft and lifter removal.

12. Note the position of the chain sprocket dowel pin for reassembly. Remove camshaft being careful not to damage the camshaft or journals.

13. Remove the camshaft carrier from the cylinder head and remove the gasket. Discard the gasket.

To install:

14. Thoroughly clean the mating surfaces of the camshaft carrier and the cylinder head, bolts and bolt holes. Install a new gasket and place the housing on the head. Install 1 bolt loosely to hold in place.

15. Install the lifters into their bores. If the camshaft is being re-placed, the lifters must also be re-placed. Lubricate camshaft lobes, journals and lifters with camshaft and lifter prelube. The camshaft lobes and journals must be ade-quately lubricated or engine damage could occur upon start up.

16. Install camshaft in same posi-tion as when removed. The timing chain sprocket dowel pin should be straight up and align with the center-line of the lifter bores.

17. Install new camshaft housing-to-camshaft housing cover seals into the cover; do not use sealer. Make sure the correct color seal is placed in each groove. Install the cover to the housing.

18. Apply thread locking compound to the camshaft housing and cover at-taching bolt threads.

19. Install bolts, then tighten, in se-quence, to 11 ft. lbs. (15 Nm). Then rotate the bolts an additional 75 de-grees, in sequence.

20. Install timing chain housing and timing chain.

21. Install the transaxle fluid level indicator tube assembly to the ex-haust camshaft cover.

22. Attach the idle speed power steering pressure switch connector.

23. Clean any loose lubricant that is present on the ignition coil and module assembly to camshaft hous-ing bolts. Apply Loctite® 592 or equivalent, onto the ignition coil and module assembly to camshaft hous-ing bolts. Install the bolts and tighten to 13 ft. lbs. (18 Nm).

24. Attach the electrical connectors to ignition coil and module assembly.

25. Connect the negative battery cable, then start the engine and check for leaks.

3.1L Engine

1. Remove the engine assembly.
2. Remove intake manifold, valve lifters and timing chain cover.
3. Remove fuel pump and pump pushrod.

4. Remove camshaft sprocket bolts, sprocket and timing chain. A light blow to the lower edge of a tight sprocket should free it (use a plastic mallet).

5. Remove the timing chain tensioner.

6. If applicable, remove the oil pump drive shaft extension.

7. Install 2 bolts in cam bolt holes and pull cam from block.

To install:

8. Lubricate the camshaft journals with engine oil and reverse removal procedure aligning the sprocket tim-ing marks.

NOTE: If a new camshaft is be-ing installed, coat the camshaft lobes with GM Engine Oil Supple-ment (E.O.S.), or equivalent.

Balance Shaft

REMOVAL AND INSTALLATION

2.3L and 2.4L Engines

NOTE: The transaxle assembly must be removed from the vehi-cle to perform this procedure.

1. Disconnect the negative battery cable.

2. Remove the transaxle from the vehicle.

3. Matchmark and remove the pressure plate and clutch.

4. Remove the flywheel mounting bolts, the flywheel retainer, for auto-matic transaxle, and the flywheel.

5. Remove the engine oil pan.

6. Remove the nut and the bolts securing the balance shaft chain cover and remove the cover.

7. Loosen the pivot bolt on the chain tensioner and pivot the ten-sioner to remove the tension from the chain.

---- **WARNING** ----

The balance shaft sprocket bolt is reverse threaded and must be ro-tated clockwise to be removed. Turning the bolt in the counter-clockwise direction can result in the bolt shearing off in the bal-ance shaft.

---

8. Hold the crankshaft from turn-ing and remove the balance shaft sprocket bolt.

9. Mark the balance shaft sprocket so the same side of the sprocket is facing the engine during installation.

10. Separate the sprocket from the balance shaft.

11. Remove the 5 balance shaft as-sembly-to-engine block mounting bolts. Do not remove the 8 bolts hold-ing the balance shaft housing halves together.

12. Pry the oil pump pickup screen out of the housing using a suitable tool.

13. Support the balance shaft hous-ing and loosen the 8 mounting bolts in the reverse order of the tightening sequence.

14. Remove the 8 bolts and lift the top half of the balance shaft housing off.

15. Remove the 2 balance shafts from the housing.

16. Remove the balance shaft bearings.

17. Remove the thrust plate mount-ing bolts, then remove the thrust plate.

To install:

18. Clean all the components thoroughly.

19. Install the thrust plate and tighten the mounting bolts to 115 inch lbs. (13 Nm).

20. Carefully install the new bear-ing inserts into the housing halves. Use care not to scratch or gouge the bearings or the housing.

21. Lubricate the bearings, balance shaft and gears with GM 9985705 or equivalent.

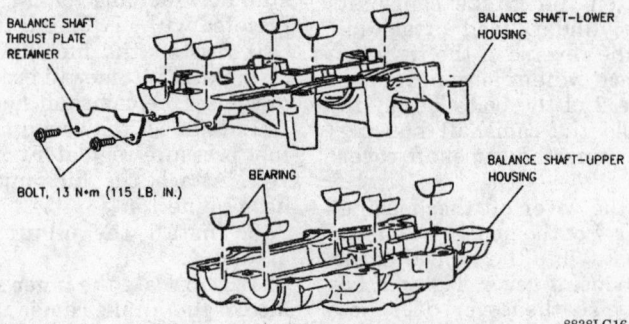

BALANCE SHAFT THRUST PLATE RETAINER

BOLT, 13 N•m (115 LB. IN.)

BEARING

BALANCE SHAFT–LOWER HOUSING

BALANCE SHAFT–UPPER HOUSING

8838LG16

Upper and lower balance shaft housings and bearing inserts — 2.3L and 2.4L engines

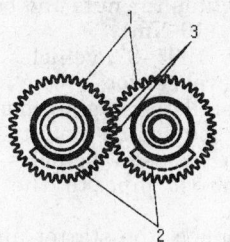

1. Balance shaft gears
2. Balance shaft counter weights
3. Balance shaft gear timing mark

8838LG27

Timing mark alignment on the balance shafts

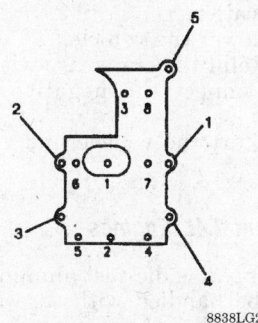

8838LG28

Balance shaft housing bolt tightening sequence

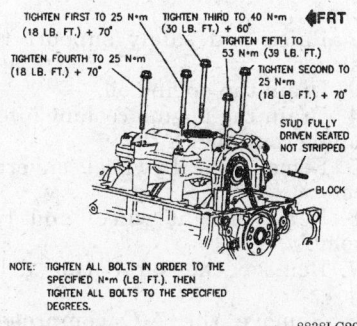

TIGHTEN FIRST TO 25 N•m (18 LB. FT.) + 70°
TIGHTEN THIRD TO 40 N•m (30 LB. FT.) + 60°
TIGHTEN FIFTH TO 53 N•m (39 LB. FT.)
TIGHTEN FOURTH TO 25 N•m (18 LB. FT.) + 70°
TIGHTEN SECOND TO 25 N•m (18 LB. FT.) + 70°
STUD FULLY DRIVEN SEATED NOT STRIPPED
BLOCK

NOTE: TIGHTEN ALL BOLTS IN ORDER TO THE SPECIFIED N•m (LB. FT.). THEN TIGHTEN ALL BOLTS TO THE SPECIFIED DEGREES.

8838LG29

Balance shaft housing-to-engine bolt tightening sequence

22. Install the balance shafts into the lower half of the housing. Make sure the timing marks on the balance shafts are lined up properly.

23. Install the upper half of the balance shaft housing and tighten the bolts in sequence to 44 inch lbs. (5 Nm). Do not tighten the bolts any tighter at this time. The final housing bolt tightening is done after the assembly is installed on the engine.

24. Do not install the oil pump pickup screen until the bolts have been tightened to their final torque.

25. Install the balance shaft assembly on the engine and after coating the mounting bolt threads with Loctite® 242 or equivalent threadlocking compound, tighten the bolts in sequence to 44 inch lbs. (5 Nm).

26. Tighten the 5 housing-to-engine bolts, as follows:

  a. Bolt 1 to 18 ft. lbs. (25 Nm).

  b. Bolt 2 to 18 ft. lbs. (25 Nm).

  c. Bolt 3 to 30 ft. lbs. (40 Nm).

  d. Bolt 4 to 18 ft. lbs. (25 Nm).

  e. Bolt 5 to 39 ft. lbs. (53 Nm).

  f. Tighten bolt 1 an additional 70 degrees.

  g. Tighten bolt 2 an additional 70 degrees.

  h. Tighten bolt 3 an additional 60 degrees.

  i. Tighten bolt 4 an additional 70 degrees.

27. Tighten the 8 upper-to-lower housing bolts in sequence to the following torques:

  a. Bolts 1 and 2 to 89 inch lbs. (10 Nm).

  b. Bolt 3 to 11 ft. lbs. (15 Nm).

  c. Bolts 4 through 7 to 89 inch lbs. (10 Nm).

  d. Bolt 8 to 11 ft. lbs. (15 Nm).

  e. Tighten all 8 bolts in sequence an additional 40 degrees.

28. Rotate the balance shaft assembly on full revolution to verify the shafts spin freely. Make sure the balance shaft assembly is returned to its original position before installation of the drive sprocket and chain.

29. Rotate the engine until the No. 1 cylinder is at TDC.

30. Rotate the engine 90 degrees.

31. Install J-41088 onto the balance shaft assembly to ensure the balance shafts don't rotate when the driven gear is installed.

32. Install the balance shaft driven gear with the mark made during disassembly facing outward and tighten the left hand threaded mounting bolt to 22 ft. lbs. (30 Nm).

33. Remove the J-41088.

34. Install the balance shaft chain guide and set the chain tension as follows:

  a. Insert 0.040 in, brass feeler gauge between the chain guide and the chain.

NOTE: A brass gauge must be used to ensure proper chain tension. Use of a steel gauge will not provide the correct flex and the chain tensioner will be incorrect.

  b. Press the chain guide against the chain with 3 lbs. of force.

  c. Tighten the chain tensioner adjuster bolt to 115 inch lbs. (13 Nm).

  d. Install the balance shaft chain cover and tighten the mounting nut and bolt to 115 inch lbs. (13 Nm).

35. Install the oil pan assembly.

36. Install the flywheel and tighten the flywheel mounting bolts to 22 ft. lbs. (30 Nm) plus a 40 degrees additional rotation.

37. Install the clutch disc and pressure plate.

38. Install the transaxle assembly.

39. Connect the negative battery cable.

40. Start the vehicle and verify the engine runs smoothly and quietly.

Piston and Connecting Rod

POSITIONING

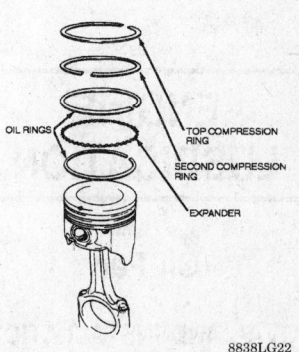

OIL RINGS
TOP COMPRESSION RING
SECOND COMPRESSION RING
EXPANDER

8838LG22

Piston ring placement — 2.0L, 2.3L, 2.4L and 3.1L engines

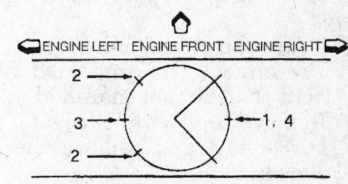

ENGINE LEFT   ENGINE FRONT   ENGINE RIGHT

1. Oil ring spacer gap (tang in hole or slot with arc)
2. Oil ring rail gaps
3. 2nd compression ring gap
4. Top compression ring gap

8838LG23

Piston ring gap locations — 2.2L and 3.1L engines

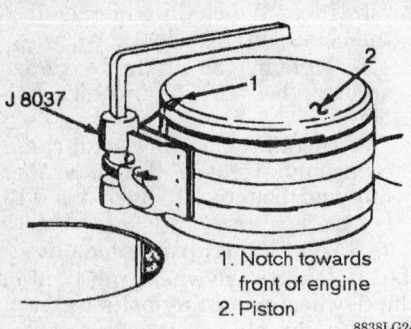

J 8037

1. Notch towards front of engine
2. Piston

8838LG24

Notch faces front of engine — 2.2L, 2.3L, 2.4L and 3.1L engines

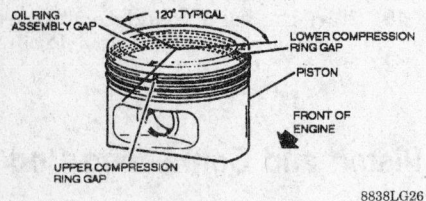

OIL RING ASSEMBLY GAP
120° TYPICAL
LOWER COMPRESSION RING GAP
PISTON
FRONT OF ENGINE
UPPER COMPRESSION RING GAP

8838LG26

Piston ring gap location — 2.3L and 2.4L engines

ENGINE LUBRICATION

Oil Pan

REMOVAL AND INSTALLATION

2.0L Engine

1. Disconnect the negative battery cable.
2. Raise and support the vehicle safely.
3. Drain the engine oil.
4. Disconnect the front exhaust pipe from the exhaust manifold.
5. Remove the flywheel cover.
6. Disconnect and remove the oil level sensor.
7. Unfasten the oil pan retaining bolts and remove the oil pan.
8. Remove the oil pump pickup tube.
9. Remove the oil deflector and gasket.
   To install:
10. Clean all the gasket surfaces completely.

11. Install the oil deflector and gasket assembly.
12. Install the oil pump pickup tube and tighten the mounting bolts to 62 inch lbs. (7 Nm).
13. Install the oil pan and tighten the mounting bolts to 97 inch lbs. (11 Nm).
14. Install the drain plug in the oil pan and tighten to 33 ft. lbs. (45 Nm).
15. Install the oil level sensor.
16. Install the flywheel cover.
17. Connect the exhaust pipe to the manifold and tighten the mounting nuts to 26 ft. lbs. (33 Nm).
18. Lower the vehicle.
19. Refill the crankcase with engine oil.
20. Connect the negative battery cable.
21. Start the engine and check for leaks.

2.2L Engine

1. Disconnect the negative terminal from the battery.
2. Raise and safely support the vehicle.
3. Drain the engine oil.
4. Remove the right front tire and wheel assembly.
5. Remove the right inner fender well splash shield.
6. Remove the starter and starter bracket.
7. Remove the flywheel cover bolts and flywheel cover.
8. Remove the engine support strut and support strut bracket.
9. On 1993–94 vehicles, disconnect the exhaust pipe from the exhaust manifold and position out of the way.
10. On 1995–97 vehicles, disconnect the oil level sensor.
11. Unfasten the oil pan mounting nuts and bolts, then remove the oil pan.
    To install:
12. Clean all the gasket surfaces completely.
13. On 1993–94 vehicles, apply a thin bead of sealer around the outside edge of the oil pan and install the oil pan gasket onto the sealer.
14. On 1995–97 vehicles, place a 2mm bead of GM 1052914 or equivalent RTV sealer to the oil pan sealing surface except at the rear seal mounting surface. Using a new oil pan rear seal, apply a thin coat RTV sealer on the end down to the ears. Position the oil pan into place and install the fasteners.
15. Install the oil pan onto the engine and loosely install all the fasteners.

16. Tighten the nuts and bolts to 89 inch lbs. (10 Nm).
17. On 1995–97 vehicles, connect the oil level sensor.
18. Install the engine mount strut bracket and engine mount strut.
19. On 1993–94 vehicles, connect the exhaust pipe to the exhaust manifold.
20. Connect the starter and install the starter and support bracket.
21. Install the flywheel cover and cover mounting bolts.
22. Install the right fender well splash shield.
23. Install the right front tire and wheel assembly and tighten to specification.
24. Lower the vehicle.
25. Refill the crankcase with oil.
26. Connect the negative battery cable.
27. Start the vehicle and verify no leaks.

2.3L and 2.4L Engines

The oil pan is die cast aluminum and must be handled with care to avoid damage. The oil pan includes an attachment to the transaxle to provide additional structural support.

1. Disconnect the negative battery cable.
2. Raise and safely support the vehicle.
3. Drain the engine oil.
4. Drain the engine coolant into a suitable container.
5. Remove the flywheel/converter cover.
6. Remove right wheel and tire assembly.
7. Remove the serpentine drive belt.
8. Remove the A/C compressor from the mounting bracket, lay it aside leaving the hoses attached.
9. Remove the engine mount strut bracket.
10. Remove the radiator outlet pipe bolts.
11. Remove the air conditioning and radiator outlet pipes from the oil pan.
12. Remove the exhaust manifold brace.
13. Remove the oil pan-to-flywheel cover bolt and nut.
14. Remove the flywheel cover stud for clearance.
15. Disconnect the radiator outlet pipe from the lower radiator hose and oil pan.
16. Disconnect the oil level sensor connector.
17. Remove the oil pan bolts and remove the oil pan.

To install:

18. Inspect the oil pan gasket for damage. The oil pan gasket is reusable, if it is not damaged.

19. With the gasket in place, position the pan to the engine and install oil pan bolts.

20. Install the flywheel cover stud, spacer and nut. Tighten this nut to 19 ft. lbs. (26 Nm).

21. Connect the oil level sensor connector.

22. Connect the radiator outlet pipe to lower radiator hose and oil pan.

23. Install the exhaust manifold brace.

24. Install the air conditioning and radiator outlet pipes to the oil pan.

25. Install the radiator outlet pipe bolts and tighten them to 124 inch lbs. (14 Nm).

26. Install the engine mount strut bracket and tighten bolts to 55 ft. lbs. (75 Nm).

27. Install the A/C compressor into the mounting bracket.

28. Install the serpentine drive belt.

29. Install the right splash shield.

30. Install the right front wheel and tire assembly.

31. Install the flywheel/converter cover.

32. Lower the vehicle.

33. Fill the crankcase with clean oil.

34. Fill the coolant system.

35. Connect the negative battery cable.

36. Start the vehicle and verify no leaks.

3.1L Engine

1. Disconnect the battery ground.
2. Raise and support the vehicle safely.
3. Drain the engine oil into a suitable container.
4. Remove the flywheel shield.

5. Disconnect and remove the starter.

6. Unfasten the oil pan retaining bolts and nuts, then remove the oil pan.

To install:

7. Clean all the gasket surfaces completely.

8. Apply a thin bead of sealer around the gasket contact are on the oil pan and install the oil pan gasket on the oil pan.

9. Install the oil pan in the vehicle and install the mounting nuts and bolts loosely. With all the fasteners installed tighten as follows:

- Oil pan mounting nuts: 89 inch lbs. (10 Nm)
- Rear oil pan mounting bolts: 18 ft. lbs. (25 Nm)
- Remaining oil pan mounting bolts: 89 inch lbs. (10 Nm)

10. Connect and install the starter.

11. Install the flywheel shield.

12. Lower the vehicle.

13. Refill the crankcase with engine oil.

14. Connect the negative battery cable.

15. Start the vehicle and verify that there are no oil leaks.

Oil Pump

REMOVAL AND INSTALLATION

2.0L Engine

1. Disconnect the negative battery cable.

2. Remove the timing belt cover.

3. Rotate the crankshaft until the timing marks on the camshaft and crankshaft sprockets are aligned with the timing marks. The crankshaft pulley mark should be in line with mark on the cylinder block and the camshaft sprocket mark should

line up with the mark on the camshaft carrier.

4. Remove the crankshaft pulley.

5. Release the tension on the timing belt and remove the timing belt from the crankshaft sprocket.

6. Remove the bolt and washer retaining the crankshaft sprocket and remove the crankshaft sprocket.

7. Remove the crankshaft key and rear thrust washer.

8. Remove the timing belt rear cover.

9. Detach the oil pressure switch electrical connector.

10. Remove the oil pan.

11. Remove the oil filter.

12. Remove the pickup tube and oil pump.

To install:

13. Clean all the gasket surfaces completely.

14. Install the oil pump with a new gasket and tighten the mounting bolts to 5 ft. lbs. (7 Nm).

15. Install a new O-ring on the pickup tube. Install the oil pan pickup tube and support.

16. Install the oil pan.

17. Coat the seal on the oil filter with clean engine oil and install the filter.

18. Connect the oil pressure sender switch.

19. Install the rear timing belt cover.

20. Install the rear thrust washer and key on crankshaft.

21. Install the crankshaft sprocket, washer and mounting bolt. Tighten the mounting bolt to 115 ft. lbs. (155 Nm).

22. With all the timing marks in alignment, install the timing belt and adjust to the proper tension.

23. Install the crankshaft pulley and tighten the mounting bolts to 20 ft. lbs. (27 Nm).

24. Install the timing belt cover.

25. Connect the negative battery cable.

26. Start the vehicle and verify no leaks.

2.2L Engine

1. Disconnect the negative battery cable.

2. Raise and safely support the vehicle.

3. Drain the engine oil into a suitable container.

4. Remove the oil pan-to-engine bolts and the oil pan.

5. Remove the oil pump-to-rear main bearing cap bolt, the oil pump and extension shaft.

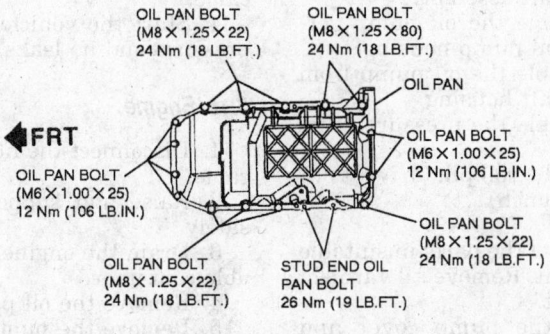

OIL PAN BOLT
(M8 × 1.25 × 22)
24 Nm (18 LB.FT.)

OIL PAN BOLT
(M8 × 1.25 × 80)
24 Nm (18 LB.FT.)

OIL PAN

◄FRT

OIL PAN BOLT
(M6 × 1.00 × 25)
12 Nm (106 LB.IN.)

OIL PAN BOLT
(M6 × 1.00 × 25)
12 Nm (106 LB.IN.)

OIL PAN BOLT
(M8 × 1.25 × 22)
24 Nm (18 LB.FT.)

OIL PAN BOLT
(M8 × 1.25 × 22)
24 Nm (18 LB.FT.)

STUD END OIL
PAN BOLT
26 Nm (19 LB.FT.)

324276

Oil pan fasteners torque specifications — 2.3L and 2.4L engines

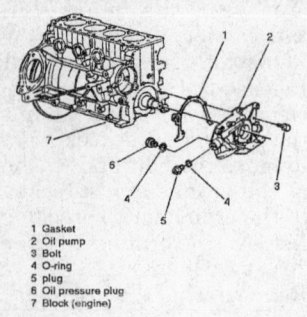

1 Gasket
2 Oil pump
3 Bolt
4 O-ring
5 plug
6 Oil pressure plug
7 Block (engine)

201936

Oil pump mounting to engine block — 2.0L engine

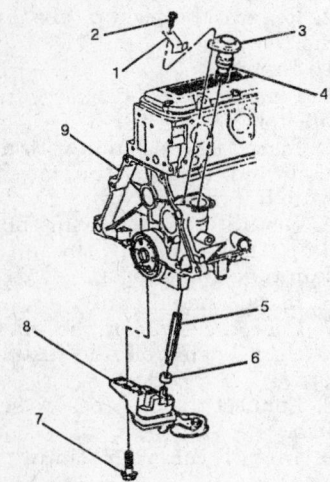

1 Bracket
2 Bolt
3 Oil pump drive assembly
4 O-ring
5 Shaft
6 Retainer: Heat and water soak prior to installation
7 Bolt
8 Oil pump
9 Cylinder block

324721

Oil pump mounting to engine block — 2.2L engines

To install:

— WARNING —

A plastic sleeve called the extension shaft retainer connects the oil pump drive shaft to the oil pump. Heat the extension shaft retainer in hot water prior to assembly. Be sure the retainer does not crack upon installation.

6. Install the extension shaft, oil pump and pump-to-rear main cap bolt. Tighten the oil pump-to-bearing cap bolt to 32 ft. lbs. (43 Nm) and the upper oil pump drive bolt to 18 ft. lbs. (25 Nm).

— WARNING —

To avoid engine damage, all oil pump cavities must be filled with petroleum jelly before installing the gears into the pump body. Also, use only original equipment gaskets. Gasket thickness is critical to proper oil pump operation.

7. Install the oil pan and attaching bolts.
8. Lower the vehicle.
9. Fill the crankcase with clean engine oil.
10. It is good practice to install a reliable mechanical oil pressure gauge so actual pump pressure can be read after startup. If no oil pressure is seen within a short time of startup, shut down the engine to determine why there is no oil pressure.
11. Connect the negative battery cable.
12. Start the engine and check oil pressure and check for leaks.
13. Turn the engine **OFF** and allow to stand. Check oil level, add as necessary.

2.3L and 2.4L Engines

Please note that the transaxle must be removed from the vehicle to service the oil pump.

1. Disconnect the negative battery cable.
2. Install engine support fixture J-28467-A or equivalent.
3. Drain the engine oil.
4. Remove the oil pan.
5. Remove the transaxle.
6. Remove the flywheel.
7. Remove the balance shaft chain cover and guide.
8. Remove the oil pump bolts, then remove the oil pump cover.
9. Pull the housing to disconnect pump gear from the balance shaft. Remove the housing assembly from the balance shaft assembly.
10. Disassemble the oil pump gerotor from the oil pump housing.
11. Disassemble the oil pump from the balance shaft housing.
12. Disassemble the pressure relief valve.
13. Remove the roll pin (drive it out with a small punch).
To install:
14. Clean all of the parts in suitable cleaning solvent. Remove all varnish, sludge and dirt.
15. Inspect the pump cover and housing for cracks and excessive wear, replace as necessary.

16. Lubricate the gears with clean engine oil.
17. Assemble the gerotor gear into the housing.

NOTE: Fill oil pump cavities with petroleum jelly prior to installation. This will ensure that there is oil pressure immediately on start-up and will prevent engine damage.

18. Install the pressure relief valve, use a ⁹/₁₆ inch deep well socket to seat the valve.
19. Install the roll pin.
20. Install the pump housing to the balance shaft assembly.
21. Install the pump cover to the oil pump housing.
22. Install the oil pump to block bolts and tighten 40 ft. lbs. (54 Nm).
23. Install the balance shaft chain guide and chain. Adjust the chain tension by inserting a 0.40 inch (1mm) brass feeler, between the chain guide and the chain.

NOTE: A brass feeler gauge must be used to ensure that correct measurements are obtained. If a steel gauge is used, it will not bend to conform to the guide and will allow for incorrect measurements.

24. Press the guide against the chain using about 3 pounds of force.
25. Tighten the chain tensioner fastener to 115 inch lbs. (13 Nm).
26. Install the balance shaft chain cover and tighten the nut and bolt to 115 inch lbs. (13 Nm).
27. Install the flywheel. Tighten the flywheel bolts to 22 ft. lbs. (30 Nm) plus an additional 45 degrees.
28. Install the transaxle.
29. Install the oil pan.
30. Lower the vehicle.
31. Fill the crankcase with clean oil.
32. Remove the engine support fixture.
33. Connect the negative battery cable.
34. Start the vehicle and verify oil pressure and no leaks.

3.1L Engine

1. Disconnect the negative battery cable.
2. Raise and support the vehicle safely.
3. Drain the engine oil into a suitable container.
4. Remove the oil pan.
5. Remove the pump-to-rear main bearing cap bolt and remove the pump and extension shaft.

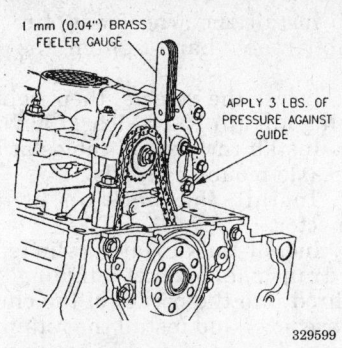

Checking chain tension — 2.3L and 2.4L engines

To install:

NOTE: Whenever the oil pump is removed or replaced it must be primed prior to installation in the vehicle to prevent oil starvation on start up.

6. Remove the 4 cover attaching screws and the cover from the oil pump assembly.
7. Pack the space around the oil pump gears completely full of petroleum jelly. There must be no air space left inside the pump.
8. Install the pump cover and 4 mounting screws and tighten the screws to 80 ft. lbs. (10 Nm).

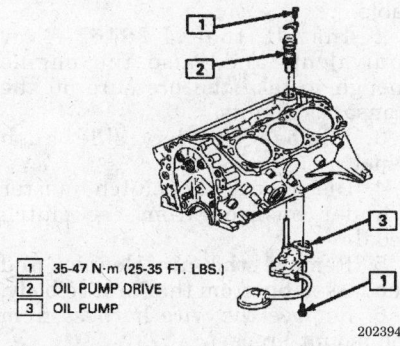

1	35-47 N·m (25-35 FT. LBS.)
2	OIL PUMP DRIVE
3	OIL PUMP

Oil pump, extension shaft and drive assembly — 3.1L engine

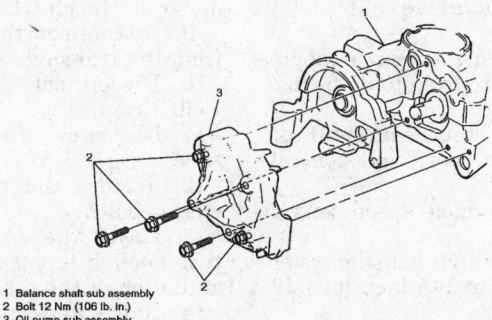

1 Balance shaft sub assembly
2 Bolt 12 Nm (106 lb. in.)
3 Oil pump sub assembly

Oil pump assembly — 2.3L and 2.4L engines

9. Assemble the pump and extension shaft with retainer to rear main bearing cap, aligning the top end of the extension shaft with the lower end of the drive gear.
10. Install the pump-to-rear bearing cap bolt. Tighten to 30 ft. lbs. (40 Nm).
11. Install the oil pan.
12. Lower the vehicle to the floor.
13. Fill the crankcase with oil to the proper level.
14. Connect the negative battery cable.

TRANSAXLE

Manual Transaxle Assembly

REMOVAL AND INSTALLATION

1993–94 Vehicles

Except Isuzu Transaxle

1. Disconnect the negative battery cable.
2. Install an engine holding support fixture (bar), J-28467-A or equivalent.

NOTE: If a lifting bar and hook is not available, a chain hoist can be used, however, during the procedure the vehicle must be raised, at which time the chain hoist must be adjusted to keep tension on the engine/transaxle assembly.

3. Remove the left sound insulator.
4. Disconnect the clutch master cylinder pushrod from the clutch pedal.
5. Remove the air cleaner and air intake duct assembly.

6. Remove the clutch actuator cylinder from the transaxle support bracket and lay it aside.
7. Remove the transaxle mount through bolt.
8. Raise and support the vehicle safely.
9. Remove the exhaust crossover bolts at the right hand manifold.
10. Lower the vehicle.
11. Disconnect the transaxle mount bracket.
12. Disconnect the shift cables and linkage.
13. Remove the transaxle mount tube.
14. Remove the upper transaxle-to-engine bolts.
15. Raise and support the vehicle safely.
16. Remove the left front tire and wheel.
17. Remove the left front inner splash shield.
18. Disconnect the transaxle strut and bracket.
19. Drain the transaxle.
20. Remove the clutch housing cover bolts.
21. Disconnect the speedometer cable.
22. Disconnect the stabilizer shaft at the left suspension support and control arm.
23. Remove the left suspension support attaching bolts and swing the suspension support aside.
24. Install drive axle boot protector J-34754 or equivalent and remove the left drive axle from the transaxle.
25. Attach the transaxle case to a jack.
26. Remove the remaining transaxle-to-engine bolts.
27. Remove the transaxle by sliding it away from the engine and carefully lowering the jack while guiding the intermediate shaft out of the transaxle.

NOTE: The engine may need to be lowered for transaxle to body clearance.

To install:
28. Reposition the transaxle and guide the intermediate shaft into the transaxle.

NOTE: The intermediate shaft cannot be easily installed after the transaxle is connected to the engine.

29. Install the transaxle-to-engine mounting bolts to 55 ft. lbs. (75 Nm).
30. Install the left drive axle into its bore at the transaxle and seat the driveaxle at the transaxle.
31. Remove the drive axle boot protector.

32. Install the suspension support to body bolts and tighten to 65 ft. lbs. (88 Nm).

33. Install the stabilizer shaft at the left suspension support and control arm.

34. Install the speedometer cable.

35. Install the clutch housing cover bolts and tighten to 115 inch lbs. (13 Nm).

36. Install the strut bracket to the transaxle.

37. Install the strut.

38. Install the inner splash shield.

39. Install the wheel and tire assembly.

40. Lower the vehicle and tighten the wheel lug nuts to 100 ft. lbs. (136 Nm).

41. Install the upper engine-to-transaxle bolts and tighten to 55 ft. lbs. (75 Nm).

42. Install the transaxle vent tube.

43. Install the shift cables.

44. Install the transaxle mount bracket.

45. Install the left exhaust crossover bolts.

46. Raise and support the vehicle safely.

47. Install the exhaust crossover bolts at the right hand manifold.

48. Lower the vehicle.

49. Install the transaxle mount through bolt.

50. Install the clutch actuator cylinder to the transaxle support bracket.

51. Install the air cleaner and air intake duct assembly.

52. Remove the engine holding support fixture (bar), J-28467-A or equivalent.

53. Connect the clutch master cylinder pushrod to the clutch pedal.

54. Install the left sound insulator.

55. Check the transaxle fluid level and add as necessary.

56. Connect the negative battery cable.

Isuzu Transaxle

1. Disconnect the negative battery cable.

2. Install an engine holding support fixture (bar), J-28467-A or equivalent and raise the engine enough to take the pressure off the motor mounts.

3. Remove the left sound insulator.

4. Disconnect the clutch master cylinder pushrod from the clutch pedal.

5. Remove the clutch actuator cylinder from the transaxle support bracket and lay it aside.

6. Disconnect the mount harness at the mount bracket.

7. Remove the transaxle mount attaching bolts.

8. Remove the transaxle mount bracket attaching bolts and nuts.

9. Remove the shift cables and retaining clamp at the transaxle.

10. Disconnect the back-up switch connector.

11. Raise and support the vehicle safely.

12. Drain the transaxle.

13. Remove the left front wheel assembly.

14. Remove the left front inner splash shield.

15. Disconnect the transaxle strut and bracket.

16. Remove the clutch housing cover bolts.

17. Disconnect the vehicle speed sensor at the transaxle.

18. Disconnect the stabilizer shaft at the left suspension support and control arm.

19. Remove the left suspension support attaching bolts and swing the suspension support aside.

20. Install drive axle boot protector J-34754 or equivalent, disconnect the drive axles and remove the left shaft from the transaxle.

21. Attach the transaxle case to a jack.

22. Remove the transaxle-to-engine bolts.

23. Remove the transaxle by sliding it away from the engine and carefully lowering the jack while guiding the right drive axle out of the transaxle.

To install:

24. Reposition the transaxle and guide the right drive axle into its bore as the transaxle is being raised.

NOTE: The right drive axle cannot be easily installed after the transaxle is connected to the engine.

25. Install the transaxle-to-engine mounting bolts to 55 ft. lbs. (75 Nm).

26. Install the left drive axle into its bore at the transaxle and seat both drive axles at the transaxle.

27. Remove the drive axle boot protector.

28. Install the suspension support to body bolts and tighten to 65 ft. lbs. (88 Nm).

29. Install the stabilizer shaft at the suspension support and control arm.

30. Install the vehicle speed sensor at the transaxle.

31. Install the clutch housing cover bolts and tighten to 115 inch lbs. (13 Nm).

32. Install the strut bracket to the transaxle, then install the strut.

33. Install the inner splash shield.

34. Install the wheel and tire assembly, then hand-tighten the lug nuts.

35. Lower the vehicle, then tighten the lug nuts to 100 ft. lbs. (136 Nm).

36. Install the ground cables at the transaxle mounting studs.

37. Install the back-up switch connector.

38. Install the actuator cylinder to the transaxle bracket aligning the pushrod into the pocket of the clutch release lever and install the retaining nuts and tighten evenly to prevent damage to the cylinder.

39. Install the transaxle mount bracket.

40. Install the transaxle mount to side frame and install the bolts.

41. Install the wiring harness at the transaxle mount bracket.

42. Install the bolt attaching the mount to the transaxle bracket and tighten to 88 ft. lbs. (120 Nm).

43. Remove the engine holding support fixture (bar), J-28467-A or equivalent.

44. Install the shift cables.

45. Check the transaxle fluid level and add as necessary.

46. Connect the negative battery cable.

1995–97 Vehicles

1. Disconnect the negative battery cable.

2. Install tool J-28467-A or equivalent, and raise the engine enough to take the pressure off the transaxle mounts.

3. Remove the left side hush panel.

4. Disconnect the clutch master cylinder pushrod from the clutch pedal.

5. Remove the air cleaner and duct assembly from the throttle body.

6. Remove the wire harness from the mount bracket.

7. Remove the upper transaxle mount-to-transaxle bolts.

8. Remove the clutch master cylinder from the clutch actuator.

9. Disconnect the ground cables from the transaxle mounting studs.

10. Disconnect the backup light switch connector.

11. Disconnect the transaxle vent tube.

12. Remove the rear transaxle-to-engine bolts.

13. Lower the engine support fixture enough to ease removal and installation of the transaxle.

14. Raise and safely support the vehicle.

15. Drain the transaxle into a suitable container.

16. Remove the tire and wheel assemblies.

17. Remove the left side splash shield.

18. Disconnect both front ABS wheel speed sensor harness connectors and move the harness out of the way.

19. Remove the flywheel cover.

20. Disconnect the vehicle speed sensor at the transaxle.

21. Remove the left and right ball joint nuts and separate them from the steering knuckle.

22. Remove the left stabilizer link pin.

23. Remove the left side U-bolt from the stabilizer bar.

24. Remove the left side suspension support attaching bolts.

25. Remove the drive axles from the transaxle.

26. Remove the front lower transaxle mount.

27. Position a suitable jack under the transaxle.

28. Remove the transaxle-to-engine mounting bolts (noting their location).

29. Remove the transaxle away from the engine by carefully lowering the jack.

To install:

30. Position the transaxle on the jack and move it into place.

31. Install the transaxle-to-engine mounting bolts and tighten them to 55 ft. lbs. (75 Nm).

32. Install the front transaxle mount.

33. Install the flywheel cover.

34. Install the drive axles into the transaxle assembly.

35. Install the left side suspension support and attaching bolts.

36. Install the left side U-bolt to the stabilizer bar.

37. Connect the ball joints to the steering knuckle and install the nuts, tighten to 48 ft. lbs. (65 Nm).

38. Install the left side stabilizer link pin assembly.

39. Route the left side ABS wheel speed sensor wiring harness and connect both front wheel speed sensor connectors.

40. Install the inner splash shield.

41. Connect the vehicle speed sensor to the transaxle.

42. Install both tire and wheel assemblies.

43. Lower the vehicle.

44. Install the ground cables to the transaxle mounting studs.

45. Install the vent tube to the transaxle.

46. Connect the backup light switch connector.

47. Install the upper transaxle mounting bolts and tighten to 55 ft. lbs. (75 Nm).

48. Install the clutch master cylinder to clutch actuator cylinder.

49. Install the rear transaxle mount. Tighten the bolts to 55 ft. lbs. (75 Nm).

50. Clip the wiring harness to the mount bracket.

51. Remove the engine support fixture.

52. Connect the shift cables clamp and nut. Tighten the nut to 89 inch lbs. (10 Nm).

53. Install the air cleaner and duct assembly to the throttle body.

54. Connect the pushrod to the clutch pedal.

55. Install the left side hush panel.

56. Connect the negative battery cable.

57. Fill the transaxle with Synchromesh Transaxle Fluid.

58. Road test the vehicle and verify proper operation.

Clutch Assembly

REMOVAL AND INSTALLATION

The manual transaxle assembly must be removed from the vehicle to service the clutch assembly.

NOTE: Prior to any vehicle service that requires the removal of the actuator cylinder, the master cylinder pushrod must be disconnected from the clutch pedal. If not disconnected, permanent damage to the actuator cylinder will occur if the clutch pedal is depressed while the actuator cylinder is disconnected.

1. Disconnect the negative battery cable.

2. Remove the clutch master cylinder pushrod from the clutch pedal.

3. Remove the transaxle.

4. If any of the parts are to be reused, mark the pressure plate assembly and the flywheel so they can be assembled in the same position. They were balanced as an assembly at the factory.

5. Loosen the attaching bolts one turn at a time until spring tension is relieved.

6. Support the pressure plate and remove the bolts. Remove the pressure plate and clutch disc. Do not disassemble the pressure plate assembly. Replace it if defective.

7. Inspect the flywheel, clutch disc, pressure plate, throwout bearing and the clutch fork and pivot shaft assembly for wear. Replace the parts as required. If the flywheel shows any signs of overheating or if it is badly grooved or scored, it should be resurfaced or replaced.

8. Clean the pressure plate and flywheel mating surfaces thoroughly.

To install:

9. Clean all parts well. Apply a small amount of high temperature grease to the pilot bearing inside the end of the crankshaft.

10. Position the clutch disc and pressure plate into the installed position, and support with clutch aligning tool J-29074 or equivalent. The clutch plate is assembled with the damper springs offset toward the transaxle. One side of the factory supplied clutch disc should be stamped "Flywheel Side."

11. Install the pressure plate-to-flywheel bolts. Tighten them gradually in a cross pattern as follows:

a. Install lightly seat all bolts.

b. Tighten bolts 1, 2, 3 then 4, 5, and 6 to 12 ft. lbs. (16 Nm).

c. Final torque bolts 1, 2, 3 then 4, 5, 6 to 15 ft. lbs. (20 Nm).

12. Lubricate the outside groove and the inside recess of the release bearing with high temperature grease. Wipe off any excess. Install the release bearing.

a. On the NVG-T550 transaxle, lubricate the inside diameter of the bearing with clutch bearing lubricant.

b. On the Isuzu transaxle, pack the inside recess of the release bearing completely full of chassis grease.

NOTE: On the Isuzu transaxle, be sure the bearing pads are located on the fork ends and both spring ends are in the fork holes with the spring completely seated in bearing groove.

13. Install the transaxle.

14. Install clutch master cylinder pushrod to the clutch pedal and install the retaining clip.

15. If equipped with cruise control, check switch adjustment at clutch pedal bracket.

NOTE: When adjusting the cruise control switch, do not exert an upward force on the clutch pedal pad of more than 20 ft. lbs. (27 Nm) or damage to the master cylinder pushrod retaining ring can result.

16. Connect the negative battery cable.

17. Bleed clutch system as necessary and road test vehicle.

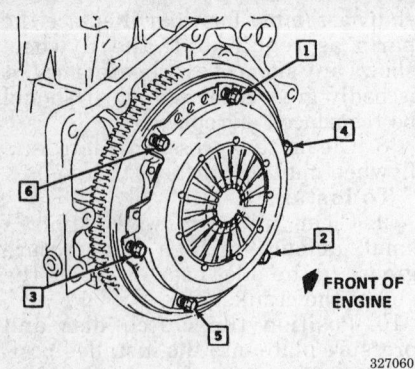

FRONT OF ENGINE

327060

Clutch cover bolt tightening sequence

Clutch Master/Actuator Cylinder Assembly

REMOVAL AND INSTALLATION

1993–94 Vehicles

NOTE: The clutch hydraulic system is serviced as a complete unit. It has been bled of air and filled with fluid. Individual components of the system are not available separately.

2.0L and 3.1L Engines

1. Disconnect the negative battery cable.
2. If equipped with the 3.1L engine, perform the following procedures:
   a. Remove the air cleaner, mass air flow sensor and the air intake duct as an assembly.
   b. Disconnect the electrical lead at the washer bottle. Remove the attaching bolts and washer bottle from the vehicle.
   c. If equipped with cruise control, remove the mounting bracket retaining nuts from the strut tower.
3. Remove the hush panel from inside the vehicle.
4. Disconnect the clutch master cylinder pushrod from the clutch pedal.
5. Remove the clutch master cylinder retaining nuts at the front of the dash.
6. Remove the actuator retaining nuts at the transaxle.
7. Remove the hydraulic system as a unit from the vehicle.
To install:
8. Install the new slave cylinder or actuator to the transaxle support bracket aligning the pushrod into the pocket on the clutch fork outer lever. Tighten the retaining nuts evenly to

prevent damage to the slave cylinder. Tighten to 14–20 ft. lbs. (19–27 Nm).

NOTE: On new replacement units, do not remove the plastic pushrod retainer from the slave or actuator cylinder. The straps will break on the first clutch pedal application.

9. Position the clutch master cylinder to the front of the dash. Install the retaining nuts and tighten the nuts evenly to prevent damage to the master cylinder. Tighten to 15–25 ft. lbs. (20–27 Nm).
10. Remove the pedal restrictor from the pushrod. Lube the pushrod bushing on the clutch pedal. Connect the pushrod to the clutch pedal and install the retaining clip.
11. If equipped with cruise control, check the switch adjustment at the pedal bracket.

NOTE: When adjusting the cruise control switch, do not exert an upward force on the clutch pedal pad of more than 20 lbs. or damage to the master cylinder pushrod retaining ring can result.

12. Install the hush panel.
13. Press the clutch pedal down several times. This will break the plastic retaining straps on the slave or actuator cylinder pushrod. Do not remove the plastic button on the end of the pushrod.
14. If equipped with the 3.1L engine, install the air cleaner, mass air flow sensor and the air intake duct as an assembly.
15. Connect the negative battery cable.

1995–97 Vehicles

Service this vehicle with Hydraulic Clutch Fluid GM P/N 12345347 or equivalent.
1. Disconnect the negative battery cable.
2. Remove the sound insulator from inside the vehicle.
3. Disconnect the clutch master cylinder pushrod from the clutch pedal.
4. Remove the clutch master cylinder retaining nuts at the front of the dash.
5. Remove the reservoir.
6. Remove the clutch master cylinder assembly from the clutch actuator cylinder assembly.

NOTE: If necessary to service the clutch actuator inside the bellhousing, note that the system uses quick-connect fittings. A washer-like release must be

pushed IN (toward the coupler) and while holding the release IN, disconnect the line. To connect, pull the washer-like release OUT, make the connection then allow the release to seat. A click should be heard.

To install:
7. Install the clutch master cylinder assembly to the clutch actuator cylinder assembly.
8. Install the reservoir.
9. Install the retaining nuts and tighten to 18 ft. lbs. (25 Nm).
10. Connect the pushrod to the clutch pedal.
11. Adjust cruise control switch (if equipped).
12. Install the sound insulator.
13. Bleed the hydraulic system.
14. Connect the negative battery cable.
15. Road test the vehicle and verify proper operation.

Hydraulic Clutch System Bleeding

PROCEDURE

1993–94 Vehicles

There are 2 possible methods of bleeding the hydraulic clutch system depending on whether or not the system contains a bleed screw. To determine which method to use, locate the slave cylinder located on the transaxle and look for a bleed screw next to the hydraulic inlet connection. If no bleed screw is present, the slave cylinder must be removed from the transaxle, follow the appropriate procedure.

Without Bleeder Screw

——— **WARNING** ———
Whenever the clutch slave cylinder is removed from the transaxle, the pushrod must first be disconnected from the clutch pedal. This is necessary to prevent the cylinder from becoming permanently damaged should the pedal be depressed with the cylinder removed from the transaxle.

1. Remove the left sound insulator and disconnect the actuator pushrod from the clutch pedal.
2. Remove the slave cylinder from the transaxle.
3. Loosen the clutch master cylinder mounting nuts out to the end of the studs, but do not remove them or the master cylinder.

4. Remove the clutch fluid reservoir cap and diaphragm. Be sure the clean the cap and surrounding area before removal in order to prevent fluid contamination.

5. Measure the slave cylinder pushrod, then depress the pushrod 0.787 in. (20mm) into the bore. Hold the pushrod in the depressed position and reinstall the reservoir cap with diaphragm.

6. Hold the slave cylinder vertically with the pushrod end facing down and at a level lower than the master cylinder and reservoir, then release the pushrod.

7. Depress the pushrod into the bore with short 0.39 in. (10mm) strokes while peering through the reservoir sides for bubbles. Continue to cycle the pushrod until air bubbles fail to appear in the reservoir

8. Install the slave cylinder to the transaxle.

9. Secure the master cylinder to the cowl bracket by evenly tightening the nuts to 15 ft. lbs. (21 Nm).

10. Inspect the clutch fluid reservoir and top-off, if necessary, using DOT 3 fluid.

11. Connect the clutch master cylinder pushrod to the clutch pedal, then install the left sound insulator.

12. Start the engine and push the clutch to the floor, then wait about 10 seconds and select the reverse gear. There should be no gear clashing. If the gears grate, the hydraulic system may still contain air and should be bled again.

With Bleeder Screw

1. Clean dirt and grease from the cap reservoir to ensure no foreign substances enter the system.

2. Remove the cap and diaphragm, then fill reservoir to the top with approved brake fluid only.

NOTE: Brake fluid must be certified to DOT 3 specification.

3. Fully loosen the bleed screw located on the slave cylinder body, next to the hydraulic inlet connection.

4. Fluid will now begin to move from the master cylinder, down the tube, to the slave cylinder. The reservoir must be kept full at all times in order to accomplish an efficient gravity fill.

5. When the slave cylinder is full, a steady stream of fluid will come from the slave outlet. At this point the system should be free of air bubbles; tighten the bleed screw to 18 inch lbs. (2 Nm).

6. Install the reservoir cap and diaphragm.

7. Start the engine and push the clutch to the floor, then wait about 10 seconds and select the reverse gear. There should be no gear clashing. If the gears grate, the hydraulic system may still contain air and should be bled again.

1995–97 Vehicles

Do not use fluid which has been bled from a system to fill the reservoir as it may be aerated, have too much moisture content or possibly be contaminated. Clean the dirt and grease from the cap to ensure that no foreign substances enter the system. It is also important to maintain the fluid level in the clutch reservoir to the top step with hydraulic clutch fluid GM part number 12345347 or equivalent.

1. Attach a hose to the bleeder screw on the clutch actuator assembly and submerge the other end of the hose in a container of hydraulic clutch fluid.

2. Depress the clutch pedal slowly and hold.

3. Loosen the bleeder screw to purge air.

4. Tighten the bleeder screw to 18 inch lbs. (2 Nm).

5. Repeat Steps 2 through 4 until the air is purged from the system.

6. Fill the reservoir to the top step with hydraulic clutch fluid.

7. Repeat this bleeding procedure if there is a grinding noise during the clutch spin down procedure.

Automatic Transaxle Assembly

REMOVAL AND INSTALLATION

1993–94 Vehicles

1. Disconnect the negative battery cable.

2. For 1993 vehicles, drain the cooling system, then disconnect the heater core hoses.

3. Remove the air intake duct or air cleaner assembly.

4. For 6-cylinder engines only, disconnect the exhaust crossover pipe from the manifolds.

5. Disconnect the TV cable from the throttle lever and the transaxle and unroute.

6. Remove the fluid level indicator and the filler tube.

7. Install J-28467-A or equivalent engine support fixture.

8. Unfasten the nut securing the wiring harness to the transaxle.

9. Remove the wiring harness-to-transaxle nut.

10. Label and detach the wires at the TCC connector and the neutral safety/back up light switch.

----- **WARNING** -----

When servicing requires that the "T" latch type wiring connector be detached from the switch, be sure to reassemble the "T" latch and the connector. Failure to do so may result in intermittent loss of switch functions.

11. Disconnect the shift linkage from the transaxle.

12. Unfasten the top 2 transaxle-to-engine bolts.

13. Remove the left upper transaxle mount and bracket assembly.

14. Disconnect the rubber hose from the transaxle-to-vent pipe.

15. Unfasten the remaining upper engine-to-transaxle bolts.

16. Raise and safely support the vehicle.

17. Remove the front wheel and tire assemblies.

18. If equipped, remove both front ABS wheel speed sensors and the harness from the left suspension support.

19. Separate both ball joints from the control arms.

20. Remove the engine-to-transaxle brace.

21. Unfasten the left stabilizer link pin bolt and frame bushing clamp nuts.

22. Remove the left suspension support assembly and cross bar. Remove both drive axles (halfshafts) from the transaxle, then support them.

23. On the 2.0L engine, disconnect the exhaust pipe from the manifold.

24. Remove the starter motor.

25. Disconnect the coolant inlet pipe from the transaxle and manifold.

26. Remove the transaxle converter cover.

27. On the 2.0L engine, remove the lower intake manifold brace.

28. Unfasten the torque converter-to-transaxle bolts, then matchmark the converter to the flywheel for assembly.

29. Disconnect and plug the transaxle cooler pipes. Remove the transaxle cable selector bracket.

30. Detach the Vehicle Speed Sensor (VSS) connector from the sensor.

31. Remove the transaxle-to-engine support bracket.

32. Position a suitable transaxle jack under the transaxle, then remove the transaxle mounting strut at the transaxle.

33. Unfasten the remaining engine-to-transaxle bolts, then remove the transaxle from the vehicle.

NOTE: The transaxle cooler and lines should be flushed any time the transaxle is removed for overhaul or to replace the pump, case or converter.

To install:

34. Put a small amount of grease on the pilot hub of the converter and make sure the converter is properly engaged with the pump.

35. Position the transaxle in the vehicle. Install the lower engine-to-transaxle bolts, then tighten to 55 ft. lbs. (75 Nm) for all except the 3.1L engine. Remove the jack.

36. Fasten the transaxle mounting strut to the transaxle.

NOTE: Install the transaxle-to-engine support bracket. It is important to center the bolt in the left rear transaxle mount slot to the body. This will minimize idle shake and harsh engagement.

37. Install the transaxle-to-engine support bracket.

38. Connect the transaxle cooler pipes.

39. Align the converter with the marks made previously on the flywheel. Hand start and tighten the net slot bolt first, then hand start and tighten the 2 remaining bolts. Tighten the converter bolts to 46 ft. lbs. (62 Nm).

40. For the 2.0L engine, install the lower intake manifold brace.

41. Fasten the transaxle selector bracket.

42. Connect the coolant inlet pipe to the transaxle and exhaust manifold studs.

43. Install the transaxle converter cover.

44. Install the starter assembly.

45. For the 2.0L engine, connect the exhaust pipe to the manifold.

46. Install the drive axles (half-shafts) to the transaxle.

47. Fasten the left suspension support assembly and the cross bar.

48. Install the left stabilizer shaft frame bushing clamp nuts and link pin bolt.

49. Install the engine-to-transaxle brace.

50. Fasten the right and left ball joints to the control arms.

51. Attach the VSS connector to the transaxle.

52. Connect the left side ABS harness and the WSS connector to the sensor.

53. Install the wheel and tire assemblies.

54. Carefully lower the vehicle.

55. Install the upper transaxle mount bolts, tighten to 55 ft. lbs. (75 Nm), except for the 3.1L engine.

56. Fasten the left side transaxle mount.

57. Attach the shift linkage to the transaxle.

58. Fasten the wiring connectors at the TCC converter and the park/neutral switch.

59. Install the wiring harness and secure to the transaxle with the retaining nut.

60. Remove the engine support fixture.

61. On the 3.1L engine, install the crossover pipe to the exhaust manifold.

62. Install the oil level indicator and fill tube.

63. Connect the TV cable at the transaxle and throttle lever.

64. Attach the rubber hose to the transaxle vent pipe.

65. Install the air inlet duct or air cleaner assembly.

66. Connect the negative battery cable, then check the transaxle fluid level and add if necessary.

1995–97 Vehicles

1. Disconnect the negative battery cable.

2. Remove the air intake duct.

3. Disconnect the TV cable, shift cable and bracket.

4. Disconnect the vacuum lines.

5. Tag and detach all necessary electrical connections.

6. Remove the power steering pump and set it aside (leaving the hoses attached).

7. Remove the filler tube.

8. Attach engine support fixture J-28467-A or equivalent.

9. Remove the top engine to transaxle bolts.

10. Raise and safely support the vehicle.

11. Remove both front tire and wheel assemblies.

12. Remove the left side splash shield.

13. Disconnect both front ABS wheel speed sensors and harness from left suspension support.

14. Disconnect both lower ball joints.

15. Disconnect the stabilizer shaft links.

16. Remove the front air deflector.

17. Remove the left suspension support.

18. Remove both drive axles (halfshafts).

19. Remove the engine to transaxle brace.

20. Remove the transaxle converter cover.

21. Remove the starter motor.

22. Remove the converter bolts.

23. Disconnect the transaxle cooler lines and remove brace.

24. Disconnect the ground wires going to transaxle.

25. Remove the exhaust brace.

26. Remove the bolts from engine and transaxle mount.

27. Support transaxle with a jack and remove transaxle mount-to-body bolts.

28. Disconnect heater core hose brace from transaxle.

29. Remove the remaining engine to transaxle bolts, then remove the transaxle assembly.

To install:

30. Make sure to properly seat the torque converter in the oil pump.

31. Install the transaxle into position with jack while installing right drive axle.

32. Install the lower engine-to-transaxle bolts and tighten to 71 ft. lbs. (96 Nm).

33. Install the transaxle mount-to-body bolts.

34. Install the engine and transaxle mount bolts.

35. Install the exhaust brace.

36. Install the cooler line brace.

37. Attach the ground wires to transaxle bolt.

38. Connect the cooler lines.

39. Install and tighten the converter bolts to 46 ft. lbs. (62 Nm).

40. Install the transaxle converter cover.

41. Install the starter.

42. Install the engine-to-transaxle brace and tighten the bolts to 32 ft. lbs. (43 Nm).

43. Install the drive axles.

44. Install the left suspension support.

45. Install the front air deflector.

46. Install the stabilizer links.

47. Connect the lower ball joints.

48. Connect both ABS wheel speed sensors.

49. Install the left splash shield.

50. Install the heater core pipe brace nut and bolt.

51. Install the front wheel and tire assemblies.

52. Lower the vehicle.

53. Install the top engine-to-transaxle bolts and tighten to 71 ft. lbs. (96 Nm).

54. Remove the engine support fixture.

55. Install the filler tube.

56. Install the power steering pump assembly.

57. Attach the electrical connections, as tagged during removal.
58. Connect the vacuum lines.
59. Attach the shift cable and bracket.
60. Attach the TV cable and adjust as necessary.
61. Install the intake air duct.
62. Connect the negative battery cable.
63. Fill transaxle and start vehicle, verify that there are no leaks.
64. Road test the vehicle.

Throttle Valve (TV) Cable

ADJUSTMENT

Setting of the TV cable must be done by rotating the throttle lever at the throttle body. Do not use the accelerator pedal to rotate the throttle lever.
1. With the engine OFF, depress and hold the reset tab at the engine end of the TV cable.
2. Move the slider until it stops against the fitting.
3. Release the rest tab.
4. Rotate the throttle lever to its full travel.
5. The slider must move (ratchet) toward the lever when the lever is rotated to its full travel position.
6. Recheck after the engine is hot and road test the vehicle.

DRIVE AXLE

Halfshaft

REMOVAL AND INSTALLATION

Some manual transaxle applications may also use an intermediate shaft.
1. Disconnect the negative battery cable.
2. With the weight of the vehicle still on the wheels, loosen, but do not remove the front hub nut. This may require an assistant holding the brakes to keep the front halfshaft from turning. It is good practice to wire-brush the exposed threads on the outer CV-joint stub shaft and apply a generous amount of penetrating oil before attempting to loosen the hub nut.
3. Raise and safely support the vehicle.
4. Remove the tire and wheel assembly.

5. Remove the hub nut and washer.
6. Install the axle boot seal protector J-33162 or the equivalent on the right hand inner boot, if equipped.
7. Remove and support the brake caliper.
8. Remove the brake rotor.
9. Remove the lower ball joint cotter pin and nut and loosen the joint. If removing the right halfshaft, turn the wheel to the left. If removing the left halfshaft turn the wheel to the right.
10. Disconnect the ABS sensor, if equipped.
11. Disconnect the stabilizer bar link.
12. Separate the lower ball joint from the steering knuckle.
13. Disengage the halfshaft stub end from the front wheel bearing and hub assembly using a suitable press-type tool, pressing until the halfshaft splines are just loose.
14. Separate the hub and bearing assembly from the halfshaft. Move the strut and knuckle assembly rearward.
15. Separate the inner joint from the transaxle using the proper puller tools such as J-33008 and J-29764 or their equivalents.
16. Remove the halfshaft from the transaxle. Do not pull the halfshaft by the CV-joint boot or on the joint itself.

To install:
17. Prior to installation, cover all sharp edges in the area of the halfshaft with shop towels so the CV-joint boots will be protected from damage. When a halfshaft is removed for any reason, the transaxle (the halfshaft male and female shank) and knuckle sealing surfaces should be inspected for debris and corrosion. If debris or corrosion are present, clean with 320 grit crocus cloth or equivalent. Transmission fluid may be used to clean off any remaining debris. The surface should be wiped clean and dry before attempting to install the halfshaft.
18. Install the halfshaft into the transaxle (or intermediate shaft, if equipped) by placing a brass drift pin into the groove on the joint housing and tapping until seated. Be careful not to damage the axle seal or dislodge the seal garter spring when installing the axle.

NOTE: Make sure the halfshaft is fully engaged in the transaxle. Verify that the halfshaft is seated by grasping the inner joint hous-

ing and pulling outward. Do not pull on the shaft or the boot, but on the inner joint housing only.

19. Install the drive axle into the hub and bearing assembly.
20. Install the lower ball joint to the steering knuckle. Tighten the ball joint-to-steering knuckle nut to 41–48 ft. lbs. (55–65 Nm) to install the cotter pin. Do not loosen the nut at any time during installation. Install a new cotter pin.
21. Install the washer and a new hub nut. To keep the hub from turning while the hub nut is being torqued, insert a drift pin through the caliper opening into one of the ventilation openings in the brake rotor. This should lock the assembly together. Torque the hub nut to 185 ft. lbs. (260 Nm).
22. Install the tire and wheel assembly.
23. Lower the vehicle.
24. Test drive vehicle to verify no front drive noise.

CV-Joint Boot

REPLACEMENT

Outer CV-Joint Boot

1. Raise and safely support the vehicle.
2. Remove the front wheel assembly. Mark the wheel to the wheel stud for re-installation purposes.
3. Remove the front halfshaft and carefully place it in a vise using a protective covering on the vise jaws.
4. Cut the large and small clamps and discard them.
5. Slide the boot down the shaft uncovering the outer joint.
6. Clean the grease away from the joint to locate the snapring.
7. Using J-8059 or equivalent, open the snapring and slide the outer joint off the shaft.
8. Remove the outer boot from the shaft.
9. Clean the outer joint completely and inspect for damage to the cage or balls. Wipe any grease off the axle shaft.

To install:
10. Slide the small clamp onto the halfshaft and push the boot down several inches past the seal mounting area.
11. Check the snapring in the outer joint and replace as necessary. Pack the outer joint with half of the grease supplied in the boot kit and install it on the end of the halfshaft.

12. Gently pull down on the joint until the splines engage. With a brass drift lightly tap the joint down until the snapring engages.

13. Pack the remaining grease from the boot kit into the boot and then pull the boot up over the end of the joint. Seat the small end on the seal mounting area.

14. Slide the small clamp into position and using J-35910 or equivalent, crimp the clamp to 130 ft. lbs. (176 Nm).

15. Install the large clamp to the proper position and crimp to 130 ft. lbs. (176 Nm).

———— WARNING ————
The boot must not be dimpled, stretched or out of shape in any way. If boot is not shaped correctly, carefully insert a thin flat blunt tool at the large end of the boot to equalize pressure. Shape the boot properly by hand and then remove the tool.

16. Install the halfshaft. Install the wheel and tire assembly, lower the vehicle and torque the wheel lug nuts to 100 ft. lbs. (136 Nm). Road test the vehicle to check for abnormal noise or vibration.

Inner CV-Joint Boot

1. Raise and safely support the vehicle.
2. Remove the front wheel assembly.
3. Remove the front halfshaft and carefully place it in a vise using a protective covering on the vise jaws.
4. Remove the outer CV-joint and boot.
5. Cut the large and small boot clamps and discard.

———— WARNING ————
Do not cut through the boot and damage the sealing surface of the the tri-pot outer housing.

6. Separate the boot from the tri-pot bushing at the large diameter end and slide the boot away from the joint along the axle.
7. Remove the tri-pot housing from the spider and axle. Clean thoroughly and set aside.
8. Clean the grease from the spider assembly to expose the spacer ring located on the outboard spider assembly. Slide the spacer ring down the halfshaft.
9. Push the spider assembly down the shaft to uncover the snapring on the end of the shaft. Using J-8059 or equivalent snapring pliers, remove

the snapring. Slide the spider assembly off the end of the shaft. Clean completely, using care not to knock the caps off the bearings. Set aside.

10. Remove the tri-pot bushing from the tri-pot housing.
11. Remove the spacer ring from the halfshaft and slide the boot off. Clean any grease off the shaft.

To install:

12. Slide the small clamp onto the halfshaft.
13. Slide the boot onto the shaft and position the neck of the boot in the groove on the axle shaft.
14. Crimp the seal retaining clamp with J-35910 or equivalent crimping tool to 100 ft. lbs. (136 Nm).
15. Put the spacer ring on the shaft, several inches below the second spacer ring groove.
16. Install the spider assembly far enough down the shaft to expose the top snapring groove. Make sure the counterbored face of the tri-pot spider faces the end of the shaft.
17. Install the top snapring and pull the spider assembly back up into position.
18. Using J-8059 or equivalent snapring pliers, lock the spacer ring in the spacer ring groove.
19. Pack the housing with half the grease supplied in the kit and put the rest of the grease in the boot.
20. Install the tri-pot bushing into the tri-pot housing.
21. Slide the larger clamp over the boot.
22. Push the tri-pot housing over the spider assembly.
23. Slide the larger diameter of the boot, into position and clamp in place over the outside of the bushing and locate lip of the boot in the bushing groove.
24. Position the tri-pot assembly and install the large clamp in position and use J-35910 or equivalent crimping tool to tighten the clamp.

———— WARNING ————
The boot must not be dimpled, stretched or out of shape in any way. If boot is not shaped correctly, carefully insert a thin flat blunt tool at the large end of the boot to equalize pressure. Shape the boot properly by hand and then remove the tool.

25. Install the outer CV-joint and boot.
26. Install halfshaft.
27. Install the wheel and tire assembly and torque the wheel lug nuts to 100 ft. lbs. (136 Nm). Road test for proper operation. Check for abnormal noise or vibration.

STEERING

Air Bag
———— CAUTION ————
Some vehicles are equipped with an air bag system, also known as the Supplemental Inflatable Restraint (SIR) system. The system must be disabled before performing service on or around system components, steering column, instrument panel components, wiring and sensors. Failure to follow safety and disabling procedures could result in accidental air bag deployment, possible personal injury and unnecessary system repairs.

PRECAUTIONS

Several precautions must be observed when handling the inflator module to avoid accidental deployment and possible personal injury.

• Never carry the inflator module by the wires or connector on the underside of the module.
• When carrying a live inflator module, hold securely with both hands, and ensure that the bag and trim cover are pointed away.
• Place the inflator module on a bench or other surface with the bag and trim cover facing up.
• With the inflator module on the bench, never place anything on or close to the module which may be thrown in the event of an accidental deployment.

DISARMING
———— CAUTION ————
The Supplemental Inflatable Restraint (SIR) system must be disarmed before performing many in-vehicle service procedures. Failure to do so may cause accidental deployment of the air bag, resulting in unnecessary SIR system repairs and/or personal injury.

Disabling the SIR System

1. Turn the steering wheel so the vehicle's wheels are pointing straight ahead.
2. Turn the ignition switch to the **LOCK** position and remove the key.
3. Remove the AIR BAG fuse from the instrument panel fuse block.

4. Remove the left hand sound insulator.

5. Disconnect the Connector Position Assurance (CPA) and yellow 2-way connector at the base of the steering column.

6. If equipped with passenger side air bags, remove the right hand sound insulator, then disconnect the CPA and yellow 2-way connector from the passenger inflator module pigtail.

Enabling the SIR System

1. Turn the ignition switch to the **LOCK** position and remove the key.

2. If equipped with passenger side air bags, connect the yellow 2-way connector and CPA to the passenger inflator module pigtail, then install the right hand sound insulator.

3. Connect the yellow 2-way connector and CPA at the base of the steering column. After installing the CPA, clip the connector to flange on the steering column support.

4. Install the left hand insulator.

5. Install the AIR BAG fuse into the instrument panel fuse block.

6. Turn the ignition switch to the **RUN** position and verify that the AIR BAG warning lamp flashes 7–9 times and then the light turns OFF.

Steering Wheel

REMOVAL AND INSTALLATION

Vehicles Without the SIR System

1. Disconnect the negative battery cable.

2. Remove steering wheel pad and disconnect the horn connection.

3. Remove the steering wheel mounting nut.

4. Remove the steering wheel using tool J-1859-03 or equivalent.

To install:

5. Align the mark on steering wheel with the mark on the column

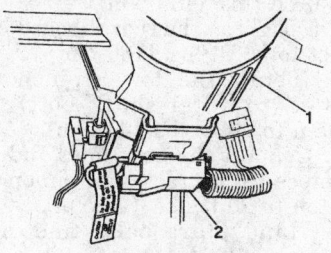

1. Steering column
2. Connector,sir(yellow)

325682

Yellow 2-way air bag CPA connector for steering column

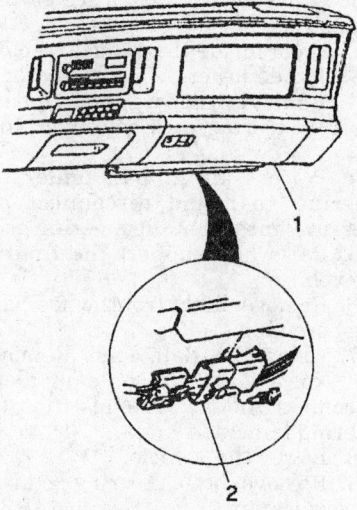

1. I/P compartment
2. Connector,sir(yellow)

325683

Yellow 2-way air bag CPA connector for passenger side — 1995–97 vehicles

shaft, then install steering wheel onto shaft.

6. Install the mounting nut and tighten to 30 ft. lbs. (41 Nm).

7. Connect the horn lead and steering wheel pad.

8. Connect the negative battery cable.

Vehicles With the SIR System

> **—— CAUTION ——**
> *The Supplemental Inflatable Restraint (SIR) system must be disarmed before removing the steering wheel. Failure to do so may cause accidental deployment of*

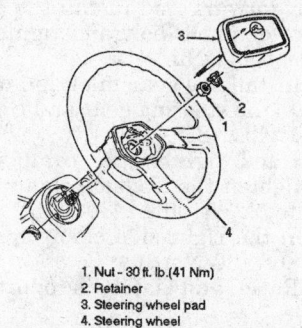

1. Nut - 30 ft. lb.(41 Nm)
2. Retainer
3. Steering wheel pad
4. Steering wheel

223416

Steering wheel view — vehicles without SIR

the air bag, resulting in unnecessary SIR system repairs and/or personal injury.

1. Properly disable the SIR (air bag) system.

2. Remove the 2 screws from the back of the steering wheel that retain the air bag module. Disconnect the module CPA and electrical connection from the rear of the air bag module and remove the module. Set module aside.

> **—— CAUTION ——**
> *When carrying a live air bag, make sure the bag and trim cover are pointed away from the body. In the unlikely event of an accidental deployment, the air bag will then deploy with minimal chance of injury. When placing a live air bag on a bench or other surface, always face the bag and trim cover up, away from the surface. This will reduce the motion of the module if it is accidentally deployed.*

3. Remove the steering wheel center nut.

4. Use steering wheel puller J-1859-03 or equivalent to prevent damage to either the steering wheel or the steering column.

To install:

5. Align the mark on the steering wheel with the mark on the shaft then install the steering wheel. Install the center nut and torque to 30 ft. lbs. (41 Nm).

6. Install the air bag module using care when making the horn and air bag connections. The air bag module retaining screws should be torqued to 89 inch lbs. (10 Nm).

7. Properly enable the air bag system.

8. Connect the negative battery cable.

9. Turn the ignition switch to the **RUN** position and make sure the INFLATABLE RESTRAINT indicator lamp flashes 7–9 times and then goes out. If the indicator lamp does not flash as indicated, SIR troubleshooting will be required.

Tie Rod Ends

REMOVAL AND INSTALLATION

1. Loosen both pinch bolts at the outer tie rod.

2. Remove the tie rod end from the strut assembly using a suitable removal tool.

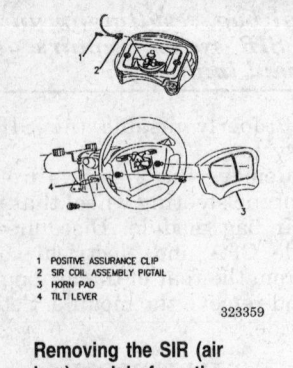

1 POSITIVE ASSURANCE CLIP
2 SIR COIL ASSEMBLY PIGTAIL
3 HORN PAD
4 TILT LEVER

323359

Removing the SIR (air bag) module from the steering wheel

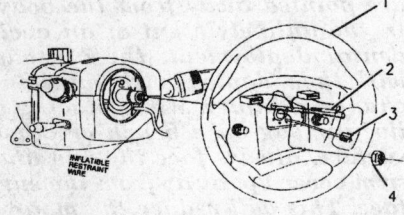

1 STEERING WHEEL
2 HORN CONTACT
3 SIR COIL ASSEMBLY PIGTAIL
4 STEERING WHEEL NUT

323360

Removing the steering wheel on vehicles with air bags

3. Unscrew the outer tie rod end from the tie rod adjuster, counting the number of revolutions required before they are disconnected.

To install:

4. Install the new tie rod end, screwing it on the same number of revolutions as counted in Step 3. When the tie rod end is installed, the tie rod adjuster must be centered between the tie rod and tie rod end, with an equal number of threads exposed on both sides of the adjuster nut. Tighten the pinch bolts to 20 ft. lbs. (27 Nm).

5. Install the tie rod end to the strut assembly and tighten to 50 ft. lbs. (68 Nm). If the cotter pin cannot be installed, tighten the nut up to 1/8 of a turn further. Never back off the nut to align the holes for the cotter pin.

6. Have the front end alignment checked or adjusted.

Power Rack and Pinion

REMOVAL AND INSTALLATION

1993–94 Vehicles

Depending on the vehicle and engine combination, it may be necessary to remove the inner and outer tie rod assemblies prior to removal of the rack and pinion unit.

1. Disconnect the negative battery cable.

2. From inside the vehicle, remove the left-side lower sound insulator.

3. Remove the upper steering shaft-to-steering rack coupling pinch bolt.

4. Place a drain pan under the steering gear and disconnect the pressure lines from the steering gear.

5. Raise and support the front of the vehicle.

6. Remove both front wheel and tire assemblies.

7. Using the Ball Joint Remover tool No. J 24319 01 or equivalent, disconnect the tie rod ends from the steering knuckles.

8. Lower the vehicle.

9. Remove both steering gear-to-chassis clamps.

10. Slide the steering gear forward and remove the lower steering shaft-to-steering rack coupling pinch bolt.

11. From the firewall, disconnect the coupling and seal from the steering gear.

12. Raise and support the front of the vehicle.

13. Through the left-wheel opening, remove the steering gear.

NOTE: If the studs were removed with the mounting clamps, reinstall the studs into the cowl panel and tighten until the studs are fully seated against the dash panel. The torque should not exceed 15 ft. lbs. (20 Nm). After a second use of the stud, a thread locking compound should be used.

To install:

14. Install the rack and pinion through the left wheel opening.

15. Install the dash seal on the rack and pinion assembly.

16. Move the rack and pinion assembly forward and install the coupling lower pinch bolt and tighten to 30 ft. lbs. (41 Nm).

17. Install the gear inlet and outlet pipes to the steering gear and tighten to 19 ft. lbs. (26 Nm).

18. Hand tighten the camp nuts, then tighten the left side clamp nuts first to 22 ft. lbs. (30 Nm), then tighten the right side clamp nuts to 22 ft. lbs. (30 Nm).

19. Raise and safely support the vehicle.

20. Install the tie rod ends to the struts and install the cotter pins after torquing the nuts to 34 ft. lbs. (47 Nm).

21. Install the wheel and tire assemblies.

22. Lower the vehicle and install the steering column upper pinch bolt to 30 ft. lbs. (41 Nm).

23. Refill power steering pump reservoir.

24. Connect the negative battery cable.

25. Bleed the power steering system.

1995–97 Vehicles

1. Disconnect the negative battery cable.

2. Remove the left side sound insulator.

3. Remove the upper pinch bolt on the inter-shaft assembly.

4. Remove the line retainer, if applicable.

5. Raise and safely support the vehicle.

6. Remove the left front tire and wheel assembly.

7. Disconnect the tie rod ends from the struts using tool J-2431-01 or equivalent.

8. Remove the left and right mounting bolts.

9. Disconnect the gear inlet and outlet hose assemblies from the rack and pinion.

10. Remove the lower pinch bolt from the flange inter-shaft assembly.

11. Remove the inter-shaft assembly.

12. Loosen the crossmember bolts to gain additional clearance for removal.

13. Remove the rack and pinion through the wheel opening.

To install:

14. Install the rack and pinion through the left wheel opening.

15. Install the crossmember bolts as follows:

a. Tighten the left rear outboard 1st to 96 ft. lbs. (130 Nm).

b. Tighten the right rear outboard 2nd to 96 ft. lbs. (130 Nm).

c. Tighten the front upper bolts to 96 ft. lbs. (130 Nm).

d. Tighten the rear inboard bolts last to 96 ft. lbs. (130 Nm).

16. Install the lower pinch bolt (flange to inter-shaft bolt) and tighten to 30 ft. lbs. (41 Nm).

17. Connect the gear inlet and outlet pipes to the rack and pinion and tighten to 20 ft. lbs. (27 Nm).

18. Hand start bolts and nuts. Tighten left side bolt to 89 ft. lbs. (120 Nm) and then tighten right side to 89 ft. lbs. (120 Nm).

19. Connect the tie rod ends to the struts and tighten to 44 ft. lbs. (60 Nm) and install new cotter pins.

20. Install left tire and wheel assembly.

21. Install line retainer, if applicable.

22. Lower the vehicle.

23. Install the upper pinch bolt and tighten to 30 ft. lbs. (41 Nm).

24. Install the left side sound insulator.

25. Connect the negative battery cable.

26. Fill with fluid and bleed air from system.

27. Check toe setting and adjust as required.

28. Road test vehicle and verify no leaks.

Power Steering Pump

BLEEDING

1. Turn the ignition switch to the **OFF** position.

2. Raise and safely support the vehicle so the front wheels are off the ground.

3. Turn the steering wheel all the way to the left.

4. Verify that the fluid in the reservoir is to the full COLD mark. Add the proper fluid as necessary. Leave the cap off the reservoir.

5. With an assistant checking the fluid level, turn the steering wheel lock-to-lock at least 20 times. The engine remains OFF. Note that on vehicles with extra long return lines or fluid coolers in the system, turn the steering wheel lock-to-lock at least 40 times. Be aware that enough trapped air may cause the fluid to overflow. Be sure to thoroughly clean any spilled fluid to allow for a leak check. Keep the fluid to the proper level.

6. Start the engine and allow to idle. Maintain the fluid level and reinstall the cap.

7. Return the wheels to center and lower the vehicle to the ground.

8. Keep the engine for running at least 2 minutes. Turn the steering wheel in both directions.

9. Check for smooth power assist, relatively quiet operation, proper fluid level, no system leaks and no bubbles or foam in the fluid. If okay, procedure is complete. If not okay, troubleshooting is required.

REMOVAL AND INSTALLATION

1993–94 Vehicles

2.0L and 2.2L Engines

1. Disconnect the negative battery cable.

2. Disconnect and cap the pressure and return hoses from the pump and drain the system into a suitable container. Use care not to damage hose fittings.

3. Cap the fittings at the pump to keep out dirt.

4. Using a 12 in. (305mm) adjustable wrench, on the tensioner casting, loosen the belt tensioner. Lift the belt off the pulley.

5. Locate the 3 pump attaching bolts through the access hole in the pulley and remove the bolts.

6. Remove the one bolt from the rear of the pump, then remove the pump assembly.

To install:

7. Install the pump assembly onto the engine block.

8. Install the one bolt from the rear of the pump and the 3 pump attaching bolts. Torque the power steering pump bolts to 18 ft. lbs. (25 Nm).

9. Install the belt onto the pulley and tighten the belt tensioner.

10. Uncap and connect the pressure and return lines to the pump.

11. Connect the negative battery cable.

12. Refill power steering pump reservoir, start engine and bleed the system. Verify no fluid leaks.

3.1L Engine

1. Disconnect the negative battery cable.

2. Remove the pump drive belt.

3. Disconnect and cap the power steering lines at the pump. Use care not to damage the hose fittings.

4. Remove the spark plug wire clip from the pump.

5. Unfasten the power steering pump bolts, then remove the pump.

To install:

6. Install the power steering pump and pump bolts. Torque the bolts to 25 ft. lbs. (34 Nm).

7. Install the spark plug wire at the pump clip.

8. Uncap and connect the power steering lines.

9. Install the drive belt.

10. Connect the negative battery cable, start the engine and bleed the system. Verify no fluid leaks.

1995–97 Vehicles

1. Disconnect the negative battery cable.

2. Remove the belt.

3. Disconnect and cap the lines at the pump.

4. Remove the mounting bolts, then remove the pump.

To install:

5. Install the pump.

6. Tighten the mounting bolts to 22 ft. lbs. (30 Nm).

7. Uncap and connect the lines to the power steering pump and tighten to 20 ft. lbs. (27 Nm).

8. Install the belt.

9. Connect the negative battery cable.

10. Fill the pump with fluid and bleed air from the power steering system.

11. Road test the vehicle and verify no leaks.

BRAKES

Anti-Lock Brake System Service

PRECAUTIONS

• Certain components within the Anti-Lock Brake System (ABS) are not intended to be serviced or repaired individually. Only those components with removal and installation procedures should be serviced.

• Do not use rubber hoses or other parts not specifically specified for an ABS system. When using repair kits, replace all parts included in the kit. Partial or incorrect repair may lead to functional problems and require the replacement of components.

• Lubricate rubber parts with clean, fresh brake fluid to ease assembly. Do not use lubricated shop air to clean parts; damage to rubber components may result.

• Use only specified brake fluid from an unopened container.

• If any hydraulic component or line is removed or replaced, it may be necessary to bleed the entire system.

• A clean repair area is essential. Always clean the reservoir and cap thoroughly before removing the cap. The slightest amount of dirt in the fluid may plug an orifice and impair the system function. Perform repairs after components have been thoroughly cleaned; use only denatured alcohol to clean components. Do not allow ABS components to come into contact with any substance containing mineral oil; this includes used shop rags.

• The Anti-Lock control unit is a microprocessor similar to other computer units in the vehicle. Ensure

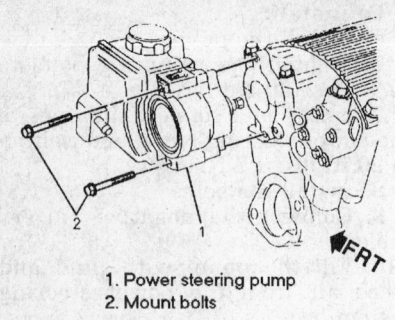

1. Power steering pump
2. Mount bolts

221848

Power steering pump view — 2.3L and 2.4L engines

that the ignition switch is **OFF** before removing or installing controller harnesses. Avoid static electricity discharge at or near the controller.

• If any arc welding is to be done on the vehicle, the control unit should be unplugged before welding operations begin.

Master Cylinder

REMOVAL AND INSTALLATION

1993–94 Vehicles

Vehicles Without ABS

1. Disconnect the negative battery cable.
2. Detach the electrical connector from the master cylinder fluid level sensor.

—————— **WARNING** ——————
When disconnecting the brake lines and removing the master cylinder from the vehicle use care not to spill brake fluid on any painted surfaces or electrical connectors.

3. Disconnect the hydraulic lines from the master cylinder. Plug each line as it is disconnected. Failure to cap the lines will result in excessive fluid loss and possible contamination of the hydraulic system.
4. Unfasten the 2 master cylinder-to-power booster mounting nuts, then remove the master cylinder.
   To install:
5. Install the master cylinder on the booster studs and tighten the mounting nuts to 28 ft. lbs.
6. Install short lengths of brake line in the master cylinder outlet ports and position them so they point back into the reservoir. Fill the reservoir with clean brake fluid until the ends of the lines are submerged.
7. Cover the reservoir with a shop towel to prevent spillage.

8. Pump the brake pedal until clear, bubble-free fluid comes out of the brake lines.
9. Remove the short brake lines and connect the vehicle brake lines to the master cylinder. Bleed each brake line at the master cylinder as follows:
   a. Have an assistant pump the brake pedal 10 times and then hold firm pressure on the pedal.
   b. Position a shop towel under the rear most brake line fitting. Loosen the fitting with a tubing wrench until a stream of brake fluid comes out. Have the assistant maintain pressure on the brake pedal until the brake line fitting is tightened again.
   c. Repeat this operation until clear, bubble free fluid comes out from around the brake line fitting.
   d. Repeat the bleeding operation at the other brake line fitting(s).
10. Final tighten the brake line fittings to 19 ft. lbs.
11. Connect the electrical connector to the fluid level sensor switch.
12. Check the master cylinder fluid level and add fluid as required. If necessary, bleed the entire brake system.
13. Road test the vehicle to verify pedal height and proper brake operation.

Vehicles With ABS

—————— **CAUTION** ——————
The hydraulic accumulator contains brake fluid and nitrogen gas at extremely high pressure. Certain portions of the hydraulic system also contain brake fluid at high pressure. It is mandatory that the gear tension be relieved before disconnecting any hoses, lines or fittings, or personal injury may result. There is gear tension within the hydraulic modulator when the displacement cylinder pistons are in the home position and each motor is under the prevailing torque of the modulator. This torque results in gear tension and makes motor pack separation difficult and dangerous. To avoid personal injury, use a GM approved hand held TECH 1® scan tool or equivalent to perform the Gear Tension Relief procedure prior to performing any portion of hydraulic brake repairs to the vehicle.

1. Disconnect the electrical connector from the brake fluid level switch.

2. Disconnect the electrical connector from the ABS system solenoids.
3. Disconnect the motor pack electrical connectors.

NOTE: When disconnecting the brake lines from the master cylinder and removing the unit, use care not to spill brake fluid on any painted surfaces or electrical connectors.

4. Disconnect and cap the brake lines from the master cylinder assembly. Plug the master cylinder ports to prevent excessive fluid loss.
5. Remove the master cylinder mounting nuts.
6. Remove the master cylinder and modulator assembly.
   To install:
7. Install the master cylinder and modulator assembly on the power booster and tighten the mounting nuts to 20 ft. lbs. (27 Nm).
8. Connect the brake lines to the master cylinder and tighten the fittings to 18 ft. lbs. (25 Nm).
9. Connect the electrical connector to the brake fluid level sensor.
10. Connect the electrical connectors to the ABS system solenoids.
11. Connect the motor pack electrical connectors.
12. Fill the master cylinder to the proper level. The proper level is to the **MAX** level indicator on the reservoir.
13. Bleed the brake system following the recommended procedure.

1995–97 Vehicles

When the ABS modulator cylinder pistons are in their uppermost position, each motor has prevailing torque due to the force necessary to ensure each piston is held firmly at the top of its travel. This torque results in "gear tension," or force on each gear that makes motor pack separation difficult. To avoid injury, or damage to the gears, the "Gear Tension Relief Sequence" briefly reverses each motor to eliminate the prevailing torque. This procedure is one of the many functions of GM's Tech 1® scan tool. Use care when using a substitute. In general, make sure the ignition switch is in the **OFF** position. Install the Tech 1® or equivalent with the correct chassis cartridge. Turn the ignition switch to the **ON** position, leaving the engine OFF. Select the proper function. The "Gear Tension Relief Sequence" is F5 on the Tech 1® scan tool. Note that this same scan tool is needed to bleed the system after repairs to the hydraulic system.

NOTE: To perform the Gear Tension Relief Sequence, a Tech 1® tool or equivalent scan tool must be used prior to removal of the ABS modulator/master cylinder assembly.

1. Perform gear tension release procedure using the Tech 1® scan tool or equivalent.
2. Remove the battery.
3. Remove the air box.
4. Detach the electrical connector from fluid level switch.
5. Disconnect the electrical connectors from both solenoids.
6. Detach the 3-pin and 6-pin motor pack electrical connectors.
7. Remove the 4 brake pipe tube nuts from master cylinder and modulator assembly. If necessary, tag for identification and placement. Place shop cloths on top of the motor pack to catch any dripping fluid. Take care not to allow brake fluid to enter the bottom of the motor pack or the electrical connectors.

NOTE: When disconnecting the brake lines from the master cylinder and removing the unit, use care not to spill brake fluid on

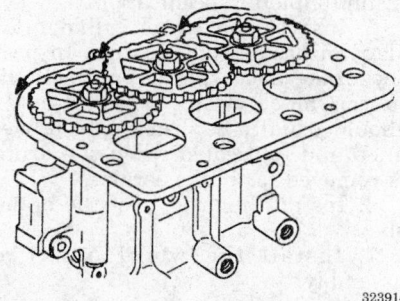

Relieve the internal gears tension when servicing the ABS unit — 1995–97 vehicles

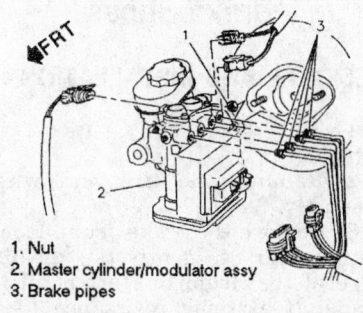

1. Nut
2. Master cylinder/modulator assy
3. Brake pipes

Master cylinder and brake lines — 1995–97 vehicles

any painted surfaces or electrical connectors. Once the brake lines are disconnected from the master cylinder, plug the lines to prevent excess brake fluid loss and contamination.

8. Remove the master cylinder mounting nuts. It may be necessary to remove the vacuum check valve from the vacuum booster to access the nut closest to the check valve.
9. Remove the master cylinder and modulator as an assembly.
10. Separate the modulator assembly from the master cylinder assembly. Use the following procedure.
   a. Turn the assembly upside down so the flat gear cover is facing up.
   b. Remove the 6 gear cover attaching screws. This should allow access to the motor pack screws.
   c. Remove the 4 motor pack screws and separate the motor pack from the ABS modulator assembly.
   d. Remove the 2 Torx head through bolts that retain the master cylinder to the motor pack assembly. Use care not to loose or damage the 2 small transfer tubes that connect between the lower part of the master cylinder assembly and the hydraulic modulator assembly. Watch for O-rings seals at both the transfer tubes and the Torx head attaching bolt openings. The transfer tubes and all O-rings must be replaced with new parts whenever the master cylinder and modulator are separated.

To install:
11. Clean all parts well. Use new O-rings and transfer tubes to assemble the master cylinder to the modulator. If the hydraulic modulator is to be replaced, install the 3 gears in the same location on the replacement modulator. Note that no repair of the hydraulic modulator is authorized.
12. Make sure 2 O-rings are properly installed on each transfer tube. Lubricate the O-rings with clean brake fluid. Install the transfer tube assemblies in the ports in the hydraulic modulator and push in by hand until they bottom.
13. Lubricate with clean brake fluid, new O-rings for the 2 Torx head through bolts and install in both the master cylinder openings and the modulator. Assemble the modulator to the master cylinder. It may be helpful to clamp just the mounting flange of the master cylinder in a soft-jaw vise. Hold the modulator and rock it into position on the master

cylinder, inserting the transfer tube assemblies into the master cylinder ports. Install the 2 Torx head through bolts and torque to 18 ft. lbs. (24 Nm).
14. With the hydraulic modulator upside down and the gears facing up, rotate each gear counterclockwise until movement stops. This procedure will position the piston very close to the top of the hydraulic modulator bore.
15. Assemble the motor pack to the hydraulic modulator. Install the gear cover.
16. Install the master cylinder and modulator assembly to the power booster assembly and tighten the mounting nuts to 20 ft. lbs. (27 Nm).
17. Connect the brake lines to the master cylinder and tighten them to 17 ft. lbs. (23 Nm).
18. Attach the electrical connector to the fluid level switch.
19. Connect the electrical connectors for both of the solenoids.
20. Attach the 3-pin and the 6-pin motor pack electrical connectors.
21. Fill the master cylinder to the proper level. The proper level is the MAX level indicator on the reservoir.
22. Bleed the hydraulic brake system.
23. Road test vehicle and verify proper operation.

Brake Caliper

REMOVAL AND INSTALLATION

1. Siphon ⅔ of the brake fluid out of the master cylinder.
2. Raise and safely support the vehicle.
3. Remove the tire and wheel assembly.
4. Compress the caliper piston back into the caliper bore using a large pair of pliers, C-clamp or special piston retracting tool.
5. If the caliper is to be completely removed from the vehicle for bench service, disconnect and cap the brake line from the caliper. Discard the old washers.
6. Remove the caliper mounting bolts and sleeves.
7. Remove the caliper from the knuckle.
8. Remove the brake pads from the caliper, if it is being replaced.
To install:
9. Install the brake pads in the caliper.
10. Install the caliper on the steering knuckle.

11. Install the mounting bolts and sleeves and tighten to 40 ft. lbs. (51 Nm).

12. If the caliper brake line was disconnected, uncap and connect the brake hose to the caliper using new washers. Tighten the mounting bolt to 35 ft. lbs. (44 Nm).

13. Refill the master cylinder and bleed the brake system.

14. Install the tire and wheel assembly and tighten to specification.

15. Lower the vehicle. Verify correct brake operation.

Disc Brake Pads

REMOVAL AND INSTALLATION

1. Siphon ⅔ of the brake fluid out of the master cylinder reservoir.

2. Raise and safely support the vehicle.

3. Remove the tire and wheel assembly.

4. Compress the caliper piston back into the caliper bore using a large pair of pliers, C-clamp or special caliper piston retracting tool.

5. Remove the caliper mounting bolts and sleeves.

6. Remove the caliper from the steering knuckle without disconnecting the brake hose. Do not allow the caliper to hang from the brake hose. Support the caliper with a piece of wire.

7. Remove the outboard pad by pushing in on the outside edge of the pad to release the mounting dowel from the hole in the caliper. When both dowels are unseated, push the pad out the bottom of the caliper.

8. Remove the inboard pad from the caliper by pulling it out of the caliper.

To install:

9. Inspect the caliper for any signs of leakage. If the caliper is leaking, new brake pads will be damaged. Also inspect the condition of the caliper support for rust and corrosion which will hinder the travel on the caliper.

10. Inspect the caliper mounting hardware. New bolts are usually recommended. Clean all parts well. Lubricate the mounting bushings and sleeves with silicone grease as required.

11. Install the inboard pad in the caliper so the spring clip on the pad back engages in the caliper piston.

12. Install the outboard pad over the caliper end until the mounting dowels snap into the mounting holes in the caliper.

13. Install the caliper over the rotor onto the steering knuckle.

14. Install the mounting bolts and sleeves and tighten to 40 ft. lbs. (51 Nm).

15. Install the tire and wheel assembly.

16. Lower the vehicle.

17. Pump the brake pedal several times to seat the pads against the rotor before attempting to move the vehicle.

18. Check the master cylinder level and add fluid as necessary.

Brake Rotor

REMOVAL AND INSTALLATION

1. Raise and safely support the vehicle.

2. Remove the tire and wheel assembly.

3. Remove the caliper mounting bolts.

4. Support the caliper on a wire out of the way. Do not disconnect the brake hose from the caliper or allow the caliper to hang from the brake hose.

5. Slide the rotor off the hub.

To install:

6. Inspect the caliper for any signs of leakage. If the caliper is leaking, new brake pads will be damaged. Also inspect the condition of the caliper support for rust and corrosion which will hinder the travel on the caliper.

7. Inspect the caliper mounting hardware. New bolts are usually recommended. Clean all parts well. Lubricate the mounting bushings and sleeves with silicone grease as required.

8. Inspect the rotor for grooves, heat cracks or excessive runout. A damaged rotor should be replaced.

9. If a new rotor is being installed the caliper piston will have to be pushed back into the caliper bore enough so the caliper will fit over the rotor. A special piston retracting tool is available. A large C-clamp may be substituted to push the piston back into the caliper bore.

10. Slide the rotor onto the hub.

11. Install the caliper and caliper mounting bolts. Tighten the bolts to 40 ft. lbs. (54 Nm).

12. Install the tire and wheel assembly.

13. Lower the vehicle. Tighten the wheel lug nuts to 100 ft. lbs. (136 Nm).

14. Pump the brake pedal several times to seat the pads against the rotor before moving the vehicle.

15. Check and add brake fluid as necessary.

Brake Drums

REMOVAL AND INSTALLATION

1. Raise and safely support the vehicle.

2. Remove the wheel and tire assembly.

3. Pull the brake drum off. It may be necessary to gently tap the rear edge of the drum to start it off the studs. If extreme resistance to removal is encountered, it will be necessary to retract the brake shoe self-adjuster screw (sometimes called the Star Wheel). Knock out the access hole in the brake drum and turn the adjuster to retract the linings from the drum. Install a replacement hole cover before reinstalling the drum.

NOTE: Do not hammer on the brake drum to remove it.

To install:

4. Inspect the inside of the brake drum. If worn, heavily grooved or if the opening is distorted, the drum should be refinished or replaced. If refinishing, observe the maximum drum diameter specification.

5. Inspect the wheel cylinder for signs of brake fluid leakage. Inspect the brake shoe springs and self-adjuster mechanism. The adjuster should usually be disassembled, cleaned and lubricated when the drum is removed for brake service.

6. Install the drum over the brake shoes.

7. Install the wheel and tire assembly.

8. Adjust the brakes.

9. Lower the vehicle. Check brake operation.

Brake Shoes

REMOVAL AND INSTALLATION

1. Raise and safely support the vehicle.

2. Remove the tire and wheel assembly.

3. Remove the brake drum. It may be necessary to gently tap the rear edge of the drum to start it off the studs. If extreme resistance to removal is encountered, it will be necessary to retract the brake shoe self-adjuster screw (sometime called the

Star Wheel). Turn the adjuster to retract the linings from the drum.

NOTE: Do not hammer on the brake drum to remove it.

4. Remove the upper return springs from the shoes using tool J-8057 or the equivalent brake spring tool.

5. Remove the hold-down springs using J-8049 or the equivalent brake spring tool.

6. Remove the shoe hold-down pins from behind the brake backing plate.

7. Lift up the actuator lever for the self-adjusting mechanism and remove the actuating link. Remove the actuator lever, pivot, and the pivot return spring.

8. Spread the shoes apart to clear the wheel cylinder pistons, then remove the parking brake strut and spring.

9. Disconnect the parking brake cable from the lever. Remove the shoes, still connected by their adjusting screw spring.

10. With the shoes removed, note the position of the adjusting spring and remove the spring and adjusting screw.

11. Remove the C-clip from the parking brake lever and remove the lever from the secondary shoe.

12. Use a damp cloth to remove all dirt and dust from the backing plate and brake parts.

To install:

13. Check the backing plate attaching bolts to make sure they are tight. Use fine emery cloth to clean all rust and dirt from the shoe contact surfaces on the plate and lubricate with brake grease. Check the wheel cylinder for signs of leakage.

14. Clean all parts completely in brake solvent and air dry.

15. Clean the backing plate shoe contact points.

16. Inspect the inside of the brake drum. If worn, heavily grooved or if the opening is distorted, the drum should be refinished or replaced.

17. Inspect the brake shoe springs and self-adjuster mechanism. Disassemble the adjuster mechanism and clean the threads and coat with grease. Make sure the adjuster assembly turns freely before installing in the vehicle.

18. Install the parking brake lever on the secondary shoe and secure with C-clip.

19. Install the adjusting screw and spring on the shoes, connecting them together. The coils of the spring must not be over the star wheel on the adjuster. The left and right hand springs are not interchangeable.

20. Spread the shoe assemblies apart and connect the parking brake cable. Install the shoes on the backing plate, engaging the shoes at the top temporarily with the wheel cylinder pistons. Make sure the star wheel on the adjuster is lined up with the adjusting hole in the backing plate, if equipped.

21. Spread the shoes apart slightly and install the parking brake strut and spring. Make sure the end of the strut without the spring engages the parking brake lever. The end with the spring engages the primary shoe (the one with the shorter lining).

22. Install the actuator pivot, lever and return spring. Install the actuating link in the shoe retainer. Lift up the actuator lever and hook the link into the lever.

23. Install the hold-down pins through the back of the plate using J-8057 or the equivalent. Install the lever pivots and hold-down springs. Install the shoe return springs using J-8049 or the equivalent. Be careful not to stretch or distort the springs.

24. Make sure the linings are in the right place, the self-adjusting mechanism is correctly installed, and the parking brake parts are hooked up.

25. Measure the distance from the edge of the primary lining to the edge of the secondary lining, then measure the inside width of the drum. Adjust the linings by means of the adjuster so the drum will fit onto the linings.

26. Install the hub and bearing assembly onto the axle if removed. Tighten the retaining bolts to 35 ft. lbs. (55 Nm.).

27. Install the drum.

28. Install the wheel and tire assembly.

29. Adjust the brakes. Install a rubber hole cover in the adjustment knock-out hole after the adjustment is complete. Adjust the parking brake.

30. Lower the car.

31. Road test the vehicle and verify proper brake operation.

Wheel Cylinder

REMOVAL AND INSTALLATION

1993–94 Vehicles

NOTE: These vehicle can have 2 different types of wheel cylinders depending on the rear brake option. Before beginning the removal procedure verify which type of wheel cylinder mounting the vehicle is equipped with.

Wheel Cylinder with Mounting Bolts

1. Raise and safely support the vehicle.

2. Remove the rear tire and wheel assembly.

3. Remove the brake drum.

4. Remove the brake shoes.

5. Disconnect and cap the brake pipe from the wheel cylinder.

6. Remove the 2 mounting bolts.

7. In order to remove the wheel cylinder from the backing plate, remove the bleeder screw from the wheel cylinder.

8. Remove the wheel cylinder.

To install:

9. Remove the bleeder screw from the new wheel cylinder.

10. Install the wheel cylinder and wheel cylinder mounting nuts. Tighten the nuts to 106 inch lbs. (12 Nm).

11. Connect the brake pipe to the wheel cylinder and tighten the fitting to 11 ft. lbs. (15 Nm)

12. Install the brake shoes.

13. Install the brake drum.

14. Install the wheel and tire assembly.

15. Adjust the brakes.

16. Lower the vehicle.

17. Refill the master cylinder and bleed the system.

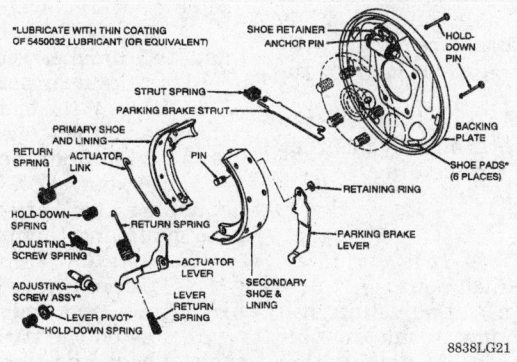

Exploded view of drum brake components

Wheel Cylinder with Retaining Ring

1. Raise and safely support the vehicle.
2. Remove the rear tire and wheel assembly.
3. Remove the brake drum.
4. Remove the brake shoes.
5. Disconnect and cap the brake pipe from the wheel cylinder.
6. In order to remove the wheel cylinder from the backing plate, it may be necessary to remove the bleeder screw from the wheel cylinder.
7. Using 2 awls or tool J-29839, release both sides of the wheel cylinder locking, work the locking ring off the wheel cylinder.
8. Remove the wheel cylinder from the backing plate.

To install:
9. Remove the bleeder screw from the new wheel cylinder.
10. Position the wheel cylinder in the backing plate.
11. Position the locking ring on the back of the wheel cylinder.
12. Place a block of wood between the wheel cylinder and the hub assembly.
13. Using a 1⅛ inch socket and a 12 inch extension, drive the retaining ring into place on the back of the wheel cylinder.
14. Connect the brake pipe to the wheel cylinder and tighten the fitting to 11 ft. lbs. (15 Nm)
15. Install the brake shoes.
16. Install the brake drum.
17. Install the wheel and tire assembly.
18. Adjust the brakes.
19. Lower the vehicle.
20. Refill the master cylinder and bleed the system.

1995–97 Vehicles

1. Raise and safely support the vehicle.
2. Remove the rear tire and wheel assembly.
3. Remove the brake drum. If it is difficult to remove the brake drum:
   a. Makes sure the parking brake is released.
   b. Back off the parking brake cable adjustment.
   c. Use a rubber mallet to tap gently around the outer rim of the drum and/or around the inner drum diameter by the spindle.
4. Remove the hub and bearing assembly.
5. Underbody components are subject to rust and corrosion. Thoroughly clean the area around the brake line connection and the retainer bolts. A

generous application of penetrating oil may make brake line removal easier. Disconnect the brake line from the back of the wheel cylinder and plug the opening in the line to prevent fluid loss and contamination.
6. Remove the 2 bolts from the back of the wheel cylinder using a No. 6 TORX® socket.
7. Remove the wheel cylinder.

To install:
8. Install wheel cylinder and tighten bolts to 15 ft. lbs. (20 Nm).
9. Install hub and bearing assembly and tighten nut to 43 ft. lbs. (58 Nm).
10. Connect the brake line to wheel cylinder and tighten to 17 ft. lbs. (23 Nm).
11. Install the brake drum.
12. Bleed wheel cylinder.
13. Install the wheel and tire assembly.
14. Lower the vehicle.
15. Check brake fluid level and fill as necessary.
16. Road test the vehicle and verify proper operation.

Parking Brake Cable

ADJUSTMENT

1993–94 Vehicles

NOTE: The rear brake shoes must be adjusted properly before parking brake adjustment can be performed. Failure to properly adjust the brakes can lead to overtightening the parking brake cable adjuster resulting in rear brake drag.

1. Pull parking lever until 5 ratchet clicks are heard.
2. Raise and safely support the vehicle.

NOTE: Clean and lubricate the exposed threads on each side of the nut before adjusting.

3. Tighten adjusting nut on the parking brake cable until the left rear wheel can turned rearward with 2 hands but locked when forward rotation is attempted.
4. Release parking brake. Rear wheels must now turn freely in either direction with no brake drag.
5. Lower the vehicle.

1995–97 Vehicles

Parking brake adjustment is not necessary. This is a self-adjusting system and damage may result from attempting to adjust or modify this system in any way. It may be nec-

essary to adjust the rear brakes to obtain proper tension in the system.

Parking Brake System Checking

1. Check the drum brake clearance and reset if required.
2. Fully apply and release the parking brake lever 4 times.
3. Apply the lever to the 3-click position.
4. Raise and safely support the vehicle.
5. Check the parking brake cable tension by attempting to turn the right rear wheel with both hands. The wheel should only rotate backward, not forward.
6. If the tire does rotate forward, readjust rear shoe to drum clearance until a light shoe-to-drum contact may be heard.
7. Release the parking brake and check for light rear brake drag. The rear wheels should turn freely and some light shoe-to-drum contact may be heard.

NOTE: The parking brake cables have the wire strand coated with a plastic material which slides against nylon seals inside the conduit end fittings. This is for corrosion protection and to reduce parking brake effort. The coated cables do not need lubrication.

Parking Brake System Adjusting

1. Fully apply and release the hand brake 4–6 times to self-adjust.
2. The parking brake is self-adjusting.

REMOVAL AND INSTALLATION

1993–94 Vehicles

Front Cable

1. Raise and safely support the vehicle.
2. Loosen the adjustment nut.
3. Remove the center console.
4. Disconnect the parking brake cable from the lever.
5. Remove the cable retaining nut and the bracket securing the front cable to the floor panel.
6. Loosen the catalytic converter shield and then remove the parking brake cable from the body.
7. Disconnect the cable from the equalizer, then remove the cable from the guide and the underbody clips.

To install:
8. Position the cable to the equalizer, guide and underbody clips.
9. Fasten the parking brake cable to the body and tighten the catalytic converter shield.

10. Install the cable retaining nut and the bracket securing the front cable to the floor panel.

11. Connect the parking brake cable to the lever.

12. Install the center console.

13. Adjust the cable.

14. Lower the vehicle.

Rear Cables

1. Raise and safely support the vehicle.

2. Back off the equalizer nut until the cable tension is eliminated.

3. Remove the tires, wheels and brake drums.

4. Insert a suitable pry tool between the brake shoe and the top part of the brake adjuster bracket. Push the bracket to the front and then release the top brake adjuster rod.

5. Remove the rear hold-down spring. Remove the actuator lever and the lever return spring.

6. Remove the adjuster screw spring.

7. Remove the top rear brake shoe return spring.

8. Unhook the parking brake cable from the parking brake lever.

9. Depress the retaining tangs on the cable housing end and remove the cable from the backing plate.

10. Remove the cable end button from the connector.

11. Depress the retaining tangs and remove the cable from the axle bracket.

To install:

12. Install the cable to the axle bracket. Make sure the locking fingers are fully engaged.

13. Install the cable end button to the connector.

14. Install the cable into the backing plate. Make sure the locking fingers are fully engaged.

15. Hook the parking brake cable to the parking brake lever.

16. Install the top rear brake shoe return spring.

17. Install the adjuster screw spring.

18. Install the rear hold-down spring. Install the actuator lever and the lever return spring.

19. Install the top brake adjuster rod.

20. Install the tires, wheels and brake drums.

21. Adjust the cable.

1995–97 Vehicles

1. Raise and safely support the vehicle.

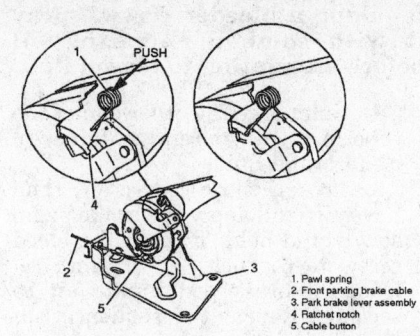

1. Pawl spring
2. Front parking brake cable
3. Park brake lever assembly
4. Ratchet notch
5. Cable button

324600

Self-adjust lockout — 1995–97 vehicles

2. Pull and hold cable towards the rear of the car to create slack in cable.

3. Bend tang on connector to allow cable removal.

4. Remove the cable from the connector.

5. Fold over the retaining clip, and remove the cable from the equalizer.

6. Lower the vehicle.

7. Remove the console (and shift boot) if equipped.

8. Move the lever to the off position.

9. Disconnect the cable conduit end fitting from handle assembly.

10. Pull the cable until the notch on the ratchet is visible through the cover plate opening.

11. Push the pawl spring downward toward notch in the ratchet.

12. Release the cable slowly to allow notch to catch leg of the spring.

13. Disconnect the front parking brake cable button from the reel assembly.

14. Remove the left rocker panel/door sill plate.

15. Remove the grommet and retainer from the floor pan.

16. Remove the cable from clip on number 2 bar under the carpet.

17. Remove the cable.

To install:

18. Push the cable through the floor pan, from the interior to exterior.

19. Install the cable conduit fitting to the handle assembly.

20. Install the front parking brake cable button-to-reel assembly.

21. Insert the cable into No. 2 bar clip (under the carpet).

22. Install the console and shifter boot, if equipped.

23. Install the left rocker panel/door sill plate.

24. Raise and safely support the vehicle.

25. Fold over the bracket to body.

26. Connect the cable to the equalizer assembly.

27. Install the cable to the connector.

28. Bend the tang on the connector to retain cable.

29. Lower the vehicle.

30. Fully apply and release the parking brake 4–6 times to activate the self-adjust system.

Brake System Bleeding

PROCEDURE

Vehicles Without ABS

As a general rule, once the master cylinder (and the brake pressure modulator valve) is bled, the remainder of the hydraulic system should be bled starting at the furthest wheel from the master cylinder and working towards the nearest wheel. Therefore, the correct bleeding sequence is:

Master cylinder
Right rear wheel cylinder
Left rear wheel cylinder
Right front caliper
Left front caliper

Most master cylinder assemblies on these vehicles are NOT equipped with bleeder valves, therefore air must be bled from the cylinders using the front brake pipe connections.

Manual Bleeding

1. Deplete the vacuum reserve by applying the brakes several times with the ignition **OFF**.

2. Clean the top of the master cylinder, remove the cover and fill the reservoirs with clean fluid. To prevent squirting fluid, and possibly damaging painted surfaces, install the cover during the procedure, but be sure to frequently check and top off the reservoirs with fresh fluid.

––––––– **WARNING** –––––––
Never reuse brake fluid which has been bled from the system because the old fluid may be contaminated with moisture and/or debris

3. The master cylinder must be bled first if it is suspected to contain air. If the master cylinder was removed and bench bled before installation it must still be bled, but it should take less time and effort. Bleed the master cylinder as follows:

a. Position a container under the master cylinder to catch the brake fluid.

WARNING

Do not allow brake fluid to spill on or come in contact with the vehicle's finish as it will remove the paint. In case of a spill, immediately flush the area with water.

---

b. Loosen the front brake line at the master cylinder and allow the fluid to flow from the front port.

c. Have a friend depress the brake pedal slowly and hold (air and/or fluid should be expelled from the loose fitting). Tighten the line, then release the brake pedal and wait 15 seconds. Loosen the fitting and repeat until all air is removed from the master cylinder bore.

d. When finished, tighten the line fitting to 20 ft. lbs. (27 Nm).

e. Repeat the sequence at the master cylinder rear pipe fitting.

NOTE: During the bleeding procedure, make sure your assistant does NOT release the brake pedal while a fitting is loosened or while a bleeder screw is opened. Otherwise, air will be drawn back into the system.

4. Check and refill the master cylinder reservoir.

NOTE: Remember, if the reservoir is allowed to empty of fluid during the procedure, air will be drawn into the system and the bleeding procedure must be restarted at the master cylinder assembly.

5. If a single line or fitting was the only hydraulic line disconnected, then only the caliper(s) or wheel cylinder(s) affected by that line must be bled. If the master cylinder required bleeding, then all calipers and wheel cylinders must be bled in the proper sequence.

6. The proper sequence is:
   a. Right rear
   b. Left rear
   c. Right front
   d. Left front

7. Bleed the individual calipers or wheel cylinders as follows:

a. Place a suitable wrench over the bleeder screw and attach a clear plastic hose over the screw end. Be sure the hose is seated snugly on the screw or you may be squirted with brake fluid.

NOTE: Be very careful when bleeding wheel cylinders and brake calipers. The bleeder screws often rust in position and may easily break off if forced. To help prevent the possibility of

breaking a bleeder screw, spray it with some penetrating oil before attempting to loosen it.

b. Submerge the other end of the tube in a transparent container of clean brake fluid.

c. Loosen the bleed screw, then have a friend apply the brake pedal slowly and hold. Tighten the bleed screw to 62 inch lbs. (7 Nm), release the brake pedal and wait 15 seconds. Repeat the sequence (including the 15 second pause) until all air is expelled from the caliper or cylinder.

d. Tighten the bleeder screw to 62 inch lbs. (7 Nm) when finished.

8. Check the pedal for a hard feeling with the engine not running. If the pedal is soft, repeat the bleeding procedure until a firm pedal is obtained.

9. If the brake warning light is ON, depress the brake pedal firmly. If there is no air in the system, the light will go out.

10. After bleeding, make sure a firm pedal is achieved before attempting to move the vehicle.

Pressure Bleeding

1. Install Pressure Bleeder Adapter Cap J-35589 or equivalent, to the master cylinder.

2. Charge Diaphragm Type Brake Bleeder J-29532 or equivalent, to 20–25 psi (140–172 kPa).

3. Connect the line to the pressure bleeder adapter cap, then open the line valve.

4. Raise and safely support the vehicle.

5. If it is necessary to bleed all of the calipers/cylinders, the following sequence should be used:
• Right rear
• Left rear
• Right front
• Left front

6. Place a proper size box end wrench (or tool J-21472) over the caliper/cylinder bleeder valve.

7. Attach a clear tube over the bleeder screw, then submerge the other end of the tube in a clear container partially filled with clean brake fluid.

8. Open the bleeder screw at least ¾ of a turn and allow flow to continue until no air is seen in the fluid.

9. Close the bleeder screw. Tighten the rear bleeder screws to 62 inch lbs. (7 Nm) and the front bleeder screws to 115 inch lbs. (13 Nm).

10. Repeat Steps 6–9 until all of the calipers and/or cylinders have been bled.

11. Carefully lower the vehicle.

12. Check the brake pedal for sponginess. If the condition is found, the entire bleeding procedure must be repeated.

13. Remove tools J-35589 and J-29532.

14. Refill the master cylinder to the proper level with brake fluid.

15. Do not attempt to move the vehicle unless a firm brake pedal is obtained.

Vehicles With ABS

Before bleeding the system, the front and rear displacement cylinder pistons must be returned to the topmost position. The preferred method uses a Tech 1® or T-100® scan tool to perform the rehoming procedure. If a Tech 1® is not available, the second procedure may be used, but it must be followed exactly.

Rehome Procedure With Tech 1® or T-100® (Preferred Method)

1. Using a Tech 1® or T-100® (CAMS), select "F5: Motor Rehome." The motor rehome function cannot be performed if current DTC's are present. If DTC's are present, the vehicle must be repaired and the codes cleared before performing the motor rehome function.

2. The entire brake system should now be bled using the pressure or manual bleeding procedures.

Rehome Procedure Without Tech 1® or T-100®

NOTE: Do not place your foot on the brake pedal through this entire procedure unless specifically instructed to do so.

This method can only be used if the ABS warning lamp is not illuminated and not DTC's are present.

1. Remove your foot from the brake pedal.

2. Start the engine and allow it to run for at least 10 seconds while observing the ABS warning lamp.

3. If the ABS warning lamp turned ON and stayed ON after about 10 seconds, the bleeding procedure must be stopped and a Tech 1® must be used to diagnose the ABS function.

4. If the ABS warning lamp turned "ON" for about 3 seconds, then turned OFF and stayed OFF, turn the ignition OFF.

5. Repeat Steps 1–4 one more time.

6. The entire brake system should now be bled by following the manual or pressure bleeding procedure.

Pressure Bleeding

NOTE: The pressure bleeding equipment must be of the diaphragm type. It must have a rubber diaphragm between the air supply and the brake fluid to prevent air, moisture and other contaminants from entering the hydraulic system.

1. Clean the master cylinder fluid reservoir cover and surrounding area, then remove the cover.
2. Add fluid, if necessary to obtain a proper fluid level.
3. Connect bleeder adapter J-35589, or equivalent, to the brake fluid reservoir, then connect the bleeder adapter to the pressure bleeding equipment.
4. Adjust the pressure bleed equipment t o 5–10 psi (35–70 kPa) and wait about 30 seconds to be sure there is no leakage.
5. Adjust the pressure bleed equipment to 30–35 psi (205–240 kPa).

——— **WARNING** ———
Use a shop rag to catch the escaping brake fluid. Be careful not to let any fluid run down the motor pack base or into the electrical connector.

6. With the pressure bleeding equipment connected and pressurized, proceed as follows:
   a. Attach a clear plastic bleeder hose to the rearward bleeder valve on the hydraulic modulator.
   b. Slowly open the bleeder valve and allow fluid to flow until no air is seen in the fluid.
   c. Close the valve when fluid flows out without any air bubbles.
   d. Repeat Steps 6b and 6c until no air bubbles are present.
   e. Relocate the bleeder hose on the forward hydraulic modulator bleed valve and repeat Steps 6a through 6d.
7. Tighten the bleeder valve to 80 inch lbs. (9 Nm).
8. Proceed to bleed the hydraulic modulator brake pipe connections as follows with the pressure bleeding equipment connected and pressurized:
   a. Slowly open the forward brake pipe tube nut on the hydraulic modulator and check for air in the escaping fluid.
   b. When the air flow ceases, immediately tighten the tube nut. Tighten the tube nut to 18 ft. lbs. (24 Nm).

9. Repeat Steps 8a and 8b for the remaining 3 brake pipe connections moving from the front to the rear.
10. Raise and safely support the vehicle.
11. Proceed, using the following steps, to bleed the wheel brakes in the following sequence: right rear, left rear, right front, then left front.
   a. Attach a clear plastic bleeder hose to the bleeder valve at the wheel, then submerge the opposite hose end in a clean container partially filled with clean brake fluid.
   b. Slowly open the bleeder valve and allow the fluid to flow.
   c. Close the valve when fluid begins to flow without any air bubbles. Tap lightly on the caliper or backing plate to dislodge any trapped air bubbles.
12. Repeat Step 11 on the other brakes using the earlier sequence.
13. Remove the pressure bleeding equipment, including bleeder adapter J-35589.
14. Carefully lower the vehicle, then check the brake fluid and add if necessary. Don't forget to put the reservoir cap back on.
15. With the ignition turned to the **RUN** position, apply the brake pedal with moderate force and hold it. Note the pedal travel and feel. If the pedal feels firm and constant and the pedal travel is not excessive, start the engine. With the engine running, recheck the pedal travel. If it's still firm and constant and pedal travel is not excessive, go to Step 17.
16. If the pedal feels soft or has excessive travel either initially or after the engine is started, the following procedure may be used:
   a. With the Tech 1® scan tool, "release" then "apply" each motor 2–3 times and cycle each solenoid 5–10 times. When finished, be sure to "apply" the front and rear motors to ensure the pistons are in the upmost position. Do not drive the vehicle.
   b. If a Tech 1® is not available, remove your foot from the brake pedal, start the engine and allow it run for at least 10 seconds to initialize the ABS; do not drive the vehicle. After 10 seconds, turn the ignition **OFF**. The initialization procedure most be repeated 5 times to ensure any trapped air has been dislodged.
   c. Repeat the bleeding procedure, starting with Step 1.
17. Road test the vehicle, and make sure the brakes are operating properly.

Manual Bleeding

1. Clean the master cylinder fluid reservoir cover and surrounding area, then remove the cover.
2. Add fluid, if necessary to obtain a proper fluid level, then put the reservoir cover back on.
3. Prime the ABS hydraulic modulator/master cylinder assembly as follows:
   a. Attach a bleeder hose to the rearward bleeder valve, then submerge the opposite hose end in a clean container partially filled with clean brake fluid.
   b. Slowly open the rearward bleeder valve.
   c. Depress and hold the brake pedal until the fluid begins to flow.
   d. Close the valve, then release the brake pedal.
   e. Repeat Steps 3b–3d until no air bubbles are present.
   f. Relocate the bleeder hose to the forward hydraulic modulator bleeder valve, then repeat Steps 3a–3e.
4. Once the fluid is seen to flow from both modulator bleeder valves, the ABS modulator/master cylinder assembly is sufficiently full of fluid. However, it may not be completely purged of air. At this point, move to the wheel brakes and bleed them. This ensures that the lowest points in the system are completely free of air and then the assembly can purged of any remaining air.
5. Remove the fluid reservoir cover. Fill to the correct level, if necessary, then fasten the cover.
6. Raise and safely support the vehicle.
7. Proceed, using the following steps, to bleed the wheel brakes in the following sequence: right rear, left rear, right front, left front.
   a. Attach a clear plastic bleeder hose to the bleeder valve at the wheel, then submerge the opposite hose end in a clean container partially filled with clean brake fluid.
   b. Open the bleeder valve.
   c. Have an assistant slowly depress the brake pedal.
   d. Close the valve and slowly release the release the brake pedal.
   e. Wait 5 seconds.
   f. Repeat Steps 7a–7e until the brake pedal feels firm at half travel and no air bubbles are observed in the bleeder hose. To assist in freeing the entrapped air, tap lightly on the caliper or braking plate to dislodge any trapped air bubbles.
8. Repeat Step 7 for the remaining brakes in the sequence given earlier.
9. Carefully lower the vehicle.

10. Remove the reservoir cover, then fill to the correct level with brake fluid and replace the cap.

11. Bleed the ABS hydraulic modulator/master cylinder assembly as follows:

a. Attach a clear plastic bleeder hose to the rearward bleeder valve on the modulator, then submerge the opposite hose end in a clean container partially filled with clean brake fluid.

b. Have an assistant depress the brake pedal with moderate force.

c. Slowly open the rearward bleeder valve and allow the fluid to flow.

d. Close the valve, then release the brake pedal.

e. Wait 5 seconds.

f. Repeat Steps 11a–11e until no air bubbles are present.

g. Relocate the bleeder hose to the forward hydraulic modulator bleeder valve, then repeat Steps 11a–11f.

12. Carefully lower the vehicle, then check the brake fluid and add if necessary. Don't forget to put the reservoir cap back on.

13. With the ignition turned to the **RUN** position, apply the brake pedal with moderate force and hold it. Note the pedal travel and feel. If the pedal feels firm and constant and the pedal travel is not excessive, start the engine. With the engine running, recheck the pedal travel. If it's still firm and constant and pedal travel is not excessive, road test the vehicle and make sure the brakes are operating properly.

14. If the pedal feels soft or has excessive travel either initially or after the engine is started, the following procedure may be used:

a. With the Tech 1® scan tool, "Release" then "Apply" each motor 2–3 times and cycle each solenoid 5–10 times. When finished, be sure to "Apply" the front and rear motors to ensure the pistons are in the upmost position. Do not drive the vehicle.

b. If a Tech 1® scan tool is not available, remove your foot from the brake pedal, start the engine and allow it run for at least 10 seconds to initialize the ABS. Do not drive the vehicle. After 10 seconds, turn the ignition **OFF**. The initialization procedure most be repeated 5 times to ensure any trapped air has been dislodged.

c. Repeat the bleeding procedure, starting with Step 1.

15. Road test the vehicle, and make sure the brakes are operating properly.

Wheel Speed Sensor

REMOVAL AND INSTALLATION

The front wheel speed sensors are of a variable reluctance type. Each sensor is attached to the knuckle assembly in close proximity to a toothed ring. An A/C voltage is generated when the toothed ring passes by the sensor. The magnitude and frequency are proportional to the speed of the wheel and both will increase with increasing speed. The sensor is not repairable, nor is the air gap adjustable. The rear wheel speed sensors operate in the same manner as the front wheel speed sensors. If and/or when the rear wheel speed sensor fails, the entire integral bearing and speed sensor assembly must be replaced.

Front Wheel Speed Sensor

1. Disconnect the negative battery cable.
2. Raise and safely support the vehicle.
3. Disconnect the front wheel speed sensor electrical connector.
4. Remove the retaining bolt.
5. Remove the front wheel speed sensor. If the sensor will not slide out of the knuckle, remove the brake rotor and use a blunt punch or equivalent tool to push the sensor from the back side of the knuckle.

NOTE: If the sensor locating pin breaks off and remains in the knuckle during removal, remove the broken pin using a blunt punch. Clean the hole using sandpaper wrapped around a screwdriver or equivalent tool. Never attempt to enlarge the hole.

To install:
6. Install the front wheel speed sensor on the mounting bracket.
7. Make sure the front wheel speed sensor is properly aligned and lays flat against the bosses on the knuckle.
8. Install the retaining bolt and tighten to 107 inch lbs. (12 Nm).
9. Attach the electrical connector to the speed sensor.
10. Lower the vehicle.
11. Connect the negative battery cable.
12. Road test the vehicle and verify proper operation.

Rear Wheel Speed Sensor

1. Disconnect the negative battery cable.
2. Raise and safely support the vehicle.
3. Remove the rear tire and wheel assembly.
4. Remove the brake drum.
5. Disconnect the rear sensor electrical connector.
6. Remove the bolts and nuts attaching the rear wheel bearing and speed sensor assembly.

NOTE: The drum brake assembly will be held in place only by the brake pipe connection after the bolts are removed. Use care not to bump or exert force on the drum brake assembly.

7. Remove the rear wheel bearing and speed sensor assembly.
To install:
8. Install the rear wheel bearing and speed sensor assembly.
9. Line up the bolt holes in bearing and sensor assembly and brake assembly and install mounting bolts.
10. Tighten the mounting bolts to 37 ft. lbs. (50 Nm).
11. Attach the electrical connector to the sensor.
12. Install the brake drum assembly.

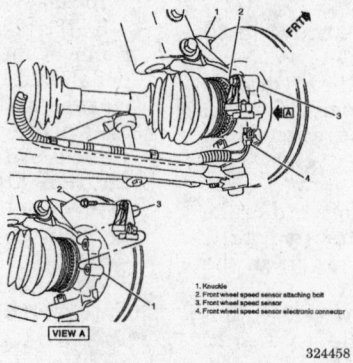

1. Knuckle
2. Front wheel speed sensor attaching bolt
3. Front wheel speed sensor
4. Front wheel speed sensor electronic connector

VIEW A

324458

Front wheel speed sensor

13. Install the tire and wheel assembly.

14. Lower the vehicle.

15. Connect the negative battery cable.

16. Road test the vehicle and verify proper operation.

FRONT SUSPENSION

Strut

REMOVAL AND INSTALLATION

1. From inside the engine compartment, pry off the cover on the strut tower, if equipped, then unfasten the upper strut-to-body nuts and/or bolts.

2. Loosen the wheel lug nuts, then raise and safely support the vehicle.

3. Place jackstands under the front crossmember. Lower the vehicle slightly so the weight of the car rests on the jackstands and NOT the control arms.

4. Remove the wheel and tire assembly.

5. Before removing front suspension components, their positions should be marked so they may assemble correctly.

-------- WARNING --------

Whenever working near the drive axles, use care to prevent damage from over extension of the drive shaft joints. When either end of the shaft is disconnected, over extension of the joint could result in separation of the internal components and possible joint failure.

6. Install a drive axle joint protective cover (modified), such as J-34754 or equivalent.

7. If necessary, remove the brake line bracket.

8. Remove the cotter pin and nut, then press the tie rod out of the strut bracket using J-24319-01 or equivalent two-armed puller. Discard the cotter pin.

9. Unfasten and remove the strut-to-steering knuckle bolts and carefully lift out the strut.

NOTE: The steering knuckle MUST be supported to prevent axle joint overextension.

10. Remove the strut assembly from the vehicle. Be careful to avoid chipping or cracking the spring coating when handling the front suspension coil spring assembly.

To install:

11. Move the strut into position, then install the nuts and/or bolts connecting the strut assembly to the body.

12. Align the steering knuckle with the strut flange scribe mark made during removal, then install the bolts and nuts. Tighten to 133 ft. lbs. (180 Nm).

13. Position the tie rod end into the strut assembly, then secure with the tie rod end bolt and new cotter pin. Tighten the tie rod end bolt to 44 ft. lbs. (60 Nm).

14. Tighten the nuts and/or bolts attaching the top of the strut to the body to 18–20 ft. lbs. (25–27 Nm).

15. Install the brake line bracket.

16. Slightly raise the vehicle, then remove the jackstands from under the suspension supports.

17. Install the tire and wheel assembly.

18. Carefully lower the vehicle, then final tighten the lug nuts to 100 ft. lbs. (140 Nm).

Lower Ball Joints

REMOVAL AND INSTALLATION

1. Raise and safely support the vehicle.

2. If suspension contact hoist is used, place safety stands under the crossmember. Lower the vehicle slightly so the weight of the vehicle rests on the crossmember.

3. Remove the tire and wheel assembly.

NOTE: Care must be exercised to prevent the axle shaft joints from being over-extended. When either end of the shaft is disconnected, over-extension of the joint could result in separation of internal components and possible joint failure. Failure to observe this can result in interior joint or boot damage and possible joint failure.

4. Remove the cotter pin and nut from the ball joint.

5. Separate the ball joint from the steering knuckle using tool J-38892 or equivalent.

6. Drill out the 3 rivets retaining ball joint to the lower control arm. Use an 1/8 inch (3mm) drill bit to make a pilot hole through the rivets. Finish drilling rivets with a 1/2 inch (13mm) drill bit.

7. Remove the nut attaching link to the stabilizer shaft.

8. Remove the ball joint from the steering knuckle and control arm.

To install:

9. Install ball joint into the control arm.

10. Install 3 new bolts and nuts (supplied with new ball joint) and tighten.

11. Position ball joint stud through the steering knuckle and tighten nut to 50 ft. lbs. (65 Nm) and install cotter pin.

12. Connect stabilizer link to stabilizer shaft and tighten nut to 13 ft. lbs. (17 Nm).

13. Install tire and wheel assembly.

14. Lower the vehicle.

15. Check front wheel alignment and adjust, if necessary.

Lower Control Arms

REMOVAL AND INSTALLATION

1. Raise and safely support vehicle.

2. If suspension contact hoist is used, place safety stands under the crossmember. Lower vehicle slightly so the weight of the vehicle rests on the crossmember and not on the control arms.

3. Remove the tire and wheel assembly.

4. Remove the nut attaching stabilizer link to stabilizer shaft.

5. Remove the cotter pin and nut.

NOTE: Care must be exercised to prevent the axle shaft joints from being over-extended. When either end of the shaft is disconnected, over-extension of the joint could result in separation of internal components and possible joint failure. Failure to observe this can result in interior joint or boot damage and possible joint failure.

6. Remove the ball joint from the steering knuckle using special tool J-38892 or equivalent ball joint press.

7. Remove the bolts attaching control arm to crossmember.

To install:

8. Install control arm into position and install bolts (leave loose to allow movement) attaching control arm to crossmember.

9. Install the ball joint stud into the steering knuckle and install nut finger-tight.

10. Tighten the ball joint nut to 41 ft. lbs. (55 Nm) and install cotter pin.

11. Remove the safety stands from under crossmember.

12. Install tire and wheel assembly.

13. Lower the vehicle. Tighten the wheel lug nuts to 100 ft. lbs. (136 Nm).

14. With vehicle at curb height, tighten control arm bolts as follows:

   a. Tighten the front control arm to crossmember bolt to 90 ft. lbs. (120 Nm).

   b. Tighten the rear control arm to crossmember bolt to 125 ft. lbs. (170 Nm).

15. Check wheel alignment and adjust as necessary.

Sway Bar

REMOVAL AND INSTALLATION

1993–94 Vehicles

1. Raise and safely support the vehicle.

2. Remove both front wheel and tire assemblies.

3. Detach the stabilizer shaft from the control arms.

4. Unfasten the retaining clamps, then disconnect the stabilizer shaft from the support assemblies.

5. Remove the rear and center bolts from the suspension support assembly, then loosen, but do not remove, the front bolts.

6. Remove the stabilizer shaft with the insulators.

To install:

7. Position the stabilizer shaft with the insulators, into the vehicle.

8. Fasten the clamps attaching the stabilizer to the suspension support assemblies and hand-tighten the retainers.

9. Install the nut attaching the stabilizer shaft to the stabilizer links.

10. Tighten the suspension support bolts, in sequence, as follows:

   a. Center bolts first, to 66 ft. lbs. (90 Nm).

   b. Front bolts second, to 65 ft. lbs. (88 Nm).

   c. Rear bolts third, to 65 ft. lbs. (88 Nm).

11. Tighten the stabilizer shaft-to-support assembly nuts to 16–22 ft. lbs. (22–30 Nm) and the stabilizer shaft-to-control arm nuts to 13–22 ft. lbs. (17–30 Nm).

12. Tighten the clamp nuts to 17 ft. lbs. (23 Nm).

13. Install the wheel and tire assemblies.

14. Carefully lower the vehicle.

1995–97 Vehicles

The sway bar (also known as the stabilizer shaft) attaches to the front suspension support (crossmember) and through a bolt and link, connects the lower control arms.

1. Raise and safely support vehicle and allow vehicle suspension to hang free.

2. Remove the front tire and wheel assemblies.

3. Remove the nuts attaching stabilizer shafts to stabilizer links.

4. Remove the clamps attaching stabilizer shaft to the crossmember.

5. Support the rear of the crossmember with adjustable safety stands.

6. Remove the rear and center bolts from crossmember assembly and loosen the front bolts.

7. Lower the crossmember 3 inches by adjusting the safety stands down.

8. Remove the stabilizer shaft with insulators.

To install:

9. Place stabilizer shaft with insulators into vehicle and hand-tighten.

10. Install the clamps attaching stabilizer shaft to crossmember assemblies and hand-tighten.

11. Install the crossmember assemblies into position and install bolts (hand-tighten).

12. Tighten crossmember bolts left rear outboard first, right rear outboard second, front upper third and rear inboard last.

   a. Tighten the left rear outboard to 96 ft. lbs. (130 Nm).

   b. Tighten the right rear outboard to 96 ft. lbs. (130 Nm).

   c. Tighten the front upper bolts to 96 ft. lbs. (130 Nm).

   d. Tighten the rear inboard bolts to 96 ft. lbs. (130 Nm).

13. Tighten the clamp bolts to crossmember to 49 ft. lbs. (66 Nm).

14. Tighten the stabilizer shaft links to control arm nuts. Tighten the nuts to 13 ft. lbs. (17 Nm).

15. Install the tire and wheel assemblies.

16. Lower the vehicle.

17. Check wheel alignment and adjust as necessary.

Front Wheel Bearings

ADJUSTMENT

These vehicles are equipped with sealed front hub and bearing assemblies. The hub and bearing assemblies are non-serviceable. If the assembly is damaged, the complete unit must be replaced.

REMOVAL AND INSTALLATION

1993–94 Vehicles

1. Safely and safely support the vehicle.

2. Remove the tire/wheel assembly.

3. Install shop towels under the outer joint to protect it from sharp edges. Inset a drift pin into the brake caliper opening and into the disc brake rotor to keep the rotor from turning. Remove the hub nut and washer.

4. Remove the caliper bolts, and support the caliper with mechanic's wire from the suspension.

5. Remove the front brake rotor.

6. Remove the ball joint cotter pin and nut and loosen the joint. If removing right axle, turn the wheel to the left. If removing the left axle, turn the wheel to the right.

7. Separate the joint by using a prybar between the suspension support and lower control arm

8. Disengage the axle (half shaft) from the hub and bearing using a suitable puller.

9. Remove the hub and bearing assembly bolts.

10. Remove the hub and bearing assembly from the steering knuckle.

11. Remove the hub and bearing seal.

To install:

12. Install the hub and bearing assembly assembly to the steering knuckle.

13. Install the hub and bearing assembly bolts and tighten bolts to 70 ft. lbs. (95 Nm).

14. Install the new hub and bearing seal into the steering knuckle.

NOTE: Lubricate the inside diameter seal lips and completely fill the cavity between the hub and bearing assembly and seal with chassis grease.

15. Install the drive axle (half shaft) into the hub and bearing assembly.

16. Install the ball joint to the steering knuckle and torque to 50 ft. lbs. (65 Nm) minimum and no more than 63 ft. lbs. (85 Nm) to install the cotter pin. Do not loosen to align cotter pin openings.

17. Install the front rotor.

18. Install the washer then a new hub nut. Inset a drift pin into the brake caliper opening and into the disc brake rotor to keep the rotor from turning. Tighten the hub nut.

Use a torque wrench to tighten to 185 ft. lbs. (260 Nm).

19. Install the brake caliper and tighten the bolts to 38 ft. lbs. (51 Nm).

20. Install the tie and wheel assembly. Safely lower the vehicle.

21. Check front end alignment after work has been completed.

1995–97 Vehicles

1. Raise and safely support the vehicle.

2. Remove the tire and wheel assembly.

3. Remove the drive axle nut.

4. Remove the brake caliper bolts and support the brake caliper to the side.

5. Remove the brake rotor.

6. Remove the 3 bolts that go through the steering knuckle from the back of the knuckle. Rust buildup may require a generous application of penetrating oil where the hub fits into the knuckle. Remove the hub and bearing assembly from the steering knuckle.

7. Remove the hub and bearing assembly from the vehicle.

To install:

8. Install hub and bearing assembly to steering knuckle.

9. Install hub and bearing bolts and tighten to 70 ft. lbs. (95 Nm).

10. Install the brake rotor.

11. Install the brake caliper and bolts.

12. Install the drive axle through hub assembly and tighten nut to 192 ft. lbs. (260 Nm).

13. Install the tire and wheel assembly.

14. Lower the vehicle.

15. Road test the vehicle and verify proper operation.

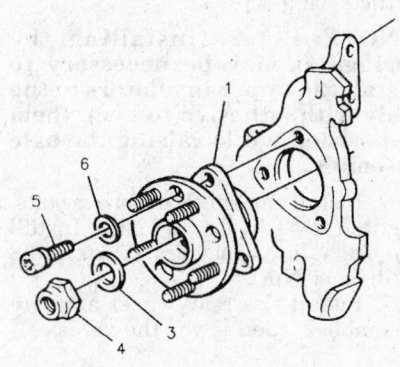

1 HUB AND BEARING ASSEMBLY
2 STEERING KNUCKLE
3 WASHER
4 DRIVE AXLE NUT – 260 N·m (192 LBS. FT.)
5 HUB AND BEARING RETAINING BOLT
6 WASHER

326537

Front hub and bearing assembly — 1995–97 vehicles

REAR SUSPENSION

Shock Absorber

REMOVAL AND INSTALLATION

1993–94 Vehicles

NOTE: Do not remove both shock absorbers at one time as suspending rear axle at full length could result in damage to brake lines and hoses.

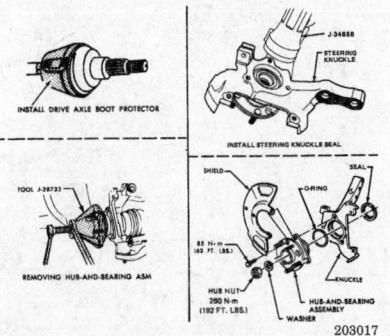

Remove and install hub and bearing assembly — 1993–94 vehicles

1. Open the deck lid.

2. Remove the shock absorber cover and remove the upper shock absorber attaching nut. Only remove one shock at a time if both shock absorbers are to be replaced.

3. Raise and safely support the vehicle.

4. Support the rear with properly positioned safety stands.

5. Remove the shock absorber mounting bolt and nut.

6. Remove the shock absorber.

To install:

7. Install the shock absorber at the lower attachment. Install bolt and nut and hand-tighten only.

8. Lower the vehicle enough to guide upper stud through the body opening and install the upper shock nut loosely.

9. Tighten the lower shock bolt and nut to 35 ft. lbs. (47 Nm).

10. Remove the axle support (safety stands, if used) and lower the vehicle.

11. Tighten the upper shock nut to 21 ft. lbs. (29 Nm).

12. Install the shock absorber upper cover.

1995–97 Vehicles

NOTE: Do not remove both shock absorbers at one time. Suspending the rear axle at full length could result in damage to brake lines and/or hoses.

1. Open the deck lid.

2. Remove the coil-over shock absorber attaching nut.

3. Raise and safely support the vehicle. Support the rear axle with safety stands.

4. Remove the bolts from the coil-over shock upper mount.

5. Remove coil-over shock mounting bolt. Remove the coil-over shock.

To install:

6. Install coil-over shock absorber at the lower attachment. Install bolt hand-tight.

7. Install bolts to the coil-over shock mount.

8. Lower the vehicle.

9. Install the coil-over shock upper mount attaching nut.

10. Tighten the lower mounting bolt to 125 ft. lbs. (170 Nm).

11. Tighten the coil-over shock upper mount bolts to 21 ft. lbs. (28 Nm).

12. Remove the safety stands and lower the vehicle.

13. Tighten the upper coil-over shock mounting nut to 15 ft. lbs. (20 Nm).

14. Road test the vehicle.

Coil Spring

REMOVAL AND INSTALLATION

——————— CAUTION ———————
The coil springs are under a considerable amount of tension. Be very careful when removing or installing them; they can exert enough force to cause very serious injuries. Make sure vehicle is safely support on the proper type equipment.

1. Raise and safely support the vehicle. Remove the rear wheel and tire assemblies.
2. Using the proper equipment, support the weight of the rear axle. Unfasten the right and left brake line bracket attaching screws, then allow the brake line to hang free.

——————— WARNING ———————
Do not suspend the rear axle by the brake hoses or damage to the hoses may result.

3. Unfasten both shock absorber lower attaching bolts from the axle.
4. Carefully lower the axle, then remove the coil spring(s) and/or insulator(s).
   To install:
5. Position the spring(s) and insulator(s) in the seats, and raise the axle. The ends of the upper coil in the spring must be positioned in the spring seat and within 9/16 in. (15mm) of the spring top.

NOTE: Before installing the springs, it may be necessary to install the upper insulators to the body with adhesive to keep them in position while raising the axle assembly.

6. Install the shock absorber bolts. Tighten to 41 ft. lbs. (55 Nm). Install the brake line brackets. Tighten to 8 ft. lbs. (11 Nm).
7. Install the rear wheel and tire assemblies, then lower the vehicle.

Rear Wheel Bearings

ADJUSTMENT

The vehicles are equipped with sealed rear hub and bearing assemblies. The assemblies are non-serviceable; if damaged, replace the complete unit.

REMOVAL AND INSTALLATION

1995–97 Vehicles

A single-unit hub and bearing assembly is bolted to both ends of the rear axle assembly or rear knuckle assembly. This hub and bearing is a sealed unit which is supposed to eliminate the need for wheel bearing adjustments and does not require periodic maintenance.

1. Raise and safely support vehicle.
2. Remove the wheel and tire assembly.
3. Remove the brake drum.
4. Remove the hub and bearing assembly mounting bolts. The top rear mounting bolt will not clear the brake shoe when removing the hub and bearing assembly. Partially remove the hub and bearing assembly prior to removing this bolt.
5. Disconnect the ABS wheel speed sensor wire connector.
6. Remove the hub and bearing assembly from the axle.
   To install:
7. Attach the rear ABS wheel speed sensor wire connector.
8. Install the hub and bearing assembly. Position the top rear mounting bolt in the hub and bearing assembly prior to installation to the axle housing.
9. Tighten the mounting bolts to 44 ft. lbs. (60 Nm).
10. Install the brake drum.
11. Install the wheel and tire assembly.
12. Lower the vehicle.
13. Road test the vehicle.

GM "L" BODY

Front Wheel Drive
CHEVROLET-Beretta • Corsica

FIRING ORDERS

NOTE: To avoid confusion, always replace spark plug wires one at a time.

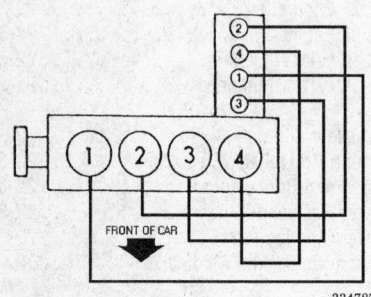

334787

2.2L (VIN 4) Engine
Engine Firing Order: 1–3–4–2
Distributorless Ignition System

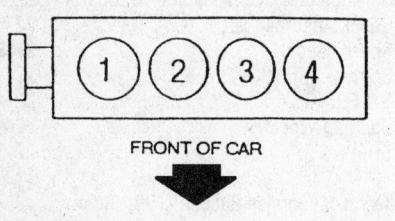

195466

2.3L (VIN A) Engine
Engine Firing Order: 1–3–4–2
Distributorless Ignition System

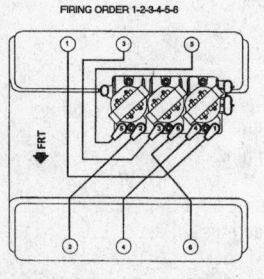

335008

3.1L (VIN T) and (VIN M) Engines
Engine Firing Order: 1–2–3–4–5–6
Distributorless Ignition System

ENGINE ELECTRICAL

NOTE: Disconnecting the negative battery cable on some vehicles may interfere with the functions of the on board computer systems and may require the computer to undergo a relearning process, once the negative battery cable is reconnected.

Ignition Timing

ADJUSTMENT

The Ignition Timing is not adjustable, and is set according to engine demand electronically. The Powertrain Control Module (PCM) controls the ignition timing for all driving conditions.

Alternator

PRECAUTIONS

Several precautions must be observed with alternator equipped vehicles to avoid damage to the unit.

• If the battery is removed for any reason, make sure it is reconnected with the correct polarity. Reversing the battery connections may result in damage to the 1-way rectifiers.

• When utilizing a booster battery as a starting aid, always connect the positive to positive terminals and the negative terminal from the booster battery to a good engine ground on the vehicle being started.

• Never use a fast charger as a booster to start vehicles.

• Disconnect the battery cables when charging the battery with a fast charger.

• Never attempt to polarize the alternator.

• Do not use test lights of more than 12 volts when checking diode continuity.

• Do not short across or ground any of the alternator terminals.

• The polarity of the battery, alternator and regulator must be matched and considered before making any electrical connections within the system.

• Never separate the alternator on an open circuit. Make sure all connections within the circuit are clean and tight.

• Disconnect the battery ground terminal when performing any service on electrical components.

• Disconnect the battery if arc welding is to be done on the vehicle.

REMOVAL AND INSTALLATION

2.2L (VIN 4) Engine

1. Disconnect the negative battery cable. Remove the serpentine belt.

2. Remove the 3 alternator mounting bolts and the 2 bracket nuts at the exhaust manifold.

3. Detach the connector from the alternator case. Remove the protective cover from the rear of the alternator and remove the battery wire-to-alternator post nut.

4. Remove the alternator.

To install:

5. Position the alternator in the mounting bracket and loosely install the mounting nuts and bolts.

6. Connect the battery wire to the alternator post and tighten the securing nut to 71 inch lbs. (8 Nm). Install the cover over the battery wire and post.

7. Attach the alternator connector.

8. Tighten the fasteners as follows:

• Upper front and rear mounting bolts: 22 ft. lbs. (30 Nm)

• Lower alternator mounting bolt (long bolt): 37 ft. lbs. (50 Nm)

• Bracket nuts at exhaust manifold: 32 ft. lbs. (43 Nm).

9. Install the serpentine belt.

10. Connect the negative battery cable.

2.3L (VIN A) Engine

1. Disconnect the negative battery cable.

2. Loosen the belt tensioner pulley bolt and rotate the pulley counterclockwise to remove the belt from the alternator pulley.

3. Remove the coolant/washer reservoir mounting screws. Detach the connector from the washer pump and position the reservoir aside.

4. Disconnect the A/C line rail clip.

5. Remove the 2 rear and one front alternator mounting bolts.

6. Remove the alternator from the mounting bracket and position so the electrical connections can be accessed.

7. Unplug the wiring harness connector from the alternator and remove the nut securing the battery wire to the rear of the alternator.

8. Remove the alternator from the vehicle.

To install:

9. Install the alternator into the engine compartment and connect the electrical connectors. Tighten the mounting nut on the battery wire to 65 inch lbs. (8 Nm).

10. Install the remaining components in the reverse order of removal. Tighten the rear mounting bolts to 19 ft. lbs. (26 Nm) and the front bolt to 37 ft. lbs. (50 Nm).

11. Connect the negative battery cable.

3.1L (VIN T) and (VIN M) Engine

1. Disconnect the negative battery cable. Detach the alternator connector.

2. Remove the serpentine belt.

3. Remove the front alternator bolt, rear rear bolt and alternator rear brace bolt.

4. Remove the alternator from the mounting bracket and tilt forward.

5. Remove the cap from the alternator post and remove the nut securing the battery wire to the post and disconnect the wire.

6. Remove the alternator from the vehicle.

7. Remove the rear alternator bracket bolt and bracket from the alternator.

To install:

8. Install the alternator rear bracket on the alternator and loosely install the mounting bolt.

9. Install the alternator assembly in the vehicle.

10. Connect the battery wire to the alternator and tighten the mounting nut to 71 inch lbs. (8 Nm). Install the protective cap over the post.

11. Install all of the mounting bolts and tighten as follows:
- Long bolt: 37 ft. lbs. (50 Nm)
- Short bolts: 18 ft. lbs. (25 Nm)
- Brace bolt: 37 ft. lbs. (50 Nm)

12. Install the serpentine belt.

13. Attach the alternator connector.

14. Connect the negative battery cable.

Drive Belt

REMOVAL AND INSTALLATION

2.2L (VIN 4) Engine

1. Disconnect negative battery cable.

2. Remove the coolant reservoir.

3. Using a 15mm wrench, pivot tensioner and remove belt from alternator.

4. Remove the serpentine belt.

To install:

5. Route the serpentine belt as follows:

  a. Under the crankshaft and A/C compressor, if equipped.

  b. Over the tensioner with the back of the belt driving the pulley.

  c. Around the water pump with the back of the belt driving the pulley.

  d. Around the power steering pump.

6. Rotate the tensioner and loop the belt around the alternator.

7. Release the tensioner.

NOTE: Do not exceed torque of 30 ft. lbs. (40 Nm) on the tensioner center bolt when installing or removing the belt. If these conditions are not followed, parts or system damage could result.

8. Verify the belt is properly seated in all the pulleys.

9. Install the coolant reservoir.

10. Connect the negative battery cable. Start the engine and check the belt alignment.

2.3L (VIN A) Engine

1. Disconnect negative battery cable.

2. Rotate the serpentine belt tensioner counterclockwise using a 13mm wrench on the tensioner center bolt. The wrench used should be at least 24 in. long.

3. With tension removed from the belt, slip the belt off the alternator pulley.

4. Slowly release the belt tensioner and remove the wrench.

5. Remove the belt from the remainder of the pulleys.

To install:

6. Install the belt under the A/C compressor and crankshaft pulley.

7. Pull the rear of the belt up and under the tensioner so the back of the belt rides on the pulley.

8. Install the wrench on the tensioner bolt and rotate the tensioner counterclockwise.

9. With the tensioner loaded, loop the belt over the alternator pulley.

10. Release the tensioner slowly and remove the wrench.

11. Make sure the belt is properly seated in all the drive belt pulleys.

12. Connect negative battery cable.

13. Start the engine and check the drive belt alignment.

3.1L (VIN T) Engine

1. Disconnect the negative battery cable. Support the oil pan with a floor jack and a block of wood.

2. Remove the right side engine mount.

3. Remove the 3 engine mount bracket bolts and remove the bracket.

4. Rotate the belt tensioner counterclockwise using a ½ inch breaker bar.

5. Remove the serpentine belt from the alternator pulley.

6. Slowly release the belt tensioner.

7. Remove the serpentine belt from the remainder of the drive belt pulleys.

To install:

8. Install the serpentine belt as follows:

  a. Loop the serpentine belt under the crankshaft and A/C compressor.

  b. Bring the belt up the front side of the engine around the water pump pulley so the back of the belt drives the pulley.

  c. Loop the belt over the power steering pump pulley.

  d. Bring the belt up the back of the engine around the tensioner pulley so the back of the belt drives the pulley.

9. Rotate the belt tensioner counterclockwise using a ½ inch breaker bar.

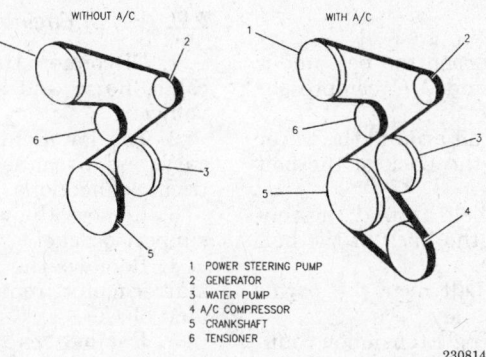

1 POWER STEERING PUMP
2 GENERATOR
3 WATER PUMP
4 A/C COMPRESSOR
5 CRANKSHAFT
6 TENSIONER

230814

Drive belt routing — 2.2L (VIN 4) engine

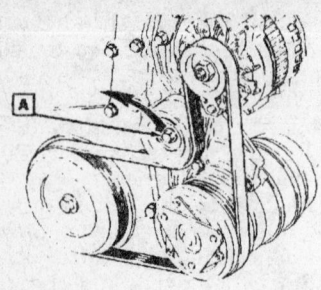

A ROTATE TENSIONER IN DIRECTION OF ARROW TO REMOVE OR INSTALL BELT.

329318

Serpentine belt routing — 2.3L (VIN A) engine

10. With the tensioner compressed loop the belt around the alternator pulley.

11. Release the belt tensioner and verify the belt is properly seated in all the drive pulleys.

12. Install the engine mount bracket and tighten the mounting bolts to 81 ft. lbs. (110 Nm).

13. Install the right engine mount.

14. Remove the floor jack.

15. Connect the negative battery cable.

16. Start the engine and check the belt alignment.

3.1L (VIN M) Engine

1. Disconnect the negative battery cable.

2. Support the engine with a floor jack. Try to place the jack near the front of the oil pan, but DO NOT support the engine by the oil pan.

3. Remove the right engine mount intermediate bracket.

4. Remove the auxiliary bracket.

5. Using a ⅜ inch breaker bar, rotate the belt tensioner counterclockwise.

6. With the tension removed from the belt, remove the serpentine belt from around the alternator pulley.

7. Slowly release the tensioner.

8. Remove the belt from the remaining pulleys.

To install:

9. Loop the serpentine belt under the crankshaft and A/C compressor pulleys.

10. Route the belt around the water pump pulley so the back of the belt drives the pulley.

11. Route the belt around the tensioner pulley so the back of the belt drives the pulley.

12. Route the belt over the power steering pump pulley.

13. Rotate the belt tensioner counterclockwise using a ⅜ inch breaker bar.

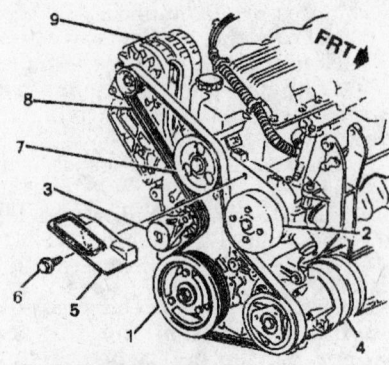

1 BALANCER, CRANKSHAFT
2 PUMP ASSEMBLY, WATER
3 TENSIONER, SERPENTINE DRIVE BELT
4 COMPRESSOR ASSEMBLY, AIR CONDITIONING
5 SHIELD, SERPENTINE DRIVE BELT
6 BOLT/SCREW, SERPENTINE DRIVE BELT SHIELD
7 PUMP ASSEMBLY, POWER STEERING
8 BELT, SERPENTINE DRIVE
9 GENERATOR ASSEMBLY

329281

Serpentine belt routing and shield — 3.1L (VIN M) engine; (VIN T) similar

14. Loop the serpentine belt around the alternator pulley.

15. Release the tensioner and verify the belt is properly seated in the drive pulleys.

16. Install the right engine mount intermediate bracket.

17. Install the auxiliary bracket.

18. Remove the floor jack.

19. Connect the negative battery cable.

20. Start the engine and check the belt for proper alignment.

Starter

REMOVAL AND INSTALLATION

2.2L (VIN 4) Engine

1. Disconnect the negative battery cable. Raise and safely support the vehicle.

2. Tag for identification, if necessary, and disconnect the starter electrical connections.

3. Remove the wiring clamp at the support bracket.

4. Remove the bolt securing the starter motor front bracket to the engine block.

5. Remove the 2 starter motor mounting bolts and remove the starter.

6. If the starter motor is being replaced with another unit, remove the 2 bracket mounting nuts and remove the bracket.

To install:

7. If removed, install the front bracket on the starter motor and tighten the nuts to 80 inch lbs. (9 Nm).

8. Install the starter motor in the vehicle and tighten the 2 starter mounting bolts to 32 ft. lbs. (43 Nm).

9. Install the starter motor bracket-to-engine bolt and tighten to 26 ft. lbs. (32 Nm).

10. Connect the wiring clamp at the support bracket.

11. Attach the connectors to the starter. Tighten the solenoid wire nut to 27 inch lbs. (3 Nm) and the battery terminal wire nut to 12 ft. lbs. (16 Nm).

12. Lower the vehicle. Connect the negative battery cable.

13. Verify vehicle starts properly without excessive noise.

2.3L (VIN A) Engine

1. Disconnect the negative battery cable.

2. Remove the air induction tube.

3. Remove the oil filter.

4. Remove the starter bolts.

5. Position the starter so the electrical wiring on the solenoid is accessible. Detach the wiring.

6. Remove the starter from the vehicle by lifting it between intake manifold and radiator.

To install:

7. Install the starter in the vehicle. Attach the electrical connections.

8. Install the starter on the engine block and tighten the mounting bolts to 74 ft. lbs. (100 Nm).

9. Coat the seal on the oil filter with clean engine oil and Install the oil filter.

10. Install the air induction tube.

11. Connect negative battery cable.

12. Top off the engine oil as necessary.

3.1L (VIN T) and (VIN M) Engine

1. Disconnect the negative battery cable. Raise and safely support the vehicle.

2. On the (VIN T), remove the flywheel cover mounting bolts and remove the flywheel cover.

3. Remove the starter motor mounting bolts. Lower the starter motor.

4. Disconnect the electrical connections from the starter motor while supporting it.

5. Remove the starter motor from the vehicle.

To install:

6. Attach the starter wiring. Tighten the battery cable nut to 71 inch lbs. (8 Nm) on the (VIN T), or to 12 ft. lbs. (16 Nm) except the (VIN T), and the solenoid wire nut to 27 inch lbs. (3 Nm).

7. Position the starter motor in the vehicle and tighten the starter mounting bolts to 32 ft. lbs. (43 Nm).

8. On the (VIN T), install the flywheel cover and mounting bolts.

9. Lower the vehicle. Connect the negative battery cable.

10. Verify that the vehicle starts properly without excessive noise.

CHASSIS ELECTRICAL

Blower Motor

REMOVAL AND INSTALLATION

1. Disconnect the negative battery cable.

2. On the 3.1L (VIN T) and (VIN M) engine, remove the serpentine belt and the alternator.

3. Disconnect the electrical connections going to the blower motor.

4. Disconnect the blower motor cooling tube.

5. Remove the retaining screws and remove the blower motor.

To install:

6. Install the blower motor into place and install the retaining screws. Install the cooling tube.

7. Attach the blower motor connections.

8. On the 3.1L (VIN T) and (VIN M) engine, install the alternator and the serpentine belt.

9. Connect the negative battery cable. Switch the blower motor ON to test operation.

Windshield Wiper Motor

REMOVAL AND INSTALLATION

1. Disconnect the negative battery cable.

2. Remove the left and right side wiper arms. Remove the air inlet screen/panel.

3. Disconnect the wiper motor drive link from the crank arm using tool J-39232 or equivalent.

4. Disconnect the electrical connectors and washer hoses.

5. Remove the wiper motor-to-chassis bolts and the wiper motor by guiding the crank arm through the hole. Remove the crank arm from the motor.

To install:

6. Install the crank arm on the new wiper motor shaft and install the attaching nut and tighten to 15 ft. lbs. (21 Nm).

7. Install the wiper motor while guiding the crank arm through cowl opening.

8. Install the wiper motor to the chassis and install the attaching bolts and tighten to 18 ft. lbs. (25 Nm).

9. Connect the electrical connectors to the wiper motor and connect the washer hoses.

10. Connect the wiper arm drive link to the crank arm using tool J-39529 or equivalent.

11. Install the air screen/panel in place onto the cowl area.

12. Install the left and right wiper arms. Connect the negative battery cable.

Headlight Switch

REMOVAL AND INSTALLATION

———— **CAUTION** ————
The Supplemental Inflatable Restraint (SIR) system must be disarmed before working around the steering column. Failure to do so may cause accidental deployment of the air bag, resulting in unnecessary SIR system repairs and/or personal injury.

1. Disconnect the negative battery cable.

2. Remove the instrument cluster bezel. Use the following procedure:

   a. Remove the screws securing the bezel to the instrument cluster.

   b. Pull the bezel to the rear to disengage the clips.

3. Squeeze the small knob at the side and pull straight out.

4. Insert a small flat blade into the slots adjacent to the center of the inner knob to disengage the knob from the switch.

5. Remove the screws attaching the switch to the bezel.

6. Remove the switch.

To install:

7. Install the switch to the bezel. Install the attaching screws.

8. Position the inner knob on the switch. Ensure the tabs are lined up with the slots and press to secure the knob.

9. Position the outer knob on the switch and align the D-shaped hole in the knob to the shaft on the switch and press to secure the knob.

10. Install the instrument cluster bezel.

11. Connect the negative battery cable.

Turn Signal Switch

REMOVAL AND INSTALLATION

1. Disable the SIR system.

2. Disconnect the negative battery cable.

3. Place the ignition in the **LOCK** position.

———— **CAUTION** ————
When carrying a live air bag, make sure the bag and trim cover are pointed away from the body. In the unlikely event of an accidental deployment, the bag will them deploy with minimal chance of injury. When placing a live air bag on a bench or other surface, always face the bag and trim cover up, away from the surface. This will reduce the motion of the module if it is accidently deployed.

4. Remove the steering wheel and air bag assembly.

5. Remove the coil assembly retaining ring.

6. Lift the coil assembly from the end of the steering shaft and allow coil to hang freely.

7. Remove the wave washer.

8. If equipped with a standard column, remove the spacer shaft lock.

9. Remove the shaft lock retaining ring using tool J-23653-C or equivalent, to compress the shaft lock.

10. Pry off the retaining ring.

11. Remove the shaft lock.

12. Remove the turn signal cancelling cam assembly.

13. Remove the upper bearing spring.

14. Position the turn signal lever to the right turn position.

15. Remove the multi-function lever by performing the following:

   a. Ensure the lever is in the center or **OFF** position.

   b. If equipped with cruise control, disconnect the cruise control connector from the steering column assembly.

   c. Pull the lever straight out of the turn signal switch.

16. Remove the hazard knob assembly.

17. Remove the screw and signal switch arm. If equipped with tilt column and cruise control, allow the switch arm to hang freely.

18. Remove the turn signal switch screws. Allow the switch to hang freely.

19. Disconnect the turn signal/hazard switch assembly terminal from the instrument panel harness.

20. If equipped with tilt column, disconnect the buzzer switch assembly terminals from the turn signal/hazard assembly connector. Remove the tan/black wire lead from cavity **E** and the light green wire from the cavity **F**.

21. Remove the upper steering column bolts.

22. Remove the wiring protector.

23. Connect a length of wire to the turn signal/hazard assembly terminal connector to aid in reassembly.

24. Gently pull the wire harness through the steering column housing shroud, steering column housing and lock assembly cover.

25. Disconnect the wire from the connector.

To install:

26. Connect the wire to the turn signal/hazard switch assembly connector.

27. Gently pull the connector through the steering column housing shroud, steering column housing and lock assembly cover.

28. Remove the wire.

29. Install the wiring protector.

30. If disconnected, connect the buzzer switch terminals to the turn signal/hazard switch assembly connector. Insert the tan/black wire lead into cavity **E** and the light green wire into cavity **F**.

31. Connect the turn signal/hazard switch assembly connector to the instrument panel harness.

32. Install the steering column support bracket bolts to the steering column. Tighten to 22 ft. lbs. (30 Nm).

33. Install the steering column upper support bolts. Tighten to 20 ft. lbs. (28 Nm).

34. Install the turn signal switch assembly and attaching screws. Tighten to 20 inch lbs. (2 Nm).

35. Install the hazard knob assembly.

36. Install the multi-function lever by performing the following:

a. Align the tab on the turn signal switch with the notch in the pivot of the turn signal switch.

b. Push the lever into the turn signal switch.

c. If equipped with cruise control, connect the connector to the steering column assembly.

37. Install the turn signal cancelling cam assembly. Lubricate with a synthetic grease.

38. Install the shaft lock.

39. Install the shaft lock retaining ring, lining up to block tooth on the shaft. Use tool J-23653-C to compress the shaft lock.

40. If equipped with a standard column, install the spacer shaft lock.

41. Install the wave washer.

42. Ensure the coil assembly is centered.

NOTE: The coil assembly will become uncentered if the steering column is separated from the steering gear and is allowed to rotate or the centering spring is pushed down, letting the hub rotate while the coil is removed from the steering column.

43. Install the coil assembly using the horn tower on the cancelling cam assembly inner ring and projections on the outer ring for alignment.

44. Install the coil assembly retaining ring. The ring must be firmly seated in the groove on the shaft. Gently pull the lower coil assembly wire to remove any wire kinks that may be inside the column.

45. Install the steering wheel and air bag, if equipped with SIR system.

46. Enable the SIR system.

47. Connect the negative battery cable.

Ignition Lock Cylinder

REMOVAL AND INSTALLATION

1. Disconnect the negative battery cable. Disable the SIR system.

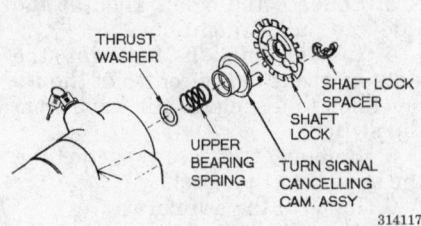

Removing components from the upper shaft

314117

2. Remove the steering wheel and air bag assembly.

3. Remove the left side sound insulators from under the dash.

4. Remove the SIR coil assembly snapring.

5. Gently pull up on the SIR coil assembly while pushing up on the SIR coil harness on the bottom side of the steering column and let the coil hang freely.

6. Remove the wave washer.

7. Remove the spacer shaft lock (standard column only).

8. Remove the shaft lock retaining ring using tool J-23653–C or equivalent, to compress the shaft lock plate.

9. Pry of the retaining ring and remove the shaft lock plate.

10. Remove the turn signal canceling cam and remove the upper bearing spring.

11. Remove the turn signal lever using the following procedure:

a. Make sure the turn signal lever is in the center, or off position.

b. If equipped with cruise control, disconnect the cruise control connector from the steering column assembly.

c. Pull the lever straight out of the turn signal switch.

12. Remove the hazard knob assembly.

13. If necessary, remove the tilt lever being careful not to damage the shaft. Grip the lever firmly and twist counterclockwise.

14. If necessary, remove the washer head screw and the switch actuator arm.

15. Remove the screws retaining the turn signal switch to the steering column and gently pull up on the wire harness, it may be necessary to disconnect the harness from the lower steering column.

16. Pull the switch out of the column and let it hang freely.

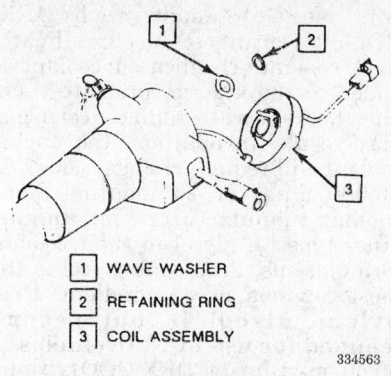

1 WAVE WASHER
2 RETAINING RING
3 COIL ASSEMBLY

334563

SIR coil assembly

17. Remove the key from the lock cylinder and gently pry out the buzzer switch.

18. Reinsert the key into the lock cylinder and turn to the lock position.

19. Remove the lock cylinder retaining screw and remove the lock cylinder.

To install:

20. Install the steering column lock cylinder.

21. Install the lock retaining screw and tighten to 40 inch lbs.

22. Remove the key from the lock cylinder and install the buzzer switch.

23. Gently grasp the turn signal switch wire harness and pull it through the column until the switch is seated, position the screws and tighten to 30 inch lbs. (4.5 Nm).

24. If necessary install the switch actuator arm tighten the washer screw 20 inch lbs. (2.3 Nm).

25. If necessary to install tilt lever grip it firmly and twist clockwise.

26. Install the hazard knob assembly.

27. Align the tab on the turn signal lever with the notch in the pivot of the turn signal switch and push the lever into the switch.

28. If equipped with cruise control, connect the lever cruise connector to the steering column assembly.

29. install the upper bearing spring and the turn signal canceling cam

30. Install the lock plate, using tool J-23653 or equivalent compress the upper bearing spring and install the retaining ring.

31. Install the spacer shaft lock (standard column only).

—————— **CAUTION** ——————

SIR coil assembly wires must be kept tight with no slack while installing the SIR coil assembly. Failure to do so may cause the

wires to be kinked near the lock plate, causing the wires to be cut when the steering wheel is turned.

32. Install the wave washer and SIR coil assembly, pulling gently on the harness on the bottom side of the steering column to remove slack from the wiring harness. Be sure to align the SIR coil assembly properly.

33. Install the snapring on top of the SIR coil.

34. If removed connect the turn signal wire harness to the steering column.

35. Install the left side sound insulators under the dash panel.

36. Properly enable the SIR system.

37. Connect the negative battery cable.

38. Check the operation of the ignition switch, turn signals, key buzzer, headlight dimmer and 4 way flashers.

Ignition Switch

REMOVAL AND INSTALLATION

1. Disconnect the negative battery cable. Remove the steering column from the vehicle.

2. Locate the ignition switch at the base of the steering column. Note that most models will also have a dimmer switch in a similar location. The ignition switch is on top of the column jacket.

3. Remove the electrical connector, remove the retainer screws and disengage the actuator rod from the ignition switch.

To install:

4. Move the ignition switch slider to the extreme right position and then move slider one detent to the left (off lock). Install the replacement ignition switch using care to make

MOVE SWITCH SLIDED TO EXTREME LEFT POSITION AND THEN MOVE SLIDER ONE DETENT TO THE RIGHT (OFF LOCK)

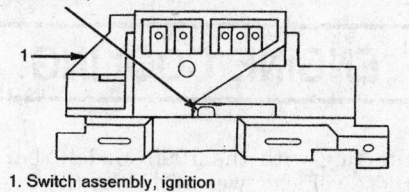

1. Switch assembly, ignition

324441

Installation position for ignition switch

sure the actuator rod engages the switch.

NOTE: Insert a 3/32 inch drill bit into the adjustment hole on the ignition switch to hold the switch slider in the proper position during installation.

5. Install and tighten the retaining screws to 35 inch lbs. (4 Nm). Use the original screws or exact replacements. Use of screws that are too long could prevent the column from collapsing on impact.

6. Connect the electrical connector to the ignition switch.

7. Raise the steering column and install the retaining bolts torquing them to 18 ft. lbs. (24 Nm).

8. Install the steering column opening filler panel under the column.

9. Connect the negative battery cable. Check operation of ignition switch.

Park/Neutral Safety Switch

REMOVAL AND INSTALLATION

1. Disconnect the negative battery cable.

2. Apply the parking brake and place the shift selector into the **N** detent.

3. Disconnect the shift linkage from the transaxle and remove the shift lever from the manual shift shaft.

4. Disconnect the park neutral switch electrical connector.

5. Remove the switch mounting bolts and remove the switch from the transaxle.

To install:

6. Make sure the transaxle is in the **N** position.

NOTE: For a new switch, if bolt holes do not align with mounting boss on transaxle, verify that the shift shaft is in the N position, do not rotate the switch. The switch is pinned in the N position. If the pin has been broken, the switch can be adjusted using the old switch installation procedure.

7. Align the flats on the manual shift shaft with the flats on the switch and install the switch over the shaft.

8. Install mounting bolts loosely.

9. If installing the original switch, insert the gage pin into the service adjustment hole and rotate the switch until the pin drops to a depth of 9/64 inch.

10. Tighten the bolts to 18 ft. lbs. (24 Nm). Attach the switch electrical connector.

11. Install the manual shift lever and connect the linkage to the lever.

12. Connect the negative battery cable. Verify that the vehicle starts in the **P** and **N** detents only.

13. Release the parking brake.

Powertrain Control Module

REMOVAL AND INSTALLATION

The Powertrain Control Module (PCM), formerly known as the Electrical Control Module (ECM) is located under the instrument panel. It is the control center of the fuel injection system and constantly monitors information supplied by the various sensors. Using this information, the PCM controls the systems that affect vehicle performance.

────── WARNING ──────
To prevent internal PCM damage, disconnect the negative battery cable before disconnecting or reconnecting power to the PCM. Do not touch the connector pins or soldered components on the circuit board as this can create electrostatic discharge damage to the PCM.

NOTE: **When replacing the production PCM with a service PCM it is important to transfer the broadcast code and production PCM number to the service PCM label.**

1. Disconnect the negative battery cable.

2. For 3.1L (VIN T) models, remove the glove box tray.

3. Remove the interior right side access panel/kick panel.

4. Disconnect the PCM harness connectors.

5. Remove the PCM from its mounting bracket and remove from vehicle.

6. Carefully remove the Programmable Read-Only Memory PROM chip (sometimes also called the Calibrator) from the PCM.

7. The PROM must be transferred to the replacement PCM. The replacement PCM is supplied without a PROM program, so the replacement PCM must be programmed by installing the original PROM chip before the vehicle will run.

To install:

────── WARNING ──────
Do not press on the ends of the PROM until the clips snap into place, the PCM circuit board and/or clips may be damaged.

8. Install the PROM by aligning the small notches in the PROM with the small notches in the socket, press inward on the clips until they snap into place. Listen for the click.

9. Install the access cover on the PCM. Attach the PCM connectors.

10. Position the PCM into its mounting location and install mounting bracket.

11. Install the interior access panel.

12. For 3.1L (VIN T) engines, install the glove box tray.

13. Connect the negative battery cable.

14. Program the PCM with a suitable scan tool.

15. Check for stored PCM codes to verify proper PROM installation.

EEPROM Programming

1. Set-up: Ensure that the following conditions have been met:
   a. The battery is fully charged.
   b. The ignition switch is in the **ON** position.
   c. The vehicle interface module cable connection at the DLC is secure.

2. Program the PCM using the latest software matching the vehicle. Refer to up-to-date dealership instructions.

3. If the PCM fails to program, proceed as follows:
   a. Ensure that all PCM connections are OK.
   b. Check the scan tool is equipped with the latest software version.
   c. Attempt again to program the PCM. If the PCM still cannot be programmed properly, replace the PCM. The replacement must be programmed or the vehicle will not run.

ENGINE COOLING

Starting with the 1996 Model Year, these vehicles were filled at the factory with a new type of antifreeze/coolant called GM Goodwrench DEX-COOL™. When adding coolant to 1996 vehicles, it is important that you use GM Goodwrench DEX-COOL™ (orange-colored, silicate-free) coolant. If silicated coolant is added to the system, premature engine, heater core or radiator corrosion may result. In addition, the engine coolant will require change sooner, at 30,000 miles or 24 months. Some coolant manufacturers are mixing other types of glycol in their coolant formulations. Propylene glycol is the most common new ingredient. **Propylene glycol is not recommended for use in GM vehicles.** A 50/50 mixture of DEX-COOL™ and clean water will provide all the recommended protection for 1996 vehicles. **DO NOT use DEX-COOL™ in pre-1996 vehicles. DO NOT mix DEX-COOL™ with any other type of antifreeze.**

Radiator

REMOVAL AND INSTALLATION

1. Disconnect the negative battery cable.

2. If equipped with a 3.1L (VIN T) engine, remove the battery.

3. Drain the cooling system.

4. Remove the air intake duct assembly.

5. Disconnect and cap the upper transaxle cooler line.

6. Disconnect the upper radiator hose and position aside.

7. Disconnect and cap the lower transaxle cooler line.

8. Detach the cooling fan connector. Remove the cooling fan assembly.

9. Remove the splash guard below the lower radiator hose.

10. Disconnect the lower radiator hose from the radiator. Remove the radiator upper mounting bolts.

11. Remove the condenser line mounting bolt, if A/C equipped.

12. Remove the radiator-to-condenser mounting bolts, if A/C equipped.

13. Disconnect the coolant surge tank hose.

14. Remove the mounting bolts, then remove the radiator from the vehicle.

To install:

15. Install the radiator in the vehicle, inserting the bottom of the radiator in the lower mounting pads. Install the radiator mounting bolts and tighten to 89 inch lbs. (10 Nm).

16. Connect the coolant surge tank hose to the radiator.

17. Install the radiator-to-condenser bolts and tighten to 89 inch lbs. (10 Nm), if A/C equipped.

18. Install the remaining components in the reverse order of removal. Tighten the transmission oil cooler line fittings to 20 ft. lbs. (27 Nm).
19. Refill the cooling system. Connect the negative battery cable.
20. Pressure test the cooling system and verify no leaks.
21. Start the vehicle and check the transmission fluid, top off as necessary.

Water Pump

REMOVAL AND INSTALLATION

2.2L (VIN 4) and 3.1L (VIN T) and (VIN M) Engines

1. Disconnect the negative battery cable. Drain the cooling system.
2. Loosen, but do not remove, the water pump pulley bolts.
3. Remove the serpentine belt.
4. Remove the alternator mounting bolts and set the alternator aside.
5. Remove the water pump pulley bolts and remove the water pump pulley.
6. Remove the 4 water pump mounting bolts and remove the water pump.
To install:
7. Clean all the gasket surfaces completely.
8. Apply a thin bead of sealer around the outer edge of the water pump gasket seating area and place he gasket on the pump.
9. Install the water pump on the engine and tighten the mounting bolts to 18 ft. lbs. (25 Nm) for the 2.2L (VIN 4) engine or to 89 inch lbs. (10 Nm) on the 3.1L (VIN T) and (VIN M) engine.
10. Install the remaining components in the reverse order of removal. Tighten the water pump pulley mounting bolts to 22 ft. lbs. (30 Nm) for the 2.2L (VIN 4) engine or to 18 ft. lbs. (25 Nm) on the 3.1L (VIN T) and (VIN M) engine.
11. Connect the negative battery cable. Refill the cooling system with the proper coolant. Bleed the cooling system and check for leaks with the engine running at idle.

2.3L (VIN A) Engine

1. Disconnect the negative battery cable. Drain the cooling system.
2. Disconnect the oxygen sensor (O_2) electrical connector.
3. Remove the upper and lower exhaust manifold heat shields.

4. Remove the exhaust manifold brace-to-manifold bolt.
5. Loosen the exhaust pipe-to-manifold spring loaded nuts.
6. Remove the radiator outlet pipe-to-water pump cover bolts.
7. Raise and safely support the vehicle.
8. Disconnect the exhaust pipe from the manifold.
9. Disconnect the radiator outlet pipe from the oil pan and transaxle.
10. If equipped with a manual transaxle, remove the exhaust manifold brace. If equipped with an automatic transaxle, leave the lower radiator hose attached and pull down on the radiator pipe to disconnect it from the water pump.
11. Lower the vehicle.
12. Remove the exhaust manifold.
13. Remove the water pump cover-to-engine bolts.
14. Remove the 3 water pump-to-timing chain housing nuts.
15. Remove the water pump and cover assembly.
16. Remove the 5 water pump cover-to-pump bolts and remove the cover.
To install:
17. Install the water pump cover and tighten the mounting bolts finger-tight.
18. Lubricate the splines of the water pump drive with chassis grease.
19. Install the water pump assembly and tighten the mounting bolts finger-tight.
20. Install the water pump-to-timing chain housing bolts finger-tight.
21. Coat the O-ring on the radiator outlet pipe with coolant and slide the pump into the pump cover and install the bolts finger-tight.
22. With all the bolts in place, tighten in the following sequence:
    a. Water pump-to-timing chain housing: 19 ft. lbs. (26 Nm)
    b. Water pump cover-to-water pump: 106 inch lbs. (12 Nm)
    c. Water pump cover-to-engine block (bottom first): 19 ft. lbs. (26 Nm)
    d. Radiator outlet pipe-to-pump cover: 10 ft. lbs. (14 Nm).
23. Install the remaining components in the reverse order of removal.
24. Refill the cooling system.
25. Connect the negative battery cable. Start the vehicle and verify no leaks.

Thermostat

REMOVAL AND INSTALLATION

1. Disconnect the negative battery cable.
2. Drain the cooling system to a level below the thermostat housing.
3. For 3.1L (VIN T) and (VIN M) engines, disconnect the air inlet from the throttle body.
4. For 3.1L (VIN T) engines, disconnect the crankcase vent tube from the air inlet duct.
5. Disconnect the upper radiator hose from the thermostat housing and position aside.
6. For the 2.3L (VIN A) engine, perform the following procedures:
    a. Disconnect the heater hose from the thermostat housing.
    b. Disconnect the electrical connector from the coolant sensor.
    c. Disconnect the throttle body heater hose from the thermostat housing.
7. Remove the thermostat housing mounting bolts and remove the thermostat housing.
8. Remove the thermostat.
To install:
9. Clean all the gasket surfaces completely.
10. Apply a thin bead of sealer around the gasket seat on the thermostat housing and install the gasket on the housing.
11. Install the thermostat in the recess in the intake manifold and install the thermostat housing and tighten the mounting bolts to 89 inch lbs. (10 Nm) for 2.2L (VIN 4) engine or 18 ft. lbs. (25 Nm) for the 2.3L (VIN A) and 3.1L (VIN T) and (VIN M) engines.
12. Install the remaining components in the reverse order of removal.
13. Refill the cooling system.
14. Connect the negative battery cable. Start the vehicle and verify no leaks.

Electric Cooling Fan

REMOVAL AND INSTALLATION

2.2L (VIN 4) and 2.3L (VIN A) Engines

1. Disconnect the negative battery cable. If equipped with the 2.2L (VIN 4) engine, remove the air intake duct.
2. Detach the connector from the fan motor and frame.
3. Remove the mounting bolt and rubber insulator, then remove the cooling fan assembly.

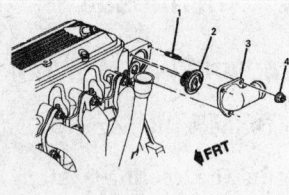

1. Water outlet stud
2. Engine coolant thermostat assembly
3. Water outlet
4. Water outlet nut

323397

Thermostat components — 2.2L (VIN 4) engine

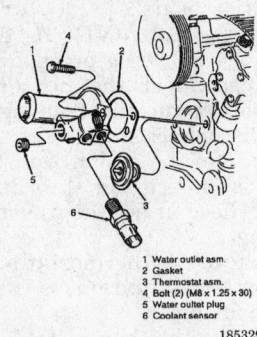

1 Water outlet asm.
2 Gasket
3 Thermostat asm.
4 Bolt (2) (M8 x 1.25 x 30)
5 Water outlet plug
6 Coolant sensor

185329

Thermostat and housing components — 2.3L (VIN A) engine

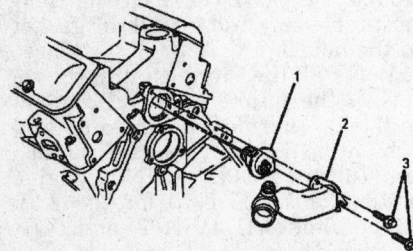

1 THERMOSTAT ASSEMBLY, ENGINE COOLANT
2 OUTLET ASSEMBLY, WATER
3 BOLT/SCREW, WATER OUTLET

226171

Thermostat components — 3.1L (VIN M) engine

To install:

4. Install the cooling fan assembly in the vehicle so the tabs on the bottom of the cooling fan legs are in the mounting notches in the radiator panel.

5. Install the insulator and mounting bolt and tighten to 89 inch lbs. (10 Nm).

6. Attach the wiring harness to the fan frame clip and the connector to the fan motor.

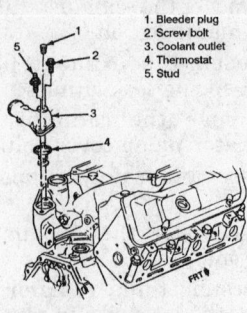

1. Bleeder plug
2. Screw bolt
3. Coolant outlet
4. Thermostat
5. Stud

183427

Exploded view of the thermostat components — 3.1L (VIN T) engine

7. If necessary, install the air intake duct. Connect the negative battery cable.

3.1L (VIN T) Engine

1. Disconnect the negative, then the positive battery cables, then remove the battery.

2. Remove the air cleaner housing and bracket assembly.

3. Disconnect the electrical connector from the cooling fan motor.

4. Disconnect the A/C line clip from the fan frame.

5. Remove the 4 cooling fan motor attaching bolts, then remove the cooling fan assembly.

To install:

6. Install the cooling fan in the vehicle. Install the mounting bolts and tighten to 89 inch lbs. (10 Nm).

7. Connect the A/C line clip to the fan frame. Attach the fan connector.

8. Install the air cleaner bracket and air cleaner housing.

9. Install the battery.

10. Connect the positive, then the negative battery cables.

3.1L (VIN M) Engine

1. Disconnect the negative battery cable. Remove the air inlet duct assembly.

2. Drain the cooling system to a level below the intake manifold.

3. Remove the cooling fan upper mounting bolt.

4. Disconnect the electrical connector from the cooling fan motor.

5. Disconnect the upper radiator hose from the radiator and position aside.

6. Remove the upper radiator mounting bolt.

7. Remove the washer fluid bottle fill tube from the the bottle.

8. Remove the vacuum tank and vacuum tank bracket.

9. Remove the cooling fan from the vehicle.

To install:

10. Install the cooling fan assembly so the tabs on the lower fan legs seat in the cutouts in the radiator support.

11. Install the remaining components in the reverse order of removal. Tighten the cooling fan mounting nut to 97 inch lbs. (97 Nm).

12. Connect the negative battery cable. Refill and bleed the cooling system.

Cooling System Bleeding

PROCEDURE

1. To bleed the system, start with the system cool, the radiator cap off and the radiator filled to about an inch below the filler neck.

2. Start the engine and run it at slightly above normal idle speed. This will insure adequate circulation. If air bubbles appear and the coolant level drops, fill the system with a mixture of anti-freeze and water to bring the level back to the proper level.

3. Run the engine this way until the thermostat opens. When this happens, the coolant will move abruptly across the top of the radiator and the temperature of the upper radiator tank and upper radiator hose will rise suddenly.

4. At this point, air is often expelled and the level may drop quite a bit. Keep refilling the system until the level is just below the neck of the radiator and remains constant.

5. Install the radiator cap and check for leaks at this time.

6. Check the level of the coolant in the reservoir and add the proper type of coolant to reach the "FULL HOT" level if needed.

7. Allow the engine to cool completely and check the coolant in the reservoir, add coolant to the "FULL COLD" level if needed.

FUEL SYSTEM

Fuel System Service Precautions

Safety is the most important factor when performing not only fuel system maintenance but any type of maintenance. Failure to conduct maintenance and repairs in a safe manner may result in serious personal injury or death. Maintenance and testing of the vehicle's fuel system components can be accomplished safely and effectively by adhering to the following rules and guidelines.

• To avoid the possibility of fire and personal injury, always disconnect the negative battery cable unless the repair or test procedure requires that battery voltage be applied.

• Always relieve the fuel system pressure prior to disconnecting any fuel system component (injector, fuel rail, pressure regulator, etc.), fitting or fuel line connection. Exercise extreme caution whenever relieving fuel system pressure to avoid exposing skin, face and eyes to fuel spray. Please be advised that fuel under pressure may penetrate the skin or any part of the body that it contacts.

• Always place a shop towel or cloth around the fitting or connection prior to loosening to absorb any excess fuel due to spillage. Ensure that all fuel spillage (should it occur) is quickly removed from engine surfaces. Ensure that all fuel soaked cloths or towels are deposited into a suitable waste container.

• Always keep a dry chemical (Class B) fire extinguisher near the work area.

• Do not allow fuel spray or fuel vapors to come into contact with a spark or open flame.

• Always use a backup wrench when loosening and tightening fuel line connection fittings. This will prevent unnecessary stress and torsion to fuel line piping. Always follow the proper torque specifications.

• Always replace worn fuel fitting O-rings with new. Do not substitute fuel hose or equivalent, where fuel pipe is installed.

Fuel System Pressure

RELIEVING

———— CAUTION ————
Fuel injection systems remain under pressure, even after the engine has been turned OFF. The fuel system pressure must be relieved before disconnecting any fuel lines. Failure to do so may result in fire and/or personal injury.

2.2L (VIN 4) and 3.1L (VIN T) and (VIN M) Engines

1. Disconnect the negative battery cable. Remove the fuel filler cap to relieve tank vapor pressure.
2. Connect a fuel gauge to the fuel pressure test fitting.

NOTE: Be sure to wrap a shop cloth around the fuel line fitting when connecting the fuel gauge tool to the fuel pressure connector.

3. Place the bleeder hose and shop cloth in an approved fuel container. Open the pressure valve to bleed the fuel pressure from the system.

———— CAUTION ————
Observe all applicable safety precautions when working around fuel. Do not allow fuel spray or fuel vapors to come in contact with a spark or open flame. Keep a dry chemical (Class B) fire extinguisher near the work area. Never drain or store fuel in an open container due to the possibility of fire or explosion.

4. After the fuel pressure is bled, retighten the fuel pressure valve.

2.3L (VIN A) Engine

1. Loosen fuel filler cap. Raise and safely support the vehicle.
2. Disconnect the fuel pump electrical connector.
3. Lower the vehicle.
4. Start the engine and run until the fuel supply remaining in the fuel lines is consumed. Engage the starter for several seconds to assure relief of any remaining pressure.
5. Raise and safely support the vehicle.
6. Connect the fuel pump electrical connector.
7. Lower the vehicle.
8. Disconnect the negative battery cable to avoid possible fuel discharge if an accidental attempt is made to start the engine.

Idle Speed

ADJUSTMENT

The idle speed is not adjustable. It is controlled by the engine control system, which consists of various sensors, switches and the Powertrain Control Module (PCM).

Fuel Filter

REMOVAL AND INSTALLATION

———— CAUTION ————
Fuel injection systems remain under pressure, even after the engine has been turned OFF. The fuel system pressure must be relieved before disconnecting any fuel lines. Failure to do so may result in fire and/or personal injury.

1. Relieve the fuel system pressure. Disconnect the negative battery cable.
2. Raise and safely support the vehicle.
3. Disconnect the fuel inlet line from the fuel filter using a backup wrench on the filter to prevent damaging the fuel line or filter.
4. Disconnect the fuel line quick-connect from the filter outlet by squeezing the tabs together and pulling the line off the filter.
5. Remove the fuel filter from the bracket.
To install:
6. Install the fuel filter in the mounting bracket.
7. Connect the quick-connect to the fuel filter. Make sure the tabs lock in place and the line can not be pulled off the filter.
8. Connect the outlet line to the filter using a backup wrench on the filter body to prevent damaging the fuel line or filter. Tighten the line fitting to 20 ft. lbs. (27 Nm).
9. Lower the vehicle.
10. Connect the negative battery cable. Pressurize the fuel system and verify no leaks.

Fuel Pump

REMOVAL AND INSTALLATION

———— CAUTION ————
Fuel injection systems remain under pressure, even after the engine has been turned OFF. The fuel system pressure must be relieved before disconnecting any fuel lines. Failure to do so may result in fire and/or personal injury.

1. Relieve the fuel system pressure. Disconnect the negative battery cable.
2. Drain, then remove the fuel tank from the vehicle.
3. While holding the modular fuel sender assembly down, remove the

snapring from the designated slots located on the retainer.

NOTE: The modular fuel sender assembly may spring up from its position. When removing the modular fuel sender from the tank, be aware that the reservoir bucket is full of fuel. It must be tipped slightly during removal to avoid damage to the float.

4. Remove the external fuel strainer.
5. Disconnect the Connector Position Assurance (CPA) piece from the electrical connector and disconnect the fuel pump electrical connector.
6. Gently release the tabs on the sides of the fuel sender at the cover assembly. Begin by squeezing the sides of the reservoir and releasing the tab opposite the fuel level sensor. Move clockwise to release the second and third tab in the same manner.
7. Lift the cover assembly out far enough to disconnect the fuel pump electrical connection.
8. Rotate the fuel pump baffle counterclockwise and remove the baffle and pump assembly from the retainer.
9. Slide the fuel pump outlet out of slot. Remove the fuel pump outlet seal.
To install:
10. Install the fuel pump outlet seal. Slide the fuel pump outlet in the slots of reservoir cover.
11. Install the fuel pump and baffle assembly onto the reservoir retainer and rotate clockwise until seated.
12. Install the lower retainer assembly partially into the reservoir. Line up all 3 sleeve tabs. Press the retainer onto the reservoir making sure all 3 tabs are firmly seated.

NOTE: Gently pull on the fuel pump reservoir from retainer to assure a secure fastening. If not secure replace entire fuel sender.

13. Connect the fuel pump connector. Attach the CPA connector to the fuel sender cover.
14. Install a new external fuel strainer.
15. Install the modular fuel sender.
16. Install the fuel tank in the vehicle.
17. Connect the negative battery cable. Pressurize the fuel system and verify no leaks.

Fuel Injector

REMOVAL AND INSTALLATION

—— CAUTION ——
Fuel injection systems remain under pressure, even after the engine has been turned OFF. The fuel system pressure must be relieved before disconnecting any fuel lines. Failure to do so may result in fire and/or personal injury.

2.2L (VIN 4) Engine

1. Disconnect the negative battery cable. Relieve the fuel system pressure.
2. Remove the upper intake manifold assembly.
3. Remove the fuel return line bracket mounting bolt and remove the bracket.
4. Position the return line away from the pressure regulator.
5. Remove the pressure regulator as follows:
   a. Disconnect the vacuum hose from the regulator.
   b. Remove the fuel return pipe clamp.
   c. Disconnect the fuel return pipe from the pressure regulator and discard the O-ring.
   d. Remove the pressure regulator bracket mounting bolt.
   e. Remove the regulator assembly and O-ring and discard the O-ring.
6. Remove the fuel injector retaining bracket mounting bolts.
7. Remove the fuel injector retaining bracket by carefully sliding it off to clear the fuel injector slots.
8. Disconnect the fuel injector electrical connector.
9. Pull the injector out.
10. Remove the old O-rings and discard.
To install:
NOTE: Each fuel injector is calibrated for a specific flow rate. Be sure to use the correct part number when ordering replacement fuel injectors.

11. Coat new O-rings with clean engine oil and install them on the injector.
12. Install the injector into the cavity in the lower intake manifold with the electrical connector facing inward.
13. Install the injector retaining bracket an mounting screws.
14. Connect the fuel injector electrical connector.

15. Install the pressure regulator as follows:
   a. Lubricate the regulator assembly and O-ring and install the O-ring on the regulator.
   b. Install the regulator onto the intake manifold.
   c. Install the pressure regulator bracket mounting bolt after coating the threads with thread locking compound. Tighten the mounting bolt to 76 inch lbs. (8.5 Nm).
   d. Connect the vacuum hose to the regulator.
   e. Connect the fuel return pipe to the pressure regulator and tighten the fitting to 13 ft. lbs. (17 Nm).
   f. Install the fuel return pipe clamp to the lower intake manifold.
16. Install the upper intake manifold assembly.
17. Connect the negative battery cable. Pressurize the fuel system and verify no leaks.
18. Start the engine and verify proper injector operation.

2.3L (VIN A) Engine

1. Disconnect the negative battery cable. Relieve the fuel system pressure.
2. Disconnect the hoses from the front side of the crankcase ventilation oil/air separator. Leave the hoses connected to the cannister purge solenoid valve.
3. Remove the oil/air separator mounting bolts.
4. Disconnect the hose from the bottom of the oil/air separator and remove the separator.
5. Position the cannister purge valve solenoid aside.
6. Remove the fuel pipe clamp bolt. Disconnect the vacuum hose at the pressure regulator.
7. Remove the fuel rail mounting bolts, then remove the fuel rail.
8. Detach the injector electrical connectors.
9. Spread the fuel injector retaining clip to release the fuel injector from the rail.
10. Remove the fuel injector.
11. Discard the injector retainer and O-rings.
To install:
12. Coat new O-rings with clean engine oil and install them on the injectors.
13. Install a new injector retainer on the injector.
14. Install the injector on the fuel rail with the electrical connector facing outward. Push the injector on until the retainer snaps in place.

15. Position the fuel rail over the cylinder head and connect the electrical connectors to the fuel injectors.

16. Install the fuel injectors into the cylinder head and tighten the mounting bolts to 19 ft. lbs. (26 Nm).

17. Install the remaining components in the reverse order of removal. Tighten the oil/air separator mounting bolts to 71 inch lbs. (8 Nm).

18. Connect the negative battery cable.

3.1L (VIN T) Engine

1. Properly relieve the fuel system pressure. Disconnect the negative battery cable.

2. Remove the upper intake manifold assembly.

3. Remove the fuel line bracket bolt.

4. Disconnect and cap the fuel lines from the fuel rail. Use a backup wrench on the fuel rail fittings to prevent them from turning.

5. Remove the O-rings from the lines and discard.

6. Disconnect the vacuum line from the pressure regulator.

7. Remove the fuel rail mounting bolts. Detach the injector connectors.

8. Remove the fuel rail assembly.

9. Remove the injector retaining clip and discard.

10. Remove the injector assembly from the fuel rail.

11. Remove the old O-rings and discard.

   To install:

12. Coat the new O-rings with clean engine oil and install the O-rings on the injector.

13. Install a new retaining clip on the fuel injector. Position the open end of the clip toward the injector electrical connector.

14. Install the injector onto the fuel rail assembly with the injector electrical connector facing outward.

15. Install the fuel rail onto the intake manifold so all the injector fit into the injector cavities.

16. Install the fuel rail mounting bolts and tighten to 89 ft. lbs. (10 Nm).

17. Connect the electrical connectors to the fuel injectors.

18. Connect the vacuum line to the pressure regulator.

19. Install new O-rings coated with engine oil on the fuel lines.

20. Connect the fuel feed and return lines and tighten the fittings to 17 ft. lbs. (23 Nm). Use a backup wrench on the fuel rail fittings to prevent them from turning.

21. Install the fuel line bracket bolt.

22. Install the upper intake manifold assembly.

23. Connect the negative battery cable. Pressurize the fuel system and verify no leaks.

3.1L (VIN M) Engine

1. Properly relieve the fuel system pressure. Disconnect the negative battery cable.

2. Remove the upper intake manifold assembly.

3. Disconnect and cap the fuel feed line from the fuel rail.

4. Remove the fuel pressure regulator as follows:

   a. Remove the fuel pressure regulator retaining screw.

   b. Place a shop towel under the pressure regulator and remove the regulator from the fuel rail.

   c. Remove the retainer and spacer bracket from the fuel rail.

   d. Disconnect the regulator from the fuel return line.

5. Disconnect the fuel injector main harness.

6. Disconnect the coolant temperature sensor electrical connector.

7. Remove the fuel rail retaining bolts and remove the fuel rail.

8. Disconnect the individual injector electrical connector.

9. Remove the injector retaining clip and discard.

10. Remove the injector assembly from the fuel rail.

11. Remove the old O-rings and discard. Save the O-ring backups for installation on the new injector.

   To install:

12. Install the O-ring backups on the fuel injector.

13. Coat the new O-rings with clean engine oil and install the O-rings on the injector.

14. Install a new retaining clip on the fuel injector. Position the open end of the clip toward the injector electrical connector.

15. Install the injector onto the fuel rail assembly with the injector electrical connector facing outward.

16. Connect the electrical connector to the fuel injector.

17. Install the fuel rail onto the intake manifold so all the injector fit into the injector cavities.

18. Install the fuel rail bolts and tighten to 89 ft. lbs. (10 Nm).

19. Connect the electrical connector to the coolant temperature switch.

20. Connect the main fuel injector harness to the fuel rail assembly.

21. Install new O-rings on the pressure regulator and fuel inlet line.

22. Connect the fuel feed line and tighten the fitting to 13 ft. lbs. (17 Nm).

23. Install the fuel pressure regulator as follows:

   a. Connect the fuel regulator to the fuel return line.

   b. Install a new retainer and spacer bracket into the slot on the fuel rail.

   c. Install the pressure regulator onto the fuel rail.

   d. Tighten the return line fitting to 13 ft. lbs. (17 Nm).

   e. Install the pressure regulator mounting bolts and tighten to 76 inch lbs. (9 Nm).

24. Install the upper intake manifold assembly.

25. Connect the negative battery cable. Pressurize the fuel system and verify no leaks.

ENGINE MECHANICAL

Engine Assembly

REMOVAL AND INSTALLATION

---- **CAUTION** ----
Fuel injection systems remain under pressure, even after the engine has been turned OFF. The fuel system pressure must be relieved before disconnecting any fuel lines. Failure to do so may result in fire and/or personal injury.

NOTE: Whenever the vehicle sub-frame is removed or lowered, the wheel alignment should be checked.

2.2L (VIN 4) Engine

1993–94 Vehicles

1. Relieve the fuel system pressure. Disconnect the negative battery cable.

2. Drain the cooling system.

3. Remove the throttle body air inlet duct. Remove the battery from the vehicle.

4. Remove the air cleaner assembly. Remove the upper radiator hose.

5. Disconnect the vacuum hose from power brake booster.

6. Remove the alternator upper brace and disconnect the wiring from the alternator.

7. Tag and disconnect the electrical wiring from the following components:
- Oxygen sensor (O_2)
- Fuel injector harness
- Idle Air Control (IAC)
- Throttle Position Sensor (TPS)
- Coolant Temperature Sender
- Park/neutral switch
- TCC solenoid
- Transaxle shift solenoid
- Transaxle ground
- MAP sensor
- EGR
- Cooling fan

8. Reclaim the refrigerant from the A/C system.

9. Disconnect the compressor-to-condenser and accumulator lines.

10. Remove the slave cylinder from the transaxle, if equipped with a manual transaxle.

11. Disconnect the fuel lines.

12. Disconnect the shift linkage from the transaxle bracket.

13. Disconnect the control cables from the throttle body lever and remove the cable bracket from the rocker arm cover and intake manifold.

14. Place a pan under the power steering pump and disconnect and cap the power steering lines.

15. Install an engine support fixture, J-28467, or the equivalent.

16. Safely raise and support the vehicle.

17. Remove the front tire and wheel assemblies.

18. Remove the right and left inner fender splash shields.

19. Disconnect the front exhaust pipe from manifold and remove the front exhaust pipe.

20. Tag and detach the ignition assembly, starter, Vehicle Speed Sensor (VSS) and the transaxle ground wire electrical connectors from the lower engine components:

21. Remove the flywheel cover.

22. Remove the lower radiator hose.

23. Disconnect the heater hoses from the heater core pipes.

24. Disconnect the transaxle cooler lines at the radiator.

25. Remove the mounting bolt from the lower engine strut at the suspension crossmember.

26. Disconnect the both front ABS wheel sensor connectors and unfasten the harnesses from the suspension crossmember.

27. Remove the cotter pins and castle nuts from the lower ball joints and separate the joints from the steering knuckles.

28. Remove the suspension crossmembers.

29. Disconnect the halfshafts from the transaxle and support aside.

30. Position a suitable table under the engine and transaxle assembly.

31. Lower the vehicle until the powertrain assembly is on the table.

32. Remove the transaxle-to-frame mount bolts.

33. Remove the intermediate bracket from the right engine mount support bracket.

34. Remove the engine support fixture.

35. Raise the vehicle, leaving the powertrain assembly on the table. When raising the vehicle take it up slowly and verify that no lines are still connected to the powertrain assembly.

36. Remove the torque converter-to-flywheel bolts and transaxle-to-engine mounting bolts and remove the transaxle from the engine.

To install:

37. Install the transaxle on the engine and tighten the mounting bolts to 68 ft. lbs. (93 Nm).

38. Install flywheel-to-converter bolts. New replacement bolts are recommended. Tighten to 46 ft. lbs. (62 Nm).

39. Position the engine on the table and lower the vehicle until the engine is in the engine bay.

40. Install an engine support fixture, J-28467, or the equivalent.

41. Raise and safely support the vehicle and remove the engine table.

42. Connect the transaxle cooler lines to the radiator.

43. Connect the heater hoses to the heater core outlet pipes.

44. Install the lower radiator hose.

45. Connect the engine mount support bracket to the intermediate bracket and tighten the bolts to 96 ft. lbs. (103 Nm).

46. Install the transaxle mount-to-body bolts and tighten to 40 ft. lbs. (54 Nm).

47. Connect the halfshafts to the transaxle.

48. Install the remaining components in the reverse order of removal. Torque the following:
- Suspension crossmember and control arm bolts to 89 ft. lbs. (120 Nm)
- Lower ball joint-to-steering knuckle bolts to 48 ft. lbs. (65 Nm)
- Engine strut-to-suspension support bolt to 89 ft. lbs. (120 Nm)
- Front exhaust pipe-to-manifold bolts to 22 ft. lbs. (30 Nm)

49. Install the battery cable and connect the positive cable.

50. Install the throttle body air inlet duct.

51. Connect the negative battery cable.

52. Fill the coolant and engine oil.

53. Refill and bleed the power steering system.

54. Start the engine and verify no leaks.

55. Check the vehicle wheel alignment and correct if necessary.

1995–96 Vehicles

Please note that on this vehicle, the engine and transaxle are removed together from under the vehicle.

1. Relieve the fuel system pressure. Disconnect the negative battery cable. Drain the cooling system.

2. Remove the battery.

3. Remove the air cleaner outlet duct. Remove the upper radiator hose.

4. Disconnect the vacuum hose from power brake booster.

5. Tag and disconnect the electrical wiring from the following components:
- Alternator
- EVAP Emission Solenoid
- Oxygen sensor (O_2)
- Fuel injector harness
- Idle Air Control (IAC)
- Throttle Position Sensor (TPS)
- Engine Coolant Temperature Sensor (ECT)
- Park/neutral switch
- TCC solenoid
- Engine grounds
- MAP sensor
- EGR

6. Remove the serpentine drive belt.

7. Remove the transaxle shift control cable from the range select lever and bracket.

8. Remove the coolant surge tank hose and surge tank.

9. Remove the vacuum line near the master cylinder.

10. Install an engine support fixture, J-28467, or the equivalent.

11. Remove the lower radiator hose.

12. Safely raise and support the vehicle. Remove the front tire and wheel assemblies.

13. Remove the right and left inner fender splash shields.

14. Disconnect the front exhaust pipe from the manifold and catalytic convertor, remove the front exhaust pipe and catalytic convertor.

15. Remove the mounting bolt from the lower engine strut at the suspension crossmember.

16. Remove the wheel speed sensor wire harness from the control arms

17. Remove the cotter pins and castle nuts from the lower ball joints and separate the joints from the steering knuckles.

18. Separate the tie rod ends from the struts using tool J-24319-01 or equivalent.

19. Remove the brake lines from the suspension support.

20. Remove the A/C compressor with the lines attached and safely support it aside.

21. Tag and disconnect the electrical connectors from the following lower engine components:
- Ignition assembly
- Starter
- Vehicle Speed Sensor (VSS)
- Transaxle ground wire
- Cooling Fan
- A/C Compressor
- Oil pressure sensor
- Oil Level Sensor

22. Disconnect the power steering lines from the rack and pinion assembly.

23. Remove the flexible coupling joint at the rack and pinion assembly.

24. Remove the accelerator control, cruise control and throttle control cables from the accelerator control bracket.

25. Remove the front suspension support assemblies.

26. Disconnect the heater hoses from the heater core pipes.

27. Disconnect the halfshafts from the transaxle and support them aside.

28. Disconnect the fuel lines.

29. Disconnect the transaxle cooler lines from the transaxle.

30. Remove the transaxle mount.

31. Remove the engine mount assembly

32. Position a suitable hydraulic support table under the engine and transaxle assembly.

33. Lower the vehicle until the powertrain assembly is on the table.

34. Remove the engine support fixture.

35. Raise the vehicle, leaving the powertrain assembly on the table. When raising the vehicle take it up slowly and verify that no lines are still connected to the powertrain assembly.

36. Remove the torque converter-to-flywheel bolts and transaxle-to-engine mounting bolts and remove the transaxle from the engine.

To install:

37. Install the transaxle on the engine and tighten the mounting bolts to 55 ft. lbs. (75 Nm).

38. Install flywheel-to-converter bolts and tighten to 55 ft. lbs. (75 Nm).

39. Position the engine on the table and lower the vehicle until the engine is in the engine bay.

40. Install an engine support fixture, J-28467, or the equivalent.

41. Raise and safely support the vehicle and remove the engine table.

42. Install the engine mount strut bracket to engine bolts, tighten to 49 ft. lbs. (66 Nm).

43. Connect the engine mount support bracket to the intermediate bracket and tighten the bolts to 125 ft. lbs. (170 Nm).

44. Install the transaxle mount.

45. Install the remaining components in the reverse order of removal. Torque the following:
- Front suspension support bolts to 89 ft. lbs. (120 Nm)
- A/C compressor bracket-to-engine bolts 67 ft. lbs. (93 Nm)
- Tie rod end-to-strut nuts to 44 ft. lbs. (60 Nm)
- Lower ball joint-to-steering knuckle bolts to 48 ft. lbs. (65 Nm) and install 2 new cotter pins
- Engine strut-to-suspension support mounting bolt to 89 ft. lbs. (120 Nm)
- Front exhaust pipe-to-manifold bolts to 22 ft. lbs. (30 Nm)

46. Install the battery cable and connect the positive cable.

47. Connect the negative battery cable.

48. Fill the coolant and engine oil.

49. Refill and bleed the power steering system.

50. Start the engine and verify no leaks.

2.3L (VIN A) Engine

1. Relieve the fuel system pressure. Disconnect the negative battery cable.

2. Drain the cooling system.

3. Recover the refrigerant from the A/C system.

4. Disconnect the heater hose from the thermostat housing.

5. Disconnect the upper radiator hose from the thermostat housing.

6. Remove the air inlet duct from the throttle body.

7. Remove the upper radiator support. Remove the cooling fan assembly.

8. Disconnect the compressor-to-condenser hose from the compressor and discard the O-ring.

9. Disconnect the vacuum hoses from the front of the engine.

10. Tag and disconnect the electrical connectors from the following components:
- Alternator
- A/C compressor
- Fuel injectors
- Idle Air Control (IAC)

- Throttle Position Sensor (TPS)
- MAP sensor
- MAT sensor
- EVAP solenoid
- Starter
- Ignition assembly
- Coolant sensors
- Oil pressure sender
- Power steering pressure switch
- Knock sensor
- Oxygen Sensor (O$_2$)
- Crankshaft sensor
- Vehicle Speed Sensor (VSS)
- Back up light switch

11. Disconnect the power brake vacuum pipe from the throttle body.

12. Disconnect the throttle cable from the throttle lever and remove the bracket and position the cable assembly aside.

13. Remove the power steering pump from the engine and set aside with the power steering lines attached.

14. Disconnect and cap the fuel lines.

15. Disconnect the shift cable from the transaxle. If equipped, remove the slave cylinder.

16. Remove the exhaust manifold.

17. Disconnect the lower radiator hose from the radiator.

18. Install an engine support fixture, J-28467 or an equivalent.

19. Remove the right engine mount.

20. Raise and safely support the vehicle.

21. Remove the front tire and wheel assemblies.

22. Remove the right lower splash shield. Remove the radiator air deflector.

23. Remove the cotter pins and castle nuts from the lower ball joints. Separate the lower ball joints from the steering knuckles.

24. Support the suspension support with a suitable jack and remove the suspension supports.

25. Disconnect the heater outlet hose from the radiator outlet pipe.

26. Remove the halfshafts and intermediate shaft from the vehicle.

27. Position a suitable engine dolly under the vehicle and lower the vehicle until the powertrain assembly is on the dolly.

28. Remove the engine mount strut and transaxle braces.

29. Using the support fixture lower the engine onto the dolly and disconnect the support fixture.

30. Raise the vehicle slowly making sure no lines or hoses are still connected to the powertrain assembly.

31. Separate the engine from the transaxle.

To install:

32. Connect the transaxle to the engine.

33. Position the engine assembly on the dolly and lower the vehicle into position over the powertrain assembly.

34. Install the engine support fixture and pull the engine into the engine bay.

35. Raise the vehicle off the dolly.

36. Install the engine support strut and transaxle braces.

37. Install the remaining components in the reverse order of removal. Torque the following:

- Suspension support rear bolts first to 66 ft. lbs. (90 Nm), the center bolts second to 66 ft. lbs. (90 Nm) and the front bolts third to 66 ft. lbs. (90 Nm)
- Ball joint-to-steering knuckle nuts to 45 ft. lbs. (62 Nm)

38. Refill the cooling system, engine crankcase, transmission and power steering fluid.

39. Connect the negative battery cable. Start the engine and verify no leaks.

40. Check the vehicle wheel alignment and correct if necessary.

3.1L (VIN T) and (VIN M) Engines

1. Relieve the fuel system pressure. Disconnect the negative battery cable.

2. Remove the air cleaner assembly and the throttle body air inlet duct.

3. Drain the cooling system into a suitable container.

4. If equipped with a VIN T engine, disconnect the transaxle cooler lines at the transaxle and remove the transaxle fill tube.

5. Remove the upper and lower radiator hoses.

6. Disconnect the coolant inlet line from the surge tank.

7. Disconnect the vacuum hoses from the vacuum modulator, EVAP canister purge and the power brake booster.

8. Disconnect the heater outlet hose from the water pump.

9. Remove the serpentine belt.

10. Disconnect the control cables from the throttle body lever.

11. Tag and disconnect the electrical wiring from the following components:

- Ignition module assembly
- Heated oxygen sensor (HO2S)
- Fuel injector harness
- Idle Air Control (IAC)

- Throttle Position Sensor (TPS)
- Coolant Temperature Sender
- Park/neutral switch
- TCC solenoid
- Transaxle shift solenoid
- Transaxle ground

12. Remove the alternator.

13. Place a pan under the power steering pump and disconnect and cap the power steering lines.

14. Disconnect the fuel lines.

15. Disconnect the shift linkage from the transaxle bracket.

16. Disconnect the vacuum hose at the vacuum reservoir.

17. Remove the cooling fan assembly.

18. Install an engine support fixture, J-28467, or the equivalent.

19. Loosen, but do not remove the 2 upper A/C compressor bolts.

20. Safely raise and support the vehicle.

21. Remove the front tire and wheel assemblies.

22. Remove the right and left inner fender splash shields.

23. Remove the mounting bolt from the lower engine strut at the suspension crossmember.

24. Disconnect the both front ABS wheel sensor connectors and unfasten the harnesses from the suspension crossmember.

25. Remove the cotter pins and castle nuts from the lower ball joints and separate the joints from the steering knuckles.

26. Remove the suspension crossmembers.

27. Disconnect the halfshafts from the transaxle and support aside.

28. Remove the oil filter.

29. Remove the oil filter adapter.

30. Remove the flywheel cover.

31. Disconnect and remove the starter.

32. Disconnect the following electrical connectors from the lower engine components:

- Knock sensor
- Front crankshaft position sensor
- Side crankshaft position sensor
- Oil level sensor
- Vehicle Speed Sensor (VSS)
- Transaxle ground wire

33. Disconnect both heater hoses.

34. Remove the lower A/C compressor mounting bolts and remove the compressor from the mounting bracket. DO NOT disconnect the refrigerant lines or allow the compressor to hang unsupported.

35. Remove the vacuum reserve tank.

36. Disconnect the front exhaust pipe from manifold and remove the front exhaust pipe.

37. Remove the engine mount strut bracket.

38. Disconnect the transaxle cooler lines at the radiator.

39. Remove the transaxle fill tube.

40. Position a suitable table under the engine and transaxle assembly.

41. Lower the vehicle until the powertrain assembly is on the table.

42. Remove the transaxle-to-frame mount bolts.

43. Remove the intermediate bracket from the right engine mount support bracket.

44. Raise the vehicle, leaving the powertrain assembly on the table. When raising the vehicle take it up slowly and verify that no lines are still connected to the powertrain assembly.

45. Remove the torque converter-to-flywheel bolts and transaxle-to-engine mounting bolts and remove the transaxle from the engine.

To install:

46. Install the transaxle on the engine and tighten the mounting bolts to 37 ft. lbs. (50 Nm).

47. Install flywheel-to-converter bolts and tighten to 47 ft. lbs. (63 Nm).

48. Position the engine on the table and lower the vehicle until the engine is in the engine bay.

49. Loosely install the serpentine belt.

50. Connect the engine mount support bracket to the intermediate bracket and tighten the bolts to 96 ft. lbs. (103 Nm).

51. Install the transaxle mount-to-body bolts and tighten to 40 ft. lbs. (54 Nm).

52. Raise and safely support the vehicle and remove the engine table.

53. Install the remaining components in the reverse order of removal. Torque the following:

- Suspension crossmembers and control arm bolts to 89 ft. lbs. (120 Nm)
- Lower ball joint-to-steering knuckle bolts to 48 ft. lbs. (65 Nm)
- Engine strut-to-suspension support bolt to 89 ft. lbs. (120 Nm)

54. Connect the negative battery cable.

55. Fill the coolant and engine oil.

56. Refill and bleed the power steering system.

57. Start the engine and verify no leaks.

Engine Mounts

REMOVAL AND INSTALLATION

2.2L (VIN 4) Engine

Right Engine Mount

1. Disconnect the negative battery cable. Install engine support fixture J-28467-A and J-35953 or their equivalents.
2. Raise the engine with the support fixture to remove the engine weight from the mount.
3. Raise and safely support the vehicle.
4. Remove 2 engine mount-to-frame bolts from the right side wheel well. Lower the vehicle.
5. Remove the upper engine mount mounting nut.
6. Remove the engine mount-to-engine bracket bolts.
7. Remove the engine mount.

To install:

8. Install over the upper stud and loosely install the mounting nut.
9. Raise and safely support the vehicle.
10. Install the 2 engine mount bolts through the fender well and tighten to 49 ft. lbs. (66 Nm).
11. Lower the vehicle.
12. Tighten the engine mount nut to 31 ft. lbs. (42 Nm).
13. Install the 3 engine mount-to-engine bracket bolts and tighten the center bolt to 31 ft. lbs. (42 Nm) and the outer bolts to 125 ft. lbs. (170 Nm).
14. Remove the engine support fixture. Connect the negative battery cable.

Lower Engine Mount Strut and Bracket

1. Disconnect the negative battery cable. Raise and safely support the vehicle.
2. Remove the strut mounting bolt at the crossmember and engine bracket and remove the strut.
3. Remove the A/C compressor from the mounting bracket and position aside with the refrigerant lines attached.
4. Remove the 4 mounting bracket bolts and remove the mounting bracket.

To install:

5. Install the mounting bracket on the engine and tighten the mounting bolts to 49 ft. lbs. (66 Nm).
6. Install the A/C compressor in the mounting bracket.
7. Install the engine support strut and tighten the strut-to-engine bracket bolt to 89 ft. lbs. (120 Nm)

and tighten the strut-to-crossmember bolt to 55 ft. lbs. (75 Nm).

8. Lower the vehicle. Connect the negative battery cable.

2.3L (VIN A) Engine

Right Engine Mount

1. Disconnect the negative battery cable.
2. Install an engine support fixture, J-28467-A or an equivalent. Use the support fixture to remove the engine weight from the engine mount.
3. Remove the 4 engine mount-to-engine bracket bolts.
4. Raise and safely support the vehicle.
5. Remove the right front tire and wheel assembly and the inner fender well splash shield..
6. Remove the 2 engine mount-to-body bolts. Lower the vehicle enough to work under the hood.
7. Remove the engine mount nut from the mounting stud.
8. Remove the engine mount.

To install:

9. Position the engine mount over the mounting stud and loosely install the mounting nut.
10. Raise and safely support the vehicle.
11. Install the 2 lower engine mount bolts and tighten to 49 ft. lbs. (66 Nm). Tighten the rear bolt first.
12. Install the inner fender splash shield.
13. Install the right front tire and wheel assembly.
14. Lower the vehicle.
15. Tighten the engine mount nut to 31 ft. lbs. (42 Nm).
16. Install the 4 engine mount-to-bracket bolts and tighten to 46 ft. lbs. (62 Nm).
17. Remove the engine support fixture. Connect the negative battery cable.

Engine Mount Strut

1. Disconnect the negative battery cable. Raise and safely support the vehicle.
2. Remove the right front tire and wheel assembly. Remove the right inner fender well splash shield.
3. Remove the engine mount strut-to-suspension support bolt.
4. Remove the engine mount strut-to-engine bracket bolt.
5. Remove the engine mount strut.

To install:

6. Position the strut in the vehicle and loosely install the mounting bolt at the suspension support.
7. Install the bolt at the engine bracket. Make sure the bolt is in-

stalled in the correct hole in the engine mount bracket. The bolt holes are marked for automatic or manual transaxle and the bolt must be installed in the correct hole.

8. Tighten the bolt at the mounting bracket to 85 ft. lbs. (115 Nm) and the suspension support bolt to 89 ft. lbs. (120 Nm).
9. Install the inner fender splash shield. Install the right front tire and wheel assembly.
10. Lower the vehicle.
11. Connect the negative battery cable.

3.1L (VIN T) Engine

Right Engine Mount

1. Disconnect the negative battery cable.
2. Install an engine support fixture, J-28467-A and J-35953 or their equivalents.
3. Raise the engine with the support fixture until the weight of the engine has been removed from the engine mount.
4. Remove the 2 bolts securing the engine mount to the engine bracket.
5. Raise and safely support the vehicle. Remove the right inner fender splash shield.
6. Remove the 2 engine mount bolts. Lower the vehicle.
7. Remove the engine mount nut under the hood and remove the engine mount.

To install:

8. Position the engine mount over the under hood stud and loosely install the mounting nut.
9. Raise and safely support the vehicle.
10. Install the 2 engine mount bolts and tighten to 49 ft. lbs. (66 Nm).
11. Install the right inner fender splash shield.
12. Lower the vehicle.
13. Tighten the engine mount nut to 31 ft. lbs. (42 Nm).
14. Install the engine mount-to-engine bracket bolts and tighten to 81 ft. lbs. (110 Nm).
15. Lower the engine and remove the support fixture.
16. Connect the negative battery cable.

Engine Mount Strut

1. Disconnect the negative battery cable. Raise and safely support the vehicle.
2. Remove the engine mount strut mounting bolts at the crossmember and engine bracket and remove the strut.

To install:

3. Install the engine strut and tighten the mounting bolts to 89 ft. lbs. (120 Nm).

4. Lower the vehicle. Connect the negative battery cable.

3.1L (VIN M) Engine

Right Engine Mount

1. Disconnect the negative battery cable. Support the engine with a floor jack and a block of wood under the oil pan.

2. Remove the 2 bolts securing the right engine mount to the intermediate bracket.

3. Remove the top nut and 2 lower bolts and remove the right engine mount.

To install:

4. Position the mount in the vehicle and tighten the mounting bolts to 49 ft. lbs. (66 Nm) and the upper mounting nut to 42 ft. lbs. (31 Nm).

5. Connect the engine mount to the intermediate bracket and tighten the mounting bolts to 96 ft. lbs. (130 Nm).

6. Remove the floor jack.

7. Connect the negative battery cable.

Engine Mount Strut

1. Disconnect the negative battery cable. Raise and safely support the vehicle.

2. Remove the right front tire and wheel assembly and the right inner fender well splash shield.

3. Remove the engine mount strut bolt at the crossmember and engine bracket and remove the strut.

To install:

4. Position the strut in the vehicle and loosely install both mounting bolts.

5. Tighten the strut-to-engine bracket bolt to 85 ft. lbs. (115 Nm) and the strut-to-crossmember bolt to 89 ft. lbs. (120 Nm).

6. Install the right inner fender well splash shield.

7. Install the right front tire and wheel assembly and tighten to specification.

8. Lower the vehicle.

9. Connect the negative battery cable.

Cylinder Head

REMOVAL AND INSTALLATION

——— CAUTION ———
Fuel injection systems remain under pressure even after the engine has been turned OFF. The fuel system pressure must be relieved before disconnecting any fuel lines. Failure to do so may result in fire and/or personal injury.

NOTE: When removing the valve train components, they must be kept in order for installation in the same locations from which they were removed.

2.2L (VIN 4) Engine

1. Relieve the fuel system pressure. Disconnect the negative battery cable. Drain the cooling system.

2. Remove the air inlet duct.

3. Tag and disconnect the vacuum lines at the intake manifold and cylinder head.

4. Tag and disconnect the following electrical connectors:
- Coolant Temperature Sensor
- Oxygen (O_2) Sensor
- Idle Air Control (IAC)
- Throttle Position Sensor (TPS)
- MAP sensor
- EVAP canister purge solenoid
- Fuel injector harness.

5. Disconnect the control cables from the throttle body and remove the cable bracket at the throttle body and rocker arm cover.

6. Remove the coolant reservoir, if necessary.

7. Install an engine support fixture, J-28467-A, or the equivalent.

8. Remove the right engine mount, if necessary.

9. Remove the serpentine belt.

10. Remove the alternator and alternator mounting bracket.

11. Disconnect and cap the power steering lines from the power steering pump and remove the power steering pump.

12. Remove the serpentine belt tensioner assembly.

13. Remove the right engine mount bracket.

14. Disconnect the spark plug wires and lay them aside.

15. Disconnect the EVAP canister purge line under the intake manifold.

16. Disconnect the upper radiator hose from the engine.

17. Disconnect the coolant inlet pipe brackets at the exhaust manifold and transaxle.

18. Remove the intake manifold brace from the power steering bracket.

19. Disconnect and cap the fuel lines at the quick disconnects.

20. Remove the rocker arm cover, rocker arm nuts, rocker arms and pushrods.

21. Remove the ignition wire bracket.

22. Remove the engine lift bracket.

23. Raise and safely support the vehicle. Disconnect the exhaust pipe from the exhaust manifold.

24. Lower the vehicle.

25. Remove the transaxle fill tube.

26. Remove the cylinder head bolts.

27. Remove the cylinder head with both manifolds. Remove the intake and exhaust manifolds from the cylinder head.

To install:

28. Clean all the gasket surfaces completely. Clean the threads on the cylinder head bolts and the block threads.

29. Install the intake and exhaust manifolds on the cylinder head.

30. Place a new cylinder head gasket in position over the dowel pins on the block. Carefully guide the cylinder head into position.

31. Install all the cylinder head bolts finger-tight. The long bolts go in bolt positions 1, 4, 5, 8 and 9. The short bolt are in positions 2, 3, 6 and 7. The stud is in position 10.

32. Tighten the bolts in sequence. The long bolts to 23 ft. lbs. (32 Nm) and the short bolts and stud to 22 ft. lbs. (29 Nm). Make second pass tightening the long bolts to 46 ft. lbs. (63 Nm) and the short bolts to 43 ft. lbs. (58 Nm). Make a final pass over all bolts tightening each an additional 90 degrees.

33. Install the engine lift bracket.

34. Install the ignition wire bracket.

35. Install the transaxle fill tube.

36. Install the pushrods, rocker arms and rocker arm nuts and tighten the nuts to 22 ft. lbs. (30 Nm).

37. Install the rocker arm cover.

38. Connect the control cables to the throttle body and install the cable brackets at the throttle body and rocker arm cover.

39. Connect the fuel lines to the throttle body. Connect the upper radiator hose.

40. Install the intake manifold brace to the power steering pump bracket.

41. Connect the spark plug wires.

42. Install the remaining components in the reverse order of removal. Tighten the serpentine belt tensioner mounting bolts to 37 ft. lbs. (50 Nm) and the exhaust pipe-to-exhaust manifold bolts to 22 ft. lbs. (30 Nm).

43. Refill the cooling system.

44. Connect the negative battery cable. Start the vehicle and verify no leaks.

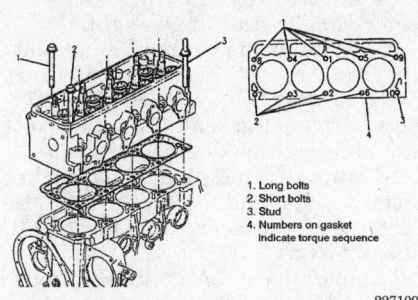

227109

Cylinder head bolt tightening sequence — 2.2L (VIN 4) engine

2.3L (VIN A) Engine

1. Disconnect the negative battery cable. Drain the cooling system.
2. Disconnect heater inlet and throttle body heater hoses from water outlet.
3. Remove the exhaust manifold.
4. Remove the intake and exhaust camshaft housings.
5. Remove the oil fill tube mounting bolt and unseat the fill tube assembly from the engine block.
6. Disconnect the injector harness electrical connector.
7. Remove the oil fill tube.
8. Disconnect the throttle body air intake duct.
9. Disconnect the power brake booster vacuum line.
10. Remove the throttle cable bracket.
11. Remove the throttle body from the intake manifold and position aside with all the lines and cables attached.
12. Disconnect the MAP sensor vacuum hose. Remove the intake manifold brace.
13. Remove the fuel rail mounting bolts and remove the fuel rail from the intake manifold with the fuel lines attached.
14. Detach the MAP sensor, MAT sensor and EVAP purge solenoid electrical connections.
15. Disconnect the upper radiator hose from the thermostat housing.
16. Disconnect the coolant sensor connectors.
17. Remove the cylinder head bolts in reverse order of the installation sequence.
18. Remove the cylinder head and gasket.
   To install:
19. Clean all gasket surfaces completely.
20. Install a new cylinder head gasket on the cylinder block and carefully position the cylinder head in place.

BOLTS 1 THROUGH 6 - 35 N•m (26 LBS. FT.)
BOLT 7 AND 8 - 20 N•m (15 LBS. FT.)
BOLTS 9 AND 10 - 30 N•m (22 LBS. FT.)

195636

Cylinder head bolt tightening sequence — 2.3L (VIN A) engine

21. Coat the cylinder head bolt threads with clean engine oil and allow the oil to drain off before installing.
22. Install the cylinder head bolts and tighten in sequence as follows:
   a. Tighten bolts 1 through 6: 26 ft. lbs. (35 Nm)
   b. Tighten bolts 7 and 8: 15 ft. lbs. (20 Nm)
   c. Tighten bolts 9 and 10: 22 ft. lbs. (30 Nm)
   d. Tighten all 10 bolts in sequence an additional 90 degrees rotation
   e. Loosen all 10 bolts in sequence 1 full turn (360 degrees)
   f. Tighten bolts 1 through 6: 26 ft. lbs. (35 Nm)
   g. Tighten bolts 7 and 8: 15 ft. lbs. (20 Nm)
   h. Tighten bolts 9 and 10: 22 ft. lbs. (30 Nm)
   i. Tighten all 10 bolts in sequence an additional 90 degrees rotation
23. Connect the heater inlet and throttle body heater hoses to the water outlet.
24. Connect the upper radiator hose to the water outlet.
25. Connect the 2 coolant sensor connections.
26. Install the fuel rail on the intake manifold.
27. Install the intake manifold brace.
28. Attach the MAP sensor, MAT sensor and EVAP purge solenoid electrical connections.
29. Connect the MAP sensor vacuum hose to the intake manifold.
30. Install the throttle body.
31. Install the throttle cable bracket. Connect the throttle body air intake duct.
32. Lubricate the O-ring on the oil fill tube with engine oil and install oil fill tube into the block and install the mounting bolt.
33. Install the exhaust and intake camshaft housings.

34. Connect the injector harness electrical connector.
35. Install the exhaust manifold.
36. Fill all fluids to their proper levels.
37. Connect the negative battery cable. Start the vehicle and verify no leaks.

3.1L (VIN T) Engine

Left Cylinder Head (Front)

1. Relieve the fuel system pressure. Remove the air cleaner assembly.
2. Disconnect the negative battery cable. Drain the cooling system.
3. Remove the rocker arm covers.
4. Remove the upper intake plenum and lower intake manifold.
5. Remove the exhaust crossover pipe. Remove the front exhaust manifold.
6. Remove the oil level indicator tube.
7. Disconnect the spark plug wires from the spark plugs and remove the wires from the wire looms. Position the wires aside.
8. Remove the rocker arm nuts, rocker arms, rocker arm balls and pushrods.
9. Remove the cylinder head bolts.
10. Remove the cylinder head.
   To install:
11. Clean all the gasket surfaces completely. Clean the cylinder head bolt threads and cylinder block threads. New replacement head bolts are recommended.
12. Install a new cylinder head gasket on the block. The gasket should fit over the dowel pins and the words **THIS SIDE UP** must be visible.
13. Install the cylinder head.
14. Coat the cylinder head bolts with sealer and install the bolts finger-tight. With all the bolts installed tighten the bolts in sequence to 33 ft. lbs. (45 Nm). Make a second pass tightening each bolt an additional 90 degrees (¼ turn).
15. Install the pushrods, rocker arms, balls and nuts.
16. Adjust the valves.
17. Install the lower intake manifold and upper intake plenum.
18. Install the rocker arm covers.
19. Install the oil level indicator tube.
20. Connect the spark plug wires to the spark plugs and the wire looms.
21. Install the front exhaust manifold and the crossover pipe.
22. Refill the cooling system. An oil and filter change is recommended since coolant can get into the oil pan during cylinder head service.

23. Connect the negative battery cable.

24. Install the air cleaner assembly.

25. Start the vehicle and verify no leaks.

Right Cylinder Head (Rear)

1. Relieve the fuel system pressure. Remove the air cleaner assembly.

2. Disconnect the negative battery cable. Drain the cooling system.

3. Raise and safely support the vehicle.

4. Disconnect the front exhaust pipe at the crossover pipe.

5. Lower the vehicle.

6. Remove the heat shield from the crossover pipe.

7. Disconnect the crossover pipe from the right manifold.

8. Remove the right exhaust manifold.

9. Disconnect the spark plug wires from the spark plugs and disconnect the wires from the looms.

10. Remove the rocker arm covers.

11. Remove the upper intake plenum and lower intake manifold.

12. Remove the rocker arm nuts, rocker arms, rocker arm balls and pushrods.

13. Remove the cylinder head bolts.

14. Remove the cylinder head.

To install:

15. Clean all the gasket surfaces completely. Clean the cylinder head bolt threads and cylinder block threads. New replacement head bolts are recommended.

16. Install a new cylinder head gasket on the block. The gasket should fit over the dowel pins and the words **THIS SIDE UP** must be visible.

17. Install the cylinder head.

18. Coat the cylinder head bolts with sealer and install the bolts finger-tight. With all the bolts installed tighten the bolts in sequence to 33 ft. lbs. (45 Nm). Make a second pass tightening each bolt an additional 90 degrees (¼ turn).

19. Install the pushrods, rocker arms, balls and nuts.

20. Adjust the valves.

21. Install the lower intake manifold and upper intake plenum.

22. Install the rocker arm covers.

23. Install the remaining components in the reverse order of removal.

24. Refill the cooling system. An oil and filter change is recommended since coolant can get into the oil pan during cylinder head service.

25. Connect the negative battery cable. Install the air cleaner.

26. Start the vehicle and verify no leaks.

3.1L (VIN M) Engine

Left Cylinder Head (Front)

1. Relieve the fuel system pressure. Disconnect the negative battery cable.

2. Drain the cooling system.

3. Remove the top half of the air cleaner assembly and remove the throttle body air inlet duct.

4. Remove the exhaust crossover pipe heat shield and crossover pipe.

5. Disconnect the spark plug wires from spark plugs and wire looms and route the wires aside.

6. Remove the rocker arm covers.

7. Remove upper intake plenum and lower intake manifold.

8. Remove the left side exhaust manifold.

9. Remove oil level indicator tube.

10. Remove rocker arms nut, rocker arms, balls and pushrods.

11. Remove the cylinder head bolts evenly. Remove the cylinder head.

To install:

12. Clean all the gasket surfaces completely. Clean the threads on the cylinder head bolts and block threads.

13. Place the cylinder head gasket in position over the dowel pins on the cylinder block so the words **THIS SIDE UP** showing.

14. Coat the bolt threads with sealer and install finger-tight.

15. Tighten the cylinder head bolts in sequence to 33 ft. lbs. (45 Nm). With all the bolts tightened make a second pass tightening all the bolts an additional 90 degrees (¼ turn).

16. Install the pushrods, rocker arms, balls and rocker arm nuts. Tighten the rocker arm nuts to 18 ft. lbs. (25 Nm).

17. Install the lower intake manifold and upper intake plenum.

18. Install the rocker arm covers.

19. Install the oil level indicator tube.

20. Connect the spark plug wires to spark plugs and wire looms.

21. Install the left side exhaust manifold.

22. Install the exhaust crossover pipe and crossover pipe heat shield.

23. Refill the cooling system.

24. Install the top half of the air cleaner assembly and the throttle body air inlet duct.

25. Connect negative battery cable.

26. Start vehicle and verify no leaks.

Right Cylinder Head (Rear)

1. Properly relieve the fuel system pressure. Disconnect the negative battery cable.

2. Drain the cooling system.

3. Remove the top half of the air cleaner assembly and remove the throttle body air inlet duct.

4. Remove the exhaust crossover pipe heat shield and crossover pipe.

5. Raise and safely support the vehicle.

6. Disconnect the Oxygen (O_2) sensor connector.

7. Disconnect the exhaust pipe from the exhaust manifold.

8. Remove the right side exhaust manifold. Lower the vehicle.

9. Disconnect the spark plug wires from spark plugs and wire looms and route the wires aside.

10. Remove the rocker arm covers.

11. Remove upper intake plenum and lower intake manifold.

12. Remove rocker arms nut, rocker arms, balls and pushrods.

13. Remove the cylinder head bolts evenly.

14. Remove the cylinder head.

To install:

15. Clean all the gasket surfaces completely. Clean the threads on the cylinder head bolts and block threads.

16. Place the cylinder head gasket in position over the dowel pins on the cylinder block so the words **THIS SIDE UP** showing.

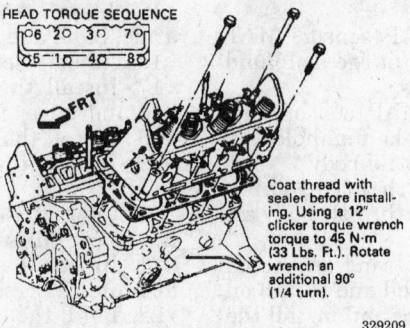

HEAD TORQUE SEQUENCE

Coat thread with sealer before installing. Using a 12" clicker torque wrench torque to 45 N·m (33 Lbs. Ft.). Rotate wrench an additional 90° (1/4 turn).

329209

Cylinder head components and bolt tightening sequence — 3.1L (VIN T) engines

17. Coat the bolt threads with sealer and install finger-tight.

18. Tighten the cylinder head bolts in sequence to 33 ft. lbs. (45 Nm). With all the bolts tightened make a second pass tightening all the bolts an additional 90 degrees (¼ turn).

19. Install the pushrods, rocker arms, balls and rocker arm nuts. Tighten the rocker arm nuts to 18 ft. lbs. (25 Nm).

20. Install the lower intake manifold and upper intake plenum.

21. Install the rocker arm covers.

22. Connect the spark plug wires to spark plugs and wire looms.

23. Raise vehicle and safely support.

24. Install the exhaust manifold.

25. Connect the exhaust pipe to the exhaust manifold.

26. Lower the vehicle.

27. Connect the Oxygen (O_2) sensor connector.

28. Install the exhaust crossover pipe and heat shield.

29. Refill the cooling system.

30. Install the top half of the air cleaner assembly and the throttle body air inlet duct.

31. Connect negative battery cable.

32. Start vehicle and verify no leaks.

Valve Lifters

BLEEDING

These vehicles are equipped with hydraulic valve lifters and do not require adjustment. Please note that if they are removed for service, they must be kept in order so they may be reinstalled in there original position.

REMOVAL AND INSTALLATION

------ CAUTION ------
Fuel injection systems remain under pressure, even after the engine has been turned OFF. The fuel system pressure must be relieved before disconnecting any fuel lines. Failure to do so may result in fire and/or personal injury.

2.2L (VIN 4) Engine

The cylinder head must be removed from the engine for this procedure.

1. Disconnect the negative battery cable. Relieve the fuel system pressure.

2. Drain the cooling system.

3. Remove the cylinder head with the intake and exhaust manifold as an assembly.

4. Remove the anti-rotation bracket bolts, the brackets and the lifters.

NOTE: Whenever new valve lifters are being installed, coat the valve lifters with camshaft assembly lube GM No. 1052365, or equivalent.

To install:

5. Coat the base of the lifters with camshaft lube or engine oil supplement prior to installation in the lifter bore.

6. If any lifters are being reused, install the lifters in the same bores from which they were removed.

7. Install the anti-rotation brackets and tighten the mounting bolts to 97 inch lbs. (11 Nm).

8. Install the cylinder head and manifolds.

9. Complete the installation by reversing the removal procedures.

10. Refill the engine with coolant. An oil and filter change is recommended.

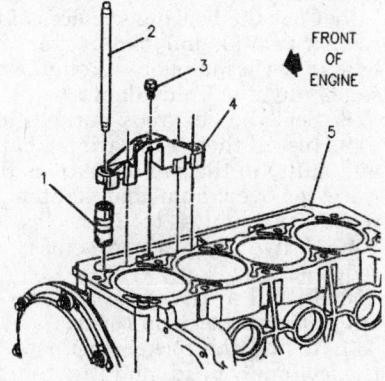

FRONT OF ENGINE

1 LIFTER, VALVE ROLLER
2 ROD, PUSH
3 BOLT – 11 N·m (97 LBS. IN.)
4 BRACKET, ANTI-ROTATION
5 BLOCK, CYLINDER

330554

Lower valve train components — 2.2L (VIN 4) engine

11. Connect the negative battery cable.

3.1L (VIN T) and (VIN M) Engines

1. Disconnect the negative battery cable. Drain the cooling system.

2. Remove the intake manifold.

3. Remove rocker arm covers.

NOTE: When removing valve train components they must be marked for installation in the same location.

4. Remove the rocker arm retaining nuts and remove the rocker arm balls, rocker arms and pushrods. Organize the components on a work bench so they can be reinstalled in their correct locations if more than one lifter is being replaced.

5. For 3.1L (VIN T) engines, remove the valve lifters from the intake valley.

6. For 3.1L (VIN M) engines, remove the 2 guide bolts from the right or left side lifter guide and remove the guide, then remove the valve lifter(s) from the lifter bores.

To install:

7. Coat the base of the valve lifter with Molykote® before installing in the lifter bore.

NOTE: Installation of a new camshaft or a wear pattern on the old valve lifter will require the replacement of the camshaft and lifters together. If camshaft replacement is not necessary, make sure to install the used valve lifters in their original position upon reinstallation.

8. For 3.1L (VIN T) engines, install the valve lifters into their bores.

9. For 3.1L (VIN M) engines, install the lifter guide and lifter guide bolts and tighten the guide bolts to 89 inch lbs. (10 Nm).

10. Install the pushrods, making sure they are properly seated in the valve lifters.

11. Install the rocker arms, rocker balls and rocker arm nuts. Tighten the rocker arm nuts to 18 ft. lbs. (25 Nm) on 1994–95 vehicles. On 1996 vehicles, torque to 89 inch lbs. (10 Nm). plus an additional 30 degrees.

12. Install the rocker arms, rocker arm balls and nuts.

13. Adjust the valve lifters.

14. Install the intake manifold.

15. Refill the cooling system.

16. Install the rocker arm covers.

17. Connect the negative battery cable.

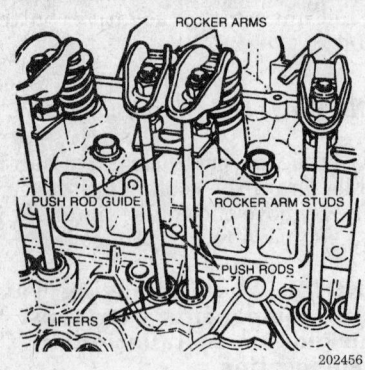

Valve train mechanism — 3.1L (VIN T) engine

Valve Lash

ADJUSTMENT

These vehicles are originally equipped with hydraulic valve lifters and do not require adjustment. Please note that if they are removed for service, they must be kept in order so they may be reinstalled in there original position.

3.1L (VIN T) and (VIN M) Engine

NOTE: These engines come from the factory with non-adjustable valve lash, BUT if the valves and the valve seats are reconditioned, adjustable rocker arm studs must be installed. The adjustment procedure is ONLY for adjustable rocker arm studs.

1. The engine should be in the No. 1 firing position. This may be determined by placing fingers on the No. 1 rocker arms as the engine assembly alignment marks on the front face of the torsional damper aligns with the arrow on the front cover. If the valves are not moving, the engine is in the No. 1 firing position. If the valves move the engine is in the No. 4 firing position and should be rotated 1 revolution to reach the No. 1 position.

2. With the engine in the No. 1 firing position, the following valves should be adjusted:

- Exhaust — 1, 2, 3
- Intake — 1, 5, 6

3. Tighten the adjusting nut until there is 0 lash + 1½ turns.

4. Crank the engine 1 revolution. This is the No. 4 firing position. With the engine in this position, the following valves should be adjusted to 0 lash + 1½ turns:

- Exhaust — 4, 5, 6
- Intake — 2, 3, 4

Rocker Arms

REMOVAL AND INSTALLATION

2.2L (VIN 4) Engine

This engine uses a simple ball pivot-type rocker arm. Motion is transmitted from the camshaft through the hydraulic roller lifters and pushrod to the rocker arm. The rocker arm pivots on its ball and transmits the camshaft motion to the valve. The rocker arm ball is located on a stud threaded onto the head and is retained by a nut. The pushrod is located by a guide plate held under the rocker arm stud.

1. Disconnect the negative battery cable. Disconnect the air intake hose from the air cleaner and throttle body.

2. Remove the control cable bracket from the rocker arm cover.

3. Disconnect the PCV hose.

4. Remove the rocker arm cover bolts, then remove the cover.

NOTE: Place components in a rack in order at disassembly so they can be reinstalled at the same location and with the same mating surface as when removed.

5. Remove the rocker arm nut(s), ball(s) and rocker arms. Pushrods may be removed by simply pulling out of the block.

To install:

6. Clean the gasket surfaces completely.

7. Coat the bearing surfaces of the rocker arm(s) and the rocker arm ball(s) with Molykote®, engine assembly lube or equivalent.

8. Seat the pushrods in the lifters.

9. Install the rocker arm(s), ball(s) and nut(s) in the same positions they were removed from and tighten the rocker arm nut(s) to 22 ft. lbs. (30 Nm). Valve lash adjustment is not required.

10. Install a new gasket in the cut out of the rocker arm cover.

11. Install the rocker arm cover on the cylinder head and tighten the rocker arm cover bolts to 89 inch lbs. (10 Nm).

12. Install the remaining components in the reverse order of removal.

13. Connect the negative battery cable. Start the engine and verify no oil leaks.

3.1L (VIN T) Engine

1. Remove the air cleaner assembly. Disconnect the negative battery cable. Drain the cooling system.

2. For the left side rocker arms (front), perform the following procedures:

a. Disconnect the ignition wire clamps from the coolant pipe.

b. Remove the coolant tube mounting bolt at the cylinder head.

c. Disconnect the coolant tube at both ends.

d. Remove the coolant tube hose from the water pump.

e. Remove the coolant tube.

f. Remove the vent pipe from the rocker arm cover to the air inlet duct.

3. For the right side rocker arms (rear), perform the following procedures:

a. Disconnect the vacuum lines at the intake plenum.

b. Remove the air inlet duct.

c. Disconnect the EGR tube at the crossover pipe.

d. Disconnect the spark plug wires from the rear plugs and disconnect the plug wire loom from the intake plenum.

e. Disconnect the coolant hoses at the throttle base.

f. Disconnect the electrical wiring from the plenum.

g. Disconnect the control cables from the throttle body.

h. Remove the bracket from the right side of the plenum.

i. Disconnect the brake booster pipe fitting from the intake plenum.

j. Remove the serpentine belt.

k. Disconnect the exhaust pipe from the crossover pipe.

l. Remove the alternator.

m. Disconnect the PCV valve from the rocker arm cover.

4. Remove the ignition wire guide.

5. Remove the 4 rocker arm cover bolts and remove the rocker arm cover.

NOTE: When removing the valve train components they must be kept in order for installation in the same locations from which they were removed.

6. Remove the rocker arm nut(s), ball(s), rocker arm(s) and pushrods.

To install:

7. Clean all the gasket surfaces completely.

8. Coat all the valve train components with engine oil prior to installation.

9. Install the pushrods making sure each is seated in the valve lifter.

10. Install the rocker arm(s) on the stud(s). Install the rocker arm ball(s) and mounting nuts. Make sure the pushrods are properly seated in the rocker arm.

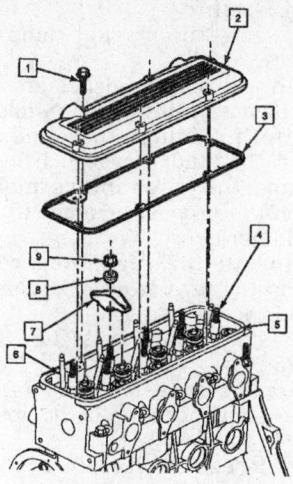

1	BOLT
2	ROCKER COVER
3	GASKET
4	ROCKER STUD
5	FLANGE, MUST BE FREE OF OIL UPON ROCKER COVER GASKET INSTALLATION.
6	PUSHROD
7	ROCKER ARM
8	BALL
9	NUT

330157

Rocker arm and cover — 2.2L (VIN 4) engine

11. Adjust the valves once all the rocker arms have been installed.

12. Install the rocker arm cover using a new gasket and tighten the rocker cover bolts to 90 inch lbs. (10 Nm).

13. Install the ignition wire guide.

14. For the left side rocker arms (front), perform the following procedures:

 a. Install the vent pipe from the rocker arm cover to the air inlet duct.

 b. Install the coolant tube and connect the coolant tube hose to the water pump.

 c. Connect the tube to the heater hose.

 d. Install the coolant tube mounting nut at the cylinder head.

 e. Connect the ignition wire clamps to the coolant tube.

15. For the right side rocker arms (rear), perform the following procedures:

 a. Connect the PCV valve to the rocker arm cover.

 b. Install the alternator.

 c. Connect the exhaust pipe to the crossover pipe.

 d. Install the serpentine belt.

 e. Connect the brake booster vacuum pipe to the intake plenum.

 f. Install the bracket to the right side of the plenum.

 g. Connect the control cables to the throttle body.

 h. Connect the electrical wiring to the plenum.

 i. Connect the coolant hoses to the throttle base.

 j. Connect the spark plug wire loom to the intake plenum and connect the spark plug wires to the plugs.

 k. Connect the EGR pipe to the crossover pipe.

 l. Install the air inlet duct.

 m. Connect the vacuum hoses to the intake plenum.

16. Refill the cooling system.

17. Connect the negative battery cable. Install the air cleaner.

18. Start the vehicle and verify no leaks.

3.1L (VIN M) Engine

1. Disconnect the negative battery cable.

2. For the left side rocker arms (front), perform the following procedures:

 a. Drain the cooling system to a level below the coolant pipe on the front of the engine.

 b. Remove the coolant bypass hose clamp at the coolant tube.

 c. Remove the 2 bolts and nut securing the coolant tube to the cylinder head and position the tube aside.

 d. Disconnect the PCV valve from the rocker arm cover.

3. For the right side rocker arms (rear), perform the following procedures:

 a. Disconnect the spark plug wires from the spark plugs and up-

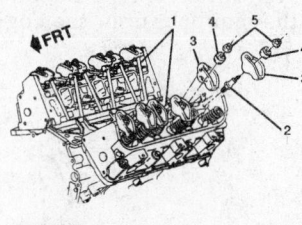

1 Pushrod
2 Valve rocker arm stud
3 Valve rocker arm
4 Valve rocker arm pivot ball
5 Valve rocker arm nut

330179

Rocker arm components — 3.1L (VIN M) engine

per intake plenum wire retainer and position aside.

 b. Disconnect the power brake booster vacuum pipe from the intake plenum.

 c. Remove the serpentine belt.

 d. Remove the alternator.

 e. Disconnect and remove the ignition assembly and EVAP canister purge solenoid as an assembly.

4. Remove the 4 rocker arm cover bolts and remove the rocker arm cover.

5. Remove the rocker arm nuts, balls, rocker arms and pushrods.

To install:

6. Clean all the gasket surfaces completely.

7. Coat all the valve train components with engine oil prior to installation.

8. Install the pushrods and install the rocker arms on the studs. Install the rocker arm balls and mounting nuts. Make sure the pushrods are properly seated in the lifter and rocker arm. Tighten the mounting nuts to 18 ft. lbs. (24 Nm).

9. Install the rocker arm cover using a new gasket and bolt grommet, tighten the rocker cover bolts to 90 inch lbs. (10 Nm).

10. For the left side rocker arms (front), perform the following procedures:

 a. Connect the PCV valve to the rocker arm cover.

 b. Position the coolant tube and connect the thermostat bypass hose.

 c. Install the coolant tube mounting nut and bolts. Tighten the screw at the water pump to 106 inch lbs. (12 Nm), the bolt at the corner of the cylinder head to 18 ft. lbs. (25 Nm) and the nut to 18 ft. lbs. (25 Nm).

11. For the right side rocker arms (rear), perform the following procedures:

 a. Install the alternator.

 b. Install the serpentine belt.

 c. Connect the power brake booster vacuum pipe to the plenum.

 d. Install the EVAP solenoid and ignition assembly.

 e. Connect the spark plug wires to the wire retainers on the plenum and the spark plugs.

12. Refill the cooling system.

13. Connect the negative battery cable. Start the engine and verify that there are no leaks.

Intake Manifold

REMOVAL AND INSTALLATION

— CAUTION —

Fuel injection systems remain under pressure, even after the engine has been turned OFF. The fuel system pressure must be relieved before disconnecting any fuel lines. Failure to do so may result in fire and/or personal injury.

2.2L (VIN 4) Engine

1993–95 Vehicles

1. Relieve the fuel system pressure. Disconnect the negative battery cable.
2. Remove the throttle body air intake duct. Drain the cooling system.
3. Identify, tag and disconnect all necessary vacuum lines.
4. Disconnect the control cables from the throttle body lever and remove the control cable bracket from the intake manifold.
5. Remove the serpentine belt.
6. Remove the power steering pump and lay it aside, without disconnecting the fluid lines.
7. Remove the transaxle fluid fill tube.
8. Tag and disconnect the MAP sensor, EGR solenoid valve, Idle Air Control (IAC) valve, Throttle Position Sensor (TPS) and fuel injector connectors.
9. Remove the MAP sensor.
10. Remove the upper intake manifold mounting bolts and the upper intake manifold.
11. Disconnect the fuel lines from the fuel rail.
12. Remove the EGR valve injector.
13. Remove the EGR valve.
14. Remove the fuel injector retainer bracket, regulator and injectors.
15. Remove the control cable bracket.
16. Remove the 6 intake manifold nuts. Remove the intake manifold.

To install:

17. Clean the gasket mounting surfaces.
18. Install a new gasket and position the lower intake manifold. Tighten the lower intake manifold nuts in the proper sequence to 24 ft. lbs. (33 Nm).
19. Install the remaining components in the reverse order of removal. Torque the following:
- Fuel injector retainer bracket bolts to 22 inch lbs. (3.5 Nm)
- Upper intake manifold nuts in the proper sequence to 22 ft. lbs. (30 Nm)
20. Connect the negative battery cable. Refill the coolant system.
21. Start the vehicle and verify no coolant or vacuum leaks.

1996 Vehicles

1. Relieve the fuel system pressure. Disconnect the negative battery cable.
2. Remove the throttle body air intake duct.
3. Remove the accelerator cable splash shield.
4. Remove the accelerator, throttle valve control and if equipped the cruise control cable.
5. Remove the vacuum harness from the top of the air inlet.
6. Remove the PCV and brake vacuum hoses at the upper intake manifold.
7. Remove the Manifold Air Pressure (MAP) sensor, the Throttle Position Sensor (TPS) and the Idle Air Control (IAC) electrical connectors.
8. Remove the 2 bolts and one nut attaching the accelerator bracket.
9. Remove the MAP sensor.
10. Remove the upper intake manifold mounting bolts and the upper intake manifold.
11. Remove the fuel feed and return quick-connect fittings.
12. Remove the bolt holding the intake brace to the power steering pump brace.
13. Remove the fuel inlet pipe retainers. Remove the transaxle fill tube attaching nut.
14. Remove the lower manifold attaching nuts.
15. Remove the 4 upper studs on the lower intake manifold and remove the manifold from the engine.

To install:

16. Clean the gasket mounting surfaces.
17. Install a new gasket and position the lower intake manifold. Install the retaining studs and nuts, tighten the studs to 89 inch lbs. (10 Nm) and the lower intake manifold nuts in the proper sequence to 24 ft. lbs. (33 Nm).
18. Install the remaining components in the reverse order of removal. Torque the following:
- Fuel inlet pipe bolts to 18 inch lbs. (25 Nm)
- Upper intake manifold nuts, in proper sequence, to 22 ft. lbs. (30 Nm)
- MAP sensor bolt to 27 inch lbs. (3 Nm)
- Accelerator cable splash shield bolts to 18 inch lbs. (25 Nm)
19. Install the air inlet duct assembly. Connect the negative battery cable.
20. Cycle the ignition switch several times to pressurize the fuel system and check for leaks.
21. Start the engine, check for proper idle and make sure there are no vacuum leaks.

2.3L (VIN A) Engine

1. Relieve the fuel system pressure. Disconnect the negative battery cable.
2. Drain the cooling system to a level below the intake manifold.
3. Disconnect the vacuum hose from the MAP sensor.
4. Detach the connectors from the MAP sensor, Intake Air Temperature (IAT) sensor, EVAP cannister purge solenoid and the fuel injectors.
5. Disconnect the vacuum hoses from the intake manifold, fuel pressure regulator and EVAP cannister purge solenoid to the canister.
6. Remove the air cleaner duct.
7. Remove the vent tube-to-air cleaner duct.

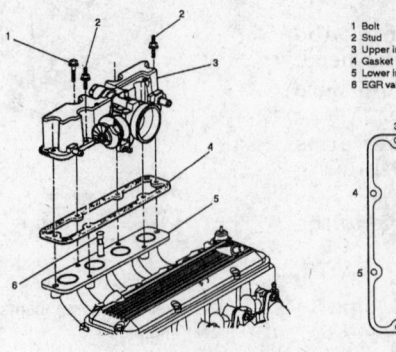
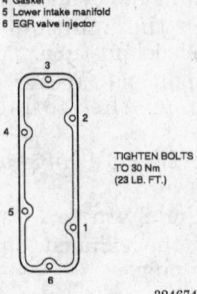

1 Bolt
2 Stud
3 Upper intake manifold assembly
4 Gasket
5 Lower intake manifold
6 EGR valve injector

TIGHTEN BOLTS
TO 30 Nm
(23 LB. FT.)

324674

Upper intake manifold — 1996 2.2L (VIN 4) engine

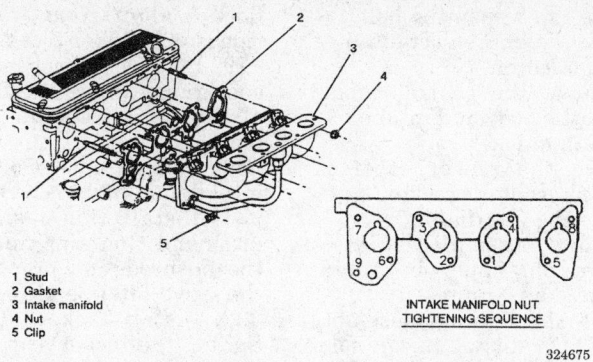

1 Stud
2 Gasket
3 Intake manifold
4 Nut
5 Clip

INTAKE MANIFOLD NUT
TIGHTENING SEQUENCE

324675

Lower intake manifold — 2.2L (VIN 4) engine

8. Remove the control cable mounting bracket from the intake manifold.

9. Disconnect the power brake vacuum line from the intake manifold and remove the vacuum line mounting bracket at the power steering pump.

10. Disconnect the coolant lines from the throttle body.

11. Remove the crankcase air/oil separator mounting bolts from the intake manifold.

12. Disconnect the hoses from the separator and remove the separator.

13. Remove the oil fill tube.

14. Remove the intake manifold support brace.

15. Remove the intake manifold mounting nuts and bolts.

16. Remove the intake manifold and discard the old gasket.

To install:

17. Clean all the gasket surfaces completely. Install the intake manifold gasket and intake manifold.

18. Tighten the intake manifold fasteners in sequence, to 18 ft. lbs. (25 Nm).

19. Install the intake manifold brace and tighten the mounting bolts to 19 ft. lbs. (26 Nm). The brace-to-block bolts must be tightened first then the brace-to-manifold bolt.

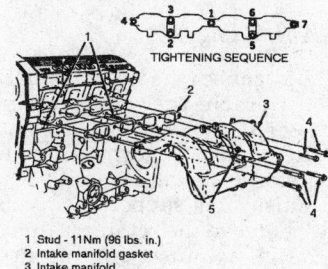

TIGHTENING SEQUENCE

1 Stud - 11Nm (96 lbs. in.)
2 Intake manifold gasket
3 Intake manifold
4 Bolt - 25 Nm (18 lbs. ft.)
5 Nut - 25 Nm (18 lbs. ft.)

327614

Intake manifold bolt tightening sequence — 2.3L (VIN A) engine

20. Connect the hoses to the air/oil separator. Position the separator on the mounting boss and install the mounting bolts.

21. Lubricate a new oil fill tube O-ring seal with engine oil and install tube between No. 1 and 2 intake tubes. Rotate as necessary to gain clearance for oil/air separator nipple on fill tube.

22. Locate the oil fill tube in its cylinder block opening. Align the fill tube so it is approximately in its installed position. Press straight down to seat fill tube and seal into cylinder block.

23. Lubricate the hoses and install the oil/air separator assembly.

24. Install the remaining components in the reverse order of removal.

25. Refill the cooling system.

26. Connect the negative battery cable.

3.1L (VIN T) Engine

1. Relieve the pressure in the fuel system. Remove the air cleaner assembly.

2. Disconnect the negative battery cable. Drain the cooling system.

3. Disconnect the accelerator and TV cables from the throttle body.

4. Disconnect the brake booster vacuum pipe from the intake manifold plenum.

5. Remove the accelerator cable and TV cable bracket from the plenum.

6. Disconnect the air inlet duct from the throttle body.

7. Disconnect and remove the throttle body at the intake plenum.

8. Disconnect and remove the EGR valve at the intake plenum.

9. Remove the upper intake plenum mounting bolts and remove the plenum.

10. Disconnect the fuel feed and return lines from the fuel rail.

11. Remove the serpentine belt.

12. Disconnect the alternator and lay it aside.

13. Disconnect the power steering lines from the alternator mounting bracket.

14. Remove the power steering pump and lay it aside.

15. Disconnect the vacuum line from the pressure regulator.

16. Remove the fuel rail mounting bolts. Detach the fuel injector connectors.

17. Remove the fuel rail assembly.

18. Disconnect the plug wires from the rear spark plugs and route over the intake manifold and aside.

19. Disconnect the heater pipe from the water pump and cylinder head.

20. Remove the front rocker arm cover. Disconnect the PCV hose from the rear rocker arm cover.

21. Remove the alternator bracket and rear brace.

22. Remove the rear rocker arm cover.

23. Disconnect the radiator hose at the thermostat housing.

24. Disconnect any remaining wiring from the intake manifold.

25. Remove the coolant sensor.

26. Remove the fuel lines from the mounting bracket.

27. Disconnect the throttle body heater hose. Disconnect the heater pipe from the intake manifold.

28. Remove the intake manifold mounting bolts. The washers on the 4 center studs must be kept in their original positions.

29. Loosen the rocker arms and remove the pushrods. The intake and exhaust pushrods are different lengths and must be kept in the same positions they were removed from. The intake pushrods are 6 in. long and the exhaust pushrods are 6⅜ in. long.

30. Remove the intake manifold.

To install:

31. Clean all the gasket surfaces completely.

32. Place a 3/16 in. (5mm) diameter bead GM sealer 1052917 or equivalent, on each ridge.

33. Install new intake manifold side gaskets on the cylinder heads.

34. Install the pushrods in their original positions and tighten the rocker arm nuts to 17 ft. lbs. (23 Nm).

35. Install the intake manifold and tighten the bolts in sequence to 15 ft. lbs. (20 Nm). After all 8 bolts have been tightened make a second pass tightening the bolts to 24 ft. lbs. (33 Nm).

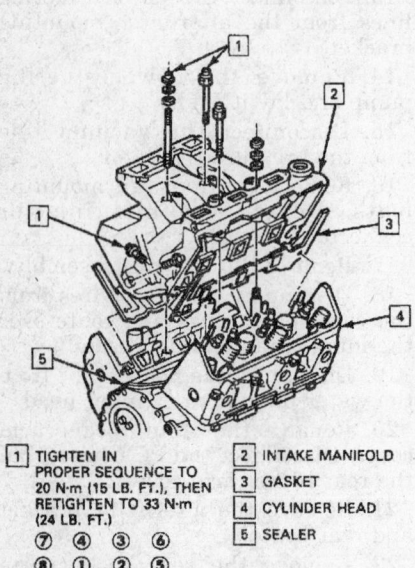

1	TIGHTEN IN PROPER SEQUENCE TO 20 N·m (15 LB. FT.), THEN RETIGHTEN TO 33 N·m (24 LB. FT.)	2	INTAKE MANIFOLD
3	GASKET
4	CYLINDER HEAD
5	SEALER

⑦ ④ ③ ⑥
⑧ ① ② ⑤

183385

Intake manifold bolt tightening sequence — 3.1L (VIN T) engine

36. Install the remaining components in the reverse order of removal. Torque the following:
- Rocker arm cover bolts to 8 ft. lbs. (11 Nm)
- Fuel rail bolts to 89 ft. lbs. (10 Nm)
- Intake plenum bolts to 16 ft. lbs. (21 Nm)
- Throttle body-to-intake plenum bolts to 18 ft. lbs. (25 Nm)

37. Refill the cooling system.
38. Connect the negative battery cable. Install the air cleaner.
39. Pressurize the fuel system before starting the vehicle to ensure all fuel line connections are tight.
40. Warm the engine to operating temperature and verify no leaks.

3.1L (VIN M) Engine

These vehicles use a two-piece intake manifold. Note that these pieces are cast aluminum. Use care when working with light alloy components.

1. Disconnect the negative battery cable. Relieve the fuel system pressure.
2. Remove the top half of the air cleaner assembly and throttle body duct.
3. Drain the cooling system.
4. Remove the EGR pipe from exhaust manifold.

5. Remove the serpentine belt.
6. Remove the brake vacuum pipe at the intake plenum.
7. Disconnect the control cables from the throttle body and intake plenum mounting bracket.
8. Remove the power steering lines at the alternator bracket.
9. Remove the alternator.
10. Disconnect the spark plug wires from the spark plugs and wire retainers on the intake plenum.
11. Remove the ignition assembly and the EVAP canister purge solenoid together.
12. Detach the connectors from the following components:
- Throttle Position Sensor (TPS)
- Idle Air Control (IAC)
- Fuel Injectors
- Coolant temperature sensor
- MAP sensor
- Camshaft Position (CMP) sensor

13. Detach the vacuum lines from the vacuum modulator, fuel pressure regulator, and the PCV valve.
14. Remove the MAP sensor from upper intake manifold.
15. Remove the upper intake plenum mounting bolts and remove the plenum.
16. Disconnect the fuel lines from the fuel rail and fuel line bracket.
17. Install engine support fixture special tool J-28467-A or an equivalent.
18. Remove the right side engine mount.
19. Remove the power steering mounting bolts and support the pump aside without disconnecting the power steering lines.
20. Disconnect the coolant inlet pipe from coolant outlet housing.
21. Remove the coolant bypass hose from the water pump and the cylinder head.
22. Disconnect the upper radiator hose at thermostat housing.
23. Remove the thermostat housing.
24. Remove both rocker arm covers.
25. Remove the lower intake manifold bolts. Make sure the washers on the 4 center bolts are installed in their original locations.

NOTE: When removing the valve train components they should be kept in order for installation the original locations.

26. Remove the rocker arm retaining nuts and remove the rocker arms and pushrods.
27. Remove the intake manifold from the engine.

To install:

28. Clean gasket material from all mating surfaces. Remove all excess

RTV sealant from front and rear ridges of cylinder block.
29. Place a 3mm bead of RTV, on each ridge, where the front and rear of the intake manifold contact the block.
30. Using a new gasket, install the intake manifold to the engine.
31. Install the pushrods, rocker arms and mounting nuts. Make sure the pushrods are properly seated in the valve lifters and rocker arms.
32. Install rocker arm nuts and tighten the rocker arm nuts to 18 ft. lbs. (24 Nm).
33. Install lower the intake manifold attaching bolts. Apply sealant PN 12345739 or equivalent to the threads of bolts, and torque bolts to 115 inch lbs. (13 Nm).
34. Install the front rocker arm cover.
35. Install the thermostat housing.
36. Connect the upper radiator hose to the thermostat housing.
37. Install the coolant inlet pipe to thermostat housing.
38. Install coolant bypass pipe at the water pump and cylinder head.
39. Install the power steering pump in the mounting bracket.
40. Connect the right side engine mount. Remove the special engine support tool.
41. Connect the fuel lines to fuel rail and bracket.
42. Install the upper intake manifold and torque the mounting bolts to 18 ft. lbs. (25 Nm).
43. Install the remaining components in the reverse order of removal.
44. Fill the cooling system. An engine oil and filter change is recommended.
45. Connect the negative battery cable. Start the vehicle and verify no leaks.

Exhaust Manifold

REMOVAL AND INSTALLATION

2.2L (VIN 4) Engine

1. Disconnect the negative battery cable. Detach the oxygen sensor connector.
2. Remove the serpentine drive belt. Remove the alternator.
3. Raise and support the vehicle safely. Remove the retainers and separate the exhaust pipe from the manifold.
4. Lower the vehicle.
5. Remove the bolt securing the oil fill tube to the engine block and pull the tube from the engine.

6. If the coolant inlet pipe interferes with removal, disconnect the mounting bracket from the transaxle stud.

7. Remove the manifold retaining nuts, then remove the manifold from the cylinder head. Remove and discard the gasket from the mating surfaces.

To install:

8. Install the manifold to the cylinder head using a new gasket. Tighten the retaining nuts to 115 inch lbs. (13 Nm).

9. Raise and support the vehicle safely, then install the exhaust pipe to the manifold. Tighten the retaining nuts to 18 ft. lbs. (25 Nm), then lower the vehicle.

10. Install the alternator and the serpentine drive belt.

11. Connect the coolant inlet pipe, if disconnected.

12. Lubricate the oil fill tube seal and install the oil fill tube.

13. Connect the wiring harness to the oxygen sensor.

14. Connect the negative battery cable. Start the engine and check for exhaust leaks.

2.3L (VIN A) Engine

1. Disconnect the negative battery cable. Disconnect the oxygen sensor (O₂) connector.

2. Remove the upper exhaust manifold heat shield mounting nuts and remove the heat shield. Remove the spacer from the center shield mounting stud.

3. Raise and safely support the vehicle.

4. Remove the exhaust manifold brace mounting nuts and remove the manifold brace.

5. Remove the spring loaded nuts securing the exhaust pipe to the manifold.

6. Pull the pipe down to disconnect it from the manifold.

7. Lower the vehicle.

8. Remove the exhaust manifold mounting nuts from the cylinder head studs.

9. Remove the exhaust manifold and gasket.

To install:

10. Clean all the gasket surfaces completely.

11. Install a new exhaust manifold gasket over the cylinder head studs and install the manifold.

12. Install the manifold mounting nuts and tighten in sequence to 31 ft. lbs. (42 Nm).

13. Raise and safely support the vehicle.

14. Install the exhaust manifold brace and tighten the upper mounting nut to 40 ft. lbs. (54 Nm) and the lower mounting nuts to 19 ft. lbs. (26 Nm).

15. Connect the exhaust pipe to the manifold and tighten the spring loaded nuts to 19 ft. lbs. (26 Nm). Tighten the nuts alternately and evenly to avoid binding the mounting nuts against the pipe.

16. Lower the vehicle.

17. Install the spacer over the center exhaust shield mounting nut.

18. Install the upper manifold heat shield and tighten the mounting nuts to 19 ft. lbs. (26 Nm).

19. Connect the the oxygen sensor (O₂) connector.

20. Connect the negative battery cable. Start the vehicle and verify no leaks.

3.1L (VIN T) Engine

Left Side Manifold (Front)

1. Disconnect the negative battery cable. Remove the air cleaner assembly.

2. Drain the cooling system.

3. Remove the engine cooling fan.

4. Remove the manifold heat shield.

5. Disconnect the exhaust crossover pipe at the left manifold.

6. Remove the exhaust manifold-to-cylinder head attaching bolts.

7. Remove the exhaust manifold.

To install:

8. Clean the gasket mounting surfaces.

9. Install the exhaust manifold and tighten the retaining bolts to 18 ft. lbs. (25 Nm).

10. Connect the exhaust crossover pipe to the left manifold and tighten the nuts to 18 ft. lbs. (25 Nm).

11. Install the manifold heat shield.

12. Install the cooling fan assembly.

13. Connect the negative battery cable. Install the air cleaner.

14. Start the vehicle and verify no exhaust leaks.

Right Side Manifold (Rear)

1. Disconnect the negative battery cable. Remove the air cleaner.

2. Raise and safely support the vehicle.

3. Remove the bolts securing the front exhaust pipe to the rear exhaust manifold.

4. Lower the vehicle.

5. Remove the heat shield from the exhaust manifold.

6. Disconnect the crossover pipe from the right manifold.

7. Disconnect the control cables from the throttle body and remove the cable bracket from the plenum.

8. Disconnect the oxygen (O₂) sensor wire.

9. Remove the exhaust manifold-to-cylinder head bolts and the exhaust manifold from the vehicle.

To install:

10. Clean the gasket mounting surfaces.

11. Install the exhaust manifold and exhaust manifold-to-cylinder head bolts. Tighten the exhaust manifold-to-cylinder head bolts to 18 ft. lbs. (25 Nm).

12. Connect the crossover pipe to the right manifold and tighten the mounting nuts to 18 ft. lbs. (25 Nm).

13. Connect the control cables to the throttle body lever and install the cable bracket on the intake plenum.

14. Connect the oxygen sensor wire.

15. Raise and safely support the vehicle.

16. Connect the exhaust pipe-to-exhaust manifold and tighten the mounting bolts to 22 ft. lbs. (30 Nm).

17. Install the manifold heat shield.

18. Lower the vehicle.

19. Connect the negative battery cable.

20. Install the air cleaner assembly.

21. Start the engine and check for leaks.

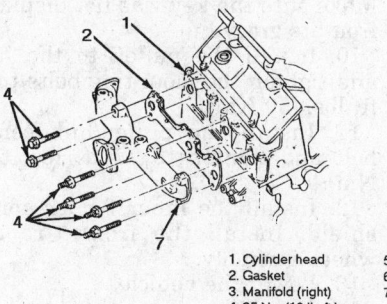
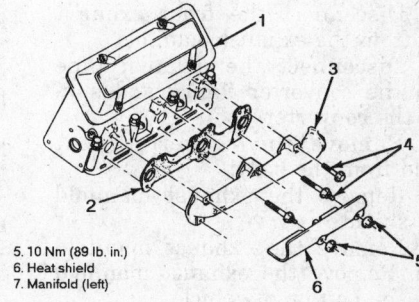

1. Cylinder head
2. Gasket
3. Manifold (right)
4. 25 Nm (18 lb. ft.)
5. 10 Nm (89 lb. in.)
6. Heat shield
7. Manifold (left)

183341

Exhaust manifold mounting — 3.1L (VIN T) engine

3.1L (VIN M) Engine

Left Side Manifold (Front)

1. Disconnect the negative battery cable.
2. Remove the top half of the air cleaner assembly and throttle cable duct.
3. Partially drain the cooling system. Remove the radiator hose from the thermostat housing.
4. Remove the coolant by-pass hose at the coolant pump and from the exhaust manifold.
5. Remove the exhaust crossover heat shield. Remove the exhaust crossover pipe from the manifold.
6. Tag and disconnect the spark plug wires.
7. Remove the exhaust manifold heat shield. Remove the exhaust manifold retaining nuts.
8. Remove the exhaust manifold.

To install:

9. Clean mating surfaces at the cylinder head and manifold.
10. Install the exhaust manifold gasket and the exhaust manifold. Torque the manifold mounting nuts to 12 ft. lbs. (16 Nm).
11. Install the remaining components in the reverse order of removal.
12. Connect the negative battery cable. Start the engine and check for leaks.

Right Side Manifold (Rear)

1. Disconnect the negative battery cable. Remove the top half of the air cleaner assembly and throttle cable duct.
2. Remove the exhaust crossover heat shield. Remove the exhaust crossover pipe from the manifold.
3. Remove the heated oxygen sensor. Remove the EGR pipe from the exhaust manifold.
4. Raise and safely support the vehicle.
5. Remove the transaxle oil fill tube and lever indicator assembly.
6. Disconnect the front exhaust pipe from the exhaust manifold.
7. Disconnect the exhaust pipe from the converter flange and support the converter.
8. Remove the converter heat shield from the body.
9. Remove the exhaust manifold heat shield.
10. Remove the exhaust manifold nuts. Remove the exhaust manifold from the bottom of vehicle.

To install:

11. Clean mating surfaces at the cylinder head and manifold.

12. Install the exhaust manifold gasket.
13. Install the exhaust manifold loosely and install heat shield at this time.
14. Install the manifold nuts and torque to 12 ft. lbs. (16 Nm).
15. Install the remaining components in the reverse order of removal.
16. Connect the negative battery cable, then start the engine and check for leaks.

Front Cover Seal

REMOVAL AND INSTALLATION

2.2L (VIN 4) Engine

1. Disconnect the negative battery cable. Remove the serpentine belt.
2. Raise and safely support the vehicle. Remove the right front tire.
3. Remove the inner fender splash shield.
4. Separate the crankshaft pulley, from the pulley hub, by removing the 3 mounting bolts.
5. Install special tool J-24420–B or equivalent damper puller on crankshaft hub, turn puller screw and remove hub. Use care not to loose the small key on the crankshaft nose.
6. Pry out oil seal with a prybar.

NOTE: Take care not to damage the seal seat in the front cover or the crankshaft surface.

To install:

7. Clean out the seal recess in the front cover.
8. Press the oil seal into the front cover, using special tool J-35468 or equivalent driver tool.
9. If removed, install the key in the slot in the crankshaft nose. A small amount of silicone sealer can be used to hold it in place. Align the pulley hub keyway (slot) with the key, and install the hub to the crankshaft. Make sure the key was not displaced from its groove.
10. Install the pulley to the hub, and tighten the mounting bolts to 37 ft. lbs. (50 Nm).
11. Install the pulley/hub center bolt and tighten to 77 ft. lbs. (105 Nm).
12. Install the inner fender splash shield. Install the front tire and wheel assembly.
13. Lower the vehicle.
14. Install the serpentine belt.
15. Connect negative battery cable.
16. Start the engine and check for leaks.

2.3L (VIN A) Engine

1. Disconnect the negative battery cable. Drain the cooling system.
2. Remove the coolant surge tank.
3. Remove the serpentine belt.
4. Remove the alternator.
5. Install an engine support fixture, J-28467–A. Install the alternator through bolt and connect the support fixture.
6. Remove the upper cover fasteners. Disconnect the front cover vent hose.
7. Remove the right engine mount and lift bracket.
8. Raise and safely support the vehicle. Remove the right front tire and wheel assembly.
9. Remove the right inner fender splash shield.
10. Remove the crankshaft balancer mounting bolt while preventing the crankshaft from turning. Using J-24420–B, or an equivalent puller, remove the crankshaft balancer.
11. Remove the lower cover fasteners. Lower the vehicle.
12. Remove the timing chain cover from the engine.
13. Support the front cover over 2 blocks of wood of equal thickness. Place the cover on the blocks so the engine side of the cover is resting on the blocks. The blocks should be as close to the front oil seal as possible without interfering with removing the oil seal.
14. Carefully drive the seal out of the front cover.

To install:

15. Install a new oil seal using J-36010 or an equivalent seal installation and alignment tool. The seal must be driven in from the engine side of the timing cover.
16. Install the front cover on the engine. Install the mounting nuts on the studs then install the mounting bolts.
17. Tighten the upper cover fasteners to 106 inch lbs. (12 Nm).
18. Raise and safely support the vehicle.
19. Tighten the lower cover fasteners to 106 inch lbs. (12 Nm).
20. Coat the seal contact area on the balancer with clean engine oil and install the balancer on the crankshaft.
21. While preventing the crankshaft from turning, tighten the balancer bolt to 129 ft. lbs. (175 Nm) plus 90 degrees additional rotation.
22. Install the right inner fender splash shield. Install the right front tire and wheel assembly.
23. Lower the vehicle.

24. Install the right engine mount bracket and engine mount. Use new replacement bolts. The engine mount bracket bolts **MUST** be replaced once they are removed.
25. Remove the engine support fixture. Install the alternator.
26. Install the serpentine belt.
27. Connect the vent hose to the front cover.
28. Install the coolant surge tank.
29. Refill the cooling system. Check the engine oil level. An oil and filter change is recommended.
30. Connect the negative battery cable. Start the engine and verify no leaks.

3.1L (VIN T) and (VIN M) Engine

1. Disconnect the negative battery cable. Remove the air cleaner assembly.
2. Remove the serpentine belt.
3. Raise and safely support the vehicle. Remove the right front tire and wheel assembly.
4. Remove the right inner fender well splash shield.
5. Remove the flywheel cover at the transaxle.
6. Remove the torsional dampener as follows:
   a. Remove the torsional damper mounting bolt while holding the crankshaft from turning.
   b. Install tool J-24420 or an equivalent puller and remove the damper from the crankshaft.
7. Pry the seal out of the front cover using a suitable prying tool.
To install:
8. Clean out the oil seal recess in the front cover.
9. Install the new seal in the front cover using J-35468, or an equivalent installation tool, to fully seat the seal in the cover.
10. Install the torsional damper as follows:
   a. Coat the seal contact area on the damper with clean engine oil.
   b. Line up the notch in the damper with the crankshaft key and slide the damper onto the crankshaft until the key is in the notch.
   c. Using tool J-29113, pull the damper onto the crankshaft.
   d. Install the damper mounting bolt and tighten to 76 ft. lbs. (102 Nm).
11. Install the flywheel cover at transaxle.
12. Install the right inner fender splash shield. Install the tire and wheel assembly.
13. Lower the vehicle.

14. Install the serpentine belt.
15. Refill the cooling system.
16. Connect the negative battery cable.
17. Install the air cleaner assembly.
18. Start the vehicle and verify no coolant leaks or oil leaks.

Timing Chain, Sprockets and Front Cover

REMOVAL AND INSTALLATION

2.2L (VIN 4) Engine

1. Disconnect the negative battery cable. Drain the engine coolant.
2. Remove the serpentine belt.
3. Remove the coolant reservoir.
4. Remove the front alternator mounting bolts.
5. Remove the 3 mounting bolts from the power steering pump. These bolts can be reached by going through the holes in the drive pulley on the pump. Lay the pump aside without disconnecting the hoses.
6. Remove the 4 tensioner mounting bolts and remove the tensioner.
7. Raise and safely support the vehicle.
8. Drain the engine oil, then remove the oil pan.
9. Remove the right front tire and wheel assembly.
10. Remove the right inner fender well splash shield.
11. Remove the crankshaft pulley.
12. Remove the crankshaft balancer.
13. Remove the front cover bolts and remove the front cover. If the cover is difficult to remove, use a soft faced mallet to lightly tap the cover to loosen it.
14. Rotate the crankshaft until the piston in No. 1 cylinder is at TDC on the compression stroke (firing position). The marks on the camshaft and crankshaft sprockets should be in alignment.
15. Loosen, but do not remove, the timing chain tensioner nut.
16. Remove the camshaft sprocket bolt and remove the sprocket and chain together. If the sprocket does not slide from the camshaft easily, a light blow with a soft mallet at the lower edge of the sprocket will dislodge it.
17. Use puller tool J-22888 or equivalent, and remove the crankshaft sprocket.
To install:
18. Install the crankshaft sprocket, using installation tool J-5590 or equivalent.

19. Install the timing chain over the camshaft sprocket and then around the crankshaft sprocket. Make sure the marks on the 2 sprockets are in alignment. Lubricate the thrust surface with Molykote® or its equivalent.
20. Align the dowel in the camshaft with the dowel hole in the sprocket and then install the sprocket onto the camshaft. Use the mounting bolt to draw the sprocket onto the camshaft and then tighten to 77 ft. lbs. (105 Nm).
21. Lubricate the timing chain with clean engine oil. Torque the bolts on the chain tensioner to 18 ft. lbs. (24 Nm).
22. Clean all gasket surfaces completely.
23. Apply a thin layer of sealer to the front cover and install a new gasket onto the cover.
24. Install the cover to the engine making sure the dowel pins line up with the holes in the front cover.
25. Install the cover mounting bolts and tighten them to 97 inch lbs. (11 Nm).
26. Install the crankshaft and pulley and tighten the pulley bolts to 37 ft. lbs. (50 Nm) and the center balancer bolt to 77 ft. lbs. (105 Nm).
27. Install the remaining components in the reverse order of removal. Torque the following:
   • Belt tensioner bolts to 37 ft. lbs. (50 Nm)
   • Power steering pump bolts to 25 ft. lbs. (34 Nm)
   • Alternator upper bolt to 22 ft. lbs. (30 Nm) and lower bolt to 37 ft. lbs. (50 Nm)
28. Fill the engine crankcase with clean oil. A filter change is recommended.
29. Refill the coolant system.
30. Connect the negative battery cable. Start the vehicle and verify no leaks.
31. Road test the vehicle and ensure proper operation.

2.3L (VIN A) Engine

Valve timing is absolutely critical. If the valve timing is even slightly incorrect, engine performance will suffer and the engine could be seriously damaged. On this engine, the timing chain removal is a complicated procedure. It is recommended that the entire procedure be reviewed before attempting to service the timing chain.

1. Disconnect the negative battery cable. Drain the cooling system.
2. Remove the coolant surge tank.

3. Remove the serpentine drive belt using a 13mm wrench that is at least 24 in. (61cm) long.

4. Disconnect and remove the alternator from the mounting bracket.

5. Install a suitable engine support, J–28467–A or the equivalent.

6. Remove the upper cover fasteners. Disconnect the upper cover vent hose.

7. Remove the right engine mount.

8. Remove the right engine mount bracket.

9. Raise and safely support the vehicle. Remove the right front tire and wheel assembly.

10. Remove the right lower splash shield from the right wheel house.

11. Remove the crankshaft balancer assembly using the following procedure:

a. Remove the torsional damper mounting bolt and washer while holding the crankshaft in place using J–38122, or the equivalent.

b. Remove the balancer using a suitable puller, J–24420–B or the equivalent.

12. Remove the lower cover fasteners. Lower the vehicle.

13. Remove the front cover, seal and gaskets.

14. Rotate the crankshaft clockwise, as viewed from front of engine (normal rotation) until the camshaft sprocket's timing dowel pin holes align with the holes in the timing chain housing. The mark on the crankshaft sprocket should align with the mark on the cylinder block. The crankshaft sprocket keyway should point upwards and align with the center line of the cylinder bores. This is the normal timed position.

15. Remove the 3 timing chain guides.

16. Raise and safely support the vehicle.

17. Gently pry off timing chain tensioner spring retainer and remove spring.

NOTE: Two styles of tensioner are used. Early production engines will have a spring post and late production ones will not. Both styles are identical in operation and are interchangeable.

18. Remove the timing chain tensioner shoe retainer.

19. Make sure all the slack in the timing chain is above the tensioner assembly. Remove the chain tensioner shoe. The timing chain must be disengaged from the wear grooves in the tensioner shoe in order to remove the shoe. Slide a prybar under the timing chain while pulling shoe outward.

20. If difficulty is encountered removing chain tensioner shoe, proceed as follows:

a. Lower the vehicle.

b. Hold the intake camshaft sprocket with a holding tool and remove the sprocket bolt and washer.

c. Remove the washer from the bolt and re-thread the bolt back into the camshaft by hand, the bolt provides a surface to push against.

d. Remove intake camshaft sprocket using a 3-jaw puller in the 3 relief holes in the sprocket. Do not attempt to pry the sprocket off the camshaft or damage to the sprocket or chain housing could occur.

21. Remove the tensioner assembly attaching bolts and the tensioner.

--- CAUTION ---

The tensioner piston is spring loaded and could fly out causing personal injury.

22. Remove the chain housing to block stud, which is actually the timing chain tensioner shoe pivot.

23. Remove the timing chain.

To install:

24. Tighten intake camshaft sprocket attaching bolt and washer, while holding the sprocket with tool J–36013, if removed.

25. Install the special tool through holes in camshaft sprockets into holes in timing chain housing. This positions the camshafts for correct timing.

26. If the camshafts are out of position and must be rotated more than ⅛ turn in order to install the alignment dowel pins:

a. The crankshaft must be rotated 90 degrees clockwise off TDC in order to give the valves adequate clearance to open.

b. Once the camshafts are in position and the dowels installed, rotate the crankshaft counterclockwise back to TDC. Do not rotate the crankshaft clockwise to TDC or valve and piston damage could occur.

27. Install the timing chain over the exhaust camshaft sprocket, around the idler sprocket and around the crankshaft sprocket.

28. Remove the alignment dowel pin from the intake camshaft. Using a dowel pin remover tool, rotate the intake camshaft sprocket counterclockwise enough to slide the timing chain over the intake camshaft sprocket. Release the camshaft sprocket wrench. The length of chain between the 2 camshaft sprockets will tighten. If properly timed, the in-

take camshaft alignment dowel pin should slide in easily. If the dowel pin does not fully index, the camshafts are not timed correctly and the procedure must be repeated.

29. Leave the alignment dowel pins installed.

30. With slack removed from chain between intake camshaft sprocket and crankshaft sprocket, the timing marks on the crankshaft and the cylinder block should be aligned. If marks are not aligned, move the chain 1 tooth forward or rearward, remove slack and recheck marks.

31. Tighten the chain housing to block stud. The stud is installed under the timing chain. Tighten to 19 ft. lbs. (26 Nm).

32. Reload timing chain tensioner assembly to its zero tension position as follows:

a. Assemble restraint cylinder, spring and nylon plug into plunger. Index slot in restraint cylinder with peg in plunger. While rotating the restraint cylinder clockwise, push the restraint cylinder into the plunger until it bottoms. Keep rotating the restraint cylinder clockwise but allow the spring to push it out of the plunger. The pin in the plunger will lock the restraint in the loaded position.

b. Install tool J–36589 or equivalent, onto plunger assembly.

c. Install plunger assembly into tensioner body with the long end toward the crankshaft when installed.

33. Install the tensioner assembly to the chain housing. Recheck plunger assembly installation. It is correctly installed when the long end is toward the crankshaft.

34. Install and tighten timing chain tensioner bolts and tighten to 10 ft. lbs. (14 Nm).

35. Install the tensioner shoe and tensioner shoe retainer. Remove special tool J–36589 and squeeze plunger assembly into the tensioner body to unload the plunger assembly.

36. Lower vehicle and remove the alignment dowel pins. Rotate crankshaft clockwise 2 full rotations. Align crankshaft timing mark with mark on cylinder block and reinstall alignment dowel pins. Alignment dowel pins will slide in easily if engine is timed correctly.

NOTE: If the engine is not correctly timed, severe engine damage could occur.

37. Install 3 timing chain guides and crankshaft oil slinger.

38. Install the front cover and gaskets on the engine and tighten the

mounting nuts and bolts to 106 inch lbs. (12 Nm).

39. Install the torsional damper as follows:

a. Coat the seal contact area on the crankshaft damper with clean engine oil.

b. Line up the damper on the crankshaft so the notch in the damper lines up with the crankshaft key.

c. Tap the balancer into place using a rubber mallet.

d. Install the damper mounting bolt and washer and tighten the bolt to 129 ft. lbs. (175 Nm) plus 90 degrees rotation.

40. Install the remaining components in the reverse order of removal.
41. Refill the cooling system.
42. Connect the negative battery cable and check for leaks.

3.1L (VIN T) Engine

1. Disconnect the negative battery cable. Drain the cooling system.
2. Remove the coolant reservoir.
3. Remove the serpentine belt and tensioner.
4. Remove the power steering pump and safely support it aside.

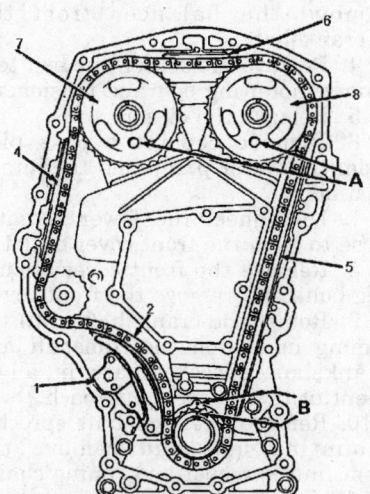

A. CAMSHAFT TIMING ALIGNMENT PIN LOCATIONS
B. CRANKSHAFT GEAR TIMING MARKS
1. SHOE ASM. TIMING CHAIN TENSIONER
2. TIMING CHAIN
3. TENSIONER, TIMING CHAIN
4. GUIDE – R.H. TIMING CHAIN
5. GUIDE – L.H. TIMING CHAIN
6. GUIDE – UPPER TIMING CHAIN
7. SPROCKET, EXHAUST CAMSHAFT
8. SPROCKET, INTAKE CAMSHAFT

330311

Timing chain and sprocket alignment positions — 2.3L (VIN A) engine

5. Remove the crankshaft balancer as follows:

a. Raise and safely support the vehicle.

b. Remove the right front tire and wheel assembly.

c. Remove the right inner fender well splash shield.

d. Remove the flywheel cover and install a flywheel holding tool.

e. Remove the balancer mounting bolt and washer.

f. Using a suitable puller, J-24420–B or the equivalent remove the balancer from the crankshaft.

6. Remove the serpentine belt idler pulley.

WARNING

Part of the oil pan removal procedure requires that the engine cradle be lowered for clearance. Failure to disconnect the intermediate steering shaft from the rack and pinion stub shaft can result in damage to the steering gear and/or intermediate shaft. This damage can cause loss of steering control which could result in personal injury.

7. Remove the engine oil pan.
8. Remove the lower timing cover bolts. Lower the vehicle.
9. Remove the radiator hose at the coolant pump. Remove the bypass pipe at the front cover.
10. Remove the canister purge hose.
11. Remove the upper timing cover bolts, then remove the cover.
12. Rotate the crankshaft until the timing marks on the camshaft and crankshaft sprockets are in alignment. The marks should be facing each other at their closest approach.

13. Remove the 3 camshaft sprocket mounting bolts and remove the camshaft sprocket and the chain.

NOTE: If the sprocket does not come off easily, a light blow on the lower edge of the sprocket with a plastic mallet should dislodge the sprocket.

14. Remove the crankshaft sprocket using tool J-5825–A, or the equivalent.
15. Remove the timing chain damper assembly bolts and damper.

To install:

16. Clean all parts well. Remove all traces of gasket material from the front cover, oil pan and engine block. Inspect the timing chain and sprockets for wear.
17. Install timing chain damper assembly and tighten the mounting bolts to 15 ft. lbs. (21 Nm).
18. Apply GM Engine Oil Supplement (EOS) P/N 1052367 or the equivalent lubricant to the sprocket thrust surface.
19. Install the camshaft sprocket in the timing chain.
20. Hold the camshaft sprocket with the timing mark pointing down and the timing chain hanging off the bottom of the sprocket.
21. Loop the chain under the crankshaft sprocket and install the camshaft sprocket over the camshaft dowel. DO NOT force the sprocket into place.
22. Align dowel in camshaft with dowel hole in camshaft sprocket.
23. Verify the timing marks are still in alignment.
24. Draw the camshaft sprocket onto camshaft using the mounting bolts. Tighten the mounting bolts to 18 ft. lbs. (25 Nm).
25. Lubricate timing chain with clean engine oil.
26. Clean all gasket surfaces completely.
27. Install a new gasket.

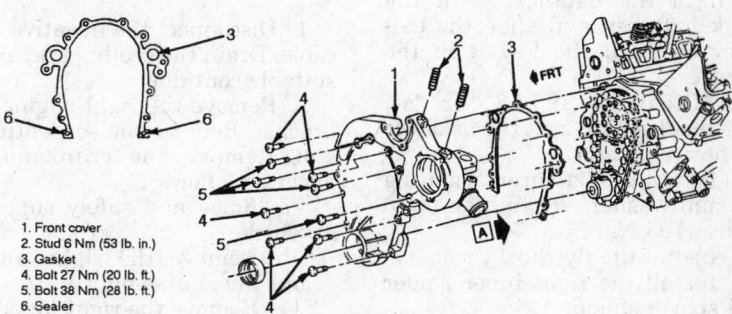

1. Front cover
2. Stud 6 Nm (53 lb. in.)
3. Gasket
4. Bolt 27 Nm (20 lb. ft.)
5. Bolt 38 Nm (28 lb. ft.)
6. Sealer

330369

Timing chain front cover assembly — 3.1L (VIN T) engine

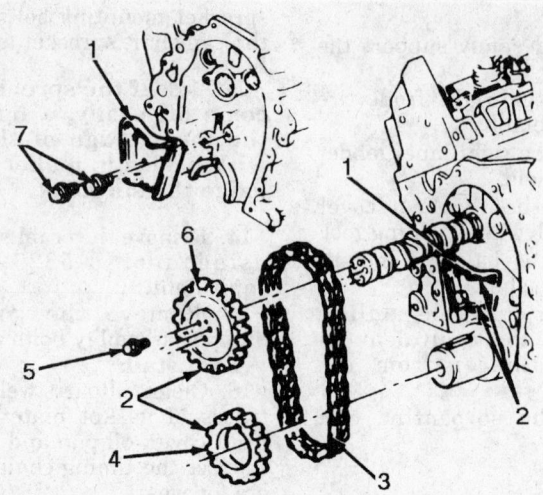

NOTE - ALIGN TIMING MARKS ON CAM
& CRANK SPROCKETS USING ALIGNMENT
MARKS ON DAMPER STAMPING OR CAST
ALIGNMENT MARKS ON CYL & CASE

1. Damper
2. Alignment marks
3. Timing chain
4. Crank sprocket
5. 28 Nm (21 lb. ft.)
6. Camshaft sprocket
7. 21 Nm (15 lb. ft.)

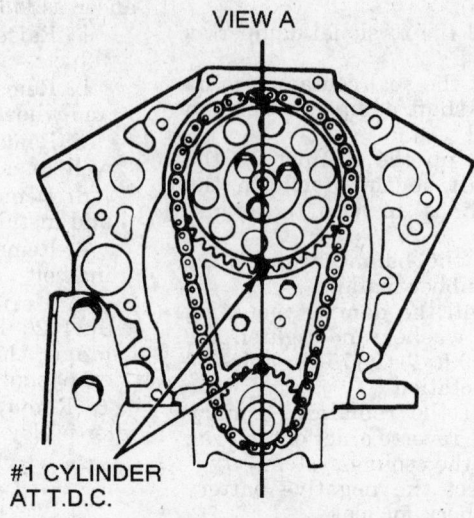

VIEW A

#1 CYLINDER
AT T.D.C.

NOTE - CAMSHAFT SPROCKET
MARK AT 6 O'CLOCK
CRANKSHAFT SPROCKET
MARK AT 12 O'CLOCK

330368

Timing chain and sprocket timing mark alignment — 3.1L (VIN T) engines

28. Apply a thin bead of sealer around the gasket sealing area of the front cover. Install a new front cover seal on the front cover.

29. Install the front cover on the engine and install the mounting bolts. Tighten the small bolts to 15 ft. lbs. (21 Nm). Tighten the large bolts to 35 ft. lbs. (45 Nm).

30. Using a new gasket, install the oil pan. Install the serpentine belt idler pulley.

31. Install crankshaft balancer as follows:

a. Coat the seal contact surface of the crankshaft balancer with clean engine oil.

b. Apply GM RTV or equivalent to the key and keyway. Line up the notch in the balancer with the crankshaft key and slide the balancer on until the key is in the balancer.

c. Using J-29113 or an equivalent puller, seat the balancer on the crankshaft.

d. Install the balancer mounting bolt and washer and tighten to 76 ft. lbs. (103 Nm).

e. Install the flywheel cover.

f. Install the right inner fender well splash shield.

g. Install the tire and wheel assembly and tighten to specification.

32. Lower the vehicle.

33. Install the radiator hose to the water pump. Install the bypass pipe at the front cover.

34. Install the canister purge hose.

35. Install the power steering pump.

36. Install the belt tensioner, tighten the bolts to 40 ft. lbs. (54 Nm). Install the serpentine drive belt.

37. Refill and bleed the cooling system.

38. Check the engine oil level. An oil and filter change is recommended.

39. Connect the negative battery cable. Start the engine and verify that there are no leaks.

3.1L (VIN M) Engine

1. Disconnect the negative battery cable. Drain the cooling system into a suitable container.

2. Remove the right engine mount bracket. Remove the serpentine belt.

3. Remove the crankshaft balancer as follows:

a. Raise and safely support the vehicle.

b. Remove the right front tire and wheel assembly.

c. Remove the right inner fender well splash shield.

d. Remove the flywheel cover and install a flywheel holding tool.

e. Remove the balancer mounting bolt and washer.

f. Using a suitable puller, J-24420–B or the equivalent remove the balancer from the crankshaft.

4. Remove the serpentine belt tensioner mounting bolt and tensioner.

5. Remove the oil pan.

6. Remove coolant bypass pipe from the water pump and the intake manifold.

7. Disconnect the lower radiator hose to from the front cover outlet.

8. Remove the front cover mounting bolts and remove the front cover.

9. Rotate the crankshaft until the timing marks on the camshaft and crankshaft sprockets are in alignment at their closest approach.

10. Remove the camshaft sprocket mounting bolt and remove the camshaft sprocket and timing chain.

11. Remove the crankshaft sprocket, using a gear puller J-5825–A or equivalent.

12. Remove the 2 mounting bolts from the timing chain damper and remove the damper.

To install:

13. Install the timing chain damper and tighten the mounting bolts to 15 ft. lbs. (21 Nm).

14. Install the crankshaft sprocket onto the crankshaft making sure the

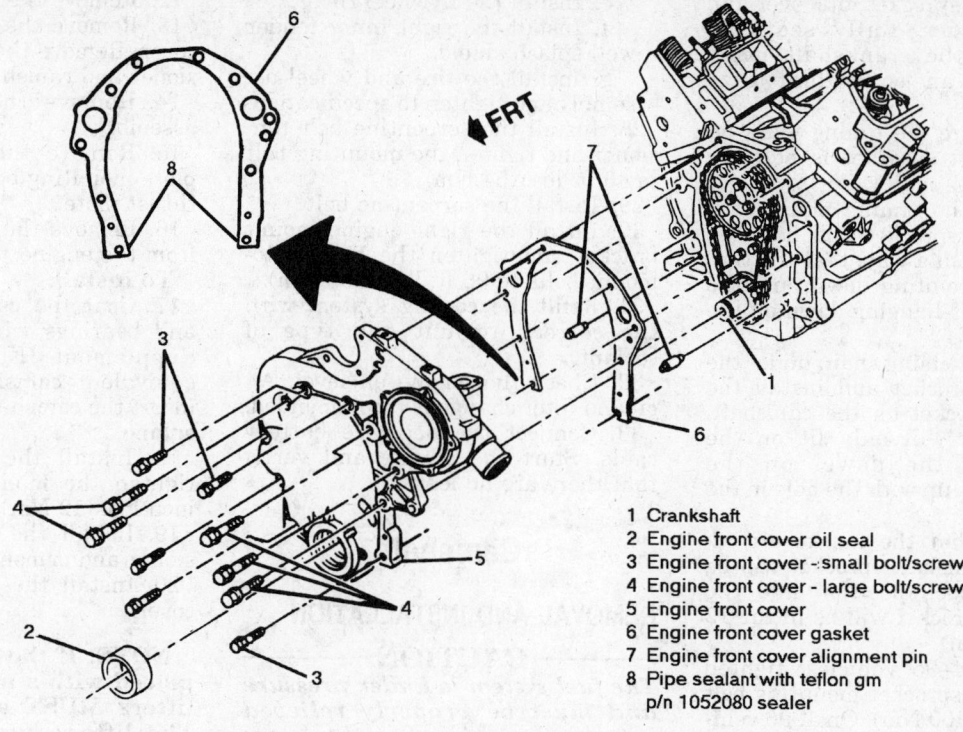

1 Crankshaft
2 Engine front cover oil seal
3 Engine front cover – small bolt/screw
4 Engine front cover - large bolt/screw
5 Engine front cover
6 Engine front cover gasket
7 Engine front cover alignment pin
8 Pipe sealant with teflon gm
  p/n 1052080 sealer

333477

Exploded view of the timing chain front cover components — 3.1L (VIN M) engine

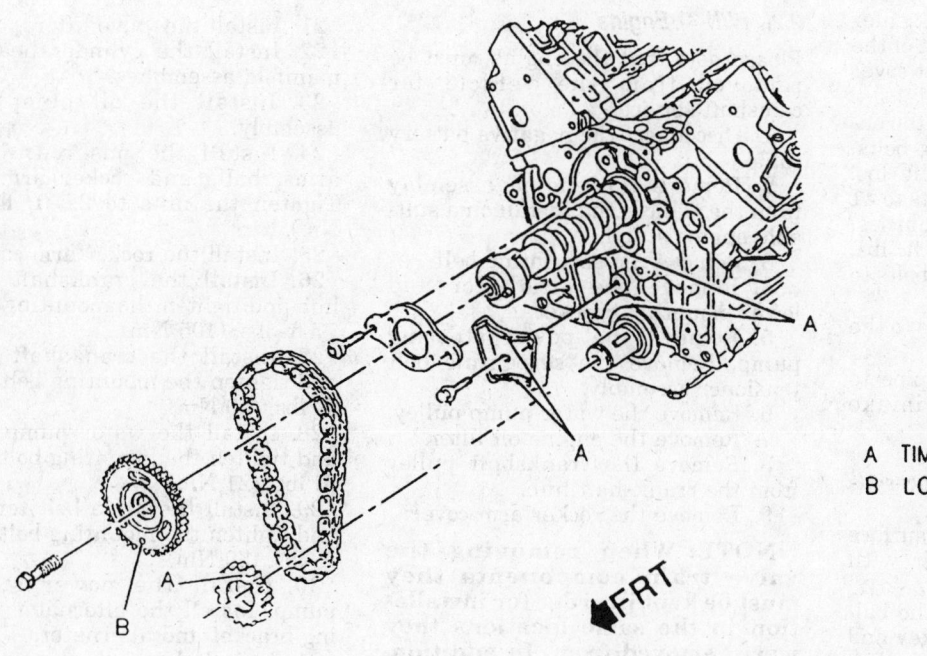

A TIMING ALIGNMENT MARKS
B LOCATOR HOLE

333813

Aligning the timing marks — 3.1L (VIN M) engine

notch in the sprocket fits over the crankshaft key. Fully seat the sprocket on the crankshaft using J-38612, or an equivalent gear installer.

15. Make sure the timing mark on the crankshaft sprocket is pointing straight up.

16. Install the timing chain over the camshaft sprocket and hold the sprocket in such a way, that the timing mark is pointing down, and the timing chain is hanging down off the sprocket.

17. Loop the timing chain under the crankshaft sprocket and install the camshaft sprocket on the camshaft. The sprocket will only fit on the camshaft if the dowel on the camshaft lines up with the hole in the sprocket.

18. Verify that the marks are aligned (the camshaft sprocket will be at the 6 o'clock position and the crankshaft sprocket will be in the 12 o'clock position).

19. On 1994–95 vehicles, tighten the camshaft sprocket mounting bolt to 74 ft. lbs. (100 Nm). On 1996 vehicles, tighten the bolt to 81 ft. lbs. (110 Nm).

20. Lubricate the timing chain components with engine oil.

21. Clean all gasket surfaces completely.

22. Apply a thin bead of sealer around the gasket sealing area of the front cover. Install a new front cover seal on the front cover.

23. Install the front cover on the engine and install the mounting bolts. Tighten the small bolts to 18 ft. lbs. (24 Nm). Tighten the large bolts to 41 ft. lbs. (55 Nm). On 1996 vehicles, tighten the small bolts to 15 ft. lbs. (21 Nm) and tighten the large bolts to 35 ft. lbs. (47 Nm).

24. Connect the radiator hose to the coolant outlet.

25. Install coolant by pass pipe to the water pump and the intake manifold.

26. Install the oil pan.

27. Install crankshaft balancer as follows:

  a. Coat the seal contact surface of the crankshaft balancer with clean engine oil.

  b. Line up the notch in the balancer with the crankshaft key and slide the balancer on until the key is in the balancer.

  c. Using J-29113 or an equivalent puller, seat the balancer on the crankshaft.

  d. Install the balancer mounting bolt and washer and tighten to 76 ft. lbs. (103 Nm).

  e. Install the flywheel cover.

  f. Install the right inner fender well splash shield.

  g. Install the tire and wheel assembly and tighten to specification.

28. Install the serpentine belt tensioner and tighten the mounting bolt to 40 ft. lbs. (54 Nm).

29. Install the serpentine belt.

30. Install the right engine mount bracket and tighten the bracket-to-mount bolts to 96 ft. lbs. (130 Nm).

31. Refill the cooling system with the correct amount and type of coolant.

32. Check the engine oil level. An oil and filter change is recommended.

33. Connect the negative battery cable. Start the engine and verify that there are no leaks.

Camshaft

REMOVAL AND INSTALLATION

—————— CAUTION ——————
The fuel system is under pressure and must be properly relieved before any service procedures are performed. Failure to properly relieve the system pressure can lead to personal injury and component damage.

2.2L (VIN 4) Engine

Please note that the engine must be removed from the vehicle for camshaft service.

1. Disconnect the negative battery cable.

2. Remove the engine assembly from the vehicle and install on a suitable engine stand.

3. Remove the serpentine belt.

4. Remove the alternator and lower mounting bracket.

5. Remove the power steering pump. Remove the serpentine belt tensioner assembly.

6. Remove the water pump pulley.

7. Remove the engine oil filter.

8. Remove the crankshaft pulley from the crankshaft hub.

9. Remove the rocker arm cover.

NOTE: When removing the valve train components they must be kept in order for installation in the same locations they were removed from. In addition, a new camshaft requires that new lifters be installed.

10. Remove the rocker arm nuts, rocker arms, balls and pushrods.

11. Remove the cylinder head with the intake and exhaust manifolds attached.

12. Remove the valve lifters.

13. Remove the timing chain front cover. Remove the timing chain, tensioner and camshaft sprocket.

14. Remove the oil pump drive assembly.

15. Remove the camshaft thrust plate mounting bolts and remove the thrust plate.

16. Remove the camshaft carefully from the engine.

To install:

17. Coat the camshaft lobes with and bearings with GM Engine Oil Supplement (E.O.S.) 1051396 or equivalent camshaft lubricant and insert the camshaft carefully into the engine.

18. Install the thrust plate and tighten the mounting bolts to 106 inch lbs. (12 Nm).

19. Install the timing chain, tensioner and camshaft sprocket.

20. Install the timing chain front cover.

NOTE: If the camshaft was replaced with a new one, the valve lifters MUST also be replaced. The lifters have already developed a wear pattern from the old camshaft and installing them on a new camshaft can cause the new camshaft to wear prematurely.

21. Install the valve lifters.

22. Install the cylinder head and manifold assemblies.

23. Install the oil pump drive assembly.

24. Install the pushrods, rocker arms, balls and rocker arm nuts. Tighten the nuts to 22 ft. lbs. (30 Nm).

25. Install the rocker arm cover.

26. Install the crankshaft pulley hub and tighten the mounting bolt to 77 ft. lbs. (105 Nm).

27. Install the crankshaft pulley and tighten the mounting bolts to 37 ft. lbs. (50 Nm).

28. Install the water pump pulley and tighten the mounting bolts to 15 ft. lbs. (21 Nm).

29. Install the drive belt tensioner and tighten the mounting bolts to 37 ft. lbs. (50 Nm).

30. Install the power steering pump. Install the alternator mounting bracket and alternator.

31. Install the engine assembly in the vehicle.

32. Coat the seal on the oil filter with clean engine oil and install the oil filter on the engine.

33. Refill the engine with oil.

34. Connect the negative battery cable. Refill the cooling system.

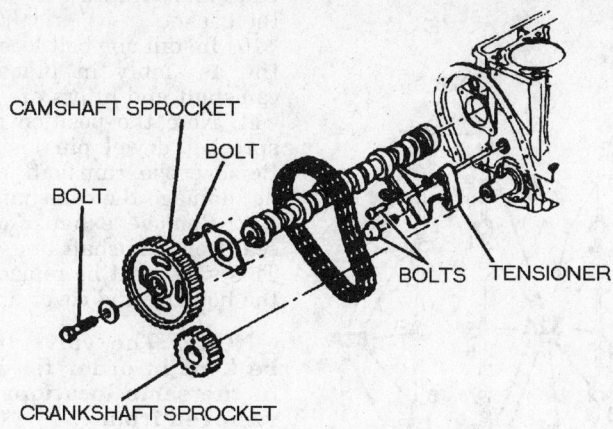

CAMSHAFT SPROCKET

BOLT

BOLT

BOLTS TENSIONER

CRANKSHAFT SPROCKET

ALIGN TABS ON TENSIONER WITH MARKS ON CAMSHAFT AND CRANKSHAFT SPROCKETS

324863

Camshaft and timing chain components — 2.2L (VIN 4) engine

35. Start the vehicle and verify no leaks.

2.3L (VIN A) Engine

Intake Camshaft

NOTE: Anytime the camshaft housing-to-cylinder head bolts are loosened or removed, the camshaft housing-to-cylinder head gasket must be replaced.

1. Disconnect the negative battery cable.
2. Disconnect the 11 pin connector from the ignition assembly cover.
3. Remove the 4 ignition assembly cover-to-camshaft housing bolts and remove assembly by pulling straight up. Use a spark plug boot wire remover to remove connector assemblies.
4. Remove the throttle lever actuator power steering pressure switch connector.
5. Loosen the power steering pump adjusting stud bolt. Loosen the power steering pump pivot bolts and remove drive belt.
6. Remove the 2 power steering pump bracket-to-transaxle bolts.
7. Remove the front power steering pump bracket-to-engine block bolt.

8. Remove the power steering pump assembly from the engine and position it aside. DO NOT disconnect the power steering lines from the pump.
9. Install J-38343-4, or an equivalent forcing pilot screw into the end of the camshaft.
10. Install J-38781, or an equivalent 3 jaw puller on the power steering drive pulley.
11. While holding the pilot screw with a wrench remove the pulley.
12. Remove oil/air separator bolts and hoses. Leave the hoses attached to the separator but disconnect from the oil fill tube, timing chain housing and intake manifold.
13. Disconnect the vacuum line from fuel pressure regulator and disconnect the fuel injector harness connector.
14. Remove the 2 bolts securing the fuel line clamps to the intake camshaft housing.
15. Remove fuel rail-to-camshaft housing mounting bolts.
16. Remove the fuel rail from the cylinder head leaving the fuel lines attached and position the fuel rail aside.
17. Remove the timing chain and camshaft sprockets.

18. Remove the timing chain housing bolts but do not remove from the engine.
19. Remove intake camshaft housing attaching bolts.
20. Remove the intake camshaft housing-to-cylinder head attaching bolts. Use the reverse of the tightening sequence when loosening camshaft housing to cylinder head attaching bolts. Leave 2 bolts loosely in place to hold the camshaft housing while separating camshaft cover from housing.
21. Push the cover off the housing by threading 4 of the housing-to-cylinder head bolts into the tapped holes in the cam housing cover. Tighten the bolts in evenly so the cover does not bind on the dowel pins.
22. Remove the 2 loosely installed camshaft housing-to-cylinder head bolts and remove the cover. Discard the gaskets.
23. Note the position of the chain sprocket dowel pin for reassembly. Remove the camshaft carefully; do not damage the camshaft oil seal.
24. Remove intake camshaft oil seal from camshaft and discard seal. This seal must be replaced any time the housing and cover are separated.

NOTE: The valve lifters must be kept in order for installation in the same locations they were removed from.

25. Remove the valve lifters from the camshaft housing.
26. Remove the camshaft carrier from the cylinder head and remove the gasket.

To install:

27. Clean all the gasket surfaces completely.
28. Install a new gasket on the cylinder head and install the camshaft housing over the dowel pins. Install one bolt loosely to hold the housing in place.

NOTE: If the camshaft was replaced the valve lifters must also be replaced.

29. Install the lifters into their original bores.
30. Lubricate camshaft lobes, journals and lifters with camshaft and lifter prelube. The camshaft lobes and journals must be adequately lubricated or engine damage could occur upon start up.
31. Install the camshaft in the same position it was in prior to removal. The timing chain sprocket dowel pin should be straight up and align with the center line of the lifter bores.

32. Install new camshaft housing-to-camshaft housing cover seals into the cover. The seals for the intake and exhaust covers are different and the correct seals must be used.

33. Remove the bolt holding the housing in place.

34. Apply thread locking compound to the camshaft housing and cover attaching bolt threads.

35. Install the camshaft housing cover.

36. Install all the mounting bolts finger-tight. With all the bolts in place tighten the bolts in sequence to 11 ft. lbs. (15 Nm) plus 75 degrees additional rotation (Long bolts) and 16 ft. lbs. (21 Nm) plus 25 degrees additional rotation (Short bolts).

37. Install the timing chain housing mounting bolts.

38. Install the timing chain and sprockets.

39. Install new fuel injector O-ring seals lubricated with oil and install the fuel rail and tighten the mounting bolts to 19 ft. lbs. (26 Nm).

40. Install the fuel line clamp mounting bolts on top of the intake camshaft housing.

41. Connect the vacuum line to the fuel pressure regulator.

42. Connect the fuel injector harness connector.

43. Install the oil/air separator assembly and connect the hoses to the oil fill tube, timing chain housing and intake manifold.

44. Lubricate the inner sealing surface of the intake camshaft seal with oil and install the seal to the housing using J-36009, or an equivalent seal installer.

45. Install the power steering pump pulley onto the intake camshaft using J-36015, or an equivalent pulley installer.

46. Install the power steering pump and install the drive belt.

47. Connect the throttle lever actuator power steering pressure switch connector.

48. Install the ignition assembly on the camshaft housing and tighten the mounting bolts to 13 ft. lbs. (18 Nm).

49. Connect the 11 pin connector to ignition assembly.

50. Connect the negative battery cable. Start the vehicle and verify no leaks.

Exhaust Camshaft

NOTE: Anytime the camshaft housing-to-cylinder head bolts are loosened or removed, the camshaft housing-to-cylinder head gasket must be replaced.

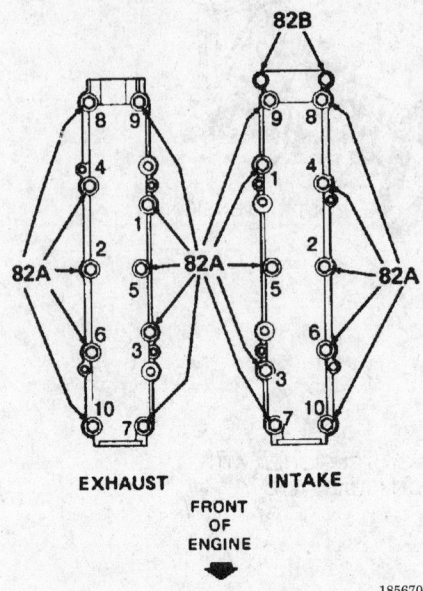

Camshaft housing bolt tightening sequence — 2.3L (VIN A) engine

1. Disconnect the negative battery cable.

2. Disconnect the 11 pin connector from the ignition assembly cover.

3. Remove the 4 ignition assembly cover-to-camshaft housing bolts and remove assembly by pulling straight up. Use a spark plug boot wire remover to remove connector assemblies.

4. Disconnect the harness from the oil pressure switch.

5. Remove the timing chain and camshaft sprockets.

6. Remove the timing chain housing bolts but do not remove from the engine.

7. Remove the exhaust camshaft housing-to-cylinder head attaching bolts. Use the reverse of the tightening sequence when loosening camshaft housing-to-cylinder head attaching bolts. Leave 2 bolts loosely in place to hold the camshaft housing while separating camshaft cover from housing.

8. Push the cover off the housing by threading 4 of the housing-to-cylinder head bolts into the tapped holes in the cam housing cover. Tighten the bolts in evenly so the cover does not bind on the dowel pins.

9. Remove the 2 loosely installed camshaft housing-to-cylinder head

bolts and remove the cover. Discard the gaskets.

10. Install one bolt loosely to retain the assembly in place while the camshaft and lifters are removed.

11. Note the position of the chain sprocket dowel pin for reassembly. Remove the camshaft carefully; do not damage the camshaft oil seal.

12. Remove exhaust camshaft oil seal from camshaft and discard seal. This seal must be replaced any time the housing and cover are separated.

NOTE: The valve lifters must be kept in order for installation in the same locations they were removed from.

13. Remove the valve lifters from the camshaft housing.

14. Remove the camshaft carrier from the cylinder head and remove the gasket.

To install:

15. Clean all the gasket surfaces completely.

16. Install a new gasket on the cylinder head and install the camshaft housing over the dowel pins. Install one bolt loosely to hold the housing in place.

NOTE: If the camshaft was replaced the valve lifters must also be replaced.

17. Install the lifters into their original bores.

18. Lubricate camshaft lobes, journals and lifters with camshaft and lifter prelube. The camshaft lobes and journals must be adequately lubricated or engine damage could occur upon start up.

19. Install the camshaft in the same position it was in prior to removal. The timing chain sprocket dowel pin should be straight up and align with the center line of the lifter bores.

20. Install new camshaft housing-to-camshaft housing cover seals into the cover. The seals for the intake and exhaust covers are different and the correct seals must be used.

21. Remove the bolt holding the housing in place.

22. Apply thread locking compound to the camshaft housing and cover attaching bolt threads.

23. Install the camshaft housing cover.

24. Install all the mounting bolts finger-tight. With all the bolts in place tighten the bolts in sequence to 11 ft. lbs. (15 Nm) plus 75 degrees additional rotation.

25. Install the timing chain housing mounting bolts.

26. Install the timing chain and sprockets.

27. Install new fuel injector O-ring seals lubricated with oil and install the fuel rail and tighten the mounting bolts to 19 ft. lbs. (26 Nm).

28. Lubricate the inner sealing surface of the exhaust camshaft seal with oil and install the seal to the housing using J-36009, or an equivalent seal installer.

29. Install the transaxle fill tube.

30. Connect the electrical connector to the oil pressure switch.

31. Install the ignition assembly on the camshaft housing and tighten the mounting bolts to 13 ft. lbs. (18 Nm).

32. Connect the 11 pin connector to ignition assembly.

33. Connect the negative battery cable.

34. Start the vehicle and verify proper operation and no leaks.

3.1L (VIN T) Engine

1. Relieve the fuel system pressure. Disconnect the negative battery cable.

2. Remove the engine assembly.

NOTE: When removing valve train components they must be marked for installation in the same location they are removed from. When the camshaft is being replaced the valve lifters should also be replaced.

3. Remove the intake manifold, valve cover, rocker arms, pushrods and valve lifters.

4. Remove the crankshaft balancer and front cover.

5. Remove the timing chain and sprockets.

6. Carefully remove the camshaft. Avoid marring the camshaft bearing surfaces.

To install:

7. Coat the camshaft with lubricant 1052365 or equivalent, and install the camshaft.

8. Install the timing chain and sprocket.

9. Install the camshaft thrust button and front cover.

10. Install the crankshaft balancer.

11. Install the intake manifold, valve cover, rocker arms, pushrods and valve lifters.

12. Install the engine assembly.

13. Connect the negative battery cable.

14. Adjust the valves, as required.

15. Start the engine and verify no oil leaks.

3.1L (VIN M) Engine

If the camshaft is being replaced, all of the valve lifters must also be replaced with new parts. Installing used lifters on a new camshaft will quickly wear the camshaft.

1. Relieve the fuel system pressure. Disconnect the negative battery cable.

2. Remove the engine assembly from the vehicle, and mount on a suitable engine stand.

3. Remove the intake manifold, valve cover, rocker arms, pushrods and valve lifters.

4. Remove the using a puller to draw the crankshaft balancer from the crankshaft nose.

5. Remove the timing chain front cover, timing chain and sprockets.

6. Remove the oil pump driven gear mounting bolt and hold-down. Remove the oil pump driven gear so the camshaft can be pulled out.

7. Remove the 2 bolts and remove the camshaft thrust plate.

8. Carefully remove the camshaft. Avoid marring the camshaft bearing surfaces.

To install:

9. Coat the camshaft with lubricant GM part No. 1052365 or equivalent camshaft break-in lubricant or quality engine oil supplement, and install the camshaft.

10. Install the camshaft thrust plate and tighten the mounting bolts to 89 inch lbs. (10 Nm).

11. Install the oil pump driven gear and tighten the mounting bolt to 27 ft. lbs. (36 Nm).

12. Install the timing chain and sprocket.

13. Verify that all timing marks are aligned. It is good practice to turn the crankshaft several complete rotations to make sure the timing marks still align when the crankshaft is returned to Top Dead Center No. 1 cylinder compression stroke (firing position). This is most important. If the valve train is not correctly timed, the engine will be damaged on start-up. When satisfied that all marks are correctly aligned, install the camshaft thrust button and front cover.

14. Install the crankshaft balancer and tighten the bolt to 75 ft. lbs. (103 Nm).

15. Install the intake manifold, valve cover, rocker arms, pushrods and valve lifters.

16. Install the engine assembly.

17. Connect the negative battery cable.

18. Verify that all fluid levels (coolant, engine oil, etc.) are full and correct. An oil and filter change is recommended.

19. Adjust the valves, as required.

20. Start the engine and verify no oil leaks.

Piston and Connecting Rod

POSITIONING

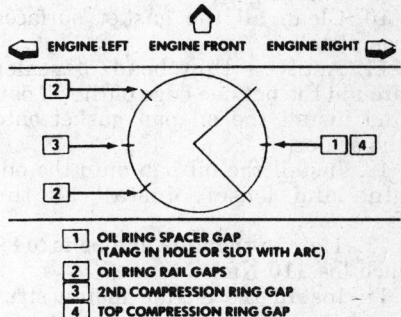

Piston ring gap location — all engines

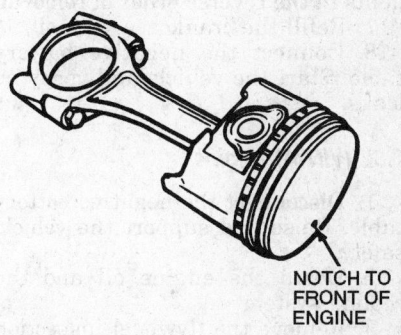

Piston position — all engines

ENGINE LUBRICATION

Oil Pan

REMOVAL AND INSTALLATION

2.2L (VIN 4) Engine

1. Disconnect the negative battery cable. Raise and safely support the vehicle.

2. Remove the right front tire and wheel assembly.

3. Remove the right inner fender well splash shield.

4. Drain the engine oil.

5. Remove the starter and starter bracket. Remove the flywheel cover bolts and flywheel cover.

6. Remove the engine support strut and support strut bracket.

7. Disconnect the exhaust pipe from the exhaust manifold and position aside.

8. Remove the oil pan mounting nuts and bolts.

9. Remove the oil pan.

To install:

10. Clean all the gasket surfaces completely.

11. Apply a thin bead of sealer around the outside edge of the oil pan and install the oil pan gasket onto the sealer.

12. Install the oil pan onto the engine and loosely install all the fasteners.

13. Tighten the nuts and bolts to 89 inch lbs. (10 Nm).

14. Install the engine mount strut bracket and engine mount strut.

15. Connect the exhaust pipe to the exhaust manifold.

16. Install the remaining components in the reverse order of removal.

17. Refill the crankcase with oil.

18. Connect the negative battery cable. Start the vehicle and verify no leaks.

2.3L (VIN A) Engine

1. Disconnect the negative battery cable. Raise and support the vehicle safely.

2. Drain the engine oil and the cooling system.

3. Remove the flywheel inspection cover.

4. Remove the right front tire and wheel assembly. Remove the right inner fender splash shield.

5. Remove the serpentine belt.

6. Disconnect the engine mount strut from the engine mount strut bracket.

7. Remove the A/C compressor from the mounting bracket and support it aside. Do not disconnect the refrigerant lines.

8. Remove the engine mount strut bracket bolts and remove the bracket.

9. Remove the radiator outlet pipe bolts.

10. Disconnect the radiator and air conditioning outlet pipes from the suspension supports.

11. Remove the exhaust manifold brace.

12. Remove the oil pan-to-flywheel cover bolt and nut and remove the flywheel cover stud and spacer.

13. Disconnect the radiator outlet pipe from the lower radiator hose and from the oil pan.

14. Disconnect the oil level sensor wire, if equipped.

15. Remove the oil pan bolts.

16. Remove the oil pan.

To install:

17. Use a new gasket and install the oil pan to the engine. Install the oil pan bolts. Tighten the chain housing and carrier seal bolts to 106 inch lbs. (12 Nm). Tighten the oil pan-to-block bolts to 17 ft. lbs. (23 Nm).

18. Install the flywheel cover spacer, stud, nut and bolt. Tighten the nut to 41 ft. lbs. (56 Nm), the stud to 115 inch lbs. (13 Nm) and the bolt to 41 ft. lbs. (56 Nm).

19. Install the oil pan-to-transaxle nut and tighten to 41 ft. lbs. (56 Nm).

20. Connect the oil level sensor wire.

21. Install the remaining components in the reverse order of removal. Tighten the engine mount strut bracket mounting bolts to 49 ft. lbs. (66 Nm).

22. Fill the crankcase with oil and the cooling system with the correct mix of coolant.

23. Connect the negative battery cable. Start the engine and check for leaks.

3.1L (VIN T) Engine

1. Disconnect the negative battery cable. Raise and support the vehicle safely.

2. Drain the engine oil, then remove the oil filter.

3. Remove the oil filter.

4. Remove the flywheel shield.

5. Disconnect and remove the starter.

6. Remove the oil pan retaining bolts and nuts, then remove the oil pan.

To install:

7. Clean all the gasket surfaces completely.

8. Apply a thin bead of sealer around the gasket contact are on the oil pan and install the oil pan gasket on the oil pan.

9. Install the oil pan in the vehicle and install the mounting nuts and bolts loosely. With all the fasteners installed tighten as follows:

- Oil pan mounting nuts: 89 inch lbs. (10 Nm)
- Rear oil pan mounting bolts: 18 ft. lbs. (25 Nm)
- Remaining oil pan mounting bolts: 89 inch lbs. (10 Nm)

10. Connect and install the starter.

11. Install the flywheel shield.

12. Coat the seal on the oil filter with engine oil and install the oil filter on the engine.

13. Lower the vehicle. Refill the crankcase with engine oil.

14. Connect the negative battery cable. Start the vehicle and verify no oil leaks.

3.1L (VIN M) Engine

1. Disconnect the negative battery cable. Remove the serpentine belt.

2. Remove the upper A/C compressor bolts, if equipped.

3. Raise and safely support the vehicle. Drain the engine oil.

4. Remove the right front tire and wheel assembly.

5. Remove the right inner fender splash shield.

6. Remove the engine mount strut from the suspension support.

7. Remove the cotter pin and castle nut from the lower ball joint and separate the joint from the steering knuckle.

8. Remove the right side sway bar link. Disconnect the ABS sensor from the right subframe.

9. Remove the right side subframe mounting bolts and remove the right side subframe and control arm as an assembly.

10. Remove the lower A/C compressor mounting bolts and position the compressor aside. DO NOT disconnect the refrigerant lines or allow the compressor to hang unsupported.

11. Remove the engine mount strut bracket from the engine.

12. Remove the engine to transaxle brace. Remove the oil filter.

13. Remove the starter.

14. Remove the flywheel cover.

15. Remove the oil pan flange retaining bolts and the oil pan side retaining bolts. Remove the oil pan.

To install:

16. Clean the gasket mating surfaces.

17. Install a new gasket on the oil pan. Apply silicon sealer to the portion of the pan that contacts the rear of the block.

18. Install the oil pan and install the mounting bolts finger-tight.

19. With all the bolts in place, tighten the oil pan flange bolts to 18 ft. lbs. (25 Nm) and the oil pan side bolts to 37 ft. lbs. (50 Nm).

20. Install the remaining components in the reverse order of removal. Torque the following:

- Engine-to-transaxle brace bolts to 68 ft. lbs. (93 Nm)
- Engine mount strut bracket bolt at the engine bracket to 85 ft. lbs. (115 Nm)
- Subframe bolts to 89 ft. lbs. (120 Nm)

- Right side sway bar link to 22 ft. lbs. (30 Nm)
- Ball joint-to-steering knuckle castle nut to 48 ft. lbs. (60 Nm)
- Engine mount strut bracket bolt to 89 ft. lbs. (120 Nm)

21. Fill the crankcase to the correct level.

22. Connect the negative battery cable. Start the vehicle and verify no leaks.

NOTE: Whenever the vehicle subframe is removed or lowered, the wheel alignment should be checked.

Oil Pump

REMOVAL AND INSTALLATION

2.2L (VIN 4) Engine

1. Disconnect the negative battery cable. Raise and safely support the vehicle.

2. Drain the engine oil, then remove the oil pan.

3. Remove the oil pump-to-rear main bearing cap bolt, the oil pump and extension shaft.

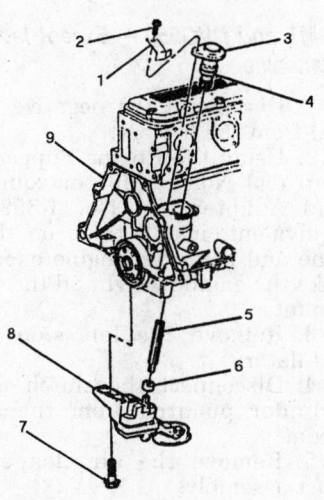

1 Bracket
2 Bolt
3 Oil pump drive assembly
4 O-ring
5 Shaft
6 Retainer; Heat and water soak prior to installation
7 Bolt
8 Oil pump
9 Cylinder block

324721

Oil pump mounting to engine block — 2.2L (VIN 4) engine

To install:

> ——— CAUTION ———
> *Heat the extension shaft retainer in hot water prior to assembly. Be sure the retainer does not crack upon installation.*

4. Install the extension shaft, oil pump and pump-to-rear main cap bolt. Tighten the oil pump-to-bearing cap bolt to 32 ft. lbs. (43 Nm) and the upper oil pump drive bolt to 18 ft. lbs. (25 Nm).

> ——— CAUTION ———
> *To avoid engine damage, all oil pump cavities must be filled with petroleum jelly before installing the gears into the pump body. This seals the gears, acts like a prime and allows the pump to start drawing oil as soon as the engine begins to crank the first time after pump service. Also use only original equipment gaskets. Gasket thickness is critical to proper oil pump operation.*

5. Install the oil pan and attaching bolts. Lower the vehicle.

6. Fill the crankcase with clean engine oil.

7. Install a mechanical oil pressure gauge so actual pump pressure can be read after startup.

8. Connect the negative battery cable.

9. Start the engine and check oil pressure and check for leaks.

10. Turn the engine **OFF** and allow to stand. Check oil level, add as necessary.

2.3L (VIN A) Engine

1. Disconnect the negative battery cable. Raise and safely support the vehicle.

2. Drain the engine oil and remove the oil pan.

3. Remove the 4 oil pump mounting bolts, then remove the oil pump assembly.

To install:

4. Install the oil pump assembly on the engine and install the 4 mounting bolts. Tighten the pump-to-block bolts to 40 ft. lbs. (54 Nm). and the oil pump screen-to-screen brace bolts to 106 inch lbs. (12 Nm).

5. Install the oil pan.

6. Fill the crankcase with clean oil to the proper level.

7. Connect the negative battery cable. Start the vehicle and verify no leaks.

3.1L (VIN T) Engine

1. Disconnect the negative battery cable. Raise and support the vehicle safely.

2. Drain the engine oil, then remove the oil pan.

3. Remove the pump-to-rear main bearing cap bolt and remove the pump and extension shaft.

To install:

NOTE: Whenever the oil pump is removed or replaced it must be primed prior to installation in the vehicle to prevent oil starvation on start up.

4. Remove the 4 cover attaching screws and the cover from the oil pump assembly.

5. Pack the space around the oil pump gears completely full of petroleum jelly. There must be no air space left inside the pump.

6. Install the pump cover and 4 mounting screws and tighten the screws to 80 ft. lbs. (10 Nm).

7. Assemble the pump and extension shaft with retainer to rear main bearing cap, aligning the top end of the extension shaft with the lower end of the drive gear.

8. Install the pump-to-the rear bearing cap bolt. Tighten to 30 ft. lbs. (40 Nm).

9. Install the oil pan.

10. Lower the vehicle to the floor.

11. Fill the crankcase with oil to the proper level.

12. Connect the negative battery cable.

3.1L (VIN M) Engine

1. Disconnect the negative battery cable. Raise and safely support the vehicle.

2. Drain the engine oil, then remove the oil pan.

3. Remove the crankshaft oil deflector bolts, then remove the deflector.

4. Remove the oil pump retaining bolts and remove the oil pump and pump driveshaft.

To install:

5. Install the oil pump and pump driveshaft. Tighten the oil pump mounting bolts to 30 ft. lbs. (41 Nm).

6. Install the crankshaft oil deflector and mounting bolts. Tighten the mounting bolts to 18 ft. lbs. (25 Nm).

7. Install the oil pan.

8. Lower the vehicle.

9. Fill the crankcase to the correct level with oil.

10. Install a mechanical oil pressure gauge so actual pump pressure can be read after startup.

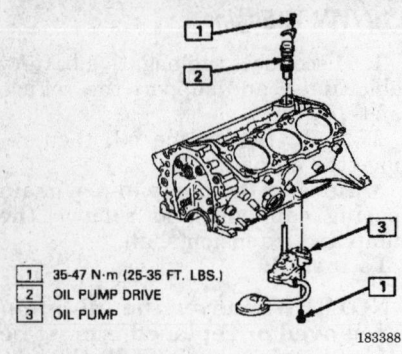

[1] 35-47 N·m (25-35 FT. LBS.)
[2] OIL PUMP DRIVE
[3] OIL PUMP

183388

Oil pump, extension shaft and drive assembly — 3.1L (VIN T) engine

11. Start the engine, check the oil pressure and check for leaks.

TRANSAXLE

Manual Transaxle Assembly

REMOVAL AND INSTALLATION

NOTE: **Before performing any maintenance that requires the removal of the slave cylinder, transaxle or clutch housing, the clutch master cylinder pushrod must first be disconnected from the clutch pedal. Failure to disconnect the pushrod will result in permanent damage to the slave cylinder if the clutch pedal is depressed with the slave cylinder disconnected.**

NVT550 — Except Isuzu Transaxle

Special lifting and support equipment is required for these procedures.

1. Disconnect the negative battery cable.
2. Remove the left side sound insulator from under the dash panel.
3. Disconnect the clutch master cylinder pushrod from the clutch pedal.
4. Remove the air cleaner and duct assembly.
5. Remove the brake booster vacuum line from the throttle body and position it aside.
6. Remove the clutch actuator cylinder quick disconnect from the clutch master cylinder line.
7. Remove the power steering bracket.
8. Remove the transaxle shift cables from the transaxle.

9. Remove the electrical connector from the backup light switch.
10. Remove the vacuum lines and tie them aside.
11. Remove the shift cable bracket.
12. Remove the upper transaxle bolts.
13. Using the Engine Support Fixture tool No. J-28467 or equivalent install it on the engine, then raise the engine enough to take the weight off the engine mounts.
14. Remove the upper transaxle mount and through bolt.
15. Raise and safely support the vehicle.
16. Drain the transaxle.
17. Remove both front ABS wheel speed sensor connectors and the left harness from the suspension support.
18. Remove both front tire and wheel assemblies.
19. Remove both front axles.
20. Remove the stabilizer link nuts.
21. Remove the left front splash shield.
22. Remove the left suspension support and attaching bolts.
23. Remove the vehicle speed sensor connector from the transaxle
24. Remove the heater hose bolt.
25. Remove the flywheel housing cover.
26. Remove the ground wire connections.
27. Remove the lower transaxle mount and through bolt.
28. Lower the vehicle.
29. Lower the engine and transaxle assembly enough to remove the transaxle from the vehicle.
30. Raise and safely support the vehicle.
31. Remove the remaining transaxle mounting bolts.
32. Remove the transaxle by carefully sliding it away from the engine and lower the jack while guiding the intermediate shaft out of the transaxle.

To install:
33. Raise the transaxle and carefully guide the intermediate shaft through the release bearing and clutch disc into the pilot bearing.
34. Install the transaxle to engine mounting bolts and studs tighten all bolt-stud positions except No. 7 to 55 ft. lbs. (75 Nm) tighten stud position No. 7 to 115 inch lbs. (13 Nm).
35. Install the lower mount and through bolt tighten to 44 ft. lbs.
36. Remove the transaxle support jack.
37. Install the backup light switch connector.
38. Install the heater hose bolt and tighten to 40 ft. (54 Nm).

39. Lower the vehicle.
40. Install the upper transaxle mount and through bolts, tighten the mount to transaxle 2 top bolts first to 96 ft. lbs. (130 Nm) the lower mount bolt last to 55 ft. lbs. (75 Nm) and the through bolt to 44 ft. lbs. (60 Nm).
41. Install the vehicle speed sensor connector.
42. Raise and safely support the vehicle.
43. Install both front drive axles.
44. Install the left suspension support and attaching bolts and nuts tighten support bolts in sequence from the rear to the front to 89 ft. lbs. (120 Nm).
45. Position the ball joint stud into the steering knuckle install the castle nut and tighten to 41 ft. lbs. (55 Nm) minimum, 48 ft. lbs. (65 Nm) maximum. Install a new cotter pin.
46. Install both stabilizer link nuts and tighten to 22 ft. lbs. (30 Nm).
47. Install the remaining components in the reverse order of removal. Tighten the shift cable bracket nuts to 89 inch lbs. (10 Nm).
48. Fill the transaxle with the recommend fluid.
49. Bleed the clutch hydraulic system. Connect the negative battery cable.
50. Apply the emergency brake, start the engine and check the transaxle and clutch operation.

5SMT and NVG550 — Except Isuzu Transaxle

1. Disconnect the negative terminal from the battery.
2. Using the Engine Support Fixture tool No. J-28467 or equivalent and Adapter tool No. J-35953 or equivalent, install them on the engine and raise the engine enough to take the engine weight off the engine mounts.
3. Remove the left side sound insulator.
4. Disconnect the clutch master cylinder pushrod from the clutch pedal.
5. Remove the air cleaner and duct assembly.
6. Disconnect the clutch slave cylinder-to-transaxle support bolts and position the cylinder aside.
7. Remove the transaxle-to-mount through bolt. Raise and support the front of the vehicle.
8. Remove the 2 exhaust crossover bolts at the right side manifold.
9. Lower the vehicle. Remove the left side exhaust manifold.
10. Disconnect the transaxle mounting bracket.
11. Disconnect the shifter cables.

12. Remove the upper transaxle-to-engine bolts.

13. Raise and support the front of the vehicle.

14. Remove the left front tire assembly and the left side inner splash shield.

15. Remove the transaxle strut and bracket.

16. Place a drain pan under the transaxle, remove the drain plug and drain the fluid from the transaxle.

17. Remove the clutch housing cover bolts.

18. Disconnect the speedometer wire.

19. From the left suspension support and control arm, disconnect the stabilizer shaft.

20. Remove the left suspension support mounting bolts and move the support aside.

21. Disconnect both halfshafts from the transaxle and remove the left halfshaft from the vehicle.

22. Using a transmission jack, attach it to and support the transaxle.

23. Remove the remaining transaxle-to-engine bolts.

24. Slide the transaxle away from the engine, lower it and remove the right side halfshaft.

To install:

25. When installing, guide the right side halfshaft into the transaxle while it is being installed in the vehicle.

26. Torque the transaxle-to-engine bolts to 60 ft. lbs., the transaxle mount-to-body bolt to 80 ft. lbs.

27. Install the left halfshaft into its bore at the transaxle then seat both halfshafts at the transaxle.

28. Install the remaining components in the reverse order of removal. Tighten the upper transaxle-to-engine bolts to 55 ft. lbs. (75 Nm).

29. Fill the transaxle with the proper type and amount of fluid.

30. Connect the negative battery cable. Start the vehicle and check for leaks.

Isuzu Transaxle

1. Disconnect the negative terminal from the battery.

2. Using the Engine Support Fixture tool No. J-28467 or equivalent and Adapter tool No. J-35953 or equivalent, install them on the engine and raise the engine enough to take the engine weight off the engine mounts.

3. Remove the left side sound insulator.

4. Disconnect the clutch master cylinder pushrod from the clutch pedal.

5. Disconnect the clutch slave cylinder-to-transaxle support bolts and position the cylinder aside.

6. Remove the wiring harness from the transaxle mount bracket and the shift wire electrical connector.

7. Remove the transaxle-to-mount bolts and the transaxle mount bracket-to-chassis nuts/bolts.

8. Disconnect the shift cables and remove the retaining clamp from the transaxle. Remove the ground cables from the transaxle mounting studs.

9. Raise and support the front of the vehicle.

10. Remove the left front tire assembly and the left side inner splash shield.

11. Remove the transaxle front strut and bracket.

12. Remove the clutch housing cover bolts. Disconnect the speedometer wire connector.

13. From the left suspension support and control arm, disconnect the stabilizer shaft.

14. Remove the left suspension support mounting bolts and move the support aside.

15. Disconnect both halfshafts from the transaxle and remove the left halfshaft from the vehicle.

16. Place a drain pan under the transaxle, remove the drain plug and drain the fluid from the transaxle.

17. Using a transmission jack, attach it to and support the transaxle.

18. Remove the transaxle-to-engine bolts.

19. Slide the transaxle away from the engine, lower it and remove the right side halfshaft.

To install:

20. When installing, guide the right side halfshaft into the transaxle while it is being installed in the vehicle.

21. Install and torque the transaxle-to-engine bolts to 55 ft. lbs. (75 Nm).

22. Install the left halfshaft into its bore at the transaxle then seat both halfshafts at the transaxle.

23. Install the remaining components in the reverse order of removal. Torque the following:

• Clutch housing cover bolts to 89 inch lbs. (10 Nm)

• Transaxle strut-to-body bolt to 40 ft. lbs. (54 Nm)

• Transaxle strut-to-transaxle to 50 ft. lbs. (68 Nm)

• Transaxle rear mount bracket-to-transaxle to 40 ft. lbs. (54 Nm)

• Transaxle mount-to-side frame to 23 ft. lbs. (30 Nm)

• Mount-to-transaxle bracket to 88 ft. lbs. (120 Nm)

24. Connect the negative cable at the battery.

MX5

1. Disconnect the negative battery cable.

2. Using the Engine Support Fixture tool No. J-28467 or equivalent and Adapter tool No. J-35953 or equivalent, install them on the engine and raise the engine enough to take the weight off the engine mounts.

3. Remove the left side sound insulator.

4. Disconnect the clutch master cylinder pushrod from the clutch pedal.

5. Disconnect the clutch slave cylinder-to-transaxle support bolts and position the cylinder aside.

6. Remove the wiring harness from the transaxle mount bracket.

7. Remove the ground cables at the transaxle mounting studs.

8. Remove the backup light switch connector.

9. Remove the transaxle vent tube.

10. Remove the transaxle-to-mount bolts and the transaxle mount bracket-to-chassis nuts/bolts.

11. Disconnect the shift cables and remove the retaining clamp from the transaxle.

12. If using the factory type engine support tooling, lower the engine with the engine support fixture at least 30 revolutions to ease the removal and installation of the transaxle. Use care if using substitute lifting and supporting equipment.

13. Raise and support the front of the vehicle.

14. Drain the transaxle.

15. Remove the left front tire assembly and the left side inner splash shield.

16. Remove both ABS wheel speed sensor harness connectors and unroute the left side harness.

17. Remove the transaxle front strut and bracket.

18. Remove the clutch housing cover bolts. Disconnect the speed sensor wire connector.

19. Disconnect the stabilizer shaft from the left suspension support and control arm.

20. Remove the left suspension support mounting bolts and move the support aside.

21. Disconnect both halfshafts from the transaxle and remove the left halfshaft from the vehicle.

22. Using a transmission jack, attach it to and support the transaxle.

23. Remove the transaxle-to-engine bolts.

24. Slide the transaxle away from the engine, lower it and remove the right side halfshaft.

To install:

25. When installing, guide the right side halfshaft into the transaxle while it is being installed in the vehicle.

26. Install and torque the transaxle-to-engine bolts to 55 ft. lbs. (75 Nm).

27. Install the remaining components in the reverse order of removal. Torque the following:

• Clutch housing cover bolts to 89 inch lbs. (10 Nm)

• Transaxle strut-to-body bolt to 40 ft. lbs. (54 Nm)

• Transaxle strut-to-transaxle to 50 ft. lbs. (68 Nm)

• Rear mount bracket to transaxle to 40 ft. lbs. (54 Nm)

• Transaxle mount to side frame to 23 ft. lbs. (30 Nm)

• Mount-to-transaxle bracket to 88 ft. lbs. (120 Nm)

28. Connect the negative cable to the battery.

29. Apply the emergency brake. Check the clutch operation and the transaxle shift pattern.

NOTE: Whenever the vehicle sub-frame is removed or lowered, the wheel alignment should be checked.

30. Check the front alignment after removing suspension components.

Clutch Assembly

REMOVAL AND INSTALLATION

1. Disconnect the negative battery cable. From inside the vehicle, remove the hush panel.

NOTE: Prior to any vehicle service that required removal of the clutch actuator cylinder (slave cylinder) such as transaxle and clutch housing removal, the master cylinder pushrod must be disconnected from the clutch pedal. If not disconnected, permanent damage to the actuator cylinder will occur if the clutch pedal is depressed while the actuator cylinder is disconnected.

2. Disconnect the clutch master cylinder pushrod from the clutch pedal.

3. Remove the transaxle.

4. With the transaxle removed and if the pressure plate is to be reused, matchmark the pressure plate and flywheel assembly to insure proper balance during reassembly.

5. Loosen the pressure plate-to-flywheel bolts (one turn at a time) until the spring pressure is removed.

6. Support the pressure plate and remove the bolts.

7. Remove the pressure plate and disc assembly. Note the flywheel side of the clutch disc.

8. Clean and inspect the clutch assembly, flywheel, release bearing, clutch fork and pivot shaft for signs of wear. Replace any necessary parts. Inspect the bearing retainer outer surface of the transaxle.

To install:

9. Clean all parts well. Apply a small amount of high temperature grease to the pilot bearing in the end of the crankshaft.

10. For the 5SMT, NVG550 and the T550 transaxles, assemble the clutch disc and pressure plate in position. Factory replacement parts should be marked when they were balanced. Align the "Heavy Side" of the flywheel assembly which should be stamped with an **X** with the clutch cover "Light Side" should be marked with paint. Support the assembly with Alignment tool No. J-29074 or equivalent clutch alignment tool.

11. For the MX5 transaxle, position the clutch disc and pressure plate into the installed position, and support with clutch aligning tool J-29074 or equivalent. The clutch plate is assembled with the damper springs offset toward the transaxle. One side of the factory supplied clutch disc should be stamped "Flywheel Side."

NOTE: The clutch disc is installed with the damper springs offset towards the transaxle. Stamped letters or some other marking on the clutch disc should identify "Flywheel Side." Make sure the clutch disc is facing the proper direction. If the same pressure plate is being reused, align the marks made during the removal.

12. Install new pressure plate-to-flywheel retaining bolts. Tighten them gradually in a cross pattern as follows:

a. Install and lightly seat bolts 1, 2, 3 then 4, 5, 6.

b. Torque bolts 1, 2, 3 to 12 ft. lbs. (16 Nm).

c. Torque bolts 4, 5, 6 to 12 ft. lbs. (16 Nm).

d. Torque bolts 1, 2, 3 then 4, 5, 6 to 15 ft. lbs. (20 Nm) plus 30 degrees additional rotation.

13. Remove the alignment tool.

14. Lightly lubricate the clutch fork ends. Fill the recess ends of the release bearing with grease. Lubricate the input shaft with a light coat of grease. Use high temperature melting-point grease for these applications. On model 5TM40 transaxles, lubricate the inside of the release bearing with clutch bearing lubricant GM PN 12345777 or equivalent. On Isuzu transaxles, pack the inside diameter recess of the release bearing completely full of GM PN 1051344 chassis grease or equivalent. In addition, on Isuzu transaxles, be sure the bearing pads are located on the fork ends (pads must be indexed) and both spring ends are in the fork holes with the spring completely seated in the bearing groove.

NOTE: The clutch lever must not be moved towards the flywheel until the transaxle is bolted to the engine. Damage to the transaxle, release bearing and clutch fork could occur if this is not followed.

15. Install the transaxle assembly.

16. Install the clutch master cylinder pushrod to the clutch pedal and install the retaining clip. If equipped with cruise control, check the switch adjustment at the clutch pedal bracket. When adjusting the cruise control switch, do not pull up on the clutch pedal pad with more than 20 pounds force or damage to the master cylinder pushrod retaining ring can result.

17. Install the sound insulator.

18. Connect the negative battery cable, then bleed the clutch system and road test the vehicle.

Clutch Master Cylinder

REMOVAL AND INSTALLATION

2.2L (VIN 4) Engine

1993–94 Vehicles

The clutch master and slave cylinders are removed from the vehicle as an assembly. After installation the clutch hydraulic system must be bled.

1. Disconnect the negative battery cable. From inside the vehicle, remove the hush panel.

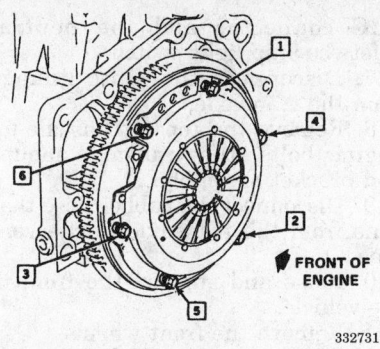

332731

Clutch cover bolt tightening sequence — MX5, 5SMT, NVG550 and T550

2. Disconnect the clutch master cylinder pushrod from the clutch master cylinder.

3. From the front of the dash, remove the trim cover.

4. Remove the clutch master cylinder-to-clutch pedal bracket nuts and the remote reservoir-to-chassis screws.

5. Remove the slave cylinder-to-transaxle nuts and the slave cylinder.

6. Remove the hydraulic system (as a unit) from the vehicle.

To install:

7. Install the slave cylinder-to-transaxle support, align the pushrod to the clutch fork outer lever pocket. Torque the slave cylinder-to-transaxle support nuts to 16 ft. lbs. (22 Nm).

NOTE: If installing a new replacement clutch hydraulic system, DO NOT break the pushrod plastic retainer. The straps will break on the first pedal application.

8. Install the master cylinder-to-clutch pedal bracket. Torque the nuts evenly (to prevent damaging the master cylinder) to 15–20 ft. lbs. (22–30 Nm) and reverse the removal procedures. Remove the pedal restrictor from the pushrod. Lubricate the pushrod bushing on the clutch pedal. If the bushing is cracked or worn, replace it.

9. If equipped with cruise control, check the switch adjustment at the clutch pedal bracket.

NOTE: When adjusting the cruise control switch, do not exert more than 20 pounds of upward force on the clutch pedal pad or damage to the master cylinder pushrod retaining rod can result.

10. Depress the clutch pedal several times to break the plastic retaining straps; DO NOT remove the plastic button from the end of the pushrod.

11. Install the hush panel.

12. Connect the negative battery cable.

1995–96 Vehicles

1. Disconnect the negative battery cable. Remove the lower left hush panel.

2. Remove the clutch master cylinder pushrod from the clutch pedal.

3. Remove the clutch master cylinder retaining nuts at the front of the dash.

4. Remove the remote reservoir.

5. Remove the clutch master cylinder assembly from the clutch actuator cylinder assembly.

To install:

6. Install the clutch master cylinder assembly to the clutch actuator cylinder assembly.

7. Install the remote reservoir.

8. Install the clutch master cylinder retaining nuts and tighten to 18 ft. lbs. (25 Nm).

9. If equipped, check the cruise control clutch switch adjustment.

10. Install the lower left hush panel.

11. Connect the negative battery cable. Bleed the hydraulic system.

12. Apply the emergency brake, start the engine and check the operation of the transaxle and clutch.

2.3L (VIN A) and 3.1L (VIN T) and (VIN M) Engines

The clutch master and slave cylinders are removed from the vehicle as an assembly. After installation the clutch hydraulic system must be bled.

1. Disconnect the negative battery cable. From inside the vehicle, remove the hush panel.

2. Remove the air cleaner duct assembly.

3. Disconnect the left fender brace.

4. Remove the battery.

5. Disconnect the MAT/IAT sensor lead at the air cleaner.

6. Disconnect the Mass Air Flow sensor lead.

7. Remove the PCV pipe retaining clamp at the air intake duct.

8. Remove the clamp retaining the air intake duct to the throttle body.

9. Remove the Mass Air Flow sensor.

10. Remove the air cleaner bracket mounting bolts at the battery tray.

11. Remove the air cleaner, Mass Air Flow sensor and the air intake duct as an assembly.

12. Disconnect the electrical lead at the washer bottle, remove the attaching bolts and remove the bottle.

13. Disconnect the cruise control mounting bracket retaining nuts from the strut tower, if equipped.

14. Disconnect the clutch master cylinder pushrod from the clutch pedal.

15. From the front of the dash, remove the trim cover, if so equipped.

16. Remove the clutch master cylinder retaining nuts at the dash.

17. Detach the clutch actuator cylinder quick disconnect from the clutch master cylinder line. Remove the actuator cylinder-to-transaxle nuts.

18. Remove the transaxle.

19. Remove the hydraulic system (as a unit) from the vehicle.

To install:

20. Install the actuator cylinder-to-transaxle support, align the pushrod to the clutch fork outer lever pocket. Torque the slave cylinder-to-transaxle support nuts to 16 ft. lbs. (22 Nm).

NOTE: If installing a new clutch hydraulic system, DO NOT break the pushrod plastic retainer. The straps will break on the first pedal application.

21. Attach the clutch actuator cylinder quick disconnect to the clutch master cylinder line.

22. Install the transaxle.

23. Install the master cylinder to the front of the dash. Torque the nuts evenly (to prevent damaging the master cylinder) to 16 ft. lbs. (22 Nm). Remove the pedal restrictor from the pushrod. Lubricate the pushrod bushing on the clutch pedal. If the bushing is cracked or worn, replace it.

24. If equipped with cruise control, check the switch adjustment at the clutch pedal bracket.

NOTE: When adjusting the cruise control switch, do not exert more than 20 pounds of upward force on the clutch pedal pad for damage to the master cylinder pushrod retaining rod can result.

25. Depress the clutch pedal several times to break the plastic retaining straps. DO NOT remove the plastic button from the end of the pushrod.

26. Install remaining components in the reverse order of removal.

27. Connect the negative terminal from the battery.

Clutch Slave Cylinder

REMOVAL AND INSTALLATION

2.2L (VIN 4) Engine

1995–96 Vehicles

1. Disconnect the negative battery cable.
2. Remove the clutch actuator cylinder (slave cylinder) from the clutch master cylinder line.
3. Remove the actuator cylinder retaining nuts at the transaxle.
4. Remove the clutch actuator cylinder assembly from the transaxle.

To install:

5. Install the clutch actuator cylinder assembly to the transaxle.
6. Install the retaining nuts and tighten to 16 ft. lbs. (22 Nm).
7. Install the clutch actuator cylinder to the clutch master cylinder line.
8. Connect the negative battery cable. Bleed the hydraulic system.
9. Apply the emergency brake, start the engine, check for proper engagement of the transaxle and clutch.

Hydraulic Clutch System

BLEEDING

Without Bleed Screw

1. Disconnect the slave cylinder from the transaxle.
2. Loosen the master cylinder mounting attaching nuts. Do not remove the master cylinder.
3. Remove any dirt or grease around the reservoir cap so dirt cannot enter the system. Fill the reservoir with an approved DOT 3 brake fluid.
4. Depress the hydraulic actuator cylinder pushrod approximately 7/8 in. (20mm) into the slave cylinder bore and hold.
5. Install the diaphragm and cap on the reservoir while holding the slave cylinder pushrod.
6. Release the slave cylinder pushrod.
7. Hold the slave cylinder vertically with the pushrod end facing the ground.

NOTE: The slave cylinder should be lower than the master cylinder.

8. Press the pushrod into the slave cylinder bore with short 3/8 in. (10mm) strokes.
9. Observe the reservoir for air bubbles. Continue until air bubbles no longer enter the reservoir.

10. Connect the slave cylinder to the transaxle. Tighten the master cylinder attaching nuts.
11. Top-up the clutch master cylinder reservoir.
12. To test the system, start the engine and push the clutch pedal to the floor. Wait 10 seconds and select reverse gear. There should be no gear clash. If clash is present, air may still be present in the system. Repeat bleeding procedure.

With Bleed Screw

1. Remove any dirt or grease around the reservoir cap so dirt cannot enter the system. Fill the reservoir with an approved DOT 3 brake fluid. Maintain the fluid level while bleeding the system.
2. Attach a hose to the bleeder screw on the clutch actuator assembly and submerge the other end of the hose in a container of hydraulic clutch fluid GM P/N 12345347 or equivalent DOT 3 brake fluid.
3. Depress the clutch pedal slowly and hold. Loosen the bleeder screw to purge air.
4. Tighten the bleeder screw to 18 inch lbs. (2 Nm).
5. Repeat the bleeding process until all the air is purged from the system.
6. Fill the master cylinder to the proper level.
7. To test the system, start the engine and push the clutch pedal to the floor. Wait 10 seconds and select reverse gear. There should be no gear clash. If clash is present, air may still be present in the system. Repeat bleeding procedure.

Automatic Transaxle Assembly

REMOVAL AND INSTALLATION

2.2L (VIN 4) Engine

1. Disconnect the negative terminal from the battery. Remove the air cleaner and air intake assembly.
2. Disconnect the Throttle Valve (TV) cable from the throttle lever and the transaxle.
3. Remove the fluid level indicator (dipstick) and the filler tube.
4. Install Engine Support Fixture tool No. J-28467 or equivalent and the Adapter tool No. J-35953 or equivalent, on the engine.
5. Remove the wiring harness-to-transaxle nut.
6. Label and disconnect the electrical connectors for the speed sensor,

TCC connector and the neutral safety/backup light switch.
7. Disconnect the shift linkage from the transaxle.
8. Remove the top 2 transaxle-to-engine bolts, the transaxle mount and bracket assembly.
9. Disconnect the rubber hose that runs from the transaxle to the vent pipe.
10. Raise and support the front of the vehicle.
11. Remove the front wheels.
12. Disconnect the shift linkage and bracket from the transaxle.
13. Remove the left side splash shield.
14. Using a modified Drive Axle Seal Protector tool No. J-34754 or equivalent, install one on each drive axle to protect the seal from damage and the joint from possible failure.
15. Using care not to damage the halfshaft boots, disconnect the halfshafts from the transaxle.
16. Remove the transaxle strut. Remove the left side stabilizer link pin bolt and bushing clamp nuts from the support.
17. Remove the left frame support bolts and move it aside.
18. Disconnect the speedometer wire from the transaxle.
19. Remove the transaxle converter cover and matchmark the torque converter-to-flywheel for reassembly.
20. Disconnect and plug the transaxle cooler pipes.
21. Remove the transaxle-to-engine support.
22. Using a transmission jack, position and secure the jack to the transaxle. Remove the remaining transaxle-to-engine bolts.
23. Remove the transaxle from the vehicle. Use care not to let the torque converter fall from the front of the transaxle.

NOTE: The transaxle cooler and lines should be flushed any time the transaxle is removed for overhaul or replacing the pump, case or converter.

To install:

24. Put a small amount of grease on the pilot hub of the converter and make sure the converter is properly engaged with the pump.
25. Raise the transaxle to the engine while guiding the right side halfshaft into the transaxle.
26. Install the lower transaxle mounting bolts and remove the jack.
27. Align the converter with the marks made previously on the flywheel and install the bolts hand-tight.

28. Torque the converter bolts to 45 ft. lbs. (61 Nm). Retorque the first bolt after the others.
29. Install the transaxle converter cover. Attach the speedometer wire connector.
30. Position the L.H. halfshaft into the transaxle. Install the left frame support assembly.
31. Install the left stabilizer shaft frame bushing nuts. Install the left stabilizer bar link pin bolt.
32. Install the transaxle strut.
33. Seat the drive axles in the transaxle. Remove the drive axle seal protectors.
34. Install the remaining components in the reverse order of removal.
35. Install the negative battery cable.
36. Fill with automatic transmission fluid and check for leaks.

2.3L (VIN A) Engine

1. Disconnect the negative battery cable. Drain the cooling system.
2. Disconnect the heater hoses at the heater core.
3. Disconnect the intake air duct.
4. Remove the cable control cover.
5. Disconnect the throttle and TV cable. Disconnect the shift cable and bracket.
6. Disconnect the throttle cable from the throttle body.
7. Tag and disconnect all vacuum lines and electrical connections.
8. Remove the power steering pump and set aside.
9. Remove the fluid fill tube.
10. Install engine support fixture J 28467-A or equivalent.
11. Remove the 4 top engine-to-transaxle bolts.
12. Raise and safely support the vehicle. Remove both front wheels.
13. Disconnect the left inner splash shield from the control arm assembly.
14. Disconnect both lower ball joints. Disconnect the stabilizer shaft links.
15. Remove the front air deflector.
16. Disconnect the left suspension support.
17. Install drive axle seal protectors and remove both drive axles.
18. Disconnect the engine to transaxle brace.
19. Remove the flywheel cover.
20. Remove the flywheel-to-torque converter bolts.
21. Disconnect the transaxle cooler pipes.
22. Disconnect the ground wires from the engine to transaxle bolt.
23. Disconnect the cooler pipe brace.

24. Disconnect the exhaust brace.
25. Remove the bolts from the engine and transaxle mount.
26. Remove the transaxle mount to body bolts.
27. Support the transaxle with a jack. Remove the remaining engine to transaxle bolts and remove the transaxle.

To install:
28. Put a small amount of grease on the pilot hub of the converter and make sure the converter is properly engaged with the pump.
29. Raise the transaxle to the engine while guiding the right side halfshaft into the transaxle.
30. Install the lower engine-to-transaxle bolts. and the transaxle mount-to-body bolts.
31. Install the bolts to the engine and transaxle mount.
32. Install the remaining components in the reverse order of removal.
33. Fill the cooling system.
34. Connect the negative battery cable.
35. Fill the transaxle with automatic transmission fluid.
36. Adjust the TV cable and shift linkage as necessary.

3.1L (VIN T) and (VIN M) Engine

1. Disconnect the negative terminal from the battery. Remove the air cleaner, bracket, Mass Air Flow (MAF) sensor and air tube as an assembly.
2. Disconnect the exhaust crossover from the right side manifold and remove the left side exhaust manifold. Raise and support the manifold/crossover assembly.
3. Disconnect the TV cable from the throttle lever and the transaxle.
4. Remove the vent hose and the shift cable from the transaxle.
5. Remove the fluid level indicator and the filler tube.
6. Using the Engine Support Fixture tool No. J-28467 or equivalent and the Adapter tool No. J-35953 or equivalent, install them on the engine. Use care to make sure the engine is safely supported and that the engine support fixture does not damage any bodywork.
7. Remove the wiring harness-to-transaxle nut.
8. Label and disconnect the wires for the speed sensor, TCC connector and the neutral safety/backup light switch.
9. Remove the upper transaxle-to-engine bolts.
10. Remove the transaxle-to-mount through bolt, the transaxle mount bracket and the mount.

11. Raise and safely support the vehicle. Remove the front wheels.
12. Disconnect the shift cable bracket from the transaxle.
13. Remove the left side splash shield.
14. Using a modified Drive Axle Seal Protector tool No. J-34754 or equivalent, install one on each drive axle to protect the seal from damage and the joint from possible failure.
15. Using care not to damage the halfshaft boots, disconnect the halfshafts from the transaxle.
16. Remove the torsional and lateral strut from the transaxle. Remove the left side stabilizer link pin bolt.
17. Remove the left frame support bolts and move it aside.
18. Disconnect the speedometer wire from the transaxle.
19. Remove the transaxle converter cover and matchmark the converter-to-flywheel for assembly.
20. Disconnect and plug the transaxle cooler pipes.
21. Remove the transaxle-to-engine support.
22. Using a transmission jack, position and secure it to the transaxle. Remove the remaining transaxle-to-engine bolts.
23. Remove the transaxle from the vehicle using care to make sure the torque converter does not fall out.

NOTE: The transaxle cooler and lines should be flushed any time the transaxle is removed for overhaul, to replace the pump, case or converter.

To install:
24. Put a small amount of grease on the pilot hub of the converter and make sure the converter is properly engaged with the pump.
25. Raise the transaxle to the engine while guiding the right side halfshaft into the transaxle.
26. Install the lower transaxle mounting bolts and remove the jack.
27. Install the cooler lines at the transmission.
28. Position the left side halfshaft into the transaxle.
29. Install the left frame support bolts.
30. Install the left stabilizer shaft bushing clamp nuts at the support.
31. Install the left stabilizer bar link pin bolt.
32. Seat the drive axles in the transaxle. Remove the drive axle seal protectors.
33. Install the shift linkage bracket to the transaxle.
34. Install the speedometer wire connector.

35. Install the transaxle brace bolts. Install the transaxle lateral strut.

36. Install the transaxle torsional strut.

37. Align the converter with the marks made previously on the flywheel and install the bolts hand-tight.

38. Torque the converter bolts to 45 ft. lbs. Retorque the first bolt after the others.

39. Install the transaxle converter cover. Install the splash shield.

40. Install both front wheels.

41. Lower the car.

42. Install the transaxle mount.

43. Install the transaxle mount through bolt.

44. Install the wiring harness and nut securing it to the transaxle.

45. Remove the engine support fixture.

46. Install the fluid level indicator and fill tube.

47. Install the neutral start and TCC connectors.

48. Install the shift cable.

49. Install the TV cable at the transaxle and throttle lever.

50. Install the rubber hoses to the transaxle vent pipe.

51. Install the left side exhaust manifold bolts and exhaust crossover bolts.

52. Install the air cleaner mounting bracket, MAF sensor and air tube as an assembly.

53. Install the negative battery cable.

54. Fill transaxle with automatic transmission fluid and check for leaks.

DRIVE AXLE

Halfshaft

REMOVAL AND INSTALLATION

1. Raise and safely support vehicle. Remove the wheel and tire assembly.

2. Install drive seal protector J-34754 or equivalent, on the outer joint.

3. Insert a drift pin through the opening in the caliper and into the ventilation openings in the brake rotor to prevent the rotor from turning.

4. Remove the drive shaft hub nut and washer.

5. Remove the lower ball joint cotter pin and nut and loosen the joint using tool J-38892 or equivalent. If removing the right axle, turn the wheel to the left. If removing the left axle, turn the wheel to the right.

6. Separate the joint, with a prybar between the suspension support.

7. Disengage the axle from the hub and bearing using J-28733-A or equivalent puller.

8. Separate the hub and bearing assembly from the drive axle and move the strut and knuckle assembly rearward.

9. Disconnect the inner joint from the transaxle using tool J-28468 or J-33008 attached to J-29794 and J-2619-01 or from the intermediate shaft (V6 and 2.3L (VIN A) engines), if equipped.

To install:

10. Install axle seal protector J-37292-A into the transaxle.

11. Insert the drive axle into the transaxle or intermediate shaft (3.1L (VIN T) and (VIN M), 2.3L (VIN A) engines), if equipped, by placing a suitable tool into the groove on the joint housing and tapping until seated.

———— CAUTION ————
Be careful not to damage the axle seal or dislodge the transaxle seal garter spring when installing the axle.

12. Verify that the drive axle is seated into the transaxle by grasping on the housing and pulling outward.

13. Install the drive axle into the hub and bearing assembly.

14. Install the lower ball joint to the knuckle. Tighten the ball joint to steering knuckle nut to 41 ft. lbs. (55 Nm) and install the cotter pin.

15. Install the washer and new drive shaft nut. Again insert a drift pin into the caliper and rotor to prevent the rotor from turning and tighten the drive shaft nut to 185 ft. lbs. (260 Nm). Install the brake caliper.

16. Remove both J-37292-B and J-34754 seal protectors.

17. Install the tire and wheel assembly. Lower the vehicle.

CV-Joint Boot

REPLACEMENT

These vehicles use several types of CV-Joints. Use care to use the correct procedure.

Outer CV-Joint Boot

Except 1996 2.2L (VIN 4) Engine

1. Raise and safely support the vehicle. Remove the halfshaft from the vehicle .

2. Secure the halfshaft in a suitable bench vise with protective jaw covers installed to protect the halfshaft.

3. Remove the inner and outer CV-joint boot clamps with side cutters and discard the clamps.

4. Slide the boot off the outer joint and down onto the halfshaft so the outer joint is exposed.

5. Clean the grease off the outer joint.

6. Locate the retaining ring in the cut out on the joint and using J-8059, or an equivalent pair of snapring pliers, spread the ears on the retainer ring.

7. Remove the CV-joint from the halfshaft. Remove the CV-joint boot from the shaft.

To install:

8. Clean all the CV-joint components completely.

9. Install the small clamp onto the halfshaft.

10. Slide the boot onto the shaft until it is below the boot seating area.

11. Pack the outer CV-joint with new grease and install the joint onto the halfshaft. Push the joint onto the axle until the retaining ring seats in the groove and the joint can not be pulled back off.

12. Put the remainder of the grease in the CV-joint boot and slide the boot up until the small end is over the boot seating area.

13. Position the small clamp over the boot and using J-35910, or an equivalent seal clamp tool, compress the clamp.

14. Fit the large end of the boot over the CV-joint and install the large clamp.

15. Using J-35910, or the equivalent, compress the clamp.

16. Install the halfshaft in the vehicle. Lower the vehicle.

1996 2.2L (VIN 4) Engine

1. Raise and safely support the vehicle.

2. Remove the halfshaft from the vehicle and support it in a suitable vise with protective jaw covers to protect the halfshaft.

3. Remove the inner and outer CV-boot clamps with side cutters and discard the clamps.

4. Slide the boot off the outer joint and down onto the halfshaft so the outer joint is exposed.

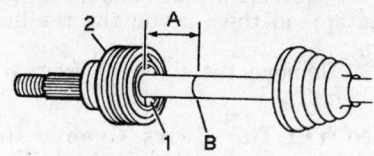

A. Measurement
B. Reference mark
1. C/V joint inner race
2. C/V joint outer race

333713

CV-joint to shaft measurement — 1996 2.2L (VIN 4) engine

5. Clean the grease off the outer joint. Scribe a reference mark on the halfshaft.

6. Measure the distance between the reference mark and the face of the CV-joint inner race. make a note of this measurement.

7. Attach tool J-46398 CV puller or equivalent, to the threaded area of the outer race.

8. Attach tool J-2619-01 slide hammer or equivalent onto the outer end of the CV puller. use the slide hammer to remove the CV-joint from the halfshaft.

9. Remove the CV puller tools from the CV-joint.

10. Remove the CV-boot from the halfshaft.

To install:

11. Clean all the CV-joint components completely.

12. Pack the CV-joint with approximately half of the grease from the service kit.

13. Install the small clamp onto the halfshaft.

14. Slide the boot onto the shaft. Expose the reference mark, which was made during disassembly, by sliding the boot until it is below the boot seating area.

15. Position the large clamp around the boot.

16. Place the axle assembly onto an arbor press. Support the tripot end and press against the CV-joint until the press cannot move any further. This ensures that the retaining ring engages in the inner race. Do not exceed 4000 pounds press load during assembly.

17. Measure the distance between the reference mark on the halfshaft and the face of the CV-joint inner race. Compare this mark with the measurement witch was made before disassembly. The distance measured after assembly must be within plus or minus 1mm of the distance measured before disassembly.

18. If the measurements do not match the ones made before disassembly the axle pressing procedure must be repeated.

19. Put the remainder of the grease in the CV-boot and slide the boot up until the small end is over the boot seating area.

20. Position the small clamp over the boot and using J-35910, or an equivalent seal clamp tool, compress the clamp.

21. Fit the large end of the boot over the CV-joint and install the large clamp.

22. Using J-35910, or the equivalent, compress the clamp.

23. Install the halfshaft in the vehicle. Lower the vehicle.

Inner CV-Joint Boot

Tripot/Free Motion Design — All Except 1996 2.2L (VIN 4) Engine

1. Raise and safely support the vehicle. Remove the halfshaft from the vehicle.

2. Secure the halfshaft in a suitable bench vise with protective jaw covers installed to protect the halfshaft.

3. Remove the inner and outer CV-joint boot clamps with side cutters and discard the clamps.

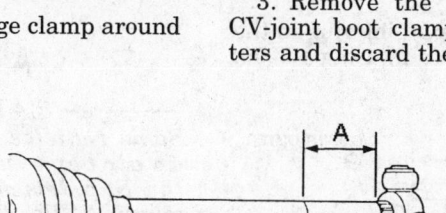

A. Measurement
B. Reference mark
C. Spider face
1. Tripot joint spider

333755

Spider joint measurement — 1996 2.2L (VIN 4) engine

4. Slide the boot off the outer joint and down onto the halfshaft so the outer joint is exposed.

5. Slide the tripot housing off the spider assembly.

6. Expand the inner retainer ring using J-8059 and work the ring down onto the halfshaft.

7. Push the spider assembly onto the axle until the top retainer ring is exposed.

8. Remove the top retainer ring using J-8059. Remove the spider assembly and inner retainer ring.

9. Remove the CV-joint boot from the shaft.

To install:

10. Clean all the CV-joint components completely.

11. Install the small clamp onto the halfshaft.

12. Slide the boot onto the shaft until it is below the boot seating area.

13. Install the inboard retainer ring until it is just below the splined portion of the axle.

14. Install the spider assembly until the top retainer ring groove is exposed.

15. Install the top retainer ring and slide the spider assembly up until it is seated over the top retainer ring.

16. Using J-8059, position the inner retainer ring in the groove below the spider assembly.

17. Pack half the grease from the service kit in the tripot housing and install the housing over the spider assembly.

18. Put the remainder of the grease in the CV-joint boot.

19. Slide the boot up the axle until the small end of the boot is seated over the boot contact area.

20. Position the small clamp over the boot and using J-35910, compress the clamp.

21. Fit the large end of the boot over the Tripot housing and install the large clamp.

22. Using J-35910, compress the clamp.

23. Install the halfshaft in the vehicle. Lower the vehicle.

Tripot Design — 1996 2.2L (VIN 4) engine

1. Raise and safely support the vehicle.

2. Remove the halfshaft from the vehicle and support it in a suitable vise with protective jaw covers to protect the halfshaft.

3. Remove the inner and outer CV-boot clamps with side cutters and discard the clamps.

4. Separate the boot from the trilobal bushing at the large diameter end of the boot and slide the boot away from the joint along the halfshaft.

JOINT SIZE	DIMENSION A	DIMENSION B
23	23.8mm (.9")	190.5mm (7.5")
27	27.05mm (1.1")	170.0mm (6.7")

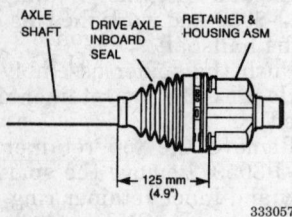

Boot dimensions for the tripot inboard joint

5. Remove the housing from the spider and shaft.

NOTE: Handle the tripot assembly with care. Tripot balls and needle rollers may separate from the tripot spider trunnions.

6. Scribe a reference mark on the halfshaft.

7. Measure the distance between the reference mark and the face of the tripot spider. Make a note of this measurement.

8. Place a brass drift on the area of the tripot spider that is next to the halfshaft.

9. Tap on the brass drift with a hammer. Remove the tripot spider assembly from the halfshaft.

10. Remove and discard the spider retaining ring from the grove on the halfshaft.

11. Remove, clean and inspect the spider assembly.

To install:

12. Install the small retaining clamp onto the neck of the boot.

13. Slide the boot onto the halfshaft toward the CV end of the axle

14. Place the large retaining ring onto the halfshaft.

15. Place the tripot spider assembly onto the halfshaft with the chamfered side of the tripot spider facing the halfshaft.

16. Place the axle assembly onto the arbor press with the tripot on the bottom plate and the CV-joint under the press head.

17. Press on the CV-joint assembly until the press cannot move any further. This ensures that the buried ring engages into the tripot spider. Do not exceed 4000 pounds press load during assembly.

18. Measure the distance between the reference mark on the halfshaft and the face of the CV-joint inner race. Compare this mark with the measurement witch was made before disassembly. The distance measured

after assembly must be within plus or minus 1mm of the distance measured before disassembly.

19. If the measurements do not match the ones made before disassembly the axle pressing procedure must be repeated.

20. Place the neck of the boot into the grove on the halfshaft and crimp the retaining clamp to 100 ft. lbs. (136 Nm) with tool J-35910 or equivalent.

21. Place approximately half of the grease from the service kit into the boot and use the remainder to repack the housing.

22. Position the large retaining clamp onto the boot.

23. Slide the tripot housing over the spider assembly and onto the halfshaft.

24. Slide the large end of the boot with the clamp onto the outside of the trilobal tripot bushing and locate the lip of the seal into the groove.

25. Seat the retaining clamp properly and latch the clamp with tool J-35566 or equivalent

26. Install the halfshaft in the vehicle. Lower the vehicle.

27. Road test the vehicle and make sure there are no vibrations.

Cross Groove Design — 1993–95 Vehicles

1. Raise and safely support the vehicle. Remove the halfshaft from the vehicle.

2. Secure the halfshaft in a suitable bench vise with protective jaw covers installed to protect the halfshaft.

3. Remove the inner and outer CV-joint boot clamps with side cutters and discard the clamps.

4. Slide the boot off the outer joint and down onto the halfshaft so the outer joint is exposed.

5. Clean the grease off the outer joint.

6. Locate the retaining ring in the cut out on the joint and using J-8059,

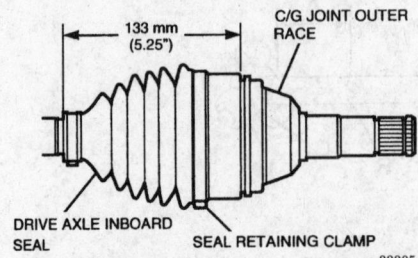

Boot dimensions for the cross groove inboard joint — 1993–95 models

or an equivalent pair of snapring pliers, spread the ears on the retainer ring.

7. Remove the CV-joint from the halfshaft.

NOTE: The Cross Groove design joint uses precision grinding and selected dimensional component fits for proper assembly and operation. Due to its complexity, DO NOT disassemble the Cross Groove joint.

8. Remove the CV-joint boot from the shaft.

To install:

9. Clean all the CV-joint components completely.

10. Install the small clamp onto the halfshaft.

11. Slide the boot onto the shaft until it is below the boot seating area.

12. Pack the outer CV-joint with new grease and install the joint onto the halfshaft. Push the joint onto the axle until the retaining ring seats in the groove and the joint can not be pulled back off.

13. Put the remainder of the grease in the CV-joint boot and slide the boot up until the small end is over the boot seating area.

14. Position the small clamp over the boot and using J-35910, or an equivalent seal clamp tool, compress the clamp.

15. Fit the large end of the boot over the CV-joint and install the large clamp.

16. Using J-35910, or the equivalent, compress the clamp.

17. Install the halfshaft in the vehicle. Lower the vehicle.

STEERING

Air Bag

— **CAUTION** —

Some vehicles are equipped with an air bag system, also known as the Supplemental Inflatable Restraint (SIR). The system must be disabled before performing service on or around system components, steering column, instrument panel components, wiring and sensors. Failure to follow safety and disabling procedures could result in accidental air bag deployment, possible personal injury and unnecessary system repairs.

PRECAUTIONS

Several precautions must be observed when handling the inflator module to avoid accidental deployment and possible personal injury.

- Never carry the inflator module by the wires or connector on the underside of the module.
- When carrying a live inflator module, hold securely with both hands, and ensure that the bag and trim cover are pointed away.
- Place the inflator module on a bench or other surface with the bag and trim cover facing up.
- With the inflator module on the bench, never place anything on or close to the module which may be thrown in the event of an accidental deployment.

DISARMING

——— CAUTION ———
The Supplemental Inflatable Restraint (SIR) system must be disarmed before performing many service functions. Failure to do so may cause accidental deployment of the air bag, resulting in unnecessary SIR system repairs and/or personal injury.

Disabling The SIR System

1. Turn the steering wheel so the vehicle's wheels are pointing straight ahead. This helps protect the clockspring mechanism under the steering wheel hub which powers the air bag module.
2. Turn the ignition switch to **LOCK** and remove the key.
3. Remove the AIR BAG fuse from the instrument panel fuse block.
4. Remove the left side sound insulator.
5. Remove the air bag connector CPA and disconnect the yellow 2-way SIR harness connector located near the base of the steering column.

NOTE: With the AIR BAG fuse removed, and if, for some reason, the negative battery cable should be connected and the ignition switch turned to the ON (not recommended), the air bag warning lamp in the instrument cluster will be ON. This is normal operation and does not indicate an SIR system malfunction.

Enabling The SIR System

1. Turn the ignition switch to the **LOCK** position and remove the key.
2. At the base of the steering column, connect the yellow 2-way con-

nector and install the CPA bar to verify that the connection is complete and tight.
3. Install the left side sound insulator.
4. Install the AIR BAG fuse into the instrument panel fuse block.
5. Turn the ignition switch to the **RUN** position and verify that AIR BAG warning lamp flashes 7 times and turns OFF. If it does not operate as described, perform SIR Diagnostic System Check.

Steering Wheel

REMOVAL AND INSTALLATION

1. Disable the SIR system.

——— CAUTION ———
When carrying a live air bag, make sure the bag and trim cover are pointed away from the body. In the unlikely event of an accidental deployment, the bag will then deploy with minimal chance of injury. When placing a live air bag on a bench or other surface, always face the bag and trim cover up, away from the surface. This will reduce the motion of the module if it is accidentally deployed.

2. Remove the air bag inflator module.
3. Detach the horn connector.
4. Disconnect the horn contact.
5. Remove the hexagon locking nut.

NOTE: When removing the steering wheel, use the specified wheel puller. Under no conditions should the end of the shaft be hammered on as this could cause internal damage.

6. Remove the steering wheel using tool J-1859-03 or equivalent puller.

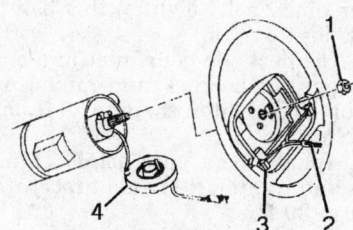

1. Steering wheel hexagon lock nut
2. Horn contact
3. Horn contact switch
4. Coil assembly

230196

Steering wheel view

To install :
7. Install the coil assembly connector through steering wheel.
8. Install the steering wheel and align steering wheel with turn signal canceling cam assembly.
9. Connect the horn contact.
10. Install the hexagon nut and tighten it to 30 ft. lbs. (41 Nm).

Tie Rod Ends

REMOVAL AND INSTALLATION

Outer Tie Rod Ends

1. Raise and safely support the vehicle. Remove the wheel and tire assembly.
2. Remove the cotter pin and castle nut from the outer tie rod end.
3. Loosen the outer tie rod pinch bolt.
4. Using steering linkage puller, J-24319–01 or equivalent, separate the tie rod from the strut steering arm.
5. Unthread the outer tie rod end from the adjuster rod keeping track of the number of turns required for removal.

To install:
6. Install the tie rod end onto the adjuster rod the same number of turns required to remove the old one.
7. Connect the tie rod end to the strut steering arm and tighten the castle nut to 35 ft. lbs. (50 Nm). If necessary to align the cotter pin holes tighten the nut a maximum of 50 ft. lbs. (75 Nm). NEVER loosen the nut to make the alignment. Install a new cotter pin.
8. Tighten the pinch bolt to 41 ft. lbs. (55 Nm).
9. Install the tire and wheel assembly and tighten the wheel nuts to 100 ft. lbs. (140 Nm).
10. Lower the vehicle.
11. Check the front toe setting and adjust as necessary.

Inner Tie Rod Ends

1. From under the hood, remove the lock plate from over the inner tie rod mounting bolts and discard the plate.
2. Remove the inner tie rod end bolt and loosen the opposite side bolt 2 turns.
3. Slide the inner tie rod out from between the plate and the steering rack.
4. Reinstall the lock plate bolt to insure proper tie rod-to-steering gear realignment.

5. Raise and safely support the vehicle. Remove the tire and wheel assembly.

6. Remove the cotter pin and castle nut from the outer tie rod end.

7. Using the steering linkage puller, J-24319-01 or equivalent, separate the tie rod from the steering knuckle.

8. Remove the inner and outer tie rod assembly from the vehicle.

9. Note the position of the inner and outer tie rods in relation to each other. Place the assembly in a vise and loosen the inner tie rod pinch bolt.

10. Unthread the inner tie rod end from the adjuster rod counting the number of turns required for removal.

To install:

11. Thread the inner tie rod end onto the adjuster the same number of turns as required for removal. Make sure the same number of threads are visible on the inboard and outboard sides of the adjuster.

12. Tighten the pinch bolt to 41 ft. lbs. (55 Nm).

13. Install the tie rod assembly in the vehicle.

14. Remove the lock plate bolt.

15. Connect the inner tie rod end to the rack and pinion unit and tighten the lock plate bolts to 65 ft. lbs. (90 Nm). Install a new lock plate retainer over the mounting bolts

16. Connect the outer tie rod end to the strut steering knuckle arm and tighten the castle nut to 35 ft. lbs. (50 Nm). If necessary to align the cotter pin holes tighten the nut a maximum of 50 ft. lbs. (75 Nm). NEVER loosen the nut to make the alignment. Install a new cotter pin.

17. Install the tire and wheel assembly and tighten the wheel nuts to 100 ft. lbs. (140 Nm).

18. Lower the vehicle.

19. Check the front toe and adjust as necessary.

Power Rack and Pinion

REMOVAL AND INSTALLATION

NOTE: The wheels of the vehicle must be straight ahead and the ignition switch must be in the LOCK position, before disconnecting the flange and coupling assembly from the steering column. Failure to do so may cause the supplemental inflatable restraint (SIR coil assembly to become off centered, which will damage the SIR coil assembly.

1. Remove the left side sound insulator.

2. Remove the upper pinch bolt on the coupling assembly.

3. Remove the line retainer (if applicable).

4. Remove the power brake assembly away from the cowl wall, leaving the master cylinder attached.

5. Raise and safely support the vehicle. Remove the front tire and wheel assemblies.

6. Disconnect the tie rod ends from the struts using tool J-24319-01 or equivalent.

7. Lower the vehicle.

8. Remove the left and right mounting clamps.

9. Disconnect the gear inlet and outlet pipes from the rack and pinion steering gear.

10. Move the rack and pinion assembly forward and remove the lower pinch bolt from flange on the coupling assembly.

11. Disconnect the coupling from the rack and pinion assembly.

12. Remove the dash seal from the rack and pinion assembly.

13. Remove the rack and pinion assembly through the left wheel opening.

NOTE: If the studs were removed with the mounting clamps, apply some thread locking compound on them and reinstall studs into cowl panel and tighten until the studs are fully seated against the dash panel. The torque should not exceed 13 ft. lbs. (18 Nm).

To install:

14. Install the rack and pinion assembly into the vehicle through the left wheel opening.

15. Attach the dash seal onto the rack and pinion assembly.

16. Move the rack and pinion assembly forward and install the coupling onto the rack and pinion.

17. Install the flange and coupling lower pinch bolt. Tighten this bolt to 30 ft. lbs. (41 Nm).

18. Connect the gear inlet and outlet pipes to the rack and pinion assembly and tighten them to 20 ft. lbs. (27 Nm).

19. Install the mounting clamp nuts and tighten (left side first) to 22 ft. lbs. (30 Nm).

20. Raise the vehicle.

21. Connect the tie rods to the struts and tighten to 44 ft. lbs. (60 Nm) and install cotter pins through the nuts.

22. Install the tire and wheel assemblies.

23. Install the line retainer (if applicable). Lower the vehicle.

24. Install the steering column pinch bolt and tighten to 30 ft. lbs. (41 Nm).

25. Install the left side sound insulator.

26. Fill the reservoir with power steering fluid and bleed air from the system.

27. Check the toe setting and adjust if necessary.

28. Road test the vehicle and verify no leaks and proper operation.

Power Steering Pump

BLEEDING

1. Turn the ignition switch to the **OFF** position.

2. Raise and safely support the vehicle so the front wheels are off the ground.

3. Turn the steering wheel all the way to the left.

4. Verify that the fluid in the reservoir is to the full COLD mark. Add the proper fluid as necessary. Leave the cap off the reservoir.

5. With an assistant checking the fluid level, turn the steering wheel lock-to-lock at least 20 times. The engine remains **OFF**. Note that on vehicles with extra long return lines or fluid coolers in the system, turn the steering wheel lock-to-lock at least 40 times. Be aware that enough trapped air may cause the fluid to overflow. Be sure to thoroughly clean any spilled fluid to allow for a leak check. Keep the fluid to the proper level.

6. Start the engine and allow to idle. Maintain the fluid level and reinstall the cap.

7. Return the wheels to center and lower the vehicle to the ground.

8. Keep the engine for running at least 2 minutes. Turn the steering wheel in both directions.

9. Check for smooth power assist, relatively quiet operation, proper fluid level, no system leaks and no bubbles or foam in the fluid. If okay, procedure is complete. If not okay, troubleshooting is required.

REMOVAL AND INSTALLATION

2.2L (VIN 4) Engine

1. Disconnect the negative battery cable. Remove the power steering pump drive belt.

2. Disconnect the power steering lines at the pump.

3. Remove the attaching bolts and remove the power steering pump.

To install:

4. Install the power steering pump and tighten the lower center bolt and upper side bolts to 22 ft. lbs. (30 Nm).

5. Install the lines to the pump and tighten inlet line to 20 ft. lbs. (27 Nm). Install the drive belt.

6. Connect the negative battery cable.

7. Fill the pump with approved power steering fluid and bleed air from power steering system.

8. Road test vehicle.

2.3L (VIN A) Engine

1. Disconnect the negative battery cable. Disconnect and cap the power steering fluid lines.

2. Back off on the adjustment stud. Remove the outer bracket.

3. Remove the drive belt and pump.

4. Unfasten the bracket to pump retainers, then remove the bracket with the pump.

5. Remove the pump from the bracket.

6. If necessary, transfer the pulley, front bracket and reservoir.

To install:

7. Position the pump with the pulley, front bracket and reservoir.

8. Install the bracket with the pump to the front of the engine.

9. Install the drive belt and the outer bracket. Adjust the drive belt tension.

10. Uncap and connect the power steering fluid lines.

11. Connect the negative battery cable.

12. Fill and bleed the power steering system. Verify no fluid leaks.

3.1L (VIN T) and (VIN M) Engines

1. Disconnect the negative battery cable.

2. Remove the power steering pump drive belt.

3. Remove the nut from bracket retaining hose on the alternator.

4. Remove the engine mount.

5. Remove the pump mounting bolts (to ease pump line removal).

6. Disconnect the lines at the pump.

7. Remove the pump.

To install:

8. Connect the lines to the pump. Tighten the inlet line to 20 ft. lbs. (27 Nm).

9. Position pump into place and install the mounting bolts. Tighten these bolts to 22 ft. lbs. (30 Nm).

10. Install the drive belt.

11. Install the engine mount.

12. Install the nut to bracket retaining hose on alternator.

13. Connect the negative battery cable.

14. Fill with approved power steering fluid and bleed air from the power steering system.

15. Road test the vehicle.

BRAKES

Anti-Lock Brake System Service

PRECAUTIONS

• Certain components within the Anti-Lock Brake System (ABS) are not intended to be serviced or repaired individually. Only those components with removal and installation procedures should be serviced.

• Do not use rubber hoses or other parts not specifically specified for and ABS system. When using repair kits, replace all parts included in the kit. Partial or incorrect repair may lead to functional problems and require the replacement of components.

• Lubricate rubber parts with clean, fresh brake fluid to ease assembly. Do not use lubricated shop air to clean parts; damage to rubber components may result.

• Use only specified brake fluid from an unopened container.

• If any hydraulic component or line is removed or replaced, it may be necessary to bleed the entire system.

• A clean repair area is essential. Always clean the reservoir and cap thoroughly before removing the cap. The slightest amount of dirt in the fluid may plug an orifice and impair the system function. Perform repairs after components have been thoroughly cleaned; use only denatured alcohol to clean components. Do not allow ABS components to come into contact with any substance containing mineral oil; this includes used shop rags.

• The Anti-Lock control unit is a microprocessor similar to other computer units in the vehicle. Ensure that the ignition switch is **OFF** before removing or installing controller harnesses. Avoid static electricity discharge at or near the controller.

• If any arc welding is to be done on the vehicle, the control unit should be unplugged before welding operations begin.

Master Cylinder

REMOVAL AND INSTALLATION

NOTE: To perform the Gear Tension Relief Sequence listed below, a Tech 1 scan tool or equivalent tool must be used prior to removal of the ABS modulator/master cylinder assembly.

————— **WARNING** —————

When servicing the master cylinder on this vehicle, the procedure below must be followed in the correct sequence. DO NOT perform any services to the master cylinder without first performing the Gear Tension Release procedure.

1. Disconnect the electrical connector from the fluid level sensor.

2. Perform gear tension release procedure.

3. Disconnect the electrical connectors from both solenoids.

4. Disconnect the 3-pin and 6 pin motor pack electrical connectors.

5. Disconnect and cap the 4 brake lines on the master cylinder and modulator.

6. Remove the master cylinder-to-power booster nuts and remove the master cylinder and modulator assembly.

To install:

7. Position the master cylinder and tighten the retaining nuts to 20 ft. lbs. (27 Nm). and the brake lines-to-master cylinder and modulator to 15 ft. lbs. (20 Nm).

8. Connect the fluid level electrical sensor wires.

9. Connect the electrical connectors to both solenoids.

10. Connect the 3-pin and 6 pin motor pack electrical connectors.

11. Refill the reservoir with an approved DOT 3 brake fluid and bleed the brake system.

Brake Caliper

REMOVAL AND INSTALLATION

1. Drain ⅔ of the brake fluid from the master cylinder.

2. Raise and safely support the vehicle. Remove the tire and wheel assembly.

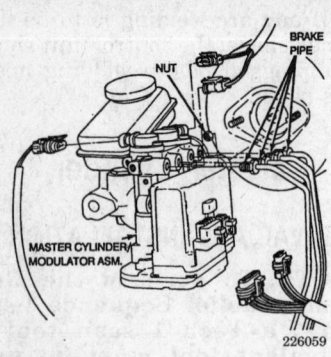

Removing the master cylinder assembly from the ABS modulator

3. Compress the caliper piston, using a suitable tool (C-clamp or large pliers).

4. If the caliper is to be removed from its brake for brake pad change or rotor service, do not disconnect the brake hose. If the caliper is to be completely removed from the vehicle, disconnect and cap the brake hose.

5. Remove the caliper mounting bolts. Remove the caliper from the vehicle.

6. If the caliper is being replaced remove the brake pads from the caliper.

To install:

7. Before installing the caliper make sure the piston is full seated in the bore and the brake pads are properly installed.

8. Install the caliper onto the steering knuckle and tighten the mounting bolts to 38 ft. lbs. (51 Nm).

9. If removed, connect the brake hose with new copper washers and tighten the brake hose bolt to 33 ft. lbs. (45 Nm).

10. Refill and bleed the brake system.

11. Install the wheel and tire assembly. Lower the vehicle.

12. Connect the negative battery cable.

13. Pump the brake pedal to verify firm and correct brake pedal feel. Road test vehicle to verify correct braking action.

Disc Brake Pads

REMOVAL AND INSTALLATION

1. Siphon ⅔ of the brake fluid out of the master cylinder reservoir.

2. Raise and safely support the vehicle. Remove the tire and wheel assembly.

3. Remove the caliper from the vehicle.

4. Remove the outboard pad by pushing in on the outside edge of the pad to release the mounting dowel from the hole in the caliper. When both dowels are unseated push the pad out the bottom of the caliper.

5. Remove the inboard pad from the caliper by pulling it out of the caliper.

To install:

6. Install the inboard pad in the caliper so the spring clip on the pad back engages in the caliper piston.

7. Install the outboard pad over the caliper end until the mounting dowels snap into the mounting holes in the caliper.

8. Install the caliper over the rotor onto the steering knuckle.

9. Install the mounting bolts and sleeves and tighten to 38 ft. lbs. (51 Nm).

10. Install the tire and wheel assembly. Lower the vehicle.

11. Pump the brake pedal several times to seat the pads against the rotor.

12. Check the master cylinder level and add fluid as necessary.

13. Verify a firm brake pedal before attempting to move the vehicle.

Brake Rotor

REMOVAL AND INSTALLATION

1. Raise and support the vehicle.

2. Remove the tire and wheel assembly.

3. Remove the caliper from the vehicle.

4. Slide the rotor off the hub.

5. If the rotor is being replaced perform the following:

   a. Siphon ⅔ of the brake fluid from the master cylinder reservoir.

   b. Compress the caliper piston into the caliper bore until it is bottomed out.

To install:

6. Slide the rotor onto the hub.

7. Install the caliper and tighten the mounting bolts to to 38 ft. lbs.

8. Torque the wheel lug nuts.

9. Lower the vehicle.

10. Pump the brake pedal several times to seat the pads against the rotor.

11. Check the brake fluid reservoir and add brake fluid as necessary.

12. Verify a firm pedal before attempting to move the vehicle.

Brake Drums

REMOVAL AND INSTALLATION

1. Raise and safely support the vehicle. Remove the tire and wheel assembly.

2. Remove the brake drum by sliding it off the hub bearing assembly. It may be necessary to back off the brake adjuster to remove the drum.

3. If the adjuster is backed off and the drum won't come off tap lightly around the edge of the drum with a soft hammer.

To install:

4. Install the drum onto the hub assembly.

5. Install the tire and wheel assembly and tighten to specification.

6. Adjust the rear brakes.

7. Lower the vehicle.

Brake Shoes

REMOVAL AND INSTALLATION

Except 1994 Vehicles With Leading/Trailing Brakes

1. Raise and safely support the vehicle. Remove the tire and wheel assembly.

2. Remove the brake drum.

3. Remove the upper return springs from the shoes with brake spring tool J-8049 or equivalent.

4. Remove the hold-down springs by gripping them with special tool J-8056 or equivalent then pressing down and turning 90 degrees (¼ turn).

5. Remove the shoe hold-down pins from behind the brake backing plate. They will simply slide out once the hold-down spring tension is relieved.

6. Lift up the actuator lever for the self-adjusting mechanism and remove the actuating link. Remove the actuator lever, pivot and the pivot return spring.

7. Spread the shoes apart to clear the wheel cylinder pistons, then remove the parking brake strut and spring.

8. Spread the shoes far enough apart to clear the hub and bearing assembly.

9. Disconnect the parking brake cable from the lever. Remove the shoes, still connected by their adjusting screw spring, from the car.

10. With the shoes removed, note the position of the adjusting spring, then remove the spring and adjusting screw.

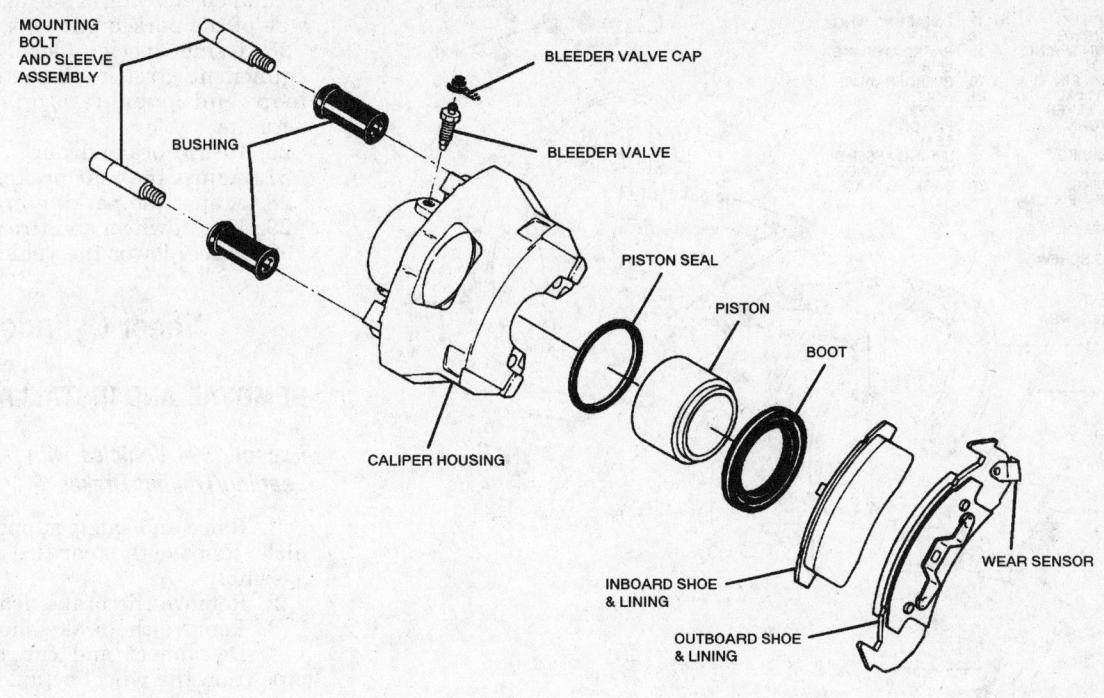

MOUNTING BOLT AND SLEEVE ASSEMBLY

BUSHING

BLEEDER VALVE CAP

BLEEDER VALVE

PISTON SEAL

PISTON

BOOT

CALIPER HOUSING

INBOARD SHOE & LINING

OUTBOARD SHOE & LINING

WEAR SENSOR

318111

Exploded view of the brake system components

11. Remove the C-clip from the parking brake lever and the lever from the secondary shoe.

To install:

12. Use a damp cloth to remove all dirt and dust from the backing plate and brake parts.

13. Lubricate the fulcrum end of the parking brake lever with brake grease specially made for the purpose. Install the lever on the secondary shoe and secure with the C-clip.

14. Install the adjusting screw and spring on the shoes, connecting them together. The coils of the spring must not be over the star wheel on the adjuster. The left and right hand springs are not interchangeable.

15. Lubricate the shoe contact surfaces on the backing plate with the brake grease.

16. Spread the shoe assemblies apart and connect the parking brake cable. Install the shoes on the backing plate, engaging the shoes at the top temporarily with the wheel cylinder pistons.

17. Spread the shoes apart slightly and install the parking brake strut and spring. Make sure the end of the strut without the spring engages the parking brake lever. The end with the spring engages the primary shoe (the one with the shorter lining).

18. Install the actuator pivot, lever and return spring. Install the actuating link in the shoe retainer. Lift up the actuator lever and hook the link into the lever.

19. Install the hold-down pins and hold-down springs. Install the shoe return springs with J-8049.

20. Install the brake drum.

21. Install the tire and wheel assembly and tighten to specification.

22. Adjust the brakes. Be sure to install a rubber hole cover in the knockout hole after the adjustment is complete.

23. Adjust the parking brake.

1994 Vehicles With Leading/Trailing Brakes

1. Raise and safely support the vehicle. Remove the wheel and tire assembly.

————— **CAUTION** —————

Brake shoes contain asbestos, which has been determined to be a cancer causing agent. Never clean the brake surfaces with compressed air. Avoid inhaling any dust from any brake surface. When cleaning brake surfaces, use a commercially available brake cleaning fluid.

2. Remove brake drum.

3. Remove actuator spring using special tool J-38400 or equivalent. Pry loop end of the spring from the adjustor actuator. Disconnect from web of the park brake shoe.

NOTE: When removing the retractor spring from either shoe and lining, do not overstretch it. Overstretching reduces the spring's effectiveness.

4. Lift the end of retractor spring from the adjuster shoe and lining. Insert hook end of special tool J-38400 or equivalent between the retractor spring and the adjuster shoe web. Pry or twist to lift end of spring out of the shoe web hole.

5. Pry end of retractor spring toward the axle with flat edge of special tool J-38400 or equivalent until spring snaps down off shoe web onto backing plate.

6. Remove the adjustor shoe and lining, adjuster actuator and adjusting screw assembly.

7. Remove park brake lever from the park brake shoe and lining. Do not remove parking brake cable from park brake lever unless lever is to replaced.

8. Lift end of retractor spring from the park brake shoe and lining. Insert hook end of J-38400 or equivalent between retractor spring

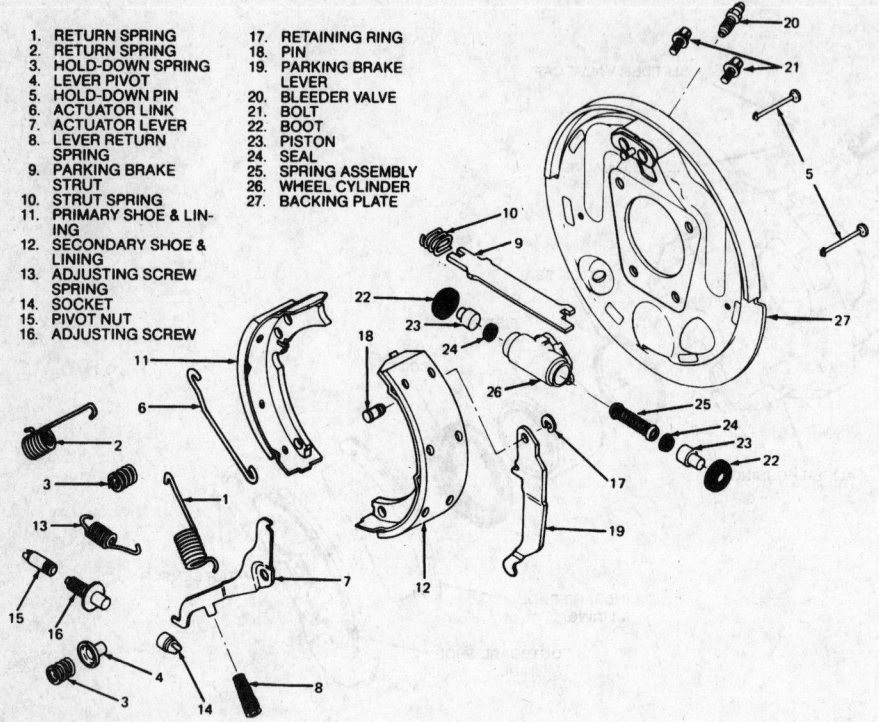

1. RETURN SPRING
2. RETURN SPRING
3. HOLD-DOWN SPRING
4. LEVER PIVOT
5. HOLD-DOWN PIN
6. ACTUATOR LINK
7. ACTUATOR LEVER
8. LEVER RETURN SPRING
9. PARKING BRAKE STRUT
10. STRUT SPRING
11. PRIMARY SHOE & LINING
12. SECONDARY SHOE & LINING
13. ADJUSTING SCREW SPRING
14. SOCKET
15. PIVOT NUT
16. ADJUSTING SCREW
17. RETAINING RING
18. PIN
19. PARKING BRAKE LEVER
20. BLEEDER VALVE
21. BOLT
22. BOOT
23. PISTON
24. SEAL
25. SPRING ASSEMBLY
26. WHEEL CYLINDER
27. BACKING PLATE

319824

Exploded view of the drum brake assembly — Except 1994 Vehicles with leading/trailing Brakes

and park brake shoe web. Pry or twist to lift end of spring out of the shoe web hole.

9. Pry end of retractor spring toward the axle with flat edge of special tool J-38400 or equivalent until the spring snaps down off the shoe web onto the backing plate.

10. Remove the brake shoe and lining. If only the shoes and linings are being replaced, the retractor spring does not have to be removed. Otherwise, remove retractor spring from the anchor plate.

To install:

11. Lubricate the 6 raised shoe pads and anchor surfaces that contact the lower ends of the brake linings on the backing plate.

--- CAUTION ---

Only a small amount of lubricant is necessary. Be careful not to contaminate the brake linings by getting any lubricant on the linings or the drum surfaces.

12. If retractor spring was removed, reinstall hooking center spring section under tab on the anchor.

13. Position shoe and lining on the backing plate.

14. Using special tool J-38400 or equivalent, pull end of retractor

spring up rest on web of the brake shoe.

15. Using special tool J-38400 or equivalent, pull end of the retractor spring over until it snaps into slot in the brake shoe.

16. Install park brake lever to the park brake shoe and lining.

17. Install parking brake cable to park brake lever, if disconnected.

18. Clean adjustor screw threads with a wire brush and apply lubricant to the threads.

19. Install the adjusting screw assembly and adjuster shoe lining. Engage pivot nut with web of park brake shoe and lining and park brake lever.

20. Position adjuster shoe and lining such that shoe web engages the deep slot in adjuster socket.

21. Install retractor spring into adjuster shoe and lining. Using special tool J-38400 or equivalent, pull end of the retractor spring up to seat on the web of brake shoe. Pull end of retractor spring over until it snaps into slot in the brake shoe.

22. Install adjustor actuator. Lubricate tab and pivot point on adjustor with brake lubricant.

23. Using special tool J-38400 or equivalent, spread brake shoes while working the adjuster actuator into position.

24. Install adjustor spring. Engage U-shaped end ring of spring in hole in web of the park brake shoe.

25. Using special tool J-38400 or equivalent, stretch spring and engage loop end over tab on adjuster actuator.

26. Install brake drum.

27. Adjust the rear brakes.

28. Adjust the parking brake.

29. Install wheel and tire assembly.

30. Safely lower the vehicle.

Wheel Cylinder

REMOVAL AND INSTALLATION

Except 1994 Vehicles with Leading/Trailing Brakes

1. Raise and safely support the vehicle. Remove the rear tire and wheel assembly.

2. Remove the brake drum.

3. Remove the brake shoes.

4. Disconnect and cap the brake pipe from the wheel cylinder.

5. In order to remove the wheel cylinder from the backing plate, remove the bleeder screw from the wheel cylinder.

6. Remove the 2 mounting bolts.

7. If necessary loosen or remove hub and bearing assembly.

8. Remove the wheel cylinder. On some vehicles, a No. 6 Torx socket may be necessary.

To install:

9. Remove the bleeder screw from the new wheel cylinder.

10. Install the wheel cylinder and wheel cylinder mounting nuts. Tighten the nuts to 106 inch lbs (12 Nm).

11. Connect the brake pipe to the wheel cylinder and tighten the fitting to 11 ft. lbs. (15 Nm).

12. If removed, install the hub and bearing assembly, torque the bolts to 43 ft. lbs. (58 Nm).

13. Install the brake shoes.

14. Install the brake drum.

15. Torque the wheel lug nuts.

16. Adjust the brakes.

17. Lower the vehicle.

18. Refill the master cylinder and bleed the system.

1994 Vehicles With Leading/Trailing Brakes

1. Raise and safely support the vehicle.

2. Remove the tire and wheel assembly. Mark the relationship between wheel to the wheel stud.

3. Remove the brake drum.

A ACCESS HOLE PLUG. NOT PART OF ASM. SERVICE ONLY ITEM.

1 ADJUSTER SOCKET

2 ADJUSTER SCREW

3 PIVOT NUT

4 RETRACTOR SPRING

5 ADJUSTER SHOE AND LINING

6 WHEEL CYLINDER

7 BLEEDER VALVE

8 BOLT

9 BACKING PLATE

10 PARK BRAKE SHOE AND LINING

11 PARK BRAKE LEVER

12 ACTUATOR SPRING

13 ADJUSTER ACTUATOR

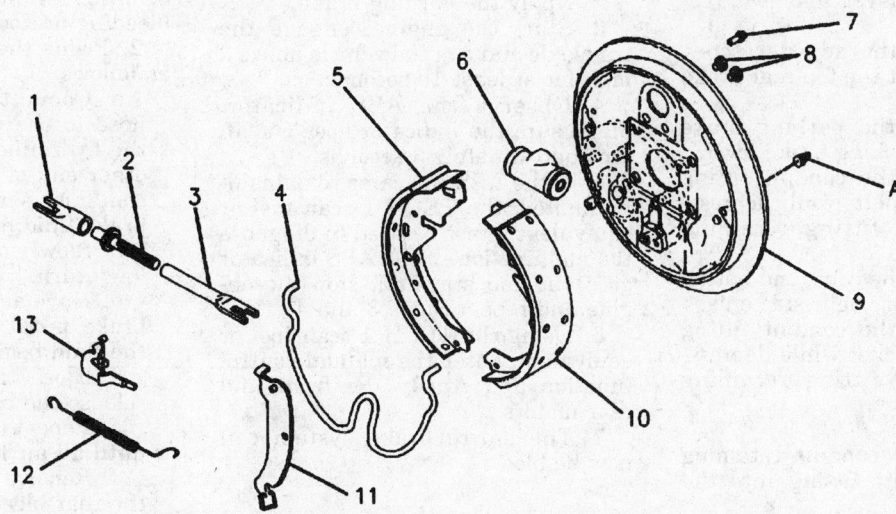

317808

Drum brake components — 1994 models with leading/trailing brakes

4. Remove the upper brake lining spring and adjuster.

5. From the rear of the backing plate, disconnect the steel brake line and cap the line opening to prevent excessive fluid loss and possible fluid contamination.

6. Remove the bleeder screw to gain clearance for removal of the wheel cylinder.

7. Remove the wheel cylinder bolts.

8. Loosen the hub assembly bolts.

9. Remove the wheel cylinder.

To install:

10. Position the wheel cylinder on the backing plate, install bolts and torque to 15 ft. lbs. (20 Nm).

11. Install the wheel cylinder bleeder screw.

12. Position the hub and bearing assembly, torque the bolts to 44 ft. lbs. (60 Nm).

13. Install the brake line to the wheel cylinder and torque the nut 17 ft. lbs. (23 Nm)

14. Install the brake shoes.

15. Install the brake drum.

16. Adjust the brakes.

17. Install the tire and wheel assembly. Lower the vehicle.

18. Refill the master cylinder, bleed the brake system and test for proper operation.

Parking Brake Cable

ADJUSTMENT

1. Adjust the rear brake shoes.

2. Pull the parking brake lever until 5 ratchet clicks are heard.

3. Raise and safely support the vehicle.

NOTE: To prevent damage to the threaded parking brake adjusting rod when servicing the parking brake, clean and lubricate the exposed threads on each side of the nut.

4. Tighten the adjusting nut on parking brake cable until the left rear wheel can be turned rearward with 2 hands but is locked when forward rotation is attempted.

5. Release the parking brake. Rear wheels must now turn freely in either direction with no brake drag.

6. Lower the vehicle.

REMOVAL AND INSTALLATION

Front

1. Raise and support the vehicle on safety stands.

2. Loosen, but do not remove, the equalizer nut to remove the cable.

3. Disconnect the cable from the equalizer and right rear cable.

4. Remove the hand grip from the parking brake lever inside the vehicle.

5. Remove the console.

6. Disconnect the cable from the parking brake lever.

7. Remove the nut holding the cable to the floor.

8. Remove the exhaust hanger bracket mounting nuts.

9. Remove the catalytic converter shield, then remove the cable.

To install:

10. Install cable to equalizer and cable to guide and underbody clips.

11. Install parking brake cable to underbody and tighten converter shield.

12. Install cable retaining nut securing front cable to the floor panel.

13. Install parking brake cable to the lever. Install the console.

14. Adjust the parking brake cable and safely lower the vehicle.

Rear

1. Raise and support the vehicle on safety stands.

2. Loosen the equalizer nut until the cable tension is released. Must be separated from the threaded rod, if removing the left cable.

3. Remove the tire and the brake drum.

4. Insert a screwdriver between the brake shoe and the top part of the brake adjuster bracket.

5. Push the bracket to the front and release the top adjuster bracket rod.

6. Disconnect the hold-down the spring, actuator lever and lever return spring.

7. Disconnect the adjuster screw spring. Disconnect the top rear brake shoe return spring.

8. Disconnect the parking brake cable from the parking brake lever.

9. Disconnect the conduit fitting from the backing plate while depressing the conduit fitting retaining tangs.

10. Disconnect the cable end button from the connector, right side only.

11. Disconnect the conduit fitting from the axle bracket while depressing the conduit fitting retaining tangs.

To install:

12. Connect the conduit retaining tangs and conduit fitting into the axle bracket.

13. Connect the cable end button to the connector, right side only.

14. Connect the conduit fitting to the backing plate.

15. Connect the parking brake cable to the parking brake lever.

16. Connect the top rear brake shoe return spring.

17. Connect the adjuster screw spring.

18. Connect the lever return spring, actuator lever, and rear hold-down spring.

19. Connect the top adjuster bracket rod.

20. Install the brake drum and tire and wheel assembly.

21. Adjust the parking brake cable and lower the vehicle.

Brake System Bleeding

PROCEDURE

NOTE: If any brake component is repaired or replaced such that air is allowed to enter the brake system, the hydraulic modulator must be bled prior to the brake calipers or wheel cylinders.

Prior to bleeding the brake system, the front and rear displacement cylinder pistons must be returned to the top-most (home) position. Using a TECH 1 scan tool or equivalent, select F5 (motor rehome). The motor rehome function cannot be performed if current Diagnostic Trouble Codes (DTC) are present. If DTCs are present, the vehicle must be repaired and the DTCs cleared before performing the motor rehome function. Proceed as follows:

1. Raise and safely support the front of the vehicle so the drive wheels are off the ground.

2. Apply the parking brake.

3. Start the engine, engage the transaxle and run the vehicle above 3 mph for at least 10 seconds.

4. Observe the ABS indicator. Make sure the indicator goes out after approximately 3 seconds.

5. If the ABS indicator remains illuminated, the TECH 1 scan tool or equivalent, must be used to diagnose the malfunction. If the ABS indicator goes out and stays off, stop the engine and repeat Steps 3 and 4.

6. Using the TECH 1 scan tool or equivalent, enter the manual control function and **Apply** the front and rear motors.

7. The entire brake system can now be bled.

Bleeding the Master Cylinder

If the master cylinder is known or suspected to have air in the bore, it must be bled first. Bleed the master cylinder as follows:

1. Remove any existing vacuum from the booster by pumping the brake pedal several times.

2. Make sure the level in the fluid reservoir is at the full or **MAX** mark. DO NOT let the fluid level drop below ½ full at any point during the bleeding procedure.

3. Disconnect the forward brake line fitting at the master cylinder, using a line wrench.

4. Allow brake fluid to fill the master cylinder bore until it starts to flow from the forward brake line port.

5. Connect the forward brake line to the master cylinder and tighten to 12 ft. lbs. (16 Nm).

6. Have an assistant depress the brake pedal slowly one time and hold down. Loosen the forward brake line fitting at the master cylinder to bleed any air from the bore. Tighten the fitting to 12 ft. lbs. (16 Nm). Slowly release the pedal.

7. Wait approximately 15 seconds and repeat the procedure until all air is purged from the master cylinder bore.

8. Repeat this procedure at the rear master cylinder bore.

9. When the procedure is completed, check the reservoir fluid level and top off, if necessary.

Bleeding the Brake System

1. Make sure the level in the master cylinder is at the **MAX** mark. DO NOT allow the fluid level to drop below ½ full at any point during the bleeding procedure.

2. Prime the hydraulic modulator as follows:

a. Connect a clear plastic bleed hose to the rear bleeder valve on the hydraulic modulator. Place the other end of the bleeder hose in a container partially filled with clean brake fluid.

b. Slowly open the bleeder valve ½–¾ turn.

c. Have an assistant depress the brake pedal and hold down until the fluid begins to flow.

d. Close the bleeder valve and release the brake pedal.

e. Repeat the bleeding procedure until all air has been purged.

f. Remove the bleeder hose from the rear bleeder valve and place it onto the forward bleeder valve of the hydraulic modulator. Repeat the bleeding procedure steps until all air has been purged from the system.

3. Bleed the wheel cylinders and calipers as follows:

a. Clean the bleeder screw at each wheel.

b. Attach a rubber hose to the bleeder screw at the right rear wheel and place the end in a clear container of brake fluid.

c. Fill the master cylinder with brake fluid. Have an assistant slowly pump up the brake pedal and hold the pressure.

d. Open the bleed screw about ¼ turn, press the brake pedal to the floor, close the bleed screw and slowly release the pedal. Continue until no more air bubbles are forced from the cylinder on application of the brake pedal.

e. Repeat the procedure on the remaining wheel cylinder and calipers in the following sequence:
- Left rear
- Right front
- Left front

f. Refill the master cylinder to the full or **MAX** mark.

4. After bleeding the wheel cylinders and calipers, the hydraulic modulator may still have remaining air. Bleed the hydraulic modulator in the following order:

a. Connect a clear bleeder hose on the rear hydraulic modulator bleed valve and submerge the other end of the hose in a clean container partially filled with brake fluid.

b. Have an assistant depress the brake pedal with medium force. Slowly open the bleeder valve ½–¾ turn and allow the brake fluid to flow.

c. Close the bleeder valve, release the brake pedal and wait approximately 5 seconds.

d. Repeat the bleeding procedure until all air has been purged from the system.

e. Remove the bleeder hose from the rear bleeder valve and connect it to the forward bleeder valve on the hydraulic modulator.

f. Repeat the hydraulic modulator bleeding procedure.

5. Check the brake fluid level and top off, if necessary.

6. Turn the ignition switch **ON**, depress the brake pedal and hold down:

a. If the brake pedal feels firm and constant, and the pedal travel is good, start the engine. Check the pedal travel while the engine is running. If the pedal feels firm and constant, and pedal travel is good, then the bleeding procedure was successful.

b. If the brake pedal feels soft or the pedal travel is excessive while the engine is **ON** or **OFF**, using the TECH 1 scanner tool, **Release** then **Apply** the motors 3 times and cycle the solenoids 10 times. Be sure to **Apply** the motors to ensure that the pistons are in the home position. **DO NOT OPERATE THE VEHICLE**. Perform the entire bleeding procedure again.

7. Road test the vehicle. Make several normal stops at a moderate speed to test proper brake system operation. Be sure to allow a reasonable cooling time between stops.

Wheel Speed Sensor

REMOVAL AND INSTALLATION

The front wheel speed sensors are attached to the steering knuckles. The sensors are not repairable, nor is the air gap adjustable. However, the air gap between the toothed ring on the outer constant velocity joint and the sensor should be between 0.020–0.070 inch (0.5–1.7mm).

The rear wheel speed sensor and toothed ring are contained within the dust cap of the integral rear wheel bearing. The sensor and toothed ring

are not repairable and no provision for air gap adjustment exists.

FRONT SUSPENSION

Strut

REMOVAL AND INSTALLATION

1. Raise and safely support the vehicle with the front wheels about a foot off the ground.

2. Remove the front tire and wheel assembly.

3. Place a floor jack under the lower control arm and raise the jack until the weight of suspension is taken up by the jack.

4. Remove the cotter pin and castle nut from the outer tie rod end and using a suitable puller, separate the tie rod end from the strut steering arm.

5. Scribe matching marks on the lower strut mounting bracket and steering knuckle for reference during installation.

6. If the suspension is equipped with a direct acting stabilizer system remove the link to strut mounting nut.

7. Remove the lower strut mounting through bolts and nuts.

8. Remove the upper strut plate mounting nuts and mounting bolt.

9. Remove the strut from the vehicle.

10. If the coil spring or strut is to be replaced, a strut compressor J-34013 or equivalent must be used for that procedure.

11. Mount the strut compressor in the holding fixture J-3289-20 or equivalent.

12. Mount the strut into the compressor and compress the spring approximately ½ its height after initial contact with the top cap.

——— CAUTION ———
Never bottom spring or dampener rod.

13. Remove the nut from the strut dampener shaft and place the guiding rod tool J-34013-27 or equivalent onto the the top of the damper shaft. Use this rod tool to guide the dampener shaft down through the bearing cap while decompressing the spring.

14. Remove the strut and any components that are necessary.

To install:

15. Install the bearing cap into the strut compressor if previously installed.

16. Mount the strut into the compressor using the bottom locking pin only. Extend the dampener shaft and install clamp J-34013-20 or equivalent on the shaft to hold it in a extended position.

17. Install the spring over the dampener and swing the assembly up so the upper locking pin can be installed.

18. Install the upper insulator, shield, bumper, and upper spring seat. Be sure the flat on the upper spring seat is facing in the proper direction. the spring seat flat should be facing the same direction as the centerline of the strut assembly.

19. Compress the coil spring while guiding the dampener shaft through the upper insulator with the guiding rod tool until the threads on the dampener shaft are visible.

20. Install the shaft nut and tighten to 52 ft. lbs. (70 Nm) while holding the shaft from turning.

21. Remove the clamp and remove the strut assembly from the compressor tool.

22. Install the strut in the vehicle and loosely install the upper mounting nuts and bolts.

23. Connect the lower strut mounting bracket to the steering knuckle and insert the through bolts. The bolts should point toward the rear of the vehicle.

24. Install the stabilizer link to the strut bracket if equipped with direct acting suspension. Install the mounting nut and tighten to 70 ft. lbs. (95 Nm).

25. Install the through bolt nuts loosely. Position the strut so the matching marks are aligned and tighten the nuts to 133 ft. lbs. (180 Nm).

26. Connect the tie rod end to the strut steering arm and tighten the castle nut to 44 ft. lbs. (60 Nm). If necessary tighten the nut up to 60 degrees (⅙ turn) additional rotation to align the cotter pin holes. NEVER loosen the nut to make the holes align.

27. Remove the floor jack from under the suspension.

28. Install the tire and wheel assembly and tighten the mounting nuts to 100 ft. lbs. (140 Nm).

29. Lower the vehicle.

30. Tighten the upper strut mounting nuts and bolt and tighten to 18 ft. lbs. (25 Nm).

31. Check the front end alignment and adjust as necessary.

Coil Spring

REMOVAL AND INSTALLATION

1. Raise and safely support the vehicle. Remove the tire and wheel assembly.
2. Remove the strut assembly.
3. Place the strut assembly into strut compressor tool J-34013-A or equivalent.
4. With the strut firmly mounted in the strut compressor tool, compress the spring and remove the center nut for strut plate. Carefully back off the strut compressor tool, after tension is fully relieved remove the coil spring from its mounting.

To install:

5. Place the coil spring into its mounting perch.
6. Place the strut mounting plate on top of the spring and attach the strut compressor.
7. Compress the coil spring enough so the mounting nut can be installed. Tighten this nut to 52 ft. lbs. (70 Nm).
8. Carefully back off the tension from strut compressor tool and remove the strut assembly.
9. Install the strut assembly into the vehicle.
10. Install the tire and wheel assembly. Lower the vehicle.
11. Check wheel alignment and road test vehicle.

Lower Ball Joints

REMOVAL AND INSTALLATION

1. Raise and safely support the vehicle.
2. If suspension contact hoist is used, place safety stands under the crossmember. Lower the vehicle slightly so the weight of the vehicle rests on the crossmember.
3. Remove the tire and wheel assembly.

NOTE: Care must be exercised to prevent the halfshaft joints from being over-extended. When either end of the shaft is disconnected, over-extension of the joint could result in separation of internal components and possible joint failure. Failure to observe this can result in interior joint or boot damage and possible joint failure.

4. Remove the cotter pin and nut from the ball joint.
5. Separate the ball joint from the steering knuckle using tool J-38892 or equivalent.
6. Drill out the 3 rivets retaining ball joint to the lower control arm. Use an 1/8 in. (3mm) drill bit to make a pilot hole through the rivets. Finish drilling rivets with a 1/2 in. (13mm) drill bit.
7. Remove the nut attaching link to the stabilizer shaft.
8. Remove the ball joint from the steering knuckle and control arm.

To install:

9. Install ball joint into the control arm.
10. Install 3 new bolts and nuts (supplied with new ball joint). Check the ball joint service kit for an instruction sheet listing bolt torque. If none is supplied, tighten the bolts to 40 ft. lbs. (55 Nm).
11. Position ball joint stud through the steering knuckle and tighten nut to 50 ft. lbs. (65 Nm) and install cotter pin.
12. Connect stabilizer link to stabilizer shaft and tighten nut to 13 ft. lbs. (17 Nm).
13. Install tire and wheel assembly.
14. Check front wheel alignment and adjust if necessary.

Lower Control Arms

REMOVAL AND INSTALLATION

1. Raise and safely support the vehicle.
2. If suspension hoist is used, place safety stands under the suspension supports and lower the vehicle slightly so weight of the vehicle rests on the suspension supports and not on the control arms.
3. Remove the tire and wheel assembly.
4. Remove the nut attaching stabilizer link to stabilizer shaft.
5. Remove the nuts attaching stabilizer shaft clamp to suspension support.

NOTE: Care must be exercised to prevent drive axle (half shaft) CV-joints from being over-extended. When either end of the shaft is disconnected, over extension of the joint could result in separation of the internal components and possible joint breakage. Failure to observe this can result in interior joint or seal damage and possible joint breakage.

6. Separate the ball joint from the steering knuckle using tool J-29330 or equivalent.
7. Remove the bolts attaching suspension support to vehicle.
8. Remove the bolts attaching the control arm to suspension support.

To install:

9. Position the control arm into place and hand-tighten attaching bolts.
10. Install suspension support into place, guiding the ball joint into the steering knuckle and loosely install bolts.
11. Install the nuts attaching stabilizer shaft clamp to suspension support.
12. Tighten these nuts to 22 ft. lbs. (30 Nm).
13. Install the ball joint nut and tighten to 48 ft. lbs. (65 Nm) and install cotter pin.
14. Remove the safety stands from under the crossmember.
15. Install the tire and wheel assembly.
16. Lower the vehicle.
17. With the vehicle at curb height, tighten control bolts as follows :
    a. Tighten the center bolts first to 89 ft. lbs. (120 Nm).
    b. Tighten the front bolts second to 89 ft. lbs. (120 Nm).
    c. Tighten the rear bolts third to 89 ft. lbs. (120 Nm).
18. Tighten the control arm attaching bolts to 61 ft. lbs. (83 Nm).
19. Check front wheel alignment and adjust as necessary.
20. Road test the vehicle.

Sway Bar

REMOVAL AND INSTALLATION

NOTE: On 1993 model years, there where 2 style front suspension systems. A Base Suspension system and the optional Direct Acting Stabilizer System. Use care to distinguish which system is used, before ordering any parts or performing any torque settings.

1. Raise and safely support the vehicle. Allow the front suspension to hang free.
2. Remove the front tire and wheel assemblies.
3. Remove the nuts that connect the sway bar links to the sway bar (also called a stabilizer bar).
4. Remove the mounting clamps from the sway bar to suspension support assemblies.

5. Remove the rear and center bolts from suspension support assemblies, and loosen the front bolts.

6. Remove the sway bar.

To install:

7. Position the sway bar with insulators onto the suspension supports.

8. Install the mounting clamps (for the sway bar) to suspension support assemblies, and hand-tighten.

9. Install the suspension support assemblies into position and install bolts, hand-tighten only.

10. For 1993 vehicles, install the nuts attaching sway bar to sway bar links. Tighten these nuts to 70 ft. lbs. (95 Nm) on Direct Acting System and 15 ft. lbs. (21 Nm) on the base systems.

11. For 1994–96 vehicles, install the nuts attaching the sway bar to the sway bar links. Tighten these nuts to 22 ft. lbs. (30 Nm).

12. Tighten the suspension support bolts as follows :

a. Center bolts to 66 ft. lbs. (90 Nm) for 1993 vehicles and to 89 ft. lbs. (120 Nm) for 1994–96 vehicles.

b. Front bolts to 66 ft. lbs. (90 Nm) for 1993 vehicles and to 89 ft. lbs. (120 Nm) for 1994–96 vehicles.

c. Rear bolts to 65 ft. lbs. (89 Nm) for 1993 vehicles and to 89 ft. lbs. (120 Nm) for 1994–96 vehicles.

13. Tighten the sway bar clamps to suspension support bolts to 22 ft. lbs. (30 Nm).

14. Install the tire and wheel assemblies.

15. Lower the vehicle.

16. Check the wheel alignment and road test the vehicle.

Front Wheel Bearings

ADJUSTMENT

The wheel bearings are not adjustable. If a wheel bearing is out of specifications, it must be replaced. Using a dial indicator, check for looseness. If it exceeds 0.005 inch (0.1270mm) on drum or disc brakes the bearing wear is excessive and the hub and bearing should be replaced.

REMOVAL AND INSTALLATION

The front wheel bearings on these vehicles are not serviced separately. The wheel bearings are pressed into the hub. If the wheel bearings are noisy or defective, the hub and bearing assembly must be replaced.

1. Loosen the axle hub nut.

2. Raise and safely support the vehicle. Remove the tire and wheel assembly.

3. Install a special boot cover J-33162 or equivalent on the L4 vehicles with automatic transaxle only. This is to protect the soft CV-joint boot.

4. Remove the axle hub nut.

5. Remove the brake caliper and rotor.

6. Remove the 3 hub and bearing mounting bolts.

7. Remove the splash shield.

NOTE: If the hub and bearing assembly is to be reused, mark the attaching bolt and corresponding hole for installation in the same position. Use of a hammer or direct heat on the halfshaft should be avoided during removal, as this could result in internal bearing damage.

8. Install special tool J-28733 or equivalent hub puller and turn the bolt to press the hub and bearing assembly off the end of the drive axle (halfshaft).

9. Disconnect the stabilizer link bolt at the lower control arm.

10. Separate the lower ball joint using special tool J-29330 or equivalent.

11. Remove the drive axle from the steering knuckle and support it aside.

12. If removing the steering knuckle use the following procedure:

a. Using a sharp tool, scribe the outline of the strut onto the steering knuckle.

b. Remove the cotter pin and castle nut from the outer tie rod end and using a steering linkage puller, separate the tie rod from the steering knuckle.

c. Remove the bolts attaching the strut to the steering knuckle and remove the knuckle.

d. Using a file, elongate the lower strut-to-knuckle hole for camber alignment adjustment.

To install:

13. Clean and inspect the hub and bearing surfaces and the steering knuckle bore.

14. If the steering knuckle was removed use the following procedure.

a. Install the steering knuckle onto the lower ball joint and tighten the ball joint nut to 48 ft. lbs. (65 Nm).

b. Connect the steering knuckle to the lower strut bracket and install the through bolts and nuts. With the matchmarks in alignment, tighten the nuts to 122 ft. lbs. (165 Nm).

c. Connect the stabilizer link bolt at the lower control arm and tighten the nut to 22 ft. lbs. (30 Nm).

15. Install the hub and bearing assembly and torque the 3 attaching bolts to 70 ft. lbs. (90 Nm).

16. If removed, connect the tie rod end to the steering knuckle and tighten the castle nut to 30 ft. lbs. (41 Nm). If necessary to align the cotter pin holes tighten the castle nut up to 60 degrees (1/6 turn) additional. Install a new cotter pin.

17. Install the drive axle (halfshaft) to the hub and bearing assembly being careful to align the axle splines to the hub and not to damage the seal.

18. Install a new hub and bearing nut on the drive axle and apply a partial torque of 75 ft. lbs. (100 Nm).

19. Install the caliper and rotor assembly.

20. Install the wheel and tire assembly.

21. Safely lower the vehicle and apply a final torque to the axle hub nut to 190 ft. lbs. (260 Nm).

22. Perform a front-end alignment check, and adjust if required. Road test vehicle.

REAR SUSPENSION

Shock Absorber

REMOVAL AND INSTALLATION

1993 Vehicles

1. Open the trunk lid, remove the trim cover if present, and remove the upper shock absorber nut. Raise and safely support the vehicle.

2. If equipped with air lift type shock absorbers, disconnect the air line. Remove the lower attaching bolt and remove the shock absorber.

3. If new shock absorbers are being installed, repeatedly compress them while inverted and extend them in their normal upright position. This will purge them of air.

To install:

4. Install the shock absorber and lower mount nuts. Torque to 38 ft. lbs. (51 Nm).

5. If equipped with air lift type shock absorbers, connect air lines.

6. Safely lower vehicle.

7. Install upper mount nuts. Torque to 13 ft. lbs. (17 Nm).

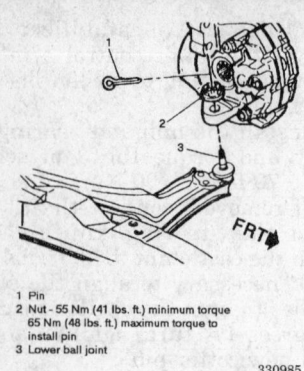

1 Pin
2 Nut - 55 Nm (41 lbs. ft.) minimum torque
  65 Nm (48 lbs. ft.) maximum torque to
  install pin
3 Lower ball joint

330985

Ball joint to knuckle installation

8. Install trim cover and close the trunk.

1994–96 Vehicles

1. Open the deck lid.
2. Remove the shock absorber upper cover and remove the upper mounting nut.

NOTE: Remove one shock absorber at a time if both shocks are to be replaced. This will ensure that the axle assembly will not be suspended by the brake hoses.

3. Raise and safely support the vehicle.
4. Support the rear axle with safety stands. When lifting vehicle with body hoist, it will be necessary to use adjustable safety stands.
5. Remove the shock absorber nut and remove the shock absorber.
To install:
6. Install the shock absorber at the lower attachment. Install the nut hand-tight.
7. Lower the vehicle enough to guide upper stud through body opening and install upper shock mounting nut.
8. Tighten the lower shock absorber mounting nuts to 35 ft. lbs. (47 Nm).
9. Remove the axle support and lower vehicle all the way then tighten shock absorber upper nut to 21 ft. lbs. (29 Nm).
10. Install the shock absorber upper cover.

Coil Spring

REMOVAL AND INSTALLATION

— CAUTION —
When removing rear springs do not use a twin-post type hoist. When certain fasteners are re-

moved from the rear axle assembly, the axle may have a tendency to cause the hoist to slip, which may cause personal injury. Perform operation on the floor if necessary.

1. Raise and safely support the vehicle. Use safety stands to support the rear axle.
2. Remove the tire and wheel assemblies.
3. Remove the right and left brake line bracket attaching screws from the body and allow brake line to hang free.
4. Remove the lower mounting bolts from the shock absorbers.

NOTE: Do not suspend rear axle by the brake hoses. Damage to hoses could result.

5. Lower the rear axle and remove the spring(s).
To install:
6. Position the coil spring in its mounting perch and raise axle. The end of the lower coil spring must be positioned in the spring seat and within 9/16 inch (15mm) of the spring stop.
7. Connect the shock absorbers to the rear axle. Tighten the nuts to 35 ft. lbs. (47 Nm).
8. Install the brake line brackets to body and tighten screws to 97 inch lbs. (11 Nm).
9. Install the tire and wheel assemblies. Remove the safety stands and lower the vehicle.

Sway Bar

REMOVAL AND INSTALLATION

1993–94 Vehicles

Stamped Axle Assembly

1. Raise and safely support the vehicle.
2. Remove the nuts and bolts at both the axle and the control arm attachments and remove the bracket, insulator and the stabilizer (sway bar) shaft.
To install:
3. Install the clamps, spacers and insulators in the axle.
4. Position the stabilizer shaft in the spacers and loosely install the lower clamp nuts.
5. Install the stabilizer shaft insulator to control arm clamps and nuts, tighten to 16 ft. lbs. (22 Nm).

6. Tighten the stabilizer shaft insulator to axle clamp nuts to 16 ft. lbs. (22 Nm).
7. Lower the vehicle.

Tubular Axle Assembly

1. Raise and safely support the vehicle.
2. Remove the nuts and bolts attaching the stabilizer shaft to the axle assembly and remove the shaft.
To install:
3. Position the stabilizer shaft.
4. Install the nuts and bolts, tighten to 16 ft. lbs. (22 Nm).
5. Lower the vehicle.

Wheel Bearings

ADJUSTMENT

The wheel bearings are not adjustable. If a wheel bearing is out of specifications, it must be replaced. Using a dial indicator, check for looseness. If it exceeds 0.005 inch (0.1270mm) on drum or disc brakes the bearing wear is excessive and the hub and bearing should be replaced.

REMOVAL AND INSTALLATION

1. Raise and safely support the vehicle. Remove the tire and wheel assembly.
2. Remove the brake drum.
3. Remove the nuts from the hub bearing mounting bolts. With the nuts removed, push the mounting bolts out of the hub bearing assembly. The top rear bolt will not clear the brake shoe and has to be left in the hub flange when removing the hub assembly.
4. Remove the rear ABS wheel speed sensor wire harness.
5. Remove the bearing assembly.
To install:
6. Connect the rear ABS wheel speed sensor wire connector.
7. Install the bearing assembly onto the backing plate with the top rear bolt already installed through the bearing bolt hole.
8. Install the remainder of the mounting bolts.
9. Install the mounting nuts and tighten to 44 ft. lbs. (60 Nm).
10. Install the brake drum.
11. Install the tire and wheel assembly and tighten the wheel nuts to 100 ft. lbs. (140 Nm).
12. Lower the vehicle.

FIRING ORDERS

NOTE: To avoid confusion, always replace spark plug wires one at a time.

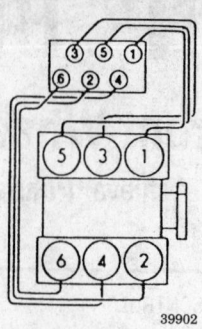

3.3L Engine
Engine Firing Order:
1–6–5–4–3–2
Distributorless
Ignition System

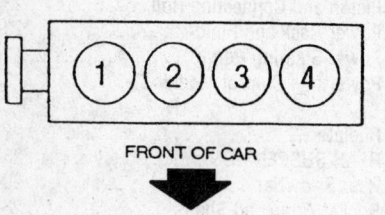

FRONT OF CAR

195466

2.3L and 2.4L Engines
Engine Firing Order: 1–3–4–2
Distributorless Ignition System

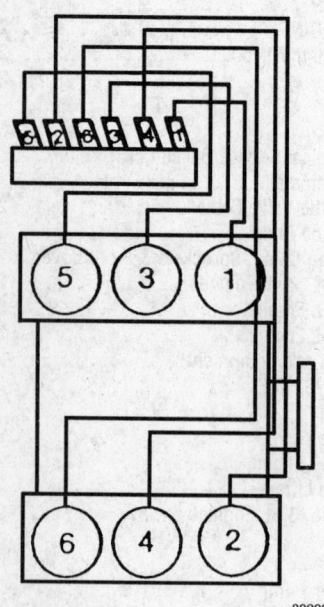

39299

3.1 Engine
Firing Order: 1–2–3–4–5–6
Distributorless Ignition System

ENGINE ELECTRICAL

NOTE: Disconnecting the negative battery cable on some vehicles may interfere with the functions of the on board computer systems and may require the computer to undergo a relearning process, once the negative battery cable is reconnected.

Ignition Timing

ADJUSTMENT

The ignition timing is not adjustable, and is set according to engine demand electronically. The Powertrain Control Module controls the ignition timing for all driving conditions. The PCM monitors input signals from the following components engine coolant sensor, intake air temperature, mass air flow sensor, park/neutral switch, throttle position sensor and the vehicle speed sensor.

Alternator

PRECAUTIONS

Several precautions must be observed with alternator equipped vehicles to avoid damage to the unit.

• If the battery is removed for any reason, make sure it is reconnected with the correct polarity. Reversing the battery connections may result in damage to the 1-way rectifiers.

• When utilizing a booster battery as a starting aid, always connect the positive to positive terminals and the negative terminal from the booster battery to a good engine ground on the vehicle being started.

• Never use a fast charger as a booster to start vehicles.

• Disconnect the battery cables when charging the battery with a fast charger.

• Never attempt to polarize the alternator.

• Do not use test lights of more than 12 volts when checking diode continuity.

• Do not short across or ground any of the alternator terminals.

• The polarity of the battery, alternator and regulator must be matched and considered before making any electrical connections within the system.

• Never separate the alternator on an open circuit. Make sure all connections within the circuit are clean and tight.

• Disconnect the battery ground terminal when performing any service on electrical components.

• Disconnect the battery if arc welding is to be done on the vehicle.

REMOVAL AND INSTALLATION

2.3L and 2.4L Engines

1. Disconnect the negative battery cable.

2. Loosen the tensioner pulley bolt, turn the pulley counterclockwise and remove the belt from the alternator pulley.

3. Disconnect the 2 vacuum lines from the oil/air separator. If necessary, disconnect the injector harness.

4. Remove the alternator mounting bolts. Partially remove the alternator until the electrical wiring is accessible.

5. Disconnect the plug in connector and battery feed wire. Lift the alternator out, being careful not to damage the A/C line or condenser.

To install:

6. Install the alternator into the engine compartment and connect the plug in connector and battery feed wire. Tighten the battery wire nut to 65 inch lbs. (8 Nm).

7. Install the upper alternator bolt and tighten hand-tight. Install the 2 rear alternator bolts and tighten the bolts to 19 ft. lbs. (26 Nm).

8. Tighten the front alternator bolt to 37 ft. lbs. (50 Nm). Install the serpentine belt.

9. Connect the injector harness. Connect the vacuum hoses to the air/oil separator. Connect the negative battery cable.

3.1L Engine

1. Disconnect the the negative battery cable. Disconnect the electrical connector from the alternator.
2. Remove the serpentine belt. If necessary, remove the alternator air inlet duct and inlet adapter from the alternator.
3. Remove bolt, stud and nuts from the alternator braces. Remove the lower alternator bolt.
4. Remove the plastic cover from the post on the back of the alternator. Disconnect battery positive lead from alternator output terminal.
5. Remove the alternator.

To install:

6. Install the alternator. Connect the battery positive lead and nut to alternator output terminal. Tighten the mounting nut to 65 inch lbs. (7 Nm). Install the protective cap.
7. Install the lower alternator bolt and tighten to 37 ft. lbs. (50 Nm).
8. Install the brace stud and bolts and nuts to alternator. Tighten the brace nuts and stud to 18 ft. lbs. (25 Nm).
9. If necessary, install the alternator air inlet adapter and air inlet duct. Install the serpentine belt.
10. Connect the electrical connector to the alternator. Connect the negative battery cable.

3.3L Engine

1. Disconnect the negative battery cable. Remove the serpentine drive belt from alternator.
2. Detach the electrical connectors. Remove the alternator bolts and the alternator.

To install:

3. Install alternator to mounting brackets with bolts, washers and nuts.
4. Install through bolts to alternator. Torque bolts to 20 ft. lbs. (27 Nm).
5. Install electrical connections. Install serpentine belt onto alternator. Connect the negative battery cable.

Drive Belt

REMOVAL AND INSTALLATION

2.3L and 2.4L Engines

1. Disconnect negative battery cable.
2. Rotate the serpentine belt tensioner counterclockwise using a 13mm wrench on the tensioner center bolt. The wrench used should be at least 24 inches in length.

3. With tension removed from the belt, slip the belt off the alternator pulley.
4. Slowly release the belt tensioner and remove the wrench. Remove the belt from the remainder of the pulleys.

To install:

5. Install the drive belt. Install the wrench on the tensioner bolt and rotate the tensioner counterclockwise.
6. With the tensioner loaded, loop the belt over the alternator pulley. Release the tensioner slowly and remove the wrench.
7. Make sure the belt is properly seated in all the drive belt pulleys. Connect negative battery cable.

3.1L Engine

1. Disconnect the negative battery cable.
2. Support the engine with a floor jack. Try to place the jack near the front of the oil pan, but DO NOT support the engine by the oil pan.
3. Remove the right engine mount intermediate bracket.
4. Remove the auxiliary bracket.
5. Rotate the belt tensioner counterclockwise using a ³/₈ inch breaker bar.

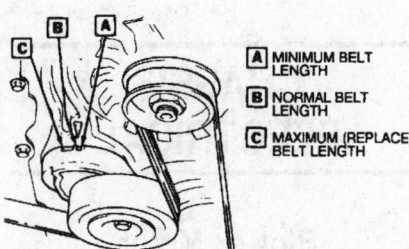

A	MINIMUM BELT LENGTH
B	NORMAL BELT LENGTH
C	MAXIMUM (REPLACE) BELT LENGTH

332057

Serpentine belt tensioner markings — 2.3L and 2.4L engines

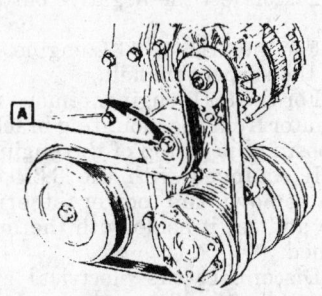

A ROTATE TENSIONER IN DIRECTION OF ARROW TO REMOVE OR INSTALL BELT.

332059

Serpentine belt routing — 2.3L and 2.4L engines

6. With the tension removed from the belt, remove the serpentine belt from around the alternator pulley.
7. Slowly release the tensioner.
8. Remove the belt from the remaining pulleys.

To install:

9. Install the serpentine belt. Rotate the belt tensioner counterclockwise using a ³/₈ inch breaker bar.
10. Loop the serpentine around the alternator pulley. Release the tensioner and verify the belt is properly seated in the the drive pulleys.
11. Install the right engine mount intermediate bracket. Install the auxiliary bracket.
12. Remove the floor jack. Connect the negative battery cable.

3.3L Engine

1. Disconnect negative battery cable.
2. Rotate the serpentine belt tensioner counterclockwise using a 18mm breaker bar.
3. With the tension removed from the belt, slip the belt off the alternator pulley.
4. Release the tensioner slowly. Remove the serpentine belt.

To install:

5. Install the serpentine belt. Rotate the belt tensioner counterclockwise using a 18mm breaker bar.
6. Install the serpentine belt over the alternator pulley.
7. Release the tensioner and verify the belt is properly seated on the drive pulleys. Connect the negative battery cable.

Starter

REMOVAL AND INSTALLATION

2.3L and 2.4L Engines

1. Disconnect the negative battery cable.
2. Remove the air induction tube.
3. On the 2.3L engine, perform the following:
   a. Remove the cooling fan assembly.
   b. Remove the oil filter.
   c. Remove the starter bolts.
4. On the 2.4L engine, perform the following:
   a. Remove the top starter bolt.
   b. Raise and safely support the vehicle.
   c. Remove the lower starter bolt.
5. Position the starter so the electrical wiring on the solenoid is accessible.

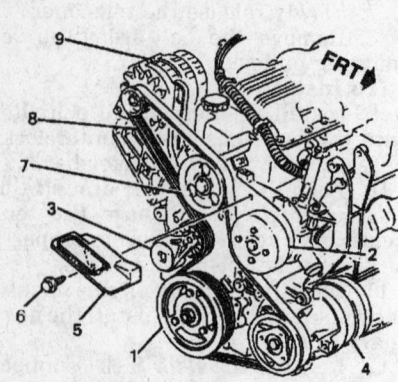

1 BALANCER, CRANKSHAFT
2 PUMP ASSEMBLY, WATER
3 TENSIONER, SERPENTINE DRIVE BELT
4 COMPRESSOR ASSEMBLY, AIR CONDITIONING
5 SHIELD, SERPENTINE DRIVE BELT
6 BOLT/SCREW, SERPENTINE DRIVE BELT SHIELD
7 PUMP ASSEMBLY, POWER STEERING
8 BELT, SERPENTINE DRIVE
9 GENERATOR ASSEMBLY

330000

Serpentine belt routing and shield — 3.1L engine

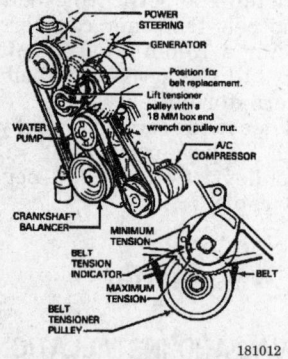

181012

Drive belt routing — 3.3L engine

6. Disconnect the electrical wiring from the starter.

7. Remove the starter from the vehicle by lifting it between intake manifold and radiator.

To install:

8. Install the starter in the vehicle.

9. Connect the electrical connectors to the starter.

10. Install the starter on the engine block and tighten the mounting bolts to 74 ft. lbs. (100 Nm) on the 1993–94 2.3L engine or to 66 ft. lbs. (90 Nm)

on the 1995 2.3L engine and 2.4L engine.

11. Install the cooling fan assembly.

12. On the 2.3L engine, coat the seal on the oil filter with clean engine oil and install the oil filter.

13. Install the air induction tube.

14. Connect negative battery cable.

15. On the 2.3L engine, top off the engine oil as necessary.

16. Start the engine and verify proper operation.

3.1L and 3.3L Engines

1. Disconnect the negative battery cable.

2. Raise and safely support the vehicle.

3. Remove the starter mounting bolts and lower the starter until the solenoid wiring is accessible.

4. Disconnect the electrical leads.

5. Remove the starter motor.

To install:

6. Install the starter motor.

7. Connect the electrical leads to the starter.

8. Install the starter mounting bolts and tighten to 32 ft. lbs. (43 Nm).

9. Lower the vehicle.

10. Connect the negative battery cable.

CHASSIS ELECTRICAL

Blower Motor

REMOVAL AND INSTALLATION

1. Disconnect the negative battery cable.

2. For the 3.1L or 3.3L engine, remove the serpentine belt.

3. For the 3.1L engine, remove the alternator from the mounting bracket and position it on top of the engine.

4. If equipped with the 3.3L engine, remove the power steering pump and set it aside with the lines attached.

5. Disconnect the electrical connectors from the blower motor.

6. Partially cut the blower case cover as indicated by the markings on the cover itself. It is not necessary to cut completely around the cover. Cut all they way around except for 2 in-

ches at the bottom and fold the cover down.

7. Disconnect the blower motor cooling tube. Remove the blower motor mounting screws and remove the blower motor.

To install:

8. Install the blower motor into the blower case and tighten the mounting screws. Connect the blower motor cooling tube.

9. Fold the blower motor cover up and secure in place with 3 mounting clips provided with the new blower motor. Attach the electrical connector to the blower motor.

10. For the 3.3L engine, install the power steering pump and serpentine belt.

11. For the 3.1L engine, install the alternator in the mounting bracket.

12. If equipped with the 3.1L or 3.3L engine, install the serpentine belt. Connect the negative battery cable.

Windshield Wiper Motor

REMOVAL AND INSTALLATION

1. Remove wiper arm and blade assemblies. Disconnect the negative battery cable. Remove the top vent grille panel.

2. Disconnect the transmission drive link socket from the wiper motor crank arm ball using J-39232 or equivalent tool.

3. Disconnect the electrical connectors from the wiper motor assembly. Remove the 3 motor retaining bolts.

4. Remove the wiper motor assembly while carefully guiding crank arm through the hole in firewall. Remove the seal from the wiper motor assembly.

To install:

5. Install the seal on the wiper motor assembly. Install the wiper motor onto the firewall while carefully guiding the crank arm through the hole.

6. Install the 3 wiper motor mounting screws and tighten to 80 inch lbs. (9 Nm).

7. Connect the electrical connectors to the wiper motor. Connect the transmission drive link socket to the wiper motor crank arm using J-39529.

8. Install the top vent grille panel. Connect the negative battery cable.

9. Cycle the wiper once and allow the system to return the the PARK position. Install the wiper arms in their proper positions.

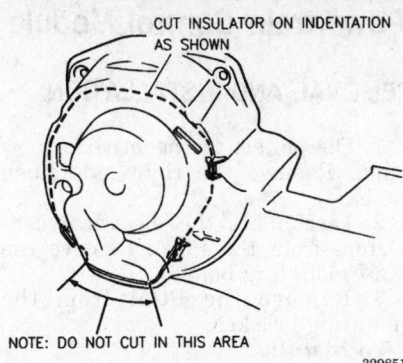

CUT INSULATOR ON INDENTATION
AS SHOWN

NOTE: DO NOT CUT IN THIS AREA

329851

Blower motor cover cut diagram

Combination Switch

REMOVAL AND INSTALLATION

——————— CAUTION ———————
The Supplemental Inflatable Restraint (SIR) system must be disarmed before working around the steering column. Failure to do so may cause accidental deployment of the air bag, resulting in unnecessary SIR system repairs and/or personal injury.

——————— CAUTION ———————
When carrying a live air bag, make sure the bag and trim cover are pointed away from the body. In the unlikely event of an accidental deployment, the bag will then deploy with minimal chance of injury. When placing a live air bag on a bench or other surface, always face the bag and trim cover up, away from the surface. This will reduce the motion of the module if it is accidentally deployed.

1993–94 Vehicles

1. If equipped with the SIR (air bag) system, disable the SIR system. Disconnect the negative battery cable.
2. Remove the steering column cover and cover screws. Remove the upper and lower steering column covers.
3. Unfasten the 2 switch mounting screws and remove the switch.
4. Remove all wire connectors from the combination switch. Push locking tabs to remove the wires.
   To install:
5. Connect all wire connectors to the combination switch. Install the combination switch to the column with 2 attaching screws.
6. Install the steering column covers.

7. Connect the negative battery cable. Enable the SIR system, if equipped.

1995–97 Vehicles

1. Properly disarm the SIR system. Disconnect the negative battery cable.
2. Remove the steering wheel. Remove the tilt wheel lever, if equipped.
3. Remove the upper and lower steering column covers. Remove the column dampener assembly.
4. Detach the electrical connector from the switch. Remove the switch mounting screws and remove the switch.
   To install:
5. Position the switch on the column and install the mounting screws. Attach the electrical connector to the switch.
6. Install the column dampener assembly. Fasten the upper and lower steering column covers.
7. Install the tilt lever. Install the steering wheel and air bag module.
8. Enable the SIR system. Connect the negative battery cable.

Ignition Lock Cylinder

REMOVAL AND INSTALLATION

——————— CAUTION ———————
The Supplemental Inflatable Restraint (SIR) system must be disarmed before removing the ignition lock cylinder. Failure to do so may cause accidental deployment of the air bag, resulting in unnecessary SIR system repairs and/or personal injury.

1. Disconnect the negative battery cable. If equipped, properly disable the SIR system.

——————— CAUTION ———————
When carrying a live air bag, make sure the bag and trim cover are pointed away from the body. In the unlikely event of an accidental deployment, the bag will then deploy with minimal chance of injury. When placing a live air bag on a bench or other surface, always face the bag and trim cover up, away from the surface. This will reduce the motion of the module if it is accidently deployed.

2. Remove the steering wheel and air bag, if equipped. Remove the left side sound insulators from under the dash.

3. If equipped, remove the SIR coil assembly snapring.
4. If equipped, gently pull up on the SIR coil assembly while pushing up on the SIR coil harness on the bottom side of the steering column and left the coil hang freely.
5. Remove the wave washer. Remove the spacer shaft lock (standard column only).
6. Remove the shaft lock retaining ring using tool J-23653-C or equivalent, to compress the shaft lock plate.
7. Pry off the retaining ring and remove the shaft lock plate. Remove the turn signal canceling cam and remove the upper bearing spring.
8. Remove the turn signal lever using the following procedure:
   a. Make sure the turn signal lever is in the center, or off position.
   b. If equipped, detach the cruise control connector from the steering column assembly.
   c. Pull the lever straight out of the turn signal switch.
9. Remove the hazard knob assembly.
10. If necessary, remove the tilt lever being careful not to damage the shaft. Grip the lever firmly and twist counterclockwise.
11. If necessary, remove the washer head screw and the switch actuator arm.
12. Remove the screws retaining the turn signal switch to the steering column and gently pull up on the wire harness; it may be necessary to disconnect the harness from the lower steering column.
13. Pull the switch out of the column and leave it hang freely. Remove the key from the lock cylinder and gently pry out the buzzer switch.
14. Reinsert the key into the lock cylinder and turn to the LOCK position. Remove the lock cylinder retaining screw and the lock cylinder.
    To install:
15. Install the steering column lock cylinder. Install the lock retaining screw and tighten to 40 inch lbs. (4.5 Nm).
16. Remove the key from the lock cylinder and install the buzzer switch.
17. Gently grasp the turn signal switch wire harness and pull it through the column until the switch is seated, position the screws and tighten to 30 inch lbs. (3.4 Nm).
18. If necessary, install the switch actuator arm, then tighten the washer screw to 20 inch lbs. (2.3 Nm).

19. If necessary, install the tilt lever by gripping it firmly and twisting clockwise. Install the hazard knob assembly.

20. Align the tab on the turn signal lever with the notch in the pivot of the turn signal switch and push the lever into the switch.

21. If equipped, connect the lever cruise connector to the steering column. Install the upper bearing spring and the turn signal cancelling cam.

22. Install the lock plate, using tool J-23653 or equivalent compress the upper bearing spring and install the retaining ring. Install the spacer shaft lock (standard column only).

— CAUTION —

SIR coil assembly wires must be kept tight with no slack while installing the SIR coil assembly. Failure to do so may cause the wires to be kinked near the lock plate, causing the wires to be cut when the steering wheel is turned.

23. Install the wave washer and SIR coil assembly, pulling gently on the harness on the bottom side of the steering column to remove slack from the wiring harness. Be sure to align the SIR coil assembly properly.

24. Install the snapring on top of the SIR coil. If removed, connect the turn signal wire harness to the steering column.

25. Install the left side sound insulators under the dash panel. Connect the negative battery cable, then properly enable the SIR system, if equipped.

26. Check the operation of the ignition switch, turn signals, key buzzer, headlight dimmer and 4 way flashers.

Ignition Switch

REMOVAL AND INSTALLATION

— CAUTION —

The Supplemental Inflatable Restraint (SIR) system must be disarmed before removing the ignition switch. Failure to do so may cause accidental deployment of the air bag, resulting in unnecessary SIR system repairs and/or personal injury.

1. Disconnect the negative battery cable. Disable the SIR system.

2. Remove the inflator module and steering wheel. Remove the upper and lower steering column covers.

3. Disconnect and remove the headlight switch. Disconnect the wiring harness from the ignition switch assembly.

4. Remove the 2 switch mounting screws. Remove ignition switch assembly.

To install:

5. Install the ignition switch on the steering column and tighten the 2 self tapping screws.

6. Connect the wiring harness to the ignition switch. Install and connect the headlight switch.

7. Install the upper and lower steering column covers. Install the steering wheel and inflator module.

8. Enable SIR system. Connect the negative battery cable.

Park/Neutral Safety Switch

REMOVAL AND INSTALLATION

1. Disconnect the negative battery cable. Disconnect the shift linkage from the shift lever on top of the transaxle.

2. Disconnect the electrical connector from the switch. Remove the mounting bolts and remove the switch from the vehicle.

To install:

3. Place the transaxle manual shift lever in **N**.

4. Align the flats on the switch with the flats on the manual shift shaft and install the switch onto the transaxle.

5. If the bolt holes in the switch do not line up with the holes in the transaxle, perform the following:

a. Verify the transaxle is in **N**.

b. Loosen the switch adjusting bolts.

c. Rotate the switch on the manual shaft until the service adjustment hole lines up with the carrier hole.

d. Insert a pin (small drill bit, piece of welding wire, etc.) in the hole to keep the switch aligned in position.

e. Tighten the mounting bolts to 18 ft. lbs. (24 Nm). Remove the alignment pin.

6. Connect the electrical connector to the switch. Connect the shift linkage to the transaxle.

7. Connect the negative battery cable. Apply the brakes firmly. Verify that the vehicle starts only in **N** and **P**.

Powertrain Control Module

REMOVAL AND INSTALLATION

1. Disconnect the negative battery cable. Remove the right side hush panel.

2. Disconnect the electrical connectors from the PCM. Remove the PCM mounting bolts.

3. Remove the PCM from the mounting bracket.

To install:

4. Install the PCM in the mounting bracket and tighten the mounting bolts. Connect the electrical connectors to the PCM.

5. Install the right side hush panel. Connect the negative battery cable.

ENGINE COOLING

Starting with the 1996 Model Year, these vehicles were filled at the factory with a new type of antifreeze/coolant called GM Goodwrench DEX-COOL™. When adding coolant to 1996 vehicles, it is important that you use GM Goodwrench DEX-COOL™ (orange-colored, silicate-free) coolant. **Propylene glycol is not recommended for use in GM vehicles.** A 50/50 mixture of DEX-COOL™ and clean water will provide all the recommended protection for 1996–97 vehicles. **DO NOT use DEX-COOL™ in pre-1996 vehicles. DO NOT mix DEX-COOL™ with any other type of antifreeze.**

Radiator

REMOVAL AND INSTALLATION

1. Disconnect the negative battery cable. Drain the cooling system into a suitable container.

2. Remove air intake duct assembly. Disconnect and cap the upper transaxle cooler line.

3. Disconnect the upper radiator hose. Disconnect and cap the lower transaxle cooler line.

4. Disconnect the cooling fan electrical connector. Remove the cooling fan mounting bolts and remove the cooling fan.

5. Raise and safely support the vehicle. Remove the splash guard below lower radiator hose.

6. Disconnect the lower radiator hose from the radiator. Remove the condenser line retaining clip.

7. Remove the condenser-to-radiator bolts. Disconnect the coolant surge tank hose.

8. Remove the radiator retaining bolts. Remove the radiator.

To install:

9. Install the radiator in the vehicle making sure the radiator fits into the lower mounts in the radiator support.

10. Install the radiator retaining bolts and tighten to 90 inch lbs. (10 Nm).

11. Connect the coolant surge tank hose. Install the condenser-to-radiator bolts.

12. Install the condenser line retaining clip. Connect the lower radiator hose to the radiator.

13. Install the splash shield under the lower radiator hose. Install the cooling fan.

14. Connect the cooling fan electrical connector. Connect the lower transaxle cooler line.

15. Connect the upper radiator hose. Connect the upper transaxle cooler line.

16. Install air intake duct assembly. Connect the negative battery cable.

17. Refill the cooling system with a 50/50 mixture of the proper antifreeze and clean water. Pressure test the cooling system and verify no leaks.

18. Start the vehicle and check the transaxle fluid level. Add fluid as necessary.

Water Pump

REMOVAL AND INSTALLATION

2.3L and 2.4L Engines

1. Disconnect the negative battery cable.

2. Drain the cooling system.

3. Disconnect the oxygen sensor (O_2) electrical connector.

4. Remove the upper and lower exhaust manifold heat shields.

5. Raise and safely support the vehicle.

6. Remove the exhaust manifold brace-to-manifold bolt.

7. Loosen the exhaust pipe-to-manifold spring loaded nuts.

8. Remove the radiator outlet pipe-to-water pump cover bolts.

9. Raise and safely support the vehicle.

10. Disconnect the exhaust pipe from the manifold.

11. Disconnect the radiator outlet pipe from the oil pan and transaxle.

12. If equipped with a manual transaxle, remove the exhaust manifold brace. If equipped with an automatic transaxle, leave the lower radiator hose attached and pull down on the radiator pipe to disconnect it from the water pump.

13. Lower the vehicle.

14. Remove the exhaust manifold.

15. Remove the water pump cover-to-engine bolts.

16. Remove the 3 water pump-to-timing chain housing nuts.

17. Remove the water pump and cover assembly.

18. Remove the 5 water pump cover-to-pump bolts and remove the cover.

To install:

19. Install the water pump-to-engine components and tighten the bolts finger-tight.

20. Lubricate the splines of the water pump drive with chassis grease.

21. To complete the installation, reverse the removal procedures.

22. With all the bolts in place tighten in the following sequence:

Water pump-to-timing chain housing: 19 ft. lbs. (26 Nm)

Water pump cover-to-water pump: 106–124 inch lbs. (12–14 Nm)

Water pump cover-to-engine block (bottom first): 19 ft. lbs. (26 Nm)

Radiator outlet pipe-to-pump cover: 10 ft. lbs. (14 Nm).

23. Refill the cooling system. Connect the negative battery cable. Start the vehicle and verify no leaks.

3.1L Engine

1. Disconnect the negative battery cable. Drain the cooling system into a suitable container.

2. Loosen, but do not remove, the water pump pulley bolts. Remove the serpentine belt.

3. Remove the water pump pulley bolts and pulley. Remove the 5 water pump mounting bolts and pump.

To install:

4. Clean all the gasket surfaces completely. Apply a thin bead of sealer around the outside edge of the water pump along the gasket sealing area and install the gasket onto the water pump.

5. Install the water pump and tighten the bolts to 89 inch lbs. (10 Nm).

6. Install the water pump pulley and tighten the pulley bolts finger-tight. Install the serpentine belt.

7. Tighten the water pump pulley bolts to 18 ft. lbs. (25 Nm). Connect the negative battery cable.

8. Refill and bleed the cooling system. Test run the engine and check for leaks.

3.3L Engine

1. Disconnect the negative battery cable. Drain the engine coolant into a suitable container.

2. Loosen, but do not remove, the water pump pulley bolts. Remove the serpentine belt.

NOTE: The bolt securing the idler pulley is reverse threaded.

3. Remove the idler pulley mounting bolt and remove the idler pulley. Disconnect the coolant hoses from the water pump.

4. Remove the water pump pulley bolts and remove the water pump pulley. Remove the water pump mounting bolts and remove the water pump.

To install:

5. Clean all the gasket surfaces completely.

6. Apply a thin bead of sealer around the gasket seating area on the water pump and install the gasket on the water pump.

7. Install the water pump on the front cover and install the mounting bolts. Tighten the water pump-to-engine block bolts (long bolts) to 22 ft. lbs. (30 Nm) and the water pump-to-front cover bolts (short bolts) to 11 ft. lbs. (15 Nm) plus 80 degrees additional rotation.

8. Install the water pump pulley and tighten the mounting bolts finger-tight. Connect the coolant hoses to the water pump.

NOTE: The bolt securing the idler pulley is reverse threaded.

9. Install the idler pulley and tighten the mounting bolt to 55 ft. lbs. (75 Nm). Install the serpentine belt.

10. Tighten the water pump pulley bolts 22 ft. lbs. (30 Nm).

11. Connect the negative battery cable. Refill and bleed the cooling system.

Thermostat

REMOVAL AND INSTALLATION

2.3L and 2.4L Engines

1993–94 Vehicles

1. Disconnect the negative battery cable.

2. Drain the cooling system to a level below the thermostat housing.

3. Disconnect the upper radiator hose from the thermostat housing and position aside.

4. Disconnect the heater hose from the thermostat housing.

5. Disconnect the electrical connector from the coolant sensor.

6. Disconnect the throttle body heater hose from the thermostat housing.

7. Remove the thermostat housing mounting bolts and remove the thermostat housing.

8. Remove the thermostat.

To install:

9. Clean all the gasket surfaces completely. Apply a thin bead of sealer around the gasket seat on the thermostat housing and install the gasket on the housing.

10. Install the thermostat, the thermostat housing and tighten the bolts to 19 ft. lbs. (26 Nm).

11. To complete the installation, reverse the removal procedures.

12. Refill the cooling system. Connect the negative battery cable. Start the vehicle and verify no leaks.

1995–97 Vehicles

1. Disconnect the negative battery cable. Drain the cooling system to a level below the thermostat.

2. Remove the coolant inlet housing bolt accessible through the exhaust manifold. Raise and safely support the vehicle.

3. Remove the radiator outlet pipe stud. Remove the second coolant inlet housing bolt.

4. Remove the coolant inlet housing. Remove the thermostat.

To install:

5. Clean all the gasket surfaces completely.

6. To complete the installation, reverse the removal procedures.

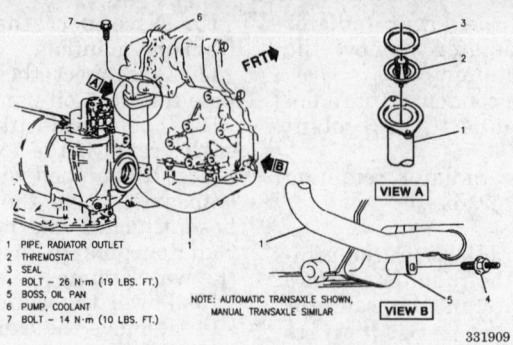

1 PIPE, RADIATOR OUTLET
2 THERMOSTAT
3 SEAL
4 BOLT – 26 N·m (19 LBS. FT.)
5 BOSS, OIL PAN
6 PUMP, COOLANT
7 BOLT – 14 N·m (10 LBS. FT.)

NOTE: AUTOMATIC TRANSAXLE SHOWN, MANUAL TRANSAXLE SIMILAR

VIEW A   VIEW B

331909

Coolant inlet pipe and mounting components — 1995–97 2.3L and 2.4L engines

7. Install the thermostat in the coolant inlet housing.

8. Torque the inlet housing bolts to 10 ft. lbs. (14 Nm) and the radiator outlet pipe stud to 19 ft. lbs. (26 Nm).

9. Lower the vehicle. Connect the negative battery cable. Refill and bleed the cooling system. Verify no leaks.

3.1L Engine

1. Disconnect the negative battery cable. Drain the cooling system to a level below the thermostat housing.

2. Remove the air cleaner assembly. Disconnect the surge tank line from the thermostat housing.

3. Remove thermostat housing-to-intake manifold attaching bolt and nut.

4. Remove the thermostat housing. Remove the thermostat.

To install:

5. Clean all gasket surfaces completely. Install the thermostat.

6. Install the thermostat housing and torque the bolt and nut to 18 ft. lbs. (25 Nm).

7. To complete the installation, reverse the removal procedures.

8. Refill the cooling system. Connect the negative battery cable. Run engine to check for leaks. Bleed the cooling system as required.

3.3L Engine

1. Disconnect the negative battery cable. Drain the cooling system to a level below the thermostat housing.

2. Disconnect the air inlet duct from the throttle body. Disconnect the upper radiator hose from the thermostat housing outlet.

3. Remove the thermostat housing bolts, the coolant pipe bracket, the thermostat housing and thermostat.

To install:

4. Clean all gasket surfaces completely. Apply a thin bead of sealer around the gasket sealing area on the thermostat housing and install the gasket on the housing.

5. Install the thermostat, thermostat housing and coolant pipe mounting bracket and tighten the mounting bolts to 20 ft. lbs. (27 Nm).

6. Connect the upper radiator hose to the thermostat housing. Connect the air inlet duct to the throttle body.

7. Refill the cooling system. Connect the negative battery cable.

Electric Cooling Fan

REMOVAL AND INSTALLATION

— **CAUTION** —

Keep hands, tools and clothing away from the engine cooling fan to avoid personal injury. This fan is electric and can come on whether or not the engine is running. The fan can start automatically in response to a heat sensor. After replacing the coolant fan, verify that the fan is moving the air in the proper direction.

2.3L and 2.4L Engines

1. Disconnect the negative battery cable. Raise the vehicle enough to remove the right front wheel assembly.

2. Remove the bolt from lower torque axis mount. Remove the fan

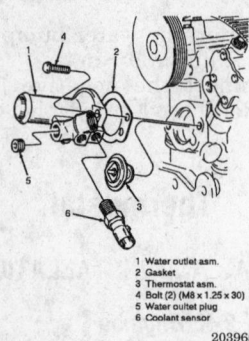

1 Water outlet asm.
2 Gasket
3 Thermostat asm.
4 Bolt (2) (M8 x 1.25 x 30)
5 Water outlet plug
6 Coolant sensor

203965

Thermostat and housing components — 1993–94 2.3L engine

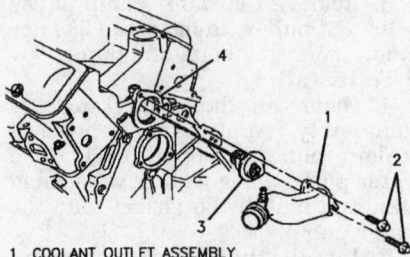

1 COOLANT OUTLET ASSEMBLY
2 BOLTS, COOLANT OUTLET 25 N·m (18 LBS. FT.)
3 THERMOSTAT
4 LOWER INTAKE MANIFOLD ASSEMBLY

331903

Thermostat assembly — 3.1L engine

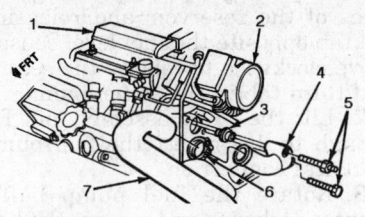

1. Intake manifold
2. Throttle body
3. Thermostat
4. Outlet asm
5. 20 ft. lb.(27 Nm)
6. Clamp 27 in. lb. (3 Nm)
7. Outlet hose

203776

Exploded view of thermostat components — 3.3L engine

mounting bolt and disconnect the wiring harness from the motor and fan frame.

3. Rock the engine rearward and remove the fan assembly.

To install:

4. Rock the engine rearward and install the fan assembly.

5. Install the fan mounting bolt and connect the electrical connector. Tighten the bolt to 97 inch lbs. (11 Nm).

6. Install the bolt to the lower torque axis mount.

7. Install the right front wheel assembly and safely lower the vehicle. Connect the negative battery cable.

3.1L Engine

1. Disconnect the negative battery cable. Drain the cooling system into a suitable container.

2. Remove the cooling fan mounting bolt. Disconnect the electrical connector at the cooling fan.

3. Remove the radiator inlet hose from the radiator. Remove the radiator mounting bolt.

4. Pull the windshield washer fluid bottle fill tube from the bottle. Remove the vacuum tank and bracket.

5. Remove the cooling fan by sliding fan leg into area left of the vacuum tank.

To install:

6. Install the fan assembly. Install the vacuum tank bracket and tank.

7. Install the windshield washer fluid bottle fill tube. Install the radiator mounting bolt and tighten to 89 inch lbs. (10 Nm).

8. Install the radiator inlet hose to the radiator. Connect the electrical connector to the fan.

9. Install the coolant fan mounting bolt and tighten to 97 inch lbs. (11 Nm).

10. Connect the negative battery cable. Fill the cooling system.

11. Connect the negative battery cable. Start the engine and verify proper operation.

Cooling System Bleeding

PROCEDURE

1. Refill the cooling system with coolant meeting GM specifications 1825 M. Use a 50/50 percent mix of coolant and clean water.

2. When refilling the engine cooling system, add engine coolant supplement sealant GM 3634621 or equivalent.

3. Fill the surge tank with the proper mix of coolant to just above the "Small Cylinder" at the base of the tank opening or until the level reaches the "Split Line" formed by the black and tan parts of the tank.

4. With the pressure cap off, start the engine and let it run until the upper radiator hose starts to get hot.

5. If coolant is lower than "Small Cylinder", add proper mix of coolant to the surge tank until the level reaches the "Split line".

6. Install the surge tank pressure cap onto the surge tank with hand-tight pressure.

FUEL SYSTEM

Fuel System Service Precautions

Safety is the most important factor when performing not only fuel system maintenance but any type of maintenance. Failure to conduct maintenance and repairs in a safe manner may result in serious personal injury or death. Maintenance and testing of the vehicle's fuel system components can be accomplished safely and effectively by adhering to the following rules and guidelines.

• To avoid the possibility of fire and personal injury, always disconnect the negative battery cable unless the repair or test procedure requires that battery voltage be applied.

• Always relieve the fuel system pressure prior to disconnecting any fuel system component (injector, fuel rail, pressure regulator, etc.), fitting or fuel line connection. Exercise extreme caution whenever relieving fuel system pressure to avoid exposing skin, face and eyes to fuel spray. Please be advised that fuel under pressure may penetrate the skin or any part of the body that it contacts.

• Always place a shop towel or cloth around the fitting or connection prior to loosening to absorb any excess fuel due to spillage. Ensure that all fuel spillage (should it occur) is quickly removed from engine surfaces. Ensure that all fuel soaked cloths or towels are deposited into a suitable waste container.

• Always keep a dry chemical (Class B) fire extinguisher near the work area.

• Do not allow fuel spray or fuel vapors to come into contact with a spark or open flame.

• Always use a backup wrench when loosening and tightening fuel line connection fittings. This will prevent unnecessary stress and torsion to fuel line piping. Always follow the proper torque specifications.

• Always replace worn fuel fitting O-rings with new. Do not substitute fuel hose or equivalent, where fuel pipe is installed.

Fuel System Pressure

RELIEVING

—— CAUTION ——

Fuel injection systems remain under pressure, even after the engine has been turned OFF. The fuel system pressure must be relieved before disconnecting any fuel lines. Failure to do so may result in fire and/or personal injury.

Except 2.3L and 2.4L Engines

1. Disconnect the negative battery cable in order to avoid possible fuel discharge if an accidental attempt is made to start the engine.

2. Loosen the fuel tank filler cap in order to relieve fuel tank pressure.

3. Connect tool J-34730-A or equivalent fuel pressure gauge to the fuel pressure test port connection. Wrap a towel around the fuel pressure connection when installing the fuel pressure gauge in order to avoid fuel spillage.

4. Install the bleed hose into an approved container and open the valve in order to bleed the fuel system pressure. The fuel pipe connections are now safe for servicing.

5. Drain any fuel remaining in the fuel pressure gauge into an approved container.

2.3L and 2.4L Engines

1. Loosen the fuel filler cap.
2. Remove the fuse marked fuel pump from the fuse block or disconnect the harness connector at the tank.
3. Start the engine and run at idle until it stalls.
4. Crank the engine for an additional 3 seconds to make sure all of the fuel pressure is exhausted from the fuel lines.
5. Turn the ignition switch **OFF**, disconnect the negative battery cable and reinstall the fuel pump fuse or connect the connector at the tank.
6. Tighten the filler cap.

Idle Speed

ADJUSTMENT

The engine idle speeds are controlled through the Powertrain Control Module (PCM) and the Idle Air Control (IAC) valve. During idle, the proper position of the IAC valve is calculated based on inputs from the battery voltage, coolant temperature, engine load, and engine RPM. There are no adjustments that can be made externally.

Mixture

ADJUSTMENT

The air/fuel mixture is controlled by the Powertrain Control Module (PCM) under various conditions the PCM will calculate from the various engine sensors and will respond to the demand of the engine conditions and engine load and will try to maintain a 14.7:1 ratio. No service adjustment is possible.

Fuel Filter

REMOVAL AND INSTALLATION

—————— CAUTION ——————
Fuel injection systems remain under pressure, even after the engine has been turned OFF. The fuel system pressure must be relieved before disconnecting any fuel lines. Failure to do so may result in fire and/or personal injury.

1. Relieve the fuel system pressure.

2. Disconnect the negative battery cable.
3. Raise and safely support the vehicle.
4. Using a backup wrench on the fuel filter disconnect the fuel line from the filter.
5. Disconnect the quick-connect fitting from the fuel filter by compressing the tabs while pulling outward on the line.
6. Remove the fuel filter from the mounting bracket.

To install:

7. Install the fuel filter in the mounting bracket.
8. Using a backup wrench on the fuel filter, tighten the fuel line fitting to 20 ft. lbs. (27 Nm).
9. Connect the quick-connect fitting to the fuel filter.
10. Lower the vehicle.
11. Connect the negative battery cable.
12. Pressurize the fuel system and verify no leaks.

Fuel Pump

REMOVAL AND INSTALLATION

—————— CAUTION ——————
Fuel injection systems remain under pressure, even after the engine has been turned OFF. The fuel system pressure must be relieved before disconnecting any fuel lines. Failure to do so may result in fire and/or personal injury.

1. Relieve the fuel system pressure. Disconnect the negative battery cable.
2. Drain fuel tank. Remove the fuel tank from the vehicle.
3. While holding the modular fuel sender assembly down, remove the snapring from the designated slots located on the retainer.

NOTE: The modular fuel sender assembly may spring up from its position. When removing the modular fuel sender from the tank, be aware that the reservoir bucket is full of fuel. It must be tipped slightly during removal to avoid damage to the float.

4. Remove the external fuel strainer.
5. Disconnect the Connector Position Assurance (CPA) piece from the electrical connector and disconnect the fuel pump electrical connector.
6. Gently release the tabs on the sides of the fuel sender at the cover

assembly. Begin by squeezing the sides of the reservoir and releasing the tab opposite the fuel level sensor. Move clockwise to release the second and third tab in the same manner.
7. Lift the cover assembly out far enough to disconnect the fuel pump electrical connection.
8. Rotate the fuel pump baffle counterclockwise and remove the baffle and pump assembly from the retainer.
9. Slide the fuel pump outlet out of slot. Remove the fuel pump outlet seal.

To install:

10. Install the fuel pump outlet seal. Slide the fuel pump outlet in the slots of reservoir cover.
11. Install the fuel pump and baffle assembly onto the reservoir retainer and rotate clockwise until seated.
12. Install the lower retainer assembly partially into the reservoir. Line up all 3 sleeve tabs. Press the retainer onto the reservoir making sure all 3 tabs are firmly seated.

NOTE: Gently pull on the fuel pump reservoir from retainer to assure a secure fastening. If not secure replace entire fuel sender.

13. Complete the installation by reverse the removal procedures.
14. Connect the negative battery cable. Pressurize the fuel system and verify no leaks.

Fuel Injector

REMOVAL AND INSTALLATION

—————— CAUTION ——————
Fuel injection systems remain under pressure, even after the engine has been turned OFF. The fuel system pressure must be relieved before disconnecting any fuel lines. Failure to do so may result in fire and/or personal injury.

2.3L and 2.4L Engines

1. Relieve the fuel system pressure. Disconnect the negative battery cable.
2. Disconnect the electrical connector from the fuel injector(s). Remove the fuel rail assembly from the cylinder head.

NOTE: It is not necessary to separate the rail from the fuel pipes.

3. Spread the injector retainer clip to release injector from the rail extru-

sion flange. Remove the fuel injector assembly.

NOTE: Each fuel injector is calibrated for a specific application, therefore it is important to order the identical part number that is inscribed on the old injector.

To install:

4. Lubricate the new injector O-ring seals with clean engine oil and install on injector assembly.

5. Position a new retainer clip on the injector (on the right side of the electrical connector).

6. Attach the fuel injector to the fuel rail with the electrical connector facing outward.

7. Install the fuel rail assembly. Connect the electrical connector.

8. Connect the negative battery cable. Turn the ignition key to the **ON** position to pressurize the fuel rail and verify no leaks.

3.1L Engine

1. Relieve the fuel system pressure. Disconnect the negative battery cable.

2. Remove the upper intake manifold assembly. Disconnect and cap the fuel feed line from the fuel rail.

3. Remove the fuel pressure regulator as follows:

a. Remove the fuel pressure regulator retaining screw.

b. Place a shop towel under the pressure regulator and remove the regulator from the fuel rail.

c. Remove the retainer and spacer bracket from the fuel rail.

d. Disconnect the regulator from the fuel return line.

4. Disconnect the fuel injector main harness. Disconnect the coolant temperature sensor electrical connector.

5. Remove the fuel rail bolts and the fuel rail. Disconnect the individual injector electrical connector.

6. Remove the injector retaining clip and discard. Remove the injector assembly from the fuel rail.

7. Remove the old O-rings and discard. Save the O-ring backups for installation on the new injector.

To install:

8. Install the O-ring backups on the fuel injector. Coat the new O-rings with clean engine oil and install the O-rings on the injector.

9. Install a new retaining clip on the fuel injector. Position the open end of the clip toward the injector electrical connector.

10. Install the injector onto the fuel rail assembly with the injector elec-

trical connector facing outward. Connect the electrical connector to the fuel injector.

11. Complete the installation by reversing the removal procedures.

12. Torque the following items to:
Fuel rail mounting bolts: 89 inch lbs. (10 Nm)
Fuel feed line fitting: 13 ft. lbs. (17 Nm)
Return line fitting: 13 ft. lbs. (17 Nm)
Pressure regulator bolts: 76 inch lbs. (9 Nm)

13. Install the upper intake manifold assembly. Connect the negative battery cable. Pressurize the fuel system and verify no leaks.

3.3L Engine

1. Relieve the fuel system pressure. Disconnect the negative battery cable.

2. Disconnect the fuel injector electrical connectors. Disconnect the vacuum line from the pressure regulator.

3. Disconnect and cap the fuel lines from the fuel rail. Use a backup wrench on the fuel rail fittings to prevent them from turning.

4. Remove the O-rings from the lines and discard. Remove the fuel rail mounting nuts.

5. Remove the fuel rail assembly. Remove the injector retaining clip and discard.

6. Remove the injector assembly from the fuel rail. Remove the old O-rings and discard.

To install:

7. Coat the new O-rings with clean engine oil and install the O-rings on the injector.

8. Complete the installation by reversing the removal procedures.

9. Torque the following items to:
Fuel rail mounting nuts: 20 ft. lbs. (27 Nm)
Fuel feed/return line fittings: 20 ft. lbs. (26 Nm)

10. Connect the negative battery cable. Pressurize the fuel system and verify no leaks.

EMISSION CONTROLS

Service Interval Lamp

RESETTING

1993 Achieva

Vehicles equipped with an **ENGINE OIL LIFE INDEX** display as a part of the **DRIVER INFORMATION SYSTEM (DIS)**, have a display that will show when to change the engine oil.

The oil change interval is determined by the driver information system and will usually fall at or between the 2 recommended alternative intervals of 3000 miles and 7500 miles, but it could be shorter than 3000 miles under some severe driving conditions. The driver information system will also signal the need for an oil change at 7500 miles or one year passed since the last oil change. If the driver information system does not indicate the need for an oil change after 7500 miles or one year or if the **ENGINE OIL LIFE INDEX** display fails to appear, the oil should be changed and the driver information system serviced.

When the **ENGINE OIL LIFE INDEX** reaches 10 percent or less, the **CHANGE OIL** light display will function as a reserve trip odometer (indicating the distance to an oil change). Until the **ENGINE OIL LIFE INDEX** reset is performed, the driver information system will display the distance to the oil change and sound a beep when the ignition switch is turned to the **ACCESSORY** or **RUN** position the first time each day.

When the distance to the next oil change reaches 0, the driver information system will display the **CHANGE OIL NOW** light. Until an **ENGINE OIL LIFE INDEX** reset is performed, the driver information system will display the **CHANGE OIL NOW** light and sound a beep when the ignition switch is turned to the **ACCESSORY** or **RUN** position at the beginning of each day.

The driver information system will not detect dusty conditions or engine malfunctions which may affect the engine oil. If driving in severe conditions exists, be sure to change the engine oil every 3000 miles or 3 months which ever comes first, unless instructed otherwise by the driver information system. The driver information center does not measure the engine oil level. It still remains the owner's responsibility to check the engine oil level. After the oil has been changed, the **ENGINE OIL LIFE INDEX** light must be reset. Resetting can be accomplished as follows:

1. The **ENGINE OIL LIFE INDEX** can be reset by pressing the **RESET** and **OIL** buttons simultaneously for at least 5 seconds while on the **ENGINE OIL LIFE INDEX** display. The driver information system will reset the **ENGINE OIL LIFE INDEX** to 100 percent and display a **ENGINE OIL LIFE INDEX** of 100 percent.

NOTE: The Engine Oil Life Index is stored on a non-volatile memory chip and will not require resetting by disconnecting the battery cables and/or fuse.

1994–95 Achieva

Vehicles that are equipped with a Drivers Information Center and have an Oil life index, will require the Oil Life indicator to be reset after each oil change.

Press the **SEL** to select **OIL**. Press **SEL** if necessary to display the oil life. The display will show a reading of the estimated oil life left. Example: **OIL LIFE 85%**. When the remaining oil life is 9 percent or less, the display will show **CHANGE OIL SOON**. Then, the vehicle is started tone will sound an the **CHANGE OIL SOON** message will display each time the vehicle is started.

When the oil life is zero, a tone will sound and the display will show, **CHANGE OIL NOW**. Then, when the vehicle is started a tone will sound and the **CHANGE OIL NOW** message will display each time the vehicle is started. Reset the Oil Life Display as follows:

1. Acknowledge all diagnostic messages in the Drivers Information Center by pressing **RESET**.

2. Press the **SEL** button on the left to select **OIL**. Press **SEL** button on the right if necessary to display oil life.

3. Press and hold the **RESET** button for about 5 seconds. Once the oil life index has been reset, a **RESET** message will be displayed and then oil life will change to 100 percent.

Be careful not reset the oil life accidentally at any time other than when the oil has just been changed. It can not be reset accurately until the next oil change.

ENGINE MECHANICAL

Engine Assembly

REMOVAL AND INSTALLATION

CAUTION

Fuel injection systems remain under pressure, even after the engine has been turned OFF. The fuel system pressure must be relieved before disconnecting any fuel lines. Failure to do so may result in fire and/or personal injury.

2.3L and 2.4L Engines

The engine and transaxle are removed as a unit from under the vehicle.

1. Relieve the fuel system pressure. Disconnect the negative battery cable.

2. Drain the cooling system into a suitable container. If equipped with A/C, recover the refrigerant.

3. Remove the left sound insulator and disconnect the clutch pushrod from pedal assembly (manual transaxle only).

4. Disconnect the heater hose and upper radiator hose from the thermostat housing. Remove the air cleaner-to-throttle body duct, the upper radiator support and cooling fan assembly.

5. If equipped with A/C, disconnect the refrigerant line and discard the O-rings.

6. Disconnect the vacuum hoses from the front of the engine. Label and disconnect the electrical connectors from the necessary components.

7. Disconnect the power brake booster vacuum lines. Disconnect the throttle cables; remove the cable bracket and position aside.

8. Remove the power steering pump rear bracket and pump; position the pump aside with the lines attached.

9. Disconnect and cap the fuel lines from the fuel rail.

10. Disconnect the shift cable from the transaxle and the TV cable on automatic transaxles.

11. Remove the exhaust manifold and heat shield.

12. Remove the lower radiator hose. Remove the coolant recovery tank and position it aside leaving the hoses attached.

13. Install an engine support fixture, J-28467-A or equivalent. Remove the right engine mount.

14. Raise and safely support the vehicle. Drain the engine crankcase into a suitable container.

15. Remove the front tire and wheel assemblies. Remove the right lower splash shield.

16. Disconnect the electrical connections from the ABS wheel speed sensors. Disconnect the ground connections from the transaxle.

17. Remove the cotter pins and castle nuts from the lower ball joints and separate the ball joints from the steering knuckles.

18. Support the suspension crossmembers; then, remove the mounting bolts and crossmembers.

19. Disconnect the heater outlet hose from the radiator outlet pipe.

20. Disconnect the halfshafts from the transaxle. If equipped with a manual transaxle, disconnect the intermediate shaft.

21. Position a suitable engine table under the power train and lower the vehicle until the powertrain is resting on the table.

22. Remove the engine strut and transaxle mount bolts. Remove the engine support fixture.

23. Raise the vehicle until it is clear of the engine. Separate the transaxle from the engine.

To install:

24. Install the transaxle onto the engine and tighten the bolts to 55 ft. lbs. (75 Nm). Position the powertrain assembly and the engine table under the vehicle.

25. Lower the vehicle and install the engine support fixture. Install the engine strut and transaxle mounts.

26. Raise the vehicle and remove the engine table. Connect the halfshafts and intermediate shaft to the transaxle.

27. Complete the installation by reversing the removal procedures.

28. Torque the ball joint-to-steering knuckle nuts to 41 ft. lbs. (55 Nm). Install new cotter pins.

29. Refill the cooling system, the crankcase and fill the remainder of the underhood fluids.

30. Connect the negative battery cable. Recharge the A/C system. Start the vehicle and verify no leaks.

NOTE: Whenever the vehicle sub-frame is removed or lowered, the wheel alignment should be checked.

3.1L Engine

Please note that the engine and transaxle are removed as an assembly from under the vehicle.

1. Relieve the fuel system pressure. Disconnect the negative battery cable.

2. Remove the top half of the air cleaner assembly and the throttle body inlet duct. Drain the cooling system.

3. Remove the upper and lower radiator hoses. Disconnect the coolant inlet line from the coolant surge tank.

4. Disconnect the vacuum hoses from the EVAP canister purge valve, vacuum modulator and power brake booster.

5. Disconnect the heater outlet hose from the water pump. Remove the serpentine drive belt.

6. Disconnect the control cables from the throttle body lever and intake manifold bracket and position the cables aside.

7. Label and disconnect the electrical connectors from the necessary components.

8. Remove the alternator. Disconnect and cap the power steering lines from the power steering pump.

9. Disconnect and cap the fuel lines from the fuel rail. Remove the cooling fan assembly.

10. Disconnect the shift control cable from the transaxle shift lever and cable bracket. Disconnect the transaxle vent tube from the transaxle.

11. Disconnect the vacuum hose from the vacuum reservoir. Install an engine support fixture, J-28467-A, or the equivalent.

12. Loosen, but do not remove the top 2 A/C compressor mounting bolts. Raise and safely support the vehicle.

13. Remove the front tire and wheel assemblies. Remove the right and left inner fender splash shields.

14. Remove the engine mount strut. Disconnect the ABS sensor wires from the wheel sensors and suspension member supports, if equipped.

15. Remove the cotter pins and castle nuts from the lower ball joints. Using a suitable tool, separate the lower ball joints from the steering knuckles.

16. Remove the lower suspension support assemblies with the lower control arms attached. Disconnect the halfshafts from the transaxle and support aside.

17. Remove the oil filter and oil filter adapter. Remove the flywheel cover. Remove the starter. Disconnect the heater hoses from the heater core.

18. Remove the A/C compressor lower mounting bolts and remove the compressor from the mounting bracket and position aside. DO NOT disconnect the refrigerant lines from the compressor or allow the lines top support the weight of the compressor.

19. Remove the vacuum reservoir tank. Disconnect the exhaust pipe from the exhaust manifold and position the pipe aside.

20. Remove the engine mount strut bracket from the engine. Disconnect the transaxle cooler lines from the radiator.

21. Remove the transaxle oil fill tube. Lower the vehicle until the power train assembly is resting on a suitable engine table.

22. Remove the transaxle mount-to-body bolts. Remove the intermediate bracket from the right engine mount.

23. Remove the engine support fixture. Raise the vehicle leaving the powertrain assembly on the engine table.

24. Separate the engine and transaxle assemblies.

To install:

25. Connect the transaxle to the engine and tighten the mounting bolts to 55 ft. lbs. (75 Nm).

26. Position the powertrain assembly under the vehicle and lower the vehicle into position. Loosely install the serpentine belt.

27. Complete the installation by reversing the removal procedures.

28. Torque the following items to:
Engine strut bracket-to-engine bolts: 44 ft. lbs. (60 Nm)
Exhaust pipe-to-exhaust manifold bolts: 18 ft. lbs. (25 Nm)
Ball joints-to-steering knuckle nuts: 41 ft. lbs. (55 Nm)

29. Connect the negative battery cable. Check and fill all the engine fluids as necessary. Start the vehicle and bleed the power steering system.

3.3L Engine

1. Relieve the fuel system pressure. Disconnect the negative battery cable.

2. Matchmark and remove the hood. Install an engine support fixture, J-28467-A, or the equivalent.

3. Drain the cooling system into a suitable container. Disconnect the heater hoses from heater core and water pump.

4. Remove the upper and lower radiator hoses. Remove the engine cooling fan assembly.

5. Remove the air inlet duct from the throttle body. Disconnect the vacuum lines from the EVAP solenoid and power brake booster.

6. Disconnect the control cables from the throttle body and intake manifold bracket. Remove the serpentine drive belt.

7. Remove the power steering pump from the mounting bracket and position aside. The power steering lines do not need to be disconnected.

8. Disconnect the electrical connectors from the engine components and route the engine harness aside. Remove the upper transaxle-to-engine bolts.

9. Raise and safely support the vehicle.

10. Remove the A/C compressor from the mounting bracket and position aside. DO NOT disconnect the refrigerant lines from the compressor.

11. Remove the right engine mount and engine strut. Remove the flywheel splash shield. Remove the flywheel-to-converter bolts.

12. Remove the lower engine-to-transaxle bolts. One of the bolts is located in between the transaxle case and the engine block and is threaded in from the opposite direction.

13. Lower the vehicle. Install a suitable engine lifting device.

14. Remove the engine support fixture. Remove the engine from the vehicle.

To install:

15. Lower the engine into the vehicle and connect the engine to the transaxle. Install 2 of the upper bolts loosely.

16. Install the engine support fixture. Remove the lifting device. Raise and safely support the vehicle.

17. Complete the installation by reversing the removal procedures.

18. Torque the following items to:
Lower transaxle-to-engine bolts: 55 ft. lbs. (75 Nm)
Converter-to-flywheel bolts: 46 ft. lbs. (62 Nm)
Upper transaxle-to-engine bolts: 55 ft. lbs. (75 Nm)

19. Refill the cooling system. Install the hood. Connect the negative battery cable.

Engine Mounts

REMOVAL AND INSTALLATION

2.3L and 2.4L Engines

Right Engine Mount

1. Disconnect the negative battery cable. Install an engine support fixture, J-28467-A or an equivalent. Use the support fixture to remove the engine weight from the engine mount.

2. Remove the 4 engine mount-to-engine bracket bolts. Raise and safely support the vehicle.

3. Remove the right front tire and wheel assembly. Remove the right inner fender well splash shield.

4. Remove the 2 engine mount-to-body bolts. Lower the vehicle enough to work under the hood.

5. Remove the engine mount nut from the mounting stud. Remove the engine mount.

To install:

6. Position the engine mount over the mounting stud and loosely install the mounting nut. Raise and safely support the vehicle.

7. Install the 2 lower engine mount bolts and tighten to 49 ft. lbs. (66 Nm). Tighten the rear bolt first.

8. Install the inner fender splash shield. Install the right front tire and wheel assembly.

9. Lower the vehicle. Tighten the engine mount nut to 31 ft. lbs. (42 Nm).

10. Install the 4 engine mount-to-bracket bolts and tighten to 46 ft. lbs. (62 Nm).

11. Remove the engine support fixture. Connect the negative battery cable.

Engine Mount Strut

1. Disconnect the negative battery cable. Raise and safely support the vehicle.

2. Remove the right front tire and wheel assembly. Remove the right inner fender well splash shield.

3. Remove the engine mount strut-to-suspension support and the engine mount strut-to-engine bracket bolt. Remove the engine mount strut.

To install:

4. Position the strut in the vehicle and loosely install the mounting bolt at the suspension support.

5. Install the bolt at the engine bracket. Make sure the bolt is installed in the correct hole in the engine mount bracket. The bolt holes are marked for automatic or manual transaxle and the bolt must be installed in the correct hole.

6. Tighten the bolt at the mounting bracket to 85 ft. lbs. (115 Nm) and the suspension support bolt to 89 ft. lbs. (120 Nm).

7. Install the inner fender splash shield. Install the right front tire and wheel assembly.

8. Lower the vehicle. Connect the negative battery cable.

3.1L Engine

Right Engine Mount

1. Disconnect the negative battery cable.

2. Support the weight of the engine. If an engine support fixture is unavailable, the engine can be supported by a hydraulic floor jack and a carefully placed block of wood between the jack and the oil pan. Lift the engine just enough to take the weight off the engine mount.

3. Remove the 2 bolts securing the right engine mount to the intermediate bracket.

4. Remove the top nut, 2 lower bolts and the right engine mount.

To install:

5. Position the mount in the vehicle and tighten the bolts to 49 ft. lbs. (66 Nm) and the upper nut to 42 ft. lbs. (31 Nm).

6. Connect the engine mount to the intermediate bracket and tighten the mounting bolts to 96 ft. lbs. (130 Nm).

7. Remove the floor jack. Connect the negative battery cable.

Engine Mount Strut

1. Disconnect the negative battery cable. Raise and safely support the vehicle.

2. Remove the right front tire and wheel assembly. Remove the right inner fender well splash shield.

3. Remove the engine mount strut bolt at the crossmember and engine bracket and remove the strut.

To install:

4. Position the strut in the vehicle and loosely install both mounting bolts.

5. Tighten the strut-to-engine bracket bolt to 85 ft. lbs. (115 Nm) and the strut-to-crossmember bolt to 89 ft. lbs. (120 Nm).

6. Install the right inner fender well splash shield. Install the right front tire and wheel assembly.

7. Lower the vehicle. Connect the negative battery cable.

3.3L Engine

Right Engine Mount

1. Disconnect the negative battery cable. Install an engine support fixture, J-28467-A or equivalent.

2. Remove the engine mount-to-engine mount bracket nuts. Raise the engine slightly with the support fixture.

3. Remove the engine mount-to-frame nuts and the engine mount.

To install:

4. Install the engine mount and tighten the engine mount-to-bracket

nuts to 31 ft. lbs. (42 Nm). Lower the engine into position.

5. Install the engine mount-to-bracket nuts and tighten to 46 ft. lbs. (62 Nm).

6. Install the engine mount-to-frame bolts and tighten to 49 ft. lbs. (66 Nm).

7. Remove the support fixture. Connect the negative battery cable.

Engine Mount Strut

1. Disconnect the negative battery cable. Raise and safely support the vehicle.

2. Remove the right front wheel assembly. Remove the right side splash shield.

3. Remove the engine mount strut-to-body bolts, the strut bracket and the mount.

To install:

4. Install the engine mount strut and tighten the bolts to 89 ft. lbs. (120 Nm).

5. Install the right side splash shield. Install the right front wheel assembly.

6. Lower the vehicle. Connect the negative battery cable.

Cylinder Head

REMOVAL AND INSTALLATION

———— CAUTION ————

Fuel injection systems remain under pressure, even after the engine has been turned OFF. The fuel system pressure must be relieved before disconnecting any fuel lines. Failure to do so may result in fire and/or personal injury.

2.3L Engine

1. Disconnect the negative battery cable. Drain the cooling system.

2. Disconnect heater inlet and throttle body heater hoses from water outlet. Remove the exhaust manifold.

3. Remove the intake and exhaust camshaft housings. Remove the oil fill tube bolt and unseat the fill tube assembly from the engine.

4. Disconnect the injector harness electrical connector. Remove the oil fill tube.

5. Disconnect the throttle body air intake duct. Disconnect the power brake booster vacuum line.

6. Remove the throttle cable bracket. Remove the throttle body and position aside with all the lines and cables attached.

7. Disconnect the MAP sensor vacuum hose. Remove the intake manifold brace.

8. Remove the fuel rail bolts and the fuel rail from the intake manifold with the fuel lines attached.

9. Disconnect the electrical connections from the following components:
- MAP sensor
- MAT sensor
- EVAP purge solenoid

10. Disconnect the upper radiator hose from the thermostat housing. Disconnect the coolant sensor connectors.

11. Remove the cylinder head bolts in reverse order of the installation sequence. Remove the cylinder head and gasket.

To install:

12. Clean all gasket surfaces completely. Install a new cylinder head gasket and the cylinder head.

13. Coat the cylinder head bolt threads with clean engine oil and allow the oil to drain off before installing.

14. For 1993–94 vehicles, install the cylinder head bolts and tighten, in sequence, as follows:

a. No's 1 thru 6: 18 ft. lbs. (25 Nm)

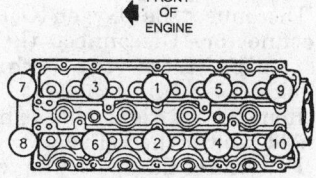

BOLTS 1 THROUGH 6 - 25 Nm (18 LBS. FT.) PLUS 90°
BOLTS 7 AND 8 - 35 Nm (26 LBS. FT.) PLUS 60°
BOLTS 9 AND 10 - 40 Nm (30 LBS. FT.) PLUS 60°

204324

Cylinder head bolt tightening sequence — 1993–94 2.3L engine

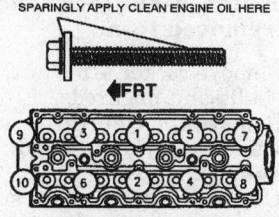

SPARINGLY APPLY CLEAN ENGINE OIL HERE

◄FRT

A. Tighten the bolts to the following N.m (lb. ft.) specification in sequence:
bolts 1 through 8: 40 N·m (30 lb. ft.)
bolts 9 and 10: 35 N·m (26 lb. ft.)
B. Then turn all 10 bolts an additional 90 degrees in sequence

223289

Cylinder head bolt tightening sequence — 1995 2.3L engine

b. No's 7 and 8: 26 ft. lbs. (35 Nm)

c. No's 9 and 10: 30 ft. lbs. (40 Nm)

d. No's 1 thru 6: plus a 90 degree turn

e. No's 7 thru 10: plus a 60 degree turn

f. Loosen all bolts: 1 full turn (360 degrees)

g. No's 1 thru 6: 18 ft. lbs. (25 Nm)

h. No's 7 and 8: 26 ft. lbs. (35 Nm)

i. No's 9 and 10: 30 ft. lbs. (40 Nm)

j. No's 1 thru 6: plus a 90 degree turn

k. No's 7 thru 10: plus a 60 degree turn

15. For 1995 vehicles, install the cylinder head bolts and tighten, in sequence, as follows:

a. No's 1 thru 8: 30 ft. lbs. (40 Nm)

b. No's 9 and 10: 26 ft. lbs. (35 Nm)

c. All 10 bolts: an additional 90 degree turn

16. Connect the heater inlet and throttle body heater hoses to the water outlet. Connect the upper radiator hose to the water outlet.

17. Connect the 2 coolant sensor connections. Install the fuel rail on the intake manifold. Install the intake manifold brace.

18. Connect the electrical connections to the following components:
- MAP sensor
- MAT sensor
- EVAP purge solenoid

19. Connect the MAP sensor vacuum hose to the intake manifold. Install the throttle body.

20. Install the throttle cable bracket. Connect the throttle body air intake duct.

21. Lubricate the O-ring on the oil fill tube with engine oil and install oil fill tube into the block and bolt.

22. Install the exhaust and intake camshaft housings. Connect the injector harness electrical connector.

23. Install the exhaust manifold. Fill all fluids to their proper levels.

24. Connect the negative battery cable. Start the vehicle and verify no leaks.

2.4L Engine

1. Disconnect the negative battery cable. Drain the cooling system.

2. Disconnect the heater inlet and throttle body heater hoses from water outlet. Remove the power brake vacuum hose from the throttle body.

3. Disconnect and tag for identification, if necessary, the electrical connections from the following components:
- MAP sensor
- MAT sensor
- EVAP purge solenoid
- Camshaft sensor

4. Remove the stud-ended bolt from the alternator. Remove the intake manifold brace. Remove the intake manifold.

5. Install the stud-ended alternator bolt back into the engine. Install engine support fixtures J-28467-400 and J-28467-A or equivalent.

6. Remove the exhaust manifold. Remove the ignition coil and module assembly.

7. Disconnect the camshaft position connector. Remove the power steering pump.

8. Disconnect the vacuum line from the fuel pressure regulator, fuel injector and fuel injector harness connector.

9. Disconnect the fuel line clamp from the bracket on top of the intake camshaft housing. Remove the fuel line rail and position it aside (leaving the fuel lines attached).

10. Disconnect the timing chain housing at the intake camshaft housing, but do not remove from the vehicle.

11. Remove the electrical connection from the oil switch.

12. If equipped with an automatic transaxle, remove the transaxle fluid level indicator tube assembly from the exhaust camshaft cover and position it aside.

NOTE: Any time the camshaft housing to cylinder head bolts are loosened or removed, the camshaft housing to cylinder head gasket must be replaced.

13. Remove the intake camshaft housing bolts, the housing and gasket.

NOTE: Turn the camshaft housing upside down as soon as it is removed from the cylinder head. The lifters will fall out of the camshaft housing if it is not turned upside down. The lifters can be damaged if they fall out and hit a hard surface.

14. Remove the exhaust camshaft housing from the cylinder head (remove the bolts in reverse of the tightening sequence).

15. Remove the radiator inlet (upper) hose from the coolant outlet. Disconnect the coolant temperature sensor electrical connections.

16. Remove the cylinder head bolts (in reverse of the tightening sequence), the cylinder head and gasket.

NOTE: Do not use abrasive pads to clean the cylinder head or block surfaces. An abrasive pad may damage the cylinder head and block. Use a razor or scraper to clean the gasket surfaces.

To install:

17. Clean all gasket surfaces completely. Make sure the threaded holes in the engine block for the cylinder head bolts are clean.

18. Install a new cylinder head gasket on the cylinder block and carefully position the cylinder head in place.

19. New replacement head bolts are recommended. Coat the cylinder head bolt threads with clean engine oil and allow the oil to drain off before installing.

20. Install the cylinder head bolts and tighten, in sequence, as follows:
   a. No's 1 thru 8: 40 ft. lbs. (65 Nm)
   b. No's 9 and 10: 30 ft. lbs. (40 Nm)
   c. All 10 bolts: an additional 90 degrees (¼ turn).

21. Install the intake and exhaust camshaft housings. Tighten the long bolts to 11 ft. lbs. (15 Nm) plus an additional 90 degrees (¼ turn). Tighten the short bolts to 11 ft. lbs. (15 Nm) plus an additional 30 degrees (⅙ turn).

22. Install the timing chain and timing chain housing. Connect the upper radiator hose to the water outlet.

23. Connect the 2 coolant sensor connections. Install the fuel rail on the intake manifold and the intake manifold brace.

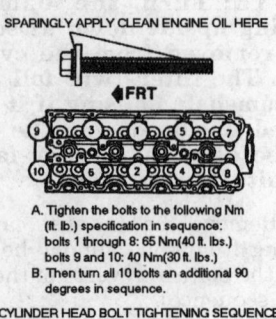

SPARINGLY APPLY CLEAN ENGINE OIL HERE

◀FRT

A. Tighten the bolts to the following Nm (ft. lb.) specification in sequence:
bolts 1 through 8: 65 Nm(40 ft. lbs.)
bolts 9 and 10: 40 Nm(30 ft. lbs.)
B. Then turn all 10 bolts an additional 90 degrees in sequence.

CYLINDER HEAD BOLT TIGHTENING SEQUENCE
330941

Cylinder head bolt tightening sequence — 2.4L engine

24. Connect the electrical connections to the following components:
   • MAP sensor
   • MAT sensor
   • EVAP purge solenoid
   • Camshaft position sensor

25. Connect the MAP sensor vacuum hose to the intake manifold. Install the power steering pump.

26. Install the throttle body and the throttle cable bracket. Connect the throttle body air intake duct.

27. Lubricate the O-ring on the oil fill tube with engine oil. Install oil fill tube into the block and the mounting.

28. Install the ignition coil and module assembly. Connect the injector harness electrical connector.

29. Remove the engine support fixture. Install the exhaust manifold.

30. Fill all fluids to their proper levels. An oil and filter change is recommended.

31. Connect the negative battery cable. Start the vehicle and verify no leaks.

3.1L Engine

Left Cylinder Head (Front)

1. Relieve the fuel system pressure. Disconnect the negative battery cable. Drain the cooling system.

2. Remove the top half of the air cleaner assembly and the throttle body air inlet duct.

3. Remove the exhaust crossover pipe heat shield and crossover pipe.

4. Disconnect the spark plug wires from spark plugs and wire looms and route the wires aside. Remove the rocker arm covers.

5. Remove upper intake plenum and lower intake manifold. Remove the left side exhaust manifold and oil level indicator tube.

NOTE: When removing the valve train components they must be kept in order for installation in the same locations they were removed from.

6. Remove rocker arms nut, rocker arms, balls and pushrods.

7. Remove the cylinder head bolts evenly and the cylinder head.

To install:

8. Clean all the gasket surfaces completely. Clean the threads on the cylinder head bolts and block threads.

9. Place the cylinder head gasket in position over the dowel pins on the cylinder block so the words **THIS SIDE UP** are showing.

10. Coat the bolt threads with sealer and install finger-tight.

11. Tighten the cylinder head bolts in sequence to 33 ft. lbs. (45 Nm).

With all the bolts tightened make a second pass tightening all the bolts an additional 90 degrees (¼ turn).

12. Install the pushrods, rocker arms, balls and rocker arm nuts. Tighten the rocker arm nuts to 18 ft. lbs. (25 Nm).

13. Install the lower intake manifold, upper intake plenum and the rocker arm covers.

14. Install the oil level indicator tube. Connect the spark plug wires to spark plugs and wire looms.

15. Install the left side exhaust manifold. Install the exhaust crossover pipe and crossover pipe heat shield.

16. Refill the cooling system. Drain the engine oil and refill with the proper type of engine oil. A filter change is recommended.

17. Install the top half of the air cleaner assembly and the throttle body air inlet duct.

18. Connect negative battery cable. Start the engine and verify no leaks.

Right Cylinder Head (Rear)

1. Relieve the fuel system pressure. Disconnect the negative battery cable. Drain the cooling system.

2. Remove the top half of the air cleaner assembly and remove the throttle body air inlet duct.

3. Remove the exhaust crossover pipe heat shield and crossover pipe. Raise and safely support the vehicle.

4. Disconnect the Oxygen (O₂) sensor connector. Disconnect the exhaust pipe from the exhaust manifold.

5. Remove the right side exhaust manifold. Lower the vehicle.

6. Disconnect the spark plug wires from spark plugs and wire looms and route the wires aside.

7. Remove the rocker arm covers, upper intake plenum and lower intake manifold.

NOTE: When removing the valve train components they must be kept in order for installation in the same locations they were removed from.

8. Remove rocker arms nut, rocker arms, balls and pushrods.

9. Remove the cylinder head bolts evenly and the cylinder head.

To install:

10. Clean all the gasket surfaces completely. Clean the threads on the cylinder head bolts and block threads.

11. Place the cylinder head gasket in position over the dowel pins on the cylinder block so the words **THIS SIDE UP** are showing.

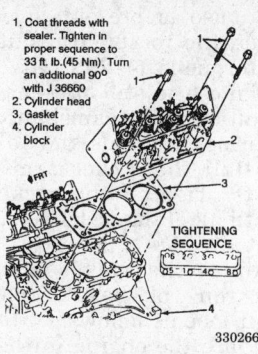

1. Coat threads with sealer. Tighten in proper sequence to 33 ft. lb. (45 Nm). Turn an additional 90° with J 36660
2. Cylinder head
3. Gasket
4. Cylinder block

TIGHTENING SEQUENCE

330266

Cylinder head bolt tightening sequence — 3.1L engine

12. Coat the bolt threads with sealer and install finger-tight.

13. Tighten the cylinder head bolts, in sequence, to 33 ft. lbs. (45 Nm). With all the bolts tightened make a second pass tightening all the bolts an additional 90 degrees (¼ turn).

14. Install the pushrods, rocker arms, balls and rocker arm nuts. Tighten the rocker arm nuts to 18 ft. lbs. (25 Nm).

15. Install the lower intake manifold and upper intake plenum. Install the rocker arm covers.

16. Connect the spark plug wires to spark plugs and wire looms. Raise vehicle and safely support.

17. Install the exhaust manifold. Connect the exhaust pipe to the exhaust manifold.

18. Lower the vehicle. Connect the Oxygen (O₂) sensor connector.

19. Install the exhaust crossover pipe and heat shield. Refill the cooling system.

20. Drain the engine oil and refill with the proper type of engine oil. A filter change is recommended.

21. Install the top half of the air cleaner assembly and the throttle body air inlet duct.

22. Connect negative battery cable. Start the engine and verify no leaks.

3.3L Engine

Right Cylinder Head (Rear)

1. Relieve the fuel system pressure. Disconnect the negative battery cable.

2. Drain the cooling system into a suitable container. Remove the intake manifold. Remove the crossover pipe.

3. Remove the nut from the heater pipe and slide the pipe out of the cover housing.

4. Remove the power steering pump mounting bolts from the mounting bracket and pull the pump forward slightly.

5. Remove the engine lift bracket and coolant pipe. Remove the exhaust manifold heat shield.

6. Remove the transaxle fill tube. Disconnect the spark plug wires from the spark plugs and wire looms.

7. Remove the spark plugs. Raise and safely support the vehicle.

8. Remove the exhaust pipe-to-exhaust manifold bolts. Lower the vehicle.

9. Remove the exhaust manifold bolts and the exhaust manifold.

10. Remove the rocker arm cover, rocker arm bolts, rocker arms, pedestals and pushrods.

11. Remove the cylinder head bolts, the cylinder head and gasket.

To install:

12. Thoroughly clean and dry all bolts, bolt holes and mating surfaces. Install the head gasket to the block and carefully position the cylinder head in place.

13. Coat the cylinder head bolts with thread lock compound and start all the cylinder head bolts finger-tight.

14. Tighten the cylinder head bolts in sequence to the following:
   a. Step 1: Tighten to 35 ft. lbs. (47 Nm).
   b. Step 2: Tighten each bolt an additional 130 degrees.
   c. Step 3: Tighten the 4 center bolts an additional 30 degrees.

15. Install the pushrods, rocker arms and pedestals and tighten the bolts to 28 ft. lbs. (38 Nm). Install the rocker arm cover.

16. Install the exhaust manifold and tighten the bolts and studs to 38 ft. lbs. (52 Nm). Raise and safely support the vehicle.

17. Connect the exhaust pipe to the exhaust manifold and tighten the bolts to 22 ft. lbs. (30 Nm). Lower the vehicle.

18. Install the spark plugs and tighten to 11 ft. lbs. (15 Nm). Connect spark plugs wires to the wire looms and spark plugs.

19. Install the transaxle fill tube and the exhaust manifold heat shield.

20. Install the coolant pipe and engine lift bracket. Install the power steering pump mounting bolts.

21. Slide the heater pipe into the front cover housing and install the mounting nut.

22. Install the crossover pipe and the intake manifold.

23. Connect the negative battery cable. Start the vehicle and verify no leaks.

Left Cylinder Head (Front)

1. Relieve the fuel system pressure. Disconnect the negative battery cable.

2. Drain the cooling system into a suitable container. Remove the intake manifold.

3. Disconnect the spark plug wires from the spark plugs and wire looms.

4. Remove the engine lift bracket and the rocker arm cover.

5. Remove the spark plugs and the exhaust crossover.

6. Remove the oil level indicator tube and the cooling fan assembly.

7. Remove the exhaust manifold heat shield and the exhaust manifold support bracket.

8. Remove the exhaust manifold bolts and the manifold. Remove the rocker arm bolts, rocker arms, pedestals and pushrods.

9. Remove the A/C bracket bolt from the cylinder head. Remove the bolts securing the alternator and ignition coil mounting bracket and position the bracket aside.

10. Disconnect the vacuum line at the rear of the cylinder head. Remove the cylinder head bolts, the cylinder head and gasket.

To install:

11. Thoroughly clean and dry all bolts, bolt holes and mating surfaces. Install the head gasket and carefully position the cylinder head in place.

12. Coat the cylinder head bolts with thread lock compound and start all the cylinder head bolts finger-tight.

13. Tighten the cylinder head bolts, in sequence, to the following:
   a. Tighten to 35 ft. lbs. (47 Nm).
   b. Tighten each bolt an additional 130 degrees.
   c. Tighten the 4 center bolts an additional 30 degrees.

14. Connect the vacuum line to the rear of the cylinder head. Install the A/C bracket bolt to the head.

15. Install the ignition assembly and alternator bracket.

16. Install the pushrods, rocker arms and pedestals and tighten the bolts to 28 ft. lbs. (38 Nm).

17. Install the exhaust manifold; tighten the bolts and studs to 38 ft. lbs. (52 Nm).

18. Install the exhaust manifold support bracket. Install the exhaust manifold heat shield and tighten the nuts to 20 ft. lbs. (27 Nm).

19. Install the cooling fan assembly. Install the oil level indicator and tighten the nut to 20 ft. lbs. (27 Nm).

20. Install the exhaust crossover pipe. Install the spark plugs and tighten to 11 ft. lbs. (15 Nm).

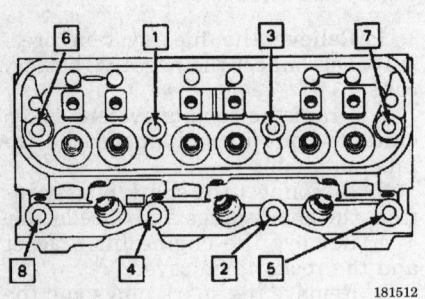

Cylinder head torque sequence — 3.3L engine

21. Install the rocker arm cover. Install the engine lift bracket.

22. Connect spark plugs wires to the wire looms and spark plugs. Install the intake manifold.

23. Connect the negative battery cable. Start the vehicle and verify no leaks.

Valve Lifters

REMOVAL AND INSTALLATION

— CAUTION —

Fuel injection systems remain under pressure, even after the engine has been turned OFF. The fuel system pressure must be relieved before disconnecting any fuel lines. Failure to do so may result in fire and/or personal injury.

NOTE: Be sure to keep all valve train parts in order so they may be reinstalled in their original locations and with the same mating surfaces as when removed.

3.1L and 3.3L Engines

1. Disconnect the negative battery cable.
2. Relieve the fuel system pressure.
3. Drain the cooling system.
4. Remove the rocker arm covers and the intake manifold.
5. On the 3.1L engine, remove the rocker arm retaining nuts, rocker arm balls, rocker arms and pushrods.
6. On the 3.3L engine, remove the rocker arm mounting bolts (make note of the rocker arm mounting studs), the rocker arms and pedestals.
7. Remove the guide bolts from the lifter guide and remove the guide.
8. Remove the valve lifter(s) from the lifter bores.

To install:

9. Lubricate the bearing surfaces with Molykote® or equivalent.

NOTE: Installation of a new camshaft or a wear pattern on the old valve lifter will require the replacement of the camshaft and lifters together. If camshaft replacement is not necessary, make sure to install the used valve lifters in their original position upon reinstallation.

10. Install the lifters in their original locations.
11. Install the lifter guide and lifter guide bolts and tighten the guide bolts to 89 inch lbs. (10 Nm) for the 3.1L engine or to 22 ft. lbs. (30 Nm) on the 3.3L engine.
12. On the 3.1L engine, perform the following:
    a. Install the pushrods, rocker arms, rocker balls and rocker arm nuts.
    b. Tighten the rocker arm nuts to 18 ft. lbs. (25 Nm) on 1994–95 engines or to 89 inch lbs. (10 Nm) plus an additional 30 degrees on 1996–97 engines.
13. On the 3.3L engine, perform the following procedure:
    a. Install rocker arms, pedestals and bolts.
    b. Apply thread lock compound to bolt threads before assembly.
    c. Tighten the mounting bolts to 11 ft. lbs. (15 Nm) plus 110 degrees additional rotation.
14. Install the intake manifold and rocker arm covers.
15. Refill the cooling system. An oil and filter change is recommended.
16. Connect the negative battery cable. Start the engine and verify that there are no leaks.

Valve Lash

ADJUSTMENT

3.1L Engine

NOTE: These engines come from the factory with non-adjustable valve lash, BUT if the valves and the valve seats are reconditioned adjustable rocker arm studs must be installed. The adjustment procedure is ONLY for adjustable rocker arm studs.

The engines are originally equipped with hydraulic valve lifters. Hydraulic valve lifters are not adjustable. The replacement rocker arm stud is adjustable and the adjustment can be performed as listed below. If valve

system noise is present, check the torque on the rocker arm nuts. The correct torque is 14–20 ft. lbs. (19–27 Nm). If noise is still present, check the condition of the camshaft, lifters, rocker arms, pushrods and valves.

1. Install the rocker arms, balls and nuts. Tighten the rocker arm nuts until all of the lash (free-play) is eliminated.
2. Adjust the valves when the lifter is on the base circle of a camshaft lobe as follows:
    a. Place the engine in the No. 1 firing position. This can be determined by placing a finger on the No. 1 rocker arm as the engine alignment mark on the front face of the of the torsional dampener pulley aligns with the arrow on the front cover. If the valves don't move freely, the engine is in the No. 1 firing position. If the valves move freely, the engine is in the No. 4 firing position and the engine should be rotated 1 full rotation to reach the No. 1 position.
    b. With the engine in the No. 1 firing order, adjust the following valves:
    • Exhaust — 1, 2, 3
    • Intake — 1, 5, 6
    c. Loosen the adjusting nut several turns backwards and then tighten until all lash (free-play) is removed. After all of the valve lash is removed turn the nut an additional 1½ turns, this will center the lifter plunger.
    d. Turn the engine 1 complete revolution until the timing tab and the alignment mark are again aligned, this will place the engine in the No. 4 firing position.
    e. With the engine in the No. 4 firing order, adjust the following valves:
    • Exhaust — 4, 5, 6
    • Intake — 2, 3, 4

Rocker Arms and Shafts

REMOVAL AND INSTALLATION

3.1L Engine

1. Disconnect the negative battery cable.
2. For the left side (front), perform the following procedures:
    a. Drain the cooling system to a level below the coolant pipe on the front of the engine.
    b. Remove the coolant bypass hose clamp at the coolant tube.
    c. Remove the 2 bolts and nut securing the coolant tube to the cyl-

inder head and position the tube aside.

d. Disconnect the PCV valve from the rocker arm cover.

3. For the right side (rear), perform the following procedures:

a. Disconnect the spark plug wires from the spark plugs and upper intake plenum wire retainer and position aside.

b. Disconnect the power brake booster vacuum pipe from the intake plenum.

c. Remove the serpentine belt.

d. Remove the alternator.

e. Disconnect and remove the ignition assembly and EVAP canister purge solenoid as an assembly.

4. Remove the 4 rocker arm cover bolts and remove the rocker arm cover.

5. Remove the rocker arm nuts, balls, rocker arms and pushrods.

To install:

6. Clean all the gasket surfaces completely.

7. Coat all the valve train components with engine oil prior to installation.

8. Install the pushrods and install the rocker arms on the studs. Install the rocker arm balls and mounting nuts. Make sure the pushrods are properly seated in the lifter and rocker arm. Tighten the mounting nuts to 18 ft. lbs. (24 Nm) for 1994–95 vehicles or to 89 inch lbs. (10 Nm) plus an additional 30 degrees for 1996–97 vehicles.

9. Install the rocker arm cover using a new gasket and tighten the rocker cover bolts to 90 inch lbs. (10 Nm).

10. For the left side (front), perform the following procedures:

a. Connect the PCV valve to the rocker arm cover.

b. Position the coolant tube and connect the thermostat bypass hose.

c. Install the coolant tube mounting nut and bolts. Tighten the screw at the water pump to 106 inch lbs. (12 Nm), the bolt at the corner of the cylinder head to 18 ft. lbs. (25 Nm) and the nut to 18 ft. lbs. (25 Nm).

11. For the right side (rear), perform the following procedures:

a. Install the alternator.

b. Install the serpentine belt.

c. Connect the power brake booster vacuum pipe to the plenum.

d. Install the EVAP solenoid and ignition assembly.

e. Connect the spark plug wires to the wire retainers on the plenum and the spark plugs.

12. Refill the cooling system.

13. Connect the negative battery cable.

14. Start the vehicle and verify no leaks.

3.3L Engine

1. Disconnect the negative battery cable.

2. Remove the serpentine belt.

3. For the left side (front), perform the following procedures:

a. Remove the bolt from the alternator brace.

b. Remove the alternator brace to intake manifold and brace.

4. For the right side (rear), perform the following procedures:

a. Loosen the power steering pump bolts and slide the pump forward.

b. Remove the power steering pump braces.

5. Tag and disconnect the spark plug wires from the plugs and move the wires away from the rocker arm cover.

6. Remove the 6 rocker arm cover bolts.

7. Remove the rocker arm cover.

8. Remove the rocker arm mounting bolt(s).

9. Remove the pedestal(s) and rocker arm(s).

10. If all 6 rocker arms are being removed from one head, remove the pushrods and pushrod guide from the cylinder head.

To install:

11. Clean all gasket surfaces completely.

12. Coat all the valve train components with engine oil prior to installation. Clean all thread locking compound from the rocker arm bolts and apply new thread locking compound.

13. Install the pushrod guide and pushrods.

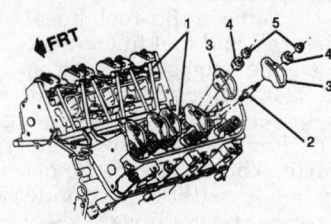

1 Pushrod
2 Valve rocker arm stud
3 Valve rocker arm
4 Valve rocker arm pivot ball
5 Valve rocker arm nut

222581

Rocker arm components — 3.1L engine

14. Install the rocker arm(s) and pedestal(s) and tighten the mounting bolt(s) to 28 ft. lbs. (38 Nm).

15. Install the rocker arm cover with a new gasket and tighten the rocker arm cover mounting bolts to 88 ft. lbs. (10 Nm).

16. Connect the spark plug wires to the plugs.

17. For the left side (front), install the rear alternator brace and tighten the bolts to 18 ft. lbs. (25 Nm).

18. For the right side (rear), install the power steering pump braces and tighten the bolts to 19 ft. lbs. (26 Nm).

19. Install the serpentine belt.

20. Connect the negative battery cable.

Intake Manifold

REMOVAL AND INSTALLATION

Starting with the 1996 Model Year, these vehicles were filled at the factory with a new type of antifreeze/coolant called GM Goodwrench DEX-COOL™. When adding coolant to 1996–97 vehicles, it is important that you use GM Goodwrench DEX-COOL™ (orange-colored, silicate-free) coolant. **Propylene glycol is not recommended for use in GM vehicles.** A 50/50 mixture of DEX-COOL™ and clean water will provide all the recommended protection for 1996–97 vehicles. **DO NOT use DEX-COOL™ in pre-1996 vehicles. DO NOT mix DEX-COOL™ with any other type of antifreeze.**

— CAUTION —

Fuel injection systems remain under pressure, even after the engine has been turned OFF. The fuel system pressure must be relieved before disconnecting any fuel lines. Failure to do so may result in fire and/or personal injury.

2.3L and 2.4L Engines

1. Relieve the fuel system pressure. Disconnect the negative battery cable.

2. Drain the cooling system to a level below the intake manifold. Disconnect the vacuum hose from the MAP sensor.

3. Disconnect the following electrical connectors from the MAP sensor, Intake Air Temperature (IAT) sensor, EVAP cannister purge solenoid and fuel injectors.

4. Disconnect the vacuum hoses from the intake manifold, fuel pres-

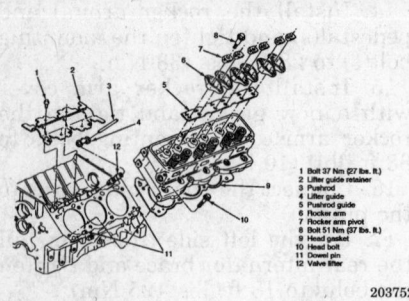

Exploded view of the valve train components — 3.3L engine

sure regulator and EVAP cannister purge solenoid to the canister.

5. Remove the air cleaner duct. Remove the vent tube-to-air cleaner duct. Remove the control cable mounting bracket from the intake manifold.

6. Disconnect the power brake vacuum line from the intake manifold and remove the vacuum line mounting bracket at the power steering pump.

7. Disconnect the coolant lines from the throttle body. Remove the crankcase air/oil separator mounting bolts from the intake manifold.

8. Disconnect the hoses from the separator and the separator. Remove the oil fill tube.

9. Remove the intake manifold support brace, the intake manifold nuts/bolts and discard the old gasket.

To install:

10. Clean all the gasket surfaces completely. Install the intake manifold gasket and intake manifold.

11. Tighten the intake manifold fasteners in sequence, to 18 ft. lbs. (25 Nm).

12. Install the intake manifold brace and tighten the mounting bolts to 19 ft. lbs. (26 Nm). The brace-to-block bolts must be tightened first then the brace-to-manifold bolt.

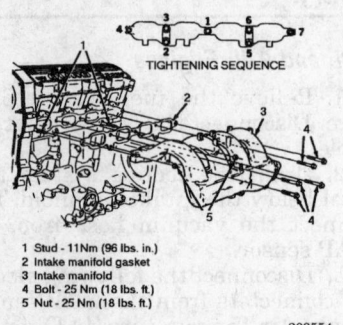

1 Stud - 11Nm (96 lbs. in.)
2 Intake manifold gasket
3 Intake manifold
4 Bolt - 25 Nm (18 lbs. ft.)
5 Nut - 25 Nm (18 lbs. ft.)

203554

Intake manifold bolt tightening sequence — 2.3L and 2.4L engine

13. Complete the installation by reversing the removal procedures.

14. Lubricate a new oil fill tube O-ring seal with engine oil and install tube between No. 1 and 2 intake tubes. Rotate as necessary to gain clearance for oil/air separator nipple on fill tube.

15. Refill the cooling system. Connect the negative battery cable.

3.1L Engine

These vehicles use a two-piece intake manifold. Note that these pieces are cast aluminum. Use care when working with light alloy components.

1. Disconnect the negative battery cable. Relieve the fuel system pressure.

2. Remove the top half of the air cleaner assembly and throttle body duct. Drain and recover the cooling system.

3. Remove the EGR pipe from exhaust manifold. Remove the serpentine belt.

4. Remove the brake vacuum pipe at the intake plenum. Disconnect the control cables from the throttle body and intake plenum mounting bracket.

5. Remove the power steering lines at the alternator bracket. Remove the alternator.

6. Disconnect the spark plug wires from the spark plugs and wire retainers on the intake plenum.

7. Remove the Ignition assembly and the EVAP canister purge solenoid together.

8. Disconnect the upper engine wiring harness connectors from the Throttle Position Sensor (TPS), Idle Air Control (IAC), fuel Injectors, coolant temperature sensor, MAP sensor and Camshaft Position (CMP) sensor.

9. Disconnect the vacuum lines from the vacuum modulator, fuel pressure regulator and PCV valve.

10. Remove the MAP sensor from upper intake manifold. Remove the upper intake plenum mounting bolts and the plenum.

11. Disconnect the fuel lines from the fuel rail and fuel line bracket. Install engine support fixture special tool J-28467-A or an equivalent.

12. Remove the right side engine mount. Remove the power steering mounting bolts and support the pump aside without disconnecting the power steering lines.

13. Disconnect the coolant inlet pipe from coolant outlet housing. Remove the coolant bypass hose from the water pump and the cylinder head.

14. Disconnect the upper radiator hose at thermostat housing.

15. Remove the thermostat housing. Remove both rocker arm covers.

16. Remove the lower intake manifold bolts. Make sure the washers on the 4 center bolts are installed in their original locations.

NOTE: When removing the valve train components they should be kept in order for installation the original locations.

17. Remove the rocker arm nuts, rocker arms and pushrods. Remove the intake manifold from the engine.

To install:

18. Clean gasket material from all mating surfaces. Remove all excess RTV sealant from front and rear ridges of cylinder block.

19. Place a 3mm bead of RTV, on each ridge, where the front and rear of the intake manifold contact the block.

20. Using a new gasket, install the intake manifold to the engine.

21. Install the pushrods, rocker arms and mounting nuts. Make sure the pushrods are properly seated in the valve lifters and rocker arms.

22. Install rocker arm nuts and tighten the rocker arm nuts to 18 ft. lbs. (24 Nm).

23. Install lower the intake manifold attaching bolts. Apply sealant PN 12345739 or equivalent to the threads of bolts, and torque bolts to 115 inch lbs. (13 Nm).

24. Complete the installation by reversing the removal procedures.

25. Install the upper intake manifold and torque the mounting bolts to 18 ft. lbs. (25 Nm).

26. Fill the cooling system. An engine oil and filter change is recommended.

27. Connect the negative battery cable. Start the vehicle and verify no leaks.

3.3L Engine

1. Relieve the fuel system pressure.

2. Disconnect the negative battery cable.

3. Drain the cooling system to a level below the intake manifold.

4. Remove the air cleaner assembly.

5. Remove the serpentine belt.

6. Disconnect and remove the alternator.

7. Remove the rear power steering pump brace.

8. Tag and disconnect the vacuum lines from the intake manifold.

9. Disconnect the upper radiator hose from the thermostat housing.

10. Disconnect the bypass and heater hoses from the intake manifold.

11. Disconnect the following electrical connectors from the Throttle Position Sensor (TPS), Idle Air Control (IAC), fuel injector, Coolant Temperature Sensor (CTS) and Mass Air Flow Sensor (MAF).

12. Disconnect the fuel line quick disconnects from the fuel rail.

13. Remove the fuel line bracket from the intake manifold.

14. Disconnect the throttle cable from the throttle body lever.

15. Disconnect the TV cable from the throttle body lever and remove the TV cable bracket from the intake manifold.

16. Remove the throttle cable bracket from the intake manifold.

17. Disconnect the rear spark plug wires from the plugs and route aside.

18. Disconnect the heater hose from the throttle body.

19. Remove the intake manifold bolt securing the power steering pump rear brace to the intake manifold.

20. Remove the manifold vacuum line clamp bolt at the alternator bracket.

21. Remove the intake manifold bolts and manifold.

To install:

22. Clean all the gasket surfaces completely.

23. Apply gasket sealer to the intake manifold end gaskets.

24. Install the intake manifold side gaskets and intake manifold.

25. Complete the installation by reversing the removal procedure.

26. Install the remainder of the intake manifold bolts after coating the threads with thread lock compound.

27. Tighten the bolts **TWICE** to 89 inch lbs. (10 Nm) in the proper sequence.

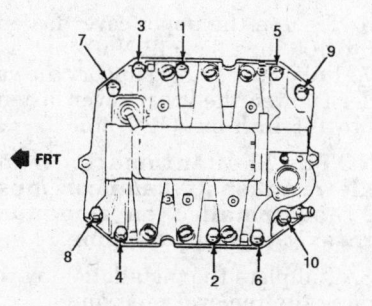

Intake manifold bolt tightening sequence — 3.3L engine

203608

28. Connect the negative battery cable.

29. Pressurize the fuel system and verify no leaks.

30. Start the vehicle and check for oil and coolant leaks.

Exhaust Manifold

REMOVAL AND INSTALLATION

2.3L and 2.4L Engines

NOTE: There are 2 different exhaust manifolds used on the Quad 4 engine. While the manifolds vary greatly in appearance, the removal and installation procedures are the same. The cast iron manifold is a single outlet design and the sheetmetal manifold is a tubular construction with a dual outlet flange.

1. Disconnect the negative battery cable. Disconnect the oxygen sensor connector. Raise and safely support the vehicle.

2. Remove the bolt that attaches the exhaust manifold brace to the manifold.

3. Remove the manifold to exhaust pipe spring loaded bolts using a

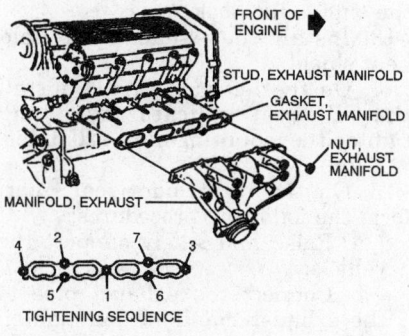

Exhaust manifold assembly — 2.3L engine

203518

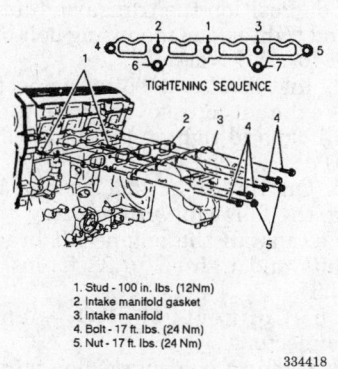

1. Stud - 100 in. lbs. (12Nm)
2. Intake manifold gasket
3. Intake manifold
4. Bolt - 17 ft. lbs. (24 Nm)
5. Nut - 17 ft. lbs. (24 Nm)

334418

Exhaust manifold assembly — 2.4L engine

13mm box wrench. The nuts should be loosened alternately and evenly to prevent the pipe from binding the nuts.

4. Pull down and back on the exhaust pipe to disengage it from the exhaust manifold bolts.

NOTE: On the 2.4L engine, do not bend the exhaust flex decoupler more than 3 degrees in any direction. Movement of more than 3 degrees will damage the flex decoupler.

5. On the 2.3L engine, remove the upper and lower exhaust manifold heat shields.

6. Lower the vehicle. Remove the exhaust manifold mounting nuts and manifold.

To install:

7. Clean all the gasket surfaces completely. Install the exhaust manifold and new gaskets.

8. Install the exhaust manifold-to-cylinder head retaining nuts and tighten in sequence to 31 ft. lbs. (42 Nm) on the 2.3L engine or to 110 inch lbs. (13 Nm) on the 2.4L engine.

9. Raise and safely support the vehicle. Install the heat shields. Install exhaust manifold brace to the manifold bolt.

10. Install the manifold-to-exhaust pipe nuts and tighten evenly to 19 ft. lbs. (26 Nm) on the 2.3L engine or to 26 ft. lbs. (35 Nm) on the 2.4L engine.

11. Lower the vehicle. Connect the oxygen sensor connector.

12. Connect the negative battery cable. Start the vehicle and verify no exhaust leaks.

3.1L Engine

1. Disconnect the negative battery cable. Remove the top half of the air cleaner assembly and throttle cable duct.

2. On the left side (front), perform the following procedures:

   a. Partially drain the cooling system.

   b. Remove the radiator hose from the thermostat housing.

   c. Remove the coolant bypass hose at the coolant pump and from the exhaust manifold.

3. Remove the exhaust crossover heat shield. Remove the exhaust crossover pipe from the manifold.

4. On the right side (rear), perform the following procedures:

   a. Remove the heated oxygen sensor.

   b. Remove the EGR pipe from the exhaust manifold.

   c. Raise and safely support the vehicle.

d. Remove the transaxle oil fill tube and lever indicator assembly.

e. Disconnect the front exhaust pipe from the exhaust manifold.

f. Disconnect the exhaust pipe from the converter flange and support the converter.

g. Remove the converter heat shield from the body.

5. Tag and disconnect the secondary ignition wires from the spark plugs.

6. Remove the exhaust manifold heat shield. Remove the exhaust manifold retaining nuts and manifold.

To install:

7. Clean mating surfaces at the cylinder head and manifold.

8. Install the exhaust manifold gasket and the exhaust manifold. Torque the manifold mounting nuts to 12 ft. lbs (16 Nm).

9. Install the exhaust manifold heat shield. Install the exhaust crossover pipe to the manifold.

10. On the right side (rear), install the converter heat shield to the body.

11. Install the exhaust crossover pipe heat shield.

12. On the left side (front), perform the following procedures:

a. Connect the secondary ignition wires to the appropriate spark plug.

b. Connect the coolant bypass pipe to the coolant pump and exhaust manifold.

c. Install the radiator hose to the coolant outlet housing.

13. On the right side (rear), perform the following procedures:

a. Connect the exhaust pipe to the exhaust manifold.

b. Install the transaxle oil level indicator and fill tube assembly.

c. Safely lower the vehicle.

d. Connect the heated oxygen sensor.

e. Connect the EGR pipe to exhaust manifold.

14. Install the top half of the air cleaner and the throttle body duct. Connect the negative battery cable.

3.3L Engine

1. Disconnect the negative battery cable.

2. On the left side (front), remove the air cleaner inlet duct.

3. Tag and disconnect the spark plug wires from the spark plugs and position aside.

4. On the right side (rear), perform the following procedures:

a. Disconnect the oxygen sensor (O_2) connector.

b. Disconnect the control cables from the throttle body lever and intake manifold bracket.

c. Disconnect the power brake booster vacuum line from the intake manifold.

d. Remove the engine lift bracket nuts from the exhaust manifold.

5. Remove the 2 bolts connecting the exhaust crossover pipe to the manifold.

6. On the left side (front), remove the oil level indicator and tube.

7. On the right side (rear), perform the following procedures:

a. Remove the transaxle fill tube bolt and reposition the tube for access.

b. Raise and safely support the vehicle.

c. Disconnect the exhaust pipe from the exhaust manifold.

d. Lower the vehicle.

8. Remove the exhaust manifold heat shield.

9. On the left side (front), remove the engine lift hook.

10. Remove the exhaust manifold mounting bolts and studs and remove the exhaust manifold.

To install:

11. Install the exhaust manifold and mounting studs and bolts and tighten to 38 ft. lbs. (52 Nm).

12. On the left side (front), install the engine lift hook.

13. Install the exhaust manifold heat shield.

14. On the left side (front), install the oil level indicator tube and tighten the mounting nut to 20 ft. lbs. (27 Nm).

15. On the right side (rear), perform the following procedures:

a. Raise and safely support the vehicle.

b. Connect the exhaust pipe to the exhaust manifold and tighten the mounting bolts to 22 ft. lbs. (30 Nm).

c. Lower the vehicle.

d. Position the transaxle fill tube and tighten the mounting bolt to 18 ft. lbs. (25 Nm).

16. Install the 2 bolts connecting the exhaust crossover pipe to the manifold and tighten to 19 ft. lbs. (26 Nm).

17. On the right side (rear), perform the following procedures:

a. Install the engine lift bracket nuts and tighten to 22 ft. lbs. (30 Nm).

b. Connect the O_2 sensor connector.

18. Connect the spark plug wires.

19. On the left side (front), install the air cleaner inlet duct.

20. On the right side (rear), perform the following procedures:

a. Connect the control cables to the throttle body lever and intake manifold bracket.

b. Connect the power booster vacuum line to the intake manifold.

21. Connect the negative battery cable. Start the vehicle and verify no exhaust leaks.

Front Cover Oil Seal

REMOVAL AND INSTALLATION

2.3L and 2.4L Engines

The front cover oil seal is installed from the engine side of the front cover. This means the front timing chain cover must be removed from the engine.

1. Disconnect the negative battery cable. Drain the cooling system until the surge tank is empty.

2. Remove the timing chain front cover from the engine.

3. To prevent distorting and damaging the front cover when removing the oil seal, use the following procedure.

a. Working on a clean benchtop, support the front cover over 2 blocks of wood of equal thickness. Place the cover on the blocks so the engine side of the cover is resting on the blocks. The blocks should be as close to the front oil seal as possible without interfering with removing the oil seal.

b. Carefully drive the seal out of the front cover.

To install:

4. Install a new oil seal using J-36010 or an equivalent seal installation driver and alignment tool. The seal must be driven in from the engine side of the timing cover.

5. Install the front cover on the engine. Install the mounting nuts on the studs then install the mounting bolts.

6. Tighten the upper cover fasteners to 106 inch lbs. (12 Nm).

7. Raise and safely support the vehicle. Tighten the lower cover fasteners to 106 inch lbs. (12 Nm).

NOTE: The automatic transaxle crankshaft balancer must NOT be installed on a manual transaxle equipped engine.

8. Complete the installation by reversing the removal procedures.

9. Refill the cooling system. An engine oil and filter change is recommended. Check the engine oil level.

10. Connect the negative battery cable. Start the vehicle and verify no leaks.

3.1L and 3.3L Engines

1. Disconnect the negative battery cable. Remove the serpentine belt.

2. Raise and safely support the vehicle. Remove the right front tire and wheel assembly.

3. Remove the right inner fender well splash shield. Remove the torsional dampener.

4. On the 3.3L engine, remove the 4 crankshaft sensor shield mounting bolts and remove the shield.

5. Pry the seal out of the front cover using a suitable prying tool. Use care not to damage the crankshaft.

To install:

6. Clean out the oil seal recess in the front cover.

7. Install the new seal in the front cover using J-34995, or an equivalent seal driver installation tool, to fully seat the seal in the cover.

8. On the 3.3L engine, install the crankshaft sensor shield and 4 mounting bolts.

9. Install the torsional damper. Install the right inner fender splash shield.

10. Install the tire and wheel assembly. Lower the vehicle.

11. Install the serpentine belt. Connect the negative battery cable.

12. Start the vehicle and verify no coolant leaks or oil leaks.

Timing Chain, Sprockets and Front Cover

REMOVAL AND INSTALLATION

2.3L and 2.4L Engines

NOTE: It is recommended that the entire procedure be reviewed before attempting to service the timing chain.

1. Disconnect the negative battery cable. Drain the cooling system into a suitable container. Remove the coolant surge tank.

2. Remove the serpentine drive belt using a 13mm wrench that is at least 24 in. (61cm) long.

3. Disconnect and remove the alternator from the mounting bracket.

4. Install a suitable engine support, J-28467-A or the equivalent.

5. Remove the upper cover fasteners. Disconnect the upper cover vent hose.

6. Remove the right engine mount and bracket. Raise and safely support the vehicle.

7. Remove the right front tire and wheel assembly. Remove the right lower splash shield from the right wheel house.

8. Remove the crankshaft balancer assembly using the following procedure:

a. Remove the torsional damper mounting bolt and washer while holding the crankshaft in place using J-38122, or the equivalent.

b. Remove the balancer using a suitable puller, J-24420-B or the equivalent.

9. Remove the lower cover fasteners. Lower the vehicle.

10. Remove the front cover and gaskets. Remove the crankshaft oil slinger.

11. Rotate the crankshaft clockwise, as viewed from front of engine (normal rotation) until the camshaft sprocket's timing dowel pin holes align with the holes in the timing chain housing. The mark on the crankshaft sprocket should align with the mark on the cylinder block. The crankshaft sprocket keyway should point upwards and align with the center line of the cylinder bores. This is the normal timed position.

12. Remove the timing chain guides. Raise and safely support the vehicle.

13. Gently pry off timing chain tensioner spring retainer and remove spring.

NOTE: Two styles of tensioner are used. Early production engines will have a spring post and late production ones will not. Both styles are identical in operation and are interchangeable.

14. Remove the timing chain tensioner shoe retainer.

15. Make sure all the slack in the timing chain is above the tensioner assembly; remove the chain tensioner shoe. The timing chain must be disengaged from the wear grooves in the tensioner shoe in order to remove the shoe. Slide a prybar under the timing chain while pulling shoe outward.

16. If difficulty is encountered removing chain tensioner shoe, proceed as follows:

a. Lower the vehicle.

b. Hold the intake camshaft sprocket with a holding tool and remove the sprocket bolt and washer.

c. Remove the washer from the bolt and re-thread the bolt back into the camshaft by hand, the bolt provides a surface to push against.

d. Remove intake camshaft sprocket using a 3-jaw puller in the 3 relief holes in the sprocket. Do not attempt to pry the sprocket off the camshaft or damage to the sprocket or chain housing could occur.

17. Remove the tensioner assembly attaching bolts and the tensioner.

CAUTION
The tensioner piston is spring loaded and could fly out causing personal injury.

18. Remove the chain housing to block stud, which is actually the timing chain tensioner shoe pivot.

19. Remove the timing chain.

To install:

20. Tighten intake camshaft sprocket attaching bolt and washer, while holding the sprocket with tool J 36013, if removed.

21. Install the special tool through holes in camshaft sprockets into holes in timing chain housing. This positions the camshafts for correct timing.

22. If the camshafts are out of position and must be rotated more than $\frac{1}{8}$ turn in order to install the alignment dowel pins:

a. The crankshaft must be rotated 90 degrees clockwise off TDC in order to give the valves adequate clearance to open.

b. Once the camshafts are in position and the dowels installed, rotate the crankshaft counterclockwise back to TDC. Do not rotate the crankshaft clockwise to TDC or valve and piston damage could occur.

23. Install the timing chain over the exhaust camshaft sprocket, around the idler sprocket and around the crankshaft sprocket.

24. Remove the alignment dowel pin from the intake camshaft. Using a dowel pin remover tool, rotate the intake camshaft sprocket counterclockwise enough to slide the timing chain over the intake camshaft sprocket. Release the camshaft sprocket wrench. The length of chain between the 2 camshaft sprockets will tighten. If properly timed, the intake camshaft alignment dowel pin should slide in easily. If the dowel pin does not fully index, the camshafts are not timed correctly and the procedure must be repeated.

25. Leave the alignment dowel pins installed.

26. With slack removed from chain between intake camshaft sprocket and crankshaft sprocket, the timing marks on the crankshaft and the cyl-

inder block should be aligned. If marks are not aligned, move the chain 1 tooth forward or rearward, remove slack and recheck marks.

27. Tighten the chain housing to block stud. The stud is installed under the timing chain. Tighten to 19 ft. lbs. (26 Nm).

28. Reload timing chain tensioner assembly to its 0 position as follows:

a. Assemble restraint cylinder, spring and nylon plug into plunger. Index slot in restraint cylinder with peg in plunger. While rotating the restraint cylinder clockwise, push the restraint cylinder into the plunger until it bottoms. Keep rotating the restraint cylinder clockwise but allow the spring to push it out of the plunger. The pin in the plunger will lock the restraint in the loaded position.

b. Install tool J 36589 or equivalent, onto plunger assembly.

c. Install plunger assembly into tensioner body with the long end toward the crankshaft when installed.

29. Install the tensioner assembly to the chain housing. Recheck plunger assembly installation. It is correctly installed when the long end is toward the crankshaft.

30. Install and tighten timing chain tensioner bolts and tighten to 10 ft. lbs. (14 Nm).

31. Install the tensioner shoe and tensioner shoe retainer. Remove special tool J 36589 and squeeze plunger assembly into the tensioner body to unload the plunger assembly.

32. Lower vehicle and remove the alignment dowel pins. Rotate crankshaft clockwise 2 full rotations. Align crankshaft timing mark with mark on cylinder block and reinstall alignment dowel pins. Alignment dowel pins will slide in easily if engine is timed correctly.

—————— WARNING ——————
If the engine is not correctly timed, severe engine damage could occur.

33. Install the timing chain guides and crankshaft oil slinger.

34. Install the front cover and gaskets on the engine and tighten the mounting nuts and bolts to 106 inch lbs. (12 Nm).

35. Install the torsional damper as follows:

a. Coat the seal contact area on the crankshaft damper with clean engine oil.

b. Line up the damper on the crankshaft so the notch in the damper lines up with the crankshaft key.

c. Tap the balancer into place using a rubber mallet.

d. Install the damper mounting bolt and washer and tighten the bolt to 129 ft. lbs. (175 Nm) plus 90 degrees rotation.

36. Install the right front lower splash shield. Install the tire and wheel assembly and tighten to specification.

37. Safely lower the vehicle. Install the right engine mount bracket.

38. Install the right engine mount. Connect the upper cover vent hose.

39. Remove the engine support. Install the alternator and connect the electrical connectors.

40. Install the serpentine belt. Install the coolant surge tank.

41. Refill the cooling system. Connect the negative battery cable and check for leaks.

3.1L Engine

1. Disconnect the negative battery cable. Drain the cooling system into a suitable container.

2. Remove the right engine mount bracket. Remove the serpentine belt.

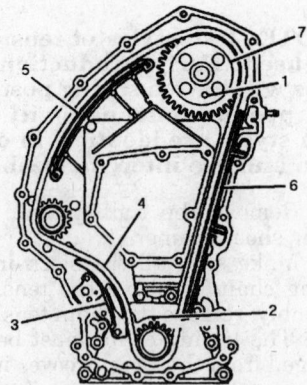

1. Camshaft timing alignment pin locations
2. Crankshaft gear timing marks
3. Shoe and tensioner assembly
4. Timing chain
5. RH timing chain guide
6. LH timing chain guide
7. Camshaft sprocket

172724

Timing chain and sprockets in the "Timed Position" — 2.3L SOHC engine

3. Remove the crankshaft balancer as follows:

a. Raise and safely support the vehicle.

b. Remove the right front tire and wheel assembly.

c. Remove the right inner fender well splash shield.

d. Remove the flywheel cover and install a flywheel holding tool.

e. Remove the balancer mounting bolt and washer.

f. Using a suitable puller, J-24420-B or the equivalent remove the balancer from the crankshaft.

4. Remove the serpentine belt tensioner mounting bolt and tensioner. Remove the oil pan.

5. Remove coolant bypass pipe from the water pump and the intake manifold.

6. Disconnect the lower radiator hose to from the front cover outlet. Remove the front cover bolts and cover.

7. Rotate the crankshaft until the timing marks on the camshaft and crankshaft sprockets are in alignment at their closest approach.

8. Remove the camshaft sprocket bolt, sprocket and timing chain.

9. Remove the crankshaft sprocket, using a gear puller J-5825-A or equivalent. Remove the timing chain damper bolts and damper.

To install:

10. Install the timing chain damper and tighten the mounting bolts to 15 ft. lbs. (21 Nm).

11. Install the crankshaft sprocket onto the crankshaft making sure the notch in the sprocket fits over the crankshaft key. Fully seat the sprocket on the crankshaft using J-38612, or an equivalent gear installer.

12. Make sure the timing mark on the crankshaft sprocket is pointing straight up.

13. Install the timing chain over the camshaft sprocket and hold the sprocket in such a way, that the timing mark is pointing down, and the timing chain is hanging down off the sprocket.

14. Loop the timing chain under the crankshaft sprocket and install the camshaft sprocket on the camshaft. The sprocket will only fit on the camshaft if the dowel on the camshaft lines up with the hole in the sprocket.

15. Verify that the marks are aligned (the camshaft sprocket will be at the 6 o'clock position and the

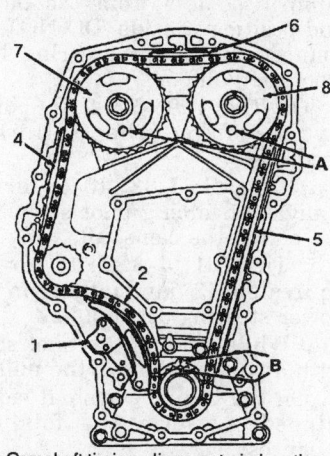

A. Camshaft timing alignment pin locations
B. Crankshaft gear timing marks
1. Shoe asm. timing chain tensioner
2. Timing chain
3. Timing chain tensioner
4. R.H. timing chain guide
5. L.H. timing chain guide
6. Upper timing chain guide
7. Exhaust camshaft sprocket
8. Intake camshaft sprocket

8838NG01

Timing chain and sprocket alignment positions — 2.3L and 2.4L DOHC engines

crankshaft sprocket will be in the 12 o'clock position).

16. On 1994–95 vehicles, tighten the camshaft sprocket mounting bolt to 74 ft. lbs. (100 Nm). On 1996–97 vehicles, tighten the bolt to 81 ft. lbs. (110 Nm).

17. Lubricate the timing chain components with engine oil. Clean all gasket surfaces completely.

18. Apply a thin bead of sealer around the gasket sealing area of the front cover. Install a new front cover seal on the front cover.

19. Install the front cover on the engine and install the mounting bolts. On 1994–95 vehicles, tighten the

small bolts to 18 ft. lbs. (24 Nm) and the large bolts to to 41 ft. lbs. (55 Nm). On 1996–97 vehicles, tighten the small bolts to 15 ft. lbs. (21 Nm) and the large bolts to 35 ft. lbs. (47 Nm).

20. Connect the radiator hose to the coolant outlet.

21. Install coolant by pass pipe to the water pump and the intake manifold. Install the oil pan.

22. Install crankshaft balancer as follows:

a. Coat the seal contact surface of the crankshaft balancer with clean engine oil.

b. Line up the notch in the balancer with the crankshaft key and slide the balancer on until the key is in the balancer.

c. Using J-29113 or an equivalent puller, seat the balancer on the crankshaft.

d. Install the balancer mounting bolt and washer and tighten to 76 ft. lbs. (103 Nm).

e. Install the flywheel cover.

f. Install the right inner fender well splash shield.

g. Install the tire and wheel assembly and tighten to specification.

23. Install the serpentine belt tensioner and tighten the mounting bolt to 40 ft. lbs. (54 Nm). Install the serpentine belt.

24. Install the right engine mount bracket and tighten the bracket-to-mount bolts to 96 ft. lbs. (130 Nm).

25. Refill the cooling system with the correct amount and type of coolant. Check the engine oil level. An oil and filter change is recommended.

26. Connect the negative battery cable. Start the engine and verify that there are no leaks.

3.3L Engine

1. Disconnect the negative battery cable. Drain the cooling system into a

suitable container. Remove the engine drive belt.

2. Disconnect the heater hose, bypass hose and lower radiator hose from the water pump and front cover.

3. Raise and safely support the vehicle. Remove the tire and wheel assembly. Remove the right inner fender well splash shield.

4. Remove the crankshaft balancer as follows:

a. Remove the crankshaft balancer bolt.

b. Using tool J-38197 or an equivalent balancer puller, remove the balancer from the crankshaft.

5. Remove the 4 crankshaft sensor shield mounting bolts and remove the shield.

6. Disconnect the electrical connectors from the crankshaft sensor, camshaft sensor and oil pressure switch.

7. Remove oil pan-to-front cover bolts. Remove front cover bolts and cover.

8. Rotate the crankshaft until the timing marks are in alignment. The crankshaft sprocket mark should point straight up at the camshaft mark which is pointing straight down. The marks should be on the closest approach.

9. Remove the timing chain damper assembly. Remove the camshaft sprocket mounting bolt.

10. Remove the camshaft sprocket, sprocket and chain assembly. Remove the crankshaft sprocket from the crankshaft using a suitable puller.

To install:

11. Install the crankshaft sprocket onto the crankshaft. It may be necessary to use a sprocket installer to fully seat the sprocket on the crankshaft. Make sure the crankshaft timing mark is pointing straight up.

12. Temporarily install the camshaft sprocket and rotate the camshaft as necessary to align the timing mark with the crankshaft timing mark.

13. Assemble timing chain on the camshaft sprocket. Hold the sprocket so the mark is pointing downward with the chain hanging off the sprocket.

14. Loop the timing chain under the crankshaft sprocket and with the marks in alignment install the camshaft sprocket on the camshaft.

15. Verify all timing marks are in alignment. If the marks are out of alignment repeat steps.

16. With the marks aligned, tighten the camshaft sprocket mounting bolts to 27 ft. lbs. (37 Nm).

VIEW A

#1 CYLINDER AT T.D.C.

NOTE - ALIGN TIMING MARKS ON CAM & CRANK SPROCKETS USING ALIGNMENT MARKS ON DAMPER STAMPING OR CAST ALIGNMENT MARKS ON CYL & CASE

1. Damper
2. Alignment marks
3. Timing chain
4. Crank sprocket
5. 28 Nm (21 lb. ft.)
6. Camshaft sprocket
7. 21 Nm (15 lb. ft.)

NOTE - CAMSHAFT SPROCKET MARK AT 6 O'CLOCK CRANKSHAFT SPROCKET MARK AT 12 O'CLOCK

333479

Timing chain and sprocket timing mark alignment — 3.1L engine

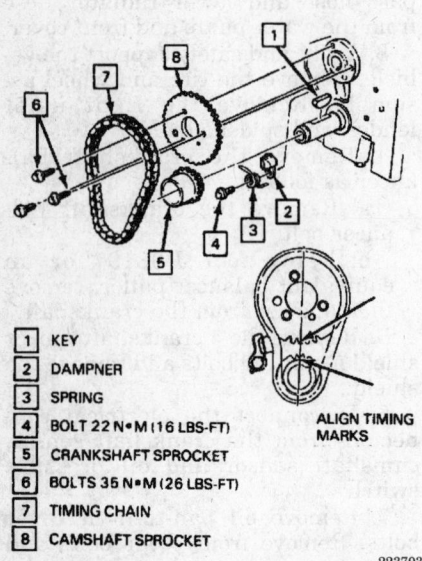

1	KEY
2	DAMPNER
3	SPRING
4	BOLT 22 N•M (16 LBS-FT)
5	CRANKSHAFT SPROCKET
6	BOLTS 35 N•M (26 LBS-FT)
7	TIMING CHAIN
8	CAMSHAFT SPROCKET

ALIGN TIMING MARKS

223703

Timing chain and sprocket components and timing mark alignment — 3.3L engine

17. Install the timing chain dampener assembly and tighten the mounting bolt to 14 ft. lbs. (19 Nm).

18. Clean all gasket surfaces completely. Apply a bead of sealer around the gasket contact area of the front cover and seat the new gasket to the cover.

19. Install the front cover on the engine and install the mounting bolts. Apply GM sealer 1052080 or equivalent to bolt threads and tighten the bolts to 22 ft. lbs. (30 Nm).

20. Install the oil pan-to-front cover bolts and tighten to 88 inch lbs. (10 Nm). Connect the electrical connectors to the camshaft sensor, crankshaft sensor and oil pressure switch.

21. Install and adjust the crankshaft sensor as follows:

a. Loosely install the crankshaft sensor on the pedestal.

b. Position the sensor with the pedestal attached on tool J-37089, or an equivalent.

c. Position the special tool on the crankshaft.

d. Install the pedestal-to-engine bolts and tighten to 21 ft. lbs. (29 Nm).

e. Tighten the sensor-to-pedestal pinch bolt to 38 inch lbs. (4.5 Nm).

f. Remove J-37089.

22. Install the crankshaft sensor shield and 4 mounting bolts.

23. Install the crankshaft balancer as follows:

a. Coat the seal contact area of the balancer with clean engine oil.

b. Line up the notch in the balancer with the crankshaft key. Slide the balancer on until the key is inside the notch.

c. Install the balancer bolt and tighten to 110 ft. lbs. (150 Nm) plus 76 degrees additional rotation. The balancer bolt being installed will fully seat the balancer on the crankshaft.

24. Install the right inner fender well splash shield. Install the tire and wheel assembly. Lower the vehicle.

25. Connect the heater hose, bypass hose and lower radiator hose to the water pump and timing cover.

26. Install the engine drive belt. Connect the negative battery cable. Check and top off the engine oil as necessary.

27. Fill and bleed the cooling system. Warm the engine to operating temperature and verify no leaks.

Camshaft

REMOVAL AND INSTALLATION

2.3L and 2.4L Engines

NOTE: Anytime the camshaft housing-to-cylinder head bolts are loosened or removed, the camshaft housing-to-cylinder head gasket must be replaced.

1. Disconnect the negative battery cable.

2. Detach the 11-pin connector from the ignition assembly cover.

3. Remove the 4 ignition assembly cover-to-camshaft housing bolts and remove assembly by pulling straight up. Use a spark plug boot wire remover to remove connector assemblies.

4. On the intake camshaft side, if equipped, perform the following procedures:

a. Remove the throttle lever actuator power steering pressure switch connector.

b. Loosen the power steering pump adjusting stud bolt. Loosen the power steering pump pivot bolts and remove drive belt.

c. Remove the 2 power steering pump bracket-to-transaxle bolts.

d. Remove the front power steering pump bracket-to-engine block bolt.

e. Remove the power steering pump assembly from the engine and position it aside. DO NOT disconnect the power steering lines from the pump.

5. On the 1993–94 2.3L engine only, perform the following procedures:

a. Install J-38343-4, or an equivalent forcing pilot screw into the end of the camshaft.

b. Install J-38781, or an equivalent 3 jaw puller on the power steering drive pulley.

c. While holding the pilot screw with a wrench remove the pulley.

6. On the intake camshaft side, if equipped, perform the following procedures:

a. Remove oil/air separator bolts and hoses. Leave the hoses attached to the separator but disconnect from the oil fill tube, timing chain housing and intake manifold.

b. Disconnect the vacuum line from fuel pressure regulator and disconnect the fuel injector harness connector.

c. Remove the 2 bolts securing the fuel line clamps to the intake camshaft housing.

d. Remove fuel rail-to-camshaft housing mounting bolts.

e. Remove the fuel rail from the cylinder head leaving the fuel lines attached and position the fuel rail aside.

7. Remove the timing chain and camshaft sprockets.

8. Remove the timing chain housing bolts but do not remove from the engine.

9. Remove camshaft housing cover-to-camshaft housing attaching bolts.

10. Remove the camshaft housing-to-cylinder head attaching bolts. Use the reverse of the tightening sequence when loosening camshaft housing to cylinder head attaching bolts. Leave 2 bolts loosely in place to hold the camshaft housing while separating camshaft cover from housing.

11. Push the cover off the housing by threading 4 of the housing-to-cylinder head bolts into the tapped holes in the cam housing cover. Tighten the bolts in evenly so the cover does not bind on the dowel pins.

12. Remove the 2 loosely installed camshaft housing-to-cylinder head bolts and remove the cover. Discard the gaskets.

13. Note the position of the chain sprocket dowel pin for reassembly. Remove the camshaft carefully; do not damage the camshaft oil seal.

14. Remove camshaft oil seal from camshaft and discard seal. This seal must be replaced any time the housing and cover are separated.

NOTE: The valve lifters must be kept in order for installation in the same locations they were removed from.

15. Remove the valve lifters from the camshaft housing.
16. Remove the camshaft carrier from the cylinder head and remove the gasket.
To install:
17. Clean all the gasket surfaces completely.
18. Install a new gasket on the cylinder head and install the camshaft housing over the dowel pins. Install one bolt loosely to hold the housing in place.

NOTE: If the camshaft was replaced the valve lifters must also be replaced.

19. Install the lifters into their original bores.
20. Lubricate camshaft lobes, journals and lifters with camshaft and lifter prelube. The camshaft lobes and journals must be adequately lubricated or engine damage could occur upon start up.
21. Install the camshaft in the same position it was in prior to removal. The timing chain sprocket dowel pin should be straight up and align with the center line of the lifter bores.
22. Install new camshaft housing-to-camshaft housing cover seals into the cover.

NOTE: The seals for the intake and exhaust covers are different and the correct seals must be used.

23. Remove the bolt holding the housing in place.
24. Apply thread locking compound to the camshaft housing and cover attaching bolt threads.
25. Install the camshaft housing cover.
26. Install all the mounting bolts finger-tight. With all the bolts in place tighten the bolts in sequence to 11 ft. lbs. (15 Nm) plus 75 degrees additional rotation (Long bolts) and 16 ft. lbs. (21 Nm) plus 25 degrees additional rotation (Short bolts).
27. Install the timing chain housing mounting bolts.
28. Install the timing chain and sprockets.
29. Install new fuel injector O-ring seals lubricated with oil and install

the fuel rail and tighten the mounting bolts to 19 ft. lbs. (26 Nm).
30. On the intake camshaft side, if equipped, perform the following procedures:
   a. Install the fuel line clamp mounting bolts on top of the intake camshaft housing.
   b. Connect the vacuum line to the fuel pressure regulator.
   c. Connect the fuel injector harness connector.
   d. Install the oil/air separator assembly and connect the hoses to the oil fill tube, timing chain housing and intake manifold.
31. Lubricate the inner sealing surface of the intake camshaft seal with oil and install the seal to the housing using J-36009, or an equivalent seal installer.
32. On the intake camshaft side, if equipped, perform the following procedures:
   a. Install the power steering pump pulley onto the intake camshaft using J-36015, or an equivalent pulley installer.
   b. Install the power steering pump and install the drive belt.
   c. Connect the throttle lever actuator power steering pressure switch connector.
33. On the exhaust camshaft side, perform the following procedures:
   a. Install the transaxle fill tube.
   b. Connect the electrical connector to the oil pressure switch.
34. Install the ignition assembly on the camshaft housing and tighten the mounting bolts to 13 ft. lbs. (18 Nm).
35. Attach the 11-pin connector to ignition assembly.
36. Connect the negative battery cable.
37. Start the vehicle and verify proper operation and no leaks.

3.1L Engine

1. Relieve the fuel system pressure. Disconnect the negative battery cable. Remove the engine assembly.

NOTE: When removing valve train components they must be marked for installation in the same location they are removed from. When the camshaft is being replaced the valve lifters should also be replaced.

2. Remove the intake manifold, valve cover, rocker arms, pushrods and valve lifters.
3. Remove the crankshaft balancer and front cover. Remove the timing chain and sprockets.

4. Remove the oil pump driven gear bolt and gear. Remove the bolts and camshaft thrust plate.
5. Carefully remove the camshaft. Avoid marring the camshaft bearing surfaces.
To install:
6. Coat the camshaft with lubricant 1052365 or equivalent, and install the camshaft.
7. Install the camshaft thrust plate and tighten bolts to 89 inch lbs. (10 Nm).
8. Install the oil pump driven gear and tighten bolt to 27 ft. lbs. (36 Nm).
9. Install the timing chain and sprocket.
10. Install the camshaft thrust button and front cover. Install the crankshaft balancer.
11. Install the intake manifold, valve cover, rocker arms, pushrods and valve lifters.
12. Install the engine assembly. Connect the negative battery cable.
13. Adjust the valves, as required. Start the engine and verify no oil leaks.

3.3L Engine

1. Relieve the fuel system pressure. Disconnect the negative battery cable.
2. Remove the intake manifold. Remove the rocker arm covers.

NOTE: When removing the valve train components they must be kept in order for installation in the same locations they were removed from.

3. Remove the rocker arm bolts, pedestals, rocker arms, pushrods and pedestal guides.
4. Remove the lifter guide retainer bolts, lifter guide retainer, lifter guides and lifters.
5. Remove the water pump pulley. Disconnect the lower radiator hose from the water pump.
6. Raise and safely support the vehicle. Remove the tire and wheel assemblies.
7. Remove the right inner fender well splash shield. Remove the flywheel covers.
8. Remove the balancer bolt and using tool J-37096, or an equivalent, to hold the crankshaft from turning.
9. Remove the balancer with a puller, J-38197, or the equivalent. Remove the crankshaft sensor shield.
10. Disconnect the crankshaft sensor and oil pressure sender connections.
11. Remove the front cover mounting bolts. Lower the vehicle.

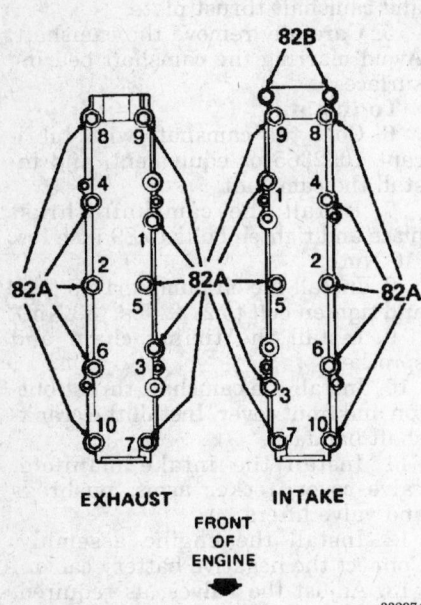

Camshaft housing bolt tightening sequence — 2.3L and 2.4L engines

12. Remove the heater pipe mounting nut and disconnect the heater pipe.

13. Remove the power steering pump brace and pump; position it aside. It is not necessary to disconnect the power steering hoses from the pump.

14. Disconnect the ground wires from the tensioner bracket. Remove the upper front cover bolts and cover.

15. Remove the timing chain and sprocket assembly.

16. Install an engine support fixture, J-28467-A and J-36462 or their equivalents.

17. Raise and safely support the vehicle. Remove the steering gear pinch bolt.

18. Remove the right ball joint pinch bolt and separate the ball joint from the steering knuckle.

19. Disconnect the wiring harness at the right front frame rail.

20. Disconnect the front exhaust pipe from the exhaust manifold.

21. Remove the cotter pin and castle nut from the tie rod end and separate the tie rod end from the right steering knuckle.

22. Install the jackstands to support the right side frame. Disconnect the knock and VSS electrical connectors.

23. Remove the right side subframe mounting bolts. Loosen the left side frame bolts.

24. Remove the right side air deflector screws and deflector. Lower the right side of the frame 6 inches.

25. Remove the camshaft thrust plate. Lower the right side of the engine and remove the camshaft.

To install:

26. Coat the camshaft with prelube, 1234551, or the equivalent. Carefully install the camshaft.

27. Install the thrust plate and tighten the mounting bolts to 11 ft. lbs. (15 Nm).

28. Raise the right side of the engine and frame into position. Install the right side frame bolt and tighten bolts to 103 ft. lbs. (150 Nm).

29. Connect the VSS and knock sensor electrical connectors. Remove the jackstands.

30. Connect the right tie rod end to the steering knuckle and tighten the castle nut to 52 ft. lbs. (70 Nm) and install a new cotter pin.

31. Connect the exhaust pipe to the exhaust manifold and tighten bolts to 22 ft. lbs. (30 Nm).

32. Connect the wiring harness at the right front frame rail.

33. Connect the ball joint to the steering knuckle and tighten the pinch bolt to 32 ft. lbs. (43 Nm).

34. Connect the steering shaft to the rack and pinion and tighten the pinch bolt to 35 ft. lbs. (48 Nm).

35. Lower the vehicle. Remove the engine support fixture.

36. Install the timing chain and sprockets. Install the front timing cover and tighten bolts to 22 ft. lbs. (30 Nm).

37. Connect the ground wire to the tensioner bracket. Install the power steering pump and tighten bolts to 21 ft. lbs. (29 Nm).

38. Install the power steering pump brace. Connect the heater pipe and install the nut.

39. Raise and safely support the vehicle. Install the oil pan-to-front cover bolts and tighten to 124 inch lbs. (14 Nm).

40. Connect the crankshaft sensor and oil sender electrical connectors. Install the crankshaft sensor shield.

41. Install the crankshaft balancer and tighten the bolt to 111 ft. lbs. (150 Nm) plus 76 degrees addition rotation, while holding the engine from turning with J-37096.

42. Install the flywheel covers. Install the right inner fender well splash shield.

43. Install the front tire and wheel assemblies. Lower the vehicle.

44. Connect the lower radiator hose to the water pump. Install the water pump pulley.

NOTE: If the camshaft was replaced the valve lifters should be replaced also. The lifters have already worn to the old camshaft and if re-used can damage the new camshaft.

45. Install the valve lifters, lifter guides and retainer; tighten the bolts to 22 ft. lbs. (30 Nm).

46. Install the pushrods, pedestal guide, rocker arms, pedestals and bolts; tighten bolts to 11 ft. lbs. (15 Nm) plus 115 degrees additional rotation.

47. Install the intake manifold. Install the rocker arm covers. Connect the negative battery cable.

48. Refill the engine with coolant and crankcase with new oil. Start the vehicle and verify no leaks.

Balance Shaft

REMOVAL AND INSTALLATION

2.3L and 2.4L Engines

The transaxle assembly must be removed from the vehicle to perform this procedure.

1. Disconnect the negative battery cable. Remove the transaxle.

2. Matchmark and remove the pressure plate and clutch. Remove the flywheel mounting bolts, the flywheel retainer on automatics and the flywheel.

3. Remove the engine oil pan. Remove the nut/bolt securing the balancer shaft chain cover and cover.

4. Loosen the pivot bolt on the chain tensioner and pivot the tensioner to remove the tension from the chain.

—— **CAUTION** ——

The balance shaft sprocket bolt is reverse threaded and must be rotated clockwise to be removed. Turning the bolt in the counterclockwise direction can result in the bolt shearing off in the balancer shaft.

5. Hold the crankshaft from turning and remove the balancer shaft sprocket bolt.

6. Mark the balancer shaft sprocket so the same side of the sprocket is facing the engine during installation.

7. Separate the sprocket from the balancer shaft.

8. Remove the 5 balance shaft assembly-to-engine block mounting bolts. DO NOT remove the 8 bolts holding the balance shaft housing halves together.

9. Pry the oil pump pickup screen out of the housing.

10. Support the balancer shaft housing and loosen the 8 mounting bolts in the reverse order of the tightening sequence.

11. Remove the 8 bolts and lift the top half of the balance shaft housing off. Remove the 2 balance shafts from the housing.

12. Remove the balance shaft bearings. Remove the thrust plate bolts and plate.

To install:

13. Clean all the components thoroughly. Install the thrust plate and tighten bolts to 115 inch lbs. (13 Nm).

14. Carefully install the new bearing inserts into the housing halves. Use care not to scratch or gouge the bearings or the housing.

15. Lubricate the bearings, balance shaft and gears with GM 9985705, or the equivalent.

16. Install the balance shafts into the lower half of the housing. Make sure the timing marks on the balance shafts are lined up properly.

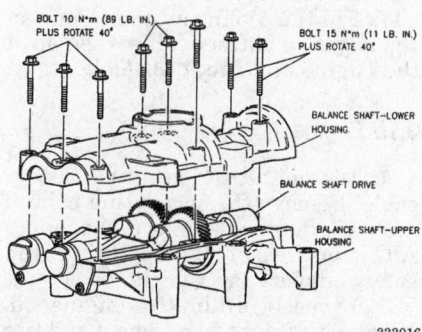

Upper and lower balance shaft housing covers and mounting bolts — 2.3L and 2.4L engines

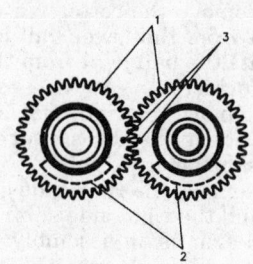

1 BALANCE SHAFT GEARS
2 BALANCE SHAFT COUNTER WEIGHTS
3 BALANCE SHAFT GEAR TIMING MARK

333018

Timing mark alignment on the balance shafts — 2.3L and 2.4L engines

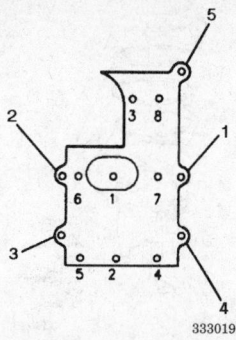

333019

Balance shaft housing bolt tightening sequence — 2.3L and 2.4L engines

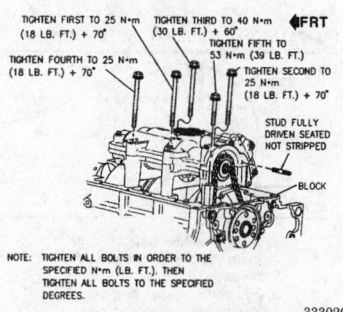

NOTE: TIGHTEN ALL BOLTS IN ORDER TO THE SPECIFIED N•m (LB. FT.). THEN TIGHTEN ALL BOLTS TO THE SPECIFIED DEGREES.

333020

Balance shaft housing-to-engine bolt tightening sequence — 2.3L and 2.4L engines

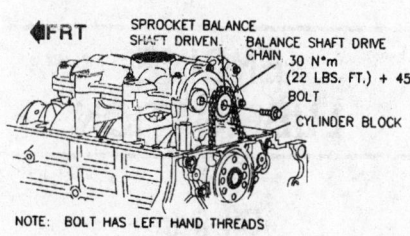

NOTE: BOLT HAS LEFT HAND THREADS

333021

Installation of the balance shaft driven gear — 2.3L and 2.4L engines

17. Install the upper half of the balance shaft housing and tighten the bolts in sequence to 44 inch lbs. (5 Nm). DO NOT tighten the bolts any tighter at this time. The final housing bolt tightening is done after the assembly is installed on the engine.

18. Do not install the oil pump pickup screen until the bolts have been tightened to their final torque.

19. Install the balance shaft assembly on the engine and after coating the mounting bolt threads with Loctite® 242 or equivalent threadlocking compound tighten the bolts in sequence to 44 inch lbs. (5 Nm).

20. Tighten the 5 housing-to-engine bolts as follows:

  a. Bolt 1 to 18 ft. lbs. (25 Nm).

  b. Bolt 2 to 18 ft. lbs. (25 Nm).

  c. Bolt 3 to 30 ft. lbs. (40 Nm).

  d. Bolt 4 to 18 ft. lbs. (25 Nm).

  e. Bolt 5 to 39 ft. lbs. (53 Nm).

  f. Tighten bolt 1 an additional 70 degrees.

  g. Tighten bolt 2 an additional 70 degrees.

  h. Tighten bolt 3 an additional 60 degrees.

  i. Tighten bolt 4 an additional 70 degrees.

21. Tighten the 8 upper-to-lower housing bolts, in sequence, to:

  a. Bolts 1 and 2 to 89 inch lbs. (10 Nm).

  b. Bolts 3 to 11 ft. lbs. (15 Nm).

  c. Bolts 4 through 7 to 89 inch lbs. (10 Nm).

  d. Bolts 8 to 11 ft. lbs. (15 Nm).

  e. Tighten all 8 bolts in sequence an additional 40 degrees.

22. Rotate the balance shaft assembly 1 full revolution to verify the shafts spin freely. Make sure the balance shaft assembly is returned to its original position before installation of the drive sprocket and chain.

23. Rotate the engine until the No. 1 cylinder is at Top Dead Center. Rotate the engine 90 degrees.

24. Install J-41088 onto the balance shaft assembly to ensure the balance shafts don't rotate when the driven gear is installed.

25. Install the balance shaft driven gear with the mark made during disassembly facing outward and tighten the left hand threaded mounting bolt to 22 ft. lbs. (30 Nm).

26. Remove J-41088.

27. Install the balance shaft chain guide and set the chain tension as follows:

  a. Insert a 0.040 inch brass feeler gage between the chain guide and the chain.

NOTE: A brass gage must be used to ensure proper chain tension. Use of a steel gage will not provide the correct flex and the chain tension will be incorrect.

  b. Press the chain guide against the chain with 3 lbs. of force.

  c. Tighten the chain tensioner adjuster bolt to 115 inch lbs. (13 Nm).

  d. Install the balance shaft chain cover and tighten the nut/bolt to 115 inch lbs. (13 Nm).

28. Install the oil pan assembly. Install the flywheel and tighten bolts to

Adjusting the balance shaft timing chain tension — 2.3L and 2.4L engines

22 ft. lbs. (30 Nm) plus 40 degrees additional rotation.

29. Install the clutch disc and pressure plate. Install the transaxle assembly.

30. Connect the negative battery cable. Start the vehicle and verify the engine runs smoothly and quietly.

Piston and Connecting Rod

POSITIONING

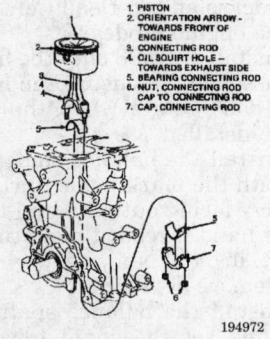

Piston and connecting rod components — 2.3L and 2.4L engines

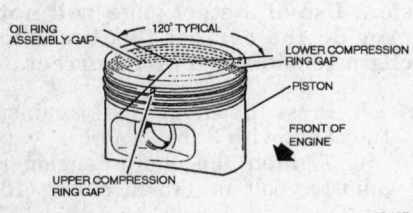

Piston ring gap location — 2.3L and 2.4L engines

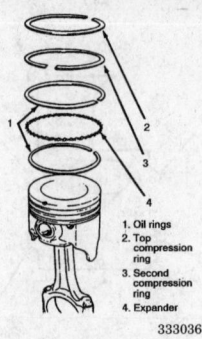

Piston, connecting rod and piston rings

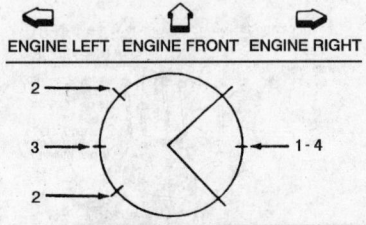

1. Oil ring spacer gap (tang in hole or slot with ARC)
2. Oil ring rail gaps
3. 2nd compression ring gap
4. Top compression ring gap

Piston ring positioning — 3.1L and 3.3L engines

ENGINE LUBRICATION

Oil Pan

REMOVAL AND INSTALLATION

2.3L and 2.4L Engines

1. Disconnect the negative battery cable. Raise and safely support the vehicle.

2. Properly drain the engine oil and cooling system. Remove the flywheel inspection cover.

3. Remove the right front wheel and tire assembly. Remove the right inner fender splash shield.

4. Remove the serpentine belt. Disconnect the engine mount strut from the engine mount strut bracket.

5. Remove the A/C compressor from the mounting bracket and support it aside. Do NOT disconnect the refrigerant lines.

6. Unfasten the engine mount strut bracket bolts, then remove the

bracket. Remove the radiator outlet pipe bolts.

7. Disconnect the radiator and A/C outlet pipes from the suspension supports. Remove the exhaust manifold brace.

8. Remove the oil pan-to-flywheel cover bolt and nut and remove the flywheel cover stud and spacer.

9. Disconnect the radiator outlet pipe from the lower radiator hose and oil pan.

10. If equipped, detach the oil level sensor wire. Unfasten the oil pan bolts, then remove the oil pan.

To install:

11. Install a new gasket, the oil pan and bolts. Tighten the chain housing and carrier seal bolts to 106 inch lbs. (12 Nm). Tighten the oil pan-to-block bolts to 17 ft. lbs. (23 Nm).

12. Install the flywheel cover spacer, stud nut and bolt. Tighten the nut to 41 ft. lbs. (56 Nm), the stud to 115 inch lbs. (13 Nm) and the bolt to 41 ft. lbs. (56 Nm).

13. Install the oil pan-to-transaxle nut and tighten to 41 ft. lbs. (56 Nm).

14. Complete the installation by reversing the removal procedures.

15. Install the engine mount strut bracket and tighten bolts to 49 ft. lbs. (66 Nm).

16. Fill the crankcase with oil. A filter change is recommended.

17. Fill the cooling system. Connect the negative battery cable, then start the engine and check for leaks.

3.1L Engine

1. Disconnect the negative battery cable. Remove the serpentine belt.

2. If equipped, remove the upper A/C compressor bolts. Raise and safely support the vehicle.

3. Properly drain the engine oil. Remove the right front wheel and tire assembly.

4. Remove the engine mount strut from the suspension support.

5. Remove the cotter pin and castle nut from the lower ball joint, the separate the ball joint from the steering knuckle.

6. Remove the right side sway bar link. Detach the ABS sensor from the right subframe.

7. Remove the right side subframe bolts and the right side subframe and control arm as an assembly.

8. If equipped, remove the lower A/C compressor bolts and position the compressor aside. Do NOT disconnect the refrigerant lines or allow the compressor to hang unsupported.

9. Remove the engine mount strut bracket from the engine. Remove the

oil filter, the starter and the flywheel cover.

10. Remove the oil pan flange bolts, the oil pan side bolts and the oil pan.

To install:

11. Clean the gasket mating surfaces. Install a new gasket on the oil pan. Apply silicone sealer to the portion of the pan that contacts the rear of the block.

12. Position the oil pan, then install the mounting bolts finger-tight.

13. With all the bolts in place, tighten the oil pan flange bolts to 18 ft. lbs. (25 Nm) and the oil pan side bolts to 34 ft. lbs. (50 Nm).

14. Complete the installation by reversing the removal procedures.

15. Torque the following item to:

a. Engine-to-transaxle brace, bolts to 68 ft. lbs. (93 Nm)

b. Engine mount strut bracket-to-engine bolts to 85 ft. lbs. (115 Nm)

c. Right side subframe/control arm assembly bolts to 89 ft. lbs. (120 Nm)

d. Right side sway bar line bolt to 22 ft. lbs. (30 Nm)

e. Ball joint-to-steering knuckle nut to 48 ft. lbs. (60 Nm). Install a new cotter pin.

f. Engine mount strut bracket bolt-to-frame bolt to 89 ft. lbs. (120 Nm)

16. Fill the crankcase to the correct level. A filter change is recommended. Connect the negative battery cable, then start the engine and check for leaks.

NOTE: Whenever the vehicle subframe is removed or lowered, the wheel alignment should be checked.

17. Check the front end alignment and adjust as required.

3.3L Engine

1. Disconnect the negative battery cable. Raise and safely support the vehicle.

2. Drain the engine oil into a suitable container. Remove the right wheel/tire assembly and the right fender well splash shield.

3. Remove the crankshaft balancer bolt and the balancer using J-38197 or equivalent puller.

4. If equipped, detach the A/C compressor electrical connector. Disconnect the A/C hose-to-suspension support clip. Remove the compressor and support it aside; do not disconnect the refrigerant lines.

5. Remove the right suspension support front bolts.

6. Loosen the remaining bolts on the suspension support so the front of the support lowers about 1½ inches.

7. Disconnect the oil level sensor from the oil pan. Remove the flywheel cover.

8. Remove the oil pan mounting bolts and the oil pan; it necessary, move the A/C line to provide clearance for oil pan removal.

To install:

9. Clean all the gasket surfaces. Install the oil pan and tighten the bolts to 124 inch lbs. (14 Nm).

10. Complete the installation by reversing the removal procedures.

11. Install the crankshaft damper and tighten the bolt to 219 ft. lbs. (297 Nm).

12. Refill the crankcase with engine oil.

13. Connect the negative battery cable, then start the engine and check for leaks.

Oil Pump

REMOVAL AND INSTALLATION

2.3L and 2.4L Engines

1993–94 Vehicles

1. Disconnect the negative battery cable. Raise and safely support the vehicle.

2. Drain the engine oil and remove the oil pan.

3. Remove the oil pump bolts and the oil pump.

To install:

4. Install the oil pump; then tighten the pump-to-block bolts to 40 ft. lbs. (54 Nm) and the oil pump screen-to-screen brace bolts to 106 inch lbs. (12 Nm).

5. Install the oil pan.

6. Carefully lower the vehicle. Fill the crankcase with clean oil to the proper level.

7. Connect the negative battery cable, then start the engine and check for leaks.

1995–97 Vehicles

NOTE: Please note that the transaxle must be removed from the vehicle to service the oil pump.

1. Disconnect the negative battery cable. Install engine support fixture J-28467-A or equivalent.

2. Properly drain the engine oil, then remove the oil pan. Remove the transaxle and the flywheel.

3. Remove the balance shaft chain cover and chain guide. Remove the oil pump bolts and the oil pump cover.

4. Pull the housing to disconnect the pump gear from the balance shaft. Remove the housing assembly from the balance shaft assembly.

To install:

5. Clean all of the parts in suitable cleaning solvent. Remove all varnish sludge and dirt.

6. Lubricate the gears with clean engine oil. Assemble the geroter gear into the housing.

NOTE: Fill oil pump cavities with petroleum jelly prior to installation. This seals the pump and acts like a "prime" so the pump will draw oil as soon as the engine begins to turn. This will ensure that there is oil pressure immediately on start-up and will prevent engine damage.

7. Install the pump housing to the balance shaft assembly, the pump cover to the oil pump housing.

8. Install the oil pump-to-block bolts and tighten to 40 ft. lbs. (54 Nm).

9. Install the balance shaft chain guide and chain. Adjust the chain tension inserting a 0.40 inch (1mm) brass feeler, between the chain guide and the chain.

NOTE: A brass feeler gauge must be used to ensure that correct measurements are obtained. If a steel gauge is used, it will not bend to conform to the guide and will allow for incorrect measurements.

10. Press the guide against the chain using about 3 lbs. of force. Tighten the chain tensioner fastener to 115 inch lbs. (13 Nm).

11. Install the balance shaft chain cover and tighten the nut and bolt to 115 inch lbs. (13 Nm).

12. Install the flywheel and the transaxle. Install the oil pan, then carefully lower the vehicle.

13. Fill the crankcase with clean engine oil. A filter change is recommended.

14. Remove the engine support fixture. Connect the negative battery cable, then start the engine and verify oil pressure and no leaks.

3.1L Engine

1. Disconnect the negative battery cable.

2. Raise and safely support the vehicle.

3. Drain the engine oil into a suitable container. Remove the oil pan.

4. Remove the crankshaft oil deflector bolts and deflector.

5. Remove the oil pump bolts, the oil pump and pump driveshaft.

To install:

6. Install the oil pump and pump driveshaft. Tighten the oil pump mounting bolts to 30 ft. lbs. (41 Nm).

7. Install the crankshaft oil deflector and tighten the nuts to 18 ft. lbs. (25 Nm).

8. Install the oil pan, then carefully lower the vehicle.

9. Fill the crankcase to the correct level with oil. An filter change is recommended.

10. Connect the negative battery cable; then, start the engine, check the oil pressure and check for leaks.

3.3L Engine

1. Disconnect the negative battery cable. Raise and safely support the vehicle.

2. Drain the engine oil into a suitable container. Remove the front cover assembly.

3. Remove the oil filter adapter, pressure regulator valve and spring.

4. Unfasten the oil pump cover screws and the cover. Remove the oil pump from the vehicle.

To install:

5. Lubricate the gears with petroleum jelly and install the gears into the housing.

6. Pack the gear cavity with petroleum jelly after the gears have been installed in the housing.

7. Install the oil pump cover and tighten screws to 97 inch lbs. (10 Nm).

8. Install the oil filter adapter with new gasket, pressure regulator valve and spring.

9. Install the front cover assembly. Carefully lower the vehicle.

10. Refill the crankcase with clean engine oil.

11. Connect the negative battery cable, then start the engine and verify correct oil pressure and that no leaks are present.

TRANSAXLE

Manual Transaxle

REMOVAL AND INSTALLATION

1993–94 Vehicles

1. Disconnect the negative battery cable. Install the engine support fixture tool J 28467 or equivalent. Raise the engine enough to take pressure off the motor mounts.

2. Remove the left sound panel. Remove the clutch master cylinder pushrod from the clutch pedal.

3. Remove the air cleaner and air intake duct assembly. Remove the clutch slave cylinder from the transaxle support bracket and lay aside.

4. Remove the transaxle mount through bolt. Raise and safely support the vehicle.

5. Remove the exhaust crossover bolts at the right manifold. Lower the vehicle.

6. Remove the left exhaust manifold. Remove the transaxle mount bracket.

7. Disconnect the shift cables. Remove the upper transaxle-to-engine bolts.

8. Raise the vehicle and support the safely. Remove the left wheel assembly.

9. Remove the left front inner splash shield. Remove the transaxle strut and bracket.

10. Drain the transaxle. Remove the clutch housing cover bolts. Disconnect the speedometer cable.

11. Disconnect the stabilizer bar at the left suspension support and control arm. Disconnect the ball joint from the steering knuckle.

12. Remove the left suspension support bolts; then, the support and control arm as an assembly.

13. Disconnect and remove the left halfshaft from the transaxle. Support the right halfshaft.

14. Attach the transaxle case to a jack. Remove the remaining transaxle-to-engine bolts.

15. Remove the transaxle by sliding toward the drive side away from the engine. Carefully lower the jack, guiding the right halfshaft from the transaxle.

To install:

16. When installing the transaxle, guide the right halfshaft into its bore as the transaxle is being raised. The right halfshaft cannot be readily installed after the transaxle is connected to the engine.

17. Install the transaxle-to-engine bolts.

18. Install the left halfshaft into its bore at the transaxle and seat both halfshafts.

19. Complete the installation by reversing the removal procedures.

20. Fill the transaxle with 5 pints (2.1L) of 5W-30 manual transaxle oil, GM part No. 1052931 or equivalent.

21. Connect the negative battery cable.

1995–97 Vehicles

1. Disconnect the negative battery cable. Install an engine support fixture, J-28467-A, or the equivalent.

2. Use the fixture to take the weight off the engine mounts. Remove the left side sound insulator from under the dash board.

3. Disconnect the clutch master cylinder pushrod from the clutch pedal stud. Remove the air cleaner and duct work from the throttle body.

4. Disconnect the shift cable from the manual shift lever on the transaxle. Disconnect the wiring harness from the transaxle mount bracket.

5. Remove the upper transaxle mount-to-transaxle bolts. Disconnect the clutch master cylinder from the slave cylinder.

6. Disconnect the ground wires from the transaxle mounting studs. Disconnect the backup light switch connector.

7. Disconnect the transaxle vent tube. Remove the upper transaxle-to-engine bolts.

8. Lower the powertrain assembly with the support fixture. Raise and safely support the vehicle.

9. Drain the transaxle fluid. Remove the front tire and wheel assemblies.

10. Remove the left inner fender well splash shield. Disconnect the front ABS sensor connectors and unroute the left side sensor wiring.

11. Remove the flywheel housing cover bolts. Disconnect the vehicle speed sensor from the transaxle.

12. Remove the cotter pins and castle nuts and using a suitable ball joint separator, disconnect the ball joints from the steering knuckles.

13. Remove the left side link kit. Remove the left side U-bolt from the sway bar.

14. Remove the left side suspension support bolts and remove the suspension support.

15. Disconnect the halfshafts from the transaxle. Remove the front lower transaxle mount.

16. Support the transaxle with a suitable jack. Remove the remainder of the engine-to-transaxle bolts.

17. Carefully slide the transaxle away from the engine. Once there is sufficient clearance, lower the transaxle from the vehicle.

To install:

18. Raise the transaxle into position with the jack and install the mounting bolts loosely. DO NOT tighten the bolts until at least 4 of the bolts are in place. Tighten the lower bolts to 55 ft. lbs. (75 Nm).

19. Install the front transaxle mount and tighten the bolts to 55 ft. lbs. (75 Nm).

20. Complete the installation by reversing the removal procedures.

21. Install the upper transaxle-to-engine bolts and tighten to 55 ft. lbs. (75 Nm).

22. Install the rear transaxle mount and tighten the bolts to 55 ft. lbs. (75 Nm) on Isuzu transaxle or 96 ft. lbs. (130 Nm) on NVG transaxle.

23. Connect the negative battery cable. Refill the transaxle with Synchromesh Transaxle Fluid.

Clutch Assembly

REMOVAL AND INSTALLATION

1. Disconnect the negative battery cable.

2. If necessary, disconnect the hydraulic line from the clutch actuator (slave cylinder).

3. Remove the transaxle from the vehicle.

4. If any clutch components are to be reused, use the following procedure:

a. Matchmark (a dots of paint or marks made with a center punch) the clutch pressure plate to the flywheel. This is to retain the balance of the original parts. If all parts are to be replaced with new, this step is not necessary.

b. If the pressure plate is to be reused, loosen the pressure plate mounting bolts by turning each bolt one full turn until all the spring pressure is removed. This helps avoid warping the pressure plate.

c. Remove the bolts the remainder of the way and remove the clutch disc and pressure plate.

To install:

5. Apply a small amount of high-temperature grease to the pilot bear-

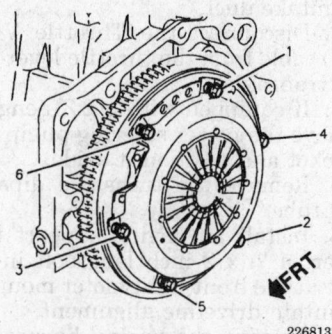

Pressure plate mounting bolt tightening sequence

226813

ing as well as the tip of the transaxle input shaft and the clutch splines.

6. Install J-29074, or an equivalent clutch alignment tool into the flywheel and install the clutch disc onto the tool. Make sure the clutch is installed in the correct direction. The light side of the disc should be visible. Many replacement discs will also be marked "Flywheel Side" as an air in getting the disc correctly positioned.

7. Install the pressure plate and loosely install the mounting bolts. New replacement bolts are recommended.

8. Make a first pass tightening the bolts in sequence to 12 ft. lbs. (16 Nm). With all the bolts tight make a second pass tightening each bolt to 15 ft. lbs. (20 Nm). Make a third pass tightening the bolts 30 degrees additional if equipped with a 3.1L engine and 45 degrees additional if equipped with a 2.3L or 2.4L engine.

9. Remove the clutch alignment tool.

10. Lubricate the inside diameter of the actuator (slave cylinder/throwout bearing) with clutch bearing lubricant.

11. Install the transaxle assembly into the vehicle.

12. If necessary, connect the clutch master cylinder hydraulic line to the actuator (slave cylinder).

13. Bleed the clutch hydraulic system.

14. Connect the negative battery cable. Road test vehicle to verify correct operation and easy shifting.

Clutch Master Cylinder

REMOVAL AND INSTALLATION

1993–94 Vehicles

1. For a 2.3L (VIN A) engine, disconnect the air intake duct from the air cleaner.

2. Disconnect the negative battery cable, then the positive battery cable.

3. For a 2.3L (VIN A) engine, perform the following:

a. Remove the left fender cross brace.

b. Remove the battery.

c. Disconnect the MAT sensor lead at the air cleaner.

d. Disconnect the MAF sensor harness.

e. Disconnect the PCV pipe retaining clamp from the air inlet duct.

f. Disconnect the air intake duct from the throttle body.

g. Remove the MAF sensor mounting bolt.

h. Remove the air cleaner bracket mounting bolts at the battery tray.

i. Remove the air inlet duct and air cleaner assembly.

j. Disconnect and remove the washer bottle.

k. Remove the cruise control mounting bracket retaining nuts from the strut tower, if equipped.

4. Remove the lower sound insulator from the under the driver's side of the dashboard.

5. Disconnect the clutch master cylinder pushrod from the clutch pedal by disconnecting the pushrod retainer from the pedal stud.

6. From under the hood, remove the 2 nuts securing the clutch master cylinder to the firewall. If equipped with a remote reservoir, disconnect and cap the fluid line from the master cylinder.

7. Remove the 2 nuts securing the slave cylinder to the transaxle.

8. Remove the hydraulic assembly from the vehicle.

To install:

9. Position the actuator assembly on the transaxle. Line up the pushrod into the pocket on the clutch lever and install the mounting nuts. Tighten the nuts alternately and evenly to 16 ft. lbs. (22 Nm).

NOTE: DO NOT remove the plastic pushrod retainer from the slave cylinder. The straps will break on the first clutch pedal application.

10. Position the clutch master cylinder on the firewall and tighten the retaining nuts to 16 ft. lbs. (22 Nm).

11. Connect the remote reservoir hose to the master cylinder, if equipped.

12. Remove the pedal restrictor from the pushrod. Lubricate the pushrod bushing on the clutch pedal.

13. Connect the pushrod to the clutch pedal and install the retaining clip.

14. If equipped with cruise control, check the adjustment of the cruise control switch.

15. Install the lower sound insulator.

16. Depress the clutch pedal several times. This will break the plastic retaining straps on the actuator pushrod. DO NOT remove the plastic button on the end of the pushrod.

17. For a 2.3L (VIN A) engine, perform the following procedure:

a. Install the washer bottle and connect the electrical connector.

b. Install the air cleaner and air duct assembly.

c. Install the air cleaner bracket mounting bolts.

d. Install the MAF sensor mounting bolt.

e. Install the clamp securing the inlet duct to the throttle body.

f. Connect the PCV pipe to the inlet duct.

g. Connect the MAF sensor electrical connector.

h. Connect the MAT sensor electrical connector.

i. Install the battery.

j. Install the left fender brace.

18. Connect the positive battery cable then the negative battery cable.

19. For a 2.3L (VIN A) engine, connect the air intake duct to the air cleaner.

1995–97 Vehicles

1. Disconnect the negative battery cable.

2. Remove the left lower sound insulator from under the dash board.

3. Disconnect the clutch master cylinder pushrod from the clutch pedal.

4. Remove the 2 master cylinder mounting nuts from the firewall.

5. Disconnect the clutch master cylinder remote reservoir hose.

6. Disconnect the hydraulic line from the master cylinder. Note that many vehicles use a quick-disconnect fitting on the hydraulic line, down near the clutch actuator (slave cylinder).

7. Remove the clutch master cylinder.

To install:

8. Position the master cylinder on the firewall studs and loosely install the mounting nuts.

9. Connect the hydraulic line to the master cylinder; then, tighten the master cylinder nuts to 18 ft. lbs. (25 Nm).

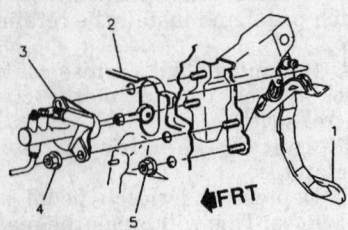

1 CLUTCH PEDAL
2 PANEL ASSEMBLY
3 MASTER CYLINDER
4 NUT, 27 N·m (20 LBS. FT.)
5 NUT, 27 N·m (20 LBS. FT.)

330576

Clutch master cylinder mounting components — 1995–97 vehicles

10. Connect the reservoir hose to the master cylinder.

11. Connect the pushrod to the clutch pedal. Install the left lower sound insulator.

12. Bleed the clutch hydraulic system. Connect the negative battery cable.

Clutch Slave Cylinder

REMOVAL AND INSTALLATION

1995–97 Vehicles

This design uses a hydraulic actuator built into the throw-out bearing. The bearing, actuator and short lengths of hydraulic tube are combined into one unit and must be replaced as a unit if service is required.

1. Disconnect the negative battery cable.

2. Disconnect the hydraulic line from the slave cylinder. Note that many vehicles use a quick-disconnect fitting on the hydraulic line, down near the clutch actuator (slave cylinder).

3. Remove the transaxle from the vehicle.

4. Remove the actuator (slave cylinder) from the transaxle.

To install:

5. Lubricate the inside diameter of the clutch actuator cylinder/throwout bearing with high temperature bearing lubricant.

6. Install the clutch actuator cylinder/throwout bearing assembly into the transaxle.

7. Install the transaxle.

8. Connect the hydraulic line to the slave cylinder. Make sure the quick-connect fitting (where used) is secure.

9. Bleed the clutch hydraulic system.

10. Connect the negative battery cable. Road test vehicle to verify correct shifting.

Hydraulic Clutch Bleeding

PROCEDURE

With Bleeder Screw

1. Make sure the reservoir is full of DOT 3 fluid and is kept topped off throughout this procedure.

2. Loosen the bleed screw, located on the actuator cylinder body next to the inlet connection.

3. When a steady stream of fluid comes out the bleeder, tighten it to 17 inch lbs. (2 Nm).

4. Refill the fluid reservoir.

5. To check the system, start the engine and wait 10 seconds.

6. Depress the clutch pedal and shift into **R**. If there is any gear clash, air may still be present.

Without Bleeder Screw

1. Remove the actuator cylinder from the transaxle.

2. Loosen the master cylinder attaching nuts to the ends of the studs.

3. Remove the reservoir cap and diaphragm.

4. Depress the actuator cylinder pushrod about ¾ in. into its bore and hold the position.

5. Install the reservoir diaphragm and cap while holding the actuator pushrod.

6. Release the pushrod when the diaphragm and cap are properly installed.

7. With the actuator lower than the master cylinder, hold the actuator vertically with the pushrod end facing the ground.

8. Press the actuator pushrod into its bore with ½ in. strokes. Check the reservoir for bubbles. Continue until no bubbles enter the reservoir.

9. Install the master cylinder and actuator. Refill the fluid reservoir.

10. To check the system, start the engine and wait 10 seconds.

11. Depress the clutch pedal and shift into reverse. If there is any gear clash, air may still be present.

Automatic Transaxle

REMOVAL AND INSTALLATION

1993 Vehicles

1. Disconnect the negative battery cable. If necessary, drain the coolant and disconnect the heater core hoses.

2. Remove the air cleaner assembly. If equipped with a 3.3L engine, remove the mass air flow sensor and air intake duct.

3. Disconnect the Throttle Valve (TV) cable from the throttle lever and the transaxle.

4. If equipped with a 2.3L engine, remove the power steering pump and bracket and position it aside.

5. Remove the transaxle dipstick and tube.

6. Install an engine support tool. Insert a ¼ x 2 inch long bolt in the hole at the front right motor mount to maintain driveline alignment.

7. Remove the wiring harness-to-transaxle nut. Disconnect the wiring connectors from the speed sensor,

TCC connector, neutral safety switch and reverse light switch.

8. Disconnect the shift linkage from the transaxle.

9. Remove the upper 2 transaxle-to-engine bolts and the upper left transaxle mount along with the bracket assembly.

10. Remove the rubber hose from the transaxle vent pipe. Remove the remaining upper engine-to-transaxle bolts.

11. Raise and safely support the vehicle. Remove both front wheels.

12. If equipped with a 2.3L engine, remove both lower ball joints and stabilizer shafts links.

13. Drain the transaxle fluid. Remove the shift linkage bracket from the transaxle.

CAUTION

On some vehicles, the sub-frame must be lowered for clearance. This means the intermediate shaft must be disconnected from the rack and pinion stub shaft. Failure to disconnect the intermediate shaft from the rack and pinion steering gear stub can result in damage to the steering gear and/or intermediate shaft. This damage can cause loss of steering control, which could result in an accident and possibly personal injury.

---

14. Install a halfshaft boot seal protector on the inner seals.

NOTE: Some vehicles may use a gray silicone boot on the inboard axle joint. Use boot protector tool on these boots. All other boots are made from a black thermo-plastic material and do not require the use of a boot seal protector.

15. Remove both ball joint-to-control arm nuts and separate the ball joints from the control arms.

16. Remove both halfshafts and support them with a cord or wire. Remove the transaxle mounting strut.

17. Remove the left stabilizer bar link pin bolt, left frame bushing clamp nuts and left frame support assembly.

18. Remove the torque converter cover. Matchmark the flexplate and torque converter for installation purposes. Remove the torque converter-to-flexplate bolts.

19. Disconnect and plug the transaxle oil cooler lines.

20. Remove the transaxle-to-engine support bracket and install the transaxle removal jack.

21. Remove the remaining transaxle-to-engine attaching bolts and the transaxle from the vehicle.

To install:

22. Securely mount the transaxle on the jack.

23. Apply a small amount of grease on the torque converter hub and seat in the oil pump.

24. Position the transaxle in the vehicle and install the lower engine to transaxle bolts.

25. Complete the installation by reversing the removal procedures.

26. Fill all fluids to their proper levels. Adjust cables as required.

27. Connect the negative battery cable and check the transaxle for proper operation and leaks.

1994 Vehicles

1. Disconnect the negative battery cable. Remove the intake air duct.

2. Disconnect the Throttle Valve (TV) cable. Remove the shift cable and bracket.

3. Tag for identification as necessary and remove vacuum lines. Remove electrical connections to transaxle.

4. Remove the power steering pump from its brackets and set aside. Remove transaxle filler tube.

5. Install engine support fixture J-28467–A or equivalent engine support.

6. Remove the top engine to transaxle bolts. Raise and safely support vehicle.

7. Remove both front tire/wheel assemblies and the left splash shield.

8. Remove both ABS wheel speed sensors and harness from the left suspension support.

9. Remove both lower ball joints and stabilizer shaft links.

10. Remove the front air deflector and left suspension support.

11. Remove both halfshafts and engine-to-transaxle brace.

12. Remove the starter and flywheel-to-torque converter bolts.

13. Remove transaxle fluid cooler pipes. Disconnect ground wires from the engine to the transaxle bolt.

14. Remove the fluid cooler pipe brace and exhaust brace. Remove engine and transaxle mount bolts.

15. Support transaxle with a transmission jack. Remove transaxle mount-to-body bolts.

16. Remove heater core hose pipe brace to transaxle nut and bolt.

17. Remove remaining engine-to-transaxle bolts and the transaxle from vehicle.

To install:

18. Place a thin film of grease on the torque converter pilot knob. Make sure to properly seat the torque converter in the oil pump.

19. Install transaxle assembly into position with jack while installing right halfaxle.

20. Install the lower engine-to-transaxle bolts into their proper location.

21. Complete the installation by reversing the removal procedures.

22. Fill transaxle. Connect the negative battery cable.

1995–97 Vehicles

With 3T40 and 4T60-E Transaxles

1. Disconnect the negative battery cable. Remove the air intake duct.

2. For 3T40 transaxle, remove the cable control cover.

3. For 4T60-E transaxle, disconnect the shift linkage from the transaxle.

4. Disconnect the vacuum modulator line from the modulator.

5. For 3T40 transaxle, disconnect the throttle cable from the throttle body.

6. Disconnect the electrical connectors from the TCC connector, Park/Neutral position switch and Shift solenoid connector.

7. For 3T40 transaxle, remove the power steering pump and set it aside leaving the hoses attached.

CAUTION

When servicing requires that the "T" latch type wiring connector be disconnected from the switch, use care to ensure proper reassembly of both the connector and the "T" latch. Failure to do so may result in intermittent loss of switch functions.

---

8. For 3T40 transaxle, remove the oil fill tube.

9. Install engine support fixture J-28467-A or equivalent. Remove the top (2) transaxle to engine bolts.

10. For 4T60-E transaxle, disconnect the rubber hose from the transaxle vent pipe.

11. Remove the remaining upper transaxle to engine bolts. Raise and safely support the vehicle.

12. Remove both front tire and wheel assemblies. Remove right and left engine splash shields.

13. Disconnect the ABS wheel speed sensor connectors and remove the harness from the left side suspension support.

14. Separate both ball joints from the control arms. Remove the left side stabilizer shaft link pin bolt.

15. Remove the left side stabilizer shaft frame bushing clamp nuts. Remove the left side suspension support assembly.

16. Remove both halfshaft. Remove the engine to transaxle brace.

17. Remove the starter motor. Remove the transaxle converter cover.

18. Remove the heater core hose pipe brace to transaxle nut and bolt. Remove the torque converter to flywheel bolts.

NOTE: Using a scribe mark the flywheel to torque converter relationship to assure proper reassembly.

19. For 4T60-E transaxle, remove the oil level indicator and fill tube.

20. Disconnect the transaxle cooler lines and plug openings to prevent excess oil leakage.

21. For 4T60-E transaxle, disconnect the vehicle speed sensor and remove the vacuum reservoir tank.

22. Position transaxle jack under transaxle. Remove the transaxle mount to body bolts.

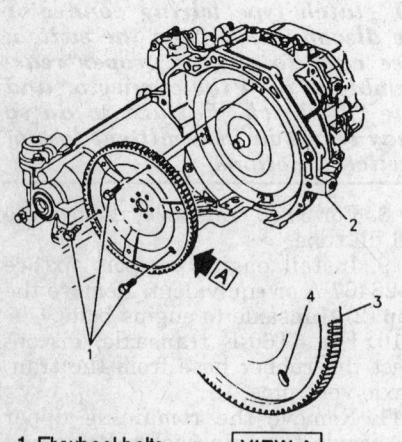

1. Flywheel bolts
2. Transaxle assy
3. Flywheel
4. Net slot

VIEW A

334079

Flywheel net slot view — 1995–97 vehicles with 4T60-E transaxle

23. Remove the remaining engine to transaxle bolts. Remove the transaxle.

NOTE: Transaxle cooler and lines should be flushed with J-35944 or equivalent whenever the transaxle has been removed for overhaul or replacement.

To install:

24. Apply a thin film of grease on the torque converter pilot hub.

NOTE: Make sure to properly seat the torque converter in the pump.

25. Position the transaxle into the vehicle. Install the lower transaxle to engine bolts.

26. Install the transaxle mount to body bolts and tighten to 66 ft. lbs. (90 Nm). Install the transaxle cooler lines.

27. Install the torque converter to flywheel bolts and tighten the bolts to 46 ft. (62 Nm).

NOTE: For 4T60-E transaxle, note that the flywheel has one oval shaped bolt opening. This is for the so-called "net slot bolt." Hand start and tighten the net slot bolt first, then tighten the remaining bolts.

28. Install the oil level indicator and fill tube.

29. For 4T60-E transaxle, install the vacuum reserve tank.

30. Install the starter. Install the transaxle converter cover.

31. For 4T60-E transaxle, connect the electrical connector to the vehicle speed sensor.

32. Install the halfshafts. Install the left side suspension support assembly.

33. Install the left stabilizer shaft frame bushing nuts. Install the stabilizer shaft link pin bolt.

34. Install the engine to transaxle brace. Connect the ball joints to the control arms.

35. Connect the ABS wheel speed sensor harness and connectors. Install the right and left side splash shields.

36. Install the heater core pipe brace to transaxle nut and bolt. Install the tire/wheel assemblies. Lower the vehicle.

37. Install the upper transaxle to engine bolts and tighten to 66 ft. lbs. (90 Nm). Connect the shift linkage to the transaxle.

38. Connect the electrical connector to the torque converter clutch.

39. Connect the electrical connector to the park/neutral and backup lamp switch.

40. For 4T60-E transaxle, install the wiring harness and nut securing it to the transaxle.

41. For 3T40 transaxle, install the throttle cable to the throttle body.

42. Remove the engine support fixture.

43. For 4T60-E transaxle, connect the rubber hose to the vent pipe.

44. Connect the vacuum line to the modulator. Install the air intake duct.

45. Connect the negative battery cable. Fill the transaxle with clean transaxle fluid.

46. Verify proper shift linkage adjustment. Start the engine and check for leaks.

47. Road test the vehicle verify proper operation and re-check the transaxle fluid level.

NOTE: Whenever the vehicle sub-frame is removed or lowered, the wheel alignment should be checked.

Throttle Valve Cable

ADJUSTMENT

Except 2.3L and 2.4L Engines

1. Disconnect the negative battery cable.

2. Depress and hold the adjustment tap at the TV cable adjuster.

3. Release the throttle lever by hand to its full travel position.

4. The slider must move toward the lever when the lever is rotated to the full travel position.

5. Inspect the cable for freedom of movement. The cable may appear to function properly with the engine stopped and cold, so recheck the cable after the engine is warm. Don't forget to connect the negative battery cable first.

6. Road test the vehicle and check for proper shifting.

2.3L and 2.4L Engines

1. Disconnect the negative battery cable.

2. Rotate the TV cable adjuster body at the transaxle 90 degrees and pull the cable conduit out until the slider mechanism contacts the stop.

3. Rotate the adjuster body back to the original position.

4. Using a torque wrench, rotate the TV cable adjuster until 75–120 inch lbs. (9–14 Nm) is reached.

5. Connect the negative battery cable, then road test the vehicle and check for proper shifting.

DRIVE AXLE

Halfshaft

REMOVAL AND INSTALLATION

1993–94 Vehicles

1. Disconnect the negative battery cable. Raise and safely support the vehicle.

2. Remove the wheels. Install halfshaft seal protectors on the outer joint, if available.

3. Insert drift a drift pin or punch through the opening in the caliper into the ventilation openings in the rake rotor to keep the rotor from turning.

4. Remove the shaft nut and washer.

5. Remove the ball joint attaching nut and separate the control arm from the steering knuckle using special tool J-38892 or equivalent. Remove the stabilizer shaft, if necessary.

6. Separate joint by using a prybar between the suspension support and lower control arm.

7. Separate the halfshaft from the hub and bearing assembly using special tool J-28733–A or equivalent puller and move the strut assembly rearward.

8. Remove the inner joint from the transaxle or intermediate shaft using special tools J-28468 or J-33008 attached to J-29794 and J-2619–01 or their equivalents.

9. To remove the intermediate shaft, remove the rear engine mount through bolt. Then remove the intermediate shaft bracket bolts and remove the assembly.

To install:

10. Seat the halfshaft into the transaxle or intermediate shaft by placing a suitable drift pin or punch into the groove on the joint housing and tapping until seated. Be careful not to damage the axle seal or spring. Verify that the axle is seated by grasping the inner joint housing and pulling outboard.

11. Install the axle to the hub and bearing assembly.

12. Install the washer and new halfshaft nut and tighten to 185 ft. lbs. (260 Nm).

13. Install the ball joint to the steering knuckle. Torque bolt to 50 ft. lbs. (65 Nm). Install the stabilizer shaft, if removed.

14. Remove the seal protectors. Install the wheels. Connect the negative battery cable.

1995–97 Vehicles

Halfshaft

1. Disconnect the negative battery cable. Raise and safely support the vehicle. Remove the wheel assembly.

2. Insert a drift or a punch through the caliper and into the rotor cooling fins to keep the axle from turning.

3. Remove the hub nut and washer and remove the drift.

4. Remove the cotter pin and castle nut from the lower ball joint. Using a suitable ball joint separator, disconnect the ball joint from the steering knuckle.

5. Disconnect the ABS sensor wire. Remove the sway bar link kit.

6. Unseat the halfshaft from the hub bearing assembly using a suitable puller, J-28733-A, or the equivalent.

7. Pivot the strut and knuckle assembly off the halfshaft and position it aside.

8. Place a pan under the transaxle.

9. Disconnect the inner joint from the transaxle using axle puller, J-28468 or J-33008 in conjuction with J-29794 and J-2619-01 or their equivalents.

To install:

10. Insert the halfshaft into the transaxle and using a non-ferrous drift positioned in the groove in the inboard joint, tap the joint in until it is seated in the transaxle.

11. Verify the joint is seated properly by grasping the inboard joint and pulling on it firmly. DO NOT pull on the axle shaft or damage to the inner joint will result.

12. Position the strut and knuckle assembly and guide the axle into the hub assembly.

13. Connect the lower ball joint to the steering knuckle and tighten the castle nut to 41 ft. lbs. (55 Nm). If necessary tighten the nut up to 60 degrees (1/6 turn) additional rotation to align the cotter pins. NEVER loosen the nut to make the holes align. Install a new cotter pin.

14. Install the washer and hub nut.

15. Insert a drift punch through the caliper and into the rotor cooling fins. Tighten the hub nut to 185 ft. lbs. (260 Nm).

16. Remove the drift. Install the link kit and tighten the mounting nut to 13 ft. lbs. (17 Nm).

17. Install the tire and wheel assembly and tighten the wheel nuts to 100 ft. lbs. (140 Nm).

18. Lower the vehicle. Connect the negative battery cable. Check the transaxle fluid level and top off as necessary.

Intermediate Shaft

1. Install an engine support fixture J-28467-A or equivalent. Raise and safely support the vehicle.

2. Remove the right side wheel assembly. Remove the sway bar link kit.

3. Remove the cotter pin and castle nut from the lower ball joint. Using a suitable ball joint separator, disconnect the ball joint from the steering knuckle.

4. Separate the inner joint from the intermediate shaft using an axle puller, J-33008 in conjunction with J-29794 and J-2619-01 or their equivalents.

5. Remove the rear engine mount through bolt. Remove the intermediate shaft support bracket-to-engine bolts.

6. Place a pan under the transaxle. Carefully disconnect the intermediate shaft from the transaxle and remove the assembly from the vehicle.

To install:

7. Insert the intermediate shaft assembly into the transaxle and position the shaft bracket so the mounting bolts can be installed finger-tight.

8. With all the bolts installed tighten them to 49 ft. lbs. (66 Nm). Coat the splines of the intermediate shaft with chassis grease.

9. Connect the halfshaft to the intermediate shaft. Install the rear engine mount through bolt.

10. Connect the lower ball joint to the steering knuckle and tighten the castle nut to 41 ft. lbs. (55 Nm). If necessary, tighten the nut up to 60 degrees (1/6 turn) additional rotation to align the cotter pins. NEVER loosen the nut to make the holes align. Install a new cotter pin.

11. Install the link kit and tighten the mounting nut to 13 ft. lbs. (17 Nm).

12. Install the tire and wheel assembly and tighten the wheel nuts to 100 ft. lbs. (140 Nm).

13. Lower the vehicle. Remove the engine support fixture.

14. Connect the negative battery cable. Check the transaxle fluid level and top off as necessary.

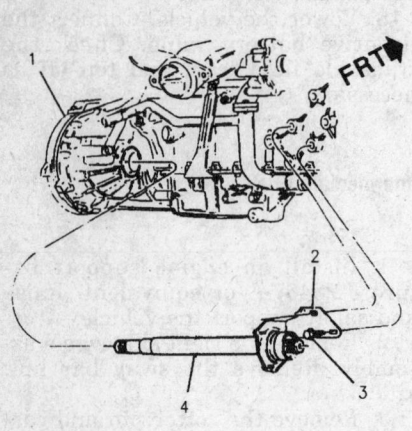

1 TRANSAXLE
2 INTERMEDIATE SHAFT SUPPORT BRACKET
3 BOLT
4 INTERMEDIATE SHAFT

286772

Intermediate shaft components — 1995–97 vehicles

CV-Joint Boot

REPLACEMENT

1. Raise and safely support the vehicle.
2. Remove the halfshaft from the vehicle.

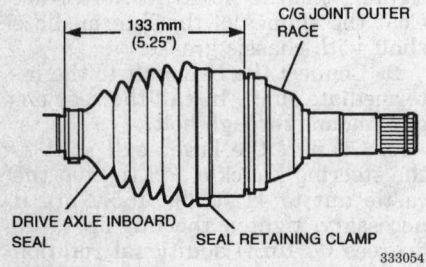

Boot dimensions for the cross groove inboard joint

3. Secure the halfshaft in a suitable bench vise with protective jaw covers installed to protect the halfshaft.
4. Remove the inner and outer CV-joint boot clamps with side cutters and discard the clamps.
5. Slide the boot off the outer joint and down onto the axle shaft so the outer joint is exposed.
6. On the Tripot/Free Motion design, perform the following procedures:
   a. Slide the tripot housing off the spider assembly.
   b. Expand the inner retainer ring using J-8059 and work the ring down onto the axle shaft.
   c. Push the spider assembly onto the axle until the top retainer ring is exposed.
   d. Remove the top retainer ring using J-8059.
   e. Remove the spider assembly and inner retainer ring.
7. On the outer CV-joint and Cross Groove design inner CV-joint, perform the following procedures:
   a. Clean the grease off the outer joint.
   b. Locate the retaining ring in the cut out on the joint and using J-8059, or an equivalent pair of snapring pliers, spread the ears on the retainer ring.
   c. Remove the CV-joint from the axle shaft.

NOTE: The Cross Groove design joint uses precision grinding and selected dimensional compo-

JOINT SIZE	DIMENSION A	DIMENSION B
23	23.8mm (.9")	190.5mm (7.5")
27	27.05mm (1.1")	170.0mm (6.7")

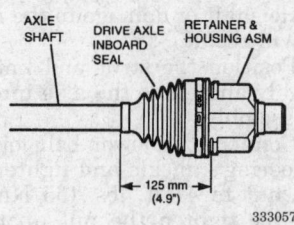

333057

Boot dimensions for the tripot inboard joint

nent fits for proper assembly and operation. Due to its complexity, DO NOT disassemble the Cross Groove joint.

8. Remove the CV-joint boot from the shaft.

To install:

9. Clean all the CV-joint components completely.
10. Install the small clamp onto the axle shaft.
11. Slide the boot onto the shaft until it is below the boot seating area.
12. On the Tripot/Free Motion Design, perform the following procedures:
   a. Install the inboard retainer ring until it is just below the splined portion of the axle.
   b. Install the spider assembly until the top retainer ring groove is exposed.
   c. Install the top retainer ring and slide the spider assembly up until it is seated over the top retainer ring.
   d. Using J-8059, position the inner retainer ring in the groove below the spider assembly.
   e. Pack half the grease from the service kit in the tripot housing and install the housing over the spider assembly.
13. On the outer CV-joint and Cross Groove design inner CV-joint, perform the following procedures:
   a. Pack the outer CV-joint with new grease and install the joint onto the axle shaft. Push the joint onto the axle until the retaining ring seats in the groove and the joint can not be pulled back off.
   b. Put the remainder of the grease in the CV-joint boot and slide the boot up until the small end is over the boot seating area.
14. Position the small clamp over the boot and using J-35910, or an equivalent seal clamp tool, compress the clamp.
15. Fit the large end of the boot over the CV joint and install the large clamp.
16. Using J-35910, or the equivalent, compress the clamp.
17. Install the halfshaft in the vehicle.
18. Lower the vehicle.

STEERING

Air Bag

---------- **CAUTION** ----------

Some vehicles are equipped with an air bag system, also known as the Supplemental Inflatable Restraint (SIR) or Supplemental Restraint System (SRS). The system must be disabled before performing service on or around system components, steering column, instrument panel components, wiring and sensors. Failure to follow safety and disabling procedures could result in accidental air bag deployment, possible personal injury and unnecessary system repairs.

PRECAUTIONS

Several precautions must be observed when handling the inflator module to avoid accidental deployment and possible personal injury.

• Never carry the inflator module by the wires or connector on the underside of the module.

• When carrying a live inflator module, hold securely with both hands, and ensure that the bag and trim cover are pointed away.

• Place the inflator module on a bench or other surface with the bag and trim cover facing up.

• With the inflator module on the bench, never place anything on or close to the module which may be thrown in the event of an accidental deployment.

DISARMING

---------- **CAUTION** ----------

The Supplemental Restraint System (SRS) must be disarmed before performing service procedures around the air bag or SRS wiring. Failure to do so may cause accidental deployment of the air bag, resulting in unnecessary SRS repairs and/or personal injury.

---

1. Disconnect the negative battery cable.

2. Turn the steering wheel so the vehicle's wheels are pointing straight ahead.

3. Turn the ignition switch to the **LOCK** position and remove the key.

4. Remove the **AIR BAG** fuse from the fuse block.

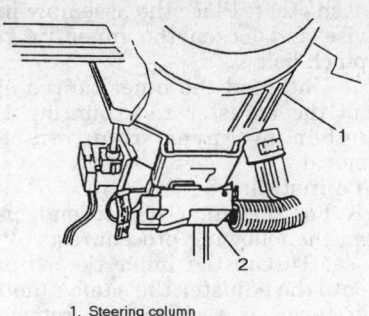

1. Steering column
2. Connector, SRS (yellow)

333823

Disconnect the SRS yellow 2-way connector

5. Remove the left sound insulator.

6. Remove the Connector Position Assurance (CPA) clip from the yellow 2-way connector at the base of the steering column, and disconnect the connector. If equipped with a passenger's side air bag, remove the CPA and disconnect the yellow 2-way connector from the passenger air bag lead.

ENABLING

1. Turn the ignition switch to the **LOCK** position and remove the key.

2. Connect the yellow 2-way connector at the base of steering column and secure it with the CPA clip. If equipped with a passenger's side air bag, connect the yellow 2-way connector at the passenger air bag lead and secure it with the CPA clip.

3. Install the left sound insulator.

4. Install the **AIR BAG** fuse in the fuse block.

5. Turn the ignition switch to the **RUN** position and verify that the **AIR BAG** warning lamp flashes 7 times and then turns OFF.

6. Connect the negative battery cable.

Steering Wheel

REMOVAL AND INSTALLATION

1993–94 Vehicles

1. Disconnect the negative battery cable.

2. If necessary, remove the 2 screws that retain the steering pad.

3. Disconnect the horn lead and remove the horn pad.

4. Remove the retainer, nut and dampener, if equipped.

5. For 1993 vehicles, matchmark the steering wheel to the shaft and remove the steering wheel from the vehicle.

6. For 1994 vehicles, remove the steering wheel using special tool J-1859–03.

To install:

7. Line up matchmark lines on the shaft and the steering wheel and install the steering wheel.

8. Install the retainer, nut and dampener, if equipped. Torque the attaching nut to 30 ft. lbs. (41 Nm).

9. Connect the horn lead and the horn pad.

10. If necessary, install the 2 screws that retain the steering pad.

11. Connect the negative battery cable.

1995–97 Vehicles

---------- **CAUTION** ----------

The Supplemental Inflatable Restraint (SIR) system must be disarmed before removing the steering wheel. Failure to do so may cause accidental deployment of the air bag, resulting in necessary SIR system repairs and/or personal injury.

---

1. Disable the SIR system.

2. Disconnect the negative battery cable.

3. Remove the 2 screws from the back of the inflator module using a No. 30 Torx® driver.

4. Disconnect the Connector Position Assurance (CPA) and inflator module electrical connector from the rear of the module assembly.

---------- **CAUTION** ----------

When carrying a live air bag, make sure the bag and trim cover are pointed away from the body. In the unlikely event of an accidental deployment, the bag will then deploy with minimal chance of injury. When placing a live air bag on a bench or other surface, always face the bag and trim cover up, away from the surface. This will reduce the motion of the module if it is accidentally deployed.

---

5. Remove the inflator module from the steering wheel.

6. Remove the retainer and nut securing the steering wheel to the steering shaft.

7. Using a suitable puller, J-1859-03 remove the steering wheel.

To install:

8. Install the steering wheel onto the steering shaft with the mark on the wheel aligned with the mark on the shaft.

9. Install the mounting nut and tighten to 30 ft. lbs. (41 Nm). Install the nut retainer.

10. Connect the CPA and inflator module electrical connectors.

11. Position the module on the steering wheel and tighten the mounting screws to 80 inch lbs. (9 Nm).

12. Enable the SIR system.

13. Connect the negative battery cable.

Tie Rod Ends

REMOVAL AND INSTALLATION

1993–94 Vehicles

1. For the inner tie rod end, perform the following procedures:

a. From under the hood, remove the lock plate from over the inner tie rod mounting bolts and discard the plate.

b. Remove the inner tie rod end bolt and loosen the opposite side bolt 2 turns.

c. Slide the inner tie rod out from between the plate and the steering rack.

d. Reinstall the lock plate bolt to insure proper tie rod-to-steering gear realignment.

2. Raise and safely support the vehicle. Remove the tire and wheel assembly.

3. Remove the cotter pin and castle nut from the outer tie rod end.

4. For the outer tie rod end, loosen the outer tie rod pinch bolt.

5. Using the steering linkage puller, J–24319–01 or equivalent, separate the tie rod from the steering knuckle.

6. For the inner tie rod end, perform the following procedures:

a. Remove the inner and outer tie rod assembly from the vehicle.

b. Note the position of the inner and outer tie rods in relation to

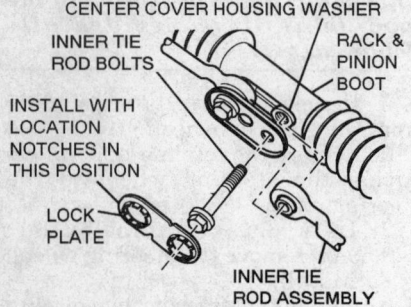

CENTER COVER HOUSING WASHER

INNER TIE ROD BOLTS

RACK & PINION BOOT

INSTALL WITH LOCATION NOTCHES IN THIS POSITION

LOCK PLATE

INNER TIE ROD ASSEMBLY

224413

Inner tie rod removal — 1993–94 vehicles

each other. Place the assembly in a vise and loosen the inner tie rod pinch bolt.

7. Unthread the inner tie rod end from the adjuster rod counting the number of turns required for removal.

To install:

8. For the inner tie rod end, perform the following procedures:

a. Thread the inner tie rod end onto the adjuster the same number of turns as required for removal. Make sure the same number of threads are visible on the inboard and outboard sides of the adjuster.

b. Tighten the pinch bolt to 41 ft. lbs. (55 Nm).

c. Install the tie rod assembly in the vehicle.

d. Remove the lock plate bolt.

e. Connect the inner tie rod end to the rack and pinion unit and tighten the lock plate bolts to 65 ft. lbs. (90 Nm). Install a new lock plate retainer over the mounting bolts

9. For the outer tie rod end, install the tie rod end onto the adjuster rod the same number of turns required to remove the old one.

10. Connect the outer tie rod end to the strut steering knuckle arm and tighten the castle nut to 41 ft. lbs. (55 Nm). If necessary to align the cotter pin holes tighten the nut up to 60 degrees additional rotation. NEVER loosen the nut to make the alignment. Install a new cotter pin.

11. For the outer tie rod end, tighten the pinch bolt temporarily to 41 ft. lbs. (55 Nm).

12. Install the tire and wheel assembly and tighten the wheel nuts to 100 ft. lbs. (140 Nm).

13. Lower the vehicle. Check the front toe and adjust as necessary.

1995–97 Vehicles

1. Raise and safely support the vehicle. Remove the tire and wheel assembly.

2. Loosen the pinch bolt at the outer tie rod end.

3. Remove the cotter pin and castle nut from the outer tie rod end and separate the tie rod end from the strut steering arm using J–24319-01 or an equivalent steering linkage puller.

4. While unthreading the tie rod end from the adjustment rod count the number of turns required to remove the end.

To install:

5. Coat the threads on the adjustment rod with grease and thread the new tie rod end on the same number

of turns required to remove the old one.

6. Connect the tie rod end to the steering knuckle and tighten the castle nut to 35 ft. lbs. (50 Nm) for 1995 vehicles, or to 44 ft. lbs. (60 Nm) for 1996–97 vehicles. If necessary tighten the nut up to 60 degrees ($\frac{1}{6}$ turn) additional rotation to align the cotter pin holes. NEVER loosen the nut to make the alignment. Install a new cotter pin.

7. Tighten the tie rod pinch bolt to 41 ft. lbs. (55 Nm) for 1995 vehicles or to 31 ft. lbs. (42 Nm) for 1996–97 vehicles.

8. Install the tire/wheel assembly and tighten the nuts to 100 ft. lbs. (140 Nm).

9. Lower the vehicle. Check the front end alignment and adjust as necessary.

Power Rack and Pinion

REMOVAL AND INSTALLATION

1. Remove the left sound insulator from under the dashboard. Remove the upper pinch bolt from the steering shaft coupling assembly.

2. Remove the power steering line retainer, if equipped. Remove the power brake booster mounting nuts and disconnect the brake pedal pushrod from the pedal stud.

3. Pull the booster away from the firewall and position away from the rack and pinion mounting clamp. DO NOT remove the master cylinder or allow the brake lines to kink.

4. Raise and safely support the vehicle. Remove the left tire and wheel assembly.

5. Remove the cotter pins and castle nuts from the outer tie rod ends and separate the tie rod ends from the strut arms using J–24319-01, or an equivalent tie rod puller.

6. Remove the 4 clamp mounting bolts and the 2 rack and pinion mounting clamps. Disconnect and cap the power steering lines from the rack and pinion unit.

7. Pull the rack away from the firewall and remove the lower steering shaft pinch bolt and separate the steering shaft from the rack and pinion stub shaft.

8. Remove the dash seal from the rack and pinion housing. Remove the rack and pinion unit out through the hood.

9. If when removing the rack and pinion mounting clamp nuts the studs came out, remove the nuts from the studs and install the studs back

into the firewall. Tighten the studs to 15 ft. lbs. (20 Nm).

To install:

10. Install the rack and pinion assembly through the engine compartment and position it on the firewall. Install the dash seal on the firewall.

11. Connect the steering shaft coupling to the stub shaft and tighten the pinch bolt to 30 ft. lbs. (41 Nm).

12. Connect the power steering lines to the rack and pinion assembly and tighten the fittings to 20 ft. lbs. (27 Nm).

13. Install the rack and pinion clamps and loosely install the mounting nuts. Tighten the left side clamps first then the right to 22 ft. lbs. (30 Nm).

14. Install the power booster on the firewall. Raise and safely support the vehicle.

15. Connect the tie rod ends to the strut arms and tighten the castle nuts. Install the left front tire and wheel assembly.

16. Connect the booster pushrod to the brake pedal and tighten the booster mounting nuts to 22 ft. lbs. (30 Nm). Install the power steering line retainer.

17. Install the steering column upper pinch bolt and tighten to 30 ft. lbs. (41 Nm). Install the left side sound insulator.

18. Refill and bleed the power steering system. Check the front end alignment and adjust as necessary.

Power Steering Pump

BLEEDING

1. Raise and safely support the vehicle so the wheels are off the ground. Turn the wheels all the way to the left. Add power steering fluid to the COLD or FULL COLD mark on the fluid lever indicator.

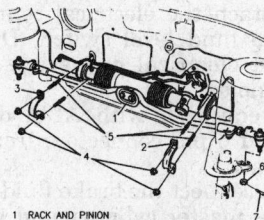

1 RACK AND PINION
2 L.H. CLAMP – HORIZONTAL SLOT AT TOP
3 R.H. CLAMP – HORIZONTAL SLOT AT TOP
4 NUT – 30 N·m (22 LBS.FT.) – HAND START ALL NUTS. TIGHTEN LEFT HAND SIDE CLAMP NUTS FIRST, THEN TIGHTEN RIGHT SIDE NUTS.
5 STUD – 18 N·m (13 LBS. FT.) – AFTER SECOND REUSE OF STUD, THREAD LOCKING KIT NO. 1052624 MUST BE USED.
6 NUT – 60 N·m (44 LBS. FT.)
7 COTTER PIN

331618

Rack and pinion mounting components

2. Start the engine and check the fluid level at fast idle. Add fluid, if necessary to bring the level up to the mark.

3. Bleed air from the system by turning the wheels from side-to-side without hitting the stops. Keep the fluid level at the COLD or FULL COLD mark. Fluid with air in it has a tan appearance.

4. Return the wheels to the center position and continue running the engine for 2–3 minutes.

5. Lower the vehicle and road test to check steering function and recheck the fluid level with the system at its normal operating temperature. Fluid should be at the HOT mark when finished.

REMOVAL AND INSTALLATION

1993–94 Vehicles

2.3L Engine — Without Variable Effort Steering (VES)

1. Disconnect the negative battery cable. Disconnect the pressure and return lines from the pump.

2. Remove the rear bracket to pump bolts. Remove the drive belt and position aside.

3. Remove the rear bracket to transaxle bolts. Remove the front bracket to engine block.

4. Remove the pump and bracket as an assembly. Transfer pulley and bracket, as necessary.

To install:

5. Install the pump and bracket as an assembly. Install front engine bracket to the engine block.

6. Install the rear bracket to transaxle bolts. Install the drive belt.

7. Install the rear bracket to pump bolts. Connect the the pressure and return lines from the pump.

8. Fill the power steering pump with fluid and bleed the system. Connect the negative battery cable and check the pump for proper operation and leaks.

2.3L Engine With Variable Effort Steering (VES)

1. Disconnect the negative battery cable. Detach the V.E.S. electrical connector.

2. Remove power steering pump lines. Back off on adjustment stud.

3. Remove brake booster hose. Remove air intake duct.

4. Remove outer bracket. Remove drive belt and pump. Remove bracket to the pump.

5. Remove the bracket with the power steering pump. Transfer pulley, front bracket and reservoir, if necessary.

To install:

6. Install the pump with pulley, front bracket and reservoir. Install the bracket with pump to the front of the engine.

7. Attach V.E.S. electrical connector, if equipped. Install serpentine drive belt.

8. Install the outer bracket. Install power steering lines.

9. Connect the negative battery cable. Fill the power steering system with fluid, then bleed the air from the system.

3.1L Engine

1. Disconnect the negative battery cable. Remove the serpentine belt.

2. Remove nut from the bracket retaining hose on the alternator.

3. Remove pump 3 bolts to ease pump hose removal.

4. Remove lines at the pump. Transfer pulley, if necessary.

To install:

5. Install lines at the pump. Install pump mounting bolts and torque bolts to 22 ft. lbs. (30 Nm).

6. Install serpentine belt. Install nut to bracket retaining hose on the alternator.

7. Fill with fluid and bleed air from power steering system.

3.3L Engine

1. Disconnect the negative battery cable. Remove the serpentine drive belt.

2. Remove the power steering pump-to-engine bolts. Pull the pump forward and disconnect the pressure tubes.

3. Remove the pump and transfer the pulley, as necessary.

To install:

4. Install the power steering pump. Connect the pressure tubes.

5. Install the power steering pump–to–engine bolts. Install the serpentine drive belt.

6. Adjust the drive belt tension. Fill the power steering pump with fluid and bleed the system.

7. Connect the negative battery cable and check the pump for proper operation and leaks.

1995–97 Vehicles

Service these vehicles with GM P/N 1050017 or equivalent power steering fluid. Do not use transmission fluid. Failure to use the proper fluid will cause hose and seal damage and fluid leaks.

2.4L Engine

1. Disconnect the negative battery cable.

2. Disconnect the Variable Effort Steering (VES) electrical connector from the power steering pump.

3. Disconnect and cap the power steering lines from the pump. Remove the power steering pump bolts and the pump.

To install:

4. Install the power steering pump and tighten the bolts to 19 ft. lbs. (26 Nm).

5. Connect the power steering lines to the pump and tighten the inlet line fitting to 20 ft. lbs. (27 Nm).

6. Connect the electrical connector to the power steering pump.

7. Connect the negative battery cable. Refill and bleed the power steering system.

3.1L Engine

1. Disconnect the negative battery cable. Remove the serpentine belt.

2. Remove the nut securing the power steering lines to the alternator bracket.

3. Remove the front engine mount. Remove the power steering bolts by working through the pulley.

4. Pivot the pump to gain access to the power steering lines. Disconnect and cap the power steering lines.

5. Remove the power steering pump.

To install:

6. Position the power steering pump under the hood in such a manner that the power steering lines can be connected. Tighten the high side line fitting to 20 ft. lbs. (27 Nm).

7. Position the pump in the mounting bracket and tighten the mounting bolts to 22 ft. lbs. (30 Nm).

8. Install the serpentine belt. Install the front engine mount.

9. Install the nut securing the power steering line bracket to the alternator bracket.

10. Connect the negative battery cable. Refill and bleed the power steering system.

BRAKES

Anti-Lock Brake System Service

PRECAUTIONS

• Certain components within the Anti-Lock Brake System (ABS) are not intended to be serviced or repaired individually. Only those components with removal and installation procedures should be serviced.

• Do not use rubber hoses or other parts not specifically specified for and ABS system. When using repair kits, replace all parts included in the kit. Partial or incorrect repair may lead to functional problems and require the replacement of components.

• Lubricate rubber parts with clean, fresh brake fluid to ease assembly. Do not use lubricated shop air to clean parts; damage to rubber components may result.

• Use only specified brake fluid from an unopened container.

• If any hydraulic component or line is removed or replaced, it may be necessary to bleed the entire system.

• A clean repair area is essential. Always clean the reservoir and cap thoroughly before removing the cap. The slightest amount of dirt in the fluid may plug an orifice and impair the system function. Perform repairs after components have been thoroughly cleaned; use only denatured alcohol to clean components. Do not allow ABS components to come into contact with any substance containing mineral oil; this includes used shop rags.

• The Anti-Lock control unit is a microprocessor similar to other computer units in the vehicle. Ensure that the ignition switch is **OFF** before removing or installing controller harnesses. Avoid static electricity discharge at or near the controller.

• If any arc welding is to be done on the vehicle, the control unit should be unplugged before welding operations begin.

Gear Tension Relief

PROCEDURE

When the ABS modulator cylinder pistons are in their uppermost position, each motor has prevailing torque due to the force necessary to en-

sure each piston is held firmly at the top of its travel. This torque results in "gear tension," or force on each gear that makes motor pack separation difficult. To avoid injury, or damage to the gears, the "Gear Tension Relief Sequence" briefly reverses each motor to eliminate the prevailing torque.

This procedure is one of the many functions of GM's Tech 1® scan tool. Use care when using a substitute. In general, make sure the ignition switch is in the **OFF** position. Install the Tech 1® or equivalent with the correct chassis cartridge. Turn the ignition switch to the **ON** position, leaving the engine OFF. Select the proper function. The "Gear Tension Relief Sequence" is **F5** on the Tech 1® scan tool. Note that the same scan tool is needed to bleed the system after repairs to the hydraulic system are completed.

Always perform the "Gear Tension Relief Sequence" prior to removing the hydraulic modulator/master cylinder assembly from the vehicle.

Each hydraulic modulator gear (large gears) should be able to be turned in one direction and then in the opposite direction when the motor pack is removed. If any gear will not move, replace the hydraulic modulator.

Master Cylinder

REMOVAL AND INSTALLATION

—————— CAUTION ——————

To perform the Gear Tension Relief Sequence, a Tech 1® scan tool or equivalent scan tool must be used prior to removal of the ABS modulator/master cylinder assembly.

1. Perform the gear tension release procedure. Disconnect the negative battery cable.

2. Detach the electrical connector from the fluid level switch. Disconnect the electrical connectors from the solenoids.

3. If equipped with ABS, detach the 3 and 6 pin connectors from the motor pack.

4. Disconnect the brake fluid pipes from the master cylinder and modulator assembly. Plug the lines and ports to prevent excessive fluid loss and possible contamination.

5. Unfasten the master cylinder mounting nuts from the power booster studs, then remove the master cylinder and modulator assembly.

To install:

6. Install the master cylinder and modulator assembly onto the power booster studs. Install the mounting nuts and tighten to 20 ft. lbs. (27 Nm).

7. Unplug and connect the brake lines to the master cylinder and modulator assembly, then tighten the fittings to 17 ft. lbs. (23 Nm).

8. Attach the electrical connector to the fluid level switch. Connect the electrical connectors to the solenoids.

9. If equipped with ABS, attach the 3 and 6 pin connectors to the motor pack. Connect the negative battery cable.

10. Refill the master cylinder with brake fluid and bleed the brake system. Note that a scan tool is needed to bleed the system after repairs to the hydraulic system are completed, if equipped with ABS.

Brake Caliper

REMOVAL AND INSTALLATION

1. Siphon ⅔ of the brake fluid out of the master cylinder.

2. Raise and safely support the vehicle. Remove the tire and wheel assembly.

3. Compress the caliper piston back into the caliper bore using a large pair of pliers, C-clamp or special piston retracting tool.

4. If the caliper is to be completely removed from the vehicle for bench service, disconnect and cap the brake line from the caliper. Discard the old washers.

5. Remove the caliper mounting bolts and sleeves. Remove the caliper from the knuckle.

6. Remove the brake pads from the caliper if the caliper is being replaced.

To install:

7. Inspect the caliper for any signs of leakage. If the caliper is leaking, new brake pads will be damaged. Also inspect the condition of the caliper support for rust and corrosion which will hinder the travel on the caliper.

8. Inspect the caliper mounting hardware. New bolts are usually recommended. Clean all parts well. Lubricate the mounting bushings and sleeves with silicone grease as required.

9. Install the brake pads in the caliper. Install the caliper on the steering knuckle.

10. Install the mounting bolts and sleeves and tighten to 40 ft. lbs. (51 Nm).

11. If the caliper brake line was disconnected, connect the brake hose to the caliper using new washers and tighten the mounting bolt to 35 ft. lbs. (44 Nm).

12. Refill the master cylinder and bleed the brake system. Install the tire and wheel assembly.

13. Lower the vehicle. Verify correct brake operation.

Disc Brake Pads

REMOVAL AND INSTALLATION

1. Siphon ⅔ of the brake fluid out of the master cylinder reservoir.

2. Raise and safely support the vehicle. Remove the tire and wheel assembly.

3. Remove the caliper from the steering knuckle without disconnecting the brake hose. DO NOT allow the caliper to hang from the brake hose. Support the caliper with a piece of wire.

4. Remove the outboard pad by pushing in on the outside edge of the pad to release the mounting dowel from the hole in the caliper. When both dowels are unseated push the pad out the bottom of the caliper.

5. Remove the inboard pad from the caliper by pulling it out of the caliper.

To install:

6. Install the inboard pad in the caliper so the spring clip on the pad back engages in the caliper piston.

7. Install the outboard pad over the caliper end until the mounting dowels snap into the mounting holes in the caliper.

8. Install the caliper over the rotor onto the steering knuckle. Install the mounting bolts and sleeves and tighten to 40 ft. lbs. (51 Nm).

9. Install the tire and wheel assembly. Lower the vehicle.

10. Pump the brake pedal several times to seat the pads against the rotor before attempting to move the vehicle.

11. Check the master cylinder level and add fluid as necessary.

Brake Rotor

REMOVAL AND INSTALLATION

1. Raise and safely support the vehicle. Remove the tire and wheel assembly.

2. Remove the caliper and support it on a wire aside. DO NOT disconnect the brake hose from the caliper or allow the caliper to hang from the brake hose.

3. Slide the rotor off the hub.

To install:

4. Inspect the rotor for grooves, heat cracks or excessive runout. A damaged rotor should be replaced.

5. If a new rotor (and/or new brake pad set) is being installed the caliper piston will have to be pushed back into the caliper bore enough so the caliper will fit over the rotor. A special piston retracting tool is available. A large C-clamp may be substituted to push the piston back into the caliper bore.

6. Slide the rotor onto the hub. Install the caliper and tighten the bolts to 40 ft. lbs. (54 Nm).

7. Torque the wheels. Lower the vehicle.

8. Pump the brake pedal several times to seat the pads against the rotor before moving the vehicle.

9. Check and add brake fluid as necessary.

Brake Drums

REMOVAL AND INSTALLATION

1. Raise and safely support the vehicle. Remove the wheel.

2. Pull the brake drum off. It may be necessary to gently tap the rear edge of the drum to start it off the studs. If extreme resistance to removal is encountered, it will be necessary to retract the brake shoe self-adjuster screw (sometimes called the Star Wheel). Knock out the access hole in the brake drum and turn the adjuster to retract the linings from the drum. Install a replacement hole cover before reinstalling the drum.

NOTE: DO NOT hammer on the brake drum to remove it.

To install:

3. Inspect the inside of the brake drum. If worn, heavily grooved or if the opening is distorted, the drum should be refinished or replaced. If refinishing, observe the maximum drum diameter specification.

4. Inspect the wheel cylinder for signs of brake fluid leakage. Inspect the brake shoe springs and self-adjuster mechanism. The adjuster should usually be disassembled, cleaned and lubricated when the drum is removed for brake service.

5. Install the drum over the brake shoes.

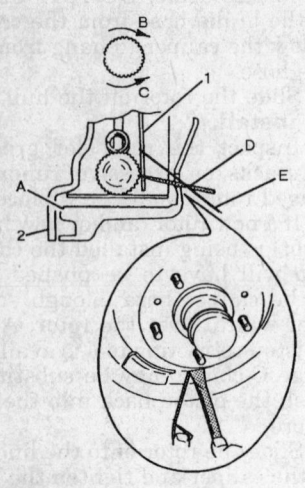

A Brake drum
B Star wheel rotation to retract brake shoes
C Star wheel rotation to expand brake shoes
D Screwdriver
E Wire hook used only when backing off adjustment
1 Parking brake lever
2 Backing plate

329960

Brake adjuster

6. Install the wheel. Adjust the brakes.

7. Lower the vehicle. Check brake operation.

Brake Shoes

REMOVAL AND INSTALLATION

1. Raise and safely support the vehicle. Remove the tire and wheel assembly. Remove the brake drum.

2. Remove the upper return springs from the shoes using tool J-8057 or the equivalent brake spring tool.

3. Remove the hold-down springs using J-8049 or the equivalent brake spring tool.

4. Remove the shoe hold-down pins from behind the brake backing plate.

5. Lift up the actuator lever for the self-adjusting mechanism and remove the actuating link. Remove the actuator lever, pivot, and the pivot return spring.

6. Spread the shoes apart to clear the wheel cylinder pistons and remove the parking brake strut and spring.

7. Disconnect the parking brake cable from the lever. Remove the shoes, still connected by their adjusting screw spring.

8. With the shoes removed, note the position of the adjusting spring and remove the spring and adjusting screw.

9. Remove the C-clip from the parking brake lever and the lever from the secondary shoe.

10. Use a damp cloth to remove all dirt and dust from the backing plate and brake parts.

To install:

11. Check the backing plate attaching bolts to make sure they are tight. Use fine emery cloth to clean all rust and dirt from the shoe contact surfaces on the plate and lubricate with brake grease. Check the wheel cylinder for signs of leakage.

12. Clean all parts completely in brake solvent and air dry. Clean the backing plate shoe contact points.

13. Inspect the inside of the brake drum. If worn, heavily grooved or if the opening is distorted, the drum should be refinished or replaced.

14. Inspect the brake shoe springs and self-adjuster mechanism. Disassemble the adjuster mechanism and clean the threads and coat with grease. Make sure the adjuster assembly turns freely before installing in the vehicle.

15. Install the parking brake lever on the secondary shoe and secure with C-clip.

16. Install the adjusting screw and spring on the shoes, connecting them together. The coils of the spring must not be over the star wheel on the adjuster. The left and right hand springs are not interchangeable.

17. Spread the shoe assemblies and connect the parking brake cable. Install the shoes on the backing plate, engaging the shoes at the top tempo-

rarily with the wheel cylinder pistons. Make sure the star wheel on the adjuster is lined up with the adjusting hole in the backing plate, if equipped.

18. Spread the shoes slightly and install the parking brake strut and spring. Make sure the end of the strut without the spring engages the parking brake lever. The end with the spring engages the primary shoe (the one with the shorter lining).

19. Install the actuator pivot, lever and return spring. Install the actuating link in the shoe retainer. Lift up the actuator lever and hook the link into the lever.

20. Install the hold-down pins through the back of the plate using J-8057 or the equivalent. Install the lever pivots and hold-down springs. Install the shoe return springs using J-8049 or the equivalent. Be careful not to stretch or distort the springs.

21. Make sure the linings are in the right place, the self-adjusting mechanism is correctly installed, and the parking brake parts are hooked up.

22. Measure the distance from the edge of the primary lining to the edge secondary lining, then measure the inside width of the drum. Adjust the linings by means of the adjuster so the drum will fit onto the linings.

23. Install the hub and bearing assembly onto the axle if removed. Tighten the retaining bolts to 35 ft. lbs. (55 Nm).

24. Install the drum and the wheels.

25. Adjust the brakes. Install a rubber hole cover in the adjustment knock-out hole after the adjustment is complete. Adjust the parking brake.

26. Lower the vehicle. Road test the vehicle and verify proper brake operation.

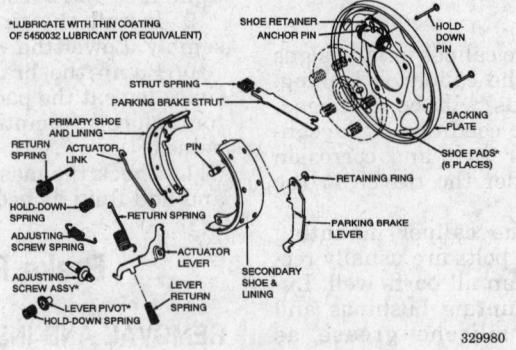

*LUBRICATE WITH THIN COATING OF 5450032 LUBRICANT (OR EQUIVALENT)

SHOE RETAINER
ANCHOR PIN
HOLD-DOWN PIN
STRUT SPRING
PARKING BRAKE STRUT
BACKING PLATE
PRIMARY SHOE AND LINING
RETURN SPRING
ACTUATOR LINK
PIN
SHOE PADS* (6 PLACES)
HOLD-DOWN SPRING
RETURN SPRING
RETAINING RING
ADJUSTING-SCREW SPRING
ACTUATOR LEVER
PARKING BRAKE LEVER
ADJUSTING-SCREW ASSY*
LEVER RETURN SPRING
SECONDARY SHOE & LINING
LEVER PIVOT*
HOLD-DOWN SPRING

329980

Drum brake components — exploded view

Wheel Cylinder

REMOVAL AND INSTALLATION

1993–95 Vehicles

NOTE: **This vehicle can have 2 different types of wheel cylinders depending on the rear brake option. Before beginning the removal procedure, verify which type of wheel cylinder mounting the vehicle is equipped with.**

1. Raise and safely support the vehicle. Remove the rear tire and wheel assembly.
2. Remove the brake drum. Remove the brake shoes.
3. Disconnect and cap the brake pipe from the wheel cylinder.
4. If wheel cylinder is equipped with mounting bolts, remove the 2 mounting bolts.
5. In order to remove the wheel cylinder from the backing plate it may be necessary to remove the bleeder screw from the wheel cylinder.
6. If wheel cylinder is equipped with a retaining ring, use 2 awls or tool J-29839, release both sides of the wheel cylinder locking, work the locking ring off the wheel cylinder.
7. Remove the wheel cylinder from the backing plate.
   To install:
8. Remove the bleeder screw from the new wheel cylinder.
9. If wheel cylinder is equipped with a retaining ring, perform the following procedure:
   a. Position the wheel cylinder in the backing plate.
   b. Position the locking ring on the back of the wheel cylinder.
   c. Place a block of wood between the wheel cylinder and the hub assembly.
   d. Using a 1⅛ inch socket and a 12 inch extension, drive the retaining ring into place on the back of the wheel cylinder.
10. If wheel cylinder is equipped with mounting bolts, install the wheel cylinder and wheel cylinder mounting nuts. Tighten the nuts to 106 inch lbs. (12 Nm).
11. Connect the brake pipe to the wheel cylinder and tighten the fitting to 11 ft. lbs. (15 Nm)
12. Install the brake shoes. Install the brake drum and the wheel.
13. Adjust the brakes. Lower the vehicle.
14. Refill the master cylinder and bleed the system.

1996–97 Vehicles

1. Raise and safely support the vehicle. Remove the tire and wheel assembly.
2. Remove the brake drum assembly. Remove the hub and bearing assembly.
3. Disconnect the inlet line at the back of the wheel cylinder and plug line opening.
4. Remove the wheel cylinder bolts and the wheel cylinder.
   To install:
5. Position the wheel cylinder into place and install the mounting bolts.
6. Tighten the mounting bolts to 15 ft. lbs. (20 Nm).
7. Install the inlet line to the back of the wheel cylinder and tighten to 17 ft. lbs. (23 Nm).
8. Install the hub and bearing assembly and tighten the bolts to 43 ft. lbs. (58 Nm).
9. Install the brake drum. Install the tire and wheel assembly.
10. Bleed the wheel cylinder. Lower the vehicle.
11. Check the master cylinder level and fill as necessary.
12. Road test the vehicle and verify proper operation.

Parking Brake Cable

ADJUSTMENT

NOTE: **The rear brake shoes must be adjusted properly before parking brake adjustment can be performed. Failure to properly adjust the brakes can lead to overtightening the parking brake cable adjuster resulting in rear brake drag.**

1993–95 Vehicles

1. Depress the parking brake pedal exactly 3 ratchet clicks. Raise and safely support the vehicle.
2. Check that the equalizer nut groove is liberally lubricated with chassis lube. Tighten the adjusting nut until the right rear wheel can just be turned to the rear with both hands but is locked when forward rotation is attempted.
3. With the mechanism totally disengaged, both rear wheels should turn freely in either direction with no brake drag. Do not adjust the parking brake so tightly as to cause brake drag.

1996–97 Vehicles

Fully apply and release the hand brake 4–6 times to self-adjust.

REMOVAL AND INSTALLATION

Except 1996–97 Grand Am

Front Cable

1. Disconnect the negative battery cable. Raise and safely support the vehicle.
2. Bend the tang on connector to allow cable removal.
3. Pull and hold cable toward the rear of car to create slack in the cable.
4. Detach the cable from the connector, equalizer assembly and guide. Lower the vehicle.
5. Disconnect the parking brake cable from the lever or reel assembly.

NOTE: **Use caution when removing spring to avoid hitting and knocking the self-adjusting spring loose.**

6. Remove the cable conduit fitting from the lever assembly while depressing the retaining tangs.
7. Remove left rocker panel/door sill plate. Remove grommet from the floor pan and cable from 2 center body clips, then remove front cable.
   To install:
8. Install parking brake cable to lever and reel assembly. Install cable conduit fitting securing cable to the lever assembly.
9. Install grommet to the floor pan, grommet retainer and 2 center body clips. Install left rocker panel/door sill plate.
10. Install cable through guide and equalizer to the connector and route under the left rear brake line.
11. Carefully lower the vehicle. Fully apply and release the foot brake 4–6 times to self adjust the system.
12. Connect the negative battery cable and check the parking brakes for proper operation.

Rear Cables

1. Disconnect the negative battery cable. Raise and safely support the vehicle.
2. Bend the tang on connector to allow cable removal. Pull and hold cable toward the rear of car to create slack in the cable.
3. Disconnect the left rear cable from the equalizer or remove the right rear cable from the connector. Remove the tire and wheel assembly and the brake drum.
4. Insert a suitable prytool between the brake shoe and the top part of the brake adjuster bracket. Push the bracket to the front and release the top adjuster bracket rod.

5. Remove the hold-down spring, actuator lever and lever return spring. Remove the adjuster screw spring.

6. Remove the top rear brake shoe return spring. Disconnect the parking brake cable from the parking brake lever.

7. Remove the conduit fitting from the backing plate while depressing the retaining tangs. Remove the conduit fitting from the axle bracket.

To install:

8. Install the conduit fitting into the backing plate Connect the cable to the parking brake lever in the drum assembly.

9. Install the top rear brake shoe return spring. Install the adjuster screw spring.

10. Install the lever return spring, actuator lever and the rear hold-down spring.

11. Install the top adjuster bracket rod. Install the brake drum(s) and the wheel and tire assembly.

12. Install the conduit fitting retaining tangs and the conduit fitting into the axle bracket.

13. Fasten the cable to the equalizer assembly (left rear cable). Attach the cable end button to the connector (front and rear cables).

14. Carefully lower the vehicle, then connect the negative battery cable. Fully apply and release the foot brake 4–6 times to self adjust the system.

1996–97 Grand Am

Front Cable

1. Raise and safely support the vehicle. Pull and hold cable toward the rear of car to create slack in the cable.

2. Bent the tang on the connector to allow cable removal. Detach the cable from the connector.

3. Fold over the retaining clip. Disconnect the cable from the equalizer. Carefully lower the vehicle.

4. Remove the center console and the shifter boot. Move the parking brake lever to the off position.

5. Detach the cable conduit end fitting from the handle assembly.

6. Pull the cable until the notch on the ratchet is visible through the cover plate opening.

7. Push the pawl spring downward toward the notch in the ratchet.

8. Release the cable slowly to allow the notch to catch the leg of the spring.

9. Detach the front parking brake cable button from the reel assembly. Remove the left rocker panel/door sill plate.

10. Remove the grommet and retainer from the floor pan. Detach the cable from the clip on the No. 2 bar, under the carpet, then remove the front cable.

To install:

11. Route the cable through the floor pan from the interior to the exterior. Attach the cable conduit fitting to the handle assembly.

12. Fasten the front parking brake cable button to the reel assembly. Install the cable into the No. 2 bar clip, under the carpet.

13. Install the console and shifter boot. Install the left rocker panel/door sill plate.

14. Raise and safely support the vehicle. Fold over bracket to the body.

15. Fasten the cable to the equalizer assembly and route under the left rear brake line.

16. Attach the cable to the connector. Bend the tang on the connector to retain the cable.

17. Carefully lower the vehicle. Fully apply and release the foot brake 4–6 times to self adjust the system.

Rear Cables

1. Remove the console and shifter boot. Move the parking brake lever to the off position.

2. Detach the cable conduit fitting from the handle assembly. Pull the cable until the notch on the ratchet is visible through the cover plate opening.

3. Push the pawl spring downward toward the notch in the ratchet. Release the cable slowly to allow the notch to catch the leg of the spring.

4. Raise and safely support the vehicle. Bend the tang on the connector to allow cable removal.

5. Pull and hold the cable towards the rear of the car to create slack in the cable. Detach the cable from the equalizer and the conduit fitting from the axle bracket.

6. Remove the tire and wheel assembly, then remove the brake drum.

7. Insert a suitable prytool between the brake shoe and the top part of the brake adjuster bracket.

8. Remove the cable from the bracket. Remove the conduit fitting from the backing plate while depressing the conduit fitting retaining tangs.

To install:

9. Install the conduit fitting into the backing plate. Connect the parking cable to the parking brake cable lever in the drum assembly.

10. Install the brake drum, then install the tire and wheel assembly. Fasten the conduit fitting retaining tangs and conduit fitting into the axle bracket.

11. Attach the cable to the equalizer, then fasten the cable to the connector. Carefully lower the vehicle

12. Install the console and shifter boot. Fully apply and release the hand brake 4–6 times to self adjust the system.

Brake System Bleeding

PROCEDURE

Use only fresh Delco supreme 11 brake fluid GM part No. 1052535 or an equivalent DOT 3 brake fluid from a sealed container. Under NO circumstances should DOT 4 or 5 be used; brake system contamination will result and every brake system component will have to be replaced.

Master Cylinder Bench Bleeding

1. Install the master cylinder in a vise, clamping at the flange. DO NOT clamp the master cylinder by the body, it may distort the body and cause a pressure loss.

2. Install short lengths of hoses/lines and fittings to the master cylinder ports, aiming the ends of the hoses/lines into the master cylinder reservoir. Fill the master cylinder with fresh DOT 3 brake fluid and make sure the ends of the hoses/lines are submerged in the brake fluid.

3. Pump the master cylinder several times until there are no air bubbles seen coming out of the hose/line ends.

4. Remove the hoses/lines and fittings and install the shipping caps to the ports, to prevent brake fluid from dripping on any painted surface.

5. Install the master cylinder cap and install the master cylinder.

6. Connect the brake lines to the master cylinder ports and snug the fittings.

7. Have an assistant depress the brake pedal slowly one time and hold. Loosen the forward brake line fitting to purge air from the master cylinder bore. Tighten the brake line fitting and have the assistant slowly release the brake pedal. Wait 15 seconds and repeat the sequence, including the 15 second wait, until all air is purged from the master cylinder bore.

8. Repeat the bleeding procedure at the rear brake line fitting.

9. Finally tighten the brake line fittings to 11 ft. lbs. (15 Nm).

10. Check the master cylinder fluid level and bleed the system. Road test and check for proper brake operation.

Master Cylinder Bleeding On-Vehicle

1. Loosen the brake line fittings at the master cylinder.
2. Pour fresh brake fluid into the master cylinder reservoir until fluid flows from the master cylinder ports, then tighten the fittings.
3. Have an assistant depress the brake pedal slowly one time and hold. Loosen the forward brake line fitting to purge air from the master cylinder bore. Tighten the brake line fitting and have the assistant slowly release the brake pedal. Wait 15 seconds and repeat the sequence, including the 15 second wait, until all air is purged from the master cylinder bore.
4. Repeat the bleeding procedure at the rear brake line fitting.
5. Final tighten the brake line fittings to 11 ft. lbs. (15 Nm).
6. Connect the electrical connector to the fluid level sensor.
7. Check the master cylinder fluid level and bleed the system. Road test and check for proper brake operation.

Brake System Bleeding

1. Have an assistant pump the brake pedal several times using smooth even strokes, pausing 10 seconds between each stroke. After the last stroke hold the pedal down.

NOTE: When pumping the brakes DO NOT force or stab the pedal down or damage to the master cylinder could result.

2. Starting with the right rear wheel, connect a clear rubber hose to the bleeder screw and submerge the other end of the hose in a container partially filled with brake fluid.
3. Open the bleeder screw about ½ turn.
4. When the brake pedal is fully depressed, close the bleeder screw.
5. Repeat the procedure as needed until no air comes out of the bleeder.
6. Repeat Steps 2–5 on the remaining wheels in the following order:
   a. Left rear
   b. Right front
   c. Left front
7. While bleeding the brakes check the fluid level in the master cylinder frequently and add fluid as necessary. DO NOT allow the master cylinder to be pumped dry.
8. After the right front wheel is done verify that all the bleeder screws are tight and not leaking. Add brake fluid to the master cylinder reservoir as required.
9. Test drive the vehicle carefully to verify correct braking operation.

Wheel Speed Sensor

REMOVAL AND INSTALLATION

The front wheel speed sensors are of a variable reluctance type. Each sensor is attached to the knuckle assembly in close proximity to a toothed ring. This results in (as the teeth pass by the sensor) an A/C voltage with a frequency proportional to the speed of the wheel. The magnitude of the voltage and frequency increase with speed. The sensor is not repairable, or adjustable. The rear wheel speed sensors operate in the same manner as the front wheel speed sensors. If a rear wheel speed sensor fails, the entire integral hub/bearing and speed sensor assembly must be replaced.

Front Wheel Speed Sensor

1. Disconnect the negative battery cable. Raise and safely support the vehicle.
2. Remove the front tire and wheel. Disconnect the wheel speed sensor electrical connector.
3. Remove the retaining bolt.
4. Remove the front wheel speed sensor. If sensor will not slide out of the knuckle, remove the brake rotor and use a blunt punch or equivalent tool to gently push the sensor from the back side of the knuckle.
   To install:
5. Install the front wheel speed sensor on the mounting bracket.

NOTE: Make sure the front wheel speed sensor is properly aligned and lays flat against the bosses on the knuckle.

6. Install the retaining bolt and tighten to 107 inch lbs. (12 Nm).
7. Connect the electrical connector to the front wheel speed sensor.
8. Install the tire and wheel. Lower the vehicle.

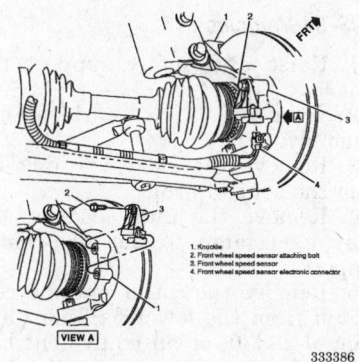

1. Knuckle
2. Front wheel speed sensor attaching bolt
3. Front wheel speed sensor
4. Front wheel speed sensor electronic connector

VIEW A

333386

Front wheel speed sensor assembly

9. Connect the negative battery cable. Road test the vehicle and verify proper operation.

Rear Wheel Speed Sensor

1. Disconnect the negative battery cable. Raise and safely support the vehicle.
2. Remove the tire and wheel and the brake drum.
3. Remove the bolts and nuts attaching the rear wheel bearing and speed sensor assembly.
4. Remove the wheel bearing and speed sensor assembly and disconnect the electrical connector from the speed sensor.

— **CAUTION** —
The brake assembly will be held in place by the brake pipe connection at this point. Use care not to bump or exert force on the brake assembly, or damage can occur to the brake pipe.

To install:
5. Connect the electrical connector to the wheel speed sensor.
6. Install the wheel bearing and speed sensor assembly into place.
7. Install the mounting bolts and nuts and tighten to 37 ft. lbs. (50 Nm).
8. Install the brake drum and the tire and wheel. Lower the vehicle.
9. Connect the negative battery cable. Road test the vehicle and verify proper operation.

FRONT SUSPENSION

Strut

REMOVAL AND INSTALLATION

1. Raise and safely support the vehicle. Remove the front tire and wheel assembly.
2. Remove the cotter pin and castle nut from the outer tie rod end and separate the ball stud from the strut arm using J-24319-01, or an equivalent tie rod puller.
3. Remove the bolt from the brake line bracket and separate the bracket from the strut.
4. Scribe reference marks on the front strut and steering knuckle for installation purposes.
5. Remove the strut lower mounting bracket nut and through bolts.

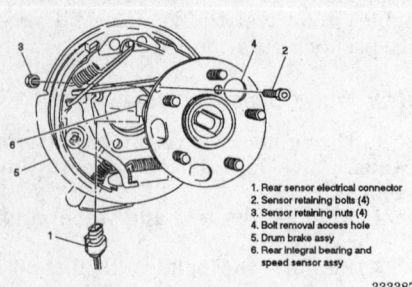

1. Rear sensor electrical connector
2. Sensor retaining bolts (4)
3. Sensor retaining nuts (4)
4. Bolt removal access hole
5. Drum brake assy
6. Rear integral bearing and speed sensor assy

333387

Rear wheel speed sensor and hub/bearing assembly

6. Lower the vehicle until the upper strut plate mounting bolts are accessible.

7. Support the lower control arm.

8. Hold the strut and remove the strut plate mounting nuts and bolt. Remove the strut from the vehicle.

WARNING

If the strut cartridge is being replaced, a strut compressor tool J-34013-A or equivalent must be used to remove the coil spring from the strut assembly. Failure to use a strut compressing tool can result in personal injury and/or part damage.

To install:

9. Install the strut assembly into the vehicle and install the upper strut plate mounting nuts.

10. Connect the lower strut bracket to the steering knuckle and install the through bolts.

11. Raise the vehicle.

12. Install the through bolt nuts and tighten to 133 ft. lbs. (180 Nm) with the reference marks in alignment.

13. Connect the tie rod end to the strut arm and tighten the castle nut to 44 ft. lbs. (60 Nm). If necessary tighten the nut up to 60 degrees (1/6 turn) additional to align the cotter pin holes. NEVER loosen the nut to make the holes align. Install a new cotter pin.

14. Connect the brake line bracket to the strut and tighten the bolt to 10 ft. lbs. (14 Nm).

15. Install the tire and wheel assembly. Lower the vehicle.

16. Tighten the upper strut plate mounting nuts and bolt to 18 ft. lbs. (25 Nm).

17. Check the front end alignment and adjust as necessary.

Lower Ball Joints

REMOVAL AND INSTALLATION

1. Raise and safely support the vehicle. Remove the tire and wheel assembly.

2. Remove the cotter pin and castle nut from the outer ball joint.

3. Separate the ball joint stud from the steering knuckle using J-29330, or an equivalent ball joint separator.

4. Drill a 1/8 inch pilot hole in the center of each of the 3 ball joint mounting rivets.

5. Using a 1/2 inch drill bit, drill the heads off the rivets.

6. With a hammer and punch, knock the rivets out of the control arm. Remove the sway bar link kit.

7. Pull the control arm down so the ball stud clears the steering knuckle. Slide the ball joint out of the control arm.

To install:

8. Position the ball joint into the lower control arm and install the bolts. The nuts must be on top of the control arm. Tighten the mounting bolts to the specification provided with the ball joint service kit.

9. Connect the ball joint to the steering knuckle and tighten the castle nut to 41 ft. lbs. (55 Nm). If necessary to align the cotter pin holes tighten the nut up to 60 degrees (1/6 turn) additional rotation. NEVER loosen the nut to make the holes align. Install a new cotter pin.

10. Install the link kit and tighten the nut to 22 ft. lbs. (30 Nm).

11. Install the tire and wheel assembly and tighten the wheel nuts to 100 ft. lbs. (140 Nm).

12. Lower the vehicle.

Lower Control Arms

REMOVAL AND INSTALLATION

1993–94 Vehicles

1. Raise and safely support the vehicle.

2. Remove tire and wheel assembly.

3. Remove the sway bar link kit from the control arm.

4. Remove the nuts securing the sway bar clamp to the suspension support.

5. Remove the cotter pin and castle nut from the lower ball joint and using J-29330, or an equivalent ball joint separator, disconnect the ball joint stud from the steering knuckle.

6. Support the suspension support with a suitable jack.

7. Remove the suspension support bolts and support assembly.

8. Remove the control arm-to-suspension support bolts and the control arm.

To install:

9. Position the control arm on the suspension support and loosely install the through bolts and nuts.

10. Raise the suspension support assembly into place and loosely install the mounting bolts.

11. Install the nuts securing the sway bar clamp to the suspension support and tighten to 30 ft. lbs. (40 Nm).

12. Connect the ball joint to the steering knuckle and tighten the castle nut to 41 ft. lbs. (55 Nm). If necessary to align the cotter pin holes tighten the nut up to 60 degrees additional rotation. NEVER loosen the nut to make the alignment.

13. Install the sway bar link kit and tighten the nut to 13 ft. lbs. (17 Nm) on vehicles with base suspension and to 70 ft. lbs. (95 Nm) on vehicles with the direct acting suspension.

14. Install the tire and wheel assembly and tighten the wheel nuts to 100 ft. lbs. (140 Nm).

15. Lower the vehicle.

16. With the vehicle at curb height tighten all suspension support bolts to 66 ft. lbs. (90 Nm).

17. Tighten the control arm bolts to 61 ft. lbs. (83 Nm).

18. Check the front end alignment and adjust as necessary.

1995–97 Vehicles

1. Raise and safely support the vehicle. Remove the tire and wheel assembly.

2. Remove the nut, washers and insulators from the link kit and remove the link kit through bolt.

3. Remove the cotter pin and castle nut from the lower ball joint and separate the ball joint from the steering knuckle using J-29330, or an equivalent ball joint separator.

4. Remove the control arm bolts and the control arm from the vehicle.

To install:

5. Install the control arm into the suspension support and loosely install the mounting bolts.

6. Connect the ball joint to the steering knuckle and tighten the castle nut to 55 ft. lbs. (75 Nm). If necessary tighten the nut up to 60 degrees (1/6 turn) additional rotation to align the cotter pin holes. NEVER loosen the nut to make the alignment. Install a new cotter pin.

7. Install the link kit and tighten the nut to 22 ft. lbs. (30 Nm).

8. Install the tire and wheel assembly. Lower the vehicle.

9. With the vehicle at the correct ride height, tighten the lower control arm mounting bolts to 61 ft. lbs. (83 Nm).

10. Check the front end alignment and adjust as necessary.

Sway Bar

REMOVAL AND INSTALLATION

1993–94 Vehicles

1. Raise and safely support the vehicle allowing the front control arms to hang free.

2. If equipped with base suspension, perform the following procedures:

   a. Remove the left front tire and wheel assembly.

   b. Remove the sway bar link kits, taking note of the positions of the washers and insulators.

   c. Remove the nuts from the sway bar center brackets.

3. If equipped with direct acting suspension, perform the following procedures:

   a. Remove the front tire and wheel assemblies.

   b. Remove the nuts securing the sway bar to the sway bar links.

   c. Disconnect the stabilizer shaft clamps from the support assemblies.

4. Loosen the front suspension support bolts and remove the rear and center bolts. Allow the rear of the suspension support assembly to hang down.

5. Remove the sway bar and sway bar bushings.

To install:

6. Position the sway bar in the vehicle with the sway bar bushings. The slits in the bushings must face toward the rear of the vehicle.

7. Install the sway bar clamps and loosely install the mounting nuts.

8. If equipped with base suspension, install the sway bar link kits and tighten the nuts to 13 ft. lbs. (17 Nm).

9. Install the suspension support bolts and tighten the center bolts to 66 ft. lbs. (90 Nm). Tighten the front bolts to 65 ft. lbs. (88 Nm) and then the rear bolts to 65 ft. lbs. (88 Nm).

10. If equipped with direct acting suspension, connect the sway bar to

the sway bar links and tighten the nuts to 70 ft. lbs. (95 Nm).

11. Make sure the sway bar is centered in the vehicle and tighten the center clamp nuts to 16 ft. lbs. (22 Nm).

12. Install the left tire and wheel assembly. Lower the vehicle.

1995–97 Vehicles

1. Raise and safely support the vehicle.

2. Remove the front tire and wheel assemblies.

3. Remove the link kits nuts, washers, insulators and through bolts.

4. Remove the nuts from the sway bar clamps at the suspension crossmember.

5. Support the rear of the suspension support and remove the rear and center support bolts and loosen but do not remove the front bolts.

6. Lower the suspension support until the sway bar clamps can be removed from the sway bar.

7. Remove the sway bar and bushings.

To install:

8. Install the sway bar bushings on the sway bar with the slits in the bushings facing toward the rear of the vehicle.

9. Position the sway bar on the suspension support and install the clamps. Install the mounting nuts loosely.

10. Raise the suspension support into position and install the mounting bolts. Tighten all the bolts to 89 ft. lbs. (120 Nm). The center bolts should be tightened first, followed by the front then rear.

11. Install the sway bar links and tighten the nuts to 22 ft. lbs. (30 Nm).

12. Make sure the sway bar is centered in the vehicle and tighten the clamp nuts to 22 ft. lbs. (30 Nm).

13. Install the tire and wheel assemblies and tighten the wheel nuts to 100 ft. lbs. (140 Nm).

14. Lower the vehicle.

Front Wheel Bearings

ADJUSTMENT

These vehicles are equipped with sealed hub and bearing assemblies. The hub and bearing assemblies are non-serviceable. If the assembly is damaged, the complete unit must be replaced.

REMOVAL AND INSTALLATION

1. Raise and safely support the vehicle. Remove the front tire and wheel assembly.

2. Insert a drift punch through the caliper and into the rotor cooling fins. This keeps the assembly from turning while the axle nut is being loosened.

3. Remove the axle nut and washer. Remove the punch.

4. Remove the caliper mounting bolts and remove the caliper from the steering knuckle. Support the caliper aside. DO NOT allow the caliper to hang unsupported from the brake hose.

5. Remove the brake rotor.

6. Remove the 3 bolts securing the hub bearing to the steering knuckle. Remove the backing plate.

7. Using J-28733-A, or an equivalent front hub remover, separate the halfshaft from the hub bearing assembly.

8. Remove the hub bearing assembly.

To install:

9. Install the hub bearing assembly over the end of the axle making sure the splines engage smoothly.

10. Install the backing plate and the 3 hub bearing mounting bolts. Tighten the bolts to 70 ft. lbs. (95 Nm).

11. Install the brake rotor.

12. Install the caliper onto the steering knuckle and tighten the caliper mounting bolts to 38 ft. lbs. (51 Nm).

13. Install a drift punch into the rotor cooling fins. This keeps the assembly from turning while the axle nut is being tightened. Torque the axle nut to 192 ft. lbs. (260 Nm).

14. Install the front tire and wheel assembly and tighten the wheel nuts to 100 ft. lbs. (140 Nm). Lower the vehicle.

REAR SUSPENSION

Shock Absorber

REMOVAL AND INSTALLATION

1993–94 Vehicles

1. Open the hatch or trunk lid, remove the trim cover if present, and upper shock absorber nut.

2. Raise and safely support the vehicle.

3. If equipped with air lift type shock absorbers, disconnect the air line. Remove the lower bolt and shock absorber.

4. If new shock absorbers are being installed, repeatedly compress them while inverted and extend them in their normal upright position. This will purge them of air.

To install:

5. Install the shock absorber and lower mount nuts. Torque to 38 ft. lbs. (51 Nm).

6. If equipped with air lift type shock absorbers, connect air lines. Safely lower vehicle.

7. Install upper mount nuts. Torque to 13 ft. lbs. (17 Nm). Install trim cover.

1995–97 Vehicles

— WARNING —

When doing this procedure, if both shocks are to be replaced, only remove one shock at a time or damage can occur to the axle assembly and/or personal injury.

1. Raise and safely support the vehicle with the wheel about 6 inches off the ground. Support the rear axle assembly with a safety stand or hydraulic floor jack.

2. From inside the trunk, remove the cap from the upper shock mounting plate. Remove the 2 shock plate mounting nuts and remove the reinforcement from the shock tower.

3. From under the suspension, remove the lower shock mounting nut. Compress the shock and remove it from the vehicle.

4. If the shock is being replaced, remove the upper shock insulator, washer and mount.

To install:

5. Install the mount, washer and insulator on the shock and tighten the mounting nut to 21 ft. lbs. (29 Nm).

6. Install the shock assembly into the vehicle. Make sure the alignment tab on the lower mount is pointing toward the rear of the vehicle.

7. Install the mounting nut and tighten to 35 ft. lbs. (47 Nm).

8. Install the reinforcement and upper shock plate mounting nuts and tighten to 18 ft. lbs. (24 Nm). Install the cap over the mounting plate.

9. Remove the safety stand or hydraulic floor jack from under the axle. Lower the vehicle.

Coil Spring

REMOVAL AND INSTALLATION

1. Raise and safely support the vehicle. Support the rear axle with a suitable screw type jack.

2. Remove the lower shock absorber mounting nuts. Remove the right and left brake line mounting bolts from the floor of the vehicle.

3. Slowly lower the rear axle assembly until the tension is removed from the coil springs. Remove the coil springs and insulators.

To install:

4. Install the springs and insulator into the vehicle. The ends of the lowest coil must be within 9/16 of the spring stop.

5. Raise the rear axle assembly until the lower shock nuts can be installed. Tighten the nuts to 35 ft. lbs. (47 Nm).

6. Connect the brake line brackets to the frame and tighten the mounting bolts to 97 inch lbs. (11 Nm).

7. Remove the jack. Lower the vehicle.

Sway Bar

REMOVAL AND INSTALLATION

1. Raise and safely support the vehicle.

2. Remove the nuts and bolts at the outboard ends of the sway bar and remove the outer sway bar mounting clamps.

3. Remove the nuts from the inboard mounting clamps. Bend the clamps open and remove the sway bar and sway bar bushings.

To install:

4. Install the sway bar and sway bar bushings into the vehicle and bend the inboard clamps around the sway bar bushings. Loosely install the mounting bolts and nuts.

5. Install the outboard sway bar clamps with the mounting nuts and bolts.

6. Make sure the sway bar is centered in the vehicle. Tighten the outboard clamp nuts to 16 ft. lbs. (22 Nm).

7. Tighten the nuts on the inboard clamps to 10 ft. lbs. (14 Nm). Lower the vehicle.

Wheel Bearings

ADJUSTMENT

These vehicles are equipped with sealed hub and bearing assemblies. The hub and bearing assemblies are non-serviceable. If the assembly is damaged, the complete unit must be replaced.

REMOVAL AND INSTALLATION

1. Raise and safely support the vehicle. Remove the wheel and tire assembly. Remove the brake drum.

2. Remove the 4 hub bearing assembly mounting nuts from behind the backing plate while holding the bolts from the front. The mounting bolts can be removed from the backing plate once the nuts have been removed. The top rear bolt will not clear the brake shoes and must be removed with the bearing assembly.

3. Disconnect the ABS speed sensor wire from the hub bearing assembly. Remove the hub bearing assembly.

To install:

4. Position the hub bearing assembly on the backing plate and insert the 4 mounting bolts. Connect the ABS wheel speed sensor, if equipped.

5. Install the hub bearing mounting nuts from behind the backing plate and tighten the nuts to 38 ft. lbs. (52 Nm) while holding the bolts.

6. Install the brake drum.

7. Install the tire and wheel assembly and tighten the wheel bolts to 100 ft. lbs. (140 Nm). Lower the vehicle.

GM "W" BODY

Front Wheel Drive

BUICK-Regal **CHEVROLET**-Lumina • Monte Carlo
OLDSMOBILE-Cutlass Supreme **PONTIAC**-Grand Prix

24

FIRING ORDERS

NOTE: To avoid confusion, always replace spark plug wires one at a time.

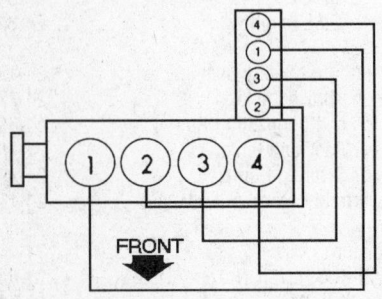

325411

2.2L (VIN 4) Engine
Engine Firing Order: 1–3–4–2
Distributorless Ignition System

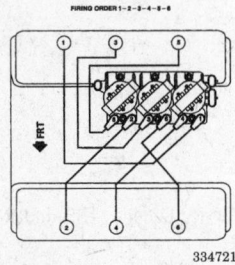

334721

3.1L (VIN T) and (VIN M)
Engine
Engine Firing Order:
1–2–3–4–5–6
Distributorless Ignition

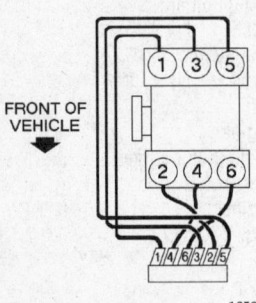

165851

3.4L (VIN X) Engine
Engine Firing Order:
1–2–3–4–5–6
Distributorless Ignition System

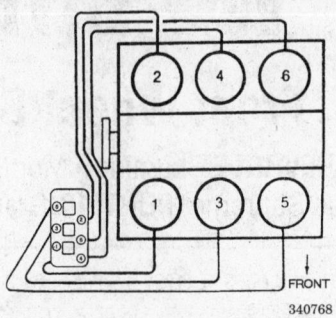

340768

3.8L (VIN L) and (VIN K) Engine
Engine Firing Order: 1–6–5–4–3–2
Distributorless Ignition System

ENGINE ELECTRICAL

NOTE: Disconnecting the negative battery cable on some vehicles may interfere with the functions of the on board computer systems and may require the computer to undergo a relearning process, once the negative battery cable is reconnected.

Ignition Timing

ADJUSTMENT

The Ignition Timing is not adjustable, and is set according to engine demand electronically. The Powertrain Control Module (PCM) controls the ignition timing for all driving conditions.

Alternator

PRECAUTIONS

Several precautions must be observed with alternator equipped vehicles to avoid damage to the unit.

- If the battery is removed for any reason, make sure it is reconnected with the correct polarity. Reversing the battery connections may result in damage to the 1-way rectifiers.
- When utilizing a booster battery as a starting aid, always connect the positive to positive terminals and the negative terminal from the booster battery to a good engine ground on the vehicle being started.
- Never use a fast charger as a booster to start vehicles.
- Disconnect the battery cables when charging the battery with a fast charger.
- Never attempt to polarize the alternator.
- Do not use test lights of more than 12 volts when checking diode continuity.
- Do not short across or ground any of the alternator terminals.
- The polarity of the battery, alternator and regulator must be matched and considered before making any electrical connections within the system.
- Never separate the alternator on an open circuit. Make sure all connections within the circuit are clean and tight.
- Disconnect the battery ground terminal when performing any service on electrical components.
- Disconnect the battery if arc welding is to be done on the vehicle.

REMOVAL AND INSTALLATION

2.2L (VIN 4) Engine

1. Disconnect the negative battery cable.
2. Remove the serpentine belt from the alternator pulley.
3. Remove the rear bracket/heat shield nuts from the exhaust manifold studs.
4. Remove the rear bracket/heat shield bolts from the block near the water pump.
5. Remove the rear bracket/heat shield bolt from the alternator.
6. Remove the rear bracket/heat shield. Detach the electrical connections from the alternator.
7. Remove the alternator front mounting bolts and remove the alternator from the mounting bracket.
   To install:
8. Position the alternator in the mounting bracket and install the front mounting bolts. With both bolts in place tighten the lower mounting bolt (long bolt) to 37 ft. lbs. (50 Nm) and the upper bolt (short bolt) to 18 ft. lbs. (25 Nm).
9. Install the remaining components in the reverse order of removal. Torque the following:
- Battery wire-to-alternator post nut to 70 inch lbs. (8 Nm)
- Mounting bolt at the alternator to 18 ft. lbs. (25 Nm)
- Bracket-to-block bolt near the water pump to 63 ft. lbs. (85 Nm)
- Bracket-to-manifold stud nuts to 37 ft. lbs. (50 Nm)
10. Connect the negative battery cable. Start the engine and verify proper charging system operation.

3.1L (VIN T) Engine

1. Remove the air cleaner assembly. Disconnect the negative battery cable.

2. Detach the alternator connector.

3. Remove the serpentine belt.

4. Remove the front alternator bolt, rear rear bolt and alternator rear brace bolt.

5. Remove the alternator from the mounting bracket and tilt forward.

6. Remove the cap from the alternator post and remove the nut securing the battery wire to the post and disconnect the wire.

7. Remove the alternator from the vehicle.

To install:

8. Install the alternator assembly in the vehicle.

9. Connect the battery wire to the alternator and tighten the mounting nut to 71 inch lbs. (8 Nm). Install the protective cap over the post.

10. Install the remaining components in the reverse order of removal. Torque the following:

• Long bolt to 35 ft. lbs. (47 Nm)

• Short bolts and brace bolt to 18 ft. lbs. (25 Nm)

11. Connect the negative battery cable. Install the air cleaner assembly.

12. Start the engine and verify proper charging system operation.

3.1L (VIN M) Engines

1. Disconnect the the negative battery cable. Remove the serpentine belt.

2. Remove the alternator mounting bolts.

3. Remove the power steering line clip mounting nut and disconnect the clip from the alternator stud.

4. Remove bolt, stud and nuts from the alternator braces.

5. Remove the lower alternator bolt. Detach the alternator connector.

6. Remove the plastic cover from the post on the back of the alternator. Remove the nut and disconnect battery positive lead from alternator output terminal.

7. Remove the alternator.

8. Remove the alternator bracket bolts.

9. Remove the alternator bracket from the alternator.

To install:

10. Position the alternator bracket on the alternator and tighten the bolt to 37 ft. lbs. (50 Nm).

11. Install the alternator.

12. Install the remaining components in the reverse order of removal. Torque the following:

• Battery positive lead nut to 15 ft. lbs. (20 Nm)

• Lower alternator bolt to 37 ft. lbs. (50 Nm)

• Brace nuts and stud to 18 ft. lbs. (25 Nm)

• Power steering line clip nut to 18 ft. lbs. (25 Nm)

13. Connect the negative battery cable. Start the engine and verify proper charging system operation.

1993–94 3.4L (VIN X) Engine

These vehicles are equipped with an alternator cooling fan behind the right side headlight, used to cool the unit through a duct hose. Make sure the duct is connected after service and is in good condition. Note that the vehicle sub-frame must be loosened to service the alternator, requiring suitable jacks and lifts.

With Automatic Transaxle

1. Remove the air cleaner assembly. Disconnect the negative battery cable.

2. Remove the serpentine belt.

3. Remove the coolant recovery bottle.

4. Remove the upper alternator stud nut and remove power steering pipe retaining clip from the stud.

5. Loosen the upper alternator stud.

6. Raise and safely support the vehicle. Remove the front tire and wheel assemblies.

7. Remove the right engine splash shield upper retaining clips and pull down the shield.

8. Remove the upper alternator stud.

9. Remove the cotter pin and lower ball joint castle nut from the right lower ball joint and separate the lower ball joint from the lower control arm.

10. Remove the metal halfshaft shield retaining screws and the shield.

11. Disconnect the right halfshaft from transaxle.

12. Remove the front exhaust pipe and converter assembly.

13. Remove the lower alternator mounting bolt.

14. Remove the alternator rear brace-to-alternator bolt. The bolt can best be accessed using a four foot extension through the left side wheel well opening.

15. Remove the alternator rear brace nut from the engine. The nut can be best accessed through the right wheel well behind the alternator.

16. Remove the alternator from the mounting bracket.

17. Disconnect the electrical connector from the alternator.

18. Remove the battery wire nut from the alternator stud and disconnect the battery positive lead from the alternator output terminal.

19. Remove the alternator.

To install:

20. Attach the alternator connector and battery wire to the alternator. Tighten the nut on the battery wire to 71 inch lbs. (8 Nm).

21. Install the alternator and loosely install the mounting bolts. If the replacement alternator does not fit into the bracket, remove the adhesive-backed shim from the rear of the alternator bracket.

22. Install the alternator rear brace, cooling duct bracket and mounting bolt. Access bolt through the left wheel opening using a 4 foot extension.

23. Tighten rear brace bolt to 18 ft. lbs., and the rear brace nut and lower mounting bolt to 37 ft. lbs. 50 (Nm).

24. Install the remaining components in the reverse order of removal. Torque the following:

• Ball joint-to-lower control nut to 63 ft. lbs. (85 Nm)

• Alternator mounting stud to 18 ft. lbs. (25 Nm)

25. Connect the negative battery cable. Install the air cleaner assembly.

26. Start the engine and verify proper charging system operation.

NOTE: Whenever the vehicle sub-frame is removed or lowered, the wheel alignment should be checked.

27. Check the vehicle wheel alignment and adjust if required.

With Manual Transaxle

1. Remove the air cleaner assembly.

2. Disconnect the negative battery cable. Remove the serpentine belt.

3. Remove the coolant recovery bottle.

4. Remove the upper alternator stud nut and remove power steering pipe retaining clip from the stud.

5. Raise and safely support the vehicle.

6. Remove the right front tire and wheel assembly.

7. Remove the right engine splash shield upper retaining clips and pull down the shield.

8. Remove the plastic outboard halfshaft shield retaining screws and the shield.

9. Remove the alternator rear brace-to-alternator bolt and nut.

10. Disconnect the electrical connector from the alternator.

11. Remove the battery wire nut from the alternator stud and disconnect the battery positive lead from the alternator output terminal.

12. Remove the cotter pin and lower ball joint castle nut from the right lower ball joint and separate the lower ball joint from the lower control arm.

13. Disconnect the right halfshaft from transaxle.

14. Remove the metal outer drive axle shield.

15. Remove the rear engine mount bracket brace mounting bolts with the metal inner halfshaft shield attached.

16. Remove the alternator mounting stud and lower mounting bolt.

17. Remove the alternator.

To install:

18. Install the alternator and loosely install the mounting bolts. If the replacement alternator does not fit into the bracket, remove the adhesive-backed shim from the rear of the alternator bracket.

19. Tighten the lower mounting bolt to 37 ft. lbs. (50 Nm) and the mounting stud to 18 ft. lbs. (25 Nm).

20. Install the remaining components in the reverse order of removal. Torque the following:

- Ball joint-to-lower control arm nut to 63 ft. lbs. (85 Nm)
- Battery wire nut to 71 inch lbs. (8 Nm)
- Rear brace bolt to 18 ft. lbs. (25 Nm)
- Rear brace nut to 37 ft. lbs. (50 Nm)

21. Connect the negative battery cable. Install the air cleaner assembly.

22. Start the engine and verify proper charging system operation.

NOTE: Whenever the vehicle sub-frame is removed or lowered, the wheel alignment should be checked.

23. Check the vehicle wheel alignment and adjust if required.

1995–97 3.4L (VIN X) Engine

1. Remove the air cleaner and duct assembly.

2. Disconnect the negative battery cable.

3. Remove the coolant recovery bottle. Remove the serpentine belt.

4. Remove the upper alternator stud nut and remove the power steering pipe retaining clip from the stud.

5. Raise and safely support the vehicle. Remove the right front tire and wheel assembly.

6. Remove the right engine splash shield upper retaining clips and pull down the shield.

7. Remove the front exhaust pipe and converter assembly.

8. Remove the front exhaust pipe heat shield.

9. Remove the alternator cooling duct bracket tab at the oil pan, if equipped.

10. Remove the alternator cooling duct, if equipped.

11. Remove the alternator rear support brace.

12. Remove the alternator rear brace-to-alternator bolt. The bolt can be accessed using a four foot long extension through the left side wheel well opening.

13. Remove the alternator rear brace nut from the engine. The nut can be accessed through the right wheel well behind the alternator.

14. Remove the intermediate shaft lower pinch bolt at the steering gear.

15. Install a jacking fixture to support the engine.

16. Remove the rear cradle bolts and lower the cradle.

17. Remove the steering gear heat shield assembly.

18. Remove the alternator lower bolt and washer.

19. Detach the alternator regulator terminal connector, if equipped with 4T60–E transaxle.

20. Remove the battery positive lead and nut from the alternator output "BAT" terminal.

21. Remove the electrical connector from the vehicle speed sensor.

22. Remove the power steering gear clip from the steering gear.

23. Remove the power steering gear inlet hose assembly from the steering gear.

24. Remove the alternator assembly.

To install:

25. Position the alternator assembly on the vehicle.

26. Install the remaining components in the reverse order of removal. Torque the following:

- Lower alternator mounting bolt to 37 ft. lbs (50 Nm)

NOTE: If the replacement alternator does not fit, remove the adhesive shim from the alternator bracket.

- Rear engine cradle bolts to 125 ft. lbs (170 Nm)

- Intermediate shaft lower pinch bolt to 35 ft. lbs. (47 Nm)
- Rear brace nut to 37 ft. lbs. (50 Nm)
- Rear brace bolt to 18 ft. lbs. (25 Nm)
- Alternator upper stud and nut to 18 ft. lbs. (25 Nm)

27. Connect the negative battery cable. Install the air cleaner and duct assembly.

28. Bleed the power steering system. Check for proper charging system operation.

NOTE: Whenever the vehicle sub-frame is removed or lowered, the wheel alignment should be checked.

29. Check the vehicle wheel alignment and adjust if required.

3.8L (VIN L) and (VIN K) Engine

1. Disconnect the negative battery cable.

2. Remove the serpentine belt from alternator pulley. It is not necessary to remove the belt from the remainder of the pulleys.

3. Remove the alternator mounting bolts.

4. Disconnect the electrical connector from the alternator and remove the nut securing the battery wire and disconnect the battery wire from the alternator stud.

5. Remove the alternator.

To install:

6. Install the alternator in the mounting bracket and loosely install the mounting bolts.

7. Tighten the alternator mounting bolts to 20 ft. lbs. (27 Nm). The lower bolt (long bolt) must be tightened first.

8. Connect the battery wire to the alternator post and tighten the mounting nut to 71 inch lbs. (8 Nm) on 1993 vehicles, and 11 ft. lbs. (15 Nm) on 1994-97 vehicles. Connect the electrical connector to the alternator.

9. Install the serpentine belt.

10. Connect the negative battery cable.

11. Start the engine and verify proper charging system operation.

Drive Belt

REMOVAL AND INSTALLATION

2.2L (VIN 4) Engine

1. Disconnect negative battery cable.

2. Remove the bolt from the reservoir bracket at the strut tower and

disconnect the bracket from the reservoir neck.

3. Disconnect and plug the reservoir hose. Remove the recovery reservoir from the side rail.

4. Using a 15mm wrench, pivot tensioner and remove belt from alternator.

5. Release the tensioner slowly.

6. Remove the serpentine belt.

To install:

7. Route the serpentine belt as follows:

a. Under the crankshaft and A/C compressor, if equipped.

b. Over the tensioner with the back of the belt driving the pulley.

c. Around the water pump with the back of the belt driving the pulley.

d. Around the power steering pump.

8. Rotate the tensioner and loop the belt around the alternator.

9. Release the tensioner.

10. Verify the belt is properly seated in all the pulleys.

11. Install the reservoir in the side rail and connect the reservoir hose.

12. Install the bracket on the reservoir neck and install the bracket-to-strut tower bolt and tighten to 18 ft. lbs. (25 Nm).

13. Connect the negative battery cable. Top off the coolant reservoir as necessary.

3.1L (VIN T) and (VIN M) Engine

1. Disconnect the negative battery cable.

2. Remove belt shield bolt and belt shield from above the water pump pulley, if equipped.

3. Rotate the belt tensioner counterclockwise using a ⅜ inch breaker bar.

4. With the tension removed from the belt, remove the serpentine belt from around the alternator pulley.

5. Slowly release the tensioner.

6. Remove the belt from the remaining pulleys.

To install:

7. Loop the serpentine belt under the crankshaft and A/C compressor pulleys.

8. Route the belt around the water pump pulley so the back of the belt drives the pulley.

9. Route the belt around the tensioner pulley so the back of the belt drives the pulley.

10. Route the belt over the power steering pump pulley.

11. Rotate the belt tensioner counterclockwise using a ⅜ inch breaker bar.

12. Loop the serpentine around the alternator pulley.

13. Release the tensioner and verify the belt is properly seated in the the the drive pulleys.

14. Install the belt shield and mounting bolt and tighten to 89 inch lbs. (10 Nm), if equipped.

15. Connect the negative battery cable.

3.4L (VIN X) Engine

1. Disconnect the negative battery cable. Remove the coolant recovery reservoir.

2. Using a box end wrench, rotate the tensioner clockwise and remove the belt from the power steering pulley.

3. Slowly release the belt tensioner.

4. Remove the serpentine belt from the remainder of the pulleys.

To install:

5. Loop the serpentine belt under the crankshaft pulley and A/C compressor. Bring the front of the belt up around the compressor and under the water pump pulley so the back of the belt drives the water pump pulley. Bring the rear of the belt over the tensioner so the rear of the belt rides on the tensioner. Loop the belt around the alternator.

6. Using a box end wrench, rotate the tensioner clockwise and install the belt around the power steering pulley last.

7. Slowly release the tensioner.

8. Verify the belt is properly seated in all the drive pulleys.

9. Install coolant recovery reservoir.

10. Connect the negative battery cable.

11. Start the engine and verify the correct routing and quiet operation of the serpentine belt.

3.8L (VIN L) and (VIN K) Engine

1. Disconnect the negative battery cable. Remove the coolant recovery bottle.

2. Rotate the tensioner assembly counterclockwise using an 18mm box end wrench on the tensioner pulley nut.

3. Slip the belt off of the alternator pulley and slowly release the tensioner.

4. Remove the wrench from the pulley and remove the belt from the remainder of the pulleys.

To install:

5. Route the serpentine belt as follows:

a. Loop the belt under the crankshaft.

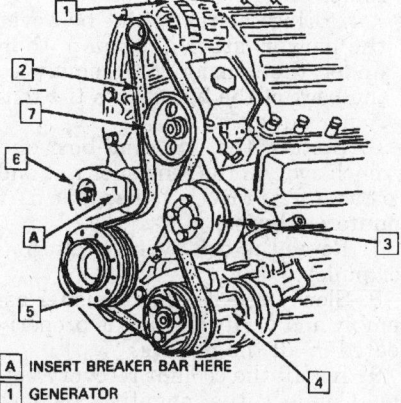

A	INSERT BREAKER BAR HERE
1	GENERATOR
2	SERPENTINE BELT
3	WATER PUMP
4	AIR CONDITIONING COMPRESSOR
5	CRANKSHAFT
6	BELT TENSIONER
7	POWER STEERING PUMP

211825

Serpentine belt routing — 3.1L (VIN T) engines

1 Generator
2 Coolant pump
3 A/C compressor
4 Engine crankshaft
5 Serpentine drive belt
6 Serpentine drive belt tensioner
7 Power steering pump

189473

Serpentine belt routing — 2.2L (VIN 4) engine

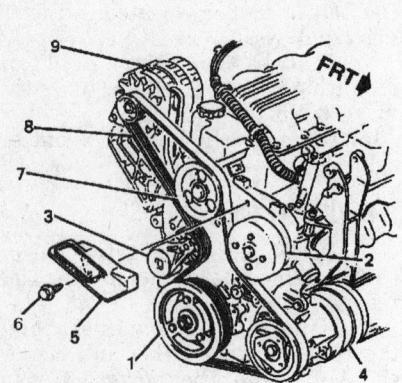

1. BALANCER, CRANKSHAFT
2. PUMP ASSEMBLY, WATER
3. TENSIONER, SERPENTINE DRIVE BELT
4. COMPRESSOR ASSEMBLY, AIR CONDITIONING
5. SHIELD, SERPENTINE DRIVE BELT
6. BOLT/SCREW, SERPENTINE DRIVE BELT SHIELD
7. PUMP ASSEMBLY, POWER STEERING
8. BELT, SERPENTINE DRIVE
9. GENERATOR ASSEMBLY

330717

Serpentine belt routing and shield — 3.1L (VIN M) engine

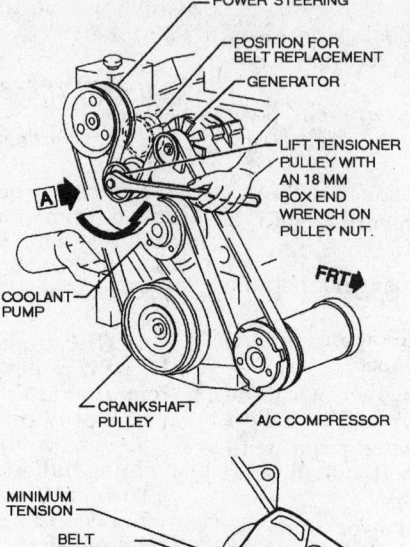

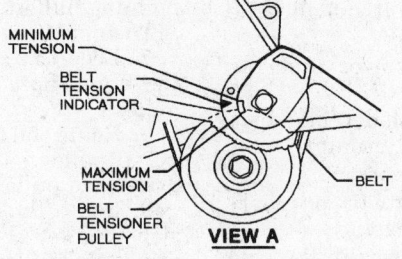

330723

Serpentine drive belt routing — 3.8L (VIN L) engine

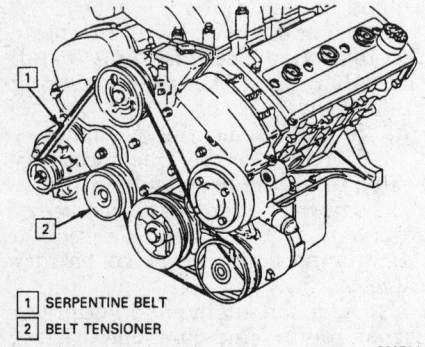

1. SERPENTINE BELT
2. BELT TENSIONER

330714

Serpentine belt routing — 3.4L (VIN X) engine

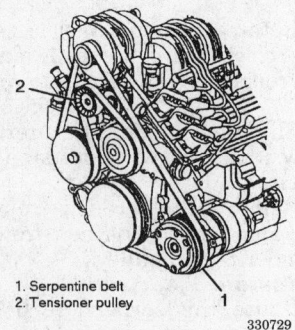

1. Serpentine belt
2. Tensioner pulley

330729

Serpentine drive belt routing — 3.8L (VIN K) engine

b. Bring the front of the belt up and around the water pump pulley so the back of the belt drives the pulley.

c. Loop the belt under the A/C compressor.

d. Bring the rear of the belt over the power steering pump then under the tensioner assembly so the back of the belt drives the tensioner pulley.

6. Install the 18 mm box end wrench over on the tensioner nut and rotate the tensioner counterclockwise.

7. Route the belt over the alternator pulley.

8. Slowly release the tensioner assembly and verify the belt is properly seated in all the pulleys.

9. Install the coolant recovery bottle. Connect the negative battery cable.

Starter

REMOVAL AND INSTALLATION

2.2L (VIN 4) Engine

1. Disconnect the negative battery cable. Raise and safely support the vehicle.

2. Remove the mounting bolts from the flywheel inspection cover and remove the cover.

3. Tag and disconnect the starter electrical connections.

4. Remove the stud securing the starter front bracket to the engine block.

5. Remove the two starter mounting bolts and remove the starter and shims if equipped.

6. If the starter is being replaced, remove the two bracket mounting nuts and remove the bracket.

To install:

7. Install the front bracket on the starter and tighten the nuts to 80 inch lbs. (9 Nm).

8. Attach the starter connectors to the starter. Tighten the solenoid wire nut to 27 inch lbs. (3 Nm) and the battery terminal wire nut to 12 ft. lbs. (16 Nm).

9. Install the starter in the vehicle and tighten the two mounting bolts to 32 ft. lbs. (43 Nm) and installing the shims if equipped.

10. Install the starter bracket-to-engine bolt and tighten to 26 ft. lbs. (32 Nm).

11. Install the flywheel cover and tighten the mounting bolts to 89 inch lbs. (10 Nm).

12. Lower the vehicle.

13. Connect the negative battery cable. Start the engine several times to ensure proper starter drive engagement into the flywheel. If a harsh engagement is encountered, it will be necessary to shim the starter motor.

3.1L Engine (VIN T) and (VIN M) Engines

1. Remove the air cleaner and air duct assembly.
2. Disconnect the negative battery cable. Raise and safely support the vehicle.
3. For 1995–97 vehicles, remove the oil filter splash shield.
4. Remove the flywheel inspection cover. Remove the two starter motor mounting bolts.
5. Disconnect the battery cable and electrical wiring from the solenoid.
6. Place a drain pan under the engine and remove the oil filter.
7. If equipped with an engine oil cooler, position the hose closest to the starter motor out of the way.
8. Disconnect the oil pressure sender electrical connector.
9. Remove the starter motor from the engine.
   To install:
10. Connect the electrical connections to the starter motor and tighten the battery nut to 84 inch lbs. (10 Nm) and the solenoid nut to 22 inch lbs. (3 Nm).
11. Install the starter motor in position in the vehicle.
12. Install the remaining components in the reverse order of removal. Tighten the starter mounting bolts to 32 ft. lbs. (43 Nm).
13. Connect the negative battery cable. Install the air cleaner and duct assembly.
14. Check the engine oil level and add as necessary. Start the engine several times to verify proper starter motor-to-flywheel engagement.

3.4L (VIN X) Engine

1. Disconnect the negative battery cable. Drain the cooling system.
2. Raise and safely support the vehicle. Remove the engine oil cooler assembly.
3. Remove the flywheel inspection cover bolts and the cover.
4. Detach the starter connectors.

5. Remove the starter motor bolts and remove the starter from the engine.
   To install:

NOTE: **Before installing the starter motor be sure the electrical terminals are secure by tightening the nut next to the cap on the solenoid battery terminal. If this terminal is NOT tight in the solenoid cap, the cap may be damaged during installation of the electrical connections and cause the starter motor to fail later.**

6. Secure the electrical terminal on the solenoid. Tighten the solenoid battery terminal inside nut to 84 inch lbs. (9.5 Nm).
7. Connect the electrical wiring to the starter.
8. Position the starter in the vehicle and tighten the starter mounting bolts to 32 ft. lbs. (43 Nm).
9. Install the remaining components in the reverse order of removal.
10. Refill with coolant. Check engine oil level and add if necessary.
11. Connect the negative battery cable. Verify starter operation as well as no oil or coolant leaks.

3.8L (VIN L) Engine

1. Disconnect the negative battery cable.
2. Remove the left side cooling fan.
3. Raise and safely support the vehicle.
4. Disconnect the oil level and knock sensor electrical connectors.
5. Disconnect the engine oil cooler pipe from the oil pipe clip.
6. Remove the flywheel cover mounting bolts and the flywheel cover. Remove the starter mounting bolts.
7. Remove the mounting nuts securing the battery cable and solenoid wire to the starter and disconnect the wires.
8. Lower the vehicle. Remove the starter from the vehicle.
   To install:
9. Position the starter on the engine. Raise and safely support the vehicle.
10. Connect the battery cable and solenoid wire to the starter and tighten the battery wire nut to 12 ft. lbs. (16 Nm) and the solenoid wire nut to 22 inch lbs. (3 Nm).

11. Install the starter mounting bolts and tighten to 32 ft. lbs. (43 Nm).
12. Install the flywheel inspection cover and tighten the mounting bolts to 89 inch lbs. (10 Nm).
13. Connect the engine oil cooler pipe to the pipe clip.
14. Attach the connectors to the oil level and knock sensors.
15. Lower the vehicle.
16. Install the left side cooling fan.
17. Connect the negative battery cable.

3.8L (VIN K) Engine

1. Disconnect the negative battery cable. Remove the upper mounting bracket above the radiator.
2. Disconnect the electrical connection at the coolant fan.
3. Remove the coolant fan assembly.
4. Remove the upper and lower oil cooler pipes from the radiator.
5. Raise and safely support the vehicle.
6. Remove the retainers from the engine harness and reposition away from the starter.
7. Disconnect both lower oil cooler hoses from the radiator.
8. Disconnect the starter electrical connections.
9. Remove the upper flywheel inspection cover.
10. Remove the starter mounting bolts and remove the starter assembly from the vehicle. There may be shims between the starter motor housing and engine block which should be saved for reuse.
    To install:
11. Install the starter to the engine with shims, if removed.
12. Install and tighten the starter mounting bolts to 32 ft. lbs. (43 Nm).
13. Connect the solenoid battery terminal outside nut to 12 ft. lbs. (16 Nm) and solenoid "S" terminal inside nut to 22 inch lbs. (2.5 Nm).
14. Install the remaining components in the reverse order of removal. Tighten the flywheel cover bolts to 89 inch lbs.(10 Nm) and the upper engine mounting bracket bolts to 32 ft. lbs. (43 Nm).
15. Check the engine oil and add as necessary. Connect the negative battery cable.
16. Start the engine several times to ensure proper starter drive to flywheel engagement. If harsh engagement is encountered the starter will have to be shimmed.

CHASSIS ELECTRICAL

Blower Motor

REMOVAL AND INSTALLATION

1. Disconnect the negative battery cable. Remove the right sound insulator panel under the instrument panel.
2. Remove the convenience center rear screws. Loosen the front screws and slide the convenience center out.
3. Grasp the carpet at top side of cowl and pull forward.
4. Remove the blower motor electrical connection.
5. Remove the harness from the clip.
6. Remove the blower motor mounting screws, then remove the motor and fan assembly.
7. If the blower motor fan (sometimes called a squirrel cage) is to be replaced, use the following procedure:
   a. Using a hot knife, cut a slot in the cage motor shaft sleeve in three places. Starting ½ inch from the base of the shaft, start the cut from the fan dome and continue to cut through the plastic material to the end of the shaft so that the fan splits from the shaft. Remove the cage from the blower motor shaft by pulling straight out.
   b. Alternative Method A. Remove the fan cage by using a small gear puller. Remove the tip of the fan at the end of the motor shaft using a soldering iron or cutting pliers. Draw the fan from the motor shaft using a small puller. Hook the legs of the puller between the fan lower rib and motor mounting plate. Thread the center post of the puller against the end of the motor shaft exposed previously.
   c. Alternative Method B. Use a drill press or arbor press to remove the fan cage. Remove the tip of the fan at the end of the motor shaft using a soldering iron or cutting pliers. Support the fan with block, angle iron, etc., while allowing the motor to be suspended. Place steady pressure in top of the motor shaft, with a drill press or small arbor press and pin arrangement. The motor should fall freely from the fan after moving the motor shaft about ½ inch.

CAUTION

Do not hammer on the motor to remove or install the fan. Do not apply force to the motor housing to seat the fan on the motor or motor/shaft bearing damage could result. Do not apply pressure to the fan rim. Be sure the correct replacement part is used.

To install:

8. If the blower motor fan is being replaced, use the following procedure:
   a. Press the new replacement fan cage onto the motor shaft. Grasp the fan by the dome and, applying hand pressure to the fan dome, insert the fan on the motor shaft until lightly seated on the shaft.
   b. While steadying the fan and motor, apply steady force to the rear of the motor shaft using a drill press or small arbor press and pin. Install the fan cage to a clearance of 0.30 inch to the motor plate.
9. Install the blower motor and fan assembly and secure with the mounting screws.
10. Install the remaining components in the reverse order of removal.
11. Connect the negative battery cable. Start the engine and verify proper blower motor operation.

Windshield Wiper Motor

REMOVAL AND INSTALLATION

1993 Vehicles

1. Disconnect the negative battery cable.
2. Disconnect and cap the washer hoses from the wiper arms.
3. Remove the wiper arms from the vehicle.
4. Remove the screws retaining the cowl cover. Lower the hood partially and remove the cowl cover and air inlet panel.
5. Disconnect the wiring harness connectors from the wiper motor and the washer hose at the firewall.

NOTE: Crank arm must be removed from the wiper motor only. Do not remove crank arm from the transmission assembly because of the factory reset adjustment.

6. Remove the 3 screws from the bellcrank housing and lower the wiper linkage.
7. Remove the wiper module assembly from the vehicle and remove the three screws securing the wiper motor.

To install:

8. Attach the wiper motor to the module assembly and install the module assembly in the vehicle.
9. Connect the electrical connectors to the wiper motor and module.
10. Install the air inlet panel and cowl panel.
11. Install the wiper arms on the wiper transmission drives and tighten the mounting nuts to 25 ft. lbs. (34 Nm).
12. Connect the washer hoses to the wiper arms.
13. Connect the negative battery cable.

1994–97 Vehicles

Except 1996–97 Lumina and Monte Carlo

1. Disconnect the negative battery cable.
2. Remove the wiper arm assemblies from the vehicle.
3. Remove the cowl panel mounting screws and remove the cowl panel.
4. Disconnect the electrical connectors from the wiper motor.
5. Remove the three screws and remove the wiper drive system module.
6. Disconnect the wiper crank arm from the wiper transmission.
7. Remove the mounting nut and remove the crank arm from the wiper motor.
8. Remove the three wiper motor mounting screws and remove the wiper motor.
9. Remove the cover from the wiper motor.

To install:

10. Install the cover on the wiper motor.
11. Install the wiper motor onto the wiper drive module and install the three mounting screws.
12. Install the remaining components in the reverse order of removal. Tighten the wiper motor crank arm mounting nut to 18 ft. lbs. (25 Nm) and the wiper drive module screws to 106 inch lbs. (12 Nm).
13. Connect the negative battery cable. Verify proper wiper operation.

1996–97 Lumina and Monte Carlo

1. Disconnect the negative battery cable.
2. Remove the wiper arm assemblies from the vehicle.
3. Remove the lower mounting reveal molding screws, then lower the hood and remove the lower reveal molding.
4. Remove the air inlet panel screws, underhood lamp switch, if equipped.

5. Remove the air inlet panel with the hood raised.

NOTE: If the wiper motor can run, place it in the inner wipe position.

6. Disconnect the 2 electrical connectors from the wiper motor and disconnect the washer hose at the firewall.

7. Remove the three screws from the bell crank housing, then lower the transmission assembly for module removal.

8. Remove the 6 screws mounting screws and remove the module assembly.

9. Remove the wiper motor crank arm from the wiper assembly using J 39232 or the equivalent.

10. Remove the nut and the wiper motor crank arm from the wiper motor assembly.

11. Remove the 3 screws, the bracket and the wiper motor assembly from the tube frame.

To install:

12. Install the cover on the wiper motor.

13. Install the wiper motor assembly, and bracket onto the tube frame with the 3 screws.

14. Tighten the screws to 106 inch lbs. (12 Nm).

15. Install the wiper motor crank arm on the wiper motor and tighten the mounting nut to 18 ft. lbs. (25 Nm).

16. Connect the wiper transmission to the wiper motor crank arm using tool J 39529 or the equivalent.

17. Position the wiper drive system module on the shroud and install the 6 mounting screws, torquing them to 89 inch lbs. (10 Nm). Make sure the body seal is in the proper position on the right side of the module.

18. Connect the electrical connectors to the wiper motor and install the connector retainers.

19. Install the washer motor hose at the firewall.

20. Install the air inlet panel and retaining screws.

21. Connect the underhood lamp, if equipped.

22. Install the lower reveal molding and retaining screws.

23. Install the wiper arm assemblies.

24. Connect the negative battery cable. Verify proper wiper operation.

Headlight Switch

REMOVAL AND INSTALLATION

—————— CAUTION ——————
The Supplemental Inflatable Restraint (SIR) system must be disarmed before working around the steering column. Failure to do so may cause accidental deployment of the air bag, resulting in unnecessary SIR system repairs and/or personal injury.

1. Disconnect the negative battery cable.

2. Disarm the SIR system.

3. Remove the steering column opening filler panel or instrument panel trim plate or pad.

4. If necessary, remove the instrument cluster assembly.

5. Remove the headlamp switch by removing the bolts or depressing the locking tabs on the switch depending on the type of switch being used.

6. Detach the connectors from the headlamp switch and remove the switch from the vehicle.

NOTE: After replacing a headlight switch due to an external short, cycle the headlight switch and the interior light dimmer switch on and off twice to activate the internal circuit protection of the switch. All interior lights will not operate if the headlight switch assembly has not been cycled.

To install:

7. Connect the electrical connections to the headlamp switch.

8. Install the headlamp switch by snapping the switch into the instrument panel or installing the retaining bolts.

9. Installing the steering column opening filler panel.

10. Install the steering column opening filler panel.

11. Enable the SIR system.

12. Connect the negative battery cable.

13. Verify the correct operation of the headlight switch assembly.

Turn Signal Switch

REMOVAL AND INSTALLATION

—————— CAUTION ——————
The Supplemental Inflatable Restraint (SIR) system must be disarmed before removing the steering wheel. Failure to do so may cause accidental deployment of the air bag, resulting in unnecessary SIR system repairs and/or personal injury.

—————— CAUTION ——————
When carrying a live air bag, make sure the bag and trim cover are pointed away from the body. In the unlikely event of an accidental deployment, the bag will then deploy with minimal chance of injury. When placing a live air bag on a bench or other surface, always face the bag and trim cover up, away from the surface. This will reduce the motion of the module if it is accidentally deployed.

1993 Vehicles And 1994 Lumina

NOTE: Tool No. J 35689-A or the equivalent, is required to remove the terminals from the connector on the turn signal switch.

1. Properly disable the SIR system. Disconnect the negative battery cable.

2. Remove the air bag module, if equipped, from the steering wheel.

3. Remove steering wheel nut retainer and jam nut.

4. Remove the steering wheel using a puller. Do not hammer on the wheel or column.

5. Pull the turn signal cancelling cam assembly from the steering shaft.

6. Remove the hazard warning knob-to-steering column screw and the knob.

NOTE: Before removing the turn signal assembly, position the turn signal lever so the turn signal assembly to steering column screws can all be removed.

7. Remove the column housing cover-to-column housing cover screw and the cover. This screw is longer and may be blue in color. If equipped with cruise control, disconnect the cruise control electrical connector.

8. Remove column cover.

9. Remove wiring protector from opening in the instrument panel bracket on the jacket and bowl assembly and separate from the wires.

10. Remove pivot and pulse switch connector from ignition and dimmer switch.

11. Remove pivot switch screw.

12. Remove pivot and pulse switch screws.

13. Remove turn signal switch connector from ignition and dimmer switch assembly connector.

14. Remove seventeen-way connector secondary lock from the turn signal connector.

15. Remove wires on buzzer switch assembly from the turn signal connector with special tool J-35689-A. Wrap wire ends with tape to protect during removal and installation.

16. Remove turn signal switch assembly from the column.

To install:

17. Install the turn signal switch to the steering column, and tighten the turn signal switch-to-steering column screws.

NOTE: Position turn signal switch so that turn signal switch screws and housing cover screw can be installed through openings in the switch.

18. Connect wires from the buzzer switch assembly to the turn signal switch connector. Connect the light green wire in location **9**. Connect tan/black wire to location **10**.

19. Install seventeen-way secondary lock to the turn signal connector and snap into place.

20. Install pivot and pulse switch assembly.

21. Install pivot switch screw.

22. Install pivot switch connector to the ignition and dimmer switch connector and snap into place.

23. Install wiring protector around all wires passing through the instrument panel bracket opening. Close protector so that interlocking grooves engage. Slide wiring the protector into the instrument panel bracket opening on the jacket and bowl assembly.

24. Install the column housing cover. Install the hazard warning knob.

25. Lubricate bottom side of the cancel cam with synthetic grease. Install turn signal cancel cam assembly.

26. Install steering wheel.

27. Install jam nut and torque to 30 ft. lbs. (41 Nm).

28. Install the air bag module.

29. Connect the negative battery cable and check the operation of the turn signal switch.

1994–97 Vehicles Except 1994 Lumina

1. Before beginning work, set the front wheels to the straight ahead position.

2. Properly disable the SIR system. Disconnect the negative battery cable.

3. Remove the steering wheel controls and the steering wheel.

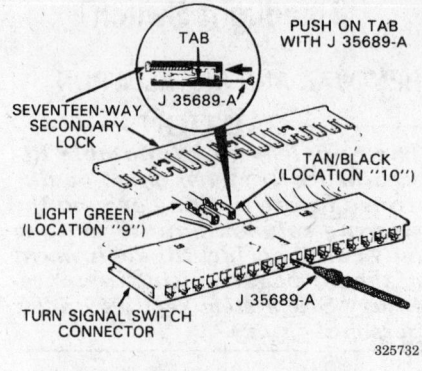

Turn signal connector special tool

4. Remove the SIR coil assembly retaining ring.

5. Remove the SIR coil assembly. Pulling gently on the wire harness while feeding the wire up the column and let the coil hang freely.

6. Remove the wave washer.

7. Remove the steering shaft lock retaining ring using special tool J 23653 SIR or the equivalent to compress the lock plate and remove the shaft lock. Dispose of shaft ring.

8. Remove the turn signal cancelling cam assembly, lock plate and the upper bearing spring.

9. Remove the upper bearing inner race seat.

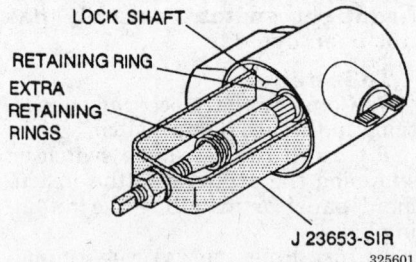

Removing the shaft lock retaining ring

10. Remove the inner race.

11. Remove the multi-function lever.

12. Remove the housing cover end cap. Tilt the end cap away from the lock housing cover and sleeve assembly and remove from the column.

13. Remove the screw with washer attached and flat head screw.

14. Remove the pivot and dimmer switch assembly. Let the switch hang freely.

15. Remove the hazard warning switch knob by using a suitable prying tool to remove the knob.

16. Position the turn signal switch to gain access to remove the screws retaining the turn signal switch to the steering column.

17. Remove the turn signal switch with the following procedure:

   a. Remove the turn signal switch connector from the bulkhead connector.

   b. Remove the wiring protector.

   c. Gently pull wire harness through the column.

18. Remove the turn signal switch from the steering column.

To install:

19. Install the turn signal switch assembly wire harness through the steering column. Connect the turn signal switch connector to the bulkhead connector.

20. Install and tighten the turn signal switch assembly three blinding head cross recess screws to 30 inch lbs. (3.4 Nm).

21. Install the pivot and dimmer switch assembly wire harness through the steering column.

22. Connect the wire harness clamp to tab inside lock housing cover of the dimmer switch.

23. Connect pivot and dimmer switch connector to bulkhead connector.

24. Place pivot and dimmer switch component to slot in the lock housing cover.

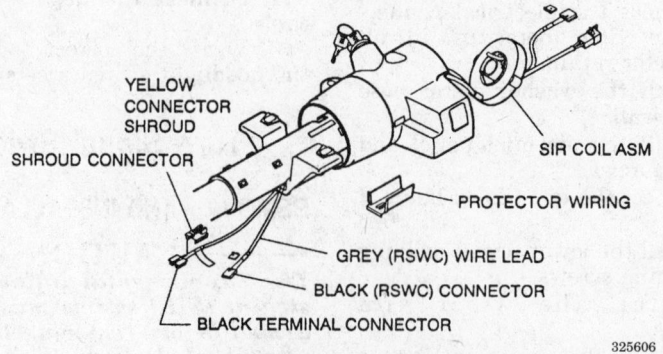

Removing the SIR coil assembly

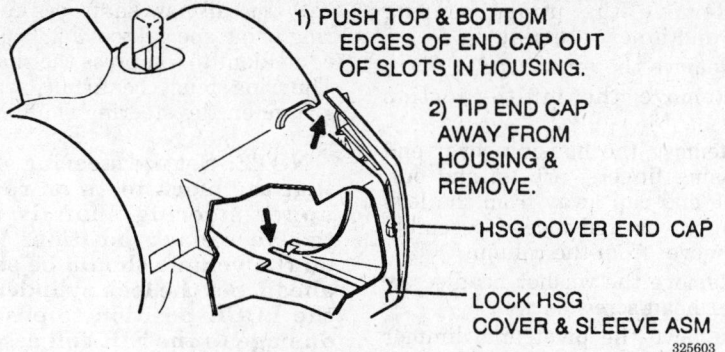

1) PUSH TOP & BOTTOM EDGES OF END CAP OUT OF SLOTS IN HOUSING.

2) TIP END CAP AWAY FROM HOUSING & REMOVE.

HSG COVER END CAP

LOCK HSG COVER & SLEEVE ASM

325603

Removing housing cover end cap

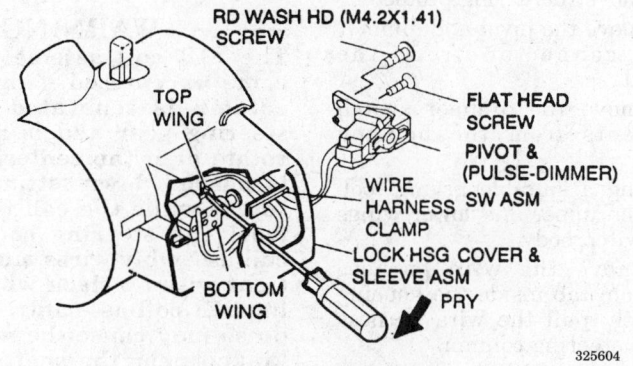

RD WASH HD (M4.2X1.41) SCREW

TOP WING

FLAT HEAD SCREW

PIVOT & (PULSE-DIMMER) SW ASM

WIRE HARNESS CLAMP

LOCK HSG COVER & SLEEVE ASM

BOTTOM WING

PRY

325604

Removing pivot and dimmer switch assembly

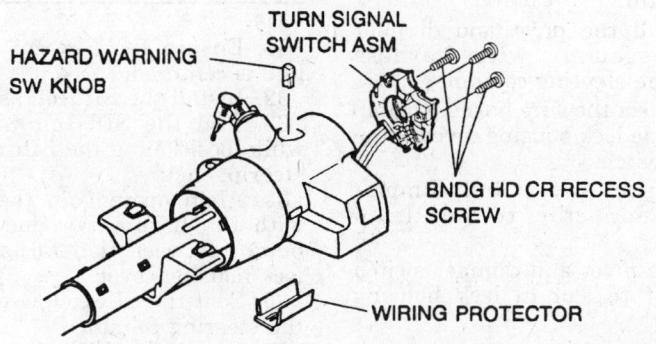

TURN SIGNAL SWITCH ASM

HAZARD WARNING SW KNOB

BNDG HD CR RECESS SCREW

WIRING PROTECTOR

325605

Removing the turn signal switch assembly

25. Install pivot and dimmer switch switch to the turn signal switch.

26. Install and tighten the round washer head screw to 27 inch lbs. (3 Nm).

27. Install and tighten the flat head screw to 18 inch lbs. (2 Nm).

28. Install the housing cover end cap. Install the multi-function and tilt lever.

29. Install the hazard warning switch knob. Position the knob through hole in the lock housing cover and snap in place.

30. Install the inner race.

31. Install the upper bearing inner race seat and spring.

32. Install the turn signal cancelling cam assembly and lubricate assembly with synthetic grease.

33. Install the shaft lock. The inner block tooth of the lock must be aligned with block tooth of race and upper shaft assembly.

34. Install new shaft lock retaining ring using special tool J 23653 SIR or the equivalent to compress the shaft lock. The ring must be firmly seated in groove on the steering shaft.

NOTE: Set the steering shaft so that the block tooth on race and upper steering shaft is at the twelve o'clock position. Wheels on the vehicle should be straight ahead. Set the lock cylinder set to the LOCK position to ensure no damage to the SIR coil assembly.

35. Install the wave washer.

NOTE: The SIR coil assembly will become uncentered if the steering column is separated from the steering gear and is allowed to rotate or if the centering spring is pushed down letting the hub rotate while the coil is removed from the steering column. In addition, the SIR coil assembly wires must be kept tight with no slack while installing SIR coil assembly. Failure to do so may cause the wires to be kinked near the shaft lock area and cut when the steering wheel is turned.

36. Ensure the SIR coil assembly hub is centered.

37. Install the SIR coil assembly.

38. Pull the SIR coil wires snug while positioning the SIR coil to the steering shaft.

39. Align opening in the SIR coil with the horn tower and "locating bump" between the two tabs on the lock housing cover.

40. Seat the SIR coil assembly into the steering column.

41. Install the SIR coil assembly retaining ring. The ring must be firmly seated in groove on the steering shaft.

NOTE: Gently pull the lower coil assembly, turn signal, pass key and pivot and dimmer switch wires to remove any kinks that may be inside the steering column assembly. Failure to do so may cause damage to the wire harness.

42. Install the wiring harness.

43. Install the steering wheel.

44. Connect the negative battery cable. Enable the SIR system.

45. Verify the correct operation of the steering column components.

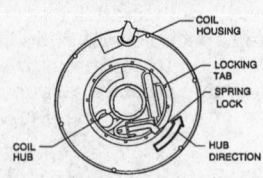

COIL HOUSING

LOCKING TAB

SPRING LOCK

COIL HUB

HUB DIRECTION

PERFORM THE FOLLOWING STEPS TO CENTER COIL ASSEMBLY

A. WHEELS STRAIGHT AHEAD.
B. REMOVE COIL ASSEMBLY.
C. HOLD COIL ASSEMBLY WITH BOTTOM UP.
D. WHILE HOLDING COIL ASSEMBLY, DEPRESS SPRING LOCK TO ROTATE HUB IN DIRECTION OF ARROW UNTIL IT STOPS.
E. THE COIL RIBBON SHOULD BE WOUND UP SNUG AGAINST CENTER HUB.
F. ROTATE COIL HUB IN OPPOSITE DIRECTION APPROXIMATELY TWO AND A HALF (2-1/2) TURNS. RELEASE SPRING LOCK BETWEEN LOCKING TABS.

325607

Centering the SIR coil assembly

Combination Switch

REMOVAL AND INSTALLATION

> **CAUTION**
> *The Supplemental Inflatable Restraint (SIR) system must be disarmed before removing the dimmer switch. Failure to do so may cause accidental deployment of the air bag, resulting in unnecessary SIR system repairs and/or personal injury.*

1. Disable the SIR system.
2. Disconnect the negative battery cable. Set the front wheels to the straight ahead position.

> **CAUTION**
> *When carrying a live air bag, make sure the bag and trim cover are pointed away from the body. In the unlikely event of an accidental deployment, the bag will then deploy with minimal chance of injury. When placing a live air bag on a bench or other surface, always face the bag and trim cover up, away from the surface. This will reduce the motion of the module if it is accidentally deployed.*

3. Remove the air bag and steering wheel using the recommended procedure.
4. Remove the SIR coil assembly retaining ring.
5. Remove the SIR coil assembly. Let the coil hang freely.
6. Remove the wave washer.
7. Remove the steering shaft lock retaining ring using special tool J 23653 SIR or the equivalent to compress the shaft lock. Dispose of shaft ring.
8. Remove the shaft lock.
9. Remove the turn signal cancelling cam assembly.

10. Remove the upper bearing spring and inner race seat.
11. Remove the inner race.
12. Remove the multi–function lever.
13. Remove the housing cover end cap. Using fingers, pry up and out. Tilt the end cap away from the lock housing cover and sleeve assembly and remove from the column.
14. Remove the washer head screw and flat head screw.
15. Remove the pivot and dimmer switch assembly using the following procedure:
   a. Remove the wiring protector.
   b. Remove the pivot and dimmer switch connector from the bulkhead.
   c. Remove the dimmer switch components from the housing cover.
   d. Using a suitable prying tool, pry on the upper and lower wings of the switch body.
   e. Remove the wire harness clamp from tab inside the housing and gently pull the wire harness from the steering column.
16. Remove the pivot and dimmer switch from the vehicle.

To install:

17. Install the pivot and dimmer switch assembly wire harness through the steering column.
18. Connect the wire harness clamp to tab inside lock housing cover of the dimmer switch.
19. Connect pivot and dimmer switch connector to bulkhead connector.
20. Place pivot and dimmer switch component to slot in lock housing cover.
21. Install pivot and dimmer switch switch to the turn signal switch.
22. Install and tighten the round washer head screw to 27 inch lbs. (3 Nm).
23. Install and tighten the flat head screw to 18 inch lbs. (2 Nm).
24. Install the housing cover end cap.
25. Install the multi–function and tilt lever. Install the inner race.
26. Install the upper bearing inner race seat. Install the upper bearing spring.
27. Install the turn signal cancelling cam assembly and lubricate assembly with synthetic grease.
28. Install the shaft lock. The inner block tooth of the lock must be aligned with block tooth of race and upper shaft assembly.

29. Install new shaft lock retaining ring using special tool J 23653 SIR or equivalent to compress the shaft lock. The ring must be firmly seated in groove on the steering shaft.

> **NOTE:** Set the steering shaft so that the block tooth on race and upper steering shaft is at the twelve o'clock position. Wheels on the vehicle should be straight ahead. Set the lock cylinder set to the LOCK position to ensure no damage to the SIR coil assembly.

30. Install the wave washer.

> **WARNING**
> **The SIR coil assembly will become uncentered if the steering column is separated from the steering gear and is allowed to rotate or if the centering spring is pushed down letting the hub rotate while the coil is removed from the steering column. SIR coil assembly wires must be kept tight with no slack while installing SIR coil assembly. Failure to do so may cause the wires to be kinked near the shaft lock area and cut when the steering wheel is turned.**

31. Ensure the SIR coil assembly hub is centered.
32. Install the SIR coil assembly.
33. Pull the SIR coil wires tight while positioning the SIR coil to the steering shaft.
34. Align opening in the SIR coil with the horn tower and "locating bump" between the two tabs on the lock housing cover.
35. Seat the SIR coil assembly into the steering column.
36. Install the SIR coil assembly retaining ring. The ring must be firmly seated in groove on the steering shaft.

> **NOTE:** Gently pull the lower coil assembly, turn signal, pass key and pivot and dimmer switch wires to remove any kinks that may be inside the steering column assembly. Failure to do so may cause damage to the wire harness.

37. Connect the wiring harness.
38. Install the steering wheel, torquing it to specifications. Install the air bag.
39. Connect the negative battery cable. Enable the SIR system.
40. Verify that all the switches are operating correctly.

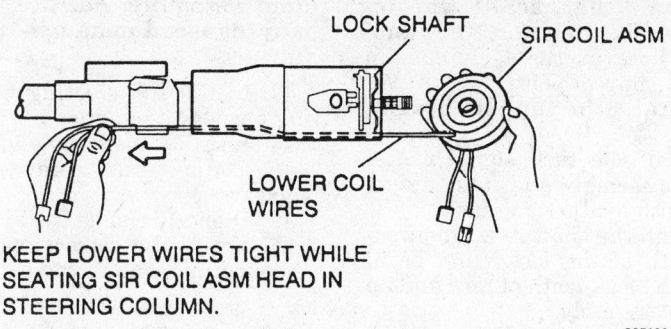

KEEP LOWER WIRES TIGHT WHILE
SEATING SIR COIL ASM HEAD IN
STEERING COLUMN.

285138

Routing the coil assembly wires

Ignition Lock Cylinder

REMOVAL AND INSTALLATION

— CAUTION —

The Supplemental Inflatable Restraint (SIR) system must be disarmed before removing the steering wheel. Failure to do so may cause accidental deployment of the air bag, resulting in unnecessary SIR system repairs and/or personal injury.

— CAUTION —

When carrying a live air bag, make sure the bag and trim cover are pointed away from the body. In the unlikely event of an accidental deployment, the bag will then deploy with minimal chance of injury. When placing a live air bag on a bench or other surface, always face the bag and trim cover up, away from the surface. This will reduce the motion of the module if it is accidentally deployed.

1993 Vehicles and 1994 Lumina

1. Remove the turn signal switch.
2. Remove the 2 lower spring retainers and discard.
3. Remove the lower bearing spring and seat.
4. Remove the adaptor screws, then remove the adaptor and the lower bearing assembly.
5. Place the lock cylinder in the **RUN** position.
6. Place opening in the retaining ring over the flat on the steering shaft, remove the retaining ring and discard.
7. Remove the thrust washer, the upper bearing spring and washer.
8. Remove the steering shaft from the lower end of the jacket and bowl assembly.

9. Remove the housing screws and remove the housing.
10. Remove the housing spacer and bearing with a drift punch and discard.
11. Remove the housing circuit bridge.
12. Place the lock cylinder in the **OFF-LOCK** position and remove the key.
13. Remove the buzzer switch and the lock retaining screw.
14. Remove the lock cylinder assembly.

To install:

15. Install the lock cylinder into the steering column. Place the lock cylinder into the **OFF-LOCK** position and remove the key. Install the retaining bolt and torque to 62 inch lbs. (7 Nm).
16. Install the buzzer switch assembly pushing down until the plastic tab covers the retaining bolt.
17. Install the housing circuit bridge to the housing and install with the clip notch over the housing lip.
18. Lubricate the bearing with lithium grease, pressing the bearing into the housing with an 1½ inch (38mm) socket until it bottoms.
19. Install the housing spacer and the column housing, torquing the screws to 88 inch lbs. (10 Nm).
20. Turn the steering column lock cylinder set to the **RUN** position and insert the steering shaft into the lower end of the jacket and bowl assembly until the shaft rests against the bearing.

NOTE: The steering shaft will extend 2.5 inches (63mm) beyond the highest surface of the steering column housing when installed properly.

21. Install the thrust washer, upper bearing spring and the other thrust washer.
22. Wrap a 2 inch (51mm) wide piece of 0.005 inch (.127mm) shim stock around the shaft and slip a new

retaining ring up to the thrust washer. Use a flat bladed tool to position the retaining ring until it seats into its proper groove. Discard the shim stock.
23. Install the adaptor and the lower bearing assembly, torquing the bolts to 26 inch lbs. (3 Nm).
24. Install the lower bearing seat and lower bearing spring.
25. Install 2 new lower spring retainers and compress the spring until the retainers are positioned 1.14 inch (29mm) from the lower end of the steering shaft.
26. Install the steering column into the vehicle.
27. Install the park lock cable into the ignition switch inhibitor and snap into place.
28. Install the turn signal switch.
29. Install steering wheel.
30. Install jam nut and torque to 30 ft. lbs. (41 Nm).
31. Install the air bag module.

1994–97 Vehicles

NOTE: When replacing the lock cylinder, the key code must be read with interrogator tool J 35628-A or equivalent. A new key must be ordered with the new lock cylinder and must be cut to the dummy key in the new lock cylinder.

1. Before beginning work, set the front wheels to the straight ahead position.
2. Disable the SIR system.
3. Disconnect the negative battery cable.
4. Remove the steering wheel controls and the steering wheel.
5. Remove the SIR coil assembly retaining ring.
6. Remove the SIR coil assembly. Pulling gently on the wire harness while feeding the wire up the column and let the coil hang freely.
7. Remove the wave washer.
8. Remove the steering shaft lock retaining ring using special tool J 23653 SIR or the equivalent to compress the lock plate and remove the shaft lock ring. Dispose of the old shaft lock ring.
9. Remove the turn signal cancelling cam assembly, lock plate and the upper bearing spring.
10. Remove the multi-function lever.
11. Remove the housing cover end cap. Tilt the end cap away from the lock housing cover and sleeve assembly and remove from the column.
12. Remove the screw with washer attached and the flat head screw.

13. Remove the pivot and dimmer switch assembly. Let the switch hang freely.

14. Remove the hazard warning switch knob by using a suitable prying tool to remove the knob.

15. Position the turn signal switch to gain access to remove the screws retaining the turn signal switch to the steering column.

16. Disconnect the ignition cylinder PASS KEY wire harness at the lower end of the steering column.

17. Attach a piece of wire to terminals for guiding the harness through the steering column.

18. Remove the retaining bolt to the ignition lock cylinder and insert the key in the lock cylinder to remove.

To install:

CAUTION

Route wires from the pass key lock cylinder carefully and make sure the retaining clip is properly located into the hole in housing. Failure to do so may result in component damage or malfunctioning of the steering column may occur.

19. Install the PASS KEY lock cylinder. Route the wires correctly and insert the terminals in the turn signal connector.

20. Install the lock cylinder fastener and torque to 22 inch lbs. (2.5 Nm).

21. Install the turn signal switch assembly to the steering column. Connect the turn signal switch connector to the bulkhead connector.

22. Install and tighten the turn signal switch assembly three blinding head cross recess screws to 30 inch lbs. (3.4 Nm).

23. Install the pivot and dimmer switch assembly wire harness through the steering column.

24. Connect the wire harness clamp to tab inside lock housing cover of the dimmer switch.

25. Connect pivot and dimmer switch connector to bulkhead connector.

26. Place pivot and dimmer switch component to slot in the lock housing cover.

27. Install pivot and dimmer switch to the turn signal switch.

28. Install and tighten the round washer head screw to 27 inch lbs. (3 Nm).

29. Install and tighten the flat head screw to 18 inch lbs. (2 Nm).

30. Install the housing cover end cap. Install the multi-function and tilt lever.

31. Install the hazard warning switch knob. Position the knob through hole in the lock housing cover and snap in place.

32. Install the upper bearing spring.

33. Install the turn signal cancelling cam assembly and lubricate assembly with synthetic grease.

34. Install the shaft lock. The inner block tooth of the lock must be aligned with block tooth of race and upper shaft assembly.

35. Install new shaft lock retaining ring using special tool J 23653 SIR or the equivalent to compress the shaft lock. The ring must be firmly seated in groove on the steering shaft.

36. Install the wave washer.

NOTE: The SIR coil assembly will become uncentered if the steering column is separated from the steering gear and is allowed to rotate or if the centering spring is pushed down letting the hub rotate while the coil is removed from the steering column. In addition, the SIR coil assembly wires must be kept tight with no slack while installing SIR coil assembly. Failure to do so may cause the wires to be kinked near the shaft lock area and cut when the steering wheel is turned.

37. Ensure the SIR coil assembly hub is centered.

38. Install the SIR coil assembly.

CAUTION

SIR coil assembly wires must be kept tight with no slack while installing the SIR coil assembly. Failure to do so may cause the wires to be kinked near the lock plate, causing the wires to be cut when the steering wheel is turned.

39. Pull the SIR coil wires snug while positioning the SIR coil to the steering shaft.

40. Align opening in the SIR coil with the horn tower and "locating bump" between the two tabs on the lock housing cover.

41. Seat the SIR coil assembly into the steering column.

42. Install the SIR coil assembly retaining ring. The ring must be firmly seated in groove on the steering shaft.

NOTE: Gently pull the lower coil assembly, turn signal, pass key and pivot and dimmer switch wires to remove any kinks that may be inside the steering col-

umn assembly. Failure to do so may cause damage to the wire harness.

43. Install the steering wheel wiring harness, if equipped.

44. Install the steering wheel.

45. Connect the negative battery cable.

46. Properly enable the SIR system.

47. Verify the correct operation of the steering column components.

Ignition Switch

REMOVAL AND INSTALLATION

1993 Vehicles and 1994 Lumina

1. Disconnect the negative battery cable. Remove the steering column from the vehicle.

2. Locate the ignition switch at the base of the steering column. Note that most models will also have a dimmer switch in a similar location. The ignition switch is on top of the column jacket.

3. Remove the electrical connector, remove the retainer screws and disengage the actuator rod from the ignition switch.

To install:

4. Move the ignition switch slider to the extreme right position and then move slider one detent to the left (off lock). Install the replacement ignition switch using care to make sure the actuator rod engages the switch.

NOTE: Insert a ⁹⁄₃₂ inch drill bit into the adjustment hole on the ignition switch to hold the switch slider in the proper position during installation.

5. Install and tighten the retaining screws to 35 inch lbs. (4 Nm). Use the original screws or exact replacements. Use of screws that are too long could prevent the column from collapsing on impact.

6. Connect the electrical connector to the ignition switch.

7. Raise the steering column and install the retaining bolts torquing them to 18 ft. lbs. (24 Nm).

8. Install the steering column opening filler panel under the column.

9. Connect the negative battery cable. Check operation of ignition switch.

1994–97 Vehicles

The mechanical key and lock cylinder switch is located in the steering column on the right side just below the

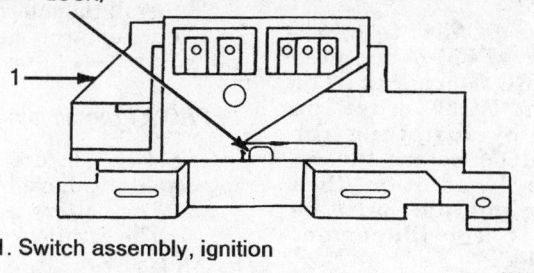

1. Switch assembly, ignition

324441

Adjustment of the ignition switch — 1993 vehicles and 1994
Lumina

steering wheel. The electrical switching portion of the assembly is separate from the key and lock cylinder, connected by a linkage rod. The electrical switching portion is located at the base of the steering column. This procedure is for removing the electrical switching portion of the ignition switch assembly. Please note that unlike many other ignition switch replacement procedures, GM recommends that the steering column be removed from the vehicle to change the ignition switch. It may, however be possible to remove and install the switch if the column can be lowered to the seat position, letting it rest on the seat cushion.

——— CAUTION ———

The Supplemental Inflatable Restraint (SIR) system must be disarmed before working around the steering column. Failure to do so may cause accidental deployment of the air bag, resulting in unnecessary SIR system repairs and/or personal injury.

1. Disable the SIR system.
2. Disconnect the negative battery cable.

NOTE: The vehicle's front wheels must be in the straight-ahead position and the key must be in the LOCK position when removing or installing the steering column. Failure to do so may cause the SIR coil assembly to become uncentered and may result in unneeded SIR system repairs. In addition, note that the steering column should never be supported by only the lower or upper support bracket alone. Damage to the column bearing adapter could result. During the steering column removal procedure, the upper and lower column bolts are removed and the

steering column lowered to rest on the seat. In this position, some repair operations may be performed.

3. Remove the steering column from the vehicle.
4. Remove the hex washer head screw. Remove the hex nut.
5. If equipped with a column shift, remove the PRNDL adjuster bracket.
6. If equipped with a floor shift, remove the column lock and ignition switch assembly from the ignition switch actuator assembly.
7. Remove the Torx® head screw.
8. Remove the dimmer switch mounting stud if necessary.
9. Remove the ignition switch assembly from the ignition switch actuator assembly.
10. Remove the turn signal switch assembly connector from the bulkhead connector.
11. Remove the pivot and dimmer switch assembly connector from the bulkhead connector.
    To install:
12. Place steering column lock cylinder in the OFF/LOCK position.

NOTE: Install the ignition switch to jacket assembly with the ignition switch in the OFF/LOCK position. The new ignition switch should be pinned in the OFF/LOCK position. Remove the plastic pin after the ignition switch is assembled to the steering column. Failure to do so may cause switch damage.

13. Adjust the ignition switch as follows:
    a. Move ignition switch slider to the extreme right position.
    b. Move switch slider one detent to left of "OFF LOCK" position.
    c. Install ³/₃₂ inch drill bit in the alignment hole on the switch to limit travel.

14. Install ignition switch assembly to the ignition switch actuator assembly.
15. Install the dimmer switch mounting stud and tighten the stud to 35 inch lbs. (4 Nm).
16. Install and tighten the Torx screw to 35 inch lbs. (4 Nm).
17. Remove the drill bit from the ignition switch.
18. If equipped with a column shift, install PRNDL adjuster bracket to stud.
19. If equipped with a floor shift, install the column lock and ignition switch assembly to the ignition switch actuator assembly.
20. Install the hex washer head screw and hex nut.
21. Install the pivot and dimmer switch assembly connector to the bulkhead connector.
22. Install the turn signal switch assembly connector to the bulkhead connector.
23. Install the steering column.
24. Connect the negative battery cable. Enable the SIR system.
25. Verify the correct operation of the steering column switches and controls.

Park/Neutral Safety Switch

REMOVAL AND INSTALLATION

1. Disconnect the negative battery cable.
2. Apply the parking brake and block the drive wheels.
3. Place the vehicles transaxle in the neutral position.
4. Remove the shift linkage lever.
5. Disconnect the electrical connector on the switch.
6. Remove the mounting bolts and remove the switch.
   To install:
7. Make sure the manual shaft is in the neutral position.
8. Align the flats of the manual shaft to the flats on the switch assembly.
9. Install the retaining bolts. DO NOT tighten at this time.

NOTE: If the bolts do not line up with the mounting boss on the transaxle, verify the manual shaft is in the neutral position. DO NOT rotate the switch, if a new switch. The switch is pinned from the factory.

10. If the switch has been rotated and the pin has been broken or when

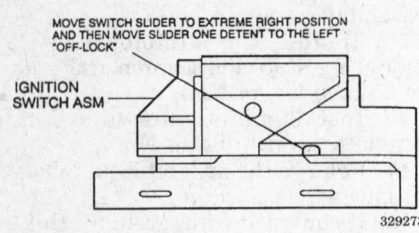

MOVE SWITCH SLIDER TO EXTREME RIGHT POSITION
AND THEN MOVE SLIDER ONE DETENT TO THE LEFT
"OFF-LOCK"

IGNITION
SWITCH ASM

329273

Ignition switch adjustment — 1994–97 vehicles, except 1994 Lumina

using the old switch, adjust as follows:

   a. Assemble the mounting bolts loosely.

   b. Align the slots using J 41545 or the equivalent.

   c. Torque the bolts to 18 ft. lbs. (25 Nm).

11. Attach the connector.

12. Install the manual shaft lever and install the retaining nut 18 ft. lbs. (25 Nm).

13. After the switch installation, verify that the engine will start in Park or Neutral position. If the engine will start in any other position, the switch needs to be readjusted.

Powertrain Control Module

REMOVAL AND INSTALLATION

——— CAUTION ———
To prevent internal PCM damage, disconnect the negative battery cable before disconnecting or reconnecting power to the PCM. Do not touch the connector pins or soldered components on the circuit board as this can create electrostatic discharge damage to the PCM.

Except 2.2L (VIN 4) Engine

1. Disconnect the negative battery cable.

2. Remove the PCM splash shield.

3. Remove the PCM mounting hardware. Detach the PCM connectors.

4. Remove the PCM from the engine compartment.

5. Remove the PCM access cover.

6. Remove the EPROM from the PCM. Using two fingers, push both retaining clips back away from the EPROM. At the same time, grasp the EPROM at both ends and lift it up

from the socket. Do not remove the cover of the EPROM.

   To install:

NOTE: If the PCM is to be replaced, the new PCM is supplied without a EPROM program. The replacement PCM must be programmed by installing the original EEPROM before the vehicle will run. In addition, then knock sensor module must be transferred if a replacement PCM is installed.

7. Remove the new PCM from its packing and check the service number to make sure it is the same as the defective PCM.

8. Remove the EPROM access cover. Install the EPROM in the EPROM socket.

NOTE: Press only the ends of the EPROM. Small notches in the EPROM must be aligned with the small notches in the EPROM socket. Press on the ends of the EPROM. Do not press on the middle of the EPROM, only on the ends.

9. Install the access cover on the PCM. Install the PCM in the engine compartment.

10. Attach the PCM connectors. Install the PCM mounting hardware.

11. Install the PCM splash panel.

12. Connect the negative battery cable.

2.2L (VIN 4) Engine

The Powertrain Control Module (PCM) is located under the instrument panel.

NOTE: When replacing the production PCM with a service PCM it is important to transfer the broadcast code and production PCM number to the service PCM label.

1. Disconnect the negative battery cable. Remove the interior access panel.

2. Detach the PCM harness connectors. Remove the PCM from its mounting bracket and remove from vehicle.

   To install:

3. If a new PCM is being installed, the new PCM Electronically Erasable Programmable Read-Only Memory or EEPROM memory chip must be programmed.

4. Install the PROM.

5. Install the access cover on the PCM. Attach the PCM connectors.

6. Position the PCM into its mounting location and install the mounting bracket.

7. Install the interior access panel.

8. Connect the negative battery cable.

EEPROM Programming

1. Set-up: Ensure that the following conditions have been met:

   a. The battery is fully charged.

   b. The ignition switch is in the **ON** position.

   c. The vehicle interface module cable connection at the DLC is secure.

2. Program the PCM using the latest software matching the vehicle. Refer to up-to-date dealership instructions.

3. If the PCM fails to program, proceed as follows:

   a. Ensure that all PCM connections are OK.

   b. Check the scan tool is equipped with the latest software version.

   c. Attempt again to program the PCM. If the PCM still cannot be programmed properly, replace the PCM. The replacement must be programmed or the vehicle will not run.

ENGINE COOLING

Starting with the 1996 Model Year, these vehicles were filled at the factory with a new type of antifreeze/coolant called GM Goodwrench DEX-COOL™. When adding coolant to 1996–97 vehicles, it is important that you use GM Goodwrench DEX-COOL™ (orange-colored, silicate-free) coolant. If silicated coolant is added to the system, premature engine, heater core or radiator corrosion may result. In addition, the engine coolant will require change sooner, at 30,000 miles or 24 months. Some coolant manufacturers are mixing other types of glycol in their coolant formulations. Propylene glycol is the most common new ingredient. **Propylene glycol is not recommended for use in GM vehicles.** A 50/50 mixture of DEX-COOL™ and clean water will provide all the recommended protection for 1996–97 vehicles. **DO NOT use DEX-COOL™ in pre-1996 vehicles. DO NOT mix DEX-COOL™ with any other type of antifreeze.**

Radiator

REMOVAL AND INSTALLATION

1. Remove the air cleaner and duct assembly.

2. Disconnect the negative battery cable. Drain the cooling system.

3. Remove the coolant recovery reservoir hose from the radiator.

4. Remove the engine strut brace bolts from the upper tie bar and rotate the struts and braces rearward. To prevent shearing of the rubber bushings, loosen the bolts on the engine struts before swinging the struts.

5. Disconnect the cooling fan electrical connectors.

6. Remove the cooling fan mounting bolts and remove the cooling fans.

7. Remove the upper radiator bracket bolts and brackets.

8. Disconnect the upper and lower hoses at the radiator.

9. Disconnect the low coolant sensor electrical connector.

10. Disconnect the transaxle fluid cooler lines from the radiator. Cap the lines to prevent the loss of transmission fluid.

11. If equipped, disconnect the engine oil cooler lines from the radiator. Cap the lines to prevent the loss of oil.

12. Remove the radiator from the vehicle.

To install:

13. Install the radiator in the vehicle. Ensure that the radiator is seated in lower insulator pads.

14. Install the remaining components in the reverse order of removal. Tighten the upper radiator bracket bolts to 89 inch lbs. (10 Nm), then cooling fan bolts to 53 inch lbs. (6 Nm) and the engine strut bolts and nuts to 32–35 ft. lbs. (43–47 Nm).

15. Connect the negative battery cable. Install the air cleaner.

16. Fill and bleed the cooling system with the proper coolant mix.

17. Start the engine and allow to come to normal operating temperature. Allow the engine to warm up sufficiently to confirm operation of cooling fan and check for coolant leaks. Also check the transaxle fluid level.

Water Pump

REMOVAL AND INSTALLATION

2.2L (VIN 4) Engine

1. Disconnect the negative battery cable. Drain the cooling system.

2. Loosen, but do not remove, the water pump pulley bolts.

3. Remove the serpentine belt.

4. Remove the alternator and alternator side bracket.

5. Remove the water pump pulley bolts and remove the pulley.

6. Remove the four mounting bolts and remove the water pump.

To install:

7. Clean all the gasket surfaces completely.

8. Apply a thin bead of sealer around the outer edge of the water pump gasket seating area and place he gasket on the pump.

9. Install the water pump on the engine and tighten the four mounting bolts to 18 ft. lbs. (25 Nm).

10. Install the remaining components in the reverse order of removal. Tighten the pulley bolts to 22 ft. lbs. (30 Nm).

11. Connect the negative battery cable. Refill and bleed the cooling system.

3.1L (VIN T) Engine

1. Disconnect the negative battery cable. Drain the cooling system.

2. Remove the nut securing the vacuum line to the top of the water pump pulley shield and position the line out of the way.

3. Remove the two shield mounting nuts and remove the pulley shield.

4. Loosen, but do not remove, the water pump pulley mounting bolts.

5. Remove the serpentine belt.

6. Remove the pulley mounting bolts and remove the pulley.

7. Remove the five attaching bolts, then remove the water pump.

To install:

8. Clean all the gasket surfaces completely.

9. Apply a thin bead of sealer around the gasket seating area on the water pump and install the gasket on the pump. The gasket is marked TOP and must be installed so the indicator is up.

10. Install the water pump and gasket on the engine and tighten to 7 ft. lbs. (10 Nm).

11. Install the remaining components in the reverse order of removal. Tighten the pulley bolts to 18 ft. lbs. (24 Nm).

12. Connect the negative battery cable. Refill and bleed the cooling system.

3.1L (VIN M) and 3.4L (VIN X) Engines

1. Disconnect the negative battery cable. Drain the cooling system.

2. Remove the coolant reservoir and lay aside.

3. Remove the serpentine belt guard, bolts and nuts.

4. Loosen the water pump pulley bolts. Remove the serpentine belt.

5. Remove the water pulley bolts and pulley.

6. Remove the water pump attaching bolts, then remove the water pump and gasket.

To install:

7. Clean all pump mating surfaces.

8. Inspect the pump. There should be a locator tab to identify the top of the pump. This locator must be in the vertical position when the pump is installed. Install the water pump and gasket. Tighten the attaching bolts to 89 inch lbs. (10 Nm).

9. Install the remaining components in the reverse order of removal. Tighten the pulley bolts to 18 ft. lbs. (25 Nm).

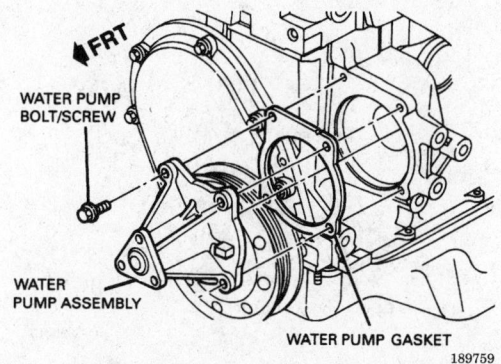

Water pump and gasket — 2.2L (VIN 4) engine

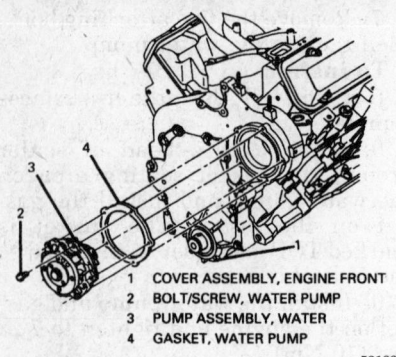

1 COVER ASSEMBLY, ENGINE FRONT
2 BOLT/SCREW, WATER PUMP
3 PUMP ASSEMBLY, WATER
4 GASKET, WATER PUMP

53183

Water pump assembly — 3.1L (VIN M) engine

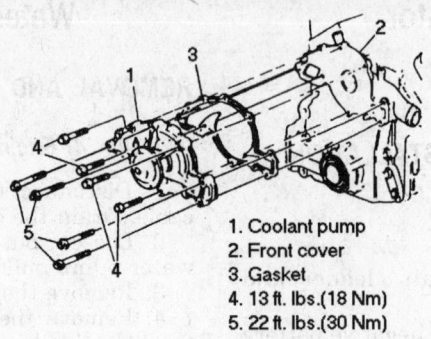

1. Coolant pump
2. Front cover
3. Gasket
4. 13 ft. lbs.(18 Nm)
5. 22 ft. lbs.(30 Nm)

328336

Water pump assembly — 3.8L (VIN L) engine

10. Connect the negative battery cable. Install the air cleaner assembly.

11. Refill the cooling system with the proper coolant. Bleed the cooling system and check for leaks with the engine running at idle.

3.8L (VIN L) and (VIN K) Engine

1. Disconnect the negative battery cable. Drain the cooling system.

2. Remove the coolant recovery reservoir.

3. Remove the serpentine belt. If additional access is needed, remove the inner fender electrical cover.

4. For VIN L engines, remove the alternator and position aside, then remove the serpentine belt tensioner pulley.

5. For VIN K engines, remove the power steering pump pulley using pulley remover J 25034-B or the equivalent.

6. Remove the water pump pulley.

7. Remove the water pump attaching bolts. Note that there are different length bolts. Use care to keep them organized for proper assembly. Remove the water pump from the vehicle.

To install:

8. Clean the water pump mating surfaces. Install the water pump using a new gasket.

9. Clean the pump bolt threads well. Install the attaching water pump bolts using care to install the proper bolts in the proper locations. Tighten long bolts to 22 ft. lbs. (30 Nm) and short bolts to 13 ft. lbs. (18 Nm) plus an additional 80 degrees using a torque angle meter.

10. Install the water pump pulley and tighten the bolts to 115 inch lbs. (13 Nm).

11. For VIN L engines, install the serpentine belt tensioner pulley, then install the alternator.

12. For VIN K engines, install the power steering pump pulley using J 25033-B or the equivalent to press the pulley onto the shaft.

13. Install the serpentine belt.

14. Install the coolant recovery reservoir.

15. Refill the cooling system with the correct amount and type of coolant.

16. Connect the negative battery cable. Start the engine and bleed the cooling system using the recommended procedure.

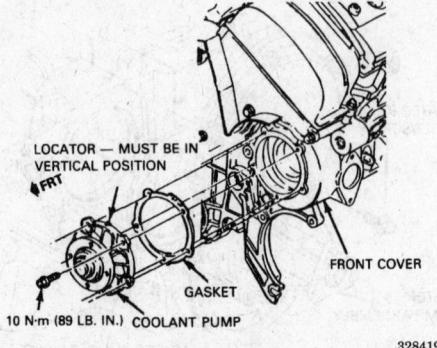

LOCATOR — MUST BE IN VERTICAL POSITION
FRT
GASKET
10 N·m (89 LB. IN.) COOLANT PUMP
FRONT COVER

328419

Water pump, gasket and mounting bolts — 3.4L (VIN X) engine

Thermostat

REMOVAL AND INSTALLATION

2.2L (VIN 4), 3.1L (VIN T) and (VIN M), and 3.8L (VIN L) and (VIN K) Engines

1. Disconnect the negative battery cable.

2. On the 3.8L, remove the air cleaner and duct.

3. Drain the cooling system to a level below the thermostat housing.

4. Disconnect the upper radiator hose from the thermostat housing.

5. Remove the thermostat housing mounting nuts and remove the thermostat housing.

6. Remove the thermostat.

To install:

7. Clean all the gasket surfaces completely. Install the thermostat in the thermostat housing.

8. Apply a thin bead of sealer around the gasket contact surface of the housing and install the gasket. The 3.8L is assembled using only RTV silicone gasket material.

9. Install the thermostat housing on the engine and tighten the mounting nuts to 89 inch lbs. (10 Nm) for the 2.2L, 18 ft. lbs. (25 Nm) for the 3.1L, or 20 ft. lbs. (27 Nm) for the 3.8L.

10. Connect the upper radiator hose to the thermostat housing.

11. Refill the cooling system.

12. Connect negative battery cable. Start engine and check for leaks.

3.4L (VIN X) Engine

— **CAUTION** —

Fuel injection systems remain under pressure, even after the engine has been turned OFF. The fuel system pressure must be relieved before disconnecting any fuel lines. Failure to do so may result in fire and/or personal injury.

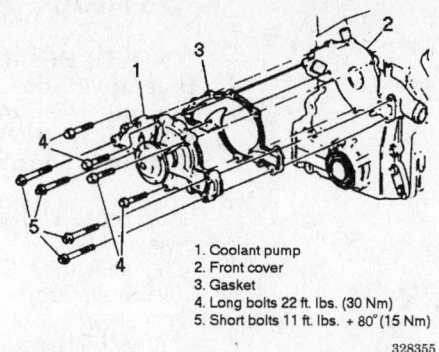

1. Coolant pump
2. Front cover
3. Gasket
4. Long bolts 22 ft. lbs. (30 Nm)
5. Short bolts 11 ft. lbs. + 80° (15 Nm)

328355

Water pump assembly — 3.8L (VIN K) engine

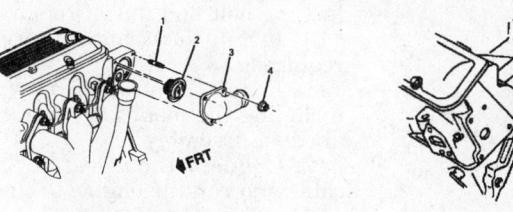

1. Water outlet stud
2. Engine coolant thermostat assembly
3. Water outlet
4. Water outlet nut

189731

Thermostat components — 2.2L (VIN 4) engine

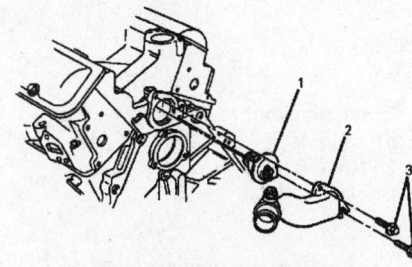

1 THERMOSTAT ASSEMBLY, ENGINE COOLANT
2 OUTLET ASSEMBLY, WATER
3 BOLT/SCREW, WATER OUTLET

191113

Thermostat components — 3.1L (VIN M) engine

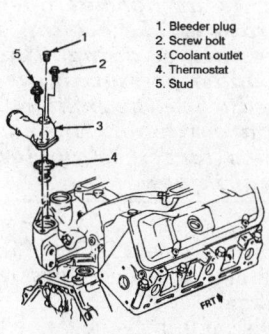

1. Bleeder plug
2. Screw bolt
3. Coolant outlet
4. Thermostat
5. Stud

207583

Thermostat components — 3.1L (VIN T) engine

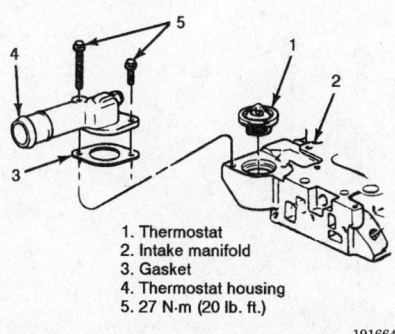

1. Thermostat
2. Intake manifold
3. Gasket
4. Thermostat housing
5. 27 N·m (20 lb. ft.)

191664

Thermostat, housing and mounting components — 3.8L (VIN L) and (VIN K) engine

1. Relieve the fuel system pressure. Remove the air cleaner assembly.
2. Disconnect the negative battery cable.
3. Partially drain the cooling system to a level below the thermostat housing.
4. Remove the torque struts.
5. Disconnect the upper radiator hose from the thermostat housing.
6. Disconnect the fuel lines from the fuel rail.

7. Remove the heater hose bracket from the thermostat housing mounting stud. Disconnect the heater hose from the throttle body.
8. Remove the thermostat housing mounting bolts, then remove the housing and the thermostat.

To install:
9. Clean all the gasket surfaces completely.
10. Apply a thin bead of sealer around the gasket contact area on the thermostat housing and install the gasket on the housing.
11. Install the thermostat in the engine and install the thermostat hous-

ing. Tighten the mounting bolts to 18 ft. lbs. (25 Nm).
12. Connect the heater hose to the throttle body. Attach the heater hose bracket to the thermostat housing mounting stud.
13. Connect the fuel lines to the fuel rail.
14. Connect the upper radiator hose to the thermostat housing.
15. Install the torque strut. Tighten the bolts to 39 ft. lbs. (53 Nm).
16. Install the air cleaner assembly.
17. Connect the negative battery cable. Refill the cooling system.
18. Pressurize the fuel system and verify no leaks. Start the vehicle and bleed the cooling system.

Electric Cooling Fan

REMOVAL AND INSTALLATION

─── **CAUTION** ───
Keep hands, tools, and clothing away from electric engine cooling fan(s) to help prevent personal injury. These fan(s) are electric and can come on whether or not the engine is running. The fan(s) can start automatically, in response to a heat sensor with the ignition in the "ON" position.

1. Disconnect the negative battery cable. Remove the air cleaner assembly, including the mounting stud and the air duct assembly.
2. Remove the coolant reservoir, as required.
3. If the engine torque strut interferes with fan removal, remove the engine strut brace bolts from the upper tie bar and rotate the strut(s) and brace(s) rearward.

NOTE: In order to prevent shearing of the rubber bushing(s), loosen the bolt(s) on the engine strut(s) before rotating.

4. Detach the connector(s) from the fan motor(s) and frame(s).
5. Remove the fan frame attaching bolts, then remove the fan assembly or assemblies.

─── **CAUTION** ───
If a fan blade assembly is bent or damaged in any way, no attempt should be made to repair and/or reuse the damaged part. A bent or damaged fan blade assembly should always be replaced with a new fan blade assembly. A bent or damaged assembly could fail and fly apart during subsequent use, possibly causing injury.

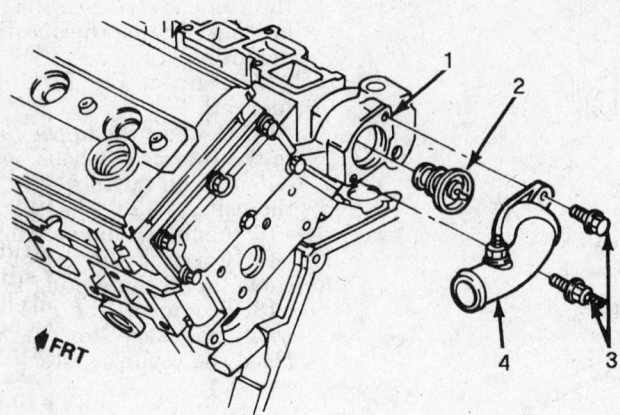

1 Intake manifold
2 Thermostat
3 25 Nm (18 lb. ft.)
4 Coolant outlet

207579

Thermostat housing and mounting components — 1993–94 3.4L (VIN X) engine

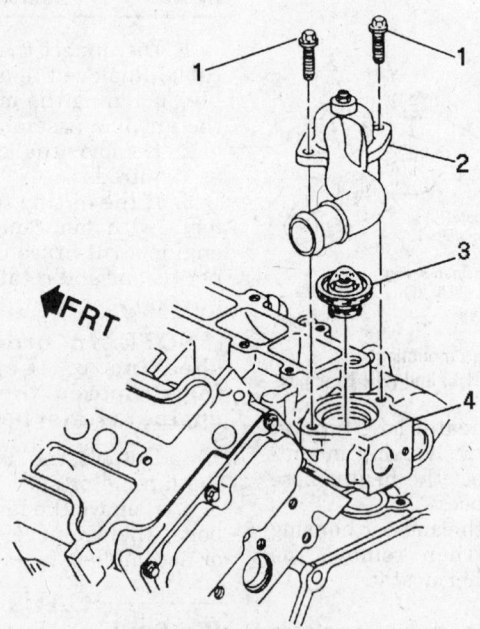

1. Mounting bolt
2. Thermostat housing
3. Thermostat
4. Lower intake manifold

328245

Thermostat housing and outlet assembly — 1995–97 3.4L (VIN X) engine

To install:

NOTE: Dirt or mud build up on the fan blades can cause an abnormal noise or vibration. Carefully clean the dirt or mud off of the fan blades, if present.

6. Install the fan assembly or assemblies.
7. Install the frame attaching bolts and tighten to 89 inch lbs. (10 Nm).
8. Attach the connector(s) to the fan motor(s) and frame(s).
9. Swing the strut(s) into the proper position and install the attaching bolt and nut, if removed.
10. Install the coolant reservoir, as required.
11. Install the air cleaner assembly, including the mounting stud and the air duct assembly.
12. Connect the negative battery cable and run the engine to check the cooling fan operation.

Cooling System Bleeding

— CAUTION —

To avoid being burned, DO NOT remove the radiator cap while the engine is at normal operating temperature. The cooling system will release scalding fluid and steam under pressure if the cap is removed while the engine and radiator are still hot. This could result in a large coolant loss and personal injury.

To ensure complete filling of the cooling system, it is necessary to bleed the system.

1. Disconnect the negative battery cable. Park the vehicle on a level surface.
2. Remove and clean the coolant recovery reservoir.
3. Remove the radiator cap when the engine is cool, as follows:
   a. Slowly rotate the cap counterclockwise to the detent. Do not pressure down while rotating the pressure cap.
   b. Wait until any residual pressure indicated by a hissing noise is relieved.
   c. After the hissing noise stops, continue to rotate the cap counterclockwise until it is removed.
4. Place a drain pan under the radiator, then open the drain valve and drain the coolant.

5. Remove the thermostat housing cap and thermostat or open bleed vents:

   a. On the 2.2L (VIN 4) engine, remove the thermostat housing cap and thermostat.

   b. On the 3.1L (VIN T) and (VIN M) engine, open the air bleed vents on the thermostat housing and the throttle body return pipe above the water pump. Open the vents 2–3 turns.

   c. On the 3.4L (VIN X) engine, open the air bleed vents on the thermostat housing and the heater coolant inlet pipe by the brake master cylinder.

   d. On the 3.8L (VIN L) and (VIN K) engine, open the air bleed vent on the thermostat housing. Open 2–3 turns.

6. Remove the engine block drain(s).

7. Fill the cooling system with coolant to the base of the radiator neck.

8. Reinstall or replace the thermostat and housing and close the air vents.

9. Close the drain valve and engine block drain(s).

10. Fill the coolant reservoir to the proper level with the proper mix of a suitable coolant and clean water.

11. Connect the negative battery cable. Start the vehicle and let the engine reach operating temperature, adding coolant as needed. Check the cooling system for leaks.

FUEL SYSTEM

Fuel System Service Precautions

Safety is the most important factor when performing not only fuel system maintenance but any type of maintenance. Failure to conduct maintenance and repairs in a safe manner may result in serious personal injury or death. Maintenance and testing of the vehicle's fuel system components can be accomplished safely and effectively by adhering to the following rules and guidelines.

• To avoid the possibility of fire and personal injury, always disconnect the negative battery cable unless the repair or test procedure requires that battery voltage be applied.

• Always relieve the fuel system pressure prior to disconnecting any fuel system component (injector, fuel rail, pressure regulator, etc.), fitting or fuel line connection. Exercise extreme caution whenever relieving fuel system pressure to avoid exposing skin, face and eyes to fuel spray. Please be advised that fuel under pressure may penetrate the skin or any part of the body that it contacts.

• Always place a shop towel or cloth around the fitting or connection prior to loosening to absorb any excess fuel due to spillage. Ensure that all fuel spillage (should it occur) is quickly removed from engine surfaces. Ensure that all fuel soaked cloths or towels are deposited into a suitable waste container.

• Always keep a dry chemical (Class B) fire extinguisher near the work area.

• Do not allow fuel spray or fuel vapors to come into contact with a spark or open flame.

• Always use a backup wrench when loosening and tightening fuel line connection fittings. This will prevent unnecessary stress and torsion to fuel line piping. Always follow the proper torque specifications.

• Always replace worn fuel fitting O-rings with new. Do not substitute fuel hose or equivalent, where fuel pipe is installed.

Fuel System Pressure

RELIEVING

—— **CAUTION** ——

Fuel injection systems remain under pressure, even after the engine has been turned OFF. The fuel system pressure must be relieved before disconnecting any fuel lines. Failure to do so may result in fire and/or personal injury. Please note that even after relieving fuel system pressure, a small amount of fuel may be released when servicing the fuel pipes or connections. In order to reduce the chance of personal injury, cover the fuel pipe fittings with a shop towel before disconnecting, to catch any fuel that may have leaked out.

1. Disconnect the negative battery cable to prevent possible discharge of fuel if an accidental attempt is made to start the engine.

2. Loosen the fuel filler cap to relieve tank pressure.

3. This procedure calls for GM special tool J 34730-1A or equivalent.

It is a fuel pressure test gauge with a line equipped with a fitting to connect to the to the fuel pressure test connection and another hose to discharge into an approved gasoline container. Wrap a shop towel around the pressure test fitting connection while connecting gauge to avoid spillage.

4. Install the bleed hose into an approved container and open the valve to bleed fuel system pressure. The fuel connections are now safe for servicing.

5. Drain any fuel remaining in the gauge into an approved container.

6. Reconnect the negative battery cable unless addition service work is being performed.

Idle Speed

ADJUSTMENT

The idle speed is controlled by the Powertrain Control Module (PCM). No adjustment is necessary or possible.

Mixture

ADJUSTMENT

The air/fuel mixture is controlled by the Powertrain Control Module (PCM); external adjustment is not possible.

Fuel Filter

REMOVAL AND INSTALLATION

—— **CAUTION** ——

Fuel injection systems remain under pressure, even after the engine has been turned OFF. The fuel system pressure must be relieved before disconnecting any fuel lines. Failure to do so may result in fire and/or personal injury.

1. Disconnect the negative battery cable. Relieve the fuel system pressure.

2. Raise and safely support the vehicle.

3. Note that on side of the fuel filter uses a quick-connect fitting. Special handling is required. To remove, grasp both sides of the fitting. Twist the female connector 1/4 turn in each direction to loosen any dirt within the fitting.

4. Using compressed air, blow dirt out of the fitting.

5. If equipped with a plastic fitting, squeeze the plastic retainer release tabs and pull the connection apart.

6. If equipped with a metal fitting, choose the correct special tool J 37088 A , J 39504 or equivalent tool set for the correct size fitting. Insert tool into the female connector, then push/pull inward to release the locking tabs. Carefully pull the connection apart.

7. Remove the fuel feed pipe nut to the outlet side of the fuel filter.

8. Remove the fuel filter from the holder.

To install:

9. Install the fuel filter into the holder.

10. Use a backup wrench on the inlet fitting while tightening the fuel filter outlet nut to 22 ft. lbs. (30 Nm).

11. Apply a few drops of clean engine oil to the male pipe end.

12. Push both sides of the fitting together to cause the retaining tabs/fingers to snap into place.

13. Once installed, pull on both sides of the fitting to make sure the connection is secure.

14. Safely lower the vehicle.

15. Tighten the fuel filler cap.

16. Turn the ignition switch to the ON position to pressurize the fuel system to check for leaks.

Fuel Pump

REMOVAL AND INSTALLATION

1. Relieve the fuel system pressure. Disconnect the negative battery cable.

2. Drain and remove the fuel tank using the recommended procedure.

3. Disconnect the quick connects from the fuel sender assembly and remove the fuel lines.

4. Using J-35731, or an equivalent spanner wrench, remove the fuel sender lock ring.

5. Lift the sender assembly carefully out of the fuel tank.

6. Remove the sender O-ring from the top of the tank and discard.

─── WARNING ───
DO NOT run the fuel pump unless it is submerged in fuel. Running the pump dry will cause serious damage to the fuel pump and may cause the pump to explode due to the oxygen in the air.

7. Note position of fuel pump strainer on fuel pump and while supporting the pump assembly in one hand twist the fuel strainer off the pump and discard the strainer.

8. Disconnect the fuel pump electrical connector.

9. Remove the clamp from the fuel line at the top of the pump.

10. Hold the fuel sender upside down on a work bench and pull the fuel pump out of the lower mounting bracket. Once the pump is clear of the lower mounting bracket tilt the pump outward and disconnect the pump from the sender assembly.

To install:

11. Install the rubber bumper and insulator on the fuel pump.

12. Hold the fuel sender upside down and install the fuel pump between fuel pulse dampener and mounting bracket.

13. Connect the fuel pump electrical connector.

14. Install the clamp on the fuel line.

15. Install a new fuel pump strainer on outer edge of ferrule until fully seated. The strainer must be facing in the same direction as prior to removal.

16. Install a new O-ring on top of the fuel tank and install the sender assembly.

17. Install the lock ring and using J-35731, or an equivalent spanner wrench.

18. Install and refill the fuel tank.

19. Refill the fuel tank.

20. Connect the negative battery cable. Turn the ignition switch to the ON position to pressurize the fuel system and check for leaks.

Fuel Injector

REMOVAL AND INSTALLATION

2.2L (VIN 4) Engine

1. Disconnect the negative battery cable. Relieve the fuel system pressure.

2. Remove the upper intake manifold assembly.

3. Remove the fuel return line bracket mounting bolt and remove the bracket.

4. Position the return line away from the pressure regulator.

5. Remove the pressure regulator as follows:

   a. Disconnect the vacuum hose from the regulator.

   b. Remove the fuel return pipe clamp.

   c. Disconnect the fuel return pipe from the pressure regulator and discard the O-ring.

   d. Remove the pressure regulator bracket mounting bolt.

   e. Remove the regulator assembly and O-ring and discard the O-ring.

6. Remove the fuel injector retaining bracket mounting bolts.

7. Remove the fuel injector retaining bracket by carefully sliding it off to clear the fuel injector slots.

8. Disconnect the fuel injector electrical connector.

9. Pull the injector out. Remove and discard the O-rings.

To install:

NOTE: Each fuel injector is calibrated for a specific flow rate. Be sure to use the correct part number when ordering replacement fuel injectors.

10. Coat new O-rings with clean engine oil, then install on the injector.

11. Install the injector into the cavity in the lower intake manifold with the electrical connector facing inward.

12. Install the remaining components in the reverse order of removal.

13. Connect the negative battery cable. Pressurize the fuel system and verify no leaks.

14. Start the engine and verify proper injector operation.

3.1L (VIN T) Engine

1. Relieve the fuel system pressure. Disconnect the negative battery cable.

2. Remove the upper intake manifold assembly. Remove the fuel line bracket bolt.

3. Disconnect and cap the fuel lines from the fuel rail. Use a backup wrench on the fuel rail fittings to prevent them from turning.

4. Remove and discard the O-rings from the lines.

5. Disconnect the vacuum line from the pressure regulator.

6. Remove the fuel rail mounting bolts, then detach the injector connectors.

7. Remove the fuel rail assembly.

8. Remove the injector retaining clip and discard.

9. Remove the injector assembly from the fuel rail. Remove and discard the O-rings.

To install:

10. Coat new O-rings with clean engine oil and install them on the injector.

11. Install a new retaining clip on the fuel injector. Position the open end of the clip toward the injector electrical connector.

12. Install the injector onto the fuel rail assembly with the injector electrical connectors facing outward.

13. Install the fuel rail onto the intake manifold so all the injectors fit into the cavities. Tighten the fuel rail mounting bolts to 19 ft. lbs. (25 Nm).

14. Install the remaining components in the reverse order of removal

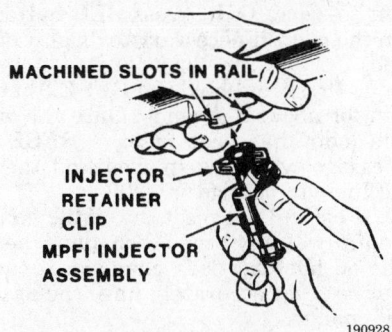

Installing the fuel injector onto the fuel rail — 3.1L (VIN T) engine shown; others are similar

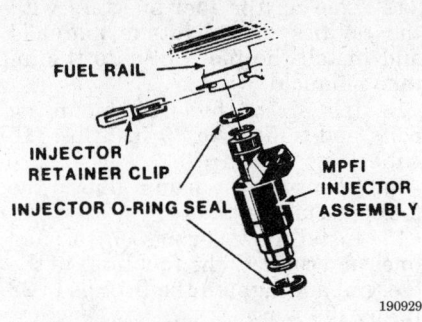

Fuel injector and mounting clips — 3.1L (VIN T) engine shown; others are similar

Tighten the fuel line fittings to 20 ft. lbs. (27 Nm), using a backup wrench.

15. Connect the negative battery cable. Pressurize the fuel system and verify no leaks.

3.1L (VIN M) Engine

1994 Vehicles

1. Disconnect the negative battery cable. Properly relieve the fuel system pressure.

2. Remove the upper intake manifold assembly.

3. Disconnect and cap the fuel feed line from the fuel rail.

4. Remove the fuel pressure regulator as follows:

  a. Remove the fuel pressure regulator retaining screw.

  b. Place a shop towel under the pressure regulator and remove the regulator from the fuel rail.

  c. Remove the retainer and spacer bracket from the fuel rail.

  d. Disconnect the regulator from the fuel return line.

5. Disconnect the fuel injector main harness.

6. Detach the coolant temperature sensor connector.

7. Remove the fuel rail retaining bolts, then remove the rail.

8. Detach the individual injector connector.

9. Remove the injector retaining clip and discard.

10. Remove the injector assembly from the fuel rail.

11. Remove and discard the O-rings. Save the O-ring backups for installation on the new injector.

To install:

12. Install the O-ring backups on the fuel injector. Coat new O-rings with clean engine oil and install them on the injector.

13. Install a new retaining clip on the fuel injector. Position the open end of the clip toward the injector electrical connector.

14. Install the injector onto the fuel rail assembly with the injector electrical connector facing outward.

15. Install the remaining components in the reverse order of removal. Install the fuel rail onto the intake manifold so all the injectors fit into the injector cavities. Tighten the rail bolts to 89 inch lbs. (10 Nm). Tighten the fuel line fittings to 22 ft. lbs. (30 Nm).

16. Connect the negative battery cable. Pressurize the fuel system and verify no leaks.

1995–97 Vehicles

1. Disconnect the negative battery cable. Relieve the fuel system pressure.

2. Remove the upper intake manifold (plenum).

3. Remove the engine fuel pipe bracket bolts. Detach the fuel pipes at the rail.

4. Remove the fuel feed and return pipe O-rings and discard.

5. Detach the fuel injector connectors.

6. Remove the fuel rail retaining bolts and lift off the fuel rail assembly with the six fuel injectors still attached.

7. Remove and discard the fuel injector retainer clip. Then remove the fuel injector by pulling out of the fuel rail. Use care in removing the injectors to prevent damage to the electrical connector pins on the injector. Also protect the nozzle. The fuel injector is serviced as a complete assembly only. Since it is an electrical component, it should not be immersed in any type of cleaner.

8. Remove the injector O-ring seal from both the spray tip end and the fuel rail end of each injector. Discard the seals. With the spray tip end O-ring removed, the O-ring backup piece may slip off of the injector. Be sure to retain the O-ring backup for reuse.

To install:

9. Clean all parts well. Use care in cleaning old gasket material from the machined aluminum surfaces on the plenum and manifold as sharp tools may damage sealing surfaces.

10. Each injector is calibrated for a specific flow rate. When replacing fuel injectors, order replacements with the identical part number as the old injectors. If fuel injectors are found to be leaking, the engine oil may be contaminated with fuel and an oil and filter change after service is complete is recommended.

11. Make sure that the O-ring backup piece is in place on the spray tip end of the injector before install-

ing a new O-ring. Lubricate new injector O-ring seals with clean engine oil and install on the injector assembly.

12. Install the new injector retainer clips on the injector assembly. Position the open end of the clip facing the injector electrical connector.

13. Install the fuel injector assembly into the opening in the fuel rail with the electrical connector facing outward. Push the injector in far enough to engage the retainer clip with the machined slots on the rail socket.

14. Install the fuel rail assembly to the intake manifold. Tilt the rail assembly to install the injectors. Install the fuel rail attaching bolts and tighten to 89 inch lbs. (10 Nm).

15. Install the remaining components in the reverse order of removal. Tighten the fuel rail nuts to 13 ft. lbs. (17 Nm). Use a backup wrench on the fittings to prevent them from turning.

16. Connect the negative battery cable. Turn the key to the **ON** position to pressurize the fuel system and check for fuel leaks.

3.4L (VIN X) Engine

1. Relieve the fuel system pressure. Disconnect the negative battery cable.

2. Remove the upper intake manifold assembly.

3. Disconnect and cap the fuel feed and return lines from the fuel rail. Remove and discard the O-rings from the fuel lines.

4. Disconnect the vacuum line from the fuel pressure regulator.

5. Remove the four fuel rail assembly mounting bolts.

6. Detach the fuel injector connectors.

7. Remove the fuel rail assembly.

8. Remove the fuel injector retainer clip and discard the clip.

9. Remove the fuel injector from the fuel rail and discard the upper O-ring.

To install:

10. Coat the new O-rings with engine oil and install the upper and lower rings on the injector.

11. Install a new injector retainer on the fuel injector and install the fuel injector on the fuel rail with the electrical connector facing outward.

12. Line up the fuel injectors with the cavities in the intake manifold and Install the fuel rail onto the intake manifold.

13. Install the fuel rail mounting bolts and tighten to 89 inch lbs. (10 Nm).

14. Connect the vacuum line to the fuel pressure regulator.

15. Install new O-rings on the fuel lines and connect the fuel lines to the fuel rail and tighten the fittings to 22 ft. lbs. (30 Nm).

16. Install the upper intake manifold plenum.

17. Connect the negative battery cable. Pressurize the fuel system and verify no leaks.

18. Start the engine and verify proper injector operation.

3.8L (VIN L) and (VIN K) Engine

1. Relieve the fuel system pressure. Disconnect the negative battery cable.

2. Remove the injector sight cover.

3. Disconnect the fuel line quick-connects at the fuel rail and cap the fuel lines.

4. Disconnect the electrical connections from the fuel injectors.

5. Disconnect the vacuum line from the pressure regulator.

6. Remove the four fuel rail-to-intake manifold mounting bolts.

7. Remove the fuel rail from the intake manifold.

8. Remove the injector retaining clip and remove the fuel injector from the fuel rail.

9. Remove the O-rings from the fuel injector and discard.

To install:

10. Coat the new O-rings with engine oil and install the O-rings on the injector.

11. Install the injector retaining clip on the fuel injector and snap the injector onto the fuel rail.

12. Line up all the injectors with the cavities in the intake manifold and push each injector into the manifold.

13. Install the fuel rail mounting bolts to 10 ft. lbs. (15 Nm).

14. Connect the fuel lines to the fuel rail. Attach the fuel injector connectors.

15. Connect the vacuum line to the pressure regulator.

16. Connect the negative battery cable.

17. Turn the ignition switch to the **ON** position to pressurize the fuel system and check for leaks.

18. Install the injector sight cover.

EMISSION CONTROLS

Service Interval Lamp

RESETTING

Vehicles that are equipped with a Drivers Information Center and have an Oil life index, will require the Oil Life indicator to be reset after each oil change.

Press the **SEL** to select **OIL**. Press **SEL** if necessary to display the oil life. The display will show a reading of the estimated oil life left. Example: **OIL LIFE 85%** . When the remaining oil life is 9% or less, the display will show **CHANGE OIL SOON**. When the vehicle is started tone will sound an the **CHANGE OIL SOON** message will display each time the vehicle is started.

When the oil life is zero, a tone will sound and the display will show, **CHANGE OIL NOW**. When then the vehicle is started a tone will sound and the **CHANGE OIL NOW** message will display each time the vehicle is started. Reset the Oil Life Display as follows:

1. Acknowledge all diagnostic messages in the Drivers Information Center by pressing **RESET**.

2. Press the **SEL** button on the left to select **OIL**. Press **SEL** button on the right if necessary to display oil life.

3. Press and hold the **RESET** button for about 5 seconds. Once the oil life index has been reset, a **RESET** message will be displayed and then oil life will change to 100% .

Be careful not reset the oil life accidentally at any time other than when the oil has just been changed. It can not be reset accurately until the next oil change.

ENGINE MECHANICAL

Engine Assembly

REMOVAL AND INSTALLATION

— CAUTION —

Fuel injection systems remain under pressure, even after the engine has been turned OFF. The fuel system pressure must be relieved before disconnecting any fuel lines. Failure to do so may result in fire and/or personal injury.

2.2L (VIN 4) Engine

1. Relieve the fuel system pressure. Disconnect the negative battery cable.

2. Mark the position of the hood hinges to aid in installation. Remove the hood hinge bolts and the hood with aid from an assistant.

3. Remove the air cleaner and the duct assembly. Drain the engine coolant.

4. Remove the engine torque strut.

5. Remove the air intake silencer assembly. Remove the coolant recovery reservoir.

6. Disconnect the upper radiator hose from the engine.

7. Remove the lower radiator hose at the water pump at the rear of the engine.

8. Disconnect the brake booster vacuum hose at the intake manifold.

9. Disconnect the throttle control cables at the throttle body.

10. Remove the serpentine drive belt.

11. Disconnect the electrical connection at the right side engine cooling fan, and remove the fan.

12. Remove the power steering pump. Properly disconnect and plug the fuel lines.

13. Remove the screws that retain the alternator heat shield and remove the heat shield.

14. Detach the alternator and O₂ sensor electrical connectors.

15. Remove the battery cable/ground wires and the bracket at the bellhousing.

16. Raise and safely support the vehicle.

17. Remove the flywheel inspection cover, and remove the flywheel to torque converter retaining bolts.

18. Remove the engine mount to frame nuts.

19. Disconnect the front exhaust pipe from the exhaust manifold.

20. Lower the vehicle.

21. To access the rear engine connections, rotate the engine as follows:

    a. Place the the transaxle in **Neutral**.

    b. Disconnect the engine torque strut from the chassis bracket.

    c. Using a prybar at the exhaust manifold, pull the engine and transaxle forward.

    d. Align the slave hole in the torque strut with the hole in the engine bracket.

    e. Install the torque strut mounting bolt, in the hole, to hold the engine in position.

22. Disconnect the electrical connections at the starter motor, including the ground wire.

23. Properly discharge the air conditioning system. Remove the bolts that retain the A/C compressor to the mount bracket and secure the compressor aside.

24. Remove the bolts that retain the engine to the transaxle bracket at the bracket.

25. Disconnect the vacuum hose at the intake manifold.

26. Disconnect the speed sensor, knock sensor, engine ground and throttle body/injector harness electrical connectors.

27. Remove the bellhousing to engine bolts/nuts and disconnect the ground wires.

28. Remove the engine torque struts.

29. Install an engine lifting/support device to the engine.

30. Begin lifting the engine out with the engine lifting device.

31. Disconnect the electrical connections at the electronic ignition module.

32. Completely remove the engine from the vehicle.

To install:

33. Lower the engine into the vehicle and attach the ignition module connections.

34. Completely lower the engine into the vehicle and remove the engine lifting device.

35. Rotate the engine forward and connect the engine torque struts.

36. Install the bellhousing to engine bolts/nuts and connect the ground wires. Tighten the mounting bolts to 55 ft. lbs. (75 Nm).

37. Install the remaining components in the reverse order of removal. Tighten the engine-to-transaxle rear bracket bolt to 63 ft. lbs. (85 Nm) and

the front and top mounting bolts to 35 ft. lbs. (48 Nm).

38. Refill the coolant to the proper level. Fill the engine oil, if not already done.

39. Install the air cleaner and the duct assembly.

40. With the aid of an assistant, install the hood while aligning the previously marked position.

41. Connect the negative battery cable. Start the engine and check for leaks.

42. Run the engine until normal operating temperature is reached, then shut the engine off and recheck all fluid levels.

3.1L (VIN T) and 1993–94 3.4L (VIN X) Engines

1. Remove the air cleaner assembly and ground wire near cleaner bracket.

2. Properly relieve the fuel system pressure, then disconnect the negative battery cable and body ground.

3. Mark the hood hinges to ensure proper reinstallation. With the help of an assistant, remove the hood retaining bolt and remove hood from the vehicle.

4. Mark and remove all necessary engine wiring; place the wiring harnesses aside.

5. Remove the throttle, TV and cruise control cables, if equipped, from the throttle body assembly.

6. Remove the fuel lines at engine.

7. Remove the AIR pump and serpentine belt.

8. Position drain pan under the radiator drain valve and drain the engine coolant. Remove coolant recovery tank. Remove the cooling fans.

9. Remove the upper and lower radiator hoses and heater hose quick connect at intake manifold.

10. Discharge the air conditioning system using the appropriate equipment. Remove the air conditioning compressor mounting bolts at the front mounting bracket.

11. Remove the power steering pump and move to the side. Attach to the body with a piece of wire or rope. Do not disconnect the pump hoses.

12. Remove the heater hoses from the engine, and position aside.

13. Remove the brake booster vacuum hose.

14. Remove the EGR hose from the exhaust manifold. Remove the pipe from the EGR valve, if equipped.

15. Raise the vehicle and support it safely.

16. Remove the A/C compressor from the engine and attach to the body with a piece of wire. The factory

recommends removal of the air conditioning manifold from compressor.

17. Remove the right front tire and wheel. Remove the right front splash shield.

18. Disconnect right ball joint nut and separate from control arm.

19. Remove halfshaft assembly. Disconnect any remaining electrical connectors at the back of the engine.

20. Remove the flywheel cover, starter motor and torque converter bolts. Matchmark the converter to the driveplate to aid installation.

21. Remove the transaxle bracket and the front engine mount nuts.

22. Disconnect the exhaust pipe and converter assembly, from the manifold.

23. Lower the vehicle.

24. Remove the torque struts.

25. Remove the exhaust crossover.

26. Detach the bulkhead electrical connector and quick connects near the PCM.

27. Disconnect the electrical connectors at the alternator assembly.

28. Support the transaxle with floor jack or equivalent.

29. Remove the remaining transaxle-to-engine bolts.

30. Attach an engine lifting device, and remove the engine from the vehicle. Check for connected wires and hoses as the engine is coming out of the body.

31. Place the engine on a workstand.

To install:

32. With an assistant, install a lifting device onto the engine and position into the vehicle.

33. Remove the lifting device.

34. Install the transaxle-to-engine bolts. Remove the transaxle support.

35. Install the remaining components in the reverse order of removal. Tighten the front engine mount retaining nuts to 32 ft. lbs. (43 Nm).

36. Align the mating marks and install the hood assembly.

37. Reconnect the battery cables.

38. Turn the ignition **ON** for 3 seconds and then return to **OFF** position. Check for fuel leaks. Repeat this procedure a second time.

39. Install the air cleaner and duct assembly.

40. Refill the engine with engine oil, coolant and transaxle fluid, if needed.

41. Inspect vehicle for fluid leaks before and after starting the engine.

3.1L (VIN M) Engine

1. Disconnect the negative battery cable. Remove the hood panel with an assistant.

2. Drain the engine coolant.

3. Remove the air cleaner and duct assembly.

4. Disconnect the transaxle fluid filler tube assembly.

5. Remove the engine mount strut brackets.

6. Raise and safely support the vehicle. Drain the engine oil.

7. Remove the front exhaust pipe and exhaust manifold heat shield.

8. Remove the lower the rear transaxle bolts.

9. Disconnect the electrical connector from the vehicle speed sensor.

10. Remove the engine mount frame side nuts.

11. Remove the flywheel inspection cover and the starter assembly.

12. Remove the torque converter bolts.

13. Remove the transaxle mount assembly. Safely lower the vehicle.

14. Remove the coolant reservoir recovery assembly.

15. Remove the serpentine belt.

16. Remove the accelerator control cable bracket and move the cables assemblies aside.

17. Remove the power brake booster vacuum hose from the upper intake manifold assembly.

18. Remove the plastic cover from the shock tower.

19. Remove the alternator front and rear braces.

20. Remove the alternator assembly from the vehicle.

21. Disconnect the fuel feed and return pipe assemblies.

22. Remove the tie straps around the ignition wiring harness assembly, engine mount strut and air conditioning compressor bracket.

23. Remove the power steering pump pulley assembly.

24. Remove the power steering pump bolts/screws and move the power steering pump aside.

25. Remove the heater outlet and inlet hose assemblies.

26. Remove the upper radiator hose assembly from the engine assembly.

27. Remove the lower radiator hose assembly from the water pump assembly.

28. Remove the engine mount strut and air conditioning compressor bracket bolts and move the compressor assembly aside.

29. Remove the automatic transaxle vacuum modulator pipe assembly.

30. Disconnect the electrical connectors from the knock sensor, oxygen sensor, coolant temperature sensor, camshaft sensor, crankshaft and wheel speed sensors.

31. Disconnect the electrical connectors from the ignition coil assembly.

32. Disconnect the electrical connectors from the idle air control valve and throttle position sensor.

33. Remove the vacuum hoses from the upper intake manifold assembly.

34. Remove the attaching transaxle bolts.

35. Install a safety bolt between the alternator bracket and the engine lift bracket assemblies.

36. Suitably support the transaxle assembly with floor stands.

37. Attach a suitable engine lifting device.

38. Remove the engine assembly from the vehicle.

To install:

39. Install the engine into the vehicle with a suitable engine lifting device.

40. Remove the engine lifting device. Remove the floor stands from the transaxle.

41. Remove the safety bolt from the alternator bracket and front engine lift bracket assemblies.

42. Install the transaxle bolts.

43. Install the remaining components in the reverse order of removal. Torque the following:
   • Torque converter bolts to 46 ft. lbs. (63 Nm)
   • Engine mount frame side nuts to drivetrain bolts to 32 ft. lbs. (43 Nm)

44. Refill the engine with clean oil.

45. Install the engine mount strut brackets.

46. Install the transaxle fluid filler tube assembly.

47. Install the air cleaner and duct assembly. Install the hood panel.

48. Refill the engine coolant and add two engine coolant sealant pellets GM 3634621 or equivalent.

49. Bleed the cooling system using the recommended procedure.

50. Start the engine and verify correct idle and performance.

1995–97 3.4L (VIN X) Engines

1. Remove the air cleaner assembly.

2. Remove the hood assembly with the help of an assistant.

3. Drain the cooling system. Remove the coolant recovery reservoir.

4. Remove the heater hoses from the engine. Remove the engine torque strut bracket and strut.

5. Remove the cooling fans.

6. Remove the radiator hoses from the engine. Detach the transaxle cooler lines.

7. Remove the radiator and attaching hoses.

8. Remove the control cables from the bracket and throttle body.

9. Remove the exhaust crossover.

10. Remove the ground straps from the bell housing.

11. Remove the fuel injector cover.

12. Remove the power steering lines at the pump and front cover.

13. Remove the serpentine drive belt. Remove the upper A/C compressor bolts.

14. Safely raise and support the vehicle. Remove the right front tire and wheel assembly.

15. Remove the right front splash shield.

16. Remove the lower A/C compressor bolts and reposition the compressor.

17. Remove the flywheel inspection cover. Remove the engine oil filter.

18. Remove the starter assembly.

19. Remove the front exhaust pipe and converter.

20. Remove the motor mount nuts from the subframe.

21. Disconnect the electrical connections from the rear of engine.

22. Remove the right ball joint nut and separate the ball joint from the control arm.

23. Remove the outer tie rod.

24. Remove the drive axle assembly.

25. Remove the transaxle shield.

26. Suitably support the transaxle with an appropriate jacking fixture.

27. Remove the motor mount bracket to the transaxle bolts and nuts.

28. Remove the transaxle support and safely lower the vehicle.

29. Remove the plastic cover from the shock tower.

30. Detach the electrical quick connectors from the PCM.

31. Disconnect all necessary vacuum lines.

32. Support the transaxle.

33. Install a suitable engine lifting device.

34. Remove the engine assembly from vehicle after disconnecting the alternator connections.

To install:

35. Install the engine assembly into the vehicle with a suitable engine lifting device.

36. Connect the electrical connections to the alternator.

37. Remove the engine lifting device. Remove the support from the transaxle.

38. Install the bell housing bolts.

39. Connect all necessary vacuum lines.

40. Attach the quick connects near the PCM.

41. Install the plastic cover to the shock tower.

42. Safely raise and support the vehicle. Safely support the transaxle with a jacking fixture.

43. Install the motor mount bracket to transaxle attaching nuts and tighten nuts to 43 ft. lbs. (58 Nm).

44. Install the remaining components in the reverse order of removal. Torque the following:

- Ball joint-to-control arm nuts to 63 ft. lbs. (85 Nm)
- Torque converter bolts to 46 ft. lbs. (63 Nm)

45. Refill the engine with clean oil.

46. Refill the cooling system.

47. Refill the power steering fluid.

48. Check and refill the transaxle fluid. Install the hood assembly.

49. Install the air cleaner assembly.

50. Connect the negative battery cable. Start the engine and check all fluids for proper level and leaks.

51. Bleed the cooling system and power steering system using the recommended procedure.

3.8L (VIN L) and (VIN K) Engine

1. Relieve the fuel system pressure. Disconnect the negative battery cable.

2. Remove the hood from the vehicle. Remove the air cleaner assembly.

3. Disconnect and cap the fuel lines from the fuel rail and mounting brackets.

4. Drain the engine coolant and remove the recovery bottle.

5. Remove the inner fender electrical cover and the fuel injector sight cover.

6. Disconnect the throttle cables from the throttle body and mounting bracket.

7. Remove the rear heat shield from the crossover pipe.

8. Remove the throttle cable mounting bracket and disconnect any vacuum lines from the bracket.

9. Disconnect the exhaust crossover from the manifolds.

10. Remove the torque strut mounting bolt and disconnect the strut from the engine bracket.

11. Remove the right side engine cooling fan.

12. Disconnect the vacuum line from the transaxle module.

13. Remove the serpentine belt.

14. Remove the power steering pump and alternator assemblies.

15. Tag and disconnect all electrical connections from the engine.

16. Disconnect the upper and lower radiator hoses as well as the heater hoses from the engine.

17. Remove the transaxle-to-engine bolts and disconnect the ground wires.

18. Raise and support the vehicle safely. Remove the right front wheel and inner splash shield.

19. Remove the flywheel cover, scribe a mark on the torque converter and flywheel and remove the flywheel-to-torque converter bolts.

20. Disconnect the wire harness clamps from the frame near the radiator.

21. Remove the A/C compressor from the bracket, lay aside and secure to the frame.

22. Remove the starter.

23. Support the transaxle with a jack and remove the transaxle-to-engine bolt, through the wheel well, using a long extension.

24. Attach a lifting device and remove the engine mount-to-frame nuts.

25. Drain the engine oil and remove the oil filter.

26. Disconnect the oil cooler pipes from the hose connections.

27. Disconnect the exhaust pipe from the manifold.

28. Lower the vehicle and remove the engine assembly from the vehicle.

To install:

29. With an assistant, install a lifting device onto the engine and position into the vehicle.

30. Support the transaxle, install the transaxle-to-engine bolts and ground wire harness and torque to 46 ft. lbs. (62 Nm).

31. Install the heater and upper and lower radiator hoses to the engine.

32. Install all electrical connections to the engine.

33. Install the alternator, power steering pump and serpentine belt.

34. Install the vacuum line to the transaxle module.

35. Install the engine torque strut and bolt and torque to 41 ft. lbs. (56 Nm).

36. Install the exhaust crossover pipe.

37. Install the throttle cable mounting bracket and vacuum lines.

38. Install the heat shield to the crossover pipe and the throttle cables to the throttle body and mounting bracket.

39. Install the inner fender electrical cover and the coolant recovery bottle.

40. Install the fuel hoses to the fuel rail and mounting brackets.

41. Raise and support the vehicle safely.

42. Connect the front exhaust pipe to the manifold.

43. Install the oil filter and oil cooler pipes.

44. Install the engine mount nuts to the frame and torque to 32 ft. lbs. (43 Nm).

45. Install the transaxle to engine bolt through the wheel well and torque to 46 ft. lbs. (62 Nm).

46. Install the starter motor assembly and connect the electrical connectors.

47. Install the air conditioner compressor to the bracket.

48. Install the wire harness clamps to the frame near the radiator.

49. Align the scribe marks, install the torque converter to flywheel bolts and torque to 46 ft. lbs. (62 Nm).

50. Install the flywheel cover and the inner fender splash shield.

51. Install the right front wheel assembly and lower the vehicle.

52. Refill the cooling system and bleed the power steering system.

53. Install the right side cooling fan.

54. Install the fuel injector sight shield and the air cleaner assembly.

55. Connect the negative battery cable and install the hood.

56. Check and add fluids as required. Test drive vehicle and recheck for leaks and correct levels.

Engine Mounts

REMOVAL AND INSTALLATION

2.2L (VIN 4) Engine

Front Engine Mount

1. Disconnect the negative battery cable. Remove the air cleaner duct assembly.

2. Remove the engine torque struts. Install engine support fixture, J-28467-A or equivalent.

3. Raise and safely support the vehicle.

4. Remove the engine mount-to-frame nuts. Lower the vehicle.

5. Raise the engine using the support fixture. Raise and safely support the vehicle.

6. Remove the engine mount-to-bracket nuts and remove the engine mount.

To install:

7. Install the engine mount onto the engine bracket and tighten the mounting nuts to 39 ft. lbs. (53 Nm).

8. Lower the vehicle.

9. Install the remaining components in the reverse order of removal.

Tighten the engine mount-to-frame nuts to 39 ft. lbs. (53 Nm).

10. Connect the negative battery cable.

Torque Strut

1. Disconnect the negative battery cable.

2. Remove the two torque strut mounting bolts and nuts.

3. Remove the torque strut from the mounting brackets.

To install:

4. Position the torque strut in the mounting brackets and install the mounting bolts and nuts.

5. Tighten the through bolts to 39 ft. lbs. (53 Nm).

6. Connect the negative battery cable.

3.1L (VIN T) Engine

Front Engine Mount

1. Disconnect the negative battery cable. Raise and safely support the vehicle.

2. Remove the two engine mount-to-frame nuts through the holes in the frame.

3. Lower the vehicle.

4. Install an engine support fixture, J-28467 or the equivalent. Raise the engine with the support fixture.

5. Raise and safely support the vehicle.

6. Remove the engine mount-to-engine bracket nuts.

7. Remove the engine mount from the vehicle.

To install:

8. Position the engine mount in the vehicle and install the engine mount-to-engine bracket nuts and tighten to 32 ft. lbs. (43 Nm).

9. Lower the vehicle.

10. Lower the engine with the support fixture until the weight of the engine is on the mount. Remove the support fixture.

11. Raise and safely support the vehicle.

12. Install the engine mount-to-frame nuts and tighten to 32 ft. lbs. (43 Nm).

13. Lower the vehicle. Connect the negative battery cable.

Rear Engine Mount

1. Disconnect the negative battery cable.

2. Install an engine support fixture, J-28467 or the equivalent.

3. Raise and safely support the vehicle.

4. Remove the transaxle mount bracket-to-frame nuts from under the

frame and the bolts from the side of the frame.

5. Lower the vehicle.

6. Remove the five transaxle mount bracket-to-transaxle mounting bolts.

7. Remove the transaxle mount assembly with bolt brackets.

8. Remove the two transaxle bracket-to-mount nuts and separate the bracket from the mount.

9. Remove the two frame bracket-to-mount nuts and separate the bracket from the mount.

To install:

10. Install the frame bracket on the mount and tighten the mounting nuts to 38 ft. lbs. (52 Nm).

11. Install the transaxle bracket onto the mount and tighten the mounting nuts to 30 ft. lbs. (47 Nm).

12. Install the assembly into the vehicle and install the transaxle bracket-to-transaxle mounting bolts and tighten to 60 ft. lbs. (82 Nm).

13. Raise and safely support the vehicle.

14. Install the frame bracket-to-frame nuts and bolts and tighten to 38 ft. lbs. (52 Nm).

15. Lower the vehicle.

16. Remove the support fixture.

17. Connect the negative battery cable.

3.1L (VIN M) Engine

Engine Mount Strut

1. Remove the engine mount strut attaching nuts.

2. Remove the radiator inlet hose clip from right hand side only on the strut assembly.

3. Remove the attaching bolt from the engine mount strut bracket.

4. Remove the engine mount strut bolt from the bracket.

5. Remove the engine mount strut assembly.

To install:

6. Position the engine mount strut assembly onto the bracket with the eyelet on the engine strut assembly facing down on dogbone style strut assembly only.

7. Install the attaching bolt to the engine mount bracket assembly.

8. Position the strut assembly to the engine mount strut bracket.

9. Install the bolt to the engine mount strut bracket.

10. Install the radiator inlet hose clip to the bolt on the right hand side only.

11. Hand start engine mount through bolt attaching nut. Tighten nut to 35 ft. lbs. (47 Nm) using a back-up wrench.

Engine Mount and Bracket

1. Remove the hood panel assembly with an assistant.
2. Remove the engine mount strut brackets. Remove the engine cooling fan assemblies.
3. Install engine support special tools J 28467-A, J 36462, J 28467-90 and J 35953 or the equivalent.
4. Raise and safely support the vehicle. Remove the front exhaust manifold pipe.
5. Remove the intermediate steering shaft bolt and move the cover aside.
6. Drain the engine oil. Remove the engine splash shield.
7. Safely support the frame assembly with jack stands.
8. Remove the transaxle mount frame side bracket nuts from the frame assembly.
9. Remove the rear drivetrain and front suspension frame bolts.
10. Safely lower the rear drivetrain and front suspension frame assembly.
11. Remove the engine mount nuts and engine mount bracket bolts. Remove the engine mount from the vehicle.

To install:

12. Position the engine mount assembly to the engine mount bracket.
13. Install and tighten the engine mount nuts to 43 ft. lbs. (58 Nm).
14. Install the transaxle mount frame side bracket nuts to the frame assembly. Tighten the nuts to 35 ft. lbs. (48 Nm).
15. Remove the jack stands from the frame assembly.
16. Install the engine splash shield.
17. Install the intermediate steering shaft and tighten the pinch bolt to 35 ft. lbs. (47 Nm).
18. Install the front exhaust manifold pipe.
19. Safely lower the vehicle. Remove the engine support tools.
20. Install the cooling fans.
21. Install the engine strut brackets.
22. Install the engine strut bracket and attaching bolts.
23. Install the hood panel assembly with an assistant.
24. Refill the engine with oil.
25. Start the engine and verify proper operation.

3.4L (VIN X) Engine

Front Engine Mount

1. Remove the hood panel assembly with an assistant.

2. Remove the air cleaner assembly. Disconnect the negative battery cable.
3. Remove the engine torque strut.
4. Install an engine support fixture J-28467-A, J-28467–90 and J-36462, or their equivalents.
5. Raise and safely support the vehicle. Remove the right front tire and wheel assembly.
6. Remove the right inner fender well splash shield.
7. Remove the oil filter.
8. Remove the engine mount nuts at the frame and the bracket.
9. Install the drive axle boot protector. Lower the vehicle.
10. Raise the engine with the support fixture, then raise and safely support the vehicle.
11. Remove the front engine mount.

To install:

12. Install the front engine mount.
13. Safely lower the vehicle.
14. Lower the engine until the weight of the engine is on the mount.
15. Raise and safely support the vehicle. Remove the drive axle boot protector.
16. Install the engine mount nuts at bracket and tighten to 35 ft. lbs. (47 Nm).
17. Install the engine mount nuts at the frame and tighten to 32 ft. lbs. (43 Nm).
18. Coat the seal on the oil filter with clean engine oil and install the oil filter.
19. Install the right inner fender well splash shield.
20. Install the right front tire and wheel assembly.
21. Lower the vehicle.
22. Remove the engine support fixture and adapters.
23. Install the torque strut.
24. Install the hood panel assembly with an assistant.
25. Connect the negative battery cable. Install the air cleaner.
26. Top off the engine oil as necessary.

Rear Engine Mount

1. Remove the hood panel assembly with an assistant.
2. Remove the air cleaner.
3. Disconnect the negative battery cable.
4. Remove the engine torque strut.
5. Install an engine support fixture, J-28467-A, J-28467–90 and J-36462 or their equivalents.
6. Raise and safely support the vehicle. Remove the right tire and wheel assembly.
7. Remove the right engine splash shield. Remove the plastic drive axle splash shield.

8. Install the drive axle boot protector.
9. Remove the cotter pin and castle nut from the lower ball joint and separate the ball joint from the control arm.
10. Remove the right axle metal shield. Disconnect the right drive axle from the transaxle.
11. Remove the rear engine mount bracket brace with the inboard metal drive axle shield attached.
12. Disconnect the front exhaust pipe from the exhaust manifold.
13. Remove the engine mount retaining nuts.
14. Lower the vehicle.
15. Raise the engine using the support fixture. Raise and safely support the vehicle.
16. Remove the rear engine mount.

To install:

17. Install the rear engine mount.
18. Lower the vehicle.
19. Lower the engine and remove the engine support fixture.
20. Raise and safely support the vehicle.
21. Install the engine mount retaining nuts and tighten to 32 ft. lbs. (43 Nm).
22. Connect the front exhaust pipe to the exhaust manifold.
23. Install the engine mount bracket brace with the inboard metal shield attached.
24. Connect the right drive axle to the transaxle.
25. Install the outboard drive axle metal shield.
26. Connect the right ball joint to the lower control arm and tighten the castle nut to 63 ft. lbs. (85 Nm). Install a new cotter pin.
27. Remove the drive axle boot protector. Install the plastic drive axle shield.
28. Install the right engine splash shield.
29. Install the right front tire and wheel assembly and tighten to specification.
30. Lower the vehicle.
31. Install the engine torque strut.
32. Connect the negative battery cable. Install the air cleaner.
33. Install the hood panel assembly with an assistant.

3.8L (VIN L) Engine

Front Engine Mounts

NOTE: There are two front engine mounts on this vehicle. The following procedure will cover removal and installation of either mount.

1. Disconnect the negative battery cable. Raise and safely support the vehicle.

2. Remove the engine mount-to-frame nuts.

3. Lower the vehicle.

4. Install an engine support fixture, J-28467-A, or the equivalent, and raise the engine about 2 inches.

5. Remove the right inner fender well splash shield.

6. Remove the engine mount-to-engine bracket nuts and remove the engine mount.

To install:

7. Position the engine mount on the mounting bracket and tighten the mounting nuts to 32 ft. lbs. (44 Nm).

8. Install the fender well splash shield. Lower the vehicle.

9. Remove the engine support fixture. Raise and safely support the vehicle.

10. Install the engine mount-to-frame nuts and tighten to 50 ft. lbs. (68 Nm).

11. Lower the vehicle. Connect the negative battery cable.

Torque Strut

1. Disconnect the negative battery cable. Remove the torque strut mounting nuts.

2. Remove the through bolts and remove the torque strut.

To install:

3. Install the torque strut and through bolts.

4. Tighten the mounting nuts to 32 ft. lbs. (43 Nm).

5. Connect the negative battery cable.

3.8L (VIN K) Engine

1. Disconnect the negative battery cable.

2. Removing the mount retaining nuts from below the frame mounting bracket.

3. Safely raise the engine to provide clearance using special engine support fixtures J 28467–A, J 28467–90 and J 35953 or equivalent.

4. Remove the drip shield and oil filter.

5. Remove the engine mount to engine bracket nuts.

6. Remove the engine mount assembly.

To install:

7. Install the engine mount and attaching engine bracket. Tighten the engine mount to frame nuts to 50 ft. lbs. (68 Nm). Tighten the engine mount to bracket nuts to 32 ft. lbs. (44 Nm). Tighten the engine bracket to engine bolts to 70 ft. lbs. (95 Nm).

8. Install the oil filter and drip shield.

9. Remove the engine support fixtures. Connect the negative battery cable.

10. Start the engine and verify proper operation.

Cylinder Head

REMOVAL AND INSTALLATION

——— CAUTION ———
Fuel injection systems remain under pressure, even after the engine has been turned OFF. The fuel system pressure must be relieved before disconnecting any fuel lines. Failure to do so may result in fire and/or personal injury.

2.2L (VIN 4) Engine

1. Properly relieve the fuel system pressure.

2. Disconnect the negative battery cable. Remove the air cleaner assembly.

3. Drain the cooling system into a suitable container.

4. Tag and disconnect the vacuum lines and electrical connectors from the intake manifold and cylinder head.

5. Disconnect the control cables from the throttle body and remove the cable bracket at the throttle body and rocker arm cover.

6. Remove the serpentine belt and serpentine belt tensioner.

7. Disconnect the spark plug wires and lay them out of the way.

8. Disconnect the canister purge line under the manifold.

9. Disconnect the upper radiator hose from the engine. Detach the upper and lower heater hoses from the intake manifold.

10. Disconnect the coolant inlet hose on the cylinder head.

11. Remove the intake manifold brace from the power steering pump bracket.

12. Disconnect and cap the fuel lines at the quick disconnects.

NOTE: When removing the valve train components they must be kept in order for installation in the same locations they were removed from.

13. Remove the rocker arm cover, rocker arm nuts, rocker arms and pushrods.

14. Remove the engine oil fill tube.

NOTE: Removal of the intake and the exhaust manifolds from the vehicle should not be necessary.

15. Remove the exhaust manifold from the cylinder head.

16. Remove the intake manifold from the cylinder head.

17. Remove the engine lift brackets.

18. Remove the cylinder head bolts.

19. Remove the cylinder head.

To install:

20. Clean all the gasket surfaces completely. Clean the threads on the cylinder head bolts and the block threads.

21. Place a new cylinder head gasket in position over the dowel pins on the block. Carefully guide the cylinder head into position.

22. Install all the cylinder head bolts finger tight. New replacement head bolts are recommended. The long bolts go in bolt positions 1, 4, 5, 8 and 9. The short bolt are in positions 2, 3, 6 and 7. The stud is in position 10.

23. Tighten the bolts in sequence. The long bolts to 23 ft. lbs. (32 Nm) and the short bolts and stud to 22 ft. lbs. (29 Nm). Make second pass tightening the long bolts to 46 ft. lbs. (63 Nm) and the short bolts to 43 ft. lbs. (58 Nm). Make a final pass over all bolts tightening each an additional 90 degrees (1/4 turn) using a torque angle meter.

24. Install the engine lift brackets. Tighten the front bracket bolt to 41 ft. lbs. (55 Nm) and the rear bracket bolt to 32 ft. lbs. (43 Nm).

25. Install the intake manifold and exhaust manifolds.

26. Install the remaining components in the reverse order of removal. Tighten the rocker arm nuts to 22 ft. lbs. (30 Nm) and the cover bolts to 89 inch lbs. (10 Nm). Tighten the serpentine belt tensioner to 37 ft. lbs. (50 Nm).

27. An oil and filter change is recommended since coolant can get into the oil system during cylinder head service.

28. Connect the negative battery cable. Refill and bleed the cooling system.

29. Start the engine and verify proper operation.

3.1L (VIN T) Engine

Left Cylinder Head (Front)

1. Remove the air cleaner assembly. Relieve the fuel system pressure.

Right Cylinder Head (Rear)

1. Remove the air cleaner assembly. Relieve the fuel system pressure.
2. Disconnect the negative battery cable.
3. Drain the cooling system into a suitable container. Raise and safely support the vehicle.
4. Remove the rocker arm covers.
5. Remove the upper intake plenum and lower intake manifold.

NOTE: When removing the valve train components they must be kept in order for installation in the same locations they were removed from.

6. Remove the rocker arm nuts, rocker arms, rocker arm balls and pushrods.
7. Disconnect the front exhaust pipe from the crossover pipe.
8. Lower the vehicle.
9. Remove the crossover pipe heat shield and crossover pipe.
10. Remove the torque strut mounting bolts at the radiator panel and loosen the bolts at the engine brackets and flip the torque struts back.
11. Rotate the engine forward.
12. Disconnect the secondary air injection pipe, if equipped.
13. Disconnect the Oxygen Sensor (O_2) electrical connector.
14. Disconnect the spark plug wires from the spark plugs and disconnect the wires from the looms.
15. Remove the cylinder head bolts.
16. Remove the cylinder head.

To install:

17. Clean all the gasket surfaces completely. Clean the cylinder head bolt threads and cylinder block threads.
18. Install a new cylinder head gasket on the block. The gasket should fit over the dowel pins and the words **THIS SIDE UP** must be visible.
19. Install the cylinder head.
20. Coat the cylinder head bolts with sealer and install the bolts finger tight. With all the bolts installed tighten the bolts in sequence to 33 ft. lbs. (45 Nm). Make a second pass tightening each bolt an additional 90°.
21. Install the remaining components in the reverse order of removal. Tighten the torque strut bolts to 41 ft. lbs. (56 Nm).
22. Refill the cooling system.
23. Connect the negative battery cable.
24. Install the air cleaner assembly.
25. Start the vehicle and verify no leaks.

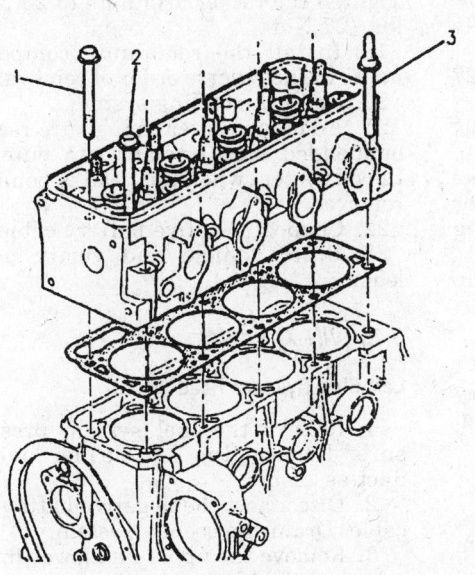

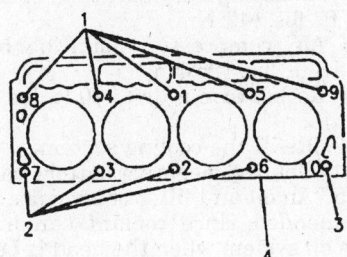

1. Long bolts
2. Short bolts
3. Stud
4. Numbers on gasket indicate torque sequence

330684

Cylinder head bolt tightening sequence — 2.2L (VIN 4) engine

2. Disconnect the negative battery cable. Drain the cooling system into a suitable container.
3. Remove the torque strut and engine-side torque strut bracket.

NOTE: When removing the valve train components they must be kept in order for installation in the same locations they were removed from.

4. Remove the rocker arm covers, rocker arm nuts, rocker arms, rocker arm balls and pushrods.
5. Remove the upper intake plenum and lower intake manifold.
6. Remove the exhaust crossover pipe.
7. Raise and safely support the vehicle. Remove the oil filter and adapter.
8. Position the oil cooler assembly aside. Lower the vehicle.
9. Remove the A/C compressor from the mounting bracket and position aside without disconnecting the refrigerant lines.
10. Disconnect the radiator hose from the water pump.
11. Remove the A/C compressor and torque strut mounting bracket.
12. Disconnect the spark plug wires from the spark plugs and remove the wires from the wire looms. Position the wires out of the way.
13. Remove the front exhaust manifold.
14. Remove the vacuum pipe nut.
15. Remove the cylinder head bolts.
16. Remove the cylinder head.

To install:

17. Clean all the gasket surfaces completely. Clean the cylinder head bolt threads and cylinder block threads.
18. Install a new cylinder head gasket on the block. The gasket should fit over the dowel pins and the words **THIS SIDE UP** must be visible.
19. Install the cylinder head.
20. Coat the cylinder head bolts with sealer and install the bolts finger tight. With all the bolts installed tighten the bolts in sequence to 33 ft. lbs. (45 Nm). Make a second pass, tightening each bolt an additional 90°.
21. Install the remaining components in the reverse order of removal.
22. Refill the cooling system.
23. Connect the negative battery cable.
24. Install the air cleaner assembly.
25. Start the vehicle and verify no leaks.

3.1L (VIN M) Engine

Left Cylinder Head (Front)

1. Disconnect the negative battery cable. Relieve the fuel system pressure.
2. Drain the cooling system into a suitable container.
3. Remove the rocker arm covers.
4. Remove upper intake plenum and lower intake manifold.
5. Remove the exhaust crossover pipe.
6. Disconnect the spark plug wires from spark plugs and wire looms and route the wires out of the way.

NOTE: When removing the valve train components use care to identify any components that will be reused. Valve train components must be kept in order for installation in the same locations from which they were removed.

7. Remove rocker arms nut, rocker arms, balls and pushrods.
8. Remove oil level indicator tube.
9. Remove any A/C compressor bolts accessible from the top.
10. Raise and safely support the vehicle. Remove the lower A/C compressor mounting bolts.
11. Disconnect the A/C compressor electrical connections and reposition the A/C compressor.
12. Remove the A/C compressor lower bracket bolts.
13. Lower the vehicle.
14. Remove the A/C compressor upper bracket bolts.
15. Remove the compressor brackets.
16. Remove the cylinder head bolts evenly.
17. Remove the cylinder head.
   To install:
18. Clean all the gasket surfaces completely. Clean the threads on the cylinder head bolts and block threads. New replacement head bolts are recommended.
19. Place the cylinder head gasket in position over the dowel pins on the cylinder block so the words **THIS SIDE UP** or other gasket identification are showing.
20. Coat the bolt threads with sealer and install finger tight.
21. Tighten the cylinder head bolts in sequence to 33 ft. lbs. (45 Nm). With all the bolts tightened make a second pass tightening all the bolts an additional 90 degrees (¼ turn).

22. Install the remaining components in the reverse order of removal. Torque the following:
   • A/C compressor bracket bolts to 35 ft. lbs. (47 Nm)
   • A/C compressor mounting bolts to 18 ft. lbs. (25 Nm)
   • Rocker arm nuts to 20 ft. lbs. (27 Nm)
23. Refill the cooling system.
24. Connect negative battery cable.
25. An oil and filter change are recommended since coolant can enter the oil system when the head is being removed.
26. Start vehicle and verify no leaks.

Right Cylinder Head (Rear)

1. Disconnect the negative battery cable. Relieve the fuel system pressure.
2. Drain the cooling system into a suitable container.
3. Remove the rocker arm covers.
4. Remove upper intake plenum and lower intake manifold.
5. Disconnect the electrical connector from the ignition assembly.
6. Remove the alternator.
7. Remove the exhaust crossover pipe. Detach the O_2 sensor connector.
8. Raise and safely support the vehicle. Disconnect the exhaust pipe from the exhaust manifold.
9. Lower the vehicle.
10. Remove the exhaust manifold.
11. Disconnect the spark plug wires from spark plugs and wire looms and route the wires out of the way.

NOTE: When removing the valve train components use care to identify any components that will be reused. Valve train components must be kept in order for installation in the same locations from which they were removed.

12. Remove rocker arms nut, rocker arms, balls and pushrods.
13. Remove the cylinder head bolts evenly. Remove the cylinder head.
   To install:
14. Clean all the gasket surfaces completely. Clean the threads on the cylinder head bolts and block threads. New replacement head bolts are recommended.
15. Place the cylinder head gasket in position over the dowel pins on the cylinder block so the words **THIS SIDE UP** or other gasket identification is showing.
16. Coat the bolt threads with sealer and install finger tight.
17. Tighten the cylinder head bolts in sequence to 33 ft. lbs. (45 Nm). With all the bolts tightened make a

second pass tightening all the bolts an additional 90 degrees.
18. Install the pushrods, rocker arms, balls and rocker arm nuts. Tighten the rocker arm nuts to 20 ft. lbs. (27 Nm).
19. Install the remaining components in the reverse order of removal.
20. Refill the cooling system.
21. An oil and filter change are recommended since coolant can enter the oil system when the head is being removed.
22. Connect negative battery cable.
23. Start vehicle and verify no leaks.

3.4L (VIN X) Engine

Left Side Cylinder Head (Front)

1. Relieve the fuel system pressure. Remove the air cleaner and duct assembly.
2. Disconnect the negative battery cable. Drain the cooling system.
3. Remove the upper and lower intake manifold components.
4. Remove the left side cam carrier cover using the following procedure.
   a. Disconnect oil/air breather hose from cam carrier cover.
   b. Disconnect the spark plug wires from the front spark plugs.
   c. Disconnect the rear spark plug wires from the clips on the front camshaft cover.
   d. Remove the four camshaft cover mounting bolts.
   e. Remove the camshaft cover and discard the old gaskets and O-rings.
5. Remove the left side cam carrier using the following procedure.
   a. Remove the camshaft timing belt.
   b. Remove the exhaust crossover pipe and reposition aside.
   c. Remove the engine torque strut.
   d. Remove the front engine lift bracket.
   e. Install six, 6-inch sections of fuel line hose under camshaft and between lifters. This will hold lifters in the carrier. For this procedure use ³/₁₆ inch (8mm) vacuum/fuel line hose for exhaust valves and ⁵/₃₂ inch (5.5mm) vacuum/fuel line hose for the intake valves.
   f. Remove the camshaft carrier bolts and remove the carrier. Discard the gasket.
6. Remove the front exhaust manifold.
7. Remove the oil level indicator tube mounting bolt and remove the tube assembly.

8. Detach the coolant temperature sensor electrical connector.

9. Remove the cylinder head mounting bolts and remove cylinder head. Discard the gasket.

To install:

10. Clean all parts well. Make sure the mating surfaces on the head and block are clean. Clean the engine block threaded holes and make sure all oil is removed from the bolt holes. Inspect the cylinder head bolt threads and make sure they are clean. New replacement head bolts are recommended. If the cylinder head is being replaced with a new head, transfer the manifold studs and temperature sensor.

11. Install a new cylinder head gasket on the engine block with the metal tabs (factory type gasket) between the cylinders facing UP.

12. Install the cylinder head on the engine block with the dowel pins lined up. Install the mounting bolts finger tight.

13. With all the bolts in place tighten the mounting bolts in sequence to 37 ft. lbs. (50 Nm) for 1993–94 vehicles and to 44 ft. lbs. (60 Nm) for 1995–97 vehicles. Make a second pass over each bolt in sequence and tighten an additional 90 degrees (¼ turn) rotation using a torque angle meter.

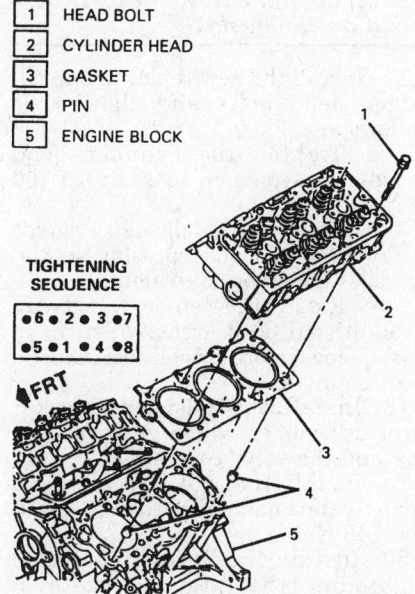

1 HEAD BOLT
2 CYLINDER HEAD
3 GASKET
4 PIN
5 ENGINE BLOCK

TIGHTENING SEQUENCE

●6 ●2 ●3 ●7
●5 ●1 ●4 ●8

◀FRT

330644

Cylinder head components and bolt tightening sequence — 3.4L (VIN X) engines

14. Connect the electrical connector to the coolant temperature sender.

15. Install the oil level tube assembly and tighten the mounting bolt to 89 inch lbs. (10 Nm).

16. Install the front exhaust manifold and tighten the attaching nuts to 18 ft. lbs. (25 Nm).

17. Clean any oil from the cam carrier to cylinder head bolt holes (bolt holes closest to the exhaust manifold).

18. Install the special camshaft holding tool J38613-A or equivalent and tighten the mounting bolt to 22 ft. lbs. (30 Nm).

NOTE: The use of petroleum jelly (never chassis grease) in the lifter bores along with the use of the lifter hold down hoses will help keep the lifters in place.

19. Install a new camshaft carrier gasket on the cylinder head and install the camshaft carrier on the cylinder head.

20. Install the camshaft carrier mounting bolts and tighten to 20 ft. lbs. (27 Nm).

21. Remove the lifter hold down hoses.

22. Install the upper and lower intake manifold assembly. If the thermostat housing was removed, install with a new gasket.

23. Install the front engine lift bracket.

24. Install the engine torque strut and tighten the bolt to 39 ft. lbs. (53 Nm).

25. Install the exhaust manifold crossover pipe.

26. Install the camshaft sprockets, if removed. Install the timing belt using great care to align all timing marks.

─────── CAUTION ───────
While installing the camshaft sprockets and timing belt, the cam timing must be set observing all timing marks. Failure to properly set the camshaft timing will result in serious engine damage.

27. Remove the special tools J-38613 camshaft holding clamps, if used.

28. Install the front engine coolant pipe. Connect the heater hose pipe to the intake plenum.

29. Connect the upper radiator hose.

30. Install the new O-rings and gasket on the camshaft carrier cover.

NOTE: Before tightening the cover mounting bolts, fully seat the bolt insulators in the camshaft cover.

31. Install the cover on the camshaft carrier and tighten the mounting bolts to 97 inch lbs. (11 Nm).

32. Connect the rear spark plug wires to the camshaft cover.

33. Connect the front spark plug wires to the spark plugs.

34. Connect the oil/air breather hose to the camshaft cover.

35. Drain the engine oil since coolant can contaminate the oil when a cylinder head is removed. Refill with the proper quantity and quality engine oil. A filter change is recommended.

36. Connect the negative battery cable. Install the air cleaner and air duct assembly.

37. Refill the cooling system.

38. Start the vehicle and verify no oil or coolant leaks. Bleed the cooling system as required.

Right Side Cylinder Head (Rear)

1. Relieve the fuel system pressure. Remove the air cleaner and duct assembly.

2. Disconnect the negative battery cable. Drain the cooling system.

3. Remove the upper intake manifold. Remove the right timing belt cover.

4. Remove the right rear spark plug wires. Disconnect oil/air breather hose from cam carrier cover.

5. Remove the four camshaft cover mounting bolts.

6. Remove the camshaft cover and discard the old gaskets and O-rings.

7. Remove the camshaft timing belt.

8. Install six 6 inch sections of fuel line hose under camshaft and between lifters. This will hold lifters in the carrier. For this procedure use ³/₁₆ inch (8mm) vacuum/fuel line hose for exhaust valves and ⁵/₃₂ inch (5.5mm) vacuum/fuel line hose for the intake valves.

9. Remove the lower intake manifold.

10. Remove the right camshaft carrier mounting bolts.

11. Remove the right camshaft carrier and gasket from the engine.

12. Remove the exhaust crossover pipe from the engine.

13. Raise and safely support the vehicle. Disconnect the front exhaust pipe at the manifold.

14. Lower the vehicle.

15. Disconnect the electrical connector from the oxygen (O_2) sensor.

16. Remove the rear timing belt tensioner bracket.

17. Remove the cylinder head mounting bolts and remove cylinder head and gasket.

To install:

18. Clean all parts well. Make sure the mating surfaces on the head and block are clean. Clean the engine block threaded holes and make sure all oil is removed from the bolt holes. Inspect the cylinder head bolt threads and make sure they are clean. New replacement head bolts are recommended. If the cylinder head is being replaced with a new head, transfer the manifold studs and temperature sensor.

19. Install a new cylinder head gasket on the engine block with the metal tabs (factory type gasket) between the cylinders facing UP.

20. Install the cylinder head on the engine block, with attached exhaust manifold, with the dowel pins lined up. Install the mounting bolts finger tight. With all the bolts in place tighten the mounting bolts in sequence to 37 ft. lbs. (50 Nm) for 1993–94 vehicles or to 44 ft. lbs. (60 Nm) for 1995–97 vehicles. Make a second pass over each bolt in sequence and tighten an additional 90 degrees (¼ turn) rotation using special tool J 36660 or equivalent angle measuring equipment.

21. Install the rear timing belt tensioner bracket.

22. Connect the electrical connector to the oxygen (O_2) sensor.

23. Raise and safely support the vehicle.

24. Connect the front exhaust pipe to the exhaust manifold.

25. Lower the vehicle.

26. Install the exhaust crossover pipe.

NOTE: Remove oil from the cam hold down tool hole in the camshaft carrier before installing and tightening the bolt.

27. Install the special camshaft hold down tool J 38613 A or equivalent and tighten to 22 ft. lbs. (30 Nm).

NOTE: The use of petroleum jelly (never chassis grease) in the lifter bores along with the use of the lifter hold down hoses, will help keep the lifters in place.

28. Install a new camshaft carrier gasket on the cylinder head and install the camshaft carrier on the cylinder head.

29. Install the mounting bolts and tighten to 20 ft. lbs. (27 Nm).

30. Remove the lifter hold down hoses.

31. Install the camshaft sprockets and timing belt.

-------- **CAUTION** --------

After installing the camshaft sprockets and timing belt the Cam Timing must be set. Failure to properly set the camshaft timing will result in serious engine damage.

32. Remove the camshaft hold down tool J-38613 or equivalent.

33. Install the lower intake manifold.

34. Install the new O-rings and gasket on the cam carrier cover.

NOTE: Before tightening the cover mounting bolts, fully seat the bolt insulators in the camshaft cover.

35. Install the cover on the camshaft carrier and tighten the mounting bolts to 97 inch lbs. (11 Nm).

36. Connect the oil/air breather hose to the camshaft cover.

37. Connect the rear spark plug wires to the camshaft cover and spark plugs.

38. Install the right timing belt cover. Install the upper intake manifold.

39. Connect the negative battery cable. Install the air cleaner and air duct assembly.

40. Drain the engine oil since coolant can contaminate the oil when a cylinder head is removed. Refill with the proper quantity and quality engine oil. A filter change is recommended,

41. Refill the cooling system.

42. Start the vehicle and verify no leaks.

3.8L (VIN L) and (VIN K) Engine

Left Side (Front)

1. Remove the air cleaner assembly. Disconnect the negative battery cable.

2. Following proper procedures, relieve the fuel system pressure.

3. Drain the cooling system and remove the intake manifold.

4. Remove the valve covers and remove the rocker arm assemblies, pedestals, valve guide plates and pushrods, keeping everything in order for reinstallation.

5. Tag and disconnect the ignition wires and remove the spark plugs.

6. Remove the the engine torque strut from the attaching bracket.

7. Remove the vacuum line from the transaxle module.

8. Remove the left exhaust manifold.

9. Remove the spark plugs.

10. Remove the alternator front mount bracket and ignition module with the mount bracket. Lay the ignition module aside.

11. Remove the cylinder head bolts and discard. Remove the cylinder head from the vehicle.

12. Clean all gasket mating surfaces and the cylinder head bolt holes in the block.

To install:

13. Clean all parts well. If the cylinder head is being replaced with a new one, transfer the torque strut bracket and exhaust support bracket with the transaxle vacuum line. If the cylinder head is to be serviced, these parts must be removed and, after servicing, installed.

14. Inspect the head gasket and mating surfaces for leaks, corrosion and blow-by.

15. **The factory specifies new replacement cylinder head bolts** when removing and installing a cylinder head. Clean the remains of sealer from all threaded openings. The cylinder block bolt holes should be cleaned using a 7/16 14 tap.

16. Place the cylinder head gasket on the engine block dowels with the note **THIS SIDE UP** facing the cylinder head and the arrow facing the front of the engine. Position the cylinder head on the engine block.

-------- **CAUTION** --------

To prevent damage to the gasket, when installing the cylinder head, do not slide the cylinder head on the gasket.

17. Install the new replacement cylinder head bolts and tighten as follows:

   a. Tighten the cylinder head bolts, in sequence, to 37 ft. lbs. (50 Nm).

   b. Rotate each bolt 120 degrees (⅓ turn) in sequence using special tool J 36660 or equivalent.

   c. Rotate the center 4 bolts an additional 30 degrees (1/12 turn) in sequence using special tool J 36660 or equivalent.

18. Install the pushrods, rocker arm assemblies, valve guide retainers and the valve covers.

19. Install the exhaust manifold and tighten manifold studs to 32 ft. lbs. (48 Nm).

20. Install the intake manifold tightening the attaching bolts in proper torque and sequence.

21. Install the alternator front mount bracket and ignition module with bracket.

22. Install the spark plugs and wires.

23. Install the engine torque strut and tighten bolt to 32 ft. lbs. (53 Nm).

24. Refill the cooling system.

25. It is recommended that the engine oil be drained and crankcase refilled with clean engine oil since coolant and debris will get into the oil pan when removing a cylinder head. A filter change is also recommended.

26. Connect the negative battery cable.

27. Install the air cleaner assembly.

28. Start the engine check for proper operation. Verify no leaks.

Right Side (Rear)

1. Remove the air cleaner assembly. Disconnect the negative battery cable.

2. Relieve the fuel system pressure.

3. Drain the cooling system and disconnect the exhaust crossover pipe.

4. Remove the intake manifold.

5. Raise and safely support the vehicle.

6. Remove the right side exhaust pipe from the manifold.

7. Safely lower the vehicle.

8. Remove the right rear valve cover. Remove the serpentine drive belt.

9. Remove the belt tensioner pulley. Remove the heater hose from the engine.

10. Remove the power steering pump mounting bracket and lay the pump aside.

11. Tag and disconnect the ignition wires and remove the spark plugs.

12. Remove the exhaust manifold from the cylinder head only.

13. Disconnect the electrical connection from the oxygen sensor.

14. Remove the rocker arm assemblies, pedestals, valve guide retainers and pushrods, keeping everything in order for reinstallation.

15. Remove the cylinder head bolts in reverse order of installation and

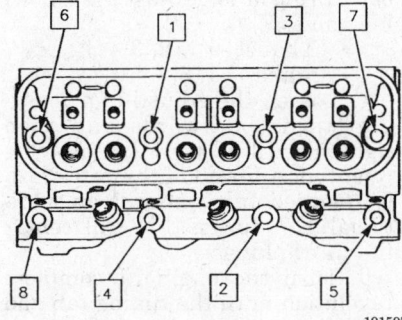

Cylinder head tightening sequence — 3.8L (VIN L) and (VIN K) engine

191597

remove the cylinder head. Discard the cylinder head bolts.

To install:

16. Clean all parts well. Inspect the head gasket and mating surfaces for leaks, corrosion and blow-by.

17. **The factory specifies new head bolts** be used when removing and installing a cylinder head. Clean the remains of sealer from all threaded openings. The cylinder block bolt holes should be cleaned using a $7/16$ 14 tap.

18. Place the cylinder head gasket on the engine block dowels with the note **THIS SIDE UP** facing the cylinder head and the arrow facing the front of the engine. Position the cylinder head on the engine block.

--- CAUTION ---

To prevent damage to the gasket, when installing the cylinder head, do not slide the cylinder head on the gasket.

19. Install the new cylinder head bolts and tighten as follows:

   a. Tighten the cylinder head bolts, in sequence, to 37 ft. lbs. (50 Nm).

   b. Rotate each bolt 120 degrees ($1/3$ turn) in sequence using special tool J 36660 or equivalent.

   c. Rotate the center 4 bolts an additional 30 degrees ($1/12$ turn) in sequence using special tool J 36660 or equivalent.

20. Connect the electrical connector to the oxygen sensor.

21. Install the exhaust manifold and intake manifold.

22. Install the pushrods, valve guide retainers and rocker arm assemblies.

23. Install the right rear valve cover. Install the spark plugs and wires in the proper order.

24. Install the power steering pump bracket and torque the bolts to 35 ft. lbs. (47 Nm).

25. Install the belt tensioner pulley and serpentine belt.

26. Install the exhaust crossover pipe. Raise and safely support the vehicle.

27. Install the front exhaust pipe to the manifold.

28. Refill the cooling system.

29. It is recommended that the engine oil be drained and crankcase refilled with clean engine oil since coolant and debris will get into the oil pan when removing a cylinder head. A filter change is also recommended.

30. Connect the negative battery cable. Install the air cleaner.

31. Start the engine check for proper operation. Verify no leaks.

Valve Lifters

REMOVAL AND INSTALLATION

--- CAUTION ---

Fuel injection systems remain under pressure, even after the engine has been turned OFF. The fuel system pressure must be relieved before disconnecting any fuel lines. Failure to do so may result in fire and/or personal injury.

2.2L (VIN 4) Engine

The cylinder head must be removed from the engine for this procedure.

1. Disconnect the negative battery cable. Relieve the fuel system pressure.

2. Drain the cooling system.

3. Remove the rocker arm cover and gasket.

NOTE: If any valvetrain components are to be reused, keep them in order so they can be reinstalled in the same locations from which they were removed.

4. Remove the rocker arm nuts. Remove the rocker arms and balls.

5. Remove the pushrods.

6. Remove the engine lift bracket from the rear of the engine.

7. Disconnect the spark plug wires and route them under the intake manifold.

8. Remove the cylinder head with the intake and exhaust manifold as an assembly.

9. Remove the bolt securing the anti-rotation brackets and remove the brackets.

10. Remove the lifters.

NOTE: Whenever new valve lifters are being installed, coat the valve lifters with camshaft assembly lube GM No. 1052365, or equivalent.

To install:

11. Coat the base of the lifters with camshaft lube or engine oil supplement prior to installation in the lifter bore.

12. If any lifters are being reused, install the lifters in the same bores from which they were removed.

13. Install the anti-rotation brackets and tighten the mounting bolts to 97 inch lbs. (11 Nm).

14. Install the cylinder head and manifolds.

15. Install the rear engine lift bracket and tighten the mounting nut to 37 ft. lbs. (50 Nm).

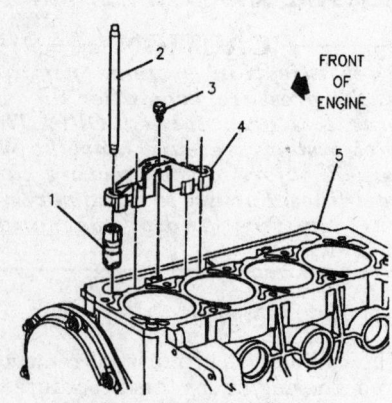

1 LIFTER, VALVE ROLLER
2 ROD, PUSH
3 BOLT – 11 N·m (97 LBS. IN.)
4 BRACKET, ANTI-ROTATION
5 BLOCK, CYLINDER

330554

Lower valve train components — 2.2L (VIN 4) engine

16. Connect the spark plug wires to the spark plugs.

17. Install the pushrods making sure they seat in the lifters correctly. Install the rocker arms, balls and mounting nuts and tighten the nuts to 22 ft. lbs. (30 Nm).

18. Install the rocker cover gasket and the rocker cover. Torque the bolts to 89 inch lbs. (10 Nm).

19. Refill the engine with coolant. An oil and filter change is recommended.

20. Connect the negative battery cable.

3.1L (VIN T) and (VIN M), and 3.8L (VIN L) and (VIN K) Engines

1. Disconnect the negative battery cable. Relieve the fuel system pressure.

2. Drain the cooling system.

3. Remove the rocker arm covers and the intake manifold.

NOTE: Be sure to keep all valve train parts in order so they may be reinstalled in their original locations and with the same mating surfaces as when removed.

4. Remove the rocker arm retaining nuts, rocker arm balls, rocker arms and pushrods.

5. If necessary, remove the 2 guide bolts from the right or left side lifter guide and remove the guide.

6. Remove the valve lifter(s) from the lifter bores.

To install:

7. Lubricate the bearing surfaces with Molykote® or equivalent.

NOTE: Installation of a new camshaft or a wear pattern on the old valve lifter will require the replacement of the camshaft and lifters together. If camshaft replacement is not necessary, make sure to install the used valve lifters in their original position upon reinstallation.

8. Install the lifters in their original locations.

9. If necessary, install the lifter guide and lifter guide bolts and tighten the guide bolts to 89 inch lbs. (10 Nm).

10. Install the pushrods, rocker arms, rocker balls and rocker arm nuts. Tighten the rocker arm nuts to 18 ft. lbs. (25 Nm) on 1993–95 vehicles. On 1996 vehicles, torque to 89 inch lbs. (10 Nm). plus an additional 30 degrees.

11. Install the intake manifold and rocker arm covers.

12. Refill the cooling system. An oil and filter change is recommended.

13. Connect the negative battery cable. Start the engine and verify that there are no leaks.

Valve Lash

ADJUSTMENT

3.1L (VIN T) and (VIN M) Engines

NOTE: These engines come from the factory with non-adjustable valve lash, BUT if the valves and the valve seats are recondi-

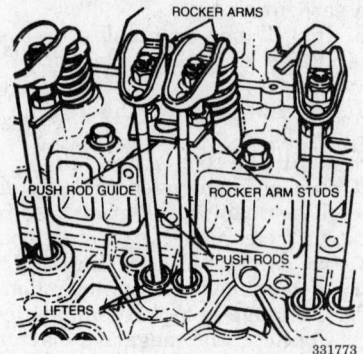

331773

Valve train mechanism — 3.1L (VIN T) engine

tioned, adjustable rocker arm studs must be installed. The adjustment procedure is ONLY for adjustable rocker arm studs.

The 3.1L (VIN T) and (VIN M) engines are originally equipped with hydraulic valve lifters. Hydraulic valve lifters are not adjustable. The 3.1L (VIN T) and (VIN M) engine replacement rocker arm stud is adjustable and the adjustment can be performed as listed below. If valve system noise is present, check the torque on the rocker arm nuts. The correct torque is 14–20 ft. lbs. (19–27 Nm) for the 3.1L (VIN T) and (VIN M) engine. If noise is still present, check the condition of the camshaft, lifters, rocker arms, pushrods and valves.

NOTE: The following adjustment procedure should be used when reconditioning a valve seat and a new adjustable type rocker arm is used.

1. Install the rocker arms, balls and nuts. Tighten the rocker arm nuts until all of the lash (free play) is eliminated.

2. Adjust the valves when the lifter is on the base circle of a camshaft lobe as follows:

a. Place the engine in the number 1 firing position. This can be determined by placing a finger on the number 1 rocker arm as the engine alignment mark on the front face of the of the torsional dampener pulley aligns with the arrow on the front cover. If the valves don't move freely, the engine is in the number 1 firing position. If the valves move freely, the engine is in the number 4 firing position and the engine should be rotated 1 full rotation to reach the number 1 position.

b. With the engine in the number 1 firing order, adjust the following valves:
- Exhaust — 1, 2, 3
- Intake — 1, 5, 6

c. Loosen the adjusting nut several turns backwards and then tighten until all lash (free play) is removed. After all of the valve lash is removed turn the nut an additional 1 ½ turns, this will center the lifter plunger.

d. Turn the engine 1 complete revolution until the timing tab and the alignment mark are again aligned, this will place the engine in the number 4 firing position.

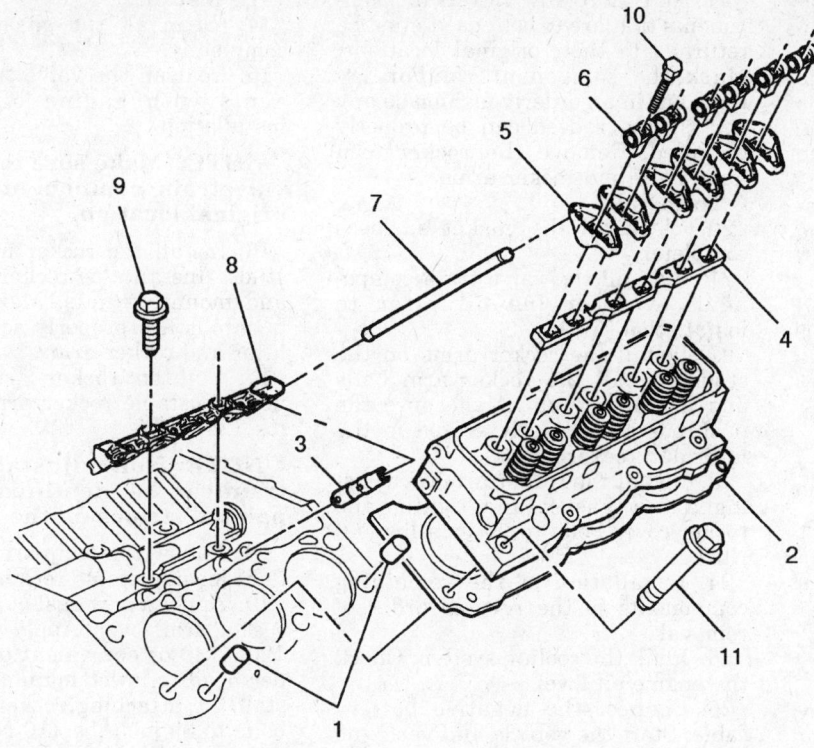

1. Dowel pin
2. Head pin
3. Valve lifter
4. Pushrod guide
5. Rocker arm
6. Rocker arm bearing
7. Pushrod
8. Lifter guide retainer
9. Bolt
10. Bolt
11. Head bolt

338522

Valve train component arrangement — 3.8L (VIN L) and (VIN K) engine

e. With the engine in the number 4 firing order, adjust the following valves:
- Exhaust — 4, 5, 6
- Intake — 2, 3, 4

2.2L (VIN 4), 3.4L (VIN X) and 3.8L (VIN L) and (VIN K) Engines

The valve clearance cannot be adjusted on these engines. The hydraulic lifters function to maintain a zero clearance when the valves are opening and closing.

Rocker Arms

REMOVAL AND INSTALLATION

2.2L (VIN 4) Engine

1. Disconnect the negative battery cable. Remove the air cleaner and duct assembly.
2. Tag and disconnect the spark plug wires from the spark plugs and disconnect the spark plug wire clips from the rocker arm cover and position aside.
3. Disconnect the PCV hose from the cover.
4. Disconnect the throttle cables from the throttle body and remove the throttle cable bracket from the intake plenum and move aside.

5. Remove the rocker arm cover bolts, then remove the cover.

NOTE: When removing the valve train components, any components that are to be reused must be kept in order for installation in the same locations from which they were removed.

6. Remove the rocker arm nut(s), ball(s) and rocker arms.
To install:
7. Clean the gasket surfaces completely.
8. Coat the bearing surfaces of the rocker arm(s) and the rocker arm ball(s) with Molykote® or equivalent engine assembly lube, engine oil supplement, etc.
9. Seat the pushrods in the lifters.
10. Install the rocker arm(s), ball(s) and nut(s). Components being reused must be installed in the same locations from which they were removed. Tighten the rocker arm nut(s) to 22 ft. lbs. (30 Nm).
11. Install the remaining components in the reverse order of removal. Tighten the rocker arm cover bolts to 89 inch lbs. (10 Nm).
12. Connect the negative battery cable. Check the engine oil level.
13. Start the engine and verify no oil leaks.

3.1L (VIN T) Engine

Left Side Rocker Arms (Front)

1. Disconnect the negative battery cable. Drain the cooling system.
2. Disconnect the ignition wire clamps from the coolant pipe.
3. Remove the coolant tube mounting bolt at the cylinder head.
4. Disconnect the coolant tube at both ends.
5. Remove the coolant tube hose from the water pump.
6. Remove the coolant tube.
7. Remove the vent pipe from the rocker arm cover to the air inlet duct.
8. Remove the ignition wire guide.
9. Remove the torque strut from the engine bracket and position out of the way.
10. Remove the four rocker arm cover bolts and remove the rocker arm cover.
11. Note that any valvetrain components that are to be reused must be returned to their original locations. Mark the components and/or lay them out in an orderly fashion so any parts being reused can be properly installed. Remove the rocker arm nuts, balls and rocker arms.
To install:
12. Clean all the gasket surfaces completely.

13. Coat all the valve train components with engine oil prior to installation.

14. Install the rocker arms on the studs. Install the rocker arm balls and mounting nuts. Make sure the pushrods are properly seated in the lifter and rocker arm.

15. Install the rocker arm cover using a new gasket and tighten the rocker cover bolts to 90 inch lbs. (10 Nm).

16. Position the torque strut and tighten the mounting bolts to 41 ft. lbs. (56 Nm).

17. Install the remaining components in the reverse order of removal.

18. Refill the cooling system.

19. Connect the negative battery cable. Start the vehicle and verify no leaks.

Right Side Rocker Arms (Rear)

1. Remove the air cleaner assembly.

2. Disconnect the negative battery cable. Drain the cooling system.

3. Remove the coolant recovery bottle.

4. Remove the torque strut mounting bolts at the engine and loosen the bolts at the radiator panel. Pull the engine forward and line up the slave hole in the torque strut with the engine bracket hole and install the mounting bolt to hold the engine in place.

5. Disconnect the vacuum lines at the intake plenum.

6. Disconnect the EGR tube at the crossover pipe.

7. Remove the ignition wire guide.

8. Disconnect the spark plug wires from the rear plugs and disconnect the plug wire loom from the intake plenum.

9. Disconnect the coolant hoses at the throttle base.

10. Disconnect the electrical wiring from the plenum.

11. Disconnect the control cables from the throttle body.

12. Remove the bracket from the right side of the plenum.

13. Disconnect the secondary air injection pump and support out of the way.

14. Disconnect the brake booster pipe fitting from the intake plenum.

15. Remove the serpentine belt.

16. Remove the alternator and alternator brace.

17. Disconnect the PCV valve from the rocker arm cover.

18. Remove the four rocker arm cover bolts and remove the rocker arm cover.

19. Note that any valvetrain components that are to be reused must be returned to their original locations. Mark the components and/or lay them out in an orderly fashion so any parts being reused can be properly installed. Remove the rocker arm nuts, balls and rocker arms.

To install:

20. Clean all the gasket surfaces completely.

21. Coat all the valve train components with engine oil prior to installation.

22. Install the rocker arms on the studs. Install the rocker arm balls and mounting nuts. Make sure the pushrods are properly seated in the lifter and rocker arm.

23. Install the rocker arm cover using a new gasket and tighten the rocker cover bolts to 90 inch lbs. (10 Nm).

24. Installation of the remaining components is the reverse order of removal.

25. Refill the cooling system. Check the engine oil level.

26. Connect the negative battery cable. Start the vehicle and verify no leaks.

3.1L (VIN M) Engine

Left Side Rocker Arms (Front)

1. Disconnect the negative battery cable. Remove the air cleaner and duct assembly.

2. Remove the thermostat bypass pipe clip nut.

3. Remove the automatic transaxle vacuum modulator pipe.

4. Remove the right engine mount strut assembly at the engine assembly.

5. Remove the upper radiator hose assembly at the engine assembly.

6. Remove the PCV valve from the rocker cover assembly.

7. Remove the heater outlet hose and thermostat bypass pipe assemblies.

8. Remove the rocker arm cover attaching screws.

9. Remove the rocker arm cover and gasket.

10. Remove the rocker arm nut and rocker arm pivot ball.

11. Remove the rocker arms.

12. Remove the pushrod assemblies.

NOTE: Use care when removing pushrods so they do not fall down into the lifter valley.

13. Place all valvetrain assemblies in a rack so they can be reinstalled in the same location from which they were removed.

To install:

14. Clean all the gasket surfaces completely.

15. Coat all the valve train components with engine oil prior to installation.

NOTE: Make sure to install all valvetrain components in their original location.

16. Install the rocker arms on the studs. Install the rocker arm balls and mounting nuts. Make sure the pushrods are properly seated in the lifter and rocker arm.

17. Tighten rocker arm nuts on non-adjustable rocker arms to 18 ft. lbs. (24 Nm).

NOTE: Non-adjustable valvetrain can be identified with two painted stripes on the pushrods.

18. Follow the adjustment procedure for adjustable rocker arms.

19. Apply a new gasket to the valve rocker arm cover. Apply sealant GM 12345739 or equivalent at the cylinder head to lower manifold joint. Install the attaching rocker arm cover bolts to 89 inch lbs. (10 Nm).

20. Install the remaining components in the reverse order of removal.

21. Fill and bleed the cooling system with the proper coolant mix for the Model Year vehicle.

22. The factory recommends that two engine coolant sealant pellets GM 3634621 or equivalent be added to the cooling system.

23. Bleed the cooling system using the recommended procedure.

24. Connect the negative battery cable. Start the vehicle and verify no leaks.

Right Side Rocker Arms (Rear)

1. Remove the air cleaner assembly. Disconnect the negative battery cable.

2. Remove the serpentine belt.

3. Remove the engine mount assemblies at the engine assembly.

4. Rotate the engine assembly.

5. Detach the connectors from the ignition coil assembly. Remove the ignition coil assembly.

6. Remove the plastic cover from the shock tower.

7. Remove the alternator front and rear braces.

8. Remove the alternator assembly.

9. Remove the EGR valve and position aside. Remove the rocker arm cover attaching screws and remove the rocker arm cover from the vehicle.

10. Remove the rocker arm nut and rocker arm pivot ball.

11. Remove the rocker arms.

12. Remove the pushrod assemblies.

NOTE: Use care when removing pushrods so they do not fall down into the lifter valley.

13. Place all valvetrain assemblies in a rack so they can be reinstalled in the same location from which they were removed.

To install:

14. Clean all the gasket surfaces completely.

15. Coat all the valve train components with engine oil prior to installation.

NOTE: Make sure to install all valvetrain components in their original location.

16. Install the rocker arms on the studs. Install the rocker arm balls and mounting nuts. Make sure the pushrods are properly seated in the lifter and rocker arm.

17. Tighten rocker arm nuts on non adjustable rocker arms to 18 ft. lbs. (24 Nm).

NOTE: Non adjustable valvetrain can be identified with two painted stripes on the pushrods.

18. Follow the adjustment procedure for adjustable rocker arms.

19. Apply a new gasket to the valve rocker arm cover. Apply sealant GM 12345739 or equivalent at the cylinder head to lower manifold joint. Install the attaching rocker arm cover and attaching bolts to 89 in. lbs. (10 Nm).

20. Install the remaining components in the reverse order of removal.

21. Connect the negative battery cable. Start the vehicle and verify no leaks.

3.8L (VIN L) and (VIN K) Engine

Left Side Rocker Arms (Front)

1. Disconnect the negative battery cable.

2. Remove the engine lift bracket from the exhaust manifold studs.

3. Remove the fuel injector sight shield.

4. Remove the serpentine belt.

5. Remove the alternator-to-brace bolt (lower bolt).

6. Remove the nut and remove the alternator brace from the intake manifold.

7. Disconnect the spark plug wires from the spark plugs. Disconnect the spark plug wire cover from the rocker arm cover and position aside.

8. Remove the rocker arm cover bolt and remove the cover.

NOTE: The rocker arms, pushrods, pedestals and bolts must be kept in order for installation in the same locations they were removed from.

9. Remove the rocker arm bolt(s), pedestal(s), rocker arm(s) and pushrod(s).

To install:

10. Clean all gasket surfaces completely.

11. Apply thread lock compound to the rocker arm bolt(s).

12. Seat the pushrod in the lifter and install the rocker arm, pedestal and mounting bolt. Tighten the mounting bolt to 28 ft. lbs. (38 Nm).

13. Install the rocker arm cover with a new gasket and tighten the mounting bolts to 89 inch lbs. (10 Nm).

14. Installation of the remaining components is the reverse of the removal procedure.

15. Connect the negative battery cable. Start the vehicle and verify no oil leaks.

Right Side Rocker Arms (Rear)

1. Disconnect the negative battery cable. Remove the coolant recovery bottle.

2. Remove the serpentine belt.

3. Remove the power steering pump mounting bolts and pull the power steering pump forward.

4. Remove the fuel injector sight shield. Remove the cannister purge valve from the bracket.

5. Remove the power steering pump support braces.

6. Remove the engine lift bracket from the exhaust manifold studs.

7. Disconnect the spark plug wires from the spark plugs. Disconnect the spark plug wire cover from the rocker arm cover and position aside.

8. Remove the rocker arm cover bolt and remove the cover.

NOTE: The rocker arms, pushrods, pedestals and bolts must be kept in order for installation in the same locations they were removed from.

9. Remove the rocker arm bolt(s), pedestal(s), rocker arm(s) and pushrod(s).

To install:

10. Clean all gasket surfaces completely.

11. Apply thread lock compound to the rocker arm bolt(s).

12. Seat the pushrod in the lifter and install the rocker arm, pedestal and mounting bolt. Tighten the mounting bolt to 28 ft. lbs. (38 Nm).

13. Install the remaining components in the reverse order of removal. Tighten the rocker arm cover bolts to 89 inch lbs. (10 Nm) and the engine lift bracket bolt to 41 ft. lbs. (55 Nm).

14. Connect the negative battery cable.

15. Start the vehicle and verify no oil leaks.

Intake Manifold

REMOVAL AND INSTALLATION

——— CAUTION ———
Fuel injection systems remain under pressure, even after the engine has been turned OFF. The fuel system pressure must be relieved before disconnecting any fuel lines. Failure to do so may result in fire and/or personal injury.

2.2L (VIN 4) Engine

1. Relieve the fuel system pressure. Disconnect the negative battery cable.

2. Remove the throttle body air intake duct. Drain the cooling system.

3. Identify, tag and disconnect all necessary vacuum lines.

4. Disconnect the control cables from the throttle body lever and remove the control cable bracket from the intake manifold.

5. Remove the serpentine belt.

6. Remove the power steering pump and lay it aside, without disconnecting the fluid lines.

7. Remove the transaxle fluid fill tube.

8. Identify, tag and disconnect the electrical connectors from the MAP sensor, EGR solenoid valve, Idle Air Control (IAC) valve, Throttle Position Sensor (TPS) and fuel injectors.

9. Remove the MAP sensor.

10. Remove the upper intake manifold mounting bolts and the upper intake manifold.

11. Disconnect the fuel lines from the fuel rail.

12. Remove the EGR valve injector.

13. Remove the EGR valve.

14. Remove the fuel injector retainer bracket, regulator and injectors.

15. Remove the control cable bracket.

16. Remove the six intake manifold nuts. Remove the intake manifold.

To install:

17. Clean the gasket mounting surfaces. Install a new gasket and posi-

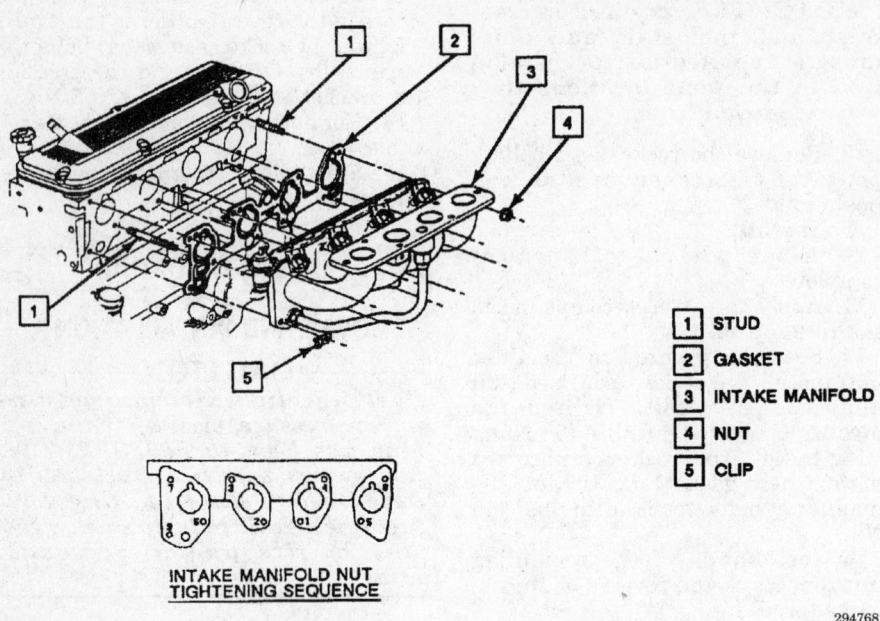

1	STUD
2	GASKET
3	INTAKE MANIFOLD
4	NUT
5	CLIP

INTAKE MANIFOLD NUT
TIGHTENING SEQUENCE

294768

Lower intake manifold torque sequence — 2.2L (VIN 4) engine

tion the lower intake manifold. Tighten the lower intake manifold nuts in the proper sequence to 24 ft. lbs. (33 Nm).

18. Install the remaining components in the reverse order of removal. Tighten the injector retainer bracket bolts to 22 inch lbs. (3.5 Nm) and the upper intake manifold bolts, in sequence, to 22 ft. lbs. (30 Nm).

19. Connect the negative battery cable. Refill the coolant system.

20. Start the vehicle and verify no coolant or vacuum leaks.

3.1L (VIN T) Engine

1. Relieve the fuel system pressure. Remove the air cleaner assembly.

2. Disconnect the negative battery cable.

3. Remove the torque strut mounting bolts from the engine bracket and loosen the bolts at the radiator panel and rotate the struts forward onto the radiator panel.

4. Drain the cooling system.

5. Disconnect the accelerator and TV cables from the throttle body.

6. Disconnect the brake booster vacuum pipe from the intake manifold plenum.

7. Remove the accelerator cable and TV cable bracket from the plenum.

8. Disconnect the air inlet duct from the throttle body.

9. Disconnect and remove the throttle body at the intake plenum.

10. Disconnect and remove the EGR valve at the intake plenum.

11. Remove the upper intake plenum mounting bolts and remove the plenum.

12. Disconnect the fuel feed and return lines from the fuel rail.

13. Remove the serpentine belt.

14. Disconnect the alternator and lay it aside.

15. Disconnect the power steering lines from the alternator mounting bracket.

16. Remove the power steering pump and lay it aside.

17. Disconnect the vacuum line from the pressure regulator.

18. Remove the fuel rail mounting bolts. Detach the fuel injector connectors, then remove the fuel rail.

19. Disconnect the plug wires from the rear spark plugs and route over the intake manifold and out of the way.

20. Disconnect the heater pipe from the water pump and cylinder head.

21. Remove the front rocker arm cover.

22. Disconnect the PCV hose from the rear rocker arm cover.

23. Remove the alternator bracket and rear brace.

24. Remove the rear rocker arm cover.

25. Disconnect the radiator hose at the thermostat housing.

26. Disconnect any remaining wiring from the intake manifold.

27. Remove the coolant sensor.

28. Remove the fuel lines from the mounting bracket.

29. Disconnect the throttle body heater hose.

30. Disconnect the heater pipe from the intake manifold.

31. Remove the intake manifold mounting bolts. The washers on the four center studs must be kept in their original positions.

32. Loosen the rocker arms and remove the pushrods. The intake and exhaust pushrods are different lengths and must be kept in the same positions they were removed from. The intake pushrods are 6 inches long and the exhaust pushrods are 6⅜ inches long.

33. Remove the intake manifold.

To install:

34. Clean all the gasket surfaces completely.

35. Place a ³/₁₆ in. (5mm) diameter bead GM sealer 1052917 or equivalent, on each ridge.

36. Install new intake manifold side gaskets on the cylinder heads.

37. Install the pushrods in their original positions and tighten the rocker arm nuts to 17 ft. lbs. (23 Nm).

38. Install the intake manifold and tighten the bolts in sequence to 15 ft. lbs. (20 Nm). After all eight bolts have been tightened make a second pass tightening the bolts to 24 ft. lbs. (33 Nm).

39. Install the remaining components in the reverse order of removal. Torque the following:

- Rocker arm cover bolts to 8 ft. lbs. (11 Nm)
- Fuel rail mounting bolts to 89 ft. lbs. (10 Nm)
- Torque strut mounting bolts to 41 ft. lbs. (56 Nm)
- Intake plenum bolts to 16 ft. lbs. (21 Nm)
- Throttle body-to-intake plenum bolts to 18 ft. lbs. (25 Nm)

40. Refill the cooling system.

41. Connect the negative battery cable. Install the air cleaner.

42. Pressurize the fuel system before starting the vehicle to ensure all fuel line connections are tight.

43. Warm the engine to operating temperature and verify no leaks.

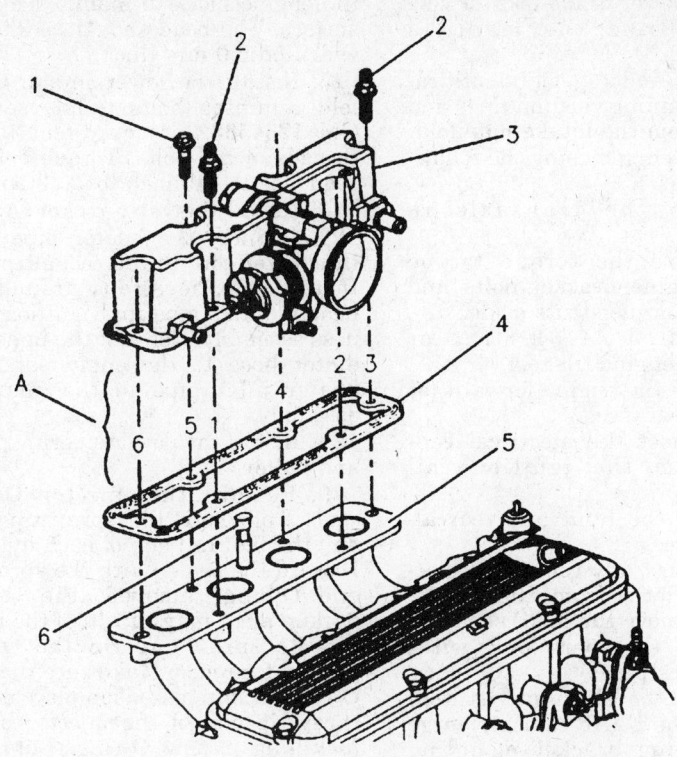

A UPPER INTAKE MANIFOLD
   ASSEMBLY TIGHTENING
   SEQUENCE
1 BOLT
2 STUD
3 UPPER INTAKE MANIFOLD ASSEMBLY
4 GASKET
5 LOWER INTAKE MANIFOLD
6 EGR VALVE INJECTOR

294769

Upper intake manifold torque sequence — 2.2L (VIN 4) engine

1994 3.1L (VIN M) Engine

TIGHTEN IN
PROPER SEQUENCE TO
20 N·m (15 LB. FT.), THEN
RETIGHTEN TO 33 N·m
(24 LB. FT.)

⑦ ④ ③ ⑥
⑧ ① ② ⑤

2 INTAKE MANIFOLD
3 GASKET
4 CYLINDER HEAD
5 SEALER

209839

Intake manifold bolt tightening sequence — 3.1L
(VIN T) engine

1. Disconnect the negative battery cable.

2. Relieve the fuel system pressure using the recommended procedure.

3. Remove the top half of the air cleaner assembly and throttle body duct.

4. Drain and recover the cooling system.

5. Remove the EGR pipe from exhaust manifold.

6. Remove the serpentine belt.

7. Remove the brake vacuum pipe at the intake plenum.

8. Disconnect the control cables from the throttle body and intake plenum mounting bracket.

9. Remove the power steering lines at the alternator bracket.

10. Remove the alternator.

11. Disconnect the spark plug wires from the spark plugs and wire retainers on the intake plenum.

12. Remove the Ignition assembly and the EVAP canister purge solenoid together.

13. Disconnect the upper engine wiring harness connectors at the following components:
- Throttle Position Sensor (TPS)
- Idle Air Control (IAC)
- Fuel Injectors
- Coolant temperature sensor
- MAP sensor
- Camshaft Position (CMP) sensor

14. Disconnect the vacuum lines from the following components:
- Vacuum modulator
- Fuel pressure regulator
- PCV valve

15. Remove the MAP sensor from upper intake manifold.

16. Remove the upper intake plenum mounting bolts and remove the plenum.

17. Disconnect the fuel lines from the fuel rail and fuel line bracket.

18. Install engine support fixture special tool J 28467-A or an equivalent.

19. Remove the right side engine mount.

20. Remove the power steering mounting bolts and support the pump out of the way without disconnecting the power steering lines.

21. Disconnect the coolant inlet pipe from coolant outlet housing.

22. Remove the coolant bypass hose from the water pump and the cylinder head.

23. Disconnect the upper radiator hose at thermostat housing.

24. Remove the thermostat housing.

25. Remove both rocker arm covers.

26. Remove the lower intake manifold bolts. Make sure the washers on the four center bolts are installed in their original locations.

NOTE: When removing the valve train components they should be kept in order for installation the original locations.

27. Remove the rocker arm retaining nuts and remove the rocker arms and pushrods.

28. Remove the intake manifold from the engine.

To install:

29. Clean gasket material from all mating surfaces. Remove all excess RTV sealant from front and rear ridges of cylinder block.

30. Place a 3mm bead of RTV, on each ridge, where the front and rear of the intake manifold contact the block.

31. Using a new gasket, install the intake manifold to the engine.

32. Install the pushrods, rocker arms and mounting nuts. Make sure the pushrods are properly seated in the valve lifters and rocker arms.

33. Install rocker arm nuts and tighten the rocker arm nuts to 18 ft. lbs. (24 Nm).

34. Install the lower intake manifold attaching bolts. Apply sealant PN 12345739 or equivalent to the threads of bolts, and torque bolts to 115 inch lbs. (13 Nm).

35. Install the remaining components in the reverse order or removal. Tighten the upper intake manifold bolts to 18 ft. lbs. (25 Nm).

36. Fill the cooling system. An engine oil and filter change is recommended.

37. Connect the negative battery cable.

38. Start the vehicle and verify no leaks.

1995–97 3.1L (VIN M) Engine

1. Disconnect the negative battery cable.

2. Drain the engine coolant. Remove the coolant recovery bottle.

3. Remove the air cleaner and duct assembly.

4. Remove the serpentine belt.

5. Disconnect the throttle and cruise control cables from the throttle body. Remove retaining brackets and set cable assemblies aside.

6. Disconnect the automatic transaxle vacuum modulator pipe. Disconnect the power brake booster vacuum hose from the manifold assembly.

7. Identify and tag for identification any remaining vacuum lines and disconnect from the intake manifold.

8. Tilt the engine using the following procedure:

   a. Place the transaxle in NEUTRAL.

   b. Remove the torque torque strut-to-engine bracket bolts and swing the torque struts aside.

   c. Install J 41131 strap or equivalent engine tilt tool.

   d. Rotate the engine forward for working clearance.

9. Disconnect the electrical connectors from the ignition coil assembly.

10. Remove the front and rear alternator braces.

11. Disconnect any remaining electrical connectors from the intake manifold. Remove the MAP sensor.

12. Remove the spark plug wires from the spark plugs.

13. Remove the electronic ignition system and the EVAP canister purge valve mounting bracket mounting bracket.

14. Remove the EGR tube assembly from the right exhaust manifold.

15. Remove the thermostat bypass pipe nut from the upper intake manifold.

16. Remove the upper intake manifold studs and bolts then remove the upper intake manifold and gaskets.

17. If the lower intake manifold needs to be removed, remove the fuel injector rail bolts and remove the fuel injector rail assembly.

18. Remove the heater inlet pipe assembly, upper radiator hose and tie straps retaining the heater outlet pipe and ignition wiring assembly. Disconnect the heater pipe from the heater core to the coolant pump.

19. Remove the power steering pump bolts and pump.

20. Remove the rocker arm covers.

21. Remove the lower intake manifold retaining bolts and remove the lower intake manifold and gasket.

To install:

22. Clean all parts well. Use care in cleaning old gasket material from the machined aluminum surfaces on the plenum and manifold as sharp tools may damage sealing surfaces.

23. Clean the mating surfaces to the intake manifold and engine block. Remove any loose pieces of RTV sealer.

24. Install the lower intake manifold to the engine block. Apply sealant GM 12345739 or equivalent at the engine block to manifold mating surface. The bead should be 3.0 mm wide and 5.0 mm thick.

25. Install the lower intake manifold retaining bolts. Apply sealant GM 12345382 or equivalent to the threads of the bolts. Torque bolts in sequence to 115 inch lbs. (13 Nm).

26. Install the valve rocker covers.

27. Connect the heater pipe from the heater core to the coolant pump. Install new tie straps around the heater outlet pipe and ignition harness assembly. Connect the upper radiator hose to the engine and the heater inlet pipe to the manifold assembly.

28. Install the power steering pump and pulley.

29. Remove the injector O-ring seals from both the spray tip ends and the fuel rail end of each injector. Discard the seals. With the spray tip end O-ring removed, the O-ring backup piece may slip off of the injector. Be sure to retain the O-ring backup for reuse. Make sure that the O-ring backup piece is in place on the spray tip end of the injector before installing a new O-ring. Lubricate new injector O-ring seals with clean engine oil and install on the injector assembly.

30. Install the fuel rail assembly to the intake manifold. Tilt the rail assembly to install the injectors. Install the fuel rail attaching bolts and tighten to 89 inch lbs. (10 Nm).

31. Connect the injector electrical connectors.

32. Install new O-rings on the fuel lines and install the fuel feed and return pipes. Tighten the fuel rail nuts to 13 ft. lbs. (17 Nm). Use a backup wrench on the fittings to prevent them from turning.

33. Using new gaskets, install the intake manifold plenum. Be sure to route the MAP sensor electrical connector to the outside of the the plenum gasket. Torque the bolts to 18 ft. lbs. (25 Nm).

34. Install the serpentine drive belt. Install the coolant recovery tank.

35. Install the MAP sensor, braces to the alternator, ignition coil front bolts and the EGR to plenum bolts. Connect the vacuum lines as noted during removal.

36. If the throttle body was removed from the upper intake manifold, inspect the throttle body before installation. Throttle body bore and valve deposits may be cleaned using carburetor cleaner and a parts cleaning brush. DO NOT use a cleaner that contains Methyl Ethyl Ketone

(MEK), an extremely strong solvent and not necessary for this type of deposit. The TP sensor and IAC valve should NOT come into contact with solvents or cleaners as they may be damaged. Verify that the gasket surfaces are clean, and, using a new flange gasket, install the throttle body. Torque the fasteners to 18 ft. lbs. (25 Nm).

37. Return the engine to its original location and install the torque strut.

38. Connect the throttle and cruise control cables.

39. Connect the IAC valve and TP sensor electrical connectors. Connect the air inlet duct. Check that the accelerator pedal is free by depressing the pedal to the floor and releasing.

40. Connect all remaining electrical connections and vacuum lines. Make sure the alternator braces are secure.

41. Refill the cooling system.

42. Since coolant can get into the engine's oil system when the intake manifold is removed, change the engine oil and filter.

43. Connect the negative battery cable.

44. Turn the key to the **ON** position several times to pressurize the fuel system and check for fuel leaks.

45. After the engine is running, bleed the cooling system and check for proper idle quality.

3.4L (VIN X) Engine

1. Relieve the fuel system pressure. Remove the air cleaner and duct work.

2. Disconnect the negative battery cable. Drain the cooling system.

3. Disconnect the control cables from the throttle body lever and the intake plenum mounting bracket.

4. Remove the fuel rail cover bolts and remove the fuel rail cover.

5. Disconnect and cap the fuel lines from the fuel rail.

6. For 1993–95 vehicles, disconnect the heater hose from the intake manifold. Disconnect the PCV valve and vacuum hose from the plenum.

7. For 1996–97 vehicles, remove the emission control purge solenoid and bracket.

8. For 1993–95 vehicles, disconnect the AIR solenoid, EGR valve and Throttle Position Sensor electrical connectors.

9. For 1996–97 vehicles, tag and disconnect the spark plug wires.

10. For 1993–95 vehicles, remove the EGR bolts and position the EGR valve away from the plenum.

11. For 1993–95 vehicles, remove the fuel line bracket from the ple-

num. Loosen the throttle body heater hose clamp at the plenum.

12. For 1993–95 vehicles, disconnect the canister purge solenoid and MAP sensor electrical connectors.

13. For 1996–97 vehicles, remove the control module mounting bolts, then remove the module.

14. Tag and disconnect the vacuum hoses from the plenum.

15. If necessary, remove the wiring loom bracket for the rear spark plug wires.

16. Remove the plenum support bracket nuts, then remove the plenum mounting bolts and remove the intake plenum.

17. Disconnect the vacuum line at the pressure regulator.

18. Remove the fuel rail assembly mounting bolts.

19. Disconnect the fuel injector electrical connectors.

20. Remove the fuel rail assembly.

21. Disconnect the radiator hose from thermostat housing.

22. Remove the heater pipe nut from the throttle body.

23. Remove the intake manifold mounting bolts and remove the intake manifold.

To install:

24. Clean all the gasket surfaces completely.

25. Install the intake manifold gasket and intake manifold.

26. For 1993–95 vehicles, insert new rubber isolators into the manifold flange. Tighten the mounting bolts to 18 ft. lbs. (25 Nm). Start with the center bolts and work outwards in a circular pattern.

27. For 1996–97 vehicles, install the M8 x 50mm bolts with the washer to the vertical holes in the the intake manifold. Tighten these bolts to 62 inch lbs. (7 Nm). Install the lower intake manifold mounting bolts. Insert the rubber insulator fully into the manifold flange before tightening any fasteners. Torque the bolts in this procedure:

 a. Draw the lower manifold into place by tightening the bolts gradually, starting with the middle bolts and working in a circular pattern. DO NOT tighten one side more than the other.

 b. Tighten these bolts to 116 inch lbs. (13 Nm).

 c. Remove the 2 bolts in the vertical holes of the intake manifold.

28. Install the remaining components in the reverse order of removal. Torque the following:

 • Fuel rail bolts to 89 inch lbs. (10 Nm)

 • Intake plenum and plenum support bracket nuts, starting in the center and working outwards in a circular pattern, to the mounting bolts to 18 ft. lbs. (25 Nm)

29. Fill and bleed the cooling system using the recommended procedure.

30. Connect the negative battery cable.

31. Install the air cleaner assembly.

32. Pressurize the fuel system and verify no leaks.

3.8L (VIN L) Engine

1. Relieve the fuel system pressure. Disconnect the negative battery cable.

2. Remove the air cleaner assembly and fuel injector sight shield.

3. Disconnect the throttle cables from the throttle body lever and intake manifold bracket.

4. Remove the coolant recovery reservoir.

5. Remove the electrical cover from the right inner fender.

6. Remove the exhaust heat shield.

7. Disconnect the EGR adapter from the upper intake manifold pipe.

8. Disconnect and cap the fuel lines from the fuel rail and remove the lines from the manifold bracket.

9. Remove the serpentine belt.

10. Remove the alternator and rear alternator brace from the intake manifold.

11. Remove the throttle body cable bracket from the intake manifold.

12. Disconnect the electrical connectors from the throttle body.

13. Disconnect the fuel injector electrical connectors.

14. Disconnect the vacuum hoses from the cannister purge solenoid, transaxle module and intake manifold connections.

15. Disconnect the vacuum hose quick connect at the upper intake manifold.

16. Remove the power steering pump mounting bolts and pull the pump forward.

17. Remove the serpentine belt tensioner pulley from the mounting bracket.

18. Remove the power steering pump support bracket.

19. Tag and disconnect the spark plug wires from the coil assembly and position them out of the way.

20. Drain the cooling system.

21. Disconnect the coolant bypass from the intake manifold.

22. Disconnect the heater pipes from the intake manifold and front cover.

23. Remove the solenoid valve mounting bracket and power steering pump brace from the intake manifold.

24. Disconnect the upper radiator hose from the thermostat housing.

25. Remove the thermostat housing and thermostat from the intake manifold.

26. Disconnect the electrical connector from the coolant sensor and remove the sensor from the intake manifold.

27. Remove the intake manifold mounting bolts and remove the intake manifold.

To install:

28. Clean all the gasket surfaces completely. Make sure all threaded opening are clean of dirt and old sealer. A thread-cutting tap can be used to clean threaded bolt holes.

29. Install the intake manifold with new gaskets.

30. Apply thread lock compound to the intake bolts and install the mounting bolts and tighten to 88 inch lbs. (10 Nm). Make a second pass to verify all bolts are tightened to 88 inch lbs. (10 Nm). Do not overtorque

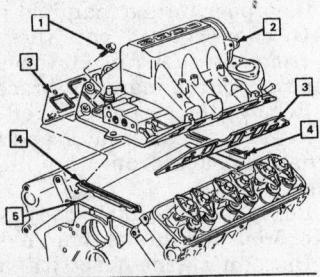

1 BOLT 10 N•M (88 LBS-IN)
TIGHTEN TWICE IN GIVEN SEQUENCE. APPLY P/N 12345493 TO BOLTS BEFORE ASSEMBLY.
2 INTAKE MANIFOLD
3 INTAKE MANIFOLD GASKET
4 INTAKE MANIFOLD SEAL
5 CLEAN SURFACE WITH SUITABLE SOLVENT AND APPLY P/N 12345336 TO THE ENDS OF INTAKE MANIFOLD SEALS.

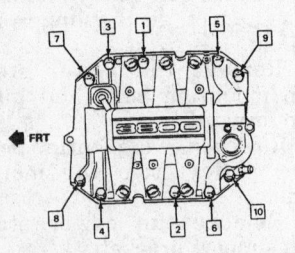

329510

Intake manifold components and bolt tightening sequence — 3.8L (VIN L) engine

or the light alloy manifold will be damaged.

31. Install the remaining components in the reverse order of removal:
- Thermostat housing bolts to 20 ft. lbs. (27 Nm)
- Power steering pump support bracket bolts to 37 ft. lbs. (50 Nm)
- Serpentine belt pulley bolt to 33 ft. lbs. (45 Nm)

32. Refill the cooling system. An oil and filter change is recommended.

33. Install the air cleaner assembly and the fuel injector sight shield.

34. Connect the negative battery cable.

35. Pressurize the fuel system by turn the ignition switch to the **ON** position to start the fuel pump and verify no fuel leaks.

36. Start the vehicle and verify no oil, vacuum or coolant leaks.

3.8L (VIN K) Engine

1. Disconnect the negative battery cable.

2. Remove the plastic engine cover clipped to the fuel rail.

3. Remove the air cleaner and duct assembly. Tag and remove the spark plug wires.

4. Properly relieve the fuel system pressure. Detach the fuel injection wiring harness.

5. Remove the canister purge electrical connector.

6. Remove the fuel rail. Remove the EGR adaptor to the upper intake pipe.

7. Remove the throttle body assembly.

8. Remove the upper manifold mounting bolts and remove the manifold from the vehicle.

9. Remove the inner fender electrical cover.

10. Remove the exhaust shield.

11. Remove the alternator and the brace. Remove the control cable bracket.

12. Disconnect the control cables, vacuum lines and electrical connectors from the throttle body and intake manifold.

13. Remove the vacuum lines from the canister purge and the transaxle module at the intake connection.

14. Remove the power steering pump and move forward.

15. Remove the belt tensioner mounting bracket.

16. Remove the steering pump support bracket.

17. Remove the coolant by-pass hose. Remove the heater pipes from the intake and the front cover.

18. Remove the alternator support brace.

19. Remove the upper radiator hose from the thermostat housing.

20. Detach the connector from the temperature sensor.

21. Remove the lower manifold mounting bolts and remove the lower manifold.

22. Remove the old sealing gaskets and clean the surfaces correctly.

To install:

23. Install the new lower intake gaskets and apply a dap of sealer to the corners of the end seals. Make sure to coat the threads of the intake bolts with P/N 12345336 or the equivalent. Torque the bolts to 89 inch lbs. (10 Nm) using the proper sequence.

24. Install the remaining components in the reverse order of removal. Tighten the upper manifold bolts, in sequence, to 11 ft. lbs. (15 Nm).

25. Fill and bleed the cooling system using the recommended procedure.

26. Install the air cleaner and duct assembly. Connect the negative battery cable.

27. Cycle the ignition key several times to ensure there are no fuel leaks before starting the vehicle.

28. Verify the proper idle quality and make sure there is no oil or coolant leaks.

Exhaust Manifold

REMOVAL AND INSTALLATION

2.2L (VIN 4) Engine

1. Disconnect the negative battery cable. Unplug the wiring harness from the oxygen sensor.

2. Remove the serpentine drive belt. Remove the alternator.

3. Raise and support the vehicle safely.

4. Remove the retainers and separate the exhaust pipe from the manifold.

5. Lower the vehicle.

6. Remove the bolt securing the oil fill tube to the engine block and pull the tube from the engine.

7. If the coolant inlet pipe interferes with removal, disconnect the mounting bracket from the transaxle stud.

8. Remove the manifold retaining nuts, then remove the manifold from the cylinder head. Remove and discard the gasket from the mating surfaces.

To install:

9. Install the manifold to the cylinder head using a new gasket.

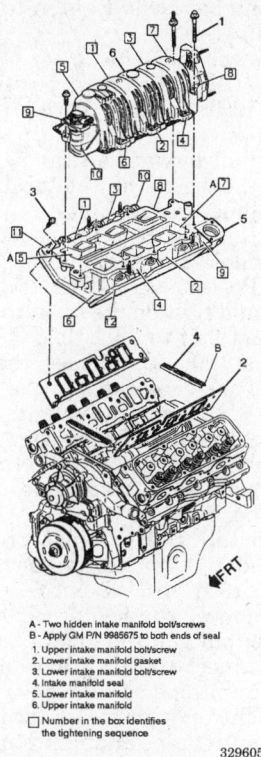

A - Two hidden intake manifold bolt/screws
B - Apply GM P/N 9985675 to both ends of seal
1. Upper intake manifold bolt/screw
2. Lower intake manifold gasket
3. Lower intake manifold bolt/screw
4. Intake manifold seal
5. Lower intake manifold
6. Upper intake manifold

☐ Number in the box identifies
the tightening sequence

329605

Upper and lower intake manifold bolt torquing sequence — 3.8L (VIN K) engine

Tighten the retaining nuts to 115 inch lbs. (13 Nm).

10. Raise and support the vehicle safely, then install the exhaust pipe to the manifold. Tighten the retaining nuts to 18 ft. lbs. (25 Nm), then lower the vehicle.

11. Install the remaining components in the reverse order of removal.

12. Connect the negative battery cable. Start the engine and check for exhaust leaks.

3.1L (VIN T) Engine

Left Side Manifold (Front)

1. Disconnect the negative battery cable. Remove the air cleaner assembly.

2. Remove the coolant recovery bottle. Remove the serpentine belt.

3. Remove the air conditioner compressor mounting bolts and remove the compressor from the mounting bracket and support aside. DO NOT disconnect the refrigerant lines from the compressor.

4. Remove the right side engine torque strut. Remove the bolts retaining the A/C compressor/torque strut bracket and remove the bracket.

5. Remove the heat shield and crossover pipe at the manifold.

6. Remove the exhaust manifold mounting bolts and remove the manifold.

To install:

7. Clean the gasket mounting surfaces. Install the exhaust manifold and a new gasket onto the cylinder head and loosely install the mounting bolts.

8. Install the exhaust crossover pipe. Tighten the exhaust manifold bolts and the crossover pipe nuts to 18 ft. lbs. (25 Nm)

9. Install the remaining components in the reverse order of removal.

10. Connect the negative battery cable. Start the engine and check for exhaust leaks.

Right Side Manifold (Rear)

1. Disconnect the negative battery cable. Remove the air cleaner assembly.

2. Raise and safely support the vehicle.

3. Disconnect the front exhaust pipe from the crossover pipe.

4. Lower the vehicle.

5. Remove the coolant recovery bottle.

6. Disconnect the torque struts from the engine mounting brackets.

7. In order to access the rear of the engine, rotate the engine as follows:

  a. Put the transaxle in **N**.

  b. Replace the passenger side torque strut to engine bracket bolt in engine bracket.

  c. Place a prybar in the bracket so it contacts the bracket and the bolt.

  d. Rotate the engine by pulling forward on the prybar. Align the slave hole in the driver side torque strut to the engine bracket hole.

  e. Retain the engine in this position using the torque strut to engine bracket bolt.

8. Remove the heat shield from the crossover pipe.

9. Remove the crossover pipe.

10. Disconnect the accelerator and TV cables from the throttle body and remove the cable mounting bracket from the intake manifold and set aside.

11. Remove the exhaust manifold mounting bolts and remove the manifold.

To install:

12. Clean the manifold mounting surfaces.

13. Install the exhaust manifold and a new gasket on the cylinder head and loosely install the mounting bolts.

14. Install the exhaust crossover pipe. Tighten the exhaust manifold bolts and the crossover pipe nuts to 18 ft. lbs. (25 Nm)

15. Install the remaining components in the reverse order of removal. Tighten the strut-to-engine bracket bolt to 32 ft. lbs. (43 Nm).

16. Connect the negative battery cable. Start the engine and check for exhaust leaks.

3.1L (VIN M) Engine

Left Side Manifold (Front)

1. Disconnect the negative battery cable. Remove the air cleaner and duct assembly.

2. Remove the coolant recovery bottle.

3. Rotate the engine as follows:

  a. Put the transaxle in **N**.

  b. Disconnect the negative battery cable.

  c. Remove the torque strut to engine bracket bolts and swing the strut assemblies aside.

  d. For 1994 vehicles, replace the passenger side torque strut to engine bracket bolt in engine bracket.

  e. For 1994 vehicles, place a prybar in the bracket so it contacts the bracket and the bolt.

  f. For 1994 vehicles, rotate the engine by pulling forward on the prybar. Align the slave hole in the driver side torque strut to the engine bracket hole.

  g. For 1995–97 vehicles, install tool J 41131 or equivalent, and rotate the engine assembly forward.

  h. Retain the engine in this position using the torque strut to engine bracket bolt.

NOTE: To prevent shearing of the rubber bushing, loosen the bolts on the engine strut before swinging the struts.

4. Relieve the accessory drive belt tension and remove the belt.

5. Disconnect the upper and lower radiator hoses from the engine assembly.

6. Remove the tie straps securing the heater outlet hose and the ignition wiring harness in position.

7. Remove the transaxle modulator vacuum pipe assembly.

8. Disconnect the heater pipe running from the water pump to the heater core.

9. Remove the heat shield and crossover pipe at the manifold.

10. Remove the engine mount strut.

11. Remove the air conditioning compressor bracket.

12. Remove the exhaust manifold mounting bolts and remove the manifold.

To install:

13. Clean the gasket mounting surfaces. Install the exhaust manifold to the engine, loosely install the mounting nuts.

14. Install the exhaust crossover pipe. Tighten the exhaust manifold nuts to 12 ft. lbs. (16 Nm) and the crossover pipe bolts to 18 ft. lbs. (25 Nm).

15. Install the exhaust manifold shield and tighten to 89 inch lbs. (10 Nm).

16. Install the engine mount strut and A/C compressor bracket.

17. Connect the heater pipe at the water pump and the heater core.

18. Connect the transaxle modulator vacuum pipe assembly.

19. Using new tie straps, secure the heater outlet pipe and ignition wiring harness in position.

20. Connect the upper and lower radiator hoses to the engine.

21. Install the serpentine belt.

22. For 1994 vehicles, position the engine in its normal resting position as follows:

a. Pull forward on the prybar to take the weight off the torque strut to engine bracket bolt.

b. Remove bolt from the strut slave hole and engine bracket.

c. Position the strut to the engine and tighten the strut to engine bracket bolt to 32 ft. lbs. (43 Nm).

23. For 1995–97 vehicles, rotate the engine backward and remove special tool J 41131 or equivalent. Install and tighten the torque strut-to-engine bracket bolts to 35 ft. lbs. (47 Nm).

24. Place the transaxle in the PARK position.

25. Install the coolant recovery bottle and refill the cooling system. Properly bleed the cooling system.

26. Install the air cleaner and duct assembly.

27. Connect the negative battery cable. Start the engine and check for exhaust leaks.

Right Side Manifold (Rear)

1. Disconnect the negative battery cable. Remove the air cleaner and duct assembly.

2. Remove the coolant recovery bottle.

3. Disconnect the upper radiator hose from the engine.

4. Remove the tie straps securing the heater outlet hose and the ignition wiring harness in position.

5. Remove the transaxle modulator vacuum pipe assembly.

6. Disconnect the heater pipe running from the water pump to the heater core.

7. Remove the heat shield and crossover pipe at the manifold.

8. Disconnect the oxygen sensor electrical connector(s).

9. In order to access the rear of the engine, rotate the engine as follows:

a. Put the transaxle in **N**.

b. Remove the torque strut to engine bracket bolts and swing the strut assemblies aside.

c. For 1994 vehicles, replace the passenger side torque strut to engine bracket bolt in engine bracket.

d. For 1994 vehicles, place a prybar in the bracket so it contacts the bracket and the bolt.

e. For 1994 vehicles, rotate the engine by pulling forward on the prybar. Align the slave hole in the driver side torque strut to the engine bracket hole.

f. For 1995–97 vehicles, install special tool J 41131 or equivalent, and rotate the engine assembly forward.

g. Retain the engine in this position using the torque strut to engine bracket bolt.

NOTE: To prevent shearing of the rubber bushing, loosen the bolts on the engine strut before swinging the struts.

10. Raise and properly support the vehicle.

11. While supporting the rear or the engine carrier assembly, remove the rear mounting bolts and lower the the rear portion of the carrier assembly.

12. Disconnect the front exhaust pipe from the manifold.

13. Disconnect the EGR tube from the manifold assembly.

14. Remove the oxygen sensor.

15. Remove the automatic transaxle dipstick and filler tube from the transaxle.

16. Remove the exhaust manifold heat shields and mounting bolts. Remove the manifold assembly from the vehicle.

To install:

17. Clean the gasket mounting surfaces.

18. Install the exhaust manifold to the engine and tighten the mounting nuts to 12 ft. lbs. (16 Nm) .

19. Install the automatic transaxle dipstick and filler tube from the transaxle.

20. Install the oxygen sensor and connect the EGR tube the manifold.

21. Raise the rear engine carrier assembly into position. Install the

mounting bolts and tighten to 107 ft. lbs. (145 Nm).

22. Lower the vehicle.

23. For 1994 vehicles, position the engine in its normal resting position as follows:

a. Pull forward on the prybar to take the weight off the torque strut to engine bracket bolt.

b. Remove bolt from the strut slave hole and engine bracket.

c. Position the strut to the engine and tighten the strut to engine bracket bolt to 32 ft. lbs. (43 Nm).

24. For 1995–97 vehicles, rotate the engine backward and remove special tool J 41131 or equivalent. Install and tighten the torque strut-to-engine bolts to 35 ft. lbs. (47 Nm).

25. Place the transaxle in the PARK position.

26. Install the exhaust crossover pipe and tighten the crossover pipe bolts to 18 ft. lbs. (25 Nm)

27. Connect the heater pipe at the water pump and the heater core.

28. Connect the transaxle modulator vacuum pipe assembly.

29. Using new tie straps, secure the heater outlet pipe and ignition wiring harness in position.

30. Connect the upper radiator hose to the engine.

31. Install the coolant recovery bottle and refill the cooling system. Properly bleed the cooling system.

32. Install the air cleaner and duct assembly.

33. Connect the negative battery cable.

34. Start the engine and check for exhaust leaks.

3.4L (VIN X) Engine

Left Side Manifold (Front)

1. Remove the air cleaner assembly. Disconnect the negative battery cable.

2. Remove the exhaust crossover pipe. remove the cooling fans.

3. If equipped with a manual transaxle, disconnect the AIR hose from the AIR pipe on the exhaust manifold.

4. Remove the exhaust manifold nuts, then remove the exhaust manifold, gasket and heat shield.

To install:

5. Clean all gasket mating surfaces completely. Install a new exhaust manifold gasket.

6. Install the exhaust manifold and heat shield and tighten the mounting nuts to 115 inch lbs. (13 Nm).

7. Install the remaining components in the reverse order of removal.

Tighten the exhaust crossover bolts to 18 ft. lbs. (25 Nm).

8. Connect the negative battery cable. Start the engine and check for exhaust leaks.

1993–94 Vehicles — Right Side Manifold (Rear)

1. Remove the air cleaner assembly. Disconnect the negative battery cable.
2. Remove the intake plenum and right side timing belt cover.
3. Tag and disconnect the right side spark plug wires.
4. Remove the air/oil separator hose at cam cover.
5. Remove the camshaft carrier cover bolts and lift the cover. Remove the gasket and O-rings from the cover.
6. Remove the secondary timing belt by removing the secondary timing belt actuator and tensioner assembly and sliding the belt from pulleys.
7. Install 6 sections of fuel line hose under the camshaft and between the lifters. This will hold lifters in carrier.
8. Remove the exhaust crossover pipe. Remove the torque strut.
9. Remove the torque strut bracket from the engine. Remove the front engine lift.
10. Remove the camshaft carrier mounting nuts and bolts and remove the cam carrier.
11. Remove camshaft carrier gasket from cylinder head.
12. Remove the exhaust manifold to crossover pipe bolts and the crossover pipe.
13. Raise and safely support the vehicle. Disconnect the front exhaust pipe from the manifold.
14. Lower the vehicle.
15. Disconnect the electrical connector from the oxygen (O_2) sensor.
16. Remove the exhaust manifold nuts, heat shield and the manifold from the engine.

To install:

17. Clean all gasket surfaces completely.
18. Install the exhaust manifold gasket, manifold and heat shields.
19. Install exhaust mounting manifold nuts and tighten to 116 inch lbs. (13 Nm).

20. Install the remaining components in the reverse order of removal. Tighten the following:
- Front exhaust pipe-to-manifold bolts to 24 ft. lbs. (32 Nm)
- Exhaust crossover pipe bolts to 18 ft. lbs. (25 Nm)
- Camshaft carrier bolts to 20 ft. lbs. (27 Nm)

21. Connect the negative battery cable. Start the engine and check for exhaust leaks.

1995–97 Vehicles — Right Side Manifold (Rear)

1. Remove the air cleaner and duct assembly.
2. Disconnect the negative battery cable. Remove the exhaust crossover pipe.
3. Remove the EGR tube from the exhaust manifold.
4. Raise and safely support the vehicle. Remove the front exhaust pipe and converter assembly.
5. Remove the oxygen sensor.
6. Remove the exhaust pipe front heat shield.
7. Remove the rear alternator brace. Remove the transmission dipstick tube.
8. Remove the intermediate shaft from the steering gear.
9. Remove the exhaust manifold attaching nuts.
10. Install a jacking fixture to support the rear cradle.
11. Remove the rear cradle bolts.
12. Safely lower the engine cradle.
13. Remove the steering gear heat shield.
14. Remove the exhaust manifold and heat shield and attaching gasket.

To install:

15. Clean all gasket surfaces completely.
16. Install the exhaust manifold gasket, manifold and heat shields.
17. Install exhaust mounting manifold nuts and tighten to 115 inch lbs. (13 Nm).
18. Install the remaining components in the reverse order of removal. Tighten the following:
- Rear cradle bolts to 125 ft. lbs. (170 Nm)
- Intermediate shaft-to-steering gear pinch bolt to 35 ft. lbs (47 Nm)
- Exhaust crossover pipe and tighten the nuts to 18 ft. lbs (25 Nm)
19. Connect the negative battery cable. Install the air cleaner and duct assembly.

20. Start the engine and check for exhaust leaks.

NOTE: Whenever the vehicle sub-frame is removed or lowered, the wheel alignment should be checked.

3.8L (VIN L) and (VIN K) Engine

Left Side Manifold (Front)

1. Remove the air cleaner and air duct assembly.
2. Disconnect the negative battery cable. Remove the torque struts.
3. Remove the right engine cooling fan.
4. Remove the heat shield from the crossover pipe and disconnect the crossover pipe at the rear manifold.
5. Drain the cooling system.
6. Disconnect the upper radiator hose from the thermostat housing.
7. Remove the fuel rail sight cover.
8. Disconnect the EGR adapter-to-intake manifold pipe.
9. Remove the EGR valve.
10. Remove the front engine lift hook.
11. Remove the torque strut bracket from the left cylinder head.
12. Remove the oil level indicator tube.
13. Disconnect the spark plug wires from the rocker arm cover and spark plugs and position out of the way.
14. Remove the A/C compressor support brace from the exhaust manifold stud.
15. Remove the manifold mounting bolts and remove the exhaust manifold.

To install:

16. Clean all the gasket surfaces completely.
17. Install the exhaust manifold and loosely install the mounting bolts. Starting with the center bolts and working outward tighten the bolts to 38 ft. lbs. (52 Nm).
18. Install the spark plugs and tighten to 11 ft. lbs. (15 Nm).
19. Install the remaining components in the reverse order of removal. Tighten the following:
- Engine lift hook nut to 22 ft. lbs. (30 Nm)
- Crossover pipe-to-rear exhaust manifold bolts to 22 ft. lbs. (30 Nm)
- Torque strut nuts to 41 ft. lbs. (56 Nm)
20. Connect the negative battery cable. Install the air cleaner and duct assembly.
21. Start the vehicle and verify no leaks.

Right Side Manifold (Rear)

1. Remove the air cleaner and air duct assembly.

2. Remove the fuel rail sight cover.

3. Disconnect the negative battery cable. Remove the coolant recovery bottle.

4. Remove the heat shield from the crossover pipe and disconnect the crossover pipe at the rear manifold.

5. Remove the transaxle oil level indicator tube.

6. Disconnect the spark plug wires from the rocker arm cover and spark plugs and position out of the way.

7. Disconnect the oxygen sensor electrical connector.

8. Remove the torque struts.

9. Remove the rear engine lift hook. Remove the spark plugs.

10. Raise and safely support the vehicle. Remove the front exhaust pipe and converter assembly.

11. Remove the rear engine mount-to-frame nuts.

12. Lower the vehicle.

13. Using a floor jack raise the right rear corner of the engine to provide clearance.

14. Remove the manifold mounting bolts and remove the exhaust manifold.

To install:

15. Clean all the gasket surfaces completely.

16. Install the exhaust manifold and loosely install the mounting bolts. Starting with the center bolts and working outward tighten the bolts to 38 ft. lbs. (52 Nm).

17. Connect the crossover pipe to the rear exhaust manifold and tighten the mounting bolts to 22 ft. lbs. (30 Nm).

18. Install the remaining components in the reverse order of removal. Tighten the following:

- Rear engine mount nuts to 50 ft. lbs. (68 Nm)

- Engine lift hook nut to 22 ft. lbs. (30 Nm)

- Spark plugs to 11 ft. lbs. (15 Nm)

- Torque strut nuts to 41 ft. lbs. (56 Nm)

19. Connect the negative battery cable. Install the air cleaner and duct assembly.

20. Start the vehicle and verify no leaks.

Front Cover Seal

REMOVAL AND INSTALLATION

2.2L (VIN 4) Engine

1. Disconnect the negative battery cable. Remove the serpentine belt.

2. Raise and properly support the vehicle. Remove the right front tire.

3. Remove the inner fender splash shield.

4. Separate the crankshaft pulley, from the pulley hub, by removing the three mounting bolts.

5. Install special tool J-24420-B or equivalent damper puller on crankshaft hub, turn puller screw and remove hub. Use care not to loose the small key on the crankshaft nose.

6. Pry out oil seal with a prybar.

NOTE: Take care not to damage the seal seat in the front cover or the crankshaft surface.

To install:

7. Clean out the seal recess in the front cover.

8. Press the oil seal into the front cover, using special tool J-35468 or equivalent driver tool.

9. If removed, install the key in the slot in the crankshaft nose. A small amount of silicone sealer can be used to hold it in place. Align the pulley hub keyway (slot) with the key, and install the hub to the crankshaft. Make sure the key was not displaced from its groove.

10. Install the pulley to the hub, and tighten the mounting bolts to 37 ft. lbs. (50 Nm).

11. Install the pulley/hub center bolt and tighten to 77 ft. lbs. (105 Nm).

12. Install the inner fender splash shield. Install the front tire and wheel assembly.

13. Lower the vehicle.

14. Install the serpentine belt.

15. Connect negative battery cable.

16. Start the engine and check for leaks.

3.1L (VIN T) and (VIN M), and 3.8L (VIN L) and (VIN K) Engines

— **CAUTION** —

The inertia weight section of the crankshaft balancer assembly is assembled to the hub with a rubber sleeve. The removal and installation procedures with the proper tools must be followed or movement of the inertia weight section on the hub will destroy the tuning of the torsional damper.

1. Disconnect the negative battery cable. Remove the serpentine belt.

2. Raise and safely support the vehicle.

3. Remove the right front tire and wheel assembly.

4. Remove the right inner fender well splash shield.

5. If necessary, remove the flywheel cover at the transaxle.

6. Remove the torsional dampener/crankshaft balancer as follows:

a. Remove the torsional damper/crankshaft balancer mounting bolt while holding the crankshaft from turning.

b. Install tool J 24420-B or an equivalent puller and remove the damper/balancer from the crankshaft. Use care not to lose the key from the crankshaft nose.

7. Pry the seal out of the front cover using a suitable prying tool. Use care not to damage the front cover or the crankshaft surface.

To install:

NOTE: Use care not to scratch or gouge the crankshaft while removing the oil seal from the front cover.

8. Clean out the oil seal recess in the front cover.

9. Install the new seal in the front cover using J-34995, or an equivalent installation tool, to fully seat the seal in the cover.

10. Install the torsional damper/crankshaft balancer as follows:

a. Coat the seal contact area on the damper with clean engine oil.

b. A small amount of silicone sealer may help hold the key in position in the crankshaft nose.

c. Line up the notch in the damper/balancer with the crankshaft key and slide the damper onto the crankshaft until the key is in the notch.

d. Using tool J-29113, pull the damper/balancer onto the crankshaft.

e. Verify that the key has not been displaced from its slot in the crankshaft nose.

f. Install the damper/balancer mounting bolt and tighten to 76 ft. lbs. (102 Nm) for the 3.1L or 111 ft. lbs. (150 Nm) plus 76 degrees turn for the 3.8L.

11. If removed, install the flywheel cover at transaxle.

12. Install the right inner fender splash shield.

13. Install the tire and wheel assembly and tighten to specification.

14. Lower the vehicle.

15. Install the serpentine belt.

16. Refill the cooling system.

17. Connect the negative battery cable. Start the vehicle and verify no coolant leaks or oil leaks.

3.4L (VIN X) Engine

1. Disconnect the negative battery cable. Remove the coolant recovery tank.

2. Remove the serpentine belt.

3. Raise and safely support the vehicle. Remove the right front tire and wheel assembly.

4. Remove the right inner fender well splash shield.

5. Remove the flywheel cover.

6. Remove the crankshaft damper bolt while holding the crankshaft from turning.

7. Remove the crankshaft pulley bolts and crankshaft pulley.

8. Remove the heater pipe mounting screws from the recovery tank mounting bracket and position the pipe out of the way.

9. Using a suitable puller, J-24420-B or the equivalent, remove the torsional damper.

10. Pry the seal out of the recess in the front cover with a suitable prying tool.

To install:

11. Clean out the seal mounting recess in the front cover.

12. Coat the seal lips with clean engine oil and install into the front cover using a suitable seal installer, J-34995 or the equivalent.

13. Coat the seal contact area on the damper with clean engine oil.

14. Install the damper onto the crankshaft with the crankshaft key in line with the notch in the damper. Seat the damper on the crankshaft using J-29113, or an equivalent puller.

15. Position the heater pipe and install the mounting screws at the recovery tank bracket.

16. Install the crankshaft pulley and tighten the mounting bolts to 37 ft. lbs. (50 Nm).

17. Install the damper mounting bolt and tighten to 78 ft. lbs. (105 Nm) while preventing the crankshaft from rotating.

18. Install the flywheel cover.

19. Install the right engine splash shield. Install the right front tire and wheel assembly.

20. Lower the vehicle.

21. Install the serpentine belt.

22. Install the coolant recovery reservoir.

23. Connect the negative battery cable. Refill the coolant in the reservoir. Start the engine and verify no leaks.

Timing Chain, Sprockets and Front Cover

REMOVAL AND INSTALLATION

2.2L (VIN 4) Engine

1. Disconnect the negative battery cable. Remove the serpentine drive belt.

2. Raise and safely support the vehicle.

3. Remove the right side tire and wheel assembly. Mark a relationship between the wheel and the wheel stud for reinstallation purposes.

4. Remove the right side inner wheel well splash shield.

5. Remove the crankshaft pulley bolts and remove the pulley.

6. Install puller J 24420-B or the equivalent and remove the hub.

7. Remove the electrical center cover. Remove the serpentine belt tensioner and bracket.

8. Remove the oil pan.

9. Remove the timing chain cover bolts and remove the cover, if the cover is difficult to remove use a rubber mallet to loosen the cover.

10. Align the marks on the timing gears before removing the timing chain and gears.

11. Remove the upper camshaft gear retaining bolt.

12. Loosen the timing chain tensioner bolt. DO NOT REMOVE THE BOLT.

13. Remove the camshaft sprocket and the timing chain.

14. If replacing the lower crankshaft gear install puller J 22888-20 or the equivalent.

To install:

15. Clean all of the sealing surfaces where the gaskets are installed.

16. Install the lower crankshaft sprocket using J 5590 or the equivalent.

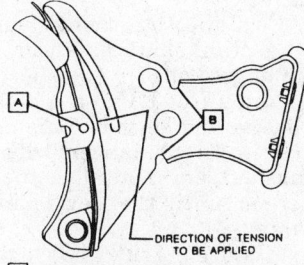

DIRECTION OF TENSION TO BE APPLIED

A INSERT PIN AFTER TENSION HAS BEEN APPLIED

B TABS, USED FOR CAMSHAFT AND CRANKSHAFT ALIGNMENT

331760

Timing chain tensioner assembly — 2.2L (VIN 4) engine

17. Compress the timing chain tensioner spring and install a cotter pin into the retaining hole.

18. Align the crankshaft and timing gears to the marks on the tensioner assembly.

19. Snug up on the tensioner bolt.

20. Align the dowel in the camshaft with the dowel in the camshaft sprocket and install the chain to the gear assemblies.

21. Tighten the timing chain tensioner retaining bolts to 18 ft. lbs. (24 Nm).

22. Tighten the camshaft sprocket retaining bolt to 77 ft. lbs. (105 Nm).

23. Remove the timing chain tensioner cotter pin.

24. Position the timing chain cover to the engine with a new gasket guiding the cover over the dowel pins.

25. Install the timing chain cover retaining bolts and tighten to 97 inch lbs. (11 Nm).

26. Install the crankshaft pulley hub using tool J 29113 or the equivalent and using RTV sealant at the keyway in the crankshaft.

27. Install the oil pan using the recommended procedure.

28. Install the crankshaft belt pulley tightening the bolts to 37 ft. lbs. (50 Nm).

29. Install the crankshaft hub bolt torquing the bolt to 77 ft. lbs. (105 Nm).

30. Install the serpentine belt and electrical cover.

31. Install the right side inner wheel well splash shield.

32. Install the wheel and tire assembly and torquing the wheel to specifications.

33. Lower the vehicle and connect the negative battery cable.

34. Fill all the fluid levels to the proper level. Start the engine and verify no leaks and proper engine performance.

3.1L (VIN T) Engine

1993–94 VIN T Engines

1. Disconnect the negative battery cable. Drain the cooling system.

2. Remove the coolant reservoir.

3. Remove the serpentine belt and tensioner.

4. Remove the power steering pump and safely support it out of the way.

5. Remove the crankshaft balancer as follows:

   a. Raise and safely support the vehicle.

   b. Remove the right front tire and wheel assembly.

   c. Remove the right inner fender well splash shield.

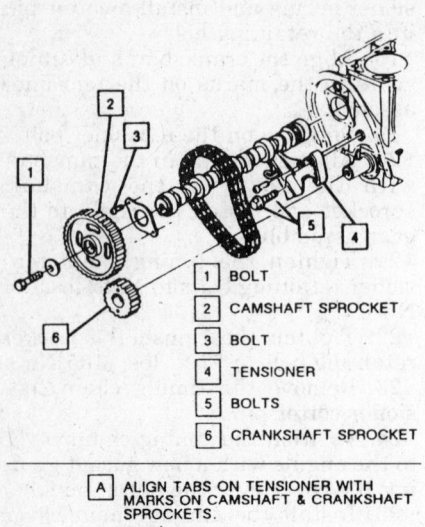

1 BOLT
2 CAMSHAFT SPROCKET
3 BOLT
4 TENSIONER
5 BOLTS
6 CRANKSHAFT SPROCKET

A ALIGN TABS ON TENSIONER WITH MARKS ON CAMSHAFT & CRANKSHAFT SPROCKETS.

331761

Timing chain and gear alignment marks — 2.2L (VIN 4) engine

d. Remove the flywheel cover and install a flywheel holding tool.

e. Remove the balancer mounting bolt and washer.

f. Using a suitable puller, J-24420-B or the equivalent remove the balancer from the crankshaft.

6. Remove the serpentine belt idler pulley.

7. Remove the engine oil pan using the recommended procedure.

8. Remove the lower timing cover bolts. Lower the vehicle.

9. Remove the radiator hose at the coolant pump. Remove the bypass pipe at the front cover.

10. Remove the canister purge hose.

11. Remove the upper timing cover bolts. Remove the timing cover.

12. Rotate the crankshaft until the timing marks on the camshaft and crankshaft sprockets are in alignment. The marks should be facing each other at their closest approach.

13. Remove the three camshaft sprocket mounting bolts and remove the camshaft sprocket and the chain.

NOTE: If the sprocket does not come off easily, a light blow on the lower edge of the sprocket with a plastic mallet should dislodge the sprocket.

14. Remove the crankshaft sprocket using tool J-5825-A, or the equivalent.

15. Remove the timing chain damper assembly bolts and damper

To install:

16. Clean all gasket surfaces completely.

17. Install timing chain damper assembly and tighten the mounting bolts to 15 ft. lbs. (21 Nm).

18. Apply GM Engine Oil Supplement (EOS) P/N 1052367 or the equivalent lubricant to the sprocket thrust surface.

19. Install the camshaft sprocket in the timing chain.

20. Hold the camshaft sprocket with the timing mark pointing down and the timing chain hanging off the bottom of the sprocket.

21. Loop the chain under the crankshaft sprocket and install the camshaft sprocket over the camshaft dowel. DO NOT force the sprocket into place.

22. Align dowel in camshaft with dowel hole in camshaft sprocket.

23. Verify the timing marks are still in alignment.

24. Draw the camshaft sprocket onto camshaft using the mounting bolts. Tighten the mounting bolts to 18 ft. lbs. (25 Nm).

25. Lubricate timing chain with clean engine oil.

26. Clean all gasket surfaces completely.

27. Install a new gasket.

28. Apply a thin bead of sealer around the gasket sealing area of the front cover. Install a new front cover seal on the front cover.

29. Install the front cover on the engine and install the mounting bolts. Tighten the small bolts to 15 ft. lbs. (21 Nm). Tighten the large bolts to 35 ft. lbs. (45 Nm).

30. Install the engine oil pan.

31. Install the serpentine belt idler pulley.

32. Install crankshaft balancer as follows:

a. Coat the seal contact surface of the crankshaft balancer with clean engine oil.

b. Apply GM RTV or equivalent to the key and keyway. Line up the notch in the balancer with the crankshaft key and slide the balancer on until the key is in the balancer.

c. Using J-29113 or an equivalent puller, seat the balancer on the crankshaft.

d. Install the balancer mounting bolt and washer and tighten to 76 ft. lbs. (103 Nm).

e. Install the flywheel cover.

f. Install the right inner fender well splash shield.

g. Install the tire and wheel assembly and tighten to specification.

33. Lower the vehicle.

34. Install the radiator hose to the coolant pump.

35. Install the bypass pipe at the front cover.

36. Install the canister purge hose.

37. Install the power steering pump.

38. Install the belt tensioner, tighten the bolts to 40 ft. lbs. (54 Nm).

39. Install the serpentine drive belt.

40. Refill and bleed the cooling system using the recommended procedure.

41. Check the engine oil level. An oil and filter change is recommended.

42. Connect the negative battery cable.

43. Start the engine and verify that there are no leaks.

1994–97 3.1L (VIN M) Engines

To remove the timing chain front cover from the 3.1L (VIN M) engine in these vehicle applications, the engine oil pan must be removed. This means the engine/drivetrain and front suspension frame must be lowered from the body assembly during this procedure. This is an involved process requiring lifts, engine support fixtures and jacking devices as various engine and transmission mounts are loosened and/or removed.

1. Disconnect the negative battery cable. Drain the engine coolant.

2. Remove the serpentine drive belt.

3. Remove the hood assembly. An assistant may be required to avoid damage to vehicle's paintwork.

NOTE: Do not allow the hood to fold back onto the windshield. Glass and paint damage may result from improper handling of the hood.

4. Remove the engine mount strut and air conditioning compressor bracket and engine mount strut bracket.

5. Remove the electric engine coolant fan assemblies.

6. Install a suitable lifting fixture that will safely support the engine with the vehicle both on the floor or raised on a lift.

7. Raise and safely support vehicle. Disconnect the front exhaust manifold pipe.

8. Remove the intermediate steering shaft bolt/screw and move the cover aside.

9. Drain the engine oil. Remove the engine splash shield.

10. Remove the engine crankshaft balancer. Use the following procedure:

NOTE: The inertia weight section of the crankshaft balancer is assembled to the hub with a rubber sleeve. The removal and installation procedures, with the proper tools MUST be followed or movement of the inertia weight section on the hub will destroy the tuning of the torsional damper.

a. Remove the right engine splash shield.

b. Remove the crankshaft balancer center bolt and washer.

c. Draw the balancer off the crankshaft using the appropriate puller.

11. Support the drivetrain and front suspension frame assembly with floor stands.

12. Remove the transaxle mount frame side bracket nuts from the drivetrain and suspension frame assembly. Use the following procedure:

a. With the vehicle safely supported, remove the left front tire and wheel assembly.

b. Remove the left front bumper fascia splash shield assembly.

c. Suitably support the transaxle assembly with floor stands.

d. Remove the transaxle mount side bracket nuts.

13. Remove the engine mount frame side nuts from the drivetrain and front suspension frame assembly.

14. Remove the rear drivetrain and front suspension frame bolts and screws.

15. Lower the drivetrain and front suspension frame assembly.

16. Remove the engine mount assembly. Remove the flywheel inspection cover.

17. Remove the starter motor assembly. Note any shims between the starter and the engine block so they can be reused at assembly.

18. Remove the transaxle mount assembly from the engine oil pan assembly.

19. Disconnect the oil level sensor. Remove the engine oil pan assembly. Do not miss the bolt going through the side of the pan and threads into the front main bearing cap.

20. Once the pan has been removed, raise the drivetrain and front suspension frame assembly back into place and install the bolts as required to temporarily secure the frame assembly. Lower the vehicle.

21. Remove the coolant recovery reservoir assembly.

22. Remove the serpentine drive belt shield.

23. Remove the power steering pump pulley, using special tools. Disconnect the ignition control wiring harness, then remove the pump mounting bolts and position the pump aside.

24. Remove the thermostat bypass pipe clip nut from the upper intake manifold. Remove the pipe from the water pump.

25. Remove the coolant pump pulley assembly. Remove the lower radiator outlet hose from the water pump.

26. Remove the serpentine drive belt tensioner.

27. If the front cover is only going to be removed and not replaced with a new cover, the 24X Crankshaft Position Sensor can be left in place on the front cover. Just disconnect the wiring harness. If the cover is going to be replaced due to damage, remove the sensor from the front cover.

NOTE: Note that this engine uses two crankshaft position sensors; one, called the 3X Crankshaft Position Sensor, is in the side of the block to read off the crankshaft. The other sensor is called the 24X Crankshaft Position Sensor and it is mounted to the bottom of the front cover. Only the 24X Crankshaft Position Sensor needs to be removed if the front cover is to be replaced.

28. Remove the bolts from the front cover, noting their locations, and separate the cover from the engine block. If the cover is being replaced, transfer the water pump to the new cover, making sure the locator on the top of the pump is installed at the top (12 o'clock) position.

29. Rotate the crankshaft until the timing marks on the camshaft and crankshaft sprockets are in alignment at their closest approach.

30. Remove the camshaft sprocket mounting bolt and remove the camshaft sprocket and timing chain.

31. Remove the crankshaft sprocket, using a gear puller J-5825-A or equivalent.

32. Remove the two mounting bolts from the timing chain damper and remove the damper.

To install:

33. Clean all parts and gasket mating surfaces well. Note that the oil pan bolts may have small O-rings under the bolt head flange which should be inspected carefully. A damaged seal will cause an oil leak.

34. Install the timing chain damper and tighten the mounting bolts to 15 ft. lbs. (21 Nm).

35. Install the crankshaft sprocket onto the crankshaft making sure the notch in the sprocket fits over the crankshaft key. Fully seat the sprocket on the crankshaft using J-38612, or equivalent.

36. Make sure the timing mark on the crankshaft sprocket is pointing straight up.

37. Install the timing chain over the camshaft sprocket and hold the sprocket in such a way, that the timing mark is pointing down, and the timing chain is hanging down off the sprocket.

38. Loop the timing chain under the crankshaft sprocket and install the camshaft sprocket on the camshaft. The sprocket will only fit on the camshaft if, the dowel on the camshaft lines up with the hole in the sprocket.

39. Verify that the marks are aligned (the camshaft sprocket will be at the 6 o'clock position and the crankshaft sprocket will be in the 12 o'clock position).

40. On 1994–95 vehicles, tighten the camshaft sprocket mounting bolt to 74 ft. lbs. (100 Nm). On 1996–97 vehicles tighten the bolt to 81 ft. lbs. (110 Nm).

41. Lubricate the timing chain components with engine oil.

42. The crankshaft front seal should be carefully removed and a replacement installed with the seal lip facing the engine. With the front cover removed, the timing chain and sprockets may be inspected. If the timing chain and sprockets are to be replaced, turn the engine crankshaft until the timing marks on the crankshaft sprocket and camshaft sprocket are aligned. There should also be timing marks on the engine block and the timing chain damper (guide). Remove the camshaft sprocket center bolt and lift off the sprocket and chain. Use a puller to remove the crankshaft sprocket. Replace the chain and sprockets as a set, using care to align all timing marks as before.

43. If the cover is being replaced with a service part or if the coolant pump is being changed out to a service replacement pump, install the coolant pump, using a new gasket, to the front cover taking care that the locator on the top of the pump is installed at the top (12 o'clock) position.

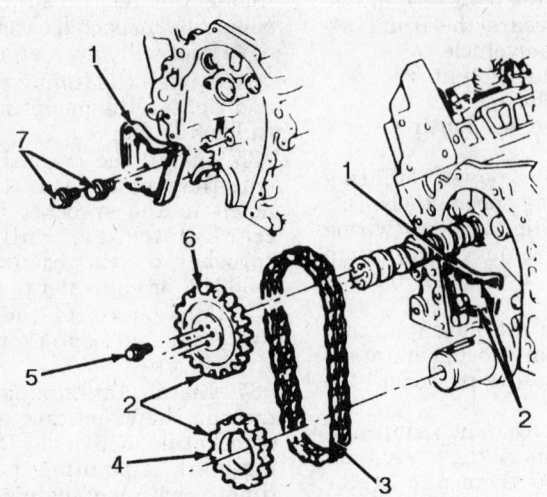

NOTE - ALIGN TIMING MARKS ON CAM
& CRANK SPROCKETS USING ALIGNMENT
MARKS ON DAMPER STAMPING OR CAST
ALIGNMENT MARKS ON CYL & CASE

1. Damper
2. Alignment marks
3. Timing chain
4. Crank sprocket
5. 28 Nm (21 lb. ft.)
6. Camshaft sprocket
7. 21 Nm (15 lb. ft.)

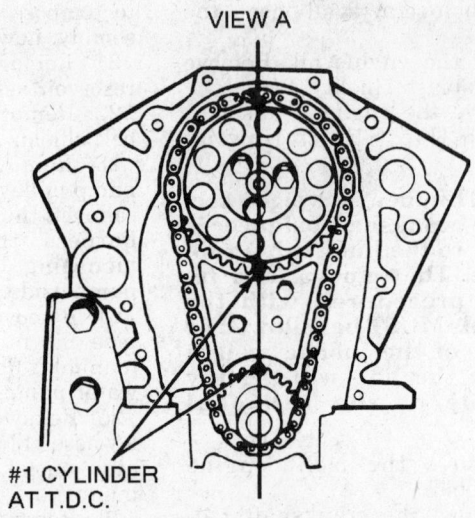

VIEW A

#1 CYLINDER
AT T.D.C.

NOTE - CAMSHAFT SPROCKET
MARK AT 6 O'CLOCK
CRANKSHAFT SPROCKET
MARK AT 12 O'CLOCK

333479

Timing chain and sprocket timing mark alignment — 3.1L (VIN M) engines

44. Coat both sides of the lower tabs of a new front cover gasket with sealer. Install the front cover assembly. Tighten the large bolts to 35 ft. lbs. (47 Nm) and the small bolts to 15 ft. lbs. (21 Nm).

45. If removed, install the 24X Crankshaft Position Sensor to the front cover and connect the wiring harness.

46. Install the serpentine drive belt tensioner. Tighten the attaching bolt to 37 ft. lbs. (50 Nm).

47. Connect the lower radiator outlet hose assembly to the coolant pump assembly. Install the pulley on the coolant pump.

48. Connect the thermostat bypass pipe to the coolant pump assembly with the pipe clip nut.

49. Position the power steering pump and install the bolts. Tighten the bolts to 25 ft. lbs. (34 Nm). Using the proper tools, install the pulley onto the power steering pump. The face of the pulley hub MUST be flush with the pump drive shaft. DO NOT use an arbor press to install the pulley.

50. Install the serpentine drive belt shield.

51. Install the coolant reservoir. Lube the reservoir hose with clean water and route it through the hole in the ECM heat shield and up to the radiator overflow fitting until the hose end butts against the filler neck.

52. Raise and safely support vehicle. Support the drivetrain and front suspension frame assembly with suitable floor stands.

53. Remove the rear drivetrain and front suspension frame bolts. Lower the drivetrain and front suspension frame assembly. This should give clearance for oil pan installation.

54. Apply a small amount of sealer on either side of the rear main bearing cap, where the seal surface on the cap meets the cylinder block (2 locations). Using a new gasket, install the oil pan and retaining bolts. Tighten the oil pan retaining bolts to 18 ft. lbs. (25 Nm). Install the oil pan side bolts to 37 ft. lbs. (50 Nm).

55. Install the transaxle mount assembly to the engine oil pan assembly.

56. Install the starter motor. Tighten the bolts to 32 ft. lbs. (43 Nm). Install the flywheel inspection cover.

NOTE: Before attaching the electrical leads, tighten the inner nuts on the solenoid terminals. If the nuts are not tight, the solenoid cap may be damaged during installation of the leads. Tighten the inner nut of BAT terminal to 84 inch lbs. (9.5 Nm).

57. Install the engine mount.

58. Raise the drivetrain and front suspension frame assembly and install the retainer bolts. Install the engine mount frame side nuts. Tighten the nuts to 32 ft. lbs. (43 Nm). Remove the floor stands from the drivetrain and front suspension frame assembly.

59. Lubricate the front cover crankshaft seal with clean engine oil. Apply a small amount of sealant to the keyway of the balancer assembly. Install the harmonic balancer using the proper tool to pull the balancer onto the crankshaft. Install the center bolt and torque to 76 ft. lbs. (103 Nm).

NOTE: The inertia weight section of the crankshaft balancer is assembled to the hub with a rubber sleeve. The removal and installation procedures, with the proper tools MUST be followed or movement of the inertia weight section on the hub will destroy the tuning of the torsional damper.

60. Install the engine splash shield.

61. Connect the intermediate steering shaft bolt. Torque to 35 ft. lbs. (48 Nm). Install the cover shield.

62. Connect the front exhaust manifold pipe. Lower vehicle.

63. Remove the engine support fixtures.

64. Install the electric cooling fan assemblies.

65. Install the engine mount strut and air conditioning compressor bracket and engine mount strut bracket.

66. Install the hood panel assembly.

NOTE: Use care when working around the hood. Glass and paint damage may result from improper handling of the hood.

67. Install the right engine splash shield and the front wheel assemblies.

68. Install the serpentine drive belt.

69. Refill the crankcase with engine oil and refill the cooling system. Connect the negative battery cable.

70. Locate the bleeder screw on the thermostat neck. Make sure it is closed but DO NOT overtorque. The air bleeder screw is made of soft brass. Start engine and allow to warm up. Open the bleeder screw as necessary to remove air from the system. Watch for oil or coolant leaks.

NOTE: The low coolant indicator lamp may come on after this procedure. After operating the vehicle so that the engine heats up and cools down three times, if the low coolant indicator lamp does NOT go out, or fails to come on at ignition check and coolant is at proper level (engine cold, coolant is level to the base of the radiator neck), electrical troubleshooting of the system is required. If at any time the TEMP warning indicator comes on, immediate action is required.

71. Road test the vehicle.

NOTE: Whenever the vehicle sub-frame is removed or lowered, the wheel alignment should be checked.

3.8L (VIN L) and (VIN K) Engine

1. Disconnect the negative battery cable. Drain the cooling system.

2. Disconnect the coolant hoses from the timing chain front cover.

3. Support the engine using support fixture J 28467-A or equivalent, then remove the engine mount.

4. Remove the drive belt(s) and belt tensioner.

5. Raise and safely support the vehicle. Remove the right front wheel.

Remove the right inner fender access panel.

6. Detach the connectors at the camshaft position sensor, crankshaft position sensor and the oil pressure sensor.

7. Keep the flywheel from turning using J 37096 or equivalent,. Remove the crankshaft balancer retaining bolts, then use puller tool J 38197 or equivalent, to remove the balancer from the crankshaft.

8. Remove the crankshaft position sensor shield and the crankshaft position sensor.

9. Remove the oil pan-to-front cover bolts.

10. Remove the front cover attaching bolts, then remove the cover.

11. Align the timing marks on the camshaft and crankshaft sprockets so they are as close together as possible.

12. Remove the timing chain damper.

13. Remove the crankshaft sprocket retaining bolts, then remove the sprocket and timing chain.

14. Remove the crankshaft sprocket.

NOTE: Do NOT rotate the camshaft or crankshaft while the timing chain and sprockets are removed.

15. Thoroughly clean all gasket mating surfaces.

To install:

16. Assembly the timing chain and sprockets with the timing marks aligned. Install the timing chain and sprockets to the camshaft and crankshaft.

17. Install the camshaft sprocket bolt and tighten to 74 ft. lbs. (100 Nm) plus an additional 90 degree turn. Recheck the camshaft and crankshaft sprocket timing marks to make sure they are still aligned.

18. Install the timing chain damper and tighten to 14 ft. lbs. (19 Nm).

19. Remove the screws and the oil pump cover from the back of the timing chain front cover. Pack the space around the oil pump gears completely full of petroleum jelly. There must be no air space left inside the pump. Reinstall the pump cover and tighten the screws to 97 inch lbs. (11 Nm).

20. Using new gaskets, install the timing chain front cover to the block. Tighten the bolts to 22 ft. lbs. (30 Nm) on 1993–95 vehicles or to 11 ft. lbs. (15 Nm) plus an additional 40 degree turn on 1996–97 vehicles. Tighten the oil pan-to-front cover bolts to 125 inch lbs. (14 Nm).

21. Install the crankshaft position sensor and tighten the bolts to 14–28

ft. lbs. (20–40 Nm). Install the crankshaft position sensor shield.

22. Keep the flywheel from turning using holder tool J 37096 or equivalent. Install the crankshaft balancer and tighten the bolt to 111 ft. lbs. (150 Nm) plus an additional 76 degree turn.

23. Attach the connectors to the camshaft position sensor, crankshaft position sensor and oil pressure sensor.

24. Install the right inner fender access panel and the right front wheel.

25. Lower the vehicle.

26. Install the tensioner assembly.

27. Install the drive belt(s).

28. Install the engine mount and remove the engine support fixture.

29. Connect the coolant hoses.

30. Connect the negative battery cable. Fill and bleed the cooling system.

31. Start the vehicle and check for leaks and proper engine operation.

Timing Belt/Chain, Sprockets, Tensioner and Front Cover

REMOVAL AND INSTALLATION

3.4L (VIN X) Engine

The 3.4L (VIN X) engine has uses a timing chain and camshaft timing belts.

1. Disconnect the negative battery cable.

2. Disconnect and remove the power steering pump from the pump mounting bracket.

3. Remove the left, right and center timing belt covers.

4. Rotate the engine clockwise to align the timing marks, TDC #1 exhaust, on the camshaft sprockets and intermediate shaft.

5. Loosely clamp the two camshaft sprockets on each side of the engine together using clamping pliers or the equivalent. Secure the belt to the right side cam sprocket with a C-clamp and a wide pad on the belt.

NOTE: When clamping the sprockets no deflection should be noticed. If any deflection is noticed, loosen the clamping devices. DO NOT mar the camshaft sprockets with the clamping device.

6. Remove the tensioner side plate retaining bolts from the tensioner and remove the side plate from the actuator and base.

7. Rotate the actuator assembly around the arm pivot and out of the base. Removal of the tensioner from the base allows it to extend to its maximum travel.

8. Set the actuator aside on a table in a vertical position to allow the oil to drain into the boot end. The tensioner should be allowed to sit for 5 minutes prior to refilling with oil.

9. Reset the timing belt actuator as follows:

a. Straighten out a paper clip or a piece of stiff wire 0.032 inch diameter to a minimum straight length of 1.85 inches. Form a double loop in the remaining end.

b. Remove the rubber end plug from the rear of the tensioner assembly. This will aid in allowing the oil in the tensioner to escape.

c. Hold the tensioner in your hand with the rubber boot end of the tensioner pointing down.

d. DO NOT remove the vent plug. Push the paper clip through the center hole in the vent plug and into the pilot hole.

e. Insert a small screwdriver into the screw slot inside the end of the tensioner.

f. Retract the tensioner by rotating the tensioner plunger in a clockwise direction while pushing the rod tip against a table top.

g. Align the screw slot to align with the vent hole, and push the straight section of the wire into the screw slot to retain the plunger in the retracted position.

h. If tensioner oil has been lost, fill the tensioner with SAE 5W30 Mobil 1®. Fill the tensioner to the bottom of the plug. The tensioner **MUST** be fully retracted before being filled with oil.

10. If the belt is being reused mark the direction of rotation on the belt.

11. Remove the timing belt tensioner pulley mounting bolt and remove the pulley.

12. Remove the timing belt after first removing the C-clamp retaining the belt to the right side camshaft sprocket.

13. Remove the Torx® head bolts securing the idler pulleys if the idlers need to be replaced.

14. Remove the intermediate shaft sprocket using the following procedure:

a. Use a suitable tool to hold the engine from turning.

b. Remove the intermediate shaft sprocket mounting bolt and washer.

c. Using J-38616, or an equivalent sprocket puller remove the sprocket from the intermediate shaft.

15. If the camshaft sprockets need to be removed proceed as follows:

a. Remove the camshaft carrier cover(s).

b. Remove the camshaft sprocket clamping pliers.

c. Rotate the camshaft being serviced so the flats on the camshaft are face up.

d. Install a camshaft hold down tool, J-38613 or the equivalent, and tighten to 22 ft. lbs. (30 Nm).

e. Remove the camshaft sprocket mounting bolt and washer while holding the camshaft from turning with J-38613 and J-38614.

f. Using J-38616, or an equivalent sprocket puller remove the sprocket from the camshaft.

g. Repeat for each camshaft sprocket as necessary.

16. Drain the cooling system.

17. Disconnect the lower radiator hose from water pump inlet pipe.

18. If equipped with a manual transaxle, disconnect the front AIR hose at the AIR pipe.

19. Disconnect the heater hose at the front cover.

20. Remove the heater pipe bracket mounting bolts at the frame.

21. Raise and safely support the vehicle. Remove right front tire and wheel assembly and the right inner fender splash shield.

22. Remove the crankshaft pulley mounting bolts and remove the pulley from the damper.

23. Remove the crankshaft damper as follows:

a. While holding the crankshaft from turning using a suitable tool, remove the damper mounting bolt and washer.

b. Install tool J-24420-B and remove the damper from the crankshaft.

24. Place an oil catch pan under the oil filter and remove the oil filter.

25. Remove the A/C compressor mounting bracket bolts.

26. Remove the lower front cover bolts.

27. On automatic transaxle vehicles, remove the halfshaft following the recommended procedure.

28. Remove the rear alternator bracket.

29. Disconnect and remove the starter following the recommended procedure.

30. Lower the vehicle.

31. Remove the intermediate shaft drive belt sprocket retaining bolt and remove the intermediate shaft drive belt sprocket using J 38616 or an equivalent puller.

32. Remove the upper alternator mounting bolts.

33. Remove the forward light relay center screws and position relay center aside.

34. Disconnect the oil cooler hose from the front cover.

35. Remove the water pump pulley.

36. Remove the upper front cover bolts and remove the front cover.

37. Mark the intermediate shaft sprocket, chain link, crankshaft sprocket and cylinder block for assembly reference. The marks should be made with paint so they won't be lost when the components are removed.

38. Retract the timing chain tensioner shoe as follows:

a. Insert J-33875 or an equivalent, on both sides of the tensioner.

b. Pull on the through pin in the tensioner arm to retract the spring located in the tensioner arm.

c. While compressing the spring, use a suitable tool, a cotter pin or nail, and insert the pin in the hole in the tensioner assembly to hold the tensioner compressed. The tool

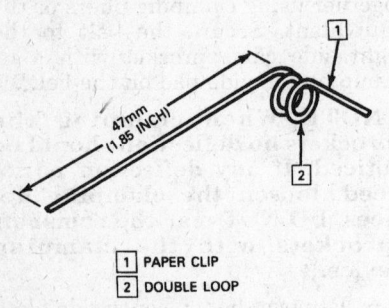

47mm (1.85 INCH)

1 PAPER CLIP
2 DOUBLE LOOP

331025

Fabricated pin for holding the actuator retracted — 3.4L (VIN X) engine

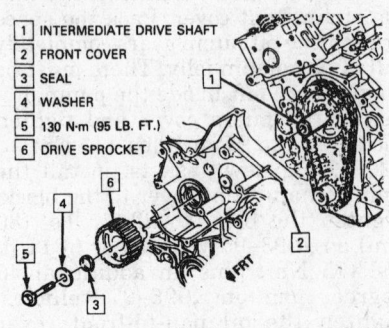

1 INTERMEDIATE DRIVE SHAFT
2 FRONT COVER
3 SEAL
4 WASHER
5 130 N·m (95 LB. FT.)
6 DRIVE SPROCKET

331027

Intermediate shaft sprocket components — 3.4L (VIN X) engine

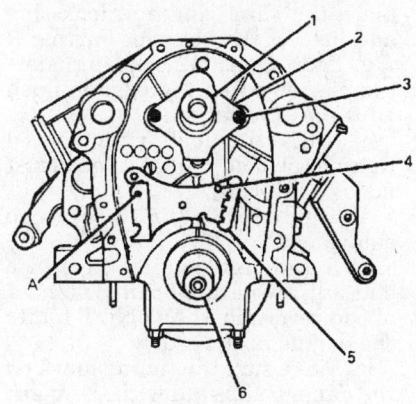

A   SPRING PIN HOLE
1   INTERMEDIATE SHAFT
2   THRUST PLATE
3   11 N•m (97 LB. IN.)
4   25 N•m (18 LB. FT.)
5   TIMING CHAIN TENSIONER
6   CRANKSHAFT

331432

Timing chain tensioner installed on engine — 3.4L (VIN X) engine

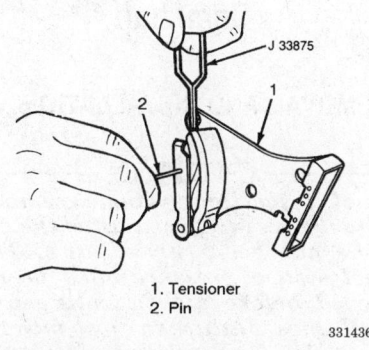

1. Tensioner
2. Pin

331436

Compressing the timing chain tensioner using J-33875 — 3.4L (VIN X) engine

used must be strong enough to hold the tensioner compressed.

NOTE: The timing chain, crankshaft sprocket and intermediate shaft sprocket will be removed at the same time. If when removing the assembly the intermediate shaft sprocket does not easily come off the intermediate shaft rotate the crankshaft back and forth to loosen the intermediate shaft sprocket.

39. Install a suitable puller, J-38611 and J-8433 or their equivalents.

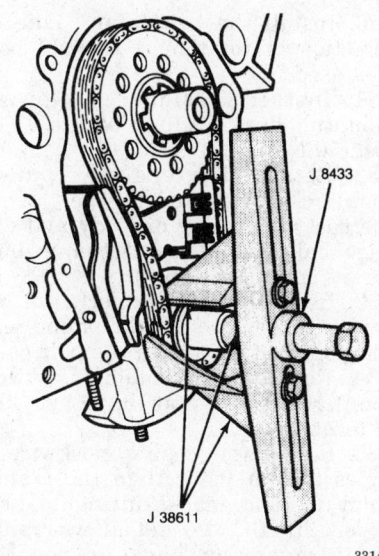

J 8433

J 38611

331434

Removing the crankshaft sprocket and timing chain using J-38611 and J-8433 — 3.4L (VIN X) engine

40. Tighten the bolt on the puller and slowly pull the crankshaft sprocket off the crankshaft. Make sure the intermediate shaft sprocket is moving along with the crankshaft sprocket.

41. Remove the timing chain and sprockets.

42. Remove the tensioner mounting bolts and remove the tensioner assembly.

To install:

43. Install the tensioner assembly and tensioner assembly mounting bolts finger tight first. Tighten the bolt in the slotted hole first to 18 ft. lbs. (25 Nm) then tighten the remainder of the bolts to 18 ft. lbs. (25 Nm).

44. Check to ensure that crankshaft key is fully seated in the crankshaft cutout and the tensioner assembly is fully retracted.

45. Assemble the timing chain, intermediate shaft sprocket and crankshaft sprocket on a work bench. The timing marks made should be in alignment. The large chamfer and counter bore of the crankshaft sprocket are installed facing toward the engine and the intermediate shaft spline sockets are installed facing away from the engine.

46. Install sprocket and chain assembly onto the engine. As the

sprockets are installed parallel alignment must be maintained.

47. The crankshaft sprocket will have to be pressed on the final 0.31 inch (8mm). This can be done using J-38612 or an equivalent puller.

48. Check to make sure timing was maintained.

49. Remove the retaining pin from tensioner. Clean all gasket surfaces completely.

50. Apply GM sealer 1052080 or equivalent, to lower edges of the sealing surface of the front cover. Install a new gasket on the front cover.

51. Install the front cover on the engine. Apply thread sealant to large bolts and tighten the bolts enough to pull the front cover against the engine block.

52. Install the water pump pulley.

53. Connect the oil cooler hose to the front cover.

54. Position the forward light relay center and install the mounting screws.

55. Install the upper alternator mounting bolt and tighten to 22 ft. lbs. (30 Nm).

56. Install the intermediate shaft drive belt sprocket. The sprocket must lock into the intermediate shaft timing chain sprocket. Install the mounting bolt and washer. While holding the engine from turning tighten the bolt to 95 ft. lbs. (130 Nm).

57. Raise and safely support vehicle.

58. Connect and install the starter. Tighten the starter mounting bolts to 32 ft. lbs. (43 Nm)

59. Install the halfshaft following the recommended procedure.

60. Install the rear alternator bracket. Tighten the mounting bolt to 22 ft. lbs. (30 Nm) and the mounting stud to 41 ft. lbs. (55 Nm) and the lower bolt to 61 ft. lbs. (83 Nm).

61. Install the lower front cover bolts and tighten the small bolts to 18 ft. lbs. (25 Nm).

62. Install the A/C compressor mounting bolts and tighten the mounting bolts to 37 ft. lbs. (50 Nm).

63. Install a new oil filter.

64. Install the crankshaft damper as follows:

a. Coat the seal contact area on the damper with clean engine oil.

b. Line up the notch inside the damper with the crankshaft key and slide the damper on until the key is started into the notch.

c. Using J-29113 or an equivalent puller, pull the damper into position on the crankshaft.

d. Install the crankshaft pulley and pulley mounting bolts. Tighten the pulley mounting bolts to 37 ft. lbs. (50 Nm).

e. Install the crankshaft damper mounting bolt and washer and tighten to 78 ft. lbs. (105 Nm).

65. Install the right side inner fender splash shield.

66. Install the tire and wheel assembly and tighten to specification.

67. Lower the vehicle.

68. Tighten the upper front cover small bolts to 18 ft. lbs. (25 Nm) and the front cover large bolts to 35 ft. lbs. (47 Nm).

69. Connect the heater hose to the front cover.

70. Install the heater pipe bracket mounting bolts.

71. If equipped with a manual transaxle, connect the front AIR hose to the AIR pipe.

72. Connect the lower radiator hose to the water pump.

73. Install the right side cooling fan. Install the upper radiator support.

74. Position the torque strut mounting bracket and install the mounting bolts and tighten to 52 ft. lbs. (70 Nm).

75. Install the front engine lift hook and tighten the mounting bolt to 52 ft. lbs. (70 Nm).

76. To install the camshaft sprockets proceed as follows:

a. Wipe the camshaft noses with clean engine oil.

b. Install the camshaft sprocket onto the nose of the camshaft.

c. Install the lock ring and shim ring.

d. Install but DO NOT tighten the camshaft sprocket mounting bolts at this time.

77. Install the intermediate shaft sprocket as follows:

a. Lubricate the seal contact area on the intermediate shaft sprocket with clean engine oil.

b. Slide the sprocket through the intermediate shaft sprocket seal and engage the locking tangs into the sockets of the chain sprocket.

c. Lightly lubricate the shaft seal and place it in position on the end of the intermediate shaft.

d. Install the intermediate shaft sprocket mounting bolt and washer. Tighten the bolt to 96 ft. lbs. (130 Nm) while holding the crankshaft from turning.

78. Install the timing belt idler pulleys and tighten the Torx® bolts to 37 ft. lbs. (50 Nm).

79. Install the actuator assembly and side plate. Tighten the actuator mounting bracket bolts to 37 ft. lbs. (50 Nm).

80. Install the belt taking note of direction of rotation if the old belt was used.

81. Install tensioner pulley to mounting base. Tighten bolt to 37 ft. lbs. (50 Nm).

82. Rotate the tensioner pulley counterclockwise into the belt using the cast square lug on body and engage ball end of the actuator into socket on pulley arm.

83. Remove tensioner lock pin allowing tensioner shaft to extend and the pulley to move into the belt.

84. Rotate the tensioner pulley counterclockwise applying 14 ft. lbs. of torque.

85. Rotate the engine clockwise 3 times to seat belt. Align the crankshaft reference marks during final rotation to TDC. Do not allow crankshaft to spring back or reverse direction of rotation.

NOTE: The timing flats on the camshafts should be 180 degrees apart from the left side to the right side. Both camshafts on the same side should be the same.

86. To perform the camshaft timing procedure proceed as follows:

a. Rotate the camshaft flats up on the right side camshafts and install a camshaft hold down tool, J-38613 or the equivalent, and tighten to 22 ft. lbs. (30 Nm).

b. Seat the lock ring on the right exhaust and intake camshaft sprockets by threading in the mounting bolt and washer.

c. Hold the sprocket from turning using tool J-38614 or the equivalent.

d. Running torque of the bolts before seating should be 55 ft. lbs. (75 Nm).

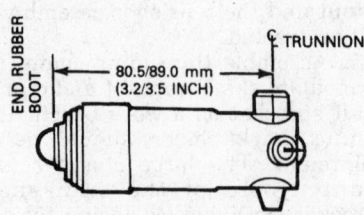

NEW BUILD ACTUATOR LENGTH
BELT INSTALLED
AFTER CAM TIMING

331024

Actuator measurement — 3.4L (VIN X) engine

e. If less torque is required replace the shim ring and lock ring.

f. If more torque is required, replace the shim ring and lock ring and inspect the bolts for burring.

g. Seating of the lock ring is accomplished when the edge is flush with the sprocket hub.

h. With the lock ring seated tighten the bolt to final torque of 81 ft. lbs. (110 Nm).

i. Remove J-38613 or the equivalent.

j. Rotate the engine clockwise one full revolution or any number of odd revolutions. **DO NOT** rotate the engine backward.

k. Make sure the timing mark on the damper lines up with the mark on the front cover.

l. Repeat steps "a" through "i" for the left side.

87. Install the camshaft carrier covers.

88. Install timing belt left, right and center covers and retaining bolts.

89. Connect the negative battery cable.

90. Start the engine and verify proper operation and engine performance.

Camshaft

REMOVAL AND INSTALLATION

--- CAUTION ---

Fuel injection systems remain under pressure, even after the engine has been turned OFF. The fuel system pressure must be relieved before disconnecting any fuel lines. Failure to do so may result in fire and/or personal injury.

2.2L (VIN 4) Engine

1. Disconnect the negative battery cable.

2. Remove the engine assembly from the vehicle and install on a suitable engine stand.

3. Remove the serpentine belt.

4. Remove the serpentine belt tensioner assembly with the alternator attached.

5. Remove the oil level indicator tube and the oil pan.

6. Remove the crankshaft balancer and front cover.

7. Remove the timing chain and camshaft sprocket.

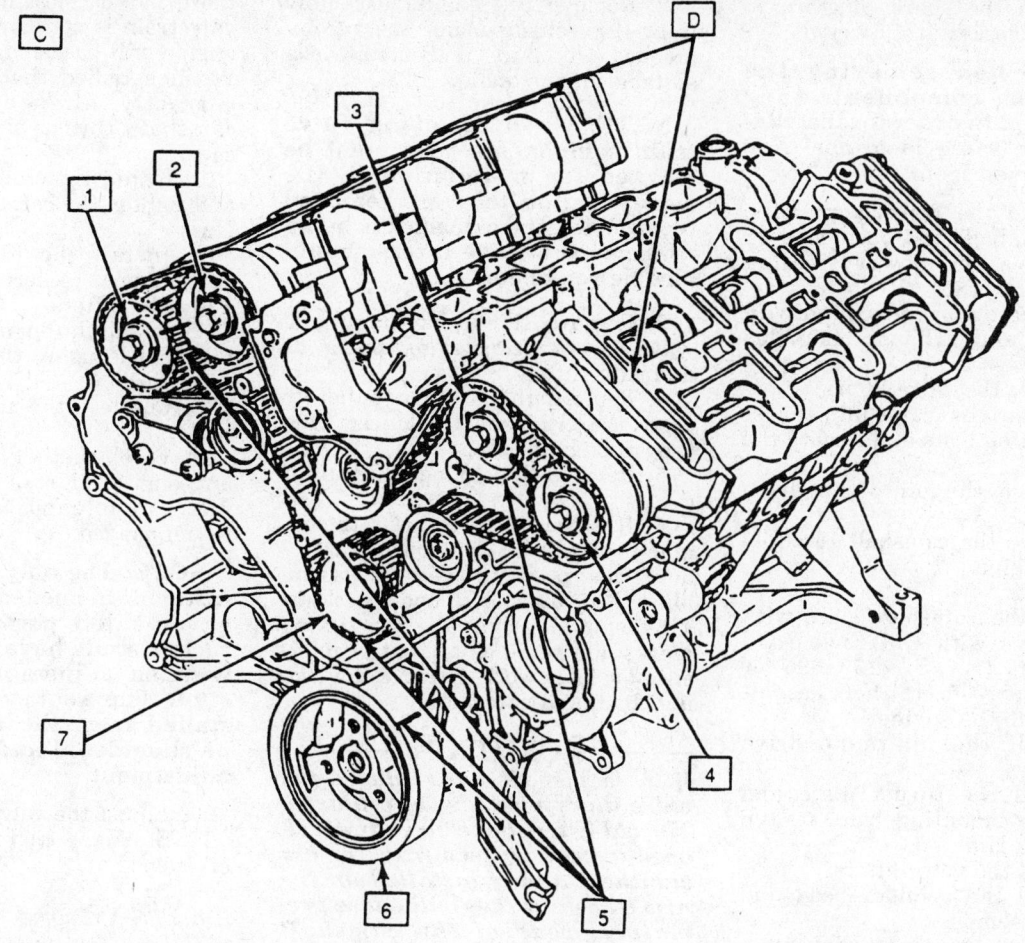

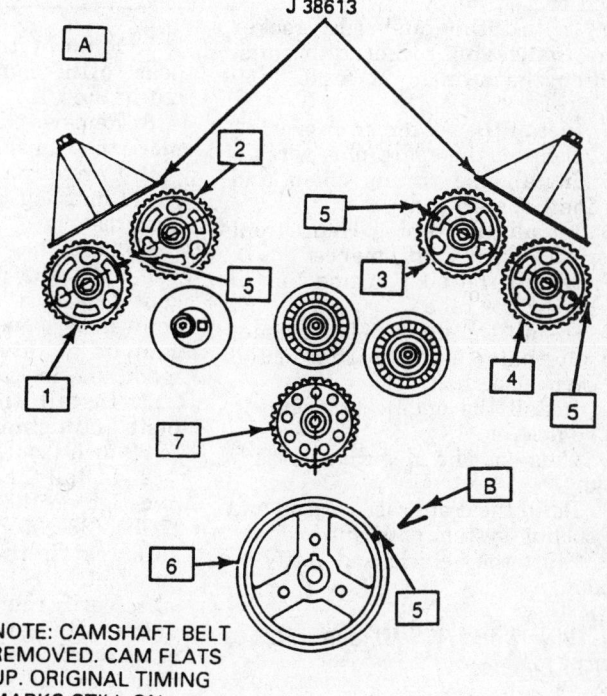

A	LOCATION OF TIMING MARKS WITH CAM HOLD DOWN TOOLS J 38613 INSTALLED (#4 TDC COMPRESSION STROKE)
B	FRONT COVER TIMING MARK
C	LOCATION OF TIMING MARKS WITH DRIVE BELT INSTALLED
D	LOCATION WHERE CAM HOLD DOWN TOOLS ARE INSTALLED
1	RH EXHAUST CAMSHAFT SPROCKET
2	RH INTAKE CAMSHAFT SPROCKET
3	LH INTAKE CAMSHAFT SPROCKET
4	LH EXHAUST CAMSHAFT SPROCKET
5	PERMANENT MARKS PAINTED DOTS REMOVE PREVIOUS MARKS IF TIMING IS BEING CHANGED AND MARK AGAIN IN THESE LOCATIONS
6	CRANKSHAFT BALANCER
7	INTERMEDIATE SHAFT SPROCKET

NOTE: CAMSHAFT BELT REMOVED. CAM FLATS UP. ORIGINAL TIMING MARKS STILL ON.

331021

Timing mark alignment — 3.4L (VIN X) engine

8. Detach the spark plug wires. Remove the rocker arm cover.

NOTE: When removing the valve train components they must be kept in order for installation in the same locations they were removed from.

9. Remove the rocker arm nuts, rocker arms, balls and pushrods.
10. Remove the power steering pump pencil brace.
11. Remove the cylinder head with the intake and exhaust manifolds attached.
12. Remove the valve lifters.
13. Remove the camshaft thrust plate mounting bolts and remove the thrust plate.
14. Remove the oil pump drive assembly.
15. Remove the camshaft carefully from the engine.

To install:
16. Coat the camshaft lobes with and bearings with GM Engine Oil Supplement (E.O.S.) 1051396 or equivalent and insert the camshaft carefully into the engine.
17. Install the oil pump drive assembly.
18. Install the thrust plate and tighten the mounting bolts to 106 inch lbs. (12 Nm).
19. Install the valve lifters.
20. Install the cylinder head and manifold assemblies.
21. Install the power steering pump pencil brace.
22. Install the pushrods, rocker arms, balls and rocker arm nuts. Tighten the nuts to 22 ft. lbs. (30 Nm).
23. Install the rocker arm cover.
24. Connect the spark plug wires.
25. Install the timing chain and camshaft sprocket.
26. Install the timing chain front cover and crankshaft balancer.
27. Install the oil pan and the oil level indicator tube.
28. Install the drive belt tensioner and alternator assembly and install the serpentine belt.
29. Install the engine assembly in the vehicle.
30. Connect the negative battery cable.
31. Refill the crankcase with oil and the cooling system with antifreeze.
32. Start the vehicle and verify no leaks.

3.1L (VIN T) and (VIN M), and 3.8L (VIN L) Engines

1. Relieve the fuel system pressure. Disconnect the negative battery cable.

2. Remove the engine assembly from the vehicle using the recommended procedure, and mount on a suitable engine stand.

NOTE: When removing valve train components they must be marked for installation in the same location they are removed from. When the camshaft is being replaced the valve lifters should also be replaced.

3. Remove the intake manifold, valve cover, rocker arms, pushrods and valve lifters.
4. Remove the using a puller to draw the crankshaft balancer from the crankshaft nose.
5. Remove the timing chain front cover.
6. It is good practice to set the engine to Top Dead Center No. 1 cylinder (firing position) before disassembling the timing chain and sprockets. This should align all the timing marks and serve as a point of reference for later work. Remove the timing chain and sprockets.

───── **CAUTION** ─────
If the camshaft was replaced the valve liters must also be replaced. The old lifters have already developed a wear pattern from the old camshaft and if installed on the new camshaft they will cause premature wear of the camshaft lobes.

7. Remove the oil pump driven gear so the camshaft can be pulled out in Step 10.
8. Remove the two bolts and remove the camshaft thrust plate.
9. Carefully remove the camshaft. Avoid marring the camshaft bearing surfaces.

To install:
10. Coat the camshaft with lubricant GM part number 1052365 or equivalent camshaft break-in lubricant or quality engine oil supplement, and install the camshaft.
11. Install the camshaft thrust plate and tighten the mounting bolts to 89 inch lbs. (10 Nm).
12. Install the oil pump driven gear and tighten the mounting bolt to 27 ft. lbs. (36 Nm).
13. Install the timing chain and sprocket.
14. Verify that all timing marks are aligned. It is good practice to turn the crankshaft several complete rotations to make sure the timing marks still align when the crankshaft is returned to Top Dead Center No. 1 cylinder compression stroke (firing posi-

tion). This is most important. If the valvetrain is not correctly timed, the engine will be damaged on start-up. When satisfied that all marks are correctly aligned, install the camshaft thrust button and front cover.
15. Install the crankshaft balancer and tighten the bolt to 75 ft. lbs. (103 Nm).
16. Install the intake manifold, valve cover, rocker arms, pushrods and valve lifters.
17. Install the engine assembly into the vehicle using the recommended procedure.
18. Connect the negative battery cable.
19. Verify that all fluid levels (coolant, engine oil, etc.) are full and correct. An oil and filter change is recommended.

NOTE: The only time valve adjustment in needed is if there was a valve job performed or the rocker studs have been replaced with an adjustable rocker arm stud. The rocker arm stud installed from the factory should be shouldered and not need any adjustment.

20. Adjust the valves, as required.
21. Start the engine and verify no oil leaks.

3.4L (VIN X) Engine

The 3.4L VIN X engine has aluminum cam carriers which house the camshafts with the actual aluminum of the carrier serving as the camshaft bearing surfaces. Use care when working with these light alloy parts. They can be damaged or broken if carelessly handled.

DOHC engines have numerous valvetrain timing marks which must be aligned or serious engine damage will result. It may be helpful to set the engine to TDC, compression stroke of No. 1 cylinder (firing position) before beginning work. This gives a known point-of-reference. In this way as the work progresses, timing marks can be observed and noted which should save time at assembly. In addition, note the routing of the timing belt so that it can eventually be properly installed.

Left Side Camshaft

1. Remove the air cleaner and duct assembly.
2. Disconnect the negative battery cable. Drain the cooling system.
3. Remove the left side camshaft cover. Remove the timing belt assembly.

4. Install six 6 inch sections of vacuum/fuel line hose under camshaft and between lifters. This will hold lifters in the carrier. For this procedure use 5/16 inch (8mm) vacuum/fuel line hose for exhaust valves and 7/32 inch (5.5mm) vacuum/fuel line hose for the intake valves.

5. Remove the exhaust crossover pipe. Remove the upper radiator hose.

6. Disconnect the heater pipe hose from the intake plenum.

7. Remove the front exhaust manifold.

8. Disconnect the engine front coolant pipe and position out of the way.

9. Remove the camshaft carrier mounting bolts.

10. Remove the camshaft carrier and gasket from the engine.

11. Place the camshaft carrier on a suitable work surface.

12. Remove the hoses securing the lifters and remove the lifters from the camshaft carrier.

13. Clean the oil out of the camshaft hold down tool mounting hole and install J-38613, or an equivalent camshaft hold down tool. Tighten the tool mounting bolt to 22 ft. lbs. (30 Nm).

14. Remove the camshaft sprocket bolts and washers, while holding the camshaft from turning using J-38614, or an equivalent camshaft sprocket holding tool.

15. Using a suitable sprocket puller, J-38616, or the equivalent, remove the camshaft sprockets.

16. Remove the six thrust plate cover screws and remove the cover and gasket.

17. Remove the two camshaft thrust plate bolts and remove the thrust plate.

18. Remove tool J-38613.

─────── CAUTION ───────
The camshaft bearing journals are all the same diameter and care must be used when removing the camshaft to prevent damage to the camshaft and/or camshaft carrier.

19. Remove the camshaft by carefully sliding it out the back of the camshaft carrier.

20. Remove the camshaft oil seal using a suitable prying tool. Do not allow the prying tool to scratch the camshaft carrier.

To install:

21. Coat the lips of the camshaft seals with clean engine oil and using J-38619, or an equivalent seal installer, seat the seals in their recesses.

22. Coat the camshaft journals and lobes with engine lube, GM 1052637 or the equivalent.

23. Carefully install the camshaft into the camshaft carrier. Make sure the camshaft does not distort the camshaft seal during installation.

24. Install the thrust plate and tighten the two mounting bolts to 89 inch lbs. (10 Nm).

25. Install the thrust plate cover and gasket and tighten the six mounting bolts to 89 inch lbs. (10 Nm).

26. Install the camshaft holding tool and tighten the mounting bolt to 22 ft. lbs. (30 Nm).

27. Install a new camshaft carrier gasket on the cylinder head and install the camshaft carrier on the cylinder head.

28. Install the mounting bolts and tighten to 18 ft. lbs. (25 Nm).

29. Install the lifters under the camshafts. Install the camshaft sprockets and timing belt.

─────── CAUTION ───────
After installing the camshaft sprockets and timing belt the Cam Timing must be set. Failure to properly set the camshaft timing will result in serious engine damage.

30. Remove the J-38613.

31. Install the front engine coolant pipe. Install the front exhaust manifold.

32. Connect the heater hose pipe to the intake plenum.

33. Connect the upper radiator hose.

34. Install the exhaust crossover pipe. Install the camshaft cover.

NOTE: Before tightening the cover mounting bolts, fully seat the bolt insulators in the camshaft cover.

35. Connect the negative battery cable. Install the air cleaner and air duct assembly.

36. Refill the cooling system.

37. Start the vehicle and verify no leaks and proper engine performance.

Right Side Camshaft

1. Remove the air cleaner and duct assembly.

2. Disconnect the negative battery cable. Drain the cooling system.

3. Remove the right side camshaft cover.

4. Remove the timing belt.

5. Install six 6 inch sections of vacuum/fuel line hose under camshaft and between lifters. This will hold lifters in the carrier. For this procedure use 5/16 inch (8mm) fuel line hose for exhaust valves and 7/32 inch (5.5mm) vacuum/fuel line hose for the intake valves.

6. Remove the front and rear engine lift brackets.

7. Remove the camshaft carrier mounting bolts, then remove the carrier and gasket from the engine.

8. Place the camshaft carrier on a suitable work surface.

9. Remove the hoses securing the lifters and remove the lifters from the camshaft carrier.

10. Clean the oil out of the camshaft hold down tool mounting hole and install J-38613, or an equivalent camshaft hold down tool. Tighten the tool mounting bolt to 22 ft. lbs. (30 Nm).

11. Remove the camshaft sprocket bolts and washers, while holding the camshaft from turning using J-38614, or an equivalent camshaft sprocket holding tool.

12. Using a suitable sprocket puller, J-38616, or the equivalent, remove the camshaft sprockets.

13. Remove the six thrust plate cover screws and remove the cover and gasket.

14. Remove the two camshaft thrust plate bolts and remove the thrust plate.

15. Remove the J-38613.

─────── CAUTION ───────
The camshaft bearing journals are all the same diameter and care must be used when removing the camshaft to prevent damage to the camshaft and/or camshaft carrier.

16. Remove the camshaft by carefully sliding it out the back of the camshaft carrier.

17. Remove the camshaft oil seal using a suitable prying tool. Do not allow the prying tool to scratch the camshaft carrier.

To install:

18. Coat the lips of the camshaft seals with clean engine oil and using J-38619, or an equivalent seal installer, seat the seals in their recesses.

19. Coat the camshaft journals and lobes with engine lube, GM 1052637 or the equivalent.

20. Carefully install the camshaft into the camshaft carrier. Make sure the camshaft does not distort the camshaft seal during installation.

21. Install the thrust plate and tighten the two mounting bolts to 89 inch lbs. (10 Nm).

22. Install the thrust plate cover and gasket and tighten the six mounting bolts to 89 inch lbs. (10 Nm).

23. Install the camshaft holding tool and tighten the mounting bolt to 22 ft. lbs. (30 Nm).

24. Install a new camshaft carrier gasket on the cylinder head and install the camshaft carrier on the cylinder head.

25. Install the mounting bolts and tighten to 18 ft. lbs. (25 Nm).

26. Install the lifters under the camshafts.

27. Install the camshaft sprockets and timing belt.

CAUTION

After installing the camshaft sprockets and timing belt the Cam Timing must be set. Failure to properly set the camshaft timing will result in serious engine damage.

28. Remove tool J-38613.

29. Install the front and rear engine lift brackets.

30. Install the camshaft cover.

NOTE: Before tightening the cover mounting bolts, fully seat the bolt insulators in the camshaft cover.

31. Connect the negative battery cable. Install the air cleaner and air duct assembly.

32. Fill and bleed the cooling system using the recommended procedure.

33. Start the vehicle and verify no leaks and proper engine performance.

3.8L (VIN K)

1. Disconnect the negative battery cable. Properly relieve the fuel system pressure.

2. Remove the engine assembly from the vehicle and mount on a suitable engine stand.

3. Remove the intake manifold.

4. Remove the rocker arm covers.

5. Remove the rocker arm assemblies, pushrods and lifters. Identify all parts as they are removed, so they can be reinstalled in their original locations.

6. Remove the crankshaft balancer center bolt and, using a suitable puller, remove the balancer from the crankshaft.

7. Remove the timing chain front cover.

8. Set the engine to Top Dead Center (TDC) No. 1 cylinder (firing position) to align the timing marks, before disassembling the timing chain and sprockets.

NOTE: Align the timing marks of the camshaft and crankshaft sprockets to avoid burring the camshaft journals by the crankshaft.

9. Remove the camshaft sprocket and timing chain.

10. Remove the camshaft thrust plate bolts and remove the thrust plate.

11. Carefully remove the camshaft from the engine block.

12. Inspect the camshaft lobes and journals for wear and/or damage; replace as necessary.

NOTE: If the camshaft was replaced the lifters must also be replaced. The old lifters have developed a wear pattern and will cause the new camshaft to wear prematurely.

To install:

13. Coat the camshaft lobes and bearings with lubricant GM part number 1052365 or equivalent camshaft break-in prelube prior to installation.

14. Carefully install the camshaft into the engine.

15. Install the camshaft thrust plate and tighten the mounting bolts to 10 ft. lbs. (14 Nm).

16. Install the camshaft sprocket and timing chain. Be sure the timing marks are aligned. Tighten the camshaft sprocket retaining bolt to 74 ft. lbs. (100 Nm) plus an additional 90 degree (¼) turn.

17. Install the timing chain front cover.

18. Install the crankshaft balancer and tighten the mounting bolt to 111 ft. lbs. (150 Nm) plus an additional 76 degree turn.

19. Coat the valve lifters with camshaft prelube and install the lifters in the lifter bores.

20. Install the lifter guides and lifter guide retainer. Tighten the retainer mounting bolts to 22 ft. lbs. (30 Nm).

21. Install the pushrods and rocker arms and tighten the rocker arm bolts to 11 ft. lbs. (15 Nm) plus an additional 90 degree turn.

22. Install the rocker arm covers.

23. Install the intake manifold.

24. Install the engine in the vehicle.

25. Verify that all fluid levels (coolant, engine oil, etc.) are full and correct.

26. Connect the negative battery cable, then start the engine and verify no leaks.

Auxiliary Shaft

REMOVAL AND INSTALLATION

3.4L (VIN X) Engine

The camshaft drive is a two-phase system. The first phase transmits power from the crankshaft to an intermediate shaft (also called an Auxiliary Shaft) by means of a chain drive. The second phase uses a belt between the intermediate shaft and the camshafts. The intermediate shaft runs in three plain bearings and is retained in the block by a thrust plate.

Although the timing chain that drives the intermediate shaft can be removed with the engine in the vehicle, if the shaft itself must be replaced, the engine will have to be removed from the vehicle.

1. Disconnect the negative battery cable. Remove the engine from vehicle.

2. Remove right side cylinder head.

3. Remove the mounting bolt and clamp and the oil pump drive assembly.

4. Remove the O-ring.

5. Remove the timing belt and timing chain assemblies. following the recommended procedures.

6. Remove the intermediate shaft (auxiliary shaft) thrust plate screws and thrust plate.

NOTE: All the intermediate shaft journals are the same diameter. Care must be taken not to damage the bearing surface when removing the shaft from the engine block.

7. Remove the intermediate shaft.

To install:

8. Inspect the intermediate shaft bearings for wear, galling, gouges and discoloration that indicates overheating. Do not attempt to repair the shaft. Replace if damaged. Bearing may be replaced in the same manner as camshaft bearings. Use care to properly align the oil feed holes when installing bearings.

9. Lubricate the intermediate shaft (auxiliary shaft) journals and gear with "Molykote" or equivalent engine assembly lube or oil supplement.

10. Install the intermediate shaft into the engine block.

11. Install the thrust plate and mounting bolts. Tighten the mounting bolts to 89 inch lbs. (10 Nm).

12. Replace the O-ring after the sprocket is installed.

13. Install the timing chain and timing belt assemblies using the recommended procedures.

14. Lubricate the oil pump drive assembly O-ring with engine oil.

15. Lubricate the oil pump drive assembly gear with engine assembly lube and install the assembly. Tighten the mounting bolt and clamp to 27 ft. lbs. (36 Nm).

16. Install the rear cylinder head following the recommended procedure.

17. Install the engine assembly following the recommended procedure.

18. Connect the negative battery cable.

Balance Shaft

REMOVAL AND INSTALLATION

3.8L (VIN L) Engine

1. Disconnect the negative battery cable. Remove the engine from the vehicle and install on a suitable engine stand.

2. Remove the flywheel.

3. Remove the intake manifold.

4. Remove the lifter guide retainer mounting bolts and remove the retainer.

5. Remove the timing chain front cover. Remove the balancer shaft gear mounting bolt.

6. Remove the camshaft sprocket mounting bolt and remove the sprocket and timing chain.

7. Remove the balance shaft drive gear from the camshaft.

8. Remove the balancer shaft retainer bolts and remove the balancer shaft retainer and gear.

9. Remove the balance shaft using J-6125-B, or the equivalent.

To install:

10. Install the balancer shaft into the block using J-36996, or the equivalent.

11. Temporarily install the balance shaft retainer and mounting bolts.

12. Install the balancer shaft gear and after coating the mounting bolt threads with thread locking compound, tighten the mounting bolt to 15 ft. lbs. (20 Nm) plus 35 degrees additional rotation.

13. Rotate the balance shaft until the timing mark is pointing straight down. Install the balance shaft drive gear on the camshaft and rotate the camshaft until the drive gear timing

mark is pointing straight up. Seat the gear fully on the camshaft.

14. Install the timing chain and camshaft sprocket.

15. Tighten the balance shaft retainer bolts to 26 ft. lbs. (35 Nm).

16. Install the timing chain front cover. Install the lifter guide retainer and tighten the mounting bolts to 27 ft. lbs. (37 Nm).

17. Install the intake manifold.

18. Install the flywheel and tighten the mounting bolts. New bolts are recommended. Torque to 132 inch lbs. (15 Nm) plus 50 degrees additional rotation.

19. Install the engine in the vehicle using the recommended procedure.

20. Connect the negative battery cable.

21. Start the vehicle and verify no leaks.

3.8L (VIN K) Engine

NOTE: The balance shaft bearing retainer bolts are designed to permanently stretch when tightened. The correct part number must be used to replace this type of fastener. DO NOT use a bolt that is stronger in this application. If the correct bolt is not used, the parts will not be tightened correctly. Part or system damage may occur. Use care to obtain correct factory replacement parts before beginning this service.

1. Disconnect the negative battery cable. Remove the engine from the vehicle and install it on a suitable engine stand.

2. Remove the flywheel-to-crankshaft bolts and the flywheel.

3. Remove the rear main carrier plate and the intake manifold.

4. Remove the lifter guide and the retainer bolts and remove the retainer.

5. Remove the timing chain front cover. Remove the camshaft sprocket and timing chain.

6. Remove the balance shaft gear-to-shaft bolt and the gear.

7. Remove the balance shaft retainer-to the-engine bolts and the retainer.

8. Using tool J 6125-B or equivalent slide hammer tool, pull the balance shaft from the front of the engine.

9. Remove the balance shaft rear plug.

10. Examine the balance shaft bearings. If in good condition, they may remain in the block. If the are worn or damaged, special tools are recommended for their removal and

installation. If using substitutes, first examine the bearings for their factory installed positions, depth and orientation to the oil feed holes. Remove the balance shaft rear bearing using the special tools J 36995–5, J 36995-3A and J 36995-2A or equivalent.

NOTE: It may take a considerable amount of force to loosen the bearing from the block bore.

To install:

11. Make sure the rear main oil seal housing gasket is removed and there is no remaining debris.

12. Dip the rear bearing in clean engine oil before the installation.

13. Using tools J 36995-2A, J 36995-3A, J 36995-1 and J 36995-8 or an equivalent, install the rear bearing. J 36995-8 assures that the rear bearing is installed in the engine block to the correct depth. The bearing will be installed properly when the tool J 36995-8 and J 36995-1 fully contact the balance shaft bore or the cylinder block/transaxle mounting flange.

14. Remove the tool J 36995 assembly or equivalent.

15. Dip the front balance shaft bearings in clean engine oil before the installation.

16. Install the balance shaft into the block using J 36996 and J 21465-13 or equivalent.

17. Temporarily install the balance shaft bearing retainer and bolts.

NOTE: This bolt is designed to permanently stretch when tightened. The correct part number must be used to replace this type of fastener. Do not use a bolt that is stronger in this application. If the correct bolt is not used, the parts will not be tightened correctly. Part or system damage may occur.

18. Install and tighten the balance shaft drive gear and the bolt with thread locking compound on the threads to 16 ft. lbs. + 70 degrees (22 Nm + 70 degrees) using the special tool J 36660 or equivalent.

19. Install the rear main seal carrier plate.

20. Measure the end play for the following applications:

   a. End play should be 0.008 inch (0.023 mm).

   b. Front radial play should be 0.0011 inch (0.028 mm).

   c. Rear radial play should be 0.0005–0.0047 inch (0.0127–0.119 mm).

21. Turn the camshaft sprocket with the camshaft gear temporarily

installed to set the timing mark pointing straight down.

22. With the camshaft sprocket and the camshaft gear removed, turn the balance shaft so that the timing mark on the gear points straight down.

23. Install the camshaft gear.

24. Align the marks on the balance shaft gear and the camshaft gear by turning the balance shaft.

25. Turn the crankshaft so that the number one piston is at Top Dead Center compression stroke (firing position).

26. Install the timing chain and camshaft sprocket. Verify that ALL timing marks are correctly aligned.

27. Measure the gear lash at four different places taking four measurements total with lash not exceeding 0.002 to 0.005 inch or (0.050mm to 0.127mm).

28. Install and tighten the balance shaft front bearing retainer and bolts to 22 ft. lbs. (30 Nm).

29. Before installing the engine front cover, verify one last time that ALL timing marks are correctly aligned. When satisfied that all timing marks are correct, install the engine front cover.

30. Install the lifter guide retainer.

31. Install the intake manifold.

32. Install and tighten the flywheel bolts to 46 ft. lbs. (62 Nm).

33. Install the engine into the vehicle. Refill and check all fluid levels.

34. Connect the negative battery cable.

35. Start the engine and check for leaks and proper engine operation.

Piston and Connecting Rod

POSITIONING

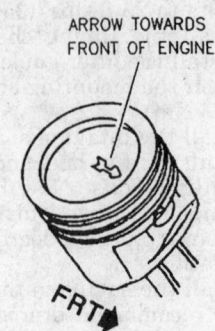

ARROW TOWARDS FRONT OF ENGINE

FRT

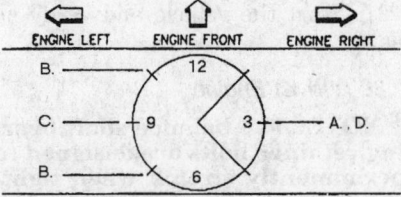

ENGINE LEFT   ENGINE FRONT   ENGINE RIGHT

B. ———— 12
C. ———— 9   3 ———— A, D.
B. ———— 6

A. OIL RING SPACER GAP (TANG IN HOLE OR SLOT WITH ARC)

B. OIL RING RAIL GAPS

C. 2ND COMPRESSION RING GAP

D. TOP COMPRESSION RING GAP

334777

Piston and ring installation — all engines similar

ENGINE LUBRICATION

Oil Pan

REMOVAL AND INSTALLATION

2.2L (VIN 4) Engine

1. Disconnect the negative terminal from the battery.

2. Raise and safely support the vehicle.

3. Disconnect and remove the starter and starter bracket.

4. Remove the flywheel cover.

5. Drain the engine oil. Remove the oil filter.

6. Remove the oil pan mounting nuts and bolts, then remove the oil pan.

To install:

7. Clean all the gasket surfaces completely.

8. Apply a thin bead of sealer around the outside edge of the oil pan and install the oil pan gasket onto the sealer.

9. Install the oil pan onto the engine and loosely install all the fasteners. Tighten the nuts and bolts to 124 inch lbs. (14 Nm).

10. Install the remaining components in the reverse order of removal.

11. Refill the crankcase with oil.

12. Connect the negative battery cable. Start the vehicle and verify no leaks.

3.1L Engine

1993–94 VIN T Engines

1. Remove the air cleaner assembly. Disconnect the negative battery cable.

2. Remove the serpentine belt.

3. Install an engine support fixture, J-28467 or the equivalent.

4. Raise and support the vehicle safely. Drain the engine oil.

5. Remove the right front tire and wheel assembly.

6. Remove the right inner fender well splash shield.

7. Remove the steering gear pinch bolt and separate the steering shaft from the rack and pinion unit.

8. Remove the transaxle mount retaining nuts.

9. Remove the engine mount-to-frame nuts.

10. Remove the engine mount bracket bolts and remove the engine mount bracket from the engine.

11. Remove the outer starter and flywheel plastic shield and the inner metal shield. Disconnect and remove the starter.

12. Place jackstands under the front and rear center crossmembers.

13. Loosen, but do not remove the rear frame bolts.

14. Remove the front frame bolts and lower the front of the frame enough to access the oil pan mounting bolts.

15. Remove the secondary air injection pipe mounting nut from the oil pan stud and position the pipe aside.

16. Remove the oil pan mounting nuts and bolts.

17. Remove the oil pan.

To install:

18. Clean all the gasket surfaces completely. Apply a thin bead of sealer around the gasket contact are on the oil pan and install the oil pan gasket on the oil pan.

19. Install the oil pan in the vehicle and install the mounting nuts and

bolts loosely. With all the fasteners installed tighten as follows:

- Oil pan mounting nuts: 89 inch lbs. (10 Nm)
- Rear oil pan mounting bolts: 18 ft. lbs. (25 Nm)
- Remaining oil pan mounting bolts: 89 inch lbs. (10 Nm)

20. Install the remaining components in the reverse order of removal. Torque the following:

- Frame mounting bolts to 103 ft. lbs. (140 Nm)
- Front engine mounting bolts to 81 ft. lbs. (110 Nm) and the side mounting bolt to 59 ft. lbs. (80 Nm)
- Engine-to-frame mounting nuts 33 ft. lbs. (45 Nm)
- Transaxle mount retaining nuts and steering shaft-to-rack and pinion pinch bolt to 35 ft. lbs. (47 Nm)

21. Refill the crankcase with engine oil. Connect the negative battery cable.

22. Start the vehicle and verify no oil leaks.

1994–97 3.1L (VIN M) Engines

1. Disconnect the negative battery cable. Remove the hood.
2. Remove the torque struts and A/C compressor/torque strut mounting bracket.
3. Remove the cooling fan assemblies.
4. Install an engine support fixture, J-28467-A, or the equivalent.
5. Raise and safely support the vehicle.
6. Disconnect the front pipe from the exhaust manifold.
7. Remove the pinch bolt from the intermediate steering shaft and disconnect the steering shaft from the rack and pinion unit.
8. Drain the engine oil.
9. Disconnect and remove the oil level indicator.
10. Remove the engine splash shield.
11. Support the frame assembly with suitable floor stands.
12. Remove the transaxle-to-frame nuts and the engine mount-to-frame nuts.
13. Remove the frame mounting bolts and lower the fame slightly.
14. Remove the front engine mount and engine bracket.
15. Remove the flywheel cover.
16. Remove the starter.
17. Remove the transaxle mount from the oil pan.
18. Remove the oil pan mounting bolts and remove the oil pan.

To install:

19. Clean all gasket surfaces completely. Install the oil pan in position with new gaskets.

20. Tighten the oil pan rail bolts to 18 ft. lbs. (25 Nm) and the oil pan side bolts to 37 ft. lbs. (50 Nm).

21. Install the remaining components in the reverse order of removal. Torque the frame mounting bolts to 103 ft. lbs. (140 Nm) and the intermediate shaft-to-rack and pinion unit pinch bolt to 35 ft. lbs. (47 Nm).

22. Refill the crankcase with oil and the power steering if disconnecting the pressure lines for clearance.

23. Connect the negative battery cable. Start the vehicle and verify no oil leaks.

NOTE: Whenever the vehicle sub-frame is removed or lowered, the wheel alignment should be checked.

3.4L (VIN X) Engine

1. Remove the air cleaner assembly. Disconnect the negative battery cable. Drain the cooling system.
2. Remove the coolant recovery tank.
3. Install an engine support fixture J-28467-A, J-28467-90, J-36462, or their equivalents.
4. Raise and safely support the vehicle. Remove the front tire and wheel assemblies.
5. Drain the engine oil.
6. Remove the steering rack and pinion gear assembly retaining bolts and hang the steering gear from the body.
7. Remove the cotter pins and castle nuts from the lower ball joints and separate the lower ball joints from the lower control arms.
8. Remove the power steering cooler line clamps at the frame.
9. Remove the engine mount nuts at the frame.
10. Support the frame with a suitable jack and remove the frame mounting bolts and remove the frame assembly.
11. Remove the oil filter.
12. Remove the engine oil cooler as follows:

   a. Disconnect the oil pressure sender and crankshaft sensor electrical connectors.

   b. Disconnect the outlet hose and position aside.

   c. Disconnect the inlet hose and position aside.

   d. Remove the mounting bolts.

   e. Remove the oil cooler.

13. Remove the starter.
14. Remove the flywheel cover.
15. Remove the oil pan retaining nuts and bolts.
16. Remove the oil pan and gasket.

To install:

17. Clean all the gasket surfaces completely. Install a new oil pan gasket and apply sealer to the gasket near the rear main bearing cap.

18. Install the oil pan and fasteners and tighten as follows:

- Oil pan nuts: 89 inch lbs. (10 Nm).
- Rear mounting bolts: 18 ft. lbs. (25 Nm).
- Remainder of bolts: 89 inch lbs. (10 Nm).

19. Install the remaining components in the reverse order of removal. Torque the following:

- Engine oil cooler bolts to 24 ft. lbs. (33 Nm)
- Frame mounting bolts to 103 ft. lbs. (140 Nm)
- Ball joint-to-control arm castle nuts to 63 ft. lbs. (85 Nm)
- Steering gear mounting bolts 59 ft. lbs. (80 Nm)

20. Install the coolant recovery tank. Refill the crankcase with oil.

21. Fill and bleed the cooling system. Connect the negative battery cable.

22. Install the air cleaner assembly.

23. Start the vehicle and verify no oil leaks.

NOTE: Whenever the vehicle sub-frame is removed or lowered, the wheel alignment should be checked.

24. Check the front end alignment and correct if necessary.

3.8L (VIN L) and (VIN K) Engine

1. Disconnect the negative battery cable. Remove the torque struts.
2. Raise and safely support the vehicle.
3. Disconnect the front exhaust pipe from the exhaust manifold.
4. Remove the right front tire and wheel assembly.
5. Remove the right inner fender well splash shield.
6. Drain the engine oil.
7. Disconnect and cap the engine oil cooler pipes.
8. Safely support the weight of the engine since the engine mounts will be disconnected. Use a suitable lifting/holding device.
9. Remove the engine mount-to-frame nuts from both front engine mounts.
10. Remove the flywheel inspection cover.
11. Raise the powertrain assembly at the transaxle to provide clearance using a suitable transmission jack.
12. Disconnect the oil level sensor indicator.

13. Remove the oil pan mounting bolts.

14. Lower the oil pan and disconnect the oil pump screen assembly and lower it into the oil pan.

15. Remove the oil pan and pump screen together.

NOTE: The oil level sensor should be removed before removing the oil pan or damage to the oil level sensor may occur.

To install:

16. Clean all parts well, especially the gasket sealing flanges on the engine block and the oil pan.

17. Install a new oil pan gasket onto the oil pan.

18. Place the oil pump screen in the oil pan and position the pan under the engine.

19. Mount the screen on the oil pump and tighten the mounting bolt to 10 ft. lbs. (15 Nm).

20. Raise the oil pan into position and tighten the mounting bolts to 10 ft. lbs. (14 Nm).

21. Install the remaining components in the reverse order of removal. Torque the following:
- Engine mount nuts and tighten to 50 ft. lbs. (68 Nm).
- Torque strut mounting bolts to 41 ft. lbs. (56 Nm)

22. Refill the crankcase with clean engine oil.

23. Connect the negative battery cable. Start the vehicle and verify no oil leaks.

Oil Pump

REMOVAL AND INSTALLATION

2.2L (VIN 4) Engine

1. Disconnect the negative battery cable. Raise and safely support the vehicle.

2. Drain the engine oil into a suitable container.

3. Remove the oil pan.

4. Remove the oil pump-to-rear main bearing cap bolt, the oil pump and extension shaft.

To install:

— CAUTION —
Heat the extension shaft retainer in hot water prior to assembly. Be sure the retainer does not crack upon installation.

5. Install the extension shaft, oil pump and pump-to-rear main cap bolt. Tighten the oil pump-to-bearing cap bolt to 32 ft. lbs. (43 Nm) and the

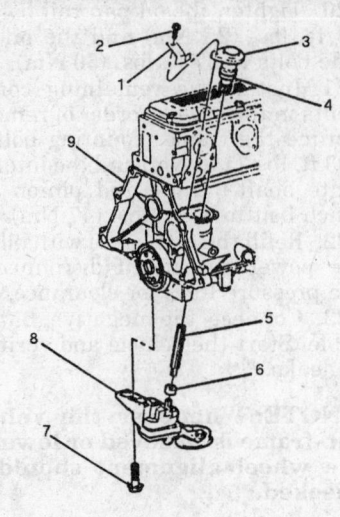

1 Bracket
2 Bolt
3 Oil pump drive assembly
4 O-ring
5 Shaft
6 Retainer; Heat and water soak prior to installation
7 Bolt
8 Oil pump
9 Cylinder block

324721

Oil pump mounting to engine block — 2.2L (VIN 4) engine

upper oil pump drive bolt to 18 ft. lbs. (25 Nm).

— CAUTION —
To avoid engine damage, all oil pump cavities must be filled with petroleum jelly before installing the gears into the pump body. This seals the gears, acts like a prime and allows the pump to start drawing oil as soon as the engine begins to crank the first time after pump service. Also use only original equipment gaskets. Gasket thickness is critical to proper oil pump operation.

6. Install the oil pan and attaching bolts. Lower the vehicle.

7. Fill the crankcase with clean engine oil.

8. It is good practice to install a reliable mechanical oil pressure gauge so that actual pump pressure can be read after startup. If no oil pressure is seen within a short time of startup, shut down the engine to determine why there is no oil pressure.

9. Connect the negative battery cable. Start the engine and check oil pressure and check for leaks.

10. Turn the engine OFF and allow to stand. Check oil level, add as necessary.

3.1L Engine

1993–94 VIN T Engines

1. Disconnect the negative battery cable. Raise and support the vehicle safely.

2. Drain the engine oil into a suitable container, then remove the oil pan.

3. Remove the pump-to-rear main bearing cap bolt and remove the pump and extension shaft.

To install:

4. Clean all parts well, especially the gasket sealing flanges on both the engine block and the oil pan.

5. Whenever the oil pump is removed or replaced it must be primed prior to installation in the vehicle to prevent oil starvation on start up. Use the following procedure.

a. Remove the 4 cover attaching screws and the cover from the oil pump assembly.

b. Pack the space around the oil pump gears completely full of petroleum jelly. This seals the pump and acts like a prime so that the pump can draw oil from the pan as soon as the engine starts to turn on initial start up after pump installation. There must be no air space left inside the pump. Do not use grease. Petroleum jelly is a low-temperature lubricant and will dissolve readily. Chassis grease must not be used since it will not melt but it will plug up the oil passages, damaging the engine.

c. Install the pump cover and four mounting screws and tighten the screws to 80 ft. lbs. (10 Nm).

6. Assemble the pump and extension shaft with retainer to rear main bearing cap, aligning the top end of the extension shaft with the lower end of the drive gear.

7. Install the pump-to-the rear bearing cap bolt. Tighten to 30 ft. lbs. (41 Nm).

8. Install the oil pan using the recommended procedure.

9. Lower the vehicle.

10. Fill the crankcase with oil to the proper level. A filter change is recommended.

11. Connect the negative battery cable. Start the engine and verify proper engine oil pressure and no leaks.

1994–97 3.1L (VIN M) Engines

1. Disconnect the negative battery cable. Raise and safely support the vehicle.

2. Drain the engine oil, then remove the oil pan.

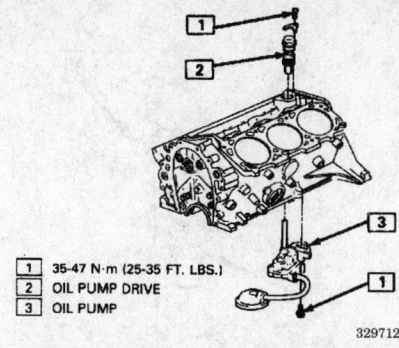

1	35-47 N·m (25-35 FT. LBS.)
2	OIL PUMP DRIVE
3	OIL PUMP

329712

Oil pump and oil pump drive assembly — 1993–94 3.1L (VIN T) engines

3. Remove the crankshaft oil deflector bolts, then remove the deflector.

4. Remove the oil pump retaining bolts and remove the oil pump and pump driveshaft.

To install:

5. Install the oil pump and pump driveshaft. Tighten the oil pump mounting bolts to 30 ft. lbs. (41 Nm).

6. Install the crankshaft oil deflector and mounting bolts. Tighten the mounting bolts to 18 ft. lbs. (25 Nm).

7. Install the oil pan.

8. Lower the vehicle.

9. Fill the crankcase to the correct level with oil.

10. It is good practice to install a reliable mechanical oil pressure gauge so that actual pump pressure can be read after startup. If no oil pressure is seen within a short time of startup, shut down the engine to determine why there is no oil pressure.

11. Start the engine, check the oil pressure and check for leaks.

3.4L (VIN X) Engine

1. Disconnect the negative battery cable. Drain the engine oil.

2. Remove the oil pan.

3. Remove the eight oil pan baffle mounting nuts and remove the oil pan baffle.

4. Remove the oil pump retaining bolt.

5. Remove the oil pump with pickup and drive shaft extension.

To install:

6. Clean all parts well. Clean the inside of the oil pan, the gasket flanges on pan and gasket rail on the block.

7. Install the oil pump with the driveshaft extension and pickup assembly onto the rear main bearing cap making sure the driveshaft extension engages in the oil pump drive.

8. Install the oil pump retaining bolt and tighten to 40 ft. lbs. 54 (Nm).

9. Install the oil pan baffle and tighten the mounting nuts to 18 ft. lbs. (25 Nm).

10. Install the oil pan following the recommended procedure.

11. Refill the crankcase with oil.

12. Connect the negative battery cable.

13. Start the vehicle and verify proper oil pressure, 15 lbs. minimum @ 1100 rpm at normal operating temperature.

3.8L (VIN L) Engine

1. Disconnect the negative battery cable. Raise and safely support the vehicle.

2. Drain the engine oil.

3. Remove the front cover assembly.

4. Remove the four bolts securing the oil filter adapter to the front cover assembly and oil filter adapter, pressure regulator valve and spring.

5. Remove the four oil pump cover attaching screws and remove the cover.

6. Remove the inner and outer pump gears.

To install:

7. Lubricate the gears with petroleum jelly and install the gears into the housing.

8. Pack the gear cavity with petroleum jelly after the gears have been installed in the housing.

9. Install the oil pump cover and screws and tighten to 97 inch lbs. (11 Nm).

10. Install the oil filter adapter with new gasket, pressure regulator valve and spring. Tighten the mounting bolts to 24 ft. lbs. (33 Nm).

11. Install the front cover assembly.

12. Refill the crankcase with clean engine oil.

13. Connect the negative battery cable. Start the vehicle and verify no leaks and proper oil pressure.

3.8L (VIN K) Engine

1. Disconnect the negative battery cable.

2. Remove the engine drive belts and tensioner assembly.

3. Remove the drive belt idler pulley and bracket, if necessary.

4. Raise and safely support the vehicle.

5. Support the engine using engine support fixture J 28467 or equivalent, and remove the torque axis mount and bracket assembly.

6. Remove the engine front cover assembly.

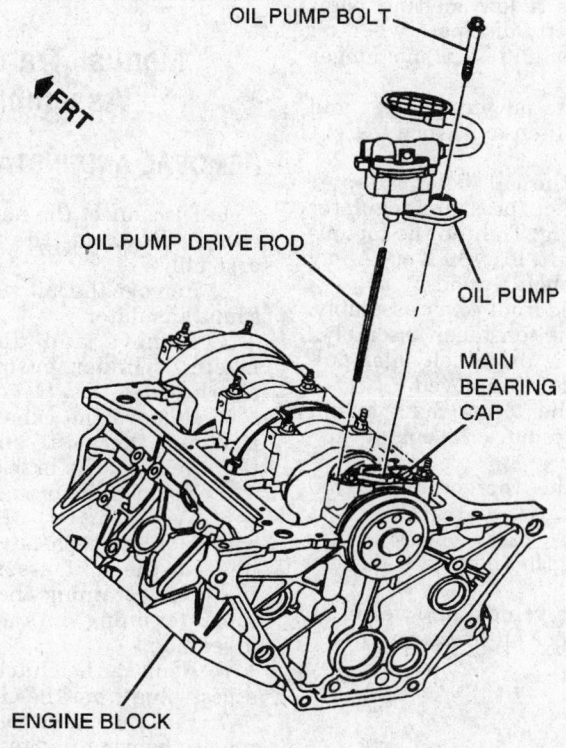

OIL PUMP BOLT

FRT

OIL PUMP DRIVE ROD

OIL PUMP

MAIN BEARING CAP

ENGINE BLOCK

324753

Oil pump and driveshaft components — 3.1L (VIN M) engine

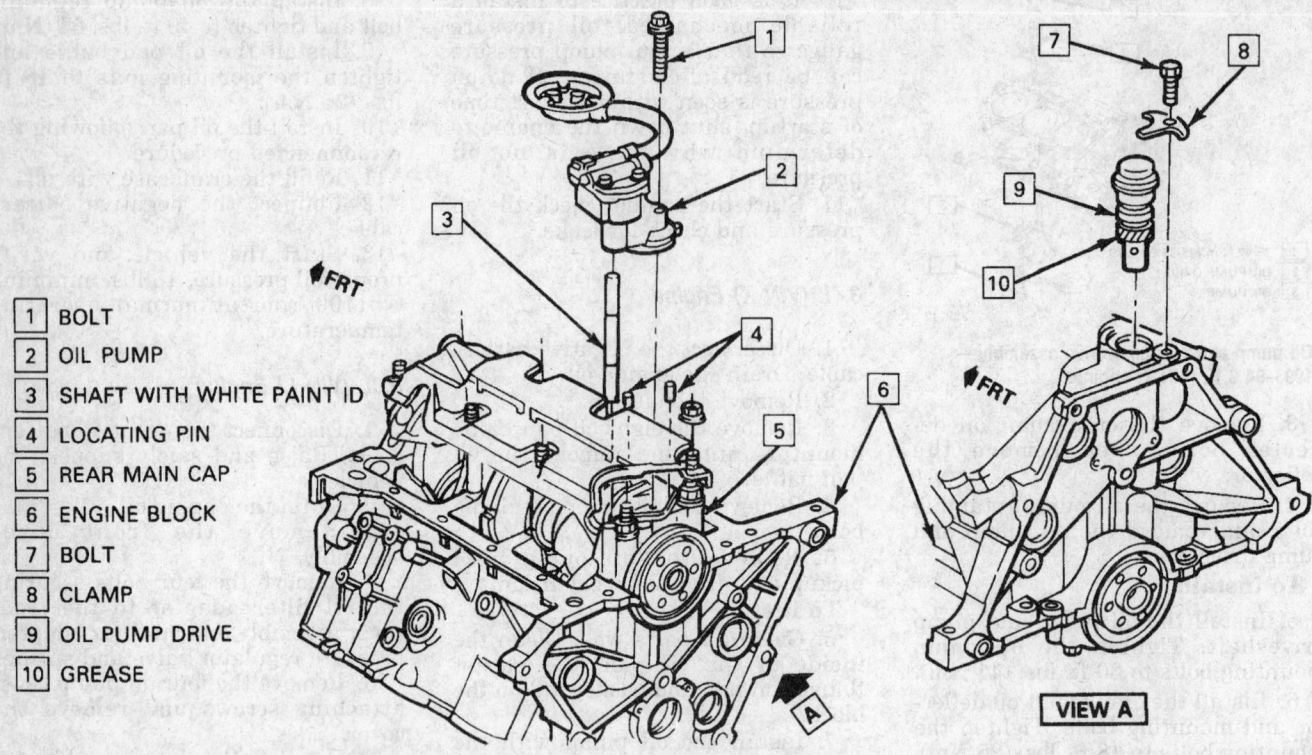

1	BOLT
2	OIL PUMP
3	SHAFT WITH WHITE PAINT ID
4	LOCATING PIN
5	REAR MAIN CAP
6	ENGINE BLOCK
7	BOLT
8	CLAMP
9	OIL PUMP DRIVE
10	GREASE

VIEW A

Oil pump components and drive gear assembly — 3.4L (VIN X) engine

7. Remove the four bolts securing the oil filter adapter to the front cover assembly and oil filter adapter, pressure regulator valve and spring.

8. Remove the four oil pump cover attaching screws and remove the cover.

9. Remove the inner and outer pump gears and inspect.

To install:

10. Lubricate the oil pump gears with petroleum jelly and install the gears into the housing.

11. Pack the gear cavity with petroleum jelly after the gears have been installed in the housing. This seals the pump and acts like a "prime" so oil will begin to drawn from the oil

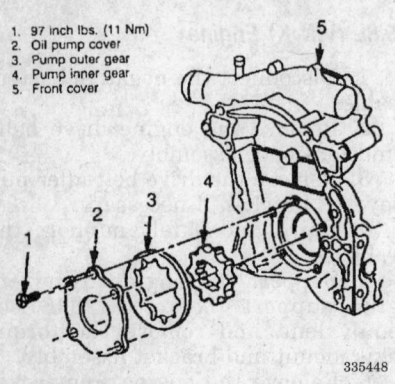

1. 97 inch lbs. (11 Nm)
2. Oil pump cover
3. Pump outer gear
4. Pump inner gear
5. Front cover

Oil pump assembly — 3.8L (VIN K) engine

pan as soon as the engine begins to turn. Do not neglect this step. DO NOT use any type of grease. Petroleum jelly has a low melting point and will correctly dissipate when oil begins to flow and it is no longer needed.

12. Install the oil pump cover and screws and tighten to 97 inch lbs. (11 Nm).

13. Install the oil filter adapter with new gasket, pressure regulator valve and spring. Tighten the mounting bolts to 24 ft. lbs. (33 Nm). Apply sealant to the bolt threads.

14. Install the front cover assembly.

15. Install the tensioner assembly.

16. Install the drive belt idler pulley and bracket, if removed.

17. Install the torque axis mount assembly and remove the engine support fixture.

18. Verify the correct engine oil level. A new oil filter is recommended.

19. Connect the negative battery cable.

20. Start the vehicle and verify no leaks and proper oil pressure.

TRANSAXLE

Manual Transaxle Assembly

REMOVAL AND INSTALLATION

1. Disconnect the negative battery cable. Remove the air cleaner assembly.

2. Remove the left side under dash sound insulator.

3. Remove and disconnect the master cylinder pushrod from the clutch pedal.

4. Remove the exhaust cross over pipe and the shift control cables at the lever and the bracket.

5. Remove the actuator and canister by using J 36221 or the equivalent to separate the hydraulic lines to the unit assembly. Remove the bolts retaining the canister and the 2 retaining nuts at the actuator assembly.

6. Remove the clutch release lever access plug from the clutch housing.

7. Pull the clutch fork lever off the release bearing flanges.

8. Install a wire through the hole in the end of the fork lever. Hold

lever out as far as possible and back towards to transaxle side of the access hole. Wrap one end of wire around the bottom attaching stud for the actuator.

9. Remove the transaxle vent hose. Remove the transaxle retaining bolts.

10. Install tool J 28467-A, J 28467-90, J 36462 engine support fixtures or their equivalents.

11. Raise and safely support the vehicle.

12. Remove the front tire and wheel assemblies, mark the position of the wheel to the wheel studs, prior to removal, for installation reference.

13. Remove the Vehicle Speed Sensor (VSS) electrical connector.

14. Remove the rack and pinion heat shield, rack and pinion retaining bolts and steering cooler line screws.

15. Remove the frame assembly from the vehicle.

16. Remove the flywheel cover and the retaining screws.

17. Remove the halfshafts from the transaxle and wire up to the body. Removal is should not be necessary.

18. Remove the left engine splash shield.

19. Remove the rear engine mount and bracket bolts.

20. Safely lower the vehicle.

21. Lower the engine using the engine support fixture.

22. Raise the vehicle and securely attach the transaxle to a suitable transmission jack.

23. Remove the intermediate shaft bracket bolts.

24. Remove the transaxle mounting studs, then remove the transaxle from the vehicle.

25. Remove the clutch release bearing from the pressure plate.

To install:

26. Install the clutch release bearing, lightly lubricate the clutch ends which contact the bearing with grease P/N 1051344 or the equivalent. Levers on the clutch fork must bear on large ears of the release bearing. Levers fit between large ears and the small tangs on the bearing. The fork lever must move freely with the bearing in place.

27. If the T-bar has been removed:

   a. Line the clutch fork lever up with the threaded stud in the clutch housing.

   b. Assemble the T-bar through the slot in the fork lever over the threaded stud.

   c. Attach the fork lever retaining nut to the threaded stud and tighten to 21 ft. lbs. (28 Nm).

28. Slide the fork lever toward the center of the transaxle over the bearing flanges.

NOTE: The clutch lever MUST NOT be moved toward the flywheel until the transaxle is bolted to the engine, or damage to the transaxle may result.

29. Install the transaxle assembly to the transaxle jack.

30. Position the transaxle to the engine assembly. Install the intermediate shaft bracket bolts.

31. Install and tighten the transaxle mounts to:

   • Transaxle to engine studs to 55 ft. lbs. (75 Nm).

   • Engine to transaxle studs to 61 ft. lbs. (83 Nm).

32. After the transaxle is assembled to the engine and all the bolts are in place, seat the clutch release bearing. This is done by pulling the fork lever away from the engine at the actuator end. In some cases in may be necessary to use a screwdriver to push the lever towards the transaxle.

33. Install the clutch release lever access plug to the clutch housing.

34. Remove the jack from the transaxle case.

35. Safely lower the vehicle.

36. Raise the engine and transaxle assembly using the engine support fixture.

37. Raise the vehicle and install the rear mount bracket brace bolts.

38. Install the engine splash shield.

39. Connect the axles into the transaxle.

40. Install the flywheel cover and retaining screws tightening to 115 inch lbs. (13 Nm).

41. Install the frame to the vehicle.

42. Install the cooler line brackets and screws, rack and pinion retaining bolts and the rack and pinion heat shield.

43. Connect the VSS connector.

44. Install the front tire and wheel assemblies, aligning the marks made during the removal.

45. Safely lower the vehicle.

46. Remove the engine support fixture. Install the transaxle retaining bolts and tighten to 55 ft. lbs. (75 Nm).

47. Install the transaxle vent hose.

48. Inspect for the actuator pushrod lever bushing, if missing it must be replaced.

49. Install the actuator and canister.

50. Install the shift control cables at the levers and the bracket assembly.

51. Install the exhaust cross over pipe.

52. Install the master cylinder push rod to the clutch pedal.

53. Press the clutch pedal down several times to assure that the effort is normal and firm. **An unusually hard pedal feel means the fork is out of position.**

54. Adjust the cruise control switch, if equipped.

NOTE: When adjusting cruise control switch, DO NOT exert an upward force on the clutch pedal pad of more than 20 pounds, or damage to the master cylinder pushrod retaining ring may result.

55. Install the left side under dash sound insulator.

56. Check the oil level of the transaxle.

57. Install the air cleaner assembly.

58. Connect the negative battery cable.

59. Road test the ensure proper operation and smooth shifting.

Clutch Assembly

REMOVAL AND INSTALLATION

The entire transaxle assembly must be removed to service the clutch assembly and its related components.

NOTE: Before performing any manual transaxle removal procedures, the clutch master cylinder pushrod MUST be disconnected from the clutch pedal and the actuator must be separated from the transaxle. The reason for this is that permanent damage may occur to the clutch actuator (once called the slave cylinder) if the clutch pedal is depressed while the system is not resisted by clutch loads. The pushrod bushing must be replaced once it has been disconnected from the master cylinder. In addition, whenever powertrain components are lowered for any reason (removal, access to other components, etc.) the clutch actuator and canister must be removed from the transaxle housing. Failure to remove the actuator and canister may cause hydraulic system damage.

1. Disconnect the negative battery cable. Some vehicles may require removal of the air cleaner assembly to access the battery and its connections.

2. Remove the sound insulator from inside the vehicle and disconnect the clutch master cylinder pushrod from the clutch pedal.

3. Remove the two actuator (slave cylinder) retaining nuts from the transaxle and remove the actuator and canister. A special tool is used to disconnect the quick-connect fitting. Position to the side and support.

4. Remove the clutch release lever access plug from the clutch housing. Pull the clutch fork lever off the release bearing flanges by pulling the lever straight out until it stops.

5. Insert a wire through the hole in the end of the fork lever. Hold the lever out as far as possible and back towards the transaxle side of the access hole. Wrap one end of the wire around the bottom attaching stud for the actuator.

6. Remove the transaxle assembly using the recommended procedure.

7. If any components are to be reused, mark the position of the clutch cover (pressure plate) to the flywheel so it can be reinstalled in the same position. This helps maintain the assembly balance.

8. Loosen the cover retaining bolts one turn at a time until all the spring pressure is released.

9. Support the clutch cover and remove all the bolts.

10. Remove the clutch cover and driven disc. To remove the clutch release bearing:

   a. Lay clutch cover assembly pressure plate-side up on a flat surface.

   b. Lightly push down on the clutch cover assembly to partially open the release bearing attaching snapring.

   c. Use snapring pliers to open the snapring and lift the assembly to allow bearing removal.

11. Inspect the clutch cover and flywheel for scoring, warpage and excessive wear. Replace the clutch cover and resurface the flywheel if damaged.

12. Inspect the clutch fork, release bearing and the bearing sleeve outer surface of the transaxle. Heavy galling on either bearing or sleeve requires replacement.

NOTE: It is generally considered good practice to replace the clutch cover (often called the pressure plate), driven disc and release bearing when servicing the clutch assembly. Also, the flywheel should be inspected for

heat cracks, scoring or other damage and be resurfaced or replaced as required.

13. Clean the clutch cover and flywheel surfaces with suitable solvent. Small marks may be removed from the flywheel friction surface with abrasives (fine sandpaper or emery cloth wrapped around a small flat block). Clean all dirt and dust from the pilot bearing in the end of the crankshaft. Apply a small amount of high temperature grease to the inside of the pilot bearing.

To install:

14. Position the clutch disc and cover onto the flywheel mating surface. The raised hub and springs should face away from the flywheel.

15. Support the clutch disc using a clutch alignment tool J 38688 or equivalent. This tool is used to align the clutch disc to the flywheel. This helps the transaxle input splines align with the clutch hub splines and also helps the end of the input shaft fit the flywheel pilot bearing, making assembly easier. Position the clutch disc and clutch cover, matching the light side of the clutch cover (usually marked with paint), with the heavy side of the flywheel (should be stamped with an **X**). The clutch disc should be marked "Flywheel" indicating the side to face the flywheel. Install the clutch cover and loosely install the clutch cover retaining bolts. **New replacement bolts are recommended.**

16. With the clutch alignment tool still installed, tighten each bolt down about ½ turn at a time. Turn the close clearance bolts first, followed by the other bolts. Continue this procedure (about 6 times) until the clutch cover stamping is touching the flywheel.

17. Torque the bolts in the same sequence to 15 ft. lbs. (20 Nm) then ro-

TIGHTENING SEQUENCE
333786

Clutch cover (pressure plate) tightening sequence. Install bolts marked "L" first

tate the bolts an additional 30 degrees in the same sequence. Use a torque angle gauge as necessary. Double check all bolt torques. Remove the clutch pilot alignment tool.

18. Install the primary part of the clutch bearing to the input shaft bearing retainer.

19. Lightly lubricate the clutch fork ends which contact the bearing with high-temperature grease, GM PN 1051344 or equivalent. Levers on the clutch fork must bear on the large ears of the release bearing. Levers fit between the large ears and small tang on bearing. The fork lever must move freely with the bearing in place.

20. If the T-bar has been removed from the clutch fork:

   a. Line the clutch fork lever up with the threaded stud in the clutch housing.

   b. Assemble the T-bar through the slot in the fork lever over the threaded stud.

   c. Attach the fork lever retaining nut to the threaded stud and tighten to 20 ft. lbs. (28 Nm).

21. Slide the fork toward the center of the transaxle over the bearing flanges. Use care. The clutch lever must not be moved toward the flywheel until the transaxle is bolted to the engine or damage to the transaxle may occur.

22. Install the transaxle assembly using the recommended procedure.

23. After the transaxle is assembled to the engine and all bolts are in place, seat the pull clutch release bearing to the clutch cover (part of the clutch assembly). This is done by pulling the fork lever away from the engine at the actuator end.

24. Install the clutch release lever access plug to the clutch housing.

25. Inspect the actuator pushrod for the lever bushing and replace if missing.

26. Install the actuator on the housing studs with the pushrod bushing centered in the pocket of the internal lever in the housing. It may require an axial load on the pushrod to compress the actuator piston spring. Tighten the actuator retaining nuts evenly to 18 ft. lbs. (25 Nm).

27. Install a new bushing in the master cylinder pushrod. Lubricate the bushing before installation. Install the master cylinder pushrod to clutch pedal with the three bushing tangs snapped into the pedal pin groove. Press the clutch pedal down several times to assure that effort is normal and firm. An unusually hard

pedal fell means the fork is out of position.

NOTE: When adjusting the cruise control switch, do not exert an upward force on the clutch pedal pad of more than 20 pounds or damage to the master cylinder pushrod retaining ring may result.

28. Adjust the cruise control switch by simply pulling back on the clutch pedal allowing the switch to self-adjust in its retainer.

29. Install the sound insulator.

30. Connect the negative battery cable and install the air cleaner assembly, as required.

31. Road test to verify proper shifting and smooth transaxle operation.

Clutch Master Cylinder

REMOVAL AND INSTALLATION

NOTE: The factory hydraulic system is serviced as a single assembly. Replacement hydraulic assemblies are pre-filled with fluid and do not require bleeding. Individual components of the system are not available separately. Check with an aftermarket parts supplier to see if individual components can be purchased separately.

Prior to any vehicle service that requires the removal of the actuator, the master cylinder pushrod must be disconnected from the clutch pedal. The reason for this is, if not disconnected, permanent damage to the actuator will occur if the clutch pedal is depressed while the system is not resisted by clutch loads. The master cylinder pushrod bushing must also be replaced whenever it has once been removed from the clutch pedal.

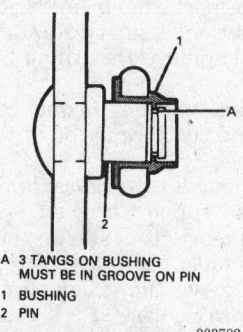

A 3 TANGS ON BUSHING
   MUST BE IN GROOVE ON PIN
1 BUSHING
2 PIN

333782

Critical bushing tang engagement, clutch cylinder pushrod to clutch pedal

Whenever any powertrain components are lowered for any reason (removal, access to other components, etc.) the clutch actuator and canister must be removed from the transaxle housing. Failure to remove actuator and canister may cause hydraulic system damage. Since the system is service by replacing the complete hydraulic system, this repair procedure covers removing the clutch hydraulic system.

1. Disconnect the negative battery cable.

2. Remove the sound insulator inside the vehicle and disconnect the master cylinder pushrod at the clutch pedal. Remove the left upper secondary cowl panel.

3. Remove the two master cylinder reservoir-to-strut tower retaining nuts.

4. Remove the anti-rotation screw located next to the master cylinder flange at the pedal support plate.

NOTE: Do not torque on the hose connection on top of the master cylinder body or damage may result.

5. Using wrench flats on the front end of the master cylinder body, twist the cylinder counterclockwise to release the twist lock attachment-to-plate. Do NOT torque on the hose connection on top of the cylinder body or damage may occur.

6. Remove the two actuator-to-transaxle retaining nuts and actuator assembly.

7. Pull the master cylinder with the pushrod attached forward out of the pedal plate. Lift the reservoir off the strut tower studs and remove the three components as a complete assembly.

To install:

8. Install the master cylinder into the opening in the pedal plate and rotate 45 degrees by applying torque on the wrench flats only.

9. Install the anti-rotation screw.

10. Install the fluid reservoir to the strut tower and torque the retaining nuts to 36 inch lbs. (4 Nm).

11. Install a new pushrod bushing and lubricate before installation.

12. Install the master cylinder pushrod to the clutch pedal.

13. Install the clutch actuator to the transaxle.

14. Press the clutch pedal down several times to ensure proper operation.

15. Install the left upper secondary cowl panel, sound insulator and connect the negative battery cable.

Hydraulic Clutch System Bleeding

1. Disconnect the negative battery cable.

2. Disconnect the quick connect fittings in clutch hydraulic line. Insert J-36221 or equivalent hydraulic line separator tool and depress plastic sleeve to separate connection.

3. Remove cap and diaphragm and fill reservoir with DOT 3 brake fluid.

4. Remove left hand upper secondary cowl.

5. Remove air from supply hose by squeezing it until no more air bubbles are seen in the reservoir.

6. Pump clutch pedal slowly until slight pressure is observed. Hold pressure on pedal and depress internal valve on quick connect fitting.

7. Repeat previous step until pedal is firm and no bubbles are seen.

8. Reconnect clutch hydraulic line. Refill clutch system and replace reservoir cap. Reconnect battery cable.

9. Verify proper clutch pedal operation.

Automatic Transaxle Assembly

REMOVAL AND INSTALLATION

1993–94 Vehicles

NOTE: These transaxles were used in a variety of General Motors vehicles. Due to model year, vehicle model and installed options, the removal and installation procedures may vary slightly. The procedures given here should suffice for most all vehicles using these transaxles.

3T40 3-Speed Automatic Transaxle

1. Disconnect the negative battery cable and remove the air cleaner assembly.

2. Remove the shift control and throttle valve (TV) cables at the transaxle.

3. Remove the throttle cable bracket and brake booster hose.

4. Remove the engine torque struts, left torque strut bracket and oil cooler lines at the transaxle.

5. Install the engine support fixture tool J28467 A and J 36462.

6. Raise and safely the vehicle with properly positioned safety stands.

7. Remove the front wheels, splash shields, calipers and rotors.

Support the caliper to the frame with wire.

8. Remove the front halfshafts (driveaxle) assemblies.

9. Disconnect both ball joints and tie rod ends from the strut assemblies.

10. Remove the engine oil filter.

11. Remove the A/C compressor and support out of the way. Do NOT disconnect the refrigerant lines.

12. Remove the rack and pinion heat shield and electrical connector.

13. Remove the rack and pinion assembly and wire to the exhaust for support.

14. Remove the power steering hoses- bracket.

15. Remove the engine and transaxle mounts at the frame.

16. Support the frame with safety stands at each end, remove the frame bolts and frame.

17. Remove the torque converter cover and bolts.

18. Remove the starter bolts and support out of the way.

19. Remove the ground cable from the transaxle case.

20. Remove the fluid fill tube bolt and transaxle mount bracket.

21. Lower the vehicle.

22. Remove the fill tube.

23. Using the engine support fixture, lower the left side of the engine about 4 in. (102mm).

24. Raise the vehicle, support with properly positioned safety stands and install a transaxle jack.

25. Remove the transaxle-to-engine bolts and transaxle.

To install:

NOTE: If the transaxle was removed due to failure, the transmission fluid cooler and fluid lines must be flushed to remove metal particles and other debris before installing the replacement unit. The torque converter should also be flushed or replaced. Neglecting to clean the complete transaxle fluid cooling system is a primary cause of repeat transaxle failures.

26. The transaxle oil cooler lines should be flushed with a converter flush kit J 35944 or equivalent.

27. Lubricate the torque converter pilot hub with chassis grease. Make sure the torque converter is properly seated in the oil pump drive. Damage to the transaxle may occur if converter is not seated completely.

28. Position the transaxle into the vehicle and guide it into place against the engine block. Make sure the alignment dowels on the block engage the openings on the front of the

transaxle case. Use care. Do not force parts in place. Light alloy components like the transaxle case are easily damaged. Double-check that the transaxle is correctly in place. Do not attempt to draw a misaligned transaxle into place by tightening the retainer bolts. This could crack the case. When satisfied with the transaxle fit, torque the transaxle-to-engine bolts to 55 ft. lbs. (75 Nm).

29. Install the transaxle mount bracket and lower the vehicle.

30. Using the engine support fixture, raise the engine into the proper position.

31. Install the transmission fluid fill tube.

32. With an assistant, position and support the frame under the vehicle. Install the frame bolts and torque to 103 ft. lbs. (140 Nm).

33. Remove the frame supports and lower the vehicle.

34. Position the engine and transaxle into the frame mounts. Remove the engine support fixture.

35. Raise the vehicle, install the torque converter-to-flywheel bolts and torque to 44 ft. lbs. (60 Nm). Install the torque converter cover.

36. Install the starter motor and transaxle ground cable.

37. Install the ball joints and tie rods.

38. Install the halfshaft (drive axle) assemblies, rotors and calipers.

39. Install the rack and pinion, lines, and heat shields.

40. Install the A/C compressor.

41. Install the engine oil filter and refill to the proper level.

42. Lower the vehicle.

43. Install the throttle cable bracket, torque strut bracket and torque strut.

44. Install the transaxle oil cooler lines to the transaxle.

45. Install the shift control and TV cables.

46. Connect the negative battery cable, install the air cleaner and recheck each operation to ensure completion of repair.

47. Adjust the shift linkage and TV cable. Check the engine and transaxle oil levels.

48. Start the engine and check for fluid leaks.

4T60 and 4T60-E 4-Speed Automatic Transaxles

1. Disconnect the negative battery cable and remove the air cleaner assembly.

2. Install the engine support fixture tool J28467 A and J 36462 or equivalent support fixtures.

3. Remove the shift control at the transaxle lever and remove the cable from its mounting bracket. Disconnect the Throttle Valve (TV) cable at the throttle linkage.

4. Remove the throttle cable bracket and brake booster hose, if required.

5. Remove the crossover pipe at the left exhaust manifold. Disconnect the EGR tube from the crossover pipe. Remove the crossover-to-exhaust pipe bolts.

6. Loosen the crossover to right exhaust manifold clamp. Swing the crossover upward to gain clearance to the top bell housing bolts.

7. Remove the four upper bell housing bolts. Leave the two lower bellhousing bolts attached for now.

8. Remove the TCC electrical connector, neutral start switch electrical connector and vacuum modulator hose at the transaxle.

9. Raise and safely support the vehicle with properly positioned safety stands.

10. Disconnect the vehicle speed sensor electrical connector.

11. Remove the front wheels, wheelhouse splash shields, calipers and rotors. Support the caliper to the frame with wire.

12. Some vehicles will require disconnecting the power steering cooler hoses from the frame, if equipped.

13. Some vehicles will require removing the power steering rack and pinion heat shield, if equipped.

14. Remove the rack and pinion assembly and wire to the exhaust for support.

15. Disconnect both ball joints and tie rod ends from the strut assemblies.

16. Remove the front halfshafts (axle) assemblies.

17. Remove the transaxle mount upper retaining nuts.

18. Remove the engine mount lower retaining nuts.

19. Support the frame with properly positioned safety stands at each end and remove the frame bolts and frame.

20. Remove the torque converter cover and the torque converter to flywheel bolts.

21. Remove the transaxle oil cooler lines and support bracket.

22. Remove the starter bolts and support out of the way.

23. Remove the ground cable from the transaxle case.

24. Support the transaxle with safety stands.

25. Remove the fluid fill tube bolt and transaxle mount bracket.

26. Remove the remaining transaxle-to-engine bolts and transaxle.

To install:

NOTE: If the transaxle was removed due to failure, the transmission fluid cooler and fluid lines must be flushed to remove metal particles and other debris before installing the replacement unit. The torque converter should also be flushed or replaced. Neglecting to clean the complete transaxle fluid cooling system is a primary cause of repeat transaxle failures.

27. The transaxle oil cooler lines should be flushed with a converter flush kit J 35944 or equivalent.

28. Lubricate the torque converter pilot hub with chassis grease. Make sure the torque converter is properly seated in the oil pump drive. Damage to the transaxle may occur if converter is not seated completely.

29. Position the transaxle into the vehicle and guide it into place against the engine block. Make sure the alignment dowels on the block engage the openings on the front of the transaxle case. Use care. Do not force parts in place. Light alloy components like the transaxle case are easily damaged. Double-check that the transaxle is correctly in place. Do not attempt to draw a misaligned transaxle into place by tightening the retainer bolts. This could crack the case. When satisfied with the transaxle fit, torque the transaxle-to-engine bolts to 55 ft. lbs. (75 Nm).

30. Install the engine to transaxle support bracket at the transaxle.

31. Install the torque converter-to-flywheel bolts and torque to 44 ft. lbs. (60 Nm).

32. Install the starter motor assembly.

33. Install the halfshafts (drive axles).

34. Install the torque converter cover.

35. Connect the oil cooler lines and support bracket.

36. Install the frame assembly and retaining bolts. Use new bolts. Torque the bolts to 103 ft. lbs. (140 Nm).

37. Install the lower engine mount retaining nuts and upper transaxle mount retaining nuts.

38. Install the ball joints-to-steering knuckle and rack and pinion assembly to the frame.

39. Install the power steering heat shield and cooler lines to frame.

40. Install the wheel house splash shields.

41. Install the rotor, calipers and front wheels.

42. Connect the vehicle speed sensor and lower the vehicle.

43. Connect the back-up/neutral safety switch.

44. Connect the vacuum modulator hose and TCC electrical connector.

45. Install the four upper bell housing bolts and torque to 55 ft. lbs. (75 Nm).

46. Install the crossover pipe to its proper position.

47. Connect the crossover pipe to right and left manifolds.

48. Connect the EGR tube to crossover.

49. Connect the TV cable at the throttle linkage and the shift cable at the transaxle.

50. Remove the engine support fixture J28467 A and J 36462.

51. Connect the negative battery cable and install the air cleaner.

52. Fill the transaxle with the proper type and quantity of transmission fluid. Adjust the TV and shift cables as required.

53. Recheck all procedures for completion of repair. Start the engine and inspect for fluid leaks.

1995–97 Vehicles

3T40 3-Speed Automatic Transaxles

1. Remove the air cleaner assembly.

2. Disconnect the negative battery cable.

3. Remove the coolant recovery reservoir.

4. Remove the shift control and T.V. cables at the transaxle.

5. Remove the throttle cable bracket and brake booster hose if equipped.

6. Remove both torque struts at the engine.

7. Remove the left torque strut bracket.

8. Remove the transaxle oil cooler lines at the transaxle and plug lines.

9. Install engine support special tools J 28467–A, J 28467 90 and J 36462 or equivalent.

10. Safely raise the vehicle.

11. Remove the tire and wheel assembly.

12. Remove the caliper bracket assemblies and rotors.

13. Remove both lower engine splash shields.

14. Remove the axle assemblies.

15. Remove the tie rods and ball joints.

16. Remove the rack and pinion heat shield and electrical connector.

17. Remove the bolts holding the main engine harness to the transaxle case.

18. Wire the rack and pinion assembly to the exhaust and remove the rack and pinion bolts from the frame.

19. Remove the bolts holding the power steering lines to the frame.

20. Remove the engine and transaxle mounts from the frame.

21. Support the frame with jackstands at each end.

22. Remove the frame bolts.

23. With an assistant, remove the frame and jackstands.

24. Remove the flywheel cover.

25. Remove the torque converter bolts.

26. Remove the starter and hang with mechanics wire.

27. Remove the ground cable at the transaxle.

28. Remove the transaxle fill tube bolt.

29. Remove the transaxle mount bracket.

30. Remove the transaxle to engine brace.

31. Safely lower the vehicle.

32. Remove the electrical connector to the transaxle.

33. Remove the transaxle fill tube.

34. Using special tool J 28467–A or equivalent and lower left side of engine approximately four inches.

35. Safely raise the vehicle.

36. Remove the fuel line bracket from the transaxle.

37. Remove the transaxle to engine bolts.

38. Remove the transaxle from the vehicle.

To install:

39. Place a small wipe of light grease on the torque converter pilot hub.

40. Position the transaxle in the vehicle.

41. Install the transaxle to engine bolts and tighten to 55 ft. lbs. (75 Nm). Remove the transaxle jack.

42. Install the fuel line bracket to the transaxle.

43. Safely lower the vehicle.

44. Using special tool J 28467–A or equivalent, raise the engine to its proper position.

45. Install the transaxle oil fill tube.

46. Connect the electrical connector to the transaxle.

47. Safely raise the vehicle.

48. Install and tighten the transaxle engine brace to 35 ft. lbs. (47 Nm).

49. Install the transaxle mount bracket.

50. Install the transaxle oil fill tube bolt.

51. With an assistant, position and support the frame under the vehicle.

52. Install new frame to body bolts and tighten bolts to 125 ft. lbs. (170 Nm).

53. Remove the frame supports.

54. Safely lower the vehicle and position engine and transaxle mount to the frame.

55. Safely raise the vehicle.

56. Install and tighten the torque converter to flywheel bolts to 46 ft. lbs. (63 Nm).

57. Install the flywheel cover.

58. Install the starter assembly.

59. Install the ground cable to the transaxle.

60. Install the ball joints and tie rod rods.

61. Install the axle assemblies.

62. Install the rotors and caliper bracket assemblies.

63. Install the rack and pinion with attaching lines to the frame.

64. Install the rack and pinion electrical connector and heat shields.

65. Install both lower engine splash shields.

66. Install the main engine harness to the transaxle case.

67. Install the tire and wheel assemblies.

68. Safely lower the vehicle.

69. Remove the engine support tools.

70. Install the throttle cable bracket and brake booster line, if equipped.

71. Install the torque strut bracket.

72. Install the transaxle oil cooler lines to the transaxle.

73. Install the torque struts.

74. Install the shift control and T.V. cables to the transaxle.

75. Install the coolant recovery reservoir.

76. Connect the negative battery cable.

77. Install the air cleaner assembly.

78. Adjust the shift linkage and T.V. cables.

79. Start engine and check engine and transaxle oil level. Add oil as necessary.

4T60-E 4-Speed Automatic Transaxles

1. Remove the hood assembly, scribe a mark at the hinge area for installation reference.

2. Remove the transaxle fluid level indicator (dipstick) assembly.

3. Remove the engine mount strut brackets.

4. Remove the electric engine cooling fan assemblies.

5. Install engine support tools J 28467–A, J 28467–90 and J 36462 or the equivalent.

6. Remove the air intake duct assembly.

7. Remove the automatic transaxle vacuum modulator pipe assembly.

8. Disconnect the electrical connections from the transaxle assembly.

9. Remove the upper transaxle to engine bolts.

10. Safely raise and support the vehicle.

11. Remove the tire and wheel assembly. Mark the position of the wheel to the wheel studs, prior to removal, for installation reference.

12. Remove the left front bumper fascia splash shield assembly.

13. Remove the front exhaust manifold pipe.

14. Remove the steering gear heat shield assembly.

15. Remove the steering gear bolts.

16. Remove the front lower control arm assemblies from the front suspension strut assemblies.

17. Remove the power steering fluid cooling pipe assembly.

18. Safely support the drivetrain and front suspension frame assembly with safety stands.

19. Remove the transaxle mount side bracket nuts.

20. Remove the engine mount frame side nuts.

21. Remove the transaxle brace.

22. Remove the drivetrain and front suspension frame assembly.

23. Disconnect the electrical connector from the vehicle speed sensor.

24. Remove the front wheel driveshaft assemblies from the transaxle.

25. Remove the transaxle converter cover assembly.

26. Remove the starter assembly.

27. Remove the transaxle torque converter bolts.

28. Remove the transaxle oil cooler upper and lower pipe assemblies.

29. Install a transaxle jack.

30. Remove the transaxle mount assembly from the engine assembly.

31. Remove the transaxle fluid filler tube assembly.

32. Remove the transaxle assembly from the vehicle.

33. Remove the transaxle torque converter assembly from the transaxle assembly.

34. Remove the transaxle assembly from the transaxle jack.

To install:

35. Install the transaxle assembly to the transaxle jack.

36. Install the transaxle torque converter assembly to the transaxle assembly.

37. Install the transaxle assembly to the engine assembly in the vehicle.

38. Install and tighten the lower rear transaxle to engine bolts to 55 ft. lbs. (75 Nm).

39. Install the transaxle fluid filler tube assembly.

40. Install the transaxle mount to engine assembly and tighten bolts to 43 ft. lbs. (58 Nm).

41. Remove the safety stand from the transaxle assembly.

42. Install the transaxle oil cooler upper and lower pipe assemblies.

43. Install and tighten the transaxle torque converter bolts to 46 ft. lbs. (63 Nm).

44. Install the starter motor assembly.

45. Install the transaxle converter cover assembly.

46. Install and tighten the transaxle converter cover bolts to 89 inch lbs. (10 Nm).

47. Install the front wheel drive shaft assemblies to the transaxle.

48. Install and tighten the drivetrain and front suspension frame bolts to 107 ft. lbs. (145 Nm).

49. Install the transaxle brace and bracket. Tighten the attaching bolts to 39 ft. lbs. (53 Nm).

50. Remove the safety stands from the drivetrain and front suspension frame assembly.

51. Install the power steering fluid cooling pipe assembly.

52. Install the lower control arm assemblies to the front suspension strut assemblies.

53. Install the steering gear bolts and heat shield assembly.

54. Install the front exhaust manifold pipe to the engine.

55. Instal the left front bumper fascia splash shield assembly.

56. Install the tire and wheel assembly, aligning the marks made on the removal.

57. Safely lower the vehicle.

58. Install and tighten the upper transaxle to engine bolts to 55 ft. lbs. (75 Nm).

59. Install the transaxle range selector assembly to the transaxle assembly.

60. Connect the electrical connectors to the transaxle assembly.

61. Install the automatic transaxle vacuum modulator pipe assembly.

62. Install the air intake duct assembly.

63. Remove the engine support fixture tools.

64. Install the engine cooling fan assemblies.

65. Install the engine mount strut brackets.

66. Install the transaxle fluid level indicator assembly.

67. Install the hood panel assembly.

68. Start the engine and check for leaks.

69. Check the fluid level in the transaxle and adjust the shift linkage.

70. Road test to verify proper transaxle shifting and smooth operation.

Throttle Valve Cable

ADJUSTMENT

1. Verify that the T.V. cable is in final routed position when the T.V. cable adjustment is performed.

2. After installation of the cable to the transaxle. engine bracket and throttle lever, check that the cable slider is in the zero or fully readjusted position. If not, the cable must be adjusted.

3. Rotate the throttle lever to the "WOT STOP" position until it reaches 62 inch lbs. (7 Nm). During adjustment, listen for several clicks of the cable adjuster.

WARNING

Do NOT use the accelerator pedal to adjust the T.V. cable, improper adjustment may result.

---

4. In case readjustment is necessary due to adjustment before or during assembly or for reprocessing, perform the following:

a. Depress and hold the readjust tab.

b. Move the slider back and forth through the fitting in direction away from the throttle lever until the slider stops against the fitting.

c. Release the readjust tab.

d. Repeat Step 3 of the procedure.

DRIVE AXLE

Halfshaft

REMOVAL AND INSTALLATION

NOTE: If vehicle is equipped with ABS brakes, use care to avoid damage to the ABS exciter ring. Damage to the ring may cause the self-diagnostic feature of the ABS system to store a system fault code.

1. Disconnect the negative battery cable.

2. Raise and safely support the vehicle. Remove the wheel and tire assembly.

3. Remove the front wheel drive axle nut.

4. Remove the brake calipers, bracket assemblies and hang caliper with mechanic's wire from the strut.

5. Remove the brake rotors.

6. Remove the four hub/bearing retaining bolts and hub.

7. Remove the ABS sensor mounting bolt and position the sensor out of the way, if equipped with ABS brakes.

8. Place a drain pan under the transaxle.

NOTE: Use care when removing the halfshaft. Tri-pot joints can be damaged if the drive axle is over-extended. It is important to handle the halfshaft in a manner to prevent over-extending.

9. Remove the halfshaft from the vehicle using the appropriate procedure for each side and transmission model:

a. To remove the right side halfshaft, use special tool J 33008, J 29794 and J 2619–01 or equivalent. Separate the halfshaft from the transaxle.

b. To remove the left side halfshaft on a 3T40 transaxle, use special tools J 33008, J 29794 and J 2619–01 or equivalent. Separate the axle from the transaxle.

c. To remove the left side halfshaft, using the frame for leverage, separate the halfshaft from the transaxle with a suitable prying tool in the groove provided on the inner joint.

10. Remove the halfshaft/bearing assembly through the knuckle.

To install:

11. Install the halfshaft/bearing assembly through the knuckle and into the transaxle. Remove special tool J 37292–A or equivalent and discard.

12. Properly position the ABS sensor and install the mounting bolt, if removed.

13. Loosely secure the bearing to the knuckle bolts.

14. Seat the halfshaft into the transaxle, using a suitable prying tool in the groove provided on the inner joint. Carefully pry against the frame or the lower control arm to seat the halfshaft.

15. Verify that the snapring is seated by tapping on the inner groove with a suitable prying tool. Grasp the inner housing of the axle shaft and pull outboard. Do not pull on the axle shaft. If the snapring is properly seated, the axle will remain in place.

16. Install the hub and bearing assembly to the axle with a new nut and washer. Tighten attaching bolts to 60 ft. lbs. (80 Nm).

17. Install the brake rotor.

18. Install the brake caliper and attaching bracket. and tighten caliper slide bolts to 80 ft. lbs. (108 Nm).

19. Install the wheel and tire assembly.

20. Safely lower the vehicle.

NOTE: Do not reuse the front wheel drive axle nut. Always use a new front wheel drive axle nut of similar design when installing the nut. Do not use a Nylock or free spinning style nut.

21. Install the front wheel drive axle nut and torque to 150 ft. lbs. (205 Nm).

22. Connect the negative battery cable.

CV-Joint Boot

REPLACEMENT

Outer CV-Joint Boot

1. Raise and safely support vehicle.

2. Remove the front tire and wheel.

3. Remove the front halfshaft and put in a vise using a protective covering on the vise jaws.

4. For swage ring (inner clamp ring) removal, use a hand grinder at cut through the ring with diagonal cutting pliers, taking care not to damage the axle shaft.

5. Slide the boot down the shaft uncovering the outer joint.

6. Clean the grease away from the joint to locate the snapring.

7. Using J 8059 or equivalent snapring pliers, open the snapring and slide outer joint off of shaft.

8. Remove the outer boot from shaft.

9. Clean the outer joint completely and inspect for damage to the cage or balls. Wipe any grease off of axle shaft.

To install:

10. Drop the small clamp on to halfshaft and push the boot down several inches past the seal mounting area.

11. Check the condition of the snapring in the outer joint, replace as necessary. Pack the outer joint with half of the grease supplied in the boot kit and install on the end of the halfshaft.

12. Gently pull down on the joint until the splines engage. With a brass drift lightly tap the joint down until the snapring engages.

13. Pack the remaining grease in the boot kit in the boot and then pull the boot up over the end of the joint. Seat the small end on the seal mounting area.

NOTE: Depending on the boot being installed, the following procedure may vary. Different aftermarket kits may use different types of clamp rings.

14. For 1993–94 vehicles, slide the small clamp into position and using J 35910 or equivalent, crimp the clamp to 130 ft. lbs. (176 Nm). Install the large clamp to the proper position and crimp to 130 ft. lbs. (176 Nm).

─────── CAUTION ───────
The boot must not be dimpled, stretched out or out of shape in any way. If the boot is not shaped correctly, carefully insert a thin flat blunt tool at the large end of the boot to equalize pressure. Shape the boot properly by hand, then remove the tool.

15. For 1995–97 vehicles, the factory recommends their J 41048 tool for inner clamp ring installation. This device mounts is a bench vise and uses two large steel blocks, tightened by two large bolts, to squeeze, or swage the clamp into place around the halfshaft. If available, mount the swage clamp tool, J 41048 or equivalent in a vise and proceed as follows:

   a. Position the outboard end of the halfshaft assembly in the special tool.

   b. Place the top half of special tool J 41048 or equivalent on the lower half of the tool and check fro proper alignment and dimension.

   c. Insert the bolts and tighten by hand until snug.

   d. Make sure that the seal, housing and swage ring all remain in alignment.

   e. Continue to tighten each bolt 180 degrees at a time, alternating until both sides are bottomed.

─────── CAUTION ───────
The boot must not be dimpled, stretched or out of shape in any way. If boot is not shaped correctly, carefully insert a thin flat blunt tool at the large end of the boot to equalize pressure. Shape the boot properly by hand and then remove the tool.

16. Install the halfshaft (a new hub nut must be used), torque the wheels to specification and road test.

Inner CV-Joint Boot

1. Raise and support vehicle. Remove the front tire and wheel.

2. Remove the front halfshaft and put in a vise using a protective covering on the vise jaws.

3. Remove the outer CV-Joint and boot.

4. Cut the large and small boot clamps and discard.

─────── CAUTION ───────
Do not cut through the boot and damage the sealing surface of the the tripot outer housing.

5. Separate boot from the trilobal tripot bushing at large diameter end and slide boot away from joint along the axle.

6. Remove the tripot housing from the spider and axle. Clean thoroughly and set aside.

7. Clean the grease from the spider assembly to expose the spacer ring located on the outboard side of the spider assembly. Using J 8059 or equivalent snapring pliers, slide the spacer ring down the half shaft.

8. Push the spider assembly down the shaft to uncover the snapring on the end of the shaft. Using J 8059 or equivalent snapring pliers, remove the snapring. Slide the spider assembly off the end of the shaft. Clean completely, using care not to knock the caps off of the bearings. Set aside.

9. Remove the trilobal pot bushing from the tripot housing.

10. Remove the spacer ring from the halfshaft and slide the boot off. Clean any grease off the shaft.

To install:

NOTE: Depending on the boot being installed, the following procedure may vary. Different aftermarket kits may use different types of clamp rings.

11. Drop the small clamp on to the halfshaft.

12. Slide the boot onto shaft and position neck of boot in groove on axle shaft.

13. Crimp seal retaining clamp with J-35910 or equivalent to 100 ft. lbs. (136 Nm).

14. Put the spacer ring on the shaft, several inches below the second spacer ring groove.

15. Install the spider assembly far enough down the shaft to expose the top snapring groove. Make sure that the counterbored face of the tripot spider faces the end of the shaft.

16. Install the top snapring and pull the spider assembly back up into position.

17. Using J 8059 or equivalent snapring pliers, lock the spacer ring in the spacer ring groove.

18. Pack the housing with half the grease supplied and put the rest in the boot.

19. Install the trilobal tripot bushing into the tripot housing.

20. Slide the larger clamp over the boot.

21. Push the tripot housing over the spider assembly.

22. Slide the larger diameter of the boot and clamp in place over the outside of the trilobal bushing and locate lip of the boot in the bushing groove.

23. Position the tripot assembly properly and install the large clamp in position. Using J 35910 or equivalent, tighten clamp.

─────── CAUTION ───────
The boot must not be dimpled, stretched or out of shape in any way. If boot is not shaped correctly, carefully insert a thin flat blunt tool at the large end of the boot to equalize pressure. Shape the boot properly by hand and then remove the tool.

24. Install the outer CV-Joint and boot.

25. Install halfshaft (a new hub nut must be used), torque wheels to specification and road test.

STEERING

Air Bag

─────── CAUTION ───────
Some vehicles are equipped with an air bag system, also known as the Supplemental Inflatable Restraint (SIR) or Supplemental Restraint System (SRS). The system must be disabled before performing service on or around system components, steering column, instrument panel components, wiring and sensors. Failure to follow safety and disabling procedures could result in accidental air bag deployment, possible personal injury and unnecessary system repairs.

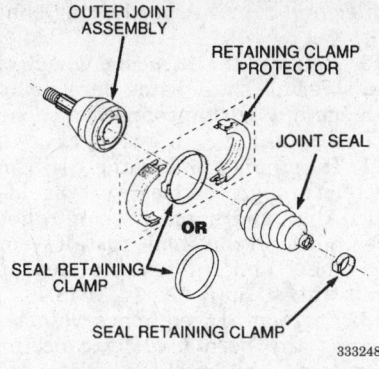

Outer CV-Joint disassembly

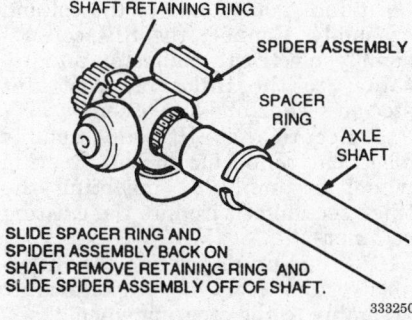

Outer CV-Joint snapring and spider assembly removal

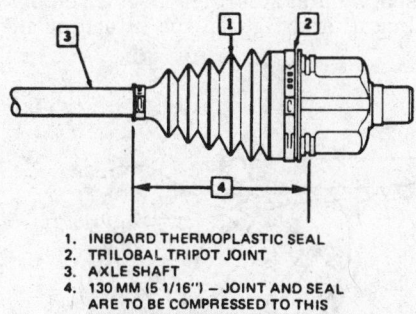

1. INBOARD THERMOPLASTIC SEAL
2. TRILOBAL TRIPOT JOINT
3. AXLE SHAFT
4. 130 MM (5 1/16") — JOINT AND SEAL ARE TO BE COMPRESSED TO THIS DIMENSION BEFORE CRIMPING CLAMPS.

333251

Inner boot collapsed dimension

PRECAUTIONS

Several precautions must be observed when handling the inflator module to avoid accidental deployment and possible personal injury.

• Never carry the inflator module by the wires or connector on the underside of the module.

• When carrying a live inflator module, hold securely with both hands, and ensure that the bag and trim cover are pointed away.

• Place the inflator module on a bench or other surface with the bag and trim cover facing up.

• With the inflator module on the bench, never place anything on or close to the module which may be thrown in the event of an accidental deployment.

DISARMING

NOTE: With the "AIR BAG" fuse removed, if for some reason the ignition switch should be switched to the ON position, the INFLATABLE RESTRAINT warning lamp will be ON. This is normal operation and does not indicate a SIR system malfunction.

1. Disconnect the negative battery cable.
2. Turn the steering wheel so that the vehicle's wheels are pointing straight ahead.
3. Turn the ignition switch to the **LOCK** position and remove the key.
4. Remove the **AIR BAG** fuse from the fuse block.
5. Remove the left sound insulator.
6. Remove the Connector Position Assurance (CPA) clip from the yellow 2-way connector at the base of the steering column, and disconnect the connector. If equipped with a passenger's side air bag, remove the CPA and disconnect the yellow 2-way connector from the passenger air bag lead.

ENABLING

1. Turn the ignition switch to the **LOCK** position and remove the key.
2. Connect the yellow 2-way connector at the base of steering column and secure it with the CPA clip. If equipped with a passenger's side air bag, connect the yellow 2-way connector at the passenger air bag lead and secure it with the CPA clip.
3. Install the left sound insulator.

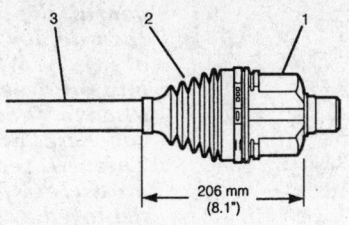

1 Retainer & housing asm
2 Drive axle inboard seal
3 Axle shaft

197744

Outboard CV-Joint boot installation measurement

4. Install the **AIR BAG** fuse in the fuse block.
5. Turn the ignition switch to the **RUN** position and verify that the **AIR BAG** warning lamp flashes 7 times and then turns off.
6. Connect the negative battery cable.

Steering Wheel

REMOVAL AND INSTALLATION

——— **CAUTION** ———

The Supplemental Inflatable Restraint (SIR) system must be disarmed before removing the steering wheel. Failure to do so may cause accidental deployment of the air bag, resulting in unnecessary SIR system repairs and/or personal injury.

NOTE: The vehicle's wheels must be in the straight ahead position and the key must be in the LOCK position when removing or installing the steering column. Failure to do so may cause the SIR coil assembly to become uncentered and may result in unneeded SIR system repairs. In addition, the steering column should never be supported by only the lower or upper support bracket alone.

1. If equipped, disable the SIR system using the recommended procedure.
2. Turn the ignition key to the **OFF** position. Disconnect the negative battery cable.
3. For non-SIR equipped vehicles, remove the horn pad. Gently pull up on horn pad to remove the horn pad.
4. For non-SIR equipped vehicles, disconnect the horn lead. Gently push down on the horn lead and turn the connector to the left and remove.

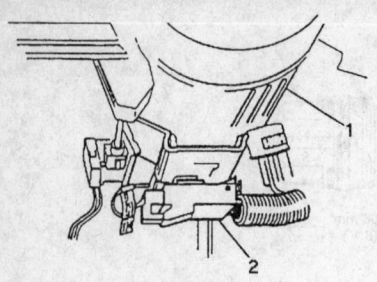

1. Steering column
2. Sir (yellow) connector

215898

Driver's side SIR connection

The wire and spring will then come out of the canceling cam tower. Remove the steering wheel retainer.

5. Loosen the inflator module attaching bolts from the back of the steering wheel assembly using No. 30 Torx™ bit until the inflatable restraint steering wheel module assembly (air bag) can be released from the steering wheel assembly.

6. Remove the inflatable restraint steering wheel module assembly from the steering wheel assembly.

CAUTION

When carrying a live air bag, make sure the bag and trim cover are pointed away from the body.

In the unlikely event of an accidental deployment, the bag will then deploy with minimal chance of injury. When placing a live air bag on a bench or other surface, always face the bag and trim cover up, away from the surface. This will reduce the motion of the module if it is accidently deployed.

---

7. Scribe an alignment on the steering mark on the steering wheel hub in line with slash mark on the steering shaft.

8. Loosen, but do not remove the steering wheel nut, positioning it flush with end of the shaft. This should protect the threaded end of the shaft when the steering wheel puller is used.

NOTE: While attaching J-1859-03 to the steering wheel, use care to prevent threading the bolts all the way through the steering wheel hub into the coil assembly and damaging the coil assembly.

9. Assemble J 1859-03 or equivalent steering wheel puller on the steering wheel hub with two $5/16$ inch bolts. Break the wheel loose from the steering shaft. Remove the

puller and remove the steering shaft nut.

10. Remove the steering wheel off the steering shaft using care not to damage any wiring connections.

   To install:

11. If equipped, route the SIR connector through the steering wheel.

12. Align mark on steering wheel with mark on the shaft. Install steering wheel and tighten the steering shaft nut to 30 ft. lbs. (41 Nm).

13. For non-SIR equipped vehicles, connect the horn lead by canceling cam tower by pushing down and turning right into the lock position.

14. For SIR vehicles, connect the horn and ground lead to the column assembly. Connect the SIR coil assembly electrical connector and retainer to the inflatable restraint steering wheel module assembly.

15. Secure the SIR coil assembly electrical connector to the steering wheel assembly by inserting the thick section of wire into the existing retainers.

16. If equipped, position the inflatable restraint steering wheel module assembly to the steering wheel.

17. If equipped, ensure the wiring is not exposed or trapped between the steering wheel module assembly and steering wheel assembly. Install and tighten the attaching in-

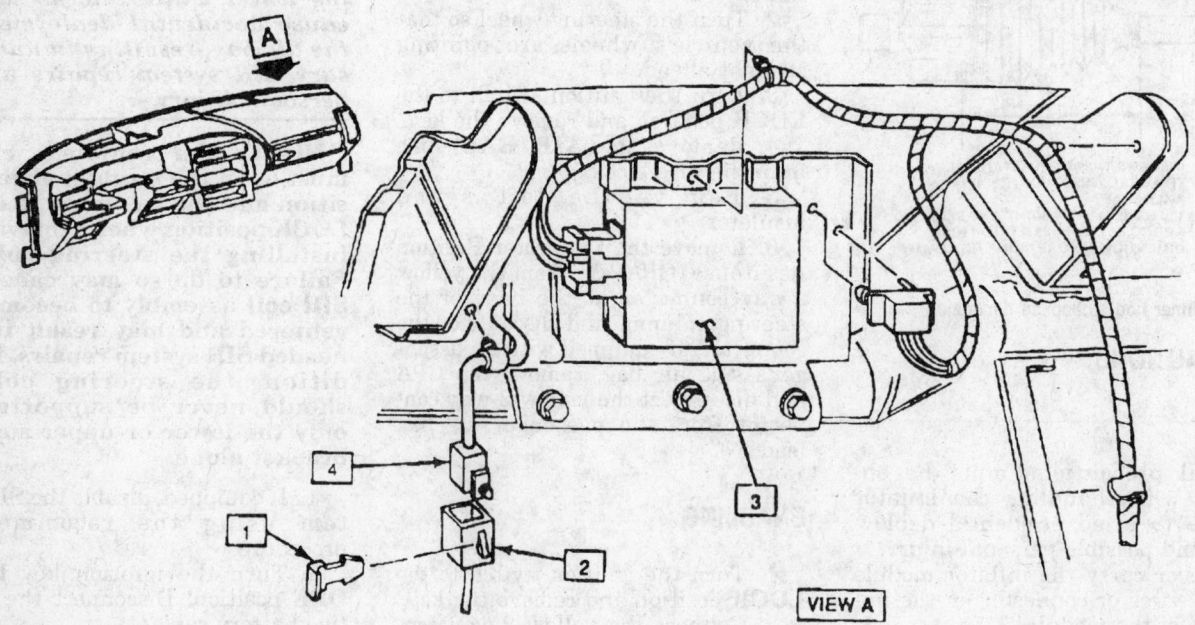

1	CONNECTOR POSITION ASSURANCE (CPA)
2	YELLOW 2-WAY CONNECTOR, SIR HARNESS
3	VEHICLE ANTI-THEFT MODULE
4	YELLOW 2-WAY CONNECTOR, INFLATOR MODULE

215899

Passenger side SIR connection

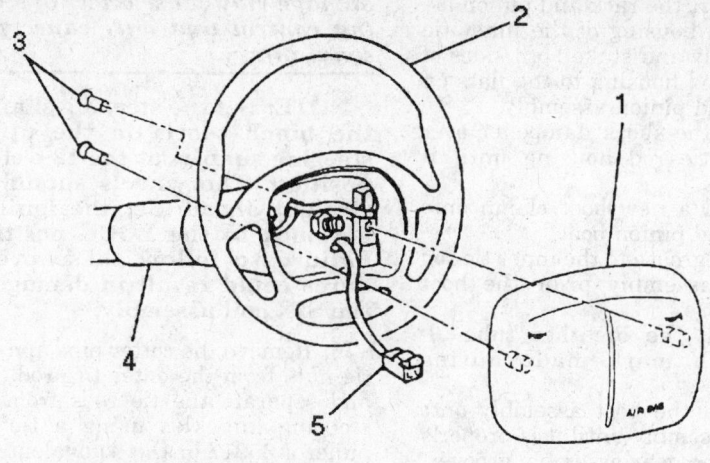

1. Inflatable restraint steering wheel module assembly
2. Steering wheel assembly
3. Module assembly bolt/screw
4. Steering column assembly
5. Sir coil assembly electrical connector
6. Position assurance (CPA) connector
7. Coil wire connector

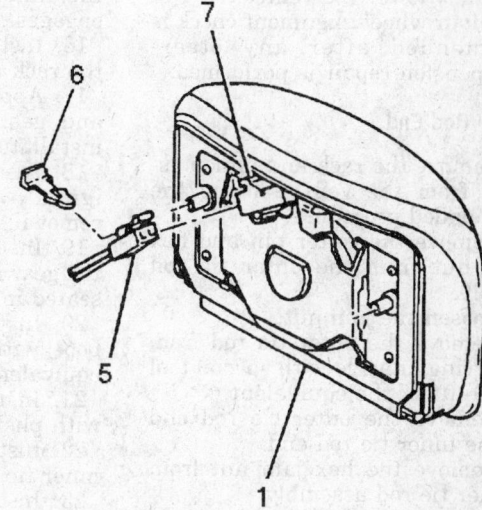

332199

Removing the inflator module

flator module screws to 25 inch lbs. (2.8 Nm).

18. For non-SIR equipped vehicles, install the horn pad and connect the steering wheel control switch connector.

19. Connect the negative battery cable. If equipped, enable the SIR system.

20. Verify correct operation of the horn and steering wheel controls, if equipped.

Tie Rod Ends

REMOVAL AND INSTALLATION

1993–94 Vehicles

1. Disconnect the negative battery cable.

2. Raise and safely support the vehicle. On some models it may be advantageous to remove the front wheel/tire assemblies.

3. Depending on the vehicle, it may be helpful to wire brush the tie rod threads. Use eye protection. Liberally coat all parts with penetrating oil. Loosen the jam nut on the inner tie rod.

4. Inspect the outer tie rod end. Most vehicles use a castellated nut and a cotter pin. Pull out and discard

the cotter pin and remove the nut. Remove the tie rod from the steering knuckle using a steering linkage removing tool J 24319-01 or equivalent. Do not use a wedge-type tool or hammer on the joint. The steering rack can be damaged.

5. Holding the inner tie rod stationary, remove the outer tie rod counting the number of turns so that the replacement may be installed close to the original's location. This speeds toe-in checking.

To install:

6. Lubricate the inner tie rod threads with anti-seize compound and install the outer tie rod the same amount of turns that it took to remove. This should approximate the original toe-in setting. Do not tighten the inner tie rod jam nut until the toe-in setting can be checked and corrected as necessary.

7. Install the outer tie rod end to the steering knuckle and install the nut. Install the castellated nut and torque the nut to 40 ft. lbs. (54 Nm). Tighten slowly to align the cotter pin slot, using care not to exceed 45 ft. lbs. Do NOT back off to align the cotter pin openings. Always use a new cotter pin.

8. Check the toe-in. Adjust if necessary, using care not to twist the rack and pinion rubber boot. When

the adjustment is satisfactory, torque the jam nut to 50 ft. lbs. (70 Nm).

9. Double check that the cotter pin is properly installed and the jam nut torqued. Many replacement tie rod ends come with grease fittings. If so, lubricate with quality chassis grease. Lower the vehicle and connect the negative battery cable.

10. Road test to verify correct toe-in angle. If necessary, check front alignment and adjust toe.

1995–97 Vehicles

Outer Tie Rod End

1. Safely raise and support the vehicle.

2. Remove the cotter pin and hex slotted nut from the outer tie rod assembly.

3. Loosen the jam nut.

4. Remove the outer tie rod from the steering knuckle with special tool J 34319-01 or equivalent.

5. Remove the outer tie rod end from the inner tie rod end.

To install:

6. Install the outer tie rod assembly to the inner tie rod and do not tighten the jam nut.

7. Install the outer tie rod to the steering knuckle. Tighten the hex slotted nut to the outer tie rod stud and tighten nut to 35 ft. lbs. (47 Nm).

Tighten nut up to ⅙ turn additional to a maximum to 52 ft. lbs. (70 Nm).

8. Install the cotter pin into hole in the tie rod strut.

9. Tighten the outer tie rod jam nut to 50 ft. lbs. (68 Nm).

10. Safely lower the vehicle.

11. A four wheel alignment check is recommended after any steering/suspension repair is performed.

Inner Tie Rod End

1. Remove the rack and pinion assembly from the vehicle using the recommended procedure.

2. Remove the cotter pin and hex slotted nut from the outer tie rod assembly.

3. Loosen the jam nut.

4. Remove the outer tie rod from the steering knuckle with special tool J 34319- 01 or the equivalent.

5. Remove the outer tie rod end from the inner tie rod end.

6. Remove the hex jam nut from the inner tie rod assembly.

7. Remove the tie rod end clamp.

8. Remove the boot clamp with side cutters and discard.

NOTE: Mark location of the breather tube on the rack and pinion assembly before removing the breather tube and rack and pinion boot.

9. Remove the rack and pinion boot and breather tube.

10. Remove the shock dampener from the inner tie rod assembly and slide back onto the rack assembly.

NOTE: The steering rack must be held during removal of the inner tie rod to prevent steering rack damage.

11. Remove the inner tie rod assembly from the rack and pinion assembly as follows:

  a. Place a wrench on flat of the rack and pinion assembly.

  b. Place another wrench on flats of the inner tie rod housing.

  c. Rotate the inner tie rod housing counterclockwise until inner tie rod separates from the rack and pinion assembly.

To install:

12. Install the shock dampener onto the rack and pinion assembly.

13. Install the inner tie rod on the rack and pinion assembly as follows:

  a. Place a wrench on the flat of the rack and pinion assembly.

  b. Place another wrench on flats of the inner tie rod housing.

  c. Tighten the inner tie rod end to 74 ft. lbs. (100 Nm).

14. Support the rack and pinion assembly and housing of the inner tie rod assembly and stake both sides of the inner rod housing to the flats on the rack and pinion assembly.

15. Slide the shock dampener over the inner tie rod housing until it engages.

16. Install a new boot clamp onto the rack and pinion boot.

17. Apply grease to the inner tie rod and gear assembly prior the boot installation.

18. Install the breather tube aligned with mark made during removal.

19. Install the boot assembly onto the gear assembly until it is properly seated in the gear assembly groove.

20. Install the boot clamp on the boot with special tool J 22610 or equivalent and crimp.

21. Install the tie rod end clamp with pliers on the boot.

22. Install the hex jam nut to the inner tie rod assembly.

23. Install the outer tie rod assembly to the inner tie rod and do not tighten the jam nut.

24. Install the outer tie rod to the steering knuckle. Tighten the hex slotted nut to the outer tie rod stud and tighten nut to 35 ft. lbs. (47 Nm). Tighten nut up to ⅙ turn additional to a maximum to 52 ft. lbs. (70 Nm).

25. Install the cotter pin into hole in the tie rod strut.

26. Tighten the outer tie rod jam nut to 50 ft. lbs. (68 Nm).

27. Safely lower the vehicle.

28. A four wheel alignment check is recommended after any steering/suspension repair is performed.

Power Rack and Pinion

REMOVAL AND INSTALLATION

Except 3.4L (VIN X) Engine

1. Disconnect the negative battery cable.

2. Raise and safely support the vehicle.

3. Remove the front wheels.

4. Remove the electrical connection from the steering gear pressure switch.

5. Remove the intermediate shaft pinch bolt at the steering gear and disconnect the intermediate shaft from the rack an pinion unit.

———— CAUTION ————
Failure to disconnect the intermediate shaft from the rack and pinion stub shaft may result in damage to the steering gear. This damage may cause a loss of steering control and may cause personal injury.

NOTE: Set the steering shaft so the block tooth on the upper steering shaft is at the 12 o'clock position. The wheels should be straight ahead. Set the ignition key lock to the LOCK position. Failure to follow these procedures could result in damage to the SIR coil assembly.

6. Remove the cotter pins and castle nuts from the outer tire rod ends and separate the tie rods from the steering knuckles using a tie rod puller J 35917 or the equivalent.

7. Support the rear of the subframe with a suitable adjustable jack.

NOTE: DO NOT lower the frame too far. Engine components near the firewall may be damaged.

8. Remove the rear frame bolts and lower the rear of the frame up to 5 inches (128mm).

9. Remove the heat shield, pipe retaining clip and the fluid pipes from the rack assembly. Use flare nut wrenches to remove the fluid pipes.

10. Remove the rack mounting bolts and nuts.

11. Remove the rack assembly out through the left wheel opening.

To install:

12. Install the rack assembly through the left wheel opening.

13. Install the mounting bolts and nuts and tighten to 59 ft. lbs. (80 Nm).

14. Connect the power steering fluid lines with new O-rings to the rack and pinion assembly. Tighten the fittings to 20 ft. lbs. (27 Nm).

15. Install the pipe retaining clips and heat shield.

16. Raise the frame and install the rear bolts and tighten to 103 ft. lbs. (140 Nm).

17. Connect the tie rod ends to the steering knuckles and tighten the castle nuts to 40 ft. lbs. (54 Nm). If necessary to align the cotter pin holes, tighten the nuts slightly until the cotter pins can be installed. NEVER loosen the nuts to align the holes.

18. Connect the intermediate shaft-to-stub shaft and tighten the lower pinch bolt to 35–40 ft. lbs. (47–54 Nm).

19. Install the front wheels and torquing to specifications.

20. Lower the vehicle.

21. Connect the negative battery cable.

22. Refill and bleed the power steering system.

NOTE: Whenever the vehicle sub-frame is removed or lowered, the wheel alignment should be checked.

23. Check the wheel alignment and adjust if required.

3.4L (VIN X) Engine

1. Disconnect the negative battery cable. Remove the air cleaner and duct assembly.
2. Install the engine support fixtures J 28467 A, J 28467 90 and J 36462 or equivalent.
3. Raise and safely support the vehicle.
4. Remove the left front wheel.
5. Loosen the right side engine splash shield.
6. Remove both left and right tie rod nuts and tie rods from the steering knuckles.
7. Remove the intermediate shaft pinch bolt at the steering gear and disconnect the intermediate shaft from the rack an pinion unit.

——————— CAUTION ———————
Failure to disconnect the intermediate shaft from the rack and pinion stub shaft may result in damage to the steering gear. This damage may cause a loss of steering control and may cause personal injury.

NOTE: Set the steering shaft so that the block tooth on the upper steering shaft is at the 12 o'clock position. The wheels should be straight ahead. Set the ignition key lock to the LOCK position. Failure to follow these procedures could result in damage to the SIR coil assembly.

8. Remove the electrical connection from the steering gear pressure switch.
9. Remove the exhaust pipe and catalytic converter assembly.
10. Support frame at the center rear using safety stands.

NOTE: DO NOT lower the frame too far. Engine components near the firewall may be damaged.

11. Remove the rear frame bolts and lower the rear of the frame up to 3 inches (76.mm).
12. Remove the heat shield, power steering line retaining clip and the fluid lines from the rack assembly. Use flare nut wrenches to remove the fluid pipes.

13. Remove the rack and pinion mounting bolts and nuts.
14. Remove the rack and pinion assembly out through the wheel opening.
15. Replace the stub shaft seals.

To install:
16. Install the rack and pinion assembly through the wheel opening.
17. Install the mounting bolts and nuts and tighten to 59 ft. lbs. (80 Nm).
18. Connect the power steering lines with new O-rings to the rack and pinion assembly. Tighten the fittings to 20 ft. lbs. (27 Nm).
19. Install the power steering line retaining clips and heat shield.
20. Raise the frame and align the steering gear stub shaft to the intermediate steering shaft. Install the rear bolts and tighten to 103 ft. lbs. (140 Nm).
21. Remove the safety stands at the frame center rear location.
22. Install the exhaust pipe and catalytic converter assembly.
23. Connect the electrical connection from the steering gear pressure switch.

——————— CAUTION ———————
When installing the intermediate shaft make sure the shaft is seated prior to the pinch bolt installation. If the pinch bolt is inserted into the coupling before the intermediate shaft installation, the two mating surfaces may disengage resulting in loss of steering.

24. Connect the intermediate shaft-to-stub shaft and tighten the lower pinch bolt to 35 ft. lbs. (47 Nm).
25. Connect the tie rod ends to the steering knuckles and tighten the castle nuts to 40 ft. lbs. (54 Nm). Additional tightening is permissible if necessary to align the cotter pin holes. NEVER loosen the nuts to align the holes.
26. Install the right side engine splash shield.
27. Install the left front wheel.
28. Lower the vehicle.
29. Remove the engine support fixtures.
30. Connect the negative battery cable. Install the air cleaner and duct assembly.
31. Refill and bleed the power steering system.

NOTE: Whenever the vehicle sub-frame is removed or lowered, the wheel alignment should be checked.

32. Check the front end alignment and adjust as necessary.

Power Steering Pump

BLEEDING

1. Fill the fluid reservoir to the **FULL COLD** mark.
2. With the engine OFF and the wheels off the ground, bleed the system by turning the wheels from side to side at least 20 times. Keep the fluid level at the **FULL COLD** mark. Continue this until the air is eliminated from the fluid.
3. Start the engine and turn the wheels lock to lock. Recheck the fluid level.
4. Return the wheels to the center position.
5. Keep the engine running for 2 minutes, turning the wheels in both directions.
6. Recheck the fluid level.

REMOVAL AND INSTALLATION

2.2L (VIN 4) Engine

1. Disconnect the negative battery cable.
2. Remove the pressure and return hoses from the pump and drain the system into a suitable container. Use care not to damage hose fittings.
3. Cap the fittings at the pump to keep out dirt.
4. Using a 12 inch (305mm) adjustable wrench, on the tensioner casting, loosen the belt tensioner. Lift the belt off the pulley.
5. Locate the three pump attaching bolts through the access hole in the pulley and remove the bolts.
6. Remove the one bolt from the rear of the pump, then remove the pump assembly.

To install:
7. Install the pump assembly onto the engine block.
8. Install the one bolt from the rear of the pump and the three pump attaching bolts. Torque the bolts to 18 ft. lbs. (25 Nm).
9. Install the belt onto the pulley and tighten belt tensioner.
10. Install the pressure and return lines into the pump.
11. Connect the negative battery cable.
12. Refill power steering pump reservoir, bleed the power steering system using the recommended procedure.

13. Start the engine and verify proper power steering assist and verify no fluid leaks.

3.1L (VIN T) Engine

1. Disconnect the negative battery cable.

2. Remove the pump drive belt.

3. Remove the power steering lines at the pump. Use care not to damage the hose fittings.

4. Remove the spark plug wire clip from the pump.

5. Remove the power steering pump bolts and pump.

To install:

6. Install the power steering pump and pump bolts. Torque bolts to 25 ft. lbs. (34 Nm).

7. Install the remaining components in the reverse order of removal.

8. Connect the negative battery cable.

9. Refill power steering pump reservoir, bleed the power steering system using the recommended procedure.

10. Start the engine and verify proper power steering assist and verify no fluid leaks.

3.1L (VIN M) Engine

1. Disconnect the negative battery cable.

2. Remove the coolant recovery reservoir.

3. Remove the serpentine drive belt at the power steering pump.

4. Remove the ignition control wiring harness near the power steering pump assembly.

5. Remove the inlet and outlet hoses from the power steering pump.

6. Remove the power steering pump attaching bolts.

7. Remove the power steering pump assembly from the vehicle

8. Remove the power steering pump pulley with special tool J 25034 B in conjunction with either J 37609 or J 37609 A or equivalent.

9. Remove the reservoir assembly from the power steering pump.

To install:

10. Install the reservoir assembly to the power steering assembly.

11. Install the power steering pulley using special tool J 25033 B or equivalent. Make sure the face of the pulley must be flush with the end of the power steering pump driveshaft.

12. Install the power steering pump onto the engine and tighten the attaching bolts to 25 ft. lbs. (34 Nm).

13. Install the remaining components in the reverse order of removal.

14. Refill power steering pump reservoir, bleed the power steering sys-

tem using the recommended procedure.

15. Start the engine and verify proper power steering assist and verify no fluid leaks.

3.4L (VIN X) Engine

1. Remove the air cleaner and duct assembly.

2. Disconnect the negative battery cable.

3. Remove the coolant recovery tank and the serpentine belt.

NOTE: Siphon the power steering fluid from the reservoir before disconnecting lines to avoid spilling fluid on the secondary timing belt cover. Use shop rags when disconnecting the lines to insure any remaining fluid does not contact the secondary timing belt cover. Power steering fluid can damage the secondary timing belt.

4. Siphon as much fluid as possible from the reservoir.

5. Remove the power steering pump from the bracket.

6. Remove the power steering lines.

To install:

7. Connect the power steering lines.

8. Install the power steering pump to the bracket and tighten the bolts to 25 ft. lbs. (34 Nm).

9. Install the remaining components in the reverse order of removal.

10. Connect the negative battery cable. Refill the power steering pump reservoir and bleed the power steering system.

11. Start the engine and verify proper power steering assist and verify no fluid leaks.

3.8L (VIN L) and (VIN K) Engine

1. Disconnect the negative battery cable.

2. Remove the coolant recovery reservoir and bracket and position aside.

3. Remove the ECM cover retaining nuts and cover.

4. Remove the serpentine belt from the pulley.

5. Remove the inlet and outlet lines at the power steering pump.

6. Remove the power steering pump mounting bolts.

7. Remove the power steering pump and reservoir.

8. Remove the reservoir from the power steering pump.

To install:

9. Install the reservoir on the power steering pump.

10. Install the power steering pump and attaching bolts to the engine. Tighten the bolts to 25 ft. lbs. (34 Nm).

11. Install the remaining components in the reverse order of removal.

12. Connect the negative battery cable.

13. Refill power steering pump reservoir, then bleed the power steering system.

14. Start the engine and verify proper power steering assist and verify no fluid leaks.

BRAKES

Anti-Lock Brake System Service

PRECAUTIONS

• Certain components within the Anti-Lock Brake System (ABS) are not intended to be serviced or repaired individually. Only those components with removal and installation procedures should be serviced.

• Do not use rubber hoses or other parts not specifically specified for and ABS system. When using repair kits, replace all parts included in the kit. Partial or incorrect repair may lead to functional problems and require the replacement of components.

• Lubricate rubber parts with clean, fresh brake fluid to ease assembly. Do not use lubricated shop air to clean parts; damage to rubber components may result.

• Use only specified brake fluid from an unopened container.

• If any hydraulic component or line is removed or replaced, it may be necessary to bleed the entire system.

• A clean repair area is essential. Always clean the reservoir and cap thoroughly before removing the cap. The slightest amount of dirt in the fluid may plug an orifice and impair the system function. Perform repairs after components have been thoroughly cleaned; use only denatured alcohol to clean components. Do not allow ABS components to come into contact with any substance containing mineral oil; this includes used shop rags.

• The Anti-Lock control unit is a microprocessor similar to other computer units in the vehicle. Ensure that the ignition switch is **OFF** before removing or installing controller harnesses. Avoid static electricity discharge at or near the controller.

• If any arc welding is to be done on the vehicle, the control unit should be unplugged before welding operations begin.

DEPRESSURIZING

With ignition **OFF** or negative cable disconnected, apply and release the service brake pedal 25 times using approximately 50 lbs on the pedal. A noticeable change in pedal feel will occur when the accumulator is completely discharged.

Master Cylinder

REMOVAL AND INSTALLATION

For vehicles with ABS, when the ABS modulator cylinder pistons are in their uppermost position, each motor has prevailing torque due to the force necessary to ensure each piston is held firmly at the top of its travel. This torque results in "gear tension," or force on each gear that makes motor pack separation difficult. To avoid injury, or damage to the gears, the "Gear Tension Relief Sequence" briefly reverses each motor to eliminate the prevailing torque. This procedure is one of the many functions of GM's Tech 1 scan tool. Use care when using a substitute. In general, make sure the ignition switch is in the **OFF** position. Install the Tech 1 or equivalent with the correct chassis cartridge. Turn the ignition switch to the **ON** position, leaving the engine OFF. Select the proper function. The "Gear Tension Relief Sequence" is F5 on the Tech 1 scan tool. Note that this same scan tool is needed to bleed the system after repairs to the hydraulic system.

For vehicles equipped with ABS, always perform the "Gear Tension Relief Sequence" prior to removing the hydraulic modulator/master cylinder assembly from the vehicle. Each hydraulic modulator gear (large gears) should be able to be turned in one direction and then in the opposite direction when the motor pack is removed. If any gear will not move, replace the hydraulic modulator.

—— CAUTION ——
To perform the Gear Tension Relief Sequence, a Tech 1 scan tool or equivalent scan tool must be used prior to removal of the ABS modulator/master cylinder assembly.

—— WARNING ——
When servicing the master cylinder on this vehicle the procedure below must be followed in the correct sequence. DO NOT perform any services to the master cylinder without first performing the gear tension release procedure.

1. If equipped with ABS, perform the gear tension release procedure.
2. Disconnect the two ABS solenoid electrical connectors.
3. Detach the connector from the brake fluid level switch.
4. If equipped with ABS, disconnect the six–way motor pack electrical connectors.

NOTE: When disconnecting the brake lines from the master cylinder and removing the unit, use care not to spill brake fluid on any painted surfaces or electrical connectors.

5. Disconnect and cap the brake lines from the master cylinder assembly. Plug the master cylinder ports to prevent excessive fluid loss.
6. Remove the master cylinder mounting nuts.
7. Remove the master cylinder or ABS master cylinder/modulator assembly.

To install:

8. Install the master cylinder or ABS master cylinder/modulator assembly on the power booster and tighten the mounting nuts to 20 ft. lbs. (27 Nm).
9. Uncap and connect the brake lines to the master cylinder and tighten the fittings to 18 ft. lbs. (24 Nm).
10. Attach the electrical connector to the brake fluid level sensor.
11. If equipped, connect the electrical connectors to the ABS system solenoids.
12. If equipped, attach the six–way motor pack electrical connectors.
13. Fill the master cylinder to the proper level. The proper level is to the **MAX** level indicator on the reservoir.
14. Bleed the brake system following the recommended procedure.

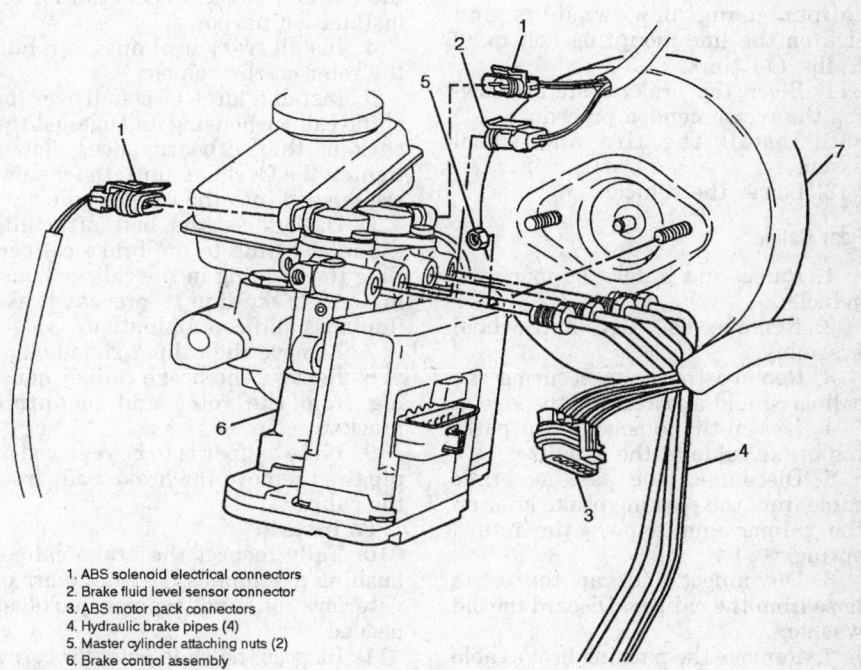

1. ABS solenoid electrical connectors
2. Brake fluid level sensor connector
3. ABS motor pack connectors
4. Hydraulic brake pipes (4)
5. Master cylinder attaching nuts (2)
6. Brake control assembly
7. Vacuum booster

250091

ABS hydraulic modulator and master cylinder assembly

Brake Caliper

REMOVAL AND INSTALLATION

1993–94 Vehicles

Front Caliper

1. Raise and safely support the vehicle.
2. Remove the tire and wheel assembly.
3. Disconnect and cap the brake hose from the caliper. Discard the old washers.
4. Remove the lower caliper mounting bolt.
5. Pivot the caliper upward until it clears the brake rotor and push it inward to disconnect the upper mounting pin from the caliper bracket.
6. If the caliper is to be replaced or repaired remove the brake pads from the caliper.

To install:

7. Before installing the caliper make sure both pistons are seated in their bores and the brake pads are correctly installed.
8. Install the caliper upper mounting bolt into the caliper mounting bracket and pivot the caliper downward to line up the bolt hole for the lower mounting bolt.
9. Install the lower mounting bolt and tighten to 79 ft. lbs. (107 Nm).
10. Connect the brake line to the caliper using new washers and tighten the line mounting bolt to 32 ft. lbs. (44 Nm).
11. Bleed the brake system following the recommended procedure.
12. Install the tire and wheel assembly.
13. Lower the vehicle.

Rear Caliper

1. Raise and safely support the vehicle.
2. Remove the tire and wheel assembly.
3. Remove the bolts securing the caliper shield and remove the shield.
4. Loosen the tension on the parking brake cable at the equalizer.
5. Disconnect the parking brake cable from the parking brake lever on the caliper and remove the return spring.
6. Disconnect and cap the brake hose from the caliper. Discard the old washers.
7. Remove the parking brake cable bracket bolt and remove the bracket.
8. Remove the caliper mounting bolts.
9. Remove the caliper from the vehicle.

To install:

10. Using a spin back tool seat the caliper piston in the caliper bore. Once the caliper piston is seated in the bore make sure the notches in the pistons are at the 6 and 12 o'clock positions.
11. Install the caliper mounting bolts and tighten to 92 ft. lbs. (125 Nm).
12. Install the parking brake cable bracket and tighten the mounting bolt to 32 ft. lbs. (44 Nm).
13. Connect the brake hose to the caliper using new washers and tighten the fitting bolt to 32 ft. lbs. (44 Nm).
14. Install the return spring for the parking brake and connect the parking brake cable to the caliper brake lever.
15. Adjust the parking brake.
16. Install the caliper heat shield.
17. Install the tire and wheel assembly and tighten to specification.
18. Bleed the brake system.
19. Lower the vehicle.

1995–97 Vehicles

Front Caliper

1. Remove ⅔ of the brake fluid from the master cylinder assembly.
2. Raise and safely support the vehicle.
3. Remove the tire and wheel assembly. Mark a relationship between the wheel and the wheel stud for reinstallation purposes.
4. Install two wheel nuts to retain the rotor on the vehicle.
5. Install a large C–clamp over top of the caliper housing and against the back of the outboard shoe. Slowly tighten the C–clamp until the pistons are pushed into the caliper bore.
6. Disconnect the bolt attaching the inlet fitting to the brake caliper. Plug the opening in the caliper housing and brake line to prevent brake fluid loss and contamination.
7. Remove the caliper slide bolt.
8. Remove the brake caliper housing from the rotor and mounting bracket.
9. If the caliper is to be replaced or repaired remove the brake pads from the caliper.

To install:

10. Fully inspect the brake caliper bushing assemblies for cuts, tears or deterioration and replace parts as needed.
11. Inspect the slide bolts for corrosion. If corrosion is found, replace the slide bolts and bushings before installing the brake caliper assembly.
12. Before installing the caliper, make sure both pistons are seated in their bores and the brake pads are correctly installed.
13. Lubricate the caliper slide bolts with silicone grease. Install the caliper slide pins through the brake caliper assembly. Tighten the slide bolts to 80 ft. lbs. (108 Nm).
14. Install the brake caliper inlet fitting and tighten to 24 ft. lbs. (32 Nm).
15. Remove wheel nuts securing the brake rotor to the hub.
16. Install the tire and wheel assembly.
17. Safely lower the vehicle.
18. Fill the master cylinder to the proper level with clean brake fluid.
19. Bleed the brake system.

Rear Caliper

1. Remove ⅔ of the brake fluid from the master cylinder assembly.
2. Raise and safely support the vehicle.
3. Remove the tire and wheel assembly. Mark a relationship between the wheel and the wheel stud for reinstallation purposes.
4. Reinstall two wheel nuts to retain rotor.
5. Remove the brake hose from caliper. Plug openings in caliper and brake hose to prevent brake fluid loss and contamination.
6. Disconnect the parking brake cable from the parking brake lever on the caliper. Lift up one end of cable spring clip free end of cable from lever.
7. Remove bolt and washer attaching cable support bracket to the caliper body assembly.
8. Remove caliper sleeve bolts.
9. Remove caliper body assembly from vehicle. Pivot caliper assembly up to clear rotor and then slide it inboard off pin sleeve.

To install:

10. Inspect caliper bolt boot, pin boot and sleeve bolt for cuts, tear or deterioration. Replace as necessary. Lubricate the mounting surfaces and the mounting sleeves.
11. Remove pin boot from caliper body assembly and install small end over pin sleeve installed on caliper support until boot seats in pin groove. This is to prevent cutting the pin boot when sliding the body assembly onto the pin sleeve.
12. Hold the caliper body assembly in position it was removed and start over end of pin sleeve. As caliper body assembly approaches the pin boot, work the large end of pin boot in caliper body groove. Push caliper body fully onto pin.
13. Pivot caliper body assembly down, using care not to dam-

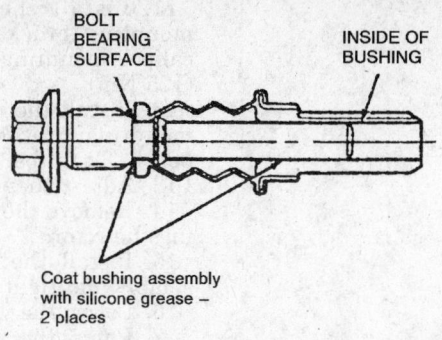

BOLT BEARING SURFACE

INSIDE OF BUSHING

Coat bushing assembly with silicone grease – 2 places

326202

Slide bolts lubrication points

age piston boot on inboard shoe. Compress sleeve boot by hand as caliper body moves into position to prevent boot damage.

14. After installing the caliper assembly into position, recheck installation of the pad clips. If necessary, use a small prying tool to reset or center the pad clips.

15. Install brake caliper sleeve bolts and tighten to 20 ft. lbs. (27 Nm).

16. Install cable support bracket with cable attached with the attaching bolt and washer. Tighten bolt to 32 ft. lbs. (43 Nm).

17. Lift up on end of the cable spring clip and work the end of the parking brake cable into notch of parking brake lever.

18. Connect the brake hose to the brake caliper. Torquing the bolt to 24 ft. lbs. (32 Nm).

19. Remove the wheel nuts securing the rotor to hub and bearing assembly.

20. Install the tire and wheel assembly and tighten to specification.

21. Safely lower the vehicle.

22. Fill master cylinder to the proper level with clean brake fluid.

23. Bleed the calipers with recommended procedure.

24. Apply approximately 175 lbs. of force three times to properly seat the brake shoe and linings against the rotor.

25. Adjust the parking brake cable as necessary.

Disc Brake Pads

REMOVAL AND INSTALLATION

Front Brake Pads

1. Siphon ⅔ of the brake fluid out of the master cylinder.

2. Raise and safely support the vehicle.

3. Mark the relationship of the wheel to the wheel stud for re-installation purposes. Remove the tire and wheel assembly.

4. Install two lug nuts to secure the rotor in place when the caliper is removed.

5. Install a large C-clamp over the top of the caliper housing and against the back of the outboard shoe. Slowly tighten the C-clamp until the caliper pistons are pushed into the caliper bore enough to slide the caliper assembly off the rotor. Use care not to tighten the C-clamp too far or the outboard shoe retaining spring will be deformed and require replacement.

6. Remove the caliper mounting bolts and remove the brake caliper from the mounting bracket.

7. DO NOT disconnect the brake hose from the caliper or allow the brake hose to support the weight of the caliper. Support the caliper on a piece of wire out of the way.

8. Remove the outer brake pad from the caliper using a suitable prying tool to lift the outboard shoe retaining spring so that it will clear the caliper center lug and pull the brake pad out of the caliper.

9. Remove the inner brake pad by by unsnapping the shoe springs from the piston.

To install:

10. Clean all parts well. If the brake pads were worn so badly that the brake rotor is damaged, it must be inspected and/or replaced. Light scoring of the rotor surfaces not exceeding 0.060 inch in depth is not harmful to brake operation and may result from normal use. Brake rotors may be refinished. Do not use a rotor that, after refinishing, will not meet the thickness specification cast in the rotor. Always replace with a new rotor.

11. If not done at removal, now use a C-clamp and clamp both pistons at the same time with a metal plate or wooden block across the face of both

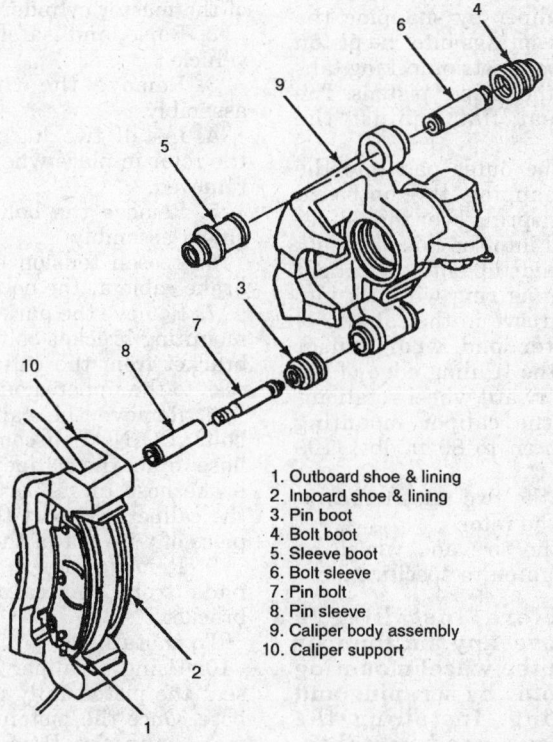

1. Outboard shoe & lining
2. Inboard shoe & lining
3. Pin boot
4. Bolt boot
5. Sleeve boot
6. Bolt sleeve
7. Pin bolt
8. Pin sleeve
9. Caliper body assembly
10. Caliper support

326203

Caliper attachments

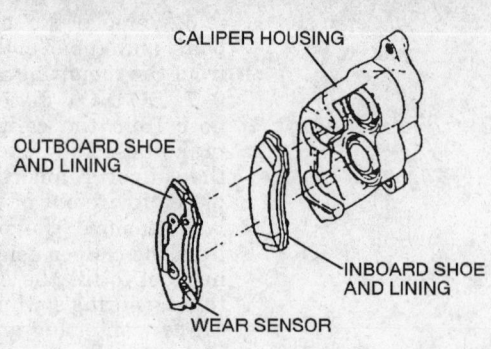

CALIPER HOUSING

OUTBOARD SHOE
AND LINING

INBOARD SHOE
AND LINING

WEAR SENSOR

326680

Front brake pads and caliper

pistons. Take care not to damage the pistons or caliper boots.

NOTE: After bottoming the pistons into the caliper bore, lift the inner edge of each caliper boot next to the piston and press out any trapped air. Make sure each boot convolution is tucked back into place. Boots must lay flat.

12. Inspect the caliper bushings for wear. Replace as necessary. Carefully inspect the slide bolts for corrosion. If corrosion if found, use new parts including the bushing assemblies when installing the caliper. Do not attempt to polish away corrosion. Lubricate caliper slide bolts with silicone grease.

13. Install the new inner disc brake pad in the caliper by snapping the shoe retainers springs into the piston making sure both sets of locking tabs are seated in the caliper pistons. The pad must seat flat against the pistons.

14. Install the outer pad into the caliper by snapping the outboard shoe retaining spring over the caliper center lug and into the housing slot.. The pad will slide up onto the caliper and the retaining ring will lock into place on the groove in the caliper.

15. The outer pad wear sensor should be at the trailing edge of the shoe during forward wheel rotation.

16. Install the caliper mounting bolts and tighten to 80 ft. lbs. (108 Nm).

17. Remove the two nuts temporarily securing the rotor.

18. Install the tire and wheel assembly and tighten to specification.

NOTE: Before installing a wheel, remove any buildup or corrosion on the wheel mounting surface or rotor by scraping and wire brushing. Installing the wheel without good metal-to-metal contact at the mounting surfaces can cause the wheel

nuts to loosen. **Torque the wheel nuts to 100 ft. lbs. (140 Nm) in a criss-cross pattern. This is important to avoid warping the brake rotor which could result in a brake pedal pulsation complaint.**

19. Lower the vehicle.

20. Pump the brake pedal several times to seat the pads against the rotor.

21. Check the brake fluid level and top off as necessary.

22. Road test the vehicle to ensure the proper brake performance.

Rear Brake Pads

1993 Vehicles and 1994 Lumina

1. Siphon 2/3 of the brake fluid out of the master cylinder.

2. Raise and safely support the vehicle.

3. Remove the tire and wheel assembly.

4. Install two lug nuts to secure the rotor in place when the caliper is removed.

5. Remove the bolts securing the shield assembly.

6. Loosen tension on the parking brake cable at the equalizer.

7. Remove the parking brake cable mounting bracket bolt and mounting bracket from the caliper to gain access to the upper mounting bolt.

8. Remove the caliper mounting bolts. DO NOT disconnect the brake hose from the caliper or allow the brake hose to support the weight of the caliper. Support the caliper on a piece of wire out of the way.

9. Remove the inner and outer pads from the caliper mounting bracket.

To install:

10. Using a caliper spin back tool, seat the piston fully into the caliper bore. Once the piston is fully seated make sure the D-shaped notch engages into the bump on the inner pad.

11. Install the caliper onto the mounting bracket and tighten the caliper mounting bolts to 92 ft. lbs. (125 Nm).

12. Install the parking brake cable mounting bracket and tighten the bolt to 32 ft. lbs. (44 Nm).

13. Adjust the parking brake.

14. Remove the two lug nuts securing the rotor.

15. Install the tire and wheel assembly and tighten to specification.

16. Lower the vehicle.

17. Pump the brake pedal several times to seat the pads against the rotor.

18. Check the brake fluid level and add fresh DOT 3 brake fluid as necessary.

19. Road test the vehicle for proper brake performance.

1994–97 Vehicles

1. Siphon 2/3 of the brake fluid out of the master cylinder.

2. Raise and safely support the vehicle.

3. Remove the tire and wheel assembly.

4. Install two lug nuts to secure the rotor in place when the caliper is removed.

5. Remove bolt and washer attaching cable support bracket to caliper body assembly. It is not necessary to disconnect the parking brake lever or disconnect the brake hose.

6. Remove the caliper retaining bolt.

7. Pivot the caliper body assembly up from the rotor and remove from the bracket. Do not completely remove the caliper assembly body.

8. Remove the outboard and inboard shoe and linings from the caliper support assembly.

9. Remove two brake lining clips from the caliper support.

To install:

NOTE: In order for the rear pads to seat in the caliper properly, the cut outs in the caliper piston must be at the 6 and 12 o'clock positions. Failure to align the piston correctly can lead to brake drag, premature brake wear and possible brake failure.

10. Using a suitable type spanner tool turn the piston in to bottom the piston fully into the caliper bore. Once the caliper is fully seated make sure the cutouts in the piston are at the 6 and 12 o'clock positions.

11. After bottoming the piston into the caliper bore, lift the inner edge of boot next to the piston assembly and press out any trapped air.

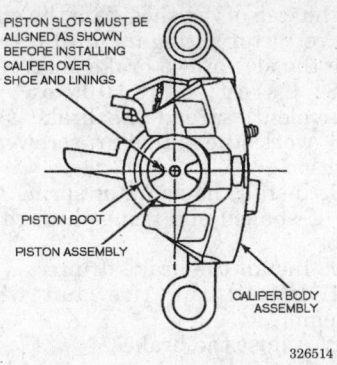

PISTON SLOTS MUST BE ALIGNED AS SHOWN BEFORE INSTALLING CALIPER OVER SHOE AND LININGS

PISTON BOOT

PISTON ASSEMBLY

CALIPER BODY ASSEMBLY

326514

Positioning rear caliper piston slots

12. Install two pad clips in the caliper support.

13. Lubricate the inner pad where it contacts the piston and mounting surfaces.

14. Install outboard and inboard shoe and linings in caliper support. Position wear sensors downward at the leading edge of the rotor during forward wheel rotation.

15. Hold the metal shoe edge against the spring end of clips in the caliper support. Push brake pad in towards the hub, bending spring ends slightly and engage shoe notches with support abutments.

16. Pivot the caliper body assembly down over the brake pad. Compress the sleeve boot by hand as the caliper body moves into position to prevent boot damage.

NOTE: After the caliper body assembly is in position, recheck installation of the brake pad clips. If necessary, use a small prying tool to reseat or center the pad clip on the support abutments.

17. Install the sleeve bolts and tighten bolt to 20 ft. lbs. (27 Nm).

18. Install cable support bracket with the cable attached and bolt washer. Tighten bolt to 32 ft. lbs. (43 Nm).

19. Remove the two lug nuts securing the rotor.

20. Install the tire and wheel assembly and tighten to specification.

21. Lower the vehicle.

22. Pump the brake pedal several times to seat the pads against the rotor.

23. Check the brake fluid level and top off as necessary.

24. Adjust parking brake as necessary.

25. Road test the vehicle for proper brake performance.

Brake Rotor

REMOVAL AND INSTALLATION

Front Rotor

1. Raise and safely support the vehicle.

2. Remove the tire and wheel assembly. Mark a relationship between the wheel and the wheel stud for installation purposes.

3. Using a T60 Torx® bit remove the two caliper mounting bracket bolts and remove the mounting bracket.

4. Remove the rotor from the front hub.

To install:

5. If a new rotor is being installed the caliper piston will have to be pushed back into the caliper bore enough so the caliper will fit over the rotor.

6. Install the rotor on the front hub and loosely install two lug nuts to hold the rotor in place while the caliper is installed.

7. Install the caliper mounting bracket and tighten the mounting bolts to 148 ft. lbs. (200 Nm).

8. Install the lower caliper bolt and tighten to 79 ft. lbs. (107 Nm).

9. Remove the two nuts securing the rotor.

10. Install the tire and wheel assembly and tighten to specification.

11. Lower the vehicle.

12. Pump the brake pedal several times to seat the pads against the rotor.

Rear Rotor

1. Raise and safely support the vehicle.

2. Remove the tire and wheel assembly. Mark a relationship between the wheel and the wheel stud for re-installation purposes.

3. Remove the parking brake cable bracket and attaching bolt.

4. Remove the disc brake caliper.

5. Mark a relationship between the brake rotor and the wheel stud for re-installation purposes.

6. Remove the brake caliper mounting bracket from the vehicle and remove the brake rotor.

To install:

7. Clean metal contact surfaces between brake rotor and hub bearing flange.

8. If a new rotor is being installed the caliper piston will have to be turned into the caliper bore enough so the caliper will fit over the rotor.

9. Install the rotor on the rear hub and loosely install two lug nuts to hold the rotor in place while the caliper is installed.

10. Install the caliper mounting bracket.

11. Install the caliper onto the mounting bracket and tighten the mounting bolts to 92 ft. lbs. (125 Nm).

12. Install the parking brake cable bracket and mounting bolt and tighten the mounting bolt to 32 ft. lbs. (44 Nm).

13. Connect the brake hose to the suspension mounting bracket and tighten the mounting bolt to 18 ft. lbs. (24 Nm).

14. Adjust the parking brake.

15. Remove the two nuts securing the rotor.

16. Install the tire and wheel assembly and tighten to specification.

17. Lower the vehicle.

18. Pump the brake pedal several times to seat the pads against the rotor.

Brake Drums

REMOVAL AND INSTALLATION

1. Raise and safely support the vehicle.

2. Mark the relationship of the wheel to the axle flange to help maintain wheel balance after assembly.

3. Remove the tire and wheel assembly.

4. Mark the relationship of the brake drum to the axle flange.

NOTE: Do not pry against the splash shield that surrounds the backing plate in an attempt to free the drum. This will bend the splash shield.

5. If difficulty is encountered in removing the brake drum, the following steps may be of assistance.

a. Make sure the parking brake is released.

b. Back off the parking brake cable adjustment.

c. Remove the access hole plug from the backing plate.

d. Using a screwdriver, back off the adjusting screw.

e. Re-install the access hole plug to prevent dirt or contamination from entering the drum brake assembly.

f. Use a small amount of penetrating oil applied around the brake drum pilot hole.

g. Carefully remove the brake drum from the vehicle.

6. After removing the brake drum it should be checked for the following:

a. Inspecting for cracks and deep grooves.

b. Inspect for out of round and taper.

c. Inspecting for hot spots (black in color).

To install:

7. Install the brake drum onto the vehicle aligning the reference marks on the axle flange.

8. Install the tire and wheel assembly, torquing it to specifications.

9. Lower the vehicle and road test for proper brake operation.

Brake Shoes

REMOVAL AND INSTALLATION

1. Raise and safely support the vehicle.

2. Remove the tire and wheel assembly.

3. Remove the brake drum.

4. Using tool J-38400, or an equivalent brake spanner and remover, remove the actuator spring from the adjuster lever. Use care not to distort the spring when removing it.

— CAUTION —

During the following steps when removing the retractor spring from either shoe and lining assembly, do not over stretch the spring. This will reduce its effectiveness. Keep fingers away from retractor spring to prevent fingers from being pinched between the spring and shoe web or spring and the backing plate.

5. Lift the end of the retractor spring from the adjuster shoe assembly. Insert the hook end of the J-38400 between the retractor spring and the shoe. Pry slightly to remove the spring end from the hole in the shoe.

6. Pry the end of the retractor spring toward the axle with the flat end of the tool until the spring snaps down off the shoe web onto the backing plate.

7. Remove the one brake shoe and remove the adjuster assembly.

8. Disconnect the parking brake lever from the shoe. DO NOT remove the parking brake lever from the cable end unless it is being replaced.

9. Using J-38400 or the equivalent, lift the end of the retractor spring from the adjuster shoe assembly. Insert the hook end of the

J-38400 or the equivalent between the retractor spring and the shoe. Pry slightly to remove the spring end from the hole in the shoe. Pry the end of the retractor spring toward the axle with the flat end of the tool until the spring snaps down off the shoe web onto the backing plate.

10. Remove the brake shoe.

To install:

11. Clean all the brake spring completely with brake solvent and allow to air dry.

12. Disassembly, clean and lubricate the adjuster screw. Once lubricated, reassemble.

13. Clean the backing plate and after it is dry apply a thin coat of brake grease to the brake shoe contact points on the backing plate.

14. Position the brake shoe, that connects to the parking brake lever, on the backing plate. Using J-38400 or the equivalent, pull the end of the retractor spring up to rest on the web of the shoe. Pull the end of the retractor spring up until it snaps into the slot in the brake shoe.

15. Connect the parking brake lever.

16. Install the remaining shoe and the adjuster screw assembly.

17. Position the brake shoe, using J-38400 or the equivalent, pull the end of the retractor spring up to rest

on the web of the shoe. Pull the end of the retractor spring up until it snaps into the slot in the brake shoe.

18. Using J-38400 or the equivalent, spread the brake shoes and work the adjuster screw into position.

19. Install the actuator spring with the U-shaped end going through the web.

20. Install the brake drum.

21. Install the tire and wheel assembly.

22. Adjust the brakes.

23. Lower the vehicle and road test for proper brake operation.

Wheel Cylinder

REMOVAL AND INSTALLATION

1. Raise and safely support the vehicle. Remove the tire and wheel assembly.

2. Remove the brake drum.

3. Using tool J-38400, or an equivalent brake spanner and remover, remove the actuator spring from the adjuster lever. Use care not to distort the spring when removing it.

NOTE: During the following steps when removing the retractor spring from either shoe and lining assembly, do not over stretch the spring. This will reduce its effectiveness.

— CAUTION —

Keep fingers away from retractor spring to prevent fingers from being pinched between the spring and shoe web or spring and the backing plate.

4. Remove the bleeder screw to gain access to the brake line.

5. Remove the wheel cylinder brake line and cap the line to prevent fluid loss and contamination.

6. Spread the upper half of the brake shoes and remove the wheel cylinder retaining bolts and remove the wheel cylinder.

To install:

7. Apply Loctite™ Master Gasket Maker or equivalent sealer to the wheel cylinder shoulder face that contacts the backing plate.

8. Position the wheel cylinder assembly on the backing plate and hold into place.

9. Install attaching bolts and tighten to 110 inch lbs. (12 Nm).

10. Connect the brake line to the wheel cylinder torquing to 12 ft. lbs. (17 Nm).

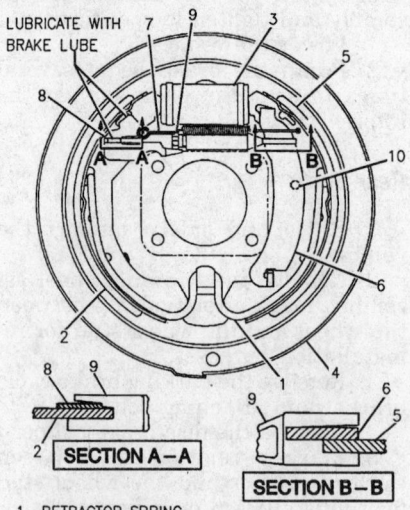

LUBRICATE WITH BRAKE LUBE

1 RETRACTOR SPRING
2 ADJUSTER SHOE AND LINING
3 WHEEL CYLINDER
4 BACKING PLATE
5 PARK BRAKE SHOE AND LINING
6 PARK BRAKE LEVER
7 ACTUATOR SPRING
8 ADJUSTER ACTUATOR
9 ADJUSTING SCREW ASSEMBLY
10 LEVER STOP

278909

Rear brake assembly component alignment

11. Bleed the wheel cylinder using the recommended procedure.

12. Install the bleeder valve and tighten to 62 inch lbs. (7 Nm).

13. Clean all the brake spring completely with brake solvent and allow to air dry.

14. Install the actuator spring and adjust the brake shoes.

15. Install the brake drum and wheel assembly, torquing the wheel to specifications.

16. Add DOT 3 brake fluid to the master cylinder if needed and road test to verify proper brake system performance.

Parking Brake Cable

ADJUSTMENT

With Rear Disc Brakes

1. Apply the service brakes (firmly) three times with pedal force approximately 175 pounds. (778 N).

2. Apply and release the parking brake three times.

3. Raise and safely support the vehicle.

4. Remove the tire and wheel assemblies. Mark the position of the wheel to the wheel studs, prior to the removal, for installation reference. This keeps spun-balanced assemblies in proper balance.

5. Check that the parking brake pedal has full release. This can be verified by turning the key to the **ON** position and monitoring the brake light on the instrument panel.

6. Install two lug nuts on each wheel to hold the rotor in position.

7. Verify that the parking brake levers on both calipers are against the caliper stops.

8. If the levers are not against the stops, loosen the adjuster nut on the equalizer until both levers are fully seated.

9. Tighten the adjuster until either of the levers is 0.020 to 0.080 inch off of the lever stop.

10. Apply the parking brake several times to verify a firm pedal. DO NOT apply the service brake at this time.

11. If the pedal is soft or the wheel can still be turned in the forward direction with the pedal fully depressed repeat Step 9.

12. Remove the lug nuts securing the rotors.

13. Install the tire and wheel assemblies, aligning the marks made at removal and tighten to specification.

14. Lower the vehicle.

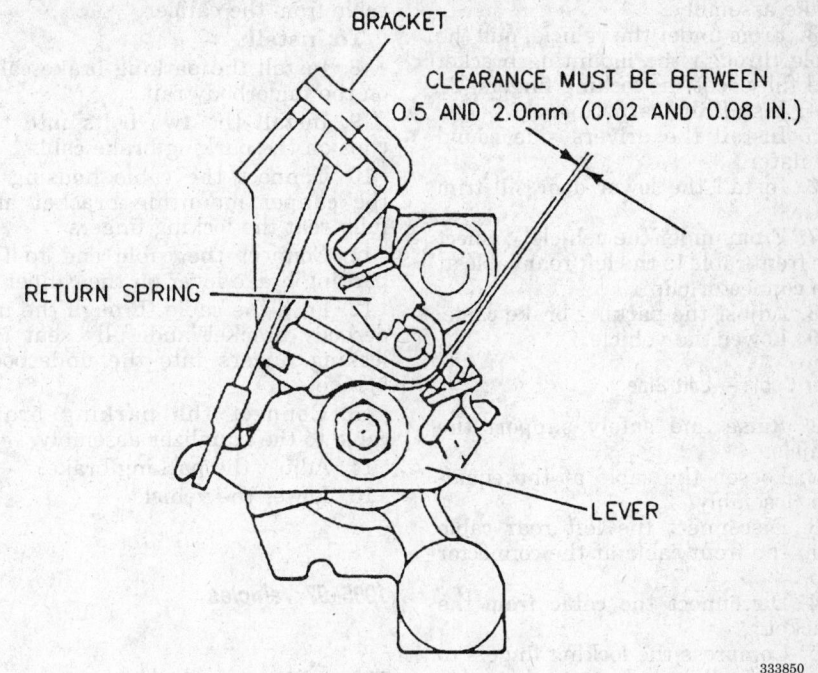

Parking brake adjustment — 1996–97 vehicles

With Rear Drum Brakes

1. Adjust the rear brake shoes before adjusting the parking brakes.

2. Apply the parking brake to 10 clicks and release. Repeat this sequence five times.

3. Check the parking brake for full release.

4. Turn the ignition to the **ON** position and check that the BRAKE indicator lamp is off.

5. If the light is on and the brake appears to be released, operate the brake pedal release lever and pull downward on the front parking brake cable to remove the slack.

6. Raise and safely support the vehicle.

7. Adjust the parking brake by turning the nut on the equalizer while spinning both rear wheels. When either rear wheel develops drag, stop adjusting and back off the equalizer nut one full turn.

8. Apply the parking brake to 4 clicks and check the rear wheel rotation.

9. Release the parking brake and check for free wheel rotation.

10. Safely lower the vehicle.

REMOVAL AND INSTALLATION

1993–94 Vehicles

Front Cable

1. Remove the lower door sill trim plate.

2. Remove the drivers side sound insulator panel.

3. Fold back the carpeting to expose the parking brake cable.

4. Raise and safely support the vehicle.

5. Loosen the parking brake cable at the equalizer.

6. Disconnect the parking brake cable from the front connector clip.

7. Compress the locking fingers and disconnect the cable housing from the bracket.

8. From inside the vehicle, disconnect the parking brake cable end from the parking brake lever assembly.

9. Compress the locking fingers and disconnect the cable housing from the lever assembly bracket.

10. Remove the cable from the vehicle.

To install:

11. Install the cable into the vehicle and from inside the vehicle connect the parking brake cable housing to the parking brake lever assembly and fully seat the locking fingers.

12. Connect the cable end to the parking brake lever on the parking brake assembly.

13. From under the vehicle, pull the cable through the mounting bracket and fully seat the locking fingers.

14. Install the carpeting.

15. Install the drivers side sound insulator.

16. Install the lower door sill trim plate.

17. From under the vehicle, connect the front cable to the left rear cable at the connector clip.

18. Adjust the parking brake cable.

19. Lower the vehicle.

Rear Cable — Left Side

1. Raise and safely support the vehicle.

2. Loosen the cable at the equalizer assembly.

3. Disconnect the left rear cable from the front cable at the connector clip.

4. Disconnect the cable from the bracket.

5. Compress the locking fingers to disconnect the parking brake cable housing from the equalizer assembly.

6. Disconnect the parking brake cable end from the parking brake lever on the caliper.

7. Disconnect the cable from the rear bracket and cable support and remove the cable from the vehicle.

To install:

8. Route the cable through the rear cable bracket and cable support.

9. Feed the cable through the equalizer and front cable bracket.

10. Fully seat the cable housing locking fingers into the equalizer assembly.

11. Connect the parking brake cable to the parking brake lever on the caliper.

12. Connect the left rear cable to the front cable at the connector clip.

13. Adjust the parking brake cable.

14. Lower the vehicle.

Rear Cable — Right Side

1. Raise and safely support the vehicle.

2. Loosen the parking brake cable at the equalizer assembly.

3. Disconnect the parking brake cable from the brake cable support.

4. Compress the locking fingers and disconnect the underbody support.

5. Remove the two bolts from the parking brake cable clips on the underbody rail.

6. Disconnect the parking brake cable end from the caliper parking brake cable lever.

7. Compress the locking fingers and disconnect the parking brake cable from the caliper.

To install:

8. Install the parking brake cable on the underbody rail.

9. Install the two bolts into the clips on the parking brake cable.

10. Connect the cable housing to the caliper mounting bracket and fully seat the locking fingers.

11. Connect the cable end to the parking brake lever on the caliper.

12. Feed the cable through the underbody bracket and fully seat the locking fingers into the underbody bracket.

13. Connect the parking brake cable to the equalizer assembly.

14. Adjust the parking brake.

15. Lower the vehicle.

1995–97 Vehicles

This vehicle is equipped with coated parking brake cable assemblies. The wire strand is coated with a clear plastic material which slides over plastic seals inside the conduit end fittings. This is for corrosion protection and reduced parking brake effort.

Handling of these cables during servicing of the parking brake system requires extra care. Damage to the plastic coating will reduce corrosion protection and if the damaged area passes through the seal, increased parking brake effort could result. Contact of the coating with sharp-edged tools, or with sharp surfaces of the vehicle underbody should be avoided.

To prevent damage to the threaded parking brake adjusting rod when servicing the parking brake, the following is recommended: before attempting to turn the adjusting nut, clean the exposed threads on each side of the nut; lubricate the threads of the adjusting rod before turning the nut.

If any one of the parking brake cables has been replaced, it is necessary to "pre-stretch" the new cable before adjusting the parking brake. To do this, apply the parking brake pedal several times. Fully release it each time. The parking brake pedal assembly, located on the dash panel left of the service brake pedal, is a "push-to-release" type mechanism which is APPLIED by depressing the pedal once and RELEASED by depressing the pedal again. No release handle is used.

NOTE: This is a parking brake NOT a emergency brake. It is designed to hold the vehicle on a flat or an incline and will not stop a vehicle in motion.

Front Cable

1. Remove the lower door sill trim plate.

2. Remove the driver's side sound insulator panel.

3. Fold back the carpeting to expose the parking brake cable.

4. Raise and safely support the vehicle.

5. Loosen the parking brake cable at the equalizer.

6. Pull the front cable at the connector clip and attach loop of the shear strap to hook of the parking brake ratcheting.

7. Remove the parking brake front cable at the connector clip.

8. Remove the parking brake front cable clip from the underbody bracket using special tool J 37043 Brake Cable Release Tool or equivalent.

9. Remove the cable button end from the lever clevis.

10. Remove the parking brake cable from the parking brake assembly using special tool J 37043 or equivalent.

To install:

11. Install the cable into the vehicle and from inside the vehicle connect the parking brake cable housing to the parking brake lever assembly and fully seat the locking fingers.

12. Install the parking brake cable end to the lever clevis.

13. Feed the parking brake cable and snap clip to the underbody.

14. Install the carpeting.

15. Install the driver's side sound insulator.

16. Install the lower door sill trim plate.

17. From under the vehicle, connect the front cable to the left rear cable at the connector clip. Tighten the nut on the equalizer to remove cable slack.

18. Adjust the parking brake cable.

19. Lower the vehicle.

Left Rear Cable

1. Raise and safely support the vehicle.

2. Loosen the cable at the equalizer assembly.

3. Disconnect the left rear cable from the front cable at the connector clip.

4. Disconnect the cable from the bracket.

5. Compress the locking fingers to disconnect the parking brake cable

housing from the equalizer assembly using special tool J 37043 or the equivalent Brake Cable Release Tool.

6. Disconnect the parking brake cable end from the parking brake lever on the caliper, if equipped with rear disc brakes. If equipped with rear drum brakes, remove the rear wheel and brake drum. Disconnect the cable end from the brake shoe lever and disengage from the backing plate.

7. Disconnect the cable from the rear bracket and cable support and remove the cable from the vehicle.

To install:

8. Route the cable through the rear cable bracket and cable support.

9. Feed the cable through the equalizer and front cable bracket.

10. Fully seat the cable housing locking fingers into the equalizer assembly.

11. Connect the parking brake cable to the parking brake lever on the caliper, if equipped with rear disc brakes. If equipped with rear drum brakes, thread the cable end through the opening in the backing plate and then connect the cable end to the brake shoe lever.

12. Connect the left rear cable to the front cable at the connector clip.

13. Tighten nut on the equalizer to remove the parking cable slack.

14. Lower the vehicle.

15. Adjust the parking brakes.

Right Rear Cable

1. Raise and safely support the vehicle.

2. Remove the parking brake cable at the equalizer.

3. Remove the parking brake cable from the brake cable support assembly.

4. Remove the parking brake cable from the rear underbody bracket using special tool J 37043 or the equivalent.

5. Remove the two bolts from clips on the underbody rail.

6. Remove the parking brake cable from the backing plate and parking brake lever for vehicle equipped with drum brakes.

7. Remove the caliper parking brake lever and bracket using special tool J 37043 or the equivalent on vehicles equipped with rear disc brakes.

To install:

8. Install the parking brake cable in position on the underbody rail.

9. Install the two bolts to support clips above the knuckle hub ward.

10. Install the parking brake cable to the backing plate and the parking brake lever on vehicle equipped with drum brakes.

11. Install the parking brake cable to the caliper parking brake lever and bracket on vehicles equipped with disc brakes. Tighten the bolts to 36 inch lbs. (4 Nm).

12. Feed the parking brake cable through the underbody bracket and snap conduit to the bracket.

13. Install the parking brake cable threaded rod to the equalizer.

14. Tighten the nut on the equalizer to remove the cable slack.

15. Safely lower the vehicle.

16. Adjust the parking brakes as required.

Parking Brake Cable With Drum Brakes

1. Raise and safely support the vehicle.

2. Loosen the parking brake cable at the equalizer.

3. Remove the cable connector at the clip if removing the left side.

4. Remove the retaining bolts if removing the right rear cable.

5. Remove the brake drum and partially remove the rear brake shoes.

6. Remove the parking brake cable from the actuator arm.

7. Remove the cable from the backing plate using tool J 37043 or the equivalent.

To install:

8. Install the brake cable into the backing plate by snapping it in.

9. Connect the opposite end to the equalizer or the connector clip.

10. Install the cable to the actuating arm and reassemble the rear brake shoes.

11. Install the brake drum and torque the wheel to specification.

12. Adjust the parking according to the recommended procedure.

13. Lower the vehicle and test the parking brake on a incline to see if will hold the weight of the vehicle.

Brake Hydraulic System

BLEEDING

NOTE: Do not move the vehicle until a firm pedal is obtained. Air in the brake system can cause loss of brakes with possible personal injury.

Vehicles Without ABS

1. If the master cylinder is known or suspected to have air in the bore, then it must be bled before bleeding any caliper in the following manner:

  a. Disconnect the forward (blind end) brake line connection at the master cylinder.

  b. Allow the brake fluid to fill the master cylinder bore until it begins to flow from the forward pipe connector port.

  c. Connect the forward brake line to the master cylinder and tighten.

  d. Depress the brake pedal slowly one time and hold. Loosen the forward brake line connection at the master cylinder to purge air from the bore. Tighten the connection and then release the brake pedal slowly. Wait fifteen seconds. Repeat the above procedure including the 15 second wait until all air is removed from the bore.

  e. Tighten the brake line to 11 ft. lbs. (15 Nm).

  f. When clear fluid flows from the forward connection, repeat the above procedure and bleed the master cylinder at the rear connection.

2. Individual calipers are bled only after all the air is removed from the master cylinder. If it is known that the calipers do not contain any air, then it will not be necessary to bleed them.

3. Place a proper size box end wrench over the caliper bleeder valve. Attach a clear tube over the bleeder valve. Submerge the other end of the tube in a clear container partially filled with brake fluid.

4. Depress the brake pedal slowly one time and hold. Loosen the bleeder valve to purge the air from the cylinder. Tighten the bleeder screw and slowly release the pedal. Wait fifteen seconds. Repeat the sequence, including the fifteen seconds wait until all air is removed. It may be necessary to repeat the sequence ten or more times to remove all air from the system.

NOTE: Do not push the pedal all the way to the floor and do not pump the pedal rapidly. Rapid pumping of the brake pedal pushes the master cylinder piston down the bore in such a way that it makes it difficult to bleed the system.

5. When all the air is expelled from the system, tighten the bleeder valves to 62 inch lbs. (7 Nm).

6. If it is necessary to bleed all the calipers, the following sequence should be used: the left front and then the right front caliper.

7. Check the pedal for a firm pedal feel. Repeat the above procedure to correct the a spongy condition.

Vehicles With ABS

Use only DOT 3 brake fluid from a clean, sealed container. Do not use fluid from an open container that may be contaminated with water. Do not use DOT (silicone) brake fluid. In the following steps, use a suitable container and shop cloths to catch fluid and prevent from contacting any painted surfaces.

1. Clean the area around the fluid reservoir cover and remove the fluid reservoir cover. Check the fluid level in the reservoir and fill to the correct level, if necessary. Install the fluid reservoir cover.

2. Prime ABS hydraulic modulator/master cylinder assembly.

a. Attach the bleeder hose to rearward bleeder valve and submerge opposite hose end in clean container partially filled with clean brake fluid.

b. Slowly open rearward bleeder valve two turns.

c. Depress the brake pedal and hold until fluid begins to flow.

d. Close the valve and release the brake pedal.

e. Repeat Steps b. through d. until no air bubbles are present.

f. Relocate the bleeder hose to the forward hydraulic modulator bleeder valve and repeat Steps b. through d.

Once fluid is seen to flow from both modulator bleeder valves, the ABS hydraulic modulator/master cylinder assembly is sufficiently full of fluid. However, it may not be completely purged of air. At this point, move to the wheel brakes and bleed them. This ensures that the lowest points in the system are completely free of air and then the ABS hydraulic modulator/master cylinder assembly can be purged of any remaining air. Bleed the wheel brakes using the following procedure.

3. Remove the fluid reservoir cover. Inspect the fluid level in reservoir and fill to the correct level if necessary. Install fluid reservoir cover.

4. Raise and safely support the vehicle.

NOTE: The proper bleeding sequence is as follows: right rear, left rear, right front and left front.

5. Proceed as follows to bleed the wheel brakes in the sequence above:

a. Attach a clear bleeder hose to the bleeder valve at the wheel and submerge the opposite end in a clean container partially filled with clean brake fluid.

b. Open the bleeder valve.

c. Slowly depress the brake pedal.

d. Close the bleeder valve and slowly release the brake pedal.

e. Wait for five seconds.

f. Repeat Steps a. through e. until the brake pedal feels firm at half travel and no air bubbles are observed in the bleeder hose. To assist in freeing entrapped air, tap lightly on the caliper or backing plate to dislodge any trapped air bubbles.

6. Repeat the above Steps on the left rear wheel brake, right front wheel brake and left front wheel brake in that order.

7. Safely lower the vehicle.

8. Remove the fluid reservoir cover. Inspect the fluid level in the reservoir and fill to the correct level, if necessary. Install the fluid reservoir cover.

9. Next, bleed the ABS hydraulic modulator/master cylinder assembly in the following manner:

a. Attach a clear plastic bleeder hose to the rearward bleeder valve on the ABS hydraulic modulator. Submerge opposite hose end in a clean container partially filled with clean brake fluid.

b. Depress the brake pedal with moderate force.

c. Slowly open the rearward bleeder valve two turns and allow the fluid to flow.

d. Close the bleeder valve and release the brake pedal.

e. Wait for 5 seconds.

f. Repeat Steps b. through e. until no air bubbles are observed in the bleeder hose.

g. Relocate the bleeder hose on the forward hydraulic modulator bleeder valve and repeat Steps a. through f.

10. Remove the fluid reservoir cover. Inspect the fluid level in the reservoir and fill to the correct level, if necessary.

11. With the ignition switch turned to the **RUN** position, apply the brake pedal with moderate force and hold. Note the pedal travel and feel.

a. If the pedal feels firm and constant and pedal travel is not excessive, start the engine. With the engine running, recheck the pedal travel. If the pedal is still firm and constant and pedal travel is not excessive, the vehicle should be ready to road test.

b. If the pedal feels soft or has excessive travel either initially or after the engine start, the following procedure may be used. With the Tech 1 or equivalent, RELEASE then APPLY each motor two to three times and cycle each solenoid five to ten times. When finished, be sure to APPLY the front and rear motors to ensure the pistons are in the upmost position. DO NOT DRIVE THE VEHICLE at this time.

c. If a Tech 1 or equivalent is not available, remove foot from the brake pedal, start the engine and allow it to run for at least ten seconds to initialize the ABS. DO NOT DRIVE THE VEHICLE at this time. After ten seconds, turn the ignition switch to the **OFF** position. This initialization procedure MUST be repeated five times to ensure any trapped air has been dislodged.

12. Now repeat the entire bleeding procedure.

13. Road test the vehicle. Make several normal non-ABS stops from moderate speed to ensure proper brake system function.

Wheel Speed Sensor

REMOVAL AND INSTALLATION

Front Wheel Speed Sensor

The front sensor is serviceable only as an assembly. Do not attempt to service the sensor harness pigtail as it is part of the sensor.

1. Disconnect the negative battery cable. Raise and safely support the vehicle.

2. Disconnect the sensor connector. Remove the bolt retaining the connector.

3. Remove the 2 bolts retaining the sensor to the knuckle.

4. Remove the sensor from the knuckle.

To install:

5. Install the sensor to the knuckle. Make sure the sensor guide pins are properly aligned and the sensor is seated.

6. Install the retaining bolts and tighten to 52 ft. lbs. (70 Nm).

7. Install the bolt retaining the electrical connector.

8. Attach the connector to the speed sensor.

9. Lower the vehicle.

10. Connect the negative battery cable. Verify proper operation of the speed sensor.

Rear Wheel Speed Sensor

The rear wheel speed sensor and the sensor ring is an integral part of the hub and bearing assembly. Should replacement be required, the hub and bearing must be replaced as a assembly.

FRONT SUSPENSION

Strut

REMOVAL AND INSTALLATION

When replacing the strut cartridge the following tools will be needed J 35668, J 35669, J 35671 and J38844 or their equivalents, it will not be necessary to remove the strut assembly from the vehicle. The strut cartridge replacement consists of removing the upper mounting plate and retainers and removing the cartridge assembly. The strut assembly is also part of the knuckle assembly.

Strut Cartridge

CAUTION

DO NOT service the strut cartridge unless the weight of the vehicle is on the suspension. The weight of the vehicle keeps the coil spring compressed. Otherwise the released coil spring could result in personal injury.

1. Disconnect the negative battery cable.
2. Scribe the strut cover to body to assure proper camber adjustment.
3. Remove the strut cover by removing the cover nuts.
4. Remove the strut shaft nut by using a no. 50 Torx® bit and J 35668 or the equivalent.
5. Remove the strut mount insulator by prying with a flat bladed tool. Use J 35668 or the equivalent to apply pressure on the strut as neces-

sary to relieve the side load (compression) on the bushing.
6. Remove the bumper by attaching J 35668 or the equivalent to the strut and pulling out the bumper.
7. Install J 38844 or the equivalent in the correct position and compress the strut down into the cartridge.
8. Remove the strut closure nut by unscrewing the closure nut using J 35671 or the equivalent.
9. Remove the cartridge and remove the oil in the strut housing using a suction pump to remove all the oil.

To install:
10. Install the self contained replacement cartridge into the strut housing.
11. Install the strut cartridge closure nut using J 35671 or the equivalent and torquing the nut to 82 ft. lbs. (110 Nm).
12. Install J 35668 or the equivalent and compress the shaft down into the cartridge.
13. Remove J 38844 or the equivalent strut alignment tool.
14. Install the strut bumper and raise the strut and remove J 35668 or the equivalent.
15. Install the strut mount insulator as follows:
   a. Use a soap solution to lubricate the bushing for ease of installation.
   b. If necessary, install J 35668 or the equivalent after the bushing is partially installed and position the strut as required to assist in the bushing installation.
16. Install the strut shaft nut using the No. 50 Torx® bit and J 35669 or the equivalent and tighten the nut to 59 ft. lbs. (80 Nm).
17. Install the strut cover mount and aligning the scribe marks then torquing the bolts to 24 ft. lbs. (33 Nm).
18. A four wheel alignment is recommended after any steer-

ing/suspension repairs are performed.

Strut Assembly

CAUTION

Strut assemblies are under high tension. DO NOT REMOVE the strut shaft nut without using the proper spring compressing tool. Failure to do so may result in personal injury.

1. Disconnect the negative battery cable.
2. Scribe the strut mount cover for reinstallation purposes.
3. Raise and safely support the vehicle.
4. Remove the tire and wheel assembly, mark a relationship between the wheel and the wheel stud for reinstallation purposes.
5. Remove the brake caliper and the bracket.
6. Remove the rotor, mark a relationship between the wheel and the wheel stud for reinstallation purposes.
7. Remove the hub and bearing attaching bolts.
8. If equipped, remove the ABS sensor mounting bolt and position sensor aside to prevent damage.
9. Remove the tie-rod to knuckle attaching nut and separate using J 35917 or the equivalent.
10. Remove the lower ball joint attaching nut and separate using J 35917 or the equivalent.
11. Remove the ball joint shield retaining bolts and the heat shield.
12. Remove the strut mount cover retaining nuts and remove the strut assembly.
13. Install the strut assembly into J 34013-A or the equivalent and using adaptor 34013-88 and its equivalent compress the spring using the forcing screw just enough to release tension from the strut shaft. Using a Torx® bit and J 35669 or the equivalent remove the strut shaft nut.
14. Relieve all of the spring pressure and remove the components from the strut assembly.
15. Inspect all components for wear and damage, replace as necessary.
   To install:
16. Install the lower spring seat, the lower spring insulator and the lower spring coil end must be visible between the step and the first retention tab of the insulator.
17. Install the spring into the the lower spring insulator and the strut shield to the spring seat.
18. Install the strut bumper and upper spring insulator, the upper

A NO 5 TORX BIT
332353
Strut shaft nut removal

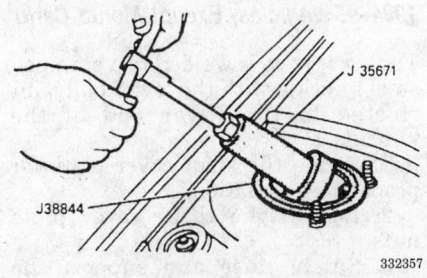

J 35671
J38844
332357
Strut closure nut removal

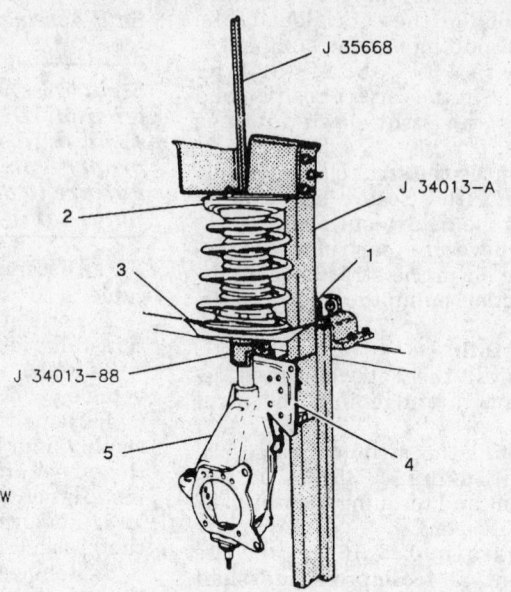

1. COMPRESSOR FORCING SCREW
2. MOUNT
3. SPRING SEAT AND BEARING
4. SPRING PLATE
5. KNUCKLE/STRUT ASSEMBLY

332401

Strut compression tool

spring coil end must be between the step and the location mark on the insulator.

19. Install the strut mount and align the strut cartridge shaft with J 35668 or the equivalent.

20. Compress the strut using J 34013-A and J 34013-88 or their equivalents. Pull the strut shaft through the strut retainer and remove J 35668 or the equivalent.

21. Install the strut mount insulator and shaft nut using J 35669 or the equivalent and Torx® bit, torquing the nut to 59 ft. lbs. (80 Nm).

22. Position the strut into the vehicle and loosely install the upper retaining nuts.

23. Install the ball joint heat shield and the retaining bolts, tighten to 62 inch lbs. (7 Nm).

24. Install the lower ball joint to control arm attaching nut and tighten to 63 ft. lbs. (85 Nm). DO NOT tighten more than 60 degrees (1 flat or 1/6 turn) to align with the hole in the ball joint stud and DO NOT loosen at any time during the installation to align and install a new cotter pin.

25. Install the tie rod to knuckle attaching nut and tightening to 63 ft. lbs. (85 Nm). DO NOT tighten more than 60 degrees (1 flat or 1/6 turn) to

align with the hole in the ball joint stud and DO NOT loosen at any time during the installation to align and install a new cotter pin.

26. Install the hub and bearing assembly, torquing to specifications.

27. Properly install the ABS speed sensor and install the retaining bolt.

28. Install the brake rotor and the brake caliper bracket, torquing it to specification.

29. Install the tire and wheel assembly, torquing to specifications.

30. Safely lower the vehicle and torque the upper retaining nuts to 24 ft. lbs. (33 Nm).

31. A four wheel alignment is recommended after any steering/suspension repairs are performed.

1994–95 Vehicles, Except Monte Carlo

The strut tube is welded to a stamped steel knuckle with the lower ball joint riveted to the lower end of the knuckle.

1. Scribe the strut cover plate for proper reinstallation.

2. Loosen the three cover plate nuts.

3. Safely raise and support the vehicle.

4. Remove the tire and wheel assembly.

5. Remove the brake caliper and bracket assembly. Support the caliper from the suspension.

6. Remove the brake rotor.

7. Remove the hub and bearing to the knuckle attaching bolts.

8. Remove the ABS sensor mounting bolt and position the ABS sensor aside to prevent damage, if equipped with ABS.

9. Separate the axle from the transaxle and remove the axle.

10. Remove the tie rod to steering knuckle attaching nut.

NOTE: Use only the recommended tools (a ball joint press or puller, not a "pickle fork") for separating the ball joints. Failure to use recommended tools may cause damage to the ball joint and seal.

11. Using special tool J 35917 or equivalent, separate the tie rod from the knuckle assembly.

12. Remove the lower ball joint to steering knuckle attaching nut.

13. Using special tool J 35917 or equivalent, separate ball joint from the lower control arm.

14. Remove the ball joint heat joint heat shield retaining bolts and heat shield assembly.

15. Remove the strut cover plate nuts.

16. Remove the steering knuckle and strut assembly.

To install:

17. Install the knuckle and strut assembly into the vehicle.

18. Install the strut mount cover plate and upper strut mount to body attaching nuts Tighten the nuts after lowering the vehicle.

19. Install the ball joint heat shield and tighten retaining bolts to 89 in. lbs. (10 Nm).

20. Install lower joint to the control arm attaching nut. Tighten ball joint nut to 63 ft. lbs. (85 Nm) with a new cotter pin. Tighten the ball joint nut to the next slot in the nut with the cotter pin hole in the stud. Do not tighten the nut more than 60° to align with hole.

21. Install the tie rod to the steering knuckle and tighten the attaching nut with a new cotter pin to 40 ft. lbs. (54 Nm).

22. Carefully install the drive axle assembly through the opening in the steering knuckle.

23. Install and tighten the hub and bearing to knuckle attaching bolts to 60 ft. lbs. (80 Nm).

24. Position the ABS sensor and install the mounting bolts, if removed.

25. Install the brake rotor.

26. Install the brake caliper and bracket assembly. Tighten the slide bolts to 80 ft. lbs. (108 Nm).

27. Install the tire and wheel assembly.

28. Safely lower the vehicle.

29. Tighten the strut cover plate nuts to 18 ft. lbs. (24 Nm) after aligning the scribe marks.

30. Road test the vehicle.

Lower Ball Joints

REMOVAL AND INSTALLATION

1. Raise and safely support the vehicle. Remove the tire and wheel assembly.

2. Remove the ball joint heat shield retaining bolts and the ball joint heat shield.

3. Remove the lower ball joint cotter pin and nut.

4. Loosen, but do not remove the stabilizer shaft bushing assembly bolts.

5. Remove the ball joint front the lower control arm using special tool J 35917 or equivalent ball joint press tool.

6. Drill out the four rivets retaining the ball joint to the steering knuckle. Use an 1/8 inch drill bit to make a pilot hole through the rivets. Finish drilling the rivets using a 1/2 inch drill bit.

NOTE: Do not damage the drive axle boots when drilling out the ball joint rivets.

7. Remove the ball joint from the steering knuckle assembly.

To install:

8. Install the ball joint in the steering knuckle assembly.

9. Install four new ball joints bolts and nuts.

10. Install the ball joint to the lower control arm and tighten the nut to 63 ft. lbs. (85 Nm). Tighten the nut to align slot in the ball joint nut with the cotter pin opening in the stud. Do not tighten the ball joint nut more than 60 degrees (one flat, or 1/6 turn) to align the hole and never loosen the nut anytime during installation. When ball joint nut installation is satisfactory, install a new cotter pin.

11. Install the ball joint heat shield.

12. Install the tire and wheel assembly.

13. Safely lower then road test the vehicle.

NOTE: A four wheel alignment is recommended after any steering/suspension repairs are performed.

Lower Control Arms

REMOVAL AND INSTALLATION

1. Raise and safely support the vehicle. Remove the tire and wheel assembly.

2. On 3.4L (VIN X) engines, remove the engine splash shields.

3. Remove the stabilizer shaft to the lower control arm insulator clamp bolts.

4. Remove the ball joint heat shield retaining bolts and the ball joint heat shield.

5. Remove the lower ball joint cotter pin and nut.

6. Loosen but do not remove the stabilizer shaft bushing assembly bolts.

7. Remove the ball joint from the lower control arm using special tool J 35917 or the equivalent.

8. Observe the direction of the mounting bolts. They must be installed in the same position. Remove the control arm to frame attaching nuts and bolts.

9. Remove the lower control arm from the vehicle.

To install:

10. Install the lower control arm to the vehicle with attaching bolts. The bolts must be installed from inside the lower control arm with the nuts on the outside.

11. Pivot the lower control arm up to the ball joint.

12. Install the ball joint to the lower control arm assembly.

13. Tighten the lower control arm to frame bolts to 52 ft. lbs. (70 Nm).

14. Install the ball joint to the lower control arm and tighten the nut to 63 ft. lbs. (85 Nm). Tighten the nut to align slot in the ball joint nut with the cotter pin opening in the stud. Do not tighten the ball joint nut more than 60° to align the hole and never loosen the nut anytime during installation. Install a new cotter pin.

15. Install the ball joint heat shield.

16. Install the stabilizer shaft to the lower control arm insulator clamp and tighten the bolts to 35 ft. lbs. (48 Nm).

17. On 3.4L (VIN X) engines, install the engine splash shields.

18. Install the tire and wheel assembly. Safely lower, then road test the vehicle.

NOTE: A four wheel alignment is recommended after any steering/suspension repairs are performed.

Sway Bar

REMOVAL AND INSTALLATION

The sway bar (also called a Stabilizer Shaft) is mounted to the top rear of the frame and the top of the lower control arm. The sway bar is attached with clamps and rubber insulators.

1. Disconnect the negative battery cable.

2. Turn the wheels to the straight ahead position and lock the wheel to prevent the Supplemental Inflatable Restraint (SIR) from becoming uncentered.

3. Raise and safely support the vehicle. Remove the front tire and wheel assemblies.

4. On 3.4L (VIN X) engines, remove the engine splash shields.

5. Move the steering shaft dust seal for access to the pinch bolt. The pinch bolt will be facing out and accessible from the left wheel opening.

CAUTION

Failure to disconnect intermediate shaft from rack and pinion stub shaft can result in damage to steering gear and/or intermediate shaft. This damage can cause loss of steering control which could result in personal injury.

6. Remove the pinch bolt from the lower intermediate steering shaft.

7. Loosen all the stabilizer insulator clamp attaching nuts and bolts.

8. Place a suitable screw type jack under the center of the rear frame crossmember.

9. Loosen the two front frame-to-body bolts (four turns only).

10. Remove the two rear frame-to-body bolts and lower the rear of the frame just enough to gain access to remove the stabilizer shaft.

11. Remove the insulators and clamps from the frame and control arms. Pull the stabilizer shaft rearward, swing down and remove from the left side of the vehicle.

To install:

12. Install the stabilizer shaft through the left side of the vehicle.

13. Coat the new insulators with rubber lubricant.

14. Loosely install the clamps to the control arms and clamps to the frame.

15. Raise the frame into position while guiding the steering gear into place.

16. Install new frame-to-body bolts and torque to 103 ft. lbs. (140 Nm).

17. Remove the jack.

18. Torque the stabilizer clamps-to-frame and control arms to 35 ft. lbs. (47 Nm).

19. Install the steering gear pinch bolt and dust seal.

20. Install the front wheels and torque the lug nuts to 100 ft. lbs. (136 Nm).

21. Connect the negative battery cable.

NOTE: Whenever the vehicle sub-frame is removed or loosened, the wheel alignment should be checked.

Front Wheel Bearings

ADJUSTMENT

The wheel bearings are not adjustable. If a wheel bearing is out of specifications, it must be replaced. Using a dial indicator, check for looseness. If it exceeds 0.005 inch (0.1270 mm) on drum or disc brakes the bearing wear is excessive and the hub and bearing should be replaced.

REMOVAL AND INSTALLATION

NOTE: Do not remove the drive axle nut at this time. Failure to follow the proper sequence of removal steps may cause permanent bearing damage.

1. Loosen the hub nut one full turn. Do not remove.

2. Raise and safely support the vehicle. Remove the tire and wheel assembly.

3. Remove the caliper mounting bracket-to-steering knuckle mounting bolts and remove the caliper and bracket assembly and support the assembly out of the way. DO NOT disconnect the brake hose from the caliper or allow the hose to support the weight of the caliper.

4. Remove the brake rotor.

5. Remove the hub nut and washer.

6. Using special tool J 28733–A or an equivalent hub puller, push the halfshaft out of the hub assembly about an inch.

7. Remove the hub assembly-to-knuckle attaching bolts.

8. If equipped, remove the ABS sensor mounting bolt and position the ABS sensor out of the way.

9. Remove the hub and bearing assembly from the vehicle.

To install:

10. Clean all parts well. Install the hub and bearing assembly onto the axle shaft splines. Make sure the

splines engage smoothly. DO NOT force the hub over the axle splines or the splines will be damaged.

11. Seat the hub and bearing assembly against the steering knuckle and install the mounting bolts. Tighten the bolts alternately and evenly to 52 ft. lbs. (70 Nm).

12. Make sure the ABS sensor is clean. The small magnet in its end will attract rust and metal chips that can degrade its performance. Install the ABS sensor and tighten the mounting bolt.

13. Install the brake rotor.

14. Install the caliper and bracket assembly on the steering knuckle and tighten the caliper bracket mounting bolts to 79 ft. lbs. (107 Nm).

15. Install the tire and wheel assembly. Lower the vehicle.

16. Install a new drive axle nut and washer. On 1993–95 vehicles, tighten the nut to 184 ft. lbs. (250 Nm). On 1996–97 vehicles, tighten the drive axle nut to 151 ft. lbs. (205 Nm).

REAR SUSPENSION

Strut

REMOVAL AND INSTALLATION

1993–94 Vehicles

1. Disconnect the negative battery cable. Raise and safely support the vehicle.

2. Remove the rear wheel assembly. Mark the position of the wheel to the wheel studs, prior to the removal, for installation reference to keep the assembly in balance.

3. Scribe the strut-to-knuckle for proper installation.

4. These vehicles are equipped with a fiberglass leaf spring. To protect the spring, rubber cushions called auxiliary springs are installed on the outboard ends of the axle. A special C-clamp shaped tool is used to compress the auxiliary spring (cushion). Remove the auxiliary spring as follows, if so equipped.

a. Remove the leaf spring retention plate.

b. Remove the front retention plate bolt just enough to rotate the plate clear of the rod.

c. Install J 37956 or the equivalent making sure the pin is in the stationary end of the clamp is inserted in the hole of the auxil-

iary spring bracket and remove the plug from the upper bracket.

d. Seat the rod in the tool channel and hand tighten and remove the rod to knuckle bolts.

e. Loosen J 37956 or the equivalent forcing screw to allow the spring to expand.

NOTE: When removing the auxiliary spring, make sure the rod/bushing clears the transverse spring on the boss on knuckle assembly.

f. Remove the auxiliary spring and J 37956 or the equivalent.

5. Remove the jack pad.

6. Remove the exhaust system, if equipped with dual exhaust.

7. Install a rear leaf spring compressor, J 35778 or the equivalent.

8. Fully compress the spring, but DO NOT remove the retention plates or the spring.

9. Remove the two strut-to-body bolts. Remove the brake hose from the strut.

10. Remove the strut and auxiliary spring upper bracket from the knuckle.

To install:

11. Position the strut in the body and knuckle bracket.

12. Install the strut-to-body bolts and torque to 34 ft. lbs. (46 Nm).

13. Install the strut-to-knuckle bolts. Install the sway bar bracket and tighten the nuts to 133 ft. lbs. (180 Nm).

14. Install the brake hose bracket and remove the spring compressing tool.

15. Install the auxiliary spring as follows:

a. Install the auxiliary spring and J 37956 or the equivalent.

b. Compress the auxiliary spring using J 37956 or the equivalent enough to install the rod to knuckle bolts.

NOTE: Make certain that the pin is in the stationary end of the clamp is inserted in the hole of the upper auxiliary spring bracket and the rod is seated in the tool channel. When compressing the auxiliary spring, make sure the rod/bushing clears the transverse spring and the boss on the knuckle.

c. Install the rod to knuckle bolt, using thread locker P/N 1052624 or equivalent and tighten the bolt to 66 ft. lbs. (90 Nm) plus an additional 90 degrees.

d. Remove J 37956 or the equivalent and install the plug.

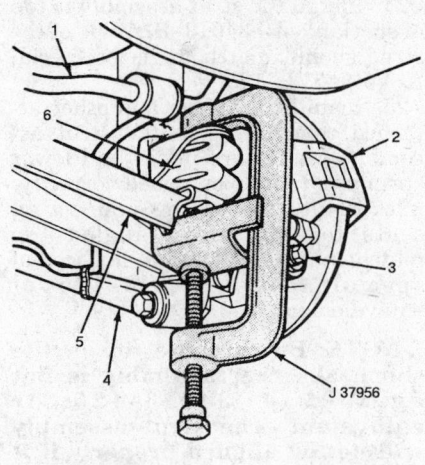

1. Stabilizer shaft
2. Caliper assembly
3. 90 Nm + 90° (66 lb. ft. + 90°)
4. Trailing arm
5. Spindle rod (rear)
6. Auxiliary spring

332708

Auxiliary spring compressing tool. Note the rubber cushion-like auxiliary spring, number 6 in the illustration

e. Properly position the retention plate and install the bolt, tighten to 15 ft. lbs. (20 Nm).

16. Install the exhaust system, if removed.

17. Install the jack pad and torque the bolts to 18 ft. lbs. (25 Nm).

18. Install the wheel and aligning the marks made on the removal, torque the lug nuts to 100 ft. lbs. (136 Nm).

19. Lower the vehicle.

20. Connect the negative battery cable.

21. A four wheel alignment check recommended after any steering/suspension repairs are performed.

1995–97 Regal, Cutlass Supreme and Grand Prix

1. Raise and safely support the vehicle.

2. Remove the tire and wheel assembly. Mark the position of the wheel to the wheel studs, prior to removal, for installation reference.

3. Scribe a reference line from the strut to the knuckle.

4. Remove the jack pad.

5. Remove the exhaust system, if equipped with dual exhausts.

6. Install special tool J 35778 or equivalent. This Y-shaped tool spans the width of the fiberglass transverse leaf spring. Fully compress the special tool to take tension off the outboard suspension components. **Do not** remove the spring or retention plates. Just relieve the tension so the strut can be removed.

NOTE: Do not use corrosive cleaning agents, silicone lubricants, engine degreasers, solvents, etc. on or near the fiberglass rear transverse spring. These materials could cause extensive damage to the spring.

7. Remove the brake hose bracket at the strut assembly.

8. Remove the strut to body bolts.

9. Remove the strut/stabilizer shaft bracket from the knuckle assembly.

To install:

10. Install the strut assembly to the body bolts. Tighten the bolts to 34 ft. lbs. (46 Nm).

11. Install the strut/stabilizer shaft bracket to the knuckle. Align the scribe marks to ensure proper alignment. Tighten the attaching bolts to 122 ft. lbs. (165 Nm).

12. Install the brake hose bracket.

13. Remove the special tool J 35778 or equivalent, if used, from the transverse leaf spring.

14. Install the exhaust system, if removed.

15. Install the jack pad and tighten the attaching bolts 18 ft. lbs. (25 Nm).

16. Install the tire and wheel assembly, aligning up the marks made during the removal.

17. Safely lower the vehicle.

18. A four wheel alignment is recommended after any steering/suspension repairs are performed.

1995–97 Lumina and Monte Carlo

These vehicles use direct double-acting struts which are attached to the body and to the rear wheel knuckle assembly. Please note that strut service requires special tools to compress and hold the components during removal and installation. This procedure calls out the GM-recommended special tool numbers. Use care if using substitutes.

1. Disconnect the negative battery cable. Raise and safely support the vehicle.

2. Remove the rear wheel assembly. Mark the position of the wheel to the wheel studs, prior to removal, for installation reference.

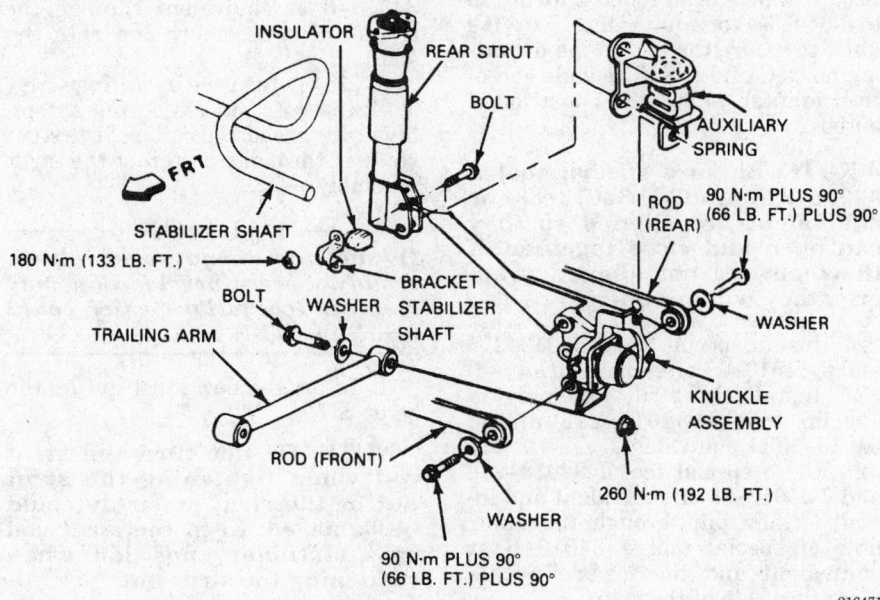

Rear suspension components

216471

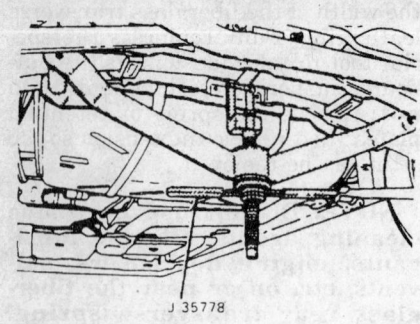

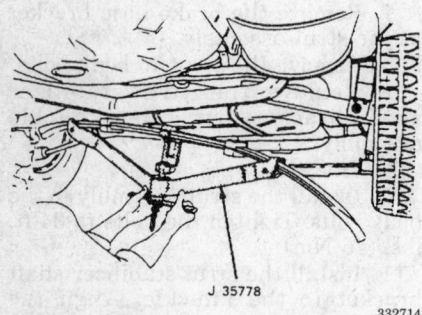

J 35778

332714

Special tool J 35778 is used to compress the fiberglass transverse rear spring during rear suspension service

3. Make a scribe mark to the strut-to-knuckle for proper installation.

4. Remove the brake hose bracket at the strut.

5. Remove the strut mount to body nuts.

6. Remove the strut stabilizer shaft bracket from the knuckle.

7. Remove the strut from the vehicle.

8. Install special tool J 34013-125 to J 34013-B or equivalent. Use wing nuts to secure it to the special tool mounting holes C and E/K. The wide end of the adapter should be positioned outward.

NOTE: Be sure that special tools J 34013-218 and J 34013-88 or equivalent are aligned so they can open and close together. If the tools are not aligned properly, they will not open.

9. Install special tool J 34013-218 and J 34013-88 or equivalent.

10. Install the strut assembly to top of special tool J 34013-B to or equivalent. Angle of the upper strut should be match the angle of special tool J 34013-125 or equivalent.

11. Install the strut assembly in to special tool J 34013-218 and J 34013-88 or equivalent.

12. Close special tool J 34013-218 and J 34013-88 and install locking pin through the lower hole of J 34013-B and lower knuckle mounting hole of the strut.

13. Turn the screw on J 34013-B or equivalent counterclockwise to raise the strut assembly up to special tool J 34013-125 or equivalent. Be sure studs go through guide holes in special tool J 34013-125 or equivalent and the top of the strut assembly is flat against the tool.

14. Compress the spring approximately 0.50 inch (13 mm) or three or four complete turns of the operating screw on special tool J 34013-125 or equivalent.

— CAUTION —

Do not over compress the spring assembly. Severe overloading may cause tool failure which could result in bodily injury.

15. Insert special tool J 35669 or equivalent on the nut then insert special tool J 34013-38 or T-50 Torx® bit into end of the shaft.

16. Discard the strut nut.

17. Turn the operating screw on special tool J 34013-B or equivalent clockwise to fully relieve spring assembly compression.

18. Remove the strut from the spring assembly.

To install:

19. Install the strut onto the spring assembly.

20. Install special tool J 34013-125 to J 34013-B or equivalent. Use wing nuts to secure the tool to the mounting holes C and E/K. The wide end of the adapter should be positioned outboard.

NOTE: Be sure special tool J 34013-218 and J 34013-88 or equivalent are aligned so they can open and close together. If the tools are not aligned properly, they will not open.

21. Install special tool J 34013-218 and J 34013-88 or equivalent.

22. Install the strut assembly to special tool J 34013-218 and J 34013-88 or equivalent.

23. Close special tool J 34013-218 and J 34013-88 or equivalent and install locking pin through the lower hole of special tool J 34013-B or equivalent and the lower knuckle mounting hole of the strut.

NOTE: Be sure the top of the shock absorber assembly is flat against special tool J 34013-125 or equivalent. The shock ab-

sorber will not be aligned correctly if it does not lay flat against the tool.

24. Install the strut assembly to top of special tool J 34013-B. Angle of the strut should match angle of special tool J 34013 -125.

25. Install the spring and other attached components to the strut assembly. Make sure upper and lower spring seats are positioned correctly.

26. Rotate the spring and spring seats together until angled top matches angle of bottom surfaces of special tool J 34013-125 or equivalent.

NOTE: Be sure the top of the shock absorber assembly is flat against J 34013-125 or equivalent. The strut assembly will not be aligned properly if it does not lay flat against the tool.

27. Turn operating screw on special tool J 34013-B or equivalent counterclockwise to raise the strut assembly up to special tool J 34013-125 or equivalent without compressing the spring assembly. Be sure studs on the strut assembly go through guide holes in special tool J 34013-125 or equivalent.

28. Fully extend strut shaft and attach special tool J 34013-20 or equivalent. This prevents shaft from retracting during compression.

29. Install plate special tool plate J 34013-38 or equivalent down through the top of special tool J 34013-B or equivalent through the top of strut assembly and onto the shaft.

30. Slowly turn the operating screw clockwise to compress spring assembly until threaded portion of the strut shaft is through the top of the strut assembly.

— CAUTION —

Do not over compress spring assembly. Severe overloading may result in tool failure which could result in bodily injury.

31. Insert the new strut nut on the strut rod.

NOTE: Do not turn the strut rod when tightening the strut nut or the strut assembly could be damaged. Keep the strut rod in a stationary position when tightening the strut nut.

32. Place special tool J 35669 or equivalent on the shock absorber nut.

33. Insert special tool J 34013-38 through J 35669 or equivalent and tighten the shock absorber nut.

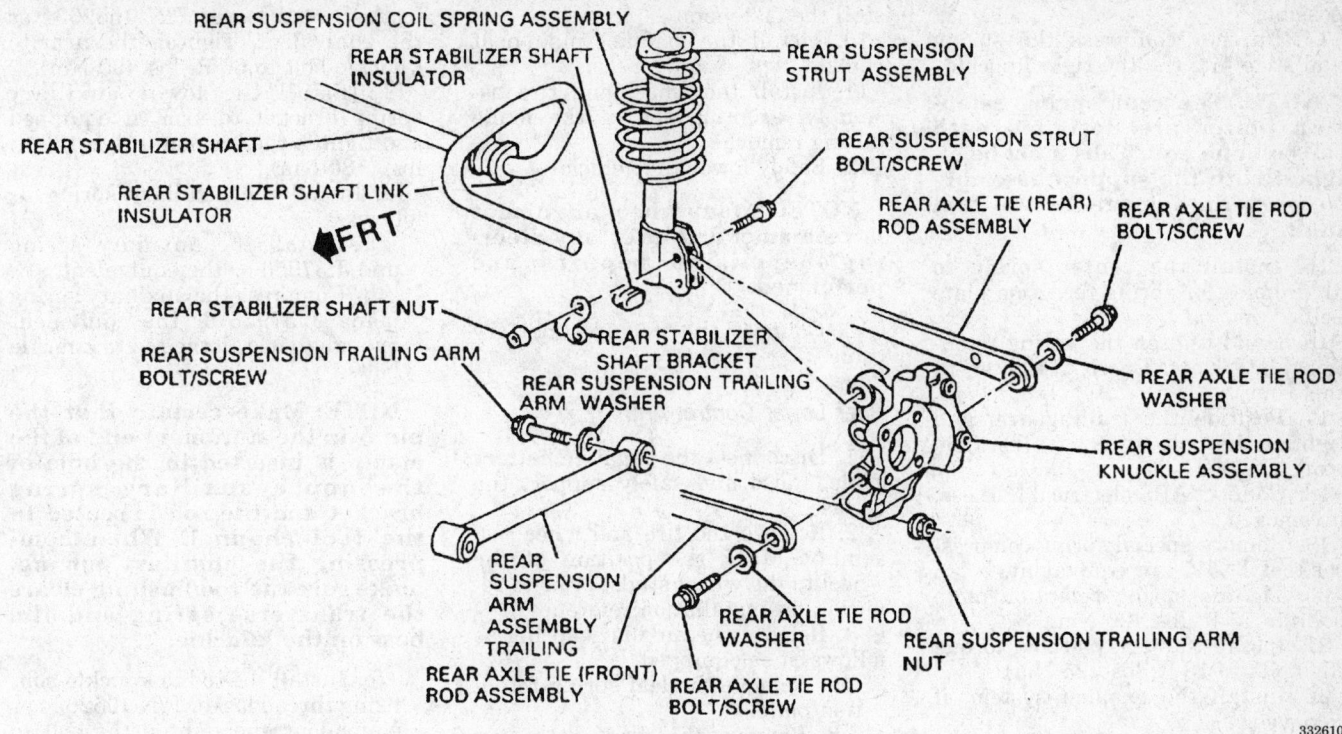

REAR SUSPENSION COIL SPRING ASSEMBLY

REAR STABILIZER SHAFT INSULATOR

REAR SUSPENSION STRUT ASSEMBLY

REAR STABILIZER SHAFT

REAR SUSPENSION STRUT BOLT/SCREW

REAR STABILIZER SHAFT LINK INSULATOR

FRT

REAR AXLE TIE (REAR) ROD ASSEMBLY

REAR AXLE TIE ROD BOLT/SCREW

REAR STABILIZER SHAFT NUT

REAR STABILIZER SHAFT BRACKET

REAR SUSPENSION TRAILING ARM BOLT/SCREW

REAR SUSPENSION TRAILING ARM WASHER

REAR AXLE TIE ROD WASHER

REAR SUSPENSION KNUCKLE ASSEMBLY

REAR SUSPENSION ARM ASSEMBLY TRAILING

REAR AXLE TIE ROD WASHER

REAR AXLE TIE (FRONT) ROD ASSEMBLY

REAR AXLE TIE ROD BOLT/SCREW

REAR SUSPENSION TRAILING ARM NUT

332610

Tri link rear suspension

34. Tighten the shock absorber nut while holding onto special tool J 35669 or equivalent. Tighten the nut to 55 ft. lbs. (75 Nm).

35. Remove special tool J 34013–20, J 34013–38, J 35669 or equivalent from J 34013–B or equivalent.

36. Install and tighten the strut mount to body nuts to 37 ft. lbs. (50 Nm).

37. Install the strut stabilizer shaft bracket to steering knuckle and tighten nuts to 122 ft. lbs. (165 Nm).

38. Install the brake hose bracket.

39. Install the tire and wheel assembly, aligning the marks made on the removal.

40. Safely lower the vehicle.

41. Four wheel alignment recommended after any steering/suspension repairs are performed.

Leaf Spring

REMOVAL AND INSTALLATION

— CAUTION —

Do NOT disconnect any rear suspension components until the transverse spring has been compressed using a rear spring compressor tool J 35778 or equivalent. Failure to follow this *procedure may result in personal injury. Wear protective eye equipment when working on the suspension.*

NOTE: Do not use corrosive cleaning agents, silicones lubricants, engine degreasers, solvents on or near the fiberglass rear transverse spring. These materials could cause extensive damage to the spring.

1. Raise and safely support the vehicle with safety stands.

2. Remove the jack pad in the middle of the spring.

3. Remove the exhaust system, if equipped with dual exhaust.

4. Remove the spring retention plates.

5. Remove the right trailing arm at the knuckle.

6. Disconnect the ABS electrical harness, if ABS equipped.

7. The factory recommends special tool J 35778 or the equivalent. This Y-shaped tool spans the width of the fiberglass transverse leaf spring. It is used to take the tension off the outboard suspension components. Separate the rear leaf spring compressor tool J 35778 or the equivalent from the center shank and hang center shank of tool at the spring center. At- tach center shank of the special tool from the front side of the vehicle only. Use care if using substitutes.

— CAUTION —

Attach the center shank of the compressor from the front side of the vehicle only.

8. Install special tool J 35778 or equivalent body to the center shank and spring. Center the spring on the rollers of the spring compressor only. Always center the spring on the rollers of special tool J 35778 or the equivalent.

9. Fully compress the spring using special tool J 35778 or equivalent.

10. Slide the spring to the left side of the vehicle. It may be necessary to pry the spring to the left using a prybar against the right knuckle and or tire and wheel assembly for leverage. When prying be careful not to damage components.

11. Relax the spring and provide removal clearance from the right side.

12. Remove the spring from the vehicle.

To install:

13. Using special tool J 35778 or the equivalent compress the spring. Install the spring to the left knuckle. Slide the spring towards to the left side as far as possible and raise the

right side of the spring as far as possible.

14. Further compress the spring and slide into the the right knuckle.

NOTE: The rear spring retention plates are designed with tabs on one end. Tabs must be aligned with the support assembly to prevent damage to the fuel tank.

15. Install the center spring to align holes for spring retention plate bolts.

16. Hand tighten the spring retention plate bolts. Do NOT tighten at this time.

17. Position the trailing arm and tighten the bolt to 192 ft. lbs. (260 Nm).

18. Connect ABS electrical harness, if removed.

19. Remove special spring compressor tool J 35778 or equivalent.

20. Tighten spring retention plate bolts to 15 ft. lbs. (20 Nm).

21. Install the jack pad and torque the bolts to 18 ft. lbs. (25 Nm).

22. Install the exhaust system, if removed.

23. Safely lower the vehicle.

24. A four wheel alignment check is recommended after any steering/suspension repairs are performed.

Lower Control Arms

REMOVAL AND INSTALLATION

Forward Rear Lower Control Arm

1. Disconnect the negative battery cable. Raise and safely support the vehicle.

2. Remove the tire and wheel assembly. Mark the position of the wheel to the wheel studs, prior to removal, for installation reference.

3. Remove the arm-to-knuckle bolt. Remove the exhaust heat shield.

4. Lower and support the fuel tank just enough to access to the bolt at the arm to the frame.

5. Remove the bolt at the frame and remove the arm.

To install:

6. Install the arm and bolt to the frame. DO NOT tighten at this time.

7. Install the bolt at the knuckle using thread locker P/N 1052624 or the equivalent.

8. Torque the following fasteners:
- Crossmember bolt to 81 ft. lbs (110 Nm) plus an additional 60 degrees.
- Knuckle bolt to 66 ft. lbs. (90 Nm) plus an additional 90 degrees.

9. Position the fuel tank and install the fasteners.

10. Install the exhaust pipe heat shield.

11. Install the wheel and tire assembly, aligning the marks made during removal.

12. Safely lower the vehicle.

NOTE: A four wheel alignment is recommended after any steering/suspension repairs are performed.

13. Connect the negative battery cable.

Rear Lower Control Arm

1. Disconnect the negative battery cable. Raise and safely support the vehicle.

2. Remove the tire and wheel assembly. Mark the position of the wheel to the wheel studs, prior to removal, for installation reference.

3. Remove the auxiliary spring as follows, if so equipped.

a. Remove the leaf spring retention plate.

b. Remove the front retention plate bolt just enough to rotate the plate clear of the rod.

c. Install J 37956 or the equivalent making sure the pin is in the stationary end of the clamp is inserted in the hole of the auxiliary spring bracket and remove the plug from the upper bracket.

d. Seat the arm in the tool channel and hand tighten and remove the arm to knuckle bolts.

e. Loosen J 37956 or the equivalent forcing screw to allow the spring to expand.

NOTE: When removing the auxiliary spring, make sure the rod /bushing clears the transverse spring on the boss on knuckle assembly.

f. Remove the auxiliary spring and J 37956 or the equivalent.

4. Remove the lower auxiliary spring bracket at the arm, if equipped.

5. Remove the arm to the knuckle bolt.

6. Remove the arm to the crossmember.

7. Push the bolt forward enough to provide arm removal clearance.

To install:

8. Position the arm to the mounting brackets and push the bolt rearward through the rod bushing.

9. Install the rod nut at the crossmember, DO NOT tighten at this time.

10. Install the arm to knuckle bolt using thread locker P/N 1052624 or the equivalent. Tighten the arm to knuckle bolt to 66 ft. lbs. (90 Nm).

11. Install the lower auxiliary spring bracket to arm, if equipped and tighten the bracket nut to 133 ft. lbs. (180 Nm).

12. Install the auxiliary spring as follows:

a. Install the auxiliary spring and J 37956 or the equivalent.

b. Compress the auxiliary spring using J 37956 or the equivalent enough to install the rod to knuckle bolts.

NOTE: Make certain that the pin is in the stationary end of the clamp is inserted in the hole of the upper auxiliary spring bracket and the rod is seated in the tool channel. When compressing the auxiliary spring, make sure the rod/bushing clears the transverse spring and the boss on the knuckle.

c. Install the rod to knuckle bolt, using thread locker P/N 1052624 or equivalent and tighten the bolt to 66 ft. lbs. (90 Nm) plus an additional 90 degrees (¼ turn).

d. Remove J 37956 or the equivalent and install the plug.

e. Properly position the retention plate and install the bolt, tighten to 15 ft. lbs. (20 Nm).

13. Install the wheel and tire assembly, aligning the marks made during removal and torquing the wheel to specifications.

14. A four-wheel alignment check is recommended after any steering/suspension repairs are performed.

15. Connect the negative battery cable.

Sway Bar

REMOVAL AND INSTALLATION

The rear suspension sway bar (also called a Stabilizer Shaft) is attached to the strut assembly and extends rearward, where it is connected to the underbody reinforcement by four rubber bushings and mounting brackets.

1. Disconnect the negative battery cable.

2. Raise and safely support the vehicle. Remove the tire and wheel assembly.

3. Scribe a matchmark on the strut noting the relation to the knuckle.

4. Remove the rear right and left stabilizer shaft link bolts and open the brackets to remove the insulators.

5. Remove the right and left strut to knuckle to stabilizer shaft nuts. DO NOT remove the strut to knuckle through bolts.

6. Remove the insulator brackets from the bolts and remove the sway bar. It may be necessary to pry the sway bar to one side for removal clearance at the strut. Use care when prying.

To install:

7. Install the sway bar by prying the shaft on one side for clearance at the strut.

8. Install the sway bar to strut brackets and install the mounting nuts on the strut through bolts.

9. Close the clamps on the link brackets and install the mounting bolts. Tighten the mounting bolts to 40 ft. lbs. (54 Nm).

10. Tighten the sway bar bracket (strut through bolt) nuts to 122 ft. lbs. (165 Nm).

11. Install the wheel and tire assembly.

12. Lower the vehicle.

13. Connect the negative battery cable.

Wheel Bearings

ADJUSTMENT

The wheel bearings are not adjustable. If a wheel bearing is out of specifications, it must be replaced. Using a dial indicator, check for looseness. If it exceeds 0.005 inch (0.1270 mm) on drum or disc brakes the bearing wear is excessive and the hub and bearing should be replaced.

REMOVAL AND INSTALLATION

Except 1995–97 Lumina and Monte Carlo

The rear suspension uses a non-serviceable wheel unit hub and bearing assembly that is bolted to the rear suspension knuckle. This hub and bearing assembly is a sealed, maintenance free unit. If the hub and/or bearing is damaged, the complete assembly must be replaced. Use the following procedure

1. Disconnect the negative battery cable. Raise and safely support the vehicle.

2. Remove the wheel and tire assembly, mark the position of the wheel to the wheel studs, prior to removal, for installation reference.

3. Scribe a mark from the strut to the knuckle for installation reference.

4. Remove the jack pad.

5. Remove the exhaust system, if equipped with dual exhaust.

6. Install tool J 35778 or the equivalent, fully compress J 35778 or the equivalent but DO NOT remove the the spring or the retention plates.

7. Remove the wheel spindle control arm to knuckle bolts.

8. Remove the caliper and brake rotor, mark the rotor to the wheel stud for installation reference.

9. If equipped, remove the ABS electrical connector.

10. Remove the hub and bearing mounting bolts and remove the bearing assembly.

11. Remove the trailing arm bolts and remove the trailing arm.

12. Remove the sway bar shaft brackets from the knuckle, if equipped.

13. Remove the spindle from the vehicle.

To install:

14. Position the spindle into the mountings.

15. Install the sway bar shaft brackets, DO NOT tighten at this time.

16. Install the control arm to spindle retaining bolt, DO NOT tighten at this time.

17. Install the trailing arm to the spindle and tighten to 66 ft. lbs. (90 Nm) plus an additional 75 degrees.

18. Install the hub and bearing assembly, torquing the bolt to 52 ft. lbs. (70 Nm).

19. If equipped, attach the ABS connector.

20. Install the brake rotor and the brake caliper.

21. Tighten the strut to knuckle bolts to 122 ft. lbs. (165 Nm).

22. Remove tool J 35778 or the equivalent.

23. Install the exhaust system, if equipped with dual exhaust.

24. Install the jack pad and tighten the bolts to 18 ft. lbs. (25 Nm).

25. Tighten the control arm bolts at this time to 177 ft. lbs. (240 Nm). Use thread locker P/N 1052624 or the equivalent.

26. Install the wheel and tire assembly, aligning the marks made during the removal.

27. Safely lower the vehicle.

28. Four wheel alignment recommended after any steering/suspension repairs are performed.

29. Connect the negative battery cable.

1995–97 Lumina and Monte Carlo

The rear suspension uses a non-serviceable wheel unit hub and bearing assembly that is bolted to the rear suspension knuckle. This hub and bearing assembly is a sealed, maintenance free unit. If the hub and/or bearing is damaged, the complete assembly must be replaced. Use the following procedure

1. Disconnect the negative battery cable. Raise and safely support the vehicle.

2. Remove the wheel and tire assembly, mark the position of the wheel to the wheel studs, prior to removal, for installation reference.

3. Scribe a mark at the strut to knuckle for installation reference.

4. Remove the rear wheel spindle to control arm knuckle bolt and washer.

5. Remove the front wheel spindle to control arm knuckle bolt and washer.

6. Remove the brake hose and drum, mark the position of the brake drum to the wheel studs, prior to removal, for installation reference. If equipped with drum brakes.

7. Remove the brake caliper and the rotor, mark the position of the brake rotor to the wheel studs, prior to removal, for installation reference. If equipped with disc brakes.

8. If equipped, remove the ABS electrical connector.

9. Remove hub and bearing mounting bolts.

10. Remove hub and bearing from the knuckle.

11. Remove the trailing arm from the knuckle.

12. Remove the strut nuts.

13. Remove the rear sway bar shaft brackets.

14. Remove the strut mounting bolts and remove the knuckle.

To install:

15. Position the knuckle into the strut assembly.

16. Install the strut mounting bolts only, do not tighten at this time.

17. Install the sway bar shaft brackets. Hand start the nuts only.

18. Install the suspension nuts and tighten to 122 ft. lbs. (165 Nm).

19. Install the front control arm spindle bolts. DO NOT tighten at this time.

20. Install the rear control arm spindle bolts. DO NOT tighten at this time.

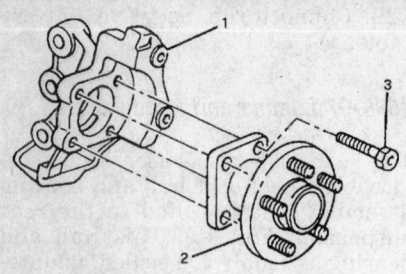

1. Rear suspension knuckle assembly
2. Hub and bearing assembly
3. Wheel bolt/screw

332865

Removing the hub and bearing

21. Install the trailing arm to the knuckle and tighten to 66 ft. lbs. (90 Nm) plus an additional 75 degrees.

22. Install the hub and bearing assembly, tightening it to 52 ft. lbs. (70 Nm).

23. If equipped, install the ABS connector.

24. Install the brake caliper and the rotor, aligning the mark made during the removal. If equipped with disc brakes.

25. Install the brake hose bracket and drum, aligning the mark made during removal. If equipped with drum brakes.

26. Tighten the control arm to knuckle bolts to 177 ft. lbs. (240 Nm).

27. Install the wheel and tire assembly, aligning the marks made during the removal.

28. Safely lower the vehicle.

29. A wheel alignment recommended after any steering/suspension repairs are performed.

GM "B" BODY

Rear Wheel Drive

BUICK-Roadmaster **CADILLAC**-Fleetwood Brougham
CHEVROLET-Caprice • Impala SS

25

FIRING ORDERS

NOTE: To avoid confusion, always replace spark plug wires one at a time.

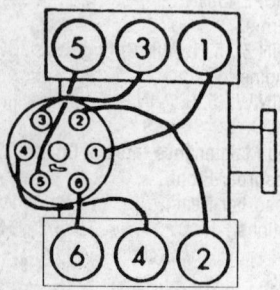

8838QG01

4.3L (VIN Z) Engine
Engine Firing Order:
1-6-5-4-3-2
Distributor Rotation: Clockwise

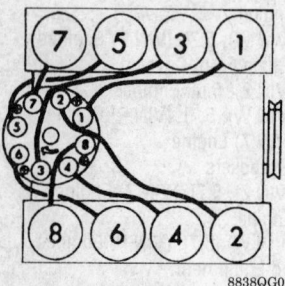

8838QG02

5.0L and 5.7L (VIN 7) Engines
Engine Firing Order:
1-8-4-3-6-5-7-2
Distributor Rotation: Clockwise

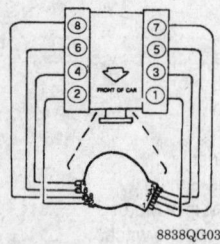

8838QG03

4.3L (VIN W), 5.7L (VIN P) Engines
Engine Firing Order:
1-8-4-3-6-5-7-2
Distributor Rotates with Camshaft

ENGINE ELECTRICAL

NOTE: Disconnecting the negative battery cable on some vehicles may interfere with the functions of the on board computer systems and may require the computer to undergo a relearning process, once the negative battery cable is reconnected.

Distributor

All 1993 vehicles use a Distributor Ignition (DI) system with electronic Ignition Control (IC). The major component of the DI system is the distributor assembly which consists of a cap, rotor, ignition module, pole piece with internal teeth and pickup coil. Vehicles use an externally mounted ignition coil. Spark timing changes are controlled electronically by the Powertrain Control Module (PCM), which monitors various engine sensors, computes the desired spark timing and signals the distributor to change the timing accordingly.

All 1994–96 vehicles, use a new distributor ignition system which was originally developed for use on the Corvette. The new system known as the Opti-Spark ignition system consists of a distributor assembly which is mounted on the front engine cover, under the water pump assembly. The system consists of a distributor assembly, control circuitry and an external coil. In the Opti-Spark system, all ignition timing is controlled by the PCM based on signals from the distributor's internally mounted optical camshaft position sensor. There is no way to bypass the PCM control or to adjust/set ignition timing on this system.

REMOVAL

1993 Vehicles

1. Disconnect the negative battery cable.
2. If necessary for access, remove the air cleaner assembly.
3. For the 4.3L (VIN Z) engine, disconnect the accelerator lever cable, TV cable and cruise control cables, as applicable, from the accelerator lever and support assembly. Then remove the retainers and the support assembly from the intake manifold.

4. Tag and disengage the wiring harness connectors from the distributor assembly.

NOTE: Use care when releasing the connector locking tabs on the distributor assembly.

5. If equipped, disconnect the external coil wire from the distributor cap.
6. Either tag and remove the spark plug wires from the distributor cap and remove the cap from the vehicle or, if clearance allows, remove the cap with the spark plug wires attached and position it out of the way.
7. Mark the position of the rotor and the distributor housing in relation to the engine.
8. Remove the distributor hold-down bolt and clamp, then pull the distributor from the engine until the rotor just stops turning. Again mark the position of rotor.
9. Tilt the distributor toward the driver's side, then remove the assembly from the vehicle. Make sure the gasket is removed with the distributor.

1994–96 Vehicles

1. Be sure the ignition is in the **OFF** or **LOCK** position, then disconnect the negative battery cable.
2. Disengage the wiring harness from the engine cooling fan assembly, then remove the cooling fan from the vehicle.
3. Remove the block drain plug and the knock sensor, then drain the engine cooling system.
4. Remove the air intake duct and the air cleaner assembly.
5. Remove the coolant and heater hoses from the water pump assembly and the throttle body.
6. Unplug the wiring harness from the Engine Coolant Temperature (ECT) sensor, then reposition the ignition coil and bracket.
7. Remove the water pump assembly.
8. Remove the serpentine drive belt from the vehicle.
9. If not done already to remove the pump or the belt, raise and safely support the vehicle for access.
10. Remove the torsional damper bolts, then remove the damper from the crankshaft hub.
11. Disconnect the spark plug wires from the distributor. Be sure to twist each boot ½ turn and pull only on the boot to remove each wire. The wire numbers should be molded into the distributor housing. If not, be sure to tag the wires before disconnection.

12. Unplug the 4-terminal PCM connector from the distributor.

13. Remove the distributor mounting bolts and pull the distributor forward until the driveshaft disengages from the engine. Mark the top of the shaft for alignment during reassembly.

INSTALLATION

Timing Not Disturbed

1993 Vehicles

NOTE: To ensure correct ignition timing when the engine has not been disturbed, the distributor must be installed with the rotor in the same final position as noted during removal.

1. Make sure the gasket is positioned on the distributor assembly.

2. Align the rotor to the last mark made, tilt distributor toward the driver's side of the vehicle and slide the distributor into the engine.

3. The rotor should turn and end up at the first mark made.

4. Install the distributor holddown bolt/clamp and tighten to 27 ft. lbs. (36 Nm), then install the distributor cap.

5. Engage all connectors or wires in the reverse order of removal.

6. Once the distributor installation has been completed, check and adjust the ignition timing.

1994–96 Vehicles

1. With the mark made on the distributor shaft earlier on top, install the distributor to the engine. Tighten the distributor bolts to 8 ft. lbs. (11 Nm).

2. Install the 4-terminal PCM connector and the spark plug wires to the distributor.

3. Position the crankshaft damper to the hub, then install the damper bolts and tighten to 60 ft. lbs. (81 Nm).

4. Install the serpentine drive belt.

5. Install the water pump assembly and engage the ECT wiring harness.

6. Reposition the ignition coil and bracket, then install the coolant and heater hoses.

7. Install the air cleaner assembly and air intake duct.

8. Install the engine block drain plug and the knock sensor.

9. Install the cooling fan assembly, then connect the wiring harness.

10. Connect the negative battery cable and fill the engine cooling system.

11. Start and run the engine, then check for leaks.

Timing Disturbed

1993 Vehicles

1. Remove the No. 1 spark plug. Place a finger over the spark plug hole and slowly rotate the engine in the normal direction of rotation, until compression is felt.

2. Align the timing mark on the crankshaft pulley to the 0 on the engine timing indicator by slowly rotating the engine in the same direction.

3. Position the rotor between No. 1 and No. 8 spark plug towers on the 4.3L (VIN W), 5.0L and 5.7L engine or the No. 1 and No. 6 spark plug towers on the 4.3L (VIN Z) engine.

4. Install the distributor, distributor cap, spark plug wiring and connectors.

5. Check the engine timing and adjust, as required.

Ignition Timing

ADJUSTMENT

When checking ignition timing NEVER pierce a secondary ignition wire. Either use a timing light with an inductive type pickup or connect an adapter between the wire and the spark plug. If the secondary wire insulation is pierced, current will eventually arc and cause engine misfiring.

NOTE: Some engines incorporate a magnetic timing probe hole for the use with electronic timing equipment. Be sure to consult the tool manufacture's instructions for the use of this equipment.

1993 Vehicles

Except 5.7L (VIN 7) Engine

1. Warm the engine to normal operating temperature. Make sure the air conditioning and all accessories are OFF.

2. With the engine running, disconnect the IC bypass connector located on the right side of the engine, near the air control valve. An Powertrain Control Module (PCM) trouble code will set when this is done.

3. Connect a timing light with the pickup lead on the No. 1 plug wire and check timing at the correct engine rpm as designated on the Vehicle Emission Information Label.

4. If the timing requires adjustment, loosen the distributor clamp bolt, the rotate the distributor, as necessary to set the timing to the specifications noted on the information label.

5. Tighten the distributor clamp bolt to 27 ft. lbs. (36 Nm) and recheck the timing to assure it was not disturbed while tightening.

6. Once the timing has been set, turn the ignition OFF, then reconnect the IC bypass harness.

7. Clear any stored trouble codes by interrupting power to the PCM for at least 30 seconds. This can be accomplished, depending on the model, by disconnecting the PCM power feed, the PCM fuse in the fuse box or by disconnecting the negative the battery cable.

5.7L (VIN 7) Engine

1. Connect a timing light and tachometer to the engine. Some tachometers currently available are not compatible with the HEI ignition system. Be sure to connect the timing light to the No. 1 spark plug wire and to connect the tachometer according to the tools instructions.

2. Start the engine and operate until normal operating temperature is reached and the chock is fully opened.

3. Turn all accessories OFF and with the engine running, ground the diagnostic terminal of the ALDL connector using a jumper.

4. Check the ignition timing at the specified rpm. If the ignition timing is not within specification, loosen the distributor clamp bolt and rotate the distributor gradually until the specified timing is obtained.

5. Tighten the distributor clamp bolt making sure the distributor does not change position. Recheck the ignition timing.

6. With the engine still running, remove the jumper from the ALDL terminal.

7. Make fuel system adjustments, as required.

8. Turn the engine OFF. Remove the tachometer and timing light.

1994–96 Vehicles

On these vehicles, base timing is preset when the engine is manufactured. All timing changes are then controlled directly by the PCM based on information from the ignition and knock sensor systems. No adjustments are necessary or possible.

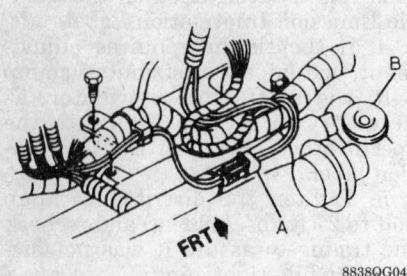

8838QG04

The IC bypass connector is located near the AIR control valve

Alternator

PRECAUTIONS

Several precautions must be observed with alternator equipped vehicles to avoid damage to the unit.

• If the battery is removed for any reason, make sure it is reconnected with the correct polarity. Reversing the battery connections may result in damage to the 1-way rectifiers.

• When utilizing a booster battery as a starting aid, always connect the positive to positive terminals and the negative terminal from the booster battery to a good engine ground on the vehicle being started.

• Never use a fast charger as a booster to start vehicles.

• Disconnect the battery cables when charging the battery with a fast charger.

• Never attempt to polarize the alternator.

• Do not use test lights of more than 12 volts when checking diode continuity.

• Do not short across or ground any of the alternator terminals.

• The polarity of the battery, alternator and regulator must be matched and considered before making any electrical connections within the system.

• Never separate the alternator on an open circuit. Make sure all connections within the circuit are clean and tight.

• Disconnect the battery ground terminal when performing any service on electrical components.

• Disconnect the battery if arc welding is to be done on the vehicle.

REMOVAL AND INSTALLATION

1993 Vehicles

1. Disconnect the negative battery cable.

2. Tag and detach the alternator electrical connections. If access to a connector is difficult, wait until the retainers are removed and the alternator is repositioned, then unplug the remaining connection.

3. With V-Belts, remove the bolt holding the slotted adjusting bracket to the alternator and remove the belt.

4. With serpentine belts, rotate the tensioner to release the drive belt, then remove the belt from the alternator.

5. Remove the fasteners to release the alternator from the engine, then unplug any remaining electrical connections and remove the alternator from the engine.

To install:

6. When reinstalling, reverse the removal procedure.

7. On V-belt engines, adjust the drive belt tension before fully tightening the fasteners.

8. Tighten the upper through-bolt and rear brace bolt to 18 ft. lbs. (28 Nm). Tighten the lower through-bolt and, if applicable, rear brace nut to 37 ft. lbs. (50 Nm).

9. Install the remaining electrical connections, then connect the negative battery cable.

1994–96 Vehicles

1. Disconnect the negative battery cable.

2. Use a breaker bar and socket to rotate the serpentine belt tensioner downward (clockwise) relieving tension from the belt, then remove the belt from the vehicle.

3. Disengage the positive battery wire and the electrical connector from the alternator assembly.

4. Remove the upper brace bolts, the lower brace bolt and the front alternator bolt all from the alternator assembly, then remove the nut retaining the upper brace to the top of the engine and remove the upper brace from the vehicle.

5. Remove the alternator from the accessory mounting bracket and from the vehicle.

To install:

6. Position the alternator to the engine and accessory mounting bracket.

7. Install the upper brace and tighten the retaining nut to 24 ft. lbs. (33 Nm), then install the alternator and brace bolts. Tighten the rear upper brace and front alternator bolts to 37 ft. lbs. (50 Nm) and tighten the rear lower brace bolt to 18 ft. lbs. (25 Nm).

8. Engage the electrical connector and the battery positive wire to the alternator assembly.

9. Install the serpentine drive belt.

10. Connect the negative battery cable.

Drive Belt

REMOVAL AND INSTALLATION

Except 4.3L (VIN W) and 5.7L (VIN P) Engines

1. Disconnect the negative battery cable.

2. Detach the 3-terminal connector and the battery charging wire (BAT lead) from the back of the alternator.

3. For V-belts, loosen the bolt holding the slotted adjusting bracket to the alternator and remove the belt.

4. For serpentine belts, loosen and rotate the tensioner to release the drive belt.

To install:

5. Position the ground strap and alternator in place.

6. For serpentine belt engines, tighten bolts and nuts to 20 ft. lbs. (27 Nm), install the drive belt over the alternator pulley, swing the belt tensioner into place and tighten.

7. Tighten the nuts and bolts to 20 ft. lbs. (27 Nm).

8. Attach the 3-terminal connector and the battery charging wire (BAT lead) to the back of the alternator.

9. Connect the negative battery cable. Check belt tensioner for the serpentine drive belt.

4.3L (VIN W) and 5.7L (VIN P) Engines

1. Disconnect the negative battery cable.

2. Use a breaker bar and a socket to rotate the serpentine belt tensioner downward (clockwise) relieving the tesnion from the belt, then remove the belt from the vehicle.

To install:

3. Route the serpentine drive belt over the pulleys, then rotate the tensioner counterclockwise to tension the belt.

4. Connect the negative battery cable.

BELT TENSION ADJUSTMENT

Serpentine Belts

The vehicles utilize a single, serpentine drive belt for all engine accessories. Serpentine belts are automati-

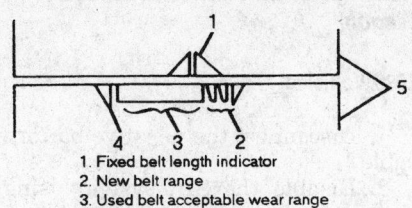

1. Fixed belt length indicator
2. New belt range
3. Used belt acceptable wear range
4. Replace belt position
5. Drive belt tensioner assembly

8838QG05

All serpentine drive belt tensioners have a belt length scale to help determine if the belt is stretched and in need of replacement

cally adjusted by the tensioner on the engine. If the belt is loose, check the condition of the belt and tensioner. Make sure the tensioner markings are within proper operating range for belt condition (new/used). The tensioner should place enough tension on the belt so it can only be twisted 90 degrees at it's longest run.

If belt slippage occurs, while the drive belt tensioner is within its operating range, and the belt does not need replacement, check the belt tension as follows:

1. Run the engine for 10 minutes, shut **OFF** the engine, then using a tension gauge between any 2 pulleys, record the belt tension.

2. Run the engine for 30 seconds and repeat Step 1.

3. Once again, run the engine for 30 seconds and repeat Step 1.

4. The belt tension is the average of the 3 readings. Serpentine belt tension should be 105–125 lbs. (467–556 Nm) for 1993 vehicles.

5. Replace the tensioner if belt length is acceptable (as measured by the tensioner scale), but the tension is below the minimum specification.

Starter

REMOVAL AND INSTALLATION

1. Disconnect the negative battery cable. Safely raise and support the vehicle.

2. Remove the retaining nuts and bolts along with the right frame bracket.

3. If necessary on police vehicles, remove the exhaust crossover pipe for access.

4. Remove the retainers and rear starter brace from the engine and starter.

5. Remove the flywheel housing cover.

6. Remove the 2 starter mounting bolts and carefully lower the starter

for access to the wiring. If present, note the position of any shims for installation purposes.

7. Tag and disconnect the wiring, then remove starter from the vehicle.

To install:

8. Position the starter and support while connecting the starter wiring as noted during removal. Be very careful not to overtighten the electrical connections and crack the solenoid cap.

9. Hold the starter in place and install the mounting bolts. If shims were removed, they must be installed in their original locations to assure proper drive pinion-to-flywheel engagement. Tighten mounting bolts to 32 ft. lbs. (43 Nm).

10. Check the flywheel-to-pinion gear clearance using a 0.20 in. (0.5mm) gauge wire between the teeth of the flywheel and starter pinion. Add or remove shims, as necessary to achieve proper clearance.

11. Install the flywheel housing cover.

12. Install the rear brace to the starter and engine. Tighten the bolt to 18 ft. lbs. (24 Nm) and the nut to 71 inch lbs. (8 Nm).

13. Install the right frame bracket and tighten the bolts to 74 ft. lbs. (100 Nm).

14. If removed, attach exhaust crossover piper.

15. Lower vehicle and connect the negative battery cable.

CHASSIS ELECTRICAL

Blower Motor

REMOVAL AND INSTALLATION

Without Air Conditioning

1. Disconnect the negative battery cable. Detach the blower motor wiring harness.

2. Remove the blower motor retaining screws and pull the blower motor and fan straight forward out of the heater module.

3. Installation is the reverse of removal. Clean and replace sealer as necessary.

With Air Conditioning

1. Disconnect the negative battery cable.

2. Remove the 4 retaining screws and remove the right side instrument panel sound insulator.

3. Disconnect the blower motor electrical connector.

4. Remove the right side hinge pillar trim finish panel by pulling it away from the front body hinge pillar.

5. Remove the screw from the secondary PCM bracket and swing the PCM module and bracket aside to provide access to the blower motor.

6. Remove the blower mounting screws, leaving the screw closest to the relay for last. Carefully, lower the blower motor fan assembly.

To install:

7. Align the blower motor and fan assembly, making sure the PCM module and retainer are out of the way, and carefully raise the assembly into place.

8. Insert and tighten the 3 mounting screws.

9. Swing the PCM module and bracket back into place and tighten the retaining screw to 17 inch lbs. (1.9 Nm).

10. Snap the right side hinge pillar trim finish panel into place on the front body hinge pillar.

11. Connect the blower motor electrical connector.

12. Insert the right side instrument panel sound insulator and attach the 4 retaining screws. Tighten the sound insulator retaining screws to 17 inch lbs. (1.9 Nm).

13. Connect the negative battery cable and check motor operation.

Windshield Wiper Motor

REMOVAL AND INSTALLATION

1. Disconnect the negative battery cable.

2. Raise the hood. Remove the right side wiper arm and hose. In order to keep the arm from pivoting, 2 holes are provided in the arm assembly, into which rivets or pins may be inserted to lock the arm in position.

3. Remove cowl screen. The left side must be removed first.

NOTE: The left side cowl screen must be removed before the right side cowl screen in order to prevent possible windshield damage.

4. If equipped, remove the retaining screws and the linkage access hole cover.

5. Loosen the transmission drive link-to-crank arm retainers, then dis-

connect link from the motor crank arm.

6. Unplug the electrical wiring and any remaining washer hoses from the motor assembly.

7. Remove the motor retaining screws, then remove the windshield wiper motor while guiding the crank arm through the hole.

To install:

8. Install the wiper motor by guiding crank arm through the hole.

9. Insert and tighten wiper motor attaching bolts to 80 inch lbs. (9 Nm).

10. Attach the electrical connectors.

11. Verify that the motor is in the **PARK** position, then attach the motor crank arm to the drive link. Tighten drive link nuts to 27 inch lbs. (3 Nm).

12. If equipped, install the wiper linkage access hole cover and tighten the screws to 18 inch lbs. (2 Nm).

13. Install right, then left side cowl screens.

NOTE: The right side cowl screen must be installed before the left side cowl screen in order to prevent possible windshield damage.

14. Attach the right side wiper arm and hose. If installed, remove the arm pivot prevention rivet or pin.

15. Connect the negative battery cable and check wiper motor operation.

Windshield Wiper Switch

REMOVAL AND INSTALLATION

NOTE: If equipped with an air bag, it is imperative that the disarming procedure is followed before repairs, and that the coil centering and rearming procedures are followed after repairs.

1. Disarm the air bag, if equipped, and disconnect the negative battery cable.

2. Remove the turn signal assembly.

3. If necessary, remove the SIR coil assembly from the column as follows:

   a. Remove wiring protector.

   b. Attach a length of mechanic's wire to the terminal connector in order to aid in reassembly.

   c. Carefully pull wire/harness connector through the column, leaving the wire in the column for assembly purposes.

4. Remove the lock cylinder set.

5. Remove lock housing cover screws.

6. Remove the tilt lever, if equipped, and remove the lock housing cover.

7. Remove the base plate and the dimmer switch rod actuator.

8. Disconnect and remove wiper switch actuator pivot pin.

9. Disengage the wiper switch connector from the vehicle wire harness, then remove the switch. Attach a piece of mechanic's wire to the connector to aid in reinstallation and gently pull the wire harness through column.

To install:

10. Connect the wiper switch assembly to the lock housing cover assembly

11. Attach the switch actuator pivot pin to the switch and cover.

12. Pull the wiper switch wire connector through the steering column with the mechanic's wire used during removal, then attach it to the vehicle wire harness.

13. Attach the dimmer switch rod actuator to the base plate and lubricate with lithium grease.

14. Connect base plate to the lock housing cover assembly. The bottom edge of the dimmer switch rod actuator must rest on the bend in the dimmer switch rod.

15. Position lock housing cover assembly in place and, if equipped, attach the tilt lever.

16. Starting with the housing cover screw in the 12 o'clock position, then 8 o'clock and finally 3 o'clock positions, tighten the screws to 80 inch lbs. (9 Nm).

17. Install the lock cylinder set.

18. Attach the SIR coil assembly, if necessary, by pulling wire connector through the steering column and attach the wire protector.

19. Install the turn signal assembly.

20. Be sure to center SIR coil assembly before reinstalling SIR inflator module and steering wheel.

21. Connect the negative battery cable and enable SIR system.

Headlight Switch

REMOVAL AND INSTALLATION

─────── **CAUTION** ───────
The Supplemental Inflatable Restraint (SIR) system must be disarmed before working around the steering column. Failure to do so may cause accidental deployment of the air bag, resulting in unnecessary SIR system repairs and/or personal injury.

Roadmaster and Roadmaster Estate Wagon

1993 Vehicles

1. Disconnect the negative battery cable.

2. Disable the SIR system using the recommended procedure.

3. Remove the bolts/screws attaching the filler assembly to the instrument panel carrier.

4. Gently pull the filler assembly assembly from the instrument panel carrier.

5. Loosen to the end of the threads, the capsule nuts attaching the steering column support bracket to the instrument panel carrier.

6. Gently lower the steering column assembly.

7. Remove the 8 bolts/screws attaching the trim plate assembly to the instrument panel carrier.

8. Remove the trim plate assembly from the instrument panel carrier.

9. Carefully unsnap the trim plate assembly and pull it away.

10. Remove the bolts/screws attaching the switch assembly to the instrument panel.

11. Pull the switch assembly away from the instrument panel.

12. Remove the electrical connectors from the headlight switch assembly.

13. Remove the bulb and socket if necessary.

To install:

14. Install the bulb and socket if removed.

15. Connect the instrument panel harness connectors to the headlight switch assembly.

16. Install the headlight switch into the instrument panel carrier.

17. Install the bolts/screws and tighten to 21 inch lbs. (2.4 Nm).

18. Align locator pins and clips on the trim plate to the instrument panel and carefully snap into place.

19. Raise the steering column into place.

20. Tighten the steering column retainer nuts to 20 ft. lbs. (27 Nm).

21. Install the 8 bolts/screws attaching the trim plate assembly to the instrument panel carrier.

22. Align the filler panel with the instrument panel carrier and snap it into place.

23. Install the 2 bolts/screws attaching the filler assembly to the instrument panel carrier.

24. Enable the SIR system.

25. Connect the negative battery cable.

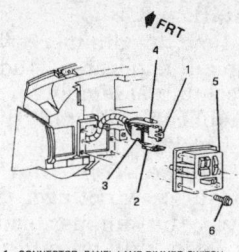

1. CONNECTOR, PANEL LAMP DIMMER SWITCH
2. CONNECTOR, TWILIGHT SENTINEL SWITCH
3. CONNECTOR, HEADLAMP SWITCH
4. LAMP, HEADLAMP SWITCH INDICATOR
5. SWITCH ASSEMBLY, HEADLAMP AND INSTRUMENT PANEL LAMP DIMMER (WITH HOUSING)
6. BOLT/SCREW, HEADLAMP AND INSTRUMENT PANEL LAMP DIMMER AND ACCESSORY SWITCH, 2.4 N·m (21 LB. IN.)

335438

Headlight switch assembly — 1993 Roadmaster

26. Check the operation of all the lights including the twilight sentinel system.

1994–96 Vehicles

1. Disconnect the negative battery cable.

2. Disable the SIR system following the recommended procedure.

3. Remove the fuse block access door.

4. Remove the instrument panel trim plate assembly.

5. Remove the headlamp switch knob with shaft from the switch assembly. To release the shaft, press button on top of the switch assembly firmly and remove shaft.

6. Remove the bolts attaching switch housing to the instrument panel assembly. Unsnap top of the switch housing.

7. Remove the lamp connector and twilight sentinel connector, if removed.

8. Unscrew the headlamp switch bezel from the switch assembly.

9. Remove the headlamp and interior lamp switch housing.

10. Remove the instrument panel harness connector from the switch assembly.

To install:

11. Install the panel harness connector to switch assembly.

12. Align the switch assembly with hole in instrument panel assembly.

13. Install the screw in headlamp switch bezel.

14. Install the lamp connector and twilight sentinel connector, if removed.

15. Align the clips on switch housing to instrument panel assembly and snap into place.

16. Install the bolts attaching switch housing to the instrument panel assembly.

17. Insert the headlamp switch knob with the shaft into lamp assembly and snap into place.

18. Install the instrument panel trim plate assembly.

19. Install the fuse block access door.

20. Enable the SIR system.

21. Connect the negative battery cable.

Fleetwood

1. Disconnect the negative battery cable.

2. Remove the left side sound insulator panel.

3. Remove the left side knee bolster and deflector.

4. Remove the headlamp switch assembly from the instrument panel using a suitable flat bladed tool to pry it out. Use thumb pressure to press the lock tang while pulling out on the Twilight Sentinel knob.

5. Disconnect the electrical connectors from the switch assembly.

To install:

6. Connect the electrical connectors to the headlamp switch assembly.

7. Install the headlamp switch assembly into the instrument panel. The lock tabs on the switch will hold the assembly in place as long as they are seated fully.

8. Install the left side knee bolster and deflector.

9. Install the left side sound insulator.

10. Connect the negative battery cable.

Caprice and Impala

1993 Vehicles

1. Disconnect the negative battery cable.

2. Disable the SIR system.

3. Remove the bolts/screws attaching the filler assembly to the instrument panel carrier.

4. Gently pull the filler assembly assembly from the instrument panel carrier.

5. Open the glove box door and gently unsnap the molding assembly from the instrument panel carrier.

6. Loosen to the end of the threads, the capsule nuts attaching the steering column support bracket to the instrument panel carrier.

7. Gently lower the steering column assembly.

8. Remove the 8 bolts/screws attaching the trim plate assembly to the instrument panel carrier.

9. Remove the trim plate assembly from the instrument panel carrier.

10. Carefully unsnap the trim plate assembly and pull it away.

11. Remove the bolts/screws attaching the switch assembly to the instrument panel.

12. Pull the switch assembly away from the instrument panel.

13. Remove the electrical connectors from the headlight switch assembly.

14. Remove the bulb and socket if necessary.

To install:

15. Install the bulb and socket if removed.

16. Connect the instrument panel harness connectors to the headlight switch assembly.

17. Install the headlight switch into the instrument panel carrier.

18. Install the bolts/screws and tighten to 21 inch lbs. (2.4 Nm).

19. Align locator pins and clips on the trim plate to the instrument panel and carefully snap into place.

20. Raise the steering column into place.

21. Tighten the capsule nuts to 20 ft. lbs. (27 Nm).

22. Align the locator pins and gently snap the molding assembly to the instrument panel carrier.

23. Install the 8 bolts/screws attaching the trim plate assembly to the instrument panel carrier.

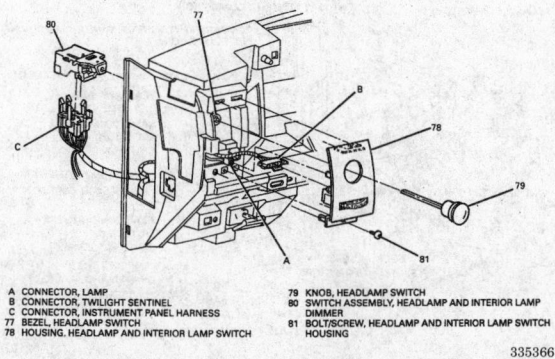

A CONNECTOR, LAMP
B CONNECTOR, TWILIGHT SENTINEL
C CONNECTOR, INSTRUMENT PANEL HARNESS
77 BEZEL, HEADLAMP SWITCH
78 HOUSING, HEADLAMP AND INTERIOR LAMP SWITCH

79 KNOB, HEADLAMP SWITCH
80 SWITCH ASSEMBLY, HEADLAMP AND INTERIOR LAMP DIMMER
81 BOLT/SCREW, HEADLAMP AND INTERIOR LAMP SWITCH HOUSING

335366

Headlight switch assembly — 1994–96 Roadmaster

24. Align the filler panel with the instrument panel carrier and snap it into place.

25. Install the 2 bolts/screws attaching the filler assembly to the instrument panel carrier.

26. Enable the SIR system

27. Connect the negative battery cable.

28. Check the operation of all the lights including the twilight sentinel system.

1994–96 Vehicles

1. Disconnect the negative battery cable.

CAUTION

The Supplemental Inflatable Restraint (SIR) system must be disarmed before working around the steering column. Failure to do so may cause accidental deployment of the air bag, resulting in unnecessary SIR system repairs and/or personal injury.

---

2. Remove instrument panel lower trim plate assembly including instrument panel cluster trim plate assembly.

3. Remove bolt attaching headlamp switch panel assembly to the instrument panel assembly.

4. Unsnap headlamp switch panel assembly from the instrument panel assembly.

5. Removal headlamp connector and dimmer connector and Twilight Sentinel connector, if equipped.

6. Remove wagon switches and accessory switch panel assembly from behind headlamp switch panel assembly, if equipped.

7. Remove instrument panel compartment assembly from behind headlamp switch panel assembly, if equipped.

8. Remove instrument panel harness connector from rear compartment lid release switch assembly, if equipped. Unsnap switch assembly from behind the housing.

To install:

9. Install instrument panel harness connector to rear compartment lid release switch assembly, if removed.

10. Install instrument panel compartment assembly to the headlamp switch assembly, if removed.

11. Install wagon switches to accessory switch panel assembly, if removed.

12. Install headlamp connector and dimmer connector and Twilight Sentinel connector, if removed.

13. Align locator pin on the instrument panel assembly with hole in bottom of headlamp switch assembly.

14. Snap clips on the switch assembly to slots on the instrument panel assembly.

15. Install bolts securing headlamp switch assembly to the instrument panel assembly.

16. Install instrument panel lower trim plate assembly including panel cluster trim plate assembly.

17. Connect the negative battery cable.

Dimmer Switch

REMOVAL AND INSTALLATION

NOTE: If equipped with an air bag system, make certain to follow the recommended disarming procedure before, and rearming procedure after, repairs.

1. Disable the SIR system, if equipped, then disconnect the negative battery cable.

2. The dimmer switch is attached to the lower steering column jacket. Remove trim panels, as necessary for access to the switch. If necessary, lower the steering column assembly and/or remove the column from the vehicle to ease access to the switch.

3. Remove the nut and screw attaching the switch and bracket to the steering column jacket.

4. If equipped, remove the shift interlock solenoid assembly and bracket from the dimmer/ignition switch mounting stud.

5. Disengage the vehicle wire harness from the switch assembly and, if applicable, remove the washer head screw, then remove the switch from the steering column.

To install:

6. Position the dimmer switch to the rod and, if applicable, stud on the steering column assembly.

7. If applicable, install the solenoid and bracket.

8. Install the screw, nut and/or washer head screw(s) finger-tight.

9. Adjust the dimmer switch by placing a $\frac{3}{32}$ in. drill bit into the hole on the switch in order to limit travel, then push the switch against the actuator rod to remove all lash. When properly adjusted, an audible click should be heard from the switch if the multifunction lever is pulled to activate the high/low beams.

10. Remove the drill bit and tighten the retainers while holding the switch in position. As applicable, tighten the washer head screw(s) and the hexagon nut to 35 inch lbs. (4.0 Nm) and the tapping screw to 22 inch lbs. (2.5 Nm).

11. Engage the vehicle wire harness to the switch assembly.

12. As applicable, install and/or raise the steering column assembly into position, then install the trim plates.

13. Connect the negative battery cable and, if equipped, properly enable the SIR system.

Turn Signal Switch

REMOVAL AND INSTALLATION

NOTE: If equipped with an air bag system, make certain to follow the recommended disarming and coil centering procedure before and after repairs.

1. Disable the SIR system, then disconnect the negative battery cable.

2. Remove the inflator module, then remove the steering wheel from the vehicle.

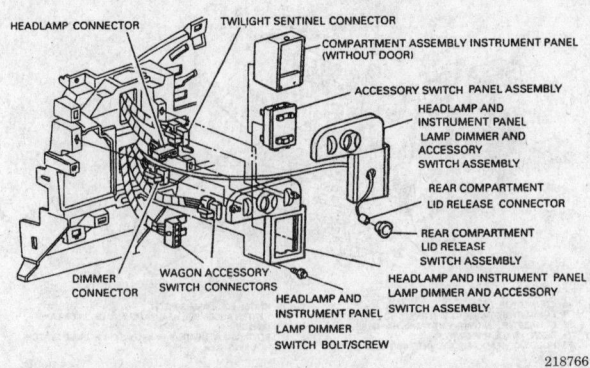

Headlamp switch assembly — 1994–96 Caprice and Impala

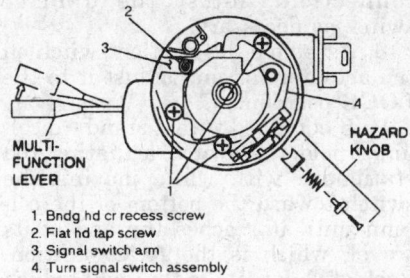

1. Bndg hd cr recess screw
2. Flat hd tap screw
3. Signal switch arm
4. Turn signal switch assembly

8838QG08

Removing the hazard switch and multifunction lever in preparation for turn signal switch removal

—————— CAUTION ——————

To avoid personal injury when carrying a live inflator module, make sure the bag and trim cover are pointed away. Always face the air bag assembly up, and never carry the inflator module by the wires or connector, otherwise personal injury may result if the module should deploy.

3. Remove the coil assembly retaining ring and allow the coil assembly to hang freely.
4. Remove the wave washer.
5. Remove the shaft lock bolt guard:
   a. Turn the ignition switch to the **RUN** position.
   b. Rotate shaft so the blocking tooth is at 7 o'clock and bolt guard screws are accessible through large slots on lock shaft.
   c. Loosen screws on lock bolt guard and remove.
   d. Return ignition to the **LOCK** position.
6. Remove and discard the shaft lock retaining ring using tool J-23653-C or equivalent.

—————— CAUTION ——————

Use a 1/2 in. wrench to hold the shaft of tool J-23653-C or equivalent, stationary when releasing the nut. Failure to do so may cause the tool to fly off and cause personal injury.

7. Remove the shaft lock, turn signal cancelling cam, upper bearing spring, upper bearing inner race seat and inner race.
8. Turn the multifunction lever to the **RIGHT TURN** position, then remove the multifunction lever and hazard knob assembly.
9. Remove the retaining screw and signal switch arm.
10. If the signal switch is being completely removed from the column assembly, remove the wiring protec-

tor from the steering column and disengage the switch connector.
11. Remove the screws retaining the turn signal switch to the steering column, using care not to drop the screws in the column.
12. If the switch is being completely removed from the column, attach a length of mechanic's wire to the harness connector to aid in reinstallation and gently pull wire harness through the steering column. Leave the wire in the column in order to pull the new harness into position.

To install:
13. Using the mechanic's wire, gently pull the turn switch connector through the steering column and attach switch connector to the vehicle wire harness.
14. Install the turn switch assembly and tighten the screws to 30 inch lbs. (3.4 Nm).
15. Install the signal switch arm and tighten the retaining screw to 20 inch lbs. (2.3 Nm).
16. Install the hazard knob assembly and multifunction lever.
17. Install the inner race, upper bearing race seat and upper bearing spring.
18. Lubricate turn signal cancelling cam with synthetic grease, then install the cancelling cam assembly.
19. Install shaft lock and new shaft lock retaining ring using tool J-23653-C or equivalent.
20. Install shaft lock bolt guard as follows:
   a. Turn ignition switch to the **RUN** position.
   b. Rotate shaft until the block tooth is at the 7 o'clock position and bolt guard screw holes are accessible though large slots on lock shaft.
   c. Tighten screws on lock bolt guard until they bottom out and torque to 20 inch lbs. (2.3 Nm).
21. Install the wave washer.
22. Install and center the coil assembly and retaining ring.
23. If removed, install the wiring protector.
24. Install steering wheel, then install the SIR inflator module.
25. Connect the negative battery cable, then properly enable the SIR system.

Ignition Lock Cylinder

REMOVAL AND INSTALLATION

—————— CAUTION ——————

When performing service on or around SIR components or wiring, follow the procedures to dis-

able the SIR system. Failure to follow procedures could result in possible air bag deployment, personal injury or unneeded SIR system repairs.

Except Fleetwood Brougham

1. Disable the SIR system, then disconnect the negative battery cable.
2. Remove the turn signal switch from the column and allow to hang freely from the wires.
3. If necessary, remove the SIR coil assembly from the column as follows:
   a. Remove wiring protector.
   b. Attach a length of mechanic's wire to the terminal connector to aid in reassembly.
   c. Carefully pull wire through the column.
4. Remove the key from the lock cylinder and remove the buzzer switch assembly.
5. Reinsert the key into the lock cylinder, be sure the key is in the **LOCK** position.
6. Remove the lock cylinder retaining screw.
7. Remove the lock cylinder from the steering column.

To install:
8. Reinstall the lock cylinder set with key inserted. Install the retaining screw and tighten to 22 inch lbs. (2.5 Nm).
9. Remove the key and install the buzzer switch with retaining clip, then insert the key.
10. If removed, gently pull the coil assembly through the steering column and allow assembly to hang freely.
11. Install the turn signal switch assembly, making sure to follow proper SIR coil centering procedures.
12. Connect the negative battery cable and enable the SIR system.

Fleetwood Brougham

1. Disable the SIR system, then disconnect the negative battery cable.
2. Remove the steering wheel.
3. Remove the spacers, the steering shaft bumper and the plastic retainer, as applicable.
4. Install a suitable lock plate tool onto the steering shaft. Tighten the tool to compress the lock plate and the spring. Remove the lock retainer.
5. Remove the lock plate, carrier assembly and the upper bearing spring from the upper steering shaft.

6. Insert the key into the ignition switch and turn the ignition switch to the **RUN** position.

NOTE: With "AIR BAG" fuse removed and ignition switch in "RUN" position, the "AIR BAG" lamp on the instrument panel cluster will light. This is normal and does not indicate an SIR fault.

7. Remove the key warning buzzer switch and retaining clip.

8. Remove the ignition cylinder retaining screw located inside the lock housing cover.

9. Remove the lock cylinder from the column.

To install:

10. Install the lock cylinder to the steering column and install the lock cylinder retaining screws.

11. Install the key warning buzzer and retaining clip, with the cylinder still in the **RUN** position.

12. Install the upper bearing spring, the carrier assembly and the lock plate, from the upper steering shaft.

13. Using a suitable lock plate tool, compress the lock plate and the spring and install the lock retainer.

14. Install the plastic retainer, the steering shaft bumper and the spacers, as applicable.

15. Turn the ignition to **LOCK**.

16. Connect the negative battery cable.

17. Enable the SIR system.

18. Turn ignition switch to **"RUN"** and verify that the "AIR BAG" lamp flashes 7 times and turns OFF.

19. Install the steering wheel.

Ignition Switch

REMOVAL AND INSTALLATION

Except Fleetwood Brougham

NOTE: If equipped with an air bag system, make certain to follow the recommended disarming procedure before, and rearming procedure after, repairs.

1. If equipped, disable the SIR system, then disconnect the negative battery cable.

2. Either lower the steering column or remove the column assembly from the vehicle, for access to the switch.

NOTE: The steering column must be supported at all times to prevent damage.

3. Remove the solenoid cable assembly from the interlock solenoid

assembly and bracket, then remove the assembly from the dimmer/ignition switch mounting bracket. Remove the ball joint spring, remove the washer head screw to remove the solenoid assembly, then remove the tapping screw and nut to remove the solenoid bracket.

4. Remove the washer head screw(s) from the dimmer switch assembly, then unplug the connector and remove the switch from the column.

5. Remove the ignition switch screws or the dimmer/ignition switch mounting stud, as applicable. Unplug the connector, then remove the ignition switch from the column.

To install:

6. Make sure the ignition key is in the **LOCK** or **OFF-LOCK** position, as applicable, then place the ignition switch in the proper position and install it to the steering column assembly and actuator rod. Move the switch slider to the extreme LEFT position, then move the slider 1 detent to the RIGHT (**OFF-LOCK**) position and insert a $^3/_{32}$ in. drill bit into the switch hole.

7. Install the ignition/dimmer switch mounting stud and tighten to 35 inch lbs. (4 Nm), then remove the drill bit from the ignition switch and connect the switch wiring harness. Install the dimmer switch followed by the interlock solenoid and bracket, hand-tightening all fasteners. Adjust the switch assembly, then tighten the fasteners.

8. Either install and/or raise the steering column into position in the vehicle.

9. Connect the negative battery cable, then enable the SIR system.

Fleetwood Brougham

1. If equipped, disable the SIR system.

2. Disconnect negative battery cable and set the ignition in the **LOCK** position.

3. Remove left sound insulator and the lower column cover as necessary to access the steering column.

4. Remove 2 nuts securing steering column to upper mounting bracket.

5. Carefully lower and support the steering column.

6. Remove the 2 screws securing the ignition switch and the dimmer switch. Detach electrical connection and remove the ignition switch.

To install:

7. Install the dimmer and ignition switches to the steering column and connect the ignition switch electrical

connection. Adjust the dimmer switch as necessary.

8. Assemble the ignition switch on the actuator rod and adjust it to the **LOCK** position.

9. If equipped with a standard column, hold the switch actuating rod stationary with while moving the switch toward the bottom of the column until it reaches the end of its travel, which is the **ACC** position. Back off 2 detents to the right, which is the **OFF/UNLOCK** position, then with the key also in the **OFF/UNLOCK** position, tighten the switch mounting screws to 35 inch lbs.

10. If equipped with a tilt wheel, hold the switch actuating rod stationary with one hand while moving the switch toward the upper end of column until it reaches the end of its travel, which is the **ACC** position. Back off 1 detent and with the key in **LOCK** position, tighten the switch mounting screws to 35 inch lbs.

11. Raise the steering column into position and attach the securing nuts.

12. Install the left sound insulator and the lower steering column support, as necessary.

13. Connect the negative battery cable and enable the SIR system.

Park/Neutral Safety Switch

OPERATION

All steering columns use a mechanical neutral start safety system. The mechanical system relies on a block which prevents starting the engine in positions other than **P** or **N**. The mechanical block is achieved by a wedge shaped finger added to the ignition switch actuator rod. The finger will only pass through the bowl plate notches when the shift lever is in the **P** and **N** positions, which then allows the lock cylinder to rotate to the **START** position.

Powertrain Control Module

LOCATION

The Powertrain Control Module (PCM) is located on the right side of the vehicle. On the Fleetwood Brougham, it is positioned in front of the right kick panel. In order to gain access to the assembly, the trim panel and, if necessary, the glove box must first be removed. Except for the

Fleetwood Brougham, it is located behind the front right side kick panel.

ENGINE COOLING

Radiator

REMOVAL AND INSTALLATION

Except Fleetwood Brougham

1. Disconnect the negative battery cable. Drain the cooling system.
2. Remove the fan shrouds.
3. Disconnect the radiator inlet and outlet hoses from the radiator assembly.
4. Disconnect and the engine oil cooler and/or transmission fluid cooler lines from the radiator. Plug all openings to prevent system contamination or excessive fluid loss.
5. If equipped, disengage the low fluid sensor connector.
6. Disconnect the coolant reservoir hose from the radiator.
7. If applicable, disconnect the heater hoses from the radiator.
8. Remove the radiator from the vehicle.

To install:

9. Position radiator in place making sure the radiator is seated on the insulators.
10. Connect the coolant recovery hose to the filler neck.
11. If equipped, connect the transmission fluid lines to the radiator and tighten to 15 ft. lbs. (21 Nm).
12. If equipped, connect the engine oil cooler lines to the radiator and tighten to 18 ft. lbs. (24 Nm).
13. Connect the radiator inlet and outlet hoses along with their clamps to the radiator.
14. If applicable, connect the heater hoses and clamp to the radiator.
15. If equipped, connect the wiring harness to the low coolant sensor.
16. Connect the upper and lower fan shrouds.
17. Connect the negative battery cable, then add coolant and properly bleed the engine cooling system.
18. Run the engine and check system for leaks.

Fleetwood Brougham

1. Disconnect the negative battery cable. Drain the cooling system.
2. Disconnect the top and bottom radiator hoses from the radiator. Re-move the reservoir hose from the radiator filler neck.
3. Disconnect and plug the transmission fluid cooler lines. Disconnect and plug the oil cooler lines, if equipped.
4. Remove the bolts retaining the engine compartment support rod to the radiator core support. Loosen each anchor bolt and position the support rods aside.
5. Remove the fan shroud retaining bolts. If without VO8 heavy duty cooling, remove the twin electric cooling fans and position the fan shroud assembly aside.
6. Remove the radiator core support cover retaining bolts. Remove the radiator core support cover.
7. Carefully lift the radiator assembly upward and out of the vehicle.

To install:

8. Lower the radiator into place making sure it is properly seated on the insulator.
9. Install the radiator core support cover and electric fans, if equipped.
10. Position the fan shroud in place and install the retaining bolts.
11. Position the engine compartment support rods to the radiator core support. Tighten the support bar bolts to 22 ft. lbs. (30 Nm).
12. Install the engine oil cooler lines, if applicable, and the transmission oil cooler lines. Tighten the fittings to 20 ft. lbs. (27 Nm).
13. Install the top, bottom and reservoir hoses and tighten the clamps to 26 inch lbs. (3 Nm).
14. Connect the negative battery cable and refill the cooling system with the proper type and quantity of coolant mixture.

Water Pump

REMOVAL AND INSTALLATION

Except 4.3L (VIN W) and 5.7L (VIN P) Engines

1. Disconnect the negative battery cable.
2. Remove cooling fan as follows:
    a. Remove upper fan shroud retaining nuts and bolts, then remove the upper fan shroud.
    b. Remove nuts and fan clutch/blade assembly.
    c. If equipped, remove the spacer. If necessary remove the connection bolts and separate the cooling fan and clutch.

NOTE: Keep the fan clutch in an upright position during repairs to prevent the silicone fluid from leaking out.

3. Relieve belt tension, then remove the serpentine belt from the water pump pulley. Remove the pulley from the hub in order to access the water pump assembly.
4. Drain the cooling system.
5. Unfasten the radiator hose.
6. Remove the bolts securing the water pump, then remove the pump from the engine.

To install:

7. Clean cylinder block and coolant pump gasket surfaces and discard old gaskets.
8. Place new gaskets on water pump and mounting bolts, then install the pump making sure the gaskets remain in position. Tighten the mounting bolts to 23 ft. lbs. (31 Nm).
9. If removed, position and install the air conditioning compressor and power steering brackets.
10. Fasten heater bypass and/or radiator hose to the pump, as applicable.
11. Install the water pump pulley, then install the serpentine drive belt.

——— WARNING ———

Inspect fan blade for bends or damage. Do not use or attempt to repair a fan blade which has been bent or damaged. It is essential that a fan blade remains in balance to prevent failure and possible injury.

12. Attach cooling fan as follows:
    a. If removed, attach the fan blade to the clutch and tighten the bolts to 18 ft. lbs. (24 Nm).
    b. Place spacer, if equipped, then attach the fan assembly to the cooling pump. Be sure to align reference marks on the fan clutch and coolant pump hub.
    c. Tighten nuts to 18 ft. lbs. (24 Nm).
    d. Install the upper fan shroud and tighten screws to 53 inch lbs. (5.8 Nm).
13. Connect the negative battery cable, then add coolant to engine.
14. Start engine and check for leaks.

4.3L (VIN W) and 5.7L (VIN P) Engines

1. Disconnect the negative battery cable.
2. Disengage the wiring harness from the cooling fan assembly, then

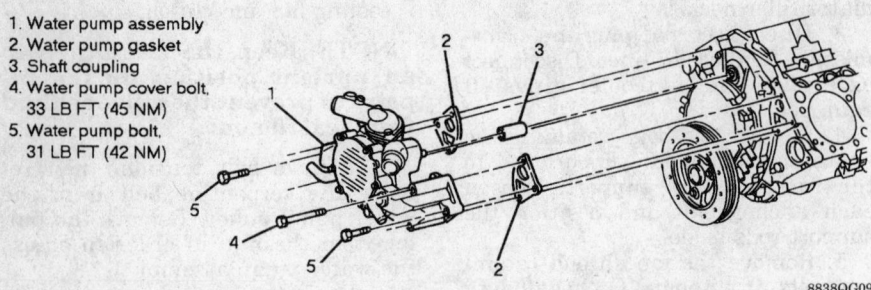

1. Water pump assembly
2. Water pump gasket
3. Shaft coupling
4. Water pump cover bolt.
   33 LB FT (45 NM)
5. Water pump bolt.
   31 LB FT (42 NM)

8838QG09

Water pump assembly removal and installation — 4.3L (VIN W) and 5.7L (VIN P) engines

remove the assembly from the vehicle.

3. Drain the engine cooling system, removing the block drain plug and the knock sensor to assure proper draining. Reinstall the drain plug and knock sensor, as soon as the system is empty.

4. Disconnect the upper and lower radiator hoses from the water pump assembly.

5. Remove the heater hose assemblies from the water pump and from the throttle body.

6. Disengage the coolant sensor wiring harness, then reposition the ignition coil and bracket assembly.

7. Remove the shorter water pump retaining bolt from the center of each pump mating flange, then remove the longer pump bolts from either side of the center bolts.

8. Carefully remove the water pump assembly and gaskets along with the pump shaft coupling.

To install:

9. Thoroughly clean the gasket mating surfaces of any remaining gasket material.

10. Install the water pump shaft coupling along with the water pump and gaskets.

11. Install the longer pump bolts and tighten 33 ft. lbs. (45 Nm), then install the shorter bolts and tighten to 31 ft. lbs. (42 Nm).

12. Reposition the ignition coil and bracket assembly.

13. Engage the coolant sensor electrical connector.

14. Install the heater and radiator hoses to the throttle body and water pump, as applicable.

15. Install the air cleaner and intake duct assemblies.

16. Install the engine cooling fan assembly and engage the wiring harness connector.

17. Connect the negative battery cable and properly fill the engine cooling system.

Thermostat

REMOVAL AND INSTALLATION

Except 4.3L (VIN W) and 5.7L (VIN P) Engines

1. Disconnect the negative battery cable.

2. Drain the coolant from the radiator into a suitable container until the level is below the thermostat housing.

3. If necessary for access, remove the air cleaner assembly from the engine.

4. Remove the radiator inlet hose from the thermostat housing assembly.

5. Remove the thermostat housing bolts, then remove the housing and the thermostat.

6. Installation is the reverse of removal. Clean the sealing surfaces, use a new gasket and tighten the housing retaining bolts to 21 ft. lbs. (28 Nm).

4.3L (VIN W) and 5.7L (VIN P) Engines

1. Open the radiator drain cock and allow the engine coolant to drain to a level just below the thermostat housing, then close the drain and tighten to 106 inch lbs. (12 Nm).

2. Remove the air cleaner assembly.

3. Disconnect the inlet hose and clamp from the thermostat assembly.

4. Remove the thermostat housing bolts, then remove the housing.

5. If equipped, remove the housing gasket.

6. Remove the thermostat assembly.

To install:

7. Make sure the flange mating surfaces are clean and free of debris, then install the thermostat assembly.

8. If equipped, install a new housing gasket.

9. Install the housing and tighten the retaining bolts to 21 ft. lbs. (28 Nm), then connect the inlet hose assembly and clamp.

10. Install the air cleaner assembly, then properly refill the engine cooling system.

Cooling System Bleeding

PROCEDURE

Except 4.3L (VIN W) and 5.7L (VIN P) Engines

1. With the cooling system completely drained, begin adding a combination of ethylene glycol antifreeze and water to achieve a mixture of at least 50 percent, but not exceeding 70 percent antifreeze.

2. Fill the radiator up to the lower portion of the filler neck.

3. Fill the coolant recovery reservoir to the **COLD FILL** or **FULL COLD** mark, then install the coolant recovery cap.

4. Start the vehicle and run the engine with the radiator cap removed until normal engine operating temperature is reached.

5. With the engine idling, add coolant to the radiator until the level reaches the bottom of the filler neck.

6. Install the radiator cap, making sure the arrow on the cap is aligned with the coolant recovery hose.

7. Inspect the system for leaks.

4.3L (VIN W) and 5.7L (VIN P) Engines

1. Make sure the vehicle is parked or supported on a level surface with the pressure cap removed. Drain the engine cooling system by opening the radiator drain cock, then opening the air bleed vents on the thermostat housing and the heater outlet hose pipe. To assure the system is fully drained, remove the engine block drain plug and the engine knock sensor from either side of the block.

2. Once the system is completely drained, close the drain cock and install the knock sensor. Cover the threads of the drain plug with sealer, then install the plug into the block. The drain plug should be tightened to 15 ft. lbs. (21 Nm), while the knock sensor should be tightened to 14 ft. lbs. (19 Nm).

3. After the repair or service has been completed, fill the cooling system with a solution which is at least 50 percent ethylene glycol antifreeze and the balance water. Do not use a solution that exceeds 70 percent antifreeze. Continue filling the radiator

until the level is just below the filler neck.

4. Once the system is filled, close the air bleeds, taking care not to overtighten and damage the brass valves.

5. Install the pressure cap, then start and run the engine to check for leaks.

FUEL SYSTEM

Fuel System Service Precautions

Safety is the most important factor when performing not only fuel system maintenance but any type of maintenance. Failure to conduct maintenance and repairs in a safe manner may result in serious personal injury or death. Maintenance and testing of the vehicle's fuel system components can be accomplished safely and effectively by adhering to the following rules and guidelines.

• To avoid the possibility of fire and personal injury, always disconnect the negative battery cable unless the repair or test procedure requires that battery voltage be applied.

• Always relieve the fuel system pressure prior to disconnecting any fuel system component (injector, fuel rail, pressure regulator, etc.), fitting or fuel line connection. Exercise extreme caution whenever relieving fuel system pressure to avoid exposing skin, face and eyes to fuel spray. Please be advised that fuel under pressure may penetrate the skin or any part of the body that it contacts.

• Always place a shop towel or cloth around the fitting or connection prior to loosening to absorb any excess fuel due to spillage. Ensure that all fuel spillage (should it occur) is quickly removed from engine surfaces. Ensure that all fuel soaked cloths or towels are deposited into a suitable waste container.

• Always keep a dry chemical (Class B) fire extinguisher near the work area.

• Do not allow fuel spray or fuel vapors to come into contact with a spark or open flame.

• Always use a backup wrench when loosening and tightening fuel line connection fittings. This will prevent unnecessary stress and torsion

to fuel line piping. Always follow the proper torque specifications.

• Always replace worn fuel fitting O-rings with new. Do not substitute fuel hose or equivalent, where fuel pipe is installed.

Fuel System Pressure

RELIEVING

1993 Vehicles With TBI

1. Disconnect the negative battery cable to prevent fuel spillage if an attempt is made to start the vehicle while fittings are still disconnected.

2. Loosen the fuel filler cap to relieve tank vapor pressure and leave the cap loosened until service is completed.

NOTE: The internal constant bleed feature of the Model 220 TBI relieves fuel pump system pressure when the engine is turned OFF. Therefore, no further relief procedure is required.

3. Be sure to tighten fuel filler cap when maintenance or repairs are finished.

1994–96 Vehicles With SFI

1. Disconnect the negative battery cable to prevent fuel discharge if the key is accidentally turned to the RUN position.

2. Loosen the fuel filler cap to relieve the tank pressure and do not tighten until service has been completed.

3. Connect J-34730-1 fuel pressure gauge or equivalent, to the fuel pressure valve. Wrap a shop cloth around the fitting while connecting the gauge to avoid spillage.

4. Place the end of the bleed hose into a suitable container and open the valve to relieve the fuel system pressure.

Idle Speed

ADJUSTMENT

The idle speed and mixture are electronically controlled by the Powertrain Control Module (PCM). All adjustments are preset at the factory and do not need periodic attention. Some throttle body units are equipped with an idle stop screw to allow adjustment of the minimum idle speed if the unit is used as a replacement. The only time the idle speed should require adjustment is

when the throttle body assembly has been replaced.

TBI Systems

1. Block the drive wheels and apply the parking brake. Remove the air cleaner assembly and/or air duct.

2. Connect a scan tool to the Data Link Connector (DLC) and select the field service mode. Turn the ignition ON and leave the engine OFF. Wait at least 45 seconds, this will allow the Idle Air Control (IAC) pintle to seat in the throttle body.

3. With the ignition switch in the ON position, the engine OFF and the scan tool in field service mode, disengage the IAC valve electrical connector and the distributor set-timing connector.

4. Connect a tachometer to the engine to monitor the engine speed.

5. Place the transmission in the P or N position and start the engine.

6. Run the engine until it reaches normal operating temperature or closed loop operation as indicated by the scan tool. It may be necessary to hold the throttle open slightly in order to maintain idle.

7. The idle speed should be 450–500 rpm, be sure the throttle and cruise control cables do not hold the throttle open. If not as specified, remove the idle speed stop screw plug and adjust as necessary.

8. Turn the ignition OFF, then reconnect the IAC valve electrical connector and the distributor set-timing connector.

9. Adjust the Idle Air Control (IAC) valve pintle position.

10. Install the air cleaner assembly, check and clear all PCM trouble codes.

SFI Systems

The idle speed and mixture are electronically controlled by the Powertrain Control Module (PCM). All adjustments are preset at the factory and do not require periodic adjustment.

Mixture

ADJUSTMENT

The idle mixture is controlled by the PCM, therefore no service adjustments are necessary. The PCM will change the air/fuel ratio by controlling the fuel injectors, based on oxygen sensor and various other outputs. A 14.7:1 ratio is required for efficient catalytic converter operation.

Fuel Filter

REMOVAL AND INSTALLATION

The fuel injection system uses an in-line filter located in the fuel feed line under the hood, attached to the frame rail or on the rear crossmember of the vehicle. The high pressure fuel system used with all fuel injection systems requires special fuel lines to contain the pressure and utilizes nylon lines with quick-connect fittings.

1. Disconnect the negative battery cable and relieve fuel system pressure.
2. Raise and support the vehicle safely.
3. Remove the filter bracket attaching bolt.
4. While grasping the fuel filter and 1 of the fuel lines, twist the line approximately ¼ turn in each direction to loosen any dirt in the fitting, then use compressed air (and safety glasses) to blow dirt out of the fitting. Squeeze the plastic tabs of the male connector on the fuel lines and the pull connection apart. Repeat for the other fitting.
5. If applicable, remove the fuel feed and return line body harness clips.
6. Remove the filter.
To install:
7. Remove the protective caps from the new filter, then position the fuel filter in the original location with the arrow pointing in correct direction.
8. Install the new connector retainers on the filter inlet and outlet tubes.
9. Apply a few drops of clean engine oil to the male ends of both fuel lines connectors.
10. Push the fuel line connectors onto the fuel filter tubes until their retaining tabs snap into place.
11. Once installed, pull on both ends of the lines to verify they are secure.
12. Secure the filter and bracket to the frame trapping the return pipe and tighten the attaching bolt to 89 inch lbs. (10 Nm).
13. Lower the vehicle.
14. Reconnect the negative battery cable.
15. Turn the ignition **ON** for 2 seconds, **OFF** for 10 seconds, then **ON** again to pressurize the fuel system. Inspect the tank and lines for leaks.

Fuel Pump

REMOVAL AND INSTALLATION

1. Disconnect the negative battery cable and relieve the fuel system pressure.
2. Drain the fuel tank using a suitable hand-operated pump, then raise and support the vehicle safely.
3. Remove the fuel tank.
4. Remove the fuel tank sending unit and pump assembly as follows: remove the assembly attaching nuts, retaining flag, assembly and O-ring from the tank. Discard the O-ring.
To install:
5. Install fuel sending unit in fuel tank as follows: position a new O-ring on fuel tank. Install fuel sender assembly, retaining flag, and attaching nuts to fuel tank. Tighten attaching nuts to 27 inch lbs. (3 Nm).
6. Install fuel tank.
7. Lower vehicle.
8. Turn the ignition **ON** for 2 seconds, **OFF** for 10 seconds, then **ON** again to pressurize the fuel system. Inspect the fuel system for leaks.

Fuel Injector

REMOVAL AND INSTALLATION

1993 Vehicles (TBI Systems)

1. Disconnect the negative battery cable and relieve fuel system pressure.
2. Remove the air cleaner assembly, then disengage the electrical connectors from the fuel injectors by squeezing the plastic tabs and pulling straight up.
3. Remove the fuel meter cover attaching screws, then remove the cover assembly.
4. Remove the fuel meter outlet passage gasket and pressure regulator dust seal. If, upon removal of the fuel meter assembly, the cover gasket is stuck to the fuel meter body, leave it in place. If it is stuck to the fuel meter cover, remove it from the cover and place it on the fuel meter body to protect the body in the next step.
5. With the cover gasket in place on the fuel meter body, carefully pry each injector from the throttle body using a small prytool and a smooth fulcrum. Carefully remove each injector and position aside.
6. Remove and discard the lower (small) O-rings from the injector nozzles.

7. Remove and discard the fuel meter cover gasket.
8. Remove and discard the upper (large) O-rings from top of each fuel injector cavity. If equipped, remove the steel backup washer, from the the top of each injector cavity.
To install:
9. Inspect the fuel injector filter for evidence of dirt and contamination. If present, check for presence of dirt in fuel lines and fuel tank.

NOTE: If replacements are required, ensure that the injector is replaced with an identical part. The model 220 TBI is capable of accepting other types of injectors but other injectors are calibrated for different flow rates and may cause driveability or emission problems.

10. If equipped, install the steel injector backup washer in the counterbore of the fuel meter body.
11. Lubricate new upper (large) O-ring with engine oil and install into the top of the fuel meter cavity, or if equipped, directly over the backup washer. Ensure the O-ring is seated properly and is flush with top of fuel meter body surface.
12. Lubricate new lower (small) O-ring with engine oil and push on nozzle end of injector until it seats against injector fuel filter.

NOTE: Backup washers and O-rings must be installed before the injectors or improper seating of large O-ring could cause fuel to leak.

13. Align the raised lug on each injector base with notch in fuel meter body cavity and install the injector. Push down with moderate pressure on injector until it is fully seated in fuel meter body. The electrical terminals of injector should be parallel with throttle shaft.
14. Install a new pressure regulator dust seal, fuel meter outlet gasket and fuel meter cover gasket.
15. Install the fuel meter cover.
16. Coat the threads of the fuel meter attaching screw with a suitable thread locking compound. Install and tighten the screws to 27 inch lbs. (3 Nm).
17. Engage the electrical connectors to their respective fuel injectors.
18. Tighten fuel filler cap, and reconnect the negative battery cable.
19. Turn the ignition **ON** for 2 seconds, **OFF** for 10 seconds, then **ON** again to pressurize the fuel system. Inspect the fuel system for leaks.

1994–96 Vehicles (SFI Systems)

1. Make sure the ignition is in the **OFF** position, then disconnect the negative battery cable.

2. Properly relieve the fuel system pressure.

3. Disengage the quick-connect fittings at the fuel rail feed and return pipes as follows:

   a. Slide the rubber dust cover from the fitting.

   b. Grasp both ends of a connection and twist ¼ turn in each direction to loosen any dirt. Repeat for other fitting.

   c. While wearing safety glasses, use compressed air to blow out dirt from the fitting.

   d. Insert a fuel line separator tool, into the female connector, then push inward to release the male connector.

   e. Repeat for the other fitting.

4. Disconnect the vacuum line at the pressure regulator, then as necessary, tag and disconnect any remaining vacuum lines which must be removed to access the fuel rail and engine fuel pipes.

5. Remove the fuel injector wiring harness from the routing clips of the fuel rail, then remove the fuel pipe attaching bolt and disengage the injector electrical connectors.

6. Remove the fuel rail attaching bolts and carefully remove the fuel rail assembly along with the injectors, from the top of the intake.

7. Rotate the injector retaining clip to the release position and remove the injector from the fuel rail assembly.

8. Remove and discard the O-ring seals from either side of the injector.

9. Remove and discard the injector retaining clip.

To install:

10. Lubricate the new injector O-rings with clean engine oil and install onto the injector.

NOTE: Always replace injectors using an identical part number as inscribed on top of the old injector.

11. Connect a new retainer clip onto the fuel injector and install the injector to the fuel rail assembly. Rotate the injector retaining clip to the lock position.

12. Install the fuel rail assembly to the intake manifold. Tighten the attaching bolts to 15 ft. lbs. (20 Nm).

13. Rotate the fuel injectors as necessary to avoid stretching the wire harnesses, then engage the injector electrical connections.

14. Install the fuel pipe retaining nut, then position the wiring harness into the routing clips at the fuel rail.

15. Connect the vacuum lines to the intake, as necessary, then connect the vacuum line to the pressure regulator.

16. Apply a few drops of clean engine oil to the male ends of the fuel line quick-connect fittings. Engage the fittings by pushing the connectors together until the retaining tabs snap into place. Pull gently on both sides of each fitting to be sure the connection is secure. When secure, slide the dust covers over the fittings.

17. Tighten the fuel filler cap and connect the negative battery cable.

18. Turn the ignition **ON** for 2 seconds, **OFF** for 10 seconds, then **ON** again and inspect the system for leaks.

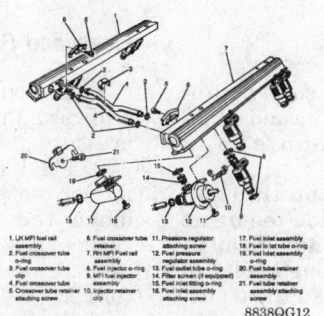

Exploded view of the fuel rail assembly — 4.3L (VIN W) and 5.7L (VIN P) engines

EMISSION CONTROLS

Emission Warning Lamps

The SERVICE ENGINE SOON Malfunction Indicator Light (MIL) located in the instrument cluster serves 2 main functions:

1. The lamp indicates to the driver when a problem has occurred and the vehicle should be taken for service as soon as reasonably possible.

2. The light may be used by technicians to monitor diagnostic trouble codes and/or open/closed loop engine operation, whenever the system is placed in the diagnostic mode.

To verify proper operation of the bulb and wiring, the lamp will illuminate when the ignition is first turned to **ON**, but the engine is not running. If the system is operating properly, the lamp will turn OFF once the engine is started.

If the MIL remains lit once the engine is started, the self-diagnostic system has detected a problem. If the problem goes away, the light will extinguish in 10 seconds (in most cases), but a diagnostic trouble code will remain in the PCM memory.

RESETTING

NOTE: In order to prevent damage to the PCM, the key must be OFF when connecting or disconnecting power to the PCM.

After repairs are made to the faulty system(s), it is necessary to make sure the PCM memory is cleared of any old diagnostic trouble codes. Removing the battery voltage to the PCM for a minimum of 30 seconds will clear all codes. This may be accomplished in various ways depending on how the vehicle is equipped. The PCM harness power feed may be disconnected at the positive battery terminal "pigtail." The fuse may be removed from the inline fuseholder which originates at the positive battery connection or from fuse block, as applicable. Also, the negative battery cable may be disconnected, but other on-board data such as the clock or radio presets will also be lost.

PCM LEARNING ABILITY

The PCM has a "learning" ability which allows it to make corrections for minor variations in the fuel system, in order to improve driveability. If the battery is disconnected to clear diagnostic codes or for safety during repairs, the "learning" process will reset and must begin again. A change may be noted in the vehicle's performance while the learning process begins. To "teach" the vehicle, make sure the engine is at normal operating temperature, then drive the vehicle at part throttle, with moderate acceleration and idle conditions, until normal performance returns.

ENGINE MECHANICAL

NOTE: Disconnecting the negative battery cable on some vehicles may interfere with the functions of the on-board computer systems and may require the computer to undergo a relearning process, once the negative battery cable is reconnected.

Engine Assembly

REMOVAL AND INSTALLATION

Except Fleetwood Brougham

1. Disconnect the negative cable, then the positive cable from the battery.
2. Relieve the fuel system, then scribe alignment marks and remove the hood from the vehicle.
3. Remove the air cleaner assembly and drain the engine cooling system into a suitable container.
4. Remove the radiator hoses and upper fan shroud.
5. Remove the engine cooling fan, then remove the radiator from the vehicle.
6. Disconnect the heater hoses at the engine
7. Disconnect the power steering pump and air conditioning compressor brackets and position out of the way. Be careful not to kink or damage the fluid or refrigerant lines.
8. Disconnect the accelerator, TV, and cruise control cables.
9. Tag and disconnect all necessary vacuum hoses.
10. Disconnect the PCM wiring harness, the engine wiring harness at the engine bulkhead, engine-to-bulkhead ground strap and any remaining wires between body and engine.

11. Remove the distributor assembly from the engine.
12. Remove the windshield wiper motor assembly.
13. Remove the MAP sensor, then for 1993 vehicles remove the EGR solenoid.
14. Remove the negative battery cable from the cylinder head.
15. If not done already, remove the brake pipe from the intake manifold.
16. Raise and support the vehicle safely.

———— WARNING ————
Never raise the engine using a jack under the oil pan, crankshaft pulley or any sheetmetal. Because there only is a small clearance between the oil pan and the oil pump screen, if the pan is bent even slightly, damage could occur to the pump screen and pickup unit.

17. Disconnect the battery positive cable and wires at the starter motor. Be sure to tag the wires for installation purposes.
18. Disconnect the crossover pipe and catalytic converter as an assembly.
19. Remove the flywheel cover and torque converter-to-flywheel bolts.
20. Remove the engine mount through-bolts.
21. Disconnect the front fuel hoses from the front fuel pipes.
22. Disconnect the transmission converter clutch wiring at the transmission and the transmission oil cooler lines at the clip on the oil pan.
23. Disconnect the equalizer rod from the transmission.
24. For 4.3L (VIN Z) engine, disconnect the catalytic converter AIR pipe at the exhaust manifold.
25. Remove the transmission-to-engine bolts.
26. Lower the vehicle.
27. Support the transmission and connect a suitable lifting device to the engine.
28. Remove the engine.

To install:
29. With the engine safely supported, lower into position and align with the motor mounts and transmission.
30. Install motor mount through-bolts and the transmission-to-engine bolts. Tighten either the nuts to 59 ft. lbs. (80 Nm) or the through-bolts to 70 ft. lbs. (95 Nm). Tighten the transmission-to-engine bolts to 35 ft. lbs. (47 Nm).
31. Raise and support the vehicle safely.

32. If applicable, connect the AIR pipe to the exhaust manifold or the equalizer rod to the transmission.
33. Connect the transmission converter clutch wiring to the transmission and the transmission oil cooler lines to the clip on the oil pan.
34. Connect the front fuel hoses to the front fuel pipes.
35. Install the torque converter-to-flywheel bolts and the flywheel housing cover.
36. Connect the crossover pipe and catalytic converter assembly.
37. Connect the battery positive cable and wires to the starter motor.
38. Lower vehicle.
39. Connect the brake pipe to the intake manifold, then connect the negative battery cable to the cylinder head.
40. As applicable, install the EGR solenoid, the MAP sensor and the windshield wiper motor.
41. Install the distributor assembly.
42. Connect the PCM wiring harness, the engine wiring harness at the engine bulkhead, engine to bulkhead ground straps and all other wires between body and engine.
43. Connect all vacuum hoses as noted during removal.
44. Connect the accelerator, TV and cruise control cables.
45. Connect the power steering pump and air conditioning compressor brackets.
46. Connect the heater hoses to the engine.
47. Install the engine cooling fan.
48. Install the radiator, hoses and fan shroud.
49. Install the air cleaner assembly.
50. Install the hood, aligning the marks made during removal.
51. Connect the negative battery cable, then fill and bleed the engine cooling system.
52. Inspect vehicle fluid levels, specifications and verify there are no fluid leaks.

Fleetwood Brougham

1. Disconnect the battery cables and properly relieve fuel system pressure.
2. Mark the hood hinge outline for proper reassembly alignment and remove the hood. Remove the air cleaner assembly.
3. Drain the cooling system. Disconnect the radiator hoses. Disconnect the heater hose from the radiator. Disconnect and plug the transmission and engine oil cooler lines.

4. Remove the radiator cover and tie struts. Disconnect the fan shroud from the radiator assembly and position it aside. Remove the radiator from the vehicle.

5. Remove the serpentine drive belt. Remove the cooling fan assembly and the fan shroud from the vehicle.

6. Disconnect the heater hose at the rear of the intake manifold. Disconnect and plug the power steering hoses at the power steering gear.

7. Remove the air conditioning compressor and position it aside.

8. Disconnect the accelerator, cruise control and throttle valve cables from their mountings and position out of the way. Remove the vacuum pipe and fuel lines from the throttle body.

9. Remove the generator assembly. Disconnect the fuel line clips at the thermostat housing and air pump. Position the fuel lines aside. As required, remove the air pump assembly.

10. Disconnect and plug all required electrical connectors. Remove the distributor cap. Remove the negative battery cable from the cylinder head.

11. Raise and support the vehicle safely. Disconnect the the crossover pipe at both manifolds.

12. Disconnect the starter electrical connectors and the positive battery cable. If necessary, remove the starter retaining bolts and remove the starter from the vehicle.

13. Remove the flywheel cover. Remove the torque converter to flywheel retaining bolts. Remove the motor mount through bolts.

14. Disconnect the transmission oil cooler lines at the clip on the oil pan. Disconnect the oil pressure, knock and oxygen sensor connectors. Remove the oil cooler hose shield.

15. Remove the ground wires from the rear of the cylinder head at both sides.

16. Remove the transmission to engine retaining bolts. Lower the vehicle.

17. Install the lifting equipment to the engine. Support the transmission properly.

18. Raise the engine slightly and pull it forward to disengage it from the transmission. Remove the engine from the vehicle.

To install:

19. Lower the engine assembly into the engine compartment; align the transmission bellhousing dowels and motor mounts.

20. Loosely install 2 transmission-to-engine bolts.

21. Remove the engine and transmission supports.

22. Raise and safely support the vehicle, install the engine mount through bolts and tighten to 70 ft. lbs. (95 Nm).

23. Route the wiring harness into its original location and reconnect the oil pressure, knock sensor and oxygen sensor connectors.

24. Reinstall the oil cooler line bracket and heat shield.

25. Connect the ground straps to the back of the cylinder heads.

26. Install and torque all transmission-to-engine bolts to 55 ft. lbs. (75 Nm).

27. Install the starter assembly, if removed and/or reconnect the wiring. Clip the transmission cooler lines to the oil pan bracket.

28. Install the flywheel-to-torque converter bolts and torque the bolts to 45 ft. lbs. (62 Nm). Install the flywheel cover.

29. Reconnect the exhaust and exhaust hangers. Tighten the crossover pipe bolts to 15 ft. lbs. (20 Nm).

30. Lower the vehicle.

31. Install heater hose to the right rear of intake manifold. Reconnect the throttle cable brackets and cables.

32. Install the distributor cap and coil wires.

33. Install the generator with wiring, but leave the rear brace disconnected. Connect the negative battery cable at the cylinder head.

34. Unplug and connect the power steering lines at the power steering gear.

35. Route the fuel lines and connect at the throttle body. Install the fuel line clips at the thermostat housing and the AIR pump.

36. Connect the rear generator brace. Connect all vacuum hoses to the throttle body. Connect all electrical connections to the intake manifold and the throttle body.

37. Connect the AIR hose from the diverter valve to the converter.

38. Install the fan and fan shroud assembly.

39. Install the radiator assembly. Connect the transmission and oil cooler lines.

40. Connect the the heater hose to the radiator tank and the radiator hoses.

41. Install the radiator cover and secure the fan shroud. Install the radiator tie struts.

42. Install the air conditioning compressor and serpentine belt. Install the air cleaner assembly.

43. Fill the cooling system and connect the battery cables.

44. Check all fluid levels, start engine and inspect for leaks.

45. Align the marks made earlier and install the hood assembly.

Engine Mounts

REMOVAL AND INSTALLATION

Except Fleetwood Brougham

1. Disconnect the negative battery cable.

2. As necessary to provide engine pivot/raising clearance:

   a. Remove the air intake duct assembly.

   b. Remove the radiator upper shroud.

   c. For 1993 vehicles, tag and disconnect the spark plug wires from the distributor cap, then remove the cap from the distributor assembly.

3. Raise and support the vehicle safely.

4. Raise the engine sufficiently to just remove the weight from the engine mount.

NOTE: Do not raise or support engine with a jack under the oil pan, crankshaft pulley or any sheetmetal. Because of the small clearance between the oil pan and oil pump screen, jacking against the oil pan may damage oil pickup assembly.

5. Remove the mount bracket bolt and nut.

6. Remove the transmission fluid cooler lines from the oil pan clip.

NOTE: For 1993 vehicles, verify the clearance between the rear of the engine and the firewall is sufficient enough to avoid possible damage to the distributor.

7. Raise the engine sufficiently to remove the motor mounts, then remove the mount bolts and nuts. Remove the engine mount from the vehicle.

To install:

8. Position the engine mount to the frame, then secure using the bolts and/or nuts. Tighten the bolts to 33 ft. lbs. (45 Nm) and the nuts to 30 ft. lbs. (41 Nm).

9. Lower engine assembly into place, then if applicable, install the

transmission fluid cooler lines to the oil pan clip.

10. Install the engine mount bracket bolt and nut. Tighten the bolt to 70 ft. lbs. (95 Nm) and/or the nut to 49 ft. lbs. (67 Nm).

11. Lower the vehicle, then install the distributor cap, radiator upper shroud and the air intake duct, as necessary.

12. Connect the negative battery cable.

Fleetwood Brougham

WARNING

Never raise the engine using a jack under the oil pan, crankshaft pulley or any sheetmetal. Because there only is a small clearance between the oil pan and the oil pump screen, if the pan is bent even slightly, damage could occur to the pump screen and pickup unit.

---

1. Disconnect the negative battery cable.

2. Remove the engine mount through bolt and nut.

3. Using a suitable lifting device, carefully raise the front of the engine far enough to remove the engine mount retaining bolts and the engine mount. Watch the clearance between the rear of the engine and the cowl panel.

4. Remove the engine mount nuts, bolts and the engine mount.

5. Installation is the reverse of the removal procedure. Tighten the engine mount bolts to 35 ft. lbs. (47 Nm) and the mount through bolts to 70 ft. lbs. (95 Nm).

Cylinder Head

REMOVAL AND INSTALLATION

1993 Vehicles

1. Disconnect the negative battery cable, then relieve the fuel system pressure.

2. Drain cooling system into a suitable container.

3. Remove the cooling fan assembly.

4. Remove the intake manifold assembly from the engine.

5. Remove the exhaust manifold assemblies from the engine.

6. Remove the rocker arm cover or valve cover.

7. For 4.3L (VIN Z) engines, remove the air conditioning pressure cycling switch, then remove the elec-

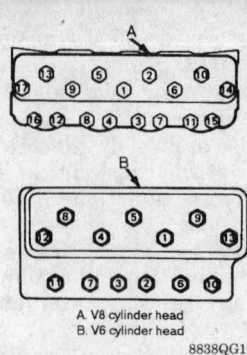

A. V8 cylinder head
B. V6 cylinder head

8838QG13

Cylinder head bolt torque sequence — 4.3L (VIN Z), 5.0L and 5.7L (VIN 7) engines

trical connectors from the wiper motor.

8. Disconnect the power steering pump, alternator, and/or air conditioning brackets, as necessary, and position aside.

9. Disconnect the ground strap and/or negative battery cable from cylinder head, as applicable.

10. Loosen cylinder head bolts gradually using at least 3 passes.

11. Clean dirt from cylinder head and adjacent area to avoid getting dirt into engine.

12. If necessary, remove rocker arm assemblies and lift out pushrods.

13. Remove cylinder head from the engine.

To install:

14. Cylinder heads using a steel gasket should have both sides of the new gasket coated with a good sealer. The coating should be thin and even. Do not use sealer on composite type gaskets.

15. Place gasket over dowel pins.

16. Place cylinder head over dowel pins and gasket.

17. Coat the threads of the cylinder head bolts with sealing compound, part 1052080 or equivalent, then install the bolts finger-tight. Following the proper torque sequence, tighten the cylinder head bolts, using multiple passes, to 68 ft. lbs. (92 Nm)

18. If removed, position pushrods and attach rocker arm assemblies.

19. As applicable, connect the ground strap and/or negative battery cable to the cylinder head.

20. Attach power steering pump, alternator and/or air conditioning brackets if removed.

21. As applicable for 4.3L (VIN Z) engines, install the AIR crossover pipe bolt and stud, then install the air conditioning pressure cycling switch, and/or engage the wiper motor electrical connectors.

22. Install the rocker arm cover or valve cover.

23. If removed, install the oil level indicator tube and/or the diverter valve.

24. Attach the intake and exhaust manifolds.

25. Install the cooling fan assembly.

26. Connect the negative battery cable.

27. Attach the fuel filler cap, add coolant and inspect the engine for leaks.

1994–96 Vehicles

Left Side

1. Disconnect the negative battery cable, then raise and support the vehicle safely.

2. Drain the engine cooling system, then disconnect the crossover pipe from the exhaust manifold.

3. Lower the vehicle.

4. Remove the intake manifold assembly.

5. Disconnect the secondary air injection hose from the check valve assembly.

6. Disconnect the coolant air bleed pipe and bolt from the left cylinder head assembly using a backup wrench on the pipe fitting.

7. Remove the ignition coil assembly.

8. Remove the left exhaust manifold assembly.

9. Remove the spark plug wire harness assembly from the clips, then disconnect the harness from the spark plugs and remove the plugs from the left cylinder head.

10. Disengage the coolant temperature sensor connector.

11. Remove the left rocker arm cover.

12. Loosen the rocker arm nuts, then remove the arms and pushrods, either tagging or arranging the components to assure installation in their original locations.

13. Remove the cylinder head bolts, then remove the cylinder head and gasket from the block.

To install:

14. Thoroughly clean the gasket mating surfaces of any remaining gasket material, then position a new cylinder head gasket on the block with the yellow tab facing upwards.

15. Install the cylinder head over the locator pins and the new gasket.

16. Coat the cylinder head bolts with a sealing compound, then install the bolts finger-tight.

17. Torque the bolts using 3 passes of the proper sequence until all bolts

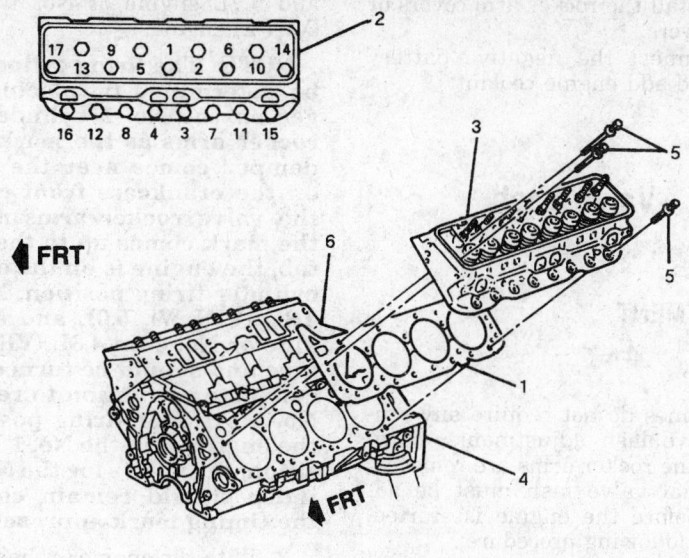

1. Gasket tab (yellow side up)
2. Bolt tightening sequence
3. Cylinder head assembly
4. Engine block
5. Cylinder head bolt
6. Cylinder head gasket

8838QG14

Cylinder head installation and bolt torque sequence — 4.3L (VIN W) and 5.7L (VIN P) engines

have been torqued to 65 ft. lbs. (88 Nm).

18. Install the pushrods and rocker arms, making sure they are in the proper locations, then adjust the valve lash.

19. Install the left rocker arm cover and tighten the retaining bolts to 100 inch lbs. (11 Nm).

20. Install the spark plugs and tighten to 11 ft. lbs. (15 Nm), then install the wiring harness assembly and secure the assembly to the clips.

21. Install the left exhaust manifold assembly.

22. Install the ignition coil assembly and tighten the bolts to 24 ft. lbs. (33 Nm).

23. Connect the engine coolant air bleed pipe and bolt to the left cylinder head assembly using a backup wrench on the pipe fitting in order to prevent component damage. Tighten the bolt to 30 ft. lbs. (40 Nm).

24. Install the secondary air injector hose to the check valve assembly.

25. Install the intake manifold assembly.

26. Raise and support the vehicle safely, then connect the crossover pipe to the exhaust manifold.

27. Lower the vehicle and fill the engine cooling system.

28. Connect then negative battery cable.

Right Side

1. Disconnect the negative battery cable.

2. Raise and support the vehicle safely, then drain the engine cooling system.

3. Remove the serpentine drive belt and the belt tensioner assembly.

4. Remove the transmission fluid level indicator tube assembly bracket from the transmission housing.

5. Remove the air conditioning compressor rear brace bolt from the engine block, then disengage the compressor connector.

6. Remove the front compressor mounting bolts, then position the compressor aside taking care not to kink or damage the lines.

7. Remove the right exhaust manifold assembly.

8. Lower the vehicle.

9. Remove the alternator assembly.

10. Remove the right rocker arm cover.

11. Remove the intake manifold assembly.

12. Remove the coolant air bleed pipe bolt from the left cylinder head assembly.

13. Disconnect the lower radiator hose and the heater hose from the water pump assembly, then position the hoses aside.

14. Remove the coolant air bleed pipe hose from the radiator.

15. Remove the power steering pump assembly.

16. Remove the engine accessory bracket bolts, then remove the bracket assembly.

17. Disconnect the wire harness assembly from the spark plugs, then remove the plugs from the right cylinder head.

18. Loosen the rocker arm nuts, then remove the arms and pushrods, either tagging or arranging the components to assure installation in their original locations.

19. Remove the cylinder head bolts, then remove the cylinder head and gasket from the block. If necessary, carefully remove the coolant air bleed pipe bolt and pipe assembly from the cylinder head.

To install:

20. If removed, loosely install the coolant air bleed pipe bolt and pipe to the cylinder head.

21. Thoroughly clean the gasket mating surfaces of any remaining gasket material, then position a new cylinder head gasket on the block with the yellow tab facing upwards.

22. Install the cylinder head over the locator pins and the new gasket.

23. Coat the cylinder head bolts with a sealing compound, then install the bolts finger-tight.

24. Torque the bolts using 3 passes of the proper sequence until all bolts have been torqued to 65 ft. lbs. (88 Nm).

25. If loosened, tighten the coolant air bleed pipe bolt to 30 ft. lbs. (40 Nm).

26. Install the pushrods and rocker arms, making sure they are in the proper locations, then adjust the valve lash.

27. Install the spark plugs and tighten to 11 ft. lbs. (15 Nm), then install the wiring harness assembly to the plugs.

28. Install the engine accessory bracket and tighten the retaining bolts to 31 ft. lbs. (42 Nm).

29. Install the right rocker arm cover and tighten the retaining bolts to 100 inch lbs. (11 Nm).

30. Install the alternator assembly.

31. Install the power steering pump assembly.

32. Connect the coolant air bleed pipe hose to the radiator, then connect the heater hose and the lower radiator hose to the water pump.

33. Connect the coolant air bleed pipe bolt to the left cylinder head and tighten to 30 ft. lbs. (41 Nm) while

using a backup wrench to prevent component damage.

34. Install the intake manifold assembly.

35. Raise and support the vehicle safely.

36. Install the right exhaust manifold assembly.

37. Position the compressor and install the front mounting bolts, then engage the electrical connector. Install the rear compressor brace bolt and tighten to 24 ft. lbs. (33 Nm).

38. Install the transmission fluid level indicator tube assembly to the transmission housing.

39. Install the serpentine drive belt, then lower the vehicle.

40. Fill the engine cooling system, then connect the negative battery cable.

Valve Lifters

REMOVAL AND INSTALLATION

1. Disconnect the negative battery cable.

2. Drain the engine cooling system.

3. Remove the intake manifold assembly.

4. Remove rocker arm covers or valve covers.

5. Remove the rocker arms and pushrods. Be sure to keep all valve train parts in order, as they must be installed in the same locations from which they were removed.

6. Remove the valve lifter retainer bolts, valve lifter retainer and/or guide.

7. Remove the valve lifters, using the a suitable valve lifter tool. If lifters are to be reinstalled, keep them in order so they may be installed in the same bores from which they were removed.

To install:

8. Coat the valve lifter rollers with Molykote®, prelube part 1052365 or equivalent.

9. Insert the valve lifters into the bores. If reinstalling used lifters, make sure they are inserted in the same location from which they were removed.

10. Install the valve lifter guide and/or retainer.

11. Tighten valve lifter retainer bolts to 12 ft. lbs. (16 Nm).

12. Install the intake manifold assembly.

13. Place the pushrods in their original positions and install the rocker arms.

14. Install the rocker arm covers or valve covers.

15. Connect the negative battery cable and add engine coolant.

Valve Lash

ADJUSTMENT

The engines do not require any routine valve lash adjustments. However, if the rocker arms are removed, the initial valve lash must be adjusted before the engine is started. Use the following procedure:.

1. With the rocker arm covers or valve covers removed and the rocker arm assemblies loosely installed, position the engine at the No. 1 cylinder Top Dead Center (TDC) position.

2. To determine TDC, slowly turn the engine until the mark on the vibration damper aligns with the center or **0** mark on the timing tab of the front cover. At this point the engine is on the No. 1 firing position or the firing position of its opposite cylinder No. 6 on 4.3L (VIN W), 5.0L

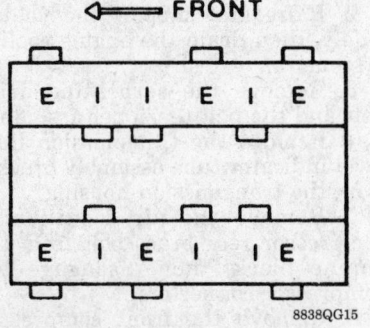

Intake and exhaust valve arrangement — 4.3L (VIN Z) engine

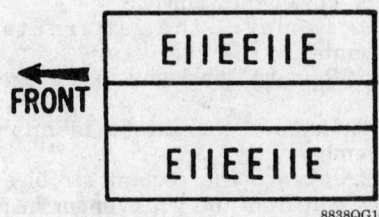

Intake and exhaust valve arrangement — 4.3L (VIN W), 5.0L and 5.7L engines

and 5.7L engine or No. 4 on 4.3L (VIN Z) engine.

NOTE: The firing cylinder may be determined by placing a finger on the No. 1 cylinder valve rocker arms as the mark on the damper comes near the 0 mark on the crankcase front cover. If the valve rocker arms move as the mark comes up to the timing tab, the engine is on the opposite cylinder firing position, No. 6 on 4.3L (VIN W), 5.0L and 5.7L engine or No. 4 on 4.3L (VIN Z) engine and should be turned over a complete revolution to reach the No. 1 cylinder firing position. If the engine is in the No. 1 TDC position, the valves for the No. 1 cylinder should remain closed as the timing mark approaches.

3. With the engine in the No. 1 firing position, adjust the following valves:

4.3L (VIN W), 5.0L and 5.7L engines
Exhaust — 1, 3, 4, 8
Intake — 1, 2, 5, 7
4.3L (VIN Z) engine
Exhaust — 1, 5, 6
Intake — 1, 2, 3

4. Adjust each valve by backing out the adjusting nut until lash is felt at the pushrod, then by tightening the adjusting nut until all lash is removed. This can be determined by rotating pushrod while turning the adjusting nut. When play has been removed, turn adjusting nut a full additional turn clockwise for 4.3L (VIN W), 5.0L and 5.7L engines or ¾ of a turn clockwise for 4.3L (VIN Z) engine. The lifter plunger will now be centered.

5. Turn the engine 1 revolution until the pointer **0** mark and the vibration damper mark are again in alignment. This is the No. 6 firing position on the 4.3L (VIN W), 5.0L and 5.7L engine or No. 4 firing position on the 4.3L (VIN Z) engine. As the timing mark approaches the pointer, the No. 1 cylinder valves should move.

6. With the engine in this position, adjust the following valves:

4.3L (VIN W), 5.0L and 5.7L engines
Exhaust — 2, 5, 6, 7
Intake — 3, 4, 6, 8
4.3L (VIN Z) engine
Exhaust — 2, 3, 4
Intake — 4, 5, 6

7. Install the rocker arm covers or valve covers.

8. Start the engine and check/adjust the minimum the idle speed, as required.

Rocker Arms

REMOVAL AND INSTALLATION

4.3L (VIN Z), 5.0L and 5.7L (VIN 7) Engines

1. Disconnect the negative battery cable.
2. Remove the valve rocker covers.
3. Remove the rocker arm assembly; nuts, balls and rocker arms. Arrange or mark each assembly to ensure installation in original positions.
4. Remove each pushrod, if necessary, and place with the appropriate assemblies, to ensure installation in original locations.

To install:

NOTE: If new rocker arms or rocker arm balls are being installed, coat the bearing surfaces with prelube part 1052365, Molykote® or equivalent.

5. Install the pushrods. Ensure that the rods are seated properly in the lifter sockets.
6. Install the rocker arms, balls and nuts. Tighten the rocker arm nuts until all the valve lash is eliminated.
7. Adjust the valves to proper specification.
8. Install the rocker arm covers, then connect the negative battery cable.
9. Start the engine and check/adjust the minimum the idle speed, as required.

4.3L (VIN W) and 5.7L (VIN P) Engines

1. Disconnect the negative battery cable.
2. Remove the left valve cover:
   a. Remove the brake booster vacuum hose.
   b. Remove the secondary AIR injection hose from the pump to check valve assembly.
   c. Remove the valve cover retaining bolts, then remove the cover and gasket.
3. Remove the right valve cover:
   a. Raise and support the vehicle safely, then remove the serpentine drive belt.
   b. Remove the transmission fluid level indicator tube assembly from the bracket on the transmission housing.
   c. Lower the vehicle, then remove the crankcase vent hose.
   d. Remove the alternator and rear alternator brace.
   e. Remove the valve cover retaining bolts, then remove the cover and gasket.

4. Remove the valve rocker arm nuts and balls, then remove the rocker arms and pushrods. Tag or arrange all valve train components to assure installation in their original locations.

To install:

5. Coat the bearing surfaces of the rocker arms, balls and pushrods with prelube.
6. Install the pushrods, making sure they are properly seated in the lifter sockets, then install the rocker arms, balls and nuts. If components are being reused, be sure they are installed in their original locations.
7. Adjust the valve lash.
8. Install the valve covers and gaskets, then tighten the retainers to 100 inch lbs. (11 Nm).
9. Install the components which were removed to access the valve covers in the reverse order of removal.
10. Connect the negative battery cable.

Intake Manifold

REMOVAL AND INSTALLATION

4.3L (VIN Z), 5.0L and 5.7L (VIN 7) Engines

1. Disconnect the negative battery cable and relieve the fuel system pressure.
2. Drain the engine coolant into a suitable container and remove the air cleaner.
3. Disconnect the fuel pipes from the throttle body, and, if necessary, remove the throttle body assembly from the manifold.
4. Disconnect the PCM engine control harness and position it aside.
5. Remove the brake tube assembly.
6. If not done already, disconnect the vacuum harness assembly.
7. Disconnect the upper radiator hose and the heater hose.

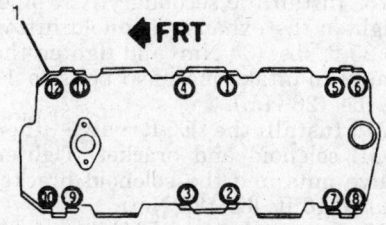

1. Intake manifold assembly

8838QG17

Intake manifold torque sequence — 5.0L and 5.7L (VIN 7) engines

8. Disengage all necessary electrical connections or remaining hose assemblies.
9. Disconnect fuel pipe clips at AIR pump bracket and at the intake manifold.
10. Disconnect accelerator and TV cables or the accelerator control cable assembly bracket, as applicable.
11. Tag and remove the spark plug wires at the distributor cap, then remove the cap.
12. Mark the position of the rotor, then remove the distributor. Remove the bracket and/or coil, as required.
13. Remove the accessory mounting brackets, as required.
14. Remove the manifold bolts and studs, then remove the intake manifold. Remove and discard the intake manifold gaskets.

To install:

15. Thoroughly clean the intake manifold and cylinder block surfaces to remove any trace of gasket material or sealant.
16. Place gasket and seals on cylinder heads and block, apply a thin bead of RTV sealer, 1052289 or equivalent, to the front and rear of cylinder block. Extend the RTV bead ½ in. up each cylinder head to seal and retain gasket.
17. Install the intake manifold, taking care not to dislodge the gaskets and seals, then install the retaining bolts and studs. Tighten the bolts and studs in proper sequence, first to 10 ft. lbs. (14 Nm) and then to 35 ft. lbs. (47 Nm).
18. Attach accessory mounting brackets, if removed. Tighten compressor brace-to-manifold nut to 18 ft. lbs. (24 Nm), compressor brace-to-compressor nut to 24 ft. lbs. (32 Nm) and/or alternator to brace nut to 37 ft. lbs. (50 Nm)
19. If removed, install the bracket and/or coil.
20. Install the distributor, aligning the rotor with the mark made during removal, then install the cap and attach the wires.
21. Install the EGR valve and EGR solenoid valve, as applicable.
22. Connect the accelerator and TV cables or the accelerator control cable assembly bracket.
23. Connect the fuel pipe clips to the AIR pump bracket and intake manifold.
24. Install the wire and hose assemblies, as applicable.
25. Connect the upper radiator hose and the heater hose.
26. Connect the PCM engine control harness.

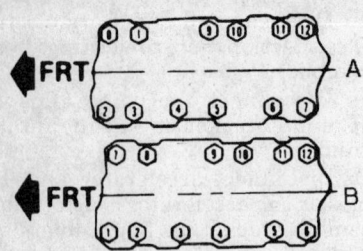

A. Initial tightening sequence
B. Final tightening sequence

8838QG18

Intake manifold torque sequence — 4.3L (VIN Z) engine

27. If removed, install the throttle body assembly.
28. Attach the fuel pipes to the TBI assembly.
29. Install the air cleaner and connect the negative battery cable.
30. Install the fuel filler cap and add engine coolant.
31. Start the engine and inspect for leaks. Check and adjust timing, as necessary.

4.3L (VIN W) and 5.7L (VIN P) Engines

1. Disconnect the negative battery cable and relieve the fuel system pressure.
2. Drain the engine cooling system into a suitable container.
3. Remove the throttle body air duct.
4. Disengage the wiring harness connectors from the fuel injectors. Disengage and reposition the left and right wiring harnesses.
5. Remove the accelerator cable bracket retainers, then disconnect the cable and bracket assembly from the throttle body.
6. Disconnect the secondary AIR diverter valve hoses.
7. Disengage the fuel pipe connectors from the fuel rail assembly.

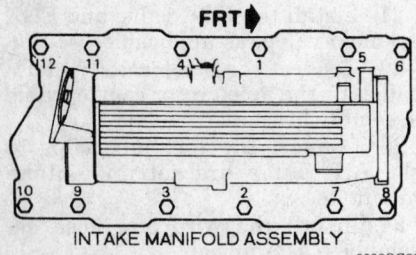

INTAKE MANIFOLD ASSEMBLY

8838QG20

Intake manifold bolt torque sequence — 4.3L (VIN W) and 5.7L (VIN P) engines

8. Remove the fuel rail bolts and disconnect the vacuum hose from the fuel pressure regulator.
9. Carefully remove the fuel rail and injector assembly from the manifold and position aside.
10. Disconnect the vacuum and crankcase vent hoses.
11. Remove the EGR solenoid assembly and the fuel EVAP canister solenoid assembly.
12. Remove the EGR valve.
13. Remove the AIR pipe from the intake and the right exhaust manifold.
14. Remove the alternator rear brace.
15. Disconnect the coolant hoses from the throttle body.
16. Remove the throttle body bolts, the throttle body and gasket from the intake.
17. Remove the intake manifold bolts and studs.
18. Remove the intake manifold and discard the old gaskets.
To install:
19. Thoroughly clean the intake manifold bolts and studs. Inspect and clean all gasket mating surfaces.
20. Apply a 3/16 in. (5mm) bead of RTV sealer to the front and rear of the cylinder block. Extend the bead 1/2 in. (13mm) up each cylinder head to seal and retain the gaskets.
21. Position the new gaskets and install the intake manifold.
22. Install the manifold bolts and studs, then tighten using 2 passes of the proper sequence. First, tighten the bolts/studs to 71 inch lbs. (8 Nm), then tighten them to 35 ft. lbs. (48 Nm).
23. Install the throttle body, gasket and retaining bolts. Tighten the throttle body bolts to 19 ft. lbs. (26 Nm).
24. Connect the coolant hoses to the throttle body, then install the alternator rear brace.
25. Install the accelerator cables and bracket, then tighten the bracket bolts to 90 inch lbs. (10 Nm).
26. Install the secondary AIR pipe. Tighten the exhaust manifold fitting to 25 ft. lbs. (34 Nm) and tighten the flange-to-intake manifold bolts to 19 ft. lbs. (26 Nm).
27. Install the EGR valve, then EGR solenoid and bracket. Tighten valve nuts and the solenoid bracket nut to 16 ft. lbs. (22 Nm).
28. Install the fuel EVAP canister purge solenoid and bracket, then tighten the bolt to 53 inch lbs. (6 Nm).
29. Connect the vacuum and crankcase vent hoses.

30. Install the fuel injector and fuel rail assembly to the intake manifold, connect the fuel pressure regulator vacuum hose and install the fuel rail bolts. Tighten the bolts to 15 ft. lbs. (20 Nm). Engage the fuel pipe connections to the fuel rail assembly.
31. Connect the secondary AIR diverter valve hoses.
32. Position the left and right wiring harnesses, then engage the fuel injector electrical connectors.
33. Install the throttle body air duct.
34. Properly fill the engine cooling system.
35. Connect the negative battery cable.

Exhaust Manifold

REMOVAL AND INSTALLATION

1993 Vehicles

1. Disconnect the negative battery cable.
2. Raise and support vehicle safely.
3. Disconnect crossover pipe at the exhaust manifold.
4. Disengage the oxygen sensor electrical connector from the left exhaust assembly.
5. Lower the vehicle.
6. Remove the air cleaner assembly, if necessary for access.
7. Remove the spark plug wires from the retainer clips.
8. Disconnect hoses, pipes, and accessory brackets, as required. If applicable, disconnect the air pipes or the hose at the check valve.
9. If necessary, remove the oil level indicator and tube.
10. Remove exhaust manifold bolts, studs, locks, washers, and if applicable, shields.
11. Remove the exhaust manifold and gasket.
To install:
12. Clean mating surfaces on manifold and cylinder head.
13. Place exhaust manifold and gasket into position on cylinder head.
14. Install shields, washers, locks, studs and bolts.
15. Tighten the center port bolts or studs to 26 ft. lbs. (35 Nm) and the front/rear port bolts or screws to 20 ft. lbs. (27 Nm).
16. Connect any hoses, pipes and accessory brackets which were removed.
17. Install the spark plug wiring to the retaining clips and, if removed, install the air cleaner assembly.

18. Raise and support vehicle safely.

19. Engage the oxygen sensor electrical connector.

20. Connect crossover pipe to the exhaust manifold and tighten nuts to 15 ft. lbs. (20 Nm).

21. Lower vehicle and connect the negative battery cable.

22. Start engine and check for leaks.

1994–96 Vehicles

Left Side

1. Disconnect the negative battery cable, then raise and support the vehicle safely.

2. Disconnect the exhaust crossover pipe from the manifold, then lower the vehicle.

3. Remove the brake booster vacuum hose.

4. Disconnect the secondary AIR pipe fitting from the exhaust manifold.

5. Disengage the oxygen sensor electrical connector.

6. Remove the exhaust manifold retaining bolts, then remove the heat shields, manifold and gasket.

To install:

7. Clean the gasket mating surfaces.

8. Position the gasket, then install the exhaust manifold and heat shields.

9. Install the manifold retaining bolts and tighten to 26 ft. lbs. (35 Nm).

10. Engage the oxygen sensor electrical connector.

11. Connect the secondary AIR pipe fitting to the exhaust manifold and tighten to 25 ft. lbs. (34 Nm).

12. Install the brake booster vacuum hose.

13. Raise and support the vehicle safely.

14. Connect the exhaust crossover pipe to the manifold, then lower the vehicle.

15. Connect the negative battery cable.

Right Side

1. Disconnect the negative battery cable, then raise and support the vehicle safely.

2. Remove the exhaust crossover pipe.

3. Remove the serpentine drive belt.

4. Remove the oil level indicator and tube assembly, then disengage the oxygen sensor electrical connector.

5. Remove the 3 rear exhaust manifold retaining bolts, then lower the vehicle.

6. Disconnect the secondary AIR pipe fitting from the exhaust manifold.

7. Remove the alternator rear lower brace.

8. Remove the remaining exhaust manifold retaining bolts, then remove the heat shields, manifold and gasket.

To install:

9. Clean the gasket mating surfaces.

10. Position the gasket, then install the exhaust manifold and heat shields.

11. Install the front 3 manifold retaining bolts and tighten to 26 ft. lbs. (35 Nm).

12. Install the alternator rear lower brace.

13. Raise and support the vehicle safely.

14. Install the remaining manifold retaining bolts and tighten to 26 ft. lbs. (35 Nm).

15. Engage the oxygen sensor electrical connector.

16. Install the oil level indicator and tube assembly.

17. Install the serpentine drive belt.

18. Install the exhaust crossover pipe.

19. Lower the vehicle.

20. Connect the secondary AIR pipe fitting to the exhaust manifold and tighten to 25 ft. lbs. (34 Nm).

21. Connect the negative battery cable.

Front Cover Seal

REMOVAL AND INSTALLATION

4.3L (VIN Z), 5.0L and 5.7L (VIN 7) Engines

The front cover oil seal may be replaced with the front cover either removed from or installed to the engine. If the cover is already removed, be sure to properly support the cover when driving in the new seal in order to prevent damage to the cover assembly.

1. Disconnect the negative battery cable.

2. Remove the air cleaner assembly, then remove the radiator upper shroud, for access.

3. Remove the serpentine drive belt and the engine cooling fan.

4. Remove the crankshaft pulley retaining bolts, then remove pulley from the damper.

5. Remove the damper retaining bolt, then using tool J-23523-E or equivalent, remove the torsional damper assembly.

6. With the torsional damper removed, carefully pry the old seal from the cover using a small prytool. Take care not to damage the front cover or the crankshaft when removing seal.

To install:

7. Position new seal with the open end toward the inside of the engine front cover and carefully drive in the new seal with tool J-35468 or equivalent.

8. After coating the damper-to-seal contact area with clean engine oil, install the torsional damper using J-23523-E or equivalent. Install the pulley and tighten the pulley retaining bolts to 43 ft. lbs. (58 Nm), then tighten the damper bolt to 70 ft. lbs. (95 Nm).

9. Install the cooling fan assembly.

10. Install the serpentine drive belt.

11. Install the radiator upper shroud and the air cleaner assembly.

12. Connect the negative battery cable and check cover for oil leaks.

4.3L (VIN W) and 5.7L (VIN P) Engines

In addition to the crankshaft oil seal, the front cover utilizes a water pump driveshaft seal and distributor driveshaft seal. All front cover oil seals on these engines are replaced in the same manner.

1. Remove the timing chain front cover from the engine.

2. Using either a suitably sized driver or a small suitable prytool, remove oil seal from the front cover. If a driver is being used, be sure to support the cover so it is not damaged. If a prytool is used, take care not to score or damage the cover sealing surfaces.

3. Use J-35468 or equivalent to install the crankshaft seal.

4. Use J-39090 or equivalent to install the distributor shaft seal.

5. Use J-39088 or equivalent to install the water pump driveshaft seal.

6. Install the timing chain front cover to the engine.

Timing Chain Front Cover

REMOVAL AND INSTALLATION

4.3L (VIN Z), 5.0L and 5.7L (VIN 7) Engines

1. Disconnect the negative battery cable.

2. Drain the cooling system into a suitable container, then remove the water pump assembly.

3. Remove the crankshaft pulley retaining bolts, then remove the pulley from the damper.

4. Remove the torsional damper retaining bolt, then using tool J-23523-E or equivalent, remove the torsional damper assembly.

5. Raise and safely support the vehicle.

6. Remove the oil pan assembly.

7. Remove the engine front cover retaining bolts, then remove the front cover and discard the gasket.

8. Clean the gasket mating surface.

To install:

9. Coat new engine front cover gasket with sealant and place into position on the engine front cover.

10. Position cover and gasket to the engine and loosely install the engine front cover-to-block upper attaching bolts. Tighten bolts alternately while carefully pressing downward on the engine front cover so the dowels in the block are aligned with the corresponding holes in the engine front cover. Be careful not to force the the front cover over the dowels to the point where the cover flange or dowels become distorted.

11. Install the remaining cover bolts and tighten all cover bolts alternately and evenly to 97 inch lbs. (11 Nm).

12. Install the oil pan.

13. After coating the damper-to-seal contact area with clean engine oil, install the torsional damper using J-23523-E, or equivalent. Install the pulley and tighten the pulley retaining bolts to 43 ft. lbs. (58 Nm), then tighten the damper bolt to 70 ft. lbs. (95 Nm).

14. Install the coolant pump assembly.

15. Connect the negative battery cable and add engine coolant.

4.3L (VIN W) and 5.7L (VIN P) Engines

1. Disconnect the negative battery cable.

2. Drain the engine oil and coolant into suitable containers.

3. Remove the throttle body air intake duct.

4. Remove the serpentine drive belt.

5. Remove the water pump assembly.

6. Remove the crankshaft balancer and hub.

   a. If not done already, raise and support the vehicle safely.

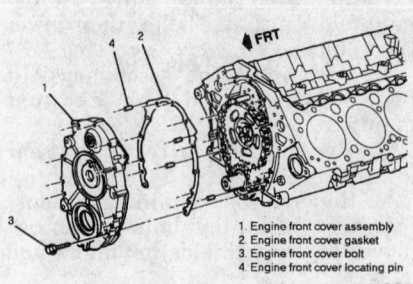

1. Engine front cover assembly
2. Engine front cover gasket
3. Engine front cover bolt
4. Engine front cover locating pin

8838QG21

Timing chain front cover — 4.3L (VIN W) and 5.7L (VIN P) Engines

   b. Remove the crankshaft balancer retaining bolts, then remove the balancer from the hub.

   c. Matchmark the crankshaft hub to the engine front cover, then remove the hub bolt and washer.

   d. Remove the crankshaft hub using J-39046 or an equivalent hub removal/installation tool. To preserve the relationship between the hub and crankshaft, DO NOT crank the engine over once the hub has been removed. If the hub is not matchmarked and installed in the original position, an engine imbalance could result.

7. Remove the distributor assembly.

8. Remove the oil pan assembly.

9. Remove the engine front cover bolts.

10. Remove the engine front cover and gasket.

To install:

11. Thoroughly clean the engine front cover and cylinder block gasket mating surfaces. Inspect the engine front cover and seals for damage, replace as necessary.

12. Using J-39087 or equivalent front cover seal protector on the water pump driveshaft, install the gasket and front cover into position over the shafts and guide pins.

13. Install the engine front cover bolts and tighten to 100 inch lbs. (11 Nm).

14. Install the oil pan and gasket.

15. Install the distributor assembly.

16. Install the crankshaft hub and torsional damper assembly.

   a. Align the matchmarks made earlier and install the crankshaft hub using the hub tool. If the engine was cranked and the matchmarks were lost, set the engine to No. 1 TDC, then install the crankshaft hub with the cast arrow in the 12 o'clock position.

   b. Install the hub washer and bolt, but do not torque at this time.

   c. Install the crankshaft balancer to the hub, then tighten the crankshaft hub bolt to 75 ft. lbs. (102 Nm) and the balancer bolts to 60 ft. lbs. (81 Nm).

NOTE: If a new balancer is installed, new balancer weights of the same size must be installed in the same hole locations as the original balancer.

17. Install the water pump assembly.

18. Install the serpentine drive belt.

19. Install the throttle body air duct.

20. Properly fill the engine crankcase with clean engine oil.

21. Properly fill the engine cooling system.

22. Connect the negative battery cable, operate the engine and check for leaks.

Timing Chain and Sprockets

REMOVAL AND INSTALLATION

4.3L (VIN Z), 5.0L and 5.7L (VIN 7) Engines

1. Disconnect the negative battery cable.

2. Remove the engine front cover.

3. Rotate the engine until the marks on the camshaft sprocket and crankshaft sprocket are aligned with the shaft centers.

4. Remove the camshaft sprocket retaining bolts, then remove the timing chain assembly, along with the camshaft sprocket.

To install:

5. Position the camshaft sprocket in the timing chain with the timing mark located as aligned during removal, then install the timing chain under the crankshaft sprocket while installing the camshaft sprocket to the engine.

6. Verify that the timing marks on the crankshaft sprocket and the camshaft sprocket are aligned with the shaft centers, then install the camshaft sprocket bolts and tighten to 21 ft. lbs. (28 Nm) Lubricate the timing chain with engine oil.

7. Install the engine front cover.

8. Connect the negative battery cable.

4.3L (VIN W) and 5.7L (VIN P) Engines

1. Remove the timing chain front cover.

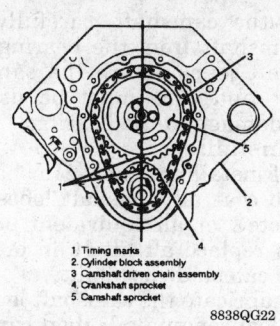

1. Timing marks
2. Cylinder block assembly
3. Camshaft driven chain assembly
4. Crankshaft sprocket
5. Camshaft sprocket

8838QG22

Timing gear alignment — 4.3L (VIN Z), 5.0L and 5.7L (VIN 7) engines

2. Rotate the crankshaft until the timing marks on the timing chain sprockets are aligned nearest each other. The camshaft sprocket mark should be at the 6 o'clock position while the mark on the crankshaft sprocket should be at the 12 o'clock position.

3. Remove the camshaft sprocket bolts.

4. Remove the camshaft sprocket and timing chain.

NOTE: To prevent piston or valve damage, do not turn the crankshaft after the timing chain has been removed.

5. Remove the water pump bearing retainer bolts, then remove the driveshaft assembly using J-39243 or equivalent driven gear assembly remover.

6. Remove the crankshaft sprocket using J-5825-A or equivalent crankshaft sprocket remover.

7. If necessary, remove the crankshaft key.

To install:

8. If removed, install the crankshaft key.

9. Install the crankshaft sprocket using J-5590 or an equivalent installation tool.

10. Install the water pump driveshaft assembly using J-39092 or an equivalent installer tool. Install the retainer bolts and tighten to 105 inch lbs. (12 Nm).

11. Align the timing marks and install the camshaft sprocket and timing chain. The gears of the camshaft sprocket and water pump driveshaft must mesh or damage to the thrust plate retainer could occur.

12. Install the camshaft sprocket bolts and tighten to 21 ft. lbs. (28 Nm).

13. Install a new O-ring to the water pump driven gear shaft using J-39089 or an equivalent seal installation tool.

14. Install the timing chain front cover.

Timing Sprockets

The camshaft sprocket is removed during the timing chain removal procedure. The following procedures should be followed if the crankshaft sprocket must also be removed.

REMOVAL AND INSTALLATION

4.3L (VIN Z), 5.0L and 5.7L (VIN 7) Engines

1. Disconnect the negative battery cable.

2. Remove the timing chain and camshaft sprocket.

3. Remove the crankshaft sprocket using tool J-5825-A or equivalent.

4. If necessary, remove crankshaft key and inspect the key/keyway for damage or wear.

To install:

5. If removed, install the crankshaft key.

6. Install the crankshaft sprocket using tool J-5590 or equivalent.

7. Install timing chain and camshaft sprocket.

8. Connect the negative battery cable.

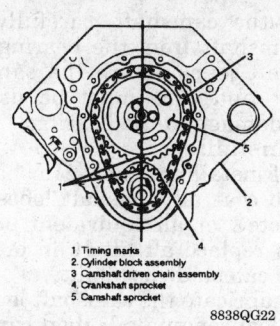

1. Timing marks
2. Keyway
3. Camshaft sprocket
4. Camshaft sprocket bolt
5. Camshaft retainer
6. Camshaft retainer bolt
7. Camshaft assembly
8. Timing chain assembly
9. Crankshaft sprocket
10. Water pump driveshaft assembly
11. Water pump driveshaft bearing retainer bolt

8838QG23

Camshaft, timing chain and sprocket alignment and installation — 4.3L (VIN W) and 5.7L (VIN P) engines

4.3L (VIN W) and 5.7L (VIN P) Engines

1. Remove the timing chain front cover.

2. Rotate the crankshaft until the timing marks on the timing chain sprockets are aligned nearest each other. The camshaft sprocket mark should be at the 6 o'clock position while the mark on the crankshaft sprocket should be at the 12 o'clock position.

3. Remove the camshaft sprocket bolts.

4. Remove the camshaft sprocket and timing chain.

NOTE: To prevent piston or valve damage, do not turn the crankshaft after the timing chain has been removed.

5. Remove the water pump bearing retainer bolts, then remove the driveshaft assembly using J-39243 or equivalent driven gear assembly remover.

6. Remove the crankshaft sprocket using J-5825-A or equivalent crankshaft sprocket remover.

7. If necessary, remove the crankshaft key.

To install:

8. If removed, install the crankshaft key.

9. Install the crankshaft sprocket using J-5590 or an equivalent installation tool.

10. Install the water pump driveshaft assembly using J-39092 or an equivalent installer tool. Install the retainer bolts and tighten to 105 inch lbs. (12 Nm).

11. Align the timing marks and install the camshaft sprocket and timing chain. The gears of the camshaft sprocket and water pump driveshaft must mesh or damage to the thrust plate retainer could occur.

12. Install the camshaft sprocket bolts and tighten to 21 ft. lbs. (28 Nm).

13. Install a new O-ring to the water pump driven gear shaft using J-39089 or an equivalent seal installation tool.

14. Install the timing chain front cover.

Camshaft

REMOVAL AND INSTALLATION

4.3L (VIN Z), 5.0L and 5.7L (VIN 7) Engines

1. Properly discharge/recover the air conditioning system, then disconnect the negative battery cable.

2. Remove the intake manifold assembly.

3. Remove the rocker arm assemblies and pushrods.

NOTE: All parts for each rocker arm/lifter valve train must be kept together and, if reused, must be installed in the same locations from which they were removed.

4. Loosen the belt tensioner and remove the serpentine drive belt.

5. Remove the upper fan shroud, radiator hoses, oil/fluid cooler lines and the radiator assembly.

6. Remove the timing chain and camshaft sprocket.

7. Disconnect the refrigerant lines from the condenser and plug to prevent system contamination, then remove the condenser from the vehicle.

8. Remove the valve lifters and arrange with the rocker arm assemblies for possible reuse, if the camshaft is not replaced.

9. Remove the camshaft retainer bolts and the camshaft retainer.

10. Install three 5/16–18 x 4 in. bolts in the camshaft bolt holes and carefully pull the camshaft from the bearings.

To install:

11. Coat camshaft lobes and journals with prelube 1052365 or equivalent.

12. Carefully slide the camshaft into the journals in the block.

13. Install the camshaft retainer and tighten the retainer bolts to 106 inch lbs. (12 Nm).

14. Install the timing chain and camshaft sprocket along with the front cover assembly.

15. Install the valve lifters. If the camshaft was replaced, use new lifters to assure durability of the camshaft lobes and lifter rollers.

16. Install the air conditioning condenser, then unplug and connect the refrigerant lines.

17. Install the radiator, oil/fluid cooler lines, radiator hoses and upper shroud assembly.

18. Install the serpentine drive belt.

19. Install the pushrods and rocker arm assemblies.

20. Install the intake manifold assembly.

21. Connect the negative battery, add coolant, and adjust valves as necessary.

22. Run the engine and check for leaks, then properly charge the air conditioning system.

4.3L (VIN W) and 5.7L (VIN P) Engines

1. Disconnect the negative battery cable.

2. Remove the intake manifold assembly.

3. Remove the rocker arms and pushrods.

4. Remove the bolt retaining the oil pump drive assembly, then lift the drive assembly from the rear of the block.

5. Remove the timing chain front cover, then remove the crankshaft shaft sprocket and timing chain.

6. Remove the valve lifters.

NOTE: If valve train components are to be reused, make sure they are tagged or arranged in order to assure installation in their original locations.

7. If necessary, properly discharge and recover the refrigerant from the air conditioning system.

8. Remove the air conditioning compressor and condenser hose from the condenser, then remove the receiver and dehydrator hose from the condenser. Plug all of the openings in order to prevent system contamination.

9. Remove the radiator and condenser assembly, then remove the air conditioning condenser support.

10. Remove the camshaft retainer bolts, then remove the camshaft retainer from the front of the block.

11. Install three 5/16–18 x 4 in. bolts or equivalent, in the camshaft bolt holes, then using the bolts to pull and rotate the camshaft, carefully pull the camshaft from the bearings. All camshaft journals are the same diameter and care must be used to avoid damage to the bearings.

To install:

12. If installing a new camshaft, be sure to coat all camshaft lobes with Molykote® or an equivalent prelube and to replace all lifters in order to assure camshaft durability.

13. Lubricate the camshaft journals with clean engine oil, then carefully insert the camshaft into the engine.

14. Install the camshaft retainer and tighten the bolts to 105 inch lbs. (12 Nm).

15. Install the condenser support, then install the radiator and condenser assembly.

16. Unplug the openings, then connect the condenser refrigerant lines.

17. Install the valve lifters. If reusing the camshaft and lifters, they must be installed into their original bores.

18. Install the camshaft sprocket and timing chain, then install the timing chain front cover.

19. Install the oil pump drive assembly and tighten the retaining bolt to 13 ft. lbs. (18 Nm).

20. Install the rocker arms and pushrods.

21. Install the intake manifold assembly.

22. Connect the negative battery cable, then if necessary, recharge the A/C system.

Piston and Connecting Rod

POSITIONING

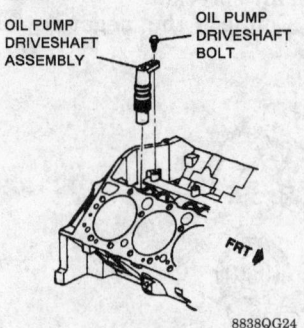

Oil pump driveshaft assembly mounting — 4.3L (VIN W) and 5.7L (VIN P) engines

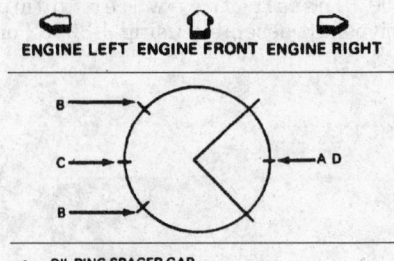

Piston ring gap locations

ENGINE LUBRICATION

Oil Pan

REMOVAL AND INSTALLATION

4.3L (VIN Z), 5.0L and 5.7L (VIN 7) Engines

1. Disconnect the negative battery cable and remove the air cleaner assembly.

2. For 5.0L and 5.7L engines, disengage the wiper motor electrical connector, then remove the fuse cover.

3. Remove the upper fan shroud.

4. Remove the transmission and engine oil dipsticks, then remove the distributor cap.

5. Raise and support the vehicle safely.

6. Drain the engine oil.

7. Disconnect the exhaust pipe at the manifolds, then remove the flywheel cover.

8. Disconnect the transmission fluid cooler lines at the clips on the oil pan.

9. Except for 4.3L (VIN Z) engine, remove the transmission dipstick tube.

10. For 5.0L and 5.7L engines, disconnect the shift linkage from the transmission and remove the right frame brace.

11. Remove the starter motor assembly.

12. If equipped with an oil level sensor, it must be disconnected and removed to prevent possible damage to the oil level sensor, oil pump pickup screen and pipe.

13. Disconnect the engine mount through-bolts, then raise the front of the engine slightly for clearance.

14. Remove the oil pan attaching nuts, bolts and reinforcement. Note the position of any bolts which may contain pipe clamps for installation purposes.

15. Place the crankshaft timing mark to the 6 o'clock position in order to move the crankshaft throw and counterbalance aside, then raise the engine sufficiently to remove the pan. Carefully remove the oil pan and discard the old gasket.

To install:

16. Clean the gasket mating surfaces. Apply a small amount of 1052914 or equivalent sealer to the front cover and rear seal retainer-to-cylinder block junctions and continue the bead 1 in. in either direction from the radius of the cavity.

17. Install the new gasket on the oil pan, and position the oil pan in place with loosely installed nuts, bolts and reinforcement. Make sure any pipe clamps, if present, are reinstalled on the bolts from which they were removed.

18. Tighten the oil pan nuts to 17 ft. lbs. (23 Nm) and bolts to 97 inch lbs. (11 Nm).

19. Lower the engine.

20. Install the engine mount through-bolts.

21. Install the right frame brace.

22. Install the starter motor assembly.

23. If removed, install the transmission fluid indicator tube.

24. For 5.0L and 5.7L engines, connect the shift linkage to the transmission.

25. Connect the transmission fluid cooler lines to the clips on the oil pan.

26. Install the flywheel cover, then connect the exhaust pipe to the manifolds.

27. If equipped, install the oil level sensor and electrical connection.

28. Lower the vehicle.

29. Install the upper fan shroud.

30. If removed, install the transmission and oil dipsticks.

31. Install the distributor cap.

32. For 5.0L and 5.7L engines, install the fuse cover, then engage the wiper motor electrical connection.

33. Install the air cleaner assembly, then connect the negative battery cable.

34. Refill the crankcase with clean engine oil, then start the engine and check engine for leaks.

4.3L (VIN W) and 5.7L (VIN P) Engines

1. Disconnect the negative battery cable.

2. Remove the air intake duct.

3. Raise and support the vehicle safely, then drain the engine crankcase of oil.

4. Drain the engine cooling system.

5. Disengage the wiring harness connector from the oil level sensor, then remove the sensor.

6. Disconnect the exhaust crossover pipe from the exhaust manifolds, then remove the pipe hanger bolts and reposition the pipe.

7. If equipped, remove the engine oil cooler hose bracket nut from the oil pan, then remove the oil cooler bolt from the oil cooler assembly. Reposition the engine oil cooler assembly.

8. Remove the transmission fluid cooler lines from the oil pan clip and remove the torque converter cover.

9. Remove the start motor assembly.

10. Remove the engine mount through-bolts, then install a suitable engine jacking fixture and carefully raise the engine, watching the clearance between engine mounted components and the firewall.

--- **WARNING** ---

Never raise the engine using a jack under the oil pan, crankshaft pulley or any sheetmetal. Because there only is a small clearance between the oil pan and the oil pump screen, if the pan is bent even slightly, damage could occur to the pump screen and pickup unit.

11. Remove the oil pan bolts, studs and nuts, then lower the oil pan assembly along with the gasket and the pan reinforcements. If necessary, rotate the crankshaft to reposition the counterweights.

To install:

12. Thoroughly clean the gasket mating surfaces of any remaining sealer and/or gasket material.

13. Apply a small amount of RTV sealer, 1052914 or equivalent, to the front cover-cylinder block junction and to the rear seal retainer-cylinder block junction. Continue the bead of sealer for 1 in. (25mm) in either direction of the radius cavity of these junctions.

14. Install the oil pan and gasket assembly, using the reinforcements, nuts, bolts and studs.

15. Tighten the corner oil pan bolts, stud or nuts to 15 ft. lbs. (20 Nm) and the remaining bolts or studs to 100 inch lbs. (11 Nm).

16. Lower the engine, remove the jacking fixture and install the engine mount through-bolts.

17. Install the oil level sensor assembly and tighten to 16 ft. lbs. (22 Nm), then engage the wiring harness connector.

18. Install the starter motor assembly.

19. Install the torque converter cover, then secure the transmission fluid cooler lines with the oil pan clip.

20. If equipped, install the engine oil cooler bolts screw to the cooler assembly and tighten to 24 ft. lbs. (33 Nm), then install the oil cooler hose bracket nut to the oil pan.

21. Install the exhaust pipe hanger bolt, then connect the crossover pipe to the exhaust manifolds.

22. Lower the vehicle and refill the engine crankcase with clean engine oil.

23. Install the air intake duct and connect the negative battery cable.

24. Start the engine and check for leaks.

Oil Pump

REMOVAL AND INSTALLATION

1. Disconnect the negative battery cable.

2. Remove the oil pan.

3. Remove the bolt(s) attaching the pump to the rear main bearing cap, then remove the pump and driveshaft extension.

4. For the 4.3L (VIN W) and 5.7L (VIN P) engines, remove the oil pump baffle retaining nuts and remove the baffle.

To install:

NOTE: The oil pump pickup should be submerged in oil and the pump primed prior to installation. Failure to prime the pump may result in oil pump failure or internal engine damage. Also, if the pickup screen and pipe assembly was removed from the pump, they must be replaced to assure a proper interference fit.

5. If the pickup screen and pipe was removed, it should be replaced as an assembly with a new part. Using a suitable tool, install a new pickup screen and pipe to the oil pump

6. Align the slot or hexagon head on the end of the shaft extension with the drive tang or the hexagon socket on the distributor shaft.

7. For the 4.3L (VIN W) and 5.7L (VIN P) engines, position the oil pump baffle before installing the pump retaining bolt.

8. Install and tighten the oil pump bolt to 77 ft. lbs. (105 Nm) for the 4.3L (VIN Z), 5.0L and 5.7L (VIN 7) engines. For the 4.3L (VIN W) and 5.7L (VIN P) engines, install the baffle nuts and tighten to 25 ft. lbs. (34 Nm).

9. Install the oil pan assembly to the engine.

10. Connect the negative battery cable, then check engine oil level and add, as necessary.

11. Start the engine while watch the indicator light or oil pressure gauge to ensure immediate oil pump operation.

TRANSMISSION

Automatic Transmission Assembly

REMOVAL AND INSTALLATION

1. Disconnect the battery negative cable and remove the air cleaner.

2. Disconnect the Throttle Valve (TV) cable at the throttle lever.

3. Remove the transmission dipstick, then remove the indicator tube from the transmission.

4. Raise and support the vehicle safely.

5. Remove the driveshaft.

6. Disconnect the shift linkage at the transmission.

7. Disengage all electrical leads at the transmission and any clips that retain the leads to the transmission case.

8. Remove the retaining bolts, then remove the flywheel cover.

9. Matchmark the flywheel and converter for installation purposes.

10. Remove and discard the torque converter-to-flywheel bolts.

11. Remove the catalytic converter support bracket.

12. Support and raise the transmission slightly using a suitable transmission jack.

13. Remove the transmission mount-to-support nut, washer, and bolt.

14. Remove the transmission support-to-frame bolts, nuts and, if used, insulators.

15. Slide the transmission support rearward.

16. Lower the transmission to gain access to the oil cooler lines and TV cable attachments.

17. Disconnect the lines and cap all openings to prevent excessive fluid loss or system contamination, then disconnect the TV cable.

18. Support the engine with a suitable tool, then and remove the transmission-to-engine bolts.

19. Install tool J-21366 or equivalent, to the torque converter or converter clutch in order to hold it in place.

20. Remove the transmission assembly from the vehicle.

To install:

21. Raise the transmission into place and remove tool J-21366.

22. Install the transmission-to-engine bolts and tighten to 35 ft. lbs. (47 Nm).

23. Unplug and install the oil cooler pipes, then connect the TV cable.

24. Install the fluid level tube using a new seal, then tighten tube retaining bolt to 35 ft. lbs. (47 Nm).

25. Install the transmission support-to-frame bolts, nuts and, if applicable, the insulators. Tighten the bolts to 41 ft. lbs. (55 Nm) for Fleetwood Brougham or to 25 ft. lbs. (34 Nm) except for Fleetwood Brougham and, if equipped the nuts to 30 ft. lbs. (41 Nm).

26. If removed or replaced, install the transmission mount bolts and tighten to 35 ft. lbs. (47 Nm). Install the transmission support nut and washer, then tighten to 30 ft. lbs. (41 Nm).

27. Remove the transmission jack, then position the converter by aligning it to the flywheel in the original position marked. Make sure the weld nuts on the converter are flush with the flywheel. Test the converter or clutch for freedom of movement.

28. Install and finger-tighten 3 new bolts, then tighten o 35 ft. lbs. (47 Nm) for Fleetwood Brougham or to 46 ft. lbs. (62 Nm) except for Fleetwood Brougham. After tightening all bolts, retorque the first bolt tightened.

29. If removed, install the floor pan reinforcement.

30. Install the catalytic converter support bracket, then install the converter cover and tighten the bolts to 89 inch lbs. (10 Nm).

31. Install the shift linkage, electrical leads, retaining clips, and if equipped, speedometer cable.

32. Install the driveshaft and lower the vehicle.

33. Install the TV cable to the throttle lever, then install the fluid level indicator.

34. Install the air cleaner, then connect the negative battery cable.

35. Adjust the shift linkage and the TV cable.

36. Flush the transmission and cooler system to prevent damage to the system components.

SHIFT LINKAGE ADJUSTMENT

The shift control linkage should be set so the engine will only start in **P** or **N**. If adjustment is necessary proceed as follows:

1. Make sure the steering column attachment and all body bolts are secure as they may affect the shift linkage.

2. Position the steering column shift lever in **N**.

3. Raise and support the vehicle safely.

4. Free the control rod and swivel.

5. Set the transmission lever to the neutral detent.

6. Hold swivel flush against the equalizer lever and finger-tighten bolt against rod. No force should be exerted in either direction on the control rod or equalizer lever while tightening bolt.

7. Tighten bolt to 21 ft. lbs. (28 Nm).

8. Lower vehicle and check that adjustment was proper.

Throttle Valve Cable

ADJUSTMENT

Setting of the TV cable must be done by rotating the throttle lever at the carburetor or throttle body. Do not use the accelerator pedal to rotate the throttle lever.

1. Turn the engine OFF and remove the air cleaner assembly for access.

2. Depress and hold the reset tab at the engine end of the TV cable.

3. Move the slider until it stops against the fitting, then release the reset tab.

4. Rotate the throttle lever to its full travel and watch the slider, it must move (ratchet) toward the lever when the lever is rotated to its full travel position.

NOTE: The TV cable assembly may appear to work correctly with the engine stopped and cold, but then not work when the engine is a normal operating temperature. Always check the TV cable adjustment again, after the engine has been warmed.

5. Recheck after the engine is hot and road test the vehicle.

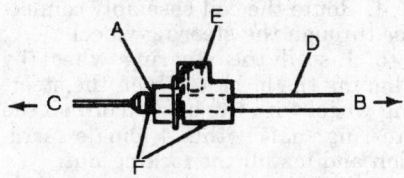

A. Slider against fitting
B. To automatic transmission assembly
C. To throttle body
D. Automatic transmission TV cable assembly
E. Automatic transmission TV cable reset tab
F. Automatic transmission TV cable

8838QG26

Throttle Valve (TV) cable adjustment

DRIVE AXLE

Driveshaft

REMOVAL AND INSTALLATION

1. Raise the vehicle and support it safely.

2. Mark the relationship of the driveshaft to the differential flange.

3. Unbolt the driveshaft-to-differential flange retaining bolts, then remove the retaining straps. Tape the bearing caps in place to prevent losing the bearing rollers. Support the driveshaft to prevent excessive strain on the universal joint.

4. Position a suitable drain pan under the transmission end to catch any fluid that may leak out when the driveshaft is removed. Pull the shaft back and remove it. Be careful not to damage the splines at the transmission end.

To install:

5. Lubricate the splines with engine 1050169 or equivalent slip yoke lubricant and slide the slip yoke into place.

6. Align the driveshaft marks, then install the connect the retaining straps and tighten the bolts to 16 ft. lbs. (22 Nm).

7. Lower the vehicle.

U-Joints

REMOVAL AND INSTALLATION

Snapring Type

1. Raise and support the vehicle safely.

2. Matchmark and remove the driveshaft.

3. Remove the snaprings from the yoke. If the snapring is difficult to remove, tap the end of the bearing cap lightly to relieve pressure from snapring.

4. Support the propeller shaft horizontally in line with the base plate of a bench vise, but never clamp the driveshaft tube.

5. Place the universal joint so the lower ear of the yoke is supported on a 1⅛ in. socket. Press 1 trunnion bearing against the socket in order to press the opposite bearing from the yoke.

6. Grasp the cap and work it out, if necessary use tool J-9522-3 and J-9522-5 or equivalents.

7. Rotate the shaft and support the other side of the yoke, then press the bearing cap from the yoke and as in previous steps.

8. Remove the trunnion from the driveshaft yoke.

9. Clean and check the condition of all parts. Use U-joint repair kits to replace all the worn parts or replace the assembly using a new U-joint.

To install:

10. Repack the bearings with chassis grease and replace the trunnion dust seals after any operation that requires disassembly of the U-joint. Be sure the lubricant reservoir at the end of the trunnion is full of lubricant. Fill the reservoirs with lubricant from the bottom.

11. Partially insert the cross into the yoke so 1 trunnion seats freely in the bearing cup, then rotate the shaft so this trunnion is on the bottom.

12. Install the opposite bearing cap part way. Be sure both trunnions are started straight into the bearing caps.

13. Press against opposite bearing caps, working the cross constantly to be sure the trunnions are free in the bearings. If binding occurs, check the needle rollers to be sure 1 or more needles have not become lodged under an end of the trunnion.

14. As soon as 1 bearing retainer groove is exposed, stop pressing and install the bearing retainer snapring.

15. Continue to press until the opposite bearing retainer can be installed. If difficulty installing the snaprings is encountered, tap the yoke with a hammer to spring the yoke ears slightly.

16. Replace the driveshaft and lower the vehicle.

Molded Retainer (Nylon Injected) Type

NOTE: Don't disassemble these joints unless replacing the complete U-joint. These factory installed joints cannot be reused and should instead be replaced by snapring type U-joints.

1. Raise and support the vehicle safely.

2. Matchmark and remove the driveshaft.

3. Support the propeller shaft horizontally in line with the base plate of a bench vise, but never clamp the driveshaft tube.

4. Place the U-joint so the lower ear of the shaft yoke is supported by a 1⅛ in. socket. Press the lower bearing cap out of the yoke ear. This will shear the nylon injected ring retaining the lower bearing cap.

5. If the bearing cup is not completely removed, lift the cross, insert J-9522-3 and J-9522-5 or equivalent separator and spacer, then press the cap completely out.

6. Rotate the driveshaft, shear the opposite plastic retainer, and press the other bearing cup out in the same manner.

7. Remove the cross from the yoke.

NOTE: Production U-joints cannot be reassembled. There are no bearing retainer grooves in the caps. Discard all parts that were removed and substitute those in the overhaul kit.

8. If the front U-joint is being removed, separate the bearing caps from the slip yoke in the same manner.

9. Remove the sheared plastic bearing retainer from the yoke. If necessary, drive a small pin or punch through the injection holes to aid in removal.

10. Install the new snapring U-joints.

11. Install the driveshaft assembly.

12. Lower the vehicle.

STEERING

Air Bag

───────── **CAUTION** ─────────
Some vehicles are equipped with an air bag system, also known as the Supplemental Inflatable Restraint (SIR) or Supplemental Restraint System (SRS). The system must be disabled before performing service on or around system components, steering column, instrument panel components, wiring and sensors. Failure to follow safety and disabling procedures could result in accidental air bag deployment, possible personal injury and unnecessary system repairs.

PRECAUTIONS

Several precautions must be observed when handling the inflator module to avoid accidental deployment and possible personal injury.

• Never carry the inflator module by the wires or connector on the underside of the module.

• When carrying a live inflator module, hold securely with both hands, and ensure that the bag and trim cover are pointed away.

• Place the inflator module on a bench or other surface with the bag and trim cover facing up.

• With the inflator module on the bench, never place anything on or close to the module which may be thrown in the event of an accidental deployment.

DISARMING

1. Align the steering wheel so the vehicle wheels are pointing in the straight-ahead position.

2. Turn the ignition switch to the **LOCK** position.

3. Remove the SIR or AIR BAG fuse from the fuse block.

4. Remove the Connector Position Assurance (CPA) device, then disengage the yellow 2-way SIR wire harness connector at the base of the steering column.

ENABLING

1. Turn the ignition switch to the **LOCK** position.

2. Engage the yellow 2-way connector at the base of the steering column, then install the CPA device.

3. Reinstall the SIR or AIR BAG fuse.

4. Turn the ignition switch to the **RUN** position.

5. Verify the SIR indicator light flashes 7–9 times, if not, inspect system for malfunction.

Air Bag Coil Assembly

NOTE: The coil assembly must remain centered in order to avoid accidental deployment of the air bag after any repair procedures to the internals of the steering column. There are 2 different styles of coil assemblies, 1 rotates clockwise and the other rotates counterclockwise. An arrow on the coil indicates the proper direction of rotation.

CENTERING THE COIL

1. With the system properly disarmed and the coil partially removed from the steering column, hold the coil assembly with the clear bottom up to see the coil ribbon.

2. While holding the coil assembly, depress the lock spring and rotate the hub in the direction of the arrow until it stops. The coil should now be wound up snug against the center hub.

3. Rotate the coil assembly in the opposite direction approximately 2½ turns and release the lock spring between the locking tabs in front of the arrow.

4. Install the coil assembly onto the steering shaft.

Steering Wheel

REMOVAL AND INSTALLATION

1. Properly disable the SIR system and disconnect the negative battery cable.

2. Remove the Torx® screws from the back of the steering wheel and lift the inflator module, then disengage the SIR coil connector and retainer from the module and the horn lead from the column.

───────── **CAUTION** ─────────
To avoid personal injury when carrying a live inflator module, make sure the bag and trim cover are pointed away. When placing a live inflator module on a bench or other surface, always face the bag and trim cover up and away from the surface. Never carry the inflator module by the wires or connector on the underside of the module, otherwise personal injury could result if the bag is deployed.

3. Remove the steering wheel lock nut, then remove the steering wheel and horn contact using a suitable puller.

NOTE: When attaching the wheel puller to the wheel, use care to prevent threading the side screws all the way through the wheel hub and into the coil, damaging the assembly.

To install:

4. Route the coil assembly connector through the steering wheel.

5. Install the steering wheel by aligning the block tooth on the steering wheel with the block tooth on the steering shaft within 1 female serration and install the locking nut.

6. Position the inflator module and connect the horn lead to the the steering column, then engage the coil assembly connector to the inflator module. Install the coil connector retainer.

7. Route the coil assembly lead around the mounting post and secure under the clip.

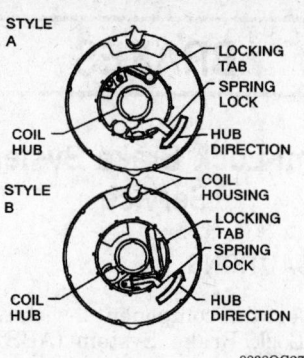

STYLE A

- LOCKING TAB
- SPRING LOCK
- COIL HUB
- HUB DIRECTION

STYLE B

- COIL HOUSING
- LOCKING TAB
- SPRING LOCK
- COIL HUB
- HUB DIRECTION

8838QG27

Centering the SIR coil assembly

8. Install the inflator module to the wheel and tighten the bolts to 53 inch lbs. (60 Nm).

9. Connect the negative battery cable and properly enable the SIR system.

Tie Rod Ends

REMOVAL AND INSTALLATION

NOTE: Because tie rod adjuster parts often become rusted in service it is recommend that a torque wrench be used to free the clamp nut from the bolts so the torque required can be measured. If a force in excess of 80 inch lbs. (9 Nm) is necessary to free the clamp nuts, they should be discarded and replaced with new nuts upon installation.

Inner

1. Raise and support the vehicle safely.
2. Remove the nut securing the inner ball stud to the intermediate or relay rod.
3. Remove the inner ball stud from the intermediate rod using J-24319-01 or an equivalent universal steering linkage puller.

4. Mark the tie rod end position before removing from the adjuster and note the position of the tie rod clamps for installation purposes. Loosen the clamp bolt and unscrew the end from the adjuster tube counting the number of turns necessary to remove the inner tie rod from the tube.

To install:

5. Lubricate the adjuster threads using chassis grease.
6. If installing a new tie rod, place the new and old components side-by-side, then copy the mark made during removal to the new component.
7. Thread the tie rod end into the adjuster using the same number of turns counted earlier, this should align the marks made earlier. If the marks do not align, make sure the same number of turns were used and/or the mark made on a new part was placed in the correct position.
8. Insert the inner ball stud to the intermediate rod, using J-29193 or an equivalent linkage installer to properly seat the tapers. Tighten the tool to 15 ft. lbs. (20 Nm), then remove the tool from the vehicle.
9. Install the retaining nut to the ball stud and tighten to 35 ft. lbs. (47 Nm).
10. Make sure the tie rod adjuster and clamps are properly positioned, then tighten the adjuster clamp bolt to 14 ft. lbs. (19 Nm).
11. Lower vehicle, then check and adjust vehicle toe, as necessary.

Outer

1. Raise and support the vehicle safely.
2. Remove the cotter pin from the outer tie rod ball stud, then remove the castellated nut.
3. Disconnect the tie rod end from the steering knuckle using a universal steering linkage puller.
4. Mark the tie rod end position before removing from the adjuster

and note the position of the tie rod clamps for installation purposes. Loosen the clamp bolt and unscrew the end from the adjuster tube counting the number of turns necessary to remove the outer tie rod from the tube.

To install:

5. Lubricate the adjuster threads using chassis grease.
6. If installing a new tie rod, place the new and old components side-by-side, then copy the mark made during removal to the new component.
7. Thread the tie rod end into the adjuster using the same number of turns counted earlier, this should align the marks made earlier. If the marks do not align, make sure the same number of turns were used and/or the mark made on a new part was placed in the correct position.
8. Connect outer ball stud to the steering knuckle, then install the attaching nut. Tighten the nut to 35 ft. lbs. (47 Nm), then install a new cotter pin. If after tightening a hole in the nut does not align with the stud hole, tighten the nut further, just enough to insert a cotter pin. Do not back off the specified torque to insert a cotter pin.
9. Make sure the tie rod adjuster and clamps are properly positioned, then tighten the adjuster clamp bolt to 14 ft. lbs. (19 Nm).
10. Lower vehicle, then check and adjust vehicle toe, as necessary.

Power Steering Gear

REMOVAL AND INSTALLATION

1. Disconnect the negative battery cable, then lock the steering wheel in the straight-ahead position to prevent damage to the SIR coil assembly.
2. Remove the shield from the steering gear return pipe nut, then remove the bolt from the intermediate shaft-to-gear coupling. Push the intermediate shaft rearward, disengaging the latch from the gear.
3. Remove the valve bracket nut from the gear.
4. Remove the steering linkage relay rod nut from the pitman arm.
5. Raise and support the vehicle safely.
6. Remove the nut from the pitman arm, then separate the arm from the steering gear using J-9172 or an equivalent puller.
7. Disconnect the power steering hoses from the gear assembly. Either

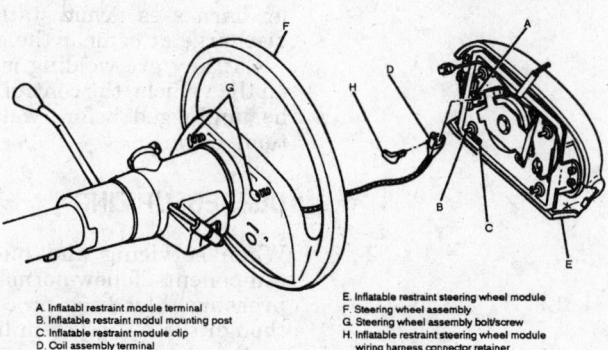

A. Inflatable restraint module terminal
B. Inflatable restraint modul mounting post
C. Inflatable restraint module clip
D. Coil assembly terminal

E. Inflatable restraint steering wheel module
F. Steering wheel assembly
G. Steering wheel assembly bolt/screw
H. Inflatable restraint steering wheel module wiring harness connector retainer

8838QG28

Removing the SIR inflator module

plug the hoses or raise and secure them to prevent excessive fluid loss.

8. Remove the steering gear mounting bolts, then remove the gear assembly.

To install:

9. Position the gear to the frame and loosely install the mounting bolts.

10. Adjust the gear to align as straight as possible with the intermediate shaft, then hold the gear in position and tighten the mounting bolts to 70 ft. lbs. (95 Nm).

11. Install the power steering hose assemblies to the gear and tighten the fittings to 21 ft. lbs. (28 Nm).

12. Install the pitman arm to the steering gear using a lock washer and a new nut. Tighten the nut to 185 ft. lbs. (250 Nm) for Fleetwood Brougham or to 179 ft. lbs. (243 Nm) except for Fleetwood Brougham.

13. Lower the vehicle, then install the steering linkage relay rod nut and tighten to 35 ft. lbs. (47 Nm).

14. Install the valve bracket nut to the gear assembly and tighten to 18 ft. lbs. (24 Nm).

15. Install the intermediate shaft coupling bolt and nut, then tighten to 30 ft. lbs. (40 Nm) for Fleetwood Brougham or to 40 ft. lbs. (54 Nm) except for Fleetwood Brougham.

16. Position the shield, making sure the latch is seated around the gear return pipe nut.

17. Connect the negative battery cable, then properly bleed the steering system.

Power Steering Pump

BLEEDING

When the power steering system has been serviced or a fitting disconnected, air must be bled from the system by using the following procedure:

1. With the engine **OFF** and the front wheels off the ground, turn wheels all the way to the left and add power steering fluid to the FULL COLD mark on the level indicator.

2. Bleed the system by turning the wheels from side to side several times, without hitting stops. Be sure to keep the level to the FULL COLD mark.

3. Start the engine and run at fast idle momentarily, then recheck the fluid level with the engine idling. If necessary add fluid to bring the level back to the FULL COLD mark.

4. Return the wheels to the center position and continue running the engine for a few minutes. Road test to check the operation of the steering.

5. Recheck the fluid level it should now be stabilized at the FULL HOT level on the indicator.

REMOVAL AND INSTALLATION

1. Disconnect the negative battery cable.

2. Loosen and remove power steering pump belt or serpentine drive belt.

3. If necessary, remove the radiator fan shroud for access.

4. Remove the power steering pump pulley using J-25034-B or equivalent puller.

5. Remove the 2 upper most pump mounting bolts, then raise and support the vehicle safely.

6. Disengage the electrical connections from the pump assembly

7. Disconnect the fluid hoses from the pump, then plug the hoses to prevent system contamination or excessive fluid loss.

8. Remove the remaining bolt and/or nut from the pump and carefully remove the pump from the vehicle.

To install:

9. Position pump assembly to bracket, then install the lower retaining bolt and tighten to 37 ft. lbs. (50 Nm).

10. Uncap and connect the inlet and outlet hoses.

11. Engage the pressure switch connector.

12. Lower the vehicle, then install the upper pump bolts and tighten to 37 ft. lbs. (50 Nm).

13. Using a suitable pump pulley installation tool, attach the pulley to the pump assembly. Make sure the pulley hub is flush with the end of the pump shaft.

14. Install the power steering pump belt or serpentine belt.

15. Fill with fluid, bleed the system and adjust belt(s) to proper tension.

BRAKES

Anti-Lock Brake System Service

PRECAUTIONS

• Certain components within the Anti-Lock Brake System (ABS) are not intended to be serviced or repaired individually. Only those components with removal and installation procedures should be serviced.

• Do not use rubber hoses or other parts not specifically specified for and ABS system. When using repair kits, replace all parts included in the kit. Partial or incorrect repair may lead to functional problems and require the replacement of components.

• Lubricate rubber parts with clean, fresh brake fluid to ease assembly. Do not use lubricated shop air to clean parts; damage to rubber components may result.

• Use only specified brake fluid from an unopened container.

• If any hydraulic component or line is removed or replaced, it may be necessary to bleed the entire system.

• A clean repair area is essential. Always clean the reservoir and cap thoroughly before removing the cap. The slightest amount of dirt in the fluid may plug an orifice and impair the system function. Perform repairs after components have been thoroughly cleaned; use only denatured alcohol to clean components. Do not allow ABS components to come into contact with any substance containing mineral oil; this includes used shop rags.

• The Anti-Lock control unit is a microprocessor similar to other computer units in the vehicle. Ensure that the ignition switch is **OFF** before removing or installing controller harnesses. Avoid static electricity discharge at or near the controller.

• If any arc welding is to be done on the vehicle, the control unit should be unplugged before welding operations begin.

DEPRESSURIZING

When servicing and bleeding ABS components, follow normal manual or pressure bleeding procedures. Although the ABS system has the ability to increase, decrease or hold brake line pressure, the hydraulic modulator cannot increase the pressure above that which is transmitted by the master cylinder. Special service procedures for bleeding the brake system with a hydraulic modulator are not required.

Master Cylinder

REMOVAL AND INSTALLATION

NOTE: Be sure to clean the area where the master cylinder is mounted, before beginning removal.

1. Disconnect the brake pipes from the master cylinder assembly. Cap or plug the openings to prevent system contamination or excessive fluid loss.
2. Remove the 2 nuts attaching the master cylinder to the brake booster assembly.
3. The combination valve bracket is mounted over the master cylinder on the power brake booster studs, pull the bracket from the studs and reposition it aside.
4. Remove the master cylinder assembly from the vehicle.

To install:

5. Install the master cylinder on the power booster studs.
6. If applicable, install the combination valve bracket on the power booster studs.
7. Install the attaching nuts and tighten to 28 ft. lbs. (38 Nm) for Fleetwood Brougham or to 20 ft. lbs. (27 Nm) except for Fleetwood Brougham.
8. Unplug the hydraulic pipes, then connect them to the master cylinder and tighten to 24 ft. lbs. (32 Nm).
9. Fill with approved brake fluid, then bleed the hydraulic brake system.

Brake Caliper

REMOVAL AND INSTALLATION

1. Remove ⅔ of the brake fluid from the master cylinder assembly.
2. Raise and support the vehicle safely.
3. Mark the relationship of the wheel to the hub for reinstallation, then remove the tire and wheel assembly.
4. Position a C-clamp over the outboard shoe and lining and the caliper housing, then slowly tighten the C-

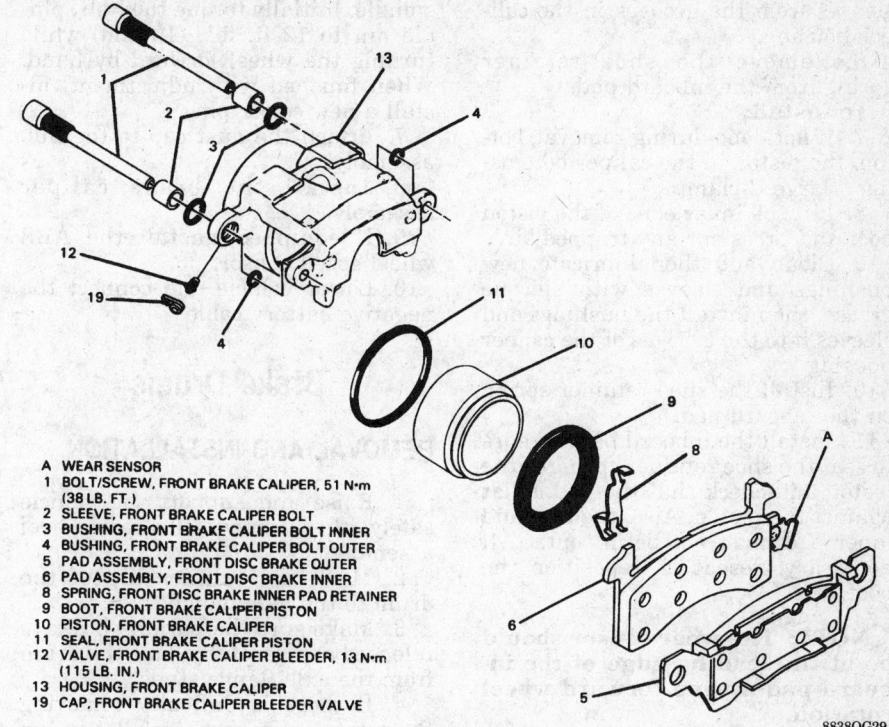

A WEAR SENSOR
1 BOLT/SCREW, FRONT BRAKE CALIPER, 51 N·m (38 LB. FT.)
2 SLEEVE, FRONT BRAKE CALIPER BOLT
3 BUSHING, FRONT BRAKE CALIPER BOLT INNER
4 BUSHING, FRONT BRAKE CALIPER BOLT OUTER
5 PAD ASSEMBLY, FRONT DISC BRAKE OUTER
6 PAD ASSEMBLY, FRONT DISC BRAKE INNER
8 SPRING, FRONT DISC BRAKE INNER PAD RETAINER
9 BOOT, FRONT BRAKE CALIPER PISTON
10 PISTON, FRONT BRAKE CALIPER
11 SEAL, FRONT BRAKE CALIPER PISTON
12 VALVE, FRONT BRAKE CALIPER BLEEDER, 13 N·m (115 LB. IN.)
13 HOUSING, FRONT BRAKE CALIPER
19 CAP, FRONT BRAKE CALIPER BLEEDER VALVE

8838QG29

Exploded view of the caliper assembly

clamp in order to bottom the piston into the caliper bore.

NOTE: If removing the caliper assembly only to access other brake parts skip Step 5. If removing the caliper entirely from the vehicle, the brake system will have to be bled.

5. If removing the caliper from the vehicle for service or replacement, remove the bolt, copper washers and inlet fitting from the caliper housing. Plug the openings to prevent system contamination or excessive fluid loss and discard the used copper washers.
6. Remove the mounting bolts and the sleeves, then remove the caliper from the rotor. If the caliper is not being completely removed from the vehicle, it must be suspended from the suspension using a hook, string or wire. Never allow the caliper to hang by the brake line or damage could occur.

To install:

7. Lubricate the sleeves and bushings with silicone grease, then insert the sleeves into caliper housing.
8. Position the caliper assembly onto the rotor and knuckle assembly.
9. Insert the mounting bolts and tighten to 38 ft. lbs. (51 Nm).
10. Measure the distance between the caliper housing and the stops on

the knuckle assembly. There should be 0.005–0.012 in. (0.13–0.30mm) of clearance. If necessary, remove the caliper assembly and file the ends of the stops on the knuckle to increase clearance. Excessive clearance requires replacement of the caliper and/or steering knuckle.

11. If removed for service or replacement, bolt new copper washers and the inlet fitting to the caliper housing, then tighten to 32 ft. lbs. (44 Nm).
12. Align the marks on the wheel and hub, then install the wheel assembly.
13. Lower the vehicle.
14. Fill the master cylinder to the proper level, and, if the line was disconnected, bleed the caliper.

Disc Brake Pads

REMOVAL AND INSTALLATION

1. Raise the vehicle and support it safely.
2. Remove the tire and wheel assembly.
3. Remove the caliper assembly and support it aside from the front suspension.
4. Remove the inner and outer brake pads from the caliper.

5. Remove the bushings and sleeves from the grooves in the caliper housing.

6. Remove the shoe retainer spring from the inboard pad.

To install:

7. If not done during removal, bottom the piston in the caliper bore using a large C-clamp.

8. Lift the inner edge of the piston boot and press out any trapped air.

9. Clean and then lubricate new bushings and sleeves with silicone grease, then install the bushings and sleeves into the grooves of the caliper housing.

10. Install the shoe retainer spring on the inboard pad.

11. Install the inboard pad. Be sure to seat the shoe retainer spring in the piston and check that the pad is flat against the piston. Also, the boot and inner pad must not be in contact. If necessary, reseat or reposition the boot.

NOTE: The wear sensor should be at the leading edge of the inboard pad during forward wheel rotation.

12. Install the outboard pad with the back of the pad flat against the housing.

13. Install the caliper assembly and clinch the pad tabs using a chisel or wedge tool between the tab and caliper casting, then using a brass punch to drive the tab down against the caliper.

14. Install the tire and wheel assembly, then lower the vehicle.

15. Apply the brake pedal 3 times to seat the pads and to achieve a firm brake pedal.

Brake Rotor

REMOVAL AND INSTALLATION

1. Disconnect the negative battery cable, then raise and support the vehicle safely.

2. If equipped, disconnect the ABS wheel speed sensor from the steering knuckle and secure.

3. Remove the brake caliper and position aside supported from the front suspension.

4. Remove the dust cap from the hub, then remove the cotter pin, nut and washer from the spindle. Carefully remove the hub and rotor from the spindle.

To install:

5. Position hub and rotor assembly onto the spindle.

6. Place the washer and nut on the spindle. Initially torque the hub spindle nut to 12 ft. lbs. (16 Nm) while turning the wheel forward by hand. When finished with adjustment, install a new cotter pin.

7. Attach the dust cap to the hub assembly.

8. Install the brake caliper assembly.

9. If equipped, install the ABS wheel speed sensor.

10. Lower vehicle and connect the negative battery cable.

Brake Drums

REMOVAL AND INSTALLATION

1. Raise and support the vehicle safely, then remove the rear wheel assembly.

2. Mark the relationship of the drum to the axle flange.

3. Make sure the parking brake is released and carefully slide the drum from the axle flange studs.

4. If there is difficulty removing the rotor, use a rubber mallet to tap gently on the outer rim of the drum and/or the inner drum diameter by the spindle. If necessary, remove the adjusting hole or knockout plate from the backing plate and back off the adjusting screw with a suitable tool.

To install:

5. Adjust the brake shoes. The outside diameter of the shoe and linings should be 0.050 in. (1.27mm) less than the inside diameter of the brake drum on each wheel.

6. Position the drum on the axle flange, aligning the marks made earlier.

7. Install the rear tire and wheel assembly.

8. Adjust the parking brake, then lower vehicle.

Brake Shoes

REMOVAL AND INSTALLATION

1. Raise and support the vehicle safely.

2. Remove the tire and wheel assemblies.

3. Remove the brake drums. If the brake drum cannot be removed, try the following:

    a. Make sure the parking brake is released.

    b. Back off the parking brake cable adjustment.

    c. Remove the adjusting hole knockout plate from the backing plate and back off the adjusting screw.

    d. Use a rubber mallet to tap on the outer rim of the drum and around the inner drum diameter by the spindle.

4. Remove the return springs.

5. Remove the hold-down pins and springs.

6. Remove the actuator lever pivot.

7. Lift up on the actuator lever to remove the actuator link. Remove the lever, pawl if equipped, and lever return spring.

8. Remove the shoe guide, the parking brake strut and spring.

9. Remove the brake shoes from the backing plate and the parking brake cable.

10. Remove the adjusting screw assembly and spring from the brake shoes.

11. Remove the parking brake lever by unhooking the lever tab from the slot in the brake shoe.

To install:

12. Clean the adjusting screw with a wire brush and then clean all components with brake cleaner or denatured alcohol.

13. Using 1052196 or equivalent, lubricate the adjusting screw threads, inside diameter of the socket and the socket face, for smooth rotation.

14. Install the parking brake lever by hooking lever tab into slot in primary or secondary shoe lining.

15. Install the adjusting screw assembly and the adjusting screw spring.

16. Connect the primary/secondary shoes to the parking brake cable and the brake backing plate.

17. Install the strut and spring by spreading the shoes and inserting the components, then install the shoe guide.

18. Install the pawl, if equipped, the actuator lever and the lever return spring.

19. Install the hold-down pins, pivot and springs.

20. Connect the actuator link to the actuator pin. Install the actuator link into actuator lever while holding up on actuator lever.

21. Install the return springs with suitable a tool.

22. Repeat the procedure for the opposite rear wheel.

23. Adjust the brake shoes. The outside diameter of both shoe and linings should be 0.050 in. (1.27mm) less than the inside diameter of the brake drum on each wheel.

24. Install the drums and tires.

25. Lower the vehicle. Be sure to adjust the parking brake cable, if the parking brake adjuster was loosened to remove the drums.

Wheel Cylinder

REMOVAL AND INSTALLATION

1. Raise and support the vehicle safely.
2. Remove the wheel and brake drum.
3. Remove the brake shoes and components.
4. Clean dirt and foreign material from around wheel cylinder assembly.
5. Disconnect the inlet tube from the cylinder and plug the opening in the line to prevent system contamination or excessive fluid loss.
6. Remove the cylinder-to-shoe pushrods.
7. Remove the 2 attaching bolts, then remove the cylinder assembly from the backing plate.
   To install:
8. Position the wheel cylinder assembly to the backing plate.
9. Connect the wheel cylinder attaching bolts and tighten to 13 ft. lbs. (18 Nm).
10. Install the cylinder shoe pushrods.
11. Connect the brake rear pipe to the wheel cylinder and tighten to 18 ft. lbs. (24 Nm).
12. Install the shoes and brake components.
13. Install the rotors and wheels.
14. Lower the vehicle and bleed the hydraulic brake system.

Parking Brake Cable

ADJUSTMENT

NOTE: Before attempting to adjust the parking brake, verify that the rear brakes are correctly adjusted. If rear brakes are adjusted properly, the parking brake cable should usually not require adjustment, unless worn or broken components were replaced.

1. Clean and lubricate the exposed threads of the adjuster rod, to either side of the nut.
2. Apply the parking brake 6 clicks, then raise and support the vehicle safely.
3. Tighten the adjusting nut until the right rear wheel can barely be turned backwards when using 2

hands, but locks up when attempts are made to move it forward.
4. With the parking brake disengaged the rear wheel should turn freely in either direction with no brake drag.
5. Lower the vehicle.

REMOVAL AND INSTALLATION

Front Cable

1. Raise the vehicle and support it safely.
2. Loosen equalizer sufficiently to gain the necessary cable slack.
3. Disconnect the front cable at the retainer.
4. Disconnect the cable casing at the frame by compressing the retainer fingers, using a ½ in. box wrench, and by pulling outward.
5. Lower the vehicle.
6. Remove driver's side wheelhouse panel screws and panel bolts. Pull panel out to gain access to the front cable.
7. Disconnect the front cable and casing at the lever assembly, by compressing retainer fingers (again using the box wrench) and by pulling outward.
8. Remove the front cable and grommet from the vehicle.
   To install:
9. Position front cable and grommet in vehicle.
10. Connect the front cable and casing at the lever assembly.
11. Install the wheelhouse panel bolts, tighten to the perimeter bolts to 18 ft. lbs. (25 Nm), and the 2 inner bolts to 89 inch lbs. (10 Nm).
12. Raise and support the vehicle safely.
13. Connect the cable casing at the frame by seating the lock fingers, then connect the cable to the retainer.
14. Adjust the parking brake and lower the vehicle.

Rear Cable

1. Raise and support the vehicle safely.
2. Loosen the equalizer enough to gain cable slack, as necessary.
3. On the left side, disengage the cable from the retainer and the equalizer.
4. On the right side, disconnect the cable from the equalizer. Disconnect the cable and casing at the frame by compressing the retainer fingers and pulling outward. Disconnect the cable and casing from the axle housing clips.

5. Remove the tire and wheel assembly, then remove the brake drum.
6. Remove the primary and secondary brake shoes, as necessary.
7. Disconnect the cable from the parking brake lever and remove the cable. Compress the retainer fingers and loosen the cable and casing from the backing plate.
   To install:
8. Connect the cable and casing into the brake backing plate, then attach to the parking brake lever.
9. Install the brake shoes and components, as necessary.
10. Adjust the brake shoes and install the drum, then install the tire and wheel assembly.
11. On the right side, connect the cable and casing to the axle housing clips and to the frame. Then connect the cable to the equalizer.
12. On the left side, connect the cable to the retainer and the equalizer.
13. Adjust the parking brake, then lower the vehicle.

Brake System Bleeding

PROCEDURE

The brake system must be bled when any brake line is disconnected or if it is suspected that there is air in the system.

NOTE: Always take extreme care to prevent fluid from touching a painted surface.

1. Clean the master cylinder of excess dirt and remove the cylinder cover and the diaphragm.
2. Fill the master cylinder to the proper level. Check the fluid level periodically during the bleeding process and replenish it, as necessary. Do not allow the master cylinder to fall below ½ full.
3. If the master cylinder is suspected or known to have air in the bore, bleed it as follows before any wheel cylinder or caliper:
   a. Disconnect the forward brake line connection at the master cylinder.
   b. Allow brake fluid to fill the master cylinder bore until it begins to flow from the forward line connector port.
   c. Connect the forward brake line to the master cylinder and tighten.
   d. Have an assistant depress the brake pedal slowly, 1 full thrust at a time and hold. Loosen the forward brake line connection at the

master cylinder to purge the air from the bore, then tighten the connection and have the assistant release the brake pedal slowly. Wait 15 seconds and repeat the sequence, including the 15 second pause, until all air is removed from the bore.

e. After all air is removed at the forward connection, repeat the procedure for the rear master cylinder connection. Both fittings should be tightened to 24 ft. lbs. (32 Nm) after the master cylinder is bled.

NOTE: Never bleed a wheel cylinder when a drum is removed.

4. Bleed the individual wheel cylinders or calipers only after all air is removed from the master cylinder. If the master cylinder was bled, then the entire system must be bled in the proper order. The correct sequence for bleeding is to work from the brake farthest from the master cylinder to the 1 closest; (right rear, left rear, right front, left front). Bleed individual components as follows:

a. Position the proper size box end wrench over the bleeder valve.

b. Attach a length of clear vinyl hose to the bleeder screw of the brake to be bled. Insert the other end of the hose into a clear jar half full of clean brake fluid so the end of the hose is beneath the level of fluid.

c. Have an assistant depress the brake pedal 1 time and hold. Loosen the bleeder valve to purge the air from the cylinder, then tighten the bleeder screw and have the assistant slowly release the pedal. Wait 15 seconds, then repeat the sequence, including the 15 second pause, until all air is removed.

NOTE: Make sure the assistant presses the brake pedal to the floor slowly. Rapid pumping of the brake pedal pushes the master cylinder secondary piston down the bore in a way that makes it difficult to bleed the rear side of the system.

5. Repeat this procedure at each of the wheels. Tighten the caliper bleeder screw to 115 inch lbs. (13 Nm) or the wheel cylinder screw to 62 inch lbs. (7 Nm) when bleeding is completed. Remember to check the master cylinder level periodically during the process. Use only fresh fluid to refill the master cylinder, not the fluid bled from the system.

6. Check the brake pedal to make sure it is free of sponginess and make sure the red BRAKE indicator lamp

is OFF indicating a proper pressure balance. If necessary, repeat the entire procedure to correct either condition.

7. When the bleeding process is complete, top off the master cylinder, install its cover and diaphragm and discard the fluid bled from the brake system.

Wheel Speed Sensor

REMOVAL AND INSTALLATION

Front Speed Sensor

1. Disconnect the negative battery cable.

2. For the right side speed sensor, unclip the connectors from the clip and separate.

3. Raise and support the vehicle safely.

4. For the left side speed sensor, with the vehicle safely supported, unclip the connectors from the clip and separate.

5. Remove the sensor wiring harness mounting bolt and bracket from the frame rail.

6. Remove the sensor retaining bolt and remove the sensor from the kunckle assembly.

To install:

7. Coat the steering knuckle with anti-corrosion compound 1052856 or equivalent, at the knuckle contact point.

8. Install the wheel speed sensor to the steering knuckle and tighten the sensor retaining bolt to 9 ft. lbs. (12 Nm) for Fleetwood Brougham or to 71 inch lbs. (8 Nm) except for Fleetwood Brougham.

NOTE: Proper installation of the wheel speed sensor cables is critical to proper operation of the ABS system. Make sure the cables are installed in the retainers. Failure to do this may result in contact with moving parts and the over extension of the cables, resulting in an open circuit.

9. Connect the sensor wiring harness mounting bolt and bracket to the frame rail and tighten to 89 inch lbs. (10 Nm).

10. For the left side speed sensor, with the vehicle safely supported, attach the connectors and position them in the clip.

11. Lower the vehicle.

12. For the right side speed sensor, attach the connectors and position them in the clip.

13. Connect the negative battery cable and road test vehicle.

Rear Speed Sensor

1. Disconnect the negative battery cable. Raise and support the vehicle safely.

2. Disconnect the rear sensor assembly from the differential sensor connector.

3. Remove the sensor wiring harness from the retainer brackets.

4. Remove the sensor retaining bolt and remove the speed sensor from the rear axle housing.

To install:

5. Install the sensor into the rear axle housing. The sensor is a tight fit but it must be pushed in by hand. Do not hammer the sensor into position.

6. Tighten the sensor retaining bolt to 71 inch lbs. (8 Nm).

7. Insert the sensor wiring harness into the retainer brackets.

8. Connect the rear sensor assembly to the differential sensor connector.

9. Lower the vehicle.

10. Connect the negative battery cable and road test the vehicle.

NOTE: Proper installation of the wheel speed sensor cables is critical to proper operation of the ABS system. Make sure the cables are installed in the retainers. Failure to do this may result in contact with moving parts and the over extension of the cables, resulting in an open circuit.

FRONT SUSPENSION

Shock Absorber

REMOVAL AND INSTALLATION

1. Raise and support the vehicle safely.

2. Hold the shock absorber upper stem from turning and remove the upper nut, then remove the retainer and grommet.

3. Remove the 2 bolts and lock washers securing the shock to the lower control arm, then lower the shock from the vehicle.

To install:

4. With the lower retainer and grommet in place over the upper stem, install the fully extended shock up through the lower control arm and spring. Make sure the upper stem passes through the mounting hole in the frame bracket.

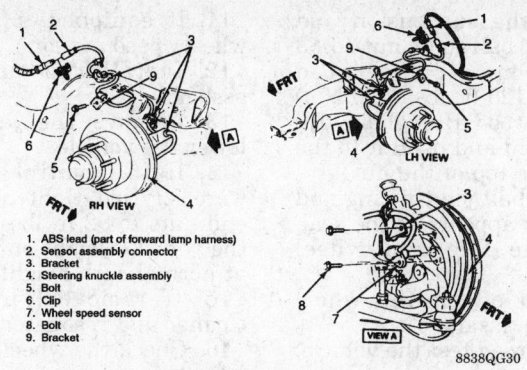

1. ABS lead (part of forward lamp harness)
2. Sensor assembly connector
3. Bracket
4. Steering knuckle assembly
5. Bolt
6. Clip
7. Wheel speed sensor
8. Bolt
9. Bracket

8838QG30

ABS front wheel speed sensor

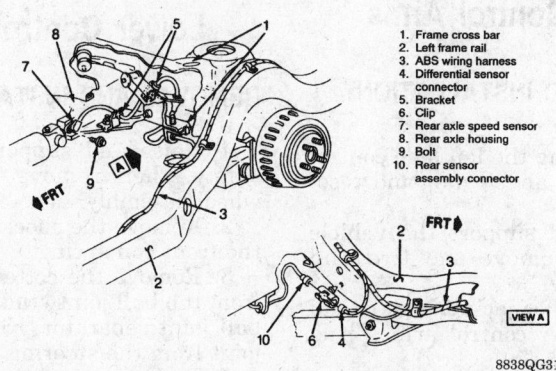

1. Frame cross bar
2. Left frame rail
3. ABS wiring harness
4. Differential sensor connector
5. Bracket
6. Clip
7. Rear axle speed sensor
8. Rear axle housing
9. Bolt
10. Rear sensor assembly connector

8838QG31

ABS rear wheel speed sensor

5. Install the upper rubber insulator, retainer and attaching nut over the shock, then tighten the nut to 97 inch lbs. (11 Nm).

6. Install the shock lower pivot to the lower control arm and tighten the 2 attaching bolts to 20 ft. lbs. (27 Nm).

7. Lower the vehicle.

Coil Spring

REMOVAL AND INSTALLATION

1. Raise and support the vehicle safely.

2. Remove the tire and wheel assembly from the vehicle.

3. If equipped, remove the ABS wheel speed sensor and secure aside.

4. Remove shock absorber.

5. Remove the stabilizer bar linkage nut, retainer and linkage from the lower control arm.

6. Remove the cotter pin and castellated nut, then separate the tie rod from the steering knuckle, using a suitable puller tool.

7. Compress the coil spring using a universal spring compressor tool.

8. Support the lower control arm using an adjustable lifting device,

then remove the lower control arm-to-frame pivot bolts.

9. Pivot the lower control arm rearward, then carefully remove the compressor and spring.

To install:

10. Position the spring onto the lower control arm making sure the insulator is in position, then install the compressor tool and compress the spring so the control arm may be repositioned.

11. Position the control arm into the frame and install the pivot bolts (install the front bolt first, positioned from the front to the rear), but wait until the suspension is supporting the vehicle's weight before tightening the control arm fasteners to specification.

12. Remove the spring compressor tool, then install the tie rod end to the steering knuckle and tighten the nut to 35 ft. lbs. (47 Nm). Tighten the nut additionally, as necessary to align the hole, then install a new cotter pin.

13. Remove the support from the lower control arm and install the stabilizer bar linkage.

14. Install the shock absorber. Tighten the lower attaching bolts to 20 ft. lbs. (27 Nm) and the upper attaching nut to 97 inch lbs. (11 Nm).

15. If equipped, install the ABS wheel speed sensor.

16. Install the wheel and lower the vehicle.

17. Tighten the lower control arm nuts to 92 ft. lbs. (125 Nm).

Torsion Bar

REMOVAL AND INSTALLATION

1. Raise and support the vehicle safely.

2. Disconnect each side of the torsion bar by removing the nut from the link bolt, then pull the bolt from the linkage and remove the retainers, grommets and spacer.

3. Remove bracket-to-frame or body bolts and remove torsion bar, rubber bushings and brackets.

To install:

4. Position the torsion bar with the shaft identification on the right side of the vehicle and the slits in the rubber bushings facing the front of the vehicle.

5. Install the rubber insulators to the stabilizer shaft and the bracket-to-frame (or body) bolts.

6. Install the link bolts, insulators, spacers, washers and nuts. Tighten the bracket bolts to 24 ft. lbs. (33 Nm), then tighten the link bolt/nut to 18 ft. lbs. (25 Nm).

7. Lower the vehicle.

Upper Ball Joints

REMOVAL AND INSTALLATION

1. Raise and safely support the vehicle; place floor stands under the lower control arm between the spring seats and the ball joints.

NOTE: Leave the jack under the spring seat during removal and installation, in order to retain the coil spring and relieve spring tension from the upper control arm. The weight of the vehicle is used to relieve spring tension on the upper control arm.

2. Remove the wheel.

3. Remove the cotter pin and nut from the upper ball joint.

4. Using a ball joint separator tool, break the stud loose and pull the stud out of the knuckle. Support the steering knuckle to prevent damage to the brake line.

5. With the control arm in a raised position, use a 1/8 in. diameter bit and drill into each of the 4 rivet heads to a depth of 1/4 in.

6. Drill off the rivet heads with a ½ in. diameter bit.

7. Punch out the rivets using a suitable driver or punch, then remove the ball joint.

To install:

8. Place the new ball joint in the upper control arm and secure it with 4 bolts and nuts in place of rivets. Torque the nuts to specifications provided with the ball joint kit.

9. Connect the ball joint to steering knuckle. Torque the nut to 61 ft. lbs. (83 Nm), then insert a new cotter pin. Do not back off the specified torque in order to install the cotter pin.

NOTE: When replacing the ball joints, use only high-quality replacement parts; bolts and nuts specified to be strong enough to endure the stress.

10. Attach the grease fitting and lubricate until grease appears at the seal.

11. Install the wheel and road test the vehicle.

Lower Ball Joints

REMOVAL AND INSTALLATION

1. Raise the vehicle and support the frame safely.

2. Remove the tire and wheel assembly.

3. Place a floor jack or axle stand under the control arm spring seat.

NOTE: Leave the jack or axle stand under the spring seat during removal and installation, in order to keep the spring and control arm positioned.

4. Remove the cotter pin and nut from the ball joint stud, then using a ball joint separator, remove the ball joint from the steering knuckle.

5. With a small putty knife or similar tool, guide the control arm from the opening in the shield to a position where the ball joint is accessible.

6. Block the steering knuckle aside using a block of wood between the frame and the upper control arm.

7. Remove the grease fittings.

8. Using a suitable ball joint remover, drive the lower ball joint from the control arm.

To install:

9. Using a ball joint installer, press in a new ball joint until it bottoms on the lower control arm.

NOTE: Make sure the grease purge on the seal faces away from the brakes.

10. Assemble the suspension and torque the lower ball joint nut to 83 ft. lbs. (113 Nm) for Fleetwood Brougham or to 125 ft. lbs. (170 Nm) except for Fleetwood Brougham. Install the cotter pin and bend it to the side, not over the top of the nut.

11. Install the ball joint fitting and lube until grease appears at the seal.

12. Install the tire and wheel assembly.

13. Check and adjust the wheel alignment, as necessary.

14. Lower and road test the vehicle.

Upper Control Arms

REMOVAL AND INSTALLATION

1. If removing the left side control ar, remove the air cleaner and resonator duct.

2. Raise and support the vehicle safely, then remove the tire and wheel assembly.

3. Place a floor jack or axle stand under the lower control arm spring seat.

NOTE: Leave the floor jack or axle stand under the spring seat during removal and installation, in order to keep the spring and control arm positioned.

4. If equipped with ABS, disconnect the wheel speed sensor and secure aside.

5. Loosen the pivot shaft-to-frame nuts and remove alignment shims. Tape the shims together and mark to assure installation in their original locations.

6. Remove the cotter pin and upper ball joint nut, then remove the ball joint from the steering knuckle, using a separator tool. Support the hub assembly to prevent damage to the brake line.

7. Remove the upper control arm shaft attaching nuts and bolts, then remove the control arm and shaft assembly.

To install:

8. Loosely install the control arm and pivot shaft assembly using the attaching bolts.

9. Install the alignment shims in the same positions from which they were removed, then install the nuts and tighten to 72 ft. lbs. (98 Nm).

10. Remove the temporary support from the hub and connect the ball joint to the steering knuckle. Tighten the upper nut to 61 ft. lbs. (83 Nm) and install a new cotter pin.

11. If equipped, install the ABS wheel speed sensor.

12. Install the tire and wheel assembly.

13. Remove the jackstands and lower the vehicle.

14. If the control arm bushings were serviced, tighten the pivot shaft end nuts to 92 ft. lbs. (125 Nm) with the vehicle weight on the suspension at normal curb height.

15. If removed, install the air cleaner and resonator.

16. Check the wheel alignment and adjust, as necessary.

Lower Control Arms

REMOVAL AND INSTALLATION

1. Raise and support the vehicle safely, then remove the tire and wheel assembly.

2. Remove the shock absorber and the front coil spring.

3. Remove the cotter pin and nut from the ball joint stud, then using a ball joint separator, remove the ball joint from the steering knuckle.

4. Remove the lower control arm assembly from the vehicle.

To install:

5. Install the lower ball joint stud into the steering knuckle.

6. Install the coil spring, and in doing so, install the shock absorber and the lower control arm.

7. Install wheel and tire assembly.

8. Lower the vehicle, then check and adjust alignment, as necessary.

Front Wheel Bearings

ADJUSTMENT

1. Raise the vehicle so the wheel can spin freely.

2. Remove the wheel cover, dust cap, cotter pin and loosen the adjusting nut.

3. Tighten the adjusting nut to 12 ft. lbs. (16 Nm) while turning the wheel, this will seat the bearings and remove any grease or burrs which could cause play later.

4. Back off the nut until it is just loose.

5. Finger-tighten the nut and install the cotter pin through the retaining ring or castle nut.

NOTE: If the cotter pin cannot be installed, back off the nut until the slot aligns with the serrations on the nut. Do not back off the nut more than ¼ of a turn.

6. Once adjusted, the front wheel bearings should have 0.001–0.005 in. (0.03–0.13mm) endplay.

7. When adjusted properly cut off any extra length from the cotter pin to prevent interference, then install the dust cap and wheel cover.

8. Lower the vehicle.

REMOVAL AND INSTALLATION

1. Raise and support the vehicle safely.

2. Remove the tire and wheel assembly.

3. Remove the rotor and hub assembly.

4. Pry the inner bearing seal from the hub, then remove the inner roller bearing assembly.

5. If necessary, remove the inner and outer bearing races using tool J-29117-A or a suitable brass punch inserted behind the races.

To install:

6. Using fresh solvent, clean all old grease from hub, spindle and bearing.

7. If the inner and outer races were removed, press or drive the races into the hub using a suitable sized driver.

8. Pack the bearings with a high temperature wheel bearing grease and reassemble the hub. Do not mix greases.

9. Install a new inner bearing seal using a flat plate to assure the seal is flush with the rotor.

10. Apply a thin coat of grease to the spindle, then install the rotor and hub assembly on the steering knuckle.

11. Adjust the wheel bearings, install a new cotter pin and replace the dust cap.

12. Install the caliper assembly.

13. Install the tire and wheel assembly.

14. Lower the vehicle.

REAR SUSPENSION

Shock Absorber

REMOVAL AND INSTALLATION

1. Raise and safely support the vehicle safely, making sure to properly support the rear axle housing.

2. If equipped with adjustable shocks, disconnect the air line by turning the spring clip 90 degrees and pulling gently on air line housing.

3. Remove the upper nuts and bolts from the shock absorber at the frame.

4. Using a wrench to hold the stud in place, remove the lower nut and washer from the shock at the rear axle housing. The stud must not be allowed to turn during this operation or damage may result in the bond between the bushing and stud.

5. Remove the shock from the vehicle.

To install:

6. Install the shock absorber and loosely connect the upper frame bolts and nuts.

7. Position the stud into the bracket on the axle housing, then install the nut and washer.

8. Either tighten the upper bolts at the frame to 20 ft. lbs. (27 Nm) or the nuts at the frame to 12 ft. lbs. (16 Nm), whichever is easier. Then while holding the stud steady with a wrench, tighten the lower shock retaining nut to 65 ft. lbs. (88 Nm) for Fleetwood Brougham or to 50 ft. lbs. (68 Nm) except for Fleetwood Brougham.

9. If equipped, connect the shock air line.

10. Remove the supports and lower the vehicle.

Coil Spring

REMOVAL AND INSTALLATION

NOTE: If both springs are to be replaced, only disconnect 1 control arm at a time in order to prevent the axle from rolling or slipping sideways.

1. Raise and support the vehicle safely, then place an adjustable support under the axle housing.

2. If equipped, disconnect the ABS rear speed sensor.

3. If equipped, disconnect the height sensor link from the upper control arm by removing the attaching nut and sliding the sensor link stud out of the hole in the upper control arm.

4. Disconnect the brake line fitting bolt at the center of the axle housing. No brake lines need to be disconnected, therefore brake bleeding should not be necessary.

5. Remove the nut and washer from the shock absorber, then disconnect the shock absorber from the axle bracket.

6. Carefully lower the axle housing sufficiently to remove the spring. Be careful not to stretch the brake hose.

7. Remove the spring, upper and lower insulator, as equipped.

To install:

8. Install the upper insulator, lower insulator, as equipped, and the rear spring to the bracket on the frame seat. Point the coil leg toward the left side of the vehicle, at a right angle from the centerline of the vehicle within 5 degrees rearward and 15 degrees forward.

NOTE: If the spring is being replaced, be sure to position the tape facing the top or spring noises could occur.

9. Raise the rear axle back into position using the adjustable lifting device.

10. Install the rear shock absorber to the bracket, and tighten the nut to 50 ft. lbs. (68 Nm).

11. Connect the rear brake line fitting bolt and tighten to 20 ft. lbs. (27 Nm).

12. If equipped, connect the height sensor link to the upper control arm and tighten the nut to 27 inch lbs. (3 Nm).

13. Remove the adjustable support from under the rear axle assembly.

14. If equipped, connect the ABS rear axle speed sensor.

15. Lower vehicle and, if necessary, adjust the height sensor.

Upper Control Arms

REMOVAL AND INSTALLATION

NOTE: If both control arms are to be replaced, only remove and install 1 control arm at a time in order to prevent the axle from rolling or slipping sideways.

1. Raise and support the vehicle safely, then place a support under the axle housing.

2. If equipped, disconnect the ABS rear speed sensor.

3. If equipped, disconnect the height sensor link from the upper control arm by removing the attaching nut and sliding the sensor link stud out of the hole in the upper control arm.

4. Remove the nut and bolt at the rear axle housing.

5. Remove the nut and bolt at the frame crossmember, then disengage the upper control arm from its mounts and remove it from the vehicle.

To install:

6. Loosely attach the upper control arm to the frame crossmember using the nut and bolt.

7. Loosely attach the upper control arm to the rear axle housing using the nut and bolt.

8. Remove the jackstand from the rear axle and place supports under the tires. Lower the vehicle sufficiently so the vehicle weight rests on the tires.

9. For Fleetwood Brougham, lower the vehicle with the while curb height, tighten the frame nut to 122 ft. lbs. (165 Nm) and the axle nut to 70 ft. lbs. (95 Nm).

10. Except for Fleetwood Brougham, with the weight of the vehicle on the tires, tighten the bolt at the frame crossmember to 114 ft. lbs. (155 Nm) or the nut to 91 ft. lbs. (123 Nm). Either the nut or the bolt must be tightened, torque whichever is easiest to access.

11. Except for Fleetwood Brougham, with the weight of the vehicle on the tires tighten either the bolt or nut at the rear axle housing, whichever is easiest to access. Tighten the bolt to 83 ft. lbs. (113 Nm) or the nut to 74 ft. lbs. (100 Nm).

12. If equipped, connect the height sensor link to the upper control arm and tighten the nut to 27 inch lbs. (3 Nm).

13. Remove the support from under the rear tires.

14. If equipped, connect the ABS rear axle speed sensor.

15. Lower vehicle and, if necessary, adjust the height sensor.

Lower Control Arm

REMOVAL AND INSTALLATION

NOTE: If both control arms are to be replaced, only remove and install 1 control arm at a time in order to prevent the axle from rolling or slipping sideways.

1. Raise and support the vehicle safely, then place a support under the axle housing.

2. If equipped, disconnect the stabilizer shaft bolts and washers from the control arm.

3. Remove the nut and bolt from the bracket on the rear axle assembly.

4. Remove the nut and bolt from the crossmember brace, if equipped, and from the bracket on the frame.

5. Remove the lower control arm from the vehicle.

To install:

6. Position the lower control arm to the vehicle.

7. Loosely install the nuts and bolts to the crossmember brace, if equipped, to the frame bracket, rear axle bracket and, if applicable, to the stabilizer shaft.

8. Remove the jackstand from the rear axle and place supports under the tires. Lower the vehicle sufficiently so the vehicle weight rests on the tires.

9. For Fleetwood Brougham, perform the following procedures:

   a. connect the stabilizer shaft bracket to the lower control arm and tighten the bolts to 21 ft. lbs. (29 Nm).

   b. For Fleetwood Brougham, lower the car and with the vehicle at normal curb height, tighten the nuts to 122 ft. lbs. (165 Nm).

10. Except for Fleetwood Brougham, perform the following procedures:

   a. With the weight of the vehicle on the tires, tighten either the nut or the bolt, whichever is easiest to access. Tighten the lower control arm bolts or nuts to 91 ft. lbs. (123 Nm).

   b. If applicable, tighten the stabilizer bolts to 63 ft. lbs. (85 Nm).

11. Remove the tire supports and lower the vehicle.

Wheel Bearings

REMOVAL AND INSTALLATION

1. Raise and support the vehicle safely.

2. Remove the tire and wheel assembly, then remove the brake drum.

3. Clean all dirt from the rear carrier cover, then loosen the bolts and remove the cover while draining the gear oil. Discard the old gasket.

4. Remove the shaft lock bolt from the differential case located in the housing, then withdraw the pinion gear shaft.

5. Push the flanged end of axle shaft toward center of the vehicle, then remove C-lock from the end of the shaft located in the housing.

6. Remove axle shaft from the housing. If replacement of the oil seal is not planned, be very careful not to damage the seal.

7. Remove the brake backing plate.

8. Remove seal from housing using a prybar behind the seal's steel case, being careful not to damage housing.

9. Insert an appropriately sized bearing remover into the bore and position it behind the bearing so tangs on tool engage bearing outer race. Remove the bearing, using a slide hammer.

To install:

10. Lubricate the new bearing with gear lubricant and install bearing using a suitable driver so the tool bottoms against the shoulder in the housing.

11. Lubricate seal lips with gear lubricant, then position the seal on a suitably sized driver and position seal into housing bore. Tap seal into place so it is flush with axle tube.

12. If removed, install the brake backing plate.

13. Insert the axle into the place while engaging the splines on the end of the shaft with the splines of the rear axle side gear. Be careful when inserting the axle not to damage the seal.

14. Install the C-lock on the bottom of the axle shaft and push the shaft outward so the lock seats in the counterbore of the rear axle side gear.

15. Install the rear axle pinion gear shaft through the differential case, thrust washers and pinions, align the hole in the shaft with the lock bolt hole. Install the lock bolt and tighten to 24 ft. lbs. (31 Nm) for 7½ in. ring gears or 20 ft. lbs. (27 Nm) for 8½ in. ring gear.

16. Position a new gasket, then install the carrier cover and tighten the bolt to 22 ft. lbs. (30 Nm) using a crosswise pattern.

17. Fill the rear assembly with the proper grade and type gear oil.

18. Install the brake drum, then install the tire and wheel assembly.

19. Lower the vehicle.

GM "F" BODY

Rear Wheel Drive

CHEVROLET-Camaro PONTIAC-Firebird

FIRING ORDERS

NOTE: To avoid confusion, always replace spark plug wires one at a time.

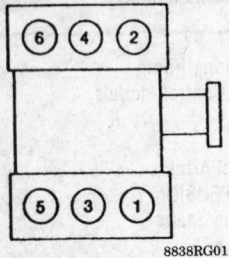

8838RG01

3.4L Engine
Engine Firing Order:
1-2-3-4-5-6
Distributorless Ignition
System

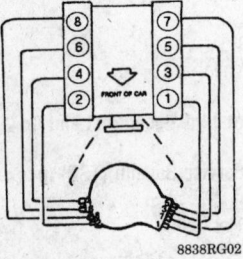

8838RG02

5.7L Engine
Engine Firing Order:
1-8-4-3-6-5-7-2
Distributor Rotates with
Camshaft

ENGINE ELECTRICAL

NOTE: Disconnecting the negative battery cable on some vehicles may interfere with the functions of the on board computer systems and may require the computer to undergo a relearning process, once the negative battery cable is reconnected.

Distributor

The 5.7L engine utilizes the new Opti-Spark distributor ignition system, which consists of a distributor assembly, control circuitry and an external coil. In the Opti-Spark system, all ignition timing is controlled by the ECM based on signals from the distributor's internally mounted optical camshaft position sensor. There is no way to bypass the ECM control or to adjust/set ignition timing on this system.

The 3.4L engine, utilizes an Electronic Ignition (EI) system and therefore has no distributor.

REMOVAL

NOTE: When making compression checks, the ignition system can be disabled by disconnecting the "BAT" or 2-terminal connector from the distributor.

5.7L Engine

1. Be sure the ignition is in the **OFF** or **LOCK** position, then disconnect the negative battery cable.
2. Disengage the wiring harness from the engine cooling fan assembly, then remove the cooling fan from the vehicle.
3. Remove the block drain plug and the knock sensor, then drain the engine cooling system.
4. Remove the air intake duct and the air cleaner assembly.
5. Remove the coolant and heater hoses from the water pump assembly and the throttle body.
6. Unplug the wiring harness from the Engine Coolant Temperature (ECT) sensor, then reposition the ignition coil and bracket.
7. Remove the water pump assembly.
8. Remove the serpentine drive belt from the vehicle.
9. If not done already to remove the pump or the belt, raise and safely support the vehicle for access.
10. Remove the torsional damper bolts, then remove the damper from the crankshaft hub.
11. Disconnect the spark plug wires from the distributor. Be sure to twist each boot ½ turn and pull only on the boot to remove each wire. The wire

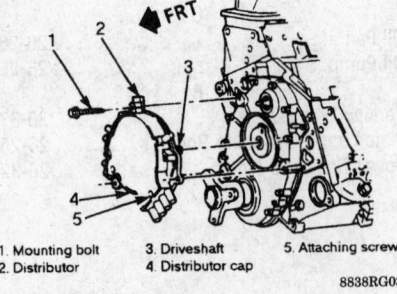

1. Mounting bolt 3. Driveshaft 5. Attaching screw
2. Distributor 4. Distributor cap

8838RG03

Distributor installation — 5.7L engine

numbers should be molded into the distributor housing. If not, be sure to tag the wires before disconnection.
12. Unplug the 4-terminal ECM connector from the distributor.
13. Remove the distributor mounting bolts and pull the distributor forward until the drive shaft disengages from the engine. Mark the top of the shaft for alignment during reassembly.

INSTALLATION

Timing Not Disturbed

5.7L Engine

1. With the mark made on the distributor shaft earlier on top, install the distributor to the engine. Tighten the distributor bolts to 8 ft. lbs. (11 Nm).
2. Install the 4-terminal ECM connector and the spark plug wires to the distributor.
3. Position the crankshaft damper to the hub, then install the damper bolts and tighten to 60 ft. lbs. (81 Nm).
4. Install the serpentine drive belt.
5. Install the water pump assembly and engage the ECT wiring harness.
6. Reposition the ignition coil and bracket, then install the coolant and heater hoses.
7. Install the air cleaner assembly and air intake duct.
8. Install the engine block drain plug and the knock sensor.
9. Install the right cooling fan assembly, then connect the wiring harness.
10. Connect the negative battery cable and fill the engine cooling system.
11. Start and run the engine, then check for leaks.

Distributorless Ignition System

The base 3.4L engine, is equipped with a distributorless Electronic Ignition (EI) system. This system uses a "waste spark" method of spark distribution. Each cylinder is paired with it's opposite in the firing order, so a cylinder on the compression stroke fires simultaneously with it's opposing cylinder on the exhaust stroke. The cylinder on the exhaust stroke requires very little voltage to fire it's plug, so most of the available voltage is used to fire the cylinder that is on the compression stroke.

The EI system consists of a coil pack, the ignition module, a camshaft position sensor, 2 crankshaft position sensors, and the Engine Control Module (ECM). The coil pack contains 3 separate and interchangeable ignition coils (1 for each pair of cylinders) attached to the module and bracket assembly. The ignition module, mounted between the coils and the bracket, is connected to the ECM. The ignition module controls the primary circuit to the ignition coils.

REMOVAL AND INSTALLATION

Camshaft Position Sensor

The camshaft position sensor is located on top of the engine block, behind the water pump, near the lower intake manifold assembly.

1. Be sure the ignition is in the **OFF** or **LOCK** position, then disconnect the negative battery cable.
2. Disengage the wiring harness connector from the sensor.
3. Remove the sensor retaining bolt, then carefully remove the sensor from the top of the engine block. If necessary, carefully pry between alternate sides of the sensor and the engine block in order to loosen the sensor from its mounting boss. Make sure the sensor is not damaged when prying.

To install:
4. Inspect the sensor O-ring and replace, if necessary.
5. Lubricate the O-ring with clean engine oil, then install the sensor into the engine block.
6. Install the sensor retaining bolt and tighten to 17 inch lbs. (1.9 Nm).
7. Engage the sensor wiring harness, then connect the negative battery cable.

24X Signal Crankshaft Position Sensor

1. Be sure the ignition is in the **OFF** or **LOCK** position, then disconnect the negative battery cable.
2. Raise and support the vehicle safely.
3. Remove the crankshaft balancer assembly from the front of the engine:
   a. Remove the serpentine drive belt.
   b. Remove the bolt and washer from the crankshaft balancer.
   c. Remove the retaining bolts from the crankshaft pulley, then remove the pulley from the balancer.
   d. Using J-24420-B or an equivalent torsional damper remover, pull the balancer from the crankshaft.
4. Remove the sensor retaining bolts.
5. Disconnect and remove the sensor from the engine.

To install:
6. Install the sensor to the engine and tighten the retaining bolts to 17 inch lbs. (1.9 Nm).
7. Use J-29113, or an equivalent installer, to pull the balancer onto the crankshaft, then install the pulley. Tighten the balancer bolt to 58 ft. lbs. (78 Nm), then tighten the pulley retaining bolts to 37 ft. lbs. (50 Nm).
8. Install the serpentine drive belt.
9. Lower the vehicle.
10. Connect the negative battery cable.

3X Signal Crankshaft Position Sensor

1. Be sure the ignition is in the **OFF** or **LOCK** position, then disconnect the negative battery cable.
2. Disengage the wiring harness connector from the sensor.
3. Remove the sensor retaining bolt, then carefully remove the sensor from the lower right side of the engine block.

To install:
4. Inspect the sensor O-ring and replace, if necessary.
5. Lubricate the O-ring with clean engine oil, then install the sensor into the side of the engine block.
6. Install the sensor retaining bolt and tighten to 89 inch lbs. (10 Nm).
7. Engage the sensor wiring harness, then connect the negative battery cable.

Ignition Module and Coils

1. Be sure the ignition is in the **OFF** or **LOCK** position, then disconnect the negative battery cable.
2. Tag and disconnect the spark plug wires from the ignition coils.
3. Disengage the electrical connectors from the assembly.
4. Remove the retaining nuts, then remove the ignition module and coils from the engine.
5. If necessary, remove the coils from the module and bracket:
   a. Remove the bolts retaining the coil(s) to the module and bracket assembly.
   b. Remove the ignition coil(s) and seal(s) from the assembly.
   c. Separate the module from the bracket.

To install:
6. If necessary, assemble the ignition module and coils:
   a. Position the module to the bracket.
   b. Install the seal(s), then position the coil(s) to the module and bracket assembly.
   c. Install the retaining bolts and tighten to 40 inch lbs. (4.5 Nm).
7. Install the ignition module and coil assembly to the engine.
8. Install the retaining nuts and tighten to 89 inch lbs. (10 Nm).
9. Engage the wiring harness connectors and the spark plug wires to the assembly.
10. Connect the negative battery cable.

Ignition Timing

ADJUSTMENT

On the 3.4L and 5.7L engines, base timing is preset when the engine is manufactured. All timing changes are then controlled directly by the ECM based on information from the ignition and knock sensor systems. No adjustments are necessary or possible.

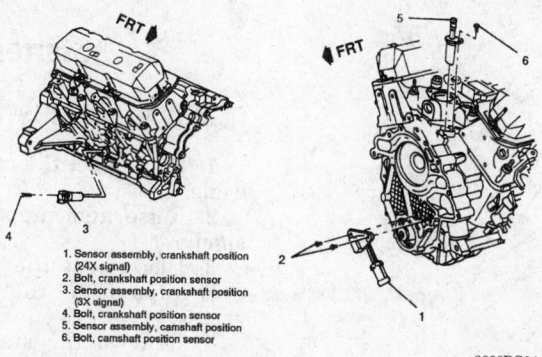

1. Sensor assembly, crankshaft position (24X signal)
2. Bolt, crankshaft position sensor
3. Sensor assembly, crankshaft position (3X signal)
4. Bolt, crankshaft position sensor
5. Sensor assembly, camshaft position
6. Bolt, camshaft position sensor

8838RG04

Ignition sensor (crankshaft/camshaft) locations — 3.4L engine

Alternator

PRECAUTIONS

Several precautions must be observed with alternator equipped vehicles to avoid damage to the unit.

- If the battery is removed for any reason, make sure it is reconnected with the correct polarity. Reversing the battery connections may result in damage to the 1-way rectifiers.
- When utilizing a booster battery as a starting aid, always connect the positive to positive terminals and the negative terminal from the booster battery to a good engine ground on the vehicle being started.
- Never use a fast charger as a booster to start vehicles.
- Disconnect the battery cables when charging the battery with a fast charger.
- Never attempt to polarize the alternator.
- Do not use test lights of more than 12 volts when checking diode continuity.
- Do not short across or ground any of the alternator terminals.
- The polarity of the battery, alternator and regulator must be matched and considered before making any electrical connections within the system.
- Never separate the alternator on an open circuit. Make sure all connections within the circuit are clean and tight.
- Disconnect the battery ground terminal when performing any service on electrical components.
- Disconnect the battery if arc welding is to be done on the vehicle.

BELT TENSION ADJUSTMENT

Serpentine belts are automatically adjusted by the tensioner on the engine. If belt slippage occurs, check the belt length scale on the drive belt

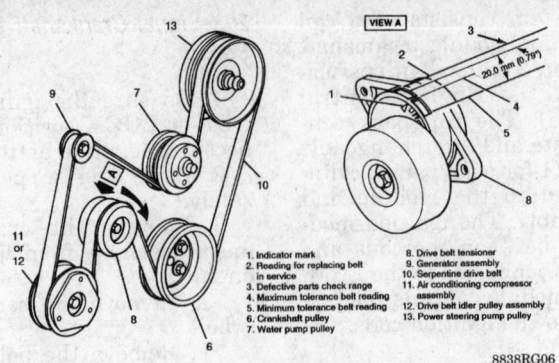

1. Indicator mark
2. Reading for replacing belt in service
3. Defective parts check range
4. Maximum tolerance belt reading
5. Minimum tolerance belt reading
6. Crankshaft pulley
7. Water pump pulley
8. Drive belt tensioner
9. Generator assembly
10. Serpentine drive belt
11. Air conditioning compressor assembly
12. Drive belt idler pulley assembly
13. Power steering pump pulley

8838RG06

Serpentine drive belt and tensioner — 3.4L engine

tensioner for the proper installed length and replace as necessary. If the drive belt tensioner is within it's operating range and belt slippage still occurs, the tensioner may need replacement.

REMOVAL AND INSTALLATION

1. Disconnect the negative battery cable.

2. Use a breaker bar and socket to rotate the serpentine belt tensioner downward (clockwise) relieving tensioner from the belt, then remove the belt from the vehicle. For 5.7L engine, it may be easier to temporarily raise and support the vehicle, then remove the belt from underneath.

3. Disengage the positive battery wire and the electrical connector from the alternator assembly.

4. For the 3.4L engine, remove the 2 rear nuts and the outer front bolt from the alternator, then remove the rear inner brace. Support the alternator, then remove the remaining front bolt and the rear brace bolt.

5. For the 5.7L engine, remove the upper brace bolts, the lower brace bolt and the front alternator bolt all from the alternator assembly, then remove the nut retaining the upper brace to the top of the engine and

remove the upper brace from the vehicle.

6. Remove the alternator from the accessory mounting bracket and from the vehicle.

To install:

7. Position the alternator to the engine and accessory mounting bracket.

8. For the 5.7L engine, install the upper brace and tighten the retaining nut to 24 ft. lbs. (33 Nm), then install the alternator and brace bolts. Tighten the rear upper brace and front alternator bolts to 37 ft. lbs. (50 Nm) and tighten the rear lower brace bolt to 18 ft. lbs. (25 Nm).

9. For the 3.4L engine, install and tighten the front inner alternator bolt and the rear upper brace-to-alternator bolt. Tighten the rear brace bolt to 18 ft. lbs. (25 Nm) and the front bolt to 37 ft. lbs. (50 Nm).

10. For the 3.4L engine, install the rear inner brace, then tighten the brace-to-engine nut to 18 ft. lbs. (25 Nm). Tighten the brace-to-alternator nut and the outer front alternator bolt to 37 ft. lbs. (50 Nm).

11. Engage the electrical connector and the battery positive wire to the alternator assembly.

12. Install the serpentine drive belt.

13. Connect the negative battery cable.

Starter

REMOVAL AND INSTALLATION

1. Disconnect the negative battery cable.

2. Raise and support the vehicle safely.

3. Disconnect the exhaust crossover pipe from the manifolds for access.

4. Remove the starter mounting bolts and lower the starter sufficiently to access the wires.

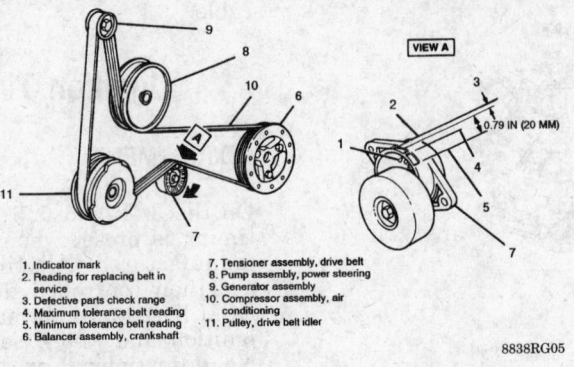

1. Indicator mark
2. Reading for replacing belt in service
3. Defective parts check range
4. Maximum tolerance belt reading
5. Minimum tolerance belt reading
6. Balancer assembly, crankshaft
7. Tensioner assembly, drive belt
8. Pump assembly, power steering
9. Generator assembly
10. Compressor assembly, air conditioning
11. Pulley, drive belt idler

8838RG05

Serpentine drive belt and tensioner — 5.7L engine

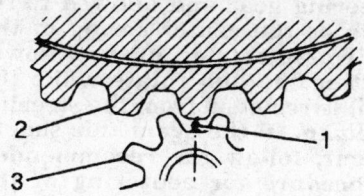

1. Insert 0.20 IN (0.5 MM) wire gage here to check
2. Flywheel assembly
3. Starter drive pinion

8838RG07

Checking flywheel-to-starter pinion clearance

5. Tag and disconnect the wiring, then remove starter from the vehicle.

To install:

NOTE: Before tightening the starter wiring, make sure the inner nuts on the solenoids are already tightened. If they are loose, the act of tightening the outer nuts could damage the solenoid cap.

6. Raise starter sufficiently and support while connecting the starter wiring. Tighten the battery terminal nut to 80 inch lbs. (9 Nm) and the ignition terminal to 22 inch lbs. (2.5 Nm).

NOTE: If shims were removed, they must be installed in their original location to assure proper drive pinion to flywheel engagement.

7. While holding the starter in place, install the mounting bolts and shims, if equipped. Tighten mounting bolts to 35 ft. lbs. (37 Nm).

8. Check that flywheel-to-pinion clearance is 0.020 inch (0.5mm) and add or subtract shims, as necessary. If removed, install the flywheel cover.

9. Position the crossover pipe and attach it to the manifolds.

10. Lower vehicle and connect the negative battery cable.

CHASSIS ELECTRICAL

Blower Motor

REMOVAL AND INSTALLATION

1. Disconnect the negative battery cable. If necessary, remove the diago-nal fender brace at the right rear corner of the engine compartment to gain access to the blower motor.

2. Disconnect the electrical wiring from the blower motor. If equipped with air conditioning, remove the blower relay and bracket as an assembly and swing them aside.

3. Remove the blower motor cooling tube.

4. Remove the blower motor retaining screws.

5. Remove the blower motor and fan as an assembly from the case. Be careful not to damage the blower fan.

To install:

6. Carefully guide the blower motor and fan into position being careful not to catch the fan on any protruding parts. Install the blower motor attaching screws and attach the blower motor cooling tube.

7. Connect the blower wiring and position the blower relay and bracket assembly into place. Install the bracket retaining screws and tighten to 13 inch lbs. (1.5 Nm).

8. If removed, install the diagonal fender brace and bolts.

9. Connect the negative battery cable and check blower operation.

Windshield Wiper Motor

REMOVAL AND INSTALLATION

1. Operate the wipers and stop them midway through the return sweep, then disconnect the negative battery cable.

2. Remove the wiper arm and blade assemblies.

3. Remove the left cowl panel and seal, then disconnect the washer hose assembly from the cowl panel and nozzle.

4. Disengage the electrical connector from the motor assembly.

5. Remove the screw and nut attaching the left linkage to the cowl.

6. Using J-39232, or an equivalent wiper linkage separator, disconnect the socket of the right hand linkage from the ball of the left hand linkage, then remove the left linkage from the vehicle.

7. Remove the screw retaining the front of the wiper motor assembly, then carefully pull the motor free from the slots of the rear retaining bracket.

8. Using the linkage separator, disconnect the right linkage from the crank arm ball of the motor assembly, then remove the motor from the vehicle.

To install:

9. Make sure the motor crank arm is in the inner wiper position and install the motor assembly to the vehicle. The crank arm drive pin must be engaged into the cam pocket.

10. Using a wiper linkage installer, press the socket of the right linkage into engagement with the motor crank arm ball.

11. Install the motor into the rear bracket by fully pressing the locator pads into the bracket slots, then install the motor retaining screw and tighten to 7.5 ft. lbs. (10 Nm).

12. Install the left linkage assembly, without the attaching parts, and press the right linkage socket onto the left linkage ball.

13. Install the left linkage screw and nut, then tighten the fasteners to 7.5 ft. lbs. (10 Nm).

14. Engage the motor wiring harness, then install the washer hose assembly to the left cowl panel and washer nozzle.

15. Install the cowl panel and hood seal, then install the wiper arm assemblies.

16. Connect then negative battery cable and verify proper operation.

Windshield Wiper Switch

REMOVAL AND INSTALLATION

NOTE: The vehicle is equipped with a SIR air bag system, it is imperative that the disarming procedure is followed before repairs, and that the coil centering and rearming procedures are followed after repairs.

1. Disarm the SIR and disconnect the negative battery cable.

2. Remove the turn signal switch assembly, but do not disconnect or remove the wiring harness. Allow the switch assembly to hang freely from the wires unless removal is necessary for switch replacement.

3. Remove the ignition lock assembly, but do not disconnect or remove the wiring harness. Allow the lock set to hang freely from the wires, unless removal is necessary for lock cylinder replacement.

4. If not done already, remove the housing cover end cap, disconnect the electrical connectors from the multi-function lever and remove the lever by pulling toward the driver's door.

5. Remove the housing cover screws, unthread and remove the tilt lever from the column assembly.

6. Remove the lock housing cover assembly.

7. Remove the base plate and the dimmer switch rod actuator and the wiper switch actuator pivot pin.

8. Remove the wire protector shield.

9. Disengage the wiper switch connector from bulkhead connector and remove the switch. Attach a piece of mechanic's wire to the connector to aid in reinstallation and gently pull the wire harness through the column. Leave the mechanic's wire routed through the column in order to pull the new switch wiring into position.

To install:

10. Connect the wiper switch to the lock housing cover assembly.

11. Attach the switch actuator pivot pin to the switch and cover.

12. Pull wiper switch wire connector through the steering column using the mechanic's wire and attach to the bulkhead harness connector. If applicable, install the wire protector shield.

13. Attach the dimmer switch rod actuator to the base plate and lubricate with lithium grease.

14. Connect base plate to lock housing cover assembly. The bottom edge of the dimmer switch rod actuator must rest on the bend in the dimmer switch rod.

15. Position lock housing cover in place and attach the tilt lever.

16. Starting with the housing cover screw in the 12 o'clock position, then 8 o'clock and finally 3 o'clock positions, tighten the screws to 80 inch lbs. (9 Nm).

17. Install the multi-function lever and engage the lever connectors on the base plate.

18. Install the housing cover end cap.

19. Install the lock cylinder assembly.

20. Install the turn signal assembly.

21. Connect the negative battery cable and enable the SIR system.

Concealed Headlights

MANUAL OPERATION

The concealed headlights used on the Firebird are electrically operated. If an electrical failure involving the headlight actuators should occur, the headlights can be operated manually. Motor knobs are located under the hood, immediately adjacent to the headlights assemblies.

To raise the headlights electrically, but without turning ON the lights,

turn ON the parking lights and lightly press the headlight switch.

Headlight Switch

REMOVAL AND INSTALLATION

1. Disconnect then negative battery cable.

2. For the Camaro, remove any instrument panel sound insulators or trimplates necessary to access the switch assembly.

3. Remove the switch assembly from the bezel. For the Firebird, the switch is removed from the panel in front of the bezel, by pulling toward the driver's seat. For the Camaro, the switch is removed from behind the bezel, by pulling toward the firewall.

4. Disengage the wiring harness connectors from the switch assembly, then remove the switch from the vehicle.

To install:

5. Engage the wiring harness connectors to the switch assembly.

6. Install the switch into the bezel assembly. Be careful not to apply pressure to the switch knob or the thumb wheel or switch damage may occur.

7. Install any trim panels or sound insulators removed for better access.

8. Connect the negative battery cable.

Turn Signal Switch

REMOVAL AND INSTALLATION

NOTE: The vehicle is equipped with a SIR air bag system, it is imperative that the disarming procedure is followed before repairs, and that the coil centering and rearming procedures are followed after repairs.

1. Properly disable the SIR system. Place the ignition switch to the **LOCK** position in order to prevent uncentering of the coil assembly.

2. Disconnect the negative battery cable.

3. Properly remove and store the inflator module and the steering wheel. Either remove the column from the vehicle or lower it, as necessary for access.

4. Remove the coil assembly retaining ring. Remove the coil assembly and allow it to hang freely from the wiring.

NOTE: The coil assembly will become uncentered if the steering column is separated from the

steering gear and allowed to rotate or the center spring of the coil assembly is pushed down, letting the hub rotate while the coil is removed from the steering column. In the event this should occur, follow the recommended procedure for centering of the coil in order to avoid accidental deployment of the air bag or damage to the internal components of the steering column.

5. Remove the wave washer.

6. Remove the shaft lock retaining ring using tool J-23653-C or equivalent shaft lock compressor. Discard the old ring.

7. Remove the shaft lock, turn signal canceling cam and upper bearing assembly.

8. Move the multi-function lever to the **RIGHT TURN** position. If equipped with cruise control, remove the column housing cover end cap by pulling toward the vehicle front and disengage the electrical harness connector. Remove the turn signal lever by pulling toward the driver's door.

9. Remove the hazard knob retaining screw and assembly.

10. Remove the turn signal switch arm and screw.

11. Remove the turn signal switch screws.

12. Disengage the switch harness connector from the bulkhead connector and remove the wiring protector.

13. Attach a length of mechanic's wire to the switch harness in order to aid in reinstallation, then gently pull the assembly up through the housing. Leave the wire routed through the column in order to pull the new harness back into position.

14. Remove the switch and harness from the vehicle.

To install:

15. Using the mechanic's wire, pull the switch harness through the column and connect to the bulkhead connector.

16. Install the harness wiring protector.

17. Position the turn signal switch assembly and install the attaching screws. Tighten the screws to 30 inch lbs. (3.4 Nm).

18. Install the switch arm and mounting screw. Tighten the screw to 20 inch lbs. (2.3 Nm).

19. Install the hazard knob assembly and the multi-function lever.

20. Install the inner race, the upper bearing race seat, and the upper bearing spring.

21. Lubricate the friction surfaces using synthetic grease, then install the turn signal canceling cam.

22. Position the shaft lock. Install the a new shaft lock retaining ring using tool J-23653-C or equivalent. Be sure the ring is firmly seated in the groove of the shaft.

23. Install the wave washer.

24. Install the coil assembly, making sure it is properly centered.

25. Position and secure the steering column, as necessary.

26. Install the steering wheel and the inflator module.

27. Connect the negative battery cable and enable the SIR system.

Ignition Lock Cylinder

REMOVAL AND INSTALLATION

NOTE: The vehicle is equipped with a SIR air bag system, it is imperative that the disarming procedure is followed before repairs, and that the coil centering and rearming procedures are followed after repairs.

1. Disable the SIR system and disconnect the negative battery cable.

2. Remove turn signal switch assembly, but do not disconnect or pull the wire harness through the column. Allow the switch assembly to hang freely from the wires.

3. If necessary, remove the coil assembly as follows:

   a. If not done already, remove the wiring protector from the vehicle.

   b. Disengage the coil terminal connector from the harness.

   c. Attach a length of mechanic's wire to the terminal connector in order to aid in reassembly.

   d. Carefully pull the wires through the column, leaving the length of mechanic's in the column for installation purposes.

4. Remove the key from the lock cylinder.

5. Remove the buzzer switch and retaining clip.

6. Reinsert the key into the lock cylinder, making sure the key is in the LOCK position.

7. Remove the lock set retaining screw.

8. Disengage the Pass Key wire harness connector from the bulkhead connector. If not done already, remove the wiring protector.

9. Attach a piece of string or mechanic's wire to the wire connector to aid in reassembly, disconnect the retaining clip from the housing cover and pull the wire up through the column. Leave the length of string or wire in the column in order to pull the new harness into position.

10. Remove the lock cylinder from the vehicle.

To install:

11. Using the length of string or the mechanic's wire, pull the lock cylinder set wire harness down through the column into the original position.

12. Install the lock cylinder set, snapping the wire retaining clip into the hole in the housing.

13. Engage the lock cylinder wiring connector to the bulkhead connector.

14. Install the lock cylinder retaining screw and tighten to 22 inch lbs. (2.5 Nm).

15. Remove the key and install the buzzer switch with retaining clip, then insert the key and leave in the LOCK position.

16. If removed, pull the turn signal switch wiring connector and/or the coil wiring connector through the steering column, connect the harnesses and install the wiring protector.

17. Install the turn signal switch assembly.

18. Connect the negative battery cable and enable the SIR system.

Ignition Switch

REMOVAL AND INSTALLATION

NOTE: The vehicle is equipped with a SIR air bag system, it is imperative that the disarming procedure is followed before repairs, and that the rearming procedure is followed after repairs.

1. Properly disable the SIR system and disconnect the negative battery cable.

2. Remove the instrument panel lower trimplate.

3. Remove the steering column mounting bolts, lower and properly support the column. If access to the switch is not sufficient, remove the column from the vehicle.

4. Remove the hex tapping screw and the washer head screw securing the switch to the column.

5. Disengage the switch assembly from the actuator rod, unplug the wiring harness connector and remove the switch assembly.

6. Remove the ignition switch screws or the dimmer/ignition switch mounting stud, as applicable. Unplug the connector, then remove the ignition switch from the column.

To install:

7. Verify that the key cylinder is in the ignition OFF and LOCKED position.

8. Make sure the switch slider is also in the OFF-LOCK position. New switches will be pinned in this position and the pin must be removed after installation or switch damage may result. To verify if a switch is in the lock position, move the slider to the extreme right position, then move the slider 1 detent to the left.

9. Install the switch to the actuator assembly and to the steering column.

10. Install and tighten the switch mounting stud or screw, as applicable. Tighten the stud to 35 inch lbs. (4.0 Nm) or the screw to 33 inch lbs. (3.7 Nm).

11. Install and adjust the dimmer switch assembly.

12. Position the steering column and secure the column in the vehicle

13. Connect the negative battery cable and enable the SIR system.

Neutral Safety Switch

ADJUSTMENT

New switches should already be set in the N position. If the bolt holes do not align, before rotating switch, verify that the shifter is in the N position. Switch adjustment is performed with the center console removed for switch access.

1. Place the transmission shifter in the N notch in the detent plate.

2. Loosen the switch attaching bolts.

3. Insert a 0.094 in. (2.34mm) drill bit or gauge pin into the adjustment hole, then rotate the switch until the gauge pin drops to a depth of 0.59 in. (15mm).

NOTE: If the proper sized gauge pin is unavailable, a 3/32 in. drill bit may be used.

4. Tighten the switch bolts to 19 inch lbs. (2.2 Nm) .

5. Remove the gauge pin and verify that the engine will only start in P or N.

REMOVAL AND INSTALLATION

1. Disconnect the negative battery cable.

2. Remove the console assembly for access to the neutral safety switch.

3. Remove the switch attaching screws, then remove the neutral safety switch from the vehicle.

To install:

4. Position shifter lever in the **N** position.

5. Insert carrier tang on the switch in the slot on the shifter.

6. If installing a new switch, install the retaining bolts and tighten the switch bolts to 19 inch lbs. (2.2 Nm). Move the shift control lever out of the **N** position to shear the factory installed plastic retaining pin.

NOTE: If installing a new switch and the holes do not align with shifter control, check that the shifter control lever is in N. Do not rotate the switch as this will shear the retaining pin. If a new switch was rotated and the pin was already broken, switch adjustment must be performed.

7. If installing an old switch, or a new switch with a sheared retaining pin, perform the switch adjustment.

8. Connect the negative battery cable and check that engine starts only in the **P** or **N** positions. If engine starts in any other position, readjust neutral switch.

9. Install center console assembly.

Powertrain Control Module

LOCATION

The Powertrain Control Module (PCM) is located under the right side the instrument panel.

ENGINE COOLING

Radiator

REMOVAL AND INSTALLATION

1. Disconnect the negative battery cable, then remove the air intake duct and air cleaner assembly.

2. Raise and support the vehicle safely, then drain the engine cooling system.

3. Disengage the cooling fan electrical connectors, then remove the fan assembly.

4. Disconnect the transmission fluid cooler lines, then plug the openings to prevent system contamination or excessive fluid loss.

5. If equipped, remove the coolant level sensor.

6. Lower the vehicle, then loosen the receiver dehydrator bracket.

7. Disconnect the lower radiator hose from the water pump assembly, then remove the upper radiator hose and the overflow tube.

8. Remove the radiator from the condenser, then carefully remove the radiator from the vehicle.

To install:

9. Install the radiator assembly into the vehicle, making sure to properly seat the radiator on the insulators, then attach the radiator to the condenser.

10. Install the overflow tube and the upper radiator hose, then connect the lower radiator hose to the water pump assembly.

11. Tighten the receiver dehydrator bracket, then raise and support the vehicle safely.

12. If equipped, install the coolant level sensor.

13. Unplug the openings, then connect the transmission fluid cooler lines.

14. Install the engine cooling fan assembly and engage the electrical connectors.

15. Lower the vehicle and install the air cleaner and intake duct assemblies.

16. Connect the negative battery cable and properly fill the engine cooling system.

Water Pump

REMOVAL AND INSTALLATION

3.4L Engine

1. Disconnect the negative battery cable and drain the engine cooling system.

2. Remove the air duct and disengage the electrical connector.

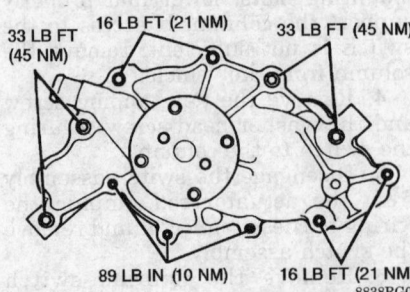

Water pump assembly torque specification — 3.4L engine

3. Disengage the electrical connector from the alternator, then remove the top ignition coil from the coil pack.

4. Loosen the tensioner pulley bolt, then remove the heater hose from the pump.

5. Loosen the pump pulley bolts, then remove the power steering bracket bolts.

6. Remove the serpentine drive belt, then remove the pump pulley.

7. Remove the power steering pump and bracket assembly, then position aside with the hoses intact.

8. Remove the water pump retaining bolts starting with the lower bolts and working alternately around to the top.

9. Remove the water pump assembly and gasket from the front of the engine.

To install:

10. Thoroughly clean the gasket mating surfaces of any remaining gasket material.

11. Position the water pump using a new gasket, then loosely install using the retaining bolts.

12. Tighten the water pump retaining bolts to specification, starting with the bolts that require the highest torque specification and working towards the bolts with the lowest specification.

13. Position the power steering pump and bracket assembly and loosely secure using the retaining bolts.

14. Install the water pump pulley and tighten the retaining bolts to 18 ft. lbs. (25 Nm), then install the serpentine drive belt.

15. Tighten the power steering pump bracket bolts to 23 ft. lbs. (31 Nm).

16. Install the heater hose to the pump assembly.

17. Install the top ignition coil to the coil pack.

18. Engage the electrical connector to the alternator.

19. Engage the electrical connector and install the air duct.

20. Connect the negative battery cable and refill the engine cooling system, then start the engine and check for leaks.

5.7L Engine

1. Disconnect the negative battery cable.

2. Disengage the wiring harness from the cooling fan assembly, then remove the assembly from the vehicle.

3. Drain the engine cooling system, removing the block drain plug

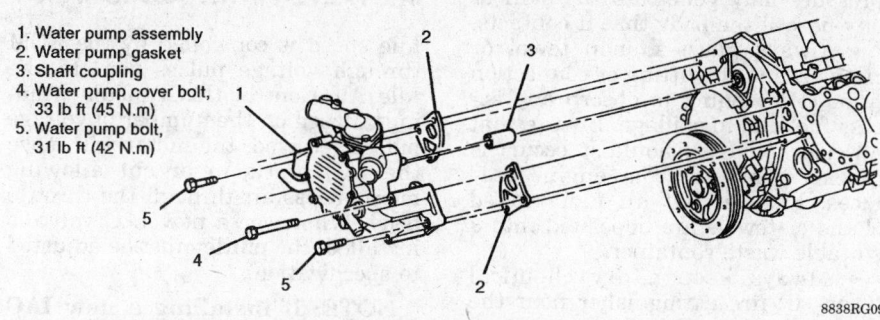

1. Water pump assembly
2. Water pump gasket
3. Shaft coupling
4. Water pump cover bolt,
   33 lb ft (45 N.m)
5. Water pump bolt,
   31 lb ft (42 N.m)

8838RG09

Water pump assembly removal and installation — 5.7L engine

and the knock sensor to assure proper draining. Reinstall the drain plug and knock sensor, as soon as the system is empty.

4. Disconnect the upper and lower radiator hoses from the water pump assembly.

5. Remove the heater hose assemblies from the water pump and from the throttle body.

6. Disengage the coolant sensor wiring harness, then reposition the ignition coil and bracket assembly.

7. Remove the shorter water pump retaining bolt from the center of each pump mating flange, then remove the longer pump bolts from either side of the center bolts.

8. Carefully remove the water pump assembly and gaskets along with the pump shaft coupling.

To install:

9. Thoroughly clean the gasket mating surfaces of any remaining gasket material.

10. Install the water pump shaft coupling along with the water pump and gaskets.

11. Install the longer pump bolts and tighten 33 ft. lbs. (45 Nm), then install the shorter bolts and tighten to 31 ft. lbs. (42 Nm).

12. Reposition the ignition coil and bracket assembly.

13. Engage the coolant sensor electrical connector.

14. Install the heater and radiator hoses to the throttle body and water pump, as applicable.

15. Install the air cleaner and intake duct assemblies.

16. Install the engine cooling fan assembly and engage the wiring harness connector.

17. Connect the negative battery cable and properly fill the engine cooling system.

Thermostat

REMOVAL AND INSTALLATION

1. Open the radiator drain cock and allow the engine coolant to drain to a level just below the thermostat housing, then close the drain and tighten to 106 inch lbs. (12 Nm).

2. Remove the air cleaner assembly.

3. Disconnect the inlet hose and clamp from the thermostat assembly.

4. Remove the thermostat housing bolts, then remove the housing.

5. If equipped, remove the housing gasket.

6. Remove the thermostat assembly.

To install:

7. Make sure the flange mating surfaces are clean and free of debris, then install the thermostat assembly.

8. If equipped, install a new housing gasket.

9. Install the housing and tighten the retaining bolts to 21 ft. lbs. (28 Nm), then connect the inlet hose assembly and clamp.

10. Install the air cleaner assembly, then properly refill the engine cooling system.

Electric Cooling Fan

REMOVAL AND INSTALLATION

1. Disconnect the negative battery cable, then remove the air cleaner assembly.

2. Disengage the fan electrical connector and clips.

3. Carefully slide the engine cooling fan assembly from the radiator and retainers.

4. If necessary, remove the fan blade nut and separate the blade from the motor and bracket.

To install:

5. If removed, install the fan blade to the motor and bracket assembly, then secure using the retaining nut.

6. Install the cooling fan assembly by sliding it into position.

7. Install the clips and engage the wiring harness connector.

8. Install the air cleaner assembly and connect the negative battery cable.

Cooling System Bleeding

PROCEDURE

3.4L Engine

1. Make sure the vehicle is parked or supported on a level surface with the pressure cap removed. Drain the engine cooling system by opening the radiator drain cock and either removing the engine block drain plug and the knock sensor (3.4L engine).

2. Once the system is completely drained, close the drain cock, then install the knock sensor and/or the drain plugs. The drain plugs should be tightened to 15 ft. lbs. (21 Nm), while the knock sensor should be tightened to 14 ft. lbs. (19 Nm).

3. After the repair or service has been completed, fill the cooling system with a solution which is at least 50 percent ethylene glycol antifreeze and the balance water. Do not use a solution that exceeds 70 percent antifreeze. Continue filling the radiator until the level is just below the filler neck.

4. Fill the coolant recovery reservoir to the **COLD** mark and install the reservoir cap.

5. Run the engine with the radiator cap removed until the normal operating temperature is reached.

—— **CAUTION** ——
Ethylene glycol in engine coolant can be flammable under some conditions. Do not spill coolant on the exhaust system or on hot engine parts.

6. With the engine idling, add coolant to the radiator until the level reaches the bottom of the filler neck.

7. Install the radiator cap. The arrows on the cap must line up with the coolant recovery reservoir hose.

5.7L Engine

1. Make sure the vehicle is parked or supported on a level surface with the pressure cap removed. Drain the engine cooling system by opening the radiator drain cock, then opening the

air bleed vents on the thermostat housing and the heater outlet hose pipe. To assure the system is fully drained, remove the engine block drain plug and the engine knock sensor from either side of the block.

2. Once the system is completely drained, close the drain cock and install the knock sensor. Cover the threads of the drain plug with sealer, then install the plug into the block. The drain plug should be tightened to 15 ft. lbs. (21 Nm), while the knock sensor should be tightened to 14 ft. lbs. (19 Nm).

3. After the repair or service has been completed, fill the cooling system with a solution which is at least 50 percent ethylene glycol antifreeze and the balance water. Do not use a solution that exceeds 70 percent antifreeze. Continue filling the radiator until the level is just below the filler neck.

4. Once the system is filled, close the air bleeds, taking care not to overtighten and damage the brass valves.

5. Install the pressure cap, then start and run the engine to check for leaks.

FUEL SYSTEM

Fuel System Service Precautions

Safety is the most important factor when performing not only fuel system maintenance but any type of maintenance. Failure to conduct maintenance and repairs in a safe manner may result in serious personal injury or death. Maintenance and testing of the vehicle's fuel system components can be accomplished safely and effectively by adhering to the following rules and guidelines.

• To avoid the possibility of fire and personal injury, always disconnect the negative battery cable unless the repair or test procedure requires that battery voltage be applied.

• Always relieve the fuel system pressure prior to disconnecting any fuel system component (injector, fuel rail, pressure regulator, etc.), fitting or fuel line connection. Exercise extreme caution whenever relieving fuel system pressure to avoid exposing skin, face and eyes to fuel spray. Please be advised that fuel under

pressure may penetrate the skin or any part of the body that it contacts.

• Always place a shop towel or cloth around the fitting or connection prior to loosening to absorb any excess fuel due to spillage. Ensure that all fuel spillage (should it occur) is quickly removed from engine surfaces. Ensure that all fuel soaked cloths or towels are deposited into a suitable waste container.

• Always keep a dry chemical (Class B) fire extinguisher near the work area.

• Do not allow fuel spray or fuel vapors to come into contact with a spark or open flame.

• Always use a backup wrench when loosening and tightening fuel line connection fittings. This will prevent unnecessary stress and torsion to fuel line piping. Always follow the proper torque specifications.

• Always replace worn fuel fitting O-rings with new. Do not substitute fuel hose or equivalent, where fuel pipe is installed.

Fuel System Pressure

RELIEVING

1. Disconnect the negative battery cable to prevent fuel discharge if the key is accidentally turned to the RUN position.

2. Loosen the fuel filler cap to relieve the tank pressure and do not tighten until service has been completed.

3. Connect J-34730-1 fuel pressure gauge or equivalent, to the fuel pressure valve. Wrap a shop cloth around the fitting while connecting the gauge to avoid spillage.

4. Place the end of the bleed hose into a suitable container and open the valve to relieve the fuel system pressure.

Idle Speed

ADJUSTMENT

The idle speed and mixture are controlled electronically by the Engine Control Module (ECM). All adjustments are preset at the factory and do require periodic attention. Some throttle body units are equipped with a idle stop screw to allow adjustment of the minimum idle speed if the unit is used as a replacement. The only time the idle speed should need adjustment is when the throttle body assembly has been replaced.

IAC VALVE PINTLE ADJUSTMENT

Idle speed is controlled by the ECM through voltage pulses sent to the Idle Air Control (IAC) motor windings. Based on the number of voltage pulses received, the motor will move the IAC pintle in or out allowing more or less air through the throttle body. Whenever a new IAC valve is installed, the pintle must be adjusted to specification.

NOTE: If installing a new IAC valve measure and adjust the valve accordingly. If reinstalling a used IAC valve, do not push or pull on the pintle to adjust pintle length or damage to the IAC worm gear threads might occur.

1. On a new IAC valve only, measure the distance between the tip of the pintle and the valve mounting surface. If greater than 1.10 inch (28mm), use light finger pressure to slowly retract the pintle. The force required to retract a new IAC valve will not damage the valve.

2. Install the IAC valve and gasket.

3. Connect the IAC valve wire connector.

4. For the 5.7L engine, reset the IAC valve pintle as follows:

   a. Depress the accelerator pedal slightly.

   b. Start the engine and run for 5 seconds.

   c. Turn the ignition OFF for 10 seconds.

   d. Restart the vehicle and check for proper idle operation.

5. For the 3.4L engine, reset the IAC valve pintle as follows:

   a. Turn the ignition ON for 5 seconds, then OFF for 10 seconds.

   b. Start the engine and check for proper idle operation.

Fuel Filter

REMOVAL AND INSTALLATION

1. Disconnect the negative battery cable and relieve the fuel system pressure.

2. Raise and support the vehicle safely.

3. Clean the fuel filter line fittings before disconnecting to prevent contamination of the fuel system.

4. Remove the fuel filter bracket, then Disengage the quick-connect fitting from the rear of the filter.

5. Loosen and disconnect the threaded outlet fitting from the from of the filter, then plug the lines to prevent system contamination.

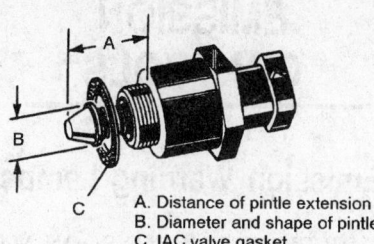

A. Distance of pintle extension
B. Diameter and shape of pintle
C. IAC valve gasket

8838RG10

Measuring the pintle position of a new IAC valve

6. Remove the filter from the vehicle.

To install:

7. Position the replacement filter and install the threaded fitting, then tighten the fitting to 20 ft. lbs. (27 Nm).

8. Install the fuel filter bracket, then engage the quick-connect fitting to the rear of the filter assembly. Remember to apply a few drops of clean engine oil to the male end of the connector in order to assure proper seal.

9. Lower the vehicle, then connect the negative battery cable and tighten the fuel filler cap.

10. Turn the ignition switch to the **ON** position for 2 seconds, **OFF** for 10 seconds, then to the **ON** position and check for fuel leaks.

Fuel Pump

REMOVAL AND INSTALLATION

The electric fuel pump is part of the fuel sender assembly located inside the fuel tank.

1. Release the fuel pressure and disconnect the negative battery cable.

2. Drain the fuel tank, then raise and support the vehicle safely.

3. Remove the fuel tank from the vehicle.

4. Clean the area surrounding the sender assembly to prevent contamination of the fuel system.

5. Remove the fuel sender assembly nuts and retaining ring, then carefully remove the sender assembly from the tank. Discard the old O-ring.

6. If necessary, separate the fuel pump from the sending assembly.

To install:

7. If removed, install the fuel pump to the sending unit. If the strainer was removed from the pump, the strainer must be replaced with a new component.

8. Inspect and clean the O-ring mating surfaces. Install the new O-ring in the groove around the tank opening, and if applicable, install a new O-ring on the fuel sender feed tube.

9. Install the fuel sender assembly; the fuel pump strainer must be in a horizontal position and when installed must not block the full travel of the float arm. Gently fold the fuel strainer over itself and slowly position the fuel sender assembly in the tank so the strainer is not damaged or trapped by sump walls.

10. Install the retaining ring and nuts, then tighten the nuts to 63 inch lbs. (7 Nm).

11. Install the fuel tank assembly to the vehicle.

12. Lower the vehicle.

13. Fill the fuel tank, tighten the fuel filler cap and connect the negative battery cable.

14. Turn the ignition switch to the **ON** position for 2 seconds, **OFF** for 10 seconds, then to the **ON** position and check for fuel leaks.

Fuel Injector

REMOVAL AND INSTALLATION

3.4L Engine

1. Make sure the ignition is **OFF**, then disconnect then negative battery cable.

2. Relieve the fuel system pressure.

3. Disengage the quick-connect fittings from the engine fuel pipes, then remove the pipe retaining bolts. Relocate the accelerator cable from the routing clip and remove the fuel tube retainer from the fuel rail, then remove the pipes from the engine.

4. Disconnect the vacuum line harness from the fuel rail assembly.

5. Disengage the wiring harness from the fuel rail assembly, then remove the fuel rail retainers.

6. Using compressed air and safety glasses, blow any dirt which may be present out of the injector bores, then remove the fuel rail assembly.

7. Rotate the fuel injector retainer clip to the release position, then carefully remove the injector from the rail assembly.

8. Remove and discard the O-ring seals from both ends of the injector, then discard the retainer clip.

To install:

9. Lubricate the new O-rings with clean engine oil, then install them onto the injector. The lower O-ring

uses a nylon collar called the O-ring backup. In order to prevent vacuum leaks and driveability problems, it is extremely important to assure that the backup O-ring is properly positioned on the injector.

10. Position a new retainer clip on the injector, then install the injector to the fuel rail assembly. Rotate the injector clip to the locked position in order to secure the injector to the rail assembly.

11. Tilt the fuel rail, as necessary to properly position the injectors, then carefully install the assembly to the intake manifold.

12. Install the fuel rail attaching bolts and tighten to 18 ft. lbs. (25 Nm).

13. Install the fuel pipes to the rail, then install the pipes to the engine. Make sure accelerator cable is properly positioned in the routing clip and that all retainers and connectors are securely fastened.

14. Rotate the injectors, as necessary, in order to avoid stretching the wire harness, then engage the harness connectors with the fuel rail assembly.

15. Install the vacuum line harness connector to the fuel rail.

16. Connect the negative battery cable, and tighten the fuel filler cap.

17. Turn the ignition switch to the **ON** position for 2 seconds, **OFF** for 10 seconds, then to the **ON** position and check for fuel leaks.

5.7L Engine

1. Make sure the ignition is in the **OFF** position, then disconnect the negative battery cable.

2. Properly relieve the fuel system pressure.

3. Disengage the quick-connect fittings at the fuel rail feed and return pipes as follows:

   a. Slide the rubber dust cover from the fitting.

   b. Grasp both ends of a connection and twist ¼ turn in each direction to loosen any dirt. Repeat for other fitting.

   c. While wearing safety glasses, use compressed air to blow out dirt from the fitting.

   d. Insert a fuel line separator tool, into the female connector, then push inward to release the male connector.

   e. Repeat for the other fitting.

4. Disconnect the vacuum line at the pressure regulator, then as necessary, tag and disconnect any remaining vacuum lines which must be removed to access the fuel rail and engine fuel pipes.

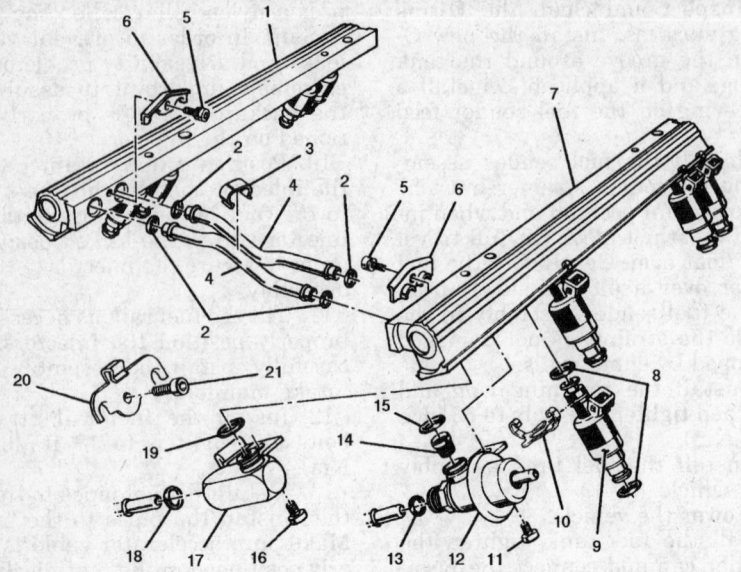

1. LH MFI fuel rail assembly
2. Fuel crossover tube o-ring
3. Fuel crossover tube clip
4. Fuel crossover tube
5. Crossover tube retainer attaching screw
6. Fuel crossover tube retainer
7. RH MFI fuel rail assembly
8. Fuel injector o-ring
9. MFI fuel injector assembly
10. Injector retainer clip
11. Pressure regulator attaching screw
12. Fuel pressure regulator assembly
13. Fuel outlet tube o-ring
14. Filter screen (if equipped)
15. Fuel inlet fitting o-ring
16. Fuel inlet assembly attaching screw
17. Fuel inlet assembly
18. Fuel inlet tube o-ring
19. Fuel inlet assembly o-ring
20. Fuel tube retainer assembly
21. Fuel tube retainer assembly attaching screw

8838RG11

Exploded view of the fuel rail assembly — 5.7L engine

5. Remove the fuel injector wiring harness from the routing clips of the fuel rail, then remove the fuel pipe attaching bolt and disengage the injector electrical connectors.

6. Remove the fuel rail attaching bolts and carefully remove the fuel rail assembly along with the injectors, from the top of the intake.

7. Rotate the injector retaining clip to the release position and remove the injector from the fuel rail assembly.

8. Remove and discard the O-ring seals from either side of the injector.

9. Remove and discard the injector retaining clip.

To install:

10. Lubricate the new injector O-rings with clean engine oil and install onto the injector.

NOTE: Always replace injectors using an identical part number as inscribed on top of the old injector.

11. Connect a new retainer clip onto the fuel injector and install the injector to the fuel rail assembly. Rotate the injector retaining clip to the lock position.

12. Install the fuel rail assembly to the intake manifold. Tighten the attaching bolts to 15 ft. lbs. (20 Nm).

13. Rotate the fuel injectors as necessary to avoid stretching the wire harnesses, then engage the injector electrical connections.

14. Install the fuel pipe retaining nut, then position the wiring harness into the routing clips at the fuel rail.

15. Connect the vacuum lines to the intake, as necessary, then connect the vacuum line to the pressure regulator.

16. Apply a few drops of clean engine oil to the male ends of the fuel line quick-connect fittings. Engage the fittings by pushing the connectors together until the retaining tabs snap into place. Pull gently on both sides of each fitting to be sure the connection is secure. When secure, slide the dust covers over the fittings.

17. Tighten the fuel filler cap and connect the negative battery cable.

18. Turn the ignition **ON** for 2 seconds, **OFF** for 10 seconds, then **ON** again and inspect the system for leaks.

EMISSION CONTROLS

Emission Warning Lamps

The SERVICE ENGINE SOON Malfunction Indicator Light (MIL) located in the instrument cluster serves 2 main functions:

1. The lamp indicates to the driver when a problem has occurred and that the vehicle should be taken for service as soon as reasonably possible.

2. The light may be used by technicians to monitor diagnostic trouble codes and/or open/closed loop engine operation, whenever the system is placed in the diagnostic mode.

To verify proper operation of the bulb and wiring, the lamp will illuminate when the ignition is first turned to **ON**, but the engine is not running. If the system is operating properly, the lamp will turn off once the engine is started.

If the MIL remains lit once the engine is started, the self-diagnostic system has detected a problem. If the problem goes away, the light will extinguish in 10 seconds (in most cases), but a diagnostic trouble code will remain in the ECM memory.

RESETTING

NOTE: In order to prevent damage to the ECM, the ignition must always be OFF when connecting or disconnecting power to the ECM.

After repairs are made to the faulty system(s), it is necessary to make sure the ECM memory is cleared of any old diagnostic trouble codes. Removing the battery voltage to the ECM for a minimum of 30 seconds will clear all codes. This may be accomplished in various ways depending on how the vehicle is equipped. The ECM harness power feed may be disconnected at the positive battery terminal "pigtail." The fuse may be removed from the inline fuseholder which originates at the positive battery connection or from fuse block, as applicable. Also, the negative battery cable may be disconnected, but other on-board data such as the clock or radio presets and, on some vehicles, IAC valve pintle position may also be lost.

ECM LEARNING ABILITY

The ECM has a "learning" ability which allows it to make corrections for minor variations in the fuel system, in order to improve driveability. If the battery is disconnected to clear diagnostic codes or for safety during repairs, the "learning" process will reset and must begin again. A change may be noted in the vehicle's performance while the learning process occurs. To "teach" the vehicle, make sure the engine is at normal operating temperature, then drive the vehicle at part throttle, with moderate acceleration and idle conditions, until normal performance returns.

ENGINE MECHANICAL

Engine Assembly

REMOVAL AND INSTALLATION

3.4L Engine

1. Make sure the ignition is **OFF**, then disconnect the negative battery cable and relieve the fuel system pressure.
2. Raise and support the vehicle safely, then remove the front wheels.
3. Drain the engine cooling system and drain the crankcase of engine oil.
4. Disconnect the exhaust crossover pipe from the intermediate pipe.
5. If equipped with an automatic transmission, remove the converter cover, then remove the converter retaining bolts.
6. Remove the front facia lower deflectors from the vehicle.
7. Remove the stabilizer bar retaining bolts.
8. Remove the serpentine drive belt, then if equipped with an automatic transmission, disconnect the transmission fluid cooler lines from the radiator.
9. Disconnect the lower radiator hose from the radiator, then disconnect the heater hoses from the pipes at the engine assembly.
10. Remove the electrical ground straps from the right side of the engine block, then disengage the electrical wiring from the starter motor assembly.

11. Disengage the wiring harness connectors from the following sensor/switches:
   - Knock sensor
   - Oxygen sensor
   - Coolant temperature sensor
   - Camshaft sensor
   - Crankshaft sensors
   - Wheel speed sensors
   - Engine oil level switch
   - Fuel pump switch/engine oil pressure gauge sensor assemblies
12. Remove the right front brake line from the caliper brake hose, then plug the openings to prevent system contamination or excessive fluid loss.
13. Remove the wiring harness and the shift linkage from the transmission assembly.
14. Matchmark, then remove the driveshaft assembly.
15. Remove the torque arm from the transmission.
16. Remove the intermediate shaft from the rack and pinion assembly.
17. Remove the electrical ground straps from the left side of the frame rail.
18. Lower the vehicle sufficiently for underhood access, then remove the air intake duct.
19. Disconnect the fuel pipe assembly, then disconnect the cruise and accelerator cables from the throttle body assembly.
20. Disconnect the upper radiator hose from the intake manifold and the lower radiator hose from the front cover assembly.
21. Disengage the electrical connectors from the fan assembly, then remove the assembly from the vehicle.
22. Disconnect the brake booster vacuum hose.
23. Remove the Y brace from the right exhaust manifold assembly.
24. Remove the alternator and air conditioning compressor bracket, then position the assembly aside, taking care not to damage the components or connections.
25. Disengage and reposition the engine wiring harness connectors.
26. Remove and reposition the brake master cylinder.
27. Remove the upper bolts and nuts from the strut assemblies, then remove the right front brake line from the modulator valve assembly and clips.
28. Raise and support the vehicle safely, then position a lift table under the engine and engine frame assembly.
29. Remove the engine frame and transmission support bolts, then carefully lift the vehicle from the engine, transmission and engine frame

assembly. Raise the vehicle slowly, stopping a few times to assure that all wiring and hoses are disconnected and free of the powertrain and/or vehicle.
30. Secure the strut assemblies to the frame, then, if applicable remove the transmission TV cable from the throttle body.
31. Remove the transmission assembly.
32. If equipped with a manual transmission, remove the clutch housing and clutch assembly.
33. Remove the power steering lines from the pump assembly. Plug the openings to prevent system contamination or excessive fluid loss.
34. Remove the engine mount through-bolts, then remove the engine from the engine frame.

To install:

35. Install the engine assembly to the engine frame, then install and tighten the mount through-bolts to 70 ft. lbs. (95 Nm).
36. Uncap the openings, then connect the lines to the power steering pump assembly.
37. If applicable, install the clutch housing and clutch assembly.
38. Install the transmission assembly, then if equipped with an automatic transmission, connect the TV cable to the throttle body.
39. Position the engine lift table under the vehicle, then align the strut assemblies and install the engine, transmission and engine frame to the vehicle.
40. Install the engine frame retaining bolts and tighten to 92 ft. lbs. (125 Nm), then remove the engine lift table and lower the vehicle as necessary for underhood access.
41. Install the upper strut assembly retaining bolts and nuts.
42. Install the right front brake line to the modulator valve assembly, then reposition and secure the master cylinder.
43. Engage the engine wiring harness connectors, then reposition and secure the alternator/A/C compressor bracket.
44. Install the Y brace to the right exhaust manifold assembly, then connect the brake booster vacuum hose.
45. Install the engine cooling fan assembly and engage the electrical connectors.
46. Connect the lower radiator hose to the front cover assembly and the upper radiator hose to the intake manifold assembly.
47. Connect the cruise and accelerator control cables to the throttle

body assembly, then install and connect the fuel pipe assembly.

48. Raise and support the vehicle safely, then connect the electrical ground straps to the left side of the frame rail.

49. Connect the intermediate shaft to the rack and pinion assembly.

50. Install the torque arm to the transmission assembly.

51. Align the matchmarks and install the driveshaft.

52. Install the shift linkage to the transmission, then engage the transmission wiring harness.

53. Connect the right front brake line to the caliper brake hose.

54. Engage all switch and sensor electrical connectors which were disengaged to remove the engine frame and powertrain assembly.

55. Install the wiring to the starter motor wiring, then install the ground straps to the right side of the engine block.

56. Connect the heater pipes to the engine assembly and the lower radiator hose to the radiator assembly.

57. If equipped with an automatic transmission, connect the fluid cooler lines to the radiator.

58. Install the serpentine drive belt.

59. Install the stabilizer bar bolts, then install the front facia lower deflectors.

60. If equipped with an automatic transmission, install the converter retaining bolts, then install the converter cover and bolts.

61. Connect the exhaust crossover pipe assembly to the intermediate pipe.

62. Bleed the hydraulic brake system, then install the wheels and lower the vehicle.

63. Fill the engine crankcase with oil and connect the negative battery cable.

64. Properly fill the engine cooling system.

65. Install the air intake duct and bleed the power steering system.

66. Check and adjust wheel alignment, as necessary.

5.7L Engine

1. Properly discharge and recover the refrigerant from the air conditioning system.

———— **WARNING** ————
The 4th generation F-body uses only the new R-134a refrigerant. R-134a is not compatible with the R-12 refrigerant formerly used in most automobiles. Service equipment and/or air conditioning sys-tem components may be contami-nated or damaged if the 2 refrigerants and oils are mixed. Never use a recovery system which was designed for and used to service R-12 equipped vehicles on a vehicle equipped with an R-134a air conditioning system.

2. Disconnect the negative battery cable and remove the air intake duct.

3. Raise and support the vehicle safely, then remove the front wheels.

4. Drain the engine cooling system and drain the crankcase of engine oil.

5. Disconnect the exhaust crossover pipe from the intermediate pipe.

6. If equipped with an automatic transmission, remove the converter cover, then remove the converter retaining bolts.

7. Remove the front facia lower deflectors from the vehicle.

8. Remove the stabilizer bar retaining bolts.

9. Remove the serpentine drive belt.

10. Remove the secondary air injection pump.

11. Disconnect the air conditioning hose assembly from the condenser and the compressor. Plug the openings to prevent system contamination and damage.

12. If equipped, disconnect the transmission fluid cooler lines from the radiator assembly. Plug the openings to prevent system contamination or excessive fluid loss.

13. Disconnect the lower radiator hose from the radiator and the lower heater hose from the water pump.

14. Remove the electrical ground straps from the right side of the engine block, then disengage the electrical wiring from the starter motor assembly.

15. Disengage the wiring harness connectors from the following sensor/switches:
• Knock sensor
• Oxygen sensor
• Coolant temperature sensor
• Wheel speed sensors
• Engine oil level switch
• Fuel pump switch/engine oil pressure gauge sensor assemblies

16. Remove the left front brake line from the caliper brake hose, then plug the openings to prevent system contamination or excessive fluid loss.

17. Remove the wiring harness and the shift linkage from the transmission assembly.

18. Matchmark, then remove the driveshaft assembly.

19. Remove the torque arm from the transmission.

20. Remove the intermediate shaft from the rack and pinion assembly.

21. Remove the electrical ground straps from the left side of the engine block and frame rail.

22. Lower the vehicle sufficiently for underhood access, then disconnect the fuel lines from the fuel rail.

23. Disconnect the cruise and accelerator cables from the throttle body assembly.

24. Disconnect the upper radiator hose, lower radiator hose and upper heater hose from the water pump assembly.

25. Disconnect the coolant air bleed pipe hose from the radiator assembly, then disconnect the brake booster vacuum hose.

26. Remove the alternator from the engine.

27. Disengage the engine wiring harness connectors, then position the harness aside.

28. Remove the air conditioning receiver and dehydrator hose from the condenser, then remove the hose from the expansion tube. Plug the openings to prevent system contamination and damage.

29. Remove the power steering reservoir from its mounting and reposition aside.

30. Remove and reposition the brake master cylinder.

31. Remove the upper bolts and nuts from the strut assemblies, then remove the right front brake line from the modulator valve assembly and clips.

32. Raise and support the vehicle safely, then position a lift table under the engine and engine frame assembly.

33. Remove the engine frame and transmission support bolts, then carefully lift the vehicle from the engine, transmission and engine frame assembly. Raise the vehicle slowly, stopping a few times to assure that all wiring and hoses are disconnected and free of the powertrain and/or vehicle.

34. Secure the strut assemblies to the frame, then if applicable, remove the transmission TV cable from the throttle body.

35. Remove the transmission assembly.

36. If equipped with a manual transmission, remove the clutch housing and clutch assembly.

37. Remove the hose assemblies from the air conditioning compressor, then loosen the retainers and remove the compressor. Plug the openings to prevent system contamination and damage.

38. Remove the power steering lines from the pump assembly. Plug the openings to prevent system contamination or excessive fluid loss.

39. Remove the engine mount through-bolts, then remove the engine from the engine frame.

To install:

40. Install the engine assembly to the engine frame, then install and tighten the mount through-bolts to 70 ft. lbs. (95 Nm).

41. Uncap the openings, then connect the lines to the power steering pump assembly.

42. Install the air conditioning compressor and retainers, then uncap the openings and install the hose assemblies to the compressor.

43. If applicable, install the clutch housing and clutch assembly.

44. Install the transmission assembly, then if equipped with an automatic transmission, connect the TV cable to the throttle body.

45. Position the engine lift table under the vehicle, then align the strut assemblies and install the engine, transmission and engine frame to the vehicle.

46. Install the engine frame retaining bolts and tighten to 92 ft. lbs. (125 Nm), then remove the engine lift table and lower the vehicle as necessary for underhood access.

47. Install the upper strut assembly retaining bolts and nuts.

48. Install the right front brake line to the modulator valve assembly and clips, then reposition and secure the master cylinder.

49. Reposition and secure the power steering reservoir.

50. Unplug the openings and install the air conditioning hose to the expansion tube, then install the receiver and dehydrator hose to the condenser.

51. Engage the engine wiring harness connectors, then install the alternator.

52. Connect the brake booster vacuum hose.

53. Connect the coolant air bleed pipe hose to the radiator assembly.

54. Connect the upper heater hose, lower radiator hose and the upper radiator hose to the water pump assembly.

55. Connect the cruise and accelerator control cables to the throttle body.

56. Connect the fuel lines to the fuel rail assembly.

57. Raise and support the vehicle safely, then install the ground straps to the left side of the engine block and the frame rail.

58. Connect the starter motor wiring.

59. Connect the intermediate shaft to the rack and pinion assembly.

60. Install the torque arm to the transmission assembly.

61. Align the matchmarks and install the driveshaft.

62. Install the shift linkage to the transmission, then engage the transmission wiring harness.

63. Connect the left front brake line to the caliper brake hose.

64. Engage all switch and sensor electrical connectors which were disengaged to remove the engine frame and powertrain assembly.

65. Connect the lower heater hose to the water pump assembly and the lower radiator hose to the radiator.

66. If applicable, connect the transmission fluid cooler lines to the radiator.

67. Unplug the openings, then connect the air conditioning hose assembly to the condenser and compressor.

68. Install the secondary air injection pump assembly.

69. Install the serpentine drive belt.

70. Install the stabilizer bar bolts, then install the front facia lower deflectors.

71. If equipped with an automatic transmission, install the converter retaining bolts, then install the converter cover and bolts.

72. Connect the exhaust crossover pipe assembly to the intermediate pipe.

73. Bleed the hydraulic brake system, then install the wheels and lower the vehicle.

74. Fill the engine crankcase with oil and connect the negative battery cable.

75. Properly fill the engine cooling system, then charge the A/C system with R-134a.

76. Install the air intake duct and bleed the power steering system.

77. Check and adjust wheel alignment, as necessary.

Engine Mounts

REMOVAL AND INSTALLATION

1. Raise and support the vehicle safely.

2. Disconnect the exhaust crossover pipe.

3. Remove the engine mount through-bolts and nuts.

— WARNING —

Never raise the engine using a jack under the oil pan, crankshaft pulley or any sheetmetal. Because there only is a small clearance between the oil pan and the oil pump screen, if the pan is bent even slightly, damage could occur to the pump screen and pickup unit.

---

4. Raise the engine using a jacking fixture.

5. Remove the engine mount-to-frame bolts and/or the engine mount bracket-to-engine bolts, as necessary.

To install:

6. If removed, install the engine mount brackets and/or the engine mounts. Tighten the retaining bolts to 43 ft. lbs. (58 Nm).

7. Lower the engine, then install the mount through-bolts.

8. Tighten either the nuts to 59 ft. lbs. (80 Nm) or the through-bolts to 70 ft. lbs. (95 Nm).

9. Connect the exhaust crossover pipe.

10. Lower the vehicle.

Cylinder Head

REMOVAL AND INSTALLATION

3.4L Engine

Left Side

1. Disconnect the negative battery cable, then raise and support the vehicle safely.

2. Drain the engine cooling system, then lower the vehicle.

3. Remove the intake manifold assembly.

4. Remove the left exhaust manifold assembly.

5. Remove the oil level indicator and tube.

6. Remove the serpentine drive belt.

7. Disengage the coolant temperature sensor connector

8. Tag and disengage the wires from the spark plugs, then remove the plugs from the left cylinder head.

9. Remove the engine lift bracket.

10. Remove the wiring harness clip from the rear of the cylinder head assembly.

11. Remove the secondary air injection pipe bracket bolt.

12. Remove the power steering pump assembly and brackets, then position aside with the hoses intact. Be careful not to kink and/or damage the hydraulic lines.

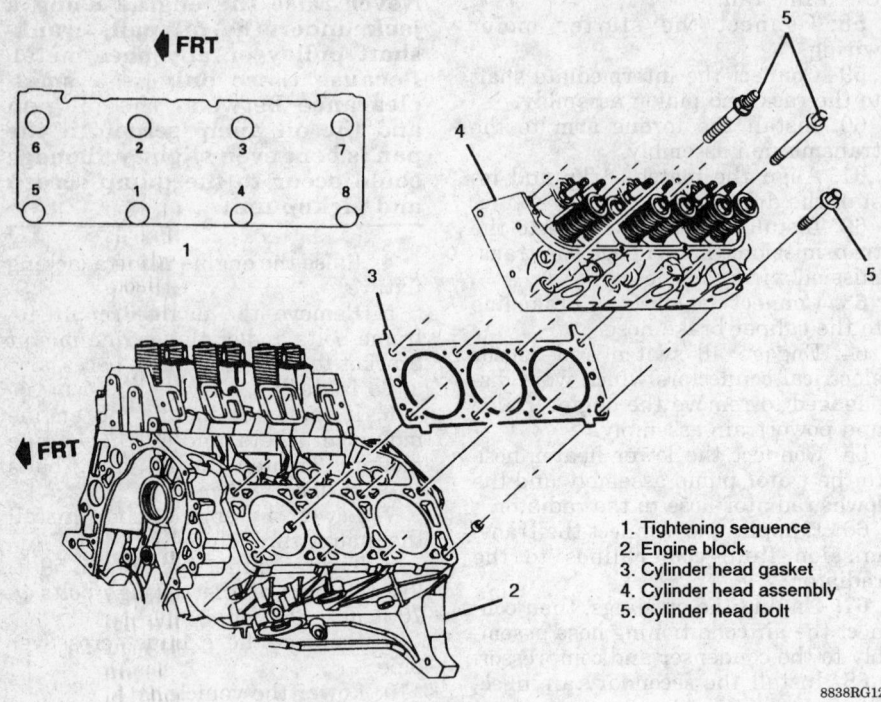

1. Tightening sequence
2. Engine block
3. Cylinder head gasket
4. Cylinder head assembly
5. Cylinder head bolt

8838RG12

Cylinder head installation and bolt torque sequence — 3.4L engine

13. Loosen the rocker arm assemblies so the pushrods may be removed. Arrange or tag the pushrods to assure installation in their original locations.

14. Remove the cylinder head bolts, then remove the cylinder head and gasket from the engine.

To install:

15. Thoroughly clean the gasket mating surfaces of any remaining gasket material, then position a new cylinder head gasket on the block.

16. Install the cylinder head over the locator pins and the new gasket.

17. Coat the bolt threads with sealant, then install the bolts finger-tight. Tighten the bolts in sequence to 41 ft. lbs. (55 Nm), then still in sequence, tighten all bolts an additional 90 degrees using a torque angle meter.

18. Install the pushrods and rocker arms, then adjust the valve lash.

19. Position and secure the power steering pump assembly and brackets.

20. Install the secondary air injection pipe bracket bolt.

21. Install and tighten the spark plugs, then connect the plug wires.

22. Install the wiring harness clip to the rear of the cylinder head assembly.

23. Install the engine lift bracket, then engage the coolant temperature sensor connector.

24. Install the serpentine drive belt.

25. Install the oil level indicator and tube.

26. Install the left exhaust manifold assembly.

27. Install the intake manifold assembly.

28. Connect the negative battery cable and fill the engine cooling system.

Right Side

1. Disconnect the negative battery cable, then raise and support the vehicle safely.

2. Drain the engine cooling system, then lower the vehicle.

3. Loosen the serpentine drive belt and remove the tensioner assembly.

4. Remove the intake manifold assembly.

5. Remove the right exhaust manifold assembly.

6. Remove the alternator and brackets.

7. Tag and disengage the wires from the spark plugs, then remove the plugs from the right cylinder head.

8. Loosen the rocker arm assemblies so the pushrods may be re-

moved. Arrange or tag the pushrods to assure installation in their original locations.

9. Remove the cylinder head bolts, then remove the cylinder head and gasket from the engine.

To install:

10. Thoroughly clean the gasket mating surfaces of any remaining gasket material, then position a new cylinder head gasket on the block.

11. Install the cylinder head over the locator pins and the new gasket.

12. Coat the bolt threads with sealant, then install the bolts finger-tight. Tighten the bolts in sequence to 41 ft. lbs. (55 Nm), then still in sequence, tighten all bolts an additional 90 degrees using a torque angle meter.

13. Install the pushrods and rocker arms, then adjust the valve lash.

14. Install and tighten the spark plugs, then connect the plug wires.

15. Install the alternator and brackets.

16. Install the right exhaust manifold assembly.

17. Install the intake manifold assembly.

18. Install the serpentine drive belt tensioner and verify that the belt is properly routed.

19. Refill the engine cooling system.

20. Connect the negative battery cable.

5.7L Engine

Left Side

1. Disconnect the negative battery cable, then raise and support the vehicle safely.

2. Drain the engine cooling system, then disconnect the crossover pipe from the exhaust manifold.

3. Lower the vehicle.

4. Remove the intake manifold assembly.

5. Disconnect the secondary air injection hose from the check valve assembly.

6. Disconnect the coolant air bleed pipe and bolt from the left cylinder head assembly using a backup wrench on the pipe fitting.

7. Remove the ignition coil assembly.

8. Remove the left exhaust manifold assembly.

9. Remove the spark plug wire harness assembly from the clips, then disconnect the harness from the spark plugs and remove the plugs from the left cylinder head.

10. Disengage the coolant temperature sensor connector.

11. Remove the left rocker arm cover.

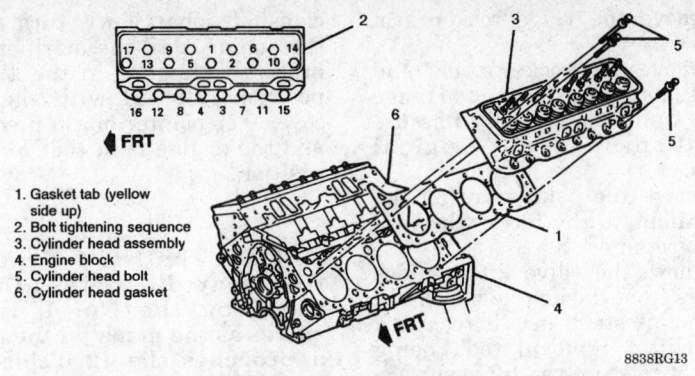

1. Gasket tab (yellow side up)
2. Bolt tightening sequence
3. Cylinder head assembly
4. Engine block
5. Cylinder head bolt
6. Cylinder head gasket

8838RG13

Cylinder head installation and bolt torque sequence — 5.7L engine

12. Loosen the rocker arm nuts, then remove the arms and pushrods, either tagging or arranging the components to assure installation in their original locations.

13. Remove the cylinder head bolts, then remove the cylinder head and old gasket from the block.

To install:

14. Thoroughly clean the gasket mating surfaces of any remaining gasket material, then position a new cylinder head gasket on the block with the yellow tab facing upwards.

15. Install the cylinder head over the locator pins and the new gasket.

16. Coat the cylinder head bolts with a sealing compound, then install the bolts finger-tight.

17. Torque the bolts using 3 passes of the proper sequence until all bolts have been torqued to 65 ft. lbs. (88 Nm).

18. Install the pushrods and rocker arms, making sure they are in the proper locations, then adjust the valve lash.

19. Install the left rocker arm cover and tighten the retaining bolts to 100 inch lbs. (11 Nm).

20. Install the spark plugs and tighten to 11 ft. lbs. (15 Nm), then install the wiring harness assembly and secure the assembly to the clips.

21. Install the left exhaust manifold assembly.

22. Install the ignition coil assembly and tighten the bolts to 24 ft. lbs. (33 Nm).

23. Connect the engine coolant air bleed pipe and bolt to the left cylinder head assembly using a backup wrench on the pipe fitting in order to prevent component damage. Tighten the bolt to 30 ft. lbs. (40 Nm).

24. Install the secondary air injector hose to the check valve assembly.

25. Install the intake manifold assembly.

26. Raise and support the vehicle safely, then connect the crossover pipe to the exhaust manifold.

27. Lower the vehicle and fill the engine cooling system.

28. Connect then negative battery cable.

Right Side

1. Disconnect the negative battery cable.

2. Raise and support the vehicle safely, then drain the engine cooling system.

3. Remove the serpentine drive belt and the belt tensioner assembly.

4. Remove the transmission fluid level indicator tube assembly bracket from the transmission housing.

5. Remove the air conditioning compressor rear brace bolt from the engine block, then disengage the compressor connector.

6. Remove the front compressor mounting bolts, then position the compressor aside taking care not to kink or damage the lines.

7. Remove the right exhaust manifold assembly.

8. Lower the vehicle.

9. Remove the alternator assembly.

10. Remove the right rocker arm cover.

11. Remove the intake manifold assembly.

12. Remove the coolant air bleed pipe bolt from the left cylinder head assembly.

13. Disconnect the lower radiator hose and the heater hose from the water pump assembly, then position the hoses aside.

14. Remove the coolant air bleed pipe hose from the radiator.

15. Remove the power steering pump assembly.

16. Remove the engine accessory bracket bolts, then remove the bracket assembly.

17. Disconnect the wire harness assembly from the spark plugs, then remove the plugs from the right cylinder head.

18. Loosen the rocker arm nuts, then remove the arms and pushrods, either tagging or arranging the components to assure installation in their original locations.

19. Remove the cylinder head bolts, then remove the cylinder head and old gasket from the block. If necessary, carefully remove the coolant air bleed pipe bolt and pipe assembly from the cylinder head.

To install:

20. If removed, loosely install the coolant air bleed pipe bolt and pipe to the cylinder head.

21. Thoroughly clean the gasket mating surfaces of any remaining gasket material, then position a new cylinder head gasket on the block with the yellow tab facing upwards.

22. Install the cylinder head over the locator pins and the new gasket.

23. Coat the cylinder head bolts with a sealing compound, then install the bolts finger-tight.

24. Torque the bolts using 3 passes of the proper sequence until all bolts have been torqued to 65 ft. lbs. (88 Nm).

25. If loosened, tighten the coolant air bleed pipe bolt to 30 ft. lbs. (40 Nm).

26. Install the pushrods and rocker arms, making sure they are in the proper locations, then adjust the valve lash.

27. Install the spark plugs and tighten to 11 ft. lbs. (15 Nm), then install the wiring harness assembly to the plugs.

28. Install the engine accessory bracket and tighten the retaining bolts to 31 ft. lbs. (42 Nm).

29. Install the right rocker arm cover and tighten the retaining bolts to 100 inch lbs. (11 Nm).

30. Install the alternator assembly.

31. Install the power steering pump assembly.

32. Connect the coolant air bleed pipe hose to the radiator, then connect the heater hose and the lower radiator hose to the water pump.

33. Connect the coolant air bleed pipe bolt to the left cylinder head and tighten to 30 ft. lbs. (41 Nm) while using a backup wrench to prevent component damage.

34. Install the intake manifold assembly.

35. Raise and support the vehicle safely.

36. Install the right exhaust manifold assembly.

37. Position the compressor and install the front mounting bolts, then engage the electrical connector. In-

stall the rear compressor brace bolt and tighten to 24 ft. lbs. (33 Nm).

38. Install the transmission fluid level indicator tube assembly to the transmission housing.

39. Install the serpentine drive belt, then lower the vehicle.

40. Fill the engine cooling system, then connect the negative battery cable.

Valve Lifters

REMOVAL AND INSTALLATION

3.4L Engine

1. Disconnect the negative battery cable.

2. Remove the intake manifold assembly.

3. Remove the rocker arms and pushrods. Be sure to tag or arrange all valve train components to assure installation in their original locations.

4. Remove the valve lifter assemblies from their bores. Be sure to tag or arrange the lifters for installation purposes.

To install:

NOTE: If lifters are replaced, be sure to use replacements that contain a narrow flat ground along the lower ¾ of the lifter. The flat is designed to provide additional oil to the cam lobe and lifter surfaces. Also, pay close attention to any lifter bores which may be marked indicating an oversized lifter. If an oversize lifter was used, the bore should be marked "0.25" and/or "O.S." in order to indicate that a 0.010 in. (0.25mm) oversize lifter was installed in that bore.

5. Coat the lifters using Molykote® or an equivalent prelube, then install the lifters into the bores. If lifters are being reused, they must be installed in their original bores.

6. Install the pushrods and the rocker arms, then adjust the valve lash.

7. Install the intake manifold.

8. Connect the negative battery cable.

5.7L Engine

1. Disconnect the negative battery cable.

2. Remove the intake manifold assembly.

3. Remove the valve rocker arm covers

4. Remove the rocker arms and pushrods. Be sure to tag or arrange all valve train components to assure installation in their original locations.

5. Remove the bolts securing the lifter retainer to the block, then remove the retainer.

6. Remove the valve lifter guides and lifters from the engine block. If a lifter remains stuck in a bore, use a suitable lifter removal tool, being careful not to score the lifter surface or the lifter bore. Tag or arrange the lifters and guides for installation in their original positions.

To install:

7. Coat the lifters with Molykote® or its equivalent before installation. If installing the old lifters, make certain each lifter is inserted into the same lifter bore from which it was removed.

8. Install the valve lifter guides and the lifter guide retainers, then tighten the bolts to 15 ft. lbs. (20 Nm).

9. Install the pushrods and rocker arms, then adjust the valve lash.

10. Install the valve rocker covers.

11. Install the intake manifold assembly.

12. Connect the negative battery cable.

Valve Lash

The valve lash on these engines must be adjusted whenever the rocker arm assemblies have been removed.

NOTE: These engines utilize hydraulic lifters which normally require very little maintenance or adjustment. These components are simple in design and are best maintained through regular, scheduled engine oil changes. If the engine is running well and no audible clicking sounds are heard from the valve train, there is no need to remove or disassemble the valve lifters.

ADJUSTMENT

1. Disconnect the negative battery cable.

2. Remove the valve rocker covers.

3. Tighten the nuts slowly until all lash is eliminated.

4. Adjust the valves when the lifter is on the base circle of the

camshaft lobe. Slowly turn or crank the engine until the mark on the vibration damper is in the 12 o'clock position (aligned with the timing cover 0 or pointer mark, if equipped) and the engine is in the No. 1 firing position.

NOTE: The No. 1 firing position may be determined by watching the No. 1 cylinder valves as the mark on the damper approaches the 12 o'clock position. If both the intake and exhaust valves are closed as the mark comes up to the timing tab, the engine is in the No. 1 firing position. If either valve opens as the timing mark approaches the top of it's travel, the engine is in the No. 6 firing position on the 5.7L engine or No. 4 firing position on the 3.4L engine and should be turned 1 full turn to reach to No. 1 firing position.

5. With the engine in the No. 1 firing position, adjust the following valves:

5.7L engine
Exhaust — 1, 3, 4, 8
Intake — 1, 2, 5, 7
3.4L engine
Exhaust — 1, 2, 3
Intake — 1, 5, 6

6. Back out the rocker arm adjusting nut until lash is felt at the pushrod, then turn the adjusting nut inward until all lash is removed. This can be determined by rotating pushrod while turning the adjusting nut. When play has been removed, the pushrod will not turn. Then, tighten the adjusting nut 1 full additional turn for 5.7L engine or 1½ additional turns for 3.4L engine.

7. Slowly turn or crank the engine 1 revolution until the vibration damper mark is at 12 o'clock again and the No. 1 cylinder valves open. This is the No. 6 firing position the No. 6 firing position on 5.7L engine or the No. 4 firing position on 3.4L engine.

8. With the engine in this position, adjust the following valves:

5.7L engine
Exhaust — 2, 5, 6, 7
Intake — 3, 4, 6, 8
3.4L engine
Exhaust — 4, 5, 6
Intake — 2, 3, 4

9. Install the valve covers and connect the negative battery cable.

10. Start the engine and check for proper operation.

Rocker Arms/Shafts

REMOVAL AND INSTALLATION

3.4L Engine

1. Disconnect the negative battery cable.

2. Remove the upper intake manifold:

a. Drain the engine cooling system and relieve the fuel system pressure, then remove the throttle body air duct.

b. Disengage the fuel rail injector wiring harness connectors, then remove the wiring harness retaining nut and position the harness aside.

c. Remove the accelerator cable bracket bolts, then remove the bracket and cables from the throttle body.

d. Disengage the fuel pipe connectors, remove the pipe bracket bolts and the pipe hold-down plate bolts, then remove and reposition the fuel pipe assembly.

e. Disconnect the coolant hoses from the throttle body.

f. Remove the fuel rail stud, then disconnect the pressure regulator vacuum hose.

g. Disengage the electrical connectors from the IAC valve, TP sensor, MAP sensor and EVAP canister purge solenoid.

h. Using compressed air and safety glasses, blow any dirt which may be present out of the injector bores, then remove the fuel rail assembly.

i. Remove the vacuum harness assembly, then remove the EGR flexible pipe bolts and reposition the pipe and gasket at the upper intake manifold assembly.

j. Remove the bolts retaining the upper manifold, then remove the manifold and gasket from the vehicle.

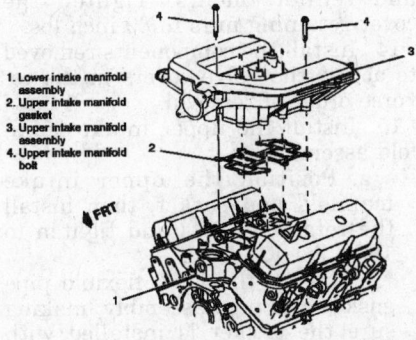

1. Lower intake manifold assembly
2. Upper intake manifold gasket
3. Upper intake manifold assembly
4. Upper intake manifold bolt

8838RG14

Upper intake manifold removal — 3.4L engine

3. To remove the left valve cover:

a. Remove the PCV valve and hose assembly.

b. Remove the spark plug wire harness assembly from the clips.

c. Remove the nuts and reinforcements, then remove the valve cover and gasket.

4. To remove the right valve cover:

a. Remove the bolts, stud and EGR valve adapter along with the EGR valve assembly from the exhaust manifold.

b. Remove the crankcase vent pipe.

c. Remove the spark plug wire harness assembly from the clips.

d. Remove the nuts and reinforcements, then remove the valve cover and gasket.

5. Remove the valve rocker arm nuts and balls, then remove the rocker arms and pushrods. Tag or arrange all valve train components to assure installation in their original locations.

To install:

6. Coat the bearing surfaces of the rocker arms, balls and pushrods with prelube.

7. Install the pushrods, making sure they are properly seated in the lifter sockets, then install the rocker arms, balls and nuts. If components are being reused, be sure they are installed in their original locations.

8. Adjust the valve lash.

9. Install the gaskets, valve covers and reinforcements. Tighten the cover retaining nuts to 89 inch lbs.

10. Install the components removed to access the valve covers in the reverse order of removal.

11. Install the upper intake manifold assembly:

a. Position the upper intake manifold and gasket, then install the retaining bolts and tighten to 18 ft. lbs. (25 Nm).

b. Install the EGR flexible pipe gasket and pipe assembly, making sure the gasket is installed with the writing TUBE SIDE facing the tube. Apply teflon sealant to the threads of the bolts, then install them and tighten to 19 ft. lbs. (26 Nm).

c. Install the vacuum harness assembly.

d. Lubricate the fuel injector O-rings using clean engine oil, then install the injectors into the manifold bore. Carefully press on the fuel rail assembly using the palms of both hands until the injectors are fully seated. Install the fuel rail retaining stud and tighten to 19 ft. lbs. (25 Nm).

e. Engage the wiring harness connectors to the IAC valve, TP sensor, MAP sensor and EVAP canister purge solenoid.

f. Connect the pressure regulator vacuum hose, then install the wiring harness retaining nut.

g. Engage the fuel pipe assembly to the fuel rail, sliding the fuel line hold-down plate over the pipe assembly. Apply teflon sealant to the hold-down plate bolt threads, then install the bolt and tighten to 18 ft. lbs. (25 Nm).

h. Install the fuel pipe bracket bolts, then position the accelerator cable bracket, bolts and cables. Tighten the cable bracket bolts to 90 inch lbs. (10 Nm), then tighten the front fuel pipe bracket bolt to 71 inch lbs. (8 Nm) and the rear bracket-to-intake bolt to 89 inch lbs. (10 Nm).

i. Engage the fuel pipe connectors.

j. Engage the fuel rail injector wiring harness connectors.

k. Connect the coolant hoses to the throttle body, then install the air duct.

l. Connect the negative battery cable and refill the engine cooling system

5.7L Engine

1. Disconnect the negative battery cable.

2. Remove the left valve cover:

a. Remove the brake booster vacuum hose.

b. Remove the secondary AIR injection hose from the pump to check valve assembly.

c. Remove the valve cover retaining bolts, then remove the cover and gasket.

3. Remove the right valve cover:

a. Raise and support the vehicle safely, then remove the serpentine drive belt.

b. Remove the transmission fluid level indicator tube assembly from the bracket on the transmission housing.

c. Lower the vehicle, then remove the crankcase vent hose.

d. Remove the alternator and rear alternator brace.

e. Remove the valve cover retaining bolts, then remove the cover and gasket.

4. Remove the valve rocker arm nuts and balls, then remove the rocker arms and pushrods. Tag or arrange all valve train components to assure installation in their original locations.

To install:

5. Coat the bearing surfaces of the rocker arms, balls and pushrods with prelube.

6. Install the pushrods, making sure they are properly seated in the lifter sockets, then install the rocker arms, balls and nuts. If components are being reused, be sure they are installed in their original locations.

7. Adjust the valve lash.

8. Install the valve covers and gaskets, then tighten the retainers to 100 inch lbs. (11 Nm).

9. Install the components which were removed to access the valve covers in the reverse order of removal.

10. Connect the negative battery cable.

Intake Manifold

REMOVAL AND INSTALLATION

3.4L Engine

1. Disconnect the negative battery cable.

2. Remove the upper intake manifold:

a. Drain the engine cooling system and relieve the fuel system pressure, then remove the throttle body air duct.

b. Disengage the fuel rail injector wiring harness connectors, then remove the wiring harness retaining nut and position the harness aside.

c. Remove the accelerator cable bracket bolts, then remove the bracket and cables from the throttle body.

d. Disengage the fuel pipe connectors, remove the pipe bracket bolts and the pipe hold-down plate bolts, then remove and reposition the fuel pipe assembly.

e. Disconnect the coolant hoses from the throttle body.

f. Remove the fuel rail stud, then disconnect the pressure regulator vacuum hose.

g. Disengage the electrical connectors from the IAC valve, TP sensor, MAP sensor and EVAP canister purge solenoid.

h. Using compressed air and safety glasses, blow any dirt which may be present out of the injector bores, then remove the fuel rail assembly.

i. Remove the vacuum harness assembly, then remove the EGR flexible pipe bolts and reposition the pipe and gasket at the upper intake manifold assembly.

j. Remove the bolts retaining the upper manifold, then remove the manifold and gasket from the vehicle.

3. To remove the left valve cover:

a. Remove the PCV valve and hose assembly.

b. Remove the spark plug wire harness assembly from the clips.

c. Remove the nuts and reinforcements, then remove the valve cover and gasket.

4. To remove the right valve cover:

a. Remove the bolts, stud and EGR valve adapter along with the EGR valve assembly from the exhaust manifold.

b. Remove the crankcase vent pipe.

c. Remove the spark plug wire harness assembly from the clips.

d. Remove the nuts and reinforcements, then remove the valve cover and gasket.

5. Disconnect the upper radiator hose and the heater hose from the intake manifold assembly.

6. Disconnect the wiring harness ground leads.

7. Remove the bolts and nuts retaining the intake manifold, then remove the intake manifold assembly and gaskets.

To install:

8. Thoroughly clean the mating surfaces of any remaining gasket material.

9. Position the intake manifold gaskets to the engine, then apply a bead of sealant at the block-to-manifold mating surfaces. The bead should be 0.08–0.012 in. (2–3mm) wide and 0.12–0.20 in. (3–5mm) thick.

10. Install the intake manifold over the gaskets, then secure using the bolts and nuts. Tighten the intake manifold fasteners to 22 ft. lbs. (30 Nm).

11. Connect the wiring harness ground leads.

12. Connect the heater and upper radiator hoses to the manifold.

13. Install the gaskets, valve covers and reinforcements. Tighten the cover retaining nuts to 89 inch lbs.

14. Install the components removed to access the valve covers in the reverse order of removal.

15. Install the upper intake manifold assembly:

a. Position the upper intake manifold and gasket, then install the retaining bolts and tighten to 18 ft. lbs. (25 Nm).

b. Install the EGR flexible pipe gasket and pipe assembly, making sure the gasket is installed with the writing TUBE SIDE facing the tube. Apply teflon sealant to the

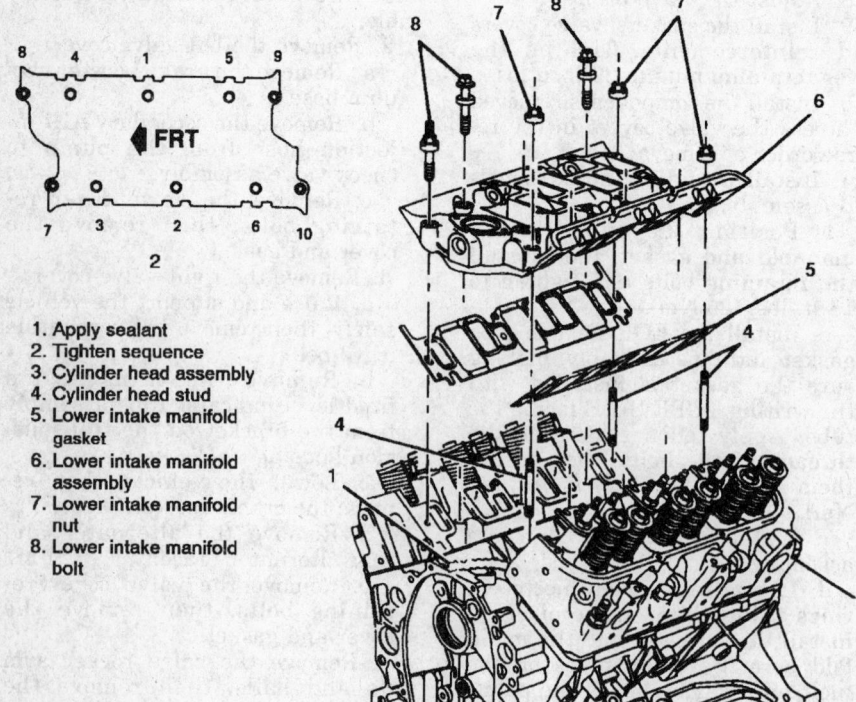

1. Apply sealant
2. Tighten sequence
3. Cylinder head assembly
4. Cylinder head stud
5. Lower intake manifold gasket
6. Lower intake manifold assembly
7. Lower intake manifold nut
8. Lower intake manifold bolt

8838RG15

Intake manifold installation and bolt torque sequence — 3.4L engine

threads of the bolts, then install them and tighten to 19 ft. lbs. (26 Nm).

c. Install the vacuum harness assembly.

d. Lubricate the fuel injector O-rings using clean engine oil, then install the injectors into the manifold bore. Carefully press on the fuel rail assembly using the palms of both hands until the injectors are fully seated. Install the fuel rail retaining stud and tighten to 19 ft. lbs. (25 Nm).

e. Engage the wiring harness connectors to the IAC valve, TP sensor, MAP sensor and EVAP canister purge solenoid.

f. Connect the pressure regulator vacuum hose, then install the wiring harness retaining nut.

g. Engage the fuel pipe assembly to the fuel rail, sliding the fuel line hold-down plate over the pipe assembly. Apply teflon sealant to the hold-down plate bolt threads, then install the bolt and tighten to 18 ft. lbs. (25 Nm).

h. Install the fuel pipe bracket bolts, then position the accelerator cable bracket, bolts and cables. Tighten the cable bracket bolts to 90 inch lbs. (10 Nm), then tighten the front fuel pipe bracket bolt to 71 inch lbs. (8 Nm) and the rear bracket-to-intake bolt to 89 inch lbs. (10 Nm).

i. Engage the fuel pipe connectors.

j. Engage the fuel rail injector wiring harness connectors.

k. Connect the coolant hoses to the throttle body, then install the air duct.

l. Connect the negative battery cable and refill the engine cooling system

5.7L Engine

1. Disconnect the negative battery cable and relieve the fuel system pressure.

2. Drain the engine cooling system into a suitable container.

3. Remove the throttle body air duct.

4. Disengage the wiring harness connectors from the fuel injectors. Disengage and reposition the left and right wiring harnesses.

5. Remove the accelerator cable bracket retainers, then disconnect the cable and bracket assembly from the throttle body.

6. Disconnect the secondary AIR diverter valve hoses.

7. Disengage the fuel pipe connectors from the fuel rail assembly.

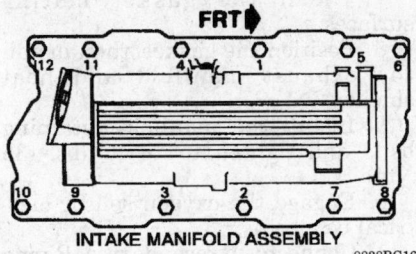

INTAKE MANIFOLD ASSEMBLY

8838RG16

Intake manifold bolt torque sequence — 5.7L engine

8. Remove the fuel rail bolts and disconnect the vacuum hose from the fuel pressure regulator.

9. Carefully remove the fuel rail and injector assembly from the manifold and position aside.

10. Disconnect the vacuum and crankcase vent hoses.

11. Remove the EGR solenoid assembly and the fuel EVAP canister solenoid assembly.

12. Remove the EGR valve.

13. Remove the AIR pipe from the intake and the right exhaust manifold.

14. Remove the alternator rear brace.

15. Disconnect the coolant hoses from the throttle body.

16. Remove the throttle body bolts, the throttle body and gasket from the intake.

17. Remove the intake manifold bolts and studs.

18. Remove the intake manifold and discard the old gaskets.

To install:

19. Thoroughly clean the intake manifold bolts and studs. Inspect and clean all gasket mating surfaces.

20. Apply a 3/16 in. (5mm) bead of RTV sealer to the front and rear of the cylinder block. Extend the bead 1/2 inch (13mm) up each cylinder head to seal and retain the gaskets.

21. Position the new gaskets and install the intake manifold.

22. Install the manifold bolts and studs, then tighten using 2 passes of the proper sequence. First, tighten the bolts/studs to 71 inch lbs. (8 Nm), then tighten them to 35 ft. lbs. (48 Nm).

23. Install the throttle body, gasket and retaining bolts. Tighten the throttle body bolts to 19 ft. lbs. (26 Nm).

24. Connect the coolant hoses to the throttle body, then install the alternator rear brace.

25. Install the accelerator cables and bracket, then tighten the bracket bolts to 90 inch lbs. (10 Nm).

26. Install the secondary AIR pipe. Tighten the exhaust manifold fitting to 25 ft. lbs. (34 Nm) and tighten the flange-to-intake manifold bolts to 19 ft. lbs. (26 Nm).

27. Install the EGR valve, then EGR solenoid and bracket. Tighten valve nuts and the solenoid bracket nut to 16 ft. lbs. (22 Nm).

28. Install the fuel EVAP canister purge solenoid and bracket, then tighten the bolt to 53 inch lbs. (6 Nm).

29. Connect the vacuum and crankcase vent hoses.

30. Install the fuel injector and fuel rail assembly to the intake manifold, connect the fuel pressure regulator vacuum hose and install the fuel rail bolts. Tighten the bolts to 15 ft. lbs. (20 Nm). Engage the fuel pipe connections to the fuel rail assembly.

31. Connect the secondary AIR diverter valve hoses.

32. Position the left and right wiring harnesses, then engage the fuel injector electrical connectors.

33. Install the throttle body air duct.

34. Properly fill the engine cooling system.

35. Connect the negative battery cable.

Exhaust Manifold

REMOVAL AND INSTALLATION

3.4L Engine

1. Disconnect the negative battery cable.

2. Raise and support the vehicle safely.

3. Remove the exhaust crossover pipe.

4. If removing the right manifold, remove the transmission filler tube, the A/C compressor rear bracket bolts and the 2 rear exhaust manifold bolts.

5. Lower the vehicle.

6. If removing the right manifold, remove the serpentine drive belt.

7. Disengage the oxygen sensor electrical connector.

8. If removing the right exhaust manifold:

a. Remove the alternator from the vehicle.

b. Disconnect the A/C compressor from the bracket and position aside, taking care not to kink or damage the refrigerant lines.

c. Remove the alternator rear Y brace.

d. Remove the EGR valve, adapter and flexible pipe.

9. If equipped with a manual transmission, disconnect the secondary AIR pipe assembly from the manifold.

10. Remove the manifold retaining bolts, then remove the manifold, heat shields and gasket.

To install:

11. Clean the gasket mating surfaces.

12. Position the gasket, then install the exhaust manifold and heat shields.

13. Install the manifold retaining bolts, except for the rear 2 bolts for the right manifold assembly, and tighten to 18 ft. lbs. (25 Nm).

14. If equipped with a manual transmission, secure the secondary AIR pipe assembly to the manifold.

15. If installing the right exhaust manifold:

a. Install the EGR valve, adapter and flexible pipe.

b. Install the alternator rear Y brace and the alternator assembly.

c. Position and secure the compressor.

16. Engage the oxygen sensor electrical connector.

17. If installing the right manifold, install the serpentine drive belt.

18. Raise and support the vehicle safely.

19. If installing the right manifold, install the transmission filler tube, the compressor rear bracket bolts and the rear 2 manifold retaining bolts.

20. Install the crossover pipe assembly.

21. Lower the vehicle and connect the negative battery cable. If applicable, check the transmission fluid level and add as necessary.

5.7L Engine

Left Side

1. Disconnect the negative battery cable, then raise and support the vehicle safely.

2. Disconnect the exhaust crossover pipe from the manifold, then lower the vehicle.

3. Remove the brake booster vacuum hose.

4. Disconnect the secondary AIR pipe fitting from the exhaust manifold.

5. Disengage the oxygen sensor electrical connector.

6. Remove the exhaust manifold retaining bolts, then remove the heat shields, manifold and gasket.

To install:

7. Clean the gasket mating surfaces.

8. Position the gasket, then install the exhaust manifold and heat shields.

9. Install the manifold retaining bolts and tighten to 26 ft. lbs. (35 Nm).

10. Engage the oxygen sensor electrical connector.

11. Connect the secondary AIR pipe fitting to the exhaust manifold and tighten to 26 ft. lbs. (35 Nm).

12. Install the brake booster vacuum hose.

13. Raise and support the vehicle safely.

14. Connect the exhaust crossover pipe to the manifold, then lower the vehicle.

15. Connect the negative battery cable.

Right Side

1. Disconnect the negative battery cable, then raise and support the vehicle safely.

2. Remove the exhaust crossover pipe.

3. Remove the serpentine drive belt.

4. Remove the oil level indicator and tube assembly, then disengage the oxygen sensor electrical connector.

5. Remove the 3 rear exhaust manifold retaining bolts, then lower the vehicle.

6. Disconnect the secondary AIR pipe fitting from the exhaust manifold.

7. Remove the alternator rear lower brace.

8. Remove the remaining exhaust manifold retaining bolts, then remove the heat shields, manifold and gasket.

To install:

9. Clean the gasket mating surfaces.

10. Position the gasket, then install the exhaust manifold and heat shields.

11. Install the front 3 manifold retaining bolts and tighten to 26 ft. lbs. (35 Nm).

12. Install the alternator rear lower brace.

13. Raise and support the vehicle safely.

14. Install the remaining manifold retaining bolts and tighten to 26 ft. lbs. (35 Nm).

15. Engage the oxygen sensor electrical connector.

16. Install the oil level indicator and tube assembly.

17. Install the serpentine drive belt.

18. Install the exhaust crossover pipe.

19. Lower the vehicle.

20. Connect the secondary AIR pipe fitting to the exhaust manifold and tighten to 25 ft. lbs. (34 Nm).

21. Connect the negative battery cable.

Front Cover Seal

REMOVAL AND INSTALLATION

In addition to the crankshaft oil seal, then 5.7L engine front cover utilizes a water pump driveshaft seal and distributor driveshaft seal. All front cover oil seals on these engines are replaced in the same manner.

1. Remove the timing chain front cover from the engine.

2. Using either a suitably sized driver or a small suitable prytool, remove oil seal from the front cover. If a driver is being used, be sure to support the cover so it is not damaged. If a prytool is used, take care not to score or damage the cover sealing surfaces.

3. For the 3.4L engine, use J-34995, or an equivalent seal installation tool to properly install the crankshaft seal to the front cover.

4. For the 5.7L engine, J-35468 or equivalent to install the crankshaft seal, J-39090 or equivalent to install the distributor shaft seal, and/or J-39088 or equivalent to install the water pump driveshaft seal.

5. Install the timing chain front cover to the engine.

Timing Chain Front Cover

REMOVAL AND INSTALLATION

3.4L Engine

1. Disconnect the negative battery cable.

2. Remove the throttle body air intake duct.

3. Remove the serpentine drive belt.

4. Remove the water pump assembly.

5. Remove the crankshaft balancer bolt and washer, then remove the pulley bolts and the pulley. Using J-24420-B, or an equivalent damper remover, pull the balancer from the end of the crankshaft assembly.

6. Remove the power steering pump and bracket assembly.

7. Remove the oil pan assembly.

8. Disconnect the lower radiator hose from the front cover assembly.

9. Remove the crankshaft sensor.

10. Remove the front cover retaining bolts, then remove the cover and gasket from the engine.

To install:

11. Thoroughly clean the mating surfaces of any remaining gasket material.

12. Coat both sides of the gasket lower tabs with sealer, then install the front cover with the gasket.

13. Install the front cover retaining bolts and tighten to 15 ft. lbs. (21 Nm).

14. Install the oil pan assembly.

15. Install the crankshaft sensor.

16. Connect the lower radiator hose to the front cover.

NOTE: If a new balancer is installed, new balancer weights of the same size must be installed in the same hole locations as the original balancer.

17. Using J-29113, or an equivalent installer tool, pull the balancer onto the end of the crankshaft. Install the crankshaft pulley and the retaining bolts, then install the crankshaft damper bolt and washer. Tighten the damper bolt to 58 ft. lbs. (78 Nm), then tighten the pulley retaining bolts to 37 ft. lbs. (50 Nm).

18. Lower the vehicle.

19. Install the water pump assembly.

20. Install the power steering pump and bracket assembly.

21. Install the serpentine drive belt, then install the throttle body air duct.

22. Connect the negative battery cable, then refill the engine cooling system.

5.7L Engine

1. Disconnect the negative battery cable.

2. Drain the engine oil and coolant into suitable containers.

3. Remove the throttle body air intake duct.

4. Remove the serpentine drive belt.

5. Remove the water pump assembly.

6. Remove the crankshaft balancer and hub.

   a. If not done already, raise and support the vehicle safely.

   b. Remove the crankshaft balancer retaining bolts, then remove the balancer from the hub.

   c. Matchmark the crankshaft hub to the engine front cover, then remove the hub bolt and washer.

   d. Remove the crankshaft hub using J-39046, or an equivalent hub removal/installation tool. To preserve the relationship between the hub and crankshaft, DO NOT crank the engine over once the hub has been removed. If the hub is not matchmarked and installed in the original position, an engine imbalance could result.

7. Remove the distributor assembly.

8. Remove the oil pan assembly.

9. Remove the engine front cover bolts.

10. Remove the engine front cover and gasket.

To install:

11. Thoroughly clean the engine front cover and cylinder block gasket mating surfaces. Inspect the engine front cover and seals for damage, replace as necessary.

12. Using J-39087 or equivalent front cover seal protector on the water pump driveshaft, install the gasket and front cover into position over the shafts and guide pins.

13. Install the engine front cover bolts and tighten to 100 inch lbs. (11 Nm).

14. Install the oil pan and gasket.

15. Install the distributor assembly.

16. Install the crankshaft hub and torsional damper assembly.

   a. Align the matchmarks made earlier and install the crankshaft hub using the hub tool. If the engine was cranked and the matchmarks were lost, set the engine to No. 1 TDC, then install the crankshaft hub with the cast arrow in the 12 o'clock position.

   b. Install the hub washer and bolt, but do not torque at this time.

   c. Install the crankshaft balancer to the hub, then tighten the crankshaft hub bolt to 75 ft. lbs. (102 Nm) and the balancer bolts to 60 ft. lbs. (81 Nm).

NOTE: If a new balancer is installed, new balancer weights of the same size must be installed in the same hole locations as the original balancer.

17. Install the water pump assembly.

18. Install the serpentine drive belt.

19. Install the throttle body air duct.

20. Properly fill the engine crankcase with clean engine oil.

21. Properly fill the engine cooling system.

22. Connect the negative battery cable, operate the engine and check for leaks.

Timing Chain and Sprockets

REMOVAL AND INSTALLATION

3.4L Engine

1. Remove the timing chain front cover.

2. Slowly rotate the crankshaft until the timing marks punched on the sprockets are aligned with the marks on the engine block or the timing chain damper. This should place the engine in the No. 4 TDC position.

3. Remove the bolts retaining the camshaft sprocket, then remove the camshaft sprocket along with the timing chain.

4. If necessary, use J-23444-A or an equivalent puller, to remove the crankshaft sprocket.

5. If necessary, remove the retaining bolts and the timing chain dampener.

To install:

6. If removed, use J-36812, or an equivalent installer, to draw the crankshaft sprocket into position.

7. Hold the camshaft sprocket vertically with the timing mark positioned as during removal, then place the timing chain over the sprocket. Position the chain under the crankshaft sprocket, then install the camshaft sprocket to the end of the camshaft.

8. Verify that all timing marks are properly aligned, then install the camshaft sprocket retaining bolts and tighten to 18 ft. lbs. (24 Nm).

9. If removed, install the timing chain dampener and tighten the retaining bolts to 15 ft. lbs. (21 Nm).

10. Install the timing chain front cover.

5.7L Engine

1. Remove the timing chain front cover.

2. Rotate the crankshaft until the timing marks on the timing chain sprockets are aligned nearest each other. The camshaft sprocket mark should be at the 6 o'clock position while the mark on the crankshaft sprocket should be at the 12 o'clock position.

3. Remove the camshaft sprocket bolts.

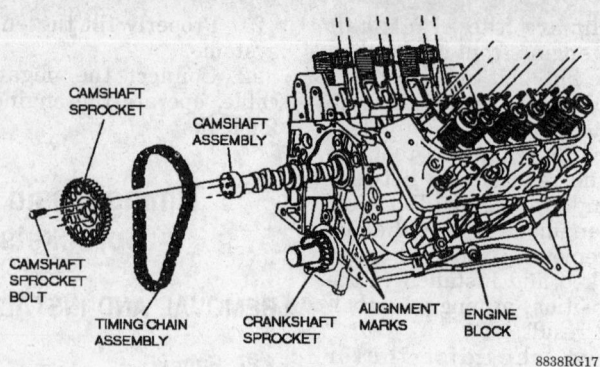

Timing chain, sprocket and camshaft installation — 3.4L engine

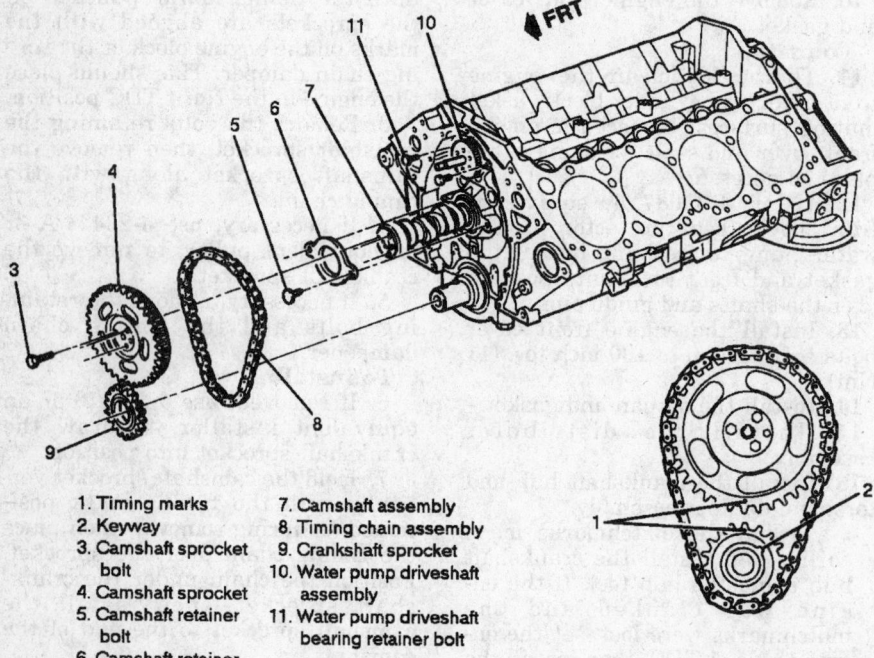

1. Timing marks
2. Keyway
3. Camshaft sprocket bolt
4. Camshaft sprocket
5. Camshaft retainer bolt
6. Camshaft retainer
7. Camshaft assembly
8. Timing chain assembly
9. Crankshaft sprocket
10. Water pump driveshaft assembly
11. Water pump driveshaft bearing retainer bolt

8838RG18

Camshaft, timing chain and sprocket alignment and installation — 5.7L engine

4. Remove the camshaft sprocket and timing chain.

NOTE: To prevent piston or valve damage, do not turn the crankshaft after the timing chain has been removed.

5. Remove the water pump bearing retainer bolts, then remove the driveshaft assembly using J-39243 or equivalent driven gear assembly remover.

6. Remove the crankshaft sprocket using J-5825-A or equivalent crankshaft sprocket remover.

7. If necessary, remove the crankshaft key.

To install:

8. If removed, install the crankshaft key.

9. Install the crankshaft sprocket using J-5590, or an equivalent installation tool.

10. Install the water pump driveshaft assembly using J-39092, or an equivalent installer tool. Install the retainer bolts and tighten to 105 inch lbs. (12 Nm).

11. Align the timing marks and install the camshaft sprocket and timing chain. The gears of the camshaft sprocket and water pump driveshaft must mesh or damage to the thrust plate retainer could occur.

12. Install the camshaft sprocket bolts and tighten to 21 ft. lbs. (28 Nm).

13. Install a new O-ring to the water pump driven gear shaft using J-39089, or an equivalent seal installation tool.

14. Install the timing chain front cover.

Camshaft

REMOVAL AND INSTALLATION

3.4L Engine

1. Disconnect the negative battery cable.

2. Remove the intake manifold assembly.

3. Remove the bolt and clamp retaining the oil pump drive assembly, then lift the drive assembly from the rear of the block.

4. Remove the timing chain front cover, then remove the crankshaft shaft sprocket and the timing chain.

5. Remove the valve lifters.

NOTE: If valve train components are to be reused, make sure they are tagged or arranged in order to assure installation in their original locations.

6. Properly discharge and recover the refrigerant from the air conditioning system.

———— **WARNING** ————

The 4th generation F-body uses only the new R-134a refrigerant. R-134a is not compatible with the R-12 refrigerant formerly used in most automobiles. Service equipment and/or air conditioning system components may be contaminated or damaged if the 2 refrigerants and oils are mixed. Never use a recovery system which was designed for and used to service R-12 equipped vehicles on a vehicle equipped with an R-134a air conditioning system.

7. Remove the air conditioning compressor and condenser hose from the condenser, then remove the receiver and dehydrator hose from the condenser. Plug all of the openings in order to prevent system contamination.

8. Remove the radiator and condenser assembly, then remove the air conditioning condenser support.

9. Install three 5/16-18 x 4 inch bolts or equivalent, in the camshaft bolt holes, then using the bolts to pull and rotate the camshaft, carefully

pull the camshaft from the bearings. All camshaft journals are the same diameter and care must be used to avoid damage to the bearings.

To install:

10. If installing a new camshaft, be sure to coat all camshaft lobes with Molykote® or an equivalent prelube and to replace all lifters in order to assure camshaft durability.

11. Lubricate the camshaft journals with clean engine oil, then carefully insert the camshaft into the engine.

12. Install the condenser support, then install the radiator and condenser assembly.

13. Unplug the openings, then connect the condenser refrigerant lines.

14. Install the valve lifters. If reusing the camshaft and lifters, they must be installed into their original bores.

15. Install the camshaft sprocket and timing chain, then install the timing chain front cover.

16. Lubricate the oil pump drive gear with prelube and the assembly with clean engine oil, then install the oil pump drive assembly and tighten the clamp bolt to 25 ft. lbs. (34 Nm).

17. Install the intake manifold assembly.

18. Connect the negative battery cable, then charge the A/C system with R-134a.

5.7L Engine

1. Disconnect the negative battery cable.

2. Remove the intake manifold assembly.

3. Remove the rocker arms and pushrods.

4. Remove the bolt retaining the oil pump drive assembly, then lift the

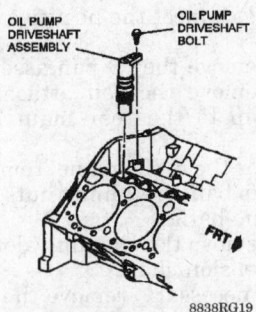

Oil pump driveshaft assembly mounting — 5.7L engine (3.4L engine similar)

drive assembly from the rear of the block.

5. Remove the timing chain front cover, then remove the crankshaft shaft sprocket and timing chain.

6. Remove the valve lifters.

NOTE: If valve train components are to be reused, make sure they are tagged or arranged in order to assure installation in their original locations.

7. Properly discharge and recover the refrigerant from the air conditioning system.

── **WARNING** ──

The 4th generation F-body uses only the new R-134a refrigerant. R-134a is not compatible with the R-12 refrigerant formerly used in most automobiles. Service equipment and/or air conditioning system components may be contaminated or damaged if the 2 refrigerants and oils are mixed. Never use a recovery system which was designed for and used to service R-12 equipped vehicles on a vehicle equipped with an R-134a air conditioning system.

8. Remove the air conditioning compressor and condenser hose from the condenser, then remove the receiver and dehydrator hose from the condenser. Plug all of the openings in order to prevent system contamination.

9. Remove the radiator and condenser assembly, then remove the air conditioning condenser support.

10. Remove the camshaft retainer bolts, then remove the camshaft retainer from the front of the block.

11. Install three ⁵/₁₆-18 x 4 inch bolts or equivalent, in the camshaft bolt holes, then using the bolts to pull and rotate the camshaft, carefully pull the camshaft from the bearings. All camshaft journals are the same diameter and care must be used to avoid damage to the bearings.

To install:

12. If installing a new camshaft, be sure to coat all camshaft lobes with Molykote® or an equivalent prelube and to replace all lifters in order to assure camshaft durability.

13. Lubricate the camshaft journals with clean engine oil, then carefully insert the camshaft into the engine.

14. Install the camshaft retainer and tighten the bolts to 105 inch lbs. (12 Nm).

15. Install the condenser support, then install the radiator and condenser assembly.

16. Unplug the openings, then connect the condenser refrigerant lines.

17. Install the valve lifters. If reusing the camshaft and lifters, they must be installed into their original bores.

18. Install the camshaft sprocket and timing chain, then install the timing chain front cover.

19. Install the oil pump drive assembly and tighten the retaining bolt to 13 ft. lbs. (18 Nm).

20. Install the rocker arms and pushrods.

21. Install the intake manifold assembly.

22. Connect the negative battery cable, then charge the A/C system with R-134a.

Piston and Connecting Rod

POSITIONING

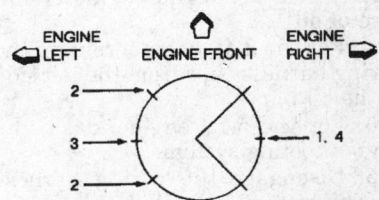

1. Oil ring spacer gap (tang in hole or slot with arc)
2. Oil ring rail gaps
3. 2nd compression ring gap
4. Top compression ring gap

8838RG20

Piston ring gap locations — all engines

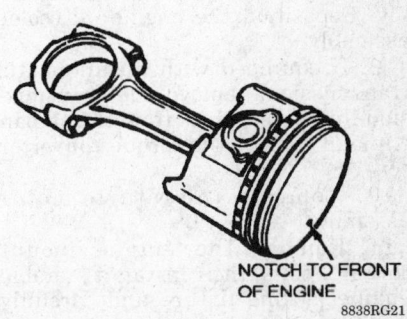

NOTCH TO FRONT OF ENGINE

8838RG21

Piston and rod positioning — 3.4L engine

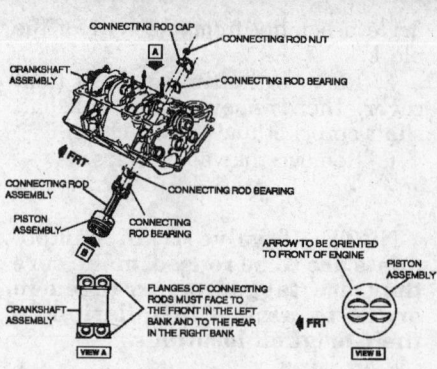

Piston and connecting rod positioning — 5.7L engine

8838RG22

ENGINE LUBRICATION

Oil Pan

REMOVAL AND INSTALLATION

1. Disconnect the negative battery cable.
2. Remove the air intake duct.
3. Raise and support the vehicle safely, then drain the engine crankcase of oil.
4. For the 3.4L engine, remove the wiring harness clips from the left side of the oil pan.
5. For the 5.7L engine, drain the engine cooling system.
6. Disengage the wiring harness connector from the oil level sensor, then remove the sensor.
7. Disconnect the exhaust crossover pipe from the exhaust manifolds, then remove the pipe hanger bolts and reposition the pipe.
8. For the 5.7L engine, remove the engine oil cooler hose bracket nut from the oil pan, then remove the oil cooler bolt from the oil cooler assembly. Reposition the engine oil cooler assembly.
9. If equipped with an automatic transmission, remove the transmission fluid cooler lines from the oil pan clip and remove the torque converter cover.
10. Remove the start motor assembly.
11. Remove the engine mount through-bolts, then install a suitable engine jacking fixture and carefully raise the engine, watching the clearance between engine mounted components and the firewall.

WARNING

Never raise the engine using a jack under the oil pan, crankshaft pulley or any sheetmetal. Because there only is a small clearance between the oil pan and the oil pump screen, if the pan is bent even slightly, damage could occur to the pump screen and pickup unit.

12. Remove the oil pan bolts, studs and nuts, then lower the oil pan assembly along with the gasket and the pan reinforcements. If necessary, rotate the crankshaft to reposition the counterweights.

To install:

13. Thoroughly clean the gasket mating surfaces of any remaining sealer and/or gasket material.
14. For the 3.4L engine, apply a small amount of sealer where the oil pan gasket tabs seat on the rear crankshaft bearing cap groove.
15. For the 5.7L engine, apply a small amount of RTV sealer, 1052914 or equivalent, to the front cover-cylinder block junction and to the rear seal retainer-cylinder block junction. Continue the bead of sealer for 1 inch (25mm) in either direction of the radius cavity of these junctions.
16. Install the oil pan and gasket assembly, using the reinforcements, nuts, bolts and studs.
17. For the 3.4L engine, tighten the front corner oil pan nuts to 24 ft. lbs. (33 Nm), the rear corner oil pan bolts to 18 ft. lbs. (25 Nm), then oil pan side rail bolts to 89 inch lbs. (10 Nm) and the oil pan studs to the 53 inch lbs. (6 Nm).
18. For the 5.7L engine, tighten corner oil pan bolts, stud or nuts to 15 ft. lbs. (20 Nm) and the remaining bolts or studs to 100 inch lbs. (11 Nm).
19. Lower the engine, remove the jacking fixture and install the engine mount through-bolts.

20. Install the oil level sensor assembly and tighten to 16 ft. lbs. (22 Nm), then engage the wiring harness connector.
21. Install the starter motor assembly.
22. If equipped with an automatic transmission, install the torque converter cover, then secure the transmission fluid cooler lines with the oil pan clip.
23. For the 5.7L engine, install the engine oil cooler bolts screw to the cooler assembly and tighten to 24 ft. lbs. (33 Nm), then install the oil cooler hose bracket nut to the oil pan.
24. Install the exhaust pipe hanger bolt, then connect the crossover pipe to the exhaust manifolds.
25. For the 3.4L engine, connect the wiring harness clips to the left side of the oil pan.
26. Lower the vehicle and refill the engine crankcase with clean engine oil.
27. Install the air intake duct and connect the negative battery cable.
28. Start the engine and check for leaks.

Oil Pump

REMOVAL AND INSTALLATION

1. Disconnect the negative battery cable.
2. Remove the oil pan assembly.
3. Remove the bolt attaching the oil pump to the rear main bearing cap.
4. For the 5.7L engine, remove the oil pump baffle retaining nuts and remove the baffle.
5. Remove the oil pump along with the extension shaft.
6. If necessary, remove the pickup screen and pipe as an assembly by placing the pump is a soft-jawed vice and extracting the pipe from the pump.

To install:

NOTE: The oil pump pickup should be submerged in oil and the pump primed prior to installation. Failure to prime the pump may result in oil pump failure or internal engine damage. Also, if the pickup screen and pipe assembly was removed from the pump, they must be replaced to assure a proper interference fit.

7. If the pickup screen and pipe were removed, it should be replaced as an assembly with a new part. Using a suitable tool, install a new pickup screen and pipe to the oil pump

8. Align the slot or hexagon head on the end of the shaft extension with the drive tang or the hexagon socket on the distributor shaft.

9. For the 5.7L engine, position the oil pump baffle before installing the pump retaining bolt.

10. Install and tighten the oil pump bolt to 30 ft. lbs. (41 Nm) on 3.4L engine or to 65 ft. lbs. (88 Nm) on 5.7L engine. For the 5.7L engine, install the baffle nuts and tighten to 25 ft. lbs. (34 Nm).

11. Install the oil pan assembly.

12. Lower the vehicle and refill the crankcase with clean engine oil.

13. Connect the negative battery cable.

14. Start the engine and check oil pressure.

TRANSMISSION

Manual Transmission Assembly

REMOVAL AND INSTALLATION

1. Disconnect the negative battery cable.

2. Remove the retainers, then unsnap the front floor console trimplate assembly.

3. For 6-speed vehicles, remove the transmission control lever handle bolts, then remove the handle assembly. Disconnect the clutch pedal pushrod in order to protect the slave cylinder during the procedure.

4. Raise and support the vehicle safely.

5. Drain the transmission fluid.

6. Matchmark and remove the driveshaft assembly.

7. Support the rear axle assembly using a jackstand, then remove the rear axle torque arm.

8. Remove the catalytic converter hanger assembly.

9. Disengage the wiring harness connector from the reverse lockout solenoid (6-speeds only) and from the backup lamp switch.

NOTE: Whenever the clutch slave cylinder is disconnected, the actuator pushrod must first be disconnected. If this is not done, the slave cylinder may become permanently damaged if the clutch pedal is depressed while the cylinder is disconnected.

10. For 6-speed vehicles, remove the actuator cylinder nuts. Remove the actuator cylinder and support aside using a length of mechanic's wire. Remove the actuator spacer, then pull the clutch fork downward to disengage from the release bearing.

11. Disengage the speed sensor harness connector.

12. Support the engine assembly with a jackstand, then position a transmission jack under the transmission.

13. Remove the transmission support and mount.

14. For 5-speed vehicles, lower the transmission sufficiently to reach the bolts on top of the housing which attach the transmission control assembly, then remove the bolts.

15. Remove the transmission to clutch housing retaining bolts, then with the aid of an assistant carefully lower the transmission from the vehicle. For 5-speed vehicles, the transmission control assembly must be held in position while lower the transmission. If this is not done, the control assembly could drop, causing damage to the component.

To install:

16. For 5-speed vehicles, thoroughly clean the sealant from the transmission extension housing and the control assembly. Support the control assembly to the installed position in the vehicle.

17. Raise the transmission into position, piloting the input shaft into the release bearing, clutch disc and pilot bearing, then support the transmission with a jackstand.

18. Install the transmission retaining bolts and tighten to 55 ft. lbs. (75 Nm) for 5-speed vehicles or to 26 ft. lbs. (35 Nm) for 6-speed vehicles.

19. For 5-speed vehicles, apply a continuous 1/8 in. (3mm) bead of RTV sealant around the housing-to-control assembly sealing surface. Raise

the transmission, engaging the control assembly, then install the bolts and tighten to 13 ft. lbs. (17 Nm).

20. Install the transmission mount and the support. Tighten the mount bolts to 40 ft. lbs. (54 Nm), then tighten the support bolts to 43 ft. lbs. (58 Nm) and the support nuts to 35 ft. lbs. (47 Nm).

21. Remove the transmission and engine jackstands or supports.

22. For 6-speed vehicles, push the clutch fork assembly upward to engage the release bearing, then install the actuator spacer, actuator cylinder and retaining nuts. Tighten the nuts to 15 ft. lbs. (20 Nm).

23. Engage the wiring harness connectors to the speed sensor assembly, backup lamp switch, and on 6-speed vehicles, to the reverse lockout solenoid.

24. Install the catalytic converter hanger assembly.

25. Install the rear axle torque arm, then remove the support from the axle.

26. Align and install the driveshaft assembly.

27. Fill the transmission assembly using Dexron®II or equivalent transmission fluid.

28. Lower the vehicle.

29. For 6-speed vehicles, install the control lever handle and tighten the retaining bolts to 18 ft. lbs. (25 Nm), then connect the clutch actuator pushrod.

30. Install the front floor console trimplate, then connect the negative battery cable.

LINKAGE ADJUSTMENT

The M49 5-speed, M28 and M29 6-speed manual transmissions fare designed with an internal shift mechanism. Shifter control adjustments are not necessary or possible.

Clutch Assembly

REMOVAL AND INSTALLATION

1. Disconnect the negative battery cable.

2. Remove the instrument panel driver knee bolster for access, then disconnect the clutch master cylinder pushrod from the clutch pedal.

3. Remove the transmission assembly.

4. For 5-speed vehicles:
   a. Remove the clutch slave cylinder actuator nuts, then support the cylinder from the vehicle using a length of mechanic's wire. Do not

allow the assembly to hang by the hydraulic line or damage could occur.

b. Remove the transmission brace nut and bolt, then remove the brace.

5. Remove the retaining bolts and the flywheel housing cover.

6. Remove the flywheel housing retaining bolts, then remove the housing from the rear of the engine.

7. Slowly and evenly loosen, then remove, the clutch pressure plate/cover bolts, then remove the pressure plate assembly along with the clutch driven disc.

To install:

8. Inspect flywheel for heat stress, cracks or other defects and repair or replace as necessary.

9. Inspect the clutch plate, disc, release bearing and fork for contamination, wear or heat stress, repair or replace as necessary.

10. Position the clutch driven disc to the flywheel, then install the pressure plate/cover assembly and finger-tighten the retaining screws.

11. Align the clutch disc with the pilot bearing and clutch pressure plate assembly using a suitable clutch alignment arbor such as J-33169 for the 5-speeds or J-38836 for 6-speeds.

12. When properly aligned, tighten the clutch pressure plate bolts to 15 ft. lbs. (20 Nm) +30 degrees of additional rotation for 5-speeds or to 22 ft. lbs. (30 Nm) for 6-speeds.

13. Install the flywheel housing assembly and tighten the retaining bolts to 75 inch lbs. (8.5 Nm).

14. For 5-speed vehicles:

a. Install the transmission brace, then tighten the retaining nut and bolt to 37 ft. lbs. (50 Nm).

b. Remove the support wire and install the slave cylinder, then tighten the retaining nuts to 15 ft. lbs. (10 Nm).

15. Install the transmission assembly.

16. Engage the clutch master cylinder pushrod to the clutch pedal assembly.

17. Install the driver side knee bolster.

18. Connect the negative battery cable.

PEDAL HEIGHT/FREE-PLAY ADJUSTMENT

The hydraulic clutch system locates the clutch pedal height and provides automatic clutch adjustment. No adjustment of clutch linkage or pedal position is required.

Clutch Hydraulic Assembly

The vehicles use a hydraulic system which is serviced only as an entire assembly.

REMOVAL AND INSTALLATION

1. Disconnect the negative battery cable.

2. Remove the instrument panel driver knee bolster for access, then disconnect the clutch master cylinder pushrod from the clutch pedal.

3. Remove the clutch master cylinder retaining nuts, then remove the master cylinder U-bolt and clutch master cylinder from the dash panel assembly.

4. Remove the clutch master cylinder reservoir retainer, then remove the reservoir from the left hood strut bracket.

5. Raise and support the vehicle safely.

6. Remove the clutch slave cylinder nuts, then remove the cylinder from the clutch housing.

7. Remove the clutch hydraulic system from the vehicle.

To install:

8. Install the clutch slave cylinder to the clutch housing, then tighten the retaining nuts to 15 ft. lbs. (20 Nm).

9. Lower the vehicle, then position the clutch master cylinder to the dash panel. Install the U-bolt and the master cylinder nuts, then tighten the nuts to 20 ft. lbs. (27 Nm).

10. Install the master cylinder reservoir to the left hood strut bracket, then install the reservoir retainer.

11. Engage the clutch master cylinder pushrod to the clutch pedal assembly.

12. Install the instrument panel driver knee bolster.

13. Connect the negative battery cable.

Hydraulic Clutch Bleeding

When bleeding the hydraulic clutch system always keep the reservoir filled with fresh clean brake fluid. NEVER use fluid which has been bled from a system as it may contain moisture, air or other contaminants.

PROCEDURE

The sealed hydraulic system used on these vehicles should not require bleeding, unless the fluid level in the reservoir has fallen so low that air is drawn into the clutch master cylinder.

1. Loosen the clutch master cylinder retaining nuts to the ends of the U-bolt threads.

2. Clean all dirt and grease from the cap to make sure no foreign substances enter the system.

3. Remove the cap and diaphragm, then fill the reservoir with the approved DOT 3 brake fluid.

4. Wrap a piece of wire around the left hood strut bracket, making sure the wire is accessible from the underside of the vehicle. The wire will be used to support the slave cylinder once it is removed from the clutch housing.

5. Raise and support the vehicle safely.

6. Remove the clutch slave cylinder retaining nuts, then remove the slave cylinder and support it using the mechanic's wire.

7. Lower the vehicle.

8. Grasp the slave cylinder and depress the cylinder pushrod approximately 0.787 in. (20.0mm) into the slave cylinder bore and hold.

9. With the pushrod held into the slave cylinder bore, install the reservoir diaphragm and cap, then release the pushrod.

10. Hold the slave cylinder vertically with the pushrod end facing downward. The slave cylinder must be held lower than the master cylinder. Press the pushrod into the slave cylinder bore with short 0.39 in. (10mm) strokes, while watching the reservoir for air bubbles.

11. Continue depressing the pushrod with short strokes until air bubbles are no longer seen entering the reservoir.

12. Raise and support the vehicle safely.

13. Remove the slave cylinder from the mechanic's wire, then install the slave cylinder and tighten the retaining nuts to 15 ft. lbs. (20 Nm), then lower the vehicle.

14. Remove the mechanic's wire from the strut bracket.

15. Install the clutch master cylinder nuts and tighten to 20 ft. lbs. (27 Nm).

16. Check the fluid level in the clutch master cylinder reservoir and add, if necessary.

17. Check the clutch pedal for proper operation and, rebleed, as necessary.

Automatic Transmission Assembly

REMOVAL AND INSTALLATION

1. Disconnect the negative battery cable.

2. Disconnect the transmission Throttle Valve (TV) cable assembly from the throttle lever.

3. Raise and support the vehicle safely.

4. Matchmark the alignment of the shaft to the axle yoke, then remove the driveshaft assembly.

5. Support the transmission assembly, then remove the nut and washer attaching the support to the transmission.

6. Remove the support bolts, then remove the support from the vehicle.

7. Remove the rear axle torque arm.

8. Remove the catalytic converter hanger assembly.

9. Remove the torque converter cover bolts, then remove the cover.

10. Remove the bolts attaching the converter to the flywheel, then disengage the flywheel and converter.

11. Move the catalytic converter heat shield, as necessary for access.

12. Disconnect the automatic transmission range selector lever cable assembly.

13. Disconnect the transmission fluid cooler pipe assemblies from the transmission oil cooler pipe clips.

14. Disengage all electrical connectors from the transmission assembly.

15. Remove the bolts attaching the transmission to the engine.

16. Remove the transmission fluid level indicator assembly, then separate the transmission from the engine block.

17. Disconnect the TV cable assembly from the transmission and remove the fluid cooler lines. Plug all openings to prevent system contamination or excessive fluid loss.

18. Carefully lower the transmission from the vehicle.

To install:

19. Raise the transmission assembly into position in the vehicle, then unplug the openings and install the fluid cooler lines.

20. Connect the TV cable assembly to the transmission assembly.

21. Align the transmission with the engine block, then install the fluid level indicator assembly.

22. Install the transmission retaining bolts and tighten to 70 ft. lbs. (95 Nm) for the 3.4L engine or to 35 ft. lbs. (47 Nm) for the 5.7L engine.

23. Engage all electrical wiring harness connectors to the transmission assembly.

24. Install the fluid cooler pipe and hose assemblies to the pipe clips.

25. Connect the transmission range selector lever cable assembly.

26. Reposition the catalytic converter heat shield.

27. Position the flywheel to the torque converter, then install the bolts and tighten to 46 ft. lbs. (63 Nm).

28. Install the torque converter cover to the transmission, then tighten the retaining bolts to 89 inch lbs. (10 Nm).

29. Install the catalytic converter hanger assembly.

30. Install the rear axle torque arm.

31. Install the support to the transmission and the rail assembly. Install the support retaining screws and tighten, then install the washer and nut to the transmission. Tighten the nut to 35 ft. lbs. (47 Nm).

32. Remove the transmission jack or adjustable support from the transmission.

33. Align and install the driveshaft assembly.

34. Lower the vehicle and connect the TV cable assembly to the throttle lever.

35. Connect the negative battery cable.

Shift Control Cable

ADJUSTMENT

1. Position the floor shifter in **N**, then raise and support the vehicle safely.

2. Unlock the cable adjustment button by turning clockwise, then separate the shift control cable from the transmission shift lever. The cable may remain in the bracket.

3. Rotate the shift lever clockwise to the **P** detent, then rotate the lever counterclockwise back through **R** to the **N** detent.

4. With the adjustment button still in the unlocked position, connect the cable to the lever.

5. Lock the cable adjustment button by turning counterclockwise.

6. Lower the vehicle and check cable adjustment.

Throttle Valve Cable

ADJUSTMENT

NOTE: Setting of the TV cable must be done by rotating the throttle lever at the throttle body. Do not use the accelerator pedal to rotate the throttle body lever.

1. Ensure the engine is **OFF**.

2. Depress and hold-down the metal reset tab at the engine end of the TV cable.

3. Move the slider until it stops against the fitting.

4. Release the reset tab.

5. Rotate the throttle lever to it's full travel position.

6. The slider must move (ratchet) toward the throttle when the lever is rotated to it's full travel position.

7. Ensure the cable moves freely. The cable may appear to function properly with the engine stopped and cold. Recheck after the engine is hot.

DRIVE AXLE

Driveshaft and U-Joints

REMOVAL AND INSTALLATION

1. Raise and support the vehicle safely.

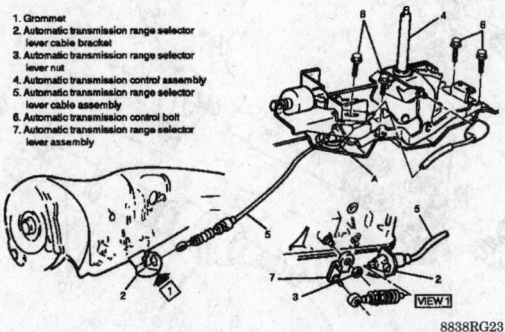

1. Grommet
2. Automatic transmission range selector lever cable bracket
3. Automatic transmission range selector lever nut
4. Automatic transmission control assembly
5. Automatic transmission range selector lever cable assembly
6. Automatic transmission control bolt
7. Automatic transmission range selector lever assembly

8838RG23

Shift control cable adjustment — automatic transmission vehicles

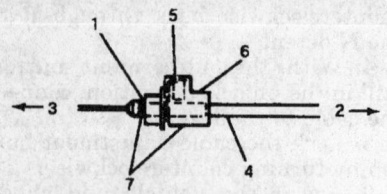

1. Slider against fitting (zero or reset position)
2. Reset direction
3. Direction of cable actuating lever
4. Automatic transmission TV cable assembly
5. Automatic transmission TV cable reset tab
6. Automatic transmission TV cable fitting
7. Automatic transmission TV cable slider

8838RG24

Throttle Valve (TV) cable adjustment

2. Matchmark the rear axle pinion flange to the driveshaft assembly for installation purposes.

3. If removing a 2-piece driveshaft, remove the bolts from the center support bearing, then remove the center support bearing and washers from the torque arm assembly.

4. Remove the 4 driveshaft strap bolts at the rear of the shaft assembly, then remove the 2 retaining straps.

5. Lower the driveshaft down slightly at the rear, then carefully pull the shaft backwards and out from the transmission housing. The transmission housing should be plugged to prevent leakage. If the bearing caps are loose, tape them together to prevent dropping and losing the bearing rollers.

6. If necessary, separate the 2-piece driveshaft assembly:

   a. Position the shaft assembly horizontally in a soft-jawed vice.

   b. Screw a suitable axle shaft remover (such as J-33008) onto a slide hammer, then position the axle remover on the shaft joint between the joint bell area and the shaft assembly.

   c. Use the slide hammer to drive the driveshaft halves apart.

   d. Remove and discard the old shaft seal from the front of the rear shaft half.

7. The U-joints will either be the nylon injected ring type (production) or the snapring type (replacement). To replace the U-joints proceed as applicable:

   a. If equipped with the snapring type, remove the snaprings from the U-joints. If a snapring does not readily come out, tap the end of the bearing cap lightly to relieve the pressure against the snapring.

   b. Support the driveshaft horizontally in line with the base plate of a press, but do not clamp the tube.

   c. Support the lower ear of the universal joint with a 1 1/8 inch socket.

   d. Remove the lower bearing cap out of the yoke ear by placing tool J-9522-3 or equivalent U-joint bearing separator, on the open horizontal bearing caps and pressing the lower bearing cap out of the yoke ear.

NOTE: If the U-joint is a nylon injected ring type, this will shear the nylon injector ring on the lower bearing cap. There are no bearing retainer grooves in the production bearing caps, therefore they cannot be reused. Replace nylon injected ring U-joints with external snapring type U-joints.

   e. If the bearing cap is not completely removed, lift tool J-9522-3 and insert tool J-9522-5 or equivalent between the bearing cap and seal, then continue pressing the U-joint out of the yoke.

   f. Repeat the procedure for the opposite side.

   g. Remove the spider from the yoke.

To install:

8. When replacing U-joints always replace the entire assembly consisting of 1 pregreased spider, 4 bearing cap assemblies with seals, needle roller bearings, round and flat derlin washers, grease and 4 snaprings. Make sure the seals are in place on the bearing caps to hold the needle roller bearings in position during service. Replace U-joints as follows:

   a. Install 1 bearing cap part way into 1 side of the yoke. Turn this yoke ear to the bottom.

   b. Using tool J-9522-3 or equivalent, seat the trunnion into the bearing cap.

   c. Install the opposite bearing cap partially onto the trunnion.

   d. Ensure that both trunnions are straight and true in the bearing caps.

   e. Press against the opposite bearing caps, while working the spider back and forth to ensure free movement of the trunnions in the bearings.

   f. If the trunnion is binding, one or more of the needle bearings may have tipped under the end of the trunnion, stop pressing and correct the situation.

   g. Stop pressing when 1 bearing cap clears the retainer groove inside the yoke.

   h. Install a snapring by pressing it into place.

A. One-piece propeller shaft assembly
B. Two-piece propeller shaft assembly
C. Trunnions
D. Spline coupling
E. Propeller shaft slip yoke assembly damper
1. Propeller shaft damper and slip
2. Propeller shaft universal joint spider
3. Propeller shaft universal joint spider bearing
4. Propeller shaft universal joint spider bearing retainer ring

5. Propeller shaft assembly
6. Front propeller shaft assembly
7. Propeller shaft center support bearing
8. Rear propeller shaft seal
9. Propeller shaft joint
10. Rear propeller shaft assembly
11. Propeller shaft front slip yoke assembly

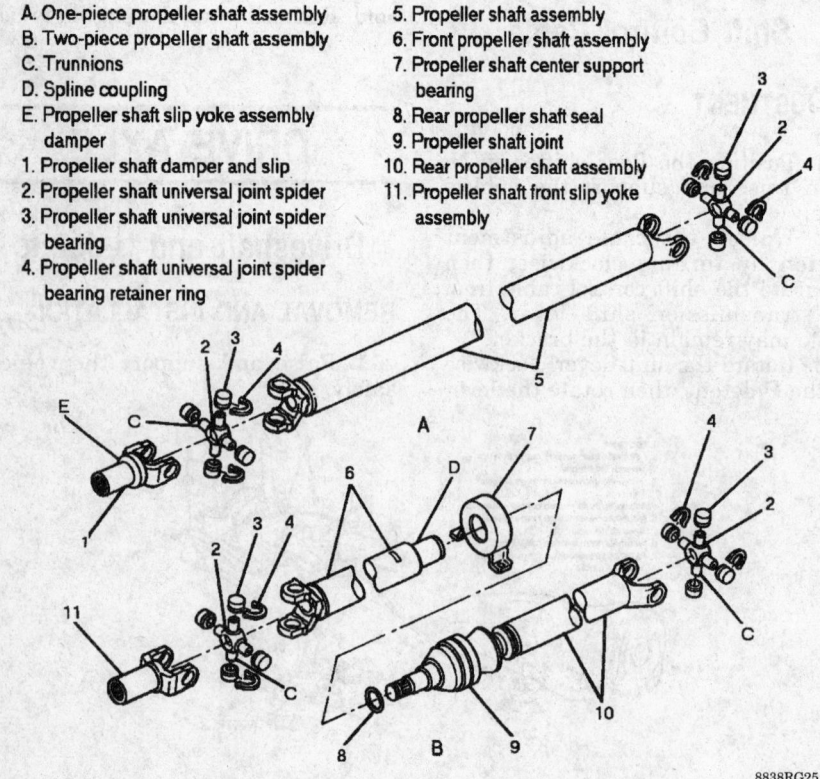

8838RG25

Exploded view of the 1 and 2-piece driveshaft assemblies

i. Repeat the procedure for the remaining bearing caps and U-joint.

9. If assembling and installing a 2-piece driveshaft:

a. Position a new shaft seal to the front of the rear half of the driveshaft assembly.

b. Align the 2 halves of the shaft assembly and push them together by hand until the snapring on the shaft joint engages in the shaft assembly. If the shafts can still be separated by hand, then snapring is not engaged, realign the shafts and try again.

10. Lubricate the spline using slip yoke lubricant , then install the slip yoke into the transmission.

11. Align the marks and install the rear of the driveshaft with the rear U-joint to the pinion yoke making sure the bearing caps are properly seated.

12. Install the retaining straps and tighten the bolts evenly to 16 ft. lbs. (22 Nm).

13. If installing a 2-piece driveshaft assembly, install the washers and the center support bearing, then install the retaining bolts and tighten to 37 ft. lbs. (50 Nm).

14. Lower the vehicle.

STEERING

Air Bag

--------------- CAUTION ---------------
Some vehicles are equipped with an air bag system, also known as the Supplemental Inflatable Restraint (SIR) or Supplemental Restraint System (SRS). The system must be disabled before performing service on or around system components, steering column, instrument panel components, wiring and sensors. Failure to follow safety and disabling procedures could result in accidental air bag deployment, possible personal injury and unnecessary system repairs.

PRECAUTIONS

Several precautions must be observed when handling the inflator module to avoid accidental deployment and possible personal injury.

• Never carry the inflator module by the wires or connector on the underside of the module.

• When carrying a live inflator module, hold securely with both hands, and ensure that the bag and trim cover are pointed away.

• Place the inflator module on a bench or other surface with the bag and trim cover facing up.

• With the inflator module on the bench, never place anything on or close to the module which may be thrown in the event of an accidental deployment.

DISARMING

1. Turn the steering wheel to align the wheels in the straight-ahead position.

2. Turn the ignition switch to the **LOCK** position.

3. Remove the SIR or AIR BAG fuse from the fuse block.

4. Remove the left side trim panel, then remove the Connector Position Assurance (CPA) device and disengage the yellow 2-way SIR harness wire connector at the base of the steering column.

5. For vehicles equipped with a passenger side inflator module, remove the glove box door, then remove the CPA device and disengage the yellow 24-way Diagnostic Energy Reserve Module (DERM) connector.

ENABLING

1. If not already done, turn the ignition switch to the **LOCK** position.

2. For vehicles equipped with a passenger side inflator module, engage the 24-way DERM connector and install the CPA device, then install the glove box door.

3. Engage the yellow 2-way connector and install the CPA device at the base of the steering column, then install the left side trim panel.

4. Install the SIR or AIR BAG fuse to the instrument panel fuse block.

5. Turn the ignition switch to the **RUN** position.

6. Verify the SIR indicator light flashes 7–9 times and then turns OFF to signify proper system operation. If light does not flash as specified, inspect system for malfunction.

Supplemental Inflatable Restraint (SIR) Coil

NOTE: After performing repairs on the internals of the steering column, the coil assembly must be centered in order to avoid coil damage or accidental air bag deployment.

ADJUSTMENT

1. Hold the coil assembly with the clear bottom up to see the coil ribbon.

2. While holding the coil assembly, depress the spring lock and rotate the hub in the direction of the arrow until it stops. The coil ribbon should now be wound snug against the center hub.

3. Rotate the coil assembly in the opposite direction approximately 2½ turns and release the lock spring between the locking tabs in front of the arrow.

4. Install the coil assembly onto the steering shaft.

Steering Wheel

REMOVAL AND INSTALLATION

--------------- CAUTION ---------------
The vehicle is equipped with a SIR air bag system, follow the recommended disarming procedures before performing any work on or around the system. Failure to do so may result in possible deployment of the air bag and/or personal injury.

1. Disable the Supplemental Inflatable Restraint (SIR) system.

2. Disconnect the negative battery cable.

3. Loosen the screws using a suitable Torx® driver, then remove the inflator module from the steering wheel.

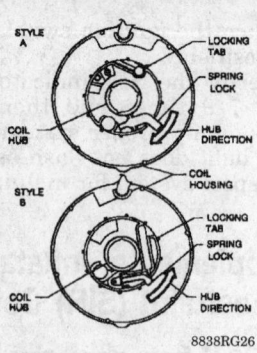

8838RG26

Centering the SIR steering
column coil assembly

CAUTION

When carrying a live inflator module, ensure the bag and trim cover are pointed away from the body. Never carry the inflator module by the wires or connector on the underside of the module. This will minimize the chance of injury should the module accidentally deploy. When placing a live inflator module on a bench or other surface, always place the bag and trim cover up, away from the surface. This is necessary so a free space is provided to allow for air bag expansion in the unlikely event of accidental deployment.

4. Disengage the coil assembly connector and Connector Position Assurance (CPA) device or clip from the inflator module terminal.

5. Remove the steering wheel locking nut.

6. Using a suitable puller, remove the steering wheel and disconnect the horn contact. When attaching the steering wheel puller, use care to prevent threading the side screws all the way through the hub, into the coil, thereby damaging the SIR coil assembly.

To install:

7. Route the coil assembly connector through the steering wheel.

8. Connect the horn contact and install the steering wheel. When installing the steering wheel, align the block tooth on the steering wheel with the block tooth on the steering shaft within 1 female serration.

9. Install the steering wheel locking nut and tighten the nut to 32 ft. lbs. (43 Nm).

10. Connect the coil assembly connector and CPA device or clip to the inflator module terminal. Secure the coil assembly connector to the steering wheel by inserting the thick section of wire into the existing retainers.

11. Install the inflator module. Ensure the wiring is not exposed or trapped between the inflator module and the steering wheel. Tighten the inflator module screws to 25 inch lbs. (2.8 Nm).

12. Connect then negative battery cable and enable the SIR system.

Tie Rod Ends

REMOVAL AND INSTALLATION

1. Disconnect battery ground cable.

2. Raise and support the vehicle safely.

3. Remove the tie rod cotter pin and hex slotted nut from the tie rod assembly.

4. Loosen tie rod jam nut.

5. Using tool J-24319-01, or an equivalent linkage puller, remove the tie rod from the steering knuckle.

6. Remove the tie rod from the steering rack assembly. To ease installation and give a point from which the alignment may be adjusted, scribe alignment marks on the rack assembly prior to tie rod end removal, and/or count the number of turns necessary to remove the tie rod end.

To install:

7. Install the tie rod to the steering rack assembly, but do not tighten the jam nut. Thread the tie rod end in the same number of turns and/or align it to the marks made during removal.

8. Install the tie rod to the steering knuckle and install the hex slotted nut to the tie rod stud.

9. Tighten the hex nut to 35 ft. lbs. (47 Nm), then insert a new cotter pin. If necessary tighten the nut additionally in order to insert the pin, but do not exceed a total torque of 52 ft. lbs. (70 Nm) and do not back off the original torque.

10. Check and adjust toe, as necessary.

NOTE: Make sure the rack and pinion boot is not twisted or puckered during installation or adjustment.

11. Adjust the toe by turning the inner tie rod, making sure the rack and pinion boot is not twisted or puckered during toe adjustment.

12. Tighten the jam nut against the tie rod to 50 ft. lbs. (68 Nm).

Power Rack and Pinion

ADJUSTMENT

1. Raise and safely support the vehicle so the front wheels are raised and the steering wheel is centered.

2. Loosen the adjuster plug locknut.

3. Turn the adjuster plug clockwise until it bottoms, then back off 50–70 degrees (approximately 1 flat).

4. Keep the adjuster plug from turning and tighten the locknut to 55 ft. lbs. (75 Nm).

5. Inspect the steering wheel returnability to center after adjustment.

REMOVAL AND INSTALLATION

1. Disconnect the negative battery cable, then raise and support the vehicle safely.

2. Remove the tire and wheel assemblies.

3. Position a drain pan under the rack assembly, then disconnect the inlet and outlet hoses from the gear assembly. Plug all openings to prevent system contamination or excessive fluid loss.

4. Disconnect the steering linkage outer tie rods from the steering knuckles.

5. Disconnect the steering rack coupling shaft assembly from the rack.

6. Remove the nuts and bolts retaining the rack to the vehicle, then remove the rack assembly.

To install:

7. Position the rack to the crossmember, then adjust the rack so it aligns as straight as possible with the rack coupling shaft assembly. Hand-start the bolts and nuts, the using a backup wrench on the nuts, tighten the retainers to 63 ft. lbs. (85 Nm).

8. Connect the steering rack coupling shaft assembly to the rack and tighten the retaining bolt to 35 ft. lbs. (47 Nm).

9. Install the steering linkage outer tie rods to the knuckles.

10. Remove the plugs, then connect the inlet and outlet hoses to the rack assembly. Make sure the hoses are properly routed and not interfering with any other components, then tighten the rack fittings to 21 ft. lbs. (28 Nm).

11. Install the tire and wheel assemblies, then lower the vehicle.

12. Connect the negative battery cable, then refill and bleed the power steering system.

Power Steering Pump

BLEEDING

1. With the engine **OFF**, raise and safely support the vehicle sufficiently high to hold the wheels off the ground.

2. Turn the wheels all the way to the left, then add power steering fluid to the FULL COLD or C mark on the fluid level indicator.

3. Bleed the system by turning the wheels from side to side without hitting the stops. It may be necessary to continue this several times in order to bleed air from the system.

4. Start the engine and allow it to idle, then recheck the fluid level. If necessary, as fluid to bring the level to the FULL COLD or C mark.

5. Return the wheels to the center position and keep the engine running for 2–3 minutes.

6. Make sure the steering functions normally and is free from noise.

7. Road test the vehicle and recheck the fluid level ensuring the level is up to the HOT or H mark.

REMOVAL AND INSTALLATION

3.4L Engine

1. Position a drain pan under the pump assembly.

2. Remove the serpentine drive belt.

3. Remove the front air intake duct.

4. Disconnect the inlet and outlet hose assemblies from the power steering pump. Plug the openings to prevent system contamination or excessive fluid loss.

5. Remove the pulley from the pump using a suitable puller.

6. Remove the pump support from the assembly, then remove the pump retaining bolts and remove the assembly from the vehicle.

To install:

7. Position the pump assembly to the brace, then install the retaining bolts and tighten to 37 ft. lbs. (50 Nm).

8. Install the pump support.

9. Install the pump pulley using a suitable installer tool. Make sure the face of the pulley hub is flush with the end of the pump shaft before a load is applied to the hub.

10. Remove the plugs, then connect the inlet and outlet hose assemblies to the pump. Tighten the hose fittings to 21 ft. lbs. (28 Nm).

11. Install the front air intake duct, then install the serpentine drive belt.

12. Refill and bleed the power steering system.

5.7L Engine

1. Disconnect the negative battery cable, then raise and support the vehicle safely.

2. Drain the engine cooling system and remove the serpentine drive belt, then lower the vehicle.

3. Position a drain pan under the pump assembly, then remove the air intake duct.

4. Remove the alternator assembly.

5. Disconnect the radiator outlet hose from the engine.

6. Disconnect the heater inlet and outlet hose assemblies from the water pump.

7. Remove the throttle body heater return hose from the heater outlet hose assembly.

8. Remove the power steering pump assembly retaining bolts.

9. Disconnect the inlet and outlet hose assemblies from the power steering pump. Plug the openings to prevent system contamination or excessive fluid loss.

10. Remove the pump assembly from the vehicle.

To install:

11. Position the pump in the vehicle and connect the hose assemblies. Tighten the hose fittings to 21 ft. lbs. (28 Nm).

12. Install the pump assembly to the bracket and tighten the retaining bolts to 18 ft. lbs. (25 Nm).

13. Connect the throttle body heater return hose to the heater outlet.

14. Connect the heater inlet and outlet hose assemblies to the water pump.

15. Connect the radiator outlet hose to the engine assembly.

16. Install the alternator assembly.

17. Raise and support the vehicle safely, then install the serpentine drive belt.

18. Lower the vehicle and connect the negative battery cable.

19. Refill and bleed the engine cooling system.

20. Refill and bleed the power steering system.

BRAKES

Anti-Lock Brake System Service

PRECAUTIONS

• Certain components within the Anti-Lock Brake System (ABS) are not intended to be serviced or repaired individually. Only those components with removal and installation procedures should be serviced.

• Do not use rubber hoses or other parts not specifically specified for and ABS system. When using repair kits, replace all parts included in the kit. Partial or incorrect repair may lead to functional problems and require the replacement of components.

• Lubricate rubber parts with clean, fresh brake fluid to ease assembly. Do not use lubricated shop air to clean parts; damage to rubber components may result.

• Use only specified brake fluid from an unopened container.

• If any hydraulic component or line is removed or replaced, it may be necessary to bleed the entire system.

• A clean repair area is essential. Always clean the reservoir and cap thoroughly before removing the cap. The slightest amount of dirt in the fluid may plug an orifice and impair the system function. Perform repairs after components have been thoroughly cleaned; use only denatured alcohol to clean components. Do not allow ABS components to come into contact with any substance containing mineral oil; this includes used shop rags.

• The Anti-Lock control unit is a microprocessor similar to other computer units in the vehicle. Ensure that the ignition switch is **OFF** before removing or installing controller harnesses. Avoid static electricity discharge at or near the controller.

• If any arc welding is to be done on the vehicle, the control unit should be unplugged before welding operations begin.

Master Cylinder

REMOVAL AND INSTALLATION

1. Drain the brake fluid from the reservoir.

2. Disconnect the 2 brake lines from the master cylinder, then plug

the openings to prevent system contamination or excessive fluid loss.

3. Remove the 2 master cylinder attaching nuts, then carefully remove the master cylinder from the vehicle. Do not allow brake fluid to come in contact with vehicle painted surfaces, wiring or electrical connections.

To install:

4. Bench bleed the master cylinder assembly:

a. Fabricate 2 plugs for the master cylinder outlet ports. This can be accomplished using 2 pieces of ³/₁₆ in. (4.75mm) brake tubing of 2 in. (50mm) in length each, along with 2 brake tube fitting nuts, one 11 x 1.5mm and the other 12 x 1.0mm. Fashion ISO flares on 1 end of each tube, then install the nuts onto the tubes. Place approximately ½ in. (13mm) of each tube unflared end in a vise (1 tube at a time), then compress the tube end using the vise, bend the tube to a 90 degree angle, then reinstall the tube in the vise and fold the tube end against itself to form an airtight seal.

b. Plug the master cylinder outlet ports using the plugs.

c. Place shop towels around the master cylinder assembly to absorb bled fluid, then mount the assembly in a vise. Make sure the front end of the cylinder assembly is mounted slightly lower than the rear.

d. Fill the master cylinder with clean brake fluid, then use a small tool with a smooth rounded end to stroke the primary piston about 1 in. (25mm) several times.

NOTE: The backpressure produced by plugging the outlet ports may not allow a full 1 in. (25mm) stroke.

e. Reposition the cylinder in the vise so the front end is positioned slightly higher than the rear and continue to stroke the primary piston.

f. Reposition the master cylinder so it is level in the vise, then loosen the plugs (1 at a time) and push the primary piston into the master cylinder to bleed air from the port. Tighten the plug before allowing the piston to withdraw, or air will be drawn back into the master cylinder.

g. Continue to loosen the plugs and push in the primary piston until all air is bleed from the master cylinder assembly.

5. Install the master cylinder assembly to the brake booster, then tighten the 2 attaching nuts to 32 ft. lbs. (43 Nm).

6. Install each of the brake pipe assemblies to the master cylinder 1 at a time by unplugging the brake line, then removing the fabricated plug and quickly installing the brake pipe fitting finger-tight.

7. Have an assistant depress the brake pedal to remove air at the loose brake pipe fittings, then tighten the fittings to 24 ft. lbs. (32 Nm).

8. Have the assistant quickly release and pump the brake pedal several times.

9. If the brake pedal feels firm, start the vehicle. If the pedal is still firm with the engine running, carefully test drive the vehicle. If the pedal feels soft or spongy, bleed the entire brake system.

Brake Caliper

REMOVAL AND INSTALLATION

Front

Single Piston

1. Remove ²/₃ of the brake fluid from the master cylinder.

2. Raise and support the vehicle safely.

3. Matchmark the relationship between the wheel and hub, then remove the wheel and tire assembly.

4. Reinstall 2 of the lug nuts in order to retain the rotor.

5. Position a C-clamp over the outboard brake pad and the caliper housing, then use the C-clamp to bottom the piston into the caliper bore.

6. If completely removing the caliper from the vehicle for replacement or service, remove the bolt, copper washers and inlet fitting from the caliper housing. Plug the openings in the inlet fitting and caliper housing to prevent system contamination or excessive fluid loss.

7. Remove the mounting bolts and sleeves.

8. Remove the caliper assembly from the rotor and bracket. If the caliper is not being completely removed, it must be suspended from the vehicle using a wire hook in order to prevent damage to the brake lines.

To install:

9. Inspect the mounting bolts and sleeves for damage or corrosion and replace as necessary. Do not attempt to polish away corrosion. Ensure all caliper-to-bracket contact points are rust free and clean. Check the inlet fitting bolt for blockage.

10. Lubricate the sleeves, bushings and slide points with a suitable silicone grease.

11. Install the sleeves into the caliper housing, then install the caliper assembly onto the rotor and bracket assembly. Install the mounting bolts and tighten to 38 ft. lbs. (51 Nm).

12. If removed, install the bolt, new copper washers and the inlet fitting to the caliper housing. Tighten the bolt to 32 ft. lbs. (44 Nm).

13. Replace the brake fluid in the master cylinder and, if the caliper was removed, bleed the caliper.

14. Align the marks made earlier and install the wheel and tire assembly.

15. Lower the vehicle, then with the engine running pump the brake pedal slowly and firmly 3 times to seat then brake pads.

Dual Piston

1. Remove ²/₃ of the brake fluid from the master cylinder.

2. Raise and support the vehicle safely.

3. Matchmark the relationship of the wheel and hub, then remove the wheel and tire assembly.

4. If completely removing the caliper from the vehicle for replacement or service, remove the bolt, inlet fitting and 2 gaskets from the caliper housing. Plug the openings in the caliper housing and inlet fitting to prevent system contamination or excessive fluid loss.

5. Remove the circlip and retainer pin.

6. Remove the caliper housing from the rotor and mounting bracket. If the caliper is not being completely removed, it must be suspended from the vehicle using a wire hook in order to prevent damage to the brake lines.

To install:

7. Check the inlet fitting bolt for blockage, clear or replace as necessary.

8. Install the caliper housing over the rotor and onto the mounting bracket. Ensure the guiding surfaces on the inboard and outboard disc brake pads and mounting bracket are seated correctly.

9. Press the caliper housing down to compress the bias springs, slide a new retainer pin into position and install a new circlip.

10. If removed, install the inlet fitting, bolt and 2 new gaskets. Tighten the bolt to 30 ft. lbs. (40 Nm).

11. Fill the master cylinder and bleed the brake system.

12. Align the matchmarks made earlier and install the wheel and tire assembly.

13. Lower the vehicle, then with the engine running pump the brake pedal slowly and firmly 3 times to seat then brake pads.

Rear

1. Raise and safely support the vehicle.
2. Disconnect the cable assembly at the equalizer.
3. Matchmark the relationship between the wheel and hub, then remove the wheel and tire assembly. Install 2 wheel nuts to retain the rotor.
4. If completely removing the caliper from the vehicle for replacement or service, remove the bolt, inlet fitting and 2 gaskets from the caliper housing. Plug the openings in the caliper housing and inlet fitting to prevent system contamination or excessive fluid loss.
5. Remove the caliper lever return spring only if it is defective. Discard the spring if the coils are opened.
6. Disconnect the parking brake cable from the caliper lever and bracket.
7. If necessary remove the vibration damper nut from the parking brake cable bracket.
8. Remove the 2 caliper guide pin bolts. Discard the pins after removal.
9. Remove the caliper housing from the rotor and mounting bracket. If the caliper is not being completely removed, it must be suspended from the vehicle using a wire hook in order to prevent damage to the brake lines.
To install:
10. Inspect the guide pins and boots and replace if corroded, worn or damaged. Check the inlet fitting bolt for blockage, clear or replace as necessary.
11. Install the caliper housing over the rotor and into the mounting bracket.
12. Install 2 new caliper guide pin bolts starting with the upper pin and tighten the bolts to 27 ft. lbs. (37 Nm).
13. If removed, install the inlet fitting, bolt and 2 new gaskets to the caliper housing. Tighten the bolt to 22 ft. lbs. (30 Nm).
14. Connect the parking brake cable to the caliper bracket and caliper lever. If removed, install the caliper lever return spring.
15. If the hose fitting was removed, bleed the caliper.
16. Lower the vehicle sufficiently and cycle the parking brake.
17. Raise and safely support the vehicle.

18. Inspect the caliper parking brake levers and ensure they are against the stops on the caliper housing. If the levers are not on their stops, check the parking brake adjustment.
19. Remove the 2 nuts securing the rotor, then align the matchmarks made earlier and install the wheel assembly.
20. Lower the vehicle, then with the engine running pump the brake pedal slowly and firmly 3 times to seat then brake pads.
21. Check the hydraulic system for leaks.

Disc Brake Pads

REMOVAL AND INSTALLATION

Front

1. Remove the caliper assembly from the rotor and mounting bracket without disconnecting the brake line, then position aside. Do not allow the caliper to hang by the brake hose, suspend it with a length of wire or a fabricated hook.
2. Remove the outer pad assembly using a small prytool, if necessary, to disengage the show buttons from the caliper assembly.
3. Remove the inner pad.
To install:
4. If not done already, bottom the piston into the caliper using a large C-clamp positioned on the housing and inside the piston well.
5. With the piston bottomed in the caliper bore, lift the inner edge of the boot next to the piston and press out any trapped air. Make sure the boot is flat.
6. Install the inner pad assembly by positioning the retainer spring into the piston. The pad must lay flat against the piston and the boot must not touch the pad. If the boot and pad are in contact, remove the pad and reposition the boot.
7. Install the outer pad assembly, making sure the back of the pad is flat against the caliper. The wear sensor should be at the trailing lower edge of the outer pad during forward wheel rotation or the outer pad has been installed on the wrong side.
8. Install the caliper assembly to the rotor and mounting bracket.

Rear

1. Remove ⅔ of the brake fluid from the master cylinder reservoir.
2. Raise and safely support the vehicle.

3. Matchmark the relationship of the wheel to the axle flange, then remove the wheel and tire assembly. Install 2 wheel nuts to retain the rotor.
4. Position a C-clamp and tighten until the piston bottoms in the base of the caliper housing. Make sure 1 end of the C-clamp rests on the inlet fitting bolt and the other against the outboard disc brake pad.

NOTE: It is not necessary to remove the parking brake caliper lever return spring to replace the disc brake pads.

5. Remove the upper caliper guide pin bolt and discard.
6. Rotate the caliper housing on the lower caliper mounting bolt. Be careful not to strain the hose or cable conduit. It may be necessary to loosen the lower caliper guide pin slightly.
7. Remove the disc brake pads and discard the shim.
To install:
8. Clean all residue from the pad guide surfaces on the mounting bracket and caliper housing. Inspect the guide pins for free movement in the mounting bracket. Replace the guide pins or boots, if they are corroded or damaged.
9. Install a new shim, then install the disc brake pads. The outboard pad with insulator is installed toward the caliper housing. The inboard pad with the wear sensor is installed nearest the caliper piston. The wear sensor must be on the leading edge with forward wheel rotation or the pad has been installed on the wrong side.
10. Rotate the caliper housing into it's operating position. The springs on the outboard brake pad must not stick through the inspection hole in the caliper housing. If the springs are sticking through the inspection hole in the caliper housing, lift the caliper housing and make the necessary corrections to the outboard brake pad positions.
11. Install a new upper caliper guide pin bolt and tighten to 27 ft. lbs. (37 Nm). Ensure that the lower caliper guide bolt is tightened to 27 ft. lbs. (37 Nm).
12. With the engine running, pump the brake pedal slowly and firmly to seat the brake pads.
13. Check the caliper parking brake levers to make sure they are against the stops on the caliper housing. If the levers are not on their stops, check the parking brake adjustment.

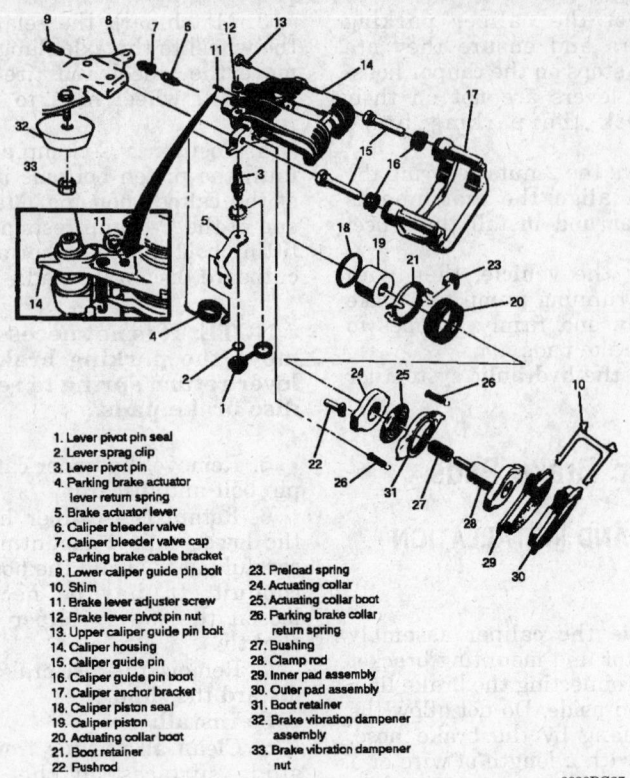

1. Lever pivot pin seal
2. Lever sprag clip
3. Lever pivot pin
4. Parking brake actuator lever return spring
5. Brake actuator lever
6. Caliper bleeder valve
7. Caliper bleeder valve cap
8. Parking brake cable bracket
9. Lower caliper guide pin bolt
10. Shim
11. Brake lever adjuster screw
12. Brake lever pivot pin nut
13. Upper caliper guide pin bolt
14. Caliper housing
15. Caliper guide pin
16. Caliper guide pin boot
17. Caliper anchor bracket
18. Caliper piston seal
19. Caliper piston
20. Actuating collar boot
21. Boot retainer
22. Pushrod
23. Preload spring
24. Actuating collar
25. Actuating collar boot
26. Parking brake collar return spring
27. Bushing
28. Clamp rod
29. Inner pad assembly
30. Outer pad assembly
31. Boot retainer
32. Brake vibration dampener assembly
33. Brake vibration dampener nut

8838RG27

Exploded view of the rear caliper assembly

14. Remove the 2 wheel nuts from the rotor, then align the marks and install the wheel assembly.

15. Lower the vehicle, check the master cylinder fluid level and roadtest the vehicle.

Brake Rotor

REMOVAL AND INSTALLATION

Front

1. Remove the caliper assembly from the rotor and mounting bracket without disconnecting the brake line, then position aside. Do not allow the caliper to hang by the brake hose, suspend it with a length of wire or a fabricated hook.

2. Slide the rotor from the hub and bearing assembly.

To install:

3. Position the rotor over the hub studs.

4. Install the caliper assembly.

Rear

The rear brake rotor may be removed, without separating the caliper from the mounting bracket.

1. Raise and support the vehicle safely.

2. Matchmark the relationship of the wheel and hub, then remove the wheel and tire assembly.

3. Remove and discard the 2 bolts retaining the caliper mounting bracket, then reposition and suspend the bracket/caliper assembly from the suspension using a length of wire or a fabricated hook.

4. Remove the rotor.

To install:

5. Install the rotor.

6. Install the caliper/mounting bracket assembly using 2 new bolts. Tighten the bolts to 74 ft. lbs. (100 Nm), then recheck the torque immediately.

7. Align the marks made earlier, then install the tire and wheel assembly.

8. Lower the vehicle.

Brake Drums

REMOVAL AND INSTALLATION

1. Raise and support the vehicle safely.

2. Matchmark the relationship of the wheel to the axle flange, then remove the wheel and tire assembly.

3. Matchmark the relationship of the drum to the axle flange and remove the brake drum. If the brake

drum is difficult to remove, try the following:

a. Ensure the parking brake is released.

b. Back off the parking brake cable adjustment.

c. Remove the adjusting hole cover or knockout plate from the backing plate. Back off the adjustment screw, using suitable brake adjusting tools.

d. Use a rubber mallet to tap gently on the outer rim of the drum and/or around the inner drum diameter by the spindle. Be careful not to deform the drum by excessive use of force.

To install:

4. Adjust the brake shoes. The outside diameter of the shoe and linings should be 0.050 inch (1.27mm) less than the inside diameter of the brake drum on each wheel.

5. Install the drum, aligning the marks on the drum and the axle flange.

6. Install the wheel and tire assembly, aligning the marks on the wheel and axle flange, then lower the vehicle.

Brake Shoes

REMOVAL AND INSTALLATION

1. Raise and support the vehicle safely.

2. Matchmark the relationship of the wheel to the axle flange, then remove the wheel and tire assembly.

3. Matchmark the relationship of the drum to the axle flange and remove the brake drum.

4. Remove the return springs using a suitable tool.

5. Remove the hold-down springs and pins. Remove the actuator pivot.

6. Remove the actuator link while lifting up on the actuator.

7. Remove the actuator lever and lever return spring.

8. Remove the shoe guide, parking brake strut and strut spring.

9. Remove the brake shoes and disconnect the parking brake lever from the appropriate shoe.

10. Remove the adjusting screw assembly and spring.

To install:

NOTE: Any part or spring which are of doubtful strength due to discoloration from heat, overstress or wear should be replaced. Clean the adjusting screw threads with a wire brush and check the threads for smooth rotations. Replace as necessary.

11. Install the parking brake lever on the appropriate shoe by hooking the lever tab into the slot.

12. Install the adjusting screw and spring. Lubricate the adjusting screw with suitable brake grease.

13. Clean and lubricate the contact points of the backing plate, then install the brake shoe assemblies and the parking brake cable to the backing plate.

14. Install the parking brake strut and strut spring by spreading the shoes apart. The strut end with the spring engages the parking brake lever and shoe, the other end engages the opposite shoe.

15. Install the shoe guide, actuator lever and lever return spring.

16. Install the hold-down pins, actuator pivot and springs.

17. Install the actuator link on the anchor pin. Install the actuator link into the actuator lever while holding up on the lever.

18. Install the shoe return springs and adjust the brakes. When properly adjusted, the brake shoe linings will be approximately 0.050 inch (1.27mm) less than the inner diameter of the brake drum.

19. Align and install the brake drum, then the wheel and tire assembly.

20. Lower the vehicle.

Wheel Cylinder

REMOVAL AND INSTALLATION

1. Raise and support the vehicle safely.

2. Matchmark and remove both the tire and wheel assemblies and the brake drums.

3. Remove the brake shoes and components, as required for access to the wheel cylinder.

4. Clean the area around the wheel cylinder and brake line. Remove the brake line from the wheel cylinder and plug the brake line to prevent system contamination or excessive fluid loss.

5. Remove the retaining bolts, then remove the wheel cylinder from the backing plate.

To install:

6. For ease of installation hold the wheel cylinder against the backing plate by inserting a block between the wheel cylinder and the axle shaft flange.

7. Install the retaining bolts and tighten to 115 inch lbs. (13 Nm).

8. Uncap and install the brake line, then tighten the fitting to 13 ft. lbs. (17 Nm).

9. Install the brake shoes and components.

10. Bleed the wheel cylinders.

11. Align and install the brake drums.

12. Install the wheel and tire assemblies, then lower the vehicle.

Parking Brake Cable

ADJUSTMENT

All vehicles feature a self-adjusting parking brake. The only adjustment possible on these vehicles is the brake shoe outer diameter adjustment when installing new brake components (drum brake vehicles) or parking brake free-travel (rear disc brake vehicles)

Parking Brake Free-Travel

Parking brake free-travel should only be adjusted if the caliper has been taken apart. This adjustment will not correct a condition where the caliper levers will not return to their stops.

NOTE: Disc brake pads must be new or parallel to within 0.006 inch (0.15mm). Parking brake free-travel adjustment is not valid with heavily tapered pads and may cause caliper/parking brake binding. Replace tapered brake pads.

1. Disconnect the parking brake cable assembly and remove the actuator lever return spring.

2. Have an assistant apply a light brake pedal load, enough to stop the rotor from turning by hand. This takes up all clearances and ensures that components are correctly aligned.

3. Apply light pressure to the caliper lever.

4. Measure the free-travel between the caliper lever and the caliper housing. The free-travel must be 0.024–0.028 inch (0.6–0.7mm).

5. If the free-travel is incorrect, do the following:

   a. Remove the adjuster screw.

   b. Clean the thread adhesive residue from the threads.

   c. Coat the threads with adhesive.

   d. Screw in the adjuster screw far enough to obtain 0.024–0.028 inch (0.6–0.7mm) free-travel between the caliper lever and the caliper housing.

6. Have an assistant release the brake pedal, then apply the brake pedal firmly 3 times. Recheck the free-travel and adjust as necessary.

7. Install the actuator lever return spring and parking brake cable assembly.

REMOVAL AND INSTALLATION

Front Cable

1. Remove the center console assembly.

2. With the parking brake lever in the up position, remove the pretension spring.

3. With the parking brake lever in the down position, rotate the arm toward the front of the vehicle until a 3mm metal pin can be inserted into the hole. It may be necessary to first remove the remaining piece of plastic shear pin, then insert the metal pin into the hole, locking out the self adjuster.

4. Pull the parking brake lever assembly all of the way back.

5. Raise and support the vehicle safely.

6. Disconnect the rear cables from the equalizer.

7. Using J-37043 or an equivalent cable release tool, disconnect the front cable fitting from the bracket.

8. Remove the grommet from the slot in the underbody.

9. Lower the vehicle, then note the retainer tab position for installation purposes and bend back the cable retainer tab on the pulley.

10. Remove the barrel-shaped front cable fitting from the adjuster track on the pulley.

11. Remove the front cable and casing from the control assembly using the cable release tool.

12. Remove the front cable from the floor pan.

To install:

13. Install the cable and seat the grommet in the floor pan. Soapy water may be used as a lubricant to ease installation, then feed the forward end of the grommet into the slot and use a small, curved, flat-ended prybar to gently press the rearward end into the slot.

14. Connect the cable casing to the control assembly.

15. Install the barrel-shaped front cable fitting into the adjuster track on the pulley, then bend the cable retainer tab to its original position on the pulley.

16. Make sure the lever is in the down position, then raise and support the vehicle safely.

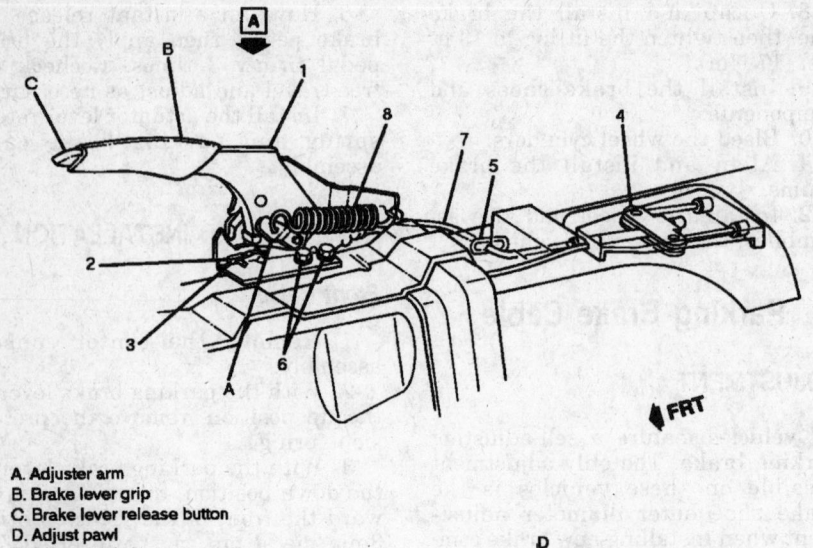

A. Adjuster arm
B. Brake lever grip
C. Brake lever release button
D. Adjust pawl
E. Adjust pawl pin
1. Parking brake lever assembly
2. Brake indiactor switch
3. Brake indicator switch bolt
4. Brake cable equalizer
5. Brake cable grommet
6. Brake lever bolt
7. Front cable assembly
8. Pretension spring

VIEW A

8838RG28

Parking brake lever assembly

17. Connect the rear cables to the equalizer and lower the vehicle.

18. With the parking brake lever in the up position, install the pretension spring.

19. Remove the metal pin from the hole, then cycle the parking brake lever 3 times.

20. Install the center console assembly.

Rear Cable

Rear Drum Brakes

1. Make sure the brake lever is fully released, then raise and support the vehicle safely.

2. Pull the equalizer rearward to gain the necessary cable slack, then insert a spacer to hold the equalizer in place and remove the left and/or right cable from the equalizer.

3. Compress the retainer fingers on the casing and pull the left and/or right rear cable out of the seat belt plate or underbody bracket.

4. Pull the cable assembly through the guides installed on the rear axle housing.

5. Matchmark and remove the left or right wheel assemblies, as applicable.

6. Matchmark and remove the brake drum(s).

7. Disconnect the left and/or right rear cable from the brake shoe operating lever, then compress the retainer fingers and pull the left and/or right cable from the backing plate.

To install:

8. Install the left and/or right cable to the backing plate and shoe actuating lever.

9. Align the marks made earlier, then install the brake drum(s) and the wheel and tire assemblies.

10. Feed the cable assembly through the guides on the rear axle housing.

11. Install the left and/or right cable into the underbody bracket or the seat belt plate and retainer.

12. Connect the left and/or right cable to the equalizer, then remove the spacer.

13. Lower the vehicle and cycle the parking brake 3 times.

Rear Disc Brakes

1. Make sure the brake lever is fully released, then raise and support the vehicle safely.

2. Pull the equalizer rearward to gain the necessary cable slack, then insert a spacer to hold the equalizer in place and remove the left and/or right cable from the equalizer.

3. Compress the retainer fingers on the casing and pull the left and/or

right rear cable out of the seat belt plate or underbody bracket.

4. Pull the cable assembly through the guides installed on the rear axle housing.

5. Matchmark and remove the left or right wheel assemblies, as applicable.

6. Pull all the cable slack to the caliper end of the cable by pushing forward on the caliper lever. Remove the left and/or right rear cable from the tang on the caliper lever, then release the lever.

7. Compress the retainer fingers on the cable casing and pull out of the bracket.

To install:

8. Push forward on the caliper lever, then install the left and/or right rear cable in the tang on the caliper lever and release the lever. Seat the finger retainers into the bracket.

9. Align the marks made earlier, then install the wheel assemblies.

10. Feed the cable assembly through the guides on the rear axle housing.

11. Install the left and/or right cable into the underbody bracket or the seat belt plate and retainer.

12. Connect the left and/or right cable to the equalizer, then remove the spacer.

13. Lower the vehicle and cycle the parking brake 3 times.

Brake System Bleeding

PROCEDURE

1. Clean the fluid reservoir cover and surrounding area to avoid system contamination, then remove the cover. Fill the reservoir to the proper level using fresh DOT 3 brake fluid from a sealed container, then reinstall the cover.

2. Prime and partially bleed the hydraulic modulator as follows:

a. Attach a length of clear plastic hose to the rearward bleeder valve on the modulator assembly, then submerge the opposite end of the hose in a container partially filled with clean brake fluid.

b. Open the bleeder valve slowly $\frac{1}{2}-\frac{3}{4}$ turn, then have an assistant depress and hold the brake pedal until fluid begins to flow. Close the bleeder valve and release the brake pedal, then repeat until no air bubbles are present in the fluid flowing from the modulator assembly.

c. Once air has been bleed from the rearward bleeder valve, move

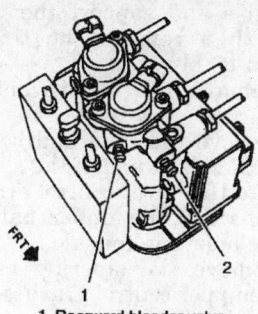

1. Rearward bleeder valve
2. Forward bleeder valve

8838RG30

ABS hydraulic modulator bleeder locations

the length of hose to the forward bleeder and repeat the bleeding procedure to further purge the modulator assembly of air.

d. At this point the modulator is sufficiently filled with fluid to bleed the wheel cylinders and/or calipers. Once the wheel points have been bled, the modulator must be rebled to assure it is completely purged of air.

3. Remove the reservoir cover and add fresh brake fluid, as necessary to fill the reservoir to the proper level, then install the cover.

4. Raise and support the vehicle safely, then bleed the wheel points in the following sequence
• Right rear
• Left rear
• Right front
• Left front

5. Bleed each of the wheel cylinders and/or calipers as follows:

a. Attach a length of clear plastic hose to the bleeder valve on the wheel cylinder or caliper assembly, then submerge the opposite end of the hose in a container partially filled with clean brake fluid.

NOTE: To help free trapped air, tap lightly on the backing plate or caliper using a rubber mallet.

b. Open the bleeder valve, then have an assistant slowly depress and hold the brake pedal. Close the bleeder valve, release the brake pedal and wait 5 seconds, then repeat the step (including the 5 second pause) until no air bubbles are present in the fluid flowing from the bleeder and a firm brake pedal is obtained.

c. Repeat the procedure for the remaining wheel points.

6. Remove the reservoir cover and add fresh brake fluid, as necessary to fill the reservoir to the proper level, then install the cover.

7. Lower the vehicle for access, then return to the hydraulic modulator and completely purge air from the bleeder valves by repeating the modulator priming and bleeding procedure. Once the modulator is bled, tighten the bleeder valves to 80 inch lbs. (9 Nm).

8. Check and add brake fluid, as necessary.

Wheel Speed Sensor

REMOVAL AND INSTALLATION

Front

The front wheel speed sensors and rings are integral with the hub and bearing assemblies. If the speed sensor or ring require replacement, the entire hub and bearing assembly must be replaced.

Rear

1. Raise and safely support the vehicle.

2. Disconnect the electrical connector from the wheel speed sensor.

3. Remove the jumper harness grommet from the retainer.

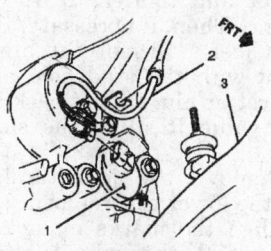

**LEFT SIDE SHOWN
RIGHT SIDE SIMILAR**

1. Front wheel speed sensor
2. Front wheel speed sensor jumper harness
3. Lower control arm

8838RG31

ABS front wheel bearing and speed sensor assembly

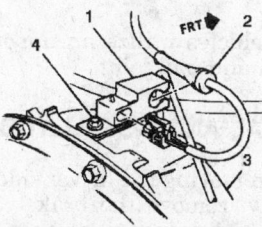

1. Rear wheel speed sensor
2. Rear wheel speed sensor jumper harness
3. Rear axle housing
4. Rear wheel speed sensor mounting bolt

8838RG32

ABS rear wheel speed sensor assembly

4. Remove the sensor-to-differential housing bolt.

5. Remove the sensor from the differential housing.

To install:

6. Lubricate the sensor O-ring with differential oil and install the O-ring with the sensor into the differential housing.

7. Torque the sensor-to-differential housing bolt to 89 inch lbs. (10 Nm).

8. Connect the electrical connector to the wheel speed sensor.

9. Install the jumper harness grommet into the retainer.

FRONT SUSPENSION

Strut

REMOVAL AND INSTALLATION

1. If removing the driver side strut assembly, remove the brake master cylinder retaining nuts, then carefully reposition the master cylinder assembly, making sure not to kink or damage the hydraulic lines.

2. Remove the bolts and screws attaching the top of the strut assembly to the upper control arm support.

3. Raise and support the vehicle safely, then remove the tire and wheel assembly.

4. Remove the stabilizer shaft link.

5. Remove the nuts and bolts retaining the lower end of the strut assembly to the lower control arm.

6. Separate the lower ball stud from the steering knuckle.

7. Using chalk or paint, match-mark the location of the strut assembly lower mount location relative to the upper mount location in case the strut is to be disassembled. Do not scribe marks or the components will be damaged.

8. Remove the strut assembly from the vehicle.

To install:

9. Install the strut assembly, positioning as noted during removal, then install the bolts and nuts retaining strut to the lower control arm and tighten to 48 ft. lbs. (65 Nm).

10. Install the lower control arm ball joint to the steering knuckle.

11. Install the stabilizer shaft link assembly.

12. Install the tire and wheel assembly, then lower the vehicle.

13. Install and hand-tighten the upper strut retaining bolts and nuts, then tighten the bolts to 37 ft. lbs. (50 Nm) and the nuts to 32 ft. lbs. (43 Nm).

14. If the driver side strut was installed, reposition the brake master cylinder to the booster assembly, then install the nuts and tighten to 32 ft. lbs. (43 Nm).

Upper Ball Joints

REMOVAL AND INSTALLATION

1. Raise and support the vehicle safely; place floor stands under the strut assembly mounting location on the lower control arm.

NOTE: Leave the jack under the strut during removal and installation, in order to retain the spring assembly and lower control arm in position.

2. Matchmark and remove the tire and wheel assembly.

3. Remove the cotter pin and nut from the upper control arm ball joint stud.

4. Support the steering knuckle with a jackstand, then using a ball joint separator tool, break the stud loose and pull the stud out of the knuckle.

5. With the control arm in a raised position, use a 1/8 inch diameter bit and drill into each of the 4 rivet heads to a depth of 1/4 inch (6mm).

6. Drill off the rivet heads with a 1/2 inch diameter bit.

7. Punch out the rivets using a suitable driver or punch, then remove the ball joint.

To install:

8. Place the new ball joint into the upper control arm and secure it with 4 bolts and nuts in place of rivets. Torque the nuts to specifications provided with the ball joint kit.

9. Remove the support from the steering knuckle, then connect the ball joint to the knuckle assembly. Torque the nut to 39 ft. lbs. (53 Nm), then insert a new cotter pin. If necessary, tighten the nut additionally in order to install the pin, but do not back off the specified torque.

NOTE: When replacing the ball joints, use only high-quality replacement parts; bolts and nuts specified to be strong enough to endure the stress.

10. Remove the floor jack from underneath the strut spring seat.

11. Install the tire and wheel assembly.

12. Lower the vehicle.

Lower Ball Joints

REMOVAL AND INSTALLATION

1. Raise and support the vehicle safely. Support the lower control arm under the spring seat or strut/spring assembly mount using a suitable jack which must remain in position throughout the procedure to retain the spring and control arm position.

2. Matchmark and remove the wheel and tire assembly.

3. Use a ball joint separator to loosen the ball stud in the steering knuckle before the cotter pin and nut is removed.

4. Remove the cotter pin and loosen the lower ball joint nut.

5. Using suitable ball joint removal tools, press the ball joint from the lower control arm.

To install:

6. Position the new ball joint into the lower control arm and press into place using suitable ball joint installation tools. If the ball joint cannot press firmly into position, either the ball joint is defective or the lower control arm must be replaced.

7. Connect the ball joint to the steering knuckle, then install the ball stud nut and tighten to 81 ft. lbs. (110 Nm). Then, if necessary, tighten the nut just sufficiently to align the nut slot with the stud hold and install a cotter pin. Do not back off the torque value to align the slot and hole.

8. Remove the jackstand from under the lower control arm.

9. Align the marks made earlier, then install the wheel and tire assembly.

10. Lower the vehicle, then check and adjust the front end alignment, as necessary.

Upper Control Arms

These vehicles utilize an upper control arm and ball joint.

REMOVAL AND INSTALLATION

1. If removing the driver side strut assembly, remove the brake master cylinder retaining nuts, then carefully reposition the master cylinder assembly, making sure not to kink or damage the hydraulic lines.

2. Remove the bolts and screws attaching the top of the strut assembly to the upper control arm support.

3. Raise and support the vehicle safely, then remove the tire and wheel assembly.

4. Remove the stabilizer shaft link.

5. Remove the nuts and bolts retaining the lower end of the strut assembly to the lower control arm.

6. Separate the upper ball stud from the steering knuckle.

7. Remove the steering knuckle from the upper control arm assembly. Remove the upper control arm and the strut assembly from the vehicle, then support the steering knuckle using jackstands.

8. Remove the nuts and bolts attaching the control arm to the control arm support, then separate the components.

To install:

9. Install the upper control arm to the control arm support, then tighten the retaining nuts and bolts to 39 ft. lbs. (53 Nm).

10. Position the upper control arm assembly and the strut assembly in the vehicle, then install the upper ball stud to the steering knuckle.

11. Remove the jackstands from underneath the steering knuckle.

12. Install the lower strut assembly fasteners, then install the stabilizer link.

13. Install the upper strut assembly fasteners.

14. Install the tire and wheel assembly, then lower the vehicle.

15. If the driver side control arm installed, reposition the brake master cylinder to the booster assembly, then install the nuts and tighten to 32 ft. lbs. (43 Nm).

Lower Control Arms

REMOVAL AND INSTALLATION

1. Raise and support the vehicle safely, then matchmark and remove the tire and wheel assembly.

2. Disconnect the steering rack outer tie rod from the steering knuckle.

3. Remove the stabilizer shaft link.

4. Remove the fasteners and disconnect the strut assembly from the lower control arm.

5. Separate and remove the lower control arm ball stud from the steering knuckle.

6. Remove the control arm retaining nuts and bolts, then remove the control arm from the vehicle.

To install:

7. Position the lower control arm to the crossmember, then install and hand-tighten the retainers. Once the retainers are installed, tighten the nuts to 96 ft. lbs. (130 Nm).

8. Position the lower control arm assembly to the steering knuckle and hand start the nut, then tighten the nut to 81 ft. lbs. (110 Nm) and install a new cotter pin. If necessary, tighten the nut further in order to align a slot with the stud hole, but do not back off from the torque specification.

9. Position the strut assembly to the lower control arm and secure using the lower strut fasteners.

10. Install the stabilizer shaft link.

11. Connect the steering rack tie rod to the steering knuckle assembly.

12. Align the matchmarks made earlier, then install the tire and wheel assembly.

13. Lower the vehicle, then check and adjust the alignment, as necessary.

Sway Bar

REMOVAL AND INSTALLATION

1. Raise and support the vehicle safely.

2. Remove the nut retaining the stabilizer shaft link to the lower control arm, then remove the link assembly and sleeve.

3. Remove the bolts and clamps retaining the stabilizer shaft to the side rail brackets.

4. Remove the sway bar/stabilizer shaft and insulators from the vehicle.

5. If necessary, remove the retaining bolts and the shaft brackets from the side rails.

To install:

6. If removed, install the brackets to the side rails and tighten the retaining bolts to 41 ft. lbs. (55 Nm).

7. Position the insulators over the shaft, making sure each insulator slit is facing the front of the vehicle.

8. Position the shaft and install the clamps. Tighten the clamp retaining bolts to 41 ft. lbs. (55 Nm).

9. Install the link sleeve between the control arm and the stabilizer shaft, then install the link assembly through the lower control arm, up through the sleeve to the stabilizer shaft. Install the retaining nut and tighten to 17 ft. lbs. (23 Nm).

10. Lower the vehicle.

Front Wheel Bearings

ADJUSTMENT

The wheel hub and bearing assembly used on these vehicles is a sealed, non-serviceable unit. No wheel bearing adjustments are necessary or possible.

REMOVAL AND INSTALLATION

1. Disconnect the negative battery cable, then raise and support the vehicle safely.

2. Matchmark and remove the tire and wheel assembly.

3. Remove the brake caliper with the brake hose attached and support aside using a hook or wire.

4. Remove the brake rotor.

5. Disengage the wheel speed sensor electrical connector.

6. Remove the 4 bolts retaining the hub and bearing assembly to the steering knuckle, then remove the assembly from the vehicle.

To install:

7. Position the hub and bearing assembly to the steering knuckle, then install the retaining bolts and tighten to 63 ft. lbs. (86 Nm).

8. Make sure the wheel speed sensor connector is properly routed in the wire bracket or system damage may occur, then engage the connector to the speed sensor.

9. Install the brake rotor.

10. Remove the support and install the caliper assembly.

11. Align and install the wheel assembly.

12. Lower the vehicle and connect the negative battery cable.

REAR SUSPENSION

Shock Absorber

REMOVAL AND INSTALLATION

1. Fold down the seatback frame assembly, then remove the quarter trim panel and pull the folding carpet back.

2. Raise and support the vehicle safely at height which will allow access to both the upper and lower shock mounts. The rear axle must be supported with a jackstand to prevent damage to the rear suspension once the shocks are disconnected.

3. Remove the upper shock attaching nut, upper washer and grommet.

4. Remove the lower shock mounting nut from the rear axle.

5. Lower the shock absorber from the vehicle along with the remaining grommet and washer.

To install:

6. Position the shock to the rear axle assembly.

7. Install the lower shock absorber mounting nut and tighten to 66 ft. lbs. (90 Nm), then position the lower pair of grommet and washer to the top of the shock and guide the shock through the underbody pan assembly.

8. With the shock properly positioned through the body mounting hole, install the upper grommet and washer over the end of the shock, then install the upper shock absorber mounting nut and tighten to 13 ft. lbs. (17 Nm).

9. Tighten the lower attaching nut to 66 ft. lbs. (90 Nm).

10. Remove the rear axle support and lower the vehicle.

11. Reposition the carpeting and trim panels, as applicable.

Coil Spring

REMOVAL AND INSTALLATION

1. Raise and support the vehicle safely so the rear axle hangs freely, then install an adjustable lifting device supporting the rear axle.

2. Disconnect the right and left shock absorber lower attaching nuts, then free the shock absorbers from the rear axle assembly.

3. Carefully lower the rear axle sufficiently to remove the upper insulators and coil springs. Make certain that at no time is the rear axle suspended by the brake lines as damage to the hydraulic brake system may occur.

To install:

4. Position the coil springs to the rear axle assembly. Be careful not to chip or damage the spring protective coatings. Should the coatings become chipped or damaged, do not used the spring until it has been replaced.

5. Position the upper insulators on the springs, then carefully raise the rear axle into position.

6. Connect the shock absorbers to the rear axle assembly, then tighten the retaining nuts to 66 ft. lbs. (90 Nm).

7. Remove the adjustable lifting device from the rear axle.

8. Lower the vehicle.

Rear Control Arms

REMOVAL AND INSTALLATION

NOTE: If both control arms are being replaced, remove and replace 1 control arm at a time to prevent the axle from rolling or slipping sideways and thus making replacement difficult or damaging components.

1. Raise and support the vehicle safely, using an adjustable jack to support the rear axle at the curb height position.
2. Remove the control arm-to-axle housing bolt and control arm-to-underbody bolt.
3. Remove the control arm assembly from the vehicle.

To install:

4. Position the control arm to the vehicle, then install the retaining nuts and bolts. The control arm-to-axle bolts should be installed from the inboard side outward.
5. With the suspension at its curb height position, tighten the control arm nuts to 60 ft. lbs. (82 Nm).
6. Remove the supports and lower the vehicle.

Track Bar (Tie Rod Assembly)

REMOVAL AND INSTALLATION

1. Raise and support the vehicle safely, then position a support to hold the rear axle at its normal curb height position.
2. Remove the bolt and nut attaching the track bar to the rear axle.
3. Remove the bolts and nuts attaching the track bar to the underbody brackets.
4. Remove the track bar and the brace from the vehicle.

To install:

5. Position the brace to the right underbody bracket, then hand-tighten 1 bolt and nut to retain it.
6. Position the track bar to the right underbody bracket then hand-tighten the other bolt and screw to retain it.
7. Install the remaining underbody bracket bolts, then tighten the bolts to 35 ft. lbs. (47 Nm) and the nuts to 75 ft. lbs. (102 Nm).
8. Install the bolt and nut retaining the track bar to the rear axle, then tighten to 75 ft. lbs. (102 Nm).
9. Remove the rear axle support, then lower the vehicle.

Torque Arm

REMOVAL AND INSTALLATION

1. Raise and support the vehicle safely, then position an adjustable lifting device under the rear axle assembly.
2. If equipped with a 2-piece driveshaft assembly, disconnect the torque arm from the driveshaft assembly.
3. Remove the bolts, nuts and washers attaching the torque arm to the rear axle housing.
4. Remove the bolts and nuts attaching the torque arm to the transmission.
5. Remove the torque arm-to-transmission bracket assemblies, then remove the torque arm from the vehicle.

To install:

6. Position the inner bracket assembly to the transmission, then install the outer bracket to the inner bracket and loosely install the fasteners.
7. Insert the torque arm into the inner and outer brackets, then tighten the fasteners. Tighten the nuts to 30 ft. lbs. (41 Nm), the bolts which thread through or into the transmission housing to 37 ft. lbs. (50 Nm) and the outer bracket-to-inner bracket bolt to 20 ft. lbs. (27 Nm).
8. Position the torque arm to the rear axle, then install the fasteners. Tighten the bolts to 96 ft. lbs. (130 Nm) and the nuts to 97 ft. lbs. (132 Nm).
9. If equipped with a 2-piece driveshaft assembly, connect the torque arm to the driveshaft.
10. Remove the rear axle support, then lower the vehicle.

Wheel Bearings

REMOVAL AND INSTALLATION

1. Raise and support the vehicle safely.
2. Remove the rear tire and wheel assembly.
3. Remove the brake rotor or the brake drum and components, as equipped.
4. Clean the carrier cover and surrounding area to prevent dirt or contamination from entering the housing, then remove the carrier cover and drain the gear oil into a suitable container.

5. Install J-39446, or an equivalent ABS exciter ring protector kit.
6. Remove the rear axle pinion shaft lockscrew and pinion shaft.
7. Push the flanged end of the axle shaft into the axle housing and remove the C-clip shaft lock from the differential case end of the shaft.
8. Remove the axle shaft from the axle housing.
9. If necessary to service the seal or bearing, use a small suitable prytool to remove the oil seal from the axle housing. Be careful not to score or damage the housing.
10. If necessary, install tool J-22813-01 or equivalent axle bearing remover, into the bore of the axle housing and position it behind the bearing, ensure the tangs of the tool engage the outer race. Remove the bearing using a slide hammer.

To install:

11. If removed, lubricate the new bearing and sealing lips with gear lubricant and install the bearing with a suitable driver so the tool bottoms against the shoulder in axle housing.
12. If removed, Position seal on suitable seal installer, then insert the seal into the housing bore. Position the seal flush with the axle tube.
13. Taking care not to damage the seal, slide the axle shaft into place so the splines engage with the splines of the side gear.
14. Insert the shaft C-lock into the bottom end of the axle shaft and push the shaft outward so the lock seats in the counterbore of the rear axle side gear.
15. Insert the rear axle pinion gear shaft through the differential case, thrust washer and pinion gears. Align the hole in the shaft with the lockscrew hole.
16. Install the lockscrew and tighten to 27 ft. lbs. (36 Nm).
17. Clean the cover gasket mating surfaces of any remaining gasket or old sealant.
18. Install the gasket onto the carrier cover, then install the carrier cover and tighten the bolts in a crosswise pattern to 22 ft. lbs. (30 Nm).

NOTE: When refilling a limited slip differential rear axle with gear oil, 4 oz (118ml) of limited slip additive should be added.

19. Fill the differential carrier with SAE 80W-90 GL-5 gear lubricant or equivalent, then install the plug.
20. Install the rear brake assemblies and rear wheels. Lower the vehicle.

FIRING ORDERS

NOTE: To avoid confusion, always replace spark plug wires one at a time.

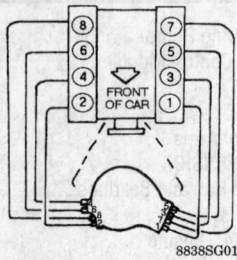

5.7L (VIN P) Engine
Engine Firing Order:
1–8–4–3–6–5–7–2
Distributor Rotates with
Camshaft

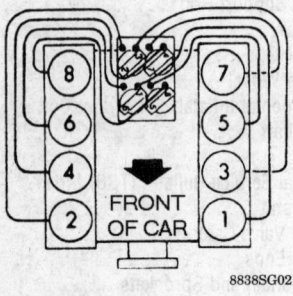

5.7L (VIN J) Engine
Engine Firing Order:
1–8–4–3–6–5–7–2
Distributorless Ignition System

ENGINE ELECTRICAL

NOTE: Disconnecting the negative battery cable on some vehicles may interfere with the functions of the on board computer systems and may require the computer to undergo a relearning process, once the negative battery cable is reconnected.

Distributor

The 5.7 (VIN P) utilizes a front engine mounted distributor assembly.

REMOVAL

5.7 (VIN P) Engine

NOTE: The ignition system may be disabled for compression checks by removing the "BAT" terminal and the 4-terminal connector from the distributor.

1. Disconnect the negative battery cable, then unplug the Intake Air Temperature (IAT) sensor harness connector.
2. Remove the air intake duct and the serpentine drive belt.
3. Drain the engine coolant into a suitable container and remove the coolant hoses from the water pump assembly.
4. Unplug the wiring harness from the Engine Coolant Temperature (ECT) sensor and remove the water pump assembly.
5. Remove the crankshaft torsional damper as follows:
   a. Raise and safely support the vehicle or lower the vehicle, as necessary.
   b. Position a suitable drain pan and remove the power steering fluid cooler. Disconnect the line from the steering gear.
   c. Remove the motor mount nuts, then carefully raise the engine sufficiently to gain tool access to the damper.
   d. Remove the torsional damper bolts, then remove the damper from the hub.
6. Remove the belt tensioner from the engine.
7. Disconnect the spark plug wires from the distributor. Be sure to twist each boot ½ turn and pull only on the boot to remove each wire. The wire numbers should be molded into the distributor housing. If not, be sure to tag the wires before disconnection.
8. Unplug the 4-terminal ECM connector from the distributor.
9. Remove the distributor mounting bolts and pull the distributor forward until the driveshaft disengages from the engine. Mark the top of the shaft for alignment during reassembly.

INSTALLATION

Timing Not Disturbed

NOTE: To ensure correct ignition timing the distributor must be installed with the rotor in the same position as it was removed.

5.7L (VIN P) Engine

1. With the mark made on the distributor shaft earlier on top, install the distributor to the engine. Tighten the distributor bolts to 8 ft. lbs. (11 Nm).
2. Install the ECM connector and the spark plug wires to the distributor.
3. Install the belt tensioner.
4. Install the torsional damper as follows:
   a. Raise and support the vehicle safely or lower the vehicle, as necessary.
   b. Position the damper to the hub and install the damper bolts. Tighten the bolts to 60 ft. lbs. (81 Nm).
   c. Connect the power steering line to the gear, then lower the engine and install the power steering fluid cooler.
   d. Install the motor mount nuts and tighten to 40 ft. lbs. (54 Nm).
5. Install the water pump assembly, the ECT wiring harness and the coolant hoses.
6. Install the serpentine drive belt and the air intake duct.
7. Install the IAT sensor connector.
8. Connect the negative battery cable, fill the engine to the proper level with coolant and bleed the power steering system, as necessary.

Distributorless Ignition System

REMOVAL AND INSTALLATION

Crankshaft Sensor

1. Disconnect the battery negative cable.
2. Raise and support vehicle safely
3. Unplug the crankshaft sensor electrical connector.
4. Remove the crankshaft sensor mounting bolt, crankshaft sensor and, if applicable, the sensor shim.

To install:

5. Coat crankshaft sensor O-ring with clean engine oil.
6. Install the sensor shim (if equipped), sensor and mounting bolt. Torque the crankshaft sensor bolt to 71 inch lbs. (8 Nm).
7. Install the wiring harness connector to the sensor terminal.
8. Lower the vehicle and connect the battery negative cable.

Ignition Module

NOTE: Before removing the ignition module, refer to the manufacturer's instructions provided with the replacement component.

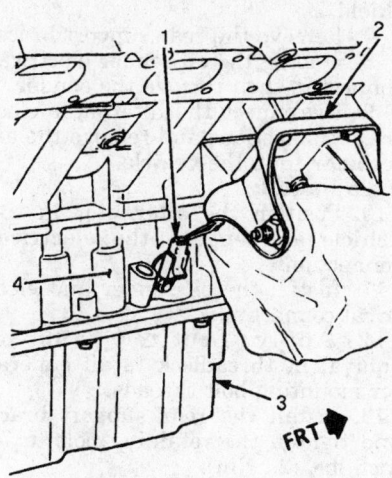

1. Crankshaft sensor
2. Right engine mount bracket
3. Oil pan
4. Engine block

8838SG03

Crankshaft position sensor location — 5.7L (VIN J) engine

1. Disconnect the battery negative cable.
2. Separate the intake plenum assembly from the top of the engine.
3. Unplug the electrical connectors from the ignition module.
4. Remove the mounting bolts, then the separate the ignition module from the underside of the intake plenum.

To install:

5. Apply a suitable dielectric grease to the back of the ignition module.
6. Position the module to the plenum and tighten the 4 mounting bolts to 89 inch lbs. (10 Nm).
7. Engage the electrical connectors nearest the front of the engine to the module.
8. Install the intake plenum assembly.
9. Connect the negative battery cable.

Ignition Coil Pack

1. Disconnect the battery negative cable.
2. Remove the intake plenum assembly.
3. Tag and disconnect the spark plug wires from the ignition coil pack.

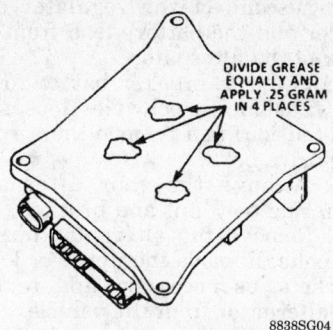

8838SG04

Ignition module dielectric grease application — 5.7L (VIN J) engine

4. Remove the 2 mounting bolts, then remove the coil pack from the ignition housing.
5. Installation is the reverse of the removal procedure.
6. Torque mounting bolts to 40 inch lbs. (4.5 Nm).

Ignition Housing

1. Disconnect battery negative cable.
2. Remove intake plenum assembly.
3. Tag and disconnect the electrical connectors and spark plug wires.
4. Remove the 4 ignition housing bracket mounting bolts.

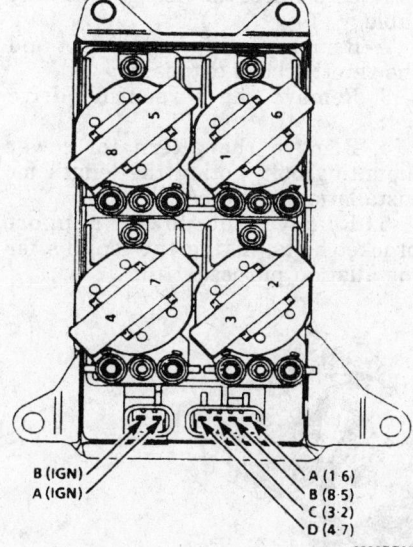

B (IGN)
A (IGN)

A (1 6)
B (8 5)
C (3 2)
D (4 7)

8838SG05

Ignition housing and coil packs — 5.7L (VIN J) engine

5. Remove the ignition coil mounting bolts. Note the position of each coil and remove the coils from the ignition housing. Remove the ignition housing from the bracket.

To install:

6. Install the ignition housing on the bracket and install the 4 seals packaged with the new housing.
7. Install the coils in their proper position on the housing and tighten the coil retaining bolts to 40 inch lbs. (4.5 Nm).
8. Install the ignition housing bracket to the engine and tighten the bracket bolts as follows: M6–16 bolts to 89 inch lbs. (10 Nm); M8–20 bolts to 19 ft. lbs. (26 Nm).
9. Engage the electrical connectors and the spark plug wires.
10. Install the intake plenum assembly.
11. Connect the negative battery cable.

Ignition Timing

ADJUSTMENT

For the 5.7L (VIN J and VIN P) engines the base engine timing is preset at the factory. The ECM then controls engine timing based on signals from the optical camshaft position sensor (VIN P) or the from crankshaft position sensor (VIN J). No adjustments are necessary or possible.

Alternator

PRECAUTIONS

Several precautions must be observed with alternator equipped vehicles to avoid damage to the unit.

• If the battery is removed for any reason, make sure it is reconnected with the correct polarity. Reversing the battery connections may result in damage to the 1-way rectifiers.
• When utilizing a booster battery as a starting aid, always connect the positive to positive terminals and the negative terminal from the booster battery to a good engine ground on the vehicle being started.
• Never use a fast charger as a booster to start vehicles.
• Disconnect the battery cables when charging the battery with a fast charger.
• Never attempt to polarize the alternator.
• Do not use test lights of more than 12 volts when checking diode continuity.

- Do not short across or ground any of the alternator terminals.
- The polarity of the battery, alternator and regulator must be matched and considered before making any electrical connections within the system.
- Never separate the alternator on an open circuit. Make sure all connections within the circuit are clean and tight.
- Disconnect the battery ground terminal when performing any service on electrical components.
- Disconnect the battery if arc welding is to be done on the vehicle.

BELT TENSION ADJUSTMENT

A single serpentine belt is used to drive all accessories. Belt tension is maintained by a spring loaded tensioner which has the ability to maintain belt tension over a broad range of belt lengths. There is an indicator to make sure the tensioner and belt are adjusted to within their operating ranges.

Belt inspection may reveal cracks in the belt ribs. These cracks will not impair belt performance. A belt should be replaced if belt slip occurs or if sections of the belt ribs are missing.

The belt tensioner can be pulled up to free the belt with the use of a ½ inch drive ratchet or breaker bar. Always disconnect the negative battery cable before servicing any of the belt driven accessories or components adjacent to the belt.

REMOVAL AND INSTALLATION

5.7L (VIN P) Engine

1. Disconnect the negative battery cable.
2. If necessary, remove the air intake duct.

3. Disconnect the regulator connector and the battery lead from the back of the alternator.
4. Use a ½ breaker bar to rotate the tensioner clockwise (loosening belt tension) and remove the serpentine drive belt.
5. Remove the rear alternator mounting bolt, nut and bracket.
6. Remove the alternator mounting bolts. Remove the upper or lower brackets, as necessary and remove the alternator from the vehicle.
   To install:
7. Position the alternator in the vehicle and install the lower mounting bolt and bracket. Be sure the bolt is finger-tight only at this time.
8. Install the alternator upper mounting bolt and bracket, but do not tighten at this time.
9. Install the rear alternator bracket, bolt and/or nut.
10. Tighten the lower and upper mounting bolts to 37 ft. lbs. (50 Nm), the rear bracket bolt to 17 ft. lbs. (23 Nm) and, if applicable, the rear bracket nut to 24 ft. lbs. (33 Nm).
11. Install the serpentine drive belt.
12. Install the regulator harness connector and the battery lead to the back of the alternator. Be careful not to overtighten the battery lead nut.
13. Install the air intake duct, if removed, and connect the negative battery cable.

5.7L (VIN J) Engine

1. Disconnect the negative battery cable.
2. Remove the air intake duct and the throttle body extension.
3. Remove the serpentine drive belt.
4. Remove the alternator lower mounting bolt, noting the length for installation purposes.
5. Remove the lower support bracket bolts, noting the lengths for installation purposes.

6. Remove the upper support bolts (noting the length) and remove the shield.
7. Remove the rear support brace.
8. Unplug the oil sender electrical connector, then remove the sender.
9. Disconnect the alternator electrical connections and remove the alternator from the vehicle.
   To install:
10. Position the alternator in the vehicle and engage the electrical connections.
11. Install the oil sender and electrical connection.
12. Apply Loctite® 565 or equivalent threadlock to all generator mounting bolt threads.
13. Install the rear support brace and tighten the retaining bolt to 70 inch lbs. (23 Nm).
14. Install the upper support shield and bolts, then the lower support bracket, spacer, support bolts and mounting bolts. Tighten the upper shield and lower mounting bolts to 38 ft. lbs. (52 Nm) and the lower support bracket bolts to 19 ft. lbs. (26 Nm).
15. Install the serpentine drive belt, then install the throttle body extension and tighten the bolts to 53 inch lbs. (6 Nm).
16. Install the air intake duct, then connect the negative battery cable.

Starter

REMOVAL AND INSTALLATION

5.7L (VIN P) Engine

1. Disconnect the negative battery cable.
2. Raise and support the vehicle safely.
3. Disconnect the wiring from the starter solenoid. Tag the wiring positions to avoid improper connections during installation.
4. Loosen the 2 starter mounting bolts, support the starter and remove the bolts. Lower the starter from the vehicle and if equipped, remove the shims, noting their locations for installation purposes.
   To install:
5. Position the shims and starter into the vehicle and insert the mounting bolts. Tighten the bolts to 34 ft. lbs. (47 Nm) and replace sealer to the front of the motor. The sealer must be applied after the motor is installed.
6. Check that flywheel-to-pinion clearance is 0.020 inch (0.5mm) and add or subtract shims, if necessary.
7. Connect the start wiring.
8. Lower vehicle and connect the battery negative cable.

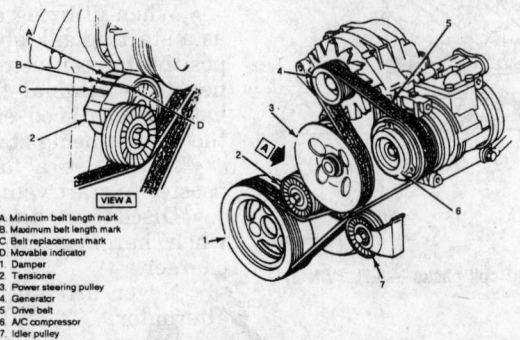

A. Minimum belt length mark
B. Maximum belt length mark
C. Belt replacement mark
D. Movable indicator
1. Damper
2. Tensioner
3. Power steering pulley
4. Generator
5. Drive belt
6. A/C compressor
7. Idler pulley

VIEW A

8838SG06

Serpentine drive belt and tensioner — 5.7L (VIN P) engine

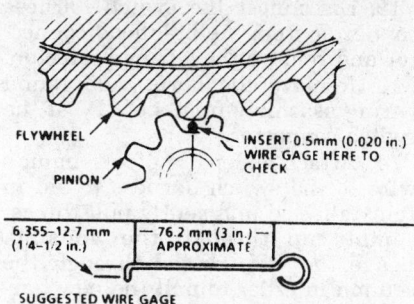

Check flywheel to starter pinion clearance using a wire gauge — 5.7L (VIN P) engine

5.7L (VIN J) Engine

1. Disconnect battery negative cable.
2. Remove the intake plenum assembly.
3. Remove the coil pack assembly.
4. Raise and support the vehicle safely.
5. Disconnect the wiring from the starter solenoid. Tag the wiring positions to avoid improper connections during installation.
6. Remove the 2 starter mounting bolts, then remove the starter from the vehicle.

To install:
7. Coat the threads of the starter mounting bolts with Loctite® 262 or equivalent threadlock.
8. Position the starter in the vehicle and install the mounting bolts. Tighten the starter mounting bolts to 38 ft. lbs. (52 Nm).
9. Connect the starter wiring and lower the vehicle.
10. Install the coil pack assembly.
11. Install the intake plenum assembly.
12. Connect the negative battery cable.

CHASSIS ELECTRICAL

Blower Motor

REMOVAL AND INSTALLATION

1. Disconnect the negative battery cable.
2. Remove the front wheel house rear panel and seal.
3. Disconnect the motor electrical connectors.

4. Remove the blower motor cooling tube.
5. Remove the motor and fan.
6. Installation is the reverse of the removal procedure.

Windshield Wiper Motor

REMOVAL AND INSTALLATION

1. Disconnect the negative battery cable.
2. Unplug the motor park switch and the circuit board electrical connectors at the motor.
3. Remove the left wiper arm and the left side plenum screen.
4. Remove the wiper transmission nuts and sockets.
5. For the VIN J engine, remove the crank nut and crank arm.
6. If equipped, disconnect the vacuum booster supply hose at the plenum.
7. Remove the wiper mounting bolts.
8. Unplug any remaining motor electrical connectors while removing wiper motor assembly.

To install:
9. Connect the wiper motor electrical connector.
10. Install the wiper motor and gasket by guiding the crank arm through the hole in the plenum panel and positioning it over the transmission.

NOTE: If installing a replacement motor on the VIN J engine, it may be necessary to file the motor shaft so the crank arm can slip over the shaft.

11. Install the motor mounting bolts and tighten to 27 inch lbs. (3 Nm).
12. For the VIN J engine, Install the crank arm nut and tighten to 30 ft. lbs. (42 Nm).
13. If applicable, connect the vacuum hose.
14. Install the transmission link sockets and nuts. Tighten the nuts to 27 inch lbs. (3 Nm).
15. Install the left plenum screen and the left wiper arm, then connect the motor upper electrical connectors.
16. Connect the negative battery cable and check motor operation.

Windshield Wiper Switch

REMOVAL AND INSTALLATION

NOTE: This vehicle is equipped with a SIR air bag system, it is imperative that the dis-

arming procedure is followed before repairs, and that the coil centering and rearming procedures are followed after repairs.

1. Disarm the SIR and disconnect the negative battery cable.
2. Remove the turn signal assembly, but do not disconnect or remove the wiring harness. Allow the switch assembly to hang freely from the wires unless removal is necessary for switch replacement.
3. Remove the ignition lock assembly, but do not disconnect or remove the wiring harness. Allow the lock set to hang freely from the wires, unless removal is necessary for lock cylinder replacement.
4. If not done already, remove the housing cover end cap, disconnect the electrical connectors from the multifunction lever and remove the lever by pulling toward the driver's door.
5. Remove the housing cover screws, unthread and remove the tilt lever from the column assembly.
6. Remove the lock housing cover assembly.
7. Remove the base plate and the dimmer switch rod actuator and the wiper switch actuator pivot pin.
8. Remove the wire protector shield.
9. Disconnect wiper switch connector from vehicle wire harness and remove switch. Attach a piece of mechanic's wire to the connector to aid in reinstallation and gently pull the wire harness through the column. Leave the mechanic's wire routed through the column to pull the new switch wiring into position.

To install:
10. Connect the wiper switch to the lock housing cover assembly.
11. Attach the switch actuator pivot pin to the switch and cover.
12. Pull wiper switch wire connector through the steering column using the mechanic's wire and attach to the vehicle wire harness. If applicable, install the wire protector shield.
13. Attach the dimmer switch rod actuator to the base plate and lubricate with lithium grease.
14. Connect base plate to lock housing cover assembly. The bottom edge of the dimmer switch rod actuator must rest on the bend in the dimmer switch rod.
15. Position lock housing cover in place and attach tilt lever.
16. Starting with the housing cover screw in the 12 o'clock position, then 8 o'clock and finally 3 o'clock positions, tighten the screws to 80 inch lbs. (9 Nm).

17. Install the multi-function lever and engage the lever connectors on the base plate.

18. Install the housing cover end cap.

19. Install the lock cylinder assembly.

20. Install the turn signal assembly.

21. Connect the negative battery cable and enable the SIR system.

Concealed Headlights

MANUAL OPERATION

The headlight doors can be opened automatically by turning the headlights switch to the ON position, then turn the switch back 1 click to the parking lights ON position and the headlight doors will stay open.

If necessary, then headlight doors may be opened manually. To open the headlight doors manually, raise the hood and turn the headlight manual control knob (located next to the headlight door) in the direction of the arrow, until the door is fully opened.

Headlight Switch

REMOVAL AND INSTALLATION

1. Disable the SIR system and disconnect the negative battery cable.

2. Remove the instrument cluster trim plate screws and reposition the trim plate to gain access to the right side of the headlight switch.

3. Remove headlight switch attaching screws and trim plate, then unplug the electrical connectors.

4. Remove the headlight switch.

5. Installation is the reverse of the removal procedure.

6. Connect the negative battery cable and properly enable the SIR system.

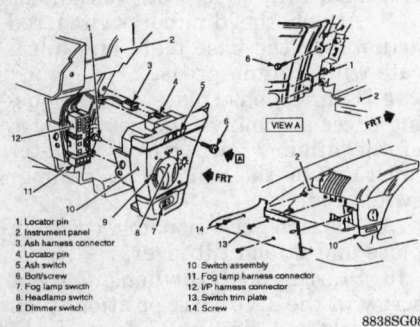

1. Locator pin
2. Instrument panel
3. Ash harness connector
4. Locator pin
5. Ash switch
6. Bolt/screw
7. Fog lamp switch
8. Headlamp switch
9. Dimmer switch
10. Switch assembly
11. Fog lamp harness connector
12. IP harness connector
13. Switch trim plate
14. Screw

8838SG08

Headlight switch mounting

Turn Signal Switch

REMOVAL AND INSTALLATION

NOTE: This vehicle is equipped with a SIR air bag system, it is imperative that the disarming procedure is followed before repairs, and that the coil centering and rearming procedures are followed after repairs. Although it may be possible to perform this procedure with the steering column installed in the vehicle, but lowered for access, the manufacturer suggests that the column be removed.

1. Properly disable the SIR system. Place the ignition switch to the **LOCK** position in order to prevent uncentering of the coil assembly.

2. Disconnect the negative battery cable.

3. Properly remove and store the inflator module and the steering wheel. Either remove the column from the vehicle or lower it for access.

4. Remove the coil assembly retaining ring. Remove the coil assembly and allow it to hang freely from the wiring.

NOTE: The coil assembly will become uncentered if the steering column is separated from the steering gear and allowed to rotate or the center spring of the coil assembly is pushed down, letting the hub rotate while the coil is removed from the steering column. In the event this should occur, follow the recommended procedure for recentering of the coil in order to avoid accidental deployment of the air bag or damage to the internal components of the steering column.

5. Remove the wave washer.

6. Remove the shaft lock retaining ring using tool J-23653-C or equivalent, shaft lock compressor. Discard the old ring.

7. Remove the shaft lock, turn signal canceling cam and upper bearing assembly.

8. Move the multi-function lever to the **RIGHT TURN** position. Remove the column housing cover end cap by pulling toward the vehicle front. Disconnect the electrical harness connector and remove the turn signal lever by pulling toward the driver door.

9. Remove the hazard knob retaining screw and assembly.

10. Remove the turn signal switch arm and screws.

11. Remove the turn signal switch screws.

12. Disconnect the switch harness connector from the bulkhead connector and remove the wiring protector.

13. Remove the horn pad ground wiring assembly from slot "D" of the switch connector.

14. Attach a length of mechanic's wire to the switch harness to aid in reinstallation and gently pull the assembly up through the housing. Leave the wire routed through the column in order to pull the new harness back into position.

15. Remove the switch and harness from the vehicle.

To install:

16. Connect the horn pad ground wiring assembly to slot "D" of the turn signal switch connector.

17. Using the mechanic's wire, pull the switch harness through the column and connect to the bulkhead connector.

18. Install the harness wiring protector.

19. Position the turn signal switch assembly and install the attaching screws. Tighten the screws to 30 inch lbs. (3.4 Nm).

20. Install the switch arm and mounting screws. Tighten the screws to 20 inch lbs. (2.3 Nm).

21. Install the hazard knob assembly and the multi-function lever.

22. Install the inner race, the upper bearing race seat, and the upper bearing spring.

23. Lubricate the friction surfaces using synthetic grease, then install the turn signal canceling cam.

24. Position the shaft lock. Install the a new shaft lock retaining ring using tool J-23653-C or equivalent. Be sure the ring is firmly seated in the groove of the shaft.

25. Install the wave washer.

26. Install the coil assembly, making sure it is properly centered.

27. Position and secure the steering column, as necessary.

28. Install the steering wheel and the inflator module.

29. Connect the negative battery cable and enable the SIR system.

Ignition Lock Cylinder

REMOVAL AND INSTALLATION

NOTE: This vehicle is equipped with a SIR air bag system, it is imperative that the disarming procedure is followed before repairs, and that the coil centering and rearming procedures are followed after repairs.

1. Hex locking nut
2. Retaining ring
3. Infl restraint coil assembly
4. Wave washer
5. Retaining ring
6. Shaft lock
7. Turn signal cancel cam assembly
8. Upper bearing spring
9. Bndg hd or recess screw
10. RD wash hd screw
11. Signal switch arm assembly
12. Turn signal switch assembly
13. Upper bearing inner race seat
14. Inner race
15. Pan head soc tap screw
16. Buzzer switch assembly
17. Lock retaining screw
18. Lock housing cover assembly
19. Steering column pass key lock cylinder set
20. Dimmer switch rod actuator
21. Switch actuator pivot pin
22. Pivot & pulse switch assembly
23. Col housing cover end base plate
24. Col housing cover end cap
25. Wiring protector
26. Connector shroud
27. Flat head tapping screw
28. Steering column housing assembly
29. Bearing assembly
30. Lock bolt
31. Lock bolt spring
32. Steering wheel lock shoe
33. Steering wheel lock shoe
34. Wire protector shield
35. Drive shaft
36. Dowel pin
37. Pivot pin
38. Shoe spring
39. Release lever spring
40. Release lever pin
41. Shoe release lever
42. Switch actuator rack
43. Rack preload spring
44. Steering column housing
45. Switch actuator sector
46. Hex washer head screw
47. Spring guide
48. Wheel tilt spring
49. Spring retainer
50. Steering column shaft assembly
51. Race & upper shaft assembly
52. Centering sphere
53. Joint preload spring
54. Lower steering shaft assembly
55. Support screw

62. Steering column housing support
71. Steering column housing shroud
72. Steering column jacket assembly
73. Cable backdrive pin spring
74. Cable backdrive pin
75. Inhibitor cross pin
76. Ignition switch actuator assembly
77. Dimmer switch rod
78. Washer head screw
79. Hexagon nut
80. Ignition switch assembly

81. Cable bracket
82. Dimmer & ignition switch mounting stud
83. Dimmer switch assembly
86. Adapter & bearing assembly
87. Hex washer head tap screw
88. Lower bearing spring
89. Lower bearing
90. Lower spring retainer
91. Retainer
92. Horn pad wire assembly

8838SG09

Exploded view of the steering column assembly

1. Disable the SIR system and disconnect the negative battery cable.

2. Remove turn signal switch assembly, but do not disconnect or pull the wire harness through the column. Allow the switch assembly to hang freely from the wires.

3. If necessary, remove the coil assembly as follows:

a. Disengage the coil terminal connector from the vehicle harness. Remove the yellow connector shroud from the black connector.

b. Remove wiring protector.

c. Attach a length of mechanic's wire to the terminal connector to aid in reassembly.

d. Carefully pull the wire harness through the column, leaving the mechanic's wire in the column for installation purposes.

4. Remove the key from the lock cylinder.

5. Remove the buzzer switch and clip.

6. Reinsert the key into the lock cylinder, making sure the key is in the **LOCK** position.

7. Remove the lock retaining screw.

8. Disengage the Pass Key wire harness connector from the bulkhead connector. If not done already, remove the wiring protector.

9. Attach a piece of string or mechanic's wire to the wire connector to aid in reassembly, disconnect the retaining clip from the housing cover and pull the wire up through the column. Leave the length of string or wire in the column in order to pull the new harness into position.

10. Remove the lock cylinder.

To install:

11. Using the length of string or the mechanic's wire, pull the PASS Key wire harness down through the column into the original position and engage the connector.

12. Install the lock cylinder set. Snap the wire retaining clip into the hole in the housing.

13. Engage the lock cylinder wiring connector to the bulkhead connector.

14. Install the lock cylinder retaining screw and tighten to 22 inch lbs. (2.5 Nm).

15. Remove the key and install the buzzer switch with retaining clip, then insert the key and leave in the **LOCK** position.

16. If removed, pull the turn signal switch wiring connector and/or the coil wiring connector through the steering column, connect the harnesses and install the wiring protector.

17. Install the turn signal switch assembly.

18. Connect the negative battery cable and enable the SIR system.

Ignition Switch

REMOVAL AND INSTALLATION

1. Disable the SIR system and disconnect the negative battery terminal.

2. Remove the column to instrument panel trim plates and attaching nuts.

3. Loosen the steering column mounting bolts.

4. Remove the steering column mounting bolts, lower and properly support the column. If access to the switch is not sufficient, remove the column from the vehicle.

NOTE: Be sure the steering column is supported at all times in order to prevent damage to the column.

5. Remove the hex nut and washer head screw securing the dimmer switch to the column.

6. If equipped, remove the horn ground strap attached to the dimmer/ignition switch mounting stud.

7. If equipped, remove the cable bracket.

8. Disengage the switch assembly from the actuator rod, unplug the wiring harness connector and remove the dimmer switch assembly.

9. Remove the dimmer and ignition switch mounting stud.

10. Remove the ignition switch from the actuating assembly and disengage the switch wire connector.

To install:

11. Verify that the key cylinder is in the **LOCK** position.

12. Move the actuator rod hole in the switch to the **LOCK** position. New switches will be pinned in this position and the pin must be removed after installation or switch damage may result.

13. Install the switch with the rod in the hole and adjust as necessary. To verify the switch is in the lock position, move the switch slider to the extreme right position and then move the slider 1 detent to the left.

14. Install the switch mounting stud and tighten to 35 inch lbs. (4.0 Nm).

15. Install the dimmer switch assembly to the actuator rod.

16. If equipped, install the cable bracket and the horn pad ground wire.

17. Install and finger-tighten the washer head screw and the hex nut.

18. Adjust the dimmer switch and tighten the screw and nut to 35 inch lbs. (4.0 Nm).

19. Position the steering column, then engage the ignition and dimmer switch connectors. Secure the column in the vehicle

20. Connect the negative battery cable and enable the SIR system.

Neutral Safety Switch

ADJUSTMENT

Switch adjustment is performed with the center console removed for switch access.

1. Place the transmission shifter in the **N** notch in the detent plate.

2. Loosen the switch attaching nuts.

3. Insert a 0.094 in. (2.34mm) drill bit or gauge pin into the adjustment hole, then rotate the switch until the gauge pin drops to a depth of 0.59 in. (15mm).

NOTE: If the proper sized gauge pin is unavailable, a ³/₃₂ in. drill bit may be used.

4. Tighten the switch nuts to 26 inch lbs. (3 Nm).

5. Remove the gauge pin and verify that the engine will only start in **P** or **N**.

REMOVAL AND INSTALLATION

1. Disconnect negative battery cable.

2. Remove the shifter knob assembly.

3. Remove the console assembly.

4. Remove the neutral switch mounting nuts.

5. Remove the switch and the gauge pin.
   To install:

6. Position shifter lever in the **N** position.

7. Insert carrier tang on the switch in the slot on the shifter.

8. If installing a new switch, install the mounting nuts and tighten the mounting nuts to 26 inch lbs. (3 Nm). Move the shift control lever out of the **N** position to shear the factory installed plastic retaining pin.

NOTE: If installing a new switch and the holes do not align with shifter control, check that the shifter control lever is in N. Do not rotate the switch as this will shear the retaining pin. If a new switch was rotated and the pin was already broken, switch adjustment must be performed.

9. If installing an old switch or a new switch with a sheared retaining pin, perform switch adjustment.

10. Connect the negative battery cable and check that engine starts only in the **P** or **N** positions. If engine starts in any other position, readjust neutral switch.

11. Install console and shifter knob assembly.

Powertrain Control Module

LOCATION

Located behind the middle of the instrument panel.

ENGINE COOLING

Radiator

REMOVAL AND INSTALLATION

1. Disconnect battery negative cable.

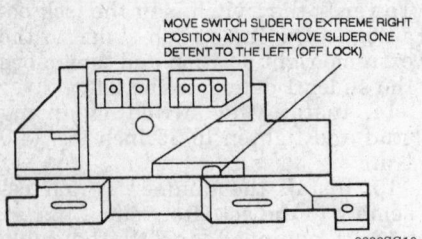

MOVE SWITCH SLIDER TO EXTREME RIGHT POSITION AND THEN MOVE SLIDER ONE DETENT TO THE LEFT (OFF LOCK)

8838SG10

Adjusting the ignition switch assembly

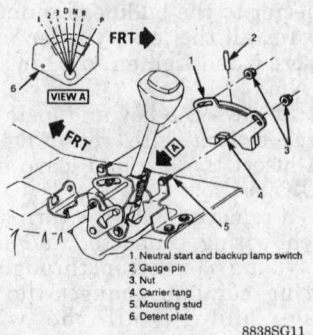

VIEW A
FRT
FRT

1. Neutral start and backup lamp switch
2. Gauge pin
3. Nut
4. Carrier tang
5. Mounting stud
6. Detent plate

8838SG11

Adjusting the neutral safety/backup light switch

2. Drain the engine coolant into a suitable container.

3. Remove the air cleaner and air duct assembly.

4. Remove the bleed hose from the radiator.

5. If equipped, disconnect the automatic transmission fluid cooler lines.

6. Remove the screws retaining the fan relays, then remove the relays from the vehicle.

7. Unplug the electrical connectors from the air pump and horn.

8. Remove the AIR pump.

9. Remove the bolts retaining the shroud to the upper support, then remove the bolts/screws retaining the upper support to the lower support.

10. Remove upper support from the vehicle.

11. Remove the radiator upper and lower hose clamps, then the disconnect the hoses from the radiator.

12. Remove the radiator from the vehicle.
    To install:

13. Install the radiator.

14. Connect and secure the inlet/outlet hoses to the radiator.

15. Install upper support to the vehicle and secure using the screws attaching the upper support to the lower support. Tighten the screws to 80 inch lbs. (9 Nm).

16. Install the nuts and bolts retaining the support to the shroud and tighten to 18 ft. lbs. (25 Nm).

17. Install the AIR pump and tighten the bolts to 89 inch lbs. (10 Nm).

18. Engage the electrical connectors to the AIR pump and horn.

19. Install the fan relays and tighten the retaining screws to 80 inch lbs. (9 Nm).

20. If equipped, connect the transmission fluid cooler lines.

21. Connect the radiator bleed hose.

22. Install the air cleaner and duct assembly.

23. Connect the negative battery cable.

24. Fill cooling system with the proper type and quantity of antifreeze and check for leaks.

Water Pump

REMOVAL AND INSTALLATION

5.7L (VIN P) Engine

1. Disconnect the negative battery cable.

2. Unplug the IAT electrical connection.

3. Remove the air cleaner and air intake duct assembly.

4. Drain the cooling system into a suitable container. Remove the knock sensors from the lower left and right side of the block to assure proper draining.

5. Remove the upper and lower radiator hoses and the heater hose from the water pump. Remove the throttle body hose from the tee fitting.

6. Unplug the coolant sensor electrical connection and remove the sensor wire harness from the retainer on the front of the coolant pump.

7. Use a box wrench or socket on the tensioner pulley bolt to rotate the tensioner and relieve belt tension, then remove the serpentine drive belt from the alternator pulley. This should create sufficient room to work, if additional room is necessary, the belt can be completely removed.

8. Remove the 6 bolts securing the water pump flanges to the engine block, then remove the water pump from the vehicle.

9. Remove and discard the old gaskets from the mating surfaces.

10. If replacing the pump, remove the coolant sensor from the old pump.

To install:

11. If replacing the pump, install the coolant sensor on the new pump and tighten to 17 ft. lbs. (23 Nm).

12. Thoroughly clean all gasket mating surfaces and apply a light coat of grease to the seals and splines before assembling the coupling to the water pump. The white band on the coupling should be positioned towards the engine.

13. Install the new gaskets with the tabs up, the coolant pump with the drive coupling and the mounting bolts. Tighten the bolts to 30 ft. lbs. (41 Nm).

14. Install the serpentine drive belt.

15. Install the coolant sensor wire harness to the retainer on the front of the pump, then engage the sensor electrical connection.

16. Connect the heater hose and the upper and lower radiator hoses to the water pump.

17. Install the throttle body hose at the tee.

18. If removed, install the knock sensors.

19. Open the bleed valves on the thermostat housing and the throttle body. Fill the cooling system through the radiator surge tank until a solid stream of coolant comes out of the bleeds.

20. Close all bleeds and continue to fill the surge tank until the coolant is level at the base of the surge tank neck.

21. Install the radiator pressure cap and check the coolant recovery reservoir for the proper level of coolant, add as necessary.

22. Install the air cleaner and intake duct assembly.

23. Engage the IAT electrical connection and clean any excess coolant from the engine compartment.

24. Connect the negative battery cable, start the engine and check for leaks.

25. If the low coolant indicator lamp is lit, the engine must be cycled from cold to normal operating temperature and back to cold 3 times. If the lamp does not go out after this and coolant is at the proper level, the indicator system must be repaired.

5.7L (VIN J) Engine

1. Disconnect the negative battery cable.

2. Drain engine coolant into a suitable container.

3. Remove the air cleaner and intake duct assembly.

4. Remove the screws attaching the throttle body extension to the throttle body, then remove the throttle body extension and gasket.

5. Remove clamps and hoses from the coolant outlets, radiator inlet and inlet pipe.

6. Remove the inlet pipe assembly and hose from the vehicle.

7. Remove the serpentine drive belt, then remove the tensioner retaining bolt and remove the tensioner from the pump. It is not necessary to remove the water pump pulley.

8. Remove the engine hose clamp, then the hose from the water pump.

9. Remove the alternator lower bracket mounting bolts, then remove the bracket from the vehicle.

10. Remove the water pump attaching bolts (noting the position and size of each bolt) and remove the bolt attaching the air conditioning compressor to the water pump. Remove the water pump from the vehicle.

To install:

11. Thoroughly clean the pump and front cover sealing surfaces.

12. Install the water pump, new gasket and bolts, finger-tight only.

13. Install and finger-tighten the bolt attaching air conditioning compressor to the pump.

14. Torque air conditioning compressor bolt and water pump attaching bolts to 20 ft. lbs. (26 Nm).

15. Install engine hose and clamp.

16. Apply Loctite® 565 to the bolt threads, then install the alternator

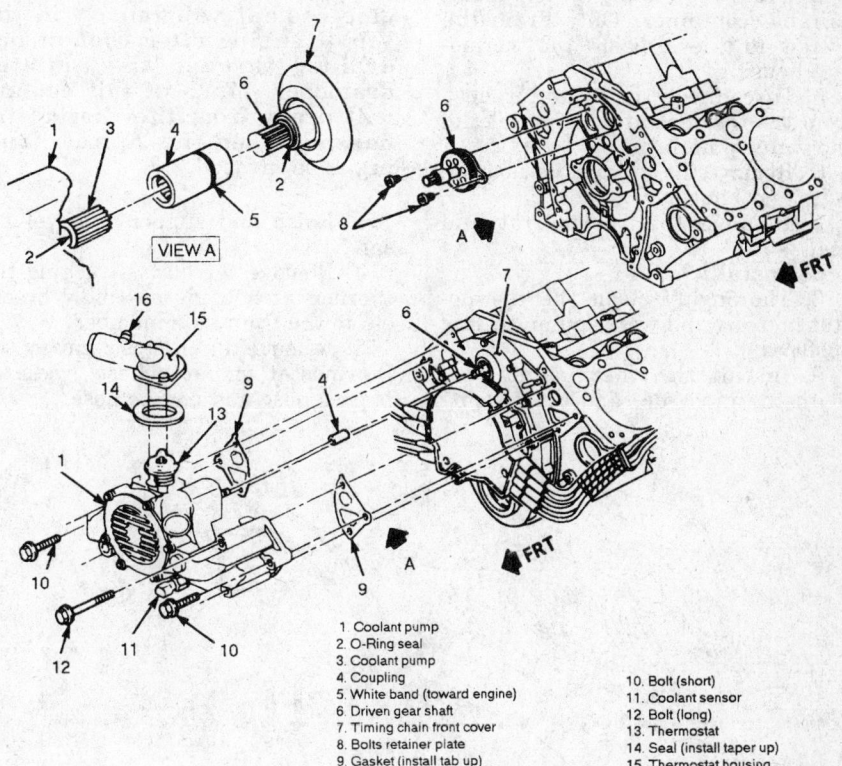

1. Coolant pump	10. Bolt (short)
2. O-Ring seal	11. Coolant sensor
3. Coolant pump	12. Bolt (long)
4. Coupling	13. Thermostat
5. White band (toward engine)	14. Seal (install taper up)
6. Driven gear shaft	15. Thermostat housing
7. Timing chain front cover	
8. Bolts retainer plate	
9. Gasket (install tab up)	

8838SG12

Exploded view of the water pump mounting — 5.7L (VIN P) engine

mounting bolts. Torque the bolts to 39 ft. lbs. (52 Nm) and the bracket bolts to 20 ft. lbs. (26 Nm).

17. Install the belt tensioner and tighten the retaining bolt to 45 ft. lbs. (60 Nm).

18. Install the serpentine drive belt.

19. Install the hose and inlet pipe assembly.

20. Install throttle body extension and gasket. Torque bolts to 53 inch lbs. (6 Nm).

21. Install air cleaner and intake duct assembly, then connect the negative battery cable.

22. Refill the cooling system with the proper type and quantity of antifreeze and inspect the system for leaks.

Thermostat

REMOVAL AND INSTALLATION

5.7L (VIN P) Engine

1. Disconnect the negative battery cable.

2. Unplug the IAT electrical connection.

3. Remove the air cleaner and intake duct assembly.

4. Drain the cooling system into a suitable container. Only drain the system to a level below the thermostat housing.

5. Disconnect the radiator hose from the thermostat housing inlet of the water pump.

6. Remove the thermostat housing bolts and housing.

7. Remove the thermostat and seal.

To install:

8. Thoroughly clean the thermostat housing and water pump sealing surfaces.

9. Install the thermostat, seal (with the taper up) and the housing

to the water pump. Tighten the bolts to 8 ft. lbs. (10 Nm).

10. Connect the radiator hose to the thermostat inlet on the water pump.

11. Open the bleed valves on the thermostat housing and the throttle body. Fill the cooling system through the radiator surge tank until a solid stream of coolant comes out of the bleeder.

12. Close all bleeds and continue to fill the surge tank until the coolant is level at the base of the surge tank neck.

13. Install the pressure cap and check the coolant recovery reservoir for the proper level of coolant, add as necessary.

14. Install the air cleaner and intake duct assembly.

15. Engage the IAT electrical connection and clean any excess coolant from the engine compartment.

16. Connect the negative battery cable, start the engine and check for leaks.

5.7L (VIN J) Engine

1. Disconnect the negative battery cable.

2. Drain the engine coolant into a suitable container.

NOTE: A large amount of engine coolant will remain in the VIN J engine after coolant has drained through the radiator draincock. Much of this coolant will drain from the thermostat housing when the 2-piece housing is separated.

3. Raise and support the vehicle safely.

4. Remove the bolts attaching the thermostat housing assembly brackets to the front side member.

5. Remove the clamps securing the thermostat radiator hose, radiator bypass hose and engine hose.

6. Remove the quick-connect fittings at the heater hose-thermostat housing junction.

7. Remove the thermostat housing assembly from the vehicle.

8. Remove the housing bolts, then remove the thermostat and seal from the housing.

To install:

9. Thoroughly clean all gasket mating surfaces.

10. Install the thermostat, seal and housing. Be sure the seal is installed with the taper towards the radiator and remains seated in housing groove when assembling the housing sections.

11. Tighten the thermostat housing bolts to 18 ft. lbs. (25 Nm).

12. Install the quick-connect fittings at the heater hose-thermostat housing junction.

13. Install the clamps to the radiator outlet hose, radiator bypass hose and the engine hose. Tighten the clamps to 35 inch lbs. (4 Nm).

14. Install the bolts retaining the housing bracket to the front side member and tighten to 18 ft. lbs. (25 Nm).

15. Lower the vehicle and connect the negative battery cable.

16. Properly fill the cooling system and check for leaks.

Electric Cooling Fans

REMOVAL AND INSTALLATION

5.7L (VIN P) Engine

Although this procedure is for both the primary (left) and auxiliary (right) fans, the primary fan does not need to be removed if only the auxiliary fan requires service.

1. Disconnect battery negative cable.

2. If removing the primary (left) fan assembly, remove the radiator upper support from the vehicle:

   a. Drain the engine coolant into a suitable container.

   b. Remove the air cleaner and air duct assembly.

   c. Remove the bleed hose from the radiator.

   d. If equipped, disconnect the automatic transmission fluid cooler lines.

   e. Remove the screws retaining the fan relays, then remove the relays from the vehicle.

   f. Unplug the electrical connectors from the air pump and horn.

   g. Remove the AIR pump.

   h. Remove the bolts retaining the shroud to the upper support,

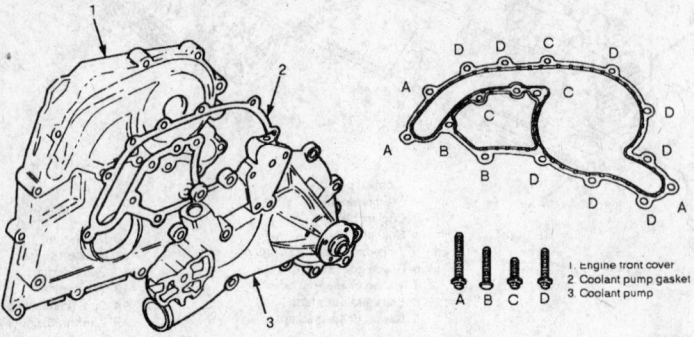

Water pump assembly mounting — 5.7L (VIN J) engine

1 Engine front cover
2 Coolant pump gasket
3 Coolant pump

8838SG13

then remove the bolts/screws retaining the upper support to the lower support.

i. Remove upper support from the vehicle.

j. If not done already, remove the shroud upper retaining bolts.

3. If removing the auxiliary (right) fan, remove the upper bolts retaining the assembly to the shroud.

4. Raise and support the vehicle safely.

5. Remove the bolts retaining the impact bar, then remove the impact bar from the vehicle.

6. Unplug the fan motor electrical connectors.

7. Remove the bolts retaining the auxiliary (right) fan assembly to the shroud, then remove the assembly from the vehicle. If necessary separate the fan blade and motor for replacement purposes.

8. If removing the primary (left) fan assembly, remove the fan shroud with the assembly still attached. With the shroud out of the vehicle, remove the assembly from the shroud. If necessary, separate the fan blade and motor for replacement purposes.

To install:

9. If separated, assemble the fan blade(s) and motor(s) for installation.

10. If removed, install the primary (left) fan assembly to the shroud and tighten the retaining bolts to 89 inch lbs. (10 Nm), then install the shroud to the vehicle. Tighten the shroud and lower mounting bolts to 80 inch lbs. (9 Nm).

11. Install the auxiliary (right) fan assembly to the shroud and tighten the retaining bolts to 89 inch lbs. (10 Nm).

12. Engage the electrical connectors to the fan assemblies.

13. Install the impact bar and tighten the retainers to 20 ft. lbs. (27 Nm), then lower the vehicle.

14. If only the auxiliary (right) fan was removed, install the upper bolts retaining the fan to the shroud and tighten to 89 inch lbs. (10 Nm).

15. If the primary (left) fan assembly was removed, install the radiator upper support.

a. Install upper support to the vehicle and secure using the screws attaching the upper support to the lower support. Tighten the screws to 80 inch lbs. (9 Nm).

b. Install the nuts and bolts retaining the support to the shroud and tighten to 18 ft. lbs. (25 Nm).

c. Install the AIR pump and tighten the bolts to 89 inch lbs. (10 Nm).

d. Engage the electrical connectors to the AIR pump and horn.

e. Install the fan relays and tighten the retaining screws to 80 inch lbs. (9 Nm).

f. If equipped, connect the transmission fluid cooler lines.

g. Connect the radiator bleed hose.

h. Install the air cleaner and duct assembly.

i. Install the shroud upper radiator bolts and tighten to 80 inch lbs. (9 Nm).

16. Connect the negative battery cable.

17. If drained, fill cooling system with the proper type and quantity of antifreeze and check for leaks.

5.7L (VIN J) Engine

Although this procedure is for both the primary and auxiliary fans, either fan assembly can be removed and serviced without removing the other.

1. Disconnect the negative battery cable.

2. If removing the auxiliary (right) fan assembly, remove the bolt retaining the upper right of the auxiliary fan assembly to the fan shroud.

3. Raise and support the vehicle safely.

4. Unplug the electrical connector from the auxiliary fan motor.

5. Remove the bolts retaining the auxiliary fan assembly to the fan shroud, then remove the assembly from the vehicle.

6. If removing the primary (left) fan assembly, lower the vehicle and drain the engine coolant into a suitable container.

7. Remove the cleaner and intake duct assembly.

8. Remove the hoses and clamps from the coolant outlets, the radiator inlet and the bypass inlet pipe.

9. Remove the hose and inlet pipe assembly from the vehicle.

10. Unplug the electrical connector from the fan motor.

11. Remove the screws retaining the fan motor to the motor support and remove the bolt retaining the air conditioning discharge line clamp to the crossmember.

12. Remove the bolts retaining the fan assembly to the fan shroud, remove the end cap from the power steering pump pulley and remove the

primary fan assembly from the vehicle.

To install:

13. If removed, install the primary fan assembly and tighten the bolts to 89 inch lbs. (10 Nm).

14. Install the screws retaining the motor to the motor support and tighten to 89 inch lbs. (10 Nm).

15. Engage the electrical connector to the primary fan motor.

16. Install the bolt retaining the air conditioning discharge line clamp to the crossmember and install the end cap to the power steering pump pulley.

17. Install the hose and inlet pipe assembly. Connect the hoses and clamps to the coolant outlets, the radiator inlet and the bypass inlet pipe.

18. Install the air cleaner and intake duct assembly

19. If the auxiliary fan assembly was removed, raise and support the vehicle safely.

20. Install the auxiliary fan assembly to the vehicle and tighten the fan-to-fan shroud bolts to 89 inch lbs. (10 Nm).

21. Engage the electrical connector to the auxiliary fan motor and lower the vehicle.

22. Install the upper right bolt retaining the assembly to the fan shroud.

23. Connect the negative battery cable.

24. If drained, fill the cooling system with the proper type and amount of coolant and check the system for leaks.

Cooling System Bleeding

If flushing is required, do not use a chemical flush. Drain and fill the cooling system using clean water until the water drained from the system in this procedure is clear.

If a flush is performed, when filling the engine with coolant begin by adding 100 percent ethylene glycol in the amount of 8.2 qts. for 5.7L (VIN P) engine or 6.9 qts. for 5.7L (VIN J) engine. Then complete the filling procedure with clean water.

PROCEDURE

5.7 (VIN J) Engine

1. With the cooling system completely drained, the engine **OFF** and radiator drain plug closed, begin adding antifreeze. Use a final mixture of 50 percent ethylene glycol anti-

freeze and 50 percent water for system refill.

NOTE: To completely drain the engine cooling system on the VIN J engine, the thermostat must be removed from the housing. The thermostat housing should temporarily be installed, without the thermostat, if system flushing is to be performed, but the thermostat must be installed before the system is filled coolant.

2. Slowly fill the cooling system through the opening in the high fill surge tank until the level is even with the base of the fill neck. Also, be sure the coolant recovery reservoir is filled to the **COLD** mark.

3. Run the engine with the pressure cap removed until normal operating temperature is reached and the upper radiator hose becomes hot.

NOTE: The coolant temperature gauge must be monitored during the running of the engine and at no time should the engine temperature be allowed to reach the 260°F mark or the engine HOT light be allowed to come ON. If this should occur, the engine should be turned OFF immediately and allowed to cool down to 80°F (27°C) before continuing with the bleeding process.

4. With the engine idling, add coolant until the level reaches the bottom of the high fill reservoir filler neck.

5. Install the pressure cap, making sure the arrows align he overflow tube.

6. Check that the coolant recovery reservoir is now at the **HOT** level, and add coolant as necessary.

7. If the low coolant indicator lamp is lit after this procedure, the engine must be cycled from cold to normal operating temperature and back to cold 3 times. If the lamp does not go out after this and engine coolant is filled to the proper level, the indicator system must be repaired.

5.7 (VIN P) Engine

1. With the cooling system completely drained, the engine **OFF** and radiator drain plug closed, open the bleed valves on the thermostat housing and the throttle body.

NOTE: To completely drain the engine cooling system on the VIN P engine, the knock sensors must be removed from the lower left and right sides of the block. The

sensors should then be reinstalled before filling the engine with coolant.

2. Begin adding antifreeze. Use a final mixture of 50 percent ethylene glycol antifreeze and 50 percent water for system refills. Fill the cooling system through the radiator surge tank until a solid stream of coolant comes out of the bleeders.

3. Close all bleeds and continue to fill the surge tank until the coolant is level at the base of the surge tank neck.

4. Install the pressure cap and check the coolant recovery reservoir for the proper level of coolant, add as necessary. If removed, install the air cleaner and intake duct assembly.

5. Clean any excess coolant from the engine compartment.

6. Connect the negative battery cable, start the engine and check for leaks.

7. Run the engine at normal operating temperature and make sure the coolant remains at the proper level.

8. If the low coolant indicator lamp is lit after this procedure, the engine must be cycled from cold to normal operating temperature and back to cold 3 times. If the lamp does not go out after this and engine coolant is filled to the proper level, the indicator system must be repaired.

FUEL SYSTEM

Fuel System Service Precautions

Safety is the most important factor when performing not only fuel system maintenance but any type of maintenance. Failure to conduct maintenance and repairs in a safe manner may result in serious personal injury or death. Maintenance and testing of the vehicle's fuel system components can be accomplished safely and effectively by adhering to the following rules and guidelines.

• To avoid the possibility of fire and personal injury, always disconnect the negative battery cable unless the repair or test procedure requires that battery voltage be applied.

• Always relieve the fuel system pressure prior to disconnecting any fuel system component (injector, fuel rail, pressure regulator, etc.), fitting

or fuel line connection. Exercise extreme caution whenever relieving fuel system pressure to avoid exposing skin, face and eyes to fuel spray. Please be advised that fuel under pressure may penetrate the skin or any part of the body that it contacts.

• Always place a shop towel or cloth around the fitting or connection prior to loosening to absorb any excess fuel due to spillage. Ensure that all fuel spillage (should it occur) is quickly removed from engine surfaces. Ensure that all fuel soaked cloths or towels are deposited into a suitable waste container.

• Always keep a dry chemical (Class B) fire extinguisher near the work area.

• Do not allow fuel spray or fuel vapors to come into contact with a spark or open flame.

• Always use a backup wrench when loosening and tightening fuel line connection fittings. This will prevent unnecessary stress and torsion to fuel line piping. Always follow the proper torque specifications.

• Always replace worn fuel fitting O-rings with new. Do not substitute fuel hose or equivalent, where fuel pipe is installed.

Fuel System Pressure

RELIEVING

1. Disconnect the negative battery cable.

2. Loosen the fuel filler cap to relieve the tank pressure.

3. Wrap a shop towel around the fuel pressure valve fitting (located on the side or end of the fuel rail assembly) to catch any fuel spray and connect J-34730-1 or an equivalent fuel gauge.

4. Install the bleed hose into a suitable container, then open the valve to bleed the fuel system pressure.

5. Close the valve and disconnect the fuel gauge. Drain any remaining fuel from the gauge into the bleed container.

Idle Speed

ADJUSTMENT

Idle speed is controlled by the ECM through the Idle Air Control (IAC) valve pintle position. The ECM is programmed to determine the correct pintle position based on various inputs, therefore idle speed is not adjustable.

When installing a new IAC valve, measure the distance between the tip of the valve pintle and the mounting flange. The distance should be no more than 1.10 inch (28mm). If the distance is greater, adjust the valve pintle by applying finger pressure to retract the pintle. Do not push the pintle of a used IAC valve, as force may be sufficient to damage the worn threads of a valve pintle which has been in service.

If the negative battery cable is disconnected and reconnected with the engine running or if the IAC valve is replaced, the idle speed may be wrong. If this is the case the IAC valve may be reset as follows:

1. Depress the accelerator pedal slightly.
2. Start the engine and run for 5 seconds.
3. Turn the ignition **OFF** for 10 seconds.
4. Restart the vehicle and check for proper idle operation.

Fuel Filter

REMOVAL AND INSTALLATION

1. Disconnect the battery negative cable and properly relieve the fuel system pressure.
2. Remove the fuel pipe retaining nut from the evaporator case.
3. Raise and support the vehicle safely.
4. For convertibles, remove the underbody brace.
5. Remove the 3 nuts retaining the fuel pipes to the chassis.
6. Clean the filter connections and surrounding areas to prevent fuel system contamination.
7. Disconnect the inlet pipe from the filter, drain any remaining fuel from the line and reposition the line for access to the filter.
8. Hold the pipe outlet nut and remove the filter by turning to unthread it from the fitting.
To install:
9. Check the fuel pipe O-rings for cuts, nicks, swelling or distortion and replace, if damaged.
10. Install the fuel filter onto the fuel outlet pipe nut.
11. Connect the fuel inlet pipe to the filter and tighten the fitting to 20 ft. lbs. (27 Nm).
12. Install the 3 chassis-to-fuel pipe retaining nuts.
13. For convertibles, install the underbody brace, tighten the retaining

nuts to 20 ft. lbs. (27 Nm) and the retaining bolts to 47 ft. lbs. (63 Nm).
14. Lower the vehicle and install the fuel pipe retaining nut to the evaporator case.
15. Tighten the fuel filler cap and connect the negative battery cable.
16. Turn the ignition **ON** for 2 seconds, **OFF** for 10 seconds, then **ON** again and inspect the system for leaks.

Fuel Pump

REMOVAL AND INSTALLATION

NOTE: Vehicles equipped with the 5.7L (VIN J) engine use a fuel sender assembly which is equipped with 2 fuel pumps. The strainers and pumps are not serviced separately from the sender and, if 1 component is damaged, the sender assembly must be replaced as a unit.

1. Disconnect the negative battery cable.
2. Properly relieve the fuel system pressure and drain the fuel tank.
3. Remove the 4 filler door bezel attaching screws, then remove the filler door bezel.
4. Lift the fuel tank filler neck housing and disconnect the drain hose from the nipple. Remove filler neck housing.
5. Clean the area around all fuel fittings to prevent system contamination, then disconnect and plug the fuel pipes and fuel vapor pipe.
6. Unplug the sending unit electrical connector, remove the attaching bolts and remove the sending unit assembly from the vehicle.
7. If equipped with the VIN J engine, replace fuel sender assembly.
8. If equipped with the VIN P engine, service the sender assembly, as necessary:
   a. Note the position of the fuel strainer on the pump.
   b. Support the pump with one hand and grasp the strainer with the other. Turn the strainer in one direction, pull the strainer off the pump and discard it.
   c. Unplug the fuel pump electrical connection.
   d. Place the fuel sender assembly upside down on a flat bench.
   e. Pull the fuel pump downward to remove it from the mounting bracket, then tilt the pump outward and remove it from the pulse dampener.
   f. Note the position of the dampener on the inlet tube, then remove

the dampener from the tube. Shake the dampener and listen for fuel, if fuel is heard inside the dampener, it must be replaced.
To install:
9. If equipped with the VIN P engine, assemble the sender for installation, as necessary:
   a. Install the fuel pulse dampener in the same position as noted during disassembly.
   b. Assemble the rear bumper and insulator onto the fuel pump.
   c. Position the fuel sender assembly upside down on a flat bench and install the fuel pump between the fuel pulse dampener and mounting bracket.
   d. Engage the pump electrical connector.
   e. Install the new fuel strainer into the same position as noted during disassembly. Push on the outer edge of ferrule until fully seated.
10. Position a new gasket on the fuel tank with the notch facing forward in the right corner of the fuel tank.
11. Carefully fold the strainer to allow it to fit through the opening in the tank. Make sure the strainer unfolds in the tank and lower the fuel sender assembly into position.
12. Install the fuel sender assembly attaching screws and tighten alternately and evenly to 45 inch lbs. (5 Nm).
13. Engage the fuel sender assembly electrical connector.
14. Connect all sender assembly fuel and vapor hoses.
15. Connect the fuel drain hose to the nipple on the rubber filler neck housing, then position the housing around the fuel tank filler neck.
16. Install the filler door bezel with the attaching screws.
17. Add fuel, tighten the filler cap and connect the negative battery cable.
18. Turn the ignition **ON** for 2 seconds, **OFF** for 10 seconds, then **ON** again and inspect the system for leaks.

Fuel Injector

REMOVAL AND INSTALLATION

5.7L (VIN P) Engine

1. Disconnect the negative battery cable and properly relieve fuel system pressure.
2. Remove the fuel rail cover.

3. Disengage the quick-connect fittings at the fuel rail feed and return pipes as follows:

a. Grasp both ends of a connection and twist ¼ turn in each direction to loosen any dirt. Repeat for other fitting.

b. While wearing safety glasses, use compressed air to blow out dirt from the fitting.

c. Insert a fuel line separator tool, into the female connector, then push inward to release the male connector and repeat for the other fitting.

4. Disconnect the vacuum line at the pressure regulator.

5. Unplug the injector electrical connectors.

6. Remove the fuel rail attaching bolts and carefully remove the fuel rail assembly along with the injectors, from the top of the intake.

7. Rotate the injector retaining clip to the release position and remove the injector from the fuel rail assembly.

8. Remove and discard the O-ring seals from either side of the injector.

9. Remove and discard the injector retaining clip.

To install:

10. Lubricate the new injector O-rings with clean engine oil and install onto the injector.

NOTE: Always replace injectors using an identical part number as inscribed in the old injector.

11. Connect a new retainer clip onto the fuel injector and install the injector to the fuel rail assembly. Rotate the injector retaining clip to the lock position.

12. Install the fuel rail assembly to the intake manifold. Tighten the attaching bolts to 15 ft. lbs. (20 Nm).

13. Rotate the fuel injectors as necessary to avoid stretching the wire harnesses and connect the injector electrical connections.

14. Connect the vacuum line to the pressure regulator.

15. Apply a few drops of clean engine oil to the male ends of the fuel line quick-connect fittings. Engage the fittings by pushing the connectors together until the retaining tabs snap into place. Pull gently on both sides of each fitting to be sure the connection is secure.

16. Tighten the fuel filler cap and connect the negative battery cable.

17. Turn the ignition **ON** for 2 seconds, **OFF** for 10 seconds, then **ON** again and inspect the system for leaks.

5.7L (VIN J) Engine

1. Disconnect the negative battery cable and properly relieve fuel system pressure.

2. Drain the cooling system into a suitable container.

3. Remove the intake plenum assembly.

4. If not done already when removing the intake plenum, disconnect the fuel feed and return lines from the fuel rail. Remove and discard the old O-rings from the fittings. Disconnect the vacuum line to the pressure regulator.

5. Unplug the fuel injector wire connectors.

6. Remove the bolts securing the fuel rail to the injector housing.

7. Carefully remove the fuel rails making sure not to damage the injector connector terminals or spray tips. Remove the spacers, if equipped. Note the routing of the vacuum hoses around the fuel rail before removing the rail.

8. Remove the injector retaining clip, then remove the injector.

9. Remove and discard the injector O-ring seals.

To install:

10. Lubricate new injector O-rings with engine oil and install the injector with retaining clip onto fuel rail. Make sure the injector wire connection is facing outward and push the injector onto the rail sufficiently to engage the clip with the machined slots on the rail socket.

NOTE: Each injector is calibrated for a specific flow rate and must be replaced with an identical part number.

11. Install the fuel rail into the injector housing, routing the vacuum lines in their previous positions around the rail.

12. If equipped, be sure the spacers are properly positioned under the rail mounting bracket.

13. Install the fuel rail bolts and tighten to 20 ft. lbs. (26 Nm).

14. Engage the injector electrical connectors, turning the injectors if necessary to avoid stretching the wire harnesses.

15. Install new O-rings to the fuel feed and return pipes.

16. Connect the fuel feed and return lines with the retaining bolts tightened to 13 ft. lbs. (18 Nm). Temporarily connect the negative battery cable and turn the ignition switch **ON** for 2 seconds, then **OFF** for 10 seconds. Cycle the ignition once again to assure proper system pressure, then disconnect the battery negative

cable and inspect the fuel system for leaks.

17. Install the plenum assembly. It may be necessary to disconnect the fuel lines for proper installation; if so be sure to relieve the system pressure before disconnecting any fuel fitting.

18. Tighten the fuel filler cap and connect the negative battery terminal.

19. Properly refill the engine cooling system.

EMISSION CONTROLS

Emission Warning Lamps

The SERVICE ENGINE SOON emission warning lamp located on the instrument panel has 2 main functions:

1. The lamp indicates to the driver that a problem has occurred and the vehicle should be taken for service as soon as reasonably possible.

2. The light may be used by technicians to monitor diagnostic trouble codes and/or open/closed loop engine operation, whenever the system is placed in the diagnostic modes.

To verify proper operation of the bulb and wiring, the lamp will illuminate when the ignition is first turned **ON**, but the engine is not running. If the system is operating properly, the lamp will turn **OFF** once the engine is started.

If the SERVICE ENGINE SOON lamp remains lit once the engine is started, the self-diagnostic system has detected a problem. If the problem goes away, the light will extinguish in 10 seconds (in most cases), but a diagnostic trouble code will remain in the ECM memory.

RESETTING

NOTE: In order to prevent damage to the ECM, the key must be OFF when connecting or disconnecting power to the ECM.

After repairs are made to the faulty system(s), it is necessary to make sure the ECM memory is cleared of any old diagnostic trouble codes. Removing the battery voltage to the ECM for a minimum of 30 seconds will clear all codes. This may be accomplished in various ways depending on how the vehicle is equipped. The ECM harness power feed may be

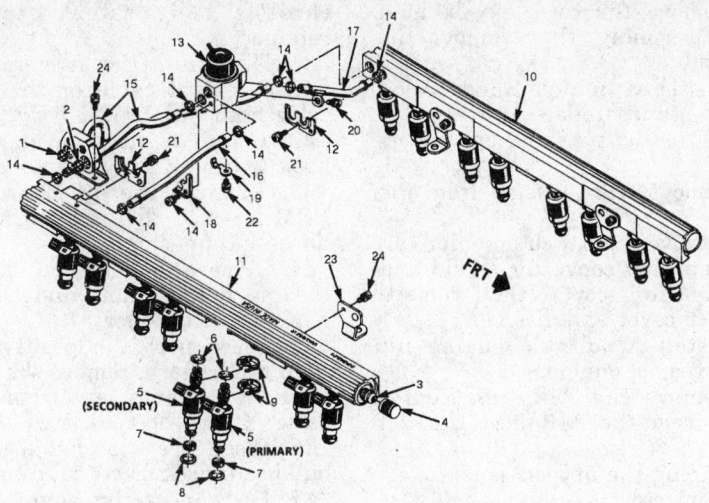

1. O-Ring fuel inlet line
2. O-Ring fuel return line
3. Connection assembly fuel pressure
4. Cap fuel pressure connection assembly
5. Injector assembly fuel: P-Primary, S-Secondary
6. O-Ring upper injector (black)
7. Backup O-Ring
8. O-Ring lower injector (brown)
9. Clip injector retainer
10. Rail assembly fuel (lh)
11. Rail assembly fuel (rh)
12. Retainer crossover tube
13. Regulator assembly fuel pressure
14. Seal O-Ring
15. Tube assembly fuel inlet and outlet
16. Tube fuel return (rh)
17. Tube fuel return (lh)
18. Retainer regulator tube
19. Retainer fuel inlet tube
20. Screw fuel return tube retainer attaching
21. Screw crossover tube retainer attaching
22. Screw fuel inlet tube retainer attaching
23. Bracket fuel rail mounting
24. Screw bracket attaching

8838SG14

Exploded view of the fuel rail assembly — 5.7L (VIN J) engine

disconnected at the positive battery terminal "pigtail." The fuse may be removed from the inline fuseholder which originates at the positive battery connection or from fuse block, as applicable. Also, the negative battery cable may be disconnected, but other on-board data such as the clock or radio presets will also be lost.

ECM LEARNING ABILITY

The ECM has a "learning" ability which allows it to make corrections for minor variations in the fuel system, in order to improve driveability. If the battery is disconnected to clear diagnostic codes or for safety during repairs, the "learning" process will reset and must begin again. A change may be noted in the vehicle's performance while the learning process begins. To "teach" the vehicle, make sure the engine is at normal operating temperature, then drive the vehicle at part throttle, with moderate acceleration and idle conditions, until normal performance returns.

ENGINE MECHANICAL

NOTE: Disconnecting the negative battery cable on some vehicles may interfere with the functions of the on board computer systems and may require the computer to undergo a relearning process.

Engine Assembly

REMOVAL AND INSTALLATION

5.7L (VIN P) Engine

1. Disconnect the negative battery cable and properly relieve fuel system pressure.
2. Drain the coolant into a suitable container.
3. Remove the air intake duct.
4. Unplug the electrical harness and vacuum connections from the top of the engine.
5. Disconnect the upper radiator, radiator hose and heater hoses from the pump, then remove the throttle body coolant hose from the radiator

tee and from the right side of the throttle body.
6. Remove the power steering pump and support aside.
7. Remove the alternator and support aside.
8. Remove the left wheel well center panel. Remove the serpentine drive belt.
9. Remove the air conditioning compressor from the bracket and position aside.
10. Unplug the electrical connector and remove the cover from the wiper motor.
11. Disconnect the AIR diverter valve hose.
12. Disconnect and plug the fuel lines at the fuel rail.
13. Remove the hoses from the power steering fluid reservoir. Plug the openings to prevent system contamination or excessive fluid loss.
14. Disconnect the accelerator cable from the throttle body.
15. Raise and support the vehicle safely.
16. Remove the starter motor.
17. Remove the left and right catalytic converters, then remove the exhaust pipe and muffler assembly.
18. Remove the transmission.
19. If equipped with a manual transmission, remove the clutch cover and plate, then remove the flywheel.
20. Remove the ground leads from the rear of the engine, then disengage the electrical connectors from the oil level, knock, oil temperature and coolant temperature sensors.
21. Remove the nuts from the engine mount studs, then lower the vehicle.
22. Install a suitable lifting device and carefully remove the engine from the vehicle.
   To install:
23. Lower the engine into position in the vehicle.
24. Remove the lifting device from the engine, then raise and support the vehicle safely.
25. Install the nuts on the engine mount studs.
26. Engage the sensor electrical connectors and then connect the ground leads to the rear of the engine.
27. If equipped with a manual transmission, install the flywheel, then install the clutch cover and plate.
28. Install the transmission.
29. Install the right and left catalytic converters, then install the exhaust pipe and muffler assembly.

30. Install the starter and lower the vehicle.

31. Connect the accelerator cable to the throttle body, then unplug the openings and install the hose to the power steering reservoir.

32. Unplug and connect the fuel lines to the fuel rail.

33. Connect the AIR diverter valve hose.

34. Install the electrical connector and cover to the wiper motor.

35. Install the air conditioning compressor.

36. Install the serpentine drive belt, then install the left wheel well center panel.

37. Install the alternator and the power steering pump.

38. Connect the coolant hose to the right side of the throttle body and connect the throttle body hose to the radiator tee. Connect the heater and radiator hoses.

39. Engage the electrical harness and all vacuum connections to the top of the engine.

40. Install the air intake duct and properly fill the engine cooling system.

41. Check all fluid levels, connect the negative battery cable and tighten the fuel filler cap.

42. Reset the CHANGE OIL indicator:

   a. Turn the ignition **ON** but do not start the engine.

   b. Depress the ENG MET button on the trip monitor, then within 5 seconds, depress the button a 2nd time. Within another 5 seconds, depress and hold the GAUGES button.

   c. While holding the GAUGES button and watch the CHANGE OIL light, it should begin to flash. Continue to hold the gauges button until the flashing stops and the light goes out indicating that the indicator is reset.

   d. If the indicator does not reset, turn the ignition **OFF** and restart the procedure.

43. Check and adjust the ASR control cables, as necessary.

44. Start the engine and check for leaks, then bleed the power steering system.

5.7L (VIN J) Engine

1. Disconnect the battery negative cable and properly relieve fuel system pressure.

2. Raise and support the vehicle safely.

3. Drain engine coolant into a suitable container and drain the engine oil.

4. Remove the complete exhaust system assembly, then remove the driveshaft.

5. Position a suitable transmission support stand under transmission and remove the transmission support beam.

6. Remove transmission from the vehicle.

7. Remove the clutch actuator cylinder, left side converter shield and clutch housing cover, then remove the clutch cover and disc.

8. Install a suitable engine lift hook to rear of engine.

9. Remove the AIR tube center section from the AIR hose and oil pan.

10. Unplug the oxygen sensor electrical connector.

11. Remove the power steering lower hose from the oil cooler.

12. Remove the negative battery cable from the cylinder case.

13. Remove the nuts attaching the engine mounts to the driveline and suspension frame, then lower the vehicle.

14. Remove the air cleaner assembly and air duct.

15. Disconnect the engine oil cooler lines from the oil filter housing.

16. Raise the rear of the engine.

17. Disconnect the fuel lines from the fuel rail.

18. Remove the evaporator housing panel and the resistor.

19. Remove the bolts attaching the right bulkhead connector.

20. Remove the engine right side wiring harness.

21. Remove the instrument panel right lower sound insulator pan.

22. Disengage the bulkhead wiring harness connectors from under the dash.

23. Remove the air bleed hose from the plenum.

24. Remove the radiator upper and lower hoses, then disconnect the power steering pump vacuum line(s).

25. Properly discharge and recover the air conditioning system.

26. Remove the air conditioning suction and discharge line flange from the compressor, then remove the air conditioning compressor-to-accumulator line from the accumulator.

27. Remove the air conditioning accumulator and position aside.

28. Remove the air conditioning accumulator bracket from the vehicle.

29. Disconnect and plug the power steering pressure line at the power steering gear.

30. Disconnect the throttle body linkage shield, then remove the throttle body cable to plenum retainer.

31. Disconnect the accelerator and cruise control cable or the control cable from the throttle body.

32. Install a suitable engine lift hook to front of the engine.

33. Remove the ECM from the ECM bracket, then disconnect ECM harness connector.

34. Remove the left front fender attaching bolts, shims and seal. Remove the left fender.

35. Disconnect the positive cable from the battery, remove the battery hold-down clamp, and remove the battery from the vehicle.

36. Disengage the engine left side bulkhead block electrical connector.

37. Disconnect the engine wiring harness fusible links at the junction block.

38. Disengage the engine harness connectors from the following:

   Secondary injector modules
   Positive battery cable at junction block
   Differential pressure switch vacuum and electrical connectors
   Air conditioning cutout relay
   Air conditioning high blower relay
   Transmission shift solenoid relay
   Fuel pump fuse
   Forward light link connector
   Positive battery lead
   Air conditioning blower resistor
   Air conditioning pressure sensors
   Air conditioning cooling fan switch
   Windshield washer pump
   Low coolant sensor
   Blower motor
   ESC knock sensor
   ESC knock sensor relay

39. Disconnect hoses from the vacuum pump, then tag and disconnect the front and rear vacuum connections.

40. Reposition engine harness aside and remove the braided ground strap from the left side frame rail.

41. Reposition the positive battery cable aside and remove the left side plenum panel screen.

42. Disconnect the brake booster vacuum hose.

43. Remove the windshield wiper motor from the vehicle.

44. Remove the MAP sensor and the MAP sensor bracket from the plenum.

45. Disconnect the AIR hose from the left exhaust manifold.

46. Using an engine lifting device, carefully remove the engine from the vehicle.

47. Transfer the following parts to the new engine, as necessary:
   Oil level indicator tube
   The exhaust manifolds
   The converter heat shields
   The wire pack heat shields
   Engine mounts
   To install:
48. Install the engine mounts to the drivetrain and to the suspension frame, finger-tighten only at this time.
49. Position the engine into the vehicle using the lifting device.
50. Install the engine mount/bracket bolts, then remove the the lifting device and lifting brackets.
51. Connect the AIR hose to the left exhaust manifold.
52. Install the MAP sensor and bracket to the plenum.
53. Install the wiper motor.
54. Install the left side plenum panel screen.
55. Route the left side wiring harness into position, then install the braided ground strap to the frame rail.
56. Install the left side bulkhead block connector.
57. Engage the engine harness fusible links and relays.
58. Install the battery and hold-down clamp.
59. Connect the battery positive cable to the battery, then install the left front fender.
60. Install the ECM to the ECM bracket, then engage the wiring harness electrical connector.
61. Remove the front engine lift hook, then connect power brake booster vacuum hose to the plenum.
62. Connect the cruise control and throttle cables or the control cable to the throttle body. Adjust the cables or the ASR cable adjuster, as applicable.
63. Install the cable shield, then install cable retainers to the plenum.
64. Connect the power steering pressure line to the power steering gear.
65. Connect the engine oil cooler lines to the engine.
66. Install the accumulator bracket and then install the accumulator.
67. Connect the air conditioning lines.
68. Attach the vacuum line(s) to the power steering pump.
69. Connect the radiator upper and lower hoses.
70. Connect the air bleed hose to the plenum.
71. Install the bulkhead wire connector to the bulkhead.
72. Engage the evaporator housing panel resistor connector.

73. Install the hose onto the vacuum pump, then install the front and rear vacuum connections.
74. Engage the engine harness connectors to the following:
   Air conditioning blower resistor
   Air conditioning pressure sensor
   Air conditioning cooling fan
   Windshield washer pump
   Low coolant sensor
   Blower motor
   ESC knock sensor
   ESC knock sensor relay
   Differential pressure switch
75. Connect the fuel lines to the fuel rail.
76. Install the engine right side wiring harness under the dash.
77. Install the instrument panel right sound insulator panel.
78. Raise and support the vehicle safely, then tighten the engine/bracket bolts and nuts to 40 ft. lbs. (54 Nm).
79. Install the power steering hose to power steering oil cooler.
80. Install the oxygen sensor wire connectors.
81. Connect the AIR tube center section to the AIR hose and oil pan.
82. Connect the negative battery cable to the engine and suitably support the engine, then remove the rear engine lift hook.
83. Install the clutch cover and disc, then install the clutch housing to the cylinder block and install the housing cover.
84. Install the left side converter shield to the housing, then position the actuator cylinder and install the retaining nuts.
85. Install the transmission and support beam.
86. Install the driveshaft.
87. Install the complete exhaust system assembly.
88. Lower the vehicle and add the proper type and amount of engine oil.
89. Tighten the fuel filler cap and connect the negative battery cable.
90. Properly fill the engine cooling system and check for leaks.
91. Recharge the air conditioning system.
92. If equipped, reset the CHANGE OIL indicator:
   a. Turn the ignition **ON** but do not start the engine.
   b. Depress the ENG MET button on the trip monitor, then within 5 seconds, depress the button a 2nd time. Within another 5 seconds, depress and hold the GAUGES button.
   c. While holding the GAUGES button and watch the CHANGE OIL light, it should begin to flash.

Continue to hold the gauges button until the flashing stops and the light goes out indicating that the indicator is reset.
   d. If the indicator does not reset, turn the ignition **OFF** and restart the procedure.

Engine Mounts

REMOVAL AND INSTALLATION

5.7L (VIN P) Engine

1. Disconnect the negative battery cable.
2. Raise and support the vehicle safely.
3. Remove the engine mount nuts from both sides.
4. Disconnect the catalytic converters from the exhaust manifolds and reposition.

NOTE: When raising and supporting the engine, NEVER place a jack under the oil pan, crankshaft pulley or any sheetmetal. There is a minimal clearance between the oil pan and the pump screen. Jacking against the pan could cause sufficient deformation to damage the oil pickup unit.

5. Raise the engine with a lifting device, sufficiently for the mount studs to clear the crossmembers.
6. Remove the engine mount through-bolt and nut.
7. Remove the engine mounts, heat shield, and any spacers, if present.
8. Remove the engine bracket bolts and remove the bracket.
   To install:
9. Install the engine bracket and bolts. Tighten the bolts to 41 ft. lbs. (56 Nm).
10. Install the engine mounts, heat shield and spacers.
11. Install the engine mount through bolt and nut.
12. Lower the engine and remove the lifting device, then tighten the mount nuts to 40 ft. lbs. (54 Nm) and the through-bolt to 77 ft. lbs. (105 Nm).
13. Install the catalytic converters and nuts to the exhaust manifolds. Tighten the nuts to 15 ft. lbs. (21 Nm).
14. Lower the vehicle and connect the negative battery cable.

5.7L (VIN J) Engine

1. Disconnect the battery negative cable.

2. Remove the exhaust manifold.

3. Remove the nut attaching the engine mount to the drivetrain and suspension frame.

4. Support and raise the engine just sufficiently for engine mount removal clearance.

5. Remove the engine mount/bracket nut and bolt from the bracket.

6. Remove the engine mount and heat shield from the vehicle.

7. Remove the bolts attaching the bracket to the cylinder case, then remove the bracket from the vehicle.

To install:

8. Install the engine mount bracket to the cylinder case and install the retaining bolts. Tighten the bolts to 38 ft. lbs. (52 Nm).

9. Install the engine mount and heat shield onto the vehicle using with the engine mount/bracket bolt and nut. Tighten the nut to 40 ft. lbs. (54 Nm).

10. Lower the engine into position and install the nut retaining the mount to the drivetrain and suspension frame. Tighten the nut to 40 ft. lbs. (54 Nm).

11. Install the exhaust manifold.

12. Connect the negative battery cable.

Cylinder Head

REMOVAL AND INSTALLATION

5.7L (VIN P) Engine

Right Side

1. Disconnect the negative battery cable and properly relieve the fuel system pressure.

2. Raise and support the vehicle safely.

3. Disconnect the catalytic converter.

4. Drain the engine cooling system, then lower the vehicle.

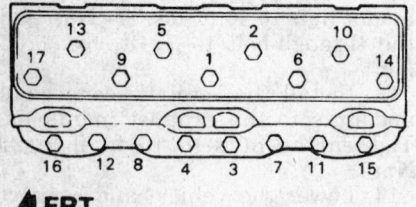

Cylinder head bolt torque sequence — 5.7L (VIN P) engine

5. Remove the lower radiator and heater hoses from the water pump.

6. Disconnect the power steering pump reservoir from the cylinder head and reposition aside.

7. Remove the coil and bracket assembly.

8. Remove the intake manifold.

9. Remove the spark plug wires from the clips, then remove the front wire bracket.

10. Remove the oil level indicator tube.

11. Disconnect the spark plug wires from the plugs, then remove the spark plugs from the cylinder head.

12. Remove the right exhaust manifold.

13. Using a backup wrench on the pipe fitting, disconnect the coolant air bleed pipe from the left cylinder head.

14. Remove the right valve rocker cover, then remove the rocker arm and pushrod assemblies.

15. Remove the cylinder head bolts.

16. Remove the cylinder head along with the coolant air bleed pipe, then remove the head gasket.

17. If necessary, remove the coolant air bleed pipe from the cylinder head.

To install:

18. Thoroughly clean the cylinder head and cylinder case mating surfaces. Make sure both surfaces are free of any foreign matter, nicks or scratches. The threads in both the bolts holes and on the bolts must be clean and free of old sealer.

19. If removed, install the coolant air bleed pipe to the cylinder head, finger-tight.

20. Position the new gasket in place on the cylinder case with the yellow tab facing up. Install the cylinder head over the dowel pins and gasket.

21. Coat the bolts with 1052080, or an equivalent sealant. Install and tighten the cylinder head bolts, using 3 passes of the proper sequence, to 65 ft. lbs. (88 Nm).

22. Install the rocker arm and pushrod assemblies.

23. Install the valve rocker cover and tighten the bolts to 100 inch lbs. (11 Nm).

24. Connect the coolant air bleed pipe to the left cylinder head and torque the pipe to both cylinder heads. Using a backup wrench, tighten the coolant air bleed pipe to 30 ft. lbs. (41 Nm).

25. Install the right exhaust manifold.

26. Install the spark plugs and tighten to 11 ft. lbs. (15 Nm).

27. Connect the spark plug wires to the plugs, then install the oil level indicator tube.

28. Install the front wire bracket, then connect the spark plug wire harness assembly to the wire bracket.

29. Install the intake manifold.

30. Install the coil and bracket assembly.

31. Position and secure the power steering pump reservoir.

32. Install the lower radiator and heater hoses to the water pump.

33. Raise and support the vehicle safely.

34. Connect the catalytic converter, then lower the vehicle.

35. Properly fill the cooling system.

36. Tighten the fuel filler cap and connect the negative battery cable.

Left Side

1. Disconnect the negative battery cable and properly relieve the fuel system pressure.

2. Raise and support the vehicle safely.

3. Disconnect the catalytic converter.

4. Drain the engine cooling system, then lower the vehicle.

5. Remove the upper radiator hose.

6. Remove the serpentine drive belt.

7. Remove the intake manifold.

8. Remove the left wheel well lower center panel.

9. Disconnect the air conditioning compressor from the bracket and position aside. Use care not to kink or damage the refrigerant lines. Remove the compressor and alternator brace.

10. Remove the spark plug wire bracket, disconnect the wires from the spark plugs and remove the spark plugs from the cylinder head.

11. Remove the left exhaust manifold.

12. Remove the remaining alternator brace, then remove the alternator.

13. Disconnect the AIR diverter valve hose.

14. Remove the left valve rocker cover.

15. Remove the drive belt idler pulley, then remove the drive belt tensioner.

16. Disconnect the power steering lines from the pump, then remove the pump. Plug the openings to prevent system contamination or excessive fluid loss.

17. Remove the spark plug and coil wires from the distributor.

18. Remove the accessory mounting bracket.

19. Remove the rocker arm and pushrod assemblies.

20. Disconnect the coolant air bleed pipe from the cylinder head.

21. Remove the cylinder head bolts, then remove the cylinder head and gasket.

To install:

22. Thoroughly clean the cylinder head and cylinder case mating surfaces. Make sure both surfaces are free of any foreign matter, nicks or scratches. The threads in both the bolts holes and on the bolts must be clean and free of old sealer.

23. Position the new gasket in place on the cylinder case with the yellow tab facing up. Install the cylinder head over the dowel pins and gasket.

24. Coat the bolts with 1052080, or an equivalent sealant. Install and tighten the cylinder head bolts, using 3 passes of the proper sequence, to 65 ft. lbs. (88 Nm).

25. Connect the coolant air bleed pipe to the cylinder head and tighten to 30 ft. lbs. (41 Nm).

26. Install the rocker arm and pushrod assemblies.

27. Install the accessory mounting bracket and bolts. Tighten the bolts to 31 ft. lbs. (42 Nm) .

28. Connect the spark plug and coil wires to the distributor.

29. Install the power steering pump, then remove the plugs from the openings and connect the lines.

30. Install the drive belt tensioner, then install the idler pulley. Tighten the tensioner and pulley bolts to 24 ft. lbs. (33 Nm).

31. Install the left valve rocker cover and bolts. Tighten the bolts to 100 inch lbs. (11 Nm).

32. Connect the AIR diverter valve hose and install the alternator lower brace.

33. Install the left exhaust manifold.

34. Install the spark plugs and tighten to 11 ft. lbs. (15 Nm). Connect the spark plug wires to the plugs and insert the wires into the brackets.

35. Install the air conditioning compressor and alternator brace, then install the compressor.

36. Install the left wheel well lower center panel.

37. Install the intake manifold.

38. Install the serpentine drive belt and the upper radiator hose.

39. Raise and safely support the vehicle, then connect the catalytic converter and lower the vehicle.

40. Properly fill the engine cooling system.

41. Tighten the fuel filler cap and connect the negative battery cable.

42. Bleed the power steering system.

5.7L (VIN J) Engine

Right Side

1. Disconnect the negative battery cable and properly relieve fuel system pressure.

2. Drain engine coolant into a suitable container.

3. Remove the intake plenum assembly.

4. Remove the right injector housing.

5. Remove the right bank valve lifters.

6. Remove the alternator assembly.

7. Disconnect the right exhaust manifold from the cylinder head. It is not necessary to completely remove the exhaust manifold from the vehicle for cylinder head removal.

8. If raised, lower the vehicle for underhood access.

9. Remove the vacuum hose from secondary port throttle valve actuator.

10. Remove the access plug from the right cylinder head.

11. Remove the top bolt attaching the right secondary timing chain fixed guide.

12. Remove cylinder head bolts, then remove the cylinder head and gasket from the vehicle.

To install:

13. Thoroughly clean the cylinder head and cylinder case mating surfaces. Make sure both surfaces are free of any foreign matter, nicks or scratches. The threads in both the bolts holes and on the bolts must be clean and free of old sealer.

NOTE: Cylinder head gaskets are not interchangeable between cylinder banks.

14. Install the cylinder head locating dowels into block, if loosened or removed, then position the new gasket in place on the cylinder case.

15. Install the cylinder head over the dowels. Coat bolt threads and washers with clean engine oil and insert.

16. Tighten the cylinder head bolts in sequence as follows:

  1st pass—45 ft. lbs. (60 Nm)
  2nd pass—74 ft. lbs. (100 Nm)
  3rd pass—118 ft. lbs. (160 Nm)

17. Apply Loctite® 262 to the fixed guide top bolt threads, install the bolt and tighten to 19 ft. lbs. (26 Nm).

18. Install the access plug into the cylinder head and torque to 15 ft. lbs. (20 Nm).

19. Connect the vacuum hose to the actuator.

20. Raise and support vehicle, drain the engine oil.

21. Install the exhaust manifold.

22. If still supported, lower the vehicle for underhood access.

23. Install the alternator.

24. Install valve lifters.

25. Install the right injector housing assembly.

26. Install the plenum assembly.

27. Fill the engine crankcase with the proper type and amount of engine oil.

28. Tighten the fuel filler cap and properly refill the cooling system.

29. Connect the negative battery cable.

30. If equipped, reset the CHANGE OIL indicator:

    a. Turn the ignition **ON** but do not start the engine.

    b. Depress the ENG MET button on the trip monitor, then within 5 seconds, depress the button a 2nd time. Within another 5 seconds, depress and hold the GAUGES button.

    c. While holding the GAUGES button and watch the CHANGE OIL light, it should begin to flash. Continue to hold the gauges button until the flashing stops and the light goes out indicating that the indicator is reset.

    d. If the indicator does not reset, turn the ignition **OFF** and restart the procedure.

Left Side

1. Disconnect the negative battery cable and properly relieve fuel system pressure.

2. Drain engine coolant into a suitable container.

3. Remove the intake plenum assembly.

4. Remove the left injector housing.

5. Remove the vacuum hose from the secondary port throttle valve actuator.

6. Remove the power brake booster assembly.

7. Remove the left bank valve lifters.

8. Remove the AIR control valve hoses, then disengage the electrical connector.

9. Remove the camshaft position sensor.

10. Disconnect the left exhaust manifold from the cylinder head. It is not necessary to completely remove the exhaust manifold from the vehicle for cylinder head removal.

11. Remove the access plug from the left cylinder head.

12. Remove the bolt attaching the left secondary timing chain guide.

13. Remove the cylinder head bolts. Remove the cylinder head and gasket from the vehicle.

To install:

14. Thoroughly clean the cylinder head and cylinder case mating surfaces. Make sure both surfaces are free of any foreign matter, nicks or scratches. The threads in both the bolts holes and on the bolts must be clean and free of old sealer.

NOTE: Cylinder head gaskets are not interchangeable between cylinder banks.

15. Install the cylinder head locating dowels into block, if loosened or removed, then position the new gasket in place on the cylinder case.

16. Install the cylinder head over the dowels. Coat bolt threads and washers with clean engine oil and insert.

17. Tighten the cylinder head bolts in sequence as follows:

    1st pass—45 ft. lbs. (60 Nm)
    2nd pass—74 ft. lbs. (100 Nm)
    3rd pass—118 ft. lbs. (160 Nm)

18. Apply Loctite® 262 to the fixed guide bolt threads, install the bolt and tighten to 19 ft. lbs. (26 Nm).

19. Install the access plug into the cylinder head and torque to 15 ft. lbs. (20 Nm).

20. Connect the vacuum hose to the actuator.

21. Raise and support vehicle, drain the engine oil and lower the vehicle.

22. Install the exhaust manifold.

23. Install the camshaft position sensor.

24. Connect the AIR control valve hoses and electrical connector.

25. Install the valve lifters.

26. Install the left injector housing assembly.

27. Install the plenum assembly.

28. Fill the engine crankcase with the proper type and amount of engine oil.

29. Tighten the fuel filler cap and properly refill the cooling system.

30. Connect the negative battery cable.

31. If equipped, reset the CHANGE OIL indicator:

    a. Turn the ignition **ON** but do not start the engine.

    b. Depress the ENG MET button on the trip monitor, then within 5 seconds, depress the button a 2nd time. Within another 5 seconds, depress and hold the GAUGES button.

    c. While holding the GAUGES button and watch the CHANGE OIL light, it should begin to flash. Continue to hold the gauges button until the flashing stops and the light goes out indicating that the indicator is reset.

    d. If the indicator does not reset, turn the ignition **OFF** and restart the procedure.

Valve Lifters

REMOVAL AND INSTALLATION

5.7L (VIN P) Engine

1. Disconnect the negative battery cable and properly relieve fuel system pressure.

2. Drain the engine cooling system into a suitable container.

3. Remove the intake manifold assembly.

4. Remove the valve rocker covers.

5. Remove the rocker arms and pushrod assemblies. Be sure to tag or arrange all parts in order to assure installation in their original locations.

6. Remove the valve lifter guide (restrictor) retainer bolts, then remove the retainer.

7. Remove the valve lifter guide (restrictor) and valve lifter assemblies. If necessary a valve lifter tool should be used to remove lifters which are stuck in their bores.

8. If lifters are to be reused, place the lifters in a rack so they may be installed in their original bores.

To install:

9. Coat the lifter rollers with 1052365 or equivalent prelube, and install the lifters to the bores. If old lifters are being reused, make sure they are installed into the same bores from which they were removed.

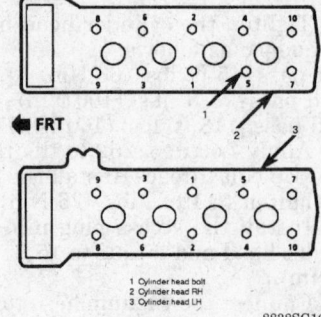

8838SG16

Cylinder head torque sequence — 5.7L (VIN J) engine

10. Install the valve lifter guides (restrictors), retainer and bolts. Tighten the bolts to 15 ft. lbs. (20 Nm).

11. Install the rocker arm and pushrod assemblies.

12. Install the valve rocker covers.

13. Install the intake manifold.

14. Tighten the fuel filler cap and connect the negative battery cable.

15. Properly fill the engine cooling system.

5.7L (VIN J) Engine

1. Disconnect the negative battery cable.

2. Remove the camshaft covers.

3. Remove the camshafts.

4. Remove lifters from bores.

NOTE: If lifters are to be reused, be sure to retain them in proper order so each lifter can be reinstalled in its original bore.

5. Installation is the reverse of the removal procedure. Lubricate lifter bores with clean engine oil.

6. Lifters should be replaced as sets with a camshaft. If new lifters are being used, be sure to pre-oil them.

Valve Lash

ADJUSTMENT

5.7L (VIN P) Engine

The valve lash on the VIN P engine is adjusted whenever the rocker arm assemblies have been removed.

NOTE: The 5.7L (VIN P) engine utilize hydraulic lifters which normally require very little maintenance or adjustment. These components are simple in design and are best maintained through regular, scheduled engine oil changes. If the engine is running well and no audible clicking sounds are heard from the valve train, there is no need to remove or disassemble the valve lifters.

1. Disconnect the negative battery cable.

2. Remove the valve rocker covers.

3. Tighten the nuts slowly until all lash is eliminated.

4. Adjust the valves when the lifter is on the base circle of the camshaft lobe. Slowly turn or crank the engine until the mark on the vibration damper is in the 12 o'clock position (aligned with the timing

cover mark, if equipped) and the engine is in the No. 1 firing position.

NOTE: The No. 1 firing position may be determined by watching the No. 1 cylinder valves as the mark on the damper approaches the 12 o'clock position. If both the intake and exhaust valves are closed as the mark comes up to the timing tab, the engine is in the No. 1 firing position. If either valve opens as the timing mark approaches the top of it's travel, the engine is in No. 6 firing position and should be turned over 1 full revolution in order to reach the No. 1 firing position.

5. With the engine in the No. 1 firing position, adjust the following valves:
- Exhaust — 1, 3, 4, 8
- Intake — 1, 2, 5, 7

6. Back out the rocker arm adjusting nut until lash is felt at the pushrod, then turn the adjusting nut inward until all lash is removed. This can be determined by rotating pushrod while turning the adjusting nut. When play has been removed, the pushrod will not turn. Then, tighten the adjusting nut 1 full additional turn.

7. Slowly turn or crank the engine 1 revolution until the vibration damper mark is at 12 o'clock again and the No. 1 cylinder valves open. This is the No. 6 firing position.

8. With the engine in this position, adjust the following valves:
- Exhaust — 2, 5, 6, 7
- Intake — 3, 4, 6, 8

9. Install the valve rocker arm covers.

10. Connect the battery negative cable.

5.7L (VIN J) Engine

This engine is equipped with hydraulic lifters which are installed in bores

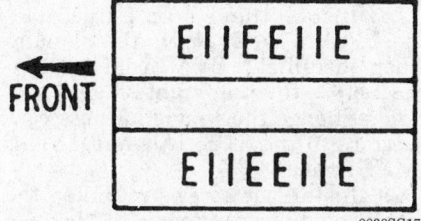

Valve arrangement — 5.7L (VIN P) engine

directly below the camshaft lobes. The lifters maintain 0 lash between the camshaft lobes and the valve stem. The lifter and installation position is non-adjustable, therefore upon failure, the lifter assembly must be replaced.

Rocker Arms

REMOVAL AND INSTALLATION

5.7L (VIN P) Engine

1. Disconnect the negative battery cable.
2. Remove the right valve rocker cover as follows:
   a. Remove the fuel rail cover and the fuel rail bolts.
   b. Disconnect the fuel pressure regulator vacuum hose.
   c. Remove the fuel injector and rail assembly from the manifold and reposition.
   d. Remove the fuel rail cover studs and position the wiring harness aside.
   e. Remove the AIR pipe and check valve from the intake and exhaust manifolds.
   f. Disconnect the crankcase vent hose.
   g. Remove the valve rocker cover bolts, cover and gasket. Replace the gasket as necessary.
3. Remove the left rocker arm cover as follows:
   a. If not done already, remove the fuel rail cover.
   b. Remove the alternator brace bolts, then remove the brace.
   c. Remove the remaining alternator bolts and position the alternator aside.
   d. Disconnect the AIR diverter valve hose from the check valve.
   e. Position the wiring harness aside.
   f. Remove the valve rocker cover bolts, cover and gasket. Replace the gasket as necessary.
4. Remove the rocker arm nuts, rocker arm balls, rocker arms and pushrods. If the valve train components are to be reused, mark or arrange the assemblies in a rack to assure installation in their original locations.

To install:
5. Coat the bearing surfaces of the rocker arms and rocker arm balls with 1052365 or equivalent pre-lube, prior to installation.
6. Install the pushrods making certain they seat in the lifter sockets.

7. Install the rocker arms, rocker arm balls and rocker arm nuts in their original positions.
8. Tighten the rocker arm nuts until all lash is eliminated.
9. Adjust the valve lash.
10. Thoroughly clean the gasket mating surfaces and install the valve rocker arm covers in the reverse order of removal. Tighten the valve rocker cover bolts to 100 inch lbs. (11 Nm). For the right valve cover, be sure to tighten the AIR pipe-to-exhaust manifold fitting to 25 ft. lbs. (34 Nm).
11. Connect the battery negative cable, start the engine and inspect for leaks.

5.7L (VIN J) Engine

This engine utilizes an overhead cam design, thus eliminating the need for any rocker arm assembly. This design improves and smoothes engine operation.

Intake Manifold

REMOVAL AND INSTALLATION

5.7L (VIN P) Engine

1. Disconnect the negative battery cable.
2. Drain engine coolant into a suitable container.
3. Remove the throttle body air duct.
4. Remove the fuel rail covers.
5. Disengage the wiring harness connectors from the fuel injectors. Disengage and reposition the left and right wiring harnesses.
6. Remove the accelerator cable bracket, then disconnect the cables from the throttle body.
7. Disconnect the AIR diverter valve hoses.
8. Remove the electrical ground strap from the intake manifold.
9. Remove the fuel rail bolts and disconnect the vacuum hose from the fuel pressure regulator.
10. Carefully remove the fuel rail and injector assembly from the manifold and position aside. Be careful not to damage the fuel lines.
11. Disconnect the vacuum and crankcase vent hoses.
12. Remove the EGR solenoid assembly and the fuel EVAP canister solenoid assembly.
13. Remove the EGR valve.
14. Remove the AIR pipe from the intake and the right exhaust manifold.
15. Remove the alternator brace.

16. Disconnect the coolant hoses from the throttle body.

17. Remove the throttle body bolts, the throttle body and gasket.

18. Remove the intake manifold bolts and studs.

19. Remove the intake manifold and gaskets.

To install:

20. Thoroughly clean the intake manifold bolts and studs. Inspect and clean all gasket mating surfaces.

21. Apply a ³/₁₆ in. (5mm) bead of RTV sealer to the front and rear of the cylinder block. Extend the bead ½ inch (13mm) up each cylinder head to seal and retain the gaskets.

22. Position the new gaskets and install the intake manifold.

23. Install the manifold bolts and studs, then tighten using 2 passes of the proper sequence. First, tighten the bolts/studs to 71 inch lbs. (8 Nm), then tighten them to 35 ft. lbs. (48 Nm).

24. Install the throttle body, gasket and retaining bolts. Tighten the throttle body bolts to 19 ft. lbs. (26 Nm).

25. Connect the coolant hoses to the throttle body.

26. Install the alternator brace.

27. Install the accelerator cables and bracket. Tighten the bracket bolts to 90 inch lbs. (10 Nm).

28. Install the AIR pipe. Tighten the exhaust manifold fitting and the bracket-to-cylinder head bolt to 25 ft. lbs. (34 Nm) and tighten the flange-to-intake manifold bolts to 19 ft. lbs. (26 Nm).

29. Install the EGR valve, then EGR solenoid and bracket. Tighten valve bolts to 16 ft. lbs. (22 Nm) and the solenoid bracket nut to 25 ft. lbs. (34 Nm).

30. Install the fuel EVAP canister purge solenoid and bracket, then tighten the nut to 15 ft. lbs. (20 Nm).

31. Connect the vacuum and crankcase vent hoses.

32. Install the fuel injector and fuel rail assembly to the intake manifold, connect the fuel pressure regulator vacuum hose and install the fuel rail bolts. Tighten the bolts to 15 ft. lbs. (20 Nm).

33. Connect the electrical ground strap to the intake manifold.

34. Connect the AIR diverter valve hoses.

35. Position the left and right wiring harnesses, then engage the fuel injector electrical connectors.

36. Install the throttle body air duct.

37. Install the fuel rail covers.

38. Properly fill the engine cooling system.

39. Connect the negative battery cable, then adjust the ASR accelerator and cruise control cables, as necessary.

Intake Plenum Assembly

The 5.7L (VIN J) engine does not utilize an intake manifold assembly like the other 5.7L Corvette engine. Instead it uses an intake plenum mated to a right and left fuel injector housing.

REMOVAL AND INSTALLATION

1. Disconnect the negative battery cable and properly relieve fuel system pressure.

2. Drain the cooling system into a suitable container.

3. Remove the air intake duct.

4. Remove the throttle cable cover and attaching hardware.

5. Remove the throttle and cruise control cables or the ASR control cable from the throttle body. Remove the cable hold-down clamp(s) and set the cables aside.

6. Remove the fresh air hose from the left and right side of the of the throttle body extension.

7. Disengage the electrical connectors from the IAC, TPS and the IAT or MAT sensors.

8. Disconnect the coolant air bleed hose from the plenum.

9. Remove the power brake booster hose, then remove the vacuum hose located between the fuel pressure regulator and the plenum.

10. Tag and remove the left and right vacuum hoses at the mid-plenum.

11. Tag and remove the MAP sensor vacuum hose.

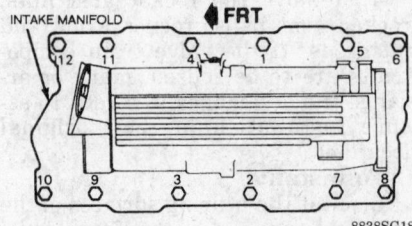

Intake manifold bolt torque sequence — 5.7L (VIN P) engine

12. Disconnect the fuel lines from the fuel rail assembly and discard the O-rings.

13. Remove the plenum assembly attaching bolts.

14. Remove the EVAP purge solenoid/PCV dual hose fitting from the plenum.

15. Remove the EVAP purge canister hose from the plenum.

16. Remove the upper EGR pipe bolts, then remove the pipe.

17. Lift the plenum and disengage the ignition module electrical connections, then remove the plenum assembly and discard the gaskets.

18. Cover the intake ports to prevent dirt or other contaminants from entering.

To install:

19. If the plenum is being replaced, transfer the MAP sensor and bracket, the throttle body, the throttle body extension and the ignition module to the new plenum.

20. Remove the tape or other cover from the intake ports and position the plenum assembly on the injector housings with the MAP sensor over the fuel pressure regulator. Engage the electrical connectors to the ignition module and MAP sensor, then install the MAP sensor vacuum hose.

21. Make sure the remaining vacuum hoses and electrical connectors are accessible, then position the new plenum gaskets between the plenum and injector housing, Install the plenum attaching bolts and tighten the bolts in their proper torque sequence to 20 ft. lbs. (26 Nm).

22. Install the vacuum hoses to mid-plenum.

23. Install new O-rings, then reconnect the fuel lines to the fuel rail assembly. Tighten the fuel line fittings to 20 ft. lbs. (26 Nm).

24. Install the vacuum hose between the pressure regulator and the plenum.

25. Connect the power brake booster vacuum hose to the plenum.

26. Install the fresh air hose onto the left and right side of the throttle body extension.

27. Install the EVAP purge solenoid/PCV hose fitting to the plenum, then install the EVAP canister connection to the rear right side.

28. Engage the wiring harness connectors to the TPS, IAC and IAT or MAT sensors.

29. Install the screws retaining the cable hold-down clamps to the plenum and tighten to 18 inch lbs. (2 Nm).

30. Install the coolant air bleed hose to the plenum.

31. Connect the throttle and cruise control cables or the ASR control cable to the throttle. Make sure the cables do not hold the throttle open and adjust, as necessary.

32. Install the cable shield, screw and nuts to the throttle body.

33. Connect the upper EGR pipe. Tighten the screw and nuts to 27 inch lbs. (3 Nm), then tighten the EGR pipe bolts to 89 inch lbs. (10 Nm).

34. Install the air intake duct, tighten the fuel filler cap and connect the negative battery terminal.

35. Properly refill the engine cooling system, then start the engine and check for leaks.

Injector Housing Assembly

The 5.7L (VIN J) engine does not utilize an intake manifold assembly like the other 5.7L Corvette engine. Instead it uses an intake plenum mated to a right and left fuel injector housing.

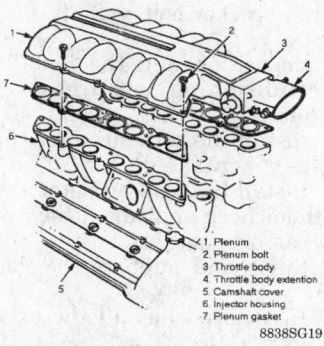

1. Plenum
2. Plenum bolt
3. Throttle body
4. Throttle body extention
5. Camshaft cover
6. Injector housing
7. Plenum gasket

8838SG19

Intake plenum installation — 5.7L (VIN J) engine

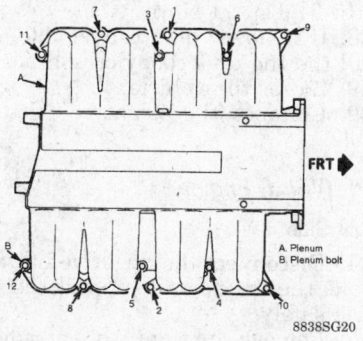

FRT ▶

A. Plenum
B. Plenum bolt

8838SG20

Intake plenum bolt torque sequence — 5.7L (VIN J) engine

REMOVAL AND INSTALLATION

Right Side

1. Disconnect the negative battery cable and properly relieve fuel system pressure.

2. Drain the cooling system into a suitable container.

3. Remove the intake plenum assembly.

4. Disengage the electrical connectors from the fuel injectors.

5. Remove the bolts attaching the fuel rail assembly to the injector housing.

6. Remove the injectors from the housing and remove the fuel rail assembly from the vehicle.

7. Disconnect the hose from the right coolant outlet pipe.

8. Remove the oil pressure sensor from the oil filter housing.

9. Remove the bolt attaching the coolant outlet pipe to the injector housing. Remove the outlet pipe and gasket from the vehicle.

10. Remove the PCV grommet from the injector housing.

11. Remove the clamp and ventilation hose from the injector housing.

12. Remove the bolt attaching the alternator rear support bracket to the alternator.

13. Remove the screws attaching the alternator rear support bracket and right side ventilation pipe to the injector housing.

14. Remove the ventilation pipe and bracket from the vehicle.

15. Remove the injector housing attaching bolts, then remove the injector housing and gasket from the vehicle.

To install:

16. Thoroughly clean all gasket mating surfaces and position the a new housing gasket.

17. Install the injector housing, rear alternator bracket, right ventilation pipe and the housing bolts. Be sure the spark plug wire harness retainer is secured by the injector housing rear bolt and tighten the fasteners to 19 ft. lbs. (26 Nm).

18. Install the ventilation hose.

19. Install PCV grommet into the injector housing.

20. Install a new gasket, the coolant outlet pipe and the retaining screws. Torque screws to 89 inch lbs. (10 Nm).

21. Install the oil pressure sensor. Apply Loctite® to sensor threads.

22. Install the hose and clamp onto the right coolant outlet pipe.

23. Install new injector lower O-rings and install the fuel rail assembly to the injector housing. Tighten

the retaining bolts to 19 ft. lbs. (26 Nm).

24. Engage the injector electrical connectors.

25. Install intake plenum assembly.

26. Connect the negative battery cable and refill the engine cooling system. Start the engine and check for leaks.

Left Side

1. Disconnect the negative battery cable and properly relieve fuel system pressure.

2. Drain the cooling system into a suitable container.

3. Remove the intake plenum assembly.

4. Disengage the electrical connectors from the fuel injectors.

5. Remove the screws attaching the fuel rail assembly to the injector housing.

6. Remove the injectors from the housing and remove the fuel rail assembly from the vehicle.

7. Disconnect the hose from the left coolant outlet pipe.

8. Remove the bolts attaching the coolant outlet pipe to the injector housing. Remove the outlet pipe and gasket from the vehicle.

9. Remove the PCV grommet from the injector housing.

10. Remove the clamp and ventilation hose from the injector housing.

11. Disengage the electrical connectors from the coolant temperature sensor and the cooling fan switch.

12. Remove the injector housing attaching bolts, then remove the injector housing and gasket from the vehicle.

To install:

13. Thoroughly clean all gasket mating surfaces and position the a new housing gasket.

14. Install the injector housing and secure using the housing retaining bolts. Be sure the spark plug wire harness retainer is secured by the injector housing rear bolt and tighten the fasteners to 19 ft. lbs. (26 Nm).

15. Install the ventilation hose and clamp.

16. Install PCV grommet into the injector housing.

17. Engage the electrical connectors to the coolant temperature sensor and the cooling fan switch.

18. Install a new gasket, the coolant outlet pipe and the retaining screws. Torque screws to 89 inch lbs. (10 Nm).

19. Install the hose and clamp onto the left coolant outlet pipe.

20. Install new injector lower O-rings and install the fuel rail assem-

bly to the injector housing. Tighten the retaining bolts to 19 ft. lbs. (26 Nm).

21. Engage the injector electrical connectors.

22. Install intake plenum assembly.

23. Connect the negative battery cable and refill the engine cooling system. Start the engine and check for leaks.

Exhaust Manifold

REMOVAL AND INSTALLATION

5.7L (VIN P) Engine

Right Side

1. Disconnect the negative battery cable.

2. Raise and support the vehicle safely.

3. For convertibles, if necessary for access, remove the underbody crossbrace.

4. Remove the nuts and disconnect the catalytic converter from the exhaust manifold. It may be necessary to loosen and remove the entire exhaust assembly.

5. Lower the vehicle.

6. Remove the fuel rail covers and disengage the fuel injector electrical connectors.

7. Remove the vacuum hose from the fuel pressure regulator.

8. Remove the fuel rail bolts, then remove the fuel injector/rail assembly from the intake manifold and position aside. Be careful not to damage the fuel lines.

9. Disconnect the spark plug wires from the plugs, then disconnect the wire clips from the supports. If necessary for clearance, or to prevent the possibility of breakage during manifold removal, remove the spark plugs.

10. Remove the front spark plug bracket and bolt.

11. Remove the oil level indicator and guide tube.

12. Remove the AIR pipe, gasket and check valve as an assembly from the intake manifold, exhaust manifolds and the cylinder head.

13. Remove the exhaust manifold studs and bolts.

14. Remove the heat shields, exhaust manifold and gasket.

To install:

15. Thoroughly clean the manifold and cylinder head gasket mating surfaces.

16. Install the exhaust manifold gasket, manifold and heat shields.

17. Install the exhaust manifold studs and bolts. Tighten the fasteners to 26 ft. lbs. (35 Nm).

18. Install the AIR pipe, gasket and check valve assembly with the retaining bolts. Tighten the AIR pipe-to-exhaust manifold fitting and the bracket bolt to 25 ft. lbs. (34 Nm) and tighten the pipe flange bolts to 19 ft. lbs. (26 Nm).

19. Apply 1052080 or equivalent, sealer to the oil level indicator guide tube ½ inch (13mm) below the bead. Install the level indicator and guide tube into the block.

20. Install the front spark plug bracket and bolt. Tighten to 108 inch lbs. (12 Nm).

21. If removed, install the spark plugs and tighten to 11 ft. lbs. (15 Nm).

22. Install the spark plug wires and clips.

23. Install the fuel injectors and fuel rail assembly to the intake manifold. Tighten the fuel rail bolts to 15 ft. lbs. (20 Nm).

24. Connect the fuel pressure regulator vacuum hose.

25. Engage the wiring harness connectors to the fuel injectors.

26. Install the fuel rail covers.

27. Raise and support the vehicle safely.

28. If removed, install the exhaust assembly.

29. Connect the catalytic converter and nuts to the exhaust manifold.

30. Tighten catalytic converter nuts to 15 ft. lbs. (21 Nm).

31. If removed on a convertible, install the the underbody crossbrace.

32. Lower the vehicle.

33. Connect the negative battery cable.

Left Side

1. Disconnect the negative battery cable.

2. Raise and support the vehicle safely.

3. For convertibles, if necessary for access, remove the underbody crossbrace.

4. Remove the nuts and disconnect the catalytic converter from the exhaust manifold. It may be necessary to loosen and remove the entire exhaust assembly.

5. Lower the vehicle.

6. Remove the air intake duct and the serpentine drive belt.

7. Remove the ASR adjuster assembly from the wheel well center panel and reposition out of the way.

8. Remove the mounting bolts and reposition the alternator and the A/C compressor.

9. Remove the AIR pipe, check valve and hose as an assembly from the exhaust manifold.

10. Remove the spark plug wires from the plugs and the clips from the supports, then position the wires aside.

11. Remove the spark plug wire supports. If necessary for clearance, or to prevent the possibility of breakage during manifold removal, remove the spark plugs.

12. Remove the exhaust manifold studs and bolts.

13. Remove the heat shields, exhaust manifold and gasket.

To install:

14. Thoroughly clean the manifold and cylinder head gasket mating surfaces.

15. Install the exhaust manifold gasket, manifold and heat shields.

16. Install the exhaust manifold studs and bolts. Tighten the fasteners to 26 ft. lbs. (35 Nm).

17. Install the spark plug wire supports and tighten to 108 inch lbs. (12 Nm).

18. If removed, install the spark plugs and tighten to 11 ft. lbs. (15 Nm).

19. Connect the spark plug wires and clips.

20. Install the AIR pipe, check valve and hose assembly. Tighten the AIR pipe-to-exhaust manifold fitting and the bracket bolt to 25 ft. lbs. (34 Nm).

21. Reposition and the install the air conditioning compressor and alternator.

22. Install the serpentine drive belt and the air intake duct.

23. Install the ASR adjuster assembly, then check and adjust the control cable, as necessary.

24. Raise and support the vehicle safely.

25. If removed, install the exhaust assembly.

26. Connect the catalytic converter and nuts to the exhaust manifold.

27. Tighten catalytic converter nuts to 15 ft. lbs. (21 Nm).

28. If removed on a convertible, install the the underbody crossbrace.

29. Lower the vehicle.

30. Connect the negative battery cable.

5.7L (VIN J) Engine

Right Side

1. Disconnect the negative battery cable, then raise and support the vehicle safely.

2. Remove the right tire and wheel assembly, then remove the wheel house lower rear and center panels.

3. Disconnect the exhaust system assembly from the catalytic converter.

4. If equipped, remove the engine block heat shield.

5. Disconnect the catalytic converter from the manifold.

6. Disengage the oxygen sensor wiring harness connector.

7. Remove the rear exhaust manifold bolts, spacers and nut.

8. Disconnect the lower EGR pipe from the manifold.

9. Lower the vehicle.

10. Disconnect the AIR check valve and hose from the manifold.

11. Remove the retaining bolt, then remove the oil level indicator and guide tube from the vehicle.

12. Remove the remaining exhaust manifold attaching bolts and spacers.

13. Remove the exhaust manifold and gasket from the vehicle. If the manifold is being replaced, transfer the oxygen sensor and heat shields to the new manifold, as necessary.

To install:

14. Thoroughly clean the manifold and cylinder head gasket mating surfaces.

15. Install the gasket and manifold to the engine using the front and center manifold bolts and spacers.

16. Install the oil level indicator and guide tube, then tighten the manifold bolts to 18 ft. lbs. (24 Nm).

17. Install the AIR check valve and hose.

18. Raise and support the vehicle safely.

19. Install the rear manifold bolts, spacers and nut. Tighten the bolts and nut to 18 ft. lbs. (24 Nm).

20. Install the wiring harness connector to the oxygen sensor.

21. Connect the catalytic converter and bolts to the manifold. Tighten the bolts to 17 ft. lbs. (23 Nm).

22. If equipped, install the engine block heat shield.

23. Connect the exhaust system assembly.

24. Install the manifold outer heat shields.

25. Install the lower EGR pipe to the manifold.

26. Lower the vehicle sufficiently for access.

27. Install the wheelhouse lower rear and center panels.

28. Install the tire and wheel assembly, then lower the vehicle completely.

29. Connect the negative battery cable.

Left Side

1. Disconnect the negative battery cable, then raise and support the vehicle safely.

2. Remove the right tire and wheel assembly, then remove the wheel house lower rear and center panels.

3. Disconnect the exhaust assembly from the catalytic converter.

4. Remove the left floor pan heat shield, the left heat shield from the frame and the engine block heat shield.

5. Disengage the converter oxygen sensor electrical connector.

6. Disconnect the AIR check valves, hoses and pipes from the manifold.

7. Remove the manifold outer heat shield and remove the catalytic converter from the exhaust manifold.

8. Remove the exhaust manifold bolts, spacers and nut.

9. If applicable, remove the center stud nut.

10. Remove the manifold and gasket from the vehicle.

To install:

11. Thoroughly clean the manifold and cylinder head gasket mating surfaces.

12. Install the gasket and manifold to the engine.

13. Install the manifold bolts, spacer, nut, and if applicable, center stud nut. Tighten the bolts and nut(s) to 18 ft. lbs. (24 Nm).

14. Install the catalytic converter and bolts to the manifold. Tighten to 17 ft. lbs. (23 Nm) and install the manifold outer heat shield. Install the AIR check valve, hoses and pipe.

15. Engage the oxygen sensor wiring harness connector.

16. Install the engine block heat shield, the left side heat shield to the frame and the floor pan heat shield.

17. Install the exhaust system assembly to the catalytic converter.

18. Install the wheelhouse lower rear and center panels, then install the tire and wheel assembly.

19. Lower the vehicle and connect the negative battery cable.

Front Cover Seal

REMOVAL AND INSTALLATION

5.7L (VIN P) Engine

1. Disconnect the battery negative cable.

2. Remove the engine front cover.

3. Using a suitable tool, remove the crankshaft, distributor shaft and/or water pump driven gear shaft seals, as necessary.

4. As applicable; use tool J-35468 or equivalent aligner and installer, to install the crankshaft seal, tool J-39090 or equivalent, to install the distributor shaft seal and/or tool J-39088 or equivalent, to install the water pump shaft seal.

5. Install the engine front cover.

6. Connect the negative battery cable.

5.7L (VIN J) Engine

1. Disconnect the battery negative cable.

2. Remove the timing chain front cover assembly.

3. Remove the seal from the front cover using J-29077-A or equivalent seal remover tool.

4. Thoroughly clean the cylinder case, front cover and water pump sealing surfaces.

5. Apply Loctite® 262 to studs and Loctite® 565 to the bolt threads.

6. Install a new cover gasket and the cover, nuts and bolts.

7. Install the new seal coated with engine oil using tool J-37309 or equivalent.

NOTE: Do not remove seal installing tool J-37309, until the front cover bolts are torqued.

8. Tighten the front cover attaching bolts to 19 inch lbs. (26 Nm) and the stud nuts to 21 ft. lbs. (28 Nm).

9. Complete the front cover installation procedure and connect the negative battery cable.

Timing Chain Front Cover

REMOVAL AND INSTALLATION

5.7L (VIN P) Engine

1. Disconnect the negative battery cable.

2. Drain the engine oil and coolant into suitable containers.

3. Remove the throttle body air intake duct.

4. Remove the serpentine drive belt.

5. Remove the water pump assembly.

6. Remove the crankshaft balancer and hub.

    a. If not done already, raise and support the vehicle safely, then remove the motor mount nuts.

    b. Remove the power steering fluid cooler, then raise the engine sufficiently for tool access to the crankshaft balancer.

NOTE: When raising and supporting the engine, NEVER place a jack under the oil pan, crankshaft pulley or any sheetmetal. There is a minimal clearance between the oil pan and the pump screen. Jacking against the pan could cause sufficient deformation to damage the oil pickup unit.

c. Remove the balancer bolts, then remove the balancer from the hub.

d. Disconnect the power steering line from the steering gear.

e. Matchmark the crankshaft hub to the engine front cover, then remove the hub bolt and washer.

f. Remove the crankshaft hub using J-39046, or an equivalent hub removal/installation tool. To preserve the relationship between the hub and crankshaft, DO NOT crank the engine over once the hub has been removed. If the hub is not matchmarked and installed in the original position, an engine imbalance could result.

7. Remove the distributor assembly.

8. Remove the oil pan assembly.

9. Remove the engine front cover bolts.

10. Remove the engine front cover and gasket.

To install:

11. Thoroughly clean the engine front cover and cylinder block gasket mating surfaces. Inspect the engine front cover and seals for damage, replace as necessary.

12. Using J-39087 or equivalent shaft gear front cover seal protector, on the water pump driveshaft, install the gasket and front cover into position over the shafts and guide pins.

13. Install the engine front cover bolts and tighten to 100 inch lbs. (11 Nm).

14. Install the oil pan and gasket.

15. Install the distributor assembly.

16. Install the and the torsional damper.

a. Align the matchmarks made earlier and install the crankshaft hub. If the engine was cranked and the matchmarks were lost, set the engine to No. 1 TDC, then install the crankshaft hub with the cast arrow in the 12 o'clock position.

b. Install the hub washer and bolt, but do not torque at this time.

c. Raise the engine, as necessary for access, then install the crank-

shaft balancer assembly. Tighten the crankshaft hub bolt to 75 ft. lbs. (102 Nm) and the balancer bolts to 60 ft. lbs. (81 Nm).

NOTE: If a new balancer is installed, new balancer weights of the same size must be installed in the same hole locations as the original balancer.

d. Install the power steering line to the gear, then lower the engine into position.

e. Install the power steering fluid cooler.

f. Install the motor mount nuts and tighten to 40 ft. lbs. (50 Nm).

17. Install the water pump assembly.

18. Install the serpentine drive belt and the throttle body air duct.

19. Properly fill the engine crankcase with oil.

20. Tighten the fuel filler cap and properly fill the engine cooling system.

21. Connect the negative battery cable, operate the engine and check for leaks.

22. Bleed the power steering hydraulic system, as necessary.

23. If equipped, reset the CHANGE OIL indicator:

a. Turn the ignition **ON** but do not start the engine.

b. Depress the ENG MET button on the trip monitor, then within 5 seconds, depress the button a 2nd time. Within another 5 seconds, depress and hold the GAUGES button.

c. While holding the GAUGES button and watch the CHANGE OIL light, it should begin to flash. Continue to hold the gauges button until the flashing stops and the light goes out indicating that the indicator is reset.

d. If the indicator does not reset, turn the ignition **OFF** and restart the procedure.

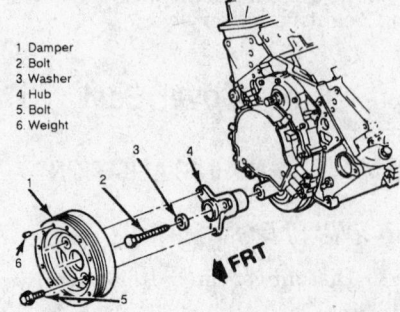

1. Damper
2. Bolt
3. Washer
4. Hub
5. Bolt
6. Weight

8838SG21

Exploded view of the crankshaft balancer and hub assembly — 5.7L (VIN P) engine

5.7L (VIN J) Engine

1. Disconnect the negative battery cable and drain the engine coolant into a suitable container.

2. Remove the water pump assembly.

3. Remove the air conditioning compressor as follows:

a. Properly discharge and recover the air conditioning system.

b. Remove the throttle body.

c. Remove the serpentine drive belt.

d. Remove the engine oil temperature sensor.

e. Remove the alternator.

f. Remove the refrigerant hose from the A/C compressor, then immediately cap or plug the openings to prevent system contamination and damage.

g. Remove the compressor mounting bolts and electrical connection.

h. Remove the compressor from the vehicle.

4. Remove the steering gear for access to the damper.

5. Remove the bolt and washer attaching the torsional damper to the crankshaft.

6. Using tool J-24420-C, or an equivalent torsional damper puller, remove the damper from the crankshaft.

7. Remove the drift key, from the crankshaft.

8. Remove the nuts and/or bolts attaching the front cover to the engine.

9. Remove the front cover and gasket from the vehicle. If necessary, remove the old seal from the front cover using J-29077-A or an equivalent oil seal remover.

To install:

10. Thoroughly clean the cylinder case, front cover and water pump sealing surfaces.

11. Apply Loctite® 262 to the stud threads and Loctite® 565 to the bolt threads.

12. Install a new cover gasket and the cover, nuts and bolts.

13. If removed, install a new front cover oil seal using J-37309 or equivalent front cover seal installer.

14. Tighten the front cover attaching bolts to 19 inch lbs. (26 Nm) and the stud nuts to 21 ft. lbs. (28 Nm).

15. Install the water pump assembly.

16. Install the air conditioning compressor in the reverse order of the removal procedure. Replace the refrigerant line seal washers and coat the new washers with 525 refrigerant oil prior to assembly. When installed,

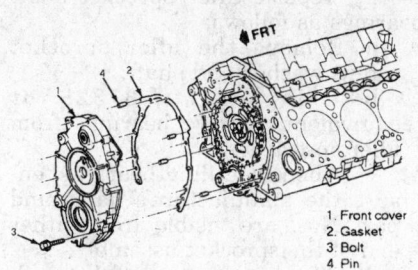

Timing chain front cover and gasket installation — 5.7L (VIN P) engine

1. Front cover
2. Gasket
3. Bolt
4. Pin

8838SG22

properly evacuate and charge the A/C system.

17. Install the key to the crankshaft and, then install the torsional damper using J-38463, or an equivalent torsional damper installer. Check for proper key seating during installation.

NOTE: If a new balancer is installed, new balancer weights of the same size must be installed in the same hole locations as the original balancer.

18. Remove the tool apply Loctite® 262 to the damper bolt threads. Install the washer and damper bolt, then tighten the bolt to 148 ft. lbs. (200 Nm).

19. Install the serpentine drive belt.

20. Install the steering gear.

21. Connect the negative battery cable, properly fill the engine cooling system and check for leaks.

22. If the engine oil was changed, and if equipped, reset the CHANGE OIL indicator:

a. Turn the ignition **ON** but do not start the engine.

b. Depress the ENG MET button on the trip monitor, then within 5 seconds, depress the button a 2nd time. Within another 5 seconds, depress and hold the GAUGES button.

c. While holding the GAUGES button and watch the CHANGE OIL light, it should begin to flash. Continue to hold the gauges button until the flashing stops and the light goes out indicating that the indicator is reset.

d. If the indicator does not reset, turn the ignition **OFF** and restart the procedure.

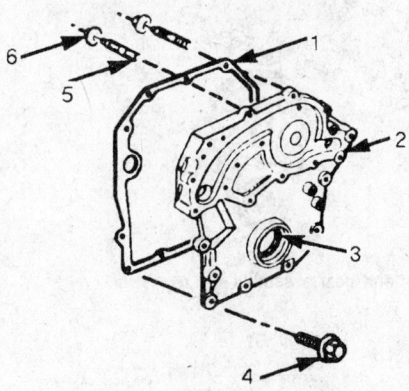

1. Engine front cover gasket
2. Engine front cover
3. Engine front cover seal
4. Engine front cover bolt
5. Engine front cover stud
6. Engine front cover stud nut

8838SG23

Timing chain front cover and gasket installation — 5.7L (VIN J) engine

Timing Chain and Sprockets

REMOVAL AND INSTALLATION

5.7L (VIN P) Engine

1. Disconnect the negative battery cable.

2. Remove the timing chain front cover.

3. Rotate the crankshaft until the timing marks on the timing chain sprockets are aligned nearest each other. The camshaft sprocket mark should be at the 6 o'clock position while the mark on the crankshaft sprocket should be at the 12 o'clock position.

4. Remove the camshaft sprocket bolts.

5. Remove the camshaft sprocket and timing chain.

NOTE: To prevent piston or valve damage, do not turn the crankshaft after the timing chain has been removed.

6. Remove the water pump bearing retainer bolts, then remove the driveshaft assembly using J-39243 or equivalent driven gear assembly remover.

7. Remove the crankshaft sprocket using J-5825-A or equivalent crankshaft sprocket remover.

8. If necessary, remove the crankshaft key.

To install:

9. If removed, install the crankshaft key.

10. Install the crankshaft sprocket using a suitable installation tool.

11. Install the water pump driveshaft assembly using a suitable tool. Install the retainer bolts and tighten to 108 inch lbs. (12 Nm).

12. Align the timing marks and install the camshaft sprocket and timing chain. The gears of the camshaft sprocket and water pump driveshaft must mesh or damage to the thrust plate retainer could occur.

13. Install the camshaft sprocket bolts and tighten to 21 ft. lbs. (28 Nm).

14. Install a new O-ring to the water pump driven gear shaft using a suitable seal installation tool.

15. Install the timing chain front cover and connect the negative battery cable.

5.7L (VIN J) Engine

Primary Timing Chain and Crankshaft Sprocket

1. Disconnect battery negative cable.

2. Remove the timing chain front cover assembly.

3. Remove the left and right intake camshafts.

4. Remove the bolts attaching the primary chain guide to the oil pump, then remove the guide from the vehicle.

5. Remove the idler sprocket assembly attaching bolts, then disengage the primary timing chain from the idler and crankshaft sprockets. Remove the chain from the vehicle.

6. Using the crankshaft torsional damper puller along with J-38211 or equivalent sprocket tool, remove the crankshaft sprocket. Note which side of the sprocket faces forward for installation purposes.

7. Remove the key and oil pump seal seat from the crankshaft.

To install:

8. Inspect the primary chain guide for excessive wear. Wear groove should not exceed a depth of 0.040 inch (1.0mm). If necessary, replace wear strip.

9. Install oil pump seal seat and key onto the crankshaft.

10. Install the crankshaft sprocket using J-38132 or equivalent sprocket installer. Make sure sprocket is in-

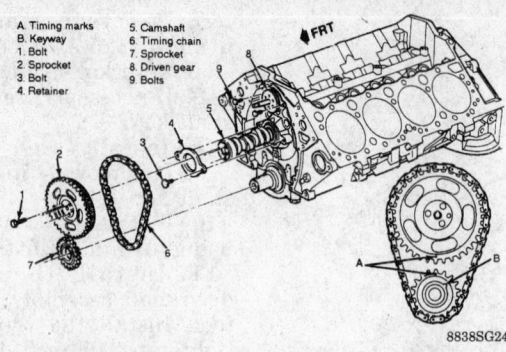

A. Timing marks
B. Keyway
1. Bolt
2. Sprocket
3. Bolt
4. Retainer
5. Camshaft
6. Timing chain
7. Sprocket
8. Driven gear
9. Bolts

8838SG24

Exploded view of the timing chain and gear assembly —
5.7L (VIN P) engine

stalled with same side to the front as noted during removal, this should be the wide shoulder.

11. Engage the primary chain onto the idler and crankshaft sprocket.

12. Apply Loctite® 262 or equivalent, to the idler sprocket assembly bolts and tighten to 19 ft. lbs. (26 Nm).

13. Apply Loctite® 262 or equivalent, to the primary chain guide bolts. Install the guide and bolts. Push the guide so the slack is removed from the chain and tighten the bolts to 89 inch lbs. (10 Nm).

NOTE: When installing guide, do not use any leverage tools, finger pressure is sufficient.

14. Install the left and right intake camshafts.

15. Install the timing chain front cover.

16. Connect the negative battery cable.

Secondary Timing Chains and Idler Sprocket Assembly

1. Disconnect battery negative cable.

2. Remove the camshafts.

3. Remove the primary timing chain and crankshaft sprocket.

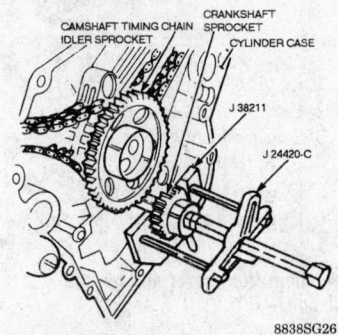

CAMSHAFT TIMING CHAIN IDLER SPROCKET
CRANKSHAFT SPROCKET
CYLINDER CASE
J 38211
J 24420-C

8838SG26

Crankshaft sprocket removal — 5.7L (VIN J) engine

4. Disengage the left and right secondary chains from the idler sprocket.

5. Remove the idler sprocket assembly.

6. Remove the left and right secondary chains from the vehicle.

To install:

7. Inspect chains and sprockets for abnormal wear or damage. If abnormal wear or damage is present on either the secondary timing chain, cam sprockets or idler sprockets, the entire assembly must be replaced.

8. Inspect the idler sprocket shaft bearings for wear or damage. If nec-

essary, replace idler sprocket shaft bearings as follows:

a. Remove the idler sprocket screw, washer and shaft.

b. Using tool J-37328 or equivalent, remove bearings from idler sprocket.

c. When installing bearings, ensure the manufacture's name and part Nos. are visible from either end of the sprocket assembly.

d. Using a press, carefully push in the bearings until they are flush with idler sprocket. Apply minimum pressure to obtain a fit 0.0–1.3mm below the surface.

9. Install the shorter (inner) secondary chain through the right head and install J-38099 or equivalent timing chain retaining tool.

10. Locate the right chain onto the rear idler sprocket.

11. Install the longer (outer) secondary chain through the left head and install J-38099 or equivalent timing chain retaining tool.

12. Locate the left chain onto the middle idler sprocket.

13. Install the primary timing chain.

14. Install the camshafts.

15. Connect the negative battery cable.

Camshaft

REMOVAL AND INSTALLATION

5.7L (VIN P) Engine

1. Disconnect battery negative cable and remove the air cleaner assembly.

2. Remove the timing chain front cover.

3. Remove the intake manifold.

4. Remove the retaining bolt and lift the oil pump driveshaft assembly from the rear of the lifter valley.

5. Remove the rocker arm and pushrod assemblies.

6. Remove the camshaft sprocket from the engine.

7. Remove the valve lifters.

8. Remove the high fill reservoir hose from the radiator.

9. Remove the relay bracket from the left side of the radiator support.

10. Remove the AIR pump intake duct and bolts, then reposition the AIR pump.

11. Remove the retaining nuts and screws, then remove the upper radiator support.

12. Remove the radiator.

13. Raise and support the vehicle safely.

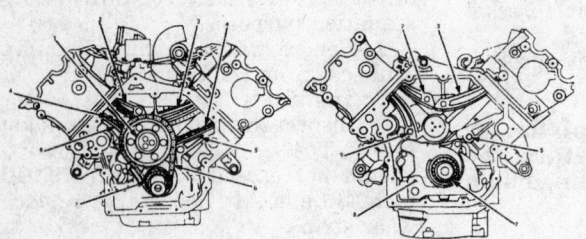

1. Camshaft timing chain idler sprocket assembly
2. Camshaft secondary timing chain fixed guide RH
3. Camshaft secondary timing chain pivot guide LH
4. Camshaft secondary timing chain
5. Camshaft secondary timing chain fixed guide LH
6. Camshaft primary timing chain
7. Crankshaft sprocket
8. Oil pump
9. Camshaft timing chain pivot guide RH

8838SG25

Primary and secondary timing chain assembly — 5.7L (VIN J) engine

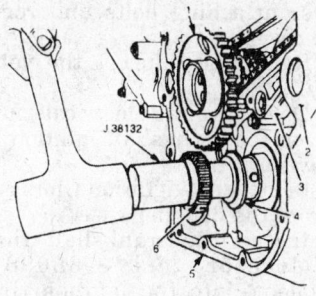

1. Camshaft timing chain idler sprocket
2. Cylinder case
3. Camshaft secondary timing chain fixed guide LH
4. Oil pump seal seat
5. Crankshaft
6. Crankshaft sprocket

8838SG27

Installing the crankshaft sprocket — 5.7L (VIN J) engine

14. Unplug the cooling fan electrical connector.
15. Remove the lower fan shroud bolts and lower the vehicle.
16. Remove the fan shroud and fan assembly.
17. Disconnect the A/C condenser line bracket at the front crossmember.
18. Raise the front of the engine with a suitable lifting device.

NOTE: When raising and supporting the engine, NEVER place a jack under the oil pan, crankshaft pulley or any sheetmetal. There is a minimal clearance between the oil pan and the pump screen. Jacking against the pan

could cause sufficient deformation to damage the oil pickup unit.

19. Remove the camshaft retainer bolts and retainer.
20. Install three ⁵⁄₁₆–18 x 4 inch bolts into the camshaft bolt holes.
21. Using the bolts, carefully rotate the camshaft and pull from the bearings. All camshaft journals are the same diameter so care must be used to avoid damaging the bearings. Remove the camshaft from the vehicle.

To install:

22. Inspect the camshaft and bearings, replace as necessary.
23. If installing a new camshaft, coat the lobes with Molykote® or equivalent pre-lube and be sure to replace all lifters to assure camshaft durability.
24. Lubricate all camshaft journals with clean engine oil and carefully insert the camshaft into the engine block.
25. Install the camshaft retainer and tighten the bolts to 108 inch lbs. (12 Nm).
26. Lower the front of the engine and connect the A/C condenser line bracket to the front crossmember.
27. Install the fan and shroud assembly.
28. Raise and support the vehicle safely, then install the lower fan shroud bolts.
29. Engage the cooling fan electrical connections and lower the vehicle.
30. Install the radiator, followed by the upper radiator support, nuts and screws.
31. Install the AIR pump, bolts and intake duct.
32. Install the relay bracket to the left side of the radiator support.
33. Connect the high fill reservoir hose to the radiator.
34. Install the valve lifters.
35. Install the camshaft sprocket.
36. Install the valve rocker arm and pushrod assemblies.

37. Install the oil pump driveshaft assembly and bolt. Tighten the bolt to 13 ft. lbs. (18 Nm).
38. Install the intake manifold.
39. Install the timing chain front cover.
40. Install the air cleaner assembly and connect the negative battery cable.

5.7L (VIN J) Engine

The VIN J engine utilizes 4 overhead camshafts. Certain shafts will have identifying bands between the first journal and lobe to distinguish between the right and left, intake and exhaust camshafts. The right intake has 1 flat band. The right exhaust has 1 raised band. The left intake has 1 flat and 1 raised band. The left exhaust has 2 raised bands.

1. Disconnect battery negative cable and drain the engine coolant into a suitable container.
2. To gain access to the right camshafts, remove the oil filter housing and right camshaft cover as follows:
 a. Remove the air intake duct.
 b. Remove the hoses and clamps from the coolant outlets, radiator inlet and inlet pipe.
 c. Remove the hoses and inlet pipe assembly from the vehicle.
 d. Remove the water pump pulley.
 e. Release the belt tensioner and remove the serpentine belt.
 f. Remove the retaining bolt and the belt tensioner from the engine.
 g. Remove the oil filter.
 h. Disengage the electrical connectors from the oil pressure sensor, oil temperature sensor and the low oil pressure switch.
 i. Remove the oil pressure sensor from the oil filter housing.
 j. Remove the alternator bracket from the oil filter housing.
 k. Disconnect and plug the oil cooler lines from the filter housing.
 l. Remove the oil filter housing mounting bolts and remove the assembly.

NOTE: If equipped with a 1 piece front cover/oil filter housing gasket, cut the old gasket along the front cover.

 m. Remove spark plug wires from plugs in the right cylinder head.
 n. Disengage the electrical connector from the blower motor resistor block.
 o. Remove the screws attaching the evaporator housing quarter panel, then remove the panel.
 p. Remove the bolts attaching the coolant outlet pipe bracket to

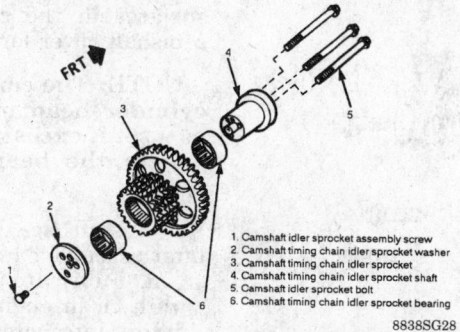

1. Camshaft idler sprocket assembly screw
2. Camshaft timing chain idler sprocket washer
3. Camshaft timing chain idler sprocket
4. Camshaft timing chain idler sprocket shaft
5. Camshaft idler sprocket bolt
6. Camshaft timing chain idler sprocket bearing

8838SG28

Timing chain idler sprocket assembly — 5.7L (VIN J) engine

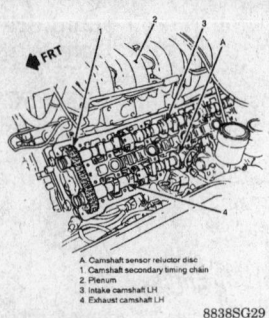

A. Camshaft sensor reluctor disc
1. Camshaft secondary timing chain
2. Plenum
3. Intake camshaft LH
4. Exhaust camshaft LH

8838SG29

Left cylinder head camshaft assembly — 5.7L (VIN J) engine

the alternator bracket and the coolant outlet to the injector housing, then position aside.

q. Remove the upper EGR pipe bolts and pipe.

r. Remove the bolt attaching the fresh air pipe bracket to the injector housing.

s. Remove the camshaft cover attaching bolts and the camshaft cover.

3. To gain access to the left camshafts, remove the air conditioning compressor and left valve cover as follows:

a. Properly discharge the air conditioning system.

b. Remove the throttle body assembly and the serpentine drive belt.

c. Remove the engine oil temperature sensor.

d. Remove the alternator assembly.

e. Remove the refrigerant hose from the A/C compressor, then immediately cap or plug the openings to prevent system contamination and damage.

f. Remove the compressor mounting bolts and electrical connection.

g. Remove the compressor from the vehicle.

h. Remove the power steering pump from the engine.

i. Remove the spark plug wires from the plugs in the left cylinder head.

j. Remove the ventilation breather pipe from the camshaft cover.

k. Remove the throttle and cruise control cable or control cable hold-down clamps from the plenum.

l. If not done already, remove the throttle body extension and coolant outlet pipe.

m. Remove the vacuum hose from the power brake booster and, if

necessary, remove the booster assembly.

n. Remove the left camshaft cover attaching bolts and remove the cover.

4. Raise and support the vehicle safely.

5. Disengage the electrical connector from the crankshaft ignition timing sensor.

6. Remove the ignition timing sensor from the cylinder case.

7. Install the crankshaft timing slot locator tool J-38098 or equivalent, into the ignition timing sensor opening. Make sure the tool head is fully seated with the indicating pin inserted into the deep notch of the crankshaft timing disc.

8. Lower vehicle.

9. Remove the bolts attaching the secondary timing chain tensioner housing to the cylinder head, then remove the housing, O-ring and tensioner from the cylinder case.

10. Remove the bolts and washers attaching the camshaft to the sprockets.

NOTE: Install a wrench on the rear camshaft hex when removing the sprocket bolts, to prevent the camshafts from exerting force on the crankshaft timing slot locator tool.

11. Remove the camshaft timing plates and pins.

12. Remove the camshaft retainers and thrust washers.

13. Remove the camshafts and sprockets from the vehicle. Install timing chain retainers J-38099 or equivalent, to retain secondary chain loops.

14. Remove lifters from bores and inspect. Make sure any lifters, to be reused, are retained in proper order so each one can be returned to its original bore.

To install:

15. Inspect the camshaft bearing journals for wear or damage.

16. Inspect the camshaft bearing surfaces in the cylinder head and camshaft cover for wear or damage.

NOTE: The camshaft cover and cylinder head must be replaced as a set if excessive wear or damage to the bearing surfaces is found.

17. Install the each camshaft and lifter assembly, 1 at a time:

a. Lubricate lifters and bores with clean engine oil, then install lifters into bores. If a camshaft is replaced, new lifters must also be used.

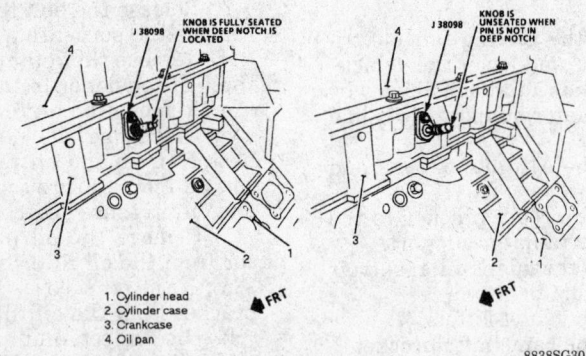

1. Cylinder head
2. Cylinder case
3. Crankcase
4. Oil pan

8838SG30

Crankshaft timing slot locator tool — 5.7L (VIN J) engine

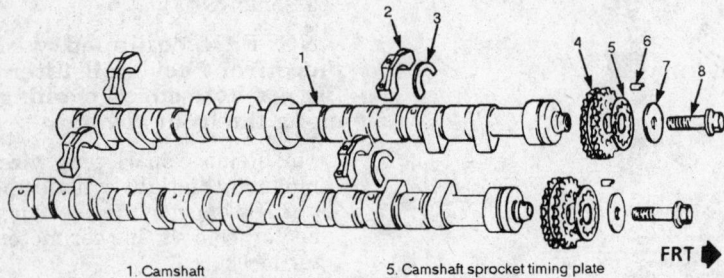

1. Camshaft
2. Camshaft retainer
3. Camshaft thrust washer
4. Camshaft sprocket
5. Camshaft sprocket timing plate
6. Camshaft sprocket pin
7. Camshaft sprocket washer
8. Camshaft sprocket bolt

8838SG31

Cylinder head camshaft assembly — 5.7L (VIN J) engine

b. Install the camshaft sprocket onto the secondary timing chain, while removing the timing chain retainer.

c. Slide the camshaft into the sprocket, noting the position of the alignment hole for timing pin tool installation. Position the camshaft in the neutral position, no valves opened.

d. Lubricate camshaft journals, lobes, thrust washers and retainers with clean engine oil.

e. Install the camshaft thrust washers, retainers and bolts. Torque bolts to 89 inch lbs. (10 Nm).

f. Repeat Steps a–e for the remaining camshafts.

18. Install timing pins J-37326 into camshaft retainers and the indexing holes in the camshafts. Camshafts can be rotated using the cast hex at the camshaft rear.

19. Install the camshaft secondary chain pre-tensioner, J-37305 or equivalent. Hand-tighten to remove slack from the timing chain, but do not overtighten.

20. Install timing plates, pins and washers. If no holes line up on the timing plate, reverse the plate.

21. Apply Loctite® 262 or equivalent, on the NEW camshaft sprocket bolts, then install and finger-tighten the bolts. New camshaft bolts should be used each time the camshaft is removed. Tighten the bolts to 18 ft. lbs. (25 Nm) and turn 80–85 degrees using a torque angle meter. A backup wrench should be used on the rear camshaft hex to prevent damaging the timing pins.

22. Remove timing pins J-37326.

23. Remove the secondary timing chain pre-tensioner tool and install the new secondary timing chain tensioner, housing, new O-ring and bolts. Lubricate tensioner with engine oil. Make sure the oil hole in the tensioner piston be installed in a vertical position and that the fork on the end of the tensioner is properly engaged onto the chain guide. After installing, use a blunt punch to release the plunger. Torque chain tensioner bolts to 89 inch lbs. (10 Nm).

24. Raise and support the vehicle safely.

25. Remove crankshaft timing slot locator J-38098 from the cylinder case.

26. Install the crankshaft position sensor into the cylinder case and tighten the retainer(s) to 71 inch lbs. (8 Nm).

27. Engage the timing sensor electrical connector and lower the vehicle.

28. Apply Permabond® A136 or equivalent, to the camshaft covers and Loctite® 565 or equivalent, to the end plugs. Install the end plugs and new spark plug bore O-rings prior to cover installation.

29. Install the camshaft covers in the reverse order of removal. The camshaft cover retainers must be tightened in the proper sequence in order to assure proper camshaft operation. Tighten the M8 bolts to 15 ft. lbs. (20 Nm), repeat 3 times. Tighten the M6 screws to 89 inch lbs. (10 Nm). Also, be sure to install a new coolant outlet cover gasket and tighten the cover screws to 89 inch lbs. (10 Nm).

30. For the right bank camshafts, install oil filter housing assembly.

31. For the left bank camshafts, install the air conditioning compressor assembly.

32. Reconnect the battery negative cable and properly fill the engine cooling system.

Piston and Connecting Rod

POSITIONING

The connecting rod chamfers must face the crankshaft counterweights on the 5.7L (VIN J) engine. For the 5.7L (VIN P) engine, make sure the chamfers face to the front on the left bank and to the rear on the right bank.

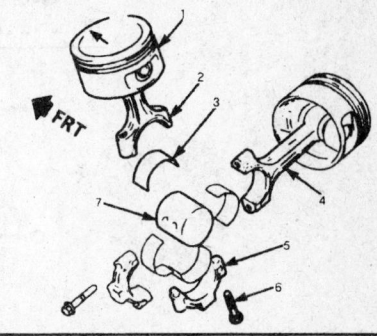

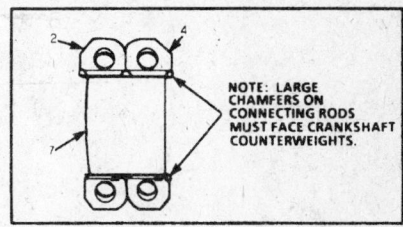

NOTE: LARGE CHAMFERS ON CONNECTING RODS MUST FACE CRANKSHAFT COUNTERWEIGHTS.

1. Piston
2. Connecting rod LH
3. Connecting rod bearing
4. Connecting rod RH
5. Connecting rod bearing cap
6. Connecting rod bearing cap bolt
7. Crankshaft

8838SG37

5.7L (VIN J) engine piston assembly — 5.7L (VIN P) engine similar

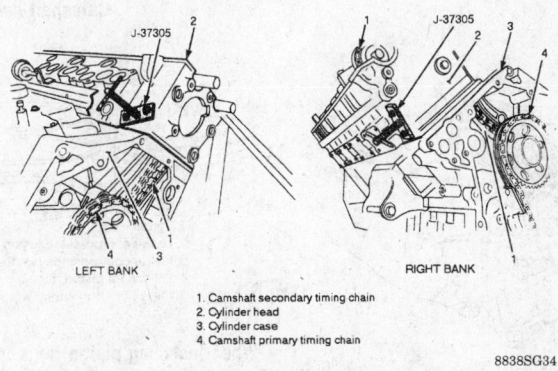

1. Camshaft secondary timing chain
2. Cylinder head
3. Cylinder case
4. Camshaft primary timing chain

8838SG34

Secondary timing chain pre-tensioner tool — 5.7L (VIN J) engine

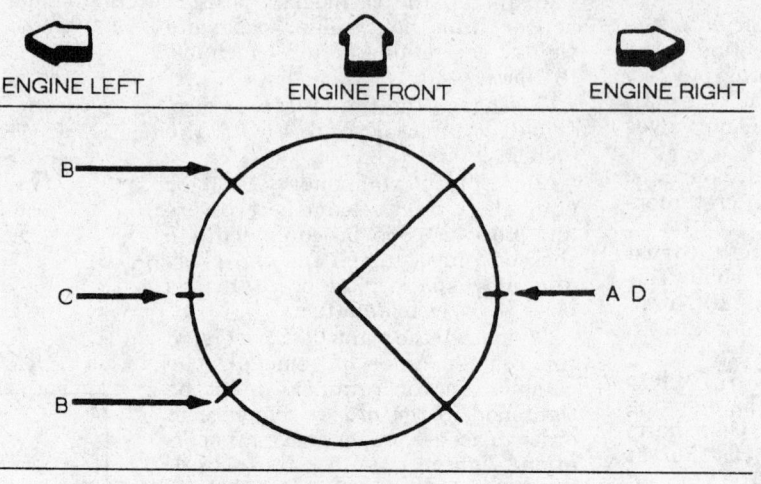

A. Oil ring spacer gap
B. Oil ring rail gaps
C. 2nd compression ring gap
D. Top compression ring gap

8838SG38

Engine ring gap locations — 5.7L (VIN P) engine

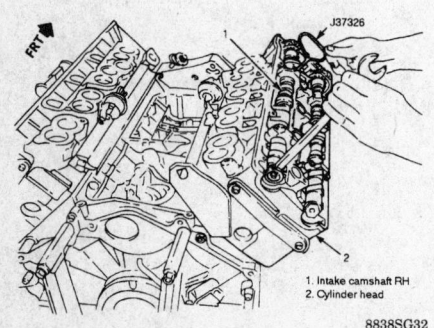

1. Intake camshaft RH
2. Cylinder head

8838SG32

Installing the camshaft timing pins — 5.7L (VIN J) engine

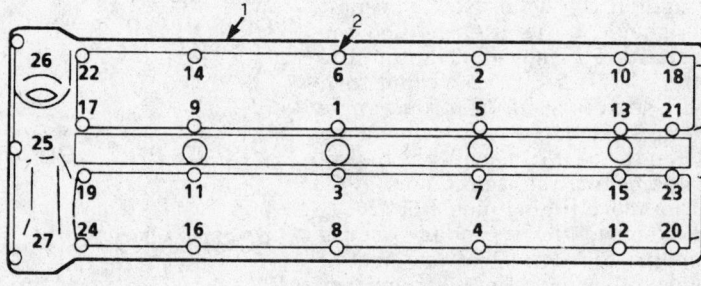

1. Camshaft cover
2. Camshaft cover bolt

8838SG35

Camshaft (valve) cover bolt torque sequence — 5.7L (VIN J) engine

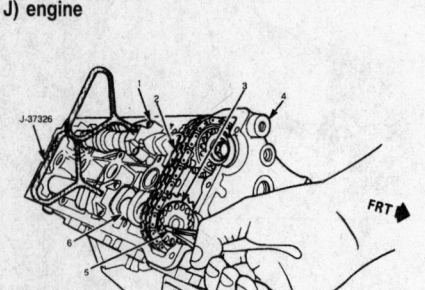

1. Camshaft retainer
2. Camshaft secondary timing chain
3. Camshaft sprocket timing plate
4. Cylinder head
5. Camshaft sprocket pin
6. Camshaft

8838SG33

Installing the camshaft sprocket pin — 5.7L (VIN J) engine

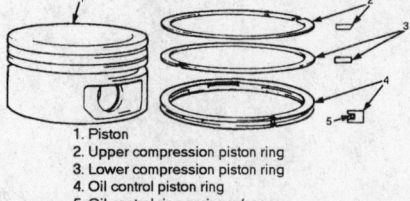

1. Piston
2. Upper compression piston ring
3. Lower compression piston ring
4. Oil control piston ring
5. Oil control ring spring w/spacer

8838SG39

When installing piston rings for the 5.7L (VIN J) engine, make sure to position the ring gaps at 120 degree intervals

ENGINE LUBRICATION

Oil Pan

REMOVAL AND INSTALLATION

5.7L (VIN P) Engine

1. Disconnect the negative battery cable.
2. Raise and support the vehicle safely, then drain the engine oil.

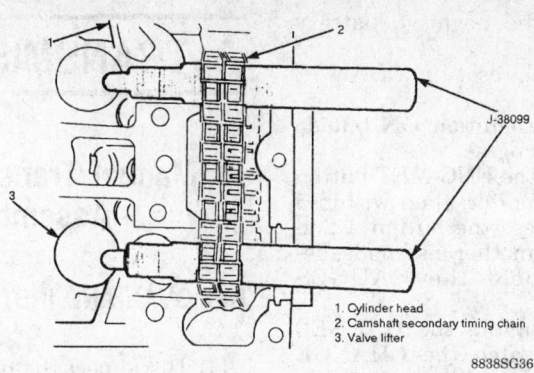

1. Cylinder head
2. Camshaft secondary timing chain
3. Valve lifter

8838SG36

Secondary timing chain retainers — 5.7L (VIN J) engine

3. Disengage the oil level sensor electrical connector and remove the sensor assembly from the side of the oil pan.

4. Remove the oil filter, then remove the oil filter adapter bolts and the adapter assembly.

5. Remove the starter motor assembly.

6. Remove the left catalytic converter.

7. Remove the flywheel cover.

8. Remove the knock sensor retaining nuts and shields.

9. Remove the oil pan bolts, nuts and studs. Be sure to note the location of stud bolts.

10. Remove the oil pan, reinforcements and gasket.

To install:

11. Thoroughly clean all gasket mating surfaces and apply a small amount of 1052914 or equivalent sealer, to the front cover and cylinder block junction and the rear seal retainer and cylinder block junction. Extend the bead of sealer approximately 1 inch (25mm) in either direction of these junctions.

12. Install the gasket onto the oil pan and reinforcements.

13. Install the gasket, pan and reinforcement assembly to the cylinder block with the bolts, studs and nuts.

14. Tighten the corner bolts or stud and nuts to 15 ft. lbs. (20 Nm). Tighten the remainder of the bolts and studs to 8 ft. lbs. (11 Nm).

15. Install the oil level sensor and tighten to 16 ft. lbs. (22 Nm).

16. Install the knock sensor shields and nuts. Tighten the nuts to 75 inch lbs. (8.5 Nm).

17. Install the flywheel cover.

18. Install the left catalytic converter.

19. Install the starter motor assembly.

20. Engage the wiring harness to the oil level sensor terminal.

21. Install the oil filter adapter and tighten the retainers to 17 ft. lbs. (23 Nm), then install the oil filter.

22. Lower the vehicle and properly fill the crankcase with clean engine oil.

23. Connect the negative battery cable.

24. Reset the CHANGE OIL indicator:

a. Turn the ignition **ON** but do not start the engine.

b. Depress the ENG MET button on the trip monitor, then within 5 seconds, depress the button a 2nd time. Within another 5 seconds, depress and hold the GAUGES button.

c. While holding the GAUGES button and watch the CHANGE OIL light, it should begin to flash. Continue to hold the gauges button until the flashing stops and the light goes out indicating that the indicator is reset.

d. If the indicator does not reset, turn the ignition **OFF** and restart the procedure.

5.7L (VIN J) Engine

1. Disconnect negative battery cable and remove the oil lever indicator from the guide tube.

2. Raise and support the vehicle safely, then drain the engine oil.

3. Remove the clutch housing cover attaching bolts, then remove the cover from the vehicle.

4. If equipped, remove the left and right wiring harness heat shields from the oil pan.

5. Disconnect the low oil sensor connection and remove the sensor from the pan.

6. Remove the bolts attaching the AIR pipe bracket to the oil pan, then remove the left and right converter heat shields.

7. Remove the nuts attaching the engine mounts at the front cross-

member rear brace on the left and right sides. Remove the bolts attaching the front crossmember to the rear braces.

8. Remove the bolts attaching the left front crossmember rear brace to the left front side member, then remove the brace from the vehicle.

9. Remove the bolts attaching the right front crossmember rear brace to the right front side member, then remove the brace from the vehicle.

10. Remove the bolts attaching the oil pan and crankcase. Remove the oil pan and gasket from the vehicle.

To install:

11. Apply Loctite® 242 to the oil pan screw threads.

12. Install the oil pan and new gasket to the engine crankcase. Tighten the oil pan front screws to 106 inch lbs. (12 Nm). Tighten the oil pan bolts to 23 ft. lbs. (31 Nm).

13. Install the front crossmember rear braces and bolts retaining the braces to the front crossmember bolts. Finger-tighten the bolts.

14. Install the bolts retaining the left front crossmember rear brace to the left front side member, finger-tight.

15. Install the bolts retaining the right front crossmember rear brace to the left front side member, finger-tight.

16. Tighten the left and right front crossmember rear brace to front crossmember bolts to 59 ft. lbs. (80 Nm), then tighten the left and right front crossmember rear brace to front side member bolts to 46 ft. lbs. (62 Nm).

17. Install the nuts retaining the engine mounts to the front crossmember and tighten to 40 ft. lbs. (54 Nm).

18. Install the converter heat shields and screws.

19. Install the bolts retaining the AIR pipe bracket to the oil pan and tighten to 89 inch lbs. (10 Nm).

20. Install the oil level sensor in the pan and tighten to 18 ft. lbs. (25 Nm), then engage the wiring harness to the sensor.

21. Install the left and right wiring harness heat shields, if equipped, and tighten the bolts to 89 inch lbs. (10 Nm).

22. Install the clutch housing cover and tighten the bolts to 80 inch lbs. (9 Nm).

23. Lower the vehicle and insert the oil level indicator into the guide tube.

24. Properly fill the crankcase with clean engine oil.

25. Connect the negative battery cable.

26. If equipped, reset the CHANGE OIL indicator:

a. Turn the ignition **ON** but do not start the engine.

b. Depress the ENG MET button on the trip monitor, then within 5 seconds, depress the button a 2nd time. Within another 5 seconds, depress and hold the GAUGES button.

c. While holding the GAUGES button and watch the CHANGE OIL light, it should begin to flash. Continue to hold the gauges button until the flashing stops and the light goes out indicating that the indicator is reset.

d. If the indicator does not reset, turn the ignition **OFF** and restart the procedure.

Oil Pump

REMOVAL AND INSTALLATION

5.7L (VIN P) Engine

1. Disconnect the negative battery cable.
2. Raise and support the vehicle safely.
3. Drain the engine oil and remove the oil pan.
4. Remove the oil pan baffle nuts.
5. Support the oil pump by hand and remove the bolt attaching the oil pump to the main bearing cap.
6. Carefully remove the baffle, the oil pump assembly, driveshaft and retainer.

To install:

NOTE: The oil pump pickup should be submerged in oil and the pump primed prior to installation. Failure to prime the pump may result in oil pump failure or internal engine damage. Also, if the pickup screen and pipe assembly was removed from the pump, they must be replaced to assure a proper interference fit.

7. Install the oil pump assembly, shaft and retainer, aligning the slot on the top of the pump driveshaft with the drive tang on the lower end of the distributor driveshaft.
8. Install the oil pan baffle, then install the bolt to the main bearing cap, followed by the baffle nuts. Tighten the retaining bolt to 65 ft. lbs. (88 Nm) and the baffle nuts to 25 ft. lbs. (34 Nm).
9. Install the oil pan and lower the vehicle.
10. Properly fill the engine crankcase with clean engine oil.

11. Connect the negative battery cable.
12. If equipped, reset the CHANGE OIL indicator:

a. Turn the ignition **ON** but do not start the engine.

b. Depress the ENG MET button on the trip monitor, then within 5 seconds, depress the button a 2nd time. Within another 5 seconds, depress and hold the GAUGES button.

c. While holding the GAUGES button and watch the CHANGE OIL light, it should begin to flash. Continue to hold the gauges button until the flashing stops and the light goes out indicating that the indicator is reset.

d. If the indicator does not reset, turn the ignition **OFF** and restart the procedure.

5.7L (VIN J) Engine

1. Disconnect battery negative cable.
2. Remove the primary timing chain and crankshaft sprocket.
3. Remove bolts attaching the oil pump to the cylinder case, then remove the oil pump from the vehicle.
4. Remove O-rings from crankshaft and, if applicable, the oil pump.
5. Remove the oil pickup seal.

To install:

6. Install new O-rings onto the crankshaft and oil pump, as applicable.
7. If applicable, install the oil pickup assembly seal.
8. Apply Loctite® 262 to the oil pump bolts and install them along with the oil pump, finger-tight.

NOTE: Make sure the 2 flats of the pump drive gear are aligned with the 2 flats on the crankshaft. Do not force pump onto crankshaft.

9. Using oil pump aligning tool J-38135 or equivalent pump aligner/seal installer, align oil pump on the crankshaft. Tighten the oil pump bolts to 19 ft. lbs. (26 Nm).
10. Install a new oil pump shaft seal using tools J-38135 and J-38463 or equivalent aligner and seal installer.

NOTE: Install a new oil pump shaft seal whenever the pump is removed from the vehicle.

11. Install the primary timing chain and crankshaft sprocket.
12. Connect the negative battery cable.

TRANSMISSION

Manual Transmission Assembly

REMOVAL AND INSTALLATION

1. Disconnect the negative battery cable.
2. Remove the shifter button, retainer, shift knob, set screw and reverse inhibitor.
3. Remove the rear trim plate screws and the screw located underneath the cup holder mat.
4. Disengage the instrument panel harness connectors from the lighter and rear compartment lid release switch, then unclip the accessory plug harness.
5. Pry the locking tabs on the underside of the boot from the shaft groove, then remove the console trim plate and boot from the shaft.
6. Raise and support the vehicle safely.
7. Remove the complete exhaust assembly.
8. Remove the bolts retaining the driveline torque beam, then slide the beam outboard to gain access to the driveshaft.
9. Remove the parking brake cable clip, then remove the bolts retaining the support bracket.
10. To maintain drivetrain balance, matchmark the relationship between the driveshaft and the differential carrier yoke, then remove the bolts attaching the driveshaft to the yoke.
11. Slide the driveline torque beam rearward until it make contact with the rear exhaust hanger.
12. Support the transmission using an adjustable transmission jack.
13. Disengage the electrical connectors from the speed sensor, backup lamp switch and the computer aided shift solenoid.
14. Remove the transmission to clutch housing attaching bolts.
15. Carefully lower the transmission and remove the transmission assembly from the vehicle.

To install:

16. Install transmission assembly into the vehicle.
17. Install and torque the transmission to clutch housing bolts to 37 ft. lbs. (50 Nm). Make sure to torque the bolts using the proper crisscross sequence, starting at the top right bolt.

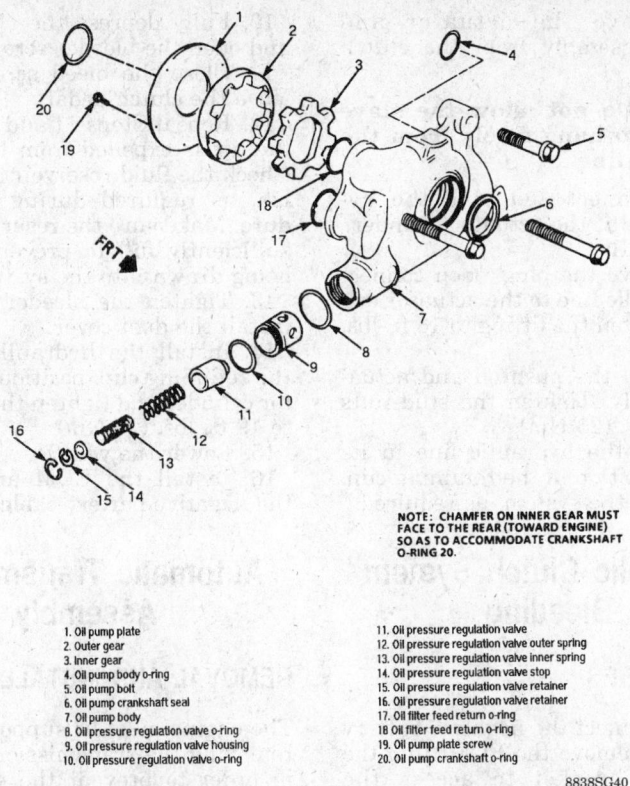

NOTE: CHAMFER ON INNER GEAR MUST
FACE TO THE REAR (TOWARD ENGINE)
SO AS TO ACCOMMODATE CRANKSHAFT
O-RING 20.

1. Oil pump plate
2. Outer gear
3. Inner gear
4. Oil pump body o-ring
5. Oil pump bolt
6. Oil pump crankshaft seal
7. Oil pump body
8. Oil pressure regulation valve o-ring
9. Oil pressure regulation valve housing
10. Oil pressure regulation valve o-ring
11. Oil pressure regulation valve
12. Oil pressure regulation valve outer spring
13. Oil pressure regulation valve inner spring
14. Oil pressure regulation valve stop
15. Oil pressure regulation valve retainer
16. Oil pressure regulation valve retainer
17. Oil filter feed return o-ring
18. Oil filter feed return o-ring
19. Oil pump plate screw
20. Oil pump crankshaft o-ring

8838SG40

Exploded view of the oil pump assembly — 5.7L (VIN J) engine

18. Engage the wiring harness connectors to the speed sensor, backup lamp switch and shift solenoid.

19. Slide the driveline torque beam forward and onto the transmission extension housing.

20. Install the driveshaft, aligning the matchmarks made on the shaft and yoke during removal. Tighten the shaft-to-yoke retaining bolts to 18 ft. lbs. (24 Nm).

21. Install the bolts retaining the support bracket and tighten to 18 ft. lbs. (25 Nm).

22. Check transmission oil level and add if necessary.

NOTE: In a horizontal position, the transmission should be filled to the point of overflow.

23. Install the parking brake cable clip.

24. Align the torque beam and install the retaining bolts. Tighten the beam-to-differential carrier bolt to 60 ft. lbs. (80 Nm) and the beam-to-transmission bolt to 37 ft. lbs. (50 Nm).

25. Install the complete exhaust system assembly.

26. Lower the vehicle.

27. Install the console trim plate and boot assembly.

28. Connect the negative battery cable.

Clutch Assembly

REMOVAL AND INSTALLATION

1. Disconnect the negative battery cable, then raise and support the vehicle safely.

2. Remove the complete exhaust system.

3. Remove the transmission assembly.

4. Except for the VIN J engine, disconnect the ground wire attached to the left clutch housing stud.

5. Remove the nuts attaching the clutch slave cylinder to the housing and support the cylinder to the side. Do not allow the cylinder to hang freely.

6. For the VIN J engine, remove the nut retaining the left converter shield to the housing.

7. Remove the clutch housing cover.

8. Remove the bolts retaining the housing to the engine block and, if applicable on the 5.7L (VIN J) engine, the right side converter heat shield.

9. Remove the housing by aligning the fork onto the 2 flats of the release bearing and push the fork away from the bearing with a twisting motion. Remove the clutch housing and, for

5.7L (VIN P) engine with magnesium housings, the aluminum spacers.

NOTE: Excessive clutch wear may require removal of the ball stud locking screw and loosening of the ball stud to disengage the fork and housing.

10. Mark the alignment of the clutch cover and flywheel for installation purposes.

11. Loosen the clutch cover bolts evenly, 1 turn at a time until spring pressure is released. Failure to properly release spring pressure may result in damage to the clutch cover assembly and the flywheel.

12. Remove the clutch plate and disc assembly.

To install:

13. Inspect flywheel, clutch plate and disc for heat stress, cracks or worn parts and replace as necessary.

14. Install the clutch assembly using a suitable universal clutch disc alignment tool.

15. Make sure the marks made earlier are in alignment, then install the cover assembly-to-flywheel bolts. Tighten the bolts in the proper sequence, 1 turn at a time, until spring pressure is properly attained and the bolts are tightened to 30 ft. lbs. (41 Nm).

16. Position the clutch housing to the engine block and engage the fork onto the release bearing. If equipped, be sure the aluminum spacer is in position.

17. Verify the housing is properly positioned on the 2 engine dowel pins and, for the 5.7L (VIN J) engine, that the right converter heat shield is installed.

18. Tighten the clutch housing bolts to 37 ft. lbs. (50 Nm) and the ball stud to 33 ft. lbs. (45 Nm). Tighten the ball stud locking screw to 11 ft. lbs. (15 Nm) for VIN P engine or to 16 ft. lbs. (22 Nm) for VIN J engine, as applicable.

19. If equipped, install the ground harness connection to the housing.

20. Install the housing cover and tighten the bolts to 80 inch lbs. (9 Nm).

21. For the VIN J engine, install the left heat shield and tighten the retaining nut to 12 inch lbs. (1.4 Nm).

22. Install the clutch slave cylinder and tighten the retaining nuts to 19 ft. lbs. (25 Nm).

23. Install the transmission assembly.

24. Install the exhaust system and lower the vehicle.

25. Connect the battery negative cable and check clutch for proper operation.

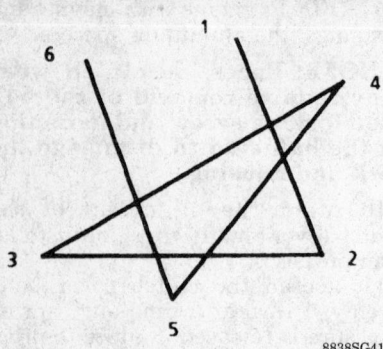

8838SG41

Clutch cover assembly torque sequence

Clutch Master Cylinder

REMOVAL AND INSTALLATION

1. Disconnect the negative battery cable, then remove the battery from the vehicle.
2. Remove the sound insulator panel from under the dash.
3. Disconnect the pushrod retaining clip and pushrod at the clutch pedal.
4. Disconnect and plug the hydraulic line at the clutch master cylinder.
5. Remove the clutch master cylinder retaining bolts at the front of the dash.
6. Remove the clutch master cylinder from the vehicle.

To install:

7. Install the master cylinder into the vehicle and tighten the mounting bolts to 12 ft. lbs. (17 Nm).
8. Remove the plug, then connect the hydraulic fitting to the master cylinder and tighten to 13 ft. lbs. (18 Nm).
9. Connect the pushrod to the pedal and install the retaining clip.
10. Install the under dash hush panel.
11. Bleed the system, as required.
12. Install the battery and connect the positive, followed by the battery negative cables.

Clutch Slave Cylinder

REMOVAL AND INSTALLATION

1. Raise and support the vehicle safely.
2. Remove the actuator cylinder stud nuts.
3. Note the position of the hydraulic line and disconnect the line from the retaining clip.

4. Remove the actuator and pushrod assembly from the clutch housing.

NOTE: Do not allow the slave cylinder to hang freely from the hydraulic line.

5. Disconnect and plug the hydraulic line at the actuator cylinder.

To install:

6. Remove the plug, then connect the hydraulic line to the actuator cylinder. Tighten the fitting to 13 ft. lbs. (18 Nm).
7. Install the pushrod and actuator assembly. Tighten the stud nuts to 19 ft. lbs. (25 Nm).
8. Place the hydraulic line in its original position in the retaining clip.
9. Bleed the system, as required.

Hydraulic Clutch System Bleeding

PROCEDURES

1. Disconnect the negative battery cable and remove the ECM from the mounting bracket to access the master cylinder for filling. Fill the master cylinder reservoir with the proper grade and type of fresh brake fluid or hydraulic clutch fluid.
2. Prior to bleeding the actuator, most of the air can be removed as follows:
    a. Remove the master cylinder cap and moisture barrier.
    b. Install the master cylinder cover.
    c. Lightly stroke the clutch pedal to release trapped air through the master cylinder.
    d. Remove the master cylinder cap and install the moisture barrier.
    e. Install the master cylinder cap.
3. Raise and support the vehicle safely.
4. Remove the actuator cylinder attaching stud nuts.
5. Remove the pushrod and actuator cylinder from the clutch housing and the hydraulic line from the retaining clip.
6. Lower cylinder slightly for access and disconnect the hydraulic hose fitting from the actuator cylinder.
7. Remove the bleed screw dust cap.
8. Position a drain pan or attach a clear plastic hose.
9. Support the slave cylinder in a horizontal position, with the bleeder screw vertical.

10. Fully depress the clutch pedal and open the bleeder screw.
11. Close the bleed screw and release the clutch pedal.
12. Repeat Steps 11 and 12 until all the air is expelled from the system. Check the fluid reservoir and replenish, as required during the procedure. Make sure the reservoir is kept sufficiently full to prevent air from being drawn into the system.
13. Tighten the bleeder screw and install the dust cover.
14. Install the hydraulic line into the retaining clip, position the actuator cylinder and tighten the stud nuts to 19 ft. lbs. (25 Nm).
15. Lower the vehicle.
16. Install the ECM and connect the negative battery cable.

Automatic Transmission Assembly

REMOVAL AND INSTALLATION

The engine must be supported before removing the transmission assembly in order to prevent the vapor blow pipe located across the rear of the engine from contacting the dash panel.

1. Disconnect the negative battery cable and remove the transmission fluid level indicator.
2. Disconnect the TV cable at the throttle lever or the adjuster assembly.
3. Raise and support the vehicle safely.
4. If equipped, remove the upper and lower underbody braces.
5. Remove the complete exhaust system.
6. Support the transmission with a suitable jack.
7. Remove the driveline support beam.
8. Matchmark and remove the driveshaft.
9. Disengage the speedometer electrical connector, then disconnect the shift control cable and the remaining electrical leads from the transmission.
10. Remove the torque converter cover and mark the relationship of the converter to the flywheel, then remove the converter-to-flywheel bolts.
11. Disconnect the oil cooler pipes at the transmission. Plug the openings to prevent system contamination or excessive fluid loss.
12. Disconnect the TV cable at the transmission.
13. Remove the transmission-to-engine mounting bolts and fasten the torque converter to the transmission

using a converter restraining tool or a length of wire.

14. Carefully move the transmission rearward, downward and out from under the vehicle. If interference is encountered with cables, cooler lines, etc., remove the component(s) before finally lowering the transmission.

To install:

15. Flush the transmission oil cooler lines using J-35944 or an equivalent transmission cooler and line flushing tool.

16. Install a converter restraint tool to hold the torque converter in place.

17. Support the transmission with a suitable jack, then raise the transmission into position and remove the torque converter holding tool.

18. Install and tighten the transmission to engine bolts to 35 ft. lbs. (47 Nm).

19. Connect the TV cable to the transmission.

20. Remove the plugs, then connect the oil cooler pipes to the transmission.

21. Align the marks made during removal and start the torque converter to flywheel bolts by hand. Tighten the bolts to 46 ft. lbs. (62 Nm).

22. Install converter cover and torque screws to 89 inch lbs. (10 Nm).

23. Engage the electrical connectors to the transmission.

24. Connect the shift control cable.

25. Engage the speedometer electrical connector.

26. Align the marks made earlier and install the driveshaft, then the driveline support beam.

27. Install the exhaust system and, if equipped, the underbody braces.

28. Lower the vehicle and install the oil level indicator.

29. Connect the TV cable to the throttle lever or to the adjuster assembly.

30. Connect the negative battery cable.

31. Check and add the proper type and amount of transmission fluid.

32. Because the driveline support beam was removed, check clearance between the air intake duct and the throttle body. If the air duct becomes dislodged from the throttle body, a driveability problem could occur.

Shift Linkage

ADJUSTMENT

1. Disconnect the negative battery cable.

2. Place the control lever in the **N** position.

3. Raise and support the vehicle safely.

4. Loosen the cable attachment at the shift lever.

5. Rotate the shift lever clockwise to **P** detent and then back to **N**.

6. Tighten the cable attachment to 15 ft. lbs. (20 Nm).

NOTE: The lever must be be held out of the P position when tightening the nut.

7. Lower the vehicle.

8. Check the cable adjustment by rotating the control lever through the detents.

9. Connect the battery negative cable.

Throttle Valve Cable

ADJUSTMENT

The Acceleration Slip Regulation (ASR) system was added to all Corvettes. This required a cable adjuster assembly which has the ability to extend cables slightly, according to commands from the control module. This extension allows the throttle close regardless of accelerator pedal position. The adjuster does not have the ability to apply throttle, it can only release it.

The cable adjuster assembly must be adjusted each time the throttle and/or TV cables are disconnect. On some models, the TV cable is also attached to a servo. The cable may be adjusted BEFORE cable adjuster assembly adjustment.

TV Cable Servo Linkage

ADJUSTMENT

1. Make sure the TV cable is installed into servo bracket.

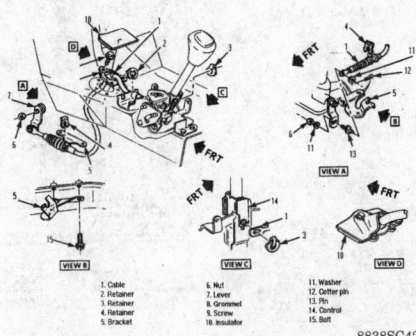

1. Cable
2. Retainer
3. Retainer
4. Retainer
5. Bracket
6. Nut
7. Lever
8. Grommet
9. Screw
10. Insulator
11. Washer
12. Cotter pin
13. Pin
14. Control
15. Bolt

8838SG42

Automatic transmission shifter cable

2. Pull servo assembly end of cable toward servo without moving the throttle lever.

3. If 1 out of the 5 holes in the servo assembly tab aligns with the cable pin, push pin through hole and connect pin to tab with retainer.

4. If the tab holes does not align with the pin, move the cable away from the servo assembly until the next closest tab hole aligns and connect the pin to the tab with the retainer.

5. Perform the adjustment procedure for the ASR accelerator and cruise control adjuster assembly.

DRIVE AXLE

Driveshaft and U-Joints

REMOVAL AND INSTALLATION

1. Raise and support the vehicle safely.

2. If equipped, remove the upper and lower underbody braces.

3. Remove the complete exhaust system as an assembly.

4. Support the transmission, then remove the bolts, washers and nuts attaching the driveline support beam at the axle and/or transmission to gain necessary clearance.

5. Mark relationship of shaft to the pinion yoke and disconnect the rear universal joint by removing trunnion bearing straps. Tape bearing cups to trunnion to prevent dropping and loss of roller bearings.

6. Place a suitable drain pan under the transmission for oil leakage, slide the slip yoke from the transmission and remove the driveshaft from the vehicle.

7. If necessary, remove the universal joints:

a. Remove the snaprings. If a snapring does not readily come out, tap the end of the bearing cap lightly to relieve pressure against the ring.

b. Place the driveshaft horizontally in line with the base plate of a press, but do not clamp the tube.

c. Support the lower ear of the universal joint with a 1⅛ inch socket.

d. Press the lower bearing cap out from the yoke using a pusher on the upper bearing cup.

e. Rotate the driveshaft, then remove the the opposite bearing cup.

f. Remove the universal joint from the yoke.

To install:

8. If removed, install the U-joints:

a. Install one bearing cap partially into 1 side of the yoke, then turn this side to the bottom.

b. Install the joint into the yoke so the trunnion seats freely in the bearing cap.

c. Install the opposite bearing cap partially into the yoke, verifying the trunnions are straight and true in the bearing caps.

d. Press against the opposite bearing caps, while working the joint in order to verify that the joint is not binding and turns freely. If the joint begins to bind, there is probably 1 or more needle bearings out of place and tipped under the trunnion.

e. When 1 bearing cap snapring retainer groove clears the inside of the yoke, stop pressing and install a snapring into place.

f. Continue to press the opposite side until a snapring can be inserted. If difficulty is encountered, strike the yoke firmly with a hammer to slightly spring the yoke ears.

g. Assemble the other half of the joint in the same manner.

9. Slide the driveshaft slip yoke into the transmission extension.

10. Align the marks made during removal and install the rear of the driveshaft to the pinion yoke. If no marks were made or the driveshaft is being replaced, align the black paint dot on the driveshaft as close to 180 degrees opposite the yellow paint dot on the axle pinion yoke.

11. Install the propeller shaft retainers and bolts. Tighten the bolts to 18 ft. lbs. (24 Nm).

12. If removed, install and align driveline support beam as follows:

a. To ensure proper alignment of the driveline, a clearance of 1.53–2.00 in. (39–51mm) must be maintained between the top of the beam to the underbody and a clearance of 0.86–1.34 in. (22–34mm) from the passenger side of the beam to the side wall.

b. Take the measurements directly above and to the right of the driveshaft yoke.

c. Apply sealer to the support sealing surfaces at the transmission extension, the differential carrier and the driveline support.

d. Install the washers, bolts and nuts then tighten the bolts at the carrier to 60 ft. lbs. (80 Nm) and the transmission bolts to 37 ft. lbs. (50 Nm).

e. Remove the transmission support.

13. Install the exhaust system assembly.

14. If equipped, install the upper and lower underbody braces.

15. Lower the vehicle.

STEERING

Air Bag

——— CAUTION ———

Some vehicles are equipped with an air bag system, also known as the Supplemental Inflatable Restraint (SIR) or Supplemental Restraint System (SRS). The system must be disabled before performing service on or around system components, steering column, instrument panel components, wiring and sensors. Failure to follow safety and disabling procedures could result in accidental air bag deployment, possible personal injury and unnecessary system repairs.

PRECAUTIONS

Several precautions must be observed when handling the inflator module to avoid accidental deployment and possible personal injury.

• Never carry the inflator module by the wires or connector on the underside of the module.

• When carrying a live inflator module, hold securely with both hands, and ensure that the bag and trim cover are pointed away.

• Place the inflator module on a bench or other surface with the bag and trim cover facing up.

• With the inflator module on the bench, never place anything on or close to the module which may be thrown in the event of an accidental deployment.

DISARMING

1. Turn the steering wheel to align the wheels in the straight-ahead position.

2. Turn the ignition switch to the **LOCK** position.

3. Remove the AIR BAG fuse from the fuse block.

4. Remove the left side lower trim panel, then unplug the Connector Position Assurance (CPA) device and the yellow 2-way SIR harness wire connector at the base of the steering column.

ENABLING

1. Turn the ignition switch to the **LOCK** position.

2. Engage the yellow 2-way connector and the CPA device at the base of the steering column.

3. Install the left side lower trim panel, then install the SIR fuse to the fuse block.

4. Turn the ignition switch to the **RUN** position.

5. Verify the SIR indicator light flashes 7–9 times and then turns OFF. If not, inspect system for malfunction.

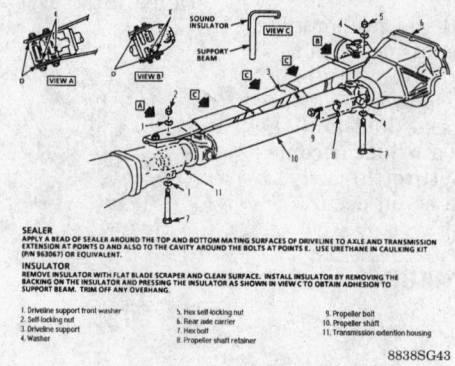

SEALER
APPLY A BEAD OF SEALER AROUND THE TOP AND BOTTOM MATING SURFACES OF DRIVELINE TO AXLE AND TRANSMISSION EXTENSION AT POINTS D AND ALSO TO THE CAVITY AROUND THE BOLTS AT POINTS E. USE URETHANE IN CAULKING KIT (PN 9630671) OR EQUIVALENT.

INSULATOR
REMOVE INSULATOR WITH FLAT BLADE SCRAPER AND CLEAN SURFACE. INSTALL INSULATOR BY REMOVING THE BACKING ON THE INSULATOR AND PRESSING THE INSULATOR AS SHOWN IN VIEW C TO OBTAIN ADHESION TO SUPPORT BEAM. WIPE OFF ANY OVERHANG.

1. Driveline support front washer	5. Hex self-locking nut	9. Propeller bolt
2. Self locking nut	6. Rear axle carrier	10. Propeller shaft
3. Driveline support	7. Hex bolt	11. Transmission extention housing
4. Washer	8. Propeller shaft retainer	

8838SG43

Driveshaft support beam alignment

Supplemental Inflatable Restraint (SIR) Coil

NOTE: After performing repairs on the internals of the steering column, the coil assembly must be centered in order to avoid coil damage or accidental air bag deployment.

ADJUSTMENT

1. Hold the coil assembly with the clear bottom up to see the coil ribbon.
2. While holding the coil assembly, depress the spring lock and rotate the hub in the direction of the arrow until it stops. The coil ribbon should now be wound up snug against the center hub.
3. Rotate the coil assembly in the opposite direction approximately 2½ turns and release the lock spring between the locking tabs in front of the arrow.
4. Install the coil assembly onto the steering shaft.

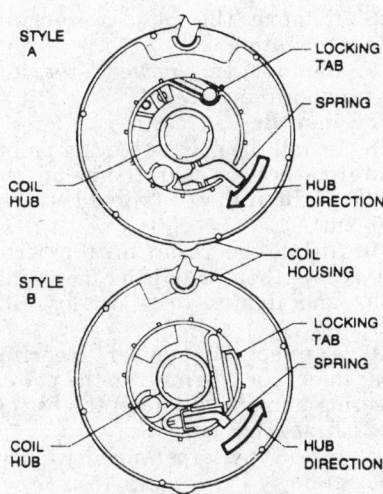

STYLE A

LOCKING TAB

SPRING

COIL HUB

HUB DIRECTION

COIL HOUSING

STYLE B

LOCKING TAB

SPRING

COIL HUB

HUB DIRECTION

PERFORM THE FOLLOWING STEPS TO CENTER COIL ASSEMBLY:
A. REMOVE COIL ASSEMBLY
B. HOLD COIL ASSEMBLY WITH CLEAR BOTTOM UP TO SEE COIL RIBBON.
C. NOTE: THERE ARE TWO DIFFERENT STYLES OF COILS. ONE ROTATES CLOCKWISE AND THE OTHER ROTATES COUNTER-CLOCKWISE.
D. WHILE HOLDING COIL ASSEMBLY, DEPRESS SPRING LOCK TO ROTATE HUB IN DIRECTION OF ARROW UNTIL IT STOPS.
E. THE COIL RIBBON SHOULD BE WOUND UP SNUG AGAINST CENTER HUB.
F. ROTATE COIL HUB IN OPPOSITE DIRECTION APPROXIMATELY TWO AND A HALF (2-1/2) TURNS. RELEASE SPRING LOCK BETWEEN LOCKING TABS IN FRONT OF ARROW.

8838SG44

Centering the SIR coil assembly

Steering Wheel

———— CAUTION ————
The Corvette is equipped with a Supplemental Inflatable Restraint system, make certain to follow the recommended disarming procedure before and the coil centering and SIR enabling procedures, after repairs.

REMOVAL AND INSTALLATION

1. Properly disable the SIR system and disconnect the negative battery cable.
2. Remove screws from the back of the steering wheel attaching the inflator module.
3. Remove the inflator module from the steering wheel.
4. Disengage the Connector Pin Assurance (CPA) device and unplug the SIR electrical connector at the inflator module.
5. Remove the steering wheel attaching nut and disengage the horn connector. Mark the relationship of the steering wheel to the column splines for installation purposes.

NOTE: To avoid damaging the SIR coil, do not use any steering wheel puller other than those recommended.

6. Using steering wheel puller tool J-1859-03 and puller screws J-38720, or equivalents, remove the steering wheel. If the steering wheel does not come off easily, proceed as follows:
   a. With the puller installed and the side screws threaded to the shoulder, tighten the puller center screw snugly against the steering shaft.
   b. Back out each side screw 1 revolution from the fully threaded position.
   c. Retighten the puller center screw.
   d. Alternately tighten each side screw ¼ turn. Tightening the screws more than ¼ turn at a time could result in damage to the steering wheel.
   e. Remove the steering wheel from the vehicle.
To install:
7. Engage the horn connector, then install the steering wheel to the column aligning the marks made earlier.
8. Install a new steering wheel retaining nut and tighten the new nut to 30 ft. lbs. (41 Nm).
9. Make sure the ignition is **OFF** and the negative battery cable is dis-

connected, the engage the SIR coil electrical connector to the inflator module. Install the CPA device to retain the connection.
10. Position the inflator module onto the steering wheel and install the module retaining screws. Tighten the screws to 87 inch lbs. (9.7 Nm).
11. Connect the negative battery cable and properly enable the SIR system.

Tie Rod Ends

REMOVAL AND INSTALLATION

1. Disconnect battery ground cable.
2. Raise and support the vehicle safely.
3. Remove the tie rod cotter pin and hex slotted nut from the tie rod assembly.
4. Loosen tie rod jam nut.
5. Using tool J-24319-01 or equivalent linkage puller, remove the tie rod from the steering knuckle.
6. Remove the tie rod from the steering rack assembly. To ease installation and give a point from which the alignment may be adjusted, scribe alignment marks on the rack assembly prior to tie rod end removal, and/or count the number of turns necessary to remove the tie rod end.
To install:
7. Install the tie rod to the steering rack assembly, but do not tighten the jam nut. Thread the tie rod end in the same number of turns and/or align it to the marks made during removal.
8. Install the tie rod to the steering knuckle and install the hex slotted nut to the tie rod stud.
9. Tighten the hex nut to 35 ft. lbs. (47 Nm), then insert a new cotter pin. If necessary tighten the nut additionally in order to insert the pin, but do not exceed a total torque of 52 ft. lbs. (70 Nm) and do not back off the original torque.
10. Check and adjust toe, as necessary.

NOTE: Make sure the rack and pinion boot is not twisted or puckered during installation or adjustment.

11. Adjust the toe by turning the inner tie rod, making sure the rack and pinion boot is not twisted or puckered during toe adjustment.
12. Tighten the jam nut against the tie rod to 50 ft. lbs. (68 Nm).

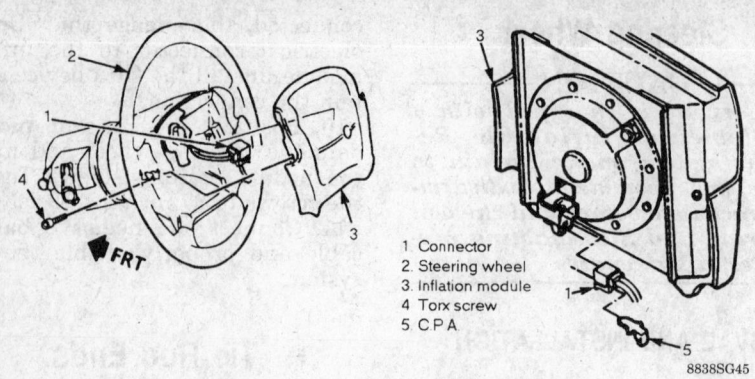

1. Connector
2. Steering wheel
3. Inflation module
4. Torx screw
5. C.P.A.

8838SG45

SIR inflator module removal and installation

Power Rack and Pinion

REMOVAL AND INSTALLATION

1. Disconnect the negative battery cable and position a drain pan under the vehicle to catch fluid.

2. For VIN P engine, remove the air intake duct, then remove the serpentine drive belt and the drive belt idler pulley.

3. Remove the power steering gear inlet hose assembly from the steering gear.

4. Remove the power steering gear outlet hose assembly from the steering gear.

NOTE: If equipped with a power steering fluid cooling pipe, disconnect fluid cooling pipe outlet hose from the fluid cooling pipe.

5. Remove the steering gear coupling shield.

6. Disconnect the intermediate shaft from the power steering gear and lower steering shaft, then position aside.

7. Raise and support the vehicle safely.

8. Remove the front tire and wheel assemblies.

9. Remove both outer tie rods from the knuckles using a suitable puller.

10. If equipped, remove the power steering cooler assembly.

11. Remove the stabilizer shaft.

12. Remove the steering gear to frame attaching clamp nuts, then remove the bolts and clamp from the vehicle.

13. Remove the power steering gear attaching attaching nut and bolt.

14. Remove the power steering gear from the vehicle.

15. If necessary, remove the outer tie rods, rack and pinion boots, and the inner tie rods from the power steering gear.

To install:

16. If removed, install the inner tie rods, boots and outer tie rods.

17. Install the power steering gear, nuts and bolts. Torque the attaching nut to 30 ft. lbs. (40 Nm). Torque the steering gear clamp nuts to 18 ft. lbs. (25 Nm).

18. Install the stabilizer shaft and, if applicable, the power steering cooler assembly.

19. Install both outer tie rods to the steering knuckle.

20. Install tire and wheel assemblies, then lower the vehicle.

21. Install the intermediate shaft and the steering gear coupling shield.

22. Install the power steering gear outlet and inlet hose assemblies to the power steering gear. Tighten fittings to 21 ft. lbs. (28 Nm).

23. For VIN P engine, install the drive belt idler pulley and the serpentine drive belt, then install the air intake duct.

24. Remove the drain and fill the power steering reservoir.

25. Connect the negative battery cable, bleed the system and check for proper operation.

Power Steering Pump

BLEEDING

1. With the engine **OFF** and wheels off the ground, turn the steering wheel all the way to the left. Add power steering fluid to the **COLD** mark on the fluid level indicator.

2. Bleed the system by turning the wheels from side-to-side without reaching the stop at either end. It may be necessary to turn the wheel from side-to-side several times. Be sure to keep the fluid full.

3. Start the engine. With engine idling, recheck the fluid level. If necessary add fluid to bring the fluid up to the **COLD** mark.

4. Return the wheels to the center position. Lower the front wheels to the ground and continue to run for 2–3 minutes.

5. Road test the vehicle to make sure the steering functions normally and without noise.

6. Check for fluid leakage. Check to make sure the fluid level is at the **HOT** mark after system is stabilized at its normal operating temperature.

REMOVAL AND INSTALLATION

5.7L (VIN P) Engine

1. Disconnect the negative battery cable and place a drain pan under the vehicle to catch fluid.

2. Remove the air intake duct, then remove the serpentine drive belt.

3. Remove the steering pump pulley hub cap, then remove the pulley using J-25034-B or an equivalent puller tool.

4. Remove the serpentine drive belt idler pulley.

5. Disconnect the gear inlet hose assembly from the pump. Remove the pump inlet pipe mounting bolts, then disconnect the pipe assembly from the pump.

6. Remove the power steering pump mounting bolts.

7. Remove the power steering pump and front bracket.

To install:

8. Install the power steering pump and front bracket with the mounting bolts. Tighten the bolts to 18 ft. lbs. (25 Nm).

9. Install the pump inlet pipe to the pump, then install the mounting bolts and tighten to 24 ft. lbs. (33 Nm).

10. Connect the power steering gear inlet hose assembly to the power steering pump and tighten the fitting to 21 ft. lbs. (28 Nm).

11. Install the serpentine drive belt idler pulley.

12. Using J-25033-B, or an equivalent installation tool, install the power steering pump pulley so the front of the pulley hub is flush with the front of the pump shaft.

13. Install the hub cap to the pulley.

14. Install the serpentine drive belt, then install the intake air duct.

15. Connect the negative battery cable and remove the drain pan.

16. Refill the power steering reservoir and properly bleed the system.

5.7L (VIN J) Engine

1. Disconnect the negative battery cable and remove the air intake duct.
2. Drain the engine cooling system into a suitable container.
3. Drain and appropriately discard the power steering fluid.
4. Remove the serpentine drive belt.
5. Remove the left coolant outlet cover and hose.
6. Disconnect the vacuum hose retainer from the pump reservoir, then remove the vacuum hose(s) and set aside.
7. Remove the bolts retaining the compressor hose clip to the crossmember.
8. Remove the pulley hub cap and separate the pulley from the pump using J-25034-B or an equivalent puller.
9. Remove the crankcase vent inlet pipe from the power steering pump bracket.
10. Remove the crankcase vent inlet hose from the throttle body extension.
11. Remove the compressor-to-power steering pump bracket bolt.
12. Remove the gear inlet pipe from the pump, then remove the outlet hose from the cooler assembly.
13. Remove the power steering pump bracket-to-cylinder head bolts, then remove the pump from the vehicle.
14. As necessary, remove the fluid reservoir hose from the assembly, remove the pump-to-mounting bracket bolts and remove the pump and/or reservoir from the mounting bracket.

To install:

15. If replacing the pump, do not use a pump with the letter **R** on the back of the pump housing.
16. If removed, install the reservoir and/or pump to the bracket. Apply Loctite® 565 or equivalent, to the pump-to-bracket mounting bolts and install. Tighten the bolts to 19 ft. lbs. (26 Nm).
17. If removed, connect the pump reservoir hose to the pump and tighten the clamp screw to 22 inch lbs. (2.5 Nm).
18. Install the pump assembly to the vehicle. Coat the bolt threads with Loctite® 565 or equivalent, then install the bolts and tighten 19 ft. lbs. (26 Nm).
19. Connect the outlet hose to the cooler assembly. Tighten the clamp screw to 22 inch lbs. (2.5 Nm).
20. Connect the power steering gear inlet pipe to the pump, then tighten the fitting to 21 ft. lbs. (28 Nm).

21. Install the compressor-to-power steering pump bracket bolt and tighten to 19 ft. lbs. (26 Nm).
22. Install the crankcase vent inlet hose to the throttle body extension and to the power steering pump bracket.
23. Install the pump pulley using J-25033-B or equivalent installation tool, then install the hub cap.
24. Install the bolt retaining the compressor hose clip to the crossmember.
25. Install the vacuum hose(s) and connect the retainer to the power steering pump reservoir. Tighten the screws to 13 inch lbs. (1.5 Nm).
26. Install the left coolant outlet cover and hose.
27. Install the serpentine drive belt.
28. Install the air intake duct and connect the negative battery cable.
29. Refill the power steering reservoir.
30. Properly fill the engine cooling system.
31. Bleed the power steering system.

BRAKES

Anti-Lock Brake System Service

PRECAUTIONS

• Certain components within the Anti-Lock Brake System (ABS) are not intended to be serviced or repaired individually. Only those components with removal and installation procedures should be serviced.

• Do not use rubber hoses or other parts not specifically specified for and ABS system. When using repair kits, replace all parts included in the kit. Partial or incorrect repair may lead to functional problems and require the replacement of components.

• Lubricate rubber parts with clean, fresh brake fluid to ease assembly. Do not use lubricated shop air to clean parts; damage to rubber components may result.

• Use only specified brake fluid from an unopened container.

• If any hydraulic component or line is removed or replaced, it may be necessary to bleed the entire system.

• A clean repair area is essential. Always clean the reservoir and cap

thoroughly before removing the cap. The slightest amount of dirt in the fluid may plug an orifice and impair the system function. Perform repairs after components have been thoroughly cleaned; use only denatured alcohol to clean components. Do not allow ABS components to come into contact with any substance containing mineral oil; this includes used shop rags.

• The Anti-Lock control unit is a microprocessor similar to other computer units in the vehicle. Ensure that the ignition switch is **OFF** before removing or installing controller harnesses. Avoid static electricity discharge at or near the controller.

• If any arc welding is to be done on the vehicle, the control unit should be unplugged before welding operations begin.

Master Cylinder

REMOVAL AND INSTALLATION

1. Disconnect the negative battery cable.
2. Unplug the electrical connector from the warning switch and, if equipped, the fluid level warning switch assemblies.
3. Disconnect the hydraulic brake lines at the master cylinder. Plug the openings to prevent fluid contamination or loss.
4. If equipped, disconnect the master cylinder prime pipe from the reservoir.
5. Remove the retaining nuts holding the cylinder to the brake booster assembly.
6. Reposition the battery cable and cruise control cable.
7. Remove the master cylinder assembly from the brake booster.

To install:

8. Position the master cylinder assembly to the power booster.
9. Clip the battery and cruise control cables into position.
10. Install the master cylinder retaining nuts and tighten to 13 ft. lbs. (18 Nm).
11. Remove the plugs, then connect the hydraulic brake lines to the master cylinder and tighten the fittings to 13 ft. lbs. (18 Nm).
12. If equipped, connect the master cylinder prime pipe.
13. Engage the electrical connections to the warning switch assemblies.
14. Fill the master cylinder and properly bleed the hydraulic brake system.

15. Connect the negative battery cable.

Brake Caliper

REMOVAL AND INSTALLATION

Front

1. Disconnect the negative battery cable and remove ⅔ of the brake fluid from the master cylinder reservoir.
2. Raise and support the vehicle safely.
3. Mark the relationship between the wheel and axle flange, then remove the tire and wheel assembly.
4. Install 2 wheel nuts to retain the brake rotor.
5. Depress the caliper pistons into the caliper bores in order to provide clearance between the pads and the rotor.
6. If the caliper is being completely removed from the vehicle for service, disconnect the brake line fitting at the caliper by removing the bolt, 2 gaskets and then the brake hose inlet fitting. Plug all openings to prevent fluid contamination or loss.

NOTE: Do not allow the fluid to come into contact with the front transverse spring as damage to the spring may occur.

7. Remove the circlip and the retainer pin, then the caliper housing from the rotor and the caliper mounting bracket. Remove the caliper from the vehicle or if the brake line is still attached, support the caliper from the control arm with a suitable hook or length of mechanic's wire.

To install:
8. Install the caliper over the brake rotor and into the caliper mounting bracket. Make sure the shoe lining guiding surfaces are correctly seated in the bracket.

NOTE: There are 2 sets of retainer pins in most repair kits. One set is for base calipers and the other is for heavy duty calipers. Make certain the correct retainer pins are installed.

9. Compress the bias springs by applying pressure to the mounting bracket, then install the new retainer pin and circlip.
10. If removed, connect the brake hose inlet fitting, 2 new gaskets and the inlet fitting bolt. Tighten the bolt to 30 ft. lbs. (40 Nm).
11. If the inlet fitting was removed, properly bleed the hydraulic brake system.

12. Remove the wheel nuts retaining the rotor, align the marks made earlier and install the tire and wheel assembly.
13. Lower the vehicle and check the brake fluid; add as necessary.
14. Connect the negative battery cable, start the engine and pump the brake pedal slowly and firmly 3 times to seat the shoe and lining assemblies.

Rear

1. Disengage the parking brake automatic adjuster as follows:
   a. Remove the drivers seat cushion.
   b. Remove the parking brake lever cover and screws.
   c. Using a suitable offset tool, disengage and hold the drive pawl from the drive sector.
   d. Insert a nail or drift through the hole in the anchor plate to retain the drive pawl in the disengaged position.
   e. Move the parking brake lever until it aligns with the lock pawl.
   f. Depress the button on the lever and move the lever to the down position.
   g. Verify the anchor plate is against the stud on the parking brake lever, if not as specified, repeat the procedure.
2. Raise and support the vehicle safely.
3. Mark the relationship between the wheel and axle flange, then remove the tire and wheel assembly.
4. Install 2 wheel nuts to retain the brake rotor.
5. If the caliper is being completely removed from the vehicle for service or replacement, disconnect the brake line fitting at the caliper by removing the bolt, 2 gaskets and the brake hose inlet fitting. Plug all openings to prevent fluid contamination or loss.

NOTE: Do not remove the lever return spring unless the parking brake cable automatic adjuster has been properly disabled.

6. Remove the lever return spring. Discard the spring if the coils are opened.
7. Disconnect the brake cable from the lever and bracket.
8. Remove the 2 guide pins bolts and discard.
9. Remove the caliper housing from the brake rotor and caliper mounting bracket.

To install:
10. Inspect the guide pins for free movement and replace the pins or boots if damaged or corroded.
11. Install the caliper over the brake rotor and into the mounting bracket.
12. Install 2 new guide pin bolts. Tighten the upper bolt to 26 ft. lbs. (35 Nm) and the lower bolt to 16 ft. lbs. (22 Nm).
13. Install the cable to the bracket and parking brake lever, then install the lever return spring.
14. If removed, connect the brake line fitting, 2 new gaskets and the inlet fitting bolt. Tighten the bolt to 30 ft. lbs. (40 Nm).
15. If the inlet fitting was removed, properly bleed the hydraulic brake system.
16. Enable the parking brake automatic adjuster in the reverse order of the disable procedure and make sure the levers are against the stops on the caliper housing.
17. Remove the 2 nuts securing the rotor to the hub, align the marks made earlier and install the tire and wheel assembly.
18. Lower the vehicle and check the brake fluid level.
19. Connect the negative battery cable, start the engine and pump the brake pedal slowly and firmly 3 times

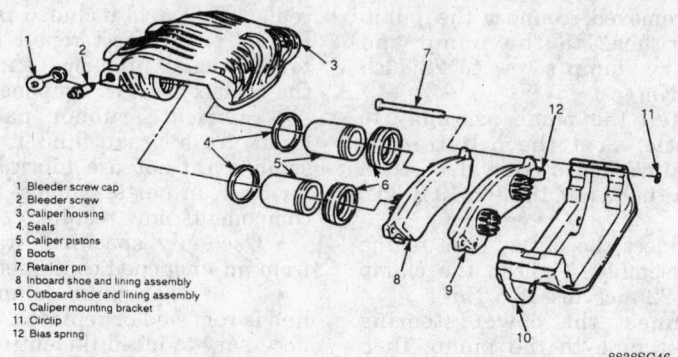

1. Bleeder screw cap
2. Bleeder screw
3. Caliper housing
4. Seals
5. Caliper pistons
6. Boots
7. Retainer pin
8. Inboard shoe and lining assembly
9. Outboard shoe and lining assembly
10. Caliper mounting bracket
11. Circlip
12. Bias spring

8838SG46

Exploded view of the front caliper assembly

to seat the shoe and lining assemblies.

Disc Brake Pads

REMOVAL AND INSTALLATION

Front

1. Disconnect the negative battery cable.
2. Remove the caliper from the mounting bracket but do not disconnect the brake hose and inlet fitting assembly.
3. Suspend the caliper from the upper control arm with wire to avoid damage to the brake hose.
4. Remove the pad and lining assemblies from the caliper.
   To install:
5. Clean all residue from the pad and lining assembly guiding surfaces on the caliper housing and the mounting bracket.
6. Install the outboard pad with the insulator to the caliper housing and the inboard pad with the wear sensor into the caliper pistons. Press the pads firmly until they are they are fully seated.
7. Remove the support and install the caliper to the mounting bracket.

8. Connect the negative battery cable.

Rear

1. Disconnect the negative battery cable and remove ⅔ of the brake fluid from the master cylinder reservoirs.
2. Raise and support the vehicle safely.
3. Mark the relationship between the wheel to the axle flange.
4. Remove the tire and wheel assembly. Install 2 wheel nuts to retain the brake rotor.
5. Use a C-clamp to depress the caliper pistons into the caliper bores to provide clearance between the pads and the rotor. Make sure 1 end of the clamp rests on the inlet fitting bolt while the other end rests on the outboard pad.
6. Remove the caliper upper guide pin bolt and discard, then rotate the caliper on the lower guide pin to access the pad linings. Be careful not to strain the cable conduit or the hoses.
7. Remove the pads from the caliper.
   To install:
8. Install the outboard pad with the insulator to the caliper housing and the inboard pad with the wear sensor nearest the caliper pistons.

The wear sensor must be in the trailing position during forward wheel rotation. Press the pads firmly until they are they are fully seated.

9. Rotate the caliper housing into position, then install a new upper guide pin bolt and tighten 26 ft. lbs. (35 Nm).
10. Remove the wheel nuts securing the rotor to the hub and install the tire and wheel assembly.
11. Lower the vehicle and fill the master cylinder to the proper level with clean brake fluid.
12. Connect the negative battery cable, start the engine and pump the brake pedal slowly and firmly 3 times to seat the shoe and lining assemblies.

Brake Rotor

REMOVAL AND INSTALLATION

Front and Rear

1. Disconnect battery negative cable.
2. Remove the caliper assembly. Either completely remove the assembly from the vehicle or suspend the assembly using a length of mechanic's wire.
3. Remove the rotor from the vehicle.
4. Installation is the reverse of removal.

Parking Brake Cable

ADJUSTMENT

The parking brake lever/cable adjustment is automatic. The adjuster must be disabled to create the necessary cable slack for certain service procedures. Following these procedures, make sure the adjuster is properly enabled to assure correct parking brake operation.

1. To disable the automatic parking brake adjuster, proceed as follows:
   a. Remove the drivers seat cushion.
   b. Remove the parking brake lever cover and screws.
   c. Using a suitable offset tool, disengage and hold the drive pawl from the drive sector.
   d. Insert a nail or drift through the hole in the anchor plate to retain the drive pawl in the disengaged position.
   e. Move the parking brake lever until it aligns with the lock pawl.

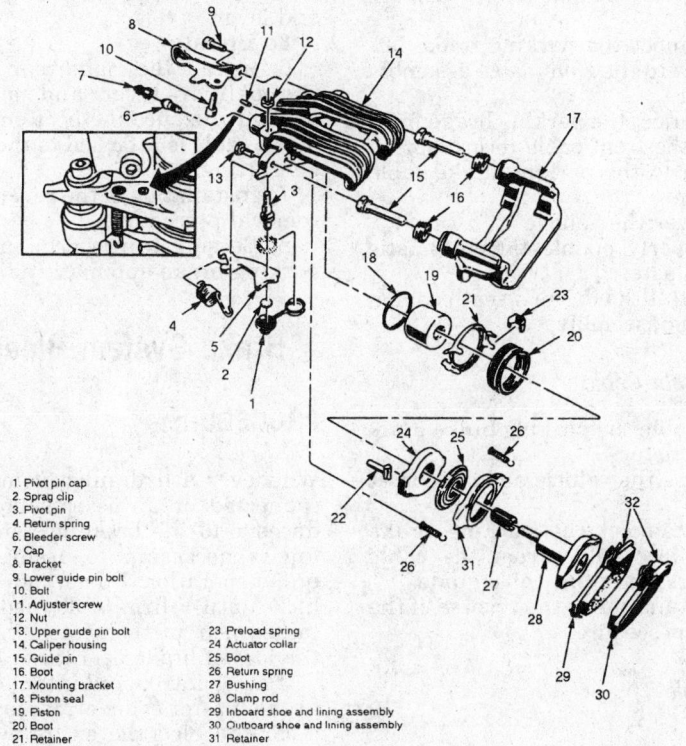

1. Pivot pin cap
2. Sprag clip
3. Pivot pin
4. Return spring
6. Bleeder screw
7. Cap
8. Bracket
9. Lower guide pin bolt
10. Bolt
11. Adjuster screw
12. Nut
13. Upper guide pin bolt
14. Caliper housing
15. Guide pin
16. Boot
17. Mounting bracket
18. Piston seal
19. Piston
20. Boot
21. Retainer
22. Pushrod
23. Preload spring
24. Actuator collar
25. Boot
26. Return spring
27. Bushing
28. Clamp rod
29. Inboard shoe and lining assembly
30. Outboard shoe and lining assembly
31. Retainer
32. Spring

8838SG47

Exploded view of the rear caliper assembly

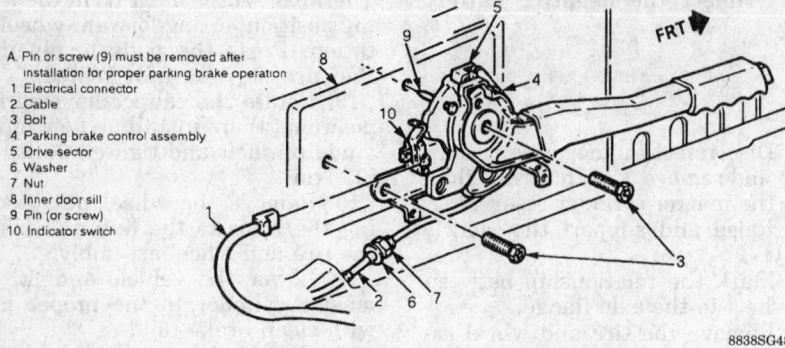

A. Pin or screw (9) must be removed after installation for proper parking brake operation
1. Electrical connector
2. Cable
3. Bolt
4. Parking brake control assembly
5. Drive sector
6. Washer
7. Nut
8. Inner door sill
9. Pin (or screw)
10. Indicator switch

8838SG48

Disabling the parking brake cable automatic adjuster

f. Depress the button on the lever and move the lever to the down position.

g. Verify the anchor plate is against the stud on the parking brake lever, if not as specified, repeat the procedure.

2. To enable the automatic parking brake adjuster, proceed as follows:

a. Remove the nail or drift pin from the anchor plate.

b. Apply and release the parking brake 3 times.

c. Pull up on the parking brake lever. Proper adjustment will result in the lever moving 3–5 ratchet clicks with a force of 61 lbs. (270 N).

d. Release the parking brake, there should be no rear brake drag and no gap between the caliper housings and caliper parking brake levers. It may be necessary to remove the tire and wheel assemblies to see the caliper housings and levers.

e. If removed, install the wheel and tire assembly.

f. Install the parking brake lever cover and screws.

g. Install the driver's seat cushion.

REMOVAL AND INSTALLATION

Front Cable

1. Remove the driver's seat cushion and frame assembly.
2. Properly disable the parking brake automatic adjuster.
3. Raise the vehicle and support it safely.
4. Disconnect the front cable from the front cable connector.
5. Disconnect the front cable from the front cable return spring.
6. Disconnect the left rear cable from the parking brake front cable assembly connector.

7. Remove the front cable attaching clip bolt and clip.
8. Lower the vehicle and remove the front cable from the automatic adjuster.
9. Remove the front cable attaching nut and washer.
10. Remove the front cable from the vehicle.

To install:
11. Install the front cable to the vehicle, then attach the front cable washer and nut. Tighten the nut 24 ft. lbs. (33 Nm).
12. Connect the front cable to the automatic adjuster.
13. Raise and support the vehicle safely.
14. Install the front cable clip and bolt. Tighten the bolt to 8 ft. lbs. (11 Nm).
15. Connect the parking brake left rear cable to the front cable assembly connector.
16. Connect the parking brake front cable to the front cable return spring and then to the parking brake cable connector.
17. Lower the vehicle.
18. Properly enable the automatic brake adjuster.
19. Install the driver's seat cushion and frame assembly.

Intermediate Cable

1. Disable the parking brake automatic adjuster.
2. Raise the vehicle and support it safely.
3. Disconnect the parking brake intermediate cable from the cable connectors and front cable guide.
4. Installation is the reverse of the removal procedure.

Rear Cable

Left

1. Disable the automatic parking brake adjuster.

2. Raise and support the vehicle safely, then remove the tire and wheel assembly.
3. Remove the front cable from the front cable return spring.
4. Disconnect the left rear cable from the front cable assembly connector.
5. Disconnect the cable from the left rear cable bracket.
6. Disconnect the left rear cable from the caliper mounting bracket and lever.

To install:
7. Install the left rear cable to the caliper lever and mounting bracket. Be sure the boot on the end of the cable is attached to the conduit end fitting.
8. Installation is the reverse of the removal procedure.
9. Be sure to properly enable the parking brake automatic adjuster.

Right

1. Disable the parking brake automatic adjuster.
2. Raise the vehicle and support it safely, then remove the tire and wheel assembly.
3. Disconnect the right rear cable from the intermediate cable.
4. Disconnect the right rear cable from the right rear cable bracket.
5. Disconnect the right rear cable from the caliper mounting bracket and lever.

To install:
6. Install the right rear cable to the caliper lever and mounting bracket. Be sure the boot on the end of the cable is attached to the conduit end fitting.
7. Installation is the reverse of the removal procedure.
8. Be sure to properly enable the parking brake automatic adjuster.

Brake System Bleeding

PROCEDURE

Whenever a hydraulic fitting is disconnected or air is somehow introduced into the brake system, bleeding is necessary to assure proper brake operation. Do not move the vehicle until a firm brake pedal is obtained. Air in the system can cause the loss of brake operation.

If air is introduced into the system at the master cylinder, it may be necessary to bled the entire system. If the disconnection of a fitting or pipe is the cause for air presence in the system, then only the caliper(s) served by that component need to be bled.

1. Fill the master cylinder reservoir with brake fluid and keep it at least ½ full of fluid at all times during the bleeding operation.

2. Deplete the brake vacuum reserve by applying and releasing the brakes several times while the engine is **OFF**.

3. If the entire system must be bled, the master cylinder prime pipe must first be bled at the hydraulic modulator located in the left rear storage compartment.

   a. Open the left rear storage compartment, then remove the sound insulator pad.

   b. Remove the cap from the modulator bleed screw, then position a box wrench and a short piece of clear tube over the screw.

   c. Position a container and rags to protect the vehicle interior from the brake fluid. Open the bleed screw and allow fluid to flow until all air is removed.

   d. Tighten the bleed screw to 106 inch lbs. (12 Nm) and remove the tubing. Wipe the screw off and make sure it has properly sealed.

   e. Install the sound insulator pad, making sure it covers the entire modulator valve or excessive noise may be heard when the system is operating. Close the rear compartment.

4. If the master cylinder is known or suspected to have air in the bore, bleed the unit before bleeding the calipers, in the following manner:

   a. Disconnect the forward (blind end) brake line connection at the master cylinder.

   b. Allow brake fluid to fill the master cylinder piston bore until it begins to flow from the forward pipe connector port at the master cylinder.

   c. Connect the forward brake line to the master cylinder and tighten.

   d. Have an assistant depress the brake pedal slowly 1 time and hold. Loosen the forward brake line connection at the master cylinder to purge air from the bore. Tighten the connection and have the assistant release the pedal slowly. Wait 15 seconds and repeat the sequence, including the 15 second pause, until all air is removed from the bore. Make sure brake fluid does not contact any painted surface.

   e. Repeat the procedure at the rear master cylinder brake line connection.

   f. If it is known that the calipers do not contain any air, it will not be necessary to bleed them.

5. If it is necessary to bleed all of the calipers, follow the proper sequence: right rear, left rear, right front, left front.

6. After all air is removed from the master cylinder, bleed the individual calipers as follows:

   a. Place a suitable sized box wrench over the bleeder valve.

   b. Attach a clear tube over the bleeder valve and allow the tube to hang, submerged in a clear container partially filled with brake fluid.

   c. Have an assistant depress the brake pedal slowly 1 time and hold. Loosen the bleeder valve to purge the air from the cylinder. Tighten the bleeder screw and have the assistant slowly release the pedal. Wait 15 seconds and repeat the sequence, including the 15 second pause, until all air is removed.

   d. It may be necessary to repeat the sequence 10 or more times to remove all of the air.

NOTE: Rapid pumping of the brake pedal pushes the master cylinder secondary piston down the bore in a way that makes it difficult to bleed the system.

7. Check the brake pedal for sponginess and the brake warning light for an indication of unbalanced pressure. Repeat the bleeding procedure to correct either of these conditions.

Wheel Speed Sensor

REMOVAL AND INSTALLATION

The front wheel speed sensors are part of the wheel hub assembly and cannot be replaced separately.

Rear Wheel

1. Disconnect the negative battery cable. Raise and safely support the vehicle. Remove the wheel and tire assembly.

2. Disconnect the sensor wiring harness from the ABS wiring harness connector. Unclip the connectors from the bracket and separate.

3. Remove the bracket and bolt from the knuckle.

4. Remove the sensor wiring harness with the grommets from the bracket. Note the position of the grommets and the harness routing for insulation purposes.

5. Remove the wheel speed sensor attaching bolt, then remove the speed sensor from the knuckle.

   To install:

6. Clean all sealant from the sensor and the sensor mounting in the knuckle.

7. Apply 12345489 or equivalent anti-corrosion sealer, to the speed sensor and install the sensor into the knuckle. The sensor is tight fit and must be installed by hand. Do not hammer the sensor into position.

8. Install the sensor retaining bolt and tighten to 86 inch lbs. (10 Nm).

9. Install the sensor wiring harness with grommets into the brackets, make sure the grommets and routing is the same as what was noted during removal.

10. Install the bracket and bolt to knuckle, tighten the bolt to 86 inch lbs. (10 Nm).

11. Connect the sensor wiring harness connector to the wiring harness connector. Make sure the connection is tight, then snap the connectors into the bracket.

12. Install the wheel and tire assembly, then lower the vehicle. Connect the negative battery cable and check for proper system operation.

FRONT SUSPENSION

Shock Absorber

REMOVAL AND INSTALLATION

Without Selective Ride Control

1. Raise and support the vehicle safely.

2. Remove the tire and wheel assemblies.

3. Disconnect the shock absorber from the lower control arm and the shock tower. If necessary, remove the front wheelhouse lower center panel to access the upper mount nut.

4. Remove the insulator and retainers from the shock absorber and the shock absorber from the vehicle.

5. Installation is the reverse of the removal procedure. Tighten the upper and lower mount nuts to 19 ft. lbs. (26 Nm).

With Selective Ride Control

1. Disconnect the negative battery cable.

2. Raise and safely support vehicle, then remove the tire and wheel assemblies.

3. Safely support the lower control arm with a jackstand.

4. Remove the actuator retaining clip, then remove the actuator from the cup retainer. Note the position of the actuator electrical leads for installation purposes.

5. Remove the shock absorber upper mounting nuts.

6. Remove the cup retainer, then the upper insulator retainer and insulator.

7. Remove the shock absorber lower mounting bolts, nuts, then compress the shock absorber and remove it from the vehicle. If necessary, remove the lower insulator from the shock.

To install:

8. If removed, install the lower insulator to the shock absorber, compress the shock and install into the vehicle.

9. Install the shock absorber lower mounting nuts and bolts, then tighten the bolts to 19 ft. lbs. (26 Nm).

10. Install the upper insulator and retainer, then install the cup assembly retainer.

11. Install the upper mounting nut and tighten the 31 ft. lbs. (42 Nm). The selector gear should be at least 0.178 inch (4.5mm) above the top of the cup assembly retainer.

12. Install and properly seat the actuator retaining clip onto the cup assembly retainer. Make sure the ends of the actuator clip protrude outward from the retainer.

13. Install the actuator onto the cup assembly retainer with the electrical leads in the same position as noted earlier. Verify that there is at least 0.315 inch (8mm) of clearance between the front wheelhouse lower center panel and the actuator electrical leads.

NOTE: Very little effort is required to snap the actuator onto the retainer, do not force it into position.

14. Remove the jackstand, then install the tire and wheel assembly.

15. Lower the vehicle and connect the negative battery cable.

Transverse Spring

REMOVAL AND INSTALLATION

1. Raise and support the vehicle safely. Position the supports so the front suspension hangs freely.

2. Remove both front tire and wheel assemblies.

NOTE: Do not use corrosive cleaning agents, engine degreasers or solvents near the fiberglass front spring, or extensive damage could occur to the spring assembly.

3. Disconnect both shock absorbers from the lower control arms, then disconnect the stabilizer shaft links from both lower control arms.

4. Disengage the wheel speed sensor electrical connectors, then remove the speed sensor wire from the bracket.

5. Remove the spring protectors.

6. Compress the front leaf springs using tool J-33432 and adapters J-33432-88 or a suitable equivalent tool, then compress the spring.

7. Disconnect the lower control arms from the steering knuckles by separating the ball joints from the knuckle bores.

8. Remove the spring retainer nuts and retainers, then carefully release the spring compression and remove the tools.

9. With the aid of an assistant, pull both lower control arms downward to release the spring ends from the lower control arms.

10. Remove the spring and retainer shims from the vehicle. Use care not to scratch or damage the spring and note the number, types (color) and positions of the shims.

To install:

11. Lubricate the spring pads with an appropriate lubricant.

12. Carefully install the retainer shims and the spring. Be careful not to scratch the spring and be sure to use the correct number and type of shims.

13. With the aid of an assistant, pull both lower control arms downward while seating the spring ends into the lower control arms.

14. Using the J-33432 and J-33432-88 or equivalents, compress the spring.

15. Install the retainers and hand-tighten the retainer nuts. Install both lower control arm ball joints into the steering knuckles. The ball joints must be positioned so the cotter pins can be inserted from the rear to the front of the vehicle.

16. Install both lower control arm ball stud washers and nuts. Tighten the hex nut to 50 ft. lbs. (68 Nm), then insert a new cotter pin. If necessary tighten the nut additionally in order to insert the pin, but do not exceed a total torque of 88 ft. lbs. (120 Nm) and do not back off the original torque.

17. Install the cotter pins from the rear to the front of the vehicle.

18. Release and remove the spring compression tools.

19. Install both spring protectors and tighten the bolts to 18 ft. lbs. (25 Nm).

20. Install the wheel speed sensor connector, cable and/or bracket, as applicable.

21. Install the stabilizer shaft links, bolts and nuts to the lower control arm. Make sure the link bolts are properly positioning, then hand-tighten the nuts.

22. Connect both shock absorbers to the lower control arms and tighten the lower mounting nuts to 19 ft. lbs. (26 Nm).

23. Use jackstands to hold the suspension at proper trim height, then tighten the spring retainer nuts to 46 ft. lbs. (63 Nm) and the stabilizer shaft link nuts to 33 ft. lbs. (45 Nm).

24. Remove the jackstand supports, then install the tire and wheel assemblies.

25. Lower the vehicle, check and adjust the front end alignment, as necessary.

Upper Ball Joints

REMOVAL AND INSTALLATION

1. Raise and support the vehicle safely.

2. Safely support the lower control arm with a jackstand.

3. Remove the tire and wheel assemblies.

4. Using J-33436 or equivalent ball joint removal tool, separate the ball joint from the knuckle.

5. Remove the upper ball joint from the control arm as follows:
   a. Center punch the rivet.
   b. Drill a pilot hole, then drill the rivet head.
   c. Punch out the rivet.

To install:

6. Install a new ball joint into the upper control arm and position so the cotter pin can be installed from the rear to the front of the vehicle.

7. Install and tighten the mounting nuts to 13 ft. lbs. (18 Nm).

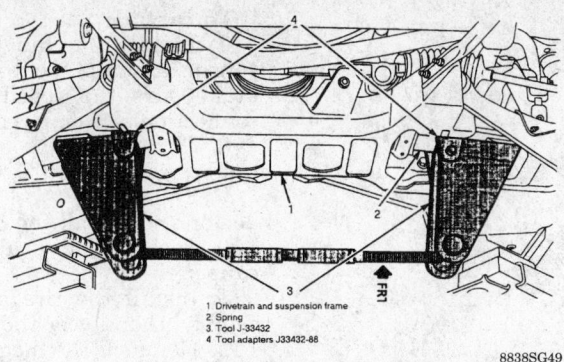

1 Drivetrain and suspension frame
2 Spring
3 Tool J-33432
4 Tool adapters J33432-88

8838SG49

Transverse spring compression tool

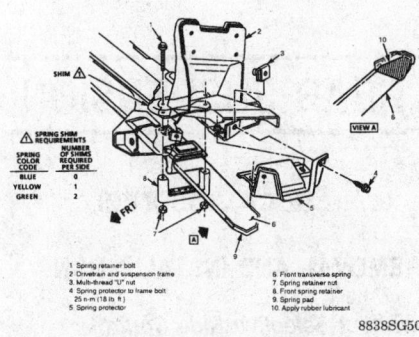

SHIM

SPRING SHIM REQUIREMENTS

SPRING COLOR CODE	NUMBER OF SHIMS REQUIRED PER SIDE
BLUE	0
YELLOW	1
GREEN	2

VIEW A

FRT

1 Spring retainer bolt
2 Drivetrain and suspension frame
3 Multi-thread "U" nut
4 Spring protector to frame bolt 25 n-m (18 lb. ft.)
5 Spring protector
6 Front transverse spring
7 Spring retainer nut
8 Front spring retainer
9 Spring pad
10 Apply rubber lubricant

8838SG50

Front transverse spring and shim installation

8. Position the ball stud into the steering knuckle, then install the upper ball joint stud washer and nut. Tighten the upper control arm ball stud nut to 33 ft. lbs. (45 Nm). Tighten the nut additionally as necessary to insert the cotter pin but do not exceed 63 ft. lbs. (85 Nm).

9. Install a new cotter pin from the rear to the front of the vehicle.

10. Remove the jackstand and lubricate the ball joint.

11. Install the tire and wheel assembly, then lower the vehicle.

Lower Ball Joints

REMOVAL AND INSTALLATION

1. Raise and support the vehicle safely.

2. Safely support the lower control arm with a jackstand.

3. Remove the tire and wheel assembly.

4. Using J-33436 or equivalent ball joint removal tool, separate the ball joint from the knuckle.

5. Press the upper ball joint from the control arm using tool J-9519-E or an equivalent removal tool.

To install:

6. Position the ball stud so the cotter pin may be installed from the rear to the front of the vehicle and press into the control arm using J-9519-E or equivalent.

7. Position the ball joint into the steering knuckle, then install the washer and nut. Tighten the lower control arm ball stud nut to 50 ft. lbs. (68 Nm). Tighten the ball stud nut additionally, as necessary to insert a cotter pin, but do not exceed 88 ft. lbs. (120 Nm) to align the cotter pin holes.

8. Install a new cotter pin from the rear to the front of the vehicle.

9. Remove the jackstand and lubricate the ball joint.

10. Install the tire and wheel assembly, then lower the vehicle.

Upper Control Arms

REMOVAL AND INSTALLATION

1. Disconnect the negative battery cable, then raise and support the vehicle safely.

2. Remove the tire and wheel assembly.

3. Remove the front wheelhouse panel seal and lower center panel.

4. If equipped, remove the shock absorber actuator wire connector.

5. Support the lower control arm with a jackstand.

6. Disengage the speed sensor electrical connector and remove the cable from the bracket.

7. Use tool J-33436 or equivalent and disconnect the upper ball joint from the knuckle.

8. Remove the upper control arm attaching bolts, shims and nuts. Note the number and position of the shims for reinstallation purposes. Remove the control arm.

To install:

9. Position the bolts through the frame, then install the upper control

arm and shims. Place the shims in the locations noted during removal. Install and tighten the control arm nuts to 37 ft. lbs. (50 Nm).

10. Position the ball stud into the steering knuckle, then install the upper ball joint stud washer and nut. Tighten the upper control arm ball stud nut to 33 ft. lbs. (45 Nm). Tighten the nut additionally as necessary to insert the cotter pin but do not exceed 63 ft. lbs. (85 Nm).

11. Install a new cotter pin from the rear to the front of the vehicle.

12. Connect the ABS speed sensor bracket, cable and/or electrical connection, as applicable.

13. If equipped, engage the shock absorber electrical actuator connection.

14. Remove the jackstand and install the front wheelhouse lower center panel and seal.

15. Install the tire and wheel assembly.

16. Lower the vehicle and connect the negative battery cable.

17. Check and adjust the front end alignment, as necessary.

Lower Control Arms

REMOVAL AND INSTALLATION

1. Disconnect the negative battery cable, then raise and support the vehicle safely.

2. Remove the tire and wheel assembly, then remove both spring protectors.

3. Using tool J-33432 and adapters J-33432-88 or equivalent, compress the spring.

4. Support the lower control arm with a jackstand.

5. Disconnect the shock absorber from the lower control arm, then disconnect the front stabilizer shaft link from the lower control arm.

6. Disengage the speed sensor electrical connector and remove the cable from the bracket.

7. Using tool J-33436 or equivalent, disconnect the lower ball joint from the knuckle.

8. Remove the engine support bracket.

9. Remove nuts, washers and bolts attaching the lower control arm to the frame.

10. Remove the jackstand and the lower control arm.

To install:

11. Install the lower control arm, bolts, washers and nuts.

12. Support the lower control arm with a jackstand.

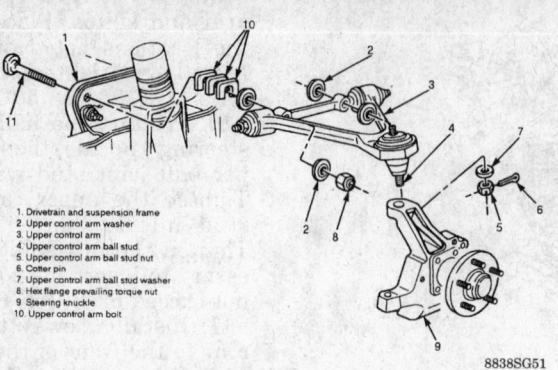

1. Drivetrain and suspension frame
2. Upper control arm washer
3. Upper control arm
4. Upper control arm ball stud
5. Upper control arm ball stud nut
6. Cotter pin
7. Upper control arm ball stud washer
8. Hex flange prevailing torque nut
9. Steering knuckle
10. Upper control arm bolt

8838SG51

Upper control arm assembly

13. Install the engine support bracket.

14. Position the ball joint into the steering knuckle, then install the washer and nut. Tighten the lower control arm ball stud nut to 50 ft. lbs. (68 Nm). Tighten the ball stud nut additionally, as necessary to insert a cotter pin, but do not exceed 88 ft. lbs. (120 Nm) to align the cotter pin holes.

15. Install a new cotter pin from the rear to the front of the vehicle.

16. Connect the ABS speed sensor bracket, cable and/or electrical connection, as applicable.

17. Connect the stabilizer shaft link to the lower control arm but hand-tighten the nuts only.

18. Remove the spring compression tool and adapters.

19. Hold the suspension at the proper trim height using jackstands and tighten the stabilizer link nuts to 35 ft. lbs. (48 Nm) and the lower control arm bolts to 82 ft. lbs. (112 Nm).

20. Connect the shock absorber to the lower control arm and tighten the nuts to 19 ft. lbs. (26 Nm).

21. Remove the jackstands and install both spring protectors. Tighten the bolts to 18 ft. lbs. (25 Nm).

22. Install the tire and wheel assembly.

23. Lower the vehicle and connect the negative battery cable.

Sway Shaft

REMOVAL AND INSTALLATION

1. Raise and support vehicle safely, then remove the tire and wheel assemblies.

2. Support the lower control arms using jackstands.

3. Remove the stabilizer shaft insulator clamp bolts and brackets from the frame.

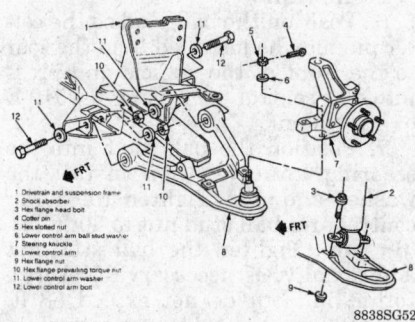

1. Drivetrain and suspension frame
2. Shock absorber
3. Hex flange head bolt
4. Cotter pin
5. Hex slotted nut
6. Lower control arm ball stud washer
7. Steering knuckle
8. Lower control arm
9. Hex flange nut
10. Hex flange prevailing torque nut
11. Lower control arm washer
12. Lower control arm bolt

8838SG52

Lower control arm assembly

4. Remove the stabilizer shaft-to-links attaching bolts. Note the bolt positioning for installation purposes.

5. Remove the stabilizer shaft from the vehicle.

6. Installation is the reverse of the removal procedure.

7. Install the shaft link bolts and nuts facing the same positions as they were removed. With the suspension held at the proper trim height, tighten the stabilizer shaft link nuts to 35 ft. lbs. (48 Nm) and the insulator clamp bolts to 40 ft. lbs. (54 Nm).

Front Wheel Bearings

REMOVAL AND INSTALLATION

1. Disconnect the negative battery cable, then raise and support the vehicle safely.

2. Remove the tire and wheel assembly.

3. Remove the caliper and support it aside, then remove the rotor.

4. Disengage the ABS speed sensor electrical connector.

5. Remove the ABS speed sensor cable bracket.

6. Remove the wheel hub/bearing/speed sensor assembly.

To install:

7. Install the hub/bearing/speed sensor assembly onto the vehicle. Make sure the speed sensor cable connection is facing rearward.

8. Tighten the assembly mounting nuts to 46 ft. lbs. (62 Nm).

9. Engage the ABS electrical connector and install the cable bracket.

10. Install the brake rotor and caliper.

11. Install the tire and wheel assembly, then lower the vehicle.

12. Connect the negative battery cable. The bearings do not require adjustment.

REAR SUSPENSION

Shock Absorber

REMOVAL AND INSTALLATION

Without Selective Ride Control

1. Raise and support the vehicle safely. Support the knuckle with a jackstand.

2. Remove the shock absorber lower mounting nut and washer.

3. Remove the shock absorber upper bracket mounting bolt.

4. Disconnect the shock absorber from the lower mounting stud.

5. If necessary, remove the shock absorber upper bracket retaining nut and remove the bracket assembly.

6. Installation is the reverse of the removal procedure.

7. Tighten the upper bracket retaining nut, if removed, to 19 ft. lbs. (26 Nm). With the suspension at proper trim height, tighten the upper bracket mounting bolts to 22 ft. lbs. (30 Nm) and the lower mounting nut to 61 ft. lbs. (83 Nm).

With Selective Ride Control

1. Disconnect the negative battery cable.

2. Raise and support the vehicle safely.

3. Support the rear knuckle with a jackstand.

4. Disconnect the shock absorber lower mounting nut and washer.

5. Remove the shock absorber upper bracket mounting bolt.

6. Disconnect the shock absorber from the mounting stud and support. Do not allow the shock to hang from the actuator harness.

7. Remove the actuator retaining clip and remove the actuator from the shock.

8. Remove the shock absorber from the vehicle.

To install:

9. Install and properly seat the actuator retaining clip onto the cup assembly. The ends of the clip should protrude from the cup.

10. Install the shock absorber electrical actuator into the cup assembly retainer. The actuator should be snapped, not be forced into position.

11. Verify that a minimum of 0.178 inch (4.5mm) of clearance exists between the selector gear and the top of the cup assembly retainer.

12. Position the shock absorber into the frame and onto the lower mounting stud.

13. Install the shock absorber upper bracket mounting bolt.

14. With the suspension held at the proper trim height. Tighten the upper bracket mounting bolts to 22 ft. lbs. (30 Nm) and the lower mounting nut to 61 ft. lbs. (83 Nm).

15. Remove the jackstands and lower the vehicle.

16. Connect the negative battery cable.

Transverse Spring

REMOVAL AND INSTALLATION

1. Raise and support the vehicle safely.

2. Remove 1 wheel and tire assembly from the vehicle.

NOTE: Do not use corrosive cleaning agents, engine degreasers or solvents near the fiberglass rear spring or extensive damage could occur to the spring assembly.

3. Install tool J-33432 or equivalent spring compressor, onto the rear transverse spring, then compress the spring.

4. Remove the cotter pins, retaining nuts, insulators and spring bolts attaching the spring to the knuckles.

5. Carefully release and remove the spring compression tool.

6. Remove the rear spring anchor plate bolts, then the anchor plate, spacers and insulator from the vehicle. Note the spacer positioning for installation purposes.

7. Remove the transverse spring from the vehicle.

To install:

8. Position the spring in the vehicle. Take care not to scratch the spring during installation.

9. Position the spacers as noted during removal, then install the insulators and anchor plates onto the differential carrier.

10. Install the anchor plate bolts and tighten to 37 ft. lbs. (50 Nm).

11. Install the spring compression tool and compress the spring.

12. Position the spring to the knuckles and install the spring bolts, insulators and nuts. Tighten the nuts until slot in nut aligns with hole in bolt and install a new cotter pin.

13. Carefully release and remove the spring compression tool.

14. Install the tire and wheel assembly.

15. Remove the jackstands and lower the vehicle.

Rear Control Arms

REMOVAL AND INSTALLATION

1. Raise and support the vehicle safely.

2. Remove the control arm nut, bolt and washers at the knuckle.

3. Remove control arm nut and bolt at the bracket.

4. Remove the control arm from the vehicle.

5. Installation is the reverse of the removal procedure.

6. With the suspension held at the proper trim height, tighten the bracket bolt to 63 ft. lbs. (85 Nm) and the knuckle nut to 140 ft. lbs. (190 Nm).

Spindle/Support Rod

REMOVAL AND INSTALLATION

1. Raise and support vehicle safely.

2. Scribe alignment marks on the wheel spindle/support rod adjustment bolt and the spindle/support rod bracket so they can be installed in the same position.

3. Remove the adjustment bolt, cam and nut, then separate the spindle/support rod from the bracket.

4. Remove the spindle/support bolt, washer and nut at the knuckle, then remove the spindle/support rod from the vehicle.

5. Installation is the reverse of the removal procedure. Be sure to align the marks made during removal.

6. With the suspension held at the proper trim height, tighten the spindle/support rod-to-knuckle nut to 107 ft. lbs. (145 Nm), then tighten the spindle/support rod adjustment nut to 186 ft. lbs. (253 Nm).

7. Check and adjust the rear suspension alignment, as necessary.

Wheel Bearings

ADJUSTMENT

REMOVAL AND INSTALLATION

1. Raise and support the vehicle safely, making sure the rear suspension hangs freely.

NOTE: Do not support the vehicle by means of the differential or the transverse leaf springs.

2. Remove the rear transverse leaf spring from the knuckle as follows:

  a. Remove 1 rear wheel assembly.

  b. Install tool J-33432 or equivalent transverse leaf com-

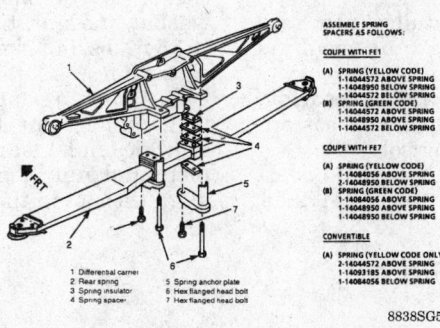

ASSEMBLE SPRING
SPACERS AS FOLLOWS:

COUPE WITH FE1
(A) SPRING (YELLOW CODE)
1-14044572 ABOVE SPRING
1-14044950 BELOW SPRING
1-14044572 BELOW SPRING
(B) SPRING (GREEN CODE)
1-14044572 ABOVE SPRING
1-14044950 ABOVE SPRING
1-14044572 BELOW SPRING

COUPE WITH FE7
(A) SPRING (YELLOW CODE)
1-14044056 ABOVE SPRING
1-14044950 BELOW SPRING
(B) SPRING (GREEN CODE)
1-14044056 ABOVE SPRING
1-14044950 BELOW SPRING

CONVERTIBLE
(A) SPRING (YELLOW CODE ONLY)
2-14044572 ABOVE SPRING
1-14093185 ABOVE SPRING
1-14084056 BELOW SPRING

1 Differential carrier   5 Spring anchor plate
2 Rear spring   6 Hex flanged head bolt
3 Spring insulator   7 Hex flanged head bolt
4 Spring spacer

8838SG53

Rear transverse spring and shim installation. Do not add extra shims to raise the trim height or the spring will be overstressed

pressor, onto the rear transverse spring and compress the spring.

c. Remove the cotter pin, nut, rubber grommets and bolt attaching the spring to the knuckle.

d. Carefully release and remove spring compressor.

3. Remove the cotter pin, nut and washer from the tie rod outer socket at knuckle. Using a suitable linkage puller, disconnect the outer tie rod from the knuckle.

4. Disconnect the spindle rod bracket at the differential carrier.

5. Remove the axle shaft universal joint straps at the both the spindle and yoke shaft ends.

6. Remove the shaft by supporting the shaft and pushing out on the knuckle assembly.

7. If necessary, remove rear axle yoke, oil seal and bearing as follows:

a. If equipped, remove the upper and lower underbody braces.

b. Remove the exhaust assembly.

c. Support the rear differential, then remove the differential carrier outer support bolts.

d. Remove the carrier cover and drain the gear oil into a suitable container.

e. Remove the snapring from the axle shaft yoke and remove the yoke.

f. If only replacing the seal, pry the axle shaft yoke seal out using a suitable tool. Be careful not to damage the yoke shaft bearing assembly.

g. If the seal cannot be removed in this manner or the bearing assembly is to be replaced as well, remove the differential assembly.

h. Using tools J-34171 for the 7.875 inch axle (automatic transmission) or J-35509 for the 8.5 inch axle (manual transmission), and driver handle J-8592 or equivalents, and a hammer, remove the seal and bearing assembly. Discard the seal and bearing.

To install:

8. If removed, install a new rear axle shaft bearing and seal as follows:

a. If installing a new bearing, clean the seal bore using a standard metal cleaning solvent.

b. Install a new rear axle bearing assembly. Use tools J-34172 for

the 7.875 inch axle (automatic transmission) or J-35510 for the 8.5 inch axle (manual transmission) with driver handle J-8592 or equivalents, and a hammer.

c. Lubricate bearings with a suitable hypoid lubricant.

d. Apply a light coat of hypoid lubricant on the lip of the axle shaft seal.

e. Install axle shaft seal using tools J-26938 for 7.875 in. axle or J-35511 for 8.5 in. axle and driver J-8592 or equivalents.

f. If removed, install the differential assembly.

g. Install the axle yoke shaft and snapring into the differential carrier.

h. If a new yoke shaft is installed, yoke shaft end play should be checked and adjusted, if necessary by using snaprings of varying thickness. Endplay should be 0.0005–0.0085 inch (0.013–0.216mm).

i. Apply a continuous ¼ inch bead of sealant to the mating surfaces, then install the differential carrier cover, with gasket, to the carrier. Tighten the bolts to specification using the proper torque sequence.

j. Install the carrier outer support retaining bolts and tighten to 60 ft. lbs. (80 Nm).

k. Remove the differential support and install the exhaust assembly or connect the crossover pipe, as applicable.

l. If applicable, install the underbody upper and lower braces.

9. Install the axle assembly shaft into the differential and spindle yoke.

10. Install the shaft U-joint retainers and tighten the bolts to 26 ft. lbs. (35 Nm).

11. Connect the spindle rod bracket to the differential carrier and tighten the spindle rod bracket bolts to 60 ft. lbs. (80 Nm).

12. Install the tie rod outer axle socket to the knuckle. Install the washer and nut, tighten the end nut to 33 ft. lbs. (45 Nm), then replace the cotter pin.

13. Using a suitable compression tool, connect the leaf spring to the knuckle and install the bolt, grommets and nut. Tighten the nut and align the slot in the nut with the hole

in the bolt, then insert a new cotter pin.

14. Lower the vehicle.

Rear Wheel Hub and Bearings

REMOVAL AND INSTALLATION

1. Disconnect the negative battery cable, then raise and support the vehicle safely.

2. Remove the tire and wheel assembly.

3. Remove the wheel speed sensor.

4. Remove the brake caliper and parking brake assembly, then remove the rotor.

5. Remove the wheel hub mounting bolts.

6. Remove the cotter pin, wheel nut retainer, spindle nut and washer.

7. Remove the wheel hub and bearing, caliper mounting plate and wheel spindle washer from the vehicle.

To install:

8. Inspect the wheel hub and bearing seal, replace if necessary. Also inspect the wheel spindle washer and replace, if damaged or excessively worn.

9. Install the wheel hub and bearing, caliper mounting plate and the wheel spindle washer. The washer flat should firmly seat against the shoulder of the wheel spindle. The lip of the washer should face the wheel spindle splines prior to hub and bearing installation.

10. Install the wheel hub mounting bolts and tighten to 66 ft. lbs. (90 Nm).

11. Install the washer and spindle nut, then tighten the nut to 164 ft. lbs. (223 Nm). The vehicle should not rest on the tires or move until the spindle nut is tightened.

12. Install the wheel retainer and a new cotter pin.

13. Install the brake rotor, then install the caliper and parking brake assembly.

14. Install the wheel speed sensor, then install the wheel and tire assembly.

15. Lower the vehicle and connect the negative battery cable.

FIRING ORDERS

NOTE: To avoid confusion, always replace spark plug wires one at a time.

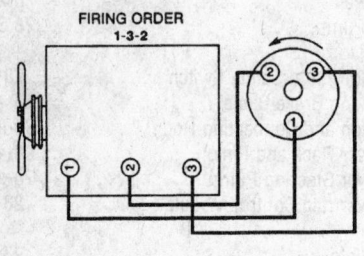

FIRING ORDER
1-3-2

322394

1.0L (VIN 6) Engine
Engine Firing Order: 1–3–2
Distributor Rotation: Counterclockwise

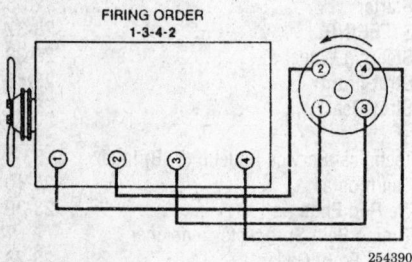

FIRING ORDER
1-3-4-2

254390

1.3L (VIN 9) Engine
Engine Firing Order: 1–3–4–2
Distributor Rotation: Counterclockwise

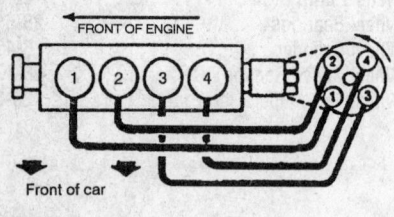

FRONT OF ENGINE

Front of car

308890

1.6L (VIN 6) and 1.8L (VIN 8) Engines
Engine Firing Order: 1–3–4–2
Distributor Rotation: Counterclockwise

ENGINE ELECTRICAL

NOTE: Disconnecting the negative battery cable on some vehicles may interfere with the functions of the on board computer systems and may require the computer to undergo a relearning process, once the negative battery cable is reconnected.

Distributor

REMOVAL AND INSTALLATION

1. Disconnect the negative battery cable.
2. Tag and disconnect the spark plug wires, wiring harness and vacuum line at the distributor, if equipped.
3. Remove the distributor cap.
4. Mark the position of the distributor rotor in relation to the distributor body. Mark the position of the distributor body in relation to the cylinder head.
5. Mark the distributor position on the housing and engine. Remove the hold-down bolt(s) and the distributor from the cylinder head. Do not rotate the engine after the distributor has been removed.
6. Remove the distributor from the engine and remove the O-ring from the distributor shaft.
7. Inspect all components of the assembly, including the cap and rotor, for cracks, terminal corrosion or wear and replace, if necessary.
To install:
8. If the engine was not rotated, proceed as follows:
   a. Align the reference marks made during removal. Position carefully and make sure the drive gear engage properly within the slot. Install the hold-down bolts but do not tighten.
   b. Reconnect the wiring and the plug wires. Connect vacuum hoses, if equipped.
   c. Install the distributor cap. Connect the battery negative cable. Check and/or adjust the ignition timing.
   d. Tighten the distributor hold-down bolt(s).
9. If the engine was rotated while the distributor was removed, the engine must be rotated until cylinder No. 1 is at TDC of the compression strike.
   a. If necessary, remove the FI Fuse from the fuse and relay box.
   b. Remove the No. 1 spark plug.
   c. Place a thumb over the spark plug hole. Have someone rotate the engine by hand, using a wrench on the crankshaft pulley until compression is felt.
   d. Align the timing mark on the crankshaft pulley with the 0 degrees mark on the timing scale attached to the front of the engine. This places the engine at TDC on the compression stroke.
   e. Turn the distributor shaft until the rotor points to the No. 1 spark plug tower on the cap.
   f. Install the distributor to the cylinder head, aligning the distributor housing-to-cylinder head marks made during the removal procedure.
   g. Connect the spark plug wires, electrical wires and vacuum advance hose, if equipped.
   h. Install the distributor cap.
   i. Install the FI fuse into the fuse and relay box, if removed.
   j. Reconnect the negative battery cable.
   k. Check and/or adjust ignition timing.
   l. Tighten the distributor hold-down bolt.

Ignition Timing

CHECKING AND ADJUSTMENT

NOTE: All 1.0L, 1.3L and 1.6L and 1.8L engines until 1995 have adjustable timing. All 1996–97 1.6L and 1.8L engines have non-adjustable timing set at 10 degrees BTDC.

1. Run the engine until it reaches normal operating temperature. Stop the engine, but keep the ignition switch in the ON position for approximately 5 seconds. Start the engine again.
2. Run the engine at 2000 RPM for approximately 5 minutes. After 5 minutes, allow it to run at idle speed.
3. Make sure all accessories are turned OFF.
4. Fully engage the parking brake.
5. Make sure the shift lever is placed in the NEUTRAL or PARK depending on transaxle type.
6. Connect a tachometer to the negative (-) terminal of the ignition coil for 1.0L and 1.3L engines, or to the IG terminal in the Data Link Connector (DLC), located next to the

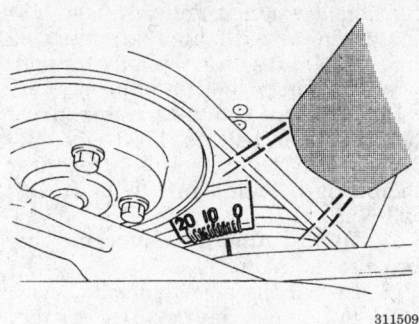

Checking ignition timing

311509

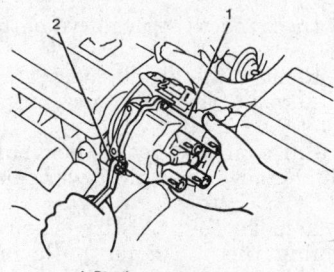

1. Distributor
2. Distributor mounting bolts

237781

Distributor mounting bolts — Prizm with 1.6L (VIN 6) or 1.8L (VIN 8) engines

left strut tower for 1.6L and 1.8L engines. Connect a timing light to the No. 1 spark plug wire. Check the engine idle speed and adjust if needed.

7. Refer to the underhood Vehicle Emission Control Information label for ignition timing specifications.

8. Remove the cap from the diagnostic check connector, located next to the ignition coil or left side strut tower, and insert a fused jumper wire between appropriate terminals.

9. For 1.0L and 1.3L engines, proceed as follows;
   a. With a 4 terminal connector, hold the connector with the locking tab at the top and jump the lower 2 terminals (**C** and **D**).
   b. With a 6 terminal connector, hold the connector with the locking tab at the top and jump the 2 terminals at the lower left (**D** and **E**).

10. For 1.6L and 1.8L engines, proceed as follows;
   a. Using a fused jumper wire, connect terminals **E1** and **TE1** together at the DLC.

11. Aim the timing light at the timing marks.

12. Loosen the distributor hold-down bolt and rotate the distributor

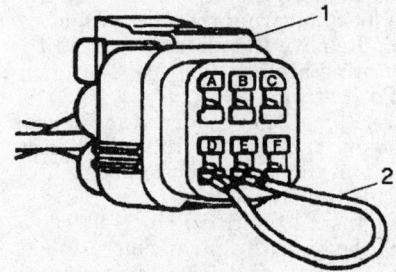

A. Blank (cavity 1)
B. Diagnostic request terminal (cavity 2)
C. Diagnostic output terminal (cavity 3)
D. Ground terminal(cavity 4)
E. Test switch terminal(cavity 5)
F. Duty check terminal(cavity 6)
1. Duty check DLC
2. Jumper

322085

Jumping terminals in Data Link Connector (DLC) — 1.0L (VIN 6) and 1.3L (VIN 9) engines

until the correct timing marks are aligned.

13. Tighten the distributor hold-down bolt to 11–15 ft. lbs. (15–20 Nm) and recheck the timing. Make sure the timing advances according to engine speed.

14. Remove the diagnostic check jumper. Remove the timing light and tachometer.

NOTE: The ignition timing is on 1996–97 1.6L and 1.8L engines NOT adjustable. If the ignition timing is out of specification a possible engine mechanical fail-

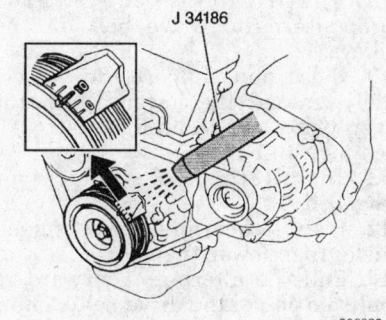

J 34186

306829

Checking ignition timing — 1993–95 Prizm with 1.6L DOHC (VIN 6) or 1.8L (VIN 8) engines

ure may have occurred. The following procedure may be used to check the timing.

15. Warm the engine to normal operating temperature. Turn all electrical accessories **OFF**.

16. Connect a tachometer or engine scan tool and check the engine idle speed and make sure it is 650–750 rpm.

17. Connect a timing light. Remove the cap on the Data Link Connector (DLC). Using a fused jumper wire, connect terminals **E1** and **TE1**.

18. Start the engine and check timing. With the jumper wire connected, the timing should be at 10 degrees BTDC.

19. Remove the jumper wire from the DLC and reinstall the cap.

20. Shut the engine **OFF** and disconnect all test equipment.

Alternator

PRECAUTIONS

Several precautions must be observed with alternator equipped vehicles to avoid damage to the unit.

• If the battery is removed for any reason, make sure it is reconnected with the correct polarity. Reversing the battery connections may result in damage to the 1-way rectifiers.

• When utilizing a booster battery as a starting aid, always connect the positive to positive terminals and the negative terminal from the booster battery to a good engine ground on the vehicle being started.

• Never use a fast charger as a booster to start vehicles.

• Disconnect the battery cables when charging the battery with a fast charger.

• Never attempt to polarize the alternator.

• Do not use test lights of more than 12 volts when checking diode continuity.

• Do not short across or ground any of the alternator terminals.

• The polarity of the battery, alternator and regulator must be matched and considered before making any electrical connections within the system.

• Never separate the alternator on an open circuit. Make sure all connections within the circuit are clean and tight.

• Disconnect the battery ground terminal when performing any service on electrical components.

• Disconnect the battery if arc welding is to be done on the vehicle.

REMOVAL AND INSTALLATION

CAUTION

Failure to disconnect the negative cable may result in injury from the positive battery lead at the alternator and may short the alternator and regulator during the removal process.

Metro

1. Disconnect the negative battery cable.

2. Remove the rubber **B** terminal cover, and unfasten the retaining nut and electrical connector.

3. Remove the alternator drive belt. Detach the adjuster bolt and loosen the pivot bolt.

4. Support the alternator while removing the pivot bolt and nut, then remove the alternator from the vehicle.

To install:

5. Install the alternator and loosely tighten the pivot bolt and nut. Loosely install the drive belt adjuster bolt.

6. Install the alternator drive belt.

7. Tighten the drive belt adjuster bolt to 17 ft. lbs. (23 Nm), then tighten the mounting bolt and nut to 17 ft. lbs. (23 Nm).

8. Fasten the electrical connector to the **B** terminal and tighten the nut to 71 inch lbs. (8 Nm). Cover the terminal with the rubber protector.

9. Reconnect the negative battery cable.

10. Run the engine and check the charging system operation.

Prizm

1. Disconnect the negative battery terminal from the battery.

2. Loosen the pivot and lock bolts.

3. Loosen the adjusting bolt. Remove the drive belt.

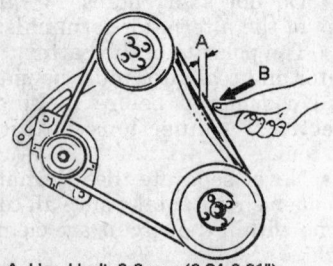

A. Used belt: 6–8 mm (0.24–0.31")
New belt: 5–7 mm (0.20–0.27")
B. 22 lbs.(10 kg)

325940

Alternator drive belt adjustment

4. Unplug the wiring connector from the alternator. Remove the nut and alternator wire.

5. Remove the pivot bolt, nut, and the lock bolt from the alternator.

6. Remove the alternator from the mounting bracket.

To install:

7. Place the alternator in the bracket and install the pivot bolt, nut, and the lock bolt. Do not tighten at this time.

8. Install the drive belt.

9. Tighten the pivot bolt to 45 ft. lbs. (61 Nm) and the adjusting lock bolt to 14 ft. lbs. (19 Nm).

10. Install the electrical connectors.

11. Connect the negative battery cable.

12. Start the engine and check the charging system operation.

Drive Belt

REMOVAL AND INSTALLATION

1.0L (VIN 6) and 1.3L (VIN 9) Engines

Alternator/Water Pump Belt

1. Disconnect the negative battery cable.

2. Remove the air cleaner.

3. Remove the A/C suction pipe bracket, if equipped.

4. Raise and support the vehicle safely. Remove the right lower splash shield.

5. If equipped, remove the A/C compressor drive belt.

6. Lower the alternator cover plate, then lower the vehicle.

7. Loosen the adjusting bolt on the upper alternator mounting bracket and remove the water pump/alternator drive belt.

8. Inspect the belt for cracking, fraying, glazing and other wear or damage; replace as necessary.

To install:

9. Install the the water pump/alternator drive belt on the pulleys.

10. Raise and support the vehicle safely. Install the lower alternator cover plate and tighten bolts to 89 inch lbs. (10 Nm).

11. Install the A/C compressor drive belt.

12. Install the right lower splash shield, then lower the vehicle.

13. Pull the alternator outward to apply tension to the drive belt. A new belt is properly tensioned when it deflects 0.20–0.27 in. (5–7mm) with approximately 22 lbs. (98 N) pressure applied at a midway point between the pulleys. A used belt is properly tensioned when it deflects 0.24–0.31 in. (6–8mm) with approximately 22 lbs. (98 N) pressure applied at a midway point between the pulleys.

14. Tighten the alternator drive belt adjuster bolt to 17 ft. lbs. (23 Nm), then tighten the alternator mounting bolt and nut to 17 ft. lbs. (23 Nm).

15. Install the A/C suction pipe bracket, as required.

16. Install the air cleaner.

17. Reconnect the negative battery cable.

Air Conditioner Compressor Belt

1. Disconnect the negative battery cable.

2. Remove the air cleaner.

3. Remove the A/C suction pipe bracket.

4. Raise and support the vehicle safely. Remove the right lower splash shield.

5. Remove the A/C drive belt by loosening the tensioner pulley nut and bolt and removing the belt from the pulleys.

6. Inspect the belt for cracking, fraying, glazing and other wear or damage; replace as necessary.

To install:

7. Install the drive belt on the pulleys, making sure it is properly aligned. Adjust the tension using the tension pulley.

8. On 1993–94 vehicles, belt tension is correct when the belt deflects 0.20–0.25 in. (5–6.5mm) with approximately 22 lbs. pressure applied to a midway point between the pulleys. On 1995–97 vehicles, belt tension is correct when the belt deflects 0.30–0.40 in. (8–10mm) with approximately 22 lbs. pressure applied to a point midway between the pulleys.

9. Tighten the tensioner pulley locknut to 20 ft. lbs. (27 Nm).

10. Install the right lower splash shield and lower the vehicle.

11. Install the suction pipe bracket.

12. Install the air cleaner.

13. Reconnect the negative battery cable.

1.6L (VIN 6) and 1.8L (VIN 8) Engines

Alternator Drive Belt

1. Disconnect the negative battery cable.

2. Loosen the pivot and adjusting lock bolts.

3. Loosen the adjusting bolt and remove the drive belt.

4. Inspect the belt for cracks, glazing and other wear and/or damage. Replace the belt, as necessary.

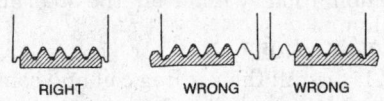

RIGHT   WRONG   WRONG

313040

Correct A/C compressor drive belt alignment — 1.0L (VIN 6) and 1.3L (VIN 9) engines

To install:

5. Install the drive belt, making sure the belt is properly aligned on the pulleys.

6. If the old belt is being reinstalled, tighten the alternator adjusting bolt until the belt deflects 0.24–0.35 in. with approximately 22 lbs. (98 N) pressure applied at a point midway between the pulleys. If a new belt has been installed, tighten the alternator adjusting bolt until the belt deflects 0.20–0.31 in. with approximately 22 lbs. (98 N) pressure applied at a point midway between the pulleys.

7. Torque the pivot bolt to 45 ft. lbs. (61 Nm) and the lock bolt to 14 ft. lbs. (19 Nm).

8. Connect the negative battery cable.

A/C Compressor Belt

1. Disconnect the negative battery cable.

2. Remove the alternator belt.

3. Remove the windshield washer fluid reservoir tank, if necessary.

4. Loosen the idler pulley locknut.

5. Loosen the adjusting bolt for the idler pulley and remove the compressor drive belt.

6. Inspect the belt for cracks, glazing and other wear and/or damage. Replace as necessary.

To install:

7. Install the drive belt, making sure it is properly aligned on the pulleys.

8. If the old belt is being reinstalled, tighten the idler pulley adjusting bolt until the belt deflects 0.33–0.37 in. with approximately 22 lbs. (98 N) pressure applied at a point midway between the pulleys. If a new belt has been installed, tighten the idler pulley adjusting bolt until the belt deflects 0.24–0.28 in. with approximately 22 lbs. (98 N) pressure applied at a point midway between the pulleys.

9. Torque the idler pulley locknut to 29 ft. lbs. (39 Nm).

10. Install the washer fluid reservoir tank, if removed.

11. Install the alternator drive belt.

12. Connect the negative battery cable.

Power Steering Belt

1. Disconnect the negative battery cable.

2. Remove the alternator drive belt.

3. If equipped, remove the A/C compressor drive belt.

4. Loosen the 2 power steering pump mounting bolts and pivot the power steering pump.

5. Remove the belt from the pulleys.

6. Inspect the belt for cracks, glazing and other wear and/or damage. Replace as necessary.

To install:

7. Install the drive belt, making sure it is properly aligned on the pulleys.

8. Adjust the belt tension by pulling back on the steering pump.

9. If the old belt is being reused, set the belt tension at 55–88 lbs. (245–391 N). If a new belt is being installed, set the belt tension at 100–121 lbs. (445–538 N).

10. Torque the power steering pump mounting bolts to 29 ft. lbs. (39 Nm).

11. If equipped, install the A/C compressor drive belt.

12. Install the alternator belt.

13. Connect the negative battery cable.

Starter

REMOVAL AND INSTALLATION

1. Disconnect the negative battery cable.

2. On 1.0L and 1.3L engines, proceed as follows;

    a. Disconnect the intake air temperature sensor from the air cleaner.

    b. Loosen the air cleaner hose clamp bolt.

    c. Disconnect the 4 air cleaner housing cover clips.

    d. Disconnect the air cleaner hose from the throttle body and remove the air cleaner housing cover together with the air cleaner hose.

3. Disconnect the wire clamp for the starter motor, if equipped.

4. Raise and safely support the vehicle.

5. Remove the retaining nut and disconnect the positive battery cable at the starter solenoid.

6. Label and disconnect the starter solenoid electrical connector.

7. Remove the mounting bolts from the starter motor, then carefully lower the assembly from the vehicle.

To install:

8. Install the starter motor assembly complete with any shims that may have been used between the engine block and the original starter.

9. Install and tighten the starter mounting bolts to 17 ft. lbs. (23 Nm).

10. Connect the starter solenoid electrical connector and positive battery cable. Tighten the retaining nut to 89 inch lbs. (10 Nm).

11. Lower the vehicle.

12. On 1.0L and 1.3L engines, proceed as follows;

    a. Connect the wire clamp for the starter motor.

    b. Connect the air cleaner hose to the throttle body. Install the air cleaner housing cover and secure with the 4 clips.

    c. Connect the IAT sensor to the air cleaner.

    d. Tighten the air cleaner hose clamp bolt.

13. Reconnect the negative battery cable.

14. Check starter operation.

CHASSIS ELECTRICAL

Blower Motor

REMOVAL AND INSTALLATION

—— **CAUTION** ——

The Air Bag system must be disarmed before working on or near the steering column. Failure to do so may cause accidental deployment, resulting in unnecessary repairs and/or personal injury.

1. Disconnect the negative battery cable.

2. If equipped, properly disable the Air Bag.

3. Remove the glove compartment to access the blower motor electrical connectors.

4. On 1995–97 models, remove the 3 mounting screws, label and disconnect the 3 electrical connectors and remove the Powertrain Control Module (PCM) from the vehicle.

5. Disconnect the blower motor and resistor wiring connectors.

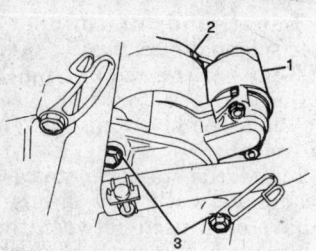

1. Solenoid
2. Starter motor assy
3. Starter motor assy mounting bolts

317240

Removing the starter motor assembly — 1.0L (VIN 6) and 1.3L (VIN 9) engines

6. Disconnect the cable from the blower case assembly.

7. Remove the blower case mounting bolts and case.

8. Remove the 3 motor retaining screws and blower motor. Disconnect the air hose.

9. Remove the blower fan retaining nut and remove the fan from the blower motor, if necessary.

To install:

10. Install the blower fan to the motor and tighten the fan retaining nut.

11. Install the blower motor assembly to the case and secure with 3 mounting screws.

12. Reconnect the air hose to the blower motor case.

13. Install the blower motor case into the vehicle and tighten the mounting bolts to 89 inch lbs. (10 Nm).

14. Reconnect the control cable. Fasten the electrical connectors to the motor and resistor.

15. On 1995–97 models, install the PCM into the vehicle and secure with 3 mounting screws. Fasten the electrical connectors.

16. Install the glove box.

17. Reconnect the negative battery cable.

18. If equipped, enable the Air Bag system.

19. Check blower motor operation.

Windshield Wiper Motor

REMOVAL AND INSTALLATION

1. Disconnect the negative battery cable.

2. Disconnect the electrical connector from the wiper motor.

3. Remove the wiper motor mounting bolts and pull the wiper motor away from the bulkhead.

4. Disconnect the wiper linkage from the motor crank arm.

5. Remove the wiper motor. Do not remove the wiper motor crank arm from the wiper motor.

To install:

6. Connect the wiper arm linkage to the wiper motor crank arm.

7. Install the wiper motor on the cowl and install the wiper motor mounting bolts. Tighten to 15 ft. lbs. (20 Nm) for Metro or 96 inch lbs. (10 Nm) for Prizm.

8. Reconnect the electrical connector to the wiper motor.

9. Reconnect the negative battery cable. Test the wiper system.

Combination Switch

REMOVAL AND INSTALLATION

Metro

——————— **CAUTION** ———————
The Air Bag system must be disarmed before removing the combination switch. Failure to do so may cause accidental deployment, resulting in unnecessary repairs and/or personal injury.

1. Place the vehicle's front wheels in the straight ahead position. Place the ignition switch into the **LOCK** position to prevent the Air Bag coil assembly from becoming uncentered.

2. Disconnect the negative battery cable.

3. Disable the Air Bag.

4. Remove the steering wheel.

——————— **CAUTION** ———————
When carrying a live Air Bag, make sure the bag and trim cover are pointed away from the body. In the unlikely event of an accidental deployment, the bag will then deploy with minimal chance of injury. When placing a live Air Bag on a bench or other surface, always face the bag and trim cover up, away from the surface. This will reduce the motion of the module if it is accidentally deployed.

5. Remove the steering column trim cover, knee bolster absorber and knee bolster panel.

6. Loosen the upper steering column mounting bolts and lower the steering column slightly.

7. Remove the upper and lower steering column covers.

8. Disconnect the Air Bag coil and combination switch electrical connectors.

9. Loosen the wiring bands for the Air Bag coil and combination switch electrical harness.

10. Remove the Air Bag coil and combination switch from the steering column.

To install:

11. Install the Air Bag coil and combination switch onto the steering column. Install and tighten the attaching screws.

12. Reconnect the electrical connectors to the combination switch and Air Bag coil. Using the wiring bands, secure the electrical harness to the steering column.

13. Install the upper and lower steering column covers and tighten the mounting screws.

14. Tighten the 2 upper steering column mounting bolts. Torque the upper column mounting bolts to 10 ft. lbs. (14 Nm).

15. Install the bolster panel and the knee bolster absorber.

16. Install the lower steering column trim cover and 3 mounting screws.

17. Install the steering wheel to the column. Torque the steering wheel retaining nut to 24 ft. lbs. (33 Nm).

18. Reconnect the negative battery cable and enable the Air Bag. Check switch operation.

Prizm

——————— **CAUTION** ———————
The Air Bag system must be disabled before performing service around the Air Bag components or wiring. Failure to do so may cause accidental deployment, resulting in unnecessary repairs and/or personal injury.

1. Make sure the front wheels are in the straight-ahead position.

2. Disconnect the negative battery cable.

3. Disable the Air Bag.

4. Remove the steering wheel.

5. Remove the upper and lower steering column trim covers.

6. Unclip the left front carpet retainer.

7. Remove the 2 screws and disconnect the hood release lever from the knee bolster.

8. Remove the 2 trim caps and unbolt and remove the knee bolster from the instrument panel.

9. Remove the tape from the Air Bag coil assembly harness and the combination switch harness.

10. Disconnect the electrical connector from the Air Bag coil.

11. Remove the 4 screws and remove the Air Bag coil assembly from the combination switch.

12. Disconnect the combination switch electrical connector.

13. Remove the 4 screws and remove the combination switch from the steering column.

To install:

14. Install the 4 screws and the combination switch to the steering column.

15. Connect the combination switch electrical connector.

16. Install the Air Bag coil and secure with the 4 screws.

17. Turn the Air Bag coil counterclockwise by hand until it reaches the stop. Turn the coil clockwise about 3 turns to align the alignment mark. The connector should be straight up.

18. Connect the Air Bag coil wiring connector.

19. Tape the Air Bag coil harness to the combination switch harness.

20. Install the knee bolster and tighten the bolts to 89 inch lbs. (10 Nm).

21. Attach the hood release lever to the knee bolster and secure with the 2 screws.

22. Position and secure the left front carpet retainer.

23. Install the steering column trim covers and secure with the 5 screws.

24. Install the steering wheel.

25. Connect the negative battery cable. Enable the Air Bag.

Ignition Lock Cylinder

REMOVAL AND INSTALLATION

——— CAUTION ———

The Air Bag system must be disarmed before performing service around the Air Bag components or wiring. Failure to do so may cause accidental deployment, resulting in unnecessary repairs and/or personal injury.

1. Disconnect the negative battery cable.

2. Disarm the Air Bag.

3. Remove the upper and lower steering column covers.

4. Insert the key and turn the ignition lock to the **ACC** position.

5. Push the lock cylinder stop in with a suitable tool (cotter pin, punch, etc.) and pull out the ignition key and the lock.

To install:

6. Install the switch with the key in the **ACC** position.

7. Install the upper and lower steering column covers.

8. Enable the Air Bag, if equipped.

9. Connect the negative battery cable.

10. Check ignition lock operation.

Ignition Switch

REMOVAL AND INSTALLATION

Metro

——— CAUTION ———

The Air Bag system must be disarmed before performing service around the Air Bag components or wiring. Failure to do so may cause accidental deployment, resulting in unnecessary repairs and/or personal injury.

1. Disconnect the negative battery cable.

2. If equipped, disarm the Air Bag.

3. Remove the steering wheel.

——— WARNING ———

If equipped with an Air Bag, the vehicle's wheels must be in the straight-ahead position and the key must be in the LOCK position.

4. Remove the steering column from the vehicle and mount in a suitable soft-jawed vise. To keep from damaging the column, wrap a shop towel around the steering shaft assembly where it contacts the vise.

5. Remove the ignition switch mounting bolts. Using a hammer and chisel, create slots on the top of the mounting bolts, then remove the bolts with a screwdriver.

——— WARNING ———

Be careful not to damage the aluminum switch body with the chisel.

6. Turn the key to **ON** or **ACC** and remove the switch from the steering column.

To install:

7. Position the oblong hole in the steering shaft so it is visible through the center of the hole in the column.

8. Install the ignition switch to the column with the key in the **ON** or **ACC** position.

9. Turn the key to the **LOCK** position and remove the key.

10. Align the ignition switch hub with the oblong hole in the steering shaft. Rotate to ensure that the steering shaft is locked.

11. Install new break-away bolts and tighten until the heads break off.

12. Turn the ignition key to **ON** or **ACC** position and check that the shaft rotates smoothly. Remove the column from the vise.

13. Install the steering column in the vehicle.

14. Install the steering wheel.

15. Connect the negative battery cable.

16. If equipped, enable the Air Bag.

17. Check switch operation.

Prizm

——— CAUTION ———

The Air Bag system must be disarmed before performing service around the Air Bag components or wiring. Failure to do so may cause accidental deployment, resulting in unnecessary repairs and/or personal injury.

1. Disconnect the negative battery cable.

2. If equipped, disarm the Air Bag.

3. Remove the upper and lower steering column covers.

4. Disconnect the electrical connector to the ignition switch.

5. Remove the 2 screws and ignition switch from the column upper bracket.

To install:

6. Install the ignition switch to the upper column bracket and install the 2 screws.

7. Connect the electrical connector to the ignition switch.

8. Install the upper and lower steering column covers.

9. If equipped, enable the Air Bag.

10. Connect the negative battery cable.

11. Check ignition switch operation.

Park/Neutral Safety Switch

REMOVAL AND INSTALLATION

Metro

1. Disconnect the negative battery cable.

2. Remove the electrical connector at the engine wiring harness.

3. Remove the harness from the retaining clamps.

4. Remove the switch from the transaxle.

To install:

5. Place the transaxle in the **N** position.

6. Using a screwdriver, turn the shift switch assembly joint clockwise or counterclockwise until a distinct click noise is heard.

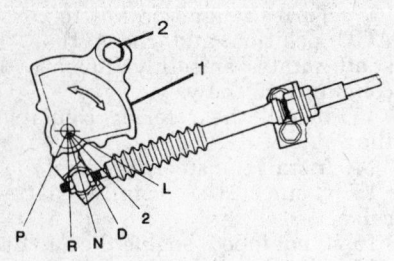

1 PARK/NEUTRAL POSITION (PNP) SWITCH
2 PNP SWITCH MOUNTING BOLT

311409

Switch mounting bolt and shift linkage — Metro

7. Install the switch and tighten the bolt to 17 ft. lbs. (23 Nm).

8. Install the harness retaining clips and connect the electrical connector.

9. Connect the negative battery cable.

10. To check the park/neutral switch for proper operation, apply the parking brake and block the wheels. Ensure the starter motor operates only when the transaxle shift lever is in the **P** or **N** positions and does not operate when in the **D**, **2**, **L** or **R** positions.

Prizm

1. Disconnect the negative battery cable.

2. Apply the parking brake and place the shifter in the **N** position.

3. Raise and safely support the vehicle.

4. Remove the left splash shield.

5. Remove the nut and washer securing the shift select cable to the switch and disconnect the cable.

6. Disconnect the electrical connector from the switch.

7. Remove the nut and washer securing the manual lever to the switch.

8. Remove the manual shaft nut and lock plate from the switch.

9. Remove the mounting bolts and remove the switch.

To install:

10. Position the switch and install the mounting bolts, finger-tight.

11. Install the lock plate and manual shaft nut onto the switch.

12. Tighten the manual shaft nut to 61 inch lbs (6.9 Nm). Stake the lock plate over the manual shaft nut.

13. Align the neutral basic line scribed in the switch with the groove in the switch sleeve, then tighten the switch bolts to 48 inch lbs. (5.4 Nm).

14. Connect the manual lever to the switch, install the washer and

tighten the nut to 106 inch lbs. (12 Nm).

15. Connect the electrical connector to the switch.

16. Connect the shift cable to the switch and install the mounting nut and tighten to 106 inch lbs. (12 Nm).

17. Install the left splash shield.

18. Lower the vehicle.

19. Connect the negative battery cable.

Powertrain Control Module

REMOVAL AND INSTALLATION

1993–94 Metro

NOTE: When removing the original Powertrain Control Module (PCM) and installing a replacement PCM, it is very important to transfer the broadcast code as well as the production PCM number to the service PCM label. This will allow positive PCM parts identification throughout the service life of the vehicle. Do not record this on the cover of the PCM.

1. Disconnect the negative battery cable.

2. Remove the bolts and the junction block from the lower left side of the instrument panel.

3. Remove the PCM mounting bolts from the lower left side of the instrument panel.

4. Disconnect the electrical connectors from the PCM and remove the PCM.

To install:

5. Connect the electrical connectors to the PCM.

6. Connect the PCM mounting bracket to the lower instrument panel and install the mounting bolts. Tighten the bolts to 11 ft. lbs. (15 Nm).

7. Install the junction block to the lower left instrument panel and

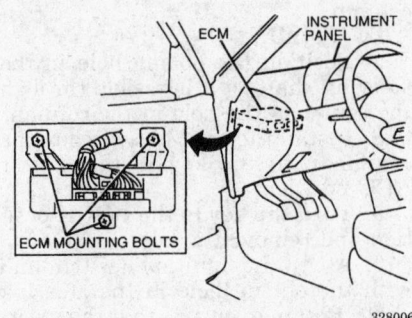

328006

ECM mounting location and mounting bolts — 1993–94 Metro

tighten the mounting bolts to 89 inch lbs. (10 Nm).

8. Connect the negative battery cable.

1995–97 Metro

NOTE: When removing the original Powertrain Control Module (PCM) and installing a replacement PCM, it is very important to transfer the broadcast code as well as the production PCM number to the service PCM label. This will allow positive PCM parts identification throughout the service life of the vehicle. Do not record this on the cover of the PCM.

1. Disconnect the negative battery cable.

2. Release the glove box clips and pull down the glove box.

3. Remove the mounting bolts to the PCM unit from behind the glove box.

4. Disconnect the electrical connectors from the PCM unit.

5. Remove the PCM from the vehicle.

To install:

6. Connect the electrical connectors to the PCM unit.

7. Install the PCM unit to the mounting bracket and install the mounting bolts. Torque the PCM mounting bolts to 11 ft. lbs. (15 Nm).

8. Push up the glove box into the dashboard and catch the clips.

9. Reconnect the negative battery cable.

Prizm

—— **WARNING** ——
To prevent internal damage to the PCM, the ignition must be OFF when disconnecting or reconnecting power to the PCM.

1. Disconnect the negative battery cable.

2. Carefully pull back the carpet from the floor under the glove box.

3. Remove the bolts securing the PCM mounting bracket to the floor bracket.

4. Disconnect the electrical connectors from the PCM.

5. Remove the PCM.

To install:

6. Connect the electrical connectors to the PCM.

7. Place the PCM and mounting bracket in position and install the 2 mounting bolts.

8. Lay the carpet back in place.

9. Connect the negative battery cable.

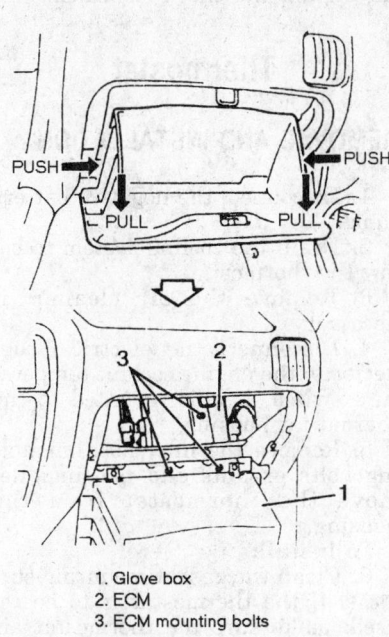

1. Glove box
2. ECM
3. ECM mounting bolts

310518

Removing the Engine Control Module — 1995–97 Metro

ENGINE COOLING

Radiator

REMOVAL AND INSTALLATION

──────── **CAUTION** ────────

Never remove the radiator cap under any conditions while the engine is operating. Failure to follow these instructions could result in personal injury and/or damage to the cooling system or engine. To avoid having scalding

hot coolant or steam blow out of the radiator, use extreme care when removing the radiator cap from a hot radiator. Wait until the engine has cooled, then wrap a thick cloth around the radiator cap and turn it slowly to the first stop. Step back while the pressure is released from the cooling system. When it is certain that all pressure has been released, press down on the radiator cap, with the cloth, turn and remove.

─────────────

1. Disconnect the negative battery cable.
2. Drain the cooling system.
3. Disconnect the cooling fan electrical connector and the air inlet hose.
4. Disconnect the upper, lower and reservoir tank hoses from the radiator.
5. If equipped with automatic transaxle, use a line wrench to disconnect the oil cooler lines from the radiator and plug them.
6. Remove the mounting bolts, then lift the radiator from the vehicle with the cooling fan attached.
7. Remove the cooling fan and shroud, if necessary.
   To install:
8. Install the cooling fan and shroud on the radiator. Tighten the lower mounting bolt(s) to 89 inch lbs. (10 Nm).
9. Place the radiator in the vehicle and install the mounting bolts. Tighten to 89 inch lbs. (10 Nm) on Metro or 115 inch lbs. (13 Nm) on Prizm.
10. If equipped with an automatic transaxle, connect the oil cooler lines to the radiator.
11. Connect the upper, lower and reservoir tank hoses to the radiator and secure with hose clamps.
12. Connect the cooling fan electrical connector and the air inlet hose.
13. Fill the cooling system and connect the negative battery cable.

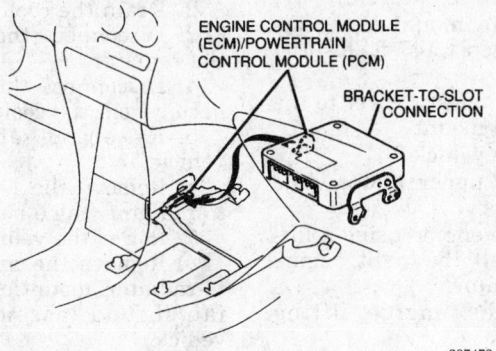

ENGINE CONTROL MODULE (ECM)/POWERTRAIN CONTROL MODULE (PCM)

BRACKET-TO-SLOT CONNECTION

237472

PCM mounting location — Prizm

14. Start the engine, allow it to reach normal operating temperature. Check the cooling system for leaks. Top off as necessary.

Water Pump

REMOVAL AND INSTALLATION

Metro

1. Disconnect the negative battery cable.
2. Drain the cooling system.
3. Remove the air cleaner.
4. Remove the suction pipe bracket for the A/C compressor, if equipped.
5. Loosen but do not remove the 4 water pump pulley bolts.
6. Raise and safely support the vehicle.
7. Remove the lower splash shield.
8. Remove the A/C compressor drive belt, if equipped.
9. Remove the alternator drive belt.
10. Remove the crankshaft and water pump pulleys.
11. Remove the timing belt.
12. Remove the oil level dipstick and tube.
13. Remove the alternator adjusting bracket from the water pump.
14. Remove the water pump rubber seals.
15. Remove the water pump mounting bolts and nuts and remove the water pump from the engine.
    To install:
16. Clean the gasket mating surfaces thoroughly.
17. Check the water pump by hand for smooth operation. If the pump does not operate smoothly or is noisy, replace it.
18. Install the pump using a new gasket. Tighten the bolts to 115 inch lbs. (13 Nm).
19. Install new rubber seals.
20. Install the upper alternator adjusting bracket and tighten the bolt to 17 ft. lbs. (23 Nm).
21. Install the oil level dipstick and tube.
22. Install the timing belt.
23. Install the water pump and crankshaft pulleys. Leave the water pump pulley bolts hand-tight.
24. Install the alternator drive belt.
25. Install the lower alternator cover plate and tighten the bolts to 89 inch lbs. (10 Nm).
26. Install the A/C compressor drive belt, if equipped.
27. Install the lower splash shield and lower the vehicle.

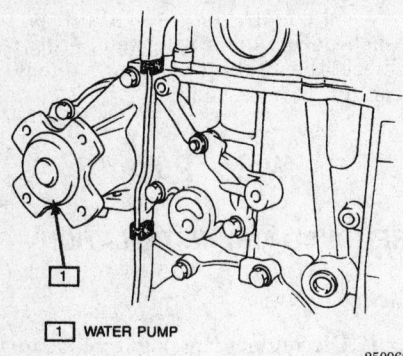

1 WATER PUMP

250960

Water pump mounting location — Metro

28. Tighten the water pump pulley mounting bolts to 18 ft. lbs. (24 Nm).

29. Adjust the water pump drive belt tension and tighten the alternator adjustment bolt to 17 ft. lbs. (23 Nm).

30. Install the suction pipe bracket for the A/C compressor, if equipped.

31. Install the air cleaner.

32. Refill the cooling system.

33. Connect the negative battery cable.

34. Start the engine and check for leaks.

Prizm

1. Disconnect the negative battery cable.

2. Drain the engine coolant into a suitable container.

3. Support the engine using a lifting device.

4. Remove the right engine mount and insulator.

5. Remove the upper and middle timing belt covers.

6. If equipped with power steering, proceed as follows:

   a. Raise and safely support the vehicle.

   b. Remove the plastic cover from the front transaxle mount.

   c. Remove the 2 mount-to-chassis bolts.

   d. Remove the nut and through-bolt, then remove the front transaxle mount from the vehicle.

   e. Lower the vehicle.

   f. Remove the coolant reservoir.

   g. Disconnect the upper hose from the radiator.

   h. Unclip the wiring connectors from the shroud.

   i. Remove the cooling fan assembly.

7. Remove the 2 nuts, one bolt and the engine wiring harness retainer.

8. If equipped, unclip the crankshaft position sensor electrical connector from the dipstick tube.

9. Remove the dipstick tube and dipstick. Immediately plug the hole in the block.

10. Disconnect the cooling fan switch electrical connector.

11. Remove the 2 nuts securing the engine coolant inlet pipe to the cylinder block.

12. Loosen the hose clamps and remove the coolant inlet pipe from the water pump.

13. Loosen the hose clamp and remove the coolant inlet hose from the water pump.

14. Remove the water pump bolts and remove the assembly. Discard the O-ring.

To install:

15. Install a new O-ring. Install the water pump, and torque the retaining bolts evenly to 10 ft. lbs. (14 Nm).

16. Connect the coolant hose to the water pump and secure the clamp.

17. Install the engine coolant inlet pipe. Secure to the hose with clamps and nuts. Tighten the nuts to 11 ft. lbs. (15 Nm).

18. Connect the cooling fan switch electrical connector.

19. Install the dipstick and tube. Torque the retaining bolt to 84 inch lbs. If equipped, clip the crankshaft position sensor electrical connector to the dipstick tube.

20. Install the engine wiring harness retainer and tighten the nuts and bolt to 89 inch lbs. (10 Nm).

21. If equipped with power steering, proceed as follows:

   a. Install the cooling fan assembly and tighten the bolts to 52 inch lbs. (6 Nm).

   b. Attach the wiring connectors to the shroud.

   c. Connect the upper hose to the radiator.

   d. Install the coolant reservoir.

   e. Raise and safely support the vehicle.

   f. Install the front transaxle mount and tighten the through-bolt nut to 64 ft. lbs. (87 Nm).

   g. Install the mount-to-chassis bolts and tighten to 47 ft. lbs. (64 Nm).

   h. Install the plastic cover to the front transaxle mount.

   i. Lower the vehicle.

22. Install the upper and middle timing belt covers.

23. Support the engine using a lifting device. Install the right engine mount and insulator.

24. Remove the engine lifting equipment.

25. Connect the negative battery cable.

26. Fill the cooling system. Start the engine and check for leaks.

Thermostat

REMOVAL AND INSTALLATION

1. Disconnect the negative battery cable.

2. Drain the cooling system to below the thermostat.

3. Remove the air cleaner, if necessary.

4. Disconnect the electrical connector to the engine coolant temperature switch, if it is mounted to the thermostat housing.

5. Remove the inlet hose, mounting bolts or nuts and housing. Remove the thermostat from the housing.

To install:

6. Clean the gasket mounting surfaces. If the thermostat is to be reused, make sure the thermostat air bleed valve is clear.

7. Install the thermostat into the housing with the spring side down.

8. Install the thermostat housing using a new gasket. Tighten bolts to 15 ft. lbs. (20 Nm), nuts to 84 inch lbs.

9. Install the upper hose to the thermostat housing and tighten the hose clamp.

10. Connect the electrical harness to the housing sensor, if necessary.

11. Refill the cooling system.

12. Connect the negative battery cable.

13. Start the engine and check for leaks.

Electric Cooling Fan

REMOVAL AND INSTALLATION

1. Disconnect the negative battery cable.

2. Drain the cooling system.

3. Disconnect the upper hose at the radiator.

4. Disconnect the electrical connector from the cooling fan motor.

5. Raise and safely support the vehicle.

6. Remove the lower fan shroud-to-radiator frame bolt(s).

7. Lower the vehicle.

8. Remove the upper fan shroud-to-radiator mounting bolts and the fan/shroud assembly from the vehicle.

9. Remove the fan blade-to-motor nut, then fan blade and washer.

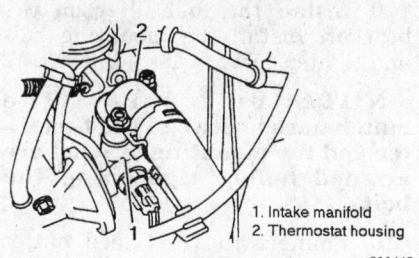

1. Intake manifold
2. Thermostat housing

311442

Thermostat housing — Metro with 1.3L (VIN 9) engine

10. Remove the fan-to-shroud bolts and the fan motor from the shroud.

To install:

11. Install the fan motor in the shroud.

12. Install the fan blade to the motor.

13. Install the cooling fan/shroud assembly to the radiator.

14. Install the upper fan shroud-to-radiator mounting bolts and tighten the bolts to 89 inch lbs. (10 Nm).

15. Raise and safely support the vehicle.

16. Install the lower cooling fan/shroud assembly mounting bolt(s). Torque the mounting bolt(s) to 89 inch lbs. (10 Nm).

17. Lower the vehicle.

18. Reconnect the fan motor electrical connector.

19. Connect the upper hose and secure with a clamp.

20. Refill the cooling system.

21. Reconnect the negative battery cable.

22. Run the engine and check the cooling fan for proper operation.

23. Turn **OFF** the engine and allow the engine to cool. Check the coolant level and top off, as necessary.

Cooling System Bleeding

PROCEDURE

———— **CAUTION** ————

Never remove the radiator cap under any conditions while the engine is operating. Failure to follow these instructions could result in personal injury and/or damage to the cooling system or engine.

To avoid having scalding coolant or steam blow out of the radiator, use extreme care when removing the radiator cap from a hot radiator. Wait until the engine has cooled, then wrap a thick cloth around the radia-

tor cap and turn it slowly to the first stop. Step back while the pressure is released from the cooling system. When certain that all pressure has been released, press down on the radiator cap, with the cloth, turn and remove.

1. When all coolant has drained, make sure the radiator drain plug is completely closed.

2. With the engine **OFF**, fill the radiator to the bottom of the filler neck with a 50/50 mix of antifreeze and water.

3. Start the engine.

4. Allow the engine to reach normal operating temperature.

5. When the thermostat opens, top off the system.

6. Install the radiator cap.

7. Fill the coolant recovery reservoir to the full hot mark.

8. Shut **OFF** the engine and allow to cool.

9. With the engine cool, check the level of coolant in the recovery bottle and top off as necessary.

GASOLINE FUEL SYSTEM

Fuel System Service Precautions

Safety is the most important factor when performing not only fuel system maintenance but any type of maintenance. Failure to conduct maintenance and repairs in a safe manner may result in serious personal injury or death. Maintenance and testing of the vehicle's fuel system components can be accomplished safely and effectively by adhering to the following rules and guidelines.

• To avoid the possibility of fire and personal injury, always disconnect the negative battery cable unless the repair or test procedure requires that battery voltage be applied.

• Always relieve the fuel system pressure prior to disconnecting any fuel system component (injector, fuel rail, pressure regulator, etc.), fitting or fuel line connection. Exercise extreme caution whenever relieving fuel system pressure to avoid exposing skin, face and eyes to fuel spray. Please be advised that fuel under pressure may penetrate the skin or any part of the body that it contacts.

• Always place a shop towel or cloth around the fitting or connection prior to loosening to absorb any excess fuel due to spillage. Ensure that all fuel spillage (should it occur) is quickly removed from engine surfaces. Ensure that all fuel soaked cloths or towels are deposited into a suitable waste container.

• Always keep a dry chemical (Class B) fire extinguisher near the work area.

• Do not allow fuel spray or fuel vapors to come into contact with a spark or open flame.

• Always use a backup wrench when loosening and tightening fuel line connection fittings. This will prevent unnecessary stress and torsion to fuel line piping. Always follow the proper torque specifications.

• Always replace worn fuel fitting O-rings with new. Do not substitute fuel hose or equivalent, where fuel pipe is installed.

Fuel System Pressure

RELIEVING

———— **CAUTION** ————

Fuel injection systems remain under pressure, even after the engine has been turned OFF. The fuel system pressure must be relieved before disconnecting any fuel lines. Failure to do so may result in fire and/or personal injury.

1. Remove the fuel filler cap.

2. On Metro, perform the following;

   a. Remove the control relay box cover from the relay box.

   b. Disconnect the fuel pump relay from the relay box connector.

3. On Prizm, perform the following;

   a. Remove the center trim bezel from the center console by gently prying around the edges.

   b. Remove the radio from the center console.

   c. Disconnect the circuit opening relay electrical connector, located in the center console.

4. Start the engine and let it run until the engine stalls due to lack of fuel.

5. Engage the starter for a few seconds to assure relief of remaining fuel pressure.

6. Disconnect the negative battery cable.

7. Continue with the required service procedure(s).

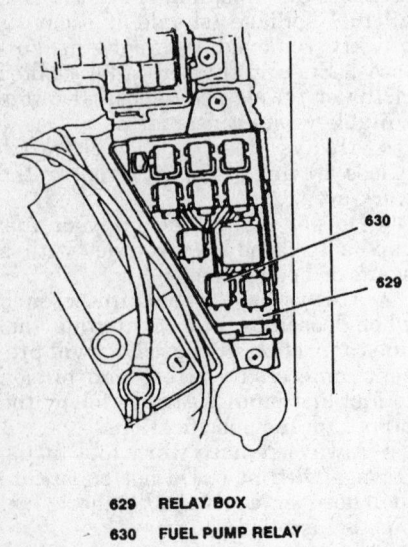

629 **RELAY BOX**

630 **FUEL PUMP RELAY**

322121

Relay box — Metro with 1.0L (VIN 6) and 1.3L (VIN 9) engines

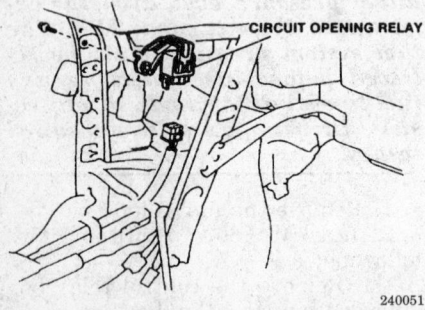

240051

Circuit opening relay — Prizm with 1.6L (VIN 6) and 1.8L (VIN 8) engines

Idle Speed

ADJUSTMENT

The idle speed is kept at the correct speed by controlling the amount of air that bypasses the throttle valve in accordance with changes in idling conditions and engine load during idling. The Powertrain Control Module (PCM) drives the Idle Air Control (IAC) motor to keep the engine running at the pre-set idle target speed in accordance with the engine coolant temperature and air conditioning

load. In addition, when the air conditioning switch is turned **OFF** and **ON** while the engine is idling, the IAC motor operates to adjust the throttle valve bypass air amount in accordance with the engine load conditions in order to avoid fluctuations in engine speed.

Mixture

ADJUSTMENT

The air/fuel mixture is not adjustable. The air/fuel mixture is controlled by the engine control system, which consists of various sensors, switches and the Powertrain Control Module (PCM). The various sensors and switches send signals to the PCM regarding engine operating conditions; the PCM then uses the data to determine the timing and opening duration of the fuel injectors.

Fuel Filter

—— CAUTION ——
Fuel injection systems remain under pressure, even after the engine has been turned OFF. The fuel system pressure must be relieved before disconnecting any fuel lines. Failure to do so may result in fire and/or personal injury.

REMOVAL AND INSTALLATION

Metro

1. Properly relieve the fuel system pressure.
2. Disconnect the negative battery cable.
3. Raise and safely support the vehicle.
4. Remove the bolt securing the parking brake cable bracket from the underbody.
5. Disconnect the fuel feed hose from the fuel filter.
6. Remove the 2 fuel filter mounting bracket bolts, and remove the fuel filter from the frame.
7. Disconnect the outlet hose from the fuel filter.
8. Remove the fuel filter from the bracket by removing the mounting bolt.

To install:
9. Install the fuel filter on the bracket. Install the mounting bolt and tighten to 11 ft. lbs. (15 Nm).

NOTE: Be sure the matchmarks between the fuel filter and the mounting bracket are aligned before tightening the bolt.

10. Connect the fuel feed outlet hose to the fuel filter. Secure with a new clamp.
11. Install the fuel filter on the frame and install the 2 bracket bolts. Tighten the bracket bolts to 11 ft. lbs. (15 Nm).
12. Connect the fuel inlet hose to the fuel filter. Secure with a new clamp.
13. Position the brake cable bracket and install the bolt. Tighten the cable bracket bolt to 11 ft. lbs. (15 Nm).
14. Lower the vehicle.
15. Reconnect the negative battery cable.
16. Turn the ignition switch to **ON** and then back to **LOCK** to pressurize the fuel system. Check for leaks and correct as necessary.

Prizm

1. Properly relieve the fuel system pressure.
2. Disconnect the negative battery cable.
3. Disconnect the Intake Air Temperature (IAT) sensor.
4. Loosen the air cleaner clamp and release the 4 air cleaner housing cover clips. Remove the hose and air cleaner cover.
5. Disconnect the hoses from the Evaporative Emissions (EVAP) canister. Remove the bolt and canister from the bracket.
6. Remove the bolt and disconnect the fuel outlet hose from the top of the fuel filter. Discard the gaskets.
7. Remove the 2 bolts and remove the fuel filter from the bracket.
8. Remove the nut and fuel inlet pipe from the bottom of the fuel filter.

To install:
9. Install the fuel filter to the bracket and tighten the bolts to 43 inch lbs. (5 Nm).
10. Connect the fuel inlet pipe to the bottom of the fuel filter and tighten the nut to 22 ft. lbs. (30 Nm).
11. Install the fuel outlet hose to the top of the fuel filter, using new gaskets. Install the bolt and tighten to 21 ft. lbs. (29 Nm).
12. Install the EVAP canister to the bracket and tighten the bolt to 21 ft. lbs. (29 Nm). Connect the hoses to the canister.

13. Install the air cleaner cover and secure with the clips. Connect the air cleaner hose.

14. Connect the IAT sensor.

15. Connect the negative battery cable. Turn the ignition switch **ON** to pressurize the fuel system. Check for fuel leaks.

Fuel Pump

———— **CAUTION** ————

Observe all applicable safety precautions when working around fuel. Do not allow fuel spray or fuel vapors to come in contact with a spark or open flame. Keep a dry chemical (Class B) fire extinguisher near the work area. Never drain or store fuel in an open container due to the possibility of fire or explosion.

REMOVAL AND INSTALLATION

Metro

1. Properly relieve the fuel system pressure.

2. Disconnect the negative battery cable.

3. Remove the filler cap. Drain the fuel tank by pumping the fuel out through the filler neck into a suitable container.

4. Reinstall the fuel filler cap.

5. Raise and safely support the vehicle.

6. Remove the fuel tank from the vehicle. Clean all dirt from the fuel pump area, to prevent fuel system contamination.

7. Remove the fuel feed and return clamps and hoses from the pump assembly.

8. Remove the screws from the pump assembly and remove with the gasket from the fuel tank.

9. Remove the screws from the pump motor assembly. Remove the fuel pump motor connectors and detach the pump motor from the pump assembly.

To install:

10. Install the 2 pump motor connectors. Install the pump motor on the pump assembly and tighten the screw.

11. Install the assembly with the gasket on the fuel tank. Install the screws and tighten.

12. Install the fuel feed and return hoses and clamps on the fuel pump assembly.

13. Install the fuel tank in the vehicle.

14. Lower the vehicle and reconnect the negative battery cable.

15. Refill the fuel tank. Turn the ignition key **ON** and allow the fuel system to pressurize. Check for leaks.

16. Start the engine and check for leaks.

Prizm

1. Properly relieve the fuel system pressure.

2. Use a pump to drain the fuel through the filler neck.

3. Remove the rear seat cushion to gain access to the service panel.

4. Remove the 4 screws, then the service access panel.

5. Disconnect the fuel pump and sending wiring.

6. Remove the nut and fuel feed hose from the sender.

7. Remove the clamp and fuel return hose from the sender.

8. Remove the bolts and remove the sender assembly from the tank.

9. Remove the pump from the bracket by pulling off the lower side of the pump from the bracket.

10. Disconnect the fuel pump electrical connector.

11. Remove the rubber cushion from the fuel pump.

12. Remove the clamp and disconnect the strainer from the pump.

To install:

13. Install the clamp and connect the strainer to the pump.

14. Install the rubber cushion to the pump.

15. Connect the fuel pump electrical connector.

16. Install the pump to the bracket.

17. Install the fuel sender assembly to the tank, using a new gasket. Secure with the bolts, tightened to 35 inch lbs. (4 Nm).

18. Install the clamp and fuel return hose to the sender.

19. Install the nut and fuel feed hose to the sender. Tighten the nut to 15 ft. lbs. (20 Nm) for 1993 vehicles

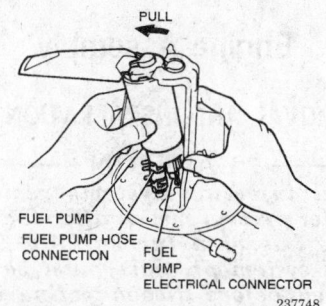

PULL

FUEL PUMP
FUEL PUMP HOSE
CONNECTION
FUEL
PUMP
ELECTRICAL CONNECTOR

237748

Removing the fuel pump from the sender assembly — Prizm with 1.6L (VIN 6) and 1.8L (VIN 8) engines

or 22 ft. lbs. (30 Nm) for 1994–97 vehicles.

20. Connect the fuel pump electrical connector.

21. Refill the fuel tank.

22. Connect the negative battery cable. Turn the ignition **ON** to pressurize the fuel system. Check for fuel leaks.

23. Install the service access panel and rear seat cushion.

Fuel Injector

REMOVAL AND INSTALLATION

———— **CAUTION** ————

Fuel injection systems remain under pressure, even after the engine has been turned OFF. The fuel system pressure must be relieved before disconnecting any fuel lines. Failure to do so may result in fire and/or personal injury.

Metro

NOTE: Do not remove the fuel pressure regulator or air valve from the throttle body. They are calibrated at the factory and are not serviceable.

1. Properly relieve the fuel system pressure, then disconnect the negative battery cable.

2. Remove the air cleaner.

3. Disconnect the intake air temperature sensor harness and A/C solenoid vacuum valve hose, if equipped.

4. Remove the air cleaner bracket from the throttle body unit.

5. Remove the injector cover and disconnect the injector electrical harness.

6. Gently pull the injector out of the throttle body.

7. Remove and discard the O-rings from the injector.

To install:

8. Install new O-rings on the injector.

9. Apply a thin coat of automatic transmission fluid to the O-rings, then install the injector into the throttle body.

NOTE: Do not twist the injector while installing, or misalignment of the O-rings may occur.

10. Install the upper insulator, cover, electrical connector and the retaining screws.

11. Install the air cleaner bracket. Torque the bracket bolts to 11 ft. lbs. (15 Nm).

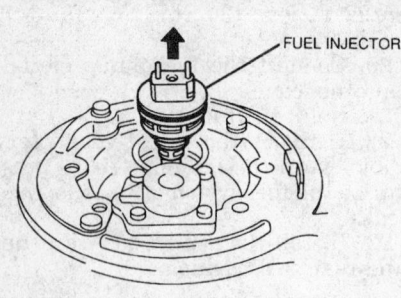

FUEL INJECTOR

311618

Removing the fuel injector — Metro with 1.0L (VIN 6) and 1.3L (VIN 9) engines

12. Reconnect the A/C solenoid vacuum valve hose and the IAT sensor, if equipped.

13. Install the remaining components.

14. Connect the negative battery cable.

15. With the ignition switch **ON** and the engine **OFF**, pressurize the fuel system and check for fuel leaks.

Prizm

1. Properly relieve the fuel system pressure.

2. Disconnect the negative battery cable.

3. Disconnect the intake air temperature sensor harness.

4. Loosen the air cleaner hose and release the housing cover clips. Remove the hose and cover.

5. Disconnect the accelerator cable from the throttle body.

6. If equipped with automatic transaxle, disconnect the TV cable from the throttle body.

7. Disconnect the vacuum hose from the fuel pressure regulator.

8. If equipped, disconnect the hoses from the EGR valve and EGR valve solenoid vacuum harness.

9. Disconnect the necessary hoses from the throttle body and intake chamber cover.

10. Label and disconnect the necessary electrical connectors from the throttle body.

11. Remove the bolts and nuts, then remove the intake chamber cover and throttle body as an assembly.

12. Disconnect the injector harnesses.

13. Remove the bolt and disconnect the fuel inlet hose from the fuel rail. Discard the gaskets.

14. Disconnect the fuel return hose from the fuel pressure regulator.

15. Remove the 2 fuel rail bolts, then remove the fuel rail with the injectors attached.

16. Remove the insulators and spacers from the intake manifold.

17. Pull the injectors free of the fuel rail.

18. Remove and discard the O-rings and grommets from the injectors.

To install:

19. Install new O-rings and grommets on each injector.

20. Coat each O-ring with a light coating of gasoline and install the injectors into the fuel rail. Make sure each injector can be smoothly rotated.

21. Install the insulators into each injector hole. Place the 2 spacers on the fuel rail mounting holes in the intake manifold.

22. Place the fuel rail and injectors on the intake manifold and check that the injectors rotate smoothly. Position the injector connector upward. Install the 2 bolts and tighten them to 11 ft. lbs. (15 Nm).

23. Connect the fuel inlet hose to the fuel rail, using new gaskets. Install and tighten the bolt to 22 ft. lbs. (29 Nm).

24. Connect the fuel return hose to the pressure regulator.

25. Connect the harnesses to each injector.

26. Install the intake chamber cover, with the throttle body attached, using a new gasket. Tighten the cover nuts and bolts evenly, to 14 ft. lbs. (19 Nm).

27. Connect the remaining components, including the TV cable, IAT harness and air cleaner hose.

28. Connect the negative battery cable. Turn the ignition **ON** to pressurize the fuel system. Check for fuel leaks.

ENGINE MECHANICAL

Engine Assembly

REMOVAL AND INSTALLATION

—————— **CAUTION** ——————
Fuel injection systems remain under pressure, even after the engine has been turned OFF. The fuel system pressure must be relieved before disconnecting any fuel lines. Failure to do so may result in fire and/or personal injury.

Metro

1. Properly relieve the fuel system pressure. Disconnect the negative battery cable.

2. Scribe a hood hinge-to-hood outline, then, with the aid of an assistant remove the hood.

3. Drain the cooling system.

4. Remove the air cleaner. Remove the radiator and cooling fan.

5. Disconnect the engine electrical connections and vacuum lines.

6. Disconnect and plug the fuel return and feed lines.

7. Disconnect the heater inlet and outlet hoses.

8. Disconnect the following cables:

   a. Accelerator cable from the throttle body.

   b. Clutch cable from manual transaxle vehicles.

   c. The gear select and the oil pressure control cable from automatic transaxle vehicles.

   d. The speedometer cable from the transaxle.

9. Support the engine assembly using tool J-28467-A and J-28467-89 or equivalents.

10. Raise and safely support the vehicle.

11. Disconnect the exhaust pipe from the manifold.

12. Disconnect the gear shift control shaft and extension from manual transaxle vehicles.

13. Drain the engine and transaxle oil.

14. Remove the left and right halfshafts. It is not necessary to remove the halfshaft from the knuckles.

15. Remove the A/C compressor and belt. Position the compressor aside, leaving the hoses connected.

16. Disconnect and plug the power steering hoses from the pump, if equipped.

17. Remove the engine rear torque rod bracket from automatic transaxle vehicles.

18. Lower the vehicle.

19. Attach suitable lifting equipment to the engine.

20. Remove the right side engine mount from the bracket.

21. On automatic transaxles, remove the transaxle rear mounting nut.

22. On manual transaxles, remove the transaxle rear mount from the body.

23. Remove the transaxle left side mounting bracket.

24. Lift the engine and transaxle assembly out of the vehicle.

To install:

25. Install the engine and transaxle assembly into the vehicle and leave the hoist connected.

26. On automatic transaxles, install the transaxle rear mounting nut.

27. On manual transaxles, install the transaxle rear mount to the body.

28. Install the transaxle left side mounting bracket. Torque the bolts to 41 ft. lbs. (55 Nm).

29. Install the transaxle right side engine mount and bracket. Torque the bolts to 41 ft. lbs. (55 Nm).

30. Remove the lifting device and support the engine assembly with tool J-28467-A and J-28467-89 or equivalents.

31. Raise and safely support the vehicle.

32. Install the rear engine torque rod bracket to the transaxle, if equipped.

33. Connect the left and right halfshafts.

34. Connect the gear shift control shaft and the extension to the transaxle, if equipped with a manual transaxle.

35. Connect the exhaust pipe to the manifold. Torque the to 37 ft. lbs. (50 Nm).

36. Install the A/C compressor and belt. Torque the compressor-to-mounting bracket bolts to 21 ft. lbs. (28 Nm).

37. Reconnect the power steering hoses to the power steering pump, if equipped.

38. Lower the vehicle.

39. Install the remaining components.

40. Install the hood. Tighten the hood mounting bolts to 20 ft. lbs. (27 Nm).

41. Adjust the clutch pedal free-play, gear select cable and accelerator cable play.

42. Refill the engine, transaxle and power steering pump, if equipped with the proper type and quantity of oil. Refill and bleed the cooling system.

43. Reconnect the negative battery cable. Start the engine and check for leaks. Make any necessary adjustments.

Prizm

1. Properly relieve the fuel system pressure.

2. Disconnect the battery cables, negative cable first. Remove the battery.

3. Scribe a hood hinge-to-hood outline. Disconnect the windshield washer fluid lines from the hood.

Support the hood and remove the hood hinge bolts and remove the hood.

4. Raise and safely support the vehicle.

5. Drain the engine oil and cooling system.

6. Remove the left and right splash shields.

7. Drain the transaxle fluid into a suitable container.

8. Lower the vehicle.

9. Disconnect the accelerator cable from the throttle lever. If equipped with an automatic transaxle, disconnect the throttle cable from the bracket.

10. Remove the radiator and cooling fan.

11. Remove the air cleaner.

12. Remove coolant reservoir support bracket.

13. Remove the windshield washer reservoir.

14. If equipped, remove the cruise control actuator and bracket as follows:

   a. Unclip the actuator cover and remove.

   b. Disconnect the accelerator cable from the actuator.

   c. Remove the 3 bolts and actuator.

   d. Remove the bolts and actuator bracket.

15. Label and disconnect the manifold absolute pressure sensor hose, brake booster hose and A/C solenoid vacuum valve hose.

16. Disconnect the A/C SV valve, MAP sensor, data link connector and A/C pressure switch wiring harnesses. Also disconnect the ground wires from the intake manifold and fenders.

17. Remove the bolts, then disconnect the 4 wiring connectors and remove the fuse and relay box.

18. Disconnect the hose from the evaporative emission canister. Loosen the canister bracket bolt and remove the canister.

19. Loosen the hose clamps and disconnect the heater hoses from the thermostat housing.

20. Disconnect the fuel feed line from the fuel rail. Discard the gaskets.

21. Disconnect the fuel return hose from fuel pressure regulator.

22. If equipped with a manual transaxle, remove the clutch slave cylinder, and disconnect the shift select and shift control cables from the transaxle.

23. If equipped with an automatic transaxle, disconnect the shift select cable from the transaxle.

24. Remove the left and right side sill plates.

25. Remove the knee bolster, glove box, center console trim bezel, and center console.

26. Remove the radio.

27. Remove the retaining bolt and set aside the left side floor carpet bracket.

28. If equipped, remove the 2 bolts from the cruise control module.

29. Disconnect the powertrain control module harnesses.

30. From the engine compartment, pull the wiring harness and grommet through the bulkhead.

31. If equipped, disconnect the A/C compressor belt and harness. Remove the compressor and suspend.

32. If equipped, remove the power steering and suspend without disconnecting the hoses.

33. Raise and safely support the vehicle.

34. Disconnect the oxygen sensor harness, then remove the front exhaust pipe.

35. Remove the left and right halfshafts. It is not necessary to remove the halfshafts from the steering knuckles.

36. Attach a suitable engine hoist to the engine/transaxle assembly.

37. Remove the front and rear transaxle mounts.

38. Lower the vehicle.

39. Remove the left transaxle mount.

40. Carefully lift the engine/transaxle assembly from the vehicle.

41. Remove the starter.

To install:

42. Install the starter.

43. Lower the engine/transaxle assembly into the vehicle.

44. Install the left transaxle mounts. Tighten the bolts to 41 ft. lbs. (56 Nm), the through-bolt to 64 ft. lbs. (87 Nm) and the reinforcement bolts to 15 ft. lbs. (21 Nm).

45. Raise and safely support the vehicle.

46. Install the front and rear transaxle mounts. Tighten the front mount bolts to 47 ft. lbs. (64 Nm) and the through-bolt to 64 ft. lbs. (87 Nm). Tighten the rear transaxle nuts to 42 ft. lbs. (52 Nm) and the through-bolt to 64 ft. lbs. (87 Nm).

47. Remove the engine hoist.

48. Raise and safely support the vehicle.

49. Install the left and right halfshafts.

50. Install the front exhaust pipe and connect the oxygen sensor.

51. If equipped, install the power steering pump and drive belt.

52. Lower the vehicle.

53. If equipped, install the A/C compressor and drive belt.

54. Insert the wiring harness through the bulkhead into the passenger compartment and install the grommet to the bulkhead.

55. Connect the wiring to the PCM.

56. If equipped, position the cruise control module and tighten the bolts to 44 inch lbs. (5 Nm).

57. Position the left side floor carpet bracket and tighten the bolt to 44 inch lbs. (5 Nm).

58. Install the radio, knee bolster, glove box, center console and trim bezel.

59. Install the left and right sill plates.

60. Connect the shift cable(s) to the transaxle.

61. If equipped with manual transaxle, install the clutch slave cylinder.

62. Connect the fuel return hose to the fuel pressure regulator.

63. Connect the fuel feed pipe to the fuel rail, using new gaskets. Tighten the bolt to 22 ft. lbs. (29 Nm).

64. Connect the heater hoses to the thermostat housing.

65. Install the EVAP canister to the bracket and tighten the bolt to 89 inch lbs. (10 Nm). Connect the hose to the canister.

66. Install the fuse and relay box. Connect the wiring connectors and tighten the bolts to 89 inch lbs. (10 Nm).

67. Connect the engine ground wires to the intake manifold and fenders. Connect the A/C pressure switch, DLC, MAP sensor and A/C SV valve wiring harnesses.

68. Connect the hoses to the A/C SV valve (if equipped), brake booster and MAP sensor.

69. If equipped, install the cruise control actuator and bracket as follows:

    a. Install the actuator bracket and tighten the bolts to 18 ft. lbs. (25 Nm).

    b. Install the actuator and tighten the bolts to 89 inch lbs. (10 Nm).

    c. Connect the accelerator cable to the actuator.

    d. Clip the actuator cover in place.

70. Install the windshield washer reservoir and secure with bolt. Connect the wiring harness and hose.

71. Install the coolant reservoir and tighten the bolts to 11 ft. lbs. (15 Nm).

72. Install the remaining components.

73. Install the battery and cables. Connect the positive cable first.

74. Install the hood.

75. Fill the engine with the proper type and quantity of oil.

76. Fill and bleed the cooling system.

77. Run the engine and check for leaks and proper operation.

Engine Mounts

REMOVAL AND INSTALLATION

Metro

Torque Rod

1. Support the engine using J-28467-A and J-28467-89, or equivalent.

2. Raise and safely support the vehicle.

3. Remove the retainer nut and torque rod-to-torque rod bracket through-bolt and washer.

4. Remove the torque rod bracket bolts and the torque rod bracket from the transaxle.

5. If equipped, remove the torque rod-to-torque rod brace bolt.

6. Remove the torque rod-to-frame through-bolt and remove from the engine.

To install:

7. Install the torque rod to the engine and loosely install the torque rod-to-frame through-bolt.

8. Install the torque rod bracket and bolts to the transaxle. Do not tighten the bolts at this time.

9. Install the torque rod-to-torque rod brace bolt, finger-tight only.

10. Install the torque rod-to-torque rod bracket through-bolt, washer and nut.

11. Tighten all torque rod nuts and bolts to 41 ft. lbs. (55 Nm).

12. Lower the vehicle.

13. Remove the engine support fixture.

Right Engine Mount

1. Support the engine using J-28467-A and J-28467-89, or equivalent.

2. Remove the 2 right engine mount-to-bracket bolts.

3. Remove the right engine mount through-bolt and mount from the bracket.

4. Remove the bracket bolts and right bracket from the vehicle.

To install:

5. Install the right bracket to the vehicle and bolts. Do not tighten at this time.

6. Install the right engine mount and through-bolt to the bracket. Install the 2 right engine mount-to-bracket bolts.

7. Tighten all mounting nuts and bolts to 41 ft. lbs. (55 Nm).

8. Remove the engine support.

Rear Engine Mount — Manual Transaxle

1. Support the engine using J-28467-A and J-28467-89, or equivalent.

2. Remove the rear engine mount nut and through-bolt.

3. Remove the 2 rear engine mount-to-bulkhead bolts, followed by the rear engine mount from the upper rear bracket.

4. Remove the retaining nuts and upper rear bracket from the lower bracket.

5. Remove the mounting bolts and bracket.

To install:

6. Install the lower mounting bracket and bolts loosely.

7. Install the upper rear bracket to the lower bracket. Install the bolts finger-tight.

8. Install the rear engine mount-to-upper rear bracket and install the bolts loosely.

1. Torque rod
2. Torque rod bracket
3. Torque rod brace

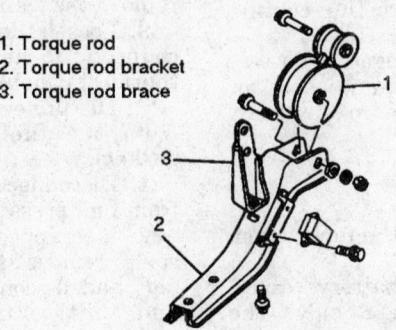

253483

Torque rod mounting components — Metro with 1.3L (VIN 9) engines

9. Install the rear through-bolt and install the nut.

10. Tighten all fasteners to 41 ft. lbs. (55 Nm).

11. Remove the engine support.

Rear Engine Mount — Automatic Transaxle

1. Support the engine using J-28467-A and J-28467-89 or equivalent.

2. Drain the engine coolant into a suitable container.

3. Raise and safely support the vehicle.

—————— CAUTION ——————

To avoid the danger of being burned, do not perform any service on the exhaust system while it is still hot. Service should only be performed after the system has cooled down.

4. Remove the front exhaust pipe/catalytic converter assembly from the exhaust manifold.

5. Remove the hose clamps, bolts and coolant return pipe from the engine.

6. Remove the nuts and rear engine bracket from the rear frame bracket.

7. Remove the bolts, nut and rear engine mount-to-engine bracket.

8. Remove the bolts and rear frame bracket from the vehicle frame.

9. Remove the nut and connecting bracket from the rear engine mount.

10. Remove the nut and rear engine mount-to-engine bracket from the rear mount.

To install:

11. Support the engine using J-28467-A and J-28467-89, or equivalent.

12. Install the rear mount bracket to the vehicle frame and install the bolts finger-tight only.

13. Install the rear mount-to-engine bracket to the engine. Install the bolts and nut finger-tight only.

14. Install the rear engine mount and nut to the bracket.

15. Install the connecting bracket to the rear mount and secure with nut finger-tight only.

16. Install the rear engine mount bracket to the rear frame bracket and install the mounting nuts.

17. Tighten all mounting nuts and bolts to 41 ft. lbs. (55 Nm).

18. Install the remaining rear frame bracket-to-vehicle bolt and torque the bolt to 41 ft. lbs. (55 Nm).

19. Reconnect the coolant return pipe to the engine. Secure with new hose clamps and bolts tightened to 15 ft. lbs. (20 Nm).

20. Install the front exhaust pipe/catalytic converter assembly. Secure with nuts and bolts. Torque the bolts to 41 ft. lbs. (55 Nm) and the nuts to 26 ft. lbs. (35 Nm).

21. Lower the vehicle.

22. Refill the engine cooling system.

23. Remove the engine support.

Prizm

Right Engine Mount

1. Disconnect the negative battery cable.

2. Remove the cruise control pump and VSV assembly, if equipped.

3. Support the engine with an engine hoist.

4. Remove the nut and bolt from the right engine mount bracket.

5. Remove the through-bolt from the engine mount.

6. Remove the mount from the vehicle.

To install:

7. Install the mount in the vehicle.

8. Install the through-bolt and the right engine mount nut and bolt. Tighten finger-tight only.

9. Tighten the through-bolt to 51 ft. lbs. (69 Nm) and the right engine mount bracket nut to 37 ft. lbs. (50 Nm), bolt to 89 ft. lbs. (121 Nm).

10. Remove the engine support fixture.

11. Install the cruise control pump and VSV assembly, if removed.

12. Connect the negative battery cable.

Torque Rod

1. Disconnect the negative battery cable.

2. Remove the support bar, if equipped.

3. Remove the right side undercover fixing bolt and center beam front fixing bolts.

4. Loosen the center beam rear fixing bolts.

5. Remove the torque rod fixing bolt from the center beam.

6. Remove the front mounting bracket fixing bolts.

7. Remove the torque rod.

To install:

8. Replace the torque rod.

9. Install the front mounting bracket. Tighten the bolt to 94 ft. lbs. (127 Nm), and the nut to 14 ft. lbs. (19 Nm).

10. Install the torque rod fixing bolt, and tighten to 51 ft. lbs. (69 Nm).

11. Install the center beam rear fixing bolt, and tighten to 56 ft. lbs. (76 Nm).

12. Install the center beam front fixing bolt, and tighten to 37 ft. lbs. (50 Nm).

13. Install the support bar, if equipped. Tighten the bolt 12 ft. lbs. (16 Nm).

14. Install the negative battery cable.

Rear Engine Mount

1. Disconnect the negative battery cable.

2. Support the engine with an engine hoist.

3. Raise and safely support the vehicle.

4. Remove the damper weight from the engine mount, if equipped.

5. Remove the rear engine mount bolts from the transmission housing.

6. Remove the mounting rubber bolt from the center beam.

7. Remove the mounting rubber assembly.

To install:

8. Replace the mounting rubber assembly.

9. Install the mounting rubber bolt, tighten to 76 ft. lbs. (103 Nm).

10. Install the rear engine mount bolt, tighten to 69 ft. lbs. (94 Nm).

11. Install the damper weight, if removed. Tighten the bolt to 37 ft. lbs. (50 Nm).

12. Lower the vehicle.

13. Remove the engine hoist.

14. Connect the negative battery cable.

Left Side Engine Mount

1. Remove the battery cables, negative first. Remove the battery and the bracket.

2. Remove the mounting rubber bracket bolts.

3. Remove the mounting rubber bolt.

4. Remove the mounting rubber.

To install:

5. Replace the mounting rubber

6. Install the mounting rubber bolt, and tighten to 51 ft. lbs. (69 Nm).

7. Install the mounting rubber bracket bolt, and tighten to 38 ft. lbs. (51 Nm).

8. Install the battery bracket and battery.

9. Install the positive battery cable, then the negative battery cable.

Cylinder Head

REMOVAL AND INSTALLATION

— CAUTION —

Fuel injection systems remain under pressure, even after the engine has been turned OFF. The fuel system pressure must be relieved before disconnecting any fuel lines. Failure to do so may result in fire and/or personal injury.

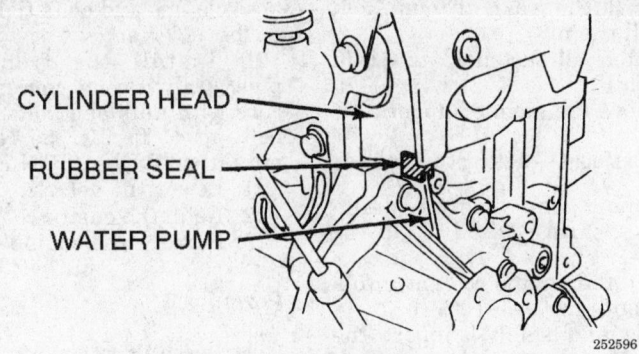

Installing the rubber seal — 1.0L (VIN 6) and 1.3L (VIN 9) engines

1.0L (VIN 6) and 1.3L (VIN 9) Engines

1. Properly relieve the fuel system pressure.
2. Disconnect the negative battery cable.
3. Drain the cooling system.
4. Remove the intake and exhaust manifolds.
5. Remove the timing belt.
6. Disconnect and label the spark plug wires. Be sure to pull firmly on the wire boot, not the plug wire.
7. Remove the distributor and case from the cylinder head.
8. Remove the cylinder head cover.
9. Loosen and remove the cylinder head bolts by reversing the tightening sequence.
10. Remove the cylinder head from the engine. Discard the cylinder head gasket and rubber seal.
11. Clean all gasket mating surfaces.
12. Check the cylinder head flatness using a straightedge and feeler gauge. Check the surface across the combustion chambers, on each side of the combustion chambers, and diagonally. If distortion exceeds 0.002 in. (0.05mm) at any location, the cylinder head should be resurfaced.

To install:

13. Install the cylinder head gasket with the TOP indicator facing upward and toward the crankshaft pulley.
14. Install the cylinder head. Lubricate the cylinder head bolts with clean engine oil and install them finger-tight.
15. Tighten the cylinder head bolts in 3 even stages, following the tightening sequence, to 54 ft. lbs. (73 Nm).
16. Install the rubber seal between the water pump and the cylinder head.
17. Install the cylinder head cover and secure with nuts and new seal washers. Tighten nuts to 44 inch lbs. (5 Nm).
18. Install the timing belt.
19. Install the distributor.

20. Connect the spark plug wires.
21. Install the intake manifold gasket, using a new gasket.
22. Install the exhaust manifold using a new gasket.
23. Refill the cooling system.
24. Connect the negative battery cable.
25. Start the engine and allow it to reach normal operating temperature. Check for leaks and adjust the ignition timing, if necessary.

1.6L (VIN 6) and 1.8L (VIN 8) Engines

1. Properly relieve the fuel system pressure.

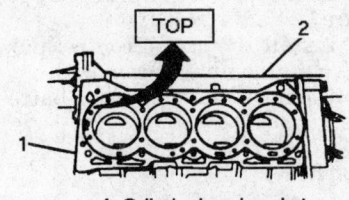

1. Cylinder head gasket
2. Cylinder block

Installing the cylinder head gasket — 1.0L (VIN 6) and 1.3L (VIN 9) engines

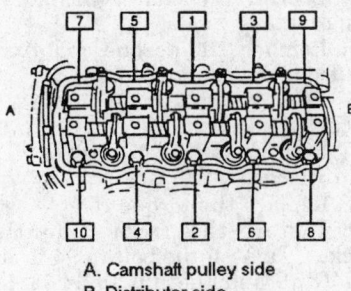

A. Camshaft pulley side
B. Distributor side

Cylinder head bolt torque sequence — 1.0L (VIN 6) and 1.3L (VIN 9) engines

2. Disconnect the negative battery cable.
3. Drain the engine coolant into a suitable container.
4. Raise and safely support the vehicle.
5. Remove the right side engine splash shield.
6. Lower the vehicle.
7. Remove the accelerator cable from the bracket and throttle body.
8. Disconnect the vacuum hose from the evaporative emissions canister.
9. Unfasten the intake air temperature sensor harness from the air cleaner cap. Loosen the air cleaner hose clamp and disconnect 4 clips to remove the cover.
10. Remove the alternator.
11. Disconnect the spark plug wires and remove the distributor case.
12. Disconnect the oxygen sensor harness, then detach the bolts and nuts and remove the front exhaust pipe.
13. Remove the exhaust manifold.
14. Remove the thermostat housing.
15. Unfasten the ground strap connector.
16. Disconnect the manifold absolute pressure sensor harness and A/C pressure switch harness, if equipped.
17. Remove the engine wiring harness from the right side fender.
18. Label and disconnect all of the hoses from the intake chamber.
19. If equipped, label and unfasten the A/C solenoid vacuum valve electrical connector.
20. Disconnect the hoses and harnesses from the EGR SV valve and vacuum modulator, then remove.
21. Remove the engine wire clamp and the intake manifold brace.
22. Disconnect the fuel return line from the pressure regulator. Remove the air pipe from the intake manifold.
23. Remove the air pipe and EGR valve.

24. Remove the throttle body assembly.

25. Label and disconnect the PCV hoses from the cylinder head cover.

26. Label and disconnect the vacuum hose from the pressure regulator. Remove the intake chamber cover.

27. Disconnect the fuel line from the fuel rail. Label and disconnect the electrical harness from the fuel injectors. Remove the fuel rail with the injectors attached.

28. Label and disconnect the A/C compressor electrical harness, if equipped. Label and disconnect the oil pressure switch, crankshaft position sensor and fan thermostat electrical harness. Remove the engine wiring harness cover. Disconnect the engine ground strap and remove the engine wire harness retainer.

29. Remove the intake manifold.

30. If equipped with cruise control, remove the actuator cover, actuator mounting bolts and actuator bracket.

31. Support the engine using a suitable jack.

32. Remove the 2 nuts from the right-hand engine mount studs.

33. If equipped, remove the 2-piece A/C line bracket from the engine mount.

34. Remove the engine mount reinforcement bracket and mount, if equipped.

─────── WARNING ───────

It may be necessary to lower the engine to gain clearance for engine mount removal. If equipped, be careful not to bend the A/C line too much or a leak may develop.

35. Remove the timing belt.

36. Remove the alternator mounting bracket.

37. Remove the oil level dipstick and tube.

38. Remove the coolant inlet pipe.

39. Remove the camshafts.

40. Uniformly loosen and remove the 10 cylinder head bolts in several passes, in reverse order of tightening.

─────── WARNING ───────

Cylinder head warpage or cracking could result from removing the bolts in the incorrect order.

41. Lift off the cylinder head assembly from the dowels.

To install:

42. Clean all gasket mating surfaces, using care not to damage the aluminum components.

43. Check the engine block and cylinder head mating surfaces, using a feeler gauge and straightedge. Check along all 4 edges and diagonally across. If distortion exceeds 0.002 in. (0.05mm) at any location, resurface the cylinder head.

44. Install a new gasket, then lower the cylinder head to the block. Make sure the dowel pins are aligned.

45. Coat the cylinder head bolt threads and the underside of each bolt head with clean engine oil and install finger-tight.

NOTE: The cylinder head bolts are in lengths of 3.54 in. (90mm) and 4.25 in. (108mm). The 3.54 in. (90mm) bolts are to be installed in the intake side of the cylinder head. The 4.25 in. (108mm) bolts are to be installed in the exhaust manifold side of the cylinder head.

46. Tighten the cylinder head bolts, in sequence, to 22 ft. lbs. (29 Nm).

47. Turn each bolt, in sequence, an additional 90 degrees. Again turn each bolt, in sequence, an additional 90 degrees.

48. Install the camshafts.

49. Install the coolant inlet pipe. Tighten the nuts to 11 ft. lbs. (15 Nm).

50. Install the engine oil level dipstick tube and tighten the bolt to 84 inch lbs. (9.5 Nm).

51. Install the alternator bracket and tighten the bolts to 19 ft. lbs. (26 Nm).

52. Install the timing belt.

53. Position the engine mount, and locate the A/C pipe, if equipped. Tighten the bolts closet to the mount to 47 ft. lbs. (64 Nm), and the bolts farthest away from the mount to 18 ft. lbs. (25 Nm).

54. Install the rubber insulator to the mount and tighten the bolt to 18 ft. lbs. (25 Nm).

55. Install the reinforcement bracket and tighten the nut and bolt to 18 ft. lbs. (25 Nm).

56. Attach the A/C pipe bracket, if equipped, to the engine mount and tighten the bolt to 89 inch lbs. (10 Nm).

57. Install the nuts to the engine mount studs and tighten to 38 ft. lbs. (52 Nm).

58. If equipped, install the cruise control actuator bracket, actuator and cover.

59. Install the intake manifold and ground strap.

60. Install the engine wire harness with the retainer and tighten the bolts and nuts to 89 inch lbs. (10 Nm). Secure the ground strap with the bolt, and tighten to 89 inch lbs. (10 Nm). Connect the fan thermostat, oil pressure switch and, if equipped, A/C compressor harnesses. Install the engine wire cover and tighten the bolts to 89 inch lbs. (10 Nm).

61. Install the fuel rail and injectors. Tighten the fuel rail bolts to 11 ft. lbs. (15 Nm) and connect the fuel injector electrical connectors.

62. Connect the fuel line to the fuel rail, using new gaskets. Tighten the union bolt to 22 ft. lbs. (29 Nm).

63. Install the intake chamber cover using a new gasket. Tighten the nuts and bolts to 14 ft. lbs. (19 Nm). Connect the vacuum hose to the pressure regulator.

64. Connect the PCV hoses to the cylinder head cover.

65. Install the throttle body assembly, using a new gasket. Tighten the bolts and nuts to 16 ft. lbs. (22 Nm). Connect the throttle position sensor harness. Install the accelerator cable bracket and tighten the bolts to 98 inch lbs. (11 Nm).

66. Install the EGR valve, using a new gasket. Tighten the nuts to 115 inch lbs. (13 Nm).

67. Connect the fuel return hose to the pressure regulator.

68. Install the intake manifold brace. Tighten the 12mm bolts to 14

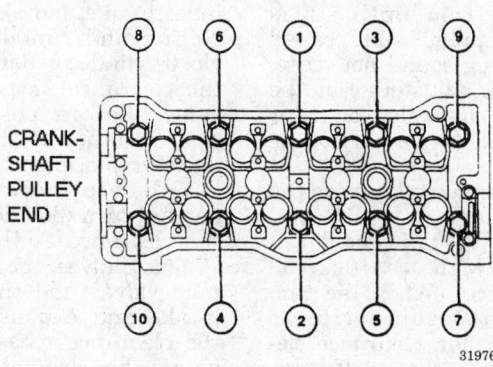

CRANK-SHAFT PULLEY END

319769

Cylinder head bolt torque sequence — 1.6L (VIN 6) and 1.8L (VIN 8) engines

ft. lbs. (19 Nm) and the 14mm bolts to 29 ft. lbs. (39 Nm).

69. Install the engine wire clamp and the EGR SV valve and vacuum modulator. Tighten the EGR SV valve bolt to 115 inch lbs. (13 Nm). Connect the hose to the EGR SV valve pipe and connect the electrical connector.

70. Install the remaining components.

71. Connect the negative battery cable.

72. Fill and bleed cooling system.

73. Run the engine and check for leaks. Check the ignition timing. Road test and check for proper operation.

Lash Adjusters

BLEEDING

1.0L (VIN 6) and 1.3L (VIN 9) Engines

Never disassemble the Hydraulic Valve Lash (HVL) adjuster. Do not apply force to the body of the adjuster or the oil in the high pressure chamber. Submerge removed adjusters in clean engine oil until installation. If the adjusters are left in air, place them with their adjuster bodies facing down. Do not place them on their side or with the adjuster bodies facing up.

Do not turn the camshaft or start the engine for about half an hour after reinstalling the HVL adjusters. This may cause valve-to piston interference.

When the engine is started, if air is trapped in the HVL adjuster, the valve may make a tapping sound when the engine is operated. Run the engine at 2000 rpm until air is purged and the tapping sound ceases. Should the tapping sound not cease, one or more HVL adjusters could be defective. If the defective adjuster cannot be located by ear, check as follows:

1. Stop the engine and remove the cylinder head cover.

2. Push the HVL adjuster downward by hand (with less than 33 lbs./147 N of force). When the cam lobe is not on the adjuster to be checked, inspect for clearance between the cam and adjuster. If clearance is found, the adjuster is defective and must be replaced.

Valve Lash

ADJUSTMENT

1.0L (VIN 6) and 1.3 (VIN 9) Engines

Hydraulic Valve Lash (HVL) adjusters, located between the camshaft and valve stems, are used to adjust the valve clearance to zero lash automatically at all times. Adjustment is not required.

1.6L (VIN 6) and 1.8L (VIN 8) Engines

NOTE: The following tools and part are needed for this procedure; Valve clearance adjustment tool set J-39871 or equivalent, a feeler gauge, a 0–1 inch micrometer and a selection of valve adjustment shims. Check and adjust the valve clearance with the engine cold.

1. Disconnect the negative battery cable.

2. Disconnect the ignition wires, then remove the spark plugs.

3. Remove the retaining nuts and remove the cylinder head cover.

4. Turn the crankshaft until the piston in No. 1 cylinder is at Top Dead Center (TDC) on the compression stroke. Align the groove in the crankshaft pulley with the **0** mark on the timing belt cover. Make sure No. 1 cylinder lash adjusters are loose and those on No. 4 are tight. If not, turn the crankshaft pulley one full revolution (360 degrees) and again align the mark on the crankshaft pulley.

5. The intake valve clearance should be 0.006–0.010 in. (0.15–0.25mm) and the exhaust valve clearance should be 0.010–0.014 in. (0.25–0.35mm).

6. Using a feeler gauge, measure the clearance between the camshaft and valve lifter shim at the No. 1 cylinder intake and exhaust valves, the No. 2 cylinder intake valves and the No. 3 cylinder exhaust valves. Record the clearance measurements for all valves that are not within specification, in order to determine the required replacement shims.

7. Rotate the crankshaft pulley one full turn (360 degrees) and check the clearance at the No. 2 cylinder exhaust valves, the No. 3 cylinder intake valves and the No. 4 cylinder intake and exhaust valves. Record the clearance measurements for all valves that are not within specification, in order to determine the required replacement shims.

8. For valves requiring adjustment, proceed as follows:

   a. Make sure the base of the camshaft lobe is directly over the valve (camshaft lobe pointing away from the valve).

   b. Insert tool J-3987-1 or equivalent, between the camshaft and lifter adjustment shim, to compress the valve spring and push the lifter down.

   c. Insert tool J-39871-2 or equivalent, between the camshaft and the lifter, to hold the lifter away from the camshaft. Position the bottom edge of the tool on the lifter.

   d. Using a small screwdriver and a magnet, remove the adjustment shim from the top of the lifter.

   e. Use the micrometer to measure the thickness of the removed shim. Determine the thickness of the new shim using the formula below. For the purposes of the following formula, T = Thickness of the shim removed; A = Valve clearance measured; N = Thickness of the required new shim.

   • For the intake camshaft valves: N = T + (A - 0.008 in. or 0.20mm)

   • For the exhaust camshaft valves: N = T + (A - 0.010 in. or 0.25mm)

   f. Select a shim closest to the calculated thickness. Shims are available in 16 sizes, in increments of 0.002 in. (0.050mm), from 0.1004 in. (2.55mm) to 0.1299 in. (3.30mm).

   g. Install the shim on the valve lifter and remove tool J-39871-2 or equivalent. Recheck the valve clearance.

9. After all valve clearances have been checked and/or adjusted, reinstall the cylinder head cover, using a new gasket. Tighten the nuts to 53 inch lbs. (6 Nm).

10. Install the spark plugs and connect the ignition wires.

11. Connect the negative battery cable.

12. Run the engine and check operation.

Rocker Arms and Shafts

REMOVAL AND INSTALLATION

1.3L (VIN 9) Engine

1. Disconnect the negative battery cable.

2. If equipped, remove the A/C compressor drive belt, then the re-

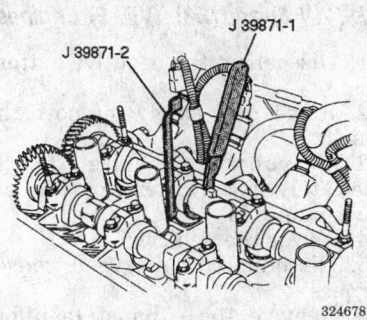

Using the valve clearance adjustment tool set to hold the lifter — 1.6L (VIN 6) and 1.8L (VIN 8) engines

move the compressor and mounting bracket. Place off to the side.

3. Disconnect the spark plug wires at the plugs.

4. Remove the air cleaner and cylinder head cover.

5. Remove the rocker shaft retaining bolts. Lift out the exhaust and intake rocker arm shafts, with the springs and rocker arms attached.

6. If necessary, remove the rocker arms, washers and springs from the rocker shafts. Note the order in which the components are removed; they must be re-assembled in the same order and position.

NOTE: The intake and exhaust rocker arm shafts are NOT the same. The 2 can be distinguished by looking at the ends of the shafts, which are different. Install the intake rocker arm shaft with the stepped end toward the distributor side.

7. Inspect the rockers arms, shafts and lash adjusters for wear and/or damage; replace as necessary.

To install:

8. If disassembled earlier, assemble the springs, washers and rocker arms on to the rocker shafts.

9. Lubricate the rocker arms and shafts with clean engine oil. Position the intake and exhaust rocker arm shafts in their original positions. Secure with the retaining bolts. Torque the shaft bolts, in the correct sequence, to 97 inch lbs. (11 Nm).

10. Apply a small amount of silicone sealer to the corners of the cylinder head cover gasket mating surface. Install a new head cover gasket and position the cylinder head cover to the head.

11. Install new seal washers and the nuts. Tighten the head cover nuts to 44 inch lbs. (5 Nm).

12. Install the remaining components.

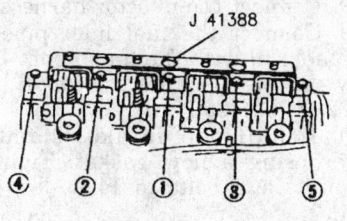

Rocker arm shaft retaining bolts torque sequence — 1.3L (VIN 9) engines

13. Reconnect the negative battery cable. Run the engine and check for leaks and proper engine operation.

Intake Manifold

REMOVAL AND INSTALLATION

— **CAUTION** —

Fuel injection systems remain under pressure, even after the engine has been turned OFF. The fuel system pressure must be relieved before disconnecting any fuel lines. Failure to do so may result in fire and/or personal injury.

1.0L (VIN 6) and 1.3L (VIN 9) Engines

1. Relieve the fuel system pressure.

2. Disconnect the negative battery cable.

3. Drain the cooling system.

4. Remove the air cleaner.

5. Disconnect the electrical harnesses to the intake manifold.

6. Disconnect the fuel return and feed hoses from the throttle body.

7. Disconnect the water hoses from the throttle body and the intake manifold.

8. Disconnect all vacuum hoses to the intake manifold.

9. Disconnect the PCV hose from the valve cover.

10. Disconnect the accelerator cable from the throttle body.

11. Remove the intake manifold fasteners and remove with the throttle body attached.

12. Remove and discard the intake manifold gasket. Clean all gasket mating surfaces.

To install:

13. Install the intake manifold to the cylinder head, using a new gasket. Install the fasteners and tighten evenly, to 17 ft. lbs. (23 Nm), starting

in the center working towards the ends.

14. Connect the PCV hose to the valve cover.

15. Connect the vacuum and coolant hoses.

16. Install the fuel feed and return hoses.

17. Reconnect the electrical harnesses to the proper locations.

18. Install the accelerator cable and adjust as follows:

a. Make sure the cable locknut and adjusting nut are loose.

b. Adjust the cable end-play to 0.4–0.6 in. (10–15mm), then tighten the lock and adjusting nuts.

c. If equipped with automatic transaxle, loosen the cable locknut and adjusting nut at the cable retaining bracket on the transaxle. Adjust the cable end-play to 0.4–0.6 in. (10–15mm), then tighten the locknut and adjusting nut.

19. Install the air cleaner.

20. Refill the cooling system and reconnect the negative battery cable.

21. Start the engine. Check for leaks.

1.6L (VIN 6) and 1.8L (VIN 9) Engines

1. Relieve the fuel system pressure.

2. Disconnect the negative battery cable.

3. Label and disconnect the following hoses:

a. Vacuum sensor hose from the fuel filter

b. Brake booster vacuum hose

c. If equipped, A/C vacuum hoses

4. If equipped, disconnect the A/C actuator electrical connector.

5. If equipped with an EGR SV valve, proceed as follows:

a. Disconnect the EGR SV electrical connector and EGR temperature sensor

b. Disconnect the vacuum hose from the EGR valve

c. Remove bolt and EGR SV valve

6. Remove the engine wire clamp.

7. Remove the intake manifold brace.

8. Disconnect the fuel return hose from the pressure regulator.

9. Remove the mounting bolt and nut and the air pipe from the intake manifold.

10. Remove the throttle body as follows:

a. Disconnect the throttle position sensor harness.

b. Loosen the locknut and adjusting nut, then remove the accelerator cable from the throttle lever.

c. If equipped with an automatic transaxle, loosen the locknut and adjusting nut and remove the throttle cable from the bracket.

d. Remove the 2 bolts and the accelerator bracket.

e. Remove the 2 bolts, 2 nuts and the throttle body with the gasket.

11. Remove the fuel inlet hose clamp bolt, intake chamber brace and gasket.

12. Disconnect the PCV hoses from the valve cover.

13. Disconnect the vacuum hose from the fuel pressure regulator.

14. Remove the bolts, nuts and the intake chamber cover with gasket.

15. Remove the union bolt and gaskets and disconnect the fuel inlet pipe from the fuel rail.

16. Disconnect the fuel injector harnesses.

17. Remove the bolts and remove the fuel rail with the injectors attached.

WARNING

Be careful not to drop the fuel injectors when removing the fuel rail. Damage could result.

---

18. Remove the insulators and spacers from the cylinder head.

19. Remove the nuts and bolt and the engine wiring harness retainer.

20. Remove the bolts, ground strap and nuts. Remove the intake manifold and gasket.

21. Clean all gasket mating surfaces.

To install:

22. Install the intake manifold using a new gasket.

23. Install the nuts, bolts and ground strap. On 1993–95 vehicles, tighten the nuts and bolts to 14 ft. lbs. (19 Nm). On 1996–97 vehicles, tighten the nuts to 9 ft. lbs. (13 Nm) and bolts to 14 ft. lbs. (19 Nm).

24. Install the engine wiring harness retainer, and tighten the nuts and bolts to 89 inch lbs. (10 Nm).

25. Install new insulators and spacers to the cylinder head.

26. Install the fuel rail with the injectors. Loosely install the retaining bolts.

NOTE: Check the fuel injectors for smooth rotation. If the injectors do not rotate smoothly, replace the O-rings and recheck. Make sure the electrical connector is pointed upward.

27. Tighten the fuel rail retaining bolts to 11 ft. lbs. (15 Nm).

28. Connect the injector harnesses.

29. Connect the fuel inlet pipe to the fuel rail with the union bolt. Use new gaskets. Tighten the union bolt to 22 ft. lbs. (29 Nm).

30. Install the intake chamber cover using a new gasket. Tighten the nuts and bolts to 14 ft. lbs. (19 Nm).

31. Connect the vacuum hose to the pressure regulator. Connect the PCV hoses to the valve cover.

32. Install the intake chamber brace with a new gasket. Tighten the bolt to 21 ft. lbs. (28 Nm).

33. Install the fuel inlet hose clamp bolt and tighten to 89 inch lbs. (10 Nm).

34. Install the throttle body using a new gasket. Tighten the nuts and bolts to 17 ft. lbs. (23 Nm).

35. Install the accelerator cable bracket and tighten the bolts to 11 ft. lbs. (15 Nm).

36. Connect the throttle cable, if equipped, and accelerator cable. Tighten the locknut and adjusting nut to 11 ft. lbs. (15 Nm).

37. If equipped, install the EGR valve, vacuum modulator and vacuum hose assembly, using new gaskets. Tighten the EGR valve-to-intake manifold bolts and nuts to 115 inch lbs. (13 Nm) and tighten the EGR valve-to-EGR pipe bolts and nuts to 43 ft. lbs. (59 Nm). Connect the 2 vacuum hoses.

38. Connect the fuel return hose to the pressure regulator.

39. Install the intake manifold brace and engine wire clamps.

40. If equipped, install the EGR SV valve and tighten the bolt to 115 inch lbs. (13 Nm). Connect the EGR SV valve and EGR TP sensors. Connect the vacuum hose to the EGR valve.

41. Connect the remaining components.

42. Connect the negative battery cable.

43. Run the engine and check for leaks.

Exhaust Manifold

REMOVAL AND INSTALLATION

CAUTION

Do not service any part of the exhaust system while it is still hot. Allow the system to cool down before performing any service.

1.0L (VIN 6) and 1.3L (VIN 9) Engines

1. Disconnect the negative battery cable.

2. Raise and safely support the vehicle.

3. Disconnect the front pipe/catalytic converter from the exhaust manifold.

4. Lower the vehicle.

5. Disconnect the oxygen sensor harness.

6. Remove the exhaust manifold heat shield.

7. Remove the manifold retaining bolts and engine hanger. Remove the manifold and gasket from the cylinder head.

8. Clean all gasket mating surfaces.

To install:

9. Install a new gasket, followed by the exhaust manifold, and engine hanger. Install the exhaust manifold bolts and nuts, torque evenly, to 17 ft. lbs. (23 Nm), working from the center towards the ends.

10. Install the manifold heat shield and tighten the retaining bolts to 11 ft. lbs. (15 Nm).

11. Raise and safely support the vehicle.

12. Connect the exhaust pipe to the manifold with a new seal. Tighten the bolts to 30–43 ft. lbs. (40–60 Nm).

13. Lower the vehicle.

14. Reconnect the oxygen sensor harness.

15. Reconnect the negative battery cable. Start the engine and check for leaks.

1.6L (VIN 6) and 1.8L (VIN 8) Engines

1. Disconnect the negative battery cable.

2. Disconnect the oxygen sensor and, if equipped, sub-oxygen sensor connectors.

3. Disconnect the exhaust pipe from the exhaust manifold or, if equipped, from the Warm Up Three Way Catalytic Converter (WU TWC).

4. Remove the 5 bolts and the upper heat shield.

5. Remove the bolts and the exhaust manifold brace.

6. Remove the nuts, then the exhaust manifold and gasket. If equipped, remove the manifold with the WU TWC attached.

7. Remove the 3 bolts and the lower heat insulator from the exhaust manifold.

8. If equipped, remove the 4 nuts and 2 WU TWC heat insulators.

9. If equipped, remove the 2 bolts, 2 nuts and WU TWC from the exhaust manifold.

To install:

10. Clean all gasket mating surfaces.

11. If equipped, install the bolts, nuts and WU TWC to the exhaust manifold. Tighten the bolts and nuts to 80 inch lbs. (9 Nm).

12. If equipped, install the nuts and WU TWC heat insulators. Tighten the nuts to 80 inch lbs. (9 Nm).

13. Install the 3 bolts and the lower heat insulator to the manifold and tighten to 80 inch lbs. (9 Nm).

14. Install a new gasket and the exhaust manifold to the cylinder head. Tighten the nuts evenly, to 25 ft. lbs. (34 Nm), working from the center towards the ends.

15. Install the exhaust manifold brace and tighten the bolts to 29 ft. lbs. (39 Nm).

16. Install the upper heat insulator and tighten the bolts to 80 inch lbs. (9 Nm).

17. Using a new gasket, connect the exhaust pipe to the exhaust manifold or WU TWC, as required. Install the nuts and tighten to 32 ft. lbs. (43 Nm).

18. Connect the oxygen sensor harness and, if equipped, sub-oxygen sensor connector.

19. Connect the negative battery cable. Run the engine and check for exhaust leaks.

Front Crankshaft Seal

REMOVAL AND INSTALLATION

1.0L (VIN 6) and 1.3L (VIN 9) Engines

1. Disconnect the negative battery cable.

2. Raise and safely support the vehicle.

3. Drain the engine oil.

4. Remove the timing belt and crankshaft sprocket.

5. Remove the oil pan and strainer.

6. Remove the oil pump bolts and pump assembly. Identify the bolts so they can be reinstalled in their proper locations.

7. Remove the crankshaft oil seal from the oil pump using a suitable removal tool.

To install:

8. Install a new crankshaft oil seal into the oil pump body using a suitable seal installer tool. Lubricate the lip of the new seal with clean engine oil.

9. Using new gaskets, install the oil pump on the engine. Use special tool J-34853 to aid installation of the oil seal over the crankshaft without

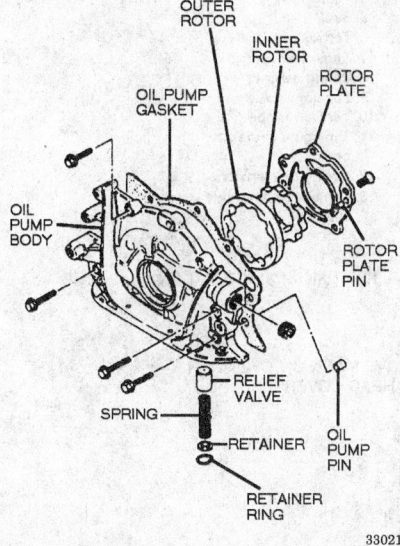

Disassembled view of the oil pump — 1.0L (VIN 6) and 1.3L (VIN 9) engines

330214

damaging the seal. The lip of the seal must **NOT** be turned out.

NOTE: Take care not to damage the front seal on the crankshaft snout.

10. Apply Loctite® pipe sealant or equivalent to the pump bolts and install. Tighten to 97 inch lbs. (11 Nm).

NOTE: After tightening, the oil pump gasket may bulge out. Trim the edge with a knife, making sure the edge is smooth with the cylinder block.

11. Install the rubber seal between the oil and coolant pumps.

12. Install the crankshaft sprocket. With the crankshaft locked, tighten the bolt to 81 ft. lbs. (110 Nm).

13. Install the oil strainer and pan.

14. Install the timing belt.

15. Lower the vehicle and fill the engine with clean engine oil.

16. Connect the negative battery cable, start the engine and check for leaks.

1.6L (VIN 6) and 1.8L (VIN 9) Engines

1. Disconnect the negative battery cable.

2. Remove the timing belt and crankshaft sprocket.

3. Cut off the oil seal lip.

WARNING

When removing the oil seal, be careful not to damage the crankshaft or the oil pump housing.

4. Carefully pry out the oil seal from the pump housing.

To install:

5. Apply clean engine oil to the new oil seal and the new oil seal with multi-purpose grease.

6. Install the new oil seal to the oil pump using a suitable seal driver.

7. Install the timing belt and crankshaft sprocket.

8. Connect the negative battery cable. Check the engine oil level.

9. Run the engine and check for leaks.

Timing Belt, Sprockets, Tensioner and Front Cover

REMOVAL AND INSTALLATION

NOTE: Timing belts must always be completely free of dirt, grease, fluids and lubricants. This includes the sprockets and contact surfaces on which the belt rides. The belt must never be crimped, twisted or bent. Never use tools to pry or wedge the belt.

1.0L (VIN 6) and 1.3 (VIN 9) Engines

1. Disconnect the negative battery cable.

2. Raise and safely support the vehicle.

3. Remove the clips and right side splash shield.

4. Remove the lower alternator cover plate.

5. Remove the alternator drive belt and, if equipped, A/C drive belt.

6. Remove the water pump pulley.

NOTE: It is not necessary to remove the crankshaft timing sprocket bolt (center bolt) to remove the crankshaft pulley.

7. Remove the crankshaft pulley bolts and crankshaft pulley.

8. Remove the retaining bolts and nut from the timing belt outside cover.

9. Remove the timing belt outside cover.

10. Turn the crankshaft to align the timing marks. The mark on the crankshaft sprocket should align with the arrow mark on the oil pump housing. The mark on the camshaft sprocket should align with the **V** mark on the timing belt inner cover or cylinder head cover.

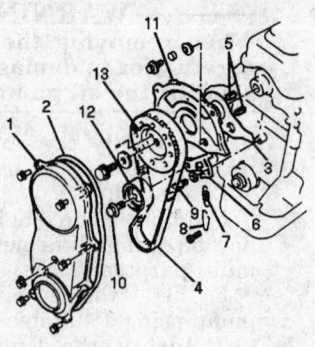

1. Timing belt cover
2. Outer timing belt cover gasket
3. Timing belt inner seal
4. Timing belt
5. Seal
6. Tensioner plate
7. Tensioner spring
8. Spring damper
9. Tensioner stud
10. Tensioner bolt
11. Timing belt inner cover
12. Timing belt tensioner
13. Camshaft timing gear sprocket

315394

Timing belt and related components — 1.0L (VIN 6) and 1.3L (VIN 9) engines

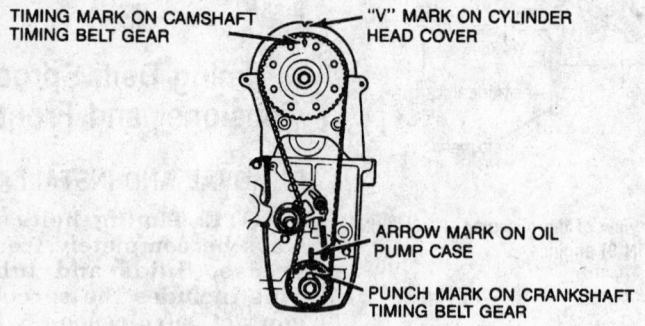

TIMING MARK ON CAMSHAFT TIMING BELT GEAR

"V" MARK ON CYLINDER HEAD COVER

ARROW MARK ON OIL PUMP CASE

PUNCH MARK ON CRANKSHAFT TIMING BELT GEAR

315398

Aligning the timing marks — Metro with 1.0L (VIN 6) and 1.3L (VIN 9) engines

11. If the timing belt is to be re-used, mark the direction of rotation on the belt.

12. Remove the timing belt tensioner, tensioner plate, tensioner spring, spring damper and timing belt.

NOTE: Never turn the camshaft or crankshaft independently after the timing belt has been removed. Interference may occur among the pistons and valves, and parts may be damaged.

13. Inspect the timing belt for wear or cracks and replace as necessary. Check the tensioner for smooth rotation.

14. If the timing belt sprockets are to be removed, proceed as follows:

a. Using a 0.39 in. (10mm) rod inserted into the camshaft, hold the camshaft and remove the retaining bolt and camshaft sprocket.

——— WARNING ———

Be careful not to damage the cylinder head or cylinder head cover mating surfaces. Place a clean shop cloth between the rod

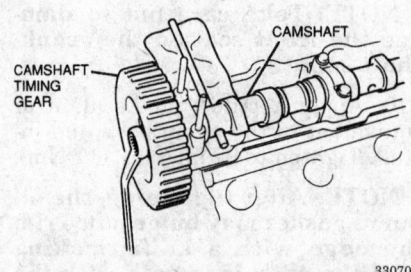

CAMSHAFT TIMING GEAR

CAMSHAFT

330702

Holding the camshaft with the rod to remove the timing belt sprocket — Metro with 1.0L (VIN 6) engines

and cylinder head. Do not bump the rod hard against the cylinder head when loosening.

b. Raise and safely support the vehicle.

c. If equipped with manual transaxle, lock the crankshaft in position by inserting a suitable flat bladed tool into the hole in the bottom of the bell housing to engage the flywheel teeth.

d. If equipped with automatic transaxle, lock the crankshaft in position by inserting a suitable flat bladed tool between the flywheel teeth and against the engine block.

e. Remove the crankshaft sprocket bolt and crankshaft sprocket.

To install:

15. If the timing belt sprockets were removed, proceed as follows:

a. Install the crankshaft sprocket, aligning the keyway. Lock the crankshaft in place and tighten the crankshaft sprocket bolt to 81 ft. lbs. (110 Nm).

b. Lower the vehicle.

c. Install the camshaft sprocket and retaining bolt. Lock the camshaft in place using the rod, and tighten the bolt to 44 ft. lbs. (60 Nm). Remove the locking rod.

16. Install the tensioner plate to the tensioner.

17. Insert the lug of the tensioner plate into the hole of the tensioner.

18. Install the tensioner, tensioner plate and spring. Do not fully tighten the tensioner bolt and stud at this time.

19. Move the tensioner plate in a counterclockwise direction. This should cause the tensioner to move in the same direction. If it does not, remove the tensioner and tensioner plate and reinsert the tensioner plate lug in the timing plate tensioner hole.

20. Check that the camshaft timing marks are aligned. If not, align the 2 marks by turning the camshaft.

21. Check that the punch mark on the crankshaft timing belt sprocket is aligned with the arrow mark on the oil pump case. If not, align the 2 marks by turning the crankshaft.

22. With the timing marks aligned, install the timing belt on the 2 sprockets. If the old belt is being re-used, be sure to install it running in the same direction of rotation.

23. Install the tensioner spring and spring damper. Turn the timing belt 2 rotations clockwise after installing the tensioner spring and damper to remove any belt slack. Tighten the tensioner stud to 8 ft. lbs. (11 Nm), then the tensioner bolt to 20 ft. lbs. (27 Nm).

NOTE: Confirm that both sets of timing marks are aligned properly.

24. Using a new seal, install the timing belt cover and tighten the bolts and nut to 97 inch lbs. (11 Nm).

25. Install the crankshaft pulley. Fit the keyway on the pulley to the crankshaft timing belt sprocket and tighten the bolts to 8–12 ft. lbs. (11–16 Nm).

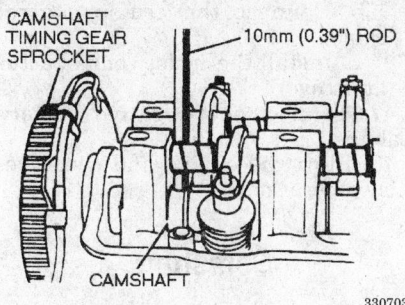

Locking the camshaft to remove the camshaft timing sprocket — Metro with 1.3L (VIN 9) engines

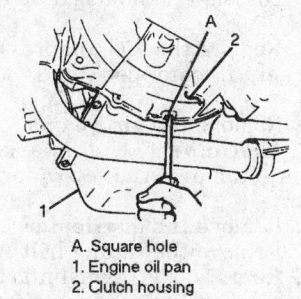

A. Square hole
1. Engine oil pan
2. Clutch housing

315508

Locking the crankshaft — Metro with manual transaxles

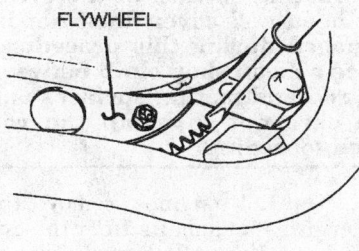

FLYWHEEL

315509

Locking the crankshaft — Metro with automatic transaxles

26. Install the alternator drive belt and, if equipped, A/C drive belt.

27. Install the lower alternator cover plate. Tighten the bolts to 89 inch lbs. (10 Nm).

28. Install the clips and right side splash shield.

29. Lower the vehicle and connect the negative battery cable.

30. Run the engine. Check for leaks.

1.6L (VIN 6) and 1.8L (VIN 8) Engines

1. Disconnect the negative battery cable.

2. Remove the windshield washer reservoir from the engine compartment.

3. If equipped with cruise control, proceed as follows:

   a. Remove the cruise control actuator cover.

   b. Disconnect the cruise control harnesses.

   c. Disconnect the control cable.

   d. Remove the bolts and actuator from the vehicle.

4. Raise and safely support the vehicle.

5. Remove the right front wheel.

6. Remove the bolts and plastic clips and right-front wheel housing.

7. Remove the alternator/water pump drive belt.

8. Lower the vehicle.

9. If equipped with A/C, proceed as follows:

   a. Remove the A/C compressor drive belt.

   b. Disconnect the compressor harness.

   c. Remove the bolts and compressor, without disconnecting the refrigerant lines. Suspend the compressor out of the way.

   d. Remove the compressor mounting bracket.

10. Remove the power steering pump drive belt.

11. Disconnect the wiring from the alternator and oil pressure switch.

12. Remove the engine wiring harness cover.

13. Remove the wiring harness from the cylinder head cover.

14. Disconnect the ignition wires from the spark plugs, then the spark plugs.

15. Remove the PCV hoses from the valve cover.

16. Remove the cap nuts, the seal washers and the cylinder head cover with gasket.

17. Turn the crankshaft to align the timing mark on the crankshaft pulley at **0**, setting the piston in No. 1 cylinder at Top Dead Center (TDC) on the compression stroke. Check that the valve lash adjusters on the No. 1 cylinder are loose. If not, turn the crankshaft pulley 1 complete revolution (360 degrees).

18. Remove the engine ground wire from the right fender apron.

19. Install a suitable support under the engine and remove the engine mount.

20. Remove the water pump pulley.

21. Remove the crankshaft pulley using a suitable puller.

22. Remove the 9 retaining bolts and the timing belt covers.

23. Slide the timing belt guide from the crankshaft.

24. Make sure the timing belt sprockets are properly aligned.

NOTE: Do not turn the crankshaft or camshaft independently after removal of the timing belt; binding or damage to engine components could result. If the timing belt is to be reused, mark the belt with an arrow showing the direction of engine revolution.

25. Remove the timing belt tensioner bolt, tensioner and tension spring.

26. Remove the timing belt from the sprockets. Inspect the timing belt for cracked or damaged teeth. Replace as necessary.

——— WARNING ———
Do not bend, twist or turn the timing belt.

27. If the camshaft sprocket is to be removed, hold the camshaft stationary using a wrench positioned on the hexagon cast into the camshaft, and remove the sprocket retaining bolt and sprocket.

——— WARNING ———
Be careful not to damage the cylinder head when holding the camshaft in place.

28. If the crankshaft sprocket is to be removed, pry it from the crankshaft using 2 flat-bladed prybars.

To install:

29. Align the camshaft key with the groove on the sprocket and slide the sprocket on. Hold the camshaft with the wrench at the wrench head portion of the camshaft, and tighten the camshaft timing sprocket bolt to 43 ft. lbs. (59 Nm).

30. Make sure the sprocket is still properly aligned.

31. Install the crankshaft timing sprocket. Align the crankshaft key with the groove on the sprocket and slide on.

32. Reinstall the timing belt tensioner and the tension spring. Pry the tensioner to the left as far as it will go and temporarily tighten the retaining bolt.

33. Install the timing belt. If installing the old belt, observe the matchmarks made during removal.

34. Loosen the retaining bolt for the timing belt tensioner and allow it to tension the belt.

35. Temporarily install the crankshaft pulley bolt and turn the crankshaft clockwise 2 full revolutions.

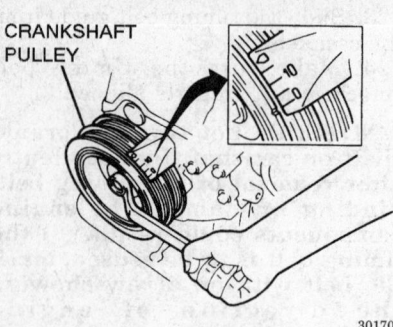

Setting No. 1 cylinder at TDC — Prizm with 1.6L (VIN 6) and 1.8L (VIN 8) Engines

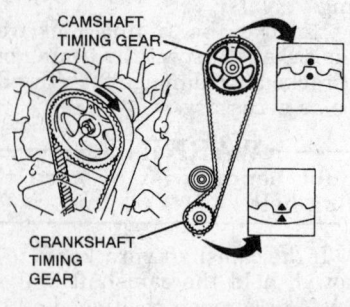

Marking the timing belt for installation reference — Prizm with 1.6L (VIN 6) and 1.8L (VIN 8) Engines

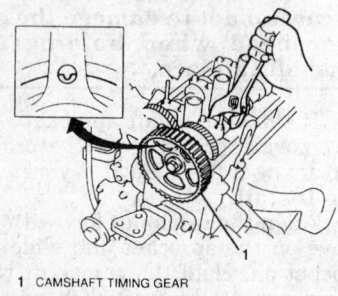

1 CAMSHAFT TIMING GEAR

301707

Aligning the camshaft timing sprocket timing marks — Prizm with 1.6L (VIN 6) and 1.8L (VIN 8) Engines

Make sure each timing mark realigns exactly.

36. Tighten the timing belt tensioner bolt to 27 ft. lbs. (37 Nm).

37. Measure the timing belt deflection. Correct deflection should be 0.20–0.24 in. (5–6mm) at 4 pounds (20 Nm) of pressure. If the deflection is not correct, adjust with the timing belt tensioner.

38. Install the timing belt guide, with the cup side facing outward.

39. Install the timing belt covers. Installing the bottom first. Tighten

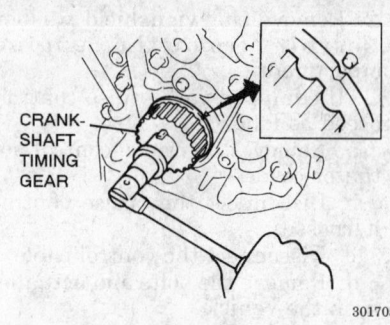

Aligning the crankshaft timing sprocket timing marks — Prizm with 1.6L (VIN 6) and 1.8L (VIN 8) Engines

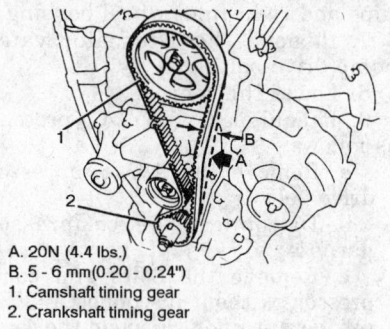

A. 20N (4.4 lbs.)
B. 5 - 6 mm(0.20 - 0.24")
1. Camshaft timing gear
2. Crankshaft timing gear

301706

Measuring timing belt deflection — Prizm with 1.6L (VIN 6) and 1.8L (VIN 8) Engines

the 9 cover bolts to 62 inch lbs. (7 Nm).

40. Install the crankshaft pulley after aligning the pulley key with the slot on the pulley. Hold the pulley with tool J-8614-01 or equivalent, and tighten the pulley bolt to 87 ft. lbs. (118 Nm).

41. Temporarily install the water pump pulley.

42. Raise and safely support the vehicle.

43. Install the engine mount.

44. Install or connect the remaining components.

45. If equipped with A/C, proceed as follows:

a. Install the compressor mounting bracket and tighten the bolts to 35 ft. lbs. (47 Nm).

b. Install the compressor and tighten the bolts to 18 ft. lbs. (25 Nm).

c. Connect the compressor wiring connector.

d. Install the compressor drive belt and adjust the tension.

46. If equipped with cruise control, proceed as follows:

a. Install the cruise control actuator and tighten the bolts to 89 inch lbs. (10 Nm).

b. Connect the cruise control cable.

c. Install the cruise control actuator cover.

47. Connect the negative battery cable.

48. Start the engine and check vehicle operation.

Camshaft

REMOVAL AND INSTALLATION

1.0L (VIN 6) and 1.3L (VIN 9) Engine

1. Disconnect the negative battery cable.

2. Remove the A/C compressor and bracket, if equipped, and position aside.

3. Remove the air cleaner.

4. Disconnect the spark plug wires. Remove the cylinder head cover.

5. Remove the distributor.

6. Remove the timing belt.

7. Remove the camshaft timing belt sprocket. Lock the camshaft with a 0.39 in. (10mm) rod inserted into the hole in the camshaft, before loosening the sprocket retaining bolt.

— WARNING —
The mating surface of the cylinder head and cover must not be damaged during this procedure. Place a clean shop cloth between the rod and mating surfaces and use care not to bump the rod when loosening.

8. On 1.3L engines, remove the rocker arms and shafts from the cylinder head. Keep all parts in order so they can be reinstalled in their original locations. The intake and exhaust rocker arm shafts are different. Be sure to identify them as they are removed.

9. Turn the crankshaft until the crankshaft sprocket timing mark is 60 degrees to the left of the arrow mark on the oil pump case.

10. Remove the camshaft housings from the cylinder head.

11. Remove the camshaft from the cylinder head.

12. On 1.3L engines remove the camshaft from the cylinder head by removing it from the flywheel end.

13. Inspect the camshaft journals and lobes for wear and/or damage and replace, as necessary.

14. Using a micrometer, measure the camshaft lobe height. If the camshaft lobe height is below the minimum limit, replace the

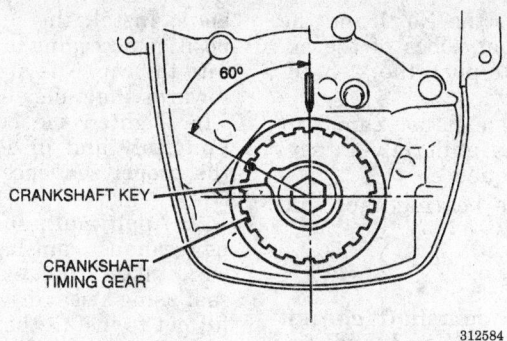

Position the crankshaft as shown before removing the camshaft — Metro with 1.0L (VIN 6) engines

camshaft. The minimum camshaft specifications are as follows:

- **Metro XFi** — 1.5562 in. (39.528mm)
- **Metro Standard and LSi** — 1.5872 in. (40.315mm)

15. On 1.3L engines, inspect the camshaft for excessive wear or scoring and replace, if necessary. Using a micrometer, measure the cam lobe height; if it is less than 1.49575 in. (38.036mm), replace the camshaft.

To install:

16. Fill the oil passage in the cylinder head with clean engine oil. Pour engine oil through the camshaft journal oil holes and check that engine oil comes out from the oil holes in the HVL adjuster bores.

17. Install the camshaft to the cylinder head. After applying engine oil to the camshaft journal and all around the cam, position the camshaft so the camshaft timing sprocket pin hole in camshaft is at the lower position.

18. On 1.0L engines, install the camshaft housings to the camshaft and the cylinder head as follows:

a. Apply clean engine oil to the sliding surface of each housing against the camshaft journal.

b. Apply silicone sealant to the mating surface of the No. 1 and No. 3 housing which will mate with the cylinder head.

c. There are marks provided on each camshaft housing indicating the position and direction for installation. Install the housing as indicated by the marks.

d. Camshaft housing No. 1 is installed first. It retains the camshaft in the proper position and thrust direction. Apply clean engine oil to the retaining bolts and install, but do not tighten fully.

e. Install the remaining camshaft housings and bolts.

f. Tighten the bolts evenly, repeating the tightening sequence

3–4 times. Tightened to 8 ft. lbs. (11 Nm).

NOTE: On 1.3L engines, exhaust and intake rocker shafts are different. To distinguish between the 2, the end dimensions are different. Install the intake rocker arm shaft with the stepped end facing the distributor side.

19. On 1.3L engines, install the intake rocker arm shaft and arms. Torque the rocker arm shaft retaining bolts to 97 inch lbs. (11 Nm).

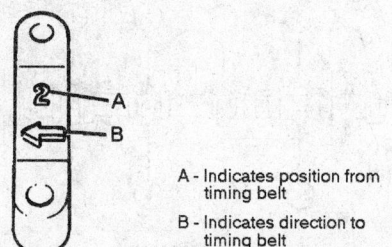

A - Indicates position from timing belt

B - Indicates direction to timing belt

Directional markings for the camshaft housings — Metro with 1.0L (VIN 6) engines

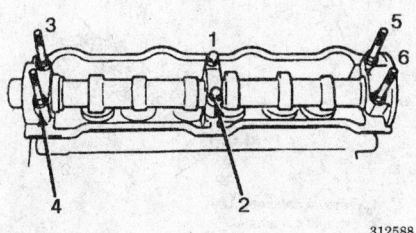

Torque sequence for the camshaft housing bolts — Metro with 1.0L (VIN 6) engines

20. On 1.3L engines, install the exhaust rocker arm shaft and arms. Torque the rocker arm shaft retaining bolts to 97 inch lbs. (11 Nm).

21. Install the camshaft oil seal. After applying engine oil to the seal lip, press-fit the camshaft oil seal until the seal surface until flush with the housing.

22. Install the camshaft timing belt sprocket after installing the dowel pin. Tighten the retaining bolt to 44 ft. lbs. (60 Nm).

23. Using a new valve cover gasket, apply a small amount of silicone sealant to the corners of the cylinder gasket, install the valve cover the cover nuts. Torque the cylinder head cover nuts to 44 in. lbs. (5 Nm).

24. Install the timing belt.

25. Install the distributor to the cylinder head. Connect the spark plug wires.

26. Install the A/C compressor and bracket, if equipped.

27. Install the air cleaner and connect the negative battery cable.

28. Start the engine and check. Adjust the ignition timing.

NOTE: When the engine is started, if air is trapped in the HVL adjuster, the valve may make a tapping sound when the engine is operated after the HVL adjuster is installed. In such a case, run the engine at 2000 rpm until the air is purged and the tapping sound ceases.

1.6L (VIN 6) and 1.8L (VIN 8) Engines

1. Disconnect the negative battery cable.

— **CAUTION** —

Work must be started after 90 seconds from the time the ignition switch is turned to the LOCK position and the negative (-) battery cable is disconnected.

2. Disconnect the spark plug wires.

3. Remove the valve cover.

4. Remove the timing belt.

5. Set the exhaust camshaft so the knock pin is slightly above the cylinder head (10 o'clock position). This angle allows the No. 1 and No. 3 cylinder cam lobes of the intake camshaft to push the lash adjusters evenly.

6. Remove the 2 bolts and the front bearing cap of the intake camshaft.

7. Secure the intake camshaft end gear to the sub gear with a service

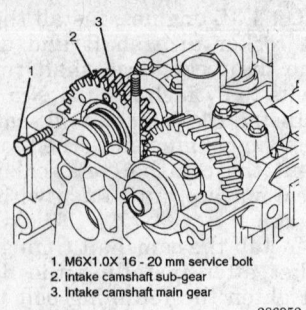

1. M6X1.0X 16 - 20 mm service bolt
2. Intake camshaft sub-gear
3. Intake camshaft main gear

286952

Fastening the sub-gear to the main gear — Prizm with 1.6L (VIN 6) and 1.8L (VIN 8) Engines

bolt. The service bolt should match the following specifications:

• Thread diameter: 6.0mm
• Thread pitch: 1.0mm
• Bolt length: 16mm

8. Uniformly loosen each intake camshaft bearing cap bolt in several passes in the proper sequence.

—————— **WARNING** ——————

The camshaft must be held level while it is being removed. If not, the portion of the cylinder head receiving the thrust may become damaged. This could cause the camshaft to bind or break.

9. Remove the bearing caps and intake camshaft.

NOTE: If the camshaft cannot be removed straight and level, install and retighten the No. 3 bearing cap. Alternately loosen the bolts on the bearing cap a little at a time while pulling upwards on the camshaft gear. DO NOT attempt to pry or force the cam loose.

10. Turn the exhaust camshaft approximately 105 degrees, so the guide pin is just past the 5 O'clock position.

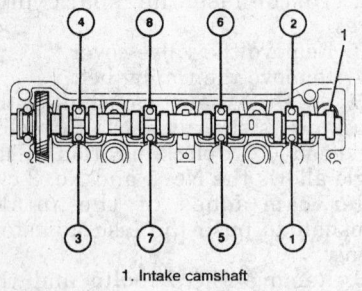

1. Intake camshaft

286950

Intake camshaft bearing cap removal sequence — Prizm with 1.6L (VIN 6) and 1.8L (VIN 8) Engines

This angle allows the No. 1 and the No. 3 cylinder cam lobes of the exhaust camshaft to push the lash adjusters evenly.

11. Loosen the exhaust camshaft bearing cap bolts uniformly in several passes in sequence.

12. Remove the bearing caps and exhaust camshaft.

NOTE: If the camshaft cannot be removed straight and level, install and retighten the No. 3 bearing cap. Alternately loosen the bolts on the bearing cap a little at a time while pulling upwards on the camshaft gear. DO NOT attempt to pry or force the cam loose.

To install:

13. Apply multi-purpose grease to the thrust portion of the camshaft.

14. Place the exhaust camshaft on the cylinder head so the cam lobes press evenly on the lash adjusters for cylinders Nos. 1 and 3. This will place the guide pin on the camshaft slightly counter clockwise at about 5 O'clock.

15. Apply a light coat of clean engine oil to the camshaft bearing cap

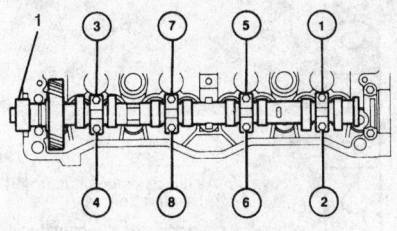

1. Exhaust camshaft

286949

Exhaust camshaft bearing cap removal sequence — Prizm with 1.6L (VIN 6) and 1.8L (VIN 8) Engines

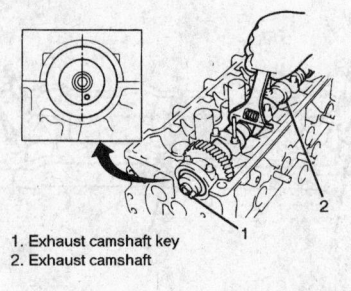

1. Exhaust camshaft key
2. Exhaust camshaft

286951

Positioning the exhaust camshaft for removal — Prizm with 1.6L (VIN 6) and 1.8L (VIN 8) Engines

bolts. Install the 5 bearing caps in position according to the number cast into the cap. The arrow should point towards the pulley end of the motor.

16. Tighten the bearing cap bolts uniformly and in several passes in the proper sequence to 9 ft. lbs. (13 Nm)

17. Apply multi-purpose grease to a new exhaust camshaft oil seal.

18. Install the exhaust camshaft seal using a seal driver. Be very careful not to install the seal on a slant.

19. Set the intake camshaft so the guide pin is slightly above the cylinder head.

20. Apply multi-purpose grease to the thrust portion of the intake camshaft.

21. Hold the intake camshaft next to the exhaust camshaft and engage the gears by matching the alignment marks.

NOTE: DO NOT use the TDC timing marks for the timing belt.

22. Keeping the gears engaged, roll the intake camshaft down and into the bearing journals. This angle allows the No. 1 and the No. 3 cylinder cam lobes of the camshaft to push the lash adjusters evenly.

23. Apply a light coat of clean engine oil to the camshaft bearing cap bolts and install the bearing caps. Observe the numbers on each cap and make certain the arrows point to the pulley end of the motor.

24. Uniformly tighten each of the bearing cap bolts in several passes in the proper sequence. Torque each bolt to 9 ft. lbs. (13 Nm)

NOTE: If the No. 1 bearing cap does not fit properly, push the camshaft gear backwards by prying apart the cylinder head and camshaft gear with a suitable tool.

25. Turn the exhaust camshaft clockwise, and set it with the guide pin facing upward. Check that the timing marks of the camshaft gears are aligned. The camshaft assembly installation marks should now be in the 12 O'clock position.

26. Secure the exhaust camshaft and install the timing belt pulley. Tighten the bolt to 43 ft. lbs. (59 Nm).

27. Install the timing belt.

28. Install the spark plug wires and connect the negative battery cable.

29. Start the engine, check for leaks. Check the ignition timing.

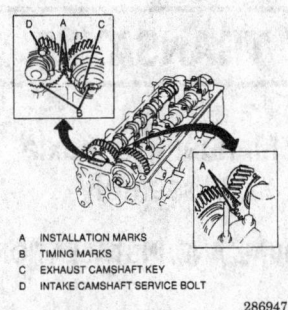

1. Exhaust camshaft

286948

Exhaust camshaft bearing cap tightening sequence — Prizm with 1.6L (VIN 6) and 1.8L (VIN 8) Engines

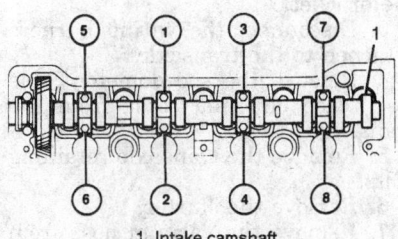

A INSTALLATION MARKS
B TIMING MARKS
C EXHAUST CAMSHAFT KEY
D INTAKE CAMSHAFT SERVICE BOLT

286947

Engaging the intake camshaft with the exhaust camshaft — Prizm with 1.6L (VIN 6) and 1.8L (VIN 8) Engines

1. Intake camshaft

286946

Intake camshaft bearing cap tightening sequence — Prizm with 1.6L (VIN 6) and 1.8L (VIN 8) Engines

Piston and Connecting Rod

POSITIONING

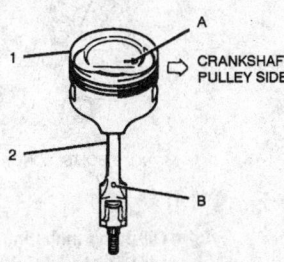

A. Arrow mark
B. Oil hole (oil hole should come on intake side)
1. Piston
2. Connecting rod

330776

Piston and connecting rod positioning — Metro with 1.0L (VIN 6) engines

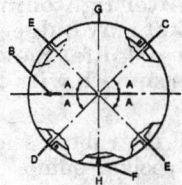

A. 45o
B. Piston arrow mark
C. First ring end gap
D. Second ring end gap
E. Oil ring end gaps
F. Oil ring spacer gap
G. Intake side
H. Exhaust side

322493

Piston ring end gap positioning — 1.0L (VIN 6) and 1.3L (VIN 9) engines

---

ENGINE LUBRICATION

Oil Pan

REMOVAL AND INSTALLATION

1. Disconnect the negative battery cable.
2. Raise and safely support the vehicle.
3. Drain the engine oil.
4. Remove the flywheel dust cover.
5. On Prizm, remove the splash shields.
6. Separate the support hanger from the catalytic converter. Remove the front exhaust pipe between the exhaust manifold and the resonator/tail pipe assembly.

7. On Prizm, perform the following;
    a. Remove the bolts and lower the front end of the center support.
    b. Remove the lower engine reinforcement brace-to-engine bolts and remove the brace.
8. Remove the crankshaft position sensor.
9. On Prizm, remove the 3 brace-to-transaxle bolts. Remove the 14 bolts from the powertrain reinforcement brace. Remove the 6 Torx® bolts and powertrain reinforcement brace from the vehicle.

NOTE: When removing the powertrain reinforcement brace, do not pry on the oil pump body or the rear main seal retainer. Be careful not to damage the contact surfaces of the powertrain reinforcement brace, cylinder block, oil pump or the rear main seal retainer.

10. Remove the oil pan bolts, nuts and oil pan from the vehicle.
11. Remove the oil pump strainer.
12. Clean all gasket material from the mating surfaces.
    To install:
13. Install the oil pump strainer with a new seal. Secure with bolt and tighten to 97 inch lbs. (11 Nm).
14. Apply a continuous bead of RTV sealant to the engine oil pan and install. Tighten the bolts and nuts to 97 inch lbs. (11 Nm) starting in the center and working outward.
15. Install the exhaust pipe with a new seal to the manifold and secure with bolts. Do not fully tighten the bolts.
16. On Prizm, apply GM 1050026 gasket paste or equivalent, to the powertrain reinforcement brace-to-cylinder block mating surface.
17. Install the powertrain reinforcement brace. Tighten the Torx® bolts to 12 ft. lbs. (16 Nm). Tighten the remaining bolts to 69 inch lbs. (7.8 Nm). Install the 3 brace-to-transaxle bolts and tighten to 17 ft. lbs. (23 Nm).
18. On Prizm, perform the following;
    a. Install the lower engine reinforcement brace to the vehicle and tighten the bolts to 47 ft. lbs. (64 Nm).
    b. Install the center support and tighten the bolts to 45 ft. lbs. (61 Nm).
19. Install the front pipe to the resonator/muffler/tail pipe assembly using a new gasket and secure with nuts.
20. Install the support hanger to the catalytic converter. If equipped,

torque the hanger bolts to 11 ft. lbs. (15 Nm).

21. Torque the front exhaust pipe-to-exhaust manifold bolts to 37 ft. lbs. (50 Nm). Torque the front exhaust pipe-to-resonator/muffler/tail pipe assembly nuts to 26 ft. lbs. (35 Nm).

22. Install the flywheel dust cover.

23. Lower the vehicle and fill the engine with oil.

24. Reconnect the negative battery cable, start the engine and check for leaks.

Oil Pump

REMOVAL AND INSTALLATION

1. Disconnect the negative battery cable.

2. Raise and safely support the vehicle.

3. Drain the engine oil.

4. Remove the water pump belt and pulley, alternator belt, body and bracket.

5. Remove the air conditioning compressor and mounting bracket, if equipped.

6. On Prizm, if equipped, unclip the crankshaft position sensor electrical connector from the dipstick tube. Remove the bolt and the dipstick tube from the oil pump. Remove the dipstick tube O-ring from the oil pump.

7. Remove the timing belt and crankshaft pulley.

8. Disconnect the engine oil level gauge.

9. Remove the crankshaft timing belt sprocket.

10. Remove the oil pan and strainer assembly.

11. Remove the oil pump bolts and the oil pump assembly. Identify the bolts so they can be reinstalled in their proper locations. If necessary, gently tap on the back side of the oil pump with a plastic or rubber faced hammer to free it from the cylinder block.

12. Remove the crankshaft oil seal from the oil pump.

13. Remove the pump rotor plate pins from the oil pump. Clean the gasket mating surfaces thoroughly.

To install:

14. Install the pump rotor plate pins into the pump body.

15. Install a new crankshaft oil seal using an oil/grease seal installer tool.

16. Lubricate the oil pump with fresh engine oil.

17. Using new gaskets, install the oil pump. Install the crankshaft seal.

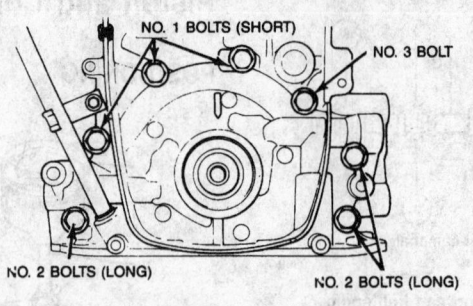

Oil pump mounting bolt identification — Metro with 1.0L (VIN 6) and 1.3L (VIN 9) engines

18. Apply Loctite® pipe sealant or equivalent to the oil pump bolts and install. On Metro, tighten the bolts to 97 inch lbs. (11 Nm) or on Prizm, tighten the bolts to 16 ft. lbs. (21 Nm).

NOTE: After tightening, the oil pump gasket may bulge. Trim the edge with a knife, making sure the edge is smooth with the cylinder block.

19. Install the rubber seal between the oil and coolant pump.

20. Install the crankshaft belt gear and guide. Tighten the pulley bolt to 81 ft. lbs. (110 Nm).

21. On Prizm, install a new dipstick tube O-ring to the oil pump. Install the dipstick tube and secure. Tighten the bolt to 84 inch lbs. (9.5 Nm). If equipped, clip the crankshaft position sensor electrical connector to the dipstick tube.

22. Install the engine oil level gauge.

23. Install the oil strainer and oil pan.

24. Install the timing belt.

25. Install the remaining components.

26. Lower the vehicle and fill the engine with clean engine oil.

27. Connect the negative battery cable. Start the engine and check for leaks.

TRANSAXLE

Manual Transaxle Assembly

REMOVAL AND INSTALLATION

Metro

1. Disconnect the negative battery cable and ground strap at the transaxle.

2. Loosen the clutch cable adjusting nuts, then remove the cable from the bracket.

3. Disconnect the wiring harness attached to the transaxle.

4. Remove the speedometer cable boot, case clip and cable from the transaxle.

5. Remove the transaxle retaining bolts.

6. Remove the starter.

7. Remove the vacuum hose from the pressure sensor.

8. Install a suitable engine support.

9. Raise and safely support the vehicle. Drain the transaxle oil.

10. Remove the gear shift control shaft bolt and nut, then detach the control shaft from the gear shift shaft.

11. Remove the extension rod nut and rod with washers.

12. Remove the exhaust pipe front and rear flange bolts.

13. Remove the clutch housing lower plate.

14. Remove the wheels.

15. Remove both the tie rod ends.

16. Remove both halfshafts at the transaxle.

17. Support the transaxle with a suitable jack and remove the transaxle retaining bolts and nuts.

18. Remove the 2 rear engine mounting bolts.

19. Remove the bolts and nuts from the transaxle left hand bracket and remove.

20. Lower the transaxle with the engine attached. Carefully pull the transaxle straight out toward the left side to disconnect the input shaft from the clutch cover. Lower and remove the transaxle assembly.

To install:

21. While the transaxle is being raised into the correct position, install the right side halfshaft.

22. Install the transaxle to the engine block using the retainer nuts and bolts. Install the left side bracket with the bolts and nuts. Torque them to 37 ft. lbs. (50 Nm).

23. Install the rear engine mounting nuts and torque to 37 ft. lbs. (50 Nm).

24. Lower the transaxle supporting jack. Torque the transaxle to engine bolt and nut to 37 ft. lbs. (50 Nm).

25. Install the left halfshaft to the transaxle.

26. Install both tie rod ends.

27. Install the front wheels.

28. Install the clutch housing lower plate.

29. Install the exhaust pipe front and rear flange nuts.

30. Install the extension rod nut and washers. Torque the nut to 24 ft. lbs. (33 Nm).

31. Install the control shaft to gear shift and the gear shift control shaft bolt and nut. Torque the gear shaft bolt and nut to 13–15 ft. lbs. (18–20 Nm).

32. Refill the transaxle with the recommended lubricant.

33. Lower the vehicle.

34. Remove the engine support fixture.

35. Install the vacuum hose to the pressure sensor.

36. Install the starter motor.

37. Install the transaxle retaining bolts. Torque the retaining bolts to 37 ft. lbs. (50 Nm).

38. Install the speedometer cable, case clip and speedometer cable boot.

39. Install the clutch cable into the bracket, Adjust the cable free-play.

40. Connect the negative battery cable and ground strap to the transaxle.

Prizm

1. Install engine support fixture J-28467 or equivalent.

2. Disconnect the battery cables, negative cable first. Remove the battery and tray.

3. Remove the air cleaner.

4. Disconnect the the reverse light and ground strap at the transaxle.

5. Remove the actuator mounting bolts and actuator line bracket.

6. Remove the shift cable retainers and end clips.

7. Remove the shift cables and place out of the way.

8. Remove the left transaxle mount cover and bracket. Remove the through-bolt from the mount.

9. Remove the 2 upper transaxle-to-engine bolts.

10. Remove the upper starter bolt. Remove the speedometer cable or disconnect the vehicle speed sensor (VSS), as applicable.

11. Raise and safely support the vehicle.

12. Remove the splash shields. Drain the transaxle oil.

13. Disconnect the starter wires. Remove the bottom starter bolt and starter.

14. Remove both halfshafts.

15. Remove the bolts holding the center crossmember to the radiator support.

16. For 1993–95 models, remove the following:

  a. Remove the front, center and rear mounting bolts.

  b. Remove the bolts holding the center crossmember to the main crossmember.

  c. Remove the exhaust hanger bracket nuts and the exhaust hanger.

17. For 1996–97 models, remove the following:

  a. Remove the bolts holding the center crossmember to the radiator support.

  b. Remove the front mount shield.

  c. Remove the front mounting bolts and rear mounting nuts.

  d. Remove the exhaust pipe support bolts.

18. Support the main crossmember. Remove the bolts holding the main crossmember to the body.

19. Remove the bolts holding the lower control arm brackets to the body.

―――――― **CAUTION** ――――――
The crossmembers are loose and free to fall. Make sure they are properly supported.

20. Slowly lower the main crossmember while holding onto the center crossmember.

21. Remove the through-bolt and mount, at the front transaxle mount

22. Remove the front mounting bracket from the transaxle.

23. Remove the center mount from the transaxle.

24. Remove the inspection cover bolt.

25. Remove the 2 lower transaxle bracket-to-transaxle bolts.

26. Lower the vehicle.

27. Remove the transaxle mount.

28. Raise and safely support the vehicle.

29. Support the transaxle with a suitable jack.

30. Remove the lower front and rear bolts holding the transaxle to the engine.

31. Remove the transaxle assembly from the engine and carefully lower.

To install:

32. Raise the transaxle into position, making sure the input shaft aligns with the clutch splines.

33. Install the lower front and rear bolts holding the transaxle to the engine. Do not tighten them fully.

34. Remove the jack from under the transaxle.

35. Install the starter and wiring.

36. Lower the vehicle.

37. Install the left transaxle mounting bolts.

38. Raise the transaxle and install the through-bolt. Tighten the through-bolt to 64 ft. lbs. (87 Nm). Tighten the transaxle mounting bolts to 45 ft. lbs. (61 Nm).

39. Install the mount cover and bracket. Tighten the cover bolts to 45 ft. lbs. (61 Nm).

40. Install the upper transaxle to engine bolts and tighten them to 34 ft. lbs. (46 Nm).

41. Connect the speedometer cable or VSS connector.

42. Position the shift cables into the brackets and connect the cable retainers and end clips.

43. Install the 2 actuator mounting bolts, then install the actuator line bracket and bolt. Tighten the line and mounting bolts to 15 ft. lbs. (20 Nm).

44. Connect the ground strap and reverse lights harnesses.

45. Install the air cleaner.

46. Raise and safely support the vehicle.

47. Tighten the lower transaxle mount bolts to 45 ft. lbs. (61 Nm).

48. Install the center mount and bolts. Tighten them to 45 ft. lbs. (61 Nm). Install the inspection cover bolt and tighten it to 45 ft. lbs. (61 Nm).

49. Install the front mount bracket on the transaxle.

50. Install the front mount and through-bolt loosely.

NOTE: When installing the front mount, weight on the mount must go toward the transaxle.

51. Position the central crossmember over the center and rear transaxle mount studs. Secure with nuts.

52. Loosely install the bolts holding the center crossmember to the radiator support.

53. Loosely install the front mount bolts.

54. Raise the main crossmember into position over the rear mount studs and align the bolts. Install the rear mount nuts loosely. Install the main crossmember to underbody bolts loosely.

55. Install the lower control arm bracket bolts loosely.

56. Loosely install the bolts holding the center crossmember to the main crossmember.

57. Install the exhaust hanger bracket and hardware.

58. Tighten the components below to the correct torque specification:

 a. Main crossmember to underbody bolts: 152 ft. lbs. (206 Nm).

 b. Lower control arm bolts: 94 ft. lbs. (127 Nm) for 1993–95 models or 161 ft. lbs. (218 Nm) for 1996–97 models.

 c. Center crossmember to radiator support bolts: 45 ft. lbs. (61 Nm).

 d. Front, center and rear mount bolts: 45 ft. lbs. (61 Nm).

 e. Exhaust bracket bolts: 115 inch lbs. (13 Nm).

 f. Front mount through-bolt: 64 ft. lbs. (87 Nm).

59. Install the remaining components.

60. Install the halfshafts.

61. Install the battery tray and battery.

62. Connect the battery cables to the battery, negative cable last.

Clutch Assembly

REMOVAL AND INSTALLATION

1. Disconnect the negative battery cable.

2. Raise and safely support the vehicle.

3. Remove the transaxle assembly.

4. Install pilot tool J-37761 or equivalent, into the pilot bearing to support the clutch.

5. Matchmark the clutch cover and flywheel.

6. On Prizm, unfasten the release fork bearing clips. Withdraw the release bearing hub and release bearing. Remove the release fork and support.

7. Loosen the clutch cover-to-flywheel bolts evenly until the spring pressure is released.

8. Remove the clutch disc and clutch cover.

To install:

9. Clean the flywheel mating surfaces of all oil, grease and metal deposits.

10. Check the diaphragm spring and pressure plate for wear or damage. If the spring or plate is excessively worn, replace the clutch cover assembly.

11. Check the pilot bearing for smooth operation. If the bearing does not spin freely, replace it.

12. Inspect the disc, pressure plate and flywheel for damage and wear using a caliper to measure depth and width and a dial indicator to measure runout.

 a. The minimum clutch disc rivet head depth is 0.012 in. (0.3mm).

 b. The maximum clutch disc runout is 0.031 in. (0.8mm).

 c. The maximum pressure plate spring depth is 0.024 in. (0.6mm).

 d. The maximum pressure plate spring width is 0.197 in. (5.0mm).

 e. The maximum flywheel runout is 0.004 in. (0.1mm).

13. When reassembling, apply a thin coating of multi-purpose grease to the release bearing hub and release fork contact points. Also, pack the groove inside the clutch hub with multi-purpose grease and lubricate the pivot points of the release fork.

14. Position the clutch disc and clutch cover with the matchmarks aligned and support with pilot tool.

15. Install the clutch cover bolts and tighten evenly to 17 ft. lbs. (23 Nm) for Metro or 14 ft. lbs. (19 Nm) for Prizm, in a crisscross pattern. Remove the pilot tool.

16. On Prizm, install the release bearing, fork, and the boot.

17. Lightly lubricate the splines, pilot bearing surface of the input shaft and release bearing with grease.

18. Install the transaxle.

19. Adjust the clutch cable.

20. Connect the negative battery cable. Check clutch operation.

Clutch Master Cylinder

REMOVAL AND INSTALLATION

Prizm

1. Drain or siphon the fluid from the master cylinder.

2. Disconnect the hydraulic line from the master cylinder.

> ——— WARNING ———
> **Do not spill brake fluid on the painted surfaces; it will destroy the finish.**

3. Working inside the vehicle, remove the underdash panel and the air duct.

4. Remove the pedal return spring.

5. Remove the spring clip and clevis pin.

6. Unfasten the bolts which secure the master cylinder to the bulkhead. Withdraw the assembly from the engine compartment side of the bulkhead.

To install:

7. Install the master cylinder to the bulkhead. Tighten the nuts to 9 ft. lbs. (12 Nm).

8. Connect the hydraulic line to the master cylinder.

9. Install the clevis pin and spring clip.

10. Install the pedal return spring.

11. Fill the reservoir with clean, fresh brake fluid and bleed the system.

12. Check the cylinder and the hose connection for leaks.

13. Reinstall the air duct and underdash cover panel.

14. Check for proper clutch operation.

Clutch Slave Cylinder

REMOVAL AND INSTALLATION

Prizm

> ——— WARNING ———
> **Do not spill brake fluid on the painted surfaces; it will destroy the finish.**

1. Raise and safely support the vehicle.

2. If necessary, remove the rear splash shield to gain access to the slave cylinder.

3. Remove the clutch fork return spring.

4. Unfasten the hydraulic line from the slave cylinder by loosening the retaining nut on the hose.

5. Remove the slave cylinder retaining nuts and remove the cylinder.

To install:

6. Install the cylinder to the clutch housing, and tighten the bolts to 106 inch lbs (12 Nm).

7. Connect the hydraulic line and tighten it to 11 ft. lbs. (15 Nm).

8. Install the clutch fork return spring.

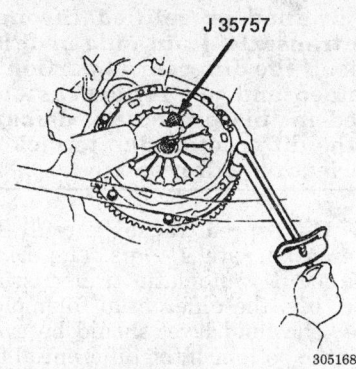

J 35757

305168

Clutch disc alignment and installation — Prizm

9. Bleed the system.
10. Install the splash shield, if removed.
11. Lower the vehicle. Check for proper clutch operation.

Hydraulic Clutch System Bleeding

PROCEDURE

NOTE: If any maintenance on the clutch system was performed or the system is suspected of containing air, bleed the system.

1. Fill the clutch reservoir with brake fluid. Check the reservoir level frequently and add fluid as needed.
2. Connect one end of a vinyl tube to the bleeder plug on the slave cylinder and submerge the other end into a clear container half-filled with clean brake fluid.
3. Slowly pump the clutch pedal several times.
4. Repeat Steps 2 and 3 until all of the air bubbles are removed from the system.
5. Tighten the bleeder plug when no more air bubbles emerge from the tube.
6. Refill the master cylinder to the proper level.
7. Check the system for leaks.

Automatic Transaxle Assembly

REMOVAL AND INSTALLATION

Metro

1. Disconnect the negative battery cable from the battery and transaxle.
2. Disconnect the throttle valve cable adjustment cover and cable from top of the transaxle.

3. Disconnect the shift cable from the manual select joint on the transaxle. Remove the retaining clip and cable from the bracket on the transaxle assembly. Loosen the adjustment nuts.
4. Remove the accelerator cable from the bracket on top of the transaxle. Remove the engine harness bracket from the rear of the transaxle.
5. Disconnect the harnesses to the vehicle speed sensor, shift solenoid and transaxle range switch.
6. Remove the speedometer cable from the gear case.
7. Disconnect and plug the inlet and outlet fluid cooler lines.
8. Remove the starter motor. Drain the transaxle fluid.
9. Remove the rear transaxle/engine mount. Remove the upper transaxle-to-engine mount bolts from the transaxle.
10. Support the engine from the top using engine support tool J-28467-A and support fixture adapters J-28467-89 or equivalents.
11. Raise and support the vehicle safely. Remove the right and left splash shields.
12. Separate both ball joints from the steering knuckles.
13. Remove the right and left halfshaft assemblies. Remove the rear engine torque rod/engine mount from the transaxle case.

NOTE: Make alignment marks on the torque converter and drive plate for assembly reference.

14. Remove the flywheel cover from the transaxle case. Lock the flywheel in place and remove the flywheel-to-torque converter bolts.
15. Remove the muffler mounting from the rear engine exhaust hanger.
16. Lower the engine slightly.
17. Support the transaxle and remove the transaxle-to-engine bolts. Slide the transaxle off the engine and lower it from the vehicle.
To install:
18. Use grease to lubricate the cup around the center of the torque converter.
19. Measure the distance between the torque converter and the edge of the transaxle housing; it should be at least 0.85 in. (21.4mm). If the distance is less than specified, the torque converter is improperly installed. Remove and reinstall it.
20. Position the transaxle to the engine. Install the lower transaxle-to-engine retaining bolt and torque to 40 ft. lbs. (55 Nm).

21. Install the left transaxle mount and secure with the bracket bolts and nuts. Torque the bolts and nuts to 40 ft. lbs. (55 Nm).
22. Remove the transaxle support jack.
23. Raise the engine assembly slightly.
24. Install the rear engine mount bracket to the transaxle. Install the bracket bolts and nuts and tighten to 40 ft. lbs. (55 Nm). Install the rear engine mount-to-bulkhead bolt and torque to 40 ft. lbs. (55 Nm).
25. Install the muffler mounting to the rear exhaust hanger.
26. Install the flywheel-to-torque converter bolts. Lock the flywheel in place and torque the flywheel-to-torque converter bolts to 14 ft. lbs. (19 Nm).
27. Install the flywheel inspection cover to the transaxle and torque the bolts to 89 inch lbs. (10 Nm).
28. Install the rear engine torque rod assembly and tighten the bolts to 40 ft. lbs. (55 Nm).
29. Install the right and left side halfshaft assemblies.
30. Install the ball joints to the steering knuckles.
31. Install the right and left splash shields.
32. Remove the engine support fixture from the vehicle.
33. Install the upper transaxle-to-engine mounting bolts and tighten to 40 ft. lbs. (55 Nm).
34. Install the through-bolt and nut to the rear engine mount and torque to 40 ft. lbs. (55 Nm).
35. Install the upper rear transaxle mount-to-bulkhead bolt and torque to 40 ft. lbs. (55 Nm).
36. Install the starter motor.
37. Connect the inlet and outlet fluid cooler lines using new hose clamps.
38. Install the speedometer cable into the gear case.
39. Connect the harness to the VSS, shift solenoid and transaxle range switch.
40. Install the remaining components.
41. Connect the negative battery cable to the transaxle and battery.
42. Refill and check the transaxle fluid level.

Prizm

1. Disconnect both battery cables, negative cable first, then remove the battery.
2. Disconnect the intake air temperature sensor, park neutral position switch, vehicle speed sensor and

solenoid wiring harness, if equipped with a 4-speed transaxle.

3. Remove the air cleaner housing cover and filter, then remove the housing from the vehicle.

4. Disconnect the ground wire from the transaxle.

5. Remove the throttle valve cable from the throttle linkage and bracket.

6. Disconnect the vehicle speed sensor harness.

7. Remove the nut and disconnect the shift select cable from the manual lever. Remove the E-clip and disconnect the cable from the cable bracket.

8. Remove the bolt and the TV cable guide bracket from the transaxle case.

9. Remove the upper transaxle-to-engine bolts. Remove the starter motor.

10. Install engine/transaxle support fixture J-28467-A or equivalent.

11. Remove the bolts and left mounting bracket.

12. Raise and safely support the vehicle.

13. Drain the transaxle fluid.

14. On 3-speed transaxles, remove the differential filler plug at the rear of the transaxle and drain the differential fluid.

15. Remove the right and left splash shields.

16. Disconnect the fluid cooler hoses from the lines at the transaxle.

17. Remove the front wheels.

18. Remove the right and left halfshafts.

19. Remove the bolts from the exhaust pipe support. Remove the nut and exhaust pipe support from the center crossmember.

20. Disconnect the oxygen sensor harness. Remove the front exhaust pipe from the vehicle.

21. Remove the plastic cover from the front transaxle mounting cavity in the center crossmember.

22. Remove the bolts/nuts from the front and rear transaxle mounts.

23. Remove the bolts and nuts, then remove the front suspension and center crossmembers.

24. Support the transaxle with a suitable jack.

25. On 3-speed transaxles, remove the bolts and the lower engine reinforcement brace.

26. Remove the flywheel access cover from the rear of the engine.

27. Remove the 6 flywheel-to-torque converter bolts.

28. Remove the drain pan from underneath the transaxle fluid pan.

29. Remove the lower transaxle-to-engine bolts from the transaxle.

30. Remove the transaxle by carefully moving the transaxle away from the engine toward the left side of the engine compartment and slowly lowering the jack.

To install:

31. Before installing the transaxle assembly, perform the following:

   a. Apply grease around the pilot shaft at the center of the torque converter.

   b. Measure the distance between the outside edge of the torque converter housing and the torque converter lug. The distance should be more than 0.906 in. (23mm). If it is less than 0.906 in. (23mm), the torque converter is improperly installed. Remove and properly seat on the input shaft.

32. Install the transaxle to the engine.

33. Install the lower transaxle-to-engine bolts and tighten to 47 ft. lbs. (64 Nm).

34. Install the flywheel-to-torque converter bolts. Tighten the bolts to 14 ft. lbs. (19 Nm).

35. Install the flywheel access cover.

36. On 3-speed transaxles, install the lower engine reinforcement brace and tighten the bolts to 47 ft. lbs. (64 Nm).

37. Remove the jack from under the transaxle.

38. Install the front suspension and center crossmembers. Tighten the front suspension crossmember bolts to 152 ft. lbs. (206 Nm) and the center crossmember bolts to 45 ft. lbs. (61 Nm).

39. Install the front and rear transaxle mount nuts. Tighten the rear nuts to 42 ft. lbs. (57 Nm), and the front to 47 ft. lbs. (64 Nm).

40. Install the plastic access cover to the center crossmember.

41. Install the front exhaust pipe and connect the oxygen sensor harness.

42. Install the halfshafts and the front wheels.

43. Install the bolts to the left transaxle bracket and tighten to 35 ft. lbs. (48 Nm).

44. Connect the fluid cooler hoses to the lines at the transaxle; secure with hose clamps.

45. Install the splash shields.

46. Install the starter.

--- **WARNING** ---

The differential portion of the 3-speed transaxle is separated from the rest of the transaxle and must be drained and refilled separately. The differential cannot be drained or refilled through the transaxle drain plug or filler tube. If the differential portion is drained and not refilled as outlined in this procedure, damage to the differential due to lack of lubrication will result.

47. On 3-speed transaxle, refill the differential with 1½ qts. (1.4 L) of Dexron®III automatic transmission fluid, into the differential filler plug hole. The fluid level should be even with the bottom of the differential filler plug hole. Install the filler plug and tighten to 29 ft. lbs. (39 Nm).

48. Lower the vehicle.

49. Install the left transaxle bracket reinforcement and tighten the bolts to 15 ft. lbs. (21 Nm).

50. Remove the engine/transaxle support fixture.

51. Install the upper transaxle-to-engine bolts and tighten to 47 ft. lbs. (64 Nm).

52. Install the TV guide cable bracket onto the transaxle and tighten the bolt to 71 inch lbs. (8 Nm).

53. Install the shift select cable into the bracket at the transaxle and secure with clip.

54. Install the shift select cable to the manual lever and secure with the nut and lockwasher. Adjust the shift select cable.

55. Connect the wiring connector to the VSS.

56. Install the TV cable to the throttle linkage. Adjust the cable as follows:

   a. Make sure the throttle valve is fully closed.

   b. Measure the distance between the end of the outer cable boot and the end of the TV cable stopper, it should be 0–0.04 in. (0–1mm).

   c. If the distance is greater than specified, loosen the TV cable locknut and tighten the adjust nut until within specification.

   d. If the distance is less than specified, loosen the TV cable adjust nut and tighten the locknut within specification.

   e. After adjustment is completed, tighten the locknut and adjust nuts to 71 inch lbs. (8 Nm).

57. Connect the ground strap to the transaxle and tighten the bolt to 115 inch lbs. (13 Nm).

58. Connect the PNP switch, IAT sensor and solenoid harness if equipped with a 4-speed transaxle.

59. Install the air cleaner and filter. Tighten the bolts to 106 inch lbs. (12 Nm). Tighten the air cleaner intake tube clamp.

60. Install the battery and connect the cables, negative cable last.

61. Refill the transaxle with the proper amount of Dexron®III automatic transmission fluid.

62. Operate the vehicle and check for leaks.

DRIVE AXLE

Halfshaft

REMOVAL AND INSTALLATION

Metro

1. Disconnect the negative battery cable.

2. Unstake and remove the halfshaft nut with the vehicle's weight on the ground.

3. Raise and safely support the vehicle.

4. Remove the front wheel.

5. If equipped with ABS, remove the wheel speed sensor.

6. Remove the ball joint from the steering knuckle.

7. Drain the transaxle fluid.

NOTE: On 4-door sedan vehicles, the right side halfshaft is stabilized by a center support bearing between the transaxle and the inner tripod joint. It is not necessary to remove the right inner drive axle and center support bearing assembly. The right side halfshaft assembly can be separated from the center support bearing by lightly tapping with a plastic mallet.

8. Using a suitable pry tool, pry on the inboard joints of the halfshaft to detach the snapring lock in the differential. On 4-door sedans, separate the right side halfshaft assembly from the center support bearing by tapping lightly with a plastic mallet.

9. Remove the stabilizer bar mounting bracket bolts.

10. Pull the halfshaft out of the transaxle, then from the steering knuckle.

To install:

11. Inspect the CV-joint boots for tears or deterioration and replace, if necessary.

12. Snap the halfshaft into the transaxle first, then into the steering knuckle.

13. Install the ball joint to the steering knuckle.

14. Install the stabilizer bar, if necessary.

15. Install the halfshaft nut and washer. Torque the halfshaft nut to 129 ft. lbs. (175 Nm) on 1993–94 vehicles or to 148 ft. lbs. (200 Nm) on 1995–97 vehicles.

16. Stake the drive axle nut with a hammer and punch.

17. If equipped with ABS, install the ABS speed sensor.

18. Install the front wheel.

19. Lower the vehicle to the ground.

20. Refill the transaxle with clean transaxle fluid.

21. Connect the negative battery cable.

22. Road test the vehicle.

Prizm

WARNING

Care must be exercised to prevent the differential-side CV-joint from being over-extended. Over-extension of the differential-side CV-joint could result in separation of internal components and possible joint failure. If the vehicle is to be lowered and moved, the front wheel bearings must be supported using a 9/16 in. (14mm) bolt, 1 3/4 in. (44.45mm) washer, 2 in. (50.8mm) washer, and a 9/16 in. (14mm) nut assembled through the shaft opening in the hub. Tighten the nut and bolt to 40 ft. lbs. (54 Nm). If equipped with Anti-lock Brake Systems (ABS), use caution not to damage the ABS speed sensor ring on the wheel-side CV-joint. DO NOT pry against the ring with metal tools. If the serrations on the speed sensor ring appear damaged, replace the wheel-side CV-joint.

---

1. Disconnect the negative battery cable.

2. Raise and safely support the vehicle.

3. Remove the front wheel.

4. Remove the 6 bolts and splash shield from the vehicle.

5. If equipped with ABS, remove the speed sensor.

6. Remove the cotter pin and lock cap and nut from the halfshaft.

7. Separate the tie rod end from the steering knuckle.

8. Separate the ball joint from the control arm.

9. Remove the wheel-side CV-joint from the steering knuckle.

10. Remove the differential-side CV-joint from the transaxle by gently prying the joint out of the transaxle.

11. Remove the halfshaft from the vehicle.

To install:

12. Inspect the CV-joint boots for tears or deterioration. Replace as necessary.

13. Inspect the front wheel bearing inner and outer oil seals for damage or deterioration. Replace as necessary.

14. Inspect the halfshaft fluid seal at the transaxle for leakage or damage. Replace as necessary.

15. Install the differential-side CV-joint into the transaxle.

16. Install the wheel-side CV-joint into the steering knuckle.

17. Install the ball joint to the control arm.

18. Install the tie rod end.

19. Install the nut onto the halfshaft and tighten to 167 ft. lbs. (228 Nm).

20. Install the lock cap onto the halfshaft and secure with a new cotter pin.

21. Install the front wheel.

22. If equipped, install the ABS speed sensor.

23. Install the splash shield and secure with the 6 bolts. Tighten the splash shield bolts to 71 inch lbs. (8 Nm). Lower the vehicle.

24. Connect the negative battery cable.

25. Check the wheel alignment.

CV-Joint Boot

REPLACEMENT

Metro

NOTE: The outboard CV-joint is not removable from the halfshaft. The outboard CV-joint boot can only be serviced by removing the inner CV-joint assembly. Outer CV-joint replacement requires replacement of the halfshaft.

1. Remove the halfshaft from the vehicle and secure in a vice equipped with jaw covers.

NOTE: During inner joint/boot service, matchmark the differential-side joint, halfshaft and tripod joint spider to ensure that all components are re-assembled in the same position.

2. Remove the large boot clamp from the joint and discard.

3. Separate the differential-side tripod joint housing from the tripod joint spider on the halfshaft.

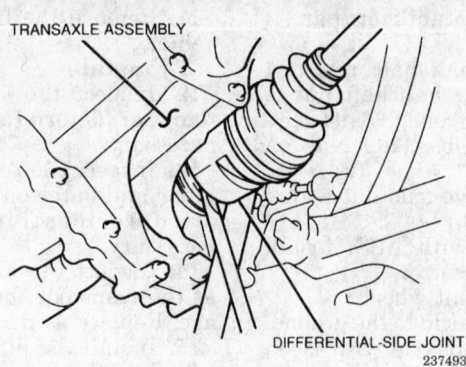

TRANSAXLE ASSEMBLY

DIFFERENTIAL-SIDE JOINT
237493

Removing the inner CV-joint from the transaxle —
Prizm

4. Remove the snapring from the inboard side of the spider joint with a suitable pair of snapring pliers, then push the spider joint down the shaft to uncover the top snapring.

5. Remove the top snapring, then remove the spider joint.

6. Remove the small inboard boot clamp and discard.

7. Remove the inboard boot.

8. Remove the outboard boot clamps and slide the boot off the inboard end of the halfshaft.

To install:

9. Inspect the tripod spider joint and/or CV-joint for excessive wear or damage and replace the joint as an assembly, if necessary.

NOTE: If the CV-joint, tripod spider joint, or boots need to be cleaned, do not wash them in solvent or degreaser. Instead, clean these components with a clean, dry, solvent-free cloth.

10. If the outer CV-joint boot was removed, install the new boot onto the halfshaft and slide down. Fill the boot with approximately 2.1–2.8 oz. (60–80g) of CV-joint lubricant provided in the boot kit.

11. Install the boot over the outer joint. Install new boot clamps.

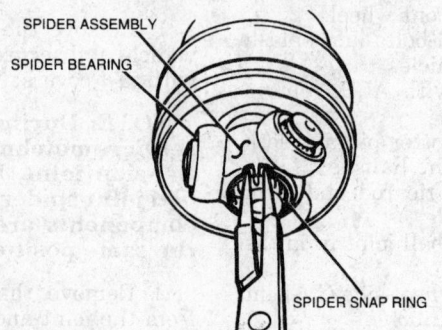

SPIDER ASSEMBLY

SPIDER BEARING

SPIDER SNAP RING

310377

Removing the outer snapring — Metro

12. Install a new small inner boot clamp onto the halfshaft. Install the inner tripod boot onto the shaft and slide down below the ring for the tripod joint.

13. Install the spider joint onto the shaft and secure with a snapring. Be sure the matchmarks line up with each other.

14. Push the spider joint into position and install the inner snapring.

15. Pack the inner tripod joint boot housing with approximately 2.8–3.5 oz. (80–100g) of CV-joint lubricant provided in the boot kit.

16. Install the tripod housing onto the spider joint.

17. Slide the boot into place and secure with new boot clamps. Be sure the boots are not distorted or dented. If this happens, pull outward on the boot in the affected area, until it is corrected. Be very careful not to separate the tripod joint housing from the axle shaft.

18. Install the halfshaft into the vehicle.

Prizm

1. Remove the halfshaft from the vehicle and secure in a vice equipped with jaw covers.

---

2. Remove the large differential boot clamp by drawing the clamp hooks together using tool J-35566 or equivalent.

3. Remove the small differential-side boot clamp. Slide the differential-side boot toward the center of the halfshaft. Place a reference mark on the differential-side joint housing and halfshaft.

4. Remove the differential joint housing from the Tripod joint spider. Place an index mark on the Tripod joint spider and halfshaft.

5. Remove the snapring and Tripod joint spider from the halfshaft.

6. Remove the differential-side boot from the halfshaft.

7. Remove the boot clamp from the wheel-side boot.

8. Remove the small clamp from the wheel-side boot.

9. Remove the wheel-side boot from the halfshaft. Place a reference mark on the wheel-side joint and the halfshaft.

10. Remove the wheel-side joint from the halfshaft by expanding the snapring.

To install:

11. Clean the boots with a clean, dry, solvent-free cloth.

12. Clean the tripod joint spider and joint with a clean, dry, solvent-free cloth.

13. Inspect the components and replace damaged or fatigued parts.

14. Align the reference marks on the wheel-side joint and halfshaft, then install the wheel-side joint onto the halfshaft by expanding the wheel-side snapring and sliding the joint onto the halfshaft. Make sure

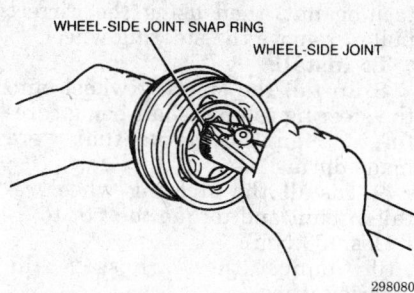

WHEEL-SIDE JOINT SNAP RING

WHEEL-SIDE JOINT

298080

Removing the outer joint snapring — Prizm

the wheel-side snapring is securely seated.

15. Install the wheel-side boot onto the halfshaft. Temporarily install the large and small wheel-side boot clamps onto the halfshaft. Do not crimp. Pack the wheel-side joint with approximately 4.6–5.3 oz. (130–150 g) of the lubricant provided in the boot kit.

16. Install the small and large boot clamps onto the wheel-side boot using tool J-22610 or equivalent.

17. Install the differential-side boot onto the halfshaft.

NOTE: When installing the Tripod joint spider onto the halfshaft, place the joint spider onto the halfshaft with the short sided splines facing toward the differential-side boot (away from transaxle).

18. Install the Tripod joint spider onto the halfshaft, aligning the reference marks; secure with snapring. Pack the differential-side joint housing with approximately 8.1–8.8 oz. (230–250 g) of the lubricant provided in the boot kit.

19. Install the differential-side joint housing onto the Tripod joint spider, aligning the reference marks.

20. Install the small differential boot clamp onto the boot using the tool J-22610 or equivalent.

NOTE: When installing the large differential-side boot clamp, use tool J-35566 or equivalent, to draw the closing hooks of the clamp together so the clamp locks into position.

21. Install the large differential-side boot clamp onto the boot using tool J-35566 or equivalent.

22. Inspect both boots for distortion or dents. Correct by pulling outward on the boot. Do not pull outward on the differential-side joint housing. If the housing is pulled, the joint may

become over-extended and detach from the halfshaft.

23. Measure the standard halfshaft length to ensure that the boots are not stretched or excessively contracted.

- Right halfshaft — 33.756 in. (857.4mm)
- Left halfshaft — 21.268 in. (540.2mm)

24. Install the halfshaft and lower the vehicle.

STEERING

Air Bag

CAUTION

Some vehicles are equipped with an Air Bag system, also known as the Supplemental Inflatable Restraint (SIR) or Supplemental Restraint System (SRS). The system must be disabled before performing service on or around system components, steering column, instrument panel components, wiring and sensors. Failure to follow safety and disabling procedures could result in accidental Air Bag deployment, possible personal injury and unnecessary system repairs.

PRECAUTIONS

Several precautions must be observed when handling the inflator module to avoid accidental deployment and possible personal injury.

- Never carry the inflator module by the wires or connector on the underside of the module.
- When carrying a live inflator module, hold securely with both hands, and ensure that the bag and trim cover are pointed away.
- Place the inflator module on a bench or other surface with the bag and trim cover facing up.
- With the inflator module on the bench, never place anything on or close to the module which may be thrown in the event of an accidental deployment.

DISARMING

Metro

CAUTION

The Air Bag system must be disarmed before performing service on components or wiring. Failure to do so may cause accidental deployment, resulting in unnecessary system repairs and/or personal injury.

1. Disconnect the negative battery cable. Wait at least 2 minutes after disconnecting the battery cable.

2. Turn the steering wheel so the front wheels are in the straight ahead position.

3. Turn the ignition switch to the **OFF** or **LOCK** position.

4. Remove the **SIR IG** fuse and the **SIR ST** fuse (if equipped) from the SIR fuse block.

5. Remove the plastic access cover from the rear of the inflator module housing.

6. Remove the Connector Position Assurance (CPA) and disconnect the yellow 2-way connector inside the inflator module housing.

7. Remove the glove box and disconnect the yellow passenger Air Bag inflator module harness, if equipped.

8. Connect the negative battery cable.

Prizm

CAUTION

The Air Bag system must be disarmed before performing service on components or wiring. Failure to do so may cause accidental deployment, resulting in unnecessary repairs and/or personal injury.

NOTE: The center sensor assembly can maintain sufficient voltage to cause deployment for up to 2 minutes after the ignition switch is turned to the LOCK position or the battery is disconnected.

1. Disconnect the negative battery cable.

2. Turn the steering wheel so the front wheels are in the straight ahead position.

3. Turn the ignition switch to **LOCK**.

4. Remove the IGN fuse and CIG and RADIO fuse from junction block 1.

5. Remove the Connector Position Assurance (CPA) and disconnect the yellow 2-way connector at the base of the steering column.

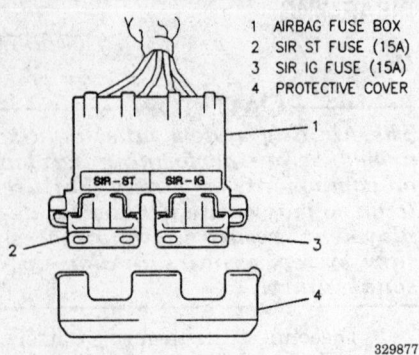

1 AIRBAG FUSE BOX
2 SIR ST FUSE (15A)
3 SIR IG FUSE (15A)
4 PROTECTIVE COVER

SIR system fuse box — 1993–94 Metro

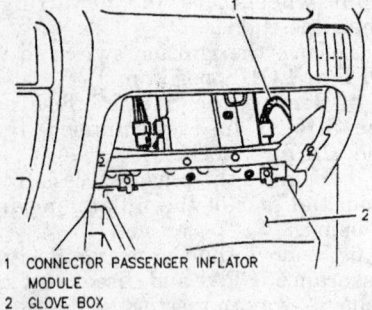

1 CONNECTOR PASSENGER INFLATOR MODULE
2 GLOVE BOX

Passenger side inflator module connector — 1993–94 Metro

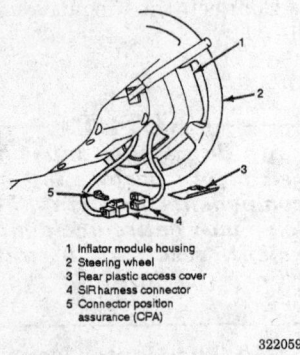

1 Inflator module housing
2 Steering wheel
3 Rear plastic access cover
4 SIR harness connector
5 Connector position assurance (CPA)

SIR harness connector — Metro

Steering Wheel

REMOVAL AND INSTALLATION

> **CAUTION**
> *The Air Bag must be disarmed before removing the steering wheel. Failure to do so may cause accidental deployment, resulting in unnecessary repairs and/or personal injury.*

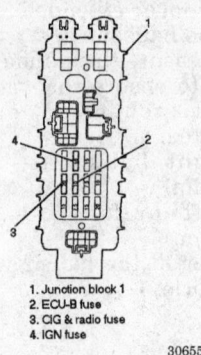

1. Junction block 1
2. ECU-B fuse
3. CIG & radio fuse
4. IGN fuse

Junction block 1 — Prizm

Metro

1. Disconnect the negative battery cable.
2. If equipped, disable the Air Bag system (SRS).
3. If equipped, remove the Air Bag inflator module attaching screws, then remove the Air Bag assembly.

> **CAUTION**
> *When carrying a live Air Bag, make sure the bag and trim cover are pointed away from the body. In the event of an accidental deployment, the bag will then deploy with minimal chance of injury. When placing a live inflator module on a bench or other surface, always place the bag and trim cover up. This will reduce the motion of the module if it is accidently deployed.*

4. If not equipped with an Air Bag, remove the steering wheel pad.
5. Scribe alignment marks across the steering wheel and shaft.
6. Disconnect the harness from the steering wheel.

> **WARNING**
> **The Air Bag system coil assembly is easily damaged if the wrong wheel pullers are used.**

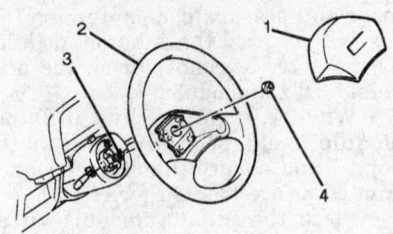

1 STEERING WHEEL PAD
2 STEERING WHEEL
3 STEERING SHAFT
4 STEERING WHEEL RETAINING NUT

Steering wheel components — Metro without an Air Bag

7. Remove the steering wheel attaching nut, then using the correct puller, remove the steering wheel.
   To install:
8. Install the steering wheel onto the steering column shaft being careful to align the marks that were made during removal.
9. Install the steering wheel retaining nut and torque the nut to 24 ft. lbs. (33 Nm).
10. Connect the harness to the steering wheel.
11. If equipped, install the steering wheel pad.
12. If equipped, install the Air Bag module and tighten the attaching screws to 89 inch lbs. (10 Nm).
13. Connect the negative battery cable.
14. Enable the Air Bag system, if equipped.

Prizm

1. Place the vehicle's front wheels in the straight ahead position.
2. Disconnect the negative battery cable.
3. Disable the Air Bag system.
4. Remove the inflator module as follows:
   a. Remove the 2 side trim covers from the steering column.
   b. Remove the 2 Torx® head screws, then release the Connector Position Assurance (CPA) and disconnect the upper steering column connector.
   c. Remove the inflator module from the steering wheel.
5. Disconnect the horn, and cruise control connectors, if equipped.
6. Place matchmarks on the steering wheel and shaft, then remove steering wheel retaining nut.

> **WARNING**
> **Do not install the puller tool bolts more than 0.5 in. (13mm) into the steering wheel. If the bolts are extended past the steering wheel, they could contact the Air Bag coil assembly and damage it.**

7. Remove the steering wheel puller tool J-1859-03 or equivalent.
   To install:
8. Install the steering wheel on the shaft, aligning the matchmarks. Tighten the retaining nut to 25 ft. lbs. (34 Nm).
9. Connect the horn, and cruise control harnesses, if equipped.
10. Install the inflator module as follows:
    a. Install the inflator module to the steering wheel and connect the

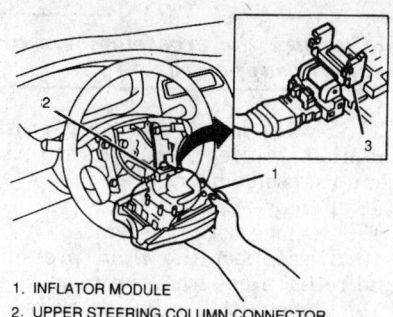

1. INFLATOR MODULE
2. UPPER STEERING COLUMN CONNECTOR
3. CONNECTOR POSITION ASSURANCE (CPA)

238850

Removing the inflator module — Prizm

upper steering column connector and the CPA.

b. Install the 2 Torx® head screws and tighten to 78 inch lbs. (8.8 Nm).

c. Install the 2 side trim covers to the steering column.

11. Enable the Air Bag system.

Tie Rod Ends

REMOVAL AND INSTALLATION

1. Raise and safely support the vehicle.

2. Remove the front wheel.

3. Remove the cotter pin and castle nut from the tie rod end ball stud.

4. Remove the tie rod end from the knuckle using ball joint separator tool J-21687-02 or the equivalent.

5. Loosen the locknut on the threaded end of the tie rod.

6. Unscrew the tie rod end from the tie rod. Note the number of turns necessary for removal.

To install:

7. Screw the tie rod end onto the tie rod the exact number of turns that was necessary to remove the tie rod.

8. Tighten the locknut to 32 ft. lbs. (43 Nm) for Metro or 35 ft. lbs. (47 Nm) for Prizm.

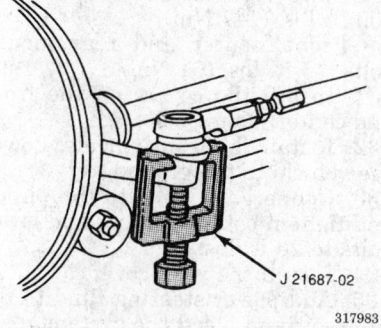

J 21687-02

317983

Separating the outer tie rod end from the steering knuckle — Metro

9. Install the tie rod ball joint stud into the steering knuckle. Install the castle nut and tighten to 32 ft. lbs. (43 Nm) for Metro or 36 ft. lbs. (49 Nm) for Prizm. Install a new cotter pin and bend the cotter pin end to secure.

10. Install the front wheel. Lower the vehicle.

11. Check and adjust the front wheel alignment.

Manual Rack and Pinion

REMOVAL AND INSTALLATION

Metro

1. Disconnect the negative battery cable.

2. Slide the driver's seat back as far as possible and pull off the front part of the floor mat on the driver's side and remove the steering shaft joint cover.

3. Loosen the steering shaft upper joint bolt, but do not remove. Remove the steering shaft lower joint bolt and disconnect the lower joint from the pinion.

4. Raise and safely support the vehicle. Remove the front wheels.

5. Remove the tie rod ends from the steering knuckles.

6. Remove the rack and pinion mounting bolts, brackets and the rack from the vehicle.

To install:

7. Install the rack, brackets and mounting bolts into the vehicle at the bulkhead. Be sure to align the pinion shaft through the floor opening. Tighten the bolts to 18 ft. lbs. (25 Nm).

8. Install the right and left tie rod ends to the steering knuckles.

9. Install the front wheels. Lower the vehicle.

10. Connect the steering shaft to the steering gear. Install the lower steering shaft-to-steering gear clinch bolt and tighten to 18 ft. lbs. (25 Nm). Tighten the steering shaft upper joint bolt to 18 ft. lbs. (25 Nm).

11. Install the steering joint cover.

12. Place the driver's side floor mat back into position.

13. Check and adjust the front wheel alignment.

Prizm

1. Disconnect the negative battery cable.

2. Remove the cover from the intermediate shaft.

3. Loosen the upper pinch bolt. Remove the lower pinch bolt at the pinion shaft.

4. Raise and safely support the vehicle. Remove both front wheels.

5. Install an engine support and tension the engine without raising it.

6. Remove the bolts holding the center crossmember to the radiator support.

7. Remove the covers from the front and center mount bolts.

8. Remove the front mount bolts, and the center mount bolts.

9. Support the crossmember and remove the rear mount bolts.

10. Remove the bolts holding the center crossmember to the main crossmember.

11. Use a floor jack and piece of wood to support the main crossmember.

12. Remove the bolts holding the main crossmember to the body.

——— CAUTION ———
When the fasteners are removed, make sure the crossmembers are properly supported.

13. Remove the bolts holding the lower control arm brackets to the body.

14. Slowly lower the main crossmember while holding onto the center crossmember.

15. Remove both tie rod joints from the steering knuckles.

16. Remove the nuts and bolts attaching the steering rack to the body.

17. Remove the rack through the right side wheel well.

To install:

18. Place the racket in position through the right side wheel well. Tighten the bracket bolts to 45 ft. lbs. (61 Nm).

19. Attach the tie rods to the steering knuckles.

20. Position the center crossmember over the center and rear transaxle mount studs; start the 2 nuts on the center mount.

21. Loosely install the bolts holding the center crossmember to the radiator support.

22. Loosely install the front mounting bolts.

23. Raise the main crossmember into position over the rear mount studs and align the underbody bolts. Install the 2 rear mount nuts loosely.

24. Install the main crossmember to underbody bolts loosely.

25. Install the lower control arm bracket bolts loosely.

26. Loosely install the bolts holding the center crossmember to the main crossmember.

27. Torque the fasteners as follows:
- Main crossmember-to-underbody bolts: 152 ft. lbs. (206 Nm)
- Center crossmember-to-radiator support bolts: 45 ft. lbs. (61 Nm)
- Lower A-frame-to-center bolts: 161 ft. lbs. (218 Nm)
- Lower A-frame-to-outer bolts: 109 ft. lbs. (147 Nm)
- Front, center and rear mount bolts: 45 ft. lbs. (61 Nm)

28. Install the covers on the front and center mount bolts.

29. Install the front wheels. Lower the vehicle.

30. Connect the yoke to the pinion and tighten both the upper and lower bolts to 26 ft. lbs.

31. Install the yoke cover.

32. Check and adjust the front end alignment as necessary. Check steering operation.

Power Rack and Pinion

REMOVAL AND INSTALLATION

Metro

1. Disconnect the negative battery cable.

2. Slide the driver's seat back. Pull off the front part of the floor mat on the driver's side and remove the steering shaft joint cover.

3. Loosen the steering shaft upper joint bolt, but do not remove. Remove the steering shaft lower joint bolt and disconnect the lower joint from the pinion.

4. Raise and safely support the vehicle. Remove the front wheels.

5. Remove the tie rod ends.

6. Disconnect the exhaust pipe at the manifold.

7. If equipped with manual transaxle, separate the shift linkage and extension rod from the transaxle.

8. If equipped with automatic transaxle, remove the engine rear torque rod with bracket from the transaxle.

9. Disconnect the power steering lines from the rack and pinion assembly. Plug the openings to prevent system contamination.

10. Remove the rack and pinion mounting bolts, brackets and the rack from the vehicle.

To install:

11. Install the rack, brackets and mounting bolts into the vehicle at the bulkhead. Be sure to align the pinion shaft through the floor opening. Tighten bolts to 18 ft. lbs. (25 Nm).

12. Unplug and reconnect the power steering lines to the rack and pinion assembly. Torque the inlet and outlet fluid lines to 25 ft. lbs. (35 Nm). Torque the remaining lines to 18 ft. lbs. (25 Nm).

13. If equipped, install the shift linkage and extension rod to the manual transaxle.

14. If equipped, install the engine rear torque rod with torque rod bracket to the automatic transaxle.

15. Connect the front exhaust pipe to the manifold using a new seal, and secure with bolts. Torque the bolts to 37 ft. lbs. (50 Nm).

16. Install the tie rod ends to the steering knuckles.

17. Install the front wheels. Lower the vehicle.

18. Connect the steering shaft to the steering gear. Install the lower steering shaft-to-steering gear clinch bolt and tighten to 18 ft. lbs. (25 Nm). Tighten the steering shaft upper joint bolt to 18 ft. lbs. (25 Nm).

19. Install the steering joint cover.

20. Place the driver's side floor mat back into the original position.

21. Bleed the system.

22. Check and adjust the front wheel alignment.

Prizm

1. Disconnect the negative battery cable.

2. Place a drain pan under the steering rack and pinion unit.

3. Remove the cover from the intermediate shaft.

4. Loosen the upper pinch bolt. Remove the lower pinch bolt at the pinion shaft.

5. Raise and safely support the vehicle.

6. Remove both front wheels.

7. Install an engine support and tension the engine without raising it.

8. Remove the bolts holding the center crossmember to the radiator support.

9. Remove the covers from the front and center mount bolts.

10. Remove the front mount bolts, then the center mount bolts and then the rear mount bolts.

11. Remove the bolts holding the center crossmember to the main crossmember.

12. Use a floor jack and piece of wood to support the main crossmember.

13. Remove the bolts holding the main crossmember to the body.

14. Remove the bolts holding the lower control arm brackets to the body.

CAUTION

Make sure the crossmembers are properly supported.

---

15. Slowly lower the main crossmember while holding onto the center crossmember.

16. Remove the tie rods.

17. Remove both ball joints.

18. Disconnect the fluid pressure and return lines from the rack.

19. Remove the nuts and bolts attaching the steering rack to the body.

20. Remove the rack and pinion unit through the right side wheel well.

To install:

21. Place the rack and pinion unit in position through the right wheel well and tighten the bracket bolts to 44 ft. lbs. (59 Nm).

22. Connect the fluid lines to the rack and tighten to 33 ft. lbs. (45 Nm). Make certain the fittings are correctly threaded before tightening.

23. Attach the tie rod ends.

24. Position the center crossmember and rear transaxle mount over the studs. Start the nuts on the center mount.

25. Loosely install the bolts holding the center crossmember to the radiator support.

26. Loosely install the front mount bolts.

27. Raise the main crossmember into position over the rear mount studs and align the underbody bolts. Install the rear mount nuts loosely.

28. Install the main crossmember to underbody bolts loosely.

29. Install the lower control arm bracket bolts loosely.

30. Tighten the components in the order listed to the following torque specification:
- Main crossmember to underbody bolts: 152 ft. lbs. (205 Nm).
- Center crossmember-to-radiator support bolts: 45 ft. lbs. (61 Nm)
- Lower A-frame-to-center bolts: 161 ft. lbs. (218 Nm)
- Lower A-frame-to-outer bolts: 109 ft. lbs. (147 Nm)
- Front, center and rear mount bolts: 45 ft. lbs. (61 Nm)

31. Install the covers on the front and center mount bolts.

32. Install the front wheels. Lower the vehicle to the ground.

33. Connect the yoke to the pinion and tighten both the upper and lower bolts to 26 ft. lbs.

34. Install the yoke cover.

35. Add power steering fluid to the reservoir and bleed the system.

36. Check and adjust the front end alignment, as necessary.

Power Steering Pump

BLEEDING

1. Raise and safely support the front of the vehicle.
2. Start the engine, and allow it to reach normal operating temperature.
3. Turn the wheels to the far left and shut the engine **OFF**.
4. Add clean power steering fluid to the **MIN** mark on the fluid level indicator.
5. Run the engine at fast idle for approximately 15 seconds and then shut **OFF** the engine.
6. Check the fluid level. If necessary, top off the fluid level to the **MIN** mark again.
7. Start the engine and bleed the system by turning the wheels from side to side 3–4 times.
8. Turn **OFF** the engine and check the fluid level. If air is still in the system, the fluid will have a light tan appearance. The bleeding procedure must be repeated until all air is purged from the system for normal steering action to be obtained.
9. Lower the vehicle. Check the fluid level and top off, if necessary.

REMOVAL AND INSTALLATION

Metro

1. Disconnect the negative battery cable.
2. Remove retaining bolt securing the high pressure line to the core support.
3. Separate the high pressure hose from the high pressure pipe at the core support.
4. Loosen the clamp and return hose from the fluid reservoir.
5. Disconnect the harness to the power steering pressure switch at the pump.
6. Raise and safely support the vehicle.
7. Remove the accessory drive belts.
8. Remove the A/C compressor mounting bolts. Hang the compressor unit off to the side, but do not disconnect the A/C hoses.
9. Remove the power steering pump/compressor mounting bracket-to-engine bolts.
10. Remove the power steering hose and bracket with power steering pump attached.
11. Remove the power steering pump from the mounting bracket.
12. Remove the nut, lock washer and high pressure power steering line from the pump.

13. Remove the return hose and hose clamp from the power steering pump.
14. Be sure to plug the power steering fluid lines/hoses to prevent system contamination.

To install:

15. Connect the fluid return hose to the pump and secure with a new clamp.
16. Connect the high pressure line to the power steering pump. Secure with a new lock washer and nut. Tighten the line nut to 18 ft. lbs. (25 Nm).
17. Install the power steering pump to the mounting bracket and tighten the mounting bolts to 25 ft. lbs. (35 Nm).
18. Install the power steering pump and A/C compressor mounting bracket, and tighten the mounting bolts to 49 ft. lbs. (65 Nm).
19. Install the A/C compressor, and tighten the mounting bolts to 25 ft. lbs. (35 Nm).
20. Lower the vehicle.
21. Reconnect the power steering pressure switch harness.
22. Install the fluid return hose to the power steering fluid reservoir.
23. Install the high pressure hose to the core support and secure with mounting bolt.
24. Reconnect the high pressure hose to the high pressure pipe. Torque the high pressure hose flare nut to 30 ft. lbs. (40 Nm).
25. Install the retaining bolt for the high pressure hose bracket and tighten to 89 inch lbs. (10 Nm).
26. Install the accessory drive belts.
27. Refill and bleed the power steering system.
28. Reconnect the negative battery cable.

Prizm

1. Disconnect the negative battery cable.
2. Remove the air cleaner.
3. Remove the power steering drive belt.
4. Remove the return hose from the power steering pump.
5. Remove the pressure hose from the pump using wrench J-29698 or equivalent.
6. Remove the upper mounting bolt from the pump.
7. Remove the vacuum hoses from the steering switch. Remove the bolts retaining the pump bracket to the engine block and mount.
8. Remove the lower mounting bolt and the power steering pump from the vehicle.

To install:

9. Install the power steering pump in the mounting bracket and secure with the lower mounting bolt.
10. Install the bolts to the engine mount, engine block and pump bracket.
11. Connect the vacuum hoses to the steering switch.
12. Install the upper mounting bolt.
13. Tighten all mounting bolts to 29 ft. lbs. (39 Nm).
14. Connect the pressure and return lines to the pump.
15. Install the drive belt.
16. Install the air cleaner.
17. Connect the negative battery cable.
18. Refill the power steering reservoir and bleed the system.

BRAKES

Anti-Lock Brake System Service

PRECAUTIONS

• Certain components within the Anti-Lock Brake System (ABS) are not intended to be serviced or repaired individually. Only those components with removal and installation procedures should be serviced.
• Do not use rubber hoses or other parts not specifically specified for and ABS system. When using repair kits, replace all parts included in the kit. Partial or incorrect repair may lead to functional problems and require the replacement of components.
• Lubricate rubber parts with clean, fresh brake fluid to ease assembly. Do not use lubricated shop air to clean parts; damage to rubber components may result.
• Use only specified brake fluid from an unopened container.
• If any hydraulic component or line is removed or replaced, it may be necessary to bleed the entire system.
• A clean repair area is essential. Always clean the reservoir and cap thoroughly before removing the cap. The slightest amount of dirt in the fluid may plug an orifice and impair the system function. Perform repairs after components have been thoroughly cleaned; use only denatured alcohol to clean components. Do not allow ABS components to come into contact with any substance contain-

ing mineral oil; this includes used shop rags.

• The Anti-Lock control unit is a microprocessor similar to other computer units in the vehicle. Ensure that the ignition switch is **OFF** before removing or installing controller harnesses. Avoid static electricity discharge at or near the controller.

• If any arc welding is to be done on the vehicle, the control unit should be unplugged before welding operations begin.

Master Cylinder

REMOVAL AND INSTALLATION

Metro

——————— **WARNING** ———————
Be careful not to spill brake fluid on the painted surfaces; it will damage the finish.

1. Disconnect the negative battery cable.
2. Clean around the reservoir cap, then remove as much brake fluid as possible.
3. Disconnect the reservoir electrical harness.
4. Disconnect and plug the brake lines.
5. Remove the mounting nuts and washers.
6. Remove the master cylinder and gasket from the vehicle.
To install:
7. Install the master cylinder, using a new gasket. Install the mounting nuts and washers and tighten to 115 inch lbs. (13 Nm).
8. Connect the brake lines to the master cylinder.
9. Reconnect the fluid reservoir electrical harness.
10. Make sure the master cylinder reservoir is filled to the **MAX** mark with clean DOT 3 brake fluid only. Install the fluid reservoir cap and bleed the brake system, if necessary.
11. Connect the negative battery cable. Check for proper brake operation.

Prizm

NOTE: Whenever the power brake booster or master cylinder is removed, the booster pushrod length must be checked using booster pushrod gauge J-34873-A or equivalent.

1. Disconnect the negative battery cable.

2. Disconnect the Intake air temperature sensor.
3. Remove the air inlet hose from the throttle body.
4. Disconnect the brake fluid level sensor.
5. Remove the cap and siphon the fluid from the reservoir.
6. Disconnect and cap the brake lines.
7. Remove the mounting nuts from the booster studs. Pull the proportioning valve bracket off the studs and position it aside.
8. Remove the master cylinder and gasket.
To install:
9. Prior to installation, check the booster pushrod length as follows:
   a. Place booster pushrod gauge J-34873-A or equivalent, on the master cylinder with the gasket in place. Lower the gauge pin until the tip slightly touches the piston.
   b. Turn the pushrod gauge upside down and set it on the booster. There must be no clearance (0 in.) between the booster pushrod and the gauge.
   c. If clearance is incorrect, adjust the pushrod length until the pushrod lightly touches the pin head.

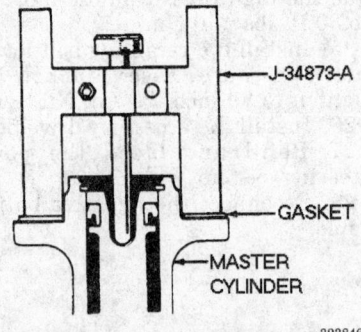

323846

Master cylinder measurement — Prizm

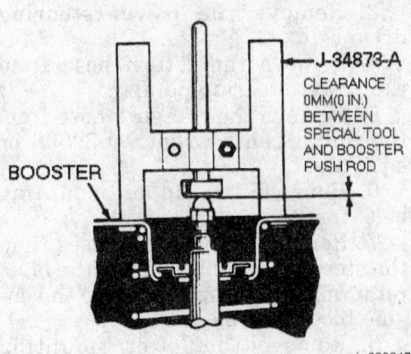

323847

Power booster measurement — Prizm

10. Prior to installation, bench bleed the master cylinder as follows:
    a. Mount the master in a soft-jawed vise, being careful not to damage the housing.
    b. Attach short lengths of brake line to the master cylinder outlet ports, with the ends of the brake lines pointing back into the master cylinder reservoir.
    c. Fill the master cylinder reservoir with fresh brake fluid, until the ends of the brake lines are submerged.
    d. Use a suitable tool to push the master cylinder piston in and out. Pump the piston until no more air bubbles appear at the ends of the submerged brake lines.
    e. Remove the brake lines from the master cylinder outlet ports. Plug the ports immediately to prevent fluid loss.
11. Install the master cylinder using a new gasket.
12. Install the proportioning valve on the booster studs and install the mounting nuts. Tighten the nuts to 115 inch lbs. (13 Nm).
13. Connect the brake lines to the master cylinder and bleed the system.
14. Connect the brake fluid level sensor electrical connector.
15. Install the air inlet hose.
16. Connect the IAT sensor.
17. Properly bleed the brake system.
18. Check the fluid level in the master cylinder reservoir and top off, as necessary.
19. Connect the negative battery cable. Road test the vehicle and check brake operation.

Brake Caliper

REMOVAL AND INSTALLATION

1. Raise and safely support the front of the vehicle. Set the parking brake and block the rear wheels.
2. Remove the front wheels.
3. Siphon a sufficient quantity of brake fluid from the master cylinder reservoir to prevent the brake fluid from overflowing the master cylinder when compressing the caliper piston. This is necessary as the piston must be forced into the cylinder bore to provide sufficient clearance to install the caliper.

NOTE: Disassemble brakes one wheel at a time. This will prevent parts confusion and also prevent the opposite caliper piston from popping out during installation.

4. Disconnect the brake hose at the caliper. Use a pan to catch any spilled fluid. Plug the disconnected hose.

5. Remove the caliper mounting bolts, then remove the caliper from the carrier.

To install:

6. Use a caliper compressor, C-clamp or large pair of pliers to slowly press the caliper piston into the caliper. When compressing the piston, be sure to place a discarded brake pad on top of the piston to prevent piston damage.

7. Install the caliper assembly to the carrier. Apply a thin, even coating of anti-seize compound to the threads and slide surfaces. Don't use grease or spray lubricants; they will not hold up under the extreme temperatures generated by the brakes. Tighten the bolts to 22 ft. lbs. (30 Nm).

8. Install the brake hose to the caliper. Secure with new washers and the union bolt. Torque the union bolt to 17 ft. lbs. (23 Nm).

9. Bleed the brake system.

10. Install the front wheels. Lower the vehicle. Check the level of the brake fluid in the master cylinder reservoir; Top off, if necessary.

11. Road test and check brake operation.

Disc Brake Pads

REMOVAL AND INSTALLATION

1. Raise and safely support the vehicle.

2. Siphon brake fluid from the master cylinder until the reservoir is ½ full. This is necessary as the piston must be forced into the cylinder bore.

3. Remove the front wheels. Reinstall 2 lug nuts finger-tight to secure the rotor in place.

NOTE: Disassemble brakes one wheel at a time. This will prevent parts confusion and also prevent the opposite caliper piston from popping out during pad installation.

4. Remove the caliper and support it with wire so it doesn't hang by the brake line.

5. Remove the brake pads, the wear indicators, the anti-squeal shims, the support plates and the anti squeal springs, if equipped.

6. Inspect the disc brake rotor for cracks, scoring, abnormal wear or other damage; machine or replace as

necessary. If machining, observe the rotor minimum thickness specification.

To install:

7. Install the pad springs or support plates onto the caliper carrier.

8. Install new pad wear indicators onto each pad, if equipped, making sure the arrow on the tab points in the direction of disc rotation.

9. Install new anti-squeal pads to the back of the pads.

10. Install the pads into the mounting bracket and install the anti-squeal springs.

11. Use a caliper compressor, C-clamp or large pair of pliers to slowly press the caliper piston back into the caliper. Be sure to place a discarded brake pad on top of the piston to prevent piston damage.

12. Install the caliper.

13. Install the front wheels. Lower the vehicle to the ground.

14. Depress the brake pedal several times and make sure the movement feels normal. The first brake pedal application may result in a very long pedal stroke due to the pistons being retracted. Always make several brake applications before starting the vehicle.

15. Check the level of the brake fluid in the master cylinder reservoir; add fluid as necessary.

16. Road test the vehicle and check brake operation.

Brake Rotor

REMOVAL AND INSTALLATION

1. Raise and safely support the vehicle.

2. Remove the wheel.

3. Remove the brake caliper and suspend out of the way on a wire. Do not allow the caliper to hang from the brake hose.

4. Inspect the rotor as follows:

NOTE: Be sure to check the front wheel bearings for looseness before taking any measurements.

a. Inspect the rotor for cracks, scoring and any abnormal wear or damage; machine or replace as necessary. If machining, observe the minimum rotor thickness specification. If the rotor passes visual inspection, proceed further and check lateral runout and rotor thickness.

b. Using a dial indicator, positioned onto the strut assembly, measure the lateral runout. If the

lateral runout exceeds 0.004 in. (0.10mm), replace the rotor or have it machined. If machining, observe the rotor minimum thickness specification.

c. Using a micrometer, measure the brake rotor thickness. Be sure to take the measurement at several points around the rotor. If the measurement is less than 0.315 in. (8mm) for 1993–94 models or 0.59 in. (15mm) for 1995–97 models replace the rotor.

5. For 1993–94 Metro, unstake and remove the halfshaft nut and washer. Discard the nut; it must not be reused. Remove the hub retaining bolts.

6. Using a slide hammer, remove the hub from the knuckle.

7. For 1995–97 Metro, remove the 2 brake rotor retaining screws.

8. Insert two 8mm bolts into the 2 opposite threaded holes on the brake rotor.

9. Slowly rotate the bolts alternately. This will separate the brake rotor from the hub assembly by pulling the rotor outward from the hub.

10. On Prizm, remove the bolts holding the caliper carrier to the steering knuckle.

11. Make a mark on the rotor indexing 1 wheel stud to 1 hole in the rotor. Remove the rotor.

12. Remove the brake rotor from the vehicle. On 1995–97 Metro, remove the two 8mm bolts from the rotor.

To install:

13. Apply a light coat of grease to the outside of the hub shaft.

14. On Prizm Models, install the caliper carrier and tighten the mounting bolts to 65 ft. lbs. (88 Nm).

15. Install the brake rotor to the knuckle.

16. Tap the hub and outer spacer onto the knuckle using a hammer. Place a block of wood on the end of the hub to absorb shock from the hammer and drive the hub into the knuckle evenly.

17. On 1993–94 models, install a new halfshaft nut and washer and tighten to 129 ft. lbs. (175 Nm). Stake the drive axle nut using a hammer and punch. Install the hub retaining bolts and tighten to 37 ft. lbs. (50 Nm).

18. On 1995–97 models, Secure the rotor using the rotor retainer screws.

19. Install the caliper.

20. Install the wheels.

21. Lower the vehicle. Pump the brake pedal several times to seat the brake pads against the rotors.

Brake Drums

REMOVAL AND INSTALLATION

1. Make sure the parking brake is fully released.
2. Raise and safely support the vehicle.
3. Remove the rear wheel.
4. If equipped without ABS, perform the following:
   a. Remove the dust cap. Using a hammer and punch, unstake the spindle nut.
   b. Remove the spindle nut and washer. Discard the spindle nut.
5. If equipped with ABS, perform the following:
   a. Remove the 2 brake drum retaining screws.
   b. Install two 8mm screws in the 2 opposite threaded holes on the brake drum.
   c. Alternately rotate each screw slowly until the 2 screws pull the drum from the hub and wheel studs.
6. Remove the drum from the vehicle. Remove the two 8mm screws from the drum.
7. Remove the rear brake drum assembly from the vehicle.
8. On Prizm, mark the position of 1 wheel lug stud in relation to a hole in the drum. This will allow the drum to be reinstalled in the original position.
9. Remove the brake drum. If the drum is frozen on the hub, tap the drum with a rubber mallet or wooden hammer handle to loosen it.

NOTE: If the brake drum is difficult to remove, remove the plug from the hole in the rear of the brake backing plate. Insert a small thin tool through the hole and hold the automatic adjusting lever away from the adjusting bolt. Use another tool to reduce the brake shoe adjustment by turning the adjuster wheel.

—— WARNING ——
Do not apply the brake pedal while the drum is removed.

10. Inspect the brake shoes for wear and/or damage; replace as necessary. Inspect the wheel cylinder for signs of brake fluid leakage; replace as necessary.
11. Inspect the drum for cracks, scoring or other wear; machine or replace as necessary. If machining, observe the maximum drum diameter specification.

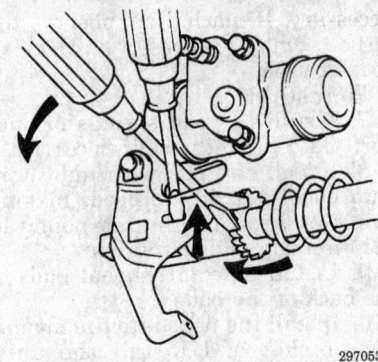

297053

Backing off the rear brake adjuster — Prizm

To install:
12. Install the brake drum assembly onto the spindle.
13. On non-ABS equipped models;
   a. Install the washer and a new spindle nut.
   b. Tighten the spindle nut to 74 ft. lbs. (100 Nm).
   c. Stake the spindle nut.
   d. Install the dust cap.
14. On ABS equipped models, install the brake drum and tighten the brake drum retaining screws.
15. On Prizm, install the drum, observing the matchmarks made earlier. Keep the drum straight while installing it, or it could damage the brake shoes.
16. Install the wheel. Lower the vehicle.
17. To adjust the brakes, depress the brake pedal 3–5 times with approximately 66 lbs. of pressure.
18. Road test and check brake operation.

Brake Shoes

REMOVAL AND INSTALLATION

Metro

1. Raise and safely support the vehicle.
2. Remove the rear wheels.
3. Remove the brake drum.
4. Remove the upper and lower return springs from the brake shoes.
5. Remove the anti-rattle spring and brake shoe adjustment strut.
6. Remove the primary and secondary brake shoe hold-down springs by pressing in and turning the hold-down pins.
7. Remove the parking brake cable.
8. Inspect all parts for wear and/or damage and replace as necessary.
9. Clean the brake backing plate and all parts to be reused with suitable brake parts cleaner and allow to dry.

To install:
10. Lubricate all contact points on the backing plate with grease.
11. Install the hold-down pins on the backing plate.
12. Install the brake cable.
13. Install the primary and secondary brake shoes to the vehicle and secure with the hold-down springs.
14. Install the brake adjustment strut and anti-rattle spring.
15. Install the upper and lower return springs to the primary and secondary brake shoes.
16. Install the brake drum.
17. Install the rear wheels.
18. Lower the vehicle.
19. Press the brake pedal 3–5 times with approximately 66 lbs. pressure to adjust the brake shoe clearance.
20. Check to ensure the brake drum is free from dragging and proper braking is obtained.

Prizm

1. Raise and safely support the vehicle.
2. Remove the rear wheels.
3. Remove the brake drum. Inspect the drum for wear and/or damage. Replace or machine the drum as necessary.

—— WARNING ——
Do not apply the brake pedal while the drum is removed.

4. Remove the upper return spring.
5. Remove the front brake shoe hold-down spring, retainers and pin.
6. Remove the anchor spring from the front brake shoe and remove the front brake shoe.
7. Remove the rear brake shoe hold-down spring, retainers and pin.
8. Remove the automatic adjusting lever spring and the adjuster.
9. Remove the parking brake cable.
10. Use a suitable tool to spread the C-clips and remove the automatic adjuster lever and parking brake lever from the rear brake shoe.
11. Clean all parts and the brake backing plate with an approved brake parts cleaner.
   To install:
12. Before reinstallation, apply grease to the contact points on the backing plate and pivot points on the adjuster lever.
13. Install the parking brake and automatic adjuster levers to the rear brake shoe.

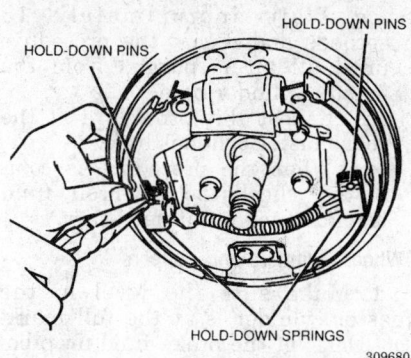

HOLD-DOWN PINS

HOLD-DOWN PINS

HOLD-DOWN SPRINGS

309680

Removing the shoe hold-down springs — Metro

14. Install new C-clips and use pliers to install them. Do not bend more than necessary to hold in place.

15. Check the clearance between the parking brake lever and the brake shoe using a feeler gauge. The clearance must be at least 0.0138 in. (0.35mm). If the clearance is not as specified, replace the shim between the lever and shoe.

16. Connect the parking brake cable.

17. Install the adjuster and the lever spring.

18. Install the rear brake shoe on the backing plate with the end of the shoe inserted in the wheel cylinder and the adjuster in place. Install the hold-down spring, retainer and pin.

— **WARNING** —

Do not allow oil or grease to get on the lining surface.

19. Install the anchor spring between the front and rear shoe.

20. Install the front brake shoe with the end of the shoe inserted in the wheel cylinder and the adjuster in place.

21. Install the hold-down spring, retainers and pin.

22. Connect the return spring.

23. Adjust the brake shoes. When adjusted properly, the outside diameter of the brake shoes will be approximately 0.024 in. (0.6mm) less than the inner diameter of the brake drum.

24. Install the brake drum and the wheel.

25. Lower the vehicle.

26. Road test and check for proper brake operation.

Wheel Cylinder

REMOVAL AND INSTALLATION

1. Raise and safely support the vehicle.

2. Remove the rear wheel.

3. Remove the brake drum and shoes.

4. Remove the bleeder screw from the wheel cylinder.

5. Using a line wrench, loosen the brake pipe flare nut and disconnect the brake pipe from the wheel cylinder. Plug the brake line opening to prevent brake system contamination.

6. Remove the wheel cylinder attaching bolts and the wheel cylinder from the backing plate. Remove the wheel cylinder-to-backing plate gasket, if equipped.

To install:

7. Install the wheel cylinder, using a new gasket if equipped, to the backing plate. If the cylinder does not have a backing plate, then apply gasket sealer to the mating surfaces. Install the retainer bolts to secure in place. Tighten the bolts to 115 inch lbs. (13 Nm) for Metro or 7.5 ft. lbs. (10 Nm) for Prism.

8. Connect the brake pipe to the wheel cylinder and tighten to 12 ft. lbs. (16 Nm) for Metro or 11 ft. lbs. for Prism.

9. Install the bleeder valve.

10. Install the brake drum and shoes.

11. Bleed the brake system.

12. Install the rear wheels.

13. Lower the vehicle. Top off the fluid reservoir to the MAX level with clean DOT 3 type brake fluid only.

14. Road test and check brake operation.

Parking Brake Cable

ADJUSTMENT

1. Make sure the rear brakes are properly adjusted. Apply and release the parking brake several times.

2. Raise and safely support the rear of the vehicle.

3. On Metro, perform the following:

a. Remove both door seal plates and the seat belt buckle bolts at the floor.

b. Disconnect the shoulder harness bolts at the floor and the bottom trim panels.

c. If equipped with a manual transaxle, remove the shift lever boot. Remove the floor console.

d. Remove the trim cover for the parking brake lever. Raise the rear seat cushion. Pull up the carpet to gain access to the parking brake lever.

4. On Prizm, perform the following;

a. Remove the center console.

b. At the rear of the parking brake lever, loosen the locknut on the brake cable.

5. Loosen the parking brake cable adjusting nuts.

6. Tighten or loosen the adjusting nuts, so when the parking brake handle is pulled, the ratcheting travel is 4–9 notches, with 44 lbs. (20 kg) of force.

7. If the rear wheels cannot turn with the parking brake fully applied, and the rear wheels turn freely without resistance when the parking brake is fully released, then the parking brake cables are properly adjusted.

8. Install the center console and any other remaining components. Lower the vehicle.

REMOVAL AND INSTALLATION

1. Disconnect the negative battery cable.

2. Remove the shift lever boot, if equipped with manual transaxle, Remove the console mounting screws and lift out the floor console.

3. On Metro, remove the rear seat cushion and parking brake lever trim cover.

4. Unfasten the harness from the parking brake switch.

5. Loosen the adjuster nut on the lever and remove the brake cables from the equalizer plate or yoke.

6. Raise and safely support the vehicle. remove the rear wheels.

7. Remove the brake cable guide brackets from the cable assembly.

8. On Prizm, remove the exhaust heat shield.

9. Remove the parking brake cable grommet from the vehicle floor pan and pull out the cable.

10. Remove the brake drums and shoes.

11. Disconnect the parking brake cable from the brake shoe lever and the backing plate by squeezing the cable retainer with pliers. Remove the cable from the vehicle.

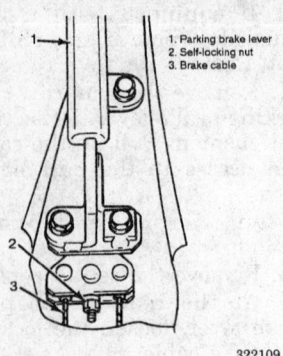

322109

Parking brake cable adjusting nut — Metro

To install:

12. Install the cable into the brake backing plate and secure with the cable retainer.

NOTE: Make certain the cable is properly routed and does not contain any sharp bends or kinks.

13. Connect the cable to the brake shoe lever.

14. Install the brake shoes and drums.

15. Route the parking brake cable and grommet into the vehicle floor pan.

16. Install the parking brake cable guide brackets and torque the guide bracket retaining bolts to 11 ft. lbs. (15 Nm).

17. On Prizm, install the exhaust heat shield.

18. Install the rear wheels. Lower the vehicle.

19. Route the parking brake cable under the interior carpeting.

20. Install the parking brake cables onto the equalizer plate or yoke.

21. Adjust the parking brake cables.

22. Connect the harness to the parking brake switch.

23. Install the remaining components.

24. Reconnect the negative battery cable.

25. Check parking brake operation.

Brake System Bleeding

PROCEDURE

Without ABS

Master Cylinder

— **WARNING** —
Do not allow brake fluid to splash or spill onto painted surfaces; the paint will be damaged. If spillage occurs, flush the area immediately with clean water.

1. If you are installing a new master cylinder, the master cylinder must be bled before installing in vehicle. To bleed a new master cylinder, proceed as follows;

a. Mount the master in a soft-jawed vise, being careful not to damage the housing.

b. Attach short lengths of brake line to the master cylinder outlet ports, with the ends of the lines pointing back into the master cylinder reservoir.

c. Fill the master cylinder reservoir with fresh brake fluid, until the ends of the brake lines are submerged.

d. Push the master cylinder piston in and out. Pump the piston until no more air bubbles appear at the ends of the submerged brake lines.

e. Remove the brake lines from the master cylinder outlet ports. Plug the ports immediately to prevent fluid loss.

2. To bleed an older master cylinder installed in the vehicle, proceed as follows:

a. Remove any existing vacuum in the booster by pumping the brake pedal several times.

b. Make sure the level in the fluid reservoir is at the full or **MAX** mark. DO NOT let the fluid level drop below ½ full at any point during the bleeding procedure.

c. Disconnect the forward brake line fitting at the master cylinder, using a line wrench.

d. Allow brake fluid to flow from the forward brake line port.

e. Connect the forward brake line to the master cylinder and tighten to 12 ft. lbs. (16 Nm).

f. Have an assistant depress the brake pedal slowly one time and hold down. Loosen the forward brake line fitting at the master cylinder to bleed any air from the bore. Tighten the fitting to 12 ft. lbs. (16 Nm). Slowly release the pedal.

g. Wait approximately 15 seconds and repeat the procedure until all air is purged from the master cylinder bore.

h. Repeat this procedure at the rear master cylinder bore.

i. When the procedure is completed, check the reservoir fluid level and top off, if necessary.

Wheel Cylinders and Calipers

1. Make sure the level in the master cylinder is at the full mark. DO NOT let the brake fluid drop below 1/2 full at any point during the process.

2. Clean the bleeder screw at each wheel.

3. Start with the wheel farthest from the master cylinder (right rear).

4. Place the correct size box-end or line wrench over the bleeder valve and attach a tight-fitting transparent hose to the bleeder. Allow the tube to hang submerged in a transparent container of clean brake fluid. The fluid must remain above the end of the hose at all times, otherwise the system will take in air instead of fluid.

5. Have an assistant slowly pump up the brake pedal and hold the pressure.

6. Open the bleed screw about ¼ turn, press the brake pedal to the floor, close the bleed screw and slowly release the pedal. Continue until no more air bubbles are forced out on application of the brake pedal.

7. Repeat procedure on remaining wheel cylinders and calipers in the following order:
- Left rear
- Right front
- Left front

8. Refill the master cylinder to the full or **MAX** mark.

— **WARNING** —
Do not reuse brake fluid which has been bled from the brake system.

9. After a firm pedal is achieved, test drive for proper operation.

With ABS

NOTE: Before bleeding, all necessary repairs must be made and any Diagnostic Trouble Codes (DTC) that may exist must be cleared.

Hydraulic Modulator

1. Using the TECH 1 TK-0-A scan tool, select the MOTOR REHOME function.

2. Make sure the level in the master cylinder is at the **MAX** mark.

DO NOT allow the fluid level to drop below ½ full at any point during the bleeding procedure.

3. Connect a clear plastic hose to the rear bleeder valve on the hydraulic modulator. Place the other end of the bleeder hose in a container partially filled with clean brake fluid.

4. Slowly open the bleeder valve ½–¾ turn.

5. Have an assistant depress the brake pedal and hold down until the fluid begins to flow.

6. Close the bleeder valve and release the brake pedal.

7. Repeat the bleeding procedure until all air has been purged.

8. Remove the bleeder hose from the rear bleeder valve and place it onto the forward bleeder valve of the hydraulic modulator. Repeat the bleeding steps until all air has been purged from the system.

Master Cylinder

NOTE: If any brake component are repaired or replaced, the hydraulic modulator must be bled prior to the brake calipers or wheel cylinders.

1. On Prizm, perform the following steps:

NOTE: Prior to bleeding the brakes, the front and rear displacement cylinder pistons must be returned to the top-most (home) position. Using a TECH 1 or T-100 scan tool or equivalent, select "F5" (motor rehome).

a. Raise and safely support the vehicle.

b. Start the engine, engage the transaxle and run the vehicle above 3 mph (5 km/h), for at least 10 seconds.

c. Observe the "ABS" indicator. Make sure the indicator goes out after approximately 3 seconds. If the "ABS" indicator remains illuminated, a TECH 1 scan tool must be used to diagnose the malfunction. If the "ABS" indicator goes out and stays off, stop the engine and proceed to Steps 2 and 3.

d. Using a TECH 1 or T-100 scan tool or equivalent, enter the manual control function and "Apply" the front and rear motors.

2. Remove any existing vacuum from the booster by pumping the brake pedal several times.

3. Make sure the level in the fluid reservoir is at the full or **MAX** mark. DO NOT let the fluid level drop below ½ full at any point during the bleeding procedure.

4. Disconnect the forward brake line fitting at the master cylinder.

5. Allow brake fluid to flow from the forward brake line port.

6. Connect the forward brake line to the master cylinder and tighten to 12 ft. lbs. (16 Nm).

7. Have an assistant depress the brake pedal slowly one time and hold down. Loosen the forward brake line fitting at the master cylinder to bleed air from the bore. Tighten the fitting to 12 ft. lbs. (16 Nm). Slowly release the pedal.

8. Wait approximately 15 seconds and repeat the procedure until all air is purged from the master cylinder bore.

9. Repeat this procedure at the rear master cylinder bore.

10. When the procedure is completed, check the reservoir fluid level and top off, if necessary.

Calipers and Wheel Cylinders

1. Clean the bleeder screw at each wheel.

2. Starting at the right rear, attach a hose to the bleeder screw and place the end in a clear container of brake fluid.

3. Fill the master cylinder with brake fluid. Have an assistant slowly pump up the brake pedal and hold the pressure.

4. Open the bleed screw about ¼ turn, press the brake pedal to the floor, close the bleed screw and slowly release the pedal. Continue until no more air bubbles are forced from the cylinder on application of the brake pedal.

5. Repeat the procedure on the remaining wheel cylinder and calipers in the following sequence:
• Left rear
• Right front
• Left front

6. Refill the master cylinder to the full or **MAX** mark.

7. Check the brake fluid level and top off, if necessary.

8. Turn the ignition switch **ON**, depress the brake pedal and hold down:
• If the brake pedal feels firm and constant, and the pedal travel is good, start the engine. Check the pedal travel while the engine is running. If the pedal feels firm and constant, and pedal travel is good, then the bleeding procedure was successful.
• If the brake pedal feels soft or the pedal travel is excessive while the engine is **ON** or **OFF**, using the TECH 1 scanner tool, **Release** then **Apply** the motors 3 times and cycle

the solenoids 10 times. Be sure to **Apply** the motors to ensure that the pistons are in the home position. **DO NOT OPERATE THE VEHICLE.** Perform the entire bleeding procedure again.

9. Road test the vehicle. Make several normal stops at a moderate speed to test the brake operation. Be sure to allow a reasonable cooling time between stops.

Wheel Speed Sensor

REMOVAL AND INSTALLATION

Front

1. Disconnect the negative battery cable.

2. Raise and safely support the vehicle.

3. Remove the front wheel and the fender liner, if equipped.

4. On Prizm, remove the 3 clamp bolts holding the sensor harness to the frame and strut.

5. On Prizm, remove the bolt holding the speed sensor to the steering knuckle. Remove the speed sensor from the vehicle.

6. Unfasten the speed sensor connector.

7. Remove the retaining bolt and the sensor from the steering knuckle.

To install:

8. Install the sensor to the knuckle and tighten the bolt to 71–106 inch lbs. (8–12 Nm) for Metro or 69 inch lbs. (8 Nm) for Prizm.

9. Connect the electrical connector to the sensor.

10. Torque the clamp bolts holding the sensor harness to the frame and strut to 48 inch lbs. (5.4 Nm).

11. Install the wheel and fender liner if equipped. Lower the vehicle.

12. Connect the negative battery cable.

Rear

1. Disconnect the negative battery cable.

2. On Prizm, remove the back seat cushion and side trim panel.

3. Raise and safely support the vehicle.

4. Remove the rear wheel.

5. On Prizm, pull out the sensor wire harness.

6. Disconnect the rear wheel speed sensor connector.

7. Remove the retaining bolt and the sensor from the knuckle.

8. On Prizm, remove the 3 clamp bolts holding the sensor wire harness to the body and strut.

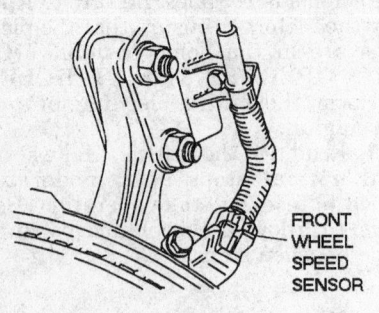

FRONT
WHEEL
SPEED
SENSOR

327710

Front wheel speed sensor — Metro

9. On Prizm, remove the lock bolt from the axle carrier and remove the speed sensor.

To install:

10. On Metro, perform the following:

a. Install the sensor to the knuckle and tighten the bolt to 71–106 inch lbs. (8–12 Nm).

b. Connect the electrical connector to the sensor.

11. On Prizm, perform the following;

a. Install the speed sensor and tighten the lock bolt to 78 inch lbs. (8 Nm).

b. Install the 2 clamp bolts to hold the sensor wire harness to the body and strut. Torque the bolts to 48 inch lbs. (5 Nm).

c. Connect the speed sensor harness and install the grommet.

d. Install the quarter trim panel and side trim and the seat cushion.

12. Install the wheel and lower the vehicle.

13. Connect the negative battery cable.

FRONT SUSPENSION

Strut

REMOVAL AND INSTALLATION

1. Disconnect the negative battery cable.

2. Remove the upper strut support nuts from the engine compartment.

3. Raise and safely support the vehicle.

4. Remove the front wheel.

5. If equipped with ABS, remove the speed sensor.

6. Remove the brake hose clip, then the hose from the strut.

7. Support the lower control arm with a floor jack.

8. Remove the strut-to-steering knuckle bolts, then remove the strut assembly from the vehicle.

To install:

9. Install the strut assembly onto the vehicle. Install the upper support nuts loosely.

10. Install the strut-to-steering knuckle bolts and tighten to 59 ft. lbs. (80 Nm).

11. Tighten the upper strut support nuts to 20–24 ft. lbs. (27–34 Nm).

12. Install the brake hose and brake hose clip. Be sure the brake hose is not twisted when installing.

13. If equipped with ABS, attach the wheel speed sensor.

14. Install the front wheels. Lower the vehicle.

15. Connect the negative battery cable. Check front wheel alignment.

Lower Ball Joints

REMOVAL AND INSTALLATION

Metro

The lower ball joint is an integral part of the lower control arm assembly and is not serviceable as an individual component. If the lower ball joint is defective, the entire lower control arm must be replaced.

Prizm

NOTE: A ball joint separator is required to complete this procedure. It is a commonly available tool which prevents damage to the joint and knuckle. Do not attempt to separate the joint without it.

1. Raise and safely support the vehicle.

———— **WARNING** ————
Do not allow the halfshaft joints to over-extend or the CV-joints may become disconnected.

2. Remove the wheel.

3. Remove the cotter pin from the ball joint nut.

4. Loosen the castle nut but do not remove it. Unscrew it just to the top of the threads and install ball joint separator tool 34754 or equivalent.

5. Use the separator to loosen the ball joint from the steering knuckle.

6. Remove the nuts and bolt holding the ball joint to the control arm.

7. Remove the ball joint from the control arm and steering knuckle.

To install:

8. Install the ball joint to the control arm and tighten the bolts and nut to 105 ft. lbs. (142 Nm).

9. Carefully install the ball joint on the steering knuckle. Use a new castle nut and tighten it to 94 ft. lbs. (128 Nm).

10. Install the cotter pin. If the holes are not in alignment, tighten the nut enough so the cotter pin can be installed. NEVER loosen the castle nut to install a cotter pin.

11. Install the wheel.

12. Lower the vehicle.

13. Check the front wheel alignment.

Lower Control Arms

REMOVAL AND INSTALLATION

Metro

1. Raise and safely support the vehicle. Remove the front wheels.

2. Remove the stabilizer bar link nut, washer and insulator from the control arm, if equipped.

3. Separate the ball joint from the knuckle.

4. Remove the control arm front mounting bracket bolts.

5. Remove the bolts and control arm rear mounting bracket.

6. Remove the control arm and the front bracket from the vehicle.

To install:

7. Install the control arm on the vehicle and secure with bolts. Do not fully tighten the bolts at this time.

8. Install the ball joint.

9. If equipped, install the stabilizer link insulator, washer and nut to the lower control arm. Do not fully tighten the nut.

10. Tighten the control arm rear mounting bolts to 32 ft. lbs. (43 Nm); front mounting bolts to 66 ft. lbs. (90 Nm) and stabilizer link nut to 20 ft. lbs. (28 Nm).

11. Install the front wheels. Lower the vehicle.

12. Check the front wheel alignment.

Prizm

NOTE: If the left side lower control arm is to be removed from a vehicle with automatic transaxle, the front suspension crossmember must first be removed.

1. Raise and safely support the vehicle.

2. Remove the front wheels.

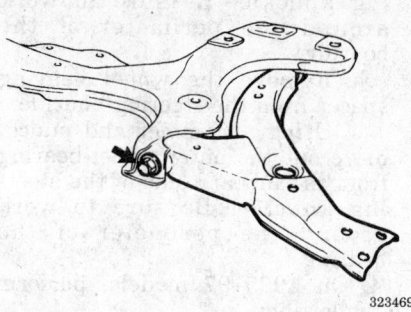

323469

Control arm-to-crossmember mounting bolt — Prizm

3. Disconnect the stabilizer bar from the lower control arm.

4. Remove the ball joint from the lower control arm.

5. If removing control arms on manual transaxle vehicles or the right side control arm on automatic transaxle vehicles, proceed as follows:

   a. Remove the nut and bolts securing the control arm bushing bracket to the vehicle.

   b. Remove the control arm-to-crossmember mounting bolt and remove the control arm.

   c. Clamp the control arm in a soft-jawed vise. Remove the retaining nut, retainer and bushing from the control arm.

6. If removing the left side control arm on automatic transaxle vehicles, proceed as follows:

   a. Remove the bolt, then separate the rear transaxle mount from the crossmember.

   b. Place a suitable jack under the engine crossmember.

   c. Remove the nuts and bolts, then separate the engine crossmember from the suspension crossmember.

   d. Remove the nut and bolts securing the right control arm bushing bracket to the vehicle.

   e. Place a suitable jack under the suspension crossmember. Use a piece of wood between the jack and crossmember to evenly distribute the load.

   f. Loosen and lower the suspension crossmember from the body.

   g. Remove the control arm-to-crossmember mounting bolt and remove the control arm.

   h. Clamp the control arm in a soft-jawed vise. Remove the retaining nut, retainer and bushing from the control arm.

To install:

7. If installing the left side control arm on automatic transaxle vehicles, proceed as follows:

   a. Install a new control arm bushing, retainer and retainer nut. Do not tighten at this time.

   b. Install the lower control arm to the suspension crossmember and tighten the control arm-to-suspension crossmember bolt to 161 ft. lbs. (218 Nm). Tighten the control arm bushing retaining nut to 101 ft. lbs. (137 Nm).

   c. Install the suspension crossmember and tighten the bolts to 152 ft. lbs. (206 Nm).

   d. Remove the support jack from under the suspension crossmember.

   e. Install the retaining bolts and nut to the control arm bushing bracket. Tighten the largest bolt to 108 ft. lbs. (147 Nm), the smaller bolt to 37 ft. lbs. (50 Nm) and the nut to 14 ft. lbs. (19 Nm).

   f. Install the nuts and one bolt securing the engine crossmember and rear transaxle mount to the suspension crossmember. Tighten the nuts to 45 ft. lbs. (61 Nm).

   g. Remove the support jack from under the engine crossmember.

8. If installing either side control arm on manual transaxle vehicles or the right side control arm on automatic transaxle vehicles, proceed as follows:

   a. Install a new control arm bushing, retainer and retainer nut. Do not tighten at this time.

   b. Install the lower control arm-to-suspension crossmember and tighten the control arm-to-suspension crossmember bolt to 159 ft. lbs. (215 Nm). Tighten the control arm bushing retaining nut to 101 ft. lbs. (137 Nm).

   c. Install the 3 retaining bolts and one nut to the control arm bushing retaining bracket. Tighten the largest bolt to 108 ft. lbs. (147 Nm), the smaller bolt to 37 ft. lbs. (50 Nm) and the nut to 14 ft. lbs. (19 Nm).

9. Install the ball joint.

10. Connect the stabilizer bar to the lower control arm.

11. Install the front wheel.

12. Lower the vehicle and bounce the vehicle up and down to stabilize the suspension. Check the wheel alignment.

Sway Bar

REMOVAL AND INSTALLATION

Metro

1. Raise and safely support the vehicle.

2. Remove the nuts, washers and cushions from the stabilizer links at the lower control arms.

3. Remove the mounting bolts, brackets and the stabilizer bar from the vehicle.

4. Remove the stabilizer links from the bar.

To install:

5. Connect the stabilizer links to the stabilizer bar and secure with retaining nuts. Do not fully tighten.

6. Install the stabilizer bar assembly to the vehicle. Install the mounting brackets and bolts. Do not fully tighten.

7. Install the stabilizer cushions, washers and nuts to the stabilizer bar links at the lower control arms.

NOTE: To make sure the stabilizer bar is centered correctly, align the color painted marks on the stabilizer bar to the outside of the mount bushings.

8. Torque the stabilizer bar bracket bolts and the stabilizer link retaining nuts to 20 ft. lbs. (28 Nm).

9. Torque the stabilizer link-to-stabilizer bar nuts to 38 ft. lbs. (50 Nm).

10. Lower the vehicle.

Prizm

1. Raise and safely support the vehicle.

2. Remove the front wheels.

3. Remove the bar links by removing the nuts.

4. Remove the stabilizer brackets by removing the nut and bolts to each bracket.

5. Disconnect the exhaust center pipe by removing the bolts and gasket.

6. Remove the stabilizer bar from the vehicle.

To install:

7. Install the stabilizer bar bushings so they align with the line painted on the stabilizer bar.

8. Install the stabilizer bar to the vehicle.

9. Using a new gasket, connect the exhaust center pipe. Torque the 2 bolts to 32 ft. lbs. (43 Nm).

NOTE: Make sure the vehicle is on the ground before torquing the nuts and bolts.

10. Install the stabilizer bar brackets and torque the bolt next to the nut to 108 ft. lbs. (147 Nm), and the bolts on the opposite side of the nut to 37 ft. lbs. (19 Nm), and the nut to 14 ft. lbs. (19 Nm).

11. Install the stabilizer bar link and torque the nuts to 33 ft. lbs. (44 Nm).

12. Install the wheels, lower the vehicle, and check the front end alignment.

Front Wheel Bearings

ADJUSTMENT

The front wheel bearings on all models cannot be adjusted. To check for a loose wheel bearing, proceed as follows:

1. Raise and safely support the vehicle.

2. Remove the front wheel.

3. Compress the caliper piston to free the caliper assembly.

4. Mount a suitable dial indicator to the knuckle or strut, with the indicator foot contacting the center of the hub next to the nut.

5. Push and pull the brake rotor by hand. If rotor movement exceeds 0.016 in. (0.04mm) for Metro or 0.020 in. (0.05mm) for Prizm, replace the wheel bearings.

6. Install the wheel and lower the vehicle.

7. Apply the brakes several times before moving the vehicle, to seat the caliper piston.

REMOVAL AND INSTALLATION

Metro

1. Raise and safely support the vehicle.

2. Remove the front wheel.

3. Remove the brake caliper and carrier and rotor. Do not let the caliper hang from the brake hose.

4. If equipped, disconnect the ABS wheel speed sensor.

5. Unstake and remove the hub nut and washer. Discard the nut.

6. Remove the hub from the steering knuckle using tools J-2619-01 and J-34866 or equivalent.

7. Separate the tie rod end from the steering knuckle.

8. Separate the ball joint from the steering knuckle.

9. Remove the strut-to-steering knuckle bolts and remove the steering knuckle from the vehicle. Support the halfshaft to prevent damage to the CV-joint.

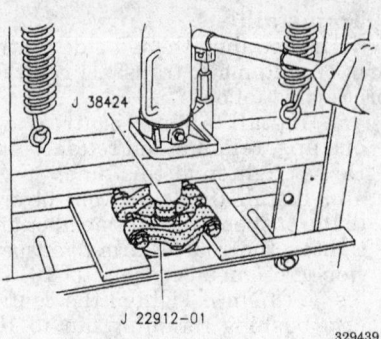

Removing the outer wheel bearing race — 1995–97 Metro

10. Remove the inner and outer oil seals using removal tools J-26941 and J-2619-01 or equivalent.

11. Remove the inner bearing race and bearing from the steering knuckle.

12. Remove the inner wheel bearing oil seal using special tools J-26941 and J-2619-01, or equivalent.

13. On 1993–94 models, perform the following;

a. Using a hammer and punch, drive out the inner wheel bearing from the outside side of the steer-

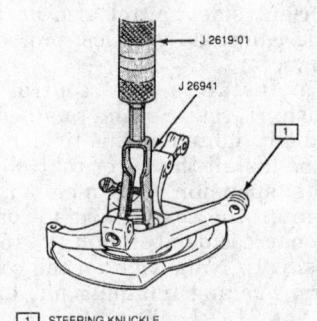

Removal of the inner wheel bearing oil seal — 1993–94 Metro

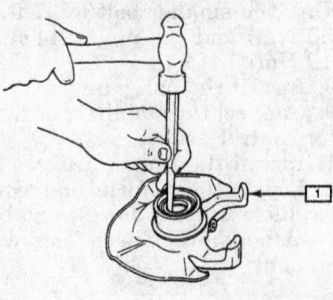

Removing the inner bearing — 1993–94 Metro

ing knuckle. Be sure to work around the perimeter of the bearing.

b. Remove the wheel bearing spacer from the steering knuckle.

c. Using a hammer and punch, drive out the outer wheel bearing from the inboard side of the steering knuckle. Be sure to work around the perimeter of the bearing.

14. On 1995–97 models, perform the following;

a. Remove the snapring from the outer side of the steering knuckle.

b. Remove the outer wheel bearing from the steering knuckle.

c. Remove the outer bearing race from the hub using split plate bearing removal tool J-22912-01, hub race puller pilot tool J-38424 or equivalents, and a press.

d. Support the steering knuckle using support tool J-41392 or equivalent, and remove the wheel bearing assembly from the steering knuckle using wheel bearing removal tool J-41391 and driver handle J-7079-2 or equivalents.

e. Inspect the wheel bearings for excessive wear, corrosion or contamination by dirt or water; replace as necessary.

To install:

15. On 1993–94 models, perform the following;

a. Install the outer wheel bearing to the outboard side of the steering knuckle using bearing installation tools J-34856 and J-7079-2, or equivalent.

b. Install the bearing spacer and inner wheel bearing to the inboard side of the steering knuckle using bearing installation tools J-34856 and J-7079-2, or equivalent.

16. On 1995–97 models, perform the following;

a. Support the steering knuckle using support tool J-41392 or equivalent.

b. Install the wheel bearing assembly into the steering knuckle using the wheel bearing race/outer seal installer tool J-39722 and driver handle J-7079-2 or equivalents.

c. Install the outer bearing into the steering knuckle.

d. Install the snapring into the outer side of the steering knuckle.

17. Install the inner wheel bearing and inner bearing race into the steering knuckle.

18. Install the outer bearing race onto the wheel hub using wheel hub race installer tool J-36050 or equivalent, and a press.

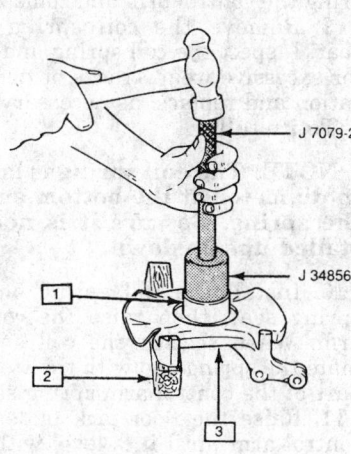

1	OUTER BEARING
2	WOOD BLOCK
3	STEERING KNUCKLE

254800

Installation of the outer wheel bearing — 1993–94 Metro

19. Apply a coating of wheel bearing grease to the lip of the inner oil seal.

20. If equipped with ABS, be sure to align the hole in the seal to the hole in the steering knuckle for the wheel speed sensor.

21. Install the inner oil seal to the inner side of the steering knuckle using installer tool J-41392 and driver handle J-7079-2 or equivalents.

22. Install the outer oil seal to the outer side of the steering knuckle using installer tool J-39722 and driver handle J-7079-2 or equivalents.

23. Apply a light coat of wheel bearing grease to the outside of the hub shaft.

24. Install the wheel hub and outer bearing race to the steering knuckle using a press.

25. Install the steering knuckle on the vehicle, inserting the CV-joint into the hub, and secure with the strut-to-steering knuckle bolts. Tighten the strut-to-steering knuckle bolts to 59 ft. lbs. (80 Nm).

26. Install the ball joint ball. Connect the tie rod end to the steering knuckle with the castle nut. Install a new cotter pin.

27. Install a new hub nut and washer. Tighten the hub nut to 129

ft. lbs. (175 Nm). Using a hammer and dull punch, stake the nut.

28. Install the brake rotor, brake caliper carrier and caliper.

29. If equipped, install the ABS wheel speed sensor. Connect the sensor electrical connector.

30. Install the front wheel and lug nuts.

31. Lower the vehicle. Check the wheel alignment.

Prizm

1. Raise and safely support the vehicle. Remove the wheel.

2. Disconnect the ABS speed sensor, if equipped.

3. Remove the halfshaft nut.

4. Remove the brake caliper and support it out of the way .Remove the disc brake rotor.

5. Separate the tie rod end from the knuckle.

6. Separate the ball joint from the knuckle.

7. Remove the 2 nuts and bolts securing the steering knuckle to the strut.

8. Remove the steering knuckle and hub assembly.

9. Clamp the knuckle in a vise with protected jaws.

10. Remove the dust deflector.

11. Use a slide hammer and extractor tool J-26941 or equivalent, to remove the inner oil seal.

12. Remove the snapring.

13. Using hub puller tools J-25287 and J-35378 or equivalents, remove the hub from the knuckle.

14. Remove the brake splash shield.

15. Using a split plate bearing remover, puller pilot and a shop press, remove the inside and outside inner bearing races from the hub.

16. Remove the outer oil seal.

17. Position the steering knuckle on support tool J-35379 or equivalent. Remove the bearing from the knuckle using a plastic hammer and remover tool J-35399 or equivalent, and driver handle J-8092 or equivalent.

To install:

18. Clean and inspect the hub and knuckle for cracks, wear or other damage; replace as necessary.

19. Press a new bearing into the steering knuckle using installation tool J-37777 or equivalent, and driver handle J-8092 or equivalent.

20. Install a new snapring into the steering knuckle.

21. Install a new inner oil seal onto the steering knuckle using seal installer J-35737-2 or equivalent.

22. Insert the side lip of a new outer oil seal into seal installer

J-35737-01 or equivalent, and drive the outer oil seal into the steering knuckle.

23. Apply multi-purpose grease to the outer oil seal lip.

24. Install the brake splash shield and tighten the bolts to 73 inch lbs. (8.3 Nm).

25. Place new inside and outside inner races on the hub bearing. Press the hub into the knuckle using installer J-35399 or equivalent, and driver handle J-8092 or equivalent.

26. Apply GM lubricant 1050109 or equivalent, to the contact surfaces of the inner oil seal lip and drive axle.

27. Drive a new dust deflector into the steering knuckle using a hammer and dull chisel.

28. Place the steering knuckle and hub assembly onto the halfshaft and install the knuckle-to-strut bolts and nuts.

29. Attach the ball joint to the control arm.

30. Attach the tie rod end to the knuckle.

31. Install the disc brake rotor and the brake caliper.

32. Install the ABS sensor, if equipped.

33. Install the nut on the end of the halfshaft.

34. Install the wheel and lower the vehicle.

35. Check the wheel alignment.

REAR SUSPENSION

Strut

REMOVAL AND INSTALLATION

Prizm

1. Raise and safely support the vehicle.

2. Remove the wheel.

3. Disconnect the brake line from the wheel cylinder (at the backing plate). Plug the lines to prevent leakage.

4. Remove the clip from the brake hose and remove the hose from the strut bracket.

5. Disconnect the sway bar link from the strut.

6. Remove the strut mounting bolts from the rear knuckle.

7. Remove the seat back side cushion (Sedan) or the rear sill side panel (Hatchback) to gain access to the upper strut mount.

8. Remove the nuts holding the upper strut mount to the body. Do not loosen the center strut piston nut.

9. Remove the strut assembly from the vehicle.

To install:

10. Place the strut in position on the vehicle and install the 3 upper retaining nuts, tightening them to 29 ft. lbs. (39 Nm).

11. Reinstall the seat back side cushion or the rear sill side panel.

12. Install the strut into the rear knuckle assembly; install the bolts and nuts and tighten them to 105 ft. lbs. (142 Nm).

13. Reconnect the sway bar link to the strut. Tighten the bolts to 32 ft. lbs. (44 Nm).

14. Connect the brake hose into the bracket and install the clip.

15. Connect the metal line to the wheel cylinder. Make certain the fittings are properly threaded before tightening them.

16. Bleed the brake system.

17. Install the wheel. Lower the vehicle.

18. Check the wheel alignment.

Shock Absorber

REMOVAL AND INSTALLATION

Metro

1. Open the vehicle hatchback or deck lid.

2. Raise and safely support the vehicle just high enough so the top and the bottom of the shock can be accessed. Remove the wheels.

3. Place a floor jack under the suspension arm for support.

4. Remove the lower shock-to-knuckle bolt.

NOTE: Do not open the knuckle slit wider than necessary. Do not lower the jack more than necessary during the strut removal or the coil spring may become unseated.

5. Remove the trim cover from the upper shock tower and remove the upper support nuts.

6. Remove the shock from the knuckle. Compress the shock and remove. If the shock is hard to remove, open the slit of the knuckle by inserting a wedge.

To install:

7. Install the shock into the vehicle. Position the bottom of the align-

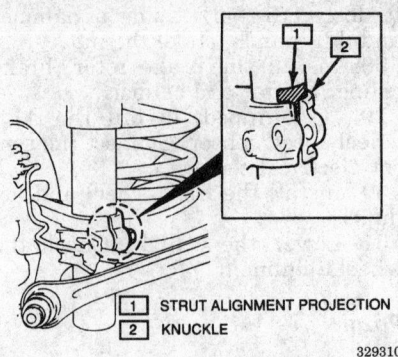

| 1 | STRUT ALIGNMENT PROJECTION |
| 2 | KNUCKLE |

329310

Shock-to-knuckle alignment — Metro

ment projection inside the knuckle opening.

8. Install the upper support nuts and tighten to 24 ft. lbs. (30 Nm). Install the trim cover to the upper shock tower.

9. Install the shock-to-knuckle bolt and tighten to 44 ft. lbs. (60 Nm). Remove the floor jack.

10. Install the rear wheels. Lower the vehicle.

11. Check the wheel alignment.

Coil Spring

REMOVAL AND INSTALLATION

Metro

1. Raise and safely support the vehicle. Remove the rear wheel.

NOTE: To facilitate the toe-in adjustment after reinstallation, confirm which one of the lines stamped on the washer is in the closest alignment with the stamped line on the control rod. If not marked, add the alignment marks.

2. Remove the brake hose from the bracket on the control rod.

3. Remove the control rod inside bolt (body center side) and outside (wheel side) stud bolts.

NOTE: The control rods are not interchangeable. Be sure the control rods are installed on the same side from which they were removed and in the same direction.

4. If equipped with ABS, disconnect the speed sensor.

5. Remove the stabilizer link from the control arm, if equipped. Support the control arm with a floor jack.

6. Remove the control arm-to-knuckle bolt and nut.

7. Manually pull down the control arm from the knuckle assembly.

8. Lower the floor jack slowly, lowering the control arm and coil spring.

9. Remove the coil spring and seat. Inspect the coil spring and seat for excessive wear, cracks or deterioration and replace, as necessary.

To install:

NOTE: The coil spring is larger in diameter at the bottom end of the spring. Be sure it is not installed upside down.

10. Install the coil spring on the spring seat, then raise the control arm. When seating the coil spring, mate the spring end with the stepped part of the control arm spring seat.

11. Raise the floor jack under the control arm until it is level with the suspension knuckle. Be sure to align the bolt holes.

12. Install the lower knuckle mount bolt.

13. If equipped with ABS, fasten the speed sensor.

14. If equipped, install the stabilizer link to the control arm.

15. Remove the jack from under the control arm.

16. Install the control rod. Install the inside and outside control rod bolts, Align the marks on the eccentric cam and body that were made before removal. Torque the control rod inside and outside nuts to 59 ft. lbs. (80 Nm).

17. Install the brake hose to the control rod bracket and secure with the E-clip.

18. Remove the floor jack.

19. Install the rear wheels. Lower the vehicle.

Control Arm

REMOVAL AND INSTALLATION

Metro

1. Raise and safely support the vehicle. Remove the rear wheel.

NOTE: To facilitate the toe-in adjustment after reinstallation, confirm which one of the lines stamped on the washer is in the closest alignment with the stamped line on the control rod. If not marked, add the alignment marks.

2. Remove the brake hose from the bracket on the control rod.

3. Remove the control rod inside bolt (body center side).

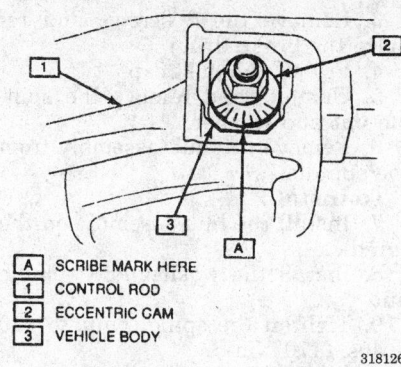

A	SCRIBE MARK HERE
1	CONTROL ROD
2	ECCENTRIC CAM
3	VEHICLE BODY

318126

Markings on the eccentric cam — Metro

4. Remove the outside (wheel side) of the control rod from the rear knuckle stud bolt.

NOTE: The control rods are not interchangeable. Be sure the control rods are installed on the same side from which they were removed and in the same direction.

5. If equipped with ABS, disconnect the speed sensor.
6. Remove the stabilizer link from the control arm, if equipped and support the control arm with a floor jack.
7. Remove the control arm-to-knuckle bolt and nut.
8. Manually pull down the control arm from the knuckle assembly.
9. Lower the floor jack slowly to lower the control arm and coil spring.
To install:
10. Install the coil spring on the spring seat of the control arm then raise the control arm. Be sure to align the bolt holes.
11. Install the lower knuckle mount bolt. Torque the bolt to 37 ft. lbs. (50 Nm).
12. If equipped with ABS, fasten the speed sensor.
13. If equipped, install the stabilizer link to the control arm and torque the stabilizer link retaining nut to 19 ft. lbs. (25 Nm).
14. Remove the jack from under the control arm.
15. Install the control rod. Install the inside and outside control rod bolt, eccentric cam, washers and nuts. Do not tighten them at this time. Align the marks on the eccentric cam and body that were made before removal.
16. To keep the eccentric cam stationary while tightening, hold the bolt head with a wrench. Torque the

control rod inside and outside nuts to 59 ft. lbs. (80 Nm).

NOTE: When tightening the nuts, it is desirable to have the vehicle off the hoist and in a non-loaded state. Also when tightening the inside nut, align the line stamped on the body with the line on the eccentric cam washer as confirmed before removal or align the matchmarks if marked.

17. Install the brake hose to the control rod bracket and secure with the bracket E-clip.
18. Remove the floor jack.
19. Install the rear wheel and lug nuts. Lower the vehicle.

Prizm

1. Raise and safely support the vehicle on jackstands. Remove the rear wheels.
2. Remove the strut rod arm nut and through-bolt at the knuckle.
3. Remove the strut rod arm nut and through-bolt at the rear crossmember.
4. Remove the strut arm.
To install:
5. Position the strut arm to the crossmember and knuckle.
6. Insert the through-bolt and nut. Hand-tighten the nuts.
7. Torque the through-bolt and nut to 67 ft. lbs. (97 Nm).
8. Install the wheels. Lower the vehicle.

Sway Bar

REMOVAL AND INSTALLATION

Metro

1. Raise and safely support the vehicle.
2. Remove the nuts, washers and cushions from the stabilizer links at the control arms.
3. Remove the mounting bracket bolts, mounting brackets and the stabilizer bar from the vehicle.
4. Remove the stabilizer links from the stabilizer bar.
To install:
5. Reconnect the stabilizer links to the stabilizer bar. Install the retaining nuts, but do not fully tighten.
6. Install the stabilizer bar onto the vehicle. Install the mounting brackets and mounting bolts. Do not fully tighten.

7. Install the link cushions, washers and retaining nuts to the stabilizer links at the control arms.

NOTE: To center the stabilizer bar correctly, align the painted marks on the stabilizer bar to the inside of the mount bushings.

8. Torque the stabilizer bar mounting bracket bolts to 20 ft. lbs. (28 Nm). Torque the stabilizer link retaining nuts to 20 ft. lbs. (28 Nm). Torque the stabilizer link-to-stabilizer bar nuts to 38 ft. lbs. (50 Nm).
9. Lower the vehicle.

Prizm

1. Raise and safely support the vehicle. Remove the rear wheels.
2. Remove the left and right stabilizer bar links.
3. Scribe matchmarks between the stabilizer bar bushings, stabilizer bar and body and remove the stabilizer bar bushings.
4. Remove the muffler from the hangers.
5. Remove the bolts and disconnect the intermediate pipe heat shield.
6. Support the fuel tank.
7. Remove the right and left side fuel tank bands.
8. Support the suspension crossmember with a jack and remove the 6 bolts attaching the crossmember to the body. Lower the suspension crossmember.
9. Remove the stabilizer bar from the vehicle.
To install:
10. Install the stabilizer bar in the vehicle.
11. Raise the suspension crossmember with the jack and install the 6 bolts attaching the crossmember to the body. Tighten the bolts to 43 ft. lbs. (58 Nm).
12. Install the fuel tank bands.
13. Install the 3 bolts and connect the intermediate pipe heat shield.
14. Install the muffler to the hanger and secure to the body with the 4 retaining bolts.
15. Install the 2 bolts and connect the resonator pipe and gasket to the muffler.
16. Install the 2 bolts and connect the resonator pipe and gasket to the catalytic converter.
17. Install the stabilizer bar bushings to the stabilizer bar aligning the matchmarks.
18. Install the left and right stabilizer bar bushing retainers and tighten to 14 ft. lbs. (19 Nm).
19. Install the left and right stabilizer bar links and tighten the stabi-

lizer bar link to strut assembly bolts to 33 ft. lbs. (44 Nm) and the stabilizer bar link to stabilizer bar bolts to 26 ft. lbs. (35 Nm).

20. Install the rear wheels.
21. Lower the vehicle.

Wheel Bearings

ADJUSTMENT

The rear wheel bearings cannot be adjusted. To check for a loose wheel bearing, proceed as follows:

1. Raise and safely support the vehicle.
2. Remove the rear wheel.
3. Mount a suitable dial indicator to the knuckle or strut, with the indicator foot contacting the center of the hub next to the hub nut.
4. Push and pull the brake drum/hub assembly by hand. If hub movement exceeds 0.012 in. (0.03mm) on Metro or 0.020 in. (0.05mm) on Prizm, replace the wheel bearings.
5. Install the wheel and lower the vehicle.

REMOVAL AND INSTALLATION

Metro

Without ABS

1. Raise and safely support the vehicle.
2. Remove the rear wheel.
3. Pry off the dust cap.
4. Unstake and remove the spindle nut and washer. Discard the spindle nut; it must not be reused.
5. Gently tap on the brake drum/hub assembly with a hammer to break loose and remove the drum/hub assembly from the spindle.
6. Place the brake drum/hub assembly on wooden blocks on a workbench.

7. Drive the inner bearing out from the drum/hub assembly using a hammer and punch. The bearing spacer can be moved from side to side to enable placement of the punch. Work around the perimeter of the inner bearing by moving this spacer.
8. Remove the bearing spacer.
9. Drive the outer bearing out from the inboard side of the drum/hub assembly using the same method.
10. Inspect the wheel bearings for wear, corrosion or contamination and replace, as necessary.

To install:

NOTE: The outer bearings is smaller in diameter than the inner one.

11. Install the outer bearing in the drum/hub assembly using installation tools J-34842 and J-7079-2 or equivalents.
12. Install the bearing spacer in the brake drum hub with the inner lip toward the outer bearing.

NOTE: Make sure the bearing spacer is properly installed. The assembly will not fit on the spindle if the spacer lip is toward the inner bearing.

13. Install the inner bearing in the drum/hub assembly using installation tools J-34842 and J-7079-2 or equivalents.
14. Install the brake drum/hub assembly on the spindle.
15. Install the washer and a new spindle nut.
16. Tighten the spindle nut to 74 ft. lbs. (100 Nm).
17. Stake the spindle nut.
18. Install the dust cap.
19. Install the wheel. Lower the vehicle.

With ABS

1. Raise and safely support the vehicle.
2. Remove the rear wheel.

3. Remove the 2 screws and remove the brake drum.
4. Pry off the dust cap.
5. Unstake and remove the spindle nut and washer.
6. Remove the hub assembly from the spindle.

To install:

7. Install the hub assembly on the spindle.
8. Install the washer and spindle nut.
9. Tighten the spindle nut to 120 ft. lbs. (170 Nm).
10. Stake the spindle nut.
11. Install the dust cap.
12. Install the wheel. Torque the lug nuts, in a cross sequence, to 44 ft. lbs. (60 Nm).
13. Lower the vehicle.

Prizm

1. Raise and safely support the vehicle.
2. Remove the wheel.
3. Remove the brake drum or rotor.
4. Check bearing end-play using a dial indicator. If end-play exceeds 0.002 in. (0.05mm), the hub must be replaced.
5. Disconnect and remove the ABS wheel speed sensor, if equipped.
6. Remove the 4 bolts securing the hub to the knuckle and remove the hub.
7. Remove the O-ring from the backing plate.

To install:

8. Install a new O-ring onto the backing plate.
9. Install the hub to the knuckle with the mounting bolts and tighten to 59 ft. lbs. (80 Nm).
10. Install the wheel speed sensor, if equipped.
11. Install the brake drum or rotor.
12. Install the wheel and lower the vehicle.
13. Check the rear wheel alignment.

FIRING ORDERS

NOTE: To avoid confusion, always replace spark plug wires one at a time.

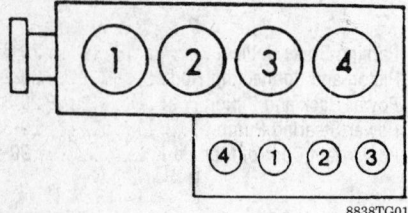

8838TG01

1.9L Engines
Engine Firing Order: 1–3–4–2
Distributorless Ignition System

ENGINE ELECTRICAL

NOTE: Disconnecting the negative battery cable on some vehicles may interfere with the functions of the on board computer systems and may require the computer to undergo a relearning process, once the negative battery cable is reconnected.

Electronic Ignition System

NOTE: The 1.9L engine is equipped with a distributorless Electronic Ignition (EI) system. Besides the Powertrain Control Module (PCM), which regulates ignition timing under normal operating conditions, the 2 major components of the EI system are an EI module/coil pack and a crankshaft position sensor.

REMOVAL AND INSTALLATION

Ignition Module and Coils

1. Properly disable the SIR system, if equipped, and disconnect the negative battery cable.
2. Tag and disconnect the spark plug wires from the EI unit. The unit is on the front of the transaxle bellhousing.
3. Disengage the electrical connectors from the ignition module.

4. Remove the 4 retaining bolts, then remove the EI unit from the vehicle.
5. If necessary, the 2 coils may be removed from the unit at this time by using a pair of needle-nose pliers to squeeze the retaining tabs while pulling the coils upward.
   To install:
6. If removed, install the coils over the retaining tabs on the ignition module.
7. Run a 6 x 1.0mm tap through the module mounting holes to remove remaining thread sealant residue and verify that the module and bellhousing mating surfaces are clean and free from grit or dirt.
8. Always use new module mounting bolts. Install the ignition module/coil assembly using the new mounting bolts with the factory applied yellow sealant and tighten the bolts to 71 inch lbs. (8 Nm). Be careful when tightening the mounting bolts, verify that each bolt head is properly seated on the module unit when tightened.
9. Engage the electrical connectors and spark plug wires to the module unit.
10. Connect the negative battery cable and, if equipped, properly enable the SIR system.
11. Start the engine and check operation.

Crankshaft Position Sensor

1. Disconnect the negative battery cable.
2. Raise the vehicle and support safely.
3. Disengage the electrical connector from the sensor located at the lower rear of the engine block.
4. Remove the retaining bolt from the sensor flange, then remove the sensor from the engine block.
5. Lubricate the sensor O-ring with clean engine oil and install in the reverse of removal. Tighten the sensor retaining bolt to 80 inch lbs. (9 Nm).

Ignition Timing

ADJUSTMENT

There is no conventional distributor for the EI system. Instead, timing is controlled by the Powertrain Control Module (PCM) through the EI module. The PCM has the ability to advance or retard ignition timing, as necessary for optimal engine performance. No timing adjustments are necessary or possible.

Alternator

PRECAUTIONS

Several precautions must be observed with alternator equipped vehicles to avoid damage to the unit.

- If the battery is removed for any reason, make sure it is reconnected with the correct polarity. Reversing the battery connections may result in damage to the 1-way rectifiers.
- When utilizing a booster battery as a starting aid, always connect the positive to positive terminals and the negative terminal from the booster battery to a good engine ground on the vehicle being started.
- Never use a fast charger as a booster to start vehicles.
- Disconnect the battery cables when charging the battery with a fast charger.
- Never attempt to polarize the alternator.
- Do not use test lights of more than 12 volts when checking diode continuity.
- Do not short across or ground any of the alternator terminals.
- The polarity of the battery, alternator and regulator must be matched and considered before making any electrical connections within the system.
- Never separate the alternator on an open circuit. Make sure all connections within the circuit are clean and tight.
- Disconnect the battery ground terminal when performing any service on electrical components.
- Disconnect the battery if arc welding is to be done on the vehicle.

BELT TENSION ADJUSTMENT

The belt is automatically adjusted using a spring loaded automatic tensioner. The marking on the tensioner arm must fall within the operating range (2 marks) on the tensioner body. If the tensioner falls outside the operating range, the serpentine drive belt must be replaced.

REMOVAL AND INSTALLATION

1. Disconnect the negative battery cable and remove the serpentine drive belt.
2. Remove the power steering pump assembly; this must be done in order to access the alternator attaching bolts.
3. Remove the alternator splash shield attaching bolt from the rear of

the alternator, then unclip the shield from the alternator.

4. Disengage the alternator electrical connections.

5. Remove the upper and lower alternator attaching bolts.

NOTE: The alternator attaching bolts can usually be reached from under the hood. If difficulty is encountered loosening the lower mounting bolt, remove the vehicle passenger side tire and splash shield. With these components removed, the alternator may be removed through the wheelwell.

6. Lift the alternator through the opening between the shock tower and the intake manifold and remove the alternator from the vehicle.

To install:

7. Position the alternator in the vehicle and install the lower attaching bolt.

NOTE: Always use a new wiring harness-to-alternator fastener to assure proper electrical contact.

8. Install the upper attaching bolt and the 2 wiring harness connectors.

9. Tighten the alternator attaching bolts to 27 ft. lbs. (37 Nm), then tighten the alternator positive terminal fastener to 89 inch lbs. (10 Nm).

10. If removed, install the vehicle passenger's side tire and splash shield.

11. Install the alternator splash shield and tighten the fastener bolt to 89 inch lbs. (10 Nm).

12. Install the power steering pump assembly.

13. Connect the negative battery cable.

Drive Belt

REMOVAL AND INSTALLATION

1. Depress the tensioner arm using Snap-On® tool S-8190A, or an equivalent 9/16 in. (14mm) wrench.

2. Remove the belt from the idler or air conditioning compressor pulley.

3. Remove the drive belt from the vehicle.

To install:

4. Route the belt around the pulleys, except for the front cover idler or air conditioning compressor pulley.

5. Depress the tensioner arm and slip the belt over the idler or air conditioning compressor pulley. Make

sure the belt ribs are properly aligned on the pulleys.

6. If the tensioner idler pulley retaining bolt is loose, remove the bolt and apply Loctite® 242 or equivalent to the bolt threads. Install the bolt and tighten to 22 ft. lbs. (30 Nm).

Starter

REMOVAL AND INSTALLATION

1. Disconnect the negative battery cable.

2. Remove the air inlet tube and fresh air hose. For the DOHC engine, the resonator must be lifted for disengagement from the engine support bracket.

3. Remove the upper starter bolts using the access hole provided next to the intake manifold support bracket. If necessary, remove the intake support bracket to provide better access and tool swing.

4. Raise and support the vehicle safely.

5. Remove the starter shield pin by pulling on it with pliers. Lift upward and carefully release the shield from the solenoid.

───── **WARNING** ─────

It is very important that the solenoid electrical connection nuts and studs are sprayed with penetrating oil prior to removal in order to avoid damage to the solenoid end cap.

6. Carefully loosen the retainers and disengage the starter electrical connectors. Be sure to tag the connections to assure proper installation.

7. Remove the lower starter bolt.

8. Remove the rear starter support bracket attaching bolt.

9. Rotate the starter until the bracket misses the axle shaft support

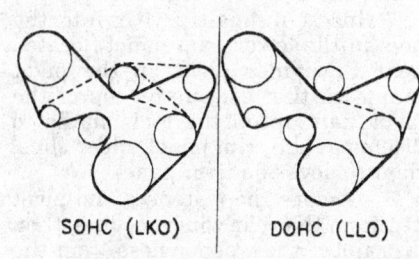

SOHC (LKO)   DOHC (LLO)

8838TG02

Accessory drive belt routing

bracket. Pull the starter rearward and toward the left side of the vehicle to remove.

To install:

10. If removed, or if the starter is being replaced, install the rear starter support bracket to the starter. Tighten the bracket nuts to 80 inch lbs. (9 Nm).

11. Guide the starter and bracket assembly into the bellhousing, then rotate the assembly until the lower bolt hole in the starter nose aligns.

12. Verify that the bracket is properly aligned, then loosely install the bracket and housing bolts.

13. If necessary, raise or lower the vehicle for access, then install the upper bolt.

14. Torque the starter mounting bolts to 27 ft. lbs. (37 Nm) and the bracket bolt to 22 ft. lbs. (30 Nm).

15. Install the electrical connectors and wiring nuts. Be careful not to overtighten the nuts and crack the solenoid end cap. Tighten the starter positive terminal to 89 inch lbs. (10 Nm) and the solenoid terminal to 44 inch lbs. (5 Nm).

16. Install the shield and push pin, making sure the pin is positioned for possible future removal.

17. Lower the vehicle, as necessary, and if removed, install the intake manifold support bracket.

18. Install the air intake tube and fresh air hose. For the DOHC engine, verify the resonator button is properly located in the service bracket.

19. Connect the negative battery cable.

CHASSIS ELECTRICAL

Blower Motor

REMOVAL AND INSTALLATION

1. Disconnect the negative battery cable.

2. Disconnect the blower motor connectors under the glove compartment.

3. Remove the blower motor mounting screws and motor assembly.

4. Install the motor in the reverse order and check operation.

Windshield Wiper Motor

REMOVAL AND INSTALLATION

1. Verify that the wipers are in the **PARK** position and disconnect the negative battery cable.

2. Remove the wiper arm finish cap and wiper arm fastening nut. Lift the blade away from the windshield and remove. Repeat for the other blade.

3. Remove the cowl trim panel fasteners at the windshield edge of the panel, then open the hood and remove the remaining fasteners. Carefully remove the cowl trim panel. If equipped with cowl mounted washer nozzles, disconnect the washer hoses from the nozzles.

4. Remove the 2 instrument panel top cover screw caps and screws. Carefully remove the cover by lifting at the rear edge to disengage the retaining clips and sliding the panel out of the windshield clips.

5. Remove the retaining screws and unsnap the windshield defroster duct from the heating and air conditioning mode valve assembly.

6. Except for SC vehicles, remove the defroster nozzle by rotating the front up and away from the windshield, exposing the wiper module fasteners. On Coupes, the windshield rake angle prevents nozzle removal.

7. Remove the wiper module fasteners and reposition the module in order to disconnect the wiring from the motor and module frame. Carefully remove the wiper module and motor assembly making sure not to contact or damage the windshield.

8. If necessary, remove the crank arm nut and disconnect the arm from the motor shaft, then remove the wiper motor attaching screws and remove the motor from the module.

To install:

9. Verify that the motor is in the **PARK** position. If necessary, temporarily connect the motor wiring and the negative battery cable, turn the wiper control ON then OFF and the motor will move to the correct position.

10. If removed, install the motor to the module and tighten the retainers to 89 inch lbs. (10 Nm).

11. Set the motor crank arm to the 9 o'clock position and install the arm onto the motor shaft. Apply Loctite® 242 or equivalent thread sealant to the crank arm retaining nut, then install the nut and tighten to 21 ft. lbs. (28 Nm).

12. Position the wiper module assembly into the vehicle and connect the wiring to the wiper motor and to the module frame.

13. Install the module retaining bolts and tighten to 89 inch lbs. (10 Nm).

14. Install the cowl trim panel and, if applicable, connect the washer hoses to the nozzles.

15. Install the wiper arm assemblies using new nuts (torque retention of the old nuts may be insufficient) and tighten the nuts to 21 ft. lbs. (28 Nm).

16. If removed, rotate the windshield defroster nozzle onto the mode valve assembly.

17. Install the screws fastening the defroster duct and make sure the duct is snapped on both sides of the heating and air conditioning module.

18. Install the instrument panel top cover and screws. Insert the panel cover screw caps.

19. Connect the negative battery cable and verify proper system operation.

Liftgate Window Wiper Motor

REMOVAL AND INSTALLATION

1. Make sure the wiper is in the **PARK** position, then disconnect the negative battery cable.

2. Remove the rear wiper arm finish cap, then remove the retaining nut.

3. Lift the wiper blade away from the liftgate window and remove the blade/arm assembly.

4. Remove the rear wiper pivot bushing.

5. Raise the liftgate, then remove the wedge blocks from either end of the liftgate assembly.

6. Remove the lower fasteners from the liftgate lower trim panel, by pushing in the of each center pin approximately ⅛ inch until it clicks, then remove the fastener.

7. Insert a small prybar into the hole in the lower trim panel (located near the wiper pivot on the pivot hump) so the tool sits on top of the pivot. Lift up on the tool handle to disengage the trim panel upper clips, then remove the trim panel.

8. Remove the fasteners and pivot nut from the rear wiper module, then disconnect the washer hose from the module check valve.

9. Disconnect the wiring from the module, then remove the module from the liftgate.

To install:

10. Position the wiper module to the liftgate, then connect the washer hose to the check valve and the wiring to the module terminals.

11. Install the module fasteners and tighten to 89 inch lbs. (10 Nm).

12. Install the grommet, washer and nut on the wiper module pivot shaft, then tighten the nut to 119 inch lbs. (14 Nm).

13. Align the upper clips on the lower trim panel to the liftgate slots, then install the panel by pushing at the clip locations.

14. Reset the trim panel push-in fasteners by spreading the center pin tabs and moving the pin so it sits approximately ¼ in. out of the fastener. Insert the fasteners into the bottom of the liftgate lower trim panel and push the center pin until flush.

15. Install the liftgate wedge blocks.

16. Install the rear wiper pivot bushing.

17. Position the arm onto the pivot shaft with the blade horizontal to the liftgate glass lower edge, then install the retaining nut and tighten to 159 inch lbs. (18 Nm).

18. Install the finish cap over the retaining nut.

19. Connect the negative battery cable, then check for proper motor operation.

Concealed Headlights

MANUAL OPERATION

If equipped with concealed headlights, the headlight doors can be opened automatically by turning the headlight switch to the ON position. Turning the switch back 1 click to the parking lights ON position or 2 clicks to the lights OFF position will leave the doors open. To close the headlight doors turn the switch 1 final click to the headlight closed position.

A headlight manual control knob is located next to each headlight door to open the doors without the aid of the electric motor. To manually open the door, raise the hood and turn the control knob in the direction of the arrow until the door is fully opened.

Combination Switch

REMOVAL AND INSTALLATION

1. Properly disable the SIR system, if equipped, and disconnect the negative battery cable.

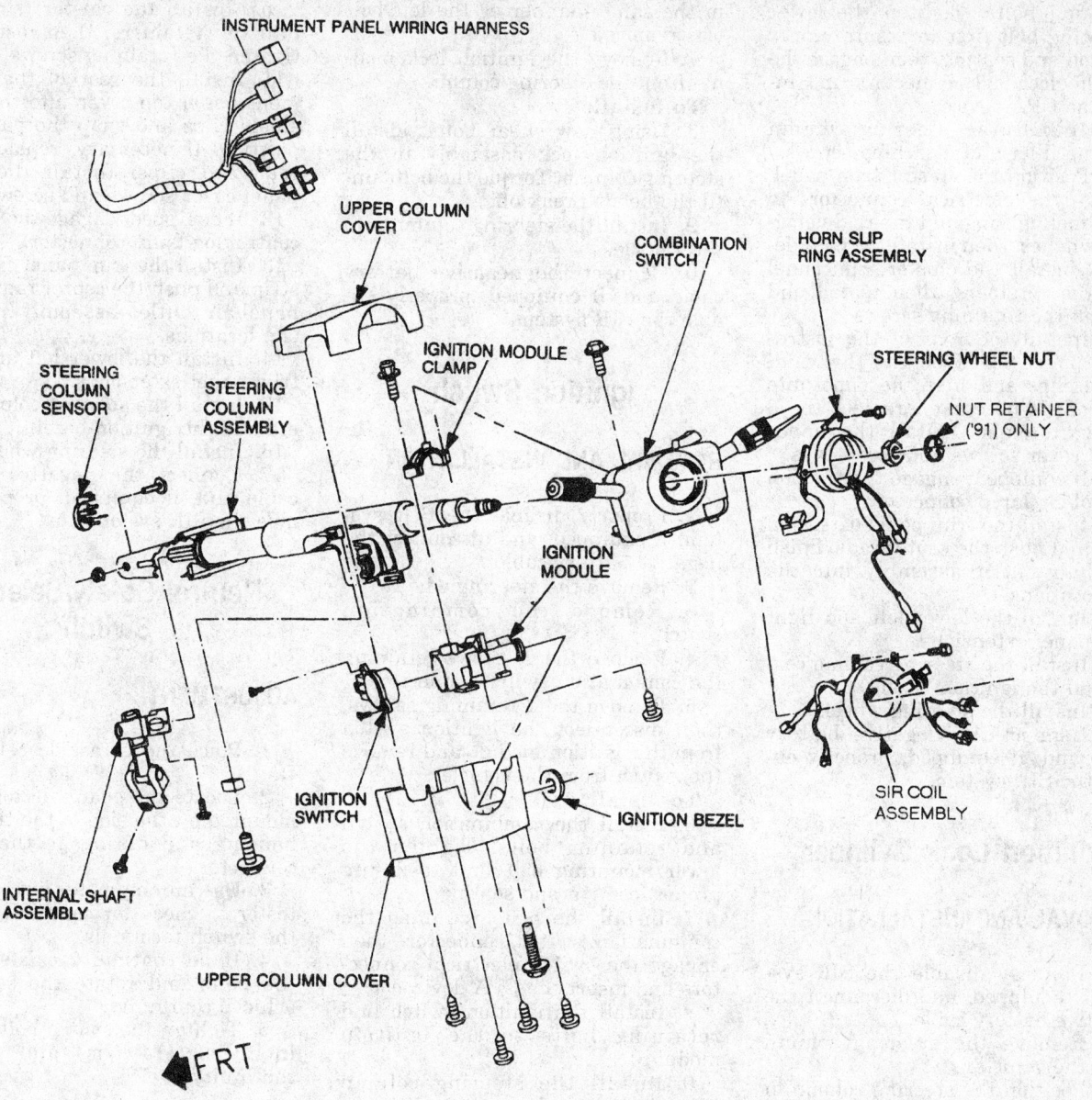

INSTRUMENT PANEL WIRING HARNESS

UPPER COLUMN COVER

COMBINATION SWITCH

HORN SLIP RING ASSEMBLY

STEERING WHEEL NUT

NUT RETAINER ('91) ONLY

IGNITION MODULE CLAMP

STEERING COLUMN SENSOR

STEERING COLUMN ASSEMBLY

IGNITION MODULE

IGNITION SWITCH

IGNITION BEZEL

SIR COIL ASSEMBLY

INTERNAL SHAFT ASSEMBLY

UPPER COLUMN COVER

FRT

8838TG03

Exploded view of the steering column assembly

2. Remove the steering wheel.

3. Remove the 2 retaining screws and the upper steering column cover from the steering column.

4. Remove the ignition lock bezel, then remove the 2 retaining screws and the lower steering column cover.

5. Disconnect the velcro fasteners, then remove the left and right lower trim panel extensions by pulling them each out of the 2 upper fasteners.

6. Carefully remove the center radio finish panel/air outlet assembly by pulling outward at the clip locations. Start at the bottom and move upward. Do not use tools that might damage the trim panel. If equipped,

disconnect the wiring from the traction control/fog lamp switch.

7. If necessary, remove the trim panel extension strip by pulling out at the fastener location.

8. Remove the 2 instrument panel top cover screw caps and screws. Carefully remove the cover by lifting at the rear edge to disengage the retaining clips and by sliding the panel out of the windshield clips.

9. Remove the 4 cluster trim panel attaching screws. Carefully pull the cluster trim panel upward/outward to disengage it from the retainers.

10. Remove the CPA devices, then disengage the electrical connectors from the instrument panel lighting

and rear window defogger switches. Remove the cluster trim panel from the vehicle.

11. Remove the 6 screws attaching the steering column opening filler, then carefully remove the assembly. Protect the console from the damage when removing the assembly.

12. Remove the CPA device and disconnect the wires from the lever control switch.

13. Remove the retaining bolts and remove the combination switch assembly from the steering column.

To install:

14. Install the combination switch aligning the bottom locating holes on the ignition module, then install the

retaining bolts. Tighten the lower mounting bolt first to assure proper location and seating, then engage the switch electrical connectors and insert the CPA device.

15. Install the steering column opening filler and attaching screws.

16. Position the cluster trim panel, engage the electrical connectors to the panel lighting and the rear defogger switches, then install the CPA devices. Install the cluster trim panel into the retainers, then install and tighten the retaining screws

17. Install the rear of the instrument panel top cover into the windshield clips and snap the panel into position. If necessary, replace the 6 rearward clips. Install the upper panel cover screws and screw caps.

18. If equipped, engage the traction control/fog lamp connector.

19. Install the trim panel extension strip and push the center radio finish panel/air outlet assembly into the clip locations.

20. Install the lower left and right trim panel extensions.

21. Install the steering column covers and the ignition bezel.

22. Install the steering wheel.

23. Connect the negative battery cable and, if equipped, properly enable the SIR system.

Ignition Lock Cylinder

REMOVAL AND INSTALLATION

1. Properly disable the SIR system, if equipped, and disconnect the negative battery cable.

2. Remove the steering column from the vehicle.

3. Position the steering column in a vise at the upper bracket.

NOTE: Always wear the proper eye protection when using drills, chisels and punches.

4. Using a center punch, mark the center of the shear bolts on the ignition lock assembly.

5. Remove the left shear bolt from the ignition module and column:
   a. Tap the chisel to create a divet on the side of the bolt.
   b. Turn the chisel and tap bolt at the divet on an angle in order to drive the bolt counterclockwise.
   c. Continue to tap the chisel and drive the bolt counterclockwise until the bolt is sufficiently loosened or removed.

6. Remove the right shear bolt from the ignition module and column

in the same manner as the left bolt was removed.

7. Remove the ignition lock module from the steering column.

To install:

8. Using new shear bolts, install the ignition lock assembly to the steering column. Torque the bolts until the heads break off.

9. Install the steering column into the vehicle.

10. Connect the negative battery cable and, if equipped, properly enable the SIR system.

Ignition Switch

REMOVAL AND INSTALLATION

1. Properly disable the SIR system, if equipped, and disconnect the negative battery cable.

2. Remove the steering wheel.

3. Remove the combination switch.

4. Remove the 2 screws retaining the combination switch connector.

5. Remove the 2 retaining screws, then disconnect the ignition switch from the ignition module and remove the switch from the vehicle.

To install:

6. Install the combination switch and retaining bolts. Tighten the lower mounting bolt first to assure proper location and seating.

7. Install the bolts retaining the combination switch connector, then engage the switch electrical connectors and insert the CPA device.

8. Install the ignition switch and retaining bolts to the ignition module.

9. Install the steering column opening filler and attaching screws.

10. Position the cluster trim panel, engage the electrical connectors to the panel lighting and the rear defogger switches, then install the CPA de-

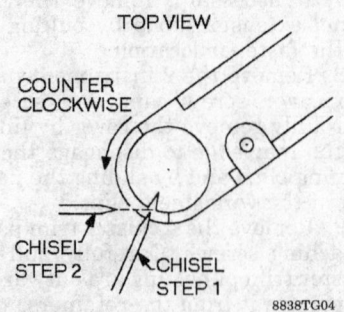

TOP VIEW

COUNTER
CLOCKWISE

CHISEL
STEP 2

CHISEL
STEP 1

8838TG04

Using a chisel to remove the shear bolts from the ignition module and steering column

vices. Install the cluster trim panel into the retainers, then install and tighten the retaining screws

11. Install the rear of the instrument panel top cover into the windshield clips and snap the panel into position. If necessary, replace the 6 rearward clips. Install the upper panel cover screws and screw caps.

12. If equipped, engage the traction control/fog lamp connector.

13. Install the trim panel extension strip and push the center radio finish panel/air outlet assembly into the clip locations.

14. Install the lower left and right trim panel extensions.

15. Install the steering column covers and the ignition bezel.

16. Install the steering wheel.

17. Connect the negative battery cable and, if equipped, properly enable the SIR system.

Neutral Safety/Selector Switch

ADJUSTMENT

1. Place the transaxle selector in **D**.

2. Locate the gear selector switch mounted on the side of the transaxle housing and disengage the switch connector.

3. Use an ohmmeter or continuity tester to check for continuity across the switch terminals.

4. If no continuity exists, loosen the bolts and rotate the switch to achieve continuity.

5. Tighten the switch bolts to 124 inch lbs. (14 Nm) and recheck continuity.

REMOVAL AND INSTALLATION

1. Disconnect the negative battery cable.

2. Remove the air induction tube. For the DOHC engine, remove the air filter box. Lift the resonator upward to disengage it from the support bracket.

3. Disengage the electrical connectors from the switch, then disconnect the shifter cable from the control lever.

4. Remove the retaining nut from the control lever shaft. Note the position of the manual lever, then remove.

5. Remove the 2 switch-to-transaxle housing retaining bolts, then remove the switch.

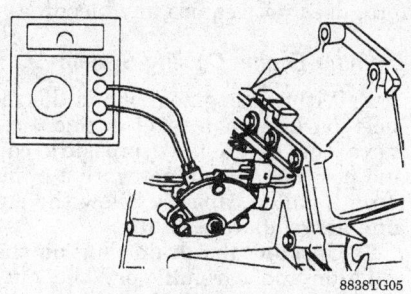

8838TG05

Adjusting the neutral safety switch

To install:

6. Install the switch to the transaxle and loosely install the retaining bolts.

7. Position the lever as noted and install the nut. Tighten the nut to 97 inch lbs. (11 Nm).

8. Adjust the switch and tighten the retaining bolts.

9. Install the shift cable to the control lever and adjust as necessary.

10. Engage the switch electrical connectors, then install the air induction tube.

11. Connect the negative battery cable and verify proper switch operation.

Powertrain Control Module

LOCATION

Located on the carrier assembly behind the left kickpanel. The PCM container the Engine Controller (EC) for all vehicles and the Transaxle Controller (TC) for vehicles with automatic transaxles.

ENGINE COOLING

Radiator

REMOVAL AND INSTALLATION

1. Disconnect the negative battery cable and drain the engine coolant.

2. Remove the air intake ducts and for the DOHC engine, remove the air cleaner housing.

3. If necessary, unplug the temperature sensor connector.

4. Remove the upper radiator hose and, if equipped with an automatic transaxle, disconnect the upper transaxle cooler line. Plug the openings to prevent fluid contamination or loss.

5. Remove the electric cooling fan assembly.

6. Remove the lower radiator hose.

7. Raise and support the vehicle safely.

8. Remove the lower splash shield, then if applicable, disconnect and plug the lower transaxle cooler line.

9. Remove the 4 condenser bracket-to-radiator bolts. Wire the condenser to the frame assembly so it stays in place, then lower the vehicle.

10. Remove the upper radiator nuts and brackets. If equipped the air conditioning, remove the upper radiator seal.

11. Remove the radiator from the vehicle.

To install:

12. Install the radiator into the vehicle.

13. Install the upper seal, if applicable, then install the brackets and retaining nuts. Be sure the L-shaped brackets do not pinch the radiator locating pins and the radiator moves freely in the grommets.

14. Raise and support the vehicle safely.

15. Install the condenser bracket bolts, and if applicable, install the lower transaxle cooler line.

16. Install the lower splash shield and lower the vehicle.

17. Install the lower radiator hose with the clamp tangs positioned at 1 o'clock.

18. Install the cooling fan assembly.

19. For vehicles with an automatic transaxle, connect the upper transaxle cooler line at a 35 degree angle inward from vertical and hold in position while tightening.

20. Install the upper radiator hose with the clamp tangs at 12 o'clock.

21. For the DOHC engine, install the air cleaner housing.

22. Install the intake air ducts and connect the air temperature sensor connector.

23. Close the radiator drain plug and install the cylinder block drain plug. Tighten the block plug to 26 ft. lbs. (35 Nm).

24. Connect the negative battery cable and properly fill the engine cooling system.

Water Pump

REMOVAL AND INSTALLATION

1. Disconnect the negative battery cable and drain the engine coolant.

2. Remove the serpentine drive belt.

3. If pump access is desired from the top, remove the air conditioning compressor bolts and position the compressor aside with the refrigerant lines intact. Be careful not to kink or damage the lines.

4. Raise the vehicle and support safely. Remove the right front tire and inner wheelwell splash shield.

5. Spray the water pump hub with a penetrating oil to loosen corrosion on the pulley and prevent damage during pump removal.

6. A 1 inch (25.4mm) block of wood may be wedged between the pump pulley and crankshaft to hold the pulley while loosening the retaining bolts. Remove the water pump pulley bolts and allow the pulley to hang freely on the hub.

7. Move the pulley outward or remove, as necessary, for access to the flange bolts.

8. Remove the 6 water pump flange bolts, then carefully pull the pump and pulley assembly away from the engine and remove the assembly from the vehicle. If necessary, a gasket scrapper may be inserted under the flange to help loosen the seal, but be careful not to damage the aluminum block sealing surface.

To install:

9. Thoroughly clean the gasket mating surface of all old gasket material. Apply a small amount of sealant at the outer edges of the bolt holes to hold that gasket in place, then position the gasket onto the water pump assembly.

10. If removed, position the pulley onto the pump hub.

11. Install the pump assembly with the small bump located next to 1 of the attaching bolts in the 11 o'clock position. Install the bolts and tighten the bolts in a criss-cross sequence (beginning with the upper left bolt) to 22 ft. lbs. (30 Nm).

12. If the pump hub exposed through the pulley is rusty, clean it with a wire brush and apply a thin coat of primer to prevent the pulley from rusting onto the hub. Install the bolts retaining the pulley to the pump hub and tighten to 19 ft. lbs. (25 Nm).

13. Install the serpentine drive belt, splash shield and tire.

14. If raised, lower the vehicle.

15. If repositioned, install the air conditioning compressor.

16. Close the radiator drain plug and install the cylinder block drain plug. Tighten the block plug to 26 ft. lbs. (35 Nm).

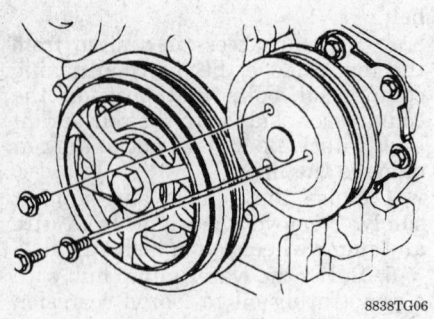

8838TG06

Water pump pulley removal

17. Connect the negative battery cable and properly fill the engine cooling system.

Thermostat

REMOVAL AND INSTALLATION

1. Drain the engine cooling system.
2. Disconnect the lower radiator hose at the thermostat housing.
3. Remove the 2 bolts at the water inlet housing, then remove the housing, thermostat and O-ring assembly.
4. Remove the thermostat from the housing using the tool provided with the replacement thermostat element.

To install:

NOTE: The thermostat will not function correctly if it has been contacted by oil. If oil is found in the cooling system, the thermostat cartridge must be replaced and the cooling system must be flushed.

5. Install the replacement thermostat using the tool provided. Make sure the element's retaining tangs are properly seated in the 2 legs and the element piston is correctly positioned in the inlet housing.
6. Install a new O-ring and position the housing assembly to the engine. Tighten the retaining bolts to 22 ft. lbs. (30 Nm).
7. Close the radiator drain plug and install the cylinder block drain plug. Tighten the block plug to 26 ft. lbs. (35 Nm).
8. Install the hose to the inlet housing.
9. Connect the negative battery cable and properly fill the engine cooling system.

Electric Cooling Fan

REMOVAL AND INSTALLATION

1. Disconnect the negative battery cable.
2. If equipped with the DOHC engine, remove the air intake ducts and unplug the temperature sensor connector.
3. Disengage the fan motor electrical connector.
4. Remove the top fan motor assembly bolts.
5. It may be necessary to loosen the top automatic transaxle cooler line, if equipped with air conditioning, and position it aside.
6. Lift the fan assembly off the lower mounting brackets. Move the assembly to the left and rotate counterclockwise (as pictured standing in front of the vehicle) lifting the right side up past the radiator hose, then remove the assembly from the vehicle.
7. If necessary, remove the retaining nut (while holding the fan), then remove the fan blade from the shaft and remove the motor from the housing.

To install:

8. If removed, install the motor and fan blade. Torque the fan nut to 27–44 inch lbs. (3–5 Nm).
9. Install the assembly with the lower left corner 1st. Rotate the assembly clockwise to place the lower left mount under the radiator hose and position the assembly onto the mounting brackets.
10. Install the upper retaining bolts.
11. If disconnected, install and tighten the transaxle oil cooler line. Position the transaxle cooler line 35 degrees inward from vertical and hold while tightening.
12. Engage the fan motor electrical connector.
13. If removed, install the intake air ducts and the temperature sensor connector.
14. Connect the negative battery cable.

Cooling System Bleeding

PROCEDURE

The cooling system uses a pressure coolant surge tank and an inlet side thermostat. Coolant is added through the pressure cap in the surge tank and fills the engine cylinder block, cylinder head, heater core and hoses and the radiator. The system therefore, does not require any bleeding.

Draining Engine Cooling System

1. With the engine cool and the vehicle parked on a level surface, remove the surge tank pressure cap and position a drain pan with a minimum 2 gallon capacity below the engine and radiator.
2. Unscrew the drain plug on the radiator and carefully pry the plug out of the housing. If necessary for replacement, pinch the housing tabs closed and remove the housing from the radiator by pulling straight out.
3. Remove the cylinder block drain plug located at the right front of the engine below the thermostat housing and allow the coolant to drain from the openings.

Refilling Engine Cooling System

1. If removed, install the drain plug housing into the radiator by pinching the tabs and inserting into the hole. Once the housing is in the hole, release the tabs and push the housing until it snaps into place. Be careful not to push the housing through the hole into the radiator.
2. Install the radiator drain plug into the housing and tighten.
3. Install the engine drain plug and tighten to 26 ft. lbs. (35 Nm).
4. Fill the system through the surge tank with a non-phosphate low silicate base ethylene glycol-based coolant mixed to the manufacturer's instructions. The coolant should be recommended for use in aluminum engines.
5. Start the engine and check for leaks.
6. Fill the surge tank to the cold line after the engine has run for 2–3 minutes and install the pressure cap.

FUEL SYSTEM

Fuel System Service Precautions

Safety is the most important factor when performing not only fuel system maintenance but any type of maintenance. Failure to conduct maintenance and repairs in a safe manner may result in serious personal injury or death. Maintenance and testing of the vehicle's fuel system components can be accomplished

safely and effectively by adhering to the following rules and guidelines.

• To avoid the possibility of fire and personal injury, always disconnect the negative battery cable unless the repair or test procedure requires that battery voltage be applied.

• Always relieve the fuel system pressure prior to disconnecting any fuel system component (injector, fuel rail, pressure regulator, etc.), fitting or fuel line connection. Exercise extreme caution whenever relieving fuel system pressure to avoid exposing skin, face and eyes to fuel spray. Please be advised that fuel under pressure may penetrate the skin or any part of the body that it contacts.

• Always place a shop towel or cloth around the fitting or connection prior to loosening to absorb any excess fuel due to spillage. Ensure that all fuel spillage (should it occur) is quickly removed from engine surfaces. Ensure that all fuel soaked cloths or towels are deposited into a suitable waste container.

• Always keep a dry chemical (Class B) fire extinguisher near the work area.

• Do not allow fuel spray or fuel vapors to come into contact with a spark or open flame.

• Always use a backup wrench when loosening and tightening fuel line connection fittings. This will prevent unnecessary stress and torsion to fuel line piping. Always follow the proper torque specifications.

• Always replace worn fuel fitting O-rings with new. Do not substitute fuel hose or equivalent, where fuel pipe is installed.

Fuel System Pressure

RELIEVING

1. Unless battery voltage is needed for testing, disconnect the negative battery cable.
2. Remove the air cleaner or air intake duct, as applicable.
3. Wrap a shop rag around the fuel test port fitting at the rear of the engine, remove the cap and connect pressure gauge tool SA9127E or equivalent.
4. Install the bleed hose into an approved container and open the valve to bleed the system pressure.
5. After the system pressure is bled, remove the gauge from the pressure test port and recap it.

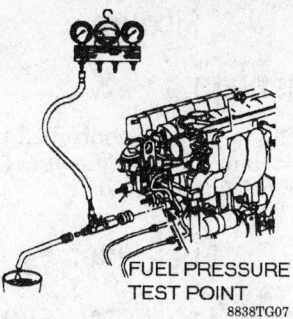

FUEL PRESSURE
TEST POINT
8838TG07

Relieving fuel system pressure — DOHC engine shown, SOHC engine similar

6. Install the air cleaner or intake duct, unless the procedure requires its removal.
7. After repairs are completed, connect the negative battery cable and prime the fuel system as follows:

   a. Turn the ignition **ON** for 5 seconds and then **OFF** for 10 seconds.

   b. Repeat the ON/OFF cycle 2 more times.

   c. Crank the engine until it starts.

   d. If it does not start, repeat Steps a–d

   e. Run the engine and check for leaks.

Idle Speed

ADJUSTMENT

The Powertrain Control Module (PCM) directly controls idle speed using the Idle Air Control (IAC) valve. Under normal circumstances and routine maintenance, tampering with the idle stop screw should NEVER occur. The screw is not used to set a engine idle speed per se, instead it is used to set the minimum position from which the PCM can use the IAC valve to control idle speed.

NOTE: The minimum idle speed adjustment is preset at the factory and requires no periodic adjustments. Adjustments should be performed ONLY when the throttle body has been replaced and/or proper idle speed cannot be obtained. The engine should be at normal operating temperature, the A/C and cooling fans should be OFF when making adjustments.

1. Before making any adjustments, clean the throttle body bore

with a shop towel and carburetor cleaner that does not contain methyl ethyl keytone. Then check the idle speed to be sure adjustment is necessary. Proper idle speeds are as follows:

SOHC

   Manual or automatic transaxle in **N** — 700–800 rpm

   Automatic transaxle in **D** — 600–700 rpm

   Automatic transaxle in **D** and A/C ON — 725–825 rpm

DOHC

   Manual transaxle in **N** — 800–900 rpm

   Automatic transaxle in **D** — 700–800 rpm

   Automatic transaxle in **D** and A/C ON — 725–825 rpm

2. If adjustment is necessary, block the wheels and apply the parking brake.

3. The IAC pintle must be properly seated in the throttle body. Connect the IAC tester SA9195E or equivalent, to the IAC valve at the throttle body. The Saturn PDT or an equivalent scan tool may attached to the ALDL and used instead of the IAC tester.

4. Remove the idle stop screw plug by piercing it with an awl and applying leverage for SOHC engines. On DOHC engines, remove the idle stop screw cover.

5. Insert the IAC air plug in the throttle body; use SA9196E for TBI or SA9106E for MFI or equivalent.

6. Connect the Saturn Portable Diagnostic Tool (PDT) or equivalent to the Assembly Line Diagnostic Link (ALDL), start the engine and check the minimum idle speed for proper setting. The minimum idle speed should be 450–650 rpm for all engines.

7. If not within specification adjust the idle screw to obtain an minimum idle speed of 500–600 rpm.

8. Turn the ignition **OFF**, then reinstall the IAC electrical connector.

9. Using the Saturn PDT or equivalent scan tool, check the TP sensor voltage. Do not replace the TP sensor unless setting is not between 0.35–0.70 volts.

10. Remove the IAC air plug and install the idle stop plug or cover.

11. Start the engine and check for proper idle operation.

12. Shut the engine **OFF** and remove the PDT or scan tool.

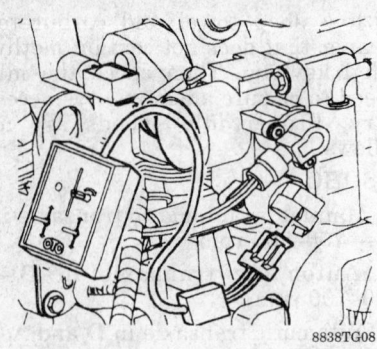

Installing the IAC connector to the valve terminals at the throttle body

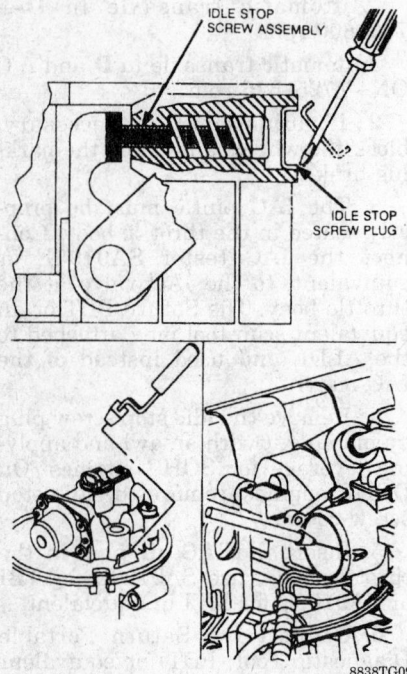

IDLE STOP
SCREW ASSEMBLY

IDLE STOP
SCREW PLUG

8838TG09

Removing the idle screw plug and installing the IAC air plug

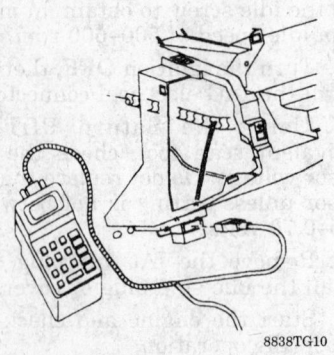

8838TG10

Connecting the Saturn diagnostic tool to the ALDL

Mixture

ADJUSTMENT

The idle mixture is controlled by the Powertrain Control Module (PCM) and is not adjustable.

Fuel Filter

REMOVAL AND INSTALLATION

1. Disconnect the negative battery cable.
2. Remove the air cleaner or air intake duct, as applicable.
3. Properly relieve the fuel system pressure.
4. Disconnect the large underhood fuel line connection located near the intake manifold support brace on the left side of the vehicle using the tool supplied with the replacement filter, SA9157E or equivalent. It may be necessary to clean the female end of the quick-connect fittings by spraying them with penetrating oil prior to disconnection.
5. Raise and support the vehicle safely.
6. Disengage the quick-connect at the fuel filter inlet by pinching the 2 plastic tangs together and pulling on the supply line.
7. Loosen the fuel filter band clamp nut, but do not completely remove.
8. Carefully push or pull the filter out of the assembly and discard the filter in an appropriate container.

To install:

9. If the band clamp was removed, clip the fuel return and vapor lines in place and install 2 new band clamp nuts. Make sure all lines are in place and will not interfere with or be damaged by filter installation and tighten the bracket nuts to 27 inch lbs. (3 Nm).
10. Clean the female end of the filter inlet quick-connect fitting by holding the line facing downward while spraying penetrating oil up into the fitting. Be careful not to bend of kink the line.
11. If not already installed, insert a new snap lock retainer into the female end of the filter inlet quick-connect fitting.
12. Route the filter's nylon outlet line through the band clamp under the A/C suction hose and engine wiring. Insert the filter far enough into the band clamp to connect the outlet line to the engine fuel line attachment. Lubricate the male end of the connector with clean engine oil, snap the connector together and pull on the line to verify proper fitting.
13. Make sure the fuel lines are properly routed and are not contacting other lines, hoses or wiring. Position the filter in the band clamp assembly with the filter's upper edge located ¼ inch (6.35mm) from the top of the band clamp.
14. Lubricate the male end of the fuel supply line with clean engine oil. Snap the line to the fuel filter and pull back to verify the fitting is secure. Tighten the band clamp nut to 89 inch lbs. (10 Nm).
15. Lower the vehicle and install the air inlet tube and/or intake duct, as applicable.
16. Connect the negative battery cable and prime the fuel system.
17. Run the engine and check for leaks.

Fuel Pump

REMOVAL AND INSTALLATION

The fuel pump module is located in the fuel tank. The tank must be removed to access the module pump/sender assembly.

1. Disconnect the negative battery cable and properly relieve fuel system pressure.
2. Drain the fuel from the tank into a suitable container and remove the tank from the vehicle.
3. Clean the area surrounding the fuel pump module and spray the cam lockring tangs with a suitable penetrating oil to loosen the fitting.
4. Using SA9156E, or an equivalent fuel module lockring removal tool, and a ½ inch breaker bar of approximately 18 in. (457mm) in length, remove the pump unit locking ring from the tank. Attempting to use a 12 in. or shorter breaker bar may cause lockring damage.
5. Lift and tilt the unit out at a 45 degree angle, being careful not to bend the sending unit filter and float arm. Remove and discard the unit-to-tank O-ring.
6. The sending unit is the only portion of the module that may be serviced. The filter may be cleaned with mineral spirits, but must be replaced as an assembly with the module if damaged. If necessary, remove the sending unit from the module as follows:
   a. Unplug the 2 electrical connections using needle-nose pliers or by pressing down the locking tab and pulling the connectors from the terminal.

b. Using a small suitable tool, push in on the sender assembly attaching tang, then lift upward and remove the sender.

c. Some floats may be serviced, check with a parts supplier for availability to determine if the vehicle's float may be replaced. If serviceable, use a ¼ in. flat-tipped prytool to carefully pry the float from the wire loop. Do not bend the float arm or deform the wire loop.

To install:

7. If removed, pinch a new float onto the float arm wire loop and/or install the sending unit to the pump module by positioning the tang in the locator slot and snapping the unit into place. Connect the 2 sending unit electrical connectors to their terminals.

8. If applicable, carefully install the fuel tank inlet check ball from inside the fuel tank by reaching through the tank module mount opening and gently pushing the ball into position.

9. Clean any debris from the O-ring mating surface and position a new O-ring onto the tank.

10. Carefully insert the pump module into the tank at a 45 degree angle to prevent sending unit and float damage. The filter and flow arm must be directed toward the front of the tank.

11. Align the pump locator tabs with the fuel tank slots, then install the cam lockring using the ring service tool.

12. Install the fuel tank assembly.

13. Lower the vehicle, connect the negative battery cable and fill the fuel tank.

14. Prime the fuel system, start the engine and check for leaks.

Fuel Injector

REMOVAL AND INSTALLATION

Throttle Body Injection (TBI)

1. Disconnect the negative battery cable.

2. Remove the air cleaner assembly.

3. Properly relieve fuel system pressure.

4. Remove the electrical connector from the injector terminal.

5. Remove the injector retaining screw and bracket.

6. Using a smooth, round fulcrum and prybar, carefully pry the injector out of the throttle body. Make sure the electrical connector and injector nozzle are protected from damage.

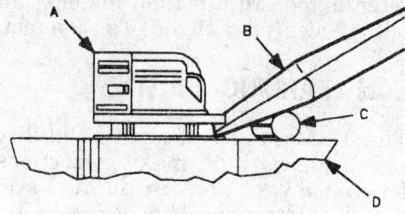

A. Fuel injector assembly
B. Prybar
C. Fulcrum
D. Fuel meter body

8838TG11

Removing the fuel injector — TBI engine

To install:

7. Remove the upper and lower O-rings and inspect the injector for dirt or contamination. The injector may be cleaned using safety glasses and compressed air, but the screen may not be removed from the injector. If injector replacement is necessary, be sure to use an identical part.

8. Install new O-rings and lubricate with clean engine oil. Be sure the upper O-ring is in the groove on the injector and the lower ring is properly installed in the fuel meter body cavity.

9. Install the injector, pushing it straight into the injector cavity with the electrical connector facing toward the fuel pressure regulator.

10. Install the retaining bracket. Coat the screw with Loctite® 242 or equivalent, then install and tighten the screw to 35 inch lbs. (3 Nm).

11. Engage the wiring harness to the injector connector, then connect the negative battery cable.

12. Prime the fuel system, start the engine and check for leaks.

13. Shut the engine **OFF** and install the air cleaner assembly.

Multi-Port Fuel Injection (MFI)

1. Disconnect the negative battery cable.

2. Remove the air intake tube with resonator and fresh air tube.

3. Properly relieve fuel system pressure.

4. Remove the fuel line bracket bolt, then disconnect the fuel pressure and return lines. Be sure to use a ¹⁵/₁₆ in. (24mm) backup wrench to prevent inlet port or bracket damage. If necessary, remove the fuel line bolts and rotate the rail slightly for wrench access. Remove the old O-rings from the lines with a suitable tool and discard.

5. Disconnect the pressure regulator vacuum hose.

6. Remove the throttle cable bracket bolts, then disconnect the cable from the throttle lever. Lay the cable over the intake manifold.

7. Remove the wiring harness connectors from the fuel injector terminals and remove the fuel rail retaining bolts.

8. Remove the fuel rail assembly by carefully pulling the rail back and upward to pull the injectors from the manifold ports. Be careful not to damage the injector spray tips and electrical connectors. Rotate the rail so the injectors point downward, then lift the rail end opposite of the fuel connections to remove the rail from between the camshaft cover and intake manifold.

9. Make sure the rails and injectors are clean and free of dirt. If injector removal is required, slide the injector retaining clip off the injector and pull the injector from the fuel rail. Remove the old injector O-rings with a suitable seal removal tool or brass seal pick and discard.

To install:

10. If removed, lubricate the new upper injector O-rings with engine oil and install onto the injector assemblies. Install the fuel injectors into the fuel rail, with the electrical connectors facing upward, then slide the injector retaining clips onto the rail. Engage the clips to retain the injectors.

11. Lubricate the new lower injector O-rings with clean engine oil, then install. Lubricate the new fuel inlet and return O-rings, then install the O-rings into the fuel outlet of the pressure regulator and inlet of the fuel rail.

12. With the pressure regulator end first and the injectors pointing downward, guide the fuel rail assembly through the passage between the camshaft cover and the intake manifold from the power steering pump side of the engine. Align the injectors with their respective ports, rotate the fuel rail and carefully push the injectors into the port.

13. Verify the injectors are properly seated in the intake manifold. Loosely connect the fuel inlet and return lines to the rail assembly, then install the fuel rail retaining bolts and tighten to 22 ft. lbs. (30 Nm).

14. Engage the wiring harness connectors to the fuel injector terminals.

15. Connect the PCV valve hose to the camshaft cover, and the vacuum line to the pressure regulator, making sure they are properly seated.

16. Install the throttle cable bracket bolts and tighten to 19 ft. lbs.

(25 Nm), then connect the throttle cable.

17. Using a backup wrench, tighten the fuel inlet and return line fittings to 133 inch lbs. (15 Nm).

18. Install the fuel line bracket bolt and tighten to 106 inch lbs. (12 Nm).

19. Connect the negative battery cable, then prime the fuel system by cycling the ignition key **ON** and **OFF**.

20. Start the engine and check for leaks.

21. Shut the engine **OFF**, then install the air intake tube and resonator assembly.

EMISSION CONTROLS

Emission Warning Lamps

The Service Engine Soon light serves 3 major functions. It informs a driver that the PCM has detected a problem and the vehicle should be taken in for service as soon as reasonably possible. It displays trouble codes stored by the PCM and it indicates if the engine is in OPEN LOOP or CLOSED LOOP operation.

The light will come ON with the key ON and engine not running. When the engine is started, the light will turn OFF. If the light stays ON, the self-diagnostic system has detected a problem. In most cases, should the condition causing a fault disappear during vehicle operation, the light will turn OFF approximately 10 seconds after the fault disappears. A code will be stored in memory even if the light extinguishes.

RESETTING

When a fault has been detected by the PCM a code will set in 2 places, general information and malfunction history. Both memories can be cleared with the use of the Saturn PDT or an equivalent scan tool. If no scan tool is available, general information can be cleared if the A and B terminals are of the ALDL are grounded 3 times in 5 seconds with the ignition ON. General information will also clear if the problem is absent for 50 ignition ON/OFF cycles or if the battery supply to the PCM is interrupted. Malfunction history can only be cleared with aid of a scan tool.

ECM LEARNING ABILITY

The PCM has a "learning" ability which allows it to make corrections for minor variations in the fuel system, in order to improve driveability. If the battery is disconnected to clear diagnostic codes or for safety during repairs, the "learning" process will reset and must begin again. A change may be noted in the vehicle's performance while the learning process begins. The following steps must be performed in order to assure the PCM will properly relearn smooth engine operation:

1. Start and warm the engine until it is at normal operating temperature.

2. Drive the vehicle at part throttle, with moderate acceleration and idle conditions, until normal performance returns.

3. Park the vehicle and engage the parking brake, but leave the engine running.

4. Place the transaxle in **D** for an automatic transaxle or neutral for a manual transaxle.

5. Allow the engine to continue idling for about 2 minutes, until the idle stabilizes. The engine must still be at normal operating temperature.

ENGINE MECHANICAL

Engine Assembly

REMOVAL AND INSTALLATION

NOTE: The manufacturer recommends that the engine and transaxle be removed as a complete unit. Instead of lifting the assembly from the vehicle, the engine cradle should be disconnected from the spaceframe and the powertrain should be lowered from the vehicle.

1. Properly disable the SIR system, if equipped.

2. Disconnect the negative and then the positive battery cable.

3. Drain the engine cooling system from the engine block and radiator drain plugs.

4. Properly relieve the fuel system pressure.

5. Unplug and label the following electrical connectors and vacuum lines:

a. The 2 coolant temperature sensors.

b. Oxygen sensor and clip at the transaxle front mount bracket.

c. Idle Air Control (IAC) valve.

d. The 2 ignition coil module connectors.

e. Throttle Position Sensor (TPS).

f. Manifold Absolute Pressure (MAP) sensor.

g. Exhaust Gas Recirculation (EGR) solenoid.

h. Brake booster vacuum hose from the booster or intake manifold.

i. Disengage the 2 ground connectors from the transaxle attachment studs at the rear side of the cylinder block.

j. Fuel injector electrical connector(s).

6. Disconnect the following automatic or manual transaxle electrical connectors. If access to any of the wires is difficult, wait until the vehicle is safely raised and supported, then unplug the connections from underneath the vehicle:

a. The 3 neutral safety switch connectors.

b. Valve body actuator connection.

c. Turbine speed sensor.

d. Temperature sensor.

e. For manual transaxles only, the reverse light switch.

f. The 2 gear shift (PRNDL switch) connectors.

7. Disconnect the accelerator cable assembly.

8. Using service tool SA9157E or equivalent, disconnect the fuel supply and return lines at the connectors. Plug the lines to prevent fuel contamination or loss. The lines may be tied to the master cylinder lines to help prevent fuel spillage and to keep them out of the way.

9. Disconnect the upper radiator hose and the cylinder head outlet and the de-aeration hose at the engine.

10. If equipped with A/C, remove serpentine drive belt, then remove the air conditioning compressor from its brackets with the hoses attached. Support the compressor from the front crossbar.

NOTE: It is not necessary to discharge the A/C compressor during engine removal, but be careful not to kink, damage or rupture the refrigerant lines.

11. If equipped, disconnect the automatic transaxle cooler lines at the

transaxle by pinching the plastic connector tabs and carefully pulling back on the lines. Plug the openings to prevent fluid loss or contamination.

12. Disconnect the automatic transaxle shifter cable or the manual shifter cables from the transaxle.

NOTE: On manual transaxle vehicles, position a block of wood under the clutch pedal to prevent accidental actuation of the slave cylinder while it is removed from the transaxle.

13. If equipped with manual transaxles, remove the 2 hydraulic slave cylinder retaining nuts from the clutch housing studs, then slide the cylinder and bracket assembly from the studs. Rotate the clutch actuator ¼ turn counterclockwise while pushing toward the housing to disengage the bayonet connector and remove it from the clutch housing. Support the clutch hydraulic system to the battery tray; being sure not to kink or pinch the hydraulic lines.

14. Using a length of an appropriate wire, tie the radiator, condenser and fan to the front crossbar. Route the wire around the 2 fan shroud supports and the crossbar.

15. Raise and support the front end of the vehicle safely.

16. Remove the front wheels and remove the fasteners connecting the side and front fender shields to the cradle.

17. Remove the brake caliper bracket attaching bolts (2 on each side) and hang the caliper assemblies from the shock tower springs using wire. Do not hang the assembly by the brake hose or damage to the brake hydraulic system may occur. The spring and shock assemblies will remain with the body when the powertrain is lowered.

18. Disconnect the struts from the knuckles on each side of the vehicle (2 bolts per side). This will allow the knuckle and hub assembly to remain with the powertrain cradle when lowered. The stabilizer bar will remain attached to the cradle and the lower control arms.

19. Disconnect the lower radiator and heater return hoses from the engine. Disconnect the heater inlet hose at the front of the dash or the engine.

20. Disengage the steering shaft and pressure switch connectors at the gear, as applicable.

21. Disconnect the front exhaust pipe at the manifold, catalytic converter and powertrain stiffening bracket.

22. If equipped with an automatic transaxle, remove the flywheel cover,

then remove the torque converter bolts.

23. Remove the alternator and starter shields.

24. Label and unplug the remaining electrical and vacuum connectors from the following components:
   a. Remove the wires from the starter solenoid and the battery feed.
   b. The alternator field and battery feed connectors.
   c. Oil pressure sensor.
   d. Knock sensor.
   e. Crankshaft position sensor.
   f. If equipped, the EVO solenoid.
   g. Vehicle speed sensor.
   h. Canister purge solenoid.
   i. Powertrain Control Module (PCM) and Oxygen sensor.
   j. If equipped, the ABS wheel sensor connector grounds.

25. Unclip the brake lines from the rear side of the cradle.

26. Carefully remove the electrical harness from the engine and transaxle, then lay the electrical harness on top of the underhood junction block and battery cover.

27. With a torque axis mount system, place a 1 inch x 1 inch x 2 inch long block of wood between the torque strut and cradle to ease removal and installation of the torque engine mount. Remove the 3 right side upper engine torque axis-to-front cover nuts and the 2 mount-to-midrail bracket nuts, allowing the powertrain to rest on the block of wood.

NOTE: Placing a block of wood under the torque axis mount prior to removing the upper mount will allow the engine to rest on the wood, thus preventing the engine from shifting. If the engine is not to be removed from the cradle, this will allow you to install the engine and the upper mount without jacking or raising the engine.

28. Place a powertrain support dolly under the cradle. Use two 4 inch x 4 inch x 36 inch pieces of wood to support the cradle on the dolly.

29. Remove the 2 right side front engine mount, torque strut brackets-to-cradle nuts.

30. Remove the 4 cradle attaching bolts and carefully lower the complete powertrain assembly from the vehicle. Verify that all necessary components are disconnected and free before complete removal.

31. Attach the 2 washers located between the cradle and body, to the cradle. They must be repositioned and installed during cradle installation.

32. Tag and disconnect the spark plug wires at the ignition module.

33. If applicable, remove the power steering pump and bracket. Support the assembly, in an upright position, from the cradle or the steering gear.

34. Install a suitable engine lifting device to the service support brackets.

35. Remove the front mount assembly and disconnect the motion restrictor bracket, if applicable.

36. Place a ½ inch x 1 inch x 3 inch block of wood under the axle shaft, then remove the starter support bracket bolt, intake manifold support brace (on DOHC engines), and 3 axle shaft bracket support bolts. Allow the bracket to rotate rearward. Lift the engine slightly for clearance, as necessary.

37. Place a 4 inch x 4 inch x 6 inch long block of wood under the transaxle housing for support. Then remove the engine strut bracket and torque strut from the front of the engine as an assembly. Lift the engine slightly as necessary for removal.

38. Remove the 4 transaxle attaching bolts/studs and separate the assembly. Manual transaxles will require the engine to be moved about 4 inches (100mm) forward in the cradle to disengage the input shaft.

39. Carefully lift the engine off the cradle.

To install:

40. If installing a manual transaxle, align the yellow dot on the clutch pressure plate near the mark on the flywheel. Use SA9145T or an equivalent clutch alignment tool to align the disk and input shaft, then tighten the pressure plate bolts to 19 ft. lbs. (25 Nm).

41. If installing an automatic transaxle, the yellow dot on the torque converter must be in the 6 o'clock position when the first flexplate-to-torque converter bolt is tightened.

42. Position the engine on the cradle aligning it with the transaxle using 2 threaded 10mm x 6 in. guide pins in the lower attachment holes. When aligned, remove the pins and install the 4 transaxle attaching bolts along with the stiffening bracket fastener. Tighten the lower bolts to 96 ft. lbs. (130 Nm) and the upper bolts to 66 ft. lbs. (90 Nm) and the stiffening bracket to powertrain fastener to 40 ft. lbs. (54 Nm). If applicable, remove the 4 inch x 4 inch x 6 inch block of wood from under the transaxle housing.

43. Install the front engine mount assembly to the engine and tighten to 41 ft. lbs. (55 Nm).

44. If removed, install the engine mount torque strut-to-cradle bracket and tighten the engine fasteners to 52 ft. lbs. (70 Nm). Hand-tighten the cradle fasteners, but do not torque until the upper midrail mount is installed.

45. Install the 1 inch x 1 inch x 2 inch long block of wood between the torque strut and cradle to ease installation of the torque mount.

46. Attach the axle shaft and starter bracket. Tighten the axle shaft bracket fasteners to 41 ft. lbs. (55 Nm) and the starter bracket to 80 inch lbs. (9 Nm).

47. Position the powertrain and cradle assembly onto the dolly, using the 2 wooden boards to support and protect the cradle assembly.

48. Carefully lift the powertrain and cradle into position. Make sure the radiator grommets are correctly aligned and that the 2 washers are reinstalled between the cradle and body at each rear cradle attachment position. If necessary, use two 9/16 in. x 18 in. long guide pins in the forward cradle holes (located next to the attaching holes) to help align the cradle. Tighten the cradle to body fasteners to 151 ft. lbs. (205 Nm).

49. Attach the brake lines to the cradle and install the steering shaft U-joint. Tighten the U-joint bolt to 35 ft. lbs. (47 Nm).

50. Position the electrical harness around the engine, then install and/or connect the following components or connectors, as applicable:

   a. Starter solenoid connector and tighten to 44 inch lbs. (5 Nm).

   b. Alternator and starter battery connectors and tighten to 89 inch lbs. (10 Nm).

   c. Oil pressure sensor, tighten to 26 ft. lbs. (35 Nm), and connector.

   d. Knock sensor and tighten to 133 inch lbs. (15 Nm), and connector.

   e. Crankshaft position sensor and tighten to 80 inch lbs. (9 Nm), and connector.

   f. Canister purge valve and tighten to 22 ft. lbs. (30 Nm) and connector/hoses.

   g. Wiring harness PCM ground and tighten to 89 inch lbs. (10 Nm).

   h. Wiring harness to the transaxle case/engine block and tighten to 18 ft. lbs. (25 Nm).

   i. If equipped, the EVO solenoid.

   j. Vehicle speed sensor.

   k. Power steering pressure switch.

   l. If equipped, the ABS wheel speed sensors.

51. Install the engine stiffening bracket bolts, as applicable and tighten to 35 ft. lbs. (47 Nm).

52. Install new gaskets and the exhaust front pipe. Tighten the pipe-to-manifold fasteners to 23 ft. lbs. (31 Nm), the pipe-to-stiffener bracket fasteners to 35 ft. lbs. (47 Nm), the pipe-to-support bracket fasteners to 23 ft. lbs. (31 Nm) and the pipe-to-catalytic converter fasteners to 35 ft. lbs. (48 Nm).

53. Connect the heater, lower radiator and coolant fill hoses, then remove the radiator assembly support wires.

54. Install the cylinder block drain plug and tighten to 27 ft. lbs. (36 Nm), then close the radiator drain.

55. Install the knuckle-to-strut attachment bolts, tighten the bolts to 148 ft. lbs. (200 Nm).

56. Install the brake caliper assemblies and tighten the bolts to 81 ft. lbs. (110 Nm).

57. Install the shift cables using new retainers.

58. Remove the supports and carefully lower the vehicle sufficiently to gain underhood access, then reposition the supports under the vehicle.

59. If equipped with a manual transaxle, install the hydraulic clutch slave cylinder, damper and shift cables, then tighten the fasteners to 19 ft. lbs. (25 Nm). Remove the wooden block from underneath the clutch pedal.

60. If equipped, install the automatic transaxle cooler lines and/or the air conditioning compressor assembly. Tighten the compressor to front bracket bolts to 35 ft. lbs. (47 Nm) and the compressor to rear bracket bolts to 22 ft. lbs. (30 Nm).

61. If removed, install the serpentine drive belt, making sure the belt is properly aligned in the grooves.

62. If equipped with a torque axis mount system, install the 2 engine mount-to-midrail bracket nuts and tighten to 52 ft. lbs. (70 Nm). Next install the 3 mount-to-front cover nuts, tighten them uniformly to 52 ft. lbs. (70 Nm) in order to prevent front cover damage. Finally, remove the block of wood from under the torque strut.

63. Tighten the strut bracket-to-cradle nuts to 52 ft. lbs. (70 Nm).

64. If equipped with an automatic transaxle, install the torque converter-to-flexplate bolts and tighten the bolts to 52 ft. lbs. (70 Nm). Install the dust cover and tighten to 89 inch lbs. (10 Nm).

65. Install the tires and splash shields, tightening the lug nuts to 103 ft. lbs. (140 Nm). Remove the supports and carefully lower the vehicle to the ground.

66. Attach the following electrical and vacuum connections:

   a. Coolant temperature sensors.

   b. Oxygen sensor.

   c. IAC valve.

   d. Fuel injector connector(s).

   e. The 2 ignition coil module connectors.

   f. TPS connector.

   g. MAP sensor.

   h. EGR solenoid.

   i. If equipped, the A/C compressor.

   j. The brake booster hose to the intake manifold or booster.

   k. The ground connectors to the transaxle attachment studs located at the rear side of the block, above the starter.

67. Attach the following transaxle connectors, as applicable:

   a. The 3 neutral safety switch connectors.

   b. Valve body actuator connector.

   c. Turbine speed sensor.

   d. On manual transaxles, the reverse light switch.

   e. The temperature sensor.

   f. The 2 gear shift (PRNDL switch) connectors.

68. Install the accelerator cable, then apply a drop of clean engine oil to the male ends of the fuel line connectors and attach the line quick-connect fittings. Make sure the lines are not kinked or damaged.

69. Install the upper radiator and de-aeration hoses, verify proper alignment of the radiator L-bracket fasteners.

NOTE: Check the upper cooling module grommets for binding or misalignment. The module retaining pins must be centered in the grommets supported by the brackets. If the grommets are pinched, loosen the brackets and reposition them. It is extremely important that the cooling module be able to move freely.

70. Install the air induction system, PCV valve and fresh air hoses.

71. Connect the battery cables and tighten to 151 inch lbs. (17 Nm).

72. If equipped, enable the SIR system.

73. Fill the engine cooling system, then check all engine and transaxle fluids, add or fill as necessary.

74. Prime the fuel system by cycling the ignition **ON** for 5 seconds and **OFF** for 10 seconds a few times without cranking the engine. Start the engine and check for leaks.

Engine Mounts

REMOVAL AND INSTALLATION

The engine mount system consists of 2 groups of interconnected components. The first group is the engine mount which bolts to the top of the front cover and to the top of the midrail bracket. The midrail bracket bolts to the mid-frame rail and supports the engine mount.

The second engine mount group consists of the engine strut which in fastened to 2 brackets, 1 mounted to the engine and the other to the cradle.

1. Properly disable the SIR system, if equipped and disconnect the negative battery cable.

2. Remove the 2 engine mount-to-midrail bracket nuts.

3. Raise and support the vehicle safely.

4. Unless only removing the engine mount, remove the right wheel, then remove the inner splash shield.

5. Position a suitable floor jack and a block of wood under the engine oil pan, then raise the powertrain slightly to unload the mount.

6. If removing the midrail bracket and/or the engine mount:

a. Remove the 3 engine mount-to-engine front cover nuts, then remove the engine mount.

b. If the engine midrail bracket removal is required, remove the 4 midrail bracket-to-midrail bolts (from the bracket side of the assembly), then remove the 3 midrail-to-bracket bolts (from the rail side of the assembly) and remove the bracket from the vehicle.

7. If removing the engine strut and/or the strut brackets:

a. Loosen the 2 strut bracket-to-cradle nuts, located under the cradle.

b. Remove the strut-to-engine bracket and strut-to-cradle bracket bolts.

c. If removing the engine strut bracket, remove the 4 bracket-to-engine bolts.

d. If removing the cradle strut bracket, remove the 2 bracket-to-engine nuts.

e. Remove the strut brackets, and/or the engine strut from the vehicle. In order free the strut and/or the cradle bracket, the floor jack must be used to raise the engine slightly for clearance.

To install:

8. If removed, install the engine strut and/or strut bracket:

a. If removed, position the strut bracket to the engine, then install the 4 retaining bolts and tighten to 40 ft. lbs. (54 Nm).

b. If removed, position the engine strut and/or the cradle bracket, then lower the jack slightly to align the fasteners and holes.

c. Loosely install the cradle bracket fasteners, then install the strut-to-bracket bolts. Tighten the strut bolts to 52 ft. lbs. (70 Nm).

d. Fully lower the powertrain and tighten the cradle fasteners to 37 ft. lbs. (50 Nm).

9. If removed, install the midrail bracket and/or engine mount:

a. If the powertrain was lowered to install the engine strut or brackets, raise it again slightly to install the engine mount.

b. If removed, position the bracket to the midrail and hand-tighten all 7 retaining bolts. After all bolts are sufficiently threaded and tightened by hand, tighten all bolts to 24 ft. lbs. (32 Nm).

c. Position the engine mount to the front cover and install the 3 nuts. Tighten the nuts to 37 ft. lbs. (50 Nm).

d. Carefully lower the powertrain, guiding the mount over the midrail studs to prevent thread damage. Remove the jack and the block of wood.

e. Install the 2 mount-to-bracket nuts and tighten to 37 ft. lbs. (50 Nm).

10. Install the inner splash shield followed by the wheel assembly.

11. Lower the vehicle and connect the negative battery cable.

12. If equipped, properly enable the SIR system.

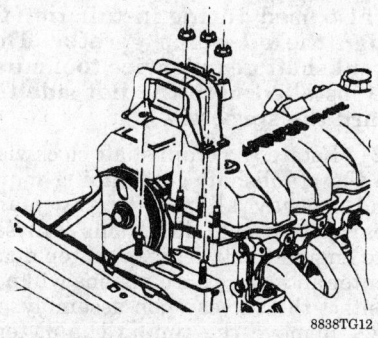

8838TG12

Removing the engine mount from the midrail bracket and front cover studs

Cylinder Head

REMOVAL AND INSTALLATION

NOTE: Remove the cylinder head when the engine is cold. Warpage may result if removed hot.

NOTE: The throttle body assembly, fuel rail, pressure regulator, injectors and intake manifold can be removed along with the cylinder head as an assembly, on all models equipped with Multi-Port Fuel Injection (MFI).

1. Disconnect the negative battery cable and properly drain the engine coolant.

2. Remove the air cleaner assembly and air inlet duct. For SOHC engines, disconnect the PCV valve and fresh air hose. For DOHC engines, disconnect the camshaft cover air hose at the cover.

3. Disconnect the accelerator cable from the throttle body and the bracket from the intake manifold.

4. Properly relieve the fuel system pressure.

5. Label and disengage the electrical connectors from the cylinder head assembly. Long nose pliers are necessary to unplug the coolant temperature sensor connectors. Position the electrical harness over the underhood junction block.

6. Label and disconnect all necessary vacuum hoses from the area around the cylinder head assembly.

7. Disconnect the upper radiator hose at the cylinder head outlet, the heater hose at the intake manifold and the deaeration hose next to the TBI assembly or at the intake manifold.

8. Remove the bolt which retains the fuel lines to the intake manifold assembly. Disconnect the fuel feed and return lines from the fuel rail and pressure regulator or from the throttle body, as applicable. For SOHC engines, remove the lower intake manifold support bracket stud. For DOHC engines, remove the upper intake manifold support bracket bolt.

9. With a torque axis mount system, unclip the lower splash shield. Place a 1 inch x 1 inch x 2 inch block of wood between the torque strut and cradle prior to mount removal. This will hold the engine in position eliminating the need for jacking to install the mount. Then, remove the 3 right side upper engine mount-to-engine front cover nuts and the 2 mount-to-

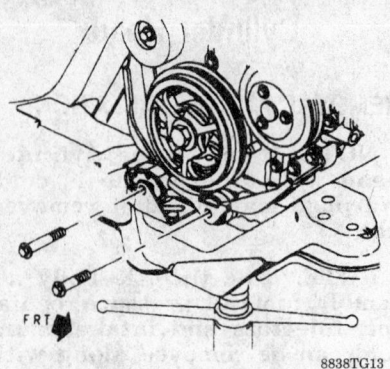

Removing the 2 strut to bracket bolts

Intake Side				
8	4	1	5	9
7	3	2	6	10
Exhaust Side				

8838TG14

Cylinder head torque sequence

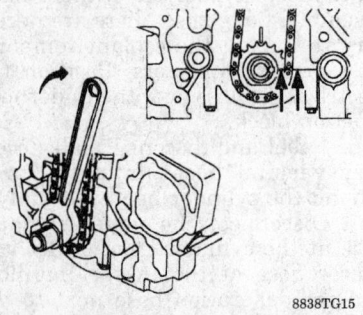

Align the crankshaft gear timing mark and keyway with the bearing cap split line to set the engine 90 degrees past TDC

midrail bracket nuts allowing the engine to rest on the block of wood.

10. Remove the serpentine drive belt and belt tensioner. It is not necessary to remove the water pump pulley, however, fit will be necessary to remove the idler pulley to access the engine front cover while the engine is still in the vehicle.

11. For SOHC engines, disconnect the deareation line at the cylinder head water outlet and from the support bracket.

12. Remove the camshaft cover, then inspect the cover silicone insula-

tors for cracks or deterioration and replace as necessary. Be sure to cover the valve train area in order to prevent foreign debris from entering the engine.

13. If equipped, remove the power steering pump bracket attaching bolts and position the assembly next to the right side front of the dash panel, away from the intake manifold and cylinder head. It is not necessary to remove the water pump pulley.

14. If equipped, remove the 3 air conditioning compressor front bracket bolts attached to the cylinder head and block, then remove the rear bracket bolts from the compressor. Do not discharge the system or disconnect the refrigerant lines. Support the compressor aside from the vehicle front support bar.

15. Raise and support vehicle safely, then drain the engine oil.

16. Remove the right side tire and splash shield.

17. For DOHC engines, remove the intake manifold support brace bolt attached to the intake manifold next to the alternator.

18. Remove the crankshaft damper/pulley assembly. Use a strap wrench or a block of wood wedged between the pulley spoke and the rear lower side of the front cover to hold the assembly while removing the bolt. Then use a 3 jaw puller on the jaw slots cast into the pulley and remove the assembly.

19. Disconnect the front exhaust pipe from the manifold.

20. Install crankshaft gear retainer tool SA9104E or equivalent to hold the gear in position, then properly remove the engine front cover. After the cover is loosened and moved approximately 1 in. (25.4mm) from the engine, the timing gear tool should be removed.

NOTE: Do not attempt to loosen the front cover without using a gear service tool which is designed to prevent damage to the timing chain guide. The tool is also used during installation to align the oil pump gerotor. The crankshaft gear service tool must be installed with the flat side toward the sprocket.

21. Rotate the crankshaft clockwise so the crankshaft gear timing mark and keyway align with the main bearing cap split line. This will set the engine to 90 degrees off top dead center and make sure pistons will not contact the valves upon assembly.

22. Remove the timing chain, tensioner, guides, camshaft sprocket(s) and chain. Use a 7/8 in. (21mm)

wrench to hold the camshaft when removing the sprocket bolts.

23. For the SOHC engine, remove the throttle body assembly and cover the intake manifold opening.

24. Use a 6-point socket to remove the 10 cylinder head bolts in several passes of the proper sequence (the reverse of the torque sequence). Failure to follow the proper sequence, or removal of the head when hot, could result in head warpage or cracking. Also, the use of a 12-point socket on the cylinder head bolts may round the bolt heads.

25. Lift the cylinder head from the dowels. Be careful not to damage the sealing surfaces if prying is necessary to remove the head from the block.

26. If necessary, remove the intake manifold or the exhaust manifold by loosening the mounting nuts in the proper sequence. If any cylinder head studs come out, the threads should be cleaned, the studs carefully installed and then tightened to 106 inch lbs. (12 Nm).

To install:

27. If removed, install the intake manifold and/or the exhaust manifold and new gasket(s). Tighten the fasteners to specification using the proper torque sequence.

28. Clean the gasket mating surfaces. Be careful not to damage the aluminum components. Make sure the block bolt holes are clean of any residual sealer, oil or foreign matter.

29. Inspect the top of the cylinder block for excessive warpage. Transverse warpage must not exceed 0.002 in. (0.05mm), while longitudinal warpage must not exceed 0.004 in. (0.10mm).

30. Check that the cylinder liners are flush or do not deviate more than 0.0005 in. (0.013mm) from the deck.

31. Make sure the crankshaft is positioned at 90 degrees past TDC to prevent valve or piston damage.

32. Install the cylinder head gasket and carefully guide the head into place over the dowels.

33. If the head bolts or the block were replaced, install the bolts and tighten in sequence to 48 ft. lbs. (65 Nm) to insure proper clamp load, then remove the bolts.

34. Coat the cylinder head bolts with clean engine oil and thread the bolts by hand until finger-tight. Tighten the bolts in sequence to 22 ft. lbs. (30 Nm).

35. Tighten the cylinder head bolts again, in sequence to 33 ft. lbs. (45 Nm) for SOHC engines or to 37 ft. lbs. (50 Nm) for DOHC engines. Install Snap-On® tool 360 or

equivalent torque angle gauge and calibrate the gauge to 0, then tighten each cylinder head bolt an additional 90 degrees, in sequence.

36. Align the camshaft(s) to TDC, then rotate the crankshaft counterclockwise 90 degrees, also to TDC. Install the timing chain, sprockets, guides and tensioner.

37. Install the front cover assembly and connect the exhaust manifold to the exhaust pipe.

38. Install the crankshaft damper/pulley assembly and tighten the bolt to 159 ft. lbs. (215 Nm).

39. For DOHC vehicles, install the intake manifold support brace bolts next to the alternator, then tighten the bracket-to-block bolt to 33 ft. lbs. (45 Nm) and tighten the bracket-to-manifold bolt to 22 ft. lbs. (30 Nm).

40. Apply a small drop of RTV across the cylinder head and front cover T-joints. Inspect the old camshaft cover gasket and replace if damaged. Install the gasket and the camshaft cover. Tighten the fasteners uniformly to 22 ft. lbs. (30 Nm) for SOHC vehicles or in proper sequence to 89 inch lbs. (10 Nm) for DOHC vehicles.

41. Install the drive belt tensioner and tighten the bolt to 22 ft. lbs. (30 Nm). Install the idler pulley and tighten the fasteners to 33 ft. lbs. (45 Nm).

42. If not done during removal, drain the engine oil. Change the filter, then install the drain plug and tighten to 26 ft. lbs. (35 Nm).

43. If removed, verify the gaps on all spark plugs and install. Tighten to 20 ft. lbs. (27 Nm).

44. For SOHC engines applicable, install a new gasket and the TBI assembly. Tighten the assembly retainers to 24 ft. lbs. (33 Nm).

45. If equipped, install the power steering pump assembly to the bracket, then tighten the bolts to 22 ft. lbs. (30 Nm).

46. If equipped, install the air conditioning compressor and bolts. Tighten the rear bracket bolts to 19 ft. lbs. (25 Nm), then tighten the front bracket bolts to 35 ft. lbs. (47 Nm).

47. Install the serpentine drive belt making sure the belt is properly aligned on the pulleys.

48. With a torque axis mounting, install the 2 mount-to-midrail bracket nuts and tighten to 52 ft. lbs. (70 Nm). Then, install the 3 upper mount-to-engine front cover nuts and tighten them uniformly to 52 ft. lbs. (70 Nm). Remove the support block of wood after the assembly is installed.

49. Install the splash shield, then install the wheel assembly.

50. Position the wiring harness and engage all wire connectors.

51. Install all vacuum hoses disconnected during removal.

52. If removed, install the accelerator cable bracket and tighten the fastener to 19 ft. lbs. (25 Nm). Connect the cable, then verify that it is properly routed and not binding.

53. Install all coolant hoses which were disconnected during removal.

54. For the SOHC engine, install the intake manifold support bracket and/or fasteners. Tighten the manifold fasteners to 21 ft. lbs. (28 Nm) and the block fasteners to 22 ft. lbs. (30 Nm).

55. Connect the fuel feed and return lines to the throttle body and tighten the fittings to 19 ft. lbs. (25 Nm) or to the fuel rail and pressure regulator and tighten the fittings to 133 inch lbs. (15 Nm). Install fuel bracket retaining bolts, as applicable.

56. Install the air cleaner and intake duct assembly.

57. Add engine oil and properly fill the engine cooling system.

58. Connect the negative battery cable.

59. Prime the fuel system, then start the engine and check for leaks.

Valve Lifters

REMOVAL AND INSTALLATION

SOHC Engine

1. Disconnect the negative battery cable.

2. Remove the rocker arm cover, then inspect the cover silicone insulators for cracks or deterioration and replace as necessary.

3. Uniformly loosen and remove the rocker arm assembly bolts.

4. Remove the 2 rocker arm shafts, rocker arm assemblies, guide

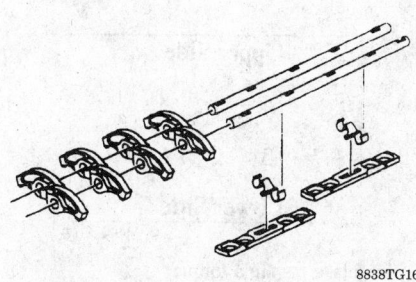

8838TG16

Lifter guide plate positioning — SOHC engine

plates and lifters. Mark all components to assure installation in their original locations.

To install:

5. Oil the lifters, then install them into the cylinder head bores.

6. Rotate the lifters until the flats are parallel with the intake and exhaust sides of cylinder head, then install the guide plates. Make sure the plates are properly seated with the retaining spring slot upwards and the lifters fitted squarely into the guide plate.

7. Install the rocker arm shaft assemblies. Make sure the rocker arm tangs are squarely seated on the lifter and the retaining spring is positioned in the guide plate slot. The retaining springs should be snapped onto the rocker arm shafts between the No. 1–2 and the No. 3–4 cylinder rocker arms.

NOTE: During installation, a flat piece of cardboard or a ratchet extension bar may be positioned on top of the rocker arm shaft assemblies to align the arms on both the valves and lifters.

8. Torque the rocker arm bolts to 19 ft. lbs. (25 Nm) in a uniform sequence.

9. Apply a small drop of RTV to each cylinder head and front cover T-joint. Inspect the rocker arm cover gasket and replace if necessary. Install the gasket and rocker arm cover, then tighten the fasteners uniformly to 22 ft. lbs. (30 Nm).

10. Connect the negative battery cable, start the engine and check for leaks.

DOHC Engine

1. Disconnect the negative battery cable.

2. Remove the camshaft cover.

3. Remove the camshafts.

NOTE: A magnet will easily pull lifters from their bores. Do not attempt to remove the lifters using pliers or other sharp tools which might score the machined surfaces.

4. Remove the valve lifters from their bores. Be sure to place all lifters in a rack or label them to assure installation in their original locations. Store lifters so the oil will not drain from the assemblies.

To install:

5. Lubricate the lifters and install them in their proper locations.

6. Properly align and install the intake and exhaust camshafts.

7. Apply a small drop of RTV across the cylinder head and front cover T-joints. Inspect the old camshaft cover gasket and replace if damaged. Install the gasket and the camshaft cover. Tighten the fasteners in the proper sequence to 89 inch lbs. (10 Nm).

8. Connect the negative battery cable, start the engine and check for leaks.

Valve Lash

ADJUSTMENT

The engines are equipped with hydraulic valve lifters which are designed to maintain zero lash. No periodic adjustments are needed or possible.

Rocker Arms/Shafts

The SOHC engine, is the only engine which utilizes rocker arms.

REMOVAL AND INSTALLATION

1. Disconnect the negative battery cable.
2. Remove the rocker arm cover, then inspect the cover silicone insulators for cracks or deterioration and replace as necessary.
3. Uniformly remove the rocker arm shaft bolts, then remove the 2 rocker arm shafts. The shafts may be unsnapped from the lifter guide plates in order to leave the guide plates and lifters in the engine.
4. If necessary, disassemble the rocker arms from the shafts.
   To install:
5. If removed, oil the rocker arm shafts, then install the arms onto the shafts.
6. Snap 1 end of each lifter guide plate retaining spring onto the rocker arm shaft between the No. 1-2 and the No. 3-4 cylinder rocker arms.

NOTE: During installation, a flat piece of cardboard or a ratchet extension bar may be positioned on top of the rocker arm shaft assemblies to align the arms on both the valves and lifters.

7. Install the rocker arm shaft assemblies. To prevent valve or piston damage, be sure the rocker arm tangs are squarely seated on the lifters and the retaining springs are positioned in the guide plate slots.

8. Verify the proper position and seating of all rocker components, then tighten the 5 rocker arm bolts on each shaft to 19 ft. lbs. (25 Nm) in a uniform sequence.

9. Apply a small drop of RTV to each cylinder head and front cover T-joint. Inspect the rocker arm cover gasket and replace if necessary. Install the gasket and rocker arm cover, then tighten the fasteners uniformly to 22 ft. lbs. (30 Nm).

10. Connect the negative battery cable, start the engine and check for leaks.

Intake Manifold

REMOVAL AND INSTALLATION

SOHC Engine

1. Disconnect the negative battery cable and drain the engine coolant.
2. Remove the air cleaner and the fresh air tube at the rocker cover. Remove the PCV tube and hose.
3. Properly relieve the fuel system pressure, then disconnect the fuel supply and return lines at the connectors using service tool SA9157E or equivalent. Plug the lines to prevent system contamination or excessive fuel spillage.
4. Disconnect the throttle cable from the throttle body, then remove the throttle cable bracket attaching nuts and position the assembly aside.
5. Label and disconnect the following wiring from the following throttle body and intake manifold components:
   - Fuel injector
   - Idle Air Control (IAC) valve
   - Throttle Position (TP) sensor
   - Exhaust Gas Recirculation (EGR) valve
   - Manifold Absolute Pressure (MAP) sensor

```
         Upper Side

      8   4   1   5

      7   3   2   6   9

         Lower Side
```
8838TG17

Intake manifold torque sequence — SOHC engine

6. Position the wiring harness away from the manifold onto the fuel relay.
7. Tag and disconnect the vacuum hoses from the TBI tube module assembly and from the brake booster.
8. Remove the heater hose from the manifold and the de-aeration line from the cylinder head, then remove the 2 clamps and position the line on the coolant bottle.
9. Remove the intake manifold support bracket bolt located next to the starter and attached to the block. If necessary, the bolt can be removed from below the vehicle.
10. Remove the serpentine drive belt, then remove the power steering pump from the bracket and support the pump next to the right side dash panel sufficiently away from the intake manifold and cylinder head.
11. On 1995–97 vehicles, disconnect the canister purge solenoid and brake booster vacuum hoses.
12. Remove the manifold retaining nuts, then remove the manifold and throttle body assembly. If necessary the lower manifold nuts can be accessed from under the vehicle.
    To install:
13. Thoroughly clean all gasket mating surfaces. Be careful not to damage or score the aluminum surface. If replaced, use Loctite® 290 or equivalent to seal a new PCV valve inlet tube into the manifold.
14. Position the new gasket, then install the manifold and retaining nuts. Tighten the nuts in sequence to 22 ft. lbs. (30 Nm).
15. Install the power steering pump and tighten the fasteners to 27 ft. lbs. (38 Nm).
16. Install the serpentine drive belt.
17. Connect the heater hose, then install the de-aeration line and clamps.
18. Install the manifold support bracket bolt and tighten to 22 ft. lbs. (30 Nm).
19. Connect the fuel supply and return lines to the TBI unit, then tighten the fittings to 19 ft. lbs. (25 Nm).
20. Reposition the wiring harness, then connect the wiring and vacuum hoses to their original locations. The harness leads to the TPS and EGR solenoid must be routed between the No. 3 and 4 intake manifold runners.
21. Install the air cleaner and fresh air tubes, then install the PCV valve hose.
22. On 1995–97 vehicles, connect the canister purge solenoid and brake booster vacuum hoses.

23. Connect the negative battery cable and properly fill the engine cooling system.

24. Prime the fuel system, start the engine and check for leaks.

DOHC Engine

1. Disconnect the negative battery cable and drain the engine coolant.

2. Remove the air inlet tube and resonator, then remove the PCV tube.

3. Properly relieve the fuel system pressure at the test port.

4. Remove the fuel line bracket bolt and disconnect the lines from the fuel rail. Disconnect the fuel lines from the connectors using service tool SA9157E or equivalent. Plug the openings to prevent fuel system contamination or excessive fuel spillage.

5. Disconnect the throttle cable from the throttle body, then remove the cable bracket assembly and position aside.

6. Label and disconnect the following wiring from the following throttle body and intake manifold components:
   • Fuel injectors
   • Idle Air Control (IAC) valve
   • Throttle Position (TP) sensor
   • Manifold Absolute Pressure (MAP) sensor

7. Disconnect the heater and de-aeration hoses from the intake manifold, then disconnect the vacuum hose from the EGR solenoid.

8. Position the wiring harness over the brake master cylinder, then remove the intake manifold support bracket bolt attached to the manifold next to the brake master cylinder.

9. Remove the serpentine drive belt. Remove the power steering pump assembly with the support bracket, then remove the upper pump bracket attachment bolts and position the pump away from the

manifold and cylinder head, near the right dash panel. Remove the lower power steering pump bracket brace.

10. Remove the 3 upper intake manifold attachment nuts, then raise and support the vehicle safely.

11. Remove the lower power steering unit support bracket. Remove the intake manifold support bracket bolt located next to the alternator, then loosen the lower bracket bolt and rotate the bracket out of the way.

12. Disconnect the canister purge solenoid and brake booster vacuum hoses.

13. Remove the intake manifold attaching stud and lower the vehicle.

14. Remove the intake manifold assembly, then remove and discard the old gasket.

To install:

15. Thoroughly clean the gasket mating surfaces. Be careful not to score or damage the aluminum sealing surfaces. If installing a new coolant de-aeration tube elbow into the manifold, use Loctite® 290 or equivalent to assure proper seal.

16. Position the new gasket, then install the intake manifold and retaining nuts. Torque the nuts in sequence to 22 ft. lbs. (30 Nm).

17. Install the power steering pump and brackets. Tighten the fasteners to 28 ft. lbs. (38 Nm).

18. Install the serpentine drive belt making sure the belt is properly aligned on the pulleys.

19. Connect the heater hose and de-aeration line to the manifold.

20. Position the manifold support brackets and install the retainers. Tighten the right bracket-to-block retainer to 41 ft. lbs. (55 Nm), then tighten the left block retainer and the support bracket-to-intake manifold bolts to 22 ft. lbs. (30 Nm).

21. Lubricate the male fuel supply and return connect fittings, then install.

22. Connect the throttle cable to the throttle body and install the support bracket. Tighten the bracket retaining bolts to 19 ft. lbs. (25 Nm). Verify that the cable locking tangs are fully engaged when assembled.

23. Position the wiring harness, then engage all electrical connectors and vacuum hoses in their original locations.

24. Install the PCV hose, the air inlet tube and resonator.

25. Connect the negative battery cable and properly fill the engine cooling system.

26. Prime the fuel system, start the engine and check for leaks.

Exhaust Manifold

REMOVAL AND INSTALLATION

1. Disconnect the negative battery cable, then raise and support the vehicle safely.

2. Remove the 2 front exhaust pipe-to-engine stiffening bracket fasteners.

3. Remove the pipe-to-manifold nuts and lower the pipe slightly, then remove the old gasket and discard.

4. Lower the vehicle.

5. If equipped, remove the the air conditioning compressor and bracket and support to the side. Do not disconnect the refrigerant lines.

6. Unplug the oxygen sensor connector. If necessary, use a 19mm, 6-point, crows foot to remove the oxygen sensor.

7. Remove the manifold retaining nuts and remove the manifold. Remove and discard the old gasket.

To install:

8. Thoroughly clean the gasket mating surfaces, be careful not to score or damage the aluminum surface.

9. Install the new gasket with the smooth side facing the manifold, then install the manifold and attaching nuts. Tighten the nuts in sequence to 16 ft. lbs. (22 Nm) for the SOHC engine or to 23 ft. lbs. (31 Nm) for the DOHC engine.

10. If replacing the oxygen sensor, coat the threads with nickel based anti-seize compound and tighten to 18 ft. lbs. (25 Nm). Install the wiring harness to the oxygen sensor electrical connector.

11. Install the air conditioning compressor and brackets. Tighten all fasteners except the front bracket-to-compressor fasteners to 19 ft. lbs. (25 Nm). Tighten the front bracket to compressor fasteners to 40 ft. lbs. (54 Nm).

12. Raise and support the vehicle safely, then install a new gasket onto the studs between the pipe and manifold.

13. Connect the pipe and manifold, then tighten the fasteners in a crosswise pattern to 23 ft. lbs. (31 Nm).

14. Install the clamp-to-front exhaust pipe and support bracket, then tighten the fasteners to 44 ft. lbs. (60 Nm).

15. Lower the vehicle, then connect the negative battery cable.

16. Start the engine and check for leaks.

Upper Side		
5	2	3
7	4	1	6
Lower Side		

8838TG18

Intake manifold torque sequence — DOHC engine

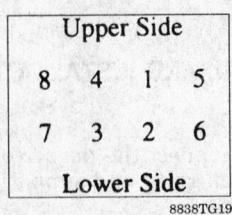

Upper Side			
8	4	1	5
7	3	2	6
Lower Side			

8838TG19

Exhaust manifold torque
sequence — SOHC engine

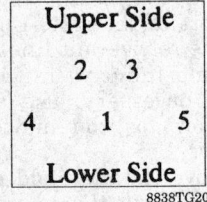

Upper Side		
	2	3
4	1	5
Lower Side		

8838TG20

Exhaust manifold
torque sequence —
DOHC engine

Front Cover Seal

REMOVAL AND INSTALLATION

Front Cover Installed

1. Disconnect the negative battery cable and drain the engine oil.
2. Raise the vehicle and support safely, then remove the right wheel and splash shield.
3. Remove the serpentine drive belt from the crankshaft damper/pulley.
4. Using a strap wrench or a piece of wood wedged between the damper spoke and the lower side of the engine front cover, hold the damper and remove the bolt. With a suitable 3 jaw puller and the slots cast into the damper, pull the crankshaft damper/pulley assembly from the crankshaft.
5. Use a small suitable tool to pry the oil seal from the front cover. Be very careful not to damage the front cover or crankshaft.
6. Clean the seal bore and oil drain back passage.
   To install:
7. Make sure the oil drain back is free of RTV or debris. Position the oil seal and thread the service seal in-

staller tool, SA9104E or equivalent, onto the end of the crankshaft. Use the tool to draw the seal into position. Never tap on the seal or the seal install with a hammer.
8. Apply a thin film of clean engine oil to the new seal lip.
9. Position the crankshaft damper/pulley assembly, then secure as accomplished during removal while tightening the bolt to 159 ft. lbs. (215 Nm).
10. Install the serpentine drive belt.
11. Install the splash shield and the wheel assembly, then lower the vehicle.
12. Properly fill the engine crankcase and connect the negative battery cable.
13. Start the engine and check for leaks.

Front Cover Removed

1. Use a suitable prytool to carefully pry the front oil seal from the front cover. Be careful not to damage the front cover or crankshaft.
2. Clean the seal bore and oil drain back passage.
3. Place the engine front cover on the base of a suitable arbor press. Support the cover in order to prevent deformation or damage.
4. Position the seal to the front cover and place tool SA9104E, or equivalent installation tool, over the seal.
5. Press the seal into the engine front cover approximately 0.04 inch (1mm) further into the engine front cover than the factory seal removed earlier.
6. Install the timing chain front cover to the vehicle.

Timing Chain Front Cover

REMOVAL AND INSTALLATION

1. Disconnect the negative battery cable and drain the engine oil.
2. Raise and support the vehicle safely, then remove the right wheel and splash shield.
3. With a torque axis mount system, place a 1 inch x 1 inch x 2 inch block of wood between the torque strut and cradle prior to mount removal. This will hold the engine in position eliminating the need for jacking to install the mount. Then, remove the 3 right side upper engine mount-to-engine front cover nuts and the 2 mount-to-midrail bracket nuts allowing the engine to rest on the block of wood.

4. Remove the serpentine drive belt and belt tensioner. It is not necessary to remove the water pump pulley, however the idler pulley must be removed to access the engine front cover.
5. Using a strap wrench or a piece of wood wedged between the damper spoke and the lower side of the engine front cover, hold the damper and remove the bolt. Using a suitable 3 jaw puller and the slots cast into the damper, pull the crankshaft damper/pulley assembly from the crankshaft.
6. Remove the power steering pump and position the assembly aside.
7. Remove the rocker or camshaft cover. Be sure to protect the valve train assemblies from foreign debris or dirt.
8. Install the special tool SA9104E or equivalent to make sure the crankshaft timing sprocket is held firmly in place and to prevent guide damage. Install with the flat side towards the crankshaft sprocket.
9. Remove the front 4 oil pan bolts, then using a suitable RTV cutting tool, cut the front seal away from the front cover.
10. Spray the 2 dowel pin holes with penetrating oil to facilitate front cover removal from the dowel pins.
11. Remove the front cover bolts. For DOHC vehicles, 1 bolt is located above the serpentine drive belt idler pulley, under the torque axis mount flange.
12. Using a small suitable tool, carefully pry the cover away from the cylinder block at the pry locations tabs which are provided. Once the cover has been loosened and moved approximately 1 in. (25.4mm), remove the crankshaft gear retaining tool, then remove the cover from the vehicle. Remove and discard the 2 oil gallery transfer seals.
    To install:
13. Make sure the oil galleys are clear. Carefully clean the gasket mating surfaces with a scraper or wire brush and carburetor solvent, brake clean or alcohol. Use a 3/16 in. drill bit and tap handle to clean the front cover holes. The non-tapped front cover holes may be cleaned using a 3/8 in. drill bit.

NOTE: If the engine front cover casting or assembly is replaced, the 3 torque axis mount studs should also be replaced. Tighten the studs to 133 inch lbs. (15 Nm).

14. If removed, install the oil pump and cover to the front cover. Be sure

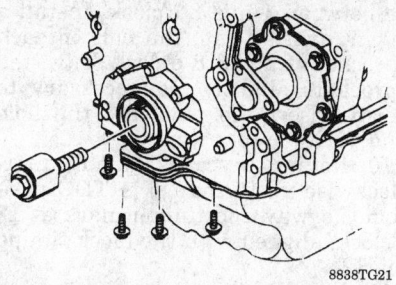

SA9104E, or an equivalent timing sprocket retainer must be used when separating the front cover in order to prevent timing chain and guide damage

to pack the pump with petroleum jelly to assure proper priming. Also, if removed, install a new front seal and/or oil pressure and suction seals.

15. Apply a 0.16 inch (4mm) bead of RTV sealer on the along the vertical sealing surfaces of the front cover to the inside of the bolt holes and a 0.08 in. (2mm) bead to the front of the oil pan. For DOHC engines, apply a thin bead around the 2 inner cover bolt holes. Be sure to assemble the front cover to the engine within 3 minutes of application.

16. In order to properly align the oil pump gerotor, install the crankshaft gear retaining tool and position the front cover to the engine. Install the bolts, then tighten the perimeter bolts starting at the center and working outwards on both sides to 19 ft. lbs. (25 Nm) for the perimeter and upper center bolts.

17. Install and tighten the front cover lower center bolt to 89 inch lbs. (10 Nm), then install the 4 oil pan front bolts and tighten to 80 inch lbs. (9 Nm).

18. After front cover installation, spray 6–12 squirts of oil though the front oil seal drain back hole to verify it is not plugged.

19. Apply a thin film of RTV between the damper/pulley assembly flange and washer only, the washer and bolt head flange are designed to prevent oil leakage.

20. Remove the crankshaft retaining tool and position the crankshaft damper/pulley assembly, then secure as accomplished during removal while tightening the bolt to 159 ft. lbs. (215 Nm).

21. Apply a small drop of RTV across the cylinder head and front cover T-joints. Inspect the old camshaft cover gasket and replace if damaged. Install the gasket and the camshaft cover. Tighten the fasteners uniformly to 22 ft. lbs. (30 Nm) for SOHC vehicles or in proper sequence

to 89 inch lbs. (10 Nm) for DOHC vehicles.

22. Install the power steering pump assembly.

23. If removed, install the idler pulley and tighten the retainer to 33 ft. lbs. (45 Nm).

24. Install the belt tensioner and the serpentine drive belt.

25. With a torque axis mounting, install the 2 mount-to-midrail bracket nuts and tighten to 52 ft. lbs. (70 Nm). Then, install the 3 upper mount-to-engine front cover nuts and tighten them uniformly to 52 ft. lbs. (70 Nm). Remove the support block of wood after the assembly is installed.

26. Install the splash shield and the wheel assembly, then lower the vehicle.

27. Properly fill the engine crankcase and connect the negative battery cable.

28. Start the engine and check for leaks.

Timing Chain and Sprockets

REMOVAL AND INSTALLATION

SOHC Engine

1. Disconnect the negative battery cable.

2. Remove the timing chain front cover.

NOTE: When removing the timing chain and sprockets, the crankshaft should be positioned 90 degrees past TDC to make sure the pistons will not contact the valves upon assembly.

3. Carefully rotate the crankshaft clockwise from TDC (timing mark at 12 o'clock) so the timing mark and keyway on the crankshaft sprocket align with the main bearing cap split line to the right of the crankshaft.

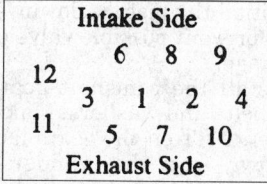

Camshaft cover torque sequence — DOHC engine

4. Remove the timing guides and tensioner.

5. Remove the camshaft sprocket bolt, using a 7/8 in. (21mm) wrench to hold the camshaft. Then remove the timing chain and camshaft sprocket. If necessary, remove the crankshaft sprocket.

To install:

6. Inspect the chain for wear and damage. Check the inside diameter of the chain, it should be no more than 16.77 in. (426mm). Inspect the chain guides for wear or cracks and the timing gears for teeth or key wear. Replace components as necessary.

7. Verify that the crankshaft is positioned 90 degrees clockwise past TDC by checking that the keyway and timing mark are at 3 o'clock, parallel with the main bearing cap split line.

8. Bring the camshaft up to the No. 1 TDC by loosely installing the sprocket and rotating the sprocket and crankshaft in the clockwise direction until the timing pin can be installed. The camshaft contains wrench flats to assist in turning the shaft. The dowel pin should be at 12 o'clock when No. 1 is at TDC and a timing pin (3/16 in. drill bit) should then install at about the 8 o'clock position.

9. With the camshaft at No. 1 TDC, rotate the crankshaft counterclockwise 90 degrees up to the No. 1 cylinder TDC position. The gear keyway and timing mark should be at 12 o'clock, aligned with the block timing mark. Remove the camshaft sprocket and timing pin so the chain can be installed.

10. Position the chain over the camshaft sprocket and under the crankshaft sprocket, then slide the camshaft sprocket into position. Install the timing chain and sprockets so the 1 silver link plate aligns with the pip mark on the top of the camshaft sprocket and the other aligns with the downward tooth at the 6 o'clock position on the crankshaft sprocket. The letters FRT on the camshaft sprocket must face forward, away from the cylinder head and excess chain slack should be located on the tensioner side of the block.

11. Install the timing pin through the camshaft sprocket to verify proper timing.

12. Install and tighten the camshaft sprocket bolt to 75 ft. lbs. (102 Nm). Again, use an wrench on the camshaft wrench flats to hold the shaft in position while tightening the bolt. Do not allow the camshaft re-

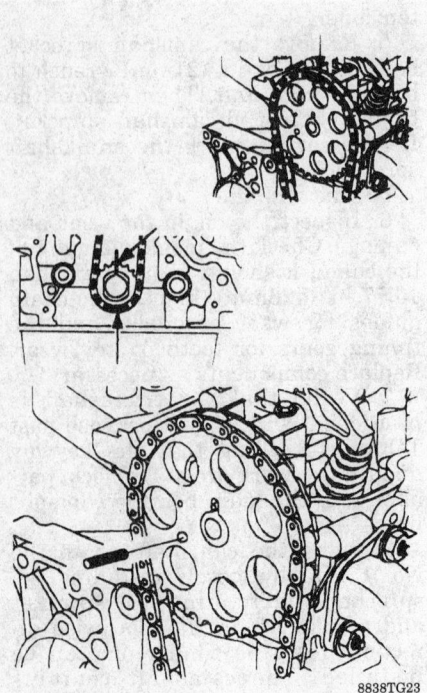

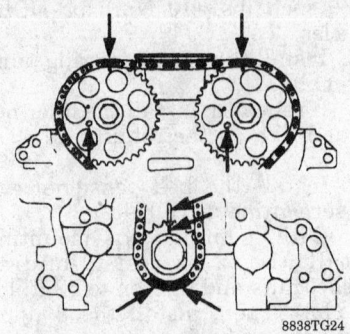

Timing chain and sprocket alignment marks — DOHC engine

8838TG24

Timing chain and sprocket alignment marks — SOHC engine

8838TG23

taining bolt to torque against the timing pin or cylinder head damage will result.

13. Install the fixed chain guides, the word FRONT on the fixed guide must be facing outward. Install the fixed guide first and verify the chain is snug against the guide, then install the pivot guide. Tighten the bolts to 19 ft. lbs. (26 Nm) and verify that the pivot guide moves freely.

14. Retract the tensioner plunger and pin the ratchet lever using a 1/8 in. No. 31 drill bit inserted in the alignment hole. Install the tensioner and tighten the bolts to 168 inch lbs. (19 Nm), then remove the drill bit.

15. Verify the proper positioning of all timing marks, then remove all alignment pins.

16. Install the timing chain front cover.

17. Connect the negative battery cable.

18. Start the engine and check for leaks.

DOHC Engine

1. Disconnect the negative battery cable.

2. Remove the timing chain front cover.

NOTE: When removing the timing chain and sprockets, the crankshaft should be positioned 90 degrees past TDC to make sure the pistons will not contact the valves upon assembly.

3. Carefully rotate the crankshaft clockwise from TDC (timing mark at 12 o'clock) so the timing mark and keyway on the crankshaft sprocket align with the main bearing cap split line to the right of the crankshaft.

4. Remove the timing guides and tensioner.

5. Remove the camshaft sprocket bolts, using a 7/8 in. (21mm) wrench to hold the camshaft from turning. Then remove the timing chain and camshaft sprockets. If necessary, remove the crankshaft sprocket.

To install:

6. Inspect the chain for wear and damage. Check the inside diameter of the chain, it should be no more than 23.15 in. (588mm). Inspect the chain guides for wear or cracks and the timing gears for teeth or key wear. Replace components as necessary.

7. If removed, install the crankshaft sprocket, then verify that the crankshaft is positioned 90 degrees clockwise past TDC. The crankshaft timing mark and keyway should be at the 3 o'clock position, aligned with the main bearing cap split line. This will position the pistons downward in order to prevent possible valve or piston damage.

8. Install the camshaft gears, retaining bolts and washers. Make sure the letters FRT on the gears face forward, away from the cylinder block. Use the wrench flats provided on the camshafts to hold the shaft and tighten the bolts to 75 ft. lbs. (102 Nm).

9. Bring the camshafts up to No. 1 TDC; the timing mark should be at 12 o'clock and the letters FRT should

be between 1 and 3 o'clock. Install a 3/16 in. drill bit into the hole in each sprocket at about 8 o'clock. Turn the sprockets slightly, as necessary to verify insert the pins into the hole and verify proper timing.

10. Rotate the crankshaft counterclockwise up to the No. 1 TDC position (keyway and timing mark at 12 o'clock, aligned with the block timing mark).

11. Position the timing chain over the camshaft sprockets so the 2 silver link plates align with the pip marks on the top center of the camshaft sprockets, then position the chain under the crankshaft sprocket so the other 2 plates align on either side of the downward tooth (at 6 o'clock position) on the crankshaft sprocket. Excess chain slack should be located on the tensioner side of the cylinder block.

12. Verify that the crankshaft pip mark aligns with the cylinder block mark at 12 o'clock and that the timing pin holes are aligned at about the 8 o'clock position. Remove the timing pins from the camshaft sprockets.

13. Install the timing chain fixed guide to the right of the block face toward the water pump. Tighten the bolts to 21 ft. lbs. (28 Nm) and verify the chain is snug against the guide.

14. Install the pivoting chain guide and check for clearance between the block and head. Tighten the bolt to 19 ft. lbs. (26 Nm) and verify that the guide pivots freely.

15. Retract the tensioner plunger and pin the ratchet lever using a 1/8 in. No. 31 drill bit inserted in the alignment hole. Install the tensioner and tighten the bolts to 168 inch lbs. (19 Nm), then remove the drill bit allowing the tensioner to extend.

16. Make sure all alignment pins are removed.

17. Install the timing chain front cover.

18. Connect the negative battery cable, then start the engine and check for leaks.

Camshaft

REMOVAL AND INSTALLATION

SOHC Engine

1. Disconnect the negative battery cable.

2. Remove the timing chain front cover.

3. Remove the timing chain and camshaft sprocket.

4. Remove the rocker arm and shaft assemblies.

5. Remove the lifters and tag or arrange to assure assembly in their original locations.

6. Remove the battery cover and battery.

7. Drive the camshaft plug (located on the battery side of the cylinder head) inward, then remove it from the cylinder head with a magnet.

8. Carefully pull the camshaft from the rear of the cylinder head though the oversized camshaft plug hole.

To install:

9. Lubricate the camshaft and install into the rear of the cylinder head.

10. Coat a new rear cylinder head plug with Loctite® 242 or equivalent and install it using a standard bushing driver.

11. Install the battery and tighten the battery hold-down nut and screw to 80 inch lbs. (9 Nm). Connect only the positive battery cable, at this time.

12. Install the valve lifters into their original bores. If a new camshaft was installed, the lifters must be replaced to assure camshaft life.

13. Install the rocker arm shaft assemblies.

14. Install the timing chain and camshaft sprocket.

15. Install the timing chain front cover.

16. Connect the negative battery cable, then start the engine and check for leaks.

DOHC Engine

1. Disconnect the negative battery cable and remove the serpentine drive belt.

2. Disconnect the spark plug wires from the plugs, remove the EGR valve solenoid attachment screw and remove the PCV fresh air hose.

3. Remove the camshaft cover, then inspect the cover silicone insulators for cracks or deterioration and replace as necessary.

4. Turn the crankshaft until the mark on the crankshaft pulley is in alignment with the front cover pointer at the damper's 12 o'clock position and No. 1 cylinder is at TDC of the compression stroke. Both camshaft dowel pins will be at the 12 o'clock position and the timing pin holes will be aligned when the No. 1 cylinder is at TDC. If necessary, the right wheel and splash shield may be

removed to help observe the timing marks.

NOTE: When removing or installing the camshaft sprocket retaining bolts, a clean shop rag should be positioned over the opening between the cylinder head and the front timing cover to prevent a bolt from being dropped between the components. If a bolt is dropped, the front cover must be removed in order to extract the bolt.

5. Carefully remove each camshaft sprocket's retaining bolts and washers. Use a ⅞ in. (21mm) open end wrench to hold each camshaft from turning while removing the bolts.

6. Position the front angled support fixture in front of the camshaft sprockets.

7. Attach the camshaft sprocket adapters to the end of each camshaft using the pilot bolts, but do not tighten the bolts. The support fixture rests between the sprockets and the sprocket adapters.

8. Remove the upper timing chain guide and both front camshaft bearing caps.

9. Secure the support fixture using the ⅜ in. bolts/blocks and align the 2 holes in each camshaft sprocket, adapter and the front support fixture. Install the 4 nuts, but do not tighten. The steel blocks should be installed against the rearward side of the camshaft sprocket. Tighten the sprocket pilot bolts to 19 ft. lbs. (25 Nm) while holding the camshafts from turning with an open end wrench.

10. Move each camshaft sprocket off the end of the camshaft by rocking the sprocket forward at the 3 and 9 o'clock positions or by carefully prying between the end of the camshaft and the sprocket. Then tighten the 4 nuts and bolts with blocks from the side of the support fixture to 19 ft. lbs. (25 Nm).

11. Install the 2 bolts retaining the support fixture to the engine front cover and tighten the bolts to 89 inch lbs. (10 Nm). Then remove each camshaft sprocket pilot bolt while holding the camshafts with a wrench.

12. Carefully pry between the sprocket and the end of the camshaft to move the camshaft rearward. Pry only sufficiently to remove its end from inside the sprocket pilot otherwise camshaft or lifter damage may occur.

13. Uniformly loosen and remove the remaining camshaft bearing cap bolts. To prevent bolt/cap damage, do not use power tools, but instead make

several passes using a hand ratchet. Then remove each camshaft. Position the caps for installation in their original locations.

To install:

14. Oil the camshaft and install with the IN camshaft on the intake side and EX camshaft on the exhaust side.

NOTE: The dowel pin in each camshaft and the crankshaft timing mark must all be located at the 12 o'clock position during installation to prevent valve and piston damage.

15. Install all bearing caps, except for the forward pair, in their original positions. Uniformly tighten the cap bolts to 124 inch lbs. (14 Nm).

16. Install 1 camshaft sprocket pilot bolt in each camshaft and tighten to 124 inch lbs. (14 Nm) in order to pull the camshaft fully forward and align the sprocket support for installation of the sprockets onto the camshafts.

17. Remove the 4 sprocket support bolt/blocks and nuts. The torque axis mount system requires the fixture to remain in place longer.

18. Verify that the camshafts are fully positioned forward and install the 2 forward bearing caps and the upper chain guide. The caps are marked E1 or I1 for exhaust or intake and must be positioned with their arrows pointing towards the sprockets. Tighten the cap bolts to 124 inch lbs. (14 Nm).

19. Make sure the camshaft dowel pin aligns with the slot in each camshaft sprocket. If necessary, rotate the camshaft slightly (1–2 degrees), then move each sprocket from the adapter onto the end of the camshaft. Fully seat each sprocket on the end of each camshaft.

20. Remove the 2 sprocket pilot bolts and adapters while using a wrench on the camshaft flats to assure the camshafts cannot move.

21. Remove the support angled fixture.

22. Install the camshaft sprocket retaining bolts and washers. Hold the camshafts and tighten the bolts to 76 ft. lbs. (103 Nm).

23. Verify all visible timing marks and holes are in alignment. If necessary, turn the crankshaft clockwise until the mark on the crankshaft pulley aligns with the mark on the front cover. Insert ³⁄₁₆ in. drill bits through the camshaft sprocket alignment holes, into the cylinder head. If the alignment pins cannot be inserted, turn the crankshaft 360 degrees clockwise and repeat. If the pins can-

not be inserted within 1–2 degrees of either TDC position, the camshafts are not properly timed. Do not start the engine until the camshafts are timed.

24. Apply a small drop of RTV across the cylinder head and front cover T-joints. Inspect the old camshaft cover gasket and replace if damaged. Install the gasket and the camshaft cover. Tighten the fasteners in proper sequence to 89 inch lbs. (10 Nm).

25. Install the right splash shield and wheel, if removed to observe the timing marks.

26. Install the PCV and fresh air hoses, the EGR valve solenoid attaching screw and the spark plug wires.

27. Install the serpentine drive belt and connect the negative battery cable.

28. Start the engine and check for leaks.

Piston and Connecting Rod

POSITIONING

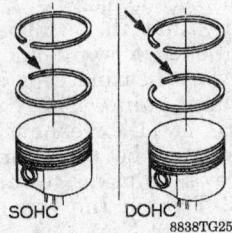

SOHC   DOHC
8838TG25

Only the top ring on the DOHC engine and the second ring on both engines have pip marks. The top ring on the SOHC engine can be installed with either side up.

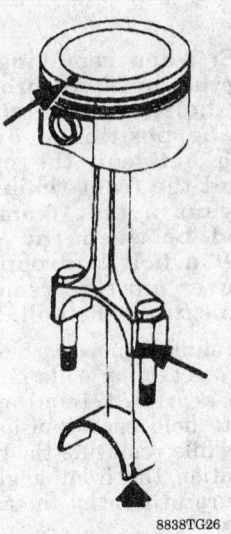

8838TG26

Piston and connecting rod positioning — align the mark on top of the piston with the front of the engine. Assemble the connecting rod to the piston with the bearing tang slots directed toward the exhaust manifold side.

ENGINE LUBRICATION

Oil Pan

REMOVAL AND INSTALLATION

1. Raise and support the vehicle safely, then drain the engine oil.

2. Remove the front exhaust pipe.

3. Remove the right wheel and splash shield, then loosen the 4 front motor mount-to-block or motor-to-front cover bolts. Back the bolts out about ½ inch (12mm).

4. Remove all the oil pan bolts. For vehicles with a manual transaxle, an 8mm flex socket may be used to access the rear oil pan bolts next to the flywheel.

5. Using SA9123E, or an equivalent RTV cutter tool, separate the oil pan from the engine. Drive the tool around the pan to shear the RTV seam, then tap the pan sideways with a rubber mallet to loosen.

6. Pry the engine mount away from the engine as necessary and remove the oil pan.

To install:

7. Carefully clean the gasket mating surfaces with a scraper and solvent.

8. Apply a 0.16 inch (4mm) bead of RTV sealer to the pan flange, towards the inside of the bolt holes.

9. Install the oil pan within 3 minutes and tighten the bolts to 80 inch lbs. (9 Nm).

10. Tighten the front mount-to-block bolts to 52 ft. lbs. (70 Nm) or the mount-to-front cover bolts to 37 ft. lbs. (50 Nm).

11. Install the right splash shield and wheel.

12. Install the exhaust pipe. Tighten the pipe-to-manifold nuts in a crosswise pattern to 23 ft. lbs. (31 Nm) and the pipe-to-converter bolts to 33 ft. lbs. (45 Nm).

13. Lower the vehicle and properly fill the engine crankcase.

14. Start the engine and check for leaks.

Oil Pump

REMOVAL AND INSTALLATION

1. Disconnect the negative battery cable and drain the engine oil.

2. Remove the timing chain front cover which contains the oil pump assembly.

3. Remove the oil pump cover Torx® bolts using a suitable impact driver. Because the pump cover screws are coated with a sealant to prevent oil leakage, they must be replaced when removed.

4. Remove the drive rotor and driven rotor.

5. If necessary, remove the relief valve using tool SA9103E or equivalent, to pull the valve from the bore. Because the puller jaws will damage the relief valve sealing seat, the valve cannot be used again when removed.

To install:

6. If removed, install a new relief valve into the cover bore. Coat the valve with clean engine oil and tap it into the bore using a hammer and SA9103E or an equivalent installer tool.

NOTE: Whenever the oil pump is installed, the assembly must be packed with petroleum jelly in order to prime the pump.

7. Install the drive and driven rotors into the pump with the chamfer toward the front oil seal.

8. Make sure the front cover bolt holes are clean, then install the pump body cover and secure using new bolts that are coated with sealant to prevent oil leakage. Tighten the bolts to 97 inch lbs. (11 Nm).

9. Install the timing chain cover and oil pump assembly to the front of the engine.

10. Properly fill the engine crankcase, then start the engine and check for leaks.

TRANSAXLE

Transaxle Assembly

REMOVAL AND INSTALLATION

1. Properly disable the SIR system, if equipped and disconnect the negative battery cable.

2. Remove the 2 air inlet duct fasteners, disengage the air temperature sensor connector and remove the air inlet duct. For the DOHC engine, loosen the flex tube-to-air box clamp, remove the 3 air box fasteners and remove the air box.

3. Remove the transaxle strut-to-cradle bracket through bolt located on the radiator side of the transaxle. Loosen the transaxle strut-to-transaxle bracket fasteners, then position the transaxle strut out of the way.

4. Disengage the backup light switch and vehicle speed sensor electrical connectors from the transaxle. Remove the vent tube retaining clip.

5. Remove the 2 ground terminals from the top 2 clutch housing studs, then unclip the oxygen sensor wire from the clutch housing.

6. Remove the top 2 clutch housing studs.

7. Remove the 4 EI coil to clutch housing bolts, then wire the coil to the cylinder head coolant outlet. Discard the old coil retaining bolts and replace with new coated fasteners upon installation.

8. Remove the shifter cables from the shift arms and clutch housing, taking care not to damage the cable boot.

───── WARNING ─────
Do not use power tools when loosening and removing the slave cylinder. The use of a power tool to remove the slave cylinder could result in breaking off the hydraulic line.

9. Rotate the clutch slave cylinder ¼ turn counterclockwise while pushing into the clutch housing, then remove the cylinder from the housing. Remove the 2 clutch hydraulic damper-to-clutch housing bolts, then wire the hydraulic assembly to the battery tray.

10. Wire the radiator to the upper radiator support in order to hold the assembly in place when the cradle is removed.

11. Install SA9105E or an equivalent engine support bar assembly.

12. Raise and support the vehicle safely, then remove the drain plug from the lower center of the housing and drain the transaxle fluid.

13. Remove the front wheels and engine splash shields from the vehicle. For coupes, remove the left and right lower facia braces.

14. Remove the front engine strut cradle bracket-to-cradle nuts from below the cradle.

15. Remove the transaxle mount-to-cradle nut from under the cradle.

16. Remove the front exhaust pipe nuts at the manifold, then disconnect the pipe from the support bracket.

17. Remove the front pipe to catalytic converter bolts and lower the pipe from the vehicle.

18. Support the steering gear with safety wire, then remove the gear-to-cradle fasteners.

19. Remove the brake line bracket push pin at the rear of the cradle.

20. Remove the engine-to-transaxle stiffening bracket bolts and remove the bracket.

21. Remove the clutch housing dust cover.

22. Remove and discard the cotter pins from the lower ball joints. Back the ball joint nut until the top of the nut is even with the top of the threads.

23. Use tool SA9132S to separate the lower control arm ball joint from the steering knuckle, then pull down on the lower control arm and remove the nut. Do not use a wedge tool or seal damage may occur.

NOTE: The outer CV-joint for vehicles equipped with ABS contains a speed sensor ring. Use of an incorrect tool to separate the control arm from the knuckle may result in damage and loss of the ABS system.

24. Use a prybar to separate the left side axle from the transaxle.

Only remove the axle sufficiently to install SA91112T or an equivalent seal protector around the axle and into the seal to prevent the seal from being cut by the shaft spline.

25. Position two 4 inch x 4 inch x 36 inch pieces of wood onto a powertrain support dolly, then position the dolly under the vehicle.

26. Remove the 4 cradle-to-body bolts, then carefully lower the cradle from the vehicle with the support dolly. Tape or wire the 2 large washers from the rear cradle-to-body attachments in position to prevent loss.

27. Support the transaxle securely with a suitable jack.

28. Remove the 2 bottom clutch housing-to-engine bolts and install a guide bolt into the bottom rear clutch housing bolt hole from the side of the engine block.

29. Carefully separate the transaxle from the engine enough to clear the intermediate shaft and lower the transaxle from the vehicle.

To install:

30. Place the transaxle assembly securely onto the jack and position under the vehicle.

31. Install axle seal protectors into seals on both sides, then place the transaxle in any gear.

32. Raise the transaxle into the vehicle guiding the unit onto the intermediate shaft. While guiding the transaxle onto the shaft, rotate the shaft back and forth to align the splines. When aligned, continue to rotate the intermediate shaft until the input shaft splines are aligned with the clutch.

33. Verify that the intermediate shaft splines are aligned with the differential side gear spline and the input shaft splines are aligned with the clutch disc splines, then install the 2 lower clutch housing-to-engine bolts and tighten to 96 ft. lbs. (130 Nm). The bolts should not be used to draw the transaxle to the engine.

34. Install the left side axle into the transaxle and remove the seal protectors. Lower the transaxle jack.

35. Clean and lubricate the ball joint threads, then raise the cradle up on the support dolly and place the ball joints into the knuckles.

36. Verify the correct positioning of the lower control arm bar studs to the knuckles, the cooling module support bushings, the engine strut bracket and the transaxle mount.

37. Insert 9/16 in. round steel rods into the cradle-to-body alignment holes near the front cradle-to-body fastener holes. Guide the cradle into

position making sure all mount studs are properly guided into their holes.

38. Make sure the washers are in place, then install the 2 rear cradle to body bolts. Verify proper cradle positioning and install the 2 front cradle bolts. Tighten the 4 cradle bolts to 151 ft. lbs. (205 Nm).

39. Remove the support dolly and lower the vehicle sufficiently for underhood access. Remove the engine support bar assembly.

40. Install the transaxle strut, and if removed, the transaxle bracket-to-transaxle bolts. Tighten the fasteners to 40 ft. lbs. (54 Nm). Install the remaining transaxle strut fasteners, as applicable, and tighten to 52 ft. lbs. (70 Nm).

41. Remove the radiator assembly support wire.

42. Use a 6 x 1.0mm tap to clean the sealant from the ignition module mounting holes in the transaxle. Install the ignition module and the new bolts with sealant. Use extreme caution to assure proper bolt installation. Tighten the bolts to 71 inch lbs. (8 Nm) and verify that the bolt heads are properly seated on the ignition module.

43. Install the 2 top clutch housing-to-engine studs and tighten to 66 ft. lbs. (90 Nm). Connect the 2 ground terminals to the studs and tighten to 18 ft. lbs. (25 Nm).

44. Engage the vehicle speed sensor and backup light switch electrical connectors to the transaxle.

45. Install the vent hose clip and the oxygen sensor wire clip to the housing.

46. Connect the shift control cables to the shift arms and the clutch housing, then install the cable retainers.

47. Remove the support wire from the battery tray, position the damper, then install the 2 slave cylinder-to-clutch housing nuts and tighten to 18 ft. lbs. (25 Nm). Push the actuator into the clutch housing and rotate ¼ turn clockwise, then install the retaining clip. Check that the master cylinder at the front of dash connection is locked in place.

48. For DOHC engines, install the air box and tighten the fasteners to 89 inch lbs. (10 Nm). Connect the flex tube to the air box, align the arrows and tighten the clamp to 18 inch lbs. (2 Nm).

49. Install the air inlet duct and fasteners, then engage the air temperature sensor electrical connector.

50. Raise and support the vehicle safely.

51. Install the transaxle mount-to-cradle fastener and tighten to 52 ft.

lbs. (70 Nm), then install the 2 engine strut cradle bracket-to-cradle fasteners to 52 ft. lbs. (70 Nm).

52. Remove the steering gear support wire and position the gear to the cradle. Install the gear bolts and nuts, then tighten the fasteners to 40 ft. lbs. (54 Nm). Connect the brake line and retainer to the cradle.

53. Install the clutch housing dust cover and tighten the fasteners to 89 inch lbs. (10 Nm).

54. Install the powertrain stiffening bracket and tighten the retainers to 35 ft. lbs. (47 Nm).

55. Position the front exhaust pipe into the vehicle, then install the gasket and the manifold retaining nuts. Tighten the nuts in a crosswise pattern to 23 ft. lbs. (31 Nm). Install the front pipe to the catalytic converter and tighten the bolts to 33 ft. lbs. (45 Nm). Finally, install the front pipe to the transaxle support bracket and tighten the fasteners to 23 ft. lbs. (31 Nm).

NOTE: If the converter flange threads are damaged use the Saturn 21010753 converter fastener kit in place of the self tapping screws in order to provide proper clamp load and prevent exhaust leaks.

56. Install the nuts onto the ball joint studs and tighten to 55 ft. lbs. (75 Nm). Continue to tighten the nuts as necessary and install new cotter pin.

57. Install the center and both wheel splash shields.

58. For coupes, install the right left lower facia braces and J-nuts. Tighten the fasteners to 89 inch lbs. (10 Nm).

59. If not done already, install the transaxle drain plug and tighten to 40 ft. lbs. (45 Nm).

60. Install the tire and wheel assemblies, then lower the vehicle.

61. Connect the negative battery cable and fill the transaxle with automatic transaxle fluid, Dexron® IIE or equivalent.

62. If equipped, properly enable the SIR system.

63. Check the vehicle alignment and adjust as necessary.

Clutch Assembly

REMOVAL AND INSTALLATION

1. Properly disable the SIR system, if equipped, and disconnect the negative battery cable.

2. Remove the transaxle from the vehicle.

3. Unsnap the release fork from the ball stud, then remove the fork and bearing from the vehicle. Slide the bearing from the fork.

NOTE: The release bearing is packed with grease and should not be washed with solvent.

4. Remove the pressure plate-to-flywheel bolts using a progressive criss-cross pattern to slowly release spring pressure and prevent cover warpage, then remove the pressure plate and clutch disc.

5. Inspect the flywheel for scores, warpage or burnt spots. Repair or replace as necessary.

To install:

6. If removed, install the flywheel and tighten the bolts in a criss-cross sequence to 59 ft. lbs. (80 Nm).

7. Install the clutch disc and pressure plate with the yellow dot on the pressure plate aligned as close as possible to the mark on the flywheel. The clutch disc is labeled FLYWHEEL SIDE in order to help correctly position the disc.

8. Install clutch alignment tool SA9145T or equivalent, in the clutch disc and push in until it bottoms out in the crankshaft, then start the pressure plate bolts.

9. Tighten the pressure plate bolts using multiple passes of a crisscross sequence to 18 ft. lbs. (25 Nm), then remove the alignment tool.

10. Lube the fork pivot point with high temperature grease, then install the release bearing to the fork. Do not lube the release bearing or bearing quill.

11. Snap the fork and bearing assembly onto the ball stud.

12. Lube the splines of the input shaft lightly with a high temperature grease.

13. Install the transaxle assembly.

14. Connect the negative battery cable and if equipped, properly enable the SIR system.

Hydraulic Clutch Assembly

REMOVAL AND INSTALLATION

NOTE: The master cylinder, pipes and slave cylinder are part of a complete fluid filled and bled assembly that must be replaced as a single unit.

1. Block the clutch pedal to prevent it from being depressed while the slave cylinder is remove from the transaxle.

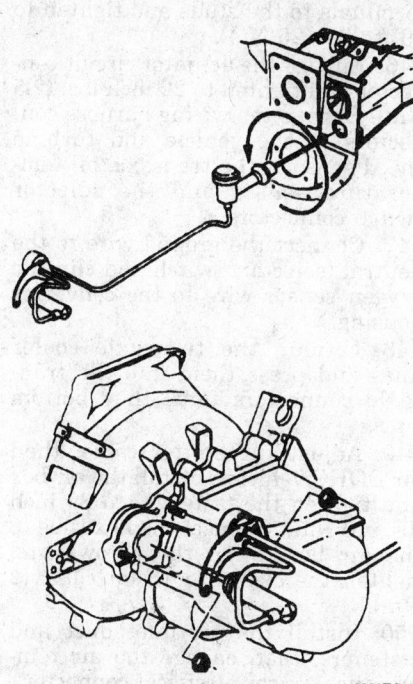

8838TG27

Hydraulic clutch release system installation — removal is in the opposite direction

2. Remove the air intake duct.

3. Check to make sure the hydraulic system has sufficient fluid and, add if necessary.

4. Rotate the slave cylinder about ¼ turn counterclockwise while pushing toward the bellhousing in order to disengage the bayonet connector and remove the cylinder from the clutch housing. Remove the slave cylinder bracket retaining nuts and pull the assembly from the studs.

5. Remove the master cylinder pushrod retaining clip and disconnect the pushrod from the clutch pedal.

6. Turn the clutch cylinder about ⅛ turn clockwise and remove from the instrument panel.

To install:

7. Position the master cylinder to the dash with the reservoir leaning toward the driver's fender. Install and turn about ⅛ turn counterclockwise to lock in position.

8. Slide the slave cylinder onto the clutch housing studs, install the nuts and tighten to 18 ft. lbs. (25 Nm).

NOTE: When installing a new assembly, the plastic retainer straps should remain in place on the slave cylinder to ensure the actuator rod seats on the release fork pocket upon installation. If reinstalling an assembly, be sure

to position a new plastic retainer strap onto the end of the pushrod and attach the straps to the cylinder.

9. Insert the slave cylinder into the housing with the hydraulic line facing downward and rotate about ¼ turn clockwise while pushing into the housing.

10. Lube the clutch pedal pin with silicone grease, then connect the pushrod to the clutch pedal and install the retaining clip.

11. Install the air inlet duct assembly and connect the negative battery cable.

12. Remove the block from behind the clutch pedal and if equipped, properly enable the SIR system.

13. Start the engine and check the pedal for proper operation.

Hydraulic Clutch System Bleeding

PROCEDURE

The clutch hydraulic assembly is serviced as a complete unit which has been filled with fluid and bled of air at the factory. The unit does not require periodic checking. The system is full when the reservoir is half full.

Only DOT 3 brake fluid should be added to the system. If fluid levels drop, inspect the system, including the slave cylinder, for leakage. A slight wetting of the slave cylinder surface is normal.

Although the slave cylinder assembly contains a bleeding screw, the manufacturer warns that the system should NOT be bleed using the bleeder. The screw is used only during the original factory fluid fill.

Automatic Transaxle Assembly

REMOVAL AND INSTALLATION

1. Properly disable the SIR system, if equipped and disconnect the negative battery cable.

2. Remove the 2 air inlet duct fasteners, disengage the air temperature sensor connector and remove the air inlet duct. For the DOHC engine, loosen the flex tube to air box clamp, remove the 3 air box fasteners and remove the air box.

3. Remove the transaxle strut-to-cradle bracket through bolt located on the radiator side of the transaxle.

4. Disengage the wiring harness connectors from the vehicle and turbine speed sensor, transaxle temperature sensor, selector switch and actuator connector from the transaxle.

5. Remove the 2 ground terminals from the top 2 converter housing studs.

6. Remove the ground wire from the neutral (selector) switch and unclip the oxygen sensor wire retainer from the converter housing.

7. Remove the top 2 converter housing studs.

8. Remove the 4 EI coil to converter housing bolts, then wire the coil to the cylinder head coolant outlet. Discard the old coil retaining bolts and replace with new bolts upon installation.

9. Wire the radiator to the upper radiator support in order to hold the assembly in place when the cradle is removed.

10. Install SA9105E or an equivalent engine support bar assembly.

11. Raise and support the vehicle safely.

12. Remove the drain plug from the transaxle housing and drain the transaxle fluid. The drain plug is on the lower cowl side of the housing and is inserted from the engine side of the vehicle.

13. Remove the front wheels and engine splash shields from the vehicle. For coupes, remove the left and right lower facia braces.

14. Remove the front engine strut cradle bracket-to-cradle nuts from below the cradle.

15. Remove the transaxle mount-to-cradle nut from under the cradle.

16. Remove and discard the cotter pin from the lower ball joints. Back the ball joint nut until the top of the nut is even with the top of the threads.

17. Use tool SA9132S to separate the lower control arm ball joint from the steering knuckle then remove the nut. Do not use a wedge tool or seal damage may occur.

NOTE: The outer CV-joint for vehicles equipped with ABS contains a speed sensor ring. Use of an incorrect tool to separate the control arm from the knuckle may result in damage and loss of the ABS system.

18. Remove the front exhaust pipe nuts at the manifold, then disconnect the pipe from the support bracket.

19. Remove the front pipe-to-catalytic converter bolts and lower the pipe from the vehicle.

20. Remove the engine-to-transaxle stiffening bracket bolts and remove the bracket.

21. Remove the steering rack-to-cradle bolts and wire the gear for support when the cradle is removed.

22. Remove the brake line from the retainer at the rear of the cradle.

23. Remove the torque converter dust cover, then remove the converter-to-flywheel bolts.

24. Position two 4 inch x 4 inch x 36 inch pieces of wood onto a powertrain support dolly and position the dolly under the vehicle.

25. Remove the 4 cradle-to-body bolts and carefully lower the cradle from the vehicle with the support dolly. Tape or wire the 2 large washers from the rear cradle-to-body attachments in position to prevent loss.

26. Squeeze the plastic tabs at the transaxle cooler line connectors and pull the lines out of the connectors. The plastic retainer should remain on the lines. Connect 1 end of a ⅜ in. rubber hose over each cooler line to prevent fluid contamination or loss.

27. If necessary for the transaxle to clear the body, lower the vehicle enough to adjust the engine support assembly and lower the transaxle side of the assembly until the valve body cover clears the frame.

28. Raise and support the vehicle safely, then support the transaxle securely with a suitable jack.

29. Use a prybar to separate the left side axle from the transaxle. Remove the axle sufficiently to install SA91112T or an equivalent seal protector around the axle and into the seal to prevent the seal from being cut by the shaft spline.

30. Remove the 2 bottom converter housing-to-engine bolts and lower the transaxle sufficiently to reach the shifter cable.

31. Separate the transaxle only sufficiently enough to install an axle seal protector on the remaining engaged axle.

32. Disconnect the transaxle shifter cable, then squeeze the retaining tabs to release the cable from the converter housing.

33. Carefully lower the transaxle from the vehicle. Use SA9165T or an equivalent transaxle cooler cleaning tool to clean the cooler and lines.

To install:

34. Place the transaxle assembly securely onto the jack and position under the vehicle. Install axle seal protectors into seals on both sides.

35. Raise the transaxle sufficiently, then connect the shifter cable to the gear selector lever and to the converter housing.

36. Raise the transaxle into the vehicle and verify that the intermediate shaft splines line up with the differential side gear splines, then install the 2 lower clutch housing-to-engine bolts and tighten to 96 ft. lbs. (130 Nm). The bolts should not be used to draw the transaxle to the engine.

37. Make sure the axle seal protectors are installed into the transaxle. Carefully install the axles to the transaxle, after the splines clear the seal, but before the axle snaps into place remove the seal protector. Push the axle all of the way into the transaxle and install the snapring. Remove the transaxle jack.

38. Clean and lubricate the ball joint threads, then raise the cradle up on the support dolly and place the ball joints into the knuckles. Verify the correct positioning of the lower control arm bar studs to the knuckles, the cooling module support bushings, the engine strut bracket and the transaxle mount.

39. Insert ⁹/₁₆ in. round steel rods into the cradle-to-body alignment holes near the front cradle to body fastener holes. Guide the cradle into position making sure all mount studs are properly guided into their holes.

40. Make sure the washers are in place, then install the 2 rear cradle to body bolts. Verify proper cradle positioning and install the 2 front cradle bolts, then tighten the 4 cradle bolts to 151 ft. lbs. (205 Nm).

41. Remove the support dolly and lower the vehicle sufficiently for underhood access. Remove the engine support bar assembly.

42. Install the transaxle strut-to-cradle bracket through bolt and nut, then tighten the fasteners to 52 ft. lbs. (70 Nm). If removed, install the strut cradle bracket-to-cradle bolt and also tighten to 52 ft. lbs. (70 Nm).

43. Remove the radiator assembly support wire.

44. Use a 6 x 1.0mm tap to clean the sealant from the ignition module mounting holes in the transaxle. Install the ignition module, then secure using the new bolts with sealant. Use extreme caution to assure proper bolt installation. Tighten the bolts to 71 inch lbs. (8 Nm) and verify that the bolt heads are properly seated on the ignition module.

45. Install the 2 top converter housing to engine studs and tighten to 66 ft. lbs. (90 Nm). Connect the 2 ground terminals to the studs and tighten to 19 ft. lbs. (25 Nm).

46. Engage the actuator circuit connector and tighten to 22 inch lbs. (2.5 Nm). Engage the wiring harness connectors to the vehicle and turbine speed sensors, the transaxle oil temperature sensor and the selector switch connectors.

47. Connect the ground wire to the neutral (selector) switch and clip the oxygen sensor wire to the converter housing.

48. Unplug the transaxle cooler lines and press them into the transaxle connectors until they bottom out.

49. Adjust the shifter cable, then for DOHC vehicles, install the air box and tighten the fasteners to 89 inch lbs. (10 Nm). Connect the flex tube to the air box, align the arrows and tighten the clamp to 18 inch lbs. (2 Nm).

50. Install the air inlet duct and fasteners, then engage the air temperature sensor electrical connector.

51. Raise and support the vehicle safely.

52. Install the transaxle mount-to-cradle nut and the transaxle strut cradle bracket-to-cradle nut, then tighten the nuts to 52 ft. lbs. (70 Nm).

53. Install the 2 engine strut cradle bracket-to-cradle fasteners from under the cradle and tighten to 52 ft. lbs. (72 Nm).

54. If applicable, install the nuts to the front transaxle-to-cradle studs and tighten to 35 ft. lbs. (48 Nm).

55. Remove the steering gear support wire and position the gear to the cradle. Install the gear bolts and nuts, then tighten the fasteners to 40 ft. lbs. (54 Nm).

56. Connect the brake line to the cradle retainer.

57. Install the torque converter-to-flexplate bolts and tighten to 52 ft. lbs. (70 Nm). Install the converter housing dust cover and tighten the bolts to 89 inch lbs. (10 Nm).

58. Install the powertrain stiffening bracket and tighten the bracket bolts to 35 ft. lbs. (47 Nm).

59. Position the exhaust manifold front pipe into the vehicle, then install the gasket and manifold retaining nuts. Tighten the nuts in a crosswise pattern to 23 ft. lbs. (31 Nm). Install the front pipe to the catalytic converter and tighten the bolts to 33 ft. lbs. (45 Nm). Finally, install the front pipe to the transaxle support

bracket and tighten the fasteners to 23 ft. lbs. (31 Nm).

NOTE: If the converter flange threads are damaged use the Saturn 21010753 converter fastener kit in place of the self tapping screws in order to provide proper clamp load and prevent exhaust leaks.

60. Install the nuts onto the ball joint studs and tighten to 55 ft. lbs. (75 Nm). Continue to tighten the nuts as necessary and install a new cotter pin.

61. Install the center and both wheel splash shields.

62. For coupes, install the right left lower facia braces and J-nuts. Tighten the fasteners to 89 inch lbs. (10 Nm).

63. Install the tire and wheel assemblies, then lower the vehicle.

64. Connect the negative battery cable and fill the transaxle with Dexron® II or equivalent fluid.

65. Properly enable the SIR system, if equipped.

66. Warm the engine and check the transaxle fluid. Check and adjust vehicle alignment, as necessary.

SHIFT CABLE ADJUSTMENT

This procedure begins with the shift cable disconnected from the transaxle lever, but routed through the converter housing.

1. Place the transaxle in the **P** position.

2. Place the transaxle shift lever in the **P** position.

3. Release the cable adjustment lock tab by lifting upward with a small prybar.

4. Connect the cable to the shift transaxle lever and install the retainer.

5. Move the cable housing back and forth in the adjuster to note end-

play, then center the cable housing in the middle of the endplay and press in the lock tab.

6. Check operation.

PARK LOCK CABLE ADJUSTMENT

1. Remove the front ashtray for access to the cable. If necessary, remove the center console assembly for additional access.

2. Turn the ignition switch to **ON**.

3. Place the shifter in **P** and make sure the brake pedal is released.

4. If adjusting a new cable, depress the lock on the cable end fitting and remove the adjustment clip (the clip is designed to leave the appropriate gap.)

5. If adjusting a cable which does not have an adjustment clip, secure the end of the cable to provide a 0.05 inch (1.25mm) gap between the cable end and the park lock connector.

6. Verify proper adjustment by the following:

   a. With the ignition **OFF**, the lever should not shift out of **P**.

   b. Turn the ignition to the **ON** position and the lever should shift.

   c. With the lever out of **P** and the ignition **OFF**, the key cannot be removed from the ignition.

   d. With the lever out of **P** and the ignition **OFF**, the lever should shift into **P**.

   e. With the ignition **OFF** and the lever in **P**, remove the key.

7. Repeat the adjustment procedure if any of the conditions are incorrect.

8. Install the ashtray and/or console assembly when properly adjusted.

DRIVE AXLE

Halfshaft

REMOVAL AND INSTALLATION

1. Remove the wheel cover or center cap, then loosen the halfshaft nut while an assistant depresses the brake pedal.

2. Raise and support the vehicle safely.

3. Remove the corresponding wheel and splash shield.

4. If removing the left side axle, drain the transaxle fluid into a suitable container.

5. Remove the halfshaft nut and washer.

6. Remove and discard the cotter pin from the lower control arm ball joint for the axle being removed. Back the ball joint nut until the top of the nut is even with the top of the threads.

7. Use tool SA9132S to separate the lower control arm ball joint from the steering knuckle, then remove the nut. Do not use a wedge tool or seal damage may occur.

NOTE: The outer CV-joint for vehicles equipped with ABS contains a speed sensor ring. Use of an incorrect tool to separate the control arm from the knuckle may result in damage and loss of the ABS system.

8. Remove the tie rod cotter pin and castle nut, then separate the tie rod end from the knuckle using a tie rod separator SA91100C or equivalent. Do not use a wedge-type tool.

9. Place a cloth over the sway bar to protect the surface, then position a prybar over the cloth with which to apply leverage to the knuckle and separate the lower control arm ball joint. Position a cloth at the prybar contact point with the cradle, then push down on the bar and separate the ball joint from the knuckle. Make sure the knuckle does not contact and damage the ball stud seal.

10. While pulling the knuckle/strut assembly away from the halfshaft, pull the end of the halfshaft from the wheel hub. If difficulty is encountered, tap on the end of the halfshaft using a block of wood and a hammer. Support the halfshaft assembly using a length of mechanic's wire or with a jack stand.

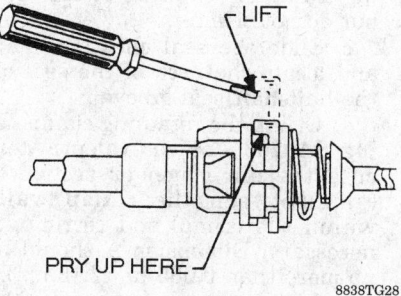

Shifter cable adjustment

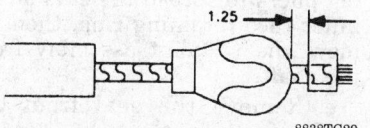

A properly adjusted park lock cable will have a 0.05 in. (1.25mm) gap between the cable end and the park lock connector

11. If removing the right halfshaft, disconnect the halfshaft from the intermediate shaft by tapping the inner joint with a hammer using a block of wood positioned to cushion the joint from the blows. Remove the halfshaft from the vehicle.

12. If removing the left halfshaft, disconnect the halfshaft by inserting a large prybar into the space between the inner joint and transaxle. Pry the halfshaft from the transaxle being careful not to contact and damage the transaxle oil seal. Remove the halfshaft from the vehicle.

To install:

13. If installing the left side halfshaft, install SA91112T or equivalent transaxle seal protector. Install the halfshaft into the transaxle, after the splines have safely passed the transaxle oil seal, remove the seal protector and fully seat the halfshaft.

14. If installing the right side halfshaft, insert the shaft onto the intermediate shaft and push firmly to engage the circlip.

15. Insert the outer end of the halfshaft into the wheel hub. Be careful not to damage the CV-joint boot.

16. Thoroughly clean and lubricate the ball joint stud threads of the lower control arm and tie rod end.

17. Install the lower control arm ball stud and install the nut, but do not tighten at this time.

18. Install the tie rod end and nut. Tighten the nut to 33 ft. lbs. (45 Nm) and install a new cotter pin. If necessary, tighten the nut additionally, but do not back off to insert the cotter pin.

19. Tighten the lower control arm ball stud nut to 55 ft. lbs. (75 Nm), tighten additionally if necessary and install a new cotter pin.

20. Install the washer and a new halfshaft nut, then tighten the nut to 145 ft. lbs. (200 Nm).

21. Install the inner splash shield and wheel.

22. Lower the vehicle and, if necessary, properly fill the transaxle.

23. Check and adjust the alignment, as necessary.

CV-Joint Boot

REPLACEMENT

1. Using soft metal or wood to protect the shaft, clamp the halfshaft to a workbench in a vise.

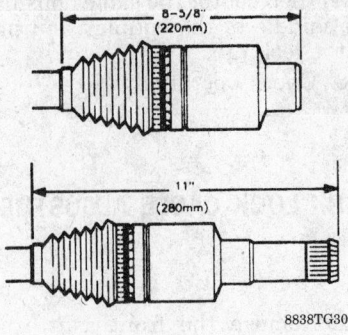

8838TG30

The tri-pot seal must be shaped to the proper dimension

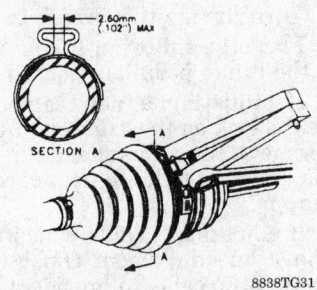

8838TG31

When a seal clamp is crimped, check dimension A (the inner gap between the walls of the crimp) — CV-joint large seal clamp shown

2. Remove and service the CV-joint from the end of the halfshaft, as follows:

a. If the halfshaft has a damaged deflector ring, use a brass drift and hammer to remove the damaged component from the CV outer race.

b. Either cut the outer seal retaining clamps using a side cutter or use a hammer and chisel to disengage the outer band from the inner band at the retaining peg on both the large and small seal clamps, then discard the clamps.

c. Separate the joint seal from the CV-joint race at the large diameter, then slide the seal away from the joint, along the halfshaft. Wipe any excess grease from the CV inner race.

d. Use a suitable pair of snapring pliers to spread the ears of the inner race retaining ring, then remove the CV-joint assembly from the shaft.

e. Remove the seal from the halfshaft.

f. If necessary, disassemble the CV-joint. Use a brass drift and gently tap on the cage until it is cocked and the first ball may be removed. Repeat this to remove the remaining balls. When the balls are removed, pivot the cage and inner

race at a 90 degree angle to the center line of outer race. Cage windows should align with the lands of the outer race, then lift the cage and inner race. Rotate the inner race up and out of the cage. Thoroughly clean and de grease all CV-joint parts and allow to dry before assembly.

3. Remove and service the tri-pot joint and joint seal from the halfshaft's other end, as follows:

a. Cut the eared seal retaining clamp on the tri-pot seal using a side cutter, then discard the clamp.

b. Remove and discard the earless clamp using a small flat blade chisel or other suitable tool.

c. Separate the seal from the tri-pot housing at the large diameter, then slide the seal away from the joint along the halfshaft. Wipe away excess grease from the face of the tri-pot spider and the inside of the housing.

d. Remove the tri-pot housing from the spider and shaft.

e. Spread the spacer ring using SA9198C, or equivalent, and slide the spacer ring along with the tri-pot spider, back on the axle towards the repositioned seal.

f. Carefully remove the spider retaining ring from the halfshaft groove, and slide the spider assembly off the shaft. Use care not to loose the tri-pot balls and needle rollers which may separate from the spider trunnions.

g. Remove the seal from the axle shaft.

h. Thoroughly clean and de grease the housing and allow to dry prior to assembly.

To install:

4. Install the tri-pot joint and seal:

a. Inspect the tri-pot joint components for unusual wear, cracks or damage and replace, as necessary. Clean the shaft; if rust is present in the seal mounting grooves, use a wire brush.

b. Install the small seal retaining clamp on the neck of the seal, but do not crimp.

c. Slide the seal onto the shaft and locate the neck of the seal on the halfshaft seal groove

d. Crimp the retaining clamp using SA9203C or equivalent. Measure the dimension or section A (gap between the clamp walls within the crimp) and recrimp, if necessary. Dimension A should be no more than 0.085 in. (2.15mm).

e. Install the spacer ring on the shaft and position it beyond the retaining groove, then install the spider, also past the retaining groove. Be sure the counterbored surface of the tri-pot spider faces the end of the shaft after installation.

f. Using SA9198C, or an equivalent tool, install the retaining ring to the halfshaft ring groove, then slide the tri-pot spider towards the end of the shaft and seat the spacer ring in the axle groove.

g. Place about ½ of the grease from the overhaul kit inside the seal and use the remainder to re-pack the tri-pot housing.

h. Install the convolute retainer (also supplied with the overhaul kit) over the seal. The retainer must be in position when the joint is assembled or seal damage may result.

i. Position the retaining clamp around the large diameter of the seal, the slide the tri-pot housing over the tri-pot assembly on the shaft.

j. Slide the large diameter of the seal over the outside of the tri-pot housing and locate the seal lip in the housing groove. The seal must not be dimpled or stretched. If necessary, carefully insert a thin, flat and blunt tool between the large seal opening and outer race to equalize pressure, then shape the seal by hand and remove the tool.

k. Position the tri-pot assembly at the proper dimensions, either 8⅝ in. or 11 in. depending on the application, then install the large seal retaining clamp around the seal. Close the clamp using SA9161C or equivalent.

5. Assemble and install the outer CV-joint and seal:

a. Inspect the CV-joint parts for any signs of unusual wear, cracks or damage and replace the joint assembly, if necessary.

b. If necessary, assemble the CV-joint for installation purposes. Put a light coat of grease on the inner and outer race grooves, then insert and rotate the inner race into the cage. Install the cage and inner race into the outer race with the windows of the cage aligned with the lines of the outer race. Use a brass drift to gently cock the cage/race and install the balls. If removed, install the race retaining ring into the inner race.

c. Pack the assembled joint using the premeasured amount of grease from the service kit.

d. Install the small retaining clamp on the neck of the new seal, then slide the seal onto the shaft and locate the seal neck in the shaft seal groove.

e. Crimp the seal retaining ring using SA9203C, or an equivalent crimping tool. A proper crimp will share the same dimensions of the tri-pot seal retaining ring crimp. The gap between the inner walls of the crimp should be no more than 0.085 in. (2.15mm).

f. Place about ½ of the provided grease inside the seal, then repack the CV-joint using the remaining grease.

g. Position the large seal retaining clamp around the seal.

h. Make sure the retaining ring side of the inner race is facing the halfshaft, then push the CV-joint onto the shaft until the ring is seated in the shaft groove.

i. Slide the seal large diameter over the outside of the CV-joint race and locate the lip of the seal in the housing groove. The seal must not be dimpled or stretched. If necessary, carefully insert a thin, flat and blunt tool between the large seal opening and outer race to equalize pressure, then shape the seal by hand and remove the tool.

j. Crimp the retaining clamp using the SA9203C or equivalent. The proper crimp gap dimension should be no larger than 0.102 in. (2.6mm). Recrimp if necessary to achieve the proper dimension.

k. If removed, position the deflecting ring at the CV-joint outer race. Use SA9160C or equivalent along with a M20 x 1.5 nut to tighten the tool until the deflector bottoms against the shoulder of the CV-joint outer race.

6. Remove the shaft assembly from the vise and install in the vehicle.

STEERING

Air Bag

—CAUTION—

Some vehicles are equipped with an air bag system, also known as the Supplemental Inflatable Restraint (SIR) or Supplemental Restraint System (SRS). The system must be disabled before performing service on or around system components, steering column, instrument panel components, wiring and sensors. Failure to follow safety and disabling procedures could result in accidental air bag deployment, possible personal injury and unnecessary system repairs.

PRECAUTIONS

Several precautions must be observed when handling the inflator module to avoid accidental deployment and possible personal injury.

• Never carry the inflator module by the wires or connector on the underside of the module.

• When carrying a live inflator module, hold securely with both hands, and ensure that the bag and trim cover are pointed away.

• Place the inflator module on a bench or other surface with the bag and trim cover facing up.

• With the inflator module on the bench, never place anything on or close to the module which may be thrown in the event of an accidental deployment.

DISARMING

1. Align the steering wheel so the vehicle wheels are pointing in the straight-ahead position.

2. Turn the ignition switch to the **OFF** position.

3. Remove the SIR fuse from the top left of the Instrument Panel Junction Block (IPJB).

4. Remove the Connector Position Assurance (CPA) device, then disengage the yellow 2-way SIR harness wire connector at the base of the steering column.

To enable system:

5. Verify that the ignition switch is in the **OFF** position.

6. Engage the yellow 2-way connector at the base of the steering column and install the CPA.

7. Install the SIR fuse.

8. Turn the ignition switch to the **RUN** position.

9. Verify the SIR indicator light flashes 7–9 times and then turns **OFF** to signify proper system operation. If light does not flash as specified, inspect system for malfunction.

Steering Wheel

REMOVAL AND INSTALLATION

Without Air Bag

1. Disconnect the negative battery cable.

2. Remove the horn pad by pulling on the edge of the pad firmly, discon-

nect the wires and remove the horn pad from the vehicle.

3. Remove the clip on the end of the steering column shaft, then remove the wheel retaining nut.

4. Note the position of the steering wheel locating notch for reassembly purposes.

5. Install a suitable steering wheel puller and remove the wheel from the steering column.

To install:

6. Route the wires through the wheel and position the steering wheel making sure to properly align the locating notch. If the locating notch is not properly positioned, any attempt to install the steering wheel will damage the wheel and column beyond repair.

7. Install a new steering wheel nut and tighten to 30 ft. lbs. (40 Nm), then install a new clip on the end of the column.

8. Connect the wires to the horn pad and press the pad firmly into position on the wheel.

9. Connect the negative battery cable.

With Air Bag

—————— CAUTION ——————
When the vehicle is equipped with a Supplemental Inflatable Restraint (SIR) system, follow the recommended disarming procedures before performing any work on or around the system. Failure to do so may result in possible deployment of the air bag and/or personal injury.

1. Properly disable the SIR system and disconnect the negative battery cable.

2. Loosen the 4 fasteners from the back of the steering wheel and lift the inflator module from the wheel.

3. Remove the CPA device and unplug the wiring harness from the module, then remove the inflator module from the steering wheel.

—————— CAUTION ——————
When carrying a live inflator module, ensure the bag and trim cover are pointed away from the body. Never carry the inflator module by the wires or connector on the underside of the module. This will minimize the chance of injury should the module accidentally deploy. When placing a live inflator module on a bench or other surface, always place the bag and trim cover up, away from the surface. This is necessary so a

free space is provided to allow for air bag expansion in the unlikely event of accidental deployment.

4. Unplug the horn connector and, if equipped, unplug the cruise control switch connector.

5. Remove the clip steering wheel retaining nut.

6. Note the position of the steering wheel locating notch for reassembly purposes.

7. Install a suitable steering wheel puller and remove the steering wheel from the steering column while extracting the wiring from the wheel.

8. Install a yellow retaining tab into the SIR coil assembly to keep it from rotating. If a retaining tab is not available, tape the coil in position to prevent coil damage.

To install:

9. Route the SIR wire and other electrical connections through the wheel, then position the steering wheel making sure to properly align the locating notch. If the locating notch is not properly positioned, any attempt to install the steering wheel will damage the wheel and column beyond repair.

10. Remove the yellow retaining tab or the tape from the SIR coil assembly.

11. Install a new steering wheel nut and tighten to 30 ft. lbs. (40 Nm).

12. Connect the wiring harness to the horn and, if equipped, to the cruise control switch.

13. Position the inflator module and connect the SIR wiring harness, then seat the module on the steering wheel.

14. Secure the module using NEW fasteners, then tighten the new fasteners to 89 inch lbs. (10 Nm).

15. Connect the negative battery cable and enable the SIR system. Check for proper system function by watching the AIR BAG indicator lamp.

Tie Rod Ends

REMOVAL AND INSTALLATION

Outer

1. Raise and support the vehicle safely.

2. Remove the front wheel and if necessary, remove the splash shield.

3. Remove the cotter pin and nut from the tie rod end.

4. Separate the tie rod end from the knuckle with separator tool SA91100C or equivalent. Do not use

a wedge-type tool or the seal may be damaged.

5. Mark the threaded portion of the steering arm for installation purposes.

6. Loosen the tie rod jam nut and thread the tie rod end off the steering shaft. Count the number of turns necessary to remove the end for installation purposes.

To install:

7. Clean the threads and grease before installation. Be sure to only lubricate the threaded portion of the stud.

8. Install the tie rod end using the same number of turns as counted during removal. Align the tie rod end to the marked location.

9. Install the tie rod to the knuckle. Install the castle nut and tighten to 33 ft. lbs. (45 Nm), then install a new cotter pin. If necessary, tighten the nut additionally, do not back off to insert the cotter pin.

10. Tighten the tie rod jam nut to 74 ft. lbs. (100 Nm).

11. Install the splash shield, if removed, then install the wheel assembly.

12. Lower the vehicle, check and adjust toe as necessary.

Inner

1. Remove the rack and pinion assembly from the vehicle.

2. Place the assembly in a holding fixture, being careful not to damage the components, then loosen the outer tie rod jam nut.

3. Unthread the outer tie rod from the inner tie rod. Count the number of turns necessary or mark for realignment.

4. Remove the outer tie rod jam nut, then remove the steering gear boot.

5. Slide the shock damper toward the steering gear and off the inner tie rod.

6. Remove the inner tie rod assembly from the steering gear. To prevent damage, place a shop cloth over the gear teeth and hold the teeth with a suitable open end wrench. If removing the right inner tie rod, the left side boot must be removed to access the teeth.

To install:

7. Remove the old Loctite® from the rack and inner tie rod threads. Then, apply an even coat of Loctite® 262 to the inner tie rod threads.

8. If removed, slide the shock damper onto the steering gear.

9. Properly hold the gear teeth and install the inner tie rod assembly

onto the steering gear. Tighten the inner rod to 70 ft. lbs. (95 Nm).

10. Slide the shock damper up against the inner tie rod assembly and install the steering boot onto the gear.

11. Thread the outer tie rod jam nut onto the inner tie rod, then install the outer tie rod to the assembly. Use the same number or turns or align the marks made earlier and tighten the jam nut to 74 ft. lbs. (100 Nm).

12. Install the rack and pinion assembly to the vehicle.

13. Align the front end and bleed the power steering system, as necessary.

Manual Rack and Pinion

REMOVAL AND INSTALLATION

1. Disconnect the negative battery cable, then raise and support the vehicle safely.

2. Remove the front tires and the left inner splash shield.

3. Remove and discard the tie rod cotter pins, then remove the castle nuts. Disconnect the tie rod ends using SA91100C or an equivalent separator tool. Do not use a wedge-type tool or seal damage may occur.

4. Loosen the intermediate shaft cover from the steering gear and move up enough to access the pinch bolt. Remove the pinch bolt.

5. Remove the steering gear-to-cradle fasteners, then remove the gear through the left fenderwell.

To install:

6. Install the steering gear and torque the steering gear retainers to 37 ft. lbs. (50 Nm). Be sure to use new nuts because the torque retention of the old nuts may be insufficient.

7. Position the intermediate steering shaft to the gear and tighten the pinch bolt to 35 ft. lbs. (47 Nm).

8. Thoroughly clean and lubricate the threads of the tie rod ends, then install the ends into the steering knuckles. Install the castle nuts and tighten to 33 ft. lbs. (45 Nm), then install new cotter pins. If necessary, tighten the nut additionally to install the pin, but do not back off the torque specification.

9. Install the left inner splash shield and install the front wheels.

10. Lower the vehicle and connect the negative battery cable.

11. Check alignment and adjust vehicle toe, as necessary.

ADJUSTMENT

Steering gear bearing preload adjustment is the same for both manual and power rack and pinion units.

Power Rack and Pinion

REMOVAL AND INSTALLATION

1. Disconnect the negative battery cable, then raise and support the vehicle safely.

2. Remove both front tires and the left inner splash shield.

3. Remove and discard the tie rod cotter pins, then remove the castle nuts. Disconnect the tie rod ends using SA91100C or an equivalent separator tool. Do not use a wedge-type tool or seal damage may occur.

4. Loosen the intermediate shaft cover from the steering gear and move up enough to access the pinch bolt. Remove the pinch bolt.

5. Place a suitable container under the steering assembly. Disconnect the pressure and return lines at the steering gear and allow the system to drain.

6. Remove the steering gear-to-cradle fasteners, then remove the gear through the left fenderwell.

To install:

7. Install the steering gear and tighten the steering gear fasteners to 37 ft. lbs. (50 Nm). Be sure to use new nuts because the torque retention of the old nuts may be insufficient.

8. Position the intermediate steering shaft to the gear and tighten the pinch bolt to 35 ft. lbs. (47 Nm).

9. Connect the pressure and return hoses, then tighten the fittings to 20 ft. lbs. (27.5 Nm).

10. Thoroughly clean and lubricate the threads of the tie rod ends, then install the ends into the steering knuckles. Install the castle nuts and tighten to 33 ft. lbs. (45 Nm), then install new cotter pins. If necessary, tighten the nut additionally to install the pin, do not back off.

11. Install the left inner splash shield and install the front wheels.

12. Lower the vehicle and connect the negative battery cable.

13. Check alignment and adjust vehicle toe, as necessary.

14. Bleed the power steering system.

ADJUSTMENT

1. Center the steering wheel, then raise and support the vehicle safely.

2. Loosen the locknut on the steering gear adjuster plug, then turn the adjuster plug clockwise until it bottoms in the gear housing.

3. Tighten the plug to 106 inch lbs. (12 Nm).

4. Back off the adjuster plug 50–70 degrees (about 1 nut flat).

5. While holding the plug steady, tighten the locknut with a crows foot wrench to 52 ft. lbs. (70 Nm).

6. Check the steering for returnability, binding or difficulty in turning.

Power Steering Pump

BLEEDING

1. Raise and support the vehicle safely.

2. Turn the steering wheels all the way to the left and fill the reservoir to the FULL mark.

3. Bleed the system by turning the wheels from side-to-side without hitting the stops. It may take several cycles to bleed the system.

4. Keep the reservoir to the FULL mark during the procedure.

5. Start the engine and check the fluid level with the engine idling. If necessary, add to bring the level to the FULL mark.

6. Return the wheels to the center position, then lower the vehicle and allow it to idle for 2–3 minutes.

7. Road test the vehicle and check for proper operation. Recheck the fluid level and make sure it is at or slightly above the full mark after the system has stabilized at normal operating temperature.

REMOVAL AND INSTALLATION

1. Disconnect the negative battery cable.

2. Remove the pump reservoir fill cap.

3. Raise and support the vehicle safely.

4. Place a suitable container under the power steering hoses, then remove the hoses from the steering gear and allow the system to drain.

NOTE: Once the hoses have been removed from the steering gear, do not rotate the steering wheel or additional fluid will be forced from the gear.

5. Lower the vehicle sufficiently for underhood access, if desired. Use a box end wrench to relieve the spring tension from the accessory drive belt tensioner and remove the

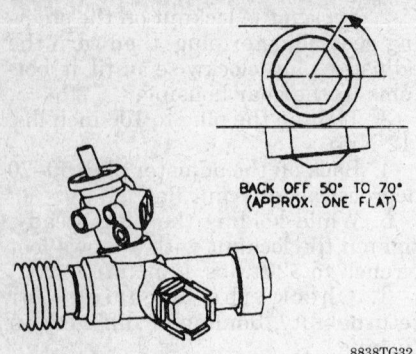

BACK OFF 50° TO 70°
(APPROX. ONE FLAT)

8838TG32

Adjusting steering gear bearing preload

serpentine belt from the steering pump pulley.

6. For DOHC engines, remove the pump-to-intake and pump-to-block fasteners and brackets.

7. Remove the 3 pump to block bolts and raise the pump sufficiently for access, then disconnect the electrical connector from the pump.

8. Remove the pump, with the hoses connected, from the vehicle. If necessary, remove the pressure and return hoses from the pump.

To install:

9. If removed, install new O-ring seals, then install the pressure and return hoses. Tighten the fittings to 20 ft. lbs. (27.5 Nm).

10. Position the pump to the block and engage the electrical connector. Install the 3 retaining bolts and tighten the bolts to 28 ft. lbs. (38 Nm).

11. For DOHC engines, install the pump-to-intake and pump-to-block brackets and fasteners. Tighten the fasteners to 22 ft. lbs. (30 Nm).

12. Install the drive belt to the steering pump pulley.

13. If lowered, raise and support the vehicle safely. Connect the pressure and return hoses to the steering gear, then tighten the fittings to 20 ft. lbs. (27.5 Nm). Route the return hose, then the pressure hose into the retaining clip.

14. Lower the vehicle and fill the power steering reservoir with clean fluid.

15. Connect the negative battery cable and bleed the power steering system.

BRAKES

Anti-Lock Brake System Service

PRECAUTIONS

• Certain components within the Anti-Lock Brake System (ABS) are not intended to be serviced or repaired individually. Only those components with removal and installation procedures should be serviced.

• Do not use rubber hoses or other parts not specifically specified for and ABS system. When using repair kits, replace all parts included in the kit. Partial or incorrect repair may lead to functional problems and require the replacement of components.

• Lubricate rubber parts with clean, fresh brake fluid to ease assembly. Do not use lubricated shop air to clean parts; damage to rubber components may result.

• Use only specified brake fluid from an unopened container.

• If any hydraulic component or line is removed or replaced, it may be necessary to bleed the entire system.

• A clean repair area is essential. Always clean the reservoir and cap thoroughly before removing the cap. The slightest amount of dirt in the fluid may plug an orifice and impair the system function. Perform repairs after components have been thoroughly cleaned; use only denatured alcohol to clean components. Do not allow ABS components to come into contact with any substance containing mineral oil; this includes used shop rags.

• The Anti-Lock control unit is a microprocessor similar to other computer units in the vehicle. Ensure that the ignition switch is **OFF** before removing or installing controller harnesses. Avoid static electricity discharge at or near the controller.

• If any arc welding is to be done on the vehicle, the control unit should be unplugged before welding operations begin.

DEPRESSURIZING

The anti-lock brake system does not operate under abnormally high hydraulic pressures that would necessitate relieving system pressure. However, if pressure relief is desired to minimize the possibility of fluid spray when servicing lines or fittings, connect 1 end of a thin plastic tube to the bleeder screw on the ABS control unit (modulator) and place the other end into a suitable container, then slowly loosen the bleeder 1/2–3/4 of a turn until the pressure is released.

Master Cylinder

REMOVAL AND INSTALLATION

Without Anti-Lock Brakes

1. Disconnect the negative battery cable, then disengage the fluid level connector from the side of the reservoir.

2. Remove the hydraulic pipes fittings from the master cylinder using a suitable wrench. Plug the openings to prevent system contamination or excessive fluid loss.

3. Remove the 2 master cylinder retaining nuts and remove the master cylinder from the vehicle.

To install:

4. Install the master cylinder onto the brake booster studs, then install the retaining nuts and tighten to 20 ft. lbs. (27 Nm).

5. Unplug the openings, then connect the hydraulic brake pipes to the master cylinder and tighten the fittings to 18 ft. lbs. (24 Nm).

6. Engage the wiring harness connector to the brake fluid level sensor terminal.

7. Connect the negative battery cable.

8. Properly bleed the hydraulic brake system.

With Anti-Lock Brakes

The master cylinder for ABS equipped vehicles is attached to, and removed with, the ABS control assembly.

Brake Caliper

REMOVAL AND INSTALLATION

Front

1. Raise and support the vehicle safely, then remove the front wheel.

2. Disconnect the brake hose from the caliper. Plug the openings to prevent system contamination or excessive fluid loss.

3. Remove the lock pin and guide pin from the caliper and support.

4. Remove the caliper from the support, being careful not to damage the pin boots. Remove the pin boots from the caliper support and inspect for damage.

To install:

5. Make sure the piston is bottomed in the bore. If necessary bottom the piston by hand or using a C-clamp.

6. If removed, install the brake pads to the caliper support.

7. Lubricate the pin boots and guide pins with silicone grease. If removed, install the pin boots into the caliper support, using the pin to assure that the boot passes all the way through the support.

8. Position the caliper onto the support and over the brake pads, then lubricate the non-threaded portion of the guide and lock pins with silicone grease. Install the pins and tighten to 27 ft. lbs. (36 Nm).

9. Make sure the brake line is properly routed with loop to the rear, then install the brake hose with new washers. Tighten the fitting to 36 ft. lbs. (49 Nm).

10. Properly bleed the hydraulic brake system.

11. Install the wheel, remove the supports and carefully lower the vehicle.

Rear

1. Raise and support the vehicle safely, then remove the rear wheel.

2. Disconnect the brake hose from the caliper. Plug the openings to prevent system contamination or excessive fluid loss.

3. Slip the end of the parking cable off the parking brake lever, then remove the cable outer housing from the cable bracket with SA9151BR or an equivalent cable release tool.

4. Remove the lock pin and guide pin.

5. Remove the caliper from the support, being careful not to damage the pin boots. If necessary, remove the pin boots from the caliper support.

To install:

6. Make sure the piston is bottomed in the bore. Do not compress the piston using a C-clamp; instead the piston must be rotated into the caliper on it's threads using a piston driver tool.

7. If removed, install the brake pads to the caliper support.

8. Lubricate the pin boots and guide pins with silicone grease. If removed, install the pin boots into the caliper support, using the pin to assure that the boot passes all the way through the support.

9. Position the caliper, then lubricate the non-threaded portion of the guide and lock pins. Install the pins and tighten to 27 ft. lbs. (36 Nm).

10. Install the brake hose using new washers, then tighten the fitting to 36 ft. lbs. (49 Nm).

11. Connect the parking brake cable.

12. Properly bleed the hydraulic brake system.

13. Install the wheel assembly and lower the vehicle.

Disc Brake Pads

REMOVAL AND INSTALLATION

Front

NOTE: Always replace the brake pads in sets, both front or both rear axle assemblies.

1. Raise and support the vehicle safely and remove the front wheels.

2. Remove the caliper lower lock pins.

NOTE: The lower caliper lock pin may be removed in order to allow pivoting of the caliper upward, away from the brake pads. Be very careful not to pivot the caliper so far as to stretch or damage the brake line. To avoid the possibility of brake line damage, remove the upper guide pin and support the caliper assembly aside.

3. Either pivot the caliper up on the guide pin or remove the upper guide pin and support the caliper from the strut using a coat hanger or length of wire.

4. Remove the 2 brake pads and the pad clips from the caliper support. Discard the old pad clips.

5. Check the caliper pins, pin boots and the piston boot for deterioration or damage.

To install:

6. By hand or using a C-clamp, bottom the piston all the way into the caliper bore.

7. Carefully lift the inner edge of the piston boot by hand to release any trapped air.

8. Install new pad clips into the caliper support.

9. Install the inner and outer brake pads into the support. If installed, remove the temporary support wire from the caliper.

10. Pivot or place the caliper body on the support and upper guide pin into position. Compress the boots by hand as the caliper is positioned onto the support.

guide and lock pins. Install the pins and tighten to 27 ft. lbs. (36 Nm).

10. Install the brake hose using new washers, then tighten the fitting to 36 ft. lbs. (49 Nm).

11. Connect the parking brake cable.

12. Properly bleed the hydraulic brake system.

13. Install the wheel assembly and lower the vehicle.

Rear

NOTE: Always replace the brake pads in sets, both front or both rear axle assemblies.

1. Raise and support the vehicle safely, then remove the rear wheels.

2. Remove the caliper lock and guide pins.

3. Remove the caliper from the support, being careful not to damage the pin boots and suspend the caliper from a wire.

4. Remove the brake pads from the support.

To install:

5. Using SA91110NE or an equivalent piston driver tool, bottom the piston by rotating it clockwise into the caliper bore; do not use a C-clamp to press the piston into the bore.

6. Align the piston slots so they are perpendicular to the brake pads.

7. Carefully lift the inner edge of the piston boot to release any trapped air. The boot must lay flat below the level of the piston face.

8. Install new pad clips into the caliper support.

9. Install the inner and outer brake pads into the clips on the support. The pad with the wear sensor should be located outboard. The piston indentation slots should be positioned to correctly accept the brake pads.

10. Position the caliper body onto the support. Lubricate the non-threaded portion of the guide and lock pins, then install the pins and tighten to 27 ft. lbs. (36 Nm).

11. Check the position of the pad clips. If necessary, use a small suitable tool to re-seat or center the pad clips on the support. Repeat the procedure for the opposite side brake pads.

12. Install the rear wheel assemblies and lower the vehicle.

13. Prior to operating the vehicle, depress the brake pedal a few times until the brake pads are seated against the rotor.

Rear

11. Lubricate the smooth ends of the removed pin(s) with silicone grease, then install the pin(s) and tighten to 27 ft. lbs. (36 Nm). Do not get grease on the pin threads.

12. Install the wheels, then lower the vehicle.

13. Prior to operating the vehicle, depress the brake pedal a few times until the brake pads are seated against the rotor.

Brake Rotor

REMOVAL AND INSTALLATION

1. Raise and support the vehicle safely, then remove the wheel assembly.
2. Remove the brake caliper from the support bracket and hang it from the suspension with wire to prevent brake line damage.
3. Remove the 2 caliper support brackets-to-knuckle mounting bolts.
4. Remove the rotor from the vehicle. If it is difficult to remove the rotor from the hub, insert two M8 x 1.25 self tapping bolts into the holes provided on the rotor and drive it from the hub.

To install:

5. Install the rotor over the hub and bearing assembly.
6. Install the caliper support bracket, then tighten the bolts to 81 ft. lbs. (110 Nm) for front caliper brackets or to 63 ft. lbs. (85 Nm) for rear caliper brackets.
7. Unwire and install the caliper.
8. Install the wheel assembly and lower the vehicle.

Brake Drums

REMOVAL AND INSTALLATION

1. Release the parking brake, then raise and support the vehicle safely.
2. Remove the rear wheel and remove the brake drum.
3. If necessary, turn the star wheel of the brake adjuster assembly to loosen the brake shoes and allow for drum removal.

Brake Shoes

REMOVAL AND INSTALLATION

NOTE: Brake shoes must be replaced as axle sets.

1. Raise and support the vehicle, then remove the wheels and brake drums.
2. Remove the lower return and adjuster springs using a universal brake spring remover. Do not over extend the springs or they will damaged and will need to be replaced.
3. Compress the leading brake shoe hold-down cup and spring while removing the pin from the rear of the backing plate. Release spring compression, then remove the hold-down cup and spring.
4. Pull the leading shoe towards the front of the vehicle and remove

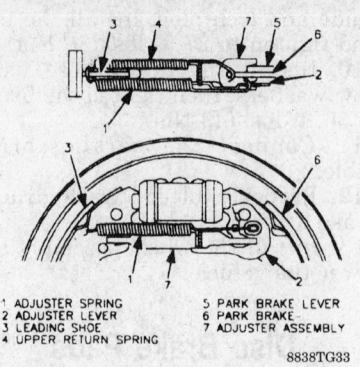

1 ADJUSTER SPRING
2 ADJUSTER LEVER
3 LEADING SHOE
4 UPPER RETURN SPRING
5 PARK BRAKE LEVER
6 PARK BRAKE
7 ADJUSTER ASSEMBLY

8838TG33

Rear brake shoe and adjuster installation

the adjuster assembly and lever. It may be necessary to turn the adjuster star wheel to shorten the adjuster's length.

5. Remove the leading shoe by twisting the shoe out of engagement with the upper return spring.
6. Remove the upper return spring from the park brake shoe, then remove the park brake shoe hold-down cup, spring and pin assembly.
7. Push the park brake shoe lever into the cable spring while disengaging the cable from the end lever and remove the parking brake shoe, lever and cable spring from the vehicle.
8. Remove the retainer and wave washer, then remove the park brake lever from the shoe.
9. Disassemble the brake adjuster socket, screw and nut, then clean the components in denatured alcohol. Inspect the assembly, making sure the screw threads smoothly into the adjusting nut over the full threaded length.
10. Inspect the wheel cylinder for signs of leakage and for cut or damaged boots. Do not attempt to repair a damaged cylinder, the assembly must be replaced.

To install:

11. Lubricate the adjuster assembly, the 6 backing plate raised shoe contact pads, the brake lever pin and surfaces which contact brake shoe webs with brake lubricant.
12. Install the park brake lever onto the pin on the brake shoe and secure with the wave washer and retainer clip. Crimp the ends of the retainer to secure the brake lever.
13. Install the cable spring into the cage on the park brake lever, then install the cable through the spring and onto the lever.
14. Install the park brake shoe using the hold-down cup assembly; use a universal spring cup remover/installer tool. Make sure the shoe is correctly engaged into the

wheel cylinder (top) and the anchor (bottom).

15. Install the long straight end of the upper return spring into the back hole in the park brake shoe, position the other brake shoe and install the other end of the spring into the back of the leading shoe.
16. Pull the lead shoe toward the front of the vehicle and install the adjuster between the park and leading brake shoes. Verify that the adjuster notches properly engage the brake shoe notches and that the shoe is properly aligned in the wheel cylinder and anchor.
17. Install the adjuster lever and adjuster spring. Make sure the notch on the lever engages the pin on the park shoe and the notch on the adjusting socket. The lower leg of the lever should engage the teeth of the star wheel adjuster assembly.
18. Secure the leading brake shoe using the hold-down cup assembly.
19. Install the adjuster spring to the upper side of the brake shoes with the short end to the lead shoe and the long end to the adjuster lever. Then install the lower return spring into the lower holes of the shoes.
20. Verify the correct location of all brake components, if necessary, use the other side brake assembly for comparison.
21. Using a suitable drum clearance gauge, measure the inner diameter of the brake drum and adjust the outside diameter of the brake shoes to 0.02 inch (0.50mm) less than the inner diameter of the drum.
22. Repeat the procedure for the opposite brake shoes and install the brake drums.
23. If the wheel cylinders have been replaced, bleed the hydraulic brake system.
24. Install the rear wheels and lower the vehicle.
25. Apply and release the brake pedal 20 times to allow the adjuster to properly position the brake shoes.
26. Check and adjust the parking brake cable, as necessary.

Wheel Cylinder

REMOVAL AND INSTALLATION

1. Raise and support the vehicle safely, then remove the wheel.
2. Remove the brake drum and shoes, as necessary for access to the cylinder.
3. Remove the bleeder valve and cap.

4. Disconnect the hydraulic brake line using a suitable wrench. Plug the line to prevent system contamination or excessive fluid loss.

5. Remove the wheel cylinder retaining bolts and remove the cylinder.

To install:

6. Install the cylinder to the backing plate, then tighten the retainers to 89 inch lbs. (10 Nm).

7. Connect the hydraulic brake line using new washers and tighten the fastener to 36 ft. lbs. (49 Nm).

8. Install the bleeder valve and cap, tighten the valve to 66 inch lbs. (7.5 Nm).

9. Install the brake shoes and drum. Be sure to properly bleed the hydraulic brake system.

10. Install the wheel assembly and lower the vehicle.

Parking Brake Cable

ADJUSTMENT

NOTE: If equipped with rear drum brakes that have been serviced, before performing parking brake adjustment procedures, apply and release the brake pedal 20 times. This allows the adjuster to position the brake shoes and prevents premature wear of the brake linings due to improper park brake adjustment.

Access to the parking brake cable adjuster may be obtained through the rear ashtray opening on the center console.

1. Remove the rear ashtray from the console. A thin stiff plastic card works well to help free the ashtray.

2. Raise and support the vehicle so the rear wheels are free to turn.

3. Pull the parking brake lever up to the 3rd click.

4. Tighten the parking brake adjuster until a light brake drag is at both rear wheels when turned by hand.

NOTE: If any of the parking brake components were replaced, the parking brake system must be set before proper adjustment can occur. To set the system, apply the parking brake lever 3–4 times with an approximate force of 100 lbs. (445 N).

5. Apply and release the lever several times.

6. Pull the lever up to the 2nd click. There should be no brake drag

at the rear wheels when turned by hand.

7. Pull the lever to the 3rd click and check for a slight drag at the rear wheels.

8. Pull the lever to the 4th click and verify the rear wheels are locked or under heavy drag.

9. Loosen or tighten the adjuster nut as necessary until these conditions are met.

10. If both rear wheels do exhibit similar drag check for damage to/incorrect installation of the drum brake components, park brake cables and/or park brake lever assembly. If any damage is found, make repairs, then return to Step 3.

11. Lower the vehicle.

REMOVAL AND INSTALLATION

1. Disconnect the negative battery cable.

2. Remove the shift knob for manual transaxles or tape the release button to the in position for automatic transaxles.

3. Remove the liner from the rear storage tray and remove the 2 screws.

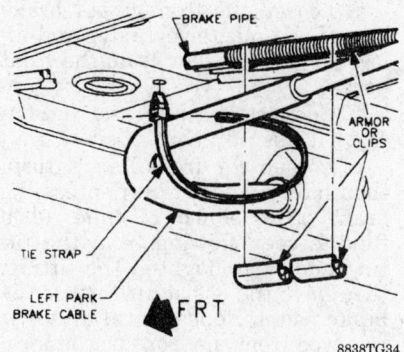

Parking brake cable tie strap

Type A parking brake equalizers must be installed with the cable entry holes facing downward

4. Remove the 2 side screws at the front of the console.

5. Apply the parking brake and remove the trim plate below the parking brake handle.

6. Remove the ashtrays and disconnect the front ashtray light.

7. If equipped, remove the power window/mirror switch by pushing the switch forward and lifting the rear edge of the assembly. Unplug the electrical connectors and remove the switch from the console.

8. Remove the side panels by disconnecting the bottom velcro fasteners and pulling the panels from the upper snaps. Lift the rear of the console and press out the seat belt bezels, then remove the console.

9. Remove the adjuster nut and, if present, the spring from the threaded rod, then remove the cables from the equalizer assembly.

10. Using SA9151BR or an equivalent release tool, remove the brake cable from the console bracket.

NOTE: Before removing the cable from the vehicle, tie a piece of wire or string to the console end of the cable. After removal, the string can be used to pull the new cable through the floor pan and into position.

11. Remove the rear seat cushion.

12. Raise the rear of the vehicle and support safely, then remove the rear wheels.

13. Remove the park brake grommet/cable assembly from the floor pan. Then remove the trailing arm/park cable-to-body fasteners.

14. If equipped with a cable retaining tie strap, cut and remove the strap.

15. If equipped with drum brakes, proceed as follows:

a. Remove the brake drum.

b. Remove the lower return spring and the adjuster spring.

c. Remove the park brake shoe hold-down cup assembly.

d. Pull the park brake shoe towards the rear of the vehicle and remove the adjuster assembly.

e. Disconnect the upper return spring from the park brake shoe.

f. Remove the park brake cable from the park brake lever, by pushing the lever into the cable spring, while disengaging the cable end from the lever.

g. Remove the park brake cable spring.

h. Use SA9151BR or an equivalent cable release tool and remove the cable from the backing plate.

16. If equipped with disc brakes, proceed as follows:

a. Remove the nut securing the cable to the floor pan stud.

b. Use SA9151BR or an equivalent cable release tool and remove the cable from the caliper lever.

17. Remove the cable from the vehicle.

To install:

18. If equipped with disc brakes, proceed as follows:

a. Install the park brake cable into the caliper bracket.

b. Attach the cable to the caliper park brake lever.

c. Install the trailing arm/park brake cable-to-body fasteners, then tighten the fasteners to 89 ft. lbs. (120 Nm).

d. Install the cable bracket to the floor pan stud. Install the nut and tighten to 25 inch lbs. (2.8 Nm).

19. If equipped with drum brakes, proceed as follows:

a. Install the cable through the brake backing plate and correctly seat the retaining fingers.

b. Install the cable spring onto the cable, then install the cable end into the actuator lever.

c. Connect the short end of the upper return spring to the leading brake shoe, then connect the other end of the spring to the park shoe.

d. Pull the park brake shoe toward the vehicle's rear and install the adjuster assembly between the shoes, then install the adjuster lever.

e. Install the adjuster spring with the short end into the leading shoe and the other end into the adjuster lever.

f. Install the lower return spring, then install the park shoe hold-down cup assembly.

g. Adjust the brake shoe outer diameter to 0.02 inch (0.50mm) less than the inner diameter of the brake drum, then install the brake drum.

h. Install the trailing arm/park brake cable-to-body fasteners, then tighten the fasteners to 89 ft. lbs. (120 Nm).

20. Pull the cable into the vehicle with the wire or string positioned during disassembly.

21. Install the grommet into the floor pan.

22. Install a park brake cable tie strap.

23. Install the rear wheels.

24. Install the cables into the console bracket, then attach the cables to the equalizer. For equalizers of the first production design, the cable entry holes (rounded outer side of the equalizer) must face downward. For the second design, the cable entry holes may face up or down.

25. Install the cable and equalizer assembly onto the threaded rod, then install the adjusting nut.

26. For rear drum vehicles, apply and release the brake pedal 20 times to allow the drum brake adjuster to position the shown.

27. Adjust the parking brake; refer to the procedure later in this section.

28. Install the center console and the rear seat cushion.

29. Lower the vehicle to the ground.

30. Connect the negative battery cable.

Brake System Bleeding

Except Anti-Lock Brakes

Make sure the master cylinder is at least ½ full with clean DOT 3 brake fluid at all times during the procedure.

1. The master cylinder must be bled first if it is suspected to contain air. Bleed the master cylinder as follows:

a. Loosen the front upper brake line at the master cylinder and allow the fluid to flow from the front port.

b. Connect the line and tighten to 24 ft. lbs. (32 Nm).

c. Loosen the front line ¼ turn, then have an assistant depress the brake pedal slowly (1 time) until fluid is seen flowing from the fitting and hold. Tighten the fitting, then have the assistant release the brake pedal. Repeat until all air is removed from the front (secondary) master cylinder bore.

d. Tighten the brake line to 24 ft. lbs. (32 Nm) when finished.

e. Repeat Steps a–d for the rear upper brake line fitting in order to bleed air from the rear (primary) master cylinder bore.

2. If a pipe or fitting was the only hydraulic line disconnected, then only the caliper(s) or wheel cylinder(s) affected by that line must be bled. If the master cylinder required bleeding, then all calipers and wheel cylinders must be bled in the proper sequence:

a. Right rear

b. Left front

c. Left rear

d. Right front

NOTE: **Calipers may be tapped lightly using a rubber mallet in order to help free trapped air.**

3. Bleed the individual calipers or wheel cylinders as follows:

a. Place a suitable wrench over the bleeder screw and attach a clear plastic hose over the screw end.

b. Submerge the other end in a transparent container of brake fluid.

c. Loosen the bleed screw, then have an assistant apply the brake pedal slowly and hold. Tighten the bleed screw to 97 inch lbs. (11 Nm) and release the brake pedal. Repeat the sequence until all air is expelled from the caliper or cylinder.

d. Tighten the bleed screw to 97 inch lbs. (11 Nm) when finished.

4. Check the pedal for a hard feeling with the engine not running. If the pedal is soft, repeat the bleeding procedure until a firm pedal is obtained.

Anti-Lock Brakes

NOTE: **Prior to bleeding the rear brakes, the rear displacement cylinder pistons must be returned to the top most or HOME position. To return the pistons to HOME, use a Scan tool to perform special test, RUN ABS PISTONS UP-HOME. This test will run the piston to the top of their travel.**

1. Fill the master cylinder with clean brake fluid and keep the reservoir at least ½ full during the bleeding operation.

2. Prime the control assembly as follows:

a. Attach a clear tube to the rear bleeder valve and allow the tube to hang in a transparent container of clean brake fluid.

b. Slowly open the valve ½–¾ of a turn, then have an assistant depress and hold the brake pedal.

c. Continue to hold the brake pedal until fluid begins to flow from the valve, then close the valve and release the pedal.

d. Tighten the valve to 62 inch lbs. (7 Nm).

e. Repeat the procedure at the front bleeder valve.

3. Once fluid flows from both control assembly valves, the master cylinder and modulator are sufficiently filled with fluid in order to bleed the system, but they may not be com-

pletely purged of air. To assure that the unit is free of air bleed the calipers to remove air from the assemblies lowest points, then return and bleed the control assembly again.

4. Bleed the calipers in the proper order:
   a. Right rear
   b. Left rear
   c. Right front
   d. Left front

NOTE: If when performing the bleed procedure on the rear calipers, brake fluid does not come out of the bleeder, the rear displacement pistons may not be at the home or top position.

5. Bleed each caliper, in the proper order, as follows:
   a. Attach a clear tube to the caliper bleeder valve and allow the tube to hang in a transparent container of clean brake fluid.
   b. Open the valve ½–¾ of a turn, then have an assistant slowly depress and hold the brake pedal.
   c. Watch for air bubbles as the fluid begins to flow from the valve, then close the valve and release the pedal.
   d. Wait 5 seconds and repeat the procedure until the pedal feels firm and no air is present in the brake line.
   e. Tighten the valve to 97 inch lbs. (11 Nm).
6. Bleed the control assembly from the 2 valves in the same fashion as the calipers are bled. Tighten the bleeder valves to 62 inch lbs. (7 Nm) when finished.
7. Check the pedal for excessive travel with the engine **OFF** and then with the engine running. If pedal feel is firm go to Step 11.
8. If the pedal feel is not firm, use a Scan tool to run the ABS motors up and down 2 times, then make sure the pistons are run up to the HOME position.
9. Start the engine, let the engine run for 2 seconds after the ABS light goes out then turn the engine **OFF**. Repeat the ignition cycle 9 more times.
10. Re-bleed the entire system.
11. With the engine running and brake applied, check the system for leaks.
12. Road test the vehicle and make several normal, non-ABS stops. Then make 1–2 ABS stops from a higher speed (about 50 MPH).
13. After road testing the vehicle it is recommended that the entire system be bled and inspected 1 final time.

Wheel Speed Sensor

REMOVAL AND INSTALLATION

Front

1. Raise and safely support the vehicle.
2. Disconnect the negative battery cable.
3. Unplug the electrical connector from the speed sensor.
4. Remove the Torx® head retaining screw from the top of the sensor.
5. Remove the wheel speed sensor from the vehicle.
6. Make sure the sensor locating pin was removed with the sensor. If the pin remained in the knuckle, pull it out with a pair of pliers. A pin which is stuck in the knuckle must be drilled out with an 8mm drill bit. Be very careful not to enlarge the locating hole in the knuckle. Incorrect sensor location would result in loss of ABS operation.
7. Use a small wooden dowel with sandpaper wrapped around it to clean the locating hole of any corrosion and allow proper sensor seating.
   To install:
8. Position the sensor to the steering knuckle, making sure that it is fully seated.
9. Install the fastener and tighten to 89 inch lbs. (10 Nm).
10. Install the wiring harness connector to the sensor.
11. Safely lower the vehicle.
12. Connect the negative battery cable.
13. Road test the vehicle to ensure proper operation.

Rear

The rear ABS speed sensor is contained within the hub and bearing assembly and is not serviceable. If the sensor is damaged and in need of replacement, the entire hub and bearing assembly must be replaced.

FRONT SUSPENSION

Strut

REMOVAL AND INSTALLATION

——— CAUTION ———
The MacPherson strut is under extreme spring pressure. Do not remove the strut shaft center support nut from the top of the shaft assembly without using an approved spring compressor. Personal injury may result if this caution is not followed.

1. If equipped with ABS, disconnect the negative battery cable, then raise and support the vehicle safely.
2. Remove the front wheel.
3. If equipped, note the ABS wiring position for assembly purposes, disconnect the ABS wiring harness from the strut wiring bracket. If the strut is being replaced, drill the rivet head retaining the ABS wiring bracket to the strut and remove the bracket.
4. Loosen the 2 steering knuckle-to-strut housing bolts, but do not remove them at this time.
5. Lower the vehicle sufficiently and remove the 3 upper strut-to-body nuts.
6. Place a rag over the CV-joint seal to protect it from damage, then remove the 2 steering knuckle-to-strut housing bolts.
7. Remove the strut assembly from the vehicle.
   To install:
8. Position the strut to the vehicle, then install 3 new upper mount nuts and tighten to 21 ft. lbs. (29 Nm). New nuts must be used because the torque retention of the old nuts may be insufficient.
9. Install the knuckle bolts, also using new nuts. Push the bottom of the strut inward while tightening the fasteners to 148 ft. lbs. (200 Nm).
10. If the strut was replaced, install the ABS wiring bracket to the strut using a new rivet.
11. If applicable, connect the ABS wiring to the bracket.
12. Install the wheel assembly and lower the vehicle.
13. Connect the negative battery cable.
14. Check and adjust the alignment as necessary.

Lower Ball Joints

REMOVAL AND INSTALLATION

The lower ball joint is an integral part of the lower control arm. If the ball joint needs replacement, the entire lower arm must be replaced as an assembly.

1. Raise and support the vehicle safely.
2. Remove the wheel and splash shield.

3. Remove and discard the cotter pin from the lower control arm ball joint. Back the ball joint nut until the top of the nut is even with the top of the threads.

4. Use tool SA9132S to separate the lower control arm ball joint from the steering knuckle, then remove the nut. Do not use a wedge tool or seal damage may occur.

NOTE: The outer CV-joint for vehicles equipped with ABS contains a speed sensor ring. Use of an incorrect tool to separate the control arm from the knuckle may result in damage and loss of the ABS system.

5. Remove the control arm-to-cradle bolt.

6. Remove the sway bar-to-control arm nut, then remove the control arm from the vehicle.

To install:

7. Position the control arm and install the arm onto the sway bar without the fastener, then place the end of the arm into the cradle. Install the cradle nut and bolt. Tighten the cradle bolt to 92 ft. lbs. (125 Nm), then tighten the cradle nut to 74 ft. lbs. (100 Nm).

8. Install the sway bar nut and tighten to 106 ft. lbs. (144 Nm).

9. Thoroughly clean and lubricate the ball joint stud threads, then install the lower control arm ball stud into the steering knuckle. Install the nut and tighten the lower control arm ball stud nut to 55 ft. lbs. (75 Nm). Tighten the nut additionally if necessary and install a new cotter pin, but do not back off the original torque specification.

10. Install the splash shield and the wheel assembly.

11. Lower the vehicle, check and adjust the alignment as necessary.

Tension Strut (Sway Bar)

REMOVAL AND INSTALLATION

1. Place the steering in the unlocked position, then raise and support the vehicle safely.

2. Remove the left wheel and splash shield.

3. Remove and discard the cotter pin from the left lower control arm ball joint. Back the ball joint nut until the top of the nut is even with the top of the threads.

4. Use tool SA9132S to separate the lower control arm ball joint from the steering knuckle, then remove

the nut. Do not use a wedge tool or seal damage may occur.

NOTE: The outer CV-joint for vehicles equipped with ABS contains a speed sensor ring. Use of an incorrect tool to separate the control arm from the knuckle may result in damage and loss of the ABS system.

5. Remove the left lower control arm-to-cradle fastener.

6. Turn the steering wheel to the left for access to and remove of the right tension strut nut and washer.

7. Remove both tension strut-to-cradle mounting brackets. If a cradle nut is damaged or broken loose from the cradle, replace the nut as follows:

a. If the nut is damaged but not broken loose and the bolt can be removed, distort the bolt threads sufficiently to lock the bolt into the nut. Insert the bolt and tighten with an air impact wrench. Continue to turn the bolt until the nut breaks free of the cradle.

b. If the nut was already broken loose and the bolt could not be extracted or if the bolt was used to break the nut loose, cut the bolt head off.

c. Retrieve the bolt shank and nut from the cradle cavity.

d. Install a new bolt 21010823 and nut 21006321 or equivalents.

8. Remove the tension strut with the left control arm from the vehicle. If necessary, remove the nut and left control arm from the strut.

To install:

9. If removed, position the left control arm to the strut, but do not tighten the fastener at this time.

10. Install the strut mounting bushings onto the strut with the bushings slits facing the front of the vehicle.

11. Position the right end of the strut into the right control arm still on the vehicle, then position the left control arm into the cradle. Do not install fasteners at this time.

12. Using new bolts or a suitable threadlock such as Loctite® 242, install the mounting brackets and tighten the bolts to 103 ft. lbs. (140 Nm). Then tighten the fasteners a second time to 103 ft. lbs. (140 Nm).

13. Temporarily install the left wheel assembly onto the vehicle and with an assistant, push the bottom of the left wheel into the vehicle in order to facilitate lower control arm-to-cradle bolt installation.

14. Tighten the lower control arm cradle bolt to 92 ft. lbs. (125 Nm), then tighten the nut to 74 ft. lbs. (100 Nm).

15. Install new nuts onto the right and left tension strut-to-control arm studs. Tighten the right nut, then left nut to 106 ft. lbs. (144 Nm).

16. Thoroughly clean and lubricate the left ball joint stud threads, then install the left lower control arm ball stud into the left steering knuckle. Install the nut and tighten the stud nut to 55 ft. lbs. (75 Nm), tighten additionally if necessary and install a new cotter pin.

17. If not done already, remove the wheel, then install the splash shield and reinstall wheel assembly.

18. Lower the vehicle, check and adjust the alignment as necessary.

Front Wheel Bearings

REMOVAL AND INSTALLATION

1. If equipped with ABS, disconnect the negative battery cable.

2. Loosen the front halfshaft nut, while an assistant depresses the brake pedal, then raise and support the vehicle safely.

3. Remove the wheel assembly.

4. Remove the brake caliper mounting bracket bolts and suspend the assembly from the strut spring with wire.

5. Loosen the strut-to-knuckle bolts, but do not remove at this time.

6. Remove the rotor, axle nut and washer.

7. Remove and discard the cotter pin from the lower control arm ball joint. Back the ball joint nut until the top of the nut is even with the top of the threads.

8. Use tool SA9132S to separate the lower control arm ball joint from the steering knuckle, then remove the nut. Do not use a wedge tool or seal damage may occur.

NOTE: The outer CV-joint for vehicles equipped with ABS contains a speed sensor ring. Use of an incorrect tool to separate the control arm from the knuckle may result in damage and loss of the ABS system.

9. Remove the tie rod cotter pin and castle nut, then separate the tie rod end from the knuckle using a tie rod separator SA91100C or equivalent. Do not use a wedge-type tool.

10. If equipped, disengage the ABS wheel speed sensor electrical connector.

11. Suspend the halfshaft from the body with wire, then remove the 2 knuckle-to-strut fasteners and re-

move the knuckle/hub assembly from the vehicle. If difficulty is encountered, position a block of wood on the end of the halfshaft and tap on the wood with a hammer to free the hub assembly.

12. If necessary, disassemble the knuckle hub assembly as follows:

a. If equipped, remove the ABS wheel speed sensor from the knuckle.

NOTE: Any time the hub or bearing is separated from the steering knuckle, a new bearing must be used upon assembly.

b. Install wheel bearing removing tools SA9159S or equivalent, to the knuckle and secure the assembly in a vice.

c. Hold the hub driver with a wrench and tighten the hub driver screw to remove the hub. If the inner bearing race is pulled out with the hub, remove the race with a bearing race remover.

d. Remove the assembly from the vice and separate the wheel hub removal tool from the knuckle.

e. Remove the bearing retainer snapring.

f. Position the knuckle in a shop press on a knuckle support tube and press the bearing from the knuckle with a suitable small driver.

To install:

13. If necessary, assemble the knuckle hub assembly as follows:

a. Use a suitable large driver and press in the new bearing until seats.

b. Use the small driver and the knuckle support tube to press in the hub assembly. The small driver must be used to support the bearing inner race with its small (pilot) side facing towards the press and away from the bearing.

c. Install the bearing retainer snapring.

d. If equipped, install the ABS wheel speed sensor into the knuckle and tighten the fastener to 6 ft. lbs. (8 Nm).

NOTE: Service knuckles may not have holes for brake dust shield mounting. The dust shield is no longer required and does not have to be reinstalled. Also, should the shield become damaged it may be removed, there is no need to repair or replace it. But, should a shield be removed and discarded, the shield should also be removed from the opposite side to maintain balance/symmetry.

14. Thoroughly clean and lubricate the ball joint stud threads of the lower control arm and tie rod end. Install the knuckle/hub assembly onto the axle shaft. Then install the washer with a new nut, but do not tighten the nut at this time.

15. Install the lower control arm ball stud through the knuckle bore and install the nut, but do not tighten at this time.

16. Install the steering knuckle-to-strut fasteners, but do not tighten at this time.

17. Install the tie rod end and nut, then tighten the nut to 33 ft. lbs. (45 Nm) and install a new cotter pin. If necessary, tighten the nut additionally, do not back off to insert the cotter pin.

18. Push inward on the bottom of the strut and tighten the knuckle fasteners to 148 ft. lbs. (200 Nm).

19. Tighten the lower control arm ball stud nut to 55 ft. lbs. (75 Nm), tighten additionally if necessary and install a new cotter pin.

20. Install the rotor onto the hub, then install the caliper mount bracket onto the knuckle. Tighten the mount bracket assembly bolts to 81 ft. lbs. (110 Nm).

21. If equipped, engage the ABS electrical connector to the wheel speed sensor.

22. While an assistant depresses the brake pedal, tighten the halfshaft nut to 148 ft. lbs. (200 Nm).

23. Install the wheel assembly and lower the vehicle.

24. Connect the negative battery cable, check and adjust the alignment, as necessary.

REAR SUSPENSION

Strut

REMOVAL AND INSTALLATION

———— **CAUTION** ————
The MacPherson strut is under extreme spring pressure. Do not remove the strut shaft center support nut from the top of the shaft assembly without using an ap- *proved spring compressor. Personal injury may result if this caution is not followed.*

1. On coupes, remove the rear seat cushion bottom, left or right rocker panel interior moldings and left or right rear sail interior panels.

2. On sedans, remove the left or right C-pillar interior moulding.

3. Fold down the rear seat backs and remove the rear seat side bolsters from the vehicle.

4. On coupes, remove the rear deck package shelf screws attaching the shelf to the side of the cargo area.

5. Remove the speaker grills and fasteners from the shelf, then remove the seatbelt bezel and separate the seat belts from the shelf. Remove the rear package shelf carpeting.

6. If equipped with ABS, disconnect the negative battery cable.

7. Raise and support the vehicle safely, then remove the appropriate rear wheel.

8. If equipped, unplug the ABS wiring from the strut wiring bracket. If the strut is being replaced, drill the rivet head retaining the ABS wiring bracket to the strut and remove the bracket.

9. Loosen the 2 strut-to-knuckle bolts; but do not remove at this time.

10. Lower the vehicle sufficiently and place a floor jack under the rear knuckle, then raise the jack enough to support the knuckle.

11. Remove and discard the 3 upper strut-to-body nuts.

12. Slowly raise the hoist, lowering the strut from the bottom.

13. Remove the strut to knuckle bolts and remove the strut assembly from the vehicle.

To install:

14. Install 3 new strut-to-body upper mount nuts and tighten to 21 ft. lbs. (29 Nm). New nuts must be used because the torque retention of the old nuts may be insufficient.

15. Install the knuckle bolts with new nuts, then push the bottom of the strut inward and tighten the fasteners to 148 ft. lbs. (200 Nm).

16. If the strut was replaced, install the ABS wiring bracket to the strut using a new rivet.

17. If applicable, connect the ABS wiring to the bracket.

18. Install the wheel assembly and lower the vehicle.

19. Install the interior components.

20. Connect the negative battery cable, then check and adjust the rear alignment as necessary.

Rear Control Arms

REMOVAL AND INSTALLATION

Lateral Links

If the front lateral link is to be replaced, the fuel tank must first be removed.

1. Raise and support the vehicle safely, then remove the rear wheel(s).
2. If removing the front lateral link, remove the fuel tank.
3. Remove the lateral link-to-knuckle bolt, then remove the crossmember bolt.
4. Remove the link from the vehicle.

To install:

5. Install the link into the crossmember using the fastener, but do not tighten at this time.
6. Install the link to the knuckle with the knuckle-to-link bolt, but do not tighten at this time.
7. Tighten the crossmember bolt, as applicable. Tighten the front link bolt to 126 ft. lbs. (170 Nm) or the rear link bolt to 89 ft. lbs. (120 Nm). Then tighten the knuckle bolt to 122 ft. lbs. (165 Nm).
8. If the front lateral link was removed, install the fuel tank.
9. Install the rear wheel(s) and lower the vehicle, then check and adjust the rear alignment as necessary.

Trailing Arm

1. Raise and support the vehicle safely, then remove the rear wheels.
2. Remove the trailing arm-to-knuckle nut.
3. Remove the trailing arm-to-body bolts.
4. Slide the trailing arm from the knuckle.

To install:

5. Install the trailing arm into the knuckle and torque the nut to 74 ft. lbs. (100 Nm).
6. Position the arm to the body, install and tighten the body bolts to 89 ft. lbs. (120 Nm).
7. Install the rear wheels and lower the vehicle.
8. Check and adjust the rear alignment, as necessary.

Wheel Bearings

Unlike the front wheel bearings, which may be removed from the hub for replacement, the rear wheel hub and bearing assembly is not serviceable. If damaged or worn, the hub and bearing assembly must be replaced as a unit.

ADJUSTMENT

The rear hub/bearing assembly is a sealed assembly, requiring no periodic maintenance. No adjustments are necessary or possible.

REMOVAL AND INSTALLATION

1. If equipped with ABS, disconnect the negative battery cable.
2. Raise and support the vehicle safely, then remove the rear wheel.
3. If equipped, unplug the ABS speed sensor connector.
4. On disc brake equipped models, remove the caliper assembly-to-knuckle mounting bolts and support it with a wire from the strut, then remove the rotor.
5. On drum brake equipped models, remove the brake drum.
6. Remove the 4 hub/bearing-to-knuckle bolts, then remove the assembly from the vehicle.

To install:

7. Install the brake backing plate, hub/bearing assembly and retaining bolts. Tighten the bolts to 63 ft. lbs. (85 Nm).
8. Install the brake drum or rotor and caliper assembly. If applicable, tighten the caliper retaining bolts to 63 ft. lbs. (85 Nm).
9. If equipped, engage the ABS speed sensor connector.
10. Install the wheel assembly and lower the vehicle. If applicable, connect the negative battery cable.

TOOLS AND EQUIPMENT 30

WHAT TOOLS ARE NEEDED

Analyzing Specific Needs

Nearly everybody needs some tools, whether they are fixing a kitchen sink, or overhauling the engine in the family car. As far as car repairs go, pliers and a can of oil will not get one very far down the path of do-it-yourself service. But, a do-it-yourselfer's garage does not have to be equipped like the local service station either. Somewhere between these two extremes is a level that suits the average do-it-yourselfer. Just where that point is depends on the home mechanic's ability and level of interest. The strategy is to match the tools and equipment to the tasks which are to be tackled.

First, things should be sorted out in an orderly manner. The do-it-yourselfer should think about his/her repair work on three levels: basic, average and advanced. Before purchasing any tools, he/she should sit down and determine the level of the expertise and cost needed to accomplish the job at hand. Knowing what repairs can, or are needed to be performed, is the most important step. Obviously, if all that is intended is changing the oil and spark plugs, many tools will not be needed. If fairly extensive repair work is planned, the do-it-yourselfer will end up with a pretty complete collection of tools. Many expensive tools can be rented from automotive parts jobbers or tool rental centers. This allows many home mechanics to do special repairs on an occasional basis.

BASIC AUTOMOTIVE TOOLS

Naturally, without the proper tools it is impossible to properly service an automobile. It would be impossible to catalog each tool which would be needed to perform every operation in this book. It would also be unwise for the amateur to rush out and buy an expensive set of tools, on the theory that one or more may be needed at sometime.

The best approach is to proceed slowly, gathering together a good quality set of those tools that are used most frequently. Don't be misled by the low cost of bargain tools. It is far better to spend a little extra money for better quality tools. Forged wrenches, 6-point sockets and fine tooth ratchets are preferable to their less expensive counterparts. As any good mechanic can concur, there are few worse experiences than trying to work on an automobile with bad tools. Any monetary savings will be far outweighed by frustration and mangled knuckles.

Certain tools, plus a basic ability to handle them, are required to get started. A basic tool set and a torque wrench are good for a start. Begin by accumulating those tools that are used most frequently (tools associated with routine maintenance/tune-up and engine repair). In addition to the normal assortment of screwdrivers and pliers, the following tools should be acquired for general routine maintenance:

• Metric and standard wrenches, sockets and combination open end/box end wrenches in sizes from 3–19mm and 1/4–7/8 in., and a spark plug socket (5/8 inch or 16mm). If possible, buy various length socket drive

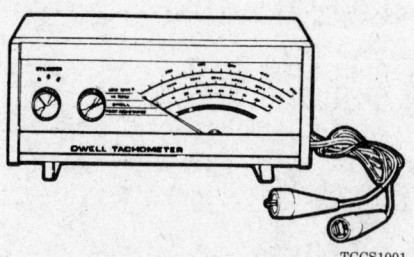

Dwell/tachometer unit (typical)

TCCS1001

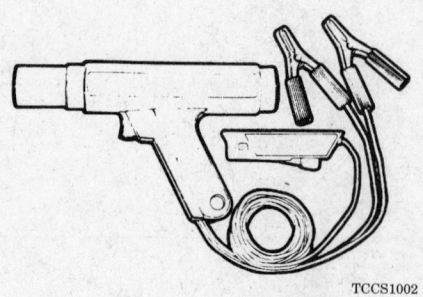

TCCS1002

Inductive type timing light

extensions. One advantage in this area is that the metric sockets available in the United States will fit SAE ratchet handles and extensions you may already have (1/4 in., 3/8 in., and 1/2 in. drive).

• Jackstands for support.
• Oil filter wrench.
• Oil filler spout or funnel.
• Grease gun for chassis lubrication.
• Hydrometer or battery tester for checking the battery.
• A low flat pan for draining oil.
• Lots of rags for wiping up the inevitable mess.

In addition to the above items, there are several other tools which, although not absolutely necessary, are handy to have around. These include oil-dry, a transmission fluid funnel and the usual supply of lubricants and fluids, all of which can be purchased on a need basis. This is a basic list for routine maintenance, but only personal needs and desires can accurately determine the final list of tools necessary.

The second list of tools is for tune-ups. While these tools are slightly more sophisticated, they need not be outrageously expensive. There are several inexpensive tach/dwell meters on the market that are every bit as good for the average mechanic as a costly professional model. Just be sure the tach/dwell meter measures to at least 1200–1500 rpm on the tach scale, and that it works on 4, 6 and 8-cylinder engines. A basic list of tune-up equipment could include:

• Tach/dwell meter.
• Spark plug wrench.
• Timing light (a DC light that works from the vehicle's battery is best).
• Wire spark plug gauge/adjusting tools.

Here again, it is best to be guided by particular needs. In addition to these basic tools, there are several other tools and gauges which may be useful. These include:

• A compression gauge. The screw-in type is slower to use, but eliminates the possibility of a faulty reading due to escaping pressure.
• A manifold vacuum gauge.
• A test light.
• A Digital Volt-Ohm Meter (DVOM). This meter allows direct testing of electrical components and grounds.

As a final note, a torque wrench will be necessary for most work. The beam type models are perfectly adequate, although the newer click (break-away) type is more precise,

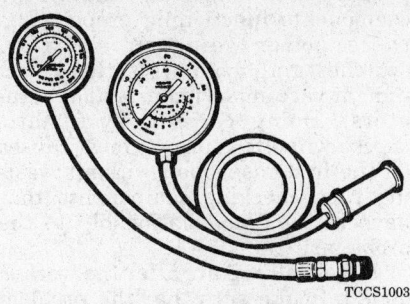

Compression gauge and a combination vacuum/fuel pressure test gauge

and does not require one to crane one's neck to see a torque reading in awkward situations. The break-away torque wrenches are more expensive and should be recalibrated periodically.

Tightening bolts to the correct torque value is extremely important on today's automobiles. The torque specification for each fastener will be given in the procedure whenever a specific torque value is required. An example of torque specifications are given (the following values are only a guide), based upon fastener size:

Bolts marked 6T

6mm bolt/nut: 5–7 ft. lbs. (7–10 Nm)

8mm bolt/nut: 12–17 ft. lbs. (16–23 Nm)

10mm bolt/nut: 23–34 ft. lbs. (31–46 Nm)

12mm bolt/nut: 41–59 ft. lbs. (56–80 Nm)

14mm bolt/nut: 56–76 ft. lbs. (76–103 Nm)

Bolts marked 8T

6mm bolt/nut: 6–9 ft. lbs. (8–12 Nm)

8mm bolt/nut: 13–20 ft. lbs. (18–27 Nm)

10mm bolt/nut: 27–40 ft. lbs. (37–54 Nm)

12mm bolt/nut: 46–69 ft. lbs. (62–94 Nm)

14mm bolt/nut: 75–101 ft. lbs. (102–137 Nm)

Special Tools

Normally, special factory tools are avoided for repair procedures, since these many not be readily available for the do-it-yourself mechanic. When it is possible to perform the job with more commonly available tools, it will be pointed out, but occasionally,

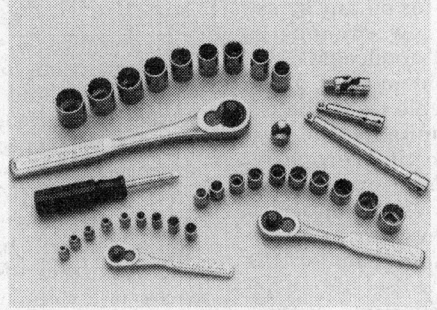

All but the most basic procedure will require an assortment of ratchets and sockets

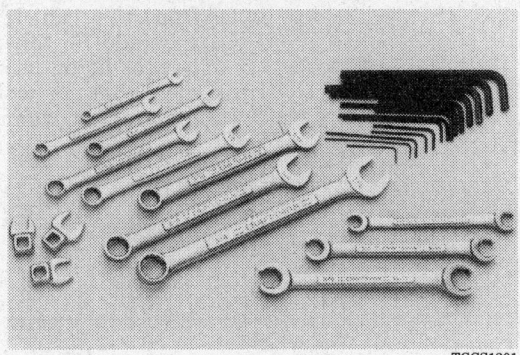

In addition to ratchets, a good set of wrenches and hex keys will be necessary

a special tool was designed to perform a specific function and should be used. Before substituting another tool, one should be convinced that neither one's safety nor the performance of the vehicle will be compromised.

Some special tools are available commercially from major tool manufacturers, automotive parts stores or a car dealership.

SPECIAL DIAGNOSTIC TOOLS

Frequent references to specific test equipment will be found in the text and in the diagnostic charts. This usually refers to scan tools used to communicate with electronic control units or special electronic testers. Among other features, scan tools combine many standard testers into a single device for quick and accurate circuit diagnosis. For many tests, a multimeter, test light, or other general test equipment can be substituted, but the technician must be aware of the risk involved. The general test equipment may not be capable of safely testing the system, or may generate incomplete or inaccurate test results. Some tests require activating system components and often this can only be done with scan tools or other special equipment.

Most test equipment is available through aftermarket tool manufacturers, but some can only be obtained through the vehicle manufacturer. Care should be taken that all test equipment being used is designed to diagnose that particular system accurately, without damaging control modules or other components.

NOTE: When using special test equipment, the manufacturer's instructions provided with the tester should be read and clearly understood before attempting any test procedures.

Electrical Test Tools

ORGANIZED TROUBLESHOOTING

When diagnosing a specific problem, there are certain troubleshooting techniques that are standard:

1. Establish when the problem occurs. Does the problem appear only

A hydraulic floor jack and a set of jackstands are essential for lifting and supporting the vehicle

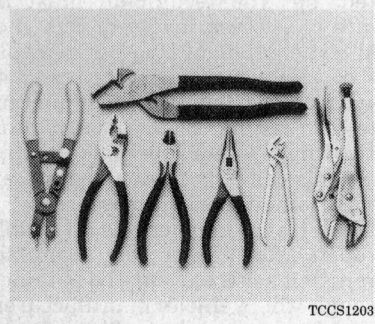

An assortment of pliers will be handy, especially for old rusted parts and stripped bolt heads

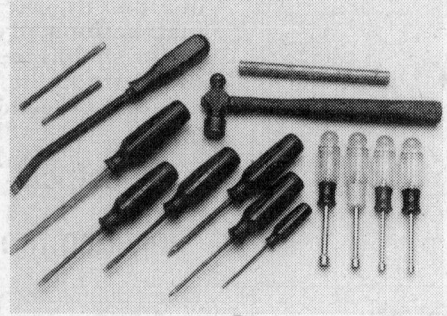

Various screwdrivers, a hammer, chisels and prybars are necessary to have in your tool box

under certain conditions? Were there any noises, odors, or other unusual symptoms? Make notes on any symptoms found, including warning lights and trouble codes, if applicable.

2. Isolate the problem area. To do this, make some simple tests and observations; then eliminate the systems that are working properly. Check for obvious problems such as broken wires, split or disconnected vacuum hoses. Always check the obvious before assuming something complicated is the cause. Be suspicious of fuses, switches and connectors; wiring itself rarely fails.

3. Test for problems systematically to determine the cause once the problem area is isolated. Are all the components functioning properly? Is there power going to electrical switches and motors? Is there vacuum at vacuum switches and/or actuators? Doing careful, systematic checks will often turn up most causes on the first inspection without wasting time checking components that have little or no relationship to the problem.

4. Test all repairs after the work is done to make sure that the problem is fixed. Some causes can be traced to more than 1 component, so a careful verification of repair work is important to pick up additional malfunctions that may cause a problem to reappear or a different problem to arise. A blown fuse, for example, is a simple problem that may require more than another fuse to repair.

The diagnostic tree charts are designed to help solve problems by leading the user through closely defined conditions and tests. Only the most likely components, vacuum and electrical circuits are checked for proper operation when troubleshooting a particular malfunction. By using the diagnostic trees to eliminate those systems and components which normally will not cause the condition described, a problem can be isolated within 1 or more systems or circuits without wasting time on unnecessary testing.

Experience has shown that most problems tend to be the result of a fairly simple and obvious cause, such as loose or corroded connectors; making careful inspection of components during testing is essential to quick and accurate troubleshooting. Frequent references to special test equipment will be found in the text and in the diagnosis charts. These devices or a compatible equivalent are necessary to perform some of the more complicated test procedures listed. Testers are available from a variety of aftermarket sources as well as from the vehicle manufacturer. Care should be taken that any test equipment being used is designed to diagnose that particular system accurately without damaging the com-

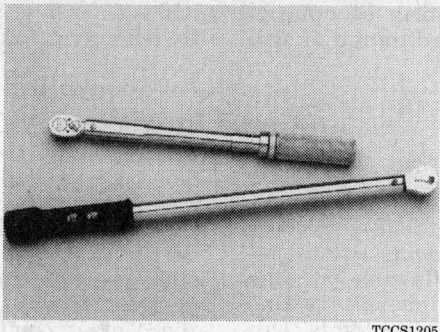

Many repairs will require the use of a torque wrench to assure the components are properly fastened

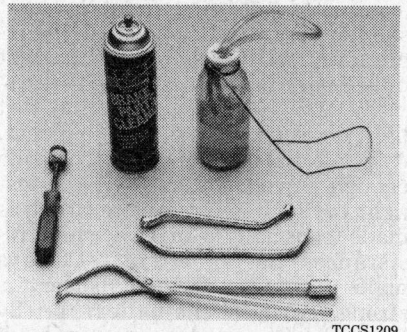

Although not always necessary, using specialized brake tools will save time

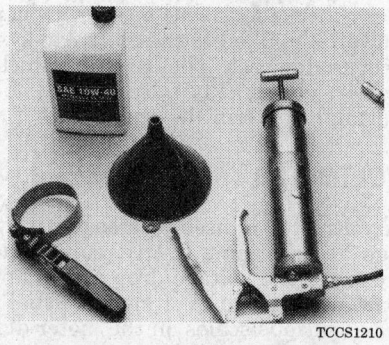

A few inexpensive lubrication tools will make regular service easier

puter control modules or components being tested.

NOTE: Pinpointing the exact cause of trouble in an electrical system can sometimes be accomplished only by the use of special test equipment. In addition to the information covered in this section, the manufacturer's instructions booklet provided with the tester should be read and clearly understood before attempting any test procedures.

Testers and Equipment

JUMPER WIRES

Jumper wires are simple, yet extremely valuable, pieces of test equipment. Jumper wires are merely wires that are used to bypass sections of a circuit. The simplest type of jumper wire is a length of multi-strand wire with an alligator clip at each end. Jumper wires are usually fabricated from lengths of standard automotive wire and whatever type of connector (alligator clip, spade connector or pin connector) is required for the vehicle being tested. Some jumper wires are made with 3 or more terminals coming from a common splice for special-purpose testing. In cramped, hard-to-reach areas it is advisable to have insulated boots over the jumper wire terminals in order to prevent accidental grounding and possible system damage.

Jumper wires are used primarily to locate open electrical circuits, on either the ground (-) side of the circuit or on the hot (+) side. If an electrical component fails to operate, connect the jumper wire between the component and a good ground. If the component operates only with the jumper installed, the ground circuit is open. If the ground circuit is good, but the component does not operate, the circuit between the power feed and component is open. Sometimes a fused jumper wire is connected directly from the battery to the hot terminal of the component, but first make sure the component uses 12 volts in operation.

By inserting an in-line fuse between a set of test leads, a fused jumper wire is created. A fused jumper wire can be used for bypas-

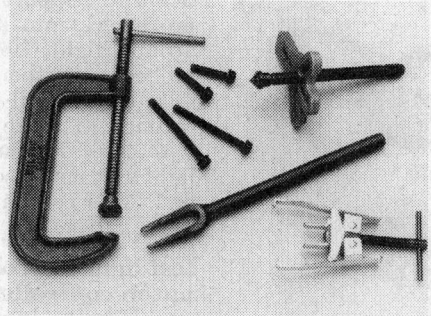

Various pullers, clamps and separator tools are needed for the repair of many components

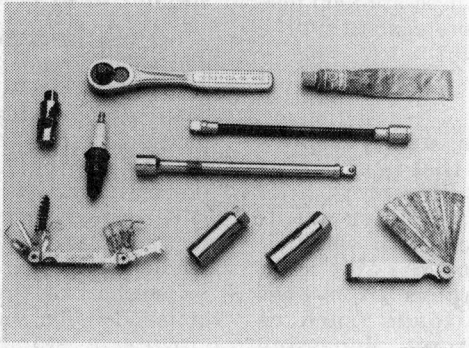

A variety of tools and gauges are needed for spark plug service

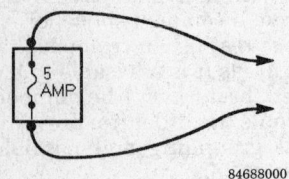

84688000

Schematic of a fused jumper wire

84688002

Fused jumper wire

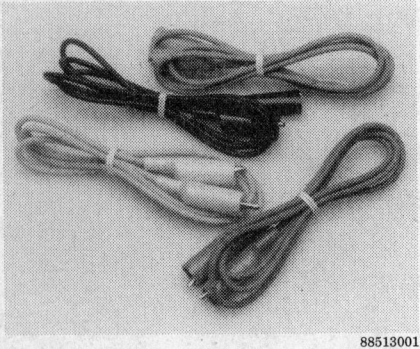

88513001

Jumper wires come in different gauges

sing open circuits. Use a 5 amp fuse to provide circuit protection.

NOTE: Never use jumpers made from wire that is of lighter gauge (smaller diameter) than used in the circuit under test. If the jumper wire is too small, it may overheat and possibly melt. Never use jumpers to bypass high-resistance loads (such as motors) in a circuit. Bypassing resistances, in effect, creates a short circuit, which may cause damage and fire. Never use a jumper for anything other than

temporary bypassing of components in a circuit, damage or fire could result.

TEST LIGHTS

12 Volt Test Light

The 12 volt test light is used to check circuits and components while electrical current is flowing through them. It is used for voltage and ground tests. 12 volt test lights come in different styles, but all have 3 main parts; a ground clip, a probe and a light.

NOTE: Avoid piercing the insulation of any wire. While most probes are designed to pierce insulation, this can lead to corrosion or broken conductors within the wire. Trace the wire to a terminal that can be probed before piercing the insulation.

The most commonly used 12 volt test lights have pick-type probes. To use a 12 volt test light, connect the ground clip to a good ground and probe wherever necessary with the pick.

The wrap-around light is handy in hard to reach areas or where it is difficult to support a wire to push a probe pick into it. To use the wrap around light, hook the wire to be probed with the hook and pull the trigger. A small pick will be forced through the wire insulation into the wire core. Only use this type of test light as a last resort and do not use it on SRS or computer data lines.

NOTE: Never use a pick-type test light to probe wiring on computer controlled systems unless specifically instructed to do so. Any wire insulation that is pierced by the test light probe should be taped and sealed with silicone after testing.

The test light does not detect specific voltage amounts; it only detects that voltage is present. It is advisable before using the test light to touch its terminals across the battery posts to make sure the light is operating properly. Do not attempt to determine voltage by how brightly the tester glows; use a voltmeter if an exact reading is needed.

Use of a LED type test light is recommended for computer controlled circuits. A standard incandescent bulb test light can load the circuit causing a high current to flow and damage the components. An LED type test light will not load the circuit

and is safer to use in a computer controlled circuit.

Self-Powered Test Light

The self-powered test light usually contains a 1.5 volt penlight battery. One type is similar in design to the 12 volt test light. This type has both the battery and the light in the handle and pick-type probe tip. The second type has the light toward the open tip, so that the light illuminates the contact point. The self-powered test light is a dual-purpose piece of equipment. It can be used to test for either open or short circuits when power is isolated from the circuit (continuity test). A powered test light should never be used on any computer controlled system or component unless specifically instructed to do so.

The 1.5 volt battery in the test light does not provide much current. A weak battery may not provide enough power to illuminate the test light even when a complete circuit is made (especially if there are high resistances in the circuit). Always make sure that the test battery is strong. To check the battery, briefly touch the ground clip to the probe; if the light glows brightly, the battery is strong enough for testing. Never use a self-powered test light to perform checks for opens or shorts when power is applied to the electrical system under test. The 12 volt vehicle power will quickly burn out the 1.5 volt light bulb in the test light.

VOLTMETER

A voltmeter is used to measure voltage at any point in a circuit, or to measure the voltage drop across any part of a circuit. Voltmeters usually have various scales on the meter dial and a selector switch to allow the selection of different test ranges. The voltmeter has a positive and a negative lead. To avoid damage to the meter, connect the negative lead, usually black, to the negative (-) side of circuit or to ground. Connect the positive lead, usually red, to the positive (+) or power side of the circuit.

A voltmeter can be connected either in parallel or in series with a circuit and has a very high resistance to current flow. When connected in parallel, only a small amount of current will flow through the voltmeter current path; the rest will flow through the normal current path and the circuit will work normally. When the voltmeter is connected in series with a circuit, only a small amount of current can flow through the circuit.

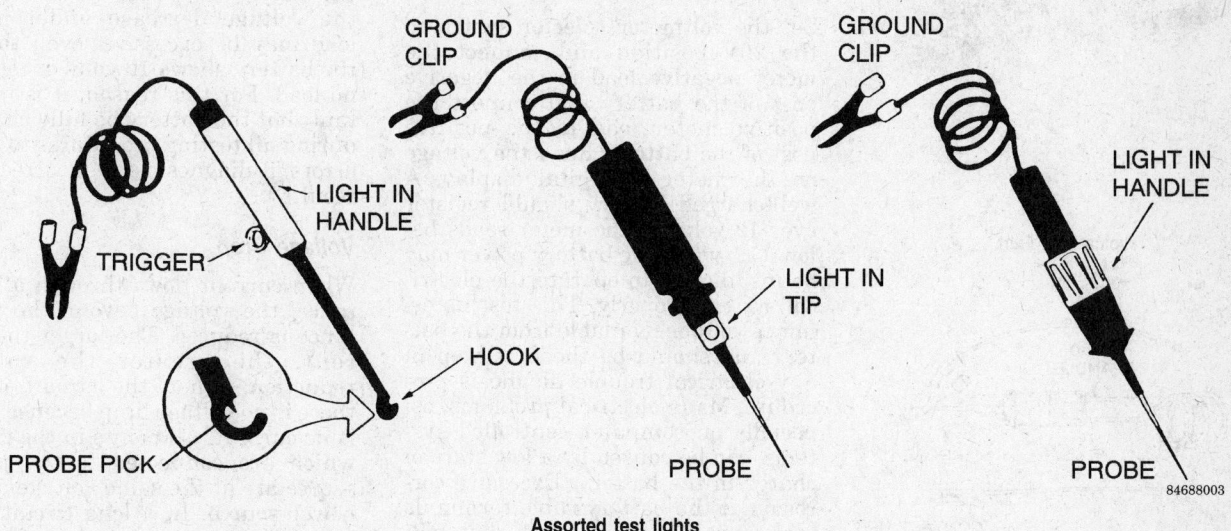

Assorted test lights

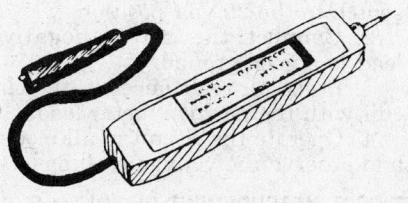

Logic probe type tester

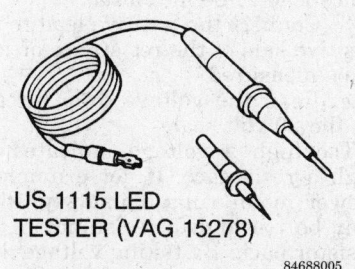

US 1115 LED TESTER (VAG 15278)

84688005

LED type test light for use on computer circuits

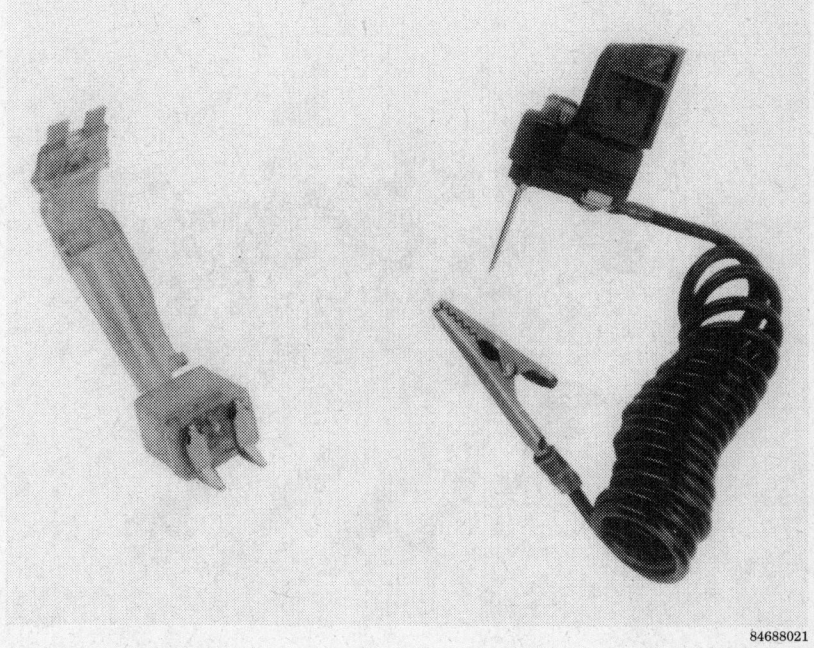

The device on the left is a fuse checker and the test light on the right is an LED type for use on computer circuits

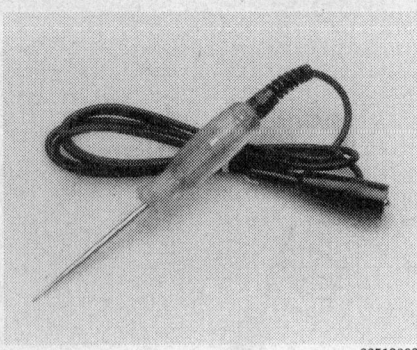

Typical test light

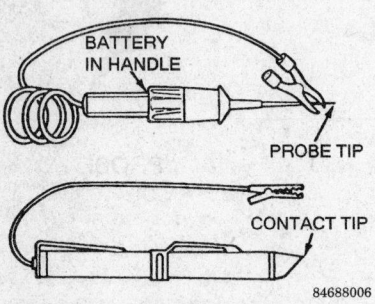

Types of self-powered test lights

The circuit will not work properly, but the voltmeter reading will show if the circuit is complete or not.

Available Voltage Measurement

Set the voltmeter selector switch to the 20V position and connect the meter negative lead to the negative post of the battery and connect the positive meter lead to the positive post of the battery. Read the voltage on the meter or digital display. A well-charged battery should register over 12 volts. If the meter reads below 11.5 volts, the battery power may be insufficient to operate the electrical system properly. This test determines voltage available from the battery and should be the first step in any electrical trouble diagnosis procedure. Many electrical problems, especially on computer controlled systems, can be caused by a low state of charge in the battery. Excessive corrosion at the battery cable terminals can cause a poor contact that will prevent proper charging and full battery current flow.

Nominal battery voltage is 12 volts, but, when fully charged, should be about 13.2 volts. When the battery is supplying current to 1 or more circuits it is said to be under load. When everything is **OFF** the electrical system is under a no-load condition. A fully charged battery showing about 12.5 volts at no load, may drop to 12 volts under medium load and will drop even lower under heavy load. If the battery is partially discharged, the voltage decrease under heavy load may be excessive, even though the battery shows 12 volts or more at no load. For this reason, it is important that the battery be fully charged during all testing procedures to avoid errors in diagnosis and incorrect test results.

Voltage Drop

When current flows through a resistance, the voltage beyond the resistance is reduced. The larger the current, the greater the voltage reduction. When the circuit is off, there is no voltage drop because there is no current. All points in the circuit which are connected to the power source are at the same voltage as the power source. In a long circuit with many connectors, a series of small, unwanted voltage drops due to corrosion at the connectors can add up to a total loss of voltage which impairs the operation of the loads in the circuit.

INDIRECT COMPUTATION OF VOLTAGE DROPS

1. Set the voltmeter selector switch to the 20 volt position.
2. Connect the meter negative lead to a good ground.
3. Probe all resistances in the circuit with the positive meter lead.
4. Operate the circuit in all modes and observe the voltage readings.

DIRECT MEASUREMENT OF VOLTAGE DROPS

1. Set the voltmeter switch to the 20 volt position.
2. Connect the voltmeter negative lead to the ground side of the resistance load to be measured.
3. Connect the positive lead to the positive side of the resistance or load to be measured.
4. Read the voltage drop directly on the 20 volt scale.

Too high a voltage indicates too high a resistance. If, for example, a blower motor runs too slowly, there may be too high a resistance in the resistor pack. By taking voltage drop readings in all parts of the circuit, the problem can be isolated. Too low a voltage drop indicates too low a resistance. If, for example, a blower motor runs too fast in the **MED** and/or **LOW** position, the problem can be isolated to the resistor pack by taking

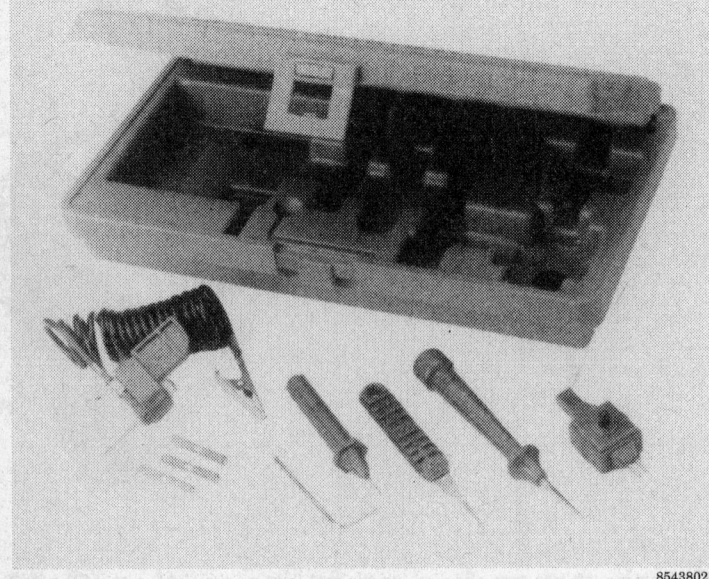

This computer circuit testing kit includes LED test lights that are safe for use on electronic circuits.

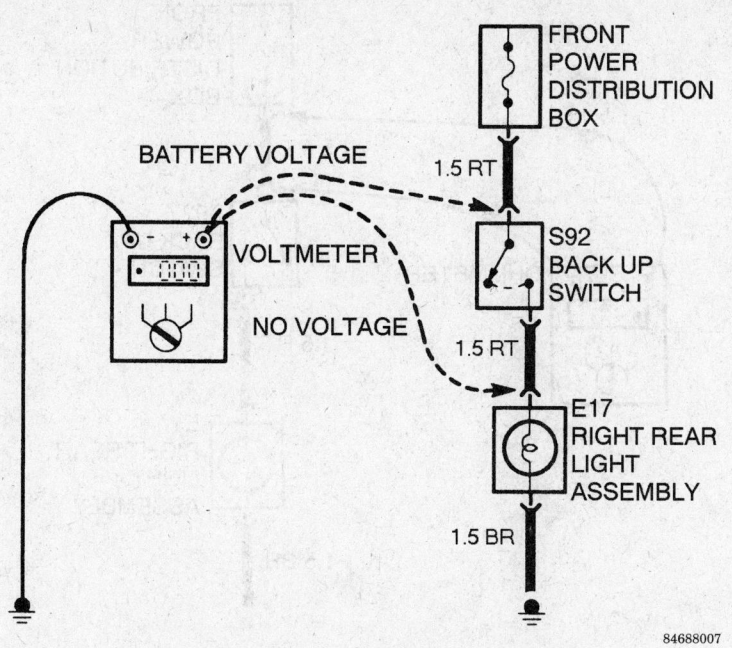

FRONT
POWER
DISTRIBUTION
BOX

BATTERY VOLTAGE

1.5 RT

VOLTMETER

S92
BACK UP
SWITCH

NO VOLTAGE

1.5 RT

E17
RIGHT REAR
LIGHT
ASSEMBLY

1.5 BR

84688007

Measuring voltage at different points in the circuit

voltage drop readings in all parts of the circuit to locate a possibly shorted resistor. The maximum allowable voltage drop under load is critical, especially if there is more than one high resistance problem in a circuit; all voltage drops are cumulative. A small drop is normal due to the resistance of the conductors.

High Resistance Testing

1. Set the voltmeter selector switch to the 2 volt position.
2. Connect the voltmeter positive lead to the positive post of the battery.
3. Turn **ON** the headlights and heater blower to provide a load.
4. Probe various points in the circuit with the negative voltmeter lead.
5. Read the voltage drop. Some average maximum allowable voltage drops are:

    Fuse panel — 0.7 volts
    Ignition switch — 0.5 volts
    Headlight switch — 0.7 volts
    Ignition coil (+) — 0.5 volts
    Any other load — 0.5–1.3 volts

NOTE: Voltage drops are all measured while a load is operating; without current flow, there will be no voltage drop.

OHMMETER

The ohmmeter is designed to read resistance (ohms) in a circuit or component. Although there are several different styles of ohmmeters, all will usually have a selector switch which permits the measurement of different ranges of resistance. Usually the selector switch allows the multiplication of the meter reading by 10, 100, 1000 or 10,000. A calibration knob allows the meter to be set at zero for accurate measurement. Since all ohmmeters are powered by an internal battery (usually 9 volts), the ohmmeter can be used as a self-powered test light. When the ohmmeter is connected, current from the ohmmeter flows through the circuit or component being tested. Since the ohmmeter's internal resistance and voltage are known values, the amount of current flow through the meter depends on the resistance of the circuit or component being tested.

The ohmmeter can be used to perform continuity tests for opens or shorts and to read actual resistance in a circuit. It should be noted that the ohmmeter is used to check the resistance of a component or wire while there is no voltage applied to the circuit. Current flow from an

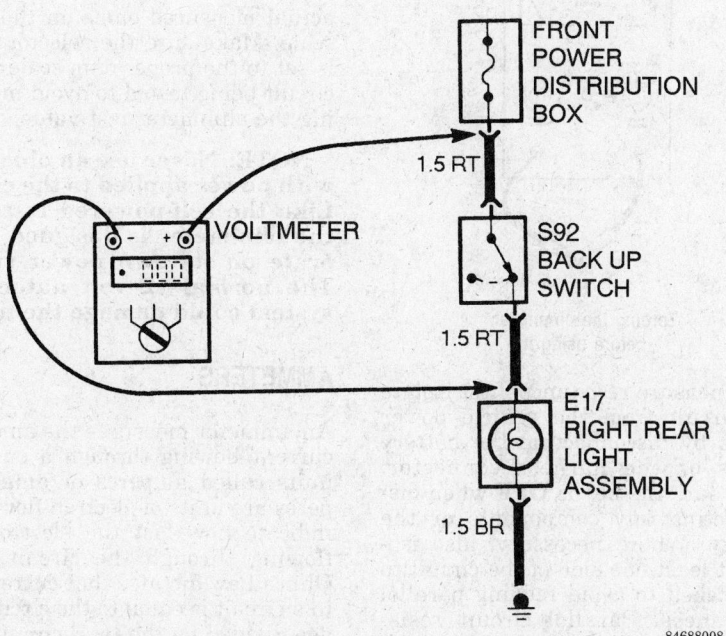

FRONT
POWER
DISTRIBUTION
BOX

1.5 RT

VOLTMETER

S92
BACK UP
SWITCH

1.5 RT

E17
RIGHT REAR
LIGHT
ASSEMBLY

1.5 BR

84688008

Checking for the voltage drop across a component in the circuit

30-9

outside voltage source (such as the vehicle battery) can damage the ohmmeter, so the circuit or component should be isolated from the vehicle electrical system before any testing is done. Since the ohmmeter uses its own voltage source, either lead can be connected to any test point.

NOTE: When checking diodes or other solid state components, the ohmmeter leads can only be connected one way in order to measure current flow in a single direction. Make sure the positive (+) and negative (-) terminal connections are as described in the test procedures to verify the one-way diode operation.

When using the meter for continuity checks, do not be concerned with the actual resistance readings. Zero resistance, or any resistance reading, indicates continuity in the circuit. Infinite resistance indicates an open in the circuit. A high resistance reading where there should be none indicates a problem in the circuit. Checks for short circuits are made in the same manner as checks for open circuits except that the circuit must be isolated from both power and normal ground. Infinite resistance indicates no continuity to ground, while zero resistance indicates a dead short to ground.

Resistance Measurement

The batteries in an ohmmeter may be affected by temperature and will weaken with age. The ohmmeter must be calibrated or "zeroed'" before taking measurements. To zero the meter, place the selector switch in its lowest range and touch the 2 leads together. Turn the calibration knob until the meter needle is exactly on zero.

NOTE: All analog (needle) type ohmmeters must be zeroed before use, but some digital ohmmeter models are automatically calibrated when the switch is turned ON. Self-calibrating digital ohmmeters do not have an adjusting knob, but it's a good idea to check for a zero readout before use by touching the leads together. All computer controlled systems require the use of a digital ohmmeter with at least 10 megohms impedance for testing. Before any test procedures are attempted, make sure the ohmmeter used is compatible with the electrical system, or damage to the on-board computer could result.

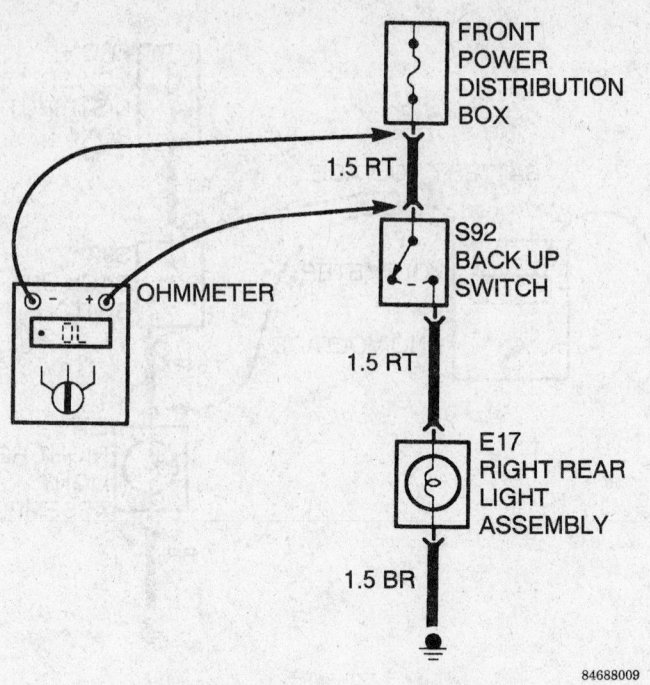

Using an ohmmeter to do a continuity test

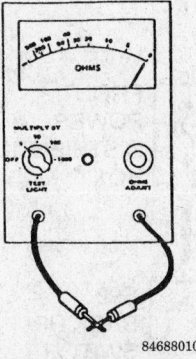

Zeroing the ohmmeter before using it

To measure resistance, first isolate the circuit from the vehicle power source by disconnecting the battery cables or the harness connector. Make sure the key is **OFF** when disconnecting any components or the battery. Where necessary, also isolate at least one side of the circuit to be checked to avoid reading parallel resistances. Parallel circuit resistances will always give a lower reading than the actual resistance of either of the branches. When measuring the resistance of parallel circuits, the total resistance will always be lower than the smallest resistance in the circuit. Connect the meter leads to both sides of the circuit (wire or component) and read the actual measured ohms on the meter scale. Make sure the selector switch is set to the proper ohm scale for the circuit being tested to avoid misreading the ohmmeter test value.

NOTE: Never use an ohmmeter with power applied to the circuit. Like the self-powered test light, the ohmmeter is designed to operate on its own power supply. The normal 12 volt automotive system could damage the meter.

AMMETERS

An ammeter measures the amount of current flowing through a circuit in units called amperes or amps. Amperes are units of electron flow which indicate how fast the electrons are flowing through the circuit. Since Ohm's Law dictates that current flow in a circuit is equal to the circuit voltage divided by the total circuit resistance, increasing voltage also increases the current level (amps). Likewise, any decrease in resistance will increase the amount of amps in a circuit. At normal operating voltage, most circuits have a characteristic amount of amperes, called "current

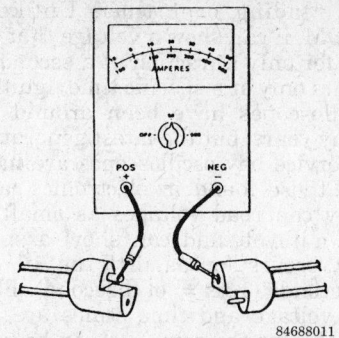

The ammeter is placed in line with the circuit to be tested

84688011

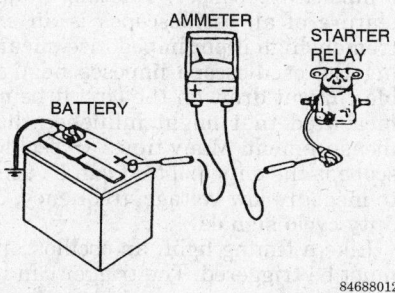

Checking the draw of the starter relay with an ammeter

84688012

draw'" which can be measured using an ammeter. By referring to a specified current draw rating, measuring the amperes, and comparing the 2 values, one can determine what is happening within the circuit to aid in diagnosis. An open circuit, for example, will not allow any current to flow so the ammeter reading will be zero. More current flows through a heavily loaded circuit or when the charging system is operating.

An ammeter is always connected in series with the circuit being tested. All of the current that normally flows through the circuit must also flow through the ammeter; if there is any other path for the current to follow, the ammeter reading will not be accurate. The ammeter itself has very little resistance to current flow and therefore will not affect the circuit, but it will measure current draw only when the circuit is closed and electricity is flowing. Excessive current draw can blow fuses and/or drain the battery; a reduced current draw can cause motors to run slowly, lights to dim and other components to operate improperly.

DIGITAL VOLT-OHM METER (DVOM)

As its name implies, this tool combines a voltmeter and an ohmmeter into a single unit that has a digital display instead of a scale and pointer. The major advantage of a fully electronic meter is that there are no moving parts that require power to operate. Analog meters with an ultra light weight needle still require some power to move the needle. This limits the range and the features that can be built into the meter. Even the most basic DVOM can read a much greater range of voltage and resistance without imposing any load on the circuit being tested. It is usually the only equipment suitable for testing computer controlled circuits and is often the only test equipment needed.

Several additional features can be built into the same unit, such as circuitry for testing diodes and measuring AC voltage, AC and DC current, temperature, duty cycle, frequency, pulse width, dwell, and rpm. Some of the more sophisticated units also have storage capability, bar graph display, automatic shut-off, and can display the difference between two readings. A top-of-the-line DVOM designed for automotive testing is prob-

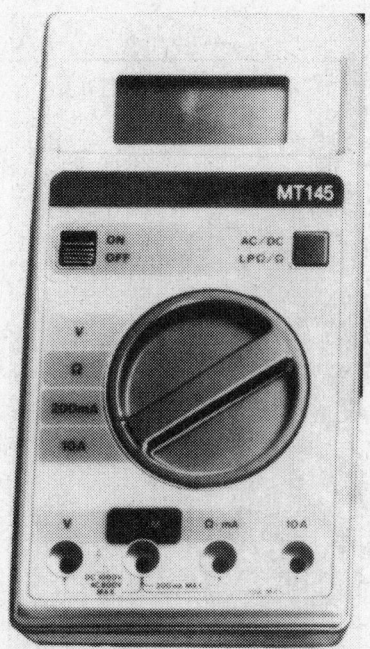

Different styles of multi-meters allow you a choice of meter functions

88513005

ably the most useful and cost effective diagnostic tool available. Be sure to buy a unit with a high impedance, usually 10 megohms or higher.

Specialty Testers

FREQUENCY PROCESSOR

Some older DVOMs are not equipped to read frequency. There is at least one unit on the market that converts frequency signals to a millivolt signal that any DVOM can read. It is a simple box with input and output jacks and a "wake-up" circuit that automatically turns the unit on when needed. Its range of 10–5000 Hz makes it useful for checking rpm sensors, mass air flow sensors, Hall effect sensors and more. Instructions provided with the processor show how to interpret the readings.

BREAK-OUT BOX

The electronic Break-Out Box (BOB) is used to tap into the wiring of a control unit. The main connector to the electronic control unit is connected to the break-out box and another wire harness is connected from the box to the control unit. The break-out box then allows the technician to access each circuit while it is operating without piercing the wire or causing damage to the connectors. All testing with the DVOM can be done safely at these terminals, eliminating the risk of damage due to backprobing at the control unit. Many times a break-out box is the only way to test a control unit function.

An Intelligent Break-Out Box (IBOB) connects to the vehicle diagnostic connector and has connector ports for a scan tool and/or a computer. On earlier electronic control units that do not generate a data stream, an IBOB will collect input/output data while the engine is running and present it to a scan tool or PC. Additionally, some manufacturers provide plastic overlays for the break-out box. This allows the box to be used on a variety of models; different overlays identify the changes in wire use or labeling. With the proper cable adaptors, an IBOB can be used with any engine, body or ABS control unit on any vehicle.

OSCILLOSCOPE

An oscilloscope is a voltmeter that presents a graphic picture of the volt-

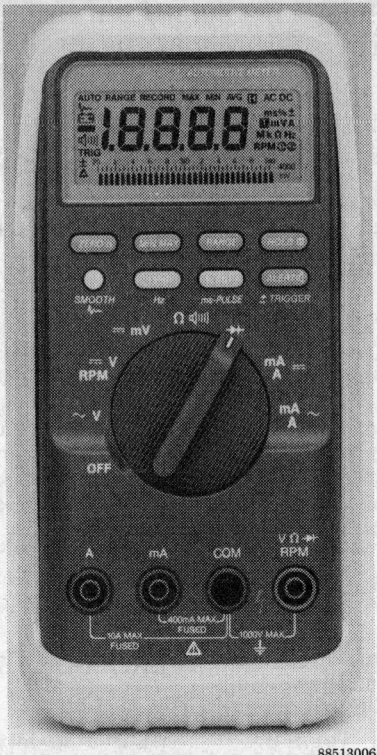

88513006

A good quality DVOM designed for automotive testing is the most useful diagnostic tool available

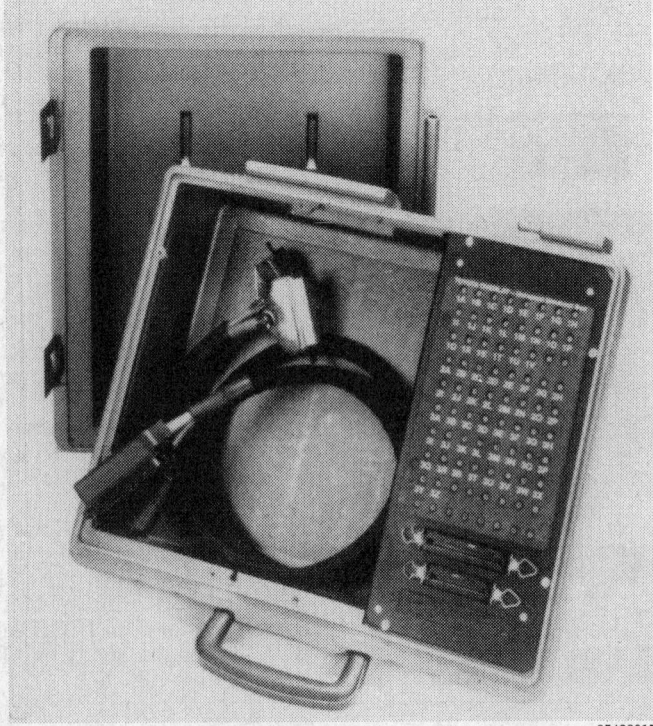

85438015

A break-out box makes it possible to T into control unit circuits

age reading over time. Unlike a DVOM, it can show a voltage that exists for only a fraction of a second or occurs only at a specific time. Ignition oscilloscopes have been around for many years, but the latest generation of service bay oscilloscopes are more like those found in electronics labs. They can read voltages as small as one millivolt and can show a spike that occurs for as little as 10 nanoseconds (1 ns = of a second). Both the voltage and time scales are adjustable, so the same tool can be used to measure the fast, high voltage signal of the secondary ignition system and slow stable signals such as a temperature sensor. Another major feature of all oscilloscopes is an extremely high input impedance, meaning the oscilloscope imposes negligible current draw on the circuit being measured that might influence that measurement. Many times an oscilloscope is the only tool that can be used to measure low voltage, frequency, or duty cycle signals.

Like a timing light, an oscilloscope must be triggered. The trigger can be internal (automatic) or can come from an external source. On a multi-channel oscilloscope, displaying the external trigger signal can show the timing of two events. For example, by taking the trigger from a suspected faulty fuel injector, it is possible to see the oxygen sensor signal only at the time of that injection event. The voltage level required to trigger the oscilloscope can also be adjusted, providing a simple method to look for low level or intermittent faults that may not set a code.

A digital oscilloscope converts the analog input signal to a digital form. A digital signal can be stored and played back by itself or along with another trace. Some units can also display the signal as numbers, minimum/maximum values, change value and average value. If the oscilloscope is equipped with a computer port, the digitized traces and other data can also be down-loaded to save and/or print out. There is computer software available to aid organization and analysis of wave forms.

With its extremely fast sampling rate and graphic display, an oscilloscope can easily show a malfunction that occurs too fast for a voltmeter to show. For example, by adjusting the time sweep of the oscilloscope to show the full up-and-down stroke of a throttle position sensor, an intermittent fault in the signal can be clearly shown. A voltmeter may also detect the fault but cannot change the dis-

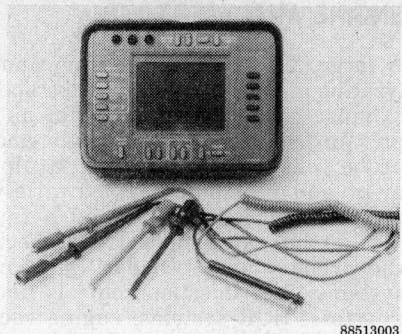

An oscilloscope shown with related testing probes

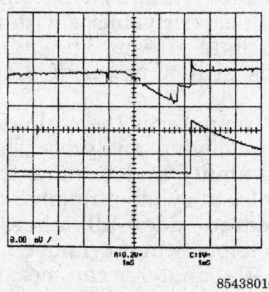

The oxygen sensor trace (top) shows a delayed cross-over coinciding with an injector pulse (bottom trace)

play fast enough to show the resistance spike. A storage oscilloscope with min/max value capability can locate intermittent faults that even other oscilloscopes cannot.

Any oscilloscope used for automotive testing must be designed for use with automobiles. Standard lab oscilloscopes are usually not able to cope with the relatively harsh automotive electronic environment. Automotive oscilloscopes are available in a variety of types with a variety of features. Some are portable hand held models that operate on batteries or vehicle

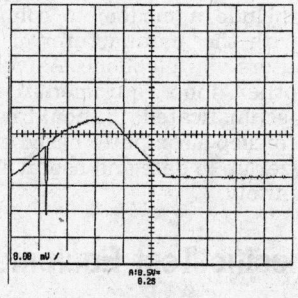

An intermittent fault in the throttle position sensor shows clearly on the scope trace

power. Even though they are small, the newest portable units include multi-trace and storage capabilities and are rugged enough to be used under the hood or on road tests. Larger more powerful models mounted in a console are often part of a top-of-the-line engine analyzer package. Most of the major diagnostic tool manufacturers produce at least one oscilloscope model.

As vehicles become more sophisticated and electronic controls become more powerful, an oscilloscope is fast becoming a necessary diagnostic tool. When the technician becomes proficient with an oscilloscope, many other diagnostic tools become unnecessary.

SCAN TOOL

This is the generic name for portable diagnostic equipment that communicates directly with an electronic control unit. The major vehicle manufacturers each have their own scan tool that is used by dealership technicians. Some of these are available through the dealer parts network or are sold outside the network under another name. Others are available only to authorized dealerships.

Scan tools are used to read and erase trouble codes stored in the control unit memory and to provide a direct data transfer link with the control unit's On Board Diagnostic (OBD) system. Reading the control unit memory through the scan tool is more complete than reading codes with the flashing light on the instrument panel. Some information is only available through the scan tool, such as the number of engine starts since the fault first appeared. Data transfer provides a real time display of control unit input/output signals. Data such as the oxygen sensor reading or idle control motor duty cycle can be displayed while the engine is

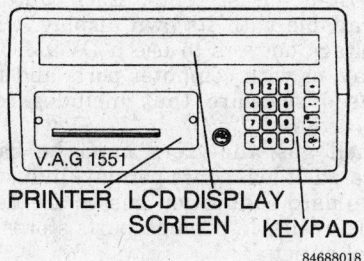

PRINTER LCD DISPLAY
SCREEN   KEYPAD

Scan tools allow the testing of electronic control units

running. The scan tool can also be used as a volt/ohmmeter to check selected circuits without disconnecting them.

Some scan tools are designed to simulate sensor inputs to test the sensor circuit, the control unit and the output device. On many vehicles, the scan tool can communicate with every control unit on the vehicle through a single diagnostic connector. Some of the more advanced scan tools are equipped with a data memory to store test data during a road test. The test data is down loaded into a PC through an RS232 computer port, greatly increasing the processing power and expanding the amount of memory and information available.

Aftermarket scan tools require adaptors to match the different vehicle diagnostic connectors. A cartridge that plugs into the scan tool contains software needed for communicating with the different control units. The software is the tool's real power and is continuously evolving to enhance its capabilities. As the tool manufacturer's data base has grown, software now includes VIN-specific information that addresses some of the most common trouble codes and driveability problems that don't always generate codes. Tests are menu-driven and many of the specifications are included right there in the program. Depending on the vehicle and the amount of computer control used, systems which may be viewed or investigated with a scan tool include:

1. Engine controllers/ECUs
2. Fuel/ignition systems
3. Electronic transmission control
4. Charging system
5. Suspension control functions
6. Anti-Lock brake system.
7. Passive restraint system
8. Anti-theft system
9. Climate Control Systems
10. Body electrical systems — including power locks, entertainment systems, sunroofs, and defoggers

Aftermarket scan tools work well with most vehicles, but no scan tool can be used on all models and all are limited in their ability to communicate with European models. The Federal On-Board Diagnostic (OBD) specification requires all vehicles to have the same diagnostic connector and diagnostic trouble codes and to use the same data transfer language. This makes it possible to use a single scan tool to communicate with all engine control units from all manufacturers. Some manufacturers began production of OBD vehicles for the

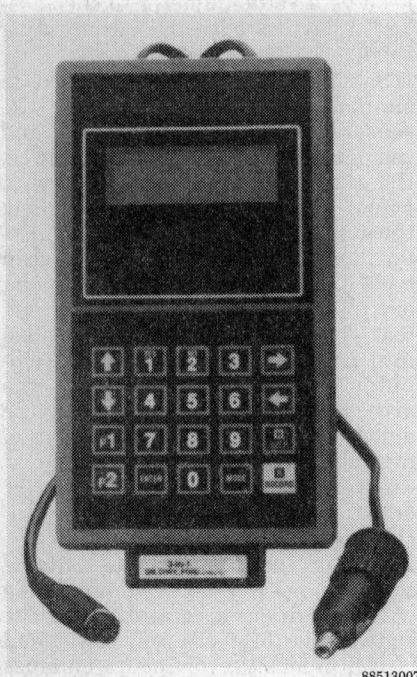

Aftermarket scan tool with a program module installed

1994 model year and all new vehicles must comply by the 1998 model year. As vehicle control units and scan tools become more powerful, data acquisition and test capabilities will also improve dramatically with each new generation of control unit and scan tool software.

EXHAUST GAS ANALYZERS

Exhaust gas analysis has long been an extremely valuable and versatile diagnostic tool. It can be used to troubleshoot fuel and ignition systems, locate vacuum leaks or EGR malfunctions, even diagnose mechanical problems such as worn valve

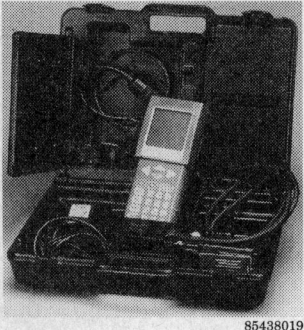

This tool can scan all OBD II-specific vehicles

guides. On most vehicles, it is the only way to accurately check air/fuel mixture.

The federal government regulates three exhaust gas components: hydrocarbons (HC), carbon monoxide (CO), and oxides of nitrogen (NOx). HC and CO are relatively easy to measure and have long been tested in states that require emissions inspections. Measuring carbon dioxide (CO_2) and oxygen (O_2) are also valuable diagnostic aids, but NOx cannot be accurately measured at idle or no-load conditions. However as states enact tighter Inspection and Maintenance programs, service bay NOx analyzers and test procedures are being developed. For complete diagnostic and certification testing, a five-gas analyzer is required.

Most gas analyzers include a single tail pipe sample probe, sample pump and filtration system and a detector cell for each of the gasses being measured. Most stand-alone four-gas analyzers used for emissions inspections are equipped with a small microprocessor that includes testing and calibration programs, a self diagnostic program and a built-in printer. Other units are designed as part of a complete diagnostic station and are connected to a PC based computer, printer and monitor screen. There is at least one portable four-gas analyzer that is used along with the same manufacturer's scan tool. The scan tool's software guides the user through test procedures based on the gas and sensor readings.

There is also a series of small, hand-held oxygen and CO monitors available that do not require a sample pump and filter system. The measuring cell is built into the tail pipe probe and the monitor is battery operated. Since there is only a wire between the probe and the monitor, these units can easily be used on a road test. They are also equipped with a memory that can record three minutes of test data. The CO monitor can be particularly useful for routine air/fuel adjustments. Each unit is available with its own display, with voltage outputs to use a DVOM display, or with computer ports and PC based software that includes test procedures.

All gas analyzers must be calibrated at least once per day. Industry standard calibration gasses are usually available through parts stores or tool outlets.

ENGINE ANALYZERS

A large, fully-equipped engine analyzer usually includes an oscilloscope, exhaust gas analyzer, vacuum and pressure sensors, a timing light and probe, electrical measuring equipment, and a computer. With a variety of electrical connections and a tail pipe probe, this analyzer can check the primary and secondary ignition systems, fuel injection controls and injectors, EGR systems, engine vacuum and compression, and the starting and charging systems. The computer can be used to read the engine control unit's data stream through the vehicle diagnostic connector. Even on earlier vehicles with no sensors or data stream, an engine analyzer is still a powerful diagnostic tool.

The computer is the real power behind an engine analyzer. The computer's ability to determine the ignition pulse at cylinder number one can be used to index all other engine events to particular cylinders. For example, the analyzer can measure the starter current needed to move each piston to TDC, indicating the relative compression of each cylinder. The analyzer can also display spark plug firing voltage and duration on the oscilloscope. A spark plug that requires more voltage with less duration could indicate a faulty injector. The analyzer can detect and clearly show all the differences between cylinders. The computer can help the technician diagnose the data, determine the necessary repairs and even provide a print-out to present a clear explanation to the vehicle owner.

These analyzers are a major investment and are well supported by the manufacturer. They are frequently updated with a computer floppy disc that includes new vehicle information and test procedures. Some machines include CD-ROM equipment to read service manuals that are available on disc. They may also include a modem to communicate with the manufacturer or other computers via telephone. As vehicles and other shop equipment become more sophisticated, it should be possible to keep a computer based engine analyzer up to date and useful almost indefinitely.

Specific Test Equipment

There are many special diagnostic tools for testing individual components or systems, such as a Hall effect sensor, idle air control motor, fuel injectors, secondary ignition systems, and others. Most are designed for use on as many vehicles as possible. Some are designed to test parts or systems on specific vehicles. Generally these devices allow the technician to quickly test components or sub-systems without going through a long diagnostic procedure. However there is a risk of incorrect diagnosis. These tools can only be dependable if the technician is familiar with their use and understands what the test results really mean. A simple vacuum leak or loose connection may produce the same test result as a faulty component.

LEAK DETECTORS

A battery powered, hand-held vacuum leak detector uses a microphone and amplifier that detects noise in the ultrasonic range. Air moving through a vacuum leak will generate sound waves in the 40 kHz range, well above the range of human hearing. The detector will sound a beeper when a leak is found. Because of the high frequency sensed by the detector, it is not generally affected by normal engine or shop noises.

Leak detectors for air conditioning systems have a vacuum pump and probe to draw an air sample into the

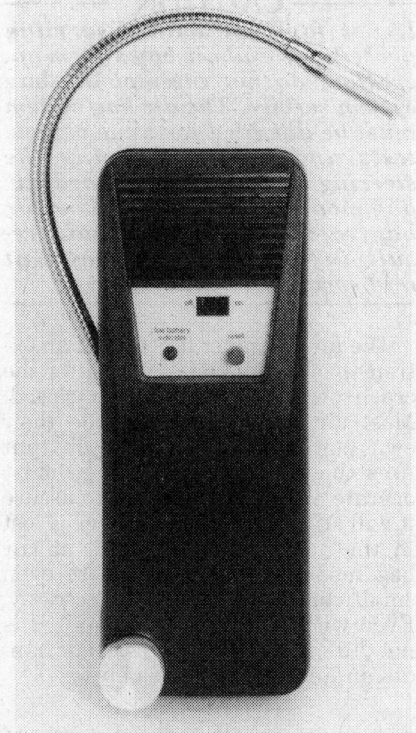

88513008

An ultrasonic vacuum leak detector is unaffected by engine or shop noise

detecting cell. The cell detects halogen gas that is common to all air conditioning refrigerants. Most are capable of indicating the type of refrigerant in the system, as well as the rate of leakage. There are battery powered hand-held models and larger AC powered units suitable for mounting on an air conditioning service cart. The newer models are capable of detecting R-134a and the sensitivity can be adjusted for possible background interference.

A combustible gas leak detector reacts to hydrocarbons present in fuels, exhaust gases, coolants, and lubricants. Models with adjustable sensitivity are typically used to look for fuel vapor leaks, head gasket leaks, and to measure the amount of exhaust leaking into the interior of a vehicle. With some imagination, this can be an extremely useful tool.

PYROMETER

A pyrometer measures a wide range of temperatures with a probe that only needs to touch the item being measured. As a general diagnostic tool, a hand held pyrometer can quickly locate hot or cold spots in a cooling system, a seized brake caliper, a dry bearing, test heater and A/C performance or even find a weak cylinder by measuring exhaust manifold runner temperatures. Most pyrometers are available with special probes for penetration and for measuring tire temperatures. There are even optical infrared non-contact pyrometers that measure temperature by the heat emission of a surface. This is useful as the surface to be measured does not have to actually be touched by a probe. They can be calibrated quickly, have a very wide temperature range and can usually be switched to display temperature in either Fahrenheit or Celsius.

IDLE AIR CONTROL TESTER

This is a kit used to isolate and test idle air control solenoids, motors, and signals. Some are made for use with a specific system, others include adaptors for use with many different vehicles. The device can activate solenoid valves and control motors to test the full range of motion with the engine not running. It can also be used to control idle speed for timing adjustment or other engine tests. Some can also check the control unit output signal to the idle air control motor. Although these functions can also be accomplished with scan tools, the idle

air control tester can be faster and easier to use for some tests.

FUEL INJECTOR TESTER

This device can quickly check the coil resistance and current draw of an electric fuel injector while it is under load. Each injector is tested individually and the results are reported on a DVOM or oscilloscope. This information makes it possible to electronically check injector balance and detect intermittent faults. When used with equipment that measures fuel pressure and injection quantity, every function of the fuel system can be tested.

OXYGEN SENSOR TESTER

This kit usually includes a propane enrichment control valve, special connectors and test instructions, and the hose and fittings needed for connecting the valve to an intake manifold. The kit allows the technician to control air/fuel mixture and check the oxygen sensor response time. When the oxygen sensor is disconnected, forcing the control unit into open loop, sensor output voltage or resistance can be read with a DVOM. The instructions also include procedures for testing the control unit's response to the oxygen sensor signal.

SENSOR SIMULATOR

This device is used to take a sensor "out of the loop" and simulate its input signals to the control unit. It can simulate every type of voltage, resistance, and frequency signal one at a time to test the control unit's response to the input. The simulator can also measure any sensor output signal by back-probing the sensor connector. In addition to displaying the reading directly, some units can also output the reading to an oscilloscope, scan tool, or other diagnostic equipment.

POWER STEERING TESTER

A power steering tester can quickly confirm that the hydraulic system is functioning properly and indicate excessive loads on the system due to mechanical malfunction in the suspension or steering linkage. The mechanical tester consists of a gauge, a heavy duty hose, and adapters for various models. Newer testers use an electronic pressure transducer that converts the pressure into an electri-

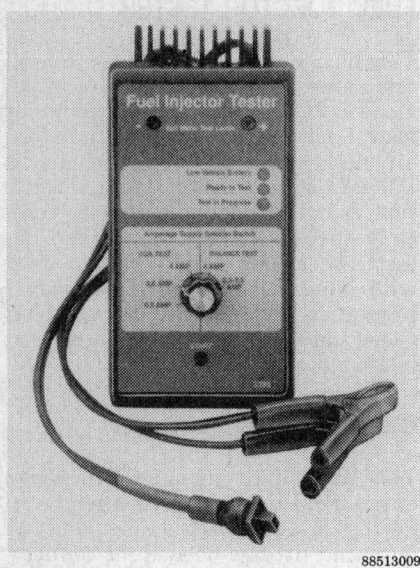

A fuel injector tester measures voltage drop while the injector is being activated

cal signal. This allows the technician to road test variable effort power steering systems with a DVOM.

ELECTRONIC SIGHT GLASS

This device, used for troubleshooting air conditioning systems, includes two transducers and a battery operated meter. The transducers attach to the outside of the metal air conditioning lines without disconnecting

85438023

A power steering system testing gauge can quickly isolate hydraulic or linkage problems

the lines. While the A/C system is operating, the device ultrasonically detects bubbles in the refrigerant and uses an LED display to simulate a sight glass, allowing the technician to actually "see" the bubbles in the system. The transducers can be fitted to any metal line at almost any point in the system, making it possible to test expansion valves and capillary tubes, or find other undesired restrictions.

ANTI-LOCK BRAKE TESTER

Anti-lock brake system control units are equipped with a self-diagnostic program that checks the system at engine start-up and de-activates the system if a fault is detected. Most control units also store diagnostic trouble codes. An ABS tester is basically a scan tool used to read and erase trouble codes and to provide a data link with the ABS control unit. The data link allows the tester to check sensor inputs and activate each solenoid and control valve to test the output system.

The testers for some of the early anti-lock braking systems are usually available only to dealers and function only with that manufacturer's vehicles. The aftermarket scan tool makers have developed the necessary adapters and software cartridges so that most scan tools used to communicate with engine control units can also be used on ABS control units. Unfortunately there are major differences in ABS designs and no scan tool is able to communicate with all ABS control units.

The most complete ABS test equipment is a software package, connector and interface pod that establishes

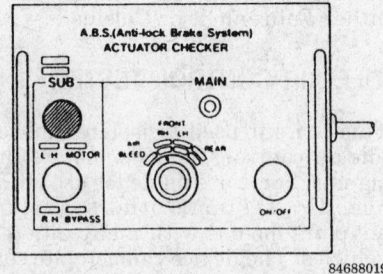

84688019

An ABS checker allows the testing of the anti-lock brake system

a data link between the control unit and a PC-based computer in an engine analyzer or alignment station. This software uses the power of the computer to completely check the control unit and "bench test" every component of the system in about one minute. It is suitable for use with every anti-lock braking system on the market and the software can be updated as required to keep the system current.

Even without a scan tool, all of the system's sensors, solenoids, and actuators can still be individually checked with a DVOM and an oscilloscope. The trouble codes and other control unit functions can only be accessed with the correct scan tool or tester.

PASSIVE RESTRAINT TESTER

While there are some differences, the air bag system in most vehicles operates basically the same way. This has made it possible to develop a unit that will test the circuits while they are fully connected and operating. It uses LEDs to read out circuit continuity, power supply, switch state and output device state. On some scan tools, a software cartridge gives the tool the ability to test the passive restraint system.

CAUTION

If not familiar with disarming procedures and air bag system operation, do not attempt air bag system service. The air bag system must be disabled for some vehicle tests and before removing the steering wheel or dashboard air bag module. Failure to follow air bag safety procedures could result in accidental deployment and serious or fatal injury.

The air bag must function in an extremely short period of time after the crash sensor switches have closed. Most air bag systems include their own power supply. The squib that fires the gas generator charge must operate at low power levels to assure it will still fire if vehicle power is lost in the accident. This makes an air bag module quite sensitive to even small currents like static electricity. Even with such a "hair trigger", it is not difficult to test the system or handle an air bag module safely.

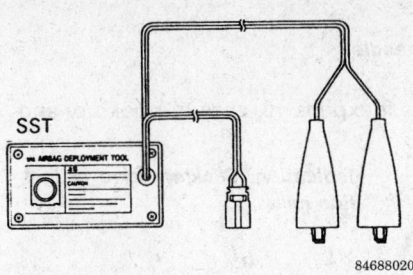

SST

84688020

An air bag deployment tool used when disposing of an air bag

Air Bag Service Precautions

—— CAUTION ——

An air bag is an explosive device. Before beginning any air bag system service, disconnect and isolate the negative battery cable and backup power supply. Follow the procedures for disarming the air bag system exactly. Do not use any type of computer memory saver device. Do not use any self powered test equipment or test lights until the system is properly disabled. Failure to follow these precautions could result in accidental deployment of the air bag and possible serious or fatal injury.

• Disconnect and isolate the battery cable and any backup power supply when servicing the air bag system.

• Do not use a memory saver.

• When re-activating an air bag system, connect the power last and make sure no one is in the vehicle.

• Do not attempt to measure the resistance across the air bag module connectors with any type of ohmmeter. The ohmmeter battery can fire the air bag module.

• When working on air bag components in the passenger compartment, try to work away from where an air bag would deploy. Accidental deployment of the air bag against a body that is not in the proper position could result in severe or fatal injury.

• A removed air bag module must be placed away from sources of heat, sparks, or electricity, including static electricity.

• A removed air bag module must be placed with the cover pad facing up, so that accidental deployment will not launch the module into the air. Also be sure to carry the module with the cover pad facing away from the body.

• When the air bag module is removed, place it away from loose objects that would be thrown in the event of accidental deployment.

• When removing a steering column with the steering wheel attached, pay attention to where the air bag is aimed. Never stand the column on the face of the steering wheel. Lock the column to avoid damage to the clockspring.

• When handling an air bag which has been deployed, there may be a powdery residue which, though mostly talc, may irritate skin, eyes, and breathing passages. Wear gloves, glasses, and a dust mask while wrapping the deployed air bag in a plastic bag for disposal.

• Sensor positioning is critical for proper system operation. If the sensor is in an area of vehicle damage, replace the sensor whether or not the air bag deployed. The proper torquing of the sensor retainers is critical.

• Any part of the air bag system found to be faulty must be replaced. No part can be repaired, not even the wiring.

• All diagnostic work is to be done with the air bag module(s) removed. Air bag simulators are available and can be installed for a full functional test of the control unit.

Mechanical Test Equipment

VACUUM GAUGE

Intake manifold vacuum is used to operate various systems and devices on all cars. To correctly diagnose and solve problems in vacuum control systems, a vacuum source is necessary for testing. In some cases, vacuum can be taken from the intake manifold while the engine is running, but vacuum is normally provided by a hand vacuum pump.

Most gauges are graduated in inches of Mercury (in. Hg) or kilopascals (kPa), although a device called a manometer reads vacuum in inches of water (in. H2O). The vacuum reading usually varies between 18–22 in. Hg (60–74 kPa) at sea level. To test engine vacuum, the vacuum gauge must be connected to a source of manifold vacuum. Many engines have a plug in the intake manifold which can be removed and replaced with an adapter fitting. Connect the vacuum gauge to the fitting with a suitable rubber hose or, if no manifold plug is available, connect the vacuum gauge to any device using manifold vacuum, such as EGR valves. The vacuum gauge can be used to determine the amount of vacuum reaching a component.

HAND VACUUM PUMP

Small, hand-held vacuum pumps come in a variety of designs and provide a source of vacuum for testing components without the engine operating. Most have a built-in vacuum gauge and allow a component to be tested without removing it from the vehicle. Operate the pump lever or plunger, applying the correct amount of vacuum required for the test. The level of vacuum in inches of Mercury (in. Hg) or kilopascals (kPa) is indicated on the pump gauge. For some testing, an additional vacuum gauge may be necessary.

COMPRESSION GAUGE

A compression gauge measures the amount of pressure in pounds per square inch (psi) or kilopascals (kPa) that a cylinder is producing. Some gauges have a hose that screws into the spark plug hole while others have a tapered rubber tip which is held by hand in the spark plug hole. Engine compression depends on the sealing ability of the rings, valves, head gasket and spark plug gaskets. If any of these parts are not sealing properly, compression will be lost and the power output of the engine will be reduced. The compression in each cylinder should be measured and the variation between cylinders should be noted. The engine should be cranked through 5 or 6 compression strokes while warm, with all plugs removed, ignition disabled and throttle valves wide open.

FUEL PRESSURE GAUGE

A fuel pressure gauge is required to test the operation of the fuel delivery and injection systems. Some systems also need a 3-way valve to check the fuel pressure in various modes of operation. Gauges may require special adapters for making fuel connections. Always observe fuel system cautions when working around any pressurized fuel system.

USING A VACUUM GAUGE

White needle = steady needle   *Dark needle = drifting needle*

The vacuum gauge is one of the most useful and easy-to-use diagnostic tools. It is inexpensive, easy to hook up, and provides valuable information about the condition of your engine.

Indication: Normal engine in good condition

Gauge reading: Steady, from 17–22 in./Hg.

Indication: Sticking valve or ignition miss

Gauge reading: Needle fluctuates from 15–20 in./Hg. at idle

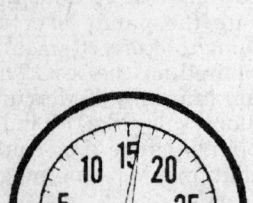

Indication: Late ignition or valve timing, low compression, stuck throttle valve, leaking carburetor or manifold gasket.

Gauge reading: Low (15–20 in./Hg.) but steady

Indication: Improper carburetor adjustment, or minor intake leak at carburetor or manifold

NOTE: Bad fuel injector O-rings may also cause this reading.

Gauge reading: Drifting needle

Indication: Weak valve springs, worn valve stem guides, or leaky cylinder head gasket (vibrating excessively at all speeds).

NOTE: A plugged catalytic converter may also cause this reading.

Gauge reading: Needle fluctuates as engine speed increases

Indication: Burnt valve or improper valve clearance. The needle will drop when the defective valve operates.

Gauge reading: Steady needle, but drops regularly

Indication: Choked muffler or obstruction in system. Speed up the engine. Choked muffler will exhibit a slow drop of vacuum to zero.

Gauge reading: Gradual drop in reading at idle

Indication: Worn valve guides

Gauge reading: Needle vibrates excessively at idle, but steadies as engine speed increases

88513010

A vacuum tool is a good gauge for diagnosing the general condition of an engine

WHERE TO START

Logical Diagnostic Procedures

Diagnosis of a driveability problem requires attention to detail and following the diagnostic procedures in the correct order. Resist the temptation to begin extensive testing before completing the preliminary diagnostic steps. The preliminary or visual inspection must be completed in detail before diagnosis begins. In many cases this will shorten diagnostic time and often cure the problem without the need for involved electronic testing.

There are two basic ways to check a vehicle engine for electronic problems. These are by symptom diagnosis and by the on-board computer self-diagnostic system. The first place to start is always the preliminary inspection. Intermittent problems are the most difficult to locate. If the problem is not present at the time of testing, the fault may not be able to be located.

PRELIMINARY INSPECTION

The visual inspection of all components is possibly the most critical step of diagnosis. A detailed examination of connectors, wiring and vacuum hoses can often lead to a repair without further diagnosis. Also, take into consideration if the vehicle has been serviced recently. Sometimes things are reconnected in the wrong place, or not at all. A careful inspector will check the undersides of hoses as well as the integrity of hard-to-reach hoses blocked by the air cleaner or other components. Correct routing for vacuum hoses can be ob-

Perform underhood inspection of all wiring and hoses

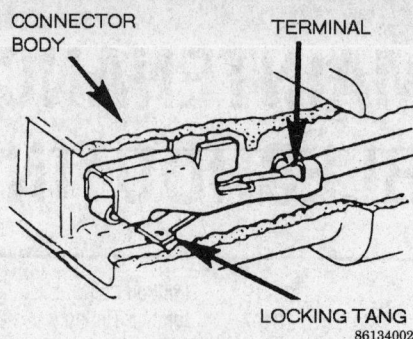

CONNECTOR BODY   TERMINAL

LOCKING TANG
86134002

Check the individual terminals and wiring connectors for damage

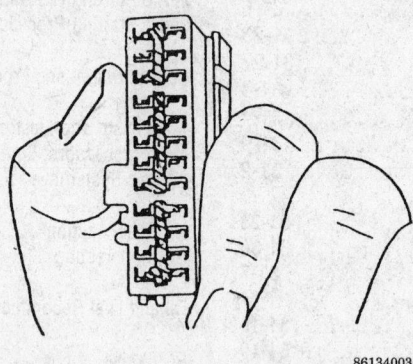

86134003

Inspect the connector terminals for damage

86134004

Check for damaged or broken wires

tained from the specific Vehicle Emission Control Information (VECI) label in the engine compartment of the vehicle. Wiring should be checked carefully for any sign of strain, burning, crimping or terminals pulled out from a connector.

Checking connectors at components or in harnesses is required; usually, pushing them together will reveal a loose fit. Also, check electrical connectors for corroded, bent, damaged, improperly seated pins, and bad wire crimps to terminals. Pay particular attention to ground circuits, making sure they are not

loose or corroded. Remember to inspect connectors and hose fittings at components not mounted on the engine, such as the evaporative canister or relays mounted on the fender aprons. Any component or wiring in the vicinity of a fluid leak or spillage should be given extra attention during inspection.

Additionally, inspect maintenance items such as belt condition and tension, battery charge and condition and the radiator cap carefully. Any of these very simple items may affect the system enough to set a fault code.

DIAGNOSIS BY SYMPTOM

Before the advent of the self-diagnostic system, diagnosis by symptom was the only method for investigation of an automotive problem. An attempt was made to solve problems by reviewing the symptoms and performing tests on suspected components until a defective component was located. The problem was then corrected and the vehicle checked for any other problems. This method is still used frequently today when a driveability complaint is made, but no code is set in the electronic control unit's memory.

When diagnosing by symptom the first step is to find out if the problem really exists. This may sound like a waste of time, but the problem must be recreated before testing is begun. This is called an "operational check". Each operational check will give either a positive or negative answer (symptom). A positive answer is found when the check gives a positive result (the horn blows when the horn button is pressed). A negative answer is found when the check gives a negative result (the radio does not play when the knob is turned on). After performing several operational checks, a pattern may develop. This pattern is used in the next step of diagnosis to determine related symptoms.

In order to determine related symptoms, perform operational checks on circuits related to the problem circuit (the radio does not work and the dash lights do not go on). These checks can be made without the use of any test equipment. Simply follow the wires in the wiring harness or, if available, obtain a copy of the vehicle's specific wiring diagram. If the radio and the dash lights are on the same circuit, first check the radio to see if it works. Then check the dash lights. If the neither the radio or dash lights work, this indicates that there is a problem in that circuit. Perform additional operational checks on that circuit and compile a list of symptoms.

When analyzing the answers, a defect will always lie between a check which gave a positive answer and one which gave a negative answer. Look at the list of symptoms and try to determine probable areas to test. If negative results are received for any related circuits, then, perhaps, the problem is at the common junction. After one has determined what the symptoms are and where one is going to look for defects, one should develop a plan for isolating the trouble. Ask a knowledgable automotive person which components frequently fail on the particular vehicle. Also notice which parts or components are easiest to reach and how can the most can be accomplished by doing the least amount of checks.

A common way of diagnosis is to use the split-in-half technique. Each test that is made essentially splits the trouble area in half. By performing this technique several times the area where a problem is located becomes smaller and smaller until the problem can be isolated in a single wire or component. This area is most commonly between the two closest check points that produced a negative answer and a positive answer.

After the problem is located, perform the repair procedure. This may involve replacing a component, repairing a component or damaged wire, or making an adjustment.

NOTE: Never assume a component is defective until it has been thoroughly tested.

The final step is to make sure the complaint is corrected. Remember, the symptoms which are uncovered may lead to several problems that require separate repairs. Repeat the diagnosis and test procedures again and again until all negative symptoms are corrected.

Diagnosis by symptom — quick reference chart

	Throttle Position Sensor	Coolant Temperature Sensor	MAP or MAF sensor	Air Temperature Sensor	Ignition Coil	Distributor	Spark Plug Wires	Fuel Filter	Air Filter	Vacuum Leak	Engine Mechanical	Knock Sensor / Spark Control	EGR System	Idle Control System	Camshaft sensor/Dist. pick-up	Oxygen Sensor	Ignition Module / Engine Computer	Torque Converter Clutch	PCV System
No Start	u	u	u		•	•		•		u	•			u	•		•		
Hard Start	•	•	•		•	•	•												u
Hesitation	•		•		•	•	•	•	•	•	•			•					
Stalling	•				•	•	•	•	•	•	•			•				•	u
Poor Idle	•				•	•	•			•				•	•				•
Dieseling								•				•		•					
Engine Lamp ON	•	•	•	•								•	•	•	•	•	•		
Knocks or Pings							•					•	•				•	•	•
High Hydrocarbons	u	u	u		•	•	•	u	•	u	•		•			•			
Black Smoke	u	•	•	•				•							•				
Blue Smoke											•								
Poor Fuel Mileage	•	•	•	•	•			•	•				•			•		•	u
Lack of Power	•	•	•	•	•	•		•	•	•	•							•	u
Back fires		•			•	•	•		•	•					•				u
Runs Poor Cold	•				•	•	•	•	•			u	•			•		u	u
Runs Poor Hot	•	•	•	•	•	•	•	•	•	•					u	•	•	u	u
High speed surging			•	•				•										•	u

u: Although possible it is unlikely this component is at fault. A totally open or shorted circuit, or severe component fault may cause this condition.

8736MW01

DIAGNOSIS BY THE STORED TROUBLE CODE

When a fault code is detected, it appears as a flash of the CHECK ENGINE light on the instrument panel. This indicates that an abnormal signal in the system has been recognized by the ECM.

When diagnosing by code, the first step is to read any fault codes from the ECM. These codes will identify the area to perform more in-depth testing. After the fault codes are read, proceed to test each of the components and component circuits indicated. Continue performing individual component tests until the failed component is located. Remember, fault codes do indicate the presence of a failure, but they do not identify the failed component directly.

SAFE VEHICLE SERVICING TIPS

It is virtually impossible to anticipate all of the hazards involved with automotive service, but care and common sense should prevent most accidents.

The rules of safety for mechanics range from "don't smoke around gasoline" to "use the proper tool for the job." The trick to avoiding injuries is to develop safe work habits and take every possible precaution.

Do's

• Do keep a fire extinguisher and first aid kit handy.
• Do wear safety glasses or safety goggles when cutting, drilling, grinding or prying. If regular glasses are worn for the sake of vision, safety goggles should be worn over the regular glasses.
• Do shield the eyes whenever working around the battery. Batteries contain sulfuric acid. In case of contact with the eyes or skin, flush the area with water or a mixture of water and baking soda, then get medical attention immediately.
• Do use safety stands for any under-vehicle service. Jacks are for raising vehicles; jackstands are for making sure the vehicle stays raised until it is to come down. Whenever the vehicle is raised, block the wheels remaining on the ground and set the parking brake.

• Do use adequate ventilation when working with chemicals. Asbestos dust from some worn brake linings can cause cancer.
• Do disconnect the negative battery cable when working on the vehicle.
• Do follow manufacturer's directions whenever working with potentially hazardous materials. Both brake fluid and most types of antifreeze are poisonous if taken internally.
• Do properly maintain all tools. Loose hammerheads, mushroomed punches and chisels, frayed or poorly grounded electrical cords, excessively worn screwdrivers, spread wrenches (open end), cracked sockets, slipping ratchets, or faulty drop light sockets can cause accidents.
• Do use the proper size and type of tool for the job.
• Do, when possible, pull on a wrench handle rather than push on it, and adjust your stance to prevent a fall.
• Do be sure that adjustable wrenches are tight on the nut or bolt and pulled so the force is on the fixed jaw.
• Do select a wrench or socket that fits the nut or bolt. The wrench or socket should sit straight, not cocked.
• Do strike squarely with a hammer. Avoid glancing blows.
• Do set the parking brake and block the wheels if work requires that the engine be running.

Don'ts

• Don't run an engine in a garage or anywhere else without proper ventilation — EVER! Carbon monoxide is poisonous and is absorbed by the body faster than oxygen. It takes a long time to leave the human body, and a deadly supply of it can be built in one's system by simply breathing in a little every day. One may not realize one is slowly poisoning oneself. Always use power vents, windows, fans or open the garage.
• Don't work around moving parts while wearing loose clothing. Short sleeves are much safer than long, loose sleeves. Hard-toed shoes with neoprene soles protect the toes and give a better grip on slippery surfaces. Jewelry such as watches, fancy belt buckles, beads, or body adornment of any kind is not safe while working around a vehicle. Long hair should be tucked under a hat or cap.
• Don't use pockets for tool boxes. A fall or bump can drive a screwdriver deep into the body. Even a

wiping cloth hanging from the back pocket can wrap around a spinning shaft or fan.
• Don't smoke around gasoline, solvent or any flammable material.
• Don't smoke around the battery. When the battery is being charged, it gives off explosive hydrogen gas.
• Don't use gasoline to wash one's hands. There are excellent soaps available. Gasoline contains compounds which are hazardous to the health and it removes natural oils from the skin so that bone dry hands will suck up oil and grease.
• Don't service the air conditioning system unless equipped with the necessary tools and training. The refrigerant is extremely cold, and when exposed to the air, will instantly freeze any surface it comes in contact with, including one's eyes. Keep refrigerant away from open flames; poisonous gas will be produced if refrigerant burns.

SERIAL NUMBER IDENTIFICATION

Vehicle Identification Number

The Vehicle Identification Number (VIN) is the number that will perhaps tell about every thing one needs to know concerning one's vehicle. The VIN number is somewhat like a Social Security Number in an automotive sense. The VIN is a standardized 17 digit number. Each digit of this number has a specific meaning or designation. Example: the 8th digit designates the engine code, the 10th digit designates the model year of the vehicle etc. The vehicle serial number is stamped on a plate fastened to the driver's side door pillar.

This number is usually located on the one of the fender aprons in the engine compartment (behind the wheel arch).

All models have the vehicle identification number stamped on a plate attached to the left side of the instrument panel. The plate is visible through the windshield.

The vehicle identification (model variation codes) may be interpreted as follows:

The serial number on all models is the new 17-digit format. The first three digits are the World Manufacturer Identification number. The next five digits are the Vehicle Description Section (same as the series identification number). The remaining nine digits are the production numbers.

NOTE: For specific identification of the particular vehicle see the Vehicle Identification Label on the vehicle being worked on. If the vehicle has be altered in some way or the engine or transmission has been changed the VIN may not coincide with the change that has been made. It a case like this, look for the serial number on the component to be sure the correct ordering of parts are made.

Engine Serial Number

The engine serial number consists of an engine series identification number followed by a six-digit production number. The number may be found in various places, depending upon the particular engine.

Vehicle Identification Number — visible through the windshield

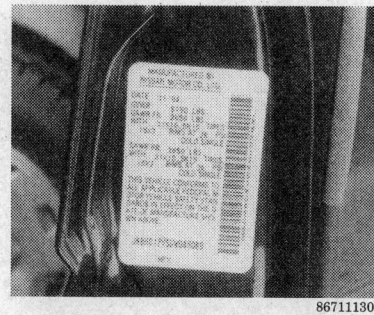

Manufacturer's label located in the door pillar area — note that the build date of vehicle is at the top

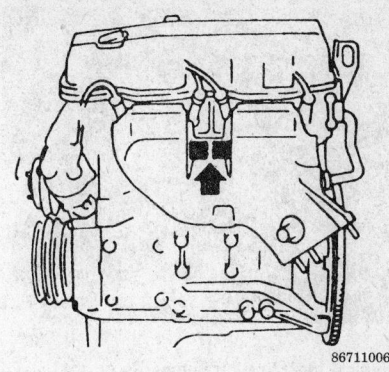

Typical engine serial number location

Transmission Serial Number

The transmission serial number is generally stamped on the front upper face of the transmission case on manual transmissions, or on the side of the transmission case on automatic transmissions.

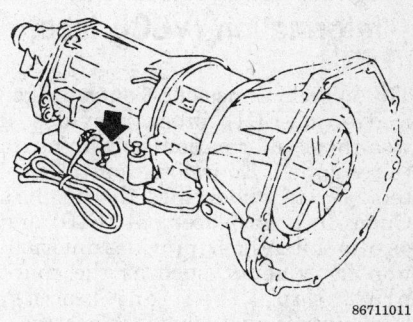

Typical automatic transmission serial number location

Transfer Case Serial Number

The transfer case serial number is generally stamped on the front upper face of the transfer case.

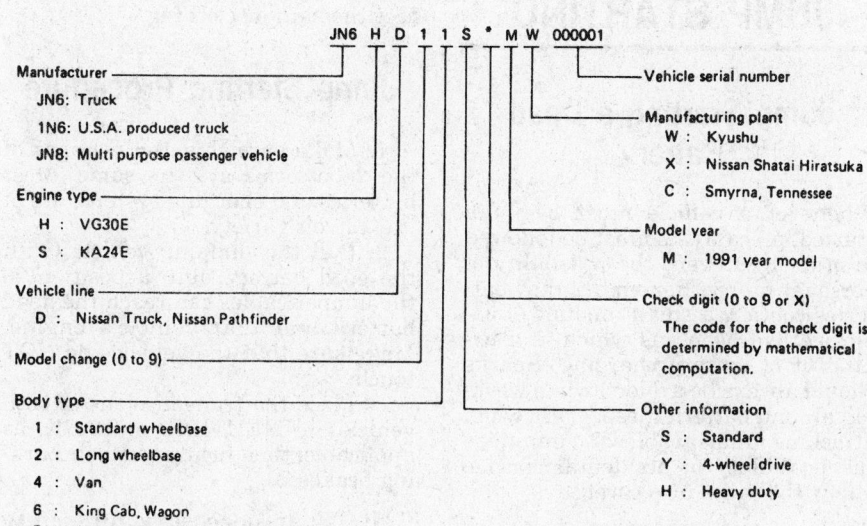

JN6 H D 1 1 S • M W 000001

Manufacturer
- JN6: Truck
- 1N6: U.S.A. produced truck
- JN8: Multi purpose passenger vehicle

Engine type
- H : VG30E
- S : KA24E

Vehicle line
- D : Nissan Truck, Nissan Pathfinder

Model change (0 to 9)

Body type
- 1 : Standard wheelbase
- 2 : Long wheelbase
- 4 : Van
- 6 : King Cab, Wagon
- 7 : 4-door wagon
- 8 : 4-door van

Vehicle serial number

Manufacturing plant
- W : Kyushu
- X : Nissan Shatai Hiratsuka
- C : Smyrna, Tennessee

Model year
- M : 1991 year model

Check digit (0 to 9 or X)
- The code for the check digit is determined by mathematical computation.

Other information
- S : Standard
- Y : 4-wheel drive
- H : Heavy duty

Vehicle Identification Number (VIN) translation

86711131

Vehicle identification number is on the firewall mounted label in the engine compartment

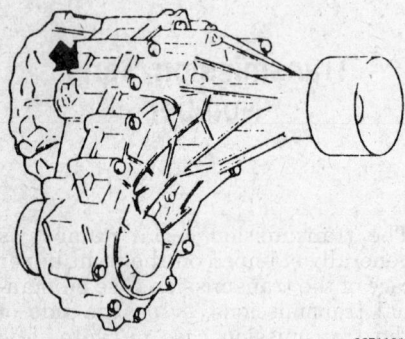

86711012

Typical transfer case serial number location

Vehicle Emissions Control Information (VECI) label

The Vehicle Emissions Control Information (VECI) label provides a wealth of information pertaining to the engine's Emission Control System. First, it identifies the engine's Cubic Inch Displacement (CID) and size in Liter(s). It provides information for tune-up, such as the spark plug gap, ignition timing, idle/mixture and valve lash specifications. In some cases, it will provide a specific adjustment procedure as required. Some labels will also incorporate the vacuum routing of the engine's emission control system. Although the VECI label is very helpful to the person working on the vehicle, it should not be used as the main source for repair information. However, if there have not been any alterations to the engine and the repair manual and sticker do not agree, use the Vehicle Emissions Control Information (VECI) sticker information, as it often reflects changes made during the production run.

Always double check for more detailed information concerning the vehicle being worked upon. Always keep in mind that the vehicle being worked on may have had an engine change. If this is the case, the year and engine code of the engine in the vehicle will have to be identified. If the vehicle is missing the Emission Control Information label, a new one can be ordered from a local dealership.

The Vehicle Emissions Control Information (VECI) label is usually located in the engine compartment. On some vehicles it will be found directly under the hood. Others may have it on the strut tower or radiator support.

86712037

Vehicle Emission Control Information (VECI) label located in the engine compartment

JUMP STARTING

Jump Starting a Dead Battery

Whenever a vehicle must be jump started, precautions must be followed in order to prevent the possibility of personal injury. Remember that batteries contain a small amount of explosive hydrogen gas which is a by-product of battery charging. Sparks should always be avoided when working around batteries, especially when attaching jumper cables. To minimize the possibility of accidental sparks, follow the procedure carefully.

——— WARNING ———
NEVER hook the batteries up in a series circuit or the entire electrical system will be severely damaged, especially the starter!

Vehicles equipped with a diesel engine utilize two 12 volt batteries, one on either side of the engine compartment. The batteries are connected in a parallel circuit (positive terminal to positive terminal, negative terminal to negative terminal). Hooking the batteries up in parallel circuit increases battery cranking power without increasing total battery voltage output. Output remains at 12 volts. On the other hand, hooking two 12 volt batteries up in a series circuit (positive terminal to negative terminal, positive terminal to negative terminal) increases total battery output to 24 volts (12 volts plus 12 volts).

Jump Starting Precautions

• Be sure that both batteries are of the same voltage. All vehicles covered by this manual and most vehicles on the road today utilize a 12 volt charging system.
• Be sure that both batteries are of the same polarity (have the same grounded terminal; in most cases NEGATIVE).
• Be sure that the vehicles are not touching or a short circuit could occur.
• On serviceable batteries, be sure the vent cap holes are not obstructed.
• Do not smoke or allow sparks anywhere near the batteries.
• In cold weather, make sure the battery electrolyte is not frozen. This can occur more readily in a battery that has been in a state of discharge.
• Do not allow electrolyte to contact the skin or clothing.

Jump Starting Procedure

1. Make sure that the voltages of the 2 batteries are the same. Most batteries and charging systems are of the 12 volt variety.
2. Pull the jumping vehicle (with the good battery) into a position so the jumper cables can reach the dead battery and that vehicle's engine. Make sure that the vehicles do NOT touch.
3. Place the transmissions of both vehicles in NEUTRAL or PARK, as applicable, then firmly set their parking brakes.

NOTE: If necessary for safety reasons, both vehicle's hazard lights may be operated throughout the entire procedure without significantly increasing the difficulty of jump starting the dead battery.

4. Turn all lights and accessories **OFF** on both vehicles. Make sure the ignition switches on both vehicles are turned to the **OFF** position.

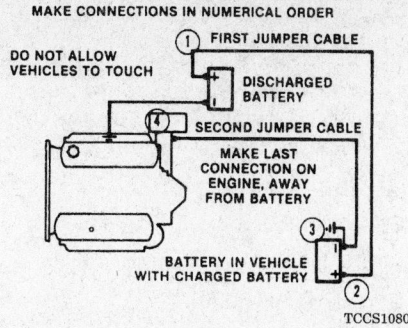

MAKE CONNECTIONS IN NUMERICAL ORDER

① FIRST JUMPER CABLE

DO NOT ALLOW VEHICLES TO TOUCH

DISCHARGED BATTERY

SECOND JUMPER CABLE

MAKE LAST CONNECTION ON ENGINE, AWAY FROM BATTERY

BATTERY IN VEHICLE WITH CHARGED BATTERY

TCCS1080

Connect the jumper cables to the batteries and engine in the order shown

5. Cover the battery cell caps with a rag, but do not cover the terminals.

6. Make sure the terminals on both batteries are clean and free of corrosion or proper electrical connection will be impeded. If necessary, clean the battery terminals before proceeding.

7. Identify the positive (+) and negative (-) terminals on both batteries.

8. Connect the first jumper cable to the positive (+) terminal of the dead battery, then connect the other end of that cable to the positive (+) terminal of the booster (good) battery.

9. Connect one end of the other jumper cable to the negative (–) terminal of the booster battery and the other cable clamp to an engine bolt head, alternator bracket or other solid, metallic point on the dead battery's engine. Try to pick a ground on the engine that is positioned away from the battery, in order to minimize the possibility of the 2 clamps touching should one loosen during the procedure. DO NOT connect this clamp to the negative (–) terminal of the bad battery.

―――― **WARNING** ――――
Be very careful to keep the jumper cables away from moving parts (cooling fan, belts, etc.) on both engines.

10. Check to make sure that the cables are routed away from any moving parts, then start the donor vehicle's engine. Run the engine at moderate speed for several minutes to allow the dead battery a chance to receive some initial charge.

11. With the donor vehicle's engine still running slightly above idle, try to start the vehicle with the dead battery. Crank the engine for no more than 10 seconds at a time and let the starter cool for at least 20 seconds

between tries. If the vehicle does not start within 3 tries, it is likely that something else is also wrong.

12. Once the vehicle is started, allow it to run at idle for a few seconds to make sure that it is properly operating.

13. Turn on the headlights, heater blower and, if equipped, the rear defroster of both vehicles in order to reduce the severity of voltage spikes and subsequent risk of damage to the vehicles' electrical systems when the cables are disconnected.

14. Carefully disconnect the cables in the reverse order of connection. Start with the negative cable that is attached to the engine ground, then the negative cable on the donor battery. Disconnect the positive cable from the donor battery, then disconnect the positive cable from the formerly dead battery. Be careful when disconnecting the cables from the positive terminals not to allow the alligator clips to touch any metal on either vehicle or a short circuit and sparks will occur.

CHARGING SYSTEM

Alternator

The alternator converts the mechanical energy supplied by the drive belt into electrical energy by a process of electromagnetic induction. When the ignition switch is turned **ON**, current flows from the battery through the charging system light (or ammeter) to the voltage regulator, and finally to the alternator. When the engine is started, the drive belt turns the rotating field (rotor) in the stationary windings (stator), inducing alternating current. This alternating current is converted into usable direct current by the diode rectifier. Most of this current is used to charge the battery and to supply power for the vehicle's electrical accessories. A small part of this current is returned to the field windings of the alternator, enabling it to increase its power output. When the current in the field windings reaches a predetermined level, the voltage regulator grounds the circuit preventing any further increase. The cycle is continued so that the voltage supply remains constant.

All models use a 12-volt alternator. Amperage ratings vary according to the year and model. All models have

an electronic, nonadjustable regulator, integral with the alternator.

ALTERNATOR PRECAUTIONS

To prevent damage to the alternator and regulator, the following precautionary measures must be taken when working with the electrical system:

• Never reverse the battery connections. Always visually check the battery polarity before any connections are made, to ensure that all connections correspond to the vehicle's battery ground polarity.

• Booster batteries must be connected properly. Make sure the positive cable of the booster battery is connected to the positive terminal of the battery which is getting the boost.

• Disconnect the battery cables before using a fast charger; the charger has a tendency to force current through the diodes in the opposite direction for which they were designed.

• Never use a fast charger as a booster for starting the vehicle.

• Never disconnect the voltage regulator while the engine is running, unless as noted for testing purposes.

• Do not ground the alternator output terminal.

• Do not operate the alternator on an open circuit with the field energized.

• Do not attempt to polarize the alternator.

• Disconnect the battery cables and remove the alternator before using an electric arc welder on the vehicle.

• Protect the alternator from excessive moisture. If the engine is to be steam cleaned, cover or remove the alternator.

REMOVAL & INSTALLATION

The procedure below is a general procedure; consult specific procedures concerning the vehicle being worked on.

NOTE: On some models, the alternator is mounted very low on the engine. On these models, it may be necessary to remove the gravel shield and work from beneath the vehicle in order to gain access to the alternator.

1. Disconnect the negative battery cable.

2. On vehicles where the alternator can only be accessed from underneath the vehicle, raise the vehicle

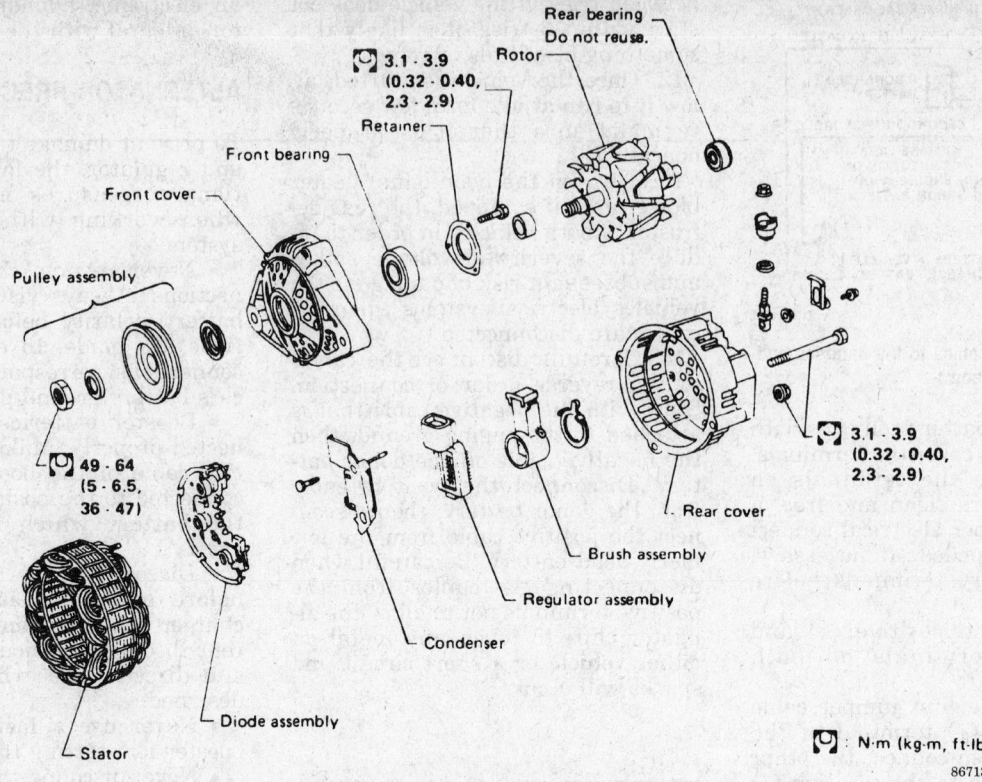

Rear bearing
Do not reuse.

3.1 - 3.9
(0.32 - 0.40,
2.3 - 2.9)

Rotor

Retainer

Front bearing

Front cover

Pulley assembly

49 - 64
(5 - 6.5,
36 - 47)

3.1 - 3.9
(0.32 - 0.40,
2.3 - 2.9)

Rear cover

Brush assembly

Regulator assembly

Condenser

Diode assembly

Stator

: N·m (kg·m, ft-lb)

86713013

Exploded view of a typical alternator assembly

and support it safely with jackstands. Make sure the jackstands are at proper locations.

3. Remove the alternator pivot bolt. Push the alternator inward and remove the drive belt.

4. Pull back the rubber boots and disconnect the wiring from the back of the alternator.

5. Remove the alternator mounting bolt, then withdraw the alternator from its bracket.

To install:

6. Position the alternator in its mounting bracket, then lightly tighten the mounting and adjusting bolts.

7. Connect the electrical leads at the rear of the alternator, and return rubber boots to the proper position.

8. Adjust the belt tension.

9. Connect the negative battery cable.

10. Start the engine and perform a charging system voltage test to insure the charging system is putting out adequately.

A basic check can be made to see if the charging system is charging by using a voltmeter. Connect the voltmeter to the battery, battery voltage is approximately 12.6 volts. Start the engine and observe the voltmeter reading, it should read between

13.2–14 volts. If it remains at 12.6 volts the system is not charging.

Regulator

Regulators may be located internally to the the alternator or externally mounted on the firewall or inner fender panel depending on the vehicle. If faulty, it must be replaced; there are no adjustments which can be made.

REMOVAL & INSTALLATION

Internal Regulator

The electronic regulator is located inside the alternator. On some alternators the regulator is a simple bolt-on procedure, others may be soldered to the brush assembly. With a little knowledge of soldering the job should be able to be accomplished. The following procedure is for a regulator that is soldered to the brush assembly. The regulator is non-adjustable, and must be replaced together with the brush assembly, if faulty.

1. Remove the alternator.

2. Remove the through-bolts and separate the front cover from the stator housing.

3. Unsolder the wire connecting the diode plate to the brush at the brush terminal.

4. Remove the bolt retaining the diode plate to the rear cover.

5. Remove the nut securing the battery terminal bolt.

6. Lift the stator slightly, together with the diode plate, to gain access to the diode plate screw. Remove the screw.

7. Separate the stator and diode, then remove the brush and regulator assembly.

8. On assembly, apply soldering heat sparingly, carrying out the operation as quickly as possible, to avoid damage to the transistors and diodes. Before assembling the alternator halves, bend a piece of wire into an L-shape, then slip it through the rear cover, next to the brushes. Use the wire to hold the brushes in a retracted position until the case halves are assembled. Remove the wire carefully, to prevent damage to the slip rings.

9. Install the alternator.

10. Start the engine and perform a charging system voltage test to insure charging system is putting out adequately.

A basic check can be made to see if the charging system is charging by using a voltmeter. Connect the voltmeter to the battery, battery voltage is approximately 12.6 volts. Start the engine and observe the voltmeter reading, it should read between 13.2–14 volts. If it remains at 12.6 volts the system is not charging.

External Regulator

Depending on the vehicle, the external regulator may be mounted on the firewall or inner fender panel. These regulators are much simpler to replace.

1. Disconnect the negative battery cable.
2. Locate the regulator.
3. Disconnect the harness connector from the regulator.
4. Remove the bolts holding the regulator to its mounting base and remove the regulator.

To install:

5. Install the regulator to the mounting base and secure it in place with the mounting screws.
6. Apply a coating of dielectric grease on the harness electrical connectors and connect.
7. Connect the negative battery cable.
8. Start the engine a perform a charging system voltage test to insure charging system is putting out adequately.

A basic check can be made to see if the charging system is charging by using a voltmeter. Connect the voltmeter to the battery, battery voltage is approximately 12.6 volts. Start the engine and observe the voltmeter reading, it should read between 13.2–14 volts. If it remains at 12.6 volts the system is not charging.

ADJUSTMENT

Voltage regulators on modern vehicles are not adjustable, most are electronic or computer controlled.

STARTING SYSTEM

Starter

REMOVAL & INSTALLATION

The procedure below is a general procedure, consult specific procedure concerning the vehicle being worked on.

1. Disconnect the negative battery cable at the battery, then disconnect

the positive battery cable at the starter.
2. On vehicles where the starter can only be accessed from underneath the vehicle, observe the following caution:

——— CAUTION ———
Raise the vehicle a support it safely with suitable jackstands. Be sure to position the jackstands at proper frame locations.

3. On some 4WD vehicles it may be necessary to perform the following procedure:
   a. Remove the front gravel shield.
   b. Detach the oil pressure switch connector.
   c. Drain the engine oil and remove the oil filter.

——— CAUTION ———
The EPA warns that prolonged contact with used engine oil may cause a number of skin disorders, including cancer! Every effort should be made to minimize exposure to used engine oil. Protective gloves should be worn when changing the oil. Wash the hands and any other exposed skin areas as soon as possible after exposure

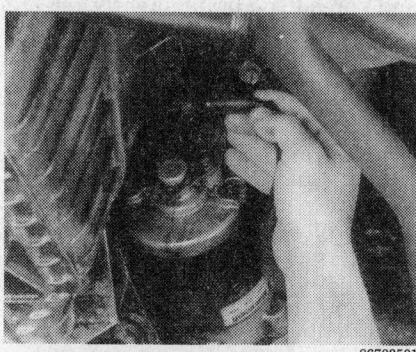

Disconnect the cable attached the starter

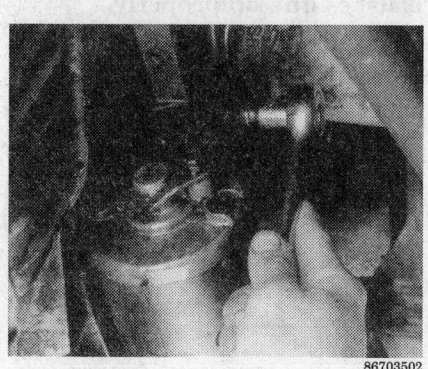

Remove the starter bracket, if equipped

to used engine oil. Soap and water, or waterless hand cleaner should be used.

   d. Remove the exhaust manifold heat insulator.
   e. Remove the fuel tube retainer bolt.
4. On some vehicles it may be necessary to perform the following procedure:
   a. Remove the front right wheel.
   b. Remove the front gravel shield.
   c. Remove the exhaust manifold heat insulator.
   d. Remove the exhaust manifold.
   e. Detach the oil pressure switch connector.
5. Unfasten the remaining electrical connections at the starter solenoid.
6. Remove the two nuts holding the starter to the bell housing, then pull the starter toward the front of the vehicle and out.

To install:

7. Insert the starter into the bell housing, being sure that the starter drive is not jammed against the flywheel.
8. Tighten the attaching nuts and secure all electrical connections to the starter assembly.
9. Install all remaining components in reverse order of removal.
10. Reconnect the battery cables. If applicable, refill and check the oil level. Check the starter assembly for proper operation.

OVERHAUL

The procedure below is a general procedure, consult specific procedures concerning the vehicle being worked on.

Solenoid Replacement

1. Remove the starter.
2. Unscrew the two solenoid switch (magnetic switch) retaining screws.
3. Remove the solenoid. In order to unhook the solenoid from the starter drive lever, lift it up at the same time that that it is being pulled out of the starter housing.
4. Installation is in the reverse order of removal. Make sure that the solenoid switch is properly engaged with the drive lever before tightening the mounting screws.

Brush Replacement

NON-REDUCTION GEAR TYPE

1. Remove the starter.

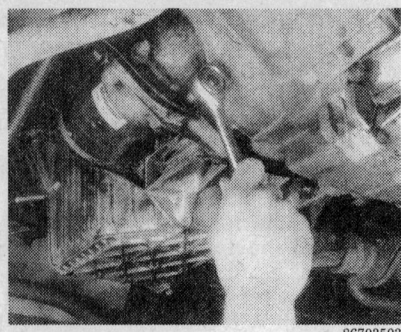

Loosen the holding the nose bracket, if equipped

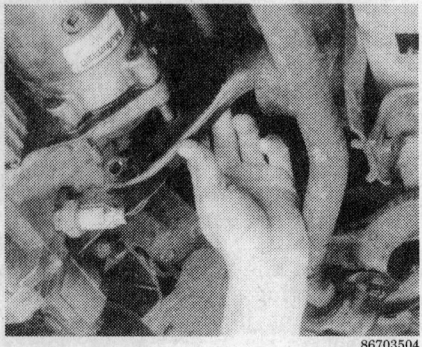

Remove the nose bracket, if equipped

Remove the bolts holding the starter in place

Remove the the starter from the engine

2. Remove the solenoid (magnetic switch).

3. Unfasten the two end frame cap mounting bolts and remove the end frame cap.

4. Remove the O-ring and lock plate from the armature shaft groove, then slide the shims off the shaft.

5. Unfasten the two long housing screws (at the front of the starter) and carefully pull off the end plate.

6. Using a screwdriver, separate the brushes from the brush holder.

7. Slide the brush holder off of the armature shaft.

8. Crush the old brushes off of the copper braid and file away any remaining solder.

9. Fit the new brushes to the braid and spread the braid slightly.

NOTE: Use a soldering iron of at least 250 watts.

10. Using a light-grade solder, solder the brush to the braid. Grip the copper braid with flat pliers to prevent the solder from flowing down its length.

11. File off any extra solder and then repeat the procedure for the remaining three brushes.

12. Installation is in the reverse order of removal.

NOTE: When installing the brush holder, make sure that the brushes line up properly.

REDUCTION GEAR TYPE

1. Remove the starter and the solenoid.

2. Remove the through-bolts and the rear cover. The rear cover can be pried off with a small pry tool, but be careful not to damage the O-ring.

3. Separate the starter housing, armature, and brush holder from the center housing. They can be removed as an assembly.

4. Remove the positive side brush from its holder. The positive brush is insulated from the brush holder, and its lead wire is connected to the field coil.

5. Carefully lift the negative brush from the commutator and remove it from the holder.

6. Installation is in the reverse order of removal.

Starter Drive Replacement

NON-REDUCTION GEAR TYPE

1. With the starter motor removed from the vehicle, separate the solenoid from the starter.

2. Remove the two through-bolts and separate the gear case from the yoke housing.

3. Remove the pinion stopper clip and the pinion stopper.

4. Slide the starter drive off the armature shaft.

5. Install the starter drive and reassemble the starter in the reverse order of removal.

REDUCTION GEAR TYPE

1. Remove the starter.

2. Unfasten the solenoid and the shift lever.

3. Remove the bolts securing the center housing to the front cover and separate the parts.

4. Remove the gears and starter drive.

5. Installation is in the reverse order of removal.

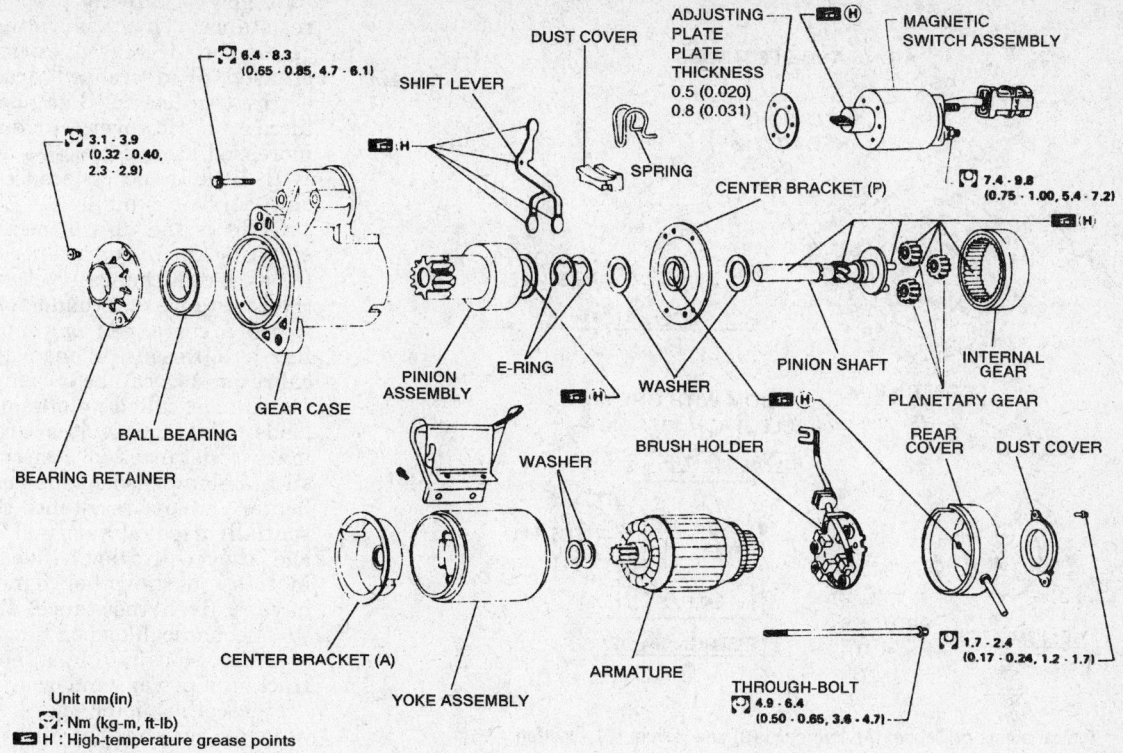

6.4 · 8.3
(0.65 · 0.85, 4.7 · 6.1)

3.1 · 3.9
(0.32 · 0.40,
2.3 · 2.9)

DUST COVER

SHIFT LEVER

ADJUSTING
PLATE
PLATE
THICKNESS
0.5 (0.020)
0.8 (0.031)

MAGNETIC
SWITCH ASSEMBLY

SPRING

CENTER BRACKET (P)

7.4 · 9.8
(0.75 · 1.00, 5.4 · 7.2)

H

PINION
ASSEMBLY

E-RING

WASHER

PINION SHAFT

INTERNAL
GEAR

GEAR CASE

PLANETARY GEAR

BALL BEARING

BRUSH HOLDER

REAR
COVER

DUST COVER

BEARING RETAINER

WASHER

1.7 · 2.4
(0.17 · 0.24, 1.2 · 1.7)

CENTER BRACKET (A)

YOKE ASSEMBLY

ARMATURE

THROUGH-BOLT
4.9 · 6.4
(0.50 · 0.65, 3.6 · 4.7)

Unit mm(in)
: Nm (kg-m, ft-lb)
H : High-temperature grease points

86713017

Exploded view of a typical reduction gear type starter

FUNDAMENTALS OF ELECTRICITY

A good understanding of basic electrical theory and how circuits work is necessary to successfully perform the service and testing outlined in this manual. Therefore, this section should be read before attempting any diagnosis and repair.

All matter is made up of tiny particles called molecules. Each molecule is made up of two or more atoms. Atoms may be divided into even smaller particles called protons, neutrons and electrons. These particles are the same in all matter and differences in materials (hard or soft, conductive or non-conductive) occur only because of the number and arrangement of these particles. In other words, the protons, neutrons and electrons in a drop of water are the same as those in an ounce of lead, there are just more of them (arranged differently) in a lead molecule than in a water molecule. Protons and neutrons packed together form the nucleus of the atom, while electrons orbit around the nucleus much the same way as the planets of the solar system orbit around the sun.

The proton is a small positive natural charge of electricity, while the neutron has no electrical charge. The electron carries a negative charge equal to the positive charge of the proton. Every electrically neutral atom contains the same number of protons and electrons, the exact number of which determines the element. The only difference between a conductor and an insulator is that a conductor possesses free electrons in large quantities, while an insulator has only a few. An element must have very few free electrons to be a good insulator and vice-versa. When we speak of electricity, we're talking about these free electrons.

In a conductor, the movement of the free electrons is hindered by collisions with the adjoining atoms of the element (matter). This hindrance to movement is called **RESISTANCE** and it varies with different materials and temperatures. As temperature increases, the movement of the free electrons increases, causing more frequent collisions and therefore increasing resistance to the movement of the electrons. The number of collisions (resistance) also increases with the number of electrons flowing (current). Current is defined as the movement of electrons through a conductor such as a wire. In a conductor (such as copper) electrons can be caused to leave their atoms and move to other atoms. This flow is continuous in that every time an atom gives up an electron, it collects another one to take its place. This movement of electrons is called electric current and is measured in amperes. When 6.28 billion, billion electrons pass a certain point in the circuit in one second, the amount of current flow is called 1 ampere.

The force or pressure which causes electrons to flow in any conductor (such as a wire) is called **VOLTAGE**. It is measured in volts and is similar to the pressure that causes water to flow in a pipe. Voltage is the difference in electrical pressure measured between 2 different points in a circuit. In a 12 volt system, for example, the force measured between the two battery posts is 12 volts. Two important concepts are voltage potential and polarity. Voltage potential is the amount of voltage or electrical pressure at a certain point in the circuit with respect to another point. For example, if the voltage potential at one post of the 12 volt battery is 0, the voltage potential at the other post is 12 volts with respect to the first post. One post of the battery is said to be

ATOMS AND ELECTRONS

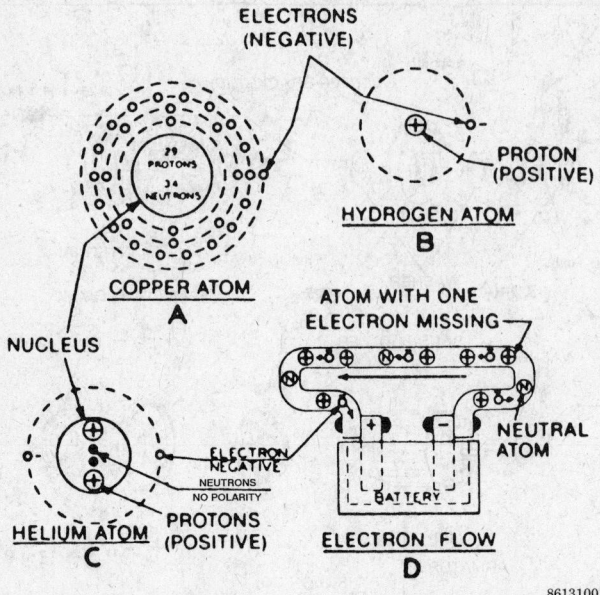

Typical atoms of Copper (A), Hydrogen (B) and Helium (C). Electron flow in a battery circuit

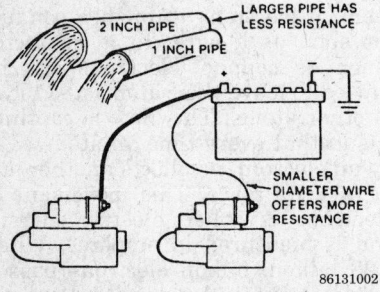

Electrical resistance can be compared to water flow through a pipe. The smaller the wire (pipe), the more resistance the flow of electrons (water)

positive (+); the other post is negative (-) and the conventional direction of current flow is from positive to negative in an electrical circuit. It should be noted that the electron flow in the wire is opposite the current flow. In other words, when the circuit is energized, the current flows from positive to negative, but the electrons actually move from negative to positive. The voltage or pressure needed to produce a current flow in a circuit must be greater than the resistance present in the circuit. In other words, if the voltage drop across the resistance is greater than or equal to the voltage input, the voltage potential will be

zero — no voltage will flow through the circuit. Resistance to the flow of electrons is measured in ohms. One volt will cause 1 ampere to flow through a resistance of 1 ohm.

Units Of Electrical Measurement

There are 3 fundamental characteristics of a direct-current electrical circuit: volts, amperes and ohms.

VOLTAGE in a circuit controls the intensity with which the loads in the circuit operate. The brightness of a lamp, the heat of an electrical defroster, the speed of a motor are all directly proportional to the voltage, if the resistance in the circuit and/or mechanical load on electric motors remains constant. Voltage available from the battery is constant (normally 12 volts), but as it operates the various loads in the circuit, voltage decreases (drops).

AMPERE is the unit of measurement of current in an electrical circuit. One ampere is the quantity of current that will flow through a resistance of 1 ohm at a pressure of 1 volt. The amount of current that flows in a circuit is controlled by the voltage and the resistance in the circuit. Cur-

rent flow is directly proportional to resistance. Thus, as voltage is increased or decreased, current is increased or decreased accordingly. Current is decreased as resistance is increased. However, current is also increased as resistance is decreased. With little or no resistance in a circuit, current is high.

OHM is the unit of measurement of resistance, represented by the Greek letter Omega (ω). One ohm is the resistance of a conductor through which a current of one ampere will flow at a pressure of one volt. Electrical resistance can be measured on an instrument called an ohmmeter. The loads (electrical devices) are the primary resistances in a circuit. Loads such as lamps, solenoids and electric heaters have a resistance that is essentially fixed; at a normal fixed voltage, they will draw a fixed current. Motors, on the other hand, do not have a fixed resistance. Increasing the mechanical load on a motor (such as might be caused by a misadjusted track in a power window system) will decrease the motor speed. The drop in motor rpm has the effect of reducing the internal resistance of the motor because the current draw of the motor varies directly with the mechanical load on the motor, although its actual resistance is unchanged. Thus, as the motor load increases, the current draw of the motor increases, and may increase up to the point where the motor stalls (cannot move the mechanical load).

Circuits are designed with the total resistance of the circuit taken into account. Troubles can arise when unwanted resistances enter into a circuit. If corrosion, dirt, grease, or any other contaminant occurs in places like switches, connectors and grounds, or if loose connections occur, resistances will develop in these areas. These resistances act like additional loads in the circuit and cause problems.

OHM'S LAW

Ohm's law is a statement of the relationship between the 3 fundamental characteristics of an electrical circuit. These rules apply to direct current (DC) only.

Ohm's law provides a means to make an accurate circuit analysis without actually seeing the circuit. If, for example, one wanted to check the condition of the rotor winding in a alternator whose specifications indicate that the field (rotor) current draw is normally 2.5 amperes at 12 volts, simply connect the rotor to a 12 volt battery and measure the current with an ammeter. If it measures about 2.5 amperes, the rotor winding can be assumed good.

An ohmmeter can be used to test components that have been removed from the vehicle in much the same manner as an ammeter. Since the voltage and the current of the rotor windings used as an earlier example are known, the resistance can be calculated using Ohms law. The formula would be ohms equals volts divided by amperes.

If the rotor resistance measures about 4.8 ohms when checked with an ohmmeter, the winding can be assumed good. By plugging in different specifications, additional circuit information can be determined such as current draw, etc.

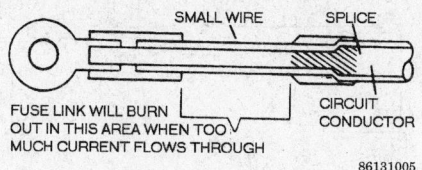

$$I = \frac{E}{R} \quad \text{or} \quad AMPERES = \frac{VOLTS}{OHMS}$$

$$R = \frac{E}{I} \quad \text{or} \quad OHMS = \frac{VOLTS}{AMPERES}$$

$$E = I \times R \quad \text{or} \quad VOLTS = AMPERES \times OHMS$$

86131003

Ohms Law is the basis for all electrical measurement. By simply plugging in two values, the third can be calculated using this formula

Electrical Circuits

An electrical circuit must start from a source of electrical supply and return to that source through a continuous path. Circuits are designed to handle a certain maximum current flow. The maximum allowable current flow is designed higher than the normal current requirements of all the loads in the circuit. Wire size, connections, insulation, etc., are designed to prevent undesirable voltage drop, overheating of conductors, arcing of contacts and other adverse effects. If the safe maximum current flow level is exceeded, damage to the circuit components will result; it is this condition that circuit protection devices are designed to prevent.

Protection devices are fuses, fusible links or circuit breakers designed to open or break the circuit quickly whenever an overload, such as a short circuit, occurs. By opening the circuit quickly, the circuit protection device prevents damage to the wiring, battery and other circuit compo-

SMALL WIRE   SPLICE

FUSE LINK WILL BURN
OUT IN THIS AREA WHEN TOO
MUCH CURRENT FLOWS THROUGH

CIRCUIT
CONDUCTOR

86131005

Typical fusible link wire

nents. Fuses and fusible links are designed to carry a preset maximum amount of current and to melt when that maximum is exceeded, while circuit breakers merely break the connection and may be manually reset. The maximum amperage rating of each fuse is marked on the fuse body and all contain a see-through portion that shows the break in the fuse element when blown. Fusible link maximum amperage rating is indicated by gauge or thickness of the wire. Never replace a blown fuse or fusible link with one of a higher amperage rating.

WARNING
Resistance wires, like fusible links, are also spliced into conductors in some areas. Do not make the mistake of replacing a fusible link with a resistance wire. Resistance wires are longer than fusible links and are stamped "RESISTOR-DO NOT CUT OR SPLICE."

Circuit breakers consist of 2 strips of metal which have different coefficients of expansion. As an overload or current flows through the bimetallic strip, the high-expansion metal will elongate due to heat and break the contact. With the circuit open, the bi-metal strip cools and shrinks, draw-

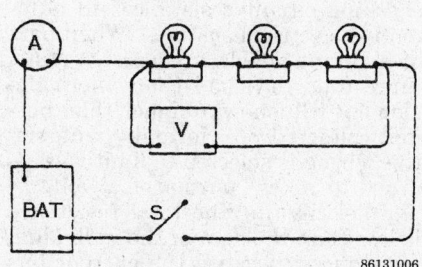

Example of a series circuit

86131006

ing the strip down until contact is re-established and current flows once again. In actual operation, the contact is broken very quickly if the overload is continuous and the circuit will be repeatedly broken and remade until the source of the overload is corrected.

The self-resetting type of circuit breaker is the one most generally used in automotive electrical systems. On manually reset circuit breakers, a button will pop up on the circuit breaker case. This button must be pushed in to reset the circuit breaker and restore power to the circuit. Always repair the source of the overload before resetting a circuit breaker or replacing a fuse or fusible link. When searching for overloads, keep in mind that the circuit protection devices protect only against overloads between the protection device and ground.

There are 2 basic types of circuit; Series and Parallel. In a series circuit, all of the elements are connected in chain fashion with the same amount of current passing through each element or load. No matter where an ammeter is connected in a series circuit, it will always read the same. The most important fact to remember about a series circuit is that the sum of the voltages across each element equals the source voltage. The total resistance of a series circuit is equal to the sum of the individual resistances within each element of the circuit. Using ohms law, one can determine the voltage drop across each element in the circuit. If the total resistance and source voltage is known, the amount of current can be calculated. Once the amount of current (amperes) is known, values can be substituted in the Ohms law formula to calculate the voltage drop across each individual element in the series circuit. The individual voltage drops must add up to the same value as the source voltage.

A parallel circuit, unlike a series circuit, contains 2 or more branches, each branch a separate path independent of the others. The total current draw from the voltage source is the sum of all the currents drawn by each branch. Each branch of a parallel circuit can be analyzed separately. The individual branches can be either simple circuits, series circuits or combinations of series-parallel circuits. Ohms law applies to parallel circuits just as it applies to series circuits, by considering each branch independently of the others. The most important thing to remember is that the

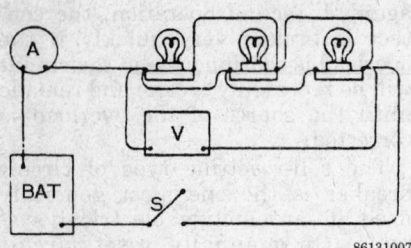

Example of a parallel circuit

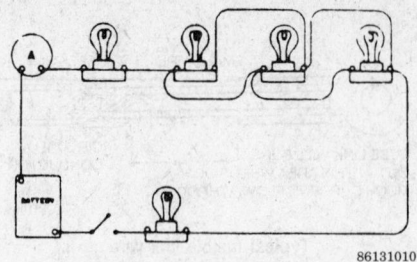

Example of a series-parallel circuit

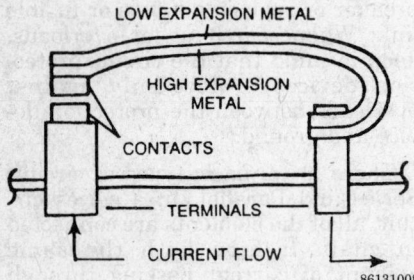

Typical circuit breaker construction

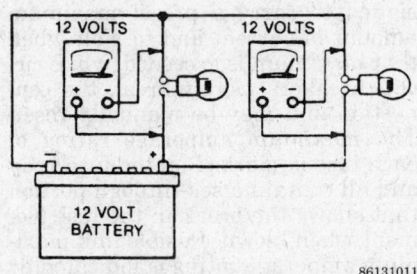

Voltage drop in a parallel circuit. Voltage drop across each lamp is 12 volts

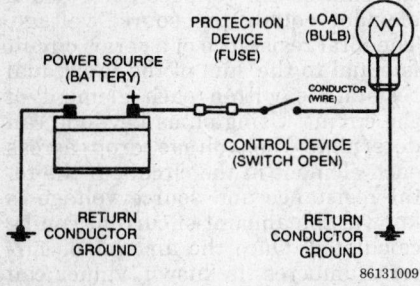

Typical circuit with all essential components

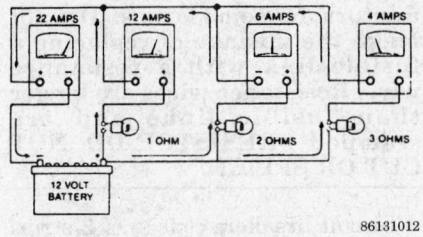

Total current in a parallel circuit: 4 + 6 +12 = 22 amps

voltage across each branch is the same as the source voltage. The current in any branch is that voltage divided by the resistance of the branch. A practical method of determining the resistance of a parallel circuit is to divide the product of the 2 resistances by the sum of 2 resistances at a time. Amperes through a parallel circuit is the sum of the amperes through the separate branches. Voltage across a parallel circuit is the same as the voltage across each branch.

By measuring the voltage drops the resistance of each element within the

circuit is being measured. The greater the voltage drop, the greater the resistance. Voltage drop measurements are a common way of checking circuit resistances in automotive electrical systems. When part of a circuit develops excessive resistance (due to a bad connection) the element will show a higher than normal voltage drop. Normally, automotive wiring is selected to limit voltage drops to a few tenths of a volt. In parallel circuits, the total resistance is less than the sum of the individual resistances; because the current has

2 paths to take, the total resistance is lower.

Magnetism and Electromagnets

Electricity and magnetism are very closely associated because when electric current passes through a wire, a magnetic field is created around the wire. When a wire carrying electric current is wound into a coil, a magnetic field with North and South poles is created just like in a bar magnet. If an iron core is placed within the coil, the magnetic field becomes stronger because iron conducts magnetic lines much easier than air. This arrangement is called an electromagnet and is the basic principle behind the operation of such components as relays, buzzers and solenoids.

A relay is basically just a remote-controlled switch that uses a small amount of current to control the flow of a large amount of current. The simplest relay contains an electromagnetic coil in series with a voltage source (battery) and a switch. A movable armature made of some magnetic material pivots at one end and is held a small distance away from the electromagnet by a spring or the spring steel of the armature itself. A contact point, made of a good conductor, is attached to the free end of the armature with another contact point a small distance away. When the relay is switched on (energized), the magnetic field created by the current flow attracts the armature, bending it until the contact points meet, closing a circuit and allowing current to flow in the second circuit through the relay to the load the circuit operates. When the relay is switched off (de-energized), the armature springs back and opens the contact points, cutting off the current flow in the secondary, or controlled, circuit. Relays can be designed to be either open or closed when energized, depending on the type of circuit control a manufacturer requires.

A buzzer is similar to a relay, but its internal connections are different. When the switch is closed, the current flows through the normally closed contacts and energizes the coil. When the coil core becomes magnetized, it bends the armature down and breaks the circuit. As soon as the circuit is broken, the spring-loaded armature remakes the circuit and again energizes the coil. This cycle

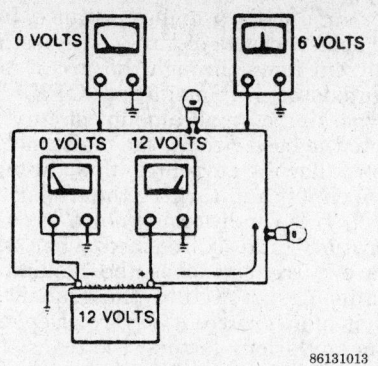

Voltage drop in a series circuit

FORCE FIELD SURROUNDING A CURRENT CARRYING COIL
(WITHOUT IRON CORE)
ALL FORCE LINES ARE COMPLETE LOOPS

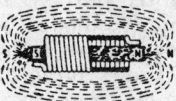

FORCE FIELD WITH SOFT IRON CORE
NOTE CONCENTRATION OF LINES IN IRON CORE

Magnetic field surrounding an electromagnet

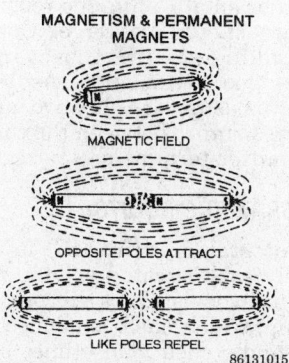

Magnetic field surrounding a bar magnet

repeats rapidly to cause the buzzing sound.

A solenoid is constructed like a relay, except that its core is allowed to move, providing mechanical motion that can be used to actuate mechanical linkage to operate a door or trunk lock or control any other mechanical function. When the switch is closed, the coil is energized and the movable core is drawn into the coil. When the switch is opened, the coil is de-energized and spring pressure returns the core to its original position.

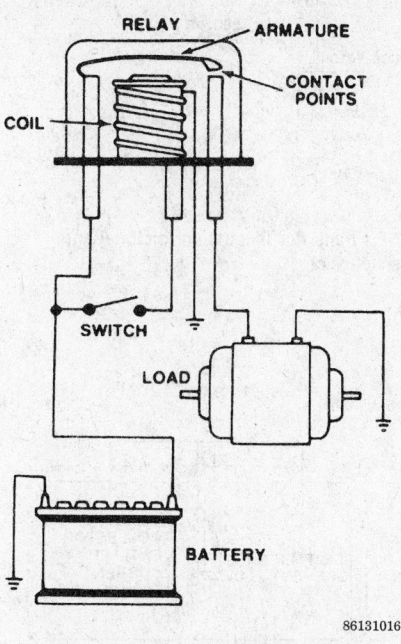

Typical relay circuit with basic components

Basic Solid State

The term "solid state" refers to devices utilizing transistors, diodes and other components which are made from materials known as semiconductors. A semiconductor is a material that is neither a good insulator nor a good conductor; principally silicon and germanium. The semiconductor material is specially treated to give it certain qualities that enhance its function, therefore becoming either P-type (positive) or N-type (negative) material. Most semiconductors are constructed of silicon and can be designed to function either as an insulator or conductor.

DIODES

The simplest semiconductor function is that of the diode or rectifier (the 2 terms mean the same thing). A diode will pass current in one direction only, like a one-way valve, because it has low resistance in one direction and high resistance on the other. Whether the diode conducts or not depends on the polarity of the voltage applied to it. A diode has 2 electrodes, an anode and a cathode. When the anode receives positive (+) voltage and the cathode receives negative (-)

voltage, current can flow easily through the diode. When the voltage is reversed, the diode becomes non-conducting and only allows a very slight amount of current to flow in the circuit. Because the semiconductor is not a perfect insulator, a small amount of reverse current leakage will occur, but the amount is usually too small to consider. The application of voltage to maintain the current flow described is called "forward bias."

A light-emitting diode (LED) is made of a particular type of crystal that glows when current is passed through it. LED's are used in display faces of many digital or electronic instrument clusters. LED's are usually arranged to display numbers (digital readout), but can be used to illuminate a variety of electronic graphic displays.

Like any other electrical device, diodes have certain ratings that must be observed and should not be exceeded. The forward current rating (or bias) indicates how much current can safely pass through the diode without causing damage or destroying it. Forward current rating is usually given in either amperes or milliamperes. The voltage drop across a diode remains constant regardless of the current flowing through it. Small diodes designed to carry low amounts of current need no special provision for dissipating the heat generated in any electrical device, but large current carrying diodes are usually mounted on heat sinks to keep the internal temperature from rising to the point where the silicon will melt and destroy the diode. When diodes are operated in a high ambient temperature environment, they must be de-rated to prevent failure.

Another diode specification is its peak inverse voltage rating. This value is the maximum amount of voltage the diode can safely handle when operating in the blocking mode. This value can be anywhere from 50–1000 volts, depending on the diode. If voltage amount is exceeded, it will damage the diode just as too much forward current will. Most semiconductor failures are caused by excessive voltage or internal heat.

One can test a diode with a small battery and a lamp with the same voltage rating. With this arrangement one can find a bad diode and determine the polarity of a good one. A diode can fail and cause either a short or open circuit, but in either case it fails to function as a diode. Testing is simply a matter of connect-

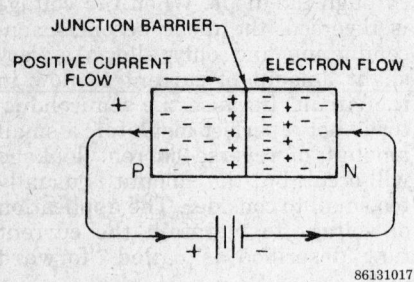

Diode with forward bias

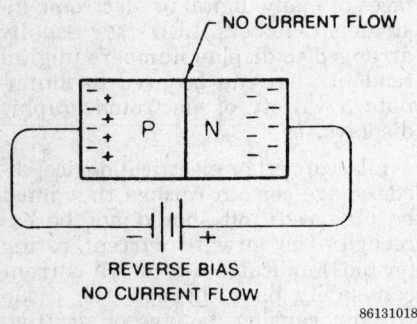

Diode with reverse bias

ing the test bulb first in one direction and then the other and making sure that current flows in one direction only. If the diode is shorted, the test bulb will remain on no matter how the light is connected.

TRANSISTORS

The transistor is an electrical device used to control voltage within a circuit. A transistor can be considered a "controllable diode" in that, in addition to passing or blocking current, the transistor can control the amount of current passing through it. Simple transistors are composed of 3 pieces of semiconductor material, P and N type, joined together and enclosed in a container. If 2 sections of P material and 1 section of N material are used, it is known as a PNP transistor; if the reverse is true, then it is known as an NPN transistor. The 2 types cannot be interchanged.

Most modern transistors are made from silicon (earlier transistors were made from germanium) and contain 3 elements; the emitter, the collector and the base. In addition to passing or blocking current, the transistor can control the amount of current passing through it and because of

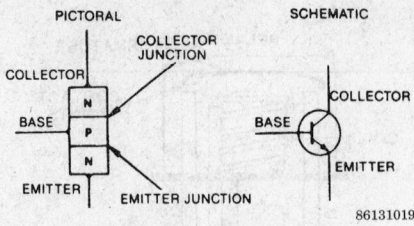

NPN transistor illustration (pictorial and schematic)

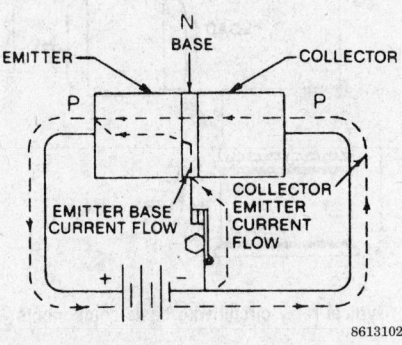

PNP transistor with base switch closed (base emitter and collector emitter current (flow)

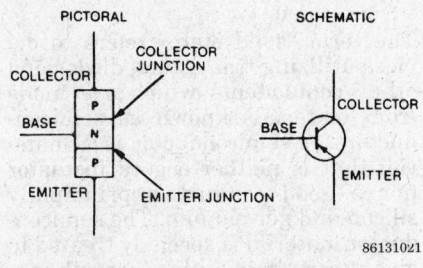

PNP transistor illustrations (pictorial and schematic)

this can function as an amplifier or a switch. The collector and emitter form the main current-carrying circuit of the transistor. The amount of current that flows through the collector-emitter junction is controlled by the amount of current in the base circuit. Only a small amount of base-emitter current is necessary to control a large amount of collector-emitter current (the amplifier effect). In automotive applications, however, the transistor is used primarily as a switch.

When no current flows in the base-emitter junction, the collector-emit-

ter circuit has a high resistance, like to open contacts of a relay. Almost no current flows through the circuit and transistor is considered **OFF**. By bypassing a small amount of current into the base circuit, the resistance is low, allowing current to flow through the circuit and turning the transistor **ON**. This condition is known as "saturation" and is reached when the base current reaches the maximum value designed into the transistor that allows current to flow. Depending on various factors, the transistor can turn on and off (go from cutoff to saturation) in less than one millionth of a second.

Much of what was said about ratings for diodes applies to transistors, since they are constructed of the same materials. When transistors are required to handle relatively high currents, such as in voltage regulators or ignition systems, they are generally mounted on heat sinks in the same manner as diodes. They can be damaged or destroyed in the same manner if their voltage ratings are exceeded. A transistor can be checked for proper operation by measuring the resistance with an ohmmeter between the base-emitter terminals and then between the base-collector terminals. The forward resistance should be small, while the reverse resistance should be large. Compare the readings with those from a known good transistor. As a final check, measure the forward and reverse resistance between the collector and emitter terminals.

INTEGRATED CIRCUITS

The integrated circuit (IC) is an extremely sophisticated solid state device that consists of a silicone wafer (or chip) which has been doped, insulated and etched many times so that it contains an entire electrical circuit with transistors, diodes, conductors and capacitors miniaturized within each tiny chip. Integrated circuits are often referred to as "computers on a chip" and are largely responsible for the current boom in electronic control technology.

Microprocessors, Computers and Logic Systems

Mechanical or electromechanical control devices lack the precision necessary to meet the requirements of modern control standards. They do not have the ability to respond to a

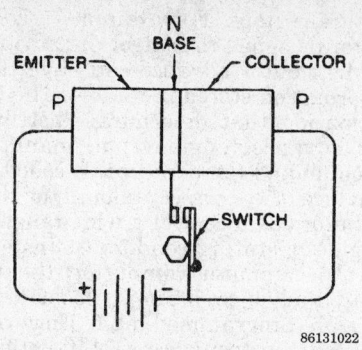

PNP transistor with base switch open (no current)

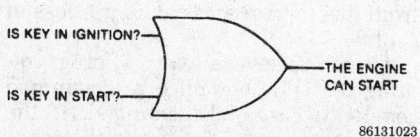

A typical two-input "OR" circuit operation

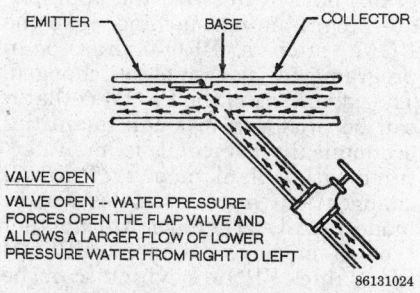

VALVE OPEN

VALVE OPEN -- WATER PRESSURE FORCES OPEN THE FLAP VALVE AND ALLOWS A LARGER FLOW OF LOWER PRESSURE WATER FROM RIGHT TO LEFT

Hydraulic analogy to transistor function is shown with the base circuit energized

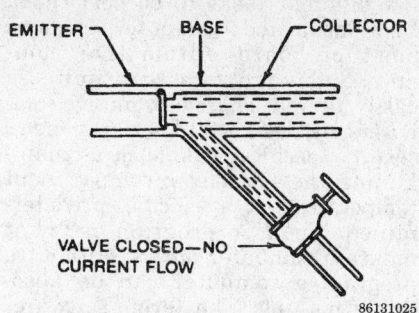

VALVE CLOSED—NO CURRENT FLOW

Hydraulic analogy to transistor function is shown with the base circuit off

variety of input conditions common to anti-lock brakes, climate control and electronic suspension operation. To meet these requirements, manufacturers have gone to solid state logic systems and microprocessors to control the basic functions of suspension, brake and temperature control, as well as other systems and accessories.

One of the more vital roles of microprocessor-based systems is their ability to perform logic functions and make decisions. Logic designers use a shorthand notation to indicate whether a voltage is present in a circuit (the number 1) or not present (the number 0). Their systems are designed to respond in different ways depending on the output signal (or the lack of it) from various control devices.

There are 3 basic logic functions or "gates" used to construct a microprocessor control system: the AND gate, the OR gate or the NOT gate. Stated simply, the AND gate works when voltage is present in 2 or more circuits which then energize a third (A and B energize C). The OR gate works when voltage is present at either circuit A or circuit B which then energizes circuit C. The NOT function is performed by a solid state device called an "inverter" which reverses the input from a circuit so that, if voltage is going in, no voltage comes out and vice versa. With these three basic building blocks, a logic designer can create complex systems easily. In actual use, a logic or decision making system may employ many logic gates and receive inputs from a number of sources (sensors), but for the most part, all utilize the basic logic gates discussed above.

Stripped to its bare essentials, a computerized decision-making system is made up of 3 subsystems:
• Input devices (sensors or switches)
• Logic circuits (computer control unit)
• Output devices (actuators or controls)

The input devices are usually nothing more than switches or sensors that provide a voltage signal to the control unit logic circuits that is read as a 1 or 0 (on or off) by the logic circuits. The output devices are anything from a warning light to solenoid-operated valves, motors, linkage, etc. In most cases, the logic circuits themselves lack sufficient output power to operate these devices directly. Instead, they operate some intermediate device such as a relay or power transistor which in turn operates the appropriate device or control. Many problems diagnosed as computer failures are really the result of a malfunctioning intermediate device like a relay. This must be kept in mind whenever troubleshooting any microprocessor-based control system.

As computer capacity is improved by the manufacturers, so does sensor technology. A few years ago, the on-board computer would receive a message from an engine sensor in a "go or no-go" form; for example the coolant temperature would either be above or below 150°F. Today's systems allow the same sensor to to pass progressively more voltage as the engine warms up. The engine computer now knows exactly what temperature the coolant is at all times. With this information the the computer can react to the changing voltage signal from the sensor instantly and control other engine functions based on engine warm-up or over heating conditions.

The logic systems discussed above are called "hardware" systems, because they consist only of the physical electronic components (gates, resistors, transistors, etc.). Hardware systems do not contain a program and are designed to perform specific or "dedicated" functions which cannot readily be changed. For many simple automotive control requirements, such dedicated logic systems are perfectly adequate. When more complex logic functions are required, or where it may be desirable to alter these functions (e.g. from one model vehicle to another) a true computer system is used. A computer can be programmed through its software to perform many different functions and, if that program is stored on a separate integrated circuit chip called a ROM (Read Only Memory), it can be easily changed simply by plugging in a different ROM with the desired program. Most on-board automotive computers are designed with this capability. The on-board computer method of engine control offers the manufacturer a flexible method of responding to data from a variety of input devices and of controlling an equally large variety of output controls. The computer response can be changed quickly and easily by simply modifying its software program.

The microprocessor is the heart of the microcomputer. It is the thinking part of the computer system through which all the data from the various sensors passes. Within the microprocessor, data is acted upon,

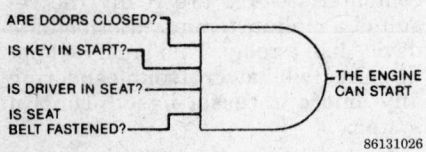

Multiple inputs "AND" operation in a typical automotive starting circuit

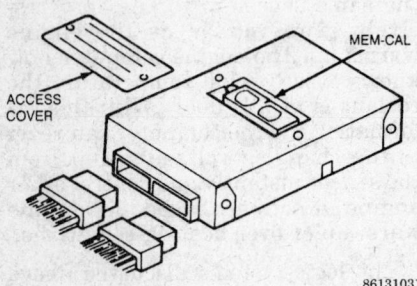

Typical General Motors engine control computer

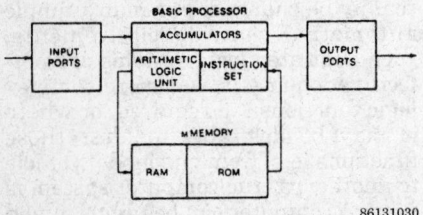

Schematic of typical microprocessor based on-board computer, showing essential common

compared, manipulated or stored for future use. A microprocessor is not necessarily a microcomputer, but the differences between the 2 are becoming very minor. Originally, a microprocessor was a major part of a microcomputer, but nowadays microprocessors are being called "single-chip microcomputers". They contain all the essential elements to make them behave as a computer, including the most important ingredient–the program.

All computers require a program. In a general purpose computer, the program can be easily changed to al-

low different tasks to be performed. In a "dedicated" computer, such as most on-board automotive computers, the program isn't quite so easily altered. These automotive computers are designed to perform one or several specific tasks, such as maintaining the passenger compartment temperature at a specific, predetermined level. A program is what makes a computer smart; without a program a computer can do absolutely nothing. The term "software" refers to the computer's program that makes the hardware preform the function needed.

The software program is simply a listing in sequential order of the steps or commands necessary to make a computer perform the desired task. Before the computer can do anything at all, the program must be fed into it by one of several possible methods. A computer can never be "smarter" than the person programming it, but it is a lot faster. Although it cannot perform any calculation or operation that the programmer himself cannot perform, its processing time is measured in millionths of a second.

Because a computer is limited to performing only those operations (instructions) programmed into its memory, the program must be broken down into a large number of very simple steps. Two different programmers can come up with 2 different programs, since there is usually more than one way to perform any task or solve a problem. In any computer, however, there is only so much memory space available, so an overly long or inefficient program may not fit into the memory. In addition to performing arithmetic functions (such as with a trip computer), a computer can also store data, look up data in a table and perform the logic functions previously discussed. A Random Access Memory (RAM) allows the computer to store bits of data temporarily while waiting to be acted upon by the program. It may also be used to store output data that is to be sent to an output device. Whatever data is stored in a RAM is lost when power is removed from the system by turning **OFF** the ignition key, for example.

Computers have another type of memory called a Read Only Memory (ROM) which is permanent. This memory is not lost when the power is removed from the system. Most programs for automotive computers are stored on a ROM memory chip. Data is usually in the form of a look-up table that saves computing time and

program steps. For example, a computer designed to control the amount of distributor advance can have this information stored in a table. The information that determines distributor advance (engine rpm, manifold vacuum and temperature) is coded to produce the correct amount of distributor advance over a wide range of engine operating conditions. Instead of the computer computing the required advance, it simply looks it up in a pre-programmed table. However, not all electronic control functions can be handled in this manner; some must be computed. On an anti-lock brake system, for example, the computer must measure the rotation of each separate wheel and then calculate how much brake pressure to apply in order to prevent one wheel from locking up and causing a loss of control.

There are several ways of programming a ROM, but once programmed the ROM cannot be changed. If the ROM is made on the same chip that contains the microprocessor, the whole computer must be altered if a program change is needed. For this reason, a ROM is usually placed on a separate chip. Another type of memory is the Programmable Read Only Memory (PROM) that has the program "burned in" with the appropriate programming machine. Like the ROM, once a PROM has been programmed, it cannot be changed. The advantage of the PROM is that it can be produced in small quantities economically, since it is manufactured with a blank memory. Program changes for various vehicles can be made readily. There is still another type of memory called an EPROM (Erasable PROM) which can be erased and programmed many times. EPROM's are used only in research and development work, not on production vehicles.

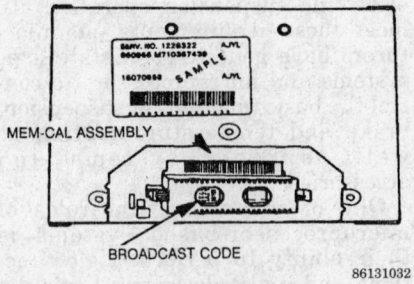

MEM-CAL ASSEMBLY

BROADCAST CODE

Typical General Motors Mem-Cal identification

General Motors refers to the engine controlling computer as an Elec-

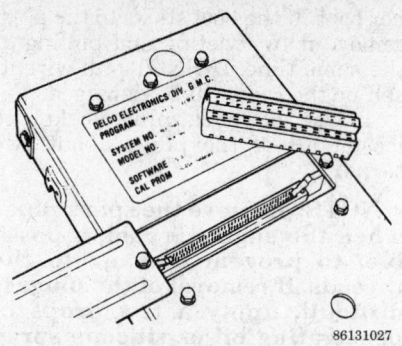

86131027

Electronic control module with Mem-Cal, used in General Motors vehicles

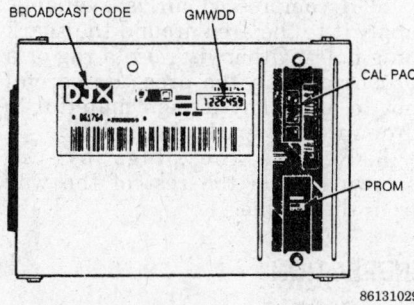

86131029

Identification of a General Motors powertrain control module (PCM)

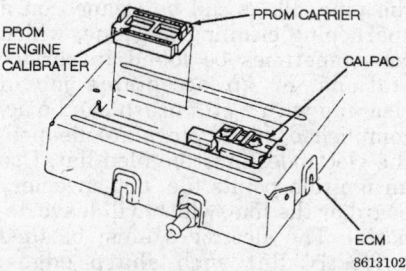

86131028

Electronic control module with PROM and CALPAK, used in General Motors vehicles

tronic Control Module (ECM). The ECM contains the PROM necessary for all engine functions, it also contains a device called a CalPak. This allows the fuel delivery function should other parts of the ECM become damaged. It has an access door in the ECM, like the PROM has. There is a third type control module used in some ECM's called a Mem-Cal. The Mem-Cal contains the function of PROM, CalPak and Electronic Spark Control (EST) module. Like the PROM, it contains the calibrations needed for a specific vehicle, as

well as the back-up fuel control circuitry required if the rest of the ECM should become damaged and the spark control. An ECM containing a PROM and CalPak can be identified by the 2 connector harnesses, while the ECM containing the Mem-Cal has 3 connector harnesses attached to it.

Engines coupled to electronically controlled transmissions employ a Powertrain Control Module (PCM) to oversee both engine and transmission operation. This unit may be referred to as the PCM, the ECM/PCM or the PCM/TCM (Transmission Control Module). The integrated functions of engine and transmission control allow accurate gear selection and improved fuel economy.

For engine diagnostics, the PCM may be considered identical to an ECM system, although the combined unit will display additional codes relating to transmission function and components.

NOTE: When the term Powertrain Control Module (PCM) is used in this manual it will refer to the engine control computer regardless that it may be a Powertrain Control Module (PCM) or Electronic Control Module (ECM).

BASIC TUNE-UP PROCEDURES

In order to extract the best performance and economy from the engine, it is essential that it be properly tuned at regular intervals. A regular tune-up/inspection will keep the engine running smoothly and will prevent the annoying minor breakdowns and poor performance associated with an untuned engine.

A complete tune-up/inspection should generally be performed every 30,000 miles (48,000 km) or 24 months, whichever comes first. This interval should be halved (as a general rule of thumb) if the vehicle is operated under severe conditions, such as trailer towing, prolonged idling, continual stop and start driving, or if starting or running problems are noticed. It is assumed that the routine maintenance has been kept up, as this will have a decided effect on the results of a tune-up.

Some 1994 and newer vehicles are specified to go up to 100,000 miles between engine tune-ups. A tune-up/inspection for all models should consists of the following items
- Inspect the drive belts.
- If necessary, check and adjust valve clearance.
- Clean the air filter housing and replacing the air filter element.
- If equipped, replace the PCV filter and Pulsed secondary air injection filter.
- Inspect or replace the fuel filter assembly.
- Inspect all fuel and vapor lines.
- Check or replace the distributor cap, rotor and ignition wires.
- Replace the spark plugs and make all necessary engine adjustments.
- Always refer to the Maintenance Interval Chart for additional service information.

NOTE: If the tune-up specifications on the Vehicle Emission Control Information sticker in the engine compartment of the vehicle disagree with the Tune-Up Specifications in the repair manual, the figures on the sticker must be used. The sticker often reflects changes made during the production run.

Spark Plugs

NOTE: The platinum type spark plug is not recommended by all manufacturers. If the vehicle has an aftermarket type platinum plug installed, these plugs are usually marked and are not to be cleaned or regapped.

Spark plugs ignite the air/fuel mixture in the cylinder as the piston reaches the top of the compression stroke. The controlled explosion that results forces the piston down, turning the crankshaft and the rest of the drive train.

The average life of a spark plug is about 30,000 miles (48,000 km). This is, however, dependent on a number of factors: the mechanical condition of the engine, the type of fuel, the driving conditions and the operator's driving style.

When the spark plugs are removed, check their condition. Plugs are a good indicator of engine condition. A small deposit of light tan or gray material on a spark plug that has been used for any period of time is considered normal. Any other color, or abnormal amounts of deposits, indi-

86712038

Because of the tangle of underhood wiring, ALWAYS tag/note wire locations before removal

cates that there may be something wrong with the engine.

When a spark plug is functioning normally or, more accurately, when the plug is installed in an engine that is functioning properly, the plugs can be taken out, cleaned, regapped, and reinstalled in the engine without causing the engine any harm.

When, and if, a plug fouls and begins to miss, the cause of the fouling and misfiring should be investigated, corrected and the plug either cleaned or replaced. There are several reasons why a spark plug will foul, and the specific reason for the malfunction of the spark plug can be ascertained simply by the appearance of the malfunctioning spark plug.

There are many spark plugs suitable for use in engines and are offered in a number of different heat ranges. The amount of heat which the plug absorbs is determined by the length of the lower insulator. The longer the insulator the hotter the plug will operate; the shorter the insulator, the cooler it will operate. A spark plug that absorbs (or retains) little heat and remains too cool will accumulate deposits of lead, oil, and carbon, because it is not hot enough to burn them off. This leads to fouling and consequent misfiring. A spark plug that absorbs too much heat will have no deposits, but the electrodes will burn away quickly and, in some cases, pre-ignition may result. Pre-ignition occurs when the spark plug tips become so heated that they ignite the air/fuel mixture before the actual spark reaches the spark plug. This premature ignition will usually cause a pinging (sounds as if marbles are in the cylinders) sound under conditions of low speed and heavy load. In severe cases, the heat may become high enough to start the air/fuel mixture burning throughout the combustion chamber rather than just to the front of the plug. In this case, the resultant explosion could be

strong enough to damage pistons, rings, and valves.

In most cases the factory recommended heat range is correct; it is chosen to perform well under a wide range of operating conditions. However, if most of the driving is long distance, high speed travel, one may want to install a spark plug one step colder than standard. If most of the driving is of the short trip variety, when the engine may not always reach operating temperature, a hotter plug may help burn off the deposits normally accumulated under those conditions.

REMOVAL

NOTE: Some engines use two spark plugs per cylinder; be sure to replace them all.

1. Disconnect the negative battery cable.

NOTE: Always keep track of the spark plug cable routing and plug wire bracket locations.

2. Number the spark plug wires so that they will be reinstalled in their respective positions.

3. Remove the wire from the end of the spark plug by grasping the rub-

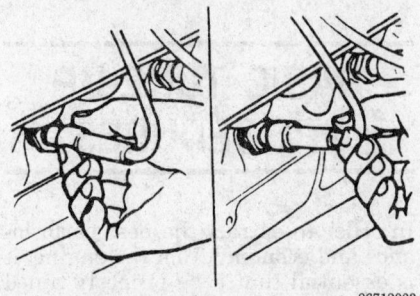

86712003

Twist and pull on the rubber boot to remove the spark plug wires; NEVER pull on the wire itself

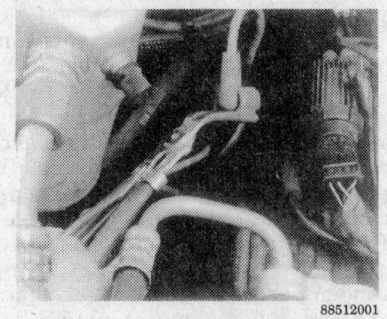

88512001

Using this special tool to remove the spark plug wire makes the job easier; NEVER pull on the wire itself

ber boot. If the boot sticks to the plug, remove it by twisting and pulling at the same time. DO NOT pull wire itself or the core will be damaged.

4. Use a spark plug socket to loosen all of the plugs about two turns.

NOTE: Remove the spark plugs when the engine is cold, if possible, to prevent damage to the threads. If removal of the plugs is difficult, apply a few drops of penetrating oil or silicone spray to the area around the base of the plug, and allow it a few minutes to work.

5. If compressed air is available, apply it to the area around the spark plug holes. Otherwise, use a rag or a brush to clean the area. Be careful not to allow any foreign material to drop into the spark plug holes.

6. Remove the plugs by unscrewing them the rest of the way from the engine.

INSPECTION

Check the plugs for deposits and wear. If new plugs are not to be installed, clean the used plugs thoroughly. Remember that any kind of deposit will decrease the efficiency of the plug. Plugs can be cleaned on a spark plug cleaning machine, which can sometimes be found in service stations, or an acceptable job of cleaning with a stiff brush can be accomplished. If the plugs are cleaned, the electrodes must be filed flat. Use an ignition points file, not an emery board or the like, which will leave deposits. The electrodes must be filed perfectly flat with sharp edges; rounded edges reduce the spark plug voltage by as much as 50%.

Check spark plug gap before installation. The ground electrode (the L-shaped electrode connected to the body of the plug) must be parallel to the center electrode and the specified size wire gauge must pass between the electrodes with a slight drag.

NOTE: NEVER adjust the gap on a used platinum type spark plug.

Always check the gap on new plugs as they are not always set correctly at the factory. Do not use a flat feeler gauge when measuring the gap on a used plug, because the reading may be inaccurate. A wire type gapping tool is the best way to check the gap. Wire gapping tools usually have a bending tool attached. Use that to adjust the side electrode until the

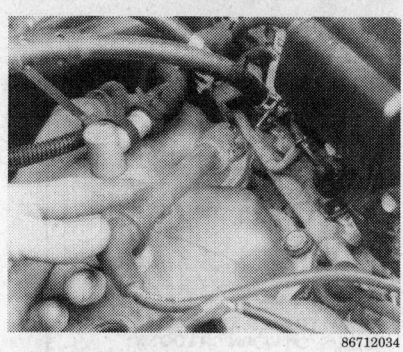

Mark and remove the spark plug wires one at a time to avoid a mix-up

Carefully unthread the spark plug from the cylinder head using the proper tools

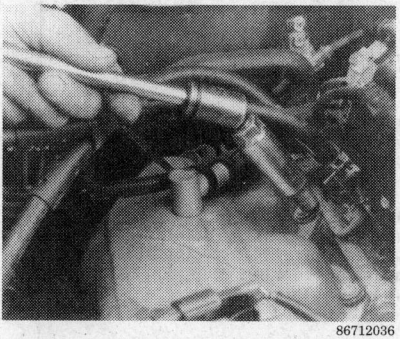

In this case, a universal joint made plug removal easier

proper distance is obtained. Absolutely never attempt to bend the center electrode. Also, be careful not to bend the side electrode too far or too often as it may weaken and break off within the engine, requiring removal of the cylinder head to retrieve it.

INSTALLATION

1. Lubricate the threads of the spark plugs with a drop of oil. Install the plugs and tighten them by hand first. Take care not to cross-thread them.

2. Tighten the spark plugs with a plug socket. Do not apply the same amount of force as one would use for a bolt; just snug them in. If a torque wrench is available, tighten to specific specifications for the vehicle being worked on.

3. Install the wires on their respective plugs. Make sure the wires are firmly connected, which will be known by feeling them click into place. Check the spark plug cable routing and always make sure the plug wires are in the correct plug wire bracket.

4. Connect the negative battery cable.

Spark Plug Wires

CHECKING & REPLACEMENT

At every tune-up/inspection, visually inspect the spark plug cables for burns cuts, or breaks in the insulation. Check the boots and the nipples on the distributor cap and coil. Replace any damaged wiring.

Every 50,000 miles (80,000 Km) or 60 months, the resistance of the wires should be checked with an ohmmeter. Wires with excessive resistance will cause misfiring, and may make the engine difficult to start in damp weather.

To check resistance, remove the distributor cap, leaving the wires attached. Connect one lead of an ohmmeter to an electrode within the cap; connect the other lead to the corresponding spark plug terminal (remove it from the plug for this test). Replace any wire which shows a resistance over 30,000 ohms. Test the high tension lead from the coil by connecting the ohmmeter between the center contact in the distributor cap and either of the primary terminals of the coil. If resistance is more than 25,000 ohms, remove the cable from the coil and check the resistance of the cable alone. Anything over 15,000 ohms is cause for replacement. It should be remembered that resistance is also a function of length; the longer the cable, the greater the resistance. Thus, if the cables on the vehicle are longer than the factory originals, resistance will be higher, and quite possibly outside these limits.

NOTE: The resistance reading given above is a general specification, consult particular specifications for the vehicle.

When installing new cables, replace them one at a time to avoid mix-ups. Start by replacing the longest one first. Install the boot firmly over the spark plug. Route the wire over the same path as the original. Insert the nipple firmly into the tower on the cap or the coil. Check the spark plug cable routing and always make sure the plug wires are in the correct plug wire bracket.

Ignition Timing

Ignition timing is the measurement in degrees of crankshaft rotation of the instant the spark plug fires, in relation to the location of the piston (while the piston is on its compression stroke).

Although no periodic service is necessary, ignition timing can be adjusted by loosening the distributor locking device and turning the distributor in the engine.

Ideally, the air/fuel mixture in the cylinder will be ignited (by the spark plug) and just begin its rapid expansion as the piston passes Top Dead Center (TDC) of the compression stroke. If this happens, the piston will be beginning the power stroke just as the compressed (by the movement of the piston) air/fuel mixture starts to expand. The expansion of the air/fuel mixture will then force the piston down on the power stroke and turn the crankshaft.

It takes a fraction of a second for the spark from the plug to completely ignite the mixture in the cylinder. Because of this, the spark plug must fire before the piston reaches TDC, if the mixture is to be completely ignited as the piston passes TDC. This measurement is given in degrees of crankshaft rotation Before the piston reaches Top Dead Center (BTDC). For example: if the ignition timing setting for the engine is seven degrees (7°) BTDC, this means that the spark plug will fire at a time when the piston for that cylinder is 7° before top dead center of its compression stroke. However, this only holds true while the engine is at idle speed.

As the engine is accelerated from idle, the speed of the engine, in revolutions per minute (rpm), increases. The increase in rpm means that the pistons are now traveling up and down much faster. Because of this, the spark plugs will have to fire even earlier if the mixture is to be completely ignited as the piston passes TDC. To accomplish this, the ECU unit incorporates devices to advance

GAP BRIDGED

IDENTIFIED BY DEPOSIT BUILD—UP CLOSING GAP BETWEEN ELECTRODES.

CAUSED BY OIL OR CARBON FOULING. REPLACE PLUG, OR, IF DEPOSITS ARE NOT EXCESSIVE THE PLUG CAN BE CLEANED.

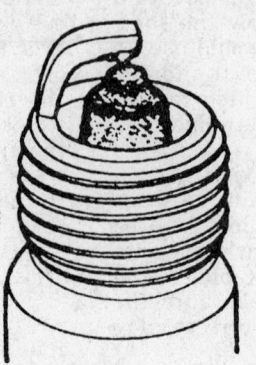

OIL FOULED

IDENTIFIED BY WET BLACK DEPOSITS ON THE INSULATOR SHELL BORE ELECTRODES.

CAUSED BY EXCESSIVE OIL ENTERING COMBUSTION CHAMBER THROUGH WORN RINGS AND PISTONS, EXCESSIVE CLEARANCE BETWEEN VALVE GUIDES AND STEMS, OR WORN OR LOOSE BEARINGS. CORRECT OIL PROBLEM. REPLACE THE PLUG.

CARBON FOULED

IDENTIFIED BY BLACK, DRY FLUFFY CARBON DEPOSITS ON INSULATOR TIPS, EXPOSED SHELL SURFACES AND ELECTRODES.

CAUSED BY TOO COLD A PLUG, WEAK IGNITION, DIRTY AIR CLEANER, DEFECTIVE FUEL PUMP, TOO RICH A FUEL MIXTURE, IMPROPERLY OPERATING HEAT RISER OR EXCESSIVE IDLING. CAN BE CLEANED.

NORMAL

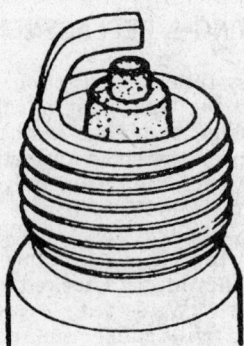

IDENTIFIED BY LIGHT TAN OR GRAY DEPOSITS ON THE FIRING TIP.

PRE-IGNITION

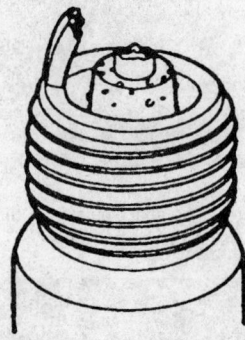

IDENTIFIED BY MELTED ELECTRODES AND POSSIBLY BLISTERED INSULATOR. METALIC DEPOSITS ON INSULATOR INDICATE ENGINE DAMAGE.

CAUSED BY WRONG TYPE OF FUEL, INCORRECT IGNITION TIMING OR ADVANCE, TOO HOT A PLUG, BURNT VALVES OR ENGINE OVERHEATING. REPLACE THE PLUG.

OVERHEATING

IDENTIFIED BY A WHITE OR LIGHT GRAY INSULATOR WITH SMALL BLACK OR GRAY BROWN SPOTS AND WITH BLUISH-BURNT APPEARANCE OF ELECTRODES.

CAUSED BY ENGINE OVER-HEATING, WRONG TYPE OF FUEL, LOOSE SPARK PLUGS, TOO HOT A PLUG, LOW FUEL PUMP PRESSURE OR INCORRECT IGNITION TIMING. REPLACE THE PLUG.

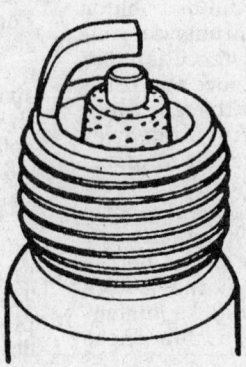

FUSED SPOT DEPOSIT

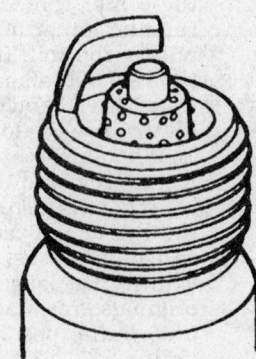

IDENTIFIED BY MELTED OR SPOTTY DEPOSITS RESEMBLING BUBBLES OR BLISTERS.

CAUSED BY SUDDEN ACCELERATION. CAN BE CLEANED IF NOT EXCESSIVE, OTHERWISE REPLACE PLUG.

TCCS2002

Inspect the spark plug to determine engine running conditions

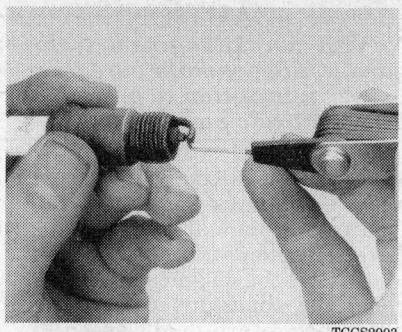

Check the gap of the spark plugs with a wire feeler gauge

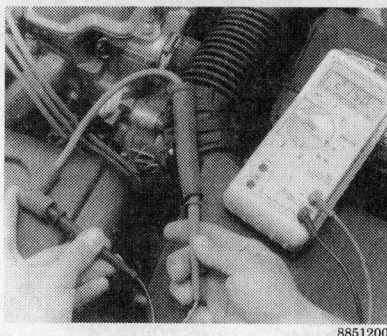

Checking individual plug wire resistance with a digital ohmmeter

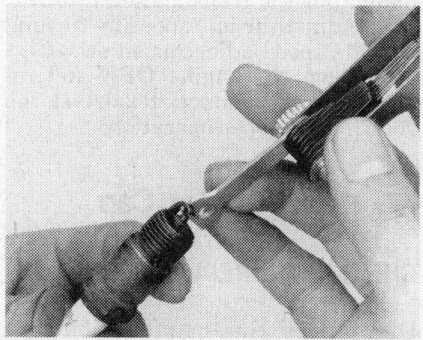

Bend the side electrode to adjust the gap

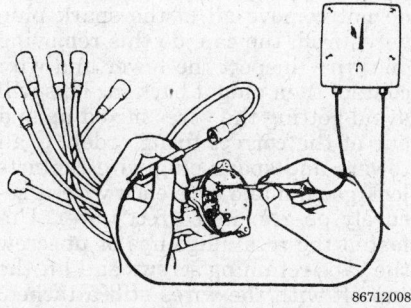

Checking plug wire resistance through the distributor cap with an ohmmeter

Checking and adjusting the ignition timing to specifications

ture ignites and expands. This will cause the piston to be forced down only a portion of its travel, resulting in poor engine performance and lack of power.

Ignition timing adjustment is checked with a timing light. This instrument is connected to the number one (No. 1) spark plug of the engine. The timing light flashes every time an electrical current is sent from the distributor, through the No. 1 spark plug wire, to the spark plug. The crankshaft pulley and the front cover of the engine are marked with a timing pointer and a timing scale. When the timing pointer is aligned with the **0** mark on the timing scale, the piston for the No. 1 cylinder is at TDC of its compression stroke. With the engine running, and the timing light aimed at the timing pointer/scale, the flashes from the timing light will allow the ignition timing to be checked. The timing light flashes every time the spark plug in the No. 1 cylinder of the engine fires. Since the flash from the timing light makes the crankshaft pulley seem stationary for a moment, the flash can be used to read the exact position of the piston in the No. 1 cylinder on the timing scale.

There are three basic types of timing lights available. The first is a simple neon bulb with two wire connections (one for the spark plug and one for the plug wire, connecting the light in series). This type of light is quite dim, and must be held closely to the marks to be seen, but it is inexpensive. The second type of light operates from the battery. Two alligator clips connect to the battery terminals, while a third wire connects to the spark plug with an adapter. This type of light is more expensive, but the xenon bulb provides a nice bright flash which can even be seen in sunlight. The third type replaces the battery source with 110 volt house current. Some timing lights have other functions built into them, such as dwell meters, tachometers, or remote starting switches. These are convenient, in that they reduce the tangle of wires under the hood, but may duplicate the functions of tools already possessed.

For most vehicles, it is best to use a timing light with an inductive pickup. This pickup simply clamps onto the No. 1 plug wire, eliminating the adapter. It is not susceptible to crossfiring or false triggering, which may occur with a conventional light, due to the greater voltages produced by electronic ignition.

the timing of the spark as engine speed increases.

If ignition timing is set too far advanced (too far BTDC), the ignition and expansion of the air/fuel mixture in the cylinder will try to force the piston down the cylinder while it is still traveling upward. This causes engine "ping", a sound which resembles marbles being dropped into an empty tin can. If the ignition timing is too far retarded (After Top Dead Center, or ATDC), the piston will have already started down on the power stroke when the air/fuel mix-

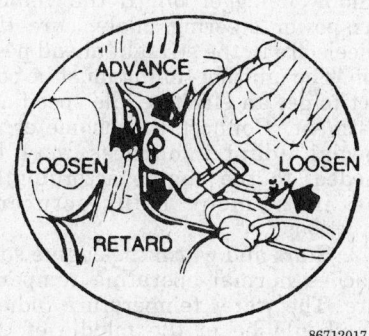

Adjust the ignition timing by rotating the distributor

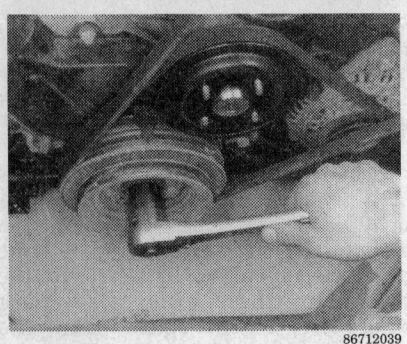

Timing marks on a 6-cylinder engine — note the fan was removed for a better view

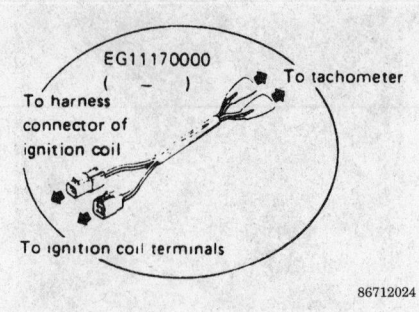

Some vehicles require a special harness for the tachometer connection to check idle speed

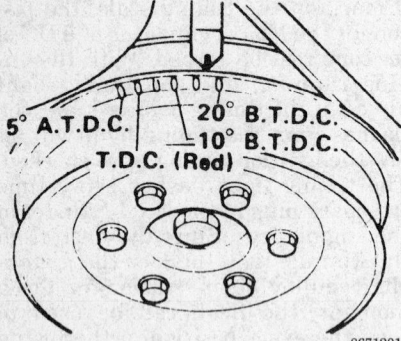

Typical timing marks

Idle Mixture Adjustment

Most vehicles today use a rather complex electronic fuel injection system which is regulated by a series of temperature, altitude (for California) and air flow sensors which feed information into an Electronic Control Unit (ECU) or Powertrain Control Module (PCM). The control unit then relays an electronic signal to the injector nozzle(s), which allow(s) a predetermined amount of fuel into the combustion chamber. In this way, all mixture control adjustments are regulated by the ECU or PCM, therefore, on these vehicles, no manual adjustments are necessary or possible.

Idle Speed Adjustment

Because of ECU or PCM control used on many vehicles today, no periodic service adjustments are necessary. If however the vehicle being worked on requires adjustment or an idle check, a general procedure is shown below. Always refer to the instructions or specifications found on the Vehicle Emission Control Information (VECI)

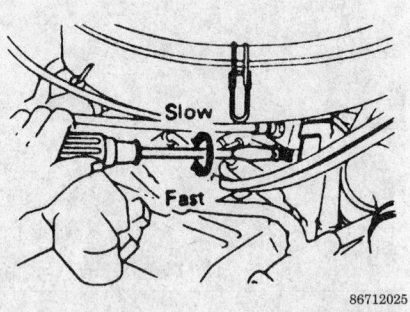

Idle speed adjustment — Carbureted engine

label found underhood for additional or updated information which is applicable to the particular vehicle.

CAUTION

For manual transmission models, set parking brake and check idle speed in N position. For automatic transmission equipped models, shifted into D for idle speed checks. When in Drive, the parking brake must be fully applied with both front and rear wheels chocked.

1. Turn the headlights, heater blower, air conditioning, and rear window defogger off. If the vehicle has power steering, make sure the wheels are in the straight ahead position. The ignition timing must be correct to get an effective idle speed adjustment. Connect a tachometer (a special adapter harness may be needed) to the engine according to the instrument manufacturer's directions.

2. Start and warm the engine so it reaches normal operating temperature. The water temperature indicator should be in the middle of the gauge.

CAUTION

NEVER run the engine in a closed garage. Always make sure there is proper ventilation to prevent carbon monoxide poisoning.

3. Run the engine at 2000 rpm for about 2 minutes under no load (in Neutral for manual transmissions or Park for automatic transmissions).

4. Race the engine to 2000–3000 rpm a few times under no load, then allow it to return to idle speed.

5. Apply the parking brake securely and block the drive wheels of the vehicle. If equipped with an automatic, put the transmission into **D**.

6. Adjust the idle speed by turning the idle speed adjusting screw.

7. Turn the engine **OFF** and remove the tachometer. Road test the vehicle for proper operation.

Distributor Cap

CHECKING & REPLACEMENT

Disconnect the negative battery cable. Individually disconnect each ignition wire (one at a time) from the distributor cap and inspect the cap towers for corrosion build-up. Note, do not remove all of the spark plug wires from the cap, do this removing one wire, inspect the tower and wire contact then plug it back in. This will avoid getting the wires mixed up and out of the correct firing order. If all towers and spark plug wire contacts look good, make sure each wire is securely plug in its correct tower. Unfasten the caps retaining clips or unscrew the caps retaining screws and lift the cap off with the wires still attached. Inspect the underside of the distributor cap for cracks or carbon streaking between the contacts. Inspect the contacts for corrosion or wear. Replace the cap if any of the signs exists.

Some times water or condensation under the distributor cap will cause the electrical current to short out between the contacts of the distributor cap or even wet ignition wires. If, when driving through a deep puddle of water, the engine stalls, chances are that either the ignition wires, or under the distributor cap, have become wet. In this case it is possible to get the engine started by using a dry cloth, and thoroughly drying the under side of the cap and the wires as well as possible.

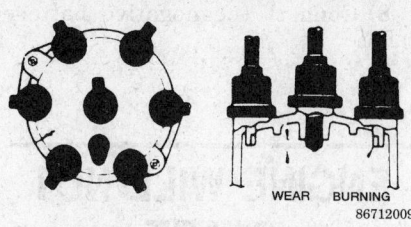

Check under the distributor cap for cracks; check the cable ends for wear

Distributor

REMOVAL

The procedure given below is a generalization and may not apply to the vehicle being worked on.

1. Disconnect the negative battery cable.

2. Unfasten the retaining clips (only remove the coil wire if necessary) and lift the distributor cap straight off. It will be easier to install the distributor if the spark plug wiring is left connected to the cap. If the plug wires must be removed from the cap, mark their positions to aid in installation.

3. Remove the dust cover and mark the position of the rotor relative to the distributor body; then mark the position of the distributor body relative to the engine block.

4. Detach the harness assembly connector.

5. Remove the pinch-bolt and lift the distributor straight out, away from the engine. The rotor and body are marked so that they can be returned to the position from which they were removed. Do not turn or disturb the engine (unless absolutely necessary) after the distributor assembly has been removed.

INSTALLATION

Timing Not Disturbed

This is a general procedure for the installation of the distributor on an engine, on which, after the distributor was removed, the crankshaft or camshaft(s) was not rotated.

1. Insert the distributor in the block and align ALL matchmarks made during removal.

2. Engage the distributor driven gear with the distributor drive.

3. Install the distributor clamp and secure it with the pinch-bolt.

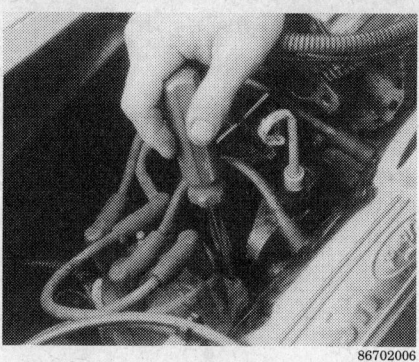

Unscrew the distributor cap retaining screws

Place the distributor cap and wires aside

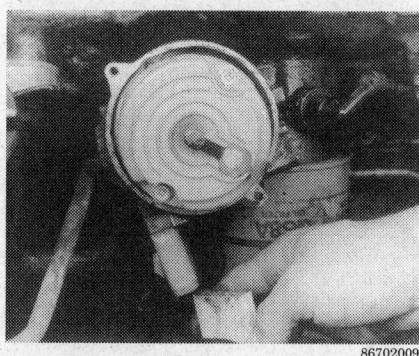

Unplug the distributor harness connector

4. Install the distributor cap and fasten the harness electrical connector.

5. If necessary, install the spark plug wires and coil wire.

6. Start the engine. Check the timing and adjust it if necessary.

Timing Disturbed

This procedure is a basic and simple way to install the distributor correctly, if the engine was cranked while the distributor was out of the engine. Another reason this procedure may be helpful is, if when the distributor was removed from the engine it was not marked for installation position as mentioned in the above procedure. However, refer to specific instructions for the vehicle being worked on.

1. It is necessary to place the No. 1 cylinder in the firing position to correctly install the distributor. To locate this position, the ignition timing marks on the crankshaft front pulley can be used.

2. Remove the No. 1 cylinder spark plug. Turn the crankshaft until the piston in the No. 1 cylinder is moving up on the compression stroke. This can be determined by placing a thumb over the spark plug hole and feeling the air being forced out of the cylinder. Stop turning the crankshaft when the timing marks indicate **TDC** or **0**.

3. Oil the distributor housing lightly where the distributor bears on the cylinder block.

4. Install the distributor so that the rotor, which is mounted on the shaft, points toward the No. 1 spark plug terminal tower position when the cap is installed. Of course, the direction in which the rotor is pointing cannot be seen if the cap is installed, so lay the cap on the top of the distributor and make a mark on the side of the distributor housing just below the No. 1 spark plug terminal. Make sure that the rotor points toward that mark when installing the distributor.

NOTE: Some engines may have an alignment mark on the distributor shaft which should be aligned with the protruding mark on the distributor housing.

5. When the distributor shaft has reached the bottom of the hole, gently move the rotor back and forth slightly until the driving lug on the end of the shaft enters the slots cut in the end of the oil pump shaft and the distributor assembly slides down into place.

6. Fasten the distributor hold-down bolt.

7. Install the spark plug into the No. 1 spark plug hole.

8. Install the distributor cap and engage the harness electrical connector.

9. If necessary, attach the spark plug wires and coil wire.

10. Start the engine. Check the timing and adjust it if necessary.

There are many variations for installing a distributor, depending on which vehicle is being worked on. Refer to specific procedures for the vehicle in question.

Paint alignment marks on both the rotor cap and engine block

Remove the distributor retaining bolt

Carefully remove the distributor

Ignition Coil

Ignition coils may be externally mounted or internally located within the distributor. The procedure given below is a basic procedure for an externally mounted ignition coil. Refer to specific procedures for the vehicle being worked on.

REMOVAL & INSTALLATION

The procedure below is a general procedure, consult specific procedures

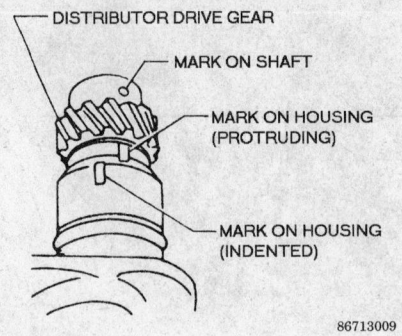

Align the mark on the housing with the mark on the shaft

concerning the vehicle being worked on.

1. Disconnect the negative battery cable.
2. Remove the two mounting bolts and lift off the ignition coil.
3. Tag and disconnect all electrical leads at the coil.
4. Disconnect the coil high tension lead and remove the coil from the engine.

To install:

5. Connect all electrical leads to the coil as tagged.
6. Plug the high tension coil wire into the coil tower.

7. Install the coil in position and tighten the mounting bolts.
8. Connect the negative battery cable.

ENGINE WILL NOT START

The procedure below is a general procedure, consult specific procedures concerning the vehicle being worked on.

No Start Test

1. Connect a voltmeter across the battery terminals. If battery voltage is not at least 12 volts, charge and test the battery before proceeding.
2. Turn the key to the **START** position and observe the voltmeter. If the engine turned over and battery voltage remained above 9.6 volts, go to next step. If the engine failed to

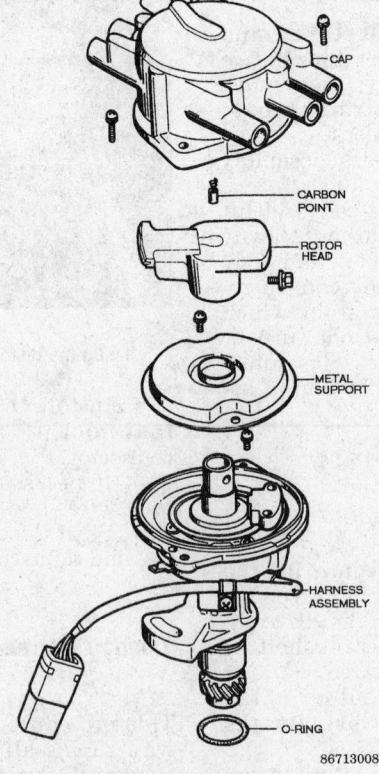

Exploded view of the distributor assembly — typical 6 cylinder engine

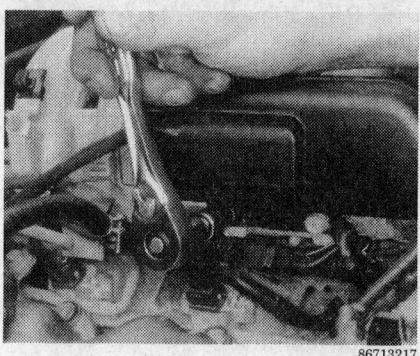

Remove the coil assembly mounting bolts

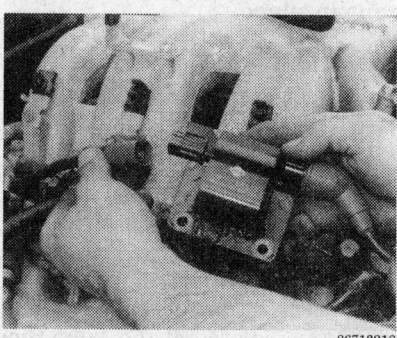

Disconnect the electrical connection from the coil assembly

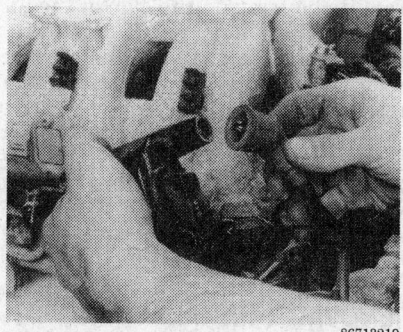

Remove the coil ignition wire from the coil assembly

crank and/or voltage was below 9.6 volts, proceed as follows:

a. If the instrument panel lights dim, load test the battery, check the battery terminals and cables, test the starter motor and verify the engine turns.

b. If the instrument panel lights do not dim, check the battery terminal connections, the ignition switch/wiring and the starter.

3. Using a spark tester, check for spark at two or more spark plugs. If

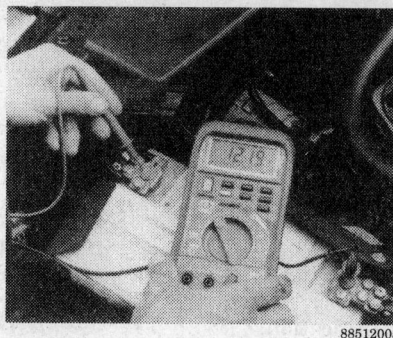

Battery voltage should remain over 9.6 volts while the engine is cranking

okay go to next step, if not okay, perform No Spark Test.

4. Cycle the ignition switch **ON** and **OFF**, several times, while listening for fuel pump operation. If fuel pump operates, proceed to next step. If fuel pump does not operate begin testing of the fuel pump circuit.

5. Verify adequate fuel in the tank, then connect a fuel pressure gauge and check fuel pressure. If fuel pressure is within specifications, proceed to next step, if not okay continue on checking the fuel pump and supply system.

6. Disconnect the fuel injector connector and connect a noid light to the wiring harness. Crank the engine, while watching the light. Perform this test on at least two injectors before proceeding. If the light does not flash, go to the next step. If the light flashes, check the engine valve timing and overall mechanical condition of the engine. If okay, items such as; poor fuel quality, faulty injectors and computer controlled devices should be checked. Although these items are less likely, a shorted TPS or faulty coolant temperature sensor, are possibilities.

7. Check and verify the Malfunction Indicator Lamp (MIL) is operating properly. If the light does not operate, check the ECM and related wiring. If the MIL lamp is operational, check the injector wiring and circuitry.

No Spark Test

1. Check for spark at two or more spark plugs. If spark does not exist go to next step, if spark is okay, check spark plugs, fuel system and engine mechanical condition.

2. Check for spark from the ignition coil wire. If spark does not exist

go to next step, if spark is okay, check distributor cap, rotor and ignition wires.

3. Check the ignition coil wire with an ohmmeter. Resistance should not exceed 1000 ohms per inch (2.5 cm) of cable. If wire resistance exceeds specification, replace the wire and retest. If wire is okay, proceed to the next step.

4. Connect a test light to the negative side of the ignition coil. Turn the key to the **ON** position and observe the test light. If the light remains brightly lit, proceed to the next step. If the light did not light, or glowed dim, check the ignition switch and power supply circuit.

5. Observe the light while cranking the engine. If the light was flashing during cranking, check the ignition coil. If the light did not flash, verify the distributor rotates smoothly, then test the ignition module, pick-up coil or hall effect switch.

RELIEVING FUEL PRESSURE

The procedure is a generic procedure for most fuel injected vehicles. Carbureted vehicles may not use an electric fuel pump. Carbureted systems are lower pressure. Refer to specific procedures for the vehicle being worked on.

1. Disable the fuel pump by one of the following methods:
- Remove the fuel pump fuse.
- Remove the fuel pump relay.
- Locate and disconnect the fuel pump wiring.

NOTE: When removing the fuel pump fuse or relay to disable the fuel pump, it is important to make certain that the fuel injectors are not part of this circuit. If the injectors do not operate the residual fuel system pressure will not be relieved.

2. Start the engine and operate it until it stalls. Once the engine has stalled, crank the starter for an additional 10 seconds.

3. Place a rag over the connection which will be disconnected, then carefully separate the connections. Use the rag to absorb any remaining fuel.

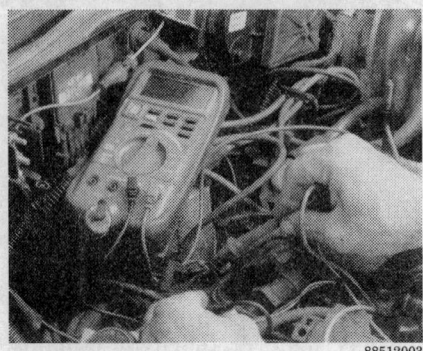

Testing the ignition coil resistance

PREVENTATIVE MAINTENANCE

Air Cleaner

The air cleaner element should be replaced at the recommended maintenance intervals. If the vehicle is operated under severely dusty conditions or severe operating conditions, more frequent changes will certainly be necessary. Inspect the element at least twice a year. Early spring and early fall are good times for an inspection. Remove the element and check for any perforations or tears in the filter. Check the cleaner housing for signs of dirt or dust that may have leaked through the filter element or in through the snorkel tube. Position a drop light on one side of the element and look through the filter at the light. If no glow of light can be seen through the element material, replace the filter. If holes in the filter element are apparent, or signs of dirt seepage through the filter are evident, replace the filter.

REMOVAL & INSTALLATION

Air cleaners come a wide selection of shapes and sizes. Most common are either round or rectangular. In any event, air filter element replacement is usually pretty simple.

If the vehicle is equipped with a round type air cleaner, it probably has one or two wing nuts and/or some clips holding the air cleaner lid in place. Just remove the wing nuts and unclip the retaining clips. Lift the lid off and remove the element.

If the vehicle is equipped with a rectangular type air cleaner, unclip the retaining clips and lift off the air cleaner lid. remove the element from the assembly. Some these units may not be as easily accessible and may require removing a hose or two, but all in all it's pretty simple to do.

Air Cleaner Assembly (Housing)

1. Disconnect all hoses, ducts and vacuum tubes from the air cleaner assembly, after tagging them for easy identification.
2. Remove the top cover wing nuts and grommet (if so equipped). Most models also utilize four to five side clips to further secure the top of the assembly. Simply pull the wire tab and release the clip. On most 1990 and later vehicles, air cleaners are secured solely by means of clips (air box-to-cleaner housing). Remove the cover and lift out the filter element.
3. Remove any side mount brackets and/or retaining bolts, then lift off the air cleaner assembly.
4. Clean or replace the filter element as detailed previously. Wipe clean all surfaces of the air cleaner housing and cover. Check the condition of the mounting gasket and replace if it appears worn or broken.
5. Reposition the air cleaner assembly, then install the mounting bracket and/or bolts.

Removing the air cleaner filter element — round type

Removing the air cleaner filter element — rectangular

6. Reposition the filter element in the case and install the cover being careful not to overtighten the wing nut(s). On round-style cleaners, be certain that the arrows on the cover lid and the snorkel match up properly.

NOTE: Filter elements on many engines have a TOP and BOTTOM side; be sure they are inserted correctly.

7. Reconnect all hoses, ductwork and vacuum lines.

NOTE: Never operate the engine without the air filter element in place.

Air Cleaner Element

The air cleaner element can be replaced by removing the wing nut(s) and/or side clips, then removing the top cover as previously detailed.

Crankcase Ventilation Filter

Certain models may also utilize an air cleaner-mounted crankcase ventilation filter. If so, it should also be cleaned or replaced at the same time as the regular air filter element. To replace the filter, remove the air cleaner top cover and pull the filter from its housing on the side of the air cleaner assembly. Push a new filter into the housing and reinstall the cover. If the filter and plastic holder need replacement, remove the clip mounting the feeder tube to the air cleaner housing, then remove the assembly from the air cleaner.

Fuel Filter

REMOVAL & INSTALLATION

— CAUTION —
NEVER SMOKE WHEN WORKING AROUND OR NEAR GASOLINE! MAKE SURE THAT THERE IS NO ACTIVE IGNITION SOURCE NEAR YOUR WORK AREA!

The procedure below is a general procedure, consult specific procedures concerning the vehicle being worked on.

— CAUTION —
Never attempt to remove the fuel filter without first relieving the fuel system pressure!

1. Release the fuel pressure from the fuel line as follows:
   a. Remove the fuel pump fuse at the fuse box.

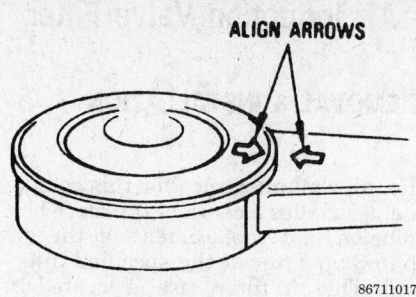

ALIGN ARROWS

86711017

Many air cleaner assemblies have arrows on the housing and lid — always make sure they align

86711018

The air filter element may be cleaned with low pressure compressed air

86711132

Loosen the air intake hose clamp before removing the filter element

b. Start the engine.

c. After the engine stalls, crank the engine two or three times to make sure that the fuel pressure is released.

d. Turn the ignition switch **OFF** and reinstall the fuel pump fuse.

2. Loosen the hose clamps at the fuel inlet and outlet lines. Wrap a shop towel or absorbent rag around the filter, then slide each line off the filter nipples.

86711133

Disconnect the air intake hose, being careful not to lose the retaining clamp

86711134

Unfasten the side retaining clamps so that the air filter housing can be opened

86711135

Remove the air filter element from the air filter housing

3. Remove the fuel filter and old hose clamps.

To install:

4. Place new hose clamps on the fuel inlet and outlet lines.

5. Connect the fuel filter, being careful to observe the correct direction of flow, then tighten the hose clamps.

6. Start the engine and check for fuel leaks.

NOTE: Always use a high pressure-type fuel filter assembly. Do not use a synthetic resinous fuel filter.

PCV Valve

The PCV valve regulates crankcase ventilation during various engine operating conditions. At high vacuum (idle speed and partial load range) it will open slightly, and at low vacuum (full throttle) it will open fully. This causes vapor to be removed from the crankcase by the engine vacuum and then be sucked into the combustion chamber where it is burned.

NOTE: The PCV system will not function properly unless the oil filler cap is tightly sealed. Check the gasket on the cap and be certain it is not leaking. Replace the cap and/or gasket, if necessary, to ensure proper sealing.

TESTING

1. Check the ventilation hoses and lines for leaks or clogging. Clean or replace as necessary.

2. With the engine running at idle, locate the PCV valve in the cylinder head cover or intake manifold and remove the ventilation hose from the valve; a strong hissing sound should be heard as air passes through the valve.

3. With the engine still idling, place a finger over the valve; a strong vacuum should be felt.

4. If the PCV valve failed either of the preceding two checks (and the ventilation hose is not clogged or broken), the valve will require replacement.

REMOVAL & INSTALLATION

1. If not already done, detach the ventilation hose from the PCV valve.

2. Remove the PCV valve. If its base is threaded, unscrew the valve; otherwise, simply pull the valve from its retaining grommet.

To install:

3. Depending on the type of valve, either screw in the replacement PCV valve or push it into its retaining grommet.

4. Slide the ventilation hose onto the end of the PCV valve.

View of the air filter element. Make sure that the element is installed in the housing properly before fastening the side clamps

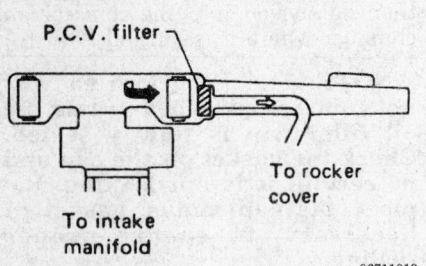

Crankcase ventilation filter replacement

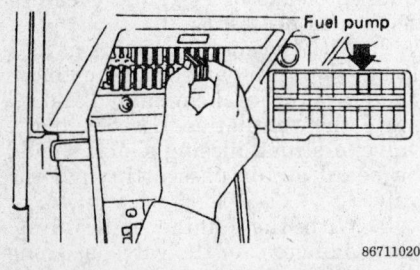

Remove the fuel pump fuse when releasing the fuel pressure — the fuse's location may vary in the box

Remove the fuel line hose clamp after releasing the fuel pressure

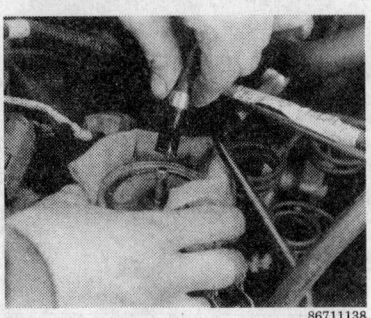

When removing the fuel line from the fuel filter, have a shop towel in position to catch any fuel that may spill from the filter

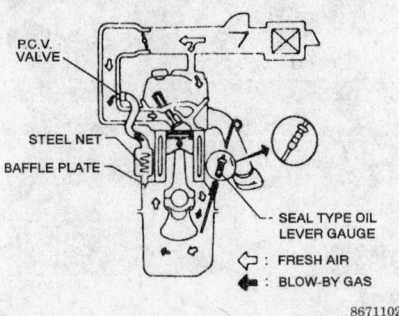

PCV valve location — typical 4-cylinder engine

Air Induction Valve Filter

REMOVAL & INSTALLATION

Regular maintenance for this component includes a check of the drive belt tension and replacement of the air pump air filter at the specified interval. The air filter case is located in the left front of the engine compartment on most models. To replace the air filter, simply unscrew the wing nut(s) securing the cover to the case, withdraw the old filter, install the new one, and reinstall the case.

Battery

NOTE: On a maintenance-free sealed battery, a built-in hydrometer or "eye" is used for checking the fluid level and specific gravity readings. If the battery is equipped with an eye, use it for checking the condition of the battery by observing the color of the eye. A green colored eye indicates good condition and a dark colored eye indicates the need for service. Replacement batteries could be either the sealed (maintenance-free) or non-sealed type.

FLUID LEVEL (EXCEPT MAINTENANCE-FREE SEALED BATTERIES)

Check the battery electrolyte level at least once a month, or more often in hot weather or during periods of extended operation. The level can be checked through the case on translucent polypropylene batteries; the cell caps must be removed on other models. The electrolyte level in each cell should be kept filled to the bottom of the split ring inside, or to the line marked on the outside of the case.

If the level is low, add only distilled water, or colorless, odorless drinking water, through the opening until the level is correct. Each cell is completely separate from the others, so each must be checked and filled individually.

If water is added in freezing weather, the vehicle should be driven several miles to allow the water to mix with the electrolyte. Otherwise, the battery could freeze.

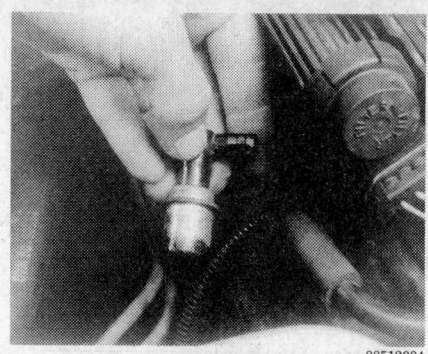

Removing the PCV valve

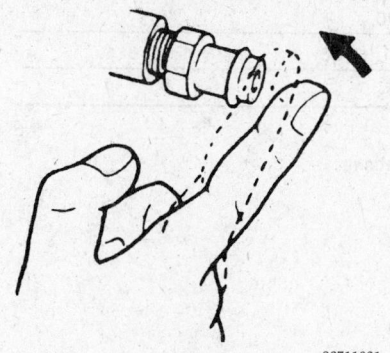

Checking the PCV valve

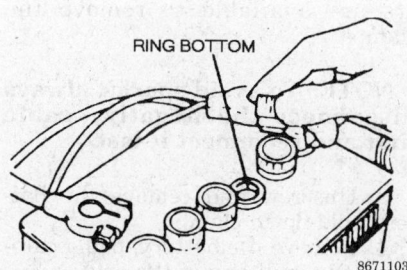

RING BOTTOM

Fill each battery cell to the bottom of the split ring with distilled water

SPECIFIC GRAVITY (EXCEPT MAINTENANCE-FREE BATTERIES)

NOTE: On a maintenance-free sealed battery, a built-in eye is used for checking the specific gravity readings. Refer to the battery case for further instructions.

At least once a year, check the specific gravity of the battery. It should be 1.26–1.28 at room temperature.

The specific gravity can be checked with the use of a hydrometer, an inexpensive instrument available from many sources, including auto parts stores. The hydrometer has a squeeze bulb at one end and a nozzle at the other. Battery electrolyte is sucked into the hydrometer until the float is lifted from its seat. The specific gravity is then read by noting the position of the float. Generally, if after charging, the specific gravity of any two cells varies more than 50 points (0.50), the battery is bad and should be replaced.

It is not possible to check the specific gravity in this manner on sealed (maintenance-free) batteries. Instead, the indicator built into the top of the case must be relied on to display any signs of battery deterioration. On most batteries if the indicator is a light color, the battery can be assumed to be OK. If the indicator is a dark color, the specific gravity is low, and the battery should be charged or replaced. There should be specific notations on the battery as to what color the indicator should be depending on the batteries state of charge.

CABLES AND CLAMPS

Once a year, the battery terminals and the cable clamps should be checked and cleaned, if necessary. Make sure that the ignition switch is turned to the **OFF** position. Loosen the clamps and remove the cables, negative cable first. On batteries with posts on top, the use of a puller specially made for this purpose is recommended. These are inexpensive, and available in most auto parts stores. Side terminal battery cables are secured with a bolt.

Clean the cable clamps and the battery terminal with a wire brush, until all corrosion, grease, etc., is removed and the metal is shiny. It is especially important to clean the inside of the clamp thoroughly, since a small deposit of foreign material or

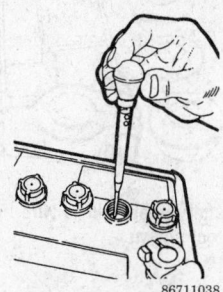

The specific gravity of the battery can be checked with a simple float-type hydrometer

oxidation there will prevent a sound electrical connection and inhibit either starting or charging. Special tools are available for cleaning these parts, one type for top post batteries and another type for side terminal batteries.

Before installing the cables, loosen the battery hold-down clamp or strap, remove the battery and check the battery tray. Clear it of any debris, and check it for soundness. Rust should be wire brushed away, and the metal given a coat of anti-rust paint. Install the battery and tighten the hold-down clamp or strap securely, but be careful not to overtighten, as doing so may crack the battery case.

After the clamps and terminals are clean, reinstall the cables, negative cable last; do not hammer on the clamps to install. Tighten the clamps securely, but do not distort them. Give the clamps and terminals a thin external coat of grease after installation, to retard corrosion.

Check the cables at the same time that the terminals are cleaned. If the cable insulation is cracked or broken, or if the ends are frayed, the cable should be replaced with a new cable of the same length and gauge.

CAUTION
Keep flame or sparks away from the battery; it gives off explosive hydrogen gas! Battery electrolyte contains sulfuric acid! If acid is splashed any on the skin or in the eyes, flush the affected area with plenty of clear water. If it lands in the eyes, get medical help immediately!

REPLACEMENT

When it becomes necessary to replace the battery, be sure to select a new battery with a cold cranking power rating equal to or greater than the battery originally installed. Deterioration, embrittlement and just plain aging of the battery cables, starter motor and associated wires makes the battery's job all the more difficult in successive years. The slow increase in electrical resistance over time makes it prudent to install a new battery with a greater capacity than the old.

REMOVAL & INSTALLATION

1. Make sure The ignition switch is turned **OFF**.
2. Disconnect the negative battery cable from the terminal, then discon-

BATTERY STATE OF CHARGE AT ROOM TEMPERATURE

Specific Gravity Reading	Charged Condition
1.260–1.280	Fully Charged
1.230–1.250	3/4 Charged
1.200–1.220	1/2 Charged
1.170–1.190	1/4 Charged
1.140–1.160	Almost no Charge
1.110–1.130	No Charge

86711039

Battery state of charge at room temperature — Generalized Specifications

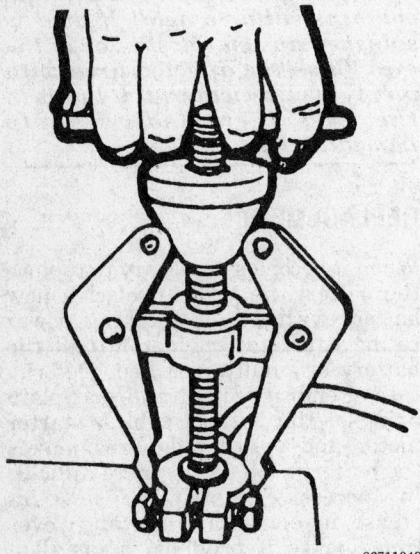

86711040

Special pullers are available to remove cable clamps

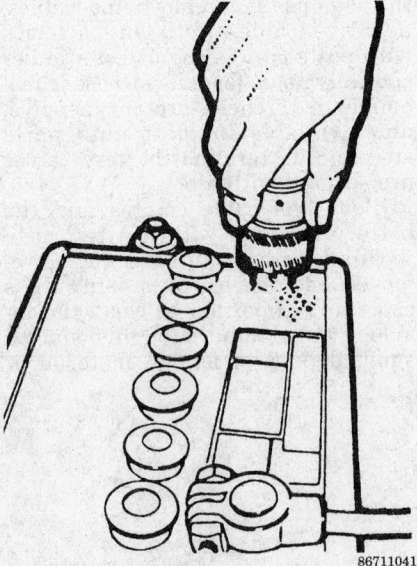

86711041

Clean the battery posts with a wire brush or the special tool shown

nect the positive cable. Special pullers are available to remove the clamps.

NOTE: To avoid sparks, always disconnect the negative cable first and reconnect it last.

3. Unscrew and remove the battery hold-down clamp.
4. Remove the battery, being careful not to spill any of the acid.

NOTE: Spilled acid can be neutralized with a baking soda and water solution. If, somehow, acid gets into the eyes, flush it out with lots of clean water and get to a doctor as quickly as possible.

To install:
5. Clean the battery posts thoroughly.
6. Clean the cable clamps using the special tools or a wire brush, both inside and out.
7. Install the battery, then fasten the hold-down clamp.
8. Connect the positive and then the negative cable. Do not hammer them into place. Coat the terminals with grease to prevent corrosion.

Clean the inside of the clamps with a wire brush or the special tool

Battery maintenance may be accomplished with household items (such as baking soda to neutralize spilled acid) or with special tools such as this post and terminal cleaner

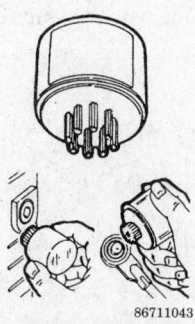

Special tools are also available for cleaning the posts and clamps of side terminal batteries

WARNING

Make absolutely sure that the battery is connected properly before turning ON the ignition switch. Reversed polarity can burn out the alternator and regulator in a matter of seconds.

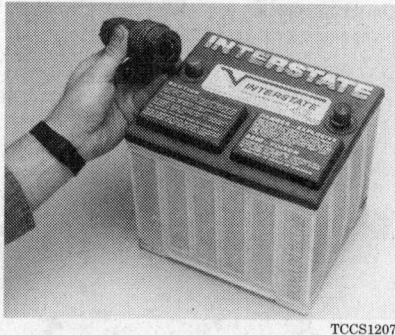

The underside of this special battery tool has a wire brush to clean post terminals

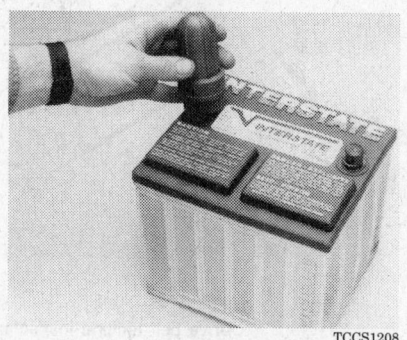

Place the tool over the terminals and twist to clean the post

Drive Belts

INSPECTION

Check the condition of the drive belts, and check the belt tension at least every 30,000 miles (48,000 km) or every 24 months.

Periodic inspection of the drive belts is important because of the following reasons; first of all, the drive belts drive various components such as the engine water pump, alternator, power steering pump and emission pump, etc.

Two of the components mention above play a vital part in keeping the engine running. They are the alternator and water pump. To give a little example of how important drive belt inspection is, picture this; suppose the alternator belt were to break due to wear or cracking, the alternator would be completely disabled and the battery would eventually go dead.

In case a like this, one just may find oneself sitting on the side of the road seeking someone to give the battery a jump to get started. Not to mention, a possible tow job, battery charge and replacement of that drive belt that could have been detected during the inspection.

The water pump drive belt could cause even more severe complications, how about an excessively overheated engine. This could result in a very expensive engine repair, like a head gasket replacement, etc. So be sure to keep a good maintenance check on the drive belts.

1. Inspect the belts for signs of glazing or cracking. A glazed belt will be perfectly smooth from slippage, while a good belt will have a slight texture of fabric visible. Cracks will generally start at the inner edge of the belt and run outward. Replace the belt at the first sign of cracking or if the glazing is severe.

2. By placing a thumb midway between the two pulleys, it should be possible to depress the belt 1/4–1/2 in. (6–13mm). If any of the belts can be depressed more than this, or cannot be depressed this much, adjust the tension. Inadequate tension will result in slippage or wear, while excessive tension will damage pulley bearings and cause belts to fray and crack.

3. It's not a bad idea to replace all drive belts at 60,000 miles (96,000 km) or 48 months, regardless of their condition.

ADJUSTMENT

Pivot Type Tensioner

This type of belt tension adjustment is commonly used in most vehicles today. This is a general procedure, consult specific adjustment and tension specifications for the vehicle being worked on.

1. Loosen the pivot and mounting bolts on the alternator.

2. Using a wooden hammer handle or broomstick, move the alternator

one way or the other until the tension is within acceptable limits.

NOTE: Never use a screwdriver or any other metal device, such as a pry bar, as a lever when adjusting the alternator belt tension!

3. Tighten the mounting bolts securely. If a new belt has been installed, always recheck the tension after a few hundred miles of driving.

Bolt Type Tensioner

Some belt tensions are adjusted by means of a tension adjusting bolt. This method of adjustment may use

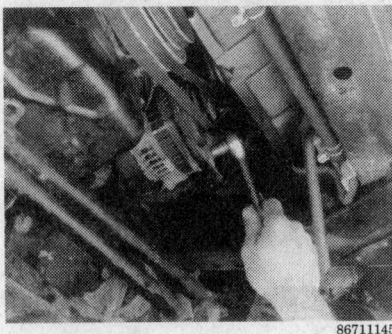
86711145

On some vehicles it is easier to access a component from underneath the vehicle

an idler pulley or the component being moved may slide on a bracket to increase or decrease belt tension. The procedure given below is is a general procedure, consult specific adjustment and tension specifications for the vehicle being worked on.

1. Loosen the pivot bolt, then turn the adjusting bolt until proper tension is achieved.

2. Tighten the mounting bolts securely. If a new belt has been installed, always recheck the tension after a few hundred miles of driving.

Serpentine Belt Type Tensioner

This type of belt tension adjustment is very commonly used in most vehicles today. Usually a serpentine type drive belt is used with the the tensioner type adjustment. The serpentine belt is one large single belt that wraps around each component's pulley. It will usually drive 3–4 components at the same time. Example: the alternator, water pump, air pump and power steering pump may be driven from this one belt. The procedure below is just a general procedure, consult specific adjustment and tension specifications for the vehicle being worked on.

1. Loosen the tensioner's pivot bolt.

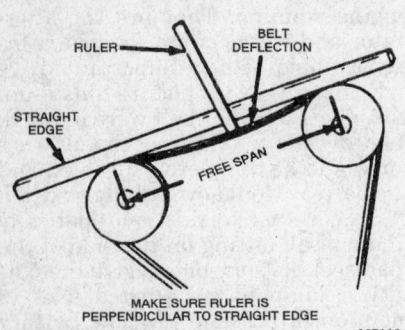
86711044

Measuring belt deflection with a straightedge and ruler

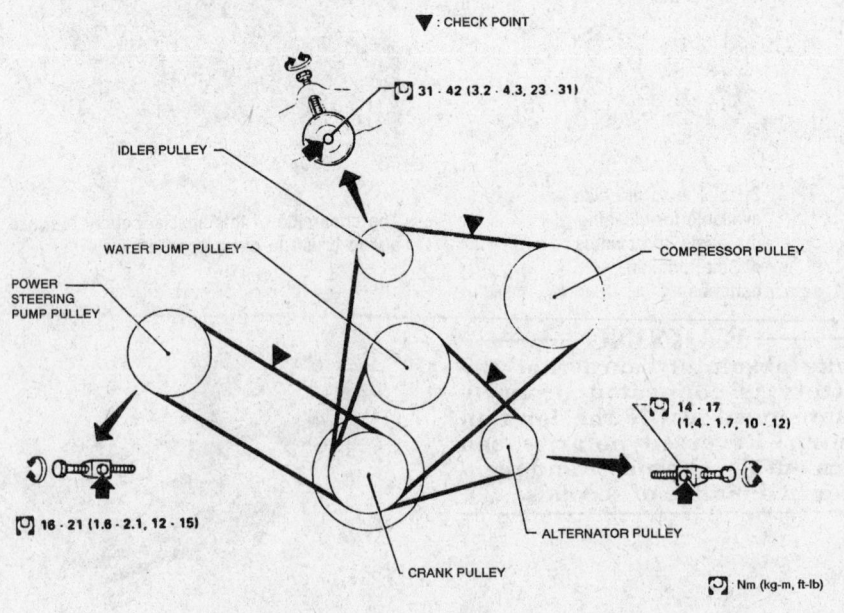
86711045

Drive belt tension inspection and adjustment points — typical

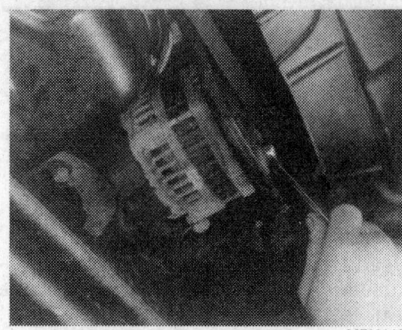

Loosen the alternator pivot bolt with a box wrench or a ratchet and socket

Use the adjusting bolt to vary tension on the belt

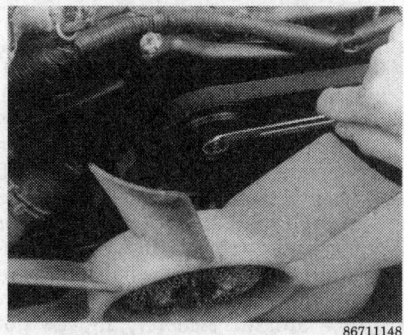

Loosen the locknut on the idler pulley before adjusting the belt

2. Usually the tensioner will have a large nut, with which a wrench can be used to relieve the tension from the belt. Moving against the tension of the tensioner will relieve the tension on the the belt. Allowing the tension of the tensioner to release will increase the tension on the belt within acceptable limits.

3. Tighten the pivot bolt securely. If a new belt has been installed, always recheck the tension after a few hundred miles of driving.

Turn the adjusting bolt until the correct belt tension is achieved

Timing Belt

INSPECTION

The timing belt is a bit more involved and a more critical service procedure. Although the do-it-yourself mechanic can service the timing belt, he/she should remember that the correct procedures must be followed exactly. Be sure to have the correct repair manual for the vehicle. If the home mechanic can enlist the aid of someone who is experienced with timing belt replacement, it would be helpful.

NOTE: Do not bend or twist the timing belt. If the timing belt breaks while driving, or the crankshaft and/or camshaft are turned separately after the timing belt is removed, valves may strike the piston heads, causing engine damage. Make sure the timing belt and tensioner are clean and free from oil and water.

As a average rule, replace the timing belt at 60,000 miles (96,000 km). These are just generalizations, always consult specific adjustment and tension specifications for the vehicle being worked on.

Evaporative Canister

SERVICING

Check the evaporation control system, if so equipped, every 15,000 miles (24,000 km) or every 12 months. Check the fuel and vapor lines/hoses for proper connections, correct routing, and condition. Replace damaged or deteriorated parts as necessary.

To check the operation of the carbon canister purge control valve, disconnect the rubber hose between the canister control valve and the T-fitting at the T-fitting. Apply vacuum to the hose leading to the control valve. The vacuum condition should be maintained indefinitely. If the control valve leaks, remove the top cover of the valve and check for a dislocated or cracked diaphragm. If the diaphragm is damaged, a repair kit containing a new diaphragm, retainer, and spring is available and should be installed.

The carbon canister has a replaceable air filter in the bottom of the canister. The filter element should be checked once a year or every 15,000 miles (24,000 km); more frequently if the vehicle is operated in dusty areas. Replace the filter by pulling it out of the bottom of the canister and installing a new one.

Hoses

INSPECTION

Inspect the condition of the radiator hoses, heater hoses and clamps periodically. Early spring and late fall are often good times to perform this, as well as other routine maintenance. Make sure the engine and cooling system are cold. Visually inspect for cracked, rotted or collapsed hoses, and replace as necessary. Run a hand along the length of the hose. If a weak or swollen spot is noted when squeezing the hose wall, replace the hose.

REPLACEMENT

1. Drain the coolant into a suitable container (if the coolant is to be reused).

— CAUTION —
When draining coolant, keep in mind that cats and dogs are attracted by ethylene glycol antifreeze, and are quite likely to drink any that is left in an uncovered container or in puddles on the ground. This will prove fatal in sufficient quantity. Always drain the coolant into a sealable container. Coolant should be reused unless it is contaminated or several years old.

2. Loosen the hose clamps at each end of the hose that requires replacement.

3. Twist, pull and slide the hose off the radiator, water pump, thermostat housing or heater connection.

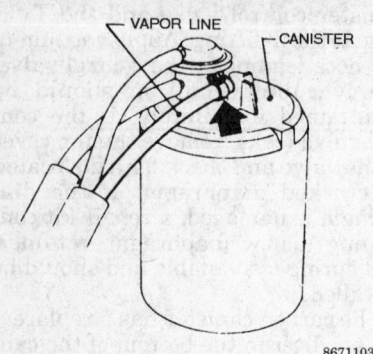

Checking the evaporative canister

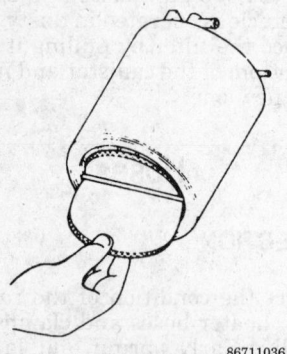

Replacing the evaporative canister filter

Remove the lines to the evaporative canister assembly before removing the canister

4. Clean the hose mounting connections. Inspect the hose clamps and replace any which are rusted or worn.

5. Position the hose clamps on the new hose.

6. Coat the connection surfaces with a water resistant sealer or equivalent and slide the hose into position. Make sure the hose clamps are located beyond the raised bead of the connector (if equipped) and centered in the clamping area of the connection.

7. Tighten the clamps evenly. Do not overtighten.

Unfasten the evaporative canister assembly retaining clamp

Remove the evaporative canister assembly from the vehicle

8. Refill the cooling system.

9. Start the engine and allow it to reach normal operating temperature. Check for coolant leaks, then top off the coolant level as necessary.

ADDITIONAL PREVENTIVE MAINTENANCE CHECKS

Antifreeze

In order to prevent heater core freeze-up during A/C operation, it is necessary to maintain permanent-type antifreeze protection of +15°F (-9°C) or lower. A reading of -15°F (-26°C) is ideal since this protection also supplies sufficient corrosion inhibitors for the protection of the engine cooling system.

NOTE: The same antifreeze should not be used longer than the manufacturer specifies.

Radiator Cap

For efficient operation of the vehicle's cooling system, the radiator cap should have a holding pressure which meets manufacturer's specifications. A cap which fails to hold the specified pressure should be replaced.

FLUIDS AND LUBRICANTS

Fluid Disposal

Used fluids such as engine oil, transmission fluid, antifreeze and brake fluid are hazardous wastes and must be disposed of properly. Before draining any fluids, consult with the local authorities; in many areas, waste oil, antifreeze, etc. are being accepted as a part of recycling programs. A number of service stations and auto parts stores are also accepting waste fluids for recycling.

Be sure of the recycling center's policies before draining any fluids, as many will not accept different fluids that have been mixed together, such as oil and antifreeze.

Oil and Fuel Recommendations

ENGINE OIL

The SAE (Society of Automotive Engineers) grade number indicates the viscosity of the engine oil (its resistance to flow at a given temperature). The lower the SAE grade number, the lighter the oil. For example, the mono-grade oils begin with SAE 5 weight, which is a thin, light oil, and continue in viscosity up to SAE 80 or 90 weight, which are heavy gear lubricants. These oils are also known as "straight weight," meaning they are of a single viscosity, and do not vary with engine temperature.

Multi-viscosity oils offer the important advantage of being adaptable to temperature extremes. These oils have designations such as 10W-40, 20W-50, etc. For example, 10W-40 means that in winter (the "W" in the designation) the oil acts like a thin 10 weight oil, allowing the engine to spin easily when cold and offering rapid lubrication. Once the engine has warmed up, however, the oil acts like a straight 40 weight, maintaining good lubrication and protection for the engine's internal components. A 20W-50 oil would therefore be slightly heavier than, and not as ideal, in cold weather as the 10W-40, but would offer better protection at higher rpm and temperatures because, when warm, it acts like a 50 weight oil. Whichever oil viscosity is

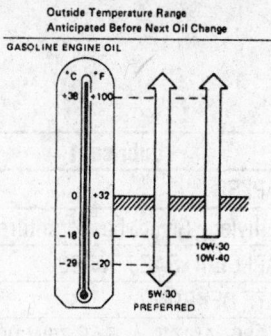

Engine oil viscosity chart — Typical

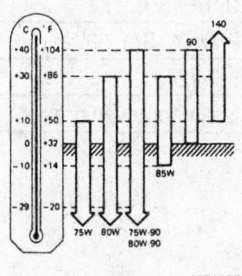

Gear oil viscosity chart — Typical

chosen when changing the oil and filter, anticipate the temperatures the engine will be operating at until the oil is changed again. Refer to the oil viscosity chart that applies to the specific vehicle for oil recommendations according to temperature.

The API (American Petroleum Institute) designation indicates the classification of engine oil used under certain given operating conditions. Only oils designated for use "Service SG" should be used. Oils of the SG type perform a variety of functions inside the engine in addition to the basic function as a lubricant. Through a balanced system of metallic detergents and polymeric dispersants, the oil prevents the formation of high and low temperature deposits, and also keeps sludge and dirt particles in suspension. Acids, particularly sulfuric acid, as well as other by-products of combustion, are neutralized. Both the SAE grade number and the API designation can be found on the oil container.

Synthetic Oil

There are many excellent synthetic and fuel-efficient oils currently available that can provide better gas mileage, longer service life and, in some cases, better engine protection. These benefits do not come without a few hitches, however, the main one being the price of synthetic oils, which is three or four times the price per quart of conventional oil.

Synthetic oil is not for every vehicle and every type of driving, so the engine's condition and type of driving the vehicle experiences should be considered. Also, check the vehicle's warranty conditions regarding the use of synthetic oils.

Brand new engines and older, high mileage engines are not good candidates for synthetic oil. The synthetic oils are so slippery that they can prevent the proper break-in of new engines; most manufacturers recommend that one should wait until the engine is properly broken in (3000 miles) before using synthetic oil. Older engines with wear have a different problem with synthetics: they "use" (consume during operation) more oil as they age. Slippery synthetic oils get past these worn parts easily. If the engine uses (burns or leaks) conventional oil, it will use synthetics much faster. Also, if the vehicle is leaking oil past old seals, it will have a much greater leak problem with synthetics.

Consider the type of driving. If most of the accumulated mileage is high speed, highway type driving, the more expensive synthetic oil may be a benefit. Extended highway driving gives the engine a chance to warm up, accumulating fewer acids in the oil, and putting less stress on the engine over the long run. Under these conditions, the oil change interval can be extended (as long as the oil filter can last the extended life of the oil) up to the advertised mileage claims of the synthetics. Vehicles with synthetic oils may show increased fuel economy in highway driving, due to less internal friction. However, many automotive experts agree that 50,000 miles (80,000 km) is too long to keep any oil in the engine.

Vehicles used under harsher circumstances (stop-and-go, city type driving, short trips, extended idling) should be serviced more frequently. For the engines in these vehicles, the much greater cost of synthetic or fuel-efficient oils may not be worth the investment. Internal wear increases much quicker on these vehicles, causing greater oil consumption and leakage.

NOTE: The mixing of conventional and synthetic oils is possible but not recommended. Nondetergent or straight mineral oils must never be used in the engine.

FUEL

It is important to use fuel of the proper octane rating in the vehicle. Octane rating is based on the quantity of anti-knock compounds added to the fuel, and also reflects the speed at which the gas will burn. The lower the octane rating, the faster it burns. The higher the octane, the slower the fuel will burn, and the greater the percentage of compounds in the fuel to prevent spark ping (knock), detonation and preignition (dieseling).

As the temperature of the engine increases, the air/fuel mixture exhibits a tendency to ignite before the spark plug is fired. If fuel of an octane rating too low for the engine is used, this will allow combustion to occur before the piston has completed its compression stroke, thereby creating a very high pressure very rapidly.

Fuel of the proper octane rating, for the compression ratio and ignition timing of the vehicle, will slow the combustion process sufficiently to allow the spark plug enough time to ignite the mixture completely and smoothly. The use of super-premium fuel is no substitution for a properly tuned and maintained engine.

Light spark knock may be noticed when accelerating or driving up hills. The slight knocking may be considered normal (with 87 octane) because the maximum fuel economy is obtained under condition of occasional light spark knock. Gasoline with an octane rating higher than 87 may be used, but it is not necessary (in most cases) for proper operation.

NOTE: The engine's fuel requirement can change with time, mainly due to carbon buildup, which changes the compression ratio. If the engine pings, knocks or runs on, switch to a higher grade of fuel. Sometimes just changing brands may cure the problem.

RECOMMENDED LUBRICANTS

Component	Lubricant
Engine oil	API SG
Coolant	Ethylene Glycol-based Antifreeze
Manual Transmission	API GL-4, SAE 75W-90
Automatic Transmission	ATF DEXRON®
Transfer Case	1989: API GL-4, SAE 75W-90
	1990-95: ATF DEXRON®
Differentials	API GL-5, SAE 80W-90
Limited Slip	Nissan-approved LSD
Master Cylinder	DOT 3, SAE J1703
Power Steering	ATF DEXRON®
Manual Steering	API GL-4, SAE 90W
Multi-Purpose Grease	NLGI #2
Free-Running Hub	Nissan-approved grease

86711059

Example of a Recommended Lubricants chart

OIL LEVEL CHECK

Every time the vehicle is refueled, the engine oil should be checked as follows:

1. Park the vehicle on level ground

NOTE: Although it is best for the engine to be at operating temperature, checking the oil immediately after stopping will lead to a false reading. Wait a few minutes after turning off the engine to allow the oil to drain back into the crankcase.

2. Open the hood and locate the dipstick. Pull the dipstick from its tube, wipe it clean and reinsert it.

NOTE: Keep in mind that this is a generalized procedure. The actual markings on the vehicle's dipstick may vary from those described here.

3. Pull the dipstick out again, and holding it horizontally, read the oil level. The oil should be between the **H** and **L** marks on the dipstick. If the oil is below the **L** mark, add oil of the proper viscosity and classification through the capped opening on top of the cylinder head cover.

4. Insert the dipstick and check the oil level again after adding any

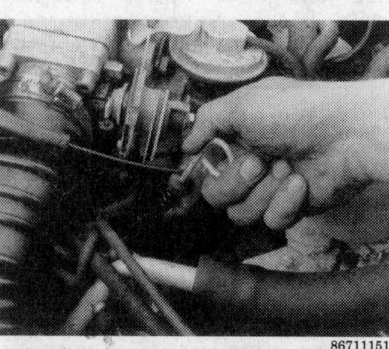

86711151

The oil dipstick in the engine compartment may be painted yellow on newer models

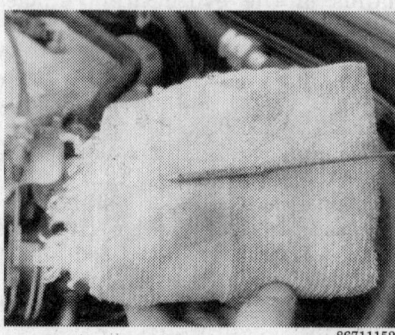

86711152

Check the oil dipstick for the correct level of engine oil — never overfill the engine oil

oil. Be careful not to overfill the crankcase. Approximately one quart of oil will raise the level from the **L** mark to the **H** mark. Excess oil will generally be consumed at an accelerated rate.

OIL AND FILTER CHANGE

NOTE: It may be a good idea to look under the vehicle, before starting any service procedure, to familiarize oneself with the necessary components and locations.

The oil should be changed at least every 7500 miles (12,000 km) or every 6 months. Some manufacturers recommend changing the oil filter with every other oil change; we suggest that the filter be changed with **every** oil change. There is approximately 1 quart of dirty oil remaining in the old oil filter if it is not changed! A few dollars more every year seems a small price to pay for extended en-

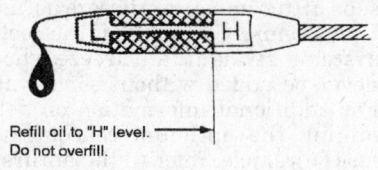

Refill oil to "H" level.
Do not overfill.

86711064

The engine oil level should be maintained between the L and H marks

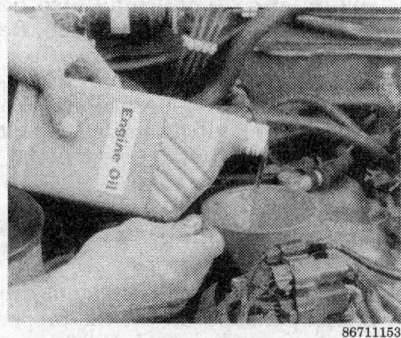

86711153

If the engine oil level is low, add engine oil, but do not overfill

gine life — so change the filter every time the oil is changed!

CAUTION

Prolonged and repeated skin contact with used engine oil, with no effort to remove the oil, may be harmful. Always follow these simple precautions when handling used motor oil.

• Avoid prolonged skin contract with used motor oil
• Remove oil from skin by washing thoroughly with soap and water, or waterless hand cleaner. Do not use gasoline, paint thinner or other solvents
• Avoid prolonged skin contact with oil-soaked clothing

The mileage figures given are sample recommended intervals assuming normal driving and conditions. If the vehicle is being used under dusty, polluted or off-road conditions, change the oil and filter more frequently than specified. The same goes for vehicles driven in stop-and-go traffic or only for short distances. Always drain the oil after the engine has been running long enough to bring it to normal operating temperature. Hot oil will flow easier and more contaminants will be removed along

with the oil than if it were drained cold. To change the oil and filter:

CAUTION

The EPA warns that prolonged contact with used engine oil may cause a number of skin disorders, including cancer! Every effort should be made to minimize personal exposure to used engine oil. Protective gloves should be worn when changing the oil. Wash the hands and any other exposed skin areas as soon as possible after exposure to used engine oil. Soap and water, or waterless hand cleaner should be used.

1. Run the engine until it reaches normal operating temperature.
2. Jack up the front of the vehicle and support it on safety stands.
3. Slide a drain pan of at least 6 quarts capacity under the oil pan.
4. Loosen the drain plug. Turn the plug out by hand. By keeping inward pressure on the plug as it is unscrewed, oil won't escape past the threads, and it can be removed without being burned by hot oil.

CAUTION

The oil will be HOT! Be careful when removing the plug, so as to avoid any painful burns.

5. Allow the oil to drain completely. Clean and inspect the drain plug and oil pan sealing surface. If the plug is equipped with a removable gasket, also clean and inspect it.
6. Using a new plug gasket, if necessary, install the drain plug and tighten to correct specifications. Do not overtighten the plug; otherwise, a new pan or a replacement plug for stripped threads will have to be purchased.

NOTE: Always consult specific service and specifications for the vehicle being worked on.

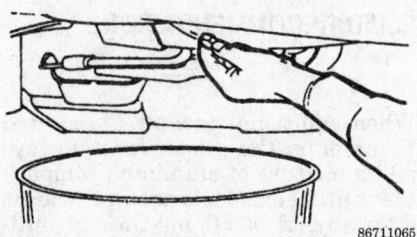

86711065

By keeping inward pressure on the drain plug as it is removed, oil won't escape past the threads

7. Some engines will require the use of a oil filter strap wrench to remove the oil filter. Others may require a cap-type filter removal tool. Keep in mind that the filter is holding about one quart of dirty, hot oil.

NOTE: If the oil filter cannot be loosened by conventional methods, punch a hole through both sides near the mounting base of the filter and insert a punch, then turn to loosen the oil filter. After the oil filter is loosened, remove it from the engine with an oil filter wrench or equivalent.

8. Empty the old filter into the drain pan and properly dispose of the filter.
9. Using a clean rag, wipe off the filter adapter on the engine block. Be sure that the rag doesn't leave any lint which could clog an oil passage.
10. Coat the rubber gasket on the filter with fresh oil. Spin it onto the engine by hand; when the gasket touches the adapter surface, give it another $1/2$–$3/4$ turn. Do not overtighten, or the gasket will be distorted and it will leak.
11. Refill the engine with the correct amount of fresh oil.
12. Check the oil level on the dipstick. It is normal for the level to be a bit above the full mark. Start the engine and allow it to idle for a few minutes.

NOTE: Do not run the engine above idle speed until it has built up oil pressure, as indicated when the oil light goes out.

13. Shut off the engine and allow the oil to drain into the crankcase for a few minutes, then check the oil level. Check around the filter and drain plug for any leaks and correct as necessary.

Power Steering Pump

Check the power steering fluid level every 6 months or 6000 miles (9600 km), whichever occurs first.

1. Park the vehicle on a level surface. Run the engine until normal operating temperature is reached.
2. Turn the steering all the way to the left and then all the way to the right several times. Center the steering wheel and shut off the engine.
3. Open the hood and check the power steering reservoir fluid level.
4. Remove the filler cap and wipe the attached dipstick clean.
5. Reinsert the dipstick and tighten the cap. Remove the dipstick

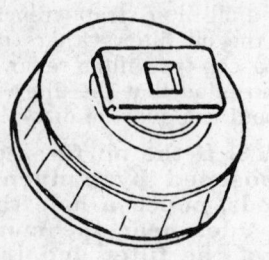

86711066

On some models, a cap-type oil filter removal tool works best

TCCS1901

Lubricate the gasket on the new filter with clean engine oil. A dry gasket may not make as good a seal, and could allow the filter to leak

86711154

Removing the oil drain plug — do not overtorque this drain plug upon installation

and note the fluid level indicated on the dipstick.

6. The level should be at any point below the upper hash mark, but not below the lower hash mark (in the HOT or COLD ranges).

7. Add fluid as necessary, but do not overfill.

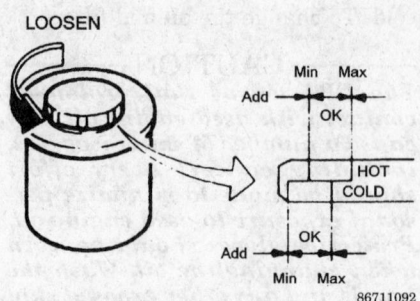

86711092

Checking the power steering fluid level

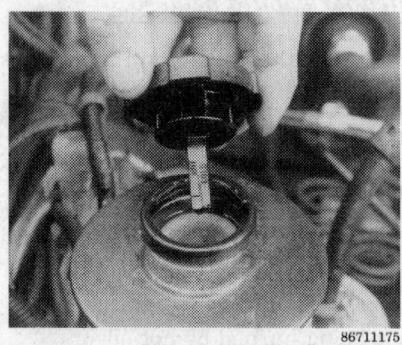

86711175

Remove the power steering cap to check the fluid level

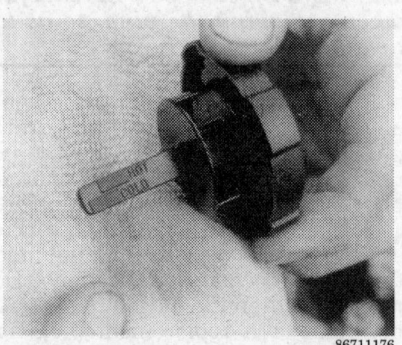

86711176

View of the power steering cap dipstick — note the hot and cold marks

Cooling System

FLUID RECOMMENDATIONS

When additional coolant is required to maintain the proper level, always add a mixture of aluminum-compatible antifreeze/coolant and water. Typically, a 50/50 mixture of antifreeze and water is recommended (even for vehicles which are not exposed to cold winter temperatures), since this mixture also imparts the necessary corrosion inhibition. A greater concentration of antifreeze may be used, but the coolant mixture's level of protection actually lessens if too much antifreeze is used. Unless simply topping off the cooling system, straight antifreeze should never be added without some water. For additional information on determining the optimum concentration for the vehicle, refer to the antifreeze manufacturer's labeling.

NOTE: **Although most manufacturers recommend ethylene glycol-based antifreeze (which has long been the prevalent type on the market), other types (such as propylene glycol) may also be suitable for use in the vehicle. Be sure to thoroughly read the alternative product's labeling to ensure compatibility before switching to a different formula. Check vehicle manufacturer's recommendations to be sure.**

FLUID LEVEL CHECK

Dealing with the cooling system can be a tricky matter unless the proper precautions are observed. It is best to check the coolant level in the radiator when the engine is cold. This is done by checking the expansion tank. If coolant is visible above the **MIN** mark on the tank, the level is satisfactory. Always be certain that the filler caps on both the radiator and the reservoir are tightly closed.

In the event that the coolant level must be checked when the engine is warm or on engines without an expansion tank, place a thick rag over the radiator cap, then slowly turn the cap counterclockwise until it reaches the first detent. Allow all the hot steam to escape. This will allow the pressure in the system to drop gradually, preventing an explosion of hot coolant. When the hissing noise stops, remove the cap the rest of the way.

It's a good idea to check the coolant every time that one stops for fuel. If the coolant level is low, add equal amounts of suitable antifreeze and clean water. Fill the expansion tank to the **MAX** level. On models without an expansion tank, add coolant through the radiator filler neck.

NOTE: **Never add cold coolant to a hot engine unless the engine is running, to avoid cracking the engine block.**

Avoid using water that is known to have a high alkaline content or is very hard, except in emergency situations. Drain and flush the cooling

Adding power steering fluid — use a funnel to avoid spills

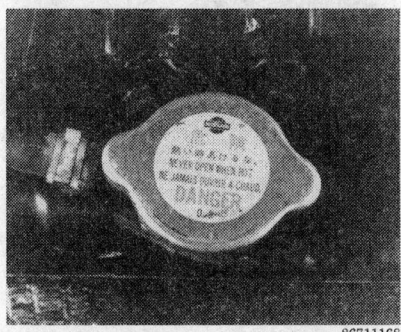

View of the radiator cap installed — never open when the engine is hot!

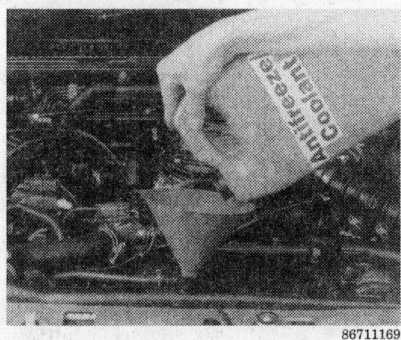

Add engine coolant to the radiator with a funnel to avoid spills

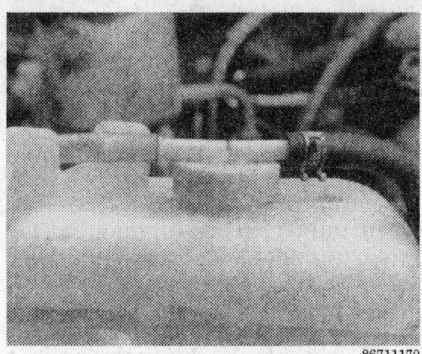

View of the coolant expansion tank

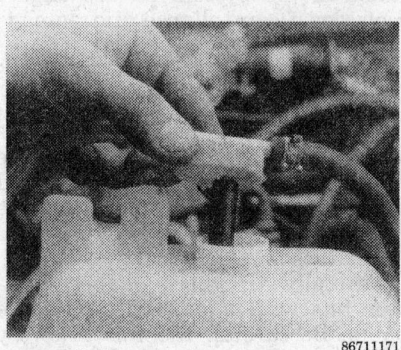

Remove the cap on the coolant expansion tank and add coolant to the proper level

Fluid level marks on the coolant recovery tank

DRAIN, REFILL AND FLUSH

system as soon as possible after using such water.

The radiator hoses and clamps and the radiator cap should be checked at the same time as the coolant level. Hoses which are brittle, cracked, or swollen should be replaced. Clamps should be checked for tightness (screwdriver-tight only)! Do not allow the clamp to cut into the hose or crush the fitting. The radiator cap gasket should be checked for any tears, cracks, swelling, or any signs of incorrect seating in the radiator neck.

CAUTION

When draining coolant, keep in mind that cats and dogs are attracted by ethylene glycol antifreeze, and are quite likely to drink any that is left in an uncovered container or in puddles on the ground. This will prove fatal in sufficient quantity. Always drain the coolant into a sealable container. Coolant should be reused unless it is contaminated or two years old.

Complete draining and refilling of the cooling system at least once every two years will remove accumulated rust, scale and other deposits.

NOTE: Use a good quality antifreeze with water pump lubricants, rust inhibitors and other corrosion inhibitors along with acid neutralizers.

1. Drain the existing coolant as follows: Position suitable drain pans beneath the radiator and engine block. Open the radiator petcock and engine drain plug(s); there may be 1 or 2 drain plugs on the engine block depending on the type of engine. Another method of draining coolant is to disconnect the bottom radiator hose at the radiator outlet.

NOTE: If it is rusted or difficult to open, spray the radiator petcock with some penetrating lubricant.

2. Set the heater temperature controls to the full HOT position.

3. Close the petcock and tighten the drain plug(s) to correct specifications or reconnect the lower hose. Open the air relief plug, if so equipped, then fill the system with water.

4. Add a can of quality radiator flush. Be sure the flush is safe to use in engines having aluminum components.

5. Idle the engine until the upper radiator hose gets hot. Race it 2 or 3 times, then shut it **OFF**. Let the engine cool down.

6. Drain the system again.

7. Repeat this process until the drained water is clear and free of scale.

8. Close the petcock and drain plug(s) or, if applicable, connect the radiator hose.

9. If equipped with a coolant recovery system, flush the reservoir with water and leave empty.

NOTE: Always open the air relief plug before filling the cooling system, in order to bleed the trapped air. Only when the cooling system is bled properly can the correct amount of coolant be added to the system.

10. Determine the capacity of the cooling system. Add the appropriate ratio of quality aluminum-compatible antifreeze and water (normally a 50/50 mix) to provide the desired protection. With the air relief plug open, add the coolant mixture through the radiator filler neck until full, then close the bleeder plug and radiator cap.

Add engine coolant to the expansion tank with a funnel to avoid spills

11. Using the same concentration of clean antifreeze and water, fill the expansion tank to the **MAX** line, then cap the tank.

SYSTEM INSPECTION

Most permanent antifreeze/coolants have a colored dye added which makes the solution an excellent leak detector. When servicing the cooling system, check for leakage at:
- All hoses and hose connections
- Radiator seams, radiator core, and radiator draincock
- All engine block and cylinder head freeze (core) plugs, and drain plugs
- Edges of all cooling system gaskets (head gaskets, thermostat gasket)
- Transmission fluid cooler
- Heating system components
- Water pump

In addition, check the engine oil dipstick for signs of coolant in the oil; also, check the coolant in the radiator for signs of oil. Investigate and correct any indication of coolant leakage.

Check the Radiator Cap

While checking the coolant level, check the radiator cap for a worn or cracked gasket. If the cap doesn't seal properly, fluid will be lost and the engine will overheat. A worn cap should be replaced with a new one. The radiator cap must maintain pressure when the engine is running, or the cooling system will "boil over". The radiator cap also has a 2-way valve design to allow coolant to be drawn into the radiator from the coolant overflow tank. If this valve is not functioning properly, the vacuum cause in the system as the engine cools down can collapse and damage the hoses.

Clean Radiator of Debris

Periodically clean any debris such as leaves, paper, insects, etc., from the radiator fins. Pick the large pieces off by hand. The smaller pieces can be washed away with water pressure from a hose.

Carefully straighten any bent radiator fins with a pair of needlenose pliers. Be careful, the fins are very soft. Don't wiggle the fins back and forth too much. Straighten them once and try not to move them again.

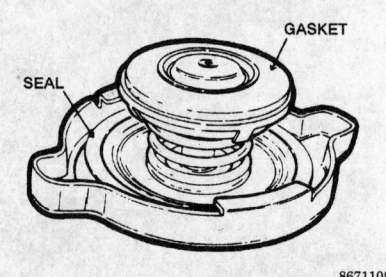

Check the radiator cap seal and gasket condition

CHECKING SYSTEM PROTECTION

A 50/50 mix of antifreeze, or coolant concentrate, and water will usually provide the necessary protection. Freeze protection may be checked by using a cooling system hydrometer. Inexpensive hydrometers (floating ball types) may be obtained from a local department store (automotive section) or an auto supply store. Follow the directions packaged with the coolant hydrometer when checking protection.

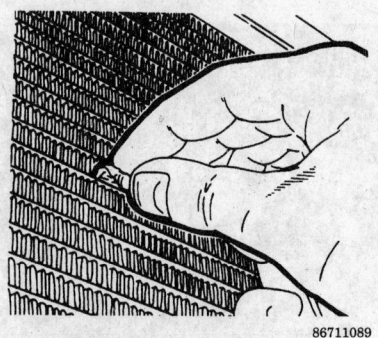

Clean the radiator fins of any debris which impedes air flow

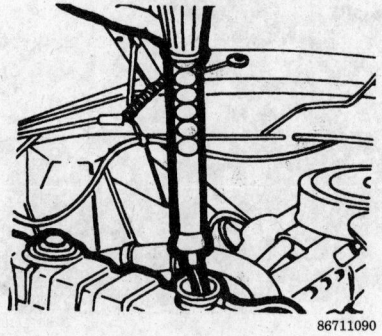

The freeze protection rating can be checked with an antifreeze tester

SPECIFICATIONS

32

CHRYSLER CORPORATION

FORD MOTOR COMPANY

GENERAL MOTORS CORPORATION

LEBARON/DAYTONA/SHADOW/SPIRIT/ACCLAIM SUNDANCE

VEHICLE IDENTIFICATION CHART

Engine Code							Model Year	
Code	Liters	Cu. In. (cc)	Cyl.	Fuel Sys.	Eng. Mfg.		Code	Year
A	2.2	135 (2212)	I4	MFI-Turbo	Chrysler		P	1993
D	2.2	135 (2212)	I4	TFI	Chrysler		R	1994
K	2.5	153 (2507)	I4	TFI	Chrysler		S	1995
V	2.5	153 (2507)	I4	MFI-M85	Chrysler			
3	3.0	181 (2972)	V6	MFI	Mitsubishi			

M85 - 85% Methanol, flexible fuel engine
MFI - Multipoint fuel injection
TFI - Throttle body fuel injection

ENGINE IDENTIFICATION
All measurements are given in inches.

Year	Model	Engine Displacement Liters (cc)	Engine Series (ID/VIN)	Fuel System	No. of Cylinders	Engine Type
1993	Acclaim	2.5 (2507)	K	TFI	4	SOHC
	Acclaim	2.5 (2507)	V	MFI-M85	4	SOHC
	Acclaim	3.0 (2972)	3	MFI	6	SOHC
	Daytona	2.2 (2212)	A	MFI-Turbo	4	DOHC
	Daytona	2.5 (2507)	K	TFI	4	SOHC
	Daytona	3.0 (2972)	3	MFI	6	SOHC
	Lebaron	2.5 (2507)	K	TFI	4	SOHC
	Lebaron	3.0 (2972)	3	MFI	4	SOHC
	Shadow	2.2 (2212)	D	TFI	4	SOHC
	Shadow	2.5 (2507)	K	TFI	4	SOHC
	Shadow	3.0 (2972)	3	MFI	6	SOHC
	Spirit	2.5 (2507)	K	TFI	4	SOHC
	Spirit	2.5 (2507)	V	MFI-M85	4	SOHC
	Spirit	3.0 (2972)	3	MFI	6	SOHC
	Sundance	2.2 (2212)	D	TFI	4	SOHC
	Sundance	2.5 (2507)	K	TFI	4	SOHC
	Sundance	3.0 (2972)	3	MFI	6	SOHC
1994	Acclaim	2.5 (2507)	K	TFI	4	SOHC
	Acclaim	2.5 (2507)	V	MFI-M85	4	SOHC
	Acclaim	3.0 (2972)	3	MFI	6	SOHC
	Lebaron	3.0 (2972)	3	MFI	6	SOHC
	Shadow	2.2 (2212)	D	TFI	4	SOHC
	Shadow	2.5 (2507)	K	TFI	4	SOHC
	Shadow	3.0 (2972)	3	MFI	6	SOHC
	Spirit	2.5 (2507)	K	TFI	4	SOHC
	Spirit	2.5 (2507)	V	MFI-M85	4	SOHC
	Spirit	3.0 (2972)	3	MFI	6	SOHC
	Sundance	2.2 (2212)	D	TFI	4	SOHC

ENGINE IDENTIFICATION

All measurements are given in inches.

Year	Model	Engine Displacement Liters (cc)	Engine Series (ID/VIN)	Fuel System	No. of Cylinders	Engine Type
1994	Sundance	2.5 (2507)	K	TFI	4	SOHC
	Sundance	3.0 (2972)	3	MFI	6	SOHC
1995	Acclaim	2.5 (2507)	K	TFI	4	SOHC
	Acclaim	2.5 (2507)	V	MFI-M85	4	SOHC
	Acclaim	3.0 (2972)	3	MFI	6	SOHC
	Lebaron	3.0 (2972)	3	MFI	6	SOHC
	Spirit	2.5 (2507)	K	TFI	4	SOHC
	Spirit	2.5 (2507)	V	MFI-M85	4	SOHC
	Spirit	3.0 (2972)	3	MFI	6	SOHC

DOHC - Double overhead camshaft
M85 - 85% Methanol flexible fuel engine
MFI - Multipoint fuel M85 - 85% Methanol flexible fuel engine

SOHC - Single overhead camshaft
TFI - Throttle body fuel injection

GENERAL ENGINE SPECIFICATIONS

Year	Engine ID/VIN	Engine Displacement Liters (cc)	Fuel System Type	Net Horsepower @ rpm	Net Torque @ rpm (ft. lbs.)	Bore x Stroke (in.)	Compression Ratio	Oil Pressure @ rpm
1993	A	2.2 (2212)	MFI-Turbo	224@6000	217@2800	3.44x3.62	7.8:1	25-80@3000
	D	2.2 (2212)	TFI	93@4800	122@3200	3.44x3.62	9.5:1	25-80@3000
	K	2.5 (2507)	TFI	100@4800	135@2800	3.44x4.09	8.9:1	25-80@3000
	V	2.5 (2507)	MFI-M85	106@4400	145@2400	3.44x4.09	9.0:1	25-80@3000
	3	3.0 (2972)	MFI	141@5000	171@2800	3.59x2.99	8.9:1	28.5@3000
1994	3	3.0 (2972)	MFI	141@5000	171@2800	3.59x2.99	8.9:1	35-75@3000
	D	2.2 (2212)	TFI	93@4800	122@3200	3.44x3.62	9.5:1	25-80@3000
	K	2.5 (2507)	TFI	100@4800	135@2800	3.44x4.09	8.9:1	25-80@3000
	V	2.5 (2507)	MFI-M85	106@4400	145@2400	3.44x4.09	9.0:1	25-80@3000
1995	3	3.0 (2972)	MFI	141@5000	170@2800	3.59x2.99	8.9:1	35-75@3000
	K	2.5 (2507)	TFI	100@4800	135@2800	3.44x4.09	8.9:1	25-80@3000
	V	2.5 (2507)	MFI-M85	106@4400	145@2400	3.44x4.09	9.0:1	25-80@3000

TFI - Throttle body fuel injection
MFI - Multiport fuel injection
M85 - 85% Methanol flexible fuel engine

GASOLINE ENGINE TUNE-UP SPECIFICATIONS

Year	Engine ID/VIN	Engine Displacement Liters (cc)	Spark Plugs Gap (in.)	Ignition Timing (deg.) MT	Ignition Timing (deg.) AT	Fuel Pump (psi)	Idle Speed (rpm) MT	Idle Speed (rpm) AT	Valve Clearance In.	Valve Clearance Ex.
1993	3	3.0 (2972)	0.040	-	12B	48	-	700	HYD	HYD
	A	2.2 (2212)	0.035	NA	-	55	850	-	HYD	HYD
	D	2.2 (2212)	0.035	12B	12B	39	900	800	HYD	HYD
	K	2.5 (2507)	0.035	12B	12B	39	850	850	HYD	HYD
	V	2.5 (2507)	0.035	12B	12B	55	850	850	HYD	HYD
1994	3	3.0 (2972)	0.035	-	①	48	-	700	HYD	HYD
	D	2.2 (2212)	0.035	①	①	39	850	800	HYD	HYD
	K	2.5 (2507)	0.035	①	①	39	900	850	HYD	HYD
	V	2.5 (2507)	0.035	12B	12B	55	850	850	HYD	HYD

GASOLINE ENGINE TUNE-UP SPECIFICATIONS

Year	Engine ID/VIN	Engine Displacement Liters (cc)	Spark Plugs Gap (in.)	Ignition Timing (deg.) MT	Ignition Timing (deg.) AT	Fuel Pump (psi)	Idle Speed (rpm) MT	Idle Speed (rpm) AT	Valve Clearance In.	Valve Clearance Ex.
1995	3	3.0 (2972)	0.035	-	①	48	-	700	HYD	HYD
	K	2.5 (2507)	0.035	-	①	39	-	850	HYD	HYD
	V	2.5 (2507)	0.035	-	①	55	-	850	HYD	HYD

NOTE: The Vehicle Emission Control Information label often reflects specification changes made during production. The label figures must be used if they differ from those in this chart.

HYD - Hydraulic

B - Before top dead center

① Refer to the Vehicle Emission Control Information label for proper timing specifications with a range of +/- 2 degrees

CAPACITIES

Year	Model	Engine ID/VIN	Engine Displacement Liters (cc)	Engine Oil with Filter (qts.)	Transmission (pts.) 4-Spd	Transmission (pts.) 5-Spd	Transmission (pts.) Auto.	Transfer Case (pts.)	Drive Axle Front (pts.)	Drive Axle Rear (pts.)	Fuel Tank (gal.)	Cooling System (qts.)
1993	Acclaim	K	2.5 (2507)	4.5	-	4.8	③	-	-	-	16.0	9.0
	Acclaim	V	2.5 (2507)	4.5	-	-	③	-	-	-	18.0	9.0
	Acclaim	3	3.0 (2972)	4.5	-	-	③	-	-	-	16.0	0.5
	Daytona	A	2.2 (2212)	4.0	-	4.8	-	-	-	-	14.0	9.0
	Daytona	K	2.5 (2507)	4.0	-	4.8	①	-	-	-	14.0	9.0
	Daytona	3	3.0 (2972)	4.0	-	4.8	①	-	-	-	16.0	9.5
	Lebaron	K	2.5 (2507)	4.5	-	5.0	①	-	-	-	14.0	9.0
	Lebaron	3	3.0 (2972)	4.5	-	5.0	①	-	-	-	14.0	9.0
	Shadow	D	2.2 (2212)	4.0	-	4.8	①	-	-	-	14.0	9.0
	Shadow	K	2.5 (2507)	4.0	-	4.8	①	-	-	-	14.0	9.0
	Shadow	3	3.0 (2972)	4.0	-	4.8	①	-	-	-	14.0	9.5
	Spirit	K	2.5 (2507)	4.0	-	4.8	①	-	-	-	16.0	9.0
	Spirit	V	2.5 (2507)	4.0	-	-	①	-	-	-	18.0	9.0
	Spirit	3	3.0 (2972)	4.0	-	-	①	-	-	-	16.0	9.5
	Sundance	D	2.2 (2212)	4.5	-	4.8	③	-	-	-	14.0	9.0
	Sundance	K	2.5 (2507)	4.5	-	4.8	③	-	-	-	14.0	9.0
	Sundance	3	3.0 (2972)	4.5	-	4.8	③	-	-	-	14.0	9.0
1994	Acclaim	K	2.5 (2507)	4.5	-	4.8	③	-	-	-	16.0	9.0
	Acclaim	V	2.5 (2507)	4.5	-	-	③	-	-	-	18.0	9.0
	Acclaim	3	3.0 (2972)	4.5	-	-	③	-	-	-	16.0	9.5
	Lebaron	3	3.0 (2972)	4.5	-	-	①	-	-	-	14.0	9.5
	Shadow	D	2.2 (2212)	4.5	-	4.8	①	-	-	-	14.0	9.0
	Shadow	K	2.5 (2507)	4.5	-	4.8	①	-	-	-	14.0	9.0
	Shadow	3	3.0 (2972)	4.5	-	4.8	①	-	-	-	14.0	9.5
	Spirit	K	2.5 (2507)	4.5	-	4.8	①	-	-	-	16.0	9.0
	Spirit	V	2.5 (2507)	4.5	-	-	①	-	-	-	18.0	9.0
	Spirit	3	3.0 (2972)	4.5	-	-	①	-	-	-	16.0	9.5
	Sundance	D	2.2 (2212)	4.5	-	4.8	③	-	-	-	14.0	9.0
	Sundance	K	2.5 (2507)	4.5	-	4.8	③	-	-	-	14.0	9.0
	Sundance	3	3.0 (2972)	4.5	-	4.8	③	-	-	-	14.0	9.5
1995	Acclaim	K	2.5 (2507)	4.5	-	4.8	③	-	-	-	16.0	9.0
	Acclaim	V	2.5 (2507)	4.5	-	-	③	-	-	-	18.0	9.0
	Acclaim	3	3.0 (2972)	4.5	-	-	③	-	-	-	16.0	9.5
	Lebaron	3	3.0 (2972)	4.5	-	-	19.8 ②	-	-	-	14.0	9.5

CAPACITIES

Year	Model	Engine ID/VIN	Engine Displacement Liters (cc)	Engine Oil with Filter (qts.)	Transmission (pts.)			Transfer Case (pts.)	Drive Axle		Fuel Tank (gal.)	Cooling System (qts.)
					4-Spd	5-Spd	Auto.		Front (pts.)	Rear (pts.)		
1995	Spirit	K	2.5 (2507)	4.5	-	4.8	①	-	-	-	16.0	9.0
	Spirit	V	2.5 (2507)	4.5	-	-	①	-	-	-	18.0	9.0
	Spirit	3	3.0 (2972)	4.5	-	-	①	-	-	-	16.0	9.5

① A413 - 17.8 pts.
   A413 (fleet) - 18.4 pts.
   A413 (lock-up) - 17.0 pts.
   A604 (electronic) - 18.2 pts.

② Overhaul fill capacity with torque converter empty

③ Non-fleet models: 8.9 qts.
   Fleet models: 9.2 qts.
   A413 with lock-up converter: 8.5 qts.
   A604 transaxle: 9.1 qts.

VALVE SPECIFICATIONS

Year	Engine ID/VIN	Engine Displacement Liters (cc)	Seat Angle (deg.)	Face Angle (deg.)	Spring Test Pressure (lbs. @ in.)	Spring Installed Height (in.)	Stem-to-Guide Clearance (in.)		Stem Diameter (in.)	
							Intake	Exhaust	Intake	Exhaust
1993	A	2.2 (2212)	45	45	225@1.34	1.650	0.0010-0.0040	0.0020-0.0040	0.2740	0.2730
	D	2.2 (2212)	45	45	114@1.65	1.650	0.0009-0.0026	0.0030-0.0047	0.3124	0.3103
	K	2.5 (2507)	45	45	114@1.65	1.650	0.0010-0.0030	0.0030-0.0047	0.3124	0.3103
	V	2.5 (2507)	45	45	114@1.65	1.650	0.0010-0.0030	0.0030-0.0047	0.3124	0.3103
	3	3.0 (2972)	44	45-45.5	73@1.59	1.990	0.0010-0.0020	0.0019-0.0030	0.3140	0.3125
1994	D	2.2 (2212)	45	45	114@1.65	1.650	0.0009-0.0026	0.0030-0.0047	0.3124	0.3103
	K	2.5 (2507)	45	45	114@1.65	1.650	0.0009-0.0026	0.0030-0.0047	0.3124	0.3103
	V	2.5 (2507)	45	45	114@1.65	1.650	0.0010-0.0030	0.0030-0.0047	0.3124	0.3103
	3	3.0 (2972)	44.5	45-45.5	73@1.59	1.960 ①	0.0010-0.0020	0.0019-0.0030	0.3140	0.3125
1995	K	2.5 (2507)	45	45	114@1.65	1.650	0.0009-0.0026	0.0030-0.0047	0.3124	0.3103
	V	2.5 (2507)	45	45	114@1.65	1.650	0.0010-0.0030	0.0030-0.0047	0.3124	0.3103
	3	3.0 (2972)	44.0-44.3	45-45.5	73@1.59	1.590	0.0010-0.0020	0.0019-0.0030	0.3140	0.3125

① Free height

TORQUE SPECIFICATIONS
All readings in ft. lbs.

Year	Engine ID/VIN	Engine Displacement Liters (cc)	Cylinder Head Bolts	Main Bearing Bolts	Rod Bearing Bolts	Crankshaft Damper Bolts	Flywheel Bolts	Manifold		Spark Plugs	Lug Nut
								Intake	Exhaust		
1993	A	2.2 (2212)	①	②	50	80	70	17	17	20	95
	D	2.2 (2212)	①	②	③	85	70	17	17	26	95
	K	2.5 (2507)	①	②	③	85	70	17	17	26	95
	V	2.5 (2507)	①	②	③	85	70	17	17	26	95
	3	3.0 (2972)	80	60	38	112	70	17	17	20	95

TORQUE SPECIFICATIONS
All readings in ft. lbs.

Year	Engine ID/VIN	Engine Displacement Liters (cc)	Cylinder Head Bolts	Main Bearing Bolts	Rod Bearing Bolts	Crankshaft Damper Bolts	Flywheel Bolts	Manifold Intake	Manifold Exhaust	Spark Plugs	Lug Nut
1994	D	2.2 (2212)	①	②	③	85	70	17	17	20	95
	K	2.5 (2507)	①	②	③	85	70	17	17	20	95
	V	2.5 (2507)	①	②	③	85	70	17	17	26	95
	3	3.0 (2972)	80	60	38	112	70	15	16	20	95
1995	K	2.5 (2507)	①	②	③	85	70	17	17	20	95
	V	2.5 (2507)	①	②	③	85	70	17	17	26	95
	3	3.0 (2972)	80	60	38	112	70	15	16	20	95

① Step 1: 45 ft. lbs.
Step 2: 65 ft. lbs.
Step 3: 65 ft. lbs.
Step 4: Plus 1/4 turn

② Step 1: 30 ft. lbs
Step 2: Plus 1/4 turn

③ Step 1: 40 ft. lbs.
Step 2: Plus 1/4 turn

BRAKE SPECIFICATIONS
All measurements in inches unless noted

Year	Model		Master Cylinder Bore	Brake Disc Original Thickness	Brake Disc Minimum Thickness	Brake Disc Maximum Runout	Brake Drum Diameter Original Inside Diameter	Brake Drum Diameter Max. Wear Limit	Brake Drum Diameter Maximum Machine Diameter	Minimum Lining Thickness Front	Minimum Lining Thickness Rear
1993	Acclaim	F	0.827	0.861	0.803	0.005	10.24	NA	NA	0.300	0.300
	①	R	-	0.856	0.797	0.005	-	-	-	-	0.280
	Daytona	F	0.827	0.940	0.882	0.006	-	-	-	0.300	-
	①	R	-	0.468	0.409	0.005	10.24	NA	NA	-	0.300
	②	R	-	0.856	0.797	0.005	10.24	NA	NA	-	0.300
	Lebaron	F	0.827	0.861	0.803	0.005	7.87	③	③	0.300	0.062
	①	R	-	0.468	0.409	0.005	-	-	-	-	0.280
	②	R	-	0.856	0.797	0.005	-	-	-	-	0.280
	Shadow	F	0.827	0.935	0.882	0.005	-	-	-	0.300	-
		R	-	0.468	0.409	0.005	7.87	NA	NA	-	0.280
	Spirit	F	0.827	0.861	0.803	0.004	-	-	-	0.300	-
		R	-	0.856	0.797	0.005	10.24	NA	NA	-	0.280
	Sundance	F	0.827	0.935	0.882	0.005	-	-	-	0.300	0.280
	①	R	-	0.468	0.409	0.005	-	-	-	-	0.280
1994	Acclaim	F	0.827	0.935	0.882	0.005	10.24	NA	NA	0.300	0.300
	⑤	R	-	0.468	0.409	0.005	-	-	-	-	0.280
	Lebaron	F	0.827	0.935	0.882	0.005	7.87	③	③	0.300	0.062
	①	R	-	0.468	0.409	0.005	-	-	-	-	0.280
	②	R	-	0.856	0.797	0.005	-	-	-	-	0.280
	Shadow	F	0.827	0.935	0.882	0.005	-	-	-	0.300	-
		R	-	0.468	0.409	0.005	7.87	NA	NA	-	0.280
	Spirit	F	0.827	0.935	0.882	0.005	-	-	-	0.300	-
	④	R	-	0.856	0.797	0.005	10.24	③	③	-	0.280
	Sundance	F	0.875	0.935	0.882	0.005	7.87	NA	NA	0.300	0.280
	①	R	-	0.468	0.409	0.005	-	-	-	-	0.280

BRAKE SPECIFICATIONS
All measurements in inches unless noted

Year	Model		Master Cylinder Bore	Brake Disc Original Thickness	Brake Disc Minimum Thickness	Brake Disc Maximum Runout	Brake Drum Diameter Original Inside Diameter	Max. Wear Limit	Maximum Machine Diameter	Minimum Lining Thickness Front	Minimum Lining Thickness Rear
1995	Acclaim	F	0.827	0.935	0.882	0.005	10.24	NA	NA	0.300	0.300
	⑤	R	-	0.468	0.409	0.005	-	-	-	-	0.280
	Lebaron	F	0.827	0.930	0.882	0.005	-	-	-	0.300	0.062
	①	R	-	0.467	0.409	0.005	8.66	③	③	-	0.280
	②	R	-	0.856	0.797	0.005	-	-	-	-	0.280
	Spirit	F	0.827	0.935	0.882	0.005	-	-	-	0.300	0.300
		R	⑤	0.468	0.409	0.005	10.24	③	③	-	0.280

F - Front
R - Rear

① Solid rear disc
② Vented rear disc
③ Maximum diameter is stamped on drum

④ Solid rear discs
Original thickness: 0.468
Minimum thickness: 0.409

⑤ Optional vented rear disc brakes:
Original thickness: 0.856
Minimum thickness: 0.797

FREQUENT MAINTENANCE LABOR
CHRYSLER LEBARON, DODGE DAYTONA

The following should be used as a guide when determining the amount of work required for a particular service if taken to a repair shop. In estimating how long a particular Frequent Maintenance Service item should take, please observe the following:
- **Factory Time** is time that is generated by the vehicle manufacturer.
- **Chilton Time** is time that is based on field research and data supplied by the vehicle manufacturer.
- All labor time operations are given in hours and tenths of an hour.
- All labor operations, are to be used as a **guide**.

COOLING

(G) Winterize Cooling System
Includes: Run engine to check for leaks, tighten all hose connections. Test radiator and pressure cap, drain radiator and engine block. Add antifreeze and refill system.
1993-95 5

(G) Belt, Drive, Renew
1993-95
V belt
Fan & Alternator (.2) 3
Power Steering (.4) 6
w/AC add 1
Air Conditioning (.3) 4
Serpentine belt (.2) 4

(G) Belt, Drive, Adjust
1993-95
one 2
each adtnl. 1

(G) Hoses, Radiator, Renew
Includes: Drain and refill cooling system as required.
1993-95
upper (.3) 4
lower (.4) 6

(G) Thermostat, Coolant, Renew
1993-95 (.4) 6

FUEL

(M) Air Cleaner, Service
1993-95 3

(G) Filter, Fuel, Renew
1993-95 in line (.3) 4
1993-95 in tank (1.0) 1.4

BRAKES

(G) Bleed Brakes (Four Wheels)
Includes: Add fluid.
1993 (.4) 6
Bleed modulator, add 6
1994-95 (.4) 6
w/Antilock 4 add 4
Bleed modulator, add 1.5

(G) Brakes, Adjust (Minor)
Includes: Adjust brakes, fill master cylinder.
1993-95, two wheels 4

(M) Parking Brake, Adjust
1993-95 (.3) 4

LUBRICATION SERVICES

(M) Engine Oil & Filter, Renew
Includes: Inspect and correct all fluid levels.
1993-95 (.2) 3

(M) Lubricate Chassis, Change Oil & Filter
Includes: Inspect and correct all fluid levels.
1993-95 6
Install grease fittings, add 1

(M) Lubricate Chassis
Includes: Inspect and correct all fluid levels.
1993-95 4
Install grease fittings, add 1

WHEELS

(G) Wheel, Renew (One)
1993-95 5

(G) Wheel, Balance
1993-95
one 3
each adtnl. 2

(G) Wheels, Rotate (All)
1993-95 5

ELECTRICAL

(G) Headlamps, Aim
1993-95
two 4
four 6

FREQUENT MAINTENANCE LABOR (cont.)
CHRYSLER LEBARON, DODGE DAYTONA

	Factory Time	Chilton Time
(G) Halogen Headlamp Bulb, Renew		
1993-95, each (.2)		.3
(G) High Mount Stop Lamp and/or Lens, Renew		
1993-95 (.7)		1.0

	Factory Time	Chilton Time
(G) License Lamp Assy., Renew		
1993-95 (.2)		.3
(G) Tail Lamp Assy., Renew		
1993-95		
Daytona, Laser (.5)		.7
LeBaron (.3)		.4

	Factory Time	Chilton Time
(G) Turn Signal & Parking Lamp Assy., Renew		
1993-95 (.2)		.4
(G) Horn, Renew		
1993-95, one electric (.2)		.4
(M) Terminals, Battery, Clean		
1993-95		.3

FREQUENT MAINTENANCE LABOR
DODGE SPIRIT, PLYMOUTH ACCLAIM

The following should be used as a guide when determining the amount of work required for a particular service if taken to a repair shop. In estimating how long a particular Frequent Maintenance Service item should take, please observe the following:
- **Factory Time** is time that is generated by the vehicle manufacturer.
- **Chilton Time** is time that is based on field research and data supplied by the vehicle manufacturer.
- All labor time operations are given in hours and tenths of an hour.
- All labor operations, are to be used as a **guide**.

	Factory Time	Chilton Time
COOLING		
(G) Winterize Cooling System		
Includes: Run engine to check for leaks, tighten all hose connections. Test radiator and pressure cap, drain radiator and engine block. Add antifreeze and refill system.		
1993-95		.5
(G) Belt, Drive, Renew		
1993-95		
V belt		
Fan & alternator		.3
Power steering		.6
w/AC add		.1
Air conditioner		.3
Serpentine		.4
(G) Belt, Drive, Adjust		
1993-95		
one		.2
each adtnl.		.1
(G) Hoses, Radiator, Renew		
Includes: Drain and refill cooling system as required.		
1993-95		
upper		.4
lower		.6
(G) Thermostat, Coolant, Renew		
1993-95		.6

	Factory Time	Chilton Time
FUEL		
(M) Air Cleaner, Service		
1993-95		.2
(G) Filter, Fuel, Renew		
1993-95		
in line		.8
in tank		1.4
BRAKES		
(G) Bleed Brakes (Four Wheels)		
Includes: Add fluid.		
1993 (.4)		.6
Bleed modulator, add		.6
1994-95 (.4)		.6
w/antilock 4 add		.6
Bleed modulator, add		1.5
(G) Brakes, Adjust (Minor)		
Includes: Adjust brakes, fill master cylinder.		
1993-95, two wheels		.4
(M) Parking Brake, Adjust		
1993-95 (.3)		.4
LUBRICATION SERVICES		
(M) Engine Oil & Filter, Renew		
Includes: Inspect and correct all fluid levels.		
1993-95		.3

	Factory Time	Chilton Time
(M) Lubricate Chassis, Change Oil & Filter		
Includes: Inspect and correct all fluid levels.		
1993-95		.6
Install grease fittings, add		.1
(M) Lubricate Chassis		
Includes: Inspect and correct all fluid levels.		
1993-95		.4
Install grease fittings, add		.1
WHEELS		
(G) Wheel, Renew (One)		
1993-95, one		.5
(G) Wheel, Balance		
1993-95		
one		.3
each adtnl.		.2
(G) Wheels, Rotate (All)		
1993-95		.5
ELECTRICAL		
(G) Headlamps, Aim		
1993-95		
two		.4
four		.6
(G) Halogen Headlamp Bulb, Renew		
1993-95, each		.3

FREQUENT MAINTENANCE LABOR (cont.)
DODGE SPIRIT, PLYMOUTH ACCLAIM

	(Factory Time)	Chilton Time
(G) High Mount Stop Lamp and/or Lens, Renew		
1993-95		.4
(G) License Lamp Assy., Renew		
1993-95		.3

	(Factory Time)	Chilton Time
(G) Tail Lamp Assy., Renew		
1993-95		.4
(G) Tail Lamp Lens or Bulb, Renew		
1993-95		.5

	(Factory Time)	Chilton Time
(G) Turn Signal & Parking Lamp Assy., Renew		
1993-95		.3
(G) Horn, Renew		
1993-95, one		.4
(M) Terminals, Battery, Clean		
1993-95		.3

FREQUENT MAINTENANCE LABOR
DODGE SHADOW, PLYMOUTH SUNDANCE

The following should be used as a guide when determining the amount of work required for a particular service if taken to a repair shop. In estimating how long a particular Frequent Maintenance Service item should take, please observe the following:
- **Factory Time** is time that is generated by the vehicle manufacturer.
- **Chilton Time** is time that is based on field research and data supplied by the vehicle manufacturer.
- All labor time operations are given in hours and tenths of an hour.
- All labor operations, are to be used as a **guide**.

COOLING

(G) Winterize Cooling System
Includes: Run engine to check for leaks, tighten all hose connections. Test radiator and pressure cap, drain radiator and engine block. Add antifreeze and refill system.
1993-94 5

(G) Belt, Drive, Renew
1993-94
  V-belt
    Fan & Alternator (.2) 3
    Power steering (.4) 6
w/AC add 1
    AC (.2) 3
  Serpentine belt (.2) 4

(G) Belt, Drive, Adjust
1993-94
  one 2
  each adtnl. 1

(G) Hoses, Radiator, Renew
Includes: Drain and refill cooling system as required.
1993-94
  upper (.3) 4
  lower (.4) 6

(G) Thermostat, Coolant, Renew
1993-94 (.4) 6

FUEL

(M) Air Cleaner, Service
1993-94 3

(G) Filter, Fuel, Renew
  in line (.3) 4
  in tank (.5) 8

1993-94
  in line (.3) 4
  in tank (1.0) 1.4

BRAKES

(G) Bleed Brakes (Four Wheels)
Includes: Add fluid.
1993 (.4) 6
Bleed modulator, add 6
1994 (.4) 6
w/antilock 4, add 4
Bleed modulator, add 1.5

(G) Brakes, Adjust (Minor)
Includes: Adjust brakes, fill master cylinder.
1993-94, two wheels 4

(M) Parking Brake, Adjust
1993-94 (.3) 4

LUBRICATION SERVICES

(M) Engine Oil & Filter, Renew
Includes: Inspect and correct all fluid levels.
1993-94 (.2) 3

(M) Lubricate Chassis, Change Oil & Filter
Includes: Inspect and correct all fluid levels.
1993-94 6
Install grease fittings, add 1

(M) Lubricate Chassis
Includes: Inspect and correct all fluid levels.
1993-94 4
Install grease fittings, add 1

WHEELS

(G) Wheel, Renew (One)
1993-94, one 5

(G) Wheel, Balance
1993-94
  one 3
  each adtnl. 2

(G) Wheels, Rotate (All)
1993-94 5

ELECTRICAL

(G) Headlamps, Aim
1993-94
  two 4
  four 6

(G) Halogen Headlamp Bulb, Renew
1993-94 each (.2) 3

(G) License Lamp Assy., Renew
1993-94 (.2) 3

(G) Tail Lamp Assy., Renew
1993-94
  2 door (.2) 4
  4 door (.3) 5

(G) Turn Signal & Parking Lamp Assy., Renew
1993-94 (.3) 4

(G) Horn, Renew
1993-94
  one (.2) 4
  each adtnl. 1

(M) Terminals, Battery, Clean
1993-94 3

SCHEDULED MAINTENANCE INTERVALS
(CHRYSLER LEBARON, DODGE DAYTONA, SHADOW, SPIRIT & PLYMOUTH ACCLAIM & SUNDANCE)

TO BE SERVICED	TYPE OF SERVICE	VEHICLE MILEAGE INTERVAL (x1000)												
		7.5	15	22.5	30	37.5	45	52.5	60	67.5	75	82.5	90	97.5
Engine oil & filter	R	✓	✓	✓	✓	✓	✓	✓	✓	✓	✓	✓	✓	✓
Exhaust system	S/I	✓	✓	✓	✓	✓	✓	✓	✓	✓	✓	✓	✓	✓
Brake hoses	S/I	✓	✓	✓	✓	✓	✓	✓	✓	✓	✓	✓	✓	✓
CV joints & front suspension components	S/I	✓	✓	✓	✓	✓	✓	✓	✓	✓	✓	✓	✓	✓
Rotate tires	S/I	✓	✓	✓	✓	✓	✓	✓	✓	✓	✓	✓	✓	✓
Coolant level, hoses & clamps	S/I	✓	✓	✓	✓	✓	✓	✓	✓	✓	✓	✓	✓	✓
Accessory drive belts	S/I		✓		✓		✓				✓		✓	
Brake linings & rear wheel bearings	S/I			✓			✓			✓			✓	
Spark plugs	R				✓				✓				✓	
Air filter element	R				✓				✓				✓	
Lubricate steering linkage, tie rod ends & ball joints	S/I				✓				✓				✓	
Engine Coolant	R							✓				✓		
Automatic transaxle fluid & filter	R				✓				✓					
Ignition cables	R								✓					
Camshaft timing belt (2.2L & 2.5L)	R												✓	
Camshaft timing belt (3.0L)	R								✓					
PCV valve	S/I								✓					

R – Replace   S/I – Service or Inspect

FREQUENT OPERATION MAINTENANCE (SEVERE SERVICE)

If a vehicle is operated under any of the following conditions it is considered severe service:
- Extremely dusty areas.
- 50% or more of the vehicle operation is in 32°C (90°F) or higher temperatures, or constant operation in temperatures below 0°C (32°F).
- Prolonged idling (vehicle operation in stop and go traffic).
- Frequent short running periods (engine does not warm to normal operating temperatures).
- Police, taxi, delivery usage or trailer towing usage.

CV joints & front suspension components - check every 3000 miles.
Oil & oil filter change – change every 3000 miles.
Rotate tires every 6000 miles.
Brake linings & rear wheel bearings– check every 9000 miles.
Air filter element – change every 15,000 miles.
Automatic transaxle fluid – change every 15,000 miles.
Tie rod ends, steering linkage & ball joints – lubricate every 15,000 miles.
PCV valve – check every 30,000 miles.

CONCORDE/LHS/NEW YORKER/INTREPID/VISION

VEHICLE IDENTIFICATION CHART

		Engine Code					Model Year	
Code	Liters	Cu. In. (cc)	Cyl.	Fuel Sys.	Eng. Mfg.		Code	Year
R	3.3	201 (3300)	V6	MFI	Chrysler		P	1993
T	3.3	201 (3301)	V6	MFI	Chrysler		R	1994
F	3.5	215 (3518)	V6	MFI	Chrysler		S	1995
L	3.8	231 (3785)	V6	MFI	Chrysler		T	1996
							V	1997

MFI - Multiport fuel injection

ENGINE IDENTIFICATION
All measurements are given in inches.

Year	Model	Engine Displacement Liters (cc)	Engine Series (ID/VIN)	Fuel System	No. of Cylinders	Engine Type
1993	Concorde	3.3 (3300)	T	MFI	6	OHV
	Concorde	3.5 (3518)	F	MFI	6	SOHC
	Intrepid	3.3 (3300)	T	MFI	6	OHV
	Intrepid	3.5 (3518)	F	MFI	6	SOHC
	New Yorker	3.3 (3300)	R	MFI	6	OHV
	New Yorker	3.8 (3785)	L	MFI	6	OHV
	Vision	3.3 (3294)	T	MFI	6	OHV
	Vision	3.5 (3518)	F	MFI	6	SOHC
1994	Concorde	3.3 (3300)	T	MFI	6	OHV
	Concorde	3.5 (3518)	F	MFI	6	SOHC
	Intrepid	3.3 (3300)	T	MFI	6	OHV
	Intrepid	3.5 (3518)	F	MFI	6	SOHC
	LHS	3.5 (3518)	F	MFI	6	SOHC
	New Yorker	3.5 (3518)	F	MFI	6	SOHC
	Vision	3.3 (3294)	T	MFI	6	OHV
	Vision	3.5 (3518)	F	MFI	6	SOHC
1995	Concorde	3.3 (3301)	T	MFI	6	OHV
	Concorde	3.5 (3518)	F	MFI	6	SOHC
	Intrepid	3.3 (3300)	T	MFI	6	OHV
	Intrepid	3.5 (3518)	F	MFI	6	SOHC
	LHS	3.5 (3518)	F	MFI	6	SOHC
	New Yorker	3.5 (3518)	F	MFI	6	SOHC
	Vision	3.3 (3301)	T	MFI	6	OHV
	Vision	3.5 (3518)	F	MFI	6	SOHC
1996-97	Concorde	3.3 (3301)	T	MFI	6	OHV
	Concorde	3.5 (3518)	F	MFI	6	SOHC
	Intrepid	3.3 (3300)	T	MFI	6	OHV
	Intrepid	3.5 (3518)	F	MFI	6	SOHC
	LHS	3.5 (3518)	F	MFI	6	SOHC
	Vision	3.3 (3301)	T	MFI	6	OHV
	Vision	3.5 (3518)	F	MFI	6	SOHC

MFI - Multiport fuel injection   SOHC - Single overhead camshaft   OHV - Overhead valve

GENERAL ENGINE SPECIFICATIONS

Year	Engine ID/VIN	Engine Displacement Liters (cc)	Fuel System Type	Net Horsepower @ rpm	Net Torque @ rpm (ft. lbs.)	Bore x Stroke (in.)	Com-pression Ratio	Oil Pressure @ rpm
1993	R	3.3 (3300)	MFI	147@4800	183@3600	3.66x3.19	8.9:1	30-80@3000
	T	3.3 (3300)	MFI	153@5300	177@2800	3.66x3.19	8.9:1	30-80@3000
	F	3.5 (3518)	MFI	214@5800	221@2800	3.78x3.19	10.5:1	25-70@3000
	L	3.8 (3785)	MFI	150@4400	203@3200	3.78x3.42	9.0:1	30-80@3000
1994	T	3.3 (3300)	MFI	161@5300	181@3200	3.66x3.19	8.9:1	30-80@3000
	F	3.5 (3518)	MFI	214@5800	221@3100	3.78x3.19	10.5:1	25-80@3000
1995	T	3.3 (3301)	MFI	161@5300	181@3200	3.66x3.19	8.9:1	30-80@3000
	F	3.5 (3518)	MFI	214@5800	221@3100	3.78x3.19	9.6:1	27-70@3000
1996-97	T	3.3 (3301)	MFI	161@5300	181@3200	3.66x3.19	8.9:1	30-80@3000
	F	3.5 (3518)	MFI	214@5800	221@3100	3.78x3.19	9.6:1	25-80@3000

MFI - Multiport fuel injection

GASOLINE ENGINE TUNE-UP SPECIFICATIONS

Year	Engine ID/VIN	Engine Displacement Liters (cc)	Spark Plugs Gap (in.)	Ignition Timing (deg.) MT	Ignition Timing (deg.) AT	Fuel Pump (psi)	Idle Speed (rpm) MT	Idle Speed (rpm) AT	Valve Clearance In.	Valve Clearance Ex.
1993	R	3.3 (3300)	0.050	-	12B	①	-	750	HYD	HYD
	T	3.3 (3300)	0.048-0.053	-	②	③	-	600-840	HYD	HYD
	F	3.5 (3518)	0.048-0.053	-	②	④	-	600-840	HYD	HYD
	L	3.8 (3785)	0.050	-	12B	①	-	750-1100	HYD	HYD
1994	T	3.3 (3300)	0.048-0.053	-	②	③	-	600-840	HYD	HYD
	F	3.5 (3518)	0.048-0.053	-	②	④	-	600-840	HYD	HYD
1995	T	3.3 (3301)	0.048-0.053	-	②	55 ①	-	600-840	HYD	HYD
	F	3.5 (3518)	0.048-0.053	-	②	48 ①	-	750-1100	HYD	HYD
1996-97	T	3.3 (3301)	0.048-0.053	-	②	55 ①	-	600-840	HYD	HYD
	F	3.5 (3518)	0.033-0.038	-	②	48 ①	-	750-1100	HYD	HYD

NOTE: The Vehicle Emission Control Information label often reflects specification changes made during production. The label figures must be used if they differ from those in this chart.

HYD - Hydraulic

B - Before top dead center

① This reading measured with vacuum hose disconnected from fuel pressure regulator

② Basic ignition timing not adjustable

③ 46 psi at idle, with vacuum applied to fuel pressure regulator

④ 39 psi at idle, with vacuum applied to fuel pressure regulator

CAPACITIES

Year	Model	Engine ID/VIN	Engine Displacement Liters (cc)	Engine Oil with Filter (qts.)	Transmission (pts.) 4-Spd	Transmission (pts.) 5-Spd	Transmission (pts.) Auto.	Transfer Case (pts.)	Drive Axle Front (pts.)	Drive Axle Rear (pts.)	Fuel Tank (gal.)	Cooling System (qts.)
1993	Concorde	T	3.3 (3300)	5.0	-	-	19.8	-	2.0	-	18.0	10.2
	Concorde	F	3.5 (3518)	5.5	-	-	19.8	-	2.0	-	18.0	11.8
	Intrepid	T	3.3 (3300)	4.0	-	-	19.8	-	2.0	-	18.0	10.2
	Intrepid	F	3.5 (3518)	4.0	-	-	19.8	-	2.0	-	18.0	11.8
	New Yorker	R	3.3 (3300)	4.5	-	-	①	-	-	-	16.0	9.0
	New Yorker	L	3.8 (3785)	4.5	-	-	①	-	-	-	16.0	9.0
	Vision	F	3.5 (3518)	5.5	-	-	19.8	-	-	-	18.0	11.8
	Vision	T	3.3 (3294)	5.0	-	-	19.8	-	-	-	18.0	10.7

CAPACITIES

Year	Model	Engine ID/VIN	Engine Displacement Liters (cc)	Engine Oil with Filter (qts.)	Transmission (pts.) 4-Spd	5-Spd	Auto.	Transfer Case (pts.)	Drive Axle Front (pts.)	Rear (pts.)	Fuel Tank (gal.)	Cooling System (qts.)
1994	Concorde	T	3.3 (3300)	5.0	-	-	19.8	-	2.0	-	18.0	10.2
	Concorde	F	3.5 (3518)	5.5	-	-	19.8	-	2.0	-	18.0	11.8
	Intrepid	T	3.3 (3300)	5.0	-	-	19.8	-	2.0	-	18.0	10.2
	Intrepid	F	3.5 (3518)	5.5	-	-	19.8	-	2.0	-	18.0	11.8
	LHS	F	3.5 (3518)	5.5	-	-	19.8	-	2.0	-	18.0	11.8
	New Yorker	F	3.5 (3518)	5.5	-	-	19.8	-	2.0	-	18.0	11.8
	Vision	T	3.3 (3301)	5.0	-	-	19.8	-	2.0	-	18.0	10.7
	Vision	F	3.5 (3518)	5.5	-	-	19.8	-	2.0	-	18.0	11.8
1995	Concorde	T	3.3 (3301)	5.0	-	-	19.8 ②	-	2.0	-	18.0	10.2
	Concorde	F	3.5 (3518)	5.5	-	-	19.8 ②	-	2.0	-	18.0	11.8
	Intrepid	T	3.3 (3300)	5.0	-	-	19.8	-	2.0	-	18.0	10.2
	Intrepid	F	3.5 (3518)	5.5	-	-	19.8	-	2.0	-	18.0	11.8
	LHS	F	3.5 (3518)	5.5	-	-	19.8 ②	-	2.0	-	18.0	11.8
	New Yorker	F	3.5 (3518)	5.5	-	-	19.8 ②	-	2.0	-	18.0	11.8
	Vision	T	3.3 (3301)	5.0	-	-	19.8	-	2.0	-	18.0	10.1
	Vision	F	3.5 (3518)	5.5	-	-	19.8	-	2.0	-	18.0	11.8
1996-97	Concorde	T	3.3 (3301)	5.0	-	-	19.8 ②	-	2.0	-	18.0	10.2
	Concorde	F	3.5 (3518)	5.5	-	-	19.7 ②	-	2.0	-	18.0	11.8
	Intrepid	T	3.3 (3300)	5.0	-	-	19.8	-	2.0	-	18.0	10.2
	Intrepid	F	3.5 (3518)	5.5	-	-	19.8	-	2.0	-	18.0	11.8
	LHS	F	3.5 (3518)	5.5	-	-	19.8 ②	-	2.0	-	18.0	11.8
	Vision	T	3.3 (3301)	5.0	-	-	19.8	-	2.0	-	18.0	10.2
	Vision	F	3.5 (3518)	5.5	-	-	19.8	-	2.0	-	18.0	11.8

① A413 - 17.8 pts.
A413 (fleet) - 18.4 pts.
A413 (lock-up) - 17.0 pts.
A604 (electronic) - 18.2 pts.

② Overhaul fill capacity with torque converter empty

VALVE SPECIFICATIONS

Year	Engine ID/VIN	Engine Displacement Liters (cc)	Seat Angle (deg.)	Face Angle (deg.)	Spring Test Pressure (lbs. @ in.)	Spring Installed Height (in.)	Stem-to-Guide Clearance (in.) Intake	Exhaust	Stem Diameter (in.) Intake	Exhaust
1993	R	3.3 (3300)	45	44.5	60@1.56	1.56	0.0010-0.0030	0.0020-0.0060	0.3110-0.3120	0.3110-0.3120
	T	3.3 (3300)	45-45.5	44.5	①	1.539-1.598	0.0010-0.0030	0.0020-0.0060	0.3120-0.3130	0.3112-0.3119
	F	3.5 (3518)	45-45.5	44.5	②	1.496	0.0009-0.0026	0.0020-0.0040	0.2730-0.2737	0.2719-0.2726
	L	3.8 (3785)	45	44.5	60@1.56	1.56	0.0010-0.0030	0.0030-0.0047	0.3120-0.3130	0.3110-0.3120
1994	T	3.3 (3300)	45-45.5	44.5-45	①	1.622-1.681	0.0010-0.0030	0.0020-0.0060	0.3120-0.3130	0.3112-0.3119
	F	3.5 (3518)	45-45.5	44.5-45	②	1.496	0.0009-0.0026	0.0020-0.0040	0.2730-0.2737	0.2719-0.2726
1995	T	3.3 (3301)	45-45.5	44.5-45	①	1.622-1.681	0.0010-0.0100	0.0020-0.0060	0.3120-0.3130	0.3112-0.3119
	F	3.5 (3518)	45-45.5	44.5-45	②	1.496	0.0009-0.0026	0.0020-0.0037	0.2730-0.2737	0.2719-0.2726

VALVE SPECIFICATIONS

Year	Engine ID/VIN	Engine Displacement Liters (cc)	Seat Angle (deg.)	Face Angle (deg.)	Spring Test Pressure (lbs. @ in.)	Spring Installed Height (in.)	Stem-to-Guide Clearance (in.) Intake	Stem-to-Guide Clearance (in.) Exhaust	Stem Diameter (in.) Intake	Stem Diameter (in.) Exhaust
1996-97	T	3.3 (3301)	45-45.5	44.5-45	①	1.622-1.681	0.0010-0.0100	0.0020-0.0060	0.3120-0.3130	0.3112-0.3119
	F	3.5 (3518)	45-45.5	44.5-45	②	1.496	0.0009-0.0026	0.0020-0.0037	0.2730-0.2737	0.2719-0.2726

① 95-100 lbs.@1.570 in. valve closed
207-229 lbs.@1.169 in. valve closed

② Intake: 201.7-218.3 lbs.@1.1752 in.
Exhaust: 158.5-171.5 lbs.@1.239 in.

TORQUE SPECIFICATIONS
All readings in ft. lbs.

Year	Engine ID/VIN	Engine Displacement Liters (cc)	Cylinder Head Bolts	Main Bearing Bolts	Rod Bearing Bolts	Crankshaft Damper Bolts	Flywheel Bolts	Manifold Intake	Manifold Exhaust	Spark Plugs	Lug Nut
1993	R	3.3 (3300)	④	②	③	50	70	17	17	26	95
	T	3.3 (3300)	④	②	③	50	70	17	17	26	95
	F	3.5 (3518)	⑤	①	③	85	75	21	17	20	95
	L	3.8 (3785)	④	②	③	50	70	17	17	26	95
1994	T	3.3 (3300)	④	②	③	40	75	17	17	20	95
	F	3.5 (3518)	⑤	①	③	85	75	21	17	20	95
1995	T	3.3 (3301)	④	②	③	40	75	17	17	20	95
	F	3.5 (3518)	⑤	①	③	85	75	21	17	20	95
1996-97	T	3.3 (3301)	④	②	③	40	75	17	17	20	100
	F	3.5 (3518)	⑤	①	③	85	75	21	17	20	100

① Main cap bolts: 30 ft. lbs. plus 1/4 turn
Main cap tie bolts: 40 ft. lbs.
② Step 1: 30 ft. lbs
Step 2: Plus 1/4 turn
③ Step 1: 40 ft. lbs.
Step 2: Plus 1/4 turn

④ Step 1: 45 ft. lbs.
Step 2: 65 ft. lbs.
Step 3: 65 ft. lbs.
Step 4: Plus 1/4 turn
Torque small bolt in rear of cylinder head to 25 ft. lbs.

⑤ Step 1: 45 ft. lbs.
Step 2: 65 ft. lbs.
Step 3: 65 ft. lbs.
Step 4: Plus 1/4 turn
Final torque should be over 90 ft. lbs.

BRAKE SPECIFICATIONS
All measurements in inches unless noted

Year	Model		Master Cylinder Bore	Brake Disc Original Thickness	Brake Disc Minimum Thickness	Brake Disc Maximum Runout	Brake Drum Diameter Original Inside Diameter	Brake Drum Diameter Max. Wear Limit	Brake Drum Diameter Maximum Machine Diameter	Minimum Lining Thickness Front	Minimum Lining Thickness Rear
1993	Concorde	F	0.937	0.945	0.803	0.003	-	-	-	0.310	0.280
		R	-	0.945	0.803	0.003	8.00	①	①	-	0.280
	Intrepid	F	0.937	0.945	0.803	0.003	-	-	-	0.310	-
		R	-	0.945	0.803	0.003	8.00	NA	NA	-	0.280
	New Yorker	F	0.827	0.861	0.803	0.005	-	-	-	0.300	-
		R	-	0.354	0.339	0.005	-	-	-	-	0.280
	Vision	F	0.937	0.945	0.803	0.003	-	-	-	0.310	-
		R	-	0.945	0.803	0.003	8.00	①	NA	-	0.280
1994	Concorde	F	0.937	0.945	0.882	0.003	-	-	-	0.310	0.280
		R	-	0.468	0.409	0.003	8.00	①	①	-	0.280
	Intrepid	F	0.937	0.945	0.882	0.003	-	-	-	0.310	-
		R	-	0.468	0.409	0.003	8.00	①	①	-	0.280

BRAKE SPECIFICATIONS
All measurements in inches unless noted

Year	Model		Master Cylinder Bore	Brake Disc Original Thickness	Brake Disc Minimum Thickness	Maximum Runout	Brake Drum Diameter Original Inside Diameter	Brake Drum Diameter Max. Wear Limit	Brake Drum Diameter Maximum Machine Diameter	Minimum Lining Thickness Front	Minimum Lining Thickness Rear
1994	LHS	F	0.937	0.945	0.882	0.003	-	-	-	0.310	-
		R	-	0.468	0.409	0.003	-	-	-	-	0.280
	New Yorker	F	0.937	0.945	0.882	0.003	-	-	-	0.300	-
		R	-	0.468	0.409	0.003	-	-	-	-	0.280
	Vision	F	0.937	0.945	0.882	0.003	-	-	-	0.310	-
		R	-	0.468	0.409	0.003	8.00	①	NA	-	0.280
1995	Concorde	F	0.937	0.945	0.882	0.003	-	-	-	0.310	0.280
		R	-	0.468	0.409	0.003	8.00	①	①	-	0.280
	Intrepid	F	0.937	0.945	0.882	0.005	-	-	-	0.310	-
		R	-	0.468	0.409	0.003	8.00	①	①	-	0.280
	LHS	F	0.937	0.945	0.882	0.003	-	-	-	0.310	-
		R	-	0.468	0.409	0.003	-	-	-	-	0.280
	New Yorker	F	0.937	0.945	0.882	0.003	-	-	-	0.300	-
		R	-	0.468	0.409	0.003	-	-	-	-	0.280
	Vision	F	0.937	0.945	0.882	0.003	-	-	-	0.312	-
		R	-	0.468	0.409	0.003	8.00	①	NA	-	0.281
1996-97	Concorde	F	0.874	0.945	0.882	0.003	-	-	-	0.310	-
		R	-	0.468	0.409	0.003	-	-	-	-	0.280
	Intrepid	F	0.937	0.945	0.882	0.005	-	-	-	0.310	-
		R	-	0.468	0.409	0.003	8.00	①	①	-	0.280
	LHS	F	0.874	0.945	0.882	0.003	-	-	-	0.310	-
		R	-	0.468	0.409	0.003	-	-	-	-	0.280
	Vision	F	0.937	0.945	0.882	0.003	-	-	-	0.312	-
		R	-	0.468	0.409	0.003	8.00	①	NA	-	0.281

F - Front
R - Rear

① Maximum diameter is stamped on drum

FREQUENT MAINTENANCE LABOR
CHRYSLER CONCORDE, LHS, NEW YORKER, DODGE INTREPID, EAGLE VISION

The following should be used as a guide when determining the amount of work required for a particular service if taken to a repair shop. In estimating how long a particular Frequent Maintenance Service item should take, please observe the following:
- **Factory Time** is time that is generated by the vehicle manufacturer.
- **Chilton Time** is time that is based on field research and data supplied by the vehicle manufacturer.
- All labor time operations are given in hours and tenths of an hour.
- All labor operations, are to be used as a **guide**.

(Factory Time)   Chilton Time

COOLING

(G) Winterize Cooling System
Includes: Run engine to check for leaks, tighten all hose connections. Test radiator and pressure cap, drain radiator and engine block. Add antifreeze and refill system.
1993-97 5

(G) Belt, Serpentine Drive, Renew
1993-97 (.2) 3

(G) Hoses, Radiator, Renew
Includes: Drain and refill cooling system as required.
1993-97
  upper (.4) 5
  lower (.5) 6
  water pump to intake (.5) 7

(G) Thermostat, Coolant, Renew
1993-97 (.5) 6

BRAKES

(G) Bleed Brakes (Four Wheels)
Includes: Add fluid.
1993-97 (.4) 5
w/ABS add 4

(G) Brakes, Adjust (Minor)
Includes: Adjust brakes, fill master cylinder.
1993-97 4

FREQUENT MAINTENANCE LABOR (cont.)
CHRYSLER CONCORDE, LHS, NEW YORKER, DODGE INTREPID, EAGLE VISION

	(Factory Time)	Chilton Time

(M) Parking Brake, Adjust
1993-97 4

LUBRICATION SERVICES

(M) Engine Oil & Filter, Renew
Includes: Inspect and correct all fluid levels.
1993-97 3

(M) Lubricate Chassis
Includes: Inspect and correct all fluid levels.
1993-97 4
Install grease fittings add 1

WHEELS

(G) Wheel, Renew (One)
1993-97 5

(G) Wheel, Balance
1993-97
one 3
each adtnl. 2

(G) Wheels, Rotate (All)
1993-97 5

ELECTRICAL

(G) Headlamps, Aim
1993-97
two 4
four 6

(G) Halogen Headlamp Bulb, Renew
1993-97 (.2) 3

(G) License Lamp Assy., Renew
1993-97 (.2) 3

(G) Park & Turn Signal Lamp Assy., Renew
1993-97 (.4) 5

(G) Tail Lamp Assy., Renew
1993-97 (.3) 4

(G) Horn, Renew
1993-97 (.3) 4

(M) Terminals, Battery, Clean
1993-97 3

SCHEDULED MAINTENANCE INTERVALS
(CHRYSLER CONCORDE, LHS, NEW YORKER, DODGE INTREPID & EAGLE VISION)

TO BE SERVICED	TYPE OF SERVICE	VEHICLE MILEAGE INTERVAL (x1000)												
		7.5	15	22.5	30	37.5	45	52.5	60	67.5	75	82.5	90	97.5
Engine oil & filter	R	✓	✓	✓	✓	✓	✓	✓	✓	✓	✓	✓	✓	✓
Exhaust system	S/I	✓	✓	✓	✓	✓	✓	✓	✓	✓	✓	✓	✓	✓
Brake hoses	S/I	✓	✓	✓	✓	✓	✓	✓	✓	✓	✓	✓	✓	✓
CV joints & front suspension components	S/I	✓	✓	✓	✓	✓	✓	✓	✓	✓	✓	✓	✓	✓
Rotate tires	S/I	✓	✓	✓	✓	✓	✓	✓	✓	✓	✓	✓	✓	✓
Coolant level, hoses & clamps	S/I	✓	✓	✓	✓	✓	✓	✓	✓	✓	✓	✓	✓	✓
Accessory drive belts	S/I		✓		✓		✓		✓		✓		✓	
Brake linings	S/I		✓	✓			✓		✓				✓	
Spark plugs	R				✓				✓				✓	
Air filter element	R				✓				✓				✓	
Lubricate steering linkage & tie rod ends	S/I				✓				✓				✓	
Engine Coolant	R						✓				✓			
PCV valve	S/I								✓				✓	

SCHEDULED MAINTENANCE INTERVALS
(CHRYSLER CONCORDE, LHS, NEW YORKER, DODGE INTREPID & EAGLE VISION)
(Cont.)

TO BE SERVICED	TYPE OF SERVICE	VEHICLE MILEAGE INTERVAL (x1000)											
		7.5	15	22.5	30	37.5	45	52.5	60	67.5	75	82.5	90	97.5
Ignition cables	R								✓				
Camshaft timing belt	R								✓				

R – Replace   S/I – Service or Inspect

FREQUENT OPERATION MAINTENANCE (SEVERE SERVICE)
If a vehicle is operated under any of the following conditions it is considered severe service:
- Extremely dusty areas.
- 50% or more of the vehicle operation is in 32°C (90°F) or higher temperatures, or constant operation in temperatures below 0°C (32°F).
- Prolonged idling (vehicle operation in stop and go traffic).
- Frequent short running periods (engine does not warm to normal operating temperatures).
- Police, taxi, delivery usage or trailer towing usage.

CV joints & front suspension components - check every 3000 miles.
Oil & oil filter change – change every 3000 miles.
Rotate tires every 3000 miles.
Brake linings – check every 9000 miles.
Air filter element – change every 15,000 miles.
Automatic transaxle fluid – change every 15,000 miles.
Differential fluid – change every 15,000 miles.
Tie rod ends & steering linkage – lubricate every 15,000 miles.
PCV valve – check every 30,000 miles.

STEALTH

VEHICLE IDENTIFICATION CHART

		Engine Code					Model Year	
Code	Liters	Cu. In. (cc)	Cyl.	Fuel Sys.	Eng. Mfg.		Code	Year
H	3.0	181 (2972)	V6	MFI	Mitsubishi		P	1993
J	3.0	181 (2972)	V6	MFI	Mitsubishi		R	1994
K	3.0	181 (2972)	V6	MFI-TT	Mitsubishi		S	1995
							T	1996
							V	1997

MFI - Multipoint fuel injection

TT - Twin Turbochargers

ENGINE IDENTIFICATION
All measurements are given in inches.

Year	Model	Engine Displacement Liters (cc)	Engine Series (ID/VIN)	Fuel System	No. of Cylinders	Engine Type
1993	Stealth	3.0 (2972)	H	MFI	6	SOHC
	Stealth	3.0 (2972)	J	MFI	6	DOHC
	Stealth	3.0 (2972)	K	MFI-TT	6	DOHC

ENGINE IDENTIFICATION
All measurements are given in inches.

Year	Model	Engine Displacement Liters (cc)	Engine Series (ID/VIN)	Fuel System	No. of Cylinders	Engine Type
1994	Stealth	3.0 (2972)	H	MFI	6	SOHC
	Stealth	3.0 (2972)	J	MFI	6	DOHC
	Stealth	3.0 (2972)	K	MFI-TT	6	DOHC
1995	Stealth	3.0 (2972)	H	MFI	6	SOHC
	Stealth	3.0 (2972)	J	MFI	6	DOHC
	Stealth	3.0 (2972)	K	MFI-TT	6	DOHC
1996-97	Stealth	3.0 (2972)	H	MFI	6	SOHC
	Stealth	3.0 (2972)	J	MFI	6	DOHC
	Stealth	3.0 (2972)	K	MFI-TT	6	DOHC

MFI - Multipoint fuel injection
TT - Twin Turbochargers
DOHC - Double overhead camshaft
SOHC - Single overhead camshaft

GENERAL ENGINE SPECIFICATIONS

Year	Engine ID/VIN	Engine Displacement Liters (cc)	Fuel System Type	Net Horsepower @ rpm	Net Torque @ rpm (ft. lbs.)	Bore x Stroke (in.)	Compression Ratio	Oil Pressure @ rpm
1993	H	3.0 (2972)	MFI	164@5500	185@4000	3.59 x 2.99	8.9:1	35-100@2000
	J	3.0 (2972)	MFI	222@6000	201@4500	3.59 x 2.99	10.0:1	35-100@2000
	K	3.0 (2972)	MFI-TT	300@6000	307@2500	3.59 x 2.99	8.0:1	35-100@2000
1994	H	3.0 (2972)	MFI	164@5500	185@4000	3.59 x 2.99	8.9:1	35-100@2000
	J	3.0 (2972)	MFI	222@6000	201@4500	3.59 x 2.99	10.0:1	35-100@2000
	K	3.0 (2972)	MFI-TT	320@6000	315@2500	3.59 x 2.99	8.0:1	35-100@2000
1995	H	3.0 (2972)	MFI	164@5500	185@4000	3.59 x 2.99	8.9:1	35-100@2000
	J	3.0 (2972)	MFI	222@6000	201@4500	3.59 x 2.99	10.0:1	35-100@2000
	K	3.0 (2972)	MFI-TT	320@6000	315@2500	3.59 x 2.99	8.0:1	35-100@2000
1996-97	H	3.0 (2972)	MFI	164@5500	185@4000	3.59 x 2.99	8.9:1	35-100@2000
	J	3.0 (2972)	MFI	222@6000	201@4500	3.59 x 2.99	10.0:1	35-100@2000
	K	3.0 (2972)	MFI-TT	320@6000	315@2500	3.59 x 2.99	8.0:1	35-100@2000

MFI - Multipoint fuel injection
TT - Twin Turbochargers

GASOLINE ENGINE TUNE-UP SPECIFICATIONS

Year	Engine ID/VIN	Engine Displacement Liters (cc)	Spark Plugs Gap (in.)	Ignition Timing (deg.) MT	Ignition Timing (deg.) AT	Fuel Pump (psi)	Idle Speed (rpm) MT	Idle Speed (rpm) AT	Valve Clearance In.	Valve Clearance Ex.
1993	H	3.0 (2972)	0.039-0.043	5B	5B	38	750	750	HYD	HYD
	J	3.0 (2972)	0.039-0.043	5B	5B	38	750	750	HYD	HYD
	K	3.0 (2972)	0.035	5B	5B	34	750	750	HYD	HYD
1994	H	3.0 (2972)	0.039-0.043	5B	5B	38	750	750	HYD	HYD
	J	3.0 (2972)	0.039-0.043	5B	5B	38	750	750	HYD	HYD
	K	3.0 (2972)	0.039-0.043	5B	5B	34	750	750	HYD	HYD
1995	H	3.0 (2972)	0.039-0.043	5B	5B	38	750	750	HYD	HYD
	J	3.0 (2972)	0.039-0.043	5B	5B	38	750	750	HYD	HYD
	K	3.0 (2972)	0.039-0.043	5B	5B	34	750	750	HYD	HYD

GASOLINE ENGINE TUNE-UP SPECIFICATIONS

Year	Engine ID/VIN	Engine Displacement Liters (cc)	Spark Plugs Gap (in.)	Ignition Timing (deg.)		Fuel Pump (psi)	Idle Speed (rpm)		Valve Clearance	
				MT	AT		MT	AT	In.	Ex.
1996-97	H	3.0 (2972)	0.039-0.043	5B	5B	38	750	750	HYD	HYD
	J	3.0 (2972)	0.039-0.043	5B	5B	38	750	750	HYD	HYD
	K	3.0 (2972)	0.039-0.043	5B	5B	34	750	750	HYD	HYD

NOTE: The Vehicle Emission Control Information label often reflects specification changes made during production. The label figures must be used if they differ from those in this chart.
B - Before top dead center
HYD - Hydraulic

CAPACITIES

Year	Model	Engine ID/VIN	Engine Displacement Liters (cc)	Engine Oil with Filter (qts.)	Transmission (pts.)			Transfer Case (pts.)	Drive Axle		Fuel Tank (gal.)	Cooling System (qts.)
					4-Spd	5-Spd	Auto.		Front (pts.)	Rear (pts.)		
1993	Stealth	H	3.0 (2972)	4.7	-	4.8	15.8	-	-	-	19.8	8.5
	Stealth	J	3.0 (2972)	4.7	-	4.8	15.8	-	-	-	19.8	8.5
	Stealth	K	3.0 (2972)	5.2	-	①	15.8	0.58	-	2.3	19.8	8.5
1994	Stealth	H	3.0 (2972)	4.7	-	4.8	15.8	-	-	-	19.8	8.5
	Stealth	J	3.0 (2972)	4.7	-	4.8	15.8	-	-	-	19.8	8.5
	Stealth	K	3.0 (2972)	5.2	-	5.0 ②	15.8	0.58	-	2.3	19.8	8.5
1995	Stealth	H	3.0 (2972)	4.7	-	4.8	15.8	-	-	-	19.8	8.5
	Stealth	J	3.0 (2972)	4.7	-	4.8	15.8	-	-	-	19.8	8.5
	Stealth	K	3.0 (2972)	5.2	-	5.0 ②	15.8	0.58	-	2.3	19.8	8.5
1996-97	Stealth	H	3.0 (2972)	4.7	-	4.8	15.8	-	-	-	19.8	8.5
	Stealth	J	3.0 (2972)	4.7	-	4.8	15.8	-	-	-	19.8	8.5
	Stealth	K	3.0 (2972)	5.2	-	5.0 ②	15.8	0.58	-	2.3	19.8	8.5

① FWD: 4.8 pts.
  AWD: 5.0 pts.
② 6 speed manual transaxle

VALVE SPECIFICATIONS

Year	Engine ID/VIN	Engine Displacement Liters (cc)	Seat Angle (deg.)	Face Angle (deg.)	Spring Test Pressure (lbs. @ in.)	Spring Installed Height (in.)	Stem-to-Guide Clearance (in.)		Stem Diameter (in.)	
							Intake	Exhaust	Intake	Exhaust
1993	H	3.0 (2972)	44-44.5	45-45.5	①	1.600-1.630	0.0012-0.0039	0.0020-0.0059	0.3140	0.3125
	J	3.0 (2972)	44-44.5	45-45.5	②	1.500-1.530	0.0008-0.0039	0.0020-0.0047	0.2600	0.2600
	K	3.0 (2972)	44-44.5	45-45.5	②	1.500-1.530	0.0008-0.0039	0.0020-0.0047	0.2600	0.2600
1994	H	3.0 (2972)	44-44.5	45-45.5	①	1.600-1.630	0.0012-0.0039	0.0020-0.0059	0.3140	0.3140
	J	3.0 (2972)	44-44.5	45-45.5	③	1.500-1.530	0.0008-0.0039	0.0020-0.0047	0.2600	0.2600
	K	3.0 (2972)	44-44.5	45-45.5	③	1.500-1.530	0.0008-0.0039	0.0020-0.0047	0.2600	0.2600
1995	H	3.0 (2972)	44-44.5	45-45.5	①	1.591-1.630	0.0012- ④ 0.0024	0.0020- ⑤ 0.0035	0.3140	0.3140
	J	3.0 (2972)	44-44.5	45-45.5	③	1.492	0.0008- ④ 0.0020	0.0020- ⑥ 0.0035	0.2600	0.2600

VALVE SPECIFICATIONS

Year	Engine ID/VIN	Engine Displacement Liters (cc)	Seat Angle (deg.)	Face Angle (deg.)	Spring Test Pressure (lbs. @ in.)	Spring Installed Height (in.)	Stem-to-Guide Clearance (in.)		Stem Diameter (in.)	
							Intake	Exhaust	Intake	Exhaust
1995	K	3.0 (2972)	44-44.5	45-45.5	③	1.492	0.0008-0.0020 ④	0.0020-0.0035 ⑥	0.2600	0.2600
1996-97	H	3.0 (2972)	45-44.5	45-45.5	①	1.591-1.630	0.0012-0.0024 ④	0.0020-0.0035 ⑤	0.3140	0.3140
	J	3.0 (2972)	45-44.5	45-45.5	③	1.492	0.0008-0.0020 ④	0.0020-0.0035 ⑥	0.2600	0.2600
	K	3.0 (2972)	45-44.5	45-45.5	③	1.492	0.0008-0.0020 ④	0.0020-0.0035 ⑥	0.2600	0.2600

① 74 @ installed height
② 62 @ installed height
③ 53 @ installed height
④ Wear limit: 0.0039
⑤ Wear limit: 0.0059
⑥ Wear limit: 0.0047

TORQUE SPECIFICATIONS
All readings in ft. lbs.

Year	Engine ID/VIN	Engine Displacement Liters (cc)	Cylinder Head Bolts	Main Bearing Bolts	Rod Bearing Bolts	Crankshaft Damper Bolts	Flywheel Bolts	Manifold		Spark Plugs	Lug Nut
								Intake	Exhaust		
1993	H	3.0 (2972)	76-83	58	38	108-116	55	13	13	18	87-101
	J	3.0 (2972)	76-83	58	38	108-116	55	13	13	18	87-101
	K	3.0 (2972)	①	58	38	130-137	55	9-11	22	18	87-101
1994	H	3.0 (2972)	76-83	58	38	108-116	55	13	13	18	87-101
	J	3.0 (2972)	①	58	38	130-137	55	13	22	18	87-101
	K	3.0 (2972)	①	58	38	130-137	55	13	22	18	87-101
1995	H	3.0 (2972)	76-83	58	38	108-116	55	13	13	15	87-101
	J	3.0 (2972)	①	67	38	130-137	55	13	22	15	87-101
	K	3.0 (2972)	①	54	38	130-137	55	13	22	15	87-101
1996-97	H	3.0 (2972)	76-83	58	38	108-116	55	13	13	15	87-101
	J	3.0 (2972)	①	67	38	130-137	55	13	22	15	87-101
	K	3.0 (2972)	①	54	38	130-137	55	13	22	15	87-101

① Step 1: 87-94 ft. lbs.
  Step 2: Fully loosen
  Step 3: 87-94 ft. lbs.

BRAKE SPECIFICATIONS
All measurements in inches unless noted

Year	Model		Master Cylinder Bore	Brake Disc			Brake Drum Diameter			Minimum Lining Thickness	
				Original Thickness	Minimum Thickness	Maximum Runout	Original Inside Diameter	Max. Wear Limit	Maximum Machine Diameter	Front	Rear
1993	Stealth	F	①	0.940	0.880	0.003	-	-	-	0.080	-
		R	-	0.710	0.650	0.003	-	-	-	-	0.080
	Stealth AWD	F	1.063	1.180	1.120	0.003	-	-	-	0.080	-
		R	-	0.790	0.720	0.003	-	-	-	-	0.080
1994	Stealth	F	①	0.940	0.880	0.003	-	-	-	0.080	-
		R	-	0.710	0.650	0.003	-	-	-	-	0.080
	Stealth AWD	F	1.063	1.180	1.120	0.003	-	-	-	0.080	-
		R	-	0.790	0.720	0.003	-	-	-	-	0.080

BRAKE SPECIFICATIONS
All measurements in inches unless noted

Year	Model		Master Cylinder Bore	Brake Disc Original Thickness	Brake Disc Minimum Thickness	Brake Disc Maximum Runout	Brake Drum Diameter Original Inside Diameter	Max. Wear Limit	Maximum Machine Diameter	Minimum Lining Thickness Front	Minimum Lining Thickness Rear
1995	Stealth	F	①	0.940	0.880	0.003	-	-	-	0.080	-
		R	-	0.710	0.650	0.003	-	-	-	-	0.080
	Stealth AWD	F	1.063	1.180	1.120	0.003	-	-	-	0.080	-
		R	-	0.790	0.720	0.003	-	-	-	-	0.080
1996-97	Stealth	F	①	0.940	0.880	0.003	-	-	-	0.080	-
		R	-	0.710	0.650	0.003	-	-	-	-	0.080
	Stealth AWD	F	1.063	1.180	1.120	0.003	-	-	-	0.080	-
		R	-	0.790	0.720	0.003	-	-	-	-	0.080

① With ABS: 1.000
  Without ABS: 1.063

FREQUENT MAINTENANCE LABOR
DODGE STEALTH

The following should be used as a guide when determining the amount of work required for a particular service if taken to a repair shop. In estimating how long a particular Frequent Maintenance Service item should take, please observe the following:

- **Factory Time** is time that is generated by the vehicle manufacturer.
- **Chilton Time** is time that is based on field research and data supplied by the vehicle manufacturer.
- All labor time operations are given in hours and tenths of an hour.
- All labor operations, are to be used as a **guide**.

COOLING

(G) Winterize Cooling System
Includes: Run engine to check for leaks, tighten all hose connections. Test radiator and pressure cap, drain radiator and engine block. Add antifreeze and refill system.
1993-97 5

(G) Belt, Drive, Renew
1993-97
SOHC 2
DOHC 7

(G) Belt, Drive, Adjust
1993-97 3

(G) Hoses, Radiator, Renew
Includes: Drain and refill cooling system as required.
1993-97
upper (.3) 4
lower (.4) 5

(G) Thermostat, Coolant, Renew
1993-97 (.4) 6

FUEL

(M) Air Cleaner, Service
1993-97 2

(G) Filter, Fuel, Renew
1993-97
in line
SOHC (.6) 1.0
DOHC Turbo (1.0) 1.5
in tank (.9) 1.3

BRAKES

(G) Bleed Brakes (Four Wheels)
Includes: Add fluid.
1993-97 5

(M) Parking Brake, Adjust
1993-97 4

LUBRICATION SERVICES

(M) Engine Oil & Filter, Renew
Includes: Inspect and correct all fluid levels.
1993-97 (.3) 3

(M) Lubricate Chassis
Includes: Inspect and correct all fluid levels.
1993-97 4
Install grease fittings add 1

WHEELS

(G) Wheel, Renew (One)
1993-97 5

(G) Wheel, Balance
1993-97
one 3
each adtnl. 2

(G) Wheels, Rotate (All)
1993-97 5

ELECTRICAL

(G) Headlamps, Aim
1993-97
two 4
four 6

(G) Parking Lamp Lens or Bulb, Renew
1993-97 (.2) 3

(G) Tail Lamp Assy., Renew
1993-97 (.5) 7

(G) Horn, Renew
1993-97 (.6) 8

(M) Terminals, Battery, Clean
1993-97 3

SCHEDULED MAINTENANCE INTERVALS
(DODGE STEALTH)

TO BE SERVICED	TYPE OF SERVICE	VEHICLE MILEAGE INTERVAL (x1000)												
		7.5	15	22.5	30	37.5	45	52.5	60	67.5	75	82.5	90	97.5
Engine oil & filter (Non-turbo) ①	R	✓	✓	✓	✓	✓	✓	✓	✓	✓	✓	✓	✓	✓
Coolant level, hoses & clamps	S/I	✓	✓	✓	✓	✓		✓	✓	✓	✓	✓	✓	✓
Rotate tires	S/I	✓	✓	✓	✓	✓	✓	✓	✓	✓	✓	✓	✓	✓
Brake hoses	S/I		✓		✓		✓		✓		✓		✓	
Drive shaft boots & front suspension components	S/I		✓		✓		✓		✓		✓		✓	
Brake linings	S/I		✓		✓		✓		✓		✓		✓	
Air filter element	R				✓				✓				✓	
Automatic transaxle fluid & filter	R				✓				✓				✓	
Differential fluid (AWD)	R				✓				✓				✓	
Engine Coolant	R				✓				✓				✓	
Spark plugs (Non-platinum)	R				✓				✓				✓	
Accessory drive belts	S/I				✓				✓				✓	
Ball joints & steering linkage seals	S/I				✓				✓				✓	
Exhaust system	S/I				✓				✓				✓	
Fuel hoses	S/I				✓				✓				✓	
Manual transaxle oil (including transfer)	S/I				✓				✓				✓	
PCV valve	S/I				✓				✓				✓	
Spark plugs (Platinum)	R								✓					
Camshaft timing belt	R								✓					
Ignition cables	R								✓					
EVAP system	S/I								✓					
Distributor cap & rotor	S/I								✓					
Fuel system	S/I								✓					

① Engine oil & filter (Turbo) - change every 5000 miles
R – Replace   S/I – Service or Inspect

SCHEDULED MAINTENANCE INTERVALS
(DODGE STEALTH) (Cont.)

FREQUENT OPERATION MAINTENANCE (SEVERE SERVICE)

    If a vehicle is operated under any of the following conditions it is considered severe service:
- Extremely dusty areas.
- 50% or more of the vehicle operation is in 32°C (90°F) or higher temperatures, or constant operation in temperatures below 0°C (32°F).
- Prolonged idling (vehicle operation in stop and go traffic).
- Frequent short running periods (engine does not warm to normal operating temperatures).
- Police, taxi, delivery usage or trailer towing usage.

CV joints & front suspension components - check every 3000 miles.
Oil & oil filter change – change every 3000 miles.
Brake linings - check every 7500 miles (1993-95) or 6000 miles (1996).
Air filter element – service or inspect every 7500 miles.
Automatic transaxle fluid – change every 15,000 miles.
Differential fluid – change every 15,000 miles.
Manual transaxle (including transfer) - change every 15,000 miles.
Spark plugs - change every 15,000 miles.
Tie rod ends & steering linkage – lubricate every 15,000 miles.

SEBRING/AVENGER

VEHICLE IDENTIFICATION CHART

Engine Code						Model Year	
Code	Liters	Cu. In. (cc)	Cyl.	Fuel Sys.	Eng. Mfg.	Code	Year
Y	2.0	122 (1996)	I4	MFI	Chrysler	S	1995
N	2.5	152 (2497)	V6	MFI	Mitsubishi	T	1996
						V	1997

MFI - Multiport fuel injection

ENGINE IDENTIFICATION
All measurements are given in inches.

Year	Model	Engine Displacement Liters (cc)	Engine Series (ID/VIN)	Fuel System	No. of Cylinders	Engine Type
1995	Avenger	2.0 (1996)	Y	MFI	4	DOHC
	Avenger	2.5 (2497)	N	MFI	6	SOHC
	Sebring	2.0 (1996)	Y	MFI	4	DOHC
	Sebring	2.5 (2497)	N	MFI	6	SOHC
1996-97	Avenger	2.0 (1996)	Y	MFI	4	DOHC
	Avenger	2.5 (2497)	N	MFI	6	SOHC
	Sebring Coupe	2.0 (1996)	Y	MFI	4	DOHC
	Sebring Coupe	2.5 (2497)	N	MFI	6	SOHC

MFI - Multiport fuel injection   DOHC - Double overhead camshaft   SOHC - Single overhead camshaft

GENERAL ENGINE SPECIFICATIONS

Year	Engine ID/VIN	Engine Displacement Liters (cc)	Fuel System Type	Net Horsepower @ rpm	Net Torque @ rpm (ft. lbs.)	Bore x Stroke (in.)	Compression Ratio	Oil Pressure @ rpm
1995	Y	2.0 (1996)	MFI	140@6000	130@4800	3.44 x 3.27	9.6:1	25-80@3000
	N	2.5 (2497)	MFI	155@5500	161@4400	3.29 x 2.99	9.4:1	35-75@3000
1996-97	Y	2.0 (1996)	MFI	140@6000	130@4800	3.44 x 3.27	9.6:1	25-80@3000
	N	2.5 (2497)	MFI	163@5500	170@4400	3.29 x 2.99	9.4:1	35-75@3000

MFI - Multiport fuel injection

GASOLINE ENGINE TUNE-UP SPECIFICATIONS

Year	Engine ID/VIN	Engine Displacement Liters (cc)	Spark Plugs Gap (in.)	Ignition Timing (deg.) MT	AT	Fuel Pump (psi)		Idle Speed (rpm) MT	AT	Valve Clearance In.	Ex.
1995	Y	2.0 (1996)	0.033-0.038	①	①	47-50		800	800	HYD	HYD
	N	2.5 (2497)	0.039-0.043	-	①	47-50	②	-	700	HYD	HYD
1996-97	Y	2.0 (1996)	0.033-0.038	①	①	47-50		800	800	HYD	HYD
	N	2.5 (2497)	0.039-0.043	-	①	47-50	②	-	750	HYD	HYD

NOTE: The Vehicle Emission Control Information label often reflects specification changes made during production. The label figures must be used if they differ from those in this chart.
HYD - Hydraulic
B - Before top dead center
① Basic ignition timing not adjustable
② This reading measured with vacuum hose disconnected from fuel pressure regulator

CAPACITIES

Year	Model	Engine ID/VIN	Engine Displacement Liters (cc)	Engine Oil with Filter (qts.)	Transmission (pts.) 4-Spd	5-Spd	Auto.	Transfer Case (pts.)	Drive Axle Front (pts.)	Rear (pts.)	Fuel Tank (gal.)	Cooling System (qts.)
1995	Avenger	Y	2.0 (1996)	4.5	-	4.2	18.0	-	-	-	16.9	7.4
	Avenger	N	2.5 (2497)	4.5	-	-	18.0	-	-	-	16.9	7.4
	Sebring	Y	2.0 (1996)	4.5	-	4.2	18.2 ①	-	-	-	16.9	7.4
	Sebring	N	2.5 (2497)	4.5	-	-	18.2 ①	-	-	-	16.9	7.4
1996-97	Avenger	Y	2.0 (1996)	4.5	-	4.2	18.0	-	-	-	16.9	7.4
	Avenger	N	2.5 (2497)	4.5	-	-	18.0	-	-	-	16.9	7.4
	Sebring Coupe	Y	2.0 (1996)	4.5	-	4.2	18.2 ①	-	-	-	16.9	7.4
	Sebring Coupe	N	2.5 (2497)	4.5	-	-	18.2 ①	-	-	-	16.9	7.4

① Overhaul fill capacity with torque converter empty

VALVE SPECIFICATIONS

Year	Engine ID/VIN	Engine Displacement Liters (cc)	Seat Angle (deg.)	Face Angle (deg.)	Spring Test Pressure (lbs. @ in.)	Spring Installed Height (in.)	Stem-to-Guide Clearance (in.) Intake	Exhaust	Stem Diameter (in.) Intake	Exhaust
1995	Y	2.0 (1996)	45	45-45.5	110-120@ ① 1.173	1.496	0.0019-0.0030	0.0029-0.0040	0.2336-0.2343	0.2325-0.2332
	N	2.5 (2497)	44-44.5	45-45.5	60@1.740	1.740	0.0008-0.0040	0.0016-0.0060	0.2360	0.2360

VALVE SPECIFICATIONS

Year	Engine ID/VIN	Engine Displacement Liters (cc)	Seat Angle (deg.)	Face Angle (deg.)	Spring Test Pressure (lbs. @ in.)	Spring Installed Height (in.)	Stem-to-Guide Clearance (in.)		Stem Diameter (in.)	
							Intake	Exhaust	Intake	Exhaust
1996-97	Y	2.0 (1996)	44.5-45	45-45.5	123-137@ ① 1.153	1.496	0.0019-0.0030	0.0029-0.0040	0.2336-0.2343	0.2325-0.2332
	N	2.5 (2497)	44-44.5	45-45.5	60@1.740	1.740	0.0008-0.0040	0.0016-0.0060	0.2360	0.2360

① With valves open

TORQUE SPECIFICATIONS
All readings in ft. lbs.

Year	Engine ID/VIN	Engine Displacement Liters (cc)	Cylinder Head Bolts	Main Bearing Bolts	Rod Bearing Bolts	Crankshaft Damper Bolts	Flywheel Bolts	Manifold		Spark Plugs	Lug Nut
								Intake	Exhaust		
1995	Y	2.0 (1996)	①	55	①	105	70	17	17	20	65-80
	N	2.5 (2497)	80	69	37	134	70	16	22	18	65-80
1996-97	Y	2.0 (1996)	①	55	①	105	70	17	17	20	88-103
	N	2.5 (2497)	80	69	37	134	70	16	22	18	88-103

① Step 1: 20 ft. lbs.
Step 2: Plus 1/4 turn

BRAKE SPECIFICATIONS
All measurements in inches unless noted

Year	Model		Master Cylinder Bore	Brake Disc			Brake Drum Diameter			Minimum Lining Thickness	
				Original Thickness	Minimum Thickness	Maximum Runout	Original Inside Diameter	Max. Wear Limit	Maximum Machine Diameter	Front	Rear
1995	Avenger	F	0.937	0.940	0.880	0.003	-	-	-	0.080	-
		R	-	0.390	0.330	0.003	9.00	②	②	-	0.080
	Sebring	F	0.937 ③	0.940	0.880	0.003	-	-	-	0.080	-
		R ①	-	0.390	0.330	0.003	⑦	②	②	-	⑧
1996-97	Avenger	F	0.937	0.940	0.880	0.003	-	-	-	0.080	-
		R	-	0.390	0.330	0.003	9.00	②	②	-	0.080
	Sebring Coupe	F	1.000 ⑥	0.940	0.880	0.003	-	-	-	0.080	-
		R ①	-	0.390 ④	0.330 ⑤	0.003	⑦	②	②	-	⑧

F - Front
R - Rear

① Solid rear disc
② Maximum diameter is stamped on drum
③ Equipped with ABS: 1.000

④ Vented rear disc: 0.790
⑤ Vented rear disc: 0.720
⑥ Equipped with standard ABS

⑦ Minimum diameter: 9.00 in.
⑧ .039 in. with drum brakes
.080 in. with disc brakes

FREQUENT MAINTENANCE LABOR
CHRYSLER SEBRING, DODGE AVENGER

The following should be used as a guide when determining the amount of work required for a particular service if taken to a repair shop. In estimating how long a particular Frequent Maintenance Service item should take, please observe the following:
- **Factory Time** is time that is generated by the vehicle manufacturer.
- **Chilton Time** is time that is based on field research and data supplied by the vehicle manufacturer.
- All labor time operations are given in hours and tenths of an hour.
- All labor operations, are to be used as a **guide**.

COOLING

(G) Winterize Cooling System
Includes: Run engine to check for leaks, tighten all hose connections. Test radiator and pressure cap, drain radiator and engine block. Add antifreeze and refill system.
1995-96 5

(G) Belt, Drive, Renew
1995-96
alternator (.5) 6
power steering (.3) 4

(G) Belt, Drive, Adjust
1995-96
one 3
each adtnl. 1

(G) Hoses, Radiator, Renew
Includes: Drain and refill cooling system as required.
1995-96
hardtop
upper (.3) 4
lower (.5) 8
convertible, each (.4) 7

FUEL

(M) Air Cleaner, Service
1995-96 3

(G) Filter, Fuel, Renew
1995-96
in line
hardtop (.7) 1.0
convertible (.4) 6
in tank
hardtop (.4) 6
convertible (1.0) 1.4

BRAKES

(G) Bleed Brakes (Four Wheels)
Includes: Add fluid.
1995-96 (.4) 5

(G) Brakes, Adjust (Minor)
Includes: Adjust brakes, fill master cylinder.
1995-96 4

LUBRICATION SERVICES

(M) Engine Oil & Filter, Renew
Includes: Inspect and correct all fluid levels.
1995-96 6

(M) Lubricate Chassis
Includes: Inspect and correct all fluid levels.
1995-96 4
Install grease fittings, add 1

WHEELS

(G) Wheel, Renew (One)
1995-96 5

(G) Wheel, Balance
1995-96
one 3
each adtnl. 2

(G) Wheels, Rotate (All)
1995-96 5

ELECTRICAL

(G) Headlamps, Aim
1995-96 4

(G) License Lamp Assy., Renew
1995-96 (.2) 3

(G) Park & Turn Signal Lamp Assy., Renew
1995-96 (.3) 4

(G) Rear Combination Lamp Assy., Renew
1995-96 (.2) 3

(G) Horn, Renew
1995-96, one (.5) 8

(M) Terminals, Battery, Clean
1995-96 3

SCHEDULED MAINTENANCE INTERVALS
(CHRYSLER SEBRING, DODGE AVENGER)

TO BE SERVICED	TYPE OF SERVICE	VEHICLE MILEAGE INTERVAL (x1000)												
		7.5	15	22.5	30	37.5	45	52.5	60	67.5	75	82.5	90	97.5
Engine oil & filter	R	✓	✓	✓	✓	✓	✓	✓	✓	✓	✓	✓	✓	✓
Coolant level, hoses & clamps	S/I	✓	✓	✓	✓	✓	✓	✓	✓	✓	✓	✓	✓	✓
Rotate tires	S/I	✓	✓	✓	✓	✓	✓	✓	✓	✓	✓	✓	✓	✓
Automatic transaxle fluid level	S/I		✓		✓		✓		✓		✓		✓	

SCHEDULED MAINTENANCE INTERVALS
(CHRYSLER SEBRING, DODGE AVENGER) (Cont.)

TO BE SERVICED	TYPE OF SERVICE	VEHICLE MILEAGE INTERVAL (x1000)												
		7.5	15	22.5	30	37.5	45	52.5	60	67.5	75	82.5	90	97.5
Brake hoses & disc brake pads	S/I		✓		✓		✓		✓		✓		✓	
Drive shaft boots & front suspension components	S/I		✓		✓		✓		✓		✓		✓	
Air filter element	R				✓				✓				✓	
Engine Coolant	R				✓				✓				✓	
Spark plugs (DOHC)	R				✓				✓				✓	
Spark plugs (SOHC 1995)①	R								✓					
Accessory drive belts	S/I				✓				✓				✓	
Ball joints & steering linkage seals	S/I				✓				✓				✓	
Exhaust system	S/I				✓				✓				✓	
Fuel hoses	S/I				✓				✓				✓	
Manual transaxle oil	S/I				✓				✓				✓	
PCV valve	S/I				✓				✓				✓	
Rear drum brake lining & rear wheel cylinders	S/I				✓				✓				✓	
Camshaft timing belt	R								✓					
Ignition cables	R								✓					
Distributor cap & rotor	S/I								✓					
EVAP system	S/I								✓					
Fuel system	S/I								✓					

① Spark plugs (SOHC 1996) - replace every 100,000 miles.
R – Replace   S/I – Service or Inspect

FREQUENT OPERATION MAINTENANCE (SEVERE SERVICE)
If a vehicle is operated under any of the following conditions it is considered severe service:
- Extremely dusty areas.
- 50% or more of the vehicle operation is in 32°C (90°F) or higher temperatures, or constant operation in temperatures below 0°C (32°F).
- Prolonged idling (vehicle operation in stop and go traffic).
- Frequent short running periods (engine does not warm to normal operating temperatures).
- Police, taxi, delivery usage or trailer towing usage.
Oil & filter change – change every 3000 miles.
Disc brake pads - check every 6000 miles.
Air filter element – change every 15,000 miles.
Automatic transaxle fluid – change every 15,000 miles.
Rear drum brake linings & rear wheel cylinders - check every 15,000 miles.
Spark plugs - change every 15,000 miles.

TALON/LASER

VEHICLE IDENTIFICATION CHART

		Engine Code					Model Year	
Code	Liters	Cu. In. (cc)	Cyl.	Fuel Sys.	Eng. Mfg.		Code	Year
B	1.8	107 (1753)	4	MFI	Mitsubishi		R	1994
E	2.0	122 (1999)	4	MFI	Mitsubishi		P	1993
F	2.0	122 (1999)	4	MFI Turbo	Mitsubsihi		S	1995
Y	2.0	122 (1996)	4	MFI	Chrysler		T	1996
							V	1997

MFI - Multiport fuel injection

ENGINE IDENTIFICATION
All measurements are given in inches.

Year	Model	Engine Displacement Liters (cc)	Engine Series (ID/VIN)	Fuel System	No. of Cylinders	Engine Type
1993	Laser	1.8 (1755)	B	MFI	4	SOHC
	Laser	2.0 (1997)	E	MFI	4	DOHC
	Laser	2.0 (1997)	F	MFI-Turbo	4	DOHC
	Talon	1.8 (1753)	B	MFI	4	SOHC
	Talon	2.0 (1999)	E	MFI	4	DOHC
	Talon	2.0 (1999)	F	MFI Turbo	4	DOHC
1994	Laser	1.8 (1755)	B	MFI	4	SOHC
	Laser	2.0 (1997)	E	MFI	4	DOHC
	Laser	2.0 (1997)	F	MFI-Turbo	4	DOHC
	Talon	1.8 (1753)	B	MFI	4	SOHC
	Talon	2.0 (1999)	E	MFI	4	DOHC
	Talon	2.0 (1999)	F	MFI Turbo	4	DOHC
1995	Talon	2.0 (1997)	F	MFI-Turbo	4	DOHC
	Talon	2.0 (1996)	Y	MFI	4	DOHC
1996-97	Talon	2.0 (1997)	F	MFI-Turbo	4	DOHC
	Talon	2.0 (1996)	Y	MFI	4	DOHC

MFI - Multiport fuel injection   DOHC - Double overhead camshaft   SOHC - Single overhead camshaft

GENERAL ENGINE SPECIFICATIONS

Year	Engine ID/VIN	Engine Displacement Liters (cc)	Fuel System Type	Net Horsepower @ rpm	Net Torque @ rpm (ft. lbs.)	Bore x Stroke (in.)	Compression Ratio	Oil Pressure @ rpm
1993	B	1.8 (1753)	MFI	92@5000	105@3500	3.17x3.39	9.0:1	41@2000
	E	2.0 (1999)	MFI	135@6000	125@5000	3.35x3.47	9.0:1	41@2000
	F	2.0 (1999)	MFI	190@6000	203@5000	3.35x3.47	7.8:1	41@2000
1994	B	1.8 (1753)	MFI	92@5000	105@3500	3.17x3.39	9.0:1	①
	E	2.0 (1999)	MFI	135@6000	125@5000	3.35x3.47	9.0:1	①
	F	2.0 (1999)	MFI	190@6000	203@5000	3.35x3.47	7.8:1	①
1995	F	2.0 (1997)	MFI	210@6000 ②	214@3000 ③	3.35x3.46	8.5:1	①
	Y	2.0 (1996)	MFI	140@6000	131@4800	3.44x3.27	9.6:1	④

GENERAL ENGINE SPECIFICATIONS

Year	Engine ID/VIN	Engine Displacement Liters (cc)	Fuel System Type	Net Horsepower @ rpm	Net Torque @ rpm (ft. lbs.)	Bore x Stroke (in.)	Compression Ratio	Oil Pressure @ rpm
1996-97	F	2.0 (1997)	MFI	210@6000 ②	214@3000 ③	3.35x3.46	8.5:1	①
	Y	2.0 (1996)	MFI	140@6000	131@4800	3.44x3.27	9.6:1	④

① 11.4 psi or more at curb idle speed
② Automatic: 205@6000

③ Automatic: 220@3000
④ 4 psi or more at curb idle speed

GASOLINE ENGINE TUNE-UP SPECIFICATIONS

Year	Engine ID/VIN	Engine Displacement Liters (cc)	Spark Plugs Gap (in.)	Ignition Timing (deg.) MT	Ignition Timing (deg.) AT	Fuel Pump (psi)	Idle Speed (rpm) MT	Idle Speed (rpm) AT	Valve Clearance In.	Valve Clearance Ex.
1993	B	1.8 (1753)	0.039-0.043	5B	5B	38	750	750	HYD	HYD
	E	2.0 (1999)	0.039-0.043	5B	5B	38	700	700	HYD	HYD
	F	2.0 (1999)	0.028-0.031	5B	5B	②	750	750	HYD	HYD
1994	B	1.8 (1753)	0.039-0.943	5B	5B	38 ③	750	750	HYD	HYD
	E	2.0 (1999)	0.039-0.043	5B	5B	38 ③	700	700	HYD	HYD
	F	2.0 (1999)	0.028-0.031	5B	5B	33 ③	750	750	HYD	HYD
1995	F	2.0 (1997)	0.028-0.030	5B	5B	33 ③	750	750	HYD	HYD
	Y	2.0 (1996)	0.033-0.038	①	①	38	700	700	HYD	HYD
1996-97	F	2.0 (1997)	0.028-0.030	5B	5B	33 ③	750	750	HYD	HYD
	Y	2.0 (1996)	0.033-0.038	①	①	38 ③	800	800	HYD	HYD

NOTE: The Vehicle Emission Control Information label often reflects specification changes made during production. The label figures must be used if they differ from those in this chart.

HYD - Hydraulic

① Basic ignition timing is not adjustable
② Manual transaxle: 27
   Automatic transaxle: 33
③ Pressure at idle with vacuum applied to fuel pressure regulator

CAPACITIES

Year	Model	Engine ID/VIN	Engine Displacement Liters (cc)	Engine Oil with Filter (qts.)	Transmission (pts.) 4-Spd	Transmission (pts.) 5-Spd	Transmission (pts.) Auto.	Transfer Case (pts.)	Drive Axle Front (pts.)	Drive Axle Rear (pts.)	Fuel Tank (gal.)	Cooling System (qts.)
1993	Laser	B	1.8 (1755)	4.1	-	1.9	13.2	-	-	-	15.9	6.6
	Laser	E	2.0 (1997)	4.6	-	1.9	13.2	-	-	-	15.9	6.6
	Laser	F	2.0 (1997)	4.8	-	①	15.2	1.2	-	1.5	15.9	7.6
	Talon	B	1.8 (1753)	4.1	-	3.8	13.0	-	-	-	16.0	6.6
	Talon	E	2.0 (1999)	4.6	-	4.6	13.0	-	-	-	16.0	7.6
	Talon	F	2.0 (1999)	4.6	-	①	13.0	1.25	-	1.5	16.0	7.6
1994	Laser	B	1.8 (1755)	4.1	-	3.8	13.0	-	-	-	15.9	6.6
	Laser	E	2.0 (1997)	4.6	-	3.8	13.2	-	-	-	15.9	7.6
	Laser	F	2.0 (1997)	4.8	-	①	14.2	1.2	-	1.5	15.9	7.6
	Talon	B	1.8 (1753)	4.1	-	3.8	13.0	-	-	-	16.0	6.6
	Talon	E	2.0 (1999)	4.6	-	4.6	13.0	-	-	-	16.0	7.6
	Talon	F	2.0 (1999)	4.6	-	①	13.0	1.25	-	1.8	16.0	7.6
1995	Talon	F	2.0 (1997)	4.6	-	①	14.2	1.06	-	1.8	16.0	7.4
	Talon	Y	2.0 (1996)	4.5	-	4.2	18.2	-	-	-	16.0	7.4
1996-97	Talon	F	2.0 (1997)	4.6	-	①	14.2	1.06	-	1.8	16.0	7.4
	Talon	Y	2.0 (1996)	4.5	-	4.2	18.2	-	-	-	16.0	7.4

① 2WD: 4.6
   4WD: 4.8

VALVE SPECIFICATIONS

Year	Engine ID/VIN	Engine Displacement Liters (cc)	Seat Angle (deg.)	Face Angle (deg.)	Spring Test Pressure (lbs. @ in.)	Spring Installed Height (in.)	Stem-to-Guide Clearance (in.)		Stem Diameter (in.)	
							Intake	Exhaust	Intake	Exhaust
1993	B	1.8 (1753)	44-44.5	45-45.5	62 ①	NA	0.0012-0.0024	0.0020-0.0035	0.3100	0.3100
	E	2.0 (1999)	44-44.5	45.45-5	66 ②	1.902 ③	0.0008-0.0019	0.0020-0.0033	0.2585-0.2591	0.2571-0.2579
	F	2.0 (1999)	44-44.5	45-45.5	66 ②	1.902 ③	0.0008-0.0019	0.0020-0.0033	0.2585-0.2591	0.2571-0.2579
1994	B	1.8 (1753)	44-44.5	45-45.5	68 ②	NA	0.0012-0.0024	0.0020-0.0035	0.3100	0.3100
	E	2.0 (1999)	44-44.5	45-45.5	66 ②	1.902 ③	0.0008-0.0019	0.0020-0.0033	0.2585-0.2591	0.2571-0.2579
	F	2.0 (1999)	44-44.5	45-45.5	66 ②	1.902 ③	0.0008-0.0019	0.0020-0.0033	0.2585-0.2591	0.2571-0.2579
1995	F	2.0 (1997)	44-44.5	45-45.5	54 ②	1.570	0.0008-0.0040	0.0020-0.0060	0.2600	0.2560
	Y	2.0 (1996)	45	45-45.5	110-120@ 1.173 ④	1.496	0.0019-0.0030	0.0029-0.0040	0.2336-0.2343	0.2325-0.2332
1996-97	F	2.0 (1997)	44-44.5	45-45.5	54 ②	1.570	0.0008-0.0040	0.0020-0.0060	0.2600	0.2560
	Y	2.0 (1996)	45	44.5-45	123-137@ 1.153 ④	1.496	0.0019-0.0030	0.0029-0.0040	0.2336-0.2343	0.2325-0.2332

① Intake: 51@1.570 (installed height)
  Exhaust: 64@1.570 (installed height)

② At installed height

③ Specification is for free length

④ With valves open

TORQUE SPECIFICATIONS
All readings in ft. lbs.

Year	Engine ID/VIN	Engine Displacement Liters (cc)	Cylinder Head Bolts	Main Bearing Bolts	Rod Bearing Bolts	Crankshaft Damper Bolts	Flywheel Bolts	Manifold		Spark Plugs	Lug Nut
								Intake	Exhaust		
1993	B	1.8 (1753)	51-54	37-39	24-25	80-94	94-101	13-18	18-22	15-21	87-101
	E	2.0 (1999)	65-72	47-51	36-38	94	94-101	18-22	18-22	15-21	87-101
	F	2.0 (1999)	65-72	47-51	36-38	94	94-101	18-22	18-22	15-21	87-101
1994	B	1.8 (1753)	51-54	37-39	24-25	80-94	94-101	13-18	18-22	15-21	65-80
	E	2.0 (1999)	①	②	14.5	94	94-101	14	18-22	15-21	65-80
	F	2.0 (1999)	①	②	14.5	94	94-101	14	18-22	18	65-80
1995	F	2.0 (1997)	④	②	14.5 ③	94	94-101	14	18-22	18	65-80
	Y	2.0 (1996)	⑤	55	⑥	105	-	17	17	20	65-80
1996-97	F	2.0 (1997)	④	②	14.5 ③	94	94-101	14	18-22	18	65-80
	Y	2.0 (1996)	⑤	55	⑥	105	-	17	17	20	65-80

① Step 1: 54 ft. lbs.
  Step 2: Fully loosen
  Step 3: 14 ft. lbs.
  Step 4: Plus 90 degrees
  Step 5: Repeat Step 4

② Step 1: 18 ft. lbs.
  Step 2: Plus 90 degrees
③ Plus 90 degrees

④ Step 1: 58 ft. lbs.
  Step 2: Fully loosen
  Step 3: 15 ft. lbs.
  Step 4: Plus 90 degrees
  Step 5: Repeat Step 4

⑤ Step 1:
  Bolts 1-6: 24 ft. lbs.
  Bolts 7-10: 20 ft. lbs.
  Step 2:
  Bolts 1-6: 49 ft. lbs.
  Bolts 7-10: 20 ft. lbs.
  Step 3: Plus 90 degrees

⑥ 20 ft. lbs. plus 90 degrees

BRAKE SPECIFICATIONS
All measurements in inches unless noted

Year	Model		Master Cylinder Bore	Brake Disc Original Thickness	Brake Disc Minimum Thickness	Maximum Runout	Brake Drum Diameter Original Inside Diameter	Brake Drum Diameter Max. Wear Limit	Brake Drum Diameter Maximum Machine Diameter	Minimum Lining Thickness Front	Minimum Lining Thickness Rear
1993	Laser		①	0.940	0.882	0.003	-	-	-	0.080	-
	①		-	0.394	0.331	0.003	-	-	-	-	0.040
	Talon	F	①	0.940	0.882	0.003	-	-	-	0.080	-
		R	-	0.390	0.331	0.003	-	-	-	-	0.080
1994	Laser		①	0.940	0.882	0.003	-	-	-	0.080	-
	①		-	0.390	0.331	0.003	-	-	-	-	0.080
	Talon	F	②	③	④	0.003	-	-	-	0.080	-
		R	-	-	-	0.003	-	-	-	-	0.080
1995	Talon	F	②	③	④	0.003	-	-	-	0.080	0.080
		R	-	-	-	0.003	-	-	-	0.080	0.080
1996-97	Talon	F	②	0.940	0.882	0.003	-	-	-	0.080	-
		R	②	0.390	0.331	0.003	-	-	-	-	0.080

① Non-turbocharged without ABS: 0.875
Non-turbocharged with ABS: 0.938
FWD turbocharged: 0.938
AWD turbocharged: 1.000

② Non-turbocharged: 0.938
FWD turbocharged: 0.938
AWD turbocharged: 1.000

③ Front: 0.940
Rear: 0.390
AWD: 0.790

④ Front: 0.882
Rear: 0.331
AWD: 0.721

FREQUENT MAINTENANCE LABOR
EAGLE TALON, PLYMOUTH LASER

The following should be used as a guide when determining the amount of work required for a particular service if taken to a repair shop. In estimating how long a particular Frequent Maintenance Service item should take, please observe the following:
  • **Factory Time** is time that is generated by the vehicle manufacturer.
  • **Chilton Time** is time that is based on field research and data supplied by the vehicle manufacturer.
  • All labor time operations are given in hours and tenths of an hour.
  • All labor operations, are to be used as a **guide**.

COOLING

(G) Winterize Cooling System
Includes: Run engine to check for leaks, tighten all hose connections. Test radiator and pressure cap, drain radiator and engine block. Add antifreeze and refill system.
1993-97 5

(G) Belt, Drive, Renew
1993-94 (.3) 4
1995-97
  Alternator
    wo/Turbo (.4) 6
    w/Turbo (.5) 7
  Power Steering (.2) 3

(G) Belt, Drive, Adjust
1993-97 3

(G) Hoses, Radiator, Renew
Includes: Drain and refill cooling system as required.
1993-97
  upper (.3) 4
  lower (.4) 5

(G) Thermostat, Coolant, Renew
1993-94 (.3) 5
1995-97 (.4) 6

FUEL

(M) Air Cleaner, Service
1993-97 2

(G) Filter, Fuel, Renew
1993-94
  in line (.5) 7
  in tank (.9) 1.3

1995-97
  in line (.7) 1.0
  in tank (.9) 1.3

BRAKES

(G) Bleed Brakes (Four Wheels)
Includes: Add fluid.
1993-97 5

(M) Parking Brake, Adjust
1993-97 4

LUBRICATION SERVICES

(M) Engine Oil & Filter, Renew
Includes: Inspect and correct all fluid levels.
1993-97 (.3) 3

FREQUENT MAINTENANCE LABOR (cont.)
EAGLE TALON, PLYMOUTH LASER

(M) Lubricate Chassis
Includes: Inspect and correct all fluid levels.
1993-97 .4
Install grease fittings, add 1

WHEELS

(G) Wheel, Renew (One)
1993-97, one 5

(G) Wheel, Balance
1993-97
  one .3
  each adtnl. 2

(G) Wheels, Rotate (All)
1993-97 5

ELECTRICAL

(G) Headlamps, Aim
1993-97
  two 4
  four 6

(G) License Lamp Assy., Renew
1993-97 (.2) 3

(G) Park & Turn Signal Lamp Assy., Renew
1993-97 (.2) 3

(G) Parking Lamp Lens or Bulb, Renew
1993-97 (.2) 3

(G) Tail Lamp Assy., Renew
1993-97 (.5) 7

(G) Horn, Renew
1993-97 (.6) 8

(M) Terminals, Battery, Clean
1993-97 3

SCHEDULED MAINTENANCE INTERVALS
(EAGLE TALON, PLYMOUTH LASER)

TO BE SERVICED	TYPE OF SERVICE	VEHICLE MILEAGE INTERVAL (x1000)												
		7.5	15	22.5	30	37.5	45	52.5	60	67.5	75	82.5	90	97.5
Engine oil & filter (Non-turbo)①	R	✓	✓	✓	✓	✓	✓	✓	✓	✓	✓	✓	✓	✓
Coolant level, hoses & clamps	S/I	✓	✓	✓	✓	✓	✓	✓	✓	✓	✓	✓	✓	✓
Rotate tires	S/I	✓	✓	✓	✓	✓	✓	✓	✓	✓	✓	✓	✓	✓
Automatic transaxle fluid level	S/I		✓		✓		✓		✓		✓		✓	
Brake hoses & disc brake pads	S/I		✓		✓		✓		✓		✓		✓	
Drive shaft boots & front suspension components	S/I		✓		✓		✓		✓		✓		✓	
Air filter element	R				✓				✓				✓	
Automatic transaxle fluid & filter②	R				✓				✓				✓	
Engine Coolant	R				✓				✓				✓	
Spark plugs	R				✓				✓				✓	
Accessory drive belts	S/I				✓				✓				✓	
Ball joints & steering linkage seals	S/I				✓				✓				✓	
Exhaust system	S/I				✓				✓				✓	
Fuel hoses	S/I				✓				✓				✓	

SCHEDULED MAINTENANCE INTERVALS
(EAGLE TALON, PLYMOUTH LASER) (Cont.)

TO BE SERVICED	TYPE OF SERVICE	VEHICLE MILEAGE INTERVAL (x1000)												
		7.5	15	22.5	30	37.5	45	52.5	60	67.5	75	82.5	90	97.5
Manual transaxle oil (including transfer)	S/I				✓				✓				✓	
Rear axle oil (AWD)	③				✓				✓				✓	
Camshaft timing belt	R								✓					
Ignition cables	R								✓					
EVAP & fuel system	S/I								✓					

① Engine oil & filter (Turbo) - change every 5000 miles. ② 1993-94: All w/A/T, 1995-97: Turbo w/A/T only.
③ 1993-94 AWD w/LSD: replace fluid at specified interval. All other w/AWD: check fluid level at specified interval.
R – Replace   S/I – Service or Inspect

FREQUENT OPERATION MAINTENANCE (SEVERE SERVICE)
If a vehicle is operated under any of the following conditions it is considered severe service:
- Extremely dusty areas.
- 50% or more of the vehicle operation is in 32°C (90°F) or higher temperatures, or constant operation in temperatures below 0°C (32°F).
- Prolonged idling (vehicle operation in stop and go traffic).
- Frequent short running periods (engine does not warm to normal operating temperatures).
- Police, taxi, delivery usage or trailer towing usage.
Oil & filter change – change every 3000 miles.
Air filter element – service or inspect every 7500 miles.
Automatic transaxle fluid – change every 15,000 miles.
Spark plugs - change every 15,000 miles.
Disc brake pads - check more frequently than every 7500 miles (1993-94, 1996-97) or 6000 miles (1995).

NEON

VEHICLE IDENTIFICATION CHART

	Engine Code						Model Year	
Code	Liters	Cu. In. (cc)	Cyl.	Fuel Sys.	Eng. Mfg.		Code	Year
C	2.0	122 (1996)	I4	MFI	Chrysler		S	1995
Y	2.0	122 (1996)	I4	MFI	Chrysler		T	1996
							V	1997

MFI - Multipoint fuel injection

ENGINE IDENTIFICATION

All measurements are given in inches.

Year	Model	Engine Displacement Liters (cc)	Engine Series (ID/VIN)	Fuel System	No. of Cylinders	Engine Type
1995	Neon	2.0 (1996)	C	MFI	4	SOHC
	Neon	2.0 (1996)	Y	MFI	4	DOHC
1996-97	Neon	2.0 (1996)	C	MFI	4	SOHC
	Neon	2.0 (1996)	Y	MFI	4	DOHC

MFI - Multipoint fuel injection

GENERAL ENGINE SPECIFICATIONS

Year	Engine ID/VIN	Engine Displacement Liters (cc)	Fuel System Type	Net Horsepower @ rpm	Net Torque @ rpm (ft. lbs.)	Bore x Stroke (in.)	Compression Ratio	Oil Pressure @ rpm
1995	C	2.0 (1996)	MFI	132@6000	129@5000	3.44x3.26	9.8:1	25-80@3000
	Y	2.0 (1996)	MFI	150@4400	NA	3.44x3.26	9.6:1	25-80@3000
1996-97	C	2.0 (1996)	MFI	132@6000	129@5000	3.44x3.26	9.8:1	25-80@3000
	Y	2.0 (1996)	MFI	150@4400	NA	3.44x3.26	9.6:1	25-80@3000

MFI - Multipoint fuel injection

GASOLINE ENGINE TUNE-UP SPECIFICATIONS

Year	Engine ID/VIN	Engine Displacement Liters (cc)	Spark Plugs Gap (in.)	Ignition Timing (deg.) MT	Ignition Timing (deg.) AT	Fuel Pump (psi)	Idle Speed (rpm) MT	Idle Speed (rpm) AT	Valve Clearance In.	Valve Clearance Ex.
1995	C	2.0 (1996)	0.035	②	②	48	③	③	HYD	HYD
	Y	2.0 (1996)	0.035	②	②	48	③	③	HYD	HYD
1996-97	C	2.0 (1996)	0.035	②	②	48	①	①	HYD	HYD
	Y	2.0 (1996)	0.035	②	②	48	①	①	HYD	HYD

NOTE: The Vehicle Emission Control Information label often reflects specification changes made during production. The label figures must be used if they differ from those in this chart.

B - Before top dead center

C - Cold

HYD - Hydraulic

NA - Not Available

① Refer to the Vehicle Emission Control Information label for correct timing specifications with a range of +/- 2 degrees

② Ignition timing cannot be adjusted. Base engine timing is set at TDC during assembly.

③ Refer to Vehicle Emissions Control Information label for proper specification

CAPACITIES

Year	Model	Engine ID/VIN	Engine Displacement Liters (cc)	Engine Oil with Filter (qts.)	Transmission (pts.) 4-Spd	Transmission (pts.) 5-Spd	Transmission (pts.) Auto.	Transfer Case (pts.)	Drive Axle Front (pts.)	Drive Axle Rear (pts.)	Fuel Tank (gal.)	Cooling System (qts.)
1995	Neon	C	2.0 (1996)	4.5	-	①	8.0	-	-	-	11.0	7.4
	Neon	Y	2.0 (1996)	4.5	-	①	8.0	-	-	-	11.0	7.4
1996-97	Neon	C	2.0 (1996)	4.5	-	①	8.0	-	-	-	11.0	7.4
	Neon	Y	2.0 (1996)	4.5	-	①	8.0	-	-	-	11.0	7.4

① Fill to bottom of fill hole

VALVE SPECIFICATIONS

Year	Engine ID/VIN	Engine Displacement Liters (cc)	Seat Angle (deg.)	Face Angle (deg.)	Spring Test Pressure (lbs. @ in.)	Spring Installed Height (in.)	Stem-to-Guide Clearance (in.)		Stem Diameter (in.)	
							Intake	Exhaust	Intake	Exhaust
1995	C	2.0 (1996)	44.5-45	45-45.5	75@1.54	1.540	0.0018-0.0025	0.0029-0.0037	0.2340	0.2330
	Y	2.0 (1996)	44.5-45	45-45.5	55-60@1.49	1.490	0.0018-0.0025	0.0029-0.0037	0.2340	0.2330
1996-97	C	2.0 (1996)	44.5-45	45-45.5	75@1.54	1.540	0.0018-0.0025	0.0029-0.0037	0.2340	0.2330
	Y	2.0 (1996)	44.5-45	45-45.5	55-60@1.49	1.490	0.0018-0.0025	0.0029-0.0037	0.2340	0.2330

TORQUE SPECIFICATIONS
All readings in ft. lbs.

Year	Engine ID/VIN	Engine Displacement Liters (cc)	Cylinder Head Bolts	Main Bearing Bolts	Rod Bearing Bolts	Crankshaft Damper Bolts	Flywheel Bolts	Manifold Intake	Manifold Exhaust	Spark Plugs	Lug Nut
1995	C	2.0 (1996)	②	①	③	105	70	17	17	20	95
	Y	2.0 (1996)	④	①	③	105	70	17	17	20	95
1996-97	C	2.0 (1996)	②	①	③	105	70	17	17	20	95
	Y	2.0 (1996)	④	①	③	105	70	17	17	20	95

① Step 1: 30 ft. lbs.
Step 2: +90 degrees

② Step 1: 25 ft. lbs.
Step 2: 50 ft. lbs.
Step 3: 50 ft. lbs.
Step 4: +90 degrees

③ Step 1: 20 ft. lbs.
Step 2: +90 degrees

④ Step 1:
Bolts 1-6: 25 ft. lbs.
Bolts 7-10: 20 ft. lbs.
Step 2:
Bolts 1-6: 50 ft. lbs.
Bolts 7-10: 20 ft. lbs.
Step 3:
Bolts 1-6: 50 ft. lbs.
Bolts 7-10: 20 ft. lbs.
Step 4: +90 degrees

BRAKE SPECIFICATIONS
All measurements in inches unless noted

Year	Model		Master Cylinder Bore	Brake Disc Original Thickness	Brake Disc Minimum Thickness	Brake Disc Maximum Runout	Brake Drum Diameter Original Inside Diameter	Brake Drum Diameter Max. Wear Limit	Brake Drum Diameter Maximum Machine Diameter	Minimum Lining Thickness Front	Minimum Lining Thickness Rear
1995	Neon	F	①	0.792	0.724	0.005	-	-	-	0.300	-
		R	-	NA	NA	NA	7.88	NA	NA	-	②
1996-97	Neon	F	①	0.792	0.724	0.005	-	-	-	0.300	-
		R	-	NA	NA	NA	7.88	NA	NA	-	②

NA - Not Available
① If equipped with rear drum brakes: 0.827
If equipped with rear disc brakes: 0.875
② Rear disc: NA
Rear drum: 0.280

FREQUENT MAINTENANCE LABOR
DODGE NEON, PLYMOUTH NEON

The following should be used as a guide when determining the amount of work required for a particular service if taken to a repair shop. In estimating how long a particular Frequent Maintenance Service item should take, please observe the following:
- **Factory Time** is time that is generated by the vehicle manufacturer.
- **Chilton Time** is time that is based on field research and data supplied by the vehicle manufacturer.
- All labor time operations are given in hours and tenths of an hour.
- All labor operations, are to be used as a **guide**.

	(Factory Time)	Chilton Time

COOLING

(G) Winterize Cooling System
Includes: Run engine to check for leaks, tighten all hose connections. Test radiator and pressure cap, drain radiator and engine block. Add antifreeze and refill system.
1995-97 5

(G) Belt, Drive, Renew
1995-97
alternator
SOHC (.4) 5
DOHC (.5) 6
power steering (.2) 3

(G) Belt, Drive, Adjust
1995-97
one (.2) 3
each adtnl. 1

(G) Hoses, Radiator, Renew
Includes: Drain and refill cooling system as required.
1995-97
upper (.3) 5
lower (.5) 6
w/AC add 2

(G) Thermostat, Coolant, Renew
1995-97 (.3) 5

FUEL

(M) Air Cleaner, Service
1995-97 2

(G) Filter, Fuel, Renew
1995-97
in line (.3) 4
in tank (.7) 1.0

BRAKES

(G) Bleed Brakes (Four Wheels)
Includes: Add fluid.
1995-97 (.4) 5
Bleed modulator, add 4

(G) Brakes, Adjust (Minor)
Includes: Adjust brakes, fill master cylinder.
1995-97 4

(M) Parking Brake, Adjust
1995-97 4

LUBRICATION SERVICES

(M) Engine Oil & Filter, Renew
Includes: Inspect and correct all fluid levels.
1995-97 3

(M) Lubricate Chassis
Includes: Inspect and correct all fluid levels.
1995-97 4
Install grease fittings, add 1

WHEELS

(G) Wheel, Renew (One)
1995-97 5

(G) Wheel, Balance
1995-97
one 3
each adtnl. 2

(G) Wheels, Rotate (All)
1995-97 5

ELECTRICAL

(G) Headlamps, Aim
1995-97
two 4
four 6

(G) License Lamp Assy., Renew
1995-97 (.2) 3

(G) Park & Turn Signal Lamp Assy., Renew
1995-97 (.2) 3

(G) Parking Lamp Lens or Bulb, Renew
1995-97 (.2) 3

(M) Terminals, Battery, Clean
1995-97 3

SCHEDULED MAINTENANCE INTERVALS
(DODGE NEON, PLYMOUTH NEON)

TO BE SERVICED	TYPE OF SERVICE	VEHICLE MILEAGE INTERVAL (x1000)												
		7.5	15	22.5	30	37.5	45	52.5	60	67.5	75	82.5	90	97.5
Engine oil & filter	R	✓	✓	✓	✓	✓	✓	✓	✓	✓	✓	✓	✓	✓
Brake hoses	S/I	✓	✓	✓	✓	✓	✓	✓	✓	✓	✓	✓	✓	✓
Coolant level, hoses & clamps	S/I	✓	✓	✓	✓	✓	✓	✓	✓	✓	✓	✓	✓	✓
CV joints & front suspension components	S/I	✓	✓	✓	✓	✓	✓	✓	✓	✓	✓	✓	✓	✓
Exhaust system	S/I	✓	✓	✓	✓	✓	✓	✓	✓	✓	✓	✓	✓	✓
Manual transaxle oil	S/I	✓	✓	✓	✓	✓	✓	✓	✓	✓	✓	✓	✓	✓
Rotate tires	S/I	✓	✓	✓	✓	✓	✓	✓	✓	✓	✓	✓	✓	✓
Accessory drive belts	S/I		✓		✓		✓		✓		✓		✓	
Brake linings	S/I			✓			✓			✓			✓	
Air filter element	R				✓				✓				✓	
Spark plugs	R				✓				✓				✓	
Lubricate ball joints	S/I				✓				✓				✓	
Engine Coolant	R						✓				✓			
PCV valve	S/I								✓				✓	
Ignition cables	R								✓					
Camshaft timing belt①	R													

① Camshaft timing belt - replace at 105,000 miles for normal service, or 102,000 miles for severe service (1995) or 105,000 miles (1996-97).

R – Replace   S/I – Service or Inspect

FREQUENT OPERATION MAINTENANCE (SEVERE SERVICE)

If a vehicle is operated under any of the following conditions it is considered severe service:
- Extremely dusty areas.
- 50% or more of the vehicle operation is in 32°C (90°F) or higher temperatures, or constant operation in temperatures below 0°C (32°F).
- Prolonged idling (vehicle operation in stop and go traffic).
- Frequent short running periods (engine does not warm to normal operating temperatures).
- Police, taxi, delivery usage or trailer towing usage.

Oil & filter change – change every 3000 miles.
Rotate tires every 6000 miles.
Brake linings - inspect every 12,000 miles.
Air filter element - service or inspect every 15,000 miles.
Automatic transaxle – change fluid & adjust bands every 15,000 miles.
Manual transaxle fluid - replace every 15,000 miles.
Engine coolant - replace at 36,000 miles, and every 30,000 miles thereafter.

STRATUS/SEBRING CONVERTIBLE/CIRRUS/BREEZE

VEHICLE IDENTIFICATION CHART

		Engine Code					Model Year	
Code	Liters	Cu. In. (cc)	Cyl.	Fuel Sys.	Eng. Mfg.		Code	Year
C	2.0	122 (1996)	I4	MFI	Chrysler		S	1995
X	2.4	148 (2429)	I4	MFI	Chrysler		T	1996
H	2.5	152 (2497)	V6	MFI	Mitsubishi		V	1997

MFI - Multiport fuel injection

ENGINE IDENTIFICATION
All measurements are given in inches.

Year	Model	Engine Displacement Liters (cc)	Engine Series (ID/VIN)	Fuel System	No. of Cylinders	Engine Type
1995	Cirrus	2.5 (2497)	H	MFI	6	SOHC
	Stratus	2.0 (1996)	C	MFI	4	SOHC
	Stratus	2.4 (2429)	X	MFI	4	DOHC
	Stratus	2.5 (2497)	H	MFI	6	SOHC
1996-97	Breeze	2.0 (1996)	C	MFI	4	SOHC
	Cirrus	2.4 (2429)	X	MFI	4	DOHC
	Cirrus	2.5 (2497)	H	MFI	6	SOHC
	Sebring Convertible	2.4 (2429)	X	MFI	4	DOHC
	Sebring Convertible	2.5 (2497)	H	MFI	6	SOHC
	Stratus	2.0 (1996)	C	MFI	4	SOHC
	Stratus	2.4 (2429)	X	MFI	4	DOHC
	Stratus	2.5 (2497)	H	MFI	6	SOHC

MFI - Multiport fuel injection   SOHC - Single overhead camshaft   DOHC - Double overhead camshaft

GENERAL ENGINE SPECIFICATIONS

Year	Engine ID/VIN	Engine Displacement Liters (cc)	Fuel System Type	Net Horsepower @ rpm	Net Torque @ rpm (ft. lbs.)	Bore x Stroke (in.)	Compression Ratio	Oil Pressure @ rpm
1995	C	2.0 (1996)	MFI	132@6000	129@5000	3.44x3.26	9.8:1	25-80@3000
	X	2.4 (2429)	MFI	140@5200	160@4000	3.44x3.98	9.4:1	25-80@3000
	H	2.5 (2497)	MFI	164@5900	163@4350	3.29x2.99	9.4:1	35-75@3000
1996-97	C	2.0 (1996)	MFI	132@6000	129@5000	3.44x3.26	9.8:1	25-80@3000
	X	2.4 (2429)	MFI	140@5200	160@4000	3.44x3.98	9.4:1	25-80@3000
	H	2.5 (2497)	MFI	164@5900	163@4350	3.29x2.99	9.4:1	35-75@3000

MFI - Multiport fuel injection

GASOLINE ENGINE TUNE-UP SPECIFICATIONS

Year	Engine ID/VIN	Engine Displacement Liters (cc)	Spark Plugs Gap (in.)	Ignition Timing (deg.)		Fuel Pump (psi)	Idle Speed (rpm)		Valve Clearance	
				MT	AT		MT	AT	In.	Ex.
1995	C	2.0 (1996)	0.035	②	②	48	①	①	HYD	HYD
	X	2.4 (2429)	0.050	-	②	49	-	①	HYD	HYD
	H	2.5 (2497)	0.038-0.043	-	②	49 ③	-	500-1100	HYD	HYD
1996-97	C	2.0 (1996)	0.035	②	②	48	①	①	HYD	HYD
	X	2.4 (2429)	0.050	-	②	49	-	①	HYD	HYD
	H	2.5 (2497)	0.038-0.043	-	②	49 ③	-	500-1100	HYD	HYD

NOTE: The Vehicle Emission Control Information label often reflects specification changes made during production. The label figures must be used if they differ from those in this chart.

HYD - Hydraulic

① Refer to the Vehicle Emission Control Information label for correct timing specifications with a range of +/- 2 degrees

② Ignition timing cannot be adjusted. Base engine timing is set at TDC during assembly.

③ This reading measured with vacuum hose disconnected from fuel pressure regulator

CAPACITIES

Year	Model	Engine ID/VIN	Engine Displacement Liters (cc)	Engine Oil with Filter (qts.)	Transmission (pts.)			Transfer Case (pts.)	Drive Axle		Fuel Tank (gal.)	Cooling System (qts.)
					4-Spd	5-Spd	Auto.		Front (pts.)	Rear (pts.)		
1995	Cirrus	H	2.5 (2497)	4.5	-	-	18.2 ①	-	-	-	16.0	10.5
	Stratus	C	2.0 (1996)	4.5	-	4.4 ②	-	-	-	-	16.0	7.4
	Stratus	X	2.4 (2429)	4.5	-	-	8.0	-	-	-	16.0	9.0
	Stratus	H	2.5 (2497)	4.5	-	-	8.0	-	-	-	16.0	10.5
1996-97	Breeze	C	2.0 (1996)	4.5	-	②	8.0	-	-	-	16.0	7.4
	Cirrus	X	2.4 (2429)	5.0	-	-	18.2 ①	-	-	-	16.0	9.0
	Cirrus	H	2.5 (2497)	4.5	-	-	18.2 ①	-	-	-	16.0	10.5
	Sebring Convertible	X	2.4 (2429)	4.5	-	-	18.2 ①	-	-	-	16.0	9.0
	Sebring Convertible	H	2.5 (2497)	4.5	-	-	18.2 ①	-	-	-	16.0	10.5
	Stratus	C	2.0 (1996)	4.5	-	4.4 ②	-	-	-	-	16.0	7.4
	Stratus	X	2.4 (2429)	4.5	-	-	8.0	-	-	-	16.0	9.0
	Stratus	H	2.5 (2497)	4.5	-	-	8.0	-	-	-	16.0	10.5

① Overhaul fill capacity with torque converter empty

② Fill to bottom of fill hole

CAMSHAFT SPECIFICATIONS
All measurements given in inches.

Year	Engine ID/VIN	Engine Displacement Liters (cc)	Journal Diameter					Elevation		Bearing Clearance	Camshaft End Play
			1	2	3	4	5	In.	Ex.		
1995	C	2.0 (1996)	1.6190-1.6200	1.6340-1.6350	1.6500-1.6510	1.6660-1.6670	1.6820-1.6830	0.3070	0.2770	0.0027-0.0030	0.0059
	X	2.4 (2429)	1.0210-1.0220	1.0210-1.0220	1.0210-1.0220	1.0210-1.0220	①	0.3240	0.2560	0.0027-0.0030	0.0019-0.0066
	H	2.5 (2497)	1.7689	1.7689	1.7689	1.7689	-	1.4800 1.4600	1.4500 1.4350	NA	0.0040-0.0160
1996-97	C	2.0 (1996)	1.6190-1.6200	1.6340-1.6350	1.6500-1.6510	1.6660-1.6670	1.6820-1.6830	0.3070	0.2770	0.0027-0.0030	0.0059
	X	2.4 (2429)	1.0210-1.0220	1.0210-1.0220	1.0210-1.0220	1.0210-1.0220	①	0.3240	0.2560	0.0027-0.0030	0.0019-0.0066
	H	2.5 (2497)	1.7689	1.7689	1.7689	1.7689	-	1.4800 1.4600	1.4500 1.4350	NA	0.0040-0.0160

NA - Not Available

① Journal Nos. 5-6: 1.021-1.022

CRANKSHAFT AND CONNECTING ROD SPECIFICATIONS

All measurements are given in inches.

Year	Engine ID/VIN	Engine Displacement Liters (cc)	Crankshaft				Connecting Rod		
			Main Brg. Journal Dia.	Main Brg. Oil Clearance	Shaft End-play	Thrust on No.	Journal Diameter	Oil Clearance	Side Clearance
1995	C	2.0 (1996)	2.0469-2.0475	0.0008-0.0024	0.0035-0.0094	3	1.8894-1.8900	0.0010-0.0023	0.0050-0.0150
	X	2.4 (2429)	2.3610-2.3630	0.0006-0.0027	0.0035-0.0094	3	1.9670-1.9685	0.0009-0.0027	0.0130-0.0150
	H	2.5 (2497)	2.3620	0.0008-0.0016	0.0020-0.0160	3	1.9690	0.0008-0.0016	0.0040-0.0160
1996-97	C	2.0 (1996)	2.0469-2.0475	0.0008-0.0024	0.0035-0.0094	3	1.8894-1.8900	0.0010-0.0023	0.0050-0.0150
	X	2.4 (2429)	2.3610-2.3625	0.0006-0.0027	0.0035-0.0095	3	1.9670-1.9685	0.0009-0.0027	0.0051-0.0150
	H	2.5 (2497)	2.3620	0.0008-0.0016	0.0020-0.0160	3	1.9690	0.0008-0.0016	0.0040-0.0160

VALVE SPECIFICATIONS

Year	Engine ID/VIN	Engine Displacement Liters (cc)	Seat Angle (deg.)	Face Angle (deg.)	Spring Test Pressure (lbs. @ in.)	Spring Installed Height (in.)	Stem-to-Guide Clearance (in.)		Stem Diameter (in.)	
							Intake	Exhaust	Intake	Exhaust
1995	C	2.0 (1996)	44.5-45	45-45.5	75@1.54	1.540	0.0018-0.0025	0.0029-0.0037	0.2340	0.2330
	X	2.4 (2429)	45-45.5	44.5-45	76@1.50	1.496	0.0018-0.0025	0.0029-0.0037	0.2340	0.2330
	H	2.5 (2497)	44-44.5	45-45.5	60@1.74	1.740	0.0008-0.0020	0.0016-0.0028	0.2360	0.2360
1996-97	C	2.0 (1996)	44.5-45	45-45.5	75@1.54	1.540	0.0018-0.0025	0.0029-0.0037	0.2340	0.2330
	X	2.4 (2429)	45-44.5	45-45.5	76@1.50	1.496	0.0018-0.0025	0.0029-0.0037	0.2340	0.2330
	H	2.5 (2497)	45-44.5	45-45.5	60@1.74	1.740	0.0008-0.0020	0.0016-0.0028	0.2360	0.2360

TORQUE SPECIFICATIONS

All readings in ft. lbs.

Year	Engine ID/VIN	Engine Displacement Liters (cc)	Cylinder Head Bolts	Main Bearing Bolts	Rod Bearing Bolts	Crankshaft Damper Bolts	Flywheel Bolts	Manifold		Spark Plugs	Lug Nut
								Intake	Exhaust		
1995	C	2.0 (1996)	①	②	③	105	70	17	17	20	95
	X	2.4 (2429)	①	②	③	100	70	17	17	20	95
	H	2.5 (2497)	80	60	37	134	70	16	22	18	95
1996-97	C	2.0 (1996)	①	②	③	105	70	17	17	20	95
	X	2.4 (2429)	①	②	③	100	70	17	17	20	95
	H	2.5 (2497)	80	60	37	134	70	16	22	18	95

① Step 1: 25 ft. lbs.
Step 2: 50 ft. lbs.
Step 3: 50 ft. lbs.
Step 4: Plus 1/4 turn

② Step 1: 30 ft. lbs
Step 2: Plus 1/4 turn

③ Step 1: 20 ft. lbs.
Step 2: +90 degrees

BRAKE SPECIFICATIONS
All measurements in inches unless noted

Year	Model	Master Cylinder Bore	Brake Disc Original Thickness	Brake Disc Minimum Thickness	Brake Disc Maximum Runout	Brake Drum Diameter Original Inside Diameter	Brake Drum Diameter Max. Wear Limit	Brake Drum Diameter Maximum Machine Diameter	Minimum Lining Thickness Front	Minimum Lining Thickness Rear
1995	Cirrus	0.874	0.911	0.843	0.005	7.88	①	①	0.035	0.062
	Stratus	0.875	0.911	0.843	0.005	7.88	①	①	0.035	0.062
1996-97	Breeze	0.875	0.911	0.843	0.003	7.88	①	①	0.035	0.062
	Cirrus	0.874	0.911	0.843	0.005	7.88	①	①	0.035	0.062
	Sebring Convertible	0.874	0.911	0.843	0.005	8.66	①	①	0.035	0.062
	Stratus	0.875	0.911	0.843	0.003	7.88	①	①	0.035	0.062

① Maximum diameter is stamped on drum

FREQUENT MAINTENANCE LABOR
CHRYSLER CIRRUS, SEBRING CONVERTIBLE, DODGE STRATUS, PLYMOUTH BREEZE

The following should be used as a guide when determining the amount of work required for a particular service if taken to a repair shop. In estimating how long a particular Frequent Maintenance Service item should take, please observe the following:
- **Factory Time** is time that is generated by the vehicle manufacturer.
- **Chilton Time** is time that is based on field research and data supplied by the vehicle manufacturer.
- All labor time operations are given in hours and tenths of an hour.
- All labor operations, are to be used as a **guide**.

COOLING

(G) Winterize Cooling System
Includes: Run engine to check for leaks, tighten all hose connections. Test radiator and pressure cap, drain radiator and engine block. Add antifreeze and refill system.
1995-97 5

(G) Belt, Drive, Renew
1995-97
  Alternator
    2.0L (.5) 6
    2.4L (.2) 3
    2.5L (.5) 6
  PS
    2.0L (.3) 4
    2.4L (.4) 5
    2.5L (.3) 4

(G) Belt, Drive, Adjust
1995-97, one 3
  each adtnl. 1

(G) Hoses, Radiator, Renew
Includes: Drain and refill cooling system as required.
1995-97
  upper (.4) 5
  lower (.5) 6

FUEL

(M) Air Cleaner, Service
1995-97 3

(G) Filter, Fuel, Renew
1995-97
  in line (.4) 5
  in tank (.9) 1.4

BRAKES

(G) Bleed Brakes (Four Wheels)
Includes: Add fluid.
1995-97 (.4) 5
Bleed antilock modulator add 4

(G) Brakes, Adjust (Minor)
Includes: Adjust brakes, fill master cylinder.
1995-97 4

LUBRICATION SERVICES

(M) Engine Oil & Filter, Renew
Includes: Inspect and correct all fluid levels.
1995-97 3

(M) Lubricate Chassis, Change Oil & Filter
Includes: Inspect and correct all fluid levels.
1995-97 6
Install grease fittings add 1

(M) Lubricate Chassis
Includes: Inspect and correct all fluid levels.
1995-97 4
Install grease fittings add 1

WHEELS

(G) Wheel, Renew (One)
1995-97 5

(G) Wheel, Balance
1995-97
  one .3
  each adtnl. 2

(G) Wheels, Rotate (All)
1995-97 5

ELECTRICAL

(G) Headlamps, Aim
1995-97
  two .4
  four .6

(G) License Lamp Assy., Renew
1995-97 (.2) 3

(G) Park & Turn Signal Lamp Assy., Renew
1995-97 (.2) 3

(G) Rear Combination Lamp Assy., Renew
1995-97 (.2) 3

(G) Horn, Renew
1995-97 (.3) 4

(M) Terminals, Battery, Clean
1995-97 3

SCHEDULED MAINTENANCE INTERVALS
(CHRYSLER STRATUS & SEBRING CONVERTIBLE, DODGE CIRRUS, PLYMOUTH BREEZE)

TO BE SERVICED	TYPE OF SERVICE	VEHICLE MILEAGE INTERVAL (x1000)												
		7.5	15	22.5	30	37.5	45	52.5	60	67.5	75	82.5	90	97.5
Engine oil & filter	R	✓	✓	✓	✓	✓	✓	✓	✓	✓	✓	✓	✓	✓
Brake hoses	S/I	✓	✓	✓	✓	✓	✓	✓	✓	✓	✓	✓	✓	✓
Coolant level, hoses & clamps	S/I	✓	✓	✓	✓	✓	✓	✓	✓	✓	✓	✓	✓	✓
CV joints & front suspension components	S/I	✓	✓	✓	✓	✓	✓	✓	✓	✓	✓	✓	✓	✓
Exhaust system	S/I	✓	✓	✓	✓	✓	✓	✓	✓	✓	✓	✓	✓	✓
Rotate tires	S/I	✓	✓	✓	✓	✓	✓	✓	✓	✓	✓	✓	✓	✓
Accessory drive belts	S/I		✓		✓		✓		✓		✓		✓	
Brake linings	S/I			✓			✓			✓			✓	
Air filter element	R				✓				✓				✓	
Spark plugs①	R				✓				✓				✓	
Lubricate front & rear ball joints	S/I				✓				✓				✓	
Engine Coolant	R						✓				✓			
PCV valve	S/I								✓				✓	
Ignition cables②	R								✓					
Camshaft timing belt③	R													

① 4 cylinder shown; 6 cylinder = 100,000 miles.
② 1995 shown. 1996-97: 4 cylinder 60,000 miles, 6 cylinder 100,000 miles.
③ Replace at 105,000 miles for normal service; replace at 102,000 miles for severe service.

R – Replace   S/I – Service or Inspect

FREQUENT OPERATION MAINTENANCE (SEVERE SERVICE)

If a vehicle is operated under any of the following conditions it is considered severe service:
- Extremely dusty areas.
- 50% or more of the vehicle operation is in 32°C (90°F) or higher temperatures, or constant operation in temperatures below 0°C (32°F).
- Prolonged idling (vehicle operation in stop and go traffic).
- Frequent short running periods (engine does not warm to normal operating temperatures).
- Police, taxi, delivery usage or trailer towing usage.

Oil & filter change – change every 3000 miles.
Rotate tires every 6000 miles.
Brake linings – check every 12,000 miles.
Air filter element – service or inspect every 15,000 miles.
Automatic transaxle – change fluid & adjust bands (if equipped) every 15,000 miles.
PCV valve – check every 30,000 miles.
Engine coolant, replace at 36,000, 51,000 & 81,000 miles.

TEMPO/TOPAZ

VEHICLE IDENTIFICATION CHART

		Engine Code					Model Year	
Code	Liters	Cu. In. (cc)	Cyl.	Fuel Sys.	Eng. Mfg.		Code	Year
X	2.3	140 (2300)	4	MFI	Ford		P	1993
X	2.3	140 (2300)	4	SFI	Ford		R	1994
U	3.0	181 (2971)	6	MFI	Ford			

MFI - Multiport fuel injection

SFI - Sequential fuel injection

ENGINE IDENTIFICATION
All measurements are given in inches.

Year	Model	Engine Displacement Liters (cc)	Engine Series (ID/VIN)	Fuel System	No. of Cylinders	Engine Type
1993	Tempo	2.3 (2300)	X	MFI	4	SOHC
	Tempo	3.0 (2971)	U	MFI	6	OHV
	Topaz	2.3 (2300)	X	MFI	4	SOHC
	Topaz	3.0 (2971)	U	MFI	6	OHV
1994	Tempo	2.3 (2300)	X	SFI	4	OHV
	Tempo	3.0 (2971)	U	SFI	6	OHV
	Topaz	2.3 (2300)	X	SFI	4	OHV
	Topaz	3.0 (2971)	U	SFI	6	OHV

MFI - Multiport fuel injection   SOHC - Single overhead camshaft

SFI - Sequential fuel injection   OHV - Overhead valve

GENERAL ENGINE SPECIFICATIONS

Year	Engine ID/VIN	Engine Displacement Liters (cc)	Fuel System Type	Net Horsepower @ rpm	Net Torque @ rpm (ft. lbs.)	Bore x Stroke (in.)	Compression Ratio	Oil Pressure @ rpm
1993	X	2.3 (2300)	MFI	98@4400	124@2200	3.70x3.30	9.0:1	55-70@2000
	U	3.0 (2971)	MFI	140@4800	160@3000	3.50x3.10	9.3:1	55-70@2000
1994	X	2.3 (2300)	SFI	98@4400	124@2200	3.68x3.30	9.0:1	55-70@2000
	U	3.0 (2971)	SFI	140@4800	160@3000	3.50x3.15	9.3:1	55-70@2500

MFI - Multiport fuel injection

SFI - Sequential fuel injection

GASOLINE ENGINE TUNE-UP SPECIFICATIONS

Year	Engine ID/VIN	Engine Displacement Liters (cc)	Spark Plugs Gap (in.)	Ignition Timing (deg.)		Fuel Pump (psi)	Idle Speed (rpm)		Valve Clearance	
				MT	AT		MT	AT	In.	Ex.
1993	X	2.3 (2300)	0.054	10B	10B	45-60	975	875	HYD	HYD
	U	3.0 (2971)	0.044	10B	10B	35-45	-	625	HYD	HYD

GASOLINE ENGINE TUNE-UP SPECIFICATIONS

Year	Engine ID/VIN	Engine Displacement Liters (cc)	Spark Plugs Gap (in.)	Ignition Timing (deg.) MT	Ignition Timing (deg.) AT	Fuel Pump (psi)	Idle Speed (rpm) MT	Idle Speed (rpm) AT	Valve Clearance In.	Valve Clearance Ex.
1994	X	2.3 (2300)	0.054	10B	10B	45-60 ②	①	①	HYD	HYD
	U	3.0 (2971)	0.044	10B	10B	30-45 ②	①	①	HYD	HYD

NOTE: The Vehicle Emission Control Information label often reflects specification changes made during production. The label figures must be used if they differ from those in this chart.

B - Before top dead center

HYD - Hydraulic

① Refer to Vehicle Emission Control Information label

② Fuel pressure with engine running, pressure regulator vacuum hose connected

CAPACITIES

Year	Model	Engine ID/VIN	Engine Displacement Liters (cc)	Engine Oil with Filter (qts.)	Transmission (pts.) 4-Spd	Transmission (pts.) 5-Spd	Transmission (pts.) Auto.	Drive Axle Front (pts.)	Drive Axle Rear (pts.)	Fuel Tank (gal.)	Cooling System (qts.)
1993	Tempo	X	2.3 (2300)	5.0	-	6.1	16.6	①	-	15.9	③
	Tempo	U	3.0 (2971)	4.5	-	6.1	16.6	①	-	15.9	②
	Topaz	X	2.3 (2300)	5.0	-	6.1	16.6	①	-	15.9	③
	Topaz	U	3.0 (2971)	4.5	-	6.1	16.6	①	-	15.9	②
1994	Tempo	X	2.3 (2300)	5.0	-	6.4	17.2 ⑤	①	-	15.9	③
	Tempo	U	3.0 (2971)	4.5	-	6.4	17.2 ⑤	①	-	15.9	④
	Topaz	X	2.3 (2300)	5.0	-	6.4	17.2 ⑤	①	-	15.9	③
	Topaz	U	3.0 (2971)	4.5	-	6.4	17.2 ⑤	①	-	15.9	④

① Included in transaxle capacity

② Manual transaxle: 7.8 qts.
Automatic transaxle: 8.4 qts.

③ Without AC: 8.3 qts.
Manual transaxle with AC: 7.3 qts.
Automatic transaxle with AC: 7.8 qts.

④ Manual transaxle: 11.8 qts.
Automatic transaxle: 7.6 qts.

⑤ Includes torque converter

VALVE SPECIFICATIONS

Year	Engine ID/VIN	Engine Displacement Liters (cc)	Seat Angle (deg.)	Face Angle (deg.)	Spring Test Pressure (lbs. @ in.)	Spring Installed Height (in.)	Stem-to-Guide Clearance (in.) Intake	Stem-to-Guide Clearance (in.) Exhaust	Stem Diameter (in.) Intake	Stem Diameter (in.) Exhaust
1993	X	2.3 (2300)	45	44	128-141@ 1.12	1.520	0.0010- 0.0027	0.0015- 0.0032	0.3416- 0.3423	0.3411- 0.3418
	U	3.0 (2971)	45	44	185@1.11	1.580	0.0001- 0.0027	0.0015- 0.0032	0.3126- 0.3129	0.3121- 0.3134
1994	X	2.3 (2300)	44-45	44-45	128-141@ 1.12	1.520	0.0018	0.0023	0.3416- 0.3423	0.3411- 0.3418
	U	3.0 (2971)	45	44	180@1.06	1.580	0.0001- 0.0027	0.0015- 0.0032	0.3126- 0.3129	0.3121- 0.3134

TORQUE SPECIFICATIONS
All readings in ft. lbs.

Year	Engine ID/VIN	Engine Displacement Liters (cc)	Cylinder Head Bolts	Main Bearing Bolts	Rod Bearing Bolts	Crankshaft Damper Bolts	Flywheel Bolts	Manifold Intake	Manifold Exhaust	Spark Plugs	Lug Nut
1993	X	2.3 (2300)	④	51-66	21-26	80-100	54-64	②	③	5-10	95
	U	3.0 (2971)	①	55-63	26	93-121	54-64	⑤	15-22	5-10	95

TORQUE SPECIFICATIONS
All readings in ft. lbs.

Year	Engine ID/VIN	Engine Displacement Liters (cc)	Cylinder Head Bolts	Main Bearing Bolts	Rod Bearing Bolts	Crankshaft Damper Bolts	Flywheel Bolts	Manifold Intake	Manifold Exhaust	Spark Plugs	Lug Nut
1994	X	2.3 (2300)	④	51-66	21-26	80-100	54-64	②	③	5-10	95
	U	3.0 (2971)	①	55-63	26	93-121	54-64	⑤	15-22	7-15	95

NOTE: Always follow proper torque patterns

NOTE: Stretch bolts are used in all procedures that require rotating the fastener a certain number of degrees. The bolts stretch and cannot be reused. For reassembly, replace with new fastners.

① Step 1: 33-41 ft. lbs.
   Step 2: 63-73 ft. lbs.
② Step 1: 5-7 ft. lbs.
   Step 2: 15-22 ft. lbs.

③ Step 1: 5-7 ft. lbs.
   Step 2: 20-30 ft. lbs.
④ Step 1: 52-59 ft. lbs.
   Step 2: 70-76 ft. lbs.

⑤ Step 1: 11 ft. lbs.
   Step 2: 18 ft. lbs.
   Step 3: 24 ft. lbs.

BRAKE SPECIFICATIONS
All measurements in inches unless noted

Year	Model	Master Cylinder Bore	Brake Disc Original Thickness	Brake Disc Minimum Thickness	Brake Disc Maximum Runout	Brake Drum Diameter Original Inside Diameter	Brake Drum Diameter Max. Wear Limit	Brake Drum Diameter Maximum Machine Diameter	Minimum Lining Thickness Front	Minimum Lining Thickness Rear
1993	Tempo	①	0.945	0.882	0.003	8.06	8.15	8.12	0.125	0.060
	Topaz	①	0.945	0.882	0.003	8.06	8.15	8.12	0.125	0.060
1994	Tempo	①	0.945	0.882	0.003	8.06	8.15	8.12	0.125	0.060
	Topaz	①	0.945	0.882	0.003	8.06	8.15	8.12	0.125	0.060

NOTE: Follow specifications stamped on rotor or drum if figures differ from those in this chart.

NA - Not Available

① Primary bore: 1.12
   Secondary bore: 0.776

FREQUENT MAINTENANCE LABOR
FORD TEMPO, MERCURY TOPAZ

The following should be used as a guide when determining the amount of work required for a particular service if taken to a repair shop. In estimating how long a particular Frequent Maintenance Service item should take, please observe the following:
- **Factory Time** is time that is generated by the vehicle manufacturer.
- **Chilton Time** is time that is based on field research and data supplied by the vehicle manufacturer.
- All labor time operations are given in hours and tenths of an hour.
- All labor operations, are to be used as a **guide**.

COOLING

(G) Winterize Cooling System

Includes: Run engine to check for leaks, tighten all hose connections. Test radiator and pressure cap, drain radiator and engine block. Add antifreeze and refill system.

1993-94 5

(G) Belt, Drive, Renew

1993-94
  AC (.4) 5
  Alternator (.3) 4

PS (.4) 5
Thermactor (.3) 4
Water Pump
  wo/AC (.4) 5
  w/AC (.5) 6

(G) Belt, Drive, Adjust

1993-94
  one (.2) 3
  each adtnl. (.1) 1

(G) Hoses, Radiator, Renew

Includes: Drain and refill cooling system as required.

1993-94
  upper (.3) 4
  lower (.4) 5
  both (.5) 6

(G) Thermostat, Coolant, Renew

1993-94, Four (.5) 7
1993-94, V6 (.7) 1.0

FREQUENT MAINTENANCE LABOR (cont.)
FORD TEMPO, MERCURY TOPAZ

	(Factory Time)	Chilton Time

FUEL

(M) Air Cleaner, Service
1993-94 2

(G) Filter, Fuel, Renew
1993-94 (.4) 5

BRAKES

(G) Bleed Brakes (Four Wheels)
Includes: Add fluid.
1993-94 (.3) 5

(G) Brakes, Adjust (Minor)
Includes: Adjust brakes, fill master cylinder.
1993-94, two wheels 4

(M) Parking Brake, Adjust
1993-94 (.4) 6

LUBRICATION SERVICES

(M) Engine Oil & Filter, Renew
Includes: Inspect and correct all fluid levels.
1993-94 5

(M) Lubricate Chassis, Change Oil & Filter
Includes: Inspect and correct all fluid levels.
1993-94 6
Install grease fittings add 1

(M) Lubricate Chassis
Includes: Inspect and correct all fluid levels.
1993-94 4
Install grease fittings add 1

WHEELS

(G) Wheel, Renew (One)
1993-94 5

(G) Wheel, Balance
1993-94
one 3
each adtnl. 2

(G) Wheels, Rotate (All)
1993-94 5

ELECTRICAL

(G) Headlamps, Aim
1993-94
two 4
four 6

(G) Halogen Headlamp Bulb, Renew
1993-94, each (.3) 3

(G) High Mount Stop Lamp and/or Lens, Renew
1993-94 (.3) 4

(G) License Lamp Assy., Renew
1993-94 (.2) 3

(G) Tail Lamp Assy., Renew
1993-94 (.4) 6

(G) Horn, Renew
1993-94 (.5) 6

(M) Terminals, Battery, Clean
1993-94 3

SCHEDULED MAINTENANCE INTERVALS
(FORD TEMPO, MERCURY TOPAZ (1993))

TO BE SERVICED	TYPE OF SERVICE	VEHICLE MILEAGE INTERVAL (x1000)												
		7.5	15	22.5	30	37.5	45	52.5	60	67.5	75	82.5	90	97.5
Engine oil & filter	R	✓	✓	✓	✓	✓	✓	✓	✓	✓	✓	✓	✓	✓
Rotate tires	S/I	✓		✓		✓		✓		✓		✓		✓
Air cleaner filter	R				✓				✓				✓	
Crankcase emission air filter	R				✓				✓				✓	
Engine coolant	R				✓				✓				✓	
Spark plugs (2.3L)	R				✓				✓				✓	
Spark plugs (3.0L)	R								✓					
Brake linings & drum (rear)	S/I				✓				✓				✓	
Disc brake pads & rotors	S/I				✓				✓				✓	
Exhaust heat shields	S/I				✓				✓				✓	

SCHEDULED MAINTENANCE INTERVALS
(FORD TEMPO, MERCURY TOPAZ (1993)) (Cont.)

TO BE SERVICED	TYPE OF SERVICE	VEHICLE MILEAGE INTERVAL (x1000)												
		7.5	15	22.5	30	37.5	45	52.5	60	67.5	75	82.5	90	97.5
Accessory drive belt(s)	S/I								✓				✓	
PCV valve	R								✓					
Repack rear wheel bearings	S/I								✓					

R – Replace   S/I – Service or Inspect

FREQUENT OPERATION MAINTENANCE (SEVERE SERVICE)

If a vehicle is operated under any of the following conditions it is considered severe service:
- Extremely dusty areas.
- 50% or more of the vehicle operation is in 32°C (90°F) or higher temperatures, or constant operation in temperatures below 0°C (32°F).
- Prolonged idling (vehicle operation in stop and go traffic).
- Frequent short running periods (engine does not warm to normal operating temperatures).
- Police, taxi, delivery usage or trailer towing usage.
Oil & oil filter – change every 3000 miles.
Automatic transaxle fluid & filter - change every 30,000 miles.

SCHEDULED MAINTENANCE INTERVALS
(FORD TEMPO, MERCURY TOPAZ (1994))

TO BE SERVICED	TYPE OF SERVICE	VEHICLE MILEAGE INTERVAL (x1000)												
		5	10	15	20	25	30	35	40	45	50	55	60	65
Engine oil & filter	R	✓	✓	✓	✓	✓	✓	✓	✓	✓	✓	✓	✓	✓
Rotate tires	S/I	✓		✓		✓		✓		✓		✓		✓
Air cleaner filter	R						✓						✓	
Crankcase emission air filter	R						✓						✓	
Engine coolant	R						✓						✓	
Spark plugs (2.3L)	R						✓						✓	
Spark plugs (3.0L)	R												✓	
Brake linings & drum (rear)	S/I						✓						✓	
Disc brake pads & rotors	S/I						✓						✓	
Exhaust heat shields	S/I						✓						✓	
Accessory drive belt(s)	S/I												✓	

SCHEDULED MAINTENANCE INTERVALS
(FORD TEMPO, MERCURY TOPAZ (1994)) (Cont.)

TO BE SERVICED	TYPE OF SERVICE	VEHICLE MILEAGE INTERVAL (x1000)												
		5	10	15	20	25	30	35	40	45	50	55	60	65
PCV valve	R												✓	
Repack rear wheel bearings	S/I												✓	

R – Replace   S/I – Service or Inspect

FREQUENT OPERATION MAINTENANCE (SEVERE SERVICE)

If a vehicle is operated under any of the following conditions it is considered severe service:
- Extremely dusty areas.
- 50% or more of the vehicle operation is in 32°C (90°F) or higher temperatures, or constant operation in temperatures below 0°C (32°F).
- Prolonged idling (vehicle operation in stop and go traffic).
- Frequent short running periods (engine does not warm to normal operating temperatures).
- Police, taxi, delivery usage or trailer towing usage.

Oil & oil filter – change every 3000 miles.
Automatic transaxle fluid & filter - change every 30,000 miles.

ASPIRE

VEHICLE IDENTIFICATION CHART

Engine Code							Model Year	
Code	Liters	Cu. In. (cc)	Cyl.	Fuel Sys.	Eng. Mfg.		Code	Year
H	1.3	81 (1319)	4	SFI	Kia Motors		R	1994
							S	1995
							T	1996
							V	1997

SFI - Sequential fuel injection

ENGINE IDENTIFICATION
All measurements are given in inches.

Year	Model	Engine Displacement Liters (cc)	Engine Series (ID/VIN)	Fuel System	No. of Cylinders	Engine Type
1994	Aspire	1.3 (1319)	H	SFI	4	SOHC
1995	Aspire	1.3 (1319)	H	SFI	4	SOHC
1996-97	Aspire	1.3 (1319)	H	SFI	4	SOHC

SFI - Sequential fuel injection   SOHC - Single overhead camshaft

GENERAL ENGINE SPECIFICATIONS

Year	Engine ID/VIN	Engine Displacement Liters (cc)	Fuel System Type	Net Horsepower @ rpm	Net Torque @ rpm (ft. lbs.)	Bore x Stroke (in.)	Compression Ratio	Oil Pressure @ rpm
1994	H	1.3 (1319)	SFI	63@5000	73@3000	2.79x3.29	9.7:1	50-64@3000
1995	H	1.3 (1319)	SFI	63@5000	73@3000	2.79x3.29	9.7:1	50-64@3000
1996-97	H	1.3 (1319)	SFI	63@5000	73@3000	2.79x3.29	9.7:1	50-64@3000

SFI - Sequential fuel injection

GASOLINE ENGINE TUNE-UP SPECIFICATIONS

Year	Engine ID/VIN	Engine Displacement Liters (cc)	Spark Plugs Gap (in.)	Ignition Timing (deg.) MT	Ignition Timing (deg.) AT	Fuel Pump (psi)	Idle Speed (rpm) MT	Idle Speed (rpm) AT	Valve Clearance In.	Valve Clearance Ex.
1994	H	1.3 (1319)	0.040	10B	10B	30-38 ①	700	750	HYD	HYD
1995	H	1.3 (1319)	0.040	10B	10B	30-38 ①	700	750	HYD	HYD
1996-97	H	1.3 (1319)	0.040	10B	10B	30-38 ①	700	750	HYD	HYD

NOTE: The Vehicle Emission Control Information label often reflects specification changes made during production. The label figures must be used if they differ from those in this chart.
B - Before top dead center
HYD - Hydraulic
① Fuel pressure with engine running, pressure regulator vacuum hose connected

CAPACITIES

Year	Model	Engine ID/VIN	Engine Displacement Liters (cc)	Engine Oil with Filter (qts.)	Transmission (pts.) 4-Spd	Transmission (pts.) 5-Spd	Transmission (pts.) Auto.	Drive Axle Front (pts.)	Drive Axle Rear (pts.)	Fuel Tank (gal.)	Cooling System (qts.)
1994	Aspire	H	1.3 (1319)	3.6	-	5.2	11.2 ②	①	-	10.0	6.3
1995	Aspire	H	1.3 (1319)	3.6	-	5.2	12.0 ②	①	-	10.0	6.3
1996-97	Aspire	H	1.3 (1319)	3.6	-	5.2	12.0 ②	①	-	10.0	6.3

① Included in transaxle capacity ② Includes torque converter

VALVE SPECIFICATIONS

Year	Engine ID/VIN	Engine Displacement Liters (cc)	Seat Angle (deg.)	Face Angle (deg.)	Spring Test Pressure (lbs. @ in.)	Spring Installed Height (in.)	Stem-to-Guide Clearance (in.) Intake	Stem-to-Guide Clearance (in.) Exhaust	Stem Diameter (in.) Intake	Stem Diameter (in.) Exhaust
1994	H	1.3 (1319)	45	45	-	1.717 ①	0.0010-0.0024	0.0012-0.0026	0.2744-0.2750	0.2742-0.2748
1995	H	1.3 (1319)	45	45	-	1.717 ①	0.0010-0.0024	0.0012-0.0026	0.2744-0.2750	0.2742-0.2748
1996-97	H	1.3 (1319)	45	45	NA	1.717 ①	0.0010-0.0024	0.0012-0.0026	0.2744-0.2750	0.2742-0.2748

① Spring height measured unloaded

TORQUE SPECIFICATIONS
All readings in ft. lbs.

Year	Engine ID/VIN	Engine Displacement Liters (cc)	Cylinder Head Bolts	Main Bearing Bolts	Rod Bearing Bolts	Crankshaft Damper Bolts	Flywheel Bolts	Manifold Intake	Manifold Exhaust	Spark Plugs	Lug Nut
1994	H	1.3 (1319)	①	40-43	②	③	71-76	14-20	12-17	15-22	85
1995	H	1.3 (1319)	①	40-43	②	③	71-76	14-20	12-17	15-22	85
1996-97	H	1.3 (1319)	①	40-43	②	③	71-76	14-20	12-17	15-22	76

NOTE: Always follow proper torque patterns
NOTE: Stretch bolts are used in all procedures that require rotating the fastener a certain number of degrees. The bolts stretch and cannot be reused. For reassembly, replace with new fastners.

① Step 1: 35-40 ft. lbs.
　Step 2: 56-60 ft. lbs.

② Step 1: 11-13 ft. lbs.
　Step 2: 22-25 ft. lbs.

③ Pulley bolts: 9-13 ft. lbs.
　Sprocket bolt: 80-87 ft. lbs.

BRAKE SPECIFICATIONS
All measurements in inches unless noted

Year	Model			Master Cylinder Bore	Brake Disc Original Thickness	Brake Disc Minimum Thickness	Brake Disc Maximum Runout	Brake Drum Diameter Original Inside Diameter	Brake Drum Diameter Max. Wear Limit	Brake Drum Diameter Maximum Machine Diameter	Minimum Lining Thickness Front	Minimum Lining Thickness Rear
1994	Aspire	①	F	②	0.710	0.630	0.004	7.87	7.93	-	0.080	0.040
		③	R	②	0.860	0.780	0.004	7.87	7.93	-	0.080	0.040
1995	Aspire	①	F	②	0.710	0.630	0.004	7.87	7.93	-	0.080	0.040
		③	R	②	0.860	0.780	0.004	7.87	7.93	-	0.080	0.040
1996-97	Aspire	①	F	②	0.710	0.630	0.004	7.87	7.93	-	0.080	0.040
		③	R	②	0.860	0.780	0.004	7.87	7.93	-	0.080	0.040

NOTE: Follow specifications stamped on rotor or drum if figures differ from those in this chart.
NA - Not Available
F - Front
R - Rear

① Manual transaxle
② Without ABS: 0.810　With ABS: 0.870
③ Automatic transaxle

FREQUENT MAINTENANCE LABOR
FORD ASPIRE

The following should be used as a guide when determining the amount of work required for a particular service if taken to a repair shop. In estimating how long a particular Frequent Maintenance Service item should take, please observe the following:
- **Factory Time** is time that is generated by the vehicle manufacturer.
- **Chilton Time** is time that is based on field research and data supplied by the vehicle manufacturer.
- All labor time operations are given in hours and tenths of an hour.
- All labor operations, are to be used as a **guide**.

COOLING

(G) Winterize Cooling System
Includes: Run engine to check for leaks, tighten all hose connections. Test radiator and pressure cap, drain radiator and engine block. Add antifreeze and refill system.
1994-97 5

(G) Belt, Drive, Renew
1994-97
　Alternator (.3) 4
　Power steering (.3) 4

(G) Belt, Drive, Adjust
1994-97
　one (.2) 3
　each adtnl. 1

(G) Hoses, Radiator, Renew
Includes: Drain and refill cooling system as required.
1994-97
　one (.4) 5
　both (.5) 7

(G) Thermostat, Coolant, Renew
1994-97 (.5) 7

FREQUENT MAINTENANCE LABOR (cont.)
FORD ASPIRE

(Factory Time)	Chilton Time

FUEL

(M) Air Cleaner, Service
1994-97 2

(G) Filter, Fuel, Renew
1994-97 (.3) 4

BRAKES

(G) Bleed Brakes (Four Wheels)
Includes: Add fluid.
1994-97 (.4) 6

(G) Brakes, Adjust (Minor)
Includes: Adjust brakes, fill master cylinder.
1994-97, two wheels 4

(M) Parking Brake, Adjust
1994-97 (.2) 3

LUBRICATION SERVICES

(M) Engine Oil & Filter, Renew
Includes: Inspect and correct all fluid levels.
1994-97 3

(M) Lubricate Chassis, Change Oil & Filter
Includes: Inspect and correct all fluid levels.
1994-97 6
Install grease fittings add 1

(M) Lubricate Chassis
Includes: Inspect and correct all fluid levels.
1994-97 4
Install grease fittings add 1

WHEELS

(G) Wheel, Renew (One)
1994-97 5

(G) Wheel, Balance
1994-97
  one .3
  each adtnl. 2

(G) Wheels, Rotate (All)
1994-97 5

ELECTRICAL

(G) Headlamps, Aim
1994-97
  two 4
  four 6

(G) Halogen Headlamp Bulb, Renew
1994-97, each (.2) 3

(G) High Mount Stop Lamp and/or Lens, Renew
1994-97 (.3) 5

(G) License Lamp Assy., Renew
1994-97 (.3) 5

(G) Tail Lamp Assy., Renew
1994-97
  one side (.3) 4
  both sides (.4) 6

(G) Horn, Renew
All models (.3) 4

(M) Terminals, Battery, Clean
1994-97 3

SCHEDULED MAINTENANCE INTERVALS
(FORD ASPIRE)

TO BE SERVICED	TYPE OF SERVICE	VEHICLE MILEAGE INTERVAL (x1000)												
		5	10	15	20	25	30	35	40	45	50	55	60	65
Engine oil & filter	R	✓	✓	✓	✓	✓	✓	✓	✓	✓	✓	✓	✓	✓
Rotate tires	S/I	✓		✓		✓		✓		✓		✓		✓
Air cleaner element & engine coolant	R						✓						✓	
Spark plugs	R						✓						✓	
Automatic transaxle fluid & filter	R						✓						✓	
Exhaust heat shields	S/I						✓						✓	
Disc brake pads & rotors, brake linings, drum, brake lines, hoses & connections	S/I						✓						✓	

SCHEDULED MAINTENANCE INTERVALS
(FORD ASPIRE) (Cont.)

TO BE SERVICED	TYPE OF SERVICE	VEHICLE MILEAGE INTERVAL (x1000)												
		5	10	15	20	25	30	35	40	45	50	55	60	65
Accessory drive belt(s)	S/I						✓						✓	
Fuel lines, hoses & idle speed	S/I						✓						✓	
Cooling system, hoses, clamps & coolant strength	S/I						✓						✓	
Clutch pedal operation	S/I						✓						✓	
Front wheel driveshaft joint boots	S/I						✓						✓	
Front suspension ball joints, steering operation & linkage	S/I						✓						✓	
Timing belt/chain & fuel filter	R												✓	
Fuel lines & tubes (emission)	S/I												✓	
Ignition timing	S/I												✓	
Repack front & rear wheel bearings	S/I												✓	

R – Replace   S/I – Service or Inspect

FREQUENT OPERATION MAINTENANCE (SEVERE SERVICE)

If a vehicle is operated under any of the following conditions it is considered severe service:
- Extremely dusty areas.
- 50% or more of the vehicle operation is in 32°C (90°F) or higher temperatures, or constant operation in temperatures below 0°C (32°F).
- Prolonged idling (vehicle operation in stop and go traffic).
- Frequent short running periods (engine does not warm to normal operating temperatures).
- Police, taxi, delivery usage or trailer towing usage.

Oil & oil filter – change every 3000 miles.
Rotate tires at 6000 miles & every 9000 miles thereafter.
Air cleaner element - service or inspect every 15,000 miles.
Automatic transaxle fluid & filter - change every 21,000 miles.

PROBE

VEHICLE IDENTIFICATION CHART

		Engine Code						Model Year	
Code	Liters	Cu. In. (cc)	Cyl.	Fuel Sys.	Eng. Mfg.			Code	Year
A	2.0	122 (1993)	4	MFI	Mazda			P	1993
B	2.5	153 (2501)	6	MFI	Mazda			R	1994
								S	1995
								T	1996
								V	1997

MFI - Multiport fuel injection

ENGINE IDENTIFICATION

All measurements are given in inches.

Year	Model	Engine Displacement Liters (cc)	Engine Series (ID/VIN)	Fuel System	No. of Cylinders	Engine Type
1993	Probe	2.0 (1993)	A	MFI	4	DOHC
	Probe	2.5 (2501)	B	MFI	6	DOHC
1994	Probe	2.0 (1993)	A	SFI	4	DOHC
	Probe	2.5 (2501)	B	SFI	6	DOHC
1995	Probe	2.0 (1993)	A	SFI	4	DOHC
	Probe	2.5 (2501)	B	SFI	6	DOHC
1996-97	Probe	2.0 (1993)	A	SFI	4	DOHC
	Probe	2.5 (2501)	B	SFI	6	DOHC

MFI - Multiport fuel injection   SFI - Sequential fuel injection   DOHC - Double overhead camshaft

GENERAL ENGINE SPECIFICATIONS

Year	Engine ID/VIN	Engine Displacement Liters (cc)	Fuel System Type	Net Horsepower @ rpm	Net Torque @ rpm (ft. lbs.)	Bore x Stroke (in.)	Compression Ratio	Oil Pressure @ rpm
1993	A	2.0 (1993)	MFI	115@5500	124@3500	3.27x3.62	9.0:1	57-71@2000
	B	2.5 (2501)	MFI	164@6000	156@4000	3.33x2.92	9.2:1	49-71@2000
1994	A	2.0 (1993)	SFI	115@5500	124@3500	3.27x3.62	9.0:1	57-71@2000
	B	2.5 (2501)	SFI	164@6000	156@4000	3.33x2.92	9.2:1	49-71@3000
1995	A	2.0 (1993)	SFI	115@5500	124@3500	3.27x3.62	9.0:1	57-71@2000
	B	2.5 (2501)	SFI	164@6000	156@4000	3.33x2.92	9.2:1	49-71@3000
1996-97	A	2.0 (1993)	SFI	118@5500	127@4500	3.27x3.62	9.0:1	57-71@2000
	B	2.5 (2501)	SFI	①	②	3.33x2.92	9.2:1	49-71@3000

MFI - Multiport fuel injection
SFI - Sequential fuel injection

① California: 160@5500
   Except California: 164@5600

② California: 156@5000
   Except California: 160@4800

GASOLINE ENGINE TUNE-UP SPECIFICATIONS

Year	Engine ID/VIN	Engine Displacement Liters (cc)	Spark Plugs Gap (in.)	Ignition Timing (deg.) MT	Ignition Timing (deg.) AT	Fuel Pump (psi)	Idle Speed (rpm) MT	Idle Speed (rpm) AT	Valve Clearance In.	Valve Clearance Ex.
1993	A	2.0 (1993)	0.040	10B	12B	64-92	700	700	HYD	HYD
	B	2.5 (2501)	0.040	10B	10B	72-92	650	650	HYD	HYD
1994	A	2.0 (1993)	0.040	10B	12B	30-38 ①	700	700	HYD	HYD
	B	2.5 (2501)	0.040	10B	10B	30-36 ①	650	650	HYD	HYD
1995	A	2.0 (1993)	0.040	10B	12B	30-38 ①	700	700	HYD	HYD
	B	2.5 (2501)	0.040	10B	10B	30-36 ①	650	650	HYD	HYD
1996-97	A	2.0 (1993)	0.041	10B	12B	30-38 ①	700	700	HYD	HYD
	B	2.5 (2501)	0.041	10B	10B	30-36 ①	700	700	HYD	HYD

NOTE: The Vehicle Emission Control Information label often reflects specification changes made during production. The label figures must be used if they differ from those in this chart.

B - Before top dead center

HYD - Hydraulic

① Fuel pressure with engine running, pressure regulator vacuum hose connected

CAPACITIES

Year	Model	Engine ID/VIN	Engine Displacement Liters (cc)	Engine Oil with Filter (qts.)	Transmission (pts.) 4-Spd	Transmission (pts.) 5-Spd	Transmission (pts.) Auto.	Drive Axle Front (pts.)	Drive Axle Rear (pts.)	Fuel Tank (gal.)	Cooling System (qts.)
1993	Probe	A	2.0 (1993)	3.7	-	5.8	18.6	①	-	15.5	7.4
	Probe	B	2.5 (2501)	4.2	-	5.8	18.6	①	-	15.5	7.9
1994	Probe	A	2.0 (1993)	3.7	-	5.8	17.6 ②	①	-	15.5	7.4
	Probe	B	2.5 (2501)	4.2	-	5.8	18.6 ②	①	-	15.5	7.9
1995	Probe	A	2.0 (1993)	3.7	-	5.8	17.6 ②	①	-	15.5	7.4
	Probe	B	2.5 (2501)	4.2	-	5.8	14.4 ②	①	-	15.5	7.9
1996-97	Probe	A	2.0 (1993)	3.7	-	5.8	17.6 ②	①	-	15.5	7.4
	Probe	B	2.5 (2501)	4.2	-	5.8	14.4 ②	①	-	15.5	7.9

① Included in transaxle capacity   ② Includes torque converter

VALVE SPECIFICATIONS

Year	Engine ID/VIN	Engine Displacement Liters (cc)	Seat Angle (deg.)	Face Angle (deg.)	Spring Test Pressure (lbs. @ in.)	Spring Installed Height (in.)	Stem-to-Guide Clearance (in.) Intake	Stem-to-Guide Clearance (in.) Exhaust	Stem Diameter (in.) Intake	Stem Diameter (in.) Exhaust
1993	A	2.0 (1993)	45	45	①	①	0.0010-0.0024	0.0012-0.0026	0.2350-0.2356	0.2348-0.2354
	B	2.5 (2501)	45	45	②	②	0.0010-0.0023	0.0012-0.0026	0.2351-0.2356	0.2349-0.2354
1994	A	2.0 (1993)	45	45	①	①	0.0010-0.0024	0.0012-0.0026	0.2350-0.2356	0.2348-0.2354
	B	2.5 (2501)	45	45	②	②	0.0010-0.0023	0.0012-0.0026	0.2351-0.2356	0.2349-0.2354
1995	A	2.0 (1993)	45	45	①	①	0.0010-0.0024	0.0012-0.0026	0.2350-0.2356	0.2348-0.2354
	B	2.5 (2501)	45	45	②	②	0.0010-0.0023	0.0012-0.0026	0.2351-0.2356	0.2349-0.2354

VALVE SPECIFICATIONS

Year	Engine ID/VIN	Engine Displacement Liters (cc)	Seat Angle (deg.)	Face Angle (deg.)	Spring Test Pressure (lbs. @ in.)	Spring Installed Height (in.)	Stem-to-Guide Clearance (in.)		Stem Diameter (in.)	
							Intake	Exhaust	Intake	Exhaust
1996-97	A	2.0 (1993)	45	45	①	①	0.0010-0.0024	0.0012-0.0026	0.2350-0.2356	0.2348-0.2354
	B	2.5 (2501)	45	45	②	②	0.0010-0.0023	0.0012-0.0026	0.2351-0.2356	0.2349-0.2354

① Measure spring free length and out of square.
   Maximum allowable out-of-square: 0.061
   Spring free length: 1.732

② Measure spring free length and out of square.
   Maximum allowable out-of-square: 0.642
   Spring free length: Intake: 1.729, Exhaust: 1.847

TORQUE SPECIFICATIONS
All readings in ft. lbs.

Year	Engine ID/VIN	Engine Displacement Liters (cc)	Cylinder Head Bolts	Main Bearing Bolts	Rod Bearing Bolts	Crankshaft Damper Bolts	Flywheel Bolts	Manifold		Spark Plugs	Lug Nut
								Intake	Exhaust		
1993	A	2.0 (1993)	①	②	②	116-123	70-75	14-19	③	11-17	85
	B	2.5 (2501)	④	⑤	⑥	116-123	70-75	14-18	14-18	11-16	85
1994	A	2.0 (1993)	⑧	③	⑥	116-123	70-75	14-19	14-21	11-17	85
	B	2.5 (2501)	⑦	⑨	⑥	116-123	45-49	14-18	14-18	11-16	85
1995	A	2.0 (1993)	⑧	③	⑥	116-123	70-75	14-19	14-21	11-17	85
	B	2.5 (2501)	⑧	⑨	⑥	116-123	45-49	14-18	14-18	11-16	85
1996-97	A	2.0 (1993)	⑧	③	⑩	116-123	70-75	14-19	14-21	11-17	85
	B	2.5 (2501)	⑧	⑨	⑩	116-123	45-49	14-18	14-18	11-16	85

NOTE: Always follow proper torque patterns
NOTE: Stretch bolts are used in all procedures that require rotating the fastener a certain number of degrees. The bolts stretch and cannot be reused. For reassembly, replace with new fastners.

① Step 1: 37-50 ft. lbs.
   Step 2: 62-68 ft. lbs.
② Step 1: 37-50 ft. lbs.
   Step 2: :58-64 ft. lbs.
③ Step 1: 12 ft. lbs.
   Step 2: Rotate each bolt 85-95 degrees
④ Step 1: 22-26 ft. lbs.
   Step 2: 33-36 ft. lbs.
⑤ Step 1: Inner bolts - 17-19 ft. lbs in two to three steps
   Step 1: Outer bolts - 13-15 ft. lbs. in two to three steps
   Step 2: Inner bolts 1-3: 70 degrees; Inner bolt 4: 80 degrees
   Step 3: Tighten outer bolts 60 degrees
   Step 4: Repeat Step 3

⑥ Step 1: 16-19 ft. lbs.
   Step 2: Rotate each bolt 90 degrees
   Step 3: Repeat Step 2
⑦ Step 1: 11 ft. lbs.
   Step 2: 18 ft. lbs.
   Step 3: 24 ft. lbs.
⑧ Step 1: 8-10 ft. lbs.
   Step 2: 13-16 ft. lbs.
   Step 3: Rotate 90 degrees
   Step 4: Repeat Step 3

⑨ Step 1: Inner main bolts: 10-12 ft. lbs.
   Step 2: Inner main bolts: 17-19 ft. lbs.
   Step 3: Outer main bolts: 6-8 ft. lbs.
   Step 4: Outer main bolts: 13-15 ft. lbs.
   Step 5: Rotate inner bolts 75 degrees
   Step 6: Rotate outer bolts 60 degrees
   Step 7: Repeat Steps 5 and 6
   Step 8: Outer cylinder block bolts: 14-15 ft. lbs.
⑩ 16-19 ft. lbs. plus 90 degrees

BRAKE SPECIFICATIONS
All measurements in inches unless noted

Year	Model			Master Cylinder Bore	Brake Disc			Brake Drum Diameter			Minimum Lining Thickness	
					Original Thickness	Minimum Thickness	Maximum Runout	Original Inside Diameter	Max. Wear Limit	Maximum Machine Diameter	Front	Rear
1993	Probe	①	F	0.937	0.940	0.890	0.004	9.00	9.06	9.06	0.040	0.040
		②	R	-	0.390	0.345	0.004	-	-	-	-	0.040
1994	Probe	①	F	0.937	0.940	0.860	0.004	9.00	9.06	9.86	0.040	0.040
		②	R	-	0.390	0.315	0.004	-	-	-	-	0.040
1995	Probe	①	F	0.937	0.890	0.860	0.004	9.00	9.06	9.86	0.040	0.040
		②	R	-	0.390	0.315	0.004	-	-	-	-	0.040

BRAKE SPECIFICATIONS
All measurements in inches unless noted

Year	Model		Master Cylinder Bore	Brake Disc			Brake Drum Diameter			Minimum Lining Thickness	
				Original Thickness	Minimum Thickness	Maximum Runout	Original Inside Diameter	Max. Wear Limit	Maximum Machine Diameter	Front	Rear
1996-97	Probe	F	0.937	0.890	0.860	0.004	-	-	-	0.040	-
		R	-	0.345	0.315	0.004	-	-	-	-	0.040

NOTE: Follow specifications stamped on rotor or drum if figures differ from those in this chart.

NA - Not Available

F - Front

R - Rear

① Except rear disc

② With rear disc

FREQUENT MAINTENANCE LABOR
FORD PROBE

The following should be used as a guide when determining the amount of work required for a particular service if taken to a repair shop. In estimating how long a particular Frequent Maintenance Service item should take, please observe the following:

- **Factory Time** is time that is generated by the vehicle manufacturer.
- **Chilton Time** is time that is based on field research and data supplied by the vehicle manufacturer.
- All labor time operations are given in hours and tenths of an hour.
- All labor operations, are to be used as a **guide**.

COOLING

(G) Winterize Cooling System
Includes: Run engine to check for leaks, tighten all hose connections. Test radiator and pressure cap, drain radiator and engine block. Add antifreeze and refill system.
1993-97 5

(G) Belt, Drive, Renew
1993-97
Four (.4) 6
V6 (.3) 5

(G) Belt, Drive, Adjust
1993-97
one (.2) 3
each adtnl. 1

(G) Hoses, Radiator, Renew
Includes: Drain and refill cooling system as required.
1993-97
upper (.3) 4
lower (.4) 5
both (.5) 6

(G) Thermostat, Coolant, Renew
1993-97
Four (.5) 7
V6 (.6) 8

FUEL

(M) Air Cleaner, Service
1993-97 (.2) 2

(G) Filter, Fuel, Renew
1993-97 (.4) 6

BRAKES

(G) Bleed Brakes (Four Wheels)
Includes: Add fluid.
1993-97 (.3) 5
w/ABS add 4

(G) Brakes, Adjust (Minor)
Includes: Adjust brakes, fill master cylinder.
1993-97 4

(M) Parking Brake, Adjust
1993-97 (.3) 3

LUBRICATION SERVICES

(M) Engine Oil & Filter, Renew
Includes: Inspect and correct all fluid levels.
1993-97 3

(M) Lubricate Chassis, Change Oil & Filter
Includes: Inspect and correct all fluid levels.
1993-97 6
Install grease fittings add 1

(M) Lubricate Chassis
Includes: Inspect and correct all fluid levels.
1993-97 4
Install grease fittings add 1

WHEELS

(G) Wheel, Renew (One)
1993-97 5

(G) Wheel, Balance
1993-97
one 3
each adtnl. 2

(G) Wheels, Rotate (All)
1993-97 5

ELECTRICAL

(G) Headlamps, Aim
1993-97
two 4
four 6

(G) License Lamp Assy., Renew
1993-97 (.3) 4

(G) Parking Lamp Lens or Bulb, Renew
1993-97 (1.1) 1.5

(G) Tail Lamp Assy., Renew
1993-94
one (.4) 6
both (.7) 1.1
1995-97
one (.3) 4
both (.5) 7

(G) Horn, Renew
1993-97 (.4) 6

(M) Terminals, Battery, Clean
1993-97 3

SCHEDULED MAINTENANCE INTERVALS
(FORD PROBE (1993))

TO BE SERVICED	TYPE OF SERVICE	VEHICLE MILEAGE INTERVAL (x1000)												
		7.5	15	22.5	30	37.5	45	52.5	60	67.5	75	82.5	90	97.5
Engine oil & filter	R	✓	✓	✓	✓	✓	✓	✓	✓	✓	✓	✓	✓	✓
Rotate tires	S/I	✓	✓	✓	✓	✓		✓		✓	✓	✓	✓	✓
Engine coolant	R				✓				✓				✓	
Air cleaner filter	R				✓				✓				✓	
Crankcase ventilation filter	R				✓				✓				✓	
Spark plugs	R				✓				✓				✓	
Exhaust heat shields	S/I				✓				✓				✓	
Front (and/or) rear disc brake pads & rotors/brake linings & drum (rear)	S/I				✓				✓				✓	
Cooling system components	S/I				✓				✓				✓	
Idle speed	S/I				✓				✓				✓	
Fuel lines	S/I				✓				✓				✓	
Halfshaft dust boots	S/I				✓				✓				✓	
Steering operation & linkage	S/I				✓				✓				✓	
Front suspension ball joints	S/I				✓				✓				✓	
Brake lines, hoses & connections	S/I				✓				✓				✓	
Bolts & nuts on chassis & body	S/I				✓				✓				✓	
Clutch pedal operation	S/I				✓				✓				✓	
Accessory drive belt(s)	S/I				✓				✓				✓	
Timing belt & fuel filter	R								✓					

R – Replace   S/I – Service or Inspect

FREQUENT OPERATION MAINTENANCE (SEVERE SERVICE)

If a vehicle is operated under any of the following conditions it is considered severe service:
- Extremely dusty areas.
- 50% or more of the vehicle operation is in 32°C (90°F) or higher temperatures, or constant operation in temperatures below 0°C (32°F).
- Prolonged idling (vehicle operation in stop and go traffic).
- Frequent short running periods (engine does not warm to normal operating temperatures).
- Police, taxi, delivery usage or trailer towing usage.

Oil & oil filter – change every 3000 miles.
Air cleaner element - service or inspect every 15,000 miles.
Front & rear brakes - check every 15,000 miles.
Rotate tires at 6000 miles & every 15,000 miles thereafter.
Automatic transaxle fluid & filter - change every 30,000 miles.

SCHEDULED MAINTENANCE INTERVALS
(FORD PROBE (1994-97))

TO BE SERVICED	TYPE OF SERVICE	VEHICLE MILEAGE INTERVAL (x1000)												
		5	10	15	20	25	30	35	40	45	50	55	60	65
Engine oil & filter	R	✓	✓	✓	✓	✓	✓	✓	✓	✓	✓	✓	✓	✓
Rotate tires	S/I	✓		✓		✓		✓		✓		✓		✓
Air cleaner element	R						✓						✓	
Spark plugs	R						✓						✓	
Automatic transmission fluid & filter	R						✓						✓	
Exhaust heat shields	S/I						✓						✓	
Front & rear brakes	S/I						✓						✓	
Accessory drive belt(s)	S/I						✓						✓	
Fuel lines & hoses	S/I						✓						✓	
Cooling system, hoses, clamps & coolant strength	S/I						✓						✓	
Front wheel driveshaft joint boots	S/I						✓						✓	
Brake lines, hoses & connections	S/I						✓						✓	
Front suspension ball joints, steering operation & linkage	S/I						✓						✓	
Idle speed	S/I						✓						✓	
Bolts & nuts on chassis & body	S/I						✓						✓	
Engine coolant	R										✓			
Timing belt/chain & fuel filter	R												✓	
Fuel lines & tubes (emission)	S/I												✓	

R – Replace   S/I – Service or Inspect

FREQUENT OPERATION MAINTENANCE (SEVERE SERVICE)

If a vehicle is operated under any of the following conditions it is considered severe service:
- Extremely dusty areas.
- 50% or more of the vehicle operation is in 32°C (90°F) or higher temperatures, or constant operation in temperatures below 0°C (32°F).
- Prolonged idling (vehicle operation in stop and go traffic).
- Frequent short running periods (engine does not warm to normal operating temperatures).
- Police, taxi, delivery usage or trailer towing usage.

Oil & oil filter – change every 3000 miles.
Air cleaner element - check every 15,000 miles.
Bolts & nuts on chassis & body - check every 15,000 miles.
Front & rear brakes - check every 15,000 miles.
Automatic transaxle fluid & filter - change every 21,000 miles.

CAPRI

VEHICLE IDENTIFICATION CHART

		Engine Code					Model Year	
Code	Liters	Cu. In. (cc)	Cyl.	Fuel Sys.	Eng. Mfg.		Code	Year
Z	1.6	98 (1597)	4	MFI	Mazda		P	1993
6	1.6 ①	98 (1597)	4	MFI	Mazda		R	1994

MFI - Multiport fuel injection   ① Turbo

ENGINE IDENTIFICATION

Year	Model		Engine Displacement Liters (cc)	Engine Series (ID/VIN)	Fuel System	No. of Cylinders	Engine Type
1993	Capri		1.6 (1597)	Z	MFI	4	DOHC
	Capri	①	1.6 (1597)	6	MFI	4	DOHC
1994	Capri		1.6 (1597)	Z	MFI	4	DOHC
	Capri	①	1.6 (1597)	6	MFI	4	DOHC

MFI - Multiport fuel injection   DOHC - Double overhead camshaft   ① Turbo

GENERAL ENGINE SPECIFICATIONS

Year	Engine ID/VIN		Engine Displacement Liters (cc)	Fuel System Type	Net Horsepower @ rpm	Net Torque @ rpm (ft. lbs.)	Bore x Stroke (in.)	Compression Ratio	Oil Pressure @ rpm
1993	Z		1.6 (1597)	MFI	100@5750	95@5500	3.07x3.29	9.4:1	45@3000
	6	①	1.6 (1597)	MFI	132@6000	136@3000	3.07x3.29	7.9:1	45@3000
1994	Z		1.6 (1597)	MFI	100@5750	95@5500	3.07x3.29	9.4:1	45@3000
	6	①	1.6 (1597)	MFI	132@6000	136@3000	3.07x3.29	7.9:1	45@3000

MFI - Multiport fuel injection   ① Turbo

GASOLINE ENGINE TUNE-UP SPECIFICATIONS

Year	Engine ID/VIN	Engine Displacement Liters (cc)	Spark Plugs Gap (in.)	Ignition Timing (deg.)		Fuel Pump (psi)	Idle Speed (rpm)		Valve Clearance	
				MT	AT		MT	AT	In.	Ex.
1993	Z	1.6 (1597)	0.041	2B	2B	①	850	850	HYD	HYD
	6	1.6 (1597)	0.041	12B	-	①	850	-	HYD	HYD
1994	Z	1.6 (1597)	0.041	2B	2B	27-34 ②	750	750	HYD	HYD
	6	1.6 (1597)	0.041	12B	-	27-34 ②	750	-	HYD	HYD

NOTE: The Vehicle Emission Control Information label often reflects specification changes made during production. The label figures must be used if they differ from those in this chart.

B - Before top dead center

HYD - Hydraulic

① Fuel pump outlet pressure: 64-85

    Engine running pressure: 37-41

② Fuel pressure with engine running, pressure regulator vacuum hose connected

CAPACITIES

Year	Model	Engine ID/VIN	Engine Displacement Liters (cc)	Engine Oil with Filter (qts.)	Transmission (pts.)			Transfer Case (pts.)	Drive Axle		Fuel Tank (gal.)	Cooling System (qts.)
					4-Spd	5-Spd	Auto.		Front (pts.)	Rear (pts.)		
1993	Capri	Z	1.6 (1597)	3.8	-	6.8	12.0	-	①	-	11.1	5.3
	Capri	6	1.6 (1597)	3.8	-	6.8	-	-	①	-	11.1	5.3
1994	Capri	Z	1.6 (1597)	3.5	-	6.8	14.4 ②	-	①	-	11.1	6.3
	Capri	6	1.6 (1597)	3.7	-	6.8	-	-	①	-	11.1	6.3

① Included in transaxle capacity   ② Includes torque converter

VALVE SPECIFICATIONS

Year	Engine ID/VIN	Engine Displacement Liters (cc)	Seat Angle (deg.)	Face Angle (deg.)	Spring Test Pressure (lbs. @ in.)	Spring Installed Height (in.)	Stem-to-Guide Clearance (in.)		Stem Diameter (in.)	
							Intake	Exhaust	Intake	Exhaust
1993	Z	1.6 (1597)	45	45	44@1.54	1.54	0.0010-0.0024	0.0012-0.0026	0.2350-0.2356	0.2348-0.2354
	6	1.6 (1597)	45	45	44@1.54	1.54	0.0010-0.0024	0.0012-0.0026	0.2350-0.2356	0.2348-0.2354
1994	Z	1.6 (1597)	45	45	44@1.54	1.54	0.0010-0.0024	0.0012-0.0026	0.2350-0.2356	0.2348-0.2354
	6	1.6 (1597)	45	45	44@1.54	1.54	0.0010-0.0024	0.0012-0.0026	0.2350-0.2356	0.2348-0.2354

7914gc02 not found

TORQUE SPECIFICATIONS

All readings in ft. lbs.

Year	Engine ID/VIN	Engine Displacement Liters (cc)	Cylinder Head Bolts	Main Bearing Bolts	Rod Bearing Bolts	Crankshaft Damper Bolts	Flywheel Bolts	Manifold		Spark Plugs	Lug Nut
								Intake	Exhaust		
1993	Z	1.6 (1597)	①	40-43	35-38	80-87	71-76	14-19	29-42	11-17	85
	6	1.6 (1597)	①	40-43	35-38	80-87	71-76	14-19	29-42	11-17	85
1994	Z	1.6 (1597)	①	40-43	35-38	80-87	71-76	14-18	29-42	11-16	85
	6	1.6 (1597)	①	40-43	35-38	80-87	71-76	14-18	29-42	11-16	85

① Step 1: 14-25 ft. lbs.
  Step 2: 56-60 ft. lbs.

BRAKE SPECIFICATIONS

All measurements in inches unless noted

Year	Model		Master Cylinder Bore	Brake Disc			Brake Drum Diameter			Minimum Lining Thickness	
				Original Thickness	Minimum Thickness	Maximum Runout	Original Inside Diameter	Max. Wear Limit	Maximum Machine Diameter	Front	Rear
1993	Capri	F	0.811	0.710	0.630	0.004	-	-	-	0.120	-
		R	-	0.390	0.350	0.004	-	-	-	0.120	-
1994	Capri	F	0.811	0.710	0.630	0.004	-	-	-	0.120	-
		R	-	0.390	0.350	0.004	-	-	-	0.120	-

NOTE: Follow specifications stamped on rotor if figures differ from those in this chart.
F - Front
R - Rear

FREQUENT MAINTENANCE LABOR
MERCURY CAPRI

The following should be used as a guide when determining the amount of work required for a particular service if taken to a repair shop. In estimating how long a particular Frequent Maintenance Service item should take, please observe the following:
- **Factory Time** is time that is generated by the vehicle manufacturer.
- **Chilton Time** is time that is based on field research and data supplied by the vehicle manufacturer.
- All labor time operations are given in hours and tenths of an hour.
- All labor operations, are to be used as a **guide**.

COOLING

(G) Winterize Cooling System
Includes: Run engine to check for leaks, tighten all hose connections. Test radiator and pressure cap, drain radiator and engine block. Add antifreeze and refill system.
1993-94 5

(G) Belt, Serpentine Drive, Renew
1993-94 (.3) 5

(G) Belt, Drive, Renew
1993-94
Alternator (.3) 4
PS (.3) 4
AC (.4) 5

(G) Belt, Drive, Adjust
1993-94
one (.2) 3
each adtnl. 1

(G) Hoses, Radiator, Renew
Includes: Drain and refill cooling system as required.
1993-94
upper (.4) 5
lower (.5) 6
both (.6) 8

(G) Thermostat, Coolant, Renew
1993-94 (.6) 8

FUEL

(M) Air Cleaner, Service
1993-94 2

(G) Filter, Fuel, Renew
1994-94 (.4) 5

BRAKES

(G) Bleed Brakes (Four Wheels)
Includes: Add fluid.
1993-94 (.3) 5

(M) Parking Brake, Adjust
1993-94 (.3) 4

LUBRICATION SERVICES

(M) Engine Oil & Filter, Renew
Includes: Inspect and correct all fluid levels.
1993-94 3

(M) Lubricate Chassis, Change Oil & Filter
Includes: Inspect and correct all fluid levels.
1993-94 6
Install grease fittings add 1

(M) Lubricate Chassis
Includes: Inspect and correct all fluid levels.
1994-94 4
Install grease fittings add 1

WHEELS

(G) Wheel, Renew (One)
1993-94 5

(G) Wheel, Balance
1993-94
one .3
each adtnl. 2

(G) Wheels, Rotate (All)
1993-94 5

ELECTRICAL

(G) Headlamps, Aim
1993-94
two .4
four .6

(G) Halogen Headlamp Bulb, Renew
1993-94, one (.3) 3

(G) License Lamp Assy., Renew
1993-94
one (.3) 4
both (.4) 6

(G) Parking Lamp Lens or Bulb, Renew
1993-94
one (.2) 3
both (.3) 4

(G) Tail Lamp Assy., Renew
1993-94
one (.3) 4
both (.4) 6

(G) Horn, Renew
1993-94 (.7) 1.0

(M) Terminals, Battery, Clean
1993-94 3

SCHEDULED MAINTENANCE INTERVALS
(MERCURY CAPRI (1993))

TO BE SERVICED	TYPE OF SERVICE	VEHICLE MILEAGE INTERVAL (x1000)												
		7.5	15	22.5	30	37.5	45	52.5	60	67.5	75	82.5	90	97.5
Engine oil & filter①	R	✓	✓	✓	✓	✓	✓	✓	✓	✓	✓	✓	✓	✓
Rotate tires	S/I	✓		✓		✓		✓		✓		✓		✓
Front and rear disc brake pads & rotors	S/I		✓		✓		✓		✓		✓		✓	
Spark plugs (Turbo)	R		✓		✓		✓		✓		✓		✓	
Spark plugs (Non-turbo)	R				✓				✓				✓	
Engine coolant	R				✓				✓				✓	
Air cleaner element	R				✓				✓				✓	
Exhaust heat shields	S/I				✓				✓				✓	
Idle speed & fuel lines	S/I				✓				✓				✓	
Steering linkage rack guides & tie rod ends/steering operations, gear housing & rack seal boots/halfshaft dust boots	S/I				✓				✓				✓	
Front suspension ball joints	S/I				✓				✓				✓	
Brake lines & connections	S/I				✓				✓				✓	
Bolts & nuts on chassis & body	S/I				✓				✓				✓	
Clutch pedal operation	S/I				✓				✓				✓	
Accessory drive belt(s)	S/I				✓				✓				✓	
Lubricate rear wheel bearings	S/I				✓				✓				✓	
Timing belt & fuel filter	R								✓					

① Engine oil & filter (Turbo) - change engine oil & filter every 5000 miles.
R – Replace   S/I – Service or Inspect

FREQUENT OPERATION MAINTENANCE (SEVERE SERVICE)
If a vehicle is operated under any of the following conditions it is considered severe service:
- Extremely dusty areas.
- 50% or more of the vehicle operation is in 32°C (90°F) or higher temperatures, or constant operation in temperatures below 0°C (32°F).
- Prolonged idling (vehicle operation in stop and go traffic).
- Frequent short running periods (engine does not warm to normal operating temperatures).
- Police, taxi, delivery usage or trailer towing usage.
Oil & oil filter – change every 3000 miles.
Rotate tires at 6000 miles & every 9000 miles thereafter.
PCV valve - change every 15,000 miles.
Automatic transaxle fluid & filter - change every 30,000 miles.

SCHEDULED MAINTENANCE INTERVALS
(MERCURY CAPRI (1994))

TO BE SERVICED	TYPE OF SERVICE	VEHICLE MILEAGE INTERVAL (x1000)												
		5	10	15	20	25	30	35	40	45	50	55	60	65
Engine oil & filter	R	✓	✓	✓	✓	✓	✓	✓	✓	✓	✓	✓	✓	✓
Rotate tires	S/I		✓			✓			✓			✓		
Front and rear disc brake pads & rotors	S/I			✓			✓			✓			✓	
Spark plugs (Turbo)	R			✓			✓			✓			✓	
Spark plugs (Non-turbo)	R						✓						✓	
Engine coolant	R						✓						✓	
Air cleaner filter	R						✓						✓	
Exhaust heat shields	S/I						✓						✓	
Idle speed	S/I						✓						✓	
Fuel lines	S/I						✓						✓	
Steering operation & linkage	S/I						✓						✓	
Halfshaft dust boots	S/I						✓						✓	
Front suspension ball joints	S/I						✓						✓	
Brake lines, hoses & connections	S/I						✓						✓	
Bolts & nuts on chassis & body	S/I						✓						✓	
Clutch pedal operation	S/I						✓						✓	
Accessory drive belt(s)	S/I						✓						✓	
Repack front & rear wheel bearings	S/I						✓						✓	
Automatic transaxle fluid & filter	R												✓	
Fuel filter	R												✓	

R – Replace   S/I – Service or Inspect

FREQUENT OPERATION MAINTENANCE (SEVERE SERVICE)
If a vehicle is operated under any of the following conditions it is considered severe service:
- Extremely dusty areas.
- 50% or more of the vehicle operation is in 32°C (90°F) or higher temperatures, or constant operation in temperatures below 0°C (32°F).
- Prolonged idling (vehicle operation in stop and go traffic).
- Frequent short running periods (engine does not warm to normal operating temperatures).
- Police, taxi, delivery usage or trailer towing usage.
Oil & oil filter - change every 3000 miles.
Rotate tires at 6000 miles, & every 15,000 miles thereafter.
Air cleaner filter - service or inspect every 15,000 miles.
Steering operation & linkage - check every 15,000 miles.

CONTOUR/MYSTIQUE

VEHICLE IDENTIFICATION CHART

Engine Code						Model Year	
Code	Liters	Cu. In. (cc)	Cyl.	Fuel Sys.	Eng. Mfg.	Code	Year
3	2.0	122 (1999)	4	SFI	Ford	S	1995
L	2.5	153 (2507)	6	SFI	Ford	T	1996
						V	1997

SFI - Sequential fuel injection

ENGINE IDENTIFICATION

All measurements are given in inches.

Year	Model	Engine Displacement Liters (cc)	Engine Series (ID/VIN)	Fuel System	No. of Cylinders	Engine Type
1995	Contour	2.0 (1999)	3	SFI	4	DOHC
	Contour	2.5 (2507)	L	SFI	6	DOHC
	Mystique	2.0 (1999)	3	SFI	4	DOHC
	Mystique	2.5 (2507)	L	SFI	6	DOHC
1996-97	Contour	2.0 (1999)	3	SFI	4	DOHC
	Contour	2.5 (2507)	L	SFI	6	DOHC
	Mystique	2.0 (1999)	3	SFI	4	DOHC
	Mystique	2.5 (2507)	L	SFI	6	DOHC

SFI - Sequential fuel injection   DOHC - Double overhead camshaft

GENERAL ENGINE SPECIFICATIONS

Year	Engine ID/VIN	Engine Displacement Liters (cc)	Fuel System Type	Net Horsepower @ rpm	Net Torque @ rpm (ft. lbs.)	Bore x Stroke (in.)	Compression Ratio	Oil Pressure @ rpm
1995	3	2.0 (1999)	SFI	125@6000	130@4500	3.39x3.46	9.6:1	20-45@1500
	L	2.5 (2507)	SFI	170@6200	165@4200	3.25x3.13	9.7:1	25-45@1500
1996-97	3	2.0 (1999)	SFI	125@5500	130@4000	3.39x3.46	9.6:1	20-45@1500
	L	2.5 (2507)	SFI	170@6200	165@4200	3.25x3.13	9.7:1	20-45@1500

SFI - Sequential fuel injection

GASOLINE ENGINE TUNE-UP SPECIFICATIONS

Year	Engine ID/VIN	Engine Displacement Liters (cc)	Spark Plugs Gap (in.)	Ignition Timing (deg.) MT	Ignition Timing (deg.) AT	Fuel Pump (psi)		Idle Speed (rpm) MT	Idle Speed (rpm) AT	Valve Clearance In.	Valve Clearance Ex.
1995	3	2.0 (1999)	0.050	10B	10B	30-38	②	880	800	HYD	HYD
	L	2.5 (2507)	0.054	10B	10B	30-36	②	①	①	HYD	HYD
1996-97	3	2.0 (1999)	0.050	10B	10B	37-41	②	①	①	HYD	HYD
	L	2.5 (2507)	0.054	10B	10B	37-41	②	①	①	HYD	HYD

NOTE: The Vehicle Emission Control Information label often reflects specification changes made during production. The label figures must be used if they differ from those in this chart.

B - Before top dead center

HYD - Hydraulic

① Refer to Vehicle Emission Control Information label

② Fuel pressure with engine running, pressure regulator vacuum hose connected

CAPACITIES

Year	Model	Engine ID/VIN	Engine Displacement Liters (cc)	Engine Oil with Filter (qts.)	Transmission (pts.) 4-Spd	5-Spd	Auto.	Drive Axle Front (pts.)	Rear (pts.)	Fuel Tank (gal.)	Cooling System (qts.)
1995	Contour	3	2.0 (1999)	4.5	-	5.5	18.0 ②	①	-	14.5	③
	Contour	L	2.5 (2507)	5.5	-	5.5	20.6 ②	①	-	14.5	④
	Mystique	3	2.0 (1999)	4.5	-	5.5	18.0 ②	①	-	14.5	③
	Mystique	L	2.5 (2507)	5.5	-	5.5	20.6 ②	①	-	14.5	④
1996-97	Contour	3	2.0 (1999)	4.5	-	5.5	18.0 ②	①	-	14.5	③
	Contour	L	2.5 (2507)	5.8	-	5.5	20.6 ②	①	-	14.5	④
	Mystique	3	2.0 (1999)	4.5	-	5.5	18.0 ②	①	-	14.5	③
	Mystique	L	2.5 (2507)	5.8	-	5.5	20.6 ②	①	-	14.5	④

① Included in transaxle capacity
② Includes torque converter

③ Automatic transaxle: 7.5 qts.
   Manual transaxle: 7.0 qts.

④ Automatic transaxle: 9.1 qts.
   Manual transaxle: 8.9 qts.

VALVE SPECIFICATIONS

Year	Engine ID/VIN	Engine Displacement Liters (cc)	Seat Angle (deg.)	Face Angle (deg.)	Spring Test Pressure (lbs. @ in.)	Spring Installed Height (in.)	Stem-to-Guide Clearance (in.) Intake	Exhaust	Stem Diameter (in.) Intake	Exhaust
1995	3	2.0 (1999)	45	45	NA	1.346	0.0007-0.0025	0.0014-0.0032	0.2373-0.2379	0.2366-0.2372
	L	2.5 (2507)	44.75	45.5	153@1.18	1.570	0.0007-0.0027	0.0017-0.0037	0.2350-0.2358	0.2343-0.2350
1996-97	3	2.0 (1999)	45	45	NA	1.346	0.0007-0.0025	0.0014-0.0032	0.2373-0.2379	0.2366-0.2372
	L	2.5 (2507)	44.75	45.5	153@1.18	1.570	0.0007-0.0027	0.0017-0.0037	0.2350-0.2358	0.2343-0.2350

TORQUE SPECIFICATIONS
All readings in ft. lbs.

Year	Engine ID/VIN	Engine Displacement Liters (cc)	Cylinder Head Bolts	Main Bearing Bolts	Rod Bearing Bolts	Crankshaft Damper Bolts	Flywheel Bolts	Manifold Intake	Exhaust	Spark Plugs	Lug Nut
1995	3	2.0 (1999)	②	55-66	①	81-89	80-87	12-15	13-16	9-13	63
	L	2.5 (2507)	③	④	⑤	⑥	54-64	6-9	13-16	7-15	63
1996-97	3	2.0 (1999)	②	55-66	①	81-89	80-87	12-15	13-16	9-13	63
	L	2.5 (2507)	③	④	⑤	⑥	54-64	6-9	13-16	7-15	63

NOTE: Always follow proper torque patterns

NOTE: Stretch bolts are used in all procedures that require rotating the fastener a certain number of degrees. The bolts stretch and cannot be reused. For reassembly, replace with new fastners.

① Step 1: 22-25 ft. lbs.
   Step 2: Rotate each bolt 85-95 degrees
② Step 1: 15-22 ft. lbs.
   Step 2: 30-37 ft. lbs.
   Step 3: Rotate 90-120 degrees
③ Step 1: 27-32 ft. lbs.
   Step 2: Rotate 85-95 degrees
   Step 3: Loosen bolts then repeat Step 1
   Step 4: Rotate 85-95 degrees
   Step 5: Repeat Step 5

④ Step 1: 2.0-3.6 ft. lbs.
   Step 2: Push crankshaft rearward.
   Lightly seat crankshaft washer forward
   Step 3: Outer cap bolts: 16-21 ft. lbs.
   Step 4: Inner cap bolts: 27-32 ft. lbs.
   Step 5: Rotate inner and outer cap bolts 85-95 degrees
   Step 6: Remaining bolts: 15-22 ft. lbs.

⑤ 26-33 ft. lbs. plus 90-120 degrees
⑥ Step 1: 89 ft. lbs.
   Step 2: Loosen bolt
   Step 3: 35-39 ft. lbs.
   Step 4: Rotate 85-95 degrees

BRAKE SPECIFICATIONS
All measurements in inches unless noted

Year	Model		Master Cylinder Bore	Brake Disc			Brake Drum Diameter			Minimum Lining Thickness	
				Original Thickness	Minimum Thickness	Maximum Runout	Original Inside Diameter	Max. Wear Limit	Maximum Machine Diameter	Front	Rear
1995	Contour	F	NA	0.950	0.870	0.006	-	-	-	0.125	-
		R	-	0.790	0.710	0.006	-	-	-	-	0.125
		R	-	-	-	-	NA	NA	8.04	-	0.125
	Mystique	F	NA	0.950	0.870	0.006	-	-	-	0.125	-
		R	-	0.790	0.710	0.006	-	-	-	-	0.125
		R	-	-	-	-	NA	NA	8.04	-	0.125
1996-97	Contour	F	NA	0.950	0.870	0.006	-	-	-	0.125	-
		R	-	0.790	0.710	0.006	8.00	NA	8.04	-	0.125
	Mystique	F	NA	0.950	0.870	0.006	-	-	-	0.125	-
		R	-	0.790	0.710	0.006	8.00	NA	8.04	-	0.125

NOTE: Follow specifications stamped on rotor or drum if figures differ from those in this chart.

NA - Not Available

F - Front

R - Rear

FREQUENT MAINTENANCE LABOR
FORD CONTOUR, MERCURY MYSTIQUE

The following should be used as a guide when determining the amount of work required for a particular service if taken to a repair shop. In estimating how long a particular Frequent Maintenance Service item should take, please observe the following:
- **Factory Time** is time that is generated by the vehicle manufacturer.
- **Chilton Time** is time that is based on field research and data supplied by the vehicle manufacturer.
- All labor time operations are given in hours and tenths of an hour.
- All labor operations, are to be used as a **guide**.

COOLING

(G) Winterize Cooling System
Includes: Run engine to check for leaks, tighten all hose connections. Test radiator and pressure cap, drain radiator and engine block. Add antifreeze and refill system.
1995-97 5

(G) Belt, Serpentine Drive, Renew
1995-97
2.0L (.4) 5
2.5L (.3) 4

(G) Hoses, Radiator, Renew
Includes: Drain and refill cooling system as required.
1995-97
upper (.7) 8
lower (.9) 1.0
both (1.2) 1.5

(G) Thermostat, Coolant, Renew
1995-97
2.0L (.8) 1.0
2.5L (1.0) 1.3

FUEL

(G) Filter, Fuel, Renew
1995-97 (.3) 4

BRAKES

(G) Bleed Brakes (Four Wheels)
Includes: Add fluid.
1995-97 (.4) 6

(G) Brakes, Adjust (Minor)
Includes: Adjust brakes, fill master cylinder.
1995-97 4

(M) Parking Brake, Adjust
1995-97 (.3) 4

LUBRICATION SERVICES

(M) Engine Oil & Filter, Renew
Includes: Inspect and correct all fluid levels.
1995-97 3

(M) Lubricate Chassis, Change Oil & Filter
Includes: Inspect and correct all fluid levels.
1995-97 6
Install grease fittings add 1

(M) Lubricate Chassis
Includes: Inspect and correct all fluid levels.
1995-97 4
Install grease fittings add 1

WHEELS

(G) Wheel, Renew (One)
1995-97 5

FREQUENT MAINTENANCE LABOR (cont.)
FORD CONTOUR, MERCURY MYSTIQUE

	(Factory Time)	Chilton Time
(G) Wheel, Balance		
1995-97		
one		.3
each adtnl.		.2
(G) Wheels, Rotate (All)		
1995-97		.5

ELECTRICAL

	(Factory Time)	Chilton Time
(G) Headlamps, Aim		
1995-97		
two		.4
four		.6
(G) Halogen Headlamp Bulb, Renew		
1995-97, each		.3

	(Factory Time)	Chilton Time
(G) License Lamp Assy., Renew		
1995-97		
one		.3
both		.4
(G) Horn, Renew		
1995-97 (.3)		.4
(M) Terminals, Battery, Clean		
1995-97		.3

SCHEDULED MAINTENANCE INTERVALS
(FORD CONTOUR, MERCURY MYSTIQUE)

TO BE SERVICED	TYPE OF SERVICE	VEHICLE MILEAGE INTERVAL (x1000)												
		5	10	15	20	25	30	35	40	45	50	55	60	65
Engine oil & filter	R	✓	✓	✓	✓	✓	✓	✓	✓	✓	✓	✓	✓	✓
Rotate tires	S/I	✓		✓		✓		✓		✓		✓		✓
Front & rear brakes	S/I		✓		✓		✓		✓		✓		✓	
Cooling system, hoses, clamps & coolant strength	S/I			✓			✓			✓			✓	
Passenger compartment air filter	R				✓				✓				✓	
Air cleaner element	R						✓						✓	
Automatic transaxle fluid & filter (1995)	R						✓						✓	
Automatic transaxle fluid & filter (1996-97)	S/I												✓	
Exhaust heat shields	S/I						✓						✓	
Accessory drive belt(s)	S/I						✓						✓	
Fuel lines & hoses	S/I						✓						✓	
Crankcase emission filter (2.0L)	R						✓							
Engine coolant①	R										✓			

SCHEDULED MAINTENANCE INTERVALS
(FORD CONTOUR, MERCURY MYSTIQUE) (Cont.)

TO BE SERVICED	TYPE OF SERVICE	VEHICLE MILEAGE INTERVAL (x1000)												
		5	10	15	20	25	30	35	40	45	50	55	60	65
Spark plugs②	R												✓	
PCV valve	R												✓	

① Change initially at 50,000 miles, & every 30,000 miles thereafter.
② 2.0L shown; 2.5L - replace every 100,000 miles.
R – Replace   S/I – Service or Inspect

FREQUENT OPERATION MAINTENANCE (SEVERE SERVICE)
If a vehicle is operated under any of the following conditions it is considered severe service:
- Extremely dusty areas.
- 50% or more of the vehicle operation is in 32°C (90°F) or higher temperatures, or constant operation in temperatures below 0°C (32°F).
- Prolonged idling (vehicle operation in stop and go traffic).
- Frequent short running periods (engine does not warm to normal operating temperatures).
- Police, taxi, delivery usage or trailer towing usage.
Oil & oil filter – change every 3000 miles.
Front & rear brakes - check every 9000 miles.
Rotate tires at 6000 miles & every 9000 miles thereafter.
Air cleaner element - check every 15,000 miles.
Passenger compartment air filter - change every 18,000 miles.
Automatic transaxle fluid & filter - change every 21,000 miles (1995), 30,000 miles (1996-97).
Spark plugs - replace every 60,000 miles.

TAURUS/CONTINENTAL/SABLE

VEHICLE IDENTIFICATION CHART

Engine Code						Model Year	
Code	Liters	Cu. In. (cc)	Cyl.	Fuel Sys.	Eng. Mfg.	Code	Year
1 ①	3.0	182 (2982)	6	SFI	Ford	P	1993
S	3.0	183 (3049)	6	SFI	Ford	R	1994
U	3.0	181 (2971)	6	MFI	Ford	S	1995
Y	3.0	182 (2980)	6	MFI	Yamaha	T	1996
P	3.2	195 (3191)	6	MFI	Yamaha	V	1997
4	3.8	232 (3802)	6	MFI	Ford		
V	4.6	281 (4593)	8	SFI	Ford		

MFI - Multiport fuel injection   SFI - Sequential fuel injection   ① Flex fuel

ENGINE IDENTIFICATION
All measurements are given in inches.

Year	Model		Engine Displacement Liters (cc)	Engine Series (ID/VIN)	Fuel System	No. of Cylinders	Engine Type
1993	Continental		3.8 (3801)	4	MFI	6	OHV
	Sable		3.0 (2971)	U	MFI	6	OHV
	Sable		3.8 (3802)	4	MFI	6	OHV
	Taurus		3.0 (2971)	U	MFI	6	OHV
	Taurus		3.8 (3802)	4	MFI	6	OHV
	Taurus SHO		3.0 (2980)	Y	MFI	6	DOHC
	Taurus SHO		3.2 (3191)	P	MFI	6	DOHC
1994	Continental		3.8 (3802)	4	SFI	6	OHV
	Sable		3.0 (2971)	U	SFI	6	OHV
	Sable		3.8 (3802)	4	SFI	6	OHV
	Taurus		3.0 (2971)	U	SFI	6	OHV
	Taurus		3.8 (3802)	4	SFI	6	OHV
	Taurus SHO		3.0 (2980)	Y	SFI	6	DOHC
	Taurus SHO		3.2 (3191)	P	SFI	6	DOHC
1995	Continental		4.6 (4593)	V	SFI	8	DOHC
	Sable		3.0 (2980)	U	SFI	6	OHV
	Sable		3.8 (3802)	4	SFI	6	OHV
	Taurus		3.0 (2980)	U	SFI	6	OHV
	Taurus		3.8 (3802)	4	SFI	6	OHV
	Taurus SHO		3.0 (2980)	Y	SFI	6	DOHC
	Taurus SHO		3.2 (3191)	P	SFI	6	DOHC
1996-97	Continental		4.6 (4593)	V	SFI	8	DOHC
	Sable		3.0 (2982)	U	SFI	6	OHV
	Sable		3.0 (2998)	S	SFI	6	DOHC
	Taurus		3.0 (2982)	U	SFI	6	OHV
	Taurus ①		3.0 (2982)	1	SFI	6	OHV
	Taurus		3.0 (3049)	S	SFI	6	DOHC

MFI - Multiport fuel injection   DOHC - Double overhead camshaft   ① Flex fuel
SFI - Sequential fuel injection   OHV - Overhead valve

GENERAL ENGINE SPECIFICATIONS

Year	Engine ID/VIN	Engine Displacement Liters (cc)	Fuel System Type	Net Horsepower @ rpm	Net Torque @ rpm (ft. lbs.)	Bore x Stroke (in.)	Compression Ratio	Oil Pressure @ rpm
1993	U	3.0 (2971)	MFI	140@4800	160@3000	3.50x3.10	9.3:1	55-70@2000
	Y	3.0 (2971)	MFI	220@6200	200@4800	3.50x3.15	9.8:1	40-65@2000
	P	3.2 (3191)	MFI	220@6200	215@4800	3.62x3.15	9.8:1	40-60@2000
	4	3.8 (3802)	MFI	140@3800	215@2400	3.81x3.39	8.2:1	40-60@2000
1994	U	3.0 (2971)	SFI	140@4800	160@3000	3.50x3.15	9.3:1	55-70@2500
	Y	3.0 (2971)	SFI	220@6200	200@4800	3.50x3.15	9.8:1	40-65@2000
	P	3.2 (3191)	SFI	220@6200	215@4800	3.62x3.15	9.8:1	40-60@2000
	4	3.8 (3802)	SFI	140@3800	215@2400	3.81x3.39	8.2:1	40-60@2500
1995	U	3.0 (2980)	SFI	140@4800	160@3000	3.50x3.15	9.3:1	55-70@2500
	Y	3.0 (2980)	SFI	220@6200	200@4800	3.50x3.15	9.8:1	40-65@2000
	P	3.2 (3191)	SFI	220@6200	215@4800	3.62x3.15	9.8:1	40-60@2000
	4	3.8 (3802)	SFI	140@3800	215@2400	3.81x3.39	9.0:1	40-60@2500
	V	4.6 (4593)	SFI	260@5750	265@4750	3.55x3.54	9.8:1	33@1500

GENERAL ENGINE SPECIFICATIONS

Year	Engine ID/VIN	Engine Displacement Liters (cc)	Fuel System Type	Net Horsepower @ rpm	Net Torque @ rpm (ft. lbs.)	Bore x Stroke (in.)	Compression Ratio	Oil Pressure @ rpm
1996-97	U	3.0 (2982)	SFI	145@5250	170@3250	3.50x3.15	9.3:1	55-70@2500
	1 ①	3.0 (2982)	SFI	145@5250	170@3250	3.50x3.15	9.3:1	55-70@2500
	S	3.0 (2998)	SFI	200@5750	200@4500	3.50x3.13	10.0:1	20-45@1515
	V	4.6 (4593)	SFI	260@5750	265@4750	3.55x3.54	9.8:1	33@1500

MFI - Multiport fuel injection   SFI - Sequential fuel injection   ① Flex fuel

GASOLINE ENGINE TUNE-UP SPECIFICATIONS

Year	Engine ID/VIN	Engine Displacement Liters (cc)	Spark Plugs Gap (in.)	Ignition Timing (deg.) MT	Ignition Timing (deg.) AT	Fuel Pump (psi)	Idle Speed (rpm) MT	Idle Speed (rpm) AT	Valve Clearance In.	Valve Clearance Ex.
1993	U	3.0 (2980)	0.044	10B	10B	35-45	-	625	HYD	HYD
	Y	3.0 (2980)	0.044	10B	-	36-39	800	-	0.006-0.010	0.010-0.014
	P	3.2 (3191)	0.044	-	10B	30-45	-	-	0.006-0.010	0.010-0.014
	4	3.8 (3802)	0.054	10B	10B	35-45	-	550	HYD	HYD
1994	U	3.0 (2980)	0.044	10B	10B	30-45 ②	①	①	HYD	HYD
	Y	3.0 (2980)	0.044	10B	-	28-33 ②	①	-	0.006-0.010	0.010-0.014
	P	3.2 (3191)	0.044	-	10B	28-33 ②	-	750	0.006-0.010	0.010-0.014
	4	3.8 (3802)	0.054	10B	10B	30-45 ②	①	①	HYD	HYD
1995	U	3.0 (2980)	0.044	10B	10B	30-45 ②	-	①	HYD	HYD
	Y	3.0 (2980)	0.044	10B	-	28-33 ②	①	-	0.006-0.010	0.010-0.014
	P	3.2 (3191)	0.044	-	10B	30-45 ②	-	800	0.006-0.010	0.010-0.014
	4	3.8 (3802)	0.054	-	10B	30-45 ②	-	①	HYD	HYD
	V	4.6 (4593)	0.054	-	10B	30-45 ②	-	①	HYD	HYD
1996-97	U	3.0 (2982)	0.044	-	10B	30-45 ②	-	①	HYD	HYD
	1 ③	3.0 (2982)	0.044	-	①	30-45 ②	-	①	HYD	HYD
	S	3.0 (2998)	0.054	-	10B	30-45 ②	-	①	HYD	HYD
	V	4.6 (4593)	0.054	-	10B	30-45 ②	-	①	HYD	HYD

NOTE: The Vehicle Emission Control Information label often reflects specification changes made during production. The label figures must be used if they differ from those in this chart.

B - Before top dead center
HYD - Hydraulic
① Refer to Vehicle Emission Control Information label
② Fuel pressure with engine running, pressure regulator vacuum hose connected
③ Flex fuel

CAPACITIES

Year	Model	Engine ID/VIN	Engine Displacement Liters (cc)	Engine Oil with Filter (qts.)	Transmission (pts.) 4-Spd	5-Spd	Auto.	Drive Axle Front (pts.)	Rear (pts.)	Fuel Tank (gal.)	Cooling System (qts.)
1993	Continental	4	3.8 (3801)	4.5	-	-	25.6	-	-	18.4	11.1
	Sable	U	3.0 (2980)	5.0	-	6.2	21.8	①	-	③	②
	Sable	4	3.8 (3801)	4.5	-	-	25.6	①	-	18.6	12.1
	Taurus	U	3.0 (2980)	5.0	-	6.2	21.8	①	-	③	②
	Taurus SHO	Y	3.0 (2980)	5.0	-	6.2	21.8	①	-	③	11.6
	Taurus	4	3.8 (3801)	4.5	-	-	25.6	①	-	18.6	12.1
	Taurus	P	3.2 (3191)	5.0	-	-	25.6	①	-	18.6	11.2
1994	Continental	4	3.8 (3802)	4.5	-	-	24.5 ④	-	-	18.4	12.1
	Sable	U	3.0 (2980)	4.5	-	-	24.5 ④	①	-	③	11.0
	Sable	4	3.8 (3801)	4.5	-	-	24.5 ④	①	-	③	12.1
	Taurus	U	3.0 (2980)	4.5	-	-	24.5 ④	①	-	③	11.0
	Taurus SHO	Y	3.0 (2980)	5.0	-	6.2	24.5 ④	①	-	18.4	11.6
	Taurus	4	3.8 (3801)	4.5	-	-	24.5 ④	①	-	③	12.1
	Taurus	P	3.2 (3191)	5.0	-	-	24.5 ④	①	-	18.4	11.4
1995	Continental	V	4.6 (4593)	6.0	-	-	26.6 ④	①	-	18.4	14.3
	Sable	U	3.0 (2980)	4.5	-	-	24.5 ④	①	-	③	11.0
	Sable	4	3.8 (3802)	4.5	-	-	24.5 ④	①	-	③	12.1
	Taurus	U	3.0 (2980)	4.5	-	-	24.5 ④	①	-	③	11.0
	Taurus SHO	Y	3.0 (2980)	5.0	-	6.2	24.5 ④	①	-	18.6	11.6
	Taurus SHO	P	3.2 (3191)	5.0	-	-	24.5 ④	①	-	18.6	11.4
	Taurus	4	3.8 (3802)	4.5	-	-	24.5 ④	①	-	③	12.1
1996-97	Continental	V	4.6 (4593)	6.0	-	-	27.4 ④	①	-	17.8	14.3
	Sable	U	3.0 (2982)	4.5	-	-	24.5 ④	①	-	③	16.0
	Sable	S	3.0 (2998)	5.8	-	-	27.0 ④	①	-	③	16.0
	Taurus	U	3.0 (2982)	4.5	-	-	24.5 ④	①	-	③	16.0
	Taurus	1	3.0 (2982)	4.5	-	-	24.5 ④	①	-	③	16.0
	Taurus	S	3.0 (2998)	5.8	-	-	27.0 ④	①	-	③	16.0

① Included in transaxle capacity

② Wagon with AC: 11.8 qts.
Except Wagon with AC: 11.0 qts.

③ Standard tank: 16.0 gals.
Optional extended range tank: 18.6 gals.

④ Includes torque converter

VALVE SPECIFICATIONS

Year	Engine ID/VIN	Engine Displacement Liters (cc)	Seat Angle (deg.)	Face Angle (deg.)	Spring Test Pressure (lbs. @ in.)	Spring Installed Height (in.)	Stem-to-Guide Clearance (in.) Intake	Exhaust	Stem Diameter (in.) Intake	Exhaust
1993	U	3.0 (2971)	45	44	185@1.11	1.850	0.0001-0.0027	0.0015-0.0032	0.3126	0.3121
	Y	3.0 (2980)	45	45.5	121@1.19	1.760	0.0010-0.0023	0.0012-0.0025	0.2346-0.2352	0.2344-0.2350
	P	3.2 (3191)	45	45.5	121@1.19	1.520	0.0010-0.0023	0.0012-0.0025	0.2346-0.2352	0.2344-0.2350
	4	3.8 (3802)	44.5	45.8	215@1.79	1.750	0.0010-0.0027	0.0015-0.0032	0.3420	0.3415
1994	U	3.0 (2971)	45	44	180@1.06	1.580	0.0001-0.0027	0.0015-0.0032	0.3126-0.3129	0.3121-0.3134
	Y	3.0 (2980)	45	45.5	121@1.19	1.520	0.0010-0.0023	0.0012-0.0025	0.2346-0.2352	0.2344-0.2350
	P	3.2 (3191)	45	45.5	121@1.19	1.520	0.0010-0.0023	0.0012-0.0025	0.2346-0.2352	0.2344-0.2350

VALVE SPECIFICATIONS

Year	Engine ID/VIN	Engine Displacement Liters (cc)	Seat Angle (deg.)	Face Angle (deg.)	Spring Test Pressure (lbs. @ in.)	Spring Installed Height (in.)	Stem-to-Guide Clearance (in.)		Stem Diameter (in.)	
							Intake	Exhaust	Intake	Exhaust
1994	4	3.8 (3802)	44.5	45.8	220@1.18	1.970	0.0010-0.0027	0.0015-0.0032	0.3415-0.3423	0.3410-0.3418
1995	U	3.0 (2980)	45	44	180@1.06	1.580	0.0001-0.0027	0.0015-0.0032	0.3126-0.3129	0.3121-0.3134
	Y	3.0 (2980)	45	45.5	121@1.19	1.520	0.0010-0.0023	0.0012-0.0025	0.2346-0.2352	0.2344-0.2350
	P	3.2 (3191)	45	45.5	121@1.19	1.520	0.0010-0.0023	0.0012-0.0025	0.2346-0.2352	0.2344-0.2350
	4	3.8 (3802)	44.5	45.8	220@1.18	1.970	0.0010-0.0027	0.0015-0.0032	0.3415-0.3423	0.3410-0.3418
	V	4.6 (4593)	45	45.5	160@1.103	1.425	0.0008-0.0027	0.0018-0.0037	0.2746-0.2754	0.2736-0.2744
1996-97	U	3.0 (2982)	45	44	180@1.06	1.580	0.0001-0.0027	0.0015-0.0032	0.3126-0.3134	0.3121-0.3129
	1	3.0 (2982)	45	44	180@1.06	1.580	0.0001-0.0027	0.0015-0.0032	0.3126-0.3134	0.3121-0.3129
	S	3.0 (2998)	44.75	45.5	153@1.18	1.570	0.0007-0.0027	0.0017-0.0037	0.2350-0.2358	0.2343-0.2350
	V	4.6 (4593)	45	45.5	160@1.103	1.425	0.0008-0.0027	0.0018-0.0037	0.2746-0.2754	0.2736-0.2744

TORQUE SPECIFICATIONS

All readings in ft. lbs.

Year	Engine ID/VIN	Engine Displacement Liters (cc)	Cylinder Head Bolts	Main Bearing Bolts	Rod Bearing Bolts	Crankshaft Damper Bolts	Flywheel Bolts	Manifold Intake	Manifold Exhaust	Spark Plugs	Lug Nut
1993	U	3.0 (2971)	⑤	65-81	①	141-169	54-64	②	③	5-10	95
	Y	3.0 (2980)	⑥	⑦	⑧	113-126	58-64	11-16	26-38	17-19	95
	P	3.2 (3191)	⑥	⑦	⑧	112-127	58-64	11-17	26-38	16-20	95
	4	3.8 (3802)	④	65-81	31-36	85-100	75-85	⑤	15-22	5-11	95
1994	U	3.0 (2971)	⑪	55-63	26	93-121	54-64	⑨	15-22	7-15	95
	Y	3.0 (2980)	⑥	⑦	⑧	113-126	51-58	11-17	26-38	15-22	95
	P	3.2 (3191)	⑥	⑦	⑧	112-127	51-58	11-17	26-38	15-22	95
	R	3.8 (3802)	⑫	65-81	31-36	103-132	54-64	⑬	15-22	7-15	95
	4	3.8 (3802)	⑩	65-81	31-36	103-132	54-64	⑬	15-22	7-15	95
1995	U	3.0 (2980)	⑪	55-63	26	93-121	54-64	⑨	15-22	7-15	95
	Y	3.0 (2980)	⑥	⑦	⑧	113-126	51-58	11-17	26-38	15-22	95
	P	3.2 (3191)	⑥	⑦	⑧	112-127	51-58	11-17	26-38	15-22	95
	4	3.8 (3802)	⑩	65-81	31-36	103-132	54-64	⑬	15-22	7-15	95
	V	4.6 (4593)	⑭	⑮	⑯	114-121	54-64	⑰	13-16	7-15	95

TORQUE SPECIFICATIONS
All readings in ft. lbs.

Year	Engine ID/VIN	Engine Displacement Liters (cc)	Cylinder Head Bolts	Main Bearing Bolts	Rod Bearing Bolts	Crankshaft Damper Bolts	Flywheel Bolts	Manifold Intake	Manifold Exhaust	Spark Plugs	Lug Nut
1996-97	U	3.0 (2982)	⑥	55-63	23-29	93-121	54-64	⑨	15-18	7-15	95
	1	3.0 (2982)	⑥	55-63	23-29	93-121	54-64	⑨	15-18	7-15	95
	S	3.0 (2998)	⑭	⑱	⑭	⑲	54-64	6-9	13-16	7-15	95
	V	4.6 (4593)	⑭	⑮	⑯	114-121	54-64	⑰	13-16	7-15	95

① Step 1: Tighten to 20-28 ft. lbs.
Step 2: Back off the nuts a minimum of two revolutions
Step 3: Apply final torque of 20-25 ft. lbs.

② Step 1: 11 ft. lbs.
Step 2: 18 ft. lbs.
Step 3: 24 ft. lbs.

③ Step 1: 5-7 ft. lbs.
Step 2: 20-30 ft. lbs.

④ Step 1: 37 ft. lbs.
Step 2: 45 ft. lbs.
Step 3: 52 ft. lbs.
Step 4: 59 ft. lbs.

⑤ Step 1: 48-54 ft. lbs.
Step 2: 63-80 ft. lbs.

⑥ Step 1: 37-50 ft. lbs.
Step 2: 62-68 ft. lbs.

⑥ Step 1: 37-50 ft. lbs.
Step 2: 58-64 ft. lbs.

⑦ Step 1: 22-26 ft. lbs.
Step 2: 33-36 ft. lbs.
Step 1: Inner bolts - 17-19 ft. lbs in two to three steps
Step 1: Outer bolts - 13-15 ft. lbs. in two to three steps
Step 2: Inner bolts 1-3: 70 degrees; Inner bolt 4: 80 degrees
Step 3: Tighten outer bolts 60 degrees
Step 4: Repeat Step 3
Step 1: 16-19 ft. lbs.
Step 2: Rotate each bolt 90 degrees
Step 3: Repeat Step 2
NOTE: Always follow the proper torque patterns

⑨ Step 1: 15-22 ft. lbs.
Step2 : 19-24 ft. lbs.

⑩ Do not reuse cylinder head bolts
Step 1: 15 ft. lbs.
Step 2: 29 ft. lbs.
Step 3: 37 ft. lbs.
Step 4: Loosen bolts one at a time and retorque as follow
Long bolts: 11-18 ft. lbs.
Short bolts: 7-15 ft. lbs.
Step 5: Rotate 85-95 degrees

⑪ Step 1: 33-41 ft. lbs.
Step 2: 63-73 ft. lbs.

⑫ Do not reuse cylinder head bolts
Step 1: 37 ft. lbs.
Step 2: 45 ft. lbs.
Step 3: 52 ft. lbs.
Step 4: 59 ft. lbs.
Step 5: Back off bolts two to three turns, one at a time
Step 6: Tighten all bolts 37-44 ft. lbs.
Step 7: Rotate bolts an additional 180-200 degrees

⑬ Step 1: 13 ft. lbs.
Step 2: 16 ft. lbs.
NOTE: Always follow proper torque patterns

⑭ Step 1: 27-32 ft. lbs.
Step 2: Rotate 85-95 degrees
Step 3: Repeat Step 2

⑮ Step 1: Main bearing cap bolts: 6-9 ft. lbs.
Step 2: Main bearing cap bolts, outer: 16-21 ft. lbs.
Step 3: Main bearing cap bolts, inner: 27-32 ft. lbs.
Step 4: Rotate main bearing cap bolts 85-95 degrees
Step 5: Main cap adjusting screws: 4 ft. lbs. then 7.5 ft. lbs.
Step 6: Main cap side bolts: 7 ft. lbs. then 14-17 ft. lbs.

⑯ Step 1: 5 ft. lbs.
Step 2: 10 ft. lbs.
Step 3: 18-25 ft. lbs.
Step 4: Rotate 85-95 degrees

⑰ Step 1: Four inside short bolts: 9-11 ft. lbs.
Step 2: All other bolts: 13-16 ft. lbs.
Step 3: Rotate 85-95 degrees

⑱ Main caps are cast into lower cylinder block and installed as one unit
Step 1: 2.0-3.5 ft. lbs.
Step 2: Push crankshaft rearward, lightly seat crankshaft thrust washer
Step 3: Outer cap bolts: 16-21 ft. lbs.
Step 4: Inner cap bolts: 27-32 ft. lbs.
Step 5: Rotate all cap bolts 85-95 degrees
Step 6: Remaining bolts: 15-22 ft. lbs.

⑲ Step 1: 89 ft. lbs.
Step 2: Loosen bolt
Step 3: 35-39 ft. lbs.
Step 4: Rotate 85-95 degrees

BRAKE SPECIFICATIONS
All measurements in inches unless noted

Year	Model		Master Cylinder Bore	Brake Disc Original Thickness	Brake Disc Minimum Thickness	Brake Disc Maximum Runout	Brake Drum Diameter Original Inside Diameter	Brake Drum Diameter Max. Wear Limit	Brake Drum Diameter Maximum Machine Diameter	Minimum Lining Thickness Front	Minimum Lining Thickness Rear
1993	Continental	F	1.000	1.024	0.974	0.003	-	-	-	0.040	-
		R	-	0.550	0.500	0.002	-	-	-	-	0.123
	Sable		1.000	③	④	⑤	①	NA	②	0.125	0.030
	Taurus		0.875	1.024	0.974	0.003	①	NA	②	0.125	0.030
	Taurus SHO	F	0.875	NA	0.972	0.002	-	-	-	0.125	-
		R	-	-	0.900	0.002	-	-	-	-	0.123
1994	Continental	F	1.000	1.020	0.974	0.003	-	-	-	0.040	-
		R	-	0.550	0.500	0.001	-	-	-	-	0.123
	Sable		1.000	③	④	⑤	①	NA	②	0.125	0.030
	Taurus		1.000	③	④	⑤	①	NA	②	0.040	⑥
	Taurus SHO	F	0.875	NA	0.972	0.002	-	-	-	0.125	-
		R	-	-	0.900	0.002	-	-	-	-	0.123

BRAKE SPECIFICATIONS
All measurements in inches unless noted

Year	Model		Master Cylinder Bore	Brake Disc Original Thickness	Brake Disc Minimum Thickness	Maximum Runout	Brake Drum Diameter Original Inside Diameter	Brake Drum Diameter Max. Wear Limit	Brake Drum Diameter Maximum Machine Diameter	Minimum Lining Thickness Front	Minimum Lining Thickness Rear
1995	Continental	F	NA	1.020	0.974	0.003	-	-	-	0.060	-
		R	-	0.550	0.502	0.001	-	-	-	-	0.123
	Sable		1.000	③	④	⑤	①	NA	②	0.040	⑥
	Taurus		1.000	③	④	⑤	①	NA	②	0.040	⑥
	Taurus SHO		1.000	③	④	⑤				0.040	0.123
1996-97	Continental	F	NA	1.020	0.974	0.003	-	-	-	0.060	-
		R	-	0.550	0.502	0.001	-	-	-	-	0.130
	Sable	F	1.000	1.020	0.974	0.002	-	-	-	0.040	-
		R	-	0.940	0.500	0.002	①	NA	②	-	⑦
	Taurus	F	1.000	1.020	0.974	0.002	-	-	-	0.040	-
		R	-	0.940	0.500	0.002	①	NA	②	-	⑦

NOTE: Follow specifications stamped on rotor or drum if figures differ from those in this chart.
NA - Not Available
F - Front
R - Rear

① Sedan: 8.85
   Wagon: 9.84
② Sedan: 8.91
   Wagon: 9.90

③ Front: 1.020
   Rear: 0.940
④ Front: 0.974
   Rear: 0.900

⑤ Front: 0.003
   Rear: 0.002
⑥ With disc brakes: 0.123
   With drum brakes: 0.030

⑦ Riveted lining: 0.031
   Bonded lining: 0.125

FREQUENT MAINTENANCE LABOR
FORD TAURUS, MERCURY SABLE

The following should be used as a guide when determining the amount of work required for a particular service if taken to a repair shop. In estimating how long a particular Frequent Maintenance Service item should take, please observe the following:
- **Factory Time** is time that is generated by the vehicle manufacturer.
- **Chilton Time** is time that is based on field research and data supplied by the vehicle manufacturer.
- All labor time operations are given in hours and tenths of an hour.
- All labor operations, are to be used as a **guide**.

COOLING

(G) Winterize Cooling System
Includes: Run engine to check for leaks, tighten all hose connections. Test radiator and pressure cap, drain radiator and engine block. Add antifreeze and refill system.
1993-97 5

(G) Belt, Drive, Renew
1993-95
  V belt, one (.3) 5
  each adtnl. 1
  Serpentine belt
    2.5L (.3) 5
    3.2L SHO (.5) 7
    3.8L (.6) 8
1996-97
  Serpentine (.3) 4

(G) Belt, Drive, Adjust
1993-95
  one (.2) 3
  each adtnl. 1

(G) Hoses, Radiator, Renew
Includes: Drain and refill cooling system as required.
1993-95
  upper (.4) 4
  lower (.5) 5
  both (.6) 6
1996-97
  upper (.8) 1.1
  lower (.7) 1.0
  both (1.1) 1.5

(G) Thermostat, Coolant, Renew
1993-95
  3.0L (.7) 1.0
  3.0L SHO, 3.2L SHO (.5) 8
  3.8L (.6) 8
1996-97
  3.0L DOHC (.9) 1.2
  3.0L FFV, 3.0L MFI (.8) 1.1

FUEL

(G) Filter, Fuel, Renew
1993-95 (.3) 4
1996-97 (.5) 6

FREQUENT MAINTENANCE LABOR (cont.)
FORD TAURUS, MERCURY SABLE

	(Factory Time)	Chilton Time
BRAKES		
(G) Bleed Brakes (Four Wheels)		
Includes: Add fluid.		
1993-95 (.3)		.5
1996-97 (1.0)		1.5
(G) Brakes, Adjust (Minor)		
Includes: Adjust brakes, fill master cylinder.		
1993-97, two wheels		.4
(M) Parking Brake, Adjust		
1993-97 (.3)		.5
LUBRICATION SERVICES		
(M) Engine Oil & Filter, Renew		
Includes: Inspect and correct all fluid levels.		
1993-97		.4
(M) Lubricate Chassis, Change Oil & Filter		
Includes: Inspect and correct all fluid levels.		
1993-97		.6
Install grease fittings add		.1

	(Factory Time)	Chilton Time
(M) Lubricate Chassis		
Includes: Inspect and correct all fluid levels.		
1993-97		.4
Install grease fittings add		.1
WHEELS		
(G) Wheel, Renew (One)		
1993-97		.5
(G) Wheel, Balance		
1993-97		
one		.3
each adtnl.		.2
(G) Wheels, Rotate (All)		
1993-97		.5
ELECTRICAL		
(G) Headlamps, Aim		
1993-97		
two		.4
four		.6

	(Factory Time)	Chilton Time
(G) Halogen Headlamp Bulb, Renew		
1993-95, one (.3)		.3
1996-97, one (.2)		.3
(G) License Lamp Assy., Renew		
1993-97 (.2)		.3
(G) Tail Lamp Assy., Renew		
1993-95		
Sedan (.5)		.7
Station Wagon (.3)		.5
1996-97		
Sedan (.3)		.5
Station Wagon (.3)		.5
(G) Horn, Renew		
1993-97 (.3)		.4
(M) Terminals, Battery, Clean		
1993-97		.3

FREQUENT MAINTENANCE LABOR
LINCOLN CONTINENTAL

The following should be used as a guide when determining the amount of work required for a particular service if taken to a repair shop. In estimating how long a particular Frequent Maintenance Service item should take, please observe the following:
- **Factory Time** is time that is generated by the vehicle manufacturer.
- **Chilton Time** is time that is based on field research and data supplied by the vehicle manufacturer.
- All labor time operations are given in hours and tenths of an hour.
- All labor operations, are to be used as a **guide**.

	(Factory Time)	Chilton Time
COOLING		
(G) Winterize Cooling System		
Includes: Run engine to check for leaks, tighten all hose connections. Test radiator and pressure cap, drain radiator and engine block. Add antifreeze and refill system.		
1993-97		.5
(G) Belt, Serpentine Drive, Renew		
1993-94 (.6)		.8
1995-97 (.8)		1.0
(G) Hoses, Radiator, Renew		
Includes: Drain and refill cooling system as required.		
1993-94		
upper (.5)		.6
lower (.6)		.7

	(Factory Time)	Chilton Time
both (.7)		.9
1995-97		
upper (.7)		.9
lower (.9)		1.1
both (1.2)		1.5
(G) Thermostat, Coolant, Renew		
1993-94 (.6)		.8
1995-97 (.7)		1.0
FUEL		
(M) Air Cleaner, Service		
1993-97		.2
(G) Filter, Fuel, Renew		
1993-97 (.3)		.4

	(Factory Time)	Chilton Time
BRAKES		
(G) Bleed Brakes (Four Wheels)		
Includes: Add fluid.		
1993-97 (.3)		.5
(M) Parking Brake, Adjust		
1993-97 (.3)		.5
LUBRICATION SERVICES		
(M) Engine Oil & Filter, Renew		
Includes: Inspect and correct all fluid levels.		
1993-97		.3
(M) Lubricate Chassis, Change Oil & Filter		
Includes: Inspect and correct all fluid levels.		
1993-97		.6
Install grease fittings add		.1

FREQUENT MAINTENANCE LABOR (cont.)
LINCOLN CONTINENTAL

	(Factory Time)	Chilton Time
(M) Lubricate Chassis		
Includes: Inspect and correct all fluid levels.		
1993-97		.4
Install grease fittings add		.1

WHEELS

	(Factory Time)	Chilton Time
(G) Wheel, Renew (One)		
1993-97		.5

	(Factory Time)	Chilton Time
(G) Wheel, Balance		
1993-97		
one		.3
each adtnl.		.2
(G) Wheels, Rotate (All)		
1993-97		.5

ELECTRICAL

	(Factory Time)	Chilton Time
(G) Headlamps, Aim		
1993-97		
two		.4
four		.6

	(Factory Time)	Chilton Time
(G) Halogen Headlamp Bulb, Renew		
1993-97, each (.3)		.3
(G) License Lamp Assy., Renew		
1993-97 (.3)		.4
(G) Horn, Renew		
1993-94 (.6)		.8
1995-97 (.3)		.4
(M) Terminals, Battery, Clean		
1993-97		.3

SCHEDULED MAINTENANCE INTERVALS
(FORD TAURUS, LINCOLN CONTINENTAL, MERCURY SABLE (1993))

TO BE SERVICED	TYPE OF SERVICE	VEHICLE MILEAGE INTERVAL (x1000)												
		7.5	15	22.5	30	37.5	45	52.5	60	67.5	75	82.5	90	97.5
Engine oil & filter	R	✓	✓	✓	✓	✓	✓	✓	✓	✓	✓	✓	✓	✓
Rotate tires	S/I	✓		✓		✓		✓		✓		✓		✓
Spark plugs (3.0L/3.8L Federal)	R				✓				✓				✓	
Spark plugs (SHO)	R								✓					
Spark plugs (3.8L Calif.)	R								✓					
Engine coolant	R				✓				✓				✓	
Air cleaner element	R				✓				✓				✓	
Front and rear disc brake pads & rotors	S/I				✓				✓				✓	
Exhaust heat shields	S/I				✓				✓				✓	
Brake lines & connections	S/I				✓				✓				✓	
Accessory drive belt(s)	S/I				✓				✓				✓	
Lubricate rear wheel bearings	S/I				✓				✓				✓	
Brake linings & drums (rear)	S/I				✓				✓				✓	
Cam belt & adjust valve lash (SHO)	R								✓					

SCHEDULED MAINTENANCE INTERVALS
(FORD TAURUS, LINCOLN CONTINENTAL, MERCURY SABLE (1993))

TO BE SERVICED	TYPE OF SERVICE	VEHICLE MILEAGE INTERVAL (x1000)												
		7.5	15	22.5	30	37.5	45	52.5	60	67.5	75	82.5	90	97.5
PCV valve	R								✓					
Battery fluid level (SHO)①	S/I													

① Check every 24,000 miles, or more often if operating in temperatures of 32°C (90°F) or higher.
R – Replace   S/I – Service or Inspect

FREQUENT OPERATION MAINTENANCE (SEVERE SERVICE)

If a vehicle is operated under any of the following conditions it is considered severe service:
- Extremely dusty areas.
- 50% or more of the vehicle operation is in 32°C (90°F) or higher temperatures, or constant operation in temperatures below 0°C (32°F).
- Prolonged idling (vehicle operation in stop and go traffic).
- Frequent short running periods (engine does not warm to normal operating temperatures).
- Police, taxi, delivery usage or trailer towing usage.

Oil & oil filter – change every 3000 miles.
Rotate tires at 6000 miles & every 9000 miles thereafter.
Automatic transaxle fluid & filter - change every 30,000 miles.

SCHEDULED MAINTENANCE INTERVALS
(FORD TAURUS, LINCOLN CONTINENTAL, MERCURY SABLE (1994-97))

TO BE SERVICED	TYPE OF SERVICE	VEHICLE MILEAGE INTERVAL (x1000)												
		5	10	15	20	25	30	35	40	45	50	55	60	65
Engine oil & filter	R	✓	✓	✓	✓	✓	✓	✓	✓	✓	✓	✓	✓	✓
Rotate tires	S/I	✓		✓		✓		✓		✓		✓		✓
Engine coolant protection, hoses & clamps	S/I			✓			✓			✓			✓	
Pass. compartment air filter (Continental)	R			✓			✓			✓			✓	
Pass. compartment air filter (Taurus, Sable)	R				✓				✓				✓	
Air cleaner filter	R						✓						✓	
Automatic transaxle fluid & filter (1995-97)	R						✓						✓	
Brake lines & connections	S/I						✓						✓	
Exhaust heat shields	S/I						✓						✓	
Front and rear disc brake pads & rotors	S/I						✓						✓	

SCHEDULED MAINTENANCE INTERVALS
(FORD TAURUS, LINCOLN CONTINENTAL, MERCURY SABLE (1994-97)) (Cont.)

TO BE SERVICED	TYPE OF SERVICE	VEHICLE MILEAGE INTERVAL (x1000)												
		5	10	15	20	25	30	35	40	45	50	55	60	65
Accessory drive belt(s)	S/I												✓	
Engine coolant (1996-97)①	R										✓			
Engine coolant (1994-95)	R						✓						✓	
Spark plugs (1994-95 exc. 3.8L Calif.)	R						✓							
Spark plugs (1994-95 3.8L Calif.)	R												✓	
Spark plugs (1996-97 exc. 3.0L FF)②	R													
Spark plugs (1996-97 3.0L FF)	R						✓						✓	
PCV valve (except 3.0L 4-valve)	R												✓	
PCV valve (3.0L 4-valve)③	R													

① Engine coolant - change initially at 50,000 miles & thereafter every 30,000 miles.
② Platinum tip spark plugs - change every 100,000 miles.
③ Replace every 100,000 miles.

R – Replace   S/I – Service or Inspect

FREQUENT OPERATION MAINTENANCE (SEVERE SERVICE)

If a vehicle is operated under any of the following conditions it is considered severe service:
- Extremely dusty areas.
- 50% or more of the vehicle operation is in 32°C (90°F) or higher temperatures, or constant operation in temperatures below 0°C (32°F).
- Prolonged idling (vehicle operation in stop and go traffic).
- Frequent short running periods (engine does not warm to normal operating temperatures).
- Police, taxi, delivery usage or trailer towing usage.

Oil & oil filter - change every 3000 miles.
Rotate tires at 6000 miles & every 9000 miles thereafter.
Air cleaner element - service or inspect every 15,000 miles.
Automatic transaxle fluid & filter - change every 21,000 miles (1995-97) or every 30,000 miles (1994).

ESCORT/TRACER

VEHICLE IDENTIFICATION CHART

		Engine Code					Model Year	
Code	Liters	Cu. In. (cc)	Cyl.	Fuel Sys.	Eng. Mfg.		Code	Year
8	1.8	112 (1844)	4	MFI	Mazda		P	1993
J	1.9	116 (1901)	4	SFI	Ford		R	1994
							S	1995
							T	1996
							V	1997

MFI - Multiport fuel injection

ENGINE IDENTIFICATION
All measurements are given in inches.

Year	Model	Engine Displacement Liters (cc)	Engine Series (ID/VIN)	Fuel System	No. of Cylinders	Engine Type
1993	Escort	1.8 (1844)	8	MFI	4	DOHC
	Escort	1.9 (1901)	J	SFI	4	SOHC
	Tracer	1.8 (1844)	8	MFI	4	DOHC
	Tracer ①	1.9 (1901)	J	SFI	4	SOHC
1994	Escort	1.8 (1844)	8	MFI	4	DOHC
	Escort	1.9 (1901)	J	SFI	4	SOHC
	Tracer	1.8 (1844)	8	MFI	4	DOHC
	Tracer	1.9 (1901)	J	SFI	4	SOHC
1995	Escort	1.8 (1844)	8	MFI	4	DOHC
	Escort	1.9 (1901)	J	SFI	4	SOHC
	Tracer	1.8 (1844)	8	MFI	4	DOHC
	Tracer	1.9 (1901)	J	SFI	4	SOHC
1996-97	Escort	1.8 (1844)	8	MFI	4	DOHC
	Escort	1.9 (1901)	J	SFI	4	SOHC
	Tracer	1.8 (1844)	8	MFI	4	DOHC
	Tracer	1.9 (1901)	J	SFI	4	SOHC

MFI - Multiport fuel injection
SFI - Sequential fuel injection
SOHC - Single overhead camshaft
DOHC - Double overhead camshaft
① High output

GENERAL ENGINE SPECIFICATIONS

Year	Engine ID/VIN	Engine Displacement Liters (cc)	Fuel System Type	Net Horsepower @ rpm	Net Torque @ rpm (ft. lbs.)	Bore x Stroke (in.)	Compression Ratio	Oil Pressure @ rpm
1993	8	1.8 (1844)	MFI	127@6500	114@4500	3.27x3.35	9.0:1	35-65@2000
	J	1.9 (1901)	SFI	88@4400	108@4000	3.23x3.46	9.0:1	35-65@2000
1994	8	1.8 (1844)	MFI	127@6500	114@4500	3.27x3.35	9.0:1	35-65@2000
	J	1.9 (1901)	SFI	88@4400	108@4000	3.23x3.46	9.0:1	35-65@2000
1995	8	1.8 (1844)	MFI	127@6500	114@4500	3.27x3.35	9.0:1	35-65@2000
	J	1.9 (1901)	SFI	88@4400	108@4000	3.23x3.46	9.0:1	35-65@2000
1996-97	8	1.8 (1844)	MFI	127@6500	114@4500	3.27x3.35	9.0:1	28-43@1000
	J	1.9 (1901)	SFI	88@4400	108@3800	3.23x3.35	9.0:1	35-65@2000

MFI - Multiport fuel injection
SFI - Sequential fuel injection

GASOLINE ENGINE TUNE-UP SPECIFICATIONS

Year	Engine ID/VIN	Engine Displacement Liters (cc)	Spark Plugs Gap (in.)	Ignition Timing (deg.) MT	Ignition Timing (deg.) AT	Fuel Pump (psi)	Idle Speed (rpm) MT	Idle Speed (rpm) AT	Valve Clearance In.	Valve Clearance Ex.
1993	8	1.8 (1844)	0.041	10B	10B	64-85	750	750	HYD	HYD
	J	1.9 (1901)	0.054	10B	10B	17-35	950	950	HYD	HYD
1994	8	1.8 (1844)	0.041	10B	10B	31-38 ②	750	750	HYD	HYD
	J	1.9 (1901)	0.054	10B	10B	30-45 ②	780	780	HYD	HYD
1995	8	1.8 (1844)	0.041	10B	10B	31-38 ②	750	750	HYD	HYD
	J	1.9 (1901)	0.054	10B	10B	30-34 ②	780	780	HYD	HYD
1996-97	8	1.8 (1844)	0.041	10B	10B	31-38 ②	750	750	HYD	HYD
	J	1.9 (1901)	0.054	10B	10B	38-45 ②	①	①	HYD	HYD

NOTE: The Vehicle Emission Control Information label often reflects specification changes made during production. The label figures must be used if they differ from those in this chart.

B - Before top dead center

HYD - Hydraulic

① Refer to Vehicle Emission Control Information label

② Fuel pressure with engine running, pressure regulator vacuum hose connected

CAPACITIES

Year	Model	Engine ID/VIN	Engine Displacement Liters (cc)	Engine Oil with Filter (qts.)	Transmission (pts.) 4-Spd	Transmission (pts.) 5-Spd	Transmission (pts.) Auto.	Drive Axle Front (pts.)	Drive Axle Rear (pts.)	Fuel Tank (gal.)	Cooling System (qts.)
1993	Escort	8	1.8 (1844)	4.0	-	7.2	13.4	①	-	13.2	7.5
	Escort	J	1.9 (1901)	4.0	-	6.2	16.6	①	-	13.0	②
	Tracer	8	1.8 (1844)	4.0	-	6.2	13.4	①	-	13.2	6.3
	Tracer	J	1.9 (1901)	4.0	5.0	6.2	16.6	①	-	13.0	6.3
1994	Escort	8	1.8 (1844)	4.0	-	7.2	13.4 ③	①	-	13.2	7.5
	Escort	J	1.9 (1901)	4.0	-	5.7	13.4 ③	①	-	11.9	5.3
	Tracer	8	1.8 (1844)	4.0	-	7.2	13.4 ③	①	-	13.2	6.3
	Tracer	J	1.9 (1901)	4.0	-	5.7	13.4 ③	①	-	11.9	5.3
1995	Escort	8	1.8 (1844)	4.0	-	7.2	13.4 ③	①	-	13.2	④
	Escort	J	1.9 (1901)	4.0	-	5.7	13.4 ③	①	-	11.9	④
	Tracer	8	1.8 (1844)	4.0	-	7.1	13.4	①	-	13.2	④
	Tracer	J	1.9 (1901)	4.0	-	5.7	13.4	①	-	11.9	④
1996-97	Escort	8	1.8 (1944)	4.0	-	6.7	13.4 ③	①	-	13.2	6.3
	Escort	J	1.9 (1901)	4.0	-	5.7	13.4 ③	①	-	11.9	⑤
	Tracer	8	1.8 (1844)	4.0	-	6.7	13.4	①	-	13.2	6.3
	Tracer	J	1.9 (1901)	4.0	-	5.7	13.4	①	-	11.9	④

① Included in transaxle capacity

② Without AC: 8.3 qts.
Manual transaxle with AC: 7.3 qts.
Automatic transaxle with AC: 7.8 qts.

③ Includes torque converter

④ Manual transaxle: 5.3 qts.
Automatic transaxle: 6.3 qts.

⑤ Without coolant recovery reservoir: 5.8 qts
With coolant recovery reservoir: 7.9 qts.

VALVE SPECIFICATIONS

Year	Engine ID/VIN	Engine Displacement Liters (cc)	Seat Angle (deg.)	Face Angle (deg.)	Spring Test Pressure (lbs. @ in.)	Spring Installed Height (in.)	Stem-to-Guide Clearance (in.) Intake	Stem-to-Guide Clearance (in.) Exhaust	Stem Diameter (in.) Intake	Stem Diameter (in.) Exhaust
1993	8	1.8 (1844)	45	45	-	①	0.0010-0.0024	0.0012-0.0026	0.2350-0.2356	0.2348-0.2354
	J	1.9 (1901)	45	45	216@1.02	1.440-1.480	0.0008-0.0027	0.0018-0.0037	0.3160	0.3150

VALVE SPECIFICATIONS

Year	Engine ID/VIN	Engine Displacement Liters (cc)	Seat Angle (deg.)	Face Angle (deg.)	Spring Test Pressure (lbs. @ in.)	Spring Installed Height (in.)	Stem-to-Guide Clearance (in.)		Stem Diameter (in.)	
							Intake	Exhaust	Intake	Exhaust
1994	8	1.8 (1844)	45	45	-	①	0.0010-0.0024	0.0012-0.0026	0.2350-0.2356	0.2348-0.2354
	J	1.9 (1901)	45	45.6	200@1.09	1.440-1.480	0.0008-0.0027	0.0018-0.0037	0.3159-0.3167	0.3149-0.3156
1995	8	1.8 (1844)	45	45	-	①	0.0010-0.0024	0.0012-0.0026	0.2350-0.2356	0.2348-0.2354
	J	1.9 (1901)	45	45.6	200@1.09	1.440-1.480	0.0008-0.0027	0.0018-0.0037	0.3159-0.3167	0.3149-0.3156
1996-97	8	1.8 (1844)	45	45	NA	①	0.0010-0.0024	0.0012-0.0026	0.2350-0.2356	0.2348-0.2354
	J	1.9 (1901)	45	45.6	200@1.09	1.440-1.480	0.0008-0.0027	0.0018-0.0037	0.3159-0.3167	0.3149-0.3156

① Spring height measured unloaded
Minimum length: 1.821

TORQUE SPECIFICATIONS
All readings in ft. lbs.

Year	Engine ID/VIN	Engine Displacement Liters (cc)	Cylinder Head Bolts	Main Bearing Bolts	Rod Bearing Bolts	Crankshaft Damper Bolts	Flywheel Bolts	Manifold		Spark Plugs	Lug Nut
								Intake	Exhaust		
1993	8	1.8 (1844)	56-60	40-43	35-37	80-87	71-76	14-19	28-34	11-17	85
	J	1.9 (1901)	①	67-80	19-25	74-90	59-69	12-15	15-20	8-15	95
1994	8	1.8 (1844)	56-60	40-43	35-37	80-87	71-76	14-19	28-34	11-17	85
	J	1.9 (1901)	①	67-80	26-30	81-96	54-67	12-15	15-20	8-15	95
1995	8	1.8 (1844)	56-60	40-43	35-37	80-87	71-76	14-19	28-34	11-17	85
	J	1.9 (1901)	①	67-80	26-30	81-96	54-67	12-15	15-20	8-15	95
1996-97	8	1.8 (1844)	50-60	40-43	35-37	80-87	71-76	14-19	28-34	11-17	76
	J	1.9 (1901)	①	67-80	26-30	81-96	54-67	12-15	15-20	8-15	76

NOTE: Always follow proper torque patterns
NOTE: Stretch bolts are used in all procedures that require rotating the fastener a certain number of degrees. The bolts stretch and cannot be reused. For reassembly, replace with new fastners.

① Do not reuse cylinder head bolts.
   Step 1: Tighten bolts, in sequence, to 44 ft. lbs.
   Step 2: Loosen bolts approx. two turns,
       retighten in sequence to 44 ft. lbs.
   Step 3: Turn all bolts, in sequence, +90 degrees
   Step 4: Repeat Step 3

BRAKE SPECIFICATIONS
All measurements in inches unless noted

Year	Model			Master Cylinder Bore	Brake Disc Original Thickness	Brake Disc Minimum Thickness	Maximum Runout	Brake Drum Diameter Original Inside Diameter	Max. Wear Limit	Maximum Machine Diameter	Minimum Lining Thickness Front	Minimum Lining Thickness Rear
1993	Escort	①	F	0.875	0.870	0.820	0.004	7.87	7.95	7.91	0.080	0.040
		②	R	-	0.350	0.310	0.004	-	-	-	-	0.040
	Tracer		F	0.875	0.870	0.790	0.004	-	-	-	0.080	-
			R	-	0.350	0.280	0.004	7.87	7.95	7.91	-	0.040

BRAKE SPECIFICATIONS
All measurements in inches unless noted

| Year | Model | | | Master Cylinder Bore | Brake Disc | | | Brake Drum Diameter | | | Minimum Lining Thickness | |
					Original Thickness	Minimum Thickness	Maximum Runout	Original Inside Diameter	Max. Wear Limit	Maximum Machine Diameter	Front	Rear
1994	Escort	①	F	0.875	0.870	0.790	0.004	7.87	7.95	7.91	0.080	0.040
		②	R	-	0.350	0.280	0.004	-	-	-	-	0.040
	Tracer		F	0.875	0.870	0.790	0.004	-	-	-	0.080	-
			R	-	0.350	0.280	0.004	7.87	7.95	7.91	-	0.040
1995	Escort	①	F	0.875	0.870	0.790	0.004	7.87	7.95	7.91	0.080	0.040
		②	R	-	0.350	0.280	0.004	-	-	-	-	0.040
	Tracer		F	0.875	0.870	0.790	0.004	-	-	-	0.080	-
			R	-	0.350	0.280	0.004	7.87	7.95	7.91	-	0.040
1996-97	Escort		F	0.875	0.870	0.790	0.004	-	-	-	0.080	-
			R	-	0.350	0.280	0.004	7.87	7.95	7.91	-	0.040
	Tracer		F	0.875	0.870	0.790	0.004	-	-	-	0.080	-
			R	-	0.350	0.280	0.004	7.87	7.95	7.91	-	0.040

NOTE: Follow specifications stamped on rotor or drum if figures differ from those in this chart.

NA - Not Available
F - Front
R - Rear
① Except rear disc
② With rear disc

FREQUENT MAINTENANCE LABOR
FORD ESCORT, MERCURY TRACER

The following should be used as a guide when determining the amount of work required for a particular service if taken to a repair shop. In estimating how long a particular Frequent Maintenance Service item should take, please observe the following:
- **Factory Time** is time that is generated by the vehicle manufacturer.
- **Chilton Time** is time that is based on field research and data supplied by the vehicle manufacturer.
- All labor time operations are given in hours and tenths of an hour.
- All labor operations, are to be used as a **guide**.

COOLING

(G) Winterize Cooling System
Includes: Run engine to check for leaks, tighten all hose connections. Test radiator and pressure cap, drain radiator and engine block. Add antifreeze and refill system.
1993-97 5

(G) Belt, Drive, Renew
1993-97
Power Steering (.5) 7
AC (.5) 7
Serpentine (.3) 4
Alternator (.5) 7

(G) Belt, Drive, Adjust
1993-97
one (.2) 3
each adtnl. (.1) 1

(G) Hoses, Radiator, Renew
Includes: Drain and refill cooling system as required.
1993-97
upper (.4) 5
lower (.6) 7
both (.7) 1.0

(G) Thermostat, Coolant, Renew
1993-96
1.8L (.5) 7
1.9L (.8) 1.0

FUEL

(M) Air Cleaner, Service
1993-97 2

(G) Filter, Fuel, Renew
1993-97 (.3) 5

BRAKES

(G) Bleed Brakes (Four Wheels)
Includes: Add fluid.
1993-97 (.3) 5

(G) Brakes, Adjust (Minor)
Includes: Adjust brakes, fill master cylinder.
1993-97, two wheels 4

(M) Parking Brake, Adjust
1993-97 (.3) 5

LUBRICATION SERVICES

(M) Engine Oil & Filter, Renew
Includes: Inspect and correct all fluid levels.
1993-97 3

FREQUENT MAINTENANCE LABOR (cont.)
FORD ESCORT, MERCURY TRACER

	(Factory Time)	Chilton Time
(M) Lubricate Chassis, Change Oil & Filter		
Includes: Inspect and correct all fluid levels.		
1993-97		.6
Install grease fittings add		.1
(M) Lubricate Chassis		
Includes: Inspect and correct all fluid levels.		
1993-97		.4
Install grease fittings add		.1

WHEELS

(G) Wheel, Renew (One)
1993-97 5

(G) Wheel, Balance
1993-97
  one 3
  each adtnl. 2
(G) Wheels, Rotate (All)
1993-97 5

ELECTRICAL

(G) Headlamps, Aim
1993-97
  two 4
  four 6

(G) Halogen Headlamp Bulb, Renew
1993-97, one (.3) 3

(G) License Lamp Assy., Renew
1993-97 (.3) 4

(G) Horn, Renew
1993-97 (.5) 7

(M) Terminals, Battery, Clean
1993-97 3

SCHEDULED MAINTENANCE INTERVALS
(FORD ESCORT, MERCURY TRACER (1993))

TO BE SERVICED	TYPE OF SERVICE	VEHICLE MILEAGE INTERVAL (x1000)												
		7.5	15	22.5	30	37.5	45	52.5	60	67.5	75	82.5	90	97.5
Engine oil & filter	R	✓	✓	✓	✓	✓	✓	✓	✓	✓	✓	✓	✓	✓
Rotate tires	S/I	✓		✓		✓		✓		✓		✓		✓
Spark plugs	R				✓				✓				✓	
Engine coolant	R				✓				✓				✓	
Air cleaner filter	R				✓				✓				✓	
Crankcase ventilation filter (1.9L)	R				✓				✓				✓	
Disc brake pads & rotors	S/I				✓				✓				✓	
Exhaust heat shields	S/I				✓				✓				✓	
Brake lines & connections/brake linings & drums (rear)	S/I				✓				✓				✓	
Accessory drive belt(s)	S/I				✓				✓				✓	
Bolts & nuts on chassis & body	S/I				✓				✓				✓	

SCHEDULED MAINTENANCE INTERVALS
(FORD ESCORT, MERCURY TRACER (1993)) (Cont.)

TO BE SERVICED	TYPE OF SERVICE	VEHICLE MILEAGE INTERVAL (x1000)												
		7.5	15	22.5	30	37.5	45	52.5	60	67.5	75	82.5	90	97.5
Clutch pedal operation	S/I				✓				✓				✓	
Steering operation & linkage & front suspension ball joints	S/I				✓				✓				✓	
PCV valve	R								✓					
Engine timing belt (1.8L)	R								✓					
Engine coolant protection, hoses & clamps	S/I								✓					
Halfshaft dust boots	S/I								✓					

R – Replace   S/I – Service or Inspect

FREQUENT OPERATION MAINTENANCE (SEVERE SERVICE)

If a vehicle is operated under any of the following conditions it is considered severe service:
- Extremely dusty areas.
- 50% or more of the vehicle operation is in 32°C (90°F) or higher temperatures, or constant operation in temperatures below 0°C (32°F).
- Prolonged idling (vehicle operation in stop and go traffic).
- Frequent short running periods (engine does not warm to normal operating temperatures).
- Police, taxi, delivery usage or trailer towing usage.

Oil & oil filter – change every 3000 miles.
Air cleaner filter (1.8L) - service or inspect every 15,000 miles.
Brake linings & drums (rear) - check every 15,000 miles.
Disc brake pads & rotors - check every 15,000 miles.
Rotate tires at 6000 miles & every 15,000 miles thereafter.
Automatic transaxle fluid & filter - change every 30,000 miles.

SCHEDULED MAINTENANCE INTERVALS
(FORD ESCORT, MERCURY TRACER (1994-97))

TO BE SERVICED	TYPE OF SERVICE	VEHICLE MILEAGE INTERVAL (x1000)												
		5	10	15	20	25	30	35	40	45	50	55	60	65
Engine oil & filter	R	✓	✓	✓	✓	✓	✓	✓	✓	✓	✓	✓	✓	✓
Rotate tires	S/I	✓		✓		✓		✓		✓		✓		✓
Cooling system, hoses, clamps & coolant strength (1.9L)	S/I			✓			✓			✓			✓	
Air cleaner element, automatic transmission fluid & filter	R						✓						✓	

SCHEDULED MAINTENANCE INTERVALS
(FORD ESCORT, MERCURY TRACER (1994-97)) (Cont.)

TO BE SERVICED	TYPE OF SERVICE	5	10	15	20	25	30	35	40	45	50	55	60	65
Crankcase emission filter (1.9L)	R						✓						✓	
Engine coolant (1.8L)	R						✓						✓	
Engine coolant (1.9L 1994)	R						✓						✓	
Engine coolant (1.9L 1995-97) ①	R										✓			
Accessory drive belt(s) (1.8L)	S/I						✓						✓	
Bolts & nuts on chassis & body	S/I						✓						✓	
Idle speed (1.8L)	S/I						✓						✓	
Clutch pedal operation, brake lines, hoses & connections	S/I						✓						✓	
Cooling system, hoses, clamps & coolant strength (1.8L)	S/I						✓						✓	
Exhaust heat shields, front & rear brakes	S/I						✓						✓	
Front suspension lower arm ball joints, steering operation & linkage, & front wheel driveshaft joint boots	S/I						✓						✓	
Fuel lines & hoses (1.8L)	S/I						✓						✓	
Fuel filter (1.8L)	R												✓	
PCV valve (1.9L)	R												✓	
Spark plugs (1.8L)	R						✓						✓	

SCHEDULED MAINTENANCE INTERVALS
(FORD ESCORT, MERCURY TRACER (1994-97)) (Cont.)

TO BE SERVICED	TYPE OF SERVICE	VEHICLE MILEAGE INTERVAL (x1000)												
		5	10	15	20	25	30	35	40	45	50	55	60	65
Spark plugs (1.9L)	R												✓	
Timing belt (1.8L)	R												✓	
Accessory drive belt(s) (1.9L)	S/I												✓	
Evaporative emission hose for emissions (1.8L)	S/I												✓	

① Change initially at 50,000 miles & every 30,000 miles thereafter.

R – Replace   S/I – Service or Inspect

FREQUENT OPERATION MAINTENANCE (SEVERE SERVICE)

If a vehicle is operated under any of the following conditions it is considered severe service:
- Extremely dusty areas.
- 50% or more of the vehicle operation is in 32°C (90°F) or higher temperatures, or constant operation in temperatures below 0°C (32°F).
- Prolonged idling (vehicle operation in stop and go traffic).
- Frequent short running periods (engine does not warm to normal operating temperatures).
- Police, taxi, delivery usage or trailer towing usage.

Oil & oil filter – change every 3000 miles.
Air cleaner element - check every 15,000 miles.
Front & rear brakes - check every 15,000 miles.
Nuts & bolts on chassis & body - check every 15,000 miles.
Automatic transaxle fluid & filter - change every 21,000 miles.

MUSTANG/THUNDERBIRD/MARK VIII/COUGAR

VEHICLE IDENTIFICATION CHART

Engine Code						Model Year	
Code	Liters	Cu. In. (cc)	Cyl.	Fuel Sys.	Eng. Mfg.	Code	Year
M	2.3	140 (2300)	4	MFI	Ford	P	1993
4	3.8	232 (3802)	6	MFI	Ford	R	1994
R ①	3.8	232 (3802)	6	MFI	Ford	S	1995
V	4.6	281 (4593)	8	SFI	Ford	T	1996
W	4.6	281 (4593)	8	SFI	Ford	V	1997
D ③	5.0	302 (4949)	8	SFI	Ford		
E ②	5.0	302 (4949)	8	MFI	Ford		
T ②	5.0	302 (4949)	8	SFI	Ford		

MFI - Multiport fuel injection   ① Supercharged   ③ Special high perfomance
SFI - Sequential fuel injection   ② High output

ENGINE IDENTIFICATION

All measurements are given in inches.

Year	Model		Engine Displacement Liters (cc)	Engine Series (ID/VIN)	Fuel System	No. of Cylinders	Engine Type
1993	Cougar		3.8 (3802)	4	MFI	6	OHV
	Cougar		5.0 (4949)	T	MFI	8	OHV
	Mark VIII		4.6 (4593)	V	MFI	8	DOHC
	Mustang		2.3 (2300)	M	MFI	4	SOHC
	Mustang	①	5.0 (4949)	E	MFI	8	OHV
	Thunderbird		3.8 (3802)	4	MFI	6	OHV
	Thunderbird	②	3.8 (3802)	R	MFI	6	OHV
	Thunderbird		5.0 (4949)	E	MFI	8	OHV
1994	Cougar		3.8 (3802)	4	SFI	6	OHV
	Cougar		4.6 (4593)	W	SFI	8	SOHC
	Mark VIII		4.6 (4593)	V	SFI	8	DOHC
	Mustang		3.8 (3802)	4	SFI	6	OHV
	Mustang	③	5.0 (4949)	D	SFI	8	OHV
	Mustang	①	5.0 (4949)	T	SFI	8	OHV
	Thunderbird		3.8 (3802)	4	SFI	6	OHV
	Thunderbird	②	3.8 (3802)	R	SFI	6	OHV
	Thunderbird		4.6 (4593)	W	SFI	8	SOHC
1995	Cougar		3.8 (3802)	4	SFI	6	OHV
	Cougar		4.5 (4593)	W	SFI	8	SOHC
	Mark VIII		4.6 (4593)	V	SFI	8	DOHC
	Mustang		3.8 (3802)	4	SFI	6	OHV
	Mustang	③	5.0 (4949)	D	SFI	8	OHV
	Mustang	①	5.0 (4949)	T	SFI	8	OHV
	Thunderbird		3.8 (3802)	4	SFI	6	OHV
	Thunderbird	②	3.8 (3802)	R	SFI	6	OHV
	Thunderbird		4.6 (4593)	W	SFI	8	SOHC
1996-97	Cougar		3.8 (3802)	4	SFI	6	OHV
	Cougar		4.6 (4593)	W	SFI	8	SOHC
	Mark VIII		4.6 (4593)	V	SFI	8	DOHC
	Mustang		3.8 (3802)	4	SFI	6	OHV
	Mustang		4.6 (4593)	W	SFI	8	SOHC
	Mustang		4.6 (4593)	V	SFI	8	DOHC
	Thunderbird		3.8 (3802)	4	SFI	6	OHV
	Thunderbird		4.6 (4593)	W	SFI	8	SOHC

MFI - Multiport fuel injection
SFI - Sequential fuel injection
SOHC - Single overhead camshaft
DOHC - Double overhead camshaft
OHV - Overhead valve

① High output
② Supercharged
③ Special high performance

GENERAL ENGINE SPECIFICATIONS

Year	Engine ID/VIN		Engine Displacement Liters (cc)	Fuel System Type	Net Horsepower @ rpm	Net Torque @ rpm (ft. lbs.)	Bore x Stroke (in.)	Compression Ratio	Oil Pressure @ rpm
1993	M		2.3 (2300)	MFI	105@4600	135@2600	3.78x3.12	9.5:1	40-60@2000
	4		3.8 (3802)	MFI	140@3800	215@2400	3.81x3.39	8.2:1	40-60@2000
	R	②	3.8 (3802)	MFI	210@2000	315@2600	3.81x3.39	8.2:1	40-60@2000
	V		4.6 (4593)	MFI	280@5500	285@4500	3.55x3.54	9.8:1	33@2000
	E	①	5.0 (4949)	MFI	235@4600	300@3200	4.00x3.39	9.0:1	40-60@2000
	T	①	5.0 (4949)	MFI	200@4000	275@3000	4.00x3.00	9.0:1	40-60@2000

GENERAL ENGINE SPECIFICATIONS

Year	Engine ID/VIN		Engine Displacement Liters (cc)	Fuel System Type	Net Horsepower @ rpm	Net Torque @ rpm (ft. lbs.)	Bore x Stroke (in.)	Compression Ratio	Oil Pressure @ rpm
1994	4		3.8 (3802)	SFI	140@3800	215@2400	3.81x3.39	8.2:1	40-60@2500
	R	②	3.8 (3802)	SFI	210@2000	315@2600	3.81x3.39	8.2:1	40-60@2500
	W		4.6 (4593)	SFI	③	④	3.55x3.54	9.0:1	20-45@2000
	V		4.6 (4593)	SFI	280@5500	285@4500	3.55x3.54	9.8:1	33@1500
	D	⑤	5.0 (4949)	SFI	225@4200	315@2600	4.00x3.00	9.0:1	40-60@2000
	T	①	5.0 (4949)	SFI	200@4000	275@3000	4.00x3.00	9.0:1	40-60@2000
1995	4		3.8 (3802)	SFI	140@3800	215@2400	3.81x3.39	9.0:1	40-60@2500
	R	②	3.8 (3802)	SFI	210@2000	315@2600	3.81x3.39	8.2:1	40-60@2500
	W		4.6 (4593)	SFI	⑥	⑦	3.55x3.54	9.0:1	20-45@2000
	V		4.6 (4593)	SFI	280@4500	285@4500	3.55x3.54	9.8:1	33@1500
	D	⑤	5.0 (4949)	SFI	225@4200	315@2600	4.00x3.00	9.0:1	40-60@2000
	T	①	5.0 (4949)	SFI	200@4000	275@3000	4.00x3.00	9.0:1	40-60@2000
1996-97	4		3.8 (3802)	SFI	145@4000	215@2750	3.81x3.39	9.0:1	40-60@2500
	V		4.6 (4593)	SFI	⑪	⑫	3.55x3.54	9.5:1	20-45@1500
	W		4.6 (4593)	SFI	⑧	⑨	3.55x3.54	⑩	20-45@1500

MFI - Multiport fuel injection
SFI - Sequential fuel injection
① High output
② Supercharged
③ Single exhaust: 190@4200
 Dual exhaust: 210@4600
④ Single exhaust: 260@3200
 Dual exhaust: 270@3400
⑤ Special high performance

⑥ Single exhaust: 190@4250
 Dual exhaust:
 Thunderbird: 205@4500
⑦ Single exhaust: 260@3250
 Dual exhaust:
 Thunderbird: 265@3200
⑧ Thunderbird: 205@4250
 Mustang: 215@4400
⑨ Thunderbird: 280@3000
 Mustang: 285@3500

⑩ Base engine: 9.0:1
⑪ Mark VIII without LSC package: 280@5500
 Mark VIII with LSC package: 290@5750
 Mustang: 305@5800
⑫ Mark VIII without LSC package: 285@4500
 Mark VIII with LSC package: 292@4500
 Mustang: 300@4800

GASOLINE ENGINE TUNE-UP SPECIFICATIONS

Year	Engine ID/VIN	Engine Displacement Liters (cc)	Spark Plugs Gap (in.)	Ignition Timing (deg.) MT	Ignition Timing (deg.) AT	Fuel Pump (psi)	Idle Speed (rpm) MT	Idle Speed (rpm) AT	Valve Clearance In.	Valve Clearance Ex.
1993	M	2.3 (2300)	0.044	10B	10B	35-40	975	975	HYD	HYD
	4	3.8 (3802)	0.054	10B	10B	35-45	-	550	HYD	HYD
	R	3.8 (3802)	0.054	10B	10B	35-45	-	550	HYD	HYD
	V	4.6 (4593)	0.054	-	10B	35-40	-	①	HYD	HYD
	E	5.0 (4949)	0.054	10B	10B	36-42	700	700	HYD	HYD
	T	5.0 (4949)	0.054	10B	10B	30-45 ②	①	①	HYD	HYD
1994	4	3.8 (3802)	0.054	10B	10B	30-45 ②	①	①	HYD	HYD
	R	3.8 (3802)	0.054	10B	10B	30-45 ②	①	①	HYD	HYD
	W	4.6 (4593)	0.054	-	10B	30-45 ②	①	①	HYD	HYD
	V	4.6 (4593)	0.054	-	10B	30-45 ②	-	①	HYD	HYD
	D	5.0 (4949)	0.054	10B	10B	30-45 ②	①	①	HYD	HYD
	T	5.0 (4949)	0.054	10B	10B	30-45 ②	①	①	HYD	HYD
1995	4	3.8 (3802)	0.054	-	10B	30-45 ②	-	①	HYD	HYD
	R	3.8 (3802)	0.054	10B	10B	30-40 ②	①	①	HYD	HYD
	W	4.6 (4593)	0.054	-	10B	30-45 ②	-	①	HYD	HYD
	V	4.6 (4593)	0.054	-	10B	30-45 ②	-	①	HYD	HYD
	D	5.0 (4949)	0.054	10B	10B	30-45 ②	①	①	HYD	HYD
	T	5.0 (4949)	0.054	10B	10B	30-45 ②	①	①	HYD	HYD

GASOLINE ENGINE TUNE-UP SPECIFICATIONS

Year	Engine ID/VIN	Engine Displacement Liters (cc)	Spark Plugs Gap (in.)	Ignition Timing (deg.) MT	Ignition Timing (deg.) AT	Fuel Pump (psi)	Idle Speed (rpm) MT	Idle Speed (rpm) AT	Valve Clearance In.	Valve Clearance Ex.
1996-97	4	3.8 (3802)	0.054	①	①	28-54 ②	①	①	HYD	HYD
	V	4.6 (4593)	0.054	10B	-	35-45 ②	①	-	HYD	HYD
	W	4.6 (4593)	0.054	10B	10B	35-45 ②	①	①	HYD	HYD

NOTE: The Vehicle Emission Control Information label often reflects specification changes made during production. The label figures must be used if they differ from those in this chart.

B - Before top dead center

HYD - Hydraulic

① Refer to Vehicle Emission Control Information label

② Fuel pressure with engine running, pressure regulator vacuum hose connected

CAPACITIES

Year	Model	Engine ID/VIN	Engine Displacement Liters (cc)	Engine Oil with Filter (qts.)	Transmission (pts.) 4-Spd	Transmission (pts.) 5-Spd	Transmission (pts.) Auto.	Drive Axle Front (pts.)	Drive Axle Rear (pts.)	Fuel Tank (gal.)	Cooling System (qts.)
1993	Cougar	4	3.8 (3802)	5.0	-	-	22.0	-	①	21.0	11.8
	Cougar	T	5.0 (4943)	5.0	-	-	24.6	-	①	19.0	14.1
	Mark VIII	V	4.6 (4593)	6.0	-	-	24.6	-	3.00	18.0	14.1
	Mustang	M	2.3 (2300)	5.0	-	5.6	19.4	-	①	15.4	10.0
	Mustang	E	5.0 (4949)	5.0	-	5.6	24.6	-	①	15.4	14.1
	Thunderbird	4	3.8 (3802)	5.0	-	-	22.0	-	①	21.0	11.8
	Thunderbird	R	3.8 (3802) ②	5.0	-	6.3	24.0	-	①	18.8	11.8
	Thunderbird	E	5.0 (4943)	5.0	-	-	24.6	-	①	19.0	14.1
1994	Cougar	4	3.8 (3802)	5.0	-	-	25.0 ④	-	①	18.0	12.6
	Cougar	W	4.6 (4593)	5.0	-	-	25.0 ④	-	①	18.0	14.1
	Mark VIII	V	4.6 (4593)	6.0	-	-	25.0 ④	-	3.00	18.0	16.0
	Mustang	4	3.8 (3802)	5.0	-	5.6	25.0 ④	-	③	15.4	11.8
	Mustang	T	5.0 (4949)	5.0	-	5.6	25.0 ④	-	③	15.4	14.1
	Mustang	D	5.0 (4949)	5.0	-	5.6	24.6 ④	-	③	15.4	14.1
	Thunderbird	4	3.8 (3802)	5.0	-	-	25.0 ④	-	①	18.0	12.6
	Thunderbird	R	3.8 (3802) ②	5.0	-	6.3	25.0 ④	-	①	18.0	12.5
	Thunderbird	W	4.6 (4593)	5.0	-	-	25.0 ④	-	①	18.0	14.1
1995	Cougar	4	3.8 (3802)	5.0	-	-	27.2 ④	-	①	18.0	12.6
	Cougar	W	4.6 (4593)	5.0	-	-	27.2 ④	-	①	18.0	14.1
	Mark VIII	V	4.6 (4593)	6.0	-	-	25.0 ④	-	3.00	18.0	16.0
	Mustang	4	3.8 (3802)	5.0	-	5.6	27.2 ④	-	③	15.4	11.8
	Mustang	T	5.0 (4949)	5.0	-	5.6	27.2 ④	-	③	15.4	14.1
	Mustang	D	5.0 (4949)	5.0	-	5.6	27.2 ④	-	③	15.4	14.1
	Thunderbird	4	3.8 (3802)	5.0	-	-	27.2 ④	-	①	18.0	12.6
	Thunderbird	R	3.8 (3802) ②	5.0	-	6.3	27.2 ④	-	①	18.0	12.5
	Thunderbird	W	4.6 (4593)	5.0	-	-	27.2 ④	-	①	18.0	14.1
1996-97	Cougar	4	3.8 (3802)	5.0	-	-	27.8 ④	-	①	18.0	12.6
	Cougar	W	4.6 (4593)	5.3	-	-	27.8 ④	-	①	18.0	14.1
	Mark VIII	V	4.6 (4593)	6.0	-	-	25.6 ④	-	3.00	18.0	16.0
	Mustang	4	3.8 (3802)	5.0	-	5.6	27.8 ④	-	3.50	15.4	11.8
	Mustang	V	4.6 (4593)	6.0	-	6.5	27.8 ④	-	3.75	15.4	14.1
	Mustang	W	4.6 (4593)	⑤	-	6.5	25.6 ④	-	3.75	15.4	14.1
	Thunderbird	4	3.8 (3802)	5.0	-	-	27.8 ④	-	⑥	18.0	12.6
	Thunderbird	W	4.6 (4593)	5.3	-	-	27.8 ④	-	⑥	18.0	14.1

① 7.50" limited slip axle: 2.75 pts.
7.50" axle: 3.0 pts.
8.80" axle: 3.25 pts.

② Supercharged

③ 7.50" axle: 3.5 pts.
8.80" axle: 3.75 pts.

④ Includes torque converter

⑤ Automatic transmission: 6.7 qts.
Manual transmission: 6.4 qts.

⑥ 7.50" axle: 3.0 pts.
8.80" axle: 3.25 pts.

VALVE SPECIFICATIONS

Year	Engine ID/VIN	Engine Displacement Liters (cc)	Seat Angle (deg.)	Face Angle (deg.)	Spring Test Pressure (lbs. @ in.)	Spring Installed Height (in.)	Stem-to-Guide Clearance (in.) Intake	Stem-to-Guide Clearance (in.) Exhaust	Stem Diameter (in.) Intake	Stem Diameter (in.) Exhaust
1993	M	2.3 (2300)	45	44	128-141@1.12	1.520	0.0010-0.0027	0.0015-0.0032	0.3416-0.3423	0.3411-0.3418
	4	3.8 (3802)	44.5	45.8	215@1.79	1.750	0.0010-0.0027	0.0015-0.0032	0.3420	0.3415
	R	3.8 (3802)	44.5	45.8	220@1.18	1.650	0.0010-0.0027	0.0015-0.0032	0.3420	0.3415
	V	4.6 (4593)	45	45.5	180@1.103	1.575	0.0008-0.0027	0.0018-0.0037	0.3415-0.3423	0.3410-0.3418
	E	5.0 (4949)	45	44	①	②	0.0010-0.0027	0.0015-0.0032	0.3420	0.3420
	T	5.0 (4949)	45	44	①	②	0.0010-0.0027	0.0015-0.0032	0.3416-0.3423	0.3411-0.3418
1994	4	3.8 (3802)	44.5	45.8	220@1.18	1.970	0.0010-0.0027	0.0015-0.0032	0.3415-0.3423	0.3410-0.3418
	R	3.8 (3802)	44.5	45.8	220@1.18	1.970	0.0010-0.0027	0.0015-0.0032	0.3415-0.3423	0.3410-0.3418
	W	4.6 (4593)	45	45.5	132@1.10	1.570	0.0008-0.0027	0.0018-0.0037	0.2746-0.2754	0.2736-0.2744
	V	4.6 (4593)	45	45.5	180@1.103	1.425	0.0008-0.0027	0.0018-0.0037	0.3415-0.3423	0.3410-0.3418
	T	5.0 (4949)	45	44	①	②	0.0010-0.0027	0.0015-0.0032	0.3416-0.3423	0.3411-0.3418
	D	5.0 (4949) ③	45	44	④	⑤	0.0010-0.0027	0.0015-0.0032	0.3416-0.3423	0.3411-0.3418
1995	4	3.8 (3802)	44.5	45.8	220@1.18	1.970	0.0010-0.0027	0.0015-0.0032	0.3415-0.3423	0.3410-0.3418
	R	3.8 (3802)	44.5	45.8	220@1.18	1.970	0.0010-0.0027	0.0015-0.0032	0.3415-0.3423	0.3410-0.3418
	W	4.6 (4593)	45	45.5	132@1.10	1.570	0.0008-0.0027	0.0018-0.0037	0.2746-0.2754	0.2736-0.2744
	V	4.6 (4593)	45	45.5	160@1.103	1.425	0.0008-0.0027	0.0018-0.0037	0.2746-0.2754	0.2736-0.2744
	T	5.0 (4949)	45	44	①	②	0.0010-0.0027	0.0015-0.0032	0.3416-0.3423	0.3411-0.3418
	D	5.0 (4949) ③	45	44	④	⑤	0.0010-0.0027	0.0015-0.0032	0.3416-0.3423	0.3411-0.3418
1996-97	4	3.8 (3802)	44.5	45.8	220@1.18	1.650	0.0010-0.0027	0.0015-0.0032	0.3415-0.3423	0.3410-0.3418
	V	4.6 (4593)	45	45.5	160@1.10	1.425	0.0008-0.0027	0.0018-0.0037	0.2746-0.2754	0.2736-0.2744
	W	4.6 (4593)	45	45.5	132@1.10	1.570	0.0008-0.0027	0.0018-0.0037	0.2746-0.2754	0.2736-0.2744

① Intake: 211-230@1.33
Exhaust: 200-226@1.15

② Intake: 1.75-1.80
Exhaust: 1.58-1.64

③ Cobra

④ Intake: 280@1.30
Exhaust: 264@1.12

⑤ Intake: 1.80
Exhaust: 1.62

TORQUE SPECIFICATIONS
All readings in ft. lbs.

Year	Engine ID/VIN	Engine Displacement Liters (cc)	Cylinder Head Bolts	Main Bearing Bolts	Rod Bearing Bolts	Crankshaft Damper Bolts	Flywheel Bolts	Manifold Intake	Manifold Exhaust	Spark Plugs	Lug Nut
1993	M	2.3 (2300)	⑤	⑥	⑦	114-151	56-64	19-28	①	5-10	95
	R	3.8 (3801)	④	65-81	31-36	103-132	54-64	⑭	15-22	5-11	95
	4	3.8 (3802)	②	65-81	31-36	85-100	75-85	⑮	15-22	5-11	95
	V	4.6 (4593)	⑧	㉔	㉑	114-121	54-64	㉓	15-22	7-15	95
	E	5.0 (4949)	65-72	60-70	19-24	70-90	75-85	23-25	18-24	10-15	95
	T	5.0 (4949)	65-72	60-70	19-24	70-90	75-85	23-25	18-24	10-15	95
1994	4	3.8 (3802)	⑱	65-81	31-36	103-132	54-64	⑰	15-22	7-15	95
	R	3.8 (3802)	⑲	65-81	31-36	103-132	54-64	⑯	15-22	7-15	95
	W	4.6 (4593)	⑧	⑨	⑩	114-121	54-64	15-22	15-22	7-15	95
	V	4.6 (4593)	⑧	㉔	③	114-121	54-64	⑬	13-16	7-15	95
	D	5.0 (4949)	⑫	60-70	19-24	110-130	75-85	⑪	26-32	10-15	95
	T	5.0 (4949)	⑫	60-70	19-24	110-130	75-85	⑪	26-32	10-15	95
1995	4	3.8 (3802)	⑱	65-81	31-36	103-132	54-64	⑰	15-22	7-15	95
	R	3.8 (3802)	⑲	65-81	31-36	103-132	54-64	⑯	15-22	7-15	95
	W	4.6 (4593)	⑧	⑨	⑩	114-121	54-64	15-22	15-22	7-15	95
	V	4.6 (4593)	⑧	㉔	③	114-121	54-64	⑬	13-16	7-15	95
	T	5.0 (4949)	⑫	60-70	19-24	110-130	75-85	⑪	26-32	10-15	95
	D	5.0 (4949)	⑫	60-70	19-24	110-130	75-85	⑪	26-32	10-15	95
1996-97	4	3.8 (3802)	⑱	65-81	31-36	103-132	54-64	⑰	15-22	7-15	95
	V	4.6 (4593)	⑧	㉔	③	114-121	54-64	⑬	13-16	7-15	95
	W	4.6 (4593)	⑧	⑨	⑩	114-121	54-64	15-22	15-22	7-15	95

① Step 1: 5-7 ft. lbs.
Step 2: 20-30 ft. lbs.

② Step 1: 37 ft. lbs.
Step 2: 45 ft. lbs.
Step 3: 52 ft. lbs.
Step 4: 59 ft. lbs.

③ Step 1: 5 ft. lbs.
Step 2: 10 ft. lbs.
Step 3: 18-25 ft. lbs.
Step 4: Rotate 85-95 degrees

④ Step 1: 37 ft. lbs.
Step 2: 45 ft. lbs.
Step 3: 52 ft. lbs.
Step 4: 59 ft. lbs.
Step 5: Back off all bolts two to three turns
Step 6: Tighten to 48-55 ft. lbs.
Step 7: Rotate bolts an additional 90-110 degrees

⑤ Step 1: 50-60 ft. lbs.
Step 2: 80-90 ft. lbs.

⑥ Step 1: 50-60 ft. lbs.
Step 2: 75-85 ft. lbs.

⑦ Step 1: 25-30 ft. lbs.
Step 2: 30-36 ft. lbs.

⑧ Do not reuse cylinder head bolts
Step 1: 27-32 ft. lbs.
Step 2: Rotate each bolt 85-95 degrees
Step 3: Repeat Step 2

⑨ Do not reuse main cap bolts
Step 1: Main bearing cap bolts: 22-25 ft. lbs.
Step 2: Rotate each bolt 85-95 degrees
Step 3: Main bearing cap adjust screws: 4 ft. lbs. then 6-8 ft. lbs
Step 4: Main bearing cap side bolts: 7 ft. lbs. then 14-17 ft. lbs.

⑩ Do not reuse rod bolts
Step 1: 12 ft. lbs.
Step 2: Rotate 85-95 degrees

⑪ Step 1: 8 ft. lbs.
Step 2: 16 ft. lbs.
Step 3: 23-25 ft. lbs.

⑫ Do not reuse cylinder head bolts
Step 1: 22-35 ft. lbs.
Step 2: 44-55 ft. lbs.
Step 3: Rotate 85-95 degrees

⑬ Step 1: Four inside short bolts: 9-11 ft. lbs.
Step 2: All other bolts: 13-16 ft. lbs.
Step 3: Rotate 85-95 degrees

⑭ Supercharger to lower intake manifold bolts
M8 x 43mm bolts: 20-28 ft.lbs
M8 x 108mm bolts: 15-22 ft. lbs.
M12 bolt: 52-70 ft. lbs.
Lower intake manifold bolts
Step 1: 8 ft. lbs.
Step 2: 11 ft. lbs.

⑮ Upper intake manifold bolts
Step 1: 8 ft. lbs.
Step 2: 15 ft. lbs.
Step 3: 24 ft. lbs
Lower intake manifold bolts
Step 1: 8 ft. lbs.
Step 2: 11 ft. lbs.

⑯ Supercharger to lower intake manifold bolts
M8 x 43mm bolts: 20-28 ft.lbs
M8 x 108mm bolts: 15-22 ft. lbs.
M12 bolt: 52-70 ft. lbs.
Lower intake manifold bolts
Step 1: 13 ft. lbs.
Step 2: 16 ft. lbs.

⑰ Upper intake manifold bolts
Step 1: 8 ft. lbs.
Step 2: 15 ft. lbs.
Step 3: 24 ft. lbs
Lower intake manifold bolts
Step 1: 13 ft. lbs.
Step 2: 16 ft. lbs.

⑱ Do not reuse cylinder head bolts
Step 1: 15 ft. lbs.
Step 2: 29 ft. lbs.
Step 3: 37 ft. lbs.
Step 4: Loosen bolts one at a time and retorque as follows:
Long bolts: 11-18 ft. lbs.
Short bolts: 7-15 ft. lbs.
Step 5: Rotate 85-95 degrees

⑲ Do not reuse cylinder head bolts
Step 1: 37 ft. lbs.
Step 2: 45 ft. lbs.
Step 3: 52 ft. lbs.
Step 4: 59 ft. lbs.
Step 5: Back off bolts two to three turns, one at a time
Step 6: Tighten all bolts 37-44 ft. lbs.
Step 7: Rotate bolts an additional 180-200 degrees

⑳ Not Used

㉑ Step 1: 18-25 ft. lbs.
Step 2: Plus 85-95 degrees

㉒ Not Used

㉓ Tighten all bolts in numerical sequence
Tighten bolts 5, 7, 9, 11: 9-11 ft. lbs. plus 85-95 degrees
Tighten all other bolts to 13-16 ft. lbs.
Then tighten all bolts an additional 85-95 degrees

㉔ Step 1: Main bearing cap bolts: 6-9 ft. lbs.
Step 2: Main bearing cap bolts, outer: 16-21 ft. lbs.
Step 3: Main bearing cap bolts, inner: 27-32 ft. lbs.
Step 4: Rotate main bearing cap bolts 85-95 degrees
Step 5: Main cap adjusting screws: 4 ft. lbs. then 7.5 ft. lbs.
Step 6: Main cap side bolts: 7 ft. lbs. then 14-17 ft. lbs.

BRAKE SPECIFICATIONS
All measurements in inches unless noted

Year	Model		F/R	Master Cylinder Bore	Brake Disc Original Thickness	Brake Disc Minimum Thickness	Maximum Runout	Brake Drum Diameter Original Inside Diameter	Max. Wear Limit	Maximum Machine Diameter	Minimum Lining Thickness Front	Rear
1993	Cougar	①	F	0.938 ⑤	1.025	0.974	0.003	-	-	-	0.125	0.030
		②	R	-	0.710	0.657	-	9.80	9.89	9.86	0.123	-
	Mark VIII		F	1.000	1.024	0.974	0.003	-	-	-	0.040	-
			R	-	0.709	0.657	0.002	-	-	-	-	0.123
	Mustang	③	F	0.875	NA	0.810	0.003	9.00	9.89	9.06	0.125	0.030
		④	R	0.875	NA	0.972	0.003	9.00	9.89	9.06	0.125	0.030
	Thunderbird	①	F	0.983 ⑤	1.025	0.974	0.003	9.84	9.89	9.86	0.125	0.030
		②	R	-	0.945	0.896	0.003	-	-	-	-	0.123
1994	Cougar	①	F	0.938 ⑤	1.025	0.974	0.003	9.84	9.89	9.86	0.040	0.030
		②	R	-	0.710	0.657	0.003	-	-	-	-	0.123
	Mark VII		F	1.000	1.024	0.974	0.003	-	-	-	0.040	-
			R	-	0.709	0.657	0.002	-	-	-	-	0.123
	Mustang		F	1.060	1.030	0.970	0.002	-	-	-	0.040	-
			R	-	0.550	0.500	0.002	-	-	-	-	0.123
	Mustang	⑥	F	1.000	1.100	1.040	0.002	-	-	-	0.040	-
			R	-	0.550	0.500	0.002	-	-	-	-	0.123
	Thunderbird	①	F	0.983 ⑤	1.025	0.974	0.003	9.84	9.89	9.86	0.040	0.030
		②	R	-	0.709	0.657	0.003	-	-	-	-	0.123
1995	Cougar		F	0.938 ⑦	1.025	0.974	0.003	9.84	9.89	9.86	0.040	0.030
			R	-	0.709	0.657	0.003	-	-	-	-	0.123
	Mark VIII		F	NA	1.024	0.974	0.002	-	-	-	0.040	-
			R	-	0.709	0.657	0.002	-	-	-	-	0.123
	Mustang		F	1.060	1.030	0.970	0.002	-	-	-	0.040	-
		⑥	F	1.000	1.100	1.040	0.002	-	-	-	0.040	-
			R	-	0.550	0.500	0.002	-	-	-	-	0.123
	Thunderbird	①	F	0.938 ⑦	1.025	0.974	0.003	9.84	9.89	9.86	0.040	0.030
		②	R	-	0.709	0.657	0.003	-	-	-	-	0.123
1996-97	Cougar		F	0.938	1.025	0.974	0.002	-	-	-	0.125	-
			R	-	0.710	0.657	-	9.80	NA	9.90	-	0.125
	Mark VIII		F	1.000	1.024	0.974	0.003	-	-	-	0.125	-
			R	-	0.709	0.657	0.002	-	-	-	-	0.125
	Mustang	⑧	F	1.060	1.030	0.970	0.001	-	-	-	0.125	-
		⑨	F	1.000	1.030	0.970	0.002	-	-	-	0.125	-
		⑥	F	1.000	1.100	1.040	0.001	-	-	-	0.125	-
			R	-	0.550	0.500	0.002	-	-	-	-	0.123
		⑥	R	-	0.710	0.660	0.002	-	-	-	-	0.123
	Thunderbird		F	0.938	1.025	0.974	0.002	-	-	-	0.125	-
			R	-	0.710	0.657	-	9.80	NA	9.90	-	0.125

NOTE: Follow specifications stamped on rotor or drum if figures differ from those in this chart.

NA - Not Available

F - Front
R - Rear

① Except rear disc
② With rear disc
③ Except 5.0L
④ 5.0L
⑤ Except ABS
⑥ Cobra
⑦ Except ABS
With drum brakes: 0.030
⑧ 3.8L engine
⑨ 4.6L except Cobra

FREQUENT MAINTENANCE LABOR
FORD THUNDERBIRD, MERCURY COUGAR

The following should be used as a guide when determining the amount of work required for a particular service if taken to a repair shop. In estimating how long a particular Frequent Maintenance Service item should take, please observe the following:
- **Factory Time** is time that is generated by the vehicle manufacturer.
- **Chilton Time** is time that is based on field research and data supplied by the vehicle manufacturer.
- All labor time operations are given in hours and tenths of an hour.
- All labor operations, are to be used as a **guide**.

	(Factory Time)	Chilton Time
COOLING		
(G) Winterize Cooling System		
Includes: Run engine to check for leaks, tighten all hose connections. Test radiator and pressure cap, drain radiator and engine block. Add antifreeze and refill system.		
1993-97		.5
(G) Belt, Serpentine Drive, Renew		
1993-97		
V6		
wo/S/C (.3)		.5
w/S/C (.5)		.8
all		1.5
V8 (.3)		.5
(G) Belt, Drive, Supercharger, Renew		
1993-95 (.6)		.9
(G) Hoses, Radiator, Renew		
Includes: Drain and refill cooling system as required.		
1993-97		
upper (.3)		.4
lower (.4)		.6
both (.5)		.7

	(Factory Time)	Chilton Time
(G) Thermostat, Coolant, Renew		
1993-97 (.5)		.7
FUEL		
(M) Air Cleaner, Service		
1993-97		.2
BRAKES		
(G) Bleed Brakes (Four Wheels)		
Includes: Add fluid.		
1993-97 (.3)		.5
(G) Brakes, Adjust (Minor)		
Includes: Adjust brakes, fill master cylinder.		
1993-97, two wheels		.4
(M) Parking Brake, Adjust		
1993-97 (.2)		.4
LUBRICATION SERVICES		
(M) Engine Oil & Filter, Renew		
Includes: Inspect and correct all fluid levels.		
1993-97		.3

	(Factory Time)	Chilton Time
(M) Lubricate Chassis, Change Oil & Filter		
Includes: Inspect and correct all fluid levels.		
1993-97		.6
Install grease fittings add		.1
(M) Lubricate Chassis		
Includes: Inspect and correct all fluid levels.		
1993-97		.4
Install grease fittings add		.1
ELECTRICAL		
(G) Headlamps, Aim		
1993-97		
two		.4
four		.6
(G) Halogen Headlamp Bulb, Renew		
1993-95, one (.2)		.3
(G) License Lamp Assy., Renew		
1993-97 (.2)		.3
(G) Horn, Renew		
1993-97 (.5)		.7

FREQUENT MAINTENANCE LABOR
FORD MUSTANG

The following should be used as a guide when determining the amount of work required for a particular service if taken to a repair shop. In estimating how long a particular Frequent Maintenance Service item should take, please observe the following:
- **Factory Time** is time that is generated by the vehicle manufacturer.
- **Chilton Time** is time that is based on field research and data supplied by the vehicle manufacturer.
- All labor time operations are given in hours and tenths of an hour.
- All labor operations, are to be used as a **guide**.

	(Factory Time)	Chilton Time
COOLING		
(G) Winterize Cooling System		
Includes: Run engine to check for leaks, tighten all hose connections. Test radiator and pressure cap, drain radiator and engine block. Add antifreeze and refill system.		
1993-97		.5

	(Factory Time)	Chilton Time
(G) Belt, Serpentine Drive, Renew		
1993-97, V6, V8 (.3)		.5
(G) Belt, Drive, Renew		
1993		
one (.3)		.4
each adtnl.		.1

	(Factory Time)	Chilton Time
(G) Belt, Drive, Adjust		
1993		
one (.2)		.3
each adtnl.		.1

FREQUENT MAINTENANCE LABOR (cont.)
FORD MUSTANG

	(Factory Time)	Chilton Time
(G) Hoses, Radiator, Renew		
Includes: Drain and refill cooling system as required.		
1993-97		
upper (.3)		.4
lower (.4)		.5
both (.5)		.8
(G) Thermostat, Coolant, Renew		
Four		
1993 (.5)		.7
V6		
1994-97 (.5)		.7
V8		
1993-95 (.9)		1.2
1996		
4.6L DOHC (.5)		.8
4.6L MFI (.6)		.9

FUEL

	(Factory Time)	Chilton Time
(M) Air Cleaner, Service		
1993-97 (.2)		.3
(G) Filter, Fuel, Renew		
1993 (.4)		.5
1994-95		
V6 (.3)		.4
V8 (.4)		.5
1996 (.3)		.4

BRAKES

	(Factory Time)	Chilton Time
(G) Bleed Brakes (Four Wheels)		
Includes: Add fluid.		
1993-97 (.3)		.5
(G) Brakes, Adjust (Minor)		
Includes: Adjust brakes, fill master cylinder.		
1993, two wheels		.4
(M) Parking Brake, Adjust		
1993 (.3)		.4
1994-97 (.2)		.3

LUBRICATION SERVICES

	(Factory Time)	Chilton Time
(M) Engine Oil & Filter, Renew		
Includes: Inspect and correct all fluid levels.		
1993-97		.3
(M) Lubricate Chassis, Change Oil & Filter		
Includes: Inspect and correct all fluid levels.		
1993-97		.6
Install grease fittings add		.1
(M) Lubricate Chassis		
Includes: Inspect and correct all fluid levels.		
1993-97		.4
Install grease fittings add		.1

ELECTRICAL

	(Factory Time)	Chilton Time
(G) Headlamps, Aim		
1993-97		
two		.4
four		.6
(G) Halogen Headlamp Bulb, Renew		
1993-97, one (.2)		.3
(G) License Lamp Assy., Renew		
1993-97 (.2)		.3
(G) Tail Lamp Assy., Renew		
1993		
one (.3)		.4
both (.4)		.6
1994-97		
one (.3)		.4
both (.5)		.7
(G) Horn, Renew		
1993-97, under fender (.5)		.7
(M) Terminals, Battery, Clean		
1993-97		.3

FREQUENT MAINTENANCE LABOR
LINCOLN MARK

The following should be used as a guide when determining the amount of work required for a particular service if taken to a repair shop. In estimating how long a particular Frequent Maintenance Service item should take, please observe the following:

- **Factory Time** is time that is generated by the vehicle manufacturer.
- **Chilton Time** is time that is based on field research and data supplied by the vehicle manufacturer.
- All labor time operations are given in hours and tenths of an hour.
- All labor operations, are to be used as a **guide**.

COOLING

	(Factory Time)	Chilton Time
(G) Winterize Cooling System		
Includes: Run engine to check for leaks, tighten all hose connections. Test radiator and pressure cap, drain radiator and engine block. Add antifreeze and refill system.		
1993-97		.5
(G) Belt, Serpentine Drive, Renew		
1993-97 (.3)		.5

	(Factory Time)	Chilton Time
(G) Hoses, Radiator, Renew		
Includes: Drain and refill cooling system as required.		
1993-97		
upper (.4)		.4
lower (.5)		.6
both (.6)		.9

	(Factory Time)	Chilton Time
(G) Thermostat, Coolant, Renew		
Mark		
1993-97 (.6)		.8
FUEL		
(M) Air Cleaner, Service		
1993-97 (.2)		.3

FREQUENT MAINTENANCE LABOR (cont.)
LINCOLN MARK

	(Factory Time)	Chilton Time
(G) Filter, Fuel, Renew		
1993-97		
Mark (.7)		1.0

BRAKES

	(Factory Time)	Chilton Time
(G) Bleed Brakes (Four Wheels)		
Includes: Add fluid.		
1993-97 (.3)		.5
w/ABS add		.3
(M) Parking Brake, Adjust		
1993 (.3)		.5
1994-97		
Mark (.3)		.5

LUBRICATION SERVICES

	(Factory Time)	Chilton Time
(M) Engine Oil & Filter, Renew		
Includes: Inspect and correct all fluid levels.		
1993-97 (.3)		.3
(M) Lubricate Chassis, Change Oil & Filter		
Includes: Inspect and correct all fluid levels.		
1993-97		.6
Install grease fittings add		.1
(M) Lubricate Chassis		
Includes: Inspect and correct all fluid levels.		
1993-97		.4
Install grease fittings add		.1

ELECTRICAL

	(Factory Time)	Chilton Time
(G) Headlamps, Aim		
1993-97		
two (.4)		.4
four (.5)		.6
(G) Halogen Headlamp Bulb, Renew		
1993-97 (.3)		.3
(G) License Lamp Assy., Renew		
1993-97 (.2)		.4
(G) Tail Lamp Assy., Renew		
1993-97, Mark (.4)		.6
(G) Horn, Renew		
1993-97, Mark (.4)		.6
(M) Terminals, Battery, Clean		
1993-97		.3

SCHEDULED MAINTENANCE INTERVALS
(FORD MUSTANG, THUNDERBIRD, LINCOLN MARK VIII, MERCURY COUGAR (1993))

TO BE SERVICED	TYPE OF SERVICE	VEHICLE MILEAGE INTERVAL (x1000)												
		7.5	15	22.5	30	37.5	45	52.5	60	67.5	75	82.5	90	97.5
Engine oil & filter	R	✓	✓	✓	✓	✓	✓	✓	✓	✓	✓	✓	✓	✓
Adjust clutch pedal by lifting pedal (Mustang)	S/I	✓	✓	✓	✓	✓	✓	✓	✓	✓	✓	✓	✓	✓
Cooling system, hoses, clamps & coolant strength	S/I		✓		✓		✓		✓		✓		✓	
Rotate tires	S/I	✓		✓		✓		✓		✓		✓		✓
Air cleaner filter	R				✓				✓				✓	
Engine coolant	R				✓				✓				✓	
Spark plugs①	R				✓				✓				✓	
Brake lines, hoses & connections	S/I				✓				✓				✓	
Disc brake pads, rotors, brake linings & drums	S/I				✓				✓				✓	
Exhaust heat shields	S/I				✓				✓				✓	
Front & rear wheel bearings	S/I				✓				✓				✓	

SCHEDULED MAINTENANCE INTERVALS
(FORD MUSTANG, THUNDERBIRD, LINCOLN MARK VIII, MERCURY COUGAR (1993)) (Cont.)

TO BE SERVICED	TYPE OF SERVICE	VEHICLE MILEAGE INTERVAL (x1000)												
		7.5	15	22.5	30	37.5	45	52.5	60	67.5	75	82.5	90	97.5
Supercharger fluid level	S/I				✓				✓				✓	
PCV filter (5.0L)	S/I				✓				✓				✓	
PCV valve	R								✓				✓	
Accessory drive belt(s)	S/I								✓				✓	
Suspension, ball joints, & tie rods	S/I				✓				✓				✓	

① Spark plugs (T-Bird/Cougar Super Coupe or 3.8L Calif.) - change every 60,000 miles.
R – Replace   S/I – Service or Inspect

FREQUENT OPERATION MAINTENANCE (SEVERE SERVICE)

If a vehicle is operated under any of the following conditions it is considered severe service:
- Extremely dusty areas.
- 50% or more of the vehicle operation is in 32°C (90°F) or higher temperatures, or constant operation in temperatures below 0°C (32°F).
- Prolonged idling (vehicle operation in stop and go traffic).
- Frequent short running periods (engine does not warm to normal operating temperatures).
- Police, taxi, delivery usage or trailer towing usage.
Oil & oil filter – change every 3000 miles.
Adjust clutch by lifting pedal (Mustang) - adjust every 6000 miles.
Rotate tires at 6000 miles & every 9000 miles thereafter.
Air cleaner filter - service or inspect every 15,000 miles.
Automatic transmission fluid & filter - change every 30,000 miles.

SCHEDULED MAINTENANCE INTERVALS
(FORD MUSTANG, THUNDERBIRD, LINCOLN MARK VIII, MERCURY COUGAR (1994-97))

TO BE SERVICED	TYPE OF SERVICE	VEHICLE MILEAGE INTERVAL (x1000)												
		5	10	15	20	25	30	35	40	45	50	55	60	65
Engine oil & filter	R	✓	✓	✓	✓	✓	✓	✓	✓	✓	✓	✓	✓	✓
Adjust clutch pedal by lifting pedal	S/I	✓	✓	✓	✓	✓	✓	✓	✓	✓	✓	✓	✓	✓
Rotate tires	S/I	✓		✓		✓		✓		✓		✓		✓
Cooling system, hoses, clamps & coolant strength	S/I			✓			✓			✓			✓	
Lubricate steering linkage (T-Bird/Cougar)	S/I			✓			✓			✓			✓	
Air cleaner element	R						✓						✓	
Automatic transmission fluid & filter	R						✓						✓	

SCHEDULED MAINTENANCE INTERVALS
(FORD MUSTANG, THUNDERBIRD, LINCOLN MARK VIII, MERCURY COUGAR (1994-97))
(Cont.)

TO BE SERVICED	TYPE OF SERVICE	VEHICLE MILEAGE INTERVAL (x1000)												
		5	10	15	20	25	30	35	40	45	50	55	60	65
Engine coolant①	R						✓						✓	
Spark plugs (1994 Mark VIII)	R						✓						✓	
Spark plugs (Mark VIII 1995-97)②	R													
Spark plugs (T-Bird & Cougar Exc. 3.8L SC, 3.8L Calif. 1994-95)	R						✓						✓	
Spark plugs (T-Bird & Cougar 1996-97)	R												✓	
Spark plugs (T-Bird & Cougar w/3.8L SC, 3.8L Calif. 1994-95)	R												✓	
Spark plugs (Mustang 1994)	R												✓	
Spark plugs (Mustang 3.8L 1995)	R												✓	
Spark plugs (Mustang 5.0L 1995)	R						✓						✓	
Spark plugs (Mustang 1996-97)②	R													
Accessory drive belt(s)	S/I						✓						✓	
Brake lines, hoses & connections	S/I						✓						✓	
Clutch fluid level (T-Bird)	S/I						✓						✓	
Exhaust heat shields	S/I						✓						✓	
Front & rear brakes	S/I						✓						✓	
PCV filter (Mustang 5.0L)	R						✓						✓	
PCV valve	R												✓	
Rear axle lubricant②	R													
Supercharger fluid level	S/I												✓	

① Engine coolant (1996-97 models) - change engine coolant at 48,000 to 50,000 miles and thereafter every 30,000 miles.
② Replace every 100,000 miles.
R – Replace   S/I – Service or Inspect

SCHEDULED MAINTENANCE INTERVALS
(FORD MUSTANG, THUNDERBIRD, LINCOLN MARK VIII, MERCURY COUGAR (1994-97))
(Cont.)

FREQUENT OPERATION MAINTENANCE (SEVERE SERVICE)

If a vehicle is operated under any of the following conditions it is considered severe service:
- Extremely dusty areas.
- 50% or more of the vehicle operation is in 32°C (90°F) or higher temperatures, or constant operation in temperatures below 0°C (32°F).
- Prolonged idling (vehicle operation in stop and go traffic).
- Frequent short running periods (engine does not warm to normal operating temperatures).
- Police, taxi, delivery usage or trailer towing usage.

Oil & oil filter – change every 3000 miles.
Adjust clutch by lifting pedal every 3000 miles.
Rotate tires at 6000 miles & every 9000 miles thereafter.
Front & rear brakes - check every 15,000 miles.
Automatic transmission fluid & filter - change every 21,000 miles.

CROWN VICTORIA/TOWN CAR/GRAND MARQUIS

VEHICLE IDENTIFICATION CHART

Engine Code					
Code	Liters	Cu. In. (cc)	Cyl.	Fuel Sys.	Eng. Mfg.
W	4.6	281 (4593)	8	SFI	Ford

SFI - Sequential fuel injection

Model Year	
Code	Year
P	1993
R	1994
S	1995
T	1996
V	1997

ENGINE IDENTIFICATION
All measurements are given in inches.

Year	Model	Engine Displacement Liters (cc)	Engine Series (ID/VIN)	Fuel System	No. of Cylinders	Engine Type
1993	Crown Victoria	4.6 (4593)	W	SFI	8	SOHC
	Grand Marquis	4.6 (4593)	W	SFI	8	SOHC
	Town Car	4.6 (4593)	W	SFI	8	SOHC
1994	Crown Victoria	4.6 (4593)	W	SFI	8	SOHC
	Grand Marquis	4.6 (4593)	W	SFI	8	SOHC
	Town Car	4.6 (4593)	W	SFI	8	SOHC
1995	Crown Victoria	4.6 (4593)	W	SFI	8	SOHC
	Grand Marquis	4.6 (4593)	W	SFI	8	SOHC
	Town Car	4.6 (4593)	W	SFI	8	SOHC

ENGINE IDENTIFICATION
All measurements are given in inches.

Year	Model	Engine Displacement Liters (cc)	Engine Series (ID/VIN)	Fuel System	No. of Cylinders	Engine Type
1996-97	Crown Victoria	4.6 (4593)	W	SFI	8	SOHC
	Grand Marquis	4.6 (4593)	W	SFI	8	SOHC
	Town Car	4.6 (4593)	W	SFI	8	SOHC

MFI - Multiport fuel injection   SFI - Sequential fuel injection   SOHC - Single overhead camshaft

GENERAL ENGINE SPECIFICATIONS

Year	Engine ID/VIN	Engine Displacement Liters (cc)	Fuel System Type	Net Horsepower @ rpm	Net Torque @ rpm (ft. lbs.)	Bore x Stroke (in.)	Compression Ratio	Oil Pressure @ rpm
1993	W	4.6 (4593)	SFI	①	②	3.55x3.54	9.0:1	20-45@2000
1994	W	4.6 (4593)	SFI	①	②	3.55x3.5.4	9.0:1	20-45@2000
1995	W	4.6 (4593)	SFI	③	④	3.55x3.54	9.0:1	20-45@2000
1996-97	W	4.6 (4593)	SFI	⑤	⑥	3.55x3.54	⑦	20-45@1500

MFI - Multiport fuel injection
SFI - Sequential fuel injection
① Single exhaust: 190@4200
  Dual exhaust: 210@4600
② Single exhaust: 260@3200
  Dual exhaust: 270@3400

③ Single exhaust: 190@4250
  Dual exhaust:: 210@4250
④ Single exhaust: 260@3250
  Dual exhaust:: 270@3250
⑤ Single exhaust: 190@4250
  Dual exhaust: 210@4250
  Crown Victoria with natural gas: 178@4500

⑥ Single exhaust: 265@3250
  Dual exhaust: 275@3250
  Crown Victoria with natural gas: 237@3500
⑦ Base engine: 9.0:1
  Crown Victoria with natural gas: 10.0:1

GASOLINE ENGINE TUNE-UP SPECIFICATIONS

Year	Engine ID/VIN	Engine Displacement Liters (cc)	Spark Plugs Gap (in.)	Ignition Timing (deg.) MT	Ignition Timing (deg.) AT	Fuel Pump (psi)	Idle Speed (rpm) MT	Idle Speed (rpm) AT	Valve Clearance In.	Valve Clearance Ex.
1993	W	4.6 (4593)	0.054	-	10B	35-40	-	560	HYD	HYD
1994	W	4.6 (4593)	0.054	-	10B	30-45 ②	①	①	HYD	HYD
1995	W	4.6 (4593)	0.054	-	10B	30-45 ②	-	①	HYD	HYD
1996-97	W	4.6 (4593)	0.054	10B	10B	35-45 ②	①	①	HYD	HYD

NOTE: The Vehicle Emission Control Information label often reflects specification changes made during production. The label figures must be used if they differ from those in this chart.
B - Before top dead center
HYD - Hydraulic
① Refer to Vehicle Emission Control Information label
② Fuel pressure with engine running, pressure regulator vacuum hose connected

CAPACITIES

Year	Model	Engine ID/VIN	Engine Displacement Liters (cc)	Engine Oil with Filter (qts.)	Transmission (pts.) 4-Spd	Transmission (pts.) 5-Spd	Transmission (pts.) Auto.	Drive Axle Front (pts.)	Drive Axle Rear (pts.)	Fuel Tank (gal.)	Cooling System (qts.)
1993	Crown Victoria	W	4.6 (4593)	5.0	-	-	24.6	-	4.0	20.0	14.1
	Grand Marquis	W	4.6 (4593)	5.0	-	-	25.6	-	4.1	20.0	14.1
	Town Car	W	4.6 (4593)	5.0	-	-	25.6	-	4.1	20.0	14.1
1994	Crown Victoria	W	4.6 (4593)	5.0	-	-	27.2 ①	-	3.75	20.0	14.1
	Grand Marquis	W	4.6 (4593)	5.0	-	-	28.2 ①	-	3.76	20.0	14.1
	Town Car	W	4.6 (4593)	5.0	-	-	28.2 ①	-	3.76	20.0	14.1

CAPACITIES

| Year | Model | Engine ID/VIN | Engine Displacement Liters (cc) | Engine Oil with Filter (qts.) | Transmission (pts.) | | | Drive Axle | | Fuel Tank (gal.) | Cooling System (qts.) |
					4-Spd	5-Spd	Auto.	Front (pts.)	Rear (pts.)		
1995	Crown Victoria	W	4.6 (4593)	5.0	-	-	27.2 ①	-	3.0	20.0	14.1
	Grand Marquis	W	4.6 (4593)	5.0	-	-	28.2 ①	-	3.1	20.0	14.1
	Town Car	W	4.6 (4593)	5.0	-	-	28.2 ①	-	3.1	20.0	14.1
1996-97	Crown Victoria	W	4.6 (4593)	5.0	-	-	27.2 ①	-	3.75 ②	20.0	14.1
	Grand Marquis	W	4.6 (4593)	5.0	-	-	28.2 ①	-	3.75 ②	20.0	15.1
	Town Car	W	4.6 (4593)	5.0	-	-	28.2 ①	-	3.75 ②	20.0	15.1

① Includes torque converter

② 7.50" axle: 3.0 pts.
8.80" axle: 3.25 pts.

VALVE SPECIFICATIONS

| Year | Engine ID/VIN | Engine Displacement Liters (cc) | Seat Angle (deg.) | Face Angle (deg.) | Spring Test Pressure (lbs. @ in.) | Spring Installed Height (in.) | Stem-to-Guide Clearance (in.) | | Stem Diameter (in.) | |
							Intake	Exhaust	Intake	Exhaust
1993	W	4.6 (4593)	45	45.5	132@1.10	1.570	0.0008-0.0027	0.0018-0.0037	0.2746-0.2754	0.2736-0.2744
1994	W	4.6 (4593)	45	45.5	132@1.10	1.570	0.0008-0.0027	0.0018-0.0037	0.2746-0.2754	0.2736-0.2744
1995	W	4.6 (4593)	45	45.5	132@1.10	1..570	0.0008-0.0027	0.0018-0.0037	0.2746-0.2754	0.2736-0.2744
1996-97	W	4.6 (4593)	45	45.5	132@1.10	1.570	0.0008-0.0027	0.0018-0.0037	0.2746-0.2754	0.2736-0.2744

TORQUE SPECIFICATIONS
All readings in ft. lbs.

| Year | Engine ID/VIN | Engine Displacement Liters (cc) | Cylinder Head Bolts | Main Bearing Bolts | Rod Bearing Bolts | Crankshaft Damper Bolts | Flywheel Bolts | Manifold | | Spark Plugs | Lug Nut |
								Intake	Exhaust		
1993	W	4.6 (4593)	①	②	③	114-121	54-64	15-22	13-16	7-15	95
1994	W	4.6 (4593)	①	②	③	114-121	54-64	15-22	13-16	7-15	95
1995	W	4.6 (4593)	①	②	③	114-121	54-64	15-22	13-16	7-15	95
1996-97	W	4.6 (4593)	①	②	③	114-121	54-64	15-22	13-16	7-15	95

NOTE: Always follow proper torque patterns
NOTE: Stretch bolts are used in all procedures that require rotating the fastener a certain number of degrees. The bolts stretch and cannot be reused. For reassembly, replace with new fasteners.

① Step 1: 22-30 ft. lbs.
Step 2: Rotate each bolt 85-95 degrees
Step 3: Repeat Step 2

② Step 1: Main bearing cap bolts: 22-25 ft. lbs.
Step 2: Rotate each bolts 85-95 degrees
Step 3: Main bearing cap adjusting screws:
44 in. lbs. then 80-97 in. lbs. lbs.
Step 4: Main bearing cap side bolts:
7 ft. lbs. then 14-17 ft. lbs.

③ Step 1: 8 ft. lbs.
Step 2: 12 ft. lbs.
Step 3: 25-34 ft. lbs.
Step 4: Rotate 85-95 degrees

BRAKE SPECIFICATIONS
All measurements in inches unless noted

Year	Model		Master Cylinder Bore	Brake Disc Original Thickness	Brake Disc Minimum Thickness	Maximum Runout	Brake Drum Diameter Original Inside Diameter	Brake Drum Diameter Max. Wear Limit	Brake Drum Diameter Maximum Machine Diameter	Minimum Lining Thickness Front	Minimum Lining Thickness Rear
1993	Crown Victoria	F	1.000	1.030	0.974	0.003	-	-	-	0.030	-
		R	-	0.500	0.440	0.003	-	-	-	-	0.030
	Grand Marquis	F	1.000	1.030	0.974	0.003	-	-	-	0.030	-
		R	-	0.500	0.440	0.003	-	-	-	-	0.030
	Town Car	F	1.000	1.030	0.974	0.003	-	-	-	0.030	-
		R	-	0.500	0.440	0.003	-	-	-	-	0.030
1994	Crown Victoria	F	1.000	1.030	0.974	0.003	-	-	-	0.125	-
		R	-	0.500	0.440	0.003	-	-	-	-	0.125
	Grand Marquis	F	1.000	1.030	0.974	0.003	-	-	-	0.125	-
		R	-	0.500	0.440	0.003	-	-	-	-	0.125
	Town Car	F	1.000	1.030	0.974	0.003	-	-	-	0.125	-
		R	-	0.500	0.440	0.003	-	-	-	-	0.125
1995	Crown Victoria	F	1.000	1.030	0.974	0.003	-	-	-	0.125	-
		R	-	0.500	0.440	0.003	-	-	-	-	0.125
	Grand Marquis	F	1.000	1.030	0.974	0.003	-	-	-	0.125	-
		R	-	0.500	0.440	0.003	-	-	-	-	0.125
	Town Car	F	1.000	1.030	0.974	0.003	-	-	-	0.125	-
		R	-	0.500	0.440	0.003	-	-	-	-	0.125
1996-97	Crown Victoria	F	1.000	1.024	0.974	0.002	-	-	-	0.125	-
		R	-	0.550	0.510	0.002	-	-	-	-	0.125
	Grand Marquis	F	1.000	1.024	0.974	0.002	-	-	-	0.125	-
		R	-	0.550	0.510	0.002	-	-	-	-	0.125
	Town Car	F	1.000	1.024	0.974	0.002	-	-	-	0.125	-
		R	-	0.550	0.510	0.002	-	-	-	-	0.125

NOTE: Follow specifications stamped on rotor or drum if figures differ from those in this chart.
F - Front
R - Rear

FREQUENT MAINTENANCE LABOR
FORD CROWN VICTORIA, MERCURY GRAND MARQUIS

The following should be used as a guide when determining the amount of work required for a particular service if taken to a repair shop. In estimating how long a particular Frequent Maintenance Service item should take, please observe the following:
- **Factory Time** is time that is generated by the vehicle manufacturer.
- **Chilton Time** is time that is based on field research and data supplied by the vehicle manufacturer.
- All labor time operations are given in hours and tenths of an hour.
- All labor operations, are to be used as a **guide**.

COOLING

(G) Winterize Cooling System
Includes: Run engine to check for leaks, tighten all hose connections. Test radiator and pressure cap, drain radiator and engine block. Add antifreeze and refill system.
1993-97	.5

(G) Belt, Serpentine Drive, Renew
1993-97 (.3)	.5

(G) Hoses, Radiator, Renew
Includes: Drain and refill cooling system as required.
1993-97
one (.5)	.6
both (.6)	.9

(G) Thermostat, Coolant, Renew
1993-97 (.5)	.7

FUEL

(M) Air Cleaner, Service
1993-97 (.2)	.3

(G) Filter, Fuel, Renew
1993-94 (.5)	.8
1995-97	
4.6L CNG (.6)	.9
4.6L MFI (.5)	.8

BRAKES

(G) Bleed Brakes (Four Wheels)
Includes: Add fluid.
1993-97
w/ABS (.6)	.8

(M) Parking Brake, Adjust
1993-97
disc brakes (.5)	.7

LUBRICATION SERVICES

(M) Engine Oil & Filter, Renew
Includes: Inspect and correct all fluid levels.
1993-97 (.3)	.3

(M) Lubricate Chassis, Change Oil & Filter
Includes: Inspect and correct all fluid levels.
1993-97	.6
Install grease fittings add	.1

(M) Lubricate Chassis
Includes: Inspect and correct all fluid levels.
1993-97	.4
Install grease fittings add	.1

ELECTRICAL

(G) Headlamps, Aim
1993-97
two	.4
four	.6

(G) License Lamp Assy., Renew
1993-97 (.3)	.4

(G) Tail Lamp Assy., Renew
1993-97
Ford Sedan
one (.4)	.5
both (.6)	.8
Mercury	
one (.5)	.6
both (.6)	.8

(G) Horn, Renew
1993-97, each (.3)	.4

(M) Terminals, Battery, Clean
1993-97	.3

FREQUENT MAINTENANCE LABOR
LINCOLN TOWN CAR

The following should be used as a guide when determining the amount of work required for a particular service if taken to a repair shop. In estimating how long a particular Frequent Maintenance Service item should take, please observe the following:
- **Factory Time** is time that is generated by the vehicle manufacturer.
- **Chilton Time** is time that is based on field research and data supplied by the vehicle manufacturer.
- All labor time operations are given in hours and tenths of an hour.
- All labor operations, are to be used as a **guide**.

COOLING

(G) Winterize Cooling System
Includes: Run engine to check for leaks, tighten all hose connections. Test radiator and pressure cap, drain radiator and engine block. Add antifreeze and refill system.
1993-97	.5

(G) Belt, Serpentine Drive, Renew
1993-97 (.3)	.5

(G) Hoses, Radiator, Renew
Includes: Drain and refill cooling system as required.
1993-97
upper (.4)	.4
lower (.5)	.6
both (.6)	.9

(G) Thermostat, Coolant, Renew
Town Car
1993-97 (.5)	.7

FREQUENT MAINTENANCE LABOR (cont.)
LINCOLN TOWN CAR

	(Factory Time)	Chilton Time

FUEL

(M) Air Cleaner, Service
1993-97 (.2) 3

(G) Filter, Fuel, Renew
1993-97
   Town Car (.5) 8

BRAKES

(G) Bleed Brakes (Four Wheels)
Includes: Add fluid.
   1993-97 (.3) 5
w/ABS add 3

(M) Parking Brake, Adjust
1993 (.3) 5
1994-97
   Town Car (.5) 7

LUBRICATION SERVICES

(M) Engine Oil & Filter, Renew
Includes: Inspect and correct all fluid levels.
   1993-97 (.3) 3

(M) Lubricate Chassis, Change Oil & Filter
Includes: Inspect and correct all fluid levels.
   1993-97 6
Install grease fittings add 1

(M) Lubricate Chassis
Includes: Inspect and correct all fluid levels.
   1993-97 4
Install grease fittings add 1

ELECTRICAL

(G) Headlamps, Aim
1993-97
   two (.4) 4
   four (.5) 6

(G) Halogen Headlamp Bulb, Renew
1993-97 (.3) 3

(G) License Lamp Assy., Renew
1993-97 (.2) 4

(G) Tail Lamp Assy., Renew
1993-97, Town Car (.6) 8

(G) Horn, Renew
1993-97, Town Car (.3) 5

(M) Terminals, Battery, Clean
1993-97 3

SCHEDULED MAINTENANCE INTERVALS
(FORD CROWN VICTORIA, LINCOLN TOWN CAR, MERCURY GRAND MARQUIS (1993))

TO BE SERVICED	TYPE OF SERVICE	VEHICLE MILEAGE INTERVAL (x1000)												
		7.5	15	22.5	30	37.5	45	52.5	60	67.5	75	82.5	90	97.5
Engine oil & filter	R	✓	✓	✓	✓	✓	✓	✓	✓	✓	✓	✓	✓	✓
Rotate tires	S/I	✓		✓		✓		✓		✓		✓		✓
Lubricate steering linkage (pitman arm socket)	S/I		✓		✓		✓		✓				✓	
Lubricate suspension	S/I		✓		✓		✓		✓		✓		✓	
Air cleaner filter	R				✓				✓				✓	
Engine coolant	R				✓				✓				✓	
Spark plugs	R				✓				✓				✓	
Brake lines & connections	S/I				✓				✓				✓	
Disc brake pads & rotors	S/I				✓				✓				✓	
Engine coolant protection, hoses & clamps	S/I				✓				✓				✓	
Exhaust heat shields	S/I					✓			✓				✓	

SCHEDULED MAINTENANCE INTERVALS
(FORD CROWN VICTORIA, LINCOLN TOWN CAR, MERCURY GRAND MARQUIS (1993))
(Cont.)

TO BE SERVICED	TYPE OF SERVICE	VEHICLE MILEAGE INTERVAL (x1000)												
		7.5	15	22.5	30	37.5	45	52.5	60	67.5	75	82.5	90	97.5
Accessory drive belt(s)	S/I								✓				✓	
Lubricate steering linkage (inner-outer tie rod ends)	S/I		✓				✓				✓			
PCV valve	R								✓					

R – Replace   S/I – Service or Inspect

FREQUENT OPERATION MAINTENANCE (SEVERE SERVICE)
   If a vehicle is operated under any of the following conditions it is considered severe service:
- Extremely dusty areas.
- 50% or more of the vehicle operation is in 32°C (90°F) or higher temperatures, or constant operation in temperatures below 0°C (32°F).
- Prolonged idling (vehicle operation in stop and go traffic).
- Frequent short running periods (engine does not warm to normal operating temperatures).
- Police, taxi, delivery usage or trailer towing usage.
Oil & oil filter – change every 3000 miles.
Rotate tires at 6000 miles & every 15,000 miles thereafter.
Automatic transaxle fluid & filter - change every 30,000 miles.

SCHEDULED MAINTENANCE INTERVALS
(FORD CROWN VICTORIA, LINCOLN TOWN CAR, MERCURY GRAND MARQUIS (1994-97))

TO BE SERVICED	TYPE OF SERVICE	VEHICLE MILEAGE INTERVAL (x1000)												
		5	10	15	20	25	30	35	40	45	50	55	60	65
Engine oil & filter	R	✓	✓	✓	✓	✓	✓	✓	✓	✓	✓	✓	✓	✓
Rotate tires	S/I	✓		✓		✓		✓		✓		✓		✓
Cooling system, hoses, clamps & coolant strength	S/I			✓			✓			✓			✓	
Lubricate steering linkage (Crown Victoria, Grand Marquis)	S/I			✓			✓			✓			✓	
Air cleaner element	R						✓						✓	
Automatic transmission fluid & filter (1995-97)	R						✓						✓	
Spark plugs (1994-95)	R						✓						✓	
Spark plugs (1996-97)②	R													
Exhaust heat shields	S/I						✓						✓	

SCHEDULED MAINTENANCE INTERVALS
(FORD CROWN VICTORIA, LINCOLN TOWN CAR, MERCURY GRAND MARQUIS (1994-97))
(Cont.)

TO BE SERVICED	TYPE OF SERVICE	VEHICLE MILEAGE INTERVAL (x1000)												
		5	10	15	20	25	30	35	40	45	50	55	60	65
Fuel filter (NGV Crown Victoria)③	R					✓					✓			
Lubricate steering linkage (Town Car)	S/I						✓						✓	
Front & rear brakes	S/I						✓						✓	
Lubricate suspension (Town Car)	S/I						✓						✓	
Engine coolant (1994)	R						✓						✓	
Engine coolant (1995-97)①	R										✓			
PCV valve	R												✓	
Accessory drive belt(s)	S/I												✓	

① Change initially at 50,000 miles, & thereafter every 30,000 miles.
② Replace every 100,000 miles.
③ Also drain coalescer. Perform every 24,000 miles for severe service.
R – Replace   S/I – Service or Inspect

FREQUENT OPERATION MAINTENANCE (SEVERE SERVICE)
If a vehicle is operated under any of the following conditions it is considered severe service:
- Extremely dusty areas.
- 50% or more of the vehicle operation is in 32°C (90°F) or higher temperatures, or constant operation in temperatures below 0°C (32°F).
- Prolonged idling (vehicle operation in stop and go traffic).
- Frequent short running periods (engine does not warm to normal operating temperatures).
- Police, taxi, delivery usage or trailer towing usage.
Oil & oil filter – change every 3000 miles.
Rotate tires at 6000 miles & every 9000 miles thereafter.
Automatic transmission fluid & filter - change every 30,000 miles (1994) or 21,000 miles (1995-97)

CENTURY/CUTLASS CIERA/CUTLASS CRUISER

VEHICLE IDENTIFICATION CHART

Engine Code							Model Year	
Code	Liters	Cu. In. (cc)	Cyl.	Fuel Sys.	Eng. Mfg.		Code	Year
4	2.2	134 (2195)	4	MFI	BOC		P	1993
M	3.1	191 (3130)	6	MFI	BOC		R	1994
N	3.3	204 (3342)	6	MFI	BOC		S	1995
							T	1996
							V	1997

MFI - Multiport fuel injection
BOC - Buick/Oldsmobile/Cadillac

ENGINE IDENTIFICATION

Year	Model	Engine Displacement Liters (cc)	Engine Series (ID/VIN)	Fuel System	No. of Cylinders	Engine Type
1993	Century	2.2 (2195)	4	MFI	4	OHV
	Century	3.3 (3342)	N	MFI	6	OHV
	Cutlass Ciera	2.2 (2195)	4	MFI	4	OHV
	Cutlass Ciera	3.3 (3342)	N	MFI	6	OHV
	Cutlass Cruiser	2.2 (2195)	4	MFI	4	OHV
	Cutlass Cruiser	3.3 (3342)	N	MFI	6	OHV
1994	Century	2.2 (2195)	4	MFI	4	OHV
	Century	3.1 (3130)	M	MFI	6	OHV
	Cutlass Ciera	2.2 (2195)	4	MFI	4	OHV
	Cutlass Ciera	3.1 (3130)	M	MFI	6	OHV
	Cutlass Cruiser	2.2 (2195)	4	MFI	4	OHV
	Cutlass Cruiser	3.1 (3130)	M	MFI	6	OHV
1995	Century	2.2 (2195)	4	MFI	4	OHV
	Century	3.1 (3130)	M	MFI	6	OHV
	Cutlass Ciera	2.2 (2195)	4	MFI	4	OHV
	Cutlass Ciera	3.1 (3130)	M	MFI	6	OHV
	Cutlass Cruiser	2.2 (2195)	4	MFI	4	OHV
	Cutlass Cruiser	3.1 (3130)	M	MFI	6	OHV
1996-97	Century	2.2 (2195)	4	MFI	4	OHV
	Century	3.1 (3130)	M	MFI	6	OHV
	Cutlass Ciera	2.2 (2195)	4	MFI	4	OHV
	Cutlass Ciera	3.1 (3130)	M	MFI	6	OHV
	Cutlass Cruiser	3.1 (3130)	M	MFI	6	OHV

MFI - Multiport fuel injection
OHV - Overhead valve

GENERAL ENGINE SPECIFICATIONS

Year	Engine ID/VIN	Engine Displacement Liters (cc)	Fuel System Type	Net Horsepower @ rpm	Net Torque @ rpm (ft. lbs.)	Bore x Stroke (in.)	Compression Ratio	Oil Pressure @ rpm
1993	4	2.2 (2195)	MFI	110@5200	130@3200	3.50x3.46	9.0:1	56@3000
	N	3.3 (3342)	MFI	160@5200	185@2000	3.70x3.16	9.0:1	60@1850
1994	4	2.2 (2195)	MFI	120@5200	130@4000	3.50x3.46	9.0:1	56@3000
	M	3.1 (3130)	MFI	160@5200	185@4000	3.50x3.31	9.5:1	15@1100
1995	4	2.2 (2195)	MFI	120@5200	130@4000	3.50x3.46	8.85:1	56@3000
	M	3.1 (3130)	MFI	160@5200	185@4000	3.50x3.31	9.5:1	15@1100
1996-97	4	2.2 (2195)	MFI	120@5200	130@4000	3.50x3.46	8.85:1	56@3000
	M	3.1 (3130)	MFI	160@5200	185@4000	3.50x3.31	9.5:1	15@1100

MFI - Multiport fuel injection

GASOLINE ENGINE TUNE-UP SPECIFICATIONS

Year	Engine ID/VIN	Engine Displacement Liters (cc)	Spark Plugs Gap (in.)	Ignition Timing (deg.) MT	Ignition Timing (deg.) AT	Fuel Pump (psi)	Idle Speed (rpm) MT	Idle Speed (rpm) AT	Valve Clearance In.	Valve Clearance Ex.
1993	4	2.2 (2195)	0.045	①	①	41-47	②	②	HYD	HYD
	N	3.3 (3342)	0.060	①	①	41-47	②	②	HYD	HYD
1994	4	2.2 (2195)	0.060	①	①	41-47	②	②	HYD	HYD
	M	3.1 (3130)	0.060	①	①	41-47	②	②	HYD	HYD

GASOLINE ENGINE TUNE-UP SPECIFICATIONS

Year	Engine ID/VIN	Engine Displacement Liters (cc)	Spark Plugs Gap (in.)	Ignition Timing (deg.)		Fuel Pump (psi)		Idle Speed (rpm)		Valve Clearance	
				MT	AT			MT	AT	In.	Ex.
1995	4	2.2 (2195)	0.060	①	①	41-47 ③		②	②	HYD	HYD
	M	3.1 (3130)	0.060	①	①	41-47 ③		②	②	HYD	HYD
1996-97	4	2.2 (2195)	0.060	①	①	41-47		②	②	HYD	HYD
	M	3.1 (3130)	0.060	①	①	41-47		②	②	HYD	HYD

NOTE: The Vehicle Emission Control Information label often reflects specification changes made during production. The label figures must be used if they differ from those in this chart.

HYD - Hydraulic

① DIS Ignition System timing not adjustable
② Idle speed maintained by ECM. There is no recommended adjustment procedure
③ Pressure at fuel pump

CAPACITIES

Year	Model	Engine ID/VIN	Engine Displacement Liters (cc)	Engine Oil with Filter (qts.)	Transmission (pts.)			Drive Axle		Fuel Tank (gal.)	Cooling System (qts.)
					4-Spd	5-Spd	Auto.	Front (pts.)	Rear (pts.)		
1993	Century	4	2.2 (2195)	4.0 ①	-	-	②	-	-	16.5	8.3
	Century	N	3.3 (3342)	4.0 ①	-	-	②	-	-	16.5	10.5
	Cutlass Ciera	4	2.2 (2195)	4.0 ①	-	-	②	-	-	16.5	8.3
	Cutlass Ciera	N	3.3 (3342)	4.0 ①	-	-	②	-	-	16.5	10.5
	Cutlass Cruiser	4	2.2 (2195)	4.0 ①	-	-	②	-	-	16.5	8.3
	Cutlass Cruiser	N	3.3 (3342)	4.0 ①	-	-	②	-	-	16.5	10.5
1994	Century	4	2.2 (2195)	4.0 ①	-	-	③	-	-	16.5	8.7
	Century	M	3.1 (3130)	4.0 ①	-	-	③	-	-	16.5	11.6
	Cutlass Ciera	4	2.2 (2195)	4.0 ①	-	-	③	-	-	16.5	8.7
	Cutlass Ciera	M	3.1 (3130)	4.0 ①	-	-	③	-	-	16.5	11.6
	Cutlass Cruiser	4	2.2 (2195)	4.0 ①	-	-	③	-	-	16.5	8.7
	Cutlass Cruiser	M	3.1 (3130)	4.0 ①	-	-	③	-	-	16.5	11.6
1995	Century	4	2.2 (2195)	3.8 ①	-	-	③	-	-	16.5	8.7
	Century	M	3.1 (3130)	3.8 ①	-	-	③	-	-	16.5	11.6
	Cutlass Ciera	4	2.2 (2195)	3.8 ①	-	-	③	-	-	16.5	8.7
	Cutlass Ciera	M	3.1 (3130)	3.8 ①	-	-	③	-	-	16.5	11.6
	Cutlass Cruiser	4	2.2 (2195)	3.8 ①	-	-	③	-	-	16.5	8.7
	Cutlass Cruiser	M	3.1 (3130)	3.8 ①	-	-	③	-	-	16.5	11.6
1996-97	Century	4	2.2 (2195)	4.0 ①	-	-	③	-	-	16.5	8.7
	Century	M	3.1 (3130)	4.0 ①	-	-	③	-	-	16.5	11.6
	Cutlass Ciera	4	2.2 (2195)	4.0 ①	-	-	③	-	-	16.5	8.7
	Cutlass Ciera	M	3.1 (3130)	3.8 ①	-	-	③	-	-	16.5	11.6
	Cutlass Cruiser	M	3.1 (3130)	4.0 ①	-	-	③	-	-	16.5	11.6

① Specification is without filter replacement; Additional oil may be required

② 3 speed: 8.0
  4 speed: 12.0

③ 3 speed: 8.0
  4 speed: 14.8

VALVE SPECIFICATIONS

Year	Engine ID/VIN	Engine Displacement Liters (cc)	Seat Angle (deg.)	Face Angle (deg.)	Spring Test Pressure (lbs. @ in.)	Spring Installed Height (in.)	Stem-to-Guide Clearance (in.) Intake	Stem-to-Guide Clearance (in.) Exhaust	Stem Diameter (in.) Intake	Stem Diameter (in.) Exhaust
1993	4	2.2 (2195)	46	45	215-233@ 1.247	1.637	0.0011- 0.0026	0.0014- 0.0031	NA	NA
	N	3.3 (3342)	45	45	210@1.315	1.690- 1.720	0.0015- 0.0035	0.0015- 0.0035	NA	NA
1994	4	2.2 (2195)	46	45	220-236@ 1.278	1.710	0.0010- 0.0027	0.0014- 0.0031	NA	NA
	M	3.1 (3130)	45	45	250@1.239	1.710	0.0010- 0.0027	0.0010- 0.0027	NA	NA
1995	4	2.2 (2195)	46	45	220-236@ 1.278	1.710	0.0010- 0.0027	0.0014- 0.0031	NA	NA
	M	3.1 (3130)	45	45	250@1.239	1.710	0.0010- 0.0027	0.0010- 0.0027	NA	NA
1996-97	4	2.2 (2195)	46	45	220-236@ 1.278	1.710	0.0010- 0.0027	0.0014- 0.0031	NA	NA
	M	3.1 (3130)	45	45	230@1.260	1.701	0.0010- 0.0027	0.0010- 0.0027	NA	NA

NA - Not Available

TORQUE SPECIFICATIONS
All readings in ft. lbs.

Year	Engine ID/VIN	Engine Displacement Liters (cc)	Cylinder Head Bolts	Main Bearing Bolts	Rod Bearing Bolts	Crankshaft Damper Bolts	Flywheel Bolts	Manifold Intake	Manifold Exhaust	Spark Plugs	Lug Nut
1993	4	2.2 (2195)	⑨	66	38	77	54	24	10	11	100
	N	3.3 (3342)	③	④	⑤	⑥	⑦	7	38	20	100
1994	4	2.2 (2195)	⑨	70	38	77	55	24	10	⑩	100
	M	3.1 (3130)	①	⑪	⑫	76	61	②	10	⑩	100
1995	4	2.2 (2195)	⑨	70	38	77	55	24	10	⑩	100
	M	3.1 (3130)	①	⑧	⑬	76	59	②	12	⑩	100
1996-97	4	2.2 (2195)	⑨	70	38	77	55	24	10	⑩	100
	M	3.1 (3130)	①	⑧	⑬	76	59	②	12	⑩	100

① Coat threads with sealer torque to 37 ft. lbs., then turn 1/4 turn (90 degrees)
② Torque all bolts to 15 ft. lbs., Retorque to 24 ft. lbs.
③ Step 1: Tighten all bolts to 35 ft. lbs.
Step 2: Turn all bolts 130 degrees
Step 3: Rotate four center bolts an additional 30 degrees

④ 26 ft. lbs. plus 50 degrees
⑤ 20 ft. lbs. plus 50 degrees
⑥ 110 ft. lbs. plus 76 degrees
⑦ 11 ft. lbs. plus 50 degrees
⑧ 37 ft. lbs. plus 75 degrees
⑨ Short bolts: 43 ft. lbs. plus 90 degrees
Long bolts: 46 ft. lbs. plus 90 degrees

⑩ New cylinder first-time installation: 21 ft. lbs
All others: 11 ft. lbs.
⑪ 37 ft. lbs. plus 77 degrees
⑫ Nos. 1-8: 30 ft. lbs.
Nos. 9-10: 26 ft. lbs.
Tighten all bolts an additional 90 degrees
⑬ 15 ft. lbs. plus 75 degrees

BRAKE SPECIFICATIONS
All measurements in inches unless noted

Year	Model		Master Cylinder Bore	Brake Disc Original Thickness	Brake Disc Minimum Thickness	Brake Disc Maximum Runout	Brake Drum Diameter Original Inside Diameter	Brake Drum Diameter Max. Wear Limit	Brake Drum Diameter Maximum Machine Diameter	Minimum Lining Thickness Front	Minimum Lining Thickness Rear
1993	Century	①	0.874	0.885	0.815	0.004	8.863	8.909	8.880	0.030	③
	Century	②	0.944	1.043	0.957	0.004	8.863	8.909	8.880	0.030	③
	Cutlass Ciera	①	0.874	0.885	0.815	0.004	8.863	8.909	8.880	0.030	③

BRAKE SPECIFICATIONS
All measurements in inches unless noted

Year	Model		Master Cylinder Bore	Brake Disc			Brake Drum Diameter			Minimum Lining Thickness	
				Original Thickness	Minimum Thickness	Maximum Runout	Original Inside Diameter	Max. Wear Limit	Maximum Machine Diameter	Front	Rear
1993	Cutlass Ciera	②	0.944	1.043	0.957	0.004	8.863	8.909	8.880	0.030	③
	Cutlass Cruiser		0.944	1.043	0.957	0.004	8.863	④	8.880	0.030	③
1994	Century		0.944	1.028	0.957	0.002	8.860	8.909	8.880	0.030	③
	Cutlass Ciera		0.944	1.028	0.957	0.002	8.860	8.909	8.880	0.030	③
	Cutlass Cruiser		0.944	1.028	0.957	0.002	8.860	④	8.880	0.030	③
1995	Century		0.944	1.028	0.957	0.002	8.863	8.909	8.880	0.030	③
	Cutlass Ciera		0.944	1.028	0.957	0.002	8.863	8.909	8.880	0.030	③
	Cutlass Cruiser		0.944	1.028	0.957	0.002	8.863	④	8.880	0.030	③
1996-97	Century		0.944	1.028	0.957	0.002	8.863	8.909	8.920	0.030	③
	Cutlass Ciera		0.944	1.028	0.957	0.002	8.863	8.909	8.920	0.030	③
	Cutlass Cruiser		0.944	1.028	0.957	0.002	8.863	8.909	8.920	0.030	③

NA - Not Available
F - Front
R - Rear

① Standard
② Heavy duty and Wagon
③ 0.030 over rivet head; If bonded lining, use 0.062 from shoe
④ Use discard diameter cast into drum

FREQUENT MAINTENANCE LABOR
BUICK CENTURY, OLDSMOBILE CUTLASS CIERA, CUTLASS CRUISER

The following should be used as a guide when determining the amount of work required for a particular service if taken to a repair shop. In estimating how long a particular Frequent Maintenance Service item should take, please observe the following:
- **Factory Time** is time that is generated by the vehicle manufacturer.
- **Chilton Time** is time that is based on field research and data supplied by the vehicle manufacturer.
- All labor time operations are given in hours and tenths of an hour.
- All labor operations, are to be used as a **guide**.

	(Factory Time)	Chilton Time

COOLING

(G) Winterize Cooling System
Includes: Run engine to check for leaks, tighten all hose connections. Test radiator and pressure cap, drain radiator and engine block. Add antifreeze and refill system.
1993-97 5

(G) Belt, Serpentine Drive, Renew
1993-97 (.3) 5

(G) Hoses, Radiator, Renew
Includes: Drain and refill cooling system as required.
1993-97
  upper (.4) 4
  lower (.5) 5
  both (.6) 7
  throttle body
    inlet (1.3) 1.7
    outlet (.7) 1.0

(G) Thermostat, Coolant, Renew
1993-97, 4 cyl. (.4) 6
1993, V6 (.6) 8
1994-97, V6 (.7) 1.0

FUEL

(M) Air Cleaner, Service
1993-97 (.2) 3

(G) Filter, Fuel, Renew
1993-97 (.3) 3

BRAKES

(G) Bleed Brakes (Four Wheels)
Includes: Add fluid.
1993-97 (.4) 5

(G) Brakes, Adjust (Minor)
Includes: Adjust brakes, fill master cylinder.
1993-97, two wheels 4
Remove knock out plugs add each . .1

(M) Parking Brake, Adjust
1993-97 (.3) 4

LUBRICATION SERVICES

(M) Engine Oil & Filter, Renew
Includes: Inspect and correct all fluid levels.
1993-97 (.3) 3

(M) Lubricate Chassis, Change Oil & Filter
Includes: Inspect and correct all fluid levels.
1993-97 6
Install grease fittings add 1

(M) Lubricate Chassis
Includes: Inspect and correct all fluid levels.
1993-97 4
Install grease fittings add 1

FREQUENT MAINTENANCE LABOR (cont.)
BUICK CENTURY, OLDSMOBILE CUTLASS CIERA, CUTLASS CRUISER

	(Factory Time)	Chilton Time
WHEELS		
(G) Wheel, Renew (One)		
1993-97 (.5)		.5
(G) Wheel, Balance		
1993-97		
one		.3
each adtnl.		.2
(G) Wheels, Rotate (All)		
1993-97		.5

	(Factory Time)	Chilton Time
ELECTRICAL		
(G) Headlamps, Aim		
1993-97		
two		.4
four		.6
(G) High Mount Stop Lamp Bulb, Renew		
1993-97 (.2)		.3
(G) License Lamp Assy., Renew		
1993-97		
one or both (.2)		.3
(G) License Lamp Bulb, Renew		
1993-97, one or all (.2)		.3

	(Factory Time)	Chilton Time
(G) Park & Turn Signal Lamp Bulb or Lens, Renew		
1993-97, each		.3
(G) Stop, Tail & Turn Signal Lamp Bulb, Renew		
1993-97		
one		.3
each adtnl.		.1
(G) Horn, Renew		
1993-97 (.3)		.4
(M) Terminals, Battery, Clean		
1993-97		.3

SCHEDULED MAINTENANCE INTERVALS
(BUICK CENTURY, OLDSMOBILE CUTLASS CIERA & CUTLASS CRUISER)

TO BE SERVICED	TYPE OF SERVICE	VEHICLE MILEAGE INTERVAL (x1000)												
		7.5	15	22.5	30	37.5	45	52.5	60	67.5	75	82.5	90	97.5
Engine oil & filter	R	✓	✓	✓	✓	✓	✓	✓	✓	✓	✓	✓	✓	✓
Brake hoses	S/I	✓	✓	✓	✓	✓	✓	✓	✓	✓	✓	✓	✓	✓
Chassis lubrication	S/I	✓	✓	✓	✓	✓	✓	✓	✓	✓	✓	✓	✓	✓
Coolant level, hoses & clamps	S/I	✓	✓	✓	✓	✓	✓	✓	✓	✓	✓	✓	✓	✓
Drive shaft boots & front suspension components	S/I	✓	✓	✓	✓	✓	✓	✓	✓	✓	✓	✓	✓	✓
Exhaust system	S/I	✓	✓	✓	✓	✓	✓	✓	✓	✓	✓	✓	✓	✓
Lubricate transaxle shift linkage, parking brake cable guides, underbody contact points & linkage	S/I	✓	✓	✓	✓	✓	✓	✓	✓	✓	✓	✓	✓	✓
Rotate tires	S/I	✓		✓		✓		✓		✓		✓		✓
Brake linings	S/I	✓		✓		✓		✓		✓		✓		✓
Air filter element	R				✓				✓				✓	
Engine Coolant②	R				✓				✓				✓	
Spark plugs①	R				✓				✓				✓	
Accessory drive belt(s)	S/I				✓				✓				✓	

SCHEDULED MAINTENANCE INTERVALS
(BUICK CENTURY, OLDSMOBILE CUTLASS CIERA & CUTLASS CRUISER) (Cont.)

TO BE SERVICED	TYPE OF SERVICE	VEHICLE MILEAGE INTERVAL (x1000)												
		7.5	15	22.5	30	37.5	45	52.5	60	67.5	75	82.5	90	97.5
EGR & fuel systems	S/I				✓				✓				✓	
PCV valve & filter	S/I				✓				✓				✓	
Ignition Cables	S/I				✓				✓				✓	
Automatic transaxle fluid & filter③	R													
Throttle body mount bolt torque	S/I	✓												

① Platinum tip spark plugs - replace every 100,000 miles.
② Engine coolant (1996-97) - replace every 100,000 miles. Use O.E. specified (DEX-COOL™) coolant only. If any silicate coolant is used, the service interval is every 30,000 miles.
③ Replace fluid & filter every 100,000 miles (1993-95).
R – Replace   S/I – Service or Inspect

FREQUENT OPERATION MAINTENANCE (SEVERE SERVICE)
If a vehicle is operated under any of the following conditions it is considered severe service:
- Extremely dusty areas.
- 50% or more of the vehicle operation is in 32°C (90°F) or higher temperatures, or constant operation in temperatures below 0°C (32°F).
- Prolonged idling (vehicle operation in stop and go traffic).
- Frequent short running periods (engine does not warm to normal operating temperatures).
- Police, taxi, delivery usage or trailer towing usage.
Engine oil & filter change – change every 3000 miles.
Inspect CV joints & front suspension components - check every 3000 miles.
Throttle body mount bolt torque - tighten at 6000 miles.
Rotate tires at 6000 miles, then every 12,000 miles.
Brake linings – check every 9000 miles.
Air filter element – service or inspect every 15,000 miles.
Automatic transaxle fluid & filter - change every 15,000 miles (1993-95) or every 50,000 miles (1996-97).

LESABRE/PARK AVENUE/PARK AVENUE ULTRA
DEVILLE (1993)/SIXTY SPECIAL
EIGHTY-EIGHT ROYALE/NINETY-EIGHT/BONNEVILLE

VEHICLE IDENTIFICATION CHART

Code	Liters	Cu. In. (cc)	Cyl.	Fuel Sys.		Eng. Mfg.
1	3.8	231 (3785)	6	MFI	①	BOC
B	4.9	300 (4917)	8	MFI		Cadillac
K	3.8	231 (3785)	6	MFI		BOC
L	3.8	231 (3785)	6	MFI		BOC

Model Year	
Code	Year
P	1993
R	1994
S	1995
T	1996
V	1997

MFI - Multiport fuel injection
BOC - Buick/Oldsmobile/Cadillac
① Supercharged engine

ENGINE IDENTIFICATION

Year	Model	Engine Displacement Liters (cc)	Engine Series (ID/VIN)	Fuel System		No. of Cylinders	Engine Type
1993	Bonneville	3.8 (3785)	L	MFI		6	OHV
	Bonneville	3.8 (3785)	1	MFI	①	6	OHV
	DeVille	4.9 (4917)	B	MFI		8	OHV
	Eighty-Eight/Royale	3.8 (3785)	L	MFI		6	OHV
	Ninety-Eight	3.8 (3785)	L	MFI		6	OHV
	Ninety-Eight	3.8 (3785)	1	MFI	①	6	OHV
	LeSabre	3.8 (3785)	L	MFI		6	OHV
	Park Avenue	3.8 (3785)	L	MFI		6	OHV
	Park Avenue Ultra	3.8 (3785)	1	MFI	①	6	OHV
	Sixty Special	4.9 (4917)	B	MFI		8	OHV
1994	Bonneville	3.8 (3785)	L	MFI		6	OHV
	Bonneville	3.8 (3785)	1	MFI	①	6	OHV
	Eighty-Eight/Royale	3.8 (3785)	L	MFI		6	OHV
	Ninety-Eight	3.8 (3785)	L	MFI		6	OHV
	Ninety-Eight	3.8 (3785)	1	MFI	①	6	OHV
	LeSabre	3.8 (3785)	L	MFI		6	OHV
	Park Avenue	3.8 (3785)	L	MFI		6	OHV
	Park Avenue Ultra	3.8 (3785)	1	MFI	①	6	OHV
1995	Bonneville	3.8 (3785)	K	MFI		6	OHV
	Bonneville	3.8 (3785)	1	MFI	①	6	OHV
	Eighty-Eight/Royale	3.8 (3785)	K	MFI		6	OHV
	Eighty-Eight/LSS	3.8 (3785)	1	MFI	①	6	OHV
	Ninety-Eight	3.8 (3785)	K	MFI		6	OHV
	Ninety-Eight	3.8 (3785)	1	MFI	①	6	OHV
	LeSabre	3.8 (3785)	L	MFI		6	OHV
	Park Avenue	3.8 (3785)	L	MFI		6	OHV
	Park Avenue Ultra	3.8 (3785)	1	MFI	①	6	OHV

ENGINE IDENTIFICATION

Year	Model	Engine Displacement Liters (cc)	Engine Series (ID/VIN)	Fuel System	No. of Cylinders	Engine Type
1996-97	Bonneville	3.8 (3785)	K	MFI	6	OHV
	Bonneville	3.8 (3785)	1	MFI ①	6	OHV
	Eighty-Eight	3.8 (3786)	K	MFI	6	OHV
	Ninety-Eight	3.8 (3786)	K	MFI	6	OHV
	LeSabre	3.8 (3785)	K	MFI	6	OHV
	Park Avenue	3.8 (3785)	K	MFI	6	OHV
	Park Avenue Ultra	3.8 (3785)	1	MFI ①	6	OHV

MFI - Multiport fuel injection
OHV - Overhead valve
① Supercharged engine

GENERAL ENGINE SPECIFICATIONS

Year	Engine ID/VIN	Engine Displacement Liters (cc)	Fuel System Type	Net Horsepower @ rpm	Net Torque @ rpm (ft. lbs.)	Bore x Stroke (in.)	Compression Ratio	Oil Pressure @ rpm
1993	B	4.9 (4917)	MFI	200@4100	275@3000	3.62x3.62	9.5:1	53@2000
	L	3.8 (3785)	MFI	170@4800	225@3200	3.80x3.40	9.0:1	60@1850
	1	3.8 (3785)	MFI	205@4400	260@2600	3.80x3.40	8.5:1	60@1850 ①
1994	L	3.8 (3785)	MFI	170@4800	225@3200	3.80x3.40	9.0:1	60@1850
	1	3.8 (3785)	MFI	225@5000	275@3200	3.80x3.40	9.0:1	60@1850 ①
1995	K	3.8 (3875)	MFI	205@5200	230@4000	3.80x3.40	9.4:1	60@1850
	L	3.8 (3875)	MFI	170@4800	225@3200	3.80x3.40	9.0:1	60@1850
	1	3.8 (3875)	MFI	225@5000	275@3200	3.80x3.40	9.0:1	60@1850 ①
1996-97	K	3.8 (3785)	MFI	205@5200	230@4000	3.80x3.40	9.4:1	60@1850
	1	3.8 (3786)	MFI	240@5200	280@3200	3.80x3.40	9.0:1	60@1850

MFI - Multiport fuel injection
① Supercharged

GASOLINE ENGINE TUNE-UP SPECIFICATIONS

Year	Engine ID/VIN	Engine Displacement Liters (cc)	Spark Plugs Gap (in.)	Ignition Timing (deg.) MT	Ignition Timing (deg.) AT	Fuel Pump (psi)	Idle Speed (rpm) MT	Idle Speed (rpm) AT	Valve Clearance In.	Valve Clearance Ex.
1993	B	4.9 (4917)	0.060	-	①	40-50	-	①	HYD	HYD
	L	3.8 (3785)	0.060	①	①	40-47	-	②	HYD	HYD
	1	3.8 (3785)	0.060	①	①	40-47	-	②	HYD	HYD
1994	L	3.8 (3785)	0.060	①	①	40-47	-	②	HYD	HYD
	1	3.8 (3785)	0.060	①	①	40-47	-	②	HYD	HYD
1995	K	3.8 (3785)	0.060	①	①	40-47 ③	-	②	HYD	HYD
	L	3.8 (3785)	0.060	①	①	40-47 ③	-	②	HYD	HYD
	1	3.8 (3785)	0.060	①	①	40-47 ③	-	②	HYD	HYD
1996-97	K	3.8 (3785)	0.060	-	①	41-47	-	②	HYD	HYD
	1	3.8 (3786)	0.060	-	①	41-47	-	②	HYD	HYD

NOTE: The Vehicle Emission Control Information label often reflects specification changes made during production. The label figures must be used if they differ from those in this chart.
HYD - Hydraulic
① DIS Ignition System timing not adjustable
② Idle speed maintained by ECM. There is no recommended adjustment procedure
③ Pressure at fuel pump

CAPACITIES

Year	Model	Engine ID/VIN	Engine Displacement Liters (cc)	Engine Oil with Filter (qts.)		Transmission (pts.)			Drive Axle		Fuel Tank (gal.)	Cooling System (qts.)
						4-Spd	5-Spd	Auto.	Front (pts.)	Rear (pts.)		
1993	Bonneville	L	3.8 (3785)	4.0	①	-	-	13.0	-	-	18.0	13.0
	Bonneville	1	3.8 (3785)	4.0	①	-	-	13.0	-	-	18.0	13.0
	DeVille	B	4.9 (4917)	5.5		-	-	13.0	-	-	18.0	12.1
	Eighty-Eight	L	3.8 (3785)	4.0	①	-	-	13.0	-	-	18.0	13.0
	Ninety-Eight	L	3.8 (3785)	4.0	①	-	-	13.0	-	-	18.0	13.0
	Ninety-Eight	1	3.8 (3785)	4.0	①	-	-	13.0	-	-	18.0	13.0
	LeSabre	L	3.8 (3785)	4.0	①	-	-	13.0	-	-	18.0	13.0
	Park Avenue	L	3.8 (3785)	4.0	①	-	-	13.0	-	-	18.0	13.0
	Park Avenue Ultra	1	3.8 (3785)	4.0	①	-	-	13.0	-	-	18.0	13.0
	Sixty Special	B	4.9 (4917)	5.5		-	-	13.0	-	-	18.0	12.1
1994	Bonneville	L	3.8 (3785)	4.0	①	-	-	13.0	-	-	18.0	13.0
	Bonneville	1	3.8 (3785)	4.0	①	-	-	13.0	-	-	18.0	13.0
	Eighty-Eight	L	3.8 (3785)	4.0	①	-	-	13.0	-	-	18.0	13.0
	Ninety-Eight	L	3.8 (3785)	4.0	①	-	-	13.0	-	-	18.0	13.0
	Ninety-Eight	1	3.8 (3785)	4.0	①	-	-	13.0	-	-	18.0	13.0
	LeSabre	L	3.8 (3785)	4.0	①	-	-	13.0	-	-	18.0	13.0
	Park Avenue	L	3.8 (3785)	4.0	①	-	-	13.0	-	-	18.0	13.0
	Park Avenue Ultra	1	3.8 (3785)	4.0	①	-	-	13.0	-	-	18.0	13.0
1995	Bonneville	K	3.8 (3785)	3.8	①	-	-	13.0	-	-	18.0	13.0
	Bonneville	1	3.8 (3785)	3.8	①	-	-	13.0	-	-	18.0	13.0
	Eighty-Eight	K	3.8 (3785)	3.8	①	-	-	13.0	-	-	18.0	13.0
	Eighty-Eight	1	3.8 (3785)	3.8	①	-	-	13.0	-	-	18.0	13.0
	Ninety-Eight	K	3.8 (3785)	3.8	①	-	-	13.0	-	-	18.0	13.0
	Ninety-Eight	1	3.8 (3785)	3.8	①	-	-	13.0	-	-	18.0	13.0
	LeSabre	L	3.8 (3785)	4.0	①	-	-	13.0	-	-	18.0	13.0
	Park Avenue	L	3.8 (3785)	4.0	①	-	-	13.0	-	-	18.0	13.0
	Park Avenue Ultra	1	3.8 (3785)	4.0	①	-	-	13.0	-	-	18.0	13.0
1996-97	Bonneville	K	3.8 (3785)	5.0	②	-	-	12.0	-	-	18.0	13.0
	Bonneville	1	3.8 (3785)	5.0	②	-	-	12.0	-	-	18.0	13.0
	Eighty-Eight	K	3.8 (3785)	5.0	①	-	-	13.0	-	-	18.0	13.0
	Ninety-Eight	K	3.8 (3785)	5.0	①	-	-	13.0	-	-	18.0	13.0
	LeSabre	K	3.8 (3785)	5.0		-	-	12.0	-	-	18.0	13.0
	Park Avenue	K	3.8 (3785)	5.0		-	-	12.0	-	-	18.0	13.0
	Park Avenue Ultra	1	3.8 (3785)	5.0		-	-	12.0	-	-	18.0	13.0

① Specification is without filter replacement; Additional oil may be required
② Fluid change with filter

VALVE SPECIFICATIONS

Year	Engine ID/VIN	Engine Displacement Liters (cc)	Seat Angle (deg.)	Face Angle (deg.)	Spring Test Pressure (lbs. @ in.)	Spring Installed Height (in.)	Stem-to-Guide Clearance (in.)		Stem Diameter (in.)	
							Intake	Exhaust	Intake	Exhaust
1993	B	4.9 (4917)	45	45	68-76@ 1.350	1.350	0.0010- 0.0030	0.0020- 0.0040	0.3413- 0.3420	0.3401- 0.3408
	L	3.8 (3785)	45	45	210@1.315	1.690- 1.720	0.0015- 0.0035	0.0015- 0.0032	NA	NA
	1	3.8 (3785)	45	45	210@1.315	1.690- 1.720	0.0015- 0.0035	0.0015- 0.0032	NA	NA

VALVE SPECIFICATIONS

Year	Engine ID/VIN	Engine Displacement Liters (cc)	Seat Angle (deg.)	Face Angle (deg.)	Spring Test Pressure (lbs. @ in.)	Spring Installed Height (in.)	Stem-to-Guide Clearance (in.)		Stem Diameter (in.)	
							Intake	Exhaust	Intake	Exhaust
1994	L	3.8 (3785)	45	45	210@1.315	1.690-1.720	0.0015-0.0035	0.0015-0.0032	NA	NA
	1	3.8 (3785)	45	45	210@1.315	1.690-1.720	0.0015-0.0035	0.0015-0.0032	NA	NA
1995	K	3.8 (3785)	46	45	210@1.315	1.690-1.720	0.0015-0.0035	0.0015-0.0032	NA	NA
	L	3.8 (3785)	45	45	210@1.315	1.690-1.720	0.0015-0.0035	0.0015-0.0032	NA	NA
	1	3.8 (3785)	46	45	210@1.315	1.690-1.720	0.0015-0.0035	0.0015-0.0032	NA	NA
1996-97	K	3.8 (3785)	45	45	80@1.750	1.690-1.720	0.0015-0.0032	0.0015-0.0032	NA	NA
	1	3.8 (3785)	45	45	80@1.750	1.690-1.720	0.0015-0.0032	0.0015-0.0032	NA	NA

NA - Not Available

TORQUE SPECIFICATIONS
All readings in ft. lbs.

Year	Engine ID/VIN	Engine Displacement Liters (cc)	Cylinder Head Bolts	Main Bearing Bolts	Rod Bearing Bolts	Crankshaft Damper Bolts	Flywheel Bolts	Manifold		Spark Plugs	Lug Nut
								Intake	Exhaust		
1993	B	4.9 (4917)	⑥	85	25	70	70	⑦	16	23	100
	L	3.8 (3785)	①	②	③	④	⑤	7	38	20	100
	1	3.8 (3785)	①	②	③	④	⑤	7	38	20	100
1994	L	3.8 (3785)	①	②	③	④	⑤	7	38	11	100
	1	3.8 (3785)	①	②	③	④	⑤	7	38	11	100
1995	K	3.7 (3785)	①	②	③	④	⑤	11	38	11	100
	L	3.8 (3785)	①	②	③	④	⑤	11	38	11	100
	1	3.7 (3785)	①	②	③	④	⑤	11	38	11	100
1996-97	K	3.8 (3785)	①	②	③	④	⑤	11	38	11	100
	1	3.8 (3786)	①	⑧	③	④	⑤	⑨	22	11	100

① Step 1: Tighten all bolts to 35 ft. lbs.
  Step 2: Turn all bolts 130 degrees
  Step 3: Rotate four center bolts an additional 30 degrees
② 26 ft. lbs. plus 50 degrees
③ 20 ft. lbs. plus 50 degrees

④ 110 ft. lbs. plus 76 degrees
⑤ 11 ft. lbs. plus 50 degrees
⑥ Step 1: 38 ft. lbs.
  Step 2: 68 ft. lbs.
  Step 3: 90 ft. lbs.
⑦ Step 1: 8 ft. lbs.
  Step 2: 12 ft. lbs.

⑧ Step 1: Tighten caps in equal increments to 52 ft. lbs.
  Step 2: Loosen 360 degrees
  Step 3: 15 ft. lbs.
  Step 4: 54 ft. lbs.
  Step 5: Plus three turns of 35 degrees for a total of 105 degrees
⑨ Upper manifold: 8 ft. lbs.
  Lower manifold: 11 ft. lbs.

BRAKE SPECIFICATIONS
All measurements in inches unless noted

Year	Model	Master Cylinder Bore	Brake Disc			Brake Drum Diameter			Minimum Lining Thickness	
			Original Thickness	Minimum Thickness	Maximum Runout	Original Inside Diameter	Max. Wear Limit	Maximum Machine Diameter	Front	Rear
1993	Bonneville	1.000	1.276	1.209	0.004	8.860	8.909	8.880	0.030	①
	DeVille	1.000	1.276	1.209	0.004	8.860	8.909	8.880	0.030	①
	Eighty-Eight	1.000	1.276	1.209	0.004	8.860	8.909	8.880	0.030	①

BRAKE SPECIFICATIONS
All measurements in inches unless noted

Year	Model	Master Cylinder Bore	Brake Disc Original Thickness	Brake Disc Minimum Thickness	Maximum Runout	Brake Drum Diameter Original Inside Diameter	Brake Drum Diameter Max. Wear Limit	Brake Drum Diameter Maximum Machine Diameter	Minimum Lining Thickness Front	Minimum Lining Thickness Rear
1993	Ninety-Eight	1.000	1.276	1.209	0.004	8.860	8.909	8.880	0.030	①
	LeSabre	1.000	1.276	1.209	0.004	8.860	8.909	8.880	0.030	①
	Park Avenue	1.000	1.276	1.209	0.004	8.860	8.909	8.880	0.030	①
	Sixty Special	1.000	1.276	1.209	0.004	8.860	8.909	8.880	0.030	0.030
1994	Bonneville	1.000	1.260	1.209	0.002	8.863	8.909	8.880	0.030	①
	Eighty-Eight	1.000	1.260	1.209	0.002	8.863	8.909	8.880	0.030	①
	Ninety-Eight	1.000	1.260	1.209	0.002	8.863	8.909	8.880	0.030	①
	LeSabre	1.000	1.260	1.209	0.002	8.863	8.909	8.880	0.030	①
	Park Avenue	1.000	1.260	1.209	0.002	8.863	8.909	8.880	0.030	①
1995	Bonneville	1.000	1.276	1.209	0.004	8.860	8.909	8.800	0.030	①
	Eighty-Eight	1.000	1.276	1.209	0.004	8.860	8.909	8.800	0.030	①
	Ninety-Eight	1.000	1.276	1.209	0.004	8.860	8.909	8.800	0.030	①
	LeSabre	1.000	1.276	1.209	0.004	8.860	8.909	8.800	0.030	①
	Park Avenue	1.000	1.276	1.209	0.004	8.860	8.909	8.800	0.030	①
1996-97	Bonneville	1.000	1.260	1.209	0.002	8.860	8.909	8.920	0.030	0.030
	Eighty-Eight	1.000	1.276	1.209	0.004	8.860	8.909	8.800	0.030	①
	Ninety-Eight	1.000	1.276	1.209	0.004	8.860	8.909	8.800	0.030	①
	LeSabre	1.000	1.260	1.209	0.002	8.863	8.909	8.920	0.030	①
	Park Avenue	1.000	1.260	1.209	0.002	8.863	8.909	8.920	0.030	①

NA - Not Available

F - Front

R - Rear

① 0.030 over rivet head; If bonded lining, use 0.062 from shoe

FREQUENT MAINTENANCE LABOR
BUICK PARK AVENUE, CADILLAC DEVILLE, FLEETWOOD
OLDSMOBILE NINETY-EIGHT

The following should be used as a guide when determining the amount of work required for a particular service if taken to a repair shop. In estimating how long a particular Frequent Maintenance Service item should take, please observe the following:
- **Factory Time** is time that is generated by the vehicle manufacturer.
- **Chilton Time** is time that is based on field research and data supplied by the vehicle manufacturer.
- All labor time operations are given in hours and tenths of an hour.
- All labor operations, are to be used as a **guide**.

	(Factory Time)	Chilton Time
COOLING		
(G) Winterize Cooling System		
Includes: Run engine to check for leaks, tighten all hose connections. Test radiator and pressure cap, drain radiator and engine block. Add antifreeze and refill system.		
1993-97		.5

	(Factory Time)	Chilton Time
(G) Belt, Drive, Renew		
Gas		
1993 (.2)		.4
1994-95		
code K (.3)		.5
code L (.3)		.5
code 1		
outer (.2)		.3
inner (.3)		.4

	(Factory Time)	Chilton Time
1996		
3.8L code K (.3)		.5
3.8L code 1		
outer (.5)		.7
inner (.6)		.8
1993		
V8 (.2)		.4

FREQUENT MAINTENANCE LABOR (cont.)
BUICK PARK AVENUE, CADILLAC DEVILLE, FLEETWOOD
OLDSMOBILE NINETY-EIGHT

	(Factory Time)	Chilton Time
(G) Hoses, Radiator, Renew		
Includes: Drain and refill cooling system as required.		
1993		
upper (.3)		.4
lower (.4)		.5
both (.5)		.7
by-pass (.5)		.7
1994-97		
upper (.7)		.9
lower (.7)		.9
by-pass (.5)		.7
(G) Thermostat, Coolant, Renew		
1993-94		
V6 (.4)		.5
V8 (.3)		.4
1995-97		
V6		
3.8L codes L, 1 (.4)		.5
3.8L code K (.6)		.8

FUEL

	(Factory Time)	Chilton Time
(M) Air Cleaner, Service		
1993-97		.2
(G) Filter, Fuel, Renew		
1993-97 (.2)		.3

BRAKES

	(Factory Time)	Chilton Time
(G) Bleed Brakes (Four Wheels)		
Includes: Add fluid.		
1993-97 (.4)		.5
w/ABS add		.3

	(Factory Time)	Chilton Time
(G) Brakes, Adjust (Minor)		
Includes: Adjust brakes, fill master cylinder.		
1993-97, two wheels		.4
Remove knock out plugs add each		.1
(M) Parking Brake, Adjust		
1993-97 (.7)		1.0

LUBRICATION SERVICES

	(Factory Time)	Chilton Time
(M) Engine Oil & Filter, Renew		
Includes: Inspect and correct all fluid levels.		
1993-97		.3
(M) Lubricate Chassis, Change Oil & Filter		
Includes: Inspect and correct all fluid levels.		
1993-97		.6
Install grease fittings add		.1
(M) Lubricate Chassis		
Includes: Inspect and correct all fluid levels.		
1993-97		.4
Install grease fittings add		.1

WHEELS

	(Factory Time)	Chilton Time
(G) Wheel, Renew (One)		
1993-97		.5
(G) Wheel, Balance		
1993-97		
one		.3
each adtnl.		.2
(G) Wheels, Rotate (All)		
1993-97		.5

ELECTRICAL

	(Factory Time)	Chilton Time
(G) Headlamps, Aim		
1993-97		
two		.4
four		.6
(G) High Mount Stop Lamp Bulb, Renew		
1993-97 (.2)		.3
(G) License Lamp Assy., Renew		
1993-97, one or all (.2)		.3
(G) License Lamp Bulb, Renew		
1993-97, one or all (.2)		.3
(G) Park & Turn Signal Lamp Assy., Renew		
1993-97, each (.2)		.4
(G) Park & Turn Signal Lamp Bulb or Lens, Renew		
1993-97		
one (.2)		.3
each adtnl.		.1
(G) Stop, Tail & Turn Signal Lamp Bulb, Renew		
1993-97		
one		.3
each adtnl.		.1
(G) Horn, Renew		
1993-97		
one (.3)		.4
each adtnl.		.1
(M) Terminals, Battery, Clean		
1993-97		.3

FREQUENT MAINTENANCE LABOR
BUICK LESABRE, OLDSMOBILE EIGHTY-EIGHT
PONTIAC BONNEVILLE

The following should be used as a guide when determining the amount of work required for a particular service if taken to a repair shop. In estimating how long a particular Frequent Maintenance Service item should take, please observe the following:
- **Factory Time** is time that is generated by the vehicle manufacturer.
- **Chilton Time** is time that is based on field research and data supplied by the vehicle manufacturer.
- All labor time operations are given in hours and tenths of an hour.
- All labor operations, are to be used as a **guide**.

COOLING

(G) Winterize Cooling System
Includes: Run engine to check for leaks, tighten all hose connections. Test radiator and pressure cap, drain radiator and engine block. Add antifreeze and refill system.

	Factory Time	Chilton Time
1993-97		.5

(G) Belt, Serpentine Drive, Renew

	Factory Time	Chilton Time
1993 (.2)		.4
1994-95		
3.8L code K (.3)		.5
3.8L code L (.3)		.5
3.8L code 1		
outer (.2)		.3
inner (.3)		.4
1996		
3.8L code K (.3)		.5
3.8L code 1		
outer (.5)		.7
inner (.6)		.8

(G) Hoses, Radiator, Renew
Includes: Drain and refill cooling system as required.

	Factory Time	Chilton Time
1993		
upper (.3)		.4
lower (.4)		.5
both (.5)		.7
by-pass (.5)		.7
1994-97		
upper (.7)		.9
lower (.7)		.9
by-pass (.5)		.7

(G) Thermostat, Coolant, Renew

	Factory Time	Chilton Time
1993 (.4)		.5
1994-97		
codes L, 1 (.4)		.5
code K (.6)		.8

FUEL

(M) Air Cleaner, Service

	Factory Time	Chilton Time
1993-97		.3

(G) Filter, Fuel, Renew

	Factory Time	Chilton Time
1993-97 (.2)		.3

BRAKES

(G) Bleed Brakes (Four Wheels)
Includes: Add fluid.

	Factory Time	Chilton Time
1993-97 (.4)		.5
w/ABS add		.3

(G) Brakes, Adjust (Minor)
Includes: Adjust brakes, fill master cylinder.

	Factory Time	Chilton Time
1993-97, two wheels		.4
Remove knock out plugs add each		.1

(M) Parking Brake, Adjust

	Factory Time	Chilton Time
1993-97 (.7)		1.0

LUBRICATION SERVICES

(M) Engine Oil & Filter, Renew
Includes: Inspect and correct all fluid levels.

	Factory Time	Chilton Time
1993-97 (.3)		.3

(M) Lubricate Chassis, Change Oil & Filter
Includes: Inspect and correct all fluid levels.

	Factory Time	Chilton Time
1993-97		.6
Install grease fittings add		.1

(M) Lubricate Chassis
Includes: Inspect and correct all fluid levels.

	Factory Time	Chilton Time
1993-97		.4
Install grease fittings add		.1

WHEELS

(G) Wheel, Renew (One)

	Factory Time	Chilton Time
1993-97		.5

(G) Wheel, Balance

	Factory Time	Chilton Time
1993-97		
one		.3
each adtnl.		.2

(G) Wheels, Rotate (All)

	Factory Time	Chilton Time
1993-97		.5

ELECTRICAL

(G) Headlamps, Aim

	Factory Time	Chilton Time
1993-97		
two		.4
four		.6

(G) High Mount Stop Lamp Bulb, Renew

	Factory Time	Chilton Time
1993-97 (.2)		.3

(G) License Lamp Assy., Renew

	Factory Time	Chilton Time
1993-97 (.2)		.3

(G) License Lamp Bulb, Renew

	Factory Time	Chilton Time
1993-97, one or all (.2)		.3

(G) Park & Turn Signal Lamp Assy., Renew

	Factory Time	Chilton Time
1993		
Delta 88 (.2)		.3
LeSabre (.2)		.3
Bonneville		
right (.8)		1.1
left (.6)		.8
1994-97		
Bonneville (.6)		.8
Delta 88, LeSabre (.2)		.3

(G) Park & Turn Signal Lamp Bulb or Lens, Renew

	Factory Time	Chilton Time
1993-97, each		.2

(G) Rear Combination Lamp Bulb, Renew

	Factory Time	Chilton Time
1993-97		
one		.2
each adtnl.		.1

(G) Horn, Renew

	Factory Time	Chilton Time
1993-97		
one (.3)		.4
each adtnl.		.1

(M) Terminals, Battery, Clean

	Factory Time	Chilton Time
1993-97		.3

SCHEDULED MAINTENANCE INTERVALS
(BUICK LESABRE, PARK AVENUE, PARK AVENUE ULTRA, CADILLAC DEVILLE (1993), SIXTY SPECIAL, OLDSMOBILE EIGHTY-EIGHT ROYALE, NINETY-EIGHT, PONTIAC BONNEVILLE)

TO BE SERVICED	TYPE OF SERVICE	VEHICLE MILEAGE INTERVAL (x1000)												
		7.5	15	22.5	30	37.5	45	52.5	60	67.5	75	82.5	90	97.5
Engine oil & filter	R	✓	✓	✓	✓	✓	✓	✓	✓	✓	✓	✓	✓	✓
Exhaust system & brake hoses	S/I	✓	✓	✓	✓	✓	✓	✓	✓	✓	✓	✓	✓	✓
Drive shaft boots & front suspension components	S/I	✓	✓	✓	✓	✓	✓	✓	✓	✓	✓	✓	✓	✓
Lubricate chassis, suspension, steering linkage, transaxle shift linkage, parking brake cable guides, underbody contact points & linkage	S/I	✓	✓	✓	✓	✓	✓	✓	✓	✓	✓	✓	✓	✓
Coolant level, hoses & clamps	S/I	✓	✓	✓	✓	✓	✓	✓	✓	✓	✓	✓	✓	✓
Throttle linkage	S/I	✓	✓	✓	✓	✓	✓	✓	✓	✓	✓	✓	✓	✓
Brake linings & rotate tires	S/I	✓		✓		✓		✓		✓		✓		✓
Accessory drive belts supercharger oil	S/I					✓			✓				✓	
Engine coolant (1993-95)②	R				✓				✓				✓	
Spark plugs①	R				✓				✓				✓	
Air filter element	R				✓				✓				✓	
PCV filter	R				✓				✓				✓	
Ignition cables	S/I				✓				✓				✓	
EGR & fuel systems	S/I				✓				✓				✓	
Automatic transaxle fluid & filter	R													✓
Throttle body mount bolt torque	S/I	✓												

① Platinum tip spark plugs - replace every 100,000 miles.
② Engine coolant (1996-97) - replace every 100,000 miles. Use O.E. specified (DEX-COOL™) only. If any silicate coolant is used, the service interval is every 30,000 miles.
R – Replace   S/I – Service or Inspect

SCHEDULED MAINTENANCE INTERVALS
(BUICK LESABRE, PARK AVENUE, PARK AVENUE ULTRA, CADILLAC DEVILLE (1993), SIXTY SPECIAL, OLDSMOBILE EIGHTY-EIGHT ROYALE, NINETY-EIGHT, PONTIAC BONNEVILLE) (Cont.)

FREQUENT OPERATION MAINTENANCE (SEVERE SERVICE)

If a vehicle is operated under any of the following conditions it is considered severe service:
- Extremely dusty areas.
- 50% or more of the vehicle operation is in 32°C (90°F) or higher temperatures, or constant operation in temperatures below 0°C (32°F).
- Prolonged idling (vehicle operation in stop and go traffic).
- Frequent short running periods (engine does not warm to normal operating temperatures).
- Police, taxi, delivery usage or trailer towing usage.

Engine oil & filter change – change every 3000 miles.
Inspect CV joints & front suspension components - check every 3000 miles.
Brake linings – check every 6000 miles.
Chassis lubrication - lubricate every 6000 miles.
Throttle body mount bolt torque - tighten at 6000 miles.
Air filter element – service or inspect every 15,000 miles.
Automatic transaxle fluid – change every 15,000 miles (1993-94) or every 50,000 miles (1995-97).
Rotate tires at 6000 miles, then every 15,000 miles.

RIVIERA (1993)/DEVILLE (1994-97)
DEVILLE CONCOURS/ELDORADO/SEVILLE

VEHICLE IDENTIFICATION CHART

Engine Code						Model Year	
Code	Liters	Cu. In. (cc)	Cyl.	Fuel Sys.	Eng. Mfg.	Code	Year
L	3.8	231 (3785)	6	MFI	BOC	P	1993
9	4.6	279 (4573)	8	MFI	Cadillac	R	1994
B	4.9	300 (4917)	8	MFI	Cadillac	S	1995
Y	4.6	279 (4573)	8	MFI	Cadillac	T	1996
						V	1997

MFI - Multiport fuel injection

BOC - Buick/Oldsmobile/Cadillac

ENGINE IDENTIFICATION

Year	Model		Engine Displacement Liters (cc)	Engine Series (ID/VIN)	Fuel System	No. of Cylinders	Engine Type
1993	Allante		4.6 (4573)	9	MFI	8	DOHC
	Eldorado	①	4.6 (4573)	Y	MFI	8	DOHC
	Eldorado	②	4.6 (4573)	9	MFI	8	DOHC
	Eldorado		4.9 (4917)	B	MFI	8	OHV
	Seville STS		4.6 (4573)	9	MFI	8	DOHC
	Seville		4.9 (4917)	B	MFI	8	OHV
	Riviera		3.8 (3785)	L	MFI	6	OHV
1994	DeVille		4.9 (4917)	B	MFI	8	OHV
	DeVille Concours		4.6 (4573)	Y	MFI	8	DOHC
	Eldorado	②	4.6 (4573)	Y	MFI	8	DOHC
	Eldorado	①	4.6 (4573)	9	MFI	8	DOHC
	Seville SLS		4.6 (4573)	Y	MFI	8	DOHC
	Seville STS		4.6 (4573)	9	MFI	8	DOHC
1995	DeVille		4.9 (4917)	B	MFI	8	OHV
	DeVille Concours	②	4.6 (4573)	Y	MFI	8	DOHC
	Eldorado ETC	②	4.6 (4573)	Y	MFI	8	DOHC
	Eldorado	①	4.6 (4573)	9	MFI	8	DOHC
	Seville SLS	①	4.6 (4573)	Y	MFI	8	DOHC
	Seville STS	②	4.6 (4573)	9	MFI	8	DOHC
1996-97	DeVille	①	4.6 (4573)	Y	MFI	8	DOHC
	DeVille Concours	②	4.6 (4573)	9	MFI	8	DOHC
	Eldorado	①	4.6 (4573)	Y	MFI	8	DOHC
	Eldorado ETC	②	4.6 (4573)	9	MFI	8	DOHC
	Seville SLS	①	4.6 (4573)	Y	MFI	8	DOHC
	Seville STS	②	4.6 (4573)	9	MFI	8	DOHC

MFI - Multiport fuel injection
OHV - Overhead valve
DOHC - DOuble overhead camshaft

① Northstar low output (270 HP)
② Northstar high output (295 HP)

GENERAL ENGINE SPECIFICATIONS

Year	Engine ID/VIN	Engine Displacement Liters (cc)	Fuel System Type	Net Horsepower @ rpm	Net Torque @ rpm (ft. lbs.)	Bore x Stroke (in.)	Compression Ratio	Oil Pressure @ rpm
1993	L	3.8 (3785)	MFI	170@4800	225@3200	3.80x3.40	9.0:1	60@1850
	9	4.6 (4573)	MFI	295@6000	290@4400	3.66x3.31	10.3:1	35@2000
	Y	4.6 (4573)	MFI	270@5600	300@4000	3.66x3.31	10.3:1	35@2000
	B	4.9 (4917)	MFI	200@4100	275@3000	3.62x3.62	9.5:1	53@2000
1994	9	4.6 (4573)	MFI	295@6000	290@4400	3.66x3.31	10.3:1	35@2000
	Y	4.6 (4573)	MFI	270@5600	300@4000	3.66x3.31	10.3:1	35@2000
	B	4.9 (4917)	MFI	200@4100	275@3000	3.62x3.62	9.5:1	53@2000
1995	9	4.6 (4573)	MFI	300@6000	290@4400	3.66x3.31	10.3:1	35@2000
	Y	4.6 (4573)	MFI	275@5600	300@4000	3.66x3.31	10.3:1	35@2000
	B	4.9 (4917)	MFI	200@4400	275@3000	3.62x3.62	9.5:1	53@2000
1996-97	9	4.6 (4573)	MFI	300@6000	290@4400	3.66x3.31	10.3:1	35@2000
	Y	4.6 (4573)	MFI	275@5600	300@4000	3.66x3.31	10.3:1	35@2000

MFI - Multiport fuel injection

GASOLINE ENGINE TUNE-UP SPECIFICATIONS

Year	Engine ID/VIN	Engine Displacement Liters (cc)	Spark Plugs Gap (in.)	Ignition Timing (deg.) MT	Ignition Timing (deg.) AT	Fuel Pump (psi)	Idle Speed (rpm) MT	Idle Speed (rpm) AT	Valve Clearance In.	Valve Clearance Ex.
1993	L	3.8 (3785)	0.060	①	①	40-47	-	②	HYD	HYD
	9	4.6 (4573)	0.050	-	③	40-50	-	③	HYD	HYD
	Y	4.6 (4573)	0.050	-	③	40-50	-	③	HYD	HYD
	B	4.9 (4917)	0.060	-	③	40-50	-	③	HYD	HYD
1994	9	4.6 (4573)	0.050	-	③	40-50	-	③	HYD	HYD
	Y	4.6 (4573)	0.050	-	③	40-50	-	③	HYD	HYD
	B	4.9 (4917)	0.060	-	③	40-50	-	③	HYD	HYD
1995	9	4.6 (4573)	0.050	-	③	40-50	-	③	HYD	HYD
	Y	4.6 (4573)	0.050	-	③	40-50	-	③	HYD	HYD
	B	4.9 (4917)	0.060	-	③	40-50	-	③	HYD	HYD
1996-97	9	4.6 (4573)	0.050	-	③	40-50	-	③	HYD	HYD
	Y	4.6 (4573)	0.050	-	③	40-50	-	③	HYD	HYD

NOTE: The Vehicle Emission Control Information label often reflects specification changes made during production. The label figures must be used if they differ from those in this chart.

HYD - Hydraulic

① DIS Ignition System timing is not adjustable
② Idle speed is maintained by the ECM. There is no recommended adjustment procedure
③ Refer to Vehicle Emission Control Information label

CAPACITIES

Year	Model	Engine ID/VIN	Engine Displacement Liters (cc)	Engine Oil with Filter (qts.)	Transmission (pts.) 4-Spd	Transmission (pts.) 5-Spd	Transmission (pts.) Auto.	Drive Axle Front (pts.)	Drive Axle Rear (pts.)	Fuel Tank (gal.)	Cooling System (qts.)
1993	Allante	9	4.6 (4573)	7.5	-	-	16.0 ④	-	-	22.5	12.3
	Riviera	L	3.8 (3785)	4.0 ①	-	-	12.0	-	-	18.0	13.0
	Eldorado	B	4.9 (4917)	5.5	-	-	12.0 ④ ⑤	-	-	20.0	12.3
	Eldorado ②	9	4.6 (4573)	7.5	-	-	12.0 ④ ⑤	-	-	20.0	12.3
	Eldorado ③	Y	4.6 (4573)	7.5	-	-	12.0 ④ ⑤	-	-	20.0	12.3
	Seville STS	9	4.6 (4573)	7.5	-	-	12.0 ④ ⑤	-	-	20.0	12.3
	Seville	B	4.9 (4917)	5.5	-	-	12.0 ④ ⑤	-	-	20.0	12.3
1994	DeVille Concours	Y	4.6 (4573)	7.5	-	-	16.0	-	-	18.0	12.1
	DeVille	B	4.9 (4917)	5.5	-	-	13.0 ④	-	-	18.0	12.1
	Eldorado	Y	4.6 (4573)	7.5	-	-	16.0	-	-	20.0	12.3
	Eldorado ②	9	4.6 (4573)	7.5	-	-	16.0	-	-	20.0	12.3
	Seville SLS	Y	4.6 (4573)	7.5	-	-	16.0	-	-	20.0	12.3
	Seville STS	9	4.6 (4573)	7.5	-	-	16.0	-	-	20.0	12.3
1995	DeVille Concours	Y	4.6 (4573)	7.5	-	-	16.0 ⑥	-	-	18.0	12.1
	DeVille	B	4.9 (4917)	5.5	-	-	13.0 ④	-	-	18.0	12.1
	Eldorado	Y	4.6 (4573)	7.5	-	-	16.0 ⑥	-	-	20.0	12.3
	Eldorado	9	4.6 (4573)	7.5	-	-	16.0 ⑥	-	-	20.0	12.3
	Seville SLS	Y	4.6 (4573)	7.5	-	-	16.0 ⑥	-	-	20.0	12.3
	Seville STS	9	4.6 (4573)	7.5	-	-	16.0 ⑥	-	-	20.0	12.3
1996-97	DeVille	Y	4.6 (4573)	7.5	-	-	16.0 ⑥	-	-	20.0	12.3 ⑦
	DeVille Concours	9	4.6 (4573)	7.5	-	-	16.0 ⑥	-	-	18.0	12.1 ⑦
	Eldorado	Y	4.6 (4573)	7.5	-	-	16.0 ⑥	-	-	20.0	12.3 ⑦

CAPACITIES

Year	Model	Engine ID/VIN	Engine Displacement Liters (cc)	Engine Oil with Filter (qts.)	Transmission (pts.) 4-Spd	5-Spd	Auto.	Drive Axle Front (pts.)	Rear (pts.)	Fuel Tank (gal.)	Cooling System (qts.)
1996-97	Eldorado	9	4.6 (4573)	7.5	-	-	16.0 ⑥	-	-	20.0	12.3 ⑦
	Seville SLS	Y	4.6 (4573)	7.5	-	-	16.0 ⑥	-	-	20.0	12.3 ⑦
	Seville STS	9	4.6 (4573)	7.5	-	-	16.0 ⑥	-	-	20.0	12.3 ⑦

① Specification is without filter replacement; Additional oil may be required
② Touring Coupe
③ Sport
④ Fluid change with filter

⑤ With 4T80 transaxle: 16.0 pts.
⑥ Bottom pan and side cover
⑦ Dex-Cool engine coolant and three pellets of P/N 1052753 or equivalent
   (Do not mix with ethylene glycol base)

VALVE SPECIFICATIONS

Year	Engine ID/VIN	Engine Displacement Liters (cc)	Seat Angle (deg.)	Face Angle (deg.)	Spring Test Pressure (lbs. @ in.)	Spring Installed Height (in.)	Stem-to-Guide Clearance (in.) Intake	Exhaust	Stem Diameter (in.) Intake	Exhaust
1993	L	3.8 (3785)	45	45	210@1.315	1.690-1.720	0.0015-0.0035	0.0015-0.0032	NA	NA
	9	4.6 (4573)	46	45	53@1.190	1.190	0.0010-0.0030	0.0020-0.0040	0.2331-0.2339	0.2331-0.2339
	Y	4.6 (4573)	46	45	53@1.190	1.190	0.0010-0.0030	0.0020-0.0040	0.2331-0.2339	0.2331-0.2339
	B	4.9 (4917)	45	45	68-76@1.730	1.730	0.0010-0.0030	0.0020-0.0040	0.3413-0.3420	0.3401-0.3408
1994	9	4.6 (4573)	46	45	53@1.190	1.190	0.0010-0.0030	0.0020-0.0040	0.2331-0.2339	0.2331-0.2339
	Y	4.6 (4573)	46	45	53@1.190	1.190	0.0010-0.0030	0.0020-0.0040	0.2331-0.2339	0.2331-0.2339
	B	4.9 (4917)	45	45	68-76@1.350	1.350	0.0010-0.0030	0.0020-0.0040	0.3413-0.3420	0.3401-0.3408
1995	9	4.6 (4573)	46	45	53@1.190	1.190	0.0010-0.0030	0.0020-0.0040	0.2331-0.2339	0.2331-0.2339
	Y	4.6 (4573)	46	45	46@1.190	1.190	0.0010-0.0030	0.0020-0.0040	0.2331-0.2339	0.2331-0.2339
	B	4.9 (4917)	45	44	68-76@1.730	1.730	0.0010-0.0030	0.0020-0.0040	0.3413-0.3420	0.3401-0.3408
1996-97	9	4.6 (4573)	46	45	53@1.190	1.190	0.0010-0.0030	0.0020-0.0040	0.2331-0.2339	0.2331-0.2339
	Y	4.6 (4573)	46	45	46@1.190	1.190	0.0010-0.0030	0.0020-0.0040	0.2331-0.2339	0.2331-0.2339

NA - Not Available

TORQUE SPECIFICATIONS
All readings in ft. lbs.

Year	Engine ID/VIN	Engine Displacement Liters (cc)	Cylinder Head Bolts	Main Bearing Bolts	Rod Bearing Bolts	Crankshaft Damper Bolts	Flywheel Bolts	Manifold Intake	Exhaust	Spark Plugs	Lug Nut
1993	L	3.8 (3785)	①	②	③	④	⑤	7	38	20	100
	9	4.6 (4573)	⑥	⑤	③	⑦	⑧	⑨	20	14	100
	Y	4.6 (4573)	⑥	⑤	③	⑦	⑧	⑨	20	14	100
	B	4.9 (4917)	⑩	85	25	⑰	70	⑪	16	23	100

TORQUE SPECIFICATIONS
All readings in ft. lbs.

Year	Engine ID/VIN	Engine Displacement Liters (cc)	Cylinder Head Bolts	Main Bearing Bolts	Rod Bearing Bolts	Crankshaft Damper Bolts	Flywheel Bolts	Manifold Intake	Exhaust	Spark Plugs	Lug Nut
1994	9	4.6 (4573)	⑥	⑧	⑥	⑦	⑤	⑨	20	14	100
	Y	4.6 (4573)	⑥	⑧	⑥	⑦	⑤	⑨	20	14	100
	B	4.9 (4917)	⑩	85	25	⑰	70	⑪	16	23	100
1995	9	4.6 (4573)	⑥	⑧	⑫	⑬	⑤	⑭	18	11	100
	Y	4.6 (4573)	⑥	⑧	⑫	⑬	⑤	⑭	18	11	100
	B	4.9 (4917)	⑩	85	25	118	70	⑪	16	11	100
1996-97	9	4.6 (4573)	⑥	⑧	⑮	⑯	⑤	⑭	18	11	100
	Y	4.6 (4573)	⑥	⑧	⑮	⑯	⑤	⑭	18	11	100

① Step 1: 36 ft. lbs.
　Step 2: 130 degrees
　Step 3: Plus 30 degrees additional on four center bolts
　(NOTE: Must use new bolts)

② 26 ft. lbs.plus 50 degrees

③ 20 ft. lbs. plus 50 degrees

④ 110 ft. lbs. plus 76 degrees

⑤ Step 1: 11 ft. lbs.
　Step 2: Plus 50 degrees

⑥ Step 1: 22 ft. lbs.
　Step 2: Plus two turns of 90 degrees

⑦ Step 1: 105 ft. lbs.
　Step 2: Plus 120 degrees

⑧ Step 1: 15 ft. lbs.
　Step 2: Plus 65 degrees

⑨ Step 1: 4 ft. lbs.
　Step 2: Plus 120 degrees

⑩ Short Bolts (39-41mm)
　(must be taper ground to prevent bottoming out)
　Step 1: 29 ft. lbs.
　Step 2: 51 ft. lbs.
　Step 3: 85 ft. lbs.
　Step 4: Tighten 3 center inboard
　　studs 1,3 and 5 to 88 ft. lbs.
　Long Bolts (47-50mm) - should not be modified
　Step 1: 29 ft. lbs.
　Step 2: 51 ft. lbs.
　Step 3: 81 ft. lbs.
　Step 4: All inboard studs to 88 ft. lbs.
　Step 5: Tighten 3 center inboard
　　studs 1,3 and 5 to 96 ft. lbs.

⑪ Step 1: 8 ft. lbs.
　Step 2: 12 ft. lbs.

⑫ Step 1: 18 ft. lbs.
　Step 2: Plus 90 degrees

⑬ Step 1: 44 ft. lbs.
　Step 2: Plus 120 degrees

⑭ 89 in. lbs.

⑮ Step 1: 18 ft. lbs.
　Step 2: Plus 110 degrees

⑯ Step 1: 37 ft. lbs.
　Step 2: Plus 120 degrees

⑰ Washer faced bolt 118 ft. lbs.
　Standard bolt 70 ft. lbs.

BRAKE SPECIFICATIONS
All measurements in inches unless noted

Year	Model	Master Cylinder Bore	Brake Disc Original Thickness	Minimum Thickness	Maximum Runout	Brake Drum Diameter Original Inside Diameter	Max. Wear Limit	Maximum Machine Diameter	Minimum Lining Thickness Front	Rear
1993	Allante	1.000	1.268	1.209	0.002	-	-	-	0.030	-
	Allante	NA	0.433	0.374	0.002	-	-	-	-	0.030
	Eldorado	1.000	1.268	1.209	0.002	-	-	-	0.030	-
	Eldorado ①	NA	0.433	0.384	0.002	-	-	-	-	0.030
	Eldorado	1.000	1.268	1.209	0.002	-	-	-	0.030	-
	Eldorado ②	NA	0.433	0.374	0.002	-	-	-	-	-
	Eldorado ETC	1.000	1.268	1.209	0.002	-	-	-	0.030	-
	Eldorado ETC ①	NA	0.433	0.374	0.002	-	-	-	-	0.030
	Riviera	1.000	1.260	1.209	0.002	-	-	-	0.030	-
		NA	0.433	0.374	0.002	-	-	-	-	0.030
	Seville	1.000	1.268	1.209	0.002	-	-	-	0.030	-
	Seville ①	NA	0.433	0.374	0.002	-	-	-	-	0.030
	Seville STS	1.000	1.268	1.209	0.002	-	-	-	0.030	-
	Seville STS ①	NA	0.433	0.374	0.002	-	-	-	-	0.030

BRAKE SPECIFICATIONS
All measurements in inches unless noted

Year	Model	Master Cylinder Bore	Brake Disc Original Thickness	Minimum Thickness	Maximum Runout	Brake Drum Diameter Original Inside Diameter	Max. Wear Limit	Maximum Machine Diameter	Minimum Lining Thickness Front	Rear
1994	Deville	NA	1.276	1.209	0.004	8.860	8.909	8.880	0.030	-
	DeVille Concours	1.000	1.276	1.209	0.004	8.860	8.909	8.880	-	0.030
	Eldorado	1.000	1.268	1.209	0.002	-	-	-	0.030	-
	Eldorado ①	NA	0.433	0.374	0.002	-	-	-	-	0.030
	Eldorado ETC	1.000	1.268	1.209	0.002	-	-	-	0.030	-
	Eldorado ETC ①	NA	0.433	0.374	0.002	-	-	-	-	0.030
	Seville SLS	1.000	1.268	1.209	0.002	-	-	-	0.030	-
	Seville SLS ①	NA	0.433	0.374	0.002	-	-	-	-	0.030
	Seville STS	1.000	1.268	1.209	0.002	-	-	-	0.030	-
	Seville STS ①	NA	0.433	0.374	0.002	-	-	-	-	0.030
1995	DeVille	1.000	1.268	1.209	0.002	-	-	-	0.030	-
		-	0.433	0.374	0.002	-	-	-	-	0.030
	DeVille Concours	1.000	1.268	1.209	0.002	-	-	-	0.030	-
		-	0.433	0.374	0.002	-	-	-	-	0.030
	Eldorado	1.000	1.268	1.209	0.002	-	-	-	0.030	-
		-	0.433	0.374	0.002	-	-	-	-	0.030
	Eldorado ETC	1.000	1.268	1.209	0.002	-	-	-	0.030	-
		-	0.433	0.374	0.002	-	-	-	-	0.030
	Seville SLS	1.000	1.268	1.209	0.002	-	-	-	0.030	-
		-	0.433	0.374	0.002	-	-	-	-	0.030
	Seville STS	1.000	1.268	1.209	0.002	-	-	-	0.030	-
		-	0.433	0.374	0.002	-	-	-	-	0.030
1996-97	DeVille	1.000	1.268	1.209	0.002	-	-	-	0.030	-
		-	0.433	0.374	0.002	-	-	-	-	0.030
	DeVille Concours	1.000	1.268	1.209	0.002	-	-	-	0.030	-
		-	0.433	0.374	0.002	-	-	-	-	0.030
	Eldorado	1.000	1.268	1.209	0.002	-	-	-	0.030	-
		-	0.433	0.374	0.002	-	-	-	-	0.030
	Eldorado ETC	1.000	1.268	1.209	0.002	-	-	-	0.030	-
		-	0.433	0.374	0.002	-	-	-	-	0.030
	Seville SLS	1.000	1.268	1.209	0.002	-	-	-	0.030	-
		-	0.433	0.374	0.002	-	-	-	-	0.030
	Seville STS	1.000	1.268	1.209	0.002	-	-	-	0.030	-
		-	0.433	0.374	0.002	-	-	-	-	0.030

NA - Not Available   ① Rear disc brakes   ② Sport

FREQUENT MAINTENANCE LABOR
BUICK RIVIERA (1993), CADILLAC DEVILLE (1994-97)
CADILLAC CONCOURS, ELDORADO, SEVILLE

The following should be used as a guide when determining the amount of work required for a particular service if taken to a repair shop. In estimating how long a particular Frequent Maintenance Service item should take, please observe the following:
- **Factory Time** is time that is generated by the vehicle manufacturer.
- **Chilton Time** is time that is based on field research and data supplied by the vehicle manufacturer.
- All labor time operations are given in hours and tenths of an hour.
- All labor operations, are to be used as a **guide**.

COOLING

(G) Winterize Cooling System
Includes: Run engine to check for leaks, tighten all hose connections. Test radiator and pressure cap, drain radiator and engine block. Add antifreeze and refill system.
1993-97 5

(G) Belt, Serpentine Drive, Renew
1993-97 (.3) 5

(G) Belt, Drive, Renew
1993-97, V8
coolant pump (.2) 3

(G) Hoses, Radiator, Renew
Includes: Drain and refill cooling system as required.
1993
  upper (.3) 4
  lower (.4) 5
  w/4.6L add 1
  both (.5) 7
  by-pass (.5) 7
1994-97
  upper (.4) 6
  lower (.6) 8
  both (.8) 1.0

(G) Thermostat, Coolant, Renew
1993
  V6 (.4) 6
  V8 (.3) 5
1994-97, V8 (.6) 9

FUEL

(M) Air Cleaner, Service
1993-97 3

(G) Filter, Fuel, Renew
1993
  in line (.4) 6
  in tank (1.3) 1.7
1994-97
  in line
    Eldorado, Seville (.5) 7
    Deville, Concours (.4) 6
  in tank (1.3) 1.7

BRAKES

(G) Bleed Brakes (Four Wheels)
Includes: Add fluid.
1993-97 (.4) 5
w/ABS add 2

(M) Parking Brake, Adjust
1993 (.3) 4
1994-97
  Eldorado, Seville (.2) 3
  Deville, Concours (.4) 6

LUBRICATION SERVICES

(M) Engine Oil & Filter, Renew
Includes: Inspect and correct all fluid levels.
1993-97 3

(M) Lubricate Chassis, Change Oil & Filter
Includes: Inspect and correct all fluid levels.
1993-97 6
Install grease fittings add 1

(M) Lubricate Chassis
Includes: Inspect and correct all fluid levels.
1993-97 4
Install grease fittings add 1

WHEELS

(G) Wheel, Renew (One)
1993-97 (.5) 5

(G) Wheel, Balance
1993-97
  one 3
  each adtnl. 2

(G) Wheels, Rotate (All)
1993-97 5

ELECTRICAL

(G) Headlamps, Aim
1993-97
  two 4
  four 6

(G) High Mount Stop Lamp Bulb, Renew
1993-97 (.3) 4

(G) License Lamp Assy., Renew
Allante
  1993 (.2) 3
Eldorado
  1993-97 (.2) 3
Seville
  1993 (.5) 7
  1994-97 (.2) 3
Deville, Concours
  1994-97 (.3) 4

(G) License Lamp Bulb, Renew
1993-97, one or all (.2) 3

(G) Park & Turn Signal Lamp Assy., Renew
1993-97 (.2) 4

(G) Park & Turn Signal Lamp Bulb or Lens, Renew
1993-97, each 2

(G) Rear Combination Lamp Bulb, Renew
1993
  Eldorado, Seville (.2) 3
  Allante (.6) 8
  Riviera, Toronado (.2) 3
  1994-97 (.2) 3
  each adtnl. 1

(G) Horn, Renew
Allante
  1993, one (.2) 3
  each adtnl. (.1) 1
Eldorado, Seville
  1993
    one (.3) 4
    each adtnl. (.4) 5
  1994-97
    one (.3) 4
    each adtnl. 4
Reatta, Riviera
  1993, one (.2) 3
  each adtnl. (.1) 1
Deville, Concours
  1994-97
    one (.3) 4
    each adtnl. 2

(M) Terminals, Battery, Clean
1993-97 3

SCHEDULED MAINTENANCE INTERVALS
(BUICK RIVIERA (1993-94), CADILLAC DEVILLE (1994-97), DEVILLE CONCOURS, ELDORADO, SEVILLE)

TO BE SERVICED	TYPE OF SERVICE	VEHICLE MILEAGE INTERVAL (x1000)												
		7.5	15	22.5	30	37.5	45	52.5	60	67.5	75	82.5	90	97.5
Engine oil & filter	R	✓	✓	✓	✓	✓	✓	✓	✓	✓	✓	✓	✓	✓
Coolant level, hoses & clamps	S/I	✓	✓	✓	✓	✓	✓	✓	✓	✓	✓	✓	✓	✓
Drive shaft boots & front suspension components	S/I	✓	✓	✓	✓	✓	✓	✓	✓	✓	✓	✓	✓	✓
Exhaust system, brake hoses & throttle linkage	S/I	✓	✓	✓	✓	✓	✓	✓	✓	✓	✓	✓	✓	✓
Lubricate chassis, suspension, steering linkage, transaxle shift linkage, parking brake cable guides, underbody contact points & linkage	S/I	✓	✓	✓	✓	✓	✓	✓	✓	✓	✓	✓	✓	✓
Throttle body mount bolt torque	S/I	✓												
Brake linings	S/I	✓		✓		✓		✓		✓		✓		✓
Rotate tires	S/I	✓		✓		✓		✓		✓		✓		✓
Inspect throttle body bore & throttle plate for deposits	S/I		✓				✓				✓		✓	
Air filter element	R					✓			✓				✓	
Engine coolant②	R					✓			✓				✓	
PCV valve	R					✓			✓				✓	
Spark plugs①	R					✓			✓				✓	
Accessory drive belt(s)	S/I					✓			✓				✓	
Automatic transaxle fluid & filter	S/I				✓				✓				✓	
EGR & fuel systems	S/I				✓				✓				✓	
Engine timing (4.9L)	S/I				✓				✓				✓	
Ignition cables	S/I				✓				✓				✓	

① Platinum tip spark plugs - replace every 100,000 miles.
② Engine coolant (1996-97) - replace every 100,000 miles. Use O.E. specified (DEX-COOL™) coolant only. If any silicate coolant is used, the service interval is every 30,000 miles.

R – Replace   S/I – Service or Inspect

SCHEDULED MAINTENANCE INTERVALS
(BUICK RIVIERA (1993), CADILLAC DEVILLE (1994-97), DEVILLE CONCOURS, ELDORADO, SEVILLE) (Cont.)

FREQUENT OPERATION MAINTENANCE (SEVERE SERVICE)

If a vehicle is operated under any of the following conditions it is considered severe service:

- Extremely dusty areas.
- 50% or more of the vehicle operation is in 32°C (90°F) or higher temperatures, or constant operation in temperatures below 0°C (32°F).
- Prolonged idling (vehicle operation in stop and go traffic).
- Frequent short running periods (engine does not warm to normal operating temperatures).
- Police, taxi, delivery usage or trailer towing usage.

CV joints & front suspension components - service or inspect every 3000 miles.
Engine oil & filter change – change every 3000 miles
Brake linings - check every 6000 miles.
Chassis lubrication - lubricate every 6000 miles.
Suspension, steering linkage, transaxle shift linkage, parking cable guides, underbody contact points - lubricate every 6000 miles.
Throttle body mount bolt torque - tighten at 6000 miles.
Air filter element – service or inspect every 15,000 miles.
Automatic transaxle fluid – change every 15,000 miles (1993-94) or every 50,000 miles (1995-97).
Inspect throttle body bore & throttle plate for deposits - clean as required every 15,000 miles.
Rotate tires at 6000 miles, then every 15,000 miles.

RIVIERA (1995-97)/AURORA

VEHICLE IDENTIFICATION CHART

Engine Code						Model Year	
Code	Liters	Cu. In. (cc)	Cyl.	Fuel Sys.	Eng. Mfg.	Code	Year
1	3.8	231 (3785)	6	MFI ①	BOC	S	1995
K	3.8	231 (3785)	6	MFI	BOC	T	1996
C	4.0	244 (3995)	8	MFI	BOC	V	1997

MFI - Multiport fuel injection
BOC - Buick/Oldsmobile/Cadillac
① Supercharged engine

ENGINE IDENTIFICATION

Year	Model	Engine Displacement Liters (cc)	Engine Series (ID/VIN)	Fuel System	No. of Cylinders	Engine Type
1995	Aurora	4.0 (3995)	C	MFI	8	DOHC
	Riviera	3.8 (3785)	1	MFI ①	6	OHV
	Riviera	3.8 (3785)	K	MFI	6	OHV
1996-97	Aurora	4.0 (3995)	C	MFI	8	DOHC
	Riviera	3.8 (3785)	1	MFI ①	6	OHV
	Riviera	3.8 (3785)	K	MFI	6	OHV

MFI - Multiport fuel injection   OHV - Overhead valve   DOHC - Double overhead camshaft   ① Supercharged engine

GENERAL ENGINE SPECIFICATIONS

Year	Engine ID/VIN	Engine Displacement Liters (cc)	Fuel System Type	Net Horsepower @ rpm	Net Torque @ rpm (ft. lbs.)	Bore x Stroke (in.)	Com-pression Ratio	Oil Pressure @ rpm
1995	1	3.8 (3785)	MFI ①	225@5000	275@3200	3.80x3.40	9.0:1	60@1850
	K	3.8 (3785)	MFI	205@5200	230@4000	3.80x3.40	9.4:1	60@1850
	C	4.0 (3995)	MFI	250@5600	245@4400	3.43x3.31	10.2:1	30@2000
1996-97	1	3.8 (3785)	MFI ①	225@5000	275@3200	3.80x3.40	8.5:1	60@1850
	K	3.8 (3785)	MFI	205@5200	230@4000	3.80x3.40	9.4:1	60@1850
	C	4.0 (3995)	MFI	250@5600	245@4400	3.43x3.31	10.2:1	20@2000

MFI - Multiport fuel injection
TFI - Throttle body fuel injection
① Supercharged engine

GASOLINE ENGINE TUNE-UP SPECIFICATIONS

Year	Engine ID/VIN	Engine Displacement Liters (cc)	Spark Plugs Gap (in.)	Ignition Timing (deg.) MT	Ignition Timing (deg.) AT	Fuel Pump (psi)	Idle Speed (rpm) MT	Idle Speed (rpm) AT	Valve Clearance In.	Valve Clearance Ex.
1995	1	3.8 (3785)	0.060	①	①	40-47	-	③	HYD	HYD
	K	3.8 (3785)	0.060	①	①	40-47	-	③	HYD	HYD
	C	4.0 (3995)	0.050	②	②	41-47 ④	-	③	HYD	HYD
1996-97	1	3.8 (3785)	0.060	②	②	41-47	-	③	HYD	HYD
	K	3.8 (3785)	0.060	②	②	41-47	-	③	HYD	HYD
	C	4.0 (3995)	0.050	-	②	41-47	-	③	HYD	HYD

NOTE: The Vehicle Emission Control Information label often reflects specification changes made during production. The label figures must be used if they differ from those in this chart.
HYD - Hydraulic
① Refer to underhood sticker
② DIS Ignition System timing is not adjustable
③ Idle speed is maintained by the ECM. There is no recommended adjustment procedure
④ Pressure at fuel pump

CAPACITIES

Year	Model	Engine ID/VIN	Engine Displacement Liters (cc)	Engine Oil with Filter (qts.)	Transmission (pts.) 4-Spd	Transmission (pts.) 5-Spd	Transmission (pts.) Auto.	Drive Axle Front (pts.)	Drive Axle Rear (pts.)	Fuel Tank (gal.)	Cooling System (qts.)
1995	Aurora	C	4.0 (3995)	7.0 ①	-	-	6.5	-	-	20.0	13.0
	Riviera	1	3.8 (3785)	5.0	-	-	13.2	-	-	20.0	13.0
	Riviera	K	3.8 (3785)	5.0	-	-	13.2	-	-	20.0	13.0
1996-97	Aurora	C	4.0 (3995)	7.5	-	-	6.5	-	-	20.0	13.0
	Riviera	1	3.8 (3785)	5.0	-	-	12.0	-	-	20.0	13.0
	Riviera	K	3.8 (3785)	4.5	-	-	12.0	-	-	20.0	13.0

① Specification is without filter replacement; Additional oil may be required

VALVE SPECIFICATIONS

Year	Engine ID/VIN	Engine Displacement Liters (cc)	Seat Angle (deg.)	Face Angle (deg.)	Spring Test Pressure (lbs. @ in.)	Spring Installed Height (in.)	Stem-to-Guide Clearance (in.)		Stem Diameter (in.)	
							Intake	Exhaust	Intake	Exhaust
1995	1	3.8 (3785)	45	45	210@1.315	1.690-1.720	0.0015-0.0035	0.0015-0.0032	NA	NA
	K	3.8 (3785)	45	45	210@1.315	1.690-1.720	0.0015-0.0035	0.0015-0.0032	NA	NA
	C	4.0 (3995)	46	45	92@0.854	1.190	0.0010-0.0030	0.0020-0.0040	NA	NA
1996-97	1	3.8 (3785)	45	45	210@1.315	1.690-1.720	0.0015-0.0035	0.0015-0.0032	NA	NA
	K	3.8 (3785)	45	45	210@1.315	1.690-1.720	0.0015-0.0035	0.0015-0.0032	NA	NA
	C	4.0 (3995)	46	45	92@0.854	1.190	0.0010-0.0030	0.0020-0.0040	0.2331-0.2339	0.2331-0.2339

TORQUE SPECIFICATIONS
All readings in ft. lbs.

Year	Engine ID/VIN	Engine Displacement Liters (cc)	Cylinder Head Bolts	Main Bearing Bolts	Rod Bearing Bolts	Crankshaft Damper Bolts	Flywheel Bolts	Manifold Intake	Manifold Exhaust	Spark Plugs	Lug Nut
1995	1	3.8 (3785)	①	②	③	④	⑤	7	38	11	100
	K	3.8 (3785)	①	②	③	④	⑤	7	38	11	100
	C	4.0 (3995)	⑦	⑧	⑨	⑩	⑤	89	18	⑥	100
1996-97	1	3.8 (3786)	①	⑪	③	④	⑤	⑬	22	11	100
	K	3.8 (3786)	①	⑫	③	④	⑤	11	38	11	100
	C	4.0 (3995)	⑦	⑧	⑨	⑩	⑤	89	18	⑥	100

① Step 1: 36 ft. lbs.
  Step 2: 130 degrees
  Step 3: Plus 30 degrees additional on four center bolts
  (NOTE: Must use new bolts)
② 26 ft. lbs. plus 50 degrees
③ 20 ft. lbs. plus 50 degrees
④ 110 ft. lbs. plus 76 degrees
⑤ Step 1: 11 ft. lbs.
  Step 2: Plus 50 degrees

⑥ New cylinder head 1st-time installation: 20 ft. lbs.
  All others: 11 ft. lbs.
⑦ Step 1: Torque to 22 ft. lbs. then 90 degrees.
  Step 2: Retorque an additional 75 degrees.
⑧ 15 ft. lbs. plus 65 degrees
⑨ 18 ft. lbs. plus 90 degrees
⑩ 44 ft. lbs. plus 120 degrees

⑪ Step 1: Tighten caps in equal increments to 52 ft. lbs.
  Step 2: Loosen 360 degrees
  Step 3: 15 ft. lbs.
  Step 4: 54 ft. lbs.
  Step 5: Plus three turns of 35 degrees-total of 105 degree
⑫ 26 ft. lbs. plus 90 degrees
⑬ Upper manifold: 8 ft. lbs.
  Lower manifold: 11 ft. lbs.

BRAKE SPECIFICATIONS
All measurements in inches unless noted

Year	Model	Master Cylinder Bore	Brake Disc Original Thickness	Brake Disc Minimum Thickness	Maximum Runout	Brake Drum Diameter Original Inside Diameter	Max. Wear Limit	Maximum Machine Diameter	Minimum Lining Thickness Front	Minimum Lining Thickness Rear
1995	Aurora	1.000	1.260	1.209	0.002	-	-	-	0.030	-
		1.000	0.433	0.374	0.002	-	-	-	-	0.030
	Riviera	1.000	1.260	1.209	0.002	-	-	-	0.030	-
		1.000	0.433	0.374	0.002	-	-	-	-	0.030
1996-97	Aurora	1.000	1.260	1.209	0.002	-	-	-	0.030	-
		1.000	0.433	0.374	0.002	-	-	-	-	0.030
	Riviera	1.000	1.260	1.209	0.002	-	-	-	0.030	-
		1.000	0.433	0.374	0.002	-	-	-	-	0.030

FREQUENT MAINTENANCE LABOR
BUICK RIVIERA (1995-97), OLDSMOBILE AURORA

The following should be used as a guide when determining the amount of work required for a particular service if taken to a repair shop. In estimating how long a particular Frequent Maintenance Service item should take, please observe the following:
- **Factory Time** is time that is generated by the vehicle manufacturer.
- **Chilton Time** is time that is based on field research and data supplied by the vehicle manufacturer.
- All labor time operations are given in hours and tenths of an hour.
- All labor operations, are to be used as a **guide**.

	(Factory Time)	Chilton Time

COOLING

(G) Winterize Cooling System
Includes: Run engine to check for leaks, tighten all hose connections. Test radiator and pressure cap, drain radiator and engine block. Add antifreeze and refill system.
1995-97 5

(G) Belt, Serpentine Drive, Renew
1995-97
3.8L (.5) 8
4.0L (.6) 9

(G) Belt, Drive, Renew
1995, 3.8L code 1
inner (.3) 4
outer (.2) 3
1996, 3.8L code 1
inner (.6) 8
outer (.5) 7
1995-97, 4.0L
water pump (.3) 4

(G) Hoses, Radiator, Renew
Includes: Drain and refill cooling system as required.
1995-97
upper or lower (.7) 9
throttle body to thermostat
Aurora (.8) 1.1
Riviera (.5) 8
intake manifold
Aurora (.4) 6

(G) Thermostat, Coolant, Renew
1995-97 (.6) 9

FUEL

(M) Air Cleaner, Service
1995-97 2

(G) Filter, Fuel, Renew
1995-97 (.3) 5

BRAKES

(G) Bleed Brakes (Four Wheels)
Includes: Add fluid.
1995-97 (.4) 5

(M) Parking Brake, Adjust
1995-97 (.3) 4

LUBRICATION SERVICES

(M) Engine Oil & Filter, Renew
Includes: Inspect and correct all fluid levels.
1995-97 3

(M) Lubricate Chassis, Change Oil & Filter
Includes: Inspect and correct all fluid levels.
1995-97 6
Install grease fittings add 1

(M) Lubricate Chassis
Includes: Inspect and correct all fluid levels.
1995-97 4
Install grease fittings add 1

WHEELS

(G) Wheel, Renew (One)
1995-97 5

(G) Wheel, Balance
1995-97
one 3
each adtnl. 2

(G) Wheels, Rotate (All)
1995-97 5

ELECTRICAL

(G) Headlamps, Aim
1995-97
two 4
four 6

(G) High Mount Stop Lamp Bulb, Renew
1995-97
Riviera (.2) 3
Aurora (.3) 4

(G) High Mount Stop Lamp and/or Lens, Renew
1995-97
Riviera (.4) 5
Aurora (.2) 3

(G) License Lamp Assy., Renew
1995-97, one or both
Riviera (.2) 3
Aurora (.3) 4

(G) License Lamp Bulb, Renew
1995-97, one or all (.2) 3

(G) Park & Turn Signal Lamp Bulb or Lens, Renew
1995-97
Riviera
one (.2) 3
all (.4) 5
Aurora, one or all (.2) 3

(G) Stop, Tail & Turn Signal Lamp Bulb, Renew
1995-97
Riviera
one (.2) 3
all (.3) 4
Aurora, one or all (.2) 3

(G) Horn, Renew
1995-97, one (.3) 4
each adtnl. 1

(M) Terminals, Battery, Clean
1995-97 3

SCHEDULED MAINTENANCE INTERVALS
(BUICK RIVIERA (1995-97), OLDSMOBILE AURORA)

TO BE SERVICED	TYPE OF SERVICE	VEHICLE MILEAGE INTERVAL (x1000)												
		7.5	15	22.5	30	37.5	45	52.5	60	67.5	75	82.5	90	97.5
Engine oil & filter	R	✓	✓	✓	✓	✓	✓	✓	✓	✓	✓	✓	✓	✓
Coolant level, hoses & clamps	S/I	✓	✓	✓	✓	✓	✓	✓	✓	✓	✓	✓	✓	✓
Drive shaft boots & front suspension components	S/I	✓	✓	✓	✓	✓	✓	✓	✓	✓	✓	✓	✓	✓
Exhaust system & brake hoses	S/I	✓	✓	✓	✓	✓	✓	✓	✓	✓	✓	✓	✓	✓
Lubricate, chassis, suspension, steering linkage, transaxle shift linkage, parking brake cable guides, underbody contact points & linkage	S/I	✓	✓	✓	✓	✓	✓	✓	✓	✓	✓	✓	✓	✓
Throttle linkage	S/I	✓	✓	✓	✓	✓	✓	✓	✓	✓	✓	✓	✓	✓
Brake linings	S/I	✓		✓		✓		✓		✓		✓		✓
Rotate tires	S/I	✓		✓		✓		✓		✓		✓		✓
Air filter element	R				✓				✓				✓	
Engine coolant (1995)②	R				✓				✓				✓	
Spark plugs①	R				✓				✓				✓	
Accessory drive belt(s)	S/I				✓				✓				✓	
Automatic transaxle fluid & filter	S/I				✓				✓				✓	
Fuel system	S/I				✓				✓				✓	
Ignition cables	S/I				✓				✓				✓	
Inspect throttle body bore & throttle plate for deposits (Oldsmobile)	S/I		✓				✓				✓			
Supercharger oil	S/I				✓				✓				✓	
Throttle body mounting torque	S/I	✓												

① Platinum tip spark plugs - replace every 100,000 miles.
② Engine coolant (1996-97) - replace every 100,000 miles. Use O.E. specified (DEX-COOL™) coolant only. If any silicate coolant is used, the service interval is every 30,000 miles.
R – Replace   S/I – Service or Inspect

SCHEDULED MAINTENANCE INTERVALS
(BUICK RIVIERA (1995-97), OLDSMOBILE AURORA) (Cont.)

FREQUENT OPERATION MAINTENANCE (SEVERE SERVICE)

If a vehicle is operated under any of the following conditions it is considered severe service:
- Extremely dusty areas.
- 50% or more of the vehicle operation is in 32°C (90°F) or higher temperatures, or constant operation in temperatures below 0°C (32°F).
- Prolonged idling (vehicle operation in stop and go traffic).
- Frequent short running periods (engine does not warm to normal operating temperatures).
- Police, taxi, delivery usage or trailer towing usage.

CV joints & front suspension components - check every 3000 miles.
Oil & oil filter – change every 3000 miles.
Brake linings – check every 6000 miles.
Chassis lubrication - lubricate every 6000 miles.
Suspension, steering linkage, transaxle shift linkage, parking cable guides, underbody contact points, & linkage - lubricate every 6000 miles.
Throttle body mounting bolt torque - check at 6000 miles.
Air filter element – service or inspect every 15,000 miles.
Rotate tires initially at 6000 miles, then every 15,000 miles.
Automatic transaxle fluid – change every 50,000 miles.

CAVALIER/SUNBIRD/SUNFIRE

VEHICLE IDENTIFICATION CHART

Engine Code						Model Year	
Code	Liters	Cu. In. (cc)	Cyl.	Fuel Sys.	Eng. Mfg.	Code	Year
H	2.0	121 (1998)	4	MFI	Pontiac	P	1993
4	2.2	133 (2180)	4	MFI	CUS	R	1994
D	2.3	138 (2262)	4	MFI	CUS	S	1995
T	2.4	146 (2392)	4	MFI	CUS	T	1996
T	3.1	191 (3130)	6	MFI	CPC	V	1997

CUS - Chevrolet/United States

CPC - Chevrolet/Pontiac/Canada

MFI - Multiport fuel injection

ENGINE IDENTIFICATION

Year	Model	Engine Displacement Liters (cc)	Engine Series (ID/VIN)	Fuel System	No. of Cylinders	Engine Type
1993	Sunbird	2.0 (1998)	H	MFI	4	SOHC
	Sunbird	3.1 (3130)	T	MFI	6	OHV
	Cavalier	2.2 (2180)	4	MFI	4	OHV
	Cavalier	3.1 (3130)	T	MFI	6	OHV
1994	Sunbird	2.0 (1998)	H	MFI	4	SOHC
	Sunbird	3.1 (3130)	T	MFI	6	OHV
	Cavalier	2.2 (2180)	4	MFI	4	OHV
	Cavalier	3.1 (3130)	T	MFI	6	OHV

ENGINE IDENTIFICATION

Year	Model	Engine Displacement Liters (cc)	Engine Series (ID/VIN)	Fuel System	No. of Cylinders	Engine Type
1995	Cavalier	2.2 (2180)	4	MFI	4	OHV
	Cavalier	2.3 (2262)	D	MFI	4	DOHC
	Sunfire	2.2 (2180)	4	MFI	4	OHV
	Sunfire	2.3 (2262)	D	MFI	4	DOHC
1996-97	Cavalier	2.2 (2180)	4	MFI	4	OHV
	Cavalier	2.4 (2392)	T	MFI	4	DOHC
	Sunfire	2.2 (2180)	4	MFI	4	OHV
	Sunfire	2.4 (2392)	T	MFI	4	DOHC

MFI - Multiport fuel injection
OHV - Overhead valve
DOHC - Double overhead camshaft
SOHC - Single overhead camshaft

GENERAL ENGINE SPECIFICATIONS

Year	Engine ID/VIN	Engine Displacement Liters (cc)	Fuel System Type	Net Horsepower @ rpm	Net Torque @ rpm (ft. lbs.)	Bore x Stroke (in.)	Compression Ratio	Oil Pressure @ rpm
1993	4	2.2 (2180)	MFI	110@5200	130@3200	3.50x3.46	9.0:1	63-77@1200
	H	2.0 (1998)	MFI	110@5200	124@3600	3.38x3.38	9.2:1	30@2000
	T	3.1 (3130)	MFI	140@4200	185@3200	3.50x3.31	8.5:1	8@600
1994	4	2.2 (2180)	MFI	110@5200	130@3200	3.50x3.46	9.0:1	63-77@1200
	H	2.0 (1998)	MFI	110@5200	124@3600	3.38x3.38	9.2:1	30@2000
	T	3.1 (3130)	MFI	140@4200	185@3200	3.50x3.31	8.5:1	8@600
1995	4	2.2 (2180)	MFI	120@5200	130@3200	3.50x3.46	9.0:1	63-77@1200
	D	2.3 (2262)	MFI	150@6100	145@4800	3.63x3.35	9.5:1	30@2000
1996-97	4	2.2 (2180)	MFI	120@5200	130@3200	3.50x3.46	9.0:1	56@3000
	T	2.4 (2392)	MFI	150@6000	155@4400	3.54x3.70	9.5:1	30@3000

MFI - Multiport fuel injection

GASOLINE ENGINE TUNE-UP SPECIFICATIONS

Year	Engine ID/VIN	Engine Displacement Liters (cc)	Spark Plugs Gap (in.)	Ignition Timing (deg.) MT	Ignition Timing (deg.) AT	Fuel Pump (psi)	Idle Speed (rpm) MT	Idle Speed (rpm) AT	Valve Clearance In.	Valve Clearance Ex.
1993	4	2.2 (2180)	0.045	①	①	41-47	①	①	HYD	HYD
	H	2.0 (1998)	0.045	②	②	41-47	①	①	HYD	HYD
	T	3.1 (3130)	0.045	10B	10B	41-47	①	①	HYD	HYD
1994	4	2.2 (2180)	0.045	①	①	41-47	①	①	HYD	HYD
	H	2.0 (1998)	0.045	②	②	41-47	①	①	HYD	HYD
	T	3.1 (3130)	0.045	①	①	41-47	①	①	HYD	HYD
1995	4	2.2 (2180)	0.045	①	①	41-47	①	①	HYD	HYD
	D	2.3 (2262)	0.035	①	①	41-47	①	①	HYD	HYD
1996-97	4	2.2 (2180)	0.045	①	①	41-47	①	①	HYD	HYD
	T	2.4 (2392)	0.035	①	①	41-47	①	①	HYD	HYD

NOTE: The Vehicle Emission Control Information label often reflects specification changes made during production. The label figures must be used if they differ from those in this chart.
B - Before top dead center
HYD - Hydraulic
① Refer to Vehicle Emission Control Information label
② Distributorless ignition system

CAPACITIES

Year	Model	Engine ID/VIN	Engine Displacement Liters (cc)	Engine Oil with Filter (qts.)	Transmission (pts.) 4-Spd	5-Spd	Auto.	Transfer Case (pts.)	Drive Axle Front (pts.)	Rear (pts.)	Fuel Tank (gal.)	Cooling System (qts.)
1993	Cavalier	4	2.2 (2180)	4.5	-	4.0	8.0 ①	-	-	-	13.6	8.5
	Cavalier	T	3.1 (3130)	4.5	-	4.0	8.0 ①	-	-	-	13.6	11.0
	Sunbird	H	2.0 (1998)	4.5	-	4.0	8.0	-	-	-	13.6	11.7
	Sunbird	T	3.1 (3130)	4.5	-	4.0	8.0	-	-	-	13.6	14.2
1994	Cavalier	4	2.2 (2180)	4.5	-	4.0	8.0 ①	-	-	-	13.6	8.5
	Cavalier	T	3.1 (3130)	4.5	-	4.0	8.0 ①	-	-	-	13.6	11.0
	Sunbird	H	2.0 (1998)	4.5	-	4.0	8.0	-	-	-	15.2	10.7
	Sunbird	T	3.1 (3130)	4.5	-	4.0	8.0	-	-	-	15.2	13.7
1995	Cavalier	4	2.2 (2180)	4.5	-	4.0	8.0 ①	-	-	-	13.6	8.5
	Cavalier	D	2.3 (2262)	4.5	-	4.0	14.0 ①	-	-	-	13.6	10.4
	Sunfire	4	2.2 (2180)	4.5	-	4.0	8.0	-	-	-	15.2	10.7
	Sunfire	D	2.3 (2262)	4.5	-	4.0	22.0	-	-	-	15.2	10.4
1996-97	Cavalier	4	2.2 (2180)	4.5	-	4.0	8.0 ①	-	-	-	13.6	8.5
	Cavalier	T	2.4 (2392)	4.5	-	4.0	14.0 ①	-	-	-	13.6	10.4
	Sunfire	4	2.2 (2180)	4.5	-	4.0	8.0	-	-	-	15.2	10.7
	Sunfire	T	2.4 (2392)	4.5	-	4.0	-	-	-	-	15.2	10.4

① 10.0 pts. if equipped with O/D

VALVE SPECIFICATIONS

Year	Engine ID/VIN	Engine Displacement Liters (cc)	Seat Angle (deg.)	Face Angle (deg.)	Spring Test Pressure (lbs. @ in.)	Spring Installed Height (in.)	Stem-to-Guide Clearance (in.) Intake	Exhaust	Stem Diameter (in.) Intake	Exhaust
1993	4	2.2 (2180)	46	45	225-233@ ① 1.25	1.640 ②	0.0011-0.0026	0.0014-0.0031	NA	NA
	H	2.0 (1998)	46	45	90@1.701	1.693	0.0008-0.0021	0.0014-0.0030	0.2753-0.2747	0.2760-0.2755
	T	3.1 (3130)	46	45	90 ①	1.600 ②	0.0010-0.0027	0.0010-0.0027	NA	NA
1994	4	2.2 (2180)	46	45	225-233@ ① 1.25	1.640 ②	0.0011-0.0026	0.0014-0.0031	NA	NA
	H	2.0 (1998)	45	46	63-71@ 1.476	1.476	0.0006-0.0017	0.0012-0.0024	0.2753-0.2747	0.2760-0.2755
	T	3.1 (3130)	46	45	90 ①	1.600 ②	0.0010-0.0027	0.0010-0.0027	NA	NA
1995	4	2.2 (2180)	46	45	225-233@ ① 1.25	1.640 ②	0.0011-0.0026	0.0014-0.0031	NA	NA
	D	2.3 (2262)	45	44	193-207@ 1.04	1.440 ②	0.0010-0.0027	0.0015-0.0032	0.2740-0.2750	0.2740-0.2750
1996-97	4	2.2 (2180)	46	45	75-81@1.71	1.710	0.0010-0.0027	0.0014-0.0031	NA	NA
	T	2.4 (2392)	45	46	50-55@1.44	1.437	0.0009-0.0025	0.0016-0.0032	0.2331-0.2339	0.2326-0.2334

NA - Not Available
① With valve open
② With valve closed

TORQUE SPECIFICATIONS
All readings in ft. lbs.

Year	Engine ID/VIN	Engine Displacement Liters (cc)	Cylinder Head Bolts	Main Bearing Bolts	Rod Bearing Bolts	Crankshaft Damper Bolts	Flywheel Bolts	Manifold		Spark Plugs	Lug Nut
								Intake	Exhaust		
1993	4	2.2 (2180)	①	77	38	85 ②	52-55	18	6-13	20	80-100
	H	2.0 (1998)	18 ⑫	44 ⑭	26 ⑭	114	48 ⑭	16	16	15	100
	T	3.1 (3130)	⑩	73	39	66-85	45-59	⑤	18	20	100
1994	4	2.2 (2180)	①	77	38	85 ②	52-55	18	6-13	20	100
	H	2.0 (1998)	18 ⑫	44 ⑬	26 ⑬	114	48	16	16	15	100
	T	3.1 (3130)	⑩	73	39	66-85	45-59	⑤	18	20	100
1995	4	2.2 (2180)	①	77	38	85 ②	52-55	18	6-13	20	100
	D	2.3 (2262)	26 ④	⑥	⑦	⑧	⑨	⑪	③	17	100
1996-97	4	2.2 (2180)	①	70	38	77 ②	52-55	24	18	11	100
	T	2.4 (2392)	⑮	⑥	⑦	⑧	⑨	⑪	③	11	100

NA - Not Available

① Step 1: 41 ft. lbs.
Step 2: Tighten an additional 45 degrees
Step 3: Tighten an additional 45 degrees
Step 4: Long bolts 1, 4-5, 8-9 an additional 20 degrees
Step 5: Short bolts 2-3, 6-7, 10 an additional 10 degrees
② Center bolt spec shown; Pulley to hub bolts: 37 ft. lbs.
③ Bolts: 27 ft. lbs.
Studs: 106 inch lbs.

④ Cylinder head bolts should be torqued 26 ft. lbs.
Long bolts: 100 degrees
Short bolts: 120 degrees
⑤ 15 ft. lbs., then 24 ft. lbs.
⑥ 15 ft. lbs. plus 90 degrees
⑦ 18 ft. lbs. plus 80 degrees
⑧ 74 ft. lbs. plus 90 degrees
⑨ 22 ft. lbs. plus 45 degrees
⑩ 33 ft. lbs. plus 90 degrees

⑪ Nuts: 18 ft. lbs.
Studs: 96 inch lbs.
⑫ Plus three turns of 60 degrees;
Plus 30-50 degrees after warm-up
⑬ Plus 45 degrees
⑭ Plus 30 degrees
⑮ Cylinder head bolts: 40 ft. lbs.
plus 90 degrees

BRAKE SPECIFICATIONS
All measurements in inches unless noted

Year	Model	Master Cylinder Bore	Brake Disc			Brake Drum Diameter			Minimum Lining Thickness	
			Original Thickness	Minimum Thickness	Maximum Runout	Original Inside Diameter	Max. Wear Limit	Maximum Machine Diameter	Front	Rear
1993	Cavalier	0.875	0.885	0.830	0.004	7.879	7.929	7.899	0.125	0.125
	Sunbird	0.874	0.806	0.736	0.002	7.874	7.929	7.899	0.030	0.030
1994	Cavalier	0.875	0.885	0.830	0.004	7.879	7.929	7.899	0.125	0.125
	Sunbird	0.874	0.806	0.736	0.002	7.874	7.929	7.899	0.030	0.030
1995	Cavalier	0.875	0.786	0.736	0.003	7.879	7.929	7.899	0.125	0.125
	Sunfire	0.874	0.786	0.736	0.003	7.870	7.930	7.900	0.030	0.030
1996-97	Cavalier	0.874	0.806	0.736	0.003	7.880	7.930	7.900	0.030	0.030
	Sunfire	0.874	0.786	0.736	0.003	7.870	7.930	7.900	0.030	0.030

FREQUENT MAINTENANCE LABOR
CHEVROLET CAVALIER, PONTIAC SUNBIRD, SUNFIRE

The following should be used as a guide when determining the amount of work required for a particular service if taken to a repair shop. In estimating how long a particular Frequent Maintenance Service item should take, please observe the following:
- **Factory Time** is time that is generated by the vehicle manufacturer.
- **Chilton Time** is time that is based on field research and data supplied by the vehicle manufacturer.
- All labor time operations are given in hours and tenths of an hour.
- All labor operations, are to be used as a **guide**.

	(Factory Time)	Chilton Time

COOLING

(G) Winterize Cooling System
Includes: Run engine to check for leaks, tighten all hose connections. Test radiator and pressure cap, drain radiator and engine block. Add antifreeze and refill system.
1993-97 5

(G) Belt, Drive, Renew
4 cyl. OHC
1993-97
  AC (.2) 3
  Serpentine (.2) 3
  w/AC add 1
4 cyl OHV
1993-97 (.2) 3
V6
1993-94 (.3) 4

(G) Hoses, Radiator, Renew
Includes: Drain and refill cooling system as required.
1993-94
  upper (.3) 4
  lower (.7) 8
  both (.8) 1.0
1995-97
  upper (.5) 7
  lower (.5) 7
  both (.8) 1.2

(G) Thermostat, Coolant, Renew
1993-94
  4 cyl.
    OHC (.2) 4
    OHV (.4) 6
  V6 (.5) 7
1995-97
  OHC (.6) 9
  OHV (.4) 6

FUEL

(M) Air Cleaner, Service
1993-97 (.2) 3

(G) Filter, Fuel, Renew
1993-94 (.3) 4
1995-97 (.4) 5

BRAKES

(G) Bleed Brakes (Four Wheels)
Includes: Add fluid.
1993-97 (.4) 5

(G) Brakes, Adjust (Minor)
Includes: Adjust brakes, fill master cylinder.
1993-97, two wheels 4
Remove knock out plugs
  add each 1

(M) Parking Brake, Adjust
1993-94 (.3) 4

LUBRICATION SERVICES

(M) Engine Oil & Filter, Renew
Includes: Inspect and correct all fluid levels.
1993-97 3

(M) Lubricate Chassis, Change Oil & Filter
Includes: Inspect and correct all fluid levels.
1993-97 6
Install grease fittings add 1

(M) Lubricate Chassis
Includes: Inspect and correct all fluid levels.
1993-97 4
Install grease fittings add 1

WHEELS

(G) Wheel, Renew (One)
1993-97 (.5) 5

(G) Wheel, Balance
1993-97
  one 3
  each adtnl. 2

(G) Wheels, Rotate (All)
1993-97 5

ELECTRICAL

(G) Headlamps, Aim
1993-97
  two 4
  four 6

(G) High Mount Stop Lamp Bulb, Renew
1993-94 (.3) 4
1995-97 (.2) 3

(G) License Lamp Bulb, Renew
1993-97, one or all (.2) 3

(G) Park & Turn Signal Lamp Assy., Renew
1993-97, each (.3) 4

(G) Park & Turn Signal Lamp Bulb or Lens, Renew
1993-97 (.2) 3

(G) Rear Combination Lamp Assy., Renew
1993-97, each (.2) 3

(G) Stop & Tail Lamp Bulb, Renew
1993-97
  one (.2) 3
  each adtnl. 1

(G) Horn, Renew
1993
  one (.3) 4
  each adtnl. 1
1994-97 (.2) 3

(M) Terminals, Battery, Clean
1993-97 3

SCHEDULED MAINTENANCE INTERVALS
(CHEVROLET CAVALIER, PONTIAC SUNBIRD & SUNFIRE)

TO BE SERVICED	TYPE OF SERVICE	VEHICLE MILEAGE INTERVAL (x1000)												
		7.5	15	22.5	30	37.5	45	52.5	60	67.5	75	82.5	90	97.5
Engine oil & filter	R	✓	✓	✓	✓	✓	✓	✓	✓	✓	✓	✓	✓	✓
Exhaust system & brake hoses	S/I	✓	✓	✓	✓	✓	✓	✓	✓	✓	✓	✓	✓	✓
Drive shaft boots & front suspension components	S/I	✓	✓	✓	✓	✓	✓	✓	✓	✓	✓	✓	✓	✓
Coolant level, hoses & clamps	S/I	✓	✓	✓	✓	✓	✓	✓	✓	✓	✓	✓	✓	✓
Throttle linkage	S/I	✓	✓	✓	✓	✓	✓	✓	✓	✓	✓	✓	✓	✓
Lubricate, chassis, suspension, steering linkage, transaxle shift linkage, parking brake cable guides, underbody contact points & linkage	S/I	✓	✓	✓	✓	✓	✓	✓	✓	✓	✓	✓	✓	✓
Brake linings & rotate tires	S/I	✓		✓		✓		✓		✓		✓		✓
Automatic transaxle fluid & filter③	S/I	✓		✓		✓		✓		✓		✓		✓
Air filter element & PCV filter	R				✓				✓				✓	
Engine coolant (1993-95)②	R				✓				✓				✓	
Spark plugs①	R				✓				✓				✓	
Accessory drive belt(s)	S/I				✓				✓				✓	
EGR & fuel systems	S/I				✓				✓				✓	
Ignition cables	S/I				✓				✓				✓	
Throttle body mount bolt torque	S/I	✓												

① Platinum tip spark plugs - replace every 100,000 miles.
② Engine coolant (1996-97) - replace every 100,000 miles. Use O.E. specified (DEX-COOL™) coolant only. If any silicate coolant is used, the service interval is every 30,000 miles.
③ Automatic transaxle fluid & filter - replace at 100,000 miles (if not changed previously).
R – Replace   S/I – Service or Inspect

SCHEDULED MAINTENANCE INTERVALS
(CHEVROLET CAVALIER, PONTIAC SUNBIRD & SUNFIRE) (Cont.)

FREQUENT OPERATION MAINTENANCE (SEVERE SERVICE)

If a vehicle is operated under any of the following conditions it is considered severe service:
- Extremely dusty areas.
- 50% or more of the vehicle operation is in 32°C (90°F) or higher temperatures, or constant operation in temperatures below 0°C (32°F).
- Prolonged idling (vehicle operation in stop and go traffic).
- Frequent short running periods (engine does not warm to normal operating temperatures).
- Police, taxi, delivery usage or trailer towing usage.

CV joints & front suspension components - check every 3000 miles.
Oil & oil filter – change every 3000 miles.
Brake linings – check every 6000 miles.
Chassis lubrication - lubricate every 6000 miles.
Lubricate suspension, steering linkage, transaxle shift linkage, parking cable guides, underbody contact points, & linkage - lubricate every 6000 miles.
Throttle body mount bolt torque- check at 6000 miles.
Rotate tires at 6000 miles, then every 12,000 miles.
Air filter element – service or inspect every 15,000 miles.
Automatic transaxle fluid – change every 15,000 miles.

CORSICA/BERETTA

VEHICLE IDENTIFICATION CHART

		Engine Code			
Code	Liters	Cu. In. (cc)	Cyl.	Fuel Sys.	Eng. Mfg.
4	2.2	133 (2180)	4	MFI	CUS
A	2.3	138 (2262)	4	MFI	CUS
M	3.1	191 (3130)	6	MFI	CPC
T	3.1	191 (3130)	6	MFI	CPC

Model Year	
Code	Year
P	1993
R	1994
S	1995
T	1996
V	1997

CUS - Chevrolet/United States
CPC - Chevrolet/Pontiac/Canada
MFI - Multiport fuel injection

ENGINE IDENTIFICATION

Year	Model	Engine Displacement Liters (cc)	Engine Series (ID/VIN)	Fuel System	No. of Cylinders	Engine Type
1993	Beretta	2.2 (2180)	4	MFI	4	OHV
	Beretta	2.3 (2262)	A	MFI	4	DOHC
	Beretta	3.1 (3130)	T	MFI	6	OHV
	Corsica	2.2 (2180)	4	MFI	4	OHV
	Corsica	3.1 (3130)	T	MFI	6	OHV

32 SPECIFICATIONS

ENGINE IDENTIFICATION

Year	Model	Engine Displacement Liters (cc)	Engine Series (ID/VIN)	Fuel System	No. of Cylinders	Engine Type
1994	Beretta	2.2 (2180)	4	MFI	4	OHV
	Beretta	2.3 (2262)	A	MFI	4	DOHC
	Beretta	3.1 (3130)	M	MFI	6	OHV
	Corsica	2.2 (2180)	4	MFI	4	OHV
	Corsica	3.1 (3130)	M	MFI	6	OHV
1995	Beretta	2.2 (2180)	4	MFI	4	OHV
	Beretta	3.1 (3130)	M	MFI	6	OHV
	Corsica	2.2 (2180)	4	MFI	4	OHV
	Corsica	3.1 (3130)	M	MFI	6	OHV
1996-97	Beretta	2.2 (2180)	4	MFI	4	OHV
	Beretta	3.1 (3130)	M	MFI	6	OHV
	Corsica	2.2 (2180)	4	MFI	4	OHV
	Corsica	3.1 (3130)	M	MFI	6	OHV

MFI - Multiport fuel injection   OHV - Overhead valve   DOHC - Double overhead camshaft

GENERAL ENGINE SPECIFICATIONS

Year	Engine ID/VIN	Engine Displacement Liters (cc)	Fuel System Type	Net Horsepower @ rpm	Net Torque @ rpm (ft. lbs.)	Bore x Stroke (in.)	Compression Ratio	Oil Pressure @ rpm
1993	4	2.2 (2180)	MFI	110@5200	130@3200	3.50x3.46	9.0:1	63-77@1200
	A	2.3 (2262)	MFI	180@6200	160@5200	3.62x3.46	10.0:1	30@2000
	T	3.1 (3130)	MFI	140@4200	185@3200	3.50x3.31	8.5:1	8@600
1994	4	2.2 (2180)	MFI	110@5200	130@3200	3.50x3.46	9.0:1	63-77@1200
	A	2.3 (2262)	MFI	180@6200	160@5200	3.62x3.46	10.0:1	30@2000
	M	3.1 (3130)	MFI	155@5200	185@4000	3.50x3.31	9.6:1	15@1100
1995	4	2.2 (2180)	MFI	120@5200	130@3200	3.50x3.46	9.0:1	63-77@1200
	M	3.1 (3130)	MFI	155@5200	185@4000	3.50x3.31	9.6:1	15@1100
1996-97	4	2.2 (2180)	MFI	120@5200	130@3200	3.50x3.46	9.0:1	56@3000
	M	3.1 (3130)	MFI	155@5200	185@4000	3.50x3.31	9.5:1	15@1100

MFI - Multiport fuel injection

GASOLINE ENGINE TUNE-UP SPECIFICATIONS

Year	Engine ID/VIN	Engine Displacement Liters (cc)	Spark Plugs Gap (in.)	Ignition Timing (deg.) MT	Ignition Timing (deg.) AT	Fuel Pump (psi)	Idle Speed (rpm) MT	Idle Speed (rpm) AT	Valve Clearance In.	Valve Clearance Ex.
1993	4	2.2 (2180)	0.045	①	①	41-47	①	①	HYD	HYD
	A	2.3 (2262)	0.035	①	①	41-47	①	①	HYD	HYD
	T	3.1 (3130)	0.045	①	①	41-47	①	①	HYD	HYD
1994	4	2.2 (2180)	0.045	①	①	41-47	①	①	HYD	HYD
	A	2.3 (2262)	0.035	①	①	41-47	①	①	HYD	HYD
	M	3.1 (3130)	0.045	①	①	41-47	①	①	HYD	HYD
1995	4	2.2 (2180)	0.045	①	①	41-47	①	①	HYD	HYD
	M	3.1 (3130)	0.045	①	①	41-47	①	①	HYD	HYD
1996-97	4	2.2 (2180)	0.045	①	①	41-47	①	①	HYD	HYD
	M	3.1 (3130)	0.045	①	①	41-47	①	①	HYD	HYD

NOTE: The Vehicle Emission Control Information label often reflects specification changes made during production. The label figures must be used if they differ from those in this chart.
B - Before top dead center
HYD - Hydraulic
① Refer to Vehicle Emission Control Information label

CAPACITIES

Year	Model	Engine ID/VIN	Engine Displacement Liters (cc)	Engine Oil with Filter (qts.)	Transmission (pts.) 4-Spd	5-Spd	Auto.	Drive Axle Front (pts.)	Rear (pts.)	Fuel Tank (gal.)	Cooling System (qts.)
1993	Beretta	4	2.2 (2180)	4.5	-	4.0	14.0 ①	-	-	15.6	9.5
	Beretta	A	2.3 (2262)	4.5	-	4.0	14.0 ①	-	-	15.6	9.5
	Beretta	T	3.1 (3130)	4.5	-	4.0	14.0 ①	-	-	15.6	②
	Corsica	4	2.2 (2180)	4.5	-	4.0	14.0 ①	-	-	15.6	9.5
	Corsica	T	3.1 (3130)	4.5	-	4.0	14.0 ①	-	-	15.6	②
1994	Beretta	4	2.2 (2180)	4.5	-	4.0	14.0 ①	-	-	15.6	9.5
	Beretta	A	2.3 (2262)	4.5	-	4.0	14.0 ①	-	-	15.6	9.5
	Beretta	M	3.1 (3130)	4.5	-	4.0	14.0 ①	-	-	15.6	②
	Corsica	4	2.2 (2180)	4.5	-	4.0	14.0 ①	-	-	15.6	9.5
	Corsica	M	3.1 (3130)	4.5	-	4.0	14.0 ①	-	-	15.6	②
1995	Beretta	4	2.2 (2180)	4.5	-	4.0	14.0 ①	-	-	15.6	9.5
	Beretta	M	3.1 (3130)	4.5	-	4.0	14.0 ①	-	-	15.6	②
	Corsica	4	2.2 (2180)	4.5	-	4.0	14.0 ①	-	-	15.6	9.5
	Corsica	M	3.1 (3130)	4.5	-	4.0	14.0 ①	-	-	15.6	②
1996-97	Beretta	4	2.2 (2180)	4.5	-	4.0	14.0 ①	-	-	15.6	9.5
	Beretta	M	3.1 (3130)	4.5	-	4.0	14.0 ①	-	-	15.6	②
	Corsica	4	2.2 (2180)	4.5	-	4.0	14.0 ①	-	-	15.6	9.5
	Corsica	M	3.1 (3130)	4.5	-	4.0	14.0 ①	-	-	15.6	②

① Drain and refill figure, overhaul: 16.0 pts.
② Automatic transmission: 12.4 qts.
Manual transmission: 11.8 qts.

VALVE SPECIFICATIONS

Year	Engine ID/VIN	Engine Displacement Liters (cc)	Seat Angle (deg.)	Face Angle (deg.)	Spring Test Pressure (lbs. @ in.)	Spring Installed Height (in.)	Stem-to-Guide Clearance (in.) Intake	Exhaust	Stem Diameter (in.) Intake	Exhaust
1993	4	2.2 (2180)	46	45	225-233@ ① 1.25	1.64 ②	0.0011-0.0026	0.0014-0.0031	NA	NA
	A	2.3 (2262)	45	44	193-207@ ① 1.04	1.44 ②	0.0010-0.0027	0.0015-0.0032	0.2740-0.2750	0.2740-0.2750
	T	3.1 (3130)	46	45	90 ①	1.60 ②	0.0010-0.0027	0.0010-0.0027	NA	NA
1994	4	2.2 (2180)	46	45	225-233@ ① 1.25	1.64 ②	0.0011-0.0026	0.0014-0.0031	NA	NA
	A	2.3 (2262)	45	44	193-207@ ① 1.04	1.44 ②	0.0010-0.0027	0.0015-0.0032	0.2740-0.2750	0.2740-0.2750
	M	3.1 (3130)	45	45	80@1.71	1.71	0.0010-0.0027	0.0010-0.0027	NA	NA
1995	4	2.2 (2180)	46	45	225-233@ ① 1.25	1.64 ②	0.0011-0.0026	0.0014-0.0031	NA	NA
	M	3.1 (3130)	45	45	80@1.71	1.71	0.0010-0.0027	0.0010-0.0027	NA	NA
1996-97	4	2.2 (2195)	46	45	75-81@1.71	1.71	0.0010-0.0027	0.0014-0.0031	NA	NA
	M	3.1 (3136)	45	45	80@1.71	1.71	0.0010-0.0027	0.0010-0.0027	NA	NA

NA - Not Available
① 1 With valve open
② 2 With valve closed

TORQUE SPECIFICATIONS
All readings in ft. lbs.

Year	Engine ID/VIN	Engine Displacement Liters (cc)	Cylinder Head Bolts	Main Bearing Bolts	Rod Bearing Bolts	Crankshaft Damper Bolts	Flywheel Bolts	Manifold Intake	Manifold Exhaust	Spark Plugs	Lug Nut
1993	4	2.2 (2180)	①	77	38	85 ②	52-55	18	6-13	20	80-100
	A	2.3 (2262)	26 ⑤	⑦	⑧	⑨	⑩	③	④	17	80-100
	T	3.1 (3130)	⑪	73	39	66-85	45-59	⑥	18	20	100
1994	4	2.2 (2180)	①	77	38	85 ②	52-55	18	6-13	20	100
	A	2.3 (2262)	26 ⑤	⑦	⑧	⑨	⑩	⑭	④	17	100
	M	3.1 (3130)	⑪	⑫	37	76	61	⑬	12	11	100
1995	4	2.2 (2180)	①	77	38	85 ②	52-55	18	6-13	20	100
	M	3.1 (3130)	⑪	⑫	37	76	61	⑬	12	11	100
1996-97	4	2.2 (2180)	①	70	38	77 ②	52-55	24	18	11	100
	M	3.1 (3130)	⑪	⑫	37	76	61	⑬	12	11	100

NA - Not Available

① Step 1: 41 ft. lbs.
  Step 2: Tighten an additional 45 degrees
  Step 3: Tighten an additional 45 degrees
  Step 4: Long bolts 1, 4-5, 8-9 an additional 20 degrees
  Step 4: Short bolts 2-3, 6-7, 10 an additional 10 degrees
② Center bolt spec shown; Pulley to hub bolts: 37 ft. lbs.
③ Nuts: 18 ft. lbs.
  Studs: 96 inch lbs.

④ Bolts: 27 ft. lbs.
  Studs: 106 inch lbs.
⑤ Cylinder head bolts should be torqued 26 ft. lbs.
  Long bolts: 100 degrees
  Short bolts: 120 degrees
⑥ 15 ft. lbs., then 24 ft. lbs.
⑦ 15 ft. lbs. plus 90 degrees
⑧ 18 ft. lbs. plus 80 degrees

⑨ 74 ft. lbs. plus 90 degrees
⑩ 22 ft. lbs. plus 45 degrees
⑪ 33 ft. lbs. plus 90 degrees
⑫ 37 ft. lbs. plus 75 degrees
⑬ 115 inch lbs.

BRAKE SPECIFICATIONS
All measurements in inches unless noted

Year	Model	Master Cylinder Bore	Brake Disc Original Thickness	Brake Disc Minimum Thickness	Brake Disc Maximum Runout	Brake Drum Diameter Original Inside Diameter	Brake Drum Diameter Max. Wear Limit	Brake Drum Diameter Maximum Machine Diameter	Minimum Lining Thickness Front	Minimum Lining Thickness Rear
1993	Beretta	0.945	0.885	0.830	0.004	7.879	7.929	7.899	0.030	0.030
	Corsica	0.945	0.885	0.830	0.004	7.879	7.929	7.899	0.030	0.030
1994	Beretta	0.945	0.885	0.830	0.004	7.879	7.929	7.899	0.030	0.030
	Corsica	0.945	0.885	0.830	0.004	7.879	7.929	7.899	0.030	0.030
1995	Beretta	0.945	0.885	0.830	0.004	7.879	7.929	7.899	0.030	0.030
	Corsica	0.945	0.885	0.830	0.004	7.879	7.929	7.899	0.030	0.030
1996-97	Beretta	0.945	0.885	0.830	0.004	7.879	7.929	7.899	0.030	0.030
	Corsica	0.945	0.885	0.830	0.004	7.879	7.929	7.899	0.030	0.030

FREQUENT MAINTENANCE LABOR
CHEVROLET CORSICA, BERETTA

The following should be used as a guide when determining the amount of work required for a particular service if taken to a repair shop. In estimating how long a particular Frequent Maintenance Service item should take, please observe the following:
- **Factory Time** is time that is generated by the vehicle manufacturer.
- **Chilton Time** is time that is based on field research and data supplied by the vehicle manufacturer.
- All labor time operations are given in hours and tenths of an hour.
- All labor operations, are to be used as a **guide**.

	(Factory Time)	Chilton Time

COOLING

(G) Winterize Cooling System
Includes: Run engine to check for leaks, tighten all hose connections. Test radiator and pressure cap, drain radiator and engine block. Add antifreeze and refill system.
1993-96 5

(G) Belt, Serpentine Drive, Renew
1993 (.3) 5
1994-96
  4 cyl. (.2) 3
  V6 (.5) 8

(G) Hoses, Radiator, Renew
Includes: Drain and refill cooling system as required.
1993-96
  Upper (.5) 6
  Lower
    Four 6
    V6 (.6) 8
    Both (.7) 1.0

(G) Thermostat, Coolant, Renew
1993-94 (.5) 8
1995-96
  4 cyl. (.4) 6
  V6 (.5) 8

FUEL

(M) Air Cleaner, Service
1993-96 2

BRAKES

(G) Bleed Brakes (Four Wheels)
Includes: Add fluid.
1993-96 (.5) 5

(G) Brakes, Adjust (Minor)
Includes: Adjust brakes, fill master cylinder.
1993-96, two wheels 4
Remove knock out plugs 1
  add each 1

(M) Parking Brake, Adjust
1993-96 (.3) 4

LUBRICATION SERVICES

(M) Engine Oil & Filter, Renew
Includes: Inspect and correct all fluid levels.
1993-96 3

(M) Lubricate Chassis, Change Oil & Filter
Includes: Inspect and correct all fluid levels.
1993-96 6
Install grease fittings add 1

(M) Lubricate Chassis
Includes: Inspect and correct all fluid levels.
1993-96 4
Install grease fittings add 1

WHEELS

(G) Wheel, Renew (One)
1993-96 5

(G) Wheel, Balance
1993-96
  one 3
  each adtnl. 2

(G) Wheels, Rotate (All)
1993-96 5

ELECTRICAL

(G) Headlamps, Aim
1993-96
  two 4
  four 6

(G) High Mount Stop Lamp Bulb, Renew
1993-96 (.2) 3

(G) License Lamp Bulb, Renew
1993-96, one or all 3

(G) Park & Turn Signal Lamp Assy., Renew
1993-96, one (.3) 4

(G) Park & Turn Signal Lamp Bulb or Lens, Renew
1993
  one 2
  all .3

(G) Rear Combination Lamp Assy., Renew
1993-96 (.3) 5

(G) Stop, Tail & Turn Signal Lamp Bulb, Renew
1993-96
  one 3
  each adtnl. 1

(G) Horn, Renew
1993 (.7) 9
1994-96 (.3) 5

(M) Terminals, Battery, Clean
1993-96 3

SCHEDULED MAINTENANCE INTERVALS
(CHEVROLET CORSICA & BERETTA)

TO BE SERVICED	TYPE OF SERVICE	VEHICLE MILEAGE INTERVAL (x1000)												
		7.5	15	22.5	30	37.5	45	52.5	60	67.5	75	82.5	90	97.5
Engine oil & filter	R	✓	✓	✓	✓	✓	✓	✓	✓	✓	✓	✓	✓	✓
Automatic transaxle fluid & filter④	S/I	✓	✓	✓	✓	✓	✓	✓	✓	✓	✓	✓	✓	✓
Brake hoses	S/I	✓	✓	✓	✓	✓	✓	✓	✓	✓	✓	✓	✓	✓
Chassis lubrication	S/I	✓	✓	✓	✓	✓	✓	✓	✓	✓	✓	✓	✓	✓
Coolant level, hoses & clamps	S/I	✓	✓	✓	✓	✓	✓	✓	✓	✓	✓	✓	✓	✓
Drive shaft boots & front suspension components	S/I	✓	✓	✓	✓	✓	✓	✓	✓	✓	✓	✓	✓	✓
Exhaust system	S/I	✓	✓	✓	✓	✓	✓	✓	✓	✓	✓	✓	✓	✓
Lubricate suspension, steering linkage, transaxle shift linkage, parking brake cable guides, underbody contact points & linkage	S/I	✓	✓	✓	✓	✓	✓	✓	✓	✓	✓	✓	✓	✓
Manual transaxle oil	S/I	✓	✓	✓	✓	✓	✓	✓	✓	✓	✓	✓	✓	✓
Throttle linkage	S/I	✓	✓	✓	✓	✓	✓	✓	✓	✓	✓	✓	✓	✓
Brake linings	S/I	✓		✓		✓		✓		✓		✓		✓
Rotate tires③	S/I	✓		✓		✓		✓		✓		✓		✓
Air filter element	R				✓				✓				✓	
Engine coolant (1993-95)②	R				✓				✓				✓	
PCV filter	R				✓				✓				✓	
Spark plugs①	R				✓				✓				✓	
Accessory drive belt(s)	S/I				✓				✓				✓	
EGR & fuel systems	S/I				✓				✓				✓	
Ignition cables	S/I				✓				✓				✓	
Throttle body mount bolt torque	S/I	✓												

① Platinum tip spark plugs - replace every 100,000 miles.
② Engine coolant (1996-97) - replace every 100,000 miles. Use O.E. specified (DEX-COOL™) coolant only. If any silicate coolant is used, the service interval is every 30,000 miles.
③ Rotate tires front-to-rear only on Beretta GTZ and Z26.
④ Automatic transaxle fluid & filter - change at 100,000 miles (unless changed previously).
R – Replace   S/I – Service or Inspect

SCHEDULED MAINTENANCE INTERVALS
(CHEVROLET CORSICA & BERETTA) (Cont.)

FREQUENT OPERATION MAINTENANCE (SEVERE SERVICE)

If a vehicle is operated under any of the following conditions it is considered severe service:
- Extremely dusty areas.
- 50% or more of the vehicle operation is in 32°C (90°F) or higher temperatures, or constant operation in temperatures below 0°C (32°F).
- Prolonged idling (vehicle operation in stop and go traffic).
- Frequent short running periods (engine does not warm to normal operating temperatures).
- Police, taxi, delivery usage or trailer towing usage.

Oil & oil filter – change every 3000 miles.
Chassis lubrication - lubricate every 6000 miles.
Throttle body mount bolt torque - tighten at 6000 miles.
Rotate tires at 6000 miles, then every 12,000 miles. ③
Air filter element - service or inspect every 15,000 miles.
Automatic transaxle fluid & filter - change every 15,000 miles.

SKYLARK/ACHIEVA/GRAND AM

VEHICLE IDENTIFICATION CHART

		Engine Code					Model Year	
Code	Liters	Cu. In. (cc)	Cyl.	Fuel Sys.	Eng. Mfg.		Code	Year
3	2.3	138 (2261)	4	MFI	BOC		P	1993
A	2.3	138 (2262)	4	MFI	BOC		R	1994
D	2.3	138 (2261)	4	MFI	BOC		S	1995
T	2.4	146 (2392)	4	MFI	CUS		T	1996
M	3.1	191 (3130)	6	MFI	BOC		V	1997
N	3.3	204 (3342)	6	MFI	BOC			

BOC - Buick/Oldsmobile/Cadillac
CUS - Chevrolet/United States
MFI - Multiport fuel injection

ENGINE IDENTIFICATION

Year	Model	Engine Displacement Liters (cc)	Engine Series (ID/VIN)	Fuel System	No. of Cylinders	Engine Type
1993	Achieva	2.3 (2261)	3	MFI	4	DOHC
	Achieva	2.3 (2262)	A	MFI	4	DOHC
	Achieva	2.3 (2261)	D	MFI	4	DOHC
	Achieva	3.3 (3342)	N	MFI	6	OHV
	Grand Am	2.3 (2261)	3	MFI	4	SOHC
	Grand Am	2.3 (2262)	A	MFI	4	DOHC
	Grand Am	2.3 (2261)	D	MFI	4	DOHC
	Grand Am	3.3 (3342)	N	MFI	6	OHV

ENGINE IDENTIFICATION

Year	Model	Engine Displacement Liters (cc)	Engine Series (ID/VIN)	Fuel System	No. of Cylinders	Engine Type
1993	Skylark	2.3 (2261)	3	MFI	4	SOHC
	Skylark	3.3 (3342)	N	MFI	6	OHV
1994	Achieva	2.3 (2261)	3	MFI	4	DOHC
	Achieva	2.3 (2262)	A	MFI	4	DOHC
	Achieva	2.3 (2261)	D	MFI	4	DOHC
	Achieva	3.1 (3130)	M	MFI	6	OHV
	Grand Am	2.3 (2261)	3	MFI	4	SOHC
	Grand Am	2.3 (2262)	A	MFI	4	DOHC
	Grand Am	2.3 (2261)	D	MFI	4	DOHC
	Grand Am	3.1 (3130)	M	MFI	6	OHV
	Skylark	2.3 (2261)	3	MFI	4	SOHC
	Skylark	3.1 (3130)	M	MFI	6	OHV
1995	Achieva	2.3 (2261)	D	MFI	4	DOHC
	Achieva	3.1 (3130)	M	MFI	6	OHV
	Grand Am	2.3 (2261)	D	MFI	4	DOHC
	Grand Am	3.1 (3130)	M	MFI	6	OHV
	Skylark	2.3 (2261)	D	MFI	4	DOHC
	Skylark	3.1 (3130)	M	MFI	6	OHV
1996-97	Achieva	2.4 (2392)	T	MFI	4	DOHC
	Achieva	3.1 (3130)	M	MFI	6	OHV
	Grand Am	2.4 (2392)	T	MFI	4	DOHC
	Grand Am	3.1 (3130)	M	MFI	6	OHV
	Skylark	2.4 (2392)	T	MFI	4	DOHC
	Skylark	3.1 (3130)	M	MFI	6	OHV

MFI - Multiport fuel injection
SOHC - Single overhead camshaft
DOHC - Double overhead camshaft
OHV - Overhead valve

GENERAL ENGINE SPECIFICATIONS

Year	Engine ID/VIN	Engine Displacement Liters (cc)	Fuel System Type	Net Horsepower @ rpm	Net Torque @ rpm (ft. lbs.)	Bore x Stroke (in.)	Compression Ratio	Oil Pressure @ rpm
1993	3	2.3 (2262)	MFI	115@5200	140@3200	3.63x3.35	9.5:1	30@2000
	A	2.3 (2262)	MFI	175@6200	155@5200	3.63x3.35	10.0:1	30@2000
	D	2.3 (2261)	MFI	155@6000	150@4800	3.63x3.35	9.5:1	30@2000
	N	3.3 (3342)	MFI	160@5200	185@2000	3.70x3.16	9.0:1	60@1850
1994	3	2.3 (2262)	MFI	115@5200	140@3200	3.63x3.35	9.5:1	30@2000
	A	2.3 (2262)	MFI	175@6200	150@5200	3.63x3.35	10.0:1	30@2000
	D	2.3 (2261)	MFI	155@6000	150@6000	3.63x3.35	9.5:1	30@2000
	M	3.1 (3130)	MFI	160@5200	185@4000	3.50x3.31	9.5:1	15@1100
1995	D	2.3 (2261)	MFI	155@6000	145@4800	3.62x3.35	9.5:1	30@2000
	M	3.1 (3130)	MFI	160@5200	185@4000	3.50x3.31	9.5:1	15@1100
1996-97	T	2.4 (2392)	MFI	150@6000	150@4400	3.54x3.70	9.5:1	30@3000
	M	3.1 (3130)	MFI	160@5200	185@4000	3.50x3.31	9.5:1	15@1100

MFI - Multiport fuel injection

GASOLINE ENGINE TUNE-UP SPECIFICATIONS

Year	Engine ID/VIN	Engine Displacement Liters (cc)	Spark Plugs Gap (in.)	Ignition Timing (deg.) MT	Ignition Timing (deg.) AT	Fuel Pump (psi)	Idle Speed (rpm) MT	Idle Speed (rpm) AT	Valve Clearance In.	Valve Clearance Ex.
1993	3	2.3 (2261)	0.035	①	①	41-47	②	②	HYD	HYD
	A	2.3 (2262)	0.035	①	①	41-47	②	②	HYD	HYD
	D	2.3 (2261)	0.035	①	①	41-47	②	②	HYD	HYD
	N	3.3 (3342)	0.060	①	①	41-47	②	②	HYD	HYD
1994	3	2.3 (2261)	0.035	①	①	41-47	②	②	HYD	HYD
	A	2.3 (2262)	0.035	①	①	41-47	②	②	HYD	HYD
	D	2.3 (2261)	0.035	①	①	41-47	②	②	HYD	HYD
	M	3.1 (3130)	0.060	①	①	41-47	②	②	HYD	HYD
1995	D	2.3 (2262)	0.035	①	①	41-47	②	②	HYD	HYD
	M	3.1 (3130)	0.060	①	①	41-47	②	②	HYD	HYD
1996-97	T	2.4 (2392)	0.035	③	③	41-47	③	③	HYD	HYD
	M	3.1 (3130)	0.060	③	③	41-47	③	③	HYD	HYD

NOTE: The Vehicle Emission Control Information label often reflects specification changes made during production. The label figures must be used if they differ from those in this chart.

HYD - Hydraulic

① DIS Ignition System timing is not adjustable
② Idle speed is maintained by the ECM. There is no recommended adjustment procedure
③ Refer to Vehicle Emission Control Information label

CAPACITIES

Year	Model	Engine ID/VIN	Engine Displacement Liters (cc)	Engine Oil with Filter (qts.)	Transmission (pts.) 4-Spd	Transmission (pts.) 5-Spd	Transmission (pts.) Auto.	Drive Axle Front (pts.)	Drive Axle Rear (pts.)	Fuel Tank (gal.)	Cooling System (qts.)
1993	Achieva	3	2.3 (2261)	4.0 ①	-	③	8.0	-	-	15.2	10.4
	Achieva	A	2.3 (2262)	4.0 ①	-	③	8.0	-	-	15.2	10.4
	Achieva	D	2.3 (2261)	4.0 ①	-	③	8.0	-	-	15.2	10.4
	Achieva	N	3.3 (3344)	4.0 ①	-	③	8.0	-	-	15.2	10.8
	Grand Am	3	2.3 (2261)	4.0 ①	-	4.0	8.0	-	-	15.2	9.5
	Grand Am	A	2.3 (2262)	4.0 ①	-	4.0	8.0	-	-	15.2	9.5
	Grand Am	D	2.3 (2261)	4.0 ①	-	4.0	8.0	-	-	15.2	9.5
	Grand Am	N	3.3 (3344)	4.0 ①	-	-	8.0	-	-	15.2	12.7
	Skylark	3	2.3 (2261)	4.0 ①	-	-	8.0	-	-	15.2	9.5
	Skylark	N	3.3 (3344)	4.0 ①	-	-	8.0	-	-	15.2	12.7
1994	Achieva	3	2.3 (2261)	4.0 ①	-	4.0	④	-	-	15.2	10.4
	Achieva	A	2.3 (2262)	4.0 ①	-	4.2	④	-	-	15.2	10.4
	Achieva	D	2.3 (2261)	4.0 ①	-	4.0	④	-	-	15.2	10.4
	Achieva	M	3.1 (3130)	4.0 ①	-	4.0	④	-	-	15.2	10.8
	Grand Am	3	2.3 (2261)	4.0 ①	-	4.0	8.0 ②	-	-	15.2	10.4
	Grand Am	A	2.3 (2262)	4.0 ①	-	4.2	8.0 ②	-	-	15.2	10.4
	Grand Am	D	2.3 (2261)	4.0 ①	-	4.0	8.0 ②	-	-	15.2	10.4
	Grand Am	M	3.1 (3130)	4.0 ①	-	4.0	8.0 ②	-	-	15.2	10.4
	Skylark	3	2.3 (2261)	4.0 ①	-	-	8.0 ②	-	-	15.2	10.4
	Skylark	M	3.1 (3130)	4.0 ①	-	-	8.0 ②	-	-	15.2	10.4
1995	Achieva	D	2.3 (2261)	4.0 ①	-	4.0	④	-	-	15.2	10.4
	Achieva	M	3.1 (3130)	3.8 ①	-	4.0	④	-	-	15.2	10.8
	Grand Am	D	2.3 (2261)	4.0 ①	-	4.0	8.0 ②	-	-	15.2	10.4
	Grand Am	M	3.1 (3130)	3.8 ①	-	4.0	8.0 ②	-	-	15.2	13.1
	Skylark	D	2.3 (2261)	4.0 ①	-	-	8.0 ②	-	-	15.2	10.4
	Skylark	M	3.1 (3130)	4.0 ①	-	-	8.0 ②	-	-	15.2	13.1

CAPACITIES

Year	Model	Engine ID/VIN	Engine Displacement Liters (cc)	Engine Oil with Filter (qts.)	Transmission (pts.) 4-Spd	5-Spd	Auto.	Drive Axle Front (pts.)	Rear (pts.)	Fuel Tank (gal.)	Cooling System (qts.)
1996-97	Achieva	T	2.4 (2392)	4.0 ①	-	4.0	④	-	-	15.2	10.4
	Achieva	M	3.1 (3130)	4.0 ①	-	4.0	④	-	-	15.2	10.8
	Grand Am	T	2.4 (2392)	4.5	-	4.0	- ②	-	-	15.2	10.4
	Grand Am	M	3.1 (3130)	4.5	-	4.0	8.0 ②	-	-	15.2	13.1
	Grand Am	T	2.4 (2392)	4.0 ①	-	4.0	④	-	-	15.2	10.4
	Grand Am	M	3.1 (3130)	4.0 ①	-	4.0	④	-	-	15.2	10.8
	Skylark	T	2.4 (2392)	4.0 ①	-	12.0	-	-	-	15.2	10.4
	Skylark	M	3.1 (3130)	4.0 ①	-	12.0	-	-	-	15.2	13.1

① Capacity is without filter replacement; Additional oil may be required
② With 4T60E transaxle: 12.0 pts.
③ With T550: 4.2
   With Isuzu: 4.0
④ 3 speed: 8.0
   4 speed: 12

VALVE SPECIFICATIONS

Year	Engine ID/VIN	Engine Displacement Liters (cc)	Seat Angle (deg.)	Face Angle (deg.)	Spring Test Pressure (lbs. @ in.)	Spring Installed Height (in.)	Stem-to-Guide Clearance (in.) Intake	Exhaust	Stem Diameter (in.) Intake	Exhaust
1993	3	2.3 (2261)	45	②	193-207@ 1.043	0.984- 1.004 ①	0.0010- 0.0027	0.0015- 0.0032	0.2751- 0.2745	0.2740- 0.2747
	A	2.3 (2262)	45	②	193-207@ 1.043	0.984- 1.004 ①	0.0010- 0.0027	0.0015- 0.0032	0.2751- 0.2745	0.2740- 0.2747
	D	2.3 (2261)	45	②	193-207@ 1.043	0.984- 1.004 ①	0.0010- 0.0027	0.0015- 0.0032	0.2751- 0.2745	0.2740- 0.2747
	N	3.3 (3342)	45	45	210@ 1.315	1.690- 1.720	0.0015- 0.0035	0.0015- 0.0032	NA	NA
1994	3	2.3 (2261)	45	②	193-207@ 1.043	0.984- 1.004 ①	0.0010- 0.0027	0.0015- 0.0032	0.2751- 0.2745	0.2740- 0.2747
	A	2.3 (2262)	45	②	193-207@ 1.043	0.984- 1.004 ①	0.0010- 0.0027	0.0015- 0.0032	0.2751- 0.2745	0.2740- 0.2747
	D	2.3 (2261)	45	②	193-207@ 1.043	0.984- 1.004 ①	0.0010- 0.0027	0.0015- 0.0032	0.2751- 0.2745	0.2740- 0.2747
	M	3.1 (3130)	45	45	250@ 1.239	1.710	0.0010- 0.0027	0.0010- 0.0027	NA	NA
1995	D	2.3 (2261)	45	②	193-207@ 1.043	0.984- 1.004 ①	0.0010- 0.0027	0.0015- 0.0032	0.2751- 0.2745	0.2740- 0.2747
	M	3.1 (3130)	45	45	250@ 1.239	1.710	0.0010- 0.0027	0.0010- 0.0027	NA	NA
1996-97	T	2.4 (2392)	45	46	50-55@ 1.437	1.437	0.0009- 0.0025	0.0016- 0.0032	0.2331- 0.2339	0.2326- 0.2334
	M	3.1 (3130)	45	45	250@1.239	1.710	0.0010- 0.0027	0.0010- 0.0027	NA	NA

NA - Not Available
① Measured from top of valve stem to top of camshaft housing
② Intake face angle: 44 degrees
   Exhaust face angle: 44.5 degrees

TORQUE SPECIFICATIONS
All readings in ft. lbs.

Year	Engine ID/VIN	Engine Displacement Liters (cc)	Cylinder Head Bolts	Main Bearing Bolts	Rod Bearing Bolts	Crankshaft Damper Bolts	Flywheel Bolts	Manifold		Spark Plugs	Lug Nut
								Intake	Exhaust		
1993	A	2.3 (2262)	⑭	②	③	⑮	⑤	18	31	16	100
	D	2.3 (2261)	⑭	②	③	⑮	⑤	18	31	16	100
	3	2.3 (2261)	20	②	③	⑮	⑤	18	31	16	100
	N	3.3 (3342)	⑧	⑨	⑩	⑪	⑫	7	38	20	100
1994	A	2.3 (2262)	⑰	②	③	④	⑤	19	31	16	100
	D	2.3 (2261)	⑰	②	③	④	⑤	19	31	16	100
	3	2.3 (2261)	⑰	②	③	④	⑤	19	31	16	100
	M	3.1 (3130)	⑥	⑬	⑰	76	61	⑦	10	⑯	100
1995	D	2.3 (2261)	⑰	②	③	④	⑤	19	31	16	100
	M	3.1 (3130)	⑥	⑬	①	76	59	⑦	12	⑯	100
1996-97	T	2.4 (2392)	⑰	②	③	④	⑤	19	31	16	100
	M	3.1 (3130)	⑥	⑬	①	76	59	⑦	12	⑯	100

NA - Not Available
① 15 ft. lbs. plus 75 degrees
② 15 ft. lbs. plus 90 degrees
③ 18 ft. lbs. plus 80 degrees
④ 129 ft. lbs. plus 90 degrees
⑤ 22 ft. lbs. plus 45 degrees
⑥ Coat threads with sealer torque to 37 ft. lbs., then turn 1/4 turn (90 degrees)
⑦ Torque all bolts to 15 ft. lbs., Retorque to 24 ft. lbs.

⑧ Step 1: Tighten all bolts to 35 ft. lbs.
   Step 2: Turn all bolts 130 degrees
   Step 3: Rotate four center bolts an additional 30 degrees
⑨ 26 ft. lbs. plus 50 degrees
⑩ 20 ft. lbs. plus 50 degrees
⑪ 110 ft. lbs. plus 76 degrees
⑫ 11 ft. lbs. plus 50 degrees
⑬ 37 ft. lbs. plus 75 degrees

⑭ Nos. 1-6: 18 ft. lbs. plus 90 degrees
   Nos. 7-8: 22 ft. lbs. plus 60 degrees
   Nos. 9-10: 26 ft. lbs. plus 60 degrees
⑮ 110 ft. lbs. plus 90 degrees
⑯ New cylinder first-time installation: 21 ft. lbs.
   All others: 11 ft. lbs.
⑰ Nos. 1-8: 30 ft. !bs.
   Nos. 9-10: 26 ft. lbs.
   Tighten all bolts an additional 90 degrees

BRAKE SPECIFICATIONS
All measurements in inches unless noted

Year	Model	Master Cylinder Bore	Brake Disc			Brake Drum Diameter			Minimum Lining Thickness	
			Original Thickness	Minimum Thickness	Maximum Runout	Original Inside Diameter	Max. Wear Limit	Maximum Machine Diameter	Front	Rear
1993	Achieva	0.874	0.806	0.736	0.003	7.874-7.890	7.929	7.899	0.030	①
	Grand Am	0.874	0.806	0.736	0.003	7.874	7.929	7.899	0.030	①
	Skylark	0.874	0.806	0.736	0.003	7.874-7.890	7.929	7.899	0.030	①
1994	Achieva	0.874	0.806	0.736	0.003	7.874-7.890	7.929	7.899	0.030	①
	Grand Am	0.874	0.806	0.736	0.003	7.874	7.929	7.899	0.030	①
	Skylark	0.874	0.806	0.736	0.003	7.874-7.890	7.929	7.899	0.030	①
1995	Achieva	0.874	0.806	0.736	0.003	7.874-7.890	7.929	7.899	0.030	①
	Grand Am	0.874	0.806	0.736	0.003	7.874	7.929	7.899	0.030	①
	Skylark	0.874	0.806	0.736	0.003	7.874-7.890	7.929	7.899	0.030	①
1996-97	Achieva	0.874	0.806	0.736	0.003	7.874-7.890	7.930	7.899	0.030	①
	Grand Am	0.874	0.806	0.736	0.003	7.874	7.930	7.900	0.030	0.030
	Skylark	0.874	0.806	0.736	0.003	7.874-7.890	7.929	7.899	0.030	①

① 0.030 over rivet head; If bonded lining, use 0.062 from shoe

FREQUENT MAINTENANCE LABOR
BUICK SKYLARK, OLDSMOBILE ACHIEVA, PONTIAC GRAND AM

The following should be used as a guide when determining the amount of work required for a particular service if taken to a repair shop.
In estimating how long a particular Frequent Maintenance Service item should take, please observe the following:
- **Factory Time** is time that is generated by the vehicle manufacturer.
- **Chilton Time** is time that is based on field research and data supplied by the vehicle manufacturer.
- All labor time operations are given in hours and tenths of an hour.
- All labor operations, are to be used as a **guide**.

COOLING

(G) Winterize Cooling System
Includes: Run engine to check for leaks, tighten all hose connections. Test radiator and pressure cap, drain radiator and engine block. Add antifreeze and refill system.
1993-97 5

(G) Belt, Drive, Renew
1993-97, Four
 PS (.3) 5
 Serpentine (.2) 3
1993, V6
 Serpentine (.7) 1.0
1994-97, V6
 Serpentine (.5) 7

(G) Belt, Drive, Adjust
1993-94, Four, one (.2) 3

(G) Hoses, Radiator, Renew
Includes: Drain and refill cooling system as required.
1993
 upper (.3) 4
 lower
  Four (.5) 6
  V6 (.8) 1.0
1994-97
 upper
  Four (.5) 7
  V6 (.3) 4
 lower
  Four (.5) 7
  V6 (.4) 5

(G) Thermostat, Coolant, Renew
1993-95
 Four (.6) 8
 V6 (.5) 7
1996-97
 Four (.8) 1.1
 V6 (.5) 7

FUEL

(M) Air Cleaner, Service
1993-97 2

(G) Filter, Fuel, Renew
1993-97 (.3) 4

BRAKES

(G) Bleed Brakes (Four Wheels)
Includes: Add fluid.
1993-97 (.4) 5
w/ABS add 3

(G) Brakes, Adjust (Minor)
Includes: Adjust brakes, fill master cylinder.
1993-97, two wheels 4
Remove knock out plugs add each . .1

(M) Parking Brake, Adjust
1993-97 (.3) 4

LUBRICATION SERVICES

(M) Engine Oil & Filter, Renew
Includes: Inspect and correct all fluid levels.
1993-97 (.3) 4

(M) Lubricate Chassis, Change Oil & Filter
Includes: Inspect and correct all fluid levels.
1993-97 6
Install grease fittings add 1

(M) Lubricate Chassis
Includes: Inspect and correct all fluid levels.
1993-97 4
Install grease fittings add 1

WHEELS

(G) Wheel, Renew (One)
1993-97 (.5) 5

(G) Wheel, Balance
1993-97
 one 3
 each adtnl. 2

(G) Wheels, Rotate (All)
1993-97 5

ELECTRICAL

(G) Headlamps, Aim
1993-97
 two 4
 four 6

(G) High Mount Stop Lamp Bulb, Renew
1993-97 (.2) 3

(G) Park & Turn Signal Lamp Assy., Renew
1993-95
 Skylark (.2) 3
 Grand Am (1.0) 1.4
1996-97 (.2) 3

(G) Park & Turn Signal Lamp Bulb or Lens, Renew
1993-97, each (.2) 2

(G) Rear Combination Lamp Assy., Renew
1993
 Skylark (.5) 7
 Grand Am (.3) 5
 Achieva (.3) 5
1994-97 (.2) 5

(G) Stop, Tail & Turn Signal Lamp Bulb, Renew
1993-97
 one (.2) 3
 each adtnl. 1

(G) Horn, Renew
1993
 one (.3) 3
 both (.4) 4
1994-95 (.7) 1.0
1996-97 (.3) 5

(M) Terminals, Battery, Clean
1993-97 3

SCHEDULED MAINTENANCE INTERVALS
(BUICK SKYLARK, OLDSMOBILE ACHIEVA, PONTIAC GRAND AM)

TO BE SERVICED	TYPE OF SERVICE	VEHICLE MILEAGE INTERVAL (x1000)												
		7.5	15	22.5	30	37.5	45	52.5	60	67.5	75	82.5	90	97.5
Engine oil & filter	R	✓	✓	✓	✓	✓	✓	✓	✓	✓	✓	✓	✓	✓
Automatic transaxle fluid & filter③	S/I	✓	✓	✓	✓	✓	✓	✓	✓	✓	✓	✓	✓	✓
Brake hoses	S/I	✓	✓	✓	✓	✓	✓	✓	✓	✓	✓	✓	✓	✓
Chassis lubrication	S/I	✓	✓	✓	✓	✓	✓	✓	✓	✓	✓	✓	✓	✓
Coolant level, hoses & clamps	S/I	✓	✓	✓	✓	✓	✓	✓	✓	✓	✓	✓	✓	✓
Drive shaft boots & front suspension components	S/I	✓	✓	✓	✓	✓	✓	✓	✓	✓	✓	✓	✓	✓
Exhaust system	S/I	✓	✓	✓	✓	✓	✓	✓	✓	✓	✓	✓	✓	✓
Lubricate suspension, steering linkage, transaxle shift linkage, parking brake cable guides, underbody contact points & linkage	S/I	✓	✓	✓	✓	✓	✓	✓	✓	✓	✓	✓	✓	✓
Manual transaxle oil	S/I	✓	✓	✓	✓	✓	✓	✓	✓	✓	✓	✓	✓	✓
Throttle linkage	S/I	✓	✓	✓	✓	✓	✓	✓	✓	✓	✓	✓	✓	✓
Brake linings	S/I	✓		✓		✓		✓		✓		✓		✓
Rotate tires	S/I	✓		✓		✓		✓		✓		✓		✓
Air filter element & PCV filter	R				✓				✓				✓	
Engine coolant (1993-95)②	R				✓				✓				✓	
Spark plugs①	R				✓				✓				✓	
Accessory drive belt(s)	S/I				✓				✓				✓	
EGR & fuel systems	S/I				✓				✓				✓	
Ignition cables	S/I				✓				✓				✓	
Throttle body mount bolt torque	S/I	✓												

① Platinum tip spark plugs - replace every 100,000 miles.
② Engine coolant (1996-97) - replace every 100,000 miles. Use O.E. specified (DEX-COOL™) coolant only. If any silicate coolant is used, the service interval is every 30,000 miles.
③ Automatic transaxle fluid & filter - replace every 100,000 miles (if not changed previously).

R – Replace   S/I – Service or Inspect

SCHEDULED MAINTENANCE INTERVALS
(BUICK SKYLARK, OLDSMOBILE ACHIEVA, PONTIAC GRAND AM) (Cont.)

FREQUENT OPERATION MAINTENANCE (SEVERE SERVICE)

If a vehicle is operated under any of the following conditions it is considered severe service:
- Extremely dusty areas.
- 50% or more of the vehicle operation is in 32°C (90°F) or higher temperatures, or constant operation in temperatures below 0°C (32°F).
- Prolonged idling (vehicle operation in stop and go traffic).
- Frequent short running periods (engine does not warm to normal operating temperatures).
- Police, taxi, delivery usage or trailer towing usage.

Oil & oil filter – change every 3000 miles.
Throttle body mount bolt torque - tighten at 6000 miles
Rotate tires at 6000 miles, & then every 12,000 miles.
Chassis lubrication - lubricate every 6000 miles.
Automatic transaxle fluid - change every 15,000 miles.
Air filter element - service or inspect every 15,000 miles.

REGAL/LUMINA/MONTE CARLO/CUTLASS SUPREME
GRAND PRIX

VEHICLE IDENTIFICATION CHART

Engine Code						Model Year	
Code	Liters	Cu. In. (cc)	Cyl.	Fuel Sys.	Eng. Mfg.	Code	Year
4	2.2	133 (2180)	4	MFI	CUS	P	1993
M	3.1	191 (3130)	6	MFI	BOC	R	1994
T	3.1	192 (3146)	6	MFI	CPC	S	1995
X	3.4	207 (3393)	6	MFI	CPC	T	1996
L	3.8	231 (3785)	6	MFI	BOC	V	1997
K	3.8	231 (3785)	6	MFI	CPC		

MFI - Multiport fuel injection
BOC - Buick/Oldsmobile/Cadillac
CUS - Chevrolet/United States
CPC - Chevrolet/Pontiac/Canada

ENGINE IDENTIFICATION

Year	Model	Engine Displacement Liters (cc)	Engine Series (ID/VIN)	Fuel System	No. of Cylinders	Engine Type	
1993	Cutlass Supreme	3.1 (3146)	T	MFI	6	OHV	
	Cutlass Supreme	3.4 (3393)	X	MFI	6	DOHC	①
	Grand Prix	3.1 (3146)	T	MFI	6	OHV	
	Grand Prix	3.4 (3393)	X	MFI	6	DOHC	
	Lumina	2.2 (2180)	4	MFI	4	OHV	

ENGINE IDENTIFICATION

Year	Model	Engine Displacement Liters (cc)	Engine Series (ID/VIN)	Fuel System	No. of Cylinders	Engine Type
1993	Lumina	3.1 (3146)	T	MFI	6	OHV
	Lumina	3.4 (3393)	X	MFI	6	DOHC
	Regal	3.1 (3146)	T	MFI	6	OHV
	Regal	3.8 (3785)	L	MFI	6	OHV
1994	Cutlass Supreme	3.1 (3130)	M	MFI	6	OHV
	Cutlass Supreme	3.4 (3393)	X	MFI	6	DOHC ①
	Grand Prix	3.1 (3136)	M	MFI	6	OHV
	Grand Prix	3.4 (3393)	X	MFI	6	DOHC
	Lumina	3.1 (3146)	T	MFI	6	OHV
	Lumina	3.4 (3393)	X	MFI	6	DOHC
	Regal	3.1 (3130)	M	MFI	6	OHV
	Regal	3.8 (3785)	L	MFI	6	OHV
1995	Cutlass Supreme	3.1 (3130)	M	MFI	6	OHV
	Cutlass Supreme	3.4 (3393)	X	MFI	6	DOHC ①
	Grand Prix	3.1 (3130)	M	MFI	6	OHV
	Grand Prix	3.4 (3393)	X	MFI	6	DOHC
	Lumina	3.1 (3130)	M	MFI	6	OHV
	Lumina	3.4 (3393)	X	MFI	6	DOHC
	Monte Carlo	3.1 (3130)	M	MFI	6	OHV
	Monte Carlo	3.4 (3393)	X	MFI	6	DOHC
	Regal	3.1 (3130)	M	MFI	6	OHV
	Regal	3.8 (3785)	L	MFI	6	OHV
1996-97	Cutlass Supreme	3.1 (3130)	M	MFI	6	OHV
	Cutlass Supreme	3.4 (3393)	X	MFI	6	DOHC
	Grand Prix	3.1 (3130)	M	MFI	6	OHV
	Grand Prix	3.4 (3393)	X	MFI	6	DOHC
	Lumina	3.1 (3130)	M	MFI	6	OHV
	Lumina	3.4 (3393)	X	MFI	6	DOHC
	Monte Carlo	3.1 (3130)	M	MFI	6	OHV
	Monte Carlo	3.4 (3393)	X	MFI	6	DOHC
	Regal	3.1 (3130)	M	MFI	6	OHV
	Regal	3.8 (3785)	K	MFI	6	OHV

MFI - Multiport fuel injection
OHV - Overhead valve
① Twin dual overhead camshaft

GENERAL ENGINE SPECIFICATIONS

Year	Engine ID/VIN	Engine Displacement Liters (cc)	Fuel System Type	Net Horsepower @ rpm	Net Torque @ rpm (ft. lbs.)	Bore x Stroke (in.)	Compression Ratio	Oil Pressure @ rpm
1993	4	2.2 (2180)	MFI	110@5200	130@3200	3.50x3.46	9.0:1	63-77@1200
	T	3.1 (3146)	MFI	140@4200	185@3200	3.50x3.31	8.9:1	15@1100
	X	3.4 (3393)	MFI	210@5200	215@4000	3.62x3.31	9.25:1	15@1100
	L	3.8 (3785)	MFI	170@4800	225@3200	3.80x3.40	9.0:1	60@1850
1994	M	3.1 (3130)	MFI	160@5200	185@4000	3.50x3.31	9.5:1	15@1100
	T	3.1 (3146)	MFI	140@4200	185@3200	3.50x3.31	8.5:1	8@600
	X	3.4 (3393)	MFI	210@5200	215@4000	3.62x3.31	9.25:1	15@1100
	L	3.8 (3785)	MFI	170@4800	225@3200	3.80x3.40	9.0:1	60@1850

GENERAL ENGINE SPECIFICATIONS

Year	Engine ID/VIN	Engine Displacement Liters (cc)	Fuel System Type	Net Horsepower @ rpm	Net Torque @ rpm (ft. lbs.)	Bore x Stroke (in.)	Compression Ratio	Oil Pressure @ rpm
1995	M	3.1 (3130)	MFI	160@5200	185@4000	3.50x3.31	9.5:1	15@1100
	X	3.4 (3393)	MFI	210@5200	215@4000	3.62x3.31	9.25:1	15@1100
	L	3.8 (3785)	MFI	170@4800	225@3200	3.80x3.40	9.0:1	60@1850
1996-97	M	3.1 (3130)	MFI	160@5200	185@4000	3.50x3.31	9.5:1	15@1100
	X	3.4 (3393)	MFI	210@5200	215@4000	3.62x3.31	9.25:1	15@1100
	K	3.8 (3785)	MFI	205@5200	230@4000	3.80x3.40	9.4:1	60@1850

MFI - Multiport fuel injection
TFI - Throttle body fuel injection

GASOLINE ENGINE TUNE-UP SPECIFICATIONS

Year	Engine ID/VIN	Engine Displacement Liters (cc)	Spark Plugs Gap (in.)	Ignition Timing (deg.) MT	Ignition Timing (deg.) AT	Fuel Pump (psi)	Idle Speed (rpm) MT	Idle Speed (rpm) AT	Valve Clearance In.	Valve Clearance Ex.
1993	4	2.2 (2180)	0.045	④	④	41-47	④	④	HYD	HYD
	T	3.1 (3146)	0.045	①	①	41-47	③	③	HYD	HYD
	X	3.4 (3393)	0.045	④	④	41-47	④	④	HYD	HYD
	L	3.8 (3785)	0.060	①	①	40-47	-	②	HYD	HYD
1994	M	3.1 (3130)	0.060	①	①	41-47	②	②	HYD	HYD
	T	3.1 (3146)	0.045	④	④	41-47	④	④	HYD	HYD
	X	3.4 (3393)	0.045	④	④	41-47	④	④	HYD	HYD
	L	3.8 (3785)	0.060	①	①	40-47	-	②	HYD	HYD
1995	M	3.1 (3130)	0.060	①	①	41-47	②	②	HYD	HYD
	X	3.4 (3393)	0.045	④	④	41-47	④	④	HYD	HYD
	L	3.8 (3785)	0.060	①	①	40-47	-	②	HYD	HYD
1996-97	M	3.1 (3130)	0.060	①	①	41-47	②	②	HYD	HYD
	X	3.4 (3393)	0.045	④	④	41-47	④	④	HYD	HYD
	K	3.8 (3785)	0.060	①	①	41-47	-	②	HYD	HYD

NOTE: The Vehicle Emission Control Information label often reflects specification changes made during production. The label figures must be used if they differ from those in this chart.

HYD - Hydraulic

① DIS Ignition System timing is not adjustable
② Idle speed is maintained by the ECM. There is no recommended adjustment procedure
③ Idle speed is controlled by ECM; Minimum air rate is adjusted by IAC centering; Refer to manual for procedure
④ Refer to Vehicle Emission Control Information label

CAPACITIES

Year	Model	Engine ID/VIN	Engine Displacement Liters (cc)	Engine Oil with Filter (qts.)	Transmission (pts.) 4-Spd	Transmission (pts.) 5-Spd	Transmission (pts.) Auto.	Drive Axle Front (pts.)	Drive Axle Rear (pts.)	Fuel Tank (gal.)	Cooling System (qts.)
1993	Cutlass Supreme	T	3.1 (3146)	4.5	-	-	③	-	-	16.0	12.6
	Cutlass Supreme	X	3.4 (3393)	5.0	-	-	③	-	-	16.5	12.7
	Grand Prix	T	3.1 (3146)	4.5	-	-	③	-	-	16.0	12.6
	Grand Prix	X	3.4 (3393)	5.0	-	-	③	-	-	16.5	12.7
	Lumina	4	2.2 (2180)	4.5	-	-	8.0 ②	-	-	16.0	9.5
	Lumina	T	3.1 (3146)	4.5	-	-	③	-	-	16.0	12.6
	Lumina	X	3.4 (3393)	5.0	-	-	③	-	-	16.5	12.7
	Regal	T	3.1 (3146)	4.0 ①	-	-	12.0	-	-	16.5	12.6
	Regal	L	3.8 (3785)	4.0 ①	-	-	12.0	-	-	16.0	11.1

CAPACITIES

Year	Model	Engine ID/VIN	Engine Displacement Liters (cc)	Engine Oil with Filter (qts.)		Transmission (pts.)			Drive Axle		Fuel Tank (gal.)	Cooling System (qts.)
						4-Spd	5-Spd	Auto.	Front (pts.)	Rear (pts.)		
1994	Cutlass Supreme	M	3.1 (3130)	4.0 ①		-	-	③	-	-	16.5	11.8
	Cutlass Supreme	X	3.4 (3393)	5.0		-	-	③	-	-	16.5	12.7
	Grand Prix	M	3.1 (3130)	4.0 ①		-	-	③	-	-	16.5	11.8
	Grand Prix	X	3.4 (3393)	5.0		-	-	③	-	-	16.5	12.7
	Lumina	T	3.1 (3146)	4.5		-	-	③	-	-	16.0	12.6
	Lumina	X	3.4 (3393)	5.0		-	-	③	-	-	16.5	12.7
	Regal	M	3.1 (3130)	4.0 ①		-	-	12.0	-	-	16.5	11.8
	Regal	L	3.8 (3785)	4.0 ①		-	-	12.0	-	-	16.5	11.1
1995	Cutlass Supreme	M	3.1 (3130)	4.5		-	-	③	-	-	16.0	12.6
	Cutlass Supreme	X	3.4 (3393)	5.0		-	-	③	-	-	16.5	12.7
	Grand Prix	M	3.1 (3130)	4.5		-	-	③	-	-	16.0	12.6
	Grand Prix	X	3.4 (3393)	5.0		-	-	③	-	-	16.5	12.7
	Lumina	M	3.1 (3130)	4.5		-	-	③	-	-	16.0	12.6
	Lumina	X	3.4 (3393)	5.0		-	-	③	-	-	16.5	12.7
	Monte Carlo	M	3.1 (3130)	4.5		-	-	③	-	-	16.5	12.6
	Monte Carlo	X	3.4 (3393)	5.0		-	-	③	-	-	16.5	12.7
	Regal	M	3.1 (3130)	4.0 ①		-	-	12.0	-	-	16.5	11.8
	Regal	L	3.8 (3785)	4.0 ①		-	-	12.0	-	-	16.5	11.1
1996-97	Cutlass Supreme	M	3.1 (3130)	4.5		-	-	③	-	-	16.0	12.6
	Cutlass Supreme	X	3.4 (3393)	5.0		-	-	③	-	-	16.5	12.7
	Grand Prix	M	3.1 (3130)	4.5		-	-	③	-	-	16.0	12.6
	Grand Prix	X	3.4 (3393)	5.0		-	-	③	-	-	16.5	12.7
	Lumina	M	3.1 (3130)	4.5		-	-	③	-	-	16.0	12.6
	Lumina	X	3.4 (3393)	5.0		-	-	③	-	-	16.5	12.7
	Monte Carlo	M	3.1 (3130)	4.5		-	-	③	-	-	16.5	12.6
	Monte Carlo	X	3.4 (3393)	5.0		-	-	③	-	-	16.5	12.7
	Regal	M	3.1 (3130)	4.0 ①		-	-	12.0	-	-	17.1	12.5
	Regal	K	3.8 (3785)	4.0 ①		-	-	12.0	-	-	17.1	11.1

① Capacity is without filter replacement; Additional oil may be required
② 10.0 pts. if equipped with O/D
③ 3T40 trans.: 8.0 pts.
   4T60 trans.: 12.0 pts.
   4T60E trans.: 14.8 pts.

VALVE SPECIFICATIONS

Year	Engine ID/VIN	Engine Displacement Liters (cc)	Seat Angle (deg.)	Face Angle (deg.)	Spring Test Pressure (lbs. @ in.)	Spring Installed Height (in.)	Stem-to-Guide Clearance (in.)		Stem Diameter (in.)	
							Intake	Exhaust	Intake	Exhaust
1993	4	2.2 (2180)	46	45	225-233@ ① 1.25	1.64 ②	0.0011-0.0026	0.0014-0.0031	NA	NA
	T	3.1 (3146)	46	45	215@1.291	1.693	0.0008-0.0021	0.0014-0.0030	NA	NA
	X	3.4 (3393)	46	45	75@1.40	1.40	0.0011-0.0026	0.0014-0.0031	NA	NA
	L	3.8 (3785)	45	45	210@1.315	1.690-1.720	0.0015-0.0035	0.0015-0.0032	NA	NA
1994	M	3.1 (3130)	45	45	250@1.239	1.710	0.0001-0.0027	0.0010-0.0027	NA	NA
	T	3.1 (3146)	46	45	215@1.291	1.60 ②	0.0010-0.0027	0.0010-0.0027	NA	NA

VALVE SPECIFICATIONS

Year	Engine ID/VIN	Engine Displacement Liters (cc)	Seat Angle (deg.)	Face Angle (deg.)	Spring Test Pressure (lbs. @ in.)	Spring Installed Height (in.)	Stem-to-Guide Clearance (in.)		Stem Diameter (in.)	
							Intake	Exhaust	Intake	Exhaust
1994	X	3.4 (3393)	46	45	75@1.40	1.40	0.0011-0.0026	0.0014-0.0031	NA	NA
	L	3.8 (3785)	45	45	210@1.315	1.690-1.720	0.0015-0.0035	0.0015-0.0032	NA	NA
1995	M	3.1 (3130)	45	45	250@1.239	1.710	0.0001-0.0027	0.0010-0.0027	NA	NA
	X	3.4 (3393)	46	45	75@1.40	1.40	0.0011-0.0026	0.0014-0.0031	NA	NA
	L	3.8 (3785)	45	45	210@1.315	1.690-1.720	0.0015-0.0035	0.0015-0.0032	NA	NA
1996-97	M	3.1 (3130)	45	45	250@1.239	1.710	0.0001-0.0027	0.0010-0.0027	NA	NA
	X	3.4 (3393)	46	45	75@1.40	1.40	0.0011-0.0026	0.0014-0.0031	NA	NA
	K	3.8 (3785)	45	45	210@1.32	1.69-1.72	0.0015-0.0035	0.0015-0.0032	NA	NA

NA - Not Available
① With valve open
② With valve closed

TORQUE SPECIFICATIONS
All readings in ft. lbs.

Year	Engine ID/VIN	Engine Displacement Liters (cc)	Cylinder Head Bolts	Main Bearing Bolts	Rod Bearing Bolts	Crankshaft Damper Bolts	Flywheel Bolts	Manifold Intake	Manifold Exhaust	Spark Plugs	Lug Nut
1993	4	2.2 (2180)	[13]	77	38	85 [14]	52-55	18	6-13	20	80-100
	T	3.1 (3146)	[3]	[8]	37	76	52	[2]	21	11	100
	X	3.4 (3393)	[11]	[8]	39	78	61	18	[12]	11	100
	L	3.8 (3785)	[6]	[1]	[7]	[4]	[5]	[7]	38	20	100
1994	M	3.1 (3130)	[3]	[8]	[9]	76	61	[2]	10	[10]	100
	T	3.1 (3146)	[3]	[8]	37	76	52	[2]	21	11	100
	X	3.4 (3393)	[11]	[8]	39	78	61	18	[12]	11	100
	L	3.8 (3785)	[6]	[1]	[7]	[4]	[5]	[7]	38	11	100
1995	M	3.1 (3130)	[3]	[8]	[9]	76	61	[2]	10	[10]	100
	X	3.4 (3393)	[11]	[8]	39	78	61	18	[12]	11	100
	L	3.8 (3785)	[6]	[1]	[7]	[4]	[5]	[7]	38	11	100
1996-97	M	3.1 (3130)	[3]	[8]	[9]	76	61	[2]	10	[10]	100
	X	3.4 (3393)	[11]	[8]	39	78	61	18	[12]	11	100
	K	3.8 (3785)	[6]	[15]	20	[4]	[5]	[16]	18	23	100

① 26 ft. lbs. plus 50 degrees
② Torque all bolts to 15 ft. lbs. Retorque to 24 ft. lbs.
③ Coat threads with sealer torque to 33 ft. lbs., then turn 1/4 turn (90 degrees)
④ 110 ft. lbs. plus 76 degrees
⑤ 11 ft. lbs., plus 50 degrees
⑥ Step 1: 35 ft. lbs.
   Step 2: 130 degrees
   Step 3: Rotate four center bolts an additional 30 degrees
⑦ 20 ft. lbs. plus 50 degrees
⑧ 37 ft. lbs. plus 77 degrees
⑨ 15 ft. lbs. plus 75 degrees
⑩ New cylinder head:
   1st-time installation: 20 ft. lbs.
   All other installations: 11 ft. lbs.
⑪ 37 ft. lbs. plus 90 degrees

⑫ 115 inch lbs.
⑬ Step 1: 41 ft. lbs.
   Step 2: Tighten an additional 45 degrees
   Step 3: Tighten an additional 45 degrees
   Step 4: Long bolts 1, 4-5, 8-9 an additional 20 degrees
   Step 4: Short bolts 2-3, 6-7, 10 an additional 10 degrees
⑭ Center bolt spec shown; Pulley to hub bolts: 37 ft. lbs.
⑮ Step 1: Tighten to 52 ft. lbs. to fully seat caps
   Step 2: Loosen bearing cap 360 degrees counter-clockwise
   Step 3: Tighten caps 15 ft. lbs., then 30 ft. lbs., then 35 degrees,
      then an additional 35 degrees plus 40 degrees – for a total of 110 degrees
⑯ Upper manifold: 18 ft. lbs.
   Lower manifold bolt/nut: 22 ft. lbs.
   Upper manifold studs: 89 in. lbs.

BRAKE SPECIFICATIONS
All measurements in inches unless noted

Year	Model		Master Cylinder Bore	Brake Disc Original Thickness	Brake Disc Minimum Thickness	Maximum Runout	Brake Drum Diameter Original Inside Diameter	Brake Drum Diameter Max. Wear Limit	Brake Drum Diameter Maximum Machine Diameter	Minimum Lining Thickness Front	Minimum Lining Thickness Rear
1993	Cutlass Supreme	F	0.945	1.040	0.972	0.004	-	-	-	0.030	0.030
		R	0.945	0.492	0.429	0.004	-	-	-	0.030	0.030
	Grand Prix		0.944	1.039	0.972	0.003	NA	NA	NA	0.030	NA
			NA	0.492	0.429	0.003	NA	NA	NA	NA	0.030
	Lumina		0.945	①	②	0.004	NA	NA	NA	0.030	0.030
	Regal	F	0.945	1.040	0.972	0.004	-	-	-	0.030	0.030
		R	0.945	0.492	0.429	0.004	-	-	-	0.030	0.030
1994	Cutlass Supreme	F	0.945	1.039	0.972	0.003	-	-	-	0.030	0.030
		R	0.945	0.492	0.429	0.003	-	-	-	0.030	0.030
	Grand Prix		0.944	1.039	0.972	0.004	NA	NA	NA	0.030	NA
			NA	0.492	0.429	0.004	NA	NA	NA	NA	0.030
	Lumina		0.945	①	②	0.004	NA	NA	NA	0.030	0.030
	Regal	F	0.945	1.039	0.972	0.003	-	-	-	0.030	0.030
		R	0.945	0.492	0.429	0.003	-	-	-	0.030	0.030
1995	Cutlass Supreme	F	1.000	1.039	0.972	0.003	-	-	-	0.030	0.030
		R	1.000	0.492	0.429	0.003	-	-	-	0.030	0.030
	Grand Prix		0.944	1.039	0.972	0.004	NA	NA	NA	0.030	NA
			NA	0.492	0.429	0.004	NA	NA	NA	NA	0.030
	Lumina		0.945	①	②	0.004	NA	NA	NA	0.030	0.030
	Monte Carlo	F	0.945	1.040	0.972	0.004	NA	NA	NA	0.030	-
		R	0.945	0.492	0.429	0.004	NA	NA	NA	-	0.030
	Regal	F	0.945	1.039	0.972	0.003	-	-	-	0.030	0.030
		R	0.945	0.492	0.429	0.003	-	-	-	0.030	0.030
1996-97	Cutlass Supreme	F	1.000	1.039	0.972	0.003	-	-	-	0.030	0.030
		R	1.000	0.492	0.429	0.003	-	-	-	0.030	0.030
	Grand Prix		0.944	1.039	0.972	0.004	NA	NA	NA	0.030	NA
			NA	0.492	0.429	0.004	NA	NA	NA	NA	0.030
	Lumina		0.945	①	②	0.004	NA	NA	NA	0.030	0.030
	Monte Carlo	F	0.945	1.040	0.972	0.004	NA	NA	NA	0.030	-
		R	0.945	0.492	0.429	0.004	NA	NA	NA	-	0.030
	Regal	F	0.945	1.039	0.972	0.003	-	-	-	0.030	0.030
		R	0.945	0.492	0.429	0.003	-	-	-	0.030	0.030

F - Front
R - Rear

① Front: 1.040; Rear: 0.492
② Front: 0.972; Rear: 0.429

FREQUENT MAINTENANCE LABOR
BUICK REGAL, CHEVROLET LUMINA, MONTE CARLO
OLDSMOBILE CUTLASS SUPREME, PONTIAC GRAND PRIX

The following should be used as a guide when determining the amount of work required for a particular service if taken to a repair shop. In estimating how long a particular Frequent Maintenance Service item should take, please observe the following:
- **Factory Time** is time that is generated by the vehicle manufacturer.
- **Chilton Time** is time that is based on field research and data supplied by the vehicle manufacturer.
- All labor time operations are given in hours and tenths of an hour.
- All labor operations, are to be used as a **guide**.

COOLING

	(Factory Time)	Chilton Time
(G) Winterize Cooling System		
Includes: Run engine to check for leaks, tighten all hose connections. Test radiator and pressure cap, drain radiator and engine block. Add antifreeze and refill system.		
1993-97		.5
(G) Belt, Drive, Renew		
1993-97, V6		
2.8L (W), 3.1L (T) (.3)		.5
w/AIR add		.1
3.4L (X) (.3)		.5
3.8L (K, L, 1) (.2)		.4
3.1L (M, T) (.2)		.4
(G) Hoses, Radiator, Renew		
Includes: Drain and refill cooling system as required.		
1993		
upper (.3)		.4
lower (.4)		.5
both (.5)		.7
1994-97		
upper (.5)		.6
lower (.4)		.5
both (.5)		.9
(G) Thermostat, Coolant, Renew		
1993-97, V6		
2.8L (W), 3.1L (T) (.5)		.6
3.4L (X) (.8)		1.2
3.8L (K, L, 1) (.5)		.6
3.1L (M, T) (.5)		.6

FUEL

	(Factory Time)	Chilton Time
(G) Filter, Fuel, Renew		
1993 (.4)		.5
1994-97 (.2)		.3

BRAKES

	(Factory Time)	Chilton Time
(G) Bleed Brakes (Four Wheels)		
Includes: Add fluid.		
1993-97 (.4)		.5
w/ABS add		.2
(M) Parking Brake, Adjust		
1993-97 (.5)		.5

LUBRICATION SERVICES

	(Factory Time)	Chilton Time
(M) Engine Oil & Filter, Renew		
Includes: Inspect and correct all fluid levels.		
1993-97		.3
(M) Lubricate Chassis, Change Oil & Filter		
Includes: Inspect and correct all fluid levels.		
1993-97		.6
Install grease fittings add		.1
(M) Lubricate Chassis		
Includes: Inspect and correct all fluid levels.		
1993-97		.6
Install grease fittings add		.1

WHEELS

	(Factory Time)	Chilton Time
(G) Wheel, Renew (One)		
1993-97		.5

(continued)

	(Factory Time)	Chilton Time
(G) Wheel, Balance		
1993-97		
one		.3
each adtnl.		.2
(G) Wheels, Rotate (All)		
1993-97		.5

ELECTRICAL

	(Factory Time)	Chilton Time
(G) Headlamps, Aim		
1993-97		
two		.4
four		.6
(G) High Mount Stop Lamp Bulb, Renew		
1993-97 (.2)		.3
(G) License Lamp Assy., Renew		
1993-97, one or both (.2)		.3
(G) License Lamp Bulb, Renew		
1993-97, one or all (.2)		.3
(G) Park & Turn Signal Lamp Bulb or Lens, Renew		
1993-97, one		.3
(G) Stop, Tail & Turn Signal Lamp Bulb, Renew		
1993-97		
one		.3
each adtnl.		.1
(G) Horn, Renew		
1993		
one (.5)		.6
each adtnl.		.1
1994-97 (.3)		.4
(M) Terminals, Battery, Clean		
1993-97		.3

SCHEDULED MAINTENANCE INTERVALS
(BUICK REGAL, CHEVROLET LUMINA & MONTE CARLO, OLDSMOBILE CUTLASS SUPREME, PONTIAC GRAND PRIX)

TO BE SERVICED	TYPE OF SERVICE	VEHICLE MILEAGE INTERVAL (x1000)												
		7.5	15	22.5	30	37.5	45	52.5	60	67.5	75	82.5	90	97.5
Engine oil & filter	R	✓	✓	✓	✓	✓	✓	✓	✓	✓	✓	✓	✓	✓
Automatic transaxle fluid & filter③	S/I	✓	✓	✓	✓	✓	✓	✓	✓	✓	✓	✓	✓	✓
Brake hoses	S/I	✓	✓	✓	✓	✓	✓	✓	✓	✓	✓	✓	✓	✓
Coolant level, hoses & clamps	S/I	✓	✓	✓	✓	✓	✓	✓	✓	✓	✓	✓	✓	✓
Drive shaft boots & front suspension components	S/I	✓	✓	✓	✓	✓	✓	✓	✓	✓	✓	✓	✓	✓
Exhaust system & throttle linkage	S/I	✓	✓	✓	✓	✓	✓	✓	✓	✓	✓	✓	✓	✓
Lubricate chassis, suspension, steering linkage, transaxle shift linkage, parking brake cable guides, underbody contact points & linkage	S/I	✓	✓	✓	✓	✓	✓	✓	✓	✓	✓	✓	✓	✓
Rotate tires	S/I	✓		✓		✓		✓		✓		✓		✓
Air filter element	R				✓				✓				✓	
Engine coolant②	R				✓				✓				✓	
PCV filter	R				✓				✓				✓	
Spark plugs①	R				✓				✓				✓	
Accessory drive belt(s)	S/I				✓				✓				✓	
Ignition cables, EGR & fuel systems	S/I				✓				✓				✓	
Camshaft timing belt	R								✓					
Throttle body mount bolt torque	S/I	✓												

① Platinum tip spark plugs - replace every 100,000 miles.
② Engine coolant (1996-97) - replace every 100,000 miles. Use O.E. specified (DEX-COOL™) coolant only. If any silicate coolant is used, the service interval is every 30,000 miles.
③ Automatic transaxle fluid & filter (1993-95) - replace every 100,000 miles.
R – Replace   S/I – Service or Inspect

SCHEDULED MAINTENANCE INTERVALS
(BUICK REGAL, CHEVROLET LUMINA & MONTE CARLO, OLDSMOBILE CUTLASS SUPREME, PONTIAC GRAND PRIX) (Cont.)

FREQUENT OPERATION MAINTENANCE (SEVERE SERVICE)

If a vehicle is operated under any of the following conditions it is considered severe service:
- Extremely dusty areas.
- 50% or more of the vehicle operation is in 32°C (90°F) or higher temperatures, or constant operation in temperatures below 0°C (32°F).
- Prolonged idling (vehicle operation in stop and go traffic).
- Frequent short running periods (engine does not warm to normal operating temperatures).
- Police, taxi, delivery usage or trailer towing usage.

Oil & oil filter – change every 3000 miles.
Chassis lubrication - lubricate every 6000 miles.
Rotate tires at 6000 miles, then every 15,000 miles (1993-94) or every 12,000 miles (1995-97).
Throttle body mount bolt torque - tighten at 6000 miles
Air filter element - service or inspect every 15,000 miles.
Automatic transaxle fluid - change every 15,000 miles (1993-95) or every 50,000 miles (1996-97).
Camshaft timing belt - change every 60,000 miles.

ROADMASTER/ROADMASTER ESTATE WAGON
FLEETWOOD/CAPRICE/ESTATE WAGON/IMPALA SS

VEHICLE IDENTIFICATION CHART

Engine Code							Model Year	
Code	Liters	Cu. In. (cc)	Cyl.	Fuel Sys.	Eng. Mfg.		Code	Year
7	5.7	350 (5737)	8	TFI	CPC		P	1993
E	5.0	305 (4999)	8	TFI	CPC		R	1994
P	5.7	350 (5737)	8	MFI	CPC		S	1995
W	4.3	265 (4294)	8	MFI	CPC		T	1996
Z	4.3	265 (4294)	6	TFI	CPC			

CPC - Chevrolet/Pontiac/Canada
MFI - Multiport fuel injection
TFI - Throttle body fuel injection

ENGINE IDENTIFICATION

Year	Model	Engine Displacement Liters (cc)	Engine Series (ID/VIN)	Fuel System	No. of Cylinders	Engine Type
1993	Caprice	4.3 (4294)	Z	TFI	6	OHV
	Caprice	5.0 (4999)	E	TFI	8	OHV
	Caprice	5.7 (5737)	7	TFI	8	OHV
	Fleetwood	5.7 (5733)	7	TFI	8	OHV
	Roadmaster	5.7 (5737)	7	TFI	8	OHV
1994	Caprice	4.3 (4294)	W	MFI	6	OHV
	Caprice	5.7 (5737)	P	MFI	8	OHV
	Fleetwood	5.7 (5733)	P	MFI	8	OHV
	Impala SS	5.7 (5737)	P	MFI	8	OHV
	Roadmaster	5.7 (5737)	P	MFI	8	OHV
1995	Caprice	4.3 (4294)	W	MFI	8	OHV
	Caprice	5.7 (5737)	P	MFI	8	OHV
	Fleetwood	5.7 (5733)	P	MFI	8	OHV
	Impala SS	5.7 (5737)	P	MFI	8	OHV
	Roadmaster	5.7 (5737)	P	MFI	8	OHV
1996	Caprice	4.3 (4294)	W	MFI	8	OHV
	Caprice	5.7 (5737)	P	MFI	8	OHV
	Fleetwood	5.7 (5733)	P	MFI	8	OHV
	Impala SS	5.7 (5737)	P	MFI	8	OHV
	Roadmaster	5.7 (5737)	P	MFI	8	OHV

TFI - Throttle body fuel injection
MFI - Multiport fuel injection
OHV - Overhead valve

GENERAL ENGINE SPECIFICATIONS

Year	Engine ID/VIN	Engine Displacement Liters (cc)	Fuel System Type	Net Horsepower @ rpm	Net Torque @ rpm (ft. lbs.)	Bore x Stroke (in.)	Compression Ratio	Oil Pressure @ rpm
1993	Z	4.3 (4294)	TFI	140@4000	225@2000	4.00x3.48	9.3:1	18@2000
	E	5.0 (4999)	TFI	170@4000	255@2400	3.74x3.48	9.3:1	18@2000
	7	5.7 (5737)	TFI	195@4200	295@2400	4.00x3.50	9.8:1	18@2000
1994	W	4.3 (4294)	MFI	200@5200	245@2400	3.74x3.48	9.93:1	18@2000
	P	5.7 (5737)	MFI	260@5000	330@3200	4.00x3.48	10.25:1	18@2000
1995	W	4.3 (4294)	MFI	200@5200	245@2400	3.74x3.48	9.93:1	18@2000
	P	5.7 (5737)	MFI	260@5000	330@3200	4.00x3.48	10.25:1	18@2000
1996	W	4.3 (4294)	MFI	200@5200	245@2400	3.74x3.48	9.93:1	18@2000
	P	5.7 (5737)	MFI	260@5000	330@3200	4.00x3.48	10.25:1	18@2000

TFI - Throttle body fuel injection
MFI - Multiport fuel injection

GASOLINE ENGINE TUNE-UP SPECIFICATIONS

Year	Engine ID/VIN	Engine Displacement Liters (cc)	Spark Plugs Gap (in.)	Ignition Timing (deg.) MT	Ignition Timing (deg.) AT	Fuel Pump (psi)	Idle Speed (rpm) MT	Idle Speed (rpm) AT	Valve Clearance In.	Valve Clearance Ex.
1993	Z	4.3 (4294)	0.035	-	①	9-13	-	①	HYD	HYD
	E	5.0 (4999)	0.035	-	①	11	-	①	HYD	HYD
	7	5.7 (5737)	0.035	-	①	9-13	-	①	HYD	HYD

GASOLINE ENGINE TUNE-UP SPECIFICATIONS

Year	Engine ID/VIN	Engine Displacement Liters (cc)	Spark Plugs Gap (in.)	Ignition Timing (deg.) MT	Ignition Timing (deg.) AT	Fuel Pump (psi)	Idle Speed (rpm) MT	Idle Speed (rpm) AT	Valve Clearance In.	Valve Clearance Ex.
1994	W	4.3 (4294)	0.050	①	①	6-24	①	①	HYD	HYD
	P	5.7 (5737)	0.035	-	①	41-47	-	①	HYD	HYD
1995	W	4.3 (4294)	0.050	①	①	6-24	①	①	HYD	HYD
	P	5.7 (5737)	0.035	-	①	41-47	-	①	HYD	HYD
1996	W	4.3 (4294)	0.050	①	①	6-24	①	①	HYD	HYD
	P	5.7 (5737)	0.035	-	①	41-47	-	①	HYD	HYD

NOTE: The Vehicle Emission Control Information label often reflects specification changes made during production. The label figures must be used if they differ from those in this chart.

HYD - Hydraulic

① Refer to Vehicle Emission Control Information label

CAPACITIES

Year	Model	Engine ID/VIN	Engine Displacement Liters (cc)	Engine Oil with Filter (qts.)	Transmission (pts.) 4-Spd	Transmission (pts.) 5-Spd	Transmission (pts.) Auto.	Transfer Case (pts.)	Drive Axle Front (pts.)	Drive Axle Rear (pts.)	Fuel Tank (gal.)	Cooling System (qts.)
1993	Caprice	Z	4.3 (4294)	4.5	-	-	10.0	-	-	③	23.0	12.6 ①
	Caprice	E	5.0 (4999)	5.0	-	-	7.0 ②	-	-	③	24.5	16.7 ①
	Caprice	7	5.7 (5737)	5.0	-	-	7.0 ②	-	-	③	22.0	14.6 ①
	Fleetwood	7	5.7 (5737)	5.0	-	-	10.0 ⑤	-	-	4.2	23.0	15.7
	Roadmaster	7	5.7 (5737)	5.0	-	-	10.0	-	-	3.50	22.0	⑥
1994	Caprice	W	4.3 (4294)	4.5	-	-	7.0 ②	-	-	③	23.0	12.6 ①
	Caprice	P	5.7 (5737)	5.0	-	-	7.0 ②	-	-	③	22.0	14.6 ①
	Impala SS	P	5.7 (5737)	5.0	-	-	7.0 ②	-	-	③	22.0	14.6 ①
	Fleetwood	P	5.7 (5737)	5.0	-	-	10.0 ⑤	-	-	4.2	23.0	14.6
	Roadmaster	P	5.7 (5737)	5.0	-	-	10.0	-	-	③	22.0	⑦
1995	Caprice	W	4.3 (4294)	4.5	-	-	7.0 ②	-	-	③	23.0	12.6 ①
	Caprice	P	5.7 (5737)	5.0	-	-	7.0 ②	-	-	③	22.0	14.6 ①
	Impala SS	P	5.7 (5737)	5.0	-	-	7.0 ②	-	-	③	22.0	14.6 ①
	Fleetwood	P	5.7 (5737)	5.0	-	-	10.0 ⑤	-	-	4.2	23.0	14.6
	Roadmaster	P	5.7 (5737)	5.0	-	-	10.0	-	-	③	22.0	⑦
1996	Caprice	W	4.3 (4294)	4.5	-	-	7.0 ②	-	-	③	23.0	12.6 ①
	Caprice	P	5.7 (5737)	5.0	-	-	7.0 ②	-	-	③	22.0	14.6 ①
	Impala SS	P	5.7 (5737)	5.0	-	-	7.0 ②	-	-	③	22.0	14.6 ①
	Fleetwood	P	5.7 (5737)	5.0	-	-	10.0 ⑤	-	-	4.2	23.0	14.6
	Roadmaster	P	5.7 (5737)	5.0	-	-	10.0	-	-	③	⑧	④

① Add 0.6 qts. for HD radiator
② 4L60 trans.: 10.0 pts.
③ With 7 5/8" ring gear: 3.50 pts.
   With 8.5" ring gear: 4.25 pts.
   With 8.75" ring gear: 5.4 pts.

④ With standard cooling: 16.4
   With heavy duty cooling: 16.9
⑤ Fluid change with filter
⑥ With std. cooling: 14.40
   With heavy duty cooling: 15.10

⑦ With std. cooling: 14.30
   With heavy duty cooling: 14.60
⑧ Sedan: 23.0 gals.
   Wagon: 21.0 gals.

VALVE SPECIFICATIONS

Year	Engine ID/VIN	Engine Displacement Liters (cc)	Seat Angle (deg.)	Face Angle (deg.)	Spring Test Pressure (lbs. @ in.)	Spring Installed Height (in.)	Stem-to-Guide Clearance (in.)		Stem Diameter (in.)	
							Intake	Exhaust	Intake	Exhaust
1993	Z	4.3 (4294)	46	45	194-206@ 1.25	1.69- 1.71	0.0011- 0.0027	0.0011- 0.0027	NA	NA
	E	5.0 (4999)	46	45	194-206@ 1.25	①	0.0011- 0.0027	0.0011- 0.0027	NA	NA
	7	5.7 (5737)	46	45	194-206@ 1.25	1.70	0.0011- 0.0027	0.0011- 0.0027	NA	NA
1994	W	4.3 (4294)	46	45	187-203@ 1.27	1.70	0.0009- 0.0027	0.0009- 0.0027	NA	NA
	P	5.7 (5737)	46	45	187-203@ 1.27	1.70	0.0009- 0.0027	0.0009- 0.0027	NA	NA
1995	W	4.3 (4294)	46	45	187-203@ 1.27	1.70	0.0009- 0.0027	0.0009- 0.0027	NA	NA
	P	5.7 (5737)	46	45	187-203@ 1.27	1.70	0.0009- 0.0027	0.0009- 0.0027	NA	NA
1996	W	4.3 (4294)	46	45	187-203@ 1.27	1.70	0.0009- 0.0027	0.0009- 0.0027	NA	NA
	P	5.7 (5733)	46	45	187-203@ 1.27	1.78	0.0009- 0.0027	0.0009- 0.0027	NA	NA

NA - Not Available
① Intake: 1.72
   Exhaust: 1.59

TORQUE SPECIFICATIONS
All readings in ft. lbs.

Year	Engine ID/VIN	Engine Displacement Liters (cc)	Cylinder Head Bolts	Main Bearing Bolts	Rod Bearing Bolts	Crankshaft Damper Bolts	Flywheel Bolts	Manifold		Spark Plugs	Lug Nut
								Intake	Exhaust		
1993	Z	4.3 (4294)	68	77	44	②	74	35	③	11	100
	E	5.0 (4999)	60-75	70-85	42-47	60	75	25-45	①	15-20	80-100
	7	5.7 (5737)	68	77	44	70	74	35	26	22	80-100
1994	W	4.3 (4294)	65	78	47	60	74	④	35	11	100
	P	5.7 (5737)	65	78	47	60	74	④	35	11	100
1995	W	4.3 (4294)	65	78	47	60	74	④	35	11	100
	P	5.7 (5737)	65	78	47	60	74	④	35	11	100
1996	W	4.3 (4294)	65	78	47	60	74	④	35	11	100
	P	5.7 (5737)	65	78	47	60	74	④	35	11	100

① Outer bolts: 14-26 ft. lbs.
Inner bolts: 20-32 ft. lbs.

② Torsional damper: 70 ft. lbs.
Crankshaft pulley: 43 ft. lbs.

③ Outer bolts: 20 ft. lbs.
Inner bolts: 26 ft. lbs.

④ Step 1: 71 inch lbs.
Step 2: 35 ft. lbs.

BRAKE SPECIFICATIONS
All measurements in inches unless noted

Year	Model			Master Cylinder Bore	Brake Disc Original Thickness	Brake Disc Minimum Thickness	Brake Disc Maximum Runout	Brake Drum Diameter Original Inside Diameter	Brake Drum Diameter Max. Wear Limit	Brake Drum Diameter Maximum Machine Diameter	Minimum Lining Thickness Front	Minimum Lining Thickness Rear
1993	Caprice			1.125	1.043	0.980	0.004	11.00	11.09	11.06	0.030	0.030
	Fleetwood			1.125	1.043	0.965	0.004	11.00	11.09	11.06	0.030	0.030
	Roadmaster			1.125	1.043	0.965	0.003	11.00	11.09	11.06	0.030	①
1994	Caprice			1.125	1.043	0.980	0.004	11.00	11.09	11.06	0.030	0.030
	Fleetwood			1.125	1.043	0.965	0.004	11.00	11.09	11.06	0.030	0.030
	Impala SS			1.125	1.043	0.980	0.004	11.00	11.09	11.06	0.030	0.030
	Roadmaster			1.125	1.043	0.965	0.003	11.00	11.09	11.06	0.030	①
	Roadmaster	②	F	1.251	1.043	0.965	0.003	-	-	-	0.030	0.030
			R	1.251	0.787	0.728	0.004	-	-	-	0.030	0.030
1995	Caprice			1.125	1.043	0.980	0.004	11.00	11.09	11.06	0.030	0.030
	Fleetwood			1.125	1.043	0.965	0.004	11.00	11.09	11.06	0.030	0.030
	Impala SS			1.125	1.043	0.980	0.004	11.00	11.09	11.06	0.030	0.030
	Roadmaster			1.125	1.043	0.965	0.003	11.00	11.09	11.06	0.030	①
	Roadmaster	②	F	1.251	1.043	0.965	0.003	-	-	-	0.030	0.030
			R	1.251	0.787	0.728	0.004	-	-	-	0.030	0.030
1996	Caprice			1.125	1.043	0.980	0.004	11.00	11.09	11.06	0.030	0.030
	Impala SS			1.125	1.043	0.980	0.004	11.00	11.09	11.06	0.030	0.030
	Fleetwood			1.125	1.125	1.043	0.004	11.00	11.09	11.06	0.030	0.030
	Roadmaster			1.125	1.043	0.965	0.004	9.500	9.590	9.560	0.030	①
	Roadmaster	②	F	1.251	1.043	0.965	0.004	-	-	-	0.030	0.030
			R	1.251	0.787	0.728	0.004	-	-	-	0.030	0.030

F - Front
R - Rear

① 0.030 over rivet head; If bonded lining, use 0.062 from shoe
② Police

FREQUENT MAINTENANCE LABOR
BUICK ROADMASTER, ESTATE WAGON
CHEVROLET CAPRICE, ESTATE WAGON, IMPALA SS

The following should be used as a guide when determining the amount of work required for a particular service if taken to a repair shop. In estimating how long a particular Frequent Maintenance Service item should take, please observe the following:
- **Factory Time** is time that is generated by the vehicle manufacturer.
- **Chilton Time** is time that is based on field research and data supplied by the vehicle manufacturer.
- All labor time operations are given in hours and tenths of an hour.
- All labor operations, are to be used as a **guide**.

COOLING

(G) Winterize Cooling System
Includes: Run engine to check for leaks, tighten all hose connections. Test radiator and pressure cap, drain radiator and engine block. Add antifreeze and refill system.
1993-96 5

(G) Belt, Drive, Renew
1993-96
  Serpentine (.2) 3

(G) Hoses, Radiator, Renew
Includes: Drain and refill cooling system as required.
1993
  upper (.3) 4
  lower (.4) 5

  both (.5) 7
  by-pass (.5) 6
1994-96
  upper (.4) 5
  lower (.6) 7
  both (.8) 1.0

(G) Thermostat, Coolant, Renew
1993-94 (.6) 7

FREQUENT MAINTENANCE LABOR (cont.)
BUICK ROADMASTER, ESTATE WAGON
CHEVROLET CAPRICE, ESTATE WAGON, IMPALA SS

The following should be used as a guide when determining the amount of work required for a particular service if taken to a repair shop. In estimating how long a particular Frequent Maintenance Service item should take, please observe the following:
- **Factory Time** is time that is generated by the vehicle manufacturer.
- **Chilton Time** is time that is based on field research and data supplied by the vehicle manufacturer.
- All labor time operations are given in hours and tenths of an hour.
- All labor operations, are to be used as a **guide**.

	(Factory Time)	Chilton Time

FUEL

(M) Air Cleaner, Service
1993-96 (.2) 3

(G) Filter, Fuel, Renew
1993-96
   in line (.3) 4
   in tank (.8) 1.2

BRAKES

(G) Bleed Brakes (Four Wheels)
Includes: Add fluid.
1993-96 (.4) 5

(G) Brakes, Adjust (Minor)
Includes: Adjust brakes, fill master cylinder.
1993-96, two wheels 4
Remove knock out plugs,
   add each 1

(M) Parking Brake, Adjust
1993-96 (.3) 4

LUBRICATION SERVICES

(M) Engine Oil & Filter, Renew
Includes: Inspect and correct all fluid levels.
1993-96 3

(M) Lubricate Chassis, Change Oil & Filter
Includes: Inspect and correct all fluid levels.
1993-96 6
Install grease fittings add 1

(M) Lubricate Chassis
Includes: Inspect and correct all fluid levels.
1993-96 4
Install grease fittings, add 1

ELECTRICAL

(G) Headlamps, Aim
1993-96
   two 4
   four 6

(G) High Mount Stop Lamp Bulb, Renew
1993-96 (.2) 3

(G) High Mount Stop Lamp and/or Lens, Renew
1993-96 (.3) 4

(G) License Lamp Assy., Renew
1993-96, one or both (.2) 3

(G) License Lamp Bulb, Renew
1993-96, one or both (.3) 3

(G) Park & Turn Signal Lamp Assy., Renew
1993-96, each (.2) 3

(G) Park & Turn Signal Lamp Bulb or Lens, Renew
1993-96, each 2

(G) Stop, Tail & Turn Signal Lamp Bulb, Renew
1993-96
   one 3
   each adtnl. 1

(G) Horn, Renew
1993-96 (.3) 3

(M) Terminals, Battery, Clean
1993-96 3

FREQUENT MAINTENANCE LABOR
CADILLAC FLEETWOOD

The following should be used as a guide when determining the amount of work required for a particular service if taken to a repair shop. In estimating how long a particular Frequent Maintenance Service item should take, please observe the following:
- **Factory Time** is time that is generated by the vehicle manufacturer.
- **Chilton Time** is time that is based on field research and data supplied by the vehicle manufacturer.
- All labor time operations are given in hours and tenths of an hour.
- All labor operations, are to be used as a **guide**.

COOLING

(G) Winterize Cooling System
Includes: Run engine to check for leaks, tighten all hose connections. Test radiator and pressure cap, drain radiator and engine block. Add antifreeze and refill system.
	Factory	Chilton
1993-96		.5

(G) Belt, Serpentine Drive, Renew
	Factory	Chilton
1993-96	(.2)	.3
w/HD cooling add	(.2)	.2

(G) Hoses, Radiator, Renew
Includes: Drain and refill cooling system as required.
1993		
upper	(.3)	.4
lower	(.7)	.8
both	(.7)	1.0
1994-96		
upper	(.3)	.5
lower	(.5)	.7
both	(.8)	1.0
w/HD cooling add	(.3)	.3

(G) Thermostat, Coolant, Renew
	Factory	Chilton
1993-96	(.5)	.7

FUEL

(M) Air Cleaner, Service
	Factory	Chilton
1993-96	(.2)	.3

(G) Filter, Fuel, Renew
1993-96		
in line	(.3)	.4
in tank	(.8)	1.2

BRAKES

(G) Bleed Brakes (Four Wheels)
Includes: Add fluid.
1993	(.4)	.5
1994-96	(.8)	1.0

(G) Brakes, Adjust (Minor)
Includes: Adjust brakes, fill master cylinder.
1993-96, two wheels		.4
Remove knock out plugs		.1
add each		.1

(M) Parking Brake, Adjust
	Factory	Chilton
1993-96	(.3)	.5

LUBRICATION SERVICES

(M) Engine Oil & Filter, Renew
Includes: Inspect and correct all fluid levels.
	Factory	Chilton
1993-96		.3

(M) Lubricate Chassis, Change Oil & Filter
Includes: Inspect and correct all fluid levels.
1993-96		.6
Install grease fittings add		.1

(M) Lubricate Chassis
Includes: Inspect and correct all fluid levels.
1993-96		.4
Install grease fittings add		.1

ELECTRICAL

(G) Headlamps, Aim
1993-96		
two		.4
four		.6

(G) License Lamp Assy., Renew
	Factory	Chilton
1993-96, one or all	(.2)	.3

(G) Park & Turn Signal Lamp Bulb or Lens, Renew
1993-96		
one	(.2)	.3
each adtnl.		.1

(G) Stop, Tail & Turn Signal Lamp Bulb, Renew
1993-96		
one		.3
each adtnl.		.1

(G) Horn, Renew
1993-96		
one	(.3)	.4
each adtnl.		.3

(M) Terminals, Battery, Clean
1993-96		.3
w/Dual batteries add		.2

SCHEDULED MAINTENANCE INTERVALS
(BUICK ROADMASTER & ROADMASTER ESTATE WAGON, CADILLAC FLEETWOOD, CHEVROLET CAPRICE, ESTATE WAGON, & IMPALA SS)

TO BE SERVICED	TYPE OF SERVICE	VEHICLE MILEAGE INTERVAL (x1000)												
		7.5	15	22.5	30	37.5	45	52.5	60	67.5	75	82.5	90	97.5
Engine oil & filter	R	✓	✓	✓	✓	✓	✓	✓	✓	✓	✓	✓	✓	✓
Automatic transmission fluid & filter③	S/I	✓	✓	✓	✓	✓	✓	✓	✓	✓	✓	✓	✓	✓
Brake hoses	S/I	✓	✓	✓	✓	✓	✓	✓	✓	✓	✓	✓	✓	✓
Engine coolant level, hoses & clamps	S/I	✓	✓	✓	✓	✓	✓	✓	✓	✓	✓	✓	✓	✓
Exhaust system & throttle linkage	S/I	✓	✓	✓	✓	✓	✓	✓	✓	✓	✓	✓	✓	✓
Front suspension components	S/I	✓	✓	✓	✓	✓	✓	✓	✓	✓	✓	✓	✓	✓
Lubricate chassis, suspension, steering linkage, transmission shift linkage, parking brake cable guides, underbody contact points & linkage	S/I	✓	✓	✓	✓	✓	✓	✓	✓	✓	✓	✓	✓	✓
Brake linings & rotate tires	S/I	✓		✓		✓		✓		✓		✓		✓
Air filter element	R				✓				✓				✓	
Engine coolant②	R				✓				✓				✓	
Spark plugs①	R				✓				✓				✓	
Engine timing (1993)	S/I				✓				✓				✓	
Front wheel bearings	S/I				✓				✓				✓	
Ignition cables, EGR & fuel systems	S/I				✓				✓				✓	
Serpentine drive belt	S/I				✓				✓				✓	
Thermostatically controlled air cleaner	S/I				✓				✓				✓	
Rear axle oil (Limited slip)	R	✓												
Throttle body mount bolt torque	S/I	✓												

① Platinum tip spark plugs - replace every 100,000 miles.
② Engine coolant (1996-97) - replace every 100,000 miles. Use O.E. specified (DEX-COOL™) coolant only. If any silicate coolant is used, the service interval is every 30,000 miles.
③ Automatic transmission fluid & filter (1993-94) - replace every 100,000 miles.
R – Replace   S/I – Service or Inspect

SCHEDULED MAINTENANCE INTERVALS
(BUICK ROADMASTER & ROADMASTER ESTATE WAGON, CADILLAC FLEETWOOD, CHEVROLET CAPRICE, ESTATE WAGON, & IMPALA SS) (Cont.)

FREQUENT OPERATION MAINTENANCE (SEVERE SERVICE)

If a vehicle is operated under any of the following conditions it is considered severe service:
- Extremely dusty areas.
- 50% or more of the vehicle operation is in 32°C (90°F) or higher temperatures, or constant operation in temperatures below 0°C (32°F).
- Prolonged idling (vehicle operation in stop and go traffic).
- Frequent short running periods (engine does not warm to normal operating temperatures).
- Police, taxi, delivery usage or trailer towing usage.

Oil & oil filter – change every 3000 miles.
Chassis lubrication - lubricate every 6000 miles.
Rear axle oil (Limited slip) - replace every 6000 miles. If vehicle is used for towing, police or taxi service, replace every 6000 miles in either type of axle.
Throttle body mount bolt torque - tighten at 6000 miles.
Air filter element - service or inspect every 15,000 miles.
Automatic transmission fluid & filter - change every 15,000 miles (1993-94) or 50,000 miles (1995-97).
Front wheel bearings - repack every 15,000 miles.
Rotate tires at 6000 miles, then every 15,000 miles.

CAMARO/FIREBIRD

VEHICLE IDENTIFICATION CHART

Engine Code						Model Year	
Code	Liters	Cu. In. (cc)	Cyl.	Fuel Sys.	Eng. Mfg.	Code	Year
T	3.1	191 (3130)	6	MFI	CPC	P	1993
S	3.4	207 (3393)	6	MFI	CPC	R	1994
K	3.8	231 (3785)	6	MFI	CPC	S	1995
E	5.0	305 (4999)	8	TFI	CPC	T	1996
8	5.7	350 (5737)	8	MFI	CPC	V	1997
P	5.7	350 (5737)	8	MFI	CPC		

MFI - Multiport fuel injection
TFI - Throttle body fuel injection
CPC - Chevrolet/Pontiac/Canada

ENGINE IDENTIFICATION

Year	Model	Engine Displacement Liters (cc)	Engine Series (ID/VIN)	Fuel System	No. of Cylinders	Engine Type
1993	Camaro	3.1 (3130)	T	MFI	6	OHV
	Camaro	3.4 (3393)	S	MFI	6	OHV
	Camaro	5.0 (4999)	E	TFI	8	OHV
	Camaro	5.7 (5737)	8	MFI	8	OHV
	Camaro	5.7 (5737)	P	MFI	8	OHV
	Firebird	3.1 (3130)	T	MFI	6	OHV
	Firebird	3.4 (3393)	S	MFI	6	OHV
	Firebird	5.0 (4999)	E	TFI	8	OHV
	Firebird	5.7 (5737)	8	MFI	8	OHV
	Firebird	5.7 (5737)	P	MFI	8	OHV
1994	Camaro	3.1 (3130)	T	MFI	6	OHV
	Camaro	3.4 (3393)	S	MFI	6	OHV
	Camaro	5.7 (5737)	P	MFI	8	OHV
	Firebird	3.1 (3130)	T	MFI	6	OHV
	Firebird	3.4 (3393)	S	MFI	6	OHV
	Firebird	5.7 (5737)	P	MFI	8	OHV
1995	Camaro	3.1 (3130)	T	MFI	6	OHV
	Camaro	3.4 (3393)	S	MFI	6	OHV
	Camaro	5.7 (5737)	P	MFI	8	OHV
	Firebird	3.1 (3130)	T	MFI	6	OHV
	Firebird	3.4 (3393)	S	MFI	6	OHV
	Firebird	5.7 (5737)	P	MFI	8	OHV
1996-97	Camaro	3.8 (3785)	K	MFI	6	OHV
	Camaro	5.7 (5737)	P	MFI	8	OHV
	Firebird	3.8 (3785)	K	MFI	6	OHV
	Firebird	5.7 (5737)	P	MFI	8	OHV

MFI - Multiport fuel injection
TFI - Throttle body fuel injection
OHV - Overhead valve

GENERAL ENGINE SPECIFICATIONS

Year	Engine ID/VIN	Engine Displacement Liters (cc)	Fuel System Type	Net Horsepower @ rpm	Net Torque @ rpm (ft. lbs.)	Bore x Stroke (in.)	Compression Ratio	Oil Pressure @ rpm
1993	T	3.1 (3130)	MFI	140@4200	185@3200	3.50x3.31	8.5:1	8@600
	S	3.4 (3393)	MFI	160@4600	200@3600	3.62x3.31	9.0:1	15@1100
	E	5.0 (4999)	TFI	170@4000	255@2400	3.74x3.48	9.3:1	18@2000
	8	5.7 (5737)	MFI	240@4000	345@3200	4.00x3.48	9.75:1	18@2000
	P	5.7 (5737)	MFI	275@5000	325@2400	4.00x3.48	10.25:1	18@2000
1994	T	3.1 (3130)	MFI	140@4200	185@3200	3.50x3.31	8.5:1	8@600
	S	3.4 (3393)	MFI	160@4600	200@3600	3.62x3.31	9.0:1	15@1100
	P	5.7 (5737)	MFI	275@5000	325@2400	4.00x3.48	10.25:1	18@2000
1995	T	3.1 (3130)	MFI	140@4200	185@3200	3.50x3.31	8.5:1	8@600
	S	3.4 (3393)	MFI	160@4600	200@3600	3.62x3.31	9.0:1	15@1100
	P	5.7 (5737)	MFI	275@5000	325@2400	4.00x3.48	10.25:1	18@2000
1996-97	K	3.8 (3785)	MFI	160@4600	200@3600	3.80x3.40	9.4:1	60@1850
	P	5.7 (5737)	MFI	275@5000	325@2400	4.00x3.48	10.25:1	18@2000

MFI - Multiport fuel injection
TFI - Throttle body fuel injection

GASOLINE ENGINE TUNE-UP SPECIFICATIONS

Year	Engine ID/VIN	Engine Displacement Liters (cc)	Spark Plugs Gap (in.)	Ignition Timing (deg.) MT	Ignition Timing (deg.) AT	Fuel Pump (psi)	Idle Speed (rpm) MT	Idle Speed (rpm) AT	Valve Clearance In.	Valve Clearance Ex.
1993	T	3.1 (3130)	0.045	①	①	41-47	①	①	HYD	HYD
	S	3.4 (3393)	0.045	①	①	41-47	①	①	HYD	HYD
	E	5.0 (4999)	0.035	-	①	9-13	-	①	HYD	HYD
	8	5.7 (5737)	0.045	8B	8B	34-47	①	①	HYD	HYD
	P	5.7 (5737)	0.035	-	①	41-47	-	①	HYD	HYD
1994	T	3.1 (3130)	0.045	①	①	41-47	①	①	HYD	HYD
	S	3.4 (3393)	0.045	①	①	41-47	①	①	HYD	HYD
	P	5.7 (5737)	0.035	-	①	41-47	-	①	HYD	HYD
1995	T	3.1 (3130)	0.045	①	①	41-47	①	①	HYD	HYD
	S	3.4 (3393)	0.045	①	①	41-47	①	①	HYD	HYD
	P	5.7 (5737)	0.035	-	①	41-47	-	①	HYD	HYD
1996-97	K	3.8 (3785)	0.045	①	①	41-47	①	①	HYD	HYD
	P	5.7 (5737)	0.035	-	①	41-47	-	①	HYD	HYD

NOTE: The Vehicle Emission Control Information label often reflects specification changes made during production. The label figures must be used if they differ from those in this chart.

B - Before top dead center

HYD - Hydraulic

① Refer to Vehicle Emission Control Information label

CAPACITIES

Year	Model	Engine ID/VIN	Engine Displacement Liters (cc)	Engine Oil with Filter (qts.)	Transmission (pts.) 4-Spd	Transmission (pts.) 5-Spd	Transmission (pts.) Auto.	Drive Axle Front (pts.)	Drive Axle Rear (pts.)	Fuel Tank (gal.)	Cooling System (qts.)
1993	Camaro	T	3.1 (3130)	4.5 ⑨	-	5.9	10.0	-	3.5	15.5	14.7 ①
	Camaro	S	3.4 (3393)	4.5 ⑨	-	5.9	10.0	-	3.5	15.5	12.3 ⑤
	Camaro	E	5.0 (4999)	5.0 ⑨	-	5.9	10.0	-	3.5	15.5	17.9 ②
	Camaro	8	5.7 (5737)	5.0 ⑨	-	5.9	10.0	-	3.5	15.5	16.6 ③
	Camaro	P	5.7 (5737)	5.0 ⑨	-	5.9 ④	10.0	-	3.5	15.5	15.1 ⑥
	Firebird	T	3.1 (3130)	4.5 ⑨	-	5.9	10.0	-	3.5	15.5	14.7 ①
	Firebird	S	3.4 (3393)	4.5 ⑨	-	5.9	10.0	-	3.5	15.5	12.3 ⑤
	Firebird	E	5.0 (4999)	5.0 ⑨	-	5.9	10.0	-	3.5	15.5	17.9 ②
	Firebird	8	5.7 (5737)	5.0 ⑨	-	5.9	10.0	-	3.5	15.5	16.6 ③
	Firebird	P	5.7 (5737)	5.0 ⑨	-	5.9 ④	10.0	-	3.5	15.5	15.1 ⑥
1994	Camaro	T	3.1 (3130)	4.5 ⑨	-	5.9	10.0	-	3.5	15.5	14.7 ①
	Camaro	S	3.4 (3393)	4.5 ⑨	-	5.9	10.0	-	3.5	15.5	12.3 ⑦
	Camaro	P	5.7 (5737)	5.0 ⑨	-	5.9 ④	10.0	-	3.5	15.5	15.1 ⑧
	Firebird	T	3.1 (3130)	4.5 ⑨	-	5.9	10.0	-	3.5	15.5	14.7 ①
	Firebird	S	3.4 (3393)	4.5 ⑨	-	5.9	10.0	-	3.5	15.5	12.3 ⑦
	Firebird	P	5.7 (5737)	5.0 ⑨	-	5.9 ④	10.0	-	3.5	15.5	15.1 ⑧
1995	Camaro	T	3.1 (3130)	4.5 ⑨	-	5.9	10.0	-	3.5	15.5	14.7 ①
	Camaro	S	3.4 (3393)	4.5 ⑨	-	5.9	10.0	-	3.5	15.5	12.3 ⑦
	Camaro	P	5.7 (5737)	5.0 ⑨	-	5.9 ④	10.0	-	3.5	15.5	15.1 ⑧
	Firebird	T	3.1 (3130)	4.5 ⑨	-	5.9	10.0	-	3.5	15.5	14.7 ①
	Firebird	S	3.4 (3393)	4.5 ⑨	-	5.9	10.0	-	3.5	15.5	12.3 ⑦
	Firebird	P	5.7 (5737)	5.0 ⑨	-	5.9 ④	10.0	-	3.5	15.5	15.1 ⑧

CAPACITIES

Year	Model	Engine ID/VIN	Engine Displacement Liters (cc)	Engine Oil with Filter (qts.)	Transmission (pts.)			Drive Axle		Fuel Tank (gal.)	Cooling System (qts.)
					4-Spd	5-Spd	Auto.	Front (pts.)	Rear (pts.)		
1996-97	Camaro	K	3.8 (3785)	4.5 ⑨	-	5.9	10.0	-	3.5	15.5	12.5
	Camaro	P	5.7 (5737)	4.5 ⑨	-	⑩	10.0	-	3.5	15.5	15.2
	Firebird	K	3.8 (3785)	4.5 ⑨	-	5.9	10.0	-	3.5	15.5	12.5
	Firebird	P	5.7 (5737)	4.5 ⑨	-	⑩	10.0	-	3.5	15.5	15.2

① With AC: 14.8 qts.
② With AC: 18.0 qts.
③ With AC: 16.7 qts.
④ With 6 speed transmission: 8.0 pts.
⑤ With AC: 12.5 qts.
⑥ With AC: 15.3 qts.
⑦ With manual transmission: 12.5 qts.
⑧ With manual transmission: 15.3 qts.
⑨ With vehicle on level surface, check oil level. Add as required to fill
⑩ ZF 6 speed trans.: 4.4 pts.

VALVE SPECIFICATIONS

Year	Engine ID/VIN	Engine Displacement Liters (cc)	Seat Angle (deg.)	Face Angle (deg.)	Spring Test Pressure (lbs. @ in.)	Spring Installed Height (in.)	Stem-to-Guide Clearance (in.)		Stem Diameter (in.)	
							Intake	Exhaust	Intake	Exhaust
1993	T	3.1 (3130)	46	45	90 ①	1.60 ②	0.0010-0.0027	0.0010-0.0027	NA	NA
	S	3.4 (3393)	46	45	190@1.20	1.61	0.0014-0.0025	0.0015-0.0029	NA	NA
	E	5.0 (4999)	46	45	194-206@1.25	③	0.0011-0.0027	0.0011-0.0027	NA	NA
	8	5.7 (5737)	46	45	194-206@1.25	③	0.0011-0.0027	0.0011-0.0027	NA	NA
	P	5.7 (5737)	46	45	252-272@1.305	1.78	0.0011-0.0027	0.0011-0.0027	NA	NA
1994	T	3.1 (3130)	46	45	90 ①	1.60 ②	0.0010-0.0027	0.0010-0.0027	NA	NA
	S	3.4 (3393)	46	45	190@1.20	1.61	0.0014-0.0025	0.0015-0.0029	NA	NA
	P	5.7 (5737)	46	45	187-203@1.27	1.70	0.0009-0.0027	0.0009-0.0027	NA	NA
1995	T	3.1 (3130)	46	45	90 ①	1.60 ②	0.0010-0.0027	0.0010-0.0027	NA	NA
	S	3.4 (3393)	46	45	190@1.20	1.61	0.0014-0.0025	0.0015-0.0029	NA	NA
	P	5.7 (5737)	46	45	187-203@1.27	1.70	0.0009-0.0027	0.0009-0.0027	NA	NA
1996-97	K	3.8 (3785)	45	45	210@1.32	1.69-1.72	0.0015-0.0035	0.0015-0.0032	NA	NA
	P	5.7 (5737)	46	45	245-265@1.33 ①	1.78	0.0009-0.0027	0.0009-0.0027	NA	NA

NA - Not Available
① With valve open
② With valve closed
③ Intake: 1.72
   Exhaust: 1.59

TORQUE SPECIFICATIONS
All readings in ft. lbs.

Year	Engine ID/VIN	Engine Displacement Liters (cc)	Cylinder Head Bolts	Main Bearing Bolts	Rod Bearing Bolts	Crankshaft Damper Bolts	Flywheel Bolts	Manifold Intake	Manifold Exhaust	Spark Plugs	Lug Nut
1993	T	3.1 (3130)	③	73	39	66-85	45-59	①	18	20	100
	S	3.4 (3393)	⑥	④	37	68	61	⑦	18	23	100
	E	5.0 (4999)	60-75	70-85	42-47	60	75	25-45	②	15-20	80-100
	8	5.7 (5737)	65	80	45	60	75	30	②	22	80-100
	P	5.7 (5737)	65	78	47	60	74	⑤	26	11	100
1994	T	3.1 (3130)	③	73	39	66-85	45-59	①	18	20	100
	S	3.4 (3393)	⑥	④	37	58	61	⑦	18	23	100
	P	5.7 (5737)	65	78	47	60	74	⑤	35	11	100
1995	T	3.1 (3130)	③	73	39	66-85	45-59	①	18	20	100
	S	3.4 (3393)	⑥	④	37	58	61	⑦	18	23	100
	P	5.7 (5737)	65	78	47	60	74	⑤	35	11	100
1996-97	K	3.8 (3785)	⑧	⑨	20	⑩	⑪	⑦	18	23	100
	P	5.7 (5737)	65	78	47	60	74	35	22	11	100

NA - Not Available

① 15 ft. lbs., then 24 ft. lbs.
② Outer bolts: 14-26 ft. lbs.
   Inner bolts: 20-32 ft. lbs.
③ 33 ft. lbs. plus 90 degrees
④ 37 ft. lbs. plus 75 degrees
⑤ Step 1: 71 inch lbs.
   Step 2: 35 ft. lbs.

⑥ 40 ft. lbs. plus 90 degrees
⑦ Upper manifold: 18 ft. lbs.
   Lower manifold bolt/nut: 22 ft. lbs.
   Upper manifold studs: 89 inch lbs.
⑧ Step 1: 37 ft. lbs. plus 130 degrees
   Step 2: Turn center bolts an additional 30 degrees

⑨ Step 1: Tighten to 52 ft. lbs. to fully seat caps
   Step 2: Loosen bearing cap 360 degrees counter-clockwise
   Step 3: Tighten caps 15 ft. lbs., then 30 ft. lbs., then 35 degrees,
   then an additional 35 degrees plus 40 degrees — for a total of 110 degrees
⑩ 111 ft. lbs. plus 76 degrees
⑪ 11 ft. lbs. plus 50 degrees

BRAKE SPECIFICATIONS
All measurements in inches unless noted

Year	Model		Master Cylinder Bore	Brake Disc Original Thickness	Brake Disc Minimum Thickness	Brake Disc Maximum Runout	Brake Drum Diameter Original Inside Diameter	Brake Drum Diameter Max. Wear Limit	Brake Drum Diameter Maximum Machine Diameter	Minimum Lining Thickness Front	Minimum Lining Thickness Rear
1993	Camaro		①	②	③	0.005	9.50	9.59	9.56	0.030	0.030
	Firebird		①	②	③	0.005	9.50	9.59	9.56	0.030	0.030
1994	Camaro		①	②	③	0.005	9.50	9.59	9.56	0.030	0.030
	Firebird		①	②	③	0.005	9.50	9.59	9.56	0.030	0.030
1995	Camaro		①	②	③	0.005	9.50	9.59	9.56	0.030	0.030
	Firebird		①	②	③	0.005	9.50	9.59	9.56	0.030	0.030
1996-97	Camaro	F	①	1.043	0.980	0.005	-	-	-	0.030	-
		R		0.795	0.744	0.005	9.50	9.59	9.56	-	0.030
	Firebird	F	①	1.043	0.980	0.005	-	-	-	0.030	-
		R		0.795	0.744	0.005	9.50	9.59	9.56	-	0.030

NA - Not Available
F - Front
R - Rear

① Rear drum: 0.945; Rear disc: 1.00
② Front: 1.043; Rear: 0.795
③ Front: 0.980; Rear: 0.744

FREQUENT MAINTENANCE LABOR
CHEVROLET CAMARO, PONTIAC FIREBIRD

The following should be used as a guide when determining the amount of work required for a particular service if taken to a repair shop. In estimating how long a particular Frequent Maintenance Service item should take, please observe the following:
- **Factory Time** is time that is generated by the vehicle manufacturer.
- **Chilton Time** is time that is based on field research and data supplied by the vehicle manufacturer.
- All labor time operations are given in hours and tenths of an hour.
- All labor operations, are to be used as a **guide**.

COOLING

(G) Winterize Cooling System
Includes: Run engine to check for leaks, tighten all hose connections. Test radiator and pressure cap, drain radiator and engine block. Add antifreeze and refill system.
1993-97 5

(G) Belt, Serpentine Drive, Renew
1993-97
V6 (.2) 3
V8 (.3) 4

(G) Hoses, Radiator, Renew
Includes: Drain and refill cooling system as required.
1993-97
V6
upper (.4) 5
lower (.5) 6
both (.9) 1.1
V8
upper (.5) 6
lower (.8) 9
both (.9) 1.5

(G) Thermostat, Coolant, Renew
1993-97
V6 (1.0) 1.4
V8 (.6) 8

FUEL

(M) Air Cleaner, Service
1993-97 (.3) 4

(G) Filter, Fuel, Renew
1993-97 (.3) 4

BRAKES

(G) Bleed Brakes (Four Wheels)
Includes: Add fluid.
1993 (.4) 5
1994-97 (.9) 1.2

(G) Brakes, Adjust (Minor)
Includes: Adjust brakes, fill master cylinder.
1993-97, two wheels 4
Remove knock out plugs
add each 1

(M) Parking Brake, Adjust
1993-97 (.6) 8

LUBRICATION SERVICES

(M) Engine Oil & Filter, Renew
Includes: Inspect and correct all fluid levels.
1993-97 3

(M) Lubricate Chassis, Change Oil & Filter
Includes: Inspect and correct all fluid levels.
1993-97 6
Install grease fittings add 1

(M) Lubricate Chassis
Includes: Inspect and correct all fluid levels.
1993-97 4
Install grease fittings add 1

ELECTRICAL

(G) Headlamps, Aim
1993-97
two 4
four 6

(G) License Lamp Assy., Renew
1993-97, one or both (.2) 3

(G) Park & Turn Signal Lamp Assy., Renew
1993-97, each (.5) 6

(G) Park & Turn Signal Lamp Bulb or Lens, Renew
1993-97
one (.4) 5
all (.6) 7

(G) Stop, Tail & Turn Signal Lamp Bulb, Renew
1993-97
one 4
each adtnl. 1

(G) Horn, Renew
1993-97
one (.4) 5
each adtnl. 1

(M) Terminals, Battery, Clean
1993-97 3

SCHEDULED MAINTENANCE INTERVALS
(CHEVROLET CAMARO, PONTIAC FIREBIRD)

TO BE SERVICED	TYPE OF SERVICE	VEHICLE MILEAGE INTERVAL (x1000)												
		7.5	15	22.5	30	37.5	45	52.5	60	67.5	75	82.5	90	97.5
Engine oil & filter	R	✓	✓	✓	✓	✓	✓	✓	✓	✓	✓	✓	✓	✓
Coolant level, hoses & clamps	S/I	✓	✓	✓	✓	✓	✓	✓	✓	✓	✓	✓	✓	✓
Exhaust system & throttle linkage	S/I	✓	✓	✓	✓	✓	✓	✓	✓	✓	✓	✓	✓	✓
Lubricate chassis, suspension, steering linkage, transmission shift linkage, parking brake cable guides, underbody contact points & linkage	S/I	✓	✓	✓	✓	✓	✓	✓	✓	✓	✓	✓	✓	✓
Brake hoses & brake lining	S/I	✓		✓		✓		✓		✓		✓		✓
Rotate tires③	S/I	✓		✓		✓		✓		✓		✓		✓
Automatic transmission fluid & filter⑤	S/I		✓		✓		✓		✓		✓		✓	
Air filter element & PCV filter	R				✓				✓				✓	
Engine coolant (1993-95)②	R				✓				✓				✓	
Spark plugs①	R				✓				✓				✓	
Ignition cables, EGR & fuel systems	S/I				✓				✓				✓	
Serpentine drive belt	S/I				✓				✓				✓	
Rear axle oil (Limited slip)④	R	✓												

① Platinum tip spark plugs - replace every 100,000 miles.
② Engine coolant (1996-97) - replace every 100,000 miles. Use O.E. specified (DEX-COOL™) coolant only. If any silicate coolant is used, the service interval is every 30,000 miles.
③ For models with P245/50ZR16 tires, rotate front-to-rear only, & be sure that the tires roll in the direction indicated by the arrows on the side walls.
④ If the vehicle is used to tow a trailer, change the rear axle fluid every 7500 miles in either type of differential.
⑤ Automatic transmission fluid & filter (1993-95) - replace every 100,000 miles.

R – Replace   S/I – Service or Inspect

FREQUENT OPERATION MAINTENANCE (SEVERE SERVICE)

  If a vehicle is operated under any of the following conditions it is considered severe service:
- **Extremely dusty areas.**
- **50% or more of the vehicle operation is in 32°C (90°F) or higher temperatures, or constant operation in temperatures below 0°C (32°F).**
- **Prolonged idling (vehicle operation in stop and go traffic).**
- **Frequent short running periods (engine does not warm to normal operating temperatures).**
- **Police, taxi, delivery usage or trailer towing usage.**

Oil & oil filter – change every 3000 miles.
Chassis lubrication - lubricate every 6000 miles.
Automatic transmission fluid & filter - change every 15,000 miles.
Air filter element - service or inspect every 15,000 miles.
Rotate tires at 6000 miles, then every 15,000 miles.③

CORVETTE

VEHICLE IDENTIFICATION CHART

		Engine Code			
Code	Liters	Cu. In. (cc)	Cyl.	Fuel Sys.	Eng. Mfg.
J	5.7	350 (5737)	8	MFI	①
P	5.7	350 (5737)	8	MFI	CPC
5	5.7	350 (5737)	8	MFI	CPC

Model Year	
Code	Year
P	1993
R	1994
S	1995
T	1996
V	1997

CPC - Chevrolet/Pontiac/Canada

MFI - Multiport fuel injection

① Manufactured by Mercury Marine

ENGINE IDENTIFICATION

Year	Model	Engine Displacement Liters (cc)	Engine Series (ID/VIN)	Fuel System	No. of Cylinders	Engine Type
1993	Corvette	5.7 (5737)	J	MFI	8	DOHC
	Corvette	5.7 (5737)	P	MFI	8	OHV
1994	Corvette	5.7 (5737)	J	MFI	8	DOHC
	Corvette	5.7 (5737)	P	MFI	8	OHV
1995	Corvette	5.7 (5737)	J	MFI	8	DOHC
	Corvette	5.7 (5737)	P	MFI	8	OHV
1996-97	Corvette	5.7 (5737)	5	MFI	8	DOHC
	Corvette	5.7 (5737)	P	MFI	8	OHV

MFI - Multiport fuel injection

OHV - Overhead valve

DOHC - Double overhead camshaft

GENERAL ENGINE SPECIFICATIONS

Year	Engine ID/VIN	Engine Displacement Liters (cc)	Fuel System Type	Net Horsepower @ rpm	Net Torque @ rpm (ft. lbs.)	Bore x Stroke (in.)	Compression Ratio	Oil Pressure @ rpm
1993	J	5.7 (5737)	MFI	375@5800	370@4800	3.90x3.66	11.0:1	40@2000
	P	5.7 (5737)	MFI	300@5000	340@3600	4.00x3.48	10.25:1	18@2000
1994	J	5.7 (5737)	MFI	375@5800	370@4800	3.90x3.66	11.0:1	40@2000
	P	5.7 (5737)	MFI	300@5000	340@3600	4.00x3.48	10.25:1	18@2000
1995	J	5.7 (5737)	MFI	375@5800	370@4800	3.90x3.66	11.0:1	40@2000
	P	5.7 (5737)	MFI	300@5000	340@3600	4.00x3.48	10.25:1	18@2000
1996-97	5	5.7 (5737)	MFI	375@5800	370@4800	3.90x3.66	11.0:1	40@2000
	P	5.7 (5737)	MFI	300@5000	340@3600	4.00x3.48	10.25:1	18@2000

MFI - Multiport fuel injection

GASOLINE ENGINE TUNE-UP SPECIFICATIONS

Year	Engine ID/VIN	Engine Displacement Liters (cc)	Spark Plugs Gap (in.)	Ignition Timing (deg.) MT	Ignition Timing (deg.) AT	Fuel Pump (psi)	Idle Speed (rpm) MT	Idle Speed (rpm) AT	Valve Clearance In.	Valve Clearance Ex.
1993	J	5.7 (5737)	0.035	①	①	48-55	①	①	HYD	HYD
	P	5.7 (5737)	0.035	-	①	41-47	-	①	HYD	HYD
1994	J	5.7 (5737)	0.035	①	①	48-55	①	①	HYD	HYD
	P	5.7 (5737)	0.035	-	①	41-47	-	①	HYD	HYD
1995	J	5.7 (5737)	0.035	①	①	48-55	①	①	HYD	HYD
	P	5.7 (5737)	0.035	-	①	41-47	-	①	HYD	HYD
1996-97	5	5.7 (5737)	0.035	①	①	48-55	①	①	HYD	HYD
	P	5.7 (5737)	0.035	-	①	41-47	-	①	HYD	HYD

NOTE: The Vehicle Emission Control Information label often reflects specification changes made during production. The label figures must be used if they differ from those in this chart.

HYD - Hydraulic

① Refer to Vehicle Emission Control Information label

CAPACITIES

Year	Model	Engine ID/VIN	Engine Displacement Liters (cc)	Engine Oil with Filter (qts.)	Transmission (pts.) 5-Spd	Transmission (pts.) 6-Spd	Transmission (pts.) Auto.	Drive Axle (pts.)	Fuel Tank (gal.)	Cooling System (qts.)
1993	Corvette	J	5.7 (5737)	8.6	-	②	-	3.75	20.0	14.7
	Corvette	P	5.7 (5737)	4.5	-	②	10.0 ①	3.75	20.0	17.8
1994	Corvette	J	5.7 (5737)	8.6	-	②	-	3.75	20.0	14.7
	Corvette	P	5.7 (5737)	4.5	-	②	10.0 ①	3.75	20.0	17.8
1995	Corvette	J	5.7 (5737)	8.6	-	②	-	3.75	20.0	14.7
	Corvette	P	5.7 (5737)	4.5	-	②	10.0 ①	3.75	20.0	17.8
1996-97	Corvette	5	5.7 (5737)	7.6	-	②	-	3.8	20.0	14.7
	Corvette	P	5.7 (5737)	5.0	-	②	10.0 ①	3.8	20.0	17.8

① 440T4 trans.: 13.0 pts.
125C trans.: 8.0 pts.

② ZF 6 speed trans.: 4.4 pts.

VALVE SPECIFICATIONS

Year	Engine ID/VIN	Engine Displacement Liters (cc)	Seat Angle (deg.)	Face Angle (deg.)	Spring Test Pressure (lbs. @ in.)	Spring Installed Height (in.)	Stem-to-Guide Clearance (in.) Intake	Stem-to-Guide Clearance (in.) Exhaust	Stem Diameter (in.) Intake	Stem Diameter (in.) Exhaust
1993	J	5.7 (5737)	44	45	147-166@ ① 0.95	1.34 ②	0.0012-0.0026	0.0014-0.0030	NA	NA
	P	5.7 (5737)	46	45	252-272@ 1.305	1.78	0.0011-0.0027	0.0011-0.0027	NA	NA
1994	J	5.7 (5737)	44	45	147-166@ ① 0.95	1.34 ②	0.0012-0.0026	0.0014-0.0030	NA	NA
	P	5.7 (5737)	46	45	187-203@ 1.27	1.70	0.0009-0.0027	0.0009-0.0027	NA	NA
1995	J	5.7 (5737)	44	45	147-166@ 0.95	1.34 ②	0.0012-0.0026	0.0014-0.0030	NA	NA
	P	5.7 (5737)	46	45	187-203@ 1.27	1.70	0.0009-0.0027	0.0009-0.0027	NA	NA

VALVE SPECIFICATIONS

Year	Engine ID/VIN	Engine Displacement Liters (cc)	Seat Angle (deg.)	Face Angle (deg.)	Spring Test Pressure (lbs. @ in.)	Spring Installed Height (in.)	Stem-to-Guide Clearance (in.) Intake	Stem-to-Guide Clearance (in.) Exhaust	Stem Diameter (in.) Intake	Stem Diameter (in.) Exhaust
1996-97	5	5.7 (5737)	44	45	147-166@ 0.95	1.34	0.0012- 0.0026	0.0014- 0.0030	NA	NA
	P	5.7 (5737)	46	45	81-89@1.78	1.78	0.0009- 0.0027	0.0009- 0.0027	NA	NA

NA - Not Available
① Inner spring: 75-82@0.79 in.
② Inner spring: 1.18 in.

TORQUE SPECIFICATIONS
All readings in ft. lbs.

Year	Engine ID/VIN	Engine Displacement Liters (cc)	Cylinder Head Bolts	Main Bearing Bolts	Rod Bearing Bolts	Crankshaft Damper Bolts	Flywheel Bolts	Manifold Intake	Manifold Exhaust	Spark Plugs	Lug Nut
1993	J	5.7 (5737)	③	②	⑤	148	66	④	①	15	100
	P	5.7 (5737)	65	78	47	60	74	⑥	26	11	100
1994	J	5.7 (5737)	③	②	⑤	148	66	④	①	15	100
	P	5.7 (5737)	65	78	47	60	74	⑥	35	11	100
1995	J	5.7 (5737)	③	②	⑤	148	66	④	①	15	100
	P	5.7 (5737)	65	78	47	60	74	⑥	35	11	100
1996-97	P	5.7 (5737)	65	78	47	60	74	35	30	11	100
	5	5.7 (5737)	③	⑦	⑤	148	66	④	①	15	100

NA - Not Available
① Studs: 22 ft. lbs.
  Bolts: 18 ft. lbs.
② Step 1: 15 ft. lbs.
  Step 2: Inner: 65-70 degrees
  Step 3: Outer: 50-55 degrees
③ Step 1: 45 ft. lbs.
  Step 2: 74 ft. lbs.
  Step 3: 118 ft. lbs.
④ Injector housing and fuel rail bolts: 20 ft. lbs.
⑤ 22 ft. lbs. plus 80-85 degrees
⑥ Step 1: 71 inch lbs.
  Step 2: 35 ft. lbs.
⑦ Step 1: 15 ft. lbs.
  Step 2: Inner: 65-70 degrees
  Step 3: Outer: 80-85 degrees

BRAKE SPECIFICATIONS
All measurements in inches unless noted

Year	Model	Master Cylinder Bore	Brake Disc Original Thickness	Brake Disc Minimum Thickness	Maximum Runout	Brake Drum Diameter Original Inside Diameter	Max. Wear Limit	Maximum Machine Diameter	Minimum Lining Thickness Front	Minimum Lining Thickness Rear
1993	Corvette	NA	②	①	0.006	NA	NA	NA	0.030	0.030
1994	Corvette	NA	②	①	0.006	NA	NA	NA	0.030	0.030
1995	Corvette	NA	②	①	0.006	NA	NA	NA	0.030	0.030
1996-97	Corvette	NA	②	①	0.006	NA	NA	NA	0.030	0.030

NA - Not Available
① Heavy duty: 1.059; Std.: 0.744
② Heavy duty: 1.110; Std.: 0.795

FREQUENT MAINTENANCE LABOR
CHEVROLET CORVETTE

The following should be used as a guide when determining the amount of work required for a particular service if taken to a repair shop. In estimating how long a particular Frequent Maintenance Service item should take, please observe the following:

- **Factory Time** is time that is generated by the vehicle manufacturer.
- **Chilton Time** is time that is based on field research and data supplied by the vehicle manufacturer.
- All labor time operations are given in hours and tenths of an hour.
- All labor operations, are to be used as a **guide**.

	(Factory Time)	Chilton Time
COOLING		
(G) Winterize Cooling System		
Includes: Run engine to check for leaks, tighten all hose connections. Test radiator and pressure cap, drain radiator and engine block. Add antifreeze and refill system.		
1993-97		.5
(G) Hoses, Radiator, Renew		
Includes: Drain and refill cooling system as required.		
1993-97		
upper	(.3)	.4
lower	(.5)	.8
both	(.6)	1.0
radiator to thermostat	(.6)	.8
thermo to w/pump	(.6)	.8
(G) Thermostat, Coolant, Renew		
1993-97		
codes 8, P, 5	(.7)	1.0
code J	(.5)	.8
FUEL		
(M) Air Cleaner, Service		
1993-97	(.2)	.3

	(Factory Time)	Chilton Time
(G) Filter, Fuel, Renew		
1993	(.3)	.4
1994-97	(.7)	1.0
w/Folding top add		.6
BRAKES		
(G) Bleed Brakes (Four Wheels)		
Includes: Add fluid.		
1993	(1.0)	1.4
1994-97	(1.4)	1.7
(M) Parking Brake, Adjust		
1993	(.6)	.8
LUBRICATION SERVICES		
(M) Engine Oil & Filter, Renew		
Includes: Inspect and correct all fluid levels.		
1993-97		.3
(M) Lubricate Chassis, Change Oil & Filter		
Includes: Inspect and correct all fluid levels.		
1993-97		.6
Install grease fittings add		.1

	(Factory Time)	Chilton Time
(M) Lubricate Chassis		
Includes: Inspect and correct all fluid levels.		
1993-97		.4
Install grease fittings add		.1
ELECTRICAL		
(G) Headlamps, Aim		
1993-97		
two		.4
four		.6
(G) High Mount Stop Lamp Bulb, Renew		
1993	(.3)	.4
1994-97	(.2)	.3
(G) License Lamp Assy., Renew		
1993-97		
one or both	(.2)	.3
(G) Park & Turn Signal Lamp Assy., Renew		
1993-97	(1.7)	2.4
(G) Horn, Renew		
1993-97, one	(.3)	.4
(M) Terminals, Battery, Clean		
1993-97		.3

SCHEDULED MAINTENANCE INTERVALS
(CHEVROLET CORVETTE)

TO BE SERVICED	TYPE OF SERVICE	VEHICLE MILEAGE INTERVAL (x1000)												
		7.5	15	22.5	30	37.5	45	52.5	60	67.5	75	82.5	90	97.5
Engine oil & filter③	R	✓	✓	✓	✓	✓	✓	✓	✓	✓	✓	✓	✓	✓
Brake hoses & brake lining	S/I	✓	✓	✓	✓	✓	✓	✓	✓	✓	✓	✓	✓	✓
Coolant level, hoses & clamps	S/I	✓	✓	✓	✓	✓	✓	✓	✓	✓	✓	✓	✓	✓
Exhaust system & throttle linkage	S/I	✓	✓	✓	✓	✓	✓	✓	✓	✓	✓	✓	✓	✓
Lubricate chassis, suspension, steering linkage, transmission shift linkage, parking brake cable guides, underbody contact points & linkage	S/I	✓	✓	✓	✓	✓	✓	✓	✓	✓	✓	✓	✓	✓
Rear axle fluid level	S/I	✓	✓	✓	✓	✓	✓	✓	✓	✓	✓	✓	✓	✓
Automatic transmission fluid & filter④	S/I		✓		✓		✓		✓		✓		✓	
Air filter element	R				✓				✓				✓	
Engine coolant (1993-95)②	R				✓				✓				✓	
Ignition cables, EGR & fuel systems	S/I				✓				✓				✓	
Serpentine drive belt	S/I				✓				✓				✓	
Spark plugs①	S/I													

① Platinum tip spark plugs - replace every 100,000 miles.
② Engine coolant (1996-97) - replace every 100,000 miles. Use O.E. specified (DEX-COOL™) coolant only. If any silicate coolant is used, the service interval is every 30,000 miles.
③ Corvette engines require a special oil meeting GM Standard 4718M.
④ Automatic transmission fluid & filter - replace every 100,000 miles (unless previously replaced).
R – Replace   S/I – Service or Inspect

FREQUENT OPERATION MAINTENANCE (SEVERE SERVICE)
If a vehicle is operated under any of the following conditions it is considered severe service:
- Extremely dusty areas.
- 50% or more of the vehicle operation is in 32°C (90°F) or higher temperatures, or constant operation in temperatures below 0°C (32°F).
- Prolonged idling (vehicle operation in stop and go traffic).
- Frequent short running periods (engine does not warm to normal operating temperatures).
- Police, taxi, delivery usage or trailer towing usage.
Engine oil & oil filter – change every 3000 miles. ③
Chassis lubrication - lubricate every 6000 miles.
Lubricate suspension, parking brake cable guides, underbody contact points & linkage - lubricate every 6000 miles.
Air filter element - service or inspect every 15,000 miles.
Automatic transmission fluid & filter - change every 15,000 miles.

METRO/PRIZM/STORM

ENGINE IDENTIFICATION

Year	Model		Engine Displacement Liters (cc)	Engine Series (ID/VIN)	Fuel System	No. of Cylinders	Engine Type
1993	Metro		1.0 (993)	6	TFI	3	SOHC
	Prizm		1.6 (1590)	6	MFI	4	DOHC
	Prizm		1.8 (1803)	8	MFI	4	DOHC
	Storm		1.6 (1590)	6	MFI	4	SOHC
	Storm GSI		1.8 (1803)	8	MFI	4	DOHC
1994	Metro		1.0 (993)	6	TFI	3	SOHC
	Prizm		1.6 (1590)	6	MFI	4	DOHC
	Prizm		1.8 (1803)	8	MFI	4	DOHC
1995	Metro		1.0 (993)	6	TFI	3	SOHC
	Metro		1.3 (1300)	9	TFI	4	SOHC
	Prizm	①	1.6 (1590)	6	MFI	4	DOHC
	Prizm		1.8 (1803)	8	MFI	4	DOHC
1996-97	Metro		1.0 (993)	6	TFI	3	SOHC
	Metro		1.3 (1300)	9	TFI	4	SOHC
	Prizm	①	1.6 (1590)	6	MFI	4	DOHC
	Prizm		1.8 (1803)	8	MFI	4	DOHC

TFI - Throttle body fuel injection
MFI - Multiport fuel injection
SOHC - Single overhead camshaft
DOHC - Double overhead camshaft
① GSi models

GENERAL ENGINE SPECIFICATIONS

Year	Engine ID/VIN		Engine Displacement Liters (cc)	Fuel System Type	Net Horsepower @ rpm	Net Torque @ rpm (ft. lbs.)	Bore x Stroke (in.)	Compression Ratio	Oil Pressure @ rpm
1993	6		1.0 (993)	TFI	55@5700	58@3300	2.91x3.03	9.5:1	50@3000
	6	①	1.6 (1590)	MFI	108@6000	105@4800	3.20x3.00	9.5:1	54@3000
	8	②	1.8 (1803)	MFI	115@5600	115@4800	3.19x3.37	9.5:1	54@3000
	6	③	1.6 (1590)	MFI	95@5800	97@4800	3.15x3.11	9.1:1	58@3000
	8	④	1.8 (1803)	MFI	140@6400	120@4600	3.15x3.54	9.7:1	71@3000
1994	6		1.0 (993)	TFI	55@5700	58@3300	2.91x3.03	9.5:1	54@3000
	6		1.6 (1590)	MFI	108@6000	105@4800	3.20x3.00	9.5:1	54@3000
	8	②	1.8 (1803)	MFI	115@6400	115@4600	3.19x3.37	9.7:1	71@3000
1995	6		1.0 (1993)	TFI	55@5700	58@3300	2.91x3.03	9.5:1	54@3000
	9		1.3 (1300)	TFI	70@5500	74@3000	2.91x3.03	9.5:1	54@3000
	6	①	1.6 (1590)	MFI	105@5800	100@4800	3.20x3.00	9.5:1	36-71@3000
	8	②	1.8 (1803)	MFI	115@5600	115@2800	3.20x3.40	9.5:1	36-71@3000
1996-97	6		1.0 (993)	TFI	55@5700	58@3300	2.91x3.03	9.5:1	54@3000
	9		1.3 (1300)	TFI	70@5500	74@3500	2.91x3.03	9.5:1	54@3000
	6	①	1.6 (1590)	MFI	105@5800	100@4800	3.20x3.00	9.5:1	36-71@3000
	8	⑤	1.8 (1803)	MFI	115@5200	117@2800	3.20x3.40	9.5:1	36-71@3000

TFI - Throttle body fuel injection
MFI - Multiport fuel injection

① Prizm
② Prizm GSi
③ Storm
④ Storm GSi
⑤ Prizm LSi

GASOLINE ENGINE TUNE-UP SPECIFICATIONS

Year	Engine ID/VIN	Engine Displacement Liters (cc)	Spark Plugs Gap (in.)	Ignition Timing (deg.) MT		Ignition Timing (deg.) AT		Fuel Pump (psi)	Idle Speed (rpm) MT	Idle Speed (rpm) AT	Valve Clearance In.	Valve Clearance Ex.
1993	6	1.0 (993)	0.041	5B	⑥	5B	⑥	23-30	800	850	HYD	HYD
	6 ①	1.6 (1590)	0.031	10B	⑧	10B	⑧	38-44	700	700	0.006-0.010	0.008-0.010
	8 ②	1.8 (1803)	0.031	10B	⑧	10B	⑧	38-44	700	700	0.006-0.010	0.008-0.012
	6 ③	1.6 (1590)	0.041	10B	⑦	10B	⑦	40-47	700	700	0.006	0.010
	8 ④	1.8 (1803)	0.041	10B	⑦	10B	⑦	40-47	700	700	HYD	HYD
1994	6 ⑤	1.0 (993)	0.041	5B	⑥	5B	⑥	23-30	800	850	HYD	HYD
	6	1.6 (1590)	0.031	10B	⑧	10B	⑧	38-44	700	700	0.006-0.010	0.008-0.010
	8	1.8 (1803)	0.031	10B	⑧	10B	⑧	38-44	700	700	0.006-0.010	0.008-0.012
1995	6 ⑤	1.0 (993)	0.041	5B	⑨	5B	⑨	23-30	800	850	HYD	HYD
	9	1.3 (1300)	0.041	5B	⑨	5B	⑨	23-30	800	850	HYD	HYD
	6	1.6 (1590)	0.031	10B	⑩	10B	⑩	31-37	700-750	700-750	0.0060-0.0100	0.0100-0.0140
	8	1.8 (1803)	0.031	10B	⑩	10B	⑩	31-37	700-750	700-750	0.0060-0.0100	0.0100-0.0140
1996-97	6 ⑤	1.0 (993)	0.041	5B	⑨	5B	⑨	23-30	800	850	HYD	HYD
	9	1.3 (1300)	0.041	5B	⑨	5B	⑨	23-30	800	850	HYD	HYD
	6	1.6 (1590)	0.031	10B	⑩	10B	⑩	31-37	700-750	700-750	0.0060-0.010	0.0100-0.0140
	8	1.8 (1803)	0.031	10B	⑩	10B	⑩	31-37	700-750	700-750	0.0060-0.010	0.0100-0.0140

NOTE: The Vehicle Emission Control Information label often reflects specification changes made during production. The label figures must be used if they differ from those in this chart.

B - Before top dead center

HYD - Hydraulic

① Prizm
② Prizm GSi
③ Storm
④ Storm GSi
⑤ Metro

⑥ Using a fused jumper wire, jump terminals 4 & 5 in the Duty Check Connector
⑦ Using a fused jumper wire, jump terminals 1 & 3 in the Data Link Connector
⑧ Using a fused jumper wire, jump terminals E1 & TE1 in the Data Link Connector
⑨ Connect a fused jumper from Duty Check cavity 4 to cavity 5 for fixed timing (DLC connector located at left strut tower)
⑩ Insert jumper wire between terminals in DLC connector E1 and TE1

CAPACITIES

Year	Model	Engine ID/VIN	Engine Displacement Liters (cc)	Engine Oil with Filter	Transmission (pts.) 4-Spd	Transmission (pts.) 5-Spd	Transmission (pts.) Auto.	Drive Axle Front (pts.)	Drive Axle Rear (pts.)	Fuel Tank (gal.)	Cooling System (qts.)
1993	Metro	6	1.0 (993)	3.7	-	4.1	10.1 ①	-	-	10.0	4.2 ⑥
	Prizm	6	1.6 (1590)	3.5	-	5.4	11.6 ①,⑤	3.0 ②	-	13.2	6.0 ③
	Prizm	8	1.8 (1803)	4.3	-	5.4	11.6 ①,⑤	3.0 ②	-	13.2	6.4 ⑦
	Storm	6	1.6 (1590)	3.2	-	4.0	14.0 ①	-	-	12.4	7.7 ④
	Storm GSi	8	1.8 (1803)	4.0	-	4.0	14.0 ①	-	-	12.4	7.8 ⑨
1994	Metro	6	1.0 (993)	3.7	-	4.1	10.1 ①	-	-	10.0	4.2 ⑥
	Prizm	6	1.6 (1590)	3.5	-	5.5	11.6 ⑦	3.0 ②	-	13.2	6.0 ③
	Prizm	8	1.8 (1803)	4.3	-	5.5	11.6 ⑦	3.0 ②	-	13.2	6.4 ⑧
1995	Metro	6	1.0 (993)	3.7	-	5.0	10.1 ①	-	-	10.6	⑩
	Metro	9	1.3 (1300)	3.7	-	5.0	10.4 ①	-	-	10.6	4.9
	Prizm	6	1.6 (1590)	3.2	-	3.0	6.6	-	3.0 ②	13.2	6.7
	Prizm GSi	8	1.8 (1803)	3.9	-	3.0	6.6	-	3.0 ②	13.2	6.7

CAPACITIES

Year	Model	Engine ID/VIN	Engine Displacement Liters (cc)	Engine Oil with Filter	Transmission (pts.)			Drive Axle		Fuel Tank (gal.)	Cooling System (qts.)
					4-Spd	5-Spd	Auto.	Front (pts.)	Rear (pts.)		
1996-97	Metro	6	1.0 (993)	3.7	-	5.0	10.1 ①	-	-	10.6	⑩
	Metro	9	1.3 (1300)	3.7	-	5.0	10.1 ①	-	-	10.6	4.9
	Prizm	6	1.6 (1590)	3.2	-	4.0	6.6	-	3.0 ②	13.2	6.7
	Prizm GSi	8	1.8 (1803)	3.9	-	4.0	6.6	-	3.0 ②	13.2	6.7

① Automatic transmission - Specification is after complete overhaul. Drain and fill will be less
② 3 speed automatic only
③ 3 speed automatic transaxle: 5.8 qts.
    4 speed automatic transaxle: 6.1 qts.
④ Manual transaxle: 7.2 qts.
⑤ 4 speed: 16.0 pts.
⑥ Manual transaxle: 4.1 qts.
⑦ 4 speed: 15.2 pts.
⑧ Manual transaxle: 6.6 qts.
⑨ Manual transaxle: 7.3 qts.
⑩ Manual transaxle: 4.1 qts.
    Automatic transaxle: 4.2 qts.

VALVE SPECIFICATIONS

Year	Engine ID/VIN	Engine Displacement Liters (cc)	Seat Angle (deg.)	Face Angle (deg.)	Spring Test Pressure (lbs. @ in.)	Spring Installed Height (in.)	Stem-to-Guide Clearance (in.)		Stem Diameter (in.)	
							Intake	Exhaust	Intake	Exhaust
1993	6	1.0 (993)	45	45	41-47.2@ 1.28	1.28	0.0008-0.0021	0.0014-0.0024	0.2148-0.2157	0.2146-0.2151
	6 ①	1.6 (1590)	45	45.5	37.3	1.25	0.0010-0.0024	0.0012-0.0026	0.2350-0.2356	0.2348-0.2354
	8 ①	1.8 (1803)	45	45.5	37.3	1.25	0.0010-0.0024	0.0012-0.0026	0.2350-0.2356	0.2348-0.2354
	6 ②	1.6 (1590)	45	45.5	-	-	0.0090	0.0018	0.2335	0.2335
	8 ②	1.8 (1803)	45	45.5	-	-	0.0009-0.0080	0.0012-0.0080	0.2320	0.2320
1994	6	1.0 (993)	45	45	46.1-51.8@ 1.28	1.28	0.0008-0.0022	0.0018-0.0028	0.2148-0.2157	0.2142-0.2148
	6	1.6 (1590)	45	45.5	37.3	1.25	0.0010-0.0024	0.0012-0.0026	0.2350-0.2356	0.2348-0.2354
	8	1.8 (1803)	45	45.5	37.3	1.25	0.0010-0.0024	0.0012-0.0026	0.2350-0.2356	0.2348-0.2354
1995	6 ③	1.0 (993)	45	45	46.1-51.8@ 1.28	1.28	0.0008-0.0022	0.0018-0.0028	0.2148-0.2157	0.2142-0.2148
	9	1.3 (1300)	45	45	54.7-64.3@ 1.63	1.63	0.0008-0.0019	0.0014-0.0025	0.2742-0.2748	0.2737-0.2742
	6 ①	1.6 (1590)	45	45.5	37.3@1.25	1.25	0.0010-0.0024	0.0012-0.0026	0.2350-0.2356	0.2348-0.2354
	8	1.8 (1803)	45	45.5	37.3@1.25	1.25	0.0010-0.0024	0.0012-0.0026	0.2350-0.2356	0.2348-0.2354
1996-97	6 ③	1.0 (993)	45	45	46.1-51.8@ 1.28	1.28	0.0008-0.0022	0.0018-0.0028	0.2148-0.2157	0.2142-0.2148
	9	1.3 (1300)	45	45	54.7-64.3@ 1.63	1.63	0.0008-0.0019	0.0014-0.0025	0.2742-0.2748	0.2737-0.2742
	6 ①	1.6 (1590)	45	45.5	37.3@1.25	1.25	0.0010-0.0024	0.0012-0.0026	0.2350-0.2356	0.2348-0.2354
	8	1.8 (1803)	45	45.5	37.3@1.25	1.25	0.0010-0.0024	0.0012-0.0026	0.2350-0.2356	0.2348-0.2354

① Prizm models
② Storm models
③ Metro

TORQUE SPECIFICATIONS
All readings in ft. lbs.

Year	Engine ID/VIN	Engine Displacement Liters (cc)	Cylinder Head Bolts	Main Bearing Bolts	Rod Bearing Bolts	Crankshaft Damper Bolts	Flywheel Bolts	Manifold Intake	Manifold Exhaust	Spark Plugs	Lug Nut
1993	6	1.0 (993)	54	40	26	81 ①	45	17	17	18	44
	6 ②	1.6 (1590)	22 ⑨	44	22 ⑦	87 ①	58 ④	14	25	13	76
	8 ②	1.8 (1803)	22 ⑨	44	22 ⑦	87 ①	58 ④	14	25	13	76
	6 ③	1.6 (1590)	58	44	11 ⑤	87 ①	⑤	17	30	21	87
	8 ③	1.8 (1803)	⑥	65 ⑧	18 ⑦	108 ①	⑩	17	30	21	87
1994	6	1.0 (993)	54	40	26	81 ①	52	17	17	18	44
	6	1.6 (1590)	22 ⑨	44	18 ⑦	87 ①	58 ④	14	25	13	76
	8	1.8 (1803)	22 ⑨	44	22 ⑦	87 ①	58 ④	14	25	13	76
1995	6	1.0 (993)	54	40	26	81 ①	52	17	17	21	44
	9	1.3 (1300)	54	40	26	79 ①	45	17	17	21	44
	6	1.6 (1590)	⑪	44	26	87 ①	④	14	29	21	76
	8	1.8 (1590)	⑫	40	26	81 ①	58	17	17	21	70
1996-97	6	1.0 (993)	54	40	26	81 ①	52	17	17	21	44
	9	1.3 (1300)	54	40	26	79 ①	45	17	17	21	44
	6	1.6 (1590)	⑪	44	⑪	87 ①	④	14	25	21	76
	8	1.8 (1803)	⑪	44	18 ⑦	87 ①	④	14	25	21	76

① Crankshaft timing belt sprocket
② Prizm models
③ Storm models
④ Manual transaxle: 58 ft. lbs.
   Automatic transaxle: 47 ft. lbs.

⑤ Automatic Transmission - 22 ft. lbs.;
   plus an additional 45-60 degrees
   Manual Transmission - 58 ft. lbs.
⑥ Step 1: 29 ft. lbs.
   Step 2: 58 ft. lbs.

⑦ Tighten an additional 90 degrees
⑧ Tighten in three steps
⑨ Plus two additional steps of 90 degrees
⑩ Automatic Transmission - 22 ft. lbs.
   Manual Transmission - 58 ft. lbs.

⑪ Step 1: 22 ft. lbs.
   Step 2: Plus 90 degrees
⑫ Step 1: 26 ft. lbs.
   Step 2: 41 ft. lbs.
   Step 3: 52 ft. lbs.

BRAKE SPECIFICATIONS
All measurements in inches unless noted

Year	Model	Master Cylinder Bore	Brake Disc Original Thickness	Brake Disc Minimum Thickness	Brake Disc Maximum Runout	Brake Drum Diameter Original Inside Diameter	Brake Drum Diameter Max. Wear Limit	Brake Drum Diameter Maximum Machine Diameter	Minimum Lining Thickness Front	Minimum Lining Thickness Rear
1993	Metro	0.825	0.394	0.315	0.004	7.09	7.16	7.16	0.310 ①	0.110 ①
	Prizm	NA	0.866	0.787	0.004	7.87	7.91	7.91	0.039	0.039
	Storm	0.810 ②	0.866	0.811	0.005	7.87	7.93	7.93	0.039	0.039
1994	Metro	0.825	0.394	0.315	0.004	7.09	7.16	7.16	0.310 ①	0.110 ①
	Prizm	NA	0.866	0.787	0.004	7.87	7.91	7.91	0.039	0.039
1995	Metro	NA	0.670	0.590	0.004	③	④	④	0.236 ①	0.111 ①
	Prizm	NA	0.866	0.787	0.003	7.87	7.91	7.91	0.039	0.039
1996-97	Metro	NA	0.670	0.590	0.004	③	④	④	0.236 ⑤	0.111 ①
	Prizm	NA	0.866	0.787	0.003	7.87	7.91	7.91	0.390	0.390

NA - Not Available
① Minimum lining thickness includes pad/shoe backing
② GSi models: 0.875 in.

③ 2 door: 7.09
   4 door: 7.87

④ 2 door: 7.16
   4 door: 7.95

⑤ 2 door: 8.66 (service limit 0.874)
   4 door: 10.00 (service limit 10.07)

FREQUENT MAINTENANCE LABOR
GEO METRO

The following should be used as a guide when determining the amount of work required for a particular service if taken to a repair shop. In estimating how long a particular Frequent Maintenance Service item should take, please observe the following:

- **Factory Time** is time that is generated by the vehicle manufacturer.
- **Chilton Time** is time that is based on field research and data supplied by the vehicle manufacturer.
- All labor time operations are given in hours and tenths of an hour.
- All labor operations, are to be used as a **guide**.

	(Factory Time)	Chilton Time

COOLING

(G) Winterize Cooling System

Includes: Run engine to check for leaks, tighten all hose connections. Test radiator and pressure cap, drain radiator and engine block. Add antifreeze and refill system.

1993-97 .5

(G) Belt, Drive, Renew

1993-97 .3
w/AC add 1

(G) Belt, Drive, Adjust

1993-97
one 3
each adtnl. 1

(G) Hoses, Radiator, Renew

Includes: Drain and refill cooling system as required.

1993-97
upper 4
lower 5
both 6

(G) Thermostat, Coolant, Renew

1993-97 .6

FUEL

(M) Air Cleaner, Service

1993-97 .2

(G) Filter, Fuel, Renew

1993 .4
1994-97 .6

BRAKES

(G) Bleed Brakes (Four Wheels)

Includes: Add fluid.

1993-97
wo/ABS 5
w/ABS 8

(G) Brakes, Adjust (Minor)

Includes: Adjust brakes, fill master cylinder.

1993-97, two wheels 4
Remove knock out plugs
add, each 1

(M) Parking Brake, Adjust

1993-97 .5

LUBRICATION SERVICES

(M) Engine Oil & Filter, Renew

Includes: Inspect and correct all fluid levels.

1993-97 .3

(M) Lubricate Chassis, Change Oil & Filter

Includes: Inspect and correct all fluid levels.

1993-97 .6
Install grease fittings add 1

(M) Lubricate Chassis

Includes: Inspect and correct all fluid levels.

1993-97 .4
Install grease fittings add 1

WHEELS

(G) Wheel, Renew (One)

1993-97 .5

(G) Wheel, Balance

1993-97
one 3
each adtnl. 2

(G) Wheels, Rotate (All)

1993-97 .5

ELECTRICAL

(G) Headlamps, Aim

1993-97
two 4
four 6

(G) High Mount Stop Lamp Bulb, Renew

1993-97 .3

(G) License Lamp Bulb, Renew

1993-97, one or all 3

(G) Park & Turn Signal Lamp Assy., Renew

1993-97, each 3

(G) Park & Turn Signal Lamp Bulb or Lens, Renew

1993-97
one 2
all 3

(G) Rear Combination Lamp Assy., Renew

1993-97 .4

(G) Stop, Tail & Turn Signal Lamp Bulb, Renew

1993-97
one 2
each adtnl. 1

(G) Horn, Renew

1993-97 .3

(M) Terminals, Battery, Clean

1993-97 .3

FREQUENT MAINTENANCE LABOR
GEO STORM

The following should be used as a guide when determining the amount of work required for a particular service if taken to a repair shop. In estimating how long a particular Frequent Maintenance Service item should take, please observe the following:
- **Factory Time** is time that is generated by the vehicle manufacturer.
- **Chilton Time** is time that is based on field research and data supplied by the vehicle manufacturer.
- All labor time operations are given in hours and tenths of an hour.
- All labor operations, are to be used as a **guide**.

COOLING

(G) Winterize Cooling System
Includes: Run engine to check for leaks, tighten all hose connections. Test radiator and pressure cap, drain radiator and engine block. Add antifreeze and refill system.
1993 5

(G) Belt, Drive, Renew
1993
  1.6L code 5
  1.6L code 6
    Fan/generator 3
    AC or PS 3
  1.8L code 8, Serpentine 7

(G) Belt, Drive, Adjust
1993
  one 3
  all 4

(G) Hoses, Radiator, Renew
Includes: Drain and refill cooling system as required.
1993
  upper or lower 4
  both 6

(G) Thermostat, Coolant, Renew
1993
  1.6L codes 5, 6 6
  1.8L code 8 9

FUEL

(M) Air Cleaner, Service
1993 2

BRAKES

(G) Bleed Brakes (Four Wheels)
Includes: Add fluid.
1993 5

(G) Brakes, Adjust (Minor)
Includes: Adjust brakes, fill master cylinder.
1993, two wheels 4
Remove knock out plugs add, each .1

(M) Parking Brake, Adjust
1993 5

LUBRICATION SERVICES

(M) Engine Oil & Filter, Renew
Includes: Inspect and correct all fluid levels.
1993 3

(M) Lubricate Chassis, Change Oil & Filter
Includes: Inspect and correct all fluid levels.
1993 6
Install grease fittings, add 1

(M) Lubricate Chassis
Includes: Inspect and correct all fluid levels.
1993 4
Install grease fittings, add 1

WHEELS

(G) Wheel, Renew (One)
1993 5

(G) Wheel, Balance
1993
  one 3
  each adtnl. 2

(G) Wheels, Rotate (All)
1993 5

ELECTRICAL

(G) Headlamps, Aim
1993
  two 4
  four 6

(G) High Mount Stop Lamp Bulb, Renew
1993 3

(G) License Lamp Assy., Renew
1993 one or both 5

(G) License Lamp Bulb, Renew
1993, one or all 2

(G) Park & Turn Signal Lamp Assy., Renew
1993, each 3

(G) Park & Turn Signal Lamp Bulb or Lens, Renew
1993
  one 2
  all 3

(G) Rear Combination Lamp Assy., Renew
1993
  exc. Hatchback 7
  Hatchback 4

(G) Stop, Tail & Turn Signal Lamp Bulb, Renew
1993
  one 2
  each adtnl. 1

(G) Horn, Renew
1993 4

(M) Terminals, Battery, Clean
1993 3

FREQUENT MAINTENANCE LABOR
GEO PRIZM

The following should be used as a guide when determining the amount of work required for a particular service if taken to a repair shop. In estimating how long a particular Frequent Maintenance Service item should take, please observe the following:
- **Factory Time** is time that is generated by the vehicle manufacturer.
- **Chilton Time** is time that is based on field research and data supplied by the vehicle manufacturer.
- All labor time operations are given in hours and tenths of an hour.
- All labor operations, are to be used as a **guide**.

COOLING

(G) Winterize Cooling System
Includes: Run engine to check for leaks, tighten all hose connections. Test radiator and pressure cap, drain radiator and engine block. Add antifreeze and refill system.
1993-97 5

(G) Belt, Drive, Renew
1993-97
  1.6L, 1.8L codes 6, 8
    AC or PS 4
    Alternator 2

(G) Belt, Drive, Adjust
1993-97
  one 3
  all 4

(G) Hoses, Radiator, Renew
Includes: Drain and refill cooling system as required.
1993-97
  upper 4
  lower 5
  both 6

(G) Thermostat, Coolant, Renew
1993-97 5

FUEL

(M) Air Cleaner, Service
1993-97 3

(G) Filter, Fuel, Renew
1993-97 6

BRAKES

(G) Bleed Brakes (Four Wheels)
Includes: Add fluid.
1993-97 5
w/ABS add 3

(G) Brakes, Adjust (Minor)
Includes: Adjust brakes, fill master cylinder.
1993-97, two wheels 4
Remove knock out plugs add
  each 1

(M) Parking Brake, Adjust
1993-97 7

LUBRICATION SERVICES

(M) Engine Oil & Filter, Renew
Includes: Inspect and correct all fluid levels.
1993-97 3

(M) Lubricate Chassis, Change Oil & Filter
Includes: Inspect and correct all fluid levels.
1993-97 6
Install grease fittings add 1

(M) Lubricate Chassis
Includes: Inspect and correct all fluid levels.
1993-97 4
Install grease fittings add 1

WHEELS

(G) Wheel, Renew (One)
1993-97 5

(G) Wheel, Balance
1993-97
  one 3
  each adtnl. 2

(G) Wheels, Rotate (All)
1993-97 5

ELECTRICAL

(G) Headlamps, Aim
1993-97
  two 4
  four 6

(G) High Mount Stop Lamp Bulb, Renew
1993-97 3

(G) License Lamp Bulb, Renew
1993-97, one or all 2

(G) Park & Turn Signal Lamp Assy., Renew
1993-97, one 3

(G) Park & Turn Signal Lamp Bulb or Lens, Renew
1993-97
  one 2
  all 3

(G) Rear Combination Lamp Assy., Renew
1993-97
  left side 5
  right side 3

(G) Stop, Tail & Turn Signal Lamp Bulb, Renew
1993-97
  one 2
  each adtnl. 1

(G) Horn, Renew
1993-97 3

(M) Terminals, Battery, Clean
1993-97 3

SCHEDULED MAINTENANCE INTERVALS
(GEO METRO, PRIZM & STORM)

TO BE SERVICED	TYPE OF SERVICE	VEHICLE MILEAGE INTERVAL (x1000)												
		7.5	15	22.5	30	37.5	45	52.5	60	67.5	75	82.5	90	97.5
Engine oil & filter	R	✓	✓	✓	✓	✓	✓	✓	✓	✓	✓	✓	✓	✓
Chassis lubrication (Metro & Storm)	S/I	✓	✓	✓	✓	✓	✓	✓	✓	✓	✓	✓	✓	✓
Locking front hubs (1996-97 Metro)	S/I	✓	✓	✓	✓	✓	✓	✓	✓	✓	✓	✓	✓	✓
Lubricate parking brake cable guides, underbody contact points & linkage	S/I	✓	✓	✓	✓	✓	✓	✓	✓	✓	✓	✓	✓	✓
Rotate tires (1994-97)	S/I	✓	✓	✓	✓	✓	✓	✓	✓	✓	✓	✓	✓	✓
Rotate tires (1993)	S/I	✓		✓		✓		✓		✓		✓		✓
Brake system	S/I	✓		✓		✓		✓		✓		✓		✓
Engine idle speed (Prizm)	S/I	✓		✓		✓		✓		✓		✓		✓
Exhaust system	S/I	✓		✓		✓		✓		✓		✓		✓
Chassis lubrication (Prizm)	S/I		✓		✓		✓		✓		✓		✓	
Fuel tank, cap & lines (Metro & Storm)	S/I		✓		✓		✓		✓		✓		✓	
Fuel tank, cap & lines (Prizm)	S/I				✓				✓				✓	
Valve clearance (Storm 1.6L)	S/I		✓		✓		✓		✓		✓		✓	
Valve clearance (Prizm)	S/I								✓					
Air cleaner filter	R				✓				✓				✓	
Engine coolant	R				✓				✓				✓	
Fuel filter (Metro)	R				✓				✓				✓	
Fuel tank cap gasket (Prizm)	R				✓				✓				✓	
Manual transaxle oil	R				✓				✓				✓	
Spark plugs	R				✓				✓				✓	
Accessory drive belt(s) (1996-97 Prizm)	S/I								✓		✓		✓	✓
Accessory drive belt(s) (except 1996-97 Prizm)	S/I			✓					✓				✓	

SCHEDULED MAINTENANCE INTERVALS
(GEO METRO, PRIZM & STORM) (Cont.)

TO BE SERVICED	TYPE OF SERVICE	VEHICLE MILEAGE INTERVAL (x1000)												
		7.5	15	22.5	30	37.5	45	52.5	60	67.5	75	82.5	90	97.5
Cooling system (Metro & Storm)	S/I				✓				✓				✓	
Cooling system (Prizm)	S/I						✓				✓			
Automatic transaxle fluid & filter	S/I				✓				✓				✓	
EGR system	S/I				✓				✓				✓	
Engine timing (1993-95 Metro)	S/I				✓				✓				✓	
Engine timing (1996-97 Metro)	S/I								✓					
Ignition cables (Storm)	S/I				✓				✓				✓	
Ignition cables (Metro & Prizm)	S/I								✓					
PCV system	S/I				✓				✓				✓	
Brake fluid (1996-97 Metro)	R								✓					
PCV valve (1996-97 Metro)①	R													
Timing belt	R								✓					
EVAP canister (1996-97 Prizm)	S/I								✓					
Throttle body unit mount bolt torque	S/I	✓												

① PCV valve (1996-97 Metro) - replace at 50,000 miles.
R – Replace   S/I – Service or Inspect

FREQUENT OPERATION MAINTENANCE (SEVERE SERVICE)

If a vehicle is operated under any of the following conditions it is considered severe service:
- Extremely dusty areas.
- 50% or more of the vehicle operation is in 32°C (90°F) or higher temperatures, or constant operation in temperatures below 0°C (32°F).
- Prolonged idling (vehicle operation in stop and go traffic).
- Frequent short running periods (engine does not warm to normal operating temperatures).
- Police, taxi, delivery usage or trailer towing usage.

Oil & oil filter (Metro, Storm & 1995-97 Prizm) – change every 3000 miles.
Oil & oil filter (1993-94 Prizm) – change every 3750 miles.
Chassis lubrication (Metro, Storm & 1995-97 Prizm) - lubricate every 6000 miles.
Chassis lubrication (1993-94 Prizm) - lubricate every 7500 miles.
Throttle body mount bolt torque - torque at 6000 miles.
Air cleaner filter - service or inspect every 15,000 miles.
Differential fluid (1996-97 Prizm) - replace every 15,000 miles.
Rotate tires - rotate at 6000 miles and then every 15,000 miles thereafter.
Automatic transaxle fluid & filter (Prizm & Storm) - replace every 15,000 miles.
Automatic transaxle fluid & filter (Metro) - replace every 50,000 miles.

SC/SC1/SC2/SL/SL1/SL2/SW1/SW2

VEHICLE IDENTIFICATION CHART

Engine Code							Model Year	
Code	Liters	Cu. In. (cc)	Cyl.	Fuel Sys.	Eng. Mfg.		Code	Year
7	1.9	116 (1901)	4	MFI	Saturn		P	1993
8	1.9	116 (1901)	4	MFI	Saturn		R	1994
9	1.9	116 (1901)	4	TFI	Saturn		S	1995
							T	1996
							V	1997

MFI - Multiport fuel injection

TFI - Throttle body fuel injection

ENGINE IDENTIFICATION

Year	Model	Engine Displacement Liters (cc)	Engine Series (ID/VIN)	Fuel System	No. of Cylinders	Engine Type
1993	Sedan	1.9 (1901)	7	MFI	4	DOHC
	Sedan	1.9 (1901)	9	TFI	4	SOHC
	Coupe	1.9 (1901)	7	MFI	4	DOHC
	Coupe	1.9 (1901)	9	TFI	4	SOHC
	Wagon	1.9 (1901)	7	MFI	4	DOHC
	Wagon	1.9 (1901)	9	TFI	4	SOHC
1994	Sedan	1.9 (1901)	7	MFI	4	DOHC
	Sedan	1.9 (1901)	9	TFI	4	SOHC
	Coupe	1.9 (1901)	7	MFI	4	DOHC
	Coupe	1.9 (1901)	9	TFI	4	SOHC
	Wagon	1.9 (1901)	7	MFI	4	DOHC
	Wagon	1.9 (1901)	9	TFI	4	SOHC
1995	Sedan	1.9 (1901)	7	MFI	4	DOHC
	Sedan	1.9 (1901)	8	MFI	4	SOHC
	Coupe	1.9 (1901)	7	MFI	4	DOHC
	Coupe	1.9 (1901)	8	MFI	4	SOHC
	Wagon	1.9 (1901)	7	MFI	4	DOHC
	Wagon	1.9 (1901)	8	MFI	4	SOHC
1996-97	Sedan	1.9 (1901)	7	MFI	4	DOHC
	Sedan	1.9 (1901)	8	MFI	4	SOHC
	Coupe	1.9 (1901)	7	MFI	4	DOHC
	Coupe	1.9 (1901)	8	MFI	4	SOHC
	Wagon	1.9 (1901)	7	MFI	4	DOHC
	Wagon	1.9 (1901)	8	MFI	4	SOHC

MFI - Multiport fuel injection

TFI - Throttle body fuel injection

DOHC - Double overhead camshaft

SOHC - Single overhead camshaft

GENERAL ENGINE SPECIFICATIONS

Year	Engine ID/VIN	Engine Displacement Liters (cc)	Fuel System Type	Net Horsepower @ rpm	Net Torque @ rpm (ft. lbs.)	Bore x Stroke (in.)	Compression Ratio	Oil Pressure @ rpm
1993	7	1.9 (1901)	MFI	124@5600	122@4800	3.23x3.54	9.5:1	29@2000
	9	1.9 (1901)	TFI	85@5000	107@2400	3.23x3.54	9.3:1	36@2000
1994	7	1.9 (1901)	MFI	124@5600	122@4800	3.23x3.54	9.5:1	29@2000
	9	1.9 (1901)	TFI	85@5000	107@2400	3.23x3.54	9.3:1	36@2000
1995	7	1.9 (1901)	MFI	124@5600	122@4800	3.23x3.54	9.5:1	29@2000
	8	1.9 (1901)	MFI	85@5000	107@2400	3.23x3.54	9.3:1	36@2000
1996-97	7	1.9 (1901)	MFI	124@5600	122@4800	3.23x3.54	9.5:1	29@2000
	8	1.9 (1901)	MFI	100@5000	114@2400	3.23x3.54	9.3:1	36@2000

MFI - Multiport fuel injection
TFI - Throttle body fuel injection

GASOLINE ENGINE TUNE-UP SPECIFICATIONS

Year	Engine ID/VIN	Engine Displacement Liters (cc)	Spark Plugs Gap (in.)	Ignition Timing (deg.) MT	AT	Fuel Pump (psi)	Idle Speed (rpm) MT	AT	Valve Clearance In.	Ex.
1993	7	1.9 (1901)	0.040	①	①	31-36 ②	③	750 ③	HYD	HYD
	9	1.9 (1901)	0.040	①	①	26-31 ②	750 ③	650 ③	HYD	HYD
1994	7	1.9 (1901)	0.040	①	①	31-36 ②	③	750 ③	HYD	HYD
	9	1.9 (1901)	0.040	①	①	26-31 ②	750 ③	650 ③	HYD	HYD
1995	7	1.9 (1901)	0.040	①	①	31-36 ②	850 ③	750 ③	HYD	HYD
	8	1.9 (1901)	0.040	①	①	26-31 ②	750 ③	650 ③	HYD	HYD
1996-97	7	1.9 (1901)	0.040	①	①	31-36 ②	850 ③	750 ③	HYD	HYD
	8	1.9 (1901)	0.040	①	①	31-36 ②	750 ③	650 ③	HYD	HYD

NOTE: The Vehicle Emission Control Information label often reflects specification changes made during production. The label figures must be used if they differ from those in this chart.
HYD - Hydraulic
① Engines equipped with Distributorless Ignition System (DIS). Ignition timing is not adjustable
② Pressure measured at idle
③ Idle speed measured with manual transmission in neutral; automatic transmission in drive

CAPACITIES

Year	Model	Engine ID/VIN	Engine Displacement Liters (cc)	Engine Oil with Filter (qts.)	Transmission (pts.) 4-Spd	5-Spd	Auto.	Drive Axle Front (pts.)	Rear (pts.)	Fuel Tank (gal.)	Cooling System (qts.)
1993	Wagon	7	1.9 (1901)	4.0	-	5.2	7.5	-	-	12.8	7.0
	Wagon	9	1.9 (1901)	4.0	-	5.2	7.5	-	-	12.8	7.0
	Sedan	7	1.9 (1901)	4.0	-	5.2	7.5	-	-	12.8	7.0
	Sedan	9	1.9 (1901)	4.0	-	5.2	7.5	-	-	12.8	7.0
	Coupe	7	1.9 (1901)	4.0	-	5.2	7.5	-	-	12.8	7.0
	Coupe	9	1.9 (1901)	4.0	-	5.2	7.5	-	-	12.8	7.0
1994	Wagon	7	1.9 (1901)	4.0	-	5.2	7.5 ①	-	-	12.8	7.0
	Wagon	9	1.9 (1901)	4.0	-	5.2	7.5 ①	-	-	12.8	7.0
	Sedan	7	1.9 (1901)	4.0	-	5.2	7.5 ①	-	-	12.8	7.0
	Sedan	9	1.9 (1901)	4.0	-	5.2	7.5 ①	-	-	12.8	7.0
	Coupe	7	1.9 (1901)	4.0	-	5.2	7.5 ①	-	-	12.8	7.0
	Coupe	9	1.9 (1901)	4.0	-	5.2	7.5 ①	-	-	12.8	7.0

CAPACITIES

Year	Model	Engine ID/VIN	Engine Displacement Liters (cc)	Engine Oil with Filter (qts.)	Transmission (pts.) 4-Spd	5-Spd	Auto.	Drive Axle Front (pts.)	Rear (pts.)	Fuel Tank (gal.)	Cooling System (qts.)
1995	Sedan	7	1.9 (1901)	4.0	-	5.2	7.5 ①	-	-	12.8	7.0
	Sedan	8	1.9 (1901)	4.0	-	5.2	7.5 ①	-	-	12.8	7.0
	Coupe	7	1.9 (1901)	4.0	-	5.2	7.5 ①	-	-	12.8	7.0
	Coupe	8	1.9 (1901)	4.0	-	5.2	7.5 ①	-	-	12.8	7.0
	Wagon	7	1.9 (1901)	4.0	-	5.2	7.5 ①	-	-	12.8	7.0
	Wagon	8	1.9 (1901)	4.0	-	5.2	7.5 ①	-	-	12.8	7.0
1996-97	Sedan	7	1.9 (1901)	4.0	-	5.2	7.5 ①	-	-	12.8	7.0
	Sedan	8	1.9 (1901)	4.0	-	5.2	7.5 ①	-	-	12.8	7.0
	Coupe	7	1.9 (1901)	4.0	-	5.2	7.5 ①	-	-	12.8	7.0
	Coupe	8	1.9 (1901)	4.0	-	5.2	7.5 ①	-	-	12.8	7.0
	Wagon	7	1.9 (1901)	4.0	-	5.2	7.5 ①	-	-	12.8	7.0
	Wagon	8	1.9 (1901)	4.0	-	5.2	7.5 ①	-	-	12.8	7.0

① Specification is for overhaul. 4.2 qts. with fluid and filter change

VALVE SPECIFICATIONS

Year	Engine ID/VIN	Engine Displacement Liters (cc)	Seat Angle (deg.)	Face Angle (deg.)	Spring Test Pressure (lbs. @ in.)	Spring Installed Height (in.)	Stem-to-Guide Clearance (in.) Intake	Exhaust	Stem Diameter (in.) Intake	Exhaust
1993	7	1.9 (1901)	44.5-45.5	45-45.5	163-180@ 0.984	①	0.0010-0.0025	0.0015-0.0032	0.2736-0.2740	0.2729-0.2736
	9	1.9 (1901)	44.5-45.5	45-45.5	202-211@ 1.280	①	0.0010-0.0025	0.0015-0.0032	0.2736-0.2741	0.2736-0.2740
1994	7	1.9 (1901)	44.5-45.5	45-45.5	163-180@ 0.984	①	0.0010-0.0025	0.0015-0.0032	0.2736-0.2740	0.2729-0.2736
	9	1.9 (1901)	44.5-45.5	45-45.5	202-211@ 1.280	①	0.0010-0.0025	0.0015-0.0032	0.2736-0.2741	0.2736-0.2740
1995	7	1.9 (1901)	44.5-45.4	45-45.5	163-180@ 0.984	①	0.0010-0.0025	0.0015-0.0032	0.2736-0.2740	0.2729-0.2736
	8	1.9 (1901)	44.5-45.4	45-45.5	202-211@ 1.280	①	0.0010-0.0025	0.0015-0.0032	0.2736-0.2741	0.2736-0.2740
1996-97	7	1.9 (1901)	44.5-45.4	45-45.5	163-180@ 0.984	①	0.0010-0.0025	0.0015-0.0032	0.2736-0.2740	0.2729-0.2736
	8	1.9 (1901)	44.5-45.4	45-45.25	202-211@ 1.280	①	0.0010-0.0025	0.0015-0.0032	0.2736-0.2741	0.2736-0.2740

① Installed height not available
Free length SOHC: 1.8898-1.9134
Free length DOHC: 1.6100

TORQUE SPECIFICATIONS
All readings in ft. lbs.

Year	Engine ID/VIN	Engine Displacement Liters (cc)	Cylinder Head Bolts	Main Bearing Bolts	Rod Bearing Bolts	Crankshaft Damper Bolts	Flywheel Bolts	Manifold Intake	Manifold Exhaust	Spark Plugs	Lug Nut
1993	7	1.9 (1901)	②	37	33	159	59 ④	22 ③	23 ③	20	103
	9	1.9 (1901)	①	37	33	159	59 ④	15 ③	16 ③	20	103
1994	7	1.9 (1901)	②	37	33	159	59 ④	22 ③	22 ③	20	103
	9	1.9 (1901)	①	37	33	159	59 ④	22 ③	22 ③	20	103
1995	7	1.9 (1901)	②	37	33	159	59 ④	22 ③	22 ③	20	103
	8	1.9 (1901)	①	37	33	159	59 ④	22 ③	22 ③	20	103
1996-97	7	1.9 (1901)	②	37	33	159	59 ④	22 ③	19 ③	20	103
	8	1.9 (1901)	①	37	33	159	59 ④	22 ③	16 ③	20	103

① Step 1: 22 ft. lbs.
  Step 2: 33 ft. lbs.
  Step 3: 90 degrees

② Step 1: 22 ft. lbs.
  Step 2: 37 ft. lbs.
  Step 3: 90 degrees

③ Studs: 106 in. lbs.
④ Flexplate specification: 44 ft. lbs.

BRAKE SPECIFICATIONS
All measurements in inches unless noted

Year	Model	Master Cylinder Bore	Brake Disc Original Thickness	Brake Disc Minimum Thickness	Brake Disc Maximum Runout	Brake Drum Diameter Original Inside Diameter	Brake Drum Diameter Max. Wear Limit	Brake Drum Diameter Maximum Machine Diameter	Minimum Lining Thickness Front	Minimum Lining Thickness Rear
1993	Sedan	NA	①	②	0.0024	7.87	7.93	7.91	0.080	0.040
	Coupe	NA	①	②	0.0024	7.87	7.93	7.91	0.080	0.040
	Wagon	NA	①	②	0.0024	7.87	7.93	7.91	0.080	0.040
1994	Sedan	NA	①	②	0.0024	7.87	7.93	7.91	0.080	0.040
	Coupe	NA	①	②	0.0024	7.87	7.93	7.91	0.080	0.040
	Wagon	NA	①	②	0.0024	7.87	7.93	7.91	0.080	0.040
1995	Sedan	NA	①	②	0.0024	7.87	7.93	7.91	0.080	0.040
	Coupe	NA	①	②	0.0024	7.87	7.93	7.91	0.080	0.040
	Wagon	NA	①	②	0.0024	7.87	7.93	7.91	0.080	0.040
1996-97	Sedan	NA	①	②	0.0024	7.87	7.93	7.91	0.080	0.040
	Coupe	NA	①	②	0.0024	7.87	7.93	7.91	0.080	0.040
	Wagon	NA	①	②	0.0024	7.87	7.93	7.91	0.080	0.040

NA - Not Available
① Front: 0.710
  Rear: 0.430
② Front: 0.633
  Rear: 0.370

FREQUENT MAINTENANCE LABOR
SATURN

The following should be used as a guide when determining the amount of work required for a particular service if taken to a repair shop. In estimating how long a particular Frequent Maintenance Service item should take, please observe the following:
- **Factory Time** is time that is generated by the vehicle manufacturer.
- **Chilton Time** is time that is based on field research and data supplied by the vehicle manufacturer.
- All labor time operations are given in hours and tenths of an hour.
- All labor operations, are to be used as a **guide**.

COOLING

(G) Winterize Cooling System
Includes: Run engine to check for leaks, tighten all hose connections. Test radiator and pressure cap, drain radiator and engine block. Add antifreeze and refill system.
1993-97 .5

(G) Belt, Serpentine Drive, Renew
1993-94 (.2) 3
1995-97 (.3) 4

(G) Hoses, Radiator, Renew
Includes: Drain and refill cooling system as required.
1993-97
  upper (.5) 6
  lower (.6) 7
  both (.7) 1.0

(G) Thermostat, Coolant, Renew
1993-97 (.6) 8

FUEL

(M) Air Cleaner, Service
1993-97 2

BRAKES

(G) Bleed Brakes (Four Wheels)
Includes: Add fluid.
1993-97
  wo/ABS (.5) 6
  w/ABS (.7) 1.0

(G) Brakes, Adjust (Minor)
Includes: Adjust brakes, fill master cylinder.
1993-97, two wheels 4

(M) Parking Brake, Adjust
1993-94 (.6) 7
1995-97 (.3) 4

LUBRICATION SERVICES

(M) Engine Oil & Filter, Renew
Includes: Inspect and correct all fluid levels.
1993-97 (.4) 4

(M) Lubricate Chassis, Change Oil & Filter
Includes: Inspect and correct all fluid levels.
1993-97 6
Install grease fittings add 1

(M) Lubricate Chassis
Includes: Inspect and correct all fluid levels.
1993-97 4
Install grease fittings add 1

WHEELS

(G) Wheel, Renew (One)
1993-97 5

(G) Wheel, Balance
1993-97
  one .3
  each adtnl. 2

(G) Wheels, Rotate (All)
1993-97 5

ELECTRICAL

(G) Headlamps, Aim
1993-97
  two .4
  four .6

(G) High Mount Stop Lamp Bulb, Renew
1993-97
  Coupe, Sedan (.2) 3
  Wagon (.3) 4

(G) License Lamp Assy., Renew
1993-97
  one or both
    Coupe, Sedan (.2) 3
    Wagon (.5) 7

(G) License Lamp Bulb, Renew
1993-97
  one or both
    Coupe, Sedan (.2) 3
    Wagon (.3) 5

(G) Park & Turn Signal Lamp Assy., Renew
1993-97, one (.7) 9

SCHEDULED MAINTENANCE INTERVALS
(SATURN SC, SC1, SC2, SL, SL1, SL2, SW1, & SW2)

TO BE SERVICED	TYPE OF SERVICE	VEHICLE MILEAGE INTERVAL (x1000)												
		3	6	9	12	15	18	21	24	27	30	33	36	39
Engine oil & filter	R		✓		✓		✓		✓		✓		✓	
Lubricate chassis, suspension, steering linkage, transmission shift linkage, parking brake cable guides, underbody contact points & linkage	S/I		✓		✓		✓		✓		✓		✓	
Drive shaft boots, suspension bushings & ball joint seals	S/I		✓		✓		✓		✓		✓		✓	
Exhaust system & throttle linkage	S/I		✓		✓		✓		✓		✓		✓	
Rotate tires	S/I		✓				✓				✓			
Brake hoses & brake lining	S/I		✓				✓				✓			
Accessory drive belt(s)	S/I						✓						✓	
Engine coolant level, hoses & clamps	S/I						✓						✓	
Air filter element	R										✓			
Engine coolant	R												✓	
Manual transaxle oil	R		✓											
Spark plugs①	R										✓			
Automatic transaxle fluid & filter	R										✓			
Ignition cables & fuel systems	S/I										✓			
Vacuum line/hose	S/I										✓			
Fuel filter②	R													

① Platinum tip spark plugs - replace every 100,000 miles.
② Replace every 60,000 miles.
R – Replace   S/I – Service or Inspect

FREQUENT OPERATION MAINTENANCE (SEVERE SERVICE)
If a vehicle is operated under any of the following conditions it is considered severe service:
- Extremely dusty areas.
- 50% or more of the vehicle operation is in 32°C (90°F) or higher temperatures, or constant operation in temperatures below 0°C (32°F).
- Prolonged idling (vehicle operation in stop and go traffic).
- Frequent short running periods (engine does not warm to normal operating temperatures).
- Police, taxi, delivery usage or trailer towing usage.
Engine oil & oil filter - change every 3000 miles.

TECHNICAL SERVICE BULLETINS 33

TECHNICAL SERVICE BULLETINS

What is a TSB?

All vehicle manufacturers experience occasional problems with one or more of their model lines, requiring that fixes be made after the vehicle is sold to the customer. Manufacturers therefore issue Technical Service Bulletins (TSBs) to inform and to suggest certain repairs or component replacements. These fixes may cover a variety of issues including: safety, general maintenance, part replacement, engine driveability improvements or general repairs. If the item at issue is a noted safety related problem, it is likely that the manufacturer will also issue SAFETY RECALL CAMPAIGN notices.

NOTE: The major difference between a TSB and a Recall Campaign is that the manufacturer wants the repairs performed to ALL vehicles affected by a recall, in order to prevent a problem (often safety related) from occurring. TSBs, on the other hand, are issued to help the dealership service facility cope with a problem (usually non-safety related) that may occur to SOME vehicles. The repair or change may not be necessary if the component or system in question never develops a problem.

The TSBs and Recall Campaigns notices are sent directly to the dealership repair facility. Safety Recall Campaign notices are also sent to the vehicle owners, but in case of a sale, transfer of title or other circumstances, the owner may not receive the actual notice.

All of these factory notifications provide a description of the problem, the vehicles which are affected by the problem, how the problem is fixed and whether or not the problem is warranty-related.

Federal law also requires that the general public have access to TSBs and Recalls. You can obtain copies of any authorized bulletins from various sources including the manufacturer service information groups or distributors, federal or state government agencies dealing in transportation or publication, electronic based professional information systems such as Chilton On Disc, or even a cooperative dealer service department.

We have provided the following examples of bulletins so that you can see the kind of information that a TSB might provide. You will also note that each bulletin is numbered, providing a valuable way to access this information. See the index for the list of bulletins applying to your vehicle which was current at time of publication.

Using the TSB Index

The TSB index in this manual is divided into groups of charts covering each manufacturer. Each group contains separate charts for the various vehicle sub-systems or service categories.

The charts we have provided contain 4 columns. The first column, **MODELS**, provides you with a coded listing of what actual vehicle nameplates from that manufacturer are affected by the bulletin. The model listing is a numerical code, explained in the footer (the bottom) of each chart.

The second column, **YEAR**, lists the model year(s) of those nameplates which are affected. For example, 92–94 would include all 1992, 1993 and 1994 models carrying the nameplate listed in the first column. Similarly, 94–94 would only affect vehicles built for the 1994 model year.

The third column, **TSB#**, was the last revised (if more than one was published) part number or code to retrieve that particular service bulletin.

Finally, the fourth column, **DESCRIPTION**, provides a brief idea of what the bulletin is about

(new components, procedures, specifications, or possibly just dealer network service policy revisions).

To determine if there are any bulletins for your vehicle within a certain category:

1. Locate the group containing charts for your vehicle's manufacturer.

2. Next, turn to the page(s) covering the specific system you are curious about (such as Heating, Air Conditioning, Ventilation, Defogger or Lighting, Horns, Turn Signals, Steering Column).

3. Once you have reached the chart for the category, the next step is to determine if any bulletins apply to the particular model on which you are working. Determine what number represents your model using the footer at the bottom of the chart. Scan the **MODELS** column for that number.

4. Whatever matches you find are service bulletins that *could* apply to your vehicle. If the **YEAR** and **DESCRIPTION** columns also match your model and problem, then you can use the **TSB#** to help attain the bulletin in question.

5. If you do not find a number match, check the chart footer again to see if any codes such as 1=All or 2=Most are used for that manufacturer. Sometimes components or procedures are shared by all or the majority of a manufacturer's model lines. In this case, a bulletin may apply to your model as well, but only if the **YEAR** and **DESCRIPTIONS** still match.

NOTE: Keep in mind that even if your model and year match a bulletin, it does not necessarily mean that the repair is required for your vehicle. A TSB (not a Safety Recall) repair should be performed ONLY if the problem (not just the symptom) exists on your vehicle. Remember that a TSB lists the probable cause of a symptom, but it is often not the only possible cause.

Service Bulletin

File In Section: 10 - Body

Bulletin No.: 63-16-07

Date: June, 1996

Subject: Door Trim Panel Insert (Cloth/Vinyl) Separates from Door Trim Panel
(Re-attach with Adhesive)

Model: 1996 Pontiac Grand Am

Condition

Some owners may comment that the door trim panel insert is separating from the door trim panel.

Cause

When vehicles are subject to heat, inserts may separate from door trim panel.

Correction

Re-attach insert to door trim panel with hot melt adhesive.

Warranty Information

For vehicles repaired warranty, use:

Labor Operation	Labor Time
C3348/C3358	0.2 hr

**WE SUPPORT
VOLUNTARY TECHNICIAN
CERTIFICATION**

7920tsb1

A typical GM TSB for interior trim pieces

CHRYSLER CORPORATION

Lighting, Horns, Turn Signals, Steering Column

Models	Year	TSB#	Description
10	95 - 95	19-03-94	CHATTER/NOISE FROM STEERING COLUMN TILT LEVER
10	95 - 95	19-05-94	CLICKING NOISE IN STEERING COLUMN - SHIM
11	94 - 94	08-40-93	DAYTIME RUNNING LAMP - WIRING DIAGRAM - S/M REV
10	95 - 95	23-26-94	ENGINE NOISE - BASE OF STEERING COLUMN
7	93 - 94	08-38-93	HEADLAMP - AMOUNT OF CONDENSATION INFO/REPAIR PROC
5	92 - 93	8/25/92	HEADLAMP - NEW LIGHT PATTERN - IMPROVES PROJECTION
7	93 - 93	8/8/93	HEADLAMP AIMING PROCEDURE
7	93 - 93	8/8/93	HEADLAMP AIMING PROCEDURE - REVISION
12	89 - 94	08-64-93	HEADLAMP CONDENSATION - INST1 VENTS
10	95 - 95	8/25/94	HEADLAMP FLICKER WITH HEAVY ELECTRICAL LOAD
7	93 - 94	08-38-94	HEADLAMP PATTERN IMPROVEMENT - REV ASSEMBLY - REV
7	93 - 94	08-38-94	HEADLAMP PATTERN IMPROVEMENT - REVISED ASSEMBLY
14	95 - 95	8/27/95	HORN ERRATIC - DRIVER AIR BAG NOT CENTERED
10	95 - 95	21-12-95	IGN KEY HARD TO REMOVE - INTERLOCK CABLE ADJUSTMENT
14	95 - 95	8/3/95	KEY HALO LAMP & DOME LAMP OPERATION - INFO
10	95 - 95	08-49-94	LAMP - IGNITION KEY - NO ILLUMINATION
7	93 - 95	08-59-94	MOISTURE IN PARK/TURN LAMP
7	93 - 94	8/22/94	REAR READING LAMP SERVICE CAUTIONS
14	95 - 95	8/11/95	RT TURN SIGNAL DOES NOT CANCEL - INTERMITTENT - REV
14	95 - 95	19-01-95	SCRAPE NOISE IN STEERING COLUMN - SPACER
3,14	96 - 96	18-40-95	SPEED CONTROL CUTOUT/OBD II PROTOCOL - PCM

Model Legend: 1= All 2= Most 3= Avenger/Sebring(Coupe) 4= Colt 5= Daytona 6= Dynasty 7= Intrepid/Vision/LHS/Concorde/New Yorker 8= Lancer 9= Monaco 10=Neon 11= ShadowSundance 12= Spirit/LeBaron 13= Stealth 14= Stratus/Cirrus 15= Viper 16= Sebring Convertible 17= Breeze 18= Laser/Talon

Instruments, Dash Cluster, Warning Lights, Mirrors

Models	Year	TSB#	Description
7	93 - 93	5/1/93	ABS WARN LAMP INTERMITTENT ON - DIAGNOSIS/REPAIR
5	93 - 93	8/31/92	COMPASS CALIBRATION - VEHICLES W/OVERHEAD CONSOLE
2	89 - 96	8/16/95	CRUISE CNTRL OVER/UNDERSHOOT DURING INITIAL SET -REV
2	89 - 96	8/16/95	CRUISE CONTROL OVER/UNDERSHOOT DURING INITIAL SET
7	95 - 95	14-07-95	ERRATIC FUEL GAUGE READINGS - REV PARTS
12,11	94 - 95	08-39-94	ERRATIC/INOP SPEEDOMETER DIAGNOSIS
18	95 - 96	14-04-96	FUEL GAUGE INACCURATE - IMPROPER INSTALLATION
5,11	92 - 93	8/18/93	GAUGE/FUEL - DOES NOT READ FULL AFTER FILL-UP/REPAIR
7	93 - 93	8/17/92	GAUGE/FUEL LEVEL - OPERATION EXPLAINED
2	89 - 93	8/15/93	GAUGE/TEMP - FLUCTUATES DURING COLD WEATHER - INFO
14	95 - 95	8/19/95	GAUGES GO TO ZERO OR FREEZE INTERMITTENTLY
14	95 - 95	08-71-94	HEATED MIRRORS & BACKLIGHT LED INOPERATIVE
4	93 - 93	23-17-93	I/P - SUNGLASS POCKET R&I - REVISED PROCEDURE
10	95 - 95	23-43-94	I/P CARE & MAINTENANCE - INFORMATION
12	92 - 93	23-02-93	I/P TO DOOR TRIM PANEL SQUEAK - DIAGNOSIS/REPAIR
2	93 - 95	8/15/95	INTERACTIVE CRUISE CONTROL OPS & 41TE/42LE
2	93 - 96	08-15-95A	INTERACTIVE CRUISE CONTROL OPS & 41TE/42LE - REV
2	93 - 93	23-08-93	MIRROR - AVOIDING SCRATCHED GLASS - INFORMATION
10	95 - 95	8/4/94	ODOMETER SERVICE PROCEDURE
16	96 - 96	8/15/96	POWER MIRROR SWITCH INOP WITH IGNITION OFF
11,12	94 - 94	8/16/94	RATTLE/LOOSE SPEED CONTROL SWITCH
7	94 - 94	08-44-93	SPEED CONTROL DIAGN - NO CODES - NEW SRVC CARTRIDGE
7	94 - 94	08-44-93	SPEED CONTROL DIAGN - NO CODES - NEW SVC CART - REV
7	96 - 96	18-32-95	SPEED CONTROL DIAGNOSTICS/IDLE QUALITY - PCM
2	92 - 93	08-28-92A	SPEED CONTROL SYS - INTERMIT DISENGAGE AT HWY SPEEDS
2	92 - 93	8/28/92	SPEED CONTROL SYS - INTERMIT DISENGAGE AT HWY SPEEDS
10	94 - 95	08-77-94	SPEEDO INOP - NO VSS SIGNAL
8	89 - 93	HL-58-92	SPEEDOMETER - CHART FOR PROPER PINION SELECTION
12,6,5,11	88 - 93	HL-58-92	SPEEDOMETER - CHART FOR PROPER PINION SELECTION
13	91 - 93	21-12-93	SPEEDOMETER READS HIGHER THAN ACTUAL SPEED - REPAIR
7	94 - 94	21-35-93	TACH OR SPEEDOMETER ERRATIC OR A/T LIMP-IN MODE
12	92 - 95	8/12/95	TRAVELER TRIP COMPUTER - ECO FUNCTION - INFO
10	95 - 95	8/15/94	TURN SIGNAL INDICATORS LIT ON IC - HAZARD BUTTON

Model Legend: 1= All 2= Most 3= Avenger/Sebring(Coupe) 4= Colt 5= Daytona 6= Dynasty 7= Intrepid/Vision/LHS/Concorde/New Yorker 8= Lancer 9= Monaco 10=Neon 11= ShadowSundance 12= Spirit/LeBaron 13= Stealth 14= Stratus/Cirrus 15= Viper 16= Sebring Convertible 17= Breeze 18= Laser/Talon

33 TECHNICAL SERVICE BULLETINS

Chassis Electrical, Wiring Harness, Fuses-Circuit Breakers, Wipers, Window Motors

Models	Year	TSB#	Description
12	95 - 95	08-35-94	AIR BAG MODULE FUSE CHANGE - S/M REVISION
16	96 - 96	8/17/96	BODY WIRING HARNESS INSTALLATION AFTER HVAC SERVICE
7	93 - 93	8/19/92	CIRCUIT/ACCESSORY - REMAINS POWERED W/KEY OFF - PROC
10	95 - 96	8/11/96	DELAYED WASHER FLUID OUTPUT
7	93 - 93	8/10/93	ELECTRICAL FAILURES - LOSS OF GROUND CIRCUIT - PROC
14	95 - 96	8/8/96	ERRATIC WIPER DELAY INTERVALS - BCM RECALIB
3	95 - 95	8/26/95	FAULT LITE ILLUMINATION OR NO START
10	94 - 95	8/28/94	FUEL PUMP/SENDING UNIT MODULE CONNECTOR SERVICE
7	93 - 93	8/22/92	FUSE BLOCK - SHIPPING MODULE ELIMINATED - INFO
6	90 - 93	8/26/93	FUSE BLOCK/RELAY BANK/RELAY MODULE LOCATIONS & ID
7	93 - 93	8/16/92	FUSES SHIPPED REMOVED FROM IP/PREVENT BATTERY DRAIN
14	95 - 95	08-37-94	HEADLINING WIRING SERVICE
7	96 - 96	8/12/96	HIGH SPEED WINDSHIELD WASHER SPRAY KNOCK-DOWN
14	95 - 95	8/17/95	INTERMIT AIR BAG LITE - POWER ACCESSORIES INOP
2	92 - 93	18-04-92	INTERMIT ELEC CONNECTION - POOR PERFORMANCE/NO START
2	92 - 93	18-04-92	INTERMIT ELEC CONNECTION - POOR PERFORMANCE/NO START
3	95 - 95	8/31/95	INTERMIT MIL ON/HARD-NO START/INOP CONDENS FAN -2.0L
10	96 - 96	8/2/96	INTERMIT OPS OF CKTS IN ENGINE WIRE HARNESS
10	95 - 95	8/30/95	INTERMIT SRS LITE/INOP ELECT COMPONENTS AFTER START
10	95 - 95	8/24/94	INTERMITTENT WIPER OPERATION - SWITCH NOT ON - RFI
7	93 - 94	08-34-94	REVISED WINDSHIELD WIPER ARMS
2	94 - 94	08-67-93	STOP LIGHT SWITCH WIRING CONNECTOR SERVICE
12	92 - 95	8/9/95	SWITCHES INTERMITTENT OR INOP - WINDOW OR CONV TOP
18	95 - 95	8/7/95	WASHER FLUID LEAKS FROM NOZZLES
14	95 - 95	08-72-94	WINDSHIELD WASHER NOZZLE AIMING - NO ADJUSTMENT
7	93 - 95	8/20/95	WINDSHIELD WASHER NOZZLE FREEZES UP
7	93 - 95	8/20/95	WINDSHIELD WASHER NOZZLE FREEZES UP - REV
14	95 - 95	8/4/95	WINDSHIELD WASHER PERFORMANCE - POOR VOLUME
7	93 - 93	08-48-93	WINDSHIELD WASHER SPRAY PATTERN TOO HIGH - NEW PART
12,6,5,11	89 - 93	23-26-92	WINDSHIELD WASHER SYSTEM - NOZZLE FREEZE UP - SERV
14	95 - 95	08-32-95	WINDSHIELD WIPER CHATTER
14	95 - 95	8/11/95	WINDSHIELD WIPER INTERMIT DELAY INCONSISTENT - REV
14	95 - 95	8/11/95	WINDSHIELD WIPER INTERMITTENT DELAY INCONSISTENT
7	93 - 94	08-49-93	WIPER BLADE FREEZE/CLOG - NEW WINTER BLADES
7	93 - 93	8/11/93	WIPER MOTOR - IMPROVE DRAINAGE SERV INFO/REPAIR PROC
7	94 - 94	8/30/93	WIRE HARNESS - EXTRA SET OF FOG LAMP CONNECTORS/PROC
10	95 - 95	08-32-94	WIRING DIAGRAMS & CONNECTORS - S/M REVISION
12	92 - 93	8/11/92	WIRING DIAGRAMS - SERVICE MANUAL REVISIONS
7	93 - 93	8/20/92	WIRING DIAGRAMS - SERVICE MANUAL REVISIONS

Model Legend: 1= All 2= Most 3= Avenger/Sebring(Coupe) 4= Colt 5= Daytona 6= Dynasty 7= Intrepid/Vision/LHS/Concorde/New Yorker 8= Lancer 9= Monaco 10=Neon 11= ShadowSundance 12= Spirit/LeBaron 13= Stealth 14= Stratus/Cirrus 15= Viper 16= Sebring Convertible 17= Breeze 18= Laser/Talon

Auxiliary Equipment, Jacks, Trailer Hitches, Towing

Models	Year	TSB#	Description
2	89 - 95	8/31/94	INST1ATION OF RADIO TRANSMITTING EQUIPMENT
2	89 - 95	8/31/94	INST1ATION OF RADIO TRANSMITTING EQUIPMENT - REV
14	95 - 95	26-02-94	JUMP STARTING - HOISTING - TOWING

Model Legend: 1= All 2= Most 3= Avenger/Sebring(Coupe) 4= Colt 5= Daytona 6= Dynasty 7= Intrepid/Vision/LHS/Concorde/New Yorker 8= Lancer 9= Monaco 10=Neon 11= ShadowSundance 12= Spirit/LeBaron 13= Stealth 14= Stratus/Cirrus 15= Viper 16= Sebring Convertible 17= Breeze 18= Laser/Talon

Heating, Air Conditioning, Ventilation, Defogger

Models	Year	TSB#	Description
7	93 - 93	24-20-92	A/C (MANUAL) - COMP/LIGHT ON DURING ENG START - INFO
7	93 - 93	24-13-92	A/C - ERRATIC OPERATION - DIAGNOSIS & REPAIR PROC
10	95 - 95	24-01-94	A/C - INST1ATION OF NON-FACTORY UNIT NOT ADVISED
7	93 - 96	24-15-95	A/C - INTERMIT OR POOR PERF - DIAGN PROCEDURE
7	93 - 93	24-07-93	A/C - INTERMITTENT OR INOPERATIVE - REPAIR PROCEDURE
12,6,5,11	91 - 93	HL-51-92	A/C - MOISTURE COMES FROM HEAT OUTLET ON BI-LEVEL
12,11	91 - 93	24-06-93	A/C - MOISTURE EXITS HEAT OUTLETS IN HIGH AMB - FIX
,6	92 - 94	24-12-92	A/C - OFFENSIVE ODOR FROM DUCTS - REPAIR REVISED
14	95 - 95	24-09-95	A/C - POOR PERFORMANCE
2	88 - 94	24-01-95	A/C - R-12 TO R-134A ADAPTATION SERVICE PROCED - REV
2	88 - 94	24-01-95	A/C - R-12 TO R-134A ADAPTATION SERVICE PROCEDURE
7	93 - 93	24-16-92	A/C - SUDDEN STOP OF COOL AIR PRODUCTION - SERVICE
14	95 - 95	24-05-95	A/C - WHITE FLAKES FROM EVAPORATOR
2	92 - 95	24-08-95	A/C - WHITE FLAKES FROM IP AND DEFROST OUTLETS
7	93 - 93	24-01-93	A/C - WIND NOISE IN RECIRCULATION MODE - REPAIR PROC
12,6,5,11	91 - 93	24-17-92	A/C AIRFLOW/COOLING LOSS W/BLOWER FAN STILL OP- INFO
2	91 - 95	24-17-92	A/C AIRFLOW/COOLING LOSS W/BLOWER FAN STILL OP- REV
7	93 - 93	24-19-92	A/C COMP DISCHARGE LINE CONTACTS EXH MANIFOLD SHIELD
7	94 - 94	08-44-93	A/C DIAGN - NO CODES - NEW SERVICE CARTRIDGE - REV
7	94 - 94	08-44-93	A/C DIAGNOSTICS - NO CODES - NEW SERVICE CARTRIDGE
10	95 - 95	24-07-95	A/C EVAPORATOR HIGH PITCHED WHISTLE
10	95 - 95	24-07-95	A/C EVAPORATOR HIGH PITCHED WHISTLE - REV
7	93 - 95	24-14-94	A/C INOP - TOTAL OR INTERMITTENT
14	95 - 96	24-07-96	A/C POOR PERF OR COMPRESSOR FAILURE ABOVE 90 DEG F
12,11	94 - 95	24-15-94	A/C POOR PERFORMANCE
10	95 - 95	24-20-94	A/C POOR PERFORMANCE - FREEZE UP - S/M REVISED
12,11	94 - 95	24-15-94	A/C POOR PERFORMANCE - REVISED
7	93 - 94	24-29-93	A/C REFRIGERANT LEAK - DIAGNOSIS & REPAIR
7	93 - 93	24-18-93	A/C- 2100 RPM COMPRESSOR MOAN/LOW WHISTLE - SERV/REV
7	93 - 93	24-18-93	A/C- 2100 RPM COMPRESSOR MOAN/LOW WHISTLE- SERV INFO
6	93 - 93	24-18-92	ATC - NEW FRESH AIR DOOR - INTERIOR WINDOW FOGS
7	93 - 93	8/15/92	ATC - RECIRCULATION DOOR ST1 FAILURE - SERV PROC
7	94 - 94	24-17-93A	ATC - TEMP SHIFTS/BLOWER MOTOR SPEED - BCM - REVISED
7	93 - 93	24-14-92	ATC - WARM AIR FROM VENTS - OPERATION & CORRECTION

Model Legend: 1= All 2= Most 3= Avenger/Sebring(Coupe) 4= Colt 5= Daytona 6= Dynasty 7= Intrepid/Vision/LHS/Concorde/New Yorker 8= Lancer 9= Monaco 10=Neon 11= ShadowSundance 12= Spirit/LeBaron 13= Stealth 14= Stratus/Cirrus 15= Viper 16= Sebring Convertible 17= Breeze 18= Laser/Talon

33 TECHNICAL SERVICE BULLETINS

Heating, Air Conditioning, Ventilation, Defogger

Models	Year	TSB#	Description
7	94 - 94	24-07-94	ATC CALIBRATION - REV BCM - BLOWER SPEED/TEMP - REV
7	94 - 94	24-17-93B	ATC CALIBRATION - REVISED BCM - BLOWER SPEED/TEMP
7	94 - 94	24-26-93	ATC CALIBRATION - TEMP SHIFTS/BLOWER MOTOR SPEED
7	93 - 93	24-09-93	ATC LACK OF COMFORT/SUNNY WEATHER - SUN SENSOR INFO
7	94 - 94	08-43-93	ATC TEST 17A REVISED - BODY DIAGNOSTIC MANUAL REV
7	93 - 93	24-17-93	ATC- CALIB AFTER POWER DISCONNECT AT BODY CONTROLLER
7	93 - 94	24-17-93	ATC- CALIB AFTER POWER DISCONNECT AT BODY CONTROLLER
7	93 - 95	24-04-94	BLOWER MOTOR INOP - WATER INGESTED INTO HOUSING -REV
7	93 - 94	24-04-94	BLOWER MOTOR INOP - WATER INGESTION INTO HOUSING
7	93 - 93	24-10-93	CLIMATE CONTROL - WHISTLE WITH ENGINE OFF - REPAIR
7	93 - 93	24-03-93	DEFROST TIME REDUCED THROUGH MODIFICATION/FOAM TAPE
7	93 - 93	24-03-93	DEFROST TIME REDUCED THROUGH MODIFICATION/FOAM TAPE
10	95 - 95	24-04-95	DEFROSTER POOR PERFORMANCE - IMPROVEMENT
7	93 - 94	24-22-93	DISTRIBUTION DUCT - DAMAGED PUSH PIN FASTENERS
7	93 - 93	24-19-93	HEATER/AC HOUSING WATER INGESTION/LEAK ON PASS SIDE
7	93 - 93	24-15-92	HVAC - RIGHT/LEFT TEMP DIFFERENTIAL FROM I/P OUTLETS
14	95 - 95	24-12-95	HVAC SYSTEM STUCK IN DEFROST MODE
10	95 - 95	24-16-94	LEAK - UNDER INSTRUMENT PANEL FROM A/C
10	95 - 95	24-02-95	LEAVES/DEBRIS IN HEATER-A/C SYSTEM
7	93 - 93	24-24-93	MANUAL A/C INOPERATIVE - RESET BCM
2	93 - 95	24-08-94	ODOR - A/C EVAPORATOR
2	92 - 95	24-06-95	ODOR - A/C EVAPORATOR - REV TWO
2	92 - 95	24-08-94	ODOR - A/C EVAPORATOR - REVISED
2	93 - 94	24-20-93	R-134A A/C - REFRIGERANT LEAK DETECTION UNIT INFO
7	94 - 94	24-10-94	RAPID CYCLING OF ATC CONTROL HEAD LED'S
7	93 - 95	24-11-95	WINDSHIELD DEFROST IMPROVEMENT

Model Legend: 1= All 2= Most 3= Avenger/Sebring(Coupe) 4= Colt 5= Daytona 6= Dynasty 7= Intrepid/Vision/LHS/Concorde/New Yorker 8= Lancer 9= Monaco 10=Neon 11= ShadowSundance 12= Spirit/LeBaron 13= Stealth 14= Stratus/Cirrus 15= Viper 16= Sebring Convertible 17= Breeze 18= Laser/Talon

Entertainment Devices, Stereo, Radio, Etc.

Models	Year	TSB#	Description
7	93 - 93	8/4/93	AM RADIO - IGNITION NOISE - REPAIR PROCEDURE
10	95 - 95	08-70-94	AM RADIO STATIC/POOR RECEPTION
14	95 - 95	8/1/95	BUZZ NOISE - FRONT DOOR SPEAKER TO WATER SHIELD
11	94 - 94	08-55-93	CLARION RADIO SUBSTITUTED FOR HUNTSVILLE - TEMPORARY
7	93 - 94	08-44-94	DISTANT AM RADIO STATIC
13	91 - 94	08-37-95	POWER ANTENNA MAST REPLACEMENT
14,10	95 - 95	CSN#612	PREMIUM RADIO DISPLAY GOES BLANK
12,6,5,11	93 - 93	8/29/93	RADIO - POOR RECEPTION AT FRINGE OF BROADCAST AREA
7	93 - 93	8/20/93	RADIO BUZZ/INTERFERENCE DURING INT LAMP FADE TO OFF
10,14	95 - 95	8/8/95	RADIO/CLOCK DISPLAY GOES BLANK - PREMIUM CASSETTE
2	94 - 94	08-45-93	STEREO RADIO LOCK UP (INOP) - PROCEDURES

Model Legend: 1= All 2= Most 3= Avenger/Sebring(Coupe) 4= Colt 5= Daytona 6= Dynasty 7= Intrepid/Vision/LHS/Concorde/New Yorker 8= Lancer 9= Monaco 10=Neon 11= ShadowSundance 12= Spirit/LeBaron 13= Stealth 14= Stratus/Cirrus 15= Viper 16= Sebring Convertible 17= Breeze 18= Laser/Talon

Seats, Belts, Interior Trim, Carpets, Air Bags

Models	Year	TSB#	Description
10	96 - 96	8/4/96	AIR BAG REPLACEMENT - CONTROL MODULE REQUIREMENTS
7	93 - 93	HL-29-92	AIRBAG DEPLOYMENT - FIELD STUDY
16	96 - 96	23-16-96	ARMREST SWITCH BEZEL LOOSE - REMOVE PROCEDURE
7	93 - 94	23-19-94	ASH RECEIVER CUP RATTLE
7	93 - 94	23-18-94	B PILLAR RATTLE NEAR SHOULDER HARNESS ADJUSTER
7	93 - 93	23-65-93	C-PILLAR APPLIQUE CHANGE
10	95 - 95	23-22-94	CARPET PILE SEPARATES FROM BACKING
10	95 - 95	23-41-95	CARPET PILE SEPARATES FROM BACKING
12	94 - 94	23-55-94	FRONT PASSENGER SEAT BACK RATTLE - CONVERTIBLE
10	95 - 95	23-12-94	FRONT SEAT ADJUSTER RATTLE
10	95 - 95	23-12-94	FRONT SEAT ADJUSTER RATTLE - REVISED
10	95 - 95	23-12-94A	FRONT SEAT ADJUSTER RATTLE - REVISED
12	94 - 94	23-09-94	FRONT SEAT BELT RETRACTION/EXTENSION - CONVERTIBLE
7	93 - 93	23-51-93	HEAD RESTRAINT - LARGE SIZE AVAILABLE FOR EARLY MDLS
3	95 - 95	23-89-94	HEADLINER SAG REPAIR PROCEDURE
3	95 - 95	23-89-94	HEADLINER SAG REPAIR PROCEDURE - REV
10	95 - 95	23-83-94	IP TOP COVER GLARE/FIT - GLOVE BOX/ASH TRAY FIT
10	95 - 95	23-83-94	IP TOP COVER GLARE/FIT - GLOVE BOX/ASH TRAY FIT -REV
7	93 - 94	23-53-93	KNEE PANEL/DOOR PANEL INTERFERENCE
7	93 - 93	23-36-93	MANUAL SEAT ADJUSTER CLICK FROM TRACK ON ACCEL/DECEL
10	95 - 95	23-87-94	NOISE - FRONT SEAT BACK
12	94 - 94	23-01-94	NOISE/CREAK/SQUEAK FROM DOOR PANEL
7	93 - 94	23-05-94	NOISE/RATTLE/POOR FIT - C-POST APPLIQUE - PROCEDURE
3	95 - 96	23-05-96	NUMBER READ THROUGH ON LEATHER FRONT SEATS
3	95 - 96	23-23-96	POWER SEAT SWITCH STICKS
12	93 - 93	23-54-93	QUARTER TRIM BOLSTER SEPARATION - CONVERTIBLE
13	91 - 94	23-21-94	RATTLE NOISE - REAR CARGO COVER - REVISED CLIPS
18	95 - 95	23-13-95	RATTLE OR HIGH EFFORT - SUNSHADE
7	93 - 94	8/22/94	REAR ASSIST HANDLE W/READING LAMP - SERVICE CAUTIONS
10	95 - 96	23-67-95	SEAT BACK RELEASE KNOB COMES LOOSE
7	93 - 93	23-29-92	SEAT BELT - RUSTY BRACKET - REPAIR PROCEDURE
10	95 - 95	23-81-94	SEAT BELT BUCKLE REPLACEMENT
10	95 - 95	23-81-94A	SEAT BELT BUCKLE REPLACEMENT - REV
12	92 - 93	23-27-92	SEAT BELT/FRONT - WEBBING STAIN FIX (CONVERTIBLE)
11	Up - 93	SUPP: SR#545	SHOULDER BELTS/MOTORIZED - INOP SYSTEM - REC1
7	94 - 94	8/18/94	SIR - HYBRID PASSENGER AIRAG SYSTEM EXPLAINED
4,13	94 - 95	8/13/95	SRS - ECU CONNECTOR LOCK CHANGE
13	91 - 94	23-25-94	SRS TROUBLESHOOTING - DTC'S
18	95 - 95	23-79-94	STRESS MARK - INTERIOR REAR LOWER QUARTER TRIM

Model Legend: 1= All 2= Most 3= Avenger/Sebring(Coupe) 4= Colt 5= Daytona 6= Dynasty 7= Intrepid/Vision/LHS/Concorde/New Yorker 8= Lancer 9= Monaco 10=Neon 11= ShadowSundance 12= Spirit/LeBaron 13= Stealth 14= Stratus/Cirrus 15= Viper 16= Sebring Convertible 17= Breeze 18= Laser/Talon

Glass, Doors, Hood, Decklid, Tailgate, Liftgate, Locks

Models	Year	TSB#	Description
14	95 - 95	23-100-94	B-PILLAR APPLIQUE COMES LOOSE - REV
14	95 - 95	23-32-95	B-PILLAR APPLIQUE COMES LOOSE - REV 2
14	95 - 95	23-32-95A	B-PILLAR APPLIQUE COMES LOOSE - REV 3
12	94 - 94	23-38-94A	BACKLITE GLASS BREAKS WHEN FOLDING CONVERT TOP - REV
12	94 - 94	23-38-94	BACKLITE GLASS BREAKS WHEN FOLDING CONVERTIBLE TOP
7	93 - 94	23-15-94	BACKLITE RATTLE - RESET PROCEDURE/PARTS
14	95 - 95	23-42-95	CHILD PROTECTION LOCK WILL NOT ENGAGE
12	90 - 94	23-63-93	CONVERTIBLE BOOT LATCH RATTLE
12	94 - 95	23-84-94	CONVERTIBLE TOP BOOT SERVICE
14	95 - 95	23-39-95	DECK LID DIFFICULT TO LIFT
14	95 - 95	23-91-94	DECK LID LATCH SYSTEM SERVICE
10	95 - 95	23-94-94	DECK LID RATTLE - TINNY CLOSING SOUND - POOR FIT
10	95 - 95	23-25-95	DECK LID SPOILER RATTLE AND/OR WATER LEAK INTO TRUNK
7	93 - 94	23-35-96	DECK LID TORSION BAR TENSION
7	93 - 93	23-38-92	DECKLID - TORSION BARS DISENGAGE FROM RETAIN BRACKET
7	93 - 93	23-11-93	DECKLID - WATER/DUST BETWEEN SEAL & INNER PANEL/FIX
7	93 - 93	23-11-93	DECKLID - WATER/DUST BETWEEN SEAL & INNER PANEL/REV
14	95 - 96	23-06-96	DECKLID REMOTE RELEASE/VALET LOCKOUT SERVICE
3	96 - 96	23-24-96	DOOR BUZZ/RATTLE
3	96 - 96	23-24-96	DOOR BUZZ/RATTLE - REV
7	93 - 93	23-25-92	DOOR CHECK STRAP - BUSHING AND CLIP SERVICE
7	93 - 93	23-25-92	DOOR CHECK STRAP - BUSHING AND CLIP SERVICE - REV
7	93 - 93	23-14-93	DOOR CHECK STRAP - BUSHING AND CLIP SERVICE - REV
18	90 - 93	23-22-93	DOOR GLASS - HARD TO ROLL UP/COMES OUT OF RUNCHANNEL
3	95 - 96	23-14-96	DOOR GLASS ADJUST PROCEDURES
14	95 - 95	23-08-95	DOOR GLASS BREAKAGE - GLASS RUN BURRS
14	95 - 95	23-08-95	DOOR GLASS BREAKAGE - GLASS RUN BURRS - REV
14	95 - 96	23-15-96	DOOR GLASS INOP
7	93 - 95	23-11-95	DOOR GLASS LIFT PLATE SERVICE
7	93 - 93	8/13/93	DOOR LOCKS/ROLLING - OPERATION INFO - S/M REVISION
18	95 - 95	23-72-94	DOOR LOWER BODY SIDE MOLDING CONTACTS FRONT FENDER
13	91 - 93	23-55-93	DOOR WINDOW GLASS ADJUSTMENT PROCEDURE
13	94 - 94	23-75-94	DOOR WINDOW WEATHERSTRIP PULLS OUT
7	94 - 94	23-39-93	DOOR/REAR - HARD TO OPEN - REPAIR PROCEDURE
10	95 - 96	23-58-95	FRONT DOOR GLASS ADJUSTMENT PROCEDURE - 2-DR
12	94 - 95	23-16-95	FRONT WINDOW WILL NOT GO UP - MOTOR OPERATES
14	95 - 95	23-02-95	FUEL FILLER DOOR UNDER FLUSH TO QUARTER PANEL
2	94 - 95	23-68-94	GLUE OOZES OUT AT BACKLIGHT OR WINDSHIELD MOLDING
10	95 - 95	23-92-94	HIGH EFFORT - FRONT MANUAL WINDOWS
14	95 - 95	23-01-95	HOOD HARD TO CLOSE - LATCH
7	94 - 96	23-51-95	HOOD MOLDING WHISTLE NOISE
10	95 - 95	23-93-94	HOOD PROP ROD RATTLE
1	95 - 96	23-74-95	INTERIOR WINDOW FILM BUILD UP

Model Legend: 1= All 2= Most 3= Avenger/Sebring(Coupe) 4= Colt 5= Daytona 6= Dynasty 7= Intrepid/Vision/LHS/Concorde/New Yorker 8= Lancer 9= Monaco 10=Neon 11= ShadowSundance 12= Spirit/LeBaron 13= Stealth 14= Stratus/Cirrus 15= Viper 16= Sebring Convertible 17= Breeze 18= Laser/Talon

Glass, Doors, Hood, Decklid, Tailgate, Liftgate, Locks

Models	Year	TSB#	Description
1	95 - 96	23-74-95	INTERIOR WINDOW FILM BUILD UP
7	93 - 94	23-53-93	KNEE PANEL/DOOR PANEL INTERFERENCE
11	93 - 93	23-28-93	LIFT GATE - SPOILER SLOSH/GIRGLE NOISE ON OPEN/CLOSE
2	93 - 94	23-60-93	LOCK CYLINDER KEY BREAKAGE - REPAIR PROCEDURES
2	93 - 93	23-60-93	LOCK CYLINDER KEY BREAKAGE - REPAIR PROCEDURES
7	93 - 93	CSN#586	NOISE/EFFORT - DOOR CHECK STRAPS - CSN 586
7	93 - 93	23-14-93A	NOISE/EFFORT - DOOR CHECK STRAPS - REVISED
12	94 - 94	23-33-94	NOISE/FLUTTER - CONVERTIBLE TOP - FOAM TAPE
7	93 - 94	23-69-93	NOISE/GLASS RUN WEATHERSTRIP STICKS - LUBRICANT
14	95 - 95	23-82-94	RATTLE - DOOR LATCH/STRIKER AREA
10	95 - 95	23-99-94	RATTLE - FRONT FENDER INNER SPLASH SHIELD
14	95 - 95	23-40-95	REAR DOOR INTERLOCK STRIKER CORROSION
7	94 - 94	23-56-94	REAR DOORS HARD TO OPEN
3	95 - 95	23-80-94	SUNROOF CLOSING - LEVER LATCH PIN
3	95 - 96	23-22-96	SUNROOF INOP OR OPENS BY ITSELF
3	95 - 96	23-25-96	SUNROOF RATCHETING NOISE OR JERKY OPS
3	95 - 95	23-30-95	SUNROOF WILL NOT FULLY OPEN OR CLOSE
7	93 - 94	23-46-94	TRUNK WATER LEAKS
7	93 - 96	23-46-94	TRUNK WATER LEAKS - REV
7	93 - 95	23-85-94	WATER DRAINS INTO TRUNK WHEN DECK LID RAISED
14	95 - 96	23-08-96	WATER LEAK BEHIND DOOR TRIM PANEL
7	93 - 94	23-28-94	WIND NOISE - AT A PILLAR AND ABOVE DOOR OPENINGS
7	93 - 93	23-15-93	WIND NOISE - DIAGNOSIS AND REPAIR PROCEDURES
10	95 - 95	23-29-94	WIND NOISE AT B POST BELT LINE - FRONT SIDE GLASS
16	96 - 96	23-33-96	WINDNOISE/RATTLE FROM CONVERTIBLE TOP
15	93 - 95	23-20-95	WINDNOISE/WATER LEAK - SOFT TOP TO WINDSHIELD AREA
14	95 - 95	23-34-95	WINDOW BELT MOLDINGS WARP AND/OR DELAMINATE
14	95 - 95	23-34-95	WINDOW BELT MOLDINGS WARP AND/OR DELAMINATE - REV
7	93 - 94	23-18-93	WINDOW/REAR - SQUEAKING/CREAKING - REPAIR/LUBE

Model Legend: 1= All 2= Most 3= Avenger/Sebring(Coupe) 4= Colt 5= Daytona 6= Dynasty 7= Intrepid/Vision/LHS/Concorde/New Yorker 8= Lancer 9= Monaco 10=Neon 11= ShadowSundance 12= Spirit/LeBaron 13= Stealth 14= Stratus/Cirrus 15= Viper 16= Sebring Convertible 17= Breeze 18= Laser/Talon

Finishes, Body Structure, Frame, Bumpers

Models	Year	TSB#	Description
11	93 - 93	23-35-93	APPLIQUE/LOWER DOOR - LOOSE/EXCESSIVE GAP - REPAIR
14	95 - 95	23-100-94	B-PILLAR APPLIQUE COMES LOOSE
10	95 - 95	23-34-94	BACKLITE MOLDING APPEARANCE - LIFTS OFF
13	93 - 93	23-10-93	BODY - AIR DAM MOUNTING REVISION
4	93 - 93	HL-45-92	BODY - PLASTIC WRAP PROTECTION DURING SHIPPING INFO
7	93 - 93	23-35-92	BODY - ROAD NOISE REDUCTION - PARTS/PROCEDURE
7	93 - 94	23-35-92	BODY - ROAD NOISE REDUCTION - PARTS/PROCEDURE - REV
7	93 - 93	23-32-92	C-PILLAR APPLIQUE - ATTACHING SCREWS RUSTY - PROC
1	93 - 93	23-21-92	COLOR INFORMATION & VEHICLE CODE PLATE LOCATION
14	95 - 95	23-17-95	COWL COVER/SCREEN PROTECTIVE FILM RESIDUE
14	95 - 96	23-78-95	COWL/PLENUM WATER LEAKS
10	95 - 95	23-25-95	DECK LID SPOILER RATTLE AND/OR WATER LEAK INTO TRUNK
7	93 - 94	13-01-93	FASCIA CHROME STRIP - SHIPPING PROTECTOR INFO
10	95 - 95	13-01-95	FRONT & REAR FACIA APPEARANCE
3	95 - 95	23-63-95	LOOSE BODY SIDE CLADDING
7	93 - 93	23-30-92	MOULDING (DOOR BELT) - LOOSE/PULLED AWAY - REPAIR
7	94 - 95	23-42-94	PAINT - PROTECTIVE SHIPPING FILM - REVISED
7	95 - 95	23-42-94A	PAINT - PROTECTIVE SHIPPING FILM - REVISED
10	95 - 95	23-97-94	PAINT ANTI-CHIP FILM & FRONT FENDER CAP REPAIR
10	95 - 95	23-65-94	PAINT FOGGING - PROCEDURES
2	89 - 94	23-37-93	PAINT/ANTI-CHIP (STONE GUARD) - REPAIR PROCEDURES
7	93 - 93	24-19-93	PLENUM DRAIN HOLE - PROC - WATER ENTERS THRU I/P
10	95 - 95	23-24-95	PROTECTIVE COATING REMOVAL
10	95 - 95	SA#95-13	PROTECTIVE COATING REMOVAL
10	95 - 96	23-24-95	PROTECTIVE COATING REMOVAL - REV
12,6	93 - 93	23-20-93	RATTLE AT VEHICLE REAR - LICENSE PLATE INST1 INFO
10	95 - 95	23-05-95	ROOF RACK WIND NOISE
10	95 - 95	23-76-94	ROOF RAIL WEATHERSTRIP REPLACEMENT PROCEDURE
2	93 - 93	23-43-93	TAPE STRIPE AND MOULDING ADHESIVE REMOVAL/PRODUCT
17	96 - 96	23-19-96	WARPED/WAVY DECKLID APPLIQUE
10	96 - 96	23-62-95	WATER LEAK AT COWL COVER SEAM
3	95 - 95	23-30-96	WATER LEAK AT RIGHT SIDE A PILLAR
10	95 - 95	23-48-94	WATER LEAK IN PASS COMPARTMENT UNDER IP
11	93 - 93	23-34-93	WEATHERSTRIP/FRONT DOOR OPENING - WATER LEAKS - FIX
7	94 - 94	23-38-93	WEATHERSTRIP/REAR DOOR GLASS - MIS-INSTALLED - SERV
7	94 - 94	23-38-93	WEATHERSTRIP/REAR DOOR GLASS - MIS-INSTALLED - SERV
10	95 - 95	13-01-94	WHITE BUMPER FASCIA APPEARANCE
7	93 - 93	23-15-93	WIND NOISE - DIAGNOSIS AND REPAIR PROCEDURES

Model Legend: 1= All 2= Most 3= Avenger/Sebring(Coupe) 4= Colt 5= Daytona 6= Dynasty 7= Intrepid/Vision/LHS/Concorde/New Yorker 8= Lancer 9= Monaco 10=Neon
11= ShadowSundance 12= Spirit/LeBaron 13= Stealth 14= Stratus/Cirrus 15= Viper 16= Sebring Convertible 17= Breeze 18= Laser/Talon

FORD MOTOR COMPANY

Lighting, Horns, Turn Signals, Steering Column

Models	Year	TSB#	Description
8	93 - 95	95-09-04	COMBO SWITCH SLOW - RETURN AFTER LEFT TURN
5	94 - 94	94-17B-18	ELECT FLASHER - HAZARD/TURN - LOCATION - S/M REV
5	94 - 94	94-10B-24	EXTERIOR LIGHTING - SCHEMATICS & DESCRIPT - S/M REV
9	94 - 94	94-15-02	HAZARD FLASHER STICKS - DOES NOT TURN OFF
8	93 - 95	95-14-03	HEADLAMP DOORS CYCLE OPEN AND CLOSED
10	94 - 95	94-26-05	HEADLAMP REMOVAL PROCEDURE - S/M REVISION
8	93 - 93	93-19B-15	HEADLAMP RETRACTOR ELECTRICAL SCHEMATIC - S/M REV
7	95 - 96	96-08-05	HORN - INTERMITTENT OPERATION
7	91 - 93	95-05-01	IGN LOCK CYLINDER STICKS/BINDS/GRABS
12	92 - 93	93-08-05	LAMP/DOME LIGHT - STAYS ON - WIRING HARNESS FIX
3	92 - 94	94-05-07	LAMPS - REAR S-P & TAIL BURN OUT EARLY
5	93 - 93	92-24B-12	LAMPS/CORNERING - ELECTRICAL SCHEMATIC - S/M REV
6	92 - 95	94-18-06	LIGHTS (SPOT) FOR POLICE VHCLS - SERVICE PARTS
4	92 - 93	93-13B-17	LIGHTS/EXT - PHOTOCELL/AMPLIFIER ASSY R&I - S/M REV
10	94 - 94	94-24B-47	TRUNK & HIGH MOUNT S-P LAMP (COBRA) - S/M REV

Model Legend: 1= All 2= Most 3= Tempo/Topaz 4= Taurus/Sable 5= Thunderbird/Cougar 6= Crown Victoria/Grand Marquis 7= Escort/Tracer 8= Probe 9= Aspire
10= Mustang 11= Contour/Mystique 12= Festiva

Instruments, Dash Cluster, Warning Lights, Mirrors

Models	Year	TSB#	Description
6	93 - 93	93-07-01	ASHTRAY CONVERSION - CUPHOLDER ASHTRAY/INSTALL PROC
8,6,5	89 - 94	95-02-03	FUEL GAUGE - ELECTRONIC - ERRATIC OR "CO" DISPLAY
8	93 - 95	95-21-06	FUEL GAUGE INACCURATE AFTER REPL OF PUMP OR SENDER
8	95 - 96	96-04-05	FUEL GAUGE INACCURATE WHEN TANK IS FULL
6,3,5	92 - 94	94-10B-26	FUEL GAUGE PINPOINT TESTS (CONNEC-R INFO) - S/M REV
10,4	93 - 93	94-10B-26	FUEL GAUGE PINPOINT TESTS (CONNEC-R INFO) - S/M REV
8	93 - 93	93-19B-20	GAUGES TEST TG9 (TACH/OIL PR/COOLANT TEMP) - S/M REV
12	93 - 93	93-26B-25	GAUGES/WARNING LAMPS - TEST STEPS CT9/CT13 - S/M REV
7,8	91 - 94	95-17-04	GLOVE BOX - IMPROVED LOCK ASSEMBLY AVAILABLE
4	94 - 94	93-24-02	GLOVE COMPARTMENT DOOR OPENS -O QUICKLY
4	94 - 94	93-24-02	GLOVE COMPARTMENT DOOR OPENS -O QUICKLY - REVISED
9	94 - 94	94-17B-03	I/C (CONV) PINPOINT TEST STEP D2 - S/M REVISION
5	92 - 93	93-26B-13	I/C (ELECTR) PINPNT TESTS TX,TE,TF,TG,SC,SD -S/M REV
5,4	92 - 93	93-26B-12	I/C (ELECTRONIC) PINPOINT TEST FC - S/M REVISION
6	92 - 93	93-26B-21	I/C (ELECTRONIC) PINPOINT TEST FC2 - S/M REVISION
5	94 - 94	94-17B-17	I/C ELECT SCHEMATIC - OIL PRESS SWITCH - S/M REV

Model Legend: 1= All 2= Most 3= Tempo/Topaz 4= Taurus/Sable 5= Thunderbird/Cougar 6= Crown Victoria/Grand Marquis 7= Escort/Tracer 8= Probe 9= Aspire
10= Mustang 11= Contour/Mystique 12= Festiva

Instruments, Dash Cluster, Warning Lights, Mirrors

Models	Year	TSB#	Description
4	94 - 94	94-10B-51	IC (CONVENTIONAL) PINPOINT TESTS - S/M REVISION
6,10	94 - 94	94-10B-30	IC (CONVENTIONAL) TEST STEP F1 - S/M REVISION
6	92 - 93	93-19B-08	IC (ELECTRONIC) SCHEMATICS & PINPOINT TESTS- S/M REV
5	92 - 93	94-10B-27	IC (ELECTRONIC) TEST STEP FA1 - S/M REVISION
4	92 - 93	94-10B-55	IC (ELECTRONIC) TEST STEP TA7 - S/M REVISION
4	92 - 94	94-10B-54	IC (ELECTRONIC) TEST STEP TB3 - S/M REVISION
5	93 - 93	94-05B-20	IC - PINPOINT TEST SA - S/M REVISION
2	83 - 94	93-24-10	IC LAMPS DIM - IGNITION SWITCH DIAGNOSIS
5	92 - 93	93-19B-05	IC(ELECTRONIC) PINPOINT TESTS REVISED - S/M REVISION
4	90 - 93	93-05B-24	IC/ELECTRONIC - PINPOINT TEST TD - S/M REVISION
6	92 - 93	93-10-01	INDICATOR LIGHT/ABS - INTERMITTENT OR STAYS ON - FIX
5	94 - 94	94-14-01	INSTRUMENT CLUSTER GLARE - REVISED PARTS
5	90 - 93	94-10B-28	INSTRUMENT ILLUMINATION SCHEMATIC - S/M REVISION
8	93 - 93	92-24B-24	INSTRUMENT PANEL R&I PROCEDURES - S/M REVISION
6	92 - 93	94-10B-32	IP ASH RECEPTACLE - REMOVE/INSTALL - S/M REVISION
2	91 - 93	92-04-04	MIL/CES/SES - LIGHT ON W/NO SELF-TEST CODES - REV
10	91 - 93	93-13B-15	MIRROR/POWER OUTSIDE REARVIEW - R&I PROC - S/M ADD
8	93 - 93	93-20-08	OIL PRESSURE GAUGE READS ON LOW SIDE OF NORMAL
5	94 - 95	94-22-05	OUTSIDE TEMP DISPLAY SHOWS -40 DEG F - INTERMITTENT
8	93 - 93	94-10B-49	POWER MIRROR SWITCH SCHEMATIC - S/M REVISION
1	90 - 95	94-17-02	REAR VIEW MIRROR DETACHES FROM 18SHIELD - SVC TIP
4	93L - 94	94-24B-64	SPEED CONTROL CKT DESIGNATIONS - 3.2L - S/M REV
4	92 - 93	93-19B-22	SPEED CONTROL DIAG CHART & TEST STEP G1 - S/M REV
4	92 - 93	93-26B-32	SPEED CONTROL PINPOINT TESTS - S/M REVISION
8	94 - 94	94-24B-54	SPEED CONTROL SYS - TEST A5 - S/M REV
10	93 - 93	94-17B-47	SPEED CONTROL SYS ELECTRICAL SCHEMATIC - S/M REV
10	92 - 93	93-19B-14	SPEED CONTROL TEST STEP D5 - S/ REVISION
6	92 - 93	93-26B-02	SPEED CONTROL TEST STEPS REVISED - S/M REVISION
4	93 - 93	93-26B-02	SPEED CONTROL TEST STEPS REVISED - S/M REVISION
11	95 - 95	95-03-07	SPEEDO INOP - CODE 452
6	92 - 93	94-10B-31	SPEEDO/ODO TEST STEP A9 - S/M REVISION
11	95 - 95	95-07-04	SPEEDOMETER NEEDLE WAIVER OR NOISE
11	95 - 95	95-07-04	SPEEDOMETER NEEDLE WAIVER OR NOISE - REV
5	92 - 93	94-05B-22	SPEEDOMETER/ODOMETER PINPOINT TEST A - S/M REVISION
2	85 - 93	91-08-14	SPEEDOMETER/ODOMETER-REPLACEMENTS W/PRE-SET MI AVAIL
2	90 - 94	95-15-03	TEMP GAUGE - FLUCTUATES AND/OR FALSE HIGH READINGS
4,3,7,5	90 - 94	95-15-03	TEMP GAUGE - FLUCTUATES OR FALSE HIGH READINGS - REV
12	92 - 93	93-08-05	WARN BUZZER/IGN KEY - STAYS ON - WIRING HARNESS FIX
8	93 - 93	93-05B-19	WARNING CHIME SYSTEM DIAGNOSTICS - S/M REVISION

Model Legend: 1= All 2= Most 3= Tempo/Topaz 4= Taurus/Sable 5= Thunderbird/Cougar 6= Crown Victoria/Grand Marquis 7= Escort/Tracer 8= Probe 9= Aspire
10= Mustang 11= Contour/Mystique 12= Festiva

Chassis Electrical, Wiring Harness, Fuses-Circuit Breakers, Wipers, Window Motors

Models	Year	TSB#	Description
5	94 - 94	94-24B-19	CONNECTOR LOCATION INDEXES - S/M REVISION
5	93 - 93	94-24B-22	CONNECTORS C220/C226 ILLUSTRATIONS - S/M REV
10	94 - 94	94-13-07	CONVERTIBLE TOP SWITCH INOP - REVISED SWITCH
4	94 - 94	94-05B-41	ELECTRICAL - CONNECTOR C122 DESCRIPTION - S/M REV
8	94 - 94	94-24B-53	ELECTRICAL SCHEMATIC - TSS ADDED - S/M REV
7	91 - 93	92-05-10	ELECTRICAL TROUBLESHOOTING - SERVICE TIPS
7	91 - 93	92-16-03	ELECTRICAL TROUBLESHOOTING - SERVICE TIPS - REVISED
7	91 - 93	93-13-06	ELECTRICAL TROUBLESHOOTING - SERVICE TIPS - REVISED
7	91 - 94	93-17-05	ELECTRICAL TROUBLESHOOTING TIPS - REVISED
10	94 - 94	94-24B-48	ENG COMP FUSE BOX INTERNALS - S/M REV
8	93 - 94	94-25-04	HARD/NO START - FUEL SHUTOFF LAMP ON - WIRING SPLICE
8	93 - 94	94-25-04	HARD/NO START/FUEL SHUTOFF LAMP ON -WIRE SPLICE- REV
1	94 - 94	94-19-07	HARNESS CONNECTOR 14401 (IGN SW) DOES NOT MATE
8	93 - 93	93-19B-15	HEADLAMP RETRACTOR ELECTRICAL SCHEMATIC - S/M REV
6	93 - 94	93-26B-16	HEATED 18SHLD DELETED (SWITCH/WIRING/HOLE)-S/M REV
5	94 - 94	94-24B-20	IC FUSE AMPERAGE RATING - S/M REVISION
2	83 - 94	93-24-10	MALFUNCTION OF 4 PIN LOW OIL LEVEL RELAY - DIAGNOSIS
5	93 - 93	93-26B-08	PINPIONT TEST D STEP D3 - S/M REVISION
5	94 - 94	94-17B-27	POWER 18OW SWITCH CONNECTOR VIEWS - S/M REV
8	93 - 94	94-16-08	POWER 18OW/LOCK INOP - HARNESS CONNECTOR CORRODED
2	94 - 94	94-07-08	REVISED ICM (EDIS) CONNECTOR - SEPARATING TIP
4	96 - 96	96-10-08	TRAILER WIRING W/ LAMP OUT WARNING - TIP
4	92 - 93	93-08-09	WASHER FLUID FORCED OUT OF NOZZLE/PRESSURE BUILD-UP
10	87 - 93	93-21-06	WIPER MOTOR ERRATIC/SLOW/INTERMITTENT - KIT
2	86 - 94	95-04-06	WIPER MOTOR REPLACEMENT - UPGRADED SERVICE KITS
2	86 - 94	95-04-06	WIPER MOTOR REPLACEMENT - UPGRADED SERVICE KITS -REV
7	91 - 96	96-08-04	WIPERS STREAK OR CLEAN UNEVENLY
5	94 - 94	94-17B-23	WIRING - C161 INFO & LOCATION - S/M REV
5	94 - 94	94-17B-19	WIRING DIAG - C253 & C341 LOCATIONS - S/M REV
2	85 - 93	92-13-05	WIRING HARNESS - TERMINAL REPAIR KIT - REVISED
2	85 - 96	93-10-05	WIRING HARNESS - TERMINAL REPAIR KIT - REVISED
2	85 - 93	91-25-08	WIRING HARNESSES - NEW KIT/REPAIR INSTEAD OF REPLACE

Model Legend: 1= All 2= Most 3= Tempo/Topaz 4= Taurus/Sable 5= Thunderbird/Cougar 6= Crown Victoria/Grand Marquis 7= Escort/Tracer 8= Probe 9= Aspire 10= Mustang 11= Contour/Mystique 12= Festiva

Auxiliary Equipment, Jacks, Trailer Hitches, Towing

Models	Year	TSB#	Description
6	93 - 93	93-09-10	CELLULAR PHONE - RUBBER BUMPER ON CRADLE BROKEN
6	93 - 93	93-26B-18	CELLULAR PHONE FUNCTIONAL TEST A - S/M REVISION
6	92 - 94	ONP/94B58	LOAD EQUALIZING HITCH INST - OWNER MANUAL REV
2	88 - 94	91-21-05	RECOMMENDATIONS WHEN TOWING THIS VHCL BEHIND ANOTHER

Model Legend: 1= All 2= Most 3= Tempo/Topaz 4= Taurus/Sable 5= Thunderbird/Cougar 6= Crown Victoria/Grand Marquis 7= Escort/Tracer 8= Probe 9= Aspire 10= Mustang 11= Contour/Mystique 12= Festiva

Heating, Air Conditioning, Ventilation, Defogger

Models	Year	TSB#	Description
2	81 - 94	93-15-05	A/C "O" RING REMOVAL FROM SPRING LOCK COUPLING/INFO
7	94 - 94	94-24B-31	A/C & HEATER ELECTRICAL SCHEMATICS - S/M REV
1	90 - 95	95-05-12	A/C - ADDING REFRIGERANT OIL - SERVICE TIP
1	85 - 95	95-08-01	A/C - APPROVED FLUSHING PROCEDURES - SERVICE TIP
1	85 - 95	95-08-01	A/C - APPROVED FLUSHING PROCEDURES - SRVC TIP - REV
6	92 - 93	92-20-04	A/C - CLUTCH CYCLING PRESSURE SWITCH TICK/PING/POP
5	89 - 93	92-17-04	A/C - COMPRESSOR MOAN AT IDLE (3.8L)
5	89 - 93	92-03-01	A/C - CONDENSATE ENTERS PASS COMPART THRU FLOOR DUCT
2	82 - 94	92-25-3 & 93-9-8	A/C - FILTER REFRIG AFTER REP'L COMPRESSOR - REVISED
2	88 - 94	94-15-06	A/C - FS-10 & FX-15 COMPRESSORS - SERVICE TIPS
2	88 - 93	92-20-05	A/C - FX-15 COMPRESSOR AIR GAP SPECIFICATION REVISED
4	91 - 93	93-01-08	A/C - FX-15 COMPRESSOR GROWL NOISE - SERV PROC - REV
4	91 - 93	93-01-08	A/C - FX-15 COMPRESSOR GROWL NOISE - SERVICE PROC
7,4,3,5	92 - 93	93-15-08	A/C - FX-15 COMPRESSOR SQUEAL - INSPECTION/SERV PROC
2	92 - 93	93-15-08	A/C - FX-15 COMPRESSOR SQUEAL - INSPECTION/SERV PROC
2	85 - 96	95-22-07	A/C - IDENT OF NON-APPROVED REFRIGERANTS - REV
2	85 - 96	95-22-07	A/C - IDENTIFICATION OF NON-APPROVED REFRIGERANTS
2	85 - 96	95-22-07	A/C - IDENTIFICATION OF NON-APPROVED REFRIGERANTS
5	93 - 93	93-05-09	A/C - INSUFFICIENT COOL/HIGH HEAD PRESSURE AT IDLE
2	80 - 93	92-25-06	A/C - MUSTY & MILDEW ODORS - NEW PRODUCT AVAIL - REV
2	80 - 93	92-25-06	A/C - MUSTY & MILDEW ODORS - NEW PRODUCT AVAILABLE
2	80 - 93	92-05-07	A/C - NEW FILTER KIT AVAILABLE - SERVICE INFO - REV
2	80 - 93	93-23-11	A/C - NO USE OF R12 REFRIG SUBSTITUTES - REVISED
2	80 - 93	93-23-11	A/C - NO USE OF R12 REFRIG SUBSTITUTES - SERVICE TIP
2	81 - 95	93-15-05	A/C - O-RING REMOVAL FROM SPRING LOCK COUPLER - TIP
2	94 - 95	94-20-06	A/C - OIL VISIBLE AT SPRING LOCK COUPLERS - TIP
2	92 - 93	93-10-04	A/C - ON-VEHICLE EVAP CORE LEAK TEST/WARRANTY - REV
2	92 - 95	94-10-07	A/C - ON-VEHICLE EVAP CORE LEAK TEST/WARRANTY - REV2
2	92 - 93	93-10-04	A/C - ON-VEHICLE EVAP CORE LEAK TEST/WARRANTY INFO
5	91 - 93	92-26-04	A/C - POOR COOLING PERFORM/MISSING CONDENSER SEALS
3	92 - 93	94-02-10	A/C - POOR COOLING/LOSS OF REFRIGERANT - CHAFED HOSE
2	92 - 95	94-26-06	A/C - R-134A TRACER DYE FOR LEAK DETECTION
1	94 - 96	94-26-06	A/C - R-134A TRACER DYE INSTALLED IN SYSTEM - REV
1	95 - 95	94-23-10	A/C - R-134A TRACER DYE INSTALLED IN SYSTEM - TIPS
3	88 - 94	92-04-06	A/C - RATTLE (IN DASH PANEL) FROM EVAP CASE - REV
2	93 - 94	94-05B-08	A/C - REFRIGERANT CAPACITIES REVISED/ADDED - S/M REV
5	89 - 93	91-22-03	A/C - REFRIGERANT LEAK/EVAP CORE TUBE WORN - REVISED

Model Legend: 1= All 2= Most 3= Tempo/Topaz 4= Taurus/Sable 5= Thunderbird/Cougar 6= Crown Victoria/Grand Marquis 7= Escort/Tracer 8= Probe 9= Aspire
10= Mustang 11= Contour/Mystique 12= Festiva

Heating, Air Conditioning, Ventilation, Defogger

Models	Year	TSB#	Description
12	93 - 93	93-26B-24	A/C - REFRIGERANT SYS PRESSURES TEST RF1 - S/M REV
5	89 - 94	92-12-08	A/C - WATER DRIPS FROM FLOOR DUCTS - REVISED CORE
5	89 - 95	95-23-03	A/C - WATER DRIPS ON FLOOR WHEN BLOWER ON HIGH
7	91 - 93	95-06-05	A/C BLOWER MO-R - LOW SPEED TICKING NOISE
2	88 - 93	92-18-04	A/C COMP (FX-15) - OIL RECOVERY/MEASURING PROC - REV
2	88 - 93	92-18-04	A/C COMP (FX-15) - OIL RECOVERY/MEASURING PROCEDURE
6	91 - 93	92-25-04	A/C COMPRESSOR (FX-15) - NOISY - CORRECTION (4.6L)
8	94 - 94	94-24-05	A/C COMPRESSOR CLUTCH INOP - HARNESS ROUTING
9	94 - 95	95-22-09	A/C COMPRESSOR INOP WITH EQUAL PRESSURE CONDITION
9	94 - 95	96-05-07	A/C COMPRESSOR NOISE
4	93 - 95	95-16-08	A/C COMPRESSOR NOISE/MOAN - 3.8L
4	93 - 93	93-26B-34	A/C COMPRESSOR REMOVE/INSTALL PROCED (SHO) - S/M REV
4	92 - 93	93-24-09	A/C COMPRESSOR TUBE FRACTURES/LEAK - NEW PARTS- 3.8L
2	91 - 93	92-25-05	A/C COMPRESSOR/FX-15 - COIL/PULLEY APPLICATION CHART
11	95 - 96	96-03-12	A/C CONDENSATE LEAK ON- FLOOR
5	89 - 94	94-11-04	A/C EVAP CORE & CASE ASSEMBLY APPLICATION
5	94 - 94	94-10B-25	A/C EVAP CORE - REMOVE/INSTALL - S/M REV
4	89 - 94	94-02-11	A/C FX-15 COMPRESSOR MANIFOLD & TUBE ASSEMBLY CHART
2	92 - 93	93-15-08	A/C FX-15 COMPRESSOR SQUEAL - INSPECT/SERV PROC REV
2	92 - 93	93-23-10	A/C FX-15 COMPRESSOR SQUEAL - INSPECT/SERV PROC REV
5	94 - 94	94-02-14	A/C INOP - CUT OFF SW WIRE RUBS SUCTION HOSE - REV
5	94 - 94	94-02-14	A/C INOP - CUT OFF SWITCH WIRING RUBS SUCTION HOSE
2	95 - 95	94-23-09	A/C NO/POOR PERFORMANCE - FS-10 COMPRESSOR
8	94 - 94	95-14-05	A/C POOR PERF - KIT - 2.0L
5	89 - 93	94-02-09	A/C POOR PERFORM - HIGH DISCHARGE TEMP - BLEND DOOR
4	93 - 93	93-21-05	A/C POOR/NO PERFORMANCE - REPLACE SERVICE PORTS
4	93 - 93	93-21-05	A/C SERVICE PORTS MAY LEAK - R134A - PROCEDURE
2	80 - 93	93-20-06	A/C USE OF CORRECT FLUORESCENT DYE - SERVICE TIP
6,4,5	95 - 95	94-24B-15	AMBIENT/SUNLOAD SENSOR TEST VOLTAGE CHECK - S/M REV
6,4,5	92 - 94	94-11-05	ATC & EATC COLD ENG/THERMAL BLOWER LOCKOUT OPERATION
6	93 - 94	94-05B-25	ATC - BLOWER MO-R SPEED CONTROL SCHEMATIC - S/M REV
6	93 - 94	94-17B-33	BLEND DOOR ACTUATOR SERVICE INFO - S/M REV
4	93 - 94	94-02-13	BLOWER MOTOR INOPERATIVE IN AU-MATIC MODE
7,10,5	93 - 94	95-05-11	CHIRP/SQUEAL FROM BLOWER MOTOR AT LOW BLOWER SPEEDS
7,10,5	93 - 94	95-05-11	CHIRP/SQUEAL FROM BLOWER MOTOR AT LOW BLOWER SPEEDS
7,10,5	93 - 94	95-05-11	CHIRP/SQUEAL FROM BLOWER MOTOR AT LOW BLOWER SPEEDS
4	86 - 94	95-05-11	CHIRP/SQUEAL FROM BLOWER MOTOR AT LOW BLOWER SPEEDS
7,10,5	93 - 94	95-05-11	CHIRP/SQUEAL FROM BLOWER MOTOR AT LOW SPEEDS - REV
7,10,5	93 - 94	95-05-11	CHIRP/SQUEAL FROM BLOWER MOTOR AT LOW SPEEDS - REV

Model Legend: 1= All 2= Most 3= Tempo/Topaz 4= Taurus/Sable 5= Thunderbird/Cougar 6= Crown Victoria/Grand Marquis 7= Escort/Tracer 8= Probe 9= Aspire 10= Mustang 11= Contour/Mystique 12= Festiva

Heating, Air Conditioning, Ventilation, Defogger

Models	Year	TSB#	Description
7,10,5	93 - 94	95-05-11	CHIRP/SQUEAL FROM BLOWER MOTOR AT LOW SPEEDS - REV
4	86 - 94	95-05-11	CHIRP/SQUEAL FROM BLOWER MOTOR AT LOW SPEEDS - REV
7,3,5,10	93 - 94	95-06-06	CHIRP/SQUEAL FROM BLOWER MOTOR AT LOW SPEEDS - REV 2
4	86 - 94	95-06-06	CHIRP/SQUEAL FROM BLOWER MO-R AT LOW SPEEDS - REV 2
7	94 - 94	94-10B-37	CLIMATE CONTROL - DIAGN/TEST/SERVICE INFO - S/M REV
6	93 - 93	93-26B-17	CLIMATE CONTROL - DIAGNOSTIC & TEST INFO - S/M REV
7	94 - 94	94-10B-36	CLIMATE CONTROL - TEST STEP C1 - S/M REVISION
2	91 - 93	93-05B-03	CLIMATE CONTROL FIXED ORIFICE TUBE INFO - S/M REV
8	94 - 94	94-24B-55	CLIMATE CONTROL SYS TEST I12 - S/M REV
3	93 - 94	93-26B-35	CLIMATE CONTROL TEST STEPS B7/B8 - S/M REVISION
4	93 - 93	93-19-04	COOLANT LEAK OR OVERHEAT - HEATER HOSE CLAMP
2	85 - 94	91-05-01	DEFROSTER - HEATED BACKLITE INOPERATIVE - REVISED
5	94 - 94	94-17B-25	EATC - BLEND DOOR ACTUATOR PIN NUMBERS - S/M REV
6	95 - 95	94-20-05	EATC - DTC'S 115 AND/OR 125 - SERVICE TIP
4	94 - 94	94-24B-61	EATC - SECTION REFERENCE - S/M REV
6	93 - 93	93-13B-08	EATC BLEND DOOR ACTUATOR FUNCTIONAL TEST - S/M REV
4	94 - 94	93-23-12	EATC BLOWER SPEED CONTROL ENHANCEMENT 17AINED
4	94 - 94	94-17B-50	EATC PINPOINT TEST B - S/M REVISION
6	93 - 94	94-17B-34	EATC/ATC BLOWER MO-R PINPOINT TESTS - S/M REV
6	93 - 94	93-05B-10	ELEC BLEND DOOR ACTUATOR/POTENTIOMETER -S/M REVISION
6	93 - 93	93-05B-10	ELEC BLEND DOOR ACTUATOR/POTENTIOMETER INFO- S/M ADD
10,4,3,5	89 - 93	93-18-07	EVAPORATOR CORE REPLACE PROCEDURE W/O REMOVING IP
10,4,3,5	89 - 94	93-18-07	EVAPORATOR CORE REPLACEMENT PROCEDURE
10,5,4,3	89 - 94	93-21-03	EVAPORATOR CORE REPLACEMENT PROCEDURE - REVISION
10,5,4,3	89 - 94	94-03-03	EVAPORATOR CORE REPLACEMENT PROCEDURE - REVISION 2
2	80 - 96	95-04-05	FILTER REFRIGERANT AFTER COMPRESSOR REPL - REV FOUR
2	80 - 95	94-15-05	FILTER REFRIGERANT AFTER COMPRESSOR REPL - REV THREE
2	80 - 94	94-03-04	FILTER REFRIGERANT AFTER COMPRESSOR REPL - REV TWO
2	80 - 94	93-26-08	FILTER REFRIGERANT AFTER COMPRESSOR REPL - TIPS REV
4	93 - 93	93-14-07	HEAT & A/C - CONTROL KNOB SEPARATES/NEW KNOB - PROC
5	94 - 94	94-20-04	HEAT & A/C - SATC METRIC CONVERSION
3	94 - 94	94-24B-65	HEAT/DEFROST - 3.0L HEATER HOSE INSTALL - S/M REV
2	83 - 94	93-14-04	HEATER/DEFROST - POOR OUTPUT/THERMOSTAT STUCK - REV
2	83 - 93	93-14-04	HEATER/DEFROSTER - POOR OUTPUT/THERMOSTAT STUCK OPEN
5	94 - 94	94-19-09	INACCURATE OR ERRATIC TEMP CONTROL - HEAT/AIR CON
5	94 - 94	94-10B-24	MANUAL A/C & HEATER SCHEMATICS & DESCRIPT - S/M REV
10	95 - 95	94-22-06	TEMP CONTROL KNOB DOES NOT REACH FULL COOL POSITION
11	95 - 95	95-11-06	WHISTLE NOISE FROM A/C-HEATER SYSTEM

Model Legend: 1= All 2= Most 3= Tempo/Topaz 4= Taurus/Sable 5= Thunderbird/Cougar 6= Crown Victoria/Grand Marquis 7= Escort/Tracer 8= Probe 9= Aspire
10= Mustang 11= Contour/Mystique 12= Festiva

Entertainment Devices, Stereo, Radio, Etc.

Models	Year	TSB#	Description
5	92 - 93	93-26B-14	ANTENNA - TEST INFORMATION REVISED - S/M REVISION
10	94 - 94	94-17B-46	AUDIO SYS - CONNECTOR VIEWS/PIN CHART - S/M REV
9	94 - 94	94-17B-04	AUDIO SYS PINPOINT TEST STEP F8 - S/M REVISION
5	94 - 94	94-24B-18	PWR ANTENNA MODULE LOCATION - S/M REVISION
4,5,11	93 - 95	95-06-07	RADIO - AM BAND - STATIC WHILE DRIVING
4,5,11	93 - 95	95-06-07	RADIO - AM BAND - STATIC WHILE DRIVING - REV
10	94 - 95	94-17B-42	RADIO - CIRCUIT 279 WIRE COLOR - S/M REV
8	93 - 94	94-11-06	RADIO - FRONT DOOR SPEAKERS INOPERATIVE
6	93 - 93	93-02-06	RADIO - POOR BASS RESPONSE/WIRING FIX (W/CELL PHONE)
10	94 - 95	95-09-03	RADIO DISTORTION/POOR RECEPTION - 5.0L
9	94 - 95	94-24B-02	RADIO SPEAKER REAR STRAP INFO - S/M REVISION
2	85 - 94	92-06-09	RADIO/2-WAY RADIO - WHINE/BUZZ NOISE - FUEL PUMP-REV
2	85 - 95	93-18-04	WHINE/BUZZ IN SPEAKER - FUEL PUMP RFI NOISE - REV
2	85 - 96	95-11-03	WHINE/BUZZ IN SPEAKER - FUEL PUMP RFI NOISE - REV2
2	85 - 94	93-15-06	WHINE/BUZZ IN SPEAKER - FUEL PUMP RFI NOISE FILTER

Model Legend: 1= All 2= Most 3= Tempo/Topaz 4= Taurus/Sable 5= Thunderbird/Cougar 6= Crown Victoria/Grand Marquis 7= Escort/Tracer 8= Probe 9= Aspire 10= Mustang 11= Contour/Mystique 12= Festiva

Seats, Belts, Interior Trim, Carpets, Air Bags

Models	Year	TSB#	Description
6	93 - 93	94-17B-32	AIR BAG (SIR) DIAGNOSTIC & TESTING INFO - S/M REV
3	93 - 94	94-17B-52	AIR BAG (SIR) DIAGNOSTIC & TESTING INFO - S/M REV
4	93 - 93	94-17B-51	AIR BAG (SIR) SERVICE INFO - S/M REV
4	92 - 95	94-19-01	AIR BAG - CODE 51 - IMPROVED DIAGN & REPAIR - REV
4	92 - 94	94-19-01	AIR BAG - CODE 51 - IMPROVED DIAGNOSTICS & REPAIR
6	95 - 95	95-22-08	AIR BAG CODE 24 - INCORRECT DIAGNOSIS - TIP
10	94 - 95	95-22-08	AIR BAG CODE 24 - INCORRECT DIAGNOSIS - TIP
8	93 - 93	93-05B-16	AIR BAG FAULT CODE 41 - SHOP MANUAL REVISION
8	93 - 93	93-15-04	AIR BAG LIGHT ON/MODULE DISPLAYS CODE 34 - SERV INFO
8	93 - 93	93-15-04	AIR BAG LIGHT ON/MODULE DISPLAYS CODE 34 - SERV INFO
3	92 - 93	92-24B-35	AIR BAG SYSTEM - PINPOINT TEST STEP E2 - S/M REV
7	94 - 94	95-17-02	AIR BAG WARNING LITE FLASHES DTC 32
7	94 - 94	95-17-02	AIR BAG WARNING LITE FLASHES DTC 32 - REV
10	93 - 93	93-15-01	ARMREST PAD TURNS YELLOW (CNVRTBL W/WHITE INTERIOR)
8	93 - 93	93-16-01	ASHTRAY COVER FALLS INTO CENTER CONSOLE -NEW CONSOLE
8	93 - 93	93-16-01	ASHTRAY COVER FALLS INTO CONSOLE - NEW CONSOLE
10	94 - 95	95-17-03	CONSOLE/PARK BRAKE HANDLE INTERFERENCE
9	94 - 95	94-24B-03	FRONT SEAT BELT RETRACTOR & TONGUE REMOVAL - S/M REV
5,6	89 - 94	94-20-01	FRONT SEAT TRACK COVERS COME OFF
4,3,5	92 - 94	94-04-01	HEADLINER ODOR (FISH OR MUSTY) - NEW HEADLINER - REV
4,3,5	92 - 94	94-04-01	HEADLINER ODOR (FISH OR MUSTY) - REVISED HEADLINER
5	89 - 93	94-02-04	I/C FINISH PANEL LOOSE AT UPPER CORNERS
5	94 - 95	95-03-02	IC TRIM - RH SIDE A/C REGISTER MISALIGNED
5	94 - 95	95-03-02	IC TRIM - RH SIDE A/C REGISTER MISALIGNED - REV
4	96 - 96	95-22-01	INTERIOR DOOR PANELS - PROPER REMOVAL PROCEDURE/TIPS
6,5	91 - 94	94-18-02	NOISE/CLUNK OR LOOSE FRONT SEAT
8	93 - 93	92-18-03	PACKAGE TRAY - REMOVE/INSTALL PROCEDURE - S/M UPDATE
7,3,5	87 - 93	93-09A-01	PASSIVE RESTRAINT SYSTEMS/MO-RIZED - DIAG/REPAIR
3	94 - 94	94-05B-44	PASSIVE SEAT BELT PINPOINT TEST STEP D3 - S/M REV
12	92 - 93	93-08-05	RESTRAINT SYS/PASSIVE - INOP OR INTERMIT - SERV PROC
9	94 - 94	94-24B-04	SEAT BELT W/ANCHOR PLATE THREAD DAMAGE - S/M REV
6	90 - 93	93-11-01	SEAT/FRONT SPLIT BENCH - CUSHION SAGS - SERV PROC
5	94 - 94	94-17B-26	SEATS/TRACKS - CONNECTOR END VIEWS - S/M REV
3	93 - 93	94-05B-45	SIR TESTING - PIN 18 CIRCUIT NUMBER/COLOR - S/M REV
2	88 - 96	92-24-01	SRS - AIR BAG MODULE COVERS - PAINTING RESTRICT -REV
3,6,10,4	88 - 93	92-24-01	SRS - AIR BAG MODULE COVERS - PAINTING RESTRICTIONS

Model Legend: 1= All 2= Most 3= Tempo/Topaz 4= Taurus/Sable 5= Thunderbird/Cougar 6= Crown Victoria/Grand Marquis 7= Escort/Tracer 8= Probe 9= Aspire
10= Mustang 11= Contour/Mystique 12= Festiva

Glass, Doors, Hood, Decklid, Tailgate, Liftgate, Locks

Models	Year	TSB#	Description
5	93 - 93	94-02-01	NOISE - POWER MOON ROOF - RESEAL JOINT GAP
7	91 - 94	95-05-03	NOISE - REAR DOORS - WEATHERSTRIP AT CORNERS
7	91 - 95	94-20-03	NOISE - SERVICE TIPS
4	92 - 93	93-02-01	NOISE - SUMMARY OF KNOWN REPAIRS
4	92 - 94	94-06-01	NOISE - SUMMARY OF KNOWN REPAIRS - REVISED
4	92 - 95	94-18-01	NOISE - SUMMARY OF KNOWN REPAIRS - REVISED
4	96 - 96	96-03-01	NOISE AT A-PILLAR OR 18SHIELD
6	95 - 96	96-10-04	LOW - HIGH ROLL UP EFFORT - RF DOOR
5	89 - 94	95-05-02	SHIELD/BACKLITE SEALER DRIPS ON INTERIOR TRIM
10	94 - 95	96-04-01	CONV -P REPLACE PARTS DIFFER FROM ORIGINAL - TIPS
7	91 - 93	92-06-03	DOOR 18OW GLASS - DIFFICULT - ROLL UP - REVISED
4	92 - 93	93-02-01	DOOR GLASS/FRONT - TIPS FORWARD/BINDS/18NOISE
4	96 - 96	95-17-01	DOOR HINGES - NEW SERVICE PROCEDURE - TIP
5	94 - 94	94-10B-20	DOOR LOCK (ELECTRONIC) PINPOINT TESTS - S/M REVISION
6	94 - 94	94-24B-28	DOOR LOCK CONTROL SYS - CONNECTOR/CIRCUITS - S/M REV
6	94 - 94	94-24B-27	DOOR LOCK CONTROL SYS - SYMPTOM CHART - S/M REV
4	94 - 94	94-10B-50	DOOR LOCK CONTROL SYS PINPOINT TESTS - S/M REVISION
6	92 - 93	92-24B-16	DOOR LOCK CYLINDER R&I PROCEDURES - S/M REVISION
8	93 - 94	94-08-06	DOOR LOCK INOP - FREEZING - PROCEDURES
4	89 - 93	93-14-02	DOOR/REAR - HIGH CLOSE EFFORT - DIAGNOSIS/SERV TIPS
6,4,5	93 - 94	94-10-02	DOORS LOCK WHEN DRIVER EXITS - SEAT SENSOR SWITCH
2	81 - 95	95-12-01	ESSEX 18SHIELD REPLACE PRODUCTS - SHELF LIFE TIP
4	93 - 95	94-23-01	FRONT DOOR CLADDING - REPLACE PROCEDURE - TIPS
9	94 - 94	95-04-02	FRONT DOOR GLASS - HIGH ROLL UP EFFORT - 3-DOOR
8	93 - 96	94-05-04	FRONT DOOR GLASS ADJUSTMENT PROCED & SPECS - REV2
8	93 - 93	93-22-01	FRONT DOOR GLASS ADJUSTMENT PROCEDURE & SPECS
8	93 - 94	93-22-01	FRONT DOOR GLASS ADJUSTMENT PROCEDURE & SPECS - REV
8	93 - 93	93-03-02	GLASS/FRONT DOOR - ADJUSTMENT PROCEDURE
8	93 - 94	94-19-03	HATCH - HYDR LIFTS DON'T SUPPORT ADDED SPOILER
8	93 - 95	95-14-03	HEADLAMP DOORS CYCLE OPEN AND CLOSED
9	94 - 95	96-08-01	HIGH EFFORT - LOCK/UNLOCK DOOR USING KEY
2	84 - 94	93-21-01	IRIDESCENCE OR MOTTLING IN TEMPERED GLASS 17AINED
5	93 - 93	94-02-03	LEAK - MOON ROOF - BETWEEN GLASS AND FRAME
4	93 - 93	93-17-01	LIFTGATE - REMOTE RELEASE RETROFIT INSTALL PROCEDURE
7	93 - 94	94-09-01	LIFTGATE DIFFICULT - OPEN/HANDLE DIS-RTS - WAGON
6,4	90 - 93	91-18-07	LOCK CYL (IGN) - KEY/ERRATIC INTERMITTENT OPERATION
6,4	90 - 93	93-01-09	LOCK CYL (IGN) - KEY/ERRATIC INTERMITTENT OPERATION
6,10,5	96 - 96	96-07-02	LOCKS - NEW DESIGN - ON TRIAL VHCLS - APPL
10	96 - 96	96-11-01	LOCKS - NEW DESIGN - SERVICE TIP

Model Legend: 1= All 2= Most 3= Tempo/Topaz 4= Taurus/Sable 5= Thunderbird/Cougar 6= Crown Victoria/Grand Marquis 7= Escort/Tracer 8= Probe 9= Aspire 10= Mustang 11= Contour/Mystique 12= Festiva

Glass, Doors, Hood, Decklid, Tailgate, Liftgate, Locks

Models	Year	TSB#	Description
2	91 - 93	93-02-02	LOCKS/POWER DOOR - INOP/ACTUA-R DISENGAGES - PROC
4,5,6	89 - 93	93-02-02	LOCKS/POWER DOOR - INOP/ACTUA-R DISENGAGES - PROC
4	95 - 95	ONP/95B66	LOWER DOOR HINGE BOLTS REPLACEMENT
4	95 - 95	ONP/95B66R	LOWER DOOR HINGE BOLTS REPLACEMENT - PARTS INFO
4	95 - 95	ONP/95B66	LOWER DOOR HINGE BOLTS REPLACEMENT - SUPPLEMENT
4	95 - 95	ONP/95B66	LOWER DOOR HINGE BOLTS REPLACEMENT - UPDATE
4	92 - 94	94-03-01	NOISE/SQUEAK FROM REAR 18OW AREA
4	92 - 95	94-03-01	NOISE/SQUEAK FROM REAR 18OW AREA - SDN - REV
8	93 - 93	92-18B-04	PACKAGE TRAY - REMOVE/INSTALL PROCEDURES - S/M REV
5	89 - 93	93-20-01	POWER MOON ROOF 18NOISE/BUFFETING WHEN FULLY OPEN
5	89 - 94	93-20-01	POWER MOON ROOF 18NOISE/BUFFETING WHEN OPEN - REV
4,5	86 - 95	91-01-07	PWR 18OW - GROANS AFTER GLASS IS POWERED UP - REV
8	93 - 95	95-25-01	RATTLE/SQUEAK NOISE FROM SLIDING ROOL PANEL
8	93 - 93	ONP/94B40	REATTACHING LUGGAGE COMP COVER - OWNER GUIDE REV
10	94 - 94	94-08-03	SQUEAK NOISE -P OF B-PILLAR - CONVERTIBLE
5	94 - 94	94-17B-28	TRUNK DOOR RELEASE ELECTRICAL SCHEMATIC - S/M REV
7	91 - 94	93-24-05	WATER LEAKS - BODY VIEW DIAGRAM REVISED
7	91 - 94	93-12-04	WATER LEAKS - DIAGNOSTIC SERVICE TIPS - REVISED
7	91 - 93	93-12-04	WATERLEAKS - DIAGNOSTIC SERVICE TIPS

Model Legend: 1= All 2= Most 3= Tempo/Topaz 4= Taurus/Sable 5= Thunderbird/Cougar 6= Crown Victoria/Grand Marquis 7= Escort/Tracer 8= Probe 9= Aspire
10= Mustang 11= Contour/Mystique 12= Festiva

Finishes, Body Structure, Frame, Bumpers

Models	Year	TSB#	Description
5	89 - 93	93-19B-06	APRON/SIDEMEMBER/FENDER - COLLISION REPAIR MAN REV
9	94 - 94	95-14-02	BUMPER - PAINT PEEL - GENERAL BUMPER PAINT REPAIR
6,10,4,3	93 - 93	93-05-03	BUMPER COVER/PLASTIC - MINOR SCRATCH - REPAIR INFO
5	93 - 93	93-05-03	BUMPER COVER/PLASTIC - MINOR SCRATCH - REPAIR INFO
2	79 - 94	93-24-01	BUMPER COVERS - IMPROVED FLEXIBLE PART REPAIR MAT'L
11,4	95 - 95	95-03-03	BUMPER COVERS - PAINT SERVICE TIP
11,4	95 - 95	95-03-03	BUMPER COVERS - PAINT SERVICE TIP
7	96 - 96	95-03-03	BUMPER COVERS - PAINT SERVICE TIP
2	80 - 94	94-08-01	BUMPER ISOLA-R & BRACKET ASSMBLY REPLCMNT GUIDELINE
4	86 - 93	87-12-01	BUMPERS - LOOSE OR UNATTACHED ENDS - REPAIR PROC
3	92 - 93	93-08-03	COWL -P PANEL - SQUEAK/RUBBING SOUND/GRUNT NOISE
6	92 - 93	93-04-02	FENDER/FRONT - WHEEL LIP COSMETIC DAMAGE (POLICE)
6	92 - 93	93-04-02	FENDER/FRONT - WHEEL LIP COSMETIC DAMAGE (POLICE)REV
2	94 - 94	93-24-03	FLEET PAINT CODES FOR DSO PAINT
2	80 - 93	91-21-07	FRAME - RIVET REPLACEMENT W/BOLTS - SERVICE TIPS
7	91 - 94	94-11-02	FRONT VALANCE PANEL MISSING OR LOOSE
4	92 - 93	93-18-02	FUEL FILLER DOOR NOT FULLY OPEN - NEW ASSIST SPRING
10	94 - 94	94-05-02	GRILLE - COWL -P VENT - CRACKS AT WASHER JET
10	94 - 94	94-05-02	GRILLE - COWL -P VENT - CRACKS AT WASHER JET - REV
8	93 - 94	94-16-01	LOOSE AIR DAM SKIRT (VALANCE) AT FRONT CROSSMEMBER
10	96 - 96	95-16-03	MYSTIC PAINT - REPAIR MAT'L - SVT COBRA ONLY
8	94 - 95	95-15-05	NOISE/RATTLE FROM ENG COMP - 2.0L CD4E
2	93 - 93	92-20-01	PAINT - 1993 COLOR CHART/AFTERMARKET SUPPLIER INFO
1	80 - 95	95-06-01	PAINT - ACID RAIN/IRON PARTICLE REMOVAL PROCEDURES
2	83 - 93	93-08-04	PAINT - EXT CLEARCOAT HAZING/PEELING/MICROCHECKING
6,3	92 - 93	92-08-03	PAINT - LOWER BODY SIDE S-NE PROTECTION - MATERIAL
2	92 - 93	92-05-02	PAINT - NEW ID LABEL FOR PAINT CODE CHANGES
10	94 - 94	94-23-03	PAINT - REDUCED CHIP RESISTANCE - LASER RED (E-9)
1	95 - 96	95-16-02	PAINT - TRANSIT COATING - RAPGARD - SERVICE TIP
1	96 - 96	95-16-02	PAINT - TRANSIT COATING - RAPGARD - SERVICE TIP -REV
9	94 - 95	95-22-04	PAINT CHIPS AT LICENSE PLATE GARNISH
1	95 - 96	95-07-02	PAINT CODES - AFTERMARKET CROSS REFERENCE LIST
2	94 - 94	93-24-04	PAINT CODES FOR REGULAR PRODUCTION - 1994 MODELS
10	93 - 93	93-23-04	PAINT PEEL - GRILLE - NEW GRILLE - COBRA ONLY
1	80 - 95	94-23-04	PRIMED SHEET METAL - PREP PROCEDURE & MSDS INFO
10	94 - 95	94-22-02	ROCKER PANEL MOLDINGS LOOSE
3	92 - 93	93-08-02	ROOF PANEL PAINT CHIPS AT UPPER REAR OF DOOR OPENING
2	87 - 95	91-04-11	RUNNING BOARDS NOISY - INCORRECTLY INSTALLED - REV2
5	93 - 93	93-19-01	UPPER QUARTER PANEL SINK MARKS IN HOT WEATHER
7	91 - 94	93-12-04	WATER LEAKS - DIAGNOSTIC SERVICE TIPS - REVISED
7	91 - 93	93-12-04	WATERLEAKS - DIAGNOSTIC SERVICE TIPS

Model Legend: 1= All 2= Most 3= Tempo/Topaz 4= Taurus/Sable 5= Thunderbird/Cougar 6= Crown Victoria/Grand Marquis 7= Escort/Tracer 8= Probe 9= Aspire 10= Mustang 11= Contour/Mystique 12= Festiva

GENERAL MOTORS CORPORATION

Lighting, Horns, Turn Signals, Steering Column

Models	Year	TSB#	Description
24,23,22,	92 - 94	438204	CARE AND CLEANING OF EXTERIOR LAMPS
11,7	94 - 95	433206	CLICK/SCRUB NOISE IN STEER COLUMN - NEW WIRE SHIELD
2	Up - 96	53-82-09	DAYTIME RUNNING LAMPS RETROFIT KITS AVAILABLE
3,7	95 - 95	53-81-07	DAYTIME RUNNING LIGHTS (DRL) - S/M REV
9	95 - 95	53-10-07	DOME LAMP LENS SERVICE - S/M REV
9	95 - 95	53-81-05	ERRATIC INTERIOR LIGHTING - DOOR JAMB SWITCH
6	94 - 94	433201	HARD HORN EFFORT
29,30	94 - 96	63-82-01	HEADLAMP - DARK SPOTS/SHADOWS ON LOW BEAM
29,30	94 - 96	63-82-01	HEADLAMP - DARK SPOTS/SHADOWS ON LOW BEAM - REV
1	Up - 96	63-81-11	HEADLAMP AIMING EQUIP DISTORTS PLASTIC LENS -CAUTION
30,29	91 - 95	53-82-01	HEADLAMP DRAIN HOSE ROUTING
24	92 - 93	381201	HEADLAMP SWITCH LED INDICATOR - OPERATION INFO
24,22,23	92 - 95	53-81-31	HEADLAMPS ON TOO SOON/TOO LONG - TWILIGHT SENTINEL
10	95 - 95	48-81-04	HEADLAMPS WILL NOT TURN OFF - ADD'L INFO/REF #438133
16,15	95 - 95	48-81-04	HEADLAMPS WILL NOT TURN OFF - ADD'L INFO/REF #438133
29,30	95 - 95	48-81-04	HEADLAMPS WILL NOT TURN OFF - ADD'L INFO/REF #438133
10	95 - 95	43-81-33	HEADLAMPS/PARKLAMPS REMAIN ON - TIPS
16,15	95 - 95	43-81-33	HEADLAMPS/PARKLAMPS REMAIN ON - TIPS
29,30	95 - 95	43-81-33	HEADLAMPS/PARKLAMPS REMAIN ON - TIPS
21	93 - 93	218109	HORN (A NOTE) - BRACKET FALLING OFF - SERV PROC
2	89 - 93	92-026/08	HORN - ERRATIC BLOWING (VEHICLES W/SIR) - REVISION
12	94 - 94	43-32-10	HORN - HIGH EFFORT OR SOUNDS WHEN ADJ TILT WHEEL
12	89 - 93	333202	HORN BLOWS INTERMITTENTLY WHEN CAR IS COLD - SERVICE
6	94 - 95	53-32-05	HORN HONKS WITHOUT DRIVER ACTIVATION
9	92 - 94	433202	HORN PAD NOT PROPERLY RETAINED - REWORK CLIPS
9	92 - 94	433202	HORN PAD NOT PROPERLY RETAINED - REWORK CLIPS - REV
9	92 - 94	433203	IGN SWITCH - DOES NOT RETURN TO RUN AFTER START
12	92 - 94	433203	IGN SWITCH - DOES NOT RETURN TO RUN AFTER START
25	92 - 94	433203	IGN SWITCH - DOES NOT RETURN TO RUN AFTER START
24	92 - 95	53-32-08	IGNITION KEY BINDS/STUCK - FLOOR SHIFT
3,7	94 - 95	53-83-13	INTERIOR DIMMER CONTROLLED LAMPS INOP
22,23,24	94 - 94	53-81-11	INTERIOR LAMPS - SYSTEM CHECK TABLE - S/M REV
10	95 - 95	43-90-19	LAMP CONTROL MODULE SERVICE & ARTWORK - S/M REV
15,16	95 - 95	43-90-19	LAMP CONTROL MODULE SERVICE & ARTWORK - S/M REV
29,30	95 - 95	43-90-19	LAMP CONTROL MODULE SERVICE & ARTWORK - S/M REV
10	95 - 95	43-90-19A	LAMP CONTROL MODULE SERVICE & ARTWORK -S/M REV (REV)
15,16	95 - 95	43-90-19	LAMP CONTROL MODULE SERVICE & ARTWORK -S/M REV (REV)
29,30	95 - 95	43-90-19	LAMP CONTROL MODULE SERVICE & ARTWORK -S/M REV (REV)

Model Legend: 1= All 2= Most 3= Beretta/Corsica 4= Celebrity/Century/Cierra/Cutlass Cruiser/6000 5= Camaro/Firebird 6= Caprice/Roadmaster/Impala SS/Fleetwood(RWD)/Estate Wagon/ Custom Cruiser 7= Monte Carlo 8= Corvette 9= Cavalier/Sunbird/Sunfire/Skyhawk 10= Bonneville, SSE, SSEi 11= Lumina 12= Grand Am 13= Gran Prix 14= LeMans 15= Electra/Park Avenue 16= LeSabre 17= Regal 18= Reatta 19= 1993 Riviera/Toronado 20= Ultra 21= Allante 22= Concours 23= DeVille 24= Eldorado, ETC, Seville, STS 25= Achieva 26= Aurora/1994 and up Riviera 27= Calais 28= Cutlass 29= 88 Royale (1988 and up) 30= 98, Touring Sedan (1989 and up) 31= Supreme

Lighting, Horns, Turn Signals, Steering Column

Models	Year	TSB#	Description
12	92 - 93	338201	LAMP/FOG - MOISTURE COLLECTS IN LAMP ASM - NEW BULB
6	92 - 93	338108	LAMP/TRUNK - MAY TOUCH LUGGAGE - CORRECTION PROC
11	90 - 93	338103R	LIGHTING/BACKGROUND - INOP/COMEBACKS - CORRECTION
9	95 - 96	53-32-10	NEW SERVICE PART NUMBER FOR HORN ASSEMBLY
1	92 - 93	333210	POP NOISE - TILT STEER COLUMN - REPLACE BEARING SEAT
2	92 - 93	333210	POP NOISE/TILT STEER COLUMN- REPL BEARING SEAT - REV
10	93 - 93	333211	RATTLE IN STEERING WHEEL - REPLACE CONTROL SWITCH
24	92 - 93	313202	SQUEAK NOISE - STEER COLUMN WHEN MOVING SHIFT LEVER
22,23,24	95 - 95	63-32-01	STEERING COLUMN SHIFT LEVER COMES LOOSE
11,7	95 - 95	43-82-07	TRUNK LAMP BULB MAY CONTACT LUGGAGE - REPLACE BULB
6	92 - 93	93-260-10	TRUNK LAMP TOUCHES LUGGAGE - REV
24	92 - 93	318305	TURN SIGNAL - FLASHER SOUND NOT LOUD ENOUGH - REPAIR
12	94 - 95	43-83-17	TURN SIGNAL FLASHER LOCATION - S/M REVISION
25	94 - 95	43-83-17	TURN SIGNAL FLASHER LOCATION - S/M REVISION
3	91 - 94	438104	TURN SIGNAL INOP - REV SWITCH & FLASHER (TILT COL)
11,7	95 - 95	433205	TURN SIGNAL LEVER WON'T STAY IN DETENT POSITION
26	95 - 95	53-82-02	VENT HOSE ROUTING WHEN REPLACING HEADLAMP
12	96 - 96	53-15-18	WATER ENTERS TAIL LAMP
12	96 - 96	53-15-18	WATER ENTERS TAIL LAMP - REV

Model Legend: 1= All 2= Most 3= Beretta/Corsica 4= Celebrity/Century/Cierra/Cutlass Cruiser/6000 5= Camaro/Firebird 6= Caprice/Roadmaster/Impala SS/Fleetwood(RWD)/Estate Wagon/ Custom Cruiser 7= Monte Carlo 8= Corvette 9= Cavalier/Sunbird/Sunfire/Skyhawk 10= Bonneville, SSE, SSEi 11= Lumina 12= Grand Am 13= Gran Prix 14= LeMans 15= Electra/Park Avenue 16= LeSabre 17= Regal 18= Reatta 19= 1993 Riviera/Toronado 20= Ultra 21= Allante 22= Concours 23= DeVille 24= Eldorado, ETC, Seville, STS 25= Achieva 26= Aurora/1994 and up Riviera 27= Calais 28= Cutlass 29= 88 Royale (1988 and up) 30= 98, Touring Sedan (1989 and up) 31= Supreme

Instruments, Dash Cluster, Warning Lights, Mirrors

Models	Year	TSB#	Description
12	92 - 95	53-83-04	BLINKING CHECK OIL & CHECK GAUGES LITE
25	92 - 95	53-83-04	BLINKING CHECK OIL & CHECK GAUGES LITE
5	94 - 94	438109	BUZZ NOISE IN IP AFTER KEY OFF - REPLACE RAP MODULE
2	91 - 93	177138R	CEL - INTERMIT OR CONSTANT ON/TCC FLUCTUATION
25	92 - 93	338304	CHECK GAUGES INDICATOR - DESCRIPTION CHANGE - S/M UP
29,30	92 - 93	93-U-19	CLIP MOUNTED REAR VIEW MIRROR R&I PROC-S/M SUPPL/REV
24	92 - 93	318301	CONSOLE ASHTRAY LATCH - IMPROPER OPERATION - REPAIR
24	96 - 96	63-62-04	COOLANT TEMP GAUGE INDICATES RUNNING HOT
24	96 - 96	63-62-04	COOLANT TEMP GAUGE INDICATES RUNNING HOT - REV
2	93 - 93	377127	CRUISE - NO 3-4 SHIFT AFTER 4-3 DOWNSHIFT- WIRE PROC
5	93 - 95	57-90-01	CRUISE CNTRL - SPEED SURGE OR SLOW RESP - 3.4L V6
6	92 - 93	238114	CRUISE CONTROL - INTERMIT/NO OP - BRAKE SWITCH INFO
26	95 - 95	51-65-55	CRUISE CONTROL CHART C-17 - 4.0L - S/M REV
9	95 - 95	53-81-21	CRUISE CONTROL DIAGN PROCEDURE CHART 1 - S/M REV
25	95 - 95	53-81-21	CRUISE CONTROL DIAGN PROCEDURE CHART 1 - S/M REV
12,9	95 - 95	53-81-21	CRUISE CONTROL DIAGN PROCEDURE CHART 1 - S/M REV
22,23,24	94 - 94	435002	CRUISE CONTROL INOP - REPLACE LOWER STOPLAMP SWITCH
5	93 - 93	375001	CRUISE CONTROL INOP/INTERMIT - DEPRESS BRAKE FULLY
12	95 - 95	43-83-15	CRUISE CONTROL MODULE LOCATION - S/M REVISION
25	95 - 95	43-83-15	CRUISE CONTROL MODULE LOCATION - S/M REVISION
15,20,16	93 - 93	216521	CRUISE VENT/VAC SOLENOIDS-DTC 61/62 CHARTS-S/M REV
26	95 - 95	54-83-01	DIC DATE/ETA INCORRECT OR 12-DISC CD CHANGER INOP
29,30	91 - 93	91-T-118	DIC DISPLAY WARN MESSAGES (CANADIAN/EXPORT)- S/M REV
10	91 - 95	53-81-13	DIC DISPLAYS - NEW DIAGN CHART - S/M REV
29,30	91 - 95	53-81-13	DIC DISPLAYS - NEW DIAGN CHART - S/M REV
26	95 - 95	53-81-17	DIC DOOR AJAR TELLTALE INTERMITTENT
26	96 - 96	63-83-02	DIC READS "DATE NOT AVAILABLE" OR "-"
5	93 - 93	476114	ELEVATED OIL PRESS GAUGE READING - POOR GROUND
23,24	96 - 96	63-83-06	ERRATIC/INACCURATE DISPLAYS AND OR CHIMES
10	92 - 94	43-63-02	FAINT BUZZ NOISE- FRONT OF DASH - FUEL LINE RETAINER
15,16	91 - 94	43-63-02	FAINT BUZZ NOISE- FRONT OF DASH - FUEL LINE RETAINER
29,30	91 - 94	43-63-02	FAINT BUZZ NOISE- FRONT OF DASH - FUEL LINE RETAINER
10	92 - 95	53-83-17	FUEL GAUGE ERRATIC DURING QUICK DRIVING MANEUVERS
23,6	89 - 93	316306	FUEL GAUGE FLUCTUATES/HESITATION/POP ON COLD START
24	93 - 93	318303	FUEL GAUGE INACCURATE - REPLACE PROM
10	92 - 93	308102	GAUGE/FUEL - READS EMPTY W/FULL TANK - REPAIR PROC
1	90 - 93	268304	GAUGE/OIL PRESSURE - ERRATIC/INCORRECT READINGS/FIX
2	92 - 93	268306	GAUGE/OIL PRESSURE - HIGH READING AT START-UP - INFO
6	94 - 96	63-83-05	GLOVE BOX LATCH/LOCK REMOVE/INSTALL PROCED - S/M REV
10	93 - 94	439013	HEAD UP DISPLAY TEST A - S/M REVISION
22,23	94 - 94	431605	HEADLINER FALLS DOWN - REAR AREA NEAR LAMP MONITOR
10	92 - 95	10-81-50	HUD & IC SPEED DIFFERENCES - DIAGN & REPAIR

Model Legend: 1= All 2= Most 3= Beretta/Corsica 4= Celebrity/Century/Cierra/Cutlass Cruiser/6000 5= Camaro/Firebird 6= Caprice/Roadmaster/Impala SS/Fleetwood(RWD)/Estate Wagon/ Custom Cruiser 7= Monte Carlo 8= Corvette 9= Cavalier/Sunbird/Sunfire/Skyhawk 10= Bonneville, SSE, SSEi 11= Lumina 12= Grand Am 13= Gran Prix 14= LeMans 15= Electra/Park Avenue 16= LeSabre 17= Regal 18= Reatta 19= 1993 Riviera/Toronado 20= Ultra 21= Allante 22= Concours 23= DeVille 24= Eldorado, ETC, Seville, STS 25= Achieva 26= Aurora/1994 and up Riviera 27= Calais 28= Cutlass 29= 88 Royale (1988 and up) 30= 98, Touring Sedan (1989 and up) 31= Supreme

Instruments, Dash Cluster, Warning Lights, Mirrors

Models	Year	TSB#	Description
1	93 - 93	93-WA-13	I/C - VEHICLES W/WRONG I/C (FROM PRODUCTION) - PROC
11	93 - 93	231075	I/P - CREAK NOISE FROM LEFT SIDE - CORRECTION PROC
11	93 - 93	231063	I/P - POP/CREAK NOISE - ENGINE FRAME BOLT PROCEDURE
12	92 - 93	338301	I/P COMPARTMENT NOT FLUSH WITH I/P - CORRECTION PROC
15,20,16	91 - 94	92-10-33	I/P CONTACTS DOOR TRIM PAD - POOR APPEAR/SQUEAK- REV
6,9,3	91 - 93	238301R	I/P CUPHOLDER - BINDING/STUCK - SERVICE PROCEDURE
25	94 - 94	53-81-04	IC (BASE) - CELL 80 - S/M REV
10	95 - 95	43-81-43	IC - VOLTS INDICATOR DIAGNOSTICS - S/M REV
16,15	95 - 95	43-81-43	IC - VOLTS INDICATOR DIAGNOSTICS - S/M REV
29,30	95 - 95	43-81-43	IC - VOLTS INDICATOR DIAGNOSTICS - S/M REV
9	95 - 95	53-83-14	IC CONNECTOR PIN-OUT LABELING - S/M REV
22,23,24	94 - 94	318112	IC EXCHANGE - INCORRECT WORDING OF DRIVER MESSAGES
1	96 - 96	48-83-07	IC IDENTIFICATION INFO
12	96 - 96	53-83-24	IC MICROPROCESSOR DIAGN - REV INFO/CHART 7 - S/M REV
25	96 - 96	53-83-24	IC MICROPROCESSOR DIAGN - REV INFO/CHART 7 - S/M REV
9	95 - 95	53-83-12	IC REPLACEMENT - 2.3L W/ 4-SPEED A/T - S/M REV
16,15	95 - 95	43-81-42	IC VOLTS INDICATOR - DIAGNOSIS - S/M REV
29,30	95 - 95	43-81-42	IC VOLTS INDICATOR - DIAGNOSIS - S/M REV
29,30	91 - 93	338110	IC/DIGITAL - C2 PINOUT TEST REVISED - S/M UPDATE
22,23,24	96 - 96	63-83-04	INACCURATE FUEL GAUGE DISPLAY - BCM/PROM
29,30	95 - 95	43-83-22	INADVERTANT LOW FUEL WARNING CHIME - WIRING
21	93 - 93	218306	INSTR PANEL/COWL - VIBRATE/SMOOTH ROAD - DIAG/REPAIR
21	93 - 93	93-027/10	INSTR PANEL/COWL - VIBRATE/SMOOTH ROAD - REVISION
1	94 - 94	488305	INSTRUMENT CLUSTER IDENTIFICATION INFORMATION
10	92 - 95	63-90-02	INTERMIT DROP OUT OF CRUISE CONTROL - 3.8L
15,16	91 - 95	63-90-02	INTERMIT DROP OUT OF CRUISE CONTROL - 3.8L
29,30	91 - 95	63-90-02	INTERMIT DROP OUT OF CRUISE CONTROL - 3.8L
9	94 - 94	438103	IP - BASE CLUSTER DIAGNOSTICS - S/M REV
10	93 - 93	208119	IP COMPARTMENT - LOCK CYLINDER R&I PROC - S/M UPDATE
22,	94 - 94	438309	IP ITCH NOISE BETWEEN TOP COVER & WOOD TRIM
5	93 - 94	438308	IP TRIM BEZEL "ITCHING" NOISE & ETCHING
26	95 - 95	438312	IP TRIM PLATE SEPARATES FROM UPPER TRIM PAD
16,15,20	91 - 94	438311	IP TRIM PLATE WARPED/WAVY AND/OR TELLTALES DISTORTED
10	92 - 95	43-83-14	LOW ILLUMINATION OF CONSOLE PRNDL INDICATOR
5	94 - 95	43-10-24	MIRROR APPLIQUE CHANGES
10	92 - 93	208120R	MIRROR/BREAKAWAY REAR VIEW - R&I PROCEDURES - S/M UP
15,20,16	92 - 93	208120	MIRROR/CLIP MOUNTED REAR VIEW - R&I PROC - ESM UP
29,30	92 - 93	208120	MIRROR/CLIP MOUNTED REAR VIEW - R&I PROC - S/M SUPPL
10	92 - 93	208120	MIRROR/CLIP MOUNTED REAR VIEW - R&I PROC - S/M UP
2	93 - 93	208120R	MIRROR/ELECTROCHROMIC REARVIEW - R&I PROC - S/M REV
15,20	93 - 93	311503	MIRROR/ELECTROCHROMIC REARVIEW - VIB - REPLACE PROC
23,24	93 - 93	311503	MIRROR/ELECTROCHROMIC REARVIEW - VIBRATION - REPAIR

Model Legend: 1= All 2= Most 3= Beretta/Corsica 4= Celebrity/Century/Cierra/Cutlass Cruiser/6000 5= Camaro/Firebird 6= Caprice/Roadmaster/Impala SS/Fleetwood(RWD)/Estate Wagon/ Custom Cruiser 7= Monte Carlo 8= Corvette 9= Cavalier/Sunbird/Sunfire/Skyhawk 10= Bonneville, SSE, SSEi 11= Lumina 12= Grand Am 13= Gran Prix 14= LeMans 15= Electra/Park Avenue 16= LeSabre 17= Regal 18= Reatta 19= 1993 Riviera/Toronado 20= Ultra 21= Allante 22= Concours 23= DeVille 24= Eldorado, ETC, Seville, STS 25= Achieva 26= Aurora/1994 and up Riviera 27= Calais 28= Cutlass 29= 88 Royale (1988 and up) 30= 98, Touring Sedan (1989 and up) 31= Supreme

Instruments, Dash Cluster, Warning Lights, Mirrors

Models	Year	TSB#	Description
15,20,16	91 - 93	306002	MONITOR/ENGINE OIL LIFE - OPERATION AND DIAGNOSIS
22,23,24	96 - 96	63-81-02	NO CRANK AFTER IP REPLACEMENT
6	94 - 94	438112	OIL PRESSURE GAUGE MAY READ ZERO - (EXCEPT SEO)
15	92 - 94	431504	OUTSIDE ELECTROCHOMATIC MIRROR INSTALL - S/M REV
26	95 - 95	431504	OUTSIDE ELECTROCHOMATIC MIRROR INSTALL - S/M REV
29	92 - 94	431504	OUTSIDE ELECTROCHOMATIC MIRROR INSTALL - S/M REV
22,23,24	95 - 95	431504	OUTSIDE ELECTROCHOMATIC MIRROR INSTALL - S/M REV
23,6	92 - 93	431504	OUTSIDE ELECTROCHOMATIC MIRROR INSTALL - S/M REV
2	92 - 94	431504	OUTSIDE ELECTROCHOMATIC MIRROR INSTALL - S/M REV 2
22,23,24	94 - 94	43-83-19	RATTLE IN ASHTRAY AREA
24	92 - 93	92-I-01 & 93-I-01	REMANUFACTURED CLUSTER (DIGITAL/ANALOG)- RETURN INFO
9	95 - 95	53-83-11	SPEEDO/TACH GAGE FLUTTER WHEN TECH 1 CONNECTED -INFO
10	91 - 93	208122	SPEEDOMETER - INACCURATE - TEST A INFO REV - S/M UP
2	88 - 93	268305R	SPEEDOMETER - REGISTERS WHEN VEHICLE IS STAIONARY
30,29	91 - 93	208122	SPEEDOMETER INACCURATE - SPEED SENSOR DIAG - S/M REV
6	93 - 93	93C03	SPEEDOMETER INACCURATE/INCORRECT AXLE RATIO - RECALL
5	93 - 95	56-90-02	STEPPER MOTOR CRUISE CNTRL - NO RESUME OR RESET -REV
26	95 - 95	43-83-13	TCS ON/OFF SWITCH FALLS INTO CENTER CONSOLE
10	95 - 95	53-83-05	TEMP GAGE READS HIGH OR FLUCTUATES - U50 IC
12	88 - 93	88-06-56	TEMP GAUGE FLUCTUATION NORMAL - 4 CYL
3	92 - 93	338118	TEMP GAUGE READS HIGH/NO OVERHEAT - HARNESS & SENSOR
5	93 - 94	438107	TEMP GAUGE READS TOO HIGH - INSTALL RESISTOR
9,3	92 - 94	478101	TEMPERATURE GAUGE FLUCTUATION - MAY BE NORMAL - 2.2L
3	92 - 93	331041	VOLT INDICATOR ON/GAUGE READS LOW - SPLASH SHIELD
8	92 - 93	335007	WARN LITES/ABS & SERV - ON/FALSE CODE 83 - CAUSE/FIX
10	94 - 94	43-81-40	WARNINGS & ALARMS - CHIME - S/M REV
29,30	94 - 94	43-81-40	WARNINGS & ALARMS - CHIME - S/M REV
29,30	94 - 94	438304	WARPING/RATTLING - IP DEFROSTER GRILLE - REPLACE

Model Legend: 1= All 2= Most 3= Beretta/Corsica 4= Celebrity/Century/Cierra/Cutlass Cruiser/6000 5= Camaro/Firebird 6= Caprice/Roadmaster/Impala SS/Fleetwood(RWD)/Estate Wagon/ Custom Cruiser 7= Monte Carlo 8= Corvette 9= Cavalier/Sunbird/Sunfire/Skyhawk 10= Bonneville, SSE, SSEi 11= Lumina 12= Grand Am 13= Gran Prix 14= LeMans 15= Electra/Park Avenue 16= LeSabre 17= Regal 18= Reatta 19= 1993 Riviera/Toronado 20= Ultra 21= Allante 22= Concours 23= DeVille 24= Eldorado, ETC, Seville, STS 25= Achieva 26= Aurora/1994 and up Riviera 27= Calais 28= Cutlass 29= 88 Royale (1988 and up) 30= 98, Touring Sedan (1989 and up) 31= Supreme

Chassis Electrical, Wiring Harness, Fuses-Circuit Breakers, Wipers, Window Motors

Models	Year	TSB#	Description
10	95 - 95	53-81-19	ABS PUMP & ABS MAIN RELAYS CLARIFIED - S/M REV
16,15	95 - 95	53-81-19	ABS PUMP & ABS MAIN RELAYS CLARIFIED - S/M REV
29,30	95 - 95	53-81-19	ABS PUMP & ABS MAIN RELAYS CLARIFIED - S/M REV
3	94 - 94	438101	BATTERY RUNDOWN - INTERIOR LIGHTS STAY ON
9	92 - 94	438101	BATTERY RUNDOWN - INTERIOR LIGHTS STAY ON
3	92 - 94	438101	BATTERY RUNDOWN - INTERIOR LIGHTS STAY ON - REV
9	92 - 94	438101	BATTERY RUNDOWN - INTERIOR LIGHTS STAY ON - REV
3	92 - 94	43-81-01	BATTERY RUNDOWN - INTERIOR LIGHTS STAY ON - REV2
9	92 - 94	43-81-01	BATTERY RUNDOWN - INTERIOR LIGHTS STAY ON - REV2
3	94 - 94	331064	BINDING/INOP DRIVER WINDOW - REPL WINDOW MOTOR - REV
3	94 - 94	331064	BINDING/INOP DRIVER WINDOW - REPLACE WINDOW MOTOR
8	95 - 95	43-82-05	CHATTER/NOISE - RIGHT HAND WIPER BLADE
10	92 - 93	338117	CKT BREAKER 4 OPENS WHEN PWR SEAT & REAR DEFOG OPER
10	94 - 94	41-65-24	DASH HARNESS TO ENG HARNESS INTERCONNECT - S/M REV
16,15	94 - 94	41-65-24	DASH HARNESS TO ENG HARNESS INTERCONNECT - S/M REV
29,30	94 - 94	41-65-24	DASH HARNESS TO ENG HARNESS INTERCONNECT - S/M REV
10	92 - 94	53-64-01	DIM LAMPS/SLOW WIPERS - LOW/NO GENERATOR OUTPUT
16,15	92 - 94	53-64-01	DIM LAMPS/SLOW WIPERS - LOW/NO GENERATOR OUTPUT
29,30	92 - 94	53-64-01	DIM LAMPS/SLOW WIPERS - LOW/NO GENERATOR OUTPUT
1	84 - 93	178201R	DIODES (ISOLATION) - USAGE AND REPLACEMENT INFO
6	94 - 94	438116	ELECT DIAGN - SCHEMATICS & ILLUSTRATIONS - S/M REV
10	94 - 95	53-81-09	ELECTRICAL DIA12S - DOOR LOCKS - S/M REV
5	93 - 94	316526	ELECTRICAL DIAGNOSIS/CHARTS/WIRE DIAG - S/M REVISION
10	93 - 93	208123	ELECTRICAL DIAGNOSTIC REVISIONS - ESM UPDATE
15,20,16	93 - 93	208123	ELECTRICAL DIAGNOSTIC REVISIONS - ESM UPDATE
30,29	93 - 93	208123	ELECTRICAL DIAGNOSTICS - S/M REVISIONS
16,15	94 - 95	53-81-09	ELECTRICAL DIAGRAMS - DOOR LOCKS - S/M REV
29,30	94 - 95	53-81-09	ELECTRICAL DIAGRAMS - DOOR LOCKS - S/M REV
1	93 - 93	92-08-07	ELECTRICAL SYS - ISOLATION DIODES - REPLACE INFO/REV
24	92 - 93	318101	ELECTRICAL SYSTEM - ERRATIC OPERATION - SERV INFO
1	Up - 94	149301	EMI/RFI DISTURBANCE - NICKEL TAPE AVAILABLE
8	94 - 94	94C21	ENGINE WIRING HARNESS ROUTING - LT1 - RECALL
23	93 - 93	319005	FALSE THEFT ALARM/DEAD BATTERY/RAP ON/LIGHTS INOP
6	91 - 93	438117	FUSE BLOWS - REAR WINDOW DEFOG - INSTALL CKT BREAKER
9	86 - 94	438118	FUSE BLOWS/NO START/BATTERY DOWN - CONVERTIBLE ONLY
10	95 - 95	63-81-05	GROUND (G201) SCHEMATICS - S/M REV
16,15	95 - 95	63-81-05	GROUND (G201) SCHEMATICS - S/M REV
29,30	95 - 95	63-81-05	GROUND (G201) SCHEMATICS - S/M REV
9	95 - 95	53-81-23	GROUND G112 LOCATION - 2.2L - S/M REV
10	94 - 94	43-81-32	HARNESS CONNECTOR FACES ILLUSTRATIONS - S/M REV
29,30	94 - 94	43-81-32	HARNESS CONNECTOR FACES ILLUSTRATIONS - S/M REV
25	92 - 93	238120R	HARNESS/ENG - REPAIR/REROUTE - INTERMIT ELEC COND

Model Legend: 1= All 2= Most 3= Beretta/Corsica 4= Celebrity/Century/Cierra/Cutlass Cruiser/6000 5= Camaro/Firebird 6= Caprice/Roadmaster/Impala SS/Fleetwood(RWD)/Estate Wagon/ Custom Cruiser 7= Monte Carlo 8= Corvette 9= Cavalier/Sunbird/Sunfire/Skyhawk 10= Bonneville, SSE, SSEi 11= Lumina 12= Grand Am 13= Gran Prix 14= LeMans 15= Electra/Park Avenue 16= LeSabre 17= Regal 18= Reatta 19= 1993 Riviera/Toronado 20= Ultra 21= Allante 22= Concours 23= DeVille 24= Eldorado, ETC, Seville, STS 25= Achieva 26= Aurora/1994 and up Riviera 27= Calais 28= Cutlass 29= 88 Royale (1988 and up) 30= 98, Touring Sedan (1989 and up) 31= Supreme

Chassis Electrical, Wiring Harness, Fuses-Circuit Breakers, Wipers, Window Motors

Models	Year	TSB#	Description
26	95 - 95	53-81-27	I/P 1 FUSE BLOWS - PWR DOOR LOCKS INOP
9	95 - 95	53-83-14	IC CONNECTOR PIN-OUT LABELING - S/M REV
29,30	95 - 95	43-83-22	INADVERTANT LOW FUEL WARNING CHIME - WIRING
10	92 - 93	53-81-28	INOP PWR SEATS OR LOCKS WITH REAR DEFOGGER ON
16,15	92 - 93	53-81-28	INOP PWR SEATS OR LOCKS WITH REAR DEFOGGER ON
29,30	92 - 93	53-81-28	INOP PWR SEATS OR LOCKS WITH REAR DEFOGGER ON
12	95 - 95	43-81-34	INOP WASHER PUMP - REV FILTER GROMMET
25	95 - 95	43-81-34	INOP WASHER PUMP - REV FILTER GROMMET
3,7	95 - 95	43-81-34	INOP WASHER PUMP - REV FILTER GROMMET
10	93 - 93	338122	INOP/ERRATIC DRIVER SIDE POWER WINDOW (EXPRESS DOWN)
16,15	93 - 93	338122	INOP/ERRATIC DRIVER SIDE POWER WINDOW (EXPRESS DOWN)
29,30	93 - 93	338122	INOP/ERRATIC DRIVER SIDE POWER WINDOW (EXPRESS DOWN)
10	93 - 93	338122A	INOP/ERRATIC DRIVER SIDE POWER WINDOW - REVISED
15,16	93 - 93	338122	INOP/ERRATIC DRIVER SIDE POWER WINDOW - REVISED
29,30	93 - 93	338122	INOP/ERRATIC DRIVER SIDE POWER WINDOW - REVISED
22,23,24	95 - 95	53-81-35	INTERMIT BLOWN FUSE (C5) OR BATTERY DRAIN
12	92 - 93	238120R	INTERMIT ELEC CONDITIONS- REPAIR/REROUTE ENG HARNESS
11,7	95 - 95	53-32-06	INTERMITTENT PROBLEMS IN WIPER/WASHER - INFO
12	96 - 96	53-81-26	LABEL/DESCRIPTION OF CRANKLOW CIRCUIT 1350 - S/M REV
25	96 - 96	53-81-26	LABEL/DESCRIPTION OF CRANKLOW CIRCUIT 1350 - S/M REV
11,7	95 - 95	57-64-05	NO START/BATTERY DISCHARGED - FAN RELAY - NEW PROM
5	93 - 94	438202	NOISE - WINDSHIELD WIPER OPERATION
24	92 - 93	311512	NOISE/WEAR - WIPER TRANSMISSION LINKAGE
12	96 - 96	53-62-03	OVERHEAT/COOL FAN INOP - UP-SIZE CLG FAN MAXI FUSE
25	96 - 96	53-62-03	OVERHEAT/COOL FAN INOP - UP-SIZE CLG FAN MAXI FUSE
10	96 - 96	53-83-21	PERIMETER LIGHTING RELAY - REPL PROCEDURE - S/M REV
16,15	96 - 96	53-83-21	PERIMETER LIGHTING RELAY - REPL PROCEDURE - S/M REV
29,30	96 - 96	53-83-21	PERIMETER LIGHTING RELAY - REPL PROCEDURE - S/M REV
24	94 - 94	438115	POWER LOSS/MIL ON - REROUTE ENGINE HARNESS
9	93 - 94	438108	PULSE WIPER MOTOR DIAGNOSTIC CHART 3 - S/M REV
12	94 - 94	438106	RELAY POSITION ILLUSTRATION - S/M REVISION
25	94 - 94	438106	RELAY POSITION ILLUSTRATION - S/M REVISION
9	95 - 96	53-82-08	REPLACEMENT OF UPPER WIPER/WASHER NOZZLE - INFO
12	95 - 95	53-81-39	SCMEMATIC FOR SPLICES S226/S227 (EXT LIGHT)- S/M REV
25	95 - 95	53-81-39	SCMEMATIC FOR SPLICES S226/S227 (EXT LIGHT)- S/M REV
5	93 - 93	331058	SIX WAY POWER SEAT SWITCH BREAKS - REPLACE SWITCH
26	96 - 96	53-83-19	STARTER ENABLE RELAY REPLACE PROCEDURE - S/M REV
3,7	95 - 95	53-81-06	VARIOUS REV TO ELECT DIAGNOSIS (8A) - S/M REV
9	90 - 94	331042	WASHER NOZZLE/TOP VENT SCREEN REMOVAL - S/M REVISION
12	92 - 93	238203	WINDSHIELD WASHER - FLUID LOSS OR INOP - SERV PROC
9	92 - 93	338202	WINDSHIELD WASHER - FLUID LOSS/CRACKED RESERVOIR
3	92 - 93	238203	WINDSHIELD WASHER INOP/FLUID LOSS - CORRECTION

Model Legend: 1= All 2= Most 3= Beretta/Corsica 4= Celebrity/Century/Cierra/Cutlass Cruiser/6000 5= Camaro/Firebird 6= Caprice/Roadmaster/Impala SS/Fleetwood(RWD)/Estate Wagon/ Custom Cruiser 7= Monte Carlo 8= Corvette 9= Cavalier/Sunbird/Sunfire/Skyhawk 10= Bonneville, SSE, SSEi 11= Lumina 12= Grand Am 13= Gran Prix 14= LeMans 15= Electra/Park Avenue 16= LeSabre 17= Regal 18= Reatta 19= 1993 Riviera/Toronado 20= Ultra 21= Allante 22= Concours 23= DeVille 24= Eldorado, ETC, Seville, STS 25= Achieva 26= Aurora/1994 and up Riviera 27= Calais 28= Cutlass 29= 88 Royale (1988 and up) 30= 98, Touring Sedan (1989 and up) 31= Supreme

Chassis Electrical, Wiring Harness, Fuses-Circuit Breakers, Wipers, Window Motors

Models	Year	TSB#	Description
25	92 - 93	238203	WINDSHIELD WASHER INOP/FLUID LOSS - SERVICE PROC
12	92 - 93	93-08-07	WINDSHIELD WASHER INOP/FLUID LOSS - SERVICE PROC REV
25	92 - 93	93-T-033	WINDSHIELD WASHER INOP/FLUID LOSS - SERVICE PROC REV
3,7	92 - 93	93-067-08E	WINDSHIELD WASHER INOP/FLUID LOSS - SERVICE PROC REV
10	94 - 94	331070	WIPER ARM ASSEMBLY REMOVAL/REPLACE PROCED - S/M REV
16,15	94 - 94	331070	WIPER ARM ASSEMBLY REMOVAL/REPLACE PROCED - S/M REV
29,30	94 - 94	331070	WIPER ARM ASSEMBLY REMOVAL/REPLACE PROCED - S/M REV
9	95 - 95	53-82-04	WIPER ARM REMOVAL PROCEDURE - S/M REV
6	91 - 94	43-82-06	WIPER BLADE CHATTER - CORRECT PARTS USAGES
6	91 - 95	43-82-06	WIPER BLADE CHATTER - CORRECT PARTS USAGES - REV
10	94 - 94	53-82-03	WIPER BLADE CHATTER - REVISED ARM
16,15	94 - 94	53-82-03	WIPER BLADE CHATTER - REVISED ARM
22,23,24	94 - 95	53-82-03	WIPER BLADE CHATTER - REVISED ARM
29,30	94 - 94	53-82-03	WIPER BLADE CHATTER - REVISED ARM
24,23,22,	92 - 94	311513	WIPER BLADE CHATTER DURING WASH/WIPE - REV PARTS
24	90 - 94	431031	WIPER WASHER ARM NOZZLES RE-INSTALL PROCEDURE
22,23	94 - 94	431031	WIPER WASHER ARM NOZZLES RE-INSTALL PROCEDURE
16,15,20	91 - 93	338107	WIPER/WASHER - PULSE SYS DIAGNOSIS REV - S/M UPDATE
12	94 - 95	53-81-37	WIPER/WASHER DIAGNOSTIC INFO - S/M REV
25	94 - 95	53-81-37	WIPER/WASHER DIAGNOSTIC INFO - S/M REV
2	94 - 94	377133	WIRE HARNESS & TRANSAXLE PIN LOCATION CHANGES -4T60E
10	95 - 95	53-81-08	WIRING DIAG - INTERIOR LAMPS - S/M REV
16,15	95 - 95	53-81-08	WIRING DIAG - INTERIOR LAMPS - S/M REV
29,30	95 - 95	53-81-08	WIRING DIAG - INTERIOR LAMPS - S/M REV
12	93 - 93	236405	WIRING/SPARK PLUG - NEW ILLUSTRATION (3.3L) - S/M UP

Model Legend: 1= All 2= Most 3= Beretta/Corsica 4= Celebrity/Century/Cierra/Cutlass Cruiser/6000 5= Camaro/Firebird 6= Caprice/Roadmaster/Impala SS/Fleetwood(RWD)/Estate Wagon/ Custom Cruiser 7= Monte Carlo 8= Corvette 9= Cavalier/Sunbird/Sunfire/Skyhawk 10= Bonneville, SSE, SSEi 11= Lumina 12= Grand Am 13= Gran Prix 14= LeMans 15= Electra/Park Avenue 16= LeSabre 17= Regal 18= Reatta 19= 1993 Riviera/Toronado 20= Ultra 21= Allante 22= Concours 23= DeVille 24= Eldorado, ETC, Seville, STS 25= Achieva 26= Aurora/1994 and up Riviera 27= Calais 28= Cutlass 29= 88 Royale (1988 and up) 30= 98, Touring Sedan (1989 and up) 31= Supreme

Auxiliary Equipment, Jacks, Trailer Hitches, Towing

Models	Year	TSB#	Description
9	95 - 96	53-01-05	FLAT TOWING (ALL 4 WHEELS ON GROUND) - 4T40E
9	95 - 96	53-01-05	FLAT TOWING (ALL 4 WHEELS ON GROUND) - 4T40E - REV
29	94 - 95	47-90-02	GUIDESTAR NAVIGATION/INFORMATION SYSTEM
1	93 - 93	219602	PHONE/MOBILE RADIO - INSTALL & TROUBLESHOOTING INFO
1	Up - 93	219602	RADIO TELEPHONE/MOBILE RADIO - GUIDELINE BOOK AVAIL
22,23,24	94 - 94	430101	TOWING MANUAL (T65) - REVISED LIMITS
15,30	93 - 93	330113	TRAILER HITCH INSTALLATION - INSTALL ENERGY ABSORBER

Model Legend: 1= All 2= Most 3= Beretta/Corsica 4= Celebrity/Century/Cierra/Cutlass Cruiser/6000 5= Camaro/Firebird 6= Caprice/Roadmaster/Impala SS/Fleetwood(RWD)/Estate Wagon/ Custom Cruiser 7= Monte Carlo 8= Corvette 9= Cavalier/Sunbird/Sunfire/Skyhawk 10= Bonneville, SSE, SSEi 11= Lumina 12= Grand Am 13= Gran Prix 14= LeMans 15= Electra/Park Avenue 16= LeSabre 17= Regal 18= Reatta 19= 1993 Riviera/Toronado 20= Ultra 21= Allante 22= Concours 23= DeVille 24= Eldorado, ETC, Seville, STS 25= Achieva 26= Aurora/1994 and up Riviera 27= Calais 28= Cutlass 29= 88 Royale (1988 and up) 30= 98, Touring Sedan (1989 and up) 31= Supreme

Heating, Air Conditioning, Ventilation, Defogger

Models	Year	TSB#	Description
3,7	93 - 93	336108	3.1L GROWL/MOAN/SQUEAL - SERPENTINE BELT/SPLASH SHLD
10	93 - 93	301101	A/C (AUTOMATIC) - DISPLAY ILLUM CNTRL TEST - S/M REV
1	83 - 93	231205R	A/C - ALTERNATE REFRIGERANT USE INFO TO REPLACE R-12
6	92 - 93	331204	A/C - CLICK/THUMP NOISE DURING OP - CAUSE/CORRECTION
15,20,16	93 - 93	201203	A/C - COMP INOP/REVERTS TO ECON MODE/CODE 48 STORED
24,22,23	94 - 94	431203	A/C - CONTINUOUS HISS NOISE - REPL ORIFICE ASSEMBLY
10	93 - 93	201203	A/C - GOES TO ECON MODE/COMPRESSOR INOP/CODE 48
15,20,16,17	91 - 93	201201R	A/C - INOP OR NO COOLING - DIAGNOSIS AND CORRECTION
10	95 - 95	53-12-11	A/C - INSUFFICENT COOLING CHART B - S/M REV
16,15	95 - 95	53-12-11	A/C - INSUFFICENT COOLING CHART B - S/M REV
29,30	95 - 95	53-12-11	A/C - INSUFFICENT COOLING CHART B - S/M REV
12	94 - 94	43-12-22	A/C - INSUFFICIENT COOLING - IMPROPERLY SEATED SEAL
25	94 - 94	43-12-22	A/C - INSUFFICIENT COOLING - IMPROPERLY SEATED SEAL
3,7	94 - 94	43-12-22	A/C - INSUFFICIENT COOLING - IMPROPERLY SEATED SEAL
10	91 - 93	201201R	A/C - NO COOLING/NO A/C - ECM CODE 66 INFO - 2ND REV
3	92 - 93	331211	A/C - NO COOLING/REFRIGERANT LOSS - CORRECTION
22,23,24	94 - 96	63-12-05	A/C - POOR PERF IN CITY TRAFFIC AT HIGH AMB TEMP
3	92 - 93	231213R	A/C - POOR PERFORMANCE W/VEHICLE IN REVERSE - SERV
1	84 - 93	331209R	A/C - R134A REFRIGERANT - INFORMATION ON USAGE
23,6	92 - 93	200106R	A/C - RECIRCULATE MODE WHISTLE NOISE - CORRECTION
10	92 - 93	331218	A/C - REDUCED PERFORMANCE - FOAM PANELS/FOAM BLOCK
10	92 - 93	200103	A/C - TOO WARM ON EXTENDED IDLE - HOOD SEAL KIT
29,30	92 - 93	200106	A/C - WHISTLE IN RECIRCULATE MODE - CORRECTION PROC
10	92 - 93	200106	A/C - WHISTLE NOISE IN RECIRCULATE MODE - CORRECTION
15,16	92 - 93	200106	A/C - WHISTLE NOISE/RECIRCULATE MODE - DIAG/REPAIR
26	95 - 95	51-65-55	A/C CLUTCH CKT DIAG CHART C-10 - 4.0L - S/M REV
10	94 - 94	41-65-24	A/C CLUTCH DIAG - CHARTS C-10 A & B - S/M REV
16,15	94 - 94	41-65-24	A/C CLUTCH DIAG - CHARTS C-10 A & B - S/M REV
29,30	94 - 94	41-65-24	A/C CLUTCH DIAG - CHARTS C-10 A & B - S/M REV
3,9	94 - 94	331225	A/C COMP & BRACKET REVISED FASTENER TORQUE - S/M REV
12	94 - 95	431217A	A/C COMP CUT OFF DURING LONG IDLE AT HIGH TEMP -REV2
25	94 - 95	431217A	A/C COMP CUT OFF DURING LONG IDLE AT HIGH TEMP -REV2
3,7	94 - 95	431217A	A/C COMP CUT OFF DURING LONG IDLE AT HIGH TEMP -REV2
3	94 - 94	431217	A/C COMP CUTS OFF DURING LONG IDLE AT HIGH TEMP -REV
12	94 - 94	431217	A/C COMP CUTS OFF DURING LONG IDLE AT HIGH TEMP -REV
25	94 - 94	431217	A/C COMP CUTS OFF DURING LONG IDLE AT HIGH TEMP -REV
12	94 - 94	431217	A/C COMP CUTS OFF DURING LONG IDLE AT HIGH TEMP-3.1L
25	94 - 94	431217	A/C COMP CUTS OFF DURING LONG IDLE AT HIGH TEMP-3.1L
3,7	94 - 94	431217	A/C COMP CUTS OFF DURING LONG IDLE AT HIGH TEMP-3.1L
23	94 - 94	43-12-25	A/C COMP MOAN NOISE UNDER LIGHT ACCEL - 4.9L
23	94 - 94	43-12-25	A/C COMP MOAN NOISE UNDER LIGHT ACCEL -4.6/4.9L-REV
24	92 - 93	43-12-25	A/C COMP MOAN NOISE UNDER LIGHT ACCEL -4.6/4.9L-REV

Model Legend: 1= All 2= Most 3= Beretta/Corsica 4= Celebrity/Century/Cierra/Cutlass Cruiser/6000 5= Camaro/Firebird 6= Caprice/Roadmaster/Impala SS/Fleetwood(RWD)/Estate Wagon/ Custom Cruiser 7= Monte Carlo 8= Corvette 9= Cavalier/Sunbird/Sunfire/Skyhawk 10= Bonneville, SSE, SSEi 11= Lumina 12= Grand Am 13= Gran Prix 14= LeMans 15= Electra/Park Avenue 16= LeSabre 17= Regal 18= Reatta 19= 1993 Riviera/Toronado 20= Ultra 21= Allante 22= Concours 23= DeVille 24= Eldorado, ETC, Seville, STS 25= Achieva 26= Aurora/1994 and up Riviera 27= Calais 28= Cutlass 29= 88 Royale (1988 and up) 30= 98, Touring Sedan (1989 and up) 31= Supreme

Heating, Air Conditioning, Ventilation, Defogger

Models	Year	TSB#	Description
2	93 - 93	111402R	A/C COMPRESSOR (HR6HE) - SHAFT NUT REMOVAL
1	91 - 93	261204R	A/C COMPRESSOR (R4 OR HR6HE) - SHAFT NUT ELIMINATED
24	93 - 93	311201	A/C COMPRESSOR - OVERHAUL PROC - S/M DESIGNATION
12	94 - 94	331224	A/C COMPRESSOR BRAP NOISE DURING HARD ACCELERATION
25	94 - 94	331224	A/C COMPRESSOR BRAP NOISE DURING HARD ACCELERATION
3,7	94 - 94	331224	A/C COMPRESSOR BRAP NOISE DURING HARD ACCELERATION
5	93 - 93	331401	A/C COMPRESSOR BUMP - PARTS & PROCEDURES
23,6	92 - 93	311404	A/C COMPRESSOR MOAN NOISE - REPLACE CLUTCH DRIVER
23,6,24	90 - 93	91-085/01	A/C COMPRESSOR MOAN NOISE - SERVICE PROCEDURE - REV
29,30	93 - 93	301101	A/C DISPLAY ILLUM CONTROL TEST DIAGNOSIS - S/M REV
10	96 - 96	63-12-10	A/C NOISES - DIAGNOSTIC PROCEDURES - 3800
16,15,19	96 - 96	63-12-10	A/C NOISES - DIAGNOSTIC PROCEDURES - 3800
29,30	96 - 96	63-12-10	A/C NOISES - DIAGNOSTIC PROCEDURES - 3800
2	85 - 93	93-075/01	A/C ODOR AT START UP - HOT/HUMID CLIMATES - REVISION
2	85 - 93	90-082/01	A/C ODOR AT START UP - HOT/HUMID CLIMATES - REVISION
2	85 - 94	93-075A/01	A/C ODOR AT START UP - HUMID CLIMATES - REVISION
2	85 - 95	911204	A/C ODOR AT START UP - HUMID CLIMATES - REVISION 2
1	93 - 96	53-12-12	A/C ODOR AT START UP IN HUMID CLIMATES
5	93 - 93	331221	A/C POOR PERF - DIAGN - LOW PRESS CUT OFF SWITCH
6	94 - 94	431211	A/C POOR PERFORMANCE - REPLACE TEMP CONTROL CABLE
10	92 - 94	431210	A/C POOR PERFORMANCE AFTER LONG IDLE - ADJ BAFFLE
9	94 - 94	331228	A/C REFRIGERANT CHANGE TO R134A
26	95 - 95	53-12-08	A/C SYSTEM FAILURE - REFRIGERANT LOSS
2	85 - 93	211209	A/C- LABOR TIMES FOR EVAP CORE & A/C ODOR OPERATIONS
5	93 - 93	331219	A/C- WHITE FLAKES FROM VENTS (EVAP SILICATE COATING)
29,30,31	93 - 93	331103R	A/C- WHITE FLAKES FROM VENTS (EVAP SILICATE COATING)
6	93 - 93	331220R	A/C- WHITE POWDER FROM VENTS DURING OP - CORRECTION
10	92 - 93	201106	ACC - TOO WARM ON COLD START/TOO COOL IN WARM AMB
24	93 - 93	318114	ACP CODE A049 - "A/C OVERHEAT/COMPRESSOR OFF" MSG
10	95 - 95	53-81-34	AIR DELIVERY/TEMP CNTRLS SCHEMATIC (ECC) - S/M REV
16,15	95 - 95	53-81-34	AIR DELIVERY/TEMP CNTRLS SCHEMATIC (ECC) - S/M REV
29,30	95 - 95	53-81-34	AIR DELIVERY/TEMP CNTRLS SCHEMATIC (ECC) - S/M REV
6	92 - 93	231210R	BI-LEVEL MODE/EXTREME TEMP DIFFERENCE IN VENTS/DUCTS
11,7	95 - 95	43-11-10	CENTER DEFROSTER GRILL LOOSE RIGHT SIDE
24	92 - 94	431106	CONDENSATION DRIPS FROM HEATER DUCT
22,23	94 - 94	431106	CONDENSATION DRIPS FROM HEATER DUCT
1	Up - 94	43-12-23	CONTAMINATED A/C REFRIGERENT - R-12 SYSTEM
1	Up - 94	53-12-05	CONTAMINATED R12 REFRIGERANT TESTING & HANDLING
2	96 - 96	63-11-08	CREAK NOISE FROM A/C DEFROST VALVE W/ CHANGING MODES
16,15,20	91 - 93	331212	DEFROSTER/SIDE WINDOW - LITTLE TO NO AIR FROM DUCTS
10	92 - 94	431201	EXPANSION TUBE ORIENTATION SHOWN INCORRECT - S/M REV
16,15	92 - 94	431201	EXPANSION TUBE ORIENTATION SHOWN INCORRECT - S/M REV

Model Legend: 1= All 2= Most 3= Beretta/Corsica 4= Celebrity/Century/Cierra/Cutlass Cruiser/6000 5= Camaro/Firebird 6= Caprice/Roadmaster/Impala SS/Fleetwood(RWD)/Estate Wagon/ Custom Cruiser 7= Monte Carlo 8= Corvette 9= Cavalier/Sunbird/Sunfire/Skyhawk 10= Bonneville, SSE, SSEi 11= Lumina 12= Grand Am 13= Gran Prix 14= LeMans 15= Electra/Park Avenue 16= LeSabre 17= Regal 18= Reatta 19= 1993 Riviera/Toronado 20= Ultra 21= Allante 22= Concours 23= DeVille 24= Eldorado, ETC, Seville, STS 25= Achieva 26= Aurora/1994 and up Riviera 27= Calais 28= Cutlass 29= 88 Royale (1988 and up) 30= 98, Touring Sedan (1989 and up) 31= Supreme

Heating, Air Conditioning, Ventilation, Defogger

Models	Year	TSB#	Description
30,29	92 - 94	431201	EXPANSION TUBE ORIENTATION SHOWN INCORRECT - S/M REV
2	91 - 93	431207A	GUIDELINES FOR RETROFIT R-12 VHCLS TO R-134A - REV
2	91 - 93	43-12-07B	GUIDELINES FOR RETROFIT R-12 VHCLS TO R-134A - REV 2
10	91 - 93	43-12-07C	GUIDELINES FOR RETROFIT R-12 VHCLS TO R-134A - REV 2
2	91 - 93	43-12-07C	GUIDELINES FOR RETROFIT R-12 VHCLS TO R-134A - REV 4
10,13,6000	91 - 93	43-12-07D	GUIDELINES FOR RETROFIT R-12 VHCLS TO R-134A - REV 4
10	91 - 93	431207A	GUIDELINES FOR RETROFITTING R-12 VEHICLES TO R-134A
2	91 - 93	431207	GUIDELINES FOR RETROFITTING R-12 VEHICLES TO R-134A
10	92 - 93	201104	HEAT - POOR DISTRIBUTION - CHECK HEAT DUCT INSTALL
2	92 - 93	331213	HEAT DISTRIB - DRIVER'S RT FOOT HOT/UPPER LEFT COLD
5	93 - 94	431102	HEATER - POOR PERFORMANCE - CORE TO CASE SEAL
22,24	94 - 94	471101	HEATER BLOWER MOTOR INOP AFTER IGNITION SYS SERVICE
12,25	92 - 93	331102	HEATER WARM UP SLOW - AIR FLOW OUTLET ASMBLY - REV
12,25	92 - 93	331102	HEATER WARM UP SLOW - REVISED AIR FLOW OUTLET ASSMBL
6	94 - 94	431214	HEATER WARMS UP SLOWLY - PLUG RADIATOR VENT HOSE
25	93 - 93	338105	HVAC - HOSE COLORS REVISED - S/M UPDATE
2	84 - 93	88-01-26	HVAC - NEW BLOWER MOTOR REPLACEMENT PARTS - INFO/REV
10	95 - 95	53-12-10	HVAC - OUTSIDE TEMP SENSOR - ADD'L INFO - S/M REV
16,15	95 - 95	53-12-10	HVAC - OUTSIDE TEMP SENSOR - ADD'L INFO - S/M REV
29,30	95 - 95	53-12-10	HVAC - OUTSIDE TEMP SENSOR - ADD'L INFO - S/M REV
5	93 - 95	53-12-04	HVAC - POOR PERFORMANCE - HEAT AND/OR A/C
29,30	94 - 94	431219	HVAC - POOR TEMP CONTROL - DUAL ZONE - DIAGNOSIS
10	93 - 93	300101	HVAC - PRO12MER R&I/SIR INFO - S/M REVISION
9	91 - 93	331222	HVAC - SLOW DEFROST - NEW NOZZLE AND/OR NOZZLE SEAL
9	91 - 93	331222	HVAC - SLOW DEFROST - NEW NOZZLE/NOZZLE SEAL - REV
2	92 - 93	331223	HVAC - UNBALANCED DEFROSTER PERFORMANCE - REPL DUCT
2	92 - 93	331223A	HVAC - UNBALANCED DEFROSTER PERFORMANCE - REVISED
12	93 - 93	338105	HVAC - VACUUM HOSE COLORS - S/M REVISION
23,24,22,	93 - 94	431212	HVAC -"SET TEMP OFFSET" FUNCTION - S/M REV - REV
23,24,22,	93 - 94	431212	HVAC -"SET TEMP OFFSET" FUNCTION EXPLAINED - S/M REV
8	93 - 96	53-81-38	HVAC BLOWER CONTROL - DIAG CHART 1 - S/M REV
9	95 - 96	53-81-29	HVAC BLOWER MOTOR DIAGNOSTIC CHART 3 - S/M REV
6	92 - 93	331206	HVAC BLOWER MOTOR WHINE IN LOW BLOWER MODE - SERVICE
26	95 - 95	43-11-09	HVAC DISPLAY ERRORS OR IMPROPER OPERATION
26	95 - 95	43-11-09	HVAC DISPLAY ERRORS OR IMPROPER OPERATION - REV
23,6	91 - 93	63-11-07	HVAC STUCK IN DEFROST MODE AFTER PROGRAMMER REPLACE
16,15	92 - 94	43-11-08	HVAC SYS TURNS ON BY DEFAULT
24	96 - 96	53-12-13	HVAC WHISTLE NOISE WHEN COMING OFF FULL HOT POSITION
16,15	92 - 96	53-12-13	HVAC WHISTLE NOISE WHEN COMING OFF FULL HOT POSITION
29,30	92 - 96	53-12-13	HVAC WHISTLE NOISE WHEN COMING OFF FULL HOT. POSITION
6	93 - 93	331203R	HVAC- HIGH BLOWER SPDS NEEDED FOR SUFFICIENT AIRFLOW
26	95 - 95	63-65-33	MIL ON - DTC 121 SET IN RAPID AMB TEMP CHANGE - 4.0L

Model Legend: 1= All 2= Most 3= Beretta/Corsica 4= Celebrity/Century/Cierra/Cutlass Cruiser/6000 5= Camaro/Firebird 6= Caprice/Roadmaster/Impala SS/Fleetwood(RWD)/Estate Wagon/ Custom Cruiser 7= Monte Carlo 8= Corvette 9= Cavalier/Sunbird/Sunfire/Skyhawk 10= Bonneville, SSE, SSEi 11= Lumina 12= Grand Am 13= Gran Prix 14= LeMans 15= Electra/Park Avenue 16= LeSabre 17= Regal 18= Reatta 19= 1993 Riviera/Toronado 20= Ultra 21= Allante 22= Concours 23= DeVille 24= Eldorado, ETC, Seville, STS 25= Achieva 26= Aurora/1994 and up Riviera 27= Calais 28= Cutlass 29= 88 Royale (1988 and up) 30= 98, Touring Sedan (1989 and up) 31= Supreme

Heating, Air Conditioning, Ventilation, Defogger

Models	Year	TSB#	Description
11	94 - 94	431206	MOISTURE CONDENSATION ON INTERIOR WINDOWS
6	94 - 94	331107	NO SHIFT OF AIR FLOW FROM DEFROST TO BLEND/DEFOG
29,30	94 - 94	431216	NOISE/HISS - A/C AT MEDIUM BLOWER SPEEDS - MUFFLER
24,23,22	94 - 94	431101	NOISE/WHINE - HVAC BLOWER MOTOR - INERTIA DISC KIT
22	94 - 96	63-11-04	POOR HEAT DIST/LACK OF AIR FLOW FROM DUCT - 4.6/4.9L
24	92 - 96	63-11-04	POOR HEAT DIST/LACK OF AIR FLOW FROM DUCT - 4.6/4.9L
24	92 - 94	431105	POOR HEAT DISTRIBUTION - DRIVER'S FEET COLD
22,23	94 - 94	431105	POOR HEAT DISTRIBUTION - DRIVER'S FEET COLD
1	93 - 95	43-12-15	R134A LEAK DETECTION WITH TRACER DYE
6	91 - 93	431213	RATTLE FROM A/C COMPRESSOR ON START-UP
3,7	94 - 96	63-81-06	REAR DEFOG POOR PERFORM - PWR DOOR LOCKS INOP
10	94 - 94	431221	RECOMMENDED REFRIGERANT CAPACITY - LABEL IS WRONG
26	95 - 95	431221	RECOMMENDED REFRIGERANT CAPACITY - LABEL IS WRONG
16,15	94 - 94	431221	RECOMMENDED REFRIGERANT CAPACITY - LABEL IS WRONG
29,30	94 - 94	431221	RECOMMENDED REFRIGERANT CAPACITY - LABEL IS WRONG
26	95 - 95	53-11-01	REDUCED HEATER PERFORMANCE ON DRIVER SIDE
26	95 - 96	53-11-01	REDUCED HEATER PERFORMANCE ON DRIVER SIDE - REV
22,23,24	94 - 95	53-12-06	REPEAT ACP CODE A047 - 4.6/4.9L
10	91 - 93	53-12-01	RETROFITTING OUT-OF-WARRANTY VHCLS FROM R12 TO R134A
16,15	91 - 93	53-12-01	RETROFITTING OUT-OF-WARRANTY VHCLS FROM R12 TO R134A
29,30	91 - 93	53-12-01	RETROFITTING OUT-OF-WARRANTY VHCLS FROM R12 TO R134A
6,23	91 - 93	53-12-01	RETROFITTING OUT-OF-WARRANTY VHCLS FROM R12 TO R134A
1	Up - 94	331226	RETROFITTING R-12 VEHICLES TO R-134A
1	88 - 96	63-12-09	SERVICE ISSUES FOR R12 OR R134A A/C SYSTEMS
22,23,24	96 - 96	63-11-05	SLOW HVAC MODE CHANGES
9	94 - 94	331105	TEMP CONTROL ADJUSTMENT DIFFICULT/IMPOSSIBLE - CABLE
9	94 - 94	331105	TEMP CONTROL ADJUSTMENT DIFFICULT/IMPOSSIBLE - REV
29,30,	91 - 93	92-T-148	TEMP VALVE - WHISTLES COMING OFF FULL HEAT - REVISED
5	95 - 95	53-12-03	VDOT A/C SYS DIAGN CHARTS - MAN A/C - S/M REV
9	94 - 94	331106	WATER LEAK - PASS FOOTWELL - AIR INLET GASKET
2	91 - 94	43-11-07	WHISTLE NOISE FROM HVAC WHEN MOVING FROM FULL HOT

Model Legend: 1= All 2= Most 3= Beretta/Corsica 4= Celebrity/Century/Cierra/Cutlass Cruiser/6000 5= Camaro/Firebird 6= Caprice/Roadmaster/Impala SS/Fleetwood(RWD)/Estate Wagon/ Custom Cruiser 7= Monte Carlo 8= Corvette 9= Cavalier/Sunbird/Sunfire/Skyhawk 10= Bonneville, SSE, SSEi 11= Lumina 12= Grand Am 13= Gran Prix 14= LeMans 15= Electra/Park Avenue 16= LeSabre 17= Regal 18= Reatta 19= 1993 Riviera/Toronado 20= Ultra 21= Allante 22= Concours 23= DeVille 24= Eldorado, ETC, Seville, STS 25= Achieva 26= Aurora/1994 and up Riviera 27= Calais 28= Cutlass 29= 88 Royale (1988 and up) 30= 98, Touring Sedan (1989 and up) 31= Supreme

Entertainment Devices, Stereo, Radio, Etc.

Models	Year	TSB#	Description
21,24,22	93 - 94	479601	AM RADIO NOISE - REPLACE ICM - 4.6L NORTHSTAR
2	92 - 94	449601	BUZZ NOISE IN AM BAND - FILTER - TRUCKS ONLY
8	91 - 94	339603	CD PLAYER SKIPS ON ROUGH ROADS (U1F RADIO)
24	93 - 93	93-061/09	IGNITION NOISE OVER SPEAKERS - REPL AMP OR RECEIVER
2	94 - 94	439601	INSTALLATION OF FIXED MAST ANTENNAS
25	95 - 95	43-90-14	MOBILE RADIOS & RFI NOISE - CRUISE CONTROL - FLEET
12,9	95 - 95	43-90-14	MOBILE RADIOS & RFI NOISE - CRUISE CONTROL - FLEET
3,9	95 - 95	43-90-14	MOBILE RADIOS & RFI NOISE - CRUISE CONTROL - FLEET
10	92 - 93	338123	NOISE/POP REAR SPEAKERS - PHONE/AMP INTERFERENCE
11,7	95 - 95	53-96-01	POOR AM RADIO RECEPTION
5	93 - 94	339601	POOR RADIO RECEPTION - REPL ANTENNA BRACKET ASSEMBLY
22,23	94 - 94	43-96-09	POP NOISE IN CENTER FRONT SPEAKER WHEN DOORS LOCKED
1	91 - 94	439604	POWER ANTENNA MAST REPLACEMENT
2	94 - 95	53-96-02	RADIO - AM STATIC W/ ENG ON
6	94 - 94	339602	RADIO - LOW FREQ DISTORTION W/ EXTENDED RANGE SPKRS
25	93 - 94	439605	RADIO ASSEMBLY - DIFFICULT OR UNABLE TO REMOVE
22,23,24	96 - 96	53-81-22	RADIO CLOCK STUCK ON 12:00 - BCM (PZM)
2	94 - 96	54-90-01	RADIO DISPLAY - "CLN" APPEARS - CLEAN/RESET
24,22,	96 - 96	54-90-01	RADIO DISPLAY - "CLN" APPEARS - CLEAN/RESET
2	95 - 96	64-90-04	RADIO DISPLAY - ERROR CODES - CAUSE/CORRECTION
1	Up - 93	349212R	RADIO FREQUENCY INTERFERENCE ID & DIAGNOSIS
1	Up - 96	93-254-09A	RADIO FREQUENCY INTERFERENCE ID & DIAGNOSIS - REV
1	Up - 96	93-09-04	RADIO FREQUENCY INTERFERENCE ID & DIAGNOSIS - REV
1	Up - 96	34-92-12	RADIO FREQUENCY INTERFERENCE ID & DIAGNOSIS - REV
1	Up - 96	SG09/93	RADIO FREQUENCY INTERFERENCE ID & DIAGNOSIS - REV
1	Up - 96	93-09-10	RADIO FREQUENCY INTERFERENCE ID & DIAGNOSIS - REV
29,30	91 - 93	338116	RADIO LOSES PRESET STATIONS OR DISPLAYS "LOC" - EXCH
24	93 - 93	319601	RADIO SPEAKERS - IGNITION NOISE - NEW AMPLIFIER
8	90 - 93	238311	RADIO/AFTERMARKET - SETS DTC'S 72/74 IN CCM - WIRING
2	93 - 94	339011	SOME TAPE CLEANERS & CD ADAPTERS NOT COMPATIBLE
29,30	92 - 93	209010	SOUND (6-SPEAKER) - DOOR MOUNTED SPEAKER BUZZ/RATTLE
5	93 - 93	331054	SPEAKER GRILLE BUZZ NOISE - ADD ADHESIVE BACKED FELT
5	93 - 93	331054	SPEAKER GRILLE BUZZ NOISE - REVISION
15,20,16	92 - 93	209010	SPEAKER/DOOR MOUNTED - BUZZ/RATTLE - CORRECTION PROC
15,20,16	92 - 93	208117	UX1/UTO OR U1A RADIO - POOR FM RECEPTION/BUZZ NOISES

Model Legend: 1= All 2= Most 3= Beretta/Corsica 4= Celebrity/Century/Cierra/Cutlass Cruiser/6000 5= Camaro/Firebird 6= Caprice/Roadmaster/Impala SS/Fleetwood(RWD)/Estate Wagon/ Custom Cruiser 7= Monte Carlo 8= Corvette 9= Cavalier/Sunbird/Sunfire/Skyhawk 10= Bonneville, SSE, SSEi 11= Lumina 12= Grand Am 13= Gran Prix 14= LeMans 15= Electra/Park Avenue 16= LeSabre 17= Regal 18= Reatta 19= 1993 Riviera/Toronado 20= Ultra 21= Allante 22= Concours 23= DeVille 24= Eldorado, ETC, Seville, STS 25= Achieva 26= Aurora/1994 and up Riviera 27= Calais 28= Cutlass 29= 88 Royale (1988 and up) 30= 98, Touring Sedan (1989 and up) 31= Supreme

Seats, Belts, Interior Trim, Carpets, Air Bags

Models	Year	TSB#	Description
24	94 - 94	43-16-11	ABSENCE OF INTENTIONAL SET SEAT BELT FEATURE
5,10	90 - 93	901611R	AC DELCO EXCHANGE PROCEDURE - SIR DERMS - REVISION
22,23,24	95 - 95	51-01-08	AIR BAG - INADVERTANT DEPLOY WHEN EXPOSED TO WATER
10	92 - 94	438302	ARMREST BIN LID MOUNTING PLATE BREAKAGE
29,30	94 - 94	339009	ASHTRAY RATTLE - HARD TO OPEN/CLOSE - POOR CLOSURE
5	93 - 93	431601	CENTER CONSOLE & CONSOLE DOOR LOOSE - REV PART
26	95 - 95	43-83-24	CENTER CONSOLE CUPHOLDER BINDS
2	Up - 95	43-16-06	CLEANING PROCEDURE FOR LEATHER SEAT COVERS
24	92 - 93	318302R	CONSOLE ASHTRAY - REPLACE PROC & SERV PARTS INFO
24	93 - 93	318308	CONSOLE CUPHOLDER - WON'T LATCH OR REMAIN CLOSED
25	96 - 96	53-10-22	COUPE REAR QUARTER ARMREST - REMOVE/INSTALL -S/M REV
10	92 - 95	53-83-09	CUP HOLDER SEPARATES AT CUP
24	96 - 96	53-83-23	CUPHOLDER WILL NOT LATCH OR OPEN - ASHTRAY GAPS
26	95 - 96	53-83-10	CUPHOLDERS TOO SMALL OR INOPERABLE
8	94 - 96	63-16-03	DOOR TRIM ARMREST LID HARD TO OPEN/CLOSE
8	94 - 94	431004	DOOR TRIM CONTACTS I/P - REPOSITION DOOR TRIM PAD
8	94 - 94	431004	DOOR TRIM CONTACTS I/P - REPOSITION TRIM PAD - REV
26	95 - 96	53-16-21	DOOR TRIM LACE - POOR RETENTION
12	95 - 96	63-16-07	DOOR TRIM PANEL INSERT (CLOTH/VINYL) SEPARATES FROM DOOR TRIM PANEL
9	95 - 96	53-16-17	DOOR TRIM PANEL LOOSE - COUPE
12	96 - 96	53-10-29	DOOR TRIM PANEL REMOVE/INSTALL - S/M REV
25	96 - 96	53-10-29	DOOR TRIM PANEL REMOVE/INSTALL - S/M REV
12	96 - 96	53-10-29	DOOR TRIM PANEL REMOVE/INSTALL - S/M REV - REV
25	96 - 96	53-10-29	DOOR TRIM PANEL REMOVE/INSTALL - S/M REV - REV
5	93 - 93	331025	FLOOR AREA - WATERLEAKS - CAUSE/SERVICE PROCEDURE
7	95 - 95	431604	FOLDING REAR BACK SEAT REMOVAL PROCEDURE
12,25	96 - 96	63-16-05	FR SEAT BELT TWISTS OR DOES NOT RETRACT COMPLETELY
10	92 - 93	201016	FRONT DOOR TRIM PANEL DISASSEMBLY/ASSEMBLY - S/M UP
5	93 - 93	331056	FRONT SEAT BACK LATERAL MOVEMENT - ADD WASHERS
24,21	92 - 93	311613	FRONT SEAT NOISE/POPPING/MOVEMENT ON ACCEL/DECEL
24,22,23	92 - 95	53-16-18	FRONT SEATS MOVE OR MAKE CLICK NOISE
24,22,23	92 - 96	53-16-18	FRONT SEATS MOVE OR MAKE CLICK NOISE - REV
25	92 - 94	43-10-26	GAP BETWEEN DOOR TRIM & WINDOW DEFOGGER GRILLE
24	94 - 94	311615	GAP BETWEEN REAR SEAT BACK & CUSHION AT BITE LINE
24	92 - 93	318307	GLOVE BOX HARD TO CLOSE - ADJUSTMENT PROCEDURES
10	93 - 95	53-16-05	HEADREST COVER LOOSE OR SPLITS
29,30	94 - 95	43-83-25	I/P - DEFROSTER GRILLE WARPS OR POOR FIT
8	93 - 94	331069	LEAK OR CARPET STAIN BELOW WINDSHIELD SIDE MOLDING
2	94 - 94	431019	LOOSE FITTING CENTER PILLAR TRIM PANEL & RETAINERS
6	94 - 95	53-10-10	LOOSE FRONT SEAT CUSHION (LEATHER)
30	91 - 93	278102	LUMBAR SUPPORT - WIRE COLOR CLARIFICATION - S/M UP
12	96 - 96	53-10-24	MIRROR SWITCH BEZEL - REMOVE/INSTALL - S/M REV

Model Legend: 1= All 2= Most 3= Beretta/Corsica 4= Celebrity/Century/Cierra/Cutlass Cruiser/6000 5= Camaro/Firebird 6= Caprice/Roadmaster/Impala SS/Fleetwood(RWD)/Estate Wagon/Custom Cruiser 7= Monte Carlo 8= Corvette 9= Cavalier/Sunbird/Sunfire/Skyhawk 10= Bonneville, SSE, SSEi 11= Lumina 12= Grand Am 13= Gran Prix 14= LeMans 15= Electra/Park Avenue 16= LeSabre 17= Regal 18= Reatta 19= 1993 Riviera/Toronado 20= Ultra 21= Allante 22= Concours 23= DeVille 24= Eldorado, ETC, Seville, STS 25= Achieva 26= Aurora/1994 and up Riviera 27= Calais 28= Cutlass 29= 88 Royale (1988 and up) 30= 98, Touring Sedan (1989 and up) 31= Supreme

Seats, Belts, Interior Trim, Carpets, Air Bags

Models	Year	TSB#	Description
25	96 - 96	53-10-24	MIRROR SWITCH BEZEL - REMOVE/INSTALL - S/M REV
12	96 - 96	53-10-24	MIRROR SWITCH BEZEL - REMOVE/INSTALL - S/M REV - REV
25	96 - 96	53-10-24	MIRROR SWITCH BEZEL - REMOVE/INSTALL - S/M REV - REV
5	93 - 93	431001	NOISE/CHUCKING FRONT SEAT - REPLACE ADJUSTER NUTS
6	91 - 94	431021	NOISE/SQUEAK - DOOR TRIM PAD
12	92 - 94	431007	NOISE/SQUEAK/RATTLE DIAGNOSTIC GUIDE
25	92 - 94	431007	NOISE/SQUEAK/RATTLE DIAGNOSTIC GUIDE
12	92 - 95	431007	NOISE/SQUEAK/RATTLE DIAGNOSTIC GUIDE - REV
25	92 - 95	431007	NOISE/SQUEAK/RATTLE DIAGNOSTIC GUIDE - REV
24	94 - 94	311607	QUARTER TRIM PANEL - HIDDEN SCREW INFORMATION
24	92 - 94	43-16-09	RATTLE AND/OR LOOSE QUARTER TRIM PANEL
3,7	93 - 96	53-83-16	RATTLE/ITCH BETWEEN DEFROSTER GRILLE & IP
5	93 - 94	438301	RATTLE/LOOSE I/P UPPER TRIM PANEL
26	95 - 95	53-16-19	REAR OF HEADLINER LOOSE OR RATTLES
10	94 - 94	43-10-52	REAR SEAT & CARPET REMOVE/INSTALL - S/M REV
24	93 - 95	43-16-08	REVISED CUPHOLDER ASSEMBLY
11	93 - 93	SUPP: 93C08	SEAT BACK MANUAL RECLINER MECHANISM - REVISED RECALL
24,21	92 - 93	311609	SEAT BACK/PASSENGER SIDE - WON'T FOLD FORWARD - SERV
12	93 - 93	231061	SEAT BELT (PASSENGER BUCKLE SIDE) - HARD TO REACH
25	93 - 93	231061	SEAT BELT (PASSENGER BUCKLE SIDE) - HARD TO REACH
3	92 - 93	331017	SEAT BELT/FRONT - REMOVING TWISTS FROM THE WEBBING
10	92 - 94	43-16-10	SEAT COVER WEAR/TEAR FROM SEAT BOLSTER - SSE/SSEi
2	92 - 94	311605	SEAT COVERS (NUANCE LEATHER) - CLEANING PROCEDURES
24	92 - 93	211617R	SEAT CUSHION - APPEARANCE CONCERNS - REPAIR INFO
24	92 - 93	93-041/10	SEAT CUSHION APPEARANCE UNSAT - REPLACE CUSH & COVER
24	92 - 93	211617	SEAT CUSHION APPEARANCE UNSAT - REVISED P/N
1	87 - 93	90-10-01	SEAT/CHILD - ANCHOR PLATE/BELT ASSY/EXTENDER INSTALL
24	92 - 93	311608	SEAT/FRONT - LUMBAR SUPPORT INADEQUATE - SERV PROC
5	93 - 93	331016	SEAT/PASSENGER - SPRING NOISE - CORRECTION PROCEDURE
29,30	91 - 93	92-T-109	SEATS/FRONT - WON'T ADJUST HIGH ENOUGH - REPAIR KIT
26	95 - 95	53-83-03	SHIFTER OPENING COVER REPLACEMENT PROCEDURE
24	94 - 94	93-I-55	SHOULDER BELT COMFORT FEATURE INTRODUCTION DELAYED
11	92 - 94	93-240-10	SHOULDER BELT/FRONT SEAT - RUBS NECK - KIT - REVISED
11	92 - 93	331030	SHOULDER BELT/FRONT SEAT - RUBS PERSON'S NECK - PROC
2	89 - 93	331052	SHOULDER BELTS - REAR SEAT CHILD/SMALL ADULT COMFORT
24	93 - 93	43-90-20	SIR (AIR BAG) DIAG CHART 1 - S/M REV
23,6	92 - 93	439003	SIR - DERM CANNOT COMMUNICATE W/ SCAN TOOL - S/M REV
1	90 - 93	92-I-06	SIR - DERM EXCHANGE PROGRAM
2	90 - 93	92-10-02	SIR - DERM EXCHANGE PROGRAM - REVISION
9	95 - 95	53-90-08	SIR - DIAGN CHARTS & SERVICE PROCEDURE - S/M REV
12	94 - 95	53-90-10	SIR - DTC 24 DIAGN CHART - S/M REV
25	94 - 95	53-90-10	SIR - DTC 24 DIAGN CHART - S/M REV

Model Legend: 1= All 2= Most 3= Beretta/Corsica 4= Celebrity/Century/Cierra/Cutlass Cruiser/6000 5= Camaro/Firebird 6= Caprice/Roadmaster/Impala SS/Fleetwood(RWD)/Estate Wagon/Custom Cruiser 7= Monte Carlo 8= Corvette 9= Cavalier/Sunbird/Sunfire/Skyhawk 10= Bonneville, SSE, SSEi 11= Lumina 12= Grand Am 13= Gran Prix 14= LeMans 15= Electra/Park Avenue 16= LeSabre 17= Regal 18= Reatta 19= 1993 Riviera/Toronado 20= Ultra 21= Allante 22= Concours 23= DeVille 24= Eldorado, ETC, Seville, STS 25= Achieva 26= Aurora/1994 and up Riviera 27= Calais 28= Cutlass 29= 88 Royale (1988 and up) 30= 98, Touring Sedan (1989 and up) 31= Supreme

Seats, Belts, Interior Trim, Carpets, Air Bags

Models	Year	TSB#	Description
3,7	94 - 96	53-90-10	SIR - DTC 24 DIAGN CHART - S/M REV
2	91 - 93	219008	SIR - INTERMITTENT DTC 24 DIAGNOSIS - S/M UPDATE
12	94 - 94	439007	SIR - PASS COMP'T DISCRIMINATING SENSOR - S/M REV
25	94 - 94	439007	SIR - PASS COMP'T DISCRIMINATING SENSOR - S/M REV
16,15,20	92 - 93	208114	SIR - SYSTEM DIAGNOSIS REVISED - ESM UPDATE
9	95 - 95	53-81-24	SIR COMPONENT - SENSORS - LOCATIONS - S/M REV
9	95 - 95	53-32-12	SIR DISABLE/ENABLE PROCEDURE - S/M REV
25	95 - 95	53-32-12	SIR DISABLE/ENABLE PROCEDURE - S/M REV
12,9	95 - 95	53-32-12	SIR DISABLE/ENABLE PROCEDURE - S/M REV
9	95 - 95	53-32-12	SIR DISABLE/ENABLE PROCEDURE - S/M REV - REV
25	95 - 95	53-32-12	SIR DISABLE/ENABLE PROCEDURE - S/M REV - REV
12,9	95 - 95	53-32-12	SIR DISABLE/ENABLE PROCEDURE - S/M REV - REV
2	93 - 94	439001	SIR DUAL AIR BAGS - DEPLOYMENT INFORMATION
2	90 - 93	92-09-03	SIR INFLATOR MODULE (WARR REPLACED) - RETURN PROC
24,6	93 - 93	209006R	SIR INFLATOR MODULE - DEALER HANDLING/RETURN INFO
1	92 - 94	319003	SIR INFLATOR MODULE SCRAPPING PROCEDURE
1	Up - 93	91-09-05	SIR INFLATOR MODULES (LIVE) - RETURN UNDER WARR INST
1	92 - 93	92-09-23	SIR MODULES - DEALER HANDLING AND RETURN INFO
10	92 - 93	209009	SIR SYS DIAGNOSIS - CHART B REVISED - S/M UPDATE
3	94 - 94	316402	SIR WARNING LAMP COMES ON STEADY - S/M UPDATE
3	94 - 94	316402	SIR WARNING LAMP COMES ON STEADY - S/M UPDATE - REV
8	93 - 93	331048	SQUEAK AND RATTLE DIAGNOSTIC GUIDE
9	93 - 94	431020	SQUEAK BETWEEN DOOR HEADER & AUXILIARY WEATHERSTRIP
15,20,16	91 - 93	201603	TRIM AREA/DOOR - NOISE - CAUSES/CORRECTIONS
24	92 - 95	92-127/10	TRIM PAD/DOOR - RATTLES/LOOSE - REPAIR PROCED - REV
24	92 - 93	211613R	TRIM PAD/DOOR - RATTLES/LOOSE - REPAIR PROCEDURE
9	95 - 96	53-16-12A	TWIST IN SEAT BELT WEBBING - SEDAN - PROCED - REV2
25	95 - 96	53-16-12A	TWIST IN SEAT BELT WEBBING - SEDAN - PROCED - REV2
9,12	95 - 96	53-16-12A	TWIST IN SEAT BELT WEBBING - SEDAN - PROCED - REV2
9	95 - 95	53-16-12	TWIST IN SEAT BELT WEBBING - SEDAN - PROCEDURE
25	96 - 96	53-16-12	TWIST IN SEAT BELT WEBBING - SEDAN - PROCEDURE - REV
9,12	96 - 96	53-16-12	TWIST IN SEAT BELT WEBBING - SEDAN - PROCEDURE - REV
9	95 - 95	53-16-24	USE OF HAIRPIN CLIP FOR HEADREST RETENTION
22,23	94 - 95	53-16-10	VERTICAL SEAT HEIGHT ADJUSTMENT - INCREASING
22,23	94 - 96	53-16-10	VERTICAL SEAT HEIGHT ADJUSTMENT - INCREASING - REV
26	95 - 96	53-16-16	WET AND/OR SMELLY CARPET FROM WATER LEAKS
24,22,23	93 - 96	53-16-16	WET AND/OR SMELLY CARPET FROM WATER LEAKS

Model Legend: 1= All 2= Most 3= Beretta/Corsica 4= Celebrity/Century/Cierra/Cutlass Cruiser/6000 5= Camaro/Firebird 6= Caprice/Roadmaster/Impala SS/Fleetwood(RWD)/Estate Wagon/ Custom Cruiser 7= Monte Carlo 8= Corvette 9= Cavalier/Sunbird/Sunfire/Skyhawk 10= Bonneville, SSE, SSEi 11= Lumina 12= Grand Am 13= Gran Prix 14= LeMans 15= Electra/Park Avenue 16= LeSabre 17= Regal 18= Reatta 19= 1993 Riviera/Toronado 20= Ultra 21= Allante 22= Concours 23= DeVille 24= Eldorado, ETC, Seville, STS 25= Achieva 26= Aurora/1994 and up Riviera 27= Calais 28= Cutlass 29= 88 Royale (1988 and up) 30= 98, Touring Sedan (1989 and up) 31= Supreme

Glass, Doors, Hood, Decklid, Tailgate, Liftgate, Locks

Models	Year	TSB#	Description
12	94 - 94	43-81-21	AUTO DOOR LOCKS - SPLICE 303 LOCATION - S/M REV
12	94 - 95	43-81-31	AUTO DOOR UNLOCK FEATURE - NEW CHART 5 - S/M REV
25	94 - 95	43-81-31	AUTO DOOR UNLOCK FEATURE - NEW CHART 5 - S/M REV
1	Up - 95	48-01-05	AUTOMATIC KEY CODING RETRIEVAL
11,7	95 - 95	43-10-60	BINDING/STICKING INSIDE LOCK RODS - FRONT DOORS
22,23	94 - 94	53-15-05	BOOM SOUND WHEN TRUNK LID OPENED
24	96 - 96	63-10-29	CLICK NOISE WHEN CYCLING DOOR GLASS TO FULL UP
26	95 - 95	431509	CLICKING NOISE FROM OUTBOARD SIDE(S) OF SUNROOF
8	94 - 94	SUPP: 94C08	CONV HARDTOP REAR GLASS DEBOND - ADD'L VEHLS- RECALL
8	94 - 94	94C08	CONVERTIBLE HARDTOP REAR GLASS DEBOND - CAMPAIGN
8	94 - 94	331062	CONVERTIBLE TOP - #5 BOW UNLATCHES
5	94 - 94	431039	CONVERTIBLE TOP BINDS WHEN RAISING/LOWERING
5	94 - 94	431039	CONVERTIBLE TOP BINDS WHEN RAISING/LOWERING - REV
5	94 - 94	43-10-56	CONVERTIBLE TOP CONCERNS - REPAIR PROCEDURES
9	92 - 94	431008	CONVERTIBLE TOP CONCERNS - TECH ASSIST AVAILABLE
21	90 - 93	311509	CONVERTIBLE TOP DOES NOT LATCH PROPERLY
9	86 - 94	431029	CONVERTIBLE TOP SIDE RAIL REMOVE/INSTALL - S/M REV
5	94 - 95	53-10-19	CONVERTIBLE TOP SLAPS ON #2 OR #3 ROOF BOW
12	92 - 93	331049	CREAK NOISE - CENTER DOOR TRIM AREA
9	95 - 95	53-16-13	CREAK NOISE - PASS SIDE OF REAR WINDOW PANEL
26	95 - 95	53-15-14	CROSS CAR WINDNOISE AT A-PILLAR ON FRONT DOORS
6	91 - 94	431006	DIRT BUILD-UP ON ROCKER AND LOWER DOOR - FOAM SEAL
6	91 - 93	331029	DOOR & REAR COMPARTMENT WATERLEAKS - CAUSE/PROC
5	93 - 93	331035	DOOR - OUTER PANEL CRACKING AT REAR - PROC/ADHESIVE
24,21	87 - 93	311611	DOOR GLASS - NOISY OP/MOTOR FAILURE - SERV INFO
24	92 - 93	311610	DOOR GLASS AREA/UPPER REAR - WATERLEAKS/WIND NOISE
11	93 - 94	53-10-18	DOOR GLASS RATTLE 2/3 DOWN
26	95 - 95	53-10-04	DOOR GLASS RATTLES WHEN CLOSING DOOR
22,23	94 - 95	53-10-20	DOOR GLASS SCRAPING NOISE OR STICKING GLASS
5	93 - 95	63-10-23	DOOR GLASS SCRATCHING
8	94 - 95	53-10-17	DOOR GLASS SCRATCHING - REPL OUTER SEAL
24	92 - 95	53-10-23	DOOR GLASS SLAP/RATTLE CLOSING DOOR W/WINDOW DOWN
8	93 - 93	331022	DOOR GLASS/RIGHT - BINDS/INSUFFICIENT TRAVEL - PROC
26	95 - 95	53-15-02	DOOR HANDLE DISCOLORED IN KEY CYLINDER OPENING
29,30	91 - 93	201017	DOOR HINGE SYSTEM - REV SERVICE PROC - S/M UPDATE
5	93 - 96	63-10-25	DOOR INNER PANEL REPAIRS ON SMC TYPE DOORS
1	Up - 93	92-041-10	DOOR LOCK CYLINDER - FREEZES/STICKS/BINDS - REVISION
1	84 - 93	92-10-15	DOOR LOCK CYLINDERS - BIND/STICK - LUBRICANTS - REV
1	93 - 93	131070R	DOOR LOCK CYLINDERS BIND/STICK - RECOMMENDED LUB
1	Up - 93	SG09/90	DOOR LOCK CYLINDERS INOP/BIND - LUBRICANT/PROC - REV
21	90 - 93	91-090/10	DOOR LOCKS - HIGH KEY EFFORTS WHEN USING GOLD KEYS
22,23,24	94 - 94	438114	DOOR LOCKS - PROC TO KEEP LOCKED W/ SHIFT TO "PARK"

Model Legend: 1= All 2= Most 3= Beretta/Corsica 4= Celebrity/Century/Cierra/Cutlass Cruiser/6000 5= Camaro/Firebird 6= Caprice/Roadmaster/Impala SS/Fleetwood(RWD)/Estate Wagon/Custom Cruiser 7= Monte Carlo 8= Corvette 9= Cavalier/Sunbird/Sunfire/Skyhawk 10= Bonneville, SSE, SSEi 11= Lumina 12= Grand Am 13= Gran Prix 14= LeMans 15= Electra/Park Avenue 16= LeSabre 17= Regal 18= Reatta 19= 1993 Riviera/Toronado 20= Ultra 21= Allante 22= Concours 23= DeVille 24= Eldorado, ETC, Seville, STS 25= Achieva 26= Aurora/1994 and up Riviera 27= Calais 28= Cutlass 29= 88 Royale (1988 and up) 30= 98, Touring Sedan (1989 and up) 31= Supreme

Glass, Doors, Hood, Decklid, Tailgate, Liftgate, Locks

Models	Year	TSB#	Description
9	95 - 95	53-15-17	DOOR OUTSIDE HANDLE SPECIAL COATED RIVET
21	93 - 93	311617	DOOR WILL NOT OPEN WITH INSIDE DOOR HANDLE
21	93 - 93	211526	DOOR WINDOW - ADJUSTMENT PROC/SUPERSEDES PRELIM S/M
12	92 - 93	92-10-71A	DOOR WINDOW/FRONT - HARD TO CYCLE/ROLL/SEAL -REVISED
25	92 - 93	93-T-034	DOOR WINDOW/FRONT - HARD TO CYCLE/ROLL/SEAL -REVISED
25	92 - 93	93-T-034	DOOR WINDOWS/FRONT - HARD TO CYCLE/ROLL/SEAL - REV
25	92 - 93	231066	DOOR WINDOWS/FRONT - HARD TO CYCLE/ROLL/SEAL - SERV
29,30	91 - 93	201013R	DOOR/WINDOWS - SERVICE PROCEDURES - S/M CORRECTIONS
15,20,16	91 - 93	201013R	DOORS/LOCKS/WINDOWS - SERV PROC REVISED - S/M UPDATE
10	92 - 93	201013R	DOORS/LOCKS/WINDOWS - SERVICE PROCS - S/M UPDATE
22,23	94 - 94	43-15-22	FLEX/TWIST OF TRUNK LID WHEN CLOSING
21	90 - 93	311606	FOLDING TOP - RATTLE FROM STOWAGE COVER LATCH - INFO
21	87 - 93	211515	FOLDING TOP ROOF BOWS - RATTLE NOISE ON TURNS
21	90 - 93	311511	FOLDING TOP STOWAGE COVER LOCK LATCH BREAKAGE
26	96 - 96	53-10-30	FRONT DOOR WATER DEFLECTOR REMOVE PROCEDURE
11	94 - 94	431017	FRONT DOOR WON'T OPEN - ADJUST BARREL NUT
5	93 - 93	331011	FRONT DOOR WON'T OPEN USING EITHER HANDLE - PROC
22,23	94 - 94	431043	FRONT DOORS WON'T STAY OPEN ON GRADES
22,23	94 - 95	431043	FRONT DOORS WON'T STAY OPEN ON GRADES - REV
22,23	94 - 96	43-10-43A	FRONT DOORS WON'T STAY OPEN ON GRADES - REV2
9	95 - 95	53-15-10	FUEL FILLER DOOR - LATCH RETAINER LEGS LOOSE
9	95 - 95	53-15-10	FUEL FILLER DOOR - LATCH RETAINER LEGS LOOSE - REV
26	95 - 95	53-15-01	FUEL FILLER DOOR INOP
26	96 - 96	63-15-06	GAP BETWEEN DOOR OPENING WEATHERSTRIP & ROOFLINE
3,7	88 - 95	89-131-10	GLASS/FRONT DOOR - POP NOISE/BINDING - REVISED
22,23	95 - 95	53-15-15	HIGH EFFORT TO LIFT REAR COMPARTMENT LID
9	95 - 95	53-15-13	HIGH EFFORT TO PULL TRUNK RELEASE CABLE HANDLE
11	95 - 95	53-10-14	HIGH PITCH WHISTLE SOUND AROUND FRONT DOORS
10	94 - 94	43-15-20	HOOD - EXCESSIVE FLUTTER OR DIFFICULT TO LATCH
25	93 - 93	231071	HOOD - REMOVAL PROCEDURE WITHOUT BREAKING MIG WELDS
8	93 - 93	331014	HOOD - WON'T OPEN/CABLE DISCONNECTED - SERVICE PROC
22,23,24	94 - 94	431016	HOOD LATCH PREVENTS FULL CLOSURE - REPLACE LATCH
12	93 - 93	231071	HOOD REMOVAL - PROCEDURE TO AVOID BREAKING MIG WELDS
9	95 - 95	43-32-09	IGNITION KEY AND LOCK CYLINDER CHANGE
25	95 - 95	43-32-09	IGNITION KEY AND LOCK CYLINDER CHANGE
12,9	95 - 95	43-32-09	IGNITION KEY AND LOCK CYLINDER CHANGE
12	92 - 93	431030	IMPRESSION IN DOOR FORWARD OF DOOR HANDLE
3,7	87 - 94	431507	INSTALL NEW HOOD PROP ROD WHEN HOOD ASMBLY REPLACED
22,23	94 - 94	431002	INTERMITTENT FUEL DOOR OPERATION
5,11	91 - 94	338309	IP (GLOVE) COMP DOOR LOCK CYL REPLACEMENT - S/M REV
25	93 - 93	333201R	KEY & LOCK CYLINDER/IGNITION - CHANGES INFORMATION

Model Legend: 1= All 2= Most 3= Beretta/Corsica 4= Celebrity/Century/Cierra/Cutlass Cruiser/6000 5= Camaro/Firebird 6= Caprice/Roadmaster/Impala SS/Fleetwood(RWD)/Estate Wagon/ Custom Cruiser 7= Monte Carlo 8= Corvette 9= Cavalier/Sunbird/Sunfire/Skyhawk 10= Bonneville, SSE, SSEi 11= Lumina 12= Grand Am 13= Gran Prix 14= LeMans 15= Electra/Park Avenue 16= LeSabre 17= Regal 18= Reatta 19= 1993 Riviera/Toronado 20= Ultra 21= Allante 22= Concours 23= DeVille 24= Eldorado, ETC, Seville, STS 25= Achieva 26= Aurora/1994 and up Riviera 27= Calais 28= Cutlass 29= 88 Royale (1988 and up) 30= 98, Touring Sedan (1989 and up) 31= Supreme

Glass, Doors, Hood, Decklid, Tailgate, Liftgate, Locks

Models	Year	TSB#	Description
12	93 - 93	333201R	KEY AND LOCK CYLINDER CHANGES - INFORMATION
1	94 - 94	310104	KEY SETS/GOLD - REDUCED TO 1 SET PER VEHICLE
1	Up - 93	93-WA-15	KEY/EMERGENCY CREDIT CARD - DEALER REMINDER
29,30	93 - 93	313201	KEY/IGN - BAR CODED ID TAG INFORMATION
1	93 - 93	313201	KEY/IGNITION - NEW BAR CODE IDENTIFICATION TAG
12	92 - 93	333207	LOCK CYLINDER/IGNITION - REPLACEMENT - S/M REVISED
25	92 - 93	333207	LOCK CYLINDER/IGNITION - REPLACEMENT PROCEDURE
5	93 - 94	431005	LOOSE DOOR OUTER PANELS - PROCEDURES
12	94 - 95	53-15-04	LOOSE/RATTLE - TRUNK LID HINGE COVERS
25	94 - 95	53-15-04	LOOSE/RATTLE - TRUNK LID HINGE COVERS
26	95 - 95	53-15-06	LOW LIFT OF TRUNK LID
5	93 - 93	331046	MOAN/CREAK - REAR WINDOW - LUBRICATE
12	92 - 94	431007	NOISE/SQUEAK/RATTLE DIAGNOSTIC GUIDE
25	92 - 94	431007	NOISE/SQUEAK/RATTLE DIAGNOSTIC GUIDE
12	92 - 95	431007	NOISE/SQUEAK/RATTLE DIAGNOSTIC GUIDE - REV
25	92 - 95	431007	NOISE/SQUEAK/RATTLE DIAGNOSTIC GUIDE - REV
26	95 - 95	53-81-27	PWR DOOR LOCKS INOP - I/P 1 FUSE BLOWS
3,7	94 - 96	63-81-06	PWR DOOR LOCKS INOP - POOR PERF REAR DEFOG
9	95 - 95	53-10-03	QUARTER WINDOW APPLIQUE & SERVICE - S/M REV
5	93 - 95	43-10-57	RATTLE - DOOR GLASS IN FULL DOWN POSITION
5	93 - 93	331050	RATTLE - LOCK ROD IN REAR COMPARTMENT (TRUNK) AREA
24	92 - 94	93-118/10	REAR COMPARTMENT - WATER LEAKS - DIAGN/PROC - REV
24	92 - 93	311504	REAR COMPARTMENT - WATER LEAKS - DIAGNOSIS/PROCEDURE
10	96 - 96	63-15-12	REAR COMPARTMENT LID - HIGH CLOSING EFFORT
29	96 - 96	63-15-12	REAR COMPARTMENT LID - HIGH CLOSING EFFORT
30	94 - 95	53-15-11	REAR COMPARTMENT LID DOES NOT STAY OPEN
22,23	94 - 94	311510	REAR COMPARTMENT WATER LEAKS - APPLY SEALER
6	91 - 93	93-235-10	REAR COMPARTMENT WATERLEAKS - CAUSE/CORRECTION - REV
9	96 - 96	63-10-31	REAR EDGE OF HOOD RUBS WINDSHIELD REVEAL MOLDING TAB
12	94 - 94	431045	RH WINDSHIELD GARNISH MOLDING DISTORTS
25	94 - 94	431045	RH WINDSHIELD GARNISH MOLDING DISTORTS
24	96 - 96	63-10-26	RUN CHANNELS/UNSTOP ASMBLY/WINDOW ADJUST - S/M REV
2	91 - 94	43-15-10	SCRAPE NOISE/HIGH EFFORT OPENING FRONT DOORS
9	95 - 95	53-01-02	SERVICING SINGLE KEY LOCK CYLINDERS
24	92 - 93	63-10-27	SIDE DOOR ROOF RAIL WEATHERSTRIP REPLACEMENT
23	96 - 96	53-10-31	SIMULATED CONV TOP SERVICE PROCEDURE - S/M REV
21	87 - 93	92-055/10	SOFT TOP - CONTINUOUS CYCLING OF PULLDOWN UNIT - REV
8	93 - 93	331048	SQUEAK AND RATTLE DIAGNOSTIC GUIDE
2	87 - 94	431018	SUNROOF (WEBASTO SYSTEM) SERVICE HOTLINE
2	87 - 94	431018	SUNROOF (WEBASTO SYSTEM) SERVICE HOTLINE - REVISED
10	91 - 93	331037	SUNROOF - WINDNOISE AT WEATHERSTRIP SEAL - NEW SEAL
30	91 - 93	331037	SUNROOF - WINDNOISE AT WEATHERSTRIP SEAL - NEW SEAL

Model Legend: 1= All 2= Most 3= Beretta/Corsica 4= Celebrity/Century/Cierra/Cutlass Cruiser/6000 5= Camaro/Firebird 6= Caprice/Roadmaster/Impala SS/Fleetwood(RWD)/Estate Wagon/ Custom Cruiser 7= Monte Carlo 8= Corvette 9= Cavalier/Sunbird/Sunfire/Skyhawk 10= Bonneville, SSE, SSEi 11= Lumina 12= Grand Am 13= Gran Prix 14= LeMans 15= Electra/Park Avenue 16= LeSabre 17= Regal 18= Reatta 19= 1993 Riviera/Toronado 20= Ultra 21= Allante 22= Concours 23= DeVille 24= Eldorado, ETC, Seville, STS 25= Achieva 26= Aurora/1994 and up Riviera 27= Calais 28= Cutlass 29= 88 Royale (1988 and up) 30= 98, Touring Sedan (1989 and up) 31= Supreme

Finishes, Body Structure, Frame, Bumpers

Models	Year	TSB#	Description
1	93 93	111703R	PAINT - FACTORY TO REFINISH PRODUCTS CROSS REFERENCE
21	93 - 93	311702	PAINT - NEW COLOR INFORMATION
5	94 - 94	331709	PAINT - NEW HIGH PERFORMANCE CLEAR COAT INFORMATION
21	93 - 93	211701	PAINT - THREE NEW COLORS - INFORMATION
8	93 - 93	93-048-10	PAINT CODE 68 - CORRECT WA # ON SERV PARTS ID LABEL
8	93 - 93	231064	PAINT CODE 69 - CORRECT WA # ON SERV PARTS ID LABEL
1	93 - 93	93-10-02	PAINT CODES AND REFINISH INFORMATION
1	94 - 94	331706	PAINT INFORMATION - 1994 MODELS
1	94 - 94	331707	PAINT INFORMATION - TRUCK MODELS ONLY
12,25	94 - 94	431702	PAINT NUMBERS ("WA") MISSING FROM SPID LABEL
9	95 - 95	63-15-03	POOR APPEAR AT QTR BELT MOLDING TO QTR PANEL - CONV
7	95 - 95	43-20-02	POOR APPEARANCE - ROOF TO QUARTER PANEL JOINTS
11,7	95 - 95	53-20-01	POP NOISE FROM LEFT SIDE DURING BRAKE/ACCEL
29	95 - 95	53-15-09	REAR BUMPER FASCIA HEAT DISTORTION
25	94 - 94	94-C-10	REAR COMPARTMENT PAN SIDE RAIL WELDS - CAMPAIGN
9	95 - 96	53-15-07	REAR QUARTER ROAD NOISE - COUPE
9	95 - 96	53-15-07	REAR QUARTER ROAD NOISE - COUPE - REV
24	92 - 95	311616A	REVEAL MOLDING WARPAGE - REVISED
10	91 - 95	30-10-03	ROOF - RAIN NOISE - INSTALL SOUND DEADENER - REVISED
16,15	91 - 95	93-10-17	ROOF - RAIN NOISE - INSTALL SOUND DEADENER - REVISED
29,30	91 - 95	93-T-045	ROOF - RAIN NOISE - INSTALL SOUND DEADENER - REVISED
6,23	91 - 93	30-10-03	ROOF - RAIN NOISE - INSTALL SOUND DEADENER - REVISED
2	91 - 93	301003	ROOF - RAIN NOISE - INSTALL SOUND DEADENER PATCHES
5	96 - 96	53-10-27	RS APPEARANCE PKG - ADD'L INFO - S/M REV
6	92 - 93	331051	SAGGING/DRIPPING FRAME COATING - REMOVE/RECOAT
8	86 - 94	431501	SCUFFED PAINT ON STORAGE COMPART LID - CONVERTIBLE
12	92 - 95	63-15-05	SIDE MOLDINGS DISCOLORED/STAINED
8	93 - 93	331048	SQUEAK AND RATTLE DIAGNOSTIC GUIDE
9	90 - 94	331042	TOP VENT SCREEN/WASHER NOZZLE REMOVAL - S/M REVISION
7	95 - 96	63-15-10	TRUNK LID APPLIQUE SCRATCHES/ABRASIONS REMOVAL
24	93 - 93	313401	UNDERBODY RATTLE/LEFT REAR CORNER - DIAG & REPAIR
1	94 - 94	321702	VINYL TRIM AND VINYL TOP REPAIR
12,25	95 - 95	53-15-03	WATER ENTERS SPOILER
12	96 - 96	53-15-18	WATER ENTERS TRUNK AT TAIL LAMP AREA - REV
3,7,12,25	91 - 94	43-16-07	WATER LEAK - RIGHT FRONT CARPET WET
10,29,30	95 - 95	53-10-34	WATERLEAK AT RIGHT REAR AREA OF TRUNK FLOOR PAN
3,7	87 - 96	53-10-21	WATERLEAK DIAGNOSTIC GUIDE
2	91 - 93	331037	WEATHERSTRIP SEAL/SUNROOF - WINDNOISE - NEW SEAL
6	91 - 93	331039R	WIND NOISE DIAG/REPAIR - TOOLS/RECOMMENDED MATERIALS
5	93 - 93	331060	WINDNOISE & WATERLEAK DIAGNOSIS & REPAIR
10	92 - 93	201008	WINDNOISE - DIAGNOSIS AND CORRECTIONS - REVISED
6	91 - 93	331005	WINDNOISE - PROPER DIAGNOSIS/MATERIALS/PROCEDURES

Model Legend: 1= All 2= Most 3= Beretta/Corsica 4= Celebrity/Century/Cierra/Cutlass Cruiser/6000 5= Camaro/Firebird 6= Caprice/Roadmaster/Impala SS/Fleetwood(RWD)/Estate Wagon/ Custom Cruiser 7= Monte Carlo 8= Corvette 9= Cavalier/Sunbird/Sunfire/Skyhawk 10= Bonneville, SSE, SSEi 11= Lumina 12= Grand Am 13= Gran Prix 14= LeMans 15= Electra/Park Avenue 16= LeSabre 17= Regal 18= Reatta 19= 1993 Riviera/Toronado 20= Ultra 21= Allante 22= Concours 23= DeVille 24= Eldorado, ETC, Seville, STS 25= Achieva 26= Aurora/1994 and up Riviera 27= Calais 28= Cutlass 29= 88 Royale (1988 and up) 30= 98, Touring Sedan (1989 and up) 31= Supreme

Finishes, Body Structure, Frame, Bumpers

Models	Year	TSB#	Description
1	Up - 94	43-10-48	ADHESIVE CAULKING KIT - REVISIONS/PRIMERS/PROCEDURES
1	90 - 95	53-17-03	ALUMINUM WHEEL REFINISHING - REVISED PROCEDURE
1	Up - 93	93-058-10	BASE COAT/CLEAR COAT - RAIL DUST DAMAGE - REPAIR
1	84 - 93	92-033-10	BASE COAT/CLEAR COAT PAINT - INDUSTRIAL DUST DAMAGE
1	93 - 96	111702	BASE COAT/CLEAR COAT PAINT - POLISHING - REV
1	85 - 93	91-10-18	BASE COAT/CLEAR COAT PAINT - POLISHING TECHNIQUES
2	85 - 93	91-084/10	BASE COAT/CLEAR COAT PAINT - POLISHING TECHNIQUES
2	85 - 93	131060R	BASE/CLEAR COAT - INDUSTRIAL DUST DAMAGE - REPAIR
6	94 - 94	432001	BODY MOUNT ASSEMBLY DESIGN CHANGE
1	92 - 93	93-WA-11	BODY SURFACE - CHEMICAL SPOTTING LABOR OPERATIONS
1	Up - 94	431701	BUMPS OR RUST COLORED SPOTS IN PAINT - REMOVE PROCED
1	94 - 95	431701	BUMPS OR RUST COLORED SPOTS IN PAINT - REVISED
12	92 - 95	43-15-13	CENTER PILLAR UPPER WATER DEFLECTOR ADDED
25	92 - 95	43-15-13	CENTER PILLAR UPPER WATER DEFLECTOR ADDED
1	Up - 94	331708	CLEARCOAT DEGRADATION/CHALKING & WHITENING - PROCED
6	91 - 93	331029	COWL AREA & DOOR WEATHERSTRIP WATERLEAKS - CAUSE/FIX
6	91 - 93	93-235-10	COWL AREA & DOOR WEATHERSTRIP WATERLEAKS/SERV - REV
24,23,22,	92 - 94	93-072/10	COWL AREA/PLENUM CHAMBER - LEAKS - REPAIR PROC - REV
24	92 - 93	311604	COWL AREA/PLENUM CHAMBER - WATERLEAKS - REPAIR PROC
8	92 - 93	331010	COWL SURFACE - CRACKS/DAMAGE - SEALING PROCEDURE
26	95 - 95	53-15-08	CREAK NOISE AT RIGHT A-PILLAR AREA
11,7	95 - 95	53-10-16	ENG NOISE/COLD AIR ENTERS PASS COMPARTMENT
5	93 - 93	331033	FASCIA/FRONT - PAINT PEELING - NO REPAIR/ORDER PROC
9	95 - 96	63-15-11	FENDER MOUNTING - RIV-NUT INSTALL PROCEDURE
12	96 - 96	63-15-07	FR LIC PLATE BRACKET INSTALL INSTRUCTIONS
24	92 - 95	43-15-16	FRONT FENDER INSULATOR FALLS OUT OF POSITION
12	93 - 93	338204	FRONT RATTLE/NOISE - HORN BRACKET HITS BUMPER SUPPOR
25	93 - 93	338204	FRONT RATTLE/NOISE - HORN BRACKET HITS BUMPER SUPPOR
9	96 - 96	53-15-20	FUEL FILLER DOOR TOO FAR INBOARD AT REAR OF DOOR
12	92 - 94	431012	GAP AT REAR WINDOW REVEAL MOLDING & QUARTER PANEL
25	92 - 94	431012	GAP AT REAR WINDOW REVEAL MOLDING & QUARTER PANEL
12	93 - 95	93-10-12	HOOD - REMOVAL PROC TO AVOID BREAKING MIG WELDS -REV
25	93 - 95	93-U-15	HOOD - REMOVAL PROC TO AVOID BREAKING MIG WELDS -REV
7	95 - 95	431502	LOOSE/SEPARATED/WARPED REAR COMPARTMENT LID APPLIQUE
24	92 - 94	93-005/10	MOULDING WARPAGE (WINDSHIELD/REAR WINDOW) - REVISED
24	92 - 93	211529R	MOULDINGS (WINDSHIELD/REAR WINDOW)-WARPING-NEW PARTS
12	92 - 94	431007	NOISE/SQUEAK/RATTLE DIAGNOSTIC GUIDE
25	92 - 94	431007	NOISE/SQUEAK/RATTLE DIAGNOSTIC GUIDE
12	92 - 95	431007	NOISE/SQUEAK/RATTLE DIAGNOSTIC GUIDE - REV
25	92 - 95	431007	NOISE/SQUEAK/RATTLE DIAGNOSTIC GUIDE - REV
9	95 - 95	95-C-48	OMITTED WELDS - BODY "B" PILLAR - 2.2L - RECALL
1	85 - 93	111702R	PAINT - CARE OF EXTERIOR FINISHES - REPAIR PROC

Model Legend: 1= All 2= Most 3= Beretta/Corsica 4= Celebrity/Century/Cierra/Cutlass Cruiser/6000 5= Camaro/Firebird 6= Caprice/Roadmaster/Impala SS/Fleetwood(RWD)/Estate Wagon/ Custom Cruiser 7= Monte Carlo 8= Corvette 9= Cavalier/Sunbird/Sunfire/Skyhawk 10= Bonneville, SSE, SSEi 11= Lumina 12= Grand Am 13= Gran Prix 14= LeMans 15= Electra/Park Avenue 16= LeSabre 17= Regal 18= Reatta 19= 1993 Riviera/Toronado 20= Ultra 21= Allante 22= Concours 23= DeVille 24= Eldorado, ETC, Seville, STS 25= Achieva 26= Aurora/1994 and up Riviera 27= Calais 28= Cutlass 29= 88 Royale (1988 and up) 30= 98, Touring Sedan (1989 and up) 31= Supreme

Glass, Doors, Hood, Decklid, Tailgate, Liftgate, Locks

Models	Year	TSB#	Description
15,20	91 - 93	331037	SUNROOF - WINDNOISE AT WEATHERSTRIP SEAL - NEW SEAL
25	92 - 93	93-T-053	SWITCH/FRONT DOOR LOCK - ASSY DISENGAGED FROM BEZEL
25	93 - 93	331012	SWITCH/FRONT DOOR LOCK - ASSY DISENGAGED FROM BEZEL
26	95 - 96	63-15-04	TRUNK LID ASSIST ROD SEPARATES FROM LID
9	88 - 94	43-15-11	WATER LEAK - DOOR GLASS/QUARTER WINDOW - CONVERTIBLE
8	92 - 93	331045	WATER LEAK - REAR HATCH - REPL WEATHERSTRIP - COUPE
5	93 - 96	53-10-32	WATER LEAK AT TRUNK LID - SEALER
2	92 - 94	431003	WATER LEAK FROM DOORS INTO PASS AREA - DIAGN/PROCED
5	93 - 95	63-10-22	WATER LEAK INTO INTERIOR DURING CAR WASH
3,7	87 - 96	53-10-21	WATERLEAK DIAGNOSTIC GUIDE
2	88 - 95	90-10-22	WEBASTO SUNROOF REPAIRS - REV
6	91 - 93	331039R	WIND NOISE DIAG/REPAIR - TOOLS/RECOMMENDED MATERIALS
2	92 - 95	53-10-33	WIND NOISE OR WATER LEAK FROM SUNROOF
11,28	88 - 94	53-15-16	WINDNOISE
5	93 - 93	331060	WINDNOISE & WATERLEAK DIAGNOSIS & REPAIR
2	92 - 93	201008	WINDNOISE - DIAGNOSIS AND CORRECTIONS - REVISED
6	91 - 93	331005	WINDNOISE - PROPER DIAGNOSIS/MATERIALS/PROCEDURES
2	96 - 96	53-10-28	WINDNOISE AND/OR RETENTION DOOR GLASS RUN CHANNEL
12	95 - 95	53-10-13	WINDNOISE OR WATER LEAK - RT QUARTER WINDOW - COUPE
25	95 - 95	53-10-13	WINDNOISE OR WATER LEAK - RT QUARTER WINDOW - COUPE
8	84 - 93	331013	WINDOW/REAR LIFT - HARD TO OPEN/HINGE LOOSE TO GLASS
11,7	95 - 95	53-10-09	WINDOWS STICK IN FULL UP POSITION
29,30	91 - 93	301004	WINDSHIELD - CREAK NOISE - SQUEAK REDUCTION TAPE
15,20,16	91 - 93	301004	WINDSHIELD AREA - CREAK FROM CONTACT W/COWL SCREEN
10	92 - 93	301004	WINDSHIELD AREA CREAK NOISE - SQUEAK REDUCTION TAPE

Model Legend: 1= All 2= Most 3= Beretta/Corsica 4= Celebrity/Century/Cierra/Cutlass Cruiser/6000 5= Camaro/Firebird 6= Caprice/Roadmaster/Impala SS/Fleetwood(RWD)/Estate Wagon/ Custom Cruiser 7= Monte Carlo 8= Corvette 9= Cavalier/Sunbird/Sunfire/Skyhawk 10= Bonneville, SSE, SSEi 11= Lumina 12= Grand Am 13= Gran Prix 14= LeMans 15= Electra/Park Avenue 16= LeSabre 17= Regal 18= Reatta 19= 1993 Riviera/Toronado 20= Ultra 21= Allante 22= Concours 23= DeVille 24= Eldorado, ETC, Seville, STS 25= Achieva 26= Aurora/1994 and up Riviera 27= Calais 28= Cutlass 29= 88 Royale (1988 and up) 30= 98, Touring Sedan (1989 and up) 31= Supreme

GEO

Instruments, Dash Cluster, Warning Lights, Mirrors

Models	Year	TSB#	Description
4	96 - 96	63-90-05	CRUISE CONTROL CHART 9B-1 - S/M REV
4	93 - 93	338303	FUEL GAUGE INOP/IP LIGHTS FLASHING OR LOSS - DIAG
4	91 - 93	238301R	I/P CUPHOLDER - BINDING/STUCK - SERVICE PROCEDURE
4	93 - 93	93C10	MIRROR/LH MANUAL REMOTE - WON'T ADJUST - RECALL

Model Legend: 1= All 2= Most 3= Metro 4= Prizm 5= Storm

Heating, Air Conditioning, Ventilation, Defogger

Models	Year	TSB#	Description
1	Up - 94	43-12-23	CONTAMINATED A/C REFRIGERENT - R-12 SYSTEM
1	Up - 94	53-12-05	CONTAMINATED R12 REFRIGERANT TESTING & HANDLING
1	88 - 96	63-12-09	SERVICE ISSUES FOR R12 OR R134A A/C SYSTEMS
1	94 - 94	431205	USE R134A PAG REFRIGERANT OIL - S/M UPDATE

Model Legend: 1= All 2= Most 3= Metro 4= Prizm 5= Storm

Entertainment Devices, Stereo, Radio, Etc.

Models	Year	TSB#	Description
2	93 - 94	439603	RADIO - NO BASS OR TREBLE CONTROL -TURN TONE SEL OFF
2	93 - 94	339011	SOME TAPE CLEANERS & CD ADAPTERS NOT COMPATIBLE

Model Legend: 1= All 2= Most 3= Metro 4= Prizm 5= Storm

Seats, Belts, Interior Trim, Carpets, Air Bags

Models	Year	TSB#	Description
4	94 - 94	439001	SIR DUAL AIR BAGS - DEPLOYMENT INFORMATION
3	96 - 96	63-90-06	SIR (AIR BAG) - S/M REV
4	93 - 93	339001	SRS - FALSE SETTING OF DTC CODE 22 - CAUSE/SERV INFO

Model Legend: 1= All 2= Most 3= Metro 4= Prizm 5= Storm

Glass, Doors, Hood, Decklid, Tailgate, Liftgate, Locks

Models	Year	TSB#	Description
1	Up - 93	92-041-10	DOOR LOCK CYLINDER - FREEZES/STICKS/BINDS - REVISION

Model Legend: 1= All 2= Most 3= Metro 4= Prizm 5= Storm

Finishes, Body Structure, Frame, Bumpers

Models	Year	TSB#	Description
1	91 - 96	53-17-03	ALUMINUM WHEEL REFINISHING - REVISED PROCEDURE
1	Up - 93	93-058-10	BASE COAT/CLEAR COAT - RAIL DUST DAMAGE - REPAIR
1	93 - 96	111702	BASE COAT/CLEAR COAT PAINT - POLISHING - REV
1	94 - 95	431701	BUMPS OR RUST COLORED SPOTS IN PAINT - REVISED

Model Legend: 1= All 2= Most 3= Metro 4= Prizm 5= Storm

SATURN

Lighting, Horns, Turn Signals, Steering Column

Models	Year	TSB#	Description
4	91 - 93	381509	HEADLAMP BIND/SLOW/INTERMITTENT - REPAIR PROCEDURE
1	94 - 94	589001	IGNITION KEY CANNOT BE REMOVED

Model Legend: 1= All 2= Most 3= Sedan 4= Sport Coupe 5= Wagon

Instruments, Dash Cluster, Warning Lights, Mirrors

Models	Year	TSB#	Description
1	95 - 95	585005R	BUZZ/RATTLE/FLUTTER NOISE IP AREA - BRAKE BOOSTER
1	95 - 95	588303	CLICK/TICK NOISE FROM IC ODOMETER
1	91 - 95	486401	CRUISE CONTROL FLUCTUATES AT HWY SPEEDS
1	91 - 94	488303	IC CLUSTER ASSEMBLY REPLACEMENT - SERVICE INFO
1	91 - 93	389002	TELLTALE LIGHT (BAT) FLASHES AT START OF WIPER PULSE

Model Legend: 1= All 2= Most 3= Sedan 4= Sport Coupe 5= Wagon

Chassis Electrical, Wiring Harness, Fuses-Circuit Breakers, Wipers, Window Motors

Models	Year	TSB#	Description
1	93 - 93	93-C-04	BATTERY POSITIVE CABLE INSPECTION/CORRECTION -RECALL
1	92 - 94	481001	BRAKE LAMP FUSE RATING INCORRECT - S/M REVISION
1	91 - 94	488201	CPA RETAINER USAGE - SERVICE INFO
1	92 - 93	288101	ELECTRICAL WIRING HARNESS DESIGN CHANGES/SERV INFO
1	91 - 93	93-C-05	GENERATOR WIRING HARNESS REPLACE - RECALL REVISED
1	91 - 93	93-C-05	GENERATOR WIRING HARNESS REPLACEMENT - RECALL
1	91 - 94	488102	INSULATION DISPLACEMENT CONN (IDC) REPAIR -PWR LOCKS
1	93 - 93	93-C-02	WINDSHIELD WIPERS - DO NOT OP PROPERLY - RECALL

Model Legend: 1= All 2= Most 3= Sedan 4= Sport Coupe 5= Wagon

Auxiliary Equipment, Jacks, Trailer Hitches, Towing

Models	Year	TSB#	Description
1	91 - 93	289006R	S INSTALLING PHONE & OTHER ELECT ACCESSORIES - INFO

Model Legend: 1= All 2= Most 3= Sedan 4= Sport Coupe 5= Wagon

Heating, Air Conditioning, Ventilation, Defogger

Models	Year	TSB#	Description
1	91 - 93	231205R	A/C - ALTERNATE TO R-12 REFRIGERANT USAGE INFO
1	91 - 94	92-T-040	A/C - ODOR ON START-UP - DISINFECT PROCED/TOOLS -REV
1	91 - 94	481101	HVAC CONTROL LEVER STUCK OR HARD TO MOVE
1	95 - 95	481102	IP CENTER AIR OUTLET CLOSES WHEN BLOWER ON HIGH
1	91 - 95	581101	NO HEAT DURING EXTENDED IDLE IN COLD WEATHER
1	95 - 95	588202	REAR DEFOG ON WHEN IGNITION TURNED ON

Model Legend: 1= All 2= Most 3= Sedan 4= Sport Coupe 5= Wagon

Entertainment Devices, Stereo, Radio, Etc.

Models	Year	TSB#	Description
3	94 - 94	489601	MUFFLED SOUND - REAR SPEAKERS - HCS ONLY
1	91 - 94	389601	NOISE/POP IN RADIO SPEAKERS - CELLULAR PHONE ANTENNA
1	95 - 95	589603	RADIO - IGN NOISE OR INTERFERENCE IN SPEAKERS
1	93 - 94	488101	RADIO - STATIC ON AM WHEN POWER WINDOWS OPERATED
1	91 - 94	481502	RADIO ANTENNA MAST - WHISTLE TYPE WINDNOISE
1	91 - 95	94-T-025	RADIO ANTENNA MAST - WHISTLE TYPE WINDNOISE - REV
1	91 - 95	94-T-025A	RADIO ANTENNA MAST - WHISTLE TYPE WINDNOISE - REV 2
1	91 - 95	94-T-025B	RADIO ANTENNA MAST - WHISTLE TYPE WINDNOISE - REV 3

Model Legend: 1= All 2= Most 3= Sedan 4= Sport Coupe 5= Wagon

Seats, Belts, Interior Trim, Carpets, Air Bags

Models	Year	TSB#	Description
1	95 - 95	588201	AIR BAG TELLTALE ON - DTC 21
2	91 - 95	581603	DISCOLORATION/SPOTTING ON DOOR TRIM PANEL INSERT
1	91 - 93	481603	FRONT SEATBACK CUSHION/COVER CUT FROM SEATBACK FRAME
3	91 - 93	92-T-039	HEADLINER WET W/CAR PARKED FACING DOWNHILL (SUNROOF)
3,5	95 - 95	581602	POP NOISE FROM FRONT SEAT BACK FRAME
1	95 - 95	588304	RATTLE/CHATTER NOISE FROM GLOVEBOX AREA
3	94 - 94	481606	REAR SEAT HEAD SERVICE INFO - HCS (HOMECOMING COMM)
2	91 - 94	481602	SCUFF MARKS OR CUTS IN FRONT SEATS
1	91 - 94	92-T-011A	SHOULDER BELT GUIDE TRACK UNIT CHATTER OR BIND/REV 2
1	91 - 93	92-T-011	SHOULDER BELT GUIDE TRACK UNIT CHATTERS OR BINDS/REV
4	93 - 95	581604	SUNROOF SUNSHADE BINDS OR CLIPS ARE BROKEN
1	91 - 93	281606	TRIM PANEL/INT DOOR - INTERMITTENT BUZZ NOISE - PROC

Model Legend: 1= All 2= Most 3= Sedan 4= Sport Coupe 5= Wagon

Glass, Doors, Hood, Decklid, Tailgate, Liftgate, Locks

Models	Year	TSB#	Description
1	91 - 93	91-T-047	DOOR LOCK/DRIVERS - UNLOCK ALL DOORS FEATURE - REV
1	91 - 93	481601	DOOR WON'T OPEN - REPLACE LATCH - PROCEDURES
1	91 - 94	481605	FRONT DOOR WINDOW CRANK - HIGH EFFORT
5	93 - 93	93-P-01	LIFTGATE/REAR - ELECTRIC RELEASE INSTALLATION
5	93 - 93	93-P-01	LIFTGATE/REAR - ELECTRIC RELEASE INSTALLATION - REV
1	93 - 93	381601	LUGGAGE COMPARTMENT - WATERLEAK - SEALER
1	91 - 94	481501	NOISE/RATTLE IN FRONT DOOR AT BELT LINE
1	93 - 93	381503	OUTER DOOR FASTENER CHANGE - USE THREADLOCK/NEW NUT
1	91 - 94	488103	POWER DOOR LOCKS INOP - REPLACE & REPOSITION RELAY
5	93 - 94	481504	REAL LIFTGATE ELECTRIC RELEASE INOP ON FIRST TRY
4	91 - 94	92-T-O38	REAR COMPART LID - WATER DRIP INTO LUGGAGE AREA -REV
3	94 - 94	481606	REAR WINDOW TRIM PANEL SRVC - HCS (HOMECOMING COMM)
5	93 - 93	381504	SPARE TIRE WELL - WATER LEAK FROM LIFTGATE AREA/SERV
5	93 - 94	481607	SQUEAK/RATTLE FROM REAR LIFTGATE
1	93 - 93	381602	SUNROOF - LEAKS WHEN BRAKING/TURNING - FOAM SEAL
2	91 - 93	93-C-01	TRUNK LOCK DEFECT - REPAIR PROCEDURE - CAMPAIGN
1	91 - 95	481506	WATER LEAK AT FRONT UPPER DOOR FRAME
1	93 - 93	381510	WINDNOISE/HOOT AT FRONT UPPER DOOR FRAME - ADJUST FR

Model Legend: 1= All 2= Most 3= Sedan 4= Sport Coupe 5= Wagon

Finishes, Body Structure, Frame, Bumpers

Models	Year	TSB#	Description
1	91 - 95	94-T-009	BUMPS OR RUST COLORED SPOTS IN PAINT - PROCED - REV
1	91 - 94	481701R	BUMPS OR RUST COLORED SPOTS IN PAINT - PROCEDURE
2	95 - 95	486402	BUZZ NOISE AT REAR - REAR COMP DAMPERS
1	91 - 93	381506	DRIP MOLDING WATER DRIP ONTO SEAT/A-PILLAR TO CARPET
1	91 - 95	93-T-021	DRIP MOLDING WATER DRIP ONTO SEAT/A-PILLAR TO CARPET
1	93 - 93	381701	EXT PAINT "WA" NUMBERS INCORRECT - MISPRINTED LABEL
1	92 - 93	381502R	FOOTWELL AREA/FRONT - WATERLEAK - CAUSE/CORRECTION
1	92 - 93	93-T-016	FOOTWELL AREA/FRONT - WATERLEAK - CAUSE/PROC - REV
1	91 - 94	92-T-055	PAINT - CLEAN PROC FOR RUBBING COMPOUND REMOVAL -REV
1	91 - 95	92-T-055A	PAINT - CLEAN PROC FOR RUBBING COMPOUND REMOVAL -REV
1	91 - 93	281508	PAINT - CLEANING PROC FOR RUBBING COMPOUND REMOVAL
1	91 - 94	91-T-019	POLISH TECHNIQUES - BASE COAT/CLEAR COAT PAINTS -REV
3,5	93 - 94	482001	POP NOISE FROM LEFT A PILLAR - INSTALL BOLTS
3,5	93 - 94	94-T-004	POP NOISE FROM LEFT A PILLAR - INSTALL BOLTS - REV
3,5	93 - 95	94-T-004A	POP NOISE FROM LEFT A PILLAR - INSTALL BOLTS - REV
1	95 - 95	583301	RATTLE/POP/CLICK NOISE FROM FRONT END - ROUGH ROAD
1	91 - 93	91-T-030A	ROOF PANEL FLANGE - WATER LEAK - SEALING PROCED -REV
3	94 - 94	481703	TRI-COAT PAINT REPAIR INFO - HCS (HOMECOMING COMM)
1	91 - 94	481503	WATER LEAK INTO FRONT FOOTWELL FROM FRONT OF DASH
1	95 - 95	481505	WATER LEAK INTO RT FRONT FOOTWELL FROM FRONT OF DASH
5	93 - 93	381504	WATERLEAKS INTO SPARE TIRE WELL FROM LIFTGATE AREA
1	91 - 94	481502	WHISTLE TYPE WINDNOISE HIGH-UP IN DASH - ANTENNA
2	94 - 94	481704	YELLOW STAINS/SPOTS ON PAINT - WHITE VEHICLES ONLY

Model Legend: 1= All 2= Most 3= Sedan 4= Sport Coupe 5= Wagon

————NOTES————

—————NOTES—————

—————NOTES—————

—————NOTES—————